THE ENGINEERING HANDBOOK

SECOND EDITION

The Electrical Engineering Handbook Series

Series Editor
Richard C. Dorf
University of California, Davis

Titles Included in the Series

The Handbook of Ad Hoc Wireless Networks, Mohammad Ilyas
The Avionics Handbook, Cary R. Spitzer
The Biomedical Engineering Handbook, Second Edition, Joseph D. Bronzino
The Circuits and Filters Handbook, Second Edition, Wai-Kai Chen
The Communications Handbook, Second Edition, Jerry Gibson
The Computer Engineering Handbook, Vojin G. Oklobdzija
The Control Handbook, William S. Levine
The CRC Handbook of Engineering Tables, Richard C. Dorf
The Digital Signal Processing Handbook, Vijay K. Madisetti and Douglas Williams
The Electrical Engineering Handbook, Second Edition, Richard C. Dorf
The Electric Power Engineering Handbook, Leo L. Grigsby
The Electronics Handbook, Jerry C. Whitaker
The Engineering Handbook, Second Edition, Richard C. Dorf
The Handbook of Formulas and Tables for Signal Processing, Alexander D. Poularikas
The Handbook of Nanoscience, Engineering, and Technology, William A. Goddard, III, Donald W. Brenner, Sergey E. Lyshevski, and Gerald J. Iafrate
The Handbook of Optical Communication Networks, Mohammad Ilyas and Hussein T. Mouftah
The Industrial Electronics Handbook, J. David Irwin
The Measurement, Instrumentation, and Sensors Handbook, John G. Webster
The Mechanical Systems Design Handbook, Osita D.I. Nwokah and Yidirim Hurmuzlu
The Mechatronics Handbook, Robert H. Bishop
The Mobile Communications Handbook, Second Edition, Jerry D. Gibson
The Ocean Engineering Handbook, Ferial El-Hawary
The RF and Microwave Handbook, Mike Golio
The Technology Management Handbook, Richard C. Dorf
The Transforms and Applications Handbook, Second Edition, Alexander D. Poularikas
The VLSI Handbook, Wai-Kai Chen

Forthcoming Titles

The Electrical Engineering Handbook, Third Edition, Richard C. Dorf
The Electronics Handbook, Second Edition, Jerry C. Whitaker

THE ENGINEERING HANDBOOK

SECOND EDITION

Editor-in-Chief

Richard C. Dorf

University of California, Davis

CRC PRESS

Boca Raton London New York Washington, D.C.

Library of Congress Cataloging-in-Publication Data

The engineering handbook / editor-in-chief, Richard C. Dorf.
p. cm.
Includes bibliographical references and index.
ISBN 0-8493-1586-7 (alk. paper)
1. Engineering—Handbooks, manuals, etc. I. Dorf, Richard C.

TA151.E424 2004
620—dc22 2003069766

Direct all inquiries to CRC Press LLC, 2000 N.W. Corporate Blvd., Boca Raton, Florida 33431.

International Standard Book Number 0-8493-1586-7
Library of Congress Card Number 2003069766
Printed in the United States of America 1 2 3 4 5 6 7 8 9 0
Printed on acid-free paper

Preface

Due to the success of the first edition of *The Engineering Handbook* published in 1995 I am pleased to provide the second edition ten years later fully updated and expanded.

Purpose

The purpose of *The Engineering Handbook, Second Edition* is to provide in a single volume a ready reference for the practicing engineer in industry, government, and academia. The book in its comprehensive format is divided into 30 sections which encompass the field of engineering and includes 19 brand new chapters and 131 fully updated contributions. The goal is to provide the most up-to-date information in the classical fields that comprise mechanical, electrical, civil, chemical, industrial, and aerospace engineering as well as the underlying fields of mathematics, materials, biomedical engineering, and nanotechnology. This book should serve the information needs of all professional engineers engaged in the practice of the profession whether in industry, education, or government. The goal of this comprehensive handbook is to replace a myriad of books with one highly informative, well-organized, definitive source of fundamental knowledge.

Organization

The fundamentals of engineering have evolved to include a wide range of knowledge, substantial empirical data, and a broad range of practice. The focus of the handbook is on the key concepts, models, and equations that enable the engineer to analyze, design, and predict the behavior of complex devices, circuits, instruments, systems, structures, plants, computers, fuels, and the environment. While data and formulae are summarized, the main focus is the provision of the underlying theories and concepts and the appropriate application of these theories to the field of engineering. Thus, the reader will find the key concepts defined, described, and illustrated in order to serve the needs of the engineer over many years.

With equal emphasis placed on materials, structures, mechanics, dynamics, fluids, thermodynamics, fuels and energy, transportation, environmental systems, circuits and systems, computers and instruments, manufacturing, aeronautical and aerospace, and economics and management as well as mathematics, the engineer should encounter a wide range of concepts and considerable depth of exploration of these concepts as they lead to application and design.

The level of conceptual development of each topic is challenging, but tutorial and relatively fundamental. Each of the more than 200 chapters is written to enlighten the expert, refresh the knowledge of the mature engineer, and educate the novice.

The information is organized into 30 major sections. The 30 sections encompass 232 chapters, and the Appendix summarizes the applicable mathematics, symbols, and physical constants.

Each chapter includes three important and useful categories: defining terms, references, and further information. *Defining terms* are key definitions, and the first occurrence of each term defined is indicated in boldface in the text. The definitions of these terms are summarized as a list at the end of each chapter. The *references* provide a list of useful books and articles for follow-up reading. Finally, *further information* provides some general and useful sources of additional information on the topic.

Locating Your Topic

Numerous avenues of access to information contained in the handbook are provided. A complete table of contents is presented at the front of the book. In addition, an individual table of contents precedes each of the 30 sections. Finally, each chapter begins with its own table of contents. The reader should look over these tables of contents to become familiar with the structure, organization, and content of the book.

The index can also be used to locate key definitions. The page on which the definition appears for each key (defining) term is clearly identified in the index.

The Engineering Handbook, Second Edition is designed to provide answers to most inquiries and direct the inquirer to further sources and references. We hope that this handbook will be referred to often and that informational requirements will be satisfied effectively.

Acknowledgments

This handbook is testimony to the dedication of the associate editors, the publishers, and my editorial associates. I particularly wish to acknowledge at CRC Press Nora Konopka, Publisher; Helena Redshaw, Project Development Manager; Liz Spangenberger, Administrative Assistant; and Susan Fox, Project Editor.

Richard C. Dorf
Editor-in-Chief

Editor-in-Chief

Richard C. Dorf, professor of electrical and computer engineering at the University of California, Davis, teaches graduate and undergraduate courses in electrical engineering in the fields of circuits and control systems. He earned a Ph.D. in electrical engineering from the U.S. Naval Postgraduate School, an M.S. from the University of Colorado, and a B.S. from Clarkson University. Highly concerned with the discipline of engineering and its wide value to social and economic needs, he has written and lectured internationally on the contributions and advances in engineering and their value to society.

Professor Dorf has extensive experience with education and industry and is professionally active in the fields of robotics, automation, electric circuits, and communications. He has served as a visiting professor at the University of Edinburgh, Scotland; the Massachusetts Institute of Technology; Stanford University; and the University of California, Berkeley.

A Fellow of The Institute of Electrical and Electronics Engineers, Dr. Dorf is widely known to the profession for his *Modern Control Systems*, Tenth Edition (Prentice Hall, 2004) and *Introduction to Electric Circuits,* Sixth Edition (Wiley, 2004). He is the Editor-in-Chief of the *Technology Management Handbook* (CRC Press, 1999), the *Engineering Handbook,* Second Edition (CRC Press, 2004), and *CRC Handbook of Engineering Tables* (CRC Press, 2004).

Contributors

Bruno Agard
École Polytechnique de Montréal
Montréal, Québec, Canada

Ramesh K. Agarwal
Wichita State University
Wichita, Kansas

C. M. Akujobi
Prairie View A & M University
Prairie View, Texas

F. Chris Alley
Clemson University (Emeritus)
Clemson, South Carolina

Willliam Ames
Georgia Institute of Technology
Atlanta, Georgia

Appiah Amirtharajah
Georgia Institute of Technology
Atlanta, Georgia

James Amrhein
Consultant
Long Beach, California

Thalia Anagnos
San Jose State University
Palo Alto, California

Ted L. Anderson
Structural Reliability Technology
Boulder, Colorado

Sandra Arlinghaus
The University of Michigan
Ann Arbor, Michigan

Roger E. A. Arndt
University of Minnesota
Ham Lake, Minnesota

John Attia
Prairie View A&M University
Prairie View, Texas

Tung Au
Carnegie Mellon University
Pittsburgh, Pennsylvania

Robert Austin
Austin Communication Education Services
Palm Harbor, Florida

Bilal M. Ayyub
Unversity of Maryland
College Park, Maryland

A. Terry Bahill
University of Arizona
Tucson, Arizona

Rex T. Baird
Silicon Laboratories, Inc.
Nashua, New Hampshire

Terrence W. Baird
Hewlett-Packard Company
Boise, Idaho

Norman Balabanian
University of Florida
Gainesville, Florida

Partha P. Banerjee
University of Dayton
Dayton, Ohio

Randall Barron
Louisiana Tech University
Ruston, Lousiana

Charles E. Baukal, Jr.
John Zink Co. LLC
Tulsa, Oklahoma

Nelson R. Bauld, Jr.
Clemson University
Clemson, South Carolina

Yildiz Bayazitoglu
Rice University
Houston, Texas

Robert Beaves
Robert Morris University
Moon Township, Pennsylvania

R. R. Beck
U.S. Army Tank Automotive Research Development and Engineering Center
Warren, Michigan

Philip B. Bedient
Rice University
Houston, Texas

Richard C.Bennett
Swenson Process Equipment, Inc.
Harvey, Illinois

Jim Bethel
Purdue University
West Lafayette, Indiana

Bharat Bhushan
Ohio State University
Columbus, Ohio

Peter J. Biltoft
University of California, Lawrence Livermore National Laboratory
Livermore, California

Stephen Birn
Moog Inc.
Chatsworth, California

Robert H. Bishop
University of Texas
Austin, Texas

Kenneth B. Black
University of Massachusetts, Amherst
Amherst, Massachusetts

Glenn Blackwell
Purdue University
West Lafayette, Indiana

Benjamin S. Blanchard
Virginia Poytechnic Institute and State University
Blacksburg, Virginia

Robert F. Boehm
University of Nevada
Las Vegas, Nevada

Bruce W. Bomar
University of Tennesse Space Institute
Tullahoma, Tennessee

Charles Borzileri
University of California, Lawrence Livermore National Laboratory
Livermore, California

Ed Braun
University of North Carolina
Charlotte, North Carolina

Donald E. Breyer
California State Polytechnic University
Pomona, California

Robert Broadwater
Virginia Polytechnic Institute and State University
Blacksburg, Virginia

Joseph D. Bronzino
Trinity College
Hartford, Connecticut

George R. Buchanan
Tennessee Technological University
Cookeville, Tennessee

R. Ben Buckner
Surveying Education Consultant
Johnson City, Tenneesee

Michael Buehrer
Virginia Polytechnic Institute and State University
Blacksburg, Virginia

George Cain
Georgia Institute of Technology
Atlanta, Georgia

William L. Chapman
Hughes Aircraft Company
Tucson, Arizona

Som Chattopadhyay
Indiana University/Purdue University
Fort Wayne, Indiana

Shiao-Hung Chiang
University of Pittsburgh
Pittsburgh, Pennsylvania

Tim Chinowsky
University of Washington
Seattle, Washington

Jonathan W. Chipman
University of Wisconsin
Madison, Wisconsin

Tony M. Cigic
University of British Columbia
Vancouver, British Columbia, Canada

Michael D. Ciletti
University of Colorado
Colorado Springs, Colorado

William Cook
Iowa State Univerisity
Ames, Iowa

C. David Cooper
University of Central Florida
Orlando, Florida

William C. Corder
CONSOL, Inc.
Pittsburgh, Pennsylvania

Harold M. Cota
California Polytechnic State University
San Luis Obispo, California

Leon W. Couch II
University of Florida
Gainesville, Florida

Dennis J. Cronin
Iowa State Univerisity
Ames, Iowa

J. B. Cropley
Union Carbide Corporate Fellow (Retired)
Scott Depot, West Virginia

John N. Daigle
University of Mississippi
University, Mississippi

Shilpa Damle
Georgia Institute of Technology
Atlanta, Georgia

Braja M. Das
California State University
Sacramento, California

Kevin A. Delin
Jet Propulsion Laboratory
Arcadia, California

Anca Deliu
Georgia Institute of Technology
Atlanta, Georgia

Bon DeWitt
University of Florida
Gainesville, Florida

Ron Dieck
Ron Dieck Associates, Inc.
Palm Beach Gardens, Florida

Henry Domingos
Clarkson University
Potsdam, New York

John F. Donovan
McDonnell Douglas Corporation
St. Louis, Missouri

Anil Doradla
Virginia Polytechnic Institute and State University
Blacksburg, Virginia

Deepak Doraiswamy
National Starch and Chemical Co.
Bridgewater, New Jersey

Nelson C. Dorny
University of Pennsylvania
Philadelphia, Pennsylvania

Tolga Duman
Arizona State University
Tempe, Arizona

Stephen A. Dyer
Kansas State University
Manhattan, Kansas

Taan ElAli
Wilberforce University
Wilberforce, Ohio

Mohammed M. El-Wakil
University of Wisconsin
Madison, Wisconsin

Halit Eren
Polytechnic University
Hong Kong

Wolter J. Fabrycky
Virginia Polytechnic Institute and State University
Blacksburg, Virginia

James R. Fair
University of Texas
Austin, Texas

Henry O. Fatoyinbo
University of Surrey
Surrey, U.K.

Charles Fazzi
Saint Vincent College
Latrobe, Pennsylvania

Chang-Xue Jack Feng
Bradley University
Peoria, Illinois

H. Scott Fogler
University of Michigan
Ann Arbor, Michigan

Samuel W. Fordyce
Consultare Technology Group
Rockville, Maryland

Wallace T. Fowler
University of Texas
Austin, Texas

A. Keith Furr
Virginia Polytechnic Institute and State University (Retired)
Blacksburg, Virginia

Anish Gaikwad
EPRI-PEAC Corporation
Knoxville, Tennessee

Richard S. Gallegher
R. S. Gallegher and Associates
Ithaca, New York

Pierre Gehlen
Seattle University
Seattle, Washington

James M. Gere
Stanford University
Stanford, California

Peter Gergely
Cornell University
Ithaca, New York

Afshin Ghajar
Oklahoma State University
Stillwater, Oklahoma

Victor W. Goldschmidt
Purdue University
West Lafayette, Indiana

Mike Golio
Golio Consulting
Mesa, Arizona

Walter J. Grantham
Washington State University
Pullman, Washington

Ling Guan
Ryerson University
Toronto, Ontario, Canada

Francis Joseph Hale
North Carolina State University
Raleigh, North Carolina

Jerry C. Hamann
University of Wyoming
Laramie, Wyoming

Simon P. Hanson
CONSOL, Inc .
Pittsburgh, Pennsylvania

Steve J. Harrison
Queen's University
Kingston, Ontario, Canada

Barbara Hauser
Bay de Noc Community College
Escanaba, Michigan

Mary Sue Haydt
Santa Clara University
Santa Clara, California

N. W. J. Hazelton
Ohio State University
Columbus, Ohio

Scott Hazelwood
University of California
Davis, California

Scott W. Heaberlin
Pacific Northwest National
Laboratory
Richland, Washington

Chris Hendrickson
Carnegie Mellon University
Pittsburgh, Pennsylvania

John B. Herbich
Texas A & M University
College Station, Texas

Ronald A. Hess
University of California
Davis, California

Russell Hibbeler
University of Louisiana at Lafayette
Lafayette, Lousiana

David J. Hills
University of California
Davis, California

Kai F. Hoettges
University of Surrey
Surrey, U.K.

Steven M. Hoffberg
Moses & Singer LLP
New York, New York

Dave Holten
University of California, Lawrence
Livermore National Laboratory
Livermore, California

Greg Hoogers
University of Applied Sciences
Birkenfeld, Germany

Stephen Horan
New Mexico State University
Las Cruces, New Mexico

Mike Hughes
University of Surrey
Surrey, U.K.

Paul J. Hurst
University of California
Davis, California

Iqbal Husain
University of Akron
Akron, Ohio

Daniel Inman
Virginia Polytechnic Institute and
State University
Blacksburg, Virginia

Rolf Johansson
Lund University
Lund, Sweden

Steven D. Johnson
Purdue University
West Lafayette, Indiana

Otakar Jonas
Jonas, Inc.
Wilmington, Delaware

S. Casey Jones
Georgia Institute of Technology
Atlanta, Georgia

Anthony J. Kalinowski
Naval Undersea Warfare Center
Newport, Rhode Island

Bruce Karnopp
University of Michigan
Ann Arbor, Michigan

Waldemar Karwowski
University of Louisville
Louisville, Kentucky

Ralph W. Kiefer
University of Wisconsin
Madison, Wisconsin

L. Kitis
University of Virginia
Charlottesville, Virginia

Joseph F. Kmec
Purdue University
West Lafayette, Indiana

Edward Knod
Western Illinois University
Macomb, Illinois

Alan A. Kornhauser
Virginia Polytechnic Institute and State University
Blacksburg, Virginia

William J. Koros
Georgia Institute of Technology
Atlanta, Georgia

William B. Krantz
University of Colorado
Boulder, Colorado

Moncef Krarti
University of Colorado
Boulder, Colorado

Jan F. Kreider
Kreider and Associates
Boulder, Colorado

Thomas R. Kurfess
Georgia Institute of Technology
Atlanta, Georgia

Andrew Kusiak
University of Iowa
Iowa City, Iowa

Benjamin Kyle
Kansas State University
Manhattan, Kansas

Fatima H. Labeed
University of Surrey
Surrey, U.K.

Richard T. Lahey
Rensselaer Polytechnic Institute
Troy, New York

Lee S. Langston
University of Connecticut
Storrs, Connecticut

Phillip Laplante
Penn State University
Malvern, Pennsylvania

Alan O. Lebeck
Mechanical Seal Technology, Inc.
Albuquerque, New Mexico

Arthur W. Leissa
Ohio State University
Columbus, Ohio

John Leonard II
Georgia State Road and Tollway Authority
Atlanta, Georgia

John B. Ligon
Michigan Technological University
Houghton, Michigan

Thomas M. Lillesand
University of Wisconsin
Madison, Wisconsin

K. H. Lin
Oak Ridge National Laboratory
Knoxville, Tennessee

Noam Lior
University of Pennsylvania
Philadelphia, Pennsylvania

Earl Livingstone
Babcock and Wilcox Company, Retired
Norton, Ohio

Gregory L. Long
University of California
Irvine, California

Eric M. Lui
Syracuse University
Syracuse, New York

Sergey Lyshevski
Rochester Institute of Technology
Rochester, New York

Arindam Maitra
EPRI PEAC Corporation
Knoxville, Tennessee

Thomas Mancini
Sandia National Laboratories
Albuquerque, New Mexico

Thomas Marlin
McMaster University
Hamilton, Ontario, Canada

Harold E. Marshall
National Institute of Standards and Technology
Gaithersburg, Maryland

Bruce R. Martin
University of California
Davis, California

E. F. Matthys
University of California
Santa Barbara, California

Michael McCarthy
University of California
Irvine, California

Alan T. McDonald
Purdue University
West Lafayette, Indiana

Ross E. McKinney
Environmental Consultant
Chapel Hill, North Carolina

Sue McNeil
University of Illinois
Chicago, Illinois

Daniel A. Mendelsohn
Ohio State University
Columbus, Ohio

J. L. Meriam
University of California (Retired)
Santa Barbara, California

Roger Messenger
Florida Atlantic University
Boca Raton, Florida

Michael D. Meyer
Georgia Institute of Technology
Atlanta, Georgia

Karsten Meyer-Waarden
University of Karlsruhe
Karlsruhe, Germany

Scott L. Miller
University of Florida
Gainesville, Florida

Jan C. Monk
National Aeronautics and Space Administration
Huntsville, Alabama

Theodore T. Moore
Georgia Institute of Technology
Atlanta, Georgia

Samiha Mourad
Santa Clara University
Santa Clara, California

Safwat M. A. Moustafa
California Polytechnic University
San Luis Obispo, California

Rias Muhamed
SBC Laboratories , Inc.
Austin, Texas

Bruce R. Munson
Iowa State Univerisity
Ames, Iowa

Mark L. Nagurka
Marquette University
Milwaukee, Wisconsin

D. Subbaram Naidu
Idaho State University
Pocatello, Idaho

Edward G. Nawy
Rutgers University
New Brunswick, New Jersey

Paul Neudorfer
Seattle University
Seattle, Washington

J. Karl C. Nieman
Utah State University
Logan, Utah

Norman Nise
California State Polytechnic University
Pomona, California

Jay M. Ochterbeck
Clemson University
Clemson, South Carolina

M. M. Ohadi
University of Maryland
College Park, Maryland

Vojin G. Oklobdzija
University of California
Davis, California

James Y. Oldshue
Oldshue Technologies International
Fairport, New York

George Opdyke, Jr.
Dykewood Enterprises
Stratford, Connecticut

Hasan Orbey (Deceased)
University of Delaware
Newark, Delaware

Terry P. Orlando
Massachusetts Institute of Technology
Cambridge, Massachusetts

Bulent Ovunc
University of Louisiana at Lafayette
Lafayette, Lousiana

Hitay Ozbay
Bilkent University
Ankara, Turkey
and
The Ohio State University
Columbus, Ohio

Joseph Palais
Arizona State University
Tempe, Arizona

Jens Palsberg
University of California
Los Angeles, California

Gordon R. Pennock
Purdue University
West Lafayette, Indiana

Kermit Phipps
EPRI PEAC Corporation
Knoxville, Tennessee

Walter Pilkey
University of Virginia
Charlottesville, Virginia

Michael J. Piovoso
Penn State University
Malvern, Pennsylvania

Greg Placencia
University of Southern California
Los Angeles, California

Bruce Poling
University of Toledo
Toledo, Ohio

Alexander D. Poularikas
University of Alabama
Huntsville, Alabama

David Rabinowitz
Moses & Singer LLP
New York, New York

Mansour Rahimi
University of Southern California
Los Angeles, California

Kaushik S. Rajashekara
Delphi Corporation
Kokomo, Indiana

Rama Ramakumar
Oklahoma State University
Stillwater, Oklahoma

Jonathon Randall
University of Sydney
Sydney, Australia

Theodore S. Rappaport
University of Texas
Austin, Texas

Muhammad H. Rashid
University of West Florida
Pensacola, Florida

Frances H. Raven
University of Notre Dame
South Bend, Indiana

Timothy A. Reinhold
Clemson University
Clemson, South Carolina

Christopher Relf
National Instruments Certified
LabVIEW Developer
New South Wales, Australia

John L. Richards
University of Pittsburgh
Pittsburgh, Pennsylvania

Everett V. Richardson
Ayres Associates
Fort Collins, Colorado

Albert J. Rosa
University of Denver
Denver, Colorado

Jennifer Russell
University of Southern California
Los Angeles, California

M. N. O. Sadiku
Prairie View A & M University
Prairie View, Texas

Richard S. Sandige
University of Wyoming
Laramie, Wyoming

Stanley I. Sandler
University of Delaware
Newark, Delaware

Albert Sargent
Arkansas Power & Light
Hot Springs, Arkansas

Nesrin Sarigul-Klijn
University of California
Davis, California

Udaya B. Sathuvalli
Rice University
Houston, Texas

Rudolph J. Scaruzzo
University of Akron
Akron, Ohio

Charles Scawthorn
Kyoto University
Kyoto, Japan

Boyd D. Schimel
Washington State University
Pullman, Washington

Richard Schonberger
Schonberger & Associates, Inc.
Bellevue, Washington

Paul Schonfeld
University of Maryland
College Park, Maryland

William T. Segui
University of Memphis
Memphis, Tennessee

Andrea Serrani
The Ohio State University
Columbus, Ohio

James F. Shackelford
University of California
Davis, California

Sherif A. Sherif
University of Florida
Gainesville, Florida

Yung C. Shin
Purdue University
West Lafayette, Indiana

Walter Short
National Renewable Energy Laboratory
Golden, Colorado

Paul W. Shuldiner
University of Massachusetts, Amherst
Amherst, Massachusetts

Ben L. Sill
Clemson University
Clemson, South Carolina

Ronald C. Sims
Utah State University
Logan, Utah

R. Paul Singh
University of California
Davis, California

Vijay P. Singh
Louisiana State University
Baton Rouge, Louisana

Shivaji Sircar
Air Products and Chemicals, Inc.
Allentown, Pennsylvania

Timothy L. Skvarenina
Purdue University
West Lafayette, Indiana

Montgomery L. Smith
University of Tennesse
Tullahoma, Tennessee

Rosemary L. Smith
University of California
Davis, California

Sid Soclof
California State University
Los Angeles, California

Michael A. Soderstrand
University of California
Davis, California and
Oklahoma State University
Stillwater, Oklahoma

Richard E. Sonntag
University of Michigan
Ann Arbor, Michigan

Frank R. Spellman
Hampton Roads Sanitation Department
Norfolk, Virginia

John P. H. Steele
Colorado School of Mines
Denver, Colorado

Ray Stefani
California State University
Long Beach, California

Yorgos J. Stephanedes
Institute of Computer Science Foundation for Research and Technology
Hellos, Greece

Matthew P. Stephens
Purdue University
West Lafayette, Indiana

P.K. Subramanyan
Glacier Clevite Heavywall Bearings
Solon, Ohio

James A. Svoboda
Clarkson University
Potsdam, New York

Hans J. Thamhain
Bentley College
Waltham, Massachusetts

James F. Thompson
J. F. Thompson, Inc.
Houston, Texas

Y. L. Tong
Georgia Institute of Technology
Atlanta, Georgia

Matt Traini
University of California, Lawrence Livermore National Laboratory
Livermore, California

Blake P. Tullis
Utah State University
Logan, Utah

J. Paul Tullis
Tullis Engineering Consultants
Logan, Utah

Vincent Van Brunt
University of South Carolina
Columbia, South Carolina

Boudewijn H. W. van Gelder
Purdue University
West Lafayette, Indiana

Peter J. Varman
Rice University
Houston, Texas

Jerry Ventre
Florida Solar Energy Center
Cocoa, Florida

Baxter E. Vieux
University of Oklahoma
Norman, Oklahoma

Thomas Vincent
University of Arizona
Tucson, Arizona

Wolf W. von Maltzahn
Rensselaer Polytechnic Institute
Troy, New York

Curtis J. Wahlberg
Purdue University
West Lafayette, Indiana

David Wallace
Georgia Institute of Technology
Atlanta, Georgia

David Wallenstein
Woodward-Clyde Consultants
Oakland, California

Yingxu Wang
University of Calgary
Calgary, Alberta, Canada

Pao-lien Wang
University of North Carolina
Charlotte, North Carolina

Kyle K. Wetzel
Wetzel Engineering, Inc.
Lawrence, Kansas

Barry Wilkinson
University of North Carolina
Charlotte, North Carolina

Wayne Wolf
Princeton University
Princeton, New Jersey

David M. Woodall
University of Idaho
Moscow, Idaho

William W. Wu
Advanced Technology Mechanization Company
Bethesda, Maryland

Chih-Kong Ken Yang
University of California at Los Angeles
Los Angeles, California

Loren W. Zachary
Iowa State Univerisity
Ames, Iowa

Ashraf A. Zeid
Army High Performance Computing Research Center and Computer Sciences Corporation
Warren, Michigan

Rodger E. Ziemer
University of Colorado
Colorado Springs, Colorado

Contents

I Statics

II Mechanics of Materials

III Dynamics and Vibrations

VI Fluid Mechanics

VII Thermodynamics and Heat Transfer

VIII Separation Processes

XI Geotechnical

XII Transportation

XIII Coastal and Ocean Engineering

XIV Environmental System and Management

XV Water Resources Engineering

XVI Linear Systems and Models

XVII Circuits

XVIII Electronics

XIX Digital Systems

XX Communications and Signal Processing

XXI Computers

XXII Measurement and Instrumentation

XXIII Surveying

XXIV Control Systems

XXV Manufacturing

XXVI Aeronautical and Aerospace

XXVII Safety

XXVIII Engineering Economics and Management

XXIX Materials Engineering

XXX Mathematics

I

Statics

1

Force-System Resultants and Equilibrium

Russell C. Hibbeler
University of Louisiana at Lafayette

Statics is a branch of mechanics that deals with the equilibrium of bodies, that is, those that are either at rest or move with constant velocity. In order to apply the laws of statics, it is first necessary to understand how to simplify force systems and compute the moment of a force. In this chapter these topics will be discussed, and some examples will be presented to show how the laws of statics are applied.

1.1 Force-System Resultants

Concurrent Force Systems

Force is a vector quantity that is characterized by its magnitude, direction, and point of application. When two forces $\mathbf{F}_1$ and $\mathbf{F}_2$ are **concurrent** they can be added together to form a resultant $\mathbf{F}_R = \mathbf{F}_1 + \mathbf{F}_2$ using the **parallelogram law,** Figure 1.1. Here $\mathbf{F}_1$ and $\mathbf{F}_2$ are referred to as components of $\mathbf{F}_R$. Successive applications of the parallelogram law can also be applied when several concurrent forces are to be added; however, it is generally simpler to first determine the two components of each force along the axes of a coordinate system and then add the respective components. For example, the *x, y, z* (or Cartesian) components of **F** are shown in Figure 1.2. Here, **i, j, k** are unit vectors used to define the direction of the positive *x, y, z* axes, and F_x, F_y, F_z are the magnitudes of each component. By vector addition, $\mathbf{F} = F_x\mathbf{i} + F_y\mathbf{j} + F_z\mathbf{k}$. When each force in a concurrent system of forces is expressed by its Cartesian components, the resultant force is therefore

$$\mathbf{F}_R = \sum F_x\mathbf{i} + \sum F_y\mathbf{j} + \sum F_z\mathbf{k} \tag{1.1}$$

where ΣF_x, ΣF_y, ΣF_z represent the scalar additions of the *x, y, z* components, respectively.

0-8493-1586-7/05/$0.00+$1.50

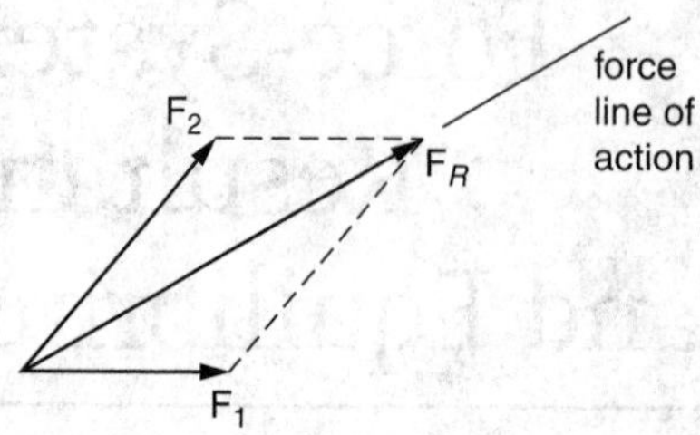

FIGURE 1.1 Addition of forces by parallelogram law.

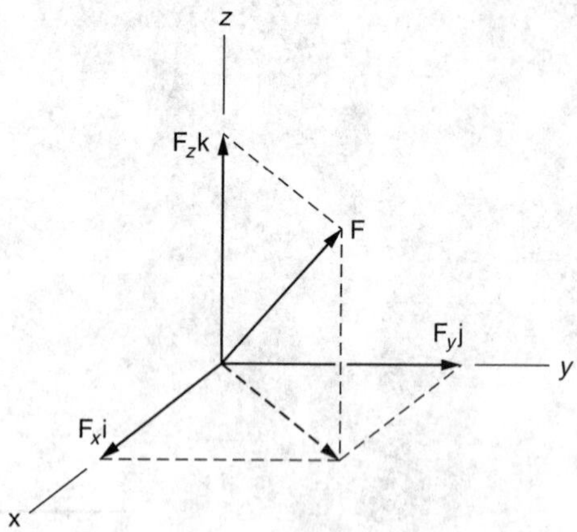

FIGURE 1.2 Resolution of a vector into its *x*, *y*, *z* components.

Moment of a Force

When a force **F** acts on a body, it will cause both external and internal effects on the body. These effects depend upon where the force is located. For example, if **F** acts at point *A* on the body in Figure 1.3, it will cause a specific translation and rotation of the body. However, if **F** is applied to some other point, *B*, which lies along the line of action of **F**, then the external effects regarding the motion of the body remain unchanged, although the body's internal effects will be different. This effect of sliding a force along its line of action is called the **principle of transmissibility**. If the force acts at point *C*, which is not along the line of action *AB*, then both the external and internal effects on the body will change. The difference in external effects — notably the difference in the rotation of the body — occurs because of the distance *d* that separates the lines of action of the two positions of the force.

This tendency for the body to rotate about a specified point *O* or axis as caused by a force is a vector quantity called a *moment*. By definition, the magnitude of the moment is

$$M_O = Fd \tag{1.2}$$

where *d* is the moment arm or perpendicular distance from the point to the line of action of the force, as in Figure 1.4. The direction of the moment is defined by the right-hand rule, whereby the curl of the right-hand fingers follows the tendency for rotation caused by the force, and the thumb specifies the directional sense of the moment. In this case, $\mathbf{M}_O$ is directed out of the page, since **F** produces counterclockwise rotation about *O*. It should be noted that the force can act at any point along its line of action and still produce the same moment about *O*.

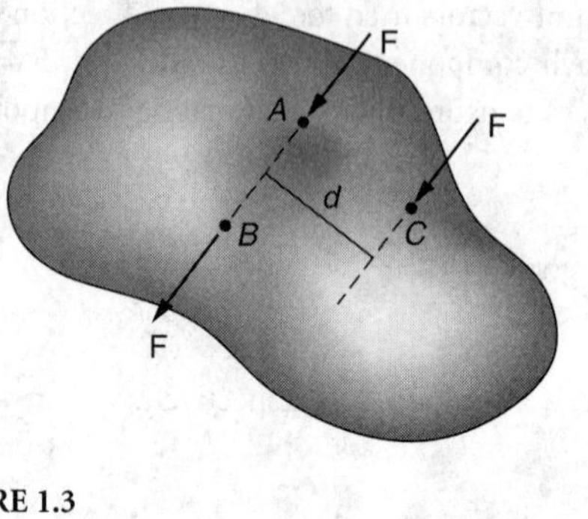

FIGURE 1.3

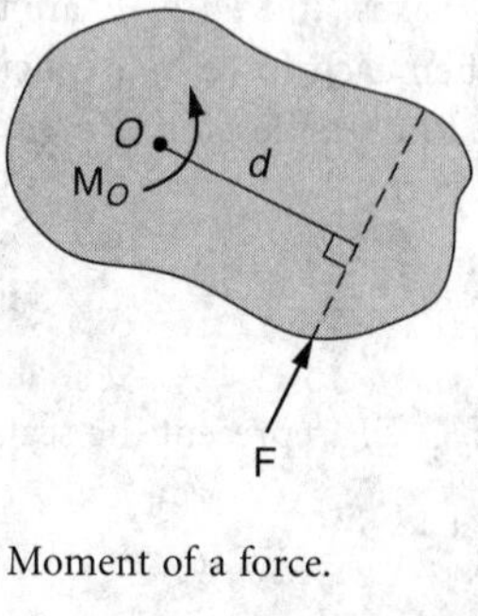

FIGURE 1.4 Moment of a force.

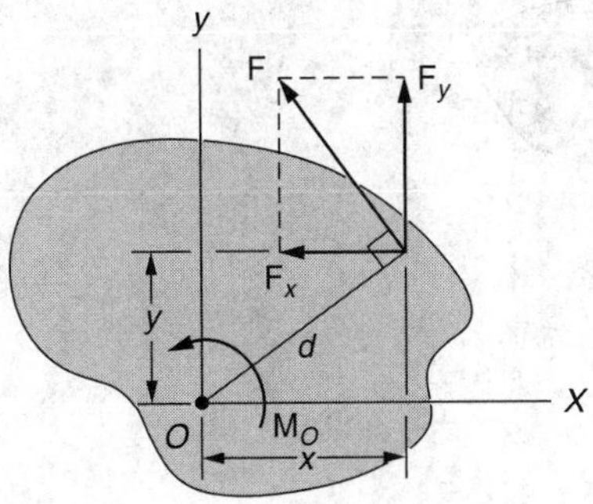

FIGURE 1.5

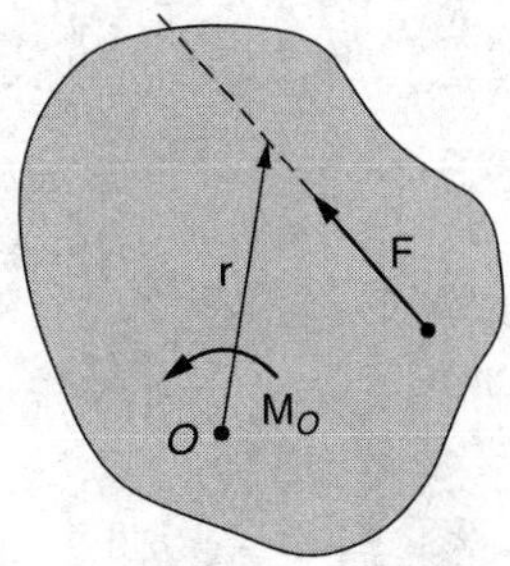

FIGURE 1.6

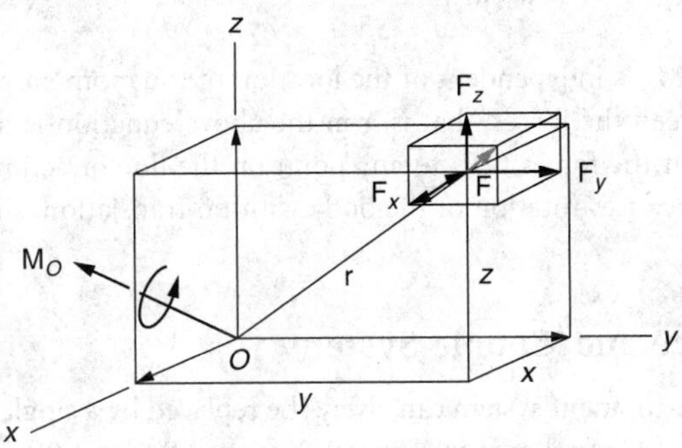

FIGURE 1.7

Sometimes the moment arm d is geometrically hard to determine. To make the calculation easier, the force is first resolved into its Cartesian components and then the moment about point O is determined using **the principle of moments**, which states that the moment of the force about O is equal to the sum of the moments of the force's components about O. Thus, as shown in Figure 1.5, we have $M_O = Fd = F_x y + F_y x$.

The moment about point O can also be expressed as a vector cross product of the position vector **r**, directed from O to any point on the line of action of the force and the force **F**, as shown in Figure 1.6. Here,

$$\mathbf{M}_O = \mathbf{r} \times \mathbf{F} \tag{1.3}$$

If **r** and **F** are expressed in terms of their Cartesian components, then as in Figure 1.7 the Cartesian components for the moment about O are

$$\begin{aligned}\mathbf{M}_O &= \mathbf{r}\times\mathbf{F} = (x\mathbf{i}+y\mathbf{j}+z\mathbf{k})\times(F_x\mathbf{i}+F_y\mathbf{j}+F_z\mathbf{k}) \\ &= (yF_z - zF_y)\mathbf{i} + (zF_x - xF_z)\mathbf{j} + (xF_y - yF_x)\mathbf{k} \qquad (1.4) \\ &= \begin{vmatrix} \mathbf{i} & \mathbf{j} & \mathbf{k} \\ x & y & z \\ F_x & F_y & F_z \end{vmatrix}\end{aligned}$$

Couple

A **couple** is defined as two parallel forces that have the same magnitude and opposite directions and are separated by a perpendicular distance d, as in Figure 1.8. The moment of a couple about the arbitrary point O is

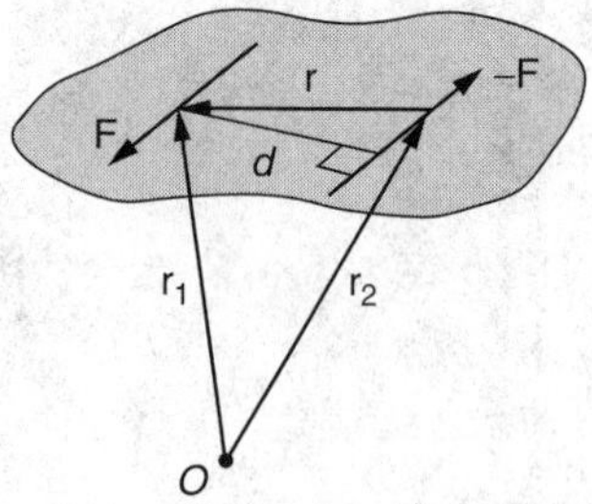

FIGURE 1.8

$$\begin{aligned}\mathbf{M}_C &= \mathbf{r}_1 \times \mathbf{F} + \mathbf{r}_2 \times (-\mathbf{F}) = (\mathbf{r}_1 - \mathbf{r}_2) \times \mathbf{F} \\ &= \mathbf{r} \times \mathbf{F}\end{aligned} \tag{1.5}$$

Here the couple moment $\mathbf{M}_C$ is independent of the location of the moment point O. Instead, it depends only on the distance between the forces; that is, **r** in the above equation is directed from any point on the line of action of one of the forces ($-\mathbf{F}$) to any point on the line of action of the other force **F**. The external effect of a couple causes rotation of the body with no translation, since the resultant force of a couple is zero.

Resultants of a Force and Couple System

A general force and couple-moment system can always be replaced by a single resultant force and couple moment acting at any point O. As shown in Figure 1.9(a) and Figure 1.9(b), these resultants are

$$\mathbf{F}_R = \sum \mathbf{F} \tag{1.6}$$

$$\mathbf{M}_{R_O} = \sum \mathbf{M}_O \tag{1.7}$$

where $\Sigma\mathbf{F} = \mathbf{F}_1 + \mathbf{F}_2 + \mathbf{F}_3$ is the vector addition of all the forces in the system, and $\Sigma\mathbf{M}_O = (\mathbf{r}_1 \times \mathbf{F}_1) + (\mathbf{r}_2 \times \mathbf{F}_2) + (\mathbf{r}_3 \times \mathbf{F}_3) + \mathbf{M}_1 + \mathbf{M}_2$ is the vector sum of the moments of all the forces about point O plus the sum of all the couple moments. This system may be further simplified by first resolving the couple moment M_{R_O} into two components — one parallel and the other perpendicular to the force $\mathbf{F}_R$, as in Figure

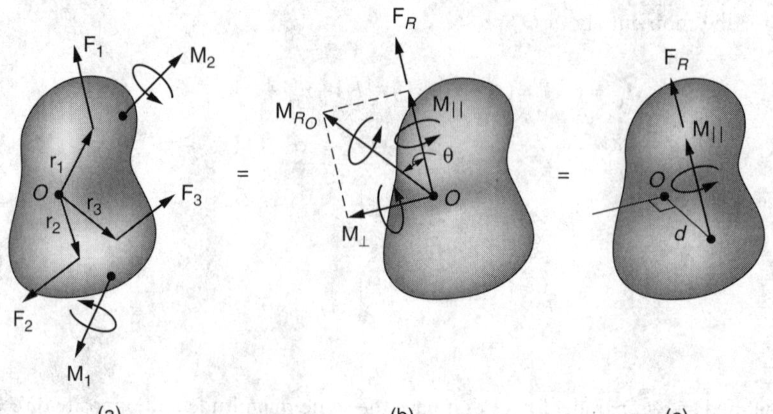

FIGURE 1.9

1.9(b). By moving the line of action of $\mathbf{F}_R$ in the plane perpendicular to $\mathbf{M}_\perp$ a distance $d = M_\perp/\mathbf{F}_R$, so that $\mathbf{F}_R$ creates the moment $M_\perp$ about O, the system can then be represented by a **wrench**, that is, a single force $\mathbf{F}_R$ and collinear moment $\mathbf{M}_\parallel$, Figure 1.9(c).

Note that in the special case of $\theta = 90°$, Figure 1.9(b), $\mathbf{M}_\parallel = \mathbf{0}$ and the system then reduces to a single resultant force $\mathbf{F}_R$ having a specified line of action. This will always be the case if the force system is either concurrent, parallel, or coplanar.

Distributed Loadings

When a body contacts another body, the loads produced are always distributed over the surface area of each body. If the area on one of the bodies is small compared to the entire surface area of the body, the loading can be represented by a single concentrated force acting at a point on the body. However, if the loading occurs over a large surface area — such as that caused by wind or a fluid — the distribution of load must be taken into account. The intensity of this surface loading at each point is represented as a pressure and its variation is defined by a load-intensity diagram. On a flat surface, the load intensity diagram is described by the loading function $p = p(x, y)$, which consists of an infinite number of parallel forces, as in Figure 1.10. Applying Equation (1.6) and Equation (1.7), the resultant of this loading and its point of application $(\bar{x}, \bar{y})$ can be determined from

$$F_R = \int p(x, y)dA \tag{1.8}$$

$$\bar{x} = \frac{\int x\, p(x, y)dA}{\int p(x, y)dA} \qquad \bar{y} = \frac{\int y\, p(x, y)dA}{\int p(x, y)dA} \tag{1.9}$$

Geometrically, F_R is equivalent to the volume under the loading diagram, and its location passes through the centroid or geometric center of this volume. Often in engineering practice, the surface loading is symmetric about an axis, in which case the loading is a function of only one coordinate, $w = w(x)$. Here the resultant is geometrically equivalent to the area under the loading curve, and the line of action of the resultant passes through the centroid of this area.

Besides surface forces as discussed above, loadings can also be transmitted to another body without direct physical contact. These body forces are distributed throughout the volume of the body. A common

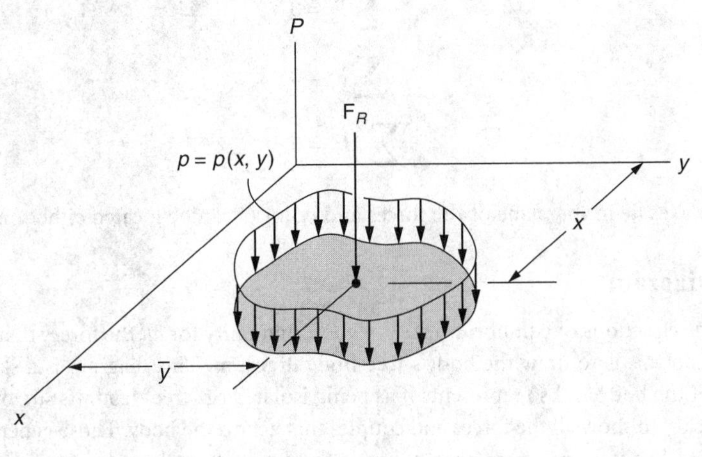

FIGURE 1.10

example is the force of gravity. The resultant of this force is termed the **weight**; it acts through the body's center of gravity and is directed towards the center of the earth.

1.2 Equilibrium

Equations of Equilibrium

A body is said to be in equilibrium when it either is at rest or moves with constant velocity. For purposes of analysis, it is assumed that the body is perfectly rigid, meaning that the particles composing the body remain at fixed distances from one another both before and after applying the load. Most engineering materials deform only slightly under load, so that moment arms and the orientation of the loading remain essentially constant. For these cases, therefore, the rigid-body model is appropriate for analysis. The necessary and sufficient conditions to maintain equilibrium of a rigid body require the resultant external force and moment acting on the body to be equal to zero. From Equation (1.6) and Equation (1.7), this can be expressed mathematically as

$$\sum \mathbf{F} = \mathbf{0} \tag{1.10}$$

$$\sum \mathbf{M}_O = \mathbf{0} \tag{1.11}$$

If the forces acting on the body are resolved into their *x, y, z* components, these equations can be written in the form of six scalar equations, namely,

$$\begin{aligned} \sum F_x &= 0 \qquad & \sum M_{Ox} &= 0 \\ \sum F_y &= 0 \qquad & \sum M_{Oy} &= 0 \\ \sum F_z &= 0 \qquad & \sum M_{Oz} &= 0 \end{aligned} \tag{1.12}$$

Actually, any set of three nonorthogonal, nonparallel axes will be suitable references for either of these force or moment summations.

If the forces on the body can be represented by a system of coplanar forces, then only three equations of equilibrium must be satisfied, namely,

$$\begin{aligned} \sum F_x &= 0 \\ \sum F_y &= 0 \\ \sum M_O &= 0 \end{aligned} \tag{1.13}$$

Here the x and y axes lie in the plane of the forces and point O can be located either on or off the body.

Free-Body Diagram

Application of the equations of equilibrium requires accountability for *all* the forces that act on the body. The best way to do this is to draw the body's **free-body diagram.** This diagram is a sketch showing an outlined shape of the body and so represents it as being isolated or "free" from its surroundings. On this sketch it is necessary to show all the forces and couples that act on the body. Those generally encountered are due to applied loadings, reactions that occur at the supports and at points of contact with other bodies, and the weight of the body. Also one should indicate the dimensions of the body necessary for

TABLE 1.1 Force Systems

Connection	Reaction	Connection	Reaction
 cable	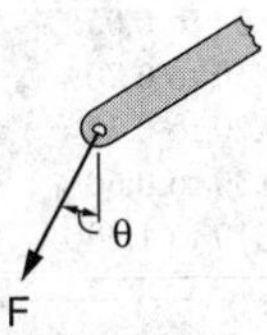	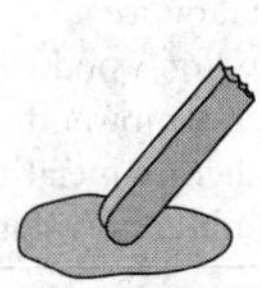 smooth surface	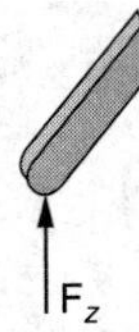
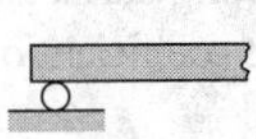 roller	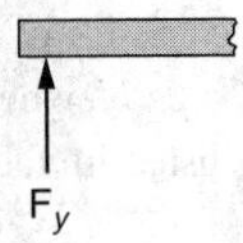	 ball and socket	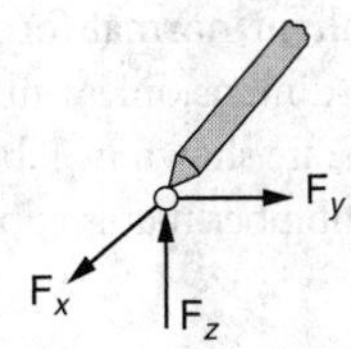
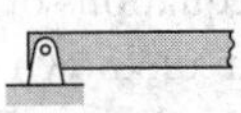 pin	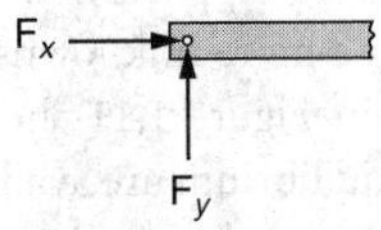	 single pin	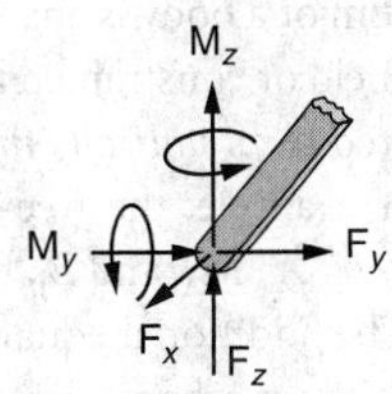
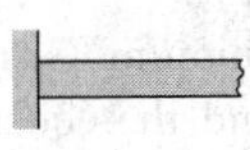 fixed support	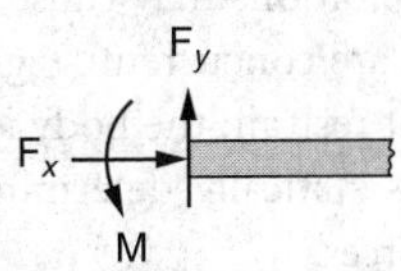	 fixed support	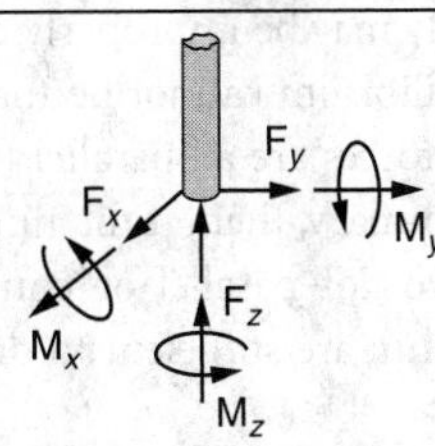

computing the moments of forces. Once the free-body diagram has been drawn and the coordinate axes established, application of the equations of equilibrium becomes a straightforward procedure.

Support Reactions

Various types of supports can be used to prevent a body from moving. Table 1.1 shows some of the most common types, along with the reactions each exerts on the body at the connection. As a general rule, if a support prevents translation in a given direction, then a force is developed on the body in that direction, whereas if rotation is prevented, a couple moment is exerted on the body.

Friction

When a body is in contact with a rough surface, a force of resistance called **friction** is exerted on the body by the surface in order to prevent or retard slipping of the body. This force always acts tangent to the surface at points of contact with the surface and is directed so as to oppose the possible or existing motion of the body. If the surface is dry, the frictional force acting on the body must satisfy the equation

$$F < \mu_s N \tag{1.14}$$

TABLE 1.2 Typical Values for Coefficients of Static Friction

Materials	μ_s
Metal on ice	0.03 to 0.05
Wood on wood	0.30 to 0.70
Leather on wood	0.20 to 0.50
Leather on metal	0.30 to 0.60
Aluminum on aluminum	1.10 to 1.70

The equality $F = \mu_s N$ applies only when motion between the contacting surfaces is impending. Here N is the resultant normal force on the body at the surface of contact, and μ_s is the coefficient of static friction, a dimensionless number that depends on the characteristics of the contacting surfaces. Typical values of μ_s are shown in Table 1.2. If the body is sliding, then $F = \mu_k N$, where μ_k is the coefficient of kinetic friction, a number that is approximately 25% smaller than those listed in Table 1.2.

Constraints

Equilibrium of a body is ensured not only by satisfying the equations of equilibrium, but also by its being properly held or constrained at its supports. If a body has more supports than are needed for equilibrium, it is referred to as *statically indeterminate,* since there will be more unknowns than equations of equilibrium. For example, the free-body diagram of the beam in Figure 1.11 shows four unknown support reactions, A_x, A_y, M_A, and B_y, but only three equations of equilibrium are available for solution [Equation (1.13)]. The additional equation needed requires knowledge of the physical properties of the body and deals with the mechanics of deformation, which is discussed in subjects such as mechanics of materials.

A body may be improperly constrained by its supports. When this occurs, the body becomes unstable and equilibrium cannot be maintained. Either of two conditions may cause this to occur — when the reactive forces are all parallel (Figure 1.12) or when they are concurrent (Figure 1.13).

In summary, then, if the number of reactive forces that restrain the body is a minimum — and these forces are not parallel or concurrent — the problem is statically determinate, and the equations of equilibrium are sufficient to determine all the reactive forces.

Internal Loadings

The equations of equilibrium can also be used to determine the internal resultant loadings in a member, provided the external loads are known. The calculation is performed using the **method of sections,** which

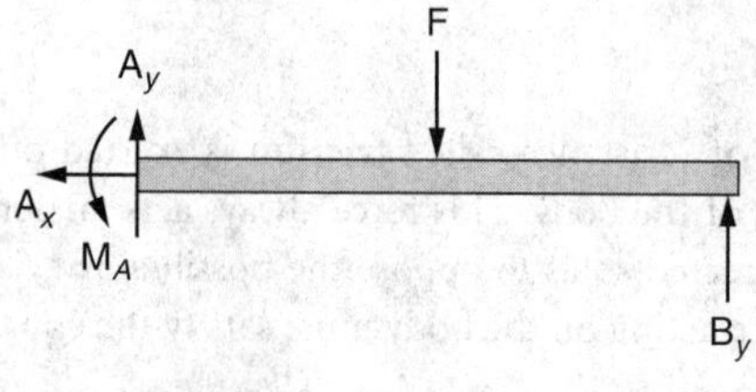

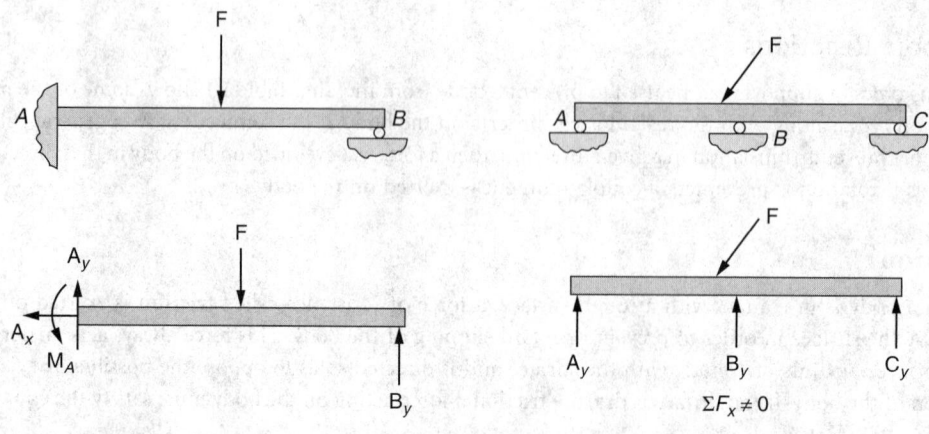

FIGURE 1.11

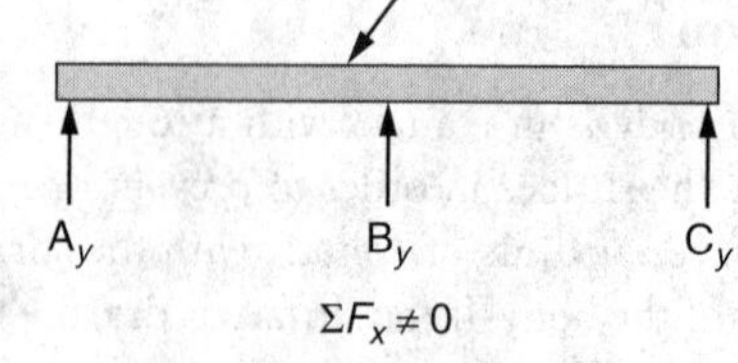

FIGURE 1.12

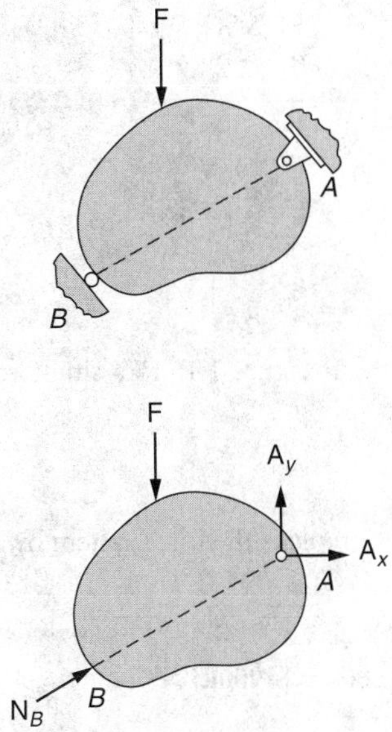

FIGURE 1.13

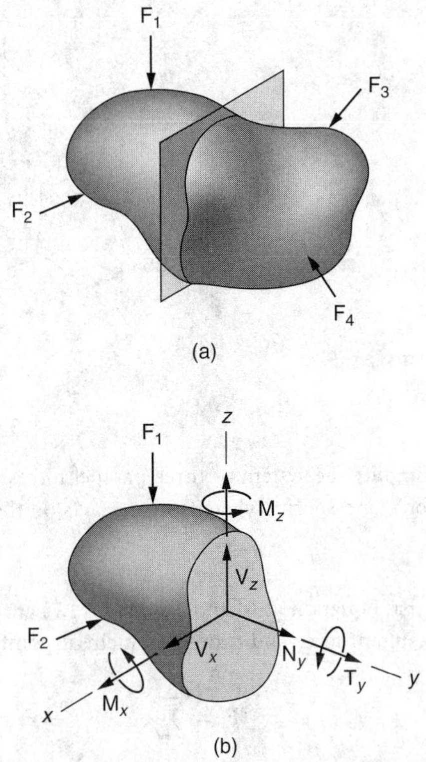

FIGURE 1.14

states that if a body is in equilibrium, then so is any segment of the body. For example, if an imaginary section is passed through the body in Figure 1.14(a), separating it into two parts, the free-body diagram of the left part is shown in Figure 1.14(b). Here the six internal resultant components are "exposed" and can be determined from the six equations of equilibrium given by Equation (1.12). These six components are referred to as the normal force, N_y, the shear-force components, V_x and V_z, the torque or twisting moment, T_y, and the bending-moment components, M_x and M_z.

If only coplanar loads act on the body [Figure 1.15(a)], then only three internal resultant loads occur [Figure 1.15(b)], namely, the normal force, *N*, the shear force, *V*, and the bending moment, *M*. Each of these loadings can be determined from Equation (1.13). Once these internal resultants have been computed, the actual load distribution over the sectioned surface, called stress, involves application of the theory related to mechanics of materials.

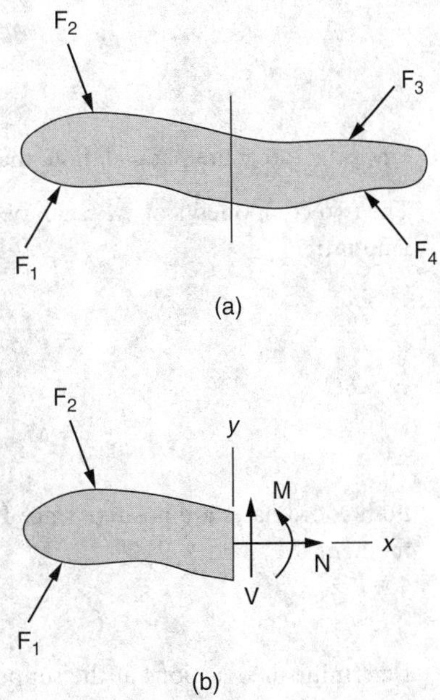

FIGURE 1.15

Numerical Applications

The following examples illustrate application of most of the principles discussed above. Solution of any problem generally requires first establishing a coordinate system, then representing the data on a diagram, and finally applying the necessary equations for solution.

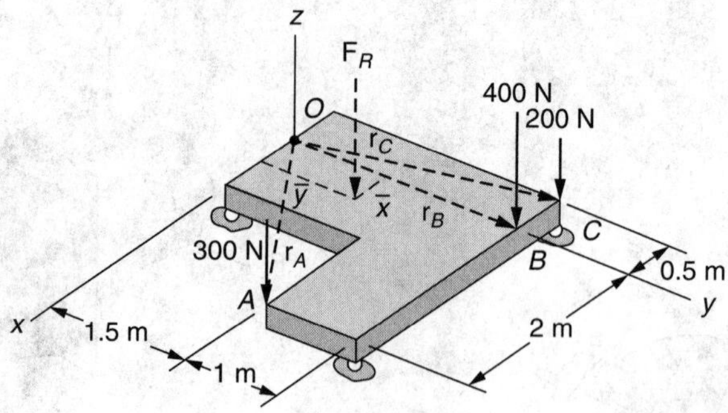

FIGURE 1.16

Example 1.1

Simplify the system of three parallel forces acting on the plate in Figure 1.16 to a single resultant force and specify where the force acts on the plate.

Solution

First, Equation (1.6) and Equation (1.7) are applied in order to replace the force system by a single resultant force and couple moment at point *O*.

$$\mathbf{F}_R = \sum \mathbf{F} \qquad \mathbf{F}_R = -300\mathbf{k} - 400\mathbf{k} - 200\mathbf{k} = \{-900\mathbf{k}\}\ \mathbf{N} \qquad \textit{Ans.}$$

$$\mathbf{M}_{R_O} = \sum \mathbf{M}_O \qquad \begin{aligned} \mathbf{M}_{R_O} &= \mathbf{r}_A \times (-300\mathbf{k}) + \mathbf{r}_B \times (-400\mathbf{k}) + \mathbf{r}_C \times (-200\mathbf{k}) \\ &= (2\mathbf{i} + 1.5\mathbf{j}) \times (-300\mathbf{j}) + (2.5\mathbf{j}) \times (-400\mathbf{k}) \\ &\quad + (-0.5\mathbf{i} + 2.5\mathbf{j}) \times (-200\mathbf{k}) \\ &= \{-1950\mathbf{i} + 500\mathbf{j}\}\mathrm{N \cdot m} \end{aligned}$$

Since the forces are parallel, note that as expected $\mathbf{F}_R$ is perpendicular to M_{R_O}.

The two components of M_{R_O} can be eliminated by moving $\mathbf{F}_R$ along the respective *y* and *x* axes an amount:

$$\bar{x} = M_{Oy} / F_R = 500\ \mathrm{N \cdot m} / 900\ \mathrm{N} = 0.556\ \mathrm{m} \qquad \textit{Ans.}$$

$$\bar{y} = M_{Ox} / F_R = 1950\ \mathrm{N \cdot m} / 900\ \mathrm{N} = 2.17\ \mathrm{m} \qquad \textit{Ans.}$$

Both coordinates are positive since $\mathbf{F}_R$, acting at $\mathbf{r} = \{0.556\mathbf{i} + 2.17\mathbf{j}\}$ m, will produce the required moment $\mathbf{M}_{R_O} = \mathbf{r} \times \mathbf{F}_R$.

Example 1.2

Determine the reactions at the supports for the beam shown in Figure 1.17(a).

Solution

Using Table 1.1, the free-body diagram for the beam is shown in Figure 1.17(b). The problem is statically determinate. The reaction $\mathbf{N}_B$ can be found by using the principle of moments and

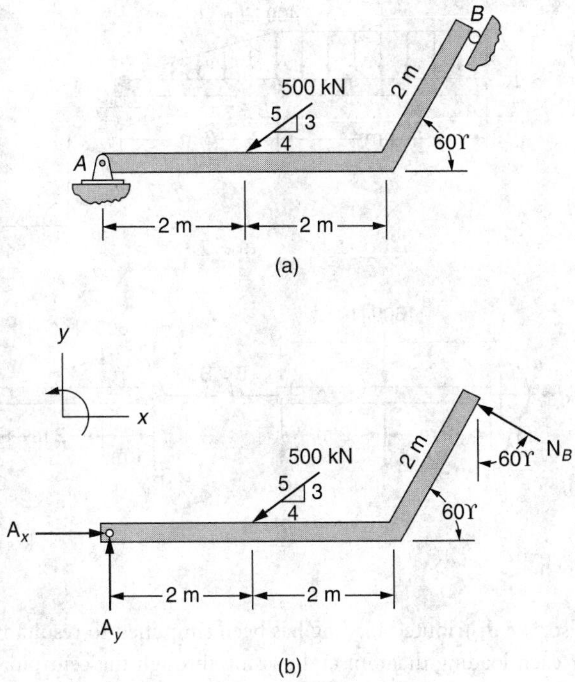

FIGURE 1.17

summing moments about point A to eliminate $\mathbf{A}_x$ and $\mathbf{A}_y$. Applying Equation (1.13) with reference to the coordinate system shown gives

$$\sum M_A = 0 \qquad -500\ \text{N}(3/5)(2\ \text{m}) + N_B \cos 60°(4\ \text{m} + 2\cos 60°\ \text{m}) + N_B \cos 60°(2 \sin 60°\ \text{m}) = 0$$

$$N_B = 150\ \text{N} \qquad \textit{Ans.}$$

$$\sum F_x = 0 \qquad A_x - 500\ \text{N}(4/5) - 150 \sin 60°\ \text{N} = 0$$

$$A_x = 530\ \text{N} \qquad \textit{Ans.}$$

$$\sum F_y = 0 \qquad A_y - 500\ \text{N}(3/5) + 150 \cos 60°\ \text{N} = 0$$

$$A_y = 225\ \text{N} \qquad \textit{Ans.}$$

Since the answers are all positive, the assumed sense of direction of the reactive forces is shown correctly on the free-body diagram.

Example 1.3

The compound beam shown in Figure 1.18(a) consists of two segments, AB and BC, which are pinned together at B. Determine the reactions on the beam at the supports.

Solution

The free-body diagrams of both segments of the beam are shown in Figure 1.18(b). Notice how the principle of action — equal but opposite reaction, Newton's third law — applies to the two force

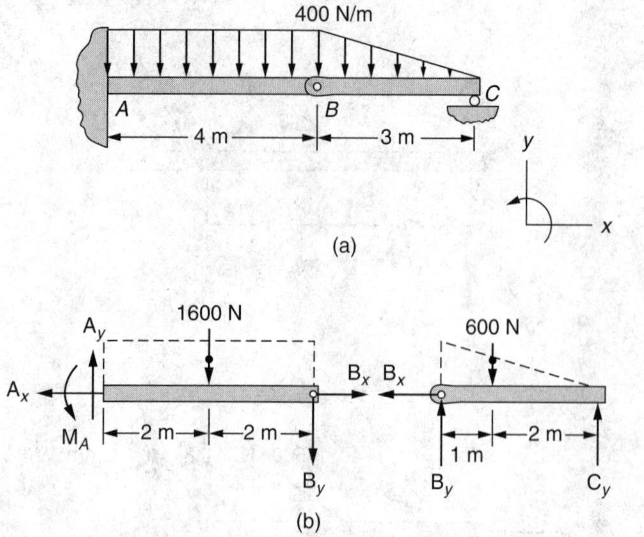

FIGURE 1.18

components at *B*. Also, the distributed loading has been simplified to resultant forces, determined from the area under each loading diagram and passing through the centroid or geometric center of each area.

The six unknowns are determined by applying Equation (1.13) to each segment. For segment *BC*:

$$\sum F_x = 0 \qquad B_x = 0 \qquad \textit{Ans.}$$

$$\sum M_B = 0 \qquad -600\ \text{N}(1\ \text{m}) + C_y(3\ \text{m}) = 0$$
$$C_y = 200\ \text{N} \qquad \textit{Ans.}$$

$$\sum F_y = 0 \qquad B_y - 600\ \text{N} + 200\ \text{N} = 0$$
$$B_y = 400\ \text{N} \qquad \textit{Ans.}$$

For segment *AB*:

$$\sum F_x = 0 \qquad A_x = 0 \qquad \textit{Ans.}$$

$$\sum F_y = 0 \qquad A_y - 1600\ \text{N} - 400\ \text{N} = 0$$
$$A_y = 2000\ \text{N} \qquad \textit{Ans.}$$

$$\sum M_A = 0 \qquad M_A = 1600\ \text{N}(2\ \text{m}) - 400\ \text{N}(4\ \text{m}) = 0$$
$$M_A - 4800\ \text{N} \cdot \text{m} \qquad \textit{Ans.}$$

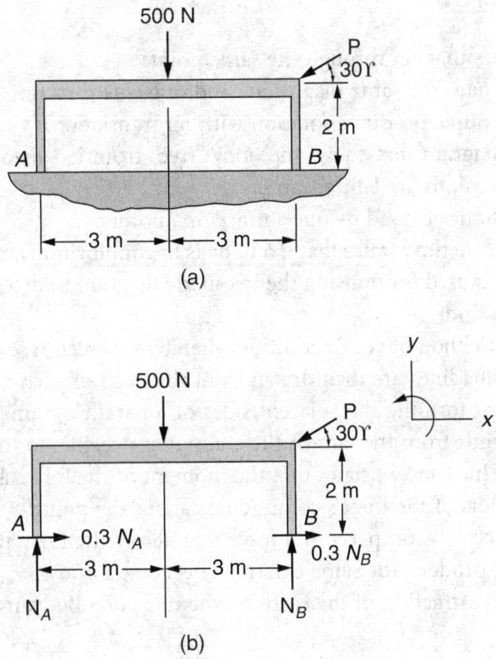

FIGURE 1.19

Example 1.4

The table in Figure 1.19(a) rests on a rough surface for which $\mu_s = 0.3$. If it supports a load of 500 N, determine the largest magnitude of force **P** that can be applied before it begins to move.

Solution

The free-body diagram is shown in Figure 1.19(b). Since the maximum force P is to be determined, slipping must impend at both A and B. Therefore, the friction equation $F = \mu_s N$ applies at these points. There are three unknowns. Applying the equations of equilibrium yields

$$\sum M_B = 0 \qquad -N_A(6 \text{ m}) + 500 \text{ N}(3 \text{ m}) + P\cos 30°(2 \text{ m}) = 0$$

$$\sum F_x = 0 \qquad 0.3N_A + 0.3N_B - P\cos 30° = 0$$

$$\sum F_y = 0 \qquad N_A + N_B - 500 \text{ N} - P\sin 30° = 0$$

Solving,

$$P = 209 \text{ N}$$

$$N_A = 310 \text{ N} \qquad \textit{Ans.}$$

$$N_B = 294 \text{ N}$$

Since N_A and N_B are both positive, the forces of the floor push up on the table as shown on the free-body diagram, and the table remains in contact with the floor.

Defining Terms

Concurrent forces — Forces that act through the same point.

Couple — Two forces that have the same magnitude and opposite directions and do not have the same line of action. A couple produces rotation with no translation.

Free-body diagram — A diagram that shows the body "free" from its surroundings. All possible loads and relevant dimensions are labeled on it.

Friction — A force of resistance caused by one surface on another.

Method of sections — This method states that if a body is in equilibrium, any sectioned part of it is also in equilibrium. It is used for drawing the free-body diagram to determine the internal loadings in any region of a body.

Parallelogram law — The method of vector addition whereby two vectors, called *components*, are joined at their tails; parallel lines are then drawn from the head of each vector so that they intersect at a common point forming the adjacent sides of a parallelogram. The resultant vector is the diagonal that extends from the tails of the component vectors to the intersection of the lines.

Principle of moments — This concept states that the moment of the force about a point is equal to the sum of the moments of the force's components about the point.

Principle of transmissibility — A property of a force that allows the force to act at any point along its line of action and produce the same external effects on a body.

Weight — The gravitational attraction of the earth on the mass of a body, usually measured at sea level and 45° latitude.

Wrench — A force and collinear moment. The effect is to produce both a push and simultaneous twist.

Reference

Hibbeler, R. C. 2004. *Engineering Mechanics: Statics,* 10th ed. Prentice Hall, Englewood Cliffs, NJ.

Further Information

Many textbooks are available for the study of statics; they can be found in any engineering library.

2

Centroids and Distributed Forces

Walter D. Pilkey
University of Virginia

L. Kitis
University of Virginia

2.1 Centroid of a Plane Area

Any set of forces acting on a rigid body is reducible to an equivalent force-couple system at any selected point *O*. This force-couple system, which consists of a force **R** equal to the vector sum of the set of forces and a couple of moment equal to the moment $\mathbf{M}_O$ of the forces about the point *O*, is equivalent to the original set of forces as far as the statics and dynamics of the entire rigid body are concerned. In particular, concurrent, coplanar, or parallel forces can always be reduced to a single equivalent force by an appropriate choice of the point *O*.

Consider, for example, a distributed load $p(x)$ acting on a straight beam (Figure 2.1). If $p(x)$ has units of force per length, the differential increment of force is $dR = p(x)\,dx$ and the total load *R* is found by integration:

$$R = \int p(x)\,dx \tag{2.1}$$

A point *O* is to be chosen such that the distributed load $p(x)$ is equivalent to a single resultant force of magnitude *R* acting at *O*. This requires that the moment of *R* about any point — say, point *A* — be the same as the moment of the load $p(x)$ about that point. Therefore the distance $\bar{x}$ between *A* and *O* is given by

$$\bar{x}R = \int x p(x)\,dx = \int x\,dR \tag{2.2}$$

The line of action of the resultant force *R* is a vertical line drawn at a distance $\bar{x}$ from point *A*. At any point on the line of action of *R*, the force-couple equivalent of the load $p(x)$ reduces to a single force.

0-8493-1586-7/05/$0.00+$1.50

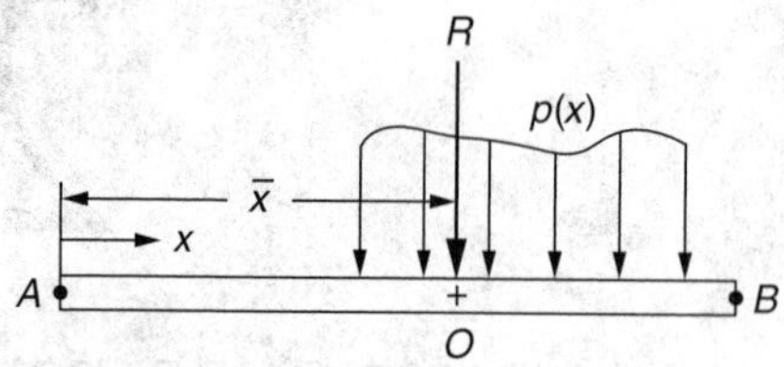

FIGURE 2.1 Distributed load on a beam.

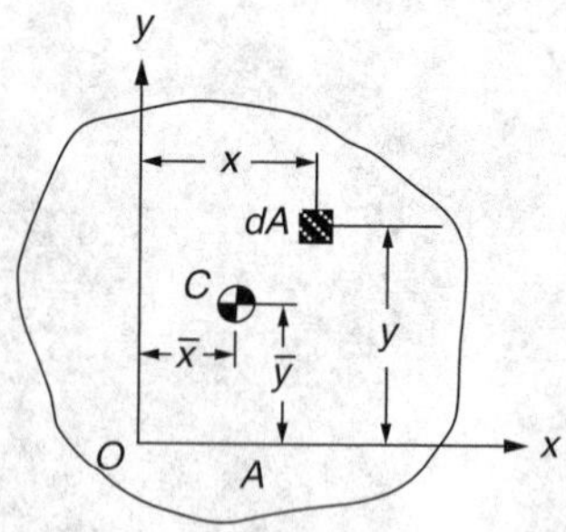

FIGURE 2.2 Centroid of an area.

This is a consequence of the *principle of transmissibility*, which states that moving a force along its line of action leaves the conditions of equilibrium or of motion of a rigid body unchanged.

Given a plane area A, the point with coordinates $\bar{x}$ and $\bar{y}$ defined by

$$\bar{x}A = \int_A x\,dA \qquad \bar{y}A = \int_A y\,dA \tag{2.3}$$

is known as the *centroid* of the area A. In the calculation of the resultant force R on a beam, the increment of force $dR = p(x)\,dx$ is a differential element of area under the load curve $p(x)$, and the length $\bar{x}$ given by Equation (2.2) is the x coordinate of the centroid of the area under the load curve. Thus, a distributed load on a beam can be replaced by a single force whose magnitude is equal to the area under the load curve $p(x)$ and whose line of action passes through the centroid of the area under $p(x)$.

Suppose that the homogeneous plate of uniform thickness t shown in Figure 2.2 is in a uniform and parallel field of force due to the earth's gravitational attraction in the negative y direction. The resultant of the gravitational forces is a single force in the same direction. The magnitude of this force is called the *weight*,

$$W = \gamma t A \tag{2.4}$$

where γ is the weight per unit volume of the material and A is the area of the plate. Since the gravitational forces and their single force equivalent must have the same moment about any point,

$$\int_A x\gamma t\,dA = M_O = \bar{x}W = \bar{x}\gamma tA \tag{2.5}$$

or, canceling the constant common factor γt on both sides,

$$\bar{x}A = \int_A x\,dA \tag{2.6}$$

Similarly, if the earth's attraction is assumed to be in the x direction, the line of action of the weight W, acting in the same direction, is specified by the coordinate $\bar{y}$, given by

$$\bar{y}A = \int_A y\,dA \tag{2.7}$$

The coordinates $\bar{x}$ and $\bar{y}$ locate the centroid of the area A of the plate. Thus, the centroid and the **center of gravity** of a homogeneous plate in a parallel and uniform gravitational field are coincident.

It can be shown that the coordinates $\bar{x}$ and $\bar{y}$ given by Equation (2.3) define the same geometric point regardless of the choice of coordinate axes. Thus, the location of the centroid is independent of any particular choice of orientations for the axes and of the choice of origin. As a consequence, the centroid is a well-defined intrinsic geometric property of any object having a rigid shape. Table 3.1 in Chapter 3 lists the centroids of some common shapes.

2.2 Centroid of a Volume

If a rigid body of volume V is subjected to a distributed force of intensity **f** (force per unit volume), the force-couple equivalent of this distributed force at an arbitrarily selected origin O of coordinates can be determined from

$$\mathbf{R} = \int_V \mathbf{f}\, dV \tag{2.8}$$

$$\mathbf{M}_O = \int_V \mathbf{r} \times \mathbf{f}\, dV \tag{2.9}$$

where **r** is the position vector of volume elements dV measured from point O. In particular, if only one component of **f** is nonzero, the resulting parallel system of distributed body forces acting on the body is reducible to a single equivalent force. For example, with $f_x = f_y = 0$ and f_z nonzero, the force-couple equivalent at O has the components

$$R_x = 0 \qquad R_y = 0 \qquad R_z = \int_V f_z\, dV \tag{2.10}$$

and

$$M_x = \int_V y f_z\, dV \qquad M_y = -\int_V x f_z\, dV \qquad M_z = 0 \tag{2.11}$$

The single force equivalent of this force-couple system is found by moving the force R_z from point O to a point in the xy plane such that the moment of R_z about O is equal to the moment of the force-couple system about O. The coordinates $\bar{x}$ and $\bar{y}$ of this point are therefore given by

$$\bar{x} R_z = \int_V x f_z\, dV \qquad \bar{y} R_z = \int_V y f_z\, dV \tag{2.12}$$

The arbitrary choice of $z = 0$ for the point of application of the resultant R_z is permissible because R_z can be slid along its line of action according to the principle of transmissibility.

If the rigid body is homogeneous and f_z is its specific weight in a uniform gravitational field, $f_z\, dV$ is an incremental weight dW and R_z is the total weight W of the body. Equation (2.12) becomes

$$\bar{x} W = \int_V x\, dW \qquad \bar{y} W = \int_V y\, dW \tag{2.13}$$

However, since the specific weight f_z is constant and

$$dW = f_z\, dV \qquad W = f_z V \tag{2.14}$$

Equation (2.13) can be written as

$$\bar{x}V = \int_V x\,dV \qquad \bar{y}V = \int_V y\,dV \tag{2.15}$$

Similarly, if the body is placed in a uniform gravitational field that exerts a force in the x direction only, then the line of action of the single equivalent force penetrates the yz plane at $\bar{y}$ defined in Equation (2.15) and $\bar{z}$ given by

$$\bar{z}V = \int_V z\,dV \tag{2.16}$$

The point with coordinates $\bar{x}, \bar{y}, \bar{z}$ defined in Equation (2.15) and Equation (2.16) is the centroid C of the volume V of the body. The centroid and the **center of gravity** of a homogeneous body in a uniform and parallel gravitational field are coincident points. If the body is not homogeneous, Equation (2.15) and Equation (2.16) cannot be used to find the center of gravity; however, they still define the centroid, which is an intrinsic geometric property of any rigid shape. The **center of mass** of a rigid body is defined by

$$\bar{\mathbf{r}}m = \int \mathbf{r}\,dm \tag{2.17}$$

where m is the mass. The center of mass depends solely on the mass distribution and is independent of the properties of the gravitational field in which the body may be placed. In a uniform and parallel gravitational field, however, $dW = g\,dm$ where g is the gravitational constant, so that the center of mass coincides with the center of gravity. If, in addition, the body is homogeneous, the x, y, z components of the vector $\bar{\mathbf{r}}$ given by Equation (2.17) are $\bar{x}, \bar{y}, \bar{z}$ defined in Eq. (2.15) and Equation (2.16). In this case, the centroid, the center of mass, and the center of gravity are coincident. The mass centers of some homogeneous solids are listed in Table 3.2, Chapter 3.

2.3 Surface Forces

Suppose the distributed load is a force per unit area $\mathbf{p}$ and let $\mathbf{p}$ be a function of the position vector $\mathbf{r}$ with respect to an origin O. For a surface of area A, the resultant of the distributed surface forces is

$$\mathbf{R} = \int_A \mathbf{p}\,dA \tag{2.18}$$

where dA is a differential surface element. The resultant moment of the distributed surface forces with respect to the reference point O is given by

$$\mathbf{M}_O = \int_A \mathbf{r} \times \mathbf{p}\,dA \tag{2.19}$$

The line of action of a single force equivalent for a parallel surface force distribution is determined as in the case of volume forces. For example, with a plane surface in the xy plane and with p_z as the only nonzero force, the line of action of the resultant force intersects the xy plane at the point whose coordinates are $\bar{x}, \bar{y}$ given by

$$\bar{x} = \frac{\int_A x p_z\,dA}{\int_A p_z\,dA} \qquad \bar{y} = \frac{\int_A y p_z\,dA}{\int_A p_z\,dA} \tag{2.20}$$

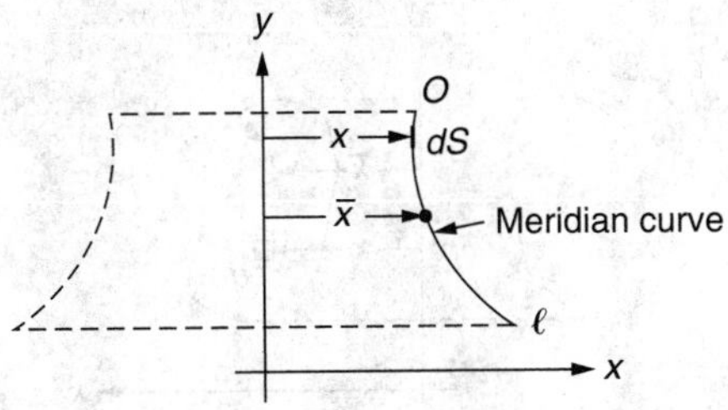

FIGURE 2.3 A surface with rotational symmetry.

2.4 Line Forces

In general, the formula for body forces can be employed for a line. For the special case of a plane curve of length l in the xy plane and distributed line force p_z (force/length) acting in the z direction, the force resultant is

$$R_z = \int_l p_z(s)\, ds \tag{2.21}$$

where s is the coordinate along the curve. The moment resultant about the reference point O is

$$\mathbf{M}_O = \int_l (y p_z \mathbf{i} - x p_z \mathbf{j})\, ds \tag{2.22}$$

where **i, j** are unit vectors in the x, y directions, respectively. The coordinates where the line of action of the single force resultant intersects the xy plane are

$$\bar{x} = \frac{\int_l x(s) p_z(s)\, ds}{\int_l p_z(s)\, ds} \qquad \bar{y} = \frac{\int_l y(s) p_z(s)\, ds}{\int_l p_z(s)\, ds} \tag{2.23}$$

2.5 Calculation of Surface Area and Volume of a Body with Rotational Symmetry

The two **Pappus–Guldin formulas**[1] take advantage of the definition of the centroid to assist in the calculation of the surface area and the volume of a body with rotational symmetry.

When a plane meridian curve (Figure 2.3) that does not intersect the y axis is rotated through ϕ radians about the y axis, a line element ds generates a surface of area

$$dS = x\phi\, ds \tag{2.24}$$

The total surface area generated by a meridian curve of length l is

[1]Pappus of Alexandria was a 3rd-century Greek geometer. Of a collection of eight mathematical books, only a portion survived. This is an informative source on ancient Greek mathematics. Included in this collection is a method for the measurement of a surface area bounded by a spiral on a sphere.

Paul Habakuk Guldin (1577–1643) was a professor of mathematics in several Italian and Austrian Jesuit colleges. In 1635 he revealed a relationship between the volume and the area of a body of revolution.

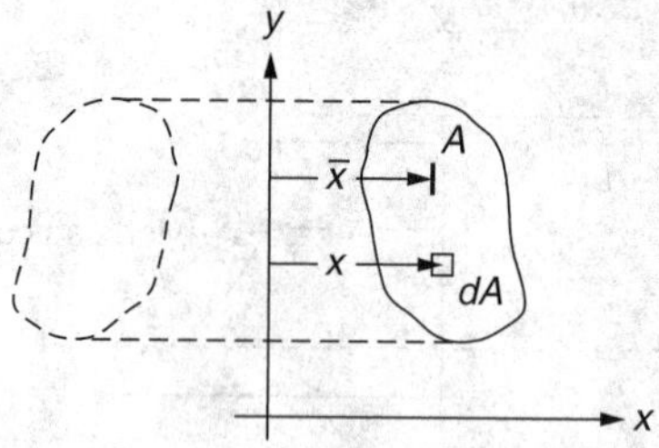

FIGURE 2.4 A volume with rotational symmetry.

$$S = \phi \int_0^l x\, ds \tag{2.25}$$

Therefore

$$S = \phi \bar{x} l \tag{2.26}$$

where $\bar{x}$ locates the centroid of the meridian curve. This is the first Pappus–Guldin formula.

If the meridian curve forms a shell of revolution, $\phi = 2\pi$ and

$$S = 2\pi \bar{x} l \tag{2.27}$$

When a plane area A (Figure 2.4) that is not cut by the y axis is rotated through an angle ϕ radians about the y axis, a volume element

$$dV = \phi x\, dA \tag{2.28}$$

is generated and the total volume is given by

$$V = \phi \bar{x} A \tag{2.29}$$

This is the second Pappus–Guldin formula.

For a solid of revolution, with $\phi = 2\pi$,

$$V = 2\pi \bar{x} A \tag{2.30}$$

If A and V are known, the centroid of A can be determined from

$$\bar{x} = \frac{V}{2\pi A} \tag{2.31}$$

2.6 Determination of Centroids

When an area or a line has an axis of symmetry, the centroid of the area or line is on that axis. If an area or line has two axes of symmetry, the centroid of the area or line is at the intersection of the axes of symmetry. Thus, the geometric centers of circles, ellipses, squares, rectangles, or lines in the shape of the perimeter of an equilateral triangle are also their centroids. When a volume has a plane of symmetry, the centroid of the volume lies on that plane. When a volume has two planes of symmetry, the centroid of the volume lies on the line of intersection of the planes of symmetry. When a volume has three planes of symmetry intersecting at a point, the point of intersection of the three planes is the centroid of the

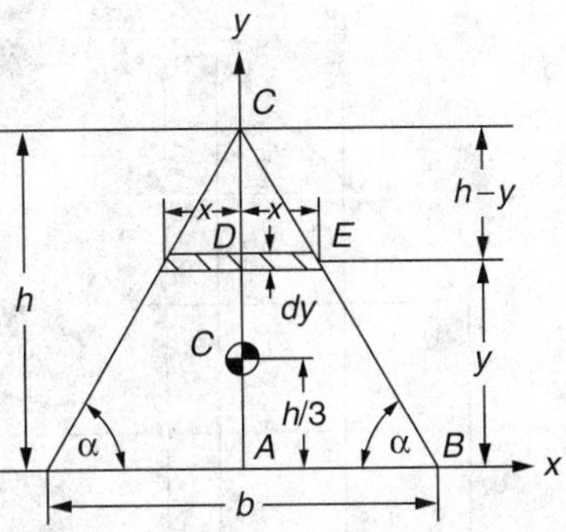

FIGURE 2.5 Centroid of a triangle.

volume. Thus, the geometric centers of spheres, ellipsoids, cubes, or rectangular parallelepipeds are also their centroids.

Centroids of unsymmetrical areas, lines, or volumes can be determined by direct integration. The Pappus–Guldin formulas can be used to determine the centroid of a plane curve when the area of the surface generated by the curve is known or to determine the centroid of a plane area when the volume generated by the area is known.

If a body can be divided into n parts, for which the volumes V_i and the centroids $\bar{x}_i, \bar{y}_i, \bar{z}_i$ are known, the centroid of the entire body has coordinates $\bar{x}, \bar{y}, \bar{z}$, given by

$$\bar{x}V = \sum_{i=1}^{n} \bar{x}_i V_i \qquad \bar{y}V = \sum_{i=1}^{n} \bar{y}_i V_i \qquad \bar{z}V = \sum_{i=1}^{n} \bar{z}_i V_i \tag{2.32}$$

where V is the total volume of the body. The same equations are applicable for a body with cutouts, provided that the volumes of the cutouts are taken as negative numbers.

Example 2.1

Determine the location of the centroid for the triangle shown in Figure 2.5. The triangle is symmetric about the y axis. Hence the centroid lies on the y axis and $\bar{x} = 0$. To determine $\bar{y}$, select an element of area dA parallel to the x axis, thus making all points in the element an equal distance from the x axis. Then $dA = 2\,x\,dy$ and, with $A = b\,h/2$,

$$\bar{y} = \frac{\int_0^h y(2x\,dy)}{bh/2} = \frac{4\int_0^h xy\,dy}{bh} \tag{2.33}$$

The expression for x in terms of y follows from the proportionality of the similar triangles ABC and DEC. This provides

$$\frac{x}{h-y} = \frac{b}{2h} \quad \text{or} \quad x = \frac{1}{2}\frac{b}{h}(h-y) \tag{2.34}$$

Finally,

$$\bar{y} = \frac{4\int_0^h \frac{1}{2}(b/h)(h-y)y\,dy}{bh} = \frac{2}{h^2}\int_0^h (hy - y^2)\,dy = \frac{2}{h^2}\left(h\frac{h^2}{2} - \frac{h^3}{3}\right) = \frac{h}{3} \tag{2.35}$$

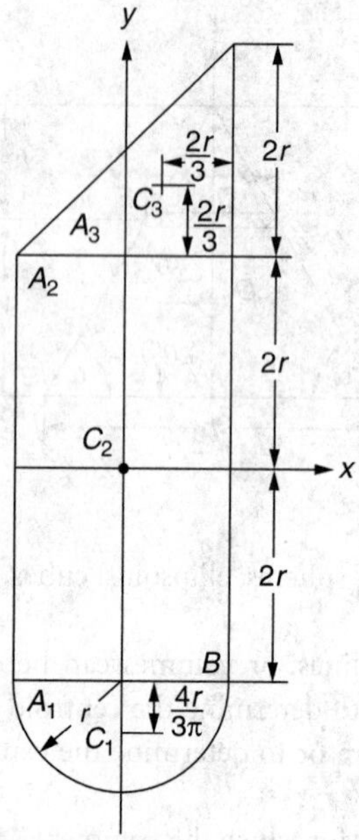

FIGURE 2.6 Centroid of a complicated area.

Thus, the centroid of the triangular area is on the y axis at a distance of one-third the altitude from the base of the triangle.

Example 2.2

Figure 2.6 shows a complicated shape that is made up of a semicircle, a rectangle, and a triangle. The areas and centroids of the semicircle, rectangle, and triangle are known:

$$A_1 = \frac{\pi r^2}{2} \qquad A_2 = 8r^2 \qquad A_3 = 2r^2 \tag{2.36}$$

$$C_1 = \left(0,\ -\frac{2(2+3\pi)r}{3\pi}\right) \qquad C_2 = (0,\ 0) \qquad C_3 = \left(\frac{r}{3},\ \frac{8r}{3}\right) \tag{2.37}$$

The x coordinate $\bar{x}$ of the centroid of the figure is given by

$$\bar{x}A = \sum_{i=1}^{3} \bar{x}_i A_i = \frac{2r^3}{3} \tag{2.38}$$

where A is the total area

$$A = A_1 + A_2 + A_3 = \frac{(\pi + 20)r^2}{2} \tag{2.39}$$

Therefore,

$$\bar{x} = \frac{4r}{3(\pi + 20)} \tag{2.40}$$

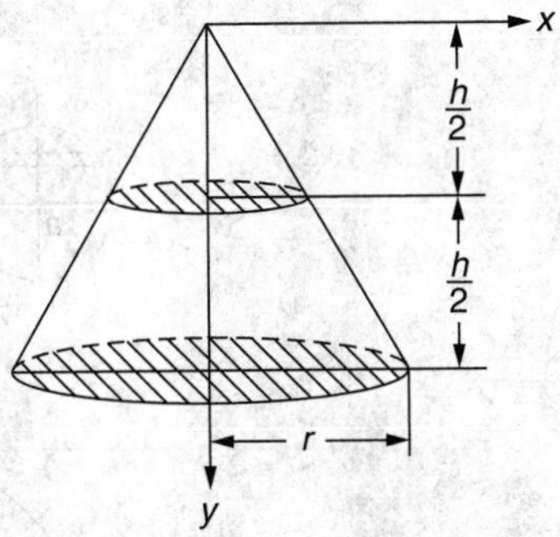

FIGURE 2.7 Frustum of a cone.

Similarly, the y coordinate $\bar{y}$ of the centroid is calculated from

$$\bar{y}A = \sum_{i=1}^{3} \bar{y}_i A_i \tag{2.41}$$

which yields

$$\bar{y} = \frac{2(14-3\pi)r}{3(\pi+20)} \tag{2.42}$$

Example 2.3

In this example, the y coordinate of the centroid of the frustum of the right circular cone shown in Figure 2.7 will be found by treating the frustum as a cone of height h with a cut-out cone of height $h/2$. From Table 3.2 of Chapter 3, the y coordinates of the centroids of these two cones are

$$y_1 = \frac{3h}{4} \qquad y_2 = \frac{3h}{8} \tag{2.43}$$

The corresponding volumes are

$$V_1 = \frac{\pi r^2 h}{3} \qquad V_2 = \frac{\pi r^2 h}{24} \tag{2.44}$$

and the y coordinate of the centroid of the frustum of the cone is

$$\bar{y} = \frac{y_1 V_1 - y_2 V_2}{V_1 - V_2} = \frac{45h}{56} \tag{2.45}$$

Example 2.4

The centroid of the area enclosed between the semicircles of radius R and r and part of the x axis (see Figure 2.8) can be found by an application of the second Pappus–Guldin formula. The volume generated by rotating this area through 2π is

$$V = \frac{4\pi}{3}(R^3 - r^3) \tag{2.46}$$

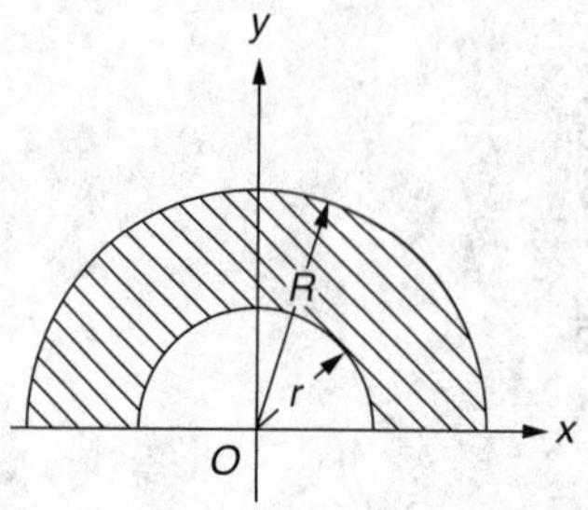

FIGURE 2.8 Plane area of Example 2.4.

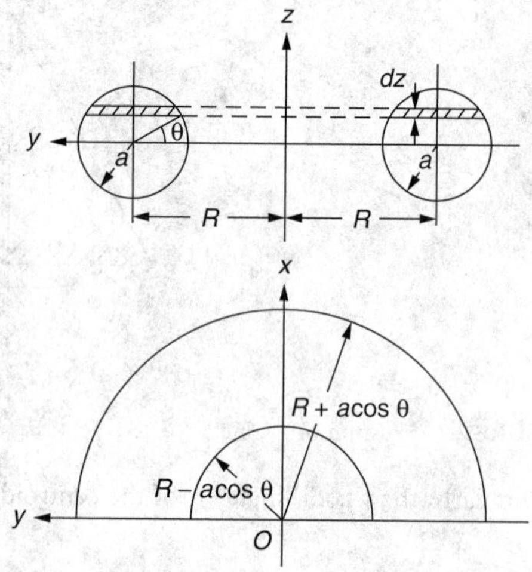

FIGURE 2.9 Example 2.5.

Hence

$$\bar{y} = \frac{V}{2\pi\pi(R^2 - r^2)/2} = \frac{4(r^2 + rR + R^2)}{3\pi(r + R)} \tag{2.47}$$

Example 2.5

Consider the problem of finding the centroid of the half torus in Table 3.2 of Chapter 3. The half torus is obtained by revolving a circle of a radius a through π radians about the z axis, and the second Pappus–Guldin formula gives its volume $V = \pi^2 a^2 R$. To calculate the x coordinate $\bar{x}$ of the centroid, a volume element obtained by taking a horizontal slice of thickness dz (see Figure 2.9) can be used. The volume of this element is given by

$$dV = \frac{\pi}{2}(r_0^2 - r_i^2)\, dz \tag{2.48}$$

where

$$\begin{aligned} r_i &= R - a\cos\theta \\ r_o &= R + a\cos\theta \end{aligned} \tag{2.49}$$

Since $dz = a \cos \theta \, d\theta$, the expression for dV simplifies to

$$dV = 2\pi a^2 R \cos^2\theta \, d\theta \tag{2.50}$$

It is known from Example 2.4 that the x coordinate of the centroid of the volume element is

$$x = \frac{4(r_i^2 + r_i r_o + r_o^2)}{3\pi(r_i + r_o)} = \frac{a^2(1 + \cos 2\theta) + 6R^2}{3\pi R} \tag{2.51}$$

Therefore, the x coordinate of the centroid of the half torus is

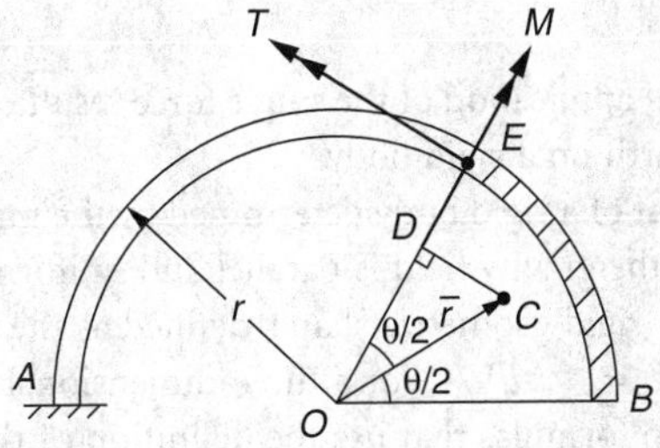

FIGURE 2.10 Semicircular cantilever beam.

$$\bar{x} = 2\frac{\int_0^{\pi/2} x\,dV}{V} = \frac{4}{3\pi^2 R}\int_0^{\pi/2}(a^2(1+\cos 2\theta)+6R^2)\cos^2\theta\,d\theta \tag{2.52}$$

in which the integration limits extend over the upper half of the volume so that the additional multiplicative factor of 2 is necessary. The result is

$$\bar{x} = \frac{a^2+4R^2}{2\pi R} \tag{2.53}$$

Example 2.6

Figure 2.10 shows a thin homogeneous semicircular cantilever beam of specific weight γ in a uniform gravitational field. Suppose that the only force acting on the beam is the gravitational force of attraction perpendicular to the plane of the figure. To determine the internal forces at any section E of the beam, the free-body diagram of the circular arc from B to E is used. The force distribution on this circular arc is a line force of constant intensity so that Equations (2.23), which give the coordinates of a point on the line of action of the single force resultant of a line force, show that the resultant passes through the centroid of the arc. The centroid C of the arc is at a distance $\bar{r}$ from the center O:

$$\bar{r} = \frac{r\sin\theta/2}{\theta/2} \tag{2.54}$$

The single force equivalent of the gravitational forces acting on the arc BE is the weight $\gamma r\theta$ acting at the centroid C in the direction perpendicular to the plane of the figure. The moment arm for the twisting moment T at E is DE and the moment arm for the bending moment M at E is DC. Hence

$$T = \gamma r\theta\left(r-\bar{r}\cos\frac{\theta}{2}\right) \tag{2.55}$$

$$M = \gamma r\theta\,\bar{r}\sin\frac{\theta}{2} \tag{2.56}$$

These expressions may be rewritten as

$$T = \gamma r^2(\theta-\sin\theta) \tag{2.57}$$

$$M = \gamma r^2(1-\cos\theta) \tag{2.58}$$

Defining Terms

Center of gravity — The point of application of the single force resultant of the distributed gravitational forces exerted by the earth on a rigid body.

Center of mass — A unique point of a rigid body determined by the mass distribution. It coincides with the center of gravity if the gravity field is parallel and uniform.

Centroid — A unique point of a rigid geometric shape, defined as the point with coordinates $\bar{x} = \int_V x \, dV/V$, $\bar{y} = \int_V y \, dV/V$, $\bar{z} = \int_V z \, dV/V$, for a three-dimensional solid of volume V.

Pappus–Guldin formulas — Two formulas that use the definition of the centroid to assist in the calculation of the area of a surface of revolution and the volume of a body of revolution.

References

Pilkey, W. D. 2004. *Formulas for Stress, Strain and Structural Matrices.* John Wiley & Sons, New York.
Pilkey, W. D. and Pilkey, O. H. 1986. *Mechanics of Solids.* Krieger, Malabar, FL.

Further Information

Consult undergraduate textbooks on statics, dynamics, and mechanics of solids.

3

Moments of Inertia

J. L. Meriam
University of California (Retired)

The **mass moment of inertia**, I, of a body is a measure of the inertial resistance of the body to rotational acceleration and is expressed by the integral $I = \int r^2\,dm$, where dm is the differential element of mass and r is the perpendicular distance from dm to the rotation axis. The **area moment of inertia** of a defined area about a given axis is expressed by the integral $I = \int s^2\,dA$, where dA is the differential element of area and s is the perpendicular distance from dA to a defined axis either in or normal to the plane of the area. The mathematical similarity to mass moment of inertia gives rise to its name. A more fitting but less used term is the *second moment of area.*

The frequent occurrence of area and mass moments of inertia in mechanics justifies establishing and tabulating their properties for commonly encountered shapes, as given in Table 3.1 and Table 3.2.

3.1 Area Moments of Inertia

Figure 3.1 illustrates the physical origin of the area moment-of-inertia integrals. In Figure 3.1(a) the surface area $ABCD$ is subject to a distributed pressure p whose intensity is proportional to the distance y from the axis AB. The moment about AB that is due to the pressure on the element of area dA is $y(p\,dA) = ky^2\,dA$. Thus the integral in question appears when the total moment $M = k\int y^2\,dA$ is evaluated.

Figure 3.1(b) shows the distribution of stress acting on a transverse section of a simple linear elastic beam bent by equal and opposite couples applied one to each end. At any section of the beam a linear distribution of force intensity or stress σ, given by $\sigma = ky$, is present, the stress being positive (tensile) below the axis O-O and negative (compressive) above the axis. The elemental moment about axis O-O is $dM = y(\sigma\,dA) = ky^2\,dA$. Thus the same integral appears when the total moment $M = k\int y^2\,dA$ is evaluated.

A third example is given in Figure 3.1(c), which shows a circular shaft subjected to a twist or torsional moment. Within the elastic limit of the material, this moment is resisted at each cross-section of the shaft by a distribution of tangential or shear stress τ that is proportional to the radial distance r from the center. Thus $\tau = kr$ and the total moment about the central axis becomes $M = \int r(\tau\,dA) = k\int r^2\,dA$. Here the integral differs from that in the preceding two examples in that the area is normal instead of parallel to the moment axis and in that r is a radial coordinate instead of a rectangular one.

0-8493-1586-7/05/$0.00+$1.50

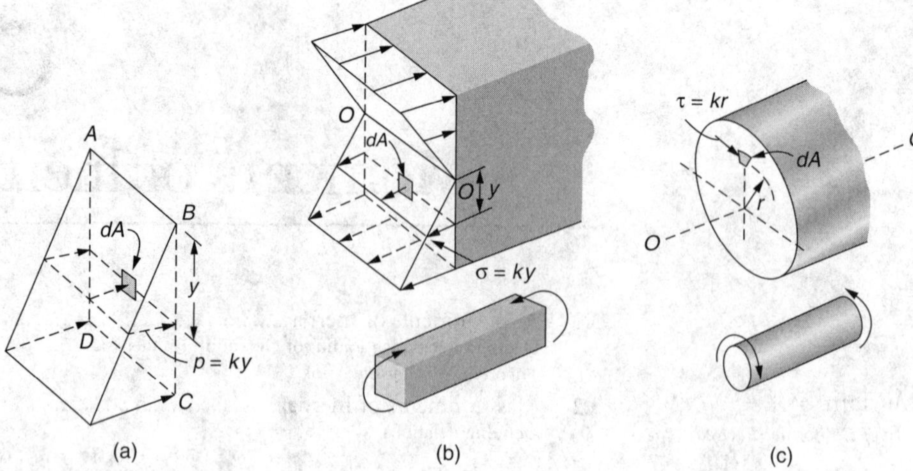

FIGURE 3.1 (*Source:* Meriam, J.L. and Kraige, L.G. 1992. *Engineering Mechanics*, 3rd ed. John Wiley & Sons, New York.)

Defining Relations

Rectangular and Polar Moments of Inertia

For area A in the xy plane, Figure 3.2, the moments of inertia of the element dA about the x and y axes are, by definition, $dI_x = y^2\, dA$ and $dI_y = x^2\, dA$, respectively, The moments of inertia of A about the same axes become

$$I_x = \int y^2 dA$$
$$I_y = \int x^2 dA \tag{3.1}$$

where the integration is carried out over the entire area.

The moment of inertia of dA about the pole O (z axis) is, by definition, $dI_z = r^2\, dA$, and the moment of inertia of the entire area about O is

$$I_z = \int r^2 dA \tag{3.2}$$

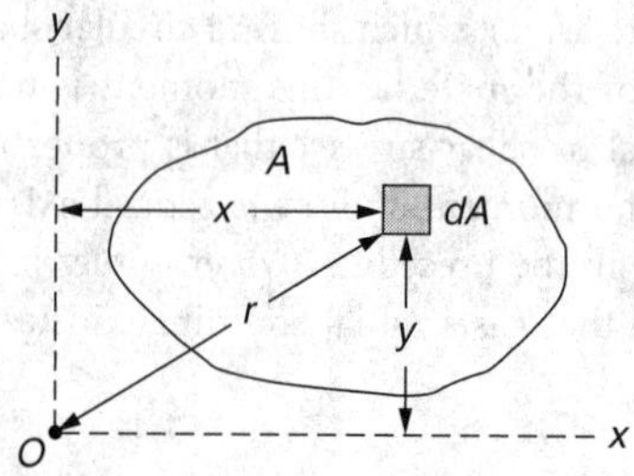

FIGURE 3.2 (*Source:* Meriam, J.L. and Kraige, L.G. 1992. *Engineering Mechanics*, 3rd ed. John Wiley & Sons, New York.)

The expressions defined by Equation (3.1) are known as *rectangular* moments of inertia, whereas the expression of Equation (3.2) is known as the *polar* moment of inertia. (In the literature the polar moment of inertia is sometimes denoted by the symbol *J*.) Because $x^2 + y^2 = r^2$, it follows that

$$I_z = I_x + I_y \tag{3.3}$$

A polar moment of inertia for an area whose boundaries are more simply described in rectangular coordinates than in polar coordinates is easily calculated using Equation (3.3).

Because the area moment of inertia involves distance squared, it is always a positive quantity for a positive area. (A hole or void may be considered a negative area.) In contrast, the first moment of area $\int y \, dA$ involves distance to the first power, so it can be positive, negative, or zero.

The dimensions of moments of inertia of areas are L^4, where L stands for the dimension of length. The SI units for moments of inertia of areas are expressed in quartic meters (m^4) or quartic millimeters (mm^4). The U.S. customary units are quartic feet (ft^4) or quartic inches ($in.^4$).

Rectangular coordinates should be used for shapes whose boundaries are most easily expressed in these coordinates. Polar coordinates will usually simplify problems where the boundaries are expressed in r and θ. The choice of an element of area that simplifies the integration as much as possible is equally important.

Radius of Gyration

Consider the area A, Figure 3.3(a), which has rectangular moments of inertia I_x and I_y and a polar moment of inertia I_z about O. If the area is visualized as being concentrated into a long narrow strip of area A a distance k_x from the x axis, Figure 3.3(b), by definition the moment of inertia of the strip about the x axis will be the same as that of the original area if $k_x^2 A = I_x$. The distance k_x is known as the **radius of gyration** of the area about the x axis. A similar relation for the y axis is found by considering the area to be concentrated into a narrow strip parallel to the y axis as shown in Figure 3.3(c). Also, by visualizing the area to be concentrated into a narrow ring of radius k_z, as shown in Figure 3.3(d), the polar moment of inertia becomes $k_z^2 A = I_z$. Summarizing,

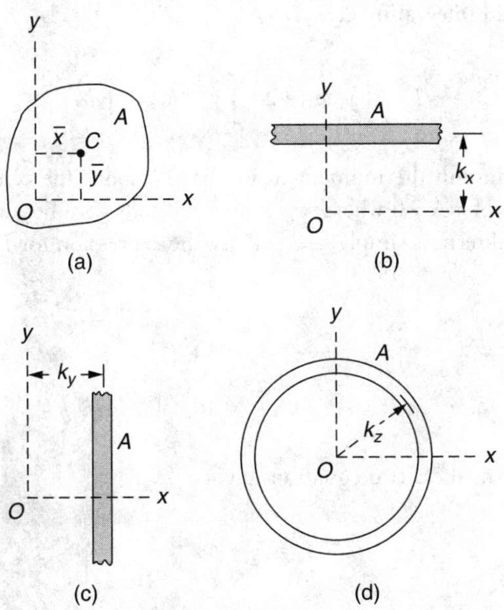

FIGURE 3.3 (*Source:* Meriam, J.L. and Kraige, L.G. 1992. *Engineering Mechanics,* 3rd ed. John Wiley & Sons, New York.)

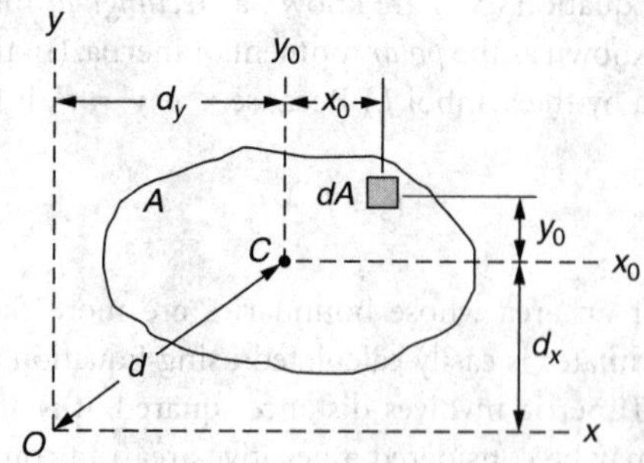

FIGURE 3.4 (*Source:* Meriam, J.L. and Kraige, L.G. 1992. *Engineering Mechanics*, 3rd ed. John Wiley & Sons, New York.)

$$I_x = k_x^2 A \qquad k_x = \sqrt{I_x / A}$$
$$I_y = k_y^2 A \qquad k_y = \sqrt{I_y / A} \tag{3.4}$$
$$I_z = k_z^2 A \qquad k_z = \sqrt{I_z / A}$$

A rectangular or polar moment of inertia may be expressed by specifying its radius of gyration and its area. Substituting Equation (3.4) into Equation (3.3) gives

$$k_z^2 = k_x^2 + k_y^2 \tag{3.5}$$

Parallel-Axis Theorem

The moment of inertia of an area about a noncentroidal axis may be easily expressed in terms of the moment of inertia about a parallel centroidal axis. In Figure 3.4 the x_0 and y_0 axes pass through the centroid C of the area. By definition the moment of inertia of the element dA about the x axis is $dI_x = (y_0 + d_x)^2\, dA$. Expanding and integrating give

$$I_x = \int y_0^2\, dA + 2d_x \int y_0\, dA + d_x^2 \int dA$$

The first integral is by definition the moment of inertia $\bar{I}_x$ about the centroidal x_0 axis. The second integral is zero because $\int y_0\, dA = A\, \bar{y}\, 0$ where $\bar{y}_0$ is automatically zero because the centroid for the area lies on the x_0 axis. The third term is simply Ad_x^2. Thus, the expression for I_x and the similar expression for I_y become

$$I_x = \bar{I}_x + Ad_x^2$$
$$I_y = \bar{I}_y + Ad_y^2 \tag{3.6}$$

By Equation (3.3), the sum of these two equations gives

$$I_z = \bar{I}_z + Ad^2 \tag{3.6a}$$

Equation (3.6) and Equation (3.6a) are the so-called *parallel-axis theorems.* It is noted that the axes between which transfer is made *must be parallel* and that one of the axes *must pass through the centroid* of the area. The parallel-axis theorems also hold for radii of gyration. Substitution of the definition of k into Equation (3.6) gives

$$k^2 = \bar{k}^2 + d^2 \tag{3.6b}$$

where $\bar{k}$ is the radius of gyration about the centroidal axis parallel to the axis about which k applies and d is the distance between the two axes. The axes may be either in or normal to the plane of the area.

A summary of formulas for area moments of inertia for various commonly encountered plane areas is given in Table 3.1.

Composite Areas

When an area is the composite of a number of distinct parts, its moment of inertia is obtained by summing the results for each of the parts in terms of its area A, its centroidal moment of inertia $\bar{I}$, the perpendicular distance d from its centroidal axis to the axis about which the moment of inertia of the composite area is being computed, and the product Ad^2. The results are easily tabulated in the form

Part	Area, A	$\bar{I}_x$	$\bar{I}_y$	d_x	d_y	Ad_x^2	Ad_y^2
Sums		$\sum \bar{I}_x$	$\sum \bar{I}_y$			$\sum Ad_x^2$	$\sum Ad_y^2$

The final results are simply

$$I_x = \sum \bar{I}_x + \sum Ad_x^2 \qquad I_y = \sum \bar{I}_y + \sum Ad_y^2$$

Products of Inertia

In certain problems involving unsymmetrical cross-sections, an expression of the form $dI_{xy} = xy\ dA$ occurs, and its integral

$$I_{xy} = \int xy\ dA \tag{3.7}$$

is known as the **product of inertia**. Unlike moments of inertia, which are always positive for positive areas, the product of inertia may be positive, negative, or zero, depending on the signs of x and y.

Rotation of Axes

It may be shown that the moments and products of inertia for the area of Figure 3.5 about the rotated axes x'- y' are given by

$$\begin{aligned} I_{x'} &= \frac{I_x + I_y}{2} + \frac{I_x - I_y}{2}\cos 2\theta - I_{xy}\sin 2\theta \\ I_{y'} &= \frac{I_x + I_y}{2} - \frac{I_x - I_y}{2}\cos 2\theta + I_{xy}\sin 2\theta \\ I_{x'y'} &= \frac{I_x - I_y}{2}\sin 2\theta + I_{xy}\cos 2\theta \end{aligned} \tag{3.8}$$

The angle that makes $I_{x'}$ and $I_{y'}$ a maximum or a minimum is determined by setting the derivative of $I_{x'}$ and $I_{y'}$ with respect to θ equal to zero. Denoting this critical angle by α gives

$$\tan 2\alpha = \frac{2I_{xy}}{I_y - I_x} \tag{3.9}$$

TABLE 3.1 Properties of Plane Areas

Figure	Centroid	Area Moments of Inertia
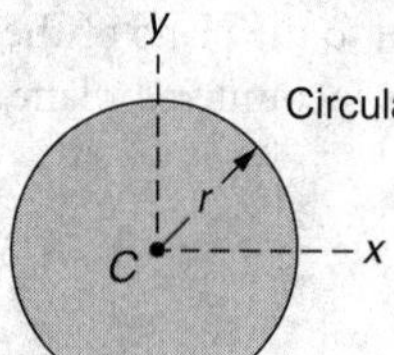 Circular area	—	$I_x = I_y = \frac{\pi r^4}{4}$ $I_z = \frac{\pi r^4}{2}$
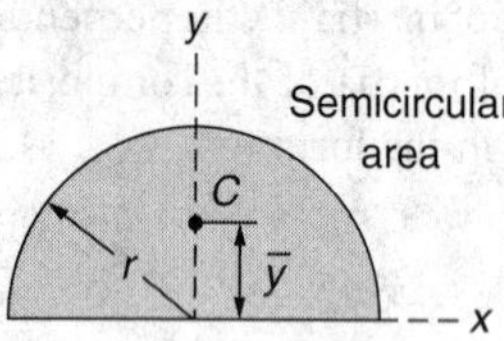 Semicircular area	$\bar{y} = \frac{4r}{3\pi}$	$I_x = I_y = \frac{\pi r^4}{8}$ $\bar{I}_x = \left(\frac{\pi}{8} - \frac{8}{9\pi}\right) r^4$ $I_z = \frac{\pi r^4}{4}$
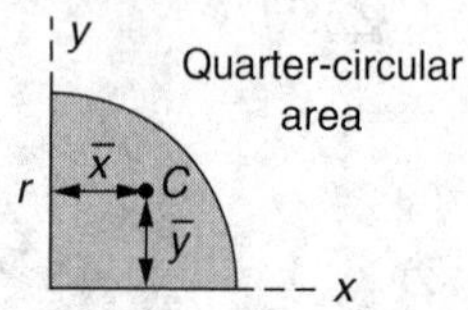 Quarter-circular area	$\bar{x} = \bar{y} = \frac{4r}{3\pi}$	$I_x = I_y = \frac{\pi r^4}{16}$ $\bar{I}_x = \bar{I}_y = \left(\frac{\pi}{16} - \frac{4}{9\pi}\right) r^4$ $I_z = \frac{\pi r^4}{8}$
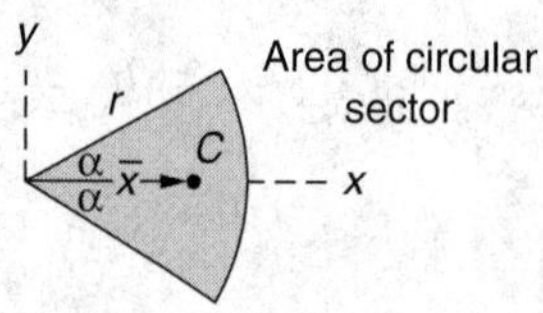 Area of circular sector	$\bar{x} = \frac{2}{3}\frac{r \sin\alpha}{\alpha}$	$I_x = \frac{r^4}{4}\left(\alpha - \frac{1}{2}\sin 2\alpha\right)$ $I_y = \frac{r^4}{4}\left(\alpha + \frac{1}{2}\sin 2\alpha\right)$ $I_z = \frac{1}{2} r^4 \alpha$
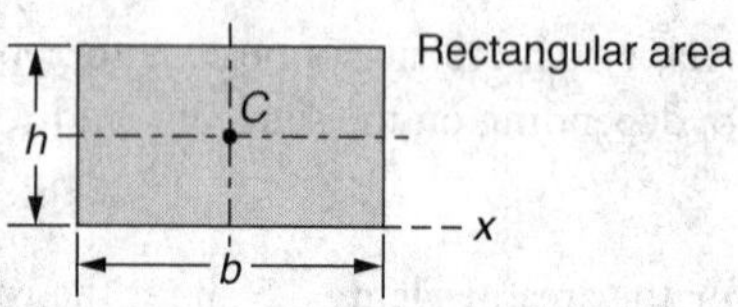 Rectangular area	—	$I_x = \frac{bh^3}{3}$ $\bar{I}_x = \frac{bh^3}{12}$ $\bar{I}_z = \frac{bh}{12}(b^2 + h^2)$
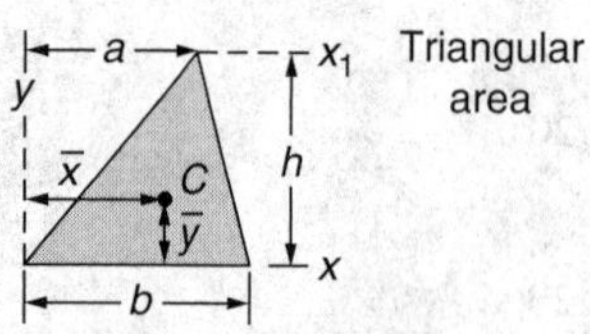 Triangular area	$\bar{x} = \frac{a+b}{3}$ $\bar{y} = \frac{h}{3}$	$I_x = \frac{bh^3}{12}$ $\bar{I}_x = \frac{bh^3}{36}$ $I_{x_1} = \frac{bh^3}{4}$
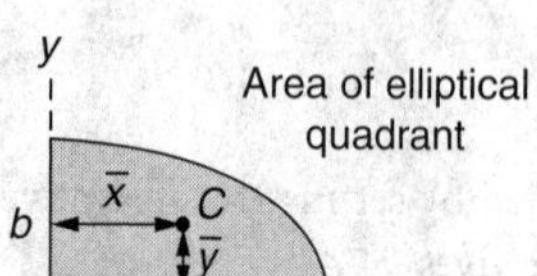 Area of elliptical quadrant	$\bar{x} = \frac{4a}{3\pi}$ $\bar{y} = \frac{4b}{3\pi}$	$I_x = \frac{\pi ab^3}{16}$, $\bar{I}_x = \left(\frac{\pi}{16} - \frac{4}{9\pi}\right) ab^3$ $I_y = \frac{\pi a^3 b}{16}$, $\bar{I}_y = \left(\frac{\pi}{16} - \frac{4}{9\pi}\right) a^3 b$ $I_z = \frac{\pi ab}{16}(a^2 + b^2)$

Source: Meriam, J.L. and Kraige, L.G. 1992. *Engineering Mechanics*, 3rd ed. John Wiley & Sons, New York.

Substitution of Equation (3.9) for 2θ in Equation (3.8) gives $I_{x'y'} = 0$ and

$$I_{max} = \frac{1}{2}\left\{ I_x + I_y + \sqrt{(I_x - I_y)^2 + 4I_{xy}^2} \right\}$$

$$I_{min} = \frac{1}{2}\left\{ I_x + I_y - \sqrt{(I_x - I_y)^2 + 4I_{xy}^2} \right\} \qquad (3.10)$$

3.2 Mass Moments of Inertia

The dynamics of bodies that rotate with angular acceleration calls for a knowledge of mass moments of inertia and is treated in the chapter on Dynamics and Vibration.

Defining Relations

Fixed Axis

The moment of inertia of the body of mass *m* in Figure 3.6 about the fixed axis *O-O* is given by

$$I_o = \int r^2 \, dm \qquad (3.11)$$

The dimensions are (mass)(length)², which are kg·m² in SI units and lb-ft-sec² in U.S. customary units. If the density ρ of the body is constant, then $dm = \rho \, dV$ and the integral becomes

$$I_o = \rho \int r^2 dV \qquad (3.12)$$

where dV is the differential volume of the mass element. To facilitate integration, coordinates that best suit the boundaries of the body should be utilized.

Radius of Gyration

The **radius of gyration** *k* of a mass *m* about an axis for which the moment of inertia is *I* is defined as

$$k = \sqrt{I/m} \quad \text{or} \quad I = k^2 m \qquad (3.13)$$

Thus *k* is a measure of the distribution of mass about the axis in question, and its definition is analogous to the similar definition of radius of gyration for area moments of inertia.

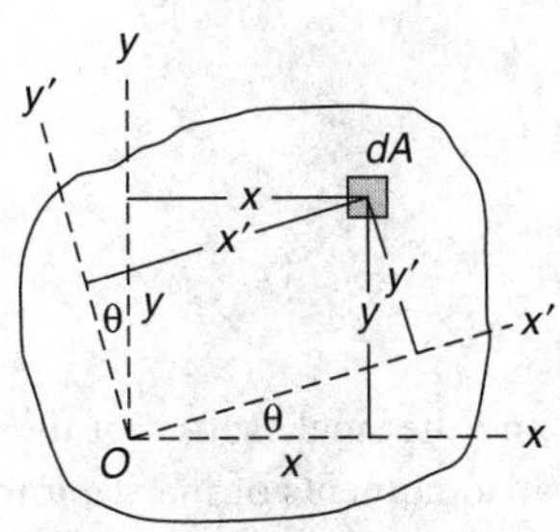

FIGURE 3.5 (*Source:* Meriam, J.L. and Kraige, L.G. 1992. *Engineering Mechanics,* 3rd ed. John Wiley & Sons, New York.)

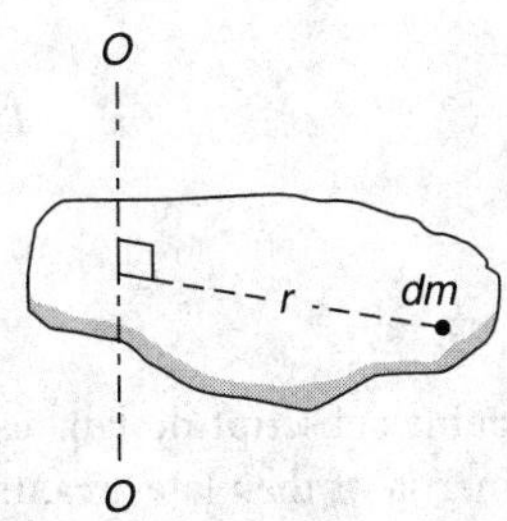

FIGURE 3.6 (*Source:* Meriam, J.L. and Kraige, L.G. 1992. *Engineering Mechanics,* 3rd ed. John Wiley & Sons, New York.)

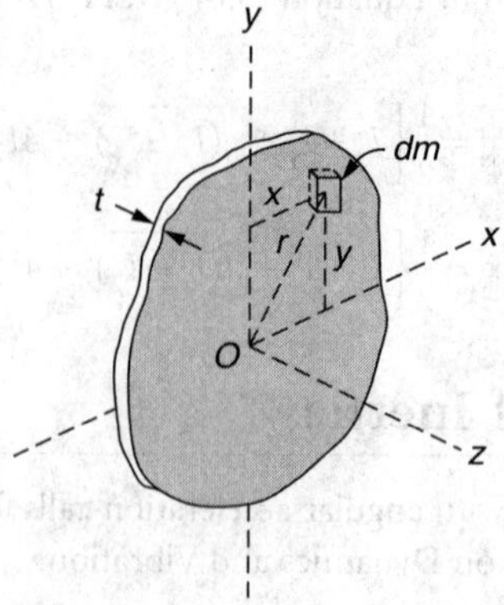

FIGURE 3.7 (*Source:* Meriam, J.L. and Kraige, L.G. 1992. *Engineering Mechanics*, 3rd ed. John Wiley & Sons, New York.)

Parallel-Axis Theorem

If the moment of inertia of a body of mass m about an axis through the mass center is known, it may easily be determined about any parallel axis by the expression

$$I = \bar{I} + md^2 \tag{3.14}$$

where $\bar{I}$ is the moment of inertia about the parallel axis through the mass center and d is the perpendicular distance between the axes. This *parallel-axis theorem* is analogous to that for area moments of inertia, Equation (3.6). It applies *only* if transfer is made to or from a parallel axis through the mass center. From the definition of Equation (3.13), it follows that

$$k^2 = \bar{k}^2 + d^2 \tag{3.15}$$

where k is the radius of gyration about an axis a distance d from the parallel axis through the mass center for which the radius of gyration is $\bar{k}$.

Flat Plates

The moments of inertia of a flat plate, Figure 3.7, about axes in the plane of and normal to the plate are frequently encountered. The elemental mass is $\rho\,(t\,dA)$, where ρ is the plate density, t is its thickness, and $dA = dx\,dy$ is the face area of dm. If ρ and t are constant, the moment of inertia about each of the axes becomes

$$\begin{aligned} I_{xx} &= \int y^2 dm = \rho t \int y^2 dA = \rho t I_x \\ I_{yy} &= \int x^2 dm = \rho t \int x^2 dA = \rho t I_y \\ I_{zz} &= \int r^2 dm = \rho t \int r^2 dA = \rho t I_z \end{aligned} \tag{3.16}$$

where the double subscript designates mass moment of inertia and the single subscript designates the moment of inertia of the plate area. Inasmuch as $I_z = I_x + I_y$ for area moments of inertia, it follows that

$$I_{zz} = I_{xx} + I_{yy} \tag{3.16a}$$

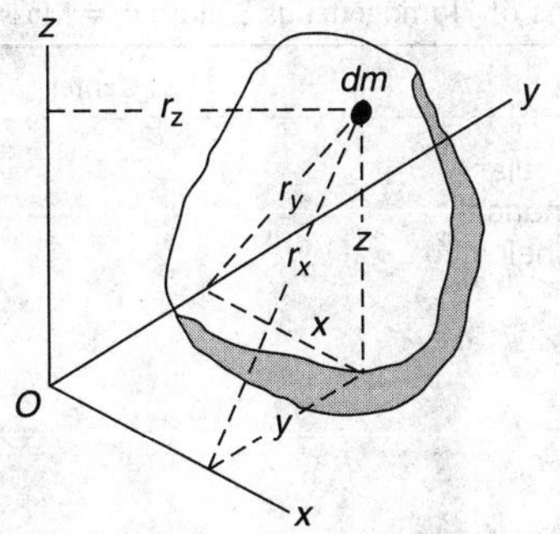

FIGURE 3.8 (*Source:* Meriam, J.L. and Kraige, L.G. 1992. *Engineering Mechanics*, 3rd ed. John Wiley & Sons, New York.)

This relation holds *only* if t is small compared with the other plate dimensions.

Composite Bodies

The mass moment of inertia of a composite body about a given axis is simply the sum of the moments of inertia of its individual components about the same axis.

A summary of formulas for mass moments of inertia for various bodies of common shape is given in Table 3.2.

General Rotation

For three-dimensional rotation of a rigid body the moments and products of inertia assume a more general form from Figure 3.8, as follows:

$$\begin{aligned} I_{xx} &= \int r_x^2 dm = \int (y^2 + z^2) dm \qquad & I_{xy} &= I_{yx} = \int xy\, dm \\ I_{yy} &= \int r_y^2 dm = \int (z^2 + x^2) dm \qquad & I_{xz} &= I_{zx} = \int xz\, dm \\ I_{zz} &= \int r_z^2 dm = \int (x^2 + y^2) dm \qquad & I_{yz} &= I_{zy} = \int yz\, dm \end{aligned} \tag{3.17}$$

Whereas the moments of inertia are always positive, the products of inertia may be positive, negative, or zero. Parallel-axis theorems for products of inertia are

$$\begin{aligned} I_{xy} &= \bar{I}_{xy} + m\bar{x}\bar{y} \\ I_{xz} &= \bar{I}_{xz} + m\bar{x}\bar{z} \\ I_{yz} &= \bar{I}_{yz} + m\bar{y}\bar{z} \end{aligned} \tag{3.18}$$

where the bar represents the product of inertia with respect to axes through the mass center and $\bar{x}$, $\bar{y}$, and $\bar{z}$ represent the coordinates of the mass center.

Defining Terms

Area moment of inertia — Defined as $\int$ (distance)^2 d(area).
Mass moment of inertia — A measure of the inertial resistance to angular acceleration.
Product of inertia — Defined as $\int xy\, dA$ for areas (xy plane) and $\int xy\, dm$, $\int xz\, dm$, and $\int yz\, dm$ for masses.
Radius of gyration — Defined as $\sqrt{I/A}$ for areas and $\sqrt{I/m}$ for masses.

TABLE 3.2 Moments of Inertia of Homogeneous Solids (m = Mass of Body Shown)

Body	Mass Center	Mass Moments of Inertia
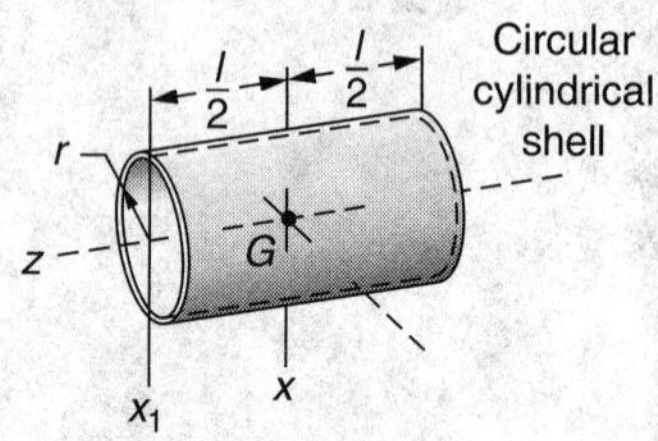	—	$I_{xx} = \frac{1}{2}mr^2 + \frac{1}{12}ml^2$ $I_{x_1x_1} = \frac{1}{2}mr^2 + \frac{1}{3}ml^2$ $I_{zz} = mr^2$
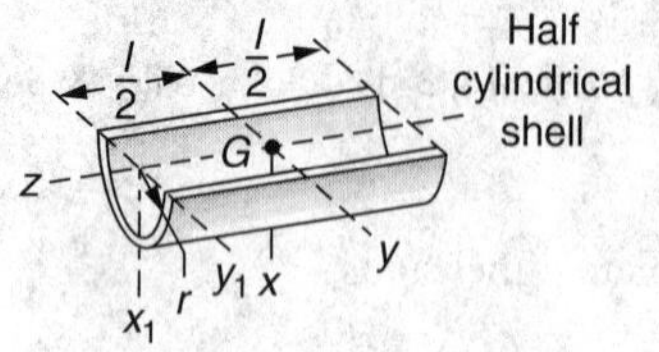	$\bar{x} = \frac{2r}{\pi}$	$I_{xx} = I_{yy} = \frac{1}{2}mr^2 + \frac{1}{12}ml^2$ $I_{x_1x_1} = I_{y_1y_1} = \frac{1}{2}mr^2 + \frac{1}{3}ml^2$ $I_{zz} = mr^2$ $\bar{I}_{zz} = \left(1 - \frac{4}{\pi^2}\right)mr^2$
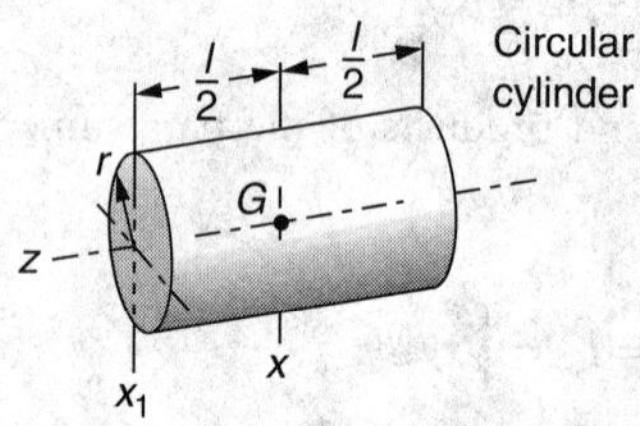	—	$I_{xx} = \frac{1}{4}mr^2 + \frac{1}{12}ml^2$ $I_{x_1x_1} = \frac{1}{4}mr^2 + \frac{1}{3}ml^2$ $I_{zz} = \frac{1}{2}mr^2$
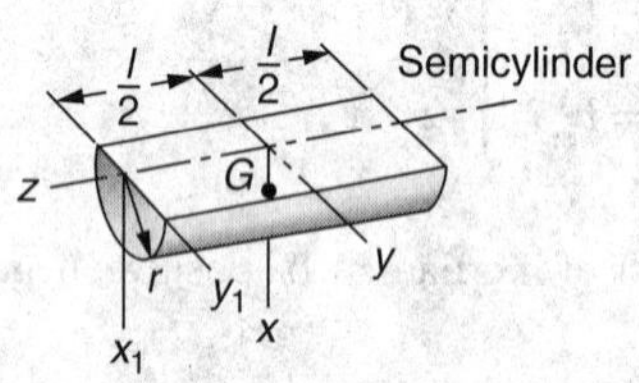	$\bar{x} = \frac{4r}{3\pi}$	$I_{xx} = I_{yy} = \frac{1}{4}mr^2 + \frac{1}{12}ml^2$ $I_{x_1x_1} = I_{y_1y_1} = \frac{1}{4}mr^2 + \frac{1}{3}ml^2$ $I_{zz} = \frac{1}{2}mr^2$ $\bar{I}_{zz} = \left(\frac{1}{2} - \frac{16}{9\pi^2}\right)mr^2$
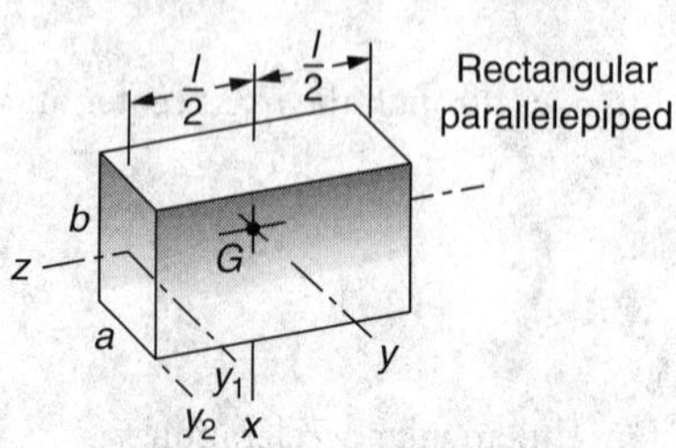	—	$I_{xx} = \frac{1}{12}m(a^2 + l^2)$ $I_{yy} = \frac{1}{12}m(b^2 + l^2)$ $I_{zz} = \frac{1}{12}m(a^2 + b^2)$ $I_{y_1y_1} = \frac{1}{12}mb^2 + \frac{1}{3}ml^2$ $I_{y_2y_2} = \frac{1}{3}m(b^2 + l^2)$

TABLE 3.2 Moments of Inertia of Homogeneous Solids (m = Mass of Body Shown) (*Continued*)

Body	Mass Center	Mass Moments of Inertia
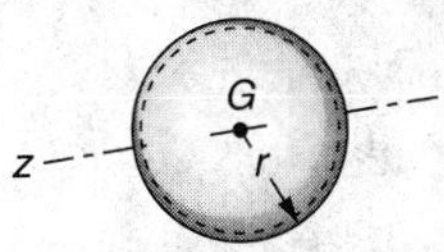 Spherical shell	—	$I_{zz} = \frac{2}{3}mr^2$
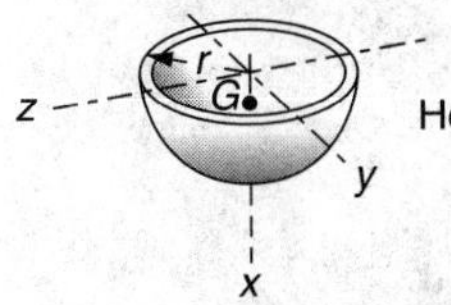 Hemispherical shell	$\bar{x} = \frac{r}{2}$	$I_{xx} = I_{yy} = I_{zz} = \frac{2}{3}mr^2$ $\bar{I}_{yy} = \bar{I}_{zz} = \frac{5}{12}mr^2$
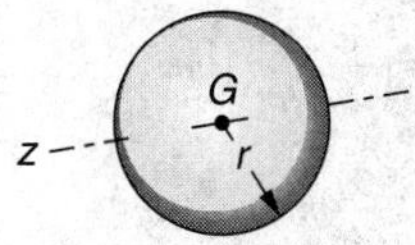 Sphere	—	$I_{zz} = \frac{2}{5}mr^2$
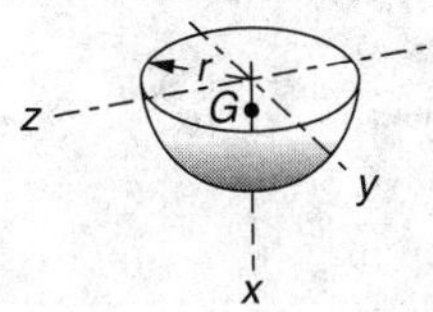 Hemisphere	$\bar{x} = \frac{3r}{8}$	$I_{xx} = I_{yy} = I_{zz} = \frac{2}{5}mr^2$ $\bar{I}_{yy} = \bar{I}_{zz} = \frac{83}{320}mr^2$
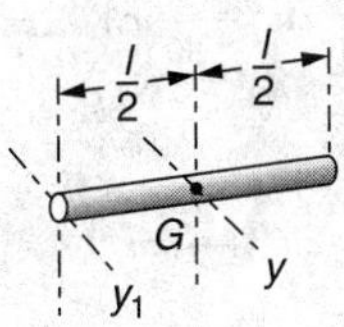 Uniform slender rod	—	$I_{yy} = \frac{1}{12}ml^2$ $I_{y_1y_1} = \frac{1}{3}ml^2$
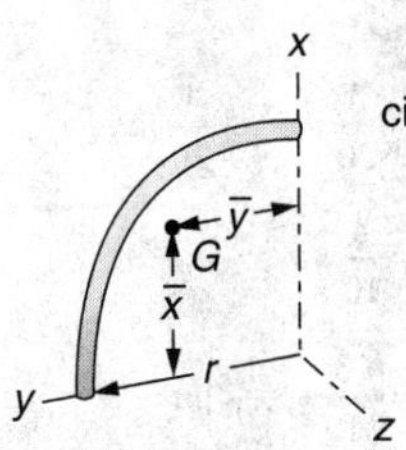 Quarter-circular rod	$\bar{x} = \bar{y}$ $= \frac{2r}{\pi}$	$I_{xx} = I_{yy} = \frac{1}{2}mr^2$ $I_{zz} = mr^2$
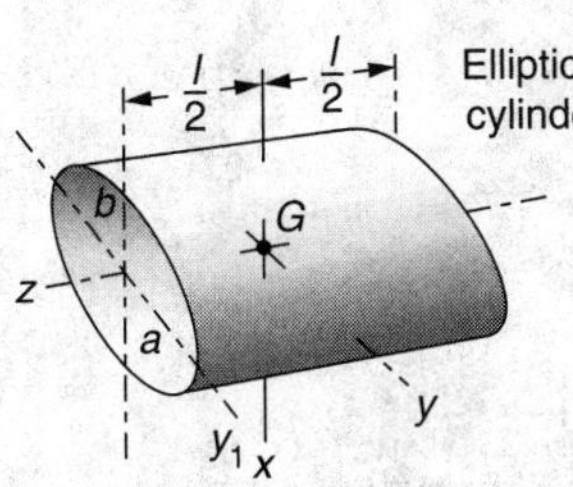 Elliptical cylinder	—	$I_{xx} = \frac{1}{4}ma^2 + \frac{1}{12}ml^2$ $I_{yy} = \frac{1}{4}mb^2 + \frac{1}{12}ml^2$ $I_{zz} = \frac{1}{4}m(a^2 + b^2)$ $I_{y_1y_1} = \frac{1}{4}mb^2 + \frac{1}{3}ml^2$
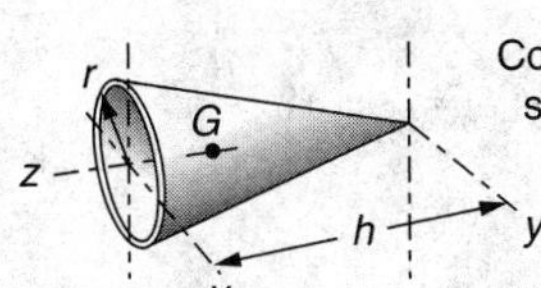 Conical shell	$\bar{z} = \frac{2h}{3}$	$I_{yy} = \frac{1}{4}mr^2 + \frac{1}{2}mh^2$ $I_{y_1y_1} = \frac{1}{4}mr^2 + \frac{1}{6}mh^2$ $I_{zz} = \frac{1}{2}mr^2$ $\bar{I}_{yy} = \frac{1}{4}mr^2 + \frac{1}{18}mh^2$

TABLE 3.2 Moments of Inertia of Homogeneous Solids (m = Mass of Body Shown) (*Continued*)

Body	Mass Center	Mass Moments of Inertia
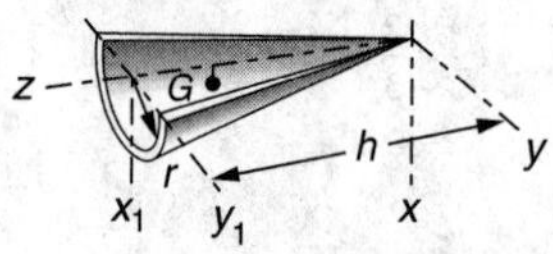 Half conical shell	$\bar{x} = \frac{4r}{3\pi}$ $\bar{z} = \frac{2h}{3}$	$I_{xx} = I_{yy}$ $= \frac{1}{4}mr^2 + \frac{1}{2}mh^2$ $I_{x_1x_1} = I_{y_1y_1}$ $= \frac{1}{4}mr^2 + \frac{1}{6}mh^2$ $I_{zz} = \frac{1}{2}mr^2$ $\bar{I}_{zz} = \left(\frac{1}{2} - \frac{16}{9\pi^2}\right)mr^2$
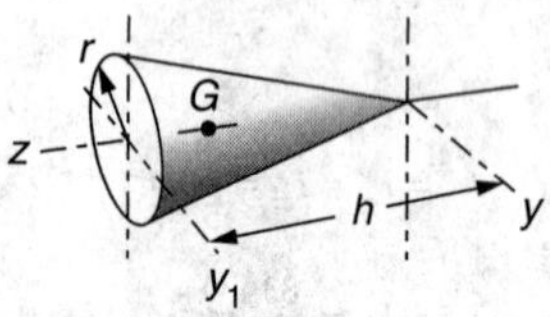 Right-circular cone	$\bar{z} = \frac{3h}{4}$	$I_{yy} = \frac{3}{20}mr^2 + \frac{3}{5}mh^2$ $I_{y_1y_1} = \frac{3}{20}mr^2 + \frac{4}{10}mh^2$ $I_{zz} = \frac{3}{10}mr^2$ $\bar{I}_{yy} = \frac{3}{20}mr^2 + \frac{3}{80}mh^2$
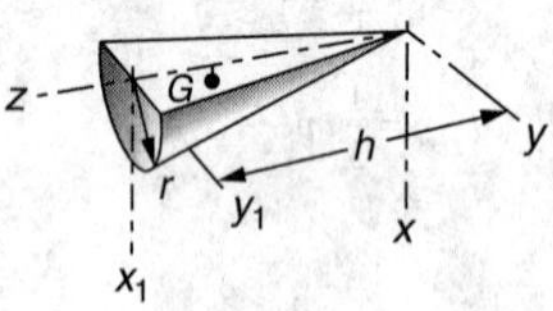 Half cone	$\bar{x} = \frac{r}{\pi}$ $\bar{z} = \frac{3h}{4}$	$I_{xx} = I_{yy}$ $= \frac{3}{20}mr^2 + \frac{3}{5}mh^2$ $I_{x_1x_1} = I_{y_1y_1}$ $= \frac{3}{20}mr^2 + \frac{1}{20}mh^2$ $I_{zz} = \frac{3}{10}mr^2$ $\bar{I}_{zz} = \left(\frac{3}{10} - \frac{1}{\pi^2}\right)mr^2$
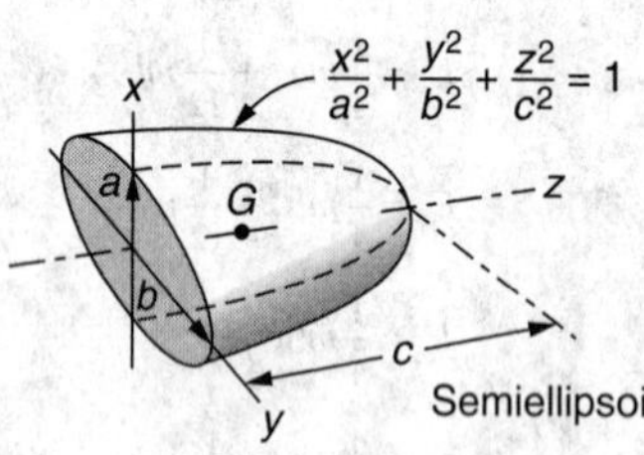 Semiellipsoid	$\bar{z} = \frac{3c}{8}$	$I_{xx} = \frac{1}{5}m(b^2 + c^2)$ $I_{yy} = \frac{1}{5}m(a^2 + c^2)$ $I_{zz} = \frac{1}{5}m(a^2 + c^2)$ $\bar{I}_{xx} = \frac{1}{5}m\left(b^2 + \frac{19}{64}c^2\right)$ $\bar{I}_{yy} = \frac{1}{5}m\left(a^2 + \frac{19}{64}c^2\right)$

TABLE 3.2 Moments of Inertia of Homogeneous Solids (m = Mass of Body Shown) (*Continued*)

Body	Mass Center	Mass Moments of Inertia
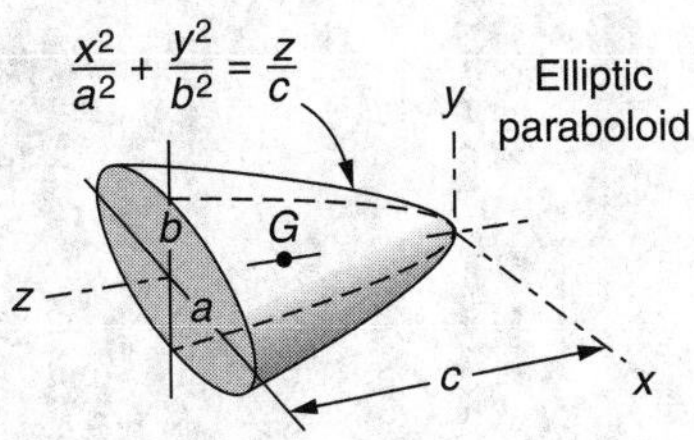	$\bar{z}=\frac{2c}{3}$	$I_{xx}=\frac{1}{6}mb^2+\frac{1}{2}mc^2$ $I_{yy}=\frac{1}{6}ma^2+\frac{1}{2}mc^2$ $I_{zz}=\frac{1}{6}m(a^2+b^2)$ $\bar{I}_{xx}=\frac{1}{6}m\left(b^2+\frac{1}{3}c^2\right)$ $\bar{I}_{yy}=\frac{1}{6}m\left(a^2+\frac{1}{3}c^2\right)$
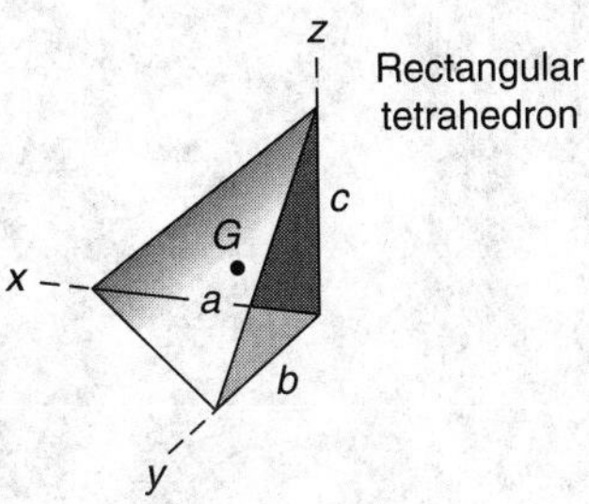	$\bar{x}=\frac{a}{4}$ $\bar{y}=\frac{b}{4}$ $\bar{z}=\frac{c}{4}$	$I_{xx}=\frac{1}{10}m(b^2+c^2)$ $I_{yy}=\frac{1}{10}m(a^2+c^2)$ $I_{zz}=\frac{1}{10}m(a^2+b^2)$ $\bar{I}_{xx}=\frac{3}{80}m(b^2+c^2)$ $\bar{I}_{yy}=\frac{3}{80}m(a^2+c^2)$ $\bar{I}_{zz}=\frac{3}{80}m(a^2+b^2)$
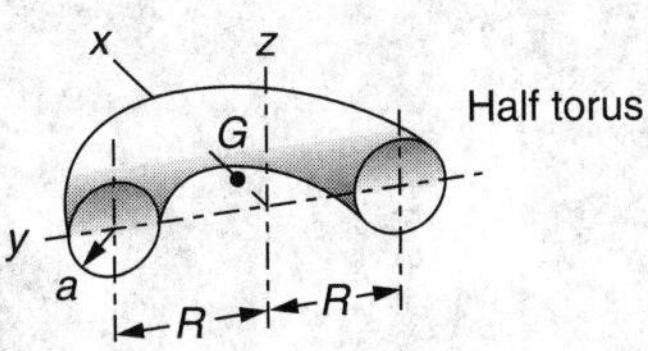	$\bar{x}=\frac{a^2+4R^2}{2\pi R}$	$I_{xx}=I_{yy}=\frac{1}{2}mR^2+\frac{5}{8}ma^2$ $I_{zz}=mR^2+\frac{3}{4}ma^2$

Reference

Meriam, J. L. and Kraige, L. G. 1992. *Engineering Mechanics*, 3rd ed. John Wiley & Sons, New York.

Further Information

Consult mechanics textbooks found in any engineering library.

II

Mechanics of Materials

4

Reactions

Thalia Anagnos
San Jose State University

For purposes of analysis, forces and moments acting on a structure or structural element can be grouped into two categories: loads and reactions. The loads acting on a structure include gravitational forces, inertial forces, friction, wind, lift, drag, hydrostatic pressure, soil pressure, and impacts. Supports are used to prevent a body from moving when subjected to these loads. **Reactions** are those moments and forces that act on the body as a consequence of the restraint provided by the supports. The magnitudes of reactions are controlled by the magnitudes of the applied loads. It should be understood that a reaction does not necessarily occur as a consequence of attaching a structure to the ground. Each structural element can have reactions due to its being connected to or supported by another structure or structural element. The type of reaction depends on the physical characteristics of the support. In order for a structure or structural element to be **stable**, there must be a sufficient number of supports to prevent it from undergoing unrestrained displacements.

4.1 Types of Supports

A variety of supports are available for restraining a structure. The type of support selected by the designer often depends on the type of structure, the material being used, the configuration of the structure, and the anticipated loads. A reaction occurs when the structure or structural element is prevented by a support from moving in a particular direction. For example, consider the fixed support shown in Table 4.1. This support prevents vertical and horizontal translation as well as rotation about an axis perpendicular to the page; thus the reactions are a vertical force, a horizontal force, and a moment.

A body at rest or moving at constant velocity is said to be in *equilibrium.* In accord with Newton's second law, equilibrium will exist if all of the forces and moments acting on the body sum to zero. To analyze the equilibrium of a structure or structural element, a free-body diagram must be drawn, on which all loads are shown and all supports have been replaced by the reactions they produce. For purposes of analysis and design many structures (but not all) subjected to two-dimensional systems of loads can be accurately modeled as **planar structures.** For these planar structures, models of support conditions can be greatly simplified. The idealized models of structural supports for planar structures are presented in Table 4.1. Table 4.2 summarizes supports and their accompanying reaction components for **space structures.** Space structures are those structures for which the members and/or the load systems are three-dimensional.

0-8493-1586-7/05/$0.00+$1.50

TABLE 4.1 Supports for Planar Structures

Support	Symbol	Reaction Components	Description of Support and Reactions
Cable	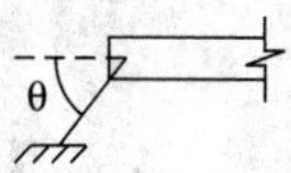	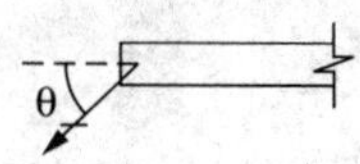	A cable can prevent translation only along its axis and can only exert tension. The unknown reaction is a tension force that acts away from the body in the direction of the cable.
Link	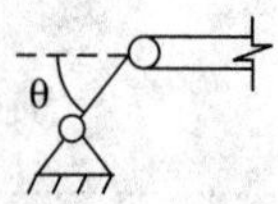	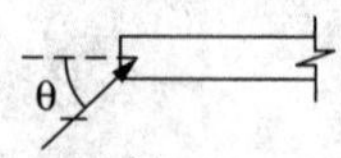	A link can prevent translation only along its axis. A link can exert tension or compression. The unknown reaction is a force that acts along the axis of the link either toward or away from the body.
Rocker	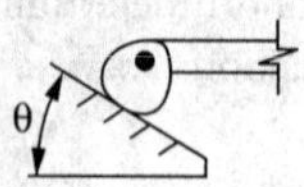	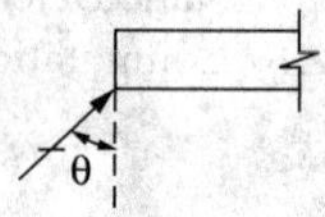	A rocker is capable only of applying a compressive force perpendicular to the surface on which it sits, and thus motion is only restrained perpendicular to that surface. The unknown reaction is a force that acts perpendicular to the surface at the point of contact.
Roller	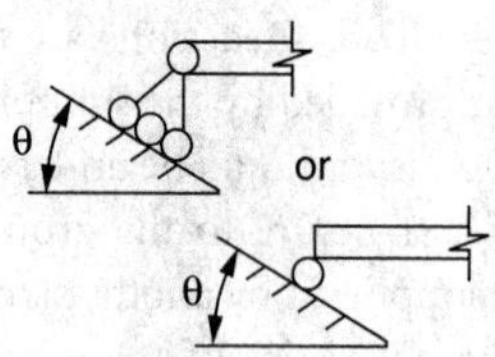	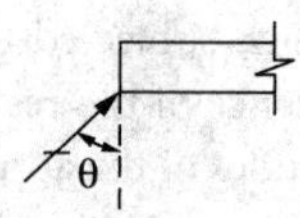	A roller is capable only of applying force (tension or compression) perpendicular to the surface on which it sits, and thus motion is only restrained perpendicular to the support surface. The body can translate parallel to the surface and can rotate. The unknown reaction is a force that acts perpendicular to the surface (in either direction) at the point of contact.
Smooth contacting surface	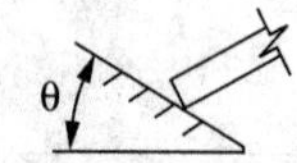	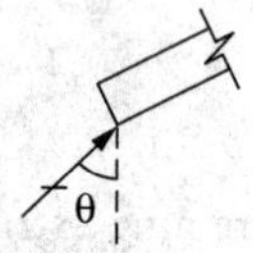	A smooth (frictionless) contacting surface is capable only of applying a compressive force and cannot provide any restraint parallel to the surface. The unknown reaction is a force that acts perpendicular to the surface at the point of contact.
Rough contacting surface	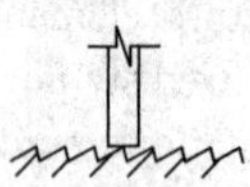	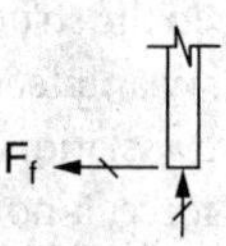	A rough contacting surface provides a frictional force (F_f) parallel to the surface. The two unknown reactions are a frictional force parallel to the surface and a compressive force that acts on the body perpendicular to the surface at the point of contact.
Frictionless pin or hinge	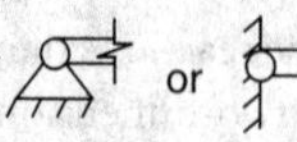	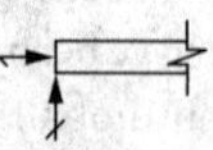	A frictionless pinned support allows a member to rotate freely but does not permit translation. The two unknown reactions are two perpendicular components of force.
Fixed support	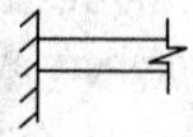	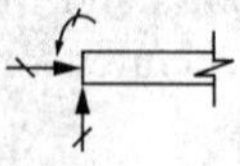	A fixed support does not permit translation or rotation. The three unknown reactions are two perpendicular force components and a moment.

TABLE 4.2 Supports for Space Structures

Support	Symbol	Reaction Components	Description of Support and Reactions
Cable		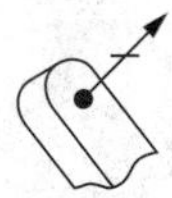	A cable can prevent translation only along its axis and can only exert tension. The unknown reaction is a tension force that acts away from the body in the direction of the cable.
Smooth contacting surface	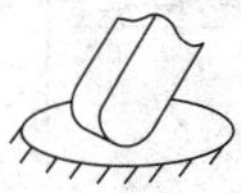		A smooth (frictionless) contacting surface cannot provide any restraint parallel to the surface. The unknown reaction is a compressive force that acts perpendicular to the surface at the point of contact.
Roller	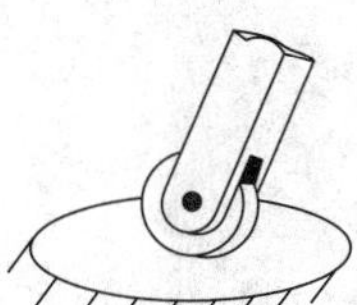		A roller is capable only of applying force (tension or compression) perpendicular to the surface on which it sits, and thus motion is only restrained perpendicular to the support surface. The body can translate parallel to the surface and can rotate. The unknown reaction is a force that acts perpendicular to the surface at the point of contact.
Ball-and-socket joint		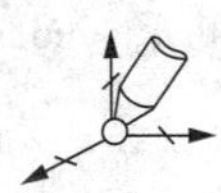	A frictionless ball-and-socket joint allows a member to rotate freely but does not permit translation. The unknown reactions are three mutually perpendicular force components.
Rough contacting surface		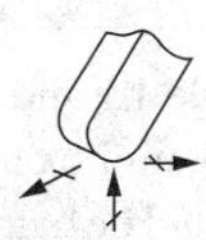	A rough contacting surface provides a frictional force parallel to the surface. The three unknown reactions are two perpendicular frictional forces parallel to the surface and a compressive force that acts perpendicular to the surface at the point of contact.
Journal bearing	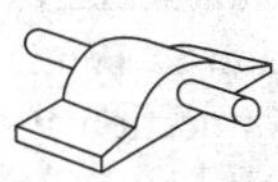	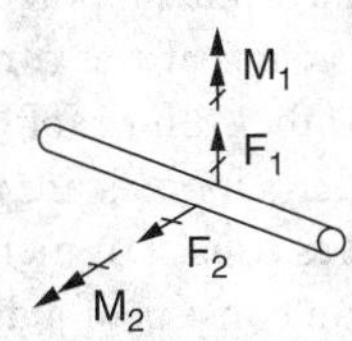	A journal bearing allows the shaft to spin freely and to translate parallel to itself. The four unknown reactions are two force components that act perpendicular to the shaft and two moments about axes perpendicular to the shaft.
Thrust bearing	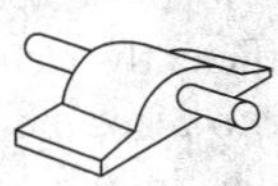	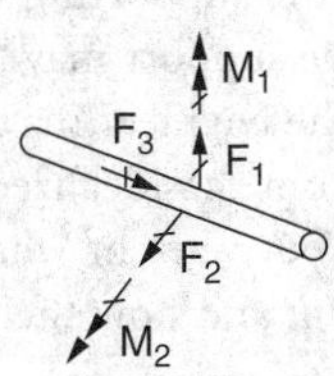	A thrust bearing allows the shaft to spin freely but does not permit any other translation or rotation. The five unknown reactions are three mutually perpendicular forces and two moments about axes perpendicular to the shaft.
Frictionless pin		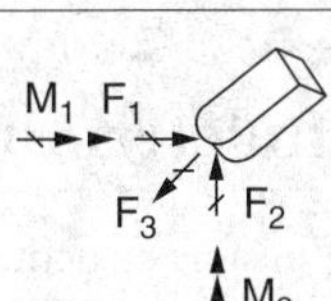	A frictionless pin permits rotation only about the axis of the pin. The five unknown reactions are three mutually perpendicular forces and two moments about axes perpendicular to the pin.
Fixed support		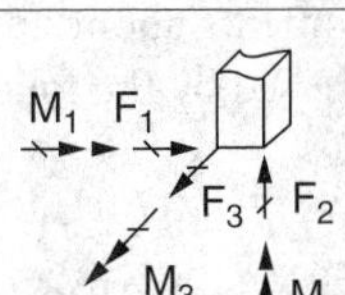	A fixed support does not permit translation or rotation. The six unknown reactions are three mutually perpendicular forces and moments about three mutually perpendicular axes.

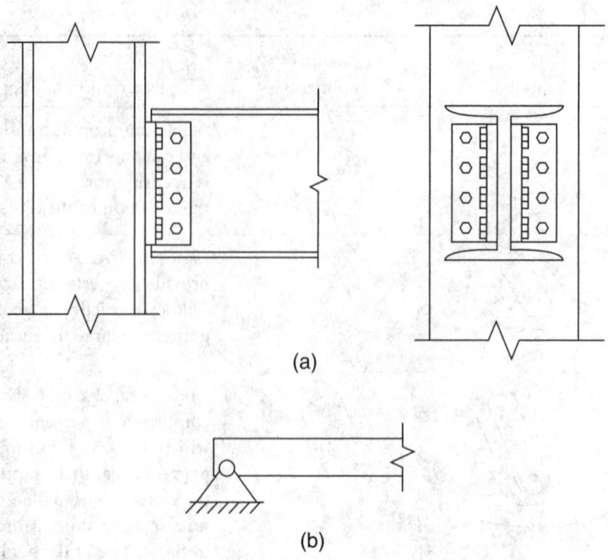

FIGURE 4.1 Connection between column and beam that is modeled as a pin support for the beam. (a) Actual connection; (b) idealized support for the beam.

4.2 Actual versus Idealized Support Conditions

The idealized models of supports presented in Table 4.1 and Table 4.2 are based on a variety of assumptions. For example, in the cases of pinned, hinged, and ball-and-socket supports, it is assumed that no friction occurs. If friction were present, rotation would be hampered and moments would result. Similarly, as shown in Figure 4.1 and Figure 4.2, a so-called pin support may not actually consist of two bodies connected by a frictionless pin. It is even possible that the connection in Figure 4.1 would be better represented by a torsional spring to account for the stiffness of the connection against rotation.

Another idealization is that hinged and roller supports are modeled by reaction forces acting at a point (see Figure 4.2). However, actual structures consist of members with finite depth or width. Thus, if a beam is resting on a wall, the actual reaction consists of a distributed force or pressure acting over the area of contact. However, if the width of the supporting wall is small compared with the length of the member, the assumption of a knife-edge support may be sufficiently accurate. In the example shown in Figure 4.2, for modeling purposes, the engineer may want to place the roller support at the center of the bearing plate. When selecting from the idealized models in Table 4.1 and Table 4.2 to represent actual supports, the engineer must be aware of what the assumptions are, how well these assumptions correspond to the actual situation, and how the idealization of the supports affects the accuracy of the analysis.

4.3 Static Determinacy and Indeterminacy

The first step in analyzing the reactions acting on any structure or structural element is to isolate the object of interest and construct a free-body diagram. Equilibrium of a body requires that the sum of the forces in any direction and the sum of the moments about any axis be zero. For a three-dimensional body, these equilibrium conditions can be expressed by the following set of equations:

$$\begin{aligned} &\Sigma F_x = 0 \qquad \Sigma F_y = 0 \qquad \Sigma F_z = 0 \\ &\Sigma M_{x\ \text{axis}} = 0 \quad \Sigma M_{y\ \text{axis}} = 0 \quad \Sigma M_{z\ \text{axis}} = 0 \end{aligned} \tag{4.1}$$

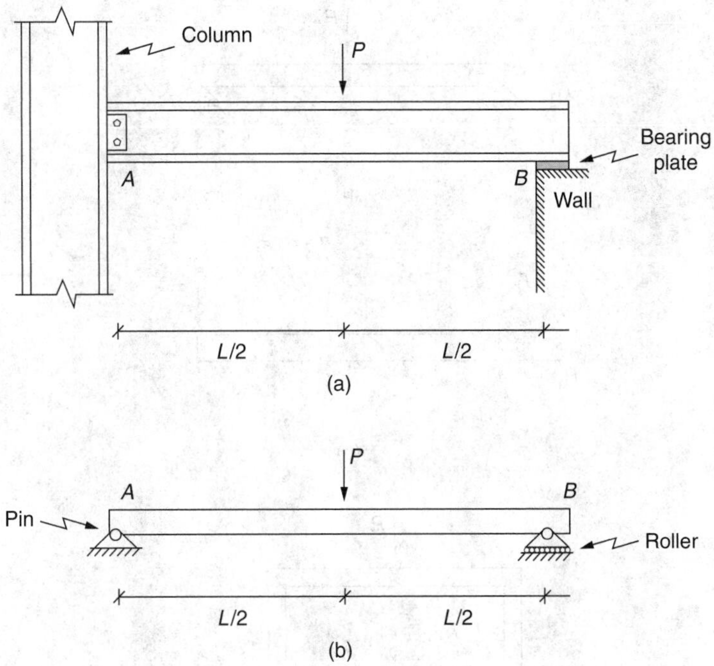

FIGURE 4.2 (a) Actual beam; (b) idealized beam.

If the body and all of the loads are in one plane (for example, the *xy* plane) the equilibrium equations reduce to

$$\Sigma F_x = 0 \quad \Sigma F_y = 0 \quad \Sigma M_{z\text{ axis}} = 0 \tag{4.2}$$

From Equations (4.1) and Equations (4.2), it can be seen that for a three-dimensional free body a maximum of six unknowns can be determined using equations of equilibrium, and for a planar free body a maximum of three unknowns. Thus, looking at an entire structure as a free body, it can be classified depending on the number and arrangement of unknown reactions. A structure for which equilibrium equations are sufficient to determine all of the unknown forces and moments is classified as **statically determinate**. If the number of unknowns exceeds the number of available equilibrium equations it is **statically indeterminate**. The number of excess unknowns defines the degree of static indeterminacy.

Examples of statically determinate planar structures are shown in Figure 4.3. The simply supported and cantilevered beams in Figure 4.3(a) and Figure 4.3(b) each have three unknown reactions. These reactions can be determined by summing forces in the vertical and horizontal directions and summing moments about an axis perpendicular to the plane of the structure [see Equations (4.2)]. The same is true for the frame shown in Figure 4.3(c). However, in this case the reaction R_3 would likely be resolved into vertical and horizontal components to simplify writing of the equilibrium equations.

A compound structure such as the one shown in Figure 4.3(d) may be made up of several rigid elements connected by pins. The procedure for analyzing a compound structure is to break it apart at the pins and draw a free-body diagram of each element. Each pin can then be replaced by the forces it applies to the element. Since no rotation can be transferred from one member to the other by a frictionless pin, the moment at a pin connection is zero. This condition provides an additional equation at each pin that can be used to solve for unknowns. Thus, while the structure in Figure 4.3(d) may at first appear to be statically indeterminate because there are four unknown reactions, it is not. The equation of condition at the pin can be combined with the three equilibrium equations to solve for all four reactions. Similarly,

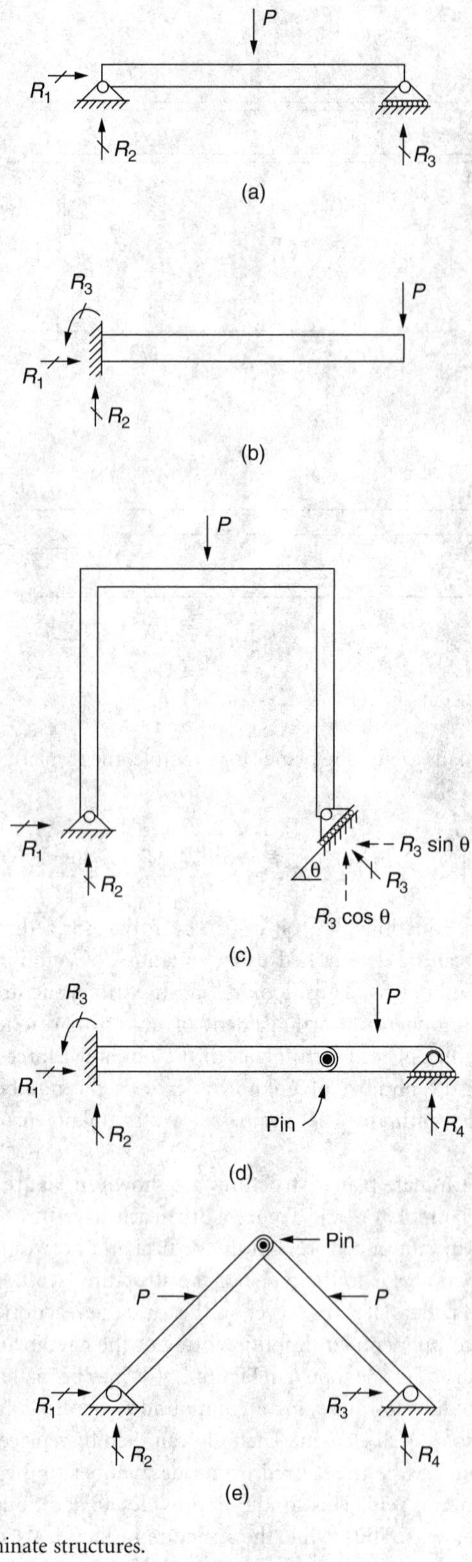

FIGURE 4.3 Statically determinate structures.

the four reactive forces on the three-hinged arch in Figure 4.3(e) can be determined by breaking the structure apart at the hinge (where the moment is zero) and applying equilibrium equations to each of the resulting free bodies.

Figure 4.4(a) and Figure 4.4(b) give examples of first-degree statically indeterminate structures. For each of these planar structures, only three equilibrium equations are available and only three reactions can be determined. The additional unknown that cannot be determined using equilibrium equations is called the *redundant.* The frame in Figure 4.4(c) has six unknown reactions and is third-degree statically indeterminate. It has three redundant reactions. The frame in Figure 4.4(d) appears to be third-degree statically indeterminate, but due to the two equations of condition provided by the two pins, it is only first-degree statically indeterminate.

A planar structure such as the one shown in Figure 4.5(a) with fewer than three reactions is **statically unstable** because there are fewer reactions than equilibrium equations. Statically unstable structures are incapable of preventing rigid body movement. It is possible, however, that while a structure has an adequate number of supports to potentially satisfy equilibrium, they can be arranged in such a way that the structure is still capable of moving. For example, in Figure 4.5(b) the beam has three reactions, but they are all parallel, so the beam can move horizontally. The reactions on the frame in Figure 4.5(c) are concurrent at point *C* and thus under certain loading conditions the frame will rotate. Structures such as these are classified as **geometrically unstable**. The reactions for a geometrically unstable structure cannot be completely determined using equations of equilibrium.

4.4 Computation of Reactions

For statically determinate structures, solving for the unknown reactions involves applying the appropriate equilibrium equations. Determination of reactions for statically indeterminate structures requires the use of additional relationships such as compatibility equations and constitutive equations involving the structural and material properties of the members. For example, either the *stiffness method* or the *flexibility method* can be used.

The procedure for determining the reactions on a statically determinate compound structure is illustrated in the following example. This requires only the use of free-body diagrams and equations of equilibrium. Example 1.2, Example 1.3, and Example 1.4 (Chapter 1) provide additional illustrations of how to determine reactions for statically determinate planar structures.

Example

The compound frame shown in Figure 4.6(a) consists of two members connected by a frictionless pin at *B*. Using the loads and dimensions shown in the figure, determine the reactions for this frame.

Solution

The unknown reactions for this frame are shown in Figure 4.6(b). The equilibrium equations [Equations (4.2)] can be combined with the equation of condition at the pin to solve for all four reactions. The frame must be separated into its two members in order to solve for the reactions. The corresponding free-body diagrams are shown in Figure 4.6(c). Starting with member *AB* because it has only three unknowns, apply Equations (4.2) with respect to the coordinate system shown.

$$\Sigma F_x = 0 \quad B_x = 0$$

$$\Sigma M_A = 0 \quad -(20\text{ k} \times 5\text{ ft}) + (B_y \times 10\text{ ft}) = 0 \quad B_y = 10\text{ k}$$

$$\Sigma F_y = 0 \quad R_{Ay} + B_y - 20\text{ k} = 0 \quad R_{Ay} = 10\text{ k}$$

Having determined the unknown forces for member AB, apply Equations (4.2) to member CBD.

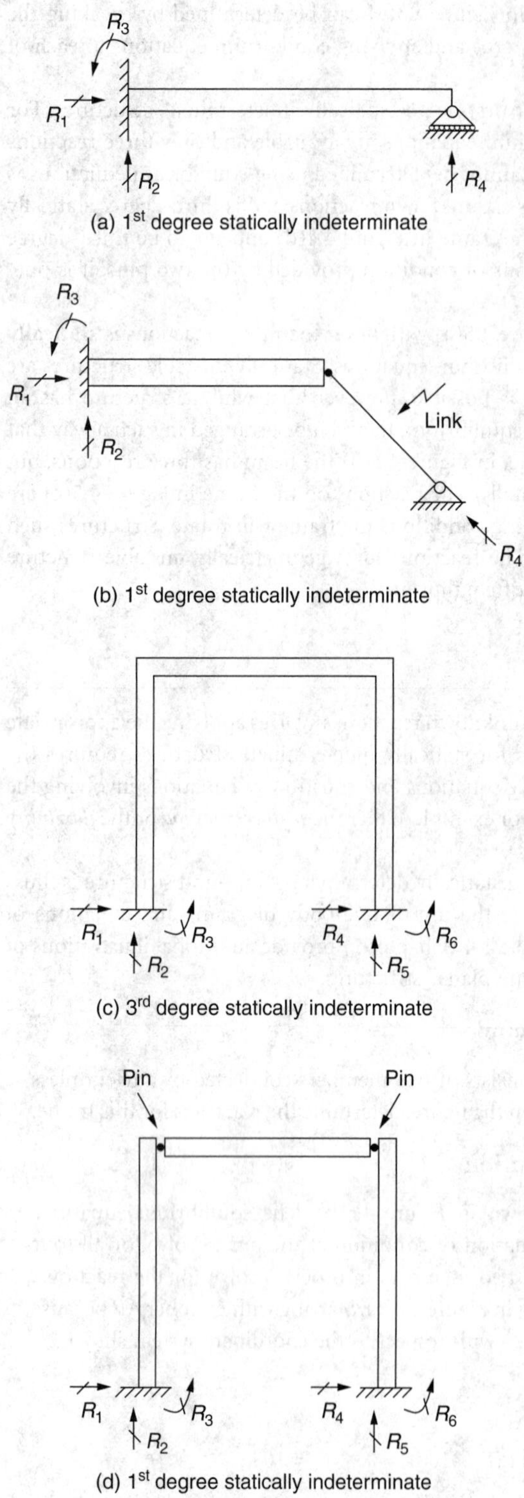

FIGURE 4.4 Statically indeterminate structures.

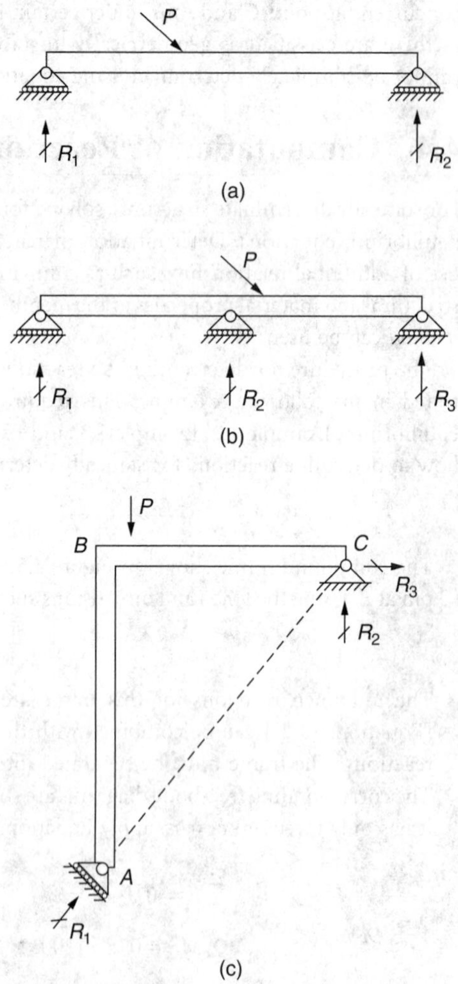

FIGURE 4.5 Examples of unstable structures: (a) statically unstable; (b) geometrically unstable; (c) geometrically unstable.

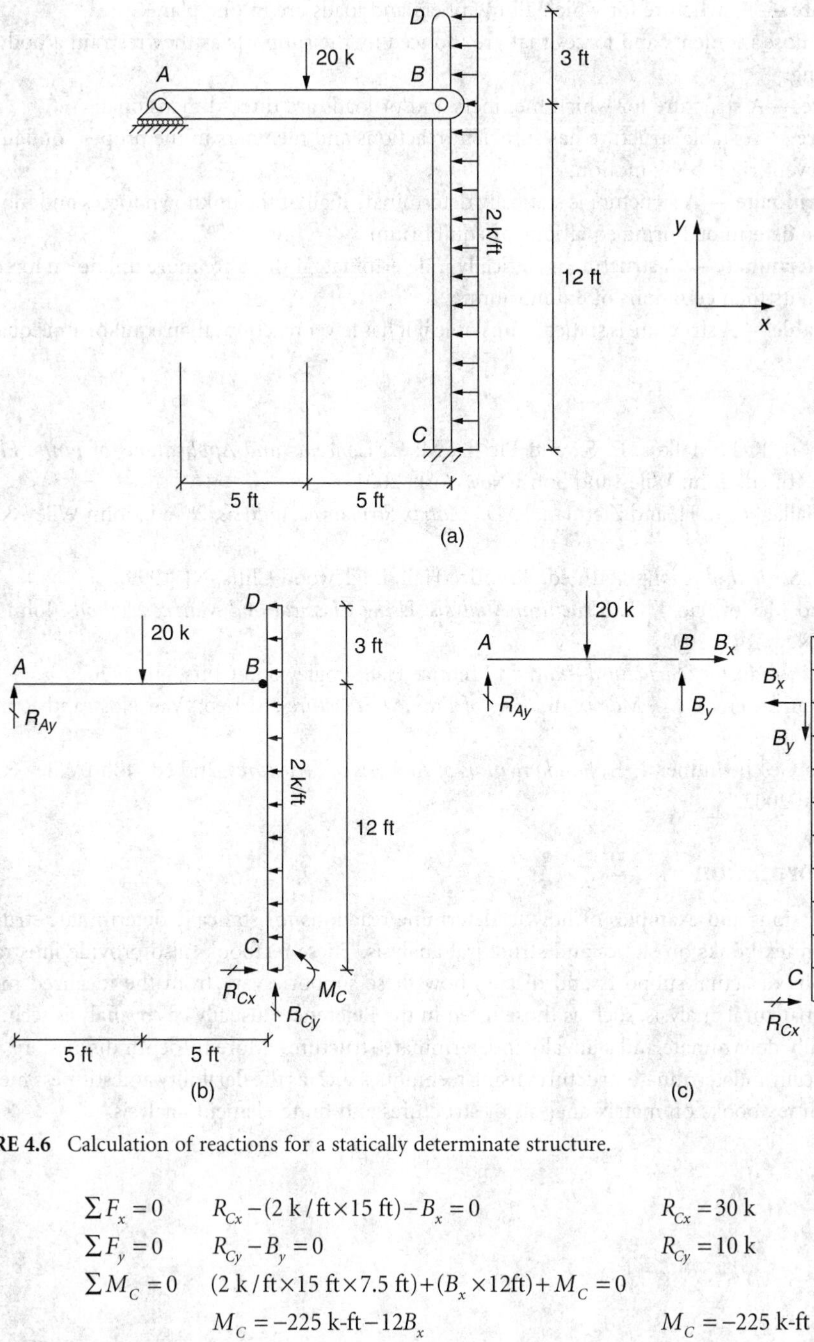

FIGURE 4.6 Calculation of reactions for a statically determinate structure.

$$\sum F_x = 0 \qquad R_{Cx} - (2\text{ k/ft}\times 15\text{ ft}) - B_x = 0 \qquad R_{Cx} = 30\text{ k}$$
$$\sum F_y = 0 \qquad R_{Cy} - B_y = 0 \qquad R_{Cy} = 10\text{ k}$$
$$\sum M_C = 0 \qquad (2\text{ k/ft}\times 15\text{ ft}\times 7.5\text{ ft}) + (B_x \times 12\text{ft}) + M_C = 0$$
$$M_C = -225\text{ k-ft} - 12B_x \qquad M_C = -225\text{ k-ft}$$

Since the value of the reaction moment at C, M_C, is negative, it is in the opposite direction to that assumed in Figure 4.6(b).

Defining Terms

Geometrically unstable — A structure is geometrically unstable if, even though it has sufficient restraints (reactions) to potentially satisfy equilibrium, it can undergo rigid-body movements due to the arrangement of the reactions.

Planar structure — A structure for which all members and loads are in one plane.
Reactions — Those moments and forces that are induced by the supports as they restrain a body from moving.
Space structure — A structure for which members and/or loads are three-dimensional.
Stable structure — A stable structure has sufficient reactions and members in the proper configuration to prevent rigid-body motion.
Statically determinate — A structure is statically determinate if all of the unknown forces and moments can be determined using equations of equilibrium.
Statically indeterminate — A structure is statically indeterminate if there are more unknown forces and moments than equations of equilibrium.
Statically unstable — A structure is statically unstable if it has fewer reactions than equilibrium equations.

References

Cook, R. D., Witt, R. J., Malkus, D. S., and Plesha, M. E., *Concepts and Applications of Finite Element Analysis,* 4th ed., John Wiley and Sons, New York, 2001.
McGuire, W., Gallagher R. H. and Ziemian, R. D., *Matrix Structural Analysis,* 2nd ed., John Wiley & Sons, New York, 1999.
Hibbeler, R. C., *Structural Analysis,* 4th ed., Prentice Hall, Englewood Cliffs, NJ, 1999.
Nelson J. K. and McCormac, J. C., *Structural Analysis: Using Classical and Matrix Methods,* John Wiley & Sons, New York, 2002.
Salmon, C. G., *Introductory Structural Analysis,* Prentice Hall, Englewood Cliffs, NJ, 1996.
Weaver, W., Jr. and Gere, J. M., *Matrix Analysis of Framed Structures,* 3rd ed., Van Nostrand Reinhold, New York, 1990.
West, H. H. and Geschwindner, L. F., *Fundamentals of Analysis of Structures,* 2nd ed., John Wiley & Sons, New York, 2002.

Further Information

Extensive discussions and examples of how to determine reactions for statically determinate structures can be found in textbooks on statics and structural analysis. These textbooks also provide illustrations and photographs of actual supports and discuss how these supports vary from the idealized models. Textbooks on structural analysis, such as those listed in the References, usually cover analysis techniques for both statically determinate and statically indeterminate structures. More in-depth discussions of the analysis of statically indeterminate structures using techniques such as the flexibility and stiffness methods can be found in textbooks on matrix analysis of structures and finite-element analysis.

5

Bending Stresses in Beams

James M. Gere
Stanford University

A *beam* is a slender structural member subjected to lateral loads. In this chapter we consider the bending stresses (i.e., normal stresses) in beams having initially straight longitudinal axes, such as the cantilever beam of Figure 5.1(a). For reference, we direct the positive x axis to the right along the longitudinal axis of the beam and the positive y axis downward (because the deflections of most beams are downward). The z axis, which is not shown in the figure, is directed away from the viewer, so that the three axes form a right-handed coordinate system. All cross-sections of the beam are assumed to be symmetric about the xy plane, and all loads are assumed to act in this plane. Consequently, the beam will deflect in this same plane [Figure 5.1(b)], which is called the **plane of bending**.

Pure bending refers to bending of a beam under a constant bending moment M, which means that the shear force V is zero (because $V = dM/dx$). **Nonuniform bending** refers to bending in the presence of shear forces, in which case the bending moment varies along the axis of the beam. The sign convention for bending moments is shown in Figure 5.2; note that positive bending moment produces tension in the lower part of the beam and compression in the upper part.

The stresses and strains in a beam are directly related to the *curvature* κ of the deflection curve. Because the x axis is positive to the right and the y axis is positive downward, the curvature is positive when the beam is bent concave downward and negative when the beam is bent concave upward (Figure 5.2).

5.1 Longitudinal Strains in Beams

Consider a segment DE of a beam subjected to pure bending by positive bending moments M [Figure 5.3(a)]. The cross-section of the beam at section mn is of arbitrary shape except that it must be symmetrical about the y axis [Figure 5.3(b)]. All cross-sections of the beam (such as mn) that were plane before bending remain plane after bending, a fact that can be proven theoretically using arguments based on symmetry. Therefore, *plane sections remain plane regardless of the material properties, whether elastic or inelastic, linear or nonlinear.* (Of course, the material properties, like the dimensions, must be symmetric about the plane of bending.)[1]

With positive bending moments, the lower part of the beam is in tension and the upper part is in compression. Therefore, longitudinal lines (i.e., line segments parallel to the x axis) in the lower part of

[1]This chapter contains selected material (text and figures) from Gere, J. M. and Timoshenko, S. P. 1990. *Mechanics of Materials*, 3rd ed. PWS, Boston. With permission.

0-8493-1586-7/05/$0.00+$1.50

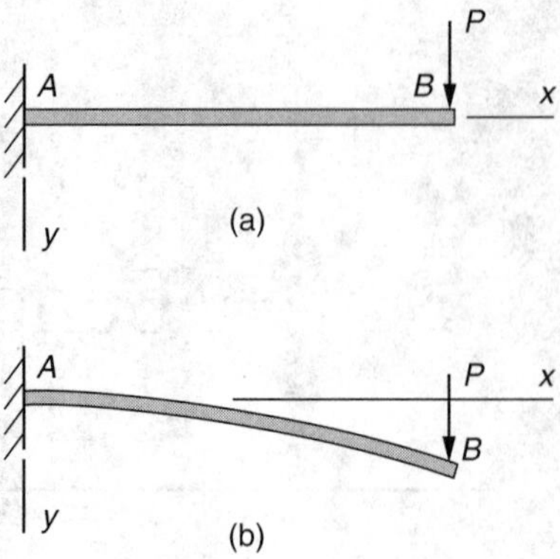

FIGURE 5.1 Bending of a cantilever beam.

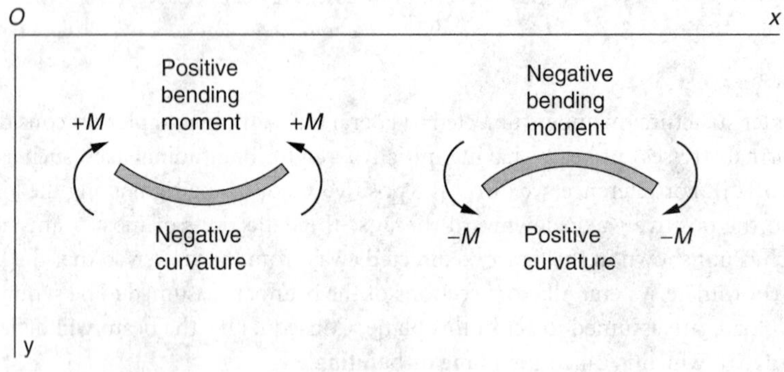

FIGURE 5.2 Sign conventions for bending moment and curvature.

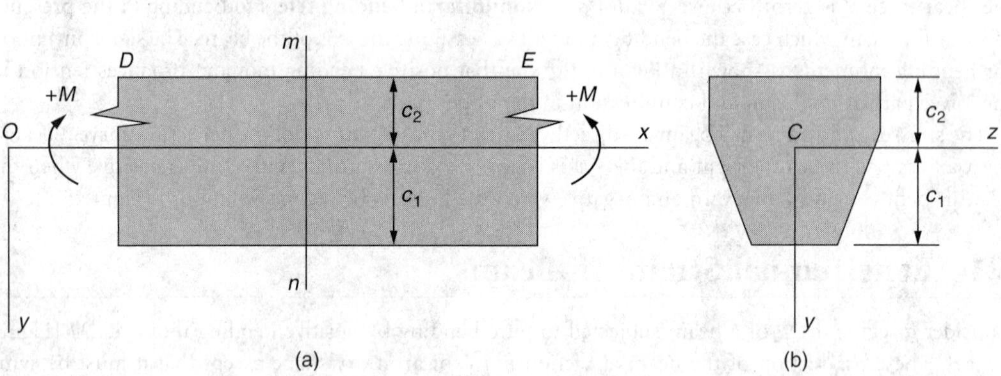

FIGURE 5.3 Beam in pure bending. (a) Side view of segment of beam showing bending moments *M* and typical section *mn.* (b) Cross-section of beam at section *mn.*

the beam are elongated and those in the upper part are shortened. The intermediate surface in which longitudinal lines do not change in length is called the **neutral surface** of the beam. We place the origin *O* of coordinates in this plane, so that the *xz* plane becomes the neutral surface. The intersection of this surface with any cross-sectional plane is called the **neutral axis of the cross-section,** for instance, the *z* axis in Figure 5.3(b).

Because plane sections remain plane, the longitudinal strains ε_x in the beam vary linearly with the distance *y* from the neutral surface, regardless of the material properties. It can also be shown that the strains are proportional to the curvature κ. Thus, the strains are given by the equation

$$\varepsilon_x = -\kappa y \tag{5.1}$$

The sign convention for ε_x is positive for elongation and negative for shortening. Note that when the curvature is positive (Figure 5.2) and y is positive (Figure 5.3), the strain is negative.

5.2 Normal Stresses in Beams (Linearly Elastic Materials)

Since longitudinal line elements in the beam are subjected only to tension or compression (elongation or shortening), they are in a state of uniaxial stress. Therefore, we can use the stress–strain diagram of the material to obtain the normal stresses σ_x from the normal strains ε_x. If the shape of the stress–strain curve can be expressed analytically, a formula can be derived for the stresses in the beam; otherwise, they must be calculated numerically.

The simplest and most common stress–strain relationship is for a linearly elastic material, in which case we can combine Hooke's law for uniaxial stress ($\sigma = E\varepsilon$) with Equation (5.1) and obtain

$$\sigma_x = E\varepsilon_x = -E\kappa y \tag{5.2}$$

in which E is the modulus of elasticity of the material. Equation (5.2) shows that the normal stresses acting on a cross-section vary linearly with the distance y from the neutral surface when the material follows Hooke's law.

Since the beam is in pure bending (Figure 5.3), the resultant of the stresses σ_x acting over the cross-section must equal the bending moment M. This observation provides two equations of statics — the first expressing that the resultant force in the x direction is equal to zero and the second expressing that the resultant moment is equal to M. The first equation of statics leads to the equation

$$\int y\, dA = 0 \tag{5.3}$$

which shows that the first moment of the cross-sectional area with respect to the z axis is zero. Therefore, the z axis must pass through the *centroid* of the cross-section. Since the z axis is also the neutral axis, we arrive at the following conclusion: The neutral axis passes through the centroid C of the cross-section provided the material follows Hooke's law and no axial force acts on the cross-section.

Since the y axis is an axis of symmetry, the y axis also passes through the centroid. Therefore, the origin of coordinates O is located at the centroid C of the cross-section. Furthermore, the symmetry of the cross-section about the y axis means that the y axis is a *principal axis*. The z axis is also a principal axis since it is perpendicular to the y axis. Therefore, when a beam of linearly elastic material is subjected to pure bending, the y and z axes are principal centroidal axes.

The second equation of statics leads to the *moment-curvature equation*

$$M = -\kappa EI \tag{5.4}$$

in which

$$I = \int y^2\, dA \tag{5.5}$$

is the moment of inertia of the cross-sectional area with respect to the z axis (that is, with respect to the neutral axis). Moments of inertia have dimensions of length to the fourth power, and typical units are in.^{4}, mm^4, and m^4 for beam calculations. The quantity EI is a measure of the resistance of the beam to bending and is called the *flexural rigidity* of the beam.

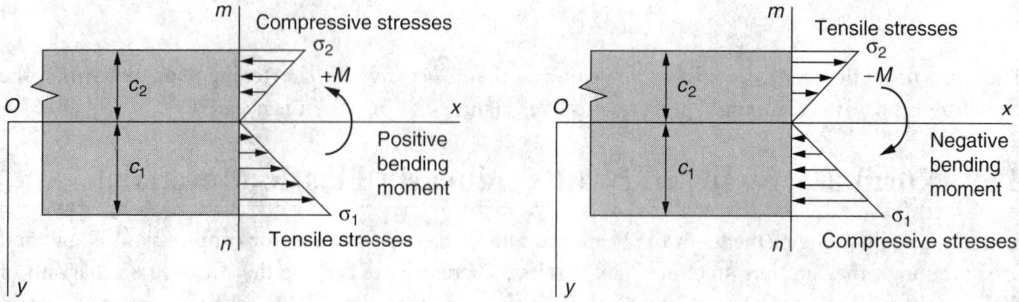

FIGURE 5.4 Bending stresses obtained from the flexure formula.

The minus sign in the moment-curvature equation is a consequence of the sign conventions we have adopted for bending moments and coordinate axes (Figure 5.2). We see that a positive bending moment produces negative curvature and a negative bending moment produces positive curvature. If the opposite sign convention for bending moments is used, or if the y axis is positive upward, then the minus sign is omitted in Equation (5.4) but a minus sign must be inserted in the flexure formula [Equation (5.6)] that follows.

The normal stresses in the beam can be related to the bending moment M by eliminating the curvature κ between Equation (5.2) and Equation (5.4), yielding

$$\sigma_x = \frac{My}{I} \tag{5.6}$$

This equation, called the **flexure formula**, shows that the stresses are directly proportional to the bending moment M and inversely proportional to the moment of inertia I of the cross-section. Furthermore, the stresses vary linearly with the distance y from the neutral axis, as shown in Figure 5.4. Stresses calculated from the flexure formula are called **bending stresses.**

The maximum tensile and compressive bending stresses occur at points located farthest from the neutral axis. Let us denote by c_1 and c_2 the distances from the neutral axis to the extreme elements in the positive and negative y directions, respectively (see Figure 5.3 and Figure 5.4). Then the corresponding maximum normal stresses σ_1 and σ_2 are

$$\sigma_1 = \frac{Mc_1}{I} = \frac{M}{S_1} \qquad \sigma_2 = -\frac{Mc_2}{I} = -\frac{M}{S_2} \tag{5.7}$$

in which

$$S_1 = \frac{I}{c_1} \qquad S_2 = \frac{I}{c_2} \tag{5.8}$$

The quantities S_1 and S_2 are known as the *section moduli* of the cross-sectional area. From Equation (5.8) we see that a section modulus has dimensions of length to the third power (for example, in.^{3}, mm^3, or m^3).

If the cross-section is symmetric with respect to the z axis, which means that it is a *doubly symmetric cross-section,* then $c_1 = c_2 = c$, and the maximum tensile and compressive stresses are equal numerically:

$$\sigma_1 = -\sigma_2 = \frac{Mc}{I} = \frac{M}{S} \tag{5.9}$$

in which

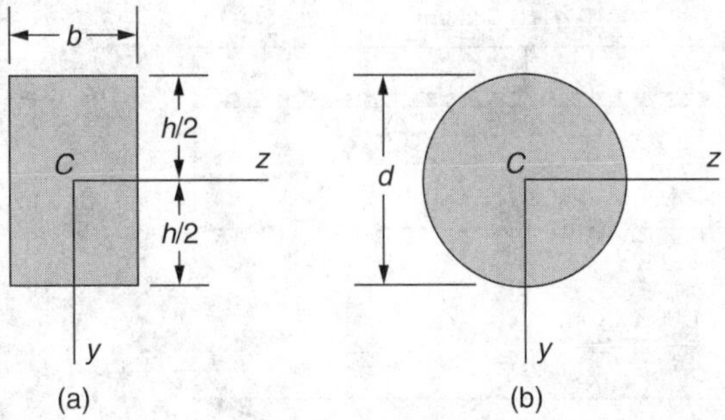

FIGURE 5.5 Doubly symmetric cross-sectional shapes.

$$S = \frac{I}{c} \tag{5.10}$$

is the section modulus. For a beam of *rectangular cross-section* with width b and height h [Figure 5.5(a)], the moment of inertia and section modulus are

$$I = \frac{bh^3}{12} \qquad S = \frac{bh^2}{6} \tag{5.11}$$

For a *circular cross-section* of diameter d [Figure 5.5(b)], these properties are

$$I = \frac{\pi d^4}{64} \qquad S = \frac{\pi d^3}{32} \tag{5.12}$$

The properties of many other plane figures are listed in textbooks and handbooks.

The preceding equations for the normal stresses apply rigorously only for pure bending, which means that no shear forces act on the cross-sections. The presence of shear forces produces warping, or out-of-plane distortion, of the cross-sections, and a cross-section that is plane before bending is no longer plane after bending. Warping due to shear greatly complicates the behavior of the beam, but detailed investigations show that the normal stresses calculated from the flexure formula are not significantly altered by the presence of the shear stresses and the associated warping. Thus, under ordinary conditions we may use the flexure formula for calculating normal stresses even when we have nonuniform bending.

The flexure formula gives results that are accurate only in regions of the beam where the stress distribution is not disrupted by abrupt changes in the shape of the beam or by discontinuities in loading. For instance, the flexure formula is not applicable at or very near the supports of a beam, where the stress distribution is irregular. Such irregularities produce localized stresses, or *stress concentrations*, that are much greater than the stresses obtained from the flexure formula. With ductile materials and static loads, we may usually disregard the effects of stress concentrations. However, they cannot be ignored when the materials are brittle or when the loads are dynamic in character.

Example

The beam *ABC* shown in Figure 5.6 has simple supports at *A* and *B* and an overhang from *B* to *C*. A uniform load of intensity $q = 3.0$ kN/m acts throughout the length of the beam. The beam is constructed of steel plates (12 mm thick) welded to form a channel section, the dimensions of which are shown in Figure 5.7(a). Calculate the maximum tensile and compressive stresses in the beam due to the uniform load.

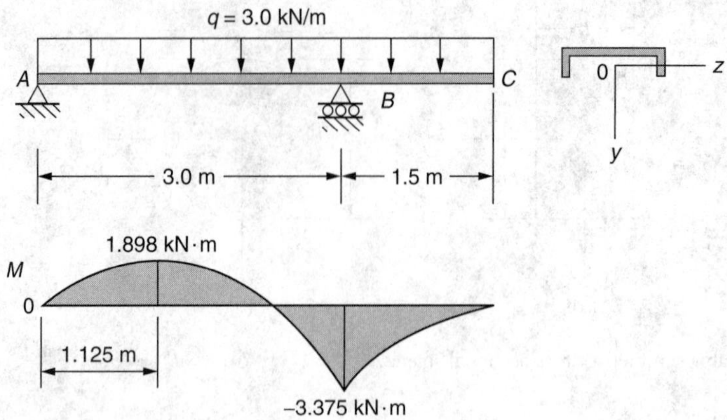

FIGURE 5.6 Beam dimensions.

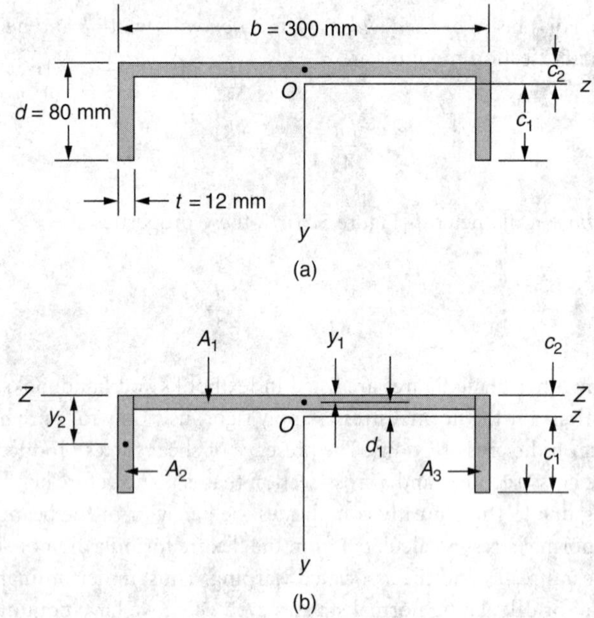

FIGURE 5.7 Cross-section of beam.

Solution

The maximum tensile and compressive stresses occur at the cross-sections where the bending moments have their maximum numerical values. Therefore, we construct the bending-moment diagram for the beam (Figure 5.6) and note that the maximum positive and negative moments equal 1.898 kN · m and −3.375 kN · m, respectively.

Next, we determine the position of the neutral axis by locating the centroid of the cross-sectional area shown in Figure 5.7(a). The results are as follows:

$$c_1 = 61.52 \text{ mm}; \; c_2 = 18.48 \text{ mm}$$

The moment of inertia of the cross-sectional area about the neutral axis (the z axis) is calculated with the aid of the parallel-axis theorem for moments of inertia; the result is

$$I = 2.469 \cdot 10^6 \text{ mm}^4$$

Also, the section moduli for the bottom and top of the beam, respectively, are

$$S_1 = \frac{I}{c_1} = 40\,100 \text{ mm}^3 \qquad S_2 = \frac{I}{c_2} = 133\,600 \text{ mm}^3$$

At the cross-section of maximum positive bending moment, the largest tensile stress occurs at the bottom of the beam (σ_1), and the largest compressive stress occurs at the top (σ_2):

$$\sigma_t = \sigma_1 = \frac{M}{S_1} = \frac{1.898 \text{ kN}\cdot\text{m}}{40\,100 \text{ mm}^3} = 47.3 \text{ MPa}$$

$$\sigma_c = \sigma_2 = -\frac{M}{S_2} = -\frac{1.898 \text{ kN}\cdot\text{m}}{133\,600 \text{ mm}^3} = -14.2 \text{ MPa}$$

Similarly, the largest stresses at the section of maximum negative moment are

$$\sigma_t = \sigma_2 = -\frac{M}{S_2} = -\frac{-3.375 \text{ kN}\cdot\text{m}}{133\,600 \text{ mm}^3} = 25.3 \text{ MPa}$$

$$\sigma_c = \sigma_1 = \frac{M}{S_1} = \frac{-3.375 \text{ kN}\cdot\text{m}}{40\,100 \text{ mm}^3} = -84.2 \text{ MPa}$$

A comparison of these four stresses shows that the maximum tensile stress due to the uniform load q is 47.3 MPa and occurs at the bottom of the beam at the section of maximum positive bending moment. The maximum compressive stress is –84.2 MPa and occurs at the bottom of the beam at the section of maximum negative moment.

Defining Terms

Bending stresses — Longitudinal normal stresses σ_x in a beam due to bending moments.

Flexure formula — The formula $\sigma_x = My/I$ for the bending stresses in a beam (linearly elastic materials only).

Neutral axis of the cross-section — The intersection of the neutral surface with a cross-sectional plane; that is, the line in the cross-section about which the beam bends and where the bending stresses are zero.

Neutral surface — The surface perpendicular to the plane of bending in which longitudinal lines in the beam do not change in length (no longitudinal strains).

Nonuniform bending — Bending in the presence of shear forces (which means that the bending moment varies along the axis of the beam).

Plane of bending — The plane of symmetry in which a beam bends and deflects.

Pure bending — Bending of a beam under a constant bending moment (no shear forces).

Section modulus — A property of the cross-section of a beam, equal to I/c [see Equation (5.8)].

References

Beer, F. P., Johnston, E. R., and DeWolf, J. T. 2001. *Mechanics of Materials,* 3rd Ed. McGraw-Hill, Inc., New York.

Gere, J. M. 2001. *Mechanics of Materials,* 5th Ed. Brooks/Cole, Pacific Grove, CA.

Hibbeler, R. C. 2000. *Mechanics of Materials,* 4th Ed., Prentice Hall, Inc., Upper Saddle River, NJ.

Lardner, T. J. and Archer, R. R. 1994. *Mechanics of Solids,* McGraw-Hill, Inc., New York.

Popov, E. P. and Balan, T. A. 1999. *Engineering Mechanics of Solids,* 2nd Ed., Prentice Hall, Inc., Upper Saddle River, NJ.

Further Information

Extensive discussions of bending, with derivations, examples, and problems, can be found in textbooks on mechanics of materials, such as those listed in the References. These books also cover many additional topics pertaining to bending stresses in beams. For instance, nonprismatic beams, fully stressed beams, beams with axial loads, stress concentrations in bending, composite beams, beams with skew loads, and stresses in inelastic beams are discussed in Gere [2001].

6

Shear Stresses in Beams

James M. Gere
Stanford University

The loads acting on a beam [Figure 6.1(a)] usually produce both bending moments M and shear forces V at cross-sections such as ab [Figure 6.1(b)]. The longitudinal normal stresses σ_x associated with the bending moments can be calculated from the flexure formula (see Chapter 5). The transverse shear stresses τ associated with the shear forces are described in this chapter.

Since the formulas for shear stresses are derived from the flexure formula, they are subject to the same limitations:

1. The beam is symmetric about the xy plane and all loads act in this plane (the *plane of bending*).
2. The beam is constructed of a linearly elastic material.
3. The stress distribution is not disrupted by abrupt changes in the shape of the beam or by discontinuities in loading (*stress concentrations*).

6.1 Shear Stresses in Rectangular Beams

A segment of a beam of rectangular cross-section (width b and height h) subjected to a vertical shear force V is shown in Figure 6.2(a). We assume that the shear stresses τ acting on the cross-section are parallel to the sides of the beam and uniformly distributed across the width (although they vary as we move up or down on the cross-section). A small element of the beam cut out between two adjacent cross-sections and between two planes that are parallel to the neutral surface is shown in Figure 6.2(a) as element $m\ n$. Shear stresses acting on one face of an element are always accompanied by complementary shear stresses of equal magnitude acting on perpendicular faces of the element, as shown in Figure 6.2(b) and Figure 6.2(c). Thus, there are horizontal shear stresses acting between horizontal layers of the beam as well as transverse shear stresses acting on the vertical cross-sections.

The equality of the horizontal and vertical shear stresses acting on element $m\ n$ leads to an interesting conclusion regarding the shear stresses at the top and bottom of the beam. If we imagine that the element $m\ n$ is located at either the top or the bottom, we see that the horizontal shear stresses vanish because there are no stresses on the outer surfaces of the beam. It follows that the vertical shear stresses also vanish at those locations; thus, $\tau = 0$ where $y = \pm h/2$. (Note that the origin of coordinates is at the centroid of the cross-section and the z axis is the neutral axis.)

0-8493-1586-7/05/$0.00+$1.50

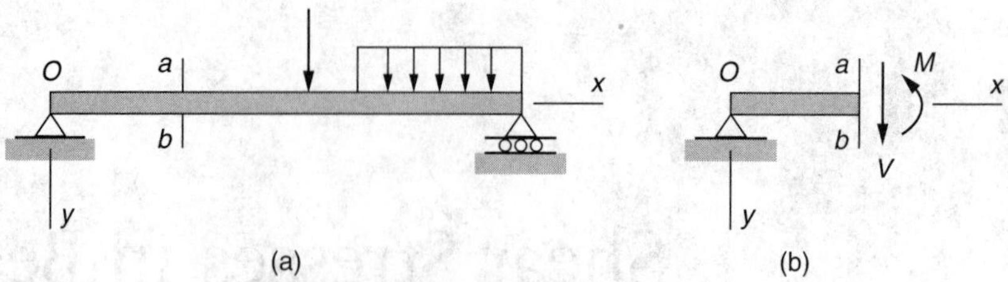

FIGURE 6.1 Beam with bending moment M and shear force V acting at cross-section *ab*.

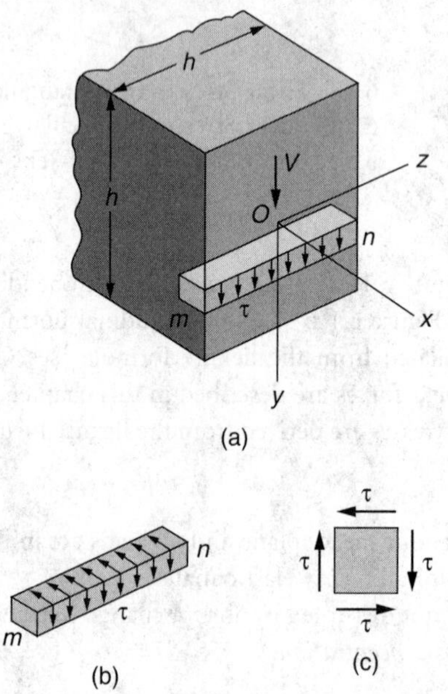

FIGURE 6.2 Shear stresses in a beam of rectangular cross-section.

The magnitude of the shear stresses can be determined by a lengthy derivation that involves only the flexure formula and static equilibrium (see References). The result is the following[1] formula for the shear stress:

$$\tau = \frac{V}{Ib}\int y\, dA \tag{6.1}$$

in which V is the shear force acting on the cross-section, I is the moment of inertia of the cross-sectional area about the neutral axis, and b is the width of the beam. The integral in Equation (6.1) is the first moment of the part of the cross-sectional area below (or above) the level at which the stress is being evaluated. Denoting this first moment by Q, that is,

$$Q = \int y\, dA \tag{6.2}$$

we can write Equation (6.1) in the simpler form

[1]Selected material (text and figures) from Chapter 5 of Gere, J. M. and Timoshenko, S. P. 1990. *Mechanics of Materials*, 3rd ed. PWS, Boston. With permission.

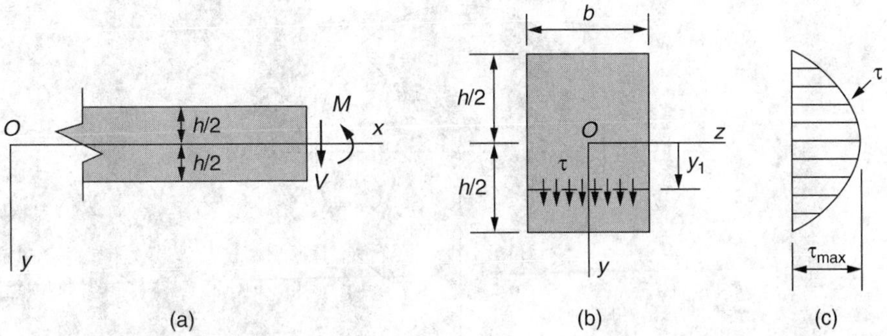

FIGURE 6.3 Distribution of shear stresses in a beam of rectangular cross-section. (a) Side view of beam showing the shear force V and bending moment M acting at a cross-section. (b) Cross-section of beam showing shear stresses τ acting at distance y_1 from the neutral axis. (c) Diagram showing the parabolic distribution of shear stresses.

$$\tau = \frac{VQ}{Ib} \tag{6.3}$$

This equation, known as the **shear formula,** can be used to determine the shear stress τ at any point in the cross-section of a rectangular beam. Note that for a specific cross-section, the shear force V, moment of inertia I, and width b are constants. However, the first moment Q (and hence the shear stress τ) varies depending upon where the stress is to be found.

To evaluate the shear stress at distance y_1 below the neutral axis (Figure 6.3), we must determine the first moment Q of the area in the cross-section below the level $y = y_1$. We can obtain this first moment by multiplying the partial area A_1 by the distance $\bar{y}_1$ from its centroid to the neutral axis:

$$Q = A_1\bar{y}_1 = b\left(\frac{h}{2} - y_1\right)\left(y_1 + \frac{h/2 - y_1}{2}\right) = \frac{b}{2}\left(\frac{h^2}{4} - y_1^2\right) \tag{6.4}$$

Of course, this same result can be obtained by integration using Equation (6.2):

$$Q = \int y\,dA = \int_{y_1}^{h/2} yb\,dy = \frac{b}{2}\left(\frac{h^2}{4} - y_1^2\right) \tag{6.5}$$

Substituting this expression for Q into the shear formula [Equation (6.3)], we get

$$\tau = \frac{V}{2I}\left(\frac{h^2}{4} - y_1^2\right) \tag{6.6}$$

This equation shows that the shear stresses in a rectangular beam vary quadratically with the distance y_1 from the neutral axis. Thus, when plotted over the height of the beam, τ varies in the manner shown by the parabolic diagram of Figure 6.3(c). Note that the shear stresses are zero when $y_1 = \pm h/2$.

The maximum value of the shear stress occurs at the neutral axis, where the first moment Q has its maximum value. Substituting $y_1 = 0$ into Equation (6.6), we get

$$\tau_{max} = \frac{Vh^2}{8I} = \frac{3V}{2A} \tag{6.7}$$

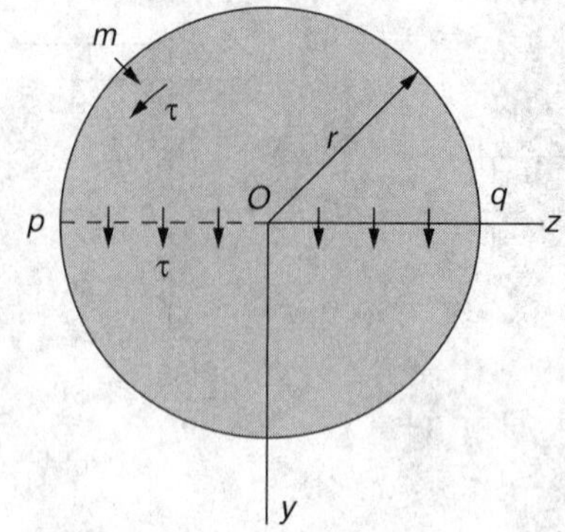

FIGURE 6.4 Shear stresses in a beam of circular cross section.

in which $A = bh$ is the cross-sectional area. Thus, the maximum shear stress is 50% larger than the average shear stress (equal to V/A). Note that the preceding equations for the shear stresses can be used to calculate either vertical shear stresses acting on a cross-section or horizontal shear stresses acting between horizontal layers of the beam.

The shear formula is valid for rectangular beams of ordinary proportions; it is exact for very narrow beams (width b much less than height h) but less accurate as b increases relative to h. For instance, when $b = h$, the true maximum shear stress is about 13% larger than the value given by Equation (6.7).

A common error is to apply the shear formula to cross-sectional shapes, such as a triangle, for which it is not applicable. The reasons it does not apply to a triangle are:

1. We assumed the cross-section had sides parallel to the y axis (so that the shear stresses acted parallel to the y axis).
2. We assumed that the shear stresses were uniform across the width of the cross-section.

These assumptions hold only in particular cases, including beams of narrow rectangular cross-section.

6.2 Shear Stresses in Circular Beams

When a beam has a circular cross-section (Figure 6.4), we can no longer assume that all of the shear stresses act parallel to the y axis. For instance, we can easily demonstrate that at a point on the boundary of the cross-section, such as point m, the shear stress τ acts tangent to the boundary. This conclusion follows from the fact that the outer surface of the beam is free of stress, and therefore the shear stress acting on the cross-section can have no component in the radial direction (because shear stresses acting on perpendicular planes must be equal in magnitude).

Although there is no simple way to find the shear stresses throughout the entire cross-section, we can readily determine the stresses at the neutral axis (where the stresses are the largest) by making some reasonable assumptions about the stress distribution. We assume that the stresses act parallel to the y axis and have constant intensity across the width of the beam (from point p to point q in Figure 6.4). Inasmuch as these assumptions are the same as those used in deriving the shear formula [Equation (6.3)], we can use that formula to calculate the shear stresses at the neutral axis. For a cross section of radius r, we obtain

$$I = \frac{\pi r^4}{4} \qquad b = 2r$$
$$Q = A_1\bar{y}_1 = \left(\frac{\pi r^2}{2}\right)\left(\frac{4r}{3\pi}\right) = \frac{2r^3}{3} \tag{6.8}$$

in which Q is the first moment of a semicircle. Substituting these expressions for I, b, and Q in the shear formula, we obtain

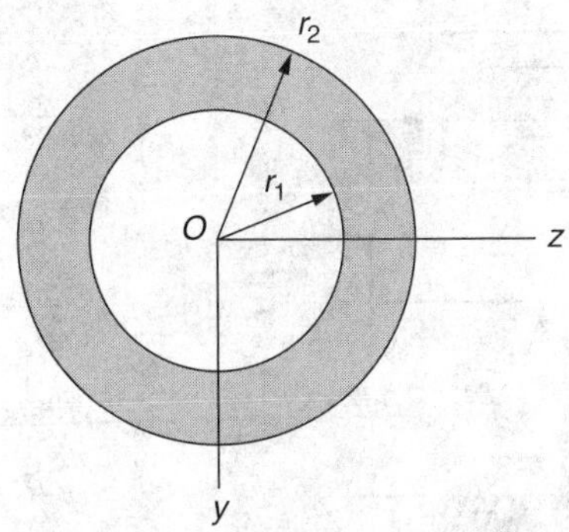

FIGURE 6.5 Shear stresses in a beam of hollow circular cross-section.

$$\tau_{max} = \frac{VQ}{Ib} = \frac{V(2r^3/3)}{(\pi r^4/4)(2r)} = \frac{4V}{3\pi r^2} = \frac{4V}{3A} \tag{6.9}$$

in which A is the area of the cross-section. This equation shows that the maximum shear stress in a circular beam is equal to 4/3 times the average shear stress V/A.

Although the preceding theory for the maximum shear stress in a circular beam is approximate, it gives results that differ by only a few percent from those obtained by more exact theories.

If a beam has a *hollow circular cross-section* (Figure 6.5), we may again assume with good accuracy that the shear stresses along the neutral axis are parallel to the y axis and uniformly distributed. Then, as before, we may use the shear formula to find the maximum shear stress. The properties of the hollow section are

$$I = \frac{\pi}{4}(r_2^4 - r_1^4) \qquad b = 2(r_2 - r_1) \qquad Q = \frac{2}{3}(r_2^3 - r_1^3) \tag{6.10}$$

and the maximum stress is

$$\tau_{max} = \frac{VQ}{Ib} = \frac{4V}{3A}\left(\frac{r_2^2 + r_2 r_1 + r_1^2}{r_2^2 + r_1^2}\right) \tag{6.11}$$

in which $A = \pi(r_2^2 - r_1^2)$ is the area of the cross-section. Note that if $r_1 = 0$, this equation reduces to Equation (6.9) for a solid circular beam.

6.3 Shear Stresses in the Webs of Beams with Flanges

When a beam of wide-flange shape [Figure 6.6(a)] is subjected to a vertical shear force, the distribution of shear stresses is more complicated than in the case of a rectangular beam. For instance, in the flanges of the beam, shear stresses act in both the vertical and horizontal directions (the y and z directions). Fortunately, the largest shear stresses occur in the web, and we can determine those stresses using the same techniques we used for rectangular beams.

Consider the shear stresses at level *ef* in the web of the beam [Figure 6.6(a)]. We assume that the shear stresses act parallel to the y axis and are uniformly distributed across the thickness of the web. Then the shear formula will still apply. However, the width b is now the thickness t of the web, and the area used in calculating the first moment Q is the area between *ef* and the bottom edge of the cross-section [that is, the shaded area of Figure 6.6(a)]. This area consists of two rectangles — the area of the flange (that is, the area below the line *abcd*) and the area *efcb* (note that we disregard the effects of the small fillets

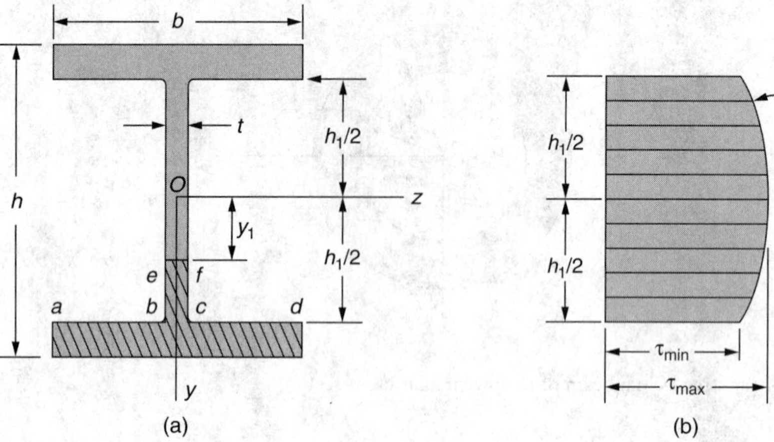

FIGURE 6.6 Shear stresses in the web of a wide-flange beam. (a) Cross-section of beam. (b) Graph showing distribution of vertical shear stresses in the web.

at the juncture of the web and flange). After evaluating the first moments of these areas and substituting into the shear formula, we get the following formula for the shear stress in the web of the beam at distance y_1 from the neutral axis:

$$\tau = \frac{VQ}{It} = \frac{V}{8It}[b(h^2 - h_1^2) + t(h_1^2 - 4y_1^2)] \tag{6.12}$$

in which I is the moment of inertia of the entire cross section, t is the thickness of the web, b is the flange width, h is the height, and h_1 is the distance between the insides of the flanges. The expression for the moment of inertia is

$$I = \frac{bh^3}{12} - \frac{(b-t)h_1^3}{12} = \frac{1}{12}(bh^3 - bh_1^3 + th_1^3) \tag{6.13}$$

Equation (6.12) is plotted in Figure 6.6(b), and we see that τ varies quadratically throughout the height of the web (from $y_1 = 0$ to $y_1 = \pm h_1/2$).

The maximum shear stress in the beam occurs in the web at the neutral axis ($y_1 = 0$), and the minimum shear stress in the web occurs where the web meets the flanges ($y_1 = \pm h_1/2$). Thus, we find

$$\tau_{max} = \frac{V}{8It}(bh^2 - bh_1^2 + th_1^2) \qquad \tau_{min} = \frac{Vb}{8It}(h^2 - h_1^2) \tag{6.14}$$

For wide-flange beams having typical cross-sectional dimensions, the maximum stress is 10 to 60% greater than the minimum stress. Also, the shear stresses in the web typically account for 90 to 98% of the total shear force; the remainder is carried by shear in the flanges.

When designing wide-flange beams, it is common practice to calculate an approximation of the maximum shear stress by dividing the total shear force by the area of the web. The result is an average shear stress in the web:

$$\tau_{ave} = \left(\frac{V}{th_1}\right) \tag{6.15}$$

For typical beams, the average stress is within 10% (plus or minus) of the actual maximum shear stress.

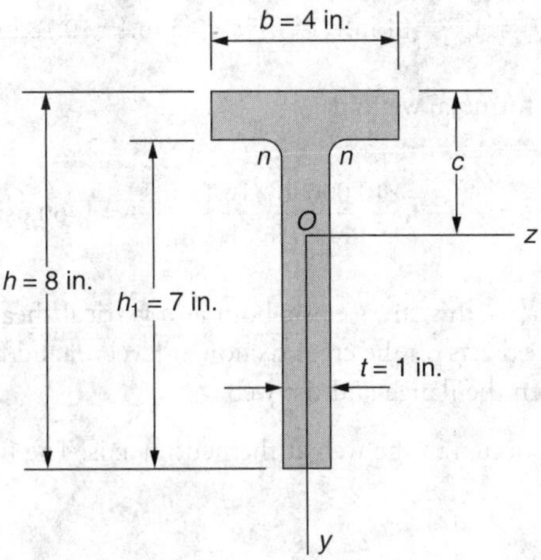

FIGURE 6.7 Example.

The elementary theory presented in the preceding paragraphs is quite satisfactory for determining shear stresses in the web. However, when investigating shear stresses in the flanges, we can no longer assume that the shear stresses are constant across the width of the section, that is, across the width b of the flanges [Figure 6.6(a)]. For instance, at the junction of the web and lower flange ($y_1 = h_1/2$), the width of the section changes abruptly from t to b. The shear stress at the free surfaces ab and cd [Figure 6.6(a)] must be zero, whereas across the web at bc the stress is τ_{min}. These observations indicate that at the junction of the web and either flange the distribution of shear stresses is more complex and cannot be investigated by an elementary analysis. The stress analysis is further complicated by the use of fillets at the reentrant corners, such as corners b and c. Without fillets, the stresses would become dangerously large. Thus, we conclude that the shear formula cannot be used to determine the vertical shear stresses in the flanges. (Further discussion of shear stresses in thin-walled beams can be found in the references.)

The method used above to find the shear stresses in the webs of wide-flange beams can also be used for certain other sections having thin webs, such as T-beams.

Example

A beam having a T-shaped cross section (Figure 6.7) is subjected to a vertical shear force $V = 10{,}000$ lb. The cross-sectional dimensions are $b = 4$ in., $t = 1$ in., $h = 8$ in., and $h_1 = 7$ in. Determine the shear stress τ_1 at the top of the web (level nn) and the maximum shear stress τ_{max}. (Disregard the areas of the fillets.)

Solution

The neutral axis is located by calculating the distance c from the top of the beam to the centroid of the cross-section. The result is

$$c = 3.045 \text{ in.}$$

The moment of inertia I of the cross-sectional area about the neutral axis (calculated with the aid of the parallel-axis theorem) is

$$I = 69.66 \text{ in.}^4$$

To find the shear stress at the top of the web we need the first moment Q_1 of the area above level nn. Thus, Q_1 is equal to the area of the flange times the distance from the neutral axis to the centroid of the flange:

$$Q_1 = A_1\bar{y}_1 = (4 \text{ in.})\,(1 \text{ in.})\,(c - 0.5 \text{ in.}) = 10.18 \text{ in.}^3$$

Substituting into the shear formula, we find

$$\tau_1 = \frac{VQ_1}{It} = \frac{(10\,000 \text{ lb})\,(10.18 \text{ in.}^3)}{(69.66 \text{ in.}^4)\,(1 \text{ in.})} = 1460 \text{ psi}$$

Like all shear stresses in beams, this stress exists both as a vertical shear stress and as a horizontal shear stress. The vertical stress acts on the cross section at level *nn* and the horizontal stress acts on the horizontal plane between the flange and the web.

The maximum shear stress occurs in the web at the neutral axis. The first moment Q_2 of the area below the neutral axis is

$$Q_2 = A_2\bar{y}_2 = (1 \text{ in.})\,(8 \text{ in.} - c)\left(\frac{8 \text{ in.} - c}{2}\right) = 12.28 \text{ in.}^3$$

Substituting into the shear formula, we obtain

$$\tau_{max} = \frac{VQ_2}{It} = \frac{(10\,000 \text{ lb})\,(12.28 \text{ in.}^3)}{(69.66 \text{ in.}^4)\,(1 \text{ in.})} = 1760 \text{ psi}$$

which is the maximum shear stress in the T-beam.

Defining Terms

Shear formula — The formula $\tau = VQ/Ib$ giving the shear stresses in a rectangular beam of linearly elastic material [Equation (6.3)].

(See also Defining Terms for Chapter 5.)

References

Beer, F. P., Johnston, E. R., and DeWolf, J. T. 2001. *Mechanics of Materials,* 3rd Ed. McGraw-Hill, Inc., New York.

Gere, J. M. 2001. *Mechanics of Materials,* 5th Ed. Brooks/Cole, Pacific Grove, CA.

Hibbeler, R. C. 2000. *Mechanics of Materials,* 4th Ed., Prentice Hall, Inc., Upper Saddle River, NJ.

Lardner, T. J. and Archer, R. R. 1994. *Mechanics of Solids,* McGraw-Hill, Inc., New York.

Popov, E. P. and Balan, T. A. 1999. *Engineering Mechanics of Solids,* 2nd Ed., Prentice Hall, Inc., Upper Saddle River, NJ.

Further Information

Extensive discussions of bending — with derivations, examples, and problems — can be found in textbooks on mechanics of materials, such as those listed in the References. These books also cover many additional topics pertaining to shear stresses in beams. For instance, built-up beams, nonprismatic beams, shear centers, and beams of thin-walled open cross-section are discussed in Gere [2001].

7

Shear and Moment Diagrams

George R. Buchanan
Tennessee Technological University

Computations for shear force and bending moment are absolutely necessary for the successful application of the theory and concepts presented in the previous chapters. The discussion presented here is an extension of Chapter 4, and the reader should already be familiar with computations for reactions. This chapter will concentrate on **statically determinate** beams. Statically indeterminate beams and frames constitute an advanced topic, and the reader is referred to the end of the chapter for further information in this area. Even though the problems that illustrate shear and moment concepts appear as structural beams, the reader should be aware that the same concepts apply to structural machine parts. A distinction should not be drawn between civil engineering and mechanical engineering problems because the methods of analysis are the same.

7.1 Sign Convention

The sign convention for moment in a beam is based on the behavior of the loaded beam. The sign convention for shear in a beam is dictated by the convenience of constructing a shear diagram using a load diagram. The sign convention is illustrated in Figure 7.1. The x axis corresponds to the longitudinal axis of the beam and must be directed from left to right. This convention dictates that shear and moment diagrams should be drawn from left to right. The direction of the positive y axis will be assumed upward, and loads that act downward on the beam will be negative. Note that it is not mandatory in shear and moment computations for positive y to be directed upward or even defined since the sign convention is independent of the vertical axis. However, for the more advanced topic of **beam deflections** positive y must be defined [Buchanan, 1988]. Positive bending moment causes compression at the top of the beam and negative bending moment causes compression at the bottom of the beam. Positive shear forces act downward on the positive face of the **free body** as shown in Figure 7.1.

7.2 Shear and Moment Diagrams

Two elementary differential equations govern the construction of shear and moment diagrams and can be derived using a free-body diagram similar to Figure 7.1, with w corresponding to a continuous load acting along the length of the beam.

0-8493-1586-7/05/$0.00+$1.50

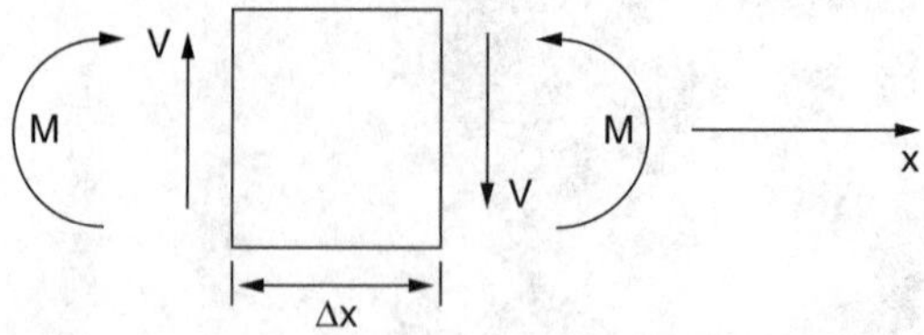

FIGURE 7.1 Beam element showing positive shear force V and positive bending moment M.

$$dV = w dx \quad \text{or} \quad \int_{V_1}^{V_2} dV = \int_{x_1}^{x_2} w dx \tag{7.1}$$

$$dM = V dx \quad \text{or} \quad \int_{M_1}^{M_2} dM = \int_{x_1}^{x_2} V dx \tag{7.2}$$

The differential equations show that the change in shear V between any two points x_1 and x_2 on a beam is equal to the area of the load diagram between those same two points and, similarly, that the change in bending moment M is equal to the area of the shear diagram. It follows that the slope of a tangent drawn at any point on the moment diagram is given by dM/dx and corresponds to the magnitude of V at that point. When the tangent has zero slope, $dM/dx = 0$, that corresponds to a **maximum or minimum moment** and can be located by examining the shear diagram for a point (x location) where $V = 0$. Locating the largest positive or negative bending moment is important for properly designing beam structures when using the equations of the previous chapters.

Shear and moment diagrams (as opposed to shear and moment equations) offer the most efficient method for analyzing beam structures for shear and moment when the beam loading can be represented as **concentrated loads** or **uniform continuous loads**. An elementary example will serve to illustrate the concept. Consider the **simply supported beam** of Figure 7.2. There is a single concentrated load with reactions as shown. The shear is obtained by directly plotting the load; the sign convention of Figure 7.1 specifically allows for this. The reaction on the left is plotted upward as the change in the shear at a point, $x = 0$. The area of the load diagram between $x = 0$ and $x = a$ is zero since the load is zero; it follows that the change in shear is zero and the shear diagram is a straight horizontal line extending from the left end of the beam to the concentrated load P. The load changes abruptly by an amount P downward and a corresponding change is noted on the shear diagram. Positive shear is above the axis of the shear diagram. There is no change in load between $x = a$ and $x = L$ and the shear remains constant. The reaction at the right end of the beam is upward and the shear is plotted upward to return to zero. The beam is simply supported, indicating that the moment must be zero at the supports. The change in moment between the left support and the point where the load is applied, $x = a$, is equal to the area of the shear diagram or a positive Pab/L. The variation in moment appears as a straight line (a line with constant slope) connecting $M = 0$ at $x = 0$ with $M = Pab/L$ at $x = a$. The area of the remaining portion of the shear diagram is $-Pab/L$ and, when plotted on the moment diagram, returns the moment to zero and satisfies the simply supported boundary condition.

Equation (7.1) and Equation (7.2) indicate that the load function is integrated to give the shear function, and similarly the shear function is integrated to give the moment function. The diagrams are a graphical illustration of the integration. An important point is that the order (power) of each function increases by one as the analyst moves from load to shear to moment. In Figure 7.2, note that when the load function is zero (1) the corresponding shear function is a constant and (2) the corresponding moment function is linear in x and is plus or minus corresponding to the sign of the area of the shear diagram.

Consider, as a second example, the beam of Figure 7.3, where a series of uniform loads and concentrated loads is applied to a beam with an overhang. The reactions are computed and shown in the figure. The shear diagram is plotted starting at the left end of the beam. Shear and moment diagrams are always

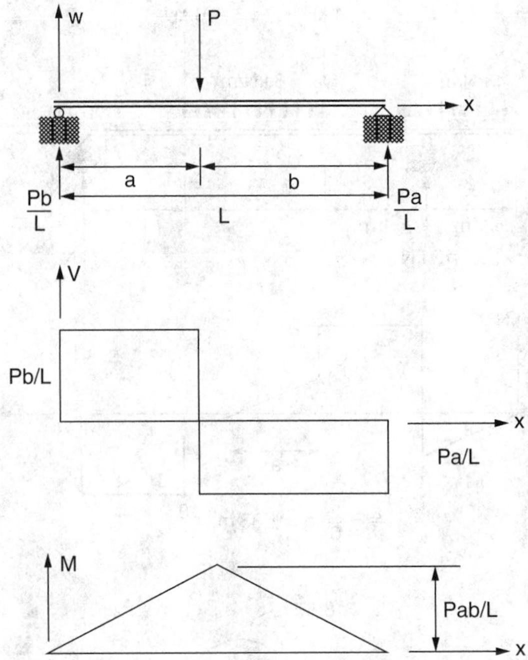

FIGURE 7.2 Shear and moment diagrams for a simple beam with a concentrated load.

constructed from left to right because Equation (7.1) and Equation (7.2) were derived in a coordinate system that is positive from left to right. The area of the load between $x = 0$ and $x = 1$ m is –4 kN; this value is the change in shear, which is plotted as a sloping line. The corresponding change in bending moment equals the area of the shear diagram, (–4 kN) (1 m)/2 = 2 kN · m. A curve with continually changing negative slope between $x = 0$ and $x = 1$ m is shown in Figure 7.3. The point of zero shear occurs in the beam section, 2 m $\leq x \leq$ 4 m, and is located using similar triangles as shown in the space between the shear and moment diagrams. For practice the reader can verify the construction of the shear and moment diagrams of Figure 7.3. A textbook on mechanics of materials or structural analysis should have a complete discussion of the topic. Again, refer to "Further Information."

Concentrated loads and uniform loads lead to shear diagrams with areas that will always be rectangles or triangles; the change in moment is easily computed. The uniformly varying load shown in Figure 7.4(e) sometimes occurs in practice; locating the point of maximum moment (point of zero shear) for some boundary conditions is not so elementary when using geometrical relationships. In such cases the use of shear and moment equations becomes a valuable analysis tool.

7.3 Shear and Moment Equations

Shear and moment equations are equations that represent the functions shown in Figure 7.2 and Figure 7.3. As with any mathematical function, they must be referenced to a coordinate origin, usually the left end of the beam. Shear and moment equations are piecewise continuous functions. Note that two separate equations are required to describe the shear diagram of Figure 7.2 and, similarly, four equations are required to describe the shear diagram of Figure 7.3. The same is true for the moment diagram.

The following procedure can be used to write shear and moment equations for almost any beam loading:

1. Choose a coordinate origin for the equation, usually, but not limited to, the left end of the beam.
2. Pass a free-body cut through the beam section where the shear and moment equations are to be written.

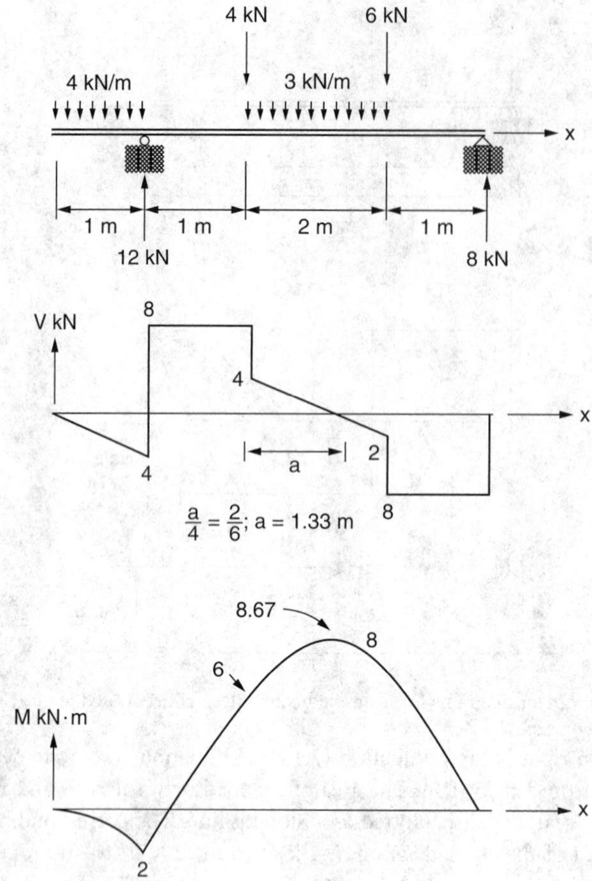

FIGURE 7.3 Shear and moment diagrams for a beam with an overhang. The concentrated loads and uniform loads illustrate the concept of maximum moment and corresponding zero shear.

3. Choose the free body that contains the coordinate origin.
4. Assume positive unknown shear and moment at the free-body cut using the sign convention defined by Figure 7.1.
5. View the free body as a **free-fixed beam** with the fixed end being at the free-body cut and the beam extending toward the coordinate origin.
6. Statically solve for the unknown shear and moment as if they were the reactions at the fixed end of any beam, that is, $\sum F = 0$ and $\sum M = 0$.
7. Always sum moments at the free-body cut such that the unknown shear passes through that point. An example should illustrate the concept.

A complete description of the beam of Figure 7.3 would require four shear and moment equations. Consider a free body of the first section — the uniformly loaded overhang. The free body is shown in Figure 7.5(a). Compare Figure 7.4(b) with Figure 7.5(a); *L* is merely replaced with *x*, the length of the free-body section. The seven steps outlined above have been followed to give

$$V_x = -wx = -4x \text{ kN}, \quad M_x = -wx^2/2 = -4x^2/2 \text{ kN}\cdot\text{m}, \quad 0 \le x \le 1 \text{ m} \tag{7.3}$$

The second segment ($1 \text{ m} \le x \le 2 \text{ m}$) is shown in Figure 7.5(b) and can be compared with Figure 7.4(c) and Figure 7.4(a): merely replace *L* with an *x*, *a* with 1 m, *P* with 12 kN, and *w* with 4 kN.

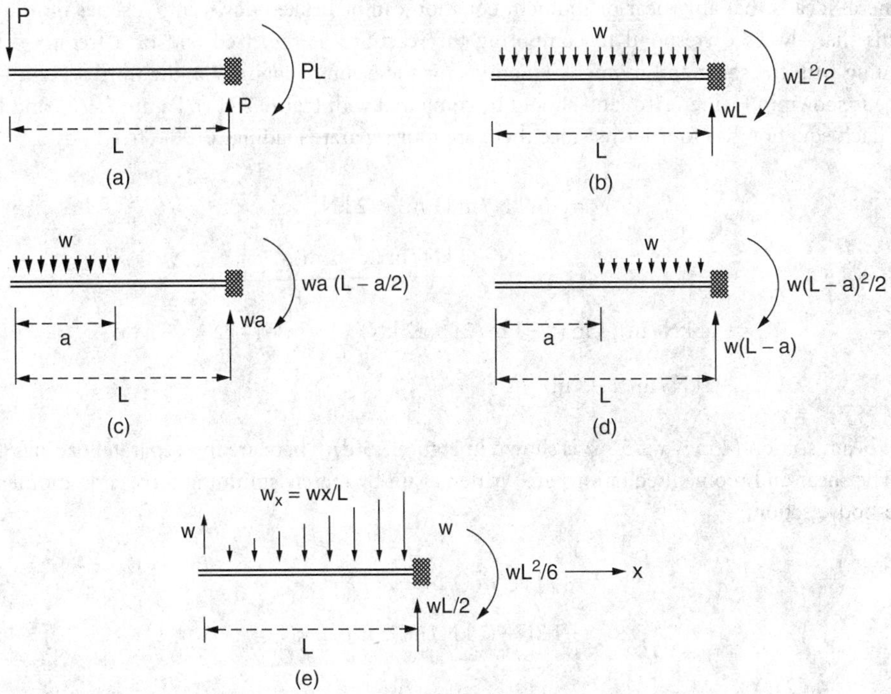

FIGURE 7.4 Shear and moment reactions for free-fixed beams.

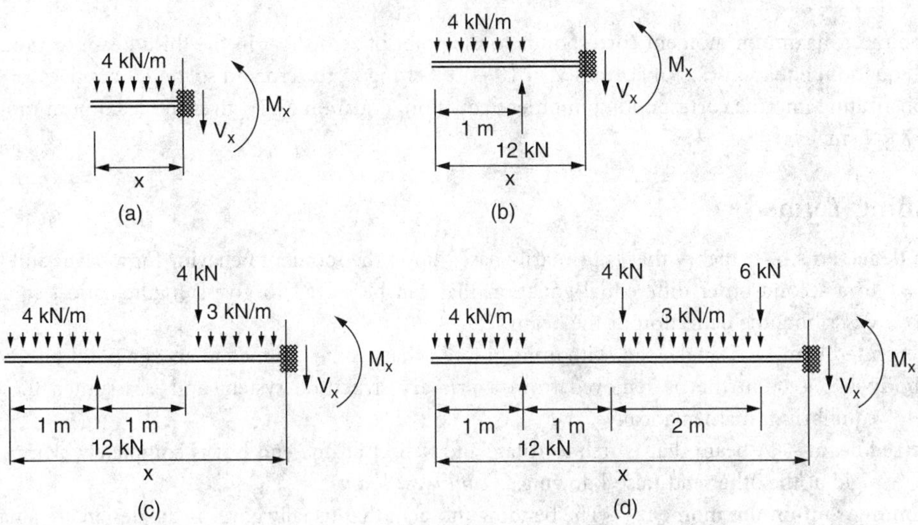

FIGURE 7.5 Free-body diagrams for the beam shown in Figure 7.3.

$$V_x = -wa + P = -(4\text{ kN/m})\,(1\text{ m}) + 12\text{ kN}, \tag{7.4}$$

$$\begin{aligned} M_x &= -wa\,(x - a/2) + P(x - a) \\ &= -(4)(1)\,(x - 1/2)\text{ kN}\cdot\text{m} + 12(x - 1)\text{ kN}\cdot\text{m} \end{aligned} \tag{7.5}$$

The general idea is that any shear or moment equation can be broken down into a series of individual problems that always correspond to computing the reactions at the fixed end of a free-fixed beam. Continuing with the shear and moment equations for the beam of Figure 7.3, the third section (2 m ≤ $x \le 4$ m) is shown in Figure 7.5(c) and should be compared with Figure 7.4(a), Figure 7.4(c), and Figure 7.4(d). Each equation has four terms since there are four separate loadings on the free body.

$$\begin{aligned} V_x &= -(4\text{ kN/m})(1\text{ m}) + 12\text{ kN} \\ &\quad -4\text{ kN} - (3\text{ kN/m})(x - 2\text{ m}), \end{aligned} \tag{7.6}$$

$$\begin{aligned} M_x &= -(4\text{ kN/m})(1\text{ m})(x - 1\text{ m}/2) + (12\text{ kN})(x - 1\text{ m}) - (4\text{ kN})(x - 2\text{ m}) \\ &\quad -(3\text{kN/m})(x - 2\text{ m})^2/2 \end{aligned} \tag{7.7}$$

The last beam section (4 m $\le x \le$ 5 m) is shown in Figure 7.5(d). There are five separate loadings on the beam. The shear and moment equations are written again by merely summing forces and moments on the free-body section.

$$\begin{aligned} V_x &= -(4\text{ kN/m})(1\text{ m}) + 12\text{ kN} \\ &\quad -4\text{ kN} - (3\text{ kN/m})(2\text{ m}) - 6\text{ kN}, \end{aligned} \tag{7.8}$$

$$\begin{aligned} M_x &= -(4\text{ kN/m})(1\text{ m})(x - 1\text{ m}/2) + (12\text{ kN})(x - 1\text{ m}) - (4\text{ kN})(x - 2\text{ m}) \\ &\quad -(3\text{ kN/m})(2\text{ m})(x - 3\text{ m}) - (6\text{ kN})(x - 4\text{ m}) \end{aligned} \tag{7.9}$$

The point of maximum moment corresponds to the point of zero shear in the third beam section. The shear equation, Equation (7.6), becomes $V_x = 10 - 3x$. Setting V_x to zero and solving for x gives $x = 3.33$ m. Substituting into the corresponding moment equation, Equation (7.7), gives the maximum moment as 8.67 kN · m.

Defining Terms

Beam deflections — A theory that is primarily based upon the moment behavior for a beam and leads to a second-order differential equation that can be solved to give a mathematical equation describing the deflection of the beam.

Concentrated load — A single load, with units of force, that can be assumed to act at a point on a beam.

Free body — A section that is removed from a primary structural system and is assumed to be in equilibrium mathematically.

Free-fixed beam — A beam that is free to rotate and deflect at one end but is completely clamped or rigid at the other end (also known as a *cantilever beam*).

Maximum or minimum moment — The bending moment that usually governs the design and analysis of beam structures.

Simply supported beam — A beam that is supported using a pin (hinge) at one end and a surface at the other end, with freedom to move along the surface

Statically determinate beam — A beam that can be analyzed for external reactions using only the equations of engineering mechanics and statics.

Uniform continuous load — A distributed beam loading of constant magnitude, with units of force per length, that acts continuously along a beam segment.

Reference

Buchanan, G. R. 1988. Shear and moment in beams, Chap. 5, and Deflection of beams, Chap. 10, in *Mechanics of Materials.* Holt, Rinehart and Winston, New York.

Further Information

Hibbeler, R. C. 1985. *Structural Analysis.* Macmillan, New York. Chapter 3 contains a discussion of shear and moment concepts. Chapters 8 and 9 cover fundamental concepts for analysis of indeterminate beams.

McCormac, J. and Elling, R. E. 1988. *Structural Analysis.* Harper and Row, New York. Chapter 3 contains a discussion of shear and moment concepts. Chapters 10, 11, and 13 cover fundamental concepts for analysis of indeterminate structures.

Gere, J. M. and Timoshenko, S. P. 1990. *Mechanics of Materials,* 3rd ed. PWS, Boston. Chapter 4 contains a discussion of shear and moment concepts. Beam deflections are covered in Chaps. 7, 8, and 10.

Nash, W. A. 1994. *Theory and Problems of Strength of Materials,* 3rd ed., McGraw-Hill, New York. Numerous solved problems for shear and moment are given in Chap. 6.

8

Columns

Loren W. Zachary
Iowa State University

John B. Ligon
Michigan Technological University

A column is an initially straight load-carrying member that is subjected to a compressive axial load. The failure of a column in compression is different from one loaded in tension. Under compression, a column can deform laterally or buckle, and this deflection can become excessive. The buckling of columns is a major cause of failure. To illustrate the fundamental aspects of the buckling of long, straight, prismatic bars, consider a thin meter stick. If a tensile axial load is applied to the meter stick, the stable equilibrium position is that of a straight line. If the stick is given a momentary side load to cause a lateral deflection, upon its release the stick immediately returns to the straight line configuration. If a compressive axial load is applied, a different result may occur. At small axial loads, the meter stick will again return to a straight line configuration after being displaced laterally. At larger loads the meter stick will remain in the displaced position. With an attempt to increase the axial load acting on the buckled column, the lateral deformations become excessive and failure occurs.

In theory, a column that is long and perfectly straight is in stable equilibrium for small loads up to a specific **critical buckling load**. At this critical buckling load, the beam will remain straight unless it is perturbed and displays large lateral deformations. This is a bifurcation point since the column can be in equilibrium with two different shapes — laterally displaced or perfectly straight. The load in this neutral equilibrium state is the critical buckling load and, for long slender columns, is referred to as the **Euler buckling load**. At loads higher than the critical load the beam is in unstable equilibrium.

8.1 Fundamentals

Buckling of Long Straight Columns

In 1757, Leonhard Euler published the solution to the problem of long slender columns buckling under compressive loads. Figure 8.1(a) shows a column that is deflected in the lateral direction. The load P_{cr} is the smallest load that will just hold the column in the laterally deflected shape. The ends of the beam are free to rotate and are commonly referred to as being *pinned, hinged,* or simply *supported.* The following assumptions are used in determining P_{cr}:

1. The beam is initially straight with a constant cross-section along its length.
2. The material is linearly elastic, isotropic, and homogeneous.
3. The load is applied axially through the centroidal axis.

0-8493-1586-7/05/$0.00+$1.50

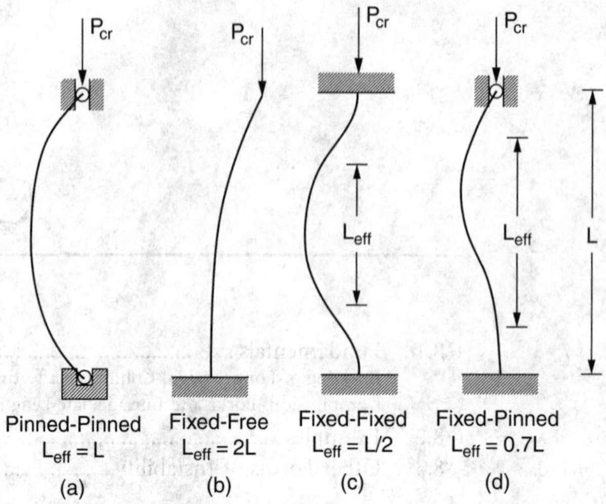

FIGURE 8.1 Effective column lengths.

4. The ends are pinned–pinned (pinned end condition at both ends).
5. No residual stresses exist in the column prior to loading.
6. No distortion or twisting of the cross-section occurs during loading.
7. The classical differential equation for the elastic curve can be used since the deflections are small.

Standard mechanics of materials textbooks [Riley and Zachary, 1989] and structural stability textbooks [Chajes, 1974] contain the derivation of the following formula:

$$\sqrt{\frac{P_{cr}}{EI}}L = n\pi, \qquad n = 1, 2, 3, \ldots \tag{8.1}$$

The smallest value for P_{cr} occurs when $n = 1$. Larger values of n give magnitudes of P_{cr} that will never be reached in practice.

$$P_{cr} = \frac{\pi^2 EI}{L^2} \tag{8.2}$$

The Euler buckling load, P_{cr}, is calculated using the moment of inertia I of the column cross-section about which axis buckling (bending) occurs. The moment of inertia can also be written in terms of the radius of gyration about the same axis:

$$I = Ar^2 \tag{8.3}$$

Table 8.1 gives some formulas that are helpful in determining the radius of gyration and moment of inertia. Using Equation (8.2) and Equation (8.3), the Euler buckling stress, σ_{cr}, in terms of the **slenderness ratio** L/r, is

$$\sigma_{cr} = \frac{P_{cr}}{A} = \frac{\pi^2 E}{(L/r)^2} \tag{8.4}$$

Table 8.1 Properties of Selected Areas

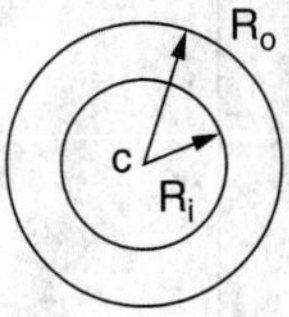

$$A = \pi(R_o^2 - R_i^2)$$

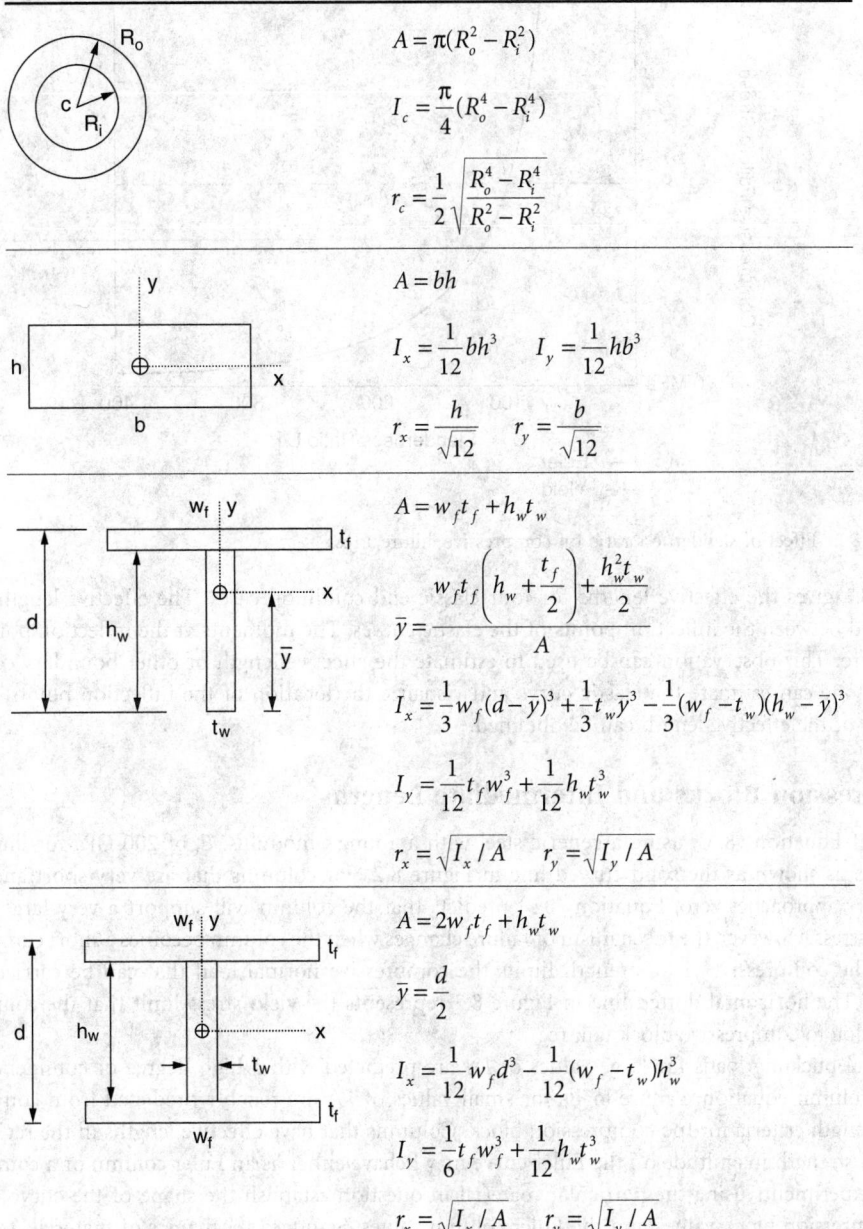

$$I_c = \frac{\pi}{4}(R_o^4 - R_i^4)$$

$$r_c = \frac{1}{2}\sqrt{\frac{R_o^4 - R_i^4}{R_o^2 - R_i^2}}$$

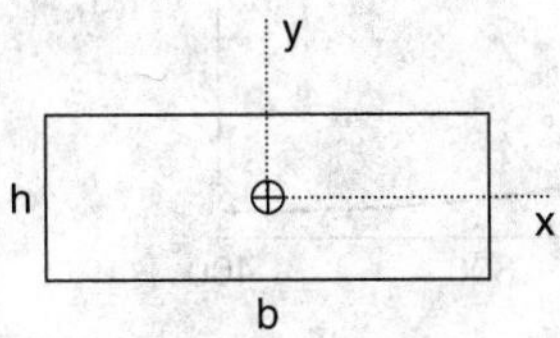

$$A = bh$$

$$I_x = \frac{1}{12}bh^3 \qquad I_y = \frac{1}{12}hb^3$$

$$r_x = \frac{h}{\sqrt{12}} \qquad r_y = \frac{b}{\sqrt{12}}$$

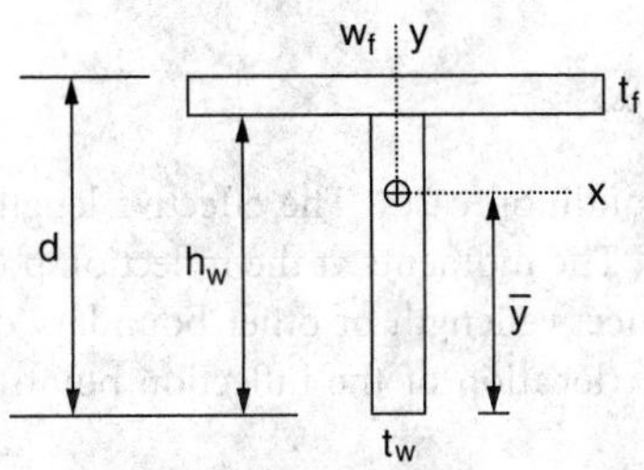

$$A = w_f t_f + h_w t_w$$

$$\bar{y} = \frac{w_f t_f\left(h_w + \frac{t_f}{2}\right) + \frac{h_w^2 t_w}{2}}{A}$$

$$I_x = \frac{1}{3}w_f(d-\bar{y})^3 + \frac{1}{3}t_w\bar{y}^3 - \frac{1}{3}(w_f - t_w)(h_w - \bar{y})^3$$

$$I_y = \frac{1}{12}t_f w_f^3 + \frac{1}{12}h_w t_w^3$$

$$r_x = \sqrt{I_x / A} \qquad r_y = \sqrt{I_y / A}$$

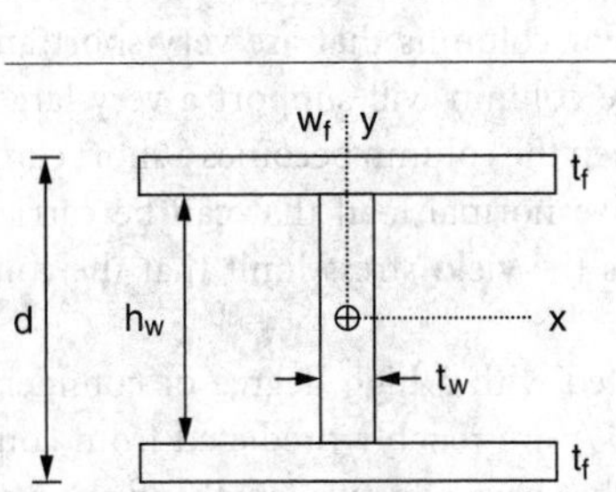

$$A = 2w_f t_f + h_w t_w$$

$$\bar{y} = \frac{d}{2}$$

$$I_x = \frac{1}{12}w_f d^3 - \frac{1}{12}(w_f - t_w)h_w^3$$

$$I_y = \frac{1}{6}t_f w_f^3 + \frac{1}{12}h_w t_w^3$$

$$r_x = \sqrt{I_x / A} \qquad r_y = \sqrt{I_y / A}$$

Effective Lengths

The development given above is for a beam with simple supports at both ends. Other boundary conditions give equations similar to Equation (8.4) if the physical length of the beam, L, is replaced by the effective length L_{eff}.

$$\sigma_{cr} = \frac{\pi^2 E}{(L_{\text{eff}} / r)^2} \tag{8.5}$$

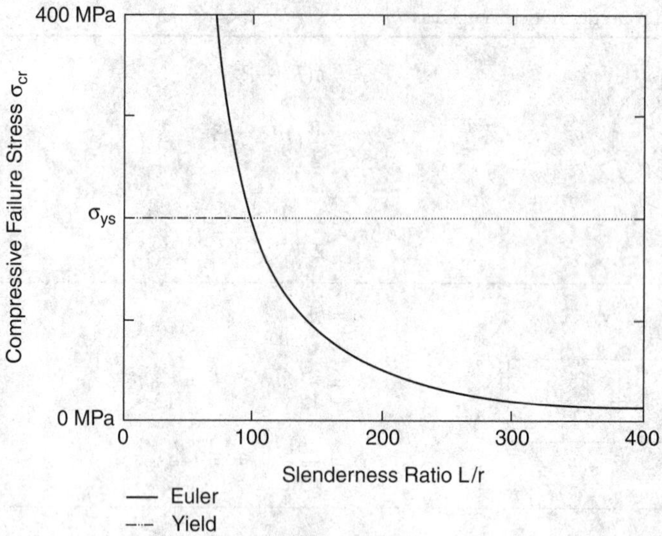

FIGURE 8.2 Effect of slenderness ratio on compressive failure stress.

Figure 8.1 gives the effective lengths for four classic end condition cases. The effective lengths are all measured between the inflection points of the elastic curves. The moments at the inflection points, $y'' = 0$, are zero. This observation can be used to estimate the effective length of other boundary condition cases. If one can estimate the elastic curve and visualize the location of the inflection points, a rough estimate of the effective length can be obtained.

Compression Blocks and Intermediate Lengths

A plot of Equation (8.5), using a generic steel with a Young's modulus, E, of 200 GPa for illustration purposes, is shown as the solid curved line in Figure 8.2. For columns that are very short and stocky where L/r approaches zero, Equation (8.5) predicts that the column will support a very large load or normal stress. However, the mechanism of failure changes when the column becomes a short compression block. The compressive yield strength limits the compressive normal load that can be carried by the column. The horizontal dotted line in Figure 8.2 represents the yield stress limit that the column can sustain due to compressive block failure.

Critical buckling loads for large values of L/r are predicted with a high degree of confidence using Euler's column equation. Failure loads for small values of L/r are reliably predicted from compressive yield strength criteria for the compression block. Columns that have effective lengths in the region near the yield strength magnitude on the Euler curve may behave either as an Euler column or a compressive block. Experiments using the particular material in question establish the shape of the curve between the compression block values and the Euler column values. Standard mechanics of materials textbooks [Riley and Zachary, 1989] give empirical formulas for this range. Using a horizontal line to cover the complete compression block and intermediate ranges usually predicts a higher critical buckling load than is found experimentally.

8.2 Examples

The following examples illustrate the procedure for determining the critical buckling load and stress for columns of several different cross-sections and effective lengths. Initially, the slenderness ratio for the two principal directions of possible buckling must be calculated to determine which direction controls. The direction with the largest slenderness ratio will give the smallest buckling load.

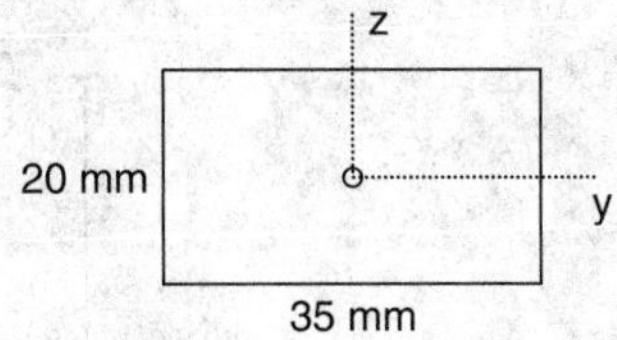

FIGURE 8.3 Cross-section used in Example 8.1.

Example 8.1: Checking Both Directions for Buckling

Consider the rectangular cross-section shown in Figure 8.3. Both ends of the column are pinned for the pinned–pinned condition in Figure 8.1(a). The physical length of the column is 1.5 m. Determine the Euler buckling load if the column is made of steel ($E = 200$ GPa). Using Table 8.1,

$$A = (20 \text{ mm})(35 \text{ mm}) = 700 \text{ mm}^2$$

$$r_y = \frac{h}{\sqrt{12}} = \frac{20 \text{ mm}}{\sqrt{12}} = 5.77 \text{ mm}$$

$$r_z = \frac{b}{\sqrt{12}} = \frac{32 \text{ mm}}{\sqrt{12}} = 10.10 \text{ mm}$$

The column tends to bend (buckle) about the y axis since r_y is smaller than r_z, producing the maximum L/r and the smallest P_{cr} and σ_{cr} for the 1.5-m length.

If the ends of the column are later restrained or fixed with respect to buckling about one of the axes — say the y axis — this changes the boundary conditions for buckling to the fixed–fixed condition [Figure 8.1(c)] for that direction. The effective length of the beam is then half the physical length or 750 mm for buckling about the y axis.

$$\frac{L_y}{r_y} = \frac{L/2}{r_y} = \frac{750 \text{ mm}}{5.77 \text{ mm}} = 130$$

$$\frac{L_z}{r_z} = \frac{L}{r_z} = \frac{1500 \text{ mm}}{10.10 \text{ mm}} = 149$$

The column will now buckle about the z axis before the load can become large enough to cause buckling about the y axis. Although the moment of inertia and radius of gyration about the y axis are smaller than about the z axis, the end conditions significantly influence the slenderness ratio and the axis about which buckling occurs.

$$P_{cr} = \frac{\pi^2 EA}{(L_z/r_z)^2} = \frac{\pi^2 200(10)^9 \text{ N/m}^2 \times 700(10)^{-6} \text{ m}^2}{(149)^2} = 62.2 \text{ kN} \qquad \textit{Ans.}$$

$$\sigma_{cr} = \frac{P_{cr}}{A} = \frac{62.2(10)^3 \text{ N}}{700(10)^{-6} \text{ m}^2} = 88.9 \text{ MPa} \qquad \textit{Ans.}$$

According to the Euler buckling formula, any load below 62.2 kN will not cause the column to buckle. No factor of safety is included in the calculations. In practice, however, such a factor should

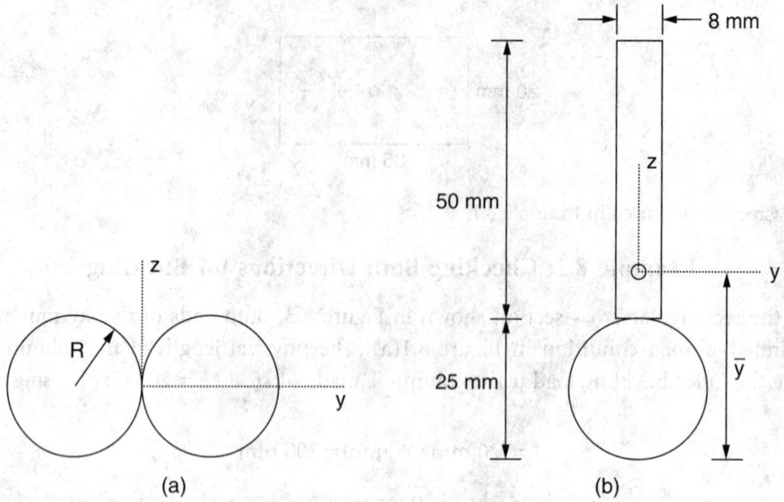

FIGURE 8.4 Cross-sections used in Example 8.2a and Example 8.2b.

be used, since real columns are never ideally straight nor is column loading purely axial. Some inadvertent bending moment or load eccentricity is always possible. The σ_{cr} calculated above is the axial compressive stress in the beam just before the beam deforms laterally. This stress is relatively small when compared to a yield strength for structural steel of approximately 250 MPa. The compressive stress in long slender beams at or below the critical buckling stress σ_{cr} can be much less than the yield strength of the material. It is imperative that failure due to buckling be checked for compressive loads.

Example 8.2a: Built-Up Section

A single 20-mm solid steel rod is being used in compression. It has been determined that the rod will buckle. It has been decided that two rods will be welded together — Figure 8.4(a) — in order to increase the buckling load. Does this increase the Euler buckling load?

The parallel axis theorem can be used to determine the combined moments of inertia and radii of gyration. The term d in the following equation is the transfer distance from the centroid of the component area to the centroid of the entire cross-section.

$$I = \sum (I_c + Ad^2)$$

$$r_z = \sqrt{I_z / A} = \sqrt{\frac{2Ar_{cz}^2 + 2Ad_z^2}{2A}} = \sqrt{r_{cz}^2 + d_z^2} = \sqrt{(R/2)^2 + R^2} = \sqrt{5}\frac{R}{2}$$

$$r_y = \sqrt{I_y / A} = \sqrt{\frac{2Ar_{cy}^2 + 2Ad_y^2}{2A}} = \sqrt{r_{cy}^2 + d_y^2} = \sqrt{(R/2)^2 + 0} = \frac{R}{2}$$

The radius of gyration is not increased for bending about the y axis compared to the single rod value of $R/2$. The buckling load will remain the same even though the column has been stiffened in one direction.

Example 8.2b: Built-Up Section

Consider the composite aluminum ($E = 70$ GPa) section in Figure 8.4(b). Determine the maximum compressive stress that can be applied. The end conditions are as follows: about the y axis, the ends are fixed–pinned [see Figure 8.1(d)], and about the z axis, the ends are fixed–fixed [see Figure 8.1(d)]. The column length is 2 m.

Compared to Example 8.2a, the radius of gyration about the y axis has the same basic definition, but the details are slightly different.

$$A_{rect} = (8\text{ mm})(50\text{ mm}) = 400\text{ mm}^2$$

$$A_{rod} = \pi(12.5\text{ mm})^2 = 490.9\text{ mm}^2$$

$$\bar{y} = \frac{\sum A\bar{y}}{\sum A} = \frac{(400\text{ mm}^2)50\text{ mm} + (490.9\text{ mm})12.5\text{ mm}}{400\text{ mm}^2 + 490.9\text{ mm}^2} = 29.34\text{ mm}$$

$$r_y = \sqrt{\frac{\sum I_y}{\sum A}} = \sqrt{\frac{\sum A(r_y^2 + d_y^2)}{\sum A}}$$

$$= \sqrt{\frac{400\left(\left(\frac{50}{\sqrt{12}}\right)^2 + (50-29.34)^2\right) + 490.9\left(\left(\frac{12.5}{2}\right)^2 + (29.34-12.5)^2\right)}{890.9}}$$

$$= 21.52\text{ mm}$$

$$r_z = \sqrt{\frac{\sum I_z}{\sum A}} = \sqrt{\frac{\sum A(r_z^2 + d_z^2)}{\sum A}}$$

$$= \sqrt{\frac{400\left(\frac{8}{\sqrt{12}}\right)^2 + 490.9\left(\frac{12.5}{2}\right)^2}{890.9}} = 4.891\text{ mm}$$

The slenderness ratio for buckling about the z axis controls since r_z is less than one-fourth of the value about the y axis:

$$\frac{L_y}{r_y} = \frac{0.7L}{r_y} = \frac{0.7(2000\text{ mm})}{21.52\text{ mm}} = 65.1$$

For many materials a slenderness ratio of 65.1 places the beam in the intermediate length range.

$$\frac{L_z}{r_z} = \frac{L/2}{r_z} = \frac{1000\text{ mm}}{4.891\text{ mm}} = 204.5$$

The beam acts as an Euler beam for buckling about the z axis:

$$\sigma_{cr} = \frac{\pi^2 E}{(L_z/r_z)^2} = \frac{\pi^2 70(10^9)\text{ N/m}^2}{(204.5)^2} = 16.52\text{ MPa} \qquad \textit{Ans.}$$

8.3 Other Forms of Instability

Beams can also fail due to local instabilities. There can be a crushing type of failure that is familiar in the crushing of thin-walled soda pop cans. The above formulas do not apply in such instances [Young, 1989]. Hollow rods subjected to torsion can buckle locally due to the compressive principal stress acting at a 45° angle to the longitudinal axis of the rod. Beams can fail in a combined bending and torsion fashion. An I-beam, with a pure moment applied, can have the flanges on the compression side buckle. When lateral loads are present in conjunction with an axial compressive load, the beam acts as a beam-column [Chen and Atsuta, 1976].

Defining Terms

Critical buckling load — The smallest compressive load at which a column will remain in the laterally displaced, buckled shape.

Critical buckling stress — The smallest compressive stress at which a column will remain in the laterally displaced, buckled shape.

E — Young's modulus of elasticity, which is the slope of the stress versus strain diagram in the initial linear region.

Euler buckling load — Same as the critical buckling load if the column is long and slender.

Euler buckling stress — Same as the critical buckling stress if the column is long and slender.

References

Chajes, A. 1974. *Principles of Structural Stability Theory.* Prentice Hall, Englewood Cliffs, NJ.

Chen, W. F. and Atsuta, T. 1976. *Theory of Beam-Columns,* vol. 1, *In-Plane Behavior and Design.* McGraw-Hill, New York.

Riley, W. F. and Zachary, L. W. 1989. *Introduction to Mechanics of Materials.* John Wiley & Sons, New York.

Young, W. C. 1989. *Roark's Formulas for Stress and Strain.* McGraw-Hill, New York.

Further Information

Journal of Structural Engineering of the American Society of Civil Engineers

Engineering Journal of American Institute of Steel Construction

Structural Stability Research Council. 1976. *Guide to Stability Design Criteria for Metal Structures,* 3rd ed. John Wiley & Sons, New York.

Salmon, C. G. and Johnson, J. E. 1990. *Steel Structures,* 3rd ed. HarperCollins, New York.

9

Pressure Vessels

Som Chattopadhyay (Second Edition)
Indiana University/Purdue University at Fort Wayne

Earl Livingston (First Edition)
Babcock and Wilcox Company, Retired

Rudolph H. Scavuzzo (First Edition)
University of Akron

The **pressure vessels** used in industry are leak-tight pressure containers, usually cylindrical or spherical in shape, with various head configurations. They are usually made from carbon steel or stainless steel and assembled by welding. The early operation of pressure vessels and boilers resulted in numerous explosions, causing loss of life and considerable property damage. In the early 1920s, the American Society of Mechanical Engineers formed a committee for the purpose of establishing minimum safety rules for boiler construction. In 1925, the committee issued a set of rules for the design and construction of unfired pressure vessels. Most states have laws mandating that these **Code** rules be met. Enforcement of these rules is accomplished via a third party employed by the state or an insurance company. These Codes are living documents in that they are constantly being revised and updated by committees comprised of individuals knowledgeable on the subject. Keeping current requires that revised Codes be published every 3 years with addenda issued every year. This chapter covers a generalized approach to pressure vessel design based on the ASME Boiler and Pressure Vessel Code, Section VIII, Division 1: Pressure Vessels.

9.1 Design Criteria

The Code design criteria consist of basic rules specifying the design method, design load, allowable stress, acceptable materials, and fabrication–inspection certification requirements for vessel **construction**. The design method, known as "design by rule," uses design pressure, allowable stress, and a design formula compatible with the geometry of the part to calculate the minimum required thickness of the part. This procedure minimizes the amount of analysis required to ensure that the vessel will not rupture or undergo excessive distortion. In conjunction with specifying the vessel thickness, the Code includes many construction details that must be followed. Where vessels are subject to complex loadings such as thermal or localized loads, and where significant discontinuities exist, the Code requires a more rigorous analysis to be performed. This method is known as the "design by analysis" method. A more complete background of both methods may be found in Chattopadhyay, 2004.

The ASME Code is included as a standard by the American National Standards Institute (ANSI). The American Petroleum Institute (API) has also developed codes for low-pressure storage tanks, and these are also part of the ANSI standards. The ASME Boiler and Pressure Vessel Code has worldwide applications, but many other industrialized countries have also developed their own boiler and pressure vessel codes. Differences in these codes sometimes cause difficulty in international trade.

0-8493-1586-7/05/$0.00+$1.50

TABLE 9.1 Acceptable Pressure Vessel Materials

Temperature Use Limit (°F)	Plate Material	Pipe Material	Forging Material
Down to –50	SA-516[a] All grades	SA 333 Gr. 1	SA 350 GR. LF1, LF2
+33 to +775	SA-285 Gr. C SA-515 Gr. 55, 60, 65 SA-516 All grades	SA-53 SA-106	SA-181 Gr. I, II
+776 to +1000	SA-204 Gr. B, C SA-387 Gr. 11, 12 Class 1	SA-335 Gr. P1, P11, P12	SA-182 Gr. F1, F11, F12

[a] Impact testing required.

Note: SA is a classification of steel used in the ASME Boiler and Pressure Vessel Code.

Design Loads

The forces that influence pressure vessel design are internal/external pressure; dead loads due to the weight of the vessel and contents; external loads from piping and attachments; wind, snow, and earthquakes; operating-type loads such as vibration and sloshing of the contents; and startup and shutdown loads. The Code considers design pressure, design temperature, and, to some extent, the influence of other loads that impact the circumferential (or hoop) and **longitudinal stresses** in shells. It is left to the designer to address the effects of the remaining loads on the vessel. Various national and local building codes must be consulted for handling wind, snow, and earthquake loadings.

Materials

The materials to be used in a pressure vessel must be selected from Code-approved material specifications. This requirement is generally not a problem, since a large database of acceptable materials is available. The factors that need to be considered in selecting a suitable material are:

Cost
Fabricability
Service condition (wear, corrosion, operating temperature)
Availability
Strength requirements

Several typical pressure vessel materials for a noncorrosive environment and for service temperatures between –50 and 1000°F are shown in Table 9.1.

Allowable Stress

The allowable stress used to determine the minimum vessel thickness is based on the tensile and yield strengths of the material at room and design temperatures. When the vessel operates at an elevated temperature (typically above 800°F), the creep properties of the material must be considered. These properties are adjusted by design factors that limit the **hoop membrane stress** to a value that precludes tensile rupture, excessive elastic and plastic deformations, and creep rupture. Table 9.2 shows typical allowable stresses for several carbon steels commonly used for unfired pressure vessels.

TABLE 9.2 Typical Allowable Stresses for Use in Pressure Vessel Design

Material Specification	Temperature Use Limit (°F)	Allowable Stress (psi)
SA-515 Gr. 60	700	14 400
	800	10 800
	900	6 500
SA 516-Gr. 70	700	16 600
	800	14 500
	900	12 000
SA-53 Gr. A	700	11 7000
	800	9 300
	900	6 500
SA-106 Gr. B	700	14 400
	800	10 800
	900	6 500
SA-181 Gr. 1	700	16 600
	800	12 000
	900	6 500

9.2 Design Formulas

The design formulas used in the "design by rule" method are based on the so-called maximum principal stress theory of failure involving the average hoop stress. The maximum principal stress theory of failure states that failure occurs when the greatest of the three principal stresses reaches the material yield strength. Ignoring the radial stress, the other two principal stresses can be determined by simplified engineering mechanics formulas.

The Code recognizes that the shell thickness may be such that the radial stresses may not be ignored; accordingly, adjustments have been made in the appropriate formulas. These formulas include a parameter, the **weld joint efficiency factor**, to address the nature of the examination performed at the welded joints during the fabrication of the vessel. Table 9.3 shows various formulas used to calculate the wall thickness for a number of pressure vessel configurations.

9.3 Opening Reinforcement

Vessel components are weakened when material is removed to provide openings for nozzles or access. High **stress concentrations** exist at the opening edge and decrease radially outward from the opening, becoming negligibly small beyond distances twice the diameter of the opening. To avoid failure in the opening region, compensation or reinforcement is required. Some ways in which this can be accomplished are:

1. Increase the vessel wall thickness.
2. Increase the nozzle wall thickness.
3. Use a combination of additional shell and nozzle thicknesses.

The Code procedure is to relocate the removed material to an area within the effective boundary around the opening. Figure 9.1 shows the steps necessary to reinforce an opening in a pressure vessel. Numerous assumptions have been made with the intent of simplifying the general approach.

The example shown in Figure 9.2 uses the design approach indicated by the Code to perform a simple sizing calculation for a typical welded carbon steel vessel. Figure 9.3 shows typical nozzle–shell and head–shell junctures that meet the Code requirements. Design specifications for the many associated vessel parts, such as bolted flanges, external attachments, and vessel supports, can be found in Chattopadhyay, 2004.

TABLE 9.3 Code Formulas for Calculation of Vessel Component Thickness

Cylindrical shell	$t = \dfrac{PR}{SE - 0.6P}$
Hemispherical head or spherical shell	$t = \dfrac{PR}{2SE - 0.2P}$
2:1 ellipsoidal head	$t = \dfrac{PD}{2SE - 0.2P}$
Flanged and dished head	$t = \dfrac{1.77PL}{2SE - 0.2P}$
Flat head	$t = d\sqrt{CP/SE}$

where
t = Mimimum required thickness (in.)
P = Deisgn pressure (psi)
R = Inside radius (in.)
S = Allowable stress (psi)
D = Inside diameter (in.)
L = Inside spherical crown radius (in.)
E = Weld joint efficiency factor, determined by joint location and degree of examination
C = Factor depending upon method of head-to-shell attachment

Defining Terms

Code — The complete rules for the construction of pressure vessels as identified in the ASME Boiler and Pressure Vessel Code, Section VIII, Division 1, Pressure Vessels.

Construction — The complete manufacturing process, including design, fabrication, inspection, examination, hydrostatic test, and certification. This applies to new construction only.

Hoop membrane stress — The average stress in a ring subjected to radial loads uniformly distributed along its circumference.

Longitudinal stress — The average stress acting on a cross-section of a vessel in the axial (lengthwise) direction.

Pressure vessel — A leak-tight pressure container, usually cylindrical or spherical in shape, subjected to internal pressure.

Stress concentration — Local high stress in the vicinity of a geometrical discontinuity such as a change in thickness or an opening in a shell.

Weld efficiency factor — A factor intended to reduce the allowable stress for welded connections, dependent on the degree of weld examination performed during the construction of the vessel.

References

American Society of Mechanical Engineers, 2004, *ASME Boiler and Pressure Vessel Code, Section VIII, Division 1, Pressure Vessels.* ASME, New York

Chattopadhyay, S. 2004, *Pressure Vessels: Design and Practice,* CRC Press, Boca Raton, FL.

Further Information

Each summer, usually in July, the Pressure Vessel and Piping Division of the American Society of Mechanical Engineers organizes an annual meeting devoted to pressure vessel technology. Over 500

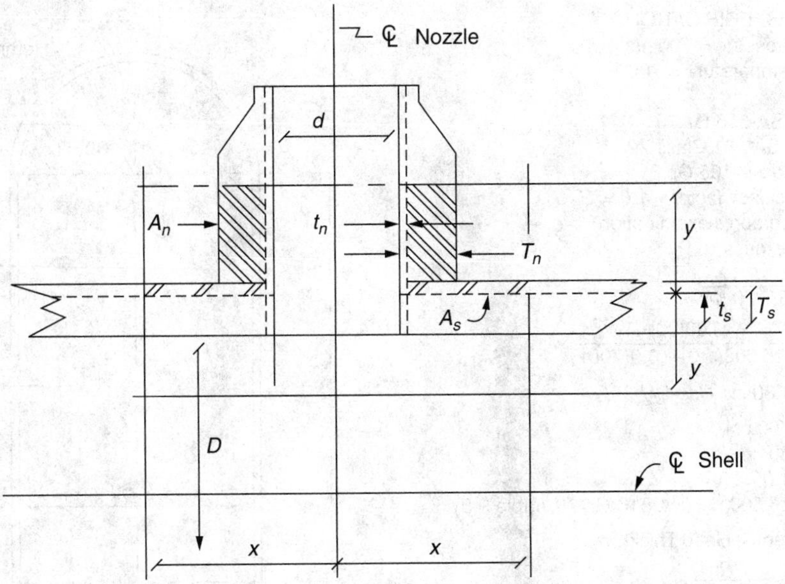

x = Larger of d or $R_n + t_n + T_n$
y = Smaller of $2\frac{1}{2}T_s$ or $2\frac{1}{2}T_n$
d = Diameter of circular opening (in.)
D = Inside diameter of shell (in.)
t_s = Required thickness of shell (in.)
T_s = Actual thickness of shell (in.)
t_n = Required thickness of nozzle (in.)
T_n = Actual thickness of nozzle (in.)
R_n = Inside radius of nozzle = $d/2$ (in.)

A_r = Area of required reinforcement (in.²)
A_s = Area available in the shell (in.²)
A_n = Area available in the nozzle (in.²)
$A_r = (d)\,(t_s)$
A_s = Larger of: $d\,(T_s - t_s) - 2T_n\,(T_s - t_s)$ or $2(T_s + t_n)(T_s - t_s) - 2t_n\,(T_s - t_s)$
A_n = Smaller of: $2\,[2\frac{1}{2}\,(T_s)(T_n - t_n)$ or $2\,[2\frac{1}{2}\,(T_n)(T_n - t_n)]$
$A_r < (A_s + A_n)$: Acceptable configuration

FIGURE 9.1 Opening reinforcement requirements.

papers are presented, many of which are published as ASME special publications in book form. Archival papers are also published in the *Journal of Pressure Vessel Technology.*

Research programs for the ASME Boiler and Pressure Vessel Code are often conducted by the Pressure Vessel Research Council (PVRC). This research is normally published in Welding Research Council (WRC) bulletins. These bulletins provide excellent documentation of major research contributions in the field of pressure vessel technology. The Electric Power Research Institute (EPRI) also conducts extensive research for the electric power industry, and this work includes some research on pressure vessels.

DESIGN SPECIFICATION
Design pressure = 700 psi
Design temperature = 700°F
Material:
Shell SA-516 Gr. 70
Head SA-181 Class 70
Nozzle SA-106 Gr. B
Weld efficiency factor = 1.0 = E
(full radiographic examination)

Shell Thickness

$$t_s = \frac{PR}{SE - 0.6P}$$

$$= \frac{700(30)}{16\,600(1.0) - 0.6(700)}$$

$= 1.30$ in. *Use* $1\frac{1}{2}" = T_s$

$P = 700$ psi
$R = 30$ in.
$E = 1.0$
$S = 16\,600$ psi (SA-516 Gr. 70, Table 9.2)

Hemispherical Head Thickness

$$t_h = \frac{PR}{2SE - 0.2P}$$

$$= \frac{700(30)}{2(16\,600)(1.0) - 0.2(200)}$$

$= 0.64$ in. *Use* $1" = T_h$

Nozzle Thickness

$$t_n = \frac{PR}{SE - 0.6P}$$

$$= \frac{700(4)}{16\,600(1.0) - 0.6(700)}$$

$= 0.17$ in. *Use* $1\frac{3}{4} = T_n$

$P = 700$ psi
$R = 4$ in.
$E = 1.0$
$S = 16\,600$ psi (SA-106 Gr. B, Table 9.2)

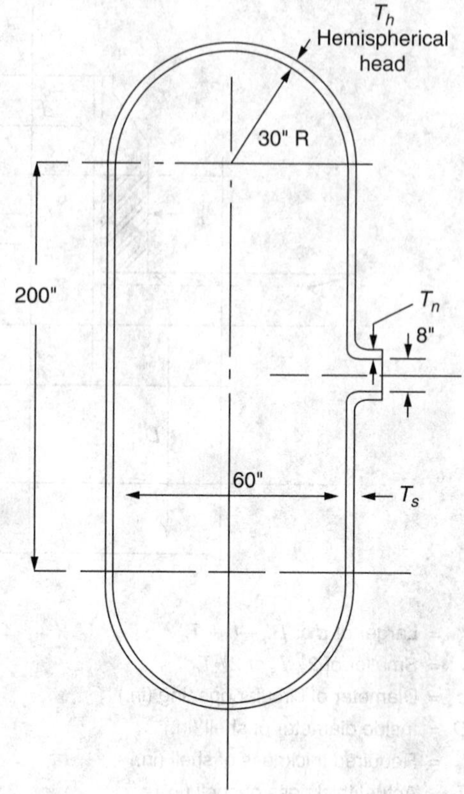

Opening Reinforcement Calculation (see Fig. 9.1)*

$A_{req'd} = (d)(t_s) = (8)(1.3) = 10.4$ in.^{2}

A_n = Smaller of: $2[2\frac{1}{2}(T_s)(T_n - t_n)]$ or $2[2\frac{1}{2}(T_n)(T_n - t_n)]$

$T_s < T_n$ *Use* T_s

$2[2\frac{1}{2}(1\frac{1}{2})(1\frac{3}{4} - 0.17)] = 11.7$ in.^{2}

$d = 8$ in.
$t_s = 1.3$ in.
$T_s = 1\frac{1}{2}$ in.
$T_n = 1\frac{3}{4}$ in.
$t_n = 0.17$ in.

*This example places all the required reinforcement in the nozzle wall since it is more economical than increasing the shell thickness.

$A_n = 11.7$ in.$^2 > A_{req'd} = 10.4$ in.^{2} — OK

FIGURE 9.2 Sample vessel calculations.

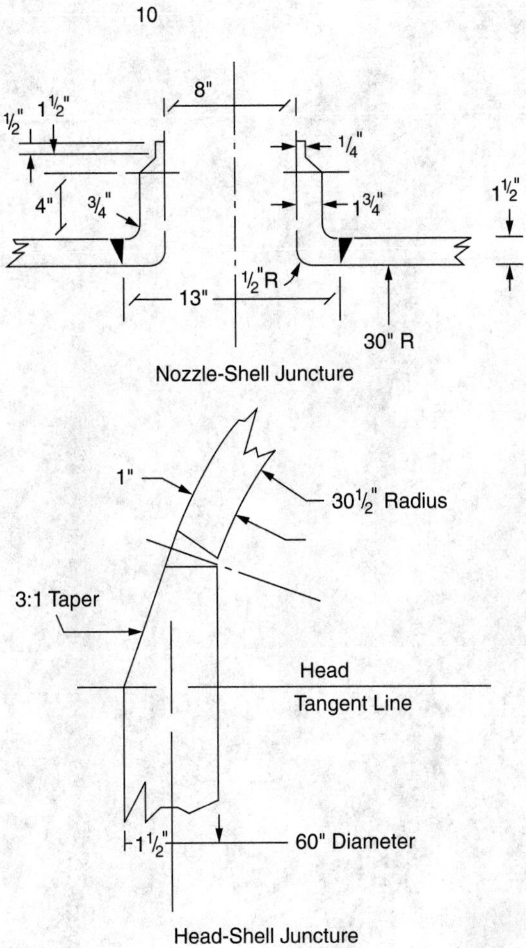

FIGURE 9.3 Fabrication details.

10

Axial Loads and Torsion

Nelson R. Bauld, Jr.
Clemson University

10.1 Axially Loaded Bars

A bar is said to be axially loaded if the action lines of all the applied forces coincide with the axis of the bar. The **bar axis** is defined as the locus of the centroids of the cross-sectional areas along the length of the bar. This locus of centroids must form a straight line, and the action lines of the applied forces must coincide with it in order for the theory of this section to apply.

Axial Strain

The axial strain in an axially loaded bar is based on the geometric assumptions that plane cross-sections in the unloaded bar, such as sections *mn* and *pq* in Figure 10.1(a), remain plane in the loaded bar as shown in Figure 10.1(b), and that they displace only axially.

The axial strain of a **line element** such as *rs* in Figure 10.1(a) is defined as the limit of the ratio of its change in length to its original length as its original length approaches zero. Thus, the axial strain ε at an arbitrary cross-section x is

$$\varepsilon(x) = \lim_{\Delta x \to 0}(\Delta x^* - \Delta x)/\Delta x = \lim_{\Delta x \to 0}[u(x+\Delta x) - u(x)]/\Delta x = du/dx \tag{10.1}$$

where $u(x)$ and $u(x + \Delta x)$ are axial displacements of the cross sections at x and $x + \Delta x$. Common units for axial strain are in./in. or mm/mm. Because axial strain is the ratio of two lengths, units for axial strain are frequently not recorded.

Axial Stress

The axial stress σ at cross-section x of an axially loaded bar is

$$\sigma(x) = N(x)/A(x) \tag{10.2}$$

0-8493-1586-7/05/$0.00+$1.50

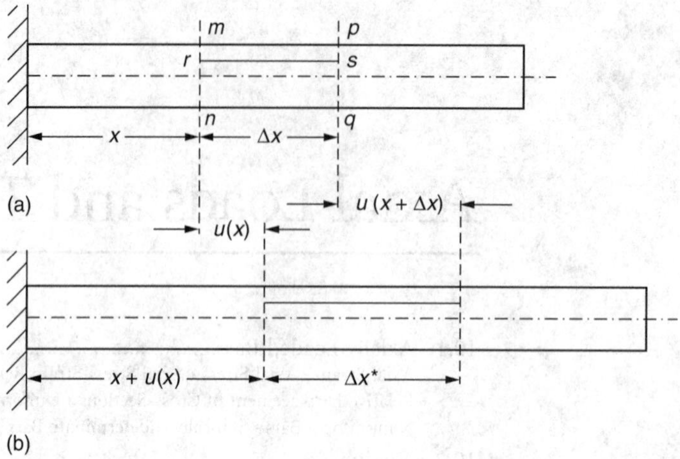

FIGURE 10.1 Axial displacements of an axially loaded bar.

where $N(x)$ is the internal force and $A(x)$ is the cross-sectional area, each at section x. Common units for axial stress are pounds per square inch (psi) or megapascals (MPa). Equation (10.2) is valid at cross-sections that satisfy the geometric assumptions stated previously. It ceases to be valid at abrupt changes in cross section and at points of load application. Cross-sections at such locations distort and therefore violate the plane cross-section assumption. Also, Equation (10.2) requires that the material at cross-section x be homogeneous; that is, the cross section cannot be made of two or more different materials.

Axial Stress-Strain Relation

The allowable stress for axially loaded bars used in most engineering structures falls within the proportional limit of the material from which they are made. Consequently, material behavior considered in this section is confined to the linearly elastic range and is given by

$$\sigma(x) = E(x)\,\varepsilon(x) \tag{10.3}$$

where $E(x)$ is the modulus of elasticity for the material at section x. Common units for the modulus of elasticity are pounds per square inch (psi) or gigapascals (GPa).

Relative Displacement of Cross-Sections

The relative displacement $e_{B/A}$ of a cross-section at x_B with respect to a cross-section at x_A is obtained by combining Equation (10.1) through Equation (10.3) and integrating from section x_A to x_B. Using Figure 10.2,

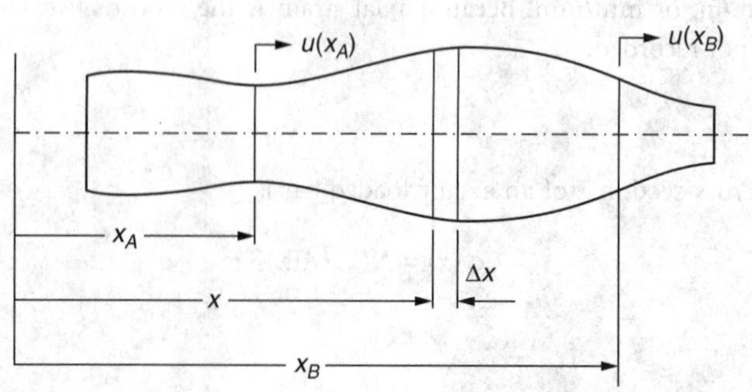

FIGURE 10.2

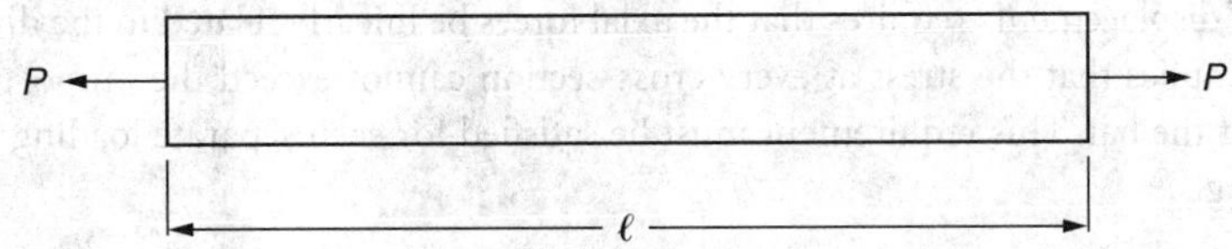

FIGURE 10.3 Uniform bar.

$$e_{B/A} = u(x_B) - u(x_A) = \int_{x_A}^{x_B} N(x)/[A(x)E(x)]dx \tag{10.4}$$

where $e_{B/A}$ denotes the change in length between the cross-sections at x_A and x_B.

Equation (10.4) must be interpreted as the sum of several integrals for a bar for which the integrand exhibits discontinuities. Discontinuities occur for cross-sections where either N, A, E, or combinations thereof change abruptly and can usually be detected by inspection.

Uniform Bar

A bar for which the internal force $N(x)$, the cross-sectional area $A(x)$, and the modulus of elasticity $E(x)$ do not change over its length is referred to as a **uniform bar**. If P denotes equilibrating forces applied to the ends of the bar and L its length, as shown in Figure 10.3, then Equation (10.4) gives the change in length of the bar as

$$e = PL/AE \tag{10.5}$$

Nonuniform Bars

A **nonuniform bar** is one for which either A, E, N, or combinations thereof change abruptly along the length of the bar. Three important methods are available to analyze axially loaded bars for which the integrand in Equation (10.4) contains discontinuities. They are as follows.

Direct Integration

Equation (10.4) is integrated directly. The internal force $N(x)$ is obtained in terms of the applied forces via the axial equilibrium equation, $A(x)$ from geometric considerations, and $E(x)$ by observing the type of material at a given section.

Discrete Elements

The bar is divided into a finite number of segments, for each of which N/AE is constant. Each segment is a uniform bar for which its change in length is given by Equation (10.5). The change in length of the nonuniform bar is the sum of the changes in length of the various segments. Accordingly, if e_i denotes the change in length of the ith segment, then the change in length e of the nonuniform bar is

$$e = \sum e_i \tag{10.6}$$

Superposition

The superposition principle applied to axially loaded bars asserts that the change in length between two cross-sections caused by several applied forces acting simultaneously is equal to the algebraic sum of the changes in length between the same two cross-sections caused by each applied force acting separately. Thus, letting $e_{B/A}$ represent the change in length caused by several applied forces acting simultaneously, and $e'_{B/A}$, $e''_{B/A}$,…represent the changes in length caused by each applied force acting separately,

$$e_{B/A} = e'_{B/A} + e''_{B/A} + \cdots \tag{10.7}$$

Superposition of displacements requires that the axial forces be linearly related to the displacements they cause, and this implies that the stress at every cross-section cannot exceed the proportional limit stress of the material of the bar. This requirement must be satisfied for each separate loading as well as for the combined loading.

Statically Indeterminate Bars

The internal force $N(x)$ in statically determinate axially loaded bars is determined via axial equilibrium alone. Subsequently, axial stress, axial strain, and axial displacements can be determined via the foregoing equations.

The internal force $N(x)$ in statically indeterminate axially loaded bars cannot be determined via axial equilibrium alone. Thus, it is necessary to augment the axial equilibrium equation with an equation (geometric compatibility equation) that accounts for any geometric constraints imposed on the bar — that is, that takes into account how the supports affect the deformation of the bar.

Three basic mechanics concepts are required to analyze statically indeterminate axially loaded bars: axial equilibrium, geometric compatibility of axial deformations, and material behavior (stress–strain relation).

Example 10.1

Determine the stresses in the aluminum and steel segments of the composite bar of Figure 10.4(a) when $P = 7000$ lb. The cross-sectional areas of the steel and aluminum segments are 2 and 4 in.^{2}, respectively, and the moduli of elasticity are $30 \cdot 10^6$ and $10 \cdot 10^6$ psi, respectively.

Solution

The bar is statically indeterminate; therefore, the solution requires the use of the three mechanics concepts discussed in the previous paragraph.

Equilibrium. The axial equilibrium equation is obtained from the free-body diagram of Figure 10.4(b) as

$$-P_{ST} + P_{AL} - 7000 = 0 \tag{10.8}$$

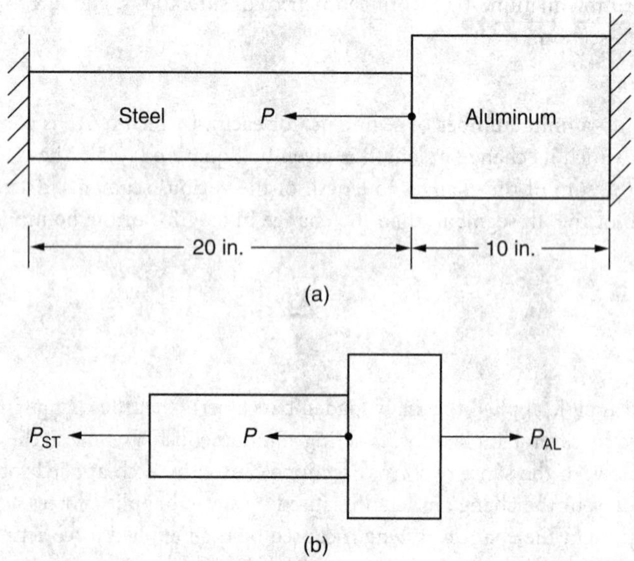

FIGURE 10.4 Statically indeterminate composite step-bar.

Geometric compatibility. The compatibility equation is obtained by noting that the total elongation of the bar is zero. Accordingly,

$$e = e_{ST} + e_{AL} = 0 \tag{10.9}$$

Material behavior. The steel and aluminum segments are assumed to behave in a linearly elastic manner, so their elongations are given by

$$e_{ST} = P_{ST}\, L_{ST}/(A_{ST}E_{ST}) \text{ and } e_{AL} = P_{AL}L_{AL}/\,(A_{AL}E_{AL}) \tag{10.10}$$

Combining Equation (10.9) and Equation (10.10) yields

$$\begin{aligned} P_{ST} &= -(L_{AL}/L_{ST})(E_{ST}/E_{AL})(A_{ST}/A_{AL})P_{AL} \\ &= -(10/20)(30/10)(2/4)P_{AL} \;= -3/4P_{AL} \end{aligned} \tag{10.11}$$

Solving Equation (10.8) and Equation (10.11) simultaneously yields

$$P_{ST} = -3000 \text{ lb} \quad \text{and} \quad P_{AL} = 4000 \text{ lb} \tag{10.12}$$

from which the stresses in the steel and aluminum are found as follows:

$$\sigma_{ST} = -3000/2 = -1500 \text{ psi} = 1500 \text{ psi (compression)}$$

$$\sigma_{AL} = 4000/4 = 1000 \text{ psi (tension)}$$

Example 10.2

Assuming that $P = 0$ in Figure 10.4(a), determine the stress in the steel and aluminum segments of the bar due to a temperature increase of 10°F. The *thermal expansion coefficients* for steel and aluminum are $\alpha_{ST} = 6.5 \cdot 10^{-6}$ and $\alpha_{AL} = 13 \cdot 10^{-6}$ inches per inch per degree Fahrenheit (in./in./°F), respectively.

Solution

Because free thermal expansion of the bar is prevented by the supports, internal stresses are induced in the two segments.

Equilibrium. The axial equilibrium equation is obtained from the free-body diagram of Figure 10.4(b). Thus,

$$-P_{ST} + P_{AL} = 0 \tag{10.13}$$

Compatibility. The compatibility equation is obtained by noting that if the bar could expand freely, its total elongation Δ would be

$$\Delta = \Delta_{ST} + \Delta_{AL} \tag{10.14}$$

where Δ_{ST} and Δ_{AL} denote the free thermal expansions of the separate segments. Because the net change in length of the bar is zero, internal strains are induced in the steel and aluminum such that the sum of the changes in lengths of the steel and aluminum segments must be equal to Δ. Therefore, the compatibility equation becomes

$$e_{ST} + e_{AL} - \Delta = 0 \tag{10.15}$$

Material behavior. Assuming linear elastic behavior for both materials

$$e_{ST} = P_{ST} L_{ST} / (A_{ST} E_{ST}) \quad \text{and} \quad e_{AL} = P_{AL} L_{AL} / (A_{AL} E_{AL}) \tag{10.16}$$

Also, because

$$\Delta_{ST} = \alpha_{ST} L_{ST} \Delta T \quad \text{and} \quad \Delta_{AL} = \alpha_{AL} L_{AL} \Delta T \tag{10.17}$$

it follows that

$$\Delta = (6.5 \cdot 10^{-6})(20)(10) + (13 \cdot 10^{-6})(10)(10) = 0.0026 \text{ in.} \tag{10.18}$$

Equation (10.13), Equation (10.15), Equation (10.16), and Equation (10.18) yield

$$P_{ST}\{1 + (E_{ST} / E_{AL})(A_{ST} / A_{AL})(L_{AL} / L_{ST})\} = (E_{ST} A_{ST} / L_{ST})\Delta$$

or

$$P_{ST}\{1 + (30/10)(2/4)(10/20)\} = \{(30 \cdot 10^6 (2)]/20\}(0.0026)$$

Thus

$$P_{ST} = P_{AL} = 4457 \text{ lb} \tag{10.19}$$

The corresponding stresses in the steel and aluminum are compression and equal to

$$\sigma_{ST} = 4457/2 = 2228 \text{ psi} \quad \text{and} \quad \sigma_{AL} = 4457/4 = 1114 \text{ psi}$$

10.2 Torsion

Torsionally loaded bars occur frequently in industrial applications such as shafts connecting motor–pump and motor–generator sets; propeller shafts in airplanes, helicopters, and ships; and torsion bars in automobile suspension systems. Many tools or tool components, such as screwdrivers and drill and router bits, possess a dominant torsional component. These tools also rely on an axial force component for their effectiveness.

Power Transmission

The specifications for a motor customarily list the power it transmits in horsepower (hp) and its angular speed in either revolutions per minute (rpm) or in cycles per second (Hz). To design or analyze a shaft, the *torque* that it is to transmit is required. Therefore, a relationship between horsepower, angular speed, and torque is required. In U.S. customary units and in the International System of Units (SI units) these relationships are

$$\text{hp} = \begin{cases} 2\pi n T / [550(12)60] = n\,T / 63\,000 \text{ (U.S. customary units)} \\ 2\pi f\,T / 745.7 = f\,T / 119 \text{ (SI units)} \end{cases} \tag{10.20}$$

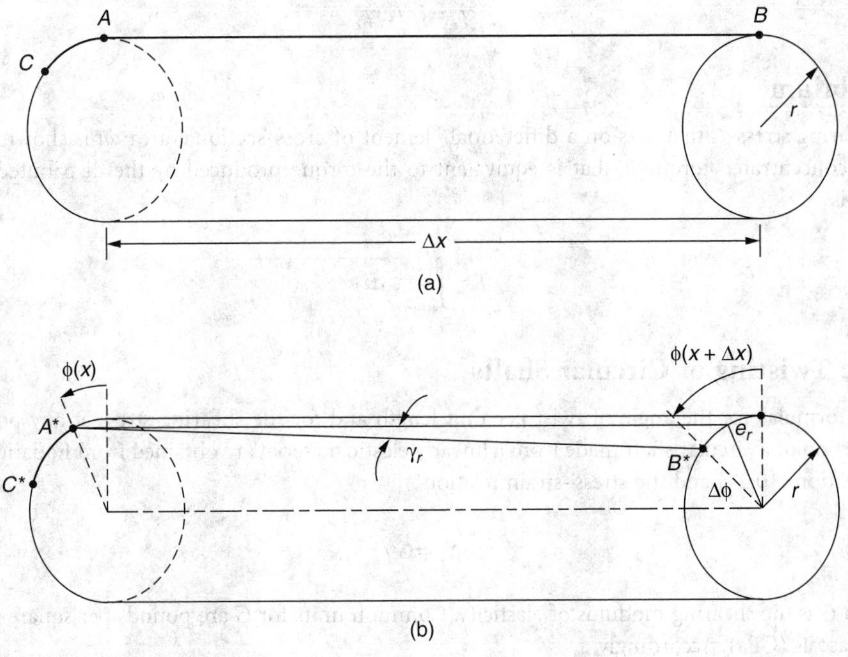

FIGURE 10.5

where f and n denote the angular speed in cycles per second and revolutions per minute, respectively, and T denotes the torque transmitted in Newton-meters (N · m) or inch-pounds (in.-lb), depending on the system of units used.

Kinematics of Circular Shafts

The theory of circular shafts is based on the geometric assumption that a plane cross-section simply rotates about the axis of the shaft and can be visualized as being composed of a series of *thin rigid disks* that rotate about the axis of the shaft.

To obtain a formula that expresses the rotation of one cross-section relative to another infinitesimally close to it, consider a shaft of radius c and examine the angular deformations of an interior segment of radius r and length Δx. This portion of the bar is indicated in Figure 10.5(a). Before twisting, line element AB is parallel to the shaft axis, and line element AC lies along a cross-sectional circle of radius r. The angle between these elements is 90 degrees. Due to twisting, AC merely moves to a new location on the circumference, but AB becomes $A* \; B*$, which is no longer parallel to the shaft axis, as in indicated in Figure 10.5(b). The shearing deformation e_r at radius r is

$$e_r = r\Delta\phi = \gamma_r \Delta x \tag{10.21}$$

where γ_r denotes the shearing strain between line elements $A* \; B*$ and $A* \; C*$, and $\Delta\phi$ represents the angular rotation of the cross-section at B relative to the cross-section at A. In the limit, as Δx becomes infinitesimal, Equation (10.21) becomes

$$\gamma_r = rd\phi / dx \tag{10.22}$$

Because a cross-section is considered rigid, Equation (10.22) indicates that the shearing strain varies linearly with distance from the center of the shaft. Consequently, because c denotes the outside radius of the shaft, the shearing strain at radius r is

$$\gamma_r = (r/c)\gamma_c \tag{10.23}$$

Equilibrium

The shearing stress τ_r that acts on a differential element of cross-sectional area *da* is shown in Figure 10.6. A concentrated torque *T* that is equivalent to the torque produced by the distributed shearing stress τ_r is

$$T = \int_{\text{area}} (\tau_r da) r \tag{10.24}$$

Elastic Twisting of Circular Shafts

Explicit formulas for the angle of twist per unit length and for the shearing stress at any point *r* in a cross-section of a circular shaft made from a linearly elastic material are obtained from Equation (10.22) and Equation (10.24) and the stress–strain relation

$$\tau_r = G\gamma_r \tag{10.25}$$

in which *G* is the shearing modulus of elasticity. Common units for *G* are pounds per square inch (psi) or gigapascals (GPa). Accordingly,

$$T = \int_{\text{area}} (G\gamma_r / r) r^2 \, da = G \, d\phi / dx \int_{\text{area}} r^2 \, da$$

or

$$d\phi / dx = T / JG \tag{10.26}$$

in which *J* is the polar moment of inertia of the cross-sectional area of the bar. Common units for *J* are inches to the fourth power (in.⁴) or meters to the fourth power (m⁴), depending on the system of units used.

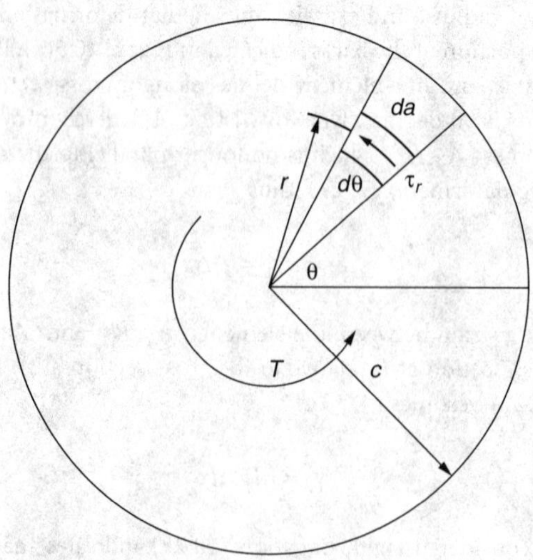

FIGURE 10.6

The shearing stress at radius r is obtained by combining Equation (10.22), Equation (10.25), and Equation (10.26). Thus,

$$\tau_r = Tr/J \tag{10.27}$$

Equation (10.26) and Equation (10.27) provide the means needed to analyze the strength and stiffness of linearly elastic shafts with circular cross-sections. These formulas remain valid for annular shafts for which the hollow and solid portions are concentric. Formulas for the polar moments of inertia J are

$$J = \begin{cases} \pi/32 d^4 & \text{(solid cross-section)} \\ \pi/32(d_o^4 - d_i^4) & \text{(annular cross-section)} \end{cases} \tag{10.28}$$

where d_o and d_i denote external and internal diameters.

Uniform Shaft

A *uniform shaft* is one for which the cross-sectional area, the shearing modulus of elasticity, and the applied torque do not change along its length. Because J, G, and T are constants over the length L, Equation (10.26) integrates to give the angle of twist of one end relative to the other end as

$$\phi = TL/JG \tag{10.29}$$

The shearing stress on any cross-section at radial distance r is

$$\tau_r = Tr/J \tag{10.30}$$

Nonuniform Shaft

A *nonuniform shaft* is one for which either J, G, T, or a combination thereof changes abruptly along the length of the shaft. Three procedures are available to determine the angle of twist for circular shafts made from linearly elastic materials.

Direct Integration

Equation (10.26) is integrated directly. Because the integrand T/JG can possess discontinuities at cross-sections for which J, G, or T changes abruptly, the integration must be interpreted as a sum of several integrations. Discontinuities in J, G, and T can usually be detected by inspection. The polar moment of inertia J is discontinuous at abrupt changes in cross-sectional area, G is discontinuous at cross-sections where the material changes abruptly, and the internal torque T is discontinuous at points where concentrated torques are applied.

Discrete Elements

The shaft is divided into a finite number of segments for each of which T/JG is constant. Consequently, the shaft is perceived to be a series of connected uniform shafts for each of which Equation (10.29) applies. Thus, if ϕ_i denotes the angle of twist of the ith segment, then the angle of twist for the shaft is

$$\phi = \sum \phi_i \tag{10.31}$$

Superposition

The superposition principle applied to the twisting of circular shafts stipulates that the relative rotation of one cross section with respect to another cross section due to several torques applied simultaneously

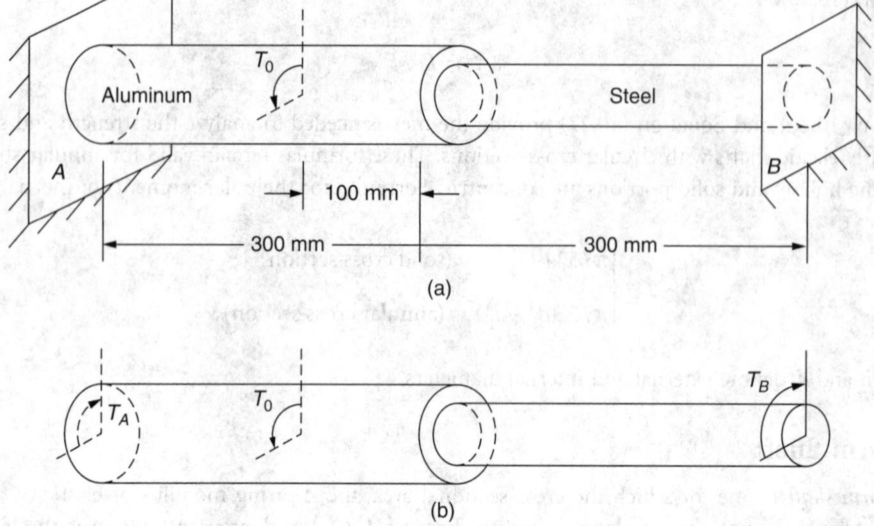

FIGURE 10.7

is equal to the algebraic sum of the relative rotations of the same cross sections due to each torque applied separately. If $\phi'_{B/A}$, $\phi''_{B/A}$,... denote relative angles of twist for each torque applied separately, then

$$\phi_{B/A} = \phi'_{B/A} + \phi''_{B/A} + \cdots \tag{10.32}$$

Superposition of angles of twist requires that the torques be linearly related to the angles of twist that they produce, which in turn implies that the shearing stress must not exceed the proportional limit stress for the material involved. This requirement must be satisfied for each separate loading, as well as for the combined loading.

Statically Indeterminate Circular Shafts

A shaft is statically indeterminate if the internal torque at a cross-section cannot be determined from moment equilibrium about the axis of the shaft. In such cases an additional equation is obtained by requiring that angles of twist be compatible with the geometric constraints imposed on the shaft. As with axially loaded bars, three basic concepts of mechanics are involved in the solution of statically indeterminate shafts: equilibrium, geometric compatibility, and material behavior.

Example 10.3

The diameters of the aluminum and steel segments of the statically indeterminate step-shaft of Figure 10.7(a) are 50 and 25 mm, respectively. Knowing that $G_{AL} = 28$ GPa, $G_{ST} = 84$ GPa, and $T_0 = 200\pi$ N · m, determine the maximum shearing stresses in the aluminum and in the steel.

Solution

Equilibrium. From Figure 10.7(b), moment equilibrium about the axis of the shaft gives

$$T_A + T_B - T_0 = 0 \tag{10.33}$$

Compatibility. The supports at the ends of the shaft prevent the cross-sections at A and B from rotating; hence, the required compatibility equation is

$$\phi_{B/A} = 0 \tag{10.34}$$

and, with the aid of the superposition principle, it can be written as

$$\phi_{B/A} = \phi'_{B/A} + \phi''_{B/A} = 0 \tag{10.35}$$

Here $\phi'_{B/A}$ and $\phi''_{B/A}$, denote the relative angular rotations of the cross-section at B with respect to the cross-section at A due to the torques T_B and T_0 acting separately.

To convert Equation (10.35) into an algebraic equation involving the torques T_B and T_0, the discrete element procedure is used. First calculate the polar moments of inertia for the two segments:

$$\begin{aligned} J_{\mathrm{AL}} &= \pi/32(0.050)^4 = 0.613\cdot 10^{-6}\,\mathrm{m}^4 \\ J_{\mathrm{ST}} &= \pi/32(0.025)^4 = 0.038\cdot 10^{-6}\,\mathrm{m}^4 \end{aligned} \tag{10.36}$$

Using Equation (10.29) for a uniform shaft, determine that

$$\begin{aligned} \phi'_{B/A} &= 0.3T_B/\{J_{\mathrm{AL}}28\cdot 10^9\} + 0.3T_B/\{J_{\mathrm{ST}}84\cdot 10^9\} = 111.47\cdot 10^{-6}\,T_B\,\mathrm{m}^4 \\ \phi''_{B/A} &= 0.3T_0/\{J_{\mathrm{AL}}28\cdot 10^9\} = 11.65\cdot 10^{-6}\,T_0\,\mathrm{m}^4 \end{aligned} \tag{10.37}$$

Consequently,

$$\phi_{B/A} = \{111.47T_B - 11.65T_0\}\cdot 10^{-6} = 0 \tag{10.38}$$

Equation (10.38) gives T_B and Equation (10.33) gives T_A. Thus,

$$T_A = 179\pi\ \mathrm{N\cdot m} \quad \text{and} \quad T_B = 21\pi\ \mathrm{N\cdot m} \tag{10.39}$$

The maximum shearing stress in each material occurs at the most remote point on a cross-section. Thus,

$$\begin{aligned} (\tau_{\mathrm{AL}})_{\max} &= T_{\mathrm{AL}}c/J_{\mathrm{AL}} = 179\pi(0.025)/0.613\cdot 10^{-6} = 22.9\ \mathrm{MPa} \\ (\tau_{\mathrm{ST}})_{\max} &= T_{\mathrm{ST}}c/J_{\mathrm{ST}} = 21\pi(0.0125)/0.038\cdot 10^{-6} = 21.7\ \mathrm{MPa} \end{aligned} \tag{10.40}$$

Defining Terms

Bar axis — Straight line locus of centroids of cross-sections along the length of a bar.

Line element — Imaginary fiber of material along a specific direction.

Nonuniform bar — A bar for which the cross-sectional area or the material composition changes abruptly along its length, or external forces are applied intermediate to its ends.

Nonuniform shaft — A bar of circular cross-section for which the diameter or material composition changes abruptly along its length, or external twisting moments are applied intermediate to its ends.

Thin rigid disk — Imaginary circular cross-section of infinitesimal thickness that is assumed to undergo no deformations in its plane.

Torque — Twisting moment.

Uniform bar — A bar of uniform cross-sectional area that is made of one material and is subjected to axial forces only at its ends.

Uniform shaft — A bar of uniform, circular cross-sectional area that is made of one material and is subjected to twisting moments only at its ends.

References

Bauld, N. R., Jr. 1986. Axially loaded members and torsion. In *Mechanics of Materials,* 2nd ed.

Beer, F. P. and Johnston, E. R., Jr. 1981. Stress and strain — axial loading and torsion. In *Mechanics of Materials.*

Gere, J. M. and Timoshenko, S. P. 1990. Axially loaded members and torsion. In *Mechanics of Materials,* 2nd ed.

Further Information

Formulas for the twisting of shafts with the following cross-sectional shapes can be found in Bauld [1986]: thin-wall, open sections of various shapes; solid elliptical, rectangular, and equilateral triangular sections; open sections composed of thin rectangles; and circular sections composed of two different concentric materials. Also available in the same reference are formulas for the twisting of circular shafts in the inelastic range.

11

Fracture Mechanics

Ted L. Anderson
Structural Reliability Technology

11.1 Introduction

Since the advent of iron and steel structures during the Industrial Revolution, a significant number of brittle fractures have occurred at stresses well below the tensile strength of the material. One of the most famous of these failures was the rupture of a molasses tank in Boston in January 1919 [Shank, 1953]. Over 2 million gallons of molasses were spilled, resulting in 12 deaths, 40 injuries, massive property damage, and several drowned horses.

The traditional strength-of-materials approach cannot explain events such as the molasses tank failure. In the first edition of his elasticity text published in 1892, Love remarked that "the conditions of rupture are but vaguely understood." Designers typically applied safety factors of 10 or more (based on the tensile strength) in an effort to avoid these seemingly random failures.

Several centuries earlier, Leonardo da Vinci had performed a series of experiments on iron wires that shed some light on the subject of brittle fracture. He found that the strength of the wires varied inversely with length. These data implied that flaws in the material controlled the strength; a longer wire corresponded to a larger sample volume and a higher probability of sampling a region containing a flaw. These results were only qualitative, however, and formal mathematical relationships between flaw size and failure stress were not developed until recently.

During World War II, a large number of Liberty ships and T2 tankers sustained brittle fractures [Bannerman and Young, 1946]. The need to understand the cause of these failures led to extensive research in the 1950s, which resulted in the engineering discipline known as **fracture mechanics**.

The field of fracture mechanics attempts to quantify the relationship between failure stress, flaw size, and material properties. Today, many segments of industry, including aerospace, oil and gas, and electric utilities, apply fracture mechanics principles in order to prevent catastrophic failures.

11.2 Fundamental Concepts

Figure 11.1 contrasts the fracture mechanics approach with the traditional approach to structural design and material selection. In the latter case, the anticipated design stress is compared to the tensile properties of candidate materials; a material is assumed to be adequate if its strength is greater than the expected applied stress. Such an approach may attempt to guard against brittle fracture by imposing a safety factor

0-8493-1586-7/05/$0.00+$1.50

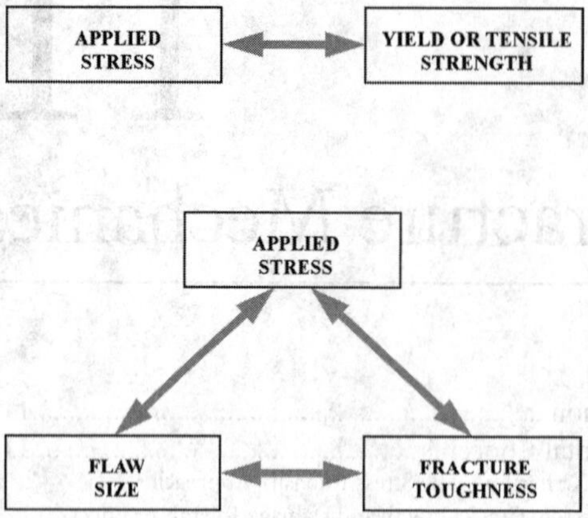

FIGURE 11.1 Comparison of the fracture mechanics approach to design with the traditional strength of materials approach. (a) The strength of materials approach. (b) The fracture mechanics approach.

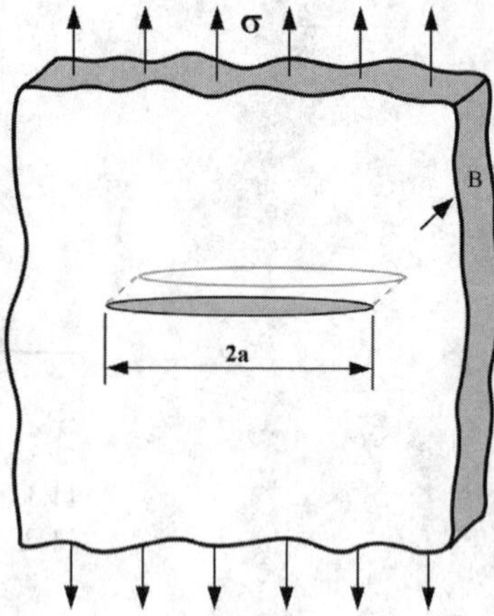

FIGURE 11.2 Through-thickness crack in an infinite plate subject to a remote tensile stress. In practical terms, "infinite" means that the width of the plate is >> 2*a*.

on stress, combined with minimum tensile elongation requirements on the material. The fracture mechanics approach [Figure 11.1(b)] has three important variables, rather than two as in Figure 11.1(a). The additional structural variable is flaw size, and **fracture toughness** replaces strength as the relevant material property. Fracture mechanics quantifies the critical combinations of these three variables.

Most fracture mechanics methodologies assume linear elastic behavior, although more advanced approaches incorporate nonlinear material behavior such as yielding. Two alternative approaches to linear elastic fracture analysis (LEFM) exist: the energy criterion and the stress intensity approaches, both of which are described below.

In addition to predicting the conditions for ultimate failure, fracture mechanics methodologies can also characterize time-dependent cracking mechanisms such as fatigue.

The Energy Criterion

The energy approach states that crack extension (i.e., fracture) occurs when the energy available for crack growth is sufficient to overcome the resistance of the material. The material resistance may include the surface energy, plastic work, or other type of energy dissipation associated with a propagating crack.

Griffith [1920] was the first to propose the energy criterion for fracture, but Irwin [1956] is primarily responsible for developing the present version of this approach: the **energy release rate**, G, which is defined as the rate of change in potential energy with crack area for a linear elastic material. At the moment of fracture, $G = G_c$, the critical energy release rate, which is a measure of fracture toughness.

For a crack of length 2*a* in an infinite plate subject to a remote tensile stress (Figure 11.2), the energy release rate is given by

$$G = \frac{\pi \sigma^2 a}{E} \tag{11.1}$$

where E is Young's modulus, σ is the remotely applied stress, and a is the half crack length. At fracture, $G = G_c$, and Equation (11.2) describes the critical combinations of stress and crack size for failure:

$$G_c = \frac{\pi \sigma_f^2 a_c}{E} \tag{11.2}$$

Note that for a constant G_c value, failure stress, σ_f, varies with $1/\sqrt{a}$. The energy release rate, G, is the driving force for fracture, while G_c is the material's resistance to fracture. To draw an analogy to the strength of materials approach of Figure 11.1(a), the applied stress can be viewed as the driving force for plastic deformation, while the yield strength is a measure of the material's resistance to deformation.

The tensile stress analogy is also useful for illustrating the concept of similitude. A yield strength value measured with a laboratory specimen should be applicable to a large structure; yield strength does not depend on specimen size, provided the material is reasonably homogeneous. One of the fundamental assumptions of fracture mechanics is that fracture toughness (G_c in this case) is independent of the size and geometry of the cracked body; a fracture toughness measurement on a laboratory specimen should be applicable to a structure. As long as this assumption is valid, all configuration effects are taken into account by the driving force, G. The similitude assumption is valid as long as the material behavior is predominantly linear elastic.

The Stress Intensity Approach

Figure 11.3 schematically shows an element near the tip of a crack in an elastic material, together with the in-plane stresses on this element. Note that each stress component is proportional to a single constant, K_I. If this constant is known, the entire stress distribution at the crack tip can be computed with the equations in Figure 11.3. This constant, which is called the **stress intensity factor**, completely characterizes the crack tip conditions in a linear elastic material [Irwin, 1957]. If one assumes that the material fails locally at some critical combination of stress and strain, then it follows that fracture must occur at a critical stress intensity, K_{IC}. Thus, K_{IC} is an alternate measure of fracture toughness.

For the plate illustrated in Figure 11.2, the stress intensity factor is given by

$$K_I = \sigma\sqrt{\pi a} \tag{11.3}$$

Failure occurs when $K_I = K_{IC}$. In this case, K_I is the driving force for fracture and K_{IC} is a measure of material resistance. As with G_c, the property of similitude should apply to K_{IC}. That is, K_{IC} is assumed to be a size-independent material property.

Comparing Equation (11.1) and Equation (11.3) results in a relationship between K_I and G:

$$G = \frac{K_I^2}{E} \tag{11.4}$$

This same relationship obviously holds for G_c and K_{IC}. Thus, the energy and stress intensity approaches to fracture mechanics are essentially equivalent for linear elastic materials.

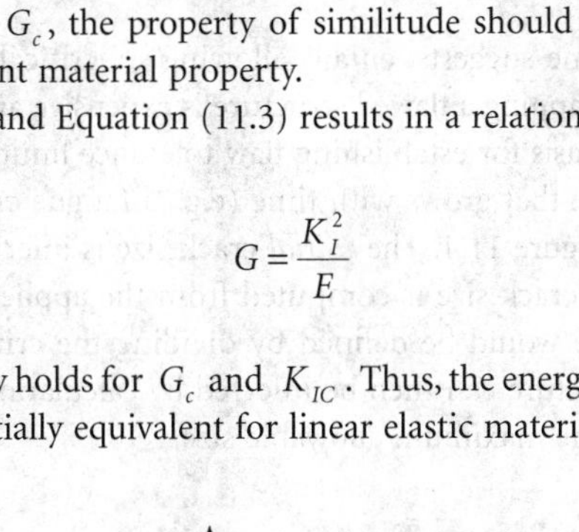

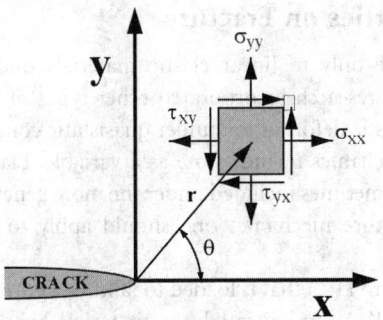

FIGURE 11.3 Stresses near the tip of a crack in an elastic material.

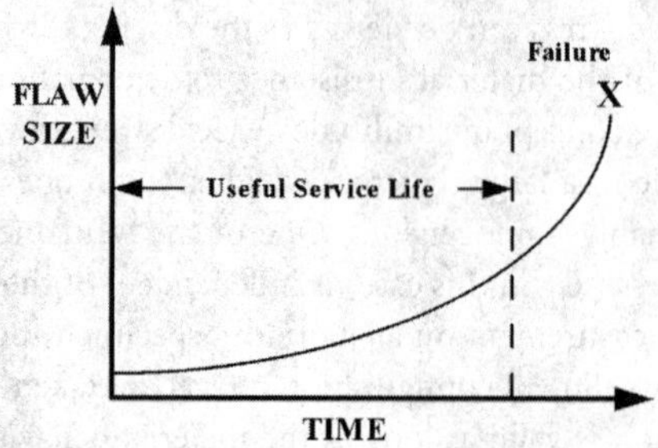

FIGURE 11.4 The damage tolerance approach to design.

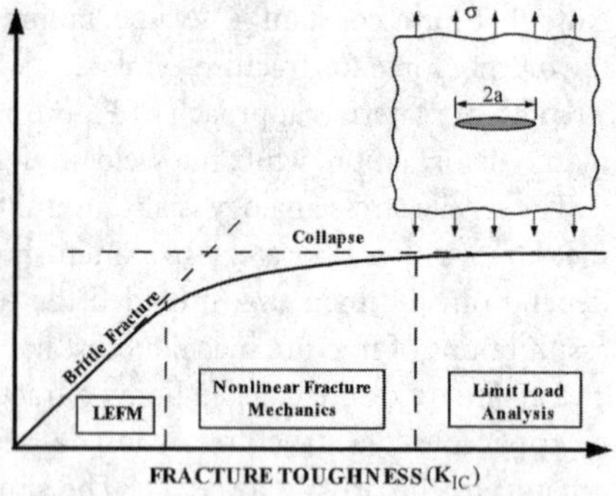

FIGURE 11.5 Effect of fracture toughness on the governing failure mechanism.

Time-Dependent Crack Growth and Damage Tolerance

Fracture mechanics often plays a role in life prediction of components that are subject to time-dependent crack growth mechanisms such as fatigue or stress corrosion cracking. The *rate* of cracking can be correlated with fracture mechanics parameters such as the stress intensity factor, and the critical crack size for failure can be computed if the fracture toughness is known. For example, Paris and Erdogan [1960] showed that the fatigue crack growth rate in metals could be described by the following empirical relationship:

$$\frac{da}{dN} = C(\Delta K)^m \tag{11.5}$$

where da/dN is the crack growth per cycle, ΔK is the stress intensity range, and C and m are material constants.

Damage tolerance, as its name suggests, entails allowing subcritical flaws to remain in a structure. Repairing flawed material or scrapping a flawed structure is expensive and is often unnecessary. Fracture mechanics provides a rational basis for establishing flaw tolerance limits.

Consider a flaw in a structure that grows with time (e.g., a fatigue crack or a stress corrosion crack) as illustrated schematically in Figure 11.4. The *initial* crack size is inferred from **nondestructive examination (NDE)**, and the *critical* crack size is computed from the applied stress and fracture toughness. Normally, an *allowable* flaw size would be defined by dividing the critical size by a safety factor. The predicted service life of the structure can then be inferred by calculating the time required for the flaw to grow from its initial size to the maximum allowable size.

Effect of Material Properties on Fracture

Most early work was applicable only to linear elastic materials under quasistatic conditions, while subsequent advances in fracture research incorporated other types of material behavior. Elastic-plastic fracture mechanics considers plastic deformation under quasistatic conditions, while dynamic, viscoelastic, and viscoplastic fracture mechanics include time as a variable. Elastic-plastic, viscoelastic, and viscoplastic fracture behavior are sometimes included under the more general heading of **nonlinear fracture mechanics**. The branch of fracture mechanics one should apply to a particular problem obviously depends on material behavior.

Consider a cracked plate (Figure 11.2) that is loaded to failure. Figure 11.5 is a schematic plot of failure stress versus fracture toughness (K_{IC}). For low toughness materials, brittle fracture is the governing failure mechanism, and critical stress varies linearly with K_{IC}, as predicted by Equation (11.3). At very high toughness values, LEFM is no longer valid, and failure is governed by the flow properties of the material.

TABLE 11.1 Typical Fracture Behavior of Selected Materials

Material	Typical Fracture Behavior
High strength steel	Linear elastic
Low- and medium-strength steel	Elastic-plastic/fully plastic
Austenitic stainless steel	Fully plastic
Precipitation-hardened aluminum	Linear elastic
Metals at high temperatures	Viscoplastic
Metals at high strain rates	Dynamic-viscoplastic
Polymers (below T_g)[a]	Linear elastic/viscoelastic
Polymers (above T_g)[a]	Viscoelastic
Monolithic ceramics	Linear elastic
Ceramic composites	Linear elastic
Ceramics at high temperatures	Viscoplastic

Note: Temperature is ambient unless otherwise specified.
[a] T_g, Glass transition temperature.

At intermediate toughness levels, there is a transition between brittle fracture under linear elastic conditions and ductile overload. Nonlinear fracture mechanics bridges the gap between LEFM and collapse. If toughness is low, LEFM is applicable to the problem, but if toughness is sufficiently high, fracture mechanics ceases to be relevant to the problem because failure stress is insensitive to toughness; a simple limit load analysis is all that is required to predict failure stress in a material with very high fracture toughness.

Table 11.1 lists various materials, together with the typical fracture regime for each material.

11.3 Concluding Remarks

Fracture is a problem that society has faced for as long as there have been manmade structures. The problem may actually be worse today than in previous centuries, because more can go wrong in our complex technological society. Major airline crashes, for instance, would not be possible without modern aerospace technology.

Fortunately, advances in the field of fracture mechanics have helped to offset some of the potential dangers posed by increasing technological complexity. Our understanding of how materials fail and our ability to prevent such failures have increased considerably since World War II. Much remains to be learned, however, and existing knowledge of fracture mechanics is not always applied when appropriate.

While catastrophic failures provide income for attorneys and consulting engineers, such events are detrimental to the economy as a whole. An economic study [Duga et al., 1983] estimated the cost of fracture in the United States in 1978 at $119 billion (in 1982 dollars), about 4% of the gross national product. Furthermore, this study estimated that the annual cost could be reduced by $35 billion if current technology were applied, and that further fracture mechanics research could reduce this figure by an additional $28 billion.

Defining Terms

Damage tolerance — A methodology that seeks to prevent catastrophic failures in components that experience time-dependent cracking. Fracture mechanics analyses are used in conjunction with nondestructive examination (NDE) to ensure that any flaws that may be present will not grow to a critical size prior to the next inspection.

Energy release rate — The rate of change in stored energy with respect to an increase in crack area. Energy release rate is a measure of the driving force for fracture. A crack will grow when the energy available for crack extension is greater than or equal to the energy required for crack extension. The latter quantity is a property of the material.

Fracture mechanics — An engineering discipline that quantifies the effect of cracks and crack-like flaws on material performance. Fracture mechanics analyses can predict both catastrophic failure and subcritical crack growth.

Fracture toughness — A measure of the ability of a material to resist crack propagation. The fracture toughness of a material can be quantified by various parameters, including a critical stress intensity factor, K_{IC}, and a critical energy release rate, G_c.

Linear elastic fracture mechanics (LEFM) — A branch of fracture mechanics that applies to materials that obey Hooke's law. LEFM is not valid when the material experiences extensive nonlinear deformation such as yielding.

Nondestructive examination (NDE) — A technology that can be used to characterize a material without altering its properties or destroying a sample. It is an indispensable tool for fracture mechanics analysis because NDE is capable of detecting and sizing crack-like flaws.

Nonlinear fracture mechanics — An extension of fracture mechanics theory to materials that experience nonlinear behavior such as yielding.

References

Bannerman, D. B. and Young, R. T., Some improvements resulting from studies of welded ship failures, *Welding Journal,* 25, 1946.

Duga, J. J., Fisher, W. H., Buxbaum, R. W., Rosenfield, A. R., Burh, A. R., Honton, E. J., and McMillan, S. C., The economic effects of fracture in the United States, NBS Special Publication 647-2, United States Department of Commerce, Washington, DC, March 1983.

Griffith, A. A., The phenomena of rupture and flow in solids, *Philosophical Transactions,* Series A, Vol. 221, 1920, pp. 163–198.

Irwin, G. R., Onset Of Fast Crack Propagation in High Strength Steel and Aluminum Alloys, *Sagamore Research Conference Proceedings, Vol. 2,* 1956, pp. 289–305.

Irwin, G. R., Analysis of stresses and strains near the end of a crack traversing a plate, *Journal of Applied Mechanics,* 24, 361–364, 1957.

Love, A. E. H., *A Treatise on the Mathematical Theory of Elasticity,* Dover Publications, New York, 1944.

Paris, P. C. and Erdogan, F., A critical analysis of crack propagation laws, *Journal of Basic Engineering,* 85, 528–534, 1960.

Shank, M. E., A Critical Review of Brittle Failure in Carbon Plate Steel Structures Other than Ships, Ship Structure Committee Report SSC-65, National Academy of Sciences–National Research Council, Washington, DC, December 1953.

Further Information

Anderson, T. L., *Fracture Mechanics: Fundamentals and Applications, Second Edition*, CRC Press, Boca Raton, FL, 1995.

Engineering Fracture Mechanics, published bi-monthly by Elsevier Science Ltd, Oxford, U.K.

International Journal of Fracture, published bi-monthly by Kluwer Academic Publishers, Dordrect, The Netherlands.

III

Dynamics and Vibrations

12

Dynamics of Particles: Kinematics and Kinetics

Bruce Karnopp
University of Michigan

Stephen Birn (Second Edition)
Moog Inc. – Aircraft Group

12.1 Dynamics of Particles

The dynamics of particles consists of five main parts:

1. **Kinematics** of a point (the geometry of a point moving through space)
2. Newton's second law
3. Moment of momentum equation
4. Momentum integrals of Newton's second law
5. Work-energy integral of Newton's second law

The concept of a **particle** is an abstraction or model of the actual physical situation. The moon in motion about the earth might be modeled as a mass point. In fact, the motion of any finite body in which the rotation effects are not important can properly be described as a particle or point mass.

Although it is possible to derive all the fundamental equations in a purely vector format, in order to describe any particular problem in dynamics, it is crucial that a specific coordinate system be employed. The coordinate systems that will be considered in this chapter are the following:

1. Cartesian coordinates
2. Natural (path) coordinates
3. Cylindrical coordinates
4. Spherical coordinates
5. Relative motion

0-8493-1586-7/05/$0.00+$1.50

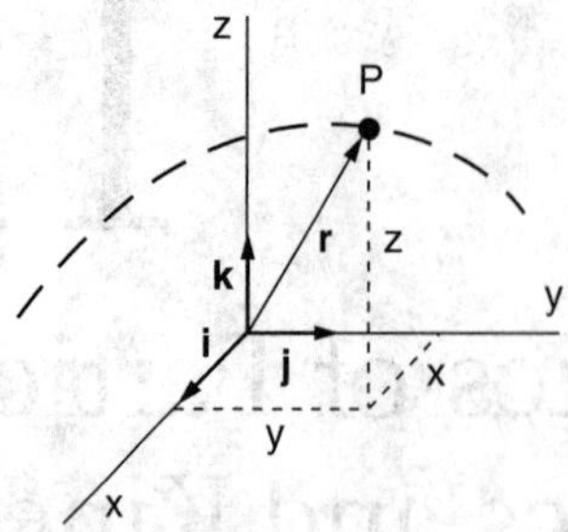

FIGURE 12.1 Cartesian coordinates.

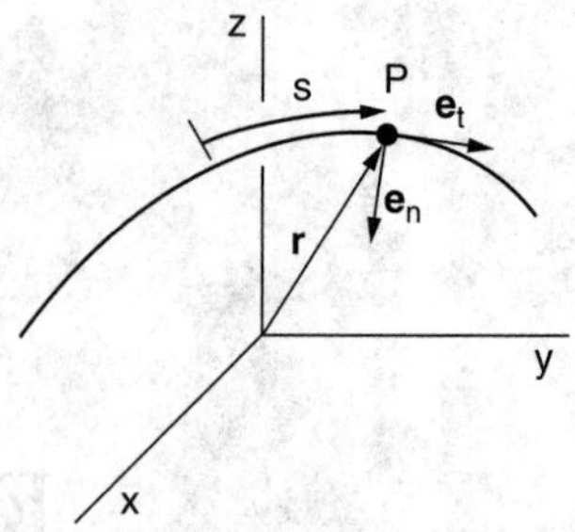

FIGURE 12.2 Natural coordinates.

The fundamental equation that is encountered in dynamics is Newton's second law, $\mathbf{F} = m\mathbf{a}$, where **F** is the total force acting on a particle and **a** is the resulting acceleration. Thus the geometric problem of dynamics consists of finding the **position**, **r**, the **velocity**, **v**, and the **acceleration**, **a**, of a point mass.

In order to use any coordinate system, the equations for the position **r**, the velocity **v**, and the acceleration **a** as expressed in that coordinate system must be known. In order to achieve these results, the derivatives of the unit vectors of the coordinate system must be determined.

Cartesian Coordinates

Consider the path of point *P* with respect to the Cartesian coordinate system shown in Figure 12.1. The position, velocity, and acceleration are expressed in Table 12.1.

Natural (Path) Coordinates

Natural or path coordinates are useful to understand the intrinsic nature of the velocity and acceleration vectors. Natural coordinates are defined by the actual trajectory of the point *P* as it moves through space. Consider the trajectory as shown in Figure 12.2.

The distance or arc length along the path (from some convenient starting position) is denoted by *s*. The velocity and acceleration of *P* are defined in terms of the path characteristics: the unit vector $\mathbf{e}_t$ tangent to the path, the unit vector $\mathbf{e}_n$ normal to the path, the radius of curvature, *R*, and the derivatives of the arc length with respect to time, $\dot{s}$ and $\ddot{s}$. The quantity *R* is the radius of curvature, and τ is the torsion of the curve. Table 12.2 lists the natural coordinate equations. Struik [1961] contains a proof of the derivatives of the unit vectors.

TABLE 12.1 Equations of Cartesian Coordinates

$\mathbf{r} = x\mathbf{i} + y\mathbf{j} + z\mathbf{k}$
$\mathbf{v} = \dot{x}\mathbf{i} + \dot{y}\mathbf{j} + \dot{z}\mathbf{k}$
$\mathbf{a} = \ddot{x}\mathbf{i} + \ddot{y}\mathbf{j} + \ddot{z}\mathbf{k}$

TABLE 12.2 Equations of Natural Coordinates

Velocity and Acceleration	Derivatives of Unit Vectors
$\mathbf{v} = \dot{s}\mathbf{e}_t$	$\frac{d\mathbf{e}_t}{dt} = \frac{\dot{s}}{R}\mathbf{e}_n$
$\mathbf{a} = \ddot{s}\mathbf{e}_t + \frac{\dot{s}^2}{R}\mathbf{e}_n$	$\frac{d\mathbf{e}_n}{dt} = -\frac{\dot{s}}{R}\mathbf{e}_t + \dot{s}\tau\mathbf{e}_b$
	$\frac{d\mathbf{e}_b}{dt} = -\dot{s}\tau\mathbf{e}_n$

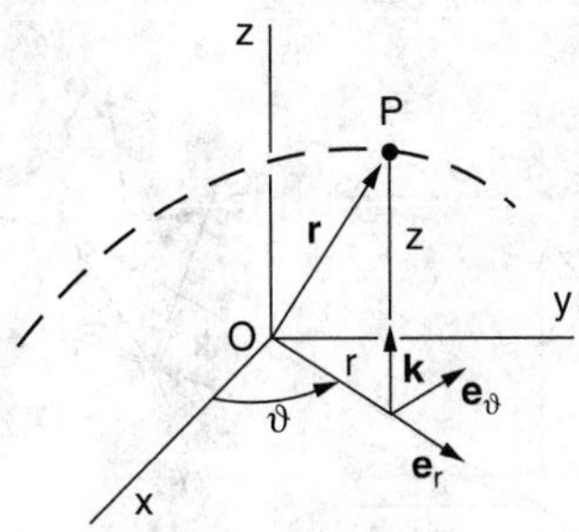

FIGURE 12.3 Cylindrical coordinates.

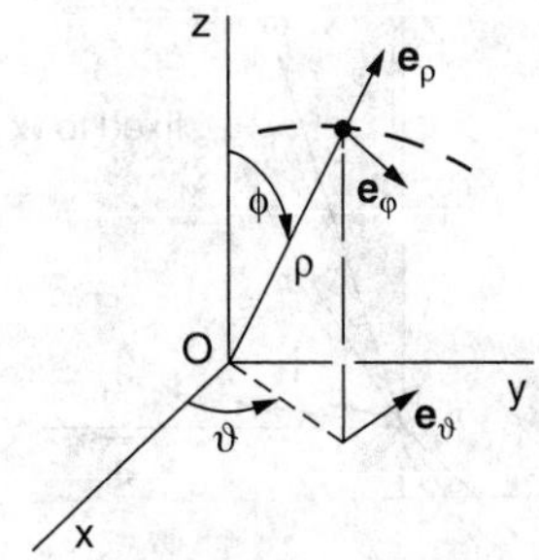

FIGURE 12.4 Spherical coordinates.

TABLE 12.3 Equations of Cylindrical Coordinates

Position, Velocity, and Acceleration	Derivatives of Unit Vectors
$\mathbf{r} = r\mathbf{e}_r + z\mathbf{k}$	$\dot{\mathbf{e}}_r = \dot{\vartheta}\mathbf{e}_\vartheta$
$\mathbf{v} = \dot{r}\mathbf{e}_r + r\dot{\vartheta}\mathbf{e}_\vartheta + \dot{z}\mathbf{k}$	$\dot{\mathbf{e}}_\vartheta = -\dot{\vartheta}\mathbf{e}_r$
$\mathbf{a} = (\ddot{r} - r\dot{\vartheta}^2)\mathbf{e}_r + (r\ddot{\vartheta} + 2\dot{r}\dot{\vartheta})\mathbf{e}_\vartheta + \ddot{z}\mathbf{k}$	$\dot{\mathbf{k}} = 0$

Cylindrical Coordinates

Cylindrical coordinates are used when there is some symmetry about a line. If this line is taken to be the z axis, the coordinates appear as in Figure 12.3. The parameters of cylindrical coordinates are introduced by dropping a line from the point P to the xy plane. The distance from the origin O to the intersection in the xy plane is denoted by the scalar r. Finally, the angle between the x axis and the line from O to the intersection is ϑ. Thus the parameters that define cylindrical coordinates are $\{r, \vartheta, z\}$, and the unit vectors are $\{\mathbf{e}_r, \mathbf{e}_\vartheta, \mathbf{k}\}$. See Table 12.3.

Spherical Coordinates

Spherical coordinates are particularly useful in problems with symmetry about a point. The coordinates are defined by the three parameters ρ, ϕ, and ϑ and the corresponding unit vectors $\mathbf{e}_\rho$, $\mathbf{e}_\phi$, and $\mathbf{e}_\vartheta$. Refer to Figure 12.4 and Table 12.4.

Kinematics of Relative Motion

The equations of relative motion are used when it is convenient to refer the motion to a coordinate system that is in motion. In general, such a coordinate system cannot be used to write the equations of

TABLE 12.4 Equations of Spherical Coordinates

Position, Velocity, and Acceleration	Derivatives of Unit Vectors
$\mathbf{r} = \rho\,\mathbf{e}_\rho$	$\dot{\mathbf{e}}_\rho = \dot{\phi}\,\mathbf{e}_\phi + \dot{\vartheta}\sin\phi\,\mathbf{e}_\vartheta$
$\mathbf{v} = \dot{\rho}\,\mathbf{e}_\rho + \rho\dot{\phi}\,\mathbf{e}_\phi + \rho\dot{\vartheta}\sin\phi\,\mathbf{e}_\vartheta$	$\dot{\mathbf{e}}_\phi = -\dot{\phi}\,\mathbf{e}_\rho + \dot{\vartheta}\cos\phi\,\mathbf{e}_\vartheta$
$\mathbf{a} = (\ddot{\rho} - \rho\dot{\phi}^2 - \rho\dot{\vartheta}^2\sin^2\phi)\mathbf{e}_\rho$	$\dot{\mathbf{e}}_\vartheta = -\dot{\vartheta}\sin\phi\,\mathbf{e}_\rho - \dot{\vartheta}\cos\phi\,\mathbf{e}_\phi$
$+(2\dot{\rho}\dot{\phi} + \rho\ddot{\phi} - \rho\dot{\vartheta}^2\sin\phi\cos\phi)\mathbf{e}_\phi$	
$+(2\dot{\rho}\dot{\vartheta}\sin\phi + 2\rho\dot{\phi}\dot{\vartheta}\cos\phi + \rho\ddot{\vartheta}\sin\phi)\mathbf{e}_\vartheta$	

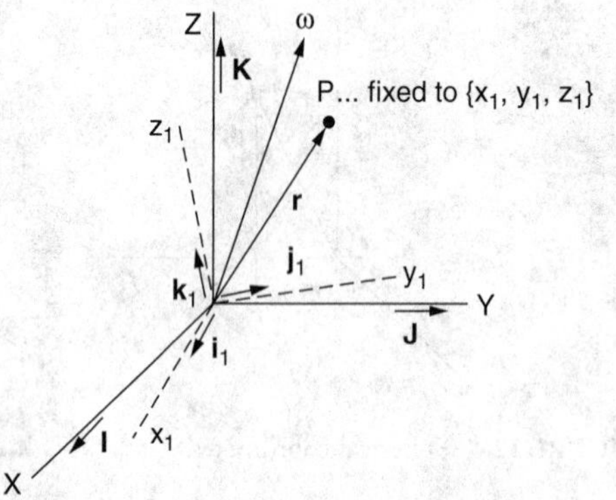

FIGURE 12.5 Angular velocity.

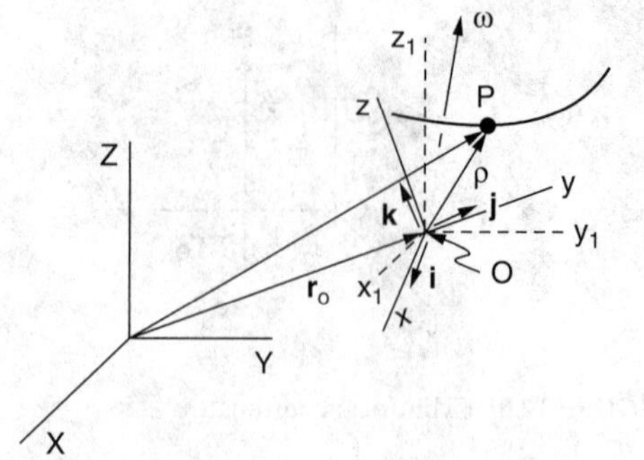

FIGURE 12.6 Relative motion.

dynamics since it will not be an inertial reference frame. The crucial concept in this regard is the **angular velocity** vector, $\boldsymbol{\omega}$. Consider Figure 12.5.

The angular velocity vector $\boldsymbol{\omega}$ is introduced through the equation

$$\mathbf{v} = \boldsymbol{\omega} \times \mathbf{r} \tag{12.1}$$

for any point P that is embedded in the moving frame $\{x_1, y_1, z_1\}$. Then

$$\frac{d\rho}{dt} = \frac{dx}{dt}\mathbf{i} + \frac{dy}{dt}\mathbf{j}\frac{dz}{dt}\mathbf{k} + x\omega \times \mathbf{i} + y\omega \times \mathbf{j} + z\omega \times \mathbf{k}$$

or

$$\frac{d\boldsymbol{\rho}}{dt} = \underbrace{\frac{dx}{dt}\mathbf{i} + \frac{dy}{dt}\mathbf{j} + \frac{dz}{dt}\mathbf{k}}_{} + \underbrace{\boldsymbol{\omega} \times (x\mathbf{i} + y\mathbf{j} + z\mathbf{k})}_{}$$

The first block of terms is the velocity as seen from the moving frame — that is, the velocity as it would appear to an observer whose feet are firmly planted in the moving frame $\{x, y, z\}$. It is convenient to denote this as $\delta\boldsymbol{\rho}/\delta t$. The second block of terms is just $\boldsymbol{\omega} \times \boldsymbol{\rho}$. Thus,

$$\frac{d\rho}{dt} = \frac{\delta\rho}{\delta t} + \omega \times \rho \tag{12.2}$$

Equation (12.2) gives an operator equation for computing time derivatives with respect to a fixed or moving frame:

$$\frac{d}{dt} = \frac{\delta}{\delta t} + \omega \times \tag{12.3}$$

Now suppose the origin of the moving frame has a motion. Then if $\mathbf{v}_o$ is the velocity of the moving origin (see Figure 12.6),

$$\mathbf{v} = \mathbf{v}_o + \frac{\delta\boldsymbol{\rho}}{\delta t} + \boldsymbol{\omega} \times \boldsymbol{\rho} \tag{12.4}$$

In order to interpret this, it is instructive to rearrange the terms:

TABLE 12.5 Equations of Relative Motion

$\mathbf{r} = \mathbf{r}_o + \boldsymbol{\rho}$
$\mathbf{v} = (\mathbf{v}_o + \boldsymbol{\omega} \times \boldsymbol{\rho}) + \dfrac{\delta \boldsymbol{\rho}}{\delta t}$
$\mathbf{a} = [\mathbf{a}_o + \omega \times (\omega \times \rho) + \dot{\omega} \times \rho] + \dfrac{\delta^2 \rho}{\delta t^2} + 2\omega \times \dfrac{\delta \rho}{\delta t}$

$$\mathbf{v} = \underbrace{(\mathbf{v}_o + \boldsymbol{\omega} \times \boldsymbol{\rho})}_{\text{Convective velocity}} + \underbrace{\frac{\delta \boldsymbol{\rho}}{\delta t}}_{\text{Relative velocity}}$$

The *relative velocity* is that which is seen by an observer fixed to the moving frame. The *convective velocity* is the velocity of a fixed point that instantaneously shares the position of the moving point.

This process is repeated to determine the acceleration. That is, the operator of Equation (12.3) is applied to Equation (12.4) to get $\mathbf{a} = d\mathbf{v}/dt$. The result is arranged as follows:

$$\mathbf{a} = \underbrace{[\mathbf{a}_o + \boldsymbol{\omega} \times (\boldsymbol{\omega} \times \boldsymbol{\rho}) + \dot{\boldsymbol{\omega}} \times \boldsymbol{\rho}]}_{\text{Convective acceleration}} + \underbrace{\frac{\delta^2 \boldsymbol{\rho}}{\delta t^2}}_{\substack{\text{Relative} \\ \text{acceleration}}} + \underbrace{2\boldsymbol{\omega} \times \frac{\delta \boldsymbol{\rho}}{\delta t}}_{\substack{\text{Coriolis} \\ \text{acceleration}}} \tag{12.5}$$

Again, the *relative acceleration* is that which a moving observer in $\{x, y, z\}$ would see. The *convective acceleration* is the acceleration of the fixed point of $\{x, y, z\}$ that shares the instantaneous position of the moving point under consideration.

The equations of position, velocity, and acceleration are summarized in Table 12.5.

12.2 Newton's Second Law

In order to write Newton's second law for a particle, m,

$$\mathbf{F} = m\mathbf{a} \tag{12.6}$$

the terms of the equation must be evaluated:

1. The force, **F**, is obtained from a free-body diagram of the particle. It should be noted that the free-body diagram will appear unbalanced because the particle is not in static equilibrium.
2. The mass, m, can be obtained from the weight of the particle:

$$\text{Weight} = mg \tag{12.7}$$

3. The acceleration is written in some convenient nonaccelerating coordinate system (from the equations in Section 12.1).

Any equation must ultimately be expressed in some unit system. The fundamental units of dynamics are force, mass, length, and time. The units for these quantities are shown in Table 12.6. Conversion of units is shown in Table 12.7.

TABLE 12.6 Unit Systems Used in Dynamics

Unit System	Type	Force	Mass	Length	Time
English (large)	Gravitational	Pound (lb)	Slug $lb \cdot s^2/ft$	Foot (ft)	Second (sec)
English (small)	Gravitational	Pound (lb)	$lb \cdot s^2/in.$	Inch (in.)	Second (sec)
MKS — metric	Absolute	Newton (N)	Kilogram (kg)	Meter (m)	Second (sec)
CGS — metric	Absolute	Dyne (dyn)	gram (g)	Centimeter (cm)	Second (sec)
Metric — large	Gravitational	Kilogram (kg)	$kg \cdot s^2/m$	Meter (m)	Second (sec)
Metric — small	Gravitational	Gram (g)	$g \cdot s^2/cm$	Centimeter (cm)	Second (sec)

TABLE 12.7 Conversion of Units

Force units	1.0 lb	= 4.448 N
	1.0 lb	$= 4.448 \cdot 10^5$ dyn
	1.0 lb	= 0.4536 (kgf)
	1.0 lb	$= 4.536 \cdot 10^2$ (g force)
Length units	1.0 in.	= 0.08333 ft
	1.0 in.	= 2.54 cm
	1.0 in.	= 0.0254 m
	1.0 ft	= 12 in.
	1.0 ft	= 30.48 cm
	1.0 ft	= 0.3048 m
Mass units	$1.0\ lb \cdot s^2/ft$	$= 12\ lb \cdot s^2/ft$
	$1.0\ lb \cdot s^2/ft$	= 1.2162 kg
	$1.0\ lb \cdot s^2/ft$	$= 1.2162 \cdot 10^3$ g
1.0 slug =	$1.0\ lb \cdot s^2/ft$	$= 0.08333\ lb \cdot s^2/in.$
	$1.0\ lb \cdot s^2/ft$	= 14.594 kg
	$1.0\ lb \cdot s^2/ft$	$= 1.4594 \cdot 10^4$ g

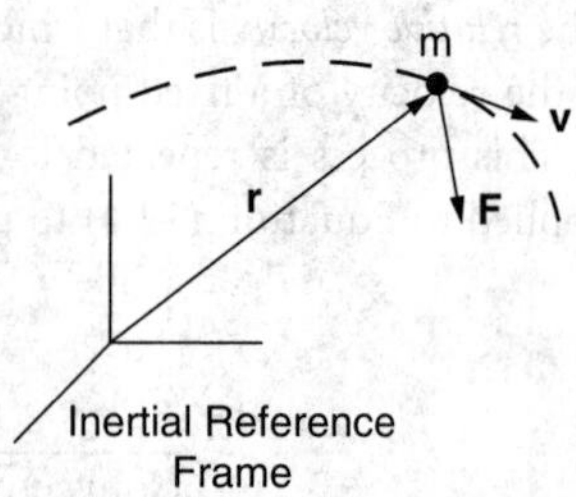

FIGURE 12.7 Moment of momentum and moment of force.

12.3 Moment of Momentum Relations

The moment of a force is determined by the (vector) *cross product:*

$$\mathbf{M}_o = \mathbf{r} \times \mathbf{F} \tag{12.8}$$

See Figure 12.7.

The *linear momentum* of *m* is

$$\mathbf{p} = m\mathbf{v} \tag{12.9}$$

and the *moment of momentum* (sometimes called the *angular momentum*) is

$$\mathbf{h}_o = \mathbf{r} \times \mathbf{p} = \mathbf{r} \times m\mathbf{v} \tag{12.10}$$

Computing the time derivative of **h** gives

$$\frac{d\mathbf{h}_o}{dt} = \frac{d\mathbf{r}}{dt} \times m\mathbf{v} + \mathbf{r} \times m\mathbf{a}$$

The first term is $\mathbf{v} \times m\mathbf{v}$. This is the cross product of two vectors in the same direction. Thus this term is zero. From Equation (12.6) and Equation (12.8), the remaining term is the moment of the force, **F**, about the point *O*. Thus:

$$M_o = \frac{d\mathbf{h}_o}{dt} \tag{12.11}$$

12.4 Momentum Integrals of Newton's Second Law

Newton's second law, Equation (12.6), can be integrated over time or space. When the former is done, the result is called an *impulse* or an *angular impulse*. When the integration is performed over space, the result is the *work*. This will be demonstrated later.

Impulse-Momentum and Angular Impulse-Moment of Momentum Relations

Recall Newton's second law, $\mathbf{F} = m\mathbf{a}$. Suppose we write $\mathbf{a} = d\mathbf{v}/dt$. Then

$$\int_{t_o}^{t_1} \mathbf{F}dt = m\int_{t_o}^{t_1} \frac{d\mathbf{v}}{dt}dt = m[\mathbf{v}(t_1) - \mathbf{v}(t_o)] \tag{12.12}$$

Equation (12.12) is called the *impulse change of linear momentum theorem.*

Similarly, taking Equation (12.11) as the basis of the time integration gives

$$\int_{r_o}^{r_1} \mathbf{M}_o dr = m\int_{t_o}^{t_1} \frac{d\mathbf{h}_o}{dt}dt = \mathbf{h}_o(t_1) - \mathbf{h}_o(t_o) \tag{12.13}$$

Equation (12.13) is called the *angular impulse change of angular momentum theorem.* Equation (12.12) and Equation (12.13) are particularly interesting when the left-hand side is zero. Then we say that linear momentum is conserved or that the moment of momentum (angular momentum) is conserved.

Two important examples that utilize these conservation laws are collision problems [Karnopp, 1974] and central force motion problems [Goldstein, 1959].

12.5 Work–Energy Integral of Newton's Second Law

In deriving the momentum laws, Newton's second law is integrated over time. In the work–energy relation, the integration takes place over space. Recall Newton's second law, $\mathbf{F} = m\mathbf{a} = m(d\mathbf{v}/dt)$. An instantaneous quantity is the *power* of the force **F**:

$$P = \mathbf{F} \cdot \mathbf{v} \tag{12.14}$$

The **power**, P, is a scalar quantity. The units of power are listed in Table 12.8.

TABLE 12.8 Units of Power

English	ft · lb/s
	in. · lb/s
	1.0 horsepower = 550 ft · lb/s
Metric	N · m/sec
	1.0 watt = 1.0 N · m/sec
Conversion	1.0 N · m/sec = 0.7376 ft · lb/s
	1.0 ft · lb/s = 1.3557 N · m/sec
	1.0 horsepower = 746 watts
	1.0 watt = 1.34048 · 10^{-3} hp

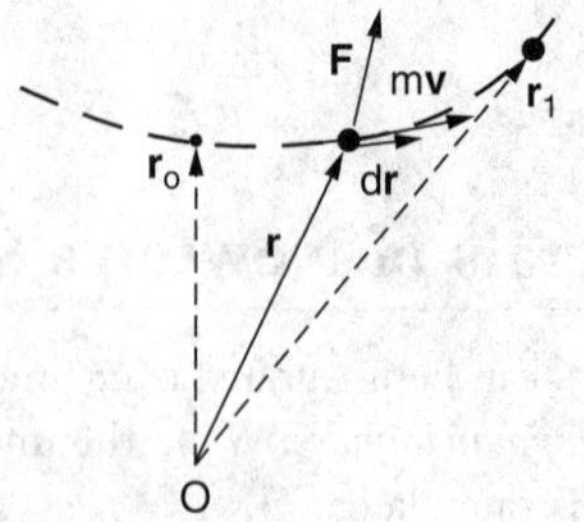

FIGURE 12.8 Work of a force.

The *work* of a force is the time integral of the power of the force. Work is also a scalar quantity:

$$W = \int_{t_o}^{t_1} P\,dt \tag{12.15}$$

With Equation (12.14), and recalling that $\mathbf{v} = d\mathbf{r}/dt$ the work W (see Figure 12.8) becomes

$$W = \int_{\mathbf{r}_o}^{\mathbf{r}_1} \mathbf{F} \cdot d\mathbf{r} \tag{12.16}$$

Equation (12.16) is what is called a *path* or *line integral.* That is, the value of the work is dependent, in general, on the particular path that is traversed between positions $\mathbf{r}_o$ and $\mathbf{r}_1$.

Recall from Equation (12.6) and Equation (12.14) that $\mathbf{F} = m\mathbf{a} = m(d\mathbf{v}/dt)$ and $P = \mathbf{F} \cdot \mathbf{v}$. Inserting Equation (12.13) into Equation (12.14) gives

$$P = m\frac{d\mathbf{v}}{dt} \cdot \mathbf{v} \tag{12.16a}$$

Now consider that

$$\frac{1}{2}\frac{d}{dt}(m\mathbf{v} \cdot \mathbf{v}) = \frac{1}{2}\left[m\frac{d\mathbf{v}}{dt} \cdot \mathbf{v} + m\mathbf{v} \cdot \frac{d\mathbf{v}}{dt}\right] = m\frac{d\mathbf{v}}{dt} \cdot \mathbf{v} \tag{12.16b}$$

Defining the *kinetic energy* of a particle to be

$$T = \frac{1}{2}m\mathbf{v} \cdot \mathbf{v} = \frac{1}{2}m\mathbf{v}^2 \tag{12.17}$$

Equation (12.16a) and Equation (12.16b) give

$$P = \frac{d}{dt}T \tag{12.18}$$

That is, the power of the force $\mathbf{F}$ equals the time rate of change of the kinetic energy T.

Finally, from Equation (12.18), the work of $\mathbf{F}$ in moving the particle from $\mathbf{r}_0$ to $\mathbf{r}_1$ equals the change in kinetic energy between $\mathbf{r}_o$ and $\mathbf{r}_1$. The work–energy theorem is derived in Table 12.9.

While Equation (12.16) gives a way to compute the work of a force, there is a very special and important class of forces that give a very simple way of computing work. These are called *conservative forces.* A force is conservative if it can be derived from a *potential energy function* through differentiation.

TABLE 12.9 The Work–Energy Theorem

The work of a force **F** is

$$W = \int_{\mathbf{r}_o}^{\mathbf{r}_1} \mathbf{F} \cdot d\mathbf{r}$$

The kinetic energy of a particle is

$$T = \frac{1}{2} m\mathbf{v} \cdot \mathbf{v} = \frac{1}{2} m\mathbf{v}^2$$

And the work–energy relation is

$$W_{\mathbf{r}_o}^{\mathbf{r}_1} = T(\mathbf{r}_1) - T(\mathbf{r}_o)$$

TABLE 12.10 Conservative Forces

	Force	Potential Energy
Gravity	$\mathbf{F} = -mg\mathbf{k}$	$V = mgz$
Universal gravitation[a]	$\mathbf{F} = -\frac{\gamma Mm}{r^2}\mathbf{e}_r$	$V = -\frac{\gamma Mm}{r}$
Spring force	$\mathbf{F} = -k\delta$	$V = \frac{1}{2}k\delta^2$

[a] For motion about the earth $\gamma M_e = 1.255 \cdot 10^3 \text{mi}^3/\text{h}^2 = 5.2277 \cdot 10^3 \text{km}^3/\text{h}^2$.

$$\mathbf{F} = -\nabla V \tag{12.19}$$

where V is the potential energy function for **F** and ∇ is the del operator. In Cartesian coordinates, Equation (12.18) becomes

$$\mathbf{F} = -\frac{\partial V}{\partial x}\mathbf{i} - \frac{\partial V}{\partial y}\mathbf{j} - \frac{\partial V}{\partial y}\mathbf{k}$$

The general form for conservative forces, Equation (12.19), is usually overly complex. Conservative forces are listed in Table 12.10.

The Work–Energy Relation for a Conservative Force

Recall the equation for the work of a force, $W = \int_{\mathbf{r}_o}^{\mathbf{r}_1} \mathbf{F} \cdot d\mathbf{r}$. Suppose **F** is conservative. Then, by Equation (12.19), $\mathbf{F} = -\nabla V$. And, finally, recall the equations of natural coordinates to write the expression for $d\mathbf{r}$: $\mathbf{v} = (d\mathbf{r}/ds)(ds/dt) = (ds/dt)\mathbf{e}_t$. Thus,

$$d\mathbf{r} = \frac{d\mathbf{r}}{ds}ds = \mathbf{e}_t ds$$

Finally, the work, by Equation (12.16), becomes

$$W = \int_{r_o}^{r_1} -(\nabla V \cdot \mathbf{e}_t) ds$$

The term inside the parentheses is just the *directional derivative* dV/ds — that is, the derivative that is taken tangent to the path. Thus the work becomes

TABLE 12.11 Units of Energy

English	ft · lb	
	in. · lb	
Metric	N · m = joule	
	dyn · cm = erg	
	Conversion	
1.0 joule =	1.0 N · m	$= 10^7$ dyn · cm $= 10^7$ erg
1.0 joule =	1.0 N · m	= 0.07376 ft · lb
1.0 joule =	1.0 N · m	= 0.88512 in. · lb
1.0 erg =	1.0 dyn · cm	$= 7.376\ 10^{-9}$ ft · lb
1.0 erg =	1.0 dyn · cm	$= 8.8512 \cdot 10^{-8}$ in. · lb
	1.0 ft · lb	= 1.3557 N · m
	1.0 ft · lb	$= 1.3557 \cdot 10^7$ dyn · cm
	1.0 in. · lb	= 0.11298 N · m
	1.0 in. · lb	$= 0.11298 \cdot 10^7$ dyn · cm

$$W = \int_{\mathbf{r}_o}^{\mathbf{r}_1} -\left(\frac{dV}{ds}\right) ds = \int_{\mathbf{r}_o}^{\mathbf{r}_1} -dV = -V(\mathbf{r}_1) + V(\mathbf{r}_o) \tag{12.20}$$

The crucial thing to note in Equation (12.20) is that the work of a conservative force depends *only* on the end positions of the path. Thus the work–energy relation derived in Table 12.9 becomes, in the case of conservative forces:

$$T(\mathbf{r}_1) + V(\mathbf{r}_1) = T(\mathbf{r}_o) + V(\mathbf{r}_o) = \text{constant}$$

For conservative and nonconservative forces:

$$[T(\mathbf{r}_o) + V(\mathbf{r}_o)] + W_{\mathbf{r}_o}^{\mathbf{r}_1} = [T(\mathbf{r}_1) + V(\mathbf{r}_1)]$$

where $W_{\mathbf{r}_o}^{\mathbf{r}_1}$ is the work of the nonconservative forces.

Units of energy are shown in Table 12.11.

The work–energy theorem is used when what is sought is the speed of a particle as a function of *position in space.* The impulse momentum theorems, on the other hand, will give the velocity as a function of *time.* Both relations are derived from Newton's second law and are called *first integrals.*

The work–energy problem solving method presented has five advantages for solving dynamics problems:

1. Coordinate system origins are arbitrary, so the problem can be simplified by choosing them carefully.
2. Accelerations do not need to be calculated.
3. Problem modifications or updates are easily made; parameterizing the dynamics problem for system layouts and case studies is also easy.
4. Scalar quantities are summed, even though the paths of motion may be complex.
5. Forces that do no work are ignored.

The main disadvantage to the energy method approach is the recognition that accelerations and work cannot be calculated for forces that do no work. In these cases, Newton's second law must be used.

12.6 Conclusion

The notion of a mass point or particle forms the basis of Newtonian mechanics. Although many systems can be modeled as a point mass, others cannot. Rigid configurations of systems, deformable systems, and so forth all require more elaborate geometrical (kinematic) description. The kinetic equations (Newton's law, momentum, moment of momentum, etc.) must be expanded in these cases. Still, the equations for particle dynamics form the basis of these discussions.

Defining Terms

Acceleration — The (vector) rate of change of velocity.
Angular velocity — The rate of change of orientation of a coordinate system.
Kinematics — The geometry of motion.
Particle — A point mass.
Position — The location of a point in space.
Power — The dot product of the force and the velocity.
Velocity — The (vector) rate of change of position.

References

Beer, F. P. and Johnston, E. R. 1984. *Vector Mechanics for Engineers: Statics and Dynamics,* 4th ed. McGraw-Hill, New York.
Goldstein, H. 1959. *Classical Mechanics.* Addison-Wesley, Reading, MA.
Hibbler, R. C. 2001. *Engineering Mechanics, Dynamics,* 9th ed. Prentice Hall, Inc.,
Englewood Cliffs, NJ.
Karnopp, B. H. 1974. *Introduction to Dynamics.* Addison-Wesley, Reading, MA.
Meriam, J. L. and Kraige, L. G. 1997. *Engineering Mechanics, Dynamics,* 4th ed. John Wiley & Sons, New York.
Struik, D. J. 1961. *Differential Geometry.* Addison-Wesley, Reading, MA.

Further Information

Synge, J. L. and Griffith, B. A. 1959. *Principles of Mechanics.* McGraw-Hill, New York.

13

Dynamics of Rigid Bodies: Kinematics and Kinetics

Ashraf A. Zeid
Army High Performance Computing Research Center and Computer Sciences Corporation

R. R. Beck
U.S. Army Tank Automotive Research Development and Engineering Center

13.1 Kinematics of Rigid Bodies

Kinematics is the study of the geometry of rigid body motion without reference to what causes the motion. **Kinematic analyses** are conducted to establish relationships between the position, **velocity,** and **acceleration** of rigid bodies or points on a rigid body.

The position and orientation of a body can be described by their distance from a perpendicular set of fixed axes called a *coordinate system.* The minimum number of independent or generalized coordinates needed to completely describe the position and orientation of a system of rigid bodies is equal to the number of *degrees of freedom* for the system. The number of degrees of freedom equals the number of nonindependent coordinates used to describe the position and orientation of each body of the system minus the number of constraints equations governing the system's motion. Therefore, the maximum number of independent coordinates needed to completely describe the position and orientation of a rigid body in space is six. Three independent equations are required to locate and describe the rigid body in translation with respect to time; the other three independent equations of motion are required to define its orientation and rotation in space with respect to time.

In general, the equations of motion of a rigid body are created relative to an inertial reference frame. The inertial reference frame is usually the rectangular set of Cartesian coordinate axes x, y, z with corresponding unit vectors i, j, k, as described previously.

A rigid body is in *rectilinear* **translation** when a line that joins any two points on the body does not rotate during motion. A rigid body is in *curvilinear translation* when all points of the body move on congruent curves. *A fixed-axis* **rotation** occurs when the line that connects any point on the body to the center of rotation rotates without any translation. When all points in a body move in parallel planes, the rigid body is in *general plane motion.* If no restriction is placed on the motion of the rigid body, it will move in *general space motion.* If the body is in general space motion and one of its points is pivoted, the body is in a *fixed-point rotation.*

0-8493-1586-7/05/$0.00+$1.50

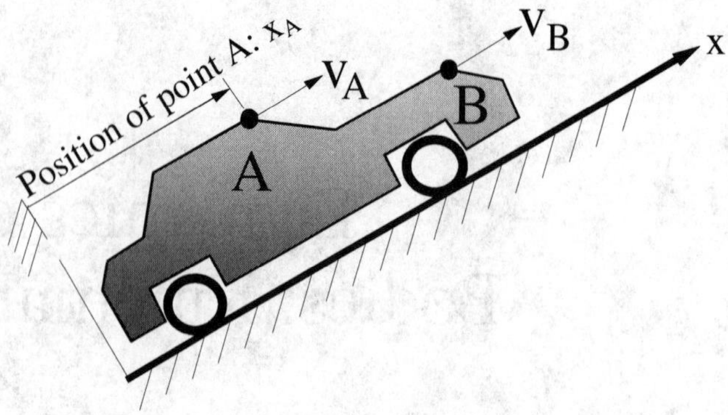

FIGURE 13.1 Rectilinear translation.

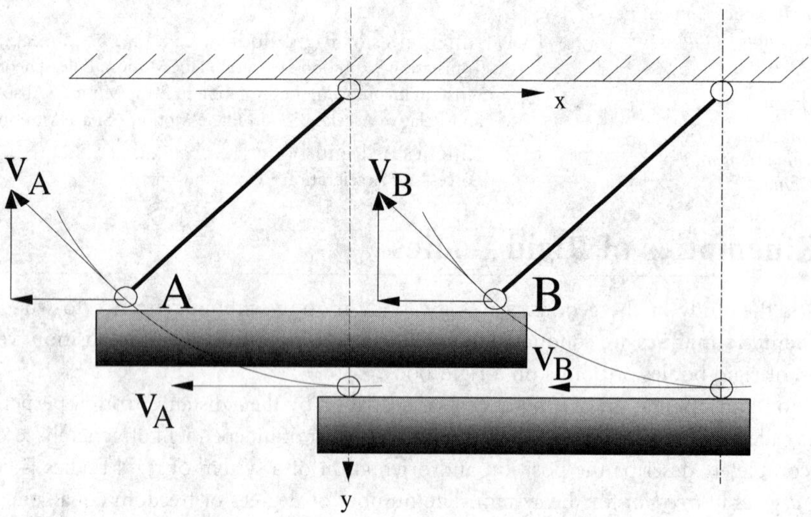

FIGURE 13.2 Curvilinear translation.

Translation

All points on a rigid body in pure translation will have the same velocity and the same acceleration at any given instant. Figure 13.1 and Figure 13.2 show examples of two different types of translational motion and a possible choice of a fixed reference frame whose axes are denoted as x and y with corresponding unit vectors i and j, respectively.

When a body is undergoing rectilinear translation, as shown in Figure 13.1, the velocities and accelerations of all points are identical in both magnitude and direction for all time.

$$\left.\begin{aligned} v_A &= v_B \\ a_A &= a_B \end{aligned}\right|_{\text{for all } t} \tag{13.1}$$

where $\{A, B, \ldots\}$ are arbitrary points on the body. In Figure 13.2, the velocities of any two points A and B on the body are identical and parallel at any instant of time; however, unlike in rectilinear translation, the velocity and acceleration directions are not constant. For curvilinear translation, the velocity equation holds at any instant of time but not necessarily throughout the entire motion:

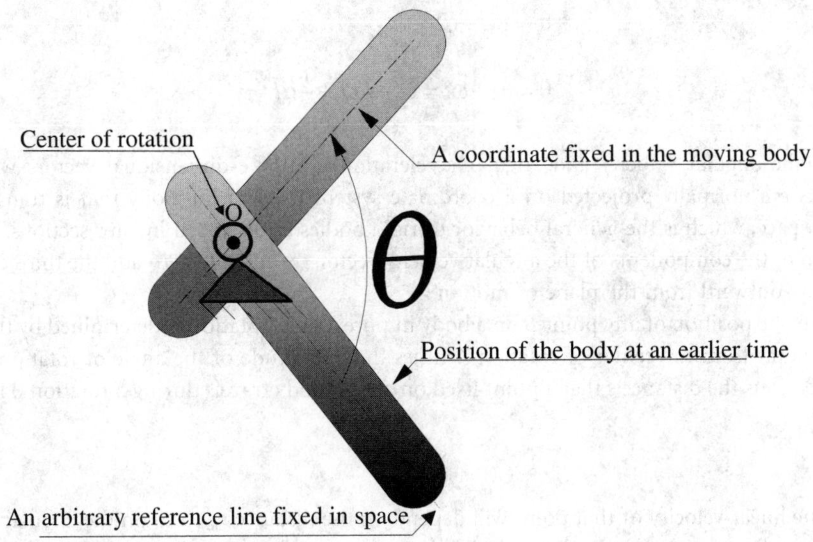

FIGURE 13.3 A body in pure rotation.

$$\left. \begin{aligned} v_A &= v_B \\ a_A &= a_B \end{aligned} \right|_{t_1 \neq t_2} \tag{13.2}$$

Rotation

The angular position of a body in pure rotation is completely defined by the angle between an arbitrary fixed reference line that passes through the center of rotation and any arbitrary line fixed to the body and passing also through the center of rotation, as shown in Figure 13.3. The rotation angle θ may be measured in degrees or radians, where

$$1 \text{ revolution} = 360 \text{ degrees} = 2\pi \text{ radians} \tag{13.3}$$

The rotation angular velocity ω is defined as the rate of change of the angular position angle θ with respect to time. It is expressed in radians per second (rps) or in revolutions per minute (rpm), as follows:

$$\omega = \frac{d\theta}{dt} \tag{13.4}$$

The rotational angular acceleration α is the time rate of change of the angular velocity resulting in the following relationship:

$$\alpha = \frac{d\omega}{dt} = \frac{d^2\theta}{dt^2} = \frac{d\omega}{d\theta}\frac{d\theta}{dt} = \omega\frac{d\omega}{d\theta} \tag{13.5}$$

In pure rotational motion, the relation between the *rotational position, velocity,* and *acceleration* are similar to pure translation. The angular velocity is the integral of the angular acceleration plus the initial velocity; the angular displacement is equal to the initial displacement added to the integral of the velocity. That is,

$$\omega = \omega_0 + \alpha t$$
$$\theta = \theta_0 + \omega t = \theta_0 + \omega_0 t + \frac{1}{2}\alpha t^2 \tag{13.6}$$

In general, the angular velocity and angular acceleration are three-dimensional vectors whose three components are normally projected on a coordinate system fixed to the body that is translating and rotating in space, which is the general behavior of rigid bodies as discussed in later sections. For planar motion, two of the components of the angular velocity vector are equal to zero and the third component points always outward from the plane of motion.

Therefore, the position of any point B on a body in pure planar rotation is determined by the distance $r_{B/A}$ of that point from the center of rotation A times the magnitude of the angle of rotation expressed in radians θ. Thus the distance s that a point fixed on a rigid body travels during a rotation θ is given by:

$$s = r_{B/A}\theta \tag{13.7}$$

Similarly, the linear velocity of that point will depend on the distance $r_{B/A}$ and on the angular velocity ω and will have a direction perpendicular to the line between the center of rotation and the point, as follows:

$$\vec{v} = \vec{\omega} \times \vec{r}_{B/A} \tag{13.8}$$

where $\times$ indicates cross product. The angular acceleration of a point on a rigid body can be decomposed into a tangential and a normal component. The tangential component is the time rate of change of the linear velocity v and is in the direction of the linear velocity, namely, along the line perpendicular to the radius of rotation $r_{B/A}$.

$$\vec{a}_t = \frac{d\vec{v}}{dt} = \vec{\alpha} \times \vec{r}_{B/A} \tag{13.9}$$

The normal acceleration depends on the time rate of change of the velocity in the tangential direction and on the angle of displacement, which gives the equation

$$\vec{a}_n = \vec{\omega} \times \vec{\omega} \times \vec{r}_{B/A} \tag{13.10}$$

General Plane Motion: Euler Theorem

General plane motion can be separated into a pure translation followed by a pure rotation about a point called the *center of rotation.* If we attach a coordinate system at point A, as shown in Figure 13.4, the position of any point on the body can be described by the position vector of point A — namely, r_A — added to the relative position of that point with respect to A — namely, the vector $r_{B/A}$ — all measured in the fixed coordinate system.

$$\vec{r}_B = \vec{r}_A + \vec{r}_{B/A} \tag{13.11}$$

Similarly, the velocity of a rigid body in general plane motion can be separated into a velocity due to pure translation v_A together with a velocity due to pure rotation $v_{A/B}$.

$$\vec{v}_B = \vec{v}_A + \vec{v}_{B/A} \tag{13.12}$$

where $\vec{v}_{B/A} = \omega \times \vec{r}_{B/A}$ and $\vec{v}_A$ is the velocity vector of point A. The velocity vector $\vec{v}_{B/A}$ is called the *relative velocity vector* of point B with respect to point A.

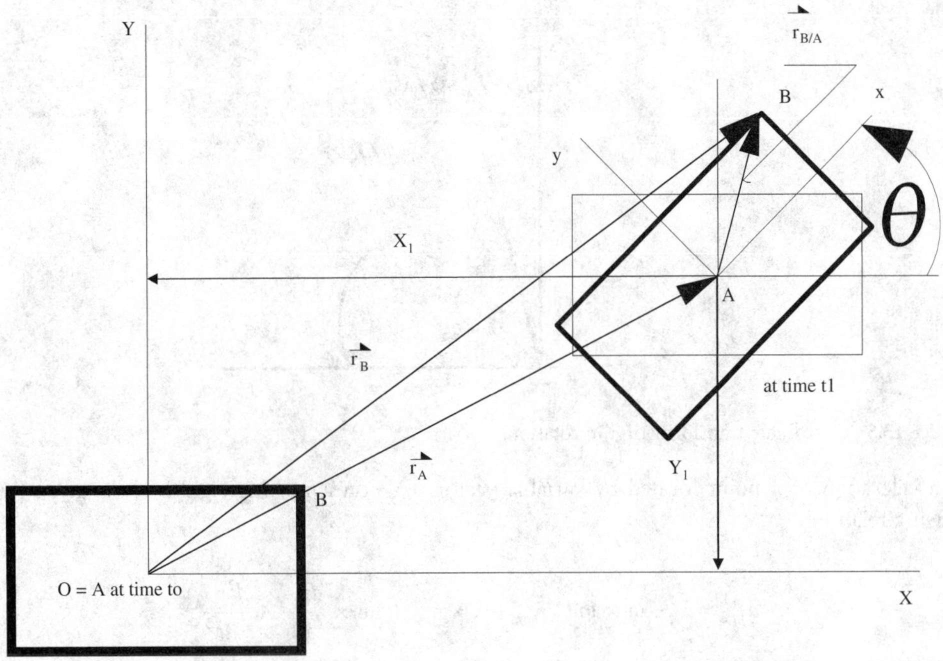

FIGURE 13.4 General motion of a rigid body in plane.

Instantaneous Center of Rotation in Plane Motion

At any instant in time, a body in general plane motion has a point — which may be either outside or on the body — around which all points of the body appear to be rotating in pure rotation. This point, called *instantaneous center of rotation*, can be found if the velocity vector of any point on the body together with the angular velocity of the body are known. The instantaneous center of rotation will lie on a line perpendicular to the velocity vector and at a distance from that point that is equal to the magnitude of the velocity of the point divided by angular velocity ω of the body.

Once the instantaneous center of rotation is found, the velocity of any point *B* on the body can be determined from the vector from that center to the point *B*, $\vec{r}_B$, as follows:

$$\vec{v}_B = \vec{\omega} \times \vec{r}_B$$

The direction of the velocity vector will be perpendicular to the vector $\vec{r}_B$.

Absolute and Relative Acceleration in Plane Motion

The angular acceleration of a point on a rigid body in plane motion also has a component due to translation and a component due to rotation; the latter component consists of a normal and a tangential component.

$$\begin{aligned} \vec{a}_B &= \vec{a}_A + \vec{a}_{B/A} \\ \vec{a}_{B/A} &= (\vec{a}_{B/A})_n + (\vec{a}_{B/A})_t \\ (\vec{a}_{B/A})_n &= \omega \times \omega \times \vec{r}_{B/A} \\ (\vec{a}_{B/A})_t &= \alpha \times \vec{r}_{B/A} \end{aligned} \tag{13.13}$$

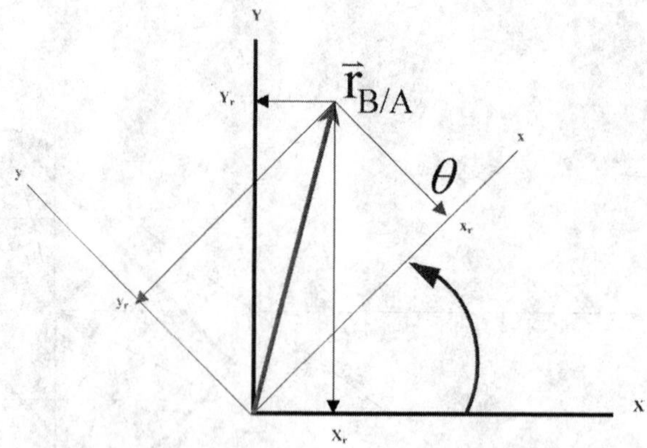

FIGURE 13.5 Coordinate transformation in rotation.

The acceleration of a point located by variable vector $\vec{r}_{B/A}$ on a moving rigid body is given by the following relation:

$$\vec{a}_{B/A} = \vec{a}_A + \omega \times \omega \times \vec{r}_{B/A} + \alpha \times \vec{r}_{B/A} + 2\omega \times \frac{d\vec{r}_{B/A}}{dt} + \frac{d^2\vec{r}_{B/A}}{dt^2} \tag{13.14}$$

where the vector $\vec{r}_{B/A}$ and its time derivative are measured in a fixed reference frame — namely, its components are $[X_r, Y_r]$ as shown in Figure 13.5. If the vector $\vec{r}_{B/A}$ is known by its components in a body-fixed coordinate $[x_r, y_r]$, then they can be transformed to the inertial coordinates as follows:

$$\begin{aligned} X_r &= x_r \cos\theta - y_r \sin\theta \\ Y_r &= x_r \sin\theta + y_r \sin\theta \end{aligned} \tag{13.15}$$

This is a coordinate transformation and is orthogonal, that is, its transpose is equal to its inverse. In matrix form the transformation of coordinates in Equation (13.15) can be written as follows:

$$\begin{bmatrix} X_r \\ Y_r \end{bmatrix} = \begin{bmatrix} \cos\theta & -\sin\theta \\ \sin\theta & \cos\theta \end{bmatrix} \begin{bmatrix} x_r \\ y_r \end{bmatrix} \quad \text{and} \quad \begin{bmatrix} x_r \\ y_r \end{bmatrix} = \begin{bmatrix} \cos\theta & \sin\theta \\ -\sin\theta & \cos\theta \end{bmatrix} \begin{bmatrix} X_r \\ Y_r \end{bmatrix} \tag{13.16}$$

If we use the prime symbol to denote that the vector components are measured in a body-fixed coordinate, then Equation (13.16) can be written in a more compact form as follows:

$$\vec{r}_{B/A} = T_z r'_{B/A} \quad \text{and} \quad r'_{B/A} = T_z^T r_{B/A} \tag{13.17}$$

where T_z^T is the transpose of the rotation matrix around the *z* axis (which would point outward from the page).

Space Motion

Three angles, called *Euler angles,* may be used to describe the orientation of a rigid body in space. These angles describe three consecutive rotations around the three coordinates of the frame fixed in a moving body with respect to an inertial fixed frame. Twelve combinations of rotation sequences can be chosen; here we choose the rotation around the *z* axis, ψ, followed by a rotation around the body-fixed *y* axis, θ, and finally a rotation around the body-fixed *x* axis, ϕ.

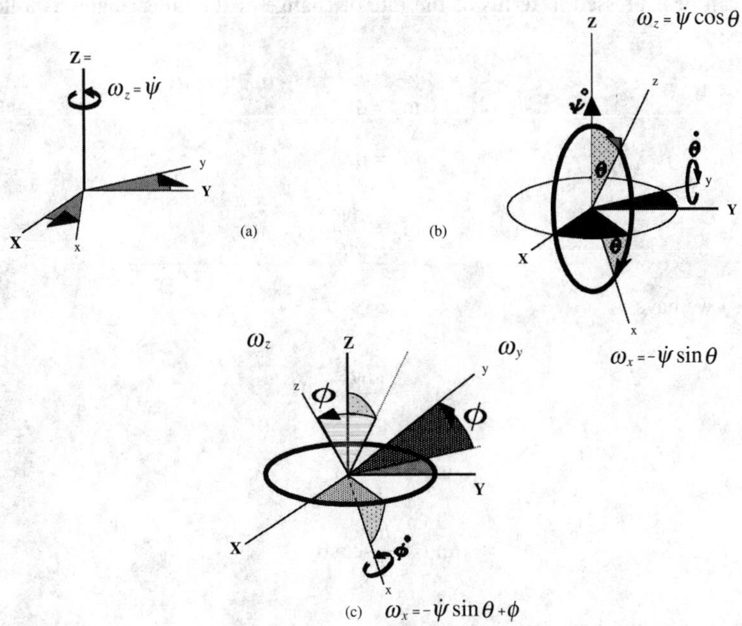

FIGURE 13.6 Three consecutive rotations of a body-fixed coordinate system (x, y, z) around the inertial-fixed axis (X, Y, Z).

Figure 13.6(c) shows the final position of a body which has a fixed coordinate system $[x, y, z]$. Originally, the body was oriented such that its fixed coordinate $[x, y, z]$ corresponded to the inertial fixed coordinate system $[X, Y, Z]$. The body was then rotated by an angle ψ around z, as shown in Figure 13.6(a), followed by a rotation of an angle θ around y, as shown in Figure 13.6(b), and finally by a rotation through an angle ϕ around z.

If the components of a vector are known in the body-fixed coordinate system, then the components of that vector can be obtained in the inertial reference frame by multiplying the vector by a transformation matrix. This transformation is obtained from the sequential product of the three successive rotation matrices around axis z, then y, and then x, respectively. As an example, the transformation matrix for the rotation in the order shown in Figure 13.6 is as follows:

$$T_{z,y,x} = \begin{bmatrix} \cos\theta\cos\psi & \cos\theta\sin\psi & -\sin\theta \\ -\cos\phi\sin\psi + \sin\phi\sin\theta\cos\psi & \cos\phi\cos\psi + \sin\phi\sin\theta\sin\psi & \sin\psi\cos\theta \\ \sin\phi\sin\psi + \cos\phi\sin\theta\cos\psi & -\sin\phi\cos\psi + \cos\phi\sin\theta\sin\psi & \cos\phi\cos\theta \end{bmatrix} \tag{13.18}$$

In order to transform any vector $r'_{B/A}$ known by its components in a body-fixed coordinate system into the corresponding vector whose components are given in inertial fixed coordinates, $r_{B/A}$, and vice versa, the vector would be multiplied by the transformation matrix as follows:

$$r_{B/A} = T_{z,y,x} r'_{B/A} \quad \text{and} \quad r'_{B/A} = T^T_{z,y,x} r_{B/A} \tag{13.19}$$

where the superscript T denotes the transpose of the matrix. Because the transformation matrix is orthogonal, its transpose is equal to its inverse, as shown by Equation (13.19).

The time derivatives of the Euler angles can be obtained from the components of the angular rotation matrix ω expressed in body coordinates. For the sequence of rotations shown in Figure 13.6, the angular velocity vector can be expressed in terms of the rate of change of the Euler angles as follows. In Figure 13.6(a) we have

$$\omega_x = 0$$

$$\omega_y = 0$$

$$\omega_z = \frac{d\psi}{dt}$$

In Figure 13.6(b) we have

$$\omega_x = -\frac{d\psi}{dt}\sin\theta$$

$$\omega_y = \frac{d\theta}{dt}$$

$$\omega_z = \frac{d\psi}{dt}\cos\theta$$

Finally, in Figure 13.6(c) we have

$$\begin{aligned}
\omega_x &= -\frac{d\psi}{dt}\sin\theta + \frac{d\phi}{dt} \\
\omega_y &= \frac{d\psi}{dt}\cos\theta\sin\phi + \frac{d\theta}{dt}\cos\phi \\
\omega_z &= \frac{d\psi}{dt}\cos\theta\cos\phi - \frac{d\theta}{dt}\sin\phi \\
\frac{d\psi}{dt} &= (\omega_y \sin\phi + \omega_z \cos\phi)/\cos\theta \\
\frac{d\theta}{dt} &= \omega_y \cos\phi - \omega_z \sin\phi \\
\frac{d\phi}{dt} &= \omega_x + (\omega_y \sin\phi - \omega_z \cos\phi)\tan\theta
\end{aligned} \tag{13.20}$$

In vector form the above equation may be written as follows:

$$\omega = E\frac{d\Sigma}{dt} \quad \text{or} \quad \frac{d\Sigma}{dt} = E^{-1}\omega \tag{13.21}$$

where the matrix E is given by:

$$E = \begin{bmatrix} -\sin\theta & 0 & 1 \\ \cos\theta\sin\phi & \cos\phi & 0 \\ \cos\theta\cos\phi & -\sin\phi & 0 \end{bmatrix}, \qquad \vec{\omega} = \begin{bmatrix} \omega_x \\ \omega_y \\ \omega_z \end{bmatrix},$$

and

$$E^{-1} = \begin{bmatrix} 0 & \dfrac{\sin\phi}{\cos\theta} & \dfrac{\cos\phi}{\cos\theta} \\ 0 & \cos\phi & -\sin\phi \\ 1 & \dfrac{\sin\phi\sin\theta}{\cos\theta} & \dfrac{\cos\phi\sin\theta}{\cos\theta} \end{bmatrix} \tag{13.22}$$

Note that the matrix E is not orthogonal, so its transpose is not equal to its inverse.

13.2 Kinetics of Rigid Bodies

Forces and Acceleration

Kinetics is the study of the relation between the forces that act on a rigid body and the resulting acceleration, velocity, and motion as a function of the body mass and geometric shape. The acceleration of a rigid body is related to its mass and to the applied forces by D'Alembert's principle, which states that the external forces acting on a rigid body are equivalent to the effective forces of the various particles of the body.

In the case of a rigid body moving in a plane motion, the D'Alembert principle amounts to the vector equation $\vec{F} = m\vec{a}$ together with the scalar equation of the moments $M = I\alpha$. In the particular case when a symmetric body is rotating around an axis that passes through its mass center — namely, centroidal rotation — the angular acceleration vector relates to the sum of moments by the equation $M = I\alpha$.

In general plane motion, the x and y components for the force vector, together with the moment equation, should be included in calculating the motion.

The **free-body diagram** is one of the essential tools for setting up the equations of motion that describe the kinetics of rigid bodies. It depicts the fundamental relation between the force vectors and the acceleration of a body by sketching the body together with all applied, reaction, and **D'Alembert force** and moment vectors drawn at the point where they are applied.

Systems of Rigid Bodies in Planar Motion

Free-body diagrams can be used to set up, and in some cases to solve, problems that involve several rigid bodies interconnected by elements forcing them to a prescribed motion — for example, a motion that follows a curve or a surface. Such elements, called *kinematical joints,* can be rigid links with negligible masses or wires in tension, such as the ones used in pulley systems.

For planar motion three equations of motion are obtained by writing down the x and y components of the forces and acceleration with the equation of moments and the angular acceleration.

For a rigid body moving under a constraint, the free-body diagram is supplemented by a *kinematical analysis,* which provides the tangential and normal component of the acceleration. Rolling of a disc on a surface, which is a noncentroidal rotation, is an example of a constrained plane motion that belongs to this class of problems. The rolling can be with no sliding, with impeding sliding, and with sliding. Rotation of a gear pair and a pulley also belong to this class of problems.

Rotation of a Three-Dimensional Body about a Fixed Axis

If several bodies each rotating in its own plane are connected by a rigid shaft then each will exert a D'Alembert force, $m\vec{a}$, on the shaft. Their combined effect will be a vector force and a couple equal to the inertia I of the body times the angular acceleration α.

If the body that rotates about a fixed axis is at rest and if the moments of the weights about the center of the rotating shaft are zero, we say that the system is *statically balanced.* When the body starts rotating, the moment due to D'Alembert forces, $m\vec{a}$, around the center of gravity of the system may not sum to zero; the system is *not dynamically balanced.* Rotating machinery strives to have its systems dynamically balanced to reduce the reaction forces at the bearings and consequently their wearing. Counterweights are added such that the total D'Alembert forces of the original bodies and the weights sum to zero.

In its most general case, the motion of a rigid body in space can be solved only through numerical integration, except for very few simple problems, such as *gyroscopic motion.*

Work and Energy

The **kinetic energy** of a particle in translation is a scalar quantity measured in joules or ft-lb and can be simply defined as $\frac{1}{2}mv^2$.

The infinitesimal element of work, Δw, is defined as the product of the projection of the force vector on the path s of the body and the infinitesimal length ds of that path: $(F \cos \beta)\, ds$, where β is the angle that the force vector makes with the path ds.

The principle of work and energy states that the energy of a body is equal to the sum, over a certain displacement path, of the work done by all external forces that acted on the body and caused that displacement plus any initial kinetic energy that the body had at the beginning of the path.

For bodies in pure rotation the work of a couple is the product of the couple and the infinitesimal angle moved due to that couple. The summation of the work over all the angular displacement caused by that couple is the rotational energy and is also defined as $\frac{1}{2}I\omega^2$, where ω is the angular velocity of the body.

The kinetic energy of a rigid body in general plane motion T is equal to the sum of kinetic energy in translation and the kinetic energy in rotation:

$$T = \frac{1}{2}(m\bar{v}^2 + \bar{I}\omega^2)$$

where $\bar{v}$ is the velocity of the mass center G of the body and $\bar{I}$ is the moment of inertia of the body about an axis through its mass center. This energy is identical to the kinetic rotational energy of the body if it is considered to be in pure rotation around its instantaneous center of rotation. In this case I would be the moment of inertia of the body around an axis that passes through the instantaneous center of rotation.

The principle of conservation of energy states that the sum of the potential and kinetic energy of a body acted upon by conservative forces — that is, nondissipative forces of friction or damping — remains constant during the time when these forces are applied.

Power is the product of the projection of the force vector on the velocity that resulted from this force. Power is measured in watt and horsepower units. The summation of power over a certain time interval is equal to the total energy stored in the body during that time.

Kinetics of Rigid Bodies in Plane Motion: Impulse and Momentum for a Rigid Body

The principle of impulse and momentum for a rigid body states that the momenta of all the particles of a rigid body at time t_1, added to the impulses of external forces acting during the time interval from time t_1 to time t_2, are equal to the system momenta at time t_2.

Momentum of a Rigid Body in Plane Motion

Translation, Rotation, and General Motion

The momenta vector of a body in plane translation motion is the product of the mass and the velocity vector. For a rigid body in plane centroidal rotation, the linear momenta vector is equal to 0 since the mass center does not have any linear velocity. The sum of the couples of forces acting on the particle of that body gives the angular momentum, which is $H_G = \bar{I}\omega$.

In a general plane motion, the momentum is a vector with components along the x, y, and ω directions. The dynamic equations of motion of a rigid body in plane motion can be obtained from D'Alembert's principle as follows:

$$\frac{dm\bar{v}_x}{dt} = (F_1)_x + (F_2)_x + \cdots$$
$$\frac{dm\bar{v}_y}{dt} = (F_1)_y + (F_2)_y + \cdots \tag{13.23}$$
$$\frac{dI\alpha}{dt} = M_1 + M_2 + \cdots + r_1 \times F_1 + r_2 \times F_2 + \cdots$$

For a system of rigid bodies, the linear momentum vector does not change in the absence of a resultant linear impulse. Similarly, the angular momentum vector does not change in the absence of an angular impulse.

Space Motion

The momentum vector of a rigid body moving in space has a linear component G and an angular component H. The linear component represents the D'Alembert principle as described by the following equations:

$$\frac{dm\bar{v}_x}{dt} = (F_1)_x + (F_2)_x + \cdots$$
$$\frac{dm\bar{v}_y}{dt} = (F_1)_y + (F_2)_y + \cdots \tag{13.24}$$
$$\frac{dm\bar{v}_z}{dt} = (F_1)_z + (F_2)_z + \cdots$$

where the velocity and the force vectors are expressed in their inertial $[X, Y, Z]$ components. If these vectors are expressed in a body-fixed coordinate system, the time derivative should include the effect of the rotation vector, as in the case of the angular momentum. The angular momentum vector H is defined as follows:

$$\vec{H} = I\vec{\omega} \tag{13.25}$$

where H and ω are expressed by their components in the body coordinate $[x, y, z]$. When a vector is expressed in a body coordinate, its time derivative should include the effect of angular rotation. For this reason — and because normally the position vector of the point of application of a force F_i from a center of rotation A, denoted by $\vec{r}_{Bi/A}$, is known by its components in a body coordinate system — the equation stating that the time rate of change of the angular momentum is equal to the sum of the moments would be written as follows:

$$\frac{dI\omega}{dt} + \omega \times I\omega = T^T_{z,y,x} M_1 + T^T_{z,y,x} M_2 + \cdots + r'_{B1/A} \times T^T_{z,y,x} F_1 + r'_{B2/A} \times T^T_{z,y,x} F_2 + \cdots \tag{13.26}$$

where I is the matrix of inertia relative to a coordinate system fixed in the body and moving with it and the center of the coordinate system is located at point A. The forces and the moments are assumed to be known by their components in an inertial fixed coordinate system, and ω is the angular velocity of the body given by its components in the body-fixed coordinates.

If the body-fixed coordinate system is chosen along the principal axis of the body, then the term $\omega \times I\omega$ can be written in matrix form as follows:

$$\omega \times I\omega = \begin{bmatrix} 0 & -\omega_z & \omega_y \\ \omega_z & 0 & -\omega_x \\ -\omega_y & \omega_x & 0 \end{bmatrix} \begin{bmatrix} I_{xx} & 0 & 0 \\ 0 & I_{yy} & 0 \\ 0 & 0 & I_{zz} \end{bmatrix} \begin{bmatrix} \omega_x \\ \omega_y \\ \omega_z \end{bmatrix} = -\begin{bmatrix} (I_{yy} - I_{zz})\omega_y\omega_z \\ (I_{zz} - I_{xx})\omega_z\omega_x \\ (I_{xx} - I_{yy})\omega_x\omega_y \end{bmatrix}$$

and the equation of motion can be reduced to the Euler's equations as follows:

$$\begin{aligned} I_{xx}\dot{\omega}_x &= (I_{yy} - I_{zz})\omega_y\omega_z + \Sigma M_x \\ I_{yy}\dot{\omega}_y &= (I_{zz} - I_{xx})\omega_z\omega_x + \Sigma M_x \\ I_{zz}\dot{\omega}_z &= (I_{xx} - I_{yy})\omega_x\omega_y + \Sigma M_z \end{aligned} \tag{13.27}$$

In general, Equation (13.25) and Equation (13.26) are solved by numerical integration, except for the cases where they are simplified, for example in gyroscopic motion. The Euler angles used in the transformation matrix T are obtained from the numerical integration of Equation (13.20).

Impulsive Motion and Eccentric Impact

The principle of conservation of momentum is useful in solving the problem of impacting bodies. If the colliding of two bodies is such that the collision point is on a line that joins their mass centers, then the collision is centroidal, the two bodies can be considered particles, and impulsive motion of particle dynamics can be used.

If the collision is noncentroidal, rotational motion will occur. In this case the projection of the velocity differential of the bodies' point of contact on the line normal to the contact surface after collision is equal to the same projection of the differential velocity prior to collision times the coefficient of restitution. This vector relation can be used to find the velocity after impact.

Rotation Around a Fixed Point and Gyroscopic Motion

When a rigid body spins at a rate ω about its axis of symmetry and is subjected to a couple of moment M about an axis perpendicular to the spin axis, then the body will precess at a rate $\mathbf{\Omega}$ about an axis that is perpendicular to both the spin and the couple axis. The rate of precession omega is equal to $M = \bar{I}\omega\Omega$.

A well-known example of gyroscopic motion is the motion of a top (see Figure 13.7), in which the couple moment M — due to gravity — is expected to force the top to fall. However, the top does not fall and rather precesses around the y axis.

Defining Terms

Acceleration — The rate of change of the velocity vector. Absolute acceleration of a rigid body is the rate of change of the velocity vector of the mass center of the body. Relative acceleration is the acceleration of a point on a body due to the angular velocity of that body only.

D'Alembert forces — Force and moment vectors due to the linear and angular accelerations of the body.

Free-body diagram — An essential sketch used to solve kinetics problems that involves sketching the rigid body together with all internal, reaction, and external force vectors.

Gyroscopic motion — Describes the motion of a rigid body that is spinning with a very large angular velocity around one axis when the couple of a moment is applied on the second axis. The resultant motion, called *precession,* is an angular velocity around the third axis.

Kinematic analysis — Starts from the geometry of constraints and uses differentiation to find the velocity and acceleration of the constrained points on a rigid body.

Kinetic energy — The accumulation of work of forces on a rigid body between two instants of time; includes kinetic energy due to translation and kinetic energy due to rotation.

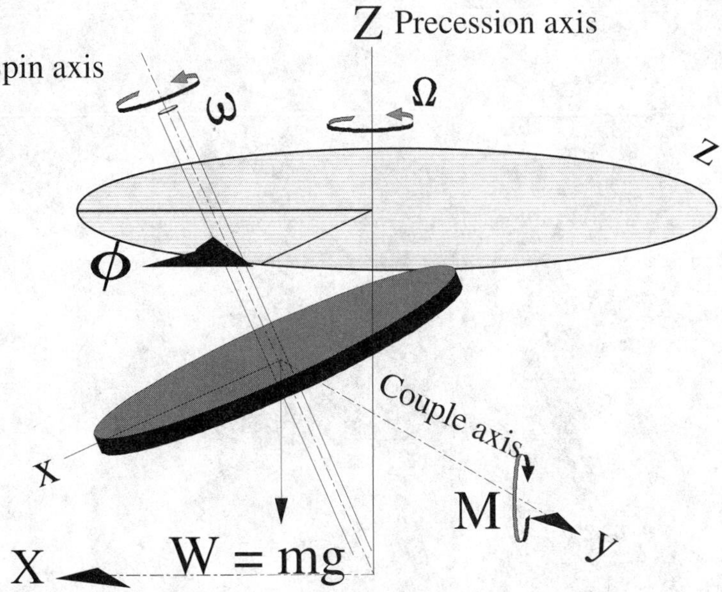

FIGURE 13.7 Gyroscopic motion.

Rotation — Centroidal rotation is the motion of a rigid body around an axis that passes through its mass center. In noncentroidal rotation the body rotates around an axis that passes through a point not corresponding to its mass center and which may not be on the body; this point is called the instantaneous center of rotation.

Translation — Rectilinear translation occurs when the velocities of any two points on the body remain equal in direction and magnitude throughout the entire duration of the motion. Curvilinear translation occurs when the velocity vectors of any two points are equal at any instant of time but change from one instant to another.

Velocity — Absolute velocity is the rate of change of the position vector of a point on a body measured from a fixed reference coordinate. Relative velocity is the rate of change of the position vector of a point on a rigid body measured from a moving reference frame.

References

Meriam, J. L. and Kraige, L. G. 1992. *Engineering Mechanics,* 3rd ed. John Wiley & Sons, New York.

Beer, F. P. and Johnston, E. R. 1987. *Mechanics for Engineers — Dynamics,* 4th ed. McGraw-Hill, New York.

Crandal, S. H., Karnopp, D. C., Kurtz., E. F., Jr., and Pridmore-Brown, D. C. 1968. *Dynamics of Mechanical and Electromechanical Systems,* McGraw-Hill, New York.

Haug, E. J. 1989. *Computer-Aided Kinematics and Dynamics of Mechanical Systems. Volume I: Basic Methods,* Allyn and Bacon, Boston.

Nikravesh, P. 1988. *Computer-Aided Analysis of Mechanical Systems,* Prentice Hall, Englewood Cliffs, NJ.

Shabana, A. A. 1994. *Computational Dynamics,* John Wiley & Sons, New York.

Further Information

Detailed treatment of the subject can be found in Meriam and Kraige [1992] and Beer and Johnston [1987].

A classical presentation of the subject can be found in Crandal *et al.* [1968].

Computer-aided analysis of the kinematics and dynamics of constrained rigid bodies in space motion can be found in Haug [1989] and Nikravesh [1988].

14

Free Vibration, Natural Frequencies, and Mode Shapes

Daniel A. Mendelsohn
Ohio State University

14.1 Basic Principles

In its simplest form, mechanical vibration is the process of a mass traveling back and forth through its position of static equilibrium under the action of a *restoring force* or *moment* that tends to return the mass to its equilibrium position. The most common restoring mechanism is a spring or elastic member that exerts a force proportional to the displacement of the mass. Gravity may also provide the restoring action, as in the case of a pendulum. The restoring mechanism of structural members is provided by the elasticity of the material of which the member is made. **Free vibration** is a condition in which there are no external forces on the system.

Cyclic or *periodic* motion in time is described by the property $x(t + \tau) = x(t)$, where t is time and τ is the *period,* that is, the time to complete one cycle of motion. The **cyclic frequency** of the motion is $f = 1/\tau$, usually measured in cycles per second (Hz). The special case of periodic motion shown in Figure 14.1 is *harmonic motion,*

$$x(t) = A\sin(\omega t) + B\cos(\omega t) \tag{14.1a}$$

$$= X\sin(\omega t + \phi) \tag{14.1b}$$

where $\omega = 2\pi f$ is the **circular frequency**, typically measured in radians (rad)/sec, $X = (A^2 + B^2)^{1/2}$ is the *amplitude* of the motion, and $\phi = \tan^{-1}(B/A)$ is the *phase angle.* Many systems exhibit harmonic motion when in free vibration but do so only at discrete **natural frequencies**. A vibrating system with n **degrees of freedom** (DOF) has n natural frequencies, and for each natural frequency there is a relationship between the amplitudes of the n independent motions, known as the **mode shape**. A structural elastic member has an infinite number of discrete natural frequencies and corresponding mode shapes. The **fundamental frequency** and associated mode shape refer to the smallest natural frequency and associated

0-8493-1586-7/05/$0.00+$1.50

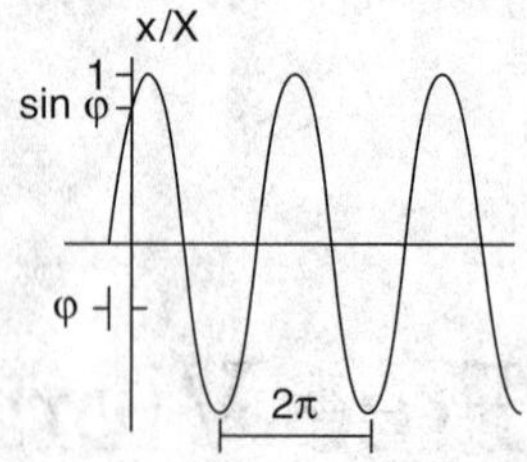

FIGURE 14.1 Time history of undamped periodic or cyclic motion.

mode shape. The study of free vibrations consists of the determination of the natural frequencies and mode shapes of a vibrating system as a function of geometry, boundary conditions, mass (density) of the components, and the strength of the restoring forces or moments. Although the natural frequencies and mode shapes are valuable to know by themselves, they have perhaps their greatest value in the analysis of forced vibrations, as discussed in detail in the following chapter.

14.2 Single-Degree-of-Freedom Systems

Equation of Motion and Fundamental Frequency

The system shown in Figure 14.2(a) consists of a mass, m, that rolls smoothly on a rigid floor and is attached to a linear spring of stiffness k. Throughout this chapter all linear (or longitudinal) springs have stiffnesses of dimension force per unit change in length from equilibrium, and all rotational (or torsional) springs have stiffnesses of dimension moment per radian of rotation from equilibrium (i.e., force times length). The distance of the mass from its equilibrium position, defined by zero stretch in the spring, is denoted by x. Applying Newton's second law to the mass in Figure 14.3(a) gives the equation of motion:

$$-kx = m\frac{d^2x}{dt^2} \Rightarrow m\frac{d^2x}{dt^2} + kx = 0 \tag{14.2}$$

Alternatively, Lagrange's equation (with only one generalized coordinate, x) may be used to find the equation of motion:

$$\frac{d}{dt}\left(\frac{\partial L}{\partial(dx/dt)}\right) - \frac{\partial L}{\partial x} = 0 \tag{14.3}$$

The Lagrangian, L, is the difference between the *kinetic energy, T*, and the *potential energy, U*, of the system. The Lagrangian for the system in Figure 14.2(a) is

$$L \equiv T - U = \frac{1}{2}m\left(\frac{dx}{dt}\right)^2 - \frac{1}{2}kx^2 \tag{14.4}$$

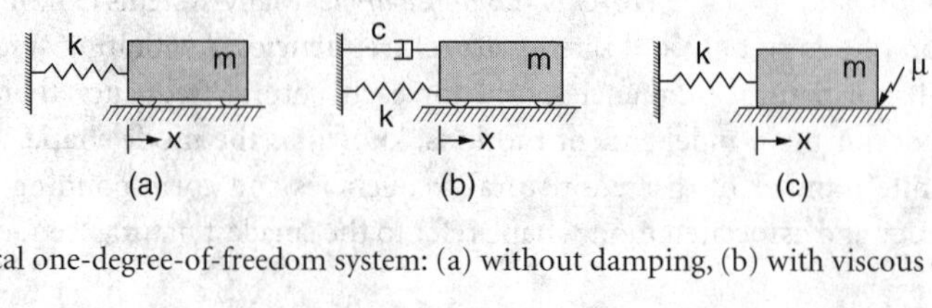

FIGURE 14.2 Typical one-degree-of-freedom system: (a) without damping, (b) with viscous damping, and (c) with frictional damping.

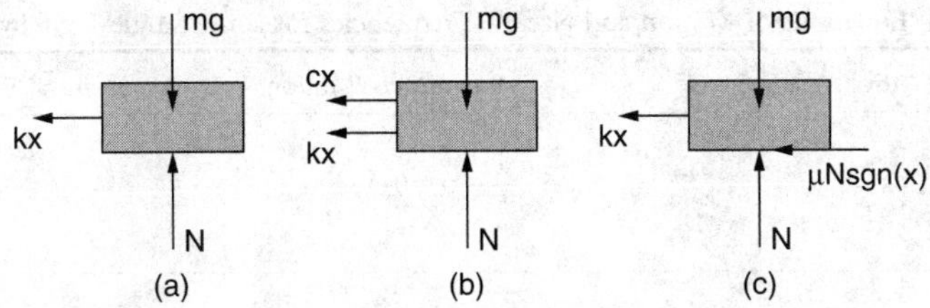

FIGURE 14.3 Free-body diagrams of the single-degree-of-freedom systems in Figure 14.2.

Substituting Equation (14.4) into Equation (14.3) gives the same equation of motion as in Equation (14.2). Using the harmonic form in Equation (14.1) for x, Equation (14.2) is satisfied if ω takes on the value

$$\omega = \sqrt{k/m} \tag{14.5}$$

which is therefore the natural frequency of the system. If the displacement and velocity are known at some time (say, $t = 0$), then the constants in Equation (14.1) may also be evaluated,

$$A = \frac{1}{\omega_0}\frac{dx}{dt}(0); \quad B = x(0) \tag{14.6}$$

and the corresponding displacement history is shown in Figure 14.1.

The natural frequency for conservative systems can also be found by the *energy method.* As the mass passes through equilibrium, $U = 0$ and $T = T_{max}$, while at its maximum displacement where the mass has zero velocity, $T = 0$ and $U = U_{max}$. Since the total energy is constant, ω is the frequency for which $T_{max} = U_{max}$. Using Equation (14.1a) and the system in Figure 14.2(a), this principle gives

$$T_{max} = \frac{1}{2}m(\omega X)^2 = \frac{1}{2}kX^2 = U_{max} \tag{14.7}$$

which in turn gives the same result for ω as in Equation (14.5).

Table 14.1 contains the equation of motion and natural frequency for some single-DOF systems. Gravity acts down, and displacements or rotations are with respect to static equilibrium. The mode shapes are of the form in Equation (14.1) with ω given in Table 14.1.

Linear Damping

Figure 14.2(b) and Figure 14.3(b) show an example of viscous damping caused by a dashpot of strength c (force per unit velocity) that acts opposite the velocity. Newton's second law then gives

$$-kx - c\frac{dx}{dt} = m\frac{d^2x}{dt^2} \Rightarrow m\frac{d^2x}{dt^2} + c\frac{dx}{dt} + kx = 0 \tag{14.8}$$

which has the solution

$$\begin{aligned} x(t) &= e^{-\zeta\omega_0 t}[A\sin(\omega_d t) + B\cos(\omega_d t)] \\ &= Xe^{-\zeta\omega_0 t}\sin(\omega_d t + \phi) \end{aligned} \tag{14.9}$$

where the *damped natural frequency,* ω_d, *damping factor,* ζ, and *critical damping coefficient,* c_c, are given by

TABLE 14.1 Equations of Motion and Natural Frequencies for some Single-DOF Systems

System	Equation of Motion	Natural Frequency
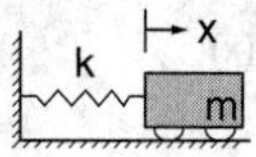	$m\frac{d^2x}{dt^2}+kx=0$	$\sqrt{\frac{k}{m}}$
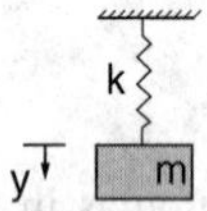	$m\frac{d^2y}{dt^2}+ky=0$	$\sqrt{\frac{k}{m}}$
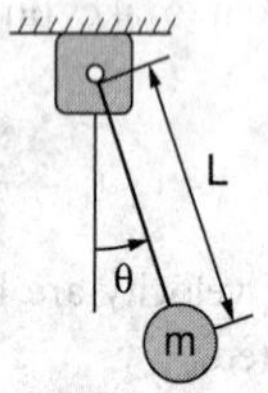	$(mL^2)\frac{d^2\theta}{dt^2}+(mgL)\theta=0$	$\sqrt{\frac{g}{L}}$
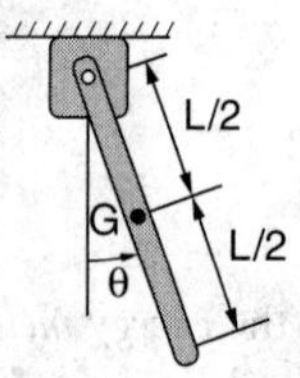 bar has mass m, center of gravity G	$\left(\frac{1}{2}mL^2\right)\frac{d^2\theta}{dt^2}+\left(mg\frac{L}{2}\right)\theta=0$	$\sqrt{\frac{3}{2}\frac{g}{L}}$
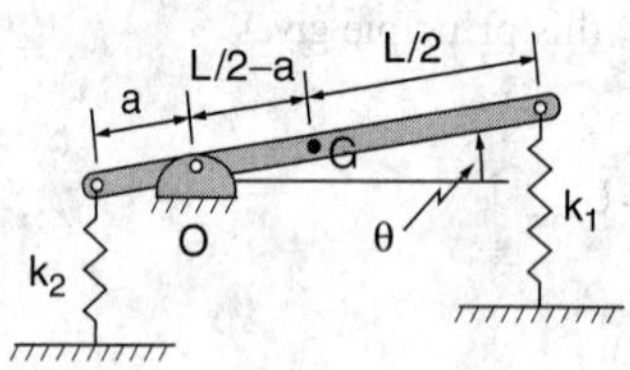	$I_0\frac{d^2\theta}{dt^2}+[k_1a^2+k_2(L-a)^2]\theta=0$ $I_0=\frac{1}{12}mL^2+m\left(\frac{L}{2}-a\right)^2$	$\sqrt{\frac{k_1}{m}\frac{a^2+\left(\frac{k_2}{k_1}\right)^2(L-a)^2}{\frac{1}{3}L^2-aL+a^2}}$
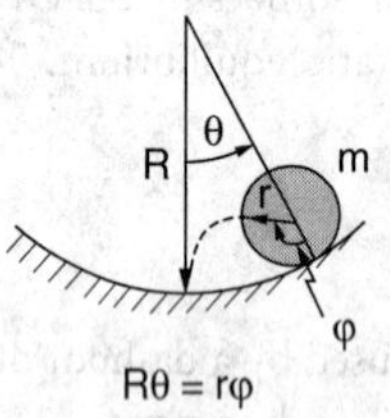	$\frac{3}{2}mr^2\left(\frac{R}{r}-1\right)\frac{d^2\theta}{dt^2}+(mgr)\theta=0$ r = radius of mass m	$\sqrt{\frac{2}{3}\frac{g}{(R-r)}}$

$$\omega_d \equiv \omega\sqrt{1-\zeta^2}; \quad \zeta \equiv \frac{c}{c_c}; \quad c_c \equiv 2m\omega = 2\sqrt{mk} \tag{14.10}$$

respectively. If $c < c_c$, then ω_d is real and Equation (14.9) represents exponentially damped oscillation, as shown in Figure 14.4. If $c \geq c_c$, then the system is *supercritically damped* and decaying motion but no vibration occurs.

The frictional effects in Figure 14.2(c) and Figure 14.3(c) are characterized by a Coulomb frictional force $F = \mu N = \mu mg$, where μ is the coefficient of sliding friction. The equation of motion is then

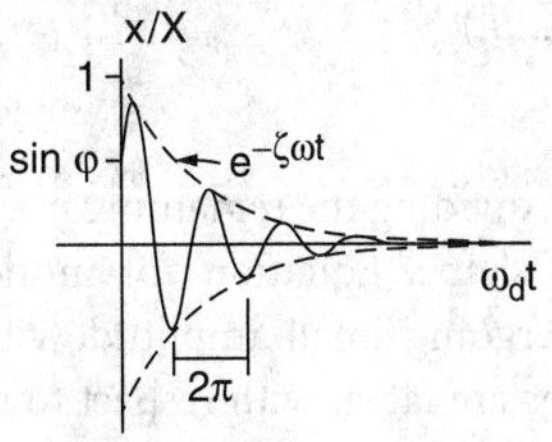

FIGURE 14.4 Time history of viscously damped vibration.

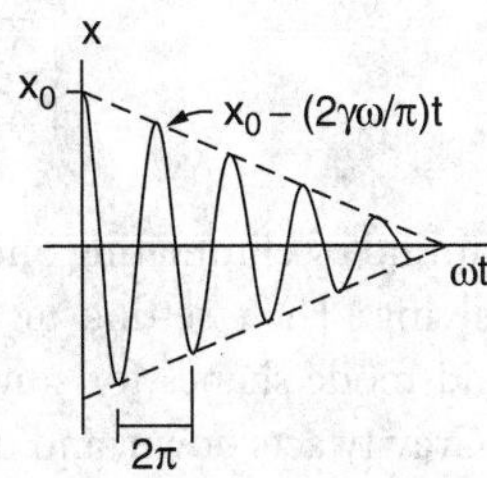

FIGURE 14.5 Time history of frictionally damped vibration.

$$m\frac{d^2x}{dt^2}+(\mu\, mg)\,\mathrm{sgn}\left(\frac{dx}{dt}\right)+kx=0 \tag{14.11}$$

where sgn(dx/dt) is equal to +1 or −1 for positive or negative values of dx/dt, respectively. This equation must be solved separately for each nth half period of the oscillation of frequency, ω,

$$x(t)=[x_0-(2n-1)\gamma]\cos(\omega t)-\mathrm{sgn}\left(\frac{dx}{dt}\right)\gamma \tag{14.12}$$

where $\gamma = (\mu mg/k)$ is the minimum initial displacement to allow motion, and ω is the undamped natural frequency, Equation (14.5). Figure 14.5 shows $x(t)$ for an initial displacement of $x_0 = 20\gamma$.

14.3 Multiple-Degree-of-Freedom Systems

For each DOF in an n-DOF system there is a *coordinate*, $x_i (i = 1,2,\ldots, n)$, which is a measure of one of the independent components of motion. The motion of the system is governed by n, generally coupled, equations of motion, which may be obtained by a Newtonian approach requiring complete free-body and acceleration diagrams for each mass. For systems with many DOFs, this approach becomes very tedious. Alternatively, applying Lagrange's equations, with no damping present,

$$\frac{d}{dt}\left(\frac{\partial L}{\partial \dot{x}_i}\right)-\frac{\partial L}{\partial x_i}=0, \quad (i=1,2,\ldots,n) \tag{14.13}$$

to the particular system yields the n equations of motion in the n unknown coordinates x_i,

$$[M]\left\{\frac{d^2x}{dt^2}\right\}+[K]\{x\}=\{0\} \tag{14.14}$$

As stated before, the Lagrangian, $L = T - U$, is the difference between the kinetic and potential energies of the system. $[M]$ and $[K]$ are the $n \times n$ *mass* and *stiffness matrices*, with elements m_{ij} and k_{ij}, which multiply the acceleration and displacement vectors of the masses, respectively. Writing x_i in the form of Equation (14.1), Equation (14.14) yields n homogeneous equations, $[A]\{X\} = \{0\}$, in the n amplitudes X_i. The elements of $[A]$ are $a_{ij} = k_{ij} - m_{ij}\omega^2$. If a solution exists, the determinant of $[A]$, an nth order polynomial in ω^2, must be zero. This yields the *frequency* or *characteristic equation*, whose n roots are the natural frequencies squared, $(\omega i)^2$. Each mode shape may written as a vector of $n - 1$ amplitude ratios:

$$\left\{\frac{X_2}{X_1}, \frac{X_3}{X_1}, \ldots, \frac{X_n}{X_1}\right\}_{(i)}, \quad (i = 1, 2, \ldots, n) \tag{14.15}$$

The ratios are found by eliminating one equation of $[A]\{X\} = \{0\}$, dividing the remaining $n - 1$ equations by X_1, and solving. Then setting $\omega = \omega_i$ gives the ith mode shape. Equations of motion, natural frequencies, and mode shapes for some two-DOF systems undergoing small amplitude vibrations are in Table 14.2. Gravity acts down, and displacements and rotations are taken with respect to the position of static equilibrium.

14.4 Continuous Systems (Infinite DOF)

The equations of motion of structural members made up of continuously distributed elastic or flexible materials are most easily obtained by a Newtonian analysis of a representative volume element. As an example, consider the longitudinal vibration of an elastic rod (Young's modulus E, density ρ) of cross-sectional area A. A free-body diagram of a volume element $A\,dx$, with normal stresses $[\sigma_x](x)$ and $[\sigma_x + (\partial\sigma_x/\partial x)dx](x)$ acting on the cross-sections is shown in Figure 14.6(a). A circular section is shown but the analysis applies to any shape of cross-section. If $u(x)$ is the displacement in the x direction of the cross-section at x, then Newton's second law gives

$$-\sigma_x A + \left(\sigma_x + \frac{\partial \sigma_x}{\partial x}dx\right)A = (\rho A dx)\frac{\partial^2 u}{\partial t^2} \tag{14.16}$$

Simplifying, letting dx go to zero, and noting uniaxial Hooke's law and the definition of the strian, ε_x,

$$\sigma_x = E\varepsilon_x = E\frac{\partial u}{\partial x} \tag{14.17}$$

Equation (14.16) can be written as

$$\frac{\partial^2 u}{\partial x^2} = \left(\frac{\rho}{E}\right)\frac{\partial^2 u}{\partial t^2} \tag{14.18}$$

which is the equation of motion for standing modes of free vibration and for wave propagation along the rod at velocity $c = \sqrt{E/p}$. If $u(x, t) = U(x)[A\sin(\omega t) + B\cos(\omega t)]$, then Equation (14.18) gives

$$\frac{d^2U}{dx^2} - \lambda^2 U = 0, \quad \lambda^2 = \frac{\rho\omega^2}{E} \tag{14.19}$$

which has solution $U(x) = C\sin(\lambda x) + D\cos(\lambda x)$. Now as an example, consider the fixed-fixed bar of length L shown in Figure 14.6(b) that has boundary conditions (BCs) $U(0) = 0$ and $U(L) = 0$, which give, respectively, $D = 0$ and either $C = 0$, which is not of interest, or

$$\sin(\lambda L) = 0 \Rightarrow \lambda = \lambda_n = \frac{n\pi}{L} \Rightarrow \omega_n = \frac{n\pi}{L}\sqrt{\frac{E}{\rho}} \quad (n = 1, 2, 3, \ldots) \tag{14.20}$$

This is the frequency equation and the resulting infinite set of discrete natural frequencies for the fixed-fixed beam of length L. The mode shapes are $U_n(x) = \sin(\lambda_n x)$.

TABLE 14.2 Equations of Motion, Natural Frequencies, and Mode Shapes for some Two-DOF Systems

Symbol	Equation of Motion	Natural Frequencies and Mode Shapes
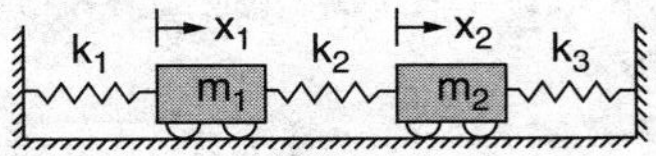	$\begin{bmatrix} m_1 & 0 \\ 0 & m_2 \end{bmatrix} \begin{bmatrix} \dfrac{d^2x_1}{dt^2} \\ \dfrac{d^2x_2}{dt^2} \end{bmatrix} + \begin{bmatrix} k_1+k_2 & -k_2 \\ -k_2 & k_2+k_3 \end{bmatrix} \begin{bmatrix} x_1 \\ x_2 \end{bmatrix} = \begin{bmatrix} 0 \\ 0 \end{bmatrix}$	$\omega_1 = \sqrt{\dfrac{k}{2m_1}}\sqrt{A-B}, \quad \omega_2 = \sqrt{\dfrac{k}{2m_2}}\sqrt{A+B}$ $A = \dfrac{m_1(k_2+k_3)+m_2(k_1+k_2)}{m_1k_2}$ $B = \sqrt{A^2 - 4\dfrac{m_2}{m_1}\left[\dfrac{(k_1+k_2)(k_2+d_3)}{k_2^2} - 1\right]}$ $\left(\dfrac{X_2}{X_1}\right)_i = 1 + \dfrac{k_1}{k_2} - \dfrac{m_1}{k_2}\omega_i^2; \quad i = 1,2$
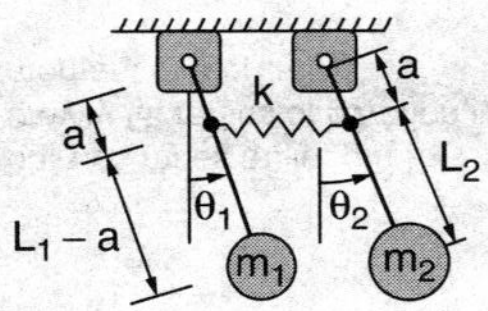	$\begin{bmatrix} m_1L_1^2 & 0 \\ 0 & m_2L_2^2 \end{bmatrix} \begin{bmatrix} \dfrac{d^2\theta_1}{dt^2} \\ \dfrac{d^2\theta_2}{dt^2} \end{bmatrix} + \begin{bmatrix} m_1gL_1+ka^2 & -ka^2 \\ -ka^2 & m_2gL_2+ka^2 \end{bmatrix} \begin{bmatrix} \theta_1 \\ \theta_2 \end{bmatrix} = \begin{bmatrix} 0 \\ 0 \end{bmatrix}$	$\omega_1 = \sqrt{\dfrac{q}{2L_1}}\sqrt{A-B}, \quad \omega_2 = \sqrt{\dfrac{q}{2L_1}}\sqrt{A+B}$ $A = 1 + \dfrac{L_1}{L_2} + \dfrac{ka^2}{m_1L_1g}\left(1 + \dfrac{m_1L_1^2}{m_2L_2^2}\right)$ $B = \sqrt{A^2 - 4\left[1 + \dfrac{ka^2}{m_1L_1g}\left(1 + \dfrac{m_2L_2}{m_1L_1}\right)\right]}$ $\left(\dfrac{\Theta_2}{\Theta_1}\right)_i = -1 + \dfrac{m_1L_1}{ka^2}(\omega_i^2L_1 - g); \quad i = 1,2$
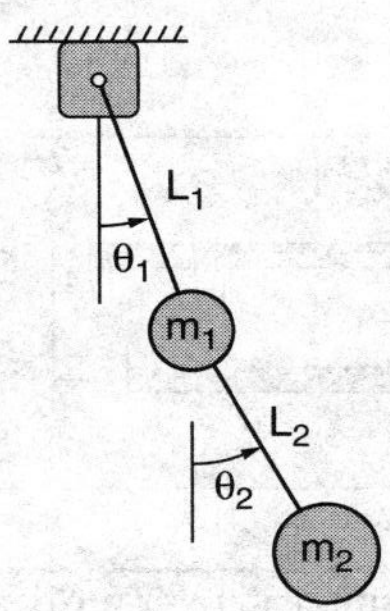	$\begin{bmatrix} m_1L_1 & 0 \\ m_2L_1 & m_2L_2 \end{bmatrix} \begin{bmatrix} \dfrac{d^2\theta_1}{dt^2} \\ \dfrac{d^2\theta_2}{dt^2} \end{bmatrix} + \begin{bmatrix} (m_1+m_2)g & -m_2g \\ -ka^2 & -m_2g \end{bmatrix} \begin{bmatrix} \theta_1 \\ \theta_2 \end{bmatrix} = \begin{bmatrix} 0 \\ 0 \end{bmatrix}$	$\omega_1 = \sqrt{\dfrac{q}{2L_1}}\sqrt{A-B}, \quad \omega_2 = \sqrt{\dfrac{q}{2L_1}}\sqrt{A+B}$ $A = \left(1 + \dfrac{m_1}{m_2}\right)\left(1 + \dfrac{L_1}{L_2}\right)$ $B = \sqrt{A^2 - 4\left(1 + \dfrac{m_2}{m_1}\right)\left(\dfrac{L_1}{L_2}\right)}$ $\left(\dfrac{\Theta_2}{\Theta_1}\right)_i = 1 + \left(\dfrac{m_1}{m_2}\right)(1 - \dfrac{\omega_i^2L_1}{g}); \quad i = 1,2$

TABLE 14.2 Equations of Motion, Natural Frequencies, and Mode Shapes for some Two-DOF Systems (*Continued*)

Symbol	Equation of Motion	Natural Frequencies and Mode Shapes
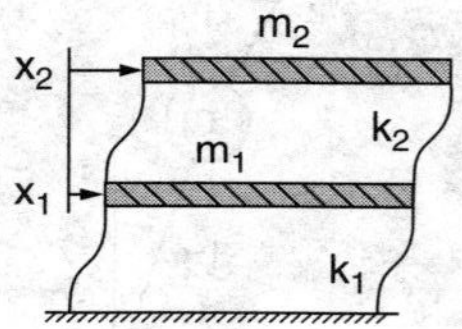	$\begin{bmatrix} m_1 & 0 \\ 0 & m_2 \end{bmatrix} \begin{bmatrix} \frac{d^2x_1}{dt^2} \\ \frac{d^2x_2}{dt^2} \end{bmatrix} + \begin{bmatrix} k_1+k_2 & -k_2 \\ -k_2 & k_2 \end{bmatrix} \begin{bmatrix} x_1 \\ x_2 \end{bmatrix} = \begin{bmatrix} 0 \\ 0 \end{bmatrix}$	$\omega_1 = \sqrt{\frac{k_1}{2m_1}}\sqrt{A-B}, \quad \omega_2 = \sqrt{\frac{k_1}{2m_1}}\sqrt{A+B}$ $A = 1 + \frac{k_2}{k_1} + \frac{k_2}{k_1}\frac{m_1}{m_2}$ $B = \sqrt{A^2 - 4\frac{k_2}{k_1}\frac{m_1}{m_2}}$ $\left(\frac{X_2}{X_1}\right)_i = 1 + \frac{k_1}{k_2} - \frac{m_1}{k_2}\omega_i^2; \quad i=1,2$
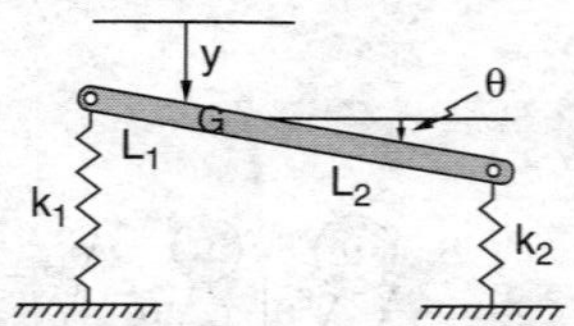bar has mass m, length $L_1 + L_2$, center of gravity G, radius of gyration r, and mass moment of inertia I_G	$\begin{bmatrix} m & 0 \\ 0 & mr^2 \end{bmatrix} \begin{bmatrix} \frac{d^2y}{dt^2} \\ \frac{d^2\theta}{dt^2} \end{bmatrix} + \begin{bmatrix} k_1+k_2 & k_2L_2-k_1L_1 \\ k_2L_2-k_1L_1 & k_1L_1^2+k_2L_2^2 \end{bmatrix} \begin{bmatrix} y \\ \theta \end{bmatrix} = \begin{bmatrix} 0 \\ 0 \end{bmatrix}$ $r = \text{Radius of gyration} = \sqrt{\frac{I_G}{m}}$	$\omega_1 = \sqrt{\frac{k_1}{2m}}\sqrt{A-B}, \quad \omega_2 = \sqrt{\frac{k_1}{2m}}\sqrt{A+B}$ $A = 1 + \frac{L_1^2}{r^2} + \frac{k_2}{k_1}\frac{L_1^2}{r^2} + \frac{k_2}{k_1}$ $B = \sqrt{A^2 - 4\frac{k_2}{k_1}\frac{(L_1+L_2)^2}{r^2}}$ $\left(\frac{Y}{\Theta}\right)_i = \frac{mr^2\omega_i^2 - (k_1L_1^2+k_2L_2^2)}{k_2L_1-k_1L_1}; \quad i=1,2$
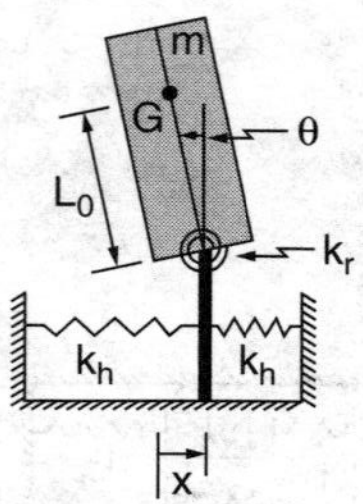	$\begin{bmatrix} m & -mL_0^2 \\ 0 & mr^2 \end{bmatrix} \begin{bmatrix} \frac{d^2x}{dt^2} \\ \frac{d^2\theta}{dt^2} \end{bmatrix} + \begin{bmatrix} k_h & 0 \\ k_hL_0 & k_r \end{bmatrix} \begin{bmatrix} x \\ \theta \end{bmatrix} = \begin{bmatrix} 0 \\ 0 \end{bmatrix}$ $r = \text{Radius of gyration} = \sqrt{\frac{I_G}{m}}$	$\omega_1 = \sqrt{\frac{k_h}{2m}}\sqrt{A-B}, \quad \omega_2 = \sqrt{\frac{k_h}{2m}}\sqrt{A+B}$ $A = 1 + \frac{L_0^2}{r^2} + \frac{k_r}{k_hr^2}$ $B = \sqrt{A^2 - 4\frac{k_r}{k_hr^2}}$ $\left(\frac{X}{\Theta}\right)_i = \frac{mr^2\omega_i^2 - k_r}{k_hL_0}; \quad i=1,2$

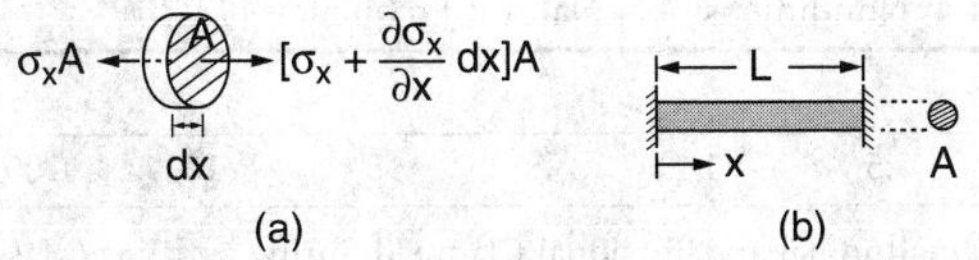

FIGURE 14.6 Longitudinal vibration of a rod of circular cross-section: (a) free-body diagram of representative volume element and (b) a clamped-clamped rod of length L.

The transverse motion $y(x, t)$ of a taut flexible string (tension T and mass per unit length ρ), the longitudinal motion $u(x, t)$ of a rod (Young's modulus E), and the torsional rotation $\phi(x, t)$ of a rod of circular or annular cross-section (shear modulus G) all share the same governing equations — Equation (14.18) and Equation (14.19) — but with different λ values: $\lambda^2 = \rho\omega^2/T$, $\rho\omega^2/E$, and $\rho\omega^2/G$, respectively. Table 14.3 and Table 14.4 contain frequency equations, nondimensional natural frequencies, and mode shapes for various combinations of BCs for a rod of length L. Only the fixed-fixed conditions apply to the string.

TABLE 14.3 Longitudinal and Torsional Vibration of a Rod

System	Frequency Equation	Natural Frequencies	Normalized Mode Shape
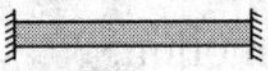	$\sin(\lambda_n L) = 0$	$(\lambda_n L) = n\pi$	$U_n(x) = \sin(\lambda_n x)$
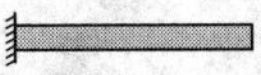	$\sin(\lambda_n L) = 0$	$(\lambda_n L) = (2n-1)\pi/2$	$U_n(x) = \sin(\lambda_n x)$
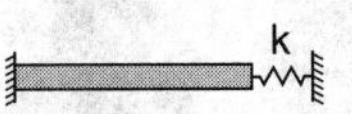	$\tan(\lambda_n L) = -\gamma(\lambda_n L)^1$	See Table 14.4	$U_n(x) = \sin(\lambda_n x)$
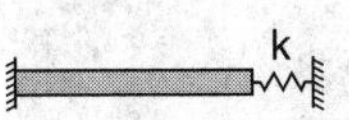	$(\lambda_n L)\tan(\lambda_n L) = \gamma^2$	See Table 14.4	$U_n(x) = \sin(\lambda_n x)$
	$\sin(\lambda_n L) = 0$	$(\lambda_n L) = n\pi$	$U_n(x) = \cos(\lambda_n x)$
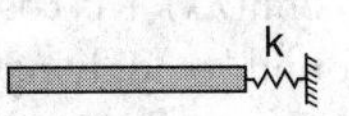	$(\lambda_n L)\tan(\lambda_n L) = -\gamma^3$	See Table 14.4	$U_n(x) = \cos(\lambda_n x)$
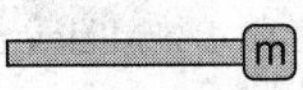	$\tan(\lambda_n L) = -\gamma(\lambda_n L)^4$	See Table 14.4	$U_n(x) = \cos(\lambda_n x)$

[1] $\gamma_{\text{long}} = \dfrac{AE}{kL}$; $\gamma_{\text{tor}} = \dfrac{I_P G}{kL}$

[2] $\gamma_{\text{long}} = \dfrac{\rho AL}{m}$; $\gamma_{\text{tor}} = \dfrac{I_P \rho L}{I_0}$

[3] $\gamma_{\text{long}} = \dfrac{kL}{AE}$; $\gamma_{\text{tor}} = \dfrac{kL}{I_P G}$

[4] $\gamma_{\text{long}} = \dfrac{m}{\rho AL}$; $\gamma_{\text{tor}} = \dfrac{I_0}{I_P \rho L}$

Note: "long" denotes longitudinal vibration and "tor" denotes torsional vibration; A = cross-sectional area, L = rod length, E = Young's modulus, G = shear modulus, k = force per length (long) or moment per radian (tor), ρ = mass per unit volume, I_P = polar moment of inertia of A about rod axis, I_0 = mass moment of inertia of attached mass; the definition of λ_n is in the text.

TABLE 14.4 Nondimensional Natural Frequencies ($\lambda_n L$)[a]

	$\|\gamma\|$							
n	0	.5	1	2	5	10	100	∞
For Longitudinal and Torsional Clamped/Spring and Free/Mass BCs[b]								
1	π	2.289	2.029	1.837	1.689	1.632	1.577	$\pi/2$
2	2π	5.087	4.913	4.814	4.754	4.734	4.715	$3\pi/2$
3	3π	8.096	7.979	7.917	7.879	7.867	7.855	$5\pi/2$
For Longitudinal and Torsional Clamped/Mass and Torsional Free/Spring BCs[c]								
1	0	0.653	0.860	1.077	1.314	1.429	1.555	$\pi/2$
2	π	3.292	3.426	3.644	4.034	4.306	4.666	$3\pi/2$
3	2π	6.362	6.437	6.578	6.910	7.228	7.776	$5\pi/2$
For Longitudinal Free/Spring BCs[d]								
1	π	2.975	2.798	2.459	1.941	1.743	1.587	$\pi/2$
2	2π	6.203	6.121	5.954	5.550	5.191	4.760	$3\pi/2$
3	3π	9.371	9.318	9.211	8.414	8.562	7.933	$5\pi/2$

[a] For the nonclassical boundary conditions in Table 14.3.
[b] See Table 14.3, cases 1 and 4.
[c] See Table 14.3, cases 2 and 3.
[d] See Table 14.3, case 3.

The transverse deflection of a beam, $w(x, t)$, is governed by the equation of motion,

$$\frac{\partial^4 w}{\partial x^4} = \left(\frac{\rho A}{EI}\right)\frac{\partial^2 w}{\partial t^2} \tag{14.21}$$

which, upon substitution of $w(x, t) = W(x)[A \sin(\omega t) + B \cos(\omega t)]$, leads to

$$\frac{d^4 w}{dx^4} - \lambda^4 W = 0, \quad \lambda^4 = \frac{\rho A \omega^2}{EI} \tag{14.22}$$

This equation has the general solution $W(x) = c_1 \sin(\lambda x) + c_2 \cos(\lambda x) + c_3 \sinh(\lambda x) + c_4 \cosh(\lambda x)$. The frequency equation, natural frequencies, and normalized mode shapes are found by applying the BCs in the same manner as above. The results for various combinations of simply supported (SS: $W = W'' = 0$), clamped (C: $W = W' = 0$), and free (F: $W'' = W''' = 0$) BCs for a beam of length L and flexural rigidity EI are given in Table 14.5.

Defining Terms

Cyclic and circular frequency — The cyclic frequency of any cyclic or periodic motion is the number of cycles of motion per second. One cycle per second is called a hertz (Hz). The circular frequency of the motion is 2π times the cyclic frequency and converts one cycle of motion into 2π radians of angular motion. The circular frequency is measured in radians per second.

Degree of freedom (DOF) — An independent motion of a moving system. A single mass rolling on a surface has one DOF, a system of two masses rolling on a surface has two DOFs, and a continuous elastic structure has an infinite number of DOFs.

Free vibration — The act of a system of masses or a structure vibrating back and forth about its position of static equilibrium in the absence of any external forces. The vibration is caused by the action of restoring forces internal to the system or by gravity.

TABLE 14.5 Transverse Vibrations of a Beam

BCs	Frequency Equation	β_1	β_2	Asymptotic to	Normalized Mode Shape
C-F	$1+\cos\beta\cosh\beta=0$	1.875	4.694	$(2n+1)\pi/2$	$(\cosh\lambda_n x-\cos\lambda_n x)-\gamma_n(\sinh\lambda_n x-\sin\lambda_n x),\quad \gamma_n=\dfrac{\cosh\beta_n+\cos\beta}{\sinh\beta_n+\sin\beta}$
SS-SS	$\sin\beta=0$	π	2π	$n\pi$	$\sin\lambda_n x$
C-SS	$\tanh\beta-\tan\beta=0$	3.927	7.069	$(4n+1)\pi/4$	$(\cosh\lambda_n x-\cos\lambda_n x)-\gamma_n(\sinh\lambda_n x-\sin\lambda_n x),\quad \gamma_n=\dfrac{\cosh\beta_n-\cos\beta}{\sinh\beta_n-\sin\beta}$
F-SS	$\tanh\beta-\tan\beta=0$	3.927	7.069	$(4n+1)\pi/4$	$(\cosh\lambda_n x+\cos\lambda_n x)-\gamma_n(\sinh\lambda_n x+\sin\lambda_n x),\quad \gamma_n=\dfrac{\cosh\beta_n-\cos\beta}{\sinh\beta_n-\sin\beta}$
C-C	$1-\cos\beta\cosh\beta=0$	4.730	7.853	$(2n+1)\pi/2$	$(\cosh\lambda_n-\cos\lambda_n x)-\gamma_n(\sinh\lambda_n x-\sin\lambda_n x),\quad \gamma_n=\dfrac{\sinh\beta_n+\sin\beta}{\cosh\beta_n-\cos\beta}$
F-F	$1-\cos\beta\cosh\beta=0$	4.730	7.853	$(2n+1)\pi/2$	$(\cosh\lambda_n+\cos\lambda_n x)-\gamma_n(\sinh\lambda_n x+\sin\lambda_n x),\quad \gamma_n=\dfrac{\sinh\beta_n+\sin\beta}{\cosh\beta_n-\cos\beta}$

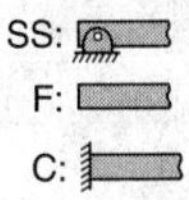

Fundamental frequency — The smallest natural frequency in a system with more than one DOF.

Mode shape — The relationship between the amplitudes (one per DOF) of the independent motions of a system in free vibration. There is one mode shape for each natural frequency, and it depends on the value of that natural frequency. For a continuous elastic structure, the mode shapes are the shapes of the structure at its maximum deformation during a cycle of vibration.

Natural frequency — The frequency or frequencies at which a system will undergo free vibration. There is one natural frequency per DOF of the system. Natural frequencies depend on the geometry, the boundary conditions (method of support or attachment), the masses of the components, and the strength of the restoring forces or moments.

References

Clark, S. K. 1972. *Dynamics of Continuous Elements,* Prentice Hall, Englewood Cliffs, NJ.

Den Hartog, J. P. 1956. *Mechanical Vibrations,* 4th ed. McGraw-Hill, New York.

Gorman, D. J. 1975. *Free Vibration Analysis of Beams and Shafts,* John Wiley & Sons, New York.

Leissa, A. W. 1993a. *Vibrations of Plates,* Acoustical Society of America, New York. (Originally issued by NASA, 1973).

Leissa, A. W. 1993b. *Vibrations of Shells,* Acoustical Society of America, New York. (Originally issued by NASA, 1973).

Magrab, E. B. 1979. *Vibrations of Elastic Structural Members,* Sijthoff and Noordhoff, Leyden, The Netherlands.

Meirovitch, L. 1967. *Analytical Methods in Vibrations,* Macmillan, New York.

Thomson, W. T. 1988. *Theory of Vibrations with Applications,* Prentice Hall, Englewood Cliffs, NJ.

Timoshenko, S. P., Young, D. H., and Weaver, J. W. 1974. *Vibration Problems in Engineering,* 4th ed. John Wiley & Sons, New York.

Further Information

Several excellent texts discuss the free vibrations of discrete systems (finite number of DOFs). In particular, the books by Den Hartog [1956], Timoshenko et al. [1974], and Thomson [1988] are recommended.

Extensive data for the natural frequencies of beams that have elastic supports (translational or rotational), end masses, multiple spans, discontinuities in cross-sections, axial tension or compression, variable thickness, or elastic foundations may be found in the monograph by Gorman.

Other important structural elements are plates and shells. Plates are flat, whereas shells have curvature (e.g., circular cylindrical, elliptic cylindrical, conical, spherical, ellipsoidal, hyperboloidal). A summary of natural frequencies for plates obtained from 500 other references is available in the book on plate vibrations by Leissa [1993a]. Extensive frequency data for various shells taken from 1000 references is also available in the book on shell vibrations by Leissa [1993b].

15

Forced Vibrations

Arthur W. Leissa
Ohio State University

Consider a mechanical system that is subjected to external forces (or moments) that are periodic in time. The forces may arise in various ways. For example, forces may be applied directly to the system (mechanical connections, fluid pressure, electromechanical), or indirectly through a foundation (which may be represented by springs and dampers). Such exciting forces always occur in rotating bodies (e.g., electric motors, internal combustion engines, gas turbines), but can also have other sources (e.g., earthquake motions, wind gusts, acoustic excitations).

The frequency (Ω) of an exciting force is typically different from the natural frequencies ($\omega_1, \omega_2, \omega_3, \ldots$) of the system. However, if Ω is close to *any* of the natural frequencies, the amplitude of the resulting motion may be very large. If Ω *equals* one of the ω_i, **resonance** exists. In this situation, if no damping were present, the amplitude would grow with time until the system failed due to excessive motion or stress. All physical systems have at least some damping, but the damping may be very small. In this situation, the amplitude of motion at resonance would remain finite, but could become very large — even excessive. When a system is excited, the responsive displacements are a combination (superposition) of all the mode shapes of free vibration. However, if Ω is close to one of the ω_i, the response is dominated by the mode shape corresponding to the ω_i.

The most important reason to know the natural frequencies of free vibration is to avoid resonant situations. One seeks to change the mass or stiffness of the system to shift the natural frequencies away from the exciting frequencies. In typical situations, the largest resonant amplitudes occur at the lowest natural frequencies. Therefore, it is particularly important to know the smallest ω_i. Free vibration mode shapes are also important because they enable one to determine *how* the system vibrates at or near resonance.

15.1 Single-Degree-of-Freedom Systems

Take the spring-mass system shown in Figure 14.2(a) of Chapter 14 and add a horizontal exciting force $F_o \sin\Omega t$ to the mass, where Ω is the *exciting frequency*. From the free-body diagram of Figure 14.3(a), the equation of motion is

$$m\ddot{x} + kx = F_o \sin\Omega t \qquad (15.1)$$

The solution of Equation (15.1) consists of the sum of two parts. One part is the *complementary* solution obtained by setting $F_o = 0$. This is the free, undamped vibration discussed in Chapter 14. The second part is the *particular* solution, due to $F_o \sin\Omega t$. This is

0-8493-1586-7/05/$0.00+$1.50

$$x = \frac{F_o/k}{1-(\Omega/\omega)^2} \sin \Omega t \tag{15.2}$$

where $\omega = \sqrt{k/m}$ is the *natural frequency.* Observing the amplitude of this motion in Equation (15.2), one sees that if excitation begins with a small frequency ($\Omega/\omega << 1$) and increases, the amplitude grows until, at $\Omega/\omega = 1$, it becomes (theoretically) infinite. This is resonance. As Ω/ω increases further, the amplitude diminishes. For large Ω/ω, it becomes very small.

If viscous damping is present, as represented in Figure 14.2(b) of Chapter 14, the equation of motion is

$$m\ddot{x} + c\dot{x} + kx = F_o \sin \Omega t \tag{15.3}$$

Again the solution has two parts, one part being the free, damped vibration, and the other part being the forced motion. The free vibration part is given by Equation (14.9) of Chapter 14. It decays with increasing time and eventually vanishes (i.e., it is *transient*). The forced vibration part is

$$x = A \sin \Omega t - B \cos \Omega t = C \sin(\Omega t - \phi) \tag{15.4a}$$

$$C = \sqrt{A^2 + B^2}, \quad \phi = \tan^{-1}(B/A) \tag{15.4b}$$

$$C = \frac{F_o/k}{[1-(\Omega/\omega)^2]^2 + [2\zeta(\Omega/\omega)]^2} \tag{15.4c}$$

where $\zeta = c/c_c$, $c_c = 2\sqrt{mk}$ as in Chapter 14. This forced vibration is called the **steady state** vibration because it remains indefinitely, even after the transient free vibration vanishes.

A graph of steady state amplitude versus forcing frequency is shown in Figure 15.1. This graph is worthy of considerable study because it shows clearly what vibratory amplitudes exist at different forcing frequencies. The *nondimensional amplitude* C/δ_{st} is used, where $\delta_{st} = F_o/k$ is the **static deflection** that the mass would have if F_o were applied. For small Ω/ω, Figure 15.1 shows that $C/\delta_{st} = 1$, regardless of the damping. The case discussed earlier with *no damping* ($\zeta = 0$) is shown, although C/δ_{st} is plotted positive for $\Omega/\omega > 1$. It is positive for all nonzero ζ (and for all Ω/ω), no matter how small. For no damping, the infinite amplitude at resonance is implied in Figure 15.1. For small damping (e.g., $\zeta = 0.1$), the peak amplitude is several times the static deflection. If ζ were only 0.01, the peak amplitude would be 50 times the static deflection.

Figure 15.1 also shows the **phase angle,** ϕ; that is, the angle by which the motion *lags* the exciting force. For small Ω/ω, it is seen that the motion is essentially **in-phase** (ϕ is nearly zero), whereas, for $\Omega/\omega >> 1$, the motion is essentially **out-of-phase** (ϕ is nearly 180°). In the vicinity of resonance ($\Omega/\omega = 1$), ϕ changes rapidly as Ω is varied, especially if the damping is small.

Suppose that, instead of applying an exciting force $F_o \sin\Omega t$ directly to the mass in Figure 14.2(b) of Chapter 14, the wall (or foundation) on the left side is given the vibratory displacement $\delta_w \sin\Omega t$. This motion causes forces to be transmitted through the spring and damper to the mass. One finds that the equation of motion is again Equation (15.3), with F_o replaced by $k\delta$. Thus, Figure 15.1 again describes the steady state vibratory amplitude of the mass, except that δ_{st} is replaced by δ_w. Now consider the *relative* displacement $x_R = x - \delta_w \sin\Omega t$ between the mass and the wall. A free-body diagram yields the equation of motion:

$$m\ddot{x}_R + c\dot{x}_R + kx_R = m\Omega^2\delta_w \sin \Omega t \tag{15.5}$$

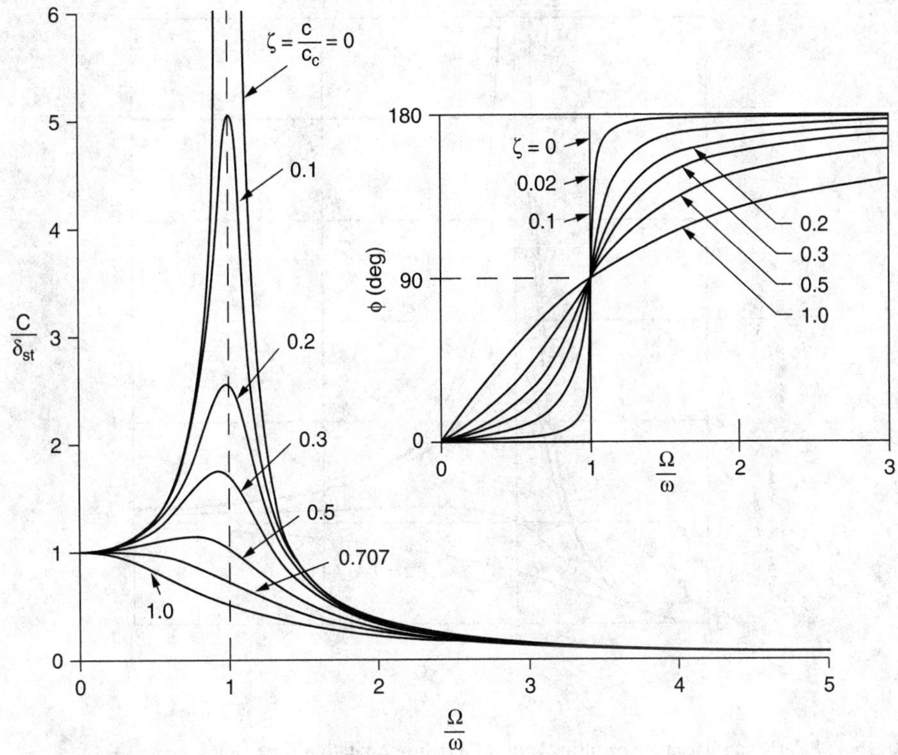

FIGURE 15.1 Displacement amplitude and phase angle resulting from an applied force versus exciting frequency for various amounts of damping (one DOF).

which has a steady state solution in the form of Equation (15.4a). The amplitude of the *relative* motion is found to be:

$$C_R = \frac{\delta_w(\Omega/\omega)^2}{[1-(\Omega/\omega)^2]^2 + [2\zeta(\Omega/\omega)]^2} \tag{15.6}$$

A graph of the ratio of the amplitudes of relative displacement and wall displacement (C_R/δ_w) is shown in Figure 15.2. The phase angle lag is the same as in Figure 15.1. In Figure 15.2, it is seen that at small excitation frequencies (Ω/ω almost zero), the relative displacement is nearly zero. But, at resonance ($\Omega/\omega = 1$), large relative motion may occur, especially for small damping (small ζ). For $\Omega/\omega >> 1$, C_R/δ_w is nearly unity, and the relative motion is 180° out of phase. This means that although the wall is shaking at a high frequency, the mass barely moves at all. This behavior is important in design when isolation from ground vibration is desired.

Other types of damping will exist in a typical mechanical system. These include:

1. Dry friction (e.g., the mass slides on a floor against opposing frictional forces)
2. Structural (or material) damping (e.g., the spring material is not perfectly elastic, but dissipates energy during each cycle of vibratory motion)
3. Aerodynamic damping (e.g., the mass vibrates in air, instead of in a vacuum as the previously described models do)

These other forms of damping may be approximated by an **equivalent viscous damping** with reasonable accuracy for many of the vibratory characteristics.

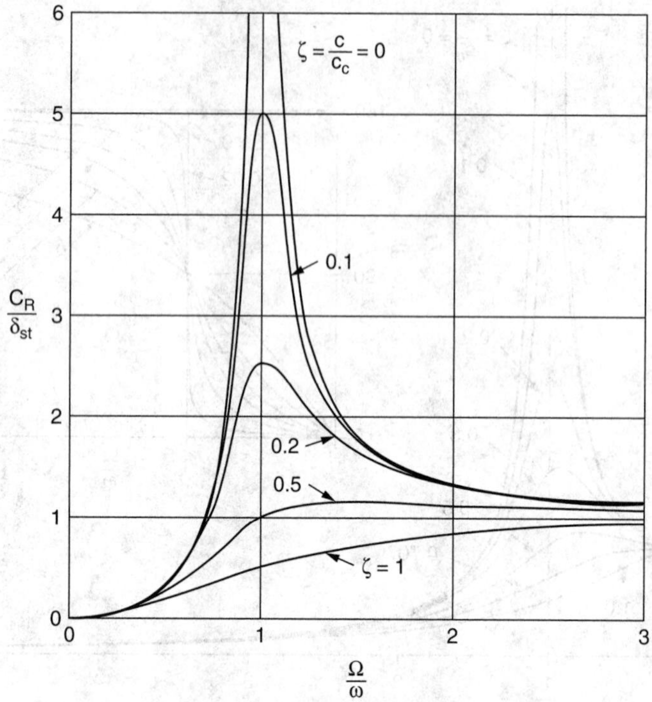

FIGURE 15.2 Relative displacement resulting from foundation excitation (one DOF).

15.2 Multiple-Degree-of-Freedom Systems

The characteristics described above (vibratory displacement and phase angle) behave similarly for systems having two or more degrees of freedom (DOF). That is, with small damping in the vicinity of a resonant frequency, the steady state displacement amplitude is large and the phase angle changes rapidly with changing Ω. The primary difference is that, instead of having a single region of resonance, there are as many regions as there are DOF. A **continuous system** (e.g., string, rod, beam, membrane, plate, shell) has infinite DOF, with an infinite number of free vibration frequencies. Thus, with small damping, large amplitudes can occur in many ranges of exciting frequency for a given exciting force or moment. Fortunately, practical applications show that, typically, only the excitations near the lowest few natural frequencies are significant (although exceptions to this can be shown).

A *two-DOF system* is depicted in Figure 15.3. The two equal masses are separated by equal springs (stiffnesses k) and equal viscous dampers (damping coefficients c). A force $F_o \sin\Omega t$ is applied to one of the masses *only.* From free-body diagrams of each mass, one may obtain two differential equations of motion in the displacements x_1 and x_2. The equations are coupled because of the spring and mass in the middle. Their steady state solution is found to be sinusoidal in time, with a common frequency Ω, but

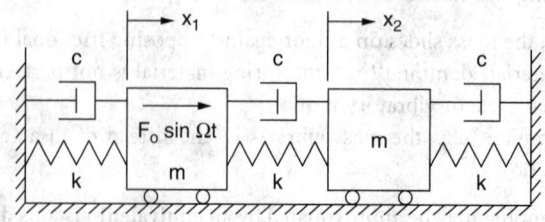

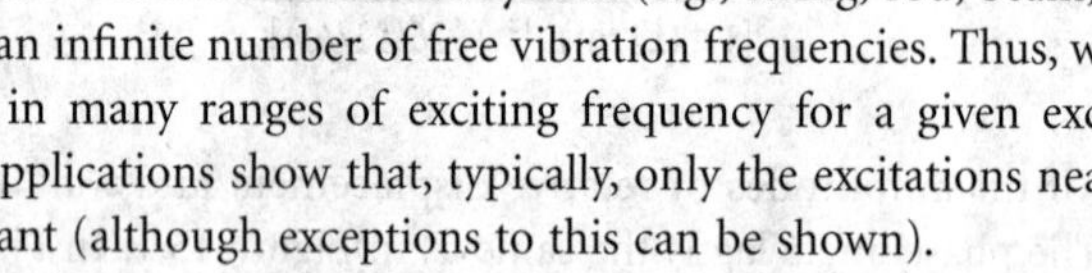

FIGURE 15.3 Two-DOF mechanical system.

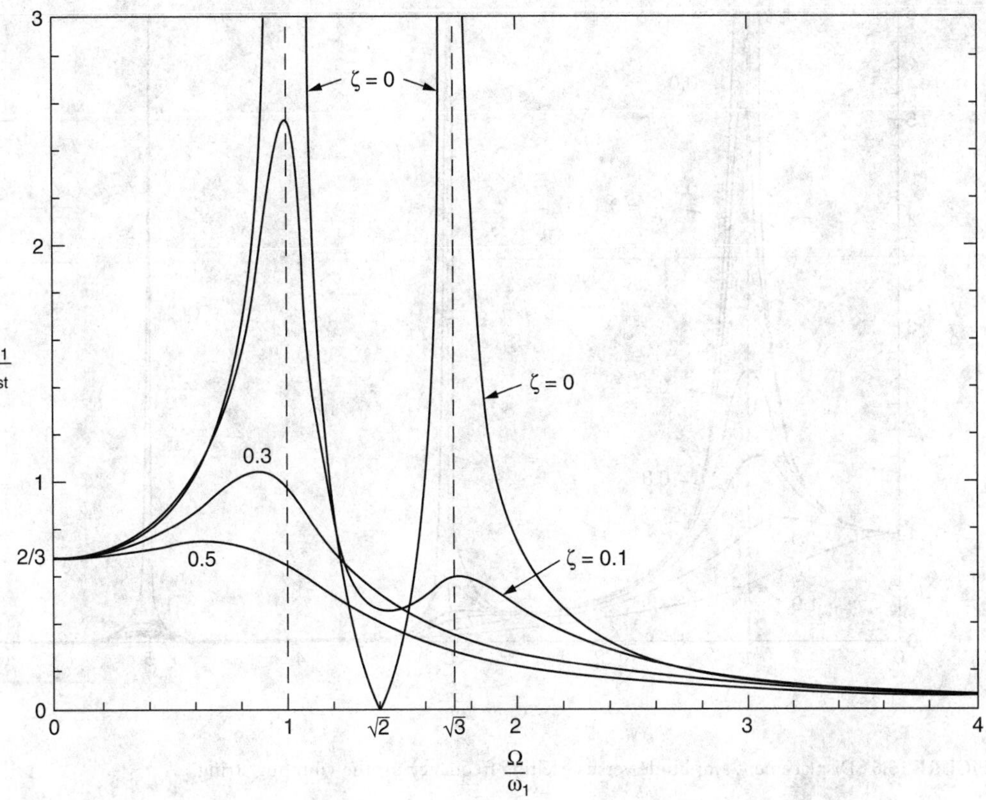

FIGURE 15.4 Displacement amplitude versus exciting frequency (two DOF).

different phase angles for each of the masses. A plot of the amplitude (C_1) for the vibratory displacement (x_1) of the mass to which the force is applied is seen in Figure 15.4 for damping ratios $\zeta = c/2\sqrt{mk} = 0$, 0.1, 0.3, and 0.5. With no damping, the amplitude is seen to become infinite at the two resonances ($\Omega/\omega_1 = 1$ and $\sqrt{3} = 1.732$, where $\omega_1 = \sqrt{k/m}$ is the smallest *natural frequency* of the system). Interestingly, for $\zeta = 0$ and $\Omega/\omega_1 = \sqrt{2} = 1.414$, there is *no* motion of the mass to which the force is applied (although the other mass vibrates). This is an example of *vibration isolation.* By adding a second mass to a single-DOF system, the vibratory motion of the first mass may be eliminated at a certain exciting frequency. The added mass need not be equal. With small damping ($\zeta = 0.1$), a large amplitude is observed in Figure 15.4 at the first resonance, but a smaller one at the second resonance. For larger damping ($\zeta = 0.3$), the second resonant peak essentially vanishes.

As an example of a **continuous system**, consider a string (or wire) of length l stretched with a tensile force (T) between two rigid walls. It has uniform thickness and mass density (ρ, mass/length), and negligible bending stiffness. Let the string be subjected to a uniformly distributed loading (p, force/length) which varies sinusoidally in time ($p = p_o \sin\Omega t$), as shown in Figure 15.5. Considering only small amplitude transverse vibrations, the equation of motion is found to be a linear, second-order, *partial*

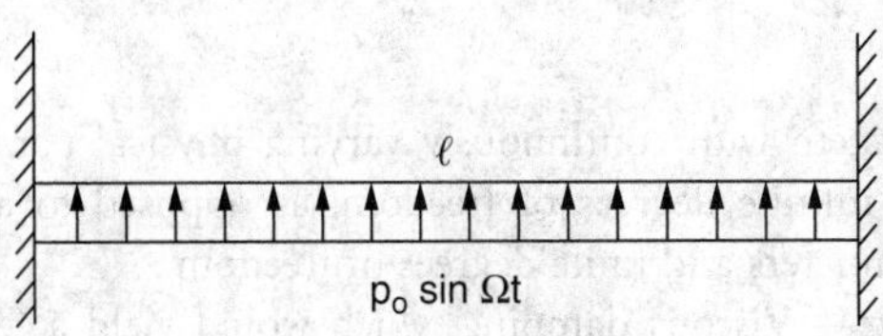

FIGURE 15.5 A string stretched between two walls (continuous system).

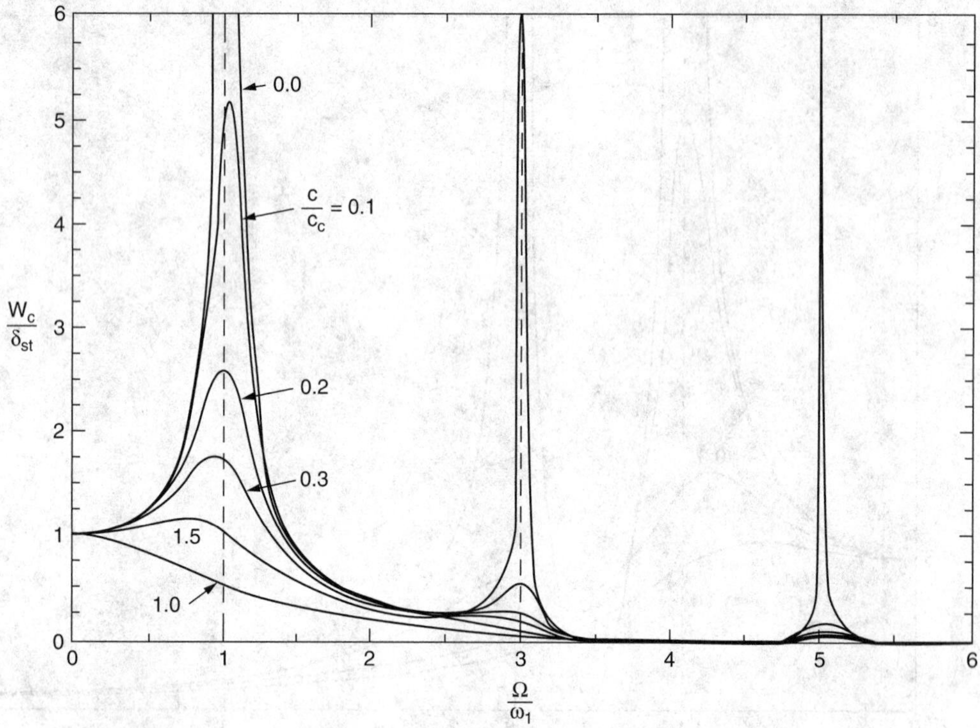

FIGURE 15.6 Displacement amplitude versus exciting frequency for the vibrating string.

differential equation. This may be solved exactly either in closed form or by taking the infinite sum of the displacement responses of the free vibration modes of the system (eigenfunction superposition). The amplitude of the transverse vibration of the center of the string (W_c) is observed in Figure 15.6. It is plotted in the nondimensional form W_c/δ_{st}, where δ_{st} would be the displacement if the pressure were *static* ($\delta_{st} = p_o l^2/8T$). The abscissa is the frequency ratio (Ω/ω_1, where ω_1 is the first natural frequency of the system). If there is no damping ($c = 0$), then infinite amplitudes (resonances) occur at the first, third, fifth, seventh, and so on natural frequencies. The natural frequencies are $\omega_m = m\pi/l$, where $m = 1,2,3,\ldots$. The free vibration mode shapes are symmetric with respect to the center of the string for $m = 1,3,5,\ldots$, and these are the modes that are excited by the symmetric loading. The antisymmetric modes ($m = 2,4,6,\ldots$) are not excited by it. In the vicinity of each resonance, the mode shape (a sine function along the length) for that natural frequency dominates. Away from resonances (e.g., $\Omega/\omega_1 = 2$), all symmetric modes are present. The width of each region of resonance decreases as the order of natural frequencies increases. Thus, for example, if W_c/δ_{st} is to be less than 3, the range of unacceptable operating frequencies (Ω/ω_1) is seen to be much smaller at the second resonance than at the first, and smaller yet at the third resonance. For small, uniformly distributed, viscous damping (e.g., $c/c_c = 0.1$, where $c_c = 2\pi\sqrt{T\rho/l^2}$ is the critical damping coefficient for the *first* mode), the amplitudes at the first three resonances are found to be $W_c/\delta_{st} = 5.18$, 0.59, and 0.21.

Defining Terms

Continuous system — A system with continuously varying physical parameters (e.g., mass, stiffness, damping), having infinite degrees of freedom; as opposed to a discrete system, which has discontinuous parameters and finite degrees of freedom.

Equivalent viscous damping — Viscous damping which would yield a forced vibratory response the same as another form of damping.

In-phase — Vibratory displacement which follows in time an exciting force (or displacement).
Out-of-phase — Vibratory displacement which is opposite to the direction of excitation.
Phase angle — The angle in a cycle of motion by which a displacement lags behind the exciting force (or displacement).
Resonance — Large amplitude motion that occurs when a forcing frequency is in the vicinity of a natural frequency of a system.
Static deflection — The limiting case of a forced vibratory displacement, when the exciting frequency is very small, so that dynamic (inertia) effects are negligible.
Steady state — The vibratory motion that persists after the transient effects die away or are neglected.

References

Den Hartog, J. P. 1956. *Mechanical Vibrations,* 4th ed. McGraw-Hill, New York.
Leissa, A. W. 1978. On a direct method for analyzing the forced vibrations of continuous systems having damping. *J. Sound Vib.* 56(3):313–324.
Leissa, A. W. 1989. Closed form exact solutions for the vibrations of continuous systems subjected to distributed exciting forces. *J. Sound Vib.* 134(3):435–454.
Leissa, A. W. and Chern, Y. T. 1992. Approximate analysis of the forced vibrations of plates. *J. Vib. Acoustics* 114:106–111.
Ruzicka, J. E. and Derby, T. F. 1971. *Influence of Damping in Vibration Isolation.* Shock and Vibration Information Center, Washington, D.C.
Snowdon, J. C. 1968. *Vibration and Shock in Damped Mechanical Systems.* John Wiley & Sons, New York.
Thomson, W. T. 1988. *Theory of Vibration with Applications.* Prentice Hall, Englewood Cliffs, NJ.
Timoshenko, S. P., Young, D. H., and Weaver, W., Jr. 1974. *Vibration Problems in Engineering,* 4th ed. John Wiley & Sons, New York.

Further Information

Discussion of forced vibrations of one-DOF systems, including nonsinusoidal exciting forces, may be found in the excellent textbooks by Den Hartog; Timoshenko, Young, and Weaver; and Thomson.

For further information on equivalent viscous damping and its representation of other forms of damping, see the textbook by Thomson. Extensive graphs of amplitude versus frequency ratio for various types of damping are in the monograph by Ruzicka and Derby.

Vibration isolation in a two-DOF system is discussed very well in the textbook by Den Hartog.

Forced vibrations of rods and beams with material damping are thoroughly discussed in the monograph by Snowdon. Closed-form exact solutions for continuous systems [Leissa, 1989] and two useful approximate methods [Leissa, 1978; Leissa and Chern, 1992] are available in individual papers.

Out-of-phase — [illegible]
Phase angle — The angle [illegible]
[illegible]

Shaft deflection — [illegible]

[illegible]
[illegible]
[illegible]
[illegible]
[illegible]
[illegible]

[illegible]

[illegible]

16

Lumped versus Distributed Parameter Systems

Bulent A. Ovunc
University of Louisiana at Lafayette

The lumped, consistent, and distributed (or continuous) mass methods are the main methods for dynamics and vibration analyses of structures. In the continuous mass method the equations of motion are satisfied at every point of the structure. In the consistent and lumped mass methods they are satisfied only at the joints of the structures. In consistent mass the displacements within members are assumed as static displacements. The lumped mass method considers the members as massless springs. The lumped and consistent mass methods are simple and fast; they are fairly approximate, but their accuracy decreases for structures subjected to the effects of the shear and rotatory inertia, member axial force, elastic medium, and so on. The continuous mass method provides accurate results under the assumptions made.

16.1 Procedure of Analysis

For frames, the general formulation is based on the vector of displacement $\{a_o(y, t)\}$ at a time t and at a point y on the center line of its constitutive members. The same formulation is valid for the above mentioned three methods, only the vector of center line displacement $\{a_o(y, t)\}$ depends on the assumptions made for each method.

For materially and geometrically linear frames, the vector $\{a_0(y, t)\}$ is made of four independent components,

0-8493-1586-7/05/$0.00+$1.50

$$\{a_o(y,t)\}^T = [u(y,t)\ v(y,t)\ w(y,t)\ \vartheta(\rho_\Theta, t)] \tag{16.1}$$

where $u(y, t)$, $v(y, t)$, and $w(y, t)$ are the displacements along the member axes x, y, z; $\vartheta(\rho_\Theta, t)$ is the twist rotation about the y axis, and $\rho_\Theta = (x^2 + y^2)^{1/2}$. It is assumed that the material properties are independent of time, and that the external disturbances applied to a structure are proportional to a same-time variable function. Thus, the displacement function $\{a_o(y, t)\}$ can be written in separable variable form. The integration of the differential equations and elimination of the integration constants gives

$$\{a_o(y,t)\} = [N(y)]\{d(t)\} = [N(y)]\{d\}f(t) \tag{16.2}$$

where $\{d(t)\} = \{d\}f(t)$, $[N(y)]$, and $\{d\}$ are time-independent matrix-of-shape functions and vector-of-member displacements, and $f(t)$ is a time-variable function of external disturbances. See Figure 16.1.

The strains $\{\varepsilon\}$ at a point within the cross-section of a member can be obtained from the center line displacements $\{a_o(y, t)\}$ [Przemieniecki, 1968] as

$$\{\varepsilon\} = [\partial]\{a_o(y,t)\} = [B(y)]\{d\}f(t) \tag{16.3}$$

where $[B(y)] = [\partial][N(y)]$, and $[\partial]$ is the matrix of differential operators.

The stresses $\{\sigma\}$ are determined from the stress–strain relationship as

$$\{\sigma\} = [E]\{\varepsilon\} = [E][B(y)]\{d\}f(t) \tag{16.4}$$

The expressions of the **strain energy**, U_i, and **kinetic energy**, $\mathcal{K}$, as well as the work done by damping forces, W_D, and by externally applied loads, W_e, are written as

$$\begin{aligned}
U_i &= \frac{1}{2}\int_v \{\varepsilon\}^T\{\sigma\}dV = \frac{1}{2}(f(t))^2\{d\}^T\left(\int_v [B(y)]^T[E][B(y)]dV\right)\{d\} \\
W_D &= -\int_v c\{\dot{a}_o\}^T\{a_o\}dV = -c\dot{f}(t)f(t)\{d\}^T\left(\int_v [N(y)]^T[N(y)]dV\right)\{d\} \\
\mathcal{K} &= \frac{1}{2}\int_v m\{\dot{a}_o\}^T\{\dot{a}_o\}dV = \frac{1}{2}(\dot{f}(t))^2\{d\}^T\left(m\int_v [N(y)]^T(N(y)]dV\right)\{d\} \\
W_e &= f_e(t)\int_y \{P_o\}^T\{a_o\}dy = f_e(t)\left(\int_y \{P_o\}^T[N(y)]dy\right)\{d\}
\end{aligned} \tag{16.5}$$

where $f_e(t)$ and $\{P_o\}$ are the time-dependent and -independent parts of the vector of externally applied joint forces.

For a member, stiffness $[k]$, mass $[m]$, and damping $[c]$ matrices are determined by substituting the strain energy U_i, **damping energy** W_D, kinetic energy $\mathcal{K}$, and **external energy** W_e into the Lagrangian dynamic equation [Ovunc, 1974],

$$\frac{\partial U_i}{\partial d_j} - \frac{\partial W_D}{\partial d_j} + \frac{d}{dt}\left(\frac{\partial K}{\partial \dot{d}_j}\right) = \frac{\partial W_e}{\partial d_j} \tag{16.6}$$

which provides the equation of forced vibration for a member as

$$[k]\{d\}f(t) - 2\nu_o[m]\{d\}\dot{f}(t) + [m]\{d\}\ddot{f}(t) = \{P_o\}f_e(t) \tag{16.7}$$

where the damping matrix $[c]$ is assumed to be proportional to mass matrix $[m]$ and ν_o is the damping coefficient.

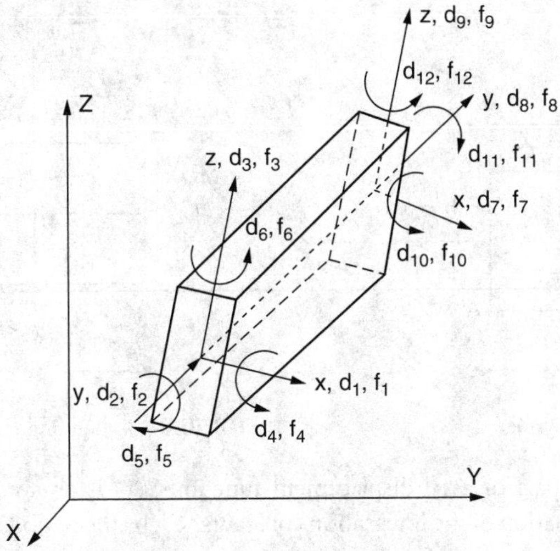

FIGURE 16.1 Coordinate axes systems.

For the free vibration, {P} = 0, the equation of motion is divided into two parts — time-independent and time-dependent:

$$([k] - \omega^2[m])\{d\} = 0 \tag{16.8}$$

$$\frac{d^2 f(t)}{dt^2} + 2\nu\omega\frac{df(t)}{dt} + \omega^2 f(t) = 0 \tag{16.9}$$

The time-dependent part, $f(t)$, Equation (16.9), is the same for all four independent cases, and $\nu = \nu_o\omega$ [Paz, 1993].

Although the lumped mass method was developed long before the continuous mass method, the continuous mass method is herein explained first. The consistent and lumped mass methods are presented as particular cases of the continuous mass method.

16.2 Continuous Mass Matrix Method

For the materially and geometrically linear frame, an arbitrary vibration of its member is obtained by combining the four independent components: axial displacement, torsional rotation, and bending in two orthogonal planes.

Member under Axial Displacement

For members with constant section, the time-independent part of the differential equation of free vibration has been given [Ovunc, 1985]. (See Figure 16.2.)

$$\frac{d^2Y(y)}{dy^2} + \alpha^2 Y(y) = 0 \tag{16.10}$$

where $\alpha^2 = (\omega^2 m - C_f p)/EA$; and where $m = (A\rho + q)/g$, p, and cf. are the mass, the peripheral area, and the friction coefficient of the elastic medium per unit length of the member, respectively.

The time-dependent part, f(t), is the same for all four independent cases.

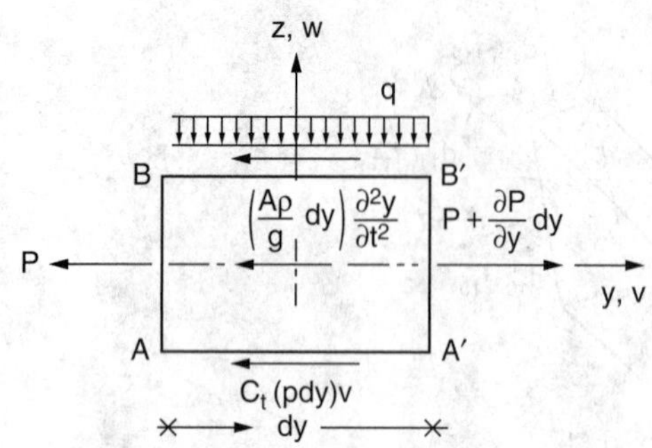

FIGURE 16.2 Axial force member.

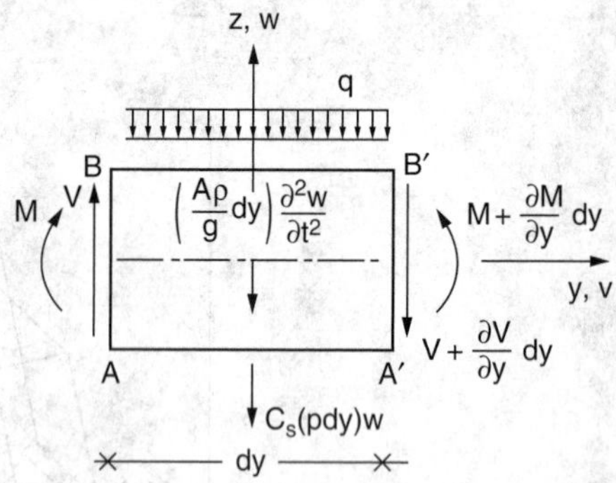

FIGURE 16.3 Bending member.

The time-independent part of axial displacement function Y(y) is obtained by integrating Equation (16.10), through the elimination of the integration constants {C}, by the boundary conditions; thus one has

$$Y(y)=\{\phi_{ax}(y)\}^{T}[L]^{-1}\{d_{ax}\}=\{N_{ax}(y)\}^{T}\{d_{ax}\} \tag{16.11}$$

The nature of the shape function $\{N_{ax}(y)\}$ depends on the sign of parameter α^2.

$$\{N_{ax}(y)\}^{T}=(1/\sin\alpha\ell)\{(\sin\alpha\ell\cos\alpha y-\cos\alpha\ell\sin\alpha y)\quad \sin\alpha y\}\text{ for }\alpha^2>0$$

$$\{N_{ax}(y)\}^{T}=(1/\sinh\alpha\ell)\{(\sinh\alpha\ell\cosh\alpha y-\cosh\alpha\ell\sinh\alpha y)\quad \sinh\alpha y\}\text{ for }\alpha^2>0$$

Member under Bending along Its Major Moment of Inertia Axes

For members with constant section, the time-independent part of the differential equation of free vibration (see Figure 16.3) has been given by Ovunc [1985] as

$$\frac{d^4Z}{dy^4}+2k^2\frac{d^2Z}{dy^2}-\beta^4 Z=0 \tag{16.12}$$

where $2k^2 = P/EI$, $\beta^4 = (\omega^2 m - C_s P')/EI$; and where I, C_s, and where p' are the moment inertia of the cross-section, the subgrade coefficient of the elastic medium, and the projected area of the cross-section, respectively.

The shape function $\{N_{bd}(y)\}$ due to deflection,

$$Z(y) = \{\phi_{bd}(y)\}^{T}[L\}^{-1}\{d_{bd}\} = \{N_{bd}(y)\}^{T}\{d_{bd}\} \tag{16.13}$$

is obtained in a similar manner as in the case of axial displacement, Equation (16.11). The nature of the shape function $\{N_{bd}(y)\}$ and its component $\{\phi_{bd}(y)\}^T$ depend on the parameters α_1^2 and α_2^2, which are expressed in terms of k^2 and β^4:

$$\alpha_1 = [(\beta^4 + k^4)^{1/2} + k^2]^{1/2},\ \alpha_2 = [(\beta^4 + k^4)^{1/2} - k^2]^{1/2}$$

Thus, for $\beta^4 > 0$ and $P > 0$ (compression is positive),

$$\{\phi_{bd}(y)\}^T = (\sin\alpha_1 y\ \cos\alpha_1 y\ \sinh\alpha_2 y\ \cosh\alpha_2 y) \tag{16.14}$$

The above expression remains the same when the axial force P is tension, except α_1 and α_2 must be interchanged. For the combination of $\beta^4 < 0$ and $P < 0$ or $P > 0$, the expression of $\{\phi_{bd}(y)\}$ can be determined in a similar manner.

The member stiffness matrices for the twist rotation and the bending in the *Oxy* plane are obtained by following similar steps as in the previous cases. The stiffness matrix for the space frame is determined by combining the stiffness matrices of all four independent cases.

Dynamic Member Stiffness Matrix for Plane and Space Frames

The dynamic member stiffness matrix $\{k_{dy}]$ for either a plane frame or a space frame is obtained by substituting either the shape functions for axial displacement and bending in *Oyz* plane [Equation (16.11) and Equation (16.13)] or the shape functions of all the four independent cases, in the Lagrangian dynamic equation [Equation (16.6)]. Integrating [Equation (16.7)] one has

$$[k_{dyn}] = [k] - \omega^2[m] \tag{16.15}$$

The continuous mass method has also been extended to frames with tapered members [Ovunc, 1990].

16.3 Consistent Mass Matrix Method

In the consistent mass matrix method, the deformations within a member are static deformations. The shape function for each independent component $\{N(y)\}$ is a static displacement due to its corresponding independent cases. Thus, the shape functions for axial displacement, $\{N_{ax}(y)\}$, and for bending in the *Oyz* plane, $\{N_{bn}(y)\}$, are given as

$$\{N_{ax}(y)\}^T = ((1-\eta)\ \eta) \tag{16.16}$$

$$\{N_{bn}(y)\}^T = ((1-3\eta^2+2\eta^3)\ (\eta-2\eta^2+\eta^3)\ell\ (3\eta^2-2\eta^3)\ (-\eta^2+\eta^3)\ell) \tag{16.17}$$

where $\eta = y/\ell$. The shape functions for twist rotation and bending in the *Oxy* plane are obtained in similar manner.

The member stiffness $[k]$ and mass $[m]$ matrices are evaluated by substituting the shape functions in the Lagrangian dynamic equation [Equation (16.6)]. Herein, the stiffness matrix $[k]$ is a static stiffness matrix and the mass matrix $[m]$ is a full matrix [Przemieniecki, 1968; Paz, 1993].

Moreover, the member stiffness matrix $[k]$ and the mass matrix $[m]$ for the consistent mass matrix method can be obtained as the first three terms of the power series expansion of the dynamic member stiffness matrix $[kdyn]$ for the continuous mass matrix [Paz, 1993].

16.4 Lumped Mass Matrix Method

The lumped mass method is obtained from the continuous mass matrix method by considering the limit when the mass of the members tends to zero. Thus, the shape functions [Equation (16.17)] and the member stiffness matrix $[k]$ are the same as in the consistent mass method. But the mass matrix $[m]$ is diagonal [Paz, 1993].

16.5 Free Vibration of Frames

The **stiffness coefficients** for the frames are evaluated from those of its members as

$$K_{i,j} = \sum k_{r,s}, \quad M_{i,j} = \sum m_{r,s}, \quad P_i = \sum p_r$$

where r and s are the member freedom numbers corresponding to the i and j of the structure freedoms.

The equation of free vibration is obtained from those of members [Equation (16.8)]

$$([K] - \omega^2[M])\{D\} = \{0\} \tag{16.18}$$

where $\{D\}$ is the vector of the structure displacements.

The natural circular frequencies, ω_i, and the corresponding modal shapes, $\{D_i\}$, are calculated from the equation of free vibration [Equation (16.18)], also called the *frequency equation.*

16.6 Forced Vibration

The second-order differential (n simultaneous equations with n unknowns)

$$[M]\{\ddot{D}(t)\} + 2\nu_o[M]\{\dot{D}(t)\} + [K]\{D(t)\} = \{P(t)\} = \{P_o\}f_e(t) \tag{16.19}$$

is converted to n separate, second-order, single-variable differential equations through the two orthogonality conditions, which proves that each modal shape vector $\{D_i\}$ is independent of the others. The ith participation factor $\mathcal{A}_i$ is defined as the component of given forced vibration on the ith modal shape vector $\{D_i\}$. Thus, any arbitrary motion can be determined by considering the summation of its components $\mathcal{A}_i$ on each modal shape vector $\{D_i\}$ [Clough and Penzien, 1993; Paz, 1993; Ovunc, 1974].

If the time variable factor $f_e(t)$ of the external disturbance [Equation (16.2)] is periodic (pulsating), the forced vibration can be directly determined without the calculation of participation factors $\mathcal{A}_i$ [Paz, 1993].

16.7 Practical Applications

For any structure, the natural circular frequency ω can be expressed in terms of a parameter C as [Paz, 1993]

$$\omega = C\sqrt{\frac{EI_j}{m_j\ell_j^4}} \tag{16.20}$$

where E is the Young's modulus, and I_j, m_j, and ℓ_j are the moment of inertia, mass, and span length of a selected member j.

Substituting the natural circular frequency ω (its expression in terms of C) into the frequency equation [Equation (16.18)] gives

$$\left|-C^2[M]+[K]\right| = 0 \tag{16.21}$$

where the general terms M_{rs} and K_{rs} of the mass and the stiffness matrix are constant and expressed as

$$M_{rs} = M_{rs}/m_j\ell_j \text{ and } K_{rs} = K_{rs}\ell_j/EI_j \tag{16.22}$$

It can be easily seen that the determinant of the frequency equation [Equation (16.21)] is independent of the member characteristics (E, m, ℓ, I) but depends on the parameter C. If, in a frame, the same characteristics of the members are multiplied by the same factor, the magnitude of the parameter C remains unchanged. However, a natural circular frequency ω_i corresponding to C_i changes [Equation (16.20)]. If the characteristics of some members change and those of the others remain constant, the parameter C_i is affected.

The advantages of one method over the others and the limits on their accuracy depend on the number and type of the members in the structures and whether the structure is subjected to additional effects.

The type of member depends on the ratio of the thickness t (of its cross-section) to its span length ℓ: $\gamma = t/\ell$.

- If the ratio $\gamma = O(1/100)$ (in the order of 1/100), the effect of bending is negligible. The member is a very thin member, called *cable.*
- If the ratio $\gamma = O(1/10)$, axial force, torsion, and bending are affecting the member. The member is *a thin member.*
- If the ratio $\gamma = O(1)$, the member is considered a *deep beam.*

The inclusion of the variation of the width or the thickness of the member, the effect of member axial force, shear, rotatory inertia, vibration of the member within an elastic medium, and so on constitutes the additional effects.

The dynamic responses of beams and frames are evaluated with or without the additional effects by lumped and consistent mass methods. The results are compared with those evaluated by continuous mass.

16.8 Linear Structures without Additional Effects

The behavior of a structure is linear when its stresses are within their elastic limits and its deformations are infinitesimal. The first one constitutes the *geometrical linearity,* the second is the *material linearity.*

Single Beams

The data related to a cantilever beam, considered as an example, composed of a single element or with two, three, or five subelements, are given in Figure 16.4. The dynamic responses of each beam have been determined by the lumped mass and the consistent mass matrix methods. The first two natural circular frequencies, ω_1 and ω_2, and the vertical displacement d_{v11}, rotation d_{r11}, and shearing force V_{11} at the free end 1 as well as the bending moment M_{21} at the fixed end 2 due to the first mode ω_1 are furnished in Table 16.1 for the lumped mass method, for the consistent mass method, and for the continuous mass method.

In the lumped mass method, increasing the number of subelements improves the accuracies of the natural circular frequencies ω_1, ω_2, and ω_i and those of the vertical displacements, d_{v11}, and rotations, d_{r11}, when their magnitudes are compared to their magnitudes obtained by the continuous mass method. For the consistent mass method, the beam has better approximations when it is subdivided into two subelements.

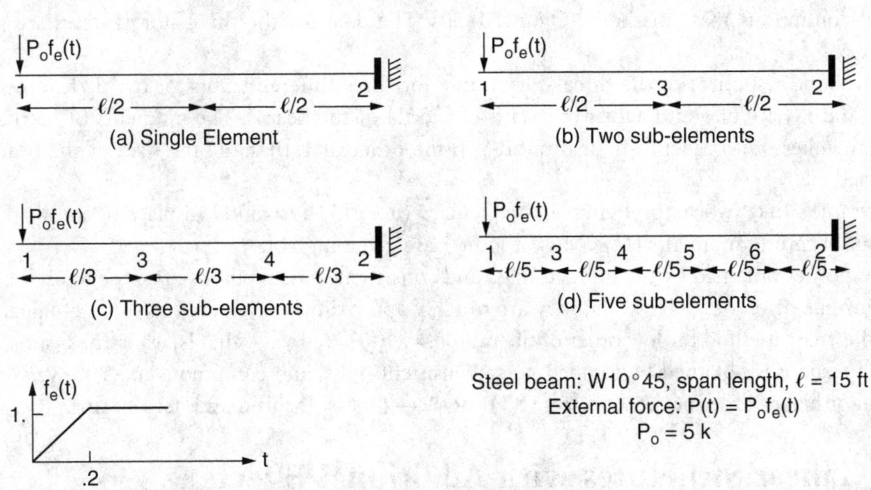

FIGURE 16.4 Cantilever beam formed by (a) a single member or (b) two, (c) three, or (d) five subelements and subjected to a force $P(t) = P_o(t)f_e(t)$.

TABLE 16.1 Comparison of Analyses

Member	ω_1	ω_2	$d_{v11} \cdot 10^{-1}$	$d_{r11} \cdot 10^{-2}$	V_{11}(k)	M_{21}(k/ft)
			Analysis by Lumped Mass Method			
Single	66.2265	0.00	0.51534	0.0	9.490	71.714
Two	85.3437	439.6537	0.39661	–0.38314	1.515	30.179
Three	90.4627	510.7243	0.37075	–0.34954	1.061	29.104
Five	93.3437	560.500	0.35681	–0.33090	0.625	28.548
			Analysis by Consistent Mass Method			
Single	95.4062	925.5823	0.35050	–0.32200	2.000	26.180
Two	94.9687	595.0385	0.34934	–0.32070	1.317	28.052
Three	94.9687	592.2191	0.34882	–0.32023	0.954	28.177
Five	94.9687	590.7352	0.34831	–0.32014	0.611	28.190
			Analysis by Continuous Mass Method			
Single	95.0762	595.8353	0.34962	–0.32080	1.312	28.025

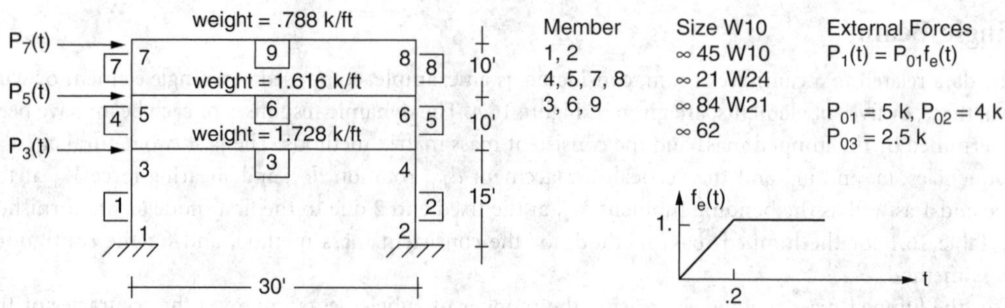

FIGURE 16.5 Three-story steel frame.

Frames

Three-story frames have been selected as an example to compare responses obtained by lumped, consistent, and continuous mass methods [Ovunc, 1980]. The data for the three-story frames are given in Figure 16.5.

The dynamic responses of the three-story frame for three different ratios of $\alpha = I_b/I_{bo}$ are selected, where I_b and I_{bo} are new and actual moments of inertia of the beams. The moments of inertia I_{bo} are for thin members with a depth-to-span ratio γ in the order of 1/10. Only the sizes of the beams have been varied.

The frame's first two natural circular frequencies (ω_1, ω_2), horizontal displacements (d_{h71}, d_{h72}) at joint 7, and bending moments (M_{11}, M_{12}) at joint 1 are given in Table 16.2.

The responses obtained by the lumped mass and consistent mass methods are close to each other for any magnitude of α (or γ). However, they are roughly approximate compared to those obtained by the continuous mass method for low order of the ratio $\alpha = O(.01)$ (or γ) — that is, when the beams are very thin. The responses obtained by lumped mass, consistent mass, and continuous mass are very close for the actual or higher order of the ratio $\alpha \leq O(1)$ (or γ) — that is, for thin and deep beams.

16.9 Linear Structures with Additional Effects

The additional effects included in the analysis of structures with linear behavior are the member axial force, soil structure interaction, and the additional mass of the members [Ovunc, 1992]. The parameters related to the member axial forces, soil–structure interaction, and additional masses of the members are

TABLE 16.2 Dynamic Responses of a Three-Story Frame

Method	$\alpha = I_b/I_{bo}$	ω_1	ω_2	d_{h71}(ft)	d_{h72}(ft)	M_{11}(k/ft)	M_{12}(k/ft)
Lumped mass	0.01	2.9275	10.9188	0.94994	−0.02215	163.871	15.2073
	1.00	7.5657	22.3703	0.11420	−0.00105	81.711	1.2817
	100.00	8.2351	23.7597	0.09400	−0.00091	78.053	1.2048
Consistent mass	0.01	2.9099	10.5168	0.95120	−0.02279	162.792	15.5520
	1.00	7.5652	22.3942	0.11440	−0.00106	81.672	1.2886
	100.00	8.2429	23.7951	0.09382	−0.00092	78.032	1.2148
Continuous mass	0.01	2.1738	10.0834	1.19336	−0.00236	228.538	39.703
	1.00	7.3964	21.8749	0.11410	−0.00106	81.849	1.317
	100.00	8.2402	23.7688	0.09370	−0.00091	78.110	2.216

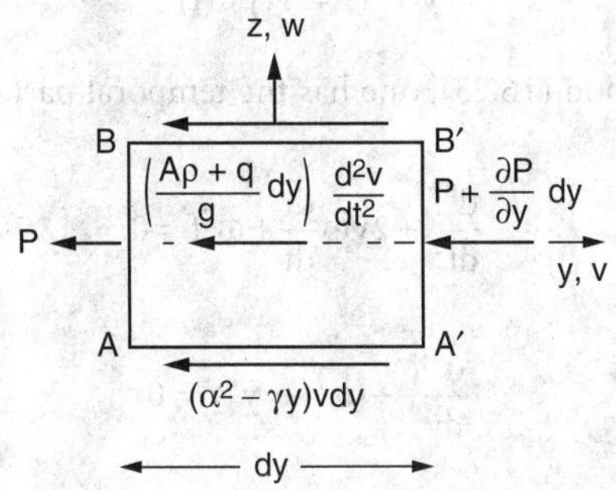

FIGURE 16.6 Axial force element.

introduced in the differential equations of motion. The stability of the structures, including the additional effects, is also analyzed by means of the continuous mass matrix method. It is assumed that the material is linear elastic and the member axial forces are static. The compaction of the soil is due to deformations and the effect of pore pressure is neglected.

Depending on the type of soil, the variation of the soil subgrade reaction is given as (Figure 16.6) [APJ RP-2A]:

- For cohesive soils,

$$C_S = k_L\,(y/L)^n \text{ k/ft}^3$$

- For cohesionless soils,

$$C_S = k\,(y/d) \text{ k/ft}^3$$

where k_L is the value of C_S at the tip of the pile ($y = L$), n and k are empirical constants that depend on the type of soil, and d is the diameter of the pile.

Under the axial displacements, the equation of motion of an infinitesimal element of the member can be written as (Figure 16.7)

$$\frac{\partial P}{dy} - \frac{Cdv}{dt} - \frac{md^2v}{dt^2} - p\left(C_{fb} + \frac{C_{ft} - C_{fb}y}{1}\right) \tag{16.23}$$

where C and p are the damping coefficient and unit peripheral area of the pile and C_{ft} and C_{fb} are the skin frictions at the top and the bottom of the pile. The mass per unit length of the pile is $m = (a\rho + q)/g$.

By expressing the axial force P in terms of displacement v(y, t), and assuming the displacement v(y, t) can be written in separable variable form,

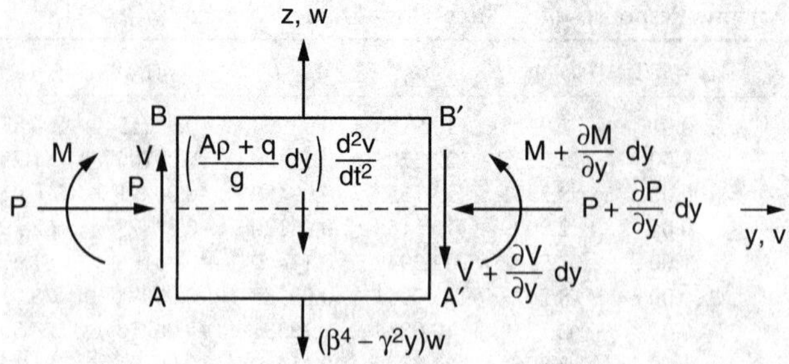

FIGURE 16.7 Bending element.

$$v(y, t) = Y(y)f(t)$$

then substituting them in Equation (16.23), one has the temporal part,

$$\frac{d^2f}{dt^2} + 2\nu\omega\frac{dt}{dt} + \omega^2 f = 0 \quad (16.24)$$

$$\frac{d^2Y}{dy^2} + (\alpha^2 - \gamma y)Y = 0 \quad (16.25)$$

where $\nu = C/2\omega m$, $\alpha^2 = (m\omega^2 - pC_{fb})/AE$, $\gamma = p(C_{ft} - C_{fb})/AEl$, and ω is the natural frequency of the vibration.

The integration of the differential equation [Equation (16.25)], for the spatial part of can be written as

$$Y(y) = \{(t)\}^T\{C\} \quad (16.26)$$

The vector of the integration constants {C} is determined from the boundary conditions. The dynamic member stiffness matrix can be determined either by writing the relations among the member end forces {f} and the member end displacements {d}, or by using the Lagrangian Dynamic Equation. Thus one has,

$$\{f\} = [k_A]\{d\} \quad (16.27)$$

where $[k_A]$, is the dynamic, axial displacement member stiffness matrix.

Noticing the similarity between the twist rotation and axial displacement functions, and the similarity between the stress–strain relationships in both cases, the dynamic member stiffness matrix for torsion can be written by analogy from that of the axial force member by simply substituting the axial force rigidity EA by the torsional rigidity GJ and the parameters α, by their expression for torsion

$$\alpha^2 = (m\omega^2/GA - pR^2C_{fb}/GJ)$$

$$\gamma = p\,R^2\,(C_{ft} - C_{fb})/GJl$$

where R is the mean radius of the cross-section.

The vibration of a member under the bending in the Oyz plane is obtained from the differential of motion as follows (Figure 16.7)

$$EI_x\frac{\partial^4 w}{dy^4} + C\frac{dw}{dt} + P\frac{d^2w}{dy^2} + m\frac{d^2w}{dt^2} + p'\left(C_{sb} - \frac{C_{sb} - C_{st}}{1}y\right) = 0 \quad (16.28)$$

where P is the member axial force, positive when it is in compression; p is the projection of the cross-section per unit length of the member on the plane subjected to soil pressure; and since the variation of the soil modula is assumed to be linear, C_{st} and C_{sb} are the soil subgrade modulus at the top and at the bottom of the pile.

Assuming that the deflection function w(y, t) can be written in separable variable form:

$$w(y, t) = Z(y)\, f(t)$$

and substituting them into the differential equation of motion (Equation 16.28), for the spatial part one has

$$\frac{d^4Z}{dy^4} + 2k^2 \frac{d^2Z}{dy^2} - (\beta^4 + \gamma^2)Z = 0 \tag{16.29}$$

where $k^2 = P/2EI_x$, $^4 = (m\omega^2 - p\, C_{sb})/EI_x$, and $\gamma^2 = p\,(C_{sb} - C_{st})/EI_x l$. The time-dependent part f(t) is the same for all four independent vibrations [Equation (16.24)].

By using the Lagrangian Dynamic Equation, the member stiffness matrix $[k_{B,\,Oyz}]$ subjected to bending in the Oyz plane can be obtained as

$$\{f\} = [k_{B,Oyz}]\{d\} \tag{16.30}$$

The nature of the shape function and the stiffness matrix depend on the values of the parameters and γ and the member axial force P, whether it is in tension or compression [Ovunc, 1985].

The dynamic member stiffness matrix for the bending in the Oxy plane is obtained following the same steps mentioned for bending in the Oyz plane.

The dynamic stiffness matrices for all four independent cases are combined, according to the type of structure: either truss or frames in planes or in space. The dynamic stiffness matrix [K] for the structure is obtained from the dynamic stiffness matrices of its constituent members as

$$\{F\} = [K]\,\{D\} \tag{16.31}$$

The dynamic stiffness matrix [K] is a transcendental function of the natural circular frequency ω of the structure.

Single Beams

The cantilever beam subdivided into three subelements is subjected to a static axial force f_{ax} at its free end 1 (Figure 16.6). The effect of member axial force f_{ax} appears in the equations of free vibration as an additional matrix — $[N_{LM}]$ and $[N_{CS}]$ for the lumped and consistent mass methods [Equation (16.18)]:

$$([K] + [N_{LM}] - \omega^2[M_{LM}])\{D\} = \{0\} \tag{16.32}$$

$$([K] + [N_{CS}] - \omega^2[M_{CS}])\{D\} = \{0\} \tag{16.33}$$

In the continuous mass method, the member axial force f_{ax} appears in the argument of the trigonometric or hyperbolic functions [Equation (16.14)].

The dynamic responses of the cantilever beam (Figure 16.6) have been computed by the lumped and consistent mass method by only changing the magnitude of the member axial force f_{ax} from zero to its critical value $(f_{ax})_{crit}$ [Equation (16.23) and Equation (16.24)]. The same computations have been performed using the continuous mass method [Equation (16.14)].

The ratios for ϕ,

$$\phi_{mi,mj} = \omega_{1,mi}/\omega_{01,mj} \tag{16.34}$$

of the first natural frequency by method *mi* (with the effect of member axial force) versus that of method *mj* (without the effect of member axial force) are plotted in Figure 16.7. The index *mi* or *mj* designates *LM, CS,* and *CT,* the lumped mass, consistent mass, and continuous mass methods.

The comparison of the variations of the ratios $\phi_{LM,LM}$ with $\phi_{LM,CT}$ [Equation (16.24)] exhibits rough approximation. The approximation involved in the variations of ratios $\phi_{CS,CS}$ and $\phi_{CS,CT}$ is very close to the actual one.

The comparison of the variations of the ratios $\phi_{LM,CT}$ and $\phi_{CS,CT}$ with $\phi_{CT,CT}$ exhibits some degree of approximation.

Frames

All the columns of the three-story steel frame are assumed to be subjected to a static axial force, f_{ax}, of the same magnitude.

The dynamic responses of the three-story frame were evaluated by lumped, consistent, and continuous mass methods. Two different cases were considered. In the first case, only the magnitudes of the weights acting on the beams have been increased by a factor m. In the second case, the magnitudes of the member axial force, f_{ax}, and the weights acting on each beam have been increased by p_x and m, in such a way that both factors have the same magnitude, $p_x = m$.

For a same-method mj, the ratio

$$\alpha_{i,mj} = (\omega_I/\omega_{oi})_{mj} \tag{16.35}$$

of the *i*th natural frequency ω_i (including the effect of member axial force and/or additional mass, only on the beams) versus the *i*th natural frequency ω_{oi} (excluding all the additional effect) is plotted in Figure 16.8. The ratio $\alpha_{i,CT}$ — for first, second, and third natural frequencies computed by the continuous mass method — is also shown in Figure 16.8.

The sways at the floor level D_{mj}, including and excluding the effect of the axial force f_{ax}, are computed by the lumped, consistent, and continuous mass matrix methods. The variations of the sway at the floor levels D_{mj} are plotted in Figure 16.9.

When the effects of member axial forces are excluded, the sways at the floor levels obtained by the lumped, consistent, and continuous mass methods are almost the same. The effect of member axial force has shown small variations in the floor sways evaluated by the lumped and consistent mass methods. But the variation in the sways at the floor levels computed by the continuous mass method is large.

Although the first buckling mode for lumped and consistent mass methods occurs by increasing sways from lower to upper floors, for the continuous mass method the first buckling mode occurs between the base and the first floor. The relative displacement of second and third floors with respect to the displacement of the first floor tends to zero.

Offshore Structures

The dynamic behavior of the plane frame of the offshore platform similar to Conoco's Main Pass 296A is selected as a structural system that includes the effects of member axial force and soil–structure interaction. Thus, the results obtained through the computations can be easily compared with those obtained previously [Ovunc, 1984 and 1985]. The static loads on the members vibrating within the plane

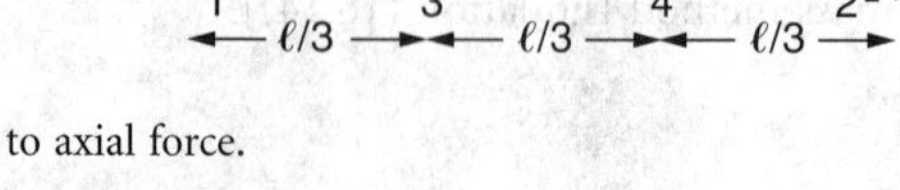

FIGURE 16.8 Beam subjected to axial force.

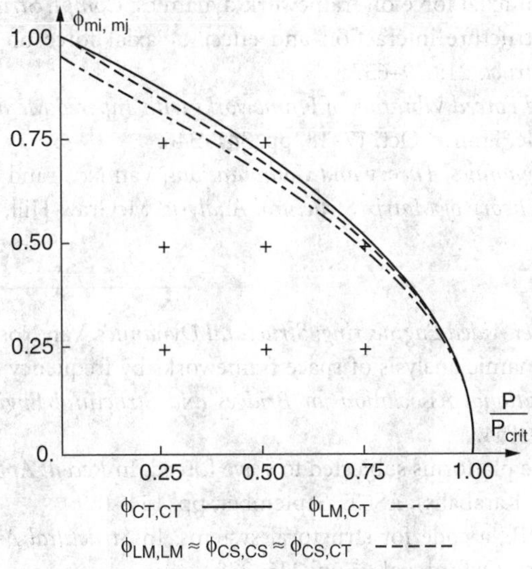

FIGURE 16.9 Effect of member axial force.

of the frame are considered as distributed on these members, whereas the static loads acting on the transversal members are lumped at the joints of the plane frame. Load factors m and l are introduced on the distributed and lumped loads. When the distributed load factor m tends to zero, the continuous mass matrix approach tends to the lumped mass matrix approach. Two different boundary conditions are taken into account at the bottom tips of the piles: fixed and free tips. The ratio of the rigidity of the beams to columns and the embedded length of the piles can be selected as additional parameters.

A dimensionless parameter α_i is defined as

$$\alpha_i = \omega_{pi}/\omega_{oi}$$

the ratio of the natural circular frequencies of the *i*th mode ω_{pi} and ω_{oi}, corresponding to the structures with and without soil–structure interaction and/or with or without the member axial force, respectively.

The data related to the structural system are assumed to be similar to those of Conoco's offshore platform. The data for the wave forces and soil characteristics are collected from information related to the Gulf of Mexico.

Defining Terms

Damping — Results from the internal friction within the material or from system vibration within another material.

Damping energy — Work done by the internal friction within the material as a result of the motion.

Kinetic energy — Work done by a mass particle as a result of its motion.

Stiffness coefficient $K_{i,j}$ — Force or moment in the direction of the first index (i) required to maintain the equilibrium of the body due to a unit displacement or rotation in the direction of the second index (j), while all the other specified displacements and rotations are equal to zero.

Strain energy — Work done by a particle due to its stress and strain.

External energy — Work done by an external force due to a displacement in its direction.

References

Clough, R. W. and Penzien, J. 1993. *Dynamic of Structures.* McGraw-Hill, New York.

Ovunc, B. A. 1974. Dynamics of frameworks by continuous mass method. *Compt. Struct.* 4:1061–1089.

Ovunc, B. A. 1980. Effect of axial force on framework dynamics. *Compt. Struct.* 11:389–395.

Ovunc, B. A. 1985. Soil-structure interaction and effect of axial force on the dynamics of offshore structures. *Compt. Struct.* 21:629–637.

Ovunc, B. A. 1990. *Free and Forced Vibration of Frameworks with Tapered Members,* Struceng and Femcad Conference, Grenoble, France, Oct. 17–18, pp. 341–346.

Paz, M. 1993. *Structural Dynamics, Theory and Computations.* Van Nostrand Reinhold, New York.

Przemieniecki, J. S. 1968. *Theory of Matrix Structural Analysis.* McGraw-Hill, New York.

Further Information

Paz, M. 1986. *Microcomputer Aided Engineering: Structural Dynamics,* Van Nostrand Reinhold, New York.

Ovunc, B. A. 1972. The dynamic analysis of space frameworks by frequency dependent stiffness matrix approach. In *International Association for Bridges and Structural Engineering,* vol. 32/2, Zurich, Switzerland, pp. 137–154.

Ovunc, B. A. 1986. Offshore platforms subjected to wave forced. In *Recent Applications in Computational Mechanics,* (ed. D. L. Karabalis), ASCE, September, pp. 154–169.

Ovunc, B. A. 1985. STDYNL, a code for structural systems. In *Structural Analysis Systems* (ed. Niku - Lari), Pergamon Press, Oxford, vol. 3, pp. 225–238.

Ovunc, B. A. 1992. *Dynamics of Offshore Structures Supported on Piles in Cohesionless Soil.* ASME, European Joint Conference on Engineering Systems Design and Analysis, Istanbul, Turkey, ESDA 1992, June 29–July 4, ASME PD, vol. 47–5, pp. 11–18.

Ovunc, B. A. 1990. Vibration of Timoshenko Frames Including Member Axial Force and Soil-Structure Interaction. FEMCAD & OPTIMIZATION Conference, Los Angeles, Nov. 5–6, pp. 359–364.

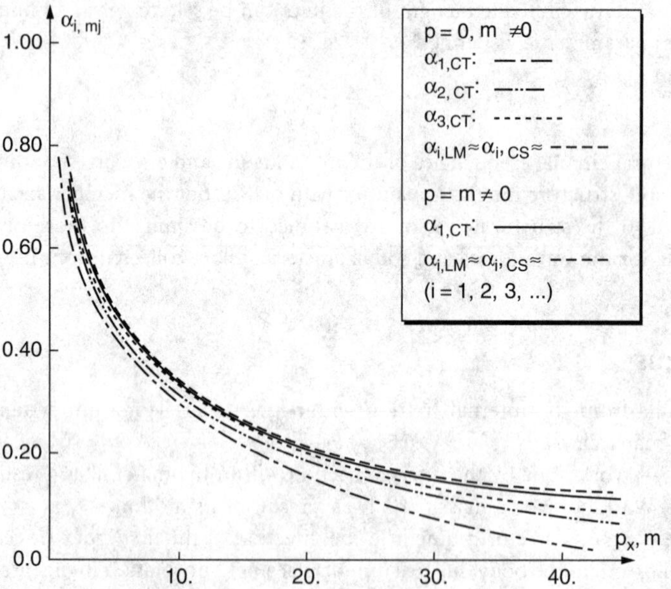

FIGURE 16.10 Additional effects of member axial force.

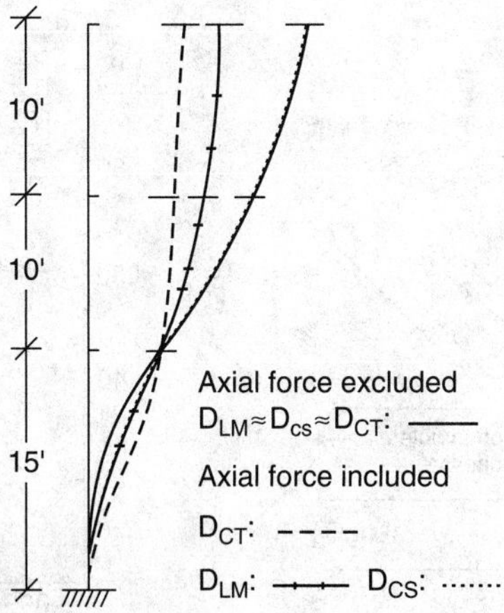

FIGURE 16.11 Sways at floor levels.

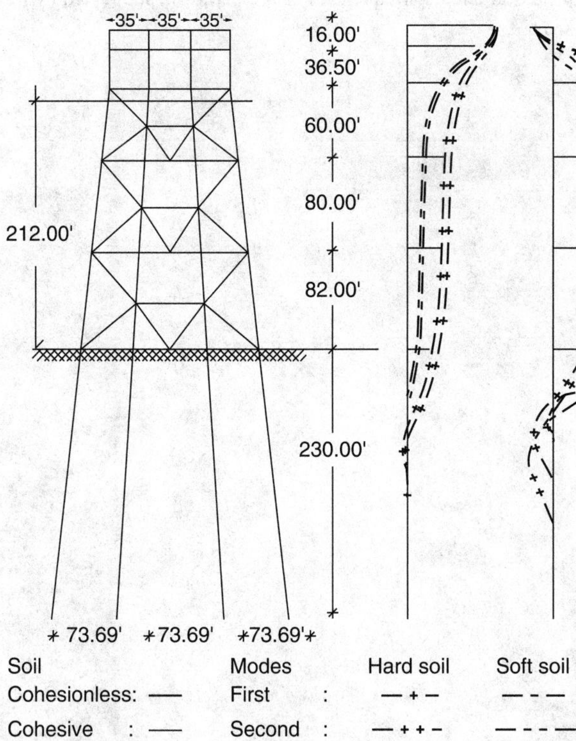

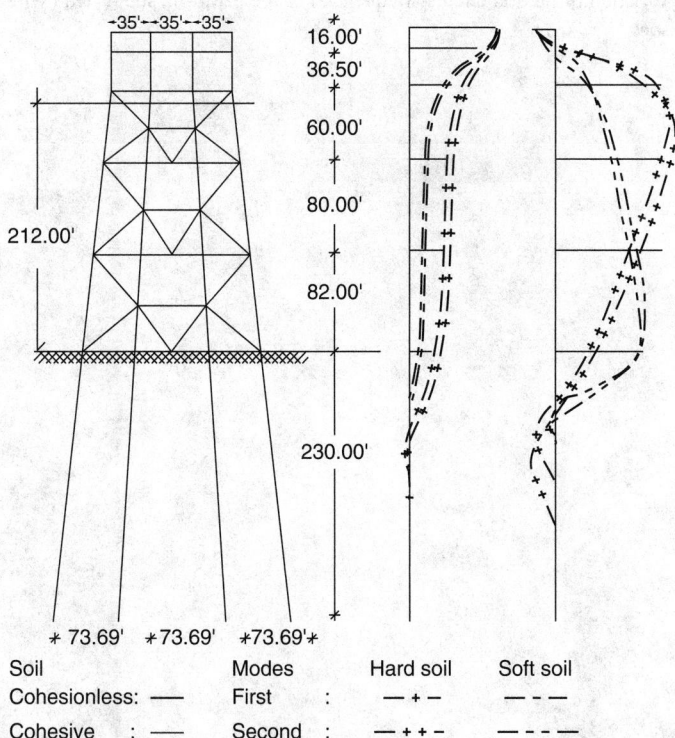

FIGURE 16.12 Offshore platform and its modal shapes.

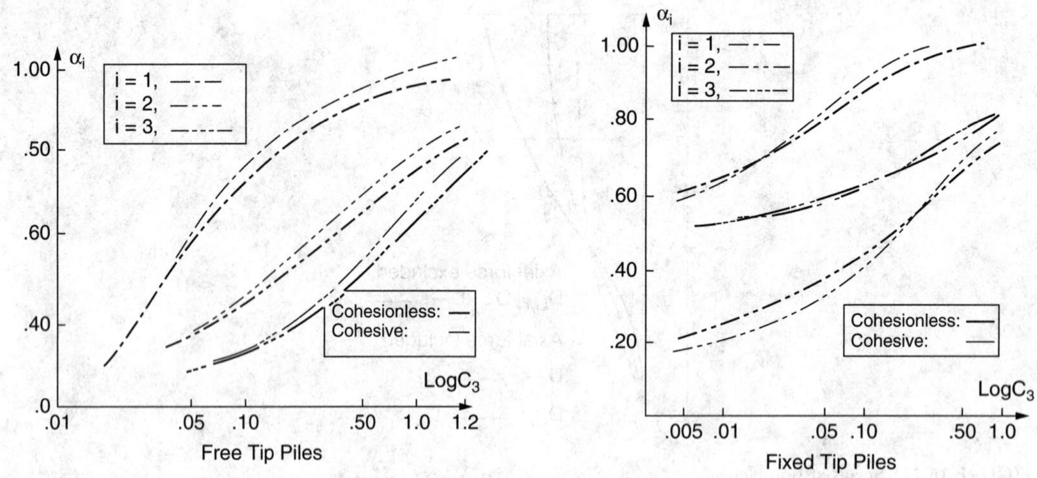

FIGURE 16.13 The variation of natural circular frequencies of the platform, supported by free and fixed tip piles vs. the soil characteristics.

17

Applications of Structural and Dynamic Principles

Anthony J. Kalinowski
Naval Undersea Warfare Center

In this chapter, we consider some practical applications of dynamics and vibrations, with specific emphasis on shock isolation and vibration isolation. In both of these isolation situations, we are concerned with the transmission of interaction forces, $u(t)$, existing between two configurations [which are referred to here as the **base configuration (BC)** and the **structural configuration (SC)**], as illustrated in Figure 17.1. This example system is a first-order representation of an idealized physical system that is general enough to represent *both* shock and vibration design situations. Most of the underlying physical principles impacting the design of either kind of isolation can be explained and illustrated with this simple one-degree-of-freedom model. The governing equations of motion can be generated from the three-part sequence shown in Figure 17.1. In Figure 17.1(a), the BC and SC are in the unloaded condition, and the two configurations are initially separated by a length L. This corresponds to the state where the model is lying horizontal relative to the vertical direction of the gravity field, or where the model is in the vertical position *before* the gravity field (acceleration of gravity $g = 386$ in./sec^2) is taken into account. For vertically oriented models, it is convenient to write the equation of motion relative to an initially gravity-loaded model, as in Figure 17.1(b), where the SC is shown at rest in a compressed state (the linear isolator has a preloaded compressive force of $u_T = -K\delta_s$, where $\delta_s = Mg/K$ is defined *positively* as the **static deflection**). Next, Figure 17.1(c) corresponds to a body in motion, and the governing equations of motion are obtained by constructing a free-body diagram of the SC and equating the sum of all vertical forces acting on the body to its mass, M, times its acceleration, $d^2z/dt^2 \equiv \ddot{z}$ (where dot notation is used to refer to time differentiation from here on), resulting in the following dynamic equilibrium equation:

$$M\ddot{z} + u_T(x, \dot{x}) + Mg - F(t) = 0 \tag{17.1}$$

The relation between the total z displacement and the isolator stretch variable x is given by

0-8493-1586-7/05/$0.00+$1.50

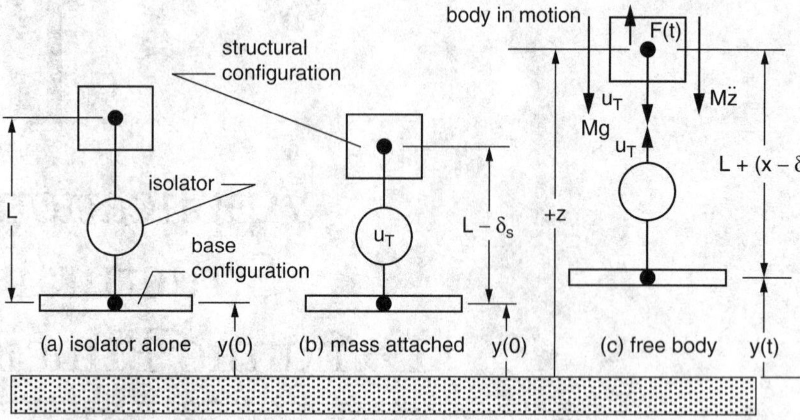

FIGURE 17.1 Multipurpose single-degree-of-freedom system.

$$z = y + x + (L - \delta_s) \tag{17.2}$$

and it follows that

$$\ddot{z} = \ddot{x} + \ddot{y} \tag{17.3}$$

In general, the isolator force, $u_T(x, \dot{x})$, could be a nonlinear function of the relative displacement x and relative velocity $\dot{x}$; however, for the purposes of this introductory development, only linear isolators will be considered. Thus, the total isolator force acting on the mass M in Figure 17.1(c) is given by

$$u_T(x, \dot{x}) = Kx + C\dot{x} - K\delta_s \equiv u(x, \dot{x}) - K\delta_s \tag{17.4}$$

where $u(x, \dot{x})$ is the dynamic portion of the isolator force not including the static deflection force. Substituting Equation (17.2) through Equation (17.4) into Equation (17.1) gives

$$M\ddot{x} + C\dot{x} + Kx = F(t) - M\ddot{y}(t) \tag{17.5}$$

subject to initial conditions

$$x(t = 0) = x_0, \quad \dot{x}(t = 0) = \dot{x}_0 \tag{17.6}$$

It is noted that the Mg term and $-K\delta_s$ cancel in the formation of Equation (17.5).

The shock and vibration will take place about the static equilibrium position shown in Figure 17.1(b); that is, when the vibrating body comes to rest (i.e., $x = 0$, $\dot{x} = 0$), the isolator is still compressed an amount equal to the static deflection δ_s. The general solution to Equation (17.5) subject to initial conditions [Equation (17.6)] can be obtained by several different methods; however, the method of Laplace transforms is used here because it readily applies to situations where:

1. The boundary conditions are of the initial-value type.
2. The right-hand side is an arbitrary function of time.
3. An equivalence between an impulse-loaded right-hand side (rapidly applied loading) and a suddenly applied initial-velocity problem with no right-hand side can be easily illustrated.

Thus, taking the Laplace transform of Equation (17.5) with respect to the Laplace transform variable s results in

$$x(s)=\frac{x_0\cdot(sM+C)}{(s^2M+sC+K)}+\frac{\dot{x}_0\cdot(M)}{(s^2M+sC+K)}+\frac{F(s)-M\ddot{y}(s)}{(s^2M+sC+K)} \tag{17.7}$$

Upon taking the inverse transform, this leads to the general solution for displacement and velocity:

$$\begin{aligned} x(t)&=e^{-\eta t}\left(x_0\cos(\omega_d t)+\frac{\dot{x}_0+\eta x_0}{\omega_d}\sin(\omega_d t)\right)\\ &+\int_{\lambda=0}^{\lambda=t}\frac{e^{-\eta(t-\lambda)}\sin[\omega_d(t-\lambda)][F(\lambda)-M\ddot{y}(\lambda)]d\lambda}{M\omega_d} \end{aligned} \tag{17.8}$$

$$\begin{aligned} \dot{x}(t)&=-\eta x(t)+e^{-\eta t}[-x_0\omega_d\sin(\omega_d t)+(\dot{x}_0+\eta x_0)\cos(\omega_d t)]\\ &+\int_{\lambda=0}^{\lambda=t}\frac{e^{-\eta(t-\lambda)}\cos[\omega_d(t-\lambda)][F(\lambda)-M\ddot{y}(\lambda)]d\lambda}{M} \end{aligned} \tag{17.9}$$

In the applications to follow, it will be more convenient to work with the variables $\omega_n\equiv\sqrt{K/M}=2\pi f_n$, corresponding to the **isolator natural frequency**, and $\zeta\equiv C/(4\pi Mf_n)$, corresponding to the **critical damping ratio**. The variables ω_d and η appearing in Equation (17.8) and Equation (17.9) are called the **damped natural frequency** and **decay constant**, respectively, and can be expressed in terms of variables ω_n and ζ using $\omega_d=\omega_n\sqrt{1-\zeta^2}$ and $\eta=\omega_n\zeta$. These new variables have the following physical meanings: ω_n corresponds to the free harmonic vibration (no driver present) of the isolator in the absence of damping; ω_d corresponds to the damped free harmonic vibration; ζ determines whether the system is *underdamped* ($\zeta<1$) or *overdamped* ($\zeta>1$), where in the former case the free motion oscillates harmonically with damped natural frequency ω_d and in the latter case the free motion does not vibrate harmonically; and finally, η corresponds to the rate at which the underdamped system exponentially decays in time.

The solutions represented by Equation (17.8) and Equation (17.9) will be used to evaluate the dynamic responses in all the example problems to follow. The forms of the solutions are general and apply to either the situation where the drivers $[F(t),\ddot{y}(t)]$ are given as an analytical expression or when they are given as a digital representation (e.g., earthquake responses). In the case of digital driver representations, the integrations in Equation (17.8) and Equation (17.9) can easily be performed by numerical integration (e.g., Simpson's rule), and in the case of analytical driver representations, closed-form integrals can be obtained with the aid of integral tables or with the aid of symbolic evaluation packages such as Maple [Redfern, 1994], MATLAB's version of Maple [Sigmon, 1994], and Mathematica [Wolfram, 1991]. A computer program using MATLAB script language [Math Works, 1992] was used to generate the results presented here.

Given the displacement and velocity versus time from Equation (17.8) and Equation (17.9), back-substituting $x(t)$ and $\dot{x}(t)$ into Equation (17.4) gives the dynamic portion of the isolator force $u(x,\dot{x})$, which serves as the major ingredient for isolating the structural configuration from the base configuration. The total acceleration $\ddot{z}$ of the SC can be obtained by substitution of $u_T(x,\dot{x})$ into Equation (17.1). We will refer to the single-degree-of-freedom model in Figure 17.1(c) for the example problems considered here, and the physical significance of the ingredients of the model will be different for each usage. We will consider two types of applications:

Base configuration loaded: Here the base configuration has a prescribed displacement motion time history, $y(t)$ (and therefore the base acceleration $d^2y/dt^2\equiv\ddot{y}$) is the basic input load, and the structural configuration has a zero-value *external forcing function* [$F(t)=0$]. Some examples of this type of problem are:

A. *Earthquake-resistant structures,* where the ground (base configuration) movement from fault slip motion excites buildings or bridges (structural configuration)
B. *Vehicle suspension,* where the ground displacement road profile [$y(\xi)$ vs. distance, ξ, in the forward direction of the vehicle traveling at constant velocity V as it moves parallel to the ground] results in the ground acting as the equivalent base configuration, with prescribed displacement road profile shape $y(\xi) = y(Vt)$ and corresponding *apparent base acceleration* $d^2y/dt^2 = V^2d^2y/d\xi^2$ measured perpendicular to the ground, and the vehicle car body (structural configuration) responds to these irregularities in the road
C. *Electronic component isolation,* where the ship superstructure (base configuration) cabinet houses electronic components mounted in cabinets (structural configuration) that are subject to waterborne shock waves that impart known base motion accelerations, $\ddot{y}$ (based on previously measured experimental data)

Structural configuration loaded: Here the structural configuration has a directly applied force time history, $F(t)$, and the base configuration has no motion (i.e., $y(t) = 0.0$). Some examples of this type of problem are:

A. *Unbalanced rotating mass,* where an eccentric mass m_e with offset r_e is rotating at constant angular speed ω (simple motor model with an unbalanced offset mass m_e) and is mounted inside the motor housing (structural configuration) and the base configuration is taken as fixed ($y(t) = 0$), where the explicit harmonic forcing function is $F(t) = m_e r_e \omega^2 \cos(\omega t)$; and
B. *Free-falling mass,* where a free-falling mass (structural configuration) is prevented from impacting a rigid surface (base configuration) by having an intermediate shock isolator break the fall of the object dropped from height H (e.g., a stunt motorcycle jumps off a ramp or a package containing fragile equipment drops), where the force on the structural configuration is $F = -mg$ for $t \geq 0$ and $F = 0$ and $t < 0$, with initial condition $dx/dt = -\sqrt{2gH}$ at $t = 0$ and $x = 0$ at $t = 0$.

17.1 Base Configuration Loaded Applications

Problem 1: Vehicle Suspension

In this example, a vehicle moving with horizontal constant velocity V passes over a roadway that has a ground profile of $y(\xi) = Y_0 \sin(k\xi)$, $0 \leq \xi \leq N_c L$, $k = 2\pi/L$, where L is the period of the ground swell. Since the relation between horizontal distance traveled and time is given by $\xi = Vt$, the vertical BC base motion can be rewritten as

$$y(t) = Y_0 \sin(\omega t), \quad \ddot{y}(t) = -\omega^2 Y_0 \sin(\omega t), \quad \omega = 2\pi V/L \tag{17.10}$$

applied over the range $0 \leq t \leq N_c L/V$ and $\ddot{y}(t) = 0$ over the rest of the time duration. If the vehicle is four-wheeled, then a higher-degree-of-freedom model is needed to represent the response due to the rotational degrees of freedom; therefore, it is assumed that the vehicle is two-wheeled and is being towed while the load is balanced over the axle. Assume that all other forces acting are negligible; therefore, $F(t) = 0.0$. Consider a vehicle weighing $W = 2000$ lb and traveling at a speed of $V = 60$ mph that encounters a two-cycle ($N_c = 2$) road swell of amplitude $Y_0 = -1.01$ in. and period $L = 20$ ft. Substituting the data into Equation (17.10) results in a peak base acceleration of $\ddot{y}_{max} = A_0 = 2$ gs and a drive frequency of $f = \omega/2\pi = 4.4$ Hz. Design a spring-damper isolator (i.e., find K and C) that limits the vehicle (structural configuration) steady state vibration **acceleration transmission ratio,** $T_A = \text{peak}|\ddot{z}/\ddot{y}_{max}|$ to a value of 0.4 and limits of maximum relative displacement of the isolator $x_{max} = \pm\frac{3}{4}\delta_s$.

Before solving for K and C, it must be noted that isolator manufacturers often use a terminology other than explicitly stating these constants. Typically, in place of K and C, a natural frequency (f_n) versus load (Mg) curve is supplied and the acceleration transmission ratio T_A is given at resonance,

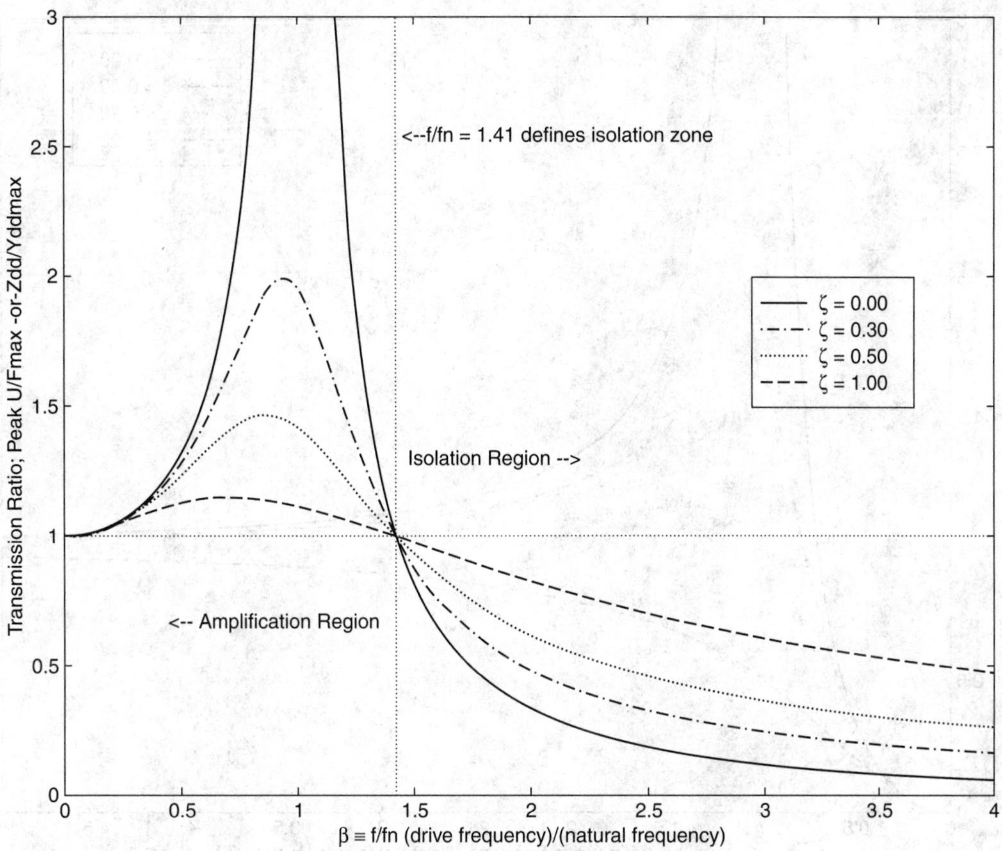

FIGURE 17.2 Transmission for harmonic structure force or base acceleration.

as in Barry, 1993. Later, it will be shown how K and C can be back-calculated from f_n and T_A. For the present, however, attention is focused on finding the desired f_n as the first step. The key ingredient in this approach is to use the transmission ratio versus drive frequency ($\beta \equiv f/f_n$) curve, as shown in Figure 17.2. This curve is obtained by substituting a solution of the form $x(t) = \bar{A}\sin(\omega t) + \bar{B}\cos(\omega t)$ into Equation (17.5), and solving for the $\bar{A}, \bar{B}$ constants, resulting in the following acceleration transmission ratio:

$$T_A = \sqrt{\frac{1+(2\zeta\beta)^2}{(1-\beta^2)^2+(2\zeta\beta)^2}}, \quad \beta \equiv \omega/\omega_n = f/f_n \tag{17.11}$$

It is of interest to note that all the curves in Figure 17.2 pass through the same frequency ratio, $\beta = \sqrt{2}$, and this special value forms the dividing line between isolation and amplification. Therefore, as a design strategy, to get isolation of the base configuration from the structural configuration, the isolator $_K$ is selected such that its natural frequency f_n results in β values $> \sqrt{2}$.

Another response ratio that applies when the relative displacements are of particular interest is the **displacement magnification ratio**, denoted as $T_D = \text{peak}\,|x/y_{max}|$, which can be derived in a manner similar to the above relation, resulting in the expression

$$T_D = \frac{\beta^2}{\sqrt{(1-\beta^2)^2+(2\zeta\beta)^2}} \tag{17.12}$$

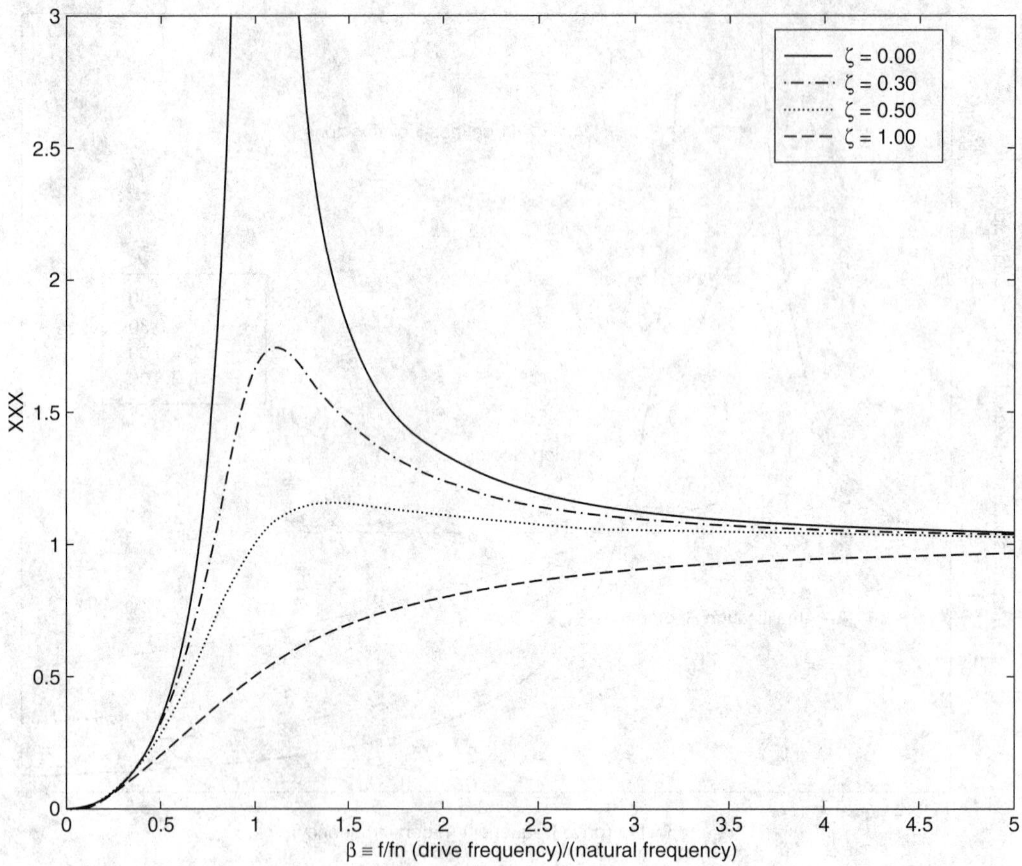

FIGURE 17.3 Magnification for harmonic base acceleration.

This is shown in Figure 17.3 plotted against the drive frequency parameter β.

At this point, the designer must decide whether to solve for the natural frequency f_n that gives the desired transmission ratio T_A, and live with the resulting displacement, *or* solve for the natural frequency f_n that gives the displacement magnification ratio T_D, and live with the resulting maximum acceleration $\ddot{z}$. It is noted that solving for β is equivalent to solving for f_n, since the drive frequency, f, is known. In this sample problem, it is decided that reducing peak acceleration has priority; therefore, with the aid of Equation (17.11), we solve for the β that gives the desired T_A. Thus,

$$\beta = \sqrt{\frac{(a+2T_A^2 - aT_A^2) + \sqrt{(a+2T_A^2 - aT_A^2)^2 - 4T_A^2(T_A^2-1)}}{2T_A^2}}, \quad a = (2\zeta)^2 \tag{17.13}$$

where for light damping (ζ < 0.05), β can be approximated with

$$\beta \approx \sqrt{1+1/T_A} \tag{17.14}$$

Before solving for β using Equation (17.13) [or Equation (17.14)], you must select a critical damping ratio ζ.

At this point, a ζ value can simply be selected according to whether light damping or heavy damping is desired, and a moderately heavy value of ζ = 0.5 is chosen in this example. Therefore, for the problem at hand, substituting $\zeta = 0.50$ and $T_A = 0.40$ into Equation (17.13) gives $\beta = 2.8131$. Upon substituting this value of β along with the drive frequency, $f = 4.4$ Hz, into the β definition [i.e., the second equation in Equation (17.11)], we can solve for the natural frequency, $f_n = 1.564$ Hz, required to limit the acceleration transmission to $T_A = 0.40$. As the final step, the actual spring constant K and damping constant C can easily be back-calculated from the f_n and ζ values in the following manner. In the sample problem, the 2000 lb load is divided equally over two isolators, so each one must carry a mass of $M = 2000 / (2 \times 386) = 2.5906$ lb-in./sec². Therefore, $K = (2\pi f_n)^2 M = 250.2$ lb/in., and $C = 4\pi M\zeta f_n = 25.46$ lb-sec/in. for each of the two isolators. The dynamic response for important variables such as $x(t)$, $\dot{x}(t)$, $\ddot{y}(t)$, and $\ddot{z}(t)/\ddot{y}_{max}$ (i.e., the SC transient acceleration transmission ratio) is computed by evaluating Equation (17.8), Equation (17.9), and Equation (17.1) over the two cycles of input and for an equally long coastdown time duration after the road profile has become flat again. Upon observing the Figure 17.4 solution, it is observed that the deflection stays within the maximum allowable constraint space of $x_{con} = \pm 0.75\,\delta_s = \pm 3.00$ in.; however, the transient portion of the peak acceleration ratio $\ddot{z}(t)/\ddot{y}_{max} = 0.62$ overshoots the target steady state solution. $T_A = 0.40$. The isolator design employed the steady state solution; therefore, it is not unusual that the transient could exceed the steady state limit. It is further noted that by the end of the second response cycle, the acceleration ratio is already approaching the 0.40 target. When the road profile turns flat, the acceleration ratio rapidly tails off toward zero, due to the large damping value. To compensate for the overshoot in transient acceleration, a larger β (and thus a smaller f_n) can be used on a second pass through the design process [e.g., enter $T_A = 0.40 \times (0.40 / 0.62)$ into Equation (17.13) and repeat the design process, where the peak $\ddot{z}(t)/\ddot{y}_{max}$ is lowered from 0.62 to 0.45].

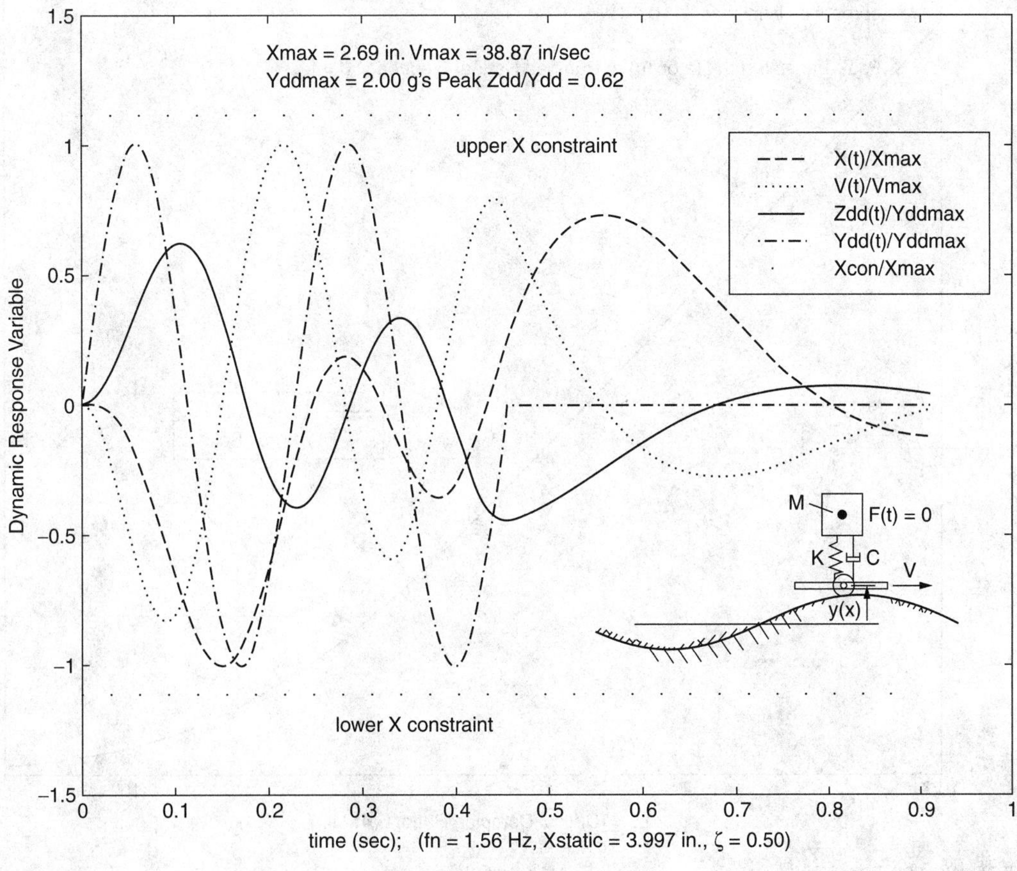

FIGURE 17.4 Roadway base acceleration transmission to structure.

In building an isolator from scratch, one can simply design it to have the physical properties K and C directly. However, if isolators are to be selected from off-the-shelf stock, a few comments are in order regarding the selection process. Manufacturers often supply a set of load (Mg) versus f_n curves for each isolator in a class of isolators. Upon entering such a set with a 1000 lb load, the isolator having a natural frequency nearest the target value, say $f_n = 1.564$ Hz, is selected. Typically there will not be an isolator corresponding to the exact f_n desired at the operating load. Thus, picking an isolator with f_n on the low side will make the steady state transmission on the low side. In some cases, the manufacturer must be contacted directly to get damping data, and in other cases damping information is given indirectly via a stated transmissibility, T_A, *at resonance.* The β location of the peak resonance T_A value in Figure 17.2 in terms of the damping ratio ζ is given by

$$\beta^2 = \frac{1}{4\zeta^2}(-1+\sqrt{1+8\zeta^2})$$

Substituting this expression into Equation (17.11) results in a relationship between T_A and ζ as plotted in Figure 17.5. For design purposes, the exact curve in inverse form can be approximated by the simpler expression

$$\zeta_{\text{res}} \approx 0.5\sqrt{1/(T_A^2-1)} \tag{17.15}$$

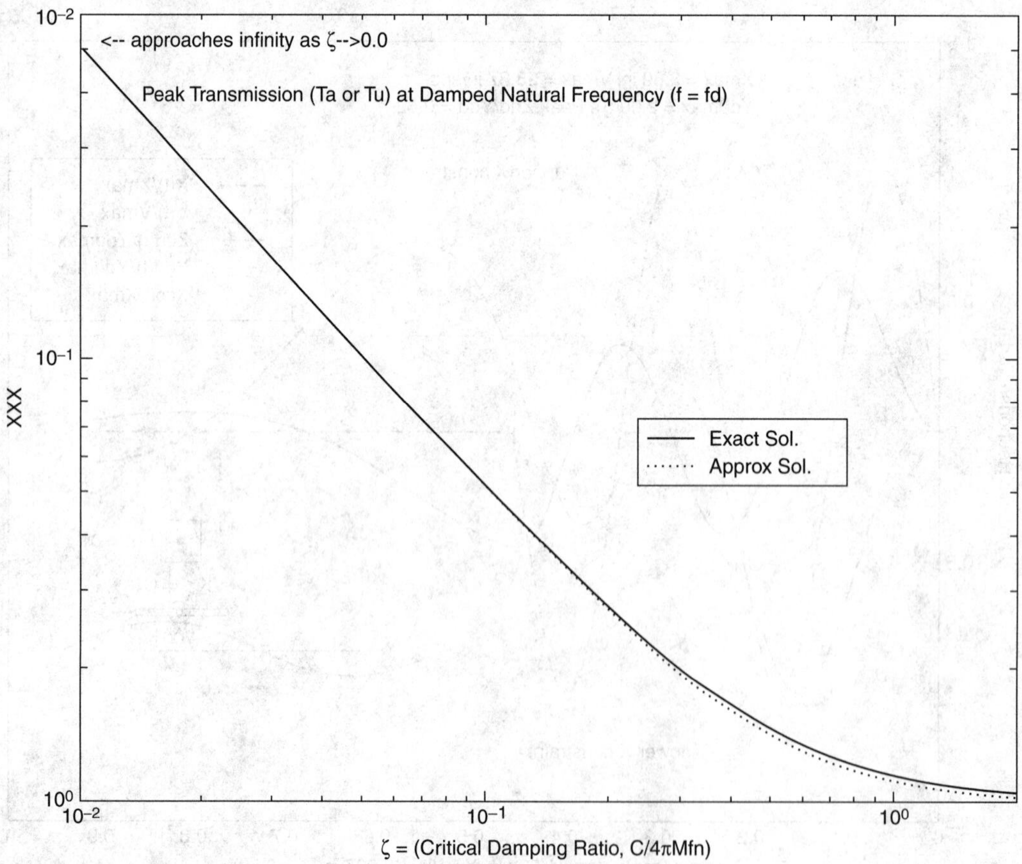

FIGURE 17.5 Transmission at resonance (harmonic force or base acceleration).

In Figure 17.5, both the exact curve and the Equation (17.15) approximation are plotted side by side. For example, if the manufacturer states that an isolator has a transmission ratio $T_A = 1.5$ *at resonance*, then substituting this value into Equation (17.15) gives a corresponding critical damping ratio of $\zeta = 0.45$. Finally, substituting $\zeta = 0.45$ into $C = 4\pi M\zeta f_n$ provides the numerical value for the isolator damping constant in question.

Problem 2: Shock Isolation of Fragile Equipment

Consider a 50 lb structural configuration, initially at rest [i.e., $x(t = 0) = 0,\ \dot{x}\,(t = 0) = 0$], that is subject to a half-sine pulse-type BC acceleration input of the form

$$\begin{aligned} \ddot{y} &= A_0 \sin(\omega t), \quad 0 \le t \le T_P; \\ \ddot{y} &= 0.0, \quad T_P < t \le \infty; \quad T_P = \pi/\omega \end{aligned} \tag{17.16}$$

and specifically the peak value of the base acceleration input is $A_0 = 16$ *gs* and the pulse duration is $T_P = 11$ msec. In this problem, the direct force $F(t)$ is taken as zero.

The design problem is to synthesize a shock isolator that will limit the acceleration transmitted to the SC to 4 *gs*; therefore, a target transmission ratio of $T_A = 4/16 = 0.25$ is sought. It is also desired that the maximum displacement be limited to $x = \pm 1.0$ in. At this point it is noted upon comparing the base motion of the previous application [Equation (17.10)] to the base motion of the current problem [Equation (17.16)] that there is a similarity between the first and second problems, except for the fact that the current shock problem has a short pulse length where only one-half of a sine wave ($N_c = 0.5$) is applied. The principle governing shock isolation is different from the corresponding vibration isolation of the previous problem in that the shock isolation process is characterized as a storage device for a sharply increasing acceleration waveform, and the design concept is to attempt to instantaneously absorb the energy and then release it at the natural frequency of the device but at a lower-level mass deceleration. One approach is, for guessed K and C values, to simply substitute these values and the input waveform of Equation (17.16) into Equation (17.7), Equation (17.8), and Equation (17.1) to get the response motion, observe the response, and reiterate the process with new K and C values until a desired response is obtained. We cannot use the steady state method of the previous problem because the steady state assumption of the input waveform would not be valid. An alternative design process is to convert the base input motion into a nearly equivalent initial, suddenly applied velocity problem, $\dot{x}(t = 0) = V_0$, with no explicit driver on the right-hand side of Equation (17.5).

The extreme case of a rapidly applied loading is a special mathematical function called a *delta function*, $\delta(t)$, whose value is ∞ at $t = 0$ and zero for $t > 0$. The right-hand-side loading can then be represented as

$$-M\ddot{y}(t) \approx -M\tilde{A}\delta(t) \tag{17.17}$$

where

$$\tilde{A} = \int_{\lambda=0}^{\lambda=T_P} \ddot{y}(\lambda)d\lambda \tag{17.18}$$

Upon substituting Equation (17.17) into Laplace-transformed Equation (17.7), with initial displacement $x_0 = 0$ and $F(s) = 0$, and noting that the Laplace transform of a delta function is 1.0, it can be seen that the $\dot{x}_0 M$ term and $-MA$ have exactly the same form (same denominator); therefore, by interchanging the roles of the driver and initial condition, we can let a problem with a zero initial velocity and pulse-type $\ddot{y}$ driver having area $\tilde{A}$ [by Equation (17.18)] be replaced with an equivalent problem having a zero $\ddot{y}$ driver but an initial velocity of $\dot{x}_0 \equiv V_0 = -\tilde{A}$. The V_0 quantity can therefore be interpreted as a suddenly applied velocity change. The advantage of this approach is that estimates of the maximum response can easily be made and used to back-calculate the isolator properties needed to achieve the desired isolator performance. Thus, for the equivalent initial velocity representation, we immediately get the solution as a special case of Equation (17.8) that simply reduces to

$$x(t) = -\tilde{A}e^{-\eta t}\sin(\omega_d t)/\omega_d \tag{17.19}$$

Upon differentiating Equation (17.19) and solving for the maximum displacement x_{max} and maximum mass acceleration $\ddot{z}_{max}$, the following result is obtained:

$$x_{max} = -\tilde{A}e^{-\alpha}/\omega_n, \quad \text{where } \alpha = (\zeta/\sqrt{1-\zeta^2})\sin^{-1}(\sqrt{1-\zeta^2}) \tag{17.20}$$

$$\ddot{z}_{max} = \tilde{A}\omega_n D_i \tag{17.21}$$

$$\text{with} \quad D_i = \exp(-\zeta\hat{t}/\sqrt{1-\zeta^2})\left(\frac{(1-2\zeta^2)}{\sqrt{1-\zeta^2}}\sin(\hat{t}) + 2\zeta\cos(\hat{t})\right) \quad 0 \le \zeta \le 0.5$$

$$D_i = 2\zeta, \quad 0.5 < \zeta \le 1.0$$

$$\text{where} \quad \hat{t} = \tan^{-1}\left(\frac{(1-4\zeta^2)\sqrt{1-\zeta^2}}{\zeta(3-4\zeta^2)}\right)$$

As in the previous example, the acceleration transmission ratio is defined as $T_A = \text{peak}\left|\ddot{z}/\ddot{y}_{max}\right|$. Substituting Equation (17.21) into this T_A expression and solving for the natural frequency gives

$$f_n = \frac{\left|\ddot{y}_{max}\right| T_A}{2\pi\tilde{A}D_i} \quad \text{(general shock)} \tag{17.22}$$

and, for very light damping (say, $\zeta < 0.1$), the approximation $D_i \approx 1.0$ can be used. The isolator selection follows the same concept as in the previous problem where the natural frequency needed to limit the transmissibility is determined. For the problem at hand, substituting the given half-sine input $\ddot{y}$ into Equation (17.18) corresponds to a velocity change of $\tilde{A} = 2A_0/\omega = 2A_0T_P/\pi$, where it is also noted that $\left|\ddot{y}_{max}\right| = A_0$. Upon using these data, Equation (17.21) reduces to

$$f_n = \frac{T_A}{4T_P D_i} \quad \text{(for half-sine pulse shock)} \tag{17.23}$$

From this point forward, the selection procedure for obtaining the isolator parameters is conceptually the same, except that the computation of the desired natural frequency is different and is governed by Equation (17.22) or Equation (17.23) as appropriate. For the problem at hand, substituting the target transmissibility $T_A = 0.25$, pulse length $T_P = 0.011$ sec, and designer-selected damping ratio $\zeta = 0.2$ (using $\zeta = 0.2$ in the D_i expression of Equation (17.21) gives $D_i = 0.8209$) results in a desired natural frequency of $f_n = 6.92$ Hz. For the 50 lb SC ($M = 50/386$), the $f_n = 6.92$

and $\zeta = 0.2$ data translate into spring and damping constants of $K = (2\pi f_n)^2 M = 244.97$ lb/in. and $C = 4\pi M\zeta f_n = 2.253$ lb-sec/in. Also, using Equation (17.20), the estimated maximum deflection is $x_{max} = 0.7520$ in.; therefore, this isolator design should meet the space constraint imposed on the problem. This x_{max} value should also be checked against the isolator manufacturer's maximum allowable spring deflection (sometimes called sway space), which is usually given in the selection catalog. Upon substituting the above K, M, and C design parameters and actual half-sine base motion $\ddot{y}(t)$ into Equation (17.8), Equation (17.9), and Equation (17.1), the dynamic response for important variables such as $x(t)$, $\dot{x}(t)$, $\ddot{y}(t)$, and $\ddot{z}(t)/\ddot{y}_{max}$ (i.e., the SC transient acceleration transmission ratio) is computed. The results are shown in Figure 17.6, where it is noted that the desired 0.25 transmissibility is achieved and the displacement constraints are not exceeded. It should be noted that the curves labeled in the legend correspond to using the actual half-sine input base motion. These results will not be exactly the same as the equivalent suddenly applied velocity (V_0) solution that was used to size the isolators, because the half-sine pulse is not exactly the idealized delta function. For illustrative purposes, the equivalent suddenly applied velocity solution is superimposed on the same plot and denoted with the unconnected symbolic markers "*" for displacement, "o" for velocity, and "+" for acceleration ratio. As expected, the velocity comparison will be different in the early time [e.g., with the actual $\ddot{y}(t)$ input the initial condition is $\dot{x}_0 = 0$, whereas in the equivalent problem the initial velocity is not zero and represents the entire input to the problem]. As a final comment, the selection of the isolators from the manufacturer's catalog follows along the same lines here and therefore will not be repeated; also check that $T_P << 1/f_n$ when using Equation (17.22) or Equation (17.23).

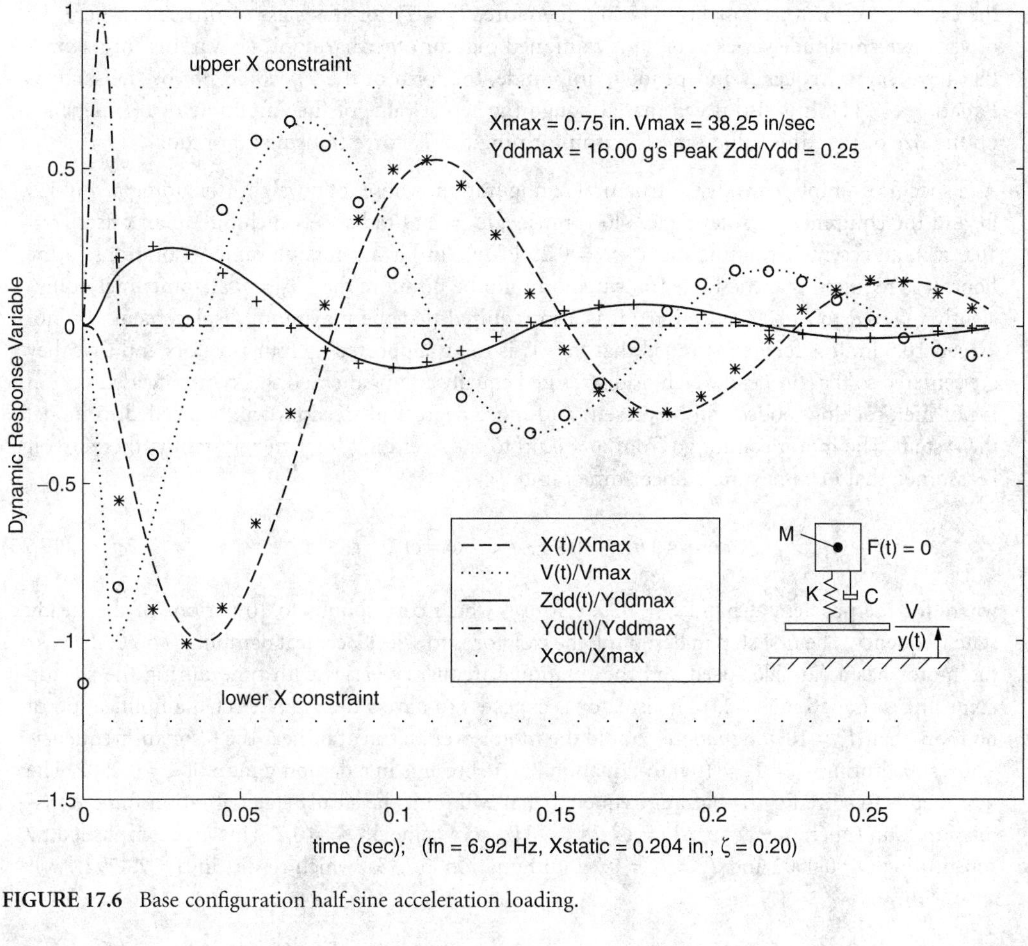

FIGURE 17.6 Base configuration half-sine acceleration loading.

17.2 Structural Configuration Loaded Applications

Problem 3: Rotating Machinery Force Transmission

In this class of problems, the base configuration is considered fixed; thus $y(t) = \dot{y}(t) = \ddot{y}(t) = 0$, and therefore the only system loading comes through the structural configuration loading $F(t)$. Perhaps one of the most common such loading problems is the situation where a piece of equipment with some sort of rotating member is spinning in a steady state mode at drive frequency ω and is resting on the structural configuration, where the total mass M of the equipment to be supported by the isolators is $M = M_s + m_e$, where M_s is the mass of the structural configuration including the rotating machinery (except for the offset mass m_e), and m_e represents the off-center eccentric mass at radius r_e. The m_e term is analogous to wet clothes clinging to the spinning drum during the spin-dry cycle of a common washing machine, where excessive vibrations are set up when the clothes are not uniformly distributed around the drum. The radial acceleration, $A_r = r_e\omega^2$, results in a reciprocating force that is represented by

$$F(t) = m_e r_e \omega^2 \cos(\omega t) \tag{17.24}$$

Therefore, by comparing Equation (17.24) to the first steady state loading example in Equation (17.10), it is seen that the $M\omega^2 Y_0 \sin(\omega t)$ driver in the differential equation is just like the current $m_e r_e \omega^2 \cos(\omega t)$ driver except for a cosine in place of a sine driver function. In fact, the steady state solutions for the transmissibility of an isolator force transmission ratio of $T_U = \text{peak}\,|u(t)/F_{\max}|$ has the exact same form as Equation (17.11); therefore, $T_U = T_A$ for this class of problem, where the driver force amplitude varies as ω^2. It is cautioned that for other harmonically varying forces, ones that have, say, a frequency-independent amplitude, the form of the T_U would not be the same as Equation (17.11). It is also noted that although the actual value of the amplitude of $F(t)$ depends on the size of $m_e r_e$, this value cancels out in forming the T_U force transmission ratio.

As a specific example, consider a structural configuration, whose total weight (including m_e) is 800 lb, and the equipment is rotating at 540 rpm (i.e., $\omega_0 = 2\pi 9$ rad/sec), which due to an offset ($r_e =$ 10.0 in.), an eccentric rotating mass ($m_e = 0.25$ lb-sec²/in.), transmits unwanted vibrations to the floor. It is required that the force transmission ratio be no more than $T_U = 0.2$ (sometimes equivalently referred to as 80% isolation). It is also required that the maximum displacement $x(t)$ not exceed ±0.5 in. It is further assumed that the SC is to be supported by four isolators and that they are centered so that the dead weight is distributed equally among them. If this centering assumption is not met, rocking modes will be present and more degrees of freedom will be needed to model the system. The motor cannot go from $\omega = 0$ up to the operating frequency instantly; therefore, it is assumed that ω has a simple linear time ramp:

$$\omega = \omega_0 t/t_c, \quad 0 \le t \le t_c; \quad \omega = \omega_0, \quad t_c \le t \tag{17.25}$$

where for the problem at hand $t_c = 10 \times (2\pi/\omega_0)$, which corresponds to 10 periods of the steady state frequency. The first step in designing the isolator is to select a critical damping ratio ζ. Because the motor has a variable speed and the rotational frequency varies with time during the startup according to Equation (17.25), the isolator is expected to have a designer-selected amplification of no more than $T_U = 10$ *at resonance*, should the motor ever be run at or near the resonant frequency. Thus, substituting $T_U = T_A = 10$ into Equation (17.15) results in a damping ratio of $\zeta = 0.0502$. The next step is to solve for the natural frequency that will limit the steady state vibration force transmission ratio (operating at speed $f = \omega_0/2\pi = 9$ Hz) to a value of $T_U = 0.2$. This is accomplished by substituting $\zeta = 0.0502$ and $T_U = T_A = 0.2$ into Equation (17.13), which results in $\beta = 2.4792$.

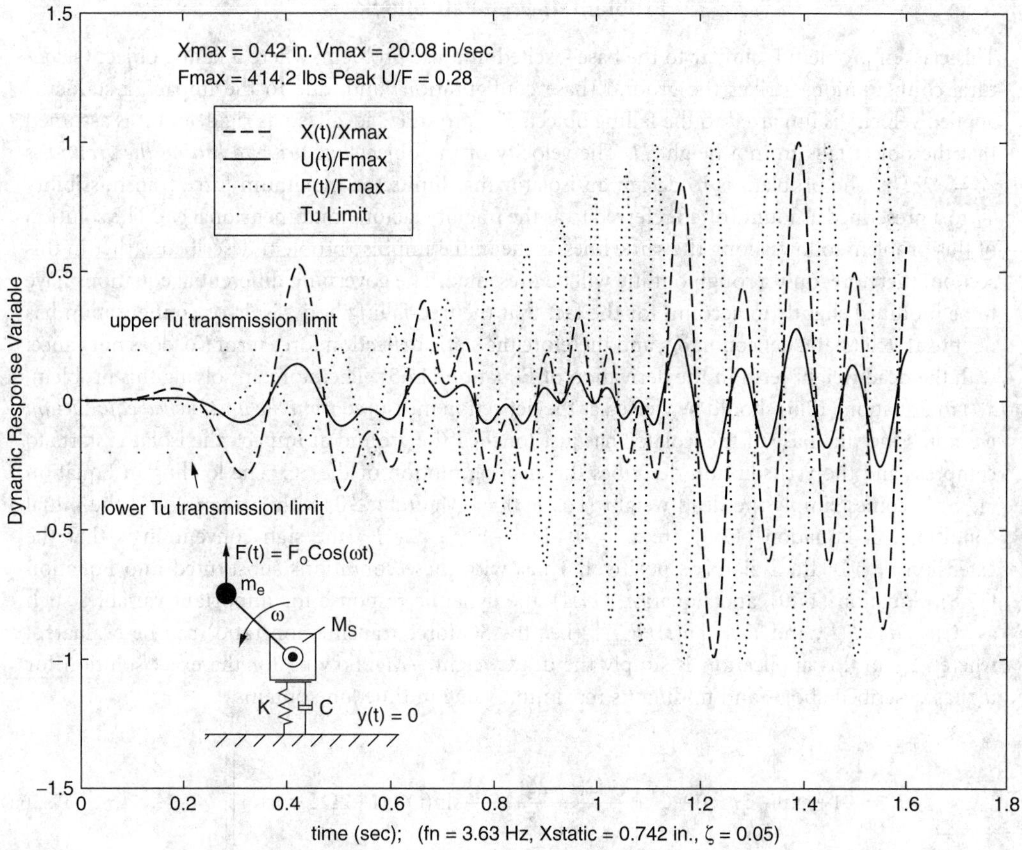

FIGURE 17.7 Transient-into-steady-state force transmission applied to foundation.

There are four isolators; therefore, the mass assigned to each isolator is determined from $M = (800/386)/4 = 0.5181$ lb-sec²/in. The natural frequency is computed from $f_n = f/\beta = 9/2.4792 = 3.63$ Hz. Finally, the spring and damping constants are computed from $K = (2\pi f_n)^2 M = 269.5$ lb/in. and $C = 4\pi M\zeta f_n = 1.186$ lb-sec/in. (Refer to the discussion on selecting isolators from manufacturers' catalogs in the first suspension isolator design example.) Upon substituting the above K, M, C design parameters and driver $F(t)$ defined by Equation (17.24) and Equation (17.25) into Equation (17.8), Equation (17.9), and Equation (17.4), the dynamic response for important variables such as $x(t)$, $\dot{x}(t)$, $F(t)$, and $|u(t)/F_{\max}|$ (i.e., the SC force transmission ratio) is computed. It is noted that during the startup phase, the variable angular velocity $\omega(t)$ results in an additional tangential acceleration component that is neglected in the analysis. The transient portion of the results is shown in Figure 17.7, where it is noted that the desired 0.20 transmissibility is exceeded ($T_A = 0.28$) during the transient (and also that the displacement constraints are not exceeded). The damping is light in this example; therefore, it takes many more cycles to reach the steady state. The horizontal dotted lines in Figure 17.7 align with the steady state limit reached for $u(t)/F_{max}$ after running the solution out to $t = 4.0$.

This demonstration example also illustrates the need to balance rotating machinery. Instead of having to live with the vibrations, it would be better to kill the source of the vibration and rebalance the equipment (e.g., make r_e smaller in this example), so that even if the vibration is still present, its magnitude will be small enough that it will have the same effect as if isolators were used to reduce the transmissibility.

Problem 4: Free-Fall Shock

This class of problem is similar to the base–excited impulse problem, where a falling object (structural configuration) strikes the ground (base configuration) and, due to the impact, a suddenly applied velocity is imparted to the falling object. The ground is idealized as rigid, and it is assumed that the object falls from a height H. The velocity of the object *just prior to hitting the ground* is $V=\sqrt{2gH}$. The problem is to design an isolator that limits the maximum force transmissibility T_U to a prescribed amount (often referred to as the fragility factor, nondimensional gs). The solution of this problem follows along the same lines as the initial impulse problem described earlier in this section; therefore, only a rough outline will be presented. The governing differential equations have to be modified slightly to account for the fact that the free-falling body–isolator configuration has no prestretch (static deflection δ_s), and therefore the static deflection force term $K\delta_s$ does not cancel with the dead weight term in the derivation of Equation (17.5). Therefore, in solving this problem, $x(t)$ in Equation (17.5) should be viewed as the deflection measured *from the unstretched equilibrium position.* Once the base of the isolator has just touched the ground at impact, the isolator starts to compress and the $x(t)$ solution describes the ensuing motion of the SC. The loading in Equation (17.5) becomes simply the dead weight (i.e., $F(t) = -Mg$ for $t \geq 0$, with $\dot{y}(t)=0$) and the initial conditions in Equation (17.6) are $x_0 = 0$, $\dot{x}_0=-V=-\sqrt{2gH}$ [the sign convention is that the extension ($+x$) of the isolator is positive]. Thus, with these conditions substituted into Equation (17.8), Equation (17.9), and Equation (17.4), the dynamic response for important variables such as $x(t)$, $\dot{x}(t)$, $F(t)$, and $T_U = |u(t)/F_{max}|$ (i.e., the SC force transmission ratio) can be evaluated, where F_{max} in this application is simply the dead weight, $-Mg$. Solving for the exact solution for $u(t)$ as described above, and finding its maximum value in time, one obtains

$$T_U=\left|-1+\exp(-\zeta\hat{t}/\sqrt{1-\zeta^2})\left[\frac{(-\Omega(1-2\zeta^2)-\zeta)}{\sqrt{1-\zeta^2}}\sin(\hat{t})+(1-2\Omega\zeta)\cos(\hat{t})\right]\right| \quad 0\leq\zeta\leq\zeta_b \tag{17.26a}$$

or

$$T_U=|-2\zeta\Omega| \quad \zeta_b\leq\zeta \tag{17.26b}$$

with

$$\zeta_b=\frac{1}{4\Omega}+0.5\sqrt{(1+1/(2\Omega)^2}, \quad \Omega=\frac{V\omega_n}{g},$$
$$\hat{t}=\tan^{-1}\left[\frac{(\Omega(1-4\zeta^2)+2\zeta)\sqrt{1-\zeta^2}}{(-1+3\Omega\zeta+2\zeta^2-4\Omega\zeta^3)}\right] \tag{17.26c}$$

For intermediate damping, Equation (17.26a) corresponds to the nondimensional time value, $\hat{t}=\omega_d t_{max}$, where the slope $du/dt = 0.0$, and Equation (17.26b) corresponds to the large damping case where the maximum force occurs at the beginning of impact, $t = 0.0$. When the damping ζ is zero (or very small), then using Equation (17.26a) with $\hat{t}\approx\tan^{-1}(\Omega/(-1))$ and solving for Ω in terms of T_U, we can solve for the natural frequency that limits the maximum transmitted force to the desired T_U value, where

$$f_n\approx\frac{g}{2\pi V}\sqrt{T_U(T_U-2)} \quad \textit{for light damping, } \zeta\approx 0.0, \text{ and } T_U>2.0 \tag{17.27}$$

The $T_U > 2.0$ limitation comes from the fact that *with zero or low damping*, fragility factors lower than 2.0 are not reachable without substantial damping. In the case of heavy damping, the maximum force occurs at the beginning, $t = 0.0$, and using Equation (17.26b), one obtains:

$$f_n \approx \frac{gT_U}{4\pi V\zeta} \quad \textit{for heavy damping, } \zeta_b \leq \zeta \tag{17.28}$$

where the breakpoint ζ_b is determined by setting the numerator of the arctan expression for $\hat{t}$ equal to zero. Finally the determination of the f_n value in terms of T_U for the intermediate damping, $0 \leq \zeta \leq \zeta_b$, is the most difficult case because the solution involves obtaining roots to a transcendental equation. It is noted that for large values $\Omega \geq 10.0$, the breakpoint value is $\zeta_b \approx 0.5$. Use the smallest positive root for $\tan^{-1}$ in Equation (17.26c) (e.g., atan2 in MATLAB).

The procedure is numerical and can easily be accomplished by substituting the desired damping ratio ζ and fragility factor T_U into Equation (17.26a) and iterating Ω until the left side equals the right side (this root, $\Omega = \Omega_{rt}$, can be found with a simple numerical root finder routine such as the "fzero" routine in MATLAB). Next, simply convert Ω_{rt} to f_n using the second of Equations (17.26c), to get

$$f_n = g\Omega_{rt}/(2\pi V) \quad \textit{for intermediate damping, } 0 \leq \zeta \leq \zeta_b \tag{17.29}$$

To illustrate the general drop application for intermediate damping, consider a 50 lb weight ($M = 50/g$) that is dropped a height $H = 6.00$ in. (i.e., $V = 68.06$ in./sec), where it is required that T_U be no bigger than 8.0, with a prescribed intermediate damping ratio $\zeta = 0.2$. Next insert these data into Equation (17.26a), solve for the root $\Omega_{rt} = 8.870$, and finally compute the isolator natural frequency $f_n = 8.0066$ Hz with Equation (17.29). The breakpoint value ζ_b must be checked to ensure that the inequality bounds of Equation (17.29) are not violated; thus, substituting $\Omega = \Omega_{rt} = 8.870$ into the first of Equations (17.26c) gives $\zeta_b = 0.5290$, which is above the $\zeta = 0.2$ value required by the constraint bounds for intermediate damping. These f_n and ζ data translate into $K = (2\pi f_n)^2 M = 327.82$ lb/in. and $C = 4\pi M\zeta f_n = 2.607$ lb-sec/in. Evaluating the dynamic response as described above with Equation (17.8), Equation (17.9), and Equation (17.4) results in the response illustrated in Figure 17.8 which, as indicated, has peak $|u(t)/F_{max}| = 8.0$. Note that the isolator force turns into tension after 0.062 sec, which implies that the structural configuration will jump up off the floor (i.e., "pogo stick effect") at a later time after the energy is absorbed. It is also noted that the $x(t)$ solution does not settle down to zero, but rather to the static deflection value; this is because Figure 17.1(a) rather than Figure 17.1(b) was the coordinate reference configuration for this free-fall problem. For comparison purposes, the solution generated with Equation (17.8), Equation (17.9), and Equation (17.4) is compared with a solution to the same problem in Church, 1963 and is indicated by the unconnected symbolic markers "*" for displacement, "o" for velocity, and "+" for the $|u(t)/F_{max}|$ ratio (after adjusting for a different sign convention for positive x). As can be seen, the agreement between the two solutions is perfect.

17.3 Additional Information

The sample problems considered here represent simple single-degree-of-freedom (SDF) models of what is often a more complicated multiple-degree-of-freedom (MDF) system. These simple example problems illustrate the main concepts involved in shock and vibration isolation; however, great care must be taken not to apply these idealized SDF-type models in situations where a MDF model is needed to represent the full picture (e.g., the four-wheel vehicle over a roadway needs a MDF model that allows the front and rear wheels to experience different parts of the roadway profile to allow for any rocking modes that may be present). Perhaps one of the most comprehensive references for shock and vibration isolation

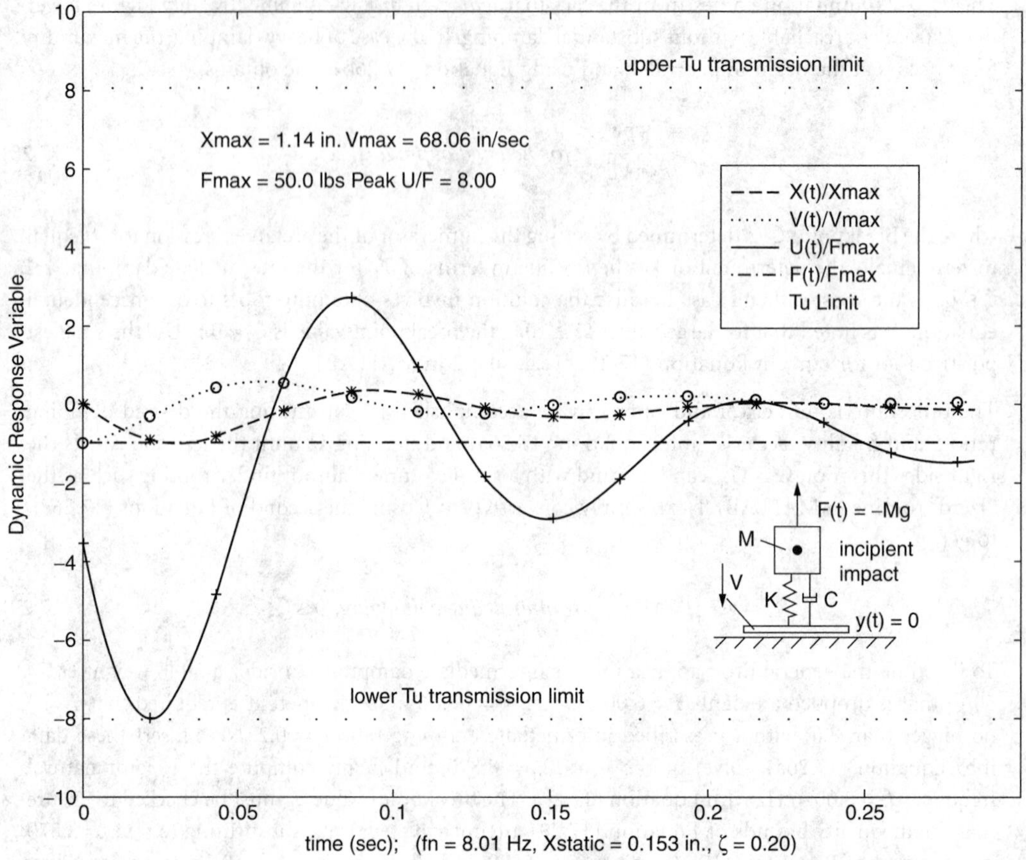

FIGURE 17.8 Free-falling structural configuration shock.

issues is Harris, 1988, which not only covers SDF models in detail for all kinds of inputs, but covers MDF models as well. Good design tips are often found directly in the manufacturer's design guides, for example, Aeroflex, 1994; Barry, 1993; Lord, 1994; and Firestone, 1994. For some more advanced ideas on optimum shock and vibration isolation concepts, see Sevin and Pilkey, 1971. The effects of nonlinearity should also be considered when necessary; for example, the simple Voigt model linear spring-damper isolators considered here eventually turn nonlinear when the deflections become large. In such cases the actual nonlinear isolator force function $u(x, \dot{x})$ could be obtained from the manufacturer and Equation (17.1) resolved as a nonlinear differential equation (which could also easily be solved with a MATLAB script file similar to the one used to generate the sample solutions).

For the analysis of MDF systems consisting of a collection of rigid bodies interconnected by any "Rube Goldberg" collection of linkages, pivot points, and so forth, a handy computer program called Working Model exists [Knowledge Revolution, 1994], enabling the user to geometrically construct the configuration to be analyzed on the computer screen. It is not unlike the object-drawing programs that come with practically all modern-day word processors. The key difference is that once the model is drawn on the screen, following the assignment of initial conditions, spring constants, damping coefficients, and loading via pulldown menus, the user is one "mouse click" away from setting the program to work and getting a graphical display of selected pertinent response variables (displacement, velocity, acceleration) as the solution is in progress. Versions exist for a variety of platforms, including Macintoshes, PCs, and workstations. Depending on the user's on-screen drawing skills, an entire analysis of a rather complex MDF system can be completed in a very short time (e.g., 5 to 10 min).

17.4 Base Motion Excitation for Multiple D.O.F. Problems

In problems 1 and 2, the base excitation was a prescribed vertical motion y(t). However, the model for the physical representation was a simple single D.O.F. For more complex structures, many degrees of freedom will of course be acting, and it is then useful to employ finite element methods to solve the problem. For example, in the Figure 17.6 inset sketch, the base could represent a building foundation, and the structural configuration above it could be a finite element representation of the building structure. The loading could still be the base excitation y(t) and would correspond to the vertical base excitation of, say, some reference earthquake data or to some prescribed shock base motions for shock and vibration testing of mounted equipment. The equations of motion look like Equation (17.5), except that the M, K, and C are now finite element method mass, stiffness, and damping matrices, respectively, and the relative displacement x(t) is replaced by a vector of unknown {Z(t)} *total* displacements and the y(t) prescribed base motion is replaced by a vector of known driver {Y(t)} nodal displacements. Following Przemieniecki [1969], for a treatment of enforced displacement problems, the full displacement vector is partitioned into an unknown and known partitions, $[\{Z(t)\}, \{Y(t)\}]^T$. We start the analysis without damping present and will add its effect on the undamped solution later. Thus the equations of motion become

$$\begin{bmatrix} M_{zz} & M_{zy} \\ M_{yz} & M_{yy} \end{bmatrix} \begin{Bmatrix} \ddot{Z} \\ \ddot{Y} \end{Bmatrix} + \begin{bmatrix} K_{zz} & K_{zy} \\ K_{yz} & K_{yy} \end{bmatrix} \begin{Bmatrix} Z \\ Y \end{Bmatrix} = \begin{Bmatrix} 0 \\ F_y \end{Bmatrix} \tag{17.30}$$

where F_y corresponds to the yet-unknown forces of constraint causing the known displacements {Y(t)} of the base structure (e.g., this motion could be an earthquake history or a base motion from a shaker table or shock machine). Expanding the top partition, we have:

$$[M_{zz}]\{\ddot{Z}\} + [K_{zz}]\{Z\} = \{\bar{F}_z\} \tag{17.31}$$

where

$$\{\bar{F}_z\} = -[M_{zy}]\{\ddot{Y}\} - [K_{zy}]\{Y\} \tag{17.32}$$

Direct Solution

The direct solution subject to the unknown structure nodal time history {Z(t)} would be to integrate this system of coupled ordinary differential equations with a standard second order in time, integration scheme (see, for example, Bathe [1996] and Cook [1988]), subject to, say, zero displacement and zero velocity initial conditions. Once having the total displacement vector {Z(t)}, the strains and stresses are then post computed. Also, the forces of constraint, $\{\bar{F}_z\}$, are post computed with the lower half of Equation (17.30). In order to account for damping, we simply add a damping term on the l.h.s. of Equation (17.31) of the form $[C_{zz}]\{\dot{Z}\}$, and correspondingly add a term - $[C_{zy}]\{\dot{Y}\}$ to the r.h.s. of Equation (17.32) and integrate the damped system of differential equations in the same manner.

Modal Solution

The goal of the modal solution is to simplify the solution so that it does not require integrating coupled equations of the entire system equations like that of the above direct approach, but rather obtains a set (one for each mode) of uncoupled second-order differential equations in time, of the form of Equation (17.5) (specifically, rewrite Equation (17.5)) by dividing it by M to obtain:

$$\ddot{x} + 2\omega_n \varsigma \dot{x} + \omega_n^2 x = F(t)/M - \ddot{y} \tag{17.33}$$

A second reason is a byproduct of the uncoupled approach and corresponds to the case where the user is given a "shock spectrum" [Cunniff, 1989; Geers, 1998; Remmers, 1983], which is the response to a single degree of freedom, subject to a base motion {Y(t)} (e.g., like Problem 1 and Problem 2 earlier).

With limited space, we can only outline the procedure; the reader is referred to Cook [1988], for detailed derivations of what is called the *mode displacement method.* The first step is to temporarily leave off the damping in Equation (17.31) and extract the normal undamped modes of vibration of Equation (17.31), with the r.h.s. set = 0, thus producing $I = 1,2,\ldots I_{max}$ normalized Eigenfunctions $\{\Phi_i\}$, stored columnwise in a matrix $[\Phi] = [\{\Phi_1\}, \{\Phi_2\}, \ldots \{\Phi_i\}]$, and Eigenvalues ω_I associated with each mode, $\{\Phi_i\}$, where I_{max} is the length of the displacement vector {Z}. The modes can be extracted using any standard finite element computer code with a structural dynamics option such as ABAQUS, or NASTRAN, or for smaller problems using MATLAB, in conjunction with m-file scripts code found in [Hwon, 2000]. Since damping is left out, these are not the actual damped modes of the system (which typically are complex), but rather modes that are used as a change in variable to uncouple the system of coupled equations Equation (17.5). We introduce a change in variable $\{Z\} = [\Phi]\{\chi\}$, where $\{\chi\}$ is defined as the new *generalized variable* unknown. Next substitute this variable change into the damped version of Equation (17.31) and into the damped version of the driver Equation (17.32). The new differential equations will now be with respect to $\{\chi\}$ and its derivatives on time, but so far the equations will still be coupled. The usual step for uncoupling the equations is to premultiply the new transformed equation by $[\Phi]^T$ and then take advantage of the orthogonality of the Eigenfunctions. When there is no damping matrix $[C_{zz}]$ or if damping is present and can be represented as *Rayleigh damping* (also known as *proportional damping*), i.e.,

$$\ddot{x} + 2\omega_n \varsigma \dot{x} + \omega_n^2 x = F(t)/M - \ddot{y} \tag{17.34}$$

(where α and β are user-defined constants), then we can simplify the equations on χ, resulting in the set of uncoupled equations for each mode "i".

$$\ddot{\chi}_i + 2\omega_i \varsigma_i \dot{\chi}_i + \omega_i^2 \chi_i = f_i(t) \qquad i = 1,2,\ldots i_{max} \tag{17.35}$$

where χ_i is the ith row variable in the transformation variable $\{\chi\}$, ζ_i = is the ith row constant in the proportional damping *diagonal matrix*

$$diag[\varsigma_i] \equiv [\Phi]^T [C_{zz}][\Phi] = [\alpha[\Phi]^T [K_{zz}][\Phi] + \beta[\Phi]^T [M_{zz}][\Phi]]$$

and the driver is given by $f_i(t) = \{\Phi_i\}^T \{F_z\}$, where the damping term $[C_{zy}]\{\dot{Y}\}$ is in place in the r.h.s. of Equation (17.32). The final step is to integrate these I_{max} equations, producing each row entry of the, χ_i, of the $\{\chi\}$ generalized column matrix. Once these have been found, the desired actual displacement vector {Z} can be found by back-substituting into the transformation equation to obtain:

$$\{Z(t)\} = \sum_{i=1}^{i=I_{cut}} \chi_i(t)\{\Phi_i\} \tag{17.36}$$

where the inequality $I_{cut} \le I_{max}$ denotes the notion that in many cases the higher modes in frequency are truncated from the final construction of the solution at a mode cutoff I_{cut}, for modes that are ordered from low to high frequency.

Integration of Uncoupled Equations

There are three options here worth noting, regarding the integration of Equations (17.35)

1. Simply integrate Equations (17.35) for each i, using a direct explicit or implicit numerical time integration scheme for ordinary differential equations.
2. Note that, remarkably, the l.h.s. of Equation (17.36) looks just like the simple single-degree-of-freedom damped harmonic motion oscillator of Equation (17.33). Therefore, by comparing like terms in these two differential equations, we can directly use Equation (17.8) and Equation (17.9) to obtain the solution for each i, where simple integration quadratures (e.g., *Simpson's rule*) can be used to obtain the solution in place of solving differential equations. Simply make the following correspondence between Equation (17.8) $\Leftrightarrow$ Equation (17.33), where: $x(t) \leftrightarrow \chi_i(t)$; $\omega_n \leftrightarrow \omega_i$; $\zeta \leftrightarrow \zeta_i$; $F(t)/M \leftrightarrow f_i(t)$; $\ddot{y} \leftrightarrow 0$ [i.e., set $\ddot{y}=0$ in Equation (17.8)].
3. Analyst is given a *shock spectrum,* which represents a single D.O.F. response to the given motion input {Y(t)}, in lieu of the input itself as in options (1) and (2). There is not enough space to treat this in any detail, but briefly the shock spectrum maxima can then be related to the maximum of each mode contribution maximum [Remmers, 1983]. The total maximum response is typically performed by adding the contributions of each mode in a root mean square sense (i.e., the square root of the sum of the squares of the contributions of each mode for displacements, and stresses). This assumes all peaks occur at the same instant in time and squaring kills any chance of cancellation. Therefore this approach (e.g., [Geers, 1998]), "is far too conservative." In contrast, however, options (1) and (2) do take proper phasing into consideration and therefore are more accurate.

Defining Terms

C: Isolator damping constant (lb-sec/in.).
K: Isolator spring constant (lb/in.).
M: Net structural configuration mass (lb-sec/in.^{2}).
g: Acceleration of gravity (386 in./sec^2).
u: Isolator force not including static deflection (lb).
$\beta = f/f_n$: Ratio of drive frequency to natural frequency (unitless).
ζ: Critical damping ratio = $C/(4\pi M f_n)$ (unitless).
δ_s: Static deflection = Mg/K, amount spring deflects under dead weight of mass *M* (in.).
ω_n: Natural frequency (rad/sec).
$\omega_n = \omega_d\sqrt{1-\zeta^2}$: damped natural frequency (rad/sec).
η: Damping decay constants (sec^{-1}).
T_U: Force transmission ratio = peak$\left| u(t)/F_{\max} \right|$ (unitless).
T_A: Acceleration transmission ratio = peak$\left|\ddot{z}/\ddot{y}_{\max}\right|$ (unitless).
T_D: Displacement magnification ratio = peak$\left| x/y_{\max} \right|$ (unitless).
BC: Base configuration, refers to the lower isolator attachment point (where base input accelerations $\ddot{y}$ are applied).
SC: Structural configuration, refers to the net vibrating mass (upper connection to the isolator).

References

Aeroflex. 1994. *Aeroflex Isolators Selection Guide,* Aeroflex International, Inc., Plainview, NY.

Barry. 1993. *Barry Controls Bulletin DOEM1,* Barry Controls, Brighton, MA.

Bathe, K. 1996. *Finite Element Procedures,* Prentice Hall, Englewood Cliffs, NJ.

Cook, R. D., Malkus, D. S., and Plesha, M. E. 1988. *Concepts and Applications of Finite Element Analysis,* 3rd ed., John Wiley & Sons, New York.

Church, A. H. 1963. *Mechanical Vibrations,* 2nd ed. John Wiley & Sons, Inc., New York.

Cunniff, P. and O'Hara, G. O. 1989. A procedure for generating shock design values, *J. Sound Vib.,* 134, 154–165.

Firestone. 1994. *Engineering Manual and Design Guide,* Firestone Industrial Products Co., Noblesville, IN.

Geers, T. L. 1998. Shock Aanalysis and design, in *Handbook of Acoustics,* Cricker, M. J., Ed., John Wiley & Sons, New York, chap. 52.

Harris, C. M. 1988. *Shock and Vibration Handbook,* 3rd ed. McGraw-Hill Book Co., New York.

Hwon, Y. W. and Bang, H. 2000. *The Finite Element Method Using MATLAB,* CRC Press, Boca Raton, FL.

Knowledge Revolution. 1994. *Working Model Demonstration Guide and Tutorial,* San Mateo, CA.

Lord. 1994. *Lord Industrial Products Catalog,* PC-2201H. Lord Industrial Products, Erie, PA.

Math Works. 1992. *The Student Edition of MATLAB.* Prentice Hall, Englewood Cliffs, NJ.

Przemieniecki, J. S. 1968, *Theory of Matrix Structural Analysis,* McGraw-Hill, New York.

Redfern, D. 1994. *The Maple Handbook: Maple V Release 3,* Springer-Verlag, New York.

Remmers, G. 1983. Maurice Biot 50th Anniversary Lecture: the Evolution of Spectral Techniques in Navy Design, *Shock and Vibration Bull.,* Part 1.

Sevin, E. and Pilkey, W. D. 1971. *Optimum Shock and Vibration,* Monogram SVM-6. Shock and Vibration Information Center, Naval Research Laboratory, Washington, D.C.

Sigmon, K. 1994. *MATLAB Primer,* 4th ed. CRC Press Inc., Boca Raton, FL.

Wolfram, S. 1991. *Mathematica: A System for Doing Mathematics by Computer,* 2nd ed. Addison-Wesley Publishing Co., Redwood City, CA.

18

Vibration Computations and Nomographs

Daniel J. Inman
Virginia Polytechnic Institute and State University

A primary concern in performing vibration analysis is just how to represent the response once the model and the various inputs of interest are known. For linear systems with a single degree of freedom, the analytical solution is closed form and can be simply plotted to illustrate the response and its important features. Historically, computation has been difficult and response vibration data have been presented in nomographs consisting of log plots of the maximum amplitudes of displacement, velocity, and acceleration versus frequency on a single two-dimensional four-axis plot. Although this approach is useful and incorporated into military and manufacturer specifications, wide availability of high-speed computing and computer codes to simulate detailed responses has produced a trend to display exact responses in both the time and frequency domains. Advances in computer algorithms and commercialization of codes has also made the computing of natural frequencies and mode shapes much easier. The following section introduces an example of a commercial computer simulation package and its use in representing the response of vibrating systems. Methods of computing natural frequencies and mode shapes are also given. In addition, the basic use of nomographs is presented.

18.1 Models for Numerical Simulation

The most common model of a vibrating system is the multiple degree of freedom (MDOF) model, which can be expressed as a vector differential equation with matrix coefficients of the form

$$M\ddot{\mathbf{x}}(t)+C\dot{\mathbf{x}}(t)+K\mathbf{x}(t)=\mathbf{f}(t) \quad \mathbf{x}(0)=\mathbf{x}_0,\ \dot{\mathbf{x}}(0)=\dot{\mathbf{x}}_0 \tag{18.1}$$

where $\mathbf{x}(t)$ is an $n \times 1$ vector of displacement coordinates, its derivative $\dot{\mathbf{x}}(t)$ is an $n \times 1$ vector of velocities, and its second derivative $\ddot{\mathbf{x}}(t)$ is an $n \times 1$ vector of accelerations. The coefficients M, C, and K are $n \times n$ matrices of mass, damping, and stiffness elements, respectively (see Chapter 14 for examples). These coefficient matrices are often symmetric and at least positive semidefinite for most common devices and

0-8493-1586-7/05/$0.00+$1.50

structures. Equation (18.1) follows from simple modeling using Newton's laws, energy methods, or dynamic finite elements. The constant vectors $\mathbf{x}_0$ and $\dot{\mathbf{x}}_0$ represent the required initial conditions. The simulation problem consists of calculating $\mathbf{x}(t)$, satisfying Equation (18.1) as time evolves, and producing a time record of each element of $\mathbf{x}(t)$, or of each degree of freedom, as opposed to a single coordinate as used in nomographs.

Currently many very well-written numerical integration codes are available commercially for less cost than is involved in writing the code and, more importantly, with less error. Codes used to solve Equation (18.1) are written based on the definition of a derivative and almost all require the equations of motion to appear in first-order form, that is, with only one instead of two time derivatives. Equation (18.1) can be easily placed into the form of a first-order vector differential equation by some simple matrix manipulations.

If the inverse of the mass matrix M exists, then the second-order vibration model of Equation (18.1) can be written as an equivalent first-order equation using new coordinates defined by the $2n \times 1$ vector:

$$\mathbf{z}(t) = \begin{bmatrix} \mathbf{x}(t) \\ \dot{\mathbf{x}}(t) \end{bmatrix}$$

Let $\mathbf{z}_1 = \mathbf{x}(t)$ and $\mathbf{z}_2 = \dot{\mathbf{x}}(t)$; then Equation (18.1) can be written as

$$\begin{aligned} \dot{\mathbf{z}}_1 &= \mathbf{z}_2 \\ \dot{\mathbf{z}}_2 &= -M^{-1}K\mathbf{z}_1 - M^{-1}C\mathbf{z}_2 + M^{-1}\mathbf{f}(t) \\ \mathbf{z}(0) &= \begin{bmatrix} \mathbf{x}_0 \\ \dot{\mathbf{x}}_0 \end{bmatrix} \end{aligned} \tag{18.2}$$

which combine to form

$$\dot{\mathbf{z}} = A\mathbf{z} + \mathbf{F}(t), \quad \mathbf{z}(0) = \mathbf{z}_0 \tag{18.3}$$

Here $\mathbf{z}$ is called a *state vector* and the *state matrix* A is defined by

$$A = \begin{bmatrix} 0 & I \\ -M^{-1}K & -M^{-1}C \end{bmatrix} \tag{18.4}$$

where I is the $n \times n$ identity matrix and 0 is an $n \times n$ matrix of zeros. The forcing function $\mathbf{F}(t)$ is "mass" scaled to be the $2n \times 1$ vector

$$\mathbf{F}(t) = \begin{bmatrix} \mathbf{0} \\ M^{-1}\mathbf{f}(t) \end{bmatrix} \tag{18.5}$$

Here $\mathbf{0}$ denotes an $n \times 1$ vector of zeros. Note that in solving vibration problems using this state space coordinate system, the first n components of the solution vector $\mathbf{z}(t)$ correspond to the individual displacements of the n degrees of freedom and the second list of n components of $\mathbf{z}(t)$ are the velocities.

18.2 Numerical Integration

The numerical solution or simulation of the system described by Equation (18.1) is easiest to discuss by first examining the scalar homogeneous (unforced) case given by

$$\dot{x}(t) = ax(t) \quad x(0) = x_0$$

where a is a simple constant. The derivative $\dot{x}(t)$ is written from its definition as

$$\frac{x(t_i + \Delta t) - x(t_i)}{\Delta t} = ax(t_i) \tag{18.6}$$

where Δt is a finite interval of time. Rewriting this expression yields

$$x(t_i + \Delta t) = x(t_i) + ax(t_i)\Delta t \tag{18.7}$$

or using a simpler notation

$$x_{i+1} = x_i + ax_i \Delta t \tag{18.8}$$

where x_i denotes $x(t_i)$. This formula gives a value of the response x_{i+1} at the "next" time interval, given the equation's coefficient a, the time increment Δt, and the previous value of the response x_i. Thus, starting with the initial value x_0, the solution is computed at each time step incrementally until the entire record over the interval of interest is calculated. This simple numerical solution is called the **Euler formula** or tangent line method and only involves addition and multiplication. Of course, the smaller Δt is, the more accurate the approximation becomes (recall the derivative is defined as the limit $\Delta t \to 0$). Unfortunately, reducing the step size Δt increases the computational time. Numerical errors (rounding and truncation) also prevent the simulations from being perfect, and users should always check their results accordingly.

The rule used to perform the simulation is often called an *algorithm*. One way to improve the accuracy of numerical simulation is to use more sophisticated algorithms. In the late 1800s, C. Runge and M.W. Kutta developed some clever formulas to improve the simple tangent or Euler methods. Essentially, these methods examine $x(t + \Delta t)$ as a Taylor series expanded in powers of Δt. The **Runge-Kutta** methods insert extra values between x_i and x_{i+1} to provide estimates of the higher-order derivative in the Taylor expansion and thus improve the accuracy of the simulation. There are several Runge-Kutta methods, thus only an example is given here.

One of the most widely used Runge-Kutta methods solves the scalar equation $\dot{x} = f(x,t)$ with initial condition $x(0) = x_0$ where $f(x, t)$ can be linear or nonlinear as well as time varying. This includes the case $\dot{x}(t) = ax(t) + g(t)$ where a is constant and $g(t)$ is an externally applied force. With x_i and Δt defined as before, the formulas for the response are

$$x_{i+1} = x_i + \frac{\Delta t}{6}(h_{i1} + 2h_{i2} + 2h_{i3} + h_{i4}) \tag{18.9}$$

where

$$h_{i1} = f(x_i, t_i) \qquad h_{i2} = f\left(x_i + \frac{\Delta t}{2}h_{i1}, t_i + \frac{\Delta t}{2}\right)$$

$$h_{i3} = f\left(x_i + \frac{\Delta t}{2}h_{i2}, t_i + \frac{\Delta t}{2}\right) \qquad h_{i4} = f\left(x_i + \Delta t h_{i3}, t_i + \Delta t\right)$$

This is referred to as a four-stage formula and represents a substantial improvement over the Euler method.

Additional improvement can be gained by adjusting the time step Δt at each interval based on how rapidly the solution $x(t)$ is changing. If the solution is not changing very rapidly, a large value of Δt_i, the ith increment of time is used. On the other hand if $x(t)$ is changing rapidly, a small Δt_i is chosen. In fact, the Δt_i can be chosen automatically as part of the algorithm.

18.3 Vibration Response by Computer Simulation

All of these methods (as well as many others not mentioned) can be applied to the simulation of the response of vibrating systems. Essentially the Runge-Kutta and Euler formulas can be applied directly to Equation (18.3) by simply enforcing a vector notation. For instance, the Euler formula applied to Equation (18.3) becomes

$$\mathbf{z}(t_{i+1}) = \mathbf{z}(t_i) + \Delta t A\mathbf{z}(t_i) + \mathbf{F}(t_i) \tag{18.10}$$

using $\mathbf{z}(0)$ as the initial value. The result will be a list of numerical values for $\mathbf{z}(t_i)$ versus the successive times t_i. Equation (18.10) can be programmed on a programmable calculator or computer system. However, many commercially available codes provide more than adequate numerical integration schemes for solving ordinary differential equations and systems of ordinary differential equations as described by Equation (18.10). Such codes are easy to use and allow studies of the effects of initial conditions and parameter changes while providing detailed solutions to complex problems.

Next, a simple example is introduced to illustrate the formulation of a vibration problem into state space form in preparation for numerical simulation. Consider then the equations of motion of a damped two-degree-of-freedom system

$$\begin{bmatrix} 9 & 0 \\ 0 & 1 \end{bmatrix}\begin{bmatrix} \ddot{x}_1(t) \\ \ddot{x}_2(t) \end{bmatrix} + \begin{bmatrix} 2.7 & -0.3 \\ -0.3 & 0.3 \end{bmatrix}\begin{bmatrix} \dot{x}_1 \\ \dot{x}_2 \end{bmatrix} + \begin{bmatrix} 27 & -3 \\ -3 & 3 \end{bmatrix}\begin{bmatrix} x_1 \\ x_2 \end{bmatrix} = \begin{bmatrix} 3 \\ 0 \end{bmatrix}\sin 3t \tag{18.11}$$

subject to the initial conditions

$$\mathbf{x}(0) = \begin{bmatrix} 0.1 \\ 0 \end{bmatrix}, \quad \dot{\mathbf{x}}(0) = \begin{bmatrix} 0 \\ 0 \end{bmatrix}$$

Here all units are SI so that $\mathbf{x}(t)$ is in meters, etc. However, any consistent set of units can be used. This is a simple example with only two degrees of freedom so chosen to fit the given space limitations. The procedure is, however, not dependent on such low order. The matrix M^{-1} in this case is simply

$$M^{-1} = \begin{bmatrix} \frac{1}{9} & 0 \\ 0 & 1 \end{bmatrix}$$

so that

$$M^{-1}K = \begin{bmatrix} 3 & -0.333 \\ -3 & 3 \end{bmatrix} \quad \text{and} \quad M^{-1}C = \begin{bmatrix} 0.3 & -0.033 \\ -0.3 & 0.3 \end{bmatrix}$$

and the state matrix becomes

$$A = \begin{bmatrix} 0 & 0 & 1 & 0 \\ 0 & 0 & 0 & 1 \\ -3 & 0.333 & 0.3 & 0.033 \\ 3 & -3 & 0.3 & -0.3 \end{bmatrix}$$

where the state vector and forcing vector are

$$\mathbf{z} = \begin{pmatrix} x_1 \\ x_2 \\ \dot{x}_1 \\ \dot{x}_2 \end{pmatrix} \quad \text{and} \quad \mathbf{F}(t) = \begin{bmatrix} 0 \\ 0 \\ \sin 3t \\ 0 \end{bmatrix}$$

respectively. Next, Equation (18.10) can be applied with these values. In the past it was required of the vibration engineer to program Equation (18.10) or some other versions of it. However, commercial software allows the matrix A to be easily assembled and allows easy simulation of the time response and plotting the frequency response.

18.4 Commercial Software for Simulation

A variety of simple-to-use, efficient, and relatively inexpensive interactive software packages are available for simulating the response. Such programs reduce by a factor of ten the amount of computer code that actually has to be written by the user. Some examples of available software containing numerical integration packages are MALAB®, Mathematica, and Mathcad. Many finite element packages also contain numerical integration routines. Here we illustrate the use of MALAB to simulate the result of the simple example above and to print the results. The MALAB code is listed in Table 18.1 and the output is plotted in Figure 18.2. The algorithm used in Table 18.1 is a modification of the formulas given in Equation (18.9), known as a Runge-Kutta-Fehlberg integration method. This method uses a fourth- and fifth-order pair of formulas and an automatic step size.

The use of high-level computational software such as MALAB, Mathcad, and Mathematica is commonplace. The slide rule has given way to the calculator, and the calculator to the personal computer. Combined with commercial software, the simulation of large and complex vibration problems can be performed without resorting to writing code in lower-level languages. This time-saving allows the vibration engineer more time to devote to design and analysis. It is, however, important to note that simulation through numerical integration is still an approximation and as such is subject to error — both **formula errors** and **round-off errors**. These should be well understood by the user. More detailed examples of these three codes used to solve vibration problems can be found in Inman [2001].

These same computer codes also contain sophisticated routines for obtaining the solutions to the eigenvalue problems associated with vibration and hence are very useful in obtaining the mode shapes and natural frequencies. These methods are discussed next.

18.5 Computing Natural Frequencies and Mode Shapes

Natural frequencies and mode shapes form the language of modern vibration analysis and are often used among practicing engineers and researchers to describe the vibration characteristics of a structure or machine. Computing natural frequencies and mode shapes use to involve many approximate techniques. The rise of modern computing and advances in numerical linear algebra and subsequent commercialization have made the computing of natural frequencies and mode shapes relatively easy. However,

TABLE 18.1 MATLAB SIMULATION Code

MATLAB code for computing and plotting the displacement versus time response of the two-degree-of-freedom system in the text. The % symbol denotes comments. Part A indicates how to input the given system and part B illustrates how to integrate and plot the response.

Part A

```
function zdot = system (t, z)
%This m-file defines the mechanical properties of the system being studied.
%The input is the current time and the previous state vector (initial conditions).
%The output is obtained by solving the state equation: zdot = A * z + f
%First, the mass, damping and stiffness matrices are defined.
M =  [9, 0; 0, 1];
C = [2, 7, -0.3; -0.3, 0.3];
K = [27, -3; -3, 3];
f = [0; 0; sin(3 * t); 0];
%The vector of external forces is defined next.
%The state matrix is assembled.
A = [zeros(2, 2) eye(2, 2); -inv(M) * K -inv(M) * C];
```

Part B

```
%This is the main file used in the simulation.
%First the initial state is defined.
z0 = [0.1; 0; 0; 0];
%Then, the solution is obtained through numerical integration using the ode command
calling the "system" input file proposed in part A.

ti = 0;                                             %Initial time of the simulation
tf = 50;                                            %Final time of the simulation
[time,solution] = ode45('system', ti, tf, z0); %Perform integration

%The displacement of each degree-of-freedom is plotted.
subplot (2,1,1);
plot (time,solution(:,1));
xlabel('t[s]');
ylabel('z1[m]');
title('First DOF;);

subplot (2,1,2);
plot(time,solution(:,2));
xlabel('t[s]');
ylabel('z2[m]');
title('Second DOF');
```

commercial codes have centered on implementing numerical linear algebra results, i.e., implementing solutions of the algebraic eigenvalue problem, rather than focusing on vibration problems directly. Thus, computations of natural frequencies and mode shapes are best done by converting the definition of these physical quantities into the algebraic eigenvalue problem. There are several ways to make this transformation and these are presented here along with their strengths and weaknesses.

Undamped systems are systems modeled by Equation (18.1) with no damping present (C = 0). Natural frequencies are defined as the frequencies at which a system will vibrate given an arbitrary initial displacement and velocity disturbance. Undamped natural frequencies (rad/sec) are denoted ω_n, and there will be one frequency for each degree of freedom in a linear system such as those successfully modeled by Equation (18.1) with $C = 0$. Natural frequencies are calculated from assuming a solution to Equation (18.1), for the undamped case of $C = 0$, of the form $\mathbf{x}(t) = \mathbf{u}e^{j\omega t}$, where $\mathbf{u}$ is an unknown vector of constants, $j = \sqrt{-1}$, t is the time, and ω is the yet-to-be-determined natural frequency. Substitution of this assumed solution into Equation (18.1) yields:

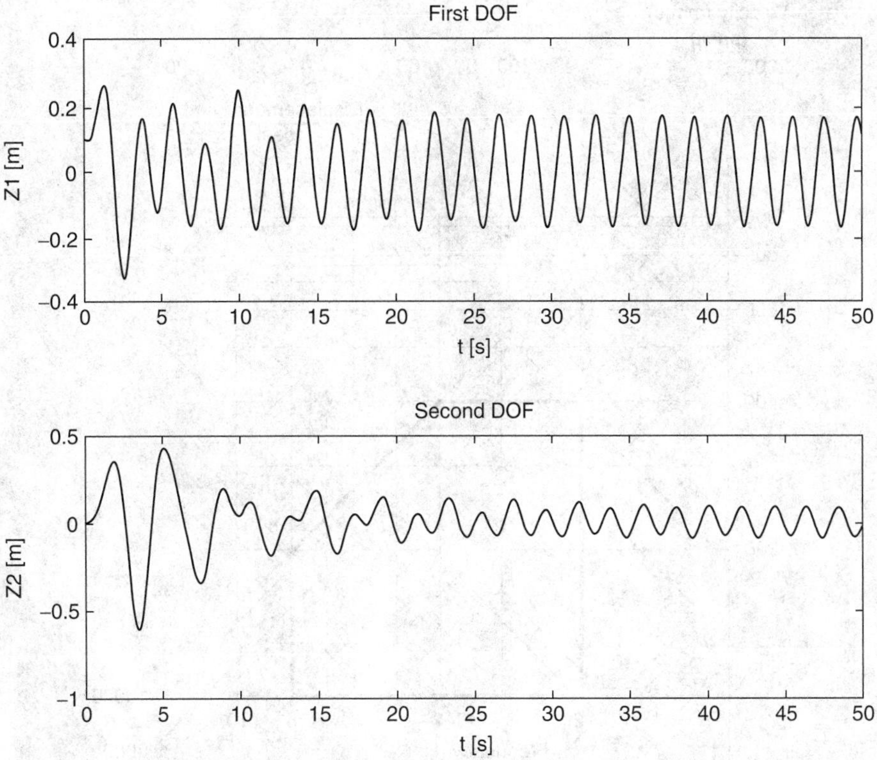

FIGURE 18.1 The output of the MALAB Code of Table 18.1, illustrating the simulation of the displacement versus time response of the system given by Equation (18.11).

$$\left(-\omega^2 M + K\right)\mathbf{u} = 0, \quad \mathbf{u} \neq \mathbf{0} \tag{18.12}$$

This last expression can only hold true if the coefficient matrix is singular, which happens if and only if the determinant of $(-\omega^2 M + K)$ vanishes. This yields an nth order polynomial in ω^2, the roots of which are the n values of the squares of the natural frequencies ω_i^2 indexed $i = 1, 2, 3, \ldots n$. The yet unknown vectors $\mathbf{u}_i$ corresponding to each ω_i are called the mode shapes, and they are found from solving the n scalar equations given by Equation (18.12) for each of the n values of ω_i^2. Since the coefficient is singular (because its determinant is zero) at each value of ω_i^2, Equation (18.12) will result in only $n - 1$ independent equations for each element of the mode shape vector $\mathbf{u}_i$. This allows for the solution of all but one value of the n elements of each mode shape vector. The nth value is computed by forcing the vector to have a unit magnitude (normalized). Rewriting Equation (18.12) reveals its relationship to the generalized eigenvalue problem:

$$K\mathbf{u}_i = \omega_i^2 M\mathbf{u}_i \Leftrightarrow A\mathbf{u}_i = \lambda_i B\mathbf{u}_i \quad \mathbf{u}_i \neq 0 \tag{18.13}$$

The generalized eigenvalue problem, stated on the right side of the above, is to compute $\mathbf{u}_i$ and λ_i given the matrices $A = K$ and $B = M$. This is exactly equivalent to computing mode shapes and natural frequencies. This is a symmetric generalized eigenvalue problem, since the matrices M and K are both symmetric.

The algorithms for computing the generalized eigenvalue problem are more costly then those of the normal eigenvalue problem, which can be obtained by multiplying Equation (18.13) by the inverse of the mass matrix M (or B) to get

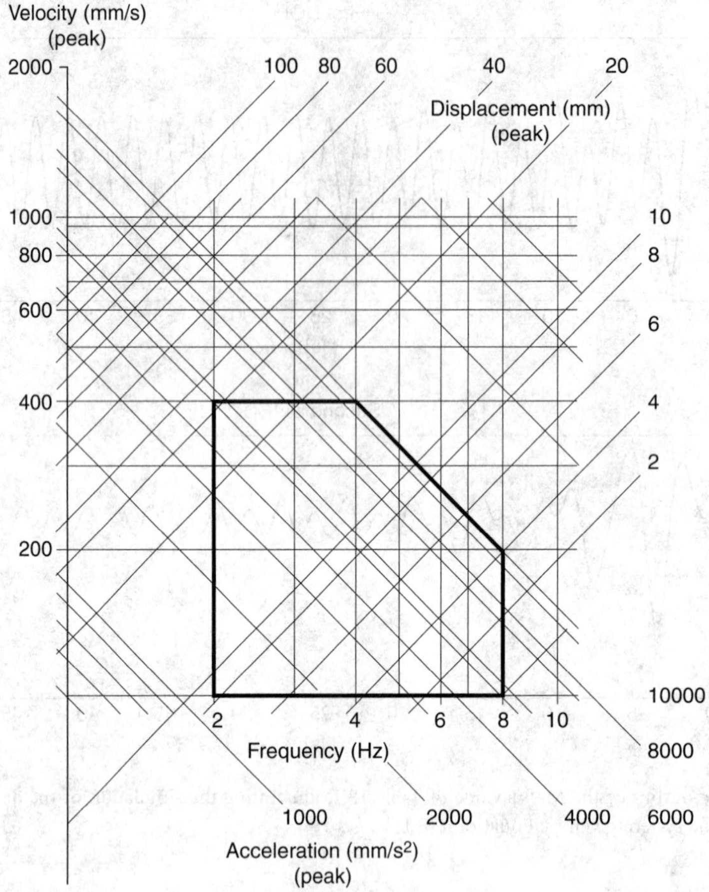

FIGURE 18.2 An example of a vibration nomograph for specifying acceptable limits of sinusoidal vibration. (*Source:* Inman, D. J. 2001. *Engineering Vibration, 2nd Edition*, Prentice Hall, Upper Saddle River, NJ. With permission.)

$$M^{-1}K\mathbf{u}_i = \omega_i^2\mathbf{u}_i \Leftrightarrow A\mathbf{u}_i = \lambda_i\mathbf{u}_i \quad \mathbf{u}_i \neq 0 \tag{18.14}$$

Here the matrix A in the statement of the eigenvalue problem is now $A = M^{-1}K$, where M^{-1} denotes the matrix inverse. The mode shapes are given directly by the vector **u** but since A is not symmetric there is no guarantee that the eigenvectors will be real valued or orthogonal (in fact they are not), nor is it obvious that the values of $\lambda_i = \omega_i^2$ will be nonnegative real numbers, as we expect for the vibration problem.

If the eigenvalue problem of Equation (18.13) is transformed by mass normalizing the stiffness matrix, a symmetric eigenvalue problem results. There are two strong advantages to using the symmetric eigenvalue problem. The first is that symmetric eigenvalue problems are known to produce orthogonal eigenvectors, even in the case of repeated frequencies, and the second is that the numerical procedure for computing the solution to the symmetric eigenvalue problem is not as subject to round-off error as the asymmetric eigenvalue problem is. To this end, assume the matrix M is positive definite and symmetric (it usually is), so that it has a positive definite matrix square root. Let $\mathbf{u} = M^{-1/2}\mathbf{v}$ in Equation (18.13) and multiply the equation by $M^{-1/2}$. This yields:

$$M^{-1/2}KM^{-1/2}\mathbf{v}_i = \omega_i^2\mathbf{v}_i \Leftrightarrow A\mathbf{v}_i = \lambda_i\mathbf{v}_i \quad \mathbf{v}_i \neq 0 \tag{18.15}$$

where the matrix $A = \tilde{K} = M^{-1/2}KM^{-1/2}$, called the mass normalized stiffness matrix, is now symmetric, and the corresponding eigenvalue problem is symmetric. From the theory of eigenvalue problems (matrix theory or linear algebra) it is well known that the solution of a symmetric eigenvalue problem results in a set of n real valued, orthogonal eigenvectors, one for each value of λ_i (regardless of repeated eigenvalues) and that the eigenvalues are all real valued and nonnegative if the matrix A is positive semidefinite (which it is in this case). This is precisely the theory needed to ensure analytical solutions such as modal expansion methods (since the $\mathbf{v}_i$ form an orthonormal basis) and in computing the time response given initial conditions. In addition, the computational algorithms for symmetric eigenvalue problems are considerably simplified compared with those needed for Equation (18.14). In particular, all the calculations assume real rather than complex arithmetic. Note that while the eigenvalues are the squares of the natural frequencies in all three eigenvalue problems listed above, the mode shapes defined from Equation (18.14) are different from the eigenvectors of Equation (18.15) and are related by $\mathbf{u} = M^{-1/2}\mathbf{v}$. In particular, the mode shapes computed by Equation (18.14) are not orthogonal as often stated in many texts, but rather are only orthogonal when weighted by the mass matrix.

This symmetric method can also proceed by using the Cholesky factors of M rather than the square root of the mass matrix. A symmetric positive definite matrix M has an upper triangular matrix (all elements below the diagonal are zero) factor L such that $L^T L = M$. If M is diagonal, then the factor L is identical to the square root of M as used in Equation (18.15). For dynamically coupled systems (i.e., those with nondiagonal mass matrices), it is more efficient to compute the mass normalized stiffness matrix in Equation (18.15) using the Cholesky factors rather than the matrix square root. The mass normalized stiffness matrix then becomes $\tilde{K} = (L^T)^{-1}KL^{-1}$. This version of the mass normalized stiffness matrix is used as the matrix A in Equation (18.15) to compute the eigenvalues and eigenvectors. (Note that some codes use a lower triangular Cholesky factor rather than upper, causing the transpose to flip sides.)

Proportionally damped systems form the most commonly used damped case. Computing natural frequencies and mode shapes when viscous damping is present, $C \neq 0$, becomes much more complicated and bifurcates into two distinct cases. The first case, often referred to as proportional damping, occurs if and only if the coefficient matrices (assumed to be symmetric) form a product $CM^{-1}K$ that is in turn symmetric. Then the eigenvectors calculated from the (undamped) symmetric eigenvalue problem of Equation (18.15) become the eigenvectors of the damped expression as well and satisfy ($\tilde{C} = M^{-1/2}CM^{-1/2}$):

$$\left(\lambda^2 I + \lambda\tilde{C} + \tilde{K}\right)\mathbf{v} = 0, \quad \mathbf{v} \neq 0 \tag{18.16}$$

where I denotes the $n \times n$ identity matrix. The matrix $\tilde{C} = M^{-1/2}CM^{-1/2}$ is called the mass normalized damping matrix. Here, λ_i are the system eigenvalues which are in general complex (for underdamped systems), and these are related to the natural frequency and damping ratio by:

$$\omega_i = \sqrt{\mathrm{Re}(\lambda_i)^2 + \mathrm{Im}(\lambda_i)^2}, \quad \zeta_i = \frac{-\mathrm{Re}(\lambda_i)}{\sqrt{\mathrm{Re}(\lambda_i)^2 + \mathrm{Im}(\lambda_i)^2}} \tag{18.17}$$

Not many numerical algorithms exist for solving Equation (18.16) for λ_i and $\mathbf{v}_i$ directly from Equation (18.16). However, because the assumption of symmetry of $CM^{-1}K$, the eigenvectors of the mass normalized stiffness matrix are also eigenvectors of the mass normalized damping matrix. Hence the eigenvectors of Equation (18.15) serve as the system eigenvectors for Equation (18.16), and the undamped natural frequencies are also computed from Equation (18.15). Hence, the natural frequencies and eigenvectors for Equation (18.16) are best calculated using Equation (18.15).

To compute the damping ratios, the system of Equation (18.16) is transformed into a diagonal system using the eigenvectors to form the orthogonal matrix P constructed by taking the columns of P to be the orthonormal vectors $\mathbf{v}_i$:

$$P = \begin{bmatrix} \mathbf{v}_1 & \mathbf{v}_2 & \cdots & \mathbf{v}_n \end{bmatrix} \tag{18.18}$$

Substitution of $\mathbf{v} = P\mathbf{r}$ into Equation (18.16) and multiplying P^T yields

$$(\lambda^2 I + \text{diag}(2\zeta_i\omega_i)\lambda + \text{diag}(\omega_i))\mathbf{r} = 0 \tag{18.19}$$

Here the vector $\mathbf{r}$ is called the modal coordinate system and the coefficient matrix of $\mathbf{r}$ is diagonal. The properties of the orthonormal matrix P for the case of proportional damping are

$$P^T P = I \quad P^T \tilde{K} P = \begin{bmatrix} \omega_1^2 & 0 & \cdots & 0 \\ 0 & \omega_2^2 & \cdots & 0 \\ \vdots & \vdots & \ddots & \vdots \\ 0 & 0 & \cdots & \omega_n^2 \end{bmatrix} \quad P^T \tilde{C} P = \begin{bmatrix} 2\zeta_1\omega_1 & 0 & \cdots & 0 \\ 0 & 2\zeta_2\omega_2 & \cdots & 0 \\ \vdots & \vdots & \ddots & \vdots \\ 0 & 0 & \cdots & 2\zeta_n\omega_n \end{bmatrix} \tag{18.20}$$

This last expression is used to compute the modal damping ratios ζ_i given the matrices M, C, and K. In practice, the modal damping ratios are guessed based on material measurements or measured using a vibration test.

The symmetry of $CM^{-1}K$ is a necessary and sufficient condition for the above analysis and results from the linear algebra theory that two matrices share the same eigenvectors if and only if they commute. However, other, simpler conditions are used in practice. These are

$$\begin{gathered} C = \alpha M + \beta K \\ C = \sum_{i=1}^{n} \alpha_{i-1} K^{i-1} \end{gathered} \tag{18.21}$$

Here the constants α, α_i, and β are arbitrary scalars chosen by the analyst so that C is positive semidefinite and to produce some reasonable amount of damping. If these conditions are satisfied, then the eigenvectors, damping ratios, and natural frequencies can be determined by Equation (18.15) and Equation (18.20). Such systems are said to possess classical normal modes (because the eigenvectors are real valued and form an orthonormal basis). They are also called normal mode systems.

General viscous damping refers to the situation when viscous damping is present but not of the proportional type. If the matrix $CM^{-1}K$ is not symmetric, then the above analysis fails and the system is said not to be proportionally damped and in general (assuming all modes to be underdamped) will have complex mode shapes. The only choice to compute the modal information is then to use the state space matrix of Equation (18.4). The associated eigenvalue problem is not symmetric and is given by

$$A\mathbf{z} = \lambda\mathbf{z} \tag{18.22}$$

Here the eigenvector $\mathbf{z}$ is related to the eigenvector in physical coordinates of Equation (18.12) by

$$\mathbf{z} = \begin{bmatrix} \mathbf{u} \\ \lambda\mathbf{u} \end{bmatrix} \tag{18.23}$$

Here both the eigenvectors $\mathbf{z}$ and the eigenvalues λ are in general complex numbers appearing in complex conjugate pairs. The relationships of these eigenvalues computed from Equation (18.22) to the natural frequencies and damping ratios are given by Equations (18.17). The mode shapes are a little more difficult to compute but are basically the first n values of the vector $\mathbf{z}$ of Equation (18.23).

TABLE 18.2 Comparison of Eigenvalue Calculation Time Using MATLAB

Method	Flops
Equation (18.13)	417
Equation (18.14)	191
Equation (18.15)	208
Cholesky	97
State matrix	2626

TABLE 18.3 MATLAB EIGENVALUE/EIGENVECTOR Codes

```
>%Enter the mass and stiffness matrices
>M=[9 0;0 1];K=[27 -3;-3 3];
>%Compute the cholesky decomposition and the mass normalized stiffness
>L=chol(M);S=inv(L);Kh=S'*K*S;
>%Compute the eigenvalues store in D and the P matrix
>[P,D]=eig(Kh)
P =
  -0.7071 -0.7071
  0.7071 -0.7071
D =
  4 0
  0 2
```

Next, the simple example of Equation (18.11) is used to illustrate the various eigenvalue problems and their solution in MATLAB. First note that this example has proportional damping since

$$CM^{-1}K = \begin{bmatrix} 2.7 & -0.3 \\ -0.3 & 3 \end{bmatrix} \begin{bmatrix} 1/9 & 0 \\ 0 & 1 \end{bmatrix} \begin{bmatrix} 27 & -3 \\ -3 & 3 \end{bmatrix} = \begin{bmatrix} 9 & -1.8 \\ -1.8 & 1 \end{bmatrix}$$

is obviously symmetric. Thus the eigenvectors and mode shapes of the undamped system serve as the eigenvectors and mode shapes of the damped system. When one proceeds to compute the solution of the eigenvalue problem (for $C = 0$) using a code, the question naturally arises: is there a numerically superior method? Table 18.2 illustrates the results of computing the eigenvalues and eigenvectors for each of the eigenvalue problems listed above applied to this simple example. The "flops" command (no longer available in MATLAB) is used to indicate how many floating-point operations are used in each calculation. Note that the most expensive choice of computing the eigenvalues and eigenvectors is using the state space formulation of Equation (18.22) followed by the generalized eigenvector problem of Equation (18.13). The cheapest computationally uses the Cholesky form of the symmetric eigenvalue problem of Equation (18.15). It should be noted that while using Equation (18.14) is relatively low, it does not result in orthogonal eigenvectors, and additional steps are required to solve the associated vibration problem. A sample code is given in Table 18.3. Detailed examples in MATLAB, Mathematica, and Mathcad for vibration analysis can be found in Inman [2001]. For simple small-order problems, the differences between eigenvalue problems is numerically not too significant. However, for practical, larger-order systems or for systems with order of magnitude differences in mass and stiffness terms, the choice of eigenvalue problem formulation could become critical.

18.6 Nomograph Fundamentals

Nomographs are graphs used to represent the relationship between displacement, velocity, acceleration, and frequency for vibrating systems. These graphs are frequently used to represent vibration limits for given parts, machines, buildings, and components. The basic premise behind a vibration nomograph is that the response of a system is harmonic of the form $x(t) = A\sin\omega_n(t)$. Here A is the amplitude of vibration and ω_n is the natural frequency of vibration in rad/sec. The velocity is the derivative $\dot{x}(t) = \omega_n A\cos\omega_n t$, and the acceleration is the second derivative $\ddot{x}(t) = -\omega_n^2 A\sin\omega_n t$. Thus, if a vibrating system has displacement amplitude A, its velocity has amplitude $\omega_n A$ and the acceleration amplitude is $\omega_n^2 A$. For a given harmonic motion, these three amplitudes can be plotted versus frequency, commonly using a log scale as illustrated in Figure 18.1. In this log scale plot, the log of the frequency (denoted f in Hz, where $f = \omega_n / 2\pi$) is plotted along the horizontal and the corresponding velocity amplitude is plotted along the vertical, also on a log scale. The lines slanting to the right, of slope +1, correspond to the log of the displacement amplitude versus frequency, whereas those slanting to the left, with slope –1, correspond to the log of the acceleration amplitude versus frequency. Thus a given point on the monograph corresponds to the amplitude of displacement, velocity, and acceleration at a specific frequency. Nomographs can be used to specify regions of acceptable vibration. Often it is not enough to specify just displacement; restrictions may also exist on velocity and acceleration amplitudes. By sketching a closed shape on a nomograph, ranges of acceptable levels of maximum displacement, velocity, and acceleration over a frequency range of interest can be easily specified. The bold lines in Figure 18.2 illustrate an example. The bold lines in the figure are used to illustrate vibration between 2 and 8 Hz with displacement amplitude limited by 30 mm, velocity amplitude limited by 400 mm/sec and acceleration limited to amplitude by 10^4 mm/sec^2.

Rather than maximum amplitude, root mean square values of displacement, velocity, and acceleration can be plotted as nomographs. As mentioned, such plots are often used in formal documents, vendor specifications, and military specifications. The International Organization for Standardization presents vibration standards for severity, which are often represented in nomograph form.

As useful as nomographs are and as frequently as they appear in codes and standards, modern computational abilities allow detailed representations of vibration data for much more complicated systems. In particular, it has become very common to simulate time responses directly.

Defining Terms

Eigenvalue problem — The algebraic process of computing the eigenvalues and eigenvectors of a given matrix.

Euler method — A simple numerical solution to a first-order ordinary differential equation based on approximating the derivative by a slope.

Formula error — Error in the computed response due to the difference between the exact solution and the approximate formula.

Mode shape — The eigenvector of the mass normalized stiffness matrix multiplied by the inverse square root of the mass matrix, indicating the shape in which a system naturally tends to vibrate.

Natural frequencies — The square roots of the eigenvalues of the mass normalized stiffness matrix and the frequencies at which a freely responding system will oscillate.

Nomogaph — A graph of displacement, velocity, and acceleration versus frequency for a single-degree-of-freedom system.

Round-off error — Error in the computed response due to numerical round-off and truncation in computer arithmetic.

Runge-Kutta method — A numerical solution to a first-order ordinary differential equation based on approximating the derivative by several estimates of the first few terms of a Taylor series expansion of the solution.

Simulation — Numerical integration to solve an (ordinary) differential equation using time steps to produce the time history of the response (of a vibrating system).

References

Boyce, W. E. and DePrima, P. C. 1986. *Elementary Differential Equations and Boundary Value Problems,* 4th ed. John Wiley & Sons, New York.

Forsythe, G. E., Malcolm, M. A., and Moler, C. B. 1977. *Computer Methods for Mathematical Computation,* Prentice Hall, Englewood Cliffs, NJ.

Golub, G. E. and Van Loan, C. R., 1996. *Matrix Computations,* 3rd ed., Johns Hopkins University Press, Baltimore, MD.

Inman, D. J. 2001. *Engineering Vibration,* 2nd ed., Prentice Hall, Upper Saddle River, NJ.

Macinante, J. A. 1984. *Seismic Mountings for Vibration Isolation,* John Wiley & Sons, New York.

Moler, C. B. 1980. *MATLAB User's Guide Technical Report CS81-1,* Department of Computer Sciences, University of New Mexico, Albuquerque.

Further Information

Further information can be found by consulting the references. More information on MALAB can be obtained from

The MathWorks, Inc.
3 Apple Hill Drive
Natick, MA 01760-2098
www.mathworks.com

Information on Mathcad can be obtained from

Headquarters
Mathsoft Engineering & Education, Inc.
101 Main Street
Cambridge, MA 02142-1521
www.mathsoft.com

Information on Mathematica can be obtained from

Corporate Headquarters
Wolfram Research, Inc.
100 Trade Center Drive
Champaign, IL 61820-7237
www.wolfram.com

19

Test Equipment and Measuring Instruments

Terrence W. Baird
Hewlett-Packard Company

Dynamic test and measurement finds application in many engineering and scientific disciplines. Although each has evolved uniquely in its types of equipment and methods employed, a degree of commonality exists in the operating principles and performance criteria of the various apparatuses used. Rather than attempting comprehensive treatment of each application or equipment type available, focus will be directed toward this commonality through a discussion of the components of a generalized test system.

The generalized model from which the discussion will be developed is as follows. A device is subjected to a specified dynamic environment produced by a test machine. The test machine input and device response motions are measured by **transducers**, whose signals are conditioned and subsequently analyzed for purposes of data reduction and test machine control. Descriptions will be limited to some commonly used test equipment and instrumentation and their key performance criteria.

Although the model described will be used as the basis for this introduction to test equipment, it is the author's intent that the technical content and principles of equipment performance be relevant to any application. The reader is encouraged to research the references and resources cited at the end of the chapter to more fully explore today's dynamic test technology on an application-specific basis.

19.1 Vibration and Shock Test Machines

Vibration Test Machines

Vibration test machines, also referred to as *shakers*, are available in several distinct designs. Two common shaker designs are electrohydraulic and electrodynamic; the names are based on the method of force generation.

Electrohydraulic shakers generate force through electrically controlled hydraulics where power is converted from the high-pressure flow of hydraulic fluid to the vibratory motion of the shaker's table. Figure 19.1 [Unholtz, 1988] illustrates a block diagram of a typical electrohydraulic shaker system and

0-8493-1586-7/05/$0.00+$1.50

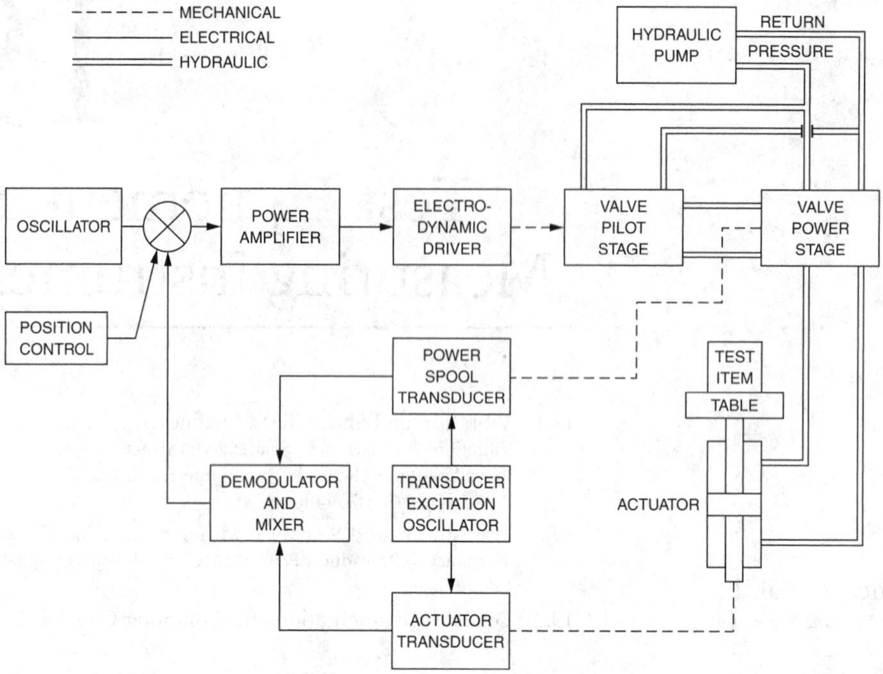

FIGURE 19.1 Block diagram of an electrohydraulic shaker. (*Source:* Harris, C. and Crede, C. 1988. *Shock and Vibration Handbook*, 3rd ed. McGraw-Hill, New York. Reproduced with permission.)

Figure 19.2 shows an actual system. Availability of large force generation, long displacement stroke, and low-frequency performance are advantages of an electrohydraulic shaker.

Electrodynamic (also called *electromagnetic*) shakers generate force by means of a coil carrying a current flow placed in a magnetic field, which passes through the coil, causing motion of the coil, moving element, and table assembly (called the *armature*). The operating principle is not unlike that of a loudspeaker system. Some electrodynamic shakers use a "double-ended" design, which incorporates both an upper and a lower field coil, resulting in reduced stray magnetic fields above the table and increased operating efficiency. A typical double-ended shaker design is shown in Figure 19.3. Higher frequency performance is a major advantage of an electrodynamic shaker.

FIGURE 19.2 Electrohydraulic shaker system including hydraulic supply, shaker, and digital controller. (Courtesy Lansmont Corp.)

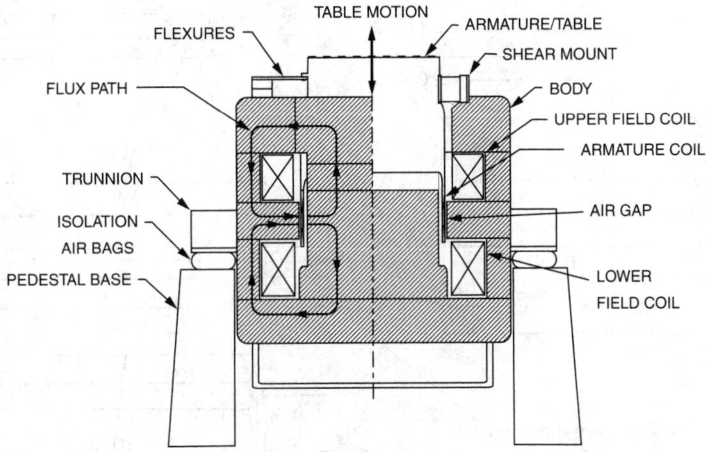

FIGURE 19.3 Schematic of double-ended electrodynamic shaker design. (Courtesy Unholtz-Dickie Corp.)

Shaker Performance Considerations

The following criteria should be evaluated relative to the application regardless of shaker type.

1. *Force rating.* The maximum force available is typically specified as a continuous rating for sine vibration through a usable frequency range. Estimated acceleration performance can be determined from:

$$A = F/W \tag{19.1}$$

   where A = maximum acceleration, g; F = force rating in pounds-force, lbf; and W = total load, lb, including armature, table, and test specimen weight.
2. *Frequency range.* Frequency versus amplitude performance is generally specified in a series of performance curves presented for various test loads. Representative ranges for typical general-purpose shakers are 1 to 500 Hz for electrohydraulic and 10 to 3000 Hz for electrodynamic, depending on test parameters.
3. *Waveform quality/harmonic distortion.* This will vary by design but should be specified.
4. *Magnetic fields.* This may be a concern for some applications, in which case it should be specified for electrodynamic shakers. It is not a concern with electrohydraulic shakers.
5. *Table or head expander* **frequency response**. The practical frequency range will depend, in part, on the table attached to the shaker. Basic design, mass, damping, and frequency response characteristics should be specified.
6. *Test orientation.* Consideration should be given to design type if independent vertical and horizontal test capability is desired. For instance, it is relatively common for an electrodynamic shaker to be supported by a base with a trunnion shaft, whereby the entire body can be rotated about its center providing for either vertical or horizontal vibration.

Shock Test Machines

Three specification types are used in defining a shock test:

1. Specification of the shock test machine (also called *shock machine*) including mounting and operating procedures. Shock machines unique to a specification will not be discussed.
2. Specification of shock motion described by simple shock pulse waveforms and parameters of peak acceleration, duration, and **velocity change.** An example is shown in Figure 19.4.
3. Specification of the **shock response spectrum (SRS)** that the test produces.

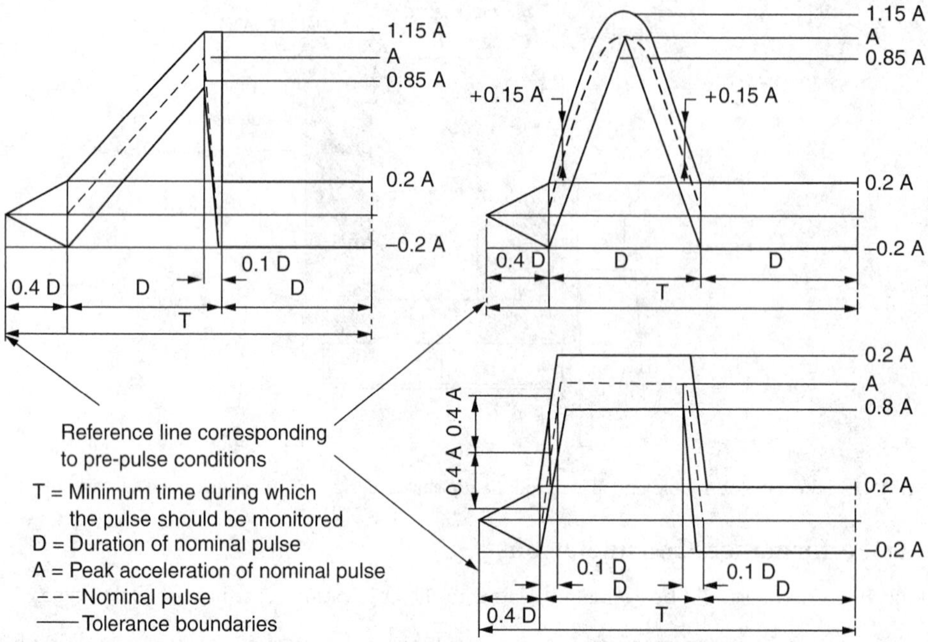

FIGURE 19.4 Preferred shock pulse waveforms of IEC 68-2-27. (*Source:* Broch, J. T. 1984. *Mechanical Vibration and Shock Measurement,* 2nd ed. Bruel and Kjaer, Naerum, Denmark. Reproduced with permission.)

Two types of machines used for shock generation are the free-fall shock machine and the previously described shakers.

A typical free-fall shock machine (sometimes called a *drop table* or *drop tester*) is shown in Figure 19.5 and is used for generating simple or classical pulse waveforms such as those in Figure 19.4. Operation is straightforward. The shock table, whose orientation and free-fall path are controlled by guide rods, is raised to a desired height and allowed to fall and impact upon a pulse programmer. Brakes are employed to prevent multiple impacts after rebound. Shock pulse velocity change is controlled by drop height. Pulse waveform, resulting peak acceleration, and duration are determined by programmer type. Significant velocity and peak acceleration capabilities are advantages of this type of shock machine.

Shakers with appropriate digital controllers are not only capable of producing classical waveforms, but can also be used for applications such as SRS programming or capturing a real-world shock pulse and using it as a control waveform for subsequent shock tests. Shakers have limitations for shock test in terms of available displacement, velocity, and peak acceleration. However, for low-level shocks of light- to medium-weight specimens, test flexibility, control, and repeatability are excellent if the test parameters are within the performance limits of the shaker.

Shock Test Machine Performance Considerations

Performance criteria that should be evaluated against the intended application include:

1. *Maximum peak acceleration.* Some shakers can achieve 100 to 200 *g*. A general-purpose free-fall shock machine is usually limited to less than 1000 *g*, whereas a high-performance shock machine may be capable of 20,000 *g* for lightweight specimens.
2. *Pulse duration,* maximum and minimum.
3. *Velocity change.* Shakers will normally be limited to less than 100 inches /second (ips). A general-purpose free-fall shock machine will achieve 300 ips, and a high-performance shock machine may be capable of 1000 ips. For very low velocity changes (0 to 50 ips), control and repeatability can be difficult with a free-fall shock machine.

FIGURE 19.5 Shock machine and associated instrumentation. (Courtesy MTS Systems Corp.)

4. *Waveform* flexibility and programmer design.
5. *Table size and weight capacity.*
6. *Table performance,* frequency response, damping, and waveform quality.
7. Potential need for *SRS control* and *waveform synthesis.*

19.2 Transducers and Signal Conditioners

Transducers

The ANSI/ISA definition of transducer is "a device which provides a usable output in response to a specified measurand" [ANSI/ISA S37.1]. The measurand is the physical quantity to be measured, and the output is an electrical quantity whose amplitude is proportional to that of the measurand. The measurand is the primary descriptor of a transducer type and for dynamic testing would likely be displacement, velocity, or acceleration. The most common measurand is acceleration, so this discussion will be limited to acceleration transducers, more commonly referred to as *accelerometers.* Two common categories of accelerometers are piezoelectric and piezoresistive, which differ fundamentally in their electrical transduction principles.

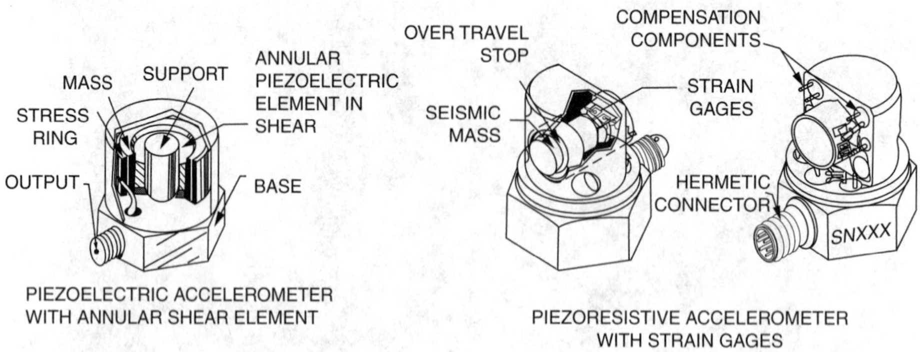

FIGURE 19.6 Accelerometer designs. (Courtesy Endevco Corp.)

Piezoelectric (PE) accelerometers incorporate sensing elements, typically quartz or ceramic crystals, that have the property of producing a charge when mechanically stressed. Specifically, when subjected to an acceleration, the PE accelerometer produces a charge proportional to the applied acceleration and is said to be *self-generating* in that the electrical output is produced without the need for auxiliary power excitation to the accelerometer. Typically, this charge then needs to be converted to a voltage in an external signal conditioner for subsequent analysis or readout. An exception to this rule is the now commonly available PE accelerometer with built-in integrated-circuit signal conditioning.

Piezoresistive (PR) accelerometers incorporate a semiconductor material such as a solid state silicon resistor, which serves as a strain-sensing or strain gage element. Arranged in pairs and typically connected electrically in a Wheatstone-bridge circuit, these PR elements exhibit a change in electrical resistance proportional to an applied acceleration. The PR accelerometer is referred to as a *passive* type accelerometer in that it does require an external power source to operate. The PR accelerometer's primary advantage is its ability to measure down to DC or steady-state acceleration, making it particularly suitable to long-duration pulses and other low-frequency applications. PE and PR accelerometer designs are shown in Figure 19.6.

Transducer Performance Considerations

The following performance criteria should be evaluated prior to an accelerometer's use. The first three are most critical and for a given design are interrelated, representing the likely trade-offs or compromises in accelerometer selection.

1. *Sensitivity.* Defined as the ratio of change in output to change in acceleration, sensitivity is expressed as coulomb/g or volt/g, depending on accelerometer type. The higher the sensitivity, the greater the system signal-to-noise ratio will be.
2. *Frequency response.* Both low- and high-frequency response may be important to an application.
3. *Mass and size.* Accelerometer weights can range from 1 to 60 g. Typically, minimizing size and mass is desirable.
4. *Mass loading effect.* Mounting an accelerometer with finite mass onto a structure changes the mechanics of the structure at that point. If the mass of the accelerometer is a significant percentage of the effective mass of the structure at the point of attachment, the structure's frequency response will be altered, resulting in a poor measurement. A simple rule of thumb or exercise to determine if mass loading is a problem is to:
   A. Measure a frequency response function of the structure using the desired accelerometer.
   B. Mount a second accelerometer of the same mass at the same point of attachment (i.e., mass is now doubled) and repeat the measurement.
   C. Compare the two measurements for amplitude changes and frequency shifts. If differences are significant, then mass loading is a problem. (Although not discussed in the text, some mea-

surement situations do exist in which mass loading effects, extreme surface temperatures, rotating structures, or other test conditions preclude the practical use of a *contact* transducer, such as an accelerometer. In these instances, a *noncontact* transducer can be employed — such as a *laser Doppler vibrometer* for motion detection — where the *electro-optical* transduction principle is employed.)

5. *Amplitude range and linearity.* Sensitivity is constant within stated tolerances over a certain amplitude range, beyond which sensitivity is nonlinear. This is usually expressed as a percentage deviation from nominal sensitivity as a function of the applied acceleration.
6. *Transverse sensitivity.* For a single-axis accelerometer, there is still a small sensitivity to transverse accelerations, which is usually expressed as a percentage of main axis sensitivity.
7. *Temperature sensitivity.* Percent deviation from the nominal sensitivity is expressed as a function of temperature.
8. *Mounting considerations.* Although not a characteristic of accelerometer design, the way in which an accelerometer is mounted to the structure for measurement has a significant influence on its effective frequency response. Figure 19.7 shows various methods used to mount accelerometers and their effect on frequency response.

Signal Conditioners

It is typical for a signal conditioner to be located in the measurement system between the transducer and the final readout or recording instruments. Signal conditioners range from simple to sophisticated and can employ internal electronics for significant signal modification or calculation of related physical quantities. Rather than describe the operating principles of signal conditioners, the following simply lists some of the key functions and available features:

1. Supply excitation voltage to a passive-circuit transducer (e.g., PR accelerometer)
2. Charge conversion to voltage from a PE accelerometer
3. "Dial-in" accelerometer sensitivity normalization
4. Gain or attenuation control to provide optimum signal-to-noise ratio in the readout instrument
5. Low-pass or high-pass filters
6. Grounding options
7. Internal electronics to perform functions such as single or double integration to obtain velocity or displacement data from acceleration signals

19.3 Digital Instrumentation and Computer Control

It is assumed the reader is familiar with the application and operating principles of time-based instruments such as the oscilloscope. Time-based instruments will be not be discussed in this text.

Today's laboratories are commonly equipped with fast Fourier transform (FFT) analysis and digital micro-processor control capabilities both for data analysis and reduction as well as test machine signal control. It is beyond the scope of this chapter to provide a comprehensive treatment of signal analysis and computer control techniques or of the many software, firmware, and hardware platforms in which these capabilities are available. The reader can find such treatments in several of the references cited at the end of this chapter. Rather, it is the author's intent to simply recognize some of the functionality and features available to the user in applying these techniques and instruments.

Dynamic signal or FFT analyzers may be used independently of test machine control for purposes of data collection and analysis. Available functionality includes:

1. Time domain or frequency domain analysis
2. Shock analysis
   - Waveform capture
   - Digital filtering

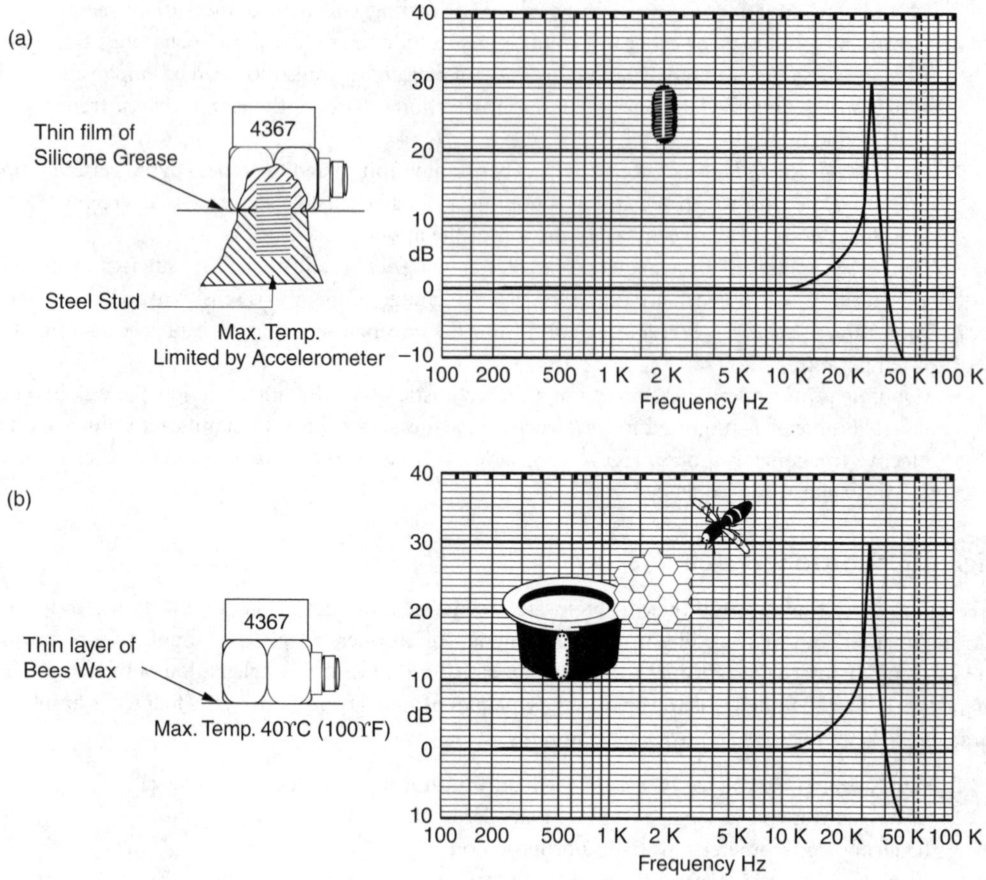

FIGURE 19.7 Typical accelerometer mounting techniques and relative frequency response characteristics. (*Source:* Broch, J. T. 1984. *Mechanical Vibration and Shock Measurement,* 2nd ed. Bruel and Kjaer, Naerum, Denmark. Reproduced with permission of Bruel and Kjaer.)

Continued.

- Math capabilities, such as integration, differentiation, and multiaxis vector resolution
- SRS computation

3. Vibration analysis
    - Power/auto spectrum analysis
    - Frequency response and transfer functions
    - Band power, harmonic power, and harmonic distortion measurements
    - Waterfall analysis
4. Programming capabilities to allow user-defined functionality

Digital control for vibration testing is generally tailored specifically to closed-loop shaker control and would normally contain fewer general analysis capabilities as compared to an independent dedicated FFT analyzer. Functions typically include:

1. Sine or random vibration control
2. Swept-sine on random
3. Narrow band random on random
4. Multiple control and response channels
5. Test control and abort limits
6. Capability of using field vibration data as shaker input signal

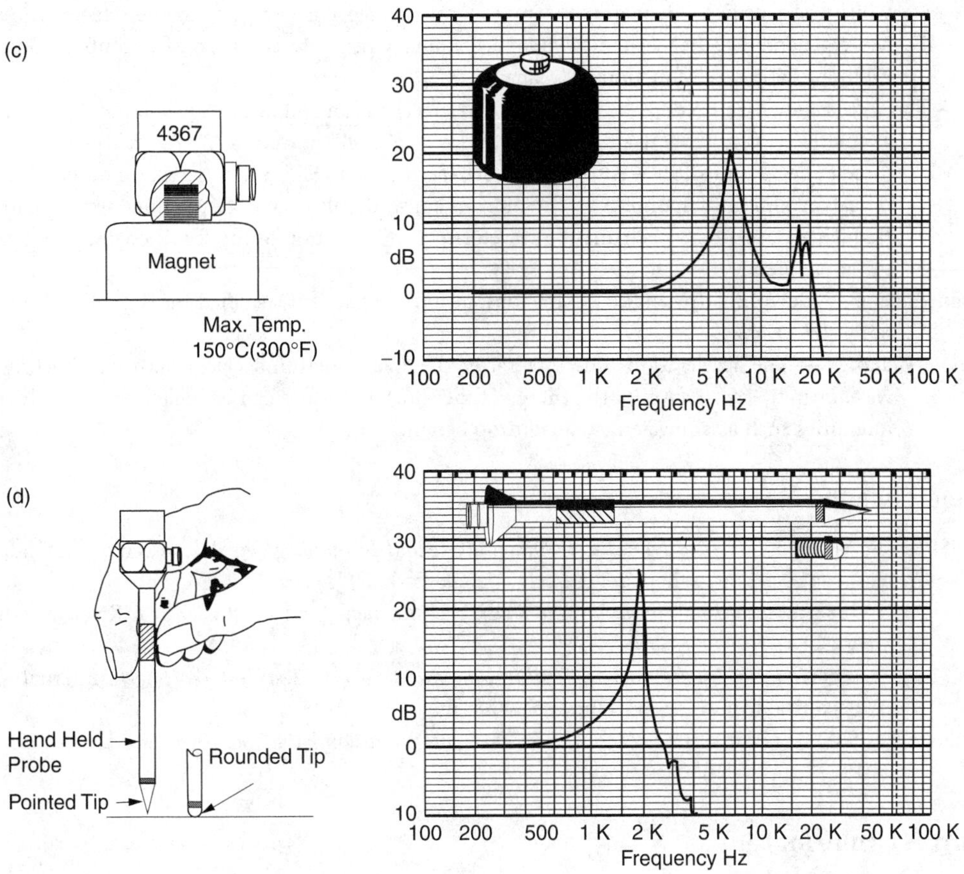

FIGURE 19.7 *Continued.*

7. Basic vibration data analysis (e.g., FFT, frequency response)

Digital control for shock-testing refers specifically to the case of controlling a shaker for shock test generation. Functions typically include:

1. Classical waveform control (e.g., half sine)
2. Direction of shock and multiple shock control
3. Transient capture of real-world shock pulses and subsequent use as shaker control
4. SRS and waveform synthesis, a function allowing the operator to specify a required SRS to the controller, which in turn synthesizes a control waveform resulting in the desired SRS
5. Basic waveform analyses such as single/double integration, vector resolution, FFT, and SRS computation

This section has provided a brief introduction to some of the test equipment, capabilities, and performance considerations associated with dynamic test and measurement. The reader is reminded that the examples cited represent only a small sample of what is available and currently in use. More thorough presentation of the concepts introduced and additional discussion on application-specific methods and equipment can be found through the resources provided at the end of the section.

Defining Terms

Frequency response — As a transducer characteristic, frequency response is the change of transducer sensitivity as a function of frequency. Normally, an operating frequency range is specified over which the sensitivity does not vary more than a stated percentage from the rated sensitivity. More generally, for a mechanical system, frequency response is a ratio of output response to input excitation as a function of frequency.

g — The acceleration produced by the force of gravity. Acceleration amplitudes are commonly described as multiples of *g*, where 1 g = 980. 665 cm/sec^2 = 386.087 in./sec^2 = 32.1739 ft/sec^2.

Shock response spectrum (SRS) — Also called *shock spectrum*, the SRS is a curve that indicates a theoretical maximum response as a function of pulse duration and responding system natural frequency. The shock spectrum of a waveform is an indication of the shock's damage potential in the frequency domain.

Transducer — A device that provides a usable output in response to a specified measurand [ANSI/ISA S37.1-1975].

Velocity change — The acceleration-time integral or the area under an acceleration-time shock pulse waveform. It is a function of the energy of the shock pulse and can be related to other physical quantities such as equivalent free-fall drop height.

References

ANSI/ISA S37.1-1975. *Electrical Transducer Nomenclature and Terminology.* American National Standard, Instrument Society of America, Research Triangle Park, NC.

Broch, J. T. 1984. *Mechanical Vibration and Shock Measurement,* 2nd ed. Bruel and Kjaer, Naerum, Denmark.

IEC 68-2-27. 1987. *Basic Environmental Testing Procedures Part 2: Test-Test Ea: Shock,* International Electrotechnical Commission, Geneva, Switzerland.

Unholtz, K. 1988. Vibration testing machines. In *Shock and Vibration Handbook,* 3rd ed., ed. C. M. Harris, p. 25–16. McGraw-Hill, New York.

Further Information

Texts

Bendat, J. S. and Piersol, A. G. 1971. *Random Data: Analysis and Measurement Procedures,* Wiley-Interscience, New York.

Gopel, W., Hesse, J., and Zemel, J. N. 1989. *Sensors: A Comprehensive Survey,* VCH Verlagsgesellschaft MBH, Weinheim, Federal Republic of Germany.

Harris, C. M. 1988. *Shock and Vibration Handbook,* 3rd ed. McGraw-Hill, New York.

Norton, H. 1989. *Handbook of Transducers,* Prentice Hall, Englewood Cliffs, NJ.

Ramirez, R. W. 1985. *The FFT, Fundamentals and Concepts,* Prentice Hall, Englewood Cliffs, NJ.

Journals

Sound and Vibration, Acoustical Publications, Inc., Bay Village, OH.

Journal of the IES, Institute of Environmental Sciences, Mount Prospect, IL.

Journal of Sound and Vibration, Academic Press Ltd., London, England.

Noise and Vibration Worldwide, IOP Publishing, Bristol, England.

The Shock and Vibration Digest, The Vibration Institute, Willowbrook, IL.

IV

Kinematics and Mechanisms

20

Linkages and Cams

J. Michael McCarthy
University of California, Irvine

Gregory L. Long
University of California, Irvine

Mechanical movement of various machine components can be coordinated using linkages and cams. These devices are assembled from hinges, ball joints, sliders, and contacting surfaces and transform an input movement such as a rotation into an output movement that may be quite complex.

20.1 Linkages

Rigid links joined together by hinges parallel to each other are constrained to move in parallel planes; the system is called a **planar linkage**. A generic value for the **degree of freedom**, or mobility, of the system is given by the formula $F = 3(n - 1) - 2j$, where n is the number of links and j is the number of hinges.

Two links and one hinge form the simplest *open chain* linkage. Open chains appear as the structure of robot manipulators. In particular, a three-degree-of-freedom planar robot is formed by four bodies joined in a series by three hinges, as in Figure 20.1(b).

If the series of links closes to form a loop, the linkage is a simple *closed chain.* The simplest case is a quadrilateral ($n = 4, j = 4$) with one degree of freedom (See Figure 20.1(a) and Figure 20.3); notice that a triangle has mobility zero. A single loop with five links has two degrees of freedom and one with six links has three degrees of freedom. This latter linkage also appears when two planar robots hold the same object.

A useful class of linkages is obtained by attaching a two-link chain to a four-link quadrilateral in various ways to obtain a one-degree-of-freedom linkage with two loops. The two basic forms of this linkage are known as the Stephenson and Watt six-bar linkages, shown in Figure 20.2.

In each of these linkages, a sliding joint, which constrains a link to a straight line rather than a circle, can replace a hinge to obtain a different movement. For example, a slider-crank linkage is a four-bar closed chain formed by three hinges and a sliding joint.

20.2 Spatial Linkages

The axes of the hinges connecting a set of links need not be parallel. In this case the system is no longer constrained to move in parallel planes and forms a **spatial linkage**. The robot manipulator with six hinged joints (denoted R for **revolute joint**) is an example of a spatial 6R open chain.

0-8493-1586-7/05/$0.00+$1.50

FIGURE 20.1 (a) Planar four-bar linkage; and (b) planar robot.

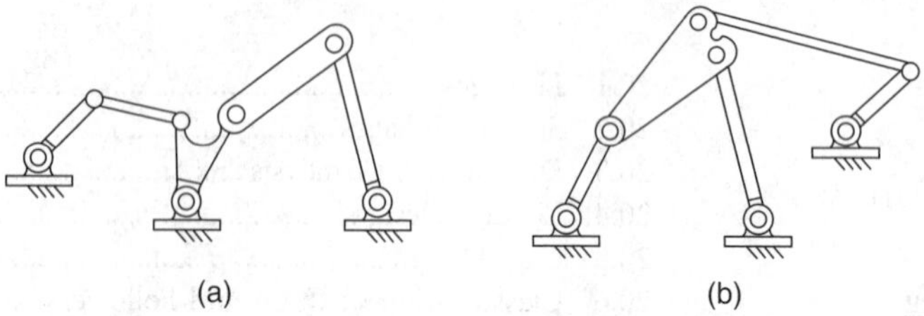

FIGURE 20.2 (a) A Watt six-bar linkage and (b) a Stephenson six-bar linkage.

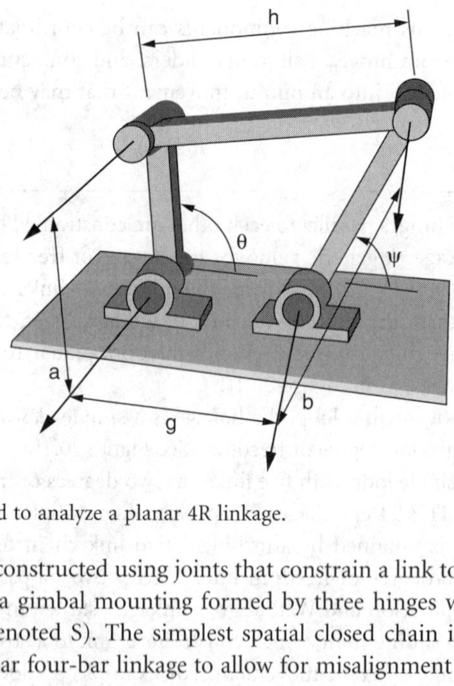

FIGURE 20.3 Dimensions used to analyze a planar 4R linkage.

Spatial linkages are often constructed using joints that constrain a link to a sphere about a point, such as a ball-in-socket joint, or a gimbal mounting formed by three hinges with concurrent axes — each termed a **spherical joint** (denoted S). The simplest spatial closed chain is the RSSR linkage, which is often used in place of a planar four-bar linkage to allow for misalignment of the cranks (Figure 20.4).

Another useful class of spatial mechanisms is produced by four hinges with concurrent axes that form a spherical quadrilateral known as a **spherical linkage.** These linkages provide a controlled reorientation movement of a body in space (Figure 20.5).

20.3 Displacement Analysis

The closed loop of the planar 4R linkage (Figure 20.3) introduces a constraint between the crank angles θ and ψ given by the equation

$$A \cos \psi + B \sin \psi = C \tag{20.1}$$

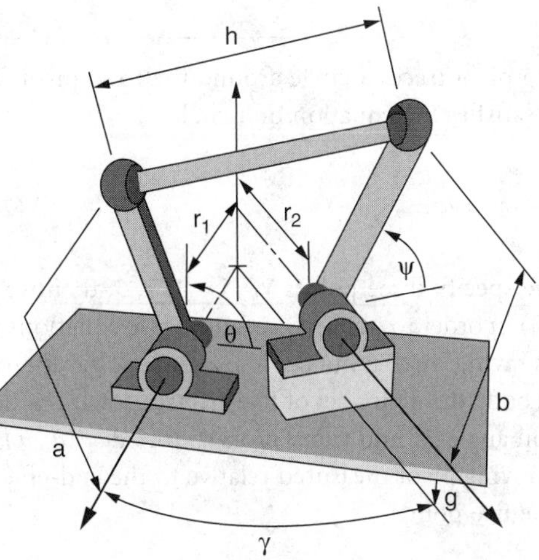

FIGURE 20.4 A spatial RSSR linkage.

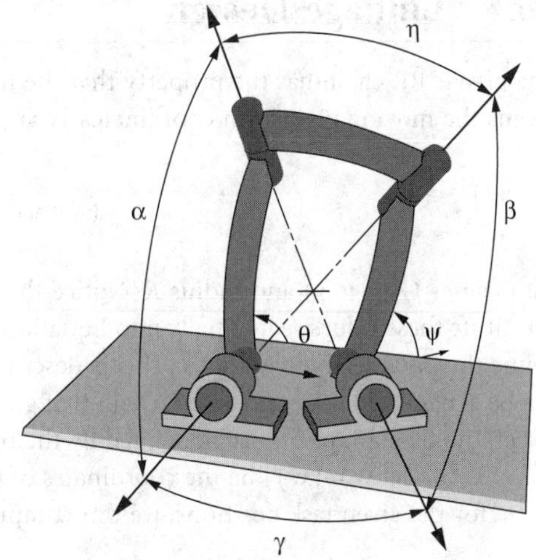

FIGURE 20.5 A spherical 4R linkage.

where $A = 2gb - 2ab\cos\theta, B = -2ab\sin\theta$, and $C = h^2 - g^2 - b^2 - a^2 + 2ga\cos\theta$

This equation can be solved to give an explicit formula for the angle ψ of the output crank in terms of the input crank rotation θ,

$$\psi(\theta) = \tan^{-1}\left(\frac{B}{A}\right) \pm \cos^{-1}\left(\frac{C}{\sqrt{A^2 + B^2}}\right). \tag{20.2}$$

This solution has two values because for a given input angle θ the output angle ψ is given by the intersection of the circles traced by the coupler and the output cranks. If the discriminant $A^2 + B^2 - C^2$ is greater than zero, then there are two solutions, if it equals zero then there is a single solution, and if it is less than zero, then there is no solution because the chain cannot close for this input angle.

The constraint equations for the spatial RSSR and spherical 4R linkages have the same form as that of the planar 4R linkage, but with coefficients as follows. For, spatial RSSR linkage (Figure 20.4),

$$A = -2ab\cos\gamma\cos\theta - 2br_1\sin\gamma,$$

$$B = 2bg - 2ab\sin\theta,$$

$$C = h^2 - g^2 - b^2 - a^2 - r_1^2 - r_2^2 + 2r_1r_2\cos\gamma + 2ar_2\sin\gamma\cos\theta + 2ga\sin\theta\,;$$

and for a spherical 4R linkage (Figure 20.5),

$$A = \sin\alpha\sin\beta\cos\gamma\cos\theta - \cos\alpha\sin\beta\sin\gamma,$$

$$B = \sin\alpha\sin\beta\sin\theta,$$

$$C = \cos\eta - \sin\alpha\cos\beta\sin\gamma\cos\theta - \cos\alpha\cos\beta\cos\gamma.$$

The formula for the output angle ψ in terms of θ for both cases is identical to that already given above for the planar 4R linkage.

20.4 Linkage Design

The planar RR chain has the property that the moving pivot traces a circle around the fixed pivot. This means the moving pivot with coordinates **W** = (x, y) satisfies the equation of a circle

$$(X-u)^2+(Y-v)^2=R^2, \tag{20.3}$$

with center **G** = (u, v) and radius R. Notice that if we specify three points $\mathbf{W}^i$, i = 1, 2, 3, then we can substitute these values sequentially into Equation (20.3) in order to obtain three quadratic equations that define the parameters u, v, and R. This is described as saying three points define a circle. This approach can be generalized to design an RR chain that guides a body through a set of five arbitrary task positions.

Let the five task positions be defined by the rotation angles θ_i and translation vectors $\mathbf{d}_i = (a_i, b_i)$, i = 1, …, 5. And let **w** = (x, y) be the coordinates of the moving pivot measured relative to the end-effector. Then, for the given task positions, we can compute the five points

$$\mathbf{W}^i=\begin{pmatrix}X\\Y\end{pmatrix}^i=\begin{bmatrix}\cos\theta_i & -\sin\theta_i\\ \sin\theta_i & \cos\theta_i\end{bmatrix}\begin{pmatrix}x\\y\end{pmatrix}+\begin{pmatrix}a_i\\b_i\end{pmatrix}, \quad \text{i} = 1, \dots, 5. \tag{20.4}$$

Substitute this into Equation (20.3) to obtain five quadratic equations in the five parameters, x, y, u, v, and R. These equations are easily solved to obtain as many as four RR chains that guide the end-effector through the specified task positions.

The planar RR chains defined in this way can be assembled into as many as six 4R linkages. The chains can individually reach the specified task positions; however, their ability to move smoothly between these positions, while connected by a coupler link, must be determined by displacement analysis of the 4R chain.

This approach to the design of serial chains can be applied to spherical RR and spatial SS chains, as well as to several other more complicated systems.

20.5 Cam Design

A *cam pair* (or *cam-follower*) consists of two primary elements called the *cam* and *follower.* The cam's motion, which is usually rotary, is transformed into either follower translation, oscillation, or combination, through direct mechanical contact. Cam pairs are found in numerous manufacturing and commercial applications requiring motion, path, and/or function generation. Cam pair mechanisms are usually simple, inexpensive, compact, and robust for the most demanding design applications. Moreover, a **cam profile** can be designed to generate virtually any desired follower motion, by either graphical or analytical methods.

20.6 Classification of Cams and Followers

The versatility of cam pairs is evidenced by the variety of shapes, forms, and motions for both cam and follower. Cams are usually classified according to their basic shape as illustrated in Figure 20.6: (a) plate cam, (b) wedge cam, (c) cylindric or barrel cam, and (d) end or face cam.

Followers are also classified according to their basic shape, with optional modifiers describing their motion characteristics. For example, a follower can oscillate [Figure 20.7 (a) and Figure 20.7 (b)] or translate [Figure 20.7 (c) through Figure 20.7 (g)]. As required by many applications, follower motion may be offset from the cam shaft's center as illustrated in Figure 20.7 (g). For all cam pairs, however, the follower must maintain constant contact with the cam surface. Constant contact can be achieved by gravity, springs, or other mechanical constraints such as grooves.

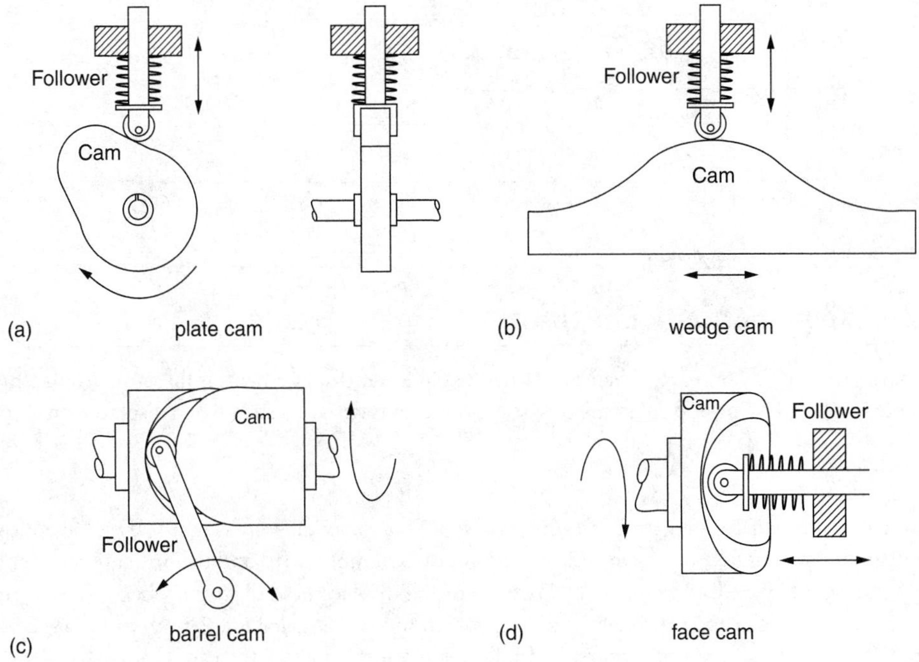

FIGURE 20.6 Basic types of cams.

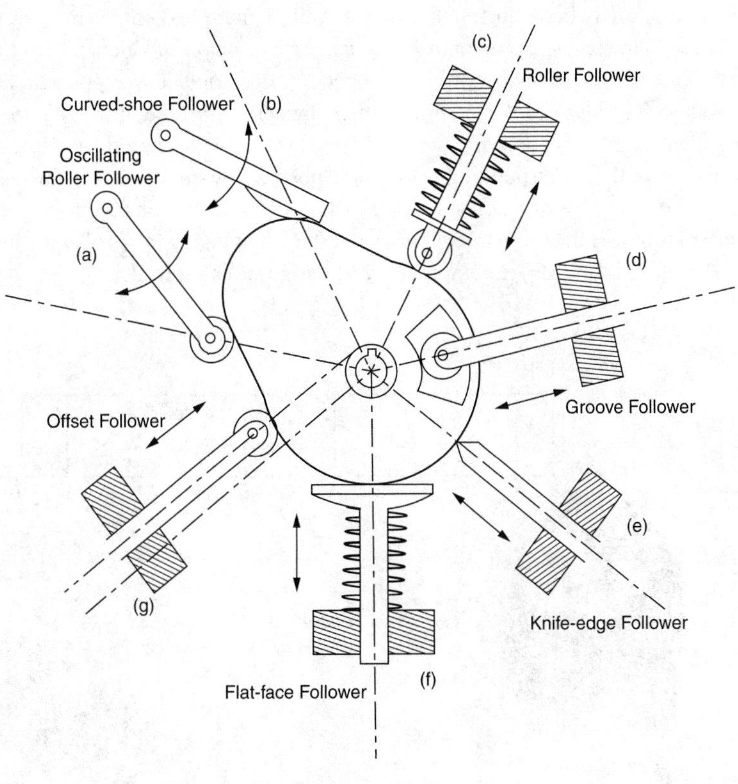

FIGURE 20.7 Basic types of followers.

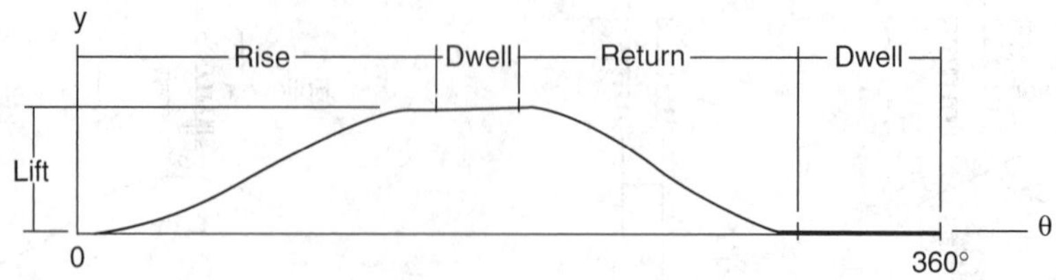

FIGURE 20.8 Displacement diagram.

20.7 Displacement Diagrams

The cam's primary function is to create a well-defined follower displacement. If the cam's displacement is designated by θ and follower displacement by y, a given cam is designed such that a displacement function

$$y = f(\theta) \tag{20.5}$$

is satisfied. A graph of y versus θ is called the *follower displacement diagram* (Figure 20.8). On a displacement diagram, the abscissa represents one revolution of cam motion (θ) and the ordinate represents the corresponding follower displacement (y). Portions of the displacement diagram, when follower motion is away from the cam's center, are called *rise*. The maximum rise is called *lift*. Periods of follower rest are referred to as *dwells*, and *returns* occur when follower motion is toward the cam's center.

The cam profile is generated from the follower displacement diagram via graphical or analytical methods that use parabolic, simple harmonic, cycloidal, and/or polynomial profiles. For many applications, the follower's velocity, acceleration, and higher time derivatives are necessary for proper cam design.

Cam profile generation is best illustrated using graphical methods where the cam profile can be constructed from the follower displacement diagram using the principle of kinematic inversion. As shown in Figure 20.9, the prime circle is divided into a number of equal angular segments and assigned station numbers. The follower displacement diagram is then divided along the abscissa into corresponding segments. Using dividers, the distances are then transferred from the displacement diagram directly onto the cam layout to locate the corresponding trace point position. A smooth curve through these points is the pitch curve. For the case of a roller follower, the roller is drawn in its proper position at each station, and the cam profile is then constructed as a smooth curve tangent to all roller positions. Analytical methods can be employed to facilitate computer-aided design of cam profiles.

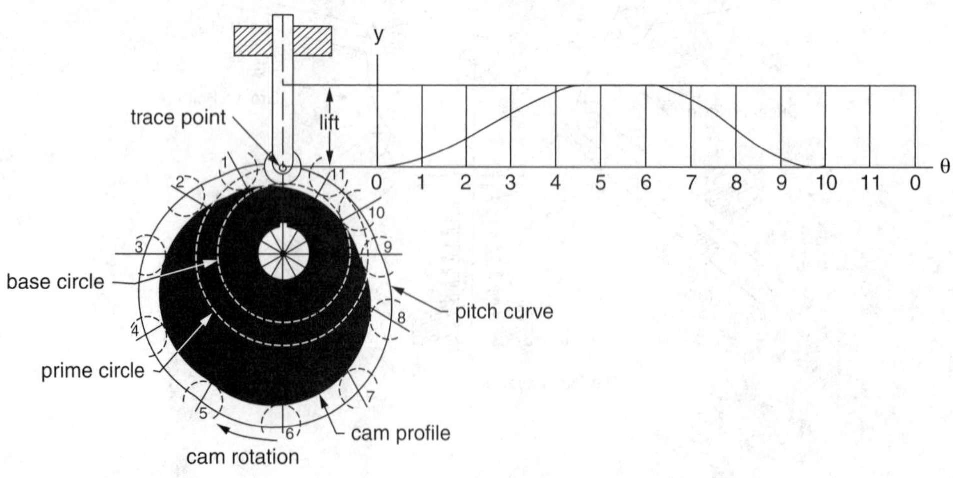

FIGURE 20.9 Cam layout.

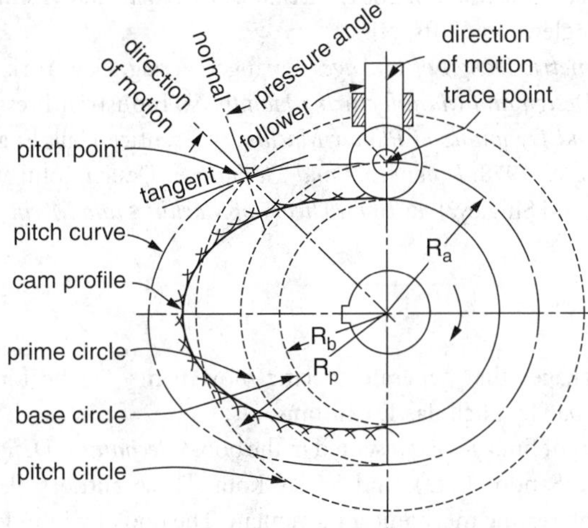

FIGURE 20.10 Cam terminology.

Defining Terms

Linkage Terminology

Standard terminology for linkages includes the following:

Degree of freedom — The number of parameters, available as input, that prescribe the configuration of a given linkage, also known as its *mobility.*

Planar linkage — A collection of links constrained to move in parallel planes.

Revolute joint — A hinged connection between two links that constrains their relative movement to the plane perpendicular to the hinge axis.

Serial chain — A series of rigid bodies connected by joints to form a chain.

Spatial linkage — A linkage with at least one link that moves out of a plane.

Spherical joint — A connection between two links that constrains their relative movement to a sphere about a point at the center of the joint.

Spherical linkage — A collection of links constrained to move on concentric spheres.

Cam Terminology

The standard cam terminology is illustrated in Figure 20.10 and defined as follows:

Base circle — The smallest circle, centered on the cam axis, that touches the cam profile (radius R_b).

Cam profile — The cam's working surface.

Pitch circle — The circle through the pitch point, centered on the cam axis (radius R_p).

Pitch curve — The path of the trace point.

Pitch point — The point on the pitch curve where pressure angle is maximum.

Pressure angle — The angle between the normal to the pitch curve and the instantaneous direction of trace point motion.

Prime circle — The smallest circle, centered on the cam axis, that touches the pitch curve (radius R_a).

Trace point — The contact point of a knife-edge follower, the center of a roller follower, or a reference point on a flat-faced follower.

References

Chironis, N. P. 1965. *Mechanisms, Linkages, and Mechanical Controls,* McGraw-Hill, New York.

Erdman, A. G, Sandor, G. N., and Kota, S. 2001. *Mechanism Design: Analysis and Synthesis,* vol. 1. (4th ed.) Prentice Hall, Englewood Cliffs, NJ.
McCarthy, J. M. 2000. *Geometric Design of Linkages,* Springer-Verlag, New York.
Norton, R. L. 2001. *CAM Design and Manufacturing Handbook,* Industrial Press.
Paul, B. 1979. *Kinematics and Dynamics of Planar Machinery,* Prentice Hall, Englewood Cliffs, NJ.
Suh, C. H. and Radcliffe, C. W. 1978. *Kinematics and Mechanism Design,* John Wiley & Sons, New York.
Uicker, J. J., Pennock, G. R., and Shigley, J. E. 2003. *Theory of Machines and Mechanisms,* Oxford University Press, New York.

Further Information

An interesting array of linkages that generate specific movements can be found in *Mechanisms and Mechanical Devices Sourcebook* by Nicholas P. Chironis.

The displacement analysis of linkages is presented in the book *Mechanism Design: Analysis and Synthesis* by Arthur Erdman, George Sandor (late), and Sridar Kota. These authors also provide a number of animations that illustrate interesting mechanical movement. The book by John Uicker, Gordon Pennock, and the late Joseph Shigley also provides the mathematical theory for linkage analysis and is particularly useful for the design of cam profiles for various applications.

A comprehensive presentation of planar, spherical and spatial linkage design can be found in the book *Kinematics and Mechanism Design* by Chung Ha Suh and Charles W. Radcliffe. The Fortran code provided in this text is still useful after many years. *Geometric Design of Linkages* is a more recent presentation of planar, spherical, and spatial linkage synthesis focusing on open chains typical to robotic systems.

Proceedings of the ASME Design Engineering Technical Conferences are published annually by the American Society of Mechanical Engineers. These proceedings document the latest developments in mechanism and machine theory.

The quarterly *ASME Journal of Mechanical Design* reports on advances in the design and analysis of linkage and cam systems. For a subscription contact American Society of Mechanical Engineers, 3 Park Ave., New York, NY 10016.

21

Tribology: Friction, Wear, and Lubrication

Bharat Bhushan
Ohio State University

In this chapter we first present the history of macrotribology and micro/nanotribology and their significance. We then describe mechanisms of friction, wear, and lubrication, followed by micro/nanotribology.

21.1 History of Tribology and its Significance to Industry

Tribology is the science and technology of two interacting surfaces in relative motion and of related subjects and practices. The popular equivalent is friction, wear, and lubrication. The word *tribology*, coined in 1966, is derived from the Greek word *tribos* meaning "rubbing," so the literal translation would be the science of rubbing [Jost, 1966]. It is only the name tribology that is relatively new, because interest in the constituent parts of tribology is older than recorded history [Dowson, 1979]. It is known that drills made during the Paleolithic period for drilling holes or producing fire were fitted with bearings made from antlers or bones, and potters' wheels or stones for grinding cereals clearly had a requirement for some form of bearings [Davidson, 1957]. A ball thrust bearing dated about 40 A.D. was found in Lake Nimi near Rome.

Records show the use of wheels from 3500 B.C., which illustrates our ancestors' concern with reducing friction in translationary motion. The transportation of large stone building blocks and monuments required the know-how of frictional devices and lubricants, such as water-lubricated sleds. Figure 21.1 illustrates the use of a sledge to transport a heavy statue by Egyptians circa 1880 B.C. [Layard, 1853]. In this transportation, 172 slaves are being used to drag a large statue weighing about 600 kN along a wooden track. One man, standing on the sledge supporting the statue, is seen pouring a liquid into the path of motion; perhaps he was one of the earliest lubrication engineers. (Dowson [1979] has estimated that

0-8493-1586-7/05/$0.00+$1.50

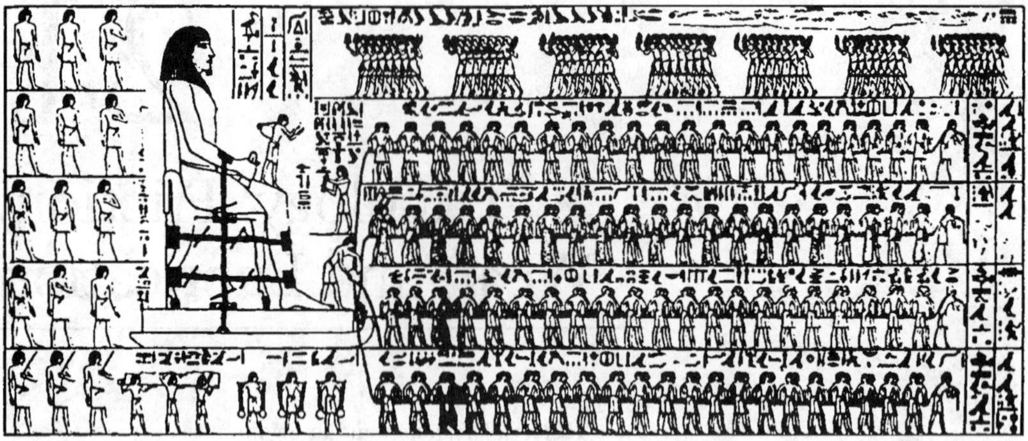

FIGURE 21.1 Egyptians using lubricant to aid movement of Colossus, El-Bersheh, ca. 1880 B.C.

each man exerted a pull of about 800 N. On this basis the total effort, which must at least equal the friction force, becomes 172 × 800 N. Thus, the coefficient of friction is about 0.23.) A tomb in Egypt that was dated several thousand years B.C. provides evidence of use of lubricants. A chariot in this tomb still contained some of the original animal-fat lubricant in its wheel bearings.

During and after the glory of the Roman empire, military engineers rose to prominence by devising both war machinery and methods of fortification, using tribological principles. It was the Renaissance engineer and artist Leonardo da Vinci (1452–1519), celebrated in his days for his genius in military construction as well as for his painting and sculpture, who first postulated a scientific approach to friction. Leonardo introduced for the first time the concept of coefficient of friction as the ratio of the friction force to normal load. In 1699, Amontons found that the friction force is directly proportional to the normal load and is independent of the apparent area of contact. These observations were verified by Coulomb in 1781, who made a clear distinction between static friction and kinetic friction.

Many other developments occurred during the 1500s, particularly in the use of improved bearing materials. In 1684, Robert Hooke suggested the combination of steel shafts and bell-metal bushes as preferable to wood shod with iron for wheel bearings. Further developments were associated with the growth of industrialization in the latter part of the eighteenth century. Early developments in the petroleum industry started in Scotland, Canada, and the U.S. in the 1850s [Parish, 1935; Dowson, 1979].

Though essential laws of viscous flow had earlier been postulated by Newton, scientific understanding of lubricated bearing operations did not occur until the end of the nineteenth century. Indeed, the beginning of our understanding of the principle of hydrodynamic lubrication was made possible by the experimental studies of Tower [1884] and the theoretical interpretations of Reynolds [1886] and related work by Petroff [1883]. Since then, developments in hydrodynamic bearing theory and practice have been extremely rapid in meeting the demand for reliable bearings in new machinery.

Wear is a much younger subject than friction and bearing development, and it was initiated on a largely empirical basis.

Since the beginning of the twentieth century, from enormous industrial growth leading to demand for better tribology, our knowledge in all areas of tribology has expanded tremendously [Holm, 1946; Bowden and Tabor, 1950, 1964; Bhushan, 1990, 1992; Bhushan and Gupta, 1991].

Tribology is crucial to modern machinery, which uses sliding and rolling surfaces. Examples of productive wear are writing with a pencil, machining, and polishing. Examples of productive friction are brakes, clutches, driving wheels on trains and automobiles, bolts, and nuts. Examples of unproductive friction and wear are internal combustion and aircraft engines, gears, cams, bearings, and seals. According to some estimates, losses resulting from ignorance of tribology amount in the U.S. to about 6% of its gross national product or about 200 billion dollars per year, and approximately one-third of the world's

Macrotribology	Micro/nanotribology
Large Mass Heavy Load	Small Mass (μg) Light Load (μg to mg)
↓	↓
Wear (Inevitable)	No Wear (Few atomic layers)
Bulk Material	Surface (few atomic layers)

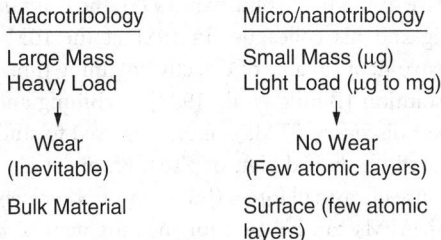

FIGURE 21.2 Comparison between macrotribology and micro/nanotribology.

energy resources in present use appear as friction in one form or another. Thus, the importance of friction reduction and wear control cannot be overemphasized for economic reasons and long-term reliability. According to Jost [1966, 1976], the United Kingdom could save approximately 500 million pounds per annum and the U.S. could save in excess of 16 billion dollars per annum by better tribological practices. The savings are both substantial and significant and could be obtained without the deployment of large capital investment.

The purpose of research in tribology is understandably the minimization and elimination of losses resulting from friction and wear at all levels of technology where the rubbing of surfaces are involved. Research in tribology leads to greater plant efficiency, better performance, fewer breakdowns, and significant savings.

21.2 Origins and Significance of Micro/Nanotribology

The advent of new techniques to measure surface topography, adhesion, friction, wear, lubricant film thickness, and mechanical properties all on micro- to nanometer scale; to image lubricant molecules; and to conduct atomic-scale simulations with the availability of supercomputers has led to development of a new field referred to as *microtribology, nanotribology, molecular tribology,* or *atomic-scale tribology.* This field deals with experimental and theoretical investigations of processes ranging from atomic and molecular scales to micro scales, occurring during adhesion, friction, wear, and thin-film lubrication at sliding surfaces. The differences between conventional or macrotribology and micro/nanotribology are contrasted in Figure 21.2. In macrotribology, tests are conducted on components with relatively large mass under heavily loaded conditions. In these tests, wear is inevitable, and the bulk properties of mating components dominate the tribological performance. In **micro/nanotribology**, measurements are made on components, at least one of the mating components with relatively small mass under lightly loaded conditions. In this situation negligible wear occurs, and the surface properties dominate the tribological performance.

Micro/nanotribological studies are needed to develop fundamental understanding of interfacial phenomena on a small scale and to study interfacial phenomena in micro- and nanostructures used in magnetic storage systems, microelectromechanical systems (MEMS), and other industrial applications [Bhushan, 1990, 1992]. The components used in micro- and nanostructures are very light (on the order of few micrograms) and operate under very light loads (on the order of few micrograms to few milligrams). As a result, friction and wear (on a nanoscale) of lightly loaded micro/nanocomponents are highly dependent on the surface interactions (few atomic layers). These structures are generally lubricated with molecularly thin films. Micro- and nanotribological techniques are ideal to study the friction and wear processes of micro- and nanostructures. Although micro/nanotribological studies are critical to study micro- and nanostructures, these studies are also valuable in fundamental understanding of interfacial phenomena in macrostructures to provide a bridge between science and engineering. Friction and wear on micro- and nanoscales have been found to be generally small compared to that at macroscales. Therefore, micro/nanotribological studies may identify the regime for ultra-low friction and near zero wear.

To give a historical perspective of the field [Bhushan, 1995], the *scanning tunneling microscope* (STM) developed by Dr. Gerd Binnig and his colleagues in 1981 at the IBM Zurich Research Laboratory, Forschungslabor, is the first instrument capable of directly obtaining three-dimensional (3-D) images of solid surfaces with atomic resolution [Binnig et al., 1982]. G. Binnig and H. Rohrer received a Nobel Prize in physics in 1986 for their discovery. STMs can only be used to study surfaces that are electrically conductive to some degree. Based on their design of STM, Binnig et al. developed, in 1985, an *atomic force microscope* (AFM) to measure ultrasmall forces (less than 1 μN) present between the AFM tip surface and the sample surface [1986]. AFMs can be used for measurement of *all engineering surfaces*, which may be either electrically conducting or insulating. AFM has become a popular surface profiler for topographic measurements on micro- to nanoscale. Mate et al. [1987] were the first to modify an AFM in order to measure both normal and friction forces; this instrument is generally called a *friction force microscope* (FFM) or *lateral force microscope* (LFM). Since then, Bhushan and other researchers have used FFM for atomic-scale and microscale friction and boundary lubrication studies [Bhushan and Ruan, 1994; Bhushan et al., 1994; Ruan and Bhushan, 1994; Bhushan, 1995; Bhushan et al., 1995]. By using a standard or a sharp diamond tip mounted on a stiff cantilever beam, Bhushan and other researchers have used AFM for scratching, wear, and measurements of elastic/plastic mechanical properties (such as indentation hardness and modulus of elasticity) [Bhushan et al., 1994; Bhushan and Koinkar, 1994a,b; Bhushan, 1995; Bhushan et al., 1995].

Surface force apparatuses (SFAs), first developed in 1969 [Tabor and Winterton, 1969], are other instruments used to study both static and dynamic properties of the molecularly thin liquid films sandwiched between two molecularly smooth surfaces [Israelachvili and Adams, 1978; Klein, 1980; Tonck et al., 1988; Georges et al., 1993, 1994]. These instruments have been used to measure the dynamic shear response of liquid films [Bhushan, 1995]. Recently, new friction attachments were developed that allow for two surfaces to be sheared past each other at varying sliding speeds or oscillating frequencies while simultaneously measuring both the friction forces and normal forces between them [Peachey et al., 1991; Bhushan, 1995]. The distance between two surfaces can also be independently controlled to within ±0.1 nm, and the force sensitivity is about 10 nN. The SFAs are used to study rheology of molecularly thin liquid films; however, the liquid under study has to be confined between molecularly smooth optically transparent surfaces with radii of curvature on the order of 1 mm (leading to poorer lateral resolution as compared to AFMs). SFAs developed by Tonck et al. [1988] and Georges et al. [1993, 1994] use an opaque and smooth ball with large radius (≈3 mm) against an opaque and smooth flat surface. Only AFMs/FFMs can be used to study *engineering surfaces* in the *dry and wet conditions* with *atomic resolution.*

21.3 Friction

Definition of Friction

Friction is the resistance to motion that is experienced whenever one solid body slides over another. The resistive force, which is parallel to the direction of motion, is called the friction force, Figure 21.3(a). If the solid bodies are loaded together and a tangential force (F) is applied, then the value of the tangential force that is required to initiate sliding is the static friction force. It may take a few milliseconds before sliding is initiated at the interface (F_{static}). The tangential force required to maintain sliding is the kinetic (or dynamic) friction force ($F_{kinetic}$). The kinetic friction force is either lower than or equal to the static friction force, Figure 21.3(b).

It has been found experimentally that there are two basic laws of intrinsic (or conventional) friction that are generally obeyed over a wide range of applications. The first law states that the friction is independent of the apparent area of contact between the contacting bodies, and the second law states that the friction force F is proportional to the normal load W between the bodies. These laws are often referred to as *Amontons laws,* after the French engineer Amontons, who presented them in 1699 [Dowson, 1979].

The second law of friction enables us to define a coefficient of friction. The law states that the friction force F is proportional to the normal load W. That is,

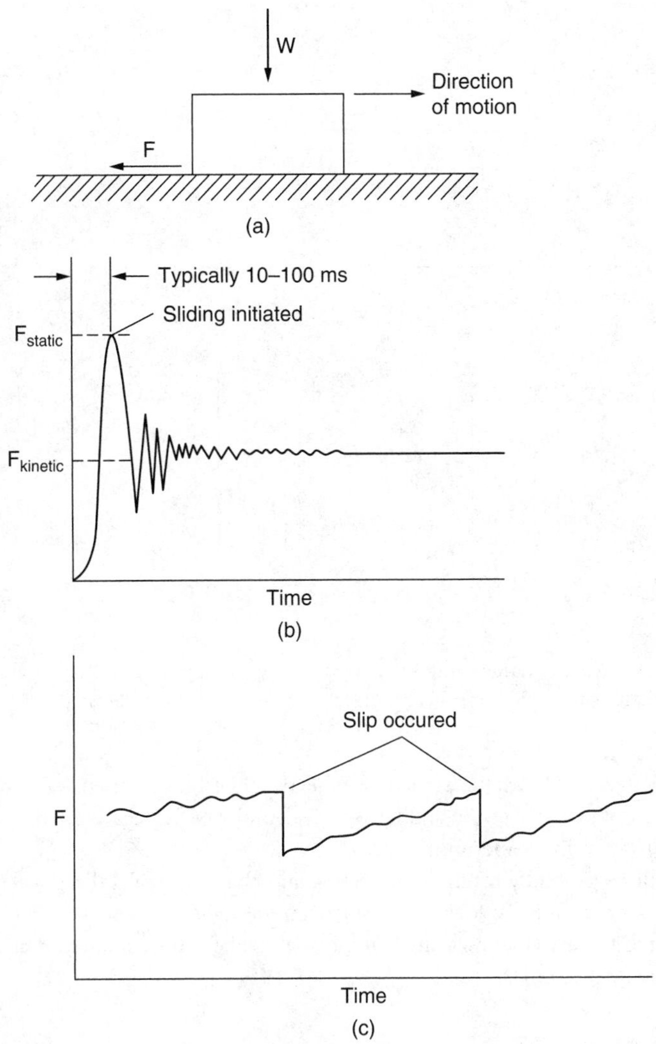

FIGURE 21.3 (a) Schematic illustration of a body sliding on a horizontal surface. *W* is the normal load and *F* is the friction force. (b) Friction force versus time or displacement. F_{static} is the force required to initiate sliding and $F_{kinetic}$ is the force required to sustain sliding. (c) Kinetic friction force versus time or displacement showing irregular stick-slip.

$$F = \mu W \tag{21.1}$$

where μ is a constant known as the *coefficient of friction*. It should be emphasized that μ is a constant only for a given pair of sliding materials under a given set of operating conditions (temperature, humidity, normal pressure, and sliding velocity). Many materials show sliding speed and normal load dependence on the coefficients of static and kinetic friction in dry and lubricated contact.

It is a matter of common experience that the sliding of one body over another under a steady pulling force proceeds sometimes at constant or nearly constant velocity and on other occasions at velocities that fluctuate widely. If the friction force (or sliding velocity) does not remain constant as a function of distance or time and produces a form of oscillation, it is generally called a *stick-slip phenomenon*, Figure 21.3(c). During the stick phase, the friction force builds up to a certain value and then slip occurs at the interface. Usually, a sawtooth pattern in the friction force–time curve [Figure 21.3(c)] is observed during the stick-slip process. Stick-slip generally arises whenever the coefficient of static friction is markedly

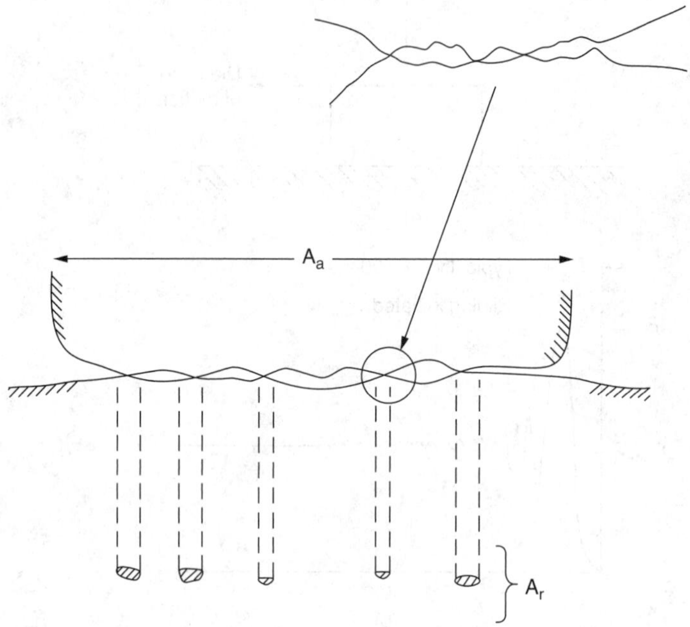

FIGURE 21.4 Schematic representation of an interface, showing the apparent (A_a) and real (A_r) areas of contact. Typical size of an asperity contact is from submicron to a few microns. Inset shows the details of a contact on a submicron scale.

greater than the coefficient of kinetic friction or whenever the rate of change of coefficient of kinetic friction as a function of velocity at the sliding velocity employed is negative. The stick-slip events can occur either repetitively or in a random manner.

The stick-slip process generally results in squealing and chattering of sliding systems. In most sliding systems the fluctuations of sliding velocity resulting from the stick-slip process and associated squeal and chatter are considered undesirable; measures are normally taken to eliminate, or at any rate to reduce, the amplitude of the fluctuations.

Theories of Friction

All engineering surfaces are rough on a microscale. When two nominally flat surfaces are placed in contact under load, the contact takes place at the tips of the asperities, the load is supported by the deformation of contacting asperities, and the discrete contact spots (junctions) are formed (Figure 21.4). The sum of the areas of all the contact spots constitutes the real (true) area of the contact (A_r) and for most materials at normal loads, this will be only a small fraction of the apparent (nominal) area of contact (A_a). The proximity of the asperities results in adhesive contacts caused by either physical or chemical interaction. When these two surfaces move relative to each other, a lateral force is required to overcome adhesion. This force is referred to as *adhesional friction force*. From classical theory of adhesion, this friction force (F_A) is defined as follows [Bowden and Tabor, 1950]. For a dry contact,

$$F_A = A_r \tau_a \tag{21.2a}$$

and for a lubricated contact,

$$F_A = A_r \left[\alpha \tau_a + (1 - \alpha)\tau_l\right] \tag{21.2b}$$

and

$$\tau_l = \eta_l \, V/h \tag{21.2c}$$

where τ_a and τ_l are the shear strengths of the dry contact and of the lubricant film, respectively; α is the fraction of unlubricated area; η_l is the dynamic viscosity of the lubricant; V is the relative sliding velocity; and h is the lubricant film thickness.

The contacts can be either elastic or plastic, depending primarily on the surface topography and the mechanical properties of the mating surfaces. The expressions for real area of contact for elastic (e) and plastic (p) contacts are as follows [Greenwood and Williamson, 1966; Bhushan, 1984, 1990]. For $\psi <$ 0.6, elastic contacts,

$$A_{re} / W \sim 3.2 / E_c (\sigma_p / R_p)^{1/2} \tag{21.3a}$$

For $\psi > 1$, plastic contacts,

$$A_{rp} / W = 1 / H \tag{21.3b}$$

Finally,

$$\psi = (E_c / H)\,(\sigma_p / R_p)^{1/2} \tag{21.3c}$$

where E_c is the composite modulus of elasticity, H is the hardness of the softer material, and σ_p and $1/R_p$ are the composite standard deviation and composite mean curvature of the summits of the mating surfaces. The real area of contact is reduced by improving the mechanical properties and in some cases by increasing the roughness (in the case of bulk of the deformation being in the elastic contact regime).

The adhesion strength depends upon the mechanical properties and the physical and chemical interaction of the contacting bodies. The adhesion strength is reduced by reducing surface interactions at the interface. For example, presence of contaminants or deliberately applied fluid film (e.g., air, water, or lubricant) would reduce the adhesion strength. Generally, most interfaces in vacuum with intimate solid–solid contact would exhibit very high values for coefficient of friction. A few pp of contaminants (air, water) may be sufficient to reduce μ dramatically. Thick films of liquids or gases would further reduce μ, as it is much easier to shear into a fluid film than to shear a solid–solid contact.

So far we have discussed theory of adhesional friction. If one of the sliding surfaces is harder than the other, the asperities of the harder surface may penetrate and plow into the softer surface. Plowing into the softer surface may also occur as a result of impacted wear debris. In addition, interaction of two rather rough surfaces may result in mechanical interlocking on micro or macro scale. During sliding, interlocking would result in plowing of one of the surfaces. In tangential motion, the plowing resistance is in addition to the adhesional friction. There is yet another mechanism of friction — deformation (or hysteresis) friction — which may be prevalent in materials with elastic hysteresis losses such as in polymers. In boundary lubricated conditions or unlubricated interfaces exposed to humid environments, the presence of some liquid may result in formation of menisci or adhesive bridges and the meniscus/viscous effects may become important; in some cases these may even dominate the overall friction force [Bhushan, 1990].

Measurements of Friction

In a friction measurement apparatus, two test specimens are loaded against each other at a desired normal load; one of the specimens is allowed to slide relative to the other at a desired sliding speed, and the tangential force required to initiate or maintain sliding is measured. Numerous apparatuses can be used to measure friction force [Benzing et al., 1976; Bhushan and Gupta, 1991]. The simplest method is an inclined-plane technique. In this method the flat test specimen of weight W is placed on top of another

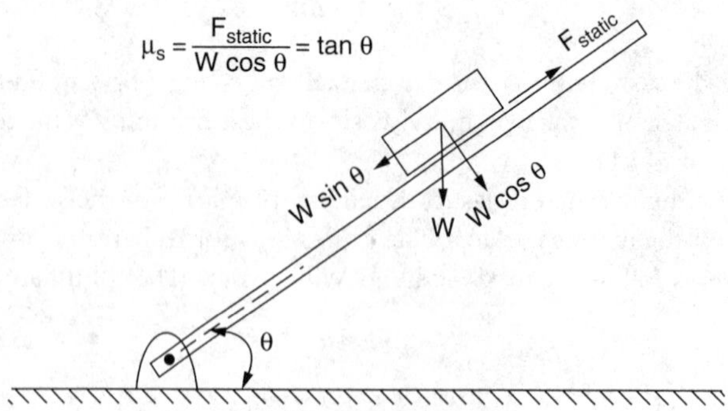

FIGURE 21.5 Inclined-plane technique to measure static friction force.

flat specimen whose inclination can be adjusted, as shown in Figure 21.5. The inclination of the lower specimen is increased from zero to an angle at which the block begins to slide. At this point, downward horizontal force being applied at the interface exceeds the static friction force, F_{static}. At the inclination angle θ, at which the block just begins to slide,

$$F_{static} = W \sin\theta$$

and the coefficient of static friction μ_s is

$$\mu_s = \frac{F_{static}}{W\cos\theta} = \tan\theta \tag{21.4}$$

The angle θ is referred to as the *friction angle.* This simple method only measures the coefficient of static friction and does not allow the measurements of the effect of sliding. However, this method demonstrates the effects of friction and provides the simplest method to measure the coefficient of static friction.

Typical values of the coefficient of friction of various material pairs are presented in Table 21.1 [Avallone and Baumeister, 1987]. It should be noted that values of the coefficient of friction depend on the operating conditions — loads, speeds, and the environment — thus, the values reported in Table 21.1 should be used with caution.

21.4 Wear

Wear is the removal of material from one or both of two solid surfaces in a solid-state contact. It occurs when solid surfaces are in a sliding, rolling, or impact motion relative to one another. Wear occurs through surface interactions at asperities, and components may need replacement after a relatively small amount of material has been removed or if the surface is unduly roughened. In well-designed tribological systems, the removal of material is usually a very slow process but it is very steady and continuous. The generation and circulation of wear debris — particularly in machine applications where the clearances are small relative to the wear particle size — may be more of a problem than the actual amount of wear.

Wear includes six principal, quite distinct phenomena that have only one thing in common: the removal of solid material from rubbing surfaces. These are [Archard, 1980; Bhushan et al., 1985a,b; Bhushan, 1990]:

1. Adhesive
2. Abrasive
3. Fatigue
4. Impact by erosion or percussion
5. Corrosive
6. Electrical arc-induced wear

TABLE 21.1 Coefficient of Friction μ for Various Material Combinations

	μ, Static		μ, Sliding (Kinetic)	
Materials	Dry	Greasy	Dry	Greasy
Hard steel on hard steel	0.78	0.11(*a*)	0.42	0.029(*h*)
		0.23(*b*)		0.081(*c*)
		0.15(*c*)		0.080(*i*)
		0.11(*d*)		0.058(*j*)
		0.0075(*p*)		0.084(*d*)
		0.0052(*h*)		0.105(*k*)
				0.096(*l*)
				0.108(*m*)
				0.12(*a*)
Mild steel on mild steel	0.74		0.57	0.09(*a*)
				0.19(*u*)
Hard steel on graphite	0.21	0.09(*a*)		
Hard steel on babbitt (ASTM 1)	0.70	0.23(*b*)	0.33	0.16(*b*)
		0.15(*c*)		0.06(*c*)
		0.08(*d*)		0.11(*d*)
		0.085(*e*)		
Hard steel on babbitt (ASTM 8)	0.42	0.17(*b*)	0.35	0.14(*b*)
		0.11(*c*)		0.065(*c*)
		0.09(*d*)		0.07(*d*)
		0.08(*e*)		0.08(*h*)
Hard steel on babbitt (ASTM 10)		0.25(*b*)		0.13(*b*)
		0.12(*c*)		0.06(*c*)
		0.10(*d*)		0.055(*d*)
		0.11(*e*)		
Mild steel on cadmium silver				0.097(*f*)
Mild steel on phosphor bronze			0.34	0.173(*f*)
Mild steel on copper lead				0.145(*f*)
Mild steel on cast iron		0.183(*c*)	0.23	0.133(*f*)
Mild steel on lead	0.95	0.5(*f*)	0.95	0.3(*f*)
Nickel on mild steel			0.64	0.178(*x*)
Aluminum on mild steel	0.61		0.47	
Magnesium on mild steel			0.42	
Magnesium on magnesium	0.6	0.08(*y*)		
Teflon on Teflon	0.04			0.04(*f*)
Teflon on steel	0.04			0.04(*f*)
Tungsten carbide on tungsten carbide	0.2	0.12(*a*)		
Tungsten carbide on steel	0.5	0.08(*a*)		
Tungsten carbide on copper	0.35			
Tungsten carbide on iron	0.8			
Bonded carbide on copper	0.35			
Bonded carbide on iron	0.8			
Cadmium on mild steel			0.46	
Copper on mild steel	0.53		0.36	0.18(*a*)
Nickel on nickel	1.10		0.53	0.12(*w*)
Brass on mild steel	0.51		0.44	
Brass on cast iron			0.30	
Zinc on cast iron	0.85		0.21	
Magnesium on cast iron			0.25	
Copper on cast iron	1.05		0.29	
Tin on cast iron			0.32	
Lead on cast iron			0.43	
Aluminum on aluminum	1.05		1.4	
Glass on glass	0.94	0.01(*p*)	0.40	0.09(*a*)
		0.005(*q*)		0.116(*v*)
Carbon on glass			0.18	
Garnet on mild steel			0.39	

TABLE 21.1 Coefficient of Friction μ for Various Material Combinations (*Continued*)

Materials	μ, Static		μ, Sliding (Kinetic)	
	Dry	Greasy	Dry	Greasy
Glass on nickel	0.78		0.56	
Copper on glass	0.68		0.53	
Cast iron on cast iron	1.10		0.15	0.070(*d*) 0.064(*n*)
Bronze on cast iron			0.22	0.077(*n*)
Oak on oak (parallel to grain)	0.62		0.48	0.164(*r*) 0.067(*s*)
Oak on oak (perpendicular)	0.54		0.32	0.072(*s*)
Leather on oak (parallel)	0.61		0.52	
Cast iron on oak			0.49	0.075(*n*)
Leather on cast iron			0.56	0.36(*t*) 0.13(*n*)
Laminated plastic on steel			0.35	0.05(*t*)
Fluted rubber bearing on steel				0.05(*t*)

Note: Reference letters indicate the lubricant used:

a = oleic acid
b = Atlantic spindle oil (light mineral)
c = castor oil
d = lard oil
e = Atlantic spindle oil plus 2% oleic acid
f = medium mineral oil
g = medium mineral oil plus 1/2% oleic acid
h = stearic acid
i = grease (zinc oxide base)
j = graphite
k = turbine oil plus 1% graphite
l = turbine oil plus 1% stearic acid
m = turbine oil (medium mineral)
n = olive oil
p = palmitic acid
q = ricinoleic acid
r = dry soap
s = lard
t = water
u = rape oil
v = 3-in-1 oil
w = octyl alcohol
x = triolein
y = 1% lauric acid in paraffin oil

Source: Adapted from Avallone, E.A. and Baumeister, T., III, 1987. *Marks' Standard Handbook for Mechanical Engineers*, 9th ed. McGraw-Hill, New York.

Other commonly encountered wear types are fretting and fretting corrosion. These are not distinct mechanisms, but rather combinations of the adhesive, corrosive, and abrasive forms of wear. According to some estimates, two-thirds of all wear encountered in industrial situations occurs because of adhesive- and abrasive-wear mechanisms.

Of the aforementioned wear mechanisms, one or more may be operating in any particular machinery. In many cases wear is initiated by one mechanism and results in other wear mechanisms, thereby complicating failure analysis.

Adhesive Wear

Adhesive wear occurs when two nominally flat solid bodies are in rubbing contact, whether lubricated or not. Adhesion (or bonding) occurs at the asperity contacts on the interface, and fragments are pulled off one surface to adhere to the other surface. Subsequently, these fragments may come off the surface on which they are formed and either be transferred back to the original surface or form loose wear particles. Severe types of adhesive wear are often called *galling, scuffing, scoring,* or *smearing*, although these terms are sometimes used loosely to describe other types of wear.

Although the adhesive wear theory can explain transferred wear particles, it does not explain how loose wear particles are formed. We now describe the actual process of formation of wear particles. Asperity contacts are sheared by sliding and a small fragment of *either surface* becomes attached to the other surface. As sliding continues, the fragment constitutes a new asperity that becomes attached once more to the original surface. This transfer element is repeatedly passed from one surface to the other

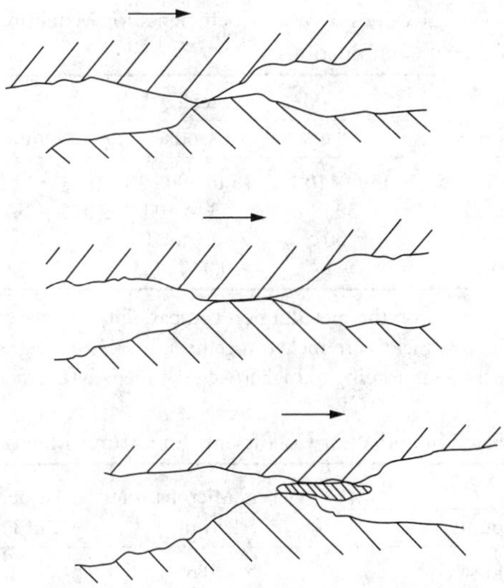

FIGURE 21.6 Schematic showing generation of wear particle as a result of adhesive wear mechanism.

and grows quickly to a large size, absorbing many of the transfer elements so as to form a flakelike particle from materials of both rubbing elements. Rapid growth of this transfer particle finally accounts for its removal as a wear particle, as shown in Figure 21.6. The occurrence of wear of the harder of the two rubbing surfaces is difficult to understand in terms of the adhesion theory. It is believed that the material transferred by adhesion to the harder surface may finally get detached by a fatigue process.

As a result of experiments carried out with various unlubricated materials — the vast majority of them metallic — it is possible to write the laws of adhesive wear, commonly referred to as Archard's law, as follows [Archard, 1953]. For plastic contacts,

$$V = kWx/H \tag{21.5}$$

where V is the volume worn away, W is the normal load, x is the sliding distance, H is the hardness of the surface being worn away, and k is a nondimensional wear coefficient dependent on the materials in contact and their exact degree of cleanliness. The term k is usually interpreted as the probability that a wear particle is formed at a given asperity encounter.

Equation (21.5) suggests that the probability of a wear-particle formation increases with an increase in the real area of contact, A_r ($A_r = W/H$ for plastic contacts), and the sliding distance. For elastic contacts occurring in materials with a low modulus of elasticity and a very low surface roughness, Equation (21.5) can be rewritten for elastic contacts (Bhushan's law of adhesive wear) as [Bhushan, 1990]

$$V = k' \, Wx / E_c (\sigma_p / R_p)^{1/2} \tag{21.6}$$

where k' is a nondimensional wear coefficient. According to this equation, elastic modulus and surface roughness govern the volume of wear. We note that in an elastic contact — though the normal stresses remain compressive throughout the entire contact — strong adhesion of some contacts can lead to generation of wear particles. Repeated elastic contacts can also fail by surface/subsurface fatigue. In addition, as the total number of contacts increases, the probability of a few plastic contacts increases, and the plastic contacts are especially detrimental from the wear standpoint.

Based on studies by Rabinowicz [1980], typical values of wear coefficients for metal on metal and nonmetal on metal combinations that are unlubricated (clean) and in various lubricated conditions are presented in Table 21.2. Wear coefficients and coefficients of friction for selected material combinations are presented in Table 21.3 [Archard, 1980].

TABLE 21.2 Typical Values of Wear Coefficients for Metal on Metal and Nonmetal on Metal Combinations

Condition	Metal on Metal		Nonmetal on Metal
	Like	Unlike[a]	
Clean (unlubricated)	$1500 \cdot 10^{-6}$	15 to $500 \cdot 10^{-6}$	$1.5 \cdot 10^{-6}$
Poorly lubricated	300	3 to 100	1.5
Average lubrication	30	0.3 to 10	0.3
Excellent lubrication	1	0.03 to 0.3	0.03

[a] The values depend on the metallurgical compatibility (degree of solid solubility when the two metals are melted together). Increasing degree of incompatibility reduces wear, leading to higher value of the wear coefficients.

TABLE 21.3 Coefficient of Friction and Wear Coefficients for Various Materials in Unlubricated Sliding

Materials		Vickers Microhardness	Coefficient	Wear Coefficient
Wearing Surface	Counter Surface	(kg/mm^2)	of Friction	(k)
Mild steel	Mild steel	186	0.62	$7.0 \cdot 10^{-3}$
60/40 leaded brass	Tool steel	95	0.24	$6.0 \cdot 10^{-4}$
Ferritic stainless steel	Tool steel	250	0.53	$1.7 \cdot 10^{-5}$
Stellite	Tool steel	690	0.60	$5.5 \cdot 10^{-5}$
PTFE	Tool steel	5	0.18	$2.4 \cdot 10^{-5}$
Polyethylene	Tool steel	17	0.53	$1.3 \cdot 10^{-7}$
Tungsten carbide	Tungsten carbide	1300	0.35	$1.0 \cdot 10^{-6}$

Note: Load = 3.9 N; speed = 1.8 m/sec. The stated value of the hardness is that of the softer (wearing) material in each example.

Source: Archard, J. F. 1980. Wear theory and mechanisms. In *Wear Control Handbook,* ed. M. B. Peterson and W. O. Winer, pp. 35–80. ASME, New York.

Abrasive Wear

Abrasive wear occurs when a rough, hard surface slides on a softer surface and plows a series of grooves in it. The surface can be plowed (plastically deformed) without removal of material. However, after the surface has been plowed several times, material removal can occur by a low-cycle fatigue mechanism. Abrasive wear is also sometimes called *plowing, scratching, scoring, gouging,* or *cutting,* depending on the degree of severity. There are two general situations for this type of wear. In the first case, the hard surface is the harder of two rubbing surfaces (two-body abrasion), for example, in mechanical operations such as grinding, cutting, and machining. In the second case the hard surface is a third body, generally a small particle of grit or abrasive, caught between the two other surfaces and sufficiently harder that it is able to abrade either one or both of the mating surfaces (three-body abrasion), for example, in lapping and polishing. In many cases the wear mechanism at the start is adhesive, which generates wear debris that gets trapped at the interface, resulting in a three-body abrasive wear.

To derive a simple quantitative expression for abrasive wear, we assume a conical asperity on the hard surface (Figure 21.7). Then the volume of wear removed is given as follows [Rabinowicz, 1965]:

$$V = k\,W x\,\overline{\tan\theta}/H \tag{21.7}$$

where $\overline{\tan\theta}$ is a weighted average of the $\tan\theta$ values of all the individual cones and k is a factor that includes the geometry of the asperities and the probability that a given asperity cuts (removes) rather than plows. Thus, the roughness effect on the volume of wear is very distinct.

Fatigue Wear

Subsurface and surface fatigue are observed during repeated rolling and sliding, respectively. For the pure rolling condition, the maximum shear stress responsible for nucleation of cracks occurs some distance

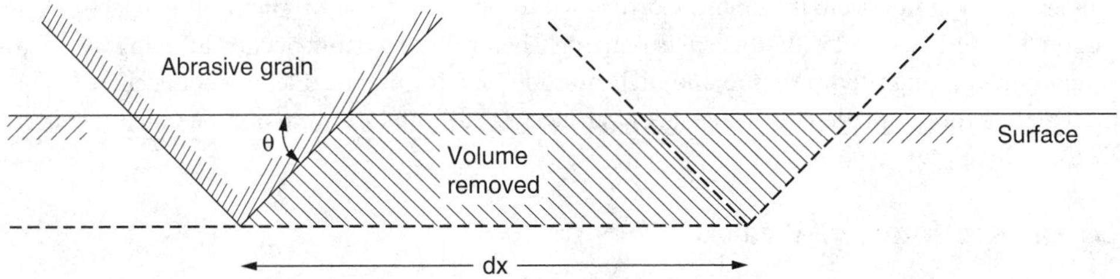

FIGURE 21.7 Abrasive wear model in which a cone removes material from a surface. (*Source:* Rabinowicz, E. 1965. *Friction and Wear of Materials.* John Wiley & Sons, New York. With permission.)

below the surface, and its location moves towards the surface with an application of the friction force at the interface. The repeated loading and unloading cycles to which the materials are exposed may induce the formation of subsurface or surface cracks, which eventually, after a critical number of cycles, will result in the breakup of the surface with the formation of large fragments, leaving large pits in the surface. Prior to this critical point, negligible wear takes place, which is in marked contrast to the wear caused by adhesive or abrasive mechanisms, where wear causes a gradual deterioration from the start of running. Therefore, the amount of material removed by fatigue wear is not a useful parameter. Much more relevant is the useful life in terms of the number of revolutions or time before fatigue failure occurs. Time to fatigue failure is dependent on the amplitude of the reversed shear stresses, the interface lubrication conditions, and the fatigue properties of the rolling materials.

Impact Wear

Two broad types of wear phenomena belong in the category of impact wear: erosive and percussive wear. Erosion can occur by jets and streams of solid particles, liquid droplets, and implosion of bubbles formed in the fluid. Percussion occurs from repetitive solid body impacts. Erosive wear by impingement of solid particles is a form of abrasion that is generally treated rather differently because the contact stress arises from the kinetic energy of a particle flowing in an air or liquid stream as it encounters a surface. The particle velocity and impact angle combined with the size of the abrasive give a measure of the kinetic energy of the erosive stream. The volume of wear is proportional to the kinetic energy of the impinging particles, that is, to the square of the velocity. Wear rate dependence on the impact angle differs between ductile and brittle materials [Bitter, 1963].

When small drops of liquid strike the surface of a solid at high speeds (as low as 300 m/sec), very high pressures are experienced, exceeding the yield strength of most materials. Thus, plastic deformation or fracture can result from a single impact, and repeated impact leads to pitting and erosive wear. Caviation erosion arises when a solid and fluid are in relative motion and bubbles formed in the fluid become unstable and implode against the surface of the solid. Damage by this process occurs in such components as ships' propellers and centrifugal pumps.

Percussion is a repetitive solid body impact, such as experienced by print hammers in high-speed electromechanical applications and high asperities of the surfaces in a gas bearing (e.g., head–medium interface in magnetic storage systems). In most practical machine applications the impact is associated with sliding; that is, the relative approach of the contacting surfaces has both normal and tangential components known as *compound impact* [Engel, 1976].

Corrosive Wear

Corrosive wear occurs when sliding takes place in a corrosive environment. In the absence of sliding, the products of the corrosion (e.g., oxides) would form a film typically less than a micrometer thick on the surfaces, which would tend to slow or even arrest the corrosion, but the sliding action wears the film away, so that the corrosive attack can continue. Thus, corrosive wear requires both corrosion and rubbing. Machinery operating in an industrial environment or near the coast generally corrode more rapidly than those operating in a clean environment. Corrosion can occur because of chemical or electrochemical interaction of the interface with the environment. Chemical corrosion occurs in a highly corrosive environment and in high-temperature and high-humidity environments. Electrochemical corrosion is a chemical reaction accompanied by the passage of an electric current; for this to occur a potential difference must exist between two regions.

Electrical Arc-Induced Wear

When a high potential is present over a thin air film in a sliding process, a dielectric breakdown results that leads to arcing. During arcing, a relatively high-power density (on the order of 1 kW/mm^2) occurs over a very short period of time (on the order of 100 μsec). The heat-affected zone is usually very shallow (on the order of 50 μm). Heating is caused by the Joule effect due to the high power density and by ion bombardment from the plasma above the surface. This heating results in considerable melting, corrosion, hardness changes, other phase changes and even the direct ablation of material. Arcing causes large craters, and any sliding or oscillation after an arc either shears or fractures the lips, leading to abrasion, corrosion, surface fatigue, and fretting. Arcing can thus initiate several modes of wear, resulting in catastrophic failures in electrical machinery [Bhushan and Davis, 1983].

Fretting and Fretting Corrosion

Fretting occurs where low-amplitude vibratory motion takes place between two metal surfaces loaded together [Anonymous, 1955]. This is a common occurrence because most machinery is subjected to vibration, both in transit and in operation. Examples of vulnerable components are shrink fits, bolted parts, and splines. Basically, fretting is a form of adhesive or abrasive wear where the normal load causes adhesion between asperities and vibrations cause ruptures, resulting in wear debris. Most commonly, fretting is combined with corrosion, in which case the wear mode is known as *fretting corrosion*.

21.5 Lubrication

Sliding between clean solid surfaces is generally characterized by a high coefficient of friction and severe wear due to the specific properties of the surfaces, such as low harness, high surface energy, reactivity, and mutual solubility. Clean surfaces readily adsorb traces of foreign substances, such as organic compounds, from the environment. The newly formed surfaces generally have a much lower coefficient of friction and wear than the clean surfaces. The presence of a layer of foreign material at an interface cannot be guaranteed during a sliding process; therefore, lubricants are deliberately applied to produce low friction and wear. The term **lubrication** is applied to two different situations: solid lubrication and fluid (liquid or gaseous) film lubrication.

Solid Lubrication

A solid lubricant is any material used in bulk or as a powder or a thin, solid film on a surface to provide protection from damage during relative movement to reduce friction and wear. Solid lubricants are used for applications in which any sliding contact occurs, for example, a bearing operative at high loads and low speeds and a hydrodynamically lubricated bearing requiring start/stop operations. The term *solid lubricants* embraces a wide range of materials that provide low friction and wear [Bhushan and Gupta, 1991]. Hard materials are also used for low wear under extreme operating conditions.

Fluid Film Lubrication

A regime of lubrication in which a thick fluid film is maintained between two sliding surfaces by an external pumping agency is called *hydrostatic lubrication*.

A summary of the lubrication regimes observed in fluid (liquid or gas) lubrication without an external pumping agency (self-acting) can be found in the familiar Stribeck curve in Figure 21.8. This plot for a

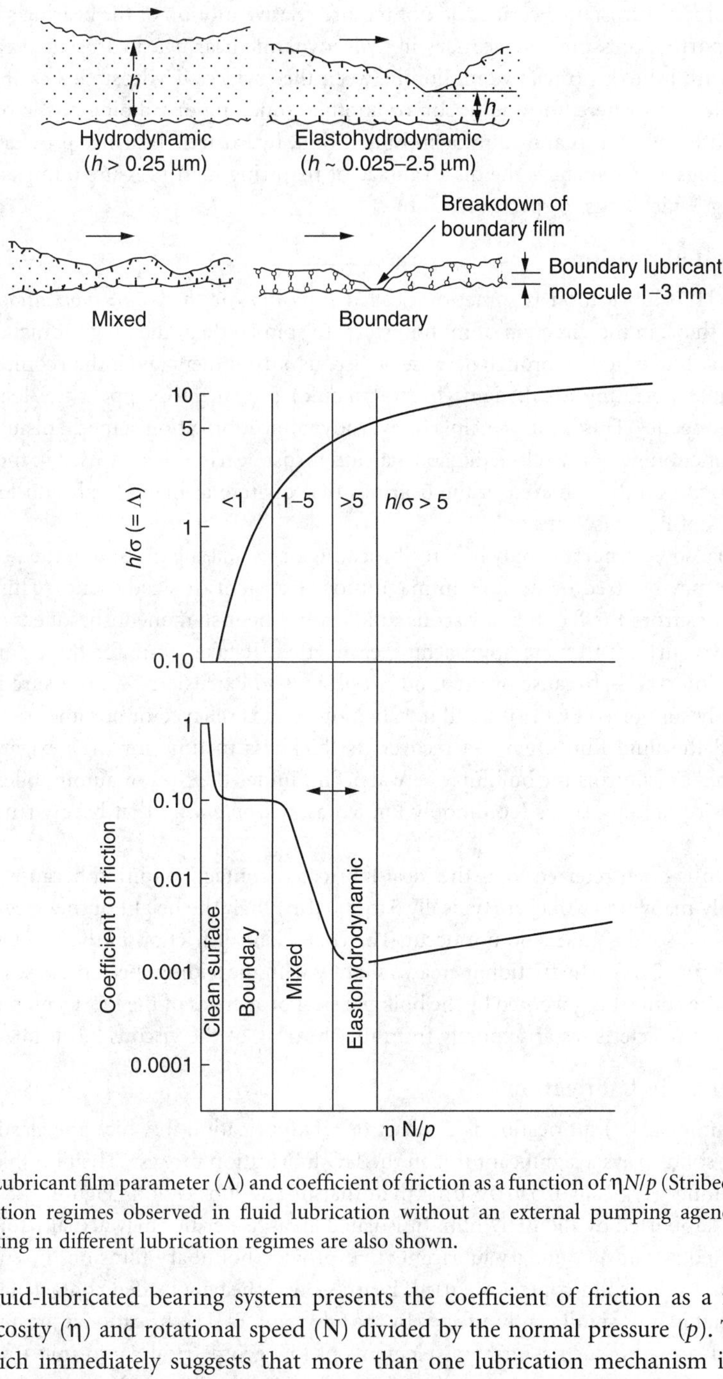

FIGURE 21.8 Lubricant film parameter (Λ) and coefficient of friction as a function of $\eta N/p$ (Stribeck curve) showing different lubrication regimes observed in fluid lubrication without an external pumping agency. Schematics of interfaces operating in different lubrication regimes are also shown.

hypothetical fluid-lubricated bearing system presents the coefficient of friction as a function of the product of viscosity (η) and rotational speed (N) divided by the normal pressure (p). The curve has a minimum, which immediately suggests that more than one lubrication mechanism is involved. The regimes of lubrication are sometimes identified by a lubricant film parameter Λ equal to h/σ, which is mean film thickness divided by composite standard deviation of surface roughnesses. Descriptions of different regimes of lubrication follow [Booser, 1984; Bhushan, 1990].

Hydrostatic Lubrication

Hydrostatic bearings support load on a thick film of fluid supplied from an external pressure source — a pump — which feeds pressurized fluid to the film. For this reason, these bearings are often called "externally pressurized." Hydrostatic bearings are designed for use with both incompressible and compressible fluids. Since hydrostatic bearings do not require relative motion of the bearing surfaces to build up the load-supporting pressures as necessary in hydrodynamic bearings, hydrostatic bearings are used in applications with little or no relative motion between the surfaces. Hydrostatic bearings may also be required in applications where, for one reason or another, touching or rubbing of the bearing surfaces cannot be permitted at startup and shutdown. In addition, hydrostatic bearings provide high stiffness. Hydrostatic bearings, however, have the disadvantage of requiring high-pressure pumps and equipment for fluid cleaning, which adds to space and cost.

Hydrodynamic Lubrication

Hydrodynamic (HD) lubrication is sometimes called *fluid-film* or *thick-film lubrication*. As a bearing with convergent shape in the direction of motion starts to spin (slide in the longitudinal direction) from rest, a thin layer of fluid is pulled through because of viscous entrainment and is then compressed between the bearing surfaces, creating a sufficient (hydrodynamic) pressure to support the load without any external pumping agency. This is the principle of hydrodynamic lubrication, a mechanism that is essential to the efficient functioning of the self-acting journal and thrust bearings widely used in modern industry. A high load capacity can be achieved in the bearings that operate at high speeds and low loads in the presence of fluids of high viscosity.

Fluid film can also be generated only by a reciprocating or oscillating motion in the normal direction (*squeeze*), which may be fixed or variable in magnitude (transient or steady state). This load-carrying phenomenon arises from the fact that a viscous fluid cannot be instantaneously squeezed out from the interface with two surfaces that are approaching each other. It takes time for these surfaces to meet, and during that interval — because of the fluid's resistance to extrusion — a pressure is built up and the load is actually supported by the fluid film. When the load is relieved or becomes reversed, the fluid is sucked in and the fluid film often can recover its thickness in time for the next application. The squeeze phenomenon controls the buildup of a water film under the tires of automobiles and airplanes on wet roadways or landing strips (commonly known as *hydroplaning*) that have virtually no relative sliding motion.

HD lubrication is often referred to as the ideal lubricated contact condition because the lubricating films are normally many times thicker (typically 5 to 500 μm) than the height of the irregularities on the bearing surface, and solid contacts do not occur. The coefficient of friction in the HD regime can be as small as 0.001 (Figure 21.8). The friction increases slightly with the sliding speed because of viscous drag. The behavior of the contact is governed by the bulk physical properties of the lubricant, notable viscosity, and the frictional characteristics arise purely from the shearing of the viscous lubricant.

Elastohydrodynamic Lubrication

Elastohydrodynamic (EHD) lubrication is a subset of HD lubrication in which the elastic deformation of the bounding solids plays a significant role in the HD lubrication process. The film thickness in EHD lubrication is thinner (typically 0.5 to 2.5 μm) than that in HD lubrication (Figure 21.8), and the load is still primarily supported by the EHD film. In isolated areas, asperities may actually touch. Therefore, in liquid lubricated systems, boundary lubricants that provide boundary films on the surfaces for protection against any solid–solid contact are used. Bearings with heavily loaded contacts fail primarily by a fatigue mode that may be significantly affected by the lubricant. EHD lubrication is most readily induced in heavily loaded contacts (such as machine elements of low geometrical conformity), where loads act over relatively small contact areas (on the order of one-thousandth of journal bearing), such as the point contacts of ball bearings and the line contacts of roller bearings and gear teeth. EHD phenomena also occur in some low elastic modulus contacts of high geometrical conformity, such as seals and conventional journal and thrust bearings with soft liners.

Mixed Lubrication

The transition between the hydrodynamic/elastohydrodynamic and boundary lubrication regimes constitutes a gray area known as *mixed lubrication*, in which two lubrication mechanisms may be functioning.

TABLE 21.4 Surface Roughness and Micro- and Macroscale Coefficients of Friction of Various Samples

Material	RMS Roughness, nm	Microscale Coefficient of Friction versus Si_3N_4 Tip[a]	Macroscale Coefficient of Friction versus Alumina Ball[b]	
			0.1 N	1 N
Si (111)	0.11	0.03	0.18	0.60
C^+-implanted Si	0.33	0.02	0.18	0.18

[a] Si_3N_4 tip (with about 50 nm radius) in the load range of 10–150 nN (1.5–3.8 Gpa), a scanning speed of 4 μm/sec and scan area of 1 × 1 μm.

[b] Alumina ball with 3-mm radius at normal loads of 0.1 and 1 N (0.23 and 0.50 Gpa) and average sliding speed of 0.8 mm/sec.

There may be more frequent solid contacts, but at least a portion of the bearing surface remains supported by a partial hydrodynamic film (Figure 21.8). The solid contacts, if between unprotected virgin metal surfaces, could lead to a cycle of adhesion, metal transfer, wear particle formation, and snowballing into seizure. However, in liquid lubricated bearings, the physi- or chemisorbed or chemically reacted films (boundary lubrication) prevent adhesion during most asperity encounters. The mixed regime is also sometimes referred to as *quasihydrodynamic, partial fluid,* or *thin-film* (typically 0.5 to 2.5 μm) *lubrication.*

Boundary Lubrication

As the load increases, speed decreases or the fluid viscosity decreases in the Stribeck curve shown in Figure 21.8; the coefficient of friction can increase sharply and approach high levels (about 0.2 or much higher). In this region, it is customary to speak of boundary lubrication. This condition can also occur in a starved contact. Boundary lubrication is that condition in which the solid surfaces are so close together that surface interaction between monomolecular or multimolecular films of lubricants (liquids or gases) and the solids dominates the contact. (This phenomenon does not apply to solid lubricants.) The concept is represented in Figure 21.8, which shows a microscopic cross-section of films on two surfaces and areas of asperity contact. In the absence of boundary lubricants and gases (no oxide films), friction may become very high (>1).

21.6 Micro/Nanotribology

AFM/FFMs are commonly used to study engineering surfaces on micro- to nanoscales. These instruments measure the normal and friction forces between a sharp tip (with a tip radius of 30 to 100 nm) and an engineering surface. Measurements can be made at loads as low as less than 1 nN and at scan rates up to about 120 Hz. A sharp AFM/FFM tip sliding on a surface simulates a single asperity contact. FFMs are used to measure coefficient of friction on micro- to nanoscales, and AFMs are used for studies of surface topography, scratching/wear and boundary lubrication, mechanical property measurements, and nanofabrication/nanomachining [Bhushan and Ruan, 1994; Bhushan et al., 1994; Bhushan and Koinkar, 1994a,b; Ruan and Bhushan, 1994; Bhushan, 1995; Bhushan et al., 1995]. For surface roughness, friction force, nanoscratching and nanowear measurements, a microfabricated square pyramidal Si_3N_4 tip with a tip radius of about 30 nm is generally used at loads ranging from 10 to 150 nN. For microscratching, microwear, nanoindentation hardness measurements, and nanofabrication, a three-sided pyramidal single-crystal natural diamond tip with a tip radius of about 100 nm is used at relatively high loads ranging from 10 to 150 μN. Friction and wear on micro- and nanoscales are found to be generally smaller compared to that at macroscales. For an example of comparison of coefficients of friction at macro- and microscales see Table 21.4.

Defining Terms

Friction — The resistance to motion whenever one solid slides over another.

Lubrication — Materials applied to the interface to produce low friction and wear in either of two situations — solid lubrication or fluid (liquid or gaseous) film lubrication.

Micro/nanotribology — The discipline concerned with experimental and theoretical investigations of processes (ranging from atomic and molecular scales to microscales) occurring during adhesion, friction, wear, and lubrication at sliding surfaces.

Tribology — The science and technology of two interacting surfaces in relative motion and of related subjects and practices.

Wear — The removal of material from one or both solid surfaces in a sliding, rolling, or impact motion relative to one another.

References

Anonymous. 1955. Fretting and fretting corrosion. *Lubrication.* 41:85–96.

Archard, J. F. 1953. Contact and rubbing of flat surfaces. *J. Appl. Phys.* 24:981–988.

Archard, J. F. 1980. Wear theory and mechanisms. *Wear Control Handbook,* ed. M. B. Peterson and W. O. Winer, pp. 35–80. ASME, New York.

Avallone, E. A. and Baumeister, T., III. 1987. *Marks' Standard Handbook for Mechanical Engineers,* 9th ed. McGraw-Hill, New York.

Benzing, R., Goldblatt, I., Hopkins, V., Jamison, W., Mecklenburg, K., and Peterson, M. 1976. *Friction and Wear Devices,* 2nd ed. ASLE, Park Ridge, IL.

Bhushan, B. 1984. Analysis of the real area of contact between a polymeric magnetic medium and a rigid surface. *ASME J. Lub. Tech.* 106:26–34.

Bhushan, B. 1990. *Tribology and Mechanics of Magnetic Storage Devices,* Springer-Verlag, New York.

Bhushan, B. 1992. *Mechanics and Reliability of Flexible Magnetic Media,* Springer-Verlag, New York.

Bhushan, B. 1995. *Handbook of Micro/Nanotribology,* CRC Press, Boca Raton, FL.

Bhushan, B. and Davis, R. E. 1983. Surface analysis study of electrical-arc-induced wear. *Thin Solid Films.* 108:135–156.

Bhushan, B., Davis, R. E., and Gordon, M. 1985a. Metallurgical re-examination of wear modes. I: Erosive, electrical arcing and fretting. *Thin Solid Films.* 123:93–112.

Bhushan, B., Davis, R. E., and Kolar, H. R. 1985b. Metallurgical re-examination of wear modes. II: Adhesive and abrasive. *Thin Solid Films.* 123:113–126.

Bhushan, B. and Gupta, B. K. 1991. *Handbook of Tribology: Materials, Coatings, and Surface Treatments,* McGraw-Hill, New York.

Bhushan, B., Israelachvili, J. N., and Landman, U. 1995. Nanotribology: friction, wear and lubrication at the atomic scale. *Nature.* 374:607–616.

Bhushan, B. and Koinkar, V. N. 1994a. Tribological studies of silicon for magnetic recording applications. *J. Appl. Phys.* 75:5741–5746.

Bhushan, B. and Koinkar, V. N. 1994b. Nanoindentation hardness measurements using atomic force microscopy. *Appl. Phys. Lett.* 64:1653–1655.

Bhushan, B., Koinkar, V. N., and Ruan, J. 1994. Microtribology of magnetic media. *Proc. Inst. Mech. Eng., Part J: J. Eng. Tribol.* 208:17–29.

Bhushan, B. and Ruan, J. 1994. Atomic-scale friction measurements using friction force microscopy: Part II — Application to magnetic media. *ASME J. Tribol.* 116:389–396.

Binnig, G., Quate, C. F., and Gerber, C. 1986. Atomic force microscope. *Phys. Rev. Lett.* 56:930–933.

Binnig, G., Rohrer, H., Gerber, C., and Weibel, E. 1982. Surface studies by scanning tunnelling microscopy. *Phys. Rev. Lett.* 49:57–61.

Bitter, J. G. A. 1963. A study of erosion phenomena. *Wear.* 6:5–21; 169–190.

Booser, E. R. 1984. *CRC Handbook of Lubrication,* vol. 2. CRC Press, Boca Raton, FL.

Bowden, F. P. and Tabor, D. 1950. *The Friction and Lubrication of Solids,* vols. I and II. Clarendon Press, Oxford.

Davidson, C. S. C. 1957. Bearing since the stone age. *Engineering.* 183:2–5.

Dowson, D. 1979. *History of Tribology,* Longman, London.

Engel, P. A. 1976. *Impact Wear of Materials,* Elsevier, Amsterdam.

Fuller, D. D. 1984. *Theory and Practice of Lubrication for Engineers,* 2nd ed. John Wiley & Sons, New York.

Georges, J. M., Millot, S., Loubet, J. L., and Tonck, A. 1993. Drainage of thin liquid films between relatively smooth surfaces. *J. Chem. Phys.* 98:7345–7360.

Georges, J. M., Tonck, A., and Mazuyer, D. 1994. Interfacial friction of wetted monolayers. *Wear.* 175:59–62.

Greenwood, J. A. and Williamson, J. B. P. 1966. Contact of nominally flat surfaces. *Proc. R. Soc. Lond.* A295:300–319.

Holm, R. 1946. *Electrical Contact.* Springer-Verlag, New York.

Israelachvili, J. N. and Adams, G. E. 1978. Measurement of friction between two mica surfaces in aqueous electrolyte solutions in the range 0–100 nm. *Chem. Soc. J., Faraday Trans. I.* 74:975–1001.

Jost, P. 1966. *Lubrication (Tribology) — A Report on the Present Position and Industry's Needs.* Department of Education and Science, H.M. Stationary Office, London.

Jost, P. 1976. Economic impact of tribology. *Proc. Mechanical Failures Prevention Group.* NBS Special Pub. 423, Gaithersburg, MD.

Klein, J. 1980. Forces between mica surfaces bearing layers of adsorbed polystyrene in cyclohexane. *Nature.* 288:248–250.

Layard, A. G. 1853. *Discoveries in the Ruins of Nineveh and Babylon,* I and II. John Murray, Albemarle Street, London.

Mate, C. M., McClelland, G. M., Erlandsson, R., and Chiang, S. 1987. Atomic-scale friction of a tungsten tip on a graphite surface. *Phys. Rev. Lett.* 59:1942–1945.

Parish, W. F. 1935. Three thousand years of progress in the development of machinery and lubricants for the hand crafts. *Mill and Factory.* Vols. 16 and 17.

Peachey, J., Van Alsten, J., and Granick, S. 1991. Design of an apparatus to measure the shear response of ultrathin liquid films. *Rev. Sci. Instrum.* 62:463–473.

Petroff, N. P. 1883. Friction in machines and the effects of the lubricant. *Eng. J.* (in Russian; St. Petersburg) 71–140, 228–279, 377–436, 535–564.

Rabinowicz, E. 1965. *Friction and Wear of Materials.* John Wiley & Sons, New York.

Rabinowicz, E. 1980. Wear coefficients — metals. *Wear Control Handbook,* ed. M. B. Peterson and W. O. Winer, pp. 475–506. ASME, New York.

Reynolds, O. O. 1886. On the theory of lubrication and its application to Mr. Beauchamp Tower's experiments. *Phil. Trans. R. Soc.* (Lond.) 177:157–234.

Ruan, J. and Bhushan, B. 1994. Atomic-scale and microscale friction of graphite and diamond using friction force microscopy. *J. Appl. Phys.* 76:5022–5035.

Tabor, D. and Winterton, R. H. S. 1969. The direct measurement of normal and retarded van der Waals forces. *Proc. R. Soc. Lond.* A312:435–450.

Tonck, A., Georges, J. M., and Loubet, J. L. 1988. Measurements of intermolecular forces and the rheology of dodecane between alumina surfaces. *J. Colloid Interf. Sci.* 126:1540–1563.

Tower, B. 1884. Report on friction experiments. *Proc. Inst. Mech. Eng.* 632.

Further Information

Societies:

Information Storage and Processing Systems Division, The American Society of Mechanical Engineers, New York.

Tribology Division, The American Society of Mechanical Engineers, New York.

Institution of Mechanical Engineers, London.

Society of Tribologists and Lubrication Engineers, Park Ridge, IL.

22

Machine Elements

John P.H. Steele
(Second Edition)
Colorado School of Mines

Gordon R. Pennock
Purdue University

Section 22.1 presents a discussion of threaded fasteners, namely, the nut and bolt, the machine screw, the cap screw, and the stud. Equations are presented for the spring stiffness of the portion of a bolt, or a cap screw, within the clamped zone, which generally consists of the unthreaded shank portion and the threaded portion. Equations for the resultant bolt load and the resultant load on the members are also included in the discussion. The section concludes with a relation that provides an estimate of the torque that is required to produce a given preload. Section 22.2 presents a discussion of clutches and brakes and the important features of these machine elements. Various types of frictional-contact clutches and brakes are included in the discussion, namely, the radial, axial, disk, and cone types. Information on positive-contact clutches and brakes is also provided. The section includes energy considerations, equations for the temperature-rise, and the characteristics of a friction material. Section 22.3 presents material on gears, including computing gear ratios for simple, compound, and epicyclic gear trains. Section 22.4 discusses roller element bearings and their selection for use in machines. Section 22.5 presents the fundamental equations for designing shafts and presents equations for constant and fluctuating torques.

22.1 Threaded Fasteners

The bolted joint with hardened steel washers is a common solution when a connection is required that can be easily disassembled (without destructive methods) and is strong enough to resist external tensile loads and shear loads. The clamping load, which is obtained by rotating the nut until the bolt is close to the elastic limit, stretches or elongates the bolt. This bolt tension will remain as the clamping force, or preload, providing the nut does not loosen. The preload induces compression in the members, which are clamped together, and exists in the connection after the nut has been properly tightened, even if there is no external load. Care must be taken to ensure that a bolted joint is properly designed and assembled [Blake, 1986]. When tightening the connection, the bolt head should be held stationary and the nut twisted. This procedure will ensure that the bolt shank will not experience the thread-friction torque. During the tightening process, the first thread on the nut tends to carry the entire load. However, yielding occurs with some strengthening due to the cold work that takes place, and the load is eventually distributed over about three nut threads. For this reason, it is recommended that nuts should not be reused; in fact, it can be dangerous if this practice is followed [Shigley and Mischke, 1989].

There are several styles of hexagonal nut, namely:

0-8493-1586-7/05/$0.00+$1.50

1. The general hexagonal nut
2. The washer-faced regular nut
3. The regular nut chamfered on both sides
4. The jam nut with washer face
5. The jam nut chamfered on both sides

Flat nuts only have a chamfered top [Shigley and Mischke, 1986]. The material of the nut must be selected carefully to match that of the bolt. Carbon steel nuts are usually made to conform to ASTM A563 Grade A specifications or to SAE Grade 2. A variety of machine screw head styles also exist; they include:

1. Fillister head
2. Flat head
3. Round head
4. Oval head
5. Truss head
6. Binding head
7. Hexagonal head (trimmed and upset)

There are also many kinds of locknuts, which have been designed to prevent a nut from loosening in service. Spring and lock washers placed beneath an ordinary nut are also common devices to prevent loosening.

Another tension-loaded connection uses cap screws threaded into one of the members. Cap screws can be used in the same applications as nuts and bolts and also in situations where one of the clamped members is threaded. The common head styles of the cap screw include:

1. Hexagonal head
2. Fillister head
3. Flat head
4. Hexagonal socket head

The head of a hexagon-head cap screw is slightly thinner than that of a hexagon-head bolt. An alternative to the cap screw is the stud, which is a rod threaded on both ends. Studs should be screwed into the lower member first, then the top member should be positioned and fastened down with hardened steel washers and nuts. The studs are regarded as permanent and the joint should be disassembled by removing only the nuts and washers. In this way, the threaded part of the lower member is not damaged by reusing the threads.

The grip of a connection is the total thickness of the clamped material [Shigley and Mischke, 1989]. In the bolted joint, the grip is the sum of the thicknesses of both the members and the washers. In a stud connection, the grip is the thickness of the top member plus that of the washer. The spring stiffness, or spring rate, of an elastic member such as a bolt is the ratio of the force applied to the member and the deflection caused by that force. The spring stiffness of the portion of a bolt, or cap screw, within the clamped zone generally consists of two parts, namely (1) that of the threaded portion and (2) that of the unthreaded shank portion. Therefore, the stiffness of a bolt is equivalent to the stiffness of two springs in series:

$$\frac{1}{k_b} = \frac{1}{k_T} + \frac{1}{k_d} \quad \text{or} \quad k_b = \frac{k_T k_d}{k_T + k_d} \tag{22.1}$$

The spring stiffnesses of the threaded and unthreaded portions of the bolt in the clamped zone, respectively, are

$$k_T = \frac{A_t E}{L_T} \quad \text{and} \quad k_d = \frac{A_d E}{L_d} \tag{22.2}$$

where A_t is the tensile-stress area, L_T is the length of the threaded portion *in the grip*, A_d is the major-diameter area of the fastener, L_d is the length of the unthreaded portion (shank), and E is the modulus of elasticity. Substituting Equation (22.2) into Equation (22.1), the estimated effective stiffness of the bolt (or cap screw) in the clamped zone can be expressed as

$$k_b = \frac{A_t A_d E}{A_t L_d + A_d L_T} \tag{22.3}$$

For short fasteners, the unthreaded area is small and so the first of the expressions in Equation (22.2) can be used to evaluate k_b. In the case of long fasteners, the threaded area is relatively small, so the second expression in Equation (22.2) can be used to evaluate the effective stiffness of the bolt. Expressions can also be obtained for the stiffness of the members in the clamped zone [Juvinall, 1983]. Both the stiffness of the fastener and the stiffness of the members in the clamped zone must be known in order to understand what happens when the connection is subjected to an external tensile load. There may be more than two members included in the grip of the fastener. Taken together the members act like compressive springs in series, and hence the total spring stiffness of the members is

$$\frac{1}{k_m} = \frac{1}{k_1} + \frac{1}{k_2} + \frac{1}{k_3} + \cdots \tag{22.4}$$

If one of the members is a soft gasket, its stiffness relative to the other members is usually so small that for all practical purposes the other members can be neglected and only the gasket stiffness need be considered. If there is no gasket, the stiffness of the members is difficult to obtain, except by experimentation, because the compression spreads out between the bolt head and the nut and hence the area is not uniform. There are, however, some cases in which this area can be determined. Ultrasonic techniques have been used to determine the pressure distribution at the member interface in a bolt-flange assembly [Ito et al., 1977]. The results show that the stress stays high out to about 1.5 times the bolt radius and then falls off farther away from the bolt. Rotsher's pressure-cone method has been suggested for stiffness calculations with a variable cone angle. This method is quite complicated; a simpler approach is to use a fixed cone angle [Little, 1967]. Both Shigley and Mischke and Juvinall and Marshek show examples of computing the material stiffness using the cone model. A cone angle of 30° has been shown to be best for standard engineering materials.

More recently, Wideman et al. (1991) used FEA (finite element analysis) to model the stiffness of clamped material for steel, aluminum, copper, and cast iron. By representing the problem in a nondimensionalized manner, they were able to develop relationships that characterized the stiffness of the material so that it could be represented by an exponential curve that could be represented by a set of parameters, A and B in the equation, i.e.,

$$\frac{k_m}{Ed} = Ae^{B(d/L)} \tag{22.5}$$

where k_m is the material stiffness, E is modulus of elasticity, d is the diameter of the bolt, L is the length of the clamped material, and A and B are the parameters that have been determined of the material being clamped, e.g., steel. Wideman et al. include a table of values as follows:

Material	Elastic Modulus [GPa]	*A*	*B*
Steel	206.8	0.78715	0.63816
Aluminum	71.0	0.79670	0.63816
Copper	118.6	0.79568	0.63553
Gray cast iron	100.0	0.77871	0.61616

Using this approach, the material stiffness can be computed in a straightforward manner. For example, for a bolted joint with 50 mm of clamper material steel and a 20-mm diameter bolt, the material stiffness would be

$$k_m = EdAe^{B(d/L)} = (206.8e9N/m^2)(0.02m)(0.078715)e^{0.62873(0.020/0.05)} = 4.187e9N/m \tag{22.6}$$

As the authors state, this approach should be used with full appreciation for the assumptions that have been made in order to arrive at this result; otherwise, its use could lead to nonconservative results. Specifically, the following assumptions were made:

1. The distance from the bolt axis to the edge of the members is several times the bolt diameter.
2. Thread friction is not excessive.
3. Shear loads are not excessive.
4. There is no slippage at the member interface.
5. The surface finish of the members is not overly rough.

The first assumption is the most important and one that designers will want to consider closely. If the design results in less than two bolt diameters between the edge and bolt hole, a larger safety factor is recommended, since these equations will compute a larger material stiffness than in fact exists. If the joint contains an unconfined gasket, then each of the materials must be considered separately and then combined using Equation (22.4), above.

Consider what happens when an external tensile load is applied to a bolted connection. Assuming that the preload has been correctly applied (by tightening the nut before the external tensile load is applied), the tensile load causes the connection to stretch through some distance. This elongation can be related to the stiffness of the bolts, or the members, by the equation

$$\delta = \frac{P_b}{k_b} = \frac{P_m}{k_m} \quad \text{or} \quad P_b = \frac{k_b}{k_m} P_m \tag{22.7}$$

where P_b is the portion of the external tensile load P taken by the bolt and P_m is the portion of P taken by the members. Since the external tensile load P is equal to $P_b + P_m$,

$$P_b = \left(\frac{k_b}{k_b + k_m}\right) P \quad \text{and} \quad P_m = \left(\frac{k_m}{k_b + k_m}\right) P \tag{22.8}$$

The resultant bolt load is $F_b = P_b + F_i$ and the resultant load on the members is $F_m = P_m - F_i$, where F_i is the preload. Therefore, the resultant bolt load can be written as

$$F_b = \left(\frac{k_b}{k_b + k_m}\right) P + F_i, \quad F_m < 0 \tag{22.9}$$

and the resultant load on the members can be written as

$$F_m = \left(\frac{k_m}{k_b + k_m}\right) P - F_i, \quad F_m < 0 \tag{22.10}$$

The best joint designs are ones in which the largest portion of the load is taken by the material, rather than the bolt. When the material carries more of the load, the bolt will see smaller stress changes when the joint is exposed to fluctuating loads. Thus one can see from Equation (22.9) and Equation (22.10) that the goal is to have the material have the larger stiffness since this will lead to the material carrying the larger load. Note, however, that Equation (22.9) and Equation (22.10) are only valid for the case when some clamping load remains in the members, which is indicated by the qualifier in the two equations. Making the grip longer causes the members to take an even greater percentage of the external load. However, if the external load is large enough to completely remove the compression, then the members will separate and the entire load will be carried by the bolts.

Since it is desirable to have a high preload in important bolted connections, methods of ensuring that the preload is actually developed when the parts are assembled must be considered. If the overall length of the bolt, L_b, can be measured (say with a micrometer) when the parts are assembled, then the bolt elongation due to the preload F_i can be computed from the relation

$$\delta = \frac{F_i L_b}{AE} \tag{22.11}$$

where A is the cross-sectional area of the bolt. The nut can then be tightened until the bolt elongates through the distance δ, which ensures that the desired preload has been obtained. In many cases, however, it is not practical or possible to measure the bolt elongation. For example, the elongation of a screw cannot be measured if the threaded end is in a blind hole. In such cases the wrench torque that is required to develop the specified preload must be estimated. Torque wrenching, pneumatic-impact wrenching, or the **turn-of-the-nut method** can be used [Blake and Kurtz, 1965]. The torque wrench has a built-in dial that indicates the proper torque. With pneumatic-impact wrenching, the air pressure is adjusted so that the wrench stalls when the proper torque is obtained or, in some cases, the air shuts off automatically at the desired torque.

The **snug-tight condition** is defined as the tightness attained by a few impacts of an impact wrench or the full effort of a person using an ordinary wrench. When the snug-tight condition is attained, all additional turning develops useful tension in the bolt. The turn-of-the-nut method requires that the fractional number of turns necessary to develop the required preload from the snug-tight condition be computed. For example, a 1-in. UNC bolt and nut that has a perfect fitup in its snug-tight condition would call for a turn of the nut of approximately 30° to bring the bolt tension up to its proof strength. A good estimate of the torque required to produce a given preload F_i can be obtained from the relation [Shigley and Mischke, 1989]

$$T = \frac{F_i d_m}{2}\left(\frac{L + \pi \mu d_m \sec\alpha}{\pi d_m - \mu L \sec\alpha}\right) + \frac{F_i \mu_c d_c}{2} \tag{22.12}$$

where d_m is the mean diameter of the bolt, L is the lead of the thread, α is half the thread angle, μ is the coefficient of thread friction, μ_c is the coefficient of collar friction, and d_c is the mean collar diameter. The coefficients of friction depend upon the surface smoothness, the accuracy, and the degree of lubrication. Although these items may vary considerably, it is interesting to note that on the average both μ and μ_c are approximately 0.15. Using the standard thread angle for SAE/ISO threads (60°) and 0.15 for friction, and approximating the mean diameter by the major diameter, we can obtain the following expression

$$T = 0.21 d F_i \tag{22.13}$$

where d is the major diameter of the bolt and F_i is the desired initial load. This expression can be used to compute the torque necessary to develop the desired clamping force in the bolt.

22.2 Clutches and Brakes

A clutch is a coupling that connects two shafts rotating at different speeds and brings the output shaft smoothly and gradually to the same speed as the input shaft. Clutches and brakes are machine elements associated with rotation and have in common the function of storing or transferring rotating energy [Remling, 1983]. When the rotating members are caused to stop by means of a brake, the kinetic energy of rotation must be absorbed by the brake. In the same way, when the members of a machine that are initially at rest are brought up to speed, slipping must occur in the clutch until the driven members have the same speed as the driver. Kinetic energy is absorbed during slippage of either a clutch or a brake, and this energy appears in the form of heat. The important features in the performance of these devices are:

1. The actuating force
2. The transmitted torque
3. The energy loss
4. The temperature rise

The torque that is transmitted is related to the actuating force, the coefficient of friction, and the geometry of the device. Essentially this is a problem in statics and can be studied separately for each geometric configuration. The rise in temperature, however, can be studied without regard to the type of device because the heat-dissipating surfaces are the geometry of interest. An approximate guide to the rise in temperature in a drum brake is the horsepower per square inch [Spotts, 1985].

The torque capacity of a clutch or brake depends upon the coefficient of friction of the material and a safe normal pressure. The character of the load may be such, however, that if this torque value is permitted, the clutch or brake may be destroyed by the generated heat. Therefore, the capacity of a clutch is limited by two factors:

1. The characteristics of the material
2. The ability of the clutch to dissipate the frictional heat

The temperature rise of a clutch or brake assembly can be approximated by the relation

$$\Delta T = \frac{H}{CW} \tag{22.11}$$

where ΔT is in °F, H is the heat generated in Btu, C is the specific heat in Btu/(lbm °F), and W is the mass of the clutch or brake assembly in lbm. If SI units are used, then

$$\Delta T = \frac{E}{Cm} \tag{22.12}$$

where ΔT is in °C, E is the total energy dissipated during the clutching operation or the braking cycle in J, C is the specific heat in J/(kg °C), and m is the mass of the clutch or brake assembly in kg. Equation (22.11) or Equation (22.12) can be used to explain what happens when a clutch or a brake is operated. However, so many variables are involved that it is most unlikely that the analytical results would approximate experimental results. For this reason, such analyses are only useful, for repetitive cycling, in pinpointing the design parameters that have the greatest effect on performance.

The friction material of a clutch or brake should have the following characteristics, to a degree that is dependent upon the severity of the service:

1. A high and uniform coefficient of friction
2. Imperviousness to environmental conditions, such as moisture
3. The ability to withstand high temperatures, as well as good heat conductivity

4. Good resiliency
5. High resistance to wear, scoring, and galling

The manufacture of friction materials is a highly specialized process, and the selection of a friction material for a specific application requires some expertise. Selection involves a consideration of all the characteristics of a friction material as well as the standard sizes that are available. The woven-cotton lining is produced as a fabric belt, which is impregnated with resins and polymerized. It is mostly used in heavy machinery and can be purchased in rolls up to 50 feet in length. The thicknesses that are available range from 0.125 to 1 in. and the width may be up to 12 in. A woven-asbestos lining is similar in construction to the cotton lining and may also contain metal particles. It is not quite as flexible as the cotton lining and comes in a smaller range of sizes. The woven-asbestos lining is also used as a brake material in heavy machinery. Due to health concerns, asbestos use has been reduced in recent years.

Molded-asbestos linings contain asbestos fiber and friction modifiers; a thermoset polymer is used, with heat, to form a rigid or a semirigid molding. The principal use is in drum brakes. Molded-asbestos pads are similar to molded linings but have no flexibility; they are used for both clutches and brakes. Sintered-metal pads are made of a mixture of copper and/or iron particles with friction modifiers, molded under high pressure and then heated to a high temperature to fuse the material. These pads are used in both brakes and clutches for heavy-duty applications. Cermet pads are similar to the sintered-metal pads and have a substantial ceramic content. Typical brake linings may consist of a mixture of asbestos fibers to provide strength and ability to withstand high temperatures; various friction particles to obtain a degree of wear resistance and higher coefficient of friction; and bonding materials. Some clutch friction materials may be run wet by allowing them to dip in oil or to be sprayed by oil. This reduces the coefficient of friction, but more heat can be transferred and higher pressure can be permitted.

The two most common methods of coupling are the frictional-contact clutch and the positive-contact clutch. Other methods include the overrunning or freewheeling clutch, the magnetic clutch, and the fluid coupling. In general, the types of frictional-contact clutches and brakes can be classified as rim type or axial type [Marks, 1987]. The analysis of all types of frictional clutches and brakes follows the same general procedure, namely:

1. Determine the pressure distribution on the frictional surfaces.
2. Find a relation between the maximum pressure and the pressure at any point.
3. Apply the conditions of static equilibrium to find the actuating force, the torque transmitted, and the support reactions.

The analysis is useful when the dimensions are known and the characteristics of the friction material are specified. In design, however, synthesis is of more interest than analysis. Here the aim is to select a set of dimensions that will provide the best device within the limitations of the frictional material that is specified by the designer [Proctor, 1961].

Rim-Type Clutches and Brakes

The rim-type brake can be designed for **self-energizing**, that is, using friction to reduce the actuating force. Self-energization is important in reducing the required braking effort; however, it also has a disadvantage. When rim-type brakes are used as vehicle brakes, a small change in the coefficient of friction will cause a large change in the pedal force required for braking. For example, it is not unusual for a 30% reduction in the coefficient of friction (due to a temperature change or moisture) to result in a 50% change in the pedal force required to obtain the same braking torque that was possible prior to the change.

The rim types may have internal expanding shoes or external contracting shoes. An internal shoe clutch consists essentially of three elements:

1. A mating frictional surface
2. A means of transmitting the torque to and from the surfaces
3. An actuating mechanism

Depending upon the operating mechanism, such clutches can be further classified as expanding-ring, centrifugal, magnetic, hydraulic, or pneumatic. The expanding-ring clutch benefits from centrifugal effects, transmits high torque even at low speeds, and requires both positive engagement and ample release force. This type of clutch is often used in textile machinery, excavators, and machine tools in which the clutch may be located within the driving pulley. The centrifugal clutch is mostly used for automatic operations. If no spring is present, the torque transmitted is proportional to the square of the speed [Beach, 1962]. This is particularly useful for electric motor drives in which, during starting, the driven machine comes up to speed without shock. Springs can be used to prevent engagement until a certain motor speed has been reached, but some shock may occur. Magnetic clutches are particularly useful for automatic and remote-control systems and are used in drives subject to complex load cycles. Hydraulic and pneumatic clutches are useful in drives with complex loading cycles, in automatic machinery, and in manipulators. Here the fluid flow can be controlled remotely using solenoid valves. These clutches are available as disk, cone, and multiple-plate clutches.

In braking systems, the internal-shoe or drum brake is used mostly for automotive applications. The actuating force of the device is applied at the end of the shoe away from the pivot. Since the shoe is usually long, the distribution of the normal forces cannot be assumed to be uniform. The mechanical arrangement permits no pressure to be applied at the heel; therefore, frictional material located at the heel contributes very little to the braking action. It is standard practice to omit the friction material for a short distance away from the heel, which also eliminates interference. In some designs the hinge pin is allowed to move to provide additional heel pressure. This gives the effect of a floating shoe. A good design concentrates as much frictional material as possible in the neighborhood of the point of maximum pressure. Typical assumptions made in an analysis of the shoe include the following:

1. The pressure at any point on the shoe is proportional to the distance from the hinge pin (zero at the heel).
2. The effect of centrifugal force is neglected (in the case of brakes, the shoes are not rotating and no centrifugal force exists; in clutch design, the effect of this force must be included in the equations of static equilibrium).
3. The shoe is rigid (in practice, some deflection will occur depending upon the load, pressure, and stiffness of the shoe; therefore, the resulting pressure distribution may be different from the assumed distribution).
4. The entire analysis is based upon a coefficient of friction that does not vary with pressure. Actually, the coefficient may vary with a number of conditions, including temperature, wear, and the environment.

For pivoted external shoe brakes and clutches, the operating mechanisms can be classified as solenoids, levers, linkages or toggle devices, linkages with spring loading, hydraulic devices, and pneumatic devices. It is common practice to concentrate on brake and clutch performance without the extraneous influences introduced by the need to analyze the statics of the control mechanisms. The moments of the frictional and normal forces about the hinge pin are the same as for the internal expanding shoes. It should be noted that when external contracting designs are used as clutches, the effect of the centrifugal force is to decrease the normal force. Therefore, as the speed increases, a larger value of the actuating force is required. A special case arises when the pivot is symmetrically located and also placed so that the moment of the friction forces about the pivot is zero.

Axial-Type Clutches and Brakes

In an axial clutch, the mating frictional members are moved in a direction parallel to the shaft. One of the earliest axial clutches was the cone clutch, which is simple in construction and yet quite powerful. Except for relatively simple installations, however, it has been largely replaced by the disk clutch, which employs one or more disks as the operating members. Advantages of the disk clutch include:

1. No centrifugal effects
2. A large frictional area that can be installed in a small space
3. More effective heat dissipation surfaces
4. A favorable pressure distribution

There are two methods in general use to obtain the axial force necessary to produce a certain torque and pressure (depending upon the construction of the clutch). The two methods are:

1. Uniform wear
2. Uniform pressure

If the disks are rigid then the greatest amount of wear will first occur in the outer areas, since the work of friction is greater in those areas. After a certain amount of wear has taken place, the pressure distribution will change so as to permit the wear to be uniform. The greatest pressure must occur at the inside diameter of the disk in order for the wear to be uniform. The second method of construction employs springs to obtain a uniform pressure over the area.

Disk Clutches and Brakes

No fundamental difference exists between a disk clutch and a disk brake [Gagne, 1953]. The disk brake has no self-energization and, hence, is not as susceptible to changes in the coefficient of friction. The axial force can be written as

$$F_a = 0.5\pi p D_1(D_2 - D_1) \tag{22.13}$$

where p is the maximum pressure, and D_1 and D_2 are the inner and outer diameters of the disk, respectively. The torque transmitted can be obtained from the relation

$$T = 0.5\mu F_a D_m \tag{22.14}$$

where μ is the coefficient of friction of the clutch material, and the mean diameter

$$D_m = 0.5(D_2 + D_1) \text{ or } D_m = \frac{2(D_2^3 - D_1^3)}{3(D_2^2 - D_1^2)} \tag{22.15}$$

for uniform wear or for uniform pressure distribution, respectively.

A common type of disk brake is the floating caliper brake. In this design the caliper supports a single floating piston actuated by hydraulic pressure. The action is much like that of a screw clamp, with the piston replacing the function of the screw. The floating action also compensates for wear and ensures an almost constant pressure over the area of the friction pads. The seal and boot are designed to obtain clearance by backing off from the piston when the piston is released.

Cone Clutches and Brakes

A cone clutch consists of:

1. A cup (keyed or splined to one of the shafts)
2. A cone that slides axially on the splines or keys on the mating shaft
3. A helical spring to hold the clutch in engagement

The clutch is disengaged by means of a fork that fits into the shifting groove on the friction cone. The axial force, in terms of the clutch dimensions, can be written as

$$F_a = \pi D_m pb \sin \alpha \tag{22.16}$$

where p is the maximum pressure, b is the face width of the cone, D_m is the mean diameter of the cone, and α is one-half the cone angle in degrees. The mean diameter can be approximated as $0.5(D_2 + D_1)$. The torque transmitted through friction can be obtained from the relation

$$T = \frac{\mu F_a D_m}{2 \sin \alpha} \tag{22.17}$$

The cone angle, the face width of the cone, and the mean diameter of the cone are the important geometric design parameters. If the cone angle is too small, say, less than about 8°, the force required to disengage the clutch may be quite large. The wedging effect lessens rapidly when larger cone angles are used. Depending upon the characteristics of the friction materials, a good compromise can usually be found using cone angles between 10° and 15°. For clutches faced with asbestos, leather, or a cork insert, a cone angle of 12.5° is recommended.

Positive-Contact Clutches

A positive-contact clutch does not slip, does not generate heat, cannot be engaged at high speeds, sometimes cannot be engaged when both shafts are at rest, and, when engaged at any speed, is accompanied by shock. The greatest differences among the various types of positive-contact clutches are concerned with the design of the jaws. To provide a longer period of time for shift action during engagement, the jaws may be ratchet shaped, spiral shaped, or gear-tooth shaped. The square-jaw clutch is another common form of a positive-contact clutch. Sometimes a great many teeth or jaws are used, and they may be cut either circumferentially, so that they engage by cylindrical mating, or on the faces of the mating elements. Positive-contact clutches are not used to the same extent as the frictional-contact clutches.

Defining Terms

Snug-tight condition — The tightness attained by a few impacts of an impact wrench or the full effort of a person using an ordinary wrench.

Turn-of-the-nut method — The fractional number of turns necessary to develop the required preload from the snug-tight condition.

Proof strength — The stress at which the bolt *begins* to take on a permanent deformation, i.e., plastic yield.

Self-energizing — A state in which friction is used to reduce the necessary actuating force. The design should make good use of the frictional material because the pressure is an allowable maximum at all points of contact.

Self-locking — When the friction moment assists in applying the brake shoe, the brake will be self-locking if the friction moment exceeds the normal moment. The designer must select the dimensions of the clutch, or the brake, to ensure that self-locking will not occur unless it is specifically desired.

Fail-safe and dead-man — These two terms are often encountered in studying the operation of clutches and brakes. Fail-safe means that the operating mechanism has been designed such that, if any element should fail to perform its function, an accident will not occur in the machine or befall the operator. Dead-man, a term from the railroad industry, refers to the control mechanism that causes the engine to come to a stop if the operator should suffer a blackout or die at the controls.

References

Beach, K. 1962. Try these formulas for centrifugal clutch design. *Product Eng.* 33(14):56–57.

Blake, A. 1986. *What Every Engineer Should Know about Threaded Fasteners: Materials and Design,* p. 202. Marcel Dekker, New York.

Blake, J. C. and Kurtz, H. J. 1965. The uncertainties of measuring fastener preload. *Machine Design.* 37(23): 128–131.

Gagne, A. F., Jr. 1953. Torque capacity and design of cone and disk clutches. *Product Eng.* 24(12): 182–187.

Ito, Y., Toyoda, J., and Nagata, S. 1977. Interface pressure distribution in a bolt-flange assembly. *Trans. ASME.* Paper No. 77-WA/DE-11, 1977.

Juvinall, R. C. 1983. *Fundamentals of Machine Component Design,* p. 761. John Wiley & Sons, New York.

Little, R. E. 1967. Bolted joints: How much give? *Machine Design.* 39(26): 173–175.

Marks, L. S. 1987. *Marks' Standard Handbook for Mechanical Engineers,* 9th ed. McGraw-Hill, New York.

Norton, R. L., 2000. *Machine Design: An Integrated Approach,* 2nd ed. Prentice Hall, Upper Saddle River, NJ.

Proctor, J. 1961. Selecting clutches for mechanical drives. *Product Eng.* 32(25): 43–58.

Remling, J. 1983. *Brakes,* 2nd ed., p. 328. John Wiley & Sons, New York.

Shigley, J. E. and Mischke, C. R. 1986. *Standard Handbook of Machine Design,* McGraw-Hill, New York.

Shigley, J. E. and Mischke, C. R. 1989. *Mechanical Engineering Design,* 5th ed., p. 779. McGraw-Hill, New York.

Spotts, M. E. 1985. *Design of Machine Elements,* 6th ed., p. 730. Prentice Hall, Englewood Cliffs, NJ.

Wideman, J, Choudhury, M., and Green, I., 1991. Computation of Member Stiffness in Bolted Connections, *Trans of ASME, Journal of Mechanical Design,* 113:432–437.

Further Information

ASME Publications Catalog. 1985. *Codes and Standards: Fasteners,* American Society of Mechanical Engineers, New York.

Bickford, J. H. 1981. *An Introduction to the Design and Behavior of Bolted Joints,* p. 443. Marcel Dekker, New York.

Burr, A. H. 1981. *Mechanical Analysis and Design,* p. 640. Elsevier Science, New York.

Crouse, W. H. 1971. *Automotive Chassis and Body,* 4th ed., pp. 262–299. McGraw-Hill, New York.

Fazekas, G. A. 1972. On circular spot brakes. *Journal of Engineering for Industry, Transactions of ASME,* 94B(3):859–863.

Ferodo, Ltd. 1968. *Friction Materials for Engineers,* Chapel-en-le-Frith, England.

Fisher, J. W. and Struik, J. H. A. 1974. *Guide to Design Criteria for Bolted and Riveted Joints,* p. 314. John Wiley & Sons, New York.

ISO Metric Screw Threads. 1981. Specifications BS 3643: Part 2, p. 10. British Standards Institute, London.

Lingaiah, K. 1994. *Machine Design Data Handbook,* McGraw-Hill, New York.

Matthews, G. P. 1964. *Art and Science of Braking Heavy Duty Vehicles,* Special Publication SP-251, Society of Automotive Engineers, Warrendale, PA.

Motosh, N. 1976. Determination of joint stiffness in bolted connections. *Journal of Engineering for Industry, Transactions of ASME,* vol. 98B(3):858–861.

Neale, M. J. (ed.), 1973. *Tribology Handbook,* John Wiley & Sons, New York.

Osgood, C. C. 1979. Saving weight in bolted joints. *Machine Design,* vol. 51:128–133.

Rodkey, E. March 1977. Making fastened joints reliable — ways to keep 'em tight. *Assemble Engineering,* pp. 24–27.

Screw Threads. 1974. ANSI Specification B1.1-1974, p. 80. American Society of Mechanical Engineers, New York.

Viglione, J. 1965. Nut design factors for long bolt life. *Machine Design,* vol. 37(18):137–141.

Wong, J. Y. 1993. *Theory of Ground Vehicles,* 2nd ed., p. 435. John Wiley & Sons, New York.

Dedication

This chapter is dedicated to the late Professor Joseph Edward Shigley who authored and coauthored several outstanding books on engineering design. The *Standard Handbook of Machine Design* and the *Mechanical Engineering Design* text (both with C. R. Mischke, see the references above) are widely used and strongly influenced the direction of this chapter.

23

Crankshaft Journal Bearings

P. K. Subramanyan
Glacier Clevite Heavywall Bearings

In modern internal combustion engines, there are two kinds of bearings in the category of crankshaft journal bearings — namely, the main bearings and the connecting rod bearings. Basically, these are wraparound, semicylindrical shell bearings. Two of them make up a set and, depending on the position in the assembly, one is called the upper and the other the lower bearing. They are of equal sizes. The main bearings support the crankshaft of the engine and the forces transmitted to the crankshaft from the cylinders. The connecting rod bearings (or simply, rod bearings) are instrumental in transferring the forces from the cylinders of the internal combustion engine to the crankshaft. These connecting rod bearings are also called big end bearings or crank pin bearings. Supporting the crankshaft and transferring the pressure-volume work from the cylinders to the pure rotational mechanical energy of the crankshaft are accomplished elegantly with minimal energy loss by shearing a suitable lubricating medium between the bearings and the journals. The segment of the crankshaft within the bounds of a set of bearings, whether main bearings or rod bearings, is called the journal. Consequently, these bearings are called journal bearings.

23.1 Role of the Journal Bearings in the Internal Combustion Engine

The crankshafts of internal combustion engines of sizes from small automotive to large slow-speed engines run at widely varying rpm (e.g., 72 to 7700). When the internal combustion engine continues to run after the startup, the crankshaft, including the crank pins, is suspended in the lubricating oil — a fluid

0-8493-1586-7/05/$0.00+$1.50

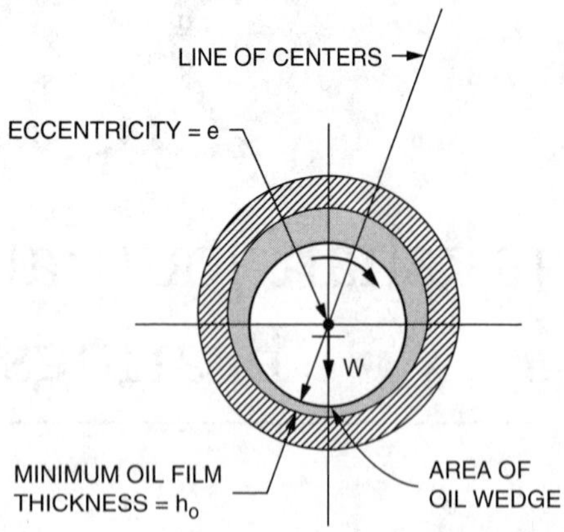

FIGURE 23.1 Schematic representation of the hydrodynamic lubricant film around a rotating journal in its bearing assembly. (*Source:* Slaymaker, R. R. 1955. *Bearing Lubrication Analysis,* John Wiley & Sons, New York. With permission.)

of very low friction. In such a condition, it is conceivable that precision-machined, semicylindrical steel shells can function as good bearings. However, there are stressful conditions, particularly in the case of automotive, truck, and medium-speed engines, when the crankshaft remains in contact with the bearings and little or no lubricating oil is present. This condition corresponds to the initial and subsequent startups. The oil pump is driven directly by the engine, and it takes several revolutions of the crankshaft before a good oil film is developed, as shown in Figure 23.1, so that the journals are completely lifted and suspended. During the revolutions prior to the formation of a sufficiently thick oil film, the journal contacts the bearing surface. In such situations, the bearings provide sufficient lubrication to avoid scuffing and **seizure**. Another stressful situation, but not as critical as the startup, is the slowing down and shutting off of the engine when the oil film reduces to a **boundary layer**.

In the case of slow-speed engines, the oil pump, which is electrically driven, is turned on to prelubricate the bearings. This provides some lubrication. Nonetheless, bearings with liners and overlays are used to avoid seizure, which can result in costly damage.

Essentially, the function of journal bearings can be stated as follows:

> Development of the **hydrodynamic lubrication oil films** in the journal bearings lifts the journals from the surfaces of the bearings and suspends the entire crankshaft on the oil films by the journals. [Theoretical aspects of this will be considered later.] The lifting of the crankshaft or, equivalently, lifting of the journals is in the range of 30 to 1000 micro-inch in the entire range of IC engines. This process allows the crankshaft to rotate with minimal energy loss. The journal bearings make it possible so that the internal combustion engine can be started, utilized, and stopped as many times as needed.

23.2 Construction of Modern Journal Bearings

The majority of modern crankshaft journal bearings have three different layers of metallic materials with distinct characteristics and functions. Conventionally, these are called trimetal bearings. The remaining bearings belong to the class of bimetal bearings and have two different metallic material layers. Bimetallic bearings are becoming very popular in the automotive industry.

All crankshaft journal bearings have a steel backing, normally of low-carbon steel. Steel backing is the thickest layer in the bearing. The next layer bonded to the steel backing is the bearing liner. This is the layer that supports the load and determines the life of the bearing. The third layer bonded to the bearing liner is the overlay. Generally, this is a precision electrodeposited layer of (1) lead, tin, and copper, (2) lead and tin, or (3) lead and indium. A very thin electrodeposited layer of nickel (0.000 05 in.) is used as a bonding layer between the liner and the lead-tin-copper overlay. This nickel layer is considered a part of the overlay, not a separate layer. Construction of a trimetal bronze bearing is illustrated in Figure 23.2.

There are two classes of bearing liners in widespread use nowadays. These are leaded bronzes and aluminum-based (frequently precipitation-strengthened) materials, such as aluminum-tin and aluminum-silicon. Bimetallic bearings have the advantage of being slightly more precise (about 0.0002 to 0.0003 in.) than trimetal bearings. The bimetal bearings have a bored or broached internal diametral (ID) surface. The electrodeposited layer in trimetal bearings is applied onto the bored or broached surface. The nickel bonding layer is applied first onto the liner, followed by the deposition of the lead-tin-copper overlay. The electrodeposited overlay introduces a certain degree of variation in the wall thickness of the bearings. In a limited application, babbitt overlays are centrifugally cast on bronze liners for slow-speed diesel engine journal bearings.

Another class of bearings is single layer solid metal bearings — namely, solid bronze and solid aluminum bearings. These bearings are not generally used as crankshaft journal bearings. However, solid aluminum is used in some medium-speed and slow-speed diesel engines.

The most popular copper-tin-based leaded bearing liner in current use has 2 to 4% tin, 23 to 27% lead, and 69 to 75% copper (all by weight). This material is applied directly on mild steel by casting or sintering. The aluminum materials are roll-bonded to steel. The material as such is produced by powder rolling as a strip or by casting and rolling.

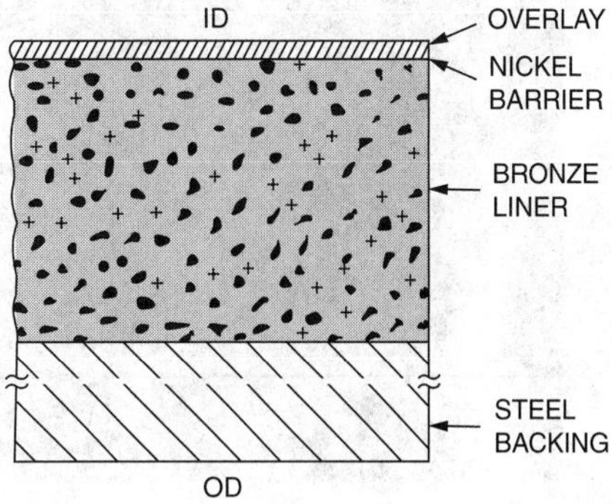

FIGURE 23.2 Schematic representation of the construction of a trimetal bearing.

23.3 The Function of the Different Material Layers in Crankshaft Journal Bearings

The bulk of modern crankshaft journal bearings is mild steel (1008 to 1026 low-carbon steels). This is the strongest of the two or three layers in the bearing. It supports the bearing liner, with or without the overlay. The bearing liner derives a certain degree of strength from the steel backing. The function of the steel backing is to carry the bearing liner, which on its own is weaker, much thinner, and less ductile. With the support of the steel backing, the bearings can be seated with a high degree of conformance and good interference fit in the housing bore (steel against steel).

The bearing liners in automotive and truck bearings have a thickness in the range of 0.006 to 0.030 in. In the case of medium-speed and slow-speed engines, the thickness of the liner ranges from 0.010 to 0.080 in. The liner material contains sufficient amounts of antifriction elements, such as lead and tin. Lead is the most valuable antifriction element in the current materials and is present as a separate phase in the matrix of copper-tin alloy in the leaded bronze materials. Similarly, tin is present as an insoluble second phase in the matrix of aluminum-based materials. Lead is also insoluble in the aluminum matrix. The liner materials play the most critical role in the bearings. Once the liner material is damaged significantly, the bearing is considered unfit for further use. In a trimetal bearing, when the overlay is lost due to wear or fatigue, the bronze liner will continue to support the load and provide adequate lubrication in times of stress. The friction coefficient of liner materials is designed to be low. Besides, the soft phases of lead (in bronze) and tin (in aluminum) function as sites for embedment of dirt particles.

The overlay, which by definition is the top layer of the bearing surface, is the softest layer in the bearing. Its functions are to provide lubrication to the journal in initial start-up situations, adjust to any misalignment or out-of-roundness of the journal, and capture dirt particles by embedment. The overlay provides sufficient lubrication during the subsequent start-up and shut-down conditions also. The journal makes a comfortable running environment in the bearing assembly during the initial runs by "bedding in." As a result of this, the wear rate of the overlay is higher in the beginning. As long as the overlay is present, the phenomenon of seizure will not occur. Once the wear progresses through the overlay, the bearing liner will provide adequate lubrication during start-up and shut-down conditions. However, if the oil supply is severely compromised or cut off for more than several seconds to a minute or so, seizure can take place once the overlay is gone, depending on the nature of the bearing liner and the load.

23.4 The Bearing Materials

All modern crankshaft journal bearing materials are mainly composed of five elements — namely, copper, aluminum, lead, tin, and silicon. These elements account for the leaded bronze and aluminum-tin,

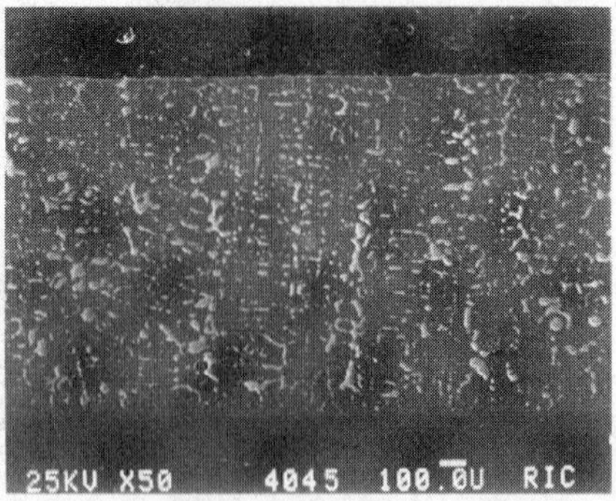

FIGURE 23.3 SEM photomicrograph of a typical cross-section of the cast leaded bronze diesel locomotive engine bearing material manufactured by Glacier Clevite Heavy-wall Bearings. The nominal composition is 3% tin, 25% lead, and 72% copper. The light gray, irregular spots represent lead in a matrix of copper-tin. This material is bonded to mild steel at the bottom. (Magnification 50×)

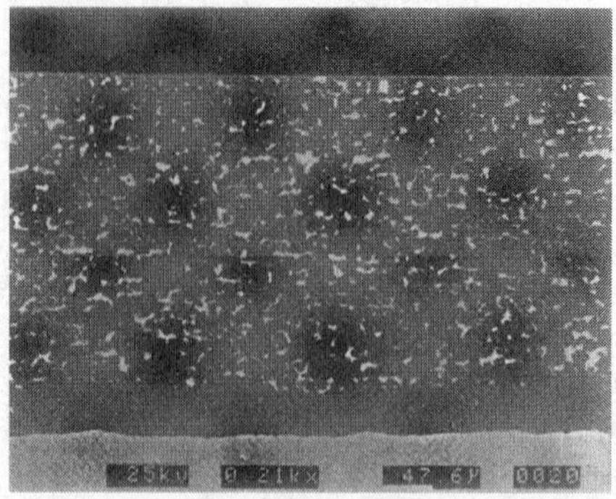

FIGURE 23.4 SEM photomicrograph of a typical cross-section of aluminum-tin material roll bonded to mild steel, manufactured by Glacier Vandervell Ltd. The nominal composition is 20% tin, 1% copper, and 79% aluminum. The light gray, irregular spots represent tin in the aluminum-copper matrix. Below the aluminum-tin layer is a layer of pure aluminum which functions as a bonding layer to the mild steel underneath. (Magnification 210×)

aluminum-lead, and aluminum-silicon materials. Indium is used as a constituent of the overlays. Antimony is used in babbitts. Silver is a bearing material with good tribological properties, but it is too expensive to use as a bearing liner in journal bearings. However, it is used in special applications in some locomotive engines. An important characteristic of a good bearing material is its ability to conduct heat. Silver, copper, and aluminum are, indeed, good conductors of heat. Silver has no affinity for iron, cobalt, and nickel [Bhushan and Gupta, 1991]. Therefore, it is expected to run very well against steel shafts. Both copper and aluminum possess a certain degree of affinity for iron. Therefore, steel journals can bond to these metals in the absence of antifriction elements, such as lead and tin, or lubricating oil. Aluminum spontaneously forms an oxide layer, which is very inert, in the presence of air or water vapor. This suppresses the seizure or the bonding tendency of aluminum. Besides, the silicon particles present in the aluminum-silicon materials keep the journals polished to reduce friction.

The microstructure of the most widely used cast leaded bronze bearing liner is shown in Figure 23.3. This has a composition of 2 to 4% tin, 23 to 27% lead, and 69 to 75% copper. Another material in widespread use, especially in automotive applications, is aluminum with 20% tin. A typical microstructure of this material is shown in Figure 23.4. It can be used as the liner for both bimetal and trimetal bearings. The copper-tin-lead material shown in Figure 23.3 is mainly used in trimetal bearings.

23.5 Basics of Hydrodynamic Journal Bearing Theory

Load-Carrying Ability

As mentioned previously, when running in good condition, the journal which was initially lying on the surface of the bearing is lifted and surrounded by the lubricant. It becomes suspended in the surrounding film of lubricating oil. If the engine keeps running, the journal will remain in its state of suspension indefinitely. The inertial load of the crankshaft and the forces transmitted from the cylinders to the crankshaft are supported by the lubricant films surrounding the main bearing journals. The oil film surrounding the rod bearing journal supports the gas forces developed in the cylinder and the inertial load of the piston and connecting rod assembly. Around each journal, a segment of the oil film develops a positive pressure to support the load, as shown in Figure 23.5. In the following brief theoretical consideration, the process that develops this load-carrying positive pressure will be illustrated.

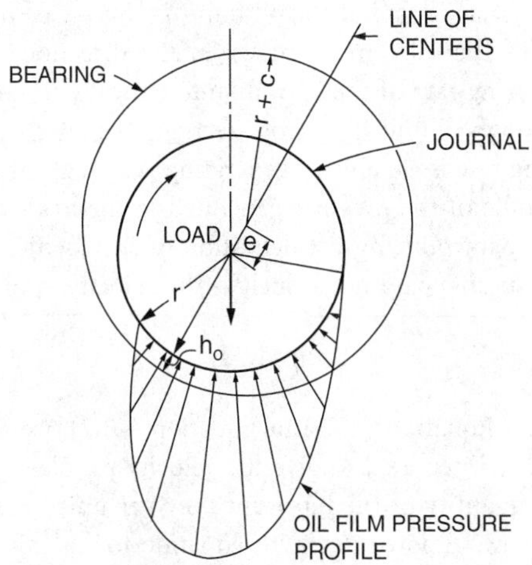

FIGURE 23.5 Schematic representation of the profile of the load supporting pressure in the oil film. (Source: Slaymaker, R. R. 1955. *Bearing Lubrication Analysis,* John Wiley & Sons, New York. By permission.)

As a background to the theoretical considerations, the following assumptions are made. The flow of the lubricating oil around the journal at all speeds is assumed to be laminar. The length of the bearing L is assumed to be infinite, or the flow of the lubricant from the edges of the bearing is negligible. The lubricant is assumed to be incompressible.

Consider a very small volume element of the lubricant moving in the direction of rotation of the journal — in this case, the x direction. The forces that act on this elemental volume and stabilize it are shown in Figure 23.6. Here, P is the pressure in the oil film at a distance x. It is independent of the thickness of the oil film or the y dimension. S is the shear stress in the oil film at a distance y above the bearing surface, which is at $y = 0$. The length L of the bearing is in the z direction. The equilibrium condition of this volume element gives us the following relationship [Slaymaker, 1955; Fuller, 1984]:

$$\left[P+\left(\frac{dP}{dx}\right)dx\right]dy\,dz+S\,dx\,dz-\left[S+\left(\frac{dS}{dy}\right)dy\right]dx\,dz-P\,dy\,dz=0 \tag{23.1}$$

Therefore,

$$\left(\frac{dS}{dy}\right)=\left(\frac{dP}{dx}\right) \tag{23.2}$$

Equation (23.2) represents a very important, fundamental relationship. It clearly shows how the load-carrying pressure P is developed. It is the rate of change of the shear stress in the direction of the oil film thickness that generates the hydrostatic pressure P. As we shall see from Equation (23.3), the shear stress

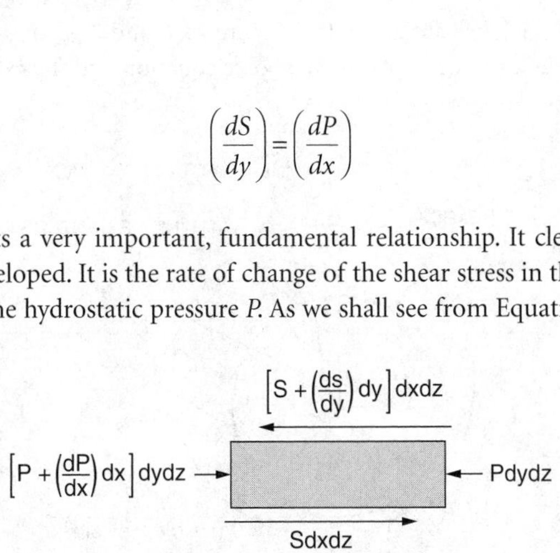

FIGURE 23.6 Schematic representation of the forces acting on a tiny volume element in the hydrodynamic lubricant film around a rotating journal.

is directly proportional to the shearing rate of the oil film (*dv/dy*) — as (*dv/dy*) increases, (*dS/dy*) must increase. Since the thickness of the oil film decreases in the direction of rotation of the journal, a progressive increase in the shearing rate of the oil film automatically occurs because the same flow rate of oil must be maintained through diminishing cross-sections (i.e., decreasing *y* dimension). This progressive increase in the shearing rate is capable of generating very high positive hydrostatic pressures to support very high loads. A profile of the pressure generated in the load-supporting segment of the oil film is shown in Figure 23.5. By introducing the definition of the coefficient of viscosity, we can relate the shear stress to a more measurable parameter, such as the velocity, *v*, of the lubricant, as

$$S = \mu(dv/dy) \tag{23.3}$$

Substituting for (*dS/dy*) from Equation (23.3) in Equation (23.2), we obtain a second-order partial differential equation in *v*. This is integrated to give the velocity profile as a function of *y*. This is then integrated to give Q, the total quantity of the lubricant flow per unit time. Applying certain boundary conditions, one can deduce the well-known Reynolds equation for the oil film pressure:

$$\left(\frac{dP}{dx}\right) = \frac{6\mu V}{h^3}(h - h_1) \tag{23.4}$$

where h is the oil film thickness, h_1 is the oil film thickness at the line of maximum oil film pressure, and V is the peripheral velocity of the journal. The variable x in the above equation can be substituted in terms of the angle of rotation θ and then integrated to obtain the Harrison equation for the oil film pressure. With reference to the diagram in Figure 23.7, the thickness of the oil film can be expressed as

$$h = c\,(1 + \varepsilon \cos\theta) \tag{23.5}$$

where c is the radial clearance and ε is the eccentricity ratio. The penultimate form of the Harrison equation can be expressed as

$$\int_0^{2\pi} dP = \int_0^{2\pi} \frac{6\mu V r \varepsilon}{c^2}\left[\frac{\cos\theta - \cos\theta_1}{(1+\varepsilon\cos\theta)^3}\right] d\theta = P - P_0 \tag{23.6}$$

where P_0 is the pressure of the lubricant at $\theta = 0$ in Figure 23.7, and θ_1 is the angle at which the oil film pressure is a maximum. Brief derivations of the Reynolds equation and the Harrison equation are given in Section 23.8.

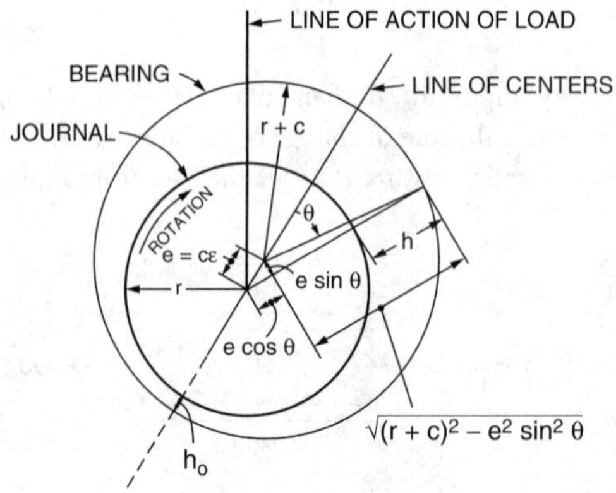

FIGURE 23.7 Illustration of the geometric relationship of a journal rotating in its bearing assembly. (*Source:* Slaymaker, R. R. 1955. *Bearing Lubrication Analysis,* John Wiley & Sons, New York. By permission.)

TABLE 23.1 Side Leakage Correction Factors for Journal Bearings

L/D	Eccentricity Ratio						
Ratio	0.80	0.90	0.92	0.94	0.96	0.98	0.99
0	1.0	1.0	1.0	1.0	1.0	1.0	1.0
2	—	0.867	0.88	0.905	0.937	0.97	0.99
1	0.605	0.72	0.745	0.79	0.843	0.91	0.958
0.5	0.33	0.50	0.56	0.635	0.732	0.84	0.908
0.3	0.17	0.30	0.355	0.435	0.551	0.705	0.81
0.1	—	0.105	0.115	0.155	0.220	0.36	0.53

For practical purposes, it is more convenient to carry out the integration of Equation (23.6) numerically rather than using Equation (23.14) in Section 23.8. This is done with good accuracy using special computer programs. The equations presented above assume that the end leakage of the lubricating oil is equal to zero. In all practical cases, there will be end leakage and hence, the oil film will not develop the maximum possible pressure profile. Therefore, its load-carrying capability will be diminished. The flow of the lubricant in the z direction needs to be taken into account. However, the Reynolds equation for this case has no general solution [Fuller, 1984]. Hence, a correction factor between zero and one is applied, depending on the length and diameter of the bearing (L/D ratio) and the eccentricity ratio of the bearing. Indeed, there are tabulated values available for the side leakage factors for bearings with various L/D ratios and eccentricity ratios [Fuller, 1984]. Some of these values are given in Table 23.1.

Booker [1965] has done considerable work in simplifying the journal center orbit calculations without loss of accuracy by introducing new concepts, such as dimensionless journal center velocity/force ratio (i.e., mobility) and maximum film pressure/specific load ratio (i.e., maximum film pressure ratio). This whole approach is called the *mobility method*. This has been developed into computer programs that are widely used in the industry to calculate film pressures and thicknesses. Further, this program calculates energy loss due to the viscous shearing of the lubricating oil. These calculations are vital for optimizing the bearing design and selecting the appropriate bearing liner with the required fatigue life. This is determined on the basis of the **peak oil film pressure** (POFP). In Booker's mobility method, the bearing assembly, including the housing, is assumed to be rigid. In reality, the bearings and housings are flexible to a certain degree, depending on the stiffness of these components. Corrections are now being made to these deviations by the elastohydrodynamic theory, which involves finite element modeling of the bearings and the housing. Also, the increase in viscosity as a function of pressure is taken into account in this calculation. The elastohydrodynamic calculations are presently done only in very special cases and have not become part of routine bearing analysis.

23.6 The Bearing Assembly

Housing

The housing into which a set of bearings is inserted and held in place is a precision-machined cylindrical bore with close tolerance. The surface finishes of the housing and the backs of the bearings must be compatible. Adequate contact between the backs of the bearings and the surface of the housing bore is a critical requirement to ensure good heat transfer through this interface. The finish of the housing bore is expected to be in the range of 60 to 90 μin. (R_a) (39.4 μin. = 1 micron). The finish on the back of the bearings is generally set at 80 μin. maximum. Nowadays, the finishes on the housing bore and the backs of the bearings are becoming finer. The finish at the parting line face of bearings of less than 12 in. gage size is expected to be less than 63 μin. For larger bearings, this is set at a maximum of 80 μin. The bearing backs may be rolled, turned, or ground. All automotive and truck bearings have rolled steel finish at the back. The housing can be bored, honed, or ground, but care must be taken to avoid circumferential and axial banding.

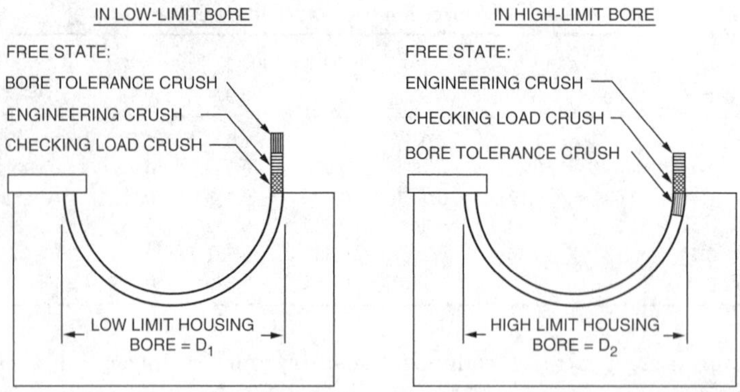

FIGURE 23.8 Schematic illustration of the components of crush of a bearing in the thinwall bearing inspection block, before application of load (i.e., in the free state). The magnitude of the crush components is exaggerated.

The Bearing Crush

The term **crush** is not used in a literal sense in this context. A quantitative measure of the crush of a bearing is equal to the excess length of the exterior circumference of the bearing over half the interior circumference of the bearing housing. Effectively, this is equal to the sum of the two parting line heights. When the bearing assembly is properly torqued, the parting line height of each bearing in the set is reduced to zero. In that state, the back of the bearing makes good contact with the housing and applies a radial pressure in the range of 800 to 1200 psi (5.5 to 8.24 Mpa). Thereby, a good interference fit is generated. If the bearings are taken out of the assembly, they are expected to spring back to their original state. Therefore, nothing is actually crushed.

The total crush or the parting line height of a bearing has three components — namely, the housing bore tolerance crush, the checking load crush, and the engineering crush. The housing bore tolerance crush is calculated as $0.5\pi(D_2 - D_1)$, where D_1 and D_2 are the lower and upper limits of the bore diameter, respectively. Suppose a bearing is inserted in its own inspection block (the diameter of which corresponds to the upper limit of the diameter of the bearing housing). The housing bore tolerance crush does not make a contribution to the actual crush, as shown in Figure 23.8 (high limit bore). If load is applied on its parting lines in increasing order and the values of these loads are plotted as a function of the cumulative decrease in parting line height, one may expect it to obey Hooke's law. Initially, however, it does not obey Hooke's law, but it does so thereafter. The initial nonlinear segment corresponds to the checking load crush. The checking load corresponds to the load required to conform the bearing properly in its housing. The final crush or the parting line height of the bearing is determined in consultation with the engine manufacturer.

Other Factors Affecting Bearing Assembly

These factors are:

1. Freespread
2. Bore distortion
3. Cap offset or twist
4. Misalignment of the crankshaft
5. Out-of-roundness of the journal
6. Deviation of the bearing clearance

The outside diameter of the bearing at the parting lines must be slightly greater than the diameter of the housing bore. This is called the freespread. It helps to snap the bearings into the housing. The required degree of freespread is determined by the wall thickness and the diameter. In the case of wall thickness, the freespread is inversely proportional to it. For a wide range of bearings, the freespread is in the range of 0.025 to 0.075 in. Bearings with negative freespread are not used because, when bolted, the side of the parting lines could rub against the journal and lead to possible seizure while running. It is possible to change the freespread from negative to positive by reforming the bearing. Bore distortion, cap offset or twist, and misalignment of the crankshaft can lead to the journal making rubbing contacts with the bearing surface. The conformability of the bearings can take care of these problems to a certain degree by local wearing of the overlay in a trimetal bearing or by melting the soft phase in a bimetal bearing, which results in the two-phase structure crushing and conforming. In severe cases, the liner materials in both cases are damaged.

By developing high oil film pressures on the peaks of the lobes, out-of-roundness in the journal can accelerate fatigue of the bearing.

If the clearance is not adequate, the bearing will suffer from oil starvation and the temperature will rise. In extreme cases, this will lead to bearing seizure and engine damage. On the other hand, if the clearance is excessive, there will be increased noise and increased peak oil film pressure, which will bring about premature fatigue of the loaded bearing.

23.7 The Design Aspects of Journal Bearings

Even though the journal bearings are of simple semicylindrical shape and apparently of unimpressive features, there are important matters to be taken into account in their design. The bearing lengths, diameters, and wall thicknesses are generally provided by the engine builder or decided in consultation with the bearing manufacturer. A journal orbit study must be done to optimize the clearance space between the journal and the bearing surface. This study also provides the **minimum oil film thickness** (MOFT) and the POFP (Figure 23.9). Values of these parameters for the optimized clearance are impor-

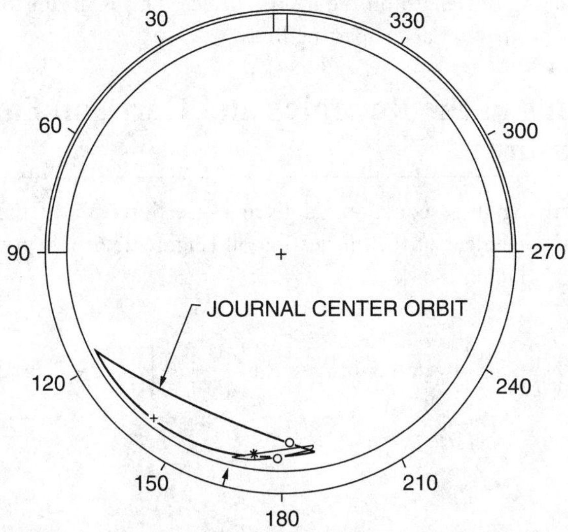

FIGURE 23.9 Journal center orbit diagram of two-stroke cycle medium-speed (900 rpm) diesel engine main bearings (no. 1 position). The inner circle represents the clearance circle of the bearings. It also represents the bearing surface. The entire cross-section of the journal is reduced to a point coinciding with the center of the journal. The upper main bearing has an oil hole at the center with a circumferential groove at the center of the bearing represented by the dark line. Maximum unit load: 1484 psi. MOFT: 151 μin. @ 70/166. POFP: 11 212 psi @ 55/171. Oil: SAE 30W. Cylinder pressure data given by the manufacturer of the engine. Clockwise rotation. The journal orbit analysis was performed at Glacier Clevite Heavywall Bearings. — * — 0–180 crank angle, — + — 180–360 crank angle, @ crank angle/bearing angle. Arrow indicates the location of MOFT.

tant factors. The MOFT is used in the calculation of the oil flow, temperature rise, and heat balance. According to Conway-Jones and Tarver [1993], about 52% of the heat generated in connecting rod bearings in automobile engines is carried away by the oil flow. Approximately 38% of the remaining heat flows into the adjacent main bearings via the crankshaft. The remaining 10% is lost by convection and radiation. In the case of main bearings, about 95% of the total heat is carried away by the oil flow, which is estimated to be more than five times the flow through the connecting rod bearings, which were fed by a single oil hole drilled in the crank pin. The POFP is the guiding factor in the selection of a bearing liner with adequate fatigue strength or fatigue life.

The bearing must be properly located in the housing bore. This is achieved by having a notch at one end of the bearing at the parting line. There must be provisions to bring in the lubricant and remove it. Therefore, appropriate grooves and holes are required. The best groove to distribute the lubricant is a circumferential groove with rounded edges, centrally placed in both bearings. If this is a square groove, the flow will be diminished by 10%. If these grooves are in the axial direction, the oil flow is decreased by 60% with respect to the circumferential ones. Having a circumferential groove in the loaded half of the bearings does increase the POFP. In the case of large slow-speed diesel engines, the POFPs are generally very low compared to the pressures in automotive, truck, and medium-speed diesel engines. Therefore, central circumferential grooves are best suited for slow-speed engines.

In the automotive, truck, and medium-speed engines, the loaded halves of the bearings do not have circumferential grooves. However, the other halves have circumferential grooves. Some of the loaded bearings have partial grooves. Otherwise, some type of oil spreader machined in the location below the parting line is desirable in the case of larger bearings. If the oil is not spread smoothly, problems of cavitation and erosion may show up. The end of the partial groove or the oil spreader must be blended.

The edges of all the bearings must be rounded or chamfered to minimize the loss of the lubricant. Edges are also chamfered to eliminate burrs. A sharp edge acts as an oil scraper and thereby enhances oil flow in the axial direction along the edges, which is harmful. Finally, bearings have a small relief just below the parting lines along the length on the inside surface. This is meant to protect the bearings in case of slight misalignment or offset at the parting lines.

23.8 Derivations of the Reynolds and Harrison Equations for Oil Film Pressure

The background for deriving these equations is given in Section 23.5 of the text. The equilibrium condition of a tiny volume element of the lubricating oil (Figure 23.6) is represented by the following equation [Slaymaker, 1955; Fuller, 1984]:

$$\left[P+\left(\frac{dP}{dx}\right)dx\right]dy\,dz+S\,dx\,dz-\left[S+\left(\frac{dS}{dy}\right)dy\right]dx\,dz-P\,dy\,dz=0 \tag{23.7}$$

Therefore,

$$\left(\frac{dS}{dy}\right)=\left(\frac{dP}{dx}\right) \tag{23.8}$$

Now, by introducing the definition of the coefficient of viscosity μ, we can relate the shear stress to a more measurable parameter, such as the velocity v of the lubricant, as

$$S=\mu\left(\frac{dv}{dy}\right) \tag{23.9}$$

Substituting for (dS/dy) from Equation (23.9) in Equation (23.8), a second-order partial differential equation in v is obtained. This is integrated to give an expression for the velocity profile as

$$v = \frac{V}{h}y - \frac{1}{2\mu}\left(\frac{dP}{dx}\right)(hy - y^2) \tag{23.10}$$

In Equation (23.10), V is the peripheral velocity of the journal and h is the oil film thickness. The boundary conditions used to derive Equation (23.10) are (1) $v = V$ when $y = h$, and (2) $v = 0$ when $y = 0$ (at the surface of the bearing). Now applying the relationship of continuity, the oil flowing past any cross-section in the z direction of the oil film around the journal must be equal. The quantity Q of oil flow per second is given by

$$Q = L\int_0^h v\,dy \tag{23.11}$$

where L is the length of the bearing that is in the z direction. Now substituting for v from Equation (23.10) in Equation (23.11) and integrating,

$$Q = L\left[\frac{Vh}{2} - \frac{h^3}{12\mu}\left(\frac{dP}{dx}\right)\right] \tag{23.12}$$

The pressure P varies as a function of x in the oil film, which is in the direction of rotation of the journal. At some point, it is expected to reach a maximum. At that point, (dP/dx) becomes zero. Let h_1 represent the oil film thickness at that point. Therefore,

$$Q = \frac{LV}{2}h_1 \tag{23.13}$$

Now we can use Equation (23.13) to eliminate Q from Equation (23.12). Hence,

$$\left(\frac{dP}{dx}\right) = \frac{6\mu V}{h^3}(h - h_1) \tag{23.14}$$

Equation (23.14) is the Reynolds equation for the oil film pressure as a function of distance in the direction of rotation of the journal. The variable x in Equation (23.14) can be substituted in terms of the angle of rotation θ and then integrated to obtain the Harrison equation for the oil film pressure. With reference to the diagram in Figure 23.7, the oil film thickness h can be expressed as

$$h = e\cos\theta + \sqrt{(r+c)^2 - e^2\sin^2\theta} - r \tag{23.15}$$

Here, e is the eccentricity, c is the radial clearance, and $e = c\varepsilon$, where ε is the eccentricity ratio. The quantity $e^2\sin^2\theta$ is much smaller compared to $(r + c)^2$. Therefore,

$$h = c(1 + \varepsilon\cos\theta) \tag{23.16}$$

Now, (dP/dx) is converted into polar coordinates by substituting $rd\theta$ for dx. Therefore, Equation (23.14) can be expressed as

$$\left(\frac{dP}{d\theta}\right)=\frac{6\mu Vr\varepsilon}{c^2}\left[\frac{\cos\theta-\cos\theta_1}{(1+\varepsilon\cos\theta)^3}\right] \tag{23.17}$$

where θ_1 is the angle at which the oil film pressure is a maximum. Integration of Equation (23.17) from $\theta = 0$ to $\theta = 2\pi$ can be expressed as

$$\int_0^{2\pi} dP=\int_0^{2\pi}\frac{6\mu Vr\varepsilon}{c^2}\left[\frac{\cos\theta-\cos\theta_1}{(1+\varepsilon\cos\theta)^3}\right]d\theta = P-P_0 \tag{23.18}$$

where P_0 is the pressure of the lubricant at the line of centers ($\theta = 0$) in Figure 23.7. If $(P - P_0)$ is assumed to be equal to zero at $\theta = 0$ and $\theta = 2\pi$, the value of $\cos\theta_1$, upon integration of Equation (23.18), is given by

$$\cos\theta_1 = -\frac{3\varepsilon}{2+\varepsilon^2} \tag{23.19}$$

and the Harrison equation for the oil film pressure for a full journal bearing by

$$P-P_0=\frac{6\mu Vr\varepsilon}{c^2}\frac{\sin\theta(2+\varepsilon\cos\theta)}{(2+\varepsilon^2)(1+\varepsilon\cos\theta)^2} \tag{23.20}$$

Acknowledgment

The author wishes to express his thanks to David Norris, President of Glacier Clevite Heavywall Bearings, for his support and interest in this article, and to J. M. Conway-Jones (Glacier Metal Company, Ltd., London), George Kingsbury (Consultant, Glacier Vandervell, Inc.), Charles Latreille (Glacier Vandervell, Inc.), and Maureen Hollander (Glacier Vandervell, Inc.) for reviewing this manuscript and offering helpful suggestions.

Defining Terms

Boundary layer lubrication — This is a marginally lubricating condition. In this case, the surfaces of two components (e.g., one sliding past the other) are physically separated by an oil film that has a thickness equal to or less than the sum of the heights of the asperities on the surfaces. Therefore, contact at the asperities can occur while running in this mode of lubrication. This is also described as "mixed lubrication." In some cases, the contacting asperities will be polished out. In other cases, they can generate enough frictional heat to destroy the two components. Certain additives can be added to the lubricating oil to reduce asperity friction drastically.

Crush — This is the property of the bearing that is responsible for producing a good interference fit in the housing bore and preventing it from spinning. A quantitative measure of the crush is equal to the excess length of the exterior circumference of the bearing over half the interior circumference of the housing. This is equal to twice the parting line height, if measured in an equalized half height measurement block.

Hydrodynamic lubrication — In this mode of lubrication, the two surfaces sliding past each other (e.g., a journal rotating in its bearing assembly) are physically separated by a liquid lubricant of suitable viscosity. The asperities do not come into contact in this case and the friction is very low.

Minimum oil film thickness (MOFT) — The hydrodynamic oil film around a rotating journal develops a continuously varying thickness. The thickness of the oil film goes through a minimum. Along this line, the journal most closely approaches the bearing. The maximum wear in the bearing is expected to occur around this line. Therefore, MOFT is an important parameter in designing bearings.

Peak oil film pressure (POFP) — The profile of pressure in the load-carrying segment of the oil film increases in the direction of rotation of the journal and goes through a maximum (Figure 23.5). This maximum pressure is a critical parameter because it determines the fatigue life of the bearing. This is also called maximum oil film pressure (MOFP).

Positive freespread — This is the excess in the outside diameter of the bearing at the parting line over the inside diameter of the housing bore. As a result of this, the bearing is clipped in position in its housing upon insertion. Bearings with negative freespread will be loose and lead to faulty assembly conditions.

Seizure — This is a critical phenomenon brought about by the breakdown of lubrication. At the core of this phenomenon is the occurrence of metal-to-metal bonding, or welding, which can develop to disastrous levels, ultimately breaking the crankshaft. With the initiation of seizure, increased generation of heat will occur, which will accelerate this phenomenon. Galling and adhesive wear are terms that refer to the same basic phenomenon. The term *scuffing* is used to describe the initial stages of seizure.

References

Bhushan, B. and Gupta, B. K. 1991. *Handbook of Tribology,* McGraw-Hill, New York.

Booker, J. F. 1965. Dynamically loaded journal bearings: Mobility method of solution. *J. Basic Eng. Trans. ASME,* series D, 87:537.

Conway-Jones, J. M. and Tarver, N. 1993. Refinement of engine bearing design techniques. *SAE Technical Paper Series, 932901, Worldwide Passenger Car Conference and Exposition,* Dearborn, MI, October 25–27.

Fuller, D. D. 1984. *Theory and Practice of Lubrication for Engineers,* 2nd ed. John Wiley & Sons, New York.

Slaymaker, R. R. 1955. *Bearing Lubrication Analysis,* John Wiley & Sons, New York.

Further Information

Yahraus, W. A. 1987. Rating sleeve bearing material fatigue life in terms of peak oil film pressure. *SAE Technical Paper Series, 871685, International Off-Highway and Powerplant Congress and Exposition,* Milwaukee, WI, September 14–17.

Booker, J. F., 1971. Dynamically loaded journal bearings: Numerical application of the mobility method. *J. Lubr. Technol. Trans. ASME,* 93:168.

Booker, J. F., 1989. Squeeze film and bearing dynamics. *Handbook of Lubrication,* ed. E. R. Booser. CRC Press, Boca Raton, FL.

Hutchings, I. M. 1992. *Tribology,* CRC Press, Boca Raton, FL.

Transactions of the ASME, Journal of Tribology

STLE Tribology Transactions

Spring and Fall Technical Conferences of the ASME/ICED

24

Fluid Sealing in Machines, Mechanical Devices, and Apparatus

Alan O. Lebeck
Mechanical Seal Technology, Inc.

The passage of fluid (leakage) between the mating parts of a machine and between other mechanical elements is prevented or minimized by a fluid seal. Commonly, a gap exists between parts formed by inherent roughness or misfit of the parts — where leakage must be prevented by a seal. One may also have of necessity gaps between parts that have relative motion, but a fluid seal is still needed. The fluid to be sealed can be any liquid or gas. Given that most machines operate with fluids and must contain fluids or exclude fluids, most mechanical devices or machines require a multiplicity of seals.

Fluid seals can be categorized as *static* or *dynamic* as follows.

Static:

- Gap to be sealed generally very small
- Accommodates imperfect surfaces, both roughness and out-of-flatness
- Subject to very small relative motions due to pressure and thermal cyclic loading
- Allows for assembly/disassembly

Dynamic:

- Gap to be sealed is much larger and exists of necessity to permit relative motion.
- Relatively large relative motions exist between surfaces to be sealed.
- Motion may be continuous (rotation) in one direction or large reciprocating or amount of motion may be limited.
- Seal must not constrain motion (usually).

0-8493-1586-7/05/$0.00+$1.50

Although some crossover exists between static and dynamic seal types, by categorizing based on the static and dynamic classification, the distinction between the various seal types is best understood.

24.1 Fundamentals of Sealing

Sealing can be accomplished by causing the gap between two surfaces to become small but defined by the geometric relationship between the parts themselves. In this case one has a fixed-clearance seal. One may also force two materials into contact with each other, and the materials may be either sliding relative to each other or static. In this case one has a surface-guided seal where the **sealing clearance** now becomes defined by the materials themselves and the dynamics of sliding in the case of a sliding seal.

There are two broad classes of surface-guided material pairs. The first and most common involves use of an **elastomeric**, plastic, or other soft material against a hard material. In this case, the soft material deforms to conform to the details of the shape of the harder surface and will usually seal off completely in the static case and nearly completely in the dynamic case. A rubber gasket on metal is an example. The second class, far less common, is where one mates a hard but wearable material to a hard material. Here the sealing gap derives from a self-lapping process plus the alignment of the faces of the material. Since both materials are relatively hard, if one material develops a roughness or grooves, the seal will leak. A mechanical face seal is an example.

24.2 Static Seals

Static seals can be categorized as follows:

- Gaskets
  - Single or composite compliant material
  - Metal encased
  - Wrapped and spiral wound
  - Solid metal
- Self-energized elastomeric rings
  - Circular cross-section (O-ring)
  - Rectangular cross-section
- Chemical compound or liquid sealants as gaskets
  - Rubbers
  - Plastics

Gaskets

Within the category of static seals, gaskets comprise the greatest fraction. The sealing principle common to gaskets is that a material is clamped between the two surfaces being sealed. The clamping force is large enough to deform the gasket material and hold it in tight contact even when the pressure attempts to open the gap between the surfaces.

A simple single-material gasket clamped between two surfaces by bolts to prevent leakage is shown in Figure 24.1. Using a compliant material, the gasket can seal even though the sealing surfaces are not flat. As shown in Figure 24.2, the gasket need not cover the entire face being sealed. A gasket can be trapped in a groove and loaded by a projection on the opposite surface as shown in Figure 24.3. Composite material gaskets or metal gaskets may be contained in grooves as in Figure 24.4. Gaskets are made in a wide variety of ways. A spiral-wound metal/fiber composite, metal or plastic clad, solid metal with sealing projections, and a solid fiber or rubber material are shown in Figure 24.5.

Gaskets can be made of relatively low-stiffness materials such as rubber or cork for applications at low pressures and where the surfaces are not very flat. For higher pressures and loads, one must utilize various composite materials and metal-encased materials as in Figure 24.5.

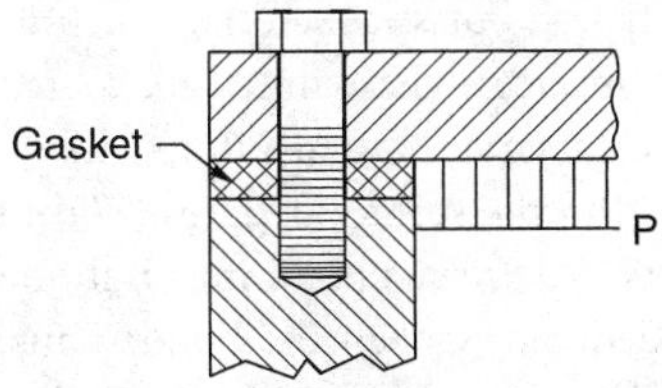

FIGURE 24.1 Gasket.

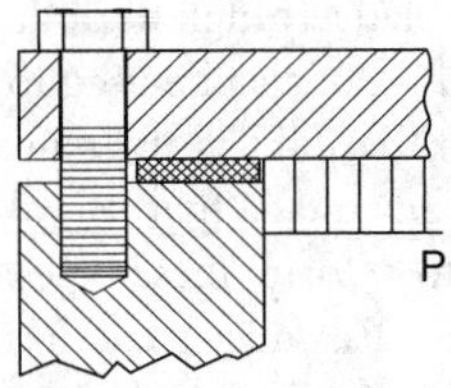

FIGURE 24.2 Gasket.

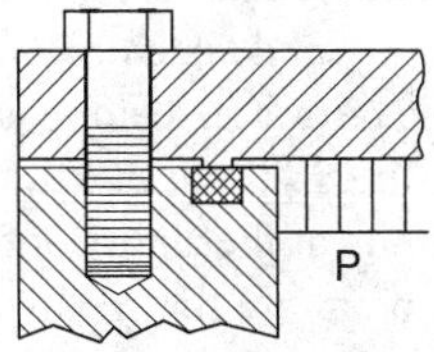

FIGURE 24.3 Loaded gasket.

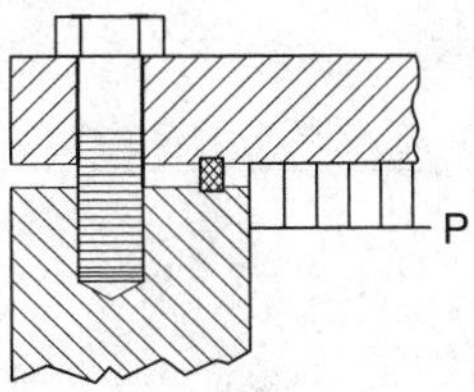

FIGURE 24.4 Hard ring gasket.

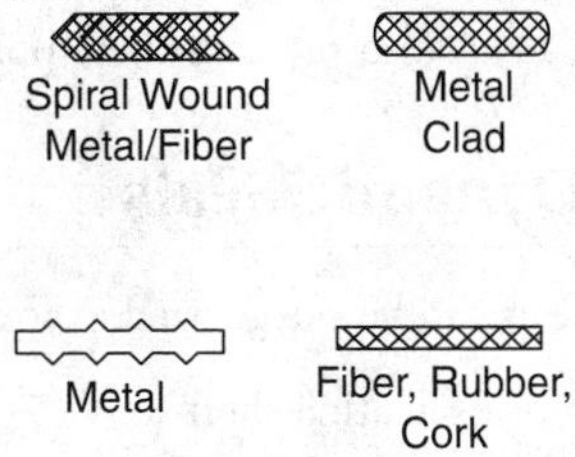

FIGURE 24.5 Varieties of gaskets.

For the highest pressures and loads a gasket may be retained in a groove and made either of very strong composite materials or even metal, as shown in Figure 24.4.

Self-Energized Seals

Elastomeric or **self-energized** rings can seal pressures to 20 MPa or even higher. As shown in Figure 24.6 and Figure 24.7, the two metal parts are clamped tightly together, and they are not supported by the elastomer. As the pressure increases, the rubber is pushed into the corner through which leakage would otherwise flow. An elastomer acts much like a fluid so that the effect of pressure on one side is to cause equal pressure on all sides. Thus, the elastomer pushes tightly against the metal walls and forms a seal. The limitation of this type of seal is that the rubber will flow or extrude out of the clearance when the pressure is high enough. This is often not a problem for static seals, since the gap can be made essentially zero as shown in Figure 24.6, which represents a typical way to utilize an elastomeric seal for static sealing.

Although the O-ring (circular cross-section) is by far the most common elastomeric seal, one can also utilize rectangular cross-sections (and even other cross-sections) as shown in Figure 24.7.

Chemical Compound or Liquid Sealants as Gaskets

Formed-in-place gaskets such as in Figure 24.8 are made by depositing a liquid-state compound on one of the surfaces before assembly. After curing, the gasket retains a thickness and flexibility, allowing it to seal very much like a separate gasket. Such gaskets are most commonly created using room-temperature vulcanizing rubbers (RTV), but other materials including epoxy can be used.

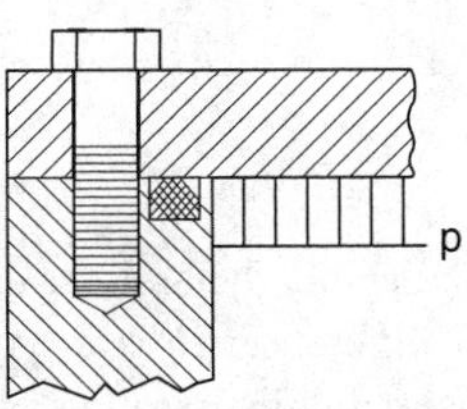

FIGURE 24.6 Elastomeric O-ring.

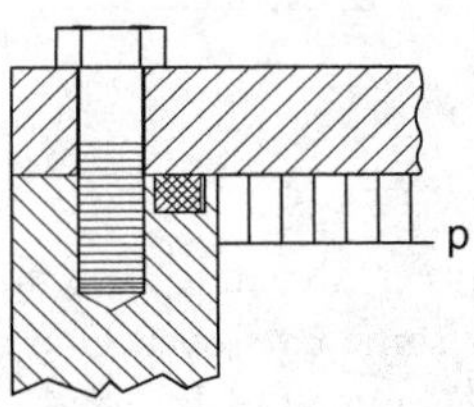

FIGURE 24.7 Elastomeric rectangular ring.

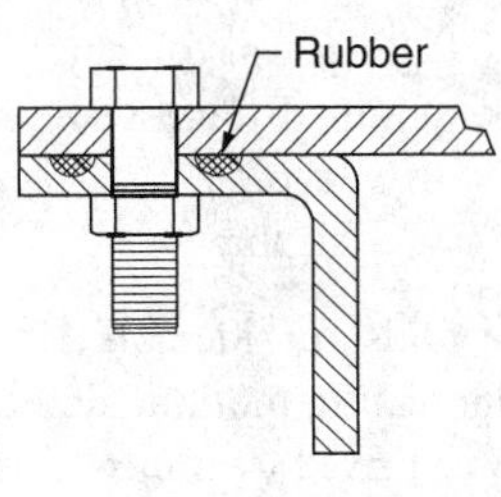

FIGURE 24.8 Formed-in-place elastomeric gasket.

While formed-in-place gaskets retain relatively high flexibility, other types of plastic materials (including epoxy and anaerobic hardening fluids) can also be used to seal two surfaces. These fluids are coated on their surfaces before assembly. Once the joint is tightened and the material hardens, it acts like a form-fitted plastic gasket, but it has the advantage that it is also bonded to the sealing surfaces. Within the limits of the ability of the materials to deform, these types of gaskets make very tight joints. But one must be aware that relative expansion of dissimilar materials so bonded can weaken the bond. Thus, such sealants are best utilized when applied to tight-fitting assemblies. These same materials are used to lock and seal threaded assemblies, including pipe fittings.

There have been many developments of chemical compounds for sealing during the past 25 years, and one is well advised to research these possibilities for sealing/assembly solutions.

24.3 Dynamic Seals

Dynamic seals can be categorized as follows:

- Rotating or oscillating shaft
  - Fixed clearance seals
    - Labyrinth
    - Clearance or bushing
    - Visco seal
    - Floating-ring seal
    - Ferrofluid seal
  - Surface-guided seals
    - Cylindrical surface
      - Circumferential seal
      - Packing
      - Lip seal
      - Elastomeric ring
    - **Annular** surface (radial face)
      - Mechanical face seal
      - Lip seal
      - Elastomeric ring
- Reciprocating
  - Fixed clearance seals
    - Bushing seal
    - Floating-ring seal
    - Clearance or bushing
  - Surface-guided seals
    - Elastomeric rings
      - Solid cross section
      - U-cups, V-rings, chevron rings
    - Split piston rings
  - Limited-travel seals
    - Bellows
    - Diaphragm

One finds considerable differences between dynamic seals for rotating shaft and dynamic seals for reciprocating motion, although there is some crossover. One of the largest differences in seal types is between fixed-clearance seals and surface-guided seals. Fixed-clearance seals maintain a sealing gap by virtue of the rigidity of the parts and purposeful creation of a fixed sealing clearance. Surface-guided seals attempt to close the sealing gap by having one of the sealing surfaces actually (or nearly) touch and rub on the other, so that the position of one surface becomes guided by the other. Fixed-clearance seals

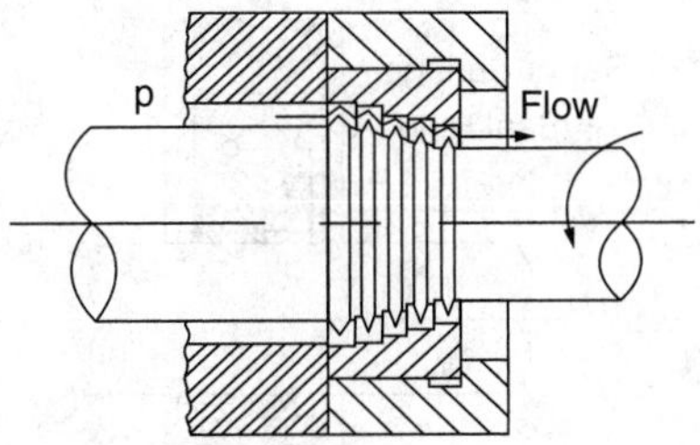

FIGURE 24.9 Labyrinth seal. (*Source*: Lebeck, A. O. 1991. *Principles and Design of Mechanical Face Seals*, John Wiley & Sons, New York. With permission.)

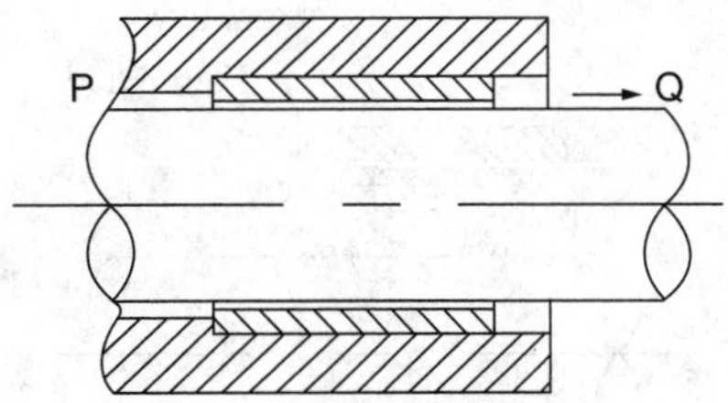

FIGURE 24.10 Bushing seal. (*Source*: Lebeck, A. O. 1991. *Principles and Design of Mechanical Face Seals*, John Wiley & Sons, New York. With permission.)

leak more than surface-guided seals as a rule, but each has its place. Finally, dynamic seals usually seal to either cylindrical surfaces or annular (radial) surfaces. Sealing to cylindrical surfaces permits easy axial freedom, whereas sealing to radial surfaces permits easy radial freedom. Many seals combine these two motions to give the needed freedom of movement in all directions.

Rotating or Oscillating Fixed-Clearance Seals

The labyrinth seal is shown in Figure 24.9. This seal has a calculable leakage depending on its exact shape, number of stages, and clearance and is commonly used in some compressors and turbomachinery as interstage seals and sometimes as seals to atmosphere. Its components can be made of readily wearable material so that a minimum initial clearance can be utilized.

The clearance or bushing seal in Figure 24.10 may leak more for the same clearance, but this represents the simplest type of clearance seal. Clearance bushings are often used as backup seals to limit flow in the event of failure of yet other seals in the system. As a first approximation, flow can be estimated using flow equations for fluid flow between parallel plates. Clearance-bushing leakage increases significantly if the bushing is eccentric.

In high-speed pumps and compressors, bushing seals interact with the shaft and bearing system dynamically. Bushing seals can utilize complex shapes and patterns of the shaft and seal surfaces to minimize leakage and to modify the dynamic stiffness and damping characteristics of the seal.

The visco seal or windback seal in Figure 24.11 is used to seal highly viscous substances, where it can be fairly effective. It acts like a screw conveyor, extruder, or spiral pump to make the fluid flow backward against sealed pressure. It can also be used at no differential pressure to retain oil within a shaft seal system by continuously pumping leaked oil back into the system.

The floating-ring seal in Figure 24.12 is used in gas compressors (it can be a series of floating rings). It can be used to seal oil where the oil serves as a barrier to gas leakage or it can seal product directly.

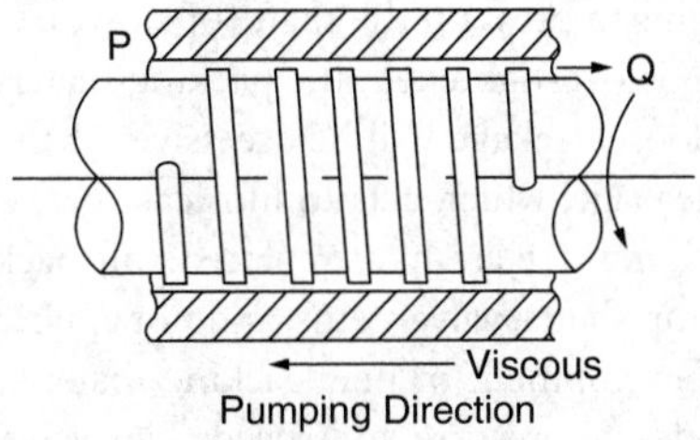

FIGURE 24.11 Visco seal. (*Source*: Lebeck, A. O. 1991. *Principles and Design of Mechanical Face Seals*, John Wiley & Sons, New York. With permission.)

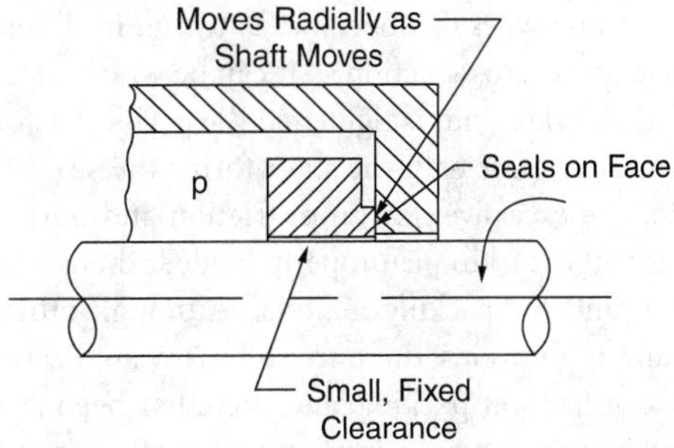

FIGURE 24.12 Floating-ring seal. (*Source*: Lebeck, A. O. 1991. *Principles and Design of Mechanical Face Seals*, John Wiley & Sons, New York. With permission.)

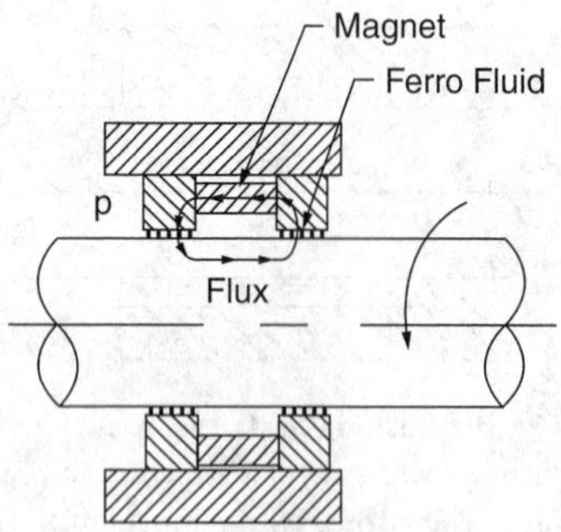

FIGURE 24.13 Ferrofluid seal. (*Source*: Lebeck, A. O. 1991. *Principles and Design of Mechanical Face Seals,* John Wiley & Sons, New York. With permission.)

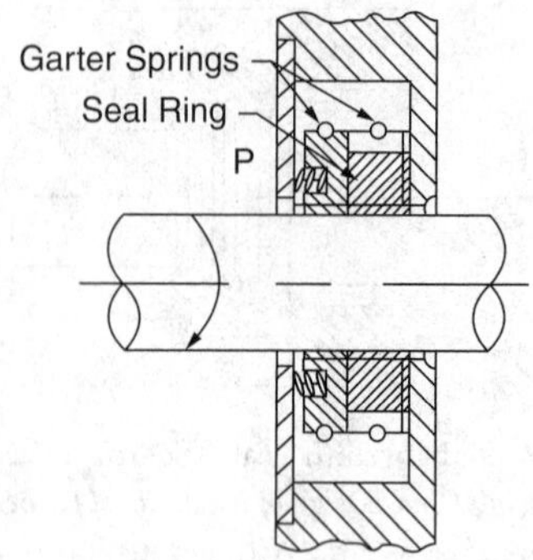

FIGURE 24.14 Circumferential seal.

This seal can be made with a very small clearance around the shaft because the seal can float radially to handle larger shaft motions. The floating-ring seal is a combination of a journal bearing where it fits around the shaft and a face seal where it is pressed against the radial face. Most of the leakage is between the shaft and the bore of the bushing, but some leakage also occurs at the face. This seal can be used in stages to reduce leakage. It can be balanced to reduce the load on the radial face. Leakage can be less than with a fixed-bushing seal.

The **ferrofluid** seal in Figure 24.13 has found application in computer disk drives where a true "positive seal" is necessary to exclude contaminants from the flying heads of the disk. The ferrofluid seal operates by retaining a ferrofluid (a suspension of iron particles in a special liquid) within the magnetic flux field, as shown. The fluid creates a continuous bridge between the rotating and nonrotating parts at all times and thus creates a positive seal. Each stage of a ferrofluid seal is capable of withstanding on the order of 20,000 Pa (3 psi), so although these seals can be staged they are usually limited to low–differential pressure applications.

Rotating Surface-Guided Seals — Cylindrical Surface

Figure 24.14 shows a segmented circumferential seal. The seal consists of angular segments with overlapping ends, and the segments are pulled radially inward by garter spring force and the sealed pressure. The seal segments are pushed against the shaft and thus are surface guided. They are also pushed against a radial face by pressure. This seal is similar to the floating-ring seal except that the seal face is pushed tight against the shaft because the segments allow for circumferential contraction. Circumferential segmented seals are commonly used in aircraft engines to seal oil and gas.

Many types of soft packing are used in the manner shown in Figure 24.15. The packing is composed of various types of fibers and is woven in different ways for various purposes. It is often formed into a rectangular cross-section so it can be wrapped around a shaft and pushed into a packing gland as shown. As the packing nut is tightened the packing deforms and begins to press on the shaft (or sleeve). Contact or near contact with the shaft forms the seal. If the packing is overtightened, the packing material will generate excessive heat from friction and burn. If it is too loose, leakage will be excessive. At the point where the packing is properly loaded, there is some small leakage, which acts to lubricate between the shaft and the packing material. Although other types of sealing devices have replaced soft packing in many applications, there are still many applications (e.g., pump shafts, valve stems, and hot applications) that utilize soft packing, and there has been a continuous development of new packing materials. Soft packing for continuously rotating shafts is restricted to moderate pressure and speeds. For valve stems and other reciprocating applications, soft packing can be used at high pressure and temperature.

The lip seal (oil seal) operating on a shaft surface represents one of the most common sealing arrangements. The lip seal is made of rubber (or, much less commonly, a plastic) or similar material that can be

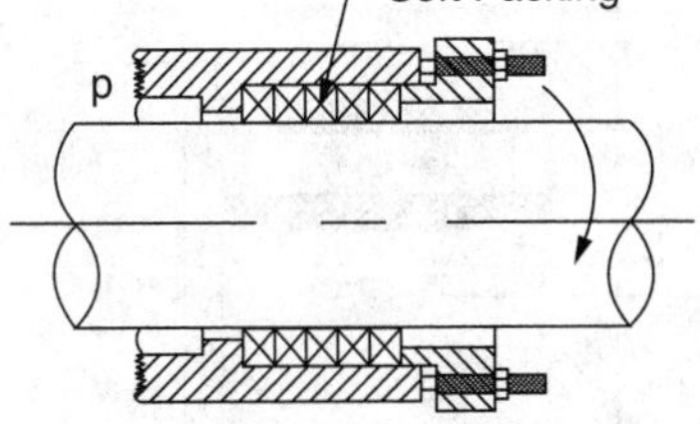

FIGURE 24.15 Soft packing.

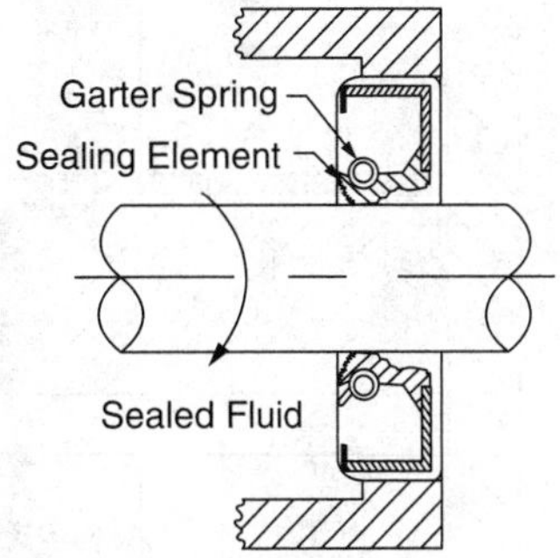

FIGURE 24.16 Lip seal.

readily deflected inward toward the shaft surface by a garter spring. The lip is very lightly loaded, and, in operation in oils with rotation, a small liquid film thickness develops between the rubber lip and the shaft. The shape of the cross-section determines which way the seal will operate. As shown in Figure 24.16, the seal will retain oil to the left. Lip seals can tolerate only moderate pressure (100,000 Pa maximum). The normal failure mechanism is deterioration (stiffening) of the rubber, so lip seals have a limited speed and temperature of service. Various elastomers are best suited for the variety of applications.

The elastomeric ring as described for static seals can also be used to seal continuous or oscillating rotary motion, given low-pressure and low-speed applications. As shown in Figure 24.17, the control of the pressure on the rubber depends on the squeeze of the rubber itself, so that compression set of the rubber will cause a loss of the seal. But if the squeeze is too high, the seal will develop too much friction heat. The use of a backup ring under high-pressure or high-gap conditions and the slipper seal to reduce friction are also shown in Figure 24.17.

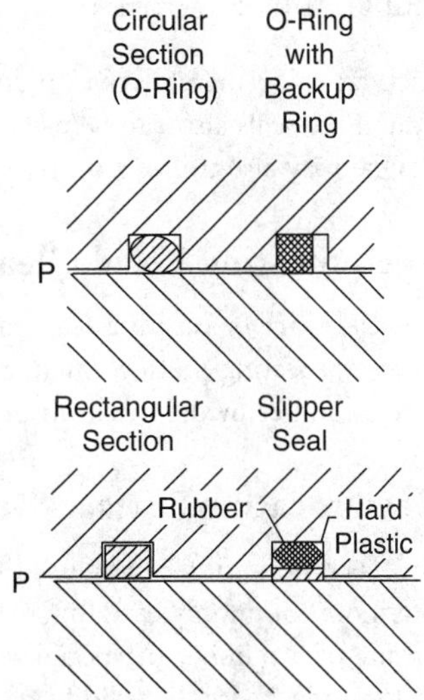

FIGURE 24.17 Elastomeric ring seals for rotating and reciprocating motion.

Rotating Surface-Guided Seals — Annular Surface

The mechanical face seal, as shown in Figure 24.18, has become widely used to seal rotating and oscillating shafts in pumps and equipment. The mechanical face seal consists of a self-aligning primary ring, a rigidly mounted mating ring, a secondary seal such as an O-ring or bellows that gives the primary ring freedom to self-align without permitting leakage, springs to provide loading of the seal faces, and a drive mechanism to flexibly provide the driving torque. It is common for the pressure to be sealed on the outside, but in some cases the pressure is on the inside. The flexibly mounted primary ring may be either the rotating or the nonrotating member.

Face seal faces are initially lapped very flat (1 micrometer or better) so that when they come into contact only a very small leakage gap results. In fact, using suitable materials,

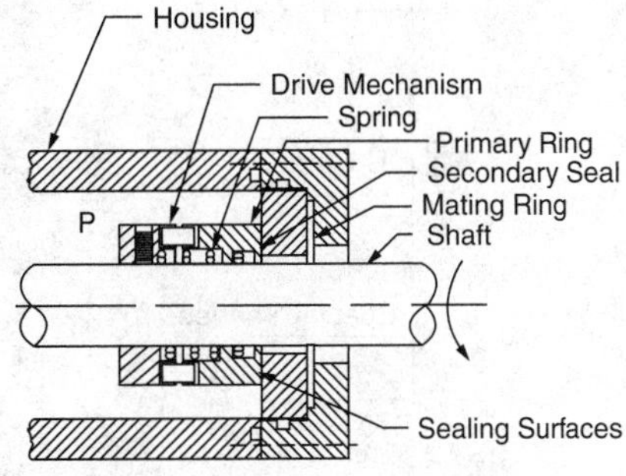

FIGURE 24.18 Mechanical face seal. (*Source*: Lebeck, A. O. 1991. *Principles and Design of Mechanical Face Seals*, John Wiley & Sons, New York. With permission.)

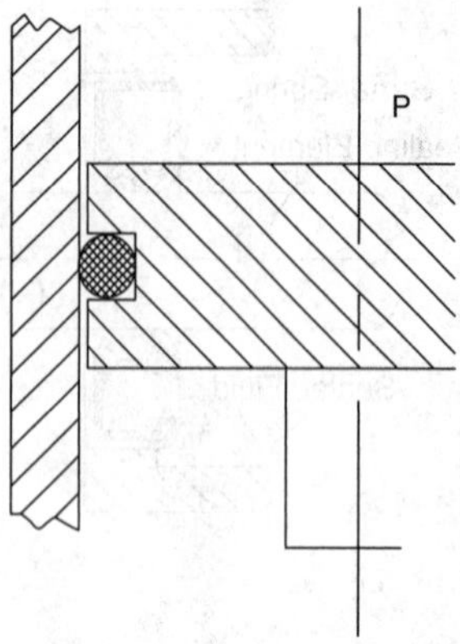

FIGURE 24.19 Elastomeric ring seal.

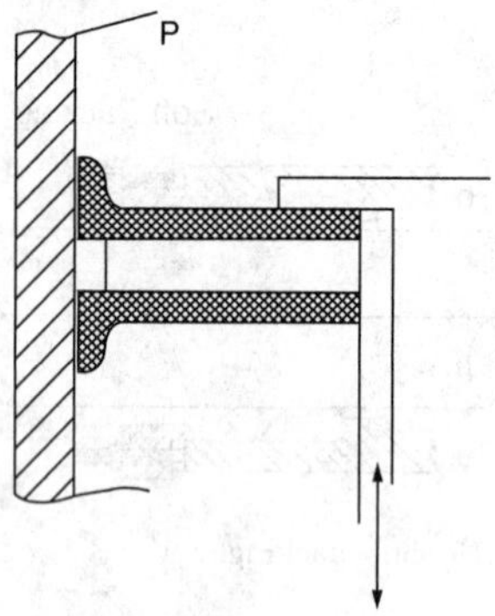

FIGURE 24.20 Cup seal.

such faces lap themselves into conformity so that such a seal can leak as little as a drop of liquid per hour. Face seals also can be used for sealing gas.

One may also utilize a lip seal or an elastomeric ring to seal rotationally on an annular face.

Reciprocating Fixed-Clearance Seals

The clearance or bushing seal (Figure 24.10) and the floating-ring seal (Figure 24.12) can also be used for reciprocating motion, such as sealing piston rods. In fact, the bushing can be made to give a near-zero clearance by deformation in such applications.

Reciprocating Surface-Guided Seals

An elastomeric ring can be used to seal the reciprocating motion of a piston, as shown in Figure 24.19. But more commonly used for such applications are cup seals (Figure 24.20), U-cups, V- or chevron rings, or any of a number of specialized shapes (Figure 24.21). Various types of these seals are used to seal piston rods, hydraulic cylinders, air cylinders, pumping rods, and pistons.

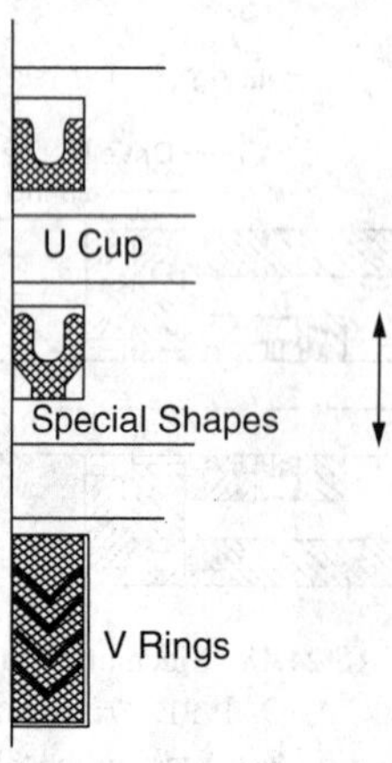

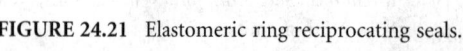

FIGURE 24.21 Elastomeric ring reciprocating seals.

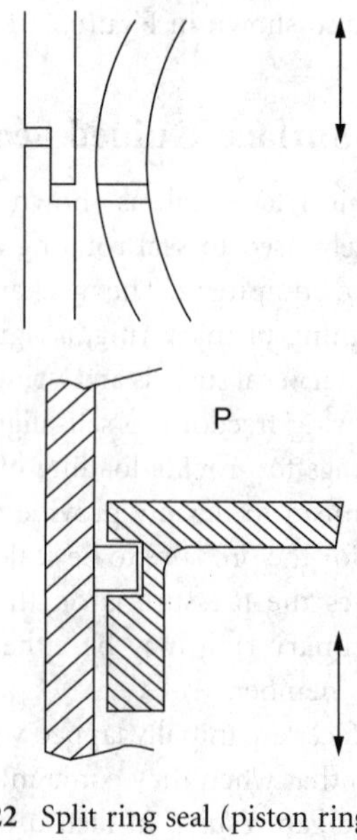

FIGURE 24.22 Split ring seal (piston ring).

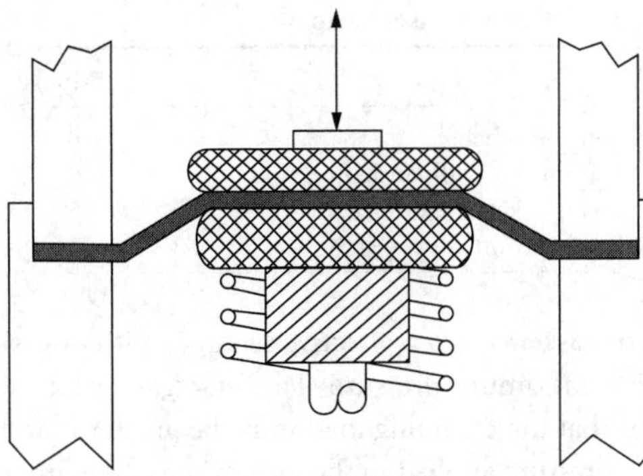

FIGURE 24.23 Diaphragm.

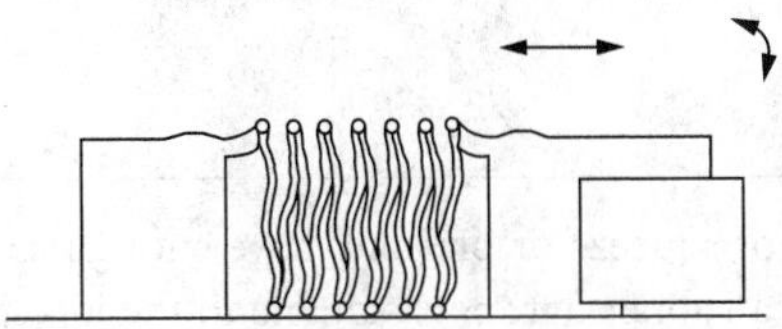

FIGURE 24.24 Welded metal bellows.

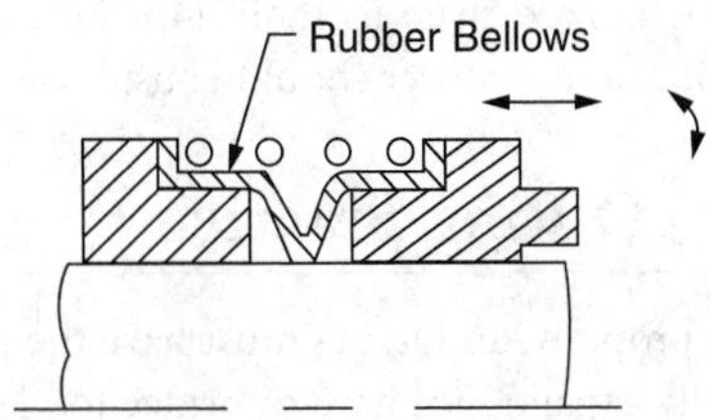

FIGURE 24.25 Rubber bellows.

Split rings such as shown in Figure 24.22 can be made of rigid materials. They are split for installation and so that they are loaded tightly against the wall by fluid pressure. Metal piston rings can be used in very hot environments. Plastic piston rings are suited to lower-temperature compressors.

Reciprocating Limited-Travel Seals

Most commonly used in pressure regulator and other limited-travel devices is the diaphragm shown in Figure 24.23. Properly designed, this seal can be absolute and have significant travel. It can also allow for angular misalignment. In Figure 24.24 is shown a metal bellows and in Figure 24.25 is a rubber bellows. Both of these permit limited axial and angular motion. They have the advantage of being absolute seals because they do not rely on a sealing interface or suffer from wear and have no significant friction. Metal bellows may be made from edge-welded disks as shown or formed from a thin metal tube.

24.4 Gasket Practice

For a gasket to seal, certain conditions must be met. There must be enough bolt or clamping force initially to seat the gasket. Then there also must be enough force to keep the gasket tightly clamped as the joint is loaded by pressure.

One may take the ASME Pressure Vessel Code [1980] formulas and simplify the gasket design procedure to illustrate the basic ideas. The clamping force, to be applied by bolts or other suitable means, must be greater than the larger of the following:

$$W_1 = \frac{\pi}{4} D^2 P + \pi 2bDmP \tag{24.1}$$

$$W_2 = \pi Dby \tag{24.2}$$

TABLE 24.1 Gasket Factors

Type	m	y (Mpa)
Soft elastometer	0.5	0
Elastometer with fabric insertion	2.5	20
Metal jacketed and filled	3.5	55
Solid flat soft copper	4.8	90

where D = effective diameter of gasket (m), b = effective seating width of gasket (m), $2b$ = effective width of gasket for pressure (m), P = maximum pressure (Pa), m = gasket factor, and y = seating load (Pa). Equation (24.1) is a statement that the clamping load must be greater than the load created by pressure plus a factor m times the same pressure applied to the area of the gasket in order to keep the gasket tight. Equation (24.2) is a statement that the initial clamping load must be greater than some load associated with a seating stress on the gasket material. To get some idea of the importance of the terms, a few m and y factors are given in Table 24.1. One should recognize that the procedure presented here is greatly simplified, and the user should consult one of the comprehensive references cited for details.

24.5 O-Ring Practice

To seal properly, an O-ring must have the proper amount of squeeze or **preload**, have enough room to thermally expand, not have to bridge too large a gap, have a rubber hardness suitable to the job, and be made of a suitable rubber. Table 24.2 shows an abbreviated version of recommendations for static O-rings and Table 24.3 for reciprocating O-rings. In many cases, one will want to span gaps larger or smaller than those recommended in the tables, so Figure 24.26 shows permissible gap as a function of pressure and hardness based on tests.

24.6 Mechanical Face Seal Practice

Figure 24.27 shows how, in general, the area on which the pressure is acting to load the primary ring may be smaller (or larger) than the area of the face. Thus, the balance ratio for a mechanical seal is defined as

$$B = \frac{r_o^2 - r_b^2}{r_o^2 - r_i^2} \tag{24.3}$$

where balance ratios less than 1.0 are considered to be "balanced" seals where in fact the face load pressure is made less than the sealed pressure. If the balance ratio is greater than 1.0, the seal is "unbalanced."

The balance radius (r_b) of a seal is used by seal designers to change the balance ratio and thus to change the load on the seal face. With reference to Figure 24.27, and noting that the face area is

$$A_f = \pi(r_o^2 - r_i^2) \tag{24.4}$$

the average **contact pressure** (load pressure not supported by fluid pressure) on the face is given by

$$p_c = (B - K)p + \frac{F_s}{A_f} \tag{24.5}$$

where the K factor represents the average value of the distribution of the fluid pressure across the face. For well-worn seals in liquid, $K = 1/2$ and, for a compressible fluid, K approaches 2/3.

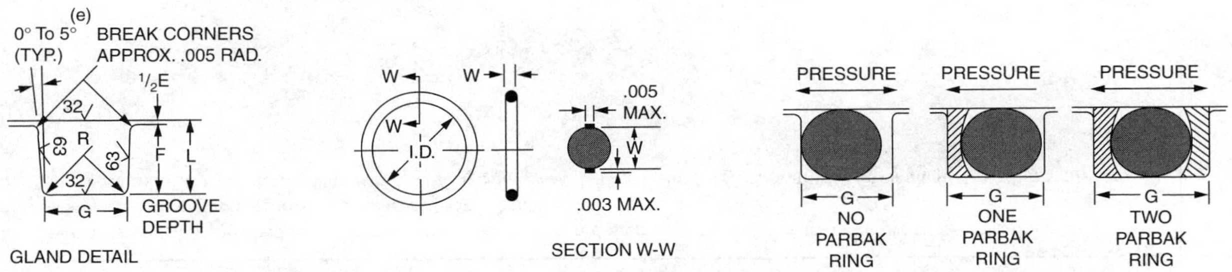

TABLE 24.2 Static O-Ring Grooves — Design Chart A5-1 for Industrial O-Ring Static Seal Glands

O-Ring Size Parker 2-	*W* Cross Section Nominal	*W* Cross Section Actual	*L* Gland Depth[a]	Squeeze[a] Actual	Squeeze[a] %	*E* Diametral Clearance[b,c]	*G* Groove Width No Parbak Rings	*G* Groove Width One Parbak Ring	*G* Groove Width Two Parbak Rings	*R* Groove Radius	Eccentricity Max.[d]
004 through 050	$\frac{1}{16}$	.070 ±.003	.050 to .052	.015 to .023	22 to 32	.002 to .005	.093 to .098	.138 to .143	.205 to .210	.005 to .015	.002
102 through 178	$\frac{3}{32}$	.103 ±.003	.081 to .083	.017 to .025	17 to 24	.002 to .005	.140 to .145	.171 to .176	.238 to .243	.005 to .015	.002
201 through 284	$\frac{1}{8}$	.139 ±.004	.111 to .113	.022 to .032	16 to 23	.003 to .006	.187 to .192	.208 to .213	.275 to .280	.010 to .025	.003
309 through 395	$\frac{3}{16}$	.210 ±.005	.170 to .173	.032 to .045	15 to 21	.003 to .006	.281 to .286	.311 to .316	.410 to .415	.020 to .035	.004
425 through 475	$\frac{1}{4}$	.275 ±.006	.226 to .229	.040 to .055	15 to 20	.004 to .007	.375 to .380	.408 to .413	.538 to .543	.020 to .035	.005

Note: (e) 0° preferred.

[a] For ease of assembly when Parbaks are used, gland depth may be increased up to 5%.

[b] Clearance gap must be held to a minimum consistent with design requirements for temperature range variation.

[c] Reduce maximum diametral clearance 50% when using silicone or fluorosilicone O-rings.

[d] Total indicator reading between groove and adjacent bearing surface.

Source: Parker Hannifin Corporation. 1990. *Parker O-Ring Handbook*, Parker Hannifin Corporation. Cleveland, OH. With permission.

TABLE 24.3 Reciprocating O-Ring Grooves — Design Chart A6-5 for Industrial Reciprocating O-Ring Packing Glands

O-Ring Size	*W* Cross Section		*L* Gland	Squeeze		*E* Diametral	*G* Groove Width			*R* Groove	Eccentricity
Parker 2-	Nominal	Actual	Depth	Actual	%	Clearance[a]	No Parbak Rings	One Parbak Ring	Two Parbak Rings	Radius	Max.[b]
006 through 012	$\frac{1}{16}$	.070 ±.003	.055 to .057	.010 to .018	15 to 25	.002 to .005	.093 to .098	.138 to .143	.205 to .210	.005 to .015	.002
104 through 116	$\frac{3}{32}$	.103 ±.003	.088 to .090	.010 to .018	10 to 17	.002 to .005	.140 to .145	.171 to .176	.238 to .243	.005 to .015	.002
201 through 222	$\frac{1}{8}$	.139 ±.004	.121 to .123	.012 to .022	9 to 16	.003 to .006	.187 to .192	.208 to .213	.275 to .280	.010 to .025	.003
309 through 349	$\frac{3}{16}$	.210 ±.005	.185 to .188	.017 to .030	8 to 14	.003 to .006	.281 to .286	.311 to .316	.410 to .415	.020 to .035	.004
425 through 460	$\frac{1}{4}$	.275 ±.006	.237 to .240	.029 to .044	11 to 16	.004 to .007	.375 to .380	.408 to .413	.538 to .543	.020 to .035	.005

[a] Clearance (extrusion gap) must be held to a minimum consistent with design requirements for temperature range variation.
[b] Total indicator reading between groove and adjacent bearing surface.
Source: Parker Hannifin Corporation. 1990. *Parker O-Ring Handbook*, Parker Hannifin Corporation. Cleveland, OH. With permission.

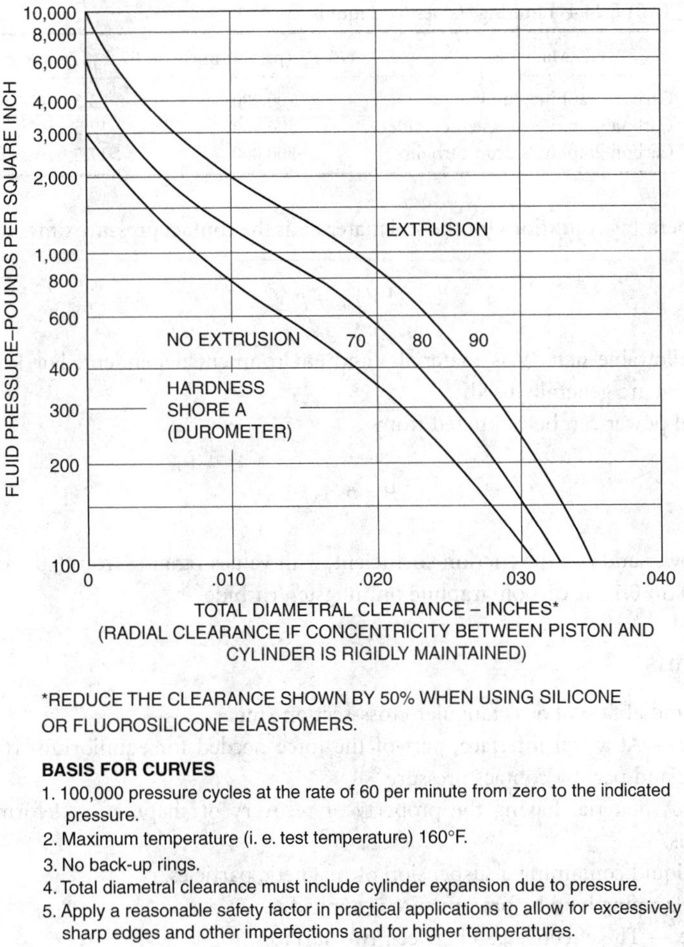

FIGURE 24.26 Limits for extrusion. (*Source*: Parker Hannifin Corporation. 1990. *Parker O-Ring Handbook*, Parker Hannifin Corporation. Cleveland, OH. With permission.)

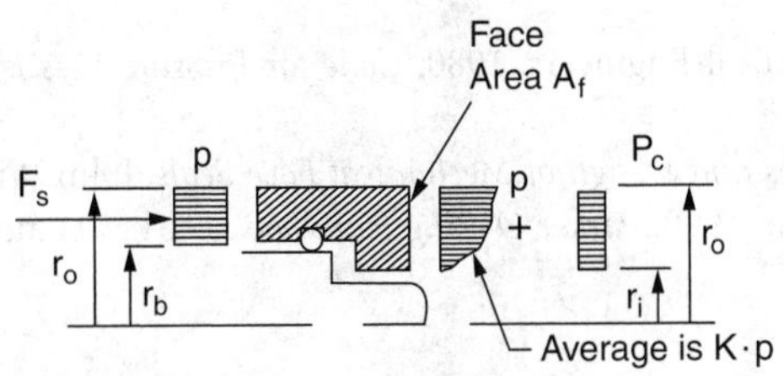

FIGURE 24.27 Mechanical seal elementary theory.

The sliding speed of the seal is based on the average face radius, or

$$V = \frac{r_o + r_i}{2}\omega \tag{24.6}$$

The severity of service for the seal is taken as the pressure times the sliding speed, or

$$(PV)_{\text{total}} = pV \tag{24.7}$$

TABLE 24.4 Limiting Values for Liquids

Materials	$(PV)_{net}$ (psi · ft/min)	$(PV)_{net}$ (Pa·m/sec)
Carbon graphite/alumina	100,000	$3.5 \cdot 10^6$
Carbon graphite/tungsten carbide	500,000	$17.5 \cdot 10^6$
Carbon graphite/silicon carbide	>500,000	$>17.5 \cdot 10^6$

The severity of operating conditions for the seal materials is the contact pressure times the sliding speed, or

$$(PV)_{net} = p_c V \tag{24.8}$$

The maximum allowable net PV is materials- and environment-dependent. For liquids, the limiting values of Table 24.4 are generally used.

Friction or seal power can be estimated from

$$P = p_c A_f f_c V \tag{24.9}$$

where P is the power and f_c is the friction coefficient, with values ranging from 0.07 for carbon graphite on silicon carbide to 0.1 for carbon graphite on tungsten carbide.

Defining Terms

Annulus — The radial face of a rectangular cross-section ring.

Contact pressure — At a seal interface, part of the force needed for equilibrium is supplied by fluid pressure and part by contact pressure.

Elastomer(ic) — A material having the property of recovery of shape after deformation; rubberlike materials.

Ferrofluid — A liquid containing a suspension of magnetic particles.

Preload — The clamping load before pressure is applied.

Sealing clearance — The effective gap between two surfaces.

Self-energized — The preload is supplied by the elastic behavior of the material itself.

References

American Society of Mechanical Engineers. 1980. *Code for Pressure Vessels,* Section VIII, Div 1. ASME, New York.

Lebeck, A. O. 1991. *Principles and Design of Mechanical Face Seals,* John Wiley & Sons, New York.

Parker Hannifin Corporation. 1990. *Parker O-Ring Handbook,* Parker Hannifin Corporation. Cleveland, OH.

Further Information

Brink, R. V., Czernik, D. E., and Horve, L. A. 1993. *Handbook of Fluid Sealing,* McGraw-Hill, New York.

Buchter, H. H. 1979. *Industrial Sealing Technology,* John Wiley & Sons, New York.

Kaydon Ring and Seals, Inc. 1987. *Engineer's Handbook — Piston Rings, Seal Rings, Mechanical Shaft Seals,* Kaydon Ring and Seals, Inc. Baltimore, MD.

Warring, R.H. 1981. *Seals and Sealing Handbook,* Gulf, Houston, TX.

V

Structures

25

Loads

Peter Gergely
Cornell University

Structures are designed to carry various loads and load combinations without collapse and with an adequate margin of safety. In addition, several **serviceability** conditions (deflections, cracking) must also be satisfied for most structures. The expected maximum values of most loads can be estimated only approximately, and building codes give only estimates of the minimum design loads, based on judgment. The design loads and load combinations rarely occur for most structures.

The two main types of loads are **dead loads** and **live loads**. Dead loads include the weight of structural and most nonstructural components: beams, columns, walls, floor slabs, bridge decks, roofing, partitions, and ceiling and flooring materials. Live loads include occupancy loads (people, building contents, traffic, movable partitions) and snow. Wind forces, earthquake forces, water pressure, and blast are similar to live loads, but they are usually considered separately. The weights of movable materials in warehouses are usually considered as live loads. In addition to these loads, temperature effects can also be considered as loads.

Most designers and building codes rely on the *Minimum Design Loads for Buildings and Other Structures* [ASCE 7-93, 1993] or the *Standard Specifications for Highway Bridges* [AASHTO, 1989] for the design of buildings and bridges, respectively. Loads for special structures, such as liquid containers, towers, cranes, and power plants, are normally specified by trade or professional organizations.

Loads are combined to produce the maximum member forces. However, codes allow reduction of combined loads if the probability of simultaneous occurrence of maximum effects is low. For example, a 0.75 factor may be applied for the combined dead load, live load, and wind or earthquake. These factors are different in the working stress design approach and in the strength (or the load and resistance factor) design approach.

25.1 Dead Loads

Dead loads are made up almost entirely of the weights of all structural elements and permanent fixtures. Therefore, it is generally easy to calculate dead loads. However, in preliminary design the sizes of structural members (beams, columns, walls, and floor slabs) are not yet known and must be estimated. The unit weights of several common materials are listed in Table 25.1.

The maximum forces in structures sometimes occur during construction — for example, during the cantilever construction of bridges. It is important to consider the loads during various stages of construction.

0-8493-1586-7/05/$0.00+$1.50

TABLE 25.1 Unit Weight of Common Construction Materials

Aluminum	165 lb/ft^3
Brick	120 lb/ft^3
Lightweight concrete	90 to 110 lb/ft^3
Normal-weight concrete	150 lb/ft^3
Steel	490 lb/ft^3
Timber	35 to 40 lb/ft^3
Roofing	5 to 10 lb/ft^2
Tile	10 to 15 lb/ft^2
6-in. hollow concrete block wall	43 lb/ft^2

25.2 Live Loads

There are many types of live loads: occupancy, weights in warehouses, traffic loads on bridges, construction or maintenance forces, automobiles in parking garages, and snow. These are much more variable than dead loads and require larger safety margins. (Wind and earthquake loads are considered separately as environmental loads.) These live loads are gravity loads and must be positioned (acting on all or part of the area) to cause maximum forces in the member being designed.

Occupancy Loads

The major type of live load in buildings is caused by occupants. The minimum specified occupancy loads depend on the use and the likelihood of congregation of many people. Typical values of distributed loads are shown in Table 25.2. In office buildings, a 20 lb/ft^2 uniform load is used to account for the weight of movable partitions.

In addition to the distributed loads, structures are also designed for concentrated loads to account for concentrations of people or furniture. Typical values are 2000 lb on office floors and on slabs in garages for passenger cars. These are assumed to be acting on a 2.5 ft^2 area.

Since it is unlikely that a very large area or most floors of a building will have the full occupancy loads, most codes allow a live load reduction factor for such cases. However, reduction is not allowed for garages and areas of public assembly. In the design of a structural member, if the influence area is more than 400 ft^2, the reduction factor is

$$0.25+\frac{15}{\sqrt{A}} \tag{25.1}$$

with a minimum value of 0.5 for one floor and 0.4 for columns receiving loads from multiple floors. For columns the influence area is four times the tributary area (thus equal to the area of all four adjoining panels), and for beams it is twice the tributary area.

TABLE 25.2 Typical Occupancy Loads (lb/ft^2)

Theaters with fixed seats	60
Theaters with movable seats	100
Corridors and lobbies	100
Garages	50
Restaurants	100
Library reading rooms	60
Offices	50
Stadium bleachers	100
Stairways	100

Bridge Live Loads

The design forces in a bridge depend on the magnitude and distribution of the vehicle load. It is not reasonable to assume that only heavy trucks will travel at close spacing, especially on a long bridge. For short bridges the actual position of the heaviest truck is important, whereas for long bridges a uniform load could be used in design. Design codes [AASHTO, 1989] specify standard loads for short and long bridges. Several standard trucks are specified — for example, the H20-44, which has a total weight of 20 tons (the 44 signifies the year of adoption). Of this weight, 8000 lb acts under the front axle and 32,000 lb under the rear wheels. Other standard truck loads are H10-44, H15-44, HS15-44, and HS20-44, where the HS designation is for semitrailers, with the weight allocation of 10, 40, and 40% for the cab, front trailer, and rear trailer wheels, respectively.

These concentrated loads must be placed on the bridge to produce maximum forces (shears and moments) in the member being designed. In addition to the individual truck loads, codes specify a uniform lane load combined with a single concentrated load. For an H10-44 loading, the distributed load is 320 lb/ft and the concentrated load is 9000 lb; the respective numbers for HS20-44 are twice as large.

25.3 Impact Loads

Moving loads on bridges and crane girders can cause vibrations and increased stresses. Simple empirical impact formulas have been developed to account for this effect, although a large number of variables, such as surface roughness, speed, and span, influence the impact effect.

In the AASHTO [1989] code the impact formula is

$$I = \frac{50}{L+125} \tag{25.2}$$

where L is in feet. The maximum value of the impact factor I is 0.3. For shorter bridges the impact effect can be high, especially when a heavy vehicle travels on the bridge at high speed.

The loads created by elevators are often increased by 100% to account for impact. Likewise, impact factors have been recommended for various other types of machinery, ranging from 20 to 50%. Craneways have three factors — 25, 20, and 10% for forces in the vertical, lateral, and longitudinal directions, respectively.

25.4 Snow Loads

The expected maximum snow accumulation in various regions is given in codes, usually for a 50-year mean recurrence interval. Values reach 50 psf in many areas but can be twice as much or more in regions with heavy snowfalls. In some regions — for example, in parts of the Rocky Mountains — local climate and topography dictate the design snow load level. The weight of snow depends on its density and typically ranges from 0.5 to 0.7 psf for 1 in. of snow after some compaction. Fresh dry snow has a specific gravity of only 0.2.

For flat roofs (outside Alaska) the snow load is

$$p_f = 0.7C_eC_tIp_g \tag{25.3}$$

C_e is the exposure factor (0.8 for windy, open areas, 0.9 for windy areas with some shelter, 1.0 if wind does not remove snow, 1.1 with little wind, and 1.2 in a forested area). However, for large roofs (greater than 400 ft in one direction), C_e should not be less than unity [Lew et al. 1987]. C_t is the thermal factor (1.0 for heated structures, 1.1 just above freezing, 1.2 for unheated structures). I is the importance factor (ranges from 0.8 to 1.2 for various occupancies). p_g is the ground snow load from maps.

The flat-roof values are corrected for sloping roofs:

$$p_s = C_s p_f \tag{25.4}$$

where C_s is the slope factor, which is given in a diagram in ASCE 7-93 [1993]. For warm roofs ($C_t = 1.0$), the slope factor is 1.0 for slopes less than about 30° and reduces linearly to zero as the slope increases to 70°. Thus roofs with slopes greater than 70° are assumed to have no snow load. Unbalanced snow load caused by a certain wind direction also must be considered.

Defining Terms

Dead load — Gravity loads produced by the weight of structural elements, permanent parts of structures such as partitions, and weight of permanent equipment.
Impact factor — Accounts for the increase in stresses caused by moving load effects.
Live load — Loads caused by occupancy and movable objects, including temporary loads.
Serviceability — Limit on behavior in service, such as on deflections, vibrations, and cracking.

References

AASHTO. 1989. *Standard Specifications for Highway Bridges,* 14th ed. The American Association of State Highway and Transportation Officials, Washington, DC.

ASCE 7-93. 1993. *Minimum Design Loads for Buildings and Other Structures,* American Society of Civil Engineers, New York.

Lew, H. S., Simiu, E., and Ellingwood, B. 1987. Loads. In *Building Structural Design Handbook,* ed. R. N. White and C. G. Salmon, pp. 9–43. John Wiley & Sons, New York.

Further Information

Uniform Building Code, International Conference of Building Officials, 5360 South Workman Mill Road, Whittier, CA 90601.

ASCE Standard, ASCE 7-93, *Minimum Design Loads for Buildings and Other Structures,* American Society of Civil Engineers, 345 East 47th Street, New York, NY 10017.

26
Wind Effects

Timothy A. Reinhold
Clemson University

Ben L. Sill
Clemson University

Wind is one of the two primary sources of lateral forces on land-based buildings and structures; the other is earthquake ground motion. Winds completely engulf the structure and generate complex distributions of pressures and, hence, loads on all exterior surfaces. Most surfaces experience negative pressures or suctions, which tend to pull the building apart. Most roofs also experience negative pressures, which act to lift the roof off of the walls and to pull roofing membranes and sheathing from the supporting structure. If the exterior surface contains openings, either because they were designed that way or because a **cladding** element fails, the interior of the structure can become exposed to some fraction of the external pressure that would have occurred at the opening. Internal pressures can also develop as a result of normal air leakage through the building skin or cladding. Internal pressures tend to be fairly uniform throughout the interior of the building and can significantly increase the loads on the walls and roof.

Wind effects on structures include the direct application of wind-induced forces, movement of the structure, and the flow of the wind around the structure, which may affect pedestrians or the function of the building. Normally, wind effects are grouped according to limit states and safety and serviceability considerations. The selection of structural systems based on their ability to resist wind-induced stresses with appropriate margins of safety is an example of a design for safety or an ultimate limit state. Limiting deflections caused by the wind loads to prevent cracking of walls or partitions and limiting the motion of the structure to prevent occupant discomfort are examples of serviceability limit state design. Following the large economic losses suffered in recent hurricanes, there have been increasing calls for protection of the building envelope against water penetration by protecting glazed openings from failure due to direct wind loads or impact loads from wind-borne debris and by reducing the penetration of wind-driven rain.

The wind effects that should be considered in the design of a particular structure vary depending on the following factors:

1. The wind climate (the expected magnitude and frequency of wind events and consideration of the types of events — that is, hurricanes, thunderstorms, tornadoes, and extra-tropical storms).
2. The local wind exposure (siting of the building or structure, including the type of terrain and terrain features surrounding the structure and the influence of neighboring structures).
3. Pressure coefficients and load factors (coefficients that depend on the exterior shape of the building or structure and factors which relate wind loads to reference wind speeds).

0-8493-1586-7/05/$0.00+$1.50

4. Dynamic effects such as resonance response and aerodynamic instabilities, which can lead to failures or to significant increases in the dynamic response and loading (these dynamic effects are not covered by normal pressure coefficients and load factors; they depend on the shape of the building or structure and properties such as mass, stiffness, damping, and modes of vibration). Typically, tall, slender structures and long, suspended structures should be evaluated for these possible effects. These include long-span bridges, stacks, towers, buildings with height-to-width ratios greater than five, and exposed long, flexible members of structures.

Within the space available, it is possible to present only a brief description of the types of wind effects that should be considered in the design of buildings and structures. Consequently, rather than reproduce a set of codelike requirements, the focus of this chapter is on describing some basic relationships that will help the engineer compare different code approaches to estimating wind loads. There are many different codes available throughout the world that use significantly different types of reference wind speeds. Often, it is not clear whether the codes will produce similar estimates of design loads if applied to the same structure (indeed, they often do produce significantly different loads). It is not the intent of this chapter to promote a particular code. The goal is to provide the tools that will allow the engineer to compare code estimates of design loads by using a consistent set of reference wind speeds, regardless of whether the code calls for a mean hourly speed, a ten-minute speed, a one-minute sustained wind speed, a fastest mile wind speed, or a gust wind speed.

In addition, the field of wind engineering remains highly empirical, which means that accurate estimates of wind loads and wind effects often require the conduct of a physical model study. The "For Further Information" section provides a list of references that can provide additional guidance on when a model study is warranted or desirable.

26.1 Wind Climate

The wind climate encompasses the range of wind events that may occur in a geographical region and the expected frequency and intensity of the event. Types of events include **extra-tropical cyclones,** thunderstorms, downbursts, microbursts, tornadoes, and hurricanes. Each type of storm has potentially different wind characteristics of importance to buildings and structures, as well as separate occurrence rate and intensity relationships. For most engineering purposes, downbursts, microbursts, and tornadoes are not considered in the establishment of design winds and loads. Thunderstorm winds are frequently buried in the historical data records and, thus, are partially built into the design winds estimated from historical data. Recent work has been conducted to extract thunderstorm winds from the historical data at selected stations. This analysis suggests that it will be important in the future to treat thunderstorms as a separate population of wind events in much the same way that current analysis considers hurricane and tornado events as separate populations for statistical analysis.

The current approach to estimating design winds in hurricane-prone regions is to conduct Monte Carlo simulations of the events using statistical information on historical tendencies of hurricanes in the area. The probabilities of experiencing hurricane winds in coastal areas are then developed from the statistics produced from the simulation of thousands of years of storms. These occurrence probabilities are then combined with probabilities for nonhurricane events to estimate design winds for various return periods ranging from 10 to 100 years. This type of analysis has been used to produce design wind speed maps for the continental U.S.; the latest edition of ASCE-7 *Minimum Design Loads for Buildings and Structures* contains the most recent map, which is generally adopted by model building codes in the U.S. This type of systematic analysis of hurricane and nonhurricane winds has not been conducted for the Hawaiian Islands or most of the rest of the world. Consequently, in hurricane-prone regions, the designer should endeavor to determine the source of the estimates for design wind speeds and the basis for the estimates. In some instances, it will be necessary to contract with a group experienced in hurricane simulations in order to produce reasonable estimates of design wind speeds.

In areas where hurricanes are not expected — which normally includes areas 100 miles inland from a hurricane coastline — a series of annual extreme wind speeds can be used to estimate the design wind speed for a specific return period using the following equation [Simiu, 1985], where it is assumed that the extreme wind climate follows a Type I extreme value distribution:

$$U_N = U_{avg} + 0.78\sigma_N(\ln N - 0.577) \tag{26.1}$$

where U_N is the design wind speed for a return period of N years, U_{avg} is the average of the annual extreme wind speeds, and σ_N is the standard deviation of the annual extreme wind speeds. It should be noted that the number of years of record greatly affects the reliability of the design wind speed estimate [Simiu, 1985].

Note also that an averaging time has not been specified in this discussion. The emphasis is placed on the wind speeds being extreme annual values that must all correspond to the same averaging time, regardless of whether it is a mean hourly, 10-min mean, or shorter-duration averaging time. These maximum speeds must also correspond to a consistent set of terrain conditions and a consistent elevation. The U.S. codes currently use a fastest mile wind speed at a height of 10 m in open terrain as the reference design wind speed and reference conditions. This type of measurement has a variable averaging time since it corresponds to the time required for one mile of wind to pass a location. Thus, any calculation of a fastest mile wind speed requires iteration. The averaging time for the fastest mile wind speed is calculated by:

$$t = 3600 / U_{FM} \tag{26.2}$$

where t is given in seconds and U_{FM} is the fastest mile wind speed expressed in miles per hour.

Other countries use similar terrain conditions and usually specify a 10 m elevation but use a wide variety of averaging times ranging from mean hourly (Canada) to peak gust (Australia), which is normally assumed to correspond to a 2- to 3-sec averaging time. The following section provides relationships for converting maximum wind speeds from one averaging time, terrain exposure, and elevation to maximum wind speeds for a different averaging time, terrain, and elevation. These equations provide a means for converting a design wind speed in an unfamiliar code to one that can be used in a code with which the designer is more familiar, if the reference conditions for the two codes are different.

26.2 Local Wind Exposure

As wind moves over the surface of the earth, the roughness of trees, buildings, and other features reduces the wind speed and creates the **atmospheric boundary layer**. The greatest reduction occurs close to the ground, with reduced effects at greater heights. There is, in fact, a height — known as the *gradient height* — at which the wind is not affected by the surface characteristics. For engineering purposes, the gradient height is generally assumed to be between 300 and 600 m, depending on the terrain. Surface roughness also affects the air flow by creating turbulent eddies (or gusts), which can have a significant effect on buildings. The gusty nature of wind is random and is analyzed using statistical approaches.

26.3 Mean Wind Speed Profile

No single analytical expression perfectly describes the mean wind speed variation with height in the atmospheric boundary layer. The two used most often are the power law profile and the logarithmic profile. The logarithmic profile is the most widely accepted, although both can give adequate descriptions of the wind speed. Each is described below, and sufficient information is given to (a) allow transfer from one profile to the other, (b) effectively convert maximum speeds in one terrain to those in another, and (c) convert from one averaging time to another.

TABLE 26.1 Parameters Used in Mean Velocity Profile Relations

Terrain	α	z_g (m)	z_0 (m)	p	β
Coastal	0.1	230	0.005	0.83	6.5
Open	0.14	275	0.07	1	6
Suburbs	0.22	375	0.3	1.15	5.25
Dense suburbs	0.26	400	1	1.33	4.85
City center	0.33	500	2.5	1.46	4

Power Law Profile

Taking z as the elevation, U as the wind speed (with the bar indicating a mean hourly value), and g as conditions at the gradient height, this profile is written as

$$\frac{\overline{U(z)}}{U_g} = \left(\frac{z}{z_g}\right)^a \tag{26.3}$$

Table 26.1 provides estimates for the gradient height and power law exponent for different terrains.

Logarithmic Profile

This expression for the variation of mean wind speed with height utilizes an aerodynamic roughness length, z_0, and a shear velocity, u_*, which is a measure of the surface drag:

$$\overline{U(z)} = \frac{u_*}{k}\ln\left(\frac{z}{z_0}\right) \tag{26.4}$$

Here, k is von Karman's constant and is usually taken as 0.4. For rough surfaces, such as dense suburban and urban conditions, a displacement height, d, should be included in Equation (26.4) by replacing z with $z - d$. For z values substantially greater than d, the correction is negligible. Writing the log law at two heights for two different terrain roughnesses (one of which is open country) and taking the ratio gives

$$\frac{\overline{U(z_1)}}{\overline{U(z_2)}_{\text{open}}} = p\frac{\ln(z_1/z_0)}{\ln(z_2/z_0)_{\text{open}}} \tag{26.5}$$

where p is the ratio of the shear velocities for the two different terrain conditions. Table 26.1 gives a summary of the parameters needed to use the profiles.

26.4 Turbulence

The wind speed can be divided into two parts — a mean or time-averaged part, $\overline{U}$, and a fluctuating or time varying part, u'. The long-term properties of the fluctuating part can be described by the variance or standard deviation, σ_u. The maximum wind speed for any averaging time, t, can be obtained by adjusting the hourly (3600 sec) average as

$$U_t(z) = \overline{U}_{3600}(z) + C(t)\sigma_u(z) \tag{26.6}$$

where the coefficient $C(t)$ is given in Table 26.2. The values of $C(t)$ for extra-tropical winds were obtained by Simiu [1981], while the values of $C(t)$ for hurricane winds reflect recent research that suggests that hurricane winds contain larger fluctuations than extra-tropical strong winds [Krayer, 1992].

TABLE 26.2 Gust Factors for Use in Calculating Maximum Wind Speeds from Mean Speeds

Conditions	Time (sec)	1	3	10	30	60	600	3600
Hurricane winds	$C(t)$	4	3.62	3.06	2.23	1.75	0.43	0
Extra-tropical winds	$C(t)$	3	2.86	2.32	1.73	1.28	0.36	0

TABLE 26.3 Example Conversions of Maximum Speeds for Different Conditions

Example	A	B	C	D	E Iteration #1	E Iteration #2	F	G
Terrain 2	Open	Open	Open	Open	Open		Open	Open
z (m)	10	10	10	10	10		10	10
u (mph)[a]					90			90
t (sec)	60	3600	60	3600	40[b]		3	40
$C(t)$	1.28	0	1.28	0	1.58		2.86	1.58
Terrain 1	Coast	Suburb	Open	Open	Suburb		Suburb	Open
z (m)	10	10	100	10	10		10	10
u (mph)[a]				90	90[c]	79	90	
t (sec)	3	3600	60	40[b]	40[b]	46	40	3
z_0 (m)	0.005	0.3	0.07	0.07	0.3		0.3	0.07
$C(t)$	2.86	0	1.58	1.58	1.58	1.49	1.58	2.86
•	6.5	5.25	6	6	5.25		5.25	6
p	0.82	1.15	1	1	1.15		1.15	1
$U_{t_1}(z_1)/U_{t_2}(z_2)$	1.39	0.81	1.42	1.31	0.88	0.86	0.56	1.19

[a] Wind speed in fastest mile.

[b] Averging time calculated from $3600/U_{FM}$.

[c] Initial guess of fastest mile speed for the first iteration selected as open country value. In the second iteration, the 79 mph value was calculated from 90 mph multiplied by the velocity ratio of 0.88.

To convert maximum wind speeds between open terrain conditions for any averaging time and elevation and another set of conditions (i.e., variations in terrain, averaging time, or elevation), the following combination of the above expressions can be used:

$$\frac{U_{t_1}(z_1)}{U_{t_2}(z_2)_{\text{open}}} = p\left[\frac{\ln(z_1/z_0)+0.4\sqrt{\beta}\,C(t_1)}{\ln(z_2/z_0)_{\text{open}}+0.98C(t_2)_{\text{open}}}\right] \tag{26.7}$$

The examples in Table 26.3 serve to illustrate the use of this expression. The first line indicates the example case, the second through fifth rows describe the open terrain wind characteristics, the sixth through thirteenth rows describe the second terrain wind characteristics, and the last row gives the ratio of maximum speeds for the stated conditions and averaging times.

26.5 Pressure Coefficients and Load Factors

Pressure coefficients and load factors, such as terrain exposure factors and gust factors, are available in building codes. These factors must be consistent with the type of reference wind used in the code. Local cladding pressures in modern codes are significantly higher than those found in earlier codes because of the improved understanding of fluctuating loads on structures. Except in areas of positive mean pressures (the windward wall), there is no direct correlation between the local pressure and the occurrence of gusts in the approaching wind.

Defining Terms

Atmospheric boundary layer — The lower part of the atmosphere where the wind flow is affected by the earth's surface.

Cladding — Parts of the exterior building surface that keep out the weather but are generally not considered part of the structural system, although they do transfer loads to the structural system.

Extra-tropical cyclones — Large-scale low-pressure systems that control most of the severe weather conditions and extreme winds in temperate regions.

References

Simiu, E. 1981. Modern developments in wind engineering: Part 1. *J. Eng. Struct.* 3:233–241.

Simiu, E. and Scanlan, R. H. 1985. *Wind Effects on Structures,* 2nd ed. John Wiley & Sons, New York.

Krayer, W. R. and Marshall, R. D. 1992. Gust factors applied to hurricane winds. *Bull. AMS.* 73.

Further Information

Information on types of physical model studies commonly performed can be obtained from *Wind Tunnel Modeling for Civil Engineering Applications,* Cambridge University Press, 1982, and ASCE Manual and Reports on Engineering Practice No. 67, *Wind Tunnel Model Studies of Buildings and Structures,* American Society of Civil Engineers, New York, 1987.

In addition to the references by Simiu listed above, a good general book on wind engineering is *The Designer's Guide to Wind Loading of Building Structures* by Cook (Butterworths, 1985). General articles, including a number of conference proceedings, are published in the *Journal of Industrial Aerodynamics and Wind Engineering.*

27

Earthquakes and Their Effects

Charles Scawthorn
Kyoto University

This chapter provides an overview of earthquakes and their effects, by first discussing the causes of earthquakes, then explaining how earthquakes are measured and how seismicity can be characterized, then discussing the effects of soils on earthquake ground motions, and concluding with an explanation of the effects of earthquakes on structures and how design codes address the earthquake issue.

27.1 Earthquakes: Causes and Faulting

Earthquakes are naturally occurring broad-banded vibratory ground motions, due to a number of causes including tectonic ground motions, volcanism, landslides, rockbursts, and manmade explosions. Of these various causes, tectonic-related earthquakes are the largest and most important. Earthquakes initiate a number of phenomena or agents, termed **seismic hazards**, that can cause significant damage to the built environment — these include fault rupture, vibratory ground motion (i.e., shaking), inundation (e.g., tsunami, seiche, dam failure), various kinds of permanent ground failure (e.g., liquefaction), fire, or hazardous materials release. It is the goal of the earthquake specialist to reduce seismic risk. For most earthquakes, shaking is the dominant and most widespread agent of damage. Typically, earthquake ground motions are powerful enough to cause damage only in the near field (i.e., within a few tens of kilometers from the causative fault); in a few instances, however, long period motions have caused significant damage at great distances to selected lightly damped structures. A prime example of this was the 1985 Mexico City earthquake, where numerous collapses of mid- and high-rise buildings were due to a magnitude 8.1 earthquake occurring at a distance of approximately 400 kilometers from Mexico City. Earthquakes killed an average of 17,000 persons per year during the twentieth century, so the study of earthquakes and the physics of the earth, as well as the design of structures to resist earthquake effects, are of vital importance for the scientific and engineering professions.

0-8493-1586-7/05/$0.00+$1.50

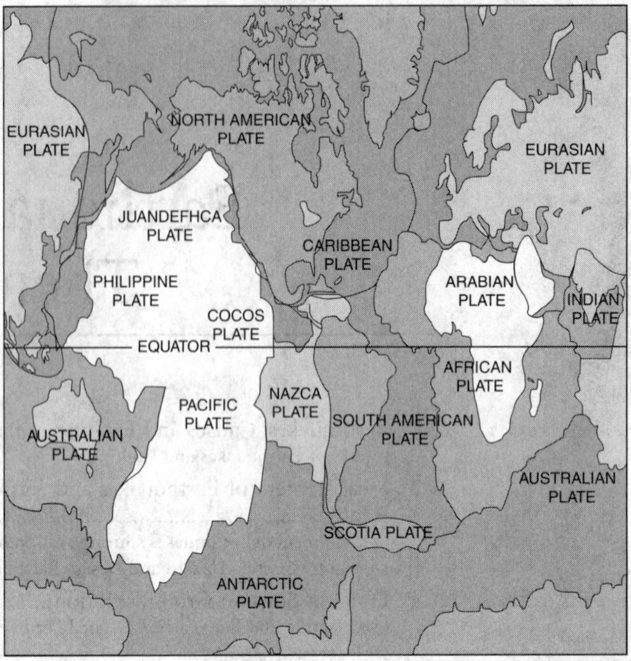

FIGURE 27.1 Global tectonic plate boundaries (Courtesy United States Geological Survey).

Causes of Earthquakes and Faulting

In a global sense, tectonic earthquakes result from motion between a number of large plates comprising the earth's crust or lithosphere (about 15 in total). See Figure 27.1.

This sudden slip releases large amounts of energy, which constitute the earthquake. While the accumulation of strain energy within the plate can cause motion (and consequent release of energy) at faults at any location, earthquakes occur with greatest frequency at the boundaries of the tectonic plates. The boundary of the Pacific plate is the source of nearly half of the world's great earthquakes. Tectonic plates move very slowly and irregularly, with occasional earthquakes. Generally, the longer a fault, the larger the earthquake it can generate. Faults are typically classified according to their sense of motion. See Figure 27.2.

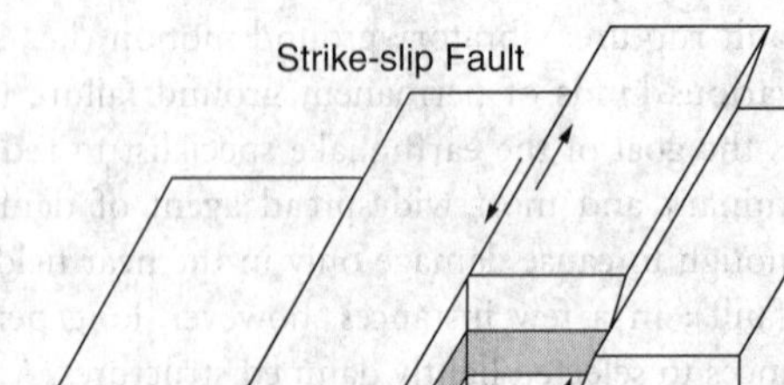

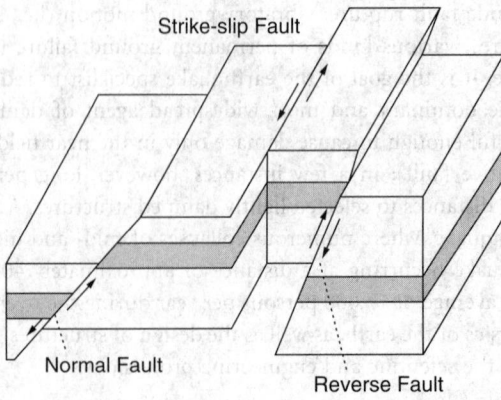

FIGURE 27.2 Earthquake fault types.

Generally, earthquakes will be concentrated in the vicinity of faults, faults that are moving more rapidly than others will tend to have higher rates of seismicity, and larger faults are more likely than others to produce a large event. Recent major earthquakes of this type have included the 1994 M_w 6.7 Northridge (California), 1995 M_w 6.9 Kobe (Japan), 1999 M_w 7.6 Kocaeli (Turkey), 1999 M_w 7.7 Ji-Ji (Taiwan), 2001 M_w 7.7 Bhuj (India), and 2003 M_w 6.6 Bam (Iran) events.

Probabilistic methods can be usefully employed to quantify the likelihood of an earthquake's occurrence, and typically form the basis for determining the **design basis earthquake**. However, the earthquake-generating process is not understood well enough to reliably predict the times, sizes, and locations of earthquakes with precision. In general, therefore, communities must be prepared for an earthquake to occur at any time.

27.2 Measurement of Earthquakes and Seismic Hazard Analysis

Measurement of Earthquakes

Engineering design requires measurement of earthquake phenomena in units such as force or displacement. An individual earthquake is a unique release of strain energy; quantification of this energy has formed the basis for measuring the earthquake event. C.F. Richter [1935] was the first to define earthquake **magnitude**, and subsequently a number of other magnitudes have been defined, the most important of which are surface wave magnitude M_S, body wave magnitude m_b and **moment magnitude** M_W. **Seismic moment** is employed to define moment magnitude M_W [Hanks and Kanamori, 1979; also denoted as boldface **M**]:

$$\log M_0 = 1.5\, M_W + 16.0 \tag{27.1}$$

where seismic moment M_0 (dyne-cm) is defined as [Lomnitz, 1974]:

$$\log M_0 = \mu A u \tag{27.2}$$

where μ is the material shear modulus, *A* is the area of fault plane rupture, and u is the mean relative displacement between the two sides of the fault (the averaged fault slip).

In general, seismic **intensity** is a metric of the effect, or the strength, of an earthquake hazard at a specific location. In the U.S., the Modified Mercalli Intensity scale (Table 27.1) is commonly employed, while other scales (MSK, EMS, JMA, etc.) are employed in other countries. It is difficult to find a reliable relationship between magnitude, which is a description of the earthquake's total energy level, and intensity, which is a subjective description of the level of shaking of the earthquake at specific sites, because shaking severity can vary with building type, design and construction practices, soil type and distance from the event.

Time histories are the actual record of the ground's motion at a site caused by an earthquake. They are recorded on strong or weak motion seismometers, typically in terms of acceleration, and can differ dramatically in duration, frequency content, and amplitude. The maximum amplitude of recorded acceleration is termed the **peak ground acceleration**, PGA (also termed the ZPA, or **zero period acceleration**); peak ground velocity (PGV) and peak ground displacement (PGD) are the maximum respective amplitudes of velocity and displacement. Recent earthquakes (1994 Northridge, M_W 6.7 and 1995 Hanshin [Kobe] M_W 6.9) have recorded PGAs of about 0.8 g and PGVs of about 100 kine[1], although it should be noted that almost 2 g was recorded in the 1992 Cape Mendocino earthquake.

If a single-degree-of-freedom mass is subjected to a time varying motion, the mass or elastic structural response can be readily calculated as a function of time, generating a structural response time history. Figure 27.3 illustrates this process, resulting in S_d, the displacement **response spectrum**, while Figure 27.4 shows:

[1] 1 kine = 1 cm/sec.

TABLE 27.1 Modified Mercalli Intensity Scale of 1931[a]

I	Not felt except by a very few under especially favorable circumstances
II	Felt only by a few persons at rest, especially on upper floors of buildings; delicately suspended objects may swing
III	Felt quite noticeably indoors, especially on upper floors of buildings, but many people do not recognize it as an earthquake; standing motor cars may rock slightly; vibration like passing truck; duration estimated
IV	During the day felt indoors by many, outdoors by few; at night some awakened; dishes, windows, and doors disturbed; walls make creaking sound; sensation like heavy truck striking building; standing motorcars rock noticeably
V	Felt by nearly everyone; many awakened; some dishes, windows, etc., broken; a few instances of cracked plaster; unstable objects overturned; disturbance of trees, poles, and other tall objects sometimes noticed; pendulum clocks may stop
VI	Felt by all; many frightened and run outdoors; some heavy furniture moved; a few instances of fallen plaster or damaged chimneys; damage slight
VII	Everybody runs outdoors; damage negligible in buildings of good design and construction, slight to moderate in well-built ordinary structures, considerable in poorly built or badly designed structures; some chimneys broken; noticed by persons driving motor cars
VIII	Damage slight in specially designed structures, considerable in ordinary substantial buildings, with partial collapse, great in poorly built structures; panel walls thrown out of frame structures; fall of chimneys, factory stacks, columns, monuments, walls; heavy furniture overturned; sand and mud ejected in small amounts; changes in well water; persons driving motor cars disturbed
IX	Damage considerable in specially designed structures, well-designed frame structures thrown out of plumb, great in substantial buildings, with partial collapse; buildings shifted off foundations; ground cracked conspicuously; underground pipes broken
X	Some well-built wooden structures destroyed, most masonry and frame structures destroyed with foundations; ground badly cracked; rails bent; landslides considerable from river banks and steep slopes; shifted sand and mud; water splashed over banks
XI	Few, if any (masonry) structures remain standing; bridges destroyed; broad fissures in ground; underground pipelines completely out of service; earth slumps and land slips in soft ground; rails bent greatly
XII	Damage total; waves seen on ground surfaces; lines of sight and level distorted; objects thrown upward into the air

[a] After Wood and Neumann, 1931.

1. S_d, the displacement response spectrum
2. S_v, the velocity response spectrum (also denoted PSV, the pseudo spectral velocity, pseudo to emphasize that this spectrum is not exactly the same as the relative velocity response spectrum [Hudson, 1979])
3. S_a, the acceleration response spectrum

Response spectra form the basis for much modern earthquake engineering structural analysis and design. They are readily calculated if the ground motion is known. Response spectra are most normally presented for 5% of critical damping. A standardized response spectrum is provided in the International Building Code [IBC, 2000] and other design codes. The spectrum is based on a smoothed average of normalized 5% damped spectrum obtained from actual ground motion records grouped by subsurface soil conditions at the location of the recording instrument, and is applicable for earthquakes characteristic of those that occur in California.

Strong Motion Attenuation and Duration

The rate at which earthquake ground motion decreases with distance, termed **attenuation**, is a function of the regional geology and inherent characteristics of the earthquake and its source. Three major factors affect the severity of ground shaking at a site:

1. Source — the size and type of the earthquake
2. Path — the distance from the source of the earthquake to the site and the geologic characteristics of the media earthquake waves pass through
3. Site-specific effects — type of soil at the site

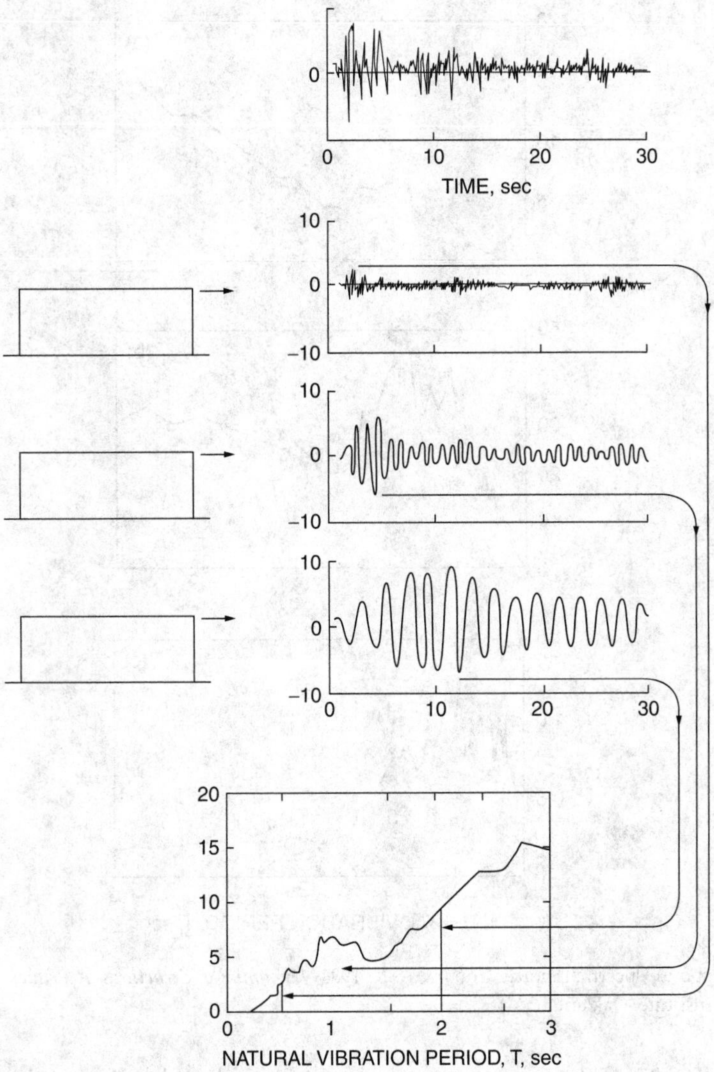

FIGURE 27.3 Computation of deformation (or displacement) response spectrum. (*Source:* Chopra, A. K. 1981. *Dynamics of Structures, A Primer,* Earthquake Engineering Research Institute, Oakland, CA.)

Figure 27.5 indicates, for alluvium, median values of the attenuation of peak horizontal acceleration with magnitude and style of faulting.

Seismic Hazard and Design Earthquake

The foregoing sections provide an overview of earthquake measures and occurrence; if an earthquake location and magnitude are specified, attenuation relations may be employed to estimate the PGA or response spectra at a site, which can then be employed for the design of a structure. Because the location and magnitude of earthquakes are typically not known beforehand, three approaches are employed in characterizing an earthquake for design purposes — they can be characterized as

1. Code approach
2. Upper-bound approach
3. Probabilistic Seismic Hazard Analysis approach

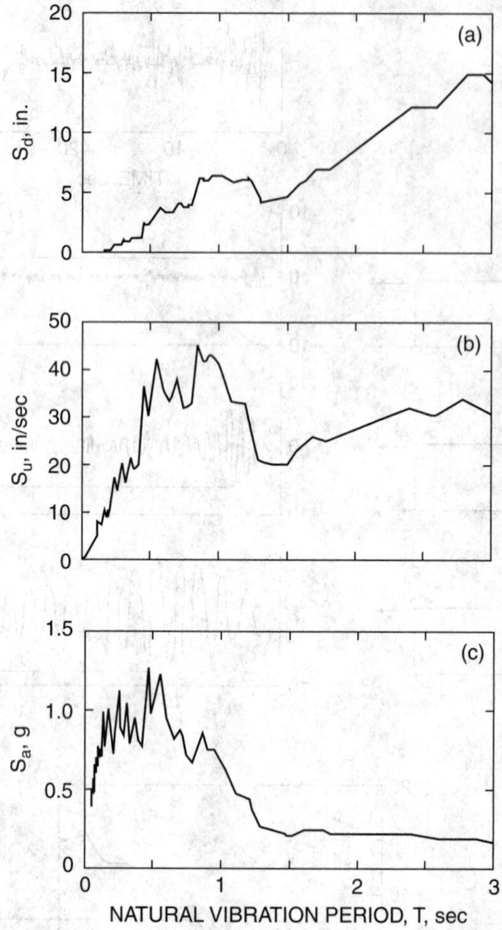

FIGURE 27.4 Response spectra. (*Source:* Chopra, A. K. 1981. *Dynamics of Structures, A Primer,* Earthquake Engineering Research Institute, Oakland, CA.)

Code approach —The code approach is to simply employ the lateral force coefficients as specified in the applicable design code. In the United States, a series of national seismic hazard maps for the United States and territories has been developed by the United States Geologic Survey (USGS) specifically for this purpose (available at http://geohazards.cr.usfs.gov/eq/index.html). Two sets of maps are available, providing contours of MCE, 5% damped, elastic spectral response acceleration at a period of (a) 0.2 sec, termed SS, and (b) 1.0 sec, termed S_1. In both cases, the spectral response acceleration values are representative of sites with subsurface conditions bordering between firm soil or soft rock. By locating a site on the maps and interpolating between the values presented for contours adjacent to the site, it is possible to rapidly estimate the MCE level shaking parameters for the site, given that it has a soft rock or firm soil profile. Figure 27.6 shows, for a portion of the western United States, contours of the 0.2-sec spectral acceleration with a 90% probability of not being exceeded in 50 years. As indicated in the figure, in zones of high seismicity these contours are quite closely spaced, making use of the maps difficult. Therefore, the USGS has furnished software, available both over the Internet (at the URL indicated above) and on a CD-ROM, that permits determination of the MCE spectral response acceleration parameters based on longitude and latitude.

Upper-bound approach — Using historic data and/or fault length–magnitude relations, a maximum magnitude event can be assigned to a fault and, using attenuation relations, a PGA or other engineering

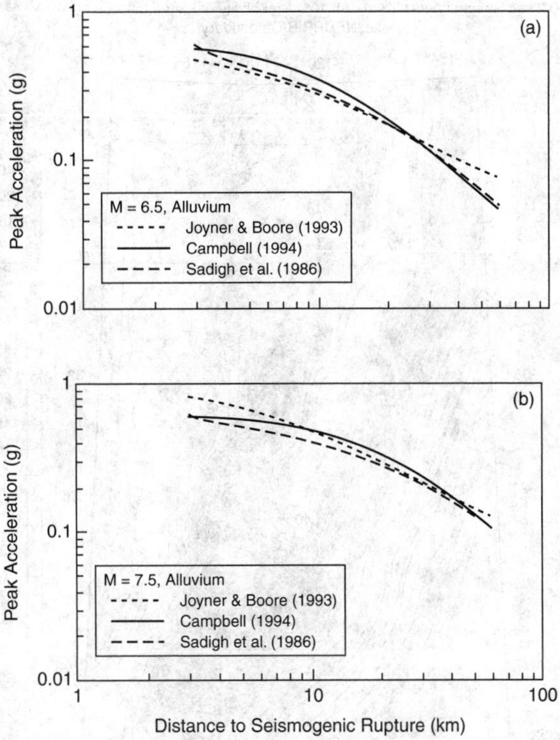

FIGURE 27.5 Campbell and Bozorgnia worldwide attenuation relationship showing (for alluvium) the scaling of peak horizontal acceleration with magnitude and style of faulting. (*Source:* Campbell, K. W. and Bozorgnia, Y. (1994) Near-Source Attenuation of Peak Horizontal Acceleration from Worldwide Accelerograms Recorded from 1957 to 1993, Proc. Fifth U.S. National Conference on Earthquake Engineering, Earthquake Engineering Research Institute, Oakland, CA.)

measure can be estimated for the site, based on the distance. The effects at a specific site are quantified on the basis of strong ground motion modeling (i.e., attenuation).

Probabilistic Seismic Hazard Analysis approach — The Probabilistic Seismic Hazard Analysis (PSHA) approach entered general practice with Cornell's [1968] seminal paper, and basically employs the theorem of total probability to formulate:

$$P(Y) = \sum_F \sum_M \sum_R p(Y \mid M, R) p(M) \tag{27.3}$$

where Y is a measure of intensity, such as PGA, response spectral parameters PSV, etc; $p(Y/M,R)$ is the probability of Y given earthquake magnitude M and distance R (i.e., attenuation); $p(M)$ is the probability of a given earthquake magnitude M; and F indicates seismic sources, whether discrete such as faults, or distributed.

Typically, various seismic sources (faults modeled as line sources and dipping planes, and various distributed or area sources, including a background source to account for miscellaneous seismicity) are identified, and their seismicity characterized on the basis of historic seismicity and/or geologic data. The effects at a specific site are quantified on the basis of strong ground motion modeling, also termed attenuation. The $p(Y/M,R)$ term represents the full probabilistic distribution of the attenuation relation; summation must occur over the full distribution, due to the significant uncertainty in attenuation. The $p(M)$ term is referred to as the **magnitude–frequency** relation, which was first characterized by Gutenberg and Richter [1954] as

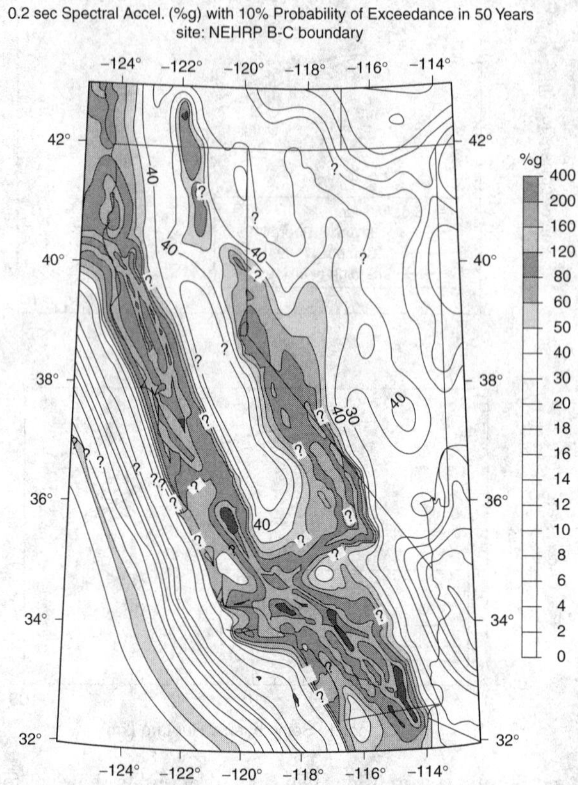

FIGURE 27.6 MCE seismic hazard map (0.2-sec spectral response acceleration) for the western United States. (Courtesy United States Geological Survey.)

$$\log \mathrm{N}(m) = \mathrm{a_N} - \mathrm{b_N}m \tag{27.4}$$

where $\mathrm{N}(m)$ = the number of earthquake events equal to or greater than magnitude m occurring on a seismic source per unit time, and $\mathrm{a_N}$ and $\mathrm{b_N}$ are regional constants ($10^{\mathrm{a_N}}$ = the total number of earthquakes with magnitude >0, and $\mathrm{b_N}$ is the rate of seismicity; $\mathrm{b_N}$ is typically 1 ± 0.3). These elements collectively are the seismotectonic model; their integration results in the seismic hazard.

A number of codes have been developed for the integration of Equation (27.3), many of which are based on the public domain FRISK (Fault Risk) code first developed by McGuire [1978].

Most probabilistic seismic hazard analysis models assume the Gutenberg–Richter exponential distribution of earthquake magnitude and that earthquakes follow a Poisson process, occurring on a seismic source zone randomly in time and space. This implies that the time between earthquake occurrences is exponentially distributed, and that the time of occurrence of the next earthquake is independent of the elapsed time since the prior earthquake.[1] The CDF for the exponential distribution is:

$$\mathrm{F}(t) = 1 - \exp(-\lambda t) \tag{27.5}$$

Note that this forms the basis for many modern building codes, in that the probabilistic seismic hazard analysis results are selected such that the seismic hazard parameter (e.g., PGA) has a "10% probability of exceedance in 50 years" [UBC, 1994] — that is, if t = 50 years and $\mathrm{F}(t)$=0.1 (i.e., only 10% probability that the event has occurred in t years), then λ = .0021 per year, or 1 per 475 years.

[1]For this aspect, the Poisson model is often termed a *memoryless* model.

Equation (27.3) is quite general and is used to develop estimates of MMI, PGA, response spectra or other measures of seismic hazard. Since probabilistic response spectra are a composite of the contributions of varying earthquake magnitudes at varying distances, the ground motions of which attenuate differently at different periods, this method has the drawback that the resulting spectra have varying (and unknown) probabilities of exceedance at different periods.

27.3 Effect of Soils on Ground Motion

The effect of different types of soils on earthquake ground motion and damage has long been noted. Quantification of the effects of soils on ground motion has generally been by either analytical or empirical methods. The current International Building Code [2000] defines seven site coefficients and provides site coefficients for the acceleration and velocity portions of response spectra as a function of expected ground motion.

Liquefaction and Liquefaction-Related Permanent Ground Displacement

Seismic liquefaction of soils has been noted in numerous earthquakes but the phenomenon was only well understood following the 1964 Alaska and, particularly, 1964 Niigata (Japan) earthquakes, where dramatic effects were observed. The process typically consists of loose granular soil with a high water table being strongly shaken during an earthquake — that is, cyclically sheared. The soil particles initially have large voids between them; due to shaking, the particles are displaced relative to each other and tend to more tightly pack, decreasing the void volume. The water, which had occupied the voids (and being incompressible), comes under increased pressure and migrates upward towards or to the surface, where the pressure is relieved. The water usually carries soil with it, and the resulting ejecta are variously termed **sand boils** or **mud volcanoes.**

Liquefaction is a major source of damage in earthquakes, since:

1. The soil's loss of shear strength results in partial or total loss of bearing capacity, resulting in foundation failure unless the structure is founded below the liquefying layer.
2. Liquefaction may result in large lateral spreads and permanent ground displacements, often measured in meters and occasionally resulting in catastrophic slides, such as occurred at Turnagain Heights in the 1964 Alaska earthquake [Seed and Wilson, 1967].
3. For both of these reasons, various lifelines, particularly buried water, wastewater, and gas pipes, typically sustain numerous breaks that can result in system failure and lead to major secondary damage, such as fire following earthquake.

Evaluating liquefaction potential requires consideration of a number of factors, including grain-size distribution. Generally, poorly graded sands (i.e., most particles about the same size) are much more susceptible to liquefaction than well-graded ones (i.e., particles of many differing sizes), since good grading results in better natural packing and better grain–grain contact. Silts and clays are generally much less susceptible to liquefaction, although the potential should not be ignored, and large-grained sands and gravels have such high permeability that pore water pressures usually dissipate before liquefaction can occur. Relative density is basically a measure of the packing; higher relative density means better packing, with more grain–grain contact, while lower relative density, or looseness, indicates a higher potential for liquefaction. Water table depth is critical, as only submerged deposits are susceptible to liquefaction. A loose sandy soil above the water table is not liquefiable by itself, although upward-flowing water from a lower liquefying layer can initiate liquefaction even above the pre-event water table. Note also that water table depths can fluctuate significantly, seasonally or over longer periods and for natural reasons or due to human intervention. Earthquake acceleration and duration are also critical, as these are the active causative agents for liquefaction.

Mitigation of liquefaction is accomplished by a number of methods, including:

- *Excavation and replacement* — that is, if the structure is sufficiently large or important, and the liquefiable layer sufficiently shallow, it may be cost-effective to place the foundation of the structure below the liquefiable layer. Note, however, that liquefaction may still occur around the structure, with possible disruption of adjacent streets, entrances and utilities.
- *Compaction* can be accomplished by a number of methods, including:
  - *Vibrostabilization*:
    - *Vibro-compaction*, in which a vibrating pile or head accompanied by water jetting is dynamically injected into the ground and then withdrawn; vibration of the head as it is withdrawn compacts an annulus of soil, and this is repeated on a closely spaced grid. Sand may be backfilled to compensate for volume reduction. This technique achieves good results in clean granular soils with less than about 20% fines.
    - *Vibro-replacement* (stone columns) — used in soils with higher fines content (>20%) or even in clay-ey soils which cannot be satisfactorily vibrated. The stone columns act as vertical drains (see below).
    - *Dynamic compaction* via the use of heavy dropped weights or small explosive charges. Weights up to 40 tons are dropped from heights up to 120 ft on a grid pattern, attaining significant compaction to depths of a maximum of 40 ft. Best used in large open areas due to vibrations, noise and flying debris.
  - *Compaction piles*
  - *Grouting*, where grout is injected at high pressure, filling voids in the soil. This technique offers the advantage of being able to be used in small, difficult to access areas (e.g., in basements of buildings, under bridges). Grouting can be used in several ways:
    - *Compaction* — A very stiff soil–cement–water mixture is injected and compacts an annulus around the borehole; this is repeated on a closely spaced grid.
    - *Chemical* — Low-viscosity chemical gels are injected, forming a strong sandstone-like material. Long-term stability of the grout should be taken into account.
    - *Jet-grouting* — Very high pressure water jets are used to cut a cylindrical hole to the desired depth, and the material is replaced or mixed with admixtures to form a stabilized column.
  - *Soil-mixing* — Large rotary augers are used to churn up and mix the soil with admixtures and form stabilized columns or walls.
- *Permeability* can be enhanced by a number of methods, including:
  - *Placement of stone columns* — Rather than by vibro-replacement, a hole is augured or clamshell-excavated, and filled with gravel or cobbles, on a closely spaced grid. If liquefaction occurs, any increase in pore water pressure is dissipated via the high permeability of the stone columns.
  - *Soil wicks* — Similar to stone columns, but wicks consisting of geotextiles are inserted into the ground on a very closely spaced grid; their high permeability similarly dissipates any increase in pore water pressure.
  - *Grouting* — Many materials are available to cement or otherwise create adhesion between soil grains, thus decreasing their cyclic mobility and any packing due to shaking.

27.4 Structures: Earthquake Effects and Seismic Design

Many different types of earthquake damage occur in structures. This section discusses general earthquake performance of buildings and selected other structures, with the emphasis on buildings, especially those typically built in the western U.S.

Buildings

In buildings, earthquake damage can be divided into two categories: structural damage and nonstructural damage, both of which can be hazardous to building occupants. Structural damage means degradation of the building's structural support systems (i.e., vertical and lateral force resisting systems), such as the

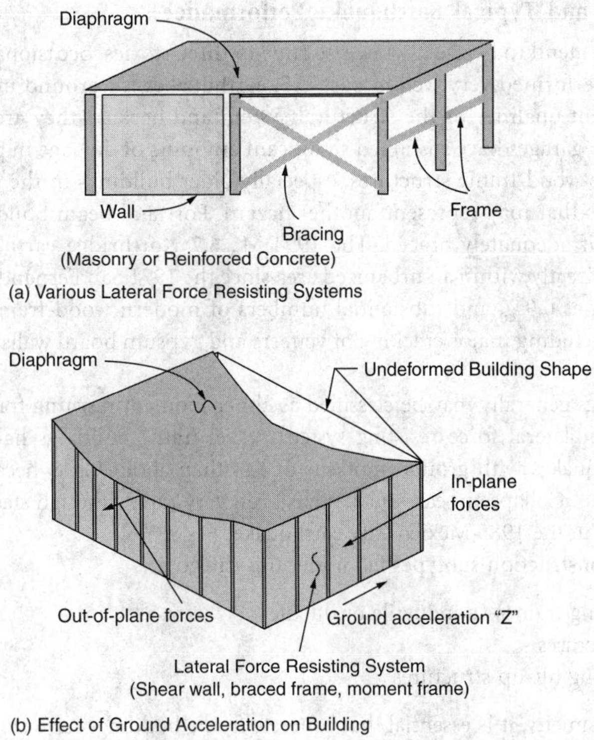

FIGURE 27.7 Resistance of earthquake forces: (a) variety of LFRS, (b) effect of ground acceleration on building.

building frames and walls, while nonstructural items include architectural features, mechanical and other equipment, and contents.

How Earthquake Forces are Resisted

Buildings experience horizontal distortion when subjected to earthquake motion (Figure 27.7). **Lateral Force Resisting Systems (LFRSs)** are usually capable of resisting only forces that result from ground motions parallel to them. However, the combined action of LFRS along the width and length of a building can typically resist earthquake motion from any direction.

In wood frame stud-wall buildings, the resistance to lateral loads is typically provided by (a) for older buildings, especially houses, wood diagonal "let-in" bracing, and (b) for newer (primarily post-World War II) buildings, plywood siding "shear walls." The earthquake resisting systems in modern buildings take many forms. Moment-resisting or braced steel frames are capable of resisting lateral loads. In concrete structures, shear walls[1] are sometimes used to provide lateral resistance, in addition to moment-resisting frames. Certain problems in earthquake resistiveness are independent of building type, and include:

Configuration, or the general vertical and/or horizontal shape of buildings, is an important factor in earthquake performance and damage. Buildings that have simple, regular, symmetric configurations generally display the best performance in earthquakes. Asymmetry can exist in the placement of bracing systems, shear walls or moment-resisting frames that are used to provide earthquake resistance in a building.

Pounding is the collision of adjacent buildings during an earthquake due to insufficient lateral clearance. Pounding has been the cause of a number of mid-rise building collapses, most notably in the 1985 Mexico City earthquake.

[1]Termed shear walls because the depth-to-width ratio is so large that deformation is primarily due to shear rather than bending.

Types of Buildings and Typical Earthquake Performance

Wood frame structures tend to be mostly low rise (one to three stories, occasionally four stories). Stud wall buildings have performed very well in past U.S. earthquakes for ground motions of about 0.5 g or less, due to inherent qualities of the structural system and because they are lightweight and low rise. Homes built over garages have sustained significant amounts of damage in past earthquakes, with many collapses. Some wood frame structures, especially older buildings in the eastern United States, have masonry veneers that may represent another hazard. Post and beam buildings tend to perform well in earthquakes, if adequately braced. The 1994 M_W 6.7 Northridge earthquake was the largest earthquake to occur directly within an urbanized area since the 1971 San Fernando earthquake; ground motions were as high as 0.9 g, and substantial numbers of modern wood-frame dwellings sustained significant damage, including major cracking of veneers and gypsum board walls and splitting of wood wall studs.

Steel frame buildings generally may be classified as either moment resisting frames (MRF) or braced frames, based on their lateral force resisting systems. Steel frame buildings have tended to perform satisfactorily in earthquakes with ground motions of less than about 0.5 g, because of their strength, flexibility, and lightness. Collapse in earthquakes has been very rare, although steel frame buildings did collapse, for example, in the 1985 Mexico City earthquake.

Concrete: Several construction subtypes fall under this category:

1. Moment resisting frames (nonductile or ductile)
2. Shear wall structures
3. Precast, including tilt-up structures

In areas of high seismicity, it is essential that concrete frame buildings have closely spaced lateral ties in the columns and beams, in order to provide ductility. This was not recognized until the 1970s, so older concrete frames are typically nonductile and a high collapse hazard. Concrete shear wall buildings consist of a concrete box or frame structural system with walls constituting the main LFRS. This building type generally tends to perform better than concrete frame buildings. Damages specific to this building type are:

1. Shear cracking and distress can occur around openings in concrete shear walls during large seismic events.
2. Shear failure can occur at wall construction joints, usually at a load level below the expected capacity.
3. Bending failures can result from insufficient chord steel lap lengths.

Tilt-up buildings are typically one story "box" systems, in which the LFRS is the precast concrete wall panels, which are anchored to the roof diaphragm. During an earthquake, weak wall anchors pull out of the walls, causing the floors or roofs to collapse. Damage to tilt-up buildings was observed again in the 1994 M_W 6.7 Northridge earthquake, where the primary problems were poor wall anchorage into the concrete and excessive forces due to flexible roof diaphragms amplifying ground motion to a greater extent than anticipated in the code.

Reinforced masonry buildings are mostly low-rise perimeter-bearing wall structures, often with wood diaphragms, although precast concrete is sometimes used. Reinforced masonry buildings can perform well in moderate earthquakes if they are adequately reinforced and grouted and if sufficient diaphragm anchorage exists. The collapse of bearing walls can lead to major building collapses. The bearing walls will be heavily damaged and collapse under large loads; wall slenderness — some of these buildings have tall story heights and thin walls.

Passive and Active Control

In about the last two decades, a number of innovative techniques have been developed and introduced to enable buildings and structures to better withstand earthquake ground motions. The general aim of these techniques has been the *avoidance* of earthquake-induced forces, rather than their resistance. These innovative techniques can be divided into two broad categories: *passive control* (base isolation, energy dissipation) and *active control*, and are increasingly being applied to the design of new structures or to the retrofit of existing structures against wind, earthquakes, and other external loads. The distinction between passive and active control is that passive systems require no active intervention or energy source, while active systems typically monitor the structure and incoming ground motion and seek to actively control masses or forces in the structure (via moving weights, variable tension tendons, etc.) so as to develop a structural response (ideally) equal and opposite to the structural response due to the incoming ground motion. Recently developed semiactive control systems appear to combine the best features of both approaches, offering the reliability of passive devices, yet maintaining the versatility and adaptability of fully active systems. Magnetorheological (MR) dampers, for example, are new semiactive control devices that use MR fluids to create controllable dampers. Initial results indicate that these devices are quite promising for civil engineering applications.

Nonbuilding

Structures

Bridges: Abutment damage rarely leads to bridge collapse. Liquefaction of saturated soils in river channels and floodplains and subsequent loss of support have caused many bridge failures in past earthquakes, notably in the 1990 M_W 7.7 Philippines and 1991 M_W 7.1 Costa Rica earthquakes. Regarding aseismic design, bridge behavior during an earthquake can be very complex. Damage in foundation systems is hard to detect, so bridge foundations should be designed to resist earthquake forces elastically.

Industrial Structures: On-ground storage tanks are subject to a variety of earthquake damage mechanisms, generally due to sliding or rocking; this can be prevented by proper detailing. Waterfront structures include quay walls, sheet-pile bulkheads, and pile-supported piers, all of which often sustain damage due to liquefaction. Many industrial structures perform well in earthquakes, but small details result in loss of service and occasionally major damage. For example, electrical power stations and transmission towers usually perform well in earthquakes, but the porcelain insulators and other elements of high voltage transformers and other substation elements are often damaged, resulting in power blackouts over wide areas.

Seismic Design Codes

Seismic design codes are a subset and are generally a section or portion of many building and other design codes. Design loading levels for earthquake and other hazards are typically set by building codes at levels that have a moderate to low probability of occurrence during the life of the structure. While buildings may be designed for earthquake shaking likely to be experienced once every 500 years, wind loads requirements are based on the anticipation, on the average, of once every 100 years, or for snow loads, on average, once every 20 years. The significant difference in recurrence intervals adopted by codes for these various hazards is a function of the hazard itself, and the adequacy of a given return period to capture a maximum, or near maximum, credible event. Building code provisions for earthquake-resistant design are unique in that, unlike the provisions for other load conditions, they do not intend that structures be capable of resisting design loading within the elastic or near elastic range of response — that is, some level of damage is permitted. Building codes intend only that buildings resist large earthquake loading without life-threatening damage and, in particular, without structural collapse or creation of large, heavy falling debris hazards.

The *2000 NEHRP Recommended Provisions for Seismic Regulation for Buildings and Other Structures (NEHRP Provisions)* represents the current state of the art in prescriptive, as opposed to performance-based, provisions for seismic-resistant design. Its provisions form the basis for earthquake design specifications contained in the 2002 edition of ASCE7, *Minimum Design Loads for Buildings and Other Structures*, either through reference or direct incorporation, the seismic regulations in the 2003 edition of the International Building Code (IBC) and also the 2002 edition of the NFPA 5000 Building Code [NFPA, 2003]. As such, the *NEHRP Provisions* will form the basis for most earthquake-resistant design in the United States, as well as other nations that base their codes on U.S. practices, throughout much of the first decade of the twenty-first century.

Defining Terms

Attenuation — The rate at which earthquake ground motion decreases with distance.

Design [basis] earthquake — The earthquake (as defined by various parameters, such as PGA, response spectra, etc.) for which the structure will be, or was, designed.

Ductile detailing — Special requirements such as, for reinforced concrete and masonry, close spacing of lateral reinforcement to attain confinement of a concrete core, appropriate relative dimensioning of beams and columns, 135-degree hooks on lateral reinforcement, hooks on main beam reinforcement within the column, etc.).

Ductile frames — Frames required to furnish satisfactory load-carrying performance under large deflections (i.e., ductility). In reinforced concrete and masonry this is achieved by **ductile detailing**.

Fault — A zone of the earth's crust within which the two sides have moved; faults may be hundreds of miles long, from 1 to over 100 miles deep, and not readily apparent on the ground surface.

Intensity — A metric of the effect, or the strength, of an earthquake hazard at a specific location, commonly measured on qualitative scales such as mmi, msk, and jma.

Lateral force resisting system (LFRS) — A structural system for resisting horizontal forces, due for example to earthquake or wind (as opposed to the vertical force resisting system, which provides support against gravity).

Liquefaction — A process resulting in a soil's loss of shear strength, due to a transient excess of pore water pressure.

Magnitude — A unique measure of an individual earthquake's release of strain energy, measured on a variety of scales, of which the **moment magnitude** M_w (derived from seismic moment) is preferred.

Mud volcanoes — see **sand boils.**

Peak ground acceleration (PGA) — The maximum amplitude of recorded acceleration (also termed the ZPA, or **zero period acceleration**).

Pounding — The collision of adjacent buildings during an earthquake due to insufficient lateral clearance.

Response spectrum — A plot of maximum amplitudes (acceleration, velocity, or displacement) of a single-degree-of-freedom oscillator (SDOF), as the natural period of the SDOF is varied across a spectrum of engineering interest (typically, for natural periods from 0.03 to 3 or more sec or frequencies of 0.3 to 30+ hz).

Sand boils — Also known as **mud volcanoes,** these are the small cone-shaped sand piles resulting from sand ejected to the surface during the liquefaction process. They have the appearance of miniature volcanoes, typically less than 1 m high, and at most several meters in diameter.

Seismic hazards — The phenomena and/or expectation of an earthquake-related agent of damage, such as fault rupture, vibratory ground motion (i.e., shaking), inundation (e.g., tsunami, seiche, dam failure), various kinds of permanent ground failure (e.g., liquefaction), fire, or hazardous materials release.

Seismic moment — The moment resulting from the forces generated on an earthquake fault during slip.

Thrust fault — Low-angle reverse faulting. (**Blind thrust faults** are faults at depth occurring under anticlinal folds; they have only subtle surface expression.)

Transform or **strike slip fault** — A fault where relative fault motion occurs in the horizontal plane, parallel to the strike of the fault.

References

American Society of Civil Engineers. 1991. *Minimum Design Loads for Buildings and Other Structures, Standard No. ASCE-7,* American Society of Civil Engineers, Reston, VA.

BSSC. 2000. NEHRP Recommended Provisions for Seismic Regulations for New Buildings, Part 1 — Provision, Part 2 — Commentary, 2000 Edition, Building Seismic Safety Council, Washington.

Campbell, K. W. and Bozorgnia, Y. (1994) Near-Source Attenuation of Peak Horizontal Acceleration from Worldwide Accelerograms Recorded from 1957 to 1993, Proc. Fifth U.S. National Conference on Earthquake Engineering, Earthquake Engineering Research Institute, Oakland, CA.

Chopra, A. K. 1981. *Dynamics of Structures, A Primer,* Earthquake Engineering Research Institute, Oakland, CA.

Hanks, T.C. and Kanamori, H. 1979. A moment magnitude scale, *J. Geophys. Res.,* 84:2348–2350.

IBC. 2000. *International Building Code 2000,* International Code Council, published by International Conference of Building Officials, Whittier, CA, and others.

Lomnitz, C. 1974. *Global Tectonics and Earthquake Risk,* Elsevier, New York.

McGuire, R. K. 1978. FRISK: computer program for seismic risk analysis using faults as earthquake sources.

McGuire, R. K. and Barnhard, T. P. 1979. The usefulness of ground motion duration in predicting the severity of seismic shaking, Proc. 2nd U.S. Natl. Conf. on Earthquake Engineering, Earthquake Engineering Research Institute, Oakland, CA.

NFPA. 2003. *NFPA 5000 Building Code,* National Fire Protection Association, Cambridge, MA.

Richter, C. F. 1935. An instrumental earthquake scale, *Bulletin of the Seismological Society of America,* 25:1–32.

Further Reading

The above material is covered in much more depth in the following publication:

Chen, W. F. and Scawthorn, C., eds. (2002) *Earthquake Engineering Handbook,* CRC Press, Boca Raton, FL.

28

Structural Analysis

Eric M. Lui
Syracuse University

Metin Oguzmert
Syracuse University

28.1 Introduction

Structural analysis deals with the evaluation of the response of a structure subjected to prescribed external loads and imposed deformations. Structural analysis often entails the computation of reactions, internal forces, deflections/deformations, stresses, and strains in a structure. This chapter presents fundamental methods of analysis for structures such as cables, arches, trusses, beams, and frames. A brief discussion of methods used for the analysis of plates and shells will also be presented.

A structure can be successfully analyzed only if it is **geometrically stable.** For a structure to be geometrically stable, the number of unknown reactions and internal forces must equal or exceed the number of *independent* equilibrium equations that can be written for the structure. In addition, the structure must be properly constrained to prevent it from undergoing rigid body motion under any perceivable loading conditions. The geometry of the structure must also be such that local or global collapse will not occur under the applied loads. As an example, the two-dimensional beam labeled Beam I in Table 28.1 is geometrically unstable because it is not properly constrained and can undergo rigid body motion in the horizontal direction under the applied load. Beam II and Beam III are both geometrically stable. Beam II is referred to as **statically determinate** because the number of unknown reactions is equal to the number of independent equilibrium equations

$$\sum F_x = 0, \quad \sum F_y = 0, \quad \sum M_z = 0 \tag{28.1}$$

that can be written for the free body diagram of the beam shown. As a result, all unknown reactions as well as any unknown internal forces and moment (see Figure 28.1) can be determined by consideration of equilibrium alone. Beam III is referred to as **statically indeterminate** because the use of equilibrium equations alone is not sufficient to solve for all the unknowns. To solve the problem, both compatibility and equilibrium equations have to be used. For instance, a compatibility equation that can be written for Beam III is

0-8493-1586-7/05/$0.00+$1.50

TABLE 28.1 Beam Examples

	Beam I	Beam II	Beam III
Original	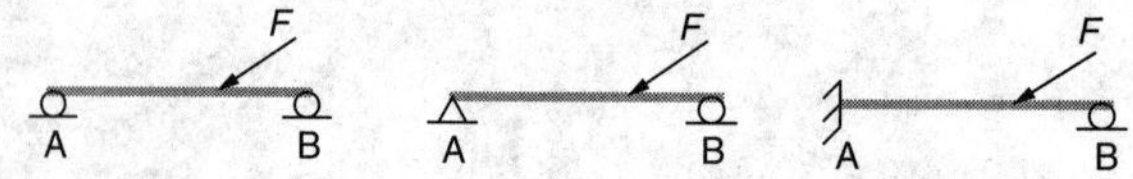		
Free body diagram	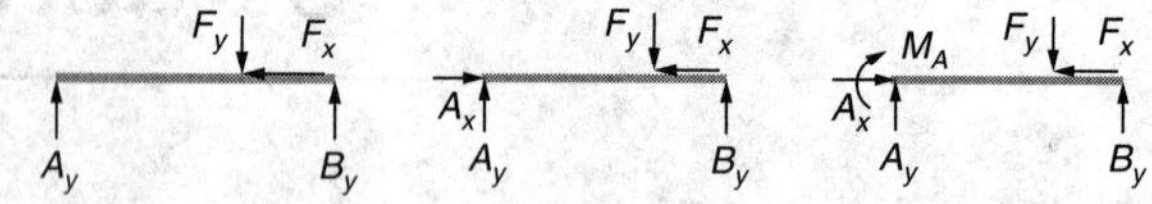		
Unknown reactions	A_y, B_y	A_x, A_y, B_y	M_A, A_x, A_y, B_y
Stability and determinacy	Unstable	Stable and statically determinate	Stable and statically indeterminate

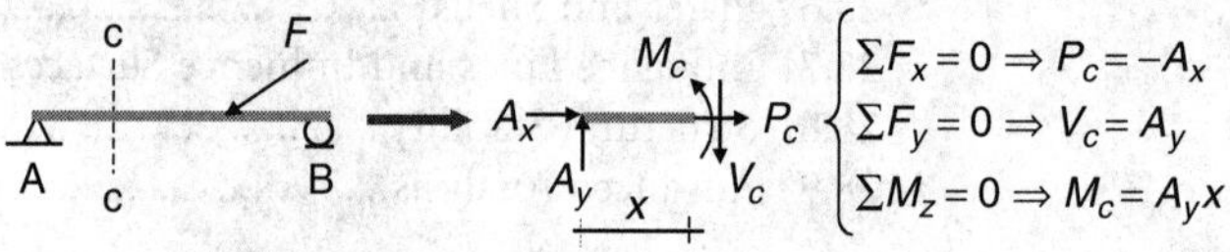

FIGURE 28.1 Computation of internal forces.

$$(\delta_{vB})_F + (\delta_{vB})_{B_y} = 0 \tag{28.2}$$

where $(\delta_{vB})_F$ and $(\delta_{vB})_{By}$ are the vertical deflection at B caused by the applied force F and the support reaction B_y, respectively, if support B is removed from the beam. The above approach of enforcing the displacement compatibility condition in a statically indeterminate structure is called the method of consistent displacements.

In the event that the applied forces are not coplanar, or the structure does not lie in a plane, the structure has to be analyzed as a three-dimensional system. In this case, Equation (28.1) becomes

$$\sum F_x = 0, \sum F_y = 0, \sum F_z = 0, \sum M_x = 0, \sum M_y = 0, \sum M_z = 0 \tag{28.3}$$

A structure is said to be linearly elastic if its force-deflection behavior is linear under both loading and unloading conditions and no permanent deformation results when the applied loads are removed. For linearly elastic structures, the principle of **superposition** can be used to facilitate the analysis. Superposition means structural responses such as displacements, internal forces, stresses and strains, etc., due to an array of forces acting simultaneously is equal to the sum of the responses due to each force acting by itself. By using superposition, the response of a complex system can be obtained by superposing the responses calculated for a subset of simpler systems derived from the original system. Equation (28.2) is an example of the use of the principle of superposition in structural analysis.

28.2 Cables

For ease of analysis, cables are often assumed to be flexible enough so the only unknown internal force acting at any given point in a cable is a (tangential) tensile force. The funicular shape of a cable is the natural shape taken by the cable under a specific set of applied loads. For the cable shown in Figure 28.2a, in which a system of concentrated forces acts, the 10 unknowns (four reactions, three internal

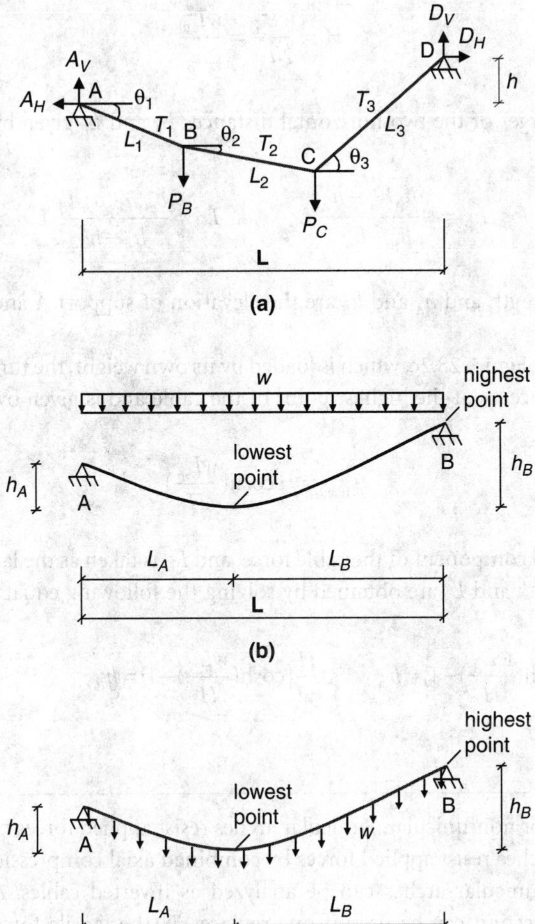

FIGURE 28.2 Cables.

forces, three angles) can be solved using eight joint equilibrium equations ($\sum F_x = 0$, $\sum F_y = 0$, written for points A, B, C, D) and two geometry equations in the form

$$\sum L_i \cos\theta_i = L, \qquad \sum L_i \sin\theta_i = h \qquad (i = 1,2,3) \tag{28.4}$$

where L_i (i = 1, 2, 3), L and h are the (known) segment lengths, span length, and difference in support elevations of the cable, respectively. The segment inclination angle, θ_i, is considered positive if it measures counterclockwise.

For the uniformly loaded cable shown in Figure 28.2b, the funicular shape is a parabola. The maximum tension occurs at the highest point of the cable and is equal to

$$T_{\max} = \sqrt{H^2 + (wL_m)^2} \tag{28.5}$$

where H is the horizontal component of the cable force, which is a constant along the entire length of the cable, given by

$$H = \frac{wL_A^2}{2h_A} = \frac{wL_B^2}{2h_B} \tag{28.6}$$

and L_m is taken as the larger of the two horizontal distances L_A and L_B given by

$$L_A = \frac{\sqrt{h_A h_B} - h_A}{h_B - h_A} L, \qquad L_B = \frac{h_B - \sqrt{h_A h_B}}{h_B - h_A} L \tag{28.7}$$

in which L is the span length and h_A and h_B are the elevation of support A and support B of the cable, respectively.

For the cable shown in Figure 28.2c, which is loaded by its own weight, the funicular shape is a catenary. The maximum tension occurs at the highest point of the cable and is given by

$$T_{\max} = H \cosh(\frac{wL_m}{H}) \tag{28.8}$$

where H is the horizontal component of the cable force, and L_m is taken as the larger of the two horizontal distances L_A and L_B. H, L_A, and L_B are obtained by solving the following equations simultaneously

$$\frac{H}{w}[\cosh(\frac{wL_A}{H}) - 1] = h_A, \qquad \frac{H}{w}[\cosh(\frac{wL_B}{H}) - 1] = h_B, \qquad L_A + L_B = L \tag{28.9}$$

28.3 Arches

Arches can be funicular or nonfunicular. Funicular arches resist applied forces by axial compression only, whereas nonfunicular arches resist applied forces by combined axial compression, shear, and bending or even twisting actions. Funicular arches can be analyzed as inverted cables. As a result, all equations presented in the above section can be used if one replaces T (the tensile force in the cable) by C (the compressive force in the arch). Nonfunicular arches have to be analyzed by other means. For instance, if the arch is statically determinate (e.g., a three-hinged arch), it can be analyzed by first separating the arch into two segments by cutting at the hinge common to both segments and exposing the internal axial and shear forces there. The problem can then be solved by applying Equation (28.1) successively to the two arch segments. The six equilibrium equations (two sets of three equilibrium equations written for the two arch segments) are sufficient to solve for the six unknowns (four unknown reaction components for the two supports and two unknown internal forces at the common hinge). If the arch is statically indeterminate (e.g., a two-hinged or a hingeless arch), the use of the method of consistent displacements [Timoshenko and Young, 1965] or other computer-based methods such as matrix [Kassimali, 1999] or finite element [Cook et al., 2001] must be used. The primary difference between the matrix and the **finite element method** is that element equilibrium is rigorously enforced in the former, but only approximately satisfied in the latter. However, for simple line-type elements such as trusses, beams, and frames, both methods will yield the same stiffness matrix. One computer-based method to analyze statically indeterminate arches is to model the curved profile of an arch by a series of straight frame elements. If the elements are kept small enough (i.e., the number of elements used to model the arch is large enough), reasonably good results can be obtained.

28.4 Trusses

Trusses are structures composed of straight and relatively slender members joined together in the form of triangles or other stable shapes. For purpose of analysis, truss members are assumed to be connected by frictionless pinned joints and the members are so arranged that loads and reactions exist only at the joints. These assumptions ensure that truss members carry only axial tension or compression forces.

A common method to analyze statically determinate trusses is the method of joints. In this method, the first two equations of Equation (28.1) for a two-dimensional truss, or the first three equations of Equation (28.3) for a three-dimensional truss, are written for each joint of the truss. These equilibrium equations can be used to solve for all unknown reactions and member forces of the truss. If the truss is statically indeterminate, the use of the matrix [Nelson and McCormac, 2003] or finite element method is preferred. In either method, an element stiffness relationship in the form

$$\mathbf{ku} + \mathbf{r}_F = \mathbf{r} \tag{28.10}$$

where **k** is the stiffness matrix, **u** is the element end displacement (referred to as degree-of-freedom) vector, **r** is the element end force vector, and $\mathbf{r}_F$ is the element equivalent end force vector, is constructed for each truss member. They are then assembled (by enforcing nodal equilibrium and compatibility conditions) into a structure stiffness relationship

$$\mathbf{KU} + \mathbf{R}_F = \mathbf{R} \tag{28.11}$$

where **K** is the structure stiffness matrix, **U** is the structural nodal displacement vector, **R** is the structural nodal force vector, and $\mathbf{R}_F$ is the structural equivalent nodal force vector.

Equation (28.11) can be rearranged and partitioned in accordance with whether the degrees of freedom (dof) are constrained (c) or unconstrained (u) into two sets of matrix equations

$$\begin{bmatrix} \mathbf{K}_{uu} & \mathbf{K}_{uc} \\ \mathbf{K}_{cu} & \mathbf{K}_{cc} \end{bmatrix} \begin{Bmatrix} \mathbf{U}_u \\ \mathbf{U}_c \end{Bmatrix} + \begin{Bmatrix} \mathbf{R}_{Fu} \\ \mathbf{R}_{Fc} \end{Bmatrix} = \begin{Bmatrix} \mathbf{R}_u \\ \mathbf{R}_c \end{Bmatrix} \tag{28.12}$$

Because the applied nodal forces $\mathbf{R}_u$ and the imposed displacements at the supports $\mathbf{U}_c$ are known, the unknown nodal displacements $\mathbf{U}_u$ can be solved using the first set of equations, and the unknown support reactions $\mathbf{R}_c$ can be solved using the second set of equations from Equation (28.12). Once $\mathbf{U}_u$ is solved, the element end displacements **u** can be extracted from $\mathbf{U}_u$, and the element end force **r** calculated from Equation (28.10).

Figure 28.3a shows a four degree-of-freedom two-dimensional truss element. The corresponding 4 × 4 element stiffness matrix **k** is

$$\mathbf{k} = \mathbf{C}^T \mathbf{k}' \mathbf{C} \tag{28.13}$$

where $\mathbf{C}^T$ is the transpose of the coordinate transformation matrix **C** given by

$$\mathbf{C} = \begin{bmatrix} \cos\theta & \sin\theta & 0 & 0 \\ -\sin\theta & \cos\theta & 0 & 0 \\ 0 & 0 & \cos\theta & \sin\theta \\ 0 & 0 & -\sin\theta & \cos\theta \end{bmatrix} \tag{28.14}$$

and **k'** is the element stiffness matrix in local coordinate given by

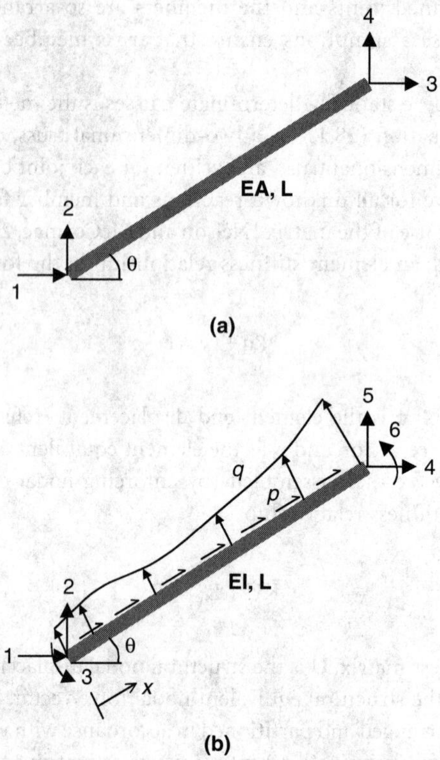

FIGURE 28.3 Truss and beam/frame elements.

$$\mathbf{k}' = \frac{EA}{L}\begin{bmatrix} 1 & 0 & -1 & 0 \\ 0 & 0 & 0 & 0 \\ -1 & 0 & 1 & 0 \\ 0 & 0 & 0 & 0 \end{bmatrix} \tag{28.15}$$

in which E is the modulus of elasticity, A is the element cross-sectional area, and L is the element length.

The equivalent end force vector $\mathbf{r}_F$ in Equation (28.10) can be used to account for the presence of any element misfit and/or temperature loading. It is equal to

$$\mathbf{r}_F = \mathbf{C}^T \mathbf{r}_F' \tag{28.16}$$

where $\mathbf{C}^T$ is the transpose of the coordinate transformation matrix expressed in Equation (28.14), and $\mathbf{r}_F'$ for the two-dimensional truss element shown in Figure 28.3a is given by

$$\mathbf{r}_F' = \left\{ \frac{EA\delta}{L} \quad 0 \quad -\frac{EA\delta}{L} \quad 0 \right\}^T \tag{28.17}$$

where δ is the amount of element misfit, taken as positive if the element is fabricated too long, and negative if the element is fabricated too short. In the case of temperature loading, $\delta = \alpha \Delta T L$, where α is the coefficient of thermal expansion and ΔT is the temperature change, taken as positive for a rise in temperature, and negative for a drop in temperature.

28.5 Beams and Frames

Beams and frames are structural elements that resist applied loads by axial, shear, bending, and twisting actions. For a two-dimensional beam or frame element experiencing small displacements subjected to a distributed load q acting through the shear center of the cross-section with flexural rigidity EI (i.e., the product of the elastic modulus E and moment of inertia I), the following differential equations relating the applied load q, internal shear V, internal bending moment M, slope θ, and deflection v can be written as

$$q = \frac{dV}{dx}, \quad V = \frac{dM}{dx}, \quad M = EI\frac{d\theta}{dx} \approx EI\frac{d(\frac{dv}{dx})}{dx} = EI\frac{d^2v}{dx^2} \tag{28.18}$$

The above equations can be used to construct shear and bending moment diagrams and to derive the equations of the elastic curve (i.e., **deflection functions**) for any beam/frame element subjected to a given set of loadings. A **shear diagram** is a plot of the internal shear force along the length of the element, and a **moment diagram** is a plot of the internal bending moment along the length of the element. By integrating the first two equations of Equation (28.18), one obtains the change in shear ΔV and change in moment ΔM from point i to j as

$$\Delta V = V_j - V_i = \int_{x_i}^{x_j} q\,dx, \quad \Delta M = M_j - M_i = \int_{x_i}^{x_j} V\,dx \tag{28.19}$$

and to account for effect of an applied concentrated load or moment, the term $\pm P_o$ is appended to the first, and the term $\pm M_o$ is appended to the second equation of Equation (28.19), respectively. The plus sign is used if the concentrated load (or moment) tends to increase the shear (or moment) at the point of application of the load (or moment). Because the integrals on the right side of the two equations expressed in Equation (28.19) represent the area under the load diagram and the area under the shear diagram between point i and point j, respectively, the equations can be used to construct shear and bending moment diagrams by starting with known values of V and M at one end of the beam and successively adding the area under the load diagram (for drawing the shear diagram) and the area under the shear diagram (for drawing the moment diagram) as one traverses from one point to another along the length of the beam. An example of a shear and moment diagram drawn for a beam is shown in Figure 28.4.

Through successive integration, the last equation of Equation (28.18) can be used to determine the slope dv/dx and deflection functions v of a segment of beam or frame element. For example,

$$\frac{dv}{dx} = \left(\int \frac{M}{EI}dx\right) + C_1, \quad v = \left(\int \frac{dv}{dx}dx\right) + C_2 \tag{28.20}$$

where C_1 and C_2 are constants of integrations that can be determined by enforcing boundary and/or continuity conditions of the beam or frame segment. **Boundary conditions** are enforced at points of supports for the structure (see Table 28.2), and continuity conditions (i.e., continuity of slopes and deflections) are enforced at points common to two adjacent beam/frame segments.

Table 28.3 gives examples of deflection equations for some basic load cases of a simply supported and a cantilever beam.

An alternative method to analyze beams and frames is the matrix or finite element method. In this method the structure is discretized into a series of elements connected together at points referred to as nodes. To perform the analysis, an element stiffness matrix **k** is formed for each element of the structural model. These element stiffness matrices are then assembled into a structure stiffness matrix **K**. The unknown nodal displacements $\mathbf{U}_u$ and reactions $\mathbf{R}_c$ are solved as described above in the section on trusses.

For the six-degree-of-freedom two-dimensional frame element shown in Figure 28.3b, the element stiffness matrix **k** can be expressed as

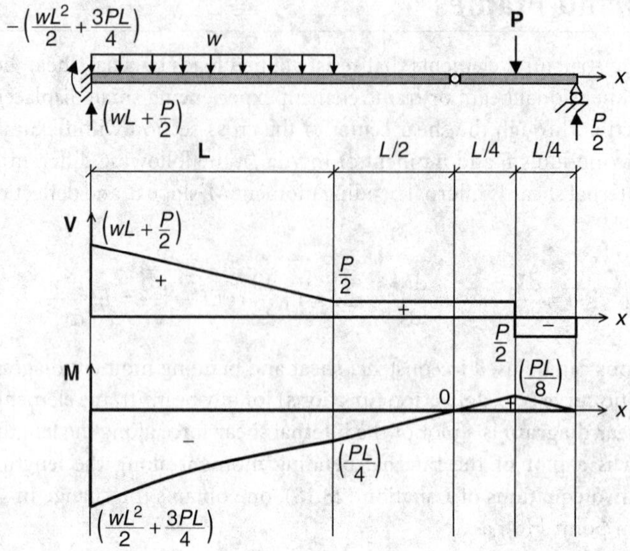

FIGURE 28.4 Shear and moment diagrams.

$$\mathbf{k} = \mathbf{C}^T\mathbf{k}'\mathbf{C} \tag{28.21}$$

where $\mathbf{C}^T$ is the transpose of the coordinate transformation matrix $\mathbf{C}$ given by

$$\mathbf{C} = \begin{bmatrix} \cos\theta & \sin\theta & 0 & 0 & 0 & 0 \\ -\sin\theta & \cos\theta & 0 & 0 & 0 & 0 \\ 0 & 0 & 1 & 0 & 0 & 0 \\ 0 & 0 & 0 & \cos\theta & \sin\theta & 0 \\ 0 & 0 & 0 & -\sin\theta & \cos\theta & 0 \\ 0 & 0 & 0 & 0 & 0 & 1 \end{bmatrix} \tag{28.22}$$

and $\mathbf{k}'$ is the element stiffness matrix in local coordinate expressed as

$$\mathbf{k}' = \frac{EI}{L}\begin{bmatrix} \frac{A}{I} & 0 & 0 & -\frac{A}{I} & 0 & 0 \\ 0 & \frac{12}{L^2} & \frac{6}{L} & 0 & -\frac{12}{L^2} & \frac{6}{L} \\ 0 & \frac{6}{L} & 4 & 0 & -\frac{6}{L} & 2 \\ -\frac{A}{I} & 0 & 0 & \frac{A}{I} & 0 & 0 \\ 0 & -\frac{12}{L^2} & -\frac{6}{L} & 0 & \frac{12}{L^2} & -\frac{6}{L} \\ 0 & \frac{6}{L} & 2 & 0 & -\frac{6}{L} & 4 \end{bmatrix} \tag{28.23}$$

It should be noted that unlike trusses in which loads are always applied to truss joints, beam and frame elements are often subjected to in-span or transverse loads. To account for these transverse loads, the equivalent element end force vector $\mathbf{r}_F$ in Equation (28.10) is now redefined as the element fixed-end force vector. For the frame element shown in Figure 28.3b, $\mathbf{r}_F$ can be calculated from the equation

TABLE 28.2 Supports and Boundary Conditions

Support	Boundary Conditions
Roller, rocker	$v = 0$
Hinged, pinned	$v = 0$
Fixed, built-in	$v = 0,\ dv/dx = 0$

TABLE 28.3 Examples of Deflection Equations

Case	Deflection Equations
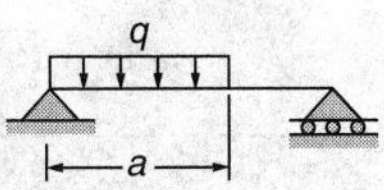	$v = -\frac{qx}{24LEI}(a^4 - 4a^3L + 4a^2L^2 + 2a^2x^2 - 4aLx^2 + Lx^3)$, for $0 \le x \le a$ $v = -\frac{qa^2}{24LEI}(-a^2L + 4L^2x + a^2x - 6Lx^2 + 2x^3)$, for $a \le x \le L$
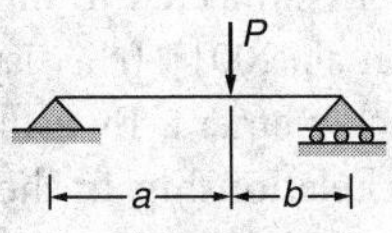	$v = -\frac{Pbx}{6LEI}(L^2 - b^2 - x^2)$, for $0 \le x \le a$
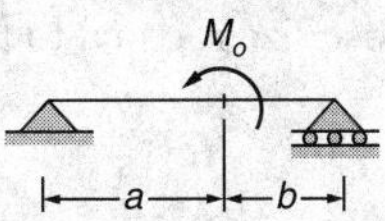	$v = -\frac{M_O x}{6LEI}(6aL - 3a^2 - 2L^2 - x^2)$, for $0 \le x \le a$
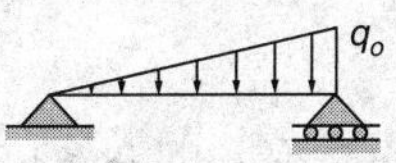	$v = -\frac{q_O x}{360LEI}(7L^4 - 10L^2x^2 + 3x^4)$
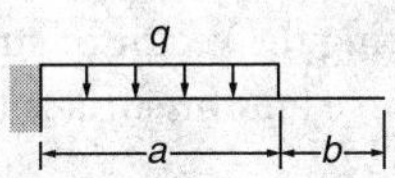	$v = -\frac{qx^2}{24EI}(6a^2 - 4ax + x^2),\ 0 \le x \le a$ $v = -\frac{qa^3}{24EI}(4x - a),\ a \le x \le L$
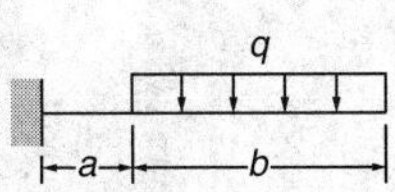	$v = -\frac{qbx^2}{12EI}(3L + 3a - 2x),\ 0 \le x \le a$ $v = -\frac{q}{24EI}(x^4 - 4Lx^3 + 6L^2x^2 - 4a^3x + a^4),\ a \le x \le L$
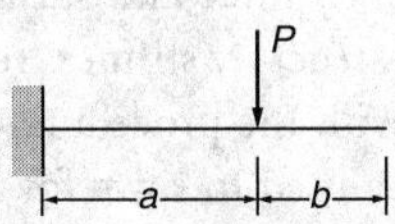	$v = -\frac{Px^2}{6EI}(3a - x),\ 0 \le x \le a$ $v = -\frac{Pa^2}{6EI}(3x - a),\ a \le x \le L$
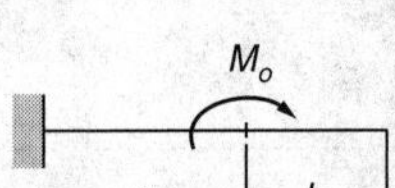	$v = -\frac{M_O x^2}{2EI},\ 0 \le x \le a$ $v = -\frac{M_O a}{2EI}(2x - a),\ a \le x \le L$
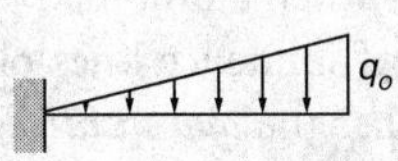	$v = -\frac{q_O x^2}{120LEI}(20L^3 - 10L^2x + x^3)$

$$\mathbf{r}_F = \mathbf{C}^T \mathbf{r}_F' \tag{28.24}$$

where $\mathbf{C}^T$ is the transpose of the coordinate transformation matrix given in Equation (28.22), and $\mathbf{r}_F'$ is the element fixed-end force vector in local coordinates whose elements are given by

$$r_{F1} = -\int_L \frac{p(L-x)}{L} dx\,, \quad r_{F4} = -\int_L \frac{px}{L} dx\,, \quad r_{F3} = -\int_L \frac{qx(L-x)^2}{L^2} dx\,, \quad r_{F6} = \int_L \frac{qx^2(L-x)}{L^2} dx \tag{28.25}$$

In the above equations, r_{F1} and r_{F4} are the fixed-end axial forces, r_{F3} and r_{F6} are the fixed-end moments of the element. The fixed-end shears r_{F2} and r_{F5} can be obtained by consideration of element equilibrium once r_{F3} and r_{F6} are calculated.

28.6 Plates and Shells

Although analytical solutions for plate and shell structures under simple loading conditions are available [Timoshenko and Woinowsky-Krieger, 1959; Gould, 1999], plates and shells subjected to more complicated load cases are often analyzed using the finite element method [Cook et al., 2001]. In a displacement-based finite element formulation, a displacement field is assumed for each element. By evaluating this assumed displacement field at each degree-of-freedom of the element, a relationship in the form $u^e = \mathbf{N}\mathbf{u}$, where u^e is the intraelement displacement, $\mathbf{N}$ is the shape matrix, and $\mathbf{u}$ is the element nodal displacement vector, can be established. An element stiffness matrix **k,** whose size is equal to the number of degrees of freedom of the element can be derived using the equation

$$\mathbf{k} = \int_V \mathbf{B}^T \mathbf{D} \mathbf{B} \, dV \tag{28.26}$$

where V is the volume of the finite element, $\mathbf{B}$ is the strain-displacement matrix (obtained by taking the appropriate derivatives of the shape function matrix **N**) relating the vector of nodal displacements **u** with the vector of strains $\boldsymbol{\varepsilon}$ (i.e., $\boldsymbol{\varepsilon} = \mathbf{B}\mathbf{u}$), $\mathbf{D}$ is the material stiffness matrix relating the vector of strains $\boldsymbol{\varepsilon}$ with the vector of stresses $\boldsymbol{\sigma}$ (i.e., $\boldsymbol{\sigma} = \mathbf{D}\boldsymbol{\varepsilon}$). To account for the effect of body force and/or surface traction, an equivalent nodal force vector can also be computed using the equation

$$\mathbf{r}_F = -\int_V \mathbf{N}^T \mathbf{f}_b \, dV - \int_S \mathbf{N}^T \mathbf{f}_s \, dS \tag{28.27}$$

where $\mathbf{N}^T$ is the transpose of the element shape matrix **N**, $\mathbf{f}_b$ is the body force, $\mathbf{f}_s$ is the surface traction, and V and S are the volume and surface of the element over which the body force and surface traction act, respectively. Once the element stiffness relationship is assembled into a structure stiffness relationship, the unknown displacements and reactions can be solved in accordance with the procedure outlined in the section Trusses. The matrix equations $\boldsymbol{\varepsilon} = \mathbf{B}\mathbf{u}$ and $\boldsymbol{\sigma} = \mathbf{D}\boldsymbol{\varepsilon}$ can then be used to solve for strains and stresses. Some examples of plate and shell finite elements are shown in Figure 28.5.

28.7 Influence Lines and Influence Surfaces

An influence line is a plot of the variation of a response function (such as a reaction, an internal shear, an internal moment, etc.) at a given point of a two-dimensional structure when a unit load is applied at different locations on the structure. An influence surface is a collection of influence lines plotted for a three-dimensional structure. Influence lines can be constructed using either the *quantitative method,* in which the desired response function at a given point of the structure is computed repeatedly when a unit

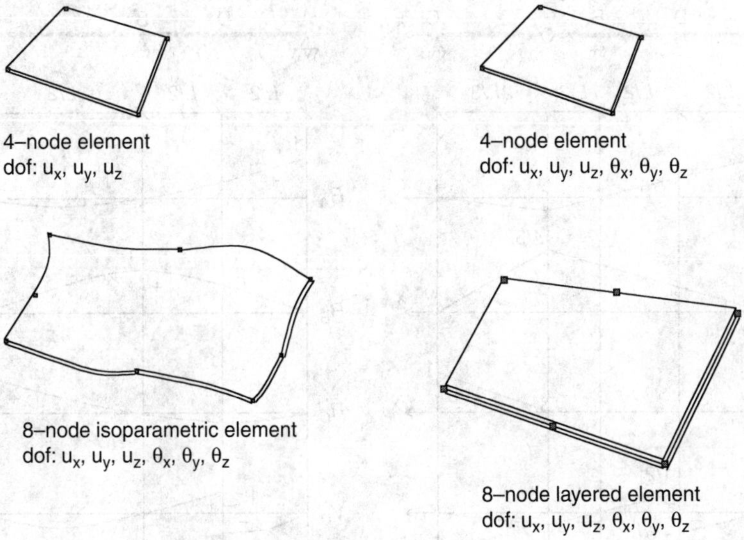

FIGURE 28.5 Plate and shell finite elements.

load is applied at different locations on the structure, or the *qualitative method* in which the Müller-Breslau principle is used [Hibbeler, 2002]. In using the Müller-Breslau principle to construct an influence line, the constraint that gives rise to the response function is removed, but its effect on the structure is retained. The influence line is obtained as the deflected shape of the structure subject to this effect. To obtain the correct values (called **influence coefficients**) for the influence line, the deflected shape of the structure is scaled to a value of unity at the location of the constraint. For instance, to draw the influence line for a support reaction, the support is first removed, but its effect on the structure is retained through the use of a concentrated load applied at the support location. The deflected shape of the structure, scaled to a value of unity at the support, is now the support reaction influence line for the structure. Figure 28.6 shows some examples of influence lines drawn for a statically determinate and a statically indeterminate beam. Note that influence lines for statically determinate beams are straight line segments but influence lines for statically indeterminate beams are curves.

The influence line for a given response function can be used to determine the locations on the structure over which a live load should be placed to educe the maximum effect of that response function on the structure. For instance, suppose the live load is a concentrated load, to obtain the maximum reaction at support B for the statically determinate beam shown in Figure 28.6a, the load should be placed at the internal hinge to the right of B where the influence coefficient has the highest value. However, to obtain the maximum support reaction at B (i.e., R_B) for the statically indeterminate beam shown in Figure 28.6b, the live concentrated load should be placed right over support B. If the maximum positive moment at point D (i.e., M_D) is desired, the live concentrated load should be placed right at point D for both beams. Now, suppose the live load is a distributed load, it should be placed over the entire beam from point A to point C for both beams to obtain the maximum reaction at support B, and over span AB only for both beams to obtain the maximum positive moment at D, and over span BC only for both beams to obtain the maximum negative moment at D. Once the live load is properly placed, its effect on the structure in terms of the magnitude of the response function can be calculated using the equations

$$Py, \qquad \text{for a single concentrated load } P$$

$$\sum_{i=1}^{n} P_i y_i, \qquad \text{for a concentrated load series } P_i\ (i = 1,2,\ldots,\text{n}) \tag{28.28}$$

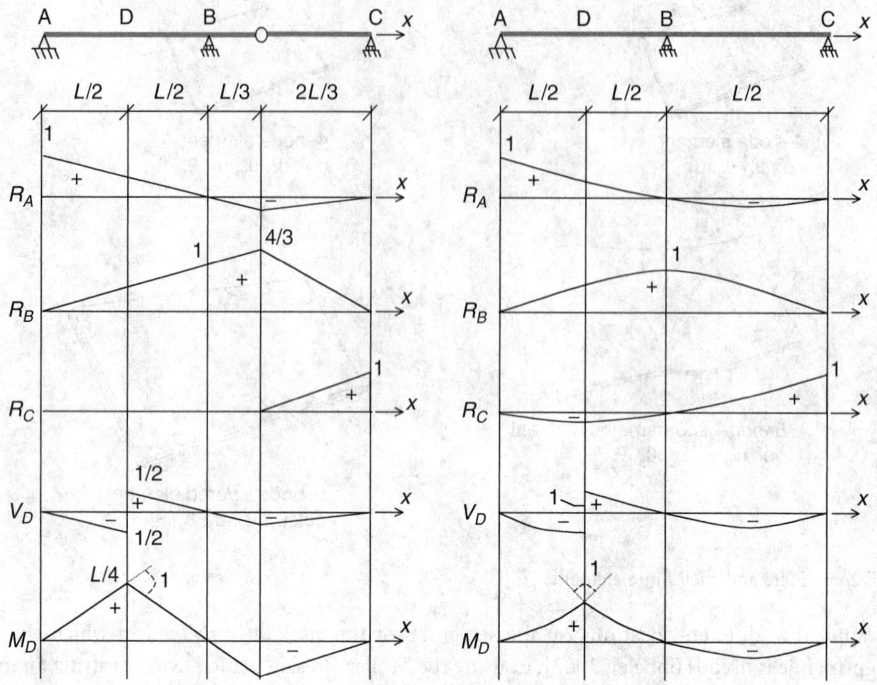

FIGURE 28.6 Influence lines.

$$\int_a^b qy\,dx, \qquad \text{for a distributed load } q \text{ applied from point a to b}$$

In the above equations, y is the influence coefficient of the influence line constructed for the specific response function. The above equations can also be used to compute magnitudes of response functions for three-dimensional structures when influence surfaces are available. In this case y will be the influence coefficient of the corresponding influence surface.

28.8 Structural Stability

Structural stability is the ability of a structural system to maintain its geometric integrity under compressive force or stress. When a structure is subjected to high compressive force/stress, it has a tendency to lose its stiffness and deflect in a direction perpendicular to the direction of the applied stress. When this occurs, the ability of the structure to carry the applied load will be compromised. For instance, a slender column buckles (i.e., becomes unstable under a compressive force) when the applied load reaches a critical value. If buckling occurs in the elastic range, this critical load can be computed as an eigenvalue problem [Chen and Lui, 1987] using the differential equation

$$EI\frac{d^4v}{dx^4} + P\frac{d^2v}{dx^2} = 0 \tag{28.29}$$

in conjunction with a set of boundary conditions that represent the nature of the supports at the two ends of the column. In the above equation, v is the lateral displacement and EI (the product of the modulus of elasticity of the material and moment of inertia of the cross-section) is the flexural rigidity of the column. The critical load is obtained as the eigenvalue of the characteristic equation, and the buckling mode shape is obtained as the eigenvector of the equation. Examples of critical loads P_{cr} for

TABLE 28.4 P_{cr} for Columns with Different Boundary Conditions

Case	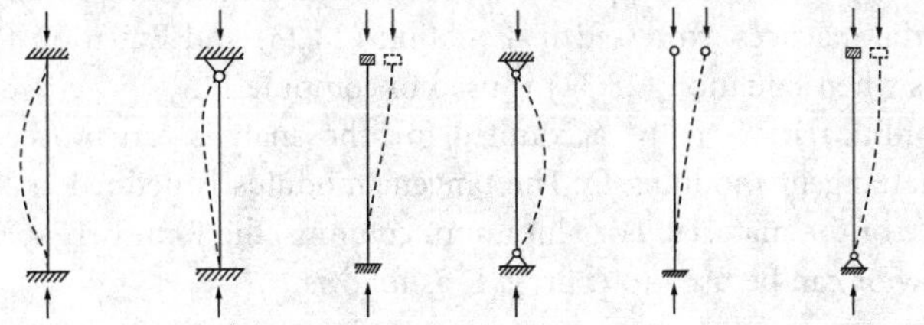					
P_{cr}	$\frac{\pi^2 EI}{(0.5L)^2}$	$\frac{\pi^2 EI}{(0.7L)^2}$	$\frac{\pi^2 EI}{L^2}$	$\frac{\pi^2 EI}{L^2}$	$\frac{\pi^2 EI}{(2.0L)^2}$	$\frac{\pi^2 EI}{(2.0L)^2}$
K	0.5	0.7	1.0	1.0	2.0	2.0

columns with different boundary conditions are given in Table 28.4. Note that P_{cr} for all these columns can be expressed in the following compact form

$$P_{cr} = \frac{\pi^2 EI}{(KL)^2} \tag{28.30}$$

where K is referred to as the effective length factor. The effective length KL of a column can be interpreted as the length between adjacent inflection points (or points of zero moment) in the column.

An alternative method to calculate P_{cr} is the finite element method [McGuire et al., 2000]. In this method P_{cr} is calculated by solving the characteristic polynomial obtained by setting the determinant of the system stiffness matrix equal to zero as follows

$$\det\left|\mathbf{K} + \mathbf{K}_G\right| = 0 \tag{28.31}$$

In the above equation, **K** is the first-order system stiffness matrix (that does not account for any geometrical nonlinear effect) and $\mathbf{K}_G$ is the geometrical system stiffness matrix (which is a function of P and accounts for the geometrical nonlinear effect).

For instance, if Equation (28.31) is used to compute P_{cr} for a column, the column is first discretized into small elements. The first-order system stiffness matrix **K** is then obtained by assembling the element stiffness matrix **k** calculated from Equation (28.21) for each element, and the geometrical system stiffness matrix $\mathbf{K}_G$ is obtained by assembling the element geometrical stiffness matrix $\mathbf{k}_G$ calculated from the equation

$$\mathbf{k} = \mathbf{C}^{T}\mathbf{k}'_G\mathbf{C} \tag{28.32}$$

in which **C** is given by Equation (28.22) and $\mathbf{k}'_G$ is given by

$$\mathbf{k}'_G = \frac{P}{L}\begin{bmatrix} 1 & 0 & 0 & -1 & 0 & 0 \\ 0 & \frac{6}{5} & \frac{L}{10} & 0 & -\frac{6}{5} & \frac{L}{10} \\ 0 & \frac{L}{10} & \frac{2L^2}{15} & 0 & -\frac{L}{10} & -\frac{L^2}{30} \\ -1 & 0 & 0 & 1 & 0 & 0 \\ 0 & -\frac{6}{5} & -\frac{L}{10} & 0 & \frac{6}{5} & -\frac{L}{10} \\ 0 & \frac{L}{10} & -\frac{L^2}{30} & 0 & -\frac{L}{10} & \frac{2L^2}{15} \end{bmatrix} \tag{28.33}$$

where P is the axial force in the element. In a column buckling analysis, the effect of initial column shortening in the direction of the applied load is usually ignored. As a result, the first and fourth rows and columns in the matrices expressed in Equation (28.23) and Equation (28.33) are usually deleted from the matrices when Equation (28.31) is used to compute P_{cr}.

If **material nonlinearity** is to be accounted for, the analysis can proceed by replacing the elastic modulus E by the tangent modulus E_t. The tangent modulus is defined as the slope of the nonlinear stress–strain curve of the material. For aluminum columns, the Ramsberg–Osgood [1943] equation for stress–strain behavior can be used to express E_t as follows

$$E_t = \frac{E}{[1+\frac{0.002nE}{\sigma_{0.2}}(\frac{\sigma}{\sigma_{0.2}})^{n-1}]} \tag{28.34}$$

where $\sigma_{0.2}$ is the 0.2% offset yield stress, and n is the hardening parameter.

For steel columns, E_t can be approximated by [Galambos, 1998]

$$E_t = \begin{cases} 4E[\frac{\sigma}{\sigma_y}(1-\frac{\sigma}{\sigma_y})] & , \text{ if } \sigma \le 0.5\sigma_y \\ E & , \text{ if } \sigma > 0.5\sigma_y \end{cases} \tag{28.35}$$

where σ is the axial stress, and σ_y is the material yield stress.

28.9 Advanced Analysis

The term advanced analysis is used to describe any analysis method that explicitly takes into consideration all important parameters that affect the behavior of a structure. For instance, in the analysis of framed structures, a plastic zone analysis is considered an advanced analysis. In a plastic zone analysis, the effects of **geometrical nonlinearity** as well as spread of plasticity from one region to another within the structure are explicitly modeled in the analysis. While the geometrical nonlinear effect can be accounted for by the use of a geometrical stiffness matrix $\mathbf{K_G}$, one method to account for the material nonlinear effect is the use of fiber elements. In this method, a frame member is modeled by bundling together groups of finite length fiber elements to form the member cross-section. The behavior of each fiber is described by predefined pre- and post-yield constitutive relationships. Needless to say, such an analysis often requires proprietary software and is extremely computationally intensive. It is therefore seldom performed. A somewhat simpler analysis method that can be used for nonlinear frame analysis is the plastic hinge (or its derivative, the modified plastic hinge) method. In a plastic hinge method of analysis, occurrences of plasticity are assumed to concentrate in regions of "zero length" plastic hinges. Regions outside of plastic hinges are assumed to be elastic. Using this assumption, the analysis can proceed with the use of an iterative process applied to the following increment equation

$$\Delta\mathrm{R} = [\mathrm{K} + \mathrm{K_G} + \mathrm{K_P}]\Delta\mathrm{U} \tag{28.36}$$

where $\mathbf{\Delta R}$ is the incremental nodal force vector, $\mathbf{\Delta U}$ is the incremental displacement vector, $\mathbf{K}$ is the first-order system stiffness matrix, $\mathbf{K_G}$ is the system geometrical stiffness matrix, and $\mathbf{K_P}$ is the system plasticity stiffness matrix obtained by assembling $\mathbf{k_P}$ given by [McGuire et al., 2000]

$$\mathrm{k_P} = -\mathrm{kG}[\mathrm{G^T kG}]^{-1}\mathrm{G^T k} \tag{28.37}$$

where **k** is the first-order (or elastic) element stiffness matrix given in Equation (28.21) for a two-dimensional frame element, **G** is the gradient matrix that satisfies the orthogonality condition

$$\mathbf{G}^{\mathrm{T}}\Delta\mathbf{R} = 0 \tag{28.38}$$

The gradient matrix **G** is obtained by taking partial derivatives of a **yield surface** Φ with respect to each force degree-of-freedom of the element. For the two-dimensional frame element shown in Figure 28.3b, **G** can be expressed as

$$\mathbf{G} = \begin{bmatrix} \dfrac{\partial \Phi}{\partial P_i} & 0 \\ \dfrac{\partial \Phi}{\partial V_i} & 0 \\ \dfrac{\partial \Phi}{\partial M_i} & 0 \\ 0 & \dfrac{\partial \Phi}{\partial P_j} \\ 0 & \dfrac{\partial \Phi}{\partial V_j} \\ 0 & \dfrac{\partial \Phi}{\partial M_j} \end{bmatrix} \tag{28.39}$$

where P_i, V_i, and M_i are the axial force, shear force, and bending moment acting on the left end of the element, and P_j, V_j, and M_j are the axial force, shear force, and bending moment acting on the right end of of the element.

There are various expressions proposed to describe the yield surface for I-shaped steel sections. One such expression is [McGuire et al., 2000]

$$\Phi = \begin{cases} p^2 + m^2 + 3.5p^2m^2, & \text{for major axis bending} \\ p^2 + m^4 + 3p^6m^2, & \text{for minor axis bending} \end{cases} \tag{28.40}$$

where $p = P/P_y$, $m = M/M_p$, in which P and M are the axial force and bending moment acting on the cross-section, respectively, P_y is the cross-section **yield load**, and M_p is the cross-section **plastic moment** capacity.

Because the problem is nonlinear, it has to be solved incrementally using an iterative technique. In each cycle of analysis $\mathbf{K_G}$ and $\mathbf{K_P}$ in Equation (28.36) are updated to reflect the effects of change in geometry (geometrical nonlinearity) and material inelasticity (material nonlinearity) on the behavior of the structure. Various iterative techniques have been proposed over the years (e.g., the Load Control Newton-Raphson Method, the Displacement Control Method, the Arc Length Method, the Work Control Method, etc.) by a number of researchers. These techniques are also well documented in the literature [see for example, Chen and Lui, 1991, McGuire et al., 2000]. Because advanced analysis can capture structural behavior more accurately, it is becoming more important in the structural engineering profession. A number of international structural design specifications already have provisions for the use of such an analysis for design.

Defining Terms

Boundary conditions — The kinematic and/or force conditions that exist at or along the boundary of a structure.

Deflection functions — Equations that describe the deflected shape of a structure under applied loads.

Finite element method — An analysis method in which a structure is modeled by an array of small elements, each with a predetermined response characteristic.

Geometrically stable — A condition in which a structure will not undergo rigid body motion nor experience local or global collapse under any configurations of applied loads.

Geometrical nonlinearity — Nonlinear response as a result of change in geometry of the structure under the applied loads.

Influence coefficient — Magnitude of a response function at a certain point on the structure due to the application of a unit load on the structure.

Material nonlinearity — Nonlinear response as a result of material yielding or nonlinear stress–strain behavior.

Moment diagram — A diagram showing the variation of internal bending moment in a structural member.

Plastic moment — The moment capacity of a cross-section that corresponds to the condition of full yielding of all fibers in the cross-section.

Shear diagram — A diagram showing the variation of internal shear force in a structural member.

Statically determinate — A condition in which the use of equilibrium equations alone is sufficient to solve for all unknown external reactions and internal forces.

Statically indeterminate — A condition in which the use of equilibrium equations alone is *not* sufficient to solve for all unknown external reactions and internal forces.

Superposition — Structural responses such as displacements, forces, stresses and strains, etc., due to an array of forces acting simultaneously can be obtained by summing the responses due to each force acting alone.

Yield load — An axial load that will cause full yielding of all fibers in a cross-section.

Yield surface — An interaction diagram that shows how two or more forces interact to cause yielding or failure in a cross-section.

References

Chen, W. F. and Lui, E. M. 1987. *Structural Stability – Theory and Implementation,* Elsevier, New York.

Chen, W. F. and Lui, E. M. 1991. *Stability Design of Steel Frames,* CRC Press, Boca Raton, FL.

Cook, R. D., Malkus, D. S., Plesha, M. E., and Witt, R. J. 2001. *Concepts and Applications of Finite Element Analysis,* 4th ed., John Wiley & Sons, New York.

Galambos, T. V. (ed.). 1998. *Guide to Stability Design Criteria for Metal Structures,* 5th ed., John Wiley & Sons, New York.

Gould, P. L. 1999. *Analysis of Plates and Shells,* Prentice Hall, Englewood Cliffs, NJ.

Hibbeler, R. C. 2002. *Structural Analysis,* 5th ed., Prentice Hall, Englewood Cliffs, NJ.

Kassimali, A. 1999. *Matrix Analysis of Structures,* Brooks/Cole, CA.

McGuire, W., Gallagher, R. H., and Ziemian, R. D. 2000. *Matrix Structural Analysis,* 2nd ed., John Wiley & Sons, New York.

Nelson, J. K. and McCormac, J. C. 2003. Structural Analysis: *Using Classical and Matrix Methods,* 3rd ed., John Wiley & Sons, New York.

Ramberg, W. and Osgood, W. R. 1943. Description of Stress-Strain Curves by Three Parameters, National Advisory Committee on Aeronautics, Technical Note No. 902.

Timoshenko, S. P. and Young, D. H. 1965. *Theory of Structures,* McGraw-Hill, New York.

Timoshenko, S. P. and Woinowsky-Krieger, S. 1959. *Theory of Plates and Shells,* McGraw-Hill, New York.

Further Information

Bathe, K.-J. 1996. *Finite Element Procedures,* Prentice Hall, New York.

Ghali, A., Neville, A. M., Brown, T. G., and Pettipiece, D. A. 2003. *Structural Analysis – A Unified Classical and Matrix Approach,* 5th ed., Routledge, London.
Young, W. C. and Budynas, R. G. 2001. *Roark's Formulas for Stress & Strains,* 7th ed., McGraw-Hill, New York.

29

Structural Steel

William T. Segui
The University of Memphis

Structural steel is used for the framework of buildings and bridges, either alone or in combination with other materials such as reinforced concrete. Steel buildings are usually constructed with standard shapes produced by hot-rolling (Figure 29.1), although custom shapes can be fabricated from plate material. Various grades of steel, as classified by the American Society for Testing and Materials [2003], are suitable for building and bridge construction. One commonly used steel is ASTM A36, with a minimum tensile yield stress F_y of 36 ksi and an ultimate tensile stress F_u between 58 and 80 ksi. (The actual yield stress of most A36 steel currently being produced is close to 50 ksi.) Other currently used steels are ASTM A572 Grade 50 and ASTM A992, both with a yield stress of 50 ksi and an ultimate stress of 65 ksi.

Several design approaches are available to the structural engineer.

In *allowable stress design* (ASD), structural members are proportioned so that the maximum computed stress is less than a permissible, or allowable, stress. This approach is also called *working stress design* or *elastic design.*

In *load and resistance factor design* (LRFD), members are proportioned so that the applied load is less than the resistance (strength) of the member. This approach can be represented by the following relationship:

$$\sum \gamma_i Q_i \leq \phi R_n \tag{29.1}$$

where γ_i is a **load factor,** Q_i is a load effect (force or moment), ϕ is a **resistance factor,** and R_n is the nominal resistance, or **nominal strength.** The summation indicates that the total factored load effect is the sum of the products of individual load effects (such as dead, live, and snow) and corresponding load factors — which are a function of not only the type of load effect, but also the combination of loads under consideration. The nominal strength is a theoretical strength, and the resistance factor reduces it to a practical value. This reduced value, ϕR_n, is called the **design strength.** Equation (29.1) states that the sum of the factored load effects must not exceed the design strength.

A third approach, *plastic design,* also uses load factors but is primarily a structural analysis method of obtaining failure loads by a consideration of collapse mechanisms.

Although all three approaches are acceptable, ASD and LRFD are the most widely used.

0-8493-1586-7/05/$0.00+$1.50

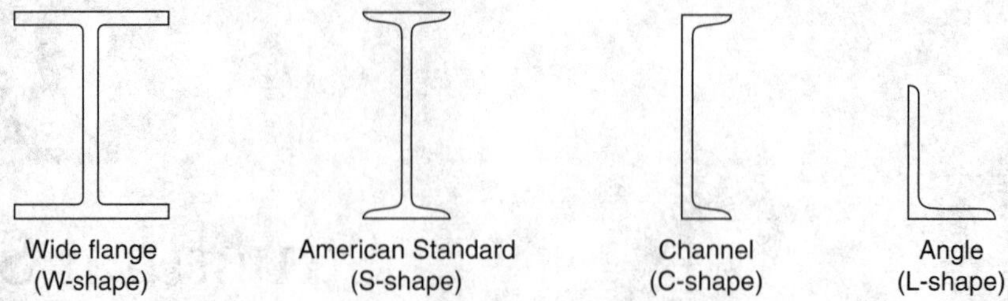

FIGURE 29.1 Examples of standard hot-rolled shapes (cross-sectional views).

The design of structural steel buildings in the U.S. is usually based on the provisions of the specifications of the American Institute of Steel Construction. For allowable stress design, the requirements are given in the 1989 Specification [American Institute of Steel Construction, 1989a] and the ninth edition ASD Manual of Steel Construction [American Institute of Steel Construction, 1989b]. For load and resistance factor design, the requirements are given in the 1999 Specification [American Institute of Steel Construction, 1999] and the third edition LRFD Manual of Steel Construction [American Institute of Steel Construction, 2001]. In 2005, AISC will publish a unified specification and manual, covering both ASD and LRFD. All of the requirements covered herein will be based on the current AISC LRFD Specification and Manual.

The loading conditions to be investigated in conjunction with Equation (29.1), along with the associated load factors, are given in ASCE 7-02, *Minimum Design Loads for Buildings and Other Structures* [American Society of Civil Engineers, 2002]. These combinations are not given in the AISC Specification, but they are shown in the LRFD Manual. The value of the resistance factor depends upon the type of member or connecting element being investigated and will be covered in the following sections.

29.1 Members

Tension Members

Tension members are used in trusses, bracing systems, and building and bridge suspension systems. The load and resistance factor relationship of Equation (29.1) can be expressed in the following way for a tension member:

$$P_u \leq \phi_t P_n \tag{29.2}$$

where P_u is the sum of the factored axial tension loads, ϕ_t is the resistance factor for tension, and P_n is the nominal tensile strength.

There are two possible failure modes, or **limit states,** for tension members: excessive deformation caused by yielding of the gross cross-section and fracture of the net cross-section. The net cross-sectional area is the gross area minus any area removed by bolt holes. For yielding of the gross section, the design strength is given by

$$\phi_t P_n = \phi_t F_y A_g = 0.90 F_y A_g \tag{29.3}$$

and the design strength based on fracture of the net section is

$$\phi_t P_n = \phi_t F_u A_e = 0.75 F_u A_e \tag{29.4}$$

where A_g is the gross cross-sectional area and A_e is the *effective* net area. Note that the resistance factor is different for each of the two limit states. Because of a phenomenon called **shear lag,** an effective net area must be used in those cases where some of the cross-sectional elements are unconnected. The effective net area is given by

$$A_e = A_n U \tag{29.5}$$

where A_n is the actual computed net area and U is a reduction factor. Shear lag can also occur in welded connections. In these cases, it is accounted for by using an effective area A_e defined as

$$A_e = A_g U \tag{29.6}$$

Block shear is another potential limit state for tension members. This failure mode must be investigated when the member is connected in such a way that a block of the material could tear out as in Figure 29.2. Two loading conditions are involved: shear and tension. In the case illustrated, the shear is along line *ab* and the tension is along line *bc*. The strength, or resistance, is the sum of two contributions — either shear yielding plus tension fracture or shear fracture plus tension yielding. The governing case will be the one that has the larger fracture component. Block shear can also occur in welded connections.

The following tension member example will illustrate the concepts of load combination and design strength.

Example

Select a single-angle tension member of A36 steel to resist a service dead load of 30 kips and a service live load of 90 kips. The member will be connected to a gusset plate at each end with longitudinal welds.

Solution

In this brief introductory example, some details will be omitted, and some AISC requirements will be used without a detailed explanation.

When dead and live loads are the only loads present, the controlling combination from ASCE 7-02 is usually

$$P_u = 1.2D + 1.6L$$

where P_u is the factored load, D is the dead load, and L is the live load. For this example,

$$P_u = 1.2(30) + 1.6(90) = 180 \text{ kips}$$

From Equation (29.2),

$$\phi_t P_n \geq P_u$$

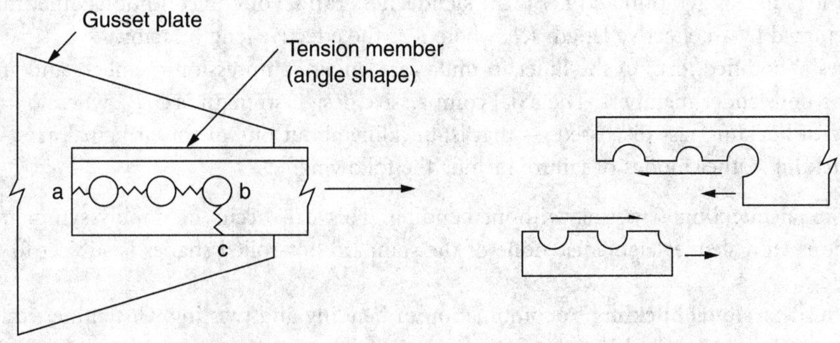

FIGURE 29.2 Block shear in a tension member.

For the limit state of yielding, from Equation (29.3), this becomes

$$0.90F_y A_g \geq P_u$$

$$A_g \geq \frac{P_u}{0.90F_y} = \frac{180}{0.90(36)} = 5.56 \text{ in.}^2$$

From the Manual of Steel Construction [American Institute of Steel Construction, 2001], an L5 × $3^1/_2 \times {}^3/_4$ has a cross-sectional area of 5.82 in.^{2} and will be investigated for effective area. For this type of welded connection, the shear lag factor U can be taken as 0.85, so

$$A_e = A_g U = 5.82(0.85) = 4.95 \text{ in.}^2$$

From Equation (29.2) and Equation (29.4), the required effective area is

$$A_e \geq \frac{P_u}{0.75F_u} = \frac{180}{0.75(58)} = 4.14 \text{ in.}^2$$

Since the furnished effective area of 4.95 in.^{2} is greater than 4.14 in.^{2}, this shape is satisfactory. Use an L5 × $3^1/_2 \times {}^3/_4$.

Compression Members

Some truss members are compression members, as are vertical supports in buildings and bridges, where they are usually referred to as columns. This discussion will be limited to *axially loaded* compression members. For a slender axially loaded compression member with pinned ends, the nominal strength is given by the Euler formula as

$$P_n = \frac{\pi^2 EI}{L^2} \tag{29.7}$$

where E is the modulus of elasticity, I is the moment of inertia about the minor principal axis of the member cross-sectional area, and L is the length. This equation can also be expressed in the following form:

$$P_n = \frac{\pi^2 E}{(L/r)^2} \tag{29.8}$$

where r is the radius of gyration and L/r is the **slenderness ratio.** For end conditions other than pinned, L can be replaced by an effective length KL, where K is the effective length factor.

AISC uses a modified form of the Euler formula for slender compression members and an empirical equation for nonslender members. The axial compressive design strength is $\phi_c P_n$, where $\phi_c = 0.85$.

The type of buckling just discussed — that is, buckling about one of the principal axes — is called **flexural buckling.** Other modes of failure include the following:

- **Torsional buckling.** Twisting without bending. This can occur in doubly-symmetrical cross-sections with slender elements (none of the standard hot-rolled shapes is subject to this failure mode).
- **Flexural-torsional buckling.** A combination of bending and twisting. Unsymmetrical cross-sections are susceptible to this type of failure.
- **Local buckling.** Localized buckling of a cross-sectional element such as a web or projecting flange.

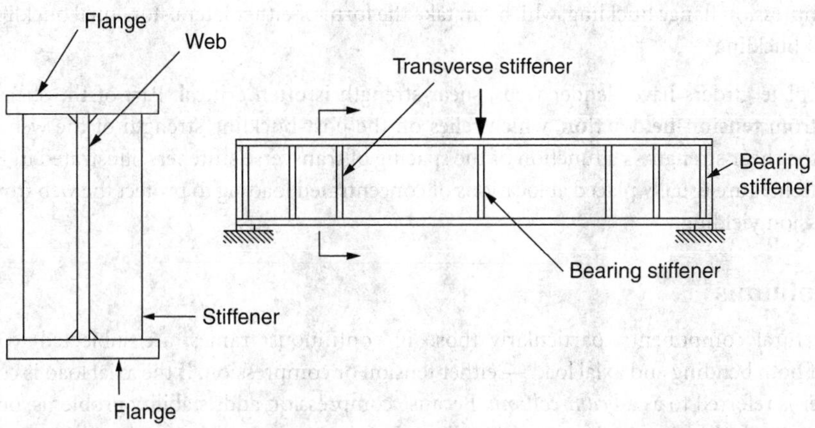

FIGURE 29.3 Plate girder details.

Beams

The flexural design strength of a beam is $\phi_b M_n$, where $\phi_b = 0.90$ and M_n is the nominal flexural strength, which is the bending moment at failure. The following discussion will be limited to hot-rolled shapes bent about the major principal axis. The nominal flexural strength is based on one of the following limit states:

- *A fully yielded cross-section.* If a beam is prevented from becoming unstable in any way, the resisting moment will be equal to the internal couple corresponding to a uniform compressive stress of F_y on one side of the neutral axis and a uniform tensile stress of F_y on the other side. This is the **plastic moment,** M_p.
- ***Lateral-torsional buckling.*** If a beam bent about its major principal axis is not adequately supported laterally (that is, in the direction perpendicular to the plane of bending), it can buckle outward, simultaneously bending about its minor principal axis and twisting about its longitudinal axis.
- *Flange local buckling.* This can occur in the *compression* flange if it is too slender.
- *Web local buckling.* This can occur in the compressed part of the web if it is too slender.

Cross-sections can be categorized as compact, noncompact, or slender, depending on the width-to-thickness ratios of their cross-sectional elements. If a shape is compact, as most of the standard hot-rolled shapes are, neither flange local buckling nor web local buckling can occur. Furthermore, if a compact beam has adequate lateral support, lateral-torsional buckling cannot occur, and the nominal strength is equal to the plastic moment; that is,

$$M_n = M_p \tag{29.9}$$

Adequate lateral support will exist when the distance between points of lateral support, called the *unbraced length*, is less than a prescribed value. If the unbraced length is too large, failure will be by either elastic lateral-torsional buckling or inelastic lateral-torsional buckling, depending on whether yielding has begun when the buckling takes place.

In addition to flexure, beams must be checked for shear strength (which usually does not control) and deflections. Deflections should be computed with service, and not factored, loads.

Flexural members with slender webs are classified by AISC as plate girders; otherwise, they are classified as beams. Plate girders (Figure 29.3) are built up from plate elements that are welded together, and their flexural strength is based on one of the following limit states:

- Tension flange yielding
- Compression flange yielding

- Compression flange buckling, which can take the form of either lateral-torsional buckling or flange local buckling

Because plate girders have slender webs, shear strength is often critical. Part of the shear resistance can come from **tension-field action,** which relies on the post-buckling strength of the web. This component of the shear strength is a function of the spacing of transverse stiffeners, illustrated in Figure 29.3. Bearing stiffeners are usually placed at locations of concentrated loading to protect the web from buckling or compression yielding.

Beam-Columns

Many structural components, particularly those in continuous frames, are subjected to significant amounts of both bending and axial load — either tension or compression. If the axial load is compressive, the member is referred to as a *beam-column.* Because compression adds stability problems, bending plus compression is usually more serious than bending plus tension.

Combined loading can be accounted for by the use of interaction equations of the form

$$\frac{P_u}{\phi_c P_n} + \frac{M_u}{\phi_b M_n} \le 1.0 \tag{29.10}$$

where each term on the left side is a ratio of a factored load effect to the corresponding design strength. The AISC specification uses two interaction equations: one for small axial loads and one for large axial loads. Furthermore, each equation has two bending terms: one for major axis bending and one for minor axis bending.

In addition to bending moment caused by transverse loads and end moments, beam-columns are subjected to secondary moments resulting from the eccentricity of the axial load with respect to the deflected member axis. This can be accounted for in an approximate, but very accurate, way through the use of **moment amplification** factors as follows:

$$M_u = B_1 M_{nt} + B_2 M_{lt} \tag{29.11}$$

where B_1 and B_2 are amplification factors, M_{nt} is the factored load moment corresponding to no joint translation, and M_{lt} is the factored load moment corresponding to lateral joint translation. If the member is part of a braced frame, then no joint translation is possible and $M_{lt} = 0$. If the member is part of an unbraced frame, M_{nt} is computed as if the member were braced against joint translation, and M_{lt} is computed as the result of only the joint translation.

If bending moments are obtained from a frame analysis computer program that performs a second-order analysis, the moments so obtained will not need to be amplified as in Equation (29.11) but can be used directly in Equation (29.10).

29.2 Connections

Modern steel structures are connected with bolts, welds, or both. Although hot-driven rivets can be found in many existing structures, they are no longer used in new construction.

Bolts

Two types of bolts are used: common and high-strength. Common bolts conform to ASTM A307 and are used in light applications. High-strength bolts are usually ASTM A325 or A490. Bolts can be loaded in tension, shear, or both. In addition, bearing stresses act between the bolts and the connected elements of shear connections. Although bearing is a problem for the connected part and not the bolt itself, the bearing load is a function of the bolt diameter and is therefore associated with the bolt strength. Thus,

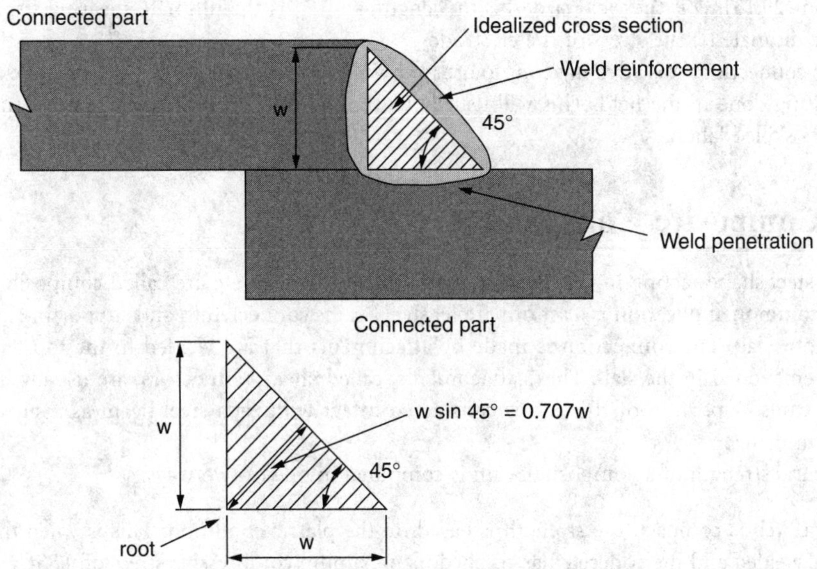

FIGURE 29.4 Fillet weld properties.

we may speak of the design strength of a bolted connection being based on shear, bearing, or tension. The shear or tension design strength of a single bolt is ϕR_n, where $\phi = 0.75$ and

$$R_n = F_n A_b \tag{29.12}$$

where F_n is the nominal ultimate shearing or tensile stress and A_b is the cross-sectional area of the bolt. Bearing strength depends upon such things as bolt spacing and edge distance and is a function of the *projected* area of contact between the bolt and the connected part.

High-strength bolts can be snug tight, where the bolts are installed by a single worker with an ordinary wrench, or fully tensioned, where they are installed to a prescribed minimum tension of about 70% of the ultimate strength. Fully tensioned bolts are used in slip-critical connections, in which slippage is not permitted. Slip-critical connections are required for members subjected to impact or fatigue loading. The AISC Specification gives the slip resistance of bolts as a function of bolt pretension and the condition of the contact surfaces.

Welds

Weld material is available in several grades, but E70 or E80 series electrodes are usually used. E70 electrodes have a tensile strength of 70 ksi and are designed to be used with steel that has a yield stress between 36 and 60 ksi. E80 electrodes have a tensile strength of 80 ksi and are used with steels that have a yield stress of 65 ksi. Welding can be performed on the job site (field welds) or in the fabricating shop (shop welds). Because of automation, shop welding is generally more economical and of higher quality. The most common types of welds are the *groove weld,* in which the weld metal is deposited into a gap, or groove, and the *fillet weld* (Figure 29.4), which is deposited into a corner. The fillet weld is the type most often used for field welding.

The strength of a fillet weld is based on the premise that failure will occur by shear on a plane through the throat of the weld. The throat is the perpendicular distance from the root to the hypotenuse on the theoretical cross-section of the weld, which is treated as an isosceles right triangle (Figure 29.4). The design strength of the weld is ϕR_n, where $\phi = 0.75$ and

$$R_n = \text{area} \times \text{ultimate shearing stress} = 0.707wLF_W \tag{29.13}$$

In Equation (29.13), w is the weld size, L is the length, and F_W is the ultimate shearing stress, equal to 60% of the ultimate tensile stress of the electrode.

In many connections it is advantageous to use both welds and bolts, with all welding done in the shop and all bolting done in the field. This will usually be the most economical arrangement, since bolting requires less skilled labor.

29.3 Composite Construction

Structural steel shapes acting in combination with reinforced concrete are called composite members. The most common application is a set of parallel steel beams connected to and supporting a reinforced concrete floor slab. The connection is made by attachments that are welded to the top flange of the beam and embedded in the slab. These attachments, called **shear connectors,** are usually in the form of headed studs. A portion of the slab is considered to act with each steel beam as a supplementary compression flange.

The flexural strength of a composite beam is computed in one of two ways:

1. If the web is compact, the strength is based on the plastic condition. This is when the steel has fully yielded and the concrete has reached its maximum compressive stress of $0.85f_c'$, where f_c' is the 28-day compressive strength.
2. If the web of the steel shape is noncompact, the strength is based on the limit state corresponding to the onset of yielding of the steel.

Deflections of composite beams are computed by the usual elastic methods but are based on the transformed section, in which a consideration of strain compatibility is used to transform the concrete into an appropriate amount of steel. The resulting cross-section can then be treated as a homogeneous steel shape.

The composite column is a combination of materials in which a structural steel shape is encased in concrete and supplemented by vertical reinforcing bars. Structural tubes or pipes filled with concrete are also used. The axial compressive strength of a composite column is computed in essentially the same way as for an ordinary steel shape but with modified values of the yield stress, modulus of elasticity, and radius of gyration.

29.4 Computer Applications

Many commercial computer programs are available to the structural steel designer. These include standard structural analysis programs for statically indeterminate structures as well as those containing an AISC "code-checking" feature. These are available for both ASD and LRFD versions of the AISC specification. If the software is capable of performing a second-order frame analysis, the bending moments can be used directly in Equation (29.10) without the need of using the moment amplification factors B_1 and B_2.

A database of standard steel shapes, as well as other software, is available from the American Institute of Steel Construction.

Defining Terms

Block shear — A limit state in which a block of material is torn from a member or connecting element such as a gusset plate. When this occurs, one or more surfaces of the block fail in shear, either by fracture or yielding, and another surface fails in tension, either by fracture or yielding.

Design strength — The nominal strength, or resistance, multiplied by a resistance factor. This is a reduced strength that accounts for uncertainties in such things as theory, material properties, and workmanship.

Flexural buckling — A mode of failure of an axially loaded compression member where buckling occurs by bending about one of the principal axes of the cross section.

Flexural-torsional buckling — A mode of failure of an axially loaded compression member in which it simultaneously buckles about one of the principal axes of its cross-section and twists about its longitudinal axis.

Lateral-torsional buckling — A limit state in which a beam buckles by deflecting laterally and twisting. This is prevented by providing lateral bracing at sufficiently close spacing.

Limit state — A failure condition upon which the strength of a member is based. Yielding of a tension member is an example of a limit state, and the axial tensile force causing the yielding is the corresponding strength.

Load factor — A multiplier of a load effect (force or moment) to bring it to a failure level. A load factor is usually greater than unity, although it can be equal to unity.

Local buckling — A localized buckling, or wrinkling, of a cross-sectional element. This form of instability is in contrast to overall buckling, as when a compression member buckles by bending.

Moment amplification — An approximate technique used to account for the secondary bending moment in beam-columns. The total moment is obtained by multiplying the primary moment by a moment amplification factor.

Nominal strength — The theoretical strength of a member before reduction by a resistance factor.

Plastic moment — The bending moment necessary to cause yielding throughout the depth of a given cross-section. There will be a uniform compressive stress equal to the yield stress on one side of the neutral axis and tension yielding on the other side. The plastic moment can be attained if the beam does not buckle prior to reaching this state.

Resistance factor — A reduction factor applied to the nominal, or theoretical, resistance (strength). Although it can be equal to unity, it is usually less than unity.

Shear connectors — Devices that are welded to the top flanges of beams and embedded in the concrete slab supported by the beams. The most common type of shear connector is the headed stud.

Shear lag — A reduction in the strength of a tension member caused by not connecting some of the cross-sectional elements. It is accounted for by reducing the actual area of the member to an effective area.

Slenderness ratio — The ratio of the effective length of a member to the radius of gyration about one of the principal axes of the cross-section.

Tension field — A condition existing in the buckled web of a plate girder in which the web cannot resist compression but is capable of resisting the diagonal tension within a panel defined by transverse web stiffeners.

Torsional buckling — A limit state for axially loaded compression members in which the member twists about its longitudinal axis.

References

American Institute of Steel Construction. 1989a. *Specification for Structural Steel Buildings: Allowable Stress Design and Plastic Design,* American Institute of Steel Construction, Chicago

American Institute of Steel Construction. 1989b. *Manual of Steel Construction: Allowable Stress Design,* 9th ed. American Institute of Steel Construction, Chicago.

American Institute of Steel Construction. 1999. *Load and Resistance Factor Design Specification for Structural Steel Buildings,* American Institute of Steel Construction, Chicago.

American Institute of Steel Construction. 2001. *Manual of Steel Construction: Load and Resistance Factor Design,* 3rd ed. American Institute of Steel Construction, Chicago.

American Society of Civil Engineers. 2002. *Minimum Design Loads for Buildings and Other Structures,* ASCE 7-02. American Society of Civil Engineers, New York.

American Society for Testing and Materials. 2003. *Annual Book of ASTM Standards,* American Society for Testing and Materials, Philadelphia.

Further Information

The American Institute of Steel Construction is a source of useful and up-to-date information on structural steel design. A monthly magazine, *Modern Steel Construction,* available in both printed and online versions, provides information on structural steel construction projects, technical issues, and AISC activities. The AISC *Engineering Journal* is a quarterly refereed journal containing technical papers with a practical orientation. Approximately every two years, AISC conducts a national lecture series on structural steel design topics of current interest, and the annual AISC North American Steel Construction Conference provides a forum for designers, fabricators, and producers.

The commentary that accompanies the AISC specification is a valuable source of background material. This document, which along with the specification is contained in the *Manual of Steel Construction,* explains and elaborates on the specification provisions. The *Manual* also contains many illustrative examples and discussions.

Several textbooks on structural steel design are available. Comprehensive works covering both allowable stress design and LRFD include *Design of Steel Structures,* by E. H. Gaylord, C. N. Gaylord, and J. E. Stallmeyer; and *Steel Structures, Design and Behavior,* by C. G. Salmon and J. E. Johnson. *Structural Steel Design: ASD Method,* by J. C. McCormac, is a current book on allowable stress design. *Structural Steel Design: LRFD Method,* by J. C. McCormac, and *LRFD Steel Design,* by W. T. Segui, treat load and resistance factor design exclusively.

30

Concrete

Edward G. Nawy
Rutgers University

30.1 Structural Concrete

Structural concrete is a product composed of a properly designed mixture comprising portland cement, coarse aggregate (stone), fine aggregate (sand), water, air, and chemical admixtures. The cement acts as the binding matrix to the aggregate and achieves its strength as a result of a process of hydration. This chemical process results in recrystallization in the form of interlocking crystals producing the cement gel, which has high compressive strength when it hardens. The aggregate could be either natural, producing normal concrete, weighing 150 lb/ft^3 (2400 kg/m^3), or artificial aggregate, such as pumice, producing lightweight concrete weighing ~110 lb/ft^3 (1750 kg/m^3).

Structural concrete should have a cylinder compressive strength of 3000 psi (20 MPa) at least, but often exceeding 4000 to 5000 psi (34.5 MPa). As the strength exceeds 6000 psi (42 MPa), such concrete is presently considered high-strength concrete. Concrete mixtures designed to produce 6000 to 12,000 psi are easily obtainable today with the use of silica fume or high-range water-reducing agents (plasticizers) to lower the water/cement ratio and hence achieve higher strength and a high performance due to the lower water content in the mix. A low water/cement or water/cementitious ratio of 30 to 25% can be achieved with these admixtures, with good workability and high slump fluidity for placing the concrete in the framework.

Concretes of compressive strengths reaching 20,000 psi (140 MPa) have been used in some buildings in the U.S. Such high strengths merit qualifying such concrete as ultra–high-strength concrete.

Admixtures in Concrete

Admixtures in concrete can be summarized as follows:

0-8493-1586-7/05/$0.00+$1.50

1. *Accelerating admixtures.* They hasten the chemical hydration process.
2. *Air-entraining admixtures.* They form minute bubbles 1 mm in diameter and smaller evenly distributed in the mix to protect the concrete from freeze and thaw cycles.
3. *Water-reducing and set-controlling admixtures.* They increase the strength of the concrete through reducing the water content but maintaining the slump (fluidity) of the concrete.
4. *Polymers.* They replace a major portion of the needed water content and can produce concretes of strength in excess of 15,000 psi.
5. *Superplasticizers.* These are high-range water-reducing chemical admixtures. A dosage not exceeding 1 to 2% by weight of cement is recommended.
6. *Silica fume admixtures.* They are pozzolanic materials as a byproduct of high-quality quartz with coal in the electric arc furnace that produces silicon and ferrosilicon alloys. They are used to attain very high-strength concrete in 3 to 7 days with relatively less increase in strength than normal concrete after 28 days. A dosage of 5 to 30% by weight of the cement can be used, depending on the strength needed. A compressive strength of 15,000 psi (105 MPa) or more can be readily achieved with good quality control.

Properties of Hardened Concrete

The mechanical properties of hardened concrete can be classified as (1) short-term or instantaneous properties, and (2) long-term properties.

The short-term properties can be enumerated as (1) strength in compression, tension, and shear, and (2) stiffness measured by the modulus of elasticity. The long-term properties can be classified in terms of creep and shrinkage:

1. *Compressive strength, f_c'.* It is based on crushing 6-in. diameter by 12-in. height standard concrete cylinders at a specified loading rate in a compression testing machine.
2. *Tensile strength, f_t'.* Tensile strength of concrete is relatively low. A good approximation of tensile strength is f_t' ranging between 10 and 15% of f_c'.
3. *Shear strength, v_c.* This value is more difficult to determine experimentally. It can vary from about $2\sqrt{f_c'}$ for normal-weight reinforced concrete beams to about 80% f_c' in direct shear combined with compression.
4. *Modulus of elasticity E_c for stiffness or ductillity determination.* The **ACI 318-05 Code** [ACI Committee 318, 2005] specifies using a secant modulus, given in psi or MPa:

$$E_c = 33W_c^{1.5}\sqrt{f_c'}\ \text{psi} \tag{30.1}$$

$$E_c = 0.043W_c^{1.5}\sqrt{f_c'}\ \text{MPa} \tag{30.2}$$

5. *Shrinkage.* There are two types of shrinkage: plastic shrinkage and drying shrinkage. *Plastic shrinkage* occurs during the first few hours after placing fresh concrete in the forms, resulting in a random map of cracks. *Drying shrinkage* occurs *after* the concrete has already attained its final set and a good portion of the chemical hydration process in the cement gel has been accomplished. Drying shrinkage results in well-defined linear cracks of larger width than plastic shrinkage cracks.
6. *Creep.* This is the lateral flow of the material under external load. The member sustains an increase in lateral strains with time due to the sustained load, hence increased stresses occur in the member and sometimes an almost 100% increase in deflection with time.

Details of all these effects and the ACI 318 Code provisions to control them in the design of reinforced and prestressed concrete structures are given in Nawy [2005a,b].

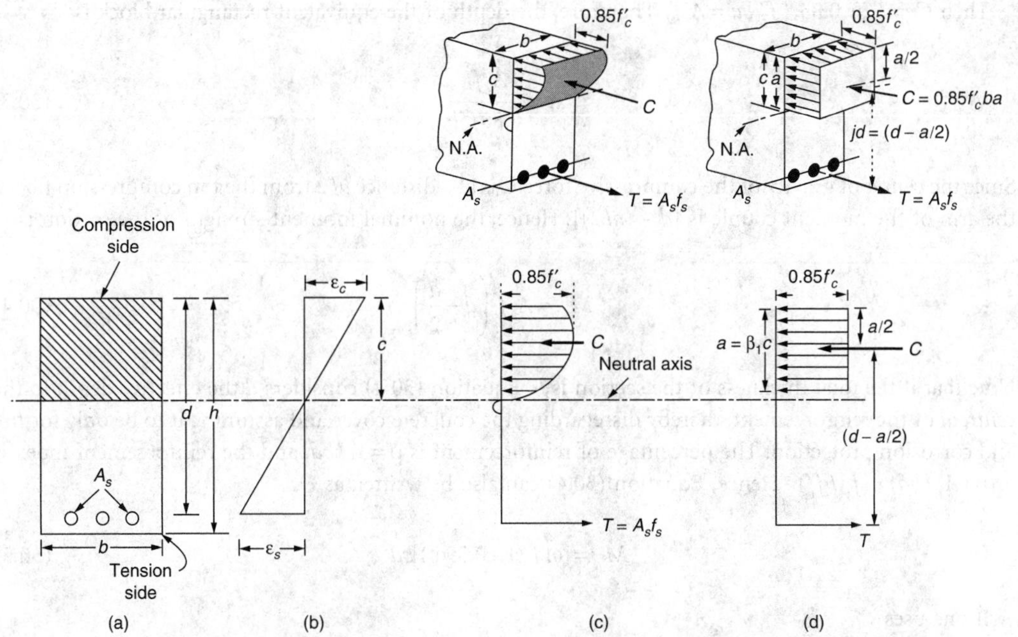

FIGURE 30.1 Stress and strain distribution across reinforced concrete beam depth: (a) beam cross-section, (b) strain distribution, (c) actual stress block, (d) assumed equivalent block. (*Source*: Nawy, E. G., 2003. *Reinforced Concrete — A Fundamental Approach*, 5th ed. Prentice Hall, Upper Saddle River, NJ.)

30.2 Flexural Design of Reinforced Concrete Members

General Principles

Concrete structural systems are generally composed of floor slabs, **beams**, **columns**, walls, and foundations. Present codes of practice (ACI, PCI, CEB, UBC, IBC) all require ultimate strength procedures in proportioning the structural elements, what is termed *strength design* by the code of the American Concrete Institute (ACI 318-05).

The strength of a particular structural unit is termed *nominal strength.* For example, in the case of a beam, the resisting moment capacity of the section calculated using the equations of equilibrium and the properties of the concrete and the steel reinforcement is called *nominal moment strength* M_n. This nominal strength is reduced using a strength reduction factor, ϕ, to account for inaccuracies in construction such as in the geometrical dimensions or position of reinforcing bars or variation in concrete properties.

The design principles are based on equilibrium of forces and moments in any section. Since concrete is weak in tension, the design assumes that the concrete in the tensile part of a beam cross section does *not* carry any load or stress, hence it is disregarded. By doing so, the tensile equilibrium force is wholly taken by the tension bars A_s in Figure 30.1(a). The actual distribution of compressive stress is parabolic, as seen in Figure 30.1(c). However, the ACI Code adopted the use of an equivalent rectangular block, as in Figure 30.1(d), in order to simplify the computations.

Singly Reinforced Beam Design

Such a beam would have reinforcement only on the tension side. If one considers the equilibrium forces in Figure 30.1(d):

C = Volume of the equivalent rectangular block = $0.85\ f'_c ba$
T = Tensile force in the reinforcement = $A_s f_y$, where f_y = yield strength of the reinforcement

Then $C = T$ or $0.85, \; f_c' ba = A_s f_y$. Therefore, the depth of the equivalent rectangular block is

$$a = \frac{A_s f_y}{0.85 f_c' b} \tag{30.3}$$

Since the center of gravity of the compressive force C is at a distance $a/2$ from the top compression fibers, the arm of the moment couple is $[d - (a/2)]$. Hence, the nominal moment strength of the section is

$$M_n = A_s f_y \left(d - \frac{a}{2} \right) \tag{30.4}$$

Note that if the total thickness of the section is h, Equation (30.4) considers d the effective depth to the *centroid* of the reinforcement, thereby disregarding the concrete cover and assuming it to be only for fire and corrosion protection. The percentage of reinforcement is $\rho = A_s / bd$, and the reinforcement index is $\omega = (A_s / bd) \times (f_y / f_c')$. Hence, Equation (30.4) can also be written as

$$M_n = [\omega f_c'(1 - 0.59\omega)] bd^2 \tag{30.5}$$

If one uses

$$R = \omega f_c'(1 - 0.59\omega) \tag{30.6}$$

then $M_n = Rbd^2$. Figure 30.2, for singly reinforced beams, can give a rapid choice of the width and depth of a beam section, as one can usually use $b = \frac{1}{2}d$. For doubly reinforced beams and for T-beams, the R value obtained from Equation 30.6 should be increased by about 50% for the first trial.

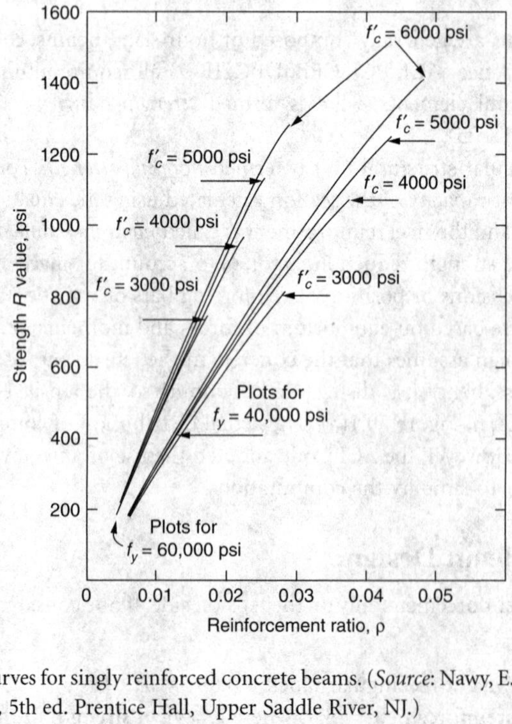

FIGURE 30.2 Strength R curves for singly reinforced concrete beams. (*Source*: Nawy, E. G., 2005. *Reinforced Concrete — A Fundamental Approach*, 5th ed. Prentice Hall, Upper Saddle River, NJ.)

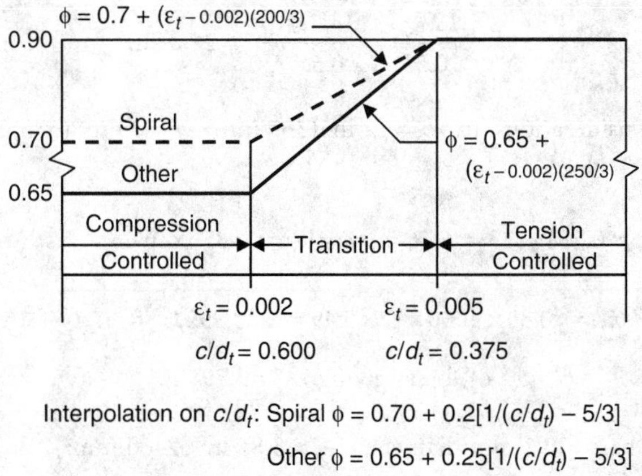

FIGURE 30.3 Strain limits zones and variation of strength reduction factor ϕ with the net tensile strain ε_t (ACI 318-05 Code).

In order to ensure ductility of the structural member and corresponding warning of failure, the ACI Code stipulates a **limit strain** in the reinforcement closest to the tensile face of the section that has to exceed $\varepsilon_t = 0.005$. It is advisable to use a strain value in the range of 0.007, which is essentially equivalent to using 50% of the balanced reinforcement ratio used in previous ACI 318 versions of the code in flexural members, such as beams and one-way and two-way slabs. One way of ensuring this higher strain value is to assume the ratio c/d_t to be about 75% of the 0.375 ratio given in Figure 30.3.

This strain state is termed as the **tension-controlled** state, shown in Figure 31.3. In sections subjected to combined flexure and compression, such as column-beams, the limit strain ranges are the **transition** state and the **compression-controlled** state, where lower strain limits are stipulated in the Code. The three strain limits zones and the strength reduction factor ϕ for these zones are shown in Figure 30.3.

The code also requires a minimum reinforcement area so that the beam can behave as a reinforced concrete section. The minimum $A_s = (3\sqrt{f_c} / f_y) b_w d$ or $200 / f_y$ whichever is larger, where b_w = width of the beam web. For the tension reinforcement in flanged sections, twice the minimum A_s in the preceding expressions is required to be used.

Design Example

A simply supported singly reinforced concrete beam is subjected to a total factored moment including its self weight, $M_u = 4.1 \times 106$ in.-lb. Given values are $f_c' = 4000$ psi (34.5 MPa), $f_y = 60{,}000$ psi (414 MPa).

Solution

Required nominal moment strength M_n equals M_u/ϕ, where $\phi = 0.90$ for flexure: $M_n = 4.1 \times 10^6/0.90 = 4.56 \times 10^6$ in.-lb. The strain ε_t has to exceed 0.005 in order to ensure ductile behavior. Hence use a c/d_t ratio $= 0.75 \times 0.375 = 0.281$.

Equation (30.5) yields

$$R = \omega f_c'(1 - 0.59\omega) = 0.215 \times 4000(1 - 0.59 \times 0.215) \simeq 750.$$

Alternatively, entering the chart in Figure 30.2 gives $R \simeq 750$. Assuming $b = \frac{1}{2}d$, Equation (30.6) yields

$$d = \sqrt[3]{\frac{M_n}{0.5R}} = \sqrt[3]{\frac{4.5\times 10^6}{0.5\times 750}} = 22.8 \text{ in.}$$

Based on practical considerations, use b = 12 in. (305 mm), d = 23 in. (585 mm), and total depth h = 26 in. (660 mm).

$$c = 0.281 \times 23 = 6.46 \text{ in. } a = \beta_1 c = 0.85 \times 6.46 = 5.49 \text{ in.}$$

$$C = 0.85 f'_c ba = 0.85\times 4000\times 12\times 5.49 = 223{,}992\, lb = A_s f_y \text{ for equilibrium}$$

$$\text{of forces giving } A_s = 3.79 \text{ in.}^2$$

Hence, try three #10 bars (32.2 mm diameter), A_s = 3.81 in^2 (2460 mm^2).

To check nominal moment strength, use Equation (30.3) to find the depth of the compression block:

$$a = \frac{A_s f_y}{0.85 f'_c b} = \frac{3.81\times 60\,000}{0.85\times 4000\times 12} = 5.60 \text{ in.}$$

From Equation 30.4, available M_n = 3.81 × 60,000[23.0 – (5.49/2)] = 4.64 × 10^6 in.-lb (5.2 × 10^5 kN-M) > Required M_n = 4.56 × 10^6 in.-lb; adopt design.

A check for minimum reinforcement and shear capacity also has to be performed, as must deflection and crack control, as detailed in [Nawy, 2003a].

$$\text{Overdesign} = \frac{4.64 - 4.56}{4.56} = 1.8\%$$

Overdesign should not be in excess of 4 to 6%.

Doubly Reinforced Sections

These are beam sections where compression reinforcement A'_s is used about 2 in. from the compression fibers. The reinforcement A'_s contributes to reducing the required depth of section where there are clearance limitations. An extra nominal moment $M' = A'_s f_y (d - d')$ is added to the section, d' being the depth from the extreme compression fibers to the centroid of the compression reinforcement A'_s. Consequently, Equation (30.4) becomes

$$M_n = (A_s - A'_s) f_y \left(d - \frac{a}{2}\right) + A'_s f_y (d - d') \tag{30.7}$$

A similar expression to Equation (30.7) can be derived for flanged sections, as in Nawy [2003a].

Columns

Columns are compression members that can fail either by material failure if they are nonslender or by buckling if they are slender. Columns designed using the material failure criteria should have a slenderness ratio kl/r not to exceed 22 for nonbraced columns. The value k is the stiffness factor at column ends, l is the effective length, and r is the radius of gyration = 0.3h for rectangular sections.

A column is essentially a doubly reinforced flexural section that is also subjected to an axial compression force P_n in addition to the forces C and T as given in Equation (30.8) and Equation (30.9), and shown in Figure 30.1. Therefore, from equilibrium of forces,

$$P_n = 0.85 f_c' ba + A_s' f_s' + A_s f_s \tag{30.8}$$

$$M_n = P_{ne} = 0.85 f_c' ba \left(\bar{y} - \frac{a}{2} \right) + A_s' f_s' (\bar{y} - d') + A_s f_s (d - \bar{y}) \tag{30.9}$$

$$e = \frac{M_n}{P_n} = \text{eccentricity}$$

where $\bar{y}$ = distance to center of gravity of section = $h/2$ for rectangular section. Notice the similarities between Equation (30.9) and Equation (30.7).

Although initial failure in beams is always by yielding of the reinforcement through limiting ρ, this is not possible in columns, as the mode of failure depends on the magnitude of eccentricity, *e*, determined by the strain level ε_t value in the range of $0.002 \le \varepsilon_t \le 0.005$. A strain state $\varepsilon \le 0.002$ would denote a concrete section in the compression-controlled region, such as columns subjected to axial compression, as the most common cases.

This subject is very extensive, particularly if buckling is also to be considered. The reader is advised to consult textbooks such as [Nawy, 2003a] and/or specialized handbooks.

Walls and Footings

The same principles for the flexural design of beams and slabs apply to walls and footings. An isolated footing is an inverted cantilever supported by a column and subjected to uniform soil pressure. It is treated as a singly reinforced beam [Equation (30.4)] subjected to a factored moment $M_u = w_u l^2/2$, where w_u is the intensity of load per unit length and l is the arm of the cantilever. In the same manner, one can design retaining walls and similar structural systems.

30.3 Shear and Torsion Design of Reinforced Concrete Members

Shear

External transverse load is resisted by internal **shear** in order to maintain section equilibrium. As concrete is weak in tension, the principal tensile stress in a beam cannot exceed the tensile strength of the concrete. The principal tensile stress is composed of two components — shear stress, v, and tensile stress, f_t — causing diagonal tension cracks at a distance d from the face of the support in beams and at a distance $d/2$ from the face of the support in two-way slabs.

Consequently, it is important that the beam web be reinforced with diagonal tension steel, called *stirrups*, in order to prevent diagonal shear cracks from opening. The resistance of the plain concrete in the web sustains part of the shear stress, and the balance has to be borne by the diagonal tension reinforcement. The shear resistance of the plain concrete in the web is termed *nominal shear strength*, V_c. A conservative general expression for V_c from the ACI 318 code is

$$V_c(\text{lb}) = 2.0\lambda \sqrt{f_c'} b_w d \tag{30.10}$$

$$V_c(\text{newton}) = \lambda \left(\sqrt{f_c'} / 6 \right) b_w d \tag{30.11}$$

where f_c' is in MPa in Equation (30.11) and λ equals

1 for stone aggregate concrete
0.85 for sand-lightweight concrete
0.75 for all lightweight concrete

If $V_n = V_u/\phi$ = the required nominal shear, the stirrups should be designed to take the difference between V_u and V_c— namely, $V_s = [(V_u/\phi) - V_c]$ — where V_u is the factored external shear and $\phi = 0.75$ in shear and torsion. The spacing of the transverse web stirrups is hence

$$s = \frac{A_v f_y d}{(V_u / \phi - V_c)} \tag{30.12}$$

where A_v = cross-sectional area of the web steel (two stirrup legs). Maximum spacing of stirrups is $d/2$ or 12 in., whichever is smaller. A concrete section designed for flexure as described earlier has to be enlarged if $V_s = (V_n - V_c) > 8\sqrt{f_c'}(b_w d)$.

Torsion

If a beam is also subjected to **torsion** combined with shear, diagonal cracks described in the previous section have to be prevented from opening. This is accomplished by use of *both* vertical closed stirrups and additional longitudinal bars evenly divided among the four faces of the beam.

The longitudinal reinforcement is required since torsion causes a three-dimensional warped surface. The ACI 318-05 code disallows utilization of the nominal torsional strength T_c of the plain concrete in the web and requires that all the torsional moment T_n be borne by the transverse closed stirrups and the longitudinal bars. It assumes that the volume of the transverse stirrups is equal to the volume of the longitudinal bars. The same equations for torsion in reinforced concrete elements are used with adjusting modifiers when applied to prestressed concrete.

Design of the particular component needs to be based on the limit state at failure. Therefore, the nonlinear behavior of a structural system after torsional cracking must be identified in one of the following two conditions

1. No redistribution of torsional stresses to other members after cracking (**equilibrium torsion**)
2. Redistribution of torsional stresses and moments after cracking to effect deformation compatibility between intersecting members (**compatibility torsion**)

For the first case, an edge beam supporting a cantilever canopy is such an example where the edge beam has to be designed to resist the *total* external factored twisting moment due to the cantilever slab; otherwise, the structure will collapse. In the second case, such as in statically indeterminate systems, stiffness assumptions, compatibility of strains at the joints, and redistribution of stresses may affect the stress resultants, leading to a **reduction** in the resulting torsional shearing stresses. The ACI Code permits a maximum factored torsional moment at the critical section *d* from the face of the supports for reinforced concrete members to be used in the design as follows:

$$T_u = \phi 4\sqrt{f_c}\left(A_{cp}^{2} \Big/ p_{cp}\right) \tag{30.13a}$$

where A_{cp} = area enclosed by outside perimeter of concrete cross-section and p_{cp} = outside perimeter of concrete cross-section A_{cp} in.

The available torsional moment strength is computed from the following expression:

$$T_n = \frac{2A_o A_l f_{yv}}{s}(\cot\theta) \tag{30.13b}$$

where A_0 = gross area enclosed by the shear flow, A_l = area of the added torsional longitudinal steel bars, s = spacing of transverse ties, and $\theta = 45^\circ$ for reinforced concrete and 37.5° for prestressed concrete.

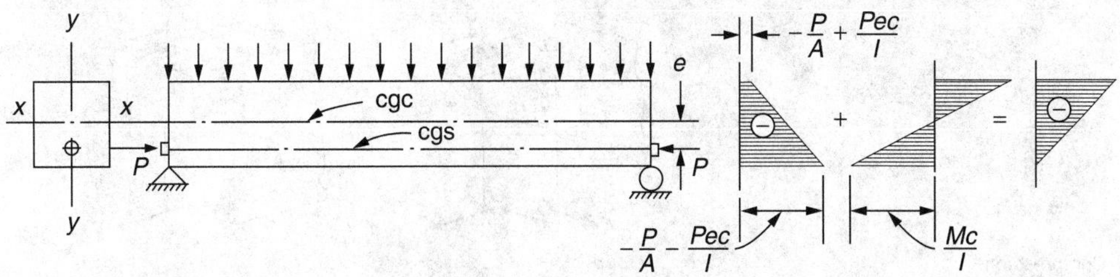

FIGURE 30.4 Stress distribution at service load in prestressed concrete beams with constant tendons. (*Source*: Nawy, E. G., 2003. *Prestressed Concrete — A Fundamental Approach,* 4th ed. Prentice Hall, Upper Saddle River, NJ.)

30.4 Prestressed Concrete

Reinforced concrete is weak in tension but strong in compression. To maximize utilization of its material properties, an internal compressive force is induced in the structural element through the use of highly stressed prestressing tendons to precompress the member prior to application of the external gravity live load and superimposed dead load. Typical effect of the prestressing action is shown in Figure 30.4, using a straight tendon, as is usually the case in precast elements. For *in situ*–cast elements, the tendon can be either harped or usually draped in a parabolic form. As can be seen from Figure 30.4, the prestressing force P alone induced a compressive stress at the bottom fibers f_b and a tensile stress at the top fibers f_t such that

$$f_b = -\frac{P_i}{A} - \frac{P_i ec}{I} \tag{30.14a}$$

$$f_t = -\frac{P_i}{A} + \frac{P_i ec}{I} \tag{30.14b}$$

where P_i is the initial prestressing force prior to losses.

With addition of stress due to self-weight of the concrete beam and external live and superimposed dead load — moment M_T — the stresses become:

$$f_b = -\frac{P_e}{A} - \frac{P_e c}{I} + \frac{M_T c}{I} \tag{30.15a}$$

$$f_t = -\frac{P_e}{A} + \frac{P_e c}{I} - \frac{M_T c}{I} \tag{30.15b}$$

where P_e is the effective prestressing force after losses in prestress.

The stress diagram at the extreme right in Figure 30.4 indicates zero tension at the bottom fibers and maximum compression at the top fibers. In this manner, prestressing the beam has resulted in full utilization of the properties of the concrete, eliminating tension cracking at the bottom fibers.

Whereas reinforced concrete members are designed only for ultimate load, prestressed concrete members are first designed for service load moments as in Equation (30.15) and then analyzed for ultimate load capacity, namely, the nominal moment strength M_n. This is necessary for determining the reserve strength available in the member between the service load level and collapse, as prestressed beams can be underreinforced (tension steel yielding) or overreinforced (compression side concrete crushing). Figure 30.5 gives the sets of equilibrium forces acting on the concrete section. Notice their similarity to

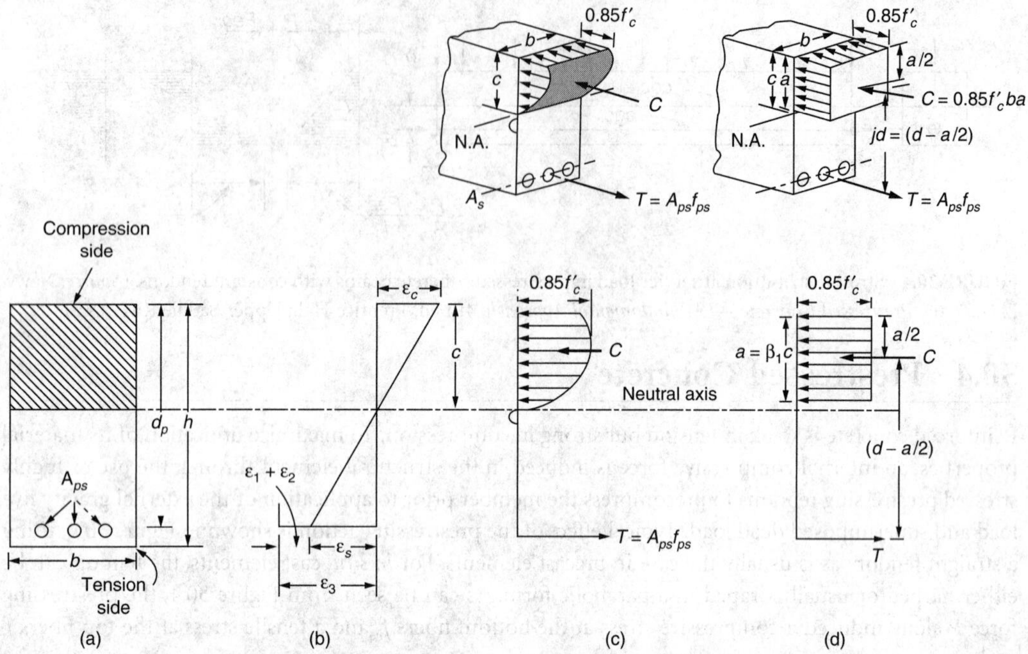

FIGURE 30.5 Stress and strain distribution across prestressed concrete beam depth: (a) beam cross-section, (b) strain distribution, (c) actual stress block, (d) assumed equivalent block. (*Source*: Nawy, E. G., 2003. *Prestressed Concrete — A Fundamental Approach,* 2nd ed. Prentice Hall, Upper Saddle River, NJ.)

those of reinforced concrete sections in Figure 30.1. If both prestressing tendons and mild steel are used in the prestressed beam, the nominal moment strength is

$$M_n = A_{ps} f_{ps}\left(d_p - \frac{a}{2}\right) + A_s f_y\left(d - \frac{a}{2}\right) + A'_s f_y (d - d') \tag{30.16}$$

It should be noted that the prestressing force P used at the service load design level should be the effective force P_e obtained from reducing the initial prestressing force, P_i, in Equation (30.14) to the *effective* service load, P_e, caused by losses due to concrete shrinkage, concrete creep, relaxation of the prestressing steel, frictional losses in posttensioned beams, and anchorage loss. A loss in the initial prestress of 20 to 25% is not unreasonable, so P_e is quite often 0.75 to 0.80 P_i.

The ACI code prescribes the following allowable stresses at service load. For concrete stresses

$$f'_{ci} \simeq 0.75 f'_c \text{ psi}$$
$$f'_{ci} \simeq 0.60 f'_c \text{ psi}$$
$$f_{ti} = 3\sqrt{f'_{ci}} \text{ psi} \quad \text{on span} \quad (\sqrt{f'_c}/4\,\text{MPa})$$
$$= 6\sqrt{f'_{ci}} \text{ psi} \quad \text{at support} \quad (\sqrt{f'_{ci}}/2\,\text{MPa})$$
$$f_c = 0.45 f'_c \text{ to } 0.60 f'_c$$
$$f_t = 6\sqrt{f'_c} \text{psi} \quad (\sqrt{f'_c}/2\,\text{MPa})$$
$$= 12\sqrt{f'_c} \text{ psi} \quad \text{if deflection verified} \quad (\sqrt{f'_c}\,\text{MPa})$$

and for reinforcing tendon stresses

Tendon jacking: $f_{ps} = 0.94 f_{py} \leq 0.80 f_{pu}$

Immediately after stress transfer: $f_{ps} = 0.82 f_{py} \leq 0.74 f_{pu}$

Posttensioned members
at anchorage immediately
after anchorage: $f_{ps} = 0.70 f_{pu}$

where

f_{ps} = Ultimate design stress allowed in tendon
f_{py} = Yield strength of tendon
f_{pu} = Ultimate strength of tendon
f_{ti} = Initial tensile stress in concrete
f_c = Service load concrete compressive strength
f_{ci} = Initial compressive stress in concrete
f_t = Service load concrete tensile strength

30.5 Serviceability Checks

Serviceability of structural components is a major factor in designing structures to sustain acceptable long-term behavior. It is controlled by limiting deflection and cracking.

For deflection computation and control, an effective moment of inertia is used. Details of design for deflection in reinforced concrete beams and slabs and for deflection and camber in prestressed concrete with design examples are given in ACI Committee 435 [1995]. Table 30.1 (from ACI) gives the allowable deflections in terms of span for reinforced concrete beams.

For crack control in beams and two-way slab floor systems, it is more effective to use smaller-diameter bars at smaller spacing for the same area of reinforcement. ACI Committee 224 [2001] gives a detailed treatment of the subject of crack control in concrete structures. Table 30.2 gives the tolerable crack widths in concrete elements.

TABLE 30.1 Minimum Thickness, *h*, of Nonprestressed Beams or One-Way Slabs (ACI 319-02 Code)

Member[a]	Simply Supported	One End Continuous	Both Ends Continuous	Cantilever
Solid one-way slabs	$l/20$	$l/24$	$l/28$	$l/10$
Beams or ribbed one-way slabs	$l/16$	$l/18.5$	$l/21$	$l/8$

Note: Span length *l* is in inches. For f_y values greater than 60,000 psi, the tabulated values should be multiplied by $(0.4 + f_y/100{,}000)$.

[a] Members not supporting or attached to partitions or other construction are likely to be damaged by large deflections.

TABLE 30.2 Tolerable Crack Widths (ACI 224-01 Report)

Exposure Condition	Tolerable Crack Width in.	mm
Dry air or protective membrane	0.016	0.40
Humidity, moist air, soil	0.012	0.30
Deicing chemicals	0.007	0.18
Seawater and seawater spray; wetting and drying	0.006	0.15
Water-retaining structures (excluding nonpressure pipes)	0.004	0.10

30.6 Computer Applications for Concrete Structures

In the design of concrete structures, several canned computer programs are available both for reinforced concrete and prestressed concrete systems. They can be either analysis or design programs. These programs are available for personal computers using MS-DOS or MS Windows operating systems. Typical general purpose programs are STRUDEL, ANSYS, SAP Professional, and ETABS. The SAP Professional program can handle in excess of 5000 nodes and requires a larger memory than the others. Except for lack of space, a long list of available programs could be compiled here. The structural engineer can without difficulty get access to all present and forthcoming programs in the market.

The specialized concrete programs are numerous. The following programs are widely used. PCA programs include

1. *ADOSS.* For two-way reinforced concrete slabs and plates
2. *PCA Columns.* For nonslender and slender regular and irregular columns
3. *PCA Frame.* For analysis of reinforced concrete frames, including second-order analysis for slender columns taking into account the $P - \Delta$ effect
4. *PCA MATS.* For the design of flexible mat foundations

Other programs include

1. *ADAPT.* A comprehensive program for the design of reinforced and prestressed concrete two-way action slabs and plates with the capability of drafting many AUTOCAD version 12.0
2. Miscellaneous design or analysis programs for proportioning sections:
   NRCPCPI. For reinforced concrete sections [Nawy, 2003a]
   NRCPCPII. For prestressed concrete sections [Nawy, 2003b]
   RCPCDH. For reinforced concrete beams, columns, and isolated footings
   RISA2D and *RISA3D.* General purpose programs for both steel and concrete and frame analysis

Among drafting programs, AUTOCAD is a comprehensive general purpose program very widely used for drafting working drawings including reinforcing details.

In summary, it should be emphasized that the computer programs discussed in this section are only representative. Other good software is available and being developed with time. Users should always strive to utilize the program that best meets their particular needs, preferences, and engineering backgrounds.

Defining Terms

Admixtures — Chemical additives to the concrete mix that change the mechanical and performance characteristics of the hardened concrete.

ACI — American Concrete Institute.

Beams — Supporting elements to floors in structural systems.

Codes — Standards governing the design and performance of constructed systems to ensure the safety and well-being of the users.

Columns — Vertical compression supports.

Ductility — Ability of the member to absorb energy and deform in response to external load.

Flexure — Bending of a structural element due to applied load.

Footings — Foundation elements within the soil supporting the superstructure.

PCI — Prestressed Concrete Institute.

Prestressed concrete — Concrete elements such as beams, columns, or piles subjected to internal (or external) compression prior to the application of external loads.

Serviceability — Cracking and deflection performance of a structural member or system.

Shear — Force due to external load acting perpendicular to the beam span to shear the section.

Torsion — Twisting moment on a section.

IBC — International Building Code.

UBC — Uniform Building Code.

References

ACI Committee 224. 2001. *Control of Cracking in Concrete Structures,* Committee Report, American Concrete Institute, Farmington Hills, MI.

ACI Committee 435. 1995. *Control of Deflection in Concrete Structures,* Committee Report, E. G., Nawy, Chairman, Publ. American Concrete Institute, Farmington Hills, MI.

ACI Committee 318. 2005. *Building Code Requirements for Reinforced Concrete: ACI 318-02 and Commentary ACI 318 R-02,* Institute Standard, American Concrete Institute, Farmington Hills, MI.

Hsu, T. C. 1993. *Unified Theory of Reinforced Concrete,* CRC Press, Boca Raton, FL.

Nawy, E. G., 2005a. *Prestressed Concrete — A Fundamental Approach,* 4th ed. Prentice Hall, Upper Saddle River, NJ.

Nawy, E. G., 2005b. *Reinforced Concrete — A Fundamental Approach,* 5th ed. Prentice Hall, Upper Saddle River, NJ.

Nawy, E. G., 2002c. *Fundamentals of High Performance Concrete,* 2nd ed., John Wiley & Sons, New York.

Further Information

A comprehensive treatment of all aspects of concrete proportioning, design, construction, and long-term performance can be found in the five-volume *Manual of Concrete Practice* published by the American Concrete Institute, Farmington Hills, MI.

The proceedings of the *Structural Journal, Materials Journal,* and *Concrete International Journal,* all published by the American Concrete Institute, are an additional source for the latest research and development in this area.

The proceedings of the *PCI Journal,* published bimonthly by the Prestressed Concrete Institute, Chicago, deals with all aspects of fabrication, design, and construction of precast and prestressed concrete beams for residential building as well as bridges.

The *Design Handbook,* in three volumes, SP-17, published annually by the American Concrete Institute, Detroit, MI, is an all-encompassing publication containing numerous charts, monograms, and tables, as well as examples pertaining to all design aspects of reinforced concrete structural members.

A comprehensive handbook, E. G., Nawy, Editor-in-Chief, *Concrete Construction Engineering Handbook,* 1998, CRC Press, Boca Raton, FL, 1250 pp. gives extensive treatment of all aspects of concrete material behavior, analysis, design, and seismic response.

31

Timber

Donald E. Breyer
California State Polytechnic University, Pomona

The four primary building materials used in the construction of civil engineering structures are reinforced concrete, reinforced masonry, structural steel, and *timber.* This chapter will give a brief introduction to *engineered wood structures.*

Wood is often used as the framing material in low-rise buildings (one to four stories), but timber has been used in taller structures. Bridges are also constructed from wood and are generally limited to relatively short-span bridges on rural and forest service roads, but glued-laminated timber framing has been used in the construction of some highway bridges. Wood is also used in foundation systems such as timber piles; utility poles and towers are other examples of wood structures. In addition to these more permanent structures, wood is commonly used for such temporary structures as concrete formwork and falsework and for shoring of trenches during construction. Although many residential structures are engineered to some extent, wood is also used as a structural material in many nonengineered homes that fall into a category known as conventional construction.

As a biological product, wood is a unique structural material. It is a renewable resource that can be obtained by growing and harvesting new trees. Proper forest management is necessary to provide a sustainable supply of wood products and to ensure that this is accomplished in an environmentally responsible way.

Wood can be used to create permanent structures. However, the proper use of wood as a structural material requires that the designer be familiar with more than stress calculations. Pound for pound, wood is stronger than many materials, but wood also has characteristics that, if used improperly, can lead to premature failure. Understanding the unique characteristics of wood is the key to its proper use.

31.1 Durability of Wood

In addition to being overstressed by some types of loading, a wood structure can be destroyed by several environmental causes, including decay, insect attack, and fire.

The moisture content of wood is defined as the weight of water in the wood expressed as a percentage of the oven dry weight of the wood. In a living tree the moisture content can be as high as 200%. In a structure, the moisture content of a wood member will be much less, and for a typical enclosed building the moisture content will range between 7 and 14%, depending on climate. However, if a structure houses a swimming pool or if there are high humidity conditions as in certain manufacturing plants, higher moisture contents will occur.

0-8493-1586-7/05/$0.00+$1.50

The best recommendation for preventing *decay* is to keep wood continuously dry. Special detailing may be required to accomplish this. However, the low moisture content of framing lumber in most enclosed buildings generally does not lead to decay. High moisture content or exposure to the weather (alternate wetting and drying) will cause decay in untreated wood products.

Wood that will be exposed to the weather or subject to other high-moisture conditions can be pressure impregnated with an approved chemical treatment to protect against decay. Pressure-treated lumber is obtained from a processing plant that specializes in treating wood products by forcing the appropriate chemicals under pressure into the wood cells. Paint-on chemicals are generally not effective.

Wood can also be destroyed by *insect attack*. Termites are the most common pest, but marine borers are found in ocean water. Termite protection may be obtained by providing a physical barrier, by maintaining a minimum clearance between the wood and soil, or by using pressure-treated wood. The same pressure-treated lumber is effective for both decay and termite protection. Marine borers require a different chemical treatment.

Fire is a threat to any structure, whether it is wood, steel, concrete, or masonry. However, wood is a combustible material, and building codes place restrictions on the height, area, and occupancy classification of a building that uses wood framing. Wood becomes harder to ignite as the cross-sectional dimensions of the framing members increase. Consequently, building codes recognize both *light-frame wood construction* (using relatively small size framing members) and *heavy-timber (HT) construction.* In a fire, a large wood member performs much better because a protective coating of char develops that helps to insulate the member.

31.2 Wood Products

The following wood products are used in structural design:

1. *Solid sawn lumber* (sawn lumber). Lumber of rectangular cross-section cut from trees. A variety of species of commercial lumber are available, and a variety of stress grades are available for each species group. Tabulated design stresses depend on the size category of a member in addition to grade. Size categories include dimension lumber, beams and stringers, posts and timbers, among others. The nominal size of a member (such as 4 × 12) is used for call-out purposes, but actual cross-sectional dimensions are less. Most grading of lumber is done by visual inspection, but some material is machine stress rated. The maximum size of sawn lumber is limited by tree size.
2. *Structural glued-laminated timber* (glu-lam). Lumber formed by gluing together small pieces (usually 2 in. nominal thickness) of wood to form virtually any size structural member. Laminating stock is usually from a western species group or southern pine. Tabulated stresses for glu-lam are generally larger than for sawn lumber because higher-quality wood can be optimized. For example, bending combinations of laminations have higher-quality laminating stock placed in the areas of higher bending stress (i.e., at the top and bottom of a beam) and lower-quality laminating stock near the neutral axis (i.e., at the center). See Figure 31.1.
3. *Structural composite lumber* (SCL). Lumber that is a reconstituted wood product. Because of its manufacturing process, SCL has even higher stress values than glu-lam.
   A. *Laminated veneer lumber* (LVL). Lumber formed by gluing together thin sheets of wood known as *veneers.*
   B. *Parallel strand lumber* (PSL). Lumber formed by gluing together thin, narrow pieces of wood known as *strands.*
4. *Round timber poles and piles.*
5. *Structural-use panels.* Usually 4 × 8 ft panels of wood with directional properties used for sheathing and other structural applications.
   A. *Plywood*
   B. *Oriented strand board* (OSB)

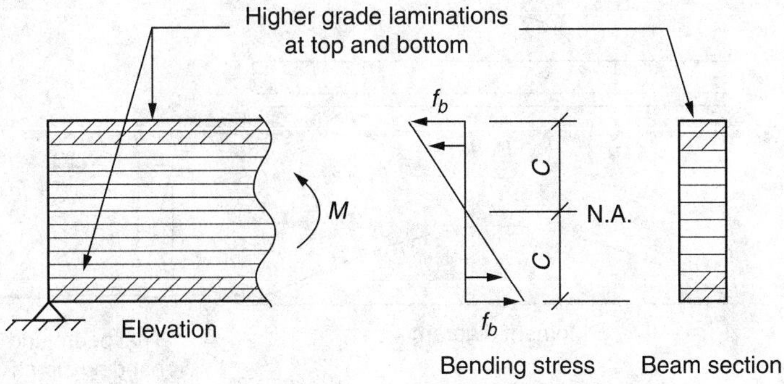

FIGURE 31.1 Layup of laminating stock in a glu-lam beam. Stronger wood is located at points of higher stress.

These products may be used individually as building components or they may be combined in a manufacturing plant to form composite structural products such as prefabricated wood trusses, wood I joists, or factory-built roof or wall panels.

The basic reference for structural design in wood is the *National Design Specification* [AF&PA, 1991]. The basic reference for plywood is the *Plywood Design Specification* [APA, 1986]. Detailed examples and additional references are given in Breyer [1993] and the *NDS Commentary* [AF&PA, 1993]. Basic building code criteria are available from such sources as the *UBC* [ICBO, 1991], which is one of the three model building codes used in the U.S.

31.3 Member Design

Wood structures are designed to resist vertical (gravity) loads and lateral (wind and earthquake) forces. Generally, a series of beams and columns is used to carry gravity loads. Thus a path is formed for the progressive transfer of gravity loads from the top of the structure down into the foundation. Design includes consideration of dead load, roof live load or snow load, floor live load, and other possible loads.

The 1991 edition of the *National Design Specification* [AF&PA, 1991] introduced sweeping changes to the design equations for engineered wood structures. Revised stresses for lumber were also introduced as a result of a 12-year study known as the *in-grade testing program*. Brief introductions to the design of a wood beam and a wood column are given in the remainder of this section.

Wood beams are designed using familiar formulas from engineering mechanics for bending, shear, deflection, and bearing. Although the basic concepts are simple, wood design can appear to be complicated because of the nature of the material. The variability of mechanical properties for different wood products is one factor. Another is the presence of natural growth characteristics such as knots, density, slope of grain, and others. The natural growth characteristics in wood have led to the development of a relatively involved lumber-grading system. However, the basic design procedure is straightforward. Basic formulas from strength of materials are used to evaluate the *actual stress* in a member. The actual stress is then checked to be less than or equal to an *allowable stress* for the species, grade, and size category. The design size is accepted or revised based on this comparison.

The allowable stress is where the unique nature of the material is taken into account. Determination of allowable stress begins by first finding the tabulated stress for a given species, stress grade, and size category. The tabulated stress applies directly to a set of base conditions (e.g., normal duration of load, dry service conditions, standard size, normal temperature, and so on). The tabulated stress is then subjected to a series of adjustment factors, which converts the base conditions for the table to the conditions for a particular design. See Figure 31.2.

For example, the bending stress in a wood beam is checked as follows:

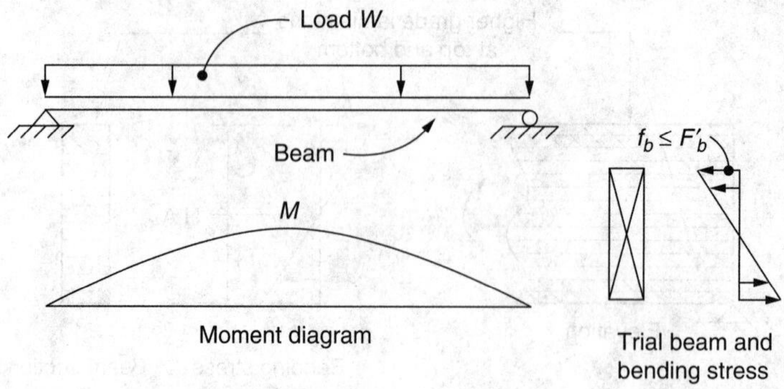

FIGURE 31.2 Bending stress in a wood beam. A trial member size is structurally safe if the actual stress f_b is less than or equal to the allowable stress F'_b.

$$f_b = \frac{Mc}{I} = \frac{M}{S} \leq F'_b$$

where

f_b = actual bending stress, psi
M = moment in beam, in.-lb
c = distance from neutral axis to extreme fiber, in.
I = moment of inertia, $in.^4$
S = section modulus, $in.^3$
F'_b = allowable bending stress, psi
= $F_b \times$ (series of adjustment factors)
= $F_b \times (C_D \times C_M \times C_L \times \text{cf.} \times C_t \times x \ldots)$
F_b = tabulated bending stress, psi
C_D = adjustment factor for duration of load (**load duration factor**)
C_M = adjustment factor for high moisture conditions (**wet service factor**)
C_L = adjustment factor for lateral torsional buckling (beam stability)
C_F = adjustment factor for size effect (**size factor**)
C_t = adjustment factor for high temperature applications (**temperature factor**)
$x \ldots$ = any other adjustment factor that may apply

For the common case of a continuously braced beam in an enclosed building (dry service at normal temperatures), a number of the adjustment factors default to unity. Thus, the complicated nature of the problem is often simplified for frequently encountered design conditions.

Wood columns are checked in a similar manner (See Figure 31.3) using the following formula for axial compressive stress:

$$f_c = \frac{P}{A} \leq F'_c$$

where

f_c = actual column stress, psi
P = axial column load, lb
A = cross-sectional area of column, $in.^2$

F'_c	=	allowable column stress, psi
	=	$F_c \times$ (series of adjustment factors)
	=	$F_c \times (C_D \times C_M \times C_P \times$ cf. $\times C_t \times x \ldots)$
C_D	=	adjustment factor for duration of load
C_M	=	adjustment factor for high-moisture conditions (wet service)
C_P	=	adjustment factor for column buckling (stability)
C_F	=	adjustment factor for size effect
C_t	=	adjustment factor for high-temperature use
$x \ldots$	=	any other adjustment factor that may apply

The column stability factor C_P is a coefficient that measures the tendency of the column to buckle between points of lateral support. C_P is based on the slenderness ratio

$$l_e/r = \text{column slenderness ratio}$$

where l_e = effective column length, in. and r = radius of gyration associated with the axis of column buckling, in.

For a column with a rectangular cross-section the radius of gyration is directly proportional to the cross-sectional dimension of the member, and the slenderness ratio becomes

$$l_e/d = \text{slenderness ratio for a rectangular column}$$

where d = cross-sectional dimension of column associated with the axis of column buckling, in inches.

In the most recent *NDS* [AF&PA, 1991] the column stability factor is defined by the Ylinen column equation, which is graphed in Figure 31.3, showing the allowable compressive stress versus column slenderness ratio. However, space does not permit a detailed review of the Ylinen equation.

Wood members that are subject to combined stress (e.g., a beam-column has both a bending moment and an axial compressive force) are handled with an interaction formula. The most recent *NDS* has a new interaction formula based on work done at the U.S. Forest Products Laboratory by Zahn.

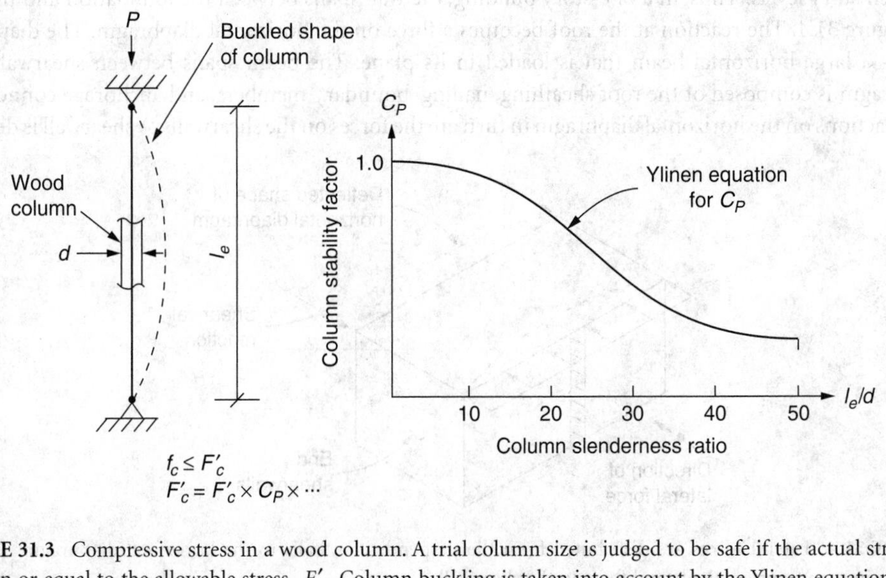

FIGURE 31.3 Compressive stress in a wood column. A trial column size is judged to be safe if the actual stress f_c is less than or equal to the allowable stress F'_c. Column buckling is taken into account by the Ylinen equation in the column stability factor C_P.

31.4 Connections

Connections in engineered wood structures may be made with a variety of fasteners and other materials. These include nails, staples, bolts, lag bolts (lag screws), wood screws, split ring connectors, shear plate connectors, nail plates, and prefabricated metal connection hardware. Fasteners may connect one wood member to another (wood-to-wood connection) or they may connect a wood member to a piece of steel connection hardware (wood-to-metal connection). The most common type of loading on a fastener is perpendicular to the axis of the fastener. This may be described as a *shear-type connection.* Connections are usually either single shear or double shear. In the past, the allowable design load on a shear connection was determined from a table. These tables for nails, bolts, and other fasteners were based on a limited series of tests conducted many years ago and were empirically determined.

The 1991 *NDS* [AF&PA, 1991] introduced a new method for obtaining the nominal design value for these types of connections. The new method is based on engineering mechanics and is referred to as the *yield limit theory for dowel-type fasteners.* In this approach an equation is evaluated for each possible mode of failure for a given type of connection. The nominal design value for the connection is defined as the smallest load capacity from the yield equations. *NDS* tables cover commonly encountered connection problems so that it is not necessary to apply the rather complicated yield limit equations for every connection design.

31.5 Lateral Force Design

Lateral forces include wind and seismic forces. Although both wind and seismic forces involve vertical components, the emphasis is on the effect of horizontal forces. The common lateral force resisting systems (*LFRSs*) are moment-resisting frames, braced frames (horizontal and vertical trusses), and horizontal diaphragms and shearwalls. Most wood frame buildings use a combination of *horizontal diaphragms* and *shearwalls.* Economy is obtained in this approach because the usual sheathing materials on roofs, floors, and walls can be designed to carry lateral forces. In order to make this happen, additional nailing for the sheathing may be required, and additional connection hardware may be necessary to tie the various elements together.

In a shearwall-type building, walls perpendicular to the lateral force are assumed to span vertically between story levels. Thus, in a one-story building, the wall spans between the foundation and the roof. See Figure 31.4. The reaction at the roof becomes a force on the horizontal diaphragm. The diaphragm acts as a large horizontal beam that is loaded in its plane. The beam spans between shearwalls. The diaphragm is composed of the roof sheathing, nailing, boundary members, and anchorage connections. The reactions on the horizontal diaphragm in turn are the forces on the shearwalls. A shearwall is designed

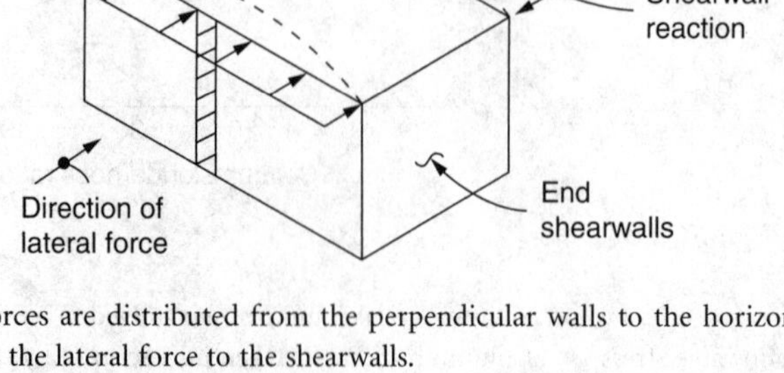

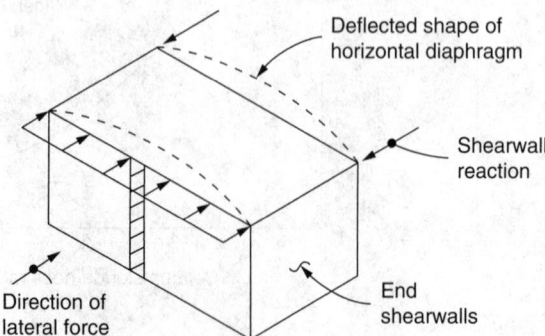

FIGURE 31.4 Lateral forces are distributed from the perpendicular walls to the horizontal diaphragm. The diaphragm in turn transfers the lateral force to the shearwalls.

as a beam that cantilevers vertically from the foundation. A shearwall includes the sheathing, nailing, boundary members, and connections to the horizontal diaphragm and to the foundation.

Defining Terms

Load duration factor, C_D — A multiplying factor used to adjust the allowable stress in a wood member or connection based on the total accumulated length of time that a load is applied to a structure. C_D ranges from 0.9 for long-term (dead) loads to 2.0 for very short-term (impact) loads.

Size factor, C_F — A multiplying factor used to adjust the allowable stress in a wood member based on the dimensions of the cross-section. Depending on the size, type of stress, and grade of lumber, the size factor may be less than, equal to, or greater than unity.

Temperature factor, C_t — A multiplying factor used to adjust the allowable stress in a wood member or connection based on the temperature conditions. C_t is 1.0 for normal temperatures and less than 1.0 for high temperatures.

Wet service factor, C_M — A multiplying factor used to adjust the allowable stress in a wood member or connection based on the moisture content of the wood. C_M is 1.0 for dry service applications and less than 1.0 for high-moisture conditions. The value of C_M depends on the type of stress and material type and may depend on other factors.

References

AF&PA. 1991. *National Design Specification for Wood Construction,* American Forest and Paper Association (formerly the National Forest Products Association), Washington, DC.

AF&PA. 1993. *Commentary on the 1991 Edition of the National Design Specification for Wood Construction,* American Forest and Paper Association (formerly the National Forest Products Association), Washington, DC.

APA. 1986. *Plywood Design Specification,* American Plywood Association, Tacoma, WA.

Breyer, D. E. 1993. *Design of Wood Structures,* 3rd ed. McGraw-Hill, New York.

ICBO. 1991. *Uniform Building Code,* International Conference of Building Officials, Whittier, CA.

Further Information

Information on engineered wood structures may be obtained from a number of sources. Several governmental and industrial organizations are listed below with areas of expertise:

General information on forest products, structural engineering, and wood research:
U.S. Forest Products Laboratory (FPL)
One Gifford Pinchot Drive
Madison, WI 53705

Sawn lumber and connection design:
American Forest and Paper Association (AF&PA)
1111 19th Street N.W., Suite 800
Washington, D.C. 20036

Glu-lam:
American Institute of Timber Construction (AITC)
7012 South Revere Parkway, Suite 140
Englewood, CO 80112

Structural-use panels and glu-lam:
American Plywood Association (APA)
P.O. Box 11700
Tacoma, WA 98411-0700

32

Masonry Design

James E. Amrhein
Consultant, Long Beach, CA

Masonry structures have been constructed since the earliest days of mankind, not only for homes but also for works of beauty and grandeur. Stone was the first masonry unit and was used for primitive but breathtaking structures, such as the 4000-year-old Stonehenge ring on England's Salisbury Plains. Stone was also used around 2500 B.C. to build the Egyptian pyramids in Giza. The 1500-mile (2400-km) Great Wall of China was constructed of brick and stone between 202 B.C. and 220 A.D.

Masonry has been used worldwide to construct impressive structures, such as St. Basil's Cathedral in Moscow and the Taj Mahal in Agra, India, as well as homes, churches, bridges, and roads. In the U.S., masonry was used from Boston to Los Angeles and has been the primary material for building construction from the 18th to the 20th centuries.

Currently, the tallest reinforced masonry structure is the 28-story Excalibur Hotel in Las Vegas, Nevada. This large high-rise complex consists of four buildings, each containing 1008 sleeping rooms. The load-bearing walls for the complex required masonry with a specified compressive strength of 4000 psi (28 MPa).

0-8493-1586-7/05/$0.00+$1.50

TABLE 32.1 Notation

a_b =	Depth of stress block for balanced strength design conditions.	F_a =	Allowable compressive stress due to axial load only, psi.
		F_b =	Allowable compressive stress due to flexure only, psi.
A =	Area of compression area for walls or columns.	F_s =	Allowable tensile or compressive stress in reinforcement, psi.
A_n =	Net cross-sectional area of masonry, in.^{2}.	F_v =	Allowable shear stress in masonry, psi.
A_s =	Area of tension steel.	h =	Effective height of column, wall, or pilaster, in.
A_{se} =	Equivalent area of tension steel considering effect of vertical load.	I =	Moment of inertia of masonry, in.^{4}.
A_v =	Cross-sectional area of shear reinforcement, in.^{2}.	j =	Ratio of distance between centroid of flexural compressive forces and centroid of tensile forces to depth, d.
b =	Width of section, in.^{2}.	k =	Ratio of depth of stress block to depth of section.
b_w =	Width of wall beam.	l =	Clear span between supports.
c_b =	Depth to neutral axis for balanced strength design conditions.	L =	Live load or related internal moments and forces.
C =	Total compression force.	M =	Maximum moment occurring simultaneously with design shear force V at the section under consideration, in.-lb.
d =	Distance from extreme compression fiber to centroid of tension reinforcement, in.	n =	Ratio of the modulus of elasticity of steel to the modulus of elasticity of masonry.
D =	Dead load or related internal moments and forces.	p =	Ratio of tensile steel area to total area of section, bd.
e =	Eccentricity of axial load, in.	P =	Design axial load, lb.
E_m =	Modulus of elasticity of masonry in compression, psi.	P_u =	Factored load on section, strength design.
E_s =	Modulus of elasticity of steel, psi.	r =	Radius of gyration, in.
f =	Calculated stress on section.	s =	Spacing of reinforcement, in.
f_a =	Calculated compressive stress in masonry due to axial load only, psi.	S =	Section modulus.
f_b =	Calculated compressive stress in masonry due to flexure only, psi.	t =	Nominal thickness of wall.
f'_m =	Specified compressive strength of masonry, psi.	T =	Total tension force on section.
f_s =	Calculated tensile or compressive stress in reinforcement, psi.	v =	Shear stress, psi.
f_t =	Calculated tension stress on masonry, psi.	V =	Design shear force.
f_v =	Calculated shear stress in masonry, psi.	w =	Load or weight per unit length or area.
f_y =	Specified yield stress of steel for reinforcement and anchors, psi.	W =	Wind load or related internal moments and forces or total uniform load.

32.1 Basis of Design

This chapter is based on the specification of materials, construction methods, and testing as given in the ASTM standards. The design parameters are in accordance and reprinted with permission from Building Code *Requirements and Commentary for Masonry Structures* (ACI 530-02, ASCE 5-02, TMS 402-02), also referred to as the Masonry Standards Joint Committee (MSJC -02).

In addition, the International Building Code, published by the International Code Council, Falls Church, VA, provides requirements and recommendations for the design and construction of masonry systems both unreinforced (plain) and reinforced. (See Table 32.1.)

32.2 Masonry Materials

The principal materials used in plain masonry are the masonry units, mortar plus grout, and reinforcing steel for reinforced masonry. These materials are assembled into homogeneous structural systems.

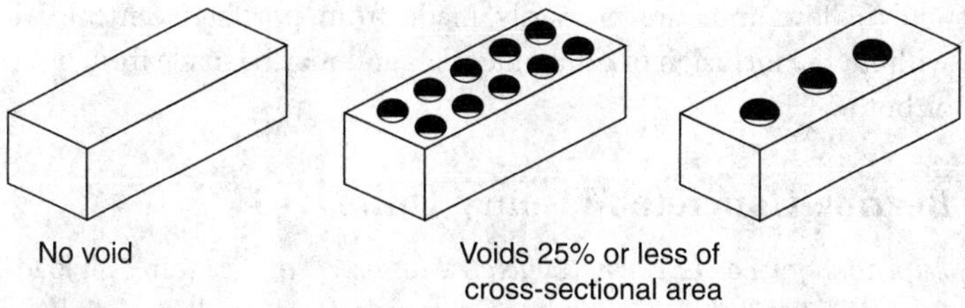

FIGURE 32.1 Solid clay brick.

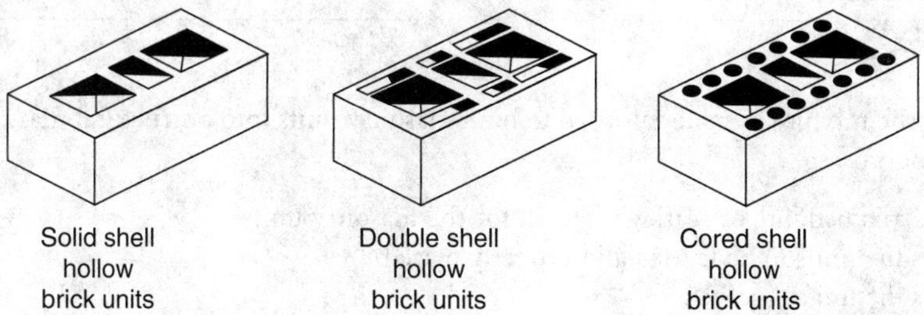

FIGURE 32.2 Hollow clay brick.

32.3 Masonry Units

The masonry units considered are clay brick, concrete brick, hollow clay bricks, and hollow concrete blocks. Masonry units are available in a variety of sizes, shapes, colors, and textures.

Clay Masonry

Clay masonry is manufactured to comply with the ASTM C 62-00, *Specification for Building Brick (Solid Masonry Units Made From Clay or Shale)*; C 216-00, *Specification for Facing Brick (Solid Masonry Units Made From Clay or Shale)*; and C 652-00a, *Specification for Hollow Brick (Hollow Masonry Units Made From Clay or Shale)*. It is made by firing clay in a kiln for 40 to 150 h, depending upon the type of kiln, size and volume of the units, and other variables. For building brick and face brick the temperature is controlled between 1600°F (870°C) and 2200°F (1200°C), whereas the temperature ranges between 2400°F (1315°C) and 2700°F (1500°C) for firebrick.

Solid Clay Units

A solid clay masonry unit, as specified in ASTM C 62-00 and C 216-00, is a unit whose net cross-sectional area, on every plane parallel to the bearing surface, is 75% or more of its gross cross-sectional area measured in the same plane. A solid brick may have a maximum coring of 25%. See Figure 32.1.

Hollow Clay Units

A hollow clay masonry unit, as specified in ASTM C 652-00a, is a unit whose net cross-sectional area in every plane parallel to the bearing surface is less than 75% of its gross cross-sectional area measured in the same plane. See Figure 32.2.

32.4 Concrete Masonry

Concrete masonry units for load-bearing systems can be either concrete brick as specified by ASTM C 55-99, *Specification for Concrete Building Brick*, or hollow load-bearing concrete masonry units as specified by ASTM C 90-00, *Specification for Hollow Load-Bearing Concrete Masonry Units*.

Concrete brick and hollow units are primarily made from portland cement, water, and suitable aggregates with or without the inclusion of other materials and may be made from lightweight or normal weight aggregates or both.

Hollow Load Bearing Concrete Masonry Units

ASTM C 90-00, *Specification for Load Bearing Concrete Masonry Units,* requires all load-bearing concrete masonry units to meet the physical requirements of dimensions, strength, and absorption.

32.5 Mortar

General

Mortar is a plastic mixture of materials used to bind masonry units into a structural mass. It is used for the following purposes:

1. It serves as a bedding or seating material for the masonry units.
2. It allows the units to be leveled and properly placed.
3. It bonds the units together.
4. It provides compressive strength.
5. It provides shear strength, particularly parallel to the wall.
6. It allows some movement and elasticity between units.
7. It seals irregularities of the masonry units.
8. It can provide color to the wall by using color additives.
9. It can provide an architectural appearance by using various types of joints.

Types of Mortar

The requirements for mortar are provided in ASTM C 270-00, *Mortar for Unit Masonry.* There are four types of mortar, which are designated M, S, N, and O. The types are identified by every other letter of the word MaSoNwOrk.

Mortar Proportions

Proportion specifications limit the amount of the constituent parts by volume. Water content, however, may be adjusted by the mason to provide proper workability under various field conditions. The most common cement-lime mortar proportions by volume are:

Type M mortar. 1 Portland cement; π lime; 3∫ sand
Type S mortar. 1 Portland cement; ∫ lime; 4∫ sand
Type N mortar. 1 Portland cement; 1 lime; 6 sand
Type O mortar. 1 Portland cement; 2 lime; 9 sand

32.6 Grout

General

Grout is a mixture of Portland cement, sand pea gravel, and water mixed to fluid consistency so that it will have a slump of 8 to 10 in. (200 to 250 mm). The minimum strength of grout is 2000 psi (13.8 MPa). Requirements for grout are given in ASTM C 476-99, *Grout for Masonry.*

Grout is placed in the cores of hollow masonry units or between wythes of solid units to bind the reinforcing steel and the masonry into a structural system. Additionally, grout provides;

1. More cross-sectional area, allowing a grouted wall to support greater vertical and lateral shear forces than a nongrouted wall
2. Added sound transmission resistance, thus reducing the sound passing through the wall
3. Increased fire resistance and an improved fire rating of the wall
4. Improved energy storage capabilities of a wall
5. Greater weight, thus improving the overturning resistance of retaining walls

Types of Grout

Fine grout is used for small grout spaces or where there is an excess of reinforcing steel. **Coarse grout is the standard grout used in the majority of grouted masonry.**

Proportions of Grout

Fine grout is proportioned 1 part portland cement and 3 parts sand with enough water for an 8- to 11-in. slump. Coarse grout is proportioned: 1 part portland cement, 3 parts sand, and 2 parts pea gravel with enough water for an 8- to 11-in. slump.

32.7 Behavior and Limits States of Structural Masonry

The design methods of structural masonry are based on a range of behavior conditions that define the limits for each of the design methods (Figure 32.3).

32.8 Unreinforced or Plain Masonry

Design of Unreinforced Masonry (MSJC 2.2)

Behavior State 1:

1. Stresses are less than the Modulus of Rupture… Masonry is uncracked.
2. Limit reached when masonry is stressed to the modulus of rupture. It cracks.

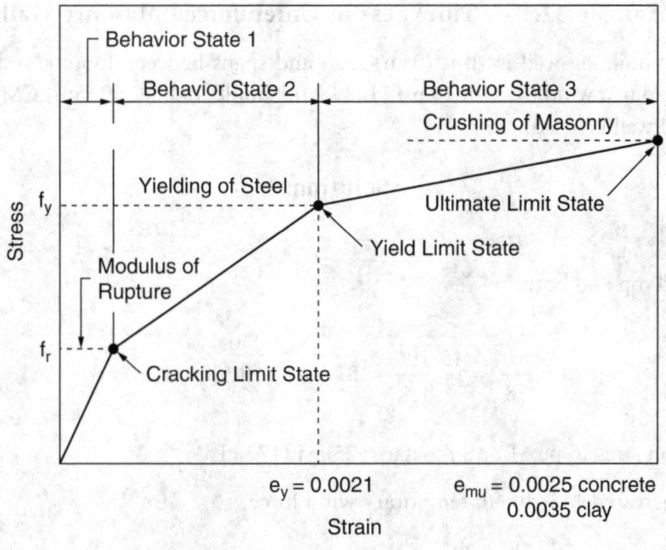

FIGURE 32.3 Behavior and limit states of structural masonry

TABLE 32.2 Allowable Flexural Tension for Clay and Concrete Masonry, psi[a] (kPa)

Masonry Type	Mortar Types: Portland Cement/Lime M or S	kPa	N	kPa	Masonry Cement and Air-Entrained Portland Cement Lime M or S	kPa	N	kPa
Normal to Bed Joints								
Solid units	40	276	30	207	24	166	15	103
Hollow units[a]								
Ungrouted	25	172	19	131	15	103	9	62
Fully grouted	68	470	58	400	41	283	26	180
Parallel to Bed Joints in Running Bond								
Solid units	80	550	60	415	48	330	30	207
Hollow units								
Ungrouted and partially grouted	50	345	38	262	30	207	19	131
Fully grouted	80	550	60	415	48	330	30	207

[a] For partially grouted masonry, allowable stresses shall be determined on the basis of linear interpolation between hollow units, which are fully grouted or ungrouted, and hollow units based on amount of grouting.

3. Masonry resists both tensile and compressive forces.
4. Tensile stress in masonry (mortar) is limited (Table 32.2).
5. Compressive stress in masonry is limited (MSJC 2.2.3.1).
6. Shear stress in masonry is limited (MSJC 2.2.5).

Unreinforced masonry considers the tensile resistance of masonry for the design of structures. The effects of stresses in reinforcement, if present, are neglected and all forces and moments are resisted by the weight of the masonry and the tension and compression capabilities of the system.

The stress due to flexural moment is $f_t = M/S$, where M is the moment on the wall and S is the section modulus. The condition is generally limited to the allowable flexural tension stress shown in Table 32.2.

Example 32.1 — Thickness of Unreinforced Masonry Wall

An unreinforced building wall is 10 ft (3 m) high and spans between footing and roof ledger. It could be subjected to a wind force of 15 psf (103 kPa). Should an 8″ (200 mm) CMU wall or a 12″ (300 mm) CMU wall be used?

Solution:

Moment on wall

Assumed pinned top and bottom

$$M = \frac{wh^2}{8} = \frac{15 \cdot 10^2}{8} = 187.5 \text{ ft} \cdot \text{lb/ft } (834 \text{ N} \cdot \text{m/m})$$

Allowable tension stress type M or S mortar = 25 psi (172 kPa)

Stress may be increased 1/3 due to temporary wind force

$$S = \frac{M}{1.33 f_t} = \frac{187 \cdot 5 \times 12}{1.33 \times 25} = 67.5 \text{ in.}^3 (1.1 \cdot 10^6 \text{ mm}^3)$$

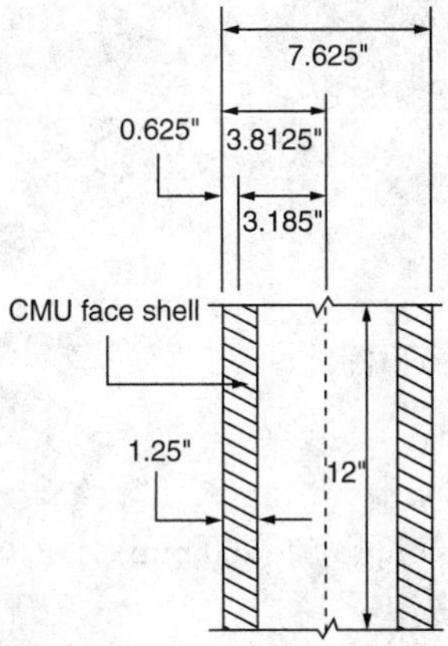

FIGURE 32.4 Vertical plane cross-section of 8" CMU wall, face shells only.

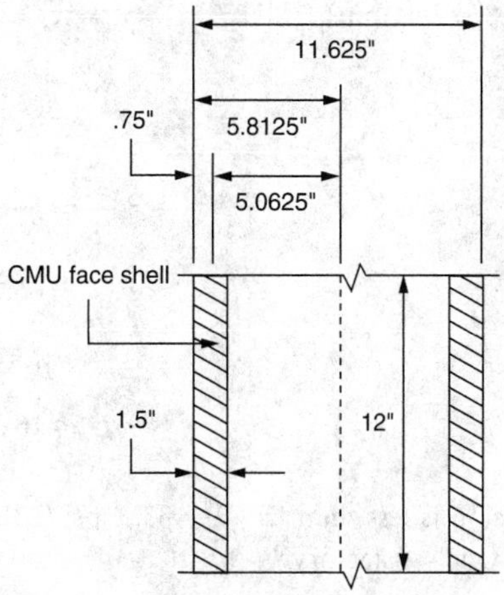

FIGURE 32.5 Vertical plane cross-section of 12" CMU wall, face shells only.

For section modulus at 8″ (200 mm) CMU (see Figure 32.4):

$$I = 2\left[\frac{bd^3}{12} + bdx^2\right] = 2\left[\frac{12\times1.25^3}{12} + 12\times1.25\times3.1875^2\right]$$

$$= 2[1.95 + 152.40] = 308.7 \text{ in.}^4 (128.5\cdot10^6 \text{ mm}^4)$$

$$S = \frac{308.7\times2}{7.625} = 81 \text{ in.}^3 (1.3\cdot10^6 \text{ mm}^3)$$

At 12″ (300 mm) CMU (see Figure 32.5):

$$I = 2\left[\frac{bd^3}{12} + bdx^2\right]$$

$$= 2\left[\frac{12\times1.5^3}{12} + 12\times1.5\times5.0625^2\right]$$

$$= 2[3.375 + 461.3] = 929.4 \text{ in.}^4 (386.8\cdot10^6 \text{ mm}^4)$$

$$S = \frac{I}{t/2} = \frac{2\times929.4}{11.625} = 159.9 \text{in.}^3 (2.6\cdot10^6 \text{mm}^3)$$

Thus, 8″ CMU = 81.0 in.^{3} (1.3 · 10^6 mm^3); 12″ CMU = 159.9 in.^{3} (2.6 · 10^6 mm^3).
Use 8″ (200 mm) CMU.

Example 32.2 — Vertical and Lateral Load on Unreinforced Masonry Wall

If a wall is subjected to a 20 psf (958 kPa) wind, and is 15 ft (4.6 m) high, and carries 2000 plf (96 kPa), what thickness concrete masonry unit should be used?

Solution:

$$M = \frac{wh^2}{8} = \frac{20\times15^2}{8} = 563 \text{ ft}\cdot\text{lb/ft} (2500 \text{ N}\cdot\text{m/m})$$

Try 8″ (200 mm) CMU.

$$S = 81 \text{ in.}^3 (1.3 \cdot 10^6 \text{ mm}^3)$$

$$\frac{P}{A} \pm \frac{M}{S} = \frac{2000}{2 \times 12 \times 1.25} \pm \frac{563 \times 12}{81}$$

$$= 66.7 \pm 83$$

$$= 150 \text{ psi compression (1.0 MPa)}$$

$$= 16.3 \text{ psi tension (115 kPa)}$$

Tension is less than $25 \times \frac{4}{3} = 33.3$ psi (230 kPa) allowable tension, so 8″ (200 mm) CMU is satisfactory.

32.9 Strength of Masonry

The ultimate compressive strength of the masonry assembly is given the symbol f'_{mu} to distinguish it from the specified compressive strength f'_m. See Table 32.3 and Table 32.4 for the allowable specified strength of masonry compared to the strength of the masonry unit.

Modulus of Elasticity

For steel reinforcement, E_s = 29,000,000 psi (199 955 MPa).

For concrete masonry, $E_m = 900 f_m'$ (see Table 32.3).

For clay masonry, $E_m = 750 f_m'$ (see Table 32.4).

TABLE 32.3 Concrete Masonry — Compressive Strength, Modulus of Elasticity, Modular Ratio

Net Area Compressive Strength of Concrete Masonry Units, psi (MPa)		Net Area Compressive Strength of Masonry, psi[a] (MPa)	Modulus of Elastciity 900 f'_m psi[a] (MPa)	Modular Ratio $n = E_s/E_m$
Type M or S Mortar	Type N Mortar			
1250 (8.62)	1300 (8.96)	1000 (6.90)	900,000 (6207)	32.2
1900 (13.10)	2150 (14.82)	1500 (10.34)	1,350,000 (9310)	21.5
2800 (19.31)	3050 (21.03)	2000 (13.79)	1,800,000 (12414)	16.1
3750 (25.86)	4050 (27.92)	2500 (17.24)	2,250,000 (15518)	12.9
4800 (33.10)	5250 (36.50)	3000 (20.69)	2,700,000 (18621)	10.7

[a] For units of less than 4 in. (102 mm) height, 85% of the values listed.

TABLE 32.4 Clay Masonry — Compressive Strength, Modulus of Elasticity, Modular Ratio

Net Area Compressive Strength for Concrete Masonry Units, psi (MPa)		Net Area Compressive Strength of Masonry, psi (MPa)	Modulus of Elasticity 750 f'_m psi (MPa)	Modular Ratio $n = E_s/E_m$
Type M or S Mortar	Type N Mortar			
1700 (11.72)	2100 (14.48)	1000 (6.90)	750,000 (5173)	38.7
3350 (23.10)	4150 (28.61)	1500 (10.34)	1,125,000 (7759)	25.8
4950 (34.13)	6200 (42.75)	2000 (13.79)	1,500,000 (10,345)	19.3
6600 (45.51)	8250 (56.88)	2500 (17.24)	1,875,000 (12,931)	15.5
8250 (56.88)	10,300 (71.02)	3000 (20.69)	2,250,000 (15,518)	12.9
9900 (68.26)	—	3500 (24.13)	2,625,000 (18,104)	11.0
13,200 (91.01)	—	4000 (27.58)	30,000,000 (20,690)	9.7

Specified Compressive Strength

For specified compressive strength of:

Concrete masonry, see Table 32.3
Clay masonry, see Table 32.4

Reinforcing Steel

Reinforcing steel in masonry has been used extensively in the West since the 1930s, revitalizing the masonry industry in earthquake-prone areas. Reinforcing steel extends the characteristics of ductility, toughness, and energy absorption that are so necessary in structures subjected to the dynamic forces of earthquakes.

Reinforcing steel may be either Grade 40, with a minimum yield strength of 40,000 psi (276 MPa), or Grade 60, minimum yield strength of 60,000 psi (414 MPa).

Allowable stresses for reinforcing steel are as follows. Tension stress in reinforcement shall not exceed the following:

Grade 40 or Grade 50 reinforcement	20,000 psi (138 MPa)
Grade 60 reinforcement	24,000 psi (165 MPa)
Wire joint reinforcement	30,000 psi (207 MPa)

Compression stress has these restrictions:

1. The compressive resistance of steel reinforcement is neglected unless lateral reinforcement is provided to tie the steel in position.
2. Compressive stress in reinforcement may not exceed the lesser of 0.4 f_y or 24,000 psi (165 MPa).

32.10 Design of Reinforced Masonry Members

Behavior State 2 — Working Stress Design (MSJC 2.3, IBC 2107):

1. Modulus of rupture is exceeded. The masonry is cracked.
2. Cracked cross-section with steel strain is less than the yield strain.
3. Reinforcing steel resists tensile forces.
4. Masonry resists compressive forces.
5. Stresses are well within the elastic range.
6. Compressive stresses are limited in masonry (MSJC 2.3.3).
7. Tensile stresses in reinforcing steel is limited (MSJC 2.3.2).

Reinforced masonry members are designed by elastic analysis using service loads and permissible stresses, which considers that the reinforcing steel resists tension forces and the masonry and grout resist compression forces.

The design and analysis of reinforced masonry structural systems have been by the straight line, elastic working stress method. In working stress design (WSD), the limits of allowable stress for the materials are established based on the properties of each material (Figure 32.6):

Yield Limit State for Behavior State 2
- Steel ratio limited to ensure ductile behavior.
- Steel yields well before masonry crushes.
- Limit reached at steel first yield.

The procedure presented is based on the working stress or straight line assumptions where all stresses are in the elastic range and:

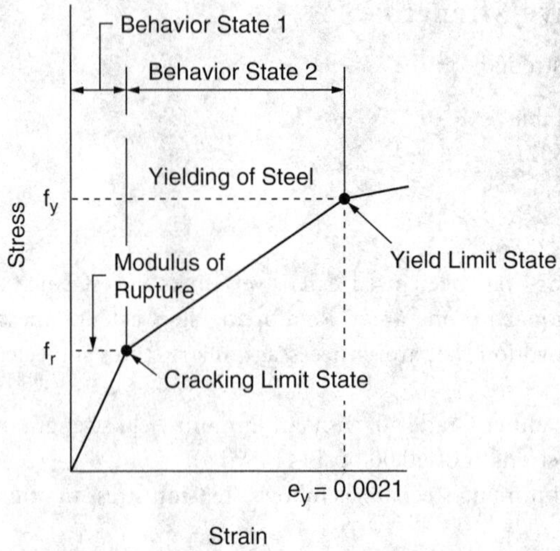

FIGURE 32.6 Range of behavior state 2.

1. Plane sections before bending remain plane during and after bending.
2. Stress is proportional to strain, which is proportional to distance from the neutral axis.
3. Modulus of elasticity is constant throughout the member.
4. Masonry carries no tensile stresses.
5. Span of the member is large compared to the depth.
6. Masonry elements combine to form a homogeneous and isotropic member.
7. External and internal moments and forces are in equilibrium.
8. Steel is stressed about the center of gravity of the bars equally.
9. The member is straight and of uniform cross-section.

Flexural Design

The basis of the flexural equations for working stress design, WSD, is the concept of the modular ratio. The modular ratio, n, is the ratio of the modulus of elasticity of steel, E_s, to the modulus of elasticity of masonry, E_m.

$$n = \frac{E_s}{E_m}$$

By use of the modular ratio, n, the steel area can be transformed into an equivalent masonry area. The strain is in proportion to the distance from the neutral axis and therefore the strain of the steel can be converted to stress in the steel. In order to establish the ratio of stresses and strains between the materials, it is necessary to locate the neutral axis.

The location of the neutral axis is defined by the dimensions, kd, which are dependent on the modular ratio, n, and the reinforcing steel ratio, $p = A_s/bd$. For a given modular ratio, n, the neutral axis can be raised by decreasing the amount of steel (reducing p) or lowered by increasing the amount of steel (increasing p). See Figure 32.7 and Figure 32.8.

Solving for k,

$$k = \sqrt{(np)^2 + 2np} - np$$

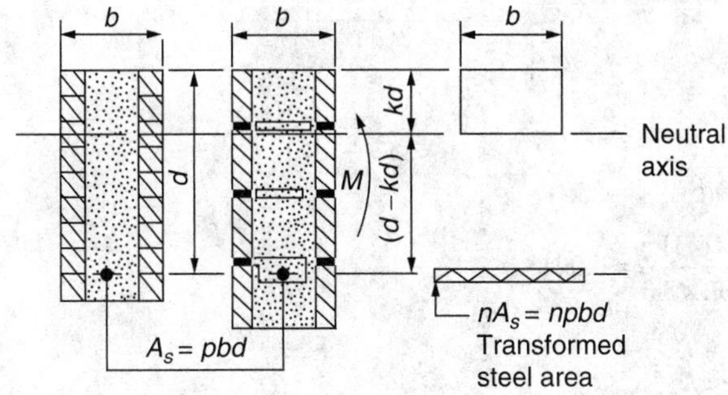

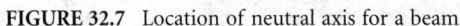

FIGURE 32.7 Location of neutral axis for a beam.

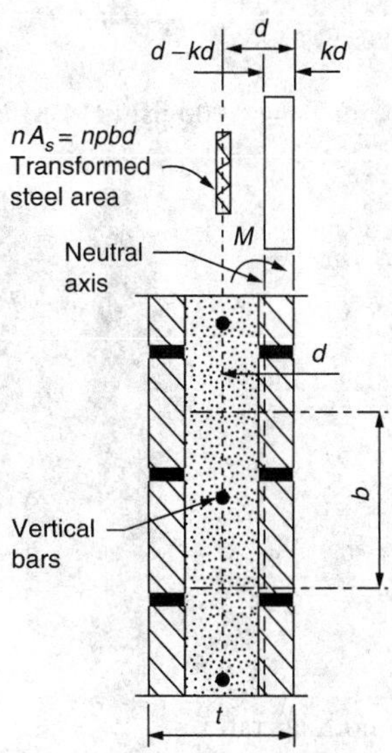

FIGURE 32.8 Location of neutral axis for a wall, in plane of wall.

Moment Capacity of a Section

The moment capacity of a reinforced structural masonry wall or beam can be limited by the allowable masonry stress (over-reinforced), allowable steel stress (under-reinforced), or both, in which case it would be a balanced design condition.

When a member is designed for the specified loads, and the masonry and reinforcing steel are stressed to their maximum allowable stresses, the design is said to be a "balanced" design. This balanced design is different from the balanced design for the strength design method. For working stresses, balanced design occurs when the masonry is stressed to its maximum allowable compressive stress and the steel is stressed to its maximum allowable tensile stress.

However, in many cases, the "balanced" design does not satisfy the conditions for the materials available or for the predetermined member size or the economy of the project. It may be advantageous to under-stress (under-reinforce) the masonry or under-stress (over-reinforce) the steel so that the size of the member can be maintained.

The moment capability of a section based on the steel stress is defined as M_s = force × moment arm, where force in the steel, $T = A_s f_s = pbdf_s$; moment arm $= jd$; $M_s = T \times jd = A_s f_s jd$; and $M_s = pbdf_s jd = f_s pjbd^2$.

The moment capability of a section based on the masonry stress is defined as M_m = force × moment arm, where force in the masonry, $C = {}^1/_2\, f_b(kd)b = {}^1/_2\, f_b kbd$; moment arm $= jd$; $M_m = C \times jd = ({}^1/_2\, f_b kbd) \times (jd)$; $M_m = {}^1/_2\, f_b kjbd^2$.

Example 32.3 — Determination of Moment Capacity of a Wall

A partially grouted 8″ (200 mm) concrete masonry wall with type S mortar is reinforced with #5 bars at 32″ (813 mm) o.c. The steel is 5.3″ (135 mm) from the compression face and is Grade 60. If t_m = 2500 psi (17.2 MPa), what is the moment capacity of the wall?

Solution

For f'_m = 2500 psi (17.2 MPa), allowable masonry stress,

$$F_b = 0.33\, f'_m = 833 \text{ psi (5.6 MPa) max. allowable compression}$$

$$E_m = 900\, f'_m = 2{,}250{,}000 \text{ psi } (15{,}500 \text{ MPa}) \text{ (see Table 32.3)}$$

Also, for $f_y = 60{,}000$ psi (414 MPa),

$$F_s = 24{,}000 \text{ psi } (165 \text{ MPa}) \text{ max. allowable tension}$$

$$E_s = 29{,}000{,}000 \text{ psi } (199\ 955 \text{ MPa})$$

For steel ratio,

$$p = \frac{A_s}{bd}$$

$$= \frac{0.31}{32 \times 5.3} = 0.0018$$

For modular ratio,

$$n = \frac{E_s}{E_m}$$

$$= \frac{29\ 000\ 000}{2\ 250\ 000} = 12.9$$

Furthermore,

$$np = 12.9 \times 0.0018 = 0.023$$

$$k = \sqrt{(np)^2 + 2n} - np$$

$$= \sqrt{(0.0023)^2 + 2 \times 0.0023} - 0.0023$$

$$= 0.193$$

$$kd = 0.193 \times 5.3 = 1.03 \text{ in. } (26 \text{ mm})$$

The neutral axis falls on the shell of CMU. Shell thickness = 1.25 inches.

$$j = 1 - \frac{k}{3} = 1 - \frac{0.193}{3} = 0.936$$

$$M_m = \frac{1}{2} f_b kjbd^2 = \frac{1}{2}(833)(0.193)(0.936)(12)(5.3)^2$$

$$= 25.362 \text{ in.-lb /ft } (9208 \text{ N}\cdot\text{m/m})$$

$$= 2.11 \text{ ft k/ft } (9128 \text{ N}\cdot\text{m/m})$$

$$M_s = f_s pjbd^2 = 24\ 000(0.0021)(0.936)(12)(5.3)^2$$

$$= 15\ 902 \text{ in.-lb /ft } (5800 \text{ N}\cdot\text{m/m})$$

$$= 1.33 \text{ ft k/ft } (5800 \text{ N}\cdot\text{m/m}) \leftarrow \text{Controls}$$

Beam Shear

Structural elements such as beams, piers, and walls are subjected to shear forces as well as flexural stresses. The unit shear stress is computed based on the formula

$$f_v = V/bd$$

When masonry members, i.e., beams, are designed to resist shear forces without the use of shear-reinforcing steel, the calculated shear stress is limited to 1.0 $(f'_m)^{1/2}$, 50 psi max. If the unit shear stress exceeds the allowable masonry shear stress, reinforcing steel must resist all the shear stress.

For flexural members, beams, with reinforcing steel resisting all the shear forces, the maximum allowable shear stress is 3.0 $(f'_m)^{1/2}$ psi with 150 psi as a maximum. The steel resists the shear by tension, and it must be anchored in the compression zone of the beam or the wall.

The unit shear, f_v, is used to determine the shear steel spacing based on the formula:

$$\text{Spacing, } s = \frac{A_v F_s}{f_v b}$$

$$\text{Unit shear stress, } f_v = \frac{A_v F_s}{b_s}$$

For continuous or fixed beams, the shear value used to determine the shear steel spacing may be taken at a distance $d/2$ from the face of the support. The shear value at the face of the support should be used to calculate the shear steel spacing in simple beams.

The maximum spacing of shear steel should not exceed $d/2$. The first shear-reinforcing bar should be located at half the calculated spacing but no more than $d/4$ from the face of support.

Vertical Load on a Wall

A load-bearing wall carries the load of the roof plus the load of floor above and is limited to the h/r of the wall. The h/r is the effective height of the wall divided by the radius of gyration, $r = (I/A)^{1/2}$

For walls with h/r not greater than 99 the load limitation is:

$$P_a = (0.25 f'_m A_n + 0.65 A_{st} F_s \left[1 - \left(\frac{h}{140r}\right)^2\right]$$

For walls with h/r greater than 99 the load limitation is:

$$P_a = (0.25 f'_m A_n + 0.65 A_{st} F_s)\left(\frac{70r}{h}\right)^2$$

Although there is a term for reinforcing steel in the equation, it is ignored because the steel is not tied to prevent buckling or displacement. Thus the maximum stress in the wall is 0.25 f'_m times the reduction factor.

Interaction of Combined Load and Moment, P/M

The direct and simple design or analysis of interaction conditions is to use the unity equation:

$$\frac{f_a}{F_a} + \frac{f_b}{F_b} \leq 1.00 \text{ or } 1.33$$

Wall Shear

For shear walls where M/Vd is less than one the masonry shear stress shall not exceed:

$$F_v = (\tfrac{1}{3})[4-(M/Vd)]\sqrt{f'_m}$$

but shall not exceed $80-45(M/Vd)$ psi

For shear walls where M/Vd is equal to or greater than one the masonry shear stress shall not exceed:

where $M/Vd \geq 1$,

$$F_v = \sqrt{f'_m}$$

but shall not exceed 35 psi (241 kPa).

When shear reinforcing is used to carry all the shear force.
For shear walls where M/Vd is less than one the shear stress shall not exceed:

where $M/Vd < 1$,

$$F_v = (\tfrac{1}{2})[4-(M/Vd)]\sqrt{f'_m}$$

but shall not exceed $120-45(M/Vd)$ psi

For shear walls where M/Vd is equal to or greater than one the shear stress shall not exceed:

where $M/Vd \geq 1$,

$$F_v = 1.5\sqrt{f'_m}$$

but shall not exceed 75 psi (517 kPa).

Columns

Columns are vertical members that basically support vertical loads. They may be plain masonry or reinforced masonry. Reduction in the load-carrying capacity is based on the h/r ratio, where h is the unbraced height and r is the minimum radius of gyration for the unbraced height.

The reduction factor for members having an h/r ratio not exceeding 99 is $[1-(h/140r)^2]$. For members with an h/r greater than 99 the factor is $(70r/h)^2$. The maximum allowable axial stress for walls or plain columns is $F_a \frac{1}{4} f'_m$ times the reduction factor.

The maximum allowable axial load on a reinforced masonry column is:

$$P_a = (0.25 f'_m A_e + 0.65 A_s F_{sc}) \text{ (reduction factor)}$$

and is limited to $P_a \leq 1/4 P_e$, where

$$P_e = \frac{\pi^2 E_m I}{h^2}\left(1-0.577\frac{e}{r}\right)^3$$

The maximum allowable unit axial stress is $F_a = P_a/A_e$.

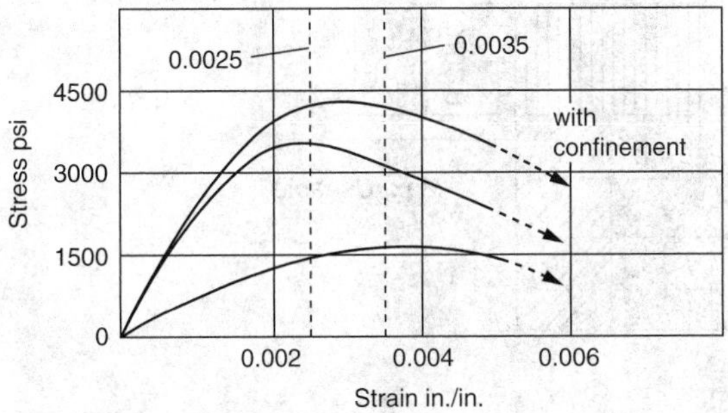

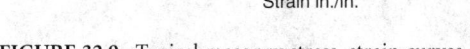

FIGURE 32.9 Typical masonry stress–strain curves.

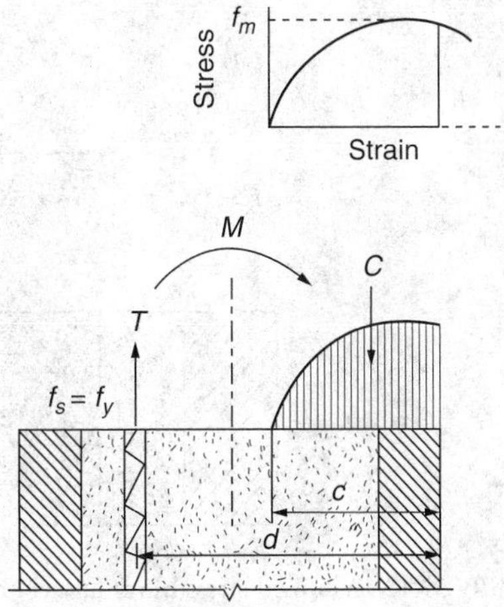

FIGURE 32.10 Stress due to flexural moment for balanced condition.

The reduction factor based on the h/r ratio is the same for reinforced columns and for walls. The same consideration is made for the determination of the effective height, h.

The effective thickness, t, is the specified thickness in the direction considered. For nonrectangular columns the effective thickness is the thickness of a square column.

32.11 Design of Structural Members

Strength Design — Behavior State 3

General

The structural design of reinforced masonry is changing from the elastic working stress method to strength design procedures.

The concept of strength design states that, when a reinforced masonry section is subjected to high flexural moments, the masonry stress from the neutral axis to the extreme compression fibers conforms to the stress–strain curve of the materials as if it were being tested in compression. See Figure 32.9 and Figure 32.10.

It also states that when the tension reinforcing reaches its yield stress, it will continue to elongate without an increase in moment or forces. This condition occurs at the yield plateau of the steel as shown on the stress–strain curve in Figure 32.11.

At the ultimate limit state, the steel is above yield strain and the masonry just reaches the maximum usable strain (0.0025 concrete; 0.0035 clay).

The compressive stress block of the masonry, as shown in Figure 32.12, is simplified from the curved or parabolic shape to a rectangular configuration. This rectangular stress block, which is called *Whitney's stress block*, is approximated as having a length of $a = 0.80\ f'_m$ and a height of $0.80\ c$.

Masonry systems have compression stress–strain curves similar to those of concrete, in that the curves are curved or parabola-shaped and that they reach the limit of usable strain.

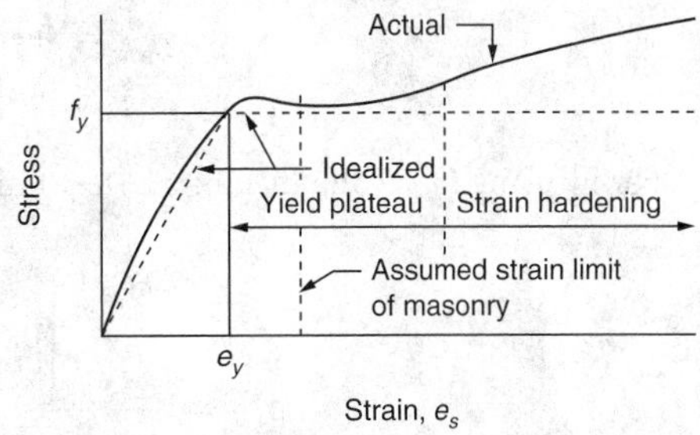

FIGURE 32.11 Idealized stress–strain diagram for reinforcing steel.

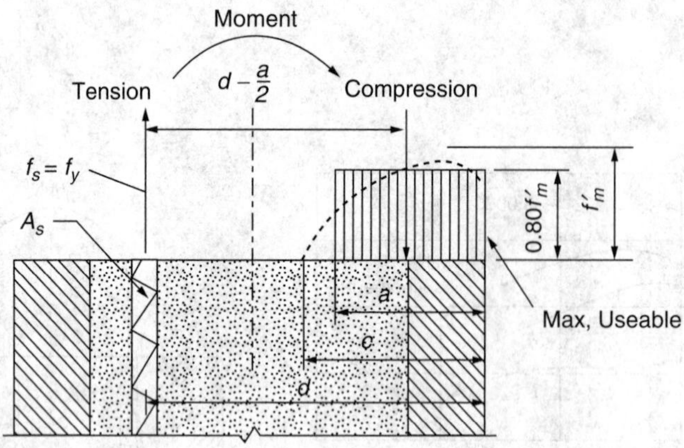

FIGURE 32.12 Assumed stress block at yield condition.

Strength Design Procedure

Two conditions are included in strength design: load parameters and design parameters.

Load Parameters

Service or actual loads are generally used for working stress design procedures. For strength design procedures, the actual or specified loads are increased by appropriate load factors. These load factors consider live and dead load, wind, earthquake, temperature, settlement, and earth pressure. (See IBC Sec. 1605.)

In addition to load factors, a capacity reduction factor, ϕ, is used to adjust for the lack of perfect materials, strength, and size. The phi factor also varies for the stress considered, whether flexural or shear.

Design Parameters

The parameters for strength design are as follows:

1. The steel is at yield stress.
2. The masonry stress block is rectangular.
3. The masonry strain is limited to 0.003 in./in.
4. The steel ratio, p, is limited to 50% of the balanced reinforcing ratio, p_b, to ensure that a ductile mechanism forms prior to brittle, crushing behavior.

For out-of-plane forces the maximum steel ratio is limited to:

$$\rho_{\max} = \frac{\left[0.64 f'_m\left(\frac{\varepsilon_{mu}}{\varepsilon_{mu}+1.3\varepsilon_y}\right)-\frac{N_u}{bd}\right]}{1.25 f_y}$$

For in-plane forces the maximum steel ratio is limited to:

$$\rho_{\max} = \frac{\left[0.64 f'_m\left(\frac{\varepsilon_{mu}}{\varepsilon_{mu}+5\varepsilon_y}\right)-\frac{N_u}{bd}\right]}{\left[1.25 f_y\left(\frac{5\varepsilon_y}{\varepsilon_{mu}+5\varepsilon_y}\right)-\frac{1}{2}\left(\frac{\varepsilon_{mu}}{\varepsilon_{mu}+5\varepsilon_y}\right)\varepsilon_{mu}E_s\right]}$$

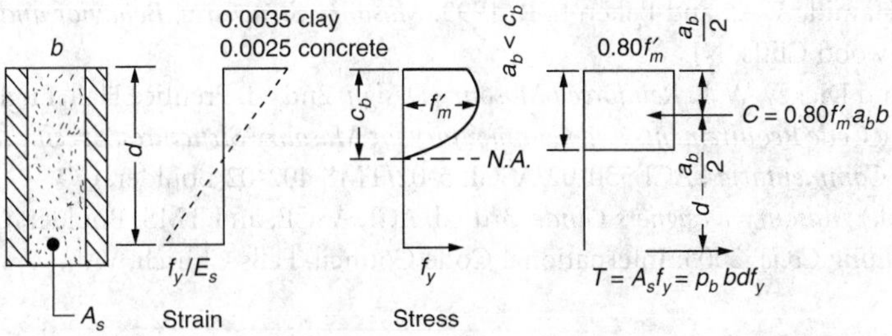

FIGURE 32.13 Strain–stress distribution on a flexural member at balanced conditions.

Strength Design for Sections with Tension Steel Only

As stated above, the limits for flexural design using strength methods are that the stress in the steel is at yield strength and that the strain in the concrete masonry is at 0.0025 and 0.0035 for clay masonry. When these conditions occur at the same moment, the section is considered to be a balanced design. See Figure 32.13.

The depth to the neutral axis, c_b, for a balanced design for concrete masonry is:

$$c_b = \frac{0.0025}{0.0025 + f_y / E_s} d = \frac{72\,500}{72\,500 + f_y} d = 0.547d$$

For clay masonry:

$$C_b = \frac{0.0035}{0.0035 + f_y / E_s} d = \frac{101500}{101500 + f_y} d = 0.629d$$

Defining Terms

Allowable work stress design or elastic design — A technique based on and limiting the stress in the structural element to a value that is always in the elastic range.

Brick — Solid unit ≤ 25% void; hollow unit > 25% and < 75% void.

Grout — Material to tie reinforcing steel and masonry units together to form a structural system.

Maximum usable strain — The limit of the masonry just before crushing.

Modular ratio — Ratio between the modulus of elasticity of steel to the modulus of elasticity of masonry.

Mortar — Plastic material between units in bed and head joints.

Steel ratio — Area of steel to area of masonry.

Strength design — A technique based on capacity of structural section considering the maximum strain in masonry, yield strength of steel, load factors for various loads considered, and phi factors for materials and workmanship.

References

Amrhein, J. E. 1994. *Reinforced Masonry Engineering Handbook,* 5th ed. Masonry Institute of America, Los Angeles, CA.

Amrhein, J. and Neville, G. *Reinforced Masonry Design, Seminar Notes No. 477i,* International Conference of Building Officials, Whittier, CA.

Amrhein, J., Chrysler, J., and Wakefield, D. *Masonry Quality Control and Field Practices, Seminar Notes No. 475i,* International Conference of Building Officials, Whittier, CA.

Beall, C. 1984. *Masonry Design and Detailing,* Prentice Hall, Englewood Cliffs, NJ.

Drysdale, R. G., Hamid, A. A., and Baker, L. R. 1993. *Masonry Structures, Behavior and Design,* Prentice Hall, Englewood Cliffs, NJ.

Schneider, R. A. and Dickey, W. L. *Reinforced Masonry Design,* 2nd ed. Prentice Hall, Englewood Cliffs, NJ.

MSJC 03. *Building Code Requirements and Commentary for Masonry Structures; Specifications for Masonry Structures; Commentaries,* ACI 530-02/ASCE 5-02/TMS 402-02 Boulder, CO

Matthys. J. H. (Ed.) *Masonry Designers Guide,* 3rd ed. ACI, ASCE, and TMS, Boulder, CO.

International Building Code, 2003. International Code Council, Falls Church, VA.

33

Nonlinear Dynamics of Continuous Mass Structural Systems

Bulent A. Ovunc
University of Louisiana at Lafayette

The nonlinearity of a structure under external disturbances may be due to large deformations, which cause geometrical nonlinearity, and/or it may be due to stresses increasing beyond their proportional limits, thus causing material nonlinearity occurring within the entire or a part of the structure. The actual stress–strain variation is represented by continuous functions, depending on the material used. For beams, the nonlinear regions appearing in their spans are determined within the analysis. The dynamic analysis can be based on a linear acceleration approach and can be performed by either an incremental or an iterative process or by using an iterative process within each increment. It can be noticed that at a section, the axial, the shearing forces, the bending moment, and the axial transversal displacements and rotation are interrelated. The effect of the nonlinearities becomes infinitesimal, negligible, when the geometrical and material nonlinearities tend to zero.

33.1 Procedure of Analysis

In the analysis it is assumed that the material is homogeneous and isotropic and the stain variation over the cross-section is linear. The internal force and moments are determined from two different sources — the equilibrium of the infinitesimal element on the deformed configuration, which is related to the geometrical nonlinearity, and from the stress distribution over the cross-section, which is related to the material nonlinearity.

0-8493-1586-7/05/$0.00+$1.50

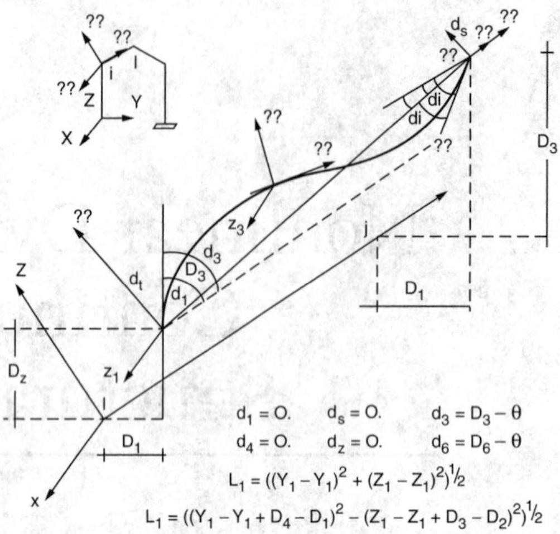

d1 = O.
ds = O.
d3 = D3 − θ
d4 = O.
dz = O.
d6 = D6 − θ
L1 = ((Y1 − Y1)2 + (Z1 − Z1)2)½
L1 = ((Y1 − Y1 + D4 − D1)2 − (Z1 − Z1 + D3 − D2)2)½

FIGURE 33.1 Axes systems in a frame.

33.2 Geometrical Behavior

The geometrical behavior of structures depends on the magnitudes of their deformations under external disturbances. The geometrical behavior is treated through three different axes systems. The joints of the structure are referred to by the Total Lagrangian axes system, XYZ. At an intermediate point within a member, the strains and the stresses are expressed with respect to the Eulerian axes system, x_e, y_e, z_e, whereas the displacements and rotations are expressed with respect to the Updated Lagrangian axes system, x_u, y_u, z_u. See Figure 33.1.

The abscissa along the Updated Lagrangian axes system, x_u, y_u, z_u (Figure 33.1), can be written in terms of the of axial displacement u(), as

$$= y_u + u() \tag{33.1}$$

The point C is located on the neutral axis of the infinitesimal element CD, of length dy_u, on the undeformed configuration. The point A, at distance z from the point C, moves to a point A′, on the deformed configuration C′D′. The point R′ is the intersection of the lines passing through the cross-sections at C′ and D′. Notice that the distances R′C′, ρ′, and R′D′, ρ″ are different than the radius of curvature ρ of the element C′D′. The point S′ is the intersection of the lines perpendicular to the cross-sections at C′and D′.

The relations among the axial and shearing force P(ξ), and V(ξ) and the increment dM/dξ on bending moment are obtained from the equilibrium of the infinitesimal element C′D′ on the deformed configuration. By considering the sum of the moments about points R′ and S′ and the sum of the forces about two axes, and by eliminating higher order differentials, one has

$$dM/d\xi = (1 + (w')^2)^{1/2} V(\xi) \tag{33.2}$$

$$dM/d\xi = -\rho \, dP/d\xi \tag{33.3}$$

$$dV/d\xi = (1/\rho)(1 + (w')^2)^{1/2} P(\xi) + (q_z w' + q_y)/(1 + (w')^2)^{1/2} \tag{33.4}$$

$$dP/d\xi = -(1/\rho)(1 + (w')^2)^{1/2} V(\xi) + (q_z - q_y w')/(1 + (w')^2)^{1/2} \tag{33.5}$$

where q_z and q_y are the distributed external loads on the infinitesimal element C′D′ on the deformed configuration.

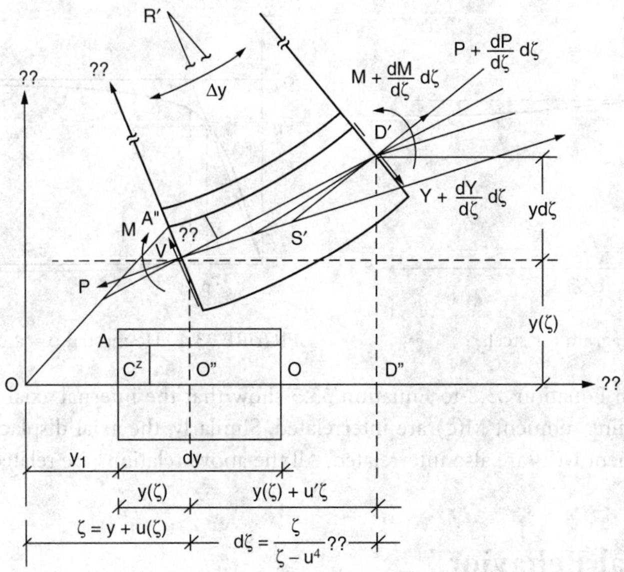

FIGURE 33.2 Infinitesimal element on deformed configuration.

For the dynamic analysis based on the continuous mass matrix method [Clough and Penzien, 1993; Ovunc, 1994],

$$q_z = m(d^2w/dt^2) \text{ and } q_y = m(d^2u/dt^2) \tag{33.6}$$

where m is the mass of the element per unit length.

The integration of the two simultaneous equations, Equation 33.4 and Equation 33.5, with $q_z = m(d^2w/dt^2)$ and $q_y = m(d^2u/dt^2)$ yields:

$$P(\xi) = ((1 + (w')^2)^{-1/2})[-C_1w' + C_2 - \xi Lm((d^2w/dt^2)w' + (d^2u/dt^2)] \tag{33.7}$$

$$V(\xi) = ((1 + (w')^2)^{-1/2})[C_1 + C_2w' - \xi Lm((d^2w/dt^2) - d^2u/dt^2)\, w')] \tag{33.8}$$

so that the increment $dM/d\xi$ on the bending moment [Equation (33.2)] can be obtained as

$$dM/d\xi = [C_1 + C_2w' - \xi Lm((d^2w/dt^2) - (d^2u/dt^2)\, w')] \tag{33.9}$$

If it is assumed that the inertia forces on the member are zero ($q_z = q_y = 0$), the above equations related to the geometrical behavior can be written as

$$dM/d\xi = [C_1 + C_2w'\,] \tag{33.10}$$

$$P(\xi) = ((1 + (w')^2)^{-1/2})[-C_1w' + C_2] \tag{33.11}$$

$$V(\xi) = ((1 + (w')^2)^{-1/2})[C_1 + C_2w'\xi] \tag{33.12}$$

The relation between the internal axial force $P(\xi)$, and shearing force $V(\xi)$ [Equation (33.11) and Equation (33.12)] can be obtained as

$$P^2 + V^2 = C^2 \tag{33.13}$$

where C is constant.

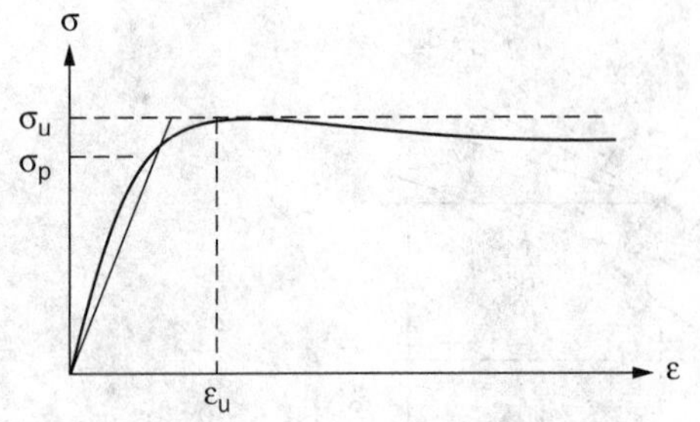

FIGURE 33.3 $\sigma - \varepsilon$ diagram for steel.

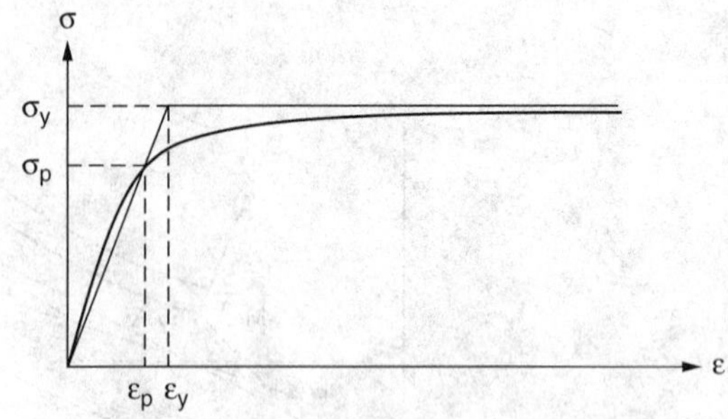

FIGURE 33.4 Hognestad $\sigma - \varepsilon$ diagram for concrete.

The relations from Equation 33.2 to Equation 33.5 show that the internal axial force $P(\xi)$, shearing force $V(\xi)$, and bending moment $M(\xi)$ are interrelated. Similarly, the axial displacement $u(\xi)$ and the transversal displacement $w(\xi)$ are also interrelated. All the above relations are related to the geometrical behavior.

33.3 Material Behavior

Material behavior depends on the variation of the stresses along the cross-section of a member. Under external disturbances, the stress in a member may vary from linear to nonlinear elastic to plastic stages. The variation of the stress versus the strain is expressed by a continuous function, corresponding to the characteristics of the material. Thus the normal stress σ_z at a distance z from the neutral axis of a section of a ductile material subjected to axial force and bending moment is expressed as [Ovunc, 1992] (Figure 33.3)

$$\sigma_z = a_\sigma \tanh(\alpha + \beta z)[1 + (\beta z)^2/(3\cosh^2(\alpha + \beta z))] \tag{33.14}$$

where $\alpha = (1/(c_\sigma L))(du/d\xi)$, $\beta = \nu\gamma$, $\nu = h/(2c_\sigma)$, $\gamma = d^2w/d\xi^2$, and a_σ and c_σ are related to the material properties.

For materials such as concrete, the expression of the stress can be approximated as (Figure 33.4)

$$\sigma_z = a_\sigma \tanh(\alpha + \beta z)[1 + (\beta z)/(2\cosh^2(\alpha + \beta z))] \tag{33.14'}$$

Additional terms can be introduced in the expression of the stress to improve the accuracy,

$$\sigma_z = a_\sigma \tanh(\alpha + \beta z)[1 + (\beta z)^2/(3\cosh^2(\alpha + \beta z)) - (\alpha + \beta z)^2/(\cosh^4(\alpha + \beta z))] \tag{33.14''}$$

but they complicate the manipulation in the derivation.

Successive Loading and Unloading

During successive loading and unloading, the stress–strain variation can be divided into four phases to describe the behavior of the material (Figure 33.5):

Phase I — The stress–strain diagram is linear up to the proportional limit. The difference between the actual and the present study is negligible.

Phase II — Between proportional and yield limits, the idealized stress–strain diagram is still linear. But in the present study, the stress–strain diagram exhibits nonlinearity.

Phase III — The yield limit does not exist for an assumed linear stress–strain diagram, where the stress increases up to failure without any change in the stiffness of the member. The yield limit exists in the stress–strain diagrams of the idealized case and present study.

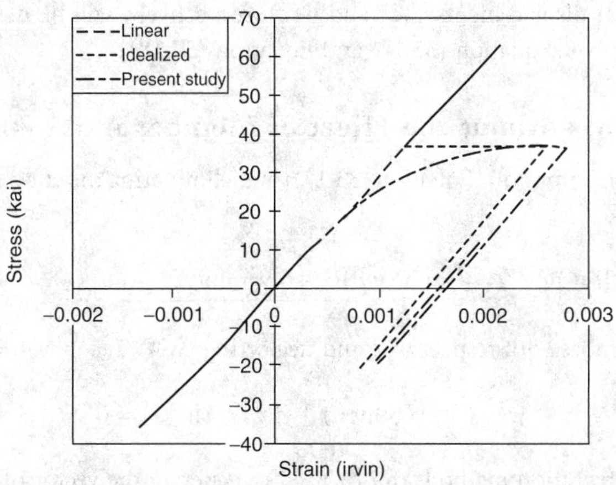

FIGURE 33.5 Successive loading and unloading.

Phase IV — During unloading, the stresses and strains decrease. For the decrease in stress, the stress–strain diagram is linear and parallel to the linear part of the loading diagram in all three cases. For the decrease beyond the lower proportional limit, the stress–strain variations in all the three cases are similar to their respective loading diagrams.

Stress Resultants

The axial force, $P(\xi)$, and the bending moment, $M(\xi)$, at an arbitrary section ξ are determined as the resultants of the stress over the cross-section, as

$$P(\xi) = a_\sigma(bh/3)\tanh\alpha[(2/\nu\gamma)\tanh(\nu\gamma) + 1 - \tanh\alpha\tanh(\nu\gamma) + 3\nu\gamma\tanh(\nu\gamma)] \tag{33.15}$$

$$M(\xi) = a_\sigma(bh/4)[\tanh(\nu\gamma) - \nu\gamma/3][1 - \tanh^2\alpha - \tanh^2(\nu\gamma)] \tag{33.16}$$

and the increment of the bending moment as

$$dM/d\xi = a_\sigma(bh^2/6)\nu w'''[1 - \tanh^2(\nu\gamma))(1 + \nu\gamma\tanh\nu\gamma)) + \tanh^2\alpha] \tag{33.16'}$$

where $\nu = h/(2c_\sigma)$ and $\gamma = d^2w/d\xi^2$.

For frames with material nonlinearity, the resisting axial force, $P(\xi)$, and the bending moment, $M(\xi)$, are no longer independent functions. The effect of the deflection $w(\xi)$ on the axial force $P(\xi)$ and the effect of the axial displacement $u(\xi)$ on the bending moment $M(\xi)$ can be seen from their expressions.

33.4 Equations of Motion

Herein, in the derivation of equations of motion, the inertia forces on the members are assumed to be zero ($q_z = q_y = 0$). Then the equations of motion are determined by equating the expressions of internal forces $P(\xi)$ and $V(\xi)$ and moment $dM/d\xi$, obtained from the geometrical behavior [Equation (33.10), Equation (33.11)], to those obtained from the material behavior [Equation (33.15) and Equation (33.16′)]

$$a_\sigma(bh^2/6)\nu w'''[1- \tanh^2(\nu\gamma))(1 + \nu\gamma\tanh\nu\gamma)) + \tanh^2\alpha] - [C_1 + C_2w'] = 0 \tag{33.17}$$

$$a_\sigma(bh/3)\tanh\alpha[(2/\nu\gamma)\tanh(\nu\gamma) + 1 - \tanh\alpha\tanh(\nu\gamma) + 3\nu\gamma\tanh(\nu\gamma)] - ((1 + (w')^2)^{-1/2)}[-C_1w' + C_2] = 0 \tag{33.18}$$

The axial and transversal displacements u(ξ) and w(ξ), respectively, will be determined from the two differential equations above [Equation (33.17) and Equation (33.18)].

Equation of Motion without the Effect of Member Axial Force

Considering the equation of motion [Equation (33.17)] and eliminating the effect of axial force (assuming $\alpha = 0$), one has

$$a_\sigma(bh^2/6)\nu w'''(1 - \tanh^2(\nu\gamma))(1 + \nu\gamma\tanh\nu\gamma)) - (C_1 + C_2 w') = 0 \tag{33.19}$$

or, differentiating once more with respect to ξ and neglecting $(w''')^2$ second order differential, one has

$$w^{iv}\,(1 + \nu\gamma\tanh\nu\gamma)) - C\,w''\cosh(\nu\gamma) = 0 \tag{33.20}$$

Various numerical integration methods can be used to generate the vector of the nonlinear member end reactions $\{f_{nl}\}$.

33.5 Linearization of the Moment and Force Displacement Relationships

To simplify the determination of displacements through the differential equation [Equation (33.20)], approximate methods can be used. As an approximate method, the expression of bending moment M(ξ) [(Equation 33.16)] can be approximated by multiplying and dividing by νγ, thus one has

$$M(\xi) = \nu\gamma([a_\sigma(bh/4)[[\tanh(\nu\gamma)/(\nu\gamma) - 1/3][1 - \tanh^2(\nu\gamma)]] \tag{33.21}$$

where γ appearing out of the bracket, say γ_{lin}, is

$$\gamma_{lin} = (d^2/d\xi^2)\,(N_1(\eta)\;N_2(\eta)\;N_3(\eta)\;N_4(\eta)\;\}^T\{d_{ben}\}$$

and $\eta = \xi/l$; $N_i(\eta)$ is the shape function and $\{d_{ben}\}$ is the vector of member end displacements and rotations.

γ appearing within the parenthesis of bending moment, M(ξ), is the summation of the displacements and rotations up to the present iteration or increment.

The shearing force–displacement relation is linearized similarly, as

$$V(\xi) = dM(\xi)/d\xi = (\nu\gamma')a_\sigma(bh/6)(1 - \tanh^2(\nu\gamma))(1 + \nu\gamma\tanh(\nu\gamma)) \tag{33.22}$$

γ′ appearing out of the bracket can be assumed, as it is in linear analysis, obtained from the shape function, for the present increment or iteration

$$\gamma' = d^3 w/d\xi^3 = (12\eta/l^3\;6\eta/l^2 - 12\eta/l^3\;6\eta)/l^2)\;\{d_{ben}\}$$

γ appearing within the parenthesis of the shearing force, V(ξ), is the summation of the displacements and rotations up to the present iteration or increment.

Substituting the linear expressions of the parameters γ and γ′ into the expressions of M(ξ) and V(ξ) provides the linearized expressions. Thus one has

$$M(\xi) = EI\;[[\tanh(\nu\gamma)/(\nu\gamma) - 1/3][1 - \tanh^2(\nu\gamma)]]\;((-6 + 12\eta)/l^2\,(-4 + 6\eta)/ \\ l\,(6 - 12\eta)/l^2\,(-2 + 6\eta)/l)\{d_{ben}\} \tag{33.23}$$

$$V(y) = EI\;(1 - \tanh^2(\nu\gamma))(1 + \nu\gamma\tanh(\nu\gamma))\;(12/l^3\;6/l^2 - 12/l^3\;6)/l^2)\{d_{ben}\} \tag{33.24}$$

where E and I are the Young's modulus and the moment of inertia of the cross-section.

Then, the equation of motion for the frame can be written as

$$([M]\{D(t)\} + [C]\{D(t)\} + [K_{nl}]\{D(t)\} = \{F(t)\} \tag{33.25}$$

where [M], [C], and $[K_{nl}]$ are the mass, damping, and nonlinear stiffness matrices for the frame, obtained from those of the members, respectively, and {D(t)} and {F(t)} are the displacements, including the rotations, and external forces, including the moments, at the joints of the frame.

Trusses

For trusses, the geometrical and material behaviors provide the equation of motion as

$$P(y) = a_c bh \tanh\alpha$$

where $\alpha = (1/(c_\sigma L))(du/d\xi)$, $u(\xi) = d_1 + (d_2 - d_1)\ \xi/L_{def}$, d_1 and d_2 are the end axial displacements at the fore and aft joints of the member with respect to the updated Lagrangian axis systems, and L_{def} is the deformed length of the member.

The linearization and the generation of the equation of motion are obtained in a similar manner to that for frames.

33.6 Consistent and Lumped Mass Matrix Methods

In the consistent and lumped mass matrix methods, the same steps as in the continuous mass matrix method are used by making the required assumptions related to their derivations.

33.7 Practical Applications

A two-span truss subjected to a suddenly applied constant load is considered as an example. The data on the geometry, material properties, external loading history, and the time steps are the same as given in Mondkar and Powell [Mondkar and Powell, 1977] (Figure 33.6).

Mondkar and Powell assumed Ramberg–Osgood continuous stress–strain variation. Their nonlinear analysis is based on Newmark's constant acceleration method. The variation of the vertical displacement

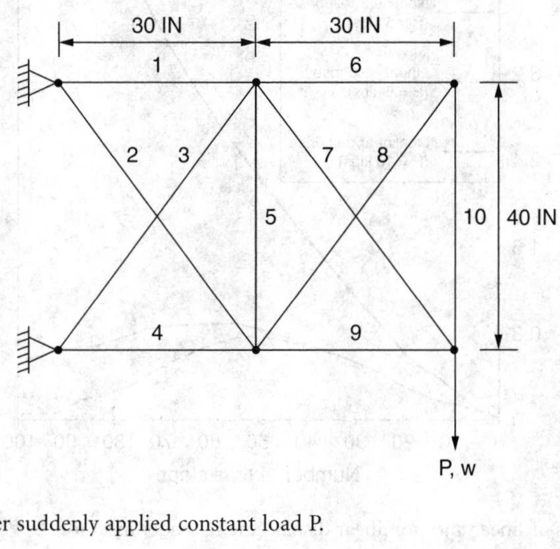

FIGURE 33.6 Truss under suddenly applied constant load P.

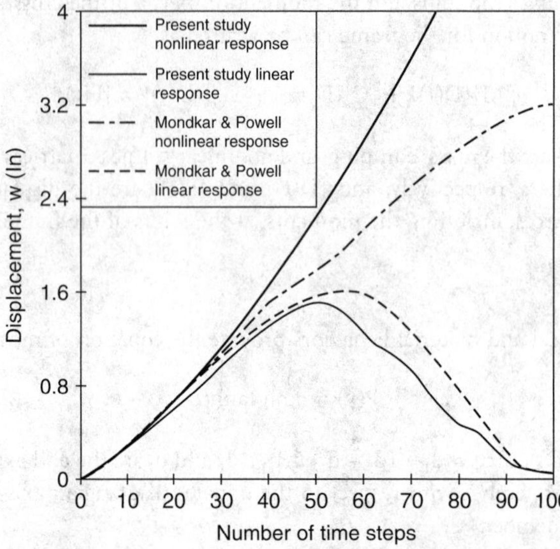

FIGURE 33.7 Linear and nonlinear responses of two-span truss under axial force P = 10 kips.

w, at the point of application of the external load P, versus the time steps is plotted in Figure 33.7. The diagrams include the linear and nonlinear responses under the Mondkar and Powell assumption and in the present study. The displacements for the linear responses using the Mondkar and Powell method and the present study have small differences up to the maximum amplitude. Beyond this point, the differences become larger. The displacements under the present study are larger than those obtained by the Mondkar and Powell assumption.

A comparison of linear and nonlinear dynamic displacements using two different external loads is shown in Figure 33.8. It can be noticed that the natural circular frequency for nonlinear response (ω_{NL}) is smaller than the natural frequency for linear response (ω_N). For external loads P = 10 k, the nonlinear displacement (w_{NL}) increases to infinity, showing the instability of the truss under P = 10 k. Since the members 7, 9, and 10, all of which are connected to the same joint, exhibit stresses at or beyond their the yield limit, local instability occurs.

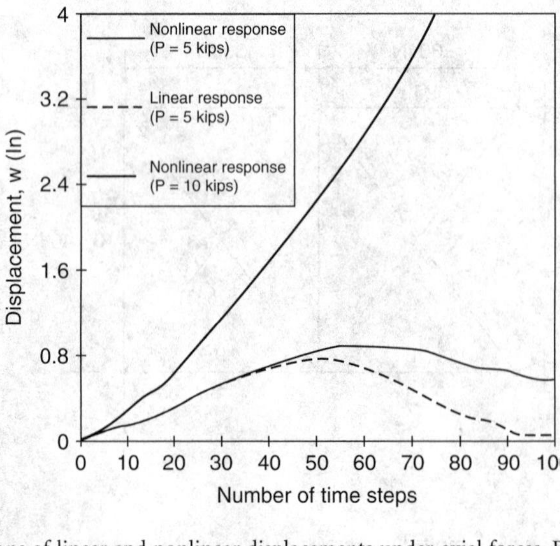

FIGURE 33.8 Comparisons of linear and nonlinear displacements under axial forces, P = 5 kips and P = 10 kips.

Defining Terms

Linear behavior — In a system, the relation between the strain or deformation and the stress or applied load is linear.

Elastic behavior — A system is elastic if it returns to its original shape when the load acting on it is removed.

Materially linear behavior — If the variation of stress versus the strain is linear within the material.

Geometrically linear behavior — If the variation of the deformation versus the applied load is linear.

Plastic behavior — If, in a system, the strain or deformation increases when the stress or applied load remains at a constant limit value.

Stability — The strength to stand or endure.

References

Clough, R. W. and Penzien, J., 1993. *Dynamics of Structures,* McGraw-Hill, New York.

Mondkar, D. P. and Powell, G. H., 1977. Finite element analysis of nonlinear static and dynamic responses, *Int. J. Numer. Meth. Eng.* 14:499–520.

Newmark, N. M., 1959. A method of computations for structural dynamics, *J. ASCE, ME* 86:67–94.

Ovunc, B. A., 1998. Nonlinear Behavior of Structures under Dynamic Disturbances, Proc., International Conference on Vibration Engineering, Aug. 6–9, Dalian, China, pp. 230–235.

Ovunc, B. A., 1985. Soil–structure interaction and effect of axial force on the dynamics of offshore structures, *Compt Struc.,* 5:629–635.

Ovunc, B. A., 1992. Frames under continuous stress–strain variation up to failure, *Bull. Technical University of Istanbul,* 45:35–52.

Ovunc, B. A., 1994. Material and Geometrical Nonlinearities in the Vibration of Frames, Proc., ESDA'94, ASME, PD-Vol. 67/7, 361–368.

Shi, J. and Atluri, S. N., 1988. Elasto-plastic large deformation analysis of space frames: a plastic hinge based, *Num. Meth. Engr.,* 26:589–615.

Wilson, E. L., Farhoomand, I., and Bathe, K. J., 1973. Nonlinear dynamic analysis of complex structures, *Int. J. Earth Eng. and Struct. Dynamics,* **1**:241–252.

Further Information

Ovunc, B. A., 1982. Geometrical Nonlinearity of Plane Frameworks, Proc., Sino-American Symposium on Bridges and Structural Engineering, Beijing, China, pp. 327–336.

Ovunc, B. A., 1990. Material Nonlinearities and Stability in the Analysis of Frames, Struceng-Femcad '90, Ed. Niku Lari, IITT, France, pp. 183–190.

Ovunc, B. A., 1996. Lumped vs Distributed Parameter Systems, Chapter 16, Dynamics and Vibration, *Engineering Handbook*, CRC Press, Boca Raton, FL, pp. 152–163.

Rossow, E., 1996. *Analysis and Behavior of Structures*, Prentice Hall, Inc., Upper Saddle River, NJ.

Veletsos, A.S. and Huang, T., 1970. Analysis of dynamic response of highway bridges, *ASCE, EM5,* 96:593–620.

34

Scour of Bridge Foundations

E. V. Richardson
Ayres Associates

Scour is the engineering term for the erosive action of flowing water, excavating and carrying away material from the bed and banks of streams and from around the piers and abutments of bridges. Scour is the most common cause of bridge failures. This chapter gives the methods and equations recommended by the U.S. Federal Highway Administration (FHWA) for the analysis and design of bridge foundations to be safe from scour. The recommendations are given in three Hydraulic Engineering Circulars (HEC-18 [Richardson and Davis, 2001], HEC 20 [Lagasse et al., 2001], and HEC 23 [Lagasse et al., 2001]) and one Hydraulic Design Series (HDS-6) [Richardson et al., 2001].

34.1 Total Scour

All streambed materials will scour. Only the rates of scour differ. Sand- and gravel-bed materials reach maximum scour depths in hours, whereas cohesive clays may take days or many storm cycles to reach maximum scour depths. However, ultimate scour in cohesive clays will be as deep as scour in sand-bed streams [Briaud et al., 1999]. Limestone and granite rock are very scour resistant (years). The total scour methods and equations given in this chapter predict maximum scour. Under some circumstances, taking the time rate of scour of clays into account may be appropriate. Briaud et al. [1999, 2001] describe equipment and methods to determine the time rate of scour in clay bed materials. Total scour at a highway crossing is comprised of three components:

1. Long-term aggradation or degradation of the river bed
2. General scour at the bridge (contraction of the flow or other flow conditions)
3. Local scour at the piers or abutments

In addition, **lateral migration** of the stream must be assessed when evaluating total scour at bridge piers and abutments.

0-8493-1586-7/05/$0.00+$1.50

These three scour components are considered to be independent of each other and are added to obtain the total scour at a pier or abutment. Considering the components independent and additive adds some conservatism to the design. Total scour is determined using a design discharge. For streams and rivers the discharge for the 20-, 50-, or 100-return period is used depending on the class of road. For bridges crossing tidal waterways, the discharge is determined using the return period of the storm surge elevation and area of the waterway.

34.2 Long-Term Aggradation or Degradation

The streambed at a bridge crossing may be aggrading, degrading, or in relative equilibrium. Aggradation involves the deposition of sediment under the bridge, whereas degradation is the lowering of the streambed under the bridge due to a deficit in sediment supply from upstream. These long-term bed elevation changes may be the natural trend of the stream or the result of some modification to the stream or watershed. Long-term aggradation and degradation do not include the cutting and filling of the streambed in the vicinity of the bridge that might occur during a runoff event (general and local scour). If the stream is aggrading, the increase in streambed elevation is not considered in the total scour. But if the stream is degrading, the estimated decrease in elevation of the streambed is included in the total scour. The engineer must assess the present state of the stream and watershed and then evaluate potential future changes in the river system. From this assessment, the long-term streambed changes are estimated.

Factors that affect long-term bed elevation changes are dams and reservoirs (up- or downstream of the bridge), changes in watershed land use (urbanization, deforestation, etc.), channelization, cutoffs of meander bends (natural or manmade), changes in the downstream channel base level (control), gravel mining from the streambed, diversion of water into or out of the stream, natural lowering of the fluvial system, movement of a bend and bridge location with respect to stream planform, and stream movement in relation to the crossing. To assess long-term bed elevation changes, a three-level fluvial system approach can be used. It consists of:

1. A qualitative determination using general geomorphic and river mechanics relationships
2. An engineering geomorphic analysis using established qualitative and quantitative relationships to estimate elevation changes resulting from various scenarios of future conditions
3. Computations using physical process computer models such as BRI-STARS [Molinas, 1990], HEC 6 [USACE, 1993], and SAM [USACE, 1988; Ayres, 2003]

The three-level approach is described in greater detail in FHWA publications HEC 20 [Lagasse et al., 2001], and HDS-6 [Richardson et al., 2001]. Available bridge inspection records, gaging station records, aerial photographs, available maps, and interviews with people knowledgeable of the stream are valuable for the analysis.

34.3 General Scour

General scour is a lowering of the streambed across the waterway at the bridge. This lowering may be uniform across the bed or it may be deeper in some parts of the cross-section. General scour is different from long-term degradation in that general scour may be cyclic and/or related to the passing of a flood. That is, the streambed lowers on the rising limb of a flood and fills on the falling limb. General scour may result from contraction of the flow or from other general scour conditions, such as flow around a bend, variable downstream control, junction of two streams, etc. Contraction scour occurs when the area of the bridge opening is smaller than the flow area of the upstream channel or channel and flood plain. Equations are given in the next section or using sediment transport computer models (see Chapter 97, Sedimentation).

34.4 Contraction Scour

Contraction may be live-bed or clear-water scour. Live-bed contraction scour occurs at a bridge when there is transport of bed material in the upstream reach into the bridge cross-section. With live-bed contraction scour, the area of the contracted section that is scoured increases until, in the limit, the transport of sediment out of the contracted section equals the sediment transported in. Clear-water contraction scour occurs when (1) there is no bed material transport from the upstream reach into the downstream reach or (2) the material being transported in the upstream reach is transported through the downstream reach mostly in suspension and at less than capacity of the flow. With clear-water contraction scour, the area of the contracted section increases until, in the limit, the velocity of the flow (V) or the shear stress (τ_o) on the bed is equal to the critical velocity (V_c) or critical shear stress (τ_c) of a certain particle size (D) in the bed material. Normally, for both live-bed and clear-water scour, the width of the contracted section is constrained, and depth increases until the limiting conditions are reached.

Live-bed contraction scour depths may be limited by sediment transport into the bridge section or by armoring of the bed by large sediment particles in the bed material. Therefore, contraction scour at a bridge can be determined by calculating the scour depths using both the clear-water and live-bed contraction scour equations and using the smaller of the two depths.

Also, to determine if the contraction scour is live bed or clear water, calculate the critical velocity for beginning of motion V_c of the D_{50} size of the bed material being considered and compare it to the mean velocity V of the flow in the main channel or overbank area upstream of the bridge opening. If the critical velocity of the bed material is larger than the mean velocity ($V_c > V$), then clear-water contraction scour will exist. If the critical velocity is less than the mean velocity ($V_c < V$), then live-bed contraction scour will exist.

There are four conditions (cases) of contraction scour at bridge sites depending on the type of contraction and whether overbank flow or relief bridges exist. Regardless of the case, contraction scour is evaluated using two basic equations: (1) the live-bed scour equation and (2) the clear-water scour equation. The four conditions (cases) of contraction scour are:

Case 1. Overbank flow on a floodplain is being forced back into the main channel by the approaches to the bridge. Case 1 conditions include:

A. The river channel width becomes narrower either due to the bridge abutments projecting into the channel or the bridge being located at a narrowing reach of the river.
B. No contraction of the main channel occurs, but the overbank flow area is completely obstructed by an embankment
C. Abutments are set back from the stream channel.

Case 2. Flow is confined to the main channel (i.e., no overbank flow is present).

Case 3. A relief bridge exists in the overbank area with little or no bed material transport.

Case 4. A relief bridge exists over a secondary stream in the overbank area with bed material transport (similar to Case 1).

Live-bed contraction scour depth is calculated using the following modified version of Laursen's [1960] equation for live-bed scour at a long contraction.

$$\frac{y_2}{y_1} = \left(\frac{Q_2}{Q_1}\right)^{6/7} \left(\frac{W_1}{W_2}\right)^{k_1} \tag{34.1}$$

$$y_s = y_2 - y_o = \text{(average contraction scour depth)} \tag{34.2}$$

where y_1 = average depth in the upstream main channel, m (ft); y_2 = average depth in the contracted section, m (ft); y_o = existing depth of flow in the contracted section before scour, m (ft); Q_1 = flow in the upstream channel transporting sediment, m^3/sec (ft^3/sec); Q_2 = flow in the contracted channel, m^3/

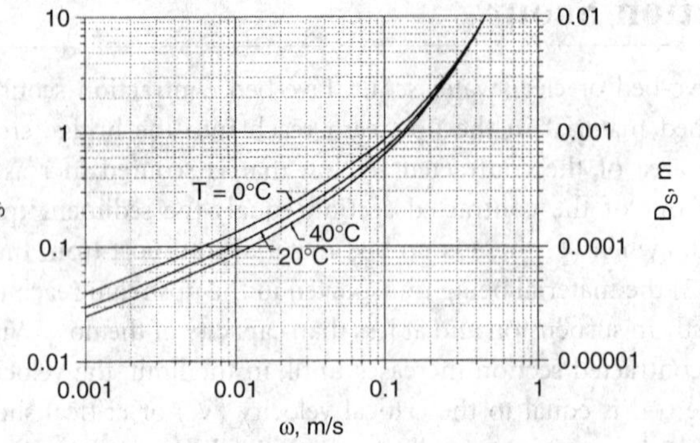

FIGURE 34.1 Fall velocity of sand-sized particles with specific gravity of 2.65.

sec (ft³/sec); W_1 = bottom width of the upstream main channel, m (ft); W_2 = bottom width of the main channel in the contracted section less pier width (s), m (ft); and k_1 = exponent determined below.

V_*/ω	k_1	Mode of Bed Material Transport
<0.50	0.59	Mostly contact bed material discharge
0.50 to 2.0	0.64	Some suspended bed material discharge
>2.0	0.69	Mostly suspended bed material discharge

$V_* = (\tau_o/\rho)^{1/2} = (gy_1\ S_1)^2$, shear velocity in the upstream section, m/sec (ft/sec); ω = fall velocity of bed material based on the D_{50}, m/sec (Figure 34.1), for fall velocity in English units (ft/sec) multiply in m/sec by 3.28; g = acceleration of gravity, 9.81 m/sec² (32.2 ft/sec²); S_1 = slope of energy grade line of main channel, m/m (ft/ft); τ_o = shear stress on the bed, Pa (N/m²) (lb/ft²); ρ = density of water, 1000 kg/m³ (1.94 slugs/ft³).

Notes

Q_2 may be the total flow going through the bridge opening as in Cases 1a and 1b. It is not the total flow for Case 1c. For Case 1c, contraction scour must be computed separately for the main channel and the left and/or right overbank areas.

Q_1 is the flow in the main channel upstream of the bridge, not including overbank flows.

Clear-water contraction scour depth is calculated using an equation based on a development suggested by Laursen. The equation is:

$$y_2 = \left[\frac{K_u Q^2}{D_m^{2/3} W^2}\right]^{3/7} \tag{34.3}$$

$$y_s = y_2 - y_o = \text{(average contraction scour depth)} \tag{34.4}$$

where y_2 = average equilibrium depth in the contracted section after contraction scour, m (ft); Q = discharge through the bridge or on the set-back overbank area at the bridge associated with the width W, m³/sec (ft³/sec); D_m = diameter of the smallest nontransportable particle in the bed material (1.25 D_{50}) in the contracted section, m (ft); D_{50} = median diameter of bed material, m (ft); W = bottom width of the contracted section less pier widths, m (ft); y_o = average existing depth in the contracted section, m (ft); K_u = 0.025 SI units; K_u = 0.0077 English units.

For stratified bed material, the depth of scour can be determined by using the clear-water scour equation sequentially with successive D_m of the bed material layers.

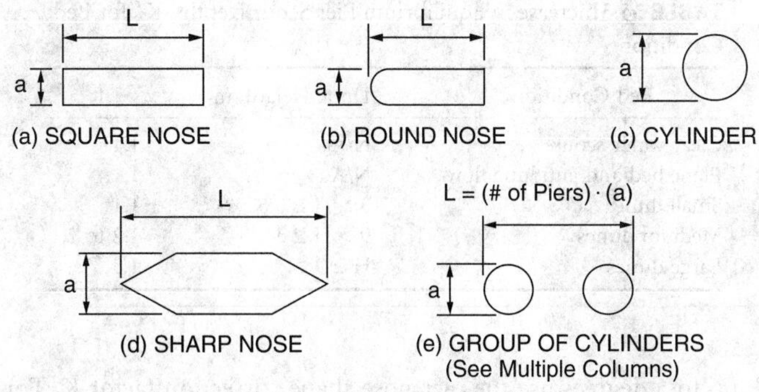

FIGURE 34.2 Common pier shapes [*Source:* Richardson, E. V. and Davis, S. R. 2001. *Evaluating Scour at Bridges, Forth Edition.* HEC 18, Pub. No. FHWA-NHI 01-001, FHWA, Washington, D.C.].

34.5 Local Scour at Bridge Piers

To determine maximum local pier scour, an equation based on an equation developed at Colorado State University [Richardson et al., 2001] is recommended for both live-bed and clear-water pier scour in HEC 18 [Richardson and Davis, 2001]. The recommendation is based on a study by Jones [1983] and a comparison by Mueller [1996] of 22 scour equations using field data collected by the U.S. Geological Survey. The data included 384 field measurements of scour at 56 bridges. The equation is:

$$\frac{y_s}{y_1} = 2.0\, K_1\, K_2\, K_3\, K_4\, Kw \left(\frac{a}{y_1}\right)^{0.65} Fr_1^{0.43} \tag{34.5}$$

where y_s = Scour depth, m (ft), y_1 = Flow depth directly upstream of the pier, m (ft), K_1 = Correction factor for pier nose shape from Figure 34.2 and Table 34.1, K_2 = Correction factor for angle of attack of flow from Equation 34.6, K_3 = Correction factor for bed condition from Table 34.3, K_4 = Correction factor for armoring by bed material size, K_w = Correction factor for pier width from Equations 34.7 and Equation 34.8, a = Pier width, m (ft), L = Length of pier, m (ft), Fr_1 = Froude Number directly upstream of the pier = $V_1/(gy_1)^{1/2}$, V_1 = Mean velocity of flow directly upstream of the pier, m/sec (ft/sec), g = Acceleration of gravity (9.81 m/sec²) (32.2 ft/sec²)

$$K_2 = (Cos\,\theta + L/a\; Sin\,\theta)^{0.65} \tag{34.6}$$

If L/a is larger than 12, use L/a = 12 as a maximum.

Table 34.2 illustrates the magnitude of the effect of the angle of attack on local pier scour.

TABLE 34.1 Correction Factor, K_1, for Pier Nose Shape

Shape of Pier Nose	K_1
(a) Square nose	1.1
(b) Round nose	1.0
(c) Circular cylinder	1.0
(d) Group of cylinders	1.0
(e) Sharp nose	0.9

TABLE 34.2 Correction Factor, K_2, for Angle of Attack, 2, of the Flow

Angle	L/a = 4	L/a = 8	L/a = 12
0	1.0	1.0	1.0
15	1.5	2.0	2.5
30	2.0	2.75	3.5
45	2.3	3.3	4.3
90	2.5	3.9	5.0

Note: Angle = skew angle of flow; L = length of pier, m.

TABLE 34.3 Increase in Equilibrium Pier Scour Depths, K_3, for Bed Condition

Bed Condition	Dune Height, m	K_3
Clear-water scour	N/A	1.1
Plane bed and antidune flow	N/A	1.1
Small dunes	$3 > H \geq 0.6$	1.1
Medium dunes	$9 > H \geq 3$	1.2 to 1.1
Large dunes	$H \geq 9$	1.3

Notes

For angles of attack up to 5 degrees use the pier nose shape correction factor K_1. For greater angles, K_2 dominates and K_1 should be considered as 1.0. K_2 should be applied using the effective length L of the pier actually subjected to the angle of attack of the flow. Plane-bed and/or antidune flow conditions are typical for most bridges over sand bed streams for the flood frequencies employed in scour design. Large dunes exist during flood flow on very large rivers, such as the Mississippi. Smaller streams with a dune bed configuration at flood flow will have smaller dunes. Piers set close to abutments (for example at the toe of a spill through abutment) must be carefully evaluated for the angle of attack and velocity of the flow coming around the abutment.

Pier Scour Armoring Factor K_4

Richardson, using the U. S. Geological Survey field data of 384 measurements at 56 bridges [Landers et al., 1999] and Mueller and Jones [1999] studies, developed the following for K_4:

- The minimum value of K_4 is 0.4.
- For $D_{50} < 2.0$ mm, $K_4 = 1.0$;
- For $D_{95} < 20.0$ mm, $K_4 = 1.0$;
- For $V_1 < V_{icD35}$, $K_4 = 0.4$.
- For $V_1 > V_{icD35}$, $K_4 = (1 - V_{icD35}/V_1)^{0.43}$. (34.7)

where V_{icDx} = the approach velocity (m/sec or ft/sec) required to initiate scour at the pier for the grain size D_x (m or ft).

$$V_{icD35} = 0.645\left(\frac{D_{35}}{a}\right)^{0.053} V_{cD35} \quad (34.8)$$

where V_{cD35} = the critical velocity (m/sec or ft/sec) for incipient motion for the grain size D_{35} (m or ft).

$$V_{cD35} = K_u y_1^{1/6} D_{35}^{1/3} \quad (34.9)$$

where y_1 = depth of flow just upstream of the pier, excluding local scour, m (ft); V_1 = velocity of the approach flow just upstream of the pier, m/sec (ft/sec); D_x = grain size for which x percent of the bed material is finer, m (ft); $K_u = 6.19$ SI units; and $K_u = 11.17$ English units.

D_{35} may be calculated using the following equation:

$$D_{35} = \frac{D_{50}}{\left(\frac{D_{84}}{D_{16}}\right)^{0.1925}} \quad (34.10)$$

If D_{84} and/or D_{16} are not measured, then take $D_{35} = 0.8\ D_{50}$.

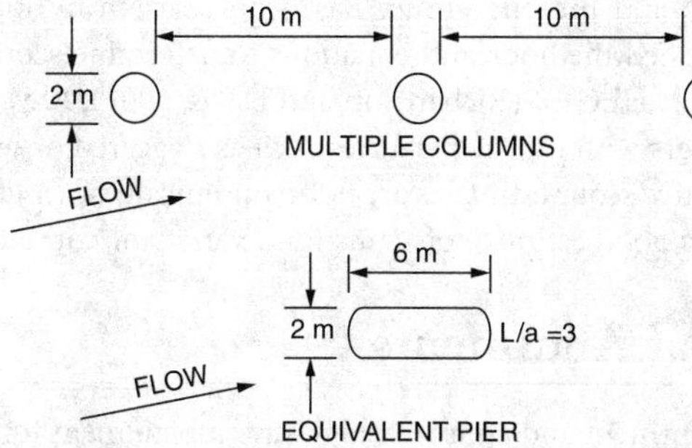

FIGURE 34.3 Multiple columns skewed to the flow [*Source:* Richardson, E. V. and Davis, S. R. 2001. *Evaluating Scour at Bridges, Fourth Edition.* HEC 18, Pub. No. FHWA-NHI 01-001, FHWA, Washington, D.C.].

K_W Correction Factor for Very Wide Piers

Johnson and Torrico [1994] suggest the following equations for a K_w factor to wide piers in shallow flow. The correction factor should be applied when the ratio of depth of flow (y) to pier width (a) is less than 0.8 (y/a < 0.8); the ratio of pier width (a) to the median diameter of the bed material (D_{50}) is greater than 50 ($a/D_{50} > 50$); and the Froude number of the flow is subcritical.

$$K_w = 2.58\left(\frac{y}{a}\right)^{0.34} Fr_1^{0.65} \quad for\ V/V_c < 1 \tag{34.11}$$

$$K_w = 1.0\left(\frac{y}{a}\right)^{0.13} Fr_1^{0.25} \quad for\ V/V_c \geq 1 \tag{34.12}$$

Care should be used in applying K_w because it is based on limited data from flume experiments. Its use should take into consideration traffic volume, class and importance of the highway, cost of a failure (potential loss of lives and dollars), and the change in cost that would occur if the K_w factor is used.

With **multiple columns** skewed to the flow (Figure 34.3) and spaced less than five pier diameters apart, the pier width "a" is a composite width of all the columns in a single bent with no space between. This composite pier width would be used in Equation 34.6 to determine K_2 in the pier scour equation. If the multiple columns are spaced five diameters or greater apart and debris is not a problem, the scour depth is taken as 1.2 times the local scour of a single column.

Pressure flow scour occurs when flow under a bridge is in contact with the underside of the bridge. The bridge may or may not be submerged. If the bridge is entirely submerged, the resulting flow is a complex combination of the plunging flow under the bridge (orifice flow) and flow over the bridge (weir flow). Pier scour with pressure flow is a combination of vertical deck contraction scour and local pier scour, which are additive [Jones et al., 1996 and Arneson and Abt, 1999]. To determine scour depths for a bridge, the reader is referred to the publications of Jones et al. [1996], Arneson and Abt [1999], and Richardson and Davis [2001].

Debris lodged on a pier can increase local scour at a pier. The scour depth can be estimated by assuming that the pier width is larger than the actual width. For additional discussion, see HEC-20 [Lagasse et al., 2001], and HEC 18 [Richardson and Davis, 2001],

34.6 Local Scour at Complex Piers

Most determinations of local pier scour depths have focused on piers with single substructural elements. In the general case, the flow could be obstructed by three substructure elements, which include the pier stem, the pile cap or footing, and the pile group. Based on research by Jones and Sheppard [2000], Richardson and Jones determined methods and equations to determine scour depths for complex pier foundations, which are given in HEC 18 (Richardson and Davis, 2001). **Physical model** studies are still recommended for complex piers with unusual features such as staggered or unevenly spaced piles or for major bridges where conservative scour estimates are not economically acceptable. However, the methods presented in HEC18 provide a good estimate of scour for a variety of complex pier situations.

34.7 Local Scour at Abutments

The flow obstructed by the abutment and approach highway embankment forms two erosion potentials. One is a horizontal vortex starting at the upstream end of the abutment and running along the toe of the abutment, and the other is a vertical wake vortex at the downstream end of the abutment. Research has only developed equations to predict scour depths caused by the horizontal vortex.. There are three general shapes of abutments:

1. Spill-through abutments
2. Vertical walls without wing walls
3. Vertical-wall abutments with wing walls

These shapes have varying angles to the flow. Maximum depth of scour from the horizontal vortex for spill-through abutments is 55% of that for vertical-wall abutments. Similarly, scour depths at vertical wall abutments with wingwalls is 82% of the scour of vertical wall abutments without wingwalls.

Almost all equations in the literature use the abutment and roadway approach length instead of the intercepted flow as one of the variables. This approach results in excessively conservative estimates of scour depth. Richardson and Richardson [1993] pointed this out in a discussion of Melville's [1992] paper. They stated, "The reason the equations in the literature predict excessively conservative abutment scour depths for the field situation is that, in the laboratory flume, the discharge intercepted by the abutment is directly related to the abutment length; whereas, in the field, this is rarely the case."

Because available equations do not estimate reliable abutment scour depths for the horizontal vortex, engineering judgment is required in designing foundations for abutments. As a minimum, abutment foundations should be designed assuming no ground support (lateral or vertical) as a result of soil loss from long-term degradation, stream instability, and contraction scour. The abutment should be protected from local scour using riprap and/or guide banks. Guidelines for the design of riprap and guide banks are given in HEC-23 [Lagasse et al., 2001]. HEC 18 [Richardson and Davis, 2001] gives two equations to provide guidance in the design of abutment foundations.

To protect the abutment and approach roadway from scour by the wake vortex, several states use a 15-meter (50-ft) guide bank extending from the downstream corner of the abutment. Otherwise, the downstream abutment and approach should be protected with riprap or other forms of bank protection.

Defining Terms

Bed form — A relief feature on the bed of a stream, such as dunes, plane bed, or antidunes. Also called *bed configuration* [Simons and Richardson, 1963; Richardson et al., 2001].

Bed material — Material found on the bed of a stream. May be transported in contact with the bed or in suspension.

Critical shear stress — The minimum amount of shear (force) exerted by the flow on a particle or group of particles that is required to initiate particle motion.

Discharge — Time rate of the movement of a quantity of water or sediment passing a given cross-section of a stream or river.

Geomorphology — The branch of physiography and geology that deals with the general configuration (form) of the earth's surface and the changes that take place as the result of the forces of nature.

Hydraulic radius — The cross-sectional area of a stream, divided by its wetted perimeter. Equals the depth of flow when the width is larger than ten times depth.

Median diameter — The particle diameter at which 50% of a sample's particles are coarser and 50% are finer (D_{50}).

Shear stress, tractive force — The force or drag on the channel boundaries that is caused by the flowing water. For uniform flow, shear stress is equal to the unit weight of water times the hydraulic radius times the slope. Usually expressed as force per unit area.

Shear velocity — The square root of the shear stress divided by the mass density of water, in units of velocity.

Slope — Fall per unit length of the channel bottom, water surface, or energy grade line.

References

Arneson, L. A. and Abt, S. R., 1999. *Vertical Contraction Scour at Bridges with Water Flowing under Pressure Conditions,* ASCE Compendium, Stream Stability and Scour at Highway Bridges, Richardson, E. V. and Lagasse, P. F. (Eds.), Reston, VA.

Ayres, 2003. *SAM, Windows Interface,* Ayres Associates, Inc., Fort Collins, CO.

Briaud, J.-L., Ting, F. C. K., Chen, H. C., Gudavalli, R., Perugu, S., and Wei, G., 1999. SRICOS: Prediction of Scour Rate in Cohesive Soils at Bridge Piers. *ASCE J. Geotech. Geoenviron. Eng.* 125.

Briaud, J-L., Ting, F. C. K., Chen, H. C., Cao, Y., Han, S. W., and Kwak, K. 2001. Erosion function apparatus for scour rate predictions. *ASCE J. Geotech. Geoenviron. Eng.* 127.

Johnson, P. A. and Torrico, E. F. 1994. Scour Around Wide Piers in Shallow Water, *Transportation Research Record* 1471, Transportation Research Board, Washington, D.C.

Jones, J. S., 1983. Comparison of Prediction Equations for Bridge Pier and Abutment Scour. *Transportation Research Record* 950, Vol. 2, Transportation Research Board, Washington, D.C.

Jones, J. S., Bertoldi, D. A., and Umbrell, E. R. 1996. Interim Procedures for Pressure Flow Scour. ASCE Hydraulic Engineering, Proc. North American Water Congress, Reston, VA.

Jones, J. S. and Sheppard, D. M. 2000. Local Scour at Complex Pier Geometries. ASCE Hydraulic Engineering Proc., Minneapolis meeting, Reston, VA.

Lagasse, P. F., Schall, J. D., and Richardson, E. V. 2001. *Stream Stability at Highway Structures, Third Edition.* HEC 20, Pub. No. FHWA NHI 01-002, Federal Highway Administration, Washington, D.C.

Lagasse, P. F., Zevenbergen, L. W., Schall, J. D., and Clopper, P. E., 2001. *Bridge Scour and Stream Instability Countermeasures — Experience, Selection, and Design Guideline,* 2nd ed. HEC 23, Pub. No. FHWA NHI 01-003, Federal Highway Administration, Washington, D.C.

Landers, M. N., Mueller, D. D., and Richardson, E. V. 1999. U.S. Geological Survey Field Measurements of Pier Scour. *ASCE Compendium, Stream Stability and Scour at Highway Bridges,* Richardson, E. V. and Lagasse, P. F. (Eds.), Reston, VA.

Laursen, E. M. 1960. Scour at bridge crossings. *ASCE J. Hydraulic Division* 86.

Melville, B. W. 1992. Local scour at bridge abutments. *ASCE J. Hydr. Eng., Hydr. Div.* 118.

Molinas, A. 1990, Bridge Stream Tube Model for Alluvial River Simulation (BRI-STARS), User's Manual, NCHRP No. HR15-11, Transportation Research Board, Washington, D.C.

Mueller, D. S., 1996. Local Scour at Bridge Piers in Nonuniform Sediment Under Dynamic Conditions. Ph.D. Dissertation, Colorado State University, Fort Collins, CO.

Mueller, D. S. and Jones, J. S. 1999. Evaluation of Recent Field and Laboratory Research at Bridge Piers in Coarse Bed Materials. *ASCE Compendium, Stream Stability and Scour at Highway Bridges,* Richardson, E. V. and Lagasse, P. F. (Eds.), Reston, VA.

Richardson, E. V. and Davis, S. R. 2001. *Evaluating Scour at Bridges, Fourth Edition.* HEC 18, Pub. No. FHWA-NHI 01-001, FHWA, Washington, D.C.

Richardson, E. V. and Richardson, J. R. 1993. Discussion of Melville, B. W., 1992 paper, Local scour at bridge abutments. *ASCE J. Hydr. Eng., Hydr. Div.* 119.

Richardson, E. V., Simons, D. B., and Lagasse, P. F. 2001. *River Engineering for Highway Encroachments — Highways in the River Environment.* Hydraulic Design Series No. 6, Pub. No. FHWA NHI 01-004, Federal Highway Administration, Washington, D.C.

Simons, D. B. and Richardson, E. V., 1963. Forms of bed roughness in alluvial channels. *ASCE Trans.* 128: .

USACE. 1988. U. S. Army Corp. of Engineers. Sam Hydraulic Design Package for Channels. WES Coastal and Hydraulic Laboratory, Vicksburg, MS.

USACE. 1993. U.S. Army Corps of Engineers, Scour and Deposition in Rivers and Reservoirs, User's Manual, HEC-6, Hydrologic Engineering Center, Davis, CA.

Further Information

Publications of the Federal Highway Administration, U.S. Department of Transportation, are available to the public through the National Technical Information Service, Springfield, VA 22161, phone 703 487 4650. The Web site is www.isddc.dot.gov.

The National Highway Institute (NHI), the technical training organization for the Federal Highway Administration (FHWA), maintains a World Wide Web site for up-to-date information on their courses and other activities. The Web site is www.nhi.fhwa.dot.gov.

The *ASCE Journal of Hydraulic Engineering, Transactions* and the annual publication titled *Hydraulic Engineering* report advances in scour. Of particular importance is the 1999 publication ASCE Compendium, Stream Stability and Scour at Highway Bridges, Richardson and Lagasse (eds.), Reston, VA. For information contact the ASCE Web site www.pubs.asce.org or e-mail to marketing@asce.org.

The Transportation Research Board, National Research Council, reports research on scour in their *Transportation Research Record* and reports on the National Cooperative Highway Research Program in *NCHRP Reports.* TRB publications may be ordered from the TRB business office, National Research Council, 2101 Constitution Avenue, NW, Washington, D.C. 20418. Telephone 202 334 3214, FAX 202 334 2519 or e-mail TRBsales@nas.edu.

VI

Fluid Mechanics

35

Incompressible Fluids

Alan T. McDonald
Purdue University, Emeritus

A fluid is a substance that cannot sustain shear stress while at rest; even a small shear stress causes a continuous rate of angular deformation within a fluid. Under typical conditions liquids and gases behave as fluids. Under extreme conditions, solids may exhibit fluid characteristics, as in the "flow" of ice in a glacier.

Fluids are characterized by the relationship between applied shear stress and rate of angular deformation (*shear rate*). **Newtonian fluids** obey a simple linear relationship. For parallel flow this may be expressed as $\tau_{yx} = \mu \, du/dy$, where τ_{yx} is shear stress applied in the direction of the velocity, y is distance perpendicular to velocity, μ is dynamic viscosity or simply **viscosity**, and du/dy is rate of angular deformation. Most gases and many liquids — such as water, gasoline, and other pure substances — are closely approximated by the Newtonian fluid model.

Viscosity depends primarily on temperature at moderate pressures. Viscosity decreases sharply with increasing temperature for liquids and increases slightly for gases. At extremely high pressures, viscosities of liquids may increase significantly.

Fluid systems are commonly encountered in engineering practice. Transportation vehicles of all types — whether immersed or floating — experience viscous and pressure drag forces caused by fluid flow around the vehicle. Pipeline transportation and human circulation are fluid systems, as are convection heating and ventilating systems.

35.1 Fundamentals of Incompressible Fluid Flow

This section covers incompressible fluid flow. *Density* is defined as mass per unit volume and denoted by ρ; fluids with constant density are *incompressible.* Liquids are nearly incompressible. Gases are compressible, but gas flow may be treated as incompressible when the maximum speed is less than one-third the speed of sound.

The objective of fluid flow analysis is to predict the pressure drop and pumping power for internal flow through conduits and the forces and moments on bodies in external flow. In principle this may be accomplished if the three components of velocity and the pressure are known. In practice it is often

0-8493-1586-7/05/$0.00+$1.50

impossible to solve problems analytically; in these cases it is necessary to rely upon experimental data from tests of models and model systems.

This section covers fundamentals of incompressible fluid mechanics. First, fluids without relative motion are considered. Details of flow fields are then considered. Dimensional analysis to help simplify design of experiments and presentation of experimental data is then treated, followed by applications of the results.

The basic equations used to analyze fluid flows are conservation of mass, Newton's second law of motion (linear momentum), and the first law of thermodynamics. These equations are derived in mechanics and physics for application to fixed masses. Special forms of the equations are required to analyze moving fluids.

A *system* is defined as a fixed mass of fluid. A *control volume* is an arbitrary boundary defined in the flow field to identify a region in space. Fluid may flow through the control volume, and exchanges of heat and work with the surroundings may occur.

The best system or control volume size for analysis depends on the information sought. Integral control volumes are used to obtain overall information such as thrust of a jet engine or force exerted by a liquid jet on a surface. Differential systems and control volumes are used to obtain detailed information about flow fields, such as point-by-point variation of velocity.

For analysis, the fluid is assumed to be a *continuum*; individual molecules are not considered. Fluid velocity is a vector that varies continuously throughout the flow; the velocity field is a vector field. It is possible to resolve the velocity into scalar components. Thus $\vec{V} = u\hat{i} + v\hat{j} + w\hat{k}$, where $\vec{V}$ is the velocity vector of the *fluid particle* (small volume of fluid surrounding point xyz); u, v, and w are scalar components of velocity; and $\hat{i}$, $\hat{j}$, $\hat{k}$ are unit vectors in the x, y, and z directions, respectively.

In the most general case, velocity is a function of three space coordinates and time, $\vec{V} = \vec{V}(x,y,z,t)$. For steady flow there is no time dependence. The number of space coordinates defines the dimensions of the flow field; $\vec{V} = \vec{V}(x)$ is a steady, one-dimensional flow field.

Stress is defined as the limiting value of force per unit area as the area is reduced to differential size. The simplest description of stress uses area elements having normals in the three coordinate directions; the infinitesimal force on each area element also may have three components. The notation τ_{yx} signifies a shear stress acting in the x direction on an area element with normal in the y direction. The *stress field* is the continuous distribution of stresses throughout the fluid; the stress field behaves as a second-order tensor.

35.2 Fluids without Relative Motion

In the absence of relative motion, no viscous stresses can be present within a fluid. The only surface stress is pressure, which acts against the surface of a fluid element. Pressure must vary continuously throughout the fluid, so it may be expanded in a Taylor series. Summing forces on an infinitesimal fluid element (Figure 35.1) leads to the expression

$$-\nabla p + \rho \vec{g} = \rho \vec{a} \tag{35.1}$$

This result shows that the *pressure gradient* ∇_p is the negative of the net body force per unit volume. (When $\vec{a} = 0$ this is the *basic equation of fluid statics.*)

When the fluid is static or the acceleration field is known, Equation (35.1) may be solved for pressure distribution. This is done most easily by expanding the pressure gradient into components (most advanced mathematics books give the *del operator* ∇ in rectangular, cylindrical, and spherical coordinates).

35.3 Basic Equations in Integral Form for Control Volumes

To formulate the basic equations for control volume application requires a limiting process [Fox et al., 2004] or derivation of the Reynolds transport theorem [Fay, 1994]. The resulting relation between the system expression and control volume variables is

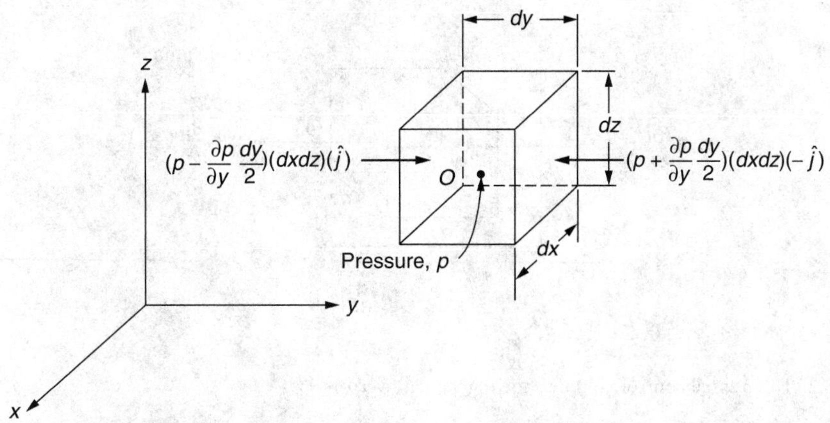

FIGURE 35.1 Differential fluid element showing pressure forces.

$$\left.\frac{dN}{dt}\right|_{\text{system}} = \frac{\partial}{\partial t}\int_{CV} \eta\rho d\forall + \int_{CS} \eta\rho \vec{V}\cdot d\vec{A} \tag{35.2}$$

To apply Equation (35.2), the system equation is formulated in terms of the rate of change of any extensive property N of the system; the corresponding intensive property is represented by η. The first integral in Equation (35.2) represents the quantity of N stored within the control volume; the second integral represents the net flux of N carried outward through the control surface.

Conservation of mass is obtained by substituting, for a system of constant mass, $dM/dt = 0$; the corresponding intensive properly is "mass per unit mass," so $\eta = 1$. Thus,

$$0 = \frac{\partial}{\partial t}\int_{CV} \rho d\forall + \int_{CS} \rho \vec{V}\cdot d\vec{A} \tag{35.3}$$

For incompressible flow, density cannot vary with time, so it is tempting to factor ρ from under the volume integral. However, parts of the control volume could be occupied by fluids having different densities at different times.

The momentum equation for control volumes is obtained by substituting the system form of Newton's second law into the left side of Equation (35.2) and setting $\eta = \vec{V}$ on the right side:

$$\vec{F}_S + \vec{F}_B = \frac{\partial}{\partial t}\int_{CV} \vec{V}\rho d\forall + \int_{CS} \vec{V}\rho \vec{V}\cdot d\vec{A} \tag{35.4}$$

The left side of Equation (35.4) represents external surface and body forces acting *on* the control volume. The first integral represents the rate of change of linear momentum contained within the control volume. The second integral accounts for the net flux of linear momentum from the control surface. Equation (35.4) is a vector equation; each of its three components must be satisfied.

The first law of thermodynamics is obtained by substituting the rate form of the system equation into Equation (35.2). The result is the scalar equation

$$\dot{Q} - \dot{W}_{\text{shaft}} = \frac{\partial}{\partial t}\int_{CV} e\rho d\forall + \int_{CS}\left(e + \frac{p}{\rho}\right)\rho \vec{V}\cdot d\vec{A} \tag{35.5}$$

In Equation (35.5) the intensive property stored energy $e = u + (V^2/2) + gz$ includes internal thermal energy u, kinetic energy $V^2/2$, and gravitational potential energy gz (all per unit mass). The rate of heat

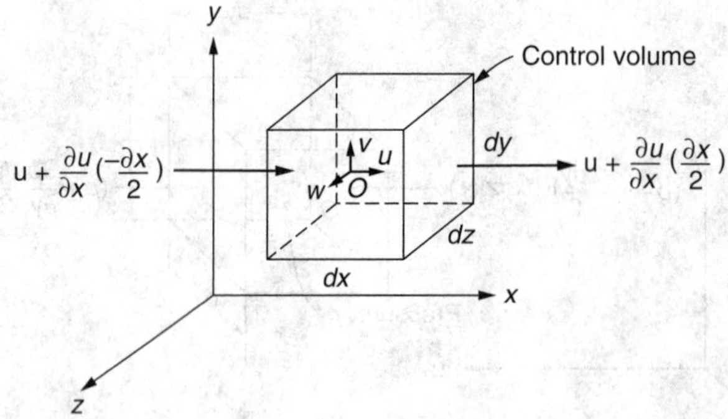

FIGURE 35.2 Differential control volume showing velocity vectors.

transfer $\dot{Q}$ is positive when into the control volume; the rate of shaft work $\dot{W}_{\text{shaft}}$ represents work done by the control volume. The first integral accounts for energy stored within the control volume; the second integral accounts for the flux of stored energy and flow work p/ρ done by pressure forces on the control surface.

35.4 Differential Analysis of Fluid Motion

Conservation of mass, Newton's second law of motion, and the first law of thermodynamics are independent physical principles that must be satisfied by any real flow. In principle it is possible to solve for three components of velocity and the pressure (four unknowns) using conservation of mass and the three components of the momentum equation. This usually is done using differential formulations to obtain detailed information about the flow field. The differential formulations may be developed using a differential system or control volume. Figure 35.2 shows a differential *CV* with velocity vectors; to set up the analysis the velocity vectors are chosen in positive coordinate directions.

Figure 35.2 shows the first term in the Taylor series expansion of each velocity component in the x direction. Similar expansions for velocity components in the other coordinate directions are summed to obtain the total flux of mass from the control volume (no mass storage term is needed since the fluid is incompressible):

$$\frac{\partial u}{\partial x}+\frac{\partial v}{\partial y}+\frac{\partial w}{\partial z}=0 \quad \text{or} \quad \nabla\cdot\vec{V}=0 \tag{35.6}$$

Equation (35.6) expresses conservation of mass in differential form. The equation was derived using an infinitesimal control volume but is valid at any point in the flow. The velocity field for incompressible flow must satisfy Equation (35.6).

Since the velocity varies from point to point, the acceleration of a fluid particle in a velocity field must be calculated using a special derivative called the *substantial derivative.* The acceleration of a fluid particle is given a special symbol D/Dt and written:

$$\frac{D\vec{V}}{Dt}=\underbrace{u\frac{\partial\vec{V}}{\partial x}+v\frac{\partial\vec{V}}{\partial y}+w\frac{\partial\vec{V}}{\partial z}}_{\text{Convective acceleration}}+\underbrace{\frac{\partial\vec{V}}{\partial t}}_{\text{Local acceleration}} \quad \text{or} \quad \frac{D\vec{V}}{Dt}=\vec{V}\cdot\nabla\vec{V}+\frac{\partial\vec{V}}{\partial t} \tag{35.7}$$

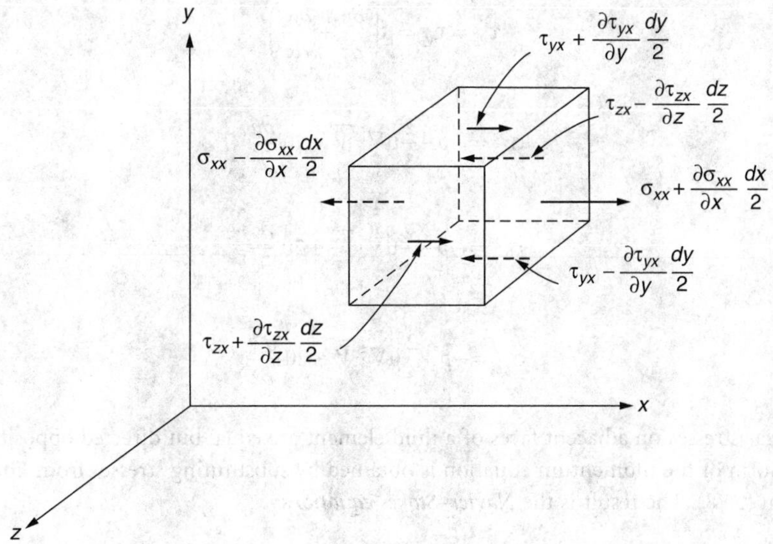

FIGURE 35.3 Differential fluid element showing stresses acting in the *x* direction.

Convective acceleration occurs when fluid particles are convected into regions of differing velocity; it may be nonzero even in a steady flow, such as steady flow through a nozzle. *Local acceleration* is caused by velocity variations with time; it is nonzero only for unsteady flow.

Forces acting on a fluid particle also may be obtained using the Taylor series expansion procedure. The results of expanding the stresses acting in the *x* direction on an infinitesimal fluid particle are shown in Figure 35.3.

Performing the same expansion in the other directions and collecting terms gives the *stress equations of motion:*

$$\rho g_x + \frac{\partial \sigma_{xx}}{\partial x} + \frac{\partial \tau_{yx}}{\partial y} + \frac{\partial \tau_{zx}}{\partial z} = \rho\left(u\frac{\partial u}{\partial x} + v\frac{\partial u}{\partial y} + w\frac{\partial u}{\partial z} + \frac{\partial u}{\partial t}\right) \tag{35.8a}$$

$$\rho g_y + \frac{\partial \tau_{xy}}{\partial x} + \frac{\partial \sigma_{yy}}{\partial y} + \frac{\partial \tau_{zy}}{\partial z} = \rho\left(u\frac{\partial v}{\partial x} + v\frac{\partial v}{\partial y} + w\frac{\partial v}{\partial z} + \frac{\partial v}{\partial t}\right) \tag{35.8b}$$

$$\rho g_z + \frac{\partial \tau_{xz}}{\partial x} + \frac{\partial \tau_{yz}}{\partial y} + \frac{\partial \sigma_{zz}}{\partial z} = \rho\left(u\frac{\partial w}{\partial x} + v\frac{\partial w}{\partial y} + w\frac{\partial w}{\partial z} + \frac{\partial w}{\partial t}\right) \tag{35.8c}$$

Before Equation (35.8a) through Equation (35.8c) may be used to solve for velocity, the stress field must be related to the velocity field. Details of this development are beyond the scope of this chapter, but are well covered by Sherman [1990]. For incompressible flow the stress components in rectangular coordinates are

$$\tau_{xy} = \tau_{yx} = \mu\left(\frac{\partial v}{\partial x} + \frac{\partial u}{\partial y}\right) \tag{35.9a}$$

$$\tau_{yz} = \tau_{zy} = \mu\left(\frac{\partial w}{\partial y} + \frac{\partial v}{\partial z}\right) \tag{35.9b}$$

$$\tau_{zx} = \tau_{xz} = \mu\left(\frac{\partial u}{\partial z} + \frac{\partial w}{\partial x}\right) \tag{35.9c}$$

$$\sigma_{xx} = -p - \frac{2}{3}\mu\nabla\cdot\vec{V} + 2\mu\frac{\partial u}{\partial x} \tag{35.9d}$$

$$\sigma_{yy} = -p - \frac{2}{3}\mu\nabla\cdot\vec{V} + 2\mu\frac{\partial v}{\partial y} \tag{35.9e}$$

$$\sigma_{zz} = -p - \frac{2}{3}\mu\nabla\cdot\vec{V} + 2\mu\frac{\partial w}{\partial z} \tag{35.9f}$$

Note that shear stresses on adjacent faces of a fluid element are equal but directed oppositely.

The final form of the momentum equation is obtained by substituting stresses from Equation (35.9) into Equation (35.8). The result is the *Navier–Stokes equations:*

$$\rho g_x - \frac{\partial p}{\partial x} + \mu\left(\frac{\partial^2 u}{\partial x^2} + \frac{\partial^2 u}{\partial y^2} + \frac{\partial^2 u}{\partial z^2}\right) = \rho\left(u\frac{\partial u}{\partial x} + v\frac{\partial u}{\partial y} + w\frac{\partial u}{\partial z} + \frac{\partial u}{\partial t}\right) \tag{35.10a}$$

$$\rho g_y - \frac{\partial p}{\partial y} + \mu\left(\frac{\partial^2 v}{\partial x^2} + \frac{\partial^2 v}{\partial y^2} + \frac{\partial^2 v}{\partial z^2}\right) = \rho\left(u\frac{\partial v}{\partial x} + v\frac{\partial v}{\partial y} + w\frac{\partial v}{\partial z} + \frac{\partial v}{\partial t}\right) \tag{35.10b}$$

$$\rho g_z - \frac{\partial p}{\partial z} + \mu\left(\frac{\partial^2 w}{\partial x^2} + \frac{\partial^2 w}{\partial y^2} + \frac{\partial^2 w}{\partial z^2}\right) = \rho\left(u\frac{\partial w}{\partial x} + v\frac{\partial w}{\partial y} + w\frac{\partial w}{\partial z} + \frac{\partial w}{\partial t}\right) \tag{35.10c}$$

These second-order, nonlinear partial differential equations are the fundamental equations of motion for viscous incompressible fluids. The Navier–Stokes equations are extremely difficult to solve analytically; only a handful of exact solutions are known [Sherman, 1990]. Some simplified cases can be solved numerically using today's advanced computers.

The Navier–Stokes equations also provide the starting point for stability analyses that predict the breakdown of laminar flow and the onset of turbulence.

35.5 Incompressible Inviscid Flow

All real fluids are viscous. However, in many situations it is reasonable to neglect viscous effects. Thus, it is useful to consider an incompressible *ideal fluid* with zero viscosity. When the fluid is inviscid there are no shear stresses; pressure is the only stress on a fluid particle.

The equations of motion for frictionless flow are called the *Euler equations.* They are obtained by substituting the acceleration of a fluid particle into Equation (35.1):

$$-\nabla p + \rho\vec{g} = \rho\frac{D\vec{V}}{Dt} \tag{35.11}$$

Equation (35.11) can be integrated to relate pressure, elevation, and velocity in a flowing fluid.

The Euler equations may be written in components using rectangular coordinates or using *streamline coordinates* defined along and normal to the flow streamlines. In streamline coordinates the components of the Euler equations are

$$\frac{\partial p}{\partial s} = -\rho V \frac{\partial V}{\partial s} \quad \text{(along a streamline)} \tag{35.12a}$$

$$\frac{\partial p}{\partial n} = \rho \frac{V^2}{R} \quad \text{(normal to a streamline)} \tag{35.12b}$$

Equation (35.12a) shows that, for frictionless flow, variations in pressure and velocity are opposite; pressure falls when velocity increases and vice versa. (Frictionless flow is an excellent model for accelerating flow; it must be used with caution for decelerating flow, in which viscous effects are likely to be important.)

Equation (35.12b) shows that pressure always increases in the direction outward from the center of curvature of streamlines. (The increasing pressure causes each fluid particle to follow a curved path along the curved streamline.) When streamlines are straight, the radius of curvature is infinite and pressure does not vary normal to the streamlines.

The *Bernoulli equation* is obtained when the Euler equation is integrated along a streamline for steady, incompressible flow without viscous effects:

$$\frac{p_1}{\rho} + \frac{V_1^2}{2} + gz_1 = \frac{p_2}{\rho} + \frac{V_2^2}{2} + gz_2 \tag{35.13}$$

The Bernoulli equation is one of the most useful equations in fluid mechanics, but it also is incorrectly applied frequently. The restrictions of steady, incompressible flow, along a streamline, without friction must be justified carefully each time the Bernoulli equation is used.

The Bernoulli equation may be used to predict pressure variations in external flow over objects and to design instrumentation for measuring pressure and velocity. *Stagnation pressure* is sensed by a total-head tube where $V = 0$. For this situation the Bernoulli equation reduces to

$$p_0 = p + \frac{1}{2}\rho V^2 \tag{35.14}$$

Equation (35.14) defines stagnation pressure p_0 as the sum of *static pressure* p and *dynamic pressure* $\frac{1}{2}\rho V^2$. A detailed discussion of fluid measurements is beyond the scope of this chapter, but these pressures can be measured using probes and a suitable instrument to sense pressure.

35.6 Dimensional Analysis and Similitude

Dimensional analysis is the process of combining key parameters of a flow situation into dimensionless groups. Several methods may be used to obtain the dimensionless groups [Fox et al., 2004], which reduce the number of variables needed to express the functional dependence of the results of an experiment or analysis. Thus, dimensionless groups simplify the presentation of data and permit analytical results to be generalized.

Each dimensionless group is a ratio of forces. Significant dimensionless groups in fluid mechanics include Reynolds number Re (ratio of inertia force to viscous force), pressure coefficient C_p (ratio of pressure force to inertia force), Froude number Fr (ratio of gravity force to inertia force), Weber number We (ratio of surface tension force to inertia force), and Mach number M (which may be interpreted as the ratio of inertia force to compressibility force).

Dynamic similarity occurs when ratios of all significant forces are the same between two flows. Dynamic similarity is required to scale model test results for use in prediction or design. Dynamic similarity is ensured for geometrically similar flows with corresponding flow patterns when all relevant dimensionless groups except one are duplicated between the two flows.

The basic differential equations also may be nondimensionalized to obtain dimensionless groups. Dynamic similarity is ensured when two flows are governed by the same differential equations with the same dimensionless coefficient values in the equations and boundary conditions. Strouhal number St is a frequency parameter that arises from boundary conditions for external flow with vortex shedding.

35.7 Internal Incompressible Viscous Flow

Laminar flow occurs at low Reynolds number; as Reynolds number increases, transition occurs and flow becomes turbulent. The numerical value corresponding to "low" Reynolds number depends on flow geometry. For circular pipes the Reynolds number at transition is $\mathrm{Re} = \rho \bar{V} D/\mu \approx 2000$, where $\bar{V}$ is the average velocity at any cross section. Transition Reynolds numbers for other geometries differ significantly.

Fully developed laminar flow cases in simple geometries can be solved by (1) using a differential control volume and Taylor series expansion to obtain an equation for shear stress variation or (2) reducing the Navier–Stokes equations to a simple form applicable to the flow field. Then the shear stress profile is integrated using appropriate boundary conditions to obtain the velocity profile. Once the velocity profile is obtained, volume flow rate, flow rate as a function of pressure drop, average velocity, and point(s) of maximum velocity can be found. Analyses of fully developed laminar flow cases are presented by Fox et al. [2004]; all known exact solutions of the Navier–Stokes equations are described in detail by Schlichting [1979] and White [1991].

Turbulent flow cannot be analyzed from first principles. Turbulence is characterized by velocity fluctuations that transport momentum across streamlines; no simple relationship exists between shear stress and strain rate in turbulent flow. Instantaneous properties cannot be predicted in a turbulent flow field; only average values can be calculated. For engineering analyses, turbulent flow is handled empirically using curve-fits to velocity profiles and experimentally determined loss coefficients.

Analysis of turbulent pipe flow is based on the first law of thermodynamics. Viscous friction causes irreversible conversion from mechanical energy to thermal energy. This conversion is regarded as a loss in mechanical energy called **head loss**:

$$\frac{p_1}{\rho} + \alpha_1 \frac{\bar{V}_1^2}{2} + g z_1 - \left(\frac{p_2}{\rho} + \alpha_2 \frac{\bar{V}_2^2}{2} + g z_2 \right) = h_{lT} \tag{35.15}$$

In Equation (35.15), total head loss h_{lT} is the difference between the mechanical energies at cross-sections 1 and 2; $\alpha \bar{V}^2/2$ is the kinetic energy flux (α is the *kinetic energy coefficient*).

To make calculations, total head loss is subdivided into "major" losses that occur in sections of constant area where flow is fully developed and "minor" losses in transitions such as entrances, fittings, valves, and exits. Major losses h_l in sections with fully developed flow are expressed in terms of the experimentally determined friction factor f:

$$h_l = f \frac{L}{D} \frac{\bar{V}^2}{2} \tag{35.16}$$

Friction factor is a function of Reynolds number Re and relative roughness e/D (equivalent roughness height e divided by tube diameter D). Results from numerous experiments were compiled and smooth curves fitted by Moody; the results are shown on the *Moody diagram* (Figure 35.4).

Minor loss data also are measured experimentally; minor losses h_{lm} may be expressed as:

$$h_{lm} = f \frac{L_e}{D} \frac{\bar{V}^2}{2} = K \frac{\bar{V}^2}{2} \tag{35.17}$$

Equivalent length ratios L_e/D and minor loss coefficients K are available from numerous sources. More details on computation of minor losses are in Chapter 40, "Valves."

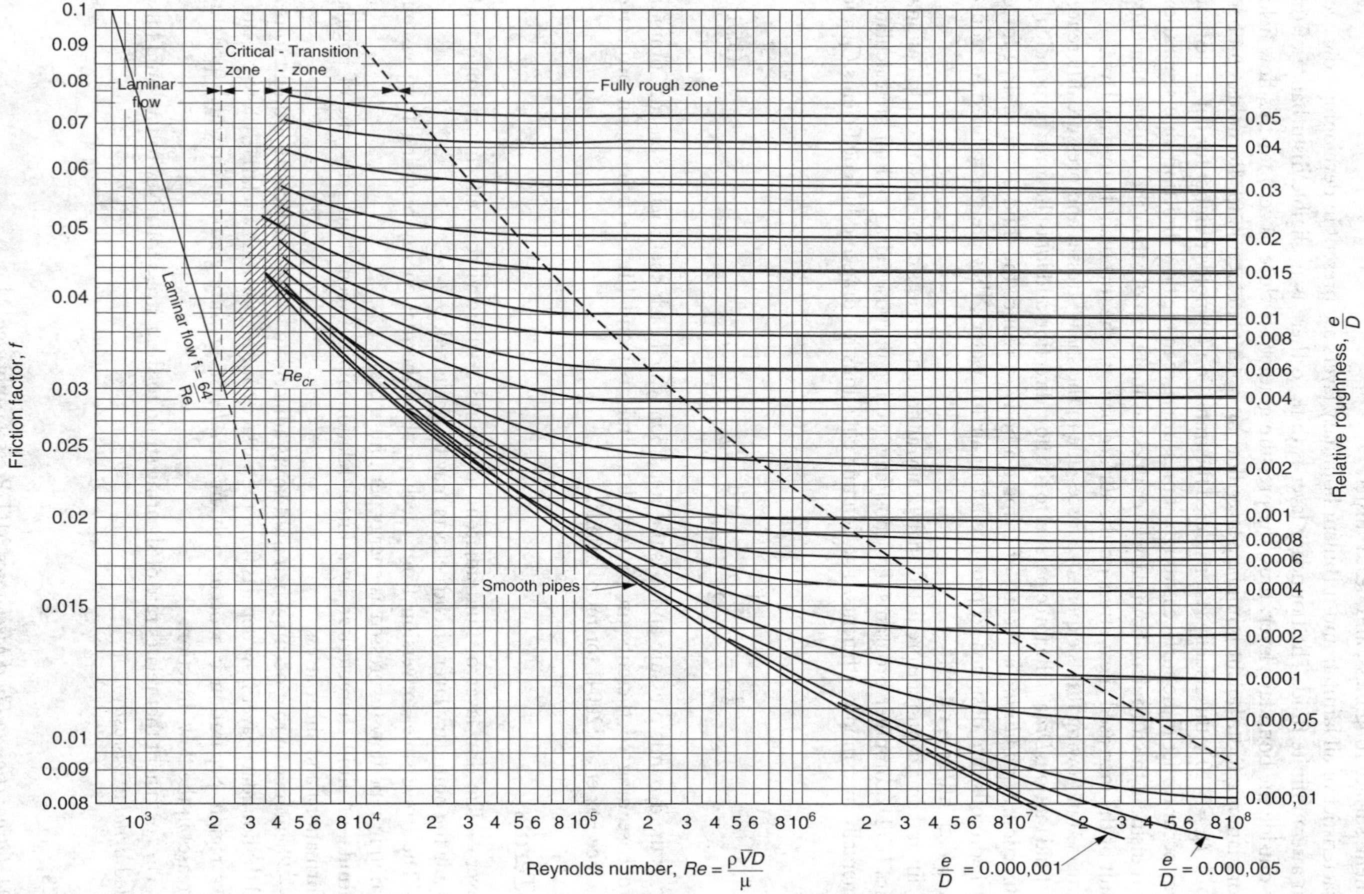

FIGURE 35.4 Moody diagram giving friction factors for pipe flow. (*Source:* Moody, L. F. 1944. *Trans. ASME.* 66(8):671–684. With permission.)

Computer programs that make calculations for pipe flow systems are commonly available. One such program accompanies the Fox et al. text [2004].

35.8 External Incompressible Viscous Flow

The **boundary layer** is the thin region near the surface of a body in which viscous effects are important. Boundary layers may be laminar or turbulent, depending on Reynolds number and factors such as pressure gradient, surface roughness, and heat transfer.

Basic characteristics of all laminar and turbulent boundary layers are present in developing flow over a flat plate in a semiinfinite fluid. The boundary layer is thin, so there is negligible disturbance of the inviscid flow outside the boundary layer; thus, the pressure gradient is close to zero for this flow field. *Transition* from laminar to turbulent boundary-layer flow on a flat plate occurs for Reynolds numbers above $\mathrm{Re} = \rho Ux/\mu \approx 5 \cdot 10^5$; this usually is considered the transition Reynolds number for flat-plate flow. Transition may occur earlier if the surface is rough, if pressure rises in the flow direction, or if **separation** occurs. After transition, the turbulent boundary layer thickens more rapidly than the laminar boundary layer as a result of the increased shear stress on the surface beneath the turbulent boundary layer.

Bodies immersed in flowing fluids experience forces due to the shear stresses and pressure differences caused by the fluid motion. *Drag* is the force parallel to the flow direction and *lift* is the force perpendicular to the flow direction. *Streamlining* is the art of shaping a body to reduce the fluid dynamic drag force. Airfoils (and hydrofoils) are designed to produce lift in air (or water); they are streamlined to reduce drag and attain high lift/drag ratios.

In general, lift and drag cannot be predicted analytically, although progress continues on computational fluid dynamics (CFD) computer programs. For most engineering purposes, drag and lift are calculated from experimentally measured coefficients. The defining equations for drag and lift coefficients C_D and C_L are

$$F_D = C_D A \frac{1}{2} \rho V^2 \quad \text{and} \quad F_L = C_L A \frac{1}{2} \rho V^2 \tag{35.18}$$

where $\frac{1}{2}\rho V^2$ is the dynamic pressure and A is the area upon which each coefficient is based. Common practice is to base drag coefficients on projected frontal area and lift coefficients on projected *planform* area. See Fox et al. [2004] for more details.

Defining Terms

Boundary layer — The thin layer of fluid adjacent to a surface where viscous effects are important; outside the boundary layer viscous effects may be neglected.

Head loss — The irreversible conversion from mechanical to thermal energy resulting from viscous friction in pipe flow (expressed as energy per unit mass).

Newtonian fluid — A fluid characterized by a linear relationship between shear rate (rate of angular deformation) and shear stress.

Separation — Phenomenon that occurs when fluid layers adjacent to a solid surface are brought to rest and the boundary-layer flow departs from the surface contour, forming a relatively low-pressure *wake* region. Separation can occur only in an *adverse pressure gradient*, in which pressure increases in the flow direction.

Viscosity — The coefficient that relates rate of shearing strain to shear stress for a Newtonian fluid (also called *dynamic viscosity*).

References

Fay, J. A. 1994. *Introduction to Fluid Mechanics,* MIT Press, Cambridge, MA.

Fox, R. W., McDonald, A. T., and Pritchard, P. J. 2004. *Introduction to Fluid Mechanics,* 6th ed. John Wiley & Sons, New York.

Moody, L. F. 1944. Friction factors for pipe flow. *Trans. ASME.* 66(8):671–684.

Schlichting, H. 1979. *Boundary-Layer Theory,* 7th ed. McGraw-Hill, New York.
Sherman, F. S. 1990. *Viscous Flow,* McGraw-Hill, New York.
White, F. M. 1991. *Viscous Fluid Flow,* McGraw-Hill, New York.

Further Information

A comprehensive source of basic information is the *Handbook of Fluid Dynamics,* edited by Victor L. Streeter (McGraw-Hill, New York, 1960).

Timely reviews of important topics are published in the *Annual Review of Fluid Mechanics* series (Annual Reviews, Palo Alto, CA). Each volume contains a cumulative index.

The *Journal of Fluids Engineering,* published quarterly (American Society of Mechanical Engineers, New York), contains articles with content ranging from fundamentals of fluid mechanics to fluid machinery.

The monthly *AIAA Journal* and semimonthly *Journal of Aircraft* (American Institute for Aeronautics and Astronautics, New York) treat aerospace applications of fluid mechanics.

Transportation aspects of fluid mechanics are covered in SAE publications (Society of Automotive Engineers, Warrendale, PA).

36

Compressible Flow

Afshin J. Ghajar
Oklahoma State University

36.1 Introduction

Compressible flow is defined as *variable density flow*; this is in contrast to incompressible flow, where the density is assumed to be constant throughout. The variation in density is mainly caused by variations in pressure and temperature. We sometimes call the study of such fluids in motion **gas dynamics.** Fluid compressibility is a very important consideration in modern engineering applications. Knowledge of compressible fluid flow theory is required in the design and operation of many devices commonly encountered in engineering practice. A few important examples are the external flow over modern high-speed aircrafts; internal flows through rocket, gas turbine, and reciprocating engines; flow through natural gas transmission pipelines; and flow in high-speed wind tunnels.

The variation of fluid density and other fluid properties for compressible flow gives rise to the occurrence of strange phenomena in compressible flow not found in incompressible flow. For example, with compressible flows we can have fluid deceleration in a convergent duct, fluid temperature decrease with heating, fluid acceleration due to friction, and discontinuous property changes in the flow.

There are many useful compressible flow references that the reader can consult, such as Anderson [2003], Oosthuizen and Carscallen [1997], Hodge and Koenig [1995], John [1984], and Zucrow and Hoffman [1976]. The objectives of this chapter are to primarily study compressibility effects by considering the steady, one-dimensional flow of an ideal gas. Although many real flows of engineering interest are more complex, these restrictions will allow us to concentrate on the effects of basic flow processes. Another aspect of this chapter is the consistent formulation of the equations in a form suitable for computer solution. The author and his associates have developed interactive software for the calculation of the properties of various compressible flows. The first version of the software called ***COMPROP*** was developed to accompany the textbook by Oosthuizen and Carscallen [1997]. A more recent version of the software called ***COMPROP2*** was developed to accompany the recent textbook by Anderson [2003]. For more detail about the development of ***COMPROP2*** and its capabilities see Tam et al. [2001].

0-8493-1586-7/05/$0.00+$1.50

36.2 The Mach Number and Flow Regimes

The single most important parameter in the analysis of the compressible fluids is the **Mach Number** (M), named after the nineteenth century Austrian physicist Ernst Mach. The Mach number (a dimensionless measure of compressibility) is defined as:

$$M = \frac{V}{a} \tag{36.1}$$

where V is the local flow velocity and a is the local **speed of sound** (the other common symbol used for the local speed of sound is "c"). For an ideal gas the speed of sound is given by [Anderson, 2003]:

$$a = \sqrt{\gamma R T} \tag{36.2}$$

where γ is the specific heat ratio (= 1.4 for air), R is the gas constant (= 287 J/kg·K for air), and T is the absolute fluid temperature. The speed of sound in a gas depends, therefore, only on the absolute temperature of the gas. For air at standard sea level conditions the speed sound is about 341 m/sec.

The Mach number can be used to characterize flow regimes as follows (the numerical values listed are only rough guides):

Incompressible flow — The Mach number is very small compared to unity ($M < 0.3$). For practical purposes the flow is treated as incompressible. For air at standard sea level conditions this assumption is good for local flow velocities of about 100 m/sec or less.

Subsonic flow — The Mach number is less than unity but large enough so that compressible flow effects are present ($0.3 < M < 1$).

Sonic flow — The Mach number is unity ($M = 1$). The significance of the point at which Mach number is equal to 1 will be demonstrated in upcoming sections.

Transonic flow — The Mach number is very close to unity ($0.8 < M < 1.2$). Modern aircrafts are mainly powered by gas turbine engines that involve transonic flows.

Supersonic flow — The Mach number is larger than unity ($M > 1$). For this flow, a shock wave is encountered. There are dramatic differences (physical and mathematical) between subsonic and supersonic flows, as will be discussed in the future sections.

Hypersonic flow — The Mach number is larger than five ($M > 5$). When a space shuttle reenters the earth's atmosphere, the flow is hypersonic. At very high Mach numbers the flowfield becomes very hot and dissociation and ionization of gases take place. In these cases the assumption of an ideal gas is no longer valid, and the flow must be analyzed by the use of kinetic theory of gases rather than continuum mechanics.

In the development of the equations of the motion of a compressible fluid, much of the analysis will appear in terms of the Mach number.

36.3 Ideal Gas Relations

Before we can proceed with the development of the equations of the motion of a compressible flow, we need to become familiar with the ideal gas fluid we will be working with. The ideal gas property changes can be evaluated from the following **equation of state** for an ideal gas:

$$p = \rho R T \tag{36.3}$$

where p is the fluid absolute pressure, ρ is the fluid density, T is the fluid absolute temperature, and R is the gas constant (= 287 J/kg·K for air). The **gas constant**, R, represents a constant for each distinct ideal gas, where

$$R = \frac{\overline{R}}{M} \tag{36.4}$$

with this notation, $\overline{R}$ is the universal gas constant (= 8314 J/kg·mol·K) and M is the molecular weight of the ideal gas (= 28.97 for air).

For an ideal gas, **internal energy**, u, and **enthalpy**, h, are considered to be functions of temperature only, and where the **specific heats at constant volume and pressure,** c_v and c_p, are also functions of temperature only. The changes in the internal energy and enthalpy of an ideal gas are computed for constant specific heats as:

$$u_2 - u_1 = c_v(T_2 - T_1) \tag{36.5}$$

$$h_2 - h_1 = c_p(T_2 - T_1) \tag{36.6}$$

For variable specific heats one must integrate du = $\int c_v dT$ and $dh = \int c_p dT$ or use the gas tables [Moran and Shapiro, 2000]. Most modern thermodynamics texts now contain software for evaluating properties of nonideal gases [Çengel and Boles, 2002].

From Equation (36.5) and Equation (36.6), we see that changes in internal energy and enthalpy are related to the changes in temperature by the values of c_v and c_p. We will now develop useful relations for determining c_v and c_p. From Equation (36.5), Equation (36.6), and the definition of enthalpy (h = u + pv = u + RT) it can be shown that [Moran and Shapiro, 2000]:

$$c_p - c_v = R \tag{36.7}$$

Equation (36.7) indicates that the difference between c_v and c_p is constant for each ideal gas regardless of temperature. Also $c_p > c_v$. If the **specific heat ratio,** γ, is defined as (the other common symbol used for specific heat ratio is "k")

$$\gamma = \frac{c_p}{c_v} \tag{36.8}$$

then combining Equation (36.7) and Equation (36.8) leads to

$$c_p = \frac{\gamma R}{\gamma - 1} \tag{36.9}$$

and

$$c_v = \frac{R}{\gamma - 1} \tag{36.10}$$

For air at standard conditions, c_p = 1005 J/kg·K and c_v = 718 J/kg·K. Equation (36.9) and Equation (36.10) will be useful in our subsequent treatment of compressible flow.

For compressible flows, changes in the thermodynamic property **entropy**, s, are also important. From the first and the second laws of thermodynamics, it can be shown that the change in entropy of an ideal gas with constant specific heat values (c_v and c_p) can be obtained from [Anderson, 2003]:

$$s_2 - s_1 = c_p \ln\frac{T_2}{T_1} - R\ln\frac{p_2}{p_1} \tag{36.11}$$

and

$$s_2 - s_1 = c_v \ln\frac{T_2}{T_1} + R\ln\frac{v_2}{v_1} \tag{36.12}$$

For variable specific heats one must integrate $\int c_p dT$ and $\int c_v dT$ or use the gas tables [Moran and Shapiro, 2000]. Equation (36.11) and Equation (36.12) allow the calculation of the change in entropy of an ideal gas between two states with constant specific heat values in terms of either the temperature and pressure, or the temperature and specific volume. Note that entropy is a function of both T and p, or T and v but not temperature alone (unlike internal energy and enthalpy).

36.4 Isentropic Flow Relations

An adiabatic flow (no heat transfer) which is frictionless (ideal or reversible) is referred to as **isentropic** (constant entropy) flow. Such flow does not occur in nature. However, the actual changes experienced by the large regions of the compressible flow field are often well approximated by this process. This is the case in internal flows such as for nozzles and external flows such as around an airfoil. In the regions adjacent to the nozzle walls or the airfoil surface, a thin boundary layer is formed and isentropic flow approximation fails. In this region flow is not adiabatic and reversible which causes the entropy to increase in the boundary layer.

Important relations for an isentropic flow of an ideal gas with constant c_v and c_p can be obtained directly from Equation (36.11) and Equation (36.12) by setting the left-hand side of these equations to zero ($s_2 = s_1$)

$$\frac{p_2}{p_1} = \left(\frac{\rho_2}{\rho_1}\right)^{\gamma} = \left(\frac{T_2}{T_1}\right)^{\gamma/(\gamma-1)} \tag{36.13}$$

Equation (36.13) relates absolute pressure, density, and absolute temperature for an isentropic process, and is very frequently used in the analysis of compressible flows.

36.5 Stagnation State and Properties

Stagnation state is defined as a state that would be reached by a fluid if it were brought to rest isentropically (reversibly and adaibatically) and without work. Figure 36.1 shows a stagnation point in compressible flow. The properties at the stagnation state are refereed to as **stagnation properties** (or total properties). The stagnation state and the stagnation properties are designated by the subscript 0 (or t). Stagnation properties are very useful and are used as a reference state for compressible flows.

Consider the steady flow of a fluid through a duct such as a nozzle, diffuser, or some other flow passage where the flow takes place adiabatically and with no shaft or electrical work. Assuming the fluid experi-

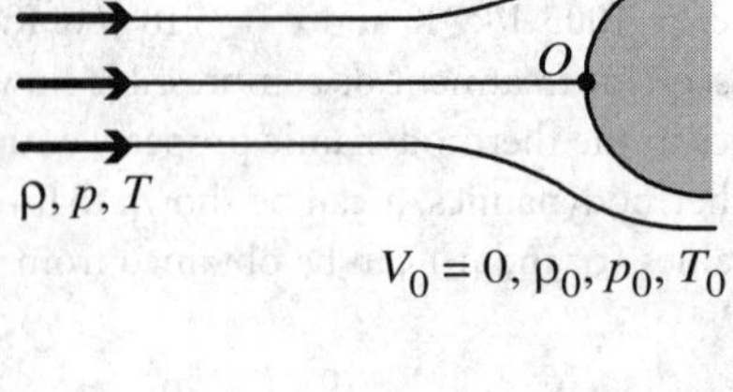

FIGURE 36.1 Stagnation point.

ences little or no change in its elevation and its potential energy, the energy equation between any two points in the flow for this single-stream steady-flow system reduces to

$$h_1 + \frac{V_1^2}{2} = h_2 + \frac{V_2^2}{2} \tag{36.14}$$

In Equation (36.14), if we let one of the points to be stagnation point (V = 0), then

$$h_0 = h + \frac{V^2}{2} \tag{36.15}$$

where h_0 is the **stagnation enthalpy** and h is the **static enthalpy** of the fluid. Combining Equation (36.14) and Equation (36.15)

$$h_{01} = h_{02} \tag{36.16}$$

That is, in the absence of any heat and work interactions and any changes in potential energy, the stagnation enthalpy of a fluid remains constant during a steady-flow process. Flows through nozzles and diffusers usually satisfy these conditions, and any increase (or decrease) in fluid velocity in these devices will create an equivalent decrease (or increase) in the static enthalpy of the fluid.

Frequently, there is difficulty in understanding the difference between stagnation and static properties. Stagnation properties are those properties experienced by a fixed observer, the fluid being brought to rest at the observer (V = 0). **Static properties** are those properties experienced by an observer moving with the same velocity as the stream. The difference between the static and stagnation properties is due to the velocity (or kinetic energy) of the flow, see Equation (36.15). We may regard the stagnation conditions as local fluid properties. Aside from analytical convenience, the definition of the stagnation state is useful experimentally, since stagnation temperature, T_0, and stagnation pressure, p_0, are relatively easily measured. It is usually much more convenient to measure stagnation temperature T_0 than the static temperature T.

36.6 Stagnation Property Relations

Recall the definition of stagnation enthalpy given by Equation (36.15), for an ideal gas with constant specific heats, its static and stagnation enthalpies can be replaced by c_pT or c_pT_0, respectively,

$$c_pT_0 = c_pT + \frac{V^2}{2}$$

or

$$T_0 = T + \frac{V^2}{2c_p} \tag{36.17}$$

In Equation (36.17), the **stagnation temperature,** T_0, represents the temperature an ideal gas will attain when it is brought to rest adiabatically. The term $V^2/2c_p$ corresponds to the temperature rise during such a process and is called the **dynamic temperature** (or impact temperature rise). Note that for low-speed flows, the stagnation and static temperatures are typically the same. But for high-speed flows, the stagnation temperature (measured by a stationary probe, for example) may be significantly higher than the static temperature of the fluid.

Introducing Equation (36.1) for M and Equation (36.9) for c_p into Equation (36.17) we obtain

$$\frac{T_0}{T} = \left(1 + \frac{\gamma - 1}{2} M^2\right) \tag{36.18}$$

Equation (36.18) gives the ratio of the stagnation to static temperature at a point in a flow as a function of the Mach number at that point. Equation (36.17) and Equation (36.18) are valid for any adiabatic flow whether thermodynamically reversible or not. They are, therefore, valid across a shock wave which is irreversible.

The ratio of the **stagnation pressure** to static pressure is obtained by substituting Equation (36.18) into the isentropic relation for pressure given by Equation (36.13) and letting state 2 be the stagnation state:

$$\frac{p_0}{p} = \left(\frac{T_0}{T}\right)^{\gamma/(\gamma-1)} = \left(1 + \frac{\gamma - 1}{2} M^2\right)^{\gamma/(\gamma-1)} \tag{36.19}$$

The ratio of the **stagnation density** to static density is obtained by substituting Equation (36.18) into the isentropic relation for density given by Equation (36.13) and letting state 2 be the stagnation state:

$$\frac{\rho_0}{\rho} = \left(\frac{T_0}{T}\right)^{1/(\gamma-1)} = \left(1 + \frac{\gamma - 1}{2} M^2\right)^{1/(\gamma-1)} \tag{36.20}$$

Equation (36.19) and Equation (36.20) give the ratios of stagnation to static pressure and density, respectively, at a point in the flow field as a function of the Mach number at that point.

Equation (36.18), Equation (36.19), and Equation (36.20) provide important relations for stagnation properties and are usually tabulated as a function of Mach number M for $\gamma = 1.4$ (corresponds to air at standard conditions) in most standard compressible flow (gas dynamics) textbooks [Anderson, 2003; John, 1984; Zucrow and Hoffman, 1976]. Figure 36.2 was developed using ***COMPROP2*** [Tam et al., 2001] and shows the variation of these stagnation properties as a function of Mach number for $\gamma = 1.4$. It is important to note that the *local* value of stagnation property depends only upon the local value of the static property and the local Mach number and is independent of the flow process. Equation (36.18), Equation (36.19), and Equation (36.20) may be used to determine these local stagnation values, even for nonisentropic flow, assuming that the local static property and local Mach number are known. These equations also allow us to relate stagnation properties between any two points (say points 1 and 2) in the compressible flowfield. For example, if the actual flow between points 1 and 2 in the flowfield is reversible and adiabatic (isentropic), then T_0, p_0, and ρ_0 have constant values at every point in the

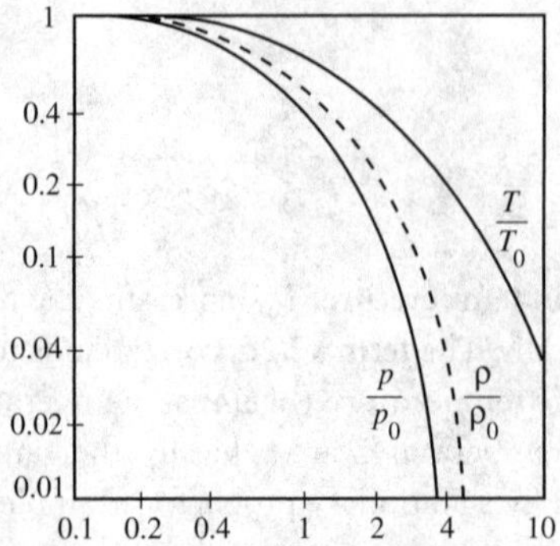

FIGURE 36.2 Stagnation property ratios for an ideal gas with $\gamma = 1.4$.

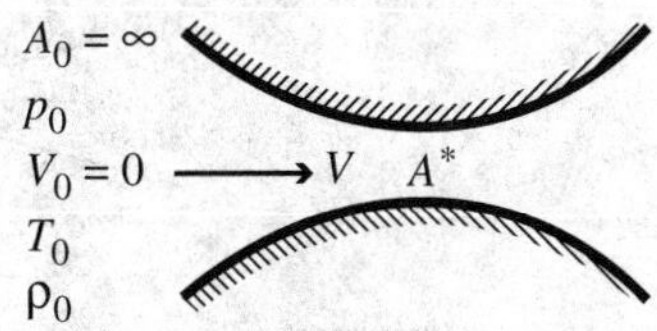

FIGURE 36.3 Converging–diverging nozzle.

flowfield. On the other hand, if the flow is irreversible and adiabatic, then only T_0 will remain constant at every point in the flow. However, for the case where the actual flow is irreversible and nonadiabatic, none of the stagnation properties stay constant between points 1 and 2 in the flowfield ($T_{01} \neq T_{02}$, $p_{01} \neq p_{02}$, and $\rho_{01} \neq \rho_{02}$).

36.7 Isentropic Flow with Area Changes

Nozzles are flow passages which accelerate the fluid to higher speeds. **Diffusers** accomplish the opposite, that is, they are used to decelerate the flow to lower speeds. These devices are quite common in gas turbines, rockets, and flow metering devices. In **incompressible flow** (ρ = constant), the volumetric flow rate (product of flow velocity, V, and the cross-sectional area, A) is constant. Thus, any passage which converges (causes A to decrease in the flow direction) is a nozzle, and any diverging passage (causes A to increase in the flow direction) is a diffuser. In fact, in any subsonic flow ($M < 1$), a converging channel accelerates and a diverging channel decelerates the flow. As will be shown in this section, just the opposite is true in supersonic flow ($M > 1$).

For the converging–diverging nozzle shown in Figure 36.3, the conservation of mass (continuity) under steady state conditions for this one-dimensional flow can be written as

$$\dot{m} = \rho V A = \text{constant} \tag{36.21}$$

The above equation can be used to relate the mass flow rate ($\dot{m}$) at different sections of the channel. Taking the logarithm of Equation (36.21) and then differentiating the resulting equation, we get

$$\frac{d\rho}{\rho} + \frac{dA}{A} + \frac{dV}{V} = 0 \tag{36.22}$$

The differential form of the frictionless momentum equation for our steady one-dimensional flow is

$$dp + \rho V dV = 0 \tag{36.23}$$

Equation (36.23) could have also been obtained from the steady one-dimensional energy equation for an isentroic flow with no work interactions and no potential energy. Combining Equation (36.23) with Equation (36.22) and introducing the definition of Mach number, Equation (36.1), one can obtain

$$\frac{dA}{A} = -\frac{dV}{V}\left(1 - \frac{V^2}{dp/d\rho}\right) = -\frac{dV}{V}\left(1 - \frac{V^2}{a^2}\right)$$

or

$$\frac{dA}{dV} = \frac{A}{V}(M^2 - 1) \tag{36.24}$$

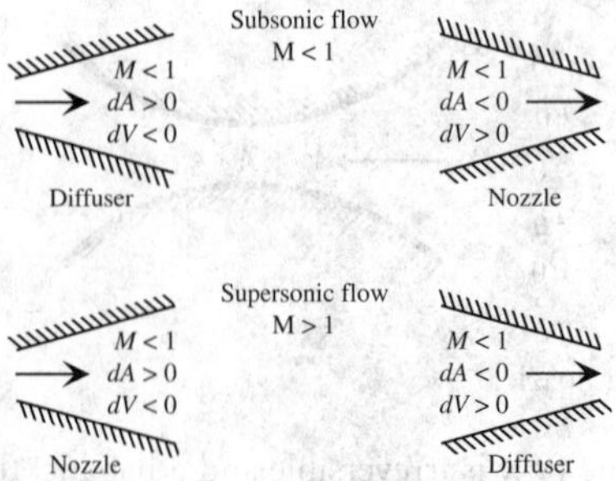

FIGURE 36.4 Area and velocity changes for subsonic and supersonic gas flow.

TABLE 36.1 Variation of Flow Properties in Converging or Diverging Channels

Type of Flow Passage	M	dA	dM	dV	dp	dT	dρ
Subsonic converging nozzle	<1	–	+	+	–	–	–
Subsonic diverging diffuser	<1	+	–	–	+	+	+
Supersonic converging diffuser	>1	–	–	–	+	+	+
Supersonic diverging nozzle	>1	+	+	+	–	–	–

where in the above equation for an isentropic flow, the speed of sound can be expressed as $a = \sqrt{dp/d\rho}$ [John, 1984]. Inspection of Equation (36.24), without actually solving it, shows a fascinating aspect of compressible flow. As mentioned at the beginning of this section, property changes are of opposite sign for subsonic and supersonic flow. This is because of the term $(M^2 - 1)$ in Equation (36.24). There are four combinations of area change and Mach number summarized in Figure 36.4. The variation of velocity, pressure, temperature, and density in converging–diverging channels in both subsonic and supersonic flow is tabulated in Table 36.1.

Equation (36.24) and Figure 36.4 have many ramifications. For M = 1 (sonic flow), Equation (36.24) yields dA/dV = 0. Mathematically this result suggests that the area associated with sonic flow (M = 1) is either a minimum or a maximum amount. The minimum in area is the only physically realistic solution. A convergent–divergent channel involves a minimum area (see Figure 36.3). These results indicate that the sonic condition (M = 1) can occur in a converging–diverging duct at the minimum area location, often referred to as the **throat** of a converging–diverging channel. Therefore, for the steady flow of an ideal gas to expand isentropically from subsonic to supersonic speeds, a convergent–divergent channel must be used. This is why rocket engines, in order to expand the exhaust gases to high-velocity, supersonic speeds, use a large bell-shaped exhaust nozzle. Conversely, for an ideal gas to compress isentropically from supersonic to subsonic speeds, it must also flow through a convergent–divergent channel, with a throat where M = 1 occurs.

The Mach number–area variation in a nozzle can be determined by combining the continuity relation [Equation (36.21)] with the ideal gas and isentropic flow relations. For this purpose, equate the mass flow rate at any section of the nozzle in Figure 36.3 to the mass flow rate under sonic conditions (at the throat, the flow is sonic and the conditions are denoted by an asterisk and are referred to as **critical conditions**):

$$\rho V A = \rho^* V^* A^*$$

or

$$\frac{A}{A^*} = \frac{\rho^*}{\rho}\frac{V^*}{V} = \frac{\rho^*}{\rho}\frac{M^*}{M}\frac{a^*}{a} = \frac{1}{M}\left(\frac{\rho^*}{\rho_0}\right)\left(\frac{\rho_0}{\rho}\right)\sqrt{\frac{T^*/T_0}{T/T_0}} \tag{36.25}$$

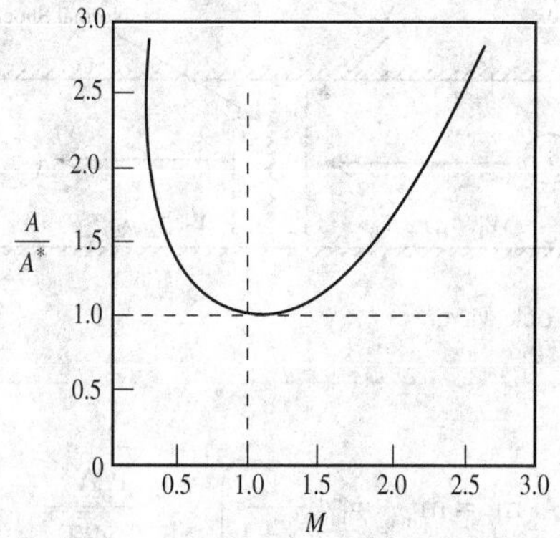

FIGURE 36.5 Variation of A/A^* with Mach number in isentropic flow for $\gamma = 1.4$.

where from Equation (36.2), $a = \sqrt{\gamma RT}$, and at the throat, the area of the throat is A^* and $M^* = 1(V^* = a^* = \sqrt{\gamma RT^*})$, and for an isentropic flow ρ_0 and T_0 is constant throughout the flow. Substituting Equation (36.18) and Equation (36.20) into Equation (36.25) and recognizing that the **critical temperature ratio** (T^*/T_0) and the **critical density ratio** (ρ^*/ρ_0) is obtained from Equation (36.18) and Equation (36.20), respectively with M = 1, we obtain

$$\frac{A}{A^*} = \frac{1}{M}\left[\frac{2}{\gamma+1}\left(1+\frac{\gamma-1}{2}M^2\right)\right]^{\frac{\gamma+1}{2(\gamma-1)}} \tag{36.26}$$

Equation (36.26) and ***COMPROP2*** [Tam et al., 2001] were used to generate the plot of A/A^* shown in Figure 36.5 for $\gamma = 1.4$. Numerical values of A/A^* versus M are also usually tabulated alongside stagnation properties given by Equation (36.18) through Equation (36.20) in most standard compressible flow textbooks, see, for example, Anderson [2003]. As can be seen from Figure 36.5, for each value of A/A^*, there are two possible isentropic solutions, one subsonic and the other supersonic. For example, from Equation (36.26) with $\gamma = 1.4$ or Figure 36.5, with $A/A^* = 2$, M = 0.3 and also M = 2.2. The minimum area (throat) occurs at M = 1. This agrees with the results of Equation (36.24), illustrated in Figure 36.4. That is, to accelerate a slow moving fluid to supersonic velocities, a converging–diverging nozzle is needed.

Under steady state conditions, the **mass flow rate** through a nozzle can be calculated from Equation (36.21) expressed in terms of M and the stagnation properties T_0 and ρ_0 from Equation (36.18) and Equation (36.19) as:

$$\dot{m} = \rho AV = \left(\frac{P}{RT}\right)A(M\sqrt{\gamma RT}) = PAM\sqrt{\frac{\gamma}{RT}} = \frac{\gamma P_0 AM}{\sqrt{\gamma RT_0}}\left(1+\frac{\gamma-1}{2}M^2\right)^{-\frac{(\gamma+1)}{2(\gamma-1)}} \tag{36.27}$$

Thus the mass flow rate of a particular fluid through a nozzle is a function of the stagnation properties of a fluid, the flow area, and the Mach number. The above relationship is valid at any location along the length of the nozzle.

For a specified flow area A and stagnation properties T_0 and ρ_0, the **maximum mass flow rate** through a nozzle can be determined by differentiating Equation (36.27) with respect to M and setting the result equal to zero. It yields M = 1. As discussed above, the only location in a nozzle where M = 1 is at the throat (minimum flow area). Therefore, the maximum possible mass flow passes through a nozzle when its throat is at the critical or sonic condition. The nozzle is then said to be **choked** and can carry no additional mass flow unless the throat is widened. If the throat is constricted further, the mass flow rate through the nozzle must decrease. We can obtain an expression for the maximum mass flow rate by substituting M = 1 in Equation (36.27):

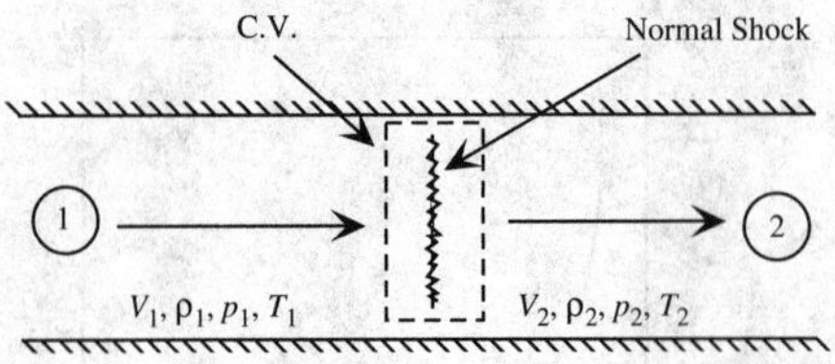

FIGURE 36.6 Stationary normal shock wave.

$$\dot{m}^* = \dot{m}_{max} = \gamma\left(\frac{2}{\gamma+1}\right)^{\frac{\gamma+1}{2(\gamma-1)}} \frac{P_0 A^*}{\sqrt{\gamma R T_0}} \tag{36.28}$$

Thus, for isentropic flow of a particular ideal gas through a nozzle, the maximum mass flow rate possible with a given throat area is fixed by the stagnation pressure and temperature of the inlet flow.

36.8 Normal and Oblique Shock Waves

When the flow velocity exceeds the speed of sound ($M > 1$), adjustments in the flow often take place through abrupt discontinuous surfaces called **shock waves.** This is one of the most interesting and unique phenomena that occurs in supersonic flow. A shock wave can be considered as a discontinuity in the properties of the flowfield. The process is irreversible. A shock wave is extremely thin, usually only a few molecular mean free paths thick (for air $\approx 10^{-5}$ cm). A shock wave is, in general, curved. However, many shock waves that occur in practical situations are straight, being either at right angles to the flow path (termed a **normal shock**) or at an angle to the flow path (termed an **oblique shock**). In case of a normal shock, the velocities both ahead (i.e., upstream) of the shock and after (i.e., downstream) the shock are at right angles to the shock wave. However, in the case of an oblique shock there is a change in the flow direction across the shock.

Attention will be given first to the changes that occur through a stationary normal shock. Fluid crossing a normal shock experiences a sudden increase in pressure, temperature, and density, accompanied by a sudden decrease in velocity from a supersonic flow to a subsonic flow. Consider an ideal gas flowing in a duct as shown in Figure 36.6. For steady state flow through a stationary normal shock, with no direction change, area change (shock is very thin), heat transfer (shock is adiabatic), or work done, the mass (continuity), momentum, and energy equations are:

Mass:
$$\rho_1 V_1 = \rho_2 V_2 \tag{36.29}$$

Momentum:
$$p_1 - p_2 = \rho_1 V_1 (V_2 - V_1) \tag{36.30}$$

Energy:
$$T_{01} = T_1 + \frac{V_1^2}{2c_p} = T_2 + \frac{V_2^2}{2c_p} = T_{02} \tag{36.31}$$

These equations, together with the definition of Mach number, Equation (36.1), the equation for speed of sound, Equation (36.2), the equation of state, Equation (36.3), and the equation for stagnation temperature, Equation (36.18), will yield two solutions. One solution, which is trivial, states that there is no change and hence no shock wave. The other solution, which corresponds to the change across a stationary normal shock wave, can be expressed in terms of the upstream Mach number:

$$M_2^2 = \frac{(\gamma-1)M_1^2+2}{2\gamma M_1^2-(\gamma-1)} \tag{36.32}$$

$$\frac{p_2}{p_1} = \frac{2\gamma M_1^2-(\gamma-1)}{(\gamma+1)} \tag{36.33}$$

$$\frac{p_{02}}{p_{01}} = \left[\frac{(\gamma+1)M_1^2}{2+(\gamma-1)M_1^2}\right]^{\gamma/(\gamma-1)}\left[\frac{\gamma+1}{2\gamma M_1^2-(\gamma-1)}\right]^{1/(\gamma-1)} \tag{36.34}$$

$$\frac{T_2}{T_1} = [2+(\gamma-1)M_1^2]\frac{2\gamma M_1^2-(\gamma-1)}{(\gamma+1)^2 M_1^2} \tag{36.35}$$

$$\frac{\rho_2}{\rho_1} = \frac{V_1}{V_2} = \frac{(\gamma+1)M_1^2}{2+(\gamma-1)M_1^2} \tag{36.36}$$

The variations of p_2/p_1, ρ_2/ρ_1, T_2/T_1, p_{02}/p_{01}, and M_2 with M_1 as obtained from Equation (36.32) through Equation (36.36) and ***COMPROP2*** [Tam et al., 2001] are plotted in Figure 36.7 for $\gamma = 1.4$, and they are normally tabulated in standard compressible flow texts in "Normal Shock Tables," see, for example, Oosthuizen and Carscallen [1997]. From Figure 36.7 we can see that rather large losses of stagnation pressure occur across the normal shock. For an adiabatic process (flow across the normal shock is adiabatic but irreversible), the stagnation pressure represents a measure of available energy of the flow in a given state. A decrease in stagnation pressure, or increase in entropy ($s_2 > s_1$), represents an energy dissipation or **loss of available energy**. The increase in entropy across the normal shock can be related to the stagnation pressure ratio across the shock by substituting Equation (36.18) and Equation (36.19) in Equation (36.11) and recognizing that stagnation temperature remains constant across the normal shock, [see energy Equation (36.31)]. The result is:

$$s_2 - s_1 = -R\ln\frac{p_{02}}{p_{01}} \tag{36.37}$$

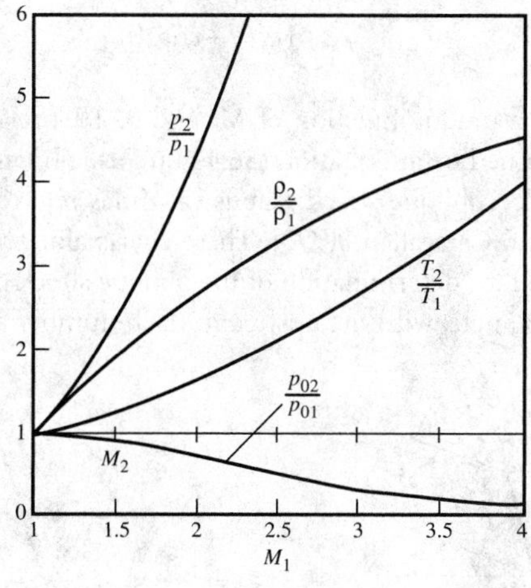

FIGURE 36.7 Variation of flow properties across a stationary normal shock wave for $\gamma = 1.4$.

From the second law of thermodynamics we must have $s_2 > s_1$. In order to minimize the loss of available energy across a normal shock, we need to have a small change in the stagnation pressure across the normal shock. Examination of Figure 36.7 shows that in order for this to happen the Mach number at the upstream of the normal shock (M_1) must be near unity.

If a plane shock is not perpendicular to the flow but inclined at an angle (termed an oblique shock), the shock will cause the fluid passing through it to change direction, in addition to increasing its pressure, temperature, and density, and decreasing its velocity. An oblique shock is illustrated in Figure 36.8; in this case the fluid flow is deflected through an angle δ, called the **deflection angle.** The angle θ shown in the figure is referred to as the **shock or wave angle** and the subscripts n and t indicate directions normal and tangent to the shock, respectively. The conservation of mass, momentum, and energy equations for the indicated control volume, see Figure 36.8, are:

Mass:
$$\rho_1 V_{1n} = \rho_2 V_{2n} \tag{36.38}$$

Momentum:
$$p_1 - p_2 = \rho_2 V_{2n}^2 - \rho_1 V_{1n}^2 \text{ (normal to the shock)} \tag{36.39}$$

$$0 = \rho_1 V_{1n}(V_{2t} - V_{1t}) \text{ or } V_{2t} = V_{1t} \text{ (tangent to the shock)} \tag{36.40}$$

Energy:
$$T_{01} = T_1 + \frac{V_{1n}^2}{2c_p} = T_2 + \frac{V_{2n}^2}{2c_p} = T_{02} \tag{36.41}$$

Since the conservation equations [Equation (36.38), Equation (36.39), and Equation (36.41)] for the oblique shock contain only the normal component of the velocity, they are identical to the normal shock conservation equations [Equation (36.29) through Equation (36.31)]. In other words, an oblique shock acts as a normal shock for the component normal to the shock, while the tangential velocity remains unchanged, [Equation (36.40)]. This fact permits the use of normal shock equations to calculate oblique shock parameters. To use normal shock equations for oblique shock calculations, in normal shock equations replace M_1 by M_{1n} and M_2 by M_{2n} where $M_{1n} = M_1 \sin\theta$ and $M_{2n} = M_2 \sin(\theta - \delta)$. Equations for M_{1n} and M_{2n} are dependent on δ and their values cannot be determined until the deflection angle is obtained. However, δ is a unique function of M_1 and θ. From the geometry of Figure 36.8 and some trigonometric manipulation, the following δ – θ – M relationship can be obtained:

$$\tan\delta = \frac{2\cot\theta(M_1^2 \sin^2\theta - 1)}{2 + M_1^2(\gamma + \cos 2\theta)} \tag{36.42}$$

Equation (36.42) specifies δ as unique function of M_1 and θ. This relation is vital to the analysis of oblique shocks. The results obtained from Equation (36.42) are usually presented in the form of a graph as shown in Figure 36.9. Detailed oblique shock graphs (at times referred to as oblique shock charts) may be found in Oosthuizen and Carscallen [1997]. These charts along with the modified form of the normal shock equations are used for determination of the oblique shock properties. Figure 36.9 is a plot of shock angle versus deflection angle, with the upstream Mach number as a parameter. It is interesting

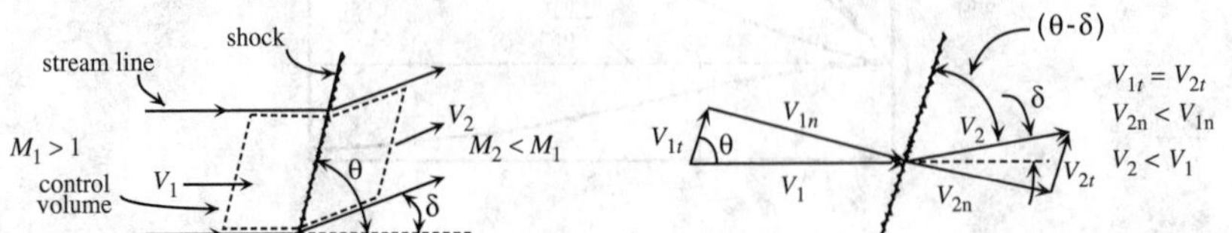

FIGURE 36.8 Oblique shock wave.

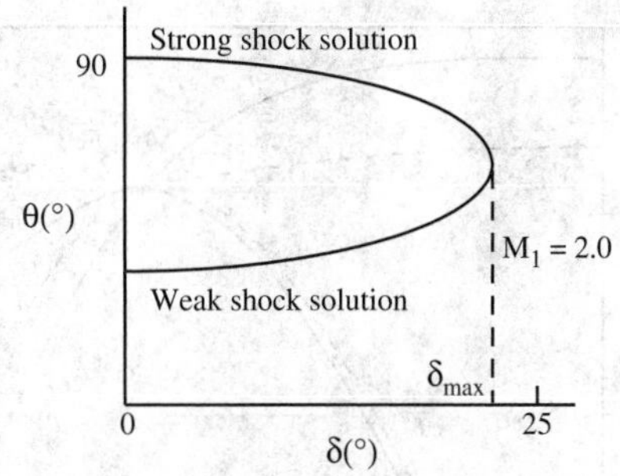

FIGURE 36.9 Oblique shock angle versus deflection angle and upstream Mach number for $\gamma = 1.4$.

to note that in the oblique shock figures, for a given initial Mach number (M_1) and a given deflection angle (δ), there are either two solutions (solid line and the dashed line in the figure) or none at all. Figure 36.9 shows that for a given M_1 (in this case $M_1 = 2.0$), a **maximum deflection angle** (δ_{max}) can be found (in this case $\delta_{max} = 23°$). This maximum varies from 0° at $M_1 = 1$ to about 45° as $M_1 \to \infty$. If δ_{max} is exceeded, the shock detaches, that is, moves ahead of the turning surface and becomes curved. For such a case, no solution exists on the oblique shock figures. In case of the attached shock, where two solutions are possible, the **weak shock** solution (solid line in Figure 36.9) is most common. A weak shock is the solution farthest removed from the normal shock case.

36.9 Rayleigh Flow

Ideal gas flow in a constant area duct with heating or cooling (stagnation temperature change) and without friction is referred to as **Rayleigh flow.** Heat can be added or removed to the gas by heat exchange through the duct walls, by radiative heat transfer, by combustion, or by evaporation and condensation. Although the frictionless (inviscid) assumption may appear unrealistic, Rayleigh flow is nevertheless useful for the analysis of jet engine combustors and flowing gaseous lasers. In these devices, the heat addition process dominates the viscous effects.

The conservation equations for the Rayleigh flow combined with the equation of state, the Mach number equation, and the definitions of the stagnation temperature and pressure, leads to the following equations; a complete analysis is given in Anderson [2003]:

$$\frac{T_0}{T_0{}^*} = \frac{(\gamma+1)M^2}{(1+\gamma M^2)^2}[2+(\gamma-1)M^2] \tag{36.43}$$

$$\frac{T}{T^*} = \frac{(1+\gamma)^2 M^2}{(1+\gamma M^2)^2} \tag{36.44}$$

$$\frac{p}{p^*} = \frac{(1+\gamma)}{(1+\gamma M^2)} \tag{36.45}$$

$$\frac{p_0}{p_0{}^*} = \left(\frac{1+\gamma}{1+\gamma M^2}\right)\left[\frac{2+(\gamma-1)M^2}{\gamma+1}\right]^{\gamma/(\gamma-1)} \tag{36.46}$$

For convenience of calculation, in Equation (36.43) through Equation (36.46) sonic flow has been used as a reference condition, where the superscript asterisk (*) signifies properties at $M = 1$. In this way, the fluid properties for Rayleigh flow have been presented as a function of a single variable, the local Mach number. Equation (36.43) through Equation (36.46) and ***COMPROP2*** [Tam et al., 2001] were used to

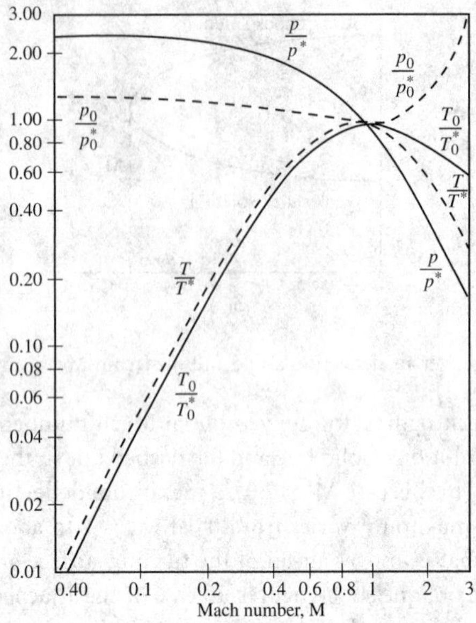

FIGURE 36.10 Rayleigh flow property variations with Mach number for γ = 1.4.

generate Figure 36.10 for γ = 1.4. These property variations are also normally tabulated in standard compressible flow texts in "Rayleigh Flow Tables," see, for example, John [1984]. Note that for a given flow no matter what the local flow properties are, the reference sonic conditions (the starred quantities) are constant values. Given particular inlet conditions to the duct (T_{01}, p_{01}, M_1) and $\dot{q}$ (the rate of heat transfer per unit mass of the flowing fluid), we can obtain the duct exit conditions after a given change in stagnation temperature (heating or cooling) as follows: The value of inlet Mach number (M_1) fixes the value of T_{01}/T_0^* from Equation (36.43) and thus the value of T_0^*, since we know T_{01}. The exit state T_{02}/T_0^* can then be determined from the following conservation of energy equation for this flow (for a given value of c_p or γ):

$$c_p(T_2 - T_1) + \frac{1}{2}(V_2^2 - V_1^2) = \dot{q} = c_p(T_{02} - T_{01}) = c_p \Delta T_0$$

or

$$\frac{T_{02}}{T_0{}^*} = \frac{T_{01}}{T_0{}^*} + \frac{\dot{q}}{c_p T_0{}^*} \tag{36.47}$$

By the value of T_{02}/T_0^*, then M_2, T/T^*, p_2/p^*, and p_{02}/p_0^* are all fixed from Equation (36.43) through Equation (36.46).

Referring to Figure 36.10 and comparing the flow properties with the T_0/T_0^* curve, several interesting facts about Rayleigh flow are evident. Considering the case of heating (increasing T_0/T_0^*), we notice that the increase in the stagnation temperature drives the Mach number toward unity for both subsonic and supersonic flow. After the Mach number has reached the sonic condition (M = 1), any further increase in heating (increase in stagnation temperature) is possible only if the initial conditions at the inlet of the duct are changed. Therefore, a maximum amount of heat can be added to flow in a duct, this maximum is determined by the attainment of Mach 1. Thus, flow in a duct can be chocked by heat addition. In this case flow is referred to as **thermally chocked.** Another interesting observation from Figure 36.10 is that the stagnation pressure always decreases for heat addition to the stream. Hence, combustion will cause a loss in stagnation pressure. Conversely, cooling tends to increase the stagnation pressure.

36.10 Fanno Flow

Flow of an ideal gas through a constant-area adiabatic duct with wall friction is referred to as **Fanno flow.** In effect, this is similar to a Moody-type pipe flow but with large changes in kinetic energy, enthalpy, and pressure in the flow.

The conservation equations for the Fanno flow combined with the definition of friction factor, equation of state, the Mach number equation, and the definitions of the stagnation temperature and pressure, leads to the following equations; a complete analysis is given in Anderson [2003]:

$$\frac{4f}{D}L^* = \left(\frac{1-M^2}{\gamma M^2}\right) + \frac{\gamma+1}{2\gamma}\ln\left[\frac{(\gamma+1)M^2}{2+(\gamma-1)M^2}\right] \tag{36.48}$$

$$\frac{T}{T^*} = \frac{(\gamma+1)}{2+(\gamma-1)M^2} \tag{36.49}$$

$$\frac{p}{p^*} = \frac{1}{M}\left[\frac{(\gamma+1)}{2+(\gamma-1)M^2}\right]^{1/2} \tag{36.50}$$

$$\frac{p_0}{p_0{}^*} = \frac{1}{M}\left[\frac{2+(\gamma-1)M^2}{(\gamma+1)}\right]^{(\gamma+1)/[2(\gamma-1)]} \tag{36.51}$$

Analogous to our discussion of Rayleigh flow, in Equation (36.48) through Equation (36.51) sonic flow ($M = 1$) has been used as a reference condition, where the flow properties are denoted by T^*, p^*, and $p_0{}^*$. L^* is defined as the length of the duct necessary to change the Mach number of the flow from M to unity and *f* is an average friction factor. Equations (36.48) through Equation (36.51) and ***COMPROP2*** [Tam et al., 2001] were used to generate Figure 36.11 for $\gamma = 1.4$. These equations are also normally tabulated in standard compressible flow texts in "Fanno Flow Tables," see, for example, John [1984].

Consider a duct of given cross-sectional area and variable length. If the inlet, mass flow rate, and average friction factor are fixed, there is a maximum length of the duct that can transmit the flow. Since the Mach number is unity at the duct exit in that case, the length is designated L^* and the flow is said to be **friction-chocked.** In other words, friction always derives the Mach number toward unity, decelerating a supersonic flow and accelerating a subsonic flow. From Equation (36.48) we can see that at any point in the duct (say point 1), the variable fL^*/D depends only on the Mach number at that point (M_1) and γ. Since the diameter (D) is constant and *f* is assumed constant, then at some other point (say 2) a distance L ($L < L^*$) downstream from point 1, we have

$$\left(\frac{4f}{D}L^*\right)_2 = \left(\frac{4f}{D}L^*\right)_1 - \left(\frac{4f}{D}L\right) \tag{36.52}$$

From Equation (36.52), we can determine M_2. If in a given situation M_2 was fixed, then Equation (36.52) can be rearranged to determine the length of duct required (L) for Mach number M_1 to change to Mach number M_2.

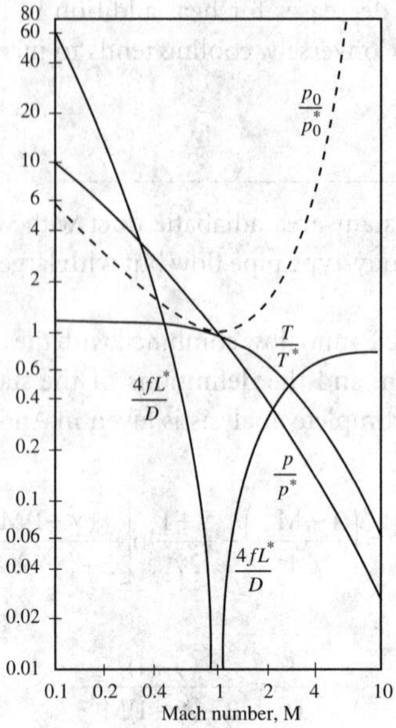

FIGURE 36.11 Fanno flow property variations with Mach number for $\gamma = 1.4$.

Defining Terms

Compressible flow — Flow in which the fluid density varies.

Isentropic flow — An adiabatic flow (no heat transfer) which is frictionless (ideal or reversible). For this flow the entropy is constant.

Stagnation state — A state that would be reached by a fluid if it were brought to rest isentropically (reversibly and adaibatically) and without work. The properties at the stagnation state are referred to as **stagnation properties** (or total properties).

Shock wave — A fully developed compression wave of large amplitude, across which density, pressure, temperature, and particle velocity change drastically. A shock wave is, in general, curved. However, many shock waves that occur in practical situations are straight, being either at right angles to the flow path (termed a **normal shock**) or at an angle to the flow path (termed an **oblique shock**).

Rayleigh flow — An idealized type of gas flow in which heat transfer may occur, satisfying the assumptions that the flow takes place in constant-area cross-section and is frictionless and steady, that the gas is ideal and has constant specific heat, that the composition of the gas does not change, and that there are no devices in the system that deliver or receive mechanical work.

Fanno flow — An ideal flow used to study the flow of fluids in long pipes; the flow obeys the same simplifying assumptions as Rayleigh flow except that the assumption there is no friction is replaced by the requirement the flow be adiabatic.

References

Anderson, J. D. 2003. *Modern Compressible Flow with Historical Perspective,* 3rd ed. McGraw-Hill, New York.

Çengel, Y. A. and Boles, M. A. 2002. *Thermodynamics: An Engineering Approach,* 4th ed. McGraw-Hill, New York.

Hodge, B. K. and Koenig, K. 1995. *Compressible Fluid Dynamics with Personal Computer Applications,* Prentice Hall, Englewood Cliffs, NJ.

John, J. E. A. 1984. *Gas Dynamics,* 2nd ed. Prentice Hall, Englewood Cliffs, NJ.

Moran, M. J. and Shapiro, H. N. 2000. *Fundamentals of Engineering Thermodynamics,* 4th ed. John Wiley & Sons, New York.

Tam, L. M., Ghajar, A. J., and Pau, C. W. 2001. Compressible Flow Software for Properties Calculations and Airfoil Analysis, Proceedings of the Eight International Conference of Enhancement and Promotion of Computing Methods in Engineering and Science (EPMESC"VIII), eds. L. Shaopei et al. July 25–28, Shanghai, China.

Oosthuizen, P. H. and Carscallen, W. E. 1997. *Compressible Fluid Flow,* McGraw-Hill, New York.

Zucrow, M. J. and Hoffman, J. D. 1976. *Gas Dynamics, Volumes I and II,* John Wiley & Sons, New York.

37

The Rheology of Non-Newtonian Fluids

Deepak Doraiswamy
National Starch & Chemical Company

Rheology may be defined as the study of the deformation and flow of matter under the influence of imposed stresses. In Newtonian fluids the stress varies linearly with the strain (or deformation) rate at constant temperature, with the constant of proportionality being the viscosity. All fluids that do not follow this simple behavior are classified as non-Newtonian. The practicing engineer is often concerned with characterizing the rheological behavior of fluids by means of rheological material functions in well-defined simple flows. This enables development of a constitutive equation that is the relationship between the stress tensor and the rate of deformation tensor. Such an equation should, in principle, enable characterization of the fluid structure and calculation of the fluid stresses for the kinematics of interest, which could involve a complex time-dependent velocity field. A physical restriction on constitutive equations, one which translates into a mathematical requirement, is that they must satisfy the principle of **material objectivity** — that is, they should be independent of the reference frame used to describe the fluid motion. An important consequence of this constraint is that all scalars associated with material functions should depend on **invariants** of the associated tensors. The following section is concerned with the rheological description and characterization of incompressible non-Newtonian liquids. A concise historical perspective on the evolution of the discipline is given in Doraiswamy [2002].

37.1 Kinematics, Flow Classification, and Material Functions

It is convenient to use two reference flows — simple shear flow and simple elongational flow — as a practical basis for evaluation of complex situations. The associated material functions in steady flow are the most widely used in rheological characterization of fluids. The kinematics of a flow field may be defined in terms of the fluid velocity field or in terms of particle displacements. It is usually more convenient to work with the deformation rate tensor $\dot{\gamma}_{ij}$, which, in Cartesian coordinates x_i, is related to the velocity gradient tensor v_{ij} as follows:

0-8493-1586-7/05/$0.00+$1.50

$$\dot{\gamma}_{ij} = v_{ij} + v_{ji} = \frac{\partial v_i}{\partial x_j} + \frac{\partial v_j}{\partial x_i} \tag{37.1}$$

It is often useful to isolate the rotational component of the fluid motion, which is defined by the vorticity tensor:

$$\omega_{ij} = \frac{\partial v_i}{\partial x_j} - \frac{\partial v_j}{\partial x_i} \tag{37.2}$$

The deformation rate tensor $\dot{\gamma}_{ij}$ is a function only of the present time t — that is, $\dot{\gamma}_{ij} = \dot{\gamma}_{ij}(t)$. It is valid for arbitrarily large deformations since it does not depend on past times t'. If the kinematics are to be described using deformations instead of deformation rates, it is necessary to consider a spatial reference state at a past time t'. The infinitesimal strain tensor $\gamma_{ij}(t,t')$, which is a measure of the strain at time t' relative to the reference state at time t, is related to the deformation rate tensor $\dot{\gamma}_{ij}(t)$ by a simple integration $\gamma_{ij}(t,t') = \int_t^{t'} \dot{\gamma}_{ij}(t'')dt'$ only in the limit of vanishingly small deformations. For finite deformations $\gamma_{ij}(t, t')$ violates the principle of material objectivity, and appropriate frame invariant measures of strain need to be defined; these are described later in the section on integral constitutive equations.

A **homogeneous flow field** is one in which the velocity gradient is constant at all points. In simple shear flow there is a nonzero component of velocity in only one direction. In a Cartesian (x, y, z) coordinate system the velocity field is defined by $v_x = \dot{\gamma} y, v_y = v_z = 0$. The scalar $\dot{\gamma}$ is termed the *shear rate* and, for the most general situation, is related to the magnitude of the second invariant of $\dot{\gamma}_{ij}$ by $\dot{\gamma} = (II_{\dot{\gamma}})^{1/2}$. The stress tensor τ_{ij} and the deformation rate tensor $\dot{\gamma}_{ij}$ enable definition of three independent material functions that are sufficient to characterize simple shear flow:

$$\eta(\dot{\gamma}) = \frac{\tau_{xy}}{\dot{\gamma}}, \quad \psi_1(\dot{\gamma}) = \frac{(\tau_{xx} - \tau_{yy})}{\dot{\gamma}^2}, \quad \text{and} \quad \psi_2(\dot{\gamma}) = \frac{(\tau_{yy} - \tau_{zz})}{\dot{\gamma}^2} \tag{37.3}$$

where $\eta(\dot{\gamma})$ is the shear viscosity and ψ_1 and ψ_2 are the first and second normal stress coefficients, respectively. Unlike most non-Newtonian systems, Newtonian fluids do not display stresses normal to the shear stress, and the coefficients ψ_1 and ψ_2 are therefore zero. The SI units of viscosity are Pascal seconds (1 Pa · s = 10 p) and of the normal stress coefficients are Pa · s^2.

For dilute solutions it is useful to define the intrinsic viscosity $[\eta]$ (which has dimensions of reciprocal concentration) as follows:

$$[\eta] = \lim_{c \to 0} \frac{\eta - \eta_s}{c\eta_s} \tag{37.4}$$

where η_s is the solvent viscosity and c is the mass concentration of the solute. At very low shear rates, the intrinsic viscosity approaches a limiting value $[\eta]_0$ known as the zero-shear rate intrinsic viscosity.

Extensional flow is defined kinematically by a rate of deformation tensor that has only diagonal components. The most common extensional flow is the simple uniaxial extension, for which the velocity field has the form $v_x = -1/2\dot{\varepsilon}x$, $v_y = -1/2\,\dot{\varepsilon}y$, and $v_z = +\,\dot{\varepsilon}z$ in Cartesian coordinates. It is approximated by the stretching of a filament by forces exerted at both ends. The scalar coefficient $\dot{\varepsilon}$ is called the principal extension rate. This flow is completely characterized by a single material function, the elongational (or extensional) viscosity $\bar{\eta}$, which, for typical non-Newtonian systems (such as polymer melts), can be several orders of magnitude higher than the shear viscosity η:

$$\bar{\eta}(\dot{\varepsilon}) = \frac{(\tau_{zz} - \tau_{xx})}{\dot{\varepsilon}} = \frac{(\tau_{zz} - \tau_{yy})}{\dot{\varepsilon}} \tag{37.5}$$

The extensional viscosity has a value three times that of the shear viscosity for a Newtonian fluid (called the Trouton relation).

Viscometric and steady extensional flows are motions of constant strain rate history in which the material properties are independent of time (although they usually have a strong dependence on strain rate). Time-dependent material properties are required for describing non-Newtonian behavior because the effect of past deformations (varying strain history) on the structure of the fluid (and consequently the stress) cannot be ignored.

A class of transient shearing flows that is widely used in rheological measurements is small-amplitude oscillatory shear flow, in which the stresses and material properties depend on time (or frequency). These involve measurement of the stress response of a fluid to an imposed sinusoidal shearing strain or shear rate of the form $\dot{\gamma}_{xy} = \gamma_0 \omega \cos \omega t$, where γ_0, ω, and t are the maximum imposed amplitude, the frequency, and the time, respectively. The shear stress response is then given by

$$\tau_{xy} = \gamma_0 G'(\omega) \sin \omega t + \gamma_0 G''(\omega) \cos \omega t \tag{37.6}$$

which provides a definition of the storage modulus, $G'(\omega)$, and the loss modulus, $G''(\omega)$ (or the analogous dynamic viscosity functions: $\eta' = G''/\omega$ and $\eta'' = G'/\omega$). For a purely elastic material the loss modulus G'' is zero, and for a purely viscous material the storage modulus G' is zero. The SI units of the elastic moduli are Pascals.

Another commonly used time-dependent material property is the relaxation modulus $G(t, \gamma_0)$, which involves measurement of the shear stress relaxation of a material after the sudden imposition of a step strain, γ_0. It is more suited to describing solid-like materials than liquid-like materials and is defined by $G(t, \gamma_0) = \tau_{xy}/\gamma_0$.

For stress growth upon inception of steady shear flow defined by

$$\begin{aligned} \dot{\gamma}_{xy} &= 0, \quad t < 0 \\ &= \dot{\gamma}_0, \quad t \geq 0 \end{aligned} \tag{37.7}$$

the associated material functions are

$$\eta^+(t, \dot{\gamma}_o) = \frac{\tau_{xy}}{\dot{\gamma}}, \quad \psi_1^+(t, \dot{\gamma}_o) = \frac{(\tau_{xx} - \tau_{yy})}{\dot{\gamma}_o^2}, \quad \text{and} \quad \psi_2^+(t, \dot{\gamma}_o) = \frac{(\tau_{yy} - \tau_{zz})}{\dot{\gamma}_o^2} \tag{37.8}$$

where the plus sign indicates that the shear rate is applied at positive times. The inverse experiment is stress relaxation after cessation of steady shear flow defined by

$$\begin{aligned} \dot{\gamma}_{xy} &= \dot{\gamma}_0, \quad t < 0 \\ &= 0, \quad t \geq 0 \end{aligned} \tag{37.9}$$

in which the material functions can be defined in a similar manner:

$$\eta^-(t, \dot{\gamma}_o) = \frac{\tau_{xy}}{\dot{\gamma}}, \quad \psi_1^-(t, \dot{\gamma}_o) = \frac{(\tau_{xx} - \tau_{yy})}{\dot{\gamma}_o^2}, \quad \text{and} \quad \psi_2^-(t, \dot{\gamma}_o) = \frac{(\tau_{yy} - \tau_{zz})}{\dot{\gamma}_o^2} \tag{37.10}$$

Analogous transient quantities, $\bar{\eta}^+(t, \dot{\varepsilon}_o)$ and $\bar{\eta}^-(t, \dot{\varepsilon}_o)$, can be defined for the stress growth or inception of simple extensional flow with an elongation rate $\dot{\varepsilon}_0$

$$\bar{\eta}^+(t, \dot{\varepsilon}_o) \quad \text{or} \quad \bar{\eta}^-(t, \dot{\varepsilon}_o) = \frac{\tau_{zz} - \tau_{xx}}{\dot{\varepsilon}_o} \tag{37.11}$$

All the above properties become independent of the imposed strain in the limit of zero strain (that is, as $\gamma_o \to 0$, $\dot{\gamma}_o \to 0$, or $\dot{\varepsilon}_o \to 0$), in which case the material functions depend only on the time. The behavior in the limit of small deformations is termed *linear viscoelasticity* and the related material properties are defined appropriately, for example, $G(t) = G(t, \gamma_o)_{\gamma_o \to 0}$. Linear viscoelasticity reveals information about material behavior in the unstrained state in which the molecular conformations and entanglements have their equilibrium values and is used more for material characterization and quality control applications than for process modeling.

In addition to the kinematics, the material properties depend on the chemical constitution of the fluid (e.g., molecular weight, molecular weight distribution, polymer branching) and the physical state of the fluid (typically measured by the temperature and concentration); consequently, they have great utility in characterization and processing operations. Some of the more commonly used correlations for material functions are discussed later. The experimental science of determining rheological material properties such as those considered in this section is termed **rheometry**; further details are provided in Dealy [1982, 1994].

37.2 Fluids

Non-Newtonian fluids may be broadly classified by their ability to retain the memory of a past deformation (which is usually reflected in a time dependence of the material properties). Fluids that display memory effects usually exhibit elasticity. A fluid is identified as viscoelastic if the stresses in it persist after the deformation has ceased (typically manifested in the decay of the shear stress and primary normal stress difference after cessation of steady shear flow). The duration of time over which appreciable stresses persist after cessation of deformation gives an estimate of the relaxation time λ of the material. A dimensionless group commonly used for evaluating the role of fluid viscoelasticity is the Deborah number, De, defined by

$$\mathrm{De} = \frac{\lambda}{T} \tag{37.12}$$

where λ is the relaxation time for the fluid (which can be estimated experimentally using an appropriate constitutive equation) and T is the characteristic time constant for the process of interest. Low values of the Deborah number therefore correspond to fluid-like behavior and high values to solid-like behavior.

The non-Newtonian materials most often encountered by the engineer are polymer melts, polymer solutions, and multiphase systems. Polymer melts and solutions are usually viscoelastic, that is, they are capable of storing elastic energy. Their viscosity–shear rate behavior typically exhibits a constant "zero shear rate viscosity" at low shear rates, followed by a shear thinning region where the viscosity decreases with shear rate. Unlike that of polymer solutions, the viscosity of melts rarely exhibits a second plateau — the infinite-shear viscosity — at the highest shear rates typically used in measurements. Also, melts usually display much higher viscosities than solutions. It is often convenient to classify solutions into dilute and concentrated systems; in the former, unlike the latter, the individual polymer chains rarely overlap. Based on simple scaling arguments it is usually assumed that the dilute solution regime ends when $[\eta]_o c \sim 1$, where c is the polymer concentration (typically 5% by weight). The source of elasticity differs between melts and solutions. In polymer solutions elasticity is mainly associated with changes in the orientation and configuration of molecules as a result of polymer–solvent interactions. In the case of polymer melts (or concentrated solutions) it is the result of mobility constraints imposed by polymer–polymer interactions. Not all fluids that retain a memory of past deformations exhibit elasticity and, in some cases (typically in concentrated suspensions), time-dependent rheological properties may be related to changes in fluid structure that are purely dissipative, as in thixotropic and rheopectic fluids. Under steady shear conditions, the viscosity decreases with time for thixotropic fluids, whereas it increases with time for rheopectic fluids. Suspensions also frequently display "yield behavior," that is, they do not

TABLE 37.1 Common Generalized Newtonian Fluids Described by the Function $\eta(\dot{\gamma})$ or $\eta(\tau)$

Model	Equation	Comments
Power-law model (Ostwald–de Waele)	$\eta = K\dot{\gamma}^{n-1}$ K = consistency index n = power law index	For "pseudoplastic" liquids, $n < 1$ For "dilatant" liquids, $n > 1$
Carreau	$\dfrac{\eta - \eta_\infty}{\eta_o - \eta_\infty} = [1 + (\lambda\dot{\gamma})^2]^{(n-1)/2}$ η_o = zero shear viscosity η_∞ = infinite shear viscosity	Describes smooth transition from η_o to η_∞ typically observed in polymer solutions.
Bingham	$\eta = \infty, \quad \tau \le \tau_y$ $\eta = \mu_0 + \dfrac{\tau_y}{\dot{\gamma}}, \quad \tau \ge \tau_y$ τ_y = yield stress μ_0 = viscosity	Describes yield behavior typically observed in suspensions.
Herschel–Bulkley	$\eta = \infty, \quad \tau \le \tau_y$ $\eta = \dfrac{\tau_y}{\dot{\gamma}} + K\dot{\gamma}^{n-1}, \quad \tau \ge \tau_y$	Describes behavior of shear-thinning suspensions.
Ellis	$\dfrac{\eta_o}{\eta} = 1 + \left(\dfrac{\tau}{\tau_{1/2}}\right)^{\alpha-1}$ $\tau_{1/2}$ = value of τ at which $\eta = \eta_o/2$	Predicts zero-shear viscosity but difficult to use since it is not explicit in shear rate.

Note: All scalar stresses refer to the magnitude or second invariant of the stress tensor τ_{ij}.

flow unless a critical (or yield) stress is applied. A number of equations describing the yield behavior of suspensions are included in Table 37.1. At low volume fractions (<5%), the viscosity η_{sp} of a rigid suspension of spheres is related to the viscosity of the suspending fluid ηs by the Einstein equation:

$$\eta_{sp} = \eta_s(1 + 2.5\phi) \tag{37.13}$$

where ϕ is the volume fraction. At higher concentrations the empirical Maron-Pierce equation is useful for estimating the viscosity of suspensions of rigid particles of narrow size distribution in viscous fluids at low deformation rates:

$$\eta_{sp} = \frac{\eta_s}{(1 - \phi/A)^2} \tag{37.14}$$

The empirical constant A may be loosely identified with the maximum packing fraction for the particulate system. It has a value of 0.68 for smooth spheres and may be approximated by the expression $A = 0.54 - 0.013a$ for aspect ratio a in the range $6 < a < 30$. The distribution of particle sizes has little effect on suspension viscosity when the volumetric loading of solids is below 20%; at higher concentration levels the effects are very pronounced. In general, the addition of suspended solids increases the shear thinning as one moves into the shear-thinning range of deformation rates and the importance of elasticity appears to decrease with increased solids content. The rheology of spherical and anisotrophic suspensions (with and without Brownian motion) is discussed by Gupta [1994].

For multiphase systems comprising fluid drops in a suspending liquid at low deformation rates and volume fractions, the Taylor analogue to the Einstein equation may be used:

$$\eta_{sp} = \eta_s \left(1 + \frac{1+2.5r}{1+r}\phi\right) \tag{37.15}$$

where r is the ratio of the viscosity of the disperse phase to that of the continuous phase. Few generalizations can be made about blends of immiscible deformable systems in view of complications arising due to droplet breakup, coalescence, and morphology effects; further details are provided in Han [1981].

The rheology of a variety of complex materials such as polymeric liquid crystals, fiber suspensions, foams and powders is discussed in Gupta [2000] and Larson [1999].

37.3 Constitutive Equations

No single constitutive equation is suitable for all purposes and the selection of one depends on the particular situation concerned. The form of the equation will reflect the type of flow (e.g., shear or elongation), the type of material (e.g., melt, concentrated solution or dilute solution), the solution scheme (differential or integral model) best suited to the problem of interest, and the particular situation to be portrayed (stress overshoot, flow instability, die-swell, etc.). The basic assumption of all constitutive equations is that the stress at any location and at any time in the flowing fluid depends on the entire flow history of the fluid element occupying that material point only (and not of adjacent elements). Constitutive equations may be broadly classified into rate (or differential) equations and integral equations (although many of the nonlinear differential forms may also have an equivalent integral expression). The development and use of a number of commonly used differential and integral models have been described by Astarita and Marrucci [1974], Bird et al. [1987a], and Larson [1988].

For a Newtonian fluid the scalar fluid viscosity η is defined by:

$$\tau_{ij} = \eta\dot{\gamma}_{ij} \tag{37.16}$$

A generalized Newtonian fluid is described by a constitutive equation in which the viscosity is only a function of the magnitude of the second invariant of the stress or the shear rate. Common empiricisms for the function $\eta(\dot{\gamma})$ or $\eta(\tau)$ (where τ is the magnitude or second invariant of τ_{ij}) are summarized in Table 37.1. The generalized Newtonian fluid is best suited to describing steady state shear flows or small deviations from such flows as long as the Deborah number is sufficiently low. Like the Newtonian fluid, it cannot describe normal stress effects or time-dependent elastic effects. In elongational flows and in rapidly changing flows, the generalized Newtonian models should not be used.

Linear viscoelasticity is primarily concerned with the description of fluid deformations that are very small or very slow. The theory of linear viscoelasticity does not satisfy the principle of frame invariance except in the zero deformation limit. Differential formulations of linear viscoelastic models combine classical ideas of Hookean solids and Newtonian fluids and are represented by the Maxwell model,

$$\tau_{ij} + \lambda_1 \frac{\partial \tau_{ij}}{\partial t} = \eta_o \dot{\gamma}_{ij} \tag{37.17}$$

and the Jeffreys model,

$$\tau_{ij} + \lambda_1 \frac{\partial \tau_{ij}}{\partial t} = \eta_o \left(\dot{\gamma}_{ij} + \lambda_2 \frac{\partial \dot{\gamma}_{ij}}{\partial t}\right) \tag{37.18}$$

in which λ_1 and λ_2 are the relaxation time and the retardation time, respectively. The most general linear viscoelastic model, which includes both the above forms, is the generalized Maxwell model. It is conve-

TABLE 37.2 Common Constitutive Equations Included in the Eight-Constant Oldroyd Model

			Time Constants				
Model Name	λ_1	λ_2	λ_3	λ_4	λ_5	λ_6	λ_7
Oldroyd six-constant						0	0
Oldroyd four-constant			0	0		0	0
Oldroyd fluid A			$2\lambda_1$	$2\lambda_2$	0	0	0
Oldroyd fluid B							0
Corotational Jeffreys			λ_1	λ_1	0	0	0
Second-order fluid	0		0		0	0	0
Upper convected Maxwell		0	0	0	0	0	0

niently represented in an integral form (which can therefore accommodate an infinite number of time constants) and is given by:

$$\tau_{ij} = \int_{-\infty}^{t} G(t-t')\dot{\gamma}_{ij}(t')dt' = \int_{-\infty}^{t} M(t-t')\gamma_{ij}(t,t')dt' \tag{37.19}$$

in which $G(t-t')$ is the relaxation modulus and $M(t-t') = dG(t-t')/dt'$ is the memory function. Various relationships between linear viscoelastic properties and material structure are provided by Ferry [1980].

The difference between the various nonlinear viscoelastic equations based on continuum mechanics are due primarily to the types of time derivative that arise as a result of rewriting them in an appropriate reference frame in order to make them objective. One of the most general frame invariant differential formulations for the stress tensor τ_{ij} is the eight-constant Oldroyd model, which may be expressed by the following equation:

$$\begin{aligned} \tau_{ij} + \lambda_1 \frac{\mathrm{D}}{\mathrm{D}t}\tau_{ij} + \frac{\lambda_3}{2}(\dot{\gamma}_{ik}\tau_{kj}\dot{\gamma}_{kj}) + \frac{\lambda_5}{2}\tau_{kk}\dot{\gamma}_{ij} + \frac{\lambda_6}{2}\tau_{kk}\dot{\gamma}_{ij}\delta_{ij} \\ = \eta_0\left[\dot{\gamma}_{ij} + \lambda_2 \frac{\mathrm{D}}{\mathrm{D}t}\tau_{ij} + \lambda_4\dot{\gamma}_{ik}\dot{\gamma}_{kj} + \frac{\lambda_7}{2}\dot{\gamma}_{ij}\dot{\gamma}_{ji}\delta_{ij}\right] \end{aligned} \tag{37.20}$$

where *D/Dt* is the corotational or Jaumann time derivative defined by:

$$\frac{\mathrm{D}}{\mathrm{D}t}\tau_{ij} = \frac{\partial \tau_{ij}}{\partial t} + v_i \frac{d\tau_{ij}}{dx_i} + \frac{1}{2}(\omega_{ik}\tau_{kj} - \tau_{ik}\omega_{kj}) \tag{37.21}$$

The eight-constant Oldroyd model includes a number of common differential models, some of which are summarized in Table 37.2 [Bird et al., 1987a]. The steady material functions for the various models can be obtained by assigning relevant values of the constants from Table 37.2 into the following expressions [Bird et al., 1987a]:

$$\frac{\eta}{\eta_o} = \frac{1 + [\lambda_2(\lambda_3+\lambda_5) + \lambda_4(\lambda_1-\lambda_3-\lambda_5) + \lambda_7(\lambda_1-\lambda_3-\frac{3}{2}\lambda_5)]\dot{\gamma}^2}{1 + [\lambda_1(\lambda_3+\lambda_5) + \lambda_3(\lambda_1-\lambda_3-\lambda_5) + \lambda_6(\lambda_1-\lambda_3-\frac{3}{2}\lambda_5)]\dot{\gamma}^2} \tag{37.22}$$

$$\frac{\psi_1}{2\eta_o\lambda_1} = \frac{\eta(\dot{\gamma})}{\eta_o} - \frac{\lambda_2}{\lambda_1} \tag{37.23}$$

$$\frac{\psi_2}{2\eta_o\lambda_1} = \frac{\psi_1}{2\eta_o\lambda_1} + \frac{(\lambda_1 - \lambda_3)\eta(\dot{\gamma})}{\lambda_1\eta_o} - \frac{(\lambda_2 - \lambda_4)}{\lambda_1} \tag{37.24}$$

$$\frac{\bar{\eta}}{3\eta_o} = \frac{1-(\lambda_2-\lambda_4)\dot{\varepsilon}+(\frac{3}{2}\lambda_5-\lambda_1+\lambda_3)(2\lambda_2-2\lambda_4-3\lambda_7)\dot{\varepsilon}^2}{1-(\lambda_1-\lambda_3)\dot{\varepsilon}+(\frac{3}{2}\lambda_5-\lambda_1+\lambda_3)(2\lambda_1-2\lambda_3-3\lambda_6)\dot{\varepsilon}^2} \tag{37.25}$$

The Oldroyd eight-parameter model illustrates how the number of constants becomes prohibitively large if only frame invariance considerations are used to formulate constitutive equations based purely on continuum mechanics considerations. It provides useful qualitative descriptions but is not quantitatively accurate. Care must be exercised in using the Oldroyd models in elongational flows since the elongational viscosity can become infinite if the parameters are not chosen carefully. Various empirical differential models have also been suggested. The White–Metzner model, for example, is obtained by making the relaxation time and the viscosity in the upper convected Maxwell model functions of the shear rate. It is useful in describing flows where the coupling of shear thinning and elasticity is significant. The Giesekus model can be obtained from molecular arguments and, unlike the Oldroyd model, is quadratic in stress. Although the model is considerably more difficult to use, it predicts decreasing viscosity and normal stress coefficients with increasing shear rate and a finite second normal stress coefficient. Differential constitutive equations are often the most convenient and practical way to determine the role of viscoelasticity in real-world systems. They are best suited to describing flows in which both shearing and extension are involved or shearing flows in which normal stresses are important.

A major advantage of integral formulations is that they are usually explicit in stress. The integral constitutive equations are conveniently represented in terms of a finite strain tensor and not the velocity field that has been considered so far. The infinitesimal strain tensor γ_{ij} described earlier is valid only for vanishingly small displacements. In order to describe the strain or displacement for finite deformations, it is necessary to consider a reference configuration. If a particle occupies position x_i at time t and position x_i' at past time t', specification of the displacement functions $x_i(x_i', t', t)$ or $x_i'(x_i, t, t')$ is equivalent to specifying the velocity field. This enables definition of the displacement gradient tensor Δ_{ij} and the inverse displacement gradient tensor Δ_{ij}^{-1} given by:

$$\Delta_{ij}(x_i, t, t') = \frac{\partial x_i'}{\partial x_j} \tag{37.26}$$

and

$$\Delta_{ij}^{-1}(x_i, t, t') = \frac{\partial x_i}{\partial x_j'} \tag{37.27}$$

The principle of material objectivity then permits definition of finite strain tensors called the *Cauchy strain tensor,* c_{ij}, and the Finger strain tensor, f_{ij}, in terms of displacement gradients rather than the velocity field:

$$c_{ij} = \Delta_{ki}\Delta_{kj} \tag{37.28}$$

$$f_{ij} = \Delta_{ik}^{-1}\Delta_{jk}^{-1} \tag{37.29}$$

Both c_{ij} and f_{ij} describe material deformation independent of superimposed rigid rotations and reduce to the unit tensor in the undeformed state. Since rheology is concerned with deviations from the undeformed state, it is sometimes more convenient to work with the relative Cauchy strain tensor C_{ij} and the relative Finger strain tensor F_{ij}, defined as follows:

$$C_{ij} = c_{ij} - \delta_{ij} = \Delta_{ki}\Delta_{kj} - \delta_{ij} \tag{37.30}$$

$$F_{ij} = \delta_{ij} - f_{ij} = \delta_{ij} - \Delta^{-1}_{ik}\Delta^{-1}_{jk} \tag{37.31}$$

Any of the finite strain tensors (f_{ij}, c_{ij}, F_{ij}, C_{ij}) or an appropriate combination of these can be used in integral formulations since they are all frame invariant measures of the fluid strain at time t' relative to that at the time t.

The most general formulation for stress, which contains a number of constitutive equations, is the memory integral expansion in which the stress is expressed as a functional of the strain history [Bird et al., 1987a]. It is obtained by integration over all past times t' following the particle that ends up at position x at present time t:

$$\begin{aligned}\tau_{ij}(x_p\ t) = &\int_{-\infty}^{t} M_{\mathrm{I}}(t-t')F'_{ij}dt' \\ &+\int_{-\infty}^{t}\int_{-\infty}^{t} M_{\mathrm{II}}(t-t',t-t'')(F'_{ik}F''_{kj}+F'_{ik}F''_{kj})dt'dt''+\cdots\end{aligned} \tag{37.32}$$

where $F'_{ij} = F_{ij}(x_p\ t, t')$, $F''_{kj} = F_{ij}(x_p\ t, t'')$, and $M_{\mathrm{I}}, M_{\mathrm{II}}, \ldots$ are kernel functions that account for memory effects. The single integral forms are the most practical and are obtained by setting all higher-dimensional integrals equal to zero. The most popular empirical integral model is the factorized K-BKZ (Kaye-Bernstein, Kearsley, Zapas) model given by:

$$\tau_{ij}(x_p, t) = \int_{-\infty}^{t} M(t-t')\left[\frac{\partial W}{\partial I_f}F'_{ij}(x_p, t, t') + \frac{\partial W}{\partial I_c}C'_{ij}(x_p, t, t')\right]dt' \tag{37.33}$$

in which $M(t-t')(\equiv M_{\mathrm{I}})$ is the linear viscoelastic memory function defined in Equation (37.19), $C'_{ij} = C_{ij}(x_p, t, t')$, and $W(I_f, I_c)$ is a potential function that needs to be experimentally determined; it is a function of the two strain invariants $I_f = f_{ii}$ and $I_c = c_{ii}$. The K-BKZ equation, which is based on rubber elasticity theory and makes no molecular assumptions, quantitatively describes important rheological phenomena such as stress growth and relaxation. The Lodge constitutive equation (which is the integral equivalent of the upper convected Maxwell model) is a special case of the K-BKZ equation and is obtained by setting $W = I_c$:

$$\tau_{ij}(x_p, t) = \int_{-\infty}^{t} M(t-t')C'_{ij}(x_p, t, t')dt' \tag{37.34}$$

In the limit of vanishingly small displacements, the general linear viscoelastic model [Equation (37.19)] is obtained from Equation (37.34) by setting C_{ij} equal to the infinitesimal strain tensor, γ_{ij}. Integral models provide a framework for including a wide class of nonlinear viscoelastic behavior. By selecting suitable empirical forms for the kernel functions, they enable description of fluid rheology using a small finite number of constants; these have the advantage of having physical meaning and can be determined from rheometric experiments. Quantitative determination of the kernel functions is, however, a nontrivial problem.

In addition to the continuum models described above, constitutive equations based on molecular theories (which result in integral forms similar to those discussed) have also been proposed. Most of these models can be classified in one of the following categories:

1. Bead-spring or random coil theories for dilute polymer solutions
2. Hydrodynamic interaction theories for moderately concentrated solutions that account for the indirect drag that one part of a polymer chain exerts on another through the solvent
3. Molecular entanglement, network, or reptation theories for concentrated solutions and melts

Although molecular models have considerable potential for material characterization and rheological flow modeling, their applicability is restricted at this time. Bird et al. [1987b] have covered such constitutive equations in great detail.

37.4 Some Useful Correlations for Material Functions

Correlations for and between the material functions are very useful in order to enable quick estimates of material properties; they are often used for material characterization and process control applications. The more commonly used correlations are described below (further details on correlations and parameter values are summarized in van Krevelen and Hoftyzer [1976], Graessley [1974], and Bird et al. [1987a]).

Equivalence of Dynamic and Steady Shear Properties

The Cox–Merz empiricism predicts that the magnitude of the complex viscosity ($|\eta^*|$) is equal to that of the viscosity at equal values of the frequency and shear rate. That is,

$$\eta(\dot{\gamma}) = |\eta^*(\omega)|_{\omega=\dot{\gamma}} = \left(\sqrt{\eta'(\omega)^2 + \eta''(\omega)^2}\right)_{\omega=\dot{\gamma}} \tag{37.35}$$

Laun's rule is another useful correlation for predicting the first normal stress difference from dynamic data:

$$\psi_1 = \frac{2\eta''(\omega)}{\omega}\left[1 + \left(\frac{\eta''}{\eta'}\right)^{0.7}\right]_{\omega=\dot{\gamma}} \tag{37.36}$$

For highly filled suspensions (~70%) that exhibit a yield stress, shear thinning, and a recoverable strain, the so-called Rutgers–Delaware rule predicts that the complex viscosity vs. maximum (or effective) shear rate ($\gamma_m\omega$) is identical to the steady shear rate vs. shear rate [Doraiswamy et al., 1991]; γ_m is the strain amplitude.

Dependence of Viscosity on Temperature

If η_o (T) is the zero-shear viscosity at the desired temperature T, and $\eta_o(T_o)$ the zero shear viscosity at an arbitrary reference temperature T_o, a shift factor a_T may be approximated as follows:

$$a_T = \frac{\eta_o(T)T_o\rho_o}{\eta_o(T_o)T\rho} \cong \frac{\eta_o(T)}{\eta_o(T_o)} \tag{37.37}$$

where ρ and ρ_o are the densities at the temperatures T and T_o, respectively. An Arrhenius-type equation is often used to determine the temperature dependence of a_T as long as the temperature is at least 100 K above the glass transition temperature, T_g:

$$a_T = \exp\left[\frac{\Delta E}{R}\left(\frac{1}{T} - \frac{1}{T_o}\right)\right] \tag{37.38}$$

where ΔE is the activation energy and R is the Boltzmann constant. For typical polymer melts $\Delta E/R$ is ~5000 K. For temperatures between T_g and $T_g + 100$, the Williams–Landel–Ferry expression can be used to estimate a_T:

$$\log a_T = \frac{-c_1^{\circ}(T - T_o)}{c_2^{\circ} + (T - T_o)} \tag{37.39}$$

If T_o is taken to be the glass-transition temperature, the following values can be used for quick estimates: $c_1^{\circ} = 17.44$ and $c_2^{\circ} = 51.6$ K.

The time–temperature superposition principle states that varying the temperature at fixed shear rate (or time) is equivalent to varying the shear rate at fixed temperature. By defining a reduced viscosity η_r and a reduced shear rate $\dot{\gamma}_r$ as follows:

$$\eta_r = \eta(\dot{\gamma}, T)\frac{\eta_o(T_o)}{\eta_o(T)} \cong \frac{\eta(\dot{\gamma}, T)}{a_T}; \quad \dot{\gamma}_r = a_T\dot{\gamma} \tag{37.40}$$

it is possible to reduce viscosity data at different temperatures and shear rates to a single master curve, provided the different isotherms are of similar shape [Bird et al., 1987a]. This approach enables extending the shear rate range of an available experimental configuration for obtaining viscometric data. Analogous procedures can be used for other material properties.

Dependence of Viscosity on Molecular Weight and Concentration

Molecular weights of polymers can be determined using solution viscosity measurements. At very low concentration or in the zero concentration limit, the intrinsic viscosity of linear, monodisperse solutions of polymers is related to the molecular weight, M, by the Mark–Houwink equation:

$$[\eta]_o = KM^a \tag{37.41}$$

where K and a are functions of the polymer, solvent, and temperature; their values are available in the literature (e.g., van Krevelen and Hoftyzer [1976]). A typical value for a is 0.7. At higher concentrations ($c > 1/[\eta]_o$), for which overlap between neighboring molecules becomes significant, the viscosity scales as the product cM. Finally, for pure melts and highly concentrated solutions beyond a critical molecular weight (typically in the range 2000–50,000), entanglements between molecules become significant and the zero shear melt viscosity is given by:

$$\eta_o = KM^{3.4} \tag{37.42}$$

For narrow distributions M can be approximated by M_w. For broad distributions, the viscosity depends on a molecular weight average between M_w and the next higher average (z average). If branching is present, M_w is replaced by gM_w, where g is a branching index.

37.5 Rheological Measurements

A detailed discussion of rheometric techniques is beyond the scope of this article; more in-depth discussions are given in Barnes et al. [1989], Gupta [2000], Dealy [1982], and Collyer and Clegg [1998], which provide original references to many of the techniques discussed. The last book also includes some relatively uncommon approaches for rheological measurements such as large-amplitude oscillatory shear and hole pressure measurements for normal stress differences. The reader is referred to these same references for a dicussion of linear viscoelastic property measurements.

Key issues that need to be considered in rheometer selection are:

1. Type of flow field (shear vs extension; non-homogeneity of the deformation rate)
2. Deformation rate range
3. Stress range
4. Transient measurements (dynamic vs steady state properties)
5. Largest size of the discrete phase
6. Viscous heating
7. Elasticity effects

Slip effects are a common occurrence in the handling of suspensions and emulsions and can be accounted for by appropriate analysis of data. Roughening the surfaces of the liquid–fixture interface is the preferred experimental route to overcoming this problem. Changes in structure during measurement (thixotropic effects) can also become a critical factor especially for concentrated suspensions and emulsions.

Shear Viscosity

The most common instruments for measuring shear viscosity are as follows:

Capillary Viscometers

The sample is pushed through a capillary using a piston or pressurized reservoir under controlled temperature conditions. The key advantages are the ease of sample loading, wide ranges of temperature and shear rates, and convenient examination of the extruded sample for texture analysis and elasticity (die swell) effects. The main disadvantages are the corrections that have to be made for shear rate variation across the capillary diameter, entrance flow, and capillary end effects. In order to avoid entrance effects, it is recommended that the ratio of the reservoir diameter to the capillary diameter be at least 10, while a capillary aspect ratio of at least 50 is usually sufficient to make end effects negligible; the so-called Bagley correction is needed for end effects with short capillaries. The adiabatic temperature rise at high shear rates can be quite large for viscous materials, and appropriate corrections need to be made; it is worth noting that the maximum adiabatic temperature rise takes place a short distance away from the wall rather than at the maximum shear at the wall. Slip effects can also become significant in some systems and are manifested by a dependence of viscosity on capillary diameter. The flow rate Q is determined as a function of the pressure drop ΔP in order to determine the viscosity $\eta(\dot{\gamma})$ as a function of shear rate $\dot{\gamma}$. If R is the tube radius and L the length, the wall shear stress is given by

$$\tau_w = \Delta P \frac{R}{2L} \tag{37.43}$$

and the wall shear rate by

$$\dot{\gamma}_w = \frac{4Q}{\pi R^3}\left(\frac{3}{4} + \frac{1}{4}\frac{d\ln Q}{d\ln \tau_w}\right) \tag{37.44}$$

The viscosity is then easily found to be

$$\eta = \frac{\tau_w}{\dot{\gamma}} \tag{37.45}$$

Equation 37.44 is the famous Weissenberg–Rabinowitsch–Mooney equation, which enables determination of the local or true value of the shear rate from the average or "apparent shear rate" $4Q/\pi R^3$.

For a power-law fluid, Equation (34.44) takes the form

$$\dot{\gamma}_w = \frac{4Q}{\pi R^3}\left(\frac{3n+1}{4n}\right) \tag{37.46}$$

where $n = d \log \tau / d \log \dot{\gamma}$. For the special case of a Newtonian liquid where n equals unity, Equation 37.43 through Equation 37.46 yield the well-known Hagen–Poiseuille equation

$$\eta = \frac{\pi \Delta P R^4}{8LQ} \tag{37.47}$$

The Bagley correction e to the wall shear stress for end effects is expressed as a multiple of the tube aspect ratio L/R and is obtained by extrapolating pressure drop–tube aspect ratio data at constant shear rate (or R) to zero pressure drop:

$$e = \frac{1}{2}\frac{\Delta P}{\tau_w} - \frac{L}{R} \tag{37.48}$$

The flow rate capacity for gas-pressure–driven instruments and the force transducer load for plunger-driven instruments typically set a limit of 10^6 l/sec for the maximum shear rate in the capillary rheometer. Viscous heating effects and extruded flow instabilities such as melt fracture are other major factors that set an upper bound on operation. Drip effects for gas-pressure–driven units and plunger-piston friction effects typically set a lower limit of 1 l/sec for the reliably attainable shear rate in the capillary device.

Most commercial capillary units are plunger-driven though gas-pressure–driven devices operating at pressures up to 200 Mpa are also available. Some capillary units have pressure transducers mounted just downstream of the entrance so that related pressure losses do not need to be accounted for.

An important variation of the capillary device is the slit rheometer where ultra-high shear rates (up to 10^6 l/sec) can be obtained under effectively isothermal conditions; slit rheometers have also been used to make measurements of the first normal stress difference for shear flow.

Coaxial Cylindrical Viscometers

Measuring viscosity using Couette-type rheometers involves filling the annular space between two concentric cylinders with the liquid and measuring the torque on the outer (or inner) cylinder while the inner (or outer) cylinder is rotated at a constant speed. In other versions of the device, one cylinder is held stationary while the other is rotated under the influence of a constant torque; the rotational speed is then measured using some form of tachometer. Such units can also be used for dynamic measurements, in which case the constant rotational speed is replaced by oscillation at a fixed frequency and small amplitude. Wide gap and infinite gap variations are also possible in order to handle suspensions and emulsions, though convenient analyses are available only for power-law fluids.

The main advantages are the effectively constant shear rate in the gap if the annulus is small (R_o/R_i = radius of outer clinder/radius of inner cylinder < 1.03), the ability to collect data at low shear rates because of the large surface area, and the relative ease of calibration. Disadvantages are sample loading and viscous heating for viscous materials and "climbing" of polymer solutions up the inner cylinder due to normal stresses. Secondary flows and turbulence effects are additional factors for low-viscosity fluids.The shear stress τ and shear rate $\dot{\gamma}$ at any radius r are related to the torque M by the following equations:

$$\tau(r)=\frac{T}{2\pi r^2 h} \quad R_i \le r \le R_o \tag{37.49}$$

$$R_i \le r \le R_o \quad R_i \le r \le R_o \tag{37.50}$$

where h is the immersed length of the cylinder and ω the angular frequency. For very narrow gaps,

$$\dot{\gamma}(r)=\frac{\omega R_i}{(R_o - R_i)} \tag{37.51}$$

An end correction is needed when the extra torque required to account for additional drag at the bottom surface is significant and is avoided by using a very large immersion length or shaping the bottom of the inner cylinder to resemble an inverted cone. The typical range of these devices is 10^{-2} to 10^{2} l/sec and is limited by the sensitivity of the torque measurement device.

A useful practical variation of the Couette device is the Brookfield viscometer, which comprises a cylindrical spindle rotating in ~ 500 ml of fluid. It is driven by a synchronous motor through a calibrated spring and the deflection of the latter serves as an empirical measure of the "viscosity" for many commercial situations.

Cone-and-Plate Viscometers

The fluid is sandwiched between a flat, circular plate and a linearly concentric cone; the former is truncated to avoid physical contact. The two are rotated relative to each other and the torque T is measured to attain a constant angular velocity ω. The typical plate radius R is a few centimeters while the cone angle θ is a couple of degrees. The downward normal force F needed to keep the relative positions of the two geometries constant is also measured. The key advantage of this rheometer is the constant shear rate throughout the fluid which also therefore enables accurate transient measurements such as dynamic properties. The device lends itself well to disposable fixtures and consequently measurements on reacting systems such as gels. Additional advantages include the small sample size and the ease of loading and cleaning. A number of limitations need to be considered. Secondary flows, centrifugal effects, and elastic instabilities become significant at high shear rates. Free surface distortion can affect the measurements as can sample drying at the outer edge; the latter problem is usually circumvented by coating the rim with a thin oil or using a solvent trap. Accurate positioning of the plates is a requirement that makes the device relatively expensive and complicated to operate.

The relevant equations for the cone-and-plate rheometer are:

$$\tau=\frac{3T}{2\pi R^3} \tag{37.52}$$

$$\gamma=\frac{\omega}{\alpha} \tag{37.53}$$

$$N_1=\frac{2F}{\pi R^2} \tag{37.54}$$

$$T_{\theta\theta}=-p_a+N_2+(N_1+2N_2)\ln\frac{r}{R} \tag{37.55}$$

where α is the cone angle in radians, p_a is the atmospheric pressure, $T_{\theta\theta}$ is the total stress exerted by the fluid normal to the plate surface, and the subscripts on T represent the θ direction in a spherical coordinate system. Determination of N_1 permits determination of N_2 through Equation (37.55) in principle if $T_{\theta\theta}$ can be determined; measurement of the latter is usually not practical to a sufficient level of accuracy.

If ω is made to depend on time, the previous set of equations still holds true for T and N; appropriate time-dependent material properties can be determined for oscillatory flow, stress growth, and relaxation. It is worth noting many modern rheometers are operated in a constant stress mode so that the shear rate becomes the dependent variable; such devices lend themselves very well to measurement of creep properties and yield stresses if present.

The cone-and-plate rheometer can be used over a wide range of shear rates varying from 10^{-6} to 10^3 l/sec. It is especially well suited to measuring the viscosity of highly viscous fluids at low shear rates and is often used to generate zero shear viscosity data for polymer melts and concentrated solutions. The typical frequency range for oscillatory measurements is 10^{-2} to 10^2 rad/sec. One of the most widely used commercial instruments is the Rheometrics unit (now owned by TA Instruments), which permits torque measurements from 0.1 to 2000 g-cm and normal forces from 0.1 to 2000 g; a special feature of the device is the so-called "force rebalance transducer," which prevents axial separation of the plates during rotation due to normal stresses.

Parallel-Plate Viscometers

Coaxial parallel disks are preferred to the cone-and-plate geometry when particles are involved whose size is comparable to the distance between the truncated cone and the plate. Other advantages are that sample preparation is easy when the initial material is a solid, and large torque readings are obtained for small amplitude oscillatory flow when the gap spacing is large. Unlike the cone-and-plate system, however, the stress and shear rates vary with radial position, and this complicates data analysis. The shear stress and shear rate at the edge of the disks (i.e., at r = R) are given by:

$$\dot{\gamma}(R) = \frac{\omega R}{h} \tag{37.56}$$

$$\tau(R) = \frac{3T}{2\pi R^3}[1 + \frac{1}{3}\frac{d\ln T}{d\ln\dot{\gamma}}] \tag{37.57}$$

$$N_1(R) - N_2(R) = \frac{2F}{\pi R^2}[1 + \frac{1}{2}\frac{d\ln F}{d\ln\dot{\gamma}(R)}] \tag{37.58}$$

where h is the gap separation. In order to determine the shear stress or normal stress at a particular shear rate, torque and normal force results are needed over a range of rim shear rates to enable differentiation of the data. The correction is seen to be analogous to the Weissenberg–Rabinowitsch–Mooney equation for tube flow. The parallel plate setup lends itself to ultra-high shear rate viscometry when sample size limitations do not permit capillary measurements. Use of very fine gaps (< 50 μ) helps overcome viscous heating problems due to the advantageous scaling of surface area with volume (gap size) and also centrifugal effects (due to the inverse scale of surface tension forces with gap size). However, care needs to be taken to account for surface nonuniformity effects and secondary flows using an appropriate calibration fluid.

The parallel-plate device enables direct determination of the dynamic moduli from measurement of the torque amplitude and the difference with the angular deformation, provided the entire sample is within the linear viscoelastic region. The unit can also be operated in the "eccentric rotating disk" mode for estimation of the dynamic properties. An offset between the axes of the two disks induces an eccentric oscillatory motion when the disks are rotated in opposing directions; an analysis of the flow field yields G' and G".

Other instruments that are used in special situations and deserve brief mention are sliding plate, torque, and squeeze flow viscometers. The sliding plate rheometer is used to measure large, transient deformations at very high shear rates, while most rheometers are used for steady-state measurement at low shear rates; the sample is typically sandwiched between two horizontal plates and is sheared by moving one plate relative to the other. Edge, friction, and gap variation effects are typically overcome by using specially designed transducers. Torque rheometers comprise two irregularly shaped screw-type mixing elements in a heated mixing chamber shaped like a figure eight.The torque needed to turn the rollers in opposing directions and the temperature are measured as a function of time and rotation speed. Although the flow field is nonhomogeneous, the torque can be approximated by the following expression if appropriate approximations are made:

$$T = C(n)N^{n} \tag{37.59}$$

where the constant C depends on the mixer geometry and power law index, and N is the rotation speed in revolutions per minute. Also, since the viscosity varies linearly with torque, the temperature dependence of the viscosity can also be obtained from the data. One of the most commonly used torque rheometers is the one made by C.W. Brabender. The squeeze flow rheometer consists of two parallel, coaxial disks that are made to approach each other at constant velocity so that the liquid contained in the gap is forced out. The applied force is measured as a function of time in constant velocity experiments or the plate separation 2H as a function of time in constant force experiments. These units are relatively simple to operate and enable handling of materials prone to fracture; squeeze flow also simulates compression molding. The main problem with these units is that the flow is time dependent and nonhomogeneous and is neither purely shear nor extensional. Also, for rapid squeeze flow, elasticity effects need to be considered. The relevant equation for power-law fluids with consistency index K and power-law index n is:

$$F = \frac{-dH}{dt}\left[\frac{2n+1}{2n}\right]\frac{\pi K}{H^{2n+1}}\frac{R^{n+3}}{(n+3)} \tag{37.60}$$

which for the special case of Newtonian fluids, where K is the viscosity and n is unity, reduces to the well known Stefan equation.

Extensional Viscometers

Measurement of extensional properties remains a daunting task because of problems related to the difficulty in clamping the test material (for mobile fluids) and applying a controlled stretch rate for a sufficient length of time to attain steady state (for both mobile and rigid materials), Extensional rheometers are broadly classified as constant-stress or constant-stretch-rate units; it is usually found that steady state is attained faster in the constant-stress mode. Some of the key categories are listed below:

Mechanical Stretching Extensiometers

These devices are usually restricted to relatively immobile systems. In these devices one end of a cylindrical sample is attached to a force transducer, and the other end is drawn out such that either the stretch rate or stress is constant. For constant stretch rate, the length L varies with time t as per the following equation:

$$\varepsilon = \ln\frac{L}{L_o} = \dot{\varepsilon}t \tag{37.61}$$

where L_o is the initial length and ε is the so-called Hencky extensional strain. The applied force is scaled with the cross-sectional area as it decreases for constant-stress devices. It is important that the experiment be carried out in an inert liquid to avoid gravitational effects and to maintain constant temperature. A major constraint is the difficulty in attaining steady state which typically corresponds to a Hencky strain of 7 or a thousand-fold increase in specimen length.

An important practical variation on this unit is the Meissner apparatus, which uses toothed wheels to stretch a constant length of sample at a constant angular velocity and a deflection transducer that measures the force needed to keep the sample length constant. Another version due to Sridhar and coworkers is the use of coaxial disks to draw the sandwiched sample apart and then determine the extensional viscosity from the measured force on the disks (using a transducer) and the filament profile (using high-resolution optics). The maximum stretch rate attained in such units is typically 1 l/sec, which is several orders of magnitude lower than those encountered in industry. Nonetheless, properties at low extension rates provide valuable insights into the origins of material properties and the development of constitutive equations.

Fiber Spinning

In this device, a pressurizing unit pumps molten polymer through a circular die and the exiting strand is pulled downwards at constant velocity using two counter-rotating knurled wheels; the drawdown force is measured using an appropriate type of transducer, and the spinline profile is determined optically. The system is usually enclosed in an isothermal chamber. In a constant pull-down velocity experiment, if it is assumed that the velocity is uniform over the cross-section, the local stretch rate can be determined from the flow rate and the diameter profile; the local liquid stress is determined from a momentum balance. The same concept is also applicable to mobile solutions which, in the unit developed by Ferguson and Hudson, involves stretching of the fluid jet emerging from a nozzle using a rotating drum and measuring the tensile force through the deflection of a thin-walled entrance tube; the analysis is similar except that fluid inertia and air drag effects do not become negligible.

A useful variation of the fiber spinning device due to Gupta and Sridhar generates stretch rates up to 1000 l/sec and involves measurement of the pressure decrease in a closed reservoir on application of suction to steady flow of exiting liquid through a capillary; the pressure decrease is related to the tensile stress and the stretch rate is determined from the spinline profile. An extension of this approach (developed by Agarwal and Gupta), which enables stretching of low viscosity and low elasticity liquids as well as miniaturization, employs intrinsic viscosity-type measuring instruments.

The determination of a true extensional viscosity is a difficult proposition in these flows due to:

1. The inherent transient nature of the process since each fluid element experiences a range of stretch rates as it moves down the spin line
2. The considerable sensitivity of the material properties to the previous history for viscoelastic properties

From a practical viewpoint, one may define an apparent extensional viscosity as the ratio of the instantaneous extensional stress to the instantaneous stretch rate. This ratio provides a convenient measure of fluid resistance to extensional deformation and may be used for parameter determination in constitutive models, which in turn may be used to simulate the processes of interest.

Converging Flow

The increase in average velocity as a liquid flows through a conical channel imparts an extensional component to the flow field. The relative amounts of shear and extension depend on the channel geometry, flow rate, fluid properties, and wall slip. According to the Cogswell analysis, the average net extensional stress in the entry region σ_E and the average stretch rate $\dot{\varepsilon}_E$ are given by

$$\sigma_E = \left(\frac{3}{8}\right)(n+1)P_{ent} \tag{37.62}$$

and

$$\dot{\varepsilon} = \frac{4\tau\dot{\gamma}}{3(n+1)P_{ent}} \tag{37.63}$$

where Q is the flow rate, n is the power-law index for shear flow, τ is the shear stress at a shear rate $\dot{\gamma}$ ($= 4Q/\pi R^3$) and P_{ent} is the total pressure drop across the abrupt contraction between two points where the flow is fully developed less the value for fully developed Poiseuelle flow.

An analogous situation occurs in the entrance region from a reservoir to a capillary as measured through the Bagley correction e, which is a measure of extensional viscosity effects.

Some other extensional rheometers are the open-syphon technique, in which liquid is sucked up from a reservoir into a capillary and the downward pull is measured; lubricated flows, in which low-viscosity Newtonian fluids are used at the walls of test sections designed to induce pure steady extensional flow; and the opposed-nozzle extensiometer due to Fuller, which involves application of a suction to fluid between two opposing nozzles to generate extension and then measuring the resulting torque on one of the nozzles that is mounted on a knife edge.

Defining Terms

Invariants of tensors — A second-order tensor T_{ij} having three scalar invariants that are independent of the coordinate system to which the components of T_{ij} are referred. They are $I_T = T_{ii}, II_T = T_{ij}T_{ji}$ and $III_T = T_{ij}T_{jk}T_{ki}$.

Material objectivity — A principle that requires that the rheological description of a material should be independent of the reference frame used to describe the fluid motion.

Relaxation time — The duration of time over which appreciable stresses persist after cessation of deformation in a fluid (e.g., the characteristic time constant for exponential decay of stress).

Rheometry — The experimental science of determining rheological material properties.

Viscometric flow — A flow field in which the deformation as seen by a fluid element is indistinguishable from simple shear flow.

References

Agarwal, S. and Gupta, R. K., 2002. An innovative extensional rheometer for low-viscosity and low-elasticity liquids, *Rheol. Acta,* 41:456.

Astarita, G. and Marrucci, G. 1974. *Principles of Non- Newtonian Fluid Mechanics.* McGraw-Hill, London.

Barnes, H. A., Hutton, J. F. and Walters, K., 1989. *An Introduction to Rheology,* Elsevier Science Publishers, Amsterdam.

Bird, R. B., Armstrong, R. C., and Hassager, O. 1987a. *Dynamics of Polymeric Liquids, Volume 1, Fluid Mechanics,* 2nd ed. John Wiley & Sons, New York.

Bird, R. B., Armstrong, R. C., and Hassager, O. 1987b. *Dynamics of Polymeric Liquids, Volume 2, Kinetic Theory,* 2nd ed. John Wiley & Sons, New York.

Collyer, A. A. and Clegg, D. W. (Eds.), 1998. *Rheological Measurement,* 2nd ed., Chapman and Hall, London.

Dealy, J. M. 1982. *Rheometers for Molten Plastics,* Van Nostrand Reinhold, New York.

Dealy, J. M. 1994. Official nomenclature for material functions describing the response of a viscoelastic fluid to various shearing and extensional deformations. *J. Rheol.* 38(1):179–191.

Doraiswamy, D., 2002. The Origins of Rheology: A Short Historical Excursion, Rheology Bulletin, 71(1), 7 (http://www.rheology.org/sor/publications/Rheology_B/Jan02/Origin_of_Rheology.pdf).

Doraiswamy, D., Majumdar, A. N., Tsao, I., Beris, A. N., Danforth, S. C., and Metzner, A. B., 1991. The Cox-Merz rule extended: a rheological model for concentrated suspensions and other materials with a yield stress, *J. Rheol.,* 35(4):647.

Ferry, J. D. 1980. *Viscoelastic Properties of Polymers,* 3rd ed. John Wiley & Sons, New York.

Graessley, W. W. 1974. The entanglement concept in polymer rheology. *Adv. Polym. Sci.* 16:1–179.

Gupta, R. K. 1994. Particulate suspensions. In *Flow and Rheology in Polymer Composites Manufacturing,* ed. S. G. Advani, pp. 9–51. Elsevier, New York.

Gupta, R. K., 2000. *Polymer and Composite Rheology,* Marcel Dekker, New York.

Han, C. D. 1981. *Multiphase Flow in Polymer Processing,* Academic, New York.
Larson, R. G. 1988. *Constitutive Equations for Polymer Melts and Solutions,* Butterworth, Boston.
Larson, R. G., 1999. *The Structure and Rheology of Complex Fluids,* Oxford University Press, New York.
van Krevelen, D. W. and Hoftyzer, P. J. 1976. *Properties of Polymers,* Elsevier, Amsterdam.

Further Information

The *Journal of Non-Newtonian Fluid Mechanics* (issued monthly), the *Journal of Rheology* (issued bimonthly), and *Rheologica Acta* (issued bimonthly) report advances in non-Newtonian fluids.

Dealy, J. M. and Wissbrun, K. F. 1990. *Melt Rheology and its Role in Plastic Processing,* Van Nostrand Reinhold, New York. Industrially relevant non-Newtonian systems (including liquid crystalline polymers) are discussed.

Tanner, R. I., 2000. *Engineering Rheology,* 2nd ed. Oxford University Press, Oxford.

38
Airfoils/Wings

Bruce R. Munson
Iowa State University

Dennis J. Cronin
Iowa State University

A simplified sketch of a wing is shown in Figure 38.1. An **airfoil** is any cross-section of the wing made by a plane parallel to the *xz* plane. The airfoil size and shape usually vary along the span.

Airfoils and wings are designed to generate a lift force, L, normal to the free stream flow that is considerably larger than the drag force, D, parallel to the free stream flow. The lift and drag are strongly dependent on the geometry (shape, size, orientation to the flow) of the wing and the speed at which it flies, V_o, as well as other parameters, including the density, ρ; viscosity, μ; and speed of sound, a; of the air. The following sections discuss some properties of airfoils and wings.

38.1 Nomenclature

The shape, size, and orientation of an airfoil can be given in terms of the following parameters (Figure 38.2): the **chord** length, c, the chord line that connects the leading and trailing edges; the **angle of attack,** α, relative to the free stream velocity, V_o; the mean **camber** line that is halfway between the upper and lower surfaces; and the thickness distribution, t, which is the distance between the upper and lower surfaces perpendicular to the camber line.

Various classes of airfoils have been developed over the years. These include the classic National Advisory Committee for Aeronautics four-, five-, and six-digit series airfoils (for example, the NACA 2412 airfoil used on the Cessna 150 or the NACA 64A109 used on the Gates Learjet [Anderson, 1991]) as well as numerous other modern airfoils [Hubin, 1992].

Performance characteristics of wings are normally given in terms of the dimensionless **lift coefficient** and **drag coefficient,**

$$C_L = \frac{L}{q_o S} \tag{38.1}$$

$$C_D = \frac{D}{q_o S} \tag{38.2}$$

where $q_o = \frac{1}{2}\rho V_o^2$ is the **dynamic pressure** and S is the planform area of the wing. The planform area is the area seen by looking into the wing from above: the span times the chord for a rectangular wing.

0-8493-1586-7/05/$0.00+$1.50

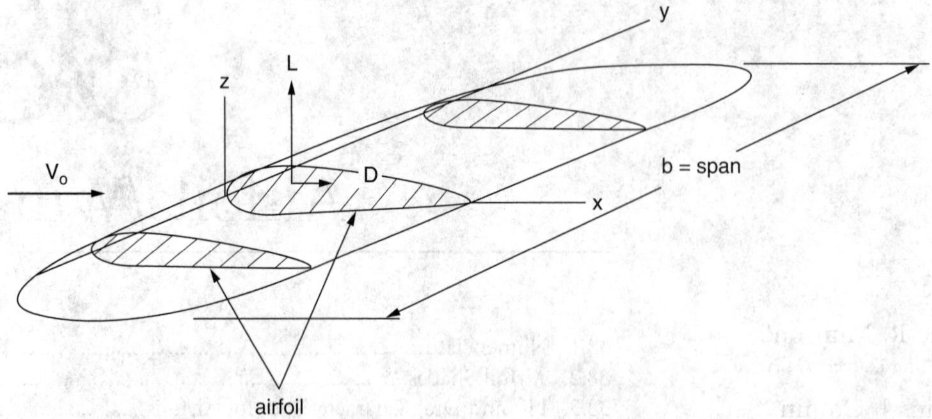

FIGURE 38.1 Wing geometry.

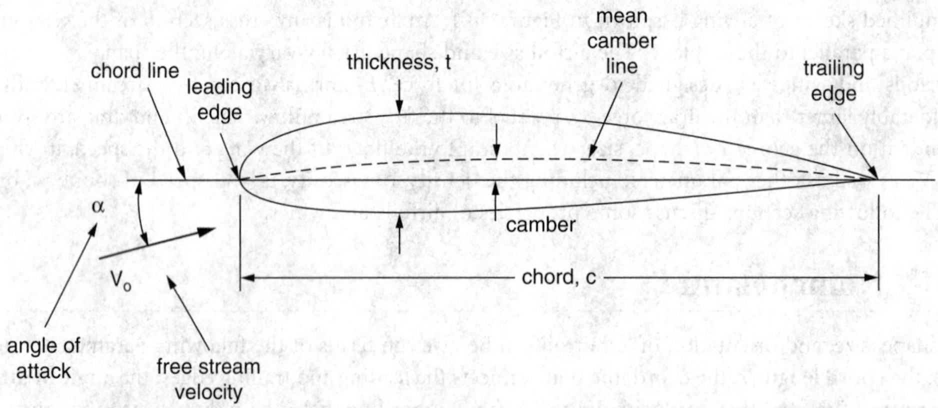

FIGURE 38.2 Airfoil geometry.

Typical characteristics for lift and drag coefficients as a function of the angle of attack are shown in Figure 38.3. An efficient wing has a large lift-to-drag ratio — that is, a large C_L/C_D.

As the angle of attack is increased from small values, the lift coefficient increases nearly linearly with α. At larger angles, there is a sudden decrease in lift and a large increase in drag. This condition indicates that the wing has stalled. The airflow has separated from the upper surface, and an area of reverse flow exists (Figure 38.3). **Stall** is a manifestation of boundary layer separation. This complex phenomenon is a result of viscous effects within a thin air layer (the boundary layer) near the upper surface of the wing in which viscous effects are important [Schlichting, 1979].

In addition to knowing the lift and drag for an airfoil, it is often necessary to know the location where these forces act. This location, the **center of pressure**, is important is determining the moments that tend to pitch the nose of the airplane up or down. Such information is often given in terms of a **moment coefficient**,

$$C_M = \frac{M}{q_o S c} \tag{38.3}$$

where M is the moment of the lift and drag forces about some specified point, often the leading edge.

For a given geometry, the lift and drag coefficients and the center of pressure (or moment coefficient) may depend on the flight speed and properties of the air. This dependence can be characterized in terms

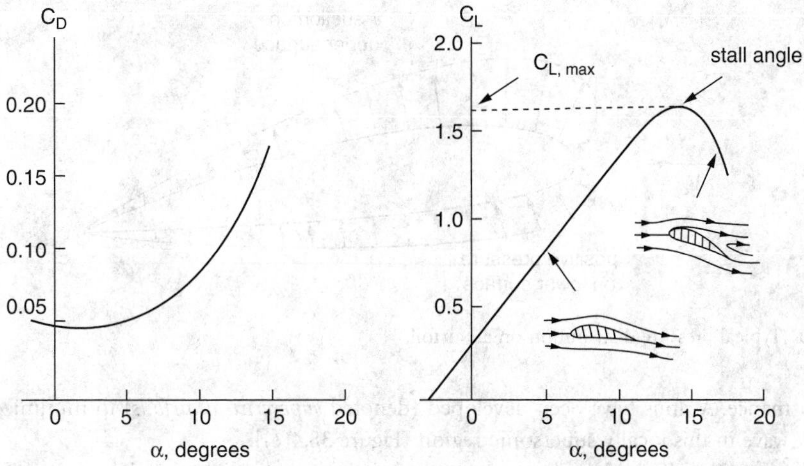

FIGURE 38.3 Typical lift and drag coefficients as a function of angle of attack.

of the Reynolds number based on chord length, $Re = \rho V_o c / \mu$, and the Mach number, $Ma = V_o / a$. For modern commercial aircraft, the Reynolds number is typically on the order of millions (10^6). Mach numbers range from less than 1 (subsonic flight) to greater than 1 (supersonic flight).

38.2 Airfoil Shapes

As shown in Figure 38.4, typical airfoil shapes have changed over the years in response to changes in flight requirements and because of increased knowledge of flow properties. Early airplanes used thin airfoils (maximum thickness 6 to 8% of the chord length), with only slight camber [Figure 38.4(a)]. Subsequently, thicker (12 to 18% maximum thickness) airfoils were developed and used successfully on a variety of low-speed aircraft [Figure 38.4(b) and Figure 38.4(c)].

A relatively recent development (ca. 1970s) has been design and construction of laminar flow airfoils that have smaller drag than previously obtainable [(Figure 38.4(d)]. This has resulted from new airfoil shapes and smooth surface construction that allow the flow over most of the airfoil to remain laminar rather than become turbulent. The performance of such airfoils can be quite sensitive to surface roughness (e.g., insects, ice, and rain) and Reynolds number effects.

With the advent of commercial and business jet aircraft, it became necessary to develop airfoils that operate properly at Mach numbers close to unity. Since air, in general, accelerates as it passes around an airfoil, the flow may be locally supersonic near portions of the upper surface, even though the flight speed is subsonic. Such supersonic flow can cause shock waves (discontinuities in the flow) that degrade

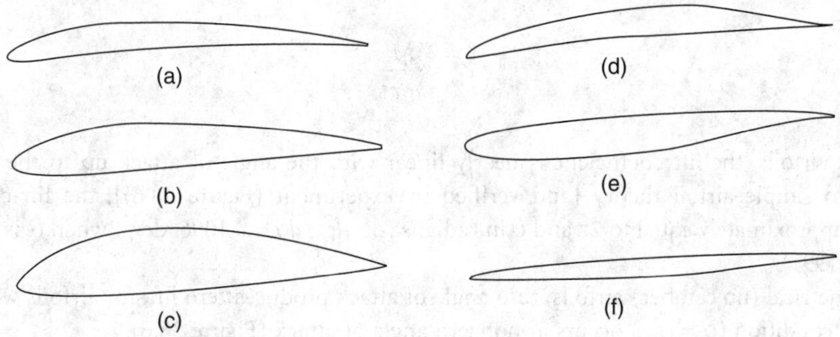

FIGURE 38.4 Various airfoil shapes.

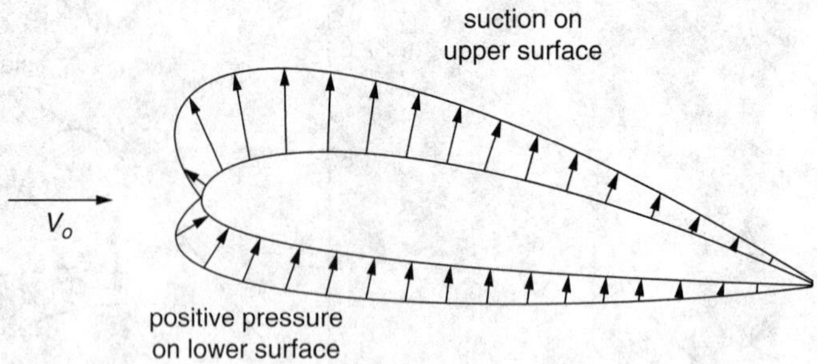

FIGURE 38.5 Typical pressure distribution on an airfoil.

airfoil performance. Airfoils have been developed (denoted *supercritical airfoils*) to minimize the effect of the shock wave in this locally supersonic region [Figure 38.4(e)].

Certain flow phenomena, such as shock waves, occur in supersonic flight that do not occur for subsonic flight [Anderson, 1991]. The result is that supersonic airfoils tend to be thinner and sharper than those for subsonic flight [see Figure 38.4(f)].

38.3 Lift and Drag Characteristics for Airfoils

Lift and drag forces on airfoils are the result of pressure and viscous forces that the air imposes on the airfoil surfaces. Pressure is the dominant factor that produces lift. A typical pressure distribution is shown in Figure 38.5. In simplistic terms, the air travels faster over the upper surface than it does over the lower surface. Hence, from **Bernoulli's principle** for steady flow, the pressure on the upper surface is lower than that on the bottom surface.

For an unstalled airfoil, most of the drag is due to viscous forces. This skin friction drag is a result of the shear stress distribution on the airfoil. For a stalled airfoil, pressure forces contribute significantly to the drag.

For a two-dimensional body such as an airfoil (a wing of infinite span), the section lift, drag, and moment coefficients (about a defined point) are based on the lift, drag, and moment per unit span, L' [force/length], D' [force/length], and M' [force · length/length], respectively. That is,

$$c_l = \frac{L'}{q_o c} \tag{38.4}$$

$$c_d = \frac{D'}{q_o c} \tag{38.5}$$

$$c_m = \frac{M'}{q_o c^2} \tag{38.6}$$

For most airfoils, the lift coefficient is nearly linear with the angle of attack up to the stall angle. According to simple airfoil theory [and verified by experiment (Figure 38.6)], the lift-curve slope, $dc_l / d\alpha$, is approximately equal to 2π and α in radians (or $dc_l / d\alpha = 0.1096$ deg^{-1} when α is in degrees) [Anderson, 1991].

For symmetrical (no camber) airfoils, zero angle of attack produces zero lift; for airfoils with camber, the zero-lift condition ($\alpha = \alpha_{0L}$) occurs at nonzero angle of attack (Figure 38.6).

The maximum lift coefficient for an airfoil ($c_l = c_{l,\max}$) is typically on the order of unity and occurs at the critical angle of attack ($\alpha = \alpha_{CR}$) (Figure 38.3). That is, the lift generated per unit span is on the

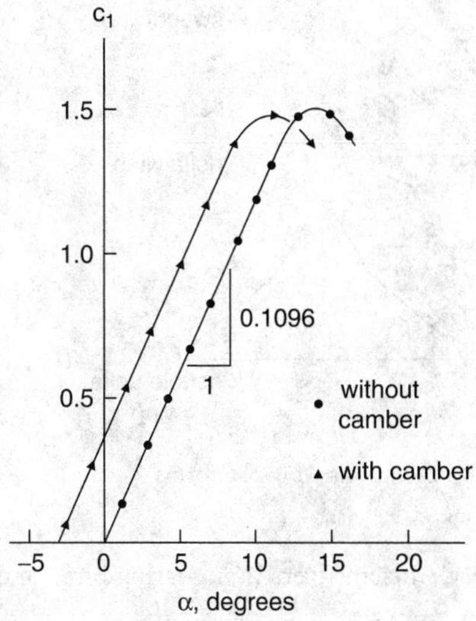

FIGURE 38.6 Effect of camber on airfoil lift coefficient.

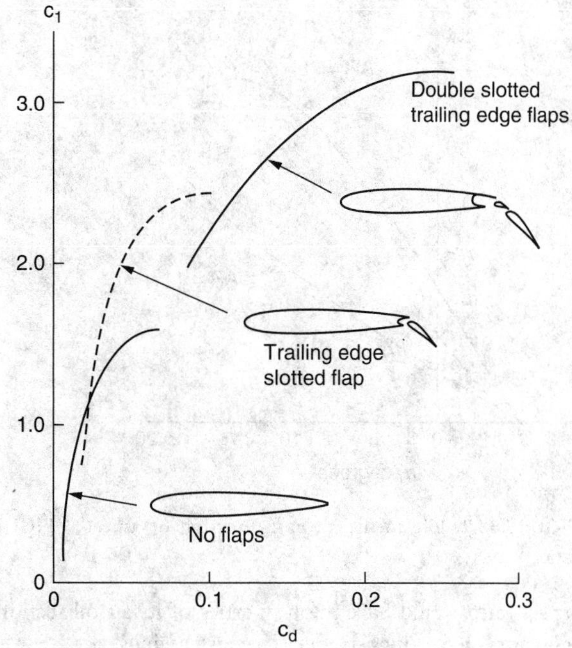

FIGURE 38.7 Effect of flaps on airfoil lift coefficient.

order of the dynamic pressure times the planform area: $L' = c_l q_o c \approx q_0 c$. The drag coefficient, on the other hand, is on the order of 0.01. Hence, the maximum lift-to-drag ratio is on the order of 100.

As illustrated in Figure 38.7, the airfoil geometry may be altered by using movable trailing or leading edge **flaps**. Such devices can significantly improve low-speed (i.e., landing or takeoff) performance by increasing the maximum lift coefficient, thereby reducing the required landing or takeoff speed.

38.4 Lift and Drag of Wings

All wings have a finite span, b, with two wing tips. The flow near the tips can greatly influence the flow characteristics over the entire wing. Hence, a wing has different lift and drag coefficients than those for the corresponding airfoil. That is, the lift and drag coefficients for a wing are a function of the aspect ratio, $AR = b^2/S$. For a wing of rectangular planform (i.e., constant chord), the aspect ratio is simply b/c.

Because of the pressure difference between the lower and upper surfaces of a wing, the air tends to "leak" around the wing tips (bottom to top) and produce a swirling flow — the trailing or wing tip vortices shown in Figure 38.8. This swirl interacts with the flow over the entire length of the wing, thereby affecting its lift and drag. The trailing vortices create a flow that makes it appear as though the wing were flying at an angle of attack different from the actual angle. This effect produces additional drag termed the *induced drag*.

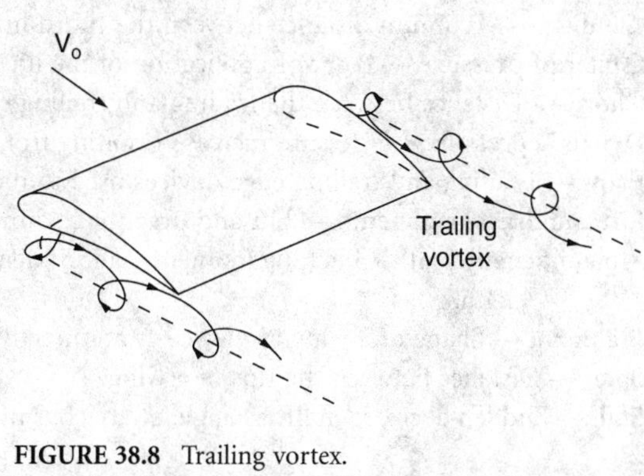

FIGURE 38.8 Trailing vortex.

As shown by theory and experiment [Anderson, 1991], the larger the aspect ratio is, the larger the lift coefficient and the smaller the drag coefficient (Figure 38.9). A wing with an infinite

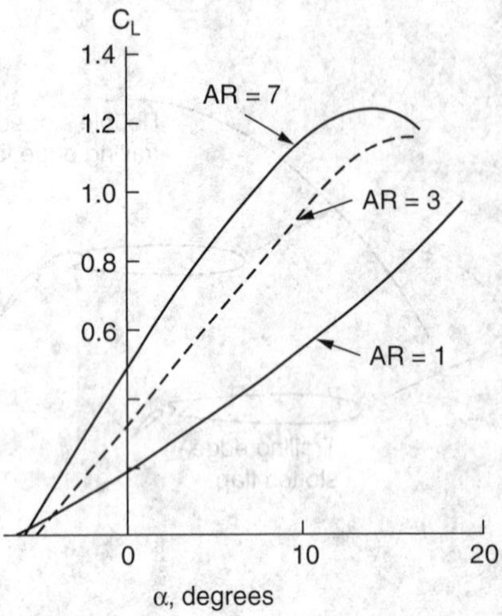

FIGURE 38.9 Lift coefficient as a function of aspect ratio.

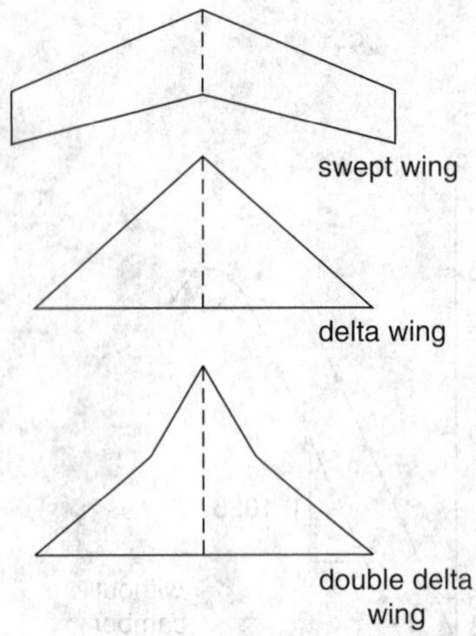

FIGURE 38.10 Various wing planforms.

aspect ratio would have the properties of its airfoil section. Very efficient flyers (i.e., soaring birds and sailplanes) have long, slender (large AR) wings.

For many wings, the chord length decreases from the wing root (next to the aircraft body) to the wing tip. In addition, the shape of the airfoil section may change from root to tip, as may the local angle of attack (i.e., the wing may have some "twist" to it).

Many modern high-speed wings are swept back with a V-shaped planform; others are delta wings with a triangular planform (Figure 38.10). Such designs take advantage of characteristics associated with high-speed compressible flow [Anderson, 1990].

Although tunnel tests of model airfoils and wings still provide valuable (and sometimes unexpected) information, modern computational fluid dynamic (CFD) techniques are widely used. Techniques involving paneling methods, finite elements, boundary elements, finite differences, and viscous-inviscid interaction are among the powerful tools currently available to the aerodynamicist [Moran, 1984].

Defining Terms

Airfoil — The cross-section of a wing, front to back.

Angle of attack — The angle between the line connecting the leading and trailing edges of an airfoil and the free stream velocity.

Bernoulli's principle — Conservation of energy principle that states that an increase in flow speed is accompanied by a decrease in pressure and vice versa.

Camber — Maximum distance between the chord line and the camber line.

Center of pressure — Point of application of the lift and drag forces.

Chord — Distance between the leading and trailing edges of an airfoil.

Dynamic pressure — Pressure increase resulting from the conversion of kinetic energy into pressure.

Flaps — Leading and trailing edge devices used to modify the geometry of an airfoil.

Lift and drag coefficients — Lift and drag made dimensionless by dynamic pressure and wing area.

Moment coefficient — Pitching moment made dimensionless by dynamic pressure, wing area, and chord length.

Planform — Shape of a wing as viewed from directly above it.

Span — Distance between the tips of a wing.

Stall — Sudden decrease in lift as angle of attack is increased to the point where flow separation occurs.

References

Anderson, J. D. 1990. *Modern Compressible Flow with Historical Perspective,* 2nd ed. McGraw-Hill, New York.

Anderson, J. D. 1991. *Fundamentals of Aerodynamics,* 2nd ed. McGraw-Hill, New York.

Hubin, W. N. 1992. *The Science of Flight: Pilot Oriented Aerodynamics,* Iowa State University Press, Ames, IA.

Moran, J. 1984. *An Introduction to Theoretical and Computational Aerodynamics,* John Wiley & Sons, New York.

Schlichting, H. 1979. *Boundary Layer Theory,* 7th ed. McGraw-Hill, New York.

Further Information

Abbott, I. H. and van Doenhoff, A. E. 1949. *Theory of Wing Sections,* McGraw-Hill, New York.

Anderson, D. A., Tannehill, J. C., and Pletcher, R. H. 1984. *Computational Fluid Mechanics and Heat Transfer,* Hemisphere, New York.

Anderson, J. D. 1985. *Introduction to Flight,* 2nd ed. McGraw-Hill, New York.

39

Boundary Layers

Edwin R. Braun
University of North Carolina, Charlotte

Pao-lien Wang
University of North Carolina, Charlotte

39.1 Theoretical Boundary Layers

In a simple model of a solid, material deformation is proportional to the strain. In a simple model of a fluid, the deformation is proportional to the rate of strain or the change in velocity over a small distance. The mathematical term describing this phenomenon is the last term in the following boundary layer equation:

$$\rho\left[\frac{\delta u}{\delta t}+u\frac{\delta u}{\delta x}+v\frac{\delta u}{\delta y}\right]=-\frac{\delta p}{\delta x}+\frac{\delta}{\delta y}\left(\mu\frac{\delta u}{\delta y}\right) \tag{39.1}$$

where u is the velocity in the x direction as a function of x, y, and t. The values ρ and μ are the density and dynamic viscosity for the fluid, respectively. This equation is good for all situations with no pressure (P) change present in the direction normal to the wall.

The left-hand side of Equation (39.1) represents time kinetic energy in flow. The pressure term is a potential energy term. The rate of strain term that represents this energy dissipates through viscous losses. When the dissipation term is significant compared to the others, a boundary layer must be considered as part of the flow analysis.

For a straight-channel, steady (not time-dependent) flow, Equation (39.1) becomes

$$\mu\frac{d^2u}{dy^2}=\frac{dp}{dx} \tag{39.2}$$

which has the solution

$$u=-\frac{1}{2\mu}\frac{dp}{dx}(D^2-y^2) \tag{39.3}$$

where d is the distance toward the wall measured from the centerline and the velocity u is zero at the walls. This velocity equation is parabola. Note that when $y = D$, Equation (39.3) becomes

0-8493-1586-7/05/$0.00+$1.50

$$u_{cL} = -\frac{D^2}{2\mu}\frac{dp}{dx} \tag{39.4}$$

Thus, if the pressure loss over a distance X is measured along with the centerline velocity (u_{cL}), the viscosity can be determined. Similarly, if the velocity is known at the centerline, the pressure loss per unit length can be calculated.

39.2 Reynolds Similarity in Test Data

As a boundary layer develops, it starts in a smooth, or laminar, state. Downstream, it transforms into a turbulent state, where the flow is irregular and contains eddies. Various physical conditions, such as wall or surface roughness or upstream turbulence, will affect the speed of this transition. In smooth-walled pipes, laminar flow occurs for Reynolds numbers (Re) of less than 2000, with fully developed turbulence for Re greater than 4000. The Reynolds number is a dimensionless number developed from dynamic similarity principles that represents the ratio of the magnitudes of the inertia forces to the friction forces in the fluid.

$$\mathrm{Re} = \frac{\text{inertia force}}{\text{friction force}}$$

where inertia force $= \rho V_c^2 L_c^2$ and friction force $= \mu V_c L_c^2$. Then,

$$\mathrm{Re} = \frac{\rho V_c L_c}{\mu} = \frac{V_c L_c}{\nu} \tag{39.5}$$

where V_c and L_c are characteristic or representative velocities and lengths, respectively. For a pipe or similar narrow channel, L_c is the internal diameter (ID) of the pipe and V_c is the average or bulk velocity obtained by dividing the mass flow rate (M) by the cross-sectional area and density of the fluid:

$$V_c = \frac{M}{\rho A} \tag{39.6}$$

Using the Reynolds number as a similarity parameter, test data can be correlated into generalized charts for frictional losses.

For the flat plate (Figure 39.1) case, V_c is taken as the free stream velocity outside the boundary layer, and L_c is the length measured along the wall standing from the leading edge.

39.3 Friction in Pipes

The energy equation for steady flow between any two points in a pipe can be written as

$$\frac{V_2^2 - V_1^2}{2g} + \frac{P_2 - P_1}{\rho g} + Z_2 - Z_1 - h_f = 0 \tag{39.7}$$

where h_f is a head loss due to friction. This equation neglects other minor losses (such as elbows, valves, exit and entrance losses, and bends). It is useful to define the head loss in terms of a friction factor (f) such that this nondimensional friction factor (f), known as the Darcy friction factor, can be determined experimentally as a function of the dimensionless Reynolds numbers and a relative roughness parameter ε/D, as shown in Figure 39.2. Rough factors, ε, are given in Table 39.1.

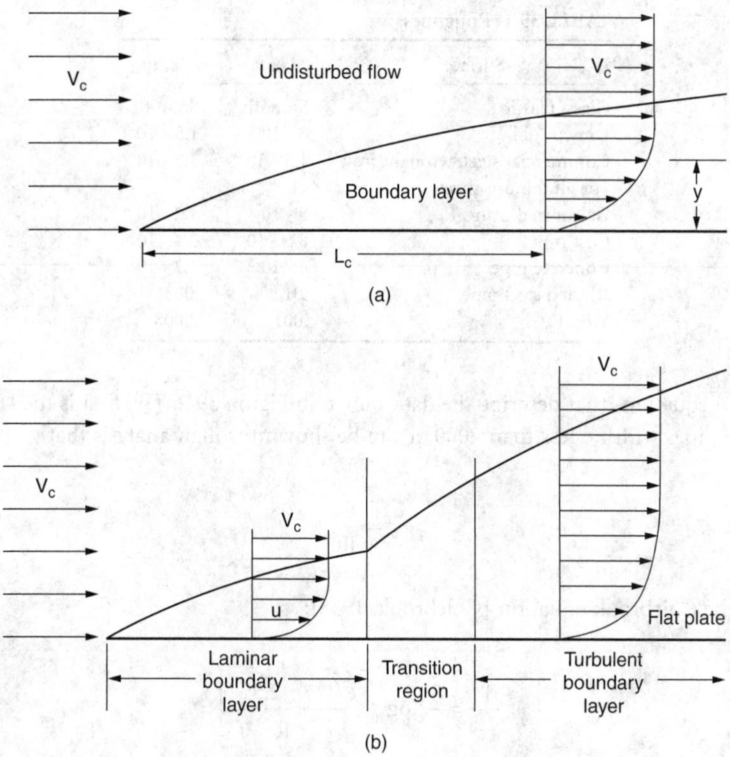

FIGURE 39.1 (a) Boundary layer along a smooth plane. (b) Laminar and turbulent boundary layers along a smooth, flat plate. (Vertical scales greatly enlarged.)

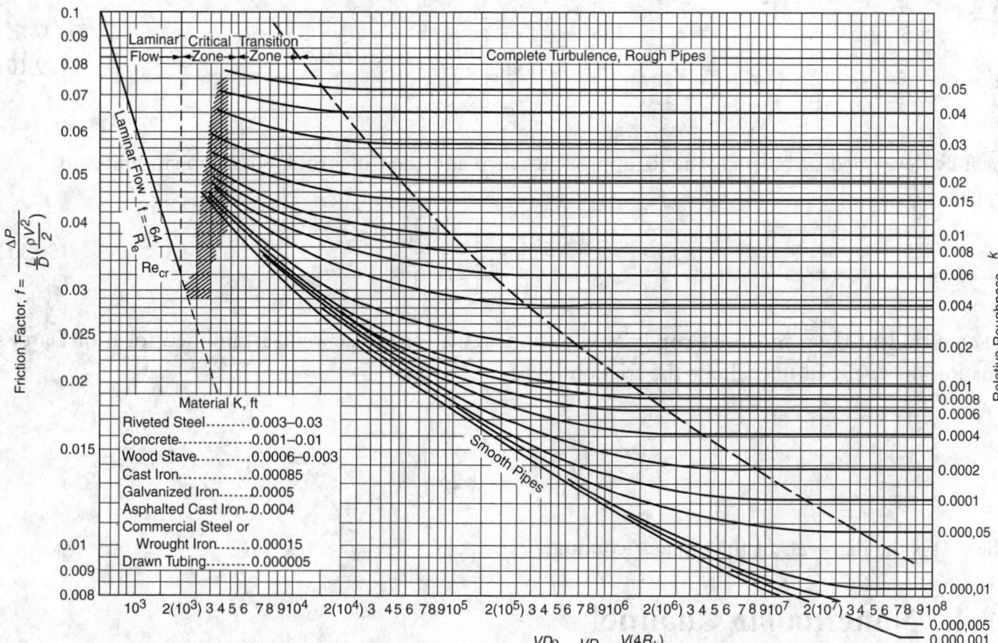

FIGURE 39.2 Friction factors for commercial pipe. (*Source*: Moody, L. F. 1944. Friction factors for pipe flow. *Trans. ASME.* 66:672. With permission.)

TABLE 39.1 Roughness

Surface	ε, ft	ε, m
Glass, plastic	Smooth	Smooth
Drawn tubing	$5 \cdot 10^{-6}$	$1.5 \cdot 10^{-6}$
Commercial steel, wrought iron or aluminum sheet	$1.5 \cdot 10^{-4}$	$4.6 \cdot 10^{-5}$
Galvanized iron	$5 \cdot 10^{-4}$	$1.2 \cdot 10^{-4}$
Cast iron	$8.5 \cdot 10^{-4}$	$2.4 \cdot 10^{-4}$
Concrete pipe	$4 \cdot 10^{-3}$	$1.2 \cdot 10^{-3}$
Riveted steel pipe	.01	.003
Wood	.001	.0003

There are two equations that describe the data shown in Figure 39.2. The first is the laminar line. For laminar flow in pipes with Re less than 2000, it can be shown through analysis that

$$f = \frac{64}{\text{Re}} \tag{39.8}$$

The second is the Colebrook equation [Colebrook, 1938]:

$$\frac{1}{\sqrt{f}} = -2\log_{10}\left[\frac{\varepsilon/D}{3.7} + \frac{2.51}{\text{Re}\sqrt{f}}\right] \tag{39.9}$$

which describes the turbulent region. Note that, as the roughness ε approaches zero, we obtain the smooth pipeline and the equation becomes

$$\frac{1}{\sqrt{f}} = 2\log_{10}\left[\frac{\text{Re}\sqrt{f}}{2.51}\right] \tag{39.10}$$

For fully developed turbulence, the Re approaches zero and the Colebrook equation simplifies to

$$\frac{1}{\sqrt{f}} = 2\log_{10}\left[\frac{3.7}{\varepsilon/D}\right] \tag{39.11}$$

For turbulent flows in closed conduits with noncircular cross-sections, a modified form of Darcy's equation may be used to evaluate the friction loss:

$$h_f = f\left(\frac{L}{D}\right)\left(\frac{v^2}{2g}\right)$$

where D is the diameter of the circular conduit.

39.4 Noncircular Channel

In the case of noncircular cross-sections, a new term, R, is introduced to replace diameter D. R is defined as hydraulic radius, which is the ratio of the cross-sectional area to the wetted perimeter (WP) of the noncircular flow section.

$$R = \frac{A}{\text{WP}}$$

For a circular pipe of diameter D the hydraulic radius R is

$$R = \frac{A}{\text{WP}} = \frac{\pi D^2/4}{\pi D} = \frac{D}{4}$$

or $D = 4R$. Substitution of $4R$ for D in Darcy's equation yields

$$h_f = \left(\frac{L}{4R}\right)\left(\frac{v^2}{2g}\right)$$

The Reynolds number can be modified as

$$\text{Re} = \frac{v(4R)\rho}{\mu} \quad \text{or} \quad \text{Re} = \frac{v(4R)}{\nu}$$

39.5 Example Solutions

Example 39.1

Refer to Figure 39.3. Water at 50°C is flowing at a rate of 0.07 m^3/sec. The pipeline is steel and has an inside diameter of 0.19 m. The length of the pipeline is 900 m. Assume the kinematic viscosity (ν) is $5.48 \cdot 10^{-7}$ m^2/sec. Find the power input to the pump if its efficiency is 82%; neglect minor losses.

Given information is as follows:

$$Q = 0.07 \text{ m}^3; \quad L = 900 \text{ m}; \quad T = 50°\text{C}$$

$$\Delta Z = 15 \text{ m}; \quad \nu = 5.48 \cdot 10^{-7} \text{ m}^2/\text{s}; \quad D = 0.2 \text{ m}$$

$$\gamma \text{ at } 50°\text{C} = 9.69 \text{ kN/m}^3$$

Find power input to pump.

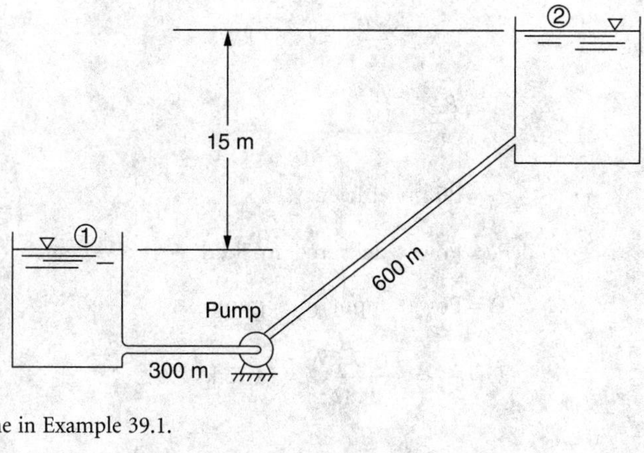

FIGURE 39.3 Pipeline in Example 39.1.

Solution

First, determine the Reynolds number:

$$\mathrm{Re} = \frac{vD}{\nu}; \quad v = \frac{Q}{A}$$

$$\mathrm{Re} = \frac{4Q}{\pi D \nu} = \frac{4(0.07)}{\pi(0.2)(5.48 \cdot 10^{-7})}$$

$$= 8.13 \cdot 10^{5}$$

Second, determine ε/D ratio and friction factor f. Roughness (ε) for steel pipe = $4.6 \cdot 10^{-5}$ m.

$$\frac{\varepsilon}{D} = \frac{4.6 \cdot 10^{-5}\ \mathrm{m}}{0.19\ \mathrm{m}} = 0.000242$$

From Moody's diagram with values of N_R and ε/D, $f = 0.0151$.

Next, determine head loss due to friction:

$$h_f = 0.0151\left(\frac{L}{D}\right)\left(\frac{v^2}{2g}\right), \quad v = \frac{Q}{A}$$

$$h_f = 0.0151\frac{8LQ^2}{\pi^2 g D^5}$$

$$= 0.0151\frac{8(900)(0.07)^2}{\pi^2(9.81)(0.02)^5}$$

$$= 0.0151\frac{35.28}{0.031} = 17.2\ \mathrm{m}$$

Finally, determine power input into pump:

$$P_A = h_A \gamma Q = 17.2\left(9.69\frac{\mathrm{kN}}{\mathrm{m}^3}\right)\left(0.07\frac{\mathrm{m}^3}{\mathrm{s}}\right)$$

$$= 11.67\frac{\mathrm{kN \cdot m}}{\mathrm{s}} = 11.67\ \mathrm{kW}$$

$$e_p = \frac{P_A}{P_1}$$

e_p = Pump efficiency

P_A = Power delivered to fluid

P_I = Power input into pump

$$P_I = \frac{P_A}{e_p} = \frac{11.67\ \mathrm{kW}}{0.82} = 14.23\ \mathrm{kW}$$

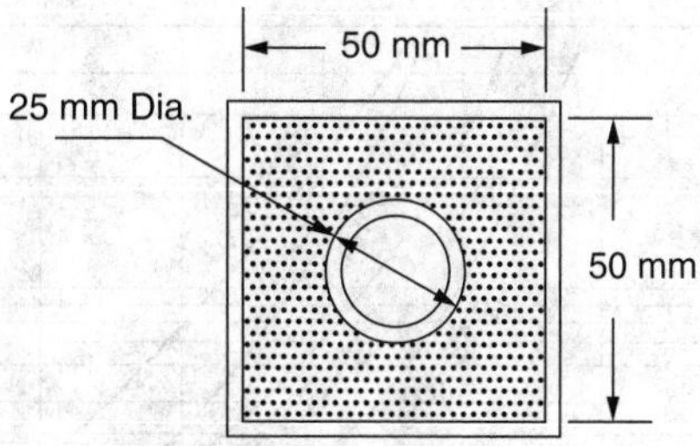

FIGURE 39.4 Duct in Example 39.2.

TABLE 39.2 Dynamic Viscosity of Liquids (μ) (mPa · sec)

Liquid	–25°C	0°C	25°C	50°C	75°C	100°C
Water		1.793	0.890	0.547	0.378	
Mercury			1.526	1.402	1.312	
Methanol	1.258	0.793	0.544			
Isobutyl acetate			0.676	0.493	0.370	0.286
Toluene	1.165	0.778	0.560	0.424	0.333	0.270
Styrene		1.050	0.695	0.507	0.390	0.310
Acetic acid			1.056	0.786	0.599	0.464
Ethanol	3.262	1.786	1.074	0.694	0.476	
Ethylene glycol			16.1	6.554	3.340	1.975

Example 39.2

Air with a specific weight of 12.5 N/m³ and dynamic viscosity of $2.0 \cdot 10^{-5}$ N · sec/m² flows through the shaded portion of the duct shown in Figure 39.4 at the rate of 0.04 m³/sec. (See Table 39.2 or Figure 39.5 for dynamic viscosities of some common liquids.) Calculate the Reynolds number of the flow, given that $\gamma = 12.5$ N/m³, $\mu = 2.0 \cdot 10^{-5}$ N · sec/m², $Q = 0.04$ m³/sec, and $L = 30$ m.

Solution

$$\rho = \gamma/g = \frac{12.5\ \mathrm{N/m^3}}{9.81\ \mathrm{m/s^2}} = 1.27\ \mathrm{N\cdot s^2/m^4}\ \text{or}\ \mathrm{kg/m^3}$$

$$\begin{aligned} A(\text{shaded}) &= (0.05\ \mathrm{m})^2 - \frac{\pi}{4}(0.025\ \mathrm{m})^2 \\ &= 0.0025\ \mathrm{m^2} - 0.00049\ \mathrm{m^2} \\ &= 0.002\ \mathrm{m^2} \end{aligned}$$

$$\begin{aligned} \text{Wet parameter (WP)} &= 4(0.05\ \mathrm{m}) + \pi(0.025\ \mathrm{m}) \\ &= 0.2\ \mathrm{m} + 0.0785\ \mathrm{m} \\ &= 0.279\ \mathrm{m} \end{aligned}$$

$$\begin{aligned} \text{Hydraulic radius } (R) &= \frac{A}{\mathrm{WP}} \\ &= \frac{0.002\ \mathrm{m^2}}{0.279\ \mathrm{m}} \\ &= 0.00717\ \mathrm{m} \end{aligned}$$

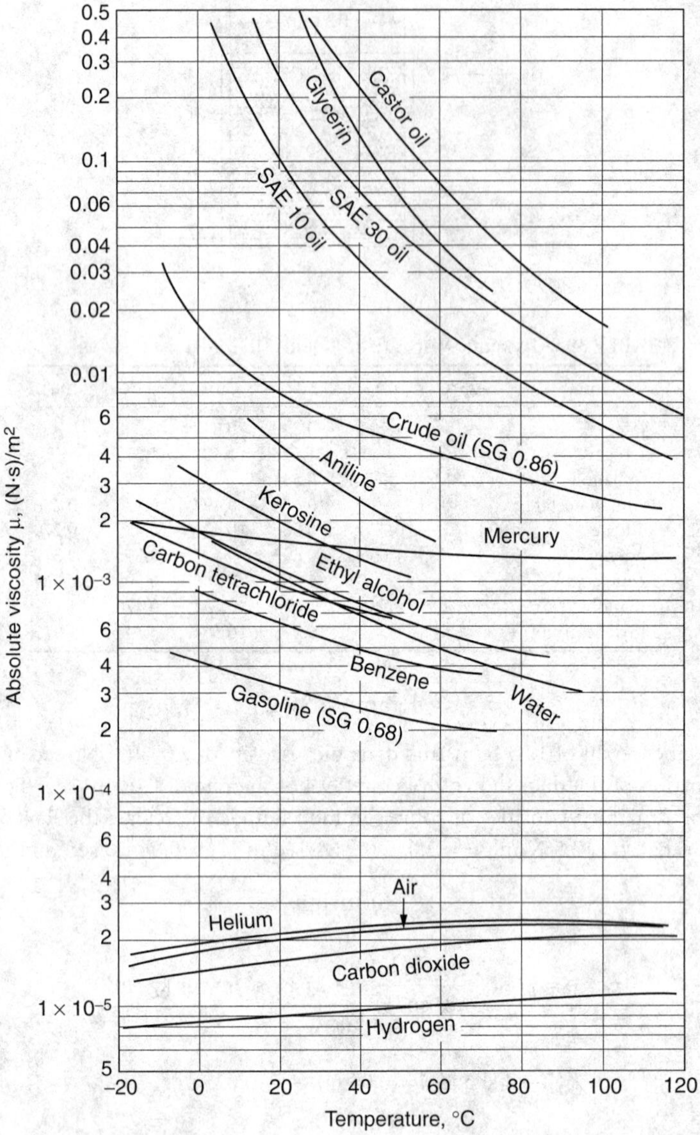

FIGURE 39.5 Absolute viscosity of common fluids at 1 atm. (*Source*: White, F. 1986. *Fluid Mechanics*, 2nd ed. McGraw-Hill, New York. With permission.)

$$v = \frac{Q}{A} = \frac{0.04\ \text{m}^3/\text{s}}{0.002\ \text{m}^2} = 20\ \text{m/s}$$

$$\text{Reynolds number (Re)} = \frac{4Rv\rho}{\mu}$$

$$N_R = \frac{4(0.00717\ \text{m})(20\ \text{m/s})(1.27\ \text{N}\cdot\text{s/m}^4)}{2.0\cdot 10^{-5}\ \text{N}\cdot\text{s/m}^2}$$

$$= 3.64\cdot 10^4$$

References

Colebrook, C. F. 1938. Turbulent flow in pipes with particular reference to the transition points between smooth and rough laws. *ICF Journal.* 2:133–156.

Moody, L. F. 1944. Friction factors for pipe flow. *Trans. ASME.* 66:672.

White, F. 1986. *Fluid Mechanics,* 2nd ed. McGraw-Hill, New York.

40 Valves

Blake P. Tullis
Utah State University

J. Paul Tullis
Tullis Engineering Consultants

Valves are mechanical devices typically installed in pipelines to control flow or pressure. The valves provide control using a variable restriction created by a rotating plug or disc; a sliding sleeve or gate; or the pinching of a flexible membrane. Since valves can be vital to the successful operation of piping systems, care should be taken to ensure that the proper valve is selected and that it is operated correctly. If not properly selected and operated, valves can cause operational problems including poor control, excessive head loss, cavitation, and hydraulic transients. Such problems can lead to accelerated wear requiring repair or replacement of the valve and, in some cases, system failure.

This chapter discusses hydraulic characteristics and function of control valves, air release/vacuum breaking valves, and check valves. Control valves are used to control flow, pressure, liquid level, cavitation, and pressure transients. Hydraulic aspects of common types of control valves are discussed, including controllability, torque requirements, cavitation, hydraulic transients, and principles of valve selection. Air valves are designed to expel large amounts of air at low pressure during filling, release small amounts of pressurized air during system operation, and admit air when the pipe is drained to prevent excessive vacuum pressures and possible pipe collapse. Guidelines are provided for selecting the proper size, type, and location of air valves to protect pipelines. Common types of check valves are described and guidelines are offered for selecting the proper check valve for different systems.

40.1 Control Valves

Hydraulic selection criteria for control valves include controllability (i.e., flow and pressure drop characteristics), torque or thrust requirements for operator sizing, cavitation performance, velocity limits, and transients generated by valve operation. Other factors that influence valve selection and performance include valve installation (flow direction and orientation), effectiveness of seating, and the influence of disturbances generated upstream from the valve.

Valve Types

Butterfly valves are popular and suitable for a variety of valve applications. They come in a wide variety of body, disc, size, and seat designs. Details of their construction can have a significant effect on their performance, especially their seating ability and the magnitude of the operating torque. All butterfly

0-8493-1586-7/05/$0.00+$1.50

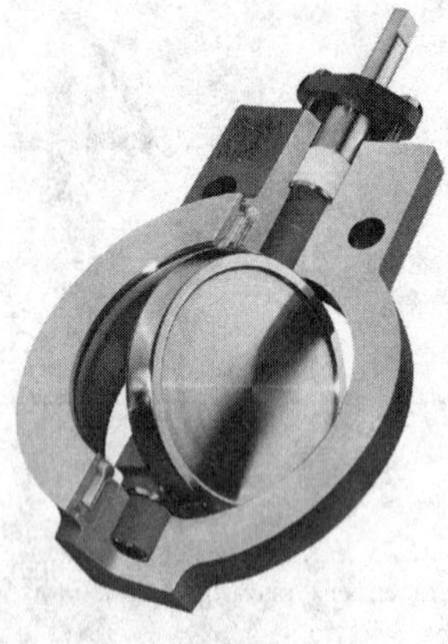

FIGURE 40.1 Metal seated butterfly valve cutaway (figure courtesy of FLOWSERVE).

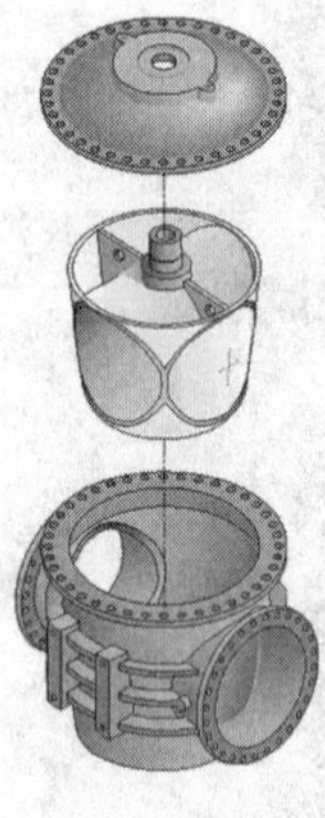

FIGURE 40.2 Typical full-port, skirted cone valve (figure courtesy of Apco Valve and Primer).

valves have a disc that rotates 90°, creating two crescent-shaped openings that discharge parallel to the wall of the downstream pipe. However, there is a wide range in valve quality, cost, and performance. A well-engineered butterfly valve can provide excellent service for many years. One type of butterfly valve is illustrated in Figure 40.1.

Cone valves have a rotating conical plug that presents no obstruction to the flow when the plug is in the fully open position. In a partially open position, these valves have two crescent-shaped throttling ports in series. There are four classifications of cone valves: full-port, reduced-port, skirted, and unskirted. Full-port means that the hole through the conical plug is the same diameter as the inlet and outlet passages of the valve body, as shown in Figure 40.2. A reduced-port valve is fabricated with the plug diameter smaller than the valve diameter, and the valve body has reducers and expanders to transition between pipe and plug diameters. The term "skirted" means that the plug is, or has the appearance of being, solid and all of the water flows through the two crescent-shaped openings in series. An "unskirted" cone valve has a fabricated plug design that allows some of the water to flow around the plug (i.e., between the plug and valve body). This design is sometimes referred to as a skeleton cone valve. The flow around the plug causes the unskirted or skeleton cone valve to have significantly worse cavitation performance than the skirted design. For unseating, the plug is lifted prior to rotation. After rotating the plug to the closed position, the plug is lowered into the seat.

Ball valves have numerous designs. One style of ball valve has a solid spherical plug with a cylindrical hole forming the flow passage (much like a skirted cone valve), as shown in Figure 40.3. For full-port designs, the flow passage is the same diameter as the valve inlet/outlet diameter. For reduced-port designs, the flow passage through the ball has a smaller diameter than the nominal valve diameter; the valve body includes reducing and expanding sections. Both of these designs have two throttling ports in series and have flow characteristics similar to cone valves. Close machining tolerances are required for the seating surfaces because the spherically shaped plug rotates into and out of the seat. Another design uses a fabricated plug essentially made of two intersecting pipes, one closed and one open, as shown in Figure 40.4. Ball valves are available with either metal or soft seats.

Plug valves are similar to cone valves, in that they have a conical rotating plug. The flow passage through the plug is typically rectangular in shape rather than circular, as shown in Figure 40.5. The eccentric plug valve design contains a control element that looks much like the visor on a helmet. This type of valve has only one throttling port.

Globe valves vary in design more than most valves do. There are far too many variations to discuss them in any detail. The conventional globe valve has a single port with a plug that moves linearly, as shown in Figure 40.6. The flow is controlled by the linear movement of the plug, relative to the seat. The internals of globe valves can vary significantly. Recent developments include a variety of single and multistage trims that suppress cavitation but reduce flow capacity. One type of trim, shown in Figure

FIGURE 40.3 Full-port, skirted, ball valve cutaway (figure courtesy of FLOWSERVE).

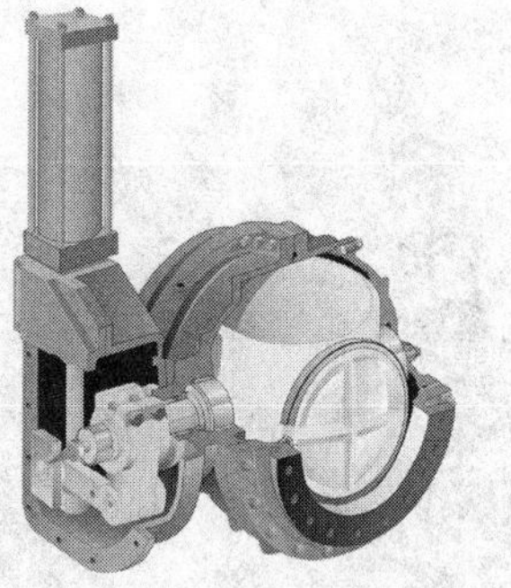

FIGURE 40.4 Unskirted ball valve cutaway (figure courtesy of Apco Valve and Primer).

40.7, has a perforated cylinder, similar to that of the sleeve valve. Some designs use multiple concentric perforated sleeves to dissipate the energy in stages. Other styles, often referred to as *stack valves,* use both parallel and series flow passages which produce very high pressure drops while suppressing cavitation.

Gate valves have a variety of fabrication differences that influence their seating performance, but their flow characteristics are similar. They have a single or double seating surface, with either soft or metal seats. The movement of a flat plate passing in and out of the flow path controls the flow through a gate valve. For most designs, the gate is completely removed from the flow path when the valve is in the full-open position. In the partially open position, all gate valves have one crescent-shaped flow opening. Gate valves can be relatively inexpensive and generate little head loss in the full-open position. As a result, gate valves are commonly used as isolation valves. Gate valves are seldom used as control valves due to vibrations and poor cavitation performance at partial valve openings.

Sleeve valves are a relatively new valve style and are excellent for controlling flow and suppressing cavitation in systems with relatively high pressure drop requirements. One style is an in-line sleeve valve that consists of a stationary sleeve with numerous holes and an external traveling sleeve, which controls the number of holes exposed for flow passage. The size and spacing of the holes can be varied to provide a variety of flow characteristics and cavitation performance. For in-line sleeve valves, the jets discharge inward, forcing the cavitation to be concentrated at the center of the discharge pipe, away from the boundaries. For use at terminal structures, sleeve valves can also be designed with the jets discharging outward into a pipe or tank. Since the perforated sleeve functions much like a strainer, sleeve valves are limited to clean water systems.

FIGURE 40.5 Full-port, skirted, plug valve cutaway (figure courtesy of FLOWSERVE).

FIGURE 40.6 Globe valve cutaway. The flow through the valve is controlled by a disc, which moves vertically up and down (figure courtesy of CLA-VAL).

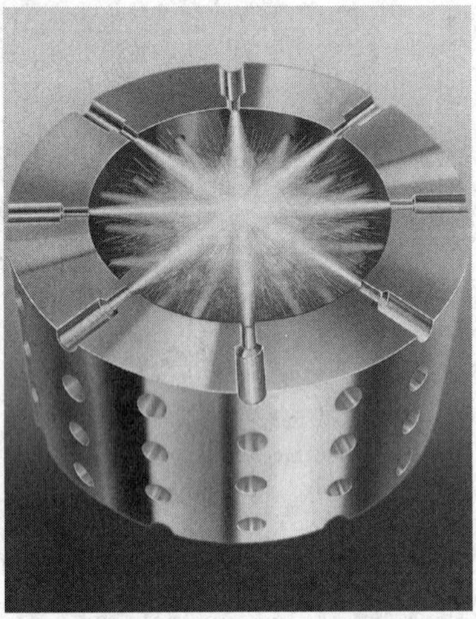

FIGURE 40.7 Globe valve cavitation trim. The figure illustrates flow passing through the trim and collecting in the center. Other trims/valves are designed for outward flow (figure courtesy of FLOWSERVE).

FIGURE 40.8 Free discharge fixed-cone (Howell–Bunger) valve (figure courtesy of Rodney Hunt).

Free-discharge, fixed-cone valves (Howell–Bunger) have a stationary cone, a traveling sleeve, and an optional discharge hood to direct the flow. A fixed-cone valve without a hood is shown in Figure 40.8. These valves are used for free-discharge releases at high pressure. Fixed-cone valves are commonly used in conjunction with low-level outlet works for dams. Since the high-velocity jet is fully aerated, cavitation is avoided. A dissipation pool is usually included to prevent erosion of the surrounding area by the jet. Care must be taken in designing the hood to limit back-splash and ensure that adequate air is supplied to prevent cavitation. Recent developments include discharge hoods designed to dissipate up to 95% of the jet's energy [Johnson et al., 2001].

For additional details about other types of control valves available in the industry and information on body styles, internal trim options, materials of construction, details on packing options, valve operators, etc., the reader is referred to Crane, 1979; Fisher, 2001; ISA, 1976; and Driskell, 1983.

Valve Coefficients

The relationship between flow and pressure drop at any valve opening can be expressed by a number of coefficients. The coefficients most commonly used are:

Discharge coefficient

$$C_d = \frac{V}{\sqrt{2\Delta P/\rho + V^2}} = \frac{V}{\sqrt{2g\Delta H + V^2}} \tag{40.1}$$

Loss coefficient

$$K = \frac{2\Delta P}{\rho V^2} = \frac{2g\Delta H}{V^2} \tag{40.2}$$

Flow coefficient

$$C_v = \frac{Q}{\sqrt{\Delta P / s.g.}} \tag{40.3}$$

Free-discharge coefficient

$$C_{df} = \frac{V}{\sqrt{2P_u/\rho + V^2}} = \frac{V}{\sqrt{2gH_u + V^2}} \tag{40.4}$$

V is the average velocity at the inlet to the valve; ρ is the fluid density; g is the acceleration due to gravity; ΔP and ΔH are the net pressure and head drop, respectively, across the valve; *s.g.* is the specific gravity of the fluid; and P_u and H_u are the gage pressure and the pressure head, respectively, at the inlet to the valve.

A few comments about these coefficients might be helpful. With the exception of Equation 40.3, the coefficients are dimensionless (as long as consistent units for ΔH, ΔP, V, g, and ρ are used). In Equation 40.3, ΔP is in pounds per square inch (psi) and Q is in U.S. gallons per minute (gpm). This makes the numeric value of C_v a function of valve size. The value of C_v has physical significance. It represents the flow rate in gpm that a valve can pass with 1 psi pressure differential. C_v data can be scaled between geometrically similar valves of different sizes using Equation 40.5. (Note: For best accuracy, the inside diameter should be used rather than the nominal diameter.)

$$C_{v1} = C_{v2}\left(\frac{d_1}{d_2}\right)^2 \tag{40.5}$$

One disadvantage of using C_v to quantify discharge capacity stems from the fact that full-port ball, cone, and gate valves have almost no pressure drop when the valve is full open. Consequently, the value of C_v approaches infinity. In contrast, the C_d value will approach 1.0 as pressure drop goes to zero. C_d is also independent of valve size and varies between 0 and 1.0 for all types of valves. This makes it convenient to compare the relative capacity of different valves.

Equation 40.4 is the recommended definition of the discharge coefficient for free discharging valves. Equation 40.4 can also be applied to predict flow capacity for a valve that is operating in choking cavitation (ch) by replacing P_u with $P_{u(ch)}$, where $P_{u(ch)} = P_u + P_{baro} - P_{vapor}$.

Figure 40.9 shows representative discharge coefficient data for a full-port skirted cone valve, a butterfly valve, and a globe valve. The data are based on actual test data, but the values only apply to the valves

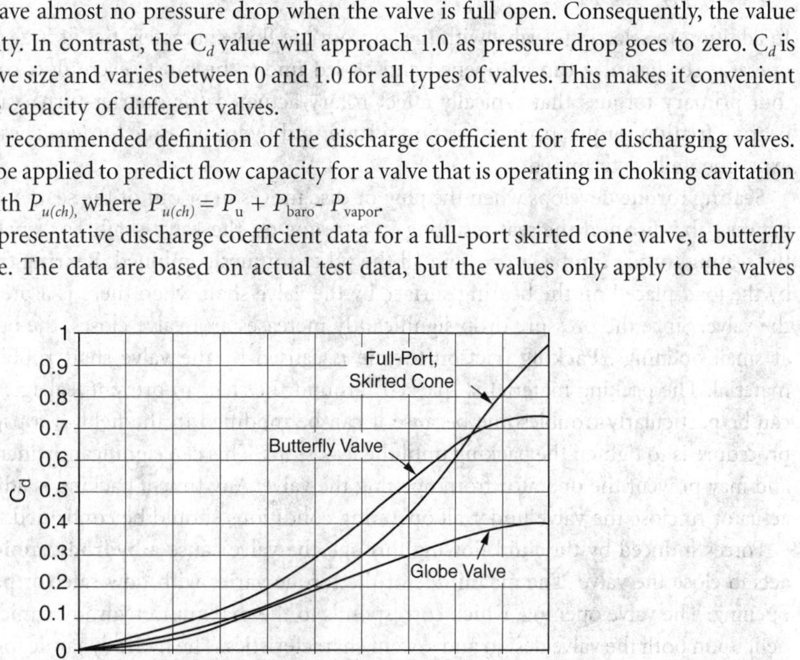

FIGURE 40.9 Discharge coefficient data for various types of control valves (data for demonstration purposes only).

tested and are not representative of all valves of the same type. When selecting a control valve for a particular application, it is necessary to obtain valve-specific performance data from the manufacturer. The data are presented only for illustration purposes.

The data in Figure 40.9 illustrate characteristic differences between valves, however, one must use caution when interpreting relative values of the discharge coefficients. At full-open, the C_d for the cone valve is 2.5 times the C_d for the globe valve. That does not necessarily mean that the cone valve will have significantly greater flow capacity in all cases. The difference in flow capacity of the valves installed in identical systems depends on the system loss (i.e., friction and minor losses). For example, if the pipeline is very long, the flow will typically be controlled almost exclusively by friction, regardless of the type of control valve used. On the other hand, if the system is very short, the control valve will produce most of the energy loss. In low-friction systems, a cone valve will provide significantly more flow than a globe valve at the same valve opening.

Note that the C_d curves for the butterfly and globe valves in Figure 40.9 flatten off near the full-open position. This means that they will not be able to provide good flow control beyond about 90% open. It is not uncommon for the C_d values for some butterfly valves to reach a maximum value at valve openings less than full-open. For the butterfly valve in Figure 40.9, the flow will actually decrease as the valve opens the last 10%. This is not a desirable flow characteristic.

When selecting a control valve, it is necessary to analyze the valve's performance as a part of the piping system and not simply consider its characteristics (Figure 40.9) alone. It is often desirable to have the flow linearly reduced (linear valve) as the control valve is closed. The linearity of a valve, however, depends more on the system losses than it does on the type of valve. A linear valve installed in a short pipe, where friction and minor losses are small, will linearly reduce the flow as the valve closes. When installed in a high friction loss system, the same valve will not reduce the flow during the first 50 to 75% of its stroke. Valve and system interaction will also influence cavitation and transient problems.

Operating Torque

In addition to selecting a valve with the appropriate flow characteristics, it is also necessary to select the operator so the valve can be opened and closed under the most severe flow conditions expected. The four primary torques that typically affect rotary-actuated (or quarter-turn) valves are seating torque, bearing friction torque, packing friction torque, and hydrodynamic torque. These torque valves must be experimentally determined.

Seating torque develops when the plug or disc moves in or out of the seat. A frictional force develops between the disc and the seat resulting in a torque on the valve shaft. For small valves with soft seats, the seating torque can be larger than all the other torques combined. **Bearing friction torque** is caused by the load placed on the bearing surface by the valve shaft when there is a pressure differential across the valve. Since the pressure drop significantly increases as a valve closes, the bearing torque is greatest at small openings. **Packing friction torque** is caused by the valve shaft rubbing against the packing material. The packing material is "packed" around the shaft to prevent leakage. Packing friction torque can be particularly troublesome because it can be modified in the field. If a packing leaks, the normal procedure is to tighten the packing until the leak stops. This can significantly increase the packing torque and may prevent the operator from opening the valve. Any time a packing is adjusted, the ability of the actuator to close the valve under all operating conditions should be confirmed.

Forces induced by the fluid flowing through the valve cause a **hydrodynamic torque**, which usually acts to close the valve. The magnitude of the torque varies with flow rate (or pressure drop) and valve opening. The valve opening, which corresponds to the maximum hydrodynamic torque value, is dependent upon both the valve design and system characteristics. The hydrodynamic forces should be evaluated over the entire valve operating range to identify the maximum operating torque.

Some rotary-actuated valves experience a torque reversal at large valve openings. At a point of torque reversal, the net torque is zero. If a valve operates at opening near a torque reversal, instabilities inherent in the flow will cause the disc to flutter. This can result in fatigue failure of the shaft and loosening or

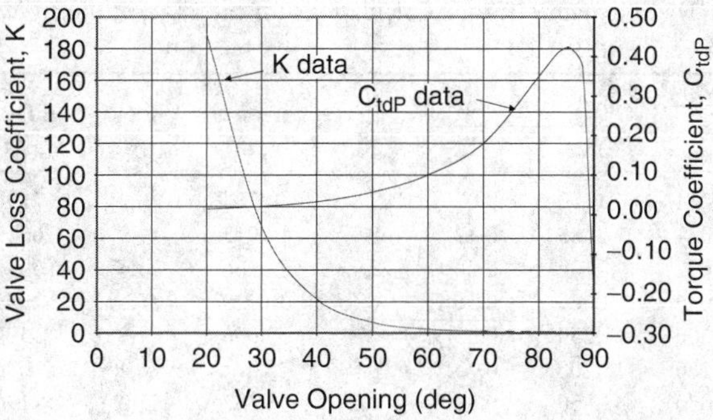

FIGURE 40.10 Sample flow and torque coefficient data for a 24-in. butterfly valve (data for demonstration purposes only).

breaking of connections. Torque reversals should be avoided by limiting the maximum opening of the valve so it will not be exposed to torque reversals.

Another factor influencing the stability and magnitude of the hydrodynamic torque is the presence of a disturbance located just upstream from the valve (i.e., another valve, elbow, pump, etc.). Turbulence and nonuniform flow generated by a disturbance can increase both the average torque and the magnitude of torque fluctuations. This will increase the possibility of fatigue failure of the shaft and connections.

Torque is typically measured by flow testing the valve in a laboratory. A properly designed testing program can identify each component of the torque. Such test data must then be scaled to valves of different sizes. Scaling the seating and packing torques is difficult but the bearing and hydrodynamic torques can be scaled fairly accurately. Scaling the hydrodynamic torque can be done using one of two common torque coefficients, C_{tv} and C_{tdP}, defined as follows:

$$C_{tv} = \frac{T}{\rho d^3 V^2} \tag{40.6}$$

$$C_{tdP} = \frac{T}{d^3 \Delta P} \tag{40.7}$$

where T is the hydrodynamic torque, ρ is the fluid density, d is the valve diameter, V is the fluid velocity, and ΔP is the net pressure differential across the valve. These coefficients are dimensionless and the standard sign convention is that a positive torque acts in the closing direction.

To illustrate how the maximum hydrodynamic torque value is determined and used, sample valve loss (K) and torque (C_{tdP}) coefficient data for a 24-inch butterfly valve are presented in Figure 40.10. For this particular valve, the torque coefficient increases with valve opening up to approximately 86° and a torque reversal occurs at approximately 89.5°. The valve loss coefficient, on the other hand, consistently decreases with increasing valve opening. Per Equation 40.7, the hydrodynamic torque is dependent upon C_{tdP} and ΔP. ΔP, like K, will increase with decreasing valve opening. As a result, the maximum torque will usually not occur at the valve opening with the highest C_{tdP} value. Table 40.1 shows hydrodynamic torque data as a function of valve opening, based on the data presented in Figure 40.10. For this example, maximum torque occurs at 50° open. This analysis should be done for several system operating conditions (if there is more than one) to determine the maximum hydrodynamic torque that the valve will be subjected to. This torque would be added to the estimated torques for seating, packing, and bearing to obtain the required operator torque.

Cavitation

Cavitation is the process of rapid vaporization and condensation of a liquid, brought about by changes in either fluid temperature or pressure. When the process occurs at a constant pressure, due to an increase

TABLE 40.1 Hydrodynamic Torque Calculations for the 24-in. Butterfly Valve Data Presented in Figure 40.10

VO (deg)	C_{tdP}	Valve Coefficients C_v	C_d	K	Flow (gpm)	Velocity (fps)	DP (lb/ft^2)	Torque (ft-lbs)
20	0.014	1250	0.07	189	8784	6.2	7106	784
30	0.021	2083	0.12	68	14,015	9.9	6515	1079
40	0.032	4100	0.21	22	21,966	15.6	5097	1305
50	0.056	6666	0.36	6.6	30,229	21.4	2959	1326
60	0.100	11,250	0.55	2.3	35,056	24.9	1397	1112
70	0.180	18,200	0.73	0.9	37,271	26.4	603	869
80	0.350	32,850	0.89	0.27	38,357	27.2	196	549
86	0.420	44,000	0.93	0.15	38,582	27.4	111	372
90	−0.200	40,000	0.92	0.18	38,522	27.3	133	−214

Note: D (in.) = 24 (pipe and valve diameter), Area (square feet) = 3.14 (pipe and valve cross-sectional area), R = 0.016 (system resistance), Δz (ft) = 120 (upstream reservoir–downstream reservoir).

System Equation: $\Delta z = (R + K_v)\dfrac{Q^2}{2g}$

in temperature, it is referred to as boiling. When the process occurs at a constant temperature, due to a decrease in pressure, it is called cavitation. Cavitation occurs when small air voids, referred to as nuclei, are subjected to vapor pressure and grow rapidly by vaporization and then are subjected to high external pressure causing collapse. All liquids contain nuclei in the form of small free air bubbles and air pockets trapped on small debris particles. The low pressure necessary for the nuclei to grow into vapor cavities occurs in zones of high velocity/low pressure, and especially at the center of eddies generated by turbulence. The pressure at the center of the eddies is lower than the ambient fluid pressure due to their high rotational velocity. When the local pressure reaches vapor pressure, the small nuclei grow rapidly by vaporization of the liquid, causing the eddies to rapidly dissipate. This subjects the vapor cavity to the ambient pressure which is greater than vapor pressure. The cavity implodes and creates extremely high local pressures and micro jets that can damage solid surfaces. For a thorough treatment of cavitation, see Knapp et al., 1970 and Tullis, 1989.

A valve exposed to excessive cavitation can experience accelerated wear; generate excessive noise, vibrations, and head loss; and even lose capacity if subjected to choking cavitation. Figure 40.11 shows

FIGURE 40.11 Cavitation damage in a globe valve (view from the downstream end of the valve).

an example of valve body damage resulting from cavitation. The acceptable cavitation level for a valve in a given system varies with valve type, valve function, details of the piping layout, and duration of operation. There are at least three cavitation limits appropriate for valve operation. These are critical, incipient damage, and choking cavitation. Critical cavitation corresponds to the onset of noise and is appropriate as a design limit for valves that need to operate essentially free of cavitation. Incipient damage corresponds to onset of pitting (material removal). It is an appropriate design limit when significant noise and vibrations can be tolerated but no damage is desired. Choked flow, choking cavitation, or flashing represents the condition where the mean pressure downstream from the valve drops to vapor pressure and the flow is at its maximum for a given upstream pressure. Between incipient damage and choking cavitation, the erosion damage and vibration levels can be severe. Using choking as a design condition is appropriate for valves intended for intermittent use, such as a pressure relief valve, where cavitation damage can be tolerated for short periods of time. It should not be used for valves intended for long-term, damage-free operation.

One unfortunate misunderstanding about valve cavitation arises from the false teaching that cavitation begins when the mean downstream pressure approaches vapor pressure, i.e., when the valve starts to choke. Choking is, in fact, the final stage of cavitation — well beyond the point where damage begins. Promoting this incorrect definition of "onset of cavitation" has resulted in many valves suffering extensive damage.

A thorough cavitation analysis is an important part of valve selection. Detailed procedures for analyzing a valve for cavitation and making adjustments for size and pressure scale effects are available [Tullis, 1989, 1993]. However, the analysis cannot be done without experimental cavitation data. Cavitation data are available for only a few valves due to the limited amount of laboratory cavitation testing performed to date. The following example demonstrates the principles involved in evaluating the cavitation intensity for a valve operating under specific flow and pressure conditions. Consider a 6-in. diameter butterfly valve that is installed in a system and operated under the following flow conditions: Pu = 44 psi, Pd = 20.8 psi, and Q = 1.29 cubic feet per second (cfs). Assume that the barometric pressure is 12.2 psia and the vapor pressure is 1.16 psia. Assume that the 6-in. butterfly valve flow characteristics are described by the butterfly valve data presented in Figure 40.9.

Determining the cavitation intensity of the valve requires comparing the value of the cavitation parameter *Sigma* for the valve at system operating conditions to *Sigma* values for the valve at different cavitation intensity levels (i.e., critical cavitation, incipient damage, and choking cavitation). *Sigma* is defined as:

$$Sigma = \frac{(Pd + Pb - Pv)}{(Pu - Pd)} \tag{40.8}$$

Pu and *Pd* are the pressures upstream and downstream from the valve, *Pb* is the barometric pressure, and *Pv* the absolute vapor pressure.

From Equation 40.1, $C_d = 0.111$, which corresponds to a valve opening of approximately 27%. From Equation 40.8, *Sigma* for the system operating conditions is:

$$Sigma_{sys} = \frac{(20.8 + 12.2 - 1.16)}{(44 - 20.8)} = 1.37$$

Values of *Sigma* corresponding to critical cavitation, incipient damage, and choking cavitation for the same valve opening must be obtained from laboratory data. Assume the following laboratory data apply to the butterfly valve at 27% open:

$$Sigma_{cr} = 1.7 \text{ (critical cavitation)}$$

$$Sigma_{id} = 1.0 \text{ (incipient damage)}$$

$$Sigma_{ch} = 0.4 \text{ (choking cavitation)}$$

Since $Sigma_{sys}$ 1.37 is between $Sigma_{cr}$ 1.7 and $Sigma_{id}$ 1.0, the valve will be operating with cavitation but not severe enough to cause cavitation damage. The specific values of the reference *Sigma* data (critical, incipient damage, and choking) vary with valve opening and should be accounted for in the analysis process.

Several options are available for suppressing cavitation such as selecting the correct valve type and size, locating valves and/or orifices in series, using cavitation-resistant materials, and injecting air. A more thorough discussion on cavitation analysis is available [Tullis, 1989, 1993].

Control Valve Selection

Some systems contain hundreds of valves and it is not reasonable to suggest that all of them receive detailed engineering attention. One of the important components of good valve selection is identifying critical valves, meaning valves that need special analysis. Critical valves include valves that will be subjected to unusual operating conditions such as high pressure drop (potential cavitation problems) and valves required to provide accurate flow control over a wide range of flow rates. Critical valves also include valves that can generate hydraulic transients if closed too quickly. In general, critical valves are those valves whose function is vital to the successful operation of the system or whose failure can lead to serious consequences.

The following hydraulic performance criteria have been suggested for selecting flow control valves [Tullis, 1989]:

1. For systems where conserving head is important, control valves should not produce excessive pressure drop in the full-open position.
2. A control valve should control the flow over at least 50% of its movement. This means that when the valve is closed from 100 to 50% open, the flow should be reduced at least 5 to 10%.
3. The maximum flow and pressure drop should be limited so the operating torque does not exceed the capacity of the operator or valve shaft and connections.
4. The valve should not be subjected to excessive cavitation.
5. Pressure transients should not exceed the safe pressure limits of the system. This requires that the valve be sized to control the flow over most of its stroke and that the closure speed be controlled. It will often require a computer simulation of the system to determine the safe valve closing time. For details on transient analysis, see Wylie and Streeter, 1993; Tullis, 1989; BHRA, 2000; and IAHR, 2000.
6. Some valves should not be operated at small openings due to potential seat damage caused by high velocity jets, cavitation near the seat, and poor flow control.
7. Some valves should not be operated near full-open, where they may have poor flow control and/or experience torque reversals leading to fatigue failures.

40.2 Air Valves

There are three types of automatic air valves: air-vacuum valves, air release valves, and combination valves. The air-vacuum valve has a large orifice to expel and admit large quantities of air at low pressure when filling or draining a pipeline. These valves contain some type of float, which rises and seals the orifice as the valve body fills with water. Once the line is pressurized, the valve cannot reopen to remove air that may subsequently accumulate. If the pressure becomes negative because of a transient or when the pipe is drained, the float drops and admits air into the line.

Air-vacuum valves should be sized based on both air release and negative pressure relief (vacuum) requirements. The larger of the two valve sizes should be used. To size the air-vacuum valve for air release (pipeline filling), the water fill rate and the allowable pressure of the air inside the pipe are needed. When filling a pipe, air released from an air valve can approach sonic velocity when the internal air pressure is approximately 7 psi. Filling rates should be controlled by the water inflow rate, not by restricting the air outflow rate. A typical safe fill rate is about 1 fps in the pipe. With the fill rate and allowable air pressure determined, the size of air-vacuum valve needed for air release can be selected using manufacturers' sizing charts.

The other criterion used to size an air-vacuum valve is pipe collapse. It is necessary to determine whether the pipe can collapse due to rapid draining from a major pipe rupture. If the pipe can collapse, the next step is to calculate the maximum draining rate and estimate the vacuum pressure the pipe can tolerate without collapse. Once those two pieces of information are determined, the air-vacuum valve can be sized using charts provided by the valve suppliers.

Since the draining rate of a ruptured pipe will typically exceed the "safe" filling rate, the size of air-vacuum valve determined for vacuum service will be larger than the valve size based on filling requirements. The larger size should be selected. There is no problem having an automatic air valve oversized for filling, provided that the water inflow rate is properly controlled.

Placement of the air-vacuum valves is critical. An analysis of the system is needed to determine where the valves should be located to protect the pipe. Generally, they are located at high points where the pipe begins its downward slope. When multiple air-vacuum valves are required at various locations along the pipe, each one must be sized independently based on a pipe rupture that will cause maximum flow in that reach of pipe.

In addition to air-vacuum valves, it is necessary to install air release valves. These valves contain a small orifice and are intended to release air that cannot be released by the air-vacuum valves during initial filling. The small orifice is controlled by a plunger attached to a float. As air accumulates in the valve body, the float drops and opens the orifice. As the air is expelled, the float rises due to buoyancy and closes off the orifice. These valves also release small amounts of pressurized air that accumulates during normal operation.

The air trapped during initial filling must be flushed to the air release valves. A flow rate equivalent to 3 to 5 fps is typically required to move pockets of trapped air to the air release valves. While doing this, the pressure in the pipeline should be kept at a minimum by controlling flow and pressure with the inlet and outlet valves.

The combination air valve has two orifices: a large one that functions as an air-vacuum valve and a small one that functions as an air release valve. The combination air valve can simply be two separate valves — one air-vacuum valve and one air release valve — or it can be one valve with both functions integrated into the design. The combination valve in Figure 40.12 is shown in the open position. As the water level in the valve rises, the float will close both the air-vacuum and the air release valves. Positive line pressures will keep the air-vacuum valve closed due to the differential pressure force. As air accumulates in the valve, the float will drop, opening the small air release orifice and expelling the air. As the air is expelled, the float will rise and close the air release orifice. Pipelines almost always require both air-vacuum and air release valves because they serve different functions.

Locating air release and air-vacuum valves depends primarily on the pipe profile. Air valves are generally placed at high points or at regular intervals if there are no significant high points.

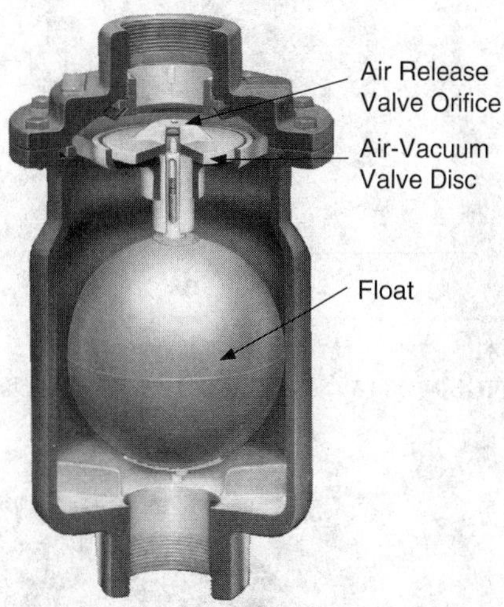

FIGURE 40.12 Combination air release and air-vacuum valve cutaway. Both the air-vacuum and air release are shown in the open position (figure used with permission from CLA-VAL).

The use of large diameter, manually operated air release valves should be avoided. If manual air valves are opened after the pipeline is pressurized, they will release huge quantities of air because the air will exit at sonic velocity. The rapid release of air causes the water to accelerate toward the air valve. When the water reaches the air valve, it will decelerate rapidly, creating a transient pressure rise that could rupture the pipeline. A similar problem occurs if a large pocket of air under high pressure is allowed to accumulate and then move through the pipe and is discharged through a control valve. To avoid both of these problems, the system should be designed to prevent the accumulation of significant amounts of air at high pressure.

40.3 Check Valves

Check Valve Types

The most common type of check valve is the **swing check**, as shown in Figure 40.13. It has a simple design, low-pressure drop at full open, and reliable sealing. It relies on gravity for valve closure, and sealing is accomplished between two flat surfaces (disc and seat). The clearances in the hinge pins and connections provide considerable freedom of movement, allowing the sealing surfaces to self-align. An advantage of a swing check, relative to other types of check valves, is that the disc will continue to seal properly with substantial wear of the hinge pin and/or connections. These valves are also relatively easy to repair, available in a wide range of sizes, and economical. The primary disadvantage of swing check valves is that they are typically the slowest closing of all the check valves due to the large mass of the disc and the long distance of travel from full open to the closed position. They are generally not appropriate for use in systems where rapid flow reversals occur. Excessive wear of the hinge pin and connections can eventually result in a valve failure.

Tilt disc check valves, like the one shown in Figure 40.14, have the hinge pin located in the flow stream, just above the centerline of the disc. They also depend on gravity for closure but close somewhat faster than swing check valves because of the shorter travel distance of the disc. Closer machining tolerances of the sealing surfaces are required, relative to the swing check, because the disc rotates into the seat. Consequently, even limited wear of the hinge pin or bushings can cause sealing problems. Due to the complexity of the sealing arrangement, field repairs to the seats are more difficult.

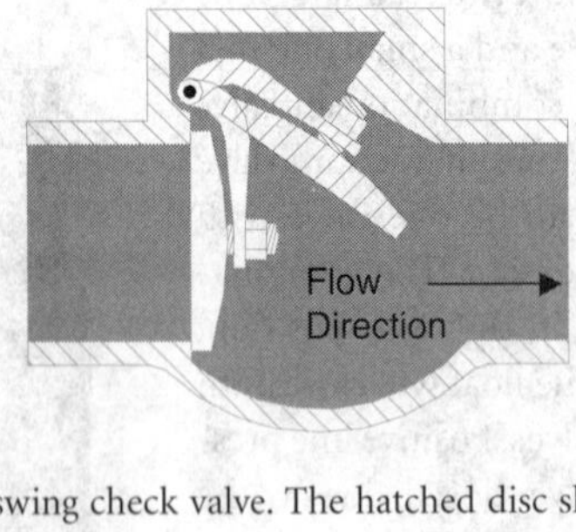

FIGURE 40.13 Schematic of typical swing check valve. The hatched disc shows the disc in the full-open position.

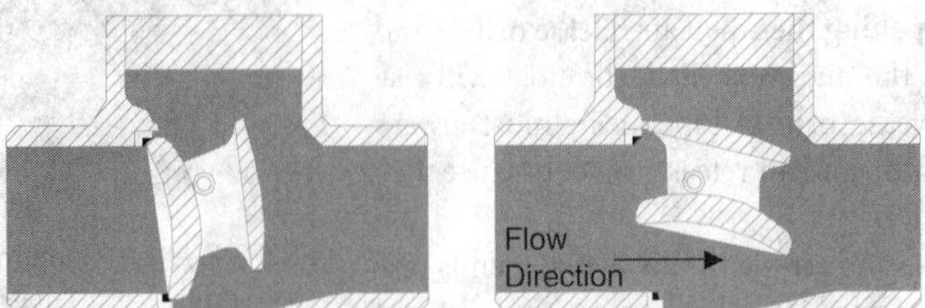

FIGURE 40.14 Schematic of typical tilt check valve. The right- and left-hand figures show the valve in the closed and open position, respectively.

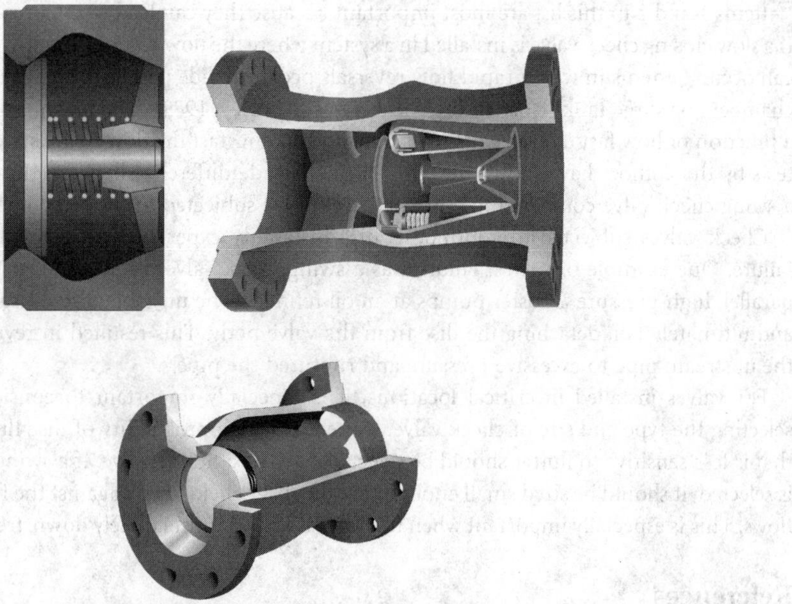

FIGURE 40.15 Three different nozzle check valve styles (figures used with permission from ENERTECH).

The body and closing elements of **lift check valves** come in a variety of configurations. They are also called by a variety of names describing either the shape of the body or disc, such as piston, ball, y-pattern, etc. The body can have the inlet and outlet in-line, or the outlet can elbow down at an angle. The disc (closing element) is generally piston shaped. Lift checks generally have larger pressure drops due to the configuration of the body. Lift checks generally rely on both gravity and spring forces for closure. Speed of closure depends primarily on the restoring force of the spring.

The **double-door check valves** have two half discs that rely on a spring force for closure. They close quickly but can experience flutter problems if the approach flow is disturbed. Restricting the maximum opening of the discs reduces their sensitivity to flutter but increases pressure drop. A broken spring is the primary cause of valve failure for double-door check valves.

Nozzle check valves are streamlined, low-pressure drop, rapidly closing check valves. The seating surface is often an annular ring machined into the valve body. The closing element is a lightweight, spring-loaded circular disc or annular ring that moves parallel to the valve centerline. Three different styles of nozzle check valves are illustrated in Figure 40.15. Because the seat diameter is large and the flow passage streamlined, a short stroke is sufficient to achieve full flow with limited pressure drop. The short stroke, the lightweight disc, and the restoring force of the spring result in a very fast disc closure. Nozzle check valves are fairly insensitive to upstream disturbances.

Check Valve Selection

The basic function of a check valve is to allow forward flow under normal conditions and avoid flow reversal, which can drain the pipe and cause reverse rotation of the pump. The characteristics of check valves, as suggested by Kalsi Engineering and Tullis Engineering Consultants, 1993, that should be considered during valve selection include:

1. Closure speed of check valves relative to the rate of flow reversal of the system
2. The stability of the disc and its sensitivity to upstream disturbances
3. The flow required to fully open and firmly backseat the disc
4. The pressure drop at maximum flow
5. Sealing effectiveness and ease of maintenance

Items 1 and 2 in this list are most important because they can lead to major system and valve failures. If a slow closing check valve is installed in a system where the flow reverses rapidly, high transient pressures can occur. Systems in which rapid flow reversals occur include parallel pumps and systems that have air chambers or surge tanks close to the check valve [Thorley, 1989]. The magnitude of the pressure rise is a function of how fast the valve disc closes relative to how fast the flow in the system reverses. Laboratory tests by the authors have shown an order-of-magnitude difference between the pressure rise caused by a swing check valve compared to a nozzle check valve subjected to the same rate of flow reversal.

Check valves subjected to continuous disc instability experience accelerated wear that can lead to failure. One example of such a failure was a swing check valve installed downstream from one of two parallel, high-pressure, booster pumps at an oil refinery. The unstable disc caused the hinge pin to wear and ultimately fail, detaching the disc from the valve body. This resulted in reverse flow that subjected the upstream pipe to excessive pressure and ruptured the pipe.

For valves installed in critical locations, it is especially important to consider disc stability when selecting the type and size of check valve. If there is a high probability of disc flutter, then a check valve that is less sensitive to flutter should be selected, such as a nozzle valve. If a swing or tilt disc check valve is selected, it should be sized small enough that the disc is held firmly against the full-open stop at normal flows. This is especially important when the valve is located immediately downstream from a disturbance.

References

BHRA. 2000. Proceedings of the 8th International Conferences on Pressure Surges, BHRA Fluid Engineering, The Hague, The Netherlands; (7th) 1996 Harrogate, England; (6th) 1989 Cambridge; (5th) 1986 Hannover, Germany; (4th) 1983 Bath, England; (3rd); 1980 Canterbury, England; (2nd) 1976 London; (1st) 1972 Canterbury, England.

Crane Co. 1979. Flow of Fluids Through Valves, Fittings, and Pipe. Technical Paper No. 410. Crane, New York.

Driskell, L. 1983. *Control-Valve Selection and Sizing,* Instrument Society of America, Pittsburgh, PA.

Fisher Controls International. 2001. *Control Valve Handbook,* 3rd ed. Fisher Controls, Marshalltown, IA.

IAHR. 2000. *Hydraulic Transients with Water Column Separation,* International Association of Hydraulic Engineering and Research. Madrid.

ISA 1976. *Handbook of Control Valves,* 2nd ed. Instrument Society of America, Pittsburgh, PA.

Johnson, M. C., Dham, R., Sager, B. T. A., and Bergquist, J. 2001. *Valves to Get Out of a Fix. International Water Power and Dam Construction*, July 2001

Kalsi Engineering and Tullis Engineering Consultants. 1993. *Application Guide for Check Valves in Nuclear Power Plants,* NP-5479, Rev. 1. Charlotte, NC. Nuclear Maintenance Applications Center (reference available only to EPRI member utilities).

Knapp, R. T., Daily, J.W., and Hammitt, F. G. 1970. *Cavitation,* McGraw-Hill, New York.

Thorley, A. R. D. 1989. *Check Valve Behavior under Transient Flow Conditions: A State-of-the-Art Review.* Vol. III. ASME.

Tullis, J. P. 1989. *Hydraulics of Pipelines — Pumps, Valves, Cavitation, Transients,* John Wiley & Sons, New York.

Tullis, J. P. 1993. *Cavitation Guide for Control Valves,* NUREG/CR-6031. U.S. Nuclear Regulatory Commission. Washington, D.C.

Wylie, E. B. and Streeter, V. L. 1993. *Fluid Transients in Systems,* Prentice Hall, Englewood Cliffs, NJ.

41

Pumps and Fans

Robert F. Boehm
University of Nevada, Las Vegas

41.1 Introduction

Pumps are devices that impart a pressure increase to a liquid. Fans are used to increase the velocity of a gas, but this is also accomplished through an increase in pressure. The pressure rise found in pumps can vary tremendously, and this is a very important design parameter along with the liquid flow rate. This pressure rise can range from simply increasing the elevation of the liquid to increasing the pressure by hundreds of atmospheres. Fan applications, on the other hand, generally deal with small pressure increases. In spite of this seemingly significant distinction between pumps and fans, there are many similarities in the fundamentals of certain types of these machines as well as with their application and theory of operation.

Pumps or fans can be used as a means of forcing flows into a region of interest or, alternatively, exhausting flows from a region of interest. For example, the inlet to an induced draft fan is attached to the region of interest. Forced draft fans are hooked with their exhaust side to the region of interest. Pumps are usually not denoted with this distinction except for vacuum pumps, which are always used to exhaust a fluid from a volume of interest.

The appropriate use of pumps and fans depends upon the satisfactory choice of device and the proper design and installation for the application. A check of sources of commercial equipment shows that many varieties of pumps and fans are available. Each of these has special characteristics that must be appreciated for achieving proper function. Preliminary design criteria for choosing among different types are given by Boehm [1987].

As is to be expected, wise applications of pumps and fans require knowledge of fluid flow fundamentals. Unless the fluid mechanics of a particular application is understood, the design could be less than desirable.

In this chapter, pump and fan types are briefly defined. In addition, typical application information is given. Also, some ideas from fluid mechanics that are especially relevant to pump and fan operation are reviewed. For more details on this latter topic, see the chapter of this book that discusses fluid mechanics fundamentals.

0-8493-1586-7/05/$0.00+$1.50

41.2 Pumps

Raising of water from wells and cisterns is the earliest form of pumping (a very detailed history of early applications is given by Ewbank [1842]). Modern applications are much broader, and these find a wide variety of machines in use. Modern pumps function on one of two principles. By far the majority of pump installations are of the *velocity head* type. In these devices, the pressure rise is achieved by giving the fluid movement. At the exit of the machine, this movement is translated into a pressure increase by slowing down the fluid. The other major type of pump is called *positive displacement.* These devices are designed to increase the pressure on the liquid while essentially trying to compress the volume. A categorization of pump types (with the exception of vacuum pumps) has been given by Krutzsch [1986], and an adaptation of this is shown below:

1. Velocity head
   - A. Centrifugal
     - i. Axial flow (single or multistage)
     - ii. Radial flow (single or double suction)
     - iii. Mixed flow (single or double suction)
     - iv. Peripheral (single or multistage)
   - B. Special effect
     - i. Gas lift
     - ii. Jet
     - iii. Hydraulic ram
     - iv. Electromagnetic
2. Positive displacement
   - A. Reciprocating
     - i. Piston, plunger
       - a. Direct acting (simplex or duplex)
       - b. Power (single or double acting, simplex, duplex, triplex, multiplex)
     - ii. Diaphragm (mechanically or fluid driven, simplex or multiplex)
   - B. Rotary
     - i. Single rotor (vane, piston, screw, flexible member, peristaltic)
     - ii. Multiple rotor (gear, lobe, screw, circumferential piston)

In the next sections, some of the more common pumps are described.

Centrifugal and Other Velocity-Head Pumps

Centrifugal pumps are used in more industrial applications than any other kind of pump. This is primarily because these pumps offer low initial and upkeep costs. Traditionally, pumps of this type have been limited to low-pressure-head applications, but modern pump designs have overcome this problem unless very high pressures are required. Some of the other good characteristics of these types of devices include smooth (nonpulsating) flow and the ability to tolerate nonflow conditions.

The most important parts of the centrifugal pump are the *impeller* and *volute.* An impeller can take on many forms, ranging from essentially a spinning disc to designs with elaborate vanes. The latter is usual. Impeller design tends to be unique to each manufacturer; a variety of designs are used for a range of applications. An example of an impeller is shown in Figure 41.1. This device imparts a radial velocity to the fluid that has entered the pump perpendicular to the impeller. The volute (there may be one or more) performs the function of slowing the fluid and increasing the pressure. A good discussion of centrifugal pumps is given by Lobanoff and Ross [1992].

The design of a pump and its performance are dependent upon a number of applicable variables. While these will be noted in more detail below, they include the pump operating speed, the flow rate,

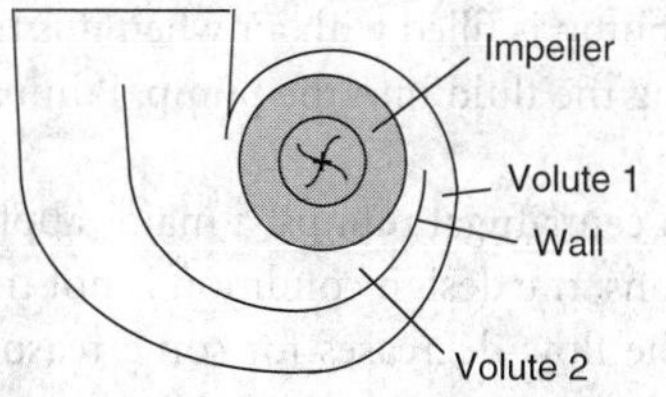

FIGURE 41.1 A schematic of a centrifugal pump is shown. The liquid enters perpendicular to the figure, and a radial velocity is imparted by clockwise spin of the impeller.

and the developed head. Pump engineers generally combine these variables together into a parameter called the *specific speed*, N_s. This is given as

$$N_s = Q^{0.5}N/H^{0.75}$$

where Q = volume rate of flow, gpm; N = operating speed of pump, rpm; and H = pump head, ft.

Historically, the efficiency is related to this parameter in a general way, showing a maximum efficiency of over 90% at a specific speed of about 2500 and very high flow rates [Yedidiah, 1996]. For lower flows, the maximum efficiency is decreased and occurs at smaller values of N_s. Rishel [2000] reported on the results of a study of actual installations of "wire-to-water" efficiencies for pumps. These ranged from 20% for small systems (20 gpm) to 84% for systems at 2000 gpm. The head was varied in the study. A plot of the data he cited is given as Figure 41.2.

An important factor in the specification of a centrifugal pump is the casing orientation and type. For example, the pump can be oriented vertically or horizontally. Horizontal mounting is most common. Vertical pumps usually offer benefits related to ease of priming and reduction in required NPSH (see discussion below). This type also requires less floor space. Submersible and immersible pumps are always of the vertical type. Another design factor is the way the casing is split, and this has implications for ease of manufacture and repair. Casings that are split perpendicular to the shaft are called radially split, while those split parallel to the shaft axis are denoted as axially split. The latter can be horizontally split or vertically split. The number of stages in the pump greatly affects the pump-output characteristics. Several stages can be incorporated into the same casing, with an associated increase in pump output. Multistage pumps are often used for applications with total developed head over (about) 50 atm.

Whether or not a pump is self-priming can be important. By this it is meant that the impeller of the pump must be immersed in the fluid to be pumped. In general, a centrifugal pump cannot begin pumping

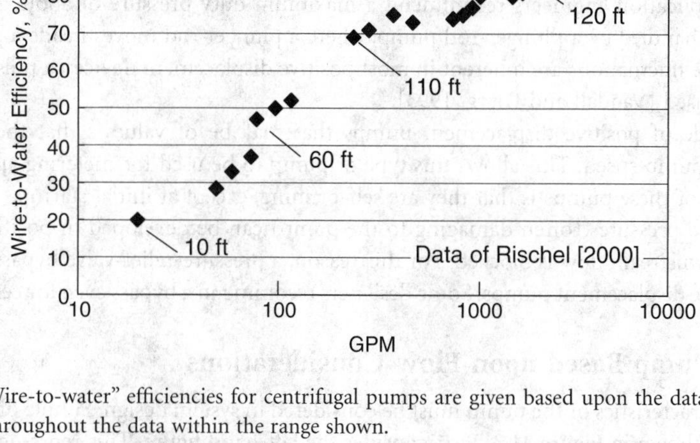

FIGURE 41.2 "Wire-to-water" efficiencies for centrifugal pumps are given based upon the data of Rishel [2000]. The head varies throughout the data within the range shown.

unless it is primed. If a centrifugal pump is filled with air when it is turned on, the initiation of pumping action may not be sufficient to bring the fluid into the pump. Pumps can be specified with features that minimize priming problems.

Although not only applicable to centrifugal pumps, a major application here is the use of a *variable speed drive* (*VSD*). In many situations, the design condition is not the only one at which the pump may have to operate. For example, if the flow decreases for some reason, the head would go up, and vice versa. This may cause problems with the particular pump installation. A VSD will allow these situations, as well as part load operation, to be accommodated. Without VSD, this would have been accomplished by mechanical devices to throttle the flow, for example. A VSD can greatly enhance the system operational efficiency.

There are other types of velocity-head pumps. *Gas lifts* accomplish a pumping action by a drag on gas bubbles that rise through a liquid.

Jet pumps (eductors) increase pressure by imparting momentum from a high-velocity liquid stream to a low-velocity or stagnant body of liquid. The resulting flow then goes through a diffuser to achieve an overall pressure increase. Related designs that use a gas or a vapor to impart the momentum are called *jet ejectors*. An excellent summary of these types of devices has been given by Power [1994]. In all cases where eductors or ejectors are used, a separate high-pressure fluid stream needs to be available to power the device.

Positive-Displacement Pumps

Positive-displacement pumps demonstrate high discharge pressures and low flow rates. Usually this is accomplished by some type of pulsating action. A piston pump is a classical example of a positive-displacement machine. Rotary pumps are one type of positive displacement device that does not impart pulsations to the exiting flow (a full description of these types of pumps is given by Turton [1994]). Several techniques are available for dealing with pulsating flows, including use of double-acting pumps (usually of the reciprocating type) and installation of pulsation dampeners.

Positive displacement pumps usually require special seals to contain the fluid. Costs are higher both initially and for maintenance compared to most pumps that operate on the velocity-head basis. Positive-displacement pumps demonstrate an efficiency that is nearly independent of flow rate, in contrast to the velocity-head type.

Reciprocating pumps offer very high efficiencies, reaching 90% in larger sizes. These types of pumps are more appropriate for pumping abrasive liquids (e.g., slurries) than are centrifugal pumps.

Piston pumps are commonly applied positive-displacement devices. The piston-cylinder arrangement is one that is familiar to most people who have ever used pumps. These pumps can be *single acting* or *double acting*. In the latter classification, the design is such that pumping takes place on both sides of the piston. Many application engineers recommend a maximum duty pressure of about 2000 psi. Higher pressures can be handled by a plunger-rod pump, where a plunger rod moves inside a pipe of the same diameter. Because fluctuations are inherent in most positive displacement devices, a pulsation dampener may have to be used [Vandall and Foerg, 1993].

A characteristic of positive displacement pumps that may be of value is that the output flow is proportional to pump speed. This allows this type of pump to be used for metering applications. Also, a positive aspect of these pumps is that they are self-priming, except at initial startup.

Very high head pressures (often damaging to the pump) can be developed in positive displacement pumps if the downstream flow is blocked. For this reason, a pressure-relief-valve bypass must always be used with positive displacement pumps. Some designers recommend a bypass even for centrifugal pumps.

Selecting a Pump Based upon Flow Considerations

Performance characteristics of the pump must be considered in system design. Simple diagrams of pump applications are shown in Figure 41.3. First consider the left-hand figure. This represents a flow circuit,

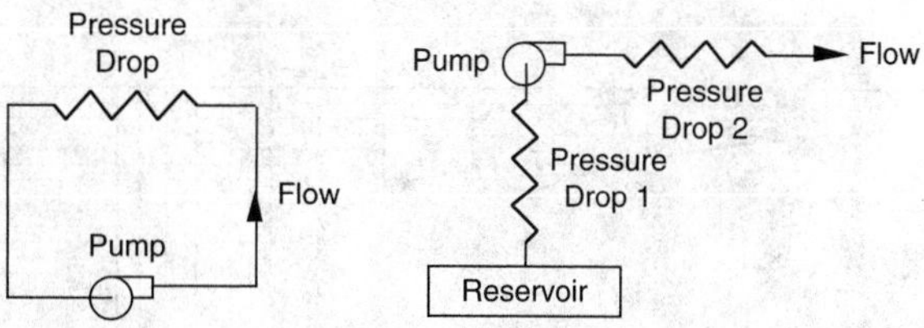

FIGURE 41.3 Typical pump applications, either in circuits or once-through arrangements, can be represented as combined fluid resistances as shown. The resistances are determined from fluid mechanics analyses.

and the pressure drops related to the piping, fittings, valves, and any other flow devices found in the circuit must be estimated using laws of fluid mechanics. Usually these resistances (pressure drops) are found to vary approximately with the square of the liquid flow rate.

Most pumps demonstrate a flow vs. pressure rise variation that is a positive value at zero flow and decreases to zero head at some larger flow. A variation typical of centrifugal pumps is shown on the left-hand side of Figure 41.4. One exception to this, and it is an important one related to these types of pumps, is the so-called "drooping head" behavior. In this situation, the head at zero flow is less than the heads that are achieved at small positive flows. This phenomenon is a result of high-efficiency design, and it is described in some detail by Paugh [1994].

Positive displacement pumps, as shown on the right-hand side of Figure 41.4, are an exception to this characteristic behavior in that these devices usually cannot tolerate a zero flow. An important aspect to note is that a closed system can presumably be pressurized by the pump to the detriment of the system and the pump.

The piping diagram shown on the right-hand side of Figure 41.3 is a once-through system, another frequently encountered installation. However, the leg of piping through "pressure drop 1" shown there can have some very important implications related to *net positive suction head*, often denoted as *NPSH*. In simple terms, NPSH indicates the difference between the local pressure and the thermodynamic saturation pressure at the fluid temperature. If $NPSH = 0$, the liquid can vaporize, and this can result in a variety of outcomes from noisy pump operation to outright failure of components. This condition is called *cavitation*, and it must be eliminated by proper design.

Cavitation, if it occurs, will first take place at the lowest pressure point within the piping arrangement. Often this point is located at, or inside, the inlet to the pump. Most manufacturers specify how much NPSH is required for satisfactory operation of their pumps. Hence, the *actual NPSH* (often times denoted as *NPSHA*) experienced by the pump must be larger than the manufacturer's required NPSH (which may be called *NPSHR*).

If a design indicates insufficient NPSH, changes should be made in the system, possibly including alternative piping layout. This might include lowering the pump location relative to the feed, changing sizes of the inlet piping and fittings, or slowing of the pump speed. Some pumps have smaller NPSH

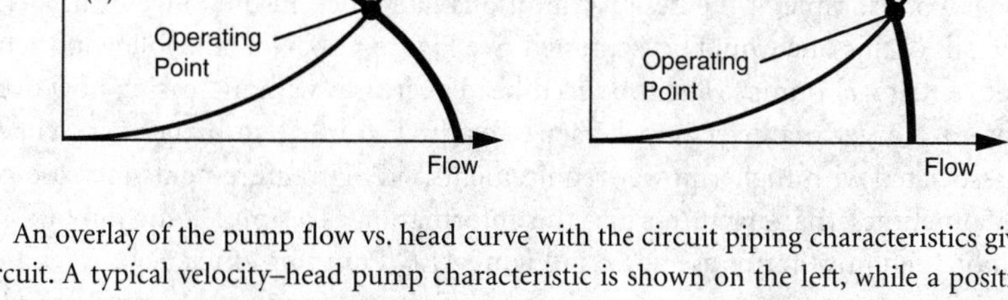

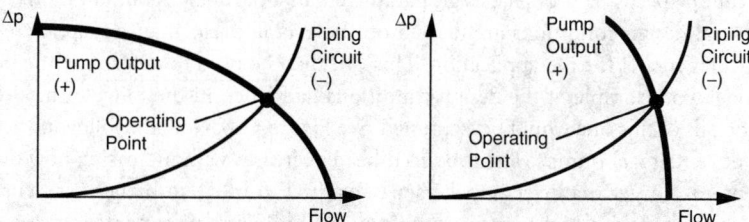

FIGURE 41.4 An overlay of the pump flow vs. head curve with the circuit piping characteristics gives the operating state of the circuit. A typical velocity–head pump characteristic is shown on the left, while a positive-displacement pump curve is shown on the right.

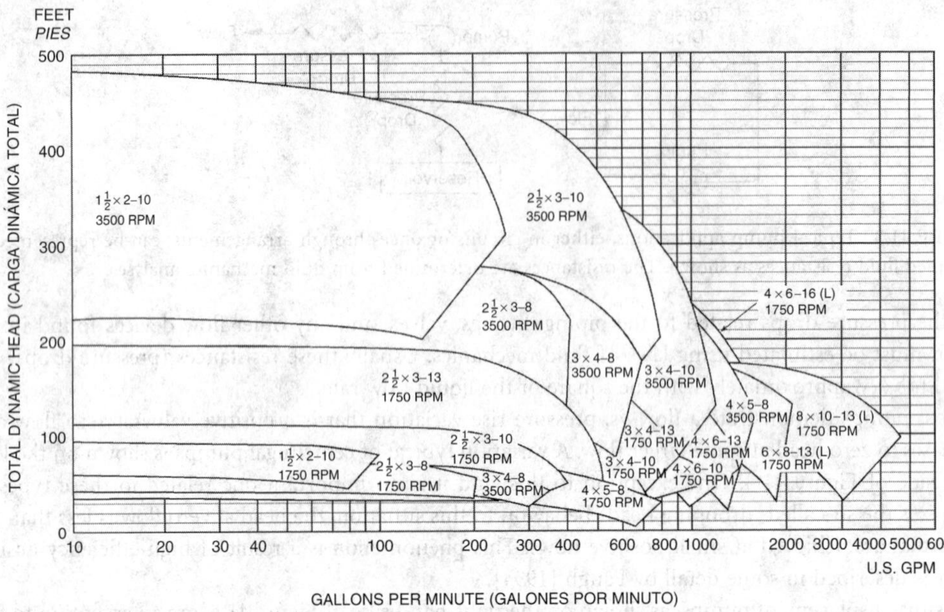

FIGURE 41.5 When selecting a pump for a given application, the starting point is a performance coverage map. From this, the model required to achieve a given performance can be selected. (Compliments of Goulds Pumps.)

requirements than others, and this might be a result of adding inducers to the inlet. Sometimes it is possible to cool the pump inlet fluid and this will increase the NPSHA. More extreme solutions could find a smaller pump installed prior to the inlet of the main pump whose purpose is to increase the NPSHA of the latter. Of course, if a pump is fed from a tank and vortices form there at typical flow conditions, this could greatly decrease the NPSHA beyond what would appear to be present.

Fluid properties can have a major impact on the choice of pump. One variable of concern is viscosity. This property can have high or extremely variable values in a given circuit. At the minimum, increases in viscosity will require more pumping power. But the value of the viscosity may influence the type of pump used. For viscosities up to slightly over 3000 cP, centrifugal pumps will work quite well. Rotary pumps can handle the less frequently encountered situations with considerably higher viscosities.

The manufacturer should be consulted for a map of operational information for a given pump. A typical set of these is shown in Figure 41.5 through Figure 41.7. This information will allow the designer to select a pump that satisfies the circuit operational requirements while meeting the necessary NPSH and most efficient operation criteria.

First consider Figure 41.5. This gives a map of overall performance for a given family of pumps. The specific operational ranges for each model with the family are shown. This range is due to use of various impeller sizes, pump speeds, and other design parameters in a particular configuration.

Once the general design conditions are located on the overall performance map, these will generally indicate a particular model for the application. This can then be used to determine the performance of the specific model. For example, if the design conditions had fallen in the range of model 1 1/2–2 – 10 3500 rpm, the details of this one would be examined. See Figure 41.6. Note the following generally typical characteristics of centrifugal pumps. The produced head decreases with increasing flow over the specific range for this pump. Larger diameter impellers are required to move to higher-level curves, and these conditions are associated with higher power requirements. NPSH requirements increase generally with increasing flow. Sometimes, manufacturers give this information as a simple function of flow, or this can be specified in more definitive terms as shown in Figure 41.6. Pumping efficiency curves are also shown. It is desirable to operate near the maximum values of efficiency if that can be accommodated between the design requirements and the characteristics of the specific pumps. This is a lower-flow pump, and,

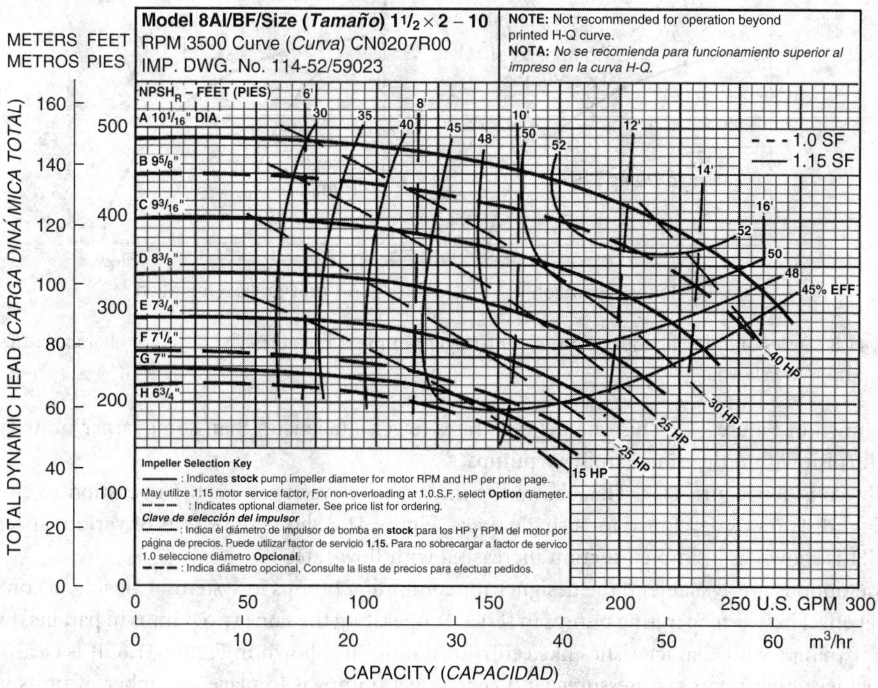

FIGURE 41.6 Once the particular model of pump is determined (see Figure 41.5), the actual performance of that model can be examined from a plot of this type. This is given here for a particular speed. (Compliments of Goulds Pumps.)

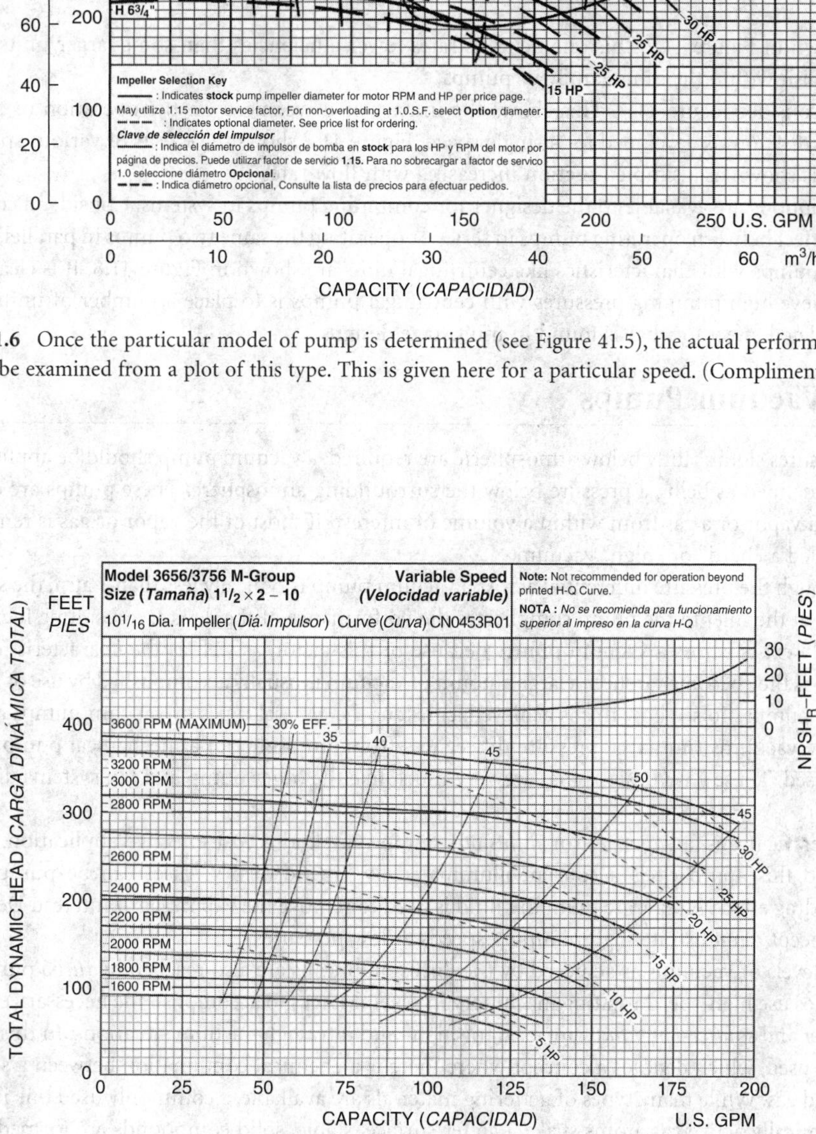

FIGURE 41.7 Impeller speed for a given model will have an impact on the pump performance (compare to Figure 41.6). (Compliments of Goulds Pumps.)

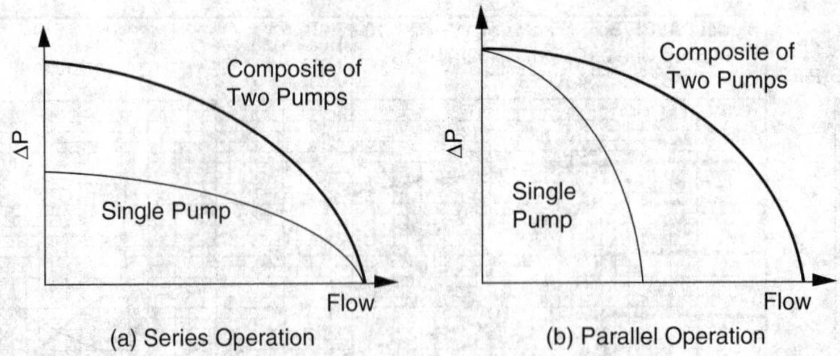

FIGURE 41.8 Series and parallel operation of centrifugal pumps is possible. The resultant characteristics for two identical pumps are shown.

as was shown in Figure 41.2, this one demonstrates lower efficiencies than some larger units would be able to furnish within the same family of pumps.

Finally, consider Figure 41.7. This shows the impact of drive speed on the operation of the pump. Where Figure 41.6 was developed for a single speed, Figure 41.7 shows the effects of various speeds. Here the NPSH is shown as a simple function increasing with flow rate.

Several options are available to the designer for combining pumps in systems. Consider a comparison of the net effect between operating pumps in series or operating the same two pumps in parallel. Examples of this for pumps with characteristics like centrifugal units are shown in Figure 41.8. It is clear that one way to achieve high pumping pressures with centrifugal pumps is to place a number of units in series. This is a related effect to what is found in multistage designs.

41.3 Vacuum Pumps

When pressures significantly below atmospheric are required, a vacuum pump should be applied. In fact, vacuum is defined as being a pressure below the surrounding atmosphere. These pumps are often used to remove a vapor or a gas from within a volume of interest. If most of the vapor or gas is removed, this is often called a “hard” or “high” vacuum.

Even though the pressure differential across a vacuum pump is typically less than 1 atm, the small value of pressure at the inlet to the pump complicates the performance of the device. A resulting high-pressure ratio usually exists across a vacuum pump. Because of this factor, as well as the characteristics of gases and vapors as they become rarefied, a large amount of vacuum is usually accomplished by use of a sequence of vacuum pumps. To see how this is achieved, consider the general types of vacuum pumps available.

At coarse vacuums (not large pressure differences from atmospheric) a mechanical pump or blower might be used. These have designs that are very much like the other pump and fan systems discussed in this chapter.

At higher vacuums, a *vapor jet* (or *diffusion*) pump may find a cost-effective application. The basic idea behind this approach is a concept illuminated by kinetic theory: gases can be pumped by the molecular drag effect. A separate pumping fluid is used to remove the vapor/gas of interest. Devices built on this concept demonstrate high pumping speeds at low pressures.

Similar levels of vacuums as achieved by the diffusion pumps are also reached by *turbo* pumps. These latter devices use high-speed rotating machinery in several stages to accomplish the necessary evacuation.

Ion-getter and *sputter-ion pumps* are used for high- and ultrahigh-vacuum situations. In this approach, *gettering* is used, which denotes a concept where there is a chemical combination between a surface and the pumped gas. While many types of gettering materials are available, a commonly used one is titanium. When chemically active gas atoms strike a getter surface, stable, solid compounds are formed there.

To achieve high- or ultrahigh-vacuums, a series of these pumps might be used. At moderate vacuums, a *roughing* pump is a good choice. As the absolute pressure decreases, then a switch is made to another type of pump, say a cryopump or a diffusion pump. Finally, the very small pressures can then be reached

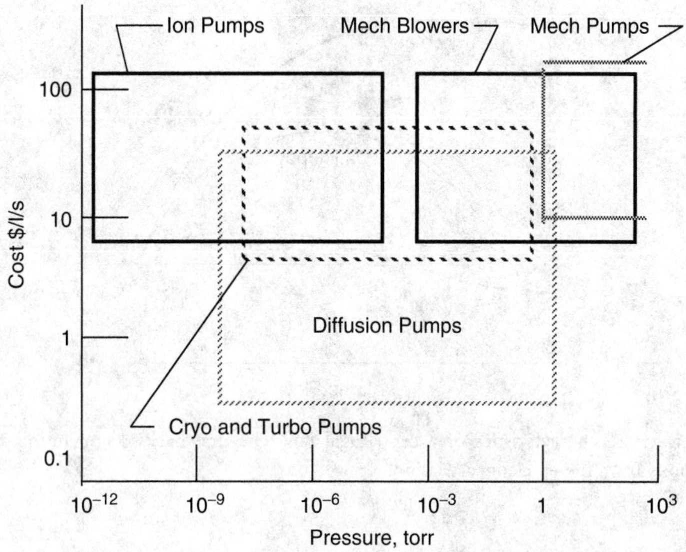

FIGURE 41.9 Approximate operational ranges where various types of vacuum pumps are most cost effective (in 1985 prices). Adapted from Hablanian, M., 1997. *High-Vacuum Technology — A Practical Guide*, 2nd ed., Marcel Dekker, New York.

using a getter or ion pump. Approximate ranges for each of these pumps are shown in Figure 41.9. Figure 41.10 shows the performance range of various types of coarse vacuum pumps. These figures were developed from information given by Hablanian [1997]. This text is an excellent source of information on all of these types of devices.

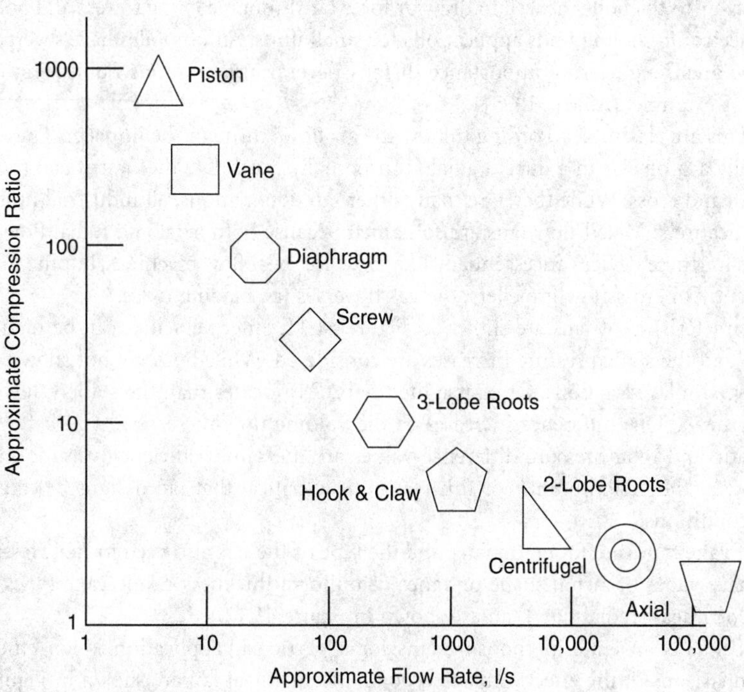

FIGURE 41.10 The operational ranges of coarse vacuum pumps are shown. Adapted from Hablanian, M., 1997. *High-Vacuum Technology — A Practical Guide*, 2nd ed., Marcel Dekker, New York.

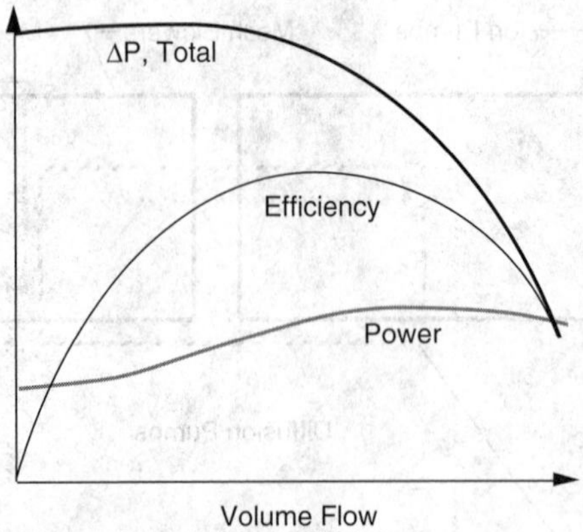

FIGURE 41.11 Shown are characteristics of a centrifugal fan. The drawbacks to operating away from optimal conditions are obvious from the efficiency variation.

41.4 Fans

As noted earlier, fans are devices that cause air to move. This definition is broad and can include a flapping palm branch, but the discussion here deals only with devices that impart air movement due to rotation of an impeller inside a fixed casing. In spite of this limiting definition, it includes a large variety of commercial designs.

Fans find application in many engineering systems. Along with chillers and boilers, they are the heart of heating, ventilating, and air conditioning (HVAC) systems. When large physical dimensions of a unit are not a design concern (usually the case), centrifugal fans are favored over axial flow units for HVAC applications. Many types of fans are found in power plants [Stultz and Kitto, 1992]. Very large fans are used to furnish air to the boiler as well to draw or force air through cooling towers and pollution control equipment. Electronic cooling finds applications for small units. Automobiles have several fans in them. Because of the great engineering importance of fans, several organizations publish rating and testing criteria (see, for example, [ASME, 1995]).

Generally, fans are classified according to how the air flows through the impeller. These flows may be axial (essentially a propeller in a duct), radial (conceptually much like the centrifugal pumps discussed earlier), mixed, and cross. While there are many other fan designations, all industrial units fall into one of these classifications. Mixed-flow fans are so named because both axial and radial flow occurs on the vanes. Casings for these devices are essentially like those for axial flow machines, but the inlet has a radial flow component. On cross-flow impellers, the gas traverses the blading twice.

Generic characteristics of fans are shown in Figure 41.11. Since velocities can be high in fans, often both the total and the static pressure increases are considered. While both are not shown on this figure, the curves have similar variations. Of course the total ΔP is greater than the static value; the difference is the velocity head. This difference increases as the volume flow increases. At zero flow (the *shut-off point*), the static and total pressure difference values are the same. Efficiency variation shows a sharp optimum value at the design point. For this reason, it is critical that fan designs be carefully tuned to the required conditions.

A variety of vane types are found on fans, and the type of these is also used for fan classification. Axial fans usually have vanes of airfoil shape or vanes of uniform thickness. Some vane types that might be found on a centrifugal (radial flow) fan are shown in Figure 41.12.

One aspect that is an issue in choosing fans for a particular application is fan efficiency. Typical efficiency comparisons of the effect of blade type on a centrifugal fan are shown in Figure 41.13. Since velocities can be high, the value of aerodynamic design is clear. Weighing against this is increased cost.

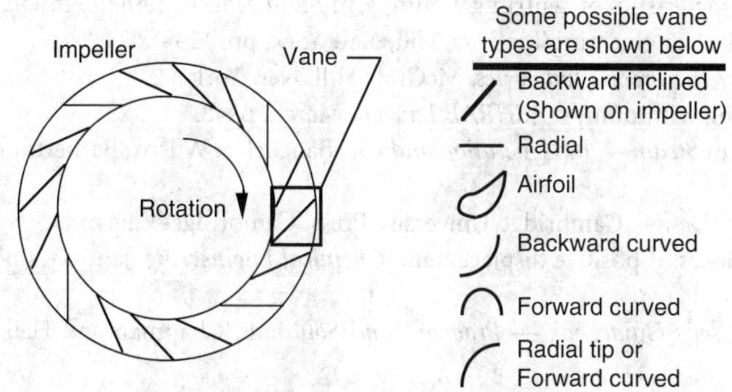

FIGURE 41.12 A variety of vane types that might be used on a centrifugal fan are shown.

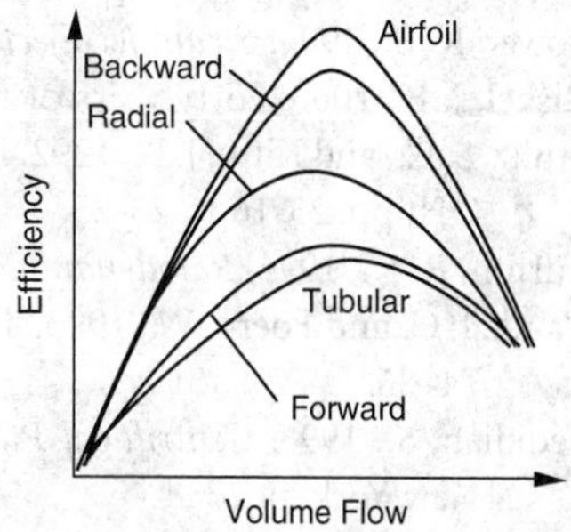

FIGURE 41.13 Efficiency variations with volume flow of centrifugal fans for a variety of vane types are shown.

An additional aspect that may be important in the choice of fans is noise generation. This may be most critical in HVAC applications. It is difficult to describe noise characteristics in brief terms because of the frequency-dependent nature of these phenomena. However, a comparison of specific sound power level (usually denoted by K_w) shows that backward-curved centrifugal fans with aerodynamic blades perform among the best designs. Details of noise characteristics are given elsewhere [ASHRAE, 1999].

While each type of fan has some specific qualities for certain applications, most installations use centrifugal (radial flow) fans. A primary exception is for very-high-flow, low-pressure-rise situations where axial (propeller) fans are used.

Similarities exist between fans and pumps because the fluid density essentially does not vary through either type of machine. Of course, in pumps this is because a liquid can be assumed to be incompressible. In fans, a gas (typically air) is moved with little pressure change. As a result, the gas density can be taken to be constant. Since most fans operate near atmospheric pressure, the ideal gas assumptions can be used in determining gas properties.

Flow control in fan applications, where needed, is a very important design concern. Methods for accomplishing this involve use of dampers (either on the inlet or on the outlet of the fan), variable pitch vanes, or variable speed control. Dampers are the least expensive to install but also the most inefficient in terms of energy use. Modern solid-state control for providing a variable frequency power to the drive motor is becoming the preferred control method, when a combination of initial and operating costs are considered.

References

ASHRAE, 1999. *ASHRAE Handbook 1999, HVAC Applications,* American Society of Heating, Refrigerating, and Air Conditioning Engineers, Atlanta, Chapter 42.

ASME, 1995. ASME Performance Test Codes, Code on Fans, ASME PTC 11-1984 (reaffirmed 1995), American Society of Mechanical Engineers, New York.

Boehm, R. F., 1987. *Design Analysis of Thermal Systems,* John Wiley & Sons, New York, pp. 17–26.

Ewbank, T., 1842. *A Description and Historical Account of Hydraulic and Other Machines for Raising Water,* 2nd ed., Greeley and McElrath, New York.

Hablanian, M., 1997. *High-Vacuum Technology — A Practical Guide*, 2nd ed., Marcel Dekker, New York.

Krutzsch, W. C., 1986, Introduction: classification and selection of pumps, Chapter 1 in *Pump Handbook*, 2nd ed. (ed. I. Karassik et al.), McGraw-Hill, New York.

Lobanoff, V. and Ross, R., 1992. *Centrifugal Pumps: Design and Application*, 2nd ed., Gulf Publishing Company, Houston.

Paugh, J. J., 1994. Head vs. capacity characteristics of centrifugal pumps, in *Fluid Movers, Second Edition,* (edited by N. P. Chopey), Chemical Engineering/McGraw Hill, New York, pp. 123-125.

Power, R. B., 1994. *Steam Jet Ejectors for the Process Industries,* McGraw Hill, New York.

Rishel, J. B., 2000. Forty years of fiddling with pumps, *ASHRAE Journal,* March, p. 48.

Stultz, S. C., and Kitto, J. B., 1992. Fans, in *Steam — Its Generation and Use,* Babcock & Wilcox, Barberton, OH, pp. 23-16 to 23-25.

Turton, R. K., 1994. *Rotodynamic Pump Design,* Cambridge University Press, Cambridge, England.

Vandell, C. and Foerg, W., 1993. The pluses of positive displacement, *Chemical Engineering,* January, pp. 74–86.

Yedidiah, S., 1996. *Centrifugal Pump User's Guidebook — Problems and Solutions,* Chapman and Hall, New York, p. 27.

42

Two-Phase Flow

Richard T. Lahey, Jr.
Rensselaer Polytechnic Institute

Multiphase flows occur in many cases of practical concern. In particular, important vapor/liquid and solid/fluid two-phase flows may occur in thermal energy production and utilization, chemical and food processing, environmental engineering, pharmaceutical manufacturing, petroleum production, and waste incineration technologies.

This chapter summarizes the essential features of two-phase flow and focuses on the engineering analysis of vapor/liquid systems. Readers interested in more in-depth treatment are referred to the work of Lahey [1992] and Roco [1993].

Two-phase flows are inherently more complicated than single-phase flows because the phases may separate and arrange themselves into distinct flow regimes. Moreover, the two phases normally do not travel at the same velocity, nor, in some situations, even in the same direction. As a consequence, important phenomena occur in two-phase flows that do not occur in single-phase flows.

The notation, conservation equations, and their associated closure relations will first be discussed. Next, flooding will be considered, and then the conservation equations will be used to analyze some situations of interest in two-phase flows.

42.1 Notation

The most important parameter that characterizes a two-phase flow is the so-called *local volume fraction* of phase k, α_k (x, t). This parameter is the time fraction that a probe at location x and time t will sense phase k during a measurement time T. That is,

$$\alpha_k = \sum_{i=1}^{N} \Delta t_i / T \tag{42.1}$$

The global volume fraction of phase k is the integral of the local volume fraction over the cross-sectional area (A_{xs}) of the conduit:

0-8493-1586-7/05/$0.00+$1.50

$$\langle \alpha_k \rangle = \iint_{A_{xs}} \alpha_k da / A_{xs} \tag{42.2}$$

By convention, in vapor/liquid two-phase flows, the volume fraction of the vapor phase, α_v, is called the *void fraction*, α.

Unlike single-phase flows, there are multiple velocities of interest in two-phase flows — in particular, the superficial velocity,

$$\langle j_k \rangle = Q_k / A_{xs} \tag{42.3}$$

and the phasic velocity,

$$\langle u_k \rangle = Q_k / A_k = \langle j_k \rangle / \langle \alpha_k \rangle \tag{42.4}$$

Also, the two phases normally do not travel with the same velocity. Indeed, in vapor/liquid systems, the vapor often travels faster than the liquid, giving rise to a local relative velocity (u_R),

$$u_R = u_v - u_l \tag{42.5a}$$

or slip ratio (S),

$$S = u_v / u_l \tag{42.5b}$$

The density of a two-phase mixture is given by

$$\langle \rho \rangle = \left[\iiint_{V_l} \rho_l dv + \iiint_{V_v} \rho_v dv \right] / (V_l + V_v) \tag{42.6a}$$

Thus, from Equation (42.2),

$$\langle \rho \rangle = \rho_l (1 - \langle \alpha \rangle) + \rho_v \langle \alpha \rangle \tag{42.6b}$$

42.2 Conservation Equations

Equations for the conservation of mass, momentum, and energy are needed to describe a flowing two-phase mixture. The appropriate one-dimensional, two-fluid (i.e., writing the conservation laws for each phase separately), and mixture conservation equations are described in the following sections. Let us consider the case of a vapor/liquid system, since this is the most complicated case.

Mass Conservation

For the vapor phase,

$$\frac{\partial}{\partial t}[\rho_v \langle \alpha \rangle A_{xs}] + \frac{\partial}{\partial z}[\rho_v \langle \alpha \rangle \langle u_v \rangle A_{xs}] = \Gamma A_{xs} \tag{42.7a}$$

For the liquid phase,

$$\frac{\partial}{\partial t}[\rho_l(1-\langle\alpha\rangle)A_{xs}]+\frac{\partial}{\partial z}[\rho_l(1-\langle\alpha\rangle)\langle u_l\rangle A_{xs}]=-\Gamma A_{xs} \tag{42.7b}$$

where Γ is the amount of liquid evaporated per unit volume per unit time.

Adding Equation (42.7a) and Equation (42.7b) together yields the one-dimensional mixture continuity equation,

$$\frac{\partial}{\partial t}[\langle\rho\rangle A_{xs}]+\frac{\partial}{\partial z}[GA_{xs}]=0 \tag{42.8}$$

where the mass flux is given by

$$G\overset{\Delta}{=}w/A_{xs}=\rho_v\langle\alpha\rangle\langle u_v\rangle+\rho_l(1-\langle\alpha\rangle)\langle u_l\rangle \tag{42.9}$$

Momentum Conservation

Let us now consider a typical flow regime in which the vapor phase does not wet the wall of the conduit. For this case we have for the vapor phase,

$$\frac{1}{g_c}\left[\frac{\partial}{\partial t}(\rho_v\langle\alpha\rangle\langle u_v\rangle)+\frac{1}{A_{xs}}\frac{\partial}{\partial z}\left(\rho_v\langle\alpha\rangle\langle u_v\rangle^2A_{xs}\right)\right]$$
$$=-\langle\alpha\rangle\frac{\partial p}{\partial z}-\frac{g}{g_c}\rho_v\langle\alpha\rangle\sin\theta-\frac{\tau_iP_i}{A_{xs}}+\frac{\Gamma u_i}{g_c} \tag{42.10a}$$

Similarly, for the liquid phase,

$$\frac{1}{g_c}\left[\frac{\partial}{\partial t}[\rho_l(1-\langle\alpha\rangle)\langle u_l\rangle]+\frac{1}{A_{xs}}\frac{\partial}{\partial z}\left[\rho_l(1-\langle\alpha\rangle)\langle u_l\rangle^2A_{xs}\right]\right]$$
$$=-(1-\langle\alpha\rangle)\frac{\partial p}{\partial z}-\frac{g}{g_c}\rho_l(1-\langle\alpha\rangle)\sin\theta-\frac{\tau_wP_f}{A_{xs}}+\frac{\tau_iP_i}{A_{xs}}-\frac{\Gamma u_i}{g_c} \tag{42.10b}$$

where the velocity of the vapor/liquid interface, u_i, the interfacial perimeter, P_i, and the interfacial shear stress, τ_i, must be constituted to achieve closure [Lahey and Drew, 1992].

Adding Equation (42.10a) and Equation (42.10b) and allowing for the possibility of N local losses, the one-dimensional mixture momentum equation can be written as

$$\frac{1}{g_c}\left[\frac{\partial G}{\partial t}+\frac{1}{A_{xs}}\frac{\partial}{\partial z}(G^2A_{xs}/\langle\rho'\rangle)\right]=-\frac{\partial p}{\partial z}-\frac{g}{g_c}\langle\rho\rangle\sin\theta-\frac{\tau_wP_f}{A_{xs}}$$
$$-\sum_{i=1}^{N}K_i\frac{G^2\delta(z-z_i)P_f}{2g_c\rho_lA_{xs}}\Phi(z) \tag{42.11}$$

where $\delta(z)$ is the Dirac delta function $\langle\rho'\rangle$ in a two-phase "density" given by:

$$\langle \rho' \rangle = \left[\frac{(1-\langle x \rangle)^2}{\rho_l (1-\langle \alpha \rangle)} + \frac{\langle x \rangle^2}{\rho_v \langle \alpha \rangle} \right]^{-1} \tag{42.12}$$

and $\langle x \rangle$ is the flow quality. It should be noted that closure models for the wall shear, τ_w, will be discussed subsequently.

It is interesting to note that one can rearrange Equation (42.11) into drift-flux form [Lahey and Moody, 1993]:

$$\frac{1}{g_c}\left[\frac{\partial G}{\partial t} + \frac{1}{A_{xs}}\frac{\partial}{\partial z}\left(\frac{G^2 A_{xs}}{\langle \rho \rangle}\right)\right] = -\frac{\partial p}{\partial z} - \frac{g}{g_c}\langle \rho \rangle \sin\theta - \frac{\tau_w P_f}{A_{xs}}$$

$$-\frac{1}{g_c A_{xs}}\frac{\partial}{\partial z}\left[A_{xs}\left(\frac{\rho_l - \langle \rho \rangle}{\langle \rho \rangle - \rho_v}\right)\frac{\rho_l \rho_v}{\langle \rho \rangle}\left(V'_{gj}\right)^2\right] + \sum_{i=1}^{N} K_i \frac{G^2 \delta(z-z_i) P_f}{2 g_c \rho_l A_{xs}} \Phi(z) \tag{42.13}$$

where, as will be discussed subsequently, $V'_{gj} = V_{gj} + (C_0 - 1)\langle j \rangle$ is the generalized drift velocity.

Energy Conservation

For the vapor phase,

$$\frac{\partial}{\partial t}[\rho_v \langle \alpha \rangle (\langle e_v \rangle - p/J\rho_v) A_{xs}] + \frac{\partial}{\partial z}[\rho_v \langle u_v \rangle \langle \alpha \rangle A_{xs} \langle e_v \rangle]$$

$$= \Gamma A_{xs} e_{v_i} - p_i A_{xs} \frac{\partial \langle \alpha \rangle}{\partial t} + q_v'''\langle \alpha \rangle A_{xs} - q_{v_i}'' P_i \tag{42.14a}$$

For the liquid phase,

$$\frac{\partial}{\partial t}[\rho_l (1-\langle \alpha \rangle)(\langle e_l \rangle - p/J\rho_l) A_{xs}] + \frac{\partial}{\partial z}[\rho_l \langle u_l \rangle (1-\langle \alpha \rangle) A_{xs} \langle \alpha \rangle]$$

$$= -\Gamma A_{xs} e_{l_i} + p_i A_{xs} \frac{\partial \langle \alpha \rangle}{\partial t} + q_w'' P_H + q_v'''(1-\langle \alpha \rangle) A_{xs} + q_{l_i}'' P \tag{42.14b}$$

where:

$$\langle e_k \rangle \triangleq h_k + \frac{\langle u_k \rangle^2}{2 g_c J} + \frac{gz \sin\theta}{g_c J} \tag{42.15}$$

As noted before, appropriate closure laws are needed for the specific phasic interfacial energy, e_{k_i}, heat fluxes, q_{ki}'', the interfacial perimeter, P_i, and the volumetric heating rate, q_k''' [Lahey and Drew, 1992]. Also, it is interesting to note that the interfacial jump condition gives

$$\Gamma = (q_{v_i}'' - q_{l_i}'') P_i / A_{xs} h_{fg} \tag{42.16}$$

If we add Equation (42.14a) and Equation (42.14b), we obtain the mixture energy equation in the form

$$\frac{\partial}{\partial t}[\rho_l(1-\langle\alpha\rangle)(\langle e_l\rangle - p/J\rho_l)+\rho_v\langle\alpha\rangle(\langle e_v\rangle - p/J\rho_v)]A_{xs}$$

$$+\frac{\partial}{\partial t}[w_l\langle e_l\rangle + w_v\langle e_v\rangle] = q''_w A_{xs} + q''' A_{xs} \tag{42.17}$$

42.3 Closure

In order to be able to evaluate the conservation equations that describe two-phase flow we must first achieve closure by constituting all parameters in these equations in terms of their state variables.

In order to demonstrate the process, let us consider the mixture conservation equations, Equation (42.8), Equation (42.13), and Equation (42.17).

We can use the Zuber–Findly drift-flux model to relate the void fraction to the superficial velocities:

$$\langle\alpha\rangle = \frac{\langle j_v\rangle}{C_0\langle j\rangle + V_{gj}} \tag{42.18}$$

where $\langle j\rangle \triangleq \langle j_v\rangle + \langle j_l\rangle, C_0$ is the void concentration parameter, and V_{gj} is the so-called *drift velocity.* Both of these drift-flux parameters are normally given by correlations that come from appropriate data [Chexal and Lellouche, 1986].

The wall shear, τ_w, is given by

$$\frac{\tau_w P_f}{A_{xs}} = \frac{fG^2\phi_{lo}^2}{2g_c D_H \rho_l} \tag{42.19}$$

where ϕ_{10}^2 is the two-phase friction pressure drop multiplier. It is just the ratio of the single-phase (liquid) to the two-phase density,

$$\phi_{lo}^2 = \rho_l / \rho_{2\phi} \tag{42.20}$$

For example, for homogeneous two-phase flow (in which the slip ratio, S, is unity) we have

$$\phi_{lo}^2 = \rho_l / \langle\rho_h\rangle = 1 + \frac{v_{fg}}{v_f}\langle x\rangle \tag{42.21}$$

For slip flows, empirical correlations, such as that of Martinelli–Nelson, shown in Figure 42.1, are often used.

For local losses (e.g., orifices, spacers, etc.) it is normal practice to assume a homogeneous multiplier; thus, as in Equation (42.21),

$$\Phi = 1 + \frac{v_{fg}}{v_f}\langle x\rangle \tag{42.22}$$

Figure 42.2 shows a typical pressure drop profile, which includes both local and distributed two-phase losses.

One of the interesting features of two-phase flows is that countercurrent flows may occur, in which the phases flow in different directions. In a vapor/liquid system this can lead to a countercurrent flow limitation (CCFL) or **flooding** condition.

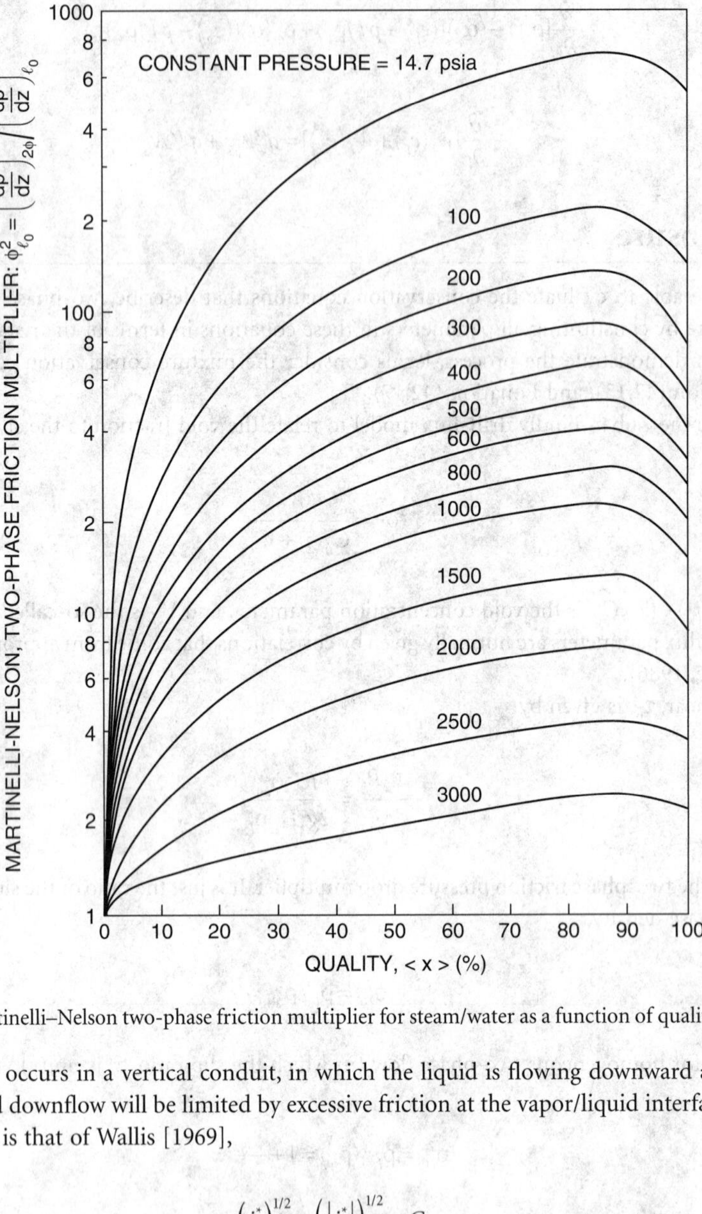

FIGURE 42.1 Martinelli–Nelson two-phase friction multiplier for steam/water as a function of quality and pressure.

When flooding occurs in a vertical conduit, in which the liquid is flowing downward and the vapor upward, the liquid downflow will be limited by excessive friction at the vapor/liquid interface. A popular CCFL correlation is that of Wallis [1969],

$$\left(j_g^*\right)^{1/2} + \left(\left|j_f^*\right|\right)^{1/2} = C \qquad (42.23)$$

where the square root of the phasic Froude number is given by

$$j_k^* = \frac{\langle j_k \rangle \rho_k^{1/2}}{[gD_H(\rho_l - \rho_v)]^{1/2}} \qquad (42.24)$$

This CCFL correlation is known to work well in small-diameter conduits.

For large-diameter conduits the Kutateladze CCFL correlation works much better:

$$K_v^{1/2} + \left|K_l\right|^{1/2} = 1.79 \qquad (42.25)$$

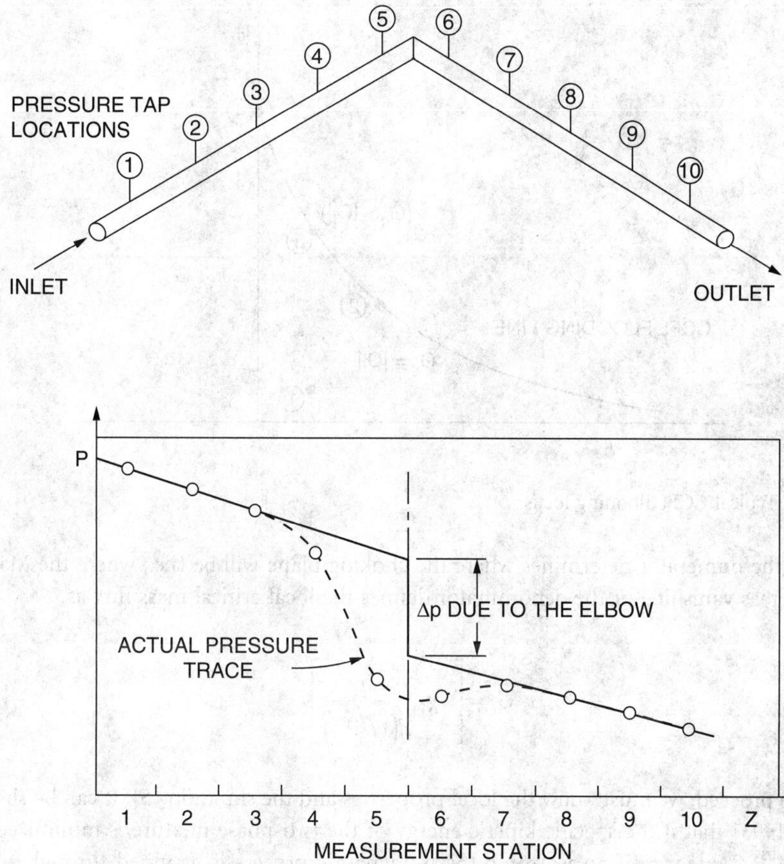

FIGURE 42.2 Two-phase pressure drop in an elbow.

where

$$K_k = \frac{\langle j_k \rangle \rho_k^{1/2}}{[\sigma g g_c(\rho_l - \rho_v)]^{1/4}} \tag{42.26}$$

A strategy for switching from one CCFL correlation to the other has been given by Henry et al. [1993] for various size conduits.

Figure 42.3 shows that both Equation (42.23) and Equation (42.25) imply that there will be no liquid downflow when the vapor upflow velocity (j_v) is large enough (i.e., at or above point 4).

Let us now use the conservation equations to analyze some two-phase flow phenomena of interest. For example, let us consider **critical flow** (i.e., the sonic discharge of a two-phase mixture).

For steady state conditions in which there are no local losses, Equation (42.11) can be expanded and rearranged to yield, for choked flow conditions,

$$-\frac{dp}{dz} = \frac{\dfrac{-G_c^2}{g_c A_{xs}\langle \rho' \rangle}\dfrac{dA_{xs}}{dx} + \dfrac{g}{g_c}\langle \rho \rangle + \dfrac{\tau_w P_f}{A_{xs}}}{\left[1 + \dfrac{G_c^2}{g_c}\dfrac{d}{dp}\left(\dfrac{1}{\langle \rho' \rangle}\right)\right]} \tag{42.27}$$

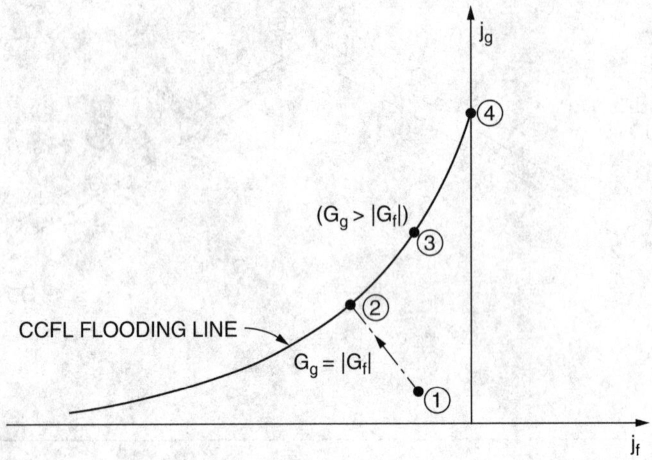

FIGURE 42.3 Typical CCFL flooding locus.

Vanishing of the numerator determines where the choking plane will be (i.e., where the Mach number is unity), whereas vanishing of the denominator defines the local critical mass flux as

$$G_c = \left[-g_c \frac{dp}{d\left(1/\langle \rho' \rangle\right)} \right]^{1/2} \tag{42.28}$$

In order to proceed, we must know the local properties and the slip ratio (S). It can be shown [Lahey and Moody, 1993] that if the specific kinetic energy of the two-phase mixture is minimized we obtain $S = (\rho_l / \rho_v)^{1/3}$. Moreover, if an isotropic thermodynamic process is assumed through a converging nozzle, Figure 42.4 is obtained. This figure is very easy to use. For example, if the stagnation pressure (p_0) and enthalpy (h_0) upstream of the nozzle are 1000 psia and 800 Btu/lbm, respectively, the critical mass flux will be about $G_c = 3995$ lbm/sec-ft^2.

42.4 Two-Phase Instabilities

It is also significant to note that important static and dynamic instabilities may occur in two-phase flows. It can be shown [Lahey and Moody, 1993] that the criterion for the occurrence of an excursive instability is

$$\frac{\partial(\Delta p_{\text{system}})}{\partial w} > \frac{\partial(\Delta p_{\text{ext}})}{\partial w} \tag{42.29}$$

where the external (ext) pressure increase (Δp_{ext}) is normally due to a pump. Figure 42.5 shows that a two-phase system having a positive displacement pump (case 1) and a centrifugal pump with a steep pump/head curve (case 4) will be stable, whereas the cases of parallel channels (case 2) and a centrifugal pump with a relatively flat pump/head curve (case 3) are unstable and have multiple operating points.

If the state variables in the mixture conservation equations are perturbed (i.e., linearized) about a steady state,

$$\delta\varphi(\eta(t)) \triangleq \varphi(\eta(t)) - \varphi_0 = \left.\frac{\partial\varphi}{\partial\eta}\right|_0 \delta\eta \tag{42.30}$$

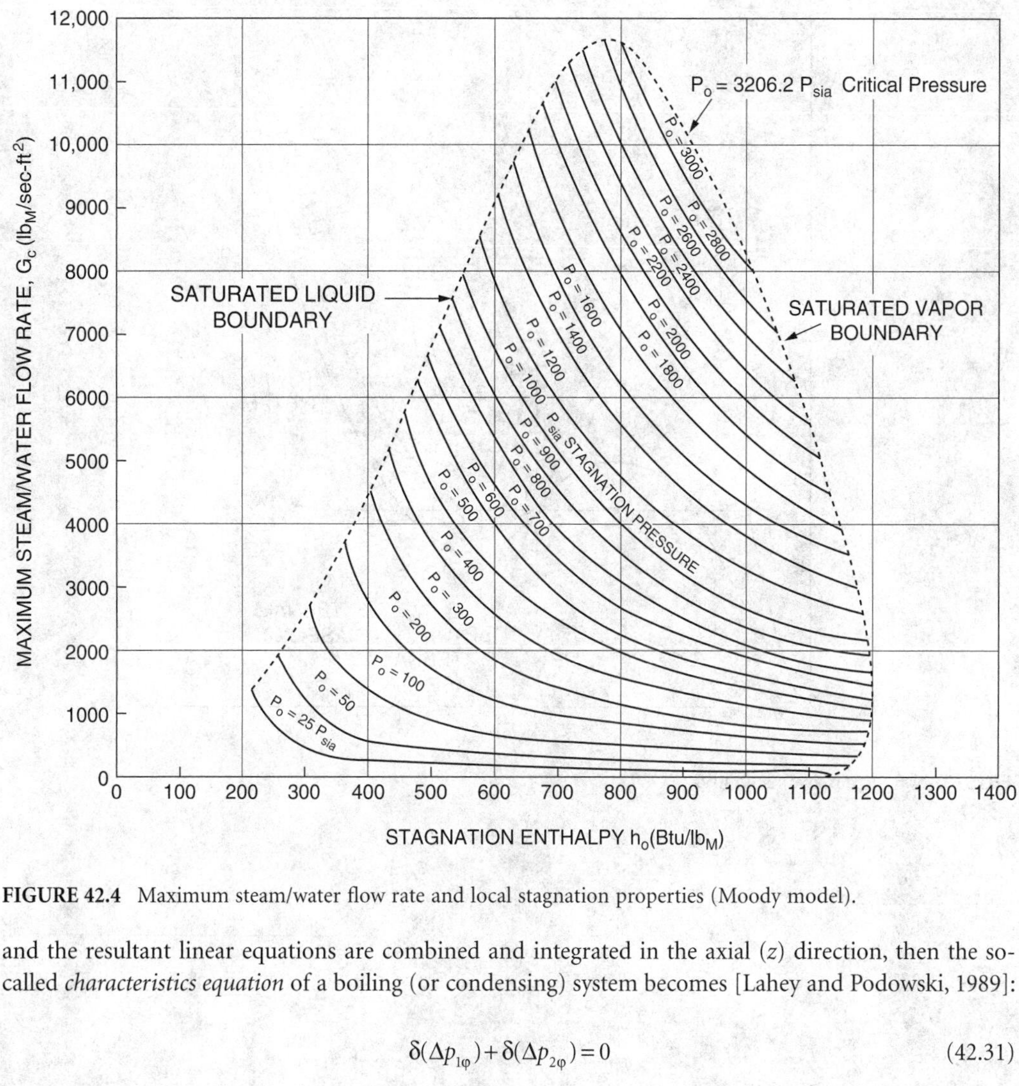

FIGURE 42.4 Maximum steam/water flow rate and local stagnation properties (Moody model).

and the resultant linear equations are combined and integrated in the axial (z) direction, then the so-called *characteristics equation* of a boiling (or condensing) system becomes [Lahey and Podowski, 1989]:

$$\delta(\Delta p_{1\varphi}) + \delta(\Delta p_{2\varphi}) = 0 \tag{42.31}$$

After Laplace-transforming Equation (42.31) — to convert it from the time domain to the frequency domain — we can apply Nyquist's stability technique to determine whether or not density-wave oscillations (DWO) are expected. Significantly, DWO are the most important and prevalent dynamic oscillations that may occur in two-phase flows. For the parallel channel case shown schematically in Figure 42.6, we see that the system is marginally stable (i.e., the Nyquist locus of $\delta(\Delta p_{2\varphi})/\delta(\Delta p_{1\varphi})$ goes through the -1 point) and thus DWO are anticipated.

A stability map of boiling/condensing systems is often given in terms of the subcooling number (N_{sub}) and the phase change number (N_{pch}). A typical plot for a typical boiling water nuclear reactor (BWR/4) is given in Figure 42.7. It can be seen that, as the inlet subcooling is increased at a given power-to-flow ratio (i.e., for N_{pch} fixed), the system may go in and out of DWO.

42.5 Conclusion

This chapter summarizes some of the methods used in the analysis of two-phase flows. The emphasis has been on two-phase fluid mechanics, although phase change heat transfer is often also very important.

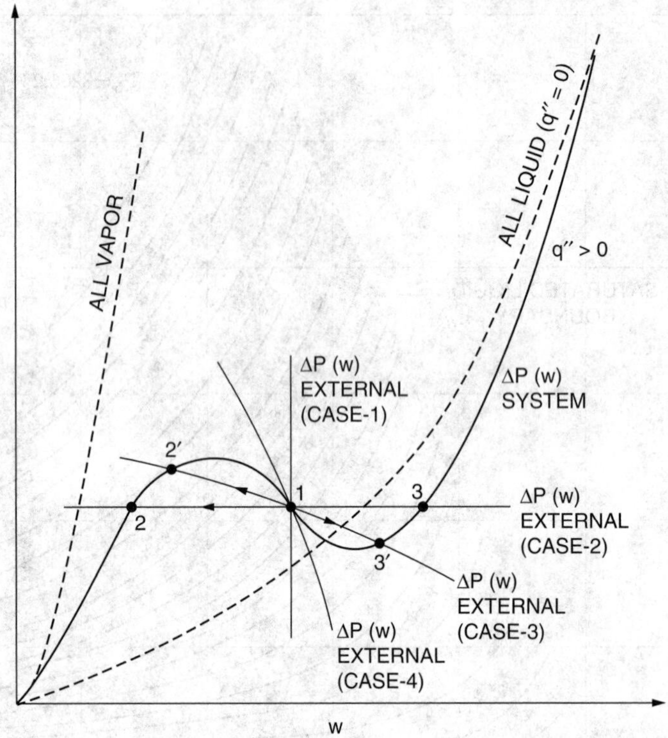

FIGURE 42.5 Excursive instability.

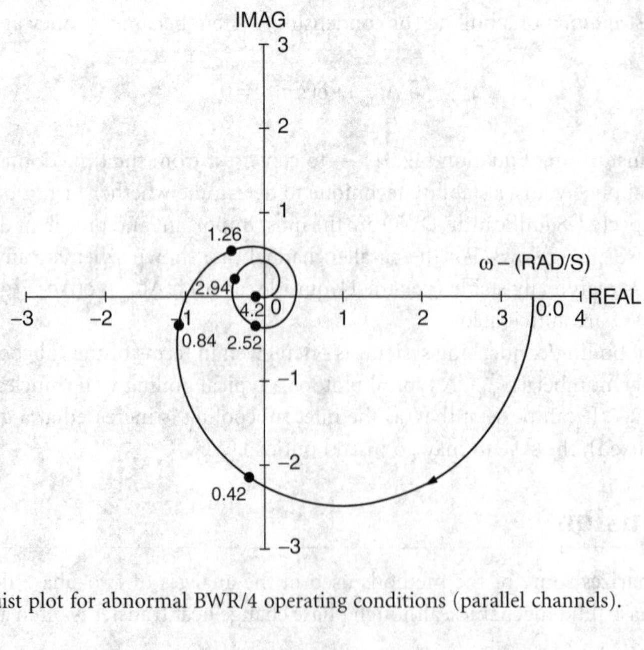

FIGURE 42.6 Nyquist plot for abnormal BWR/4 operating conditions (parallel channels).

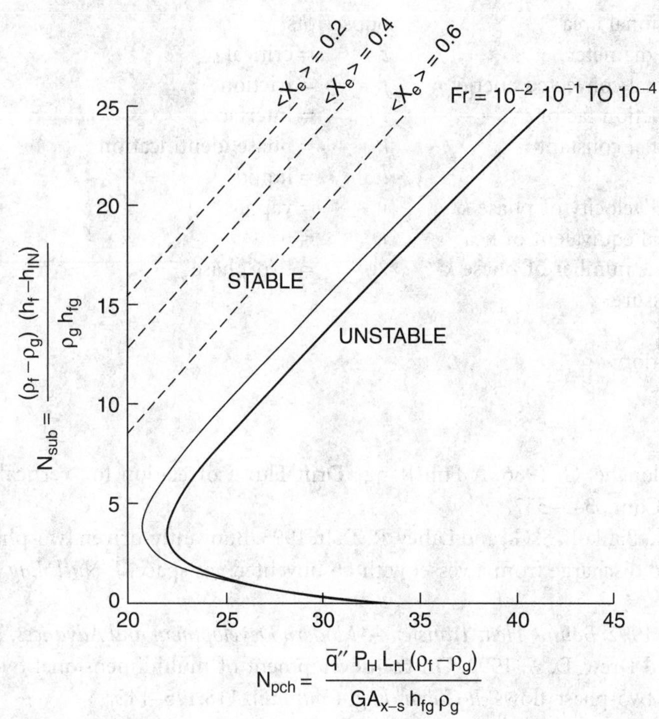

FIGURE 42.7 Typical BWR/4 stability map (K_{in} = 27.8, K_{exit} = 0.14).

The literature associated with two-phase thermal hydraulics is vast; nevertheless, it is hoped that this chapter will provide a useful road map to this literature.

Defining Terms

Critical flow — A condition in which the two-phase mixture is flowing at the local sonic velocity.
Density — The mass per unit volume of the material in question (ρ).
Enthalpy — The internal energy per unit volume plus the related flow work (h).
Flooding — A countercurrent flow limitation due to the resistance induced by cocurrent vapor/liquid flow streams.
Flow area — The cross-sectional area through which the fluid flows (A_{xs}).
Flow rate — The mass flowing per unit time (w).
Phase — The various phases of a fluid are solid, liquid, and gas.
Void fraction — The local volume fraction of the dispersed vapor phase.
Volumetric flow rate — The volume flowing per unit time of phase k (Q_k).

Nomenclature

α	= void fraction	q''	= heat flux
$\delta(z)$	= Dirac delta function	q'''	= volumetric heat generation rate
Γ	= mass of vapor generated per unit volume per unit time	Q	= volumetric flow rate
		S	= slip ratio
θ	= angle from the horizontal plane	u_k	= velocity of phase k
ρ	= density	w	= flow rate
τ	= shear stress	x	= flow quality

A_{xs}	= cross-sectional area		Subscripts	
D_H	= hydraulic diameter		c	= critical
$\langle e_k \rangle$	= specific total convected energy		f	= friction
f	= Moody friction factor		i	= interface
g_c	= gravitational constant		k	= phase identification
G	= mass flux		l	= liquid
j_k	= superficial velocity of phase k		v	= vapor
J	= mechanical equivalent of heat		w	= wall
K_k	= Kutateladze number of phase k		2ϕ	= two-phase
p	= static pressure			
P	= perimeter			
z	= axial position			

References

Chexal, B. and Lellouche, G. 1986. A Full Range Drift-Flux Correlation for Vertical Flows, Proc. Nat. Heat Trans. Conf., 343–357.

Henry, C., Henry, R., Bankoff, S. G., and Lahey, R. T., Jr. 1993. Buoyantly-driven two-phase countercurrent flow in liquid discharge from a vessel with an unvented gas space. *J. Nucl. Eng. Design.* 141(1 and 2):237–248.

Lahey, R. T., Jr., Ed. 1992. *Boiling Heat Transfer — Modern Development and Advances,* Elsevier, New York.

Lahey, R. T., Jr., and Drew, D. A. 1992. On the development of multidimensional two-fluid models for vapor/liquid two-phase flows. *J. Chem. Eng. Commun.* 118:125–142.

Lahey, R. T., Jr., and Moody, F. J. 1993. *The Thermal-Hydraulics of a Boiling Water Nuclear Reactor.* ANS Monograph. American Nuclear Society, La Grange Park, IL.

Lahey, R. T., Jr., and Podowski, M. Z. 1989. On the analysis of instabilities in two-phase flows. *Multiphase Sci. Tech.* 4:183–342.

Roco, M., Ed. 1993. *Particulate Two-Phase Flow.* Hemisphere, New York.

Wallis, G. B. 1969. *One-Dimensional Two-Phase Flow.* McGraw-Hill, New York.

Further Information

Butterworth, D. and Hewitt, G. F. 1977. *Two-Phase Flow and Heat Transfer,* Harwell Series, Oxford University Press.

Collier, J. G. 1972. *Convective Boiling and Condensation,* McGraw-Hill, New York.

Drew, D. A. and Passman, S. L. 1998. *Theory of Multicomponent Fluids,* Appl. Math. Sci.-135, Springer.

Ishii, M. 1975. *Thermo-Fluid Dynamic Theory of Two-Phase Flow,* Eyrolles, Paris.

43

Basic Mixing Principles for Various Types of Fluid Mixing Applications

James Y. Oldshue
Oldshue Technologies International, Inc.

The fluid mixing process involves three different areas of viscosity which affect flow patterns and scaleup and two different scales within the fluid itself: **macro scale** and **micro scale**. Design questions come up with regard to the performance of mixing processes in a given volume. Consideration must be given to proper impeller and tank geometry as well as to the proper speed and power for the impeller. Similar considerations arise when it is desired to scale up or scale down, and this involves another set of mixing considerations.

If the fluid discharge from an impeller is measured with a device that has a high frequency response, one can track the velocity of the fluid as a function of time. The velocity at a given point in time can then be expressed as an average velocity ($\bar{v}$) plus a fluctuating component (v'). Average velocities can be integrated across the discharge of the impeller, and the pumping capacity normal to an arbitrary discharge plane can be calculated. This arbitrary discharge plane is often defined by the boundaries of the impeller blade diameter and height. Because there is no casing, however, an additional 10 to 20% of flow typically can be considered as the primary flow of an impeller.

The velocity gradients between the average velocities operate only on larger particles. Typically, these particles are greater than 1000 μm in size. This is not a precise definition, but it does give a feel for the magnitudes involved. This defines macro-scale mixing. In the turbulent region, these macro-scale fluctuations can also arise from the finite number of impeller blades passing a finite number of baffles. These set up velocity fluctuations that can also operate on the macro scale.

Smaller particles primarily see only the fluctuating velocity component. When the particle size is much less than 100 μm, the turbulent properties of the fluid become important. This is the definition of the boundary size for micro-scale mixing.

0-8493-1586-7/05/$0.00+$1.50

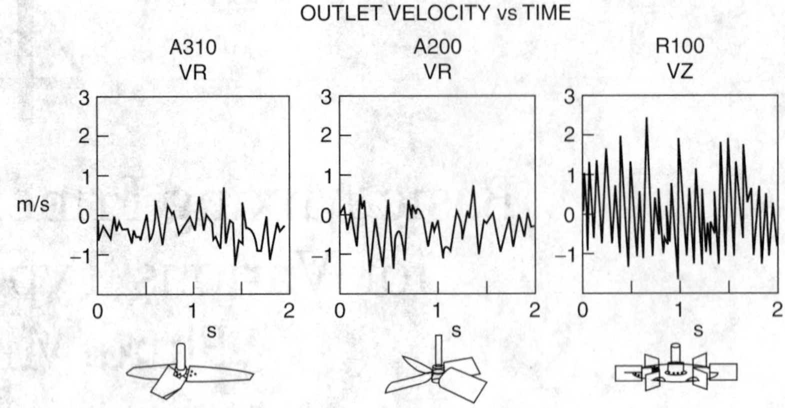

FIGURE 43.1 Typical velocity as a function of time for three different impellers, all at the same total pumping capacity. (Courtesy of LIGHTNIN.)

All of the power applied by a mixer to a fluid through the impeller appears as heat. The conversion of power to heat is through viscous shear and is 2542 Btu/h/hp. Viscous shear is present in turbulent flow only at the micro-scale level. As a result, the power per unit volume is a major component of the phenomenon of micro-scale mixing. At a 1 μm level, in fact, it does not matter what specific impeller design is used to apply the power.

Numerous experiments show that the power per unit volume in the zone of the impeller (which is about 5% of the total tank volume) is about 100 times higher than the power per unit volume in the rest of the vessel. Based on some reasonable assumptions about the fluid mechanics parameters, the root-mean-square (rms) velocity fluctuation in the zone of the impeller appears to be approximately 5 to 10 times higher than in the rest of the vessel. This conclusion has been verified by experimental measurements.

The ratio of the rms velocity fluctuation to the average velocity in the impeller zone is about 50% for many open impellers. If the rms velocity fluctuation is divided by the average velocity in the rest of the vessel, however, the ratio is on the order of 5 to 10%. This is also the ratio of rms velocity fluctuation to the mean velocity in pipeline flow. In micro-scale mixing, phenomena can occur in mixing tanks that do not occur in pipeline reactors. Whether this is good or bad depends upon the process requirements.

Figure 43.1 shows velocity versus time for three different impellers. The differences between the impellers are quite significant and can be important for mixing processes. All three impeller velocities are calculated for the same impeller flow, Q, and same diameter. The A310 (Figure 43.2) draws the least power and has the lowest velocity fluctuations. This gives the lowest micro-scale turbulence and shear rate. The A200 (Figure 43.3) displays increased velocity fluctuations and draws more power. The R100

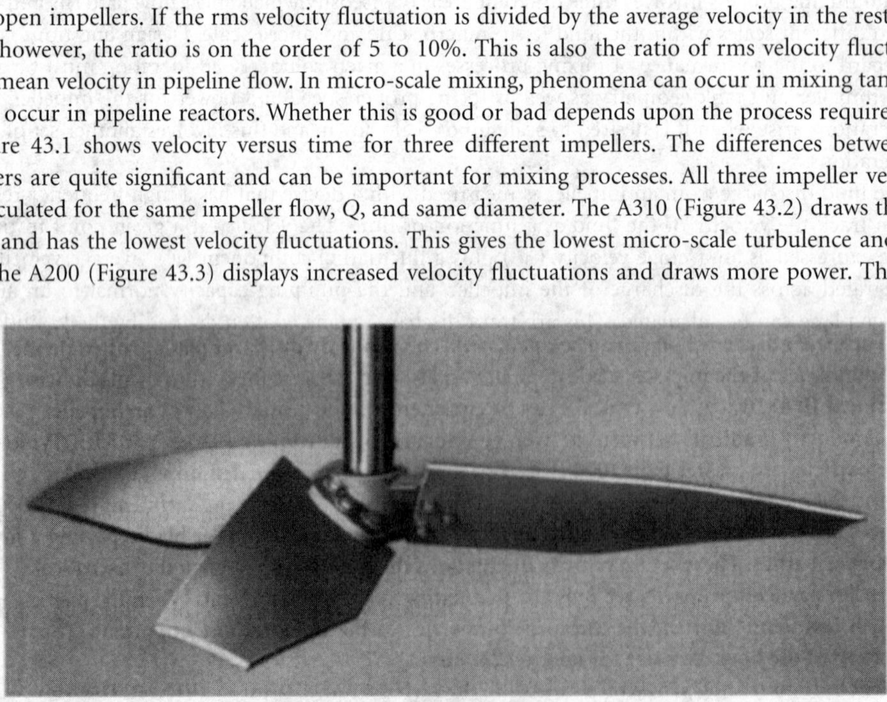

FIGURE 43.2 Fluidfoil impeller (A310). (Courtesy of LIGHTNIN.)

FIGURE 43.3 Typical axial-flow turbine (A200). (Courtesy of LIGHTNIN.)

FIGURE 43.4 Radial-flow Rushton turbine (R100). (Courtesy of LIGHTNIN.)

(Figure 43.4) draws the most power and has the highest micro-scale shear rate. The proper impeller should be used for each individual process requirement.

The velocity spectra in the axial direction for the axial-flow impeller A200 are shown in Figure 43.5. A decibel correlation has been used in this figure because of its well-known applicability in mathematical modeling as well as the practicality of putting many orders of magnitude of data in a reasonably sized chart. Other spectra of importance are the power spectra (the square of the velocity) and the Reynolds stress (the product of the *R* and *Z* velocity components), which is a measure of the momentum at a point.

The ultimate question is this: How do all these phenomena apply to process design in mixing vessels? No one today is specifying mixers for industrial processes based on meeting criteria of this type. This is largely because processes are so complex that it is not possible to define the process requirements in terms of these fluid mechanics parameters. If the process results could be defined in terms of these parameters, sufficient information probably exists to permit the calculation of an approximate mixer design. It is important to continue studying fluid mechanics parameters in both mixing and pipeline reactors to establish what is required by different processes in fundamental terms.

One of the most practical recent results of these studies has been the ability to design pilot plant experiments (and, in many cases, plant-scale experiments) that can establish the sensitivity of a process to macro-scale mixing variables (as a function of power, pumping capacity, impeller diameter, impeller tip

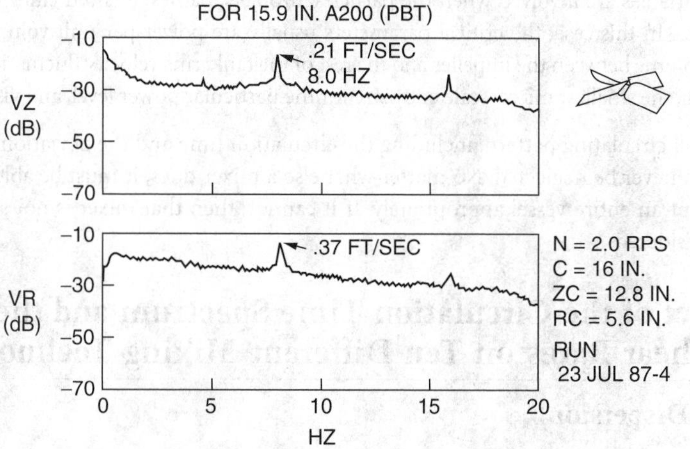

FIGURE 43.5 Typical velocity spectrum as a function of fluctuation frequency. (Courtesy of LIGHTNIN.)

speeds, and macro-scale shear rates) in contrast to micro-scale mixing variables (which are relative to power per unit volume, rms velocity fluctuations, and some estimation of the size of the micro-scale eddies).

Another useful and interesting concept is the size of the eddies at which the power of an impeller is eventually dissipated. This concept utilizes the principles of **isotropic turbulence** developed by Komolgoroff. The calculations assume some reasonable approach to the degree of isotropic turbulence, and the estimates do give some idea as to how far down in the micro scale the power per unit volume can effectively reach. The equation is

$$L = (\nu^3 / \varepsilon)^{1/4}$$

43.1 Scaleup/Scaledown

Two applications of scaleup frequently arise. One is building a model for pilot plant studies to develop an understanding of the process variables for an existing full-scale mixing installation. The other is taking a new process and studying it in the pilot plant to work out pertinent scaleup variables for a new mixing installation.

Because there are thousands of specific processes each year that involve mixing, there will be at least hundreds of different situations requiring a somewhat different pilot plant approach. Unfortunately, no set of rules states how to carry out studies for any specific program, but here are a few guidelines that can help one carry out a pilot plant program:

- For any given process, take a qualitative look at the possible role of fluid shear stresses. Try to consider pathways related to fluid shear stress that may affect the process. If none exist, then this extremely complex phenomenon can be dismissed and the process design can be based on such things as uniformity, circulation time, blend time, or velocity specifications. This is often the case in the blending of miscible fluids and the suspension of solids.
- If fluid shear stresses are likely to be involved in obtaining a process result, then one must qualitatively look at the scale at which the shear stresses influence the result. If the particles, bubbles, droplets, or fluid clumps are on the order of 1000 μm or larger, the variables are macro-scale, and average velocity at a point is the predominant variable.

When macro-scale variables are involved, every geometric design variable can affect the role of shear stresses. These variables can include power, impeller speed, impeller diameter, impeller blade shape, impeller blade width or height, thickness of the material used to make the impeller, number of blades, impeller location, baffle location, and number of impellers.

Micro-scale variables are involved when the particles, droplets, baffles, or fluid clumps are on the order of 100 μm or less. In this case, the critical parameters usually are power per unit volume, distribution of power per unit volume between the impeller and the rest of the tank, rms velocity fluctuation, energy spectra, dissipation length, the smallest micro-scale eddy size for the particular power level, and viscosity of the fluid.

- The overall circulating pattern, including the circulation time and the deviation of the circulation times, can never be neglected. No matter what else a mixer does, it must be able to circulate fluid throughout an entire vessel appropriately. If it cannot, then that mixer is not suited for the tank being considered.

43.2 Effect of the Circulation Time Spectrum and the Spectrum of Shear Rates on Ten Different Mixing Technologies

Gas–Liquid Dispersion

The macro-scale rate shear change affects the bubble size distribution in tanks of various sizes. As processes are scaled up, the linear, superficial gas velocity tends to be higher in the larger tank. This is the major contributor to the energy input of the gas stream. If the power per unit volume put in by the mixer remains relatively constant, then small tanks have a different ratio of mixing energy to gas expansion energy, which affects the flow pattern and a variety of other fluid mechanics parameters. The large tank will tend to have a larger variation of the size distribution of bubbles than will the small tank.

This phenomenon is affected by the fact that the surface tension and viscosity vary all the way from a relatively pure liquid phase to all types of situations with dissolved chemicals, either electrolytes or nonelectrolytes, and other types of surface-active agents.

Gas–Liquid Mass Transfer

If we are concerned only with the total volumetric mass transfer rate, then we can achieve very similar $K_G a$ values in large tanks and in small tanks.

Blend time enters the picture primarily for other process steps immediately preceding or following the gas-liquid mass transfer step. Blending can play an important role in the total process, of which gas–liquid mass transfer is only one components.

Solids Suspension and Dispersion

Solids suspension is not usually affected by blend time or shear rate changes in the relatively low to medium solids concentration in the range from 0 to 40% by weight. However, as solids become more concentrated, the effect of solids concentration on the power required changes the criteria from the settling velocity of the individual particles in the mixture to the apparent viscosity of the more concentrated slurry. This means that we enter an area where the blending of non-Newtonian fluid regions, the shear rates, and circulation patterns play marked roles (see Figure 43.6).

The suspension of a single solid particle should depend primarily on the upward velocity at a given point and also should be affected by the uniformity of this velocity profile across the entire tank cross-section. There are upward velocities in the tank, and there also must be corresponding downward velocities. In addition to the effect of the upward velocity on a settling particle, there is also the random motion of the micro-scale environment, which does not affect large particles very much but is a major factor in the concentration and uniformity of particles in the transition and micro-scale range.

Using a draft tube in the tank for solids suspension introduces another, different set of variables. There are other relationships that are very much affected by scaleup in this type of process. Different scaleup problems exist depending on whether the impeller is pumping up or down within the draft tube.

If the process involves the dispersion of solids in a liquid, then we may either be concerned with breaking up agglomerates or possibly physically breaking or shattering particles that have a low cohesive

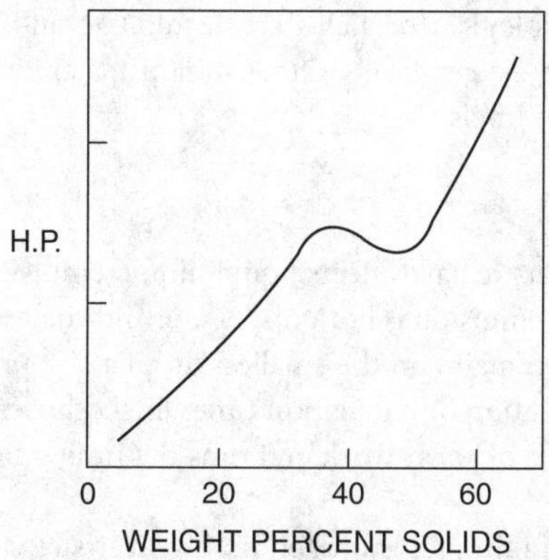

FIGURE 43.6 Effect of percent solids by weight and power required for uniformity and fluid motion. (Courtesy of LIGHTNIN.)

force between their components. Normally, we do not think of breaking ionic bonds with the shear rates available in mixing machinery.

If we know the shear stress required to break up a particle, we can determine the shear rate required of the machinery by various viscosities with the equation

$$\text{Shear stress} = \text{Viscosity} \times \text{Shear rate}$$

The shear rate available from various types of mixing and dispersion devices is known approximately, as is the range of viscosities in which they can operate. This makes the selection of the mixing equipment subject to calculation of the shear stress required for the viscosity to be used.

In the equation above, it is assumed that there is 100% transmission of the shear rate in the shear stress. However, with the slurry viscosity determined essentially by the properties of the slurry, at high slurry concentrations there is a slippage factor in which internal motion of particles in the fluids over and around each other can reduce the effective transmission of viscosity efficiencies from 100% to as low as 30%.

Animal cells in biotechnology do not normally have a tough skin as fungal cells do and are very sensitive to mixing effects. Many approaches have been tried to minimize the effect of increased shear rates on scaleup, and these include encapsulating the organism in or on micro particles and conditioning cells selectively to shear rates. In addition, traditional fermentation processes have maximum shear rate requirements in which cells become progressively more and more damaged until they become motile.

Solid–Liquid Mass Transfer

There is potentially a major effect of both shear rate and circulation time in these processes. The solids may be inorganic, in which case we are looking at the slip velocity of the particle and also whether we can break up agglomerates of particles, which may enhance mass transfer. When the particles become small enough, they tend to follow the flow pattern, so the slip velocity necessary to affect the mass transfer becomes less and available.

This shows that from the definition of off-bottom motion to complete uniformity, the effect of mixer power is much less than from going to on-bottom motion to off-bottom suspension. The initial increase in power causes more and more solids to become in active communication with the liquid and has a much greater mass transfer rate than that occurring above the power level for off-bottom suspension, in which slip velocity between the particles of fluid is the major contributor.

Since there may well be chemical or biological reactions happening on or in the solid phase, depending upon the size of the process participants, it may or may not be appropriate to consider macro- or micro-scale effects.

In the case of living organisms, their access to dissolved oxygen throughout the tank is of great concern. Large tanks in the fermentation industry often have a Z/T ratio of 2:1 to 4:1; thus, top to bottom blending can be a major factor. Some biological particles are facultative and can adapt and reestablish their metabolism at different dissolved oxygen levels. Other organisms are irreversibly destroyed by sufficient exposure to low dissolved oxygen levels.

Liquid–Liquid Emulsions

Almost every shear rate parameter we have affects liquid–liquid emulsion formation. Some of the effects are dependent upon whether the emulsion is both dispersing and coalescing in the tank, or whether there are sufficient stabilizers present to maintain the smallest droplet size produced for long periods of time. Blend time and the standard deviation of circulation times affect the length of time it takes for a particle to be exposed to the various levels of shear work and thus the time it takes to achieve the ultimate small particle size desired.

As an aside, when a large liquid droplet is broken up by shear stress, it tends to initially elongate into a dumbbell type of shape, which determines the particle size of the two large droplets formed. Then the neck in the center of the "dumbbell" may explode or shatter. This would give a debris of particle sizes that can be quite different from the two major particles produced.

Liquid–Liquid Extraction

If our main interest is in the total volumetric mass transfer between the liquids, the role of shear rate and blend time is relatively minor. However, if we are interested in the bubble size distribution — and we often are because that affects the settling time of an emulsion in a multistage cocurrent or countercurrent extraction process — then the change in macro and micro rates on scaleup is a major factor. Blend time and circulation time are usually not major factors on scaleup.

Blending

If the blending process occurs between two or more fluids with relatively low viscosity such that the blending is not affected by fluid shear rates, then the difference in blend time and circulation between small and large tanks is the only factor involved. However, if the blending involves wide disparities in the density of viscosity and surface tension between the various phases, a certain level or shear rate may be required before blending can proceed to its ultimate degree of uniformity.

The role of viscosity is a major factor in going from the turbulent regime, through the transition region, into the viscous regime, and there is a change in the rate of energy dissipation discussed previously. The role of non-Newtonian viscosity is very strong since that tends to markedly change the influence of impellers and determines the appropriate geometry.

Another factor here is the relative increase in Reynolds number on scaleup. This means that we could have pilot plants, as well as the plant, running in the turbulent region. We could have the pilot plant running in the transition region and the plant in the turbulent, or the pilot plant could be in the viscous region while the plant is in the transition region. There is no apparent way to prevent this Reynolds number change upon scaleup. In reviewing the qualitative flow pattern in a pilot scale system, it should be realized that the flow pattern in the large tank will be at an apparently much lower viscosity and therefore at a much higher Reynolds number than is observed in the pilot plant. This means that the roles of tank shape, D/T ratio, baffles, and impeller locations can be based on different criteria in the plant size unit than in the pilot size unit under observation.

Chemical Reactions

Chemical reactions are influenced by the uniformity of concentration both at the feed point and in the rest of the tank and can be markedly affected by changes in overall blend time and circulation time as well as the micro-scale environment. It is possible to keep the ratio between the power per unit volume at the impeller and that in the rest of the tank relatively similar on scaleup, but much detail needs to be considered regarding the reaction conditions, particularly where selectivity is involved. This means that reactions can take different paths depending upon chemistry and fluid mechanics, and this is a major consideration in what should be examined. The method of introducing the reagent stream can be projected in several different ways depending upon the geometry of the impeller and the feed system.

Fluid Motion

Sometimes the specification is purely in terms of pumping capacity. Obviously, the change in volume and velocity relationships depends upon the size of the two- and three-dimensional area or volume involved. The impeller flow is treated in a head/flow concept, and the head required for various types of mixing systems can be calculated or estimated.

Heat Transfer

In general, the fluid mechanics of the film on the mixer side of the heat transfer surface is a function of what happens at that surface rather than the fluid mechanics around the impeller zone. The impeller provides flow largely across and adjacent to the heat transfer surface, and that is the major consideration of the heat transfer result obtained. Many of the correlations are in terms of traditional dimensionless groups in heat transfer, while the impeller performance is often expressed as the impeller Reynolds number.

43.3 Computational Fluid Dynamics

Several software programs available to model flow patterns of mixing tanks are available. They allow the prediction of flow patterns based on certain boundary conditions. The most reliable models use accurate fluid mechanics data generated for the impellers in question and a reasonable number of modeling cells to give the overall tank flow pattern. These flow patterns can give velocities, streamlines, and localized kinetic energy values for the system. Their main use at the present time is in examining the effect of changes in mixing variables based on adjustments to the mixing process. These programs can model velocity, shear rates, and kinetic energy but probably cannot adapt to the chemistry of diffusion or mass transfer kinetics of actual industrial processes at the present time.

Relatively uncomplicated transparent tank studies using tracer fluids or particles can also give a feel for the overall flow pattern. The time and expense of calculating these flow patterns with computational fluid dynamics should be considered in relation to their applicability to an actual industrial process. The future of computational fluid dynamics looks very encouraging, and a reasonable amount of time and effort placed in this regard can yield immediate results as well as the potential for future process evaluation.

Figure 43.7 through Figure 43.9 show some approaches. Figure 43.7 shows velocity vectors for an A310 impeller. Figure 43.8 shows contours of kinetic energy of turbulence. Figure 43.9 uses a particle trajectory approach with neutral buoyancy particles.

Numerical fluid mechanics can define many of the fluid mechanics parameters for an overall reactor system. Many of the models break the mixing tank up into small microcells. Suitable material and mass transfer balances between these cells throughout the reactor are then made. This can involve massive computational requirements. Programs are available that can give reasonably acceptable models of experimental data in mixing vessels. Modeling the three-dimensional aspect of a flow pattern in a mixing tank can require a large amount of computing power.

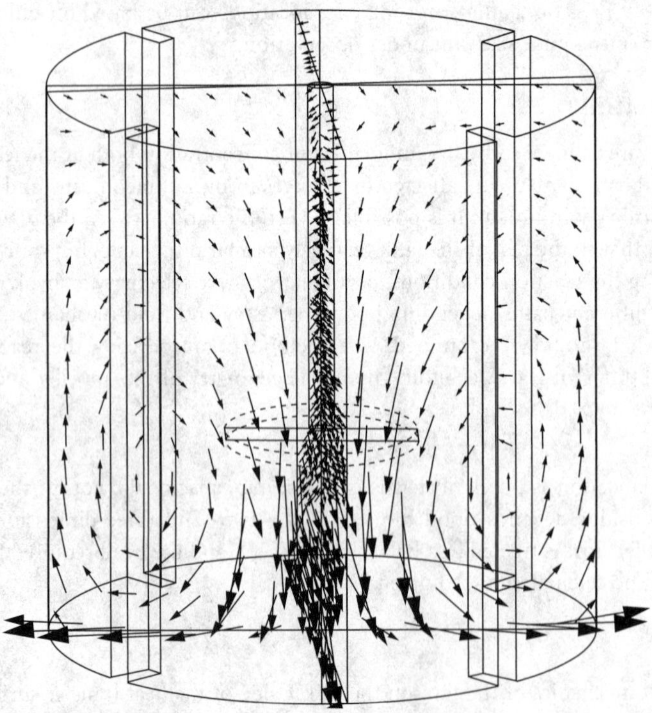

FIGURE 43.7 Typical velocity pattern for a three-dimensional model using computational fluid dynamics for an axial flow impeller (A310). (Courtesy of LIGHTNIN.)

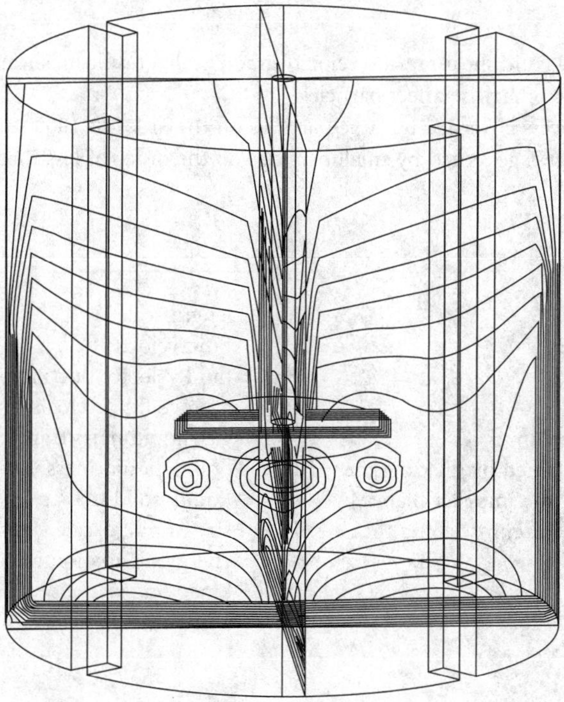

FIGURE 43.8 Typical contours of kinetic energy of turbulence using a three-dimensional model with computational fluid dynamics for an axial flow impeller (A310). (Courtesy of LIGHTNIN.)

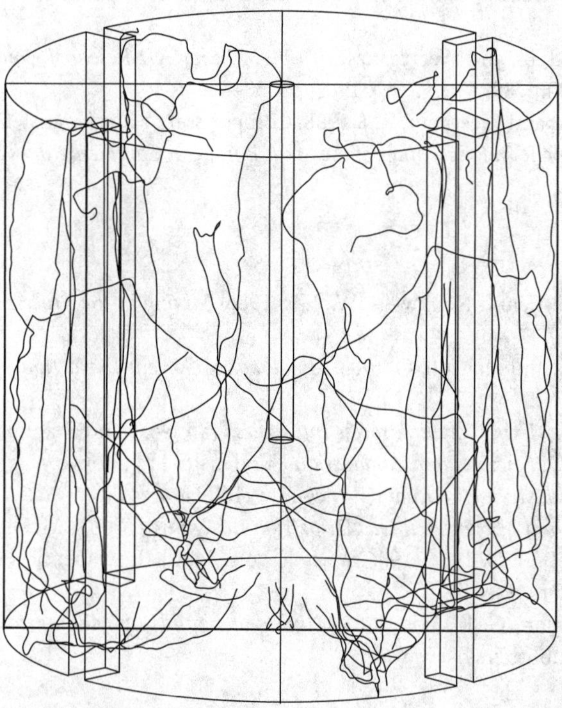

FIGURE 43.9 Typical particle trajectory using an axial flow impeller (A310) with a 100-micron particle using computational fluid dynamics. (Courtesy of LIGHTNIN.)

Defining Terms

Isotropic turbulence — Fluid shear rate is a velocity gradient that results in shear stress, which can break up, disperse, or otherwise affect particles.

Macro scale — Any process governed by large particles on the order of 1000 μm or more.

Micro scale — Any process governed by small particles on the order of less than 100 μm.

Nomenclature

N	impeller speed
D	impeller diameter
T	tank diameter
Z	liquid level
P/V	power per unit volume
SR	solidity ratio, obtained by dividing the projected area of the impeller blades by the area of a disk circumscribing the impeller blades
N_P	power number
H	velocity head, $v^2/2g$
P	power
L	length scale
v	fluid velocity
v'	fluid velocity fluctuation
$\bar{v}$	average fluid velocity
$K_G a$	gas–liquid mass transfer coefficient
$K_L a$	liquid–liquid mass transfer coefficient
k_s	liquid–solid mass transfer coefficient
L	size of microscale eddy
ε	energy dissipation rate
$\dot{u}$	kinematic viscosity
u	dynamic viscosity

References

Levich, V. 1962. *Physico-Chemical Hydrodynamics,* Prentice Hall, Englewood Cliffs, NJ.

Middleton, J. C. 1989. *Proceedings of the Third European Conference on Mixing,* BHRA, Cranfield, England, pp. 15–36.

Neinow, A. W., Buckland, B., and Weetman, R. J. 1989. *Mixing XII Research Conference,* Potosi, MO.

Oldshue, J. Y. 1989. Mixing '89. *Chem. Eng. Prog.* pp. 33–42.

Oldshue, J. Y., Post, T. A., and Weetman, R. J. 1988. Comparison of mass transfer characteristics of radial and axial flow impellers. *Proceedings of the Sixth European Conference on Mixing,* BHRA, Cranfield, England.

Further Information

Harnby, N., Edwards, M. F., and Neinow, A. W., Eds. 1986. *Mixing in the Process Industries,* Butterworth, Stoneham, MA.

Lo, T. C., Baird, M. H. I., and Hanson, C. 1983. *Handbook of Solvent Extraction,* John Wiley & Sons, New York.

McDonough, R. J. 1992. *Mixing for the Process Industries,* Van Nostrand Reinhold, New York.

Nagata, S. 1975. *Mixing: Principles and Applications,* Kodansha Ltd., Tokyo.

Oldshue, J. Y. 1983. *Fluid Mixing Technology,* McGraw-Hill, New York.

Tatterson, G. B. 1991. *Fluid Mixing and Gas Dispersion in Agitated Tanks,* McGraw-Hill, New York.

Uhl, V. W. and Gray, J. B., Eds. 1966, 1986. *Mixing,* Vols. I and II, Academic Press, New York; Vol. III, Academic Press, Orlando, FL.

Ulbrecht, J. J. and Paterson, G. K., Eds. 1985. *Mixing of Liquids by Mechanical Agitation,* Gordon and Breach Science Publishers, New York.

Proceedings

Fluid Mechanics of Mixing, ed. R. King. Kluwer Academic Publishers, Dordecht, Netherlands, 1992.

Fluid Mixing, Vol. I. Inst. Chem. Eng. Symp., Ser. No. 64 (Bradford, England). Institute of Chemical Engineers, Rugby, England, 1984.

Mixing — Theory Related to Practice, AIChE, Inst. Chem. Eng. Symp. Ser. No. 10 (London). AIChE and Institute of Chemical Engineers, London, 1965.

Proceedings of the First European Conf. on Mixing, Ed. N. G. Coles. BHRA Fluid Eng., Cranfield, England, 1974.

Proceedings of the Second European Conference on Mixing, Ed. H. S. Stephens and J. A. Clark. BHRA Fluid Eng., Cranfield, England, 1977.

Proceedings of the Third European Conference on Mixing, Ed. H. S. Stephens and C. A. Stapleton. BHRA Fluid Eng., Cranfield, England, 1979.

Proceedings of the Fourth European Conference on Mixing, Ed. H. S. Stephens and D. H. Goodes. BHRA Fluid Eng., Cranfield, England, 1982.

Proceedings of the Fifth European Conference on Mixing, Ed. S. Stanbury. BHRA Fluid Eng., Cranfield, England, 1985.

Proceedings of the Sixth European Conference on Mixing, BHRA Fluid Eng., Cranfield, England, 1988.

Process Mixing: Chemical and Biochemical Applications, Ed. G. B. Tatterson and R. V. Calabrese. AIChE Symp. Ser. No. 286, 1992.

44

Fluid Measurement Techniques

S. A. Sherif
University of Florida, Gainesville

The subject of fluid measurements covers a broad spectrum of applications. Examples include wind-tunnel experiments, turbomachinery studies, water tunnel and flume investigations, erosion studies, and meteorological research, to name a few. Many wind-tunnel investigations call for determining forces on scale models of large systems. **Similarity analysis** is then performed to permit generalization of the experimental results obtained on the laboratory model. Techniques for measuring fluid flow parameters are extremely diverse and include thermal anemometers, laser velocimeters, volume and mass flow devices, flow visualization by direction injection, and optical diagnostics. Measurements can be carried out in liquids and gases, incompressible and compressible fluids, two-phase and single-phase flows, and Newtonian and **non-Newtonian fluids.** Types of quantities measured vary greatly and include velocity, temperature, turbulence, vorticity, Reynolds stresses, turbulent heat flux, higher-order turbulence moments, volume and mass flow rates, and differential pressure. Topics covered in this chapter will be limited to differential pressure measurements, thermal anemometry, laser velocimetry, and volume and mass flow measurements. Coverage of these topics will take the form of providing a summary of fundamental principles followed by a summary of the basic equations used.

44.1 Fundamental Principles

The purpose of this section is to provide an overview of the fundamental principles involved in fluid measurements employing the most common and widely used techniques. These techniques include using differential pressure-based instruments, hot-wire and hot-film anemometers, laser Doppler velocimeters, and volume and mass flow devices. These topics will be discussed in some detail in this section.

Differential pressures may be thought of as differences between time-averaged pressures at two points in a fluid flow or between a time-averaged and an instantaneous value of pressure evaluated at a point in the flow [Blake, 1983]. This type of measurement provides an alternative to **thermal anemometry** for determining velocity magnitude and direction as well as turbulence intensity of a fluid flow. The Pitot-static tube is one of the most reliable differential pressure probes for flow measurement (see Figure 44.1). Special forms of Preston tubes have been used for measurement of wall shear stress in boundary layers of smooth walls. Preston tubes are hypodermic needles that respond to the mean velocity profile in the vicinity of the wall.

0-8493-1586-7/05/$0.00+$1.50

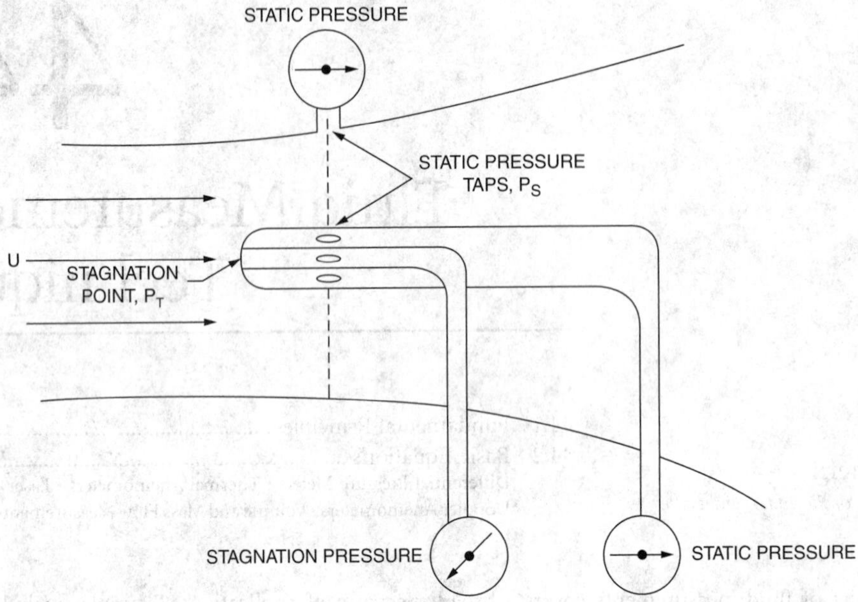

FIGURE 44.1 Typical installation of a Pitot-static tube in a duct. (*Source:* Blake, W. K. 1983. Differential pressure measurement. In *Fluid Mechanics Measurements,* R. J. Goldstein, Ed., Hemisphere, Washington, DC. pp. 61–97. Reproduced with permission. All rights reserved.)

Thermal anemometry refers to the use of a wire or film sensor for velocity, temperature, concentration, and turbulence measurements. This technique started in the late 1800s in the form of employing homemade constant-current anemometers (CCA) for velocity measurement. Three categories of anemometers now exist. The constant-temperature anemometer (CTA) supplies a sensor with heating current that changes with the flow velocity to maintain constant sensor resistance. This type is primarily used for velocity, turbulence, Reynolds stress, and vorticity measurements. The constant-current anemometer, on the other hand, supplies a constant heating current to the sensor. This can either be used for velocity measurements or for temperature and temperature fluctuation measurements. This choice is dictated by the magnitude of the probe sensor current, where the probe sensitivity to velocity fluctuations diminishes for low values of current. The third type of anemometer is the pulsed wire anemometer, which is capable of measuring the fluid velocity by momentarily heating the sensor. This causes heating of the fluid around the sensor and a subsequent convection of that fluid segment downstream to a second wire that acts as a temperature sensor. The time of flight of the heated fluid segment is inversely proportional to the fluid velocity. Typical block diagrams representing the constant-temperature and constant-current anemometers are shown in Figure 44.2 and Figure 44.3, respectively. Thermal anemometers are always connected to a probe via a probe cable. The sensor of a typical probe can either be made of wire or film. Wire probes are made of tungsten or platinum and are about 1 mm long and 5 μm in diameter. Film probes, on the other hand, are made of nickel or platinum deposited in a thin layer onto a backing material (such as quartz) and connected to the anemometer employing leads attached to the ends of the film. A thin protective coating is usually deposited over the film to prevent damage by chemical reaction or abrasion. Typical thicknesses of film probes are 70 μm.

Laser Doppler anemometry (LDA), or laser Doppler velocimetry (LDV), deals with measuring fluid velocities and higher-order turbulence quantities by detecting the Doppler frequency shift of laser light that has been scattered by small particles moving with the fluid [Adrian, 1983]. Three different types of LDA optical systems exist. These are the reference-beam system, the dual-beam system, and the dual-scatter system. The dual-beam LDA produces two types of signals — coherent and incoherent. The coherent signal occurs when at least two particles simultaneously reside in the measurement volume.

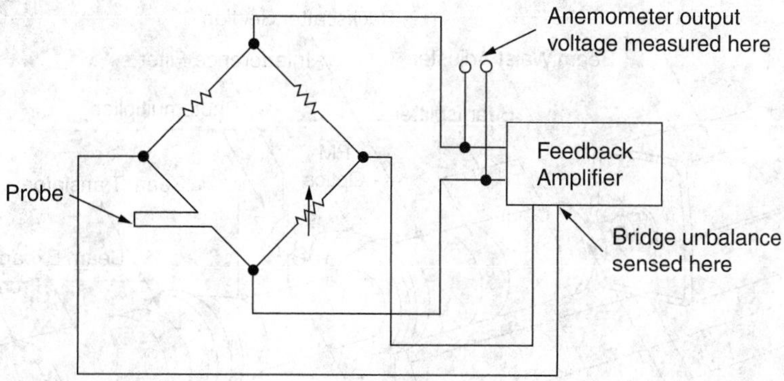

FIGURE 44.2 Block diagram of a constant-temperature anemometer.

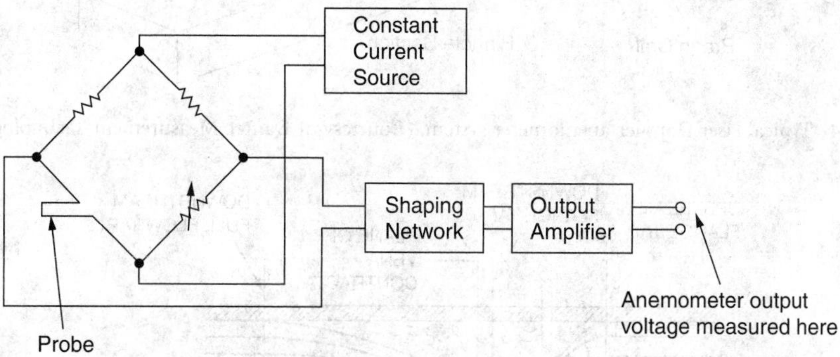

FIGURE 44.3 Block diagram of a constant-current anemometer.

The incoherent signal occurs when a single particle scatters two light waves, one from each illuminating beam. The reference-beam LDA mixes light scattered from an illuminating beam with a reference beam to detect the frequency difference. Both the reference-beam LDA and the dual-beam LDA have the disadvantage of having to satisfy a coherent aperture condition. The amplitude of the reference-beam Doppler signal is proportional to the square root of the reference-beam power, thus allowing an unlimited increase of the signal amplitude by the simple expedient of increasing the power. This feature is particularly useful when the scattered light flux is small compared to a background radiation level. A typical LDA system consists of a laser, a **beam waist adjuster**, a beamsplitter, a **bragg cell**, a backscatter section, a **photomultiplier tube**, photomultiplier optics, an interference filter, a **pinhole section**, a **beam translator**, a beam expander, and a front lens (see Figure 44.4). Other components include a **signal processor** and a **frequency tracker** or a **frequency counter**.

Fluid meters may be classified into those that determine fluid quantity and those that determine fluid flow rate. Quantity meter may be classified as weighing meters (such as weighers or tilting traps) or volumetric meters (such as calibrated tanks, reciprocating pistons, rotating disks and pistons, sliding and rotating vanes, gear and lobed impellers, bellows, and liquid sealed drums). Rate meters, on the other hand, can be classified as differential pressure meters (such as orifice, venturi, nozzle, centrifugal, Pitot-tube, or linear resistance meters), momentum meters (such as turbine, propeller, or cup anemometers), variable-area meters (such as gate, cones, floats-in-tubes, or slotted cylinder and piston meters), force meters (such as target or hydrometric pendulum meters), thermal meters (such as hot-wire and hot-film anemometers), fluid surface height or head meters (such as weirs and flumes), and miscellaneous meters (such as electromagnetic, tracers, acoustic, vortex-shedding, laser, and Coriolis meters) [Mattingly, 1983].

A typical orifice meter (see Figure 44.5) consists of four parts:

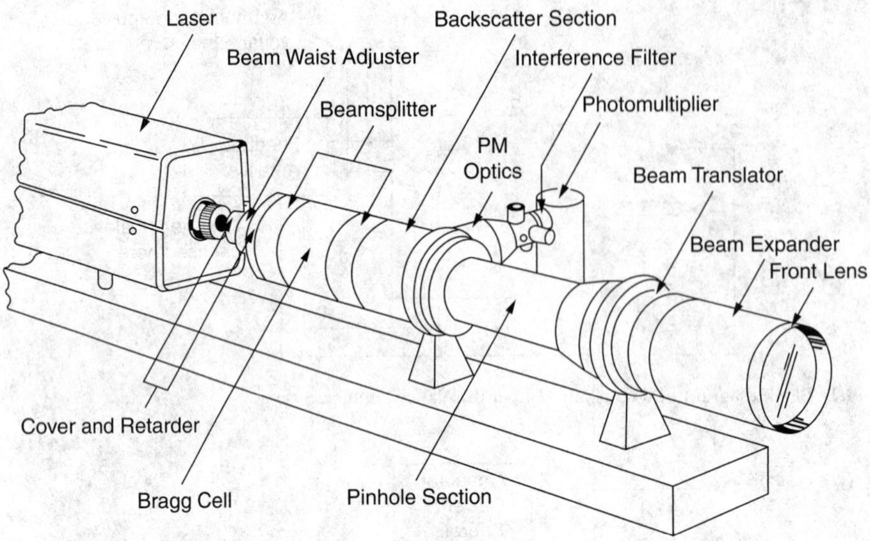

FIGURE 44.4 Typical laser Doppler anemometer system. (Courtesy of Dantec Measurement Technology A/S.)

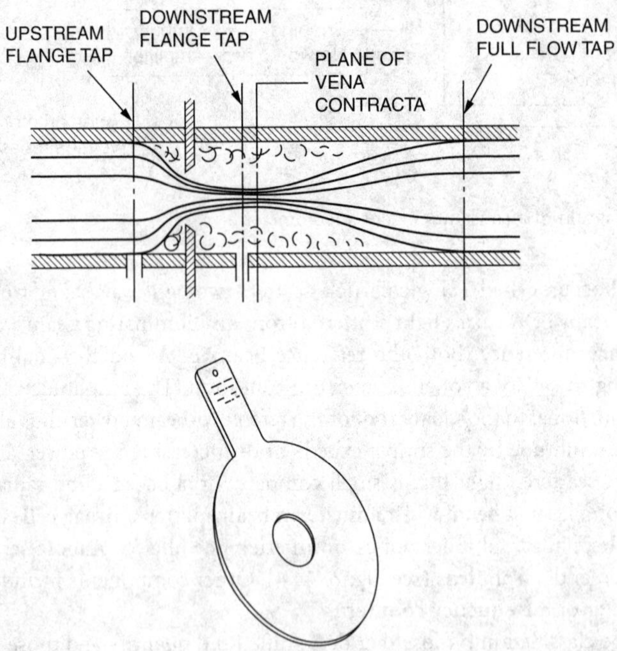

FIGURE 44.5 Orifice meter. (*Source:* Mattingly, G. E. 1983. Volume flow measurements. In *Fluid Mechanics Measurements,* R. J. Goldstein, Ed., Hemisphere, Washington, DC. pp. 245–306. Reproduced with permission. All rights reserved.)

1. The upstream section including flow-conditioning devices
2. The orifice fixture and plate assembly
3. The downstream meter-tube section
4. The secondary element (not shown)

A venturi tube typically has a shape that closely approximates the streamline pattern of the flow through a reduced cross-sectional area (see Figure 44.6). Elbow meters are considered nonintrusive since they do

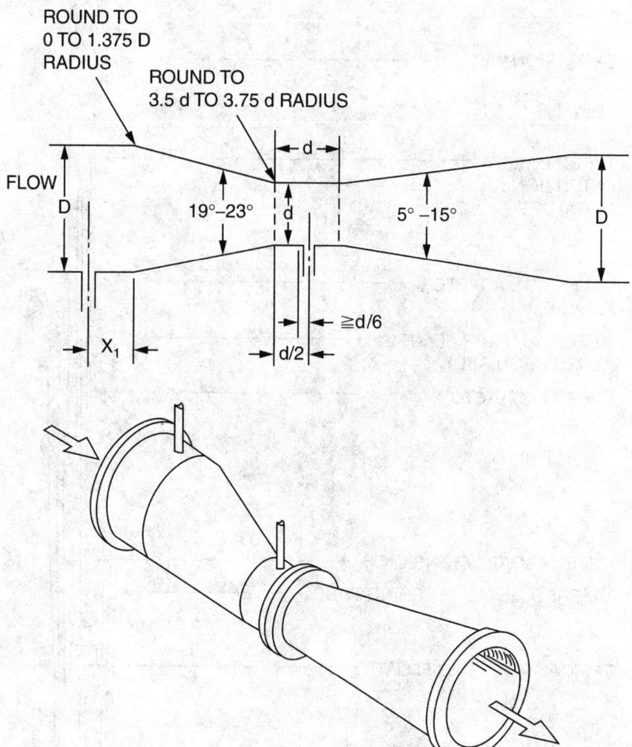

FIGURE 44.6 Venturi tube. (*Source:* Mattingly, G. E. 1983. Volume flow measurements. In *Fluid Mechanics Measurements*, R. J. Goldstein, Ed., Hemisphere, Washington, DC. pp. 245–306. Reproduced with permission. All rights reserved.)

not interfere with the flow pattern in the pipe and do not introduce structural elements in the flow [Mattingly, 1983]. Laminar flow meters are based on establishing laminar flow between the pressure taps and using the laminar flow rate relationships through a tube of a known area and with a specified pressure drop across its length. Turbine meters enable the fluid flow to spin a propeller wheel whose angular speed is related to the average flow rate in the duct. The angular speed is typically detected by the passage of the blade tips past a coil pickup in the pipe. Rotameters are vertically installed devices that operate by balancing the upward fluid drag on the float with the weight of the float in the upwardly diverging tube (see Figure 44.7). Appropriate choice of the configuration of the metering tube can allow the position of the float to be linearly proportional to the flow rate. Target meters (see Figure 44.8) operate on the principle that the average flow rate in a pipe flow is related to the fluid drag on a disk supported in the pipe. The fluid drag can be measured using secondary devices such as strain gauges and fluid-activated bellows. Target meters are particularly useful in flow metering applications involving dirty fluids so long as the suspended particles do not alter the critical geometrical arrangement. Thermal flow meters operate on the principle of sensing the increase in fluid temperature between two thermometers placed in the flow when heat is added between the thermometers. Thermal flow meters may be made to be nonintrusive and are also capable of operating on the basis of cooling rather than heating.

44.2 Basic Equations

Differential Pressure Meters

The velocity in a one-dimensional laminar flow using a Pitot tube may be expressed as

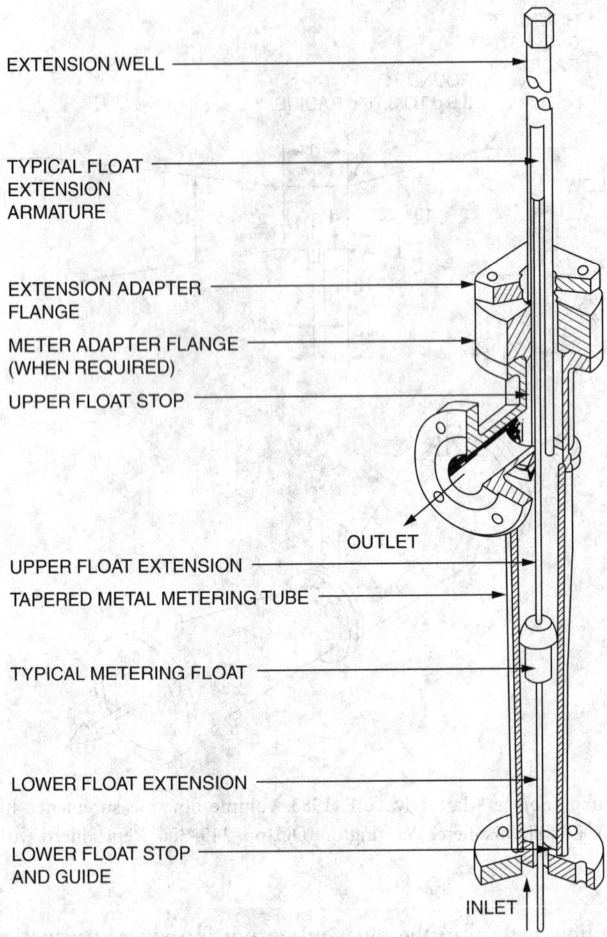

FIGURE 44.7 Rotameter. (*Source:* Mattingly, G. E. 1983. Volume flow measurements. In *Fluid Mechanics Measurements*, R. J. Goldstein, Ed., Hemisphere, Washington, DC. pp. 245–306. Reproduced with permission. All rights reserved.)

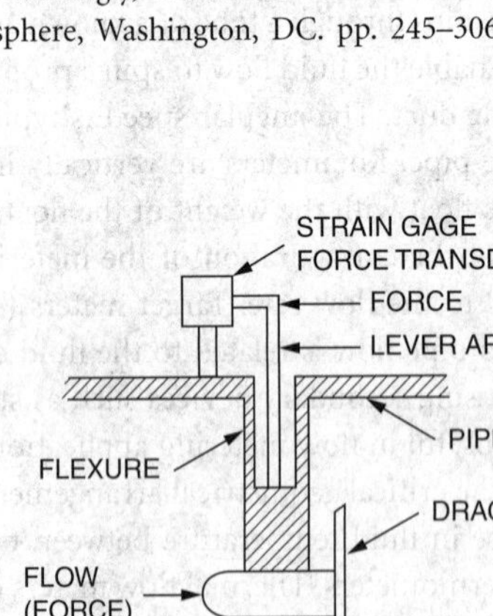

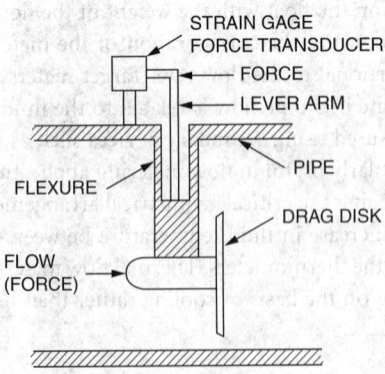

FIGURE 44.8 Target meter. (*Source:* Mattingly, G. E. 1983. Volume flow measurements. In *Fluid Mechanics Measurements*, R. J. Goldstein, Ed., Hemisphere, Washington, DC. pp. 245–306. Reproduced with permission. All rights reserved.)

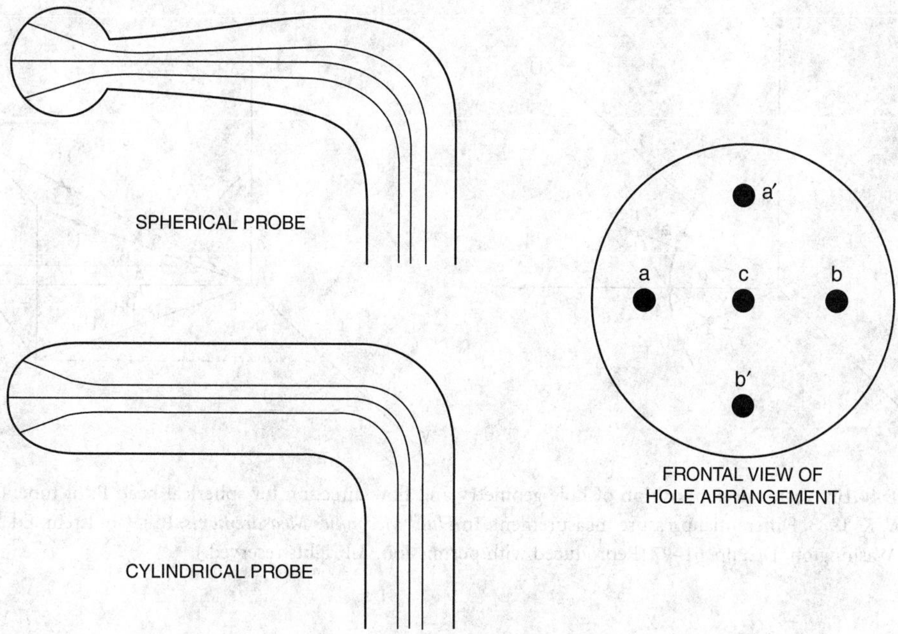

FIGURE 44.9 Some five-hole Pitot-tube geometries. (*Source:* Blake, W. K. 1983. Differential pressure measurement. In *Fluid Mechanics Measurements*, R. J. Goldstein, Ed., Hemisphere, Washington, DC. pp. 61–97. Reproduced with permission. All rights reserved.)

$$U = \sqrt{\frac{2(p_T - p_s)}{\rho}} \tag{44.1}$$

The most widely used method for measuring mean velocity in multidimensional flows is a five-hole pressure probe, which is a streamlined axisymmetric body that points into the flow (see Figure 44.9). The vector decomposition of the flow velocity U^*, which is incident on the five pressure taps a, b, c, a', and b' of a spherical probe, is shown in Figure 44.10. Pien [1958] has shown the following:

$$\frac{p_a - p_b}{2p_c - p_a - p_b} = \frac{\sin 2\alpha}{1 - \cos 2\alpha} \tan 2\beta_h \tag{44.2}$$

$$\frac{p_a - p_b}{\frac{1}{2}\rho V_h^2} = \frac{9}{4} \sin 2\alpha \sin 2\beta_h \tag{44.3}$$

Thermal Anemometers

The overheat ratio and resistance difference ratio may be expressed as

$$a_1 = \frac{R_s}{R_f} \quad \text{and} \quad a_2 = \frac{R_s - R_f}{R_f} \tag{44.4}$$

The yaw, pitch, and roll sensitivities (Figure 44.11) may be expressed by the partial derivatives $\partial E/\partial \theta$, $\partial E/\partial \phi$, and $\partial E/\partial \psi$, respectively. The effective cooling velocity, on the other hand, may be expressed in either of the following forms:

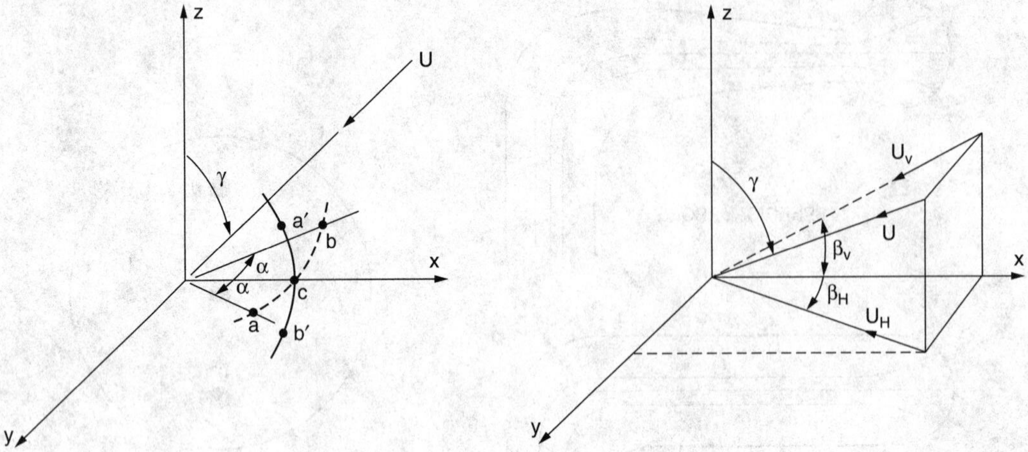

FIGURE 44.10 Vector decomposition of hole geometry and flow direction for spherical-head Pitot tube. (*Source:* Blake, W. K. 1983. Differential pressure measurement. In *Fluid Mechanics Measurements*, R. J. Goldstein, Ed., Hemisphere, Washington, DC. pp. 61–97. Reproduced with permission. All rights reserved.)

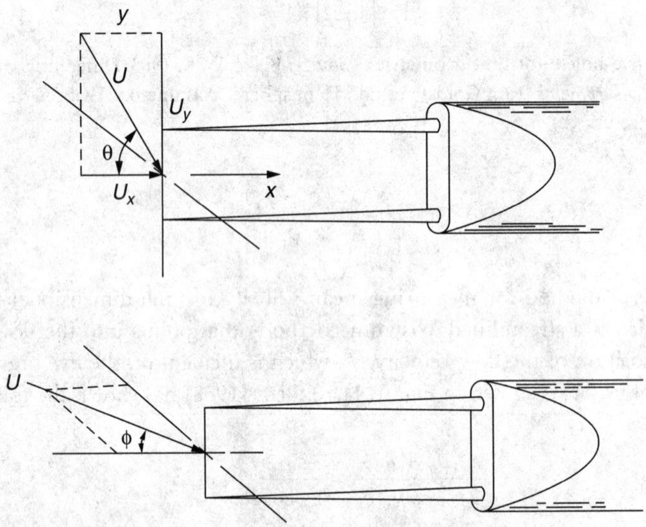

FIGURE 44.11 Yaw and pitch angles of a standard hot-wire probe.

$$U_{\text{eff}}^2 = U_x^2 + k^2U_y^2 + h^2U_z^2 \tag{44.5}$$

$$U_{\text{eff}}^2 = U^2(\cos^2\theta\cos^2\phi + k^2\sin^2\theta\cos^2\phi + h^2\sin^2\phi) \tag{44.6}$$

The resistance–temperature relationships may be expressed by

$$R = \rho_r\frac{l}{A} \quad \text{and} \quad R_s = R_o[1 + \alpha(T_s - T_o) + \alpha_1(T_s - T_o)^2 + \cdots] \tag{44.7}$$

where $\alpha = 3.5\cdot10^{-3}\,°\text{C}^{-1}$ and $\alpha_1 = -5.5\cdot10^{-7}\,°\text{C}^{-2}$ for platinum and $\alpha = 5.2\cdot10^{-3}\,°\text{C}^{-1}$ and $\alpha_1 = 7.0\cdot10^{-7}\,°\text{C}^{-2}$ for tungsten.

A number of cooling laws are commonly used in thermal anemometry. The most common of these cooling laws is King's [1914]:

$$\frac{I^2 R_s}{T_s - T_f} = A_o + B_o \sqrt{\mathrm{Re}} \tag{44.8}$$

Siddall and Davies [1972] expressed King's law in a modified form given by

$$E_b^2 = A + BU^{0.5} + CU \tag{44.9}$$

where $A = 1.273, B = 0.860,$ and $C = -0.017$. Collis and Williams [1959], on the other hand, derived the following cooling law:

$$\mathrm{Nu}\left(\frac{T_m}{T_f}\right)^{-0.17} = A_1 + B_1\,\mathrm{Re}^{n_1} \tag{44.10}$$

where $A_1, B_1,$ and n_1 take the values 0.24, 0.56, and 0.45, respectively, when the Reynolds number is between 0.02 and 44, while their values become 0, 0.48, and 0.51, respectively, for a Reynolds number larger than 44 and smaller than 140. The quantity T_m is the arithmetic average of $\overline{T}_s$ and T_f.

Kramer's cooling law [Hinze, 1959] can be expressed by

$$\mathrm{Nu} = 0.42\mathrm{Pr}^{0.20} + 0.57\mathrm{Pr}^{0.33}\mathrm{Re}^{0.50} \tag{44.11}$$

This equation is valid over the Reynolds number range of 0.1 < Re < 10,000. Other cooling laws include that of Van der Hegge Zijnen [1956], who derived the following equation:

$$\mathrm{Nu} = 0.38\mathrm{Pr}^{0.2} + (0.56\mathrm{Re}^{0.5} + 0.001\mathrm{Re})\mathrm{Pr}^{0.333} \tag{44.12}$$

The effect of ambient temperature changes may be expressed in terms of the velocity and temperature sensitivities as follows:

$$e_s = S_{\mathrm{vel}} U + S_{\mathrm{temp}} t_f \tag{44.13}$$

$$S_{\mathrm{vel}} = \left\{ \frac{n_1 \pi l k_o R_s}{2^{1.97}\overline{E}_s \overline{U} T_o^{0.80}} \left[\frac{\rho_o \overline{U} d}{\mu_o} (2T_o)^{1.76} \right]^{n_1} B_1 \right\} \cdot \frac{(T_s + \overline{T}_f)^{0.97-1.76n_1}(T_s - \overline{T}_f)}{\overline{T}_f^{0.17}} \tag{44.14}$$

$$S_{\mathrm{temp}} = -\frac{1}{2}\left\{ \frac{1.76 n_1 \pi l k_o R_s}{2^{0.97}\overline{E}_s T_o^{0.80}} \left[\frac{\rho_o \overline{U} d}{\mu_o} (2T_o)^{1.76} \right]^{n_1} \cdot B_1 \frac{T_s - \overline{T}_f}{\overline{T}_f^{0.17}(T_s + \overline{T}_f)^{0.03+1.76n_1}} \right.$$

$$\left. \cdot \overline{E}_s \left(\frac{0.17}{\overline{T}_f} + \frac{1}{T_s - \overline{T}_f} - \frac{0.97}{T_s + \overline{T}_f} \right) \right\} \tag{44.15}$$

The corrected bridge voltage is:

$$E_{bc} \simeq E_b \left[1 - \frac{T_{01} - T_{02}}{2(T_s - T_{01})} \right] \tag{44.16}$$

The frequency response of a constant-current anemometer may be expressed by the following differential equation:

$$\frac{dr_s}{dt} + \frac{\alpha R_o}{\rho c A l} \left\{ \frac{\pi l k_f}{\alpha R_o} \left[0.42 \mathrm{Pr}^{0.20} + 0.57 \mathrm{Pr}^{0.33} \left(\frac{\bar{U} d}{\nu} \right)^{0.5} \right] - \bar{I}^2 \right\} r_s$$

$$= \left(\frac{\pi k_f}{2 \rho c A \bar{U}} \right) 0.57 \mathrm{Pr}^{0.33} \left(\frac{\bar{U} d}{\nu} \right)^{0.5} (\bar{R}_s - R_f) u \tag{44.17}$$

whose time constant is

$$\tau_{\text{cca}} = \frac{\rho C A l (\bar{R}_s - R_f)}{\bar{I}^2 \alpha R_f R_o} \tag{44.18}$$

while the frequency response of a constant-temperature anemometer may be expressed by the following equation:

$$\frac{di_s}{dt} + \frac{\alpha R_o \bar{I}_s^2 [2g\bar{R}_s (\bar{R}_s - R_f) + R_f^2]}{\rho c A l (\bar{R}_s - R_f)} i_s = 0.57 \mathrm{Pr}^{0.33} \sqrt{\frac{\bar{U} d}{\nu}} \frac{\pi l k_f (\bar{R}_s - R_f)}{2 \alpha R_o \bar{U}} u \tag{44.19}$$

whose time constant is

$$\tau_{\text{cta}} = \frac{\rho c A l (\bar{R}_s - R_f)}{\alpha R_o \bar{I}_s^2 [2g R_s (\bar{R}_s - R_f) + R_f^2]} \tag{44.20}$$

The mean velocity vector may be computed using the following set of equations:

$$U_{\text{eff}_x}^2 = U^2 (\cos^2 \alpha + k^2 \sin^2 \alpha) \tag{44.21}$$

$$U_{\text{eff}_y}^2 = U^2 (\cos^2 \beta + k^2 \sin^2 \beta) \tag{44.22}$$

$$U_{\text{eff}_z}^2 = U^2 (\cos^2 \gamma + k^2 \sin^2 \gamma) \tag{44.23}$$

$$\cos \alpha = \frac{1}{U} \sqrt{\frac{U_{\text{eff}_x}^2 - k^2 U^2}{1 - k^2}} \tag{44.24}$$

$$\cos \beta = \frac{1}{U} \sqrt{\frac{U_{\text{eff}_y}^2 - k^2 U^2}{1 - k^2}} \tag{44.25}$$

$$\cos\gamma = \frac{1}{U}\sqrt{\frac{U^2_{\text{eff}_z} - k^2U^2}{1-k^2}} \tag{44.26}$$

$$\cos^2\alpha + \cos^2\beta + \cos^2\gamma = 1 \tag{44.27}$$

$$U = \sqrt{\frac{U^2_{\text{eff}_x} + U^2_{\text{eff}_y} + U^2_{\text{eff}_z}}{1+2k^2}} \tag{44.28}$$

Laser Doppler Anemometers

The Doppler frequency may be expressed in terms of half the angle between two identical laser beams with frequency f (wavelength λ) and the flow velocity component V_θ perpendicular to the two-beam bisector, as follows:

$$f_D = \frac{2\sin\theta}{\lambda}V_\theta \tag{44.29}$$

A major advantage of this equation is the fact that it is linear and does not contain any undetermined constants, thus eliminating the need for calibration. The width of the measuring volume may be expressed in terms of the transmitting length focal distance, f_T, and the diameter of the beam waist before the transmitting lens, D:

$$d = \frac{4f_T\lambda}{\pi D} \tag{44.30}$$

The length and height, on the other hand, may be expressed by

$$l = \frac{d}{\sin\theta} \quad \text{and} \quad h = \frac{d}{\cos\theta} \tag{44.31}$$

The spacing of the nearly parallel fringes (produced by the interference of the two light beams in the measuring volume) as well as the number of fringes can be expressed by

$$\delta = \frac{\lambda}{2\sin\theta} \quad \text{and} \quad n = \frac{4\Delta}{\pi D} \tag{44.32}$$

The Doppler frequency can be determined using frequency counters that time a fixed number, N, of zero crossings. This allows computing of the particle velocity by simply using the following relationship:

$$V_\theta = \frac{N\delta}{\Delta t} \tag{44.33}$$

Frequency counters are extremely accurate and have a wide dynamic range; however, their output is provided at irregular intervals. This necessitates using special statistical procedures to perform proper counting. Furthermore, the performance of frequency counters may be compromised at higher particle concentrations, when multiple particles are likely to coexist in the measuring volume.

Volume and Mass Flow Measurements

Tanks may be used to determine the flow rate for steady liquid flows by measuring the mass of liquid collected in a known period of time. For gas volume flow rate measurements, compressibility must be taken into account. Positive displacement flow meters may be used for specialized applications.

Volume or mass flow measurements may also be performed using restriction flow meters for flow in ducts. This may include flow nozzles, orifice plates, and venturis. These restriction flow meters typically produce a differential pressure across the meter.

The equations describing the ideal performance characteristics of differential pressure meters for incompressible fluids can be expressed by

$$V_2 = \left[\frac{2g(p_1 - p_2)}{\xi(1-\beta^4)}\right]^{1/2} \quad \text{and} \quad \dot{M}_I = A_2\left[\frac{2\rho(p_1 - p_2)}{(1-\beta^4)}\right]^{1/2} \tag{44.34}$$

where the quantity $1/\tilde{A}(1 - \beta^4)$ is a velocity correction factor, while β is the ratio of restriction hole to pipe diameter.

The corresponding equations for compressible fluids are:

$$V_2 = \left[\frac{2\gamma p_1(1-r^{1-1/\gamma})}{(\gamma-1)\rho_1(1-\beta^4 r^{2/\gamma})}\right]^{1/2} \quad \text{and}$$

$$\dot{M}_I = \frac{A_2 p_1}{T_1^{1/2}}\left[\frac{r^{2/\gamma}(r^{2/\gamma} - r^{1+1/\gamma})}{1-\beta^4 r^{2/\gamma}}\right]^{1/2}\left(\frac{g}{R}\frac{2\gamma}{\gamma-1}\right)^{1/2} \tag{44.35}$$

The equations describing the performance of a real compressible orifice flow, on the other hand, are:

$$\dot{Q}_h = C'\sqrt{h_w p_f} \quad \text{and} \quad C' = F_b F_r Y F_{pg} F_{tb} F_{tf} F_g F_{pv} F_m F_a F_\ell \tag{44.36}$$

The differential pressure across a paddle-type orifice plate is related to the fluid flow rate by

$$\dot{Q} = A_2 C_D\left[\frac{2\Delta p}{\rho(1-\beta^4)}\right]^{1/4} \tag{44.37}$$

The orifice plate may be clamped between pipe flanges. Its main advantages include simplicity and low cost. Disadvantages include its limited capacity and a high permanent head loss due to uncontrolled expansion downstream from the metering element.

Venturi meters can also be used to perform volume or mass flow measurements. They are typically made from castings machine to close tolerances. As a consequence, they are generally bulky, heavy, and expensive.

Defining Terms

Beam translator — Used in LDV systems to adjust the intersection angle by reducing the standard beam distance.

Beam waist adjuster — Used in LDV systems with long focal lengths to optimize the fringe pattern quality in the measuring volume.

Bragg cell — A module in LDV systems capable of providing a positive or negative optical frequency shift of the laser light.

Frequency tracker — A device capable of measuring the instantaneous frequency of the LDV signal. Two types of trackers are used in LDV: the phase-locked loop (PLL) and the frequency-locked loop (FLL).

Frequency counter — A device used in LDV systems capable of measuring the frequency of a signal by accurately timing the duration of an integral number of cycles of the signal.

Non-Newtonian fluids — Fluids in which the coefficient of viscosity is not independent of the velocity gradient.

Photomultiplier tube (PMT) — A device in LDV systems capable of using the photoelectric effect, wherein photons striking a coating of photoemissive material on the photocathode cause electrons to be emitted from the material.

Pinhole translator — Used in LDV backscatter measurements. The device can image the measuring volume on a pinhole, thereby constituting an efficient additional spatial filter, thus eliminating undesirable reflections from window surfaces and walls in the vicinity of the measuring volume.

Signal processor — A device in LDV systems designed to measure the Doppler frequency in addition to any other relevant data coming from the photomultiplier tube (PMT) signal.

Similarity analysis — One of the most powerful tools in fluid mechanics, which permits a wide generalization of experimental results.

Thermal anemometer — A device that measures fluid velocity by sensing the changes in heat transfer from a small electrically heated sensor exposed to the fluid flow.

References

Adrian, R. J. 1983. Laser velocimetry. In *Fluid Mechanics Measurements,* ed. R. J. Goldstein, Hemisphere, Washington, DC. pp. 155–244.

Blake, W. K. 1983. Differential pressure measurement. In *Fluid Mechanics Measurements,* ed. R. J. Goldstein, Hemisphere, Washington, DC. pp. 61–97.

Collis, D. C. and Williams, M. J. 1959. Two-dimensional convection from heated wires at low Reynolds numbers. *J. Fluid Mech.* 6:357–384.

Hinze, J. O. 1959. *Turbulence,* McGraw-Hill, New York.

King, L. V. 1914. On the convection of heat from small cylinders in a stream of fluid: Determination of the convection constants of small platinum wires with applications to hot-wire anemometry. *Phil. Trans. R. Soc. (London) A.* 214:373–432.

Lomas, C. G. 1986. *Fundamentals of Hot Wire Anemometry.* Cambridge University Press, Cambridge, U.K.

Mattingly, G. E. 1983. Volume flow measurements. In *Fluid Mechanics Measurements,* R. J. Goldstein, Ed., Hemisphere, Washington, DC. pp. 245–306.

Pien, P. C. 1958. *Five-Hole Spherical Pitot Tube,* DTMB Report 1229.

Siddall, R. G. and Davies, T. W. 1972. An improved response equation for hot-wire anemometry. *Int. J. Heat Mass Transfer.* 15:367–368.

Van der Hegge Zijnen, B. G. 1956. Modified correlation formulate for the heat transfer by natural and by forced convection from horizontal cylinders. *Appl. Sci. Res. A.* 6:129–140.

Further Information

A good discussion of the principles of hot-wire/film anemometry may be found in *Hot-Wire Anemometry* by A. E. Perry. The author covers all aspects of this measuring technique. A good introduction to the principles of laser Doppler velocimetry can be found in *Laser Doppler Measurements* by B. M. Watrasiewicz and M. J. Rudd. Volume flow measurements include several diverse topics, but the book by H. S. Bean on *Fluid Meters — Their Theory and Applications* may provide a good starting point for the interested reader. Other measurement techniques may be found in *Fluid Mechanics Measurements* by R. J. Goldstein. This book is an excellent reference for a broad spectrum of measurement techniques commonly used in fluid mechanics.

Fluid Measurements

List of Symbols

α	overheat
A	area
b	yaw parameter
c	specific heat
C'	orifice flow constant
C_D, C_d	dimensionless discharge coefficients
c_p	specific heat at constant pressure
c_v	specific heat at constant volume
d	diameter
D	inside pipe diameter
e	fluctuating component of voltage
E	voltage
f	frequency
F	meter factor
F_b	basic orifice factor
F_r	Reynolds-number factor
F_{pb}	pressure-base factor
F_{tb}	temperature-base factor
F_{tf}	flowing-temperature factor
F_g	specific-gravity factor
F_{pv}	supercompressibility factor
F_m	mercury-manometer factor
F_a	orifice thermal-expansion factor
F_ℓ	gauge-location factor
g	acceleration of gravity
h	coefficient of convective heat transfer, height, differential pressure
I	fluctuating component of current
I	electrical current
k	thermal conductivity, yaw factor
K	flow coefficient, flowmeter constant
l	length
$\dot{m}, \dot{M}$	mass rate of flow
n	exponent used in King's law
Nu	Nusselt number
p	steady or time-averaged pressure
Pr	Prandtl number
q	heat transfer rate
$\dot{Q}$	volume flow rate
r	fluctuating component of electrical resistance, radius, recovery factor, correlation coefficient, pressure ratio
R	electrical resistance, gas constant
Re	Reynolds number
S	sensitivity
t	fluctuating component of temperature, time
T	temperature, absolute temperature
u	fluctuating component of velocity
U	x-component of velocity, fluid velocity
V	average fluid velocity in conduit, point fluid velocity
$\dot{W}$	weight flow rate
x	characteristic length, horizontal distance
y	vertical distance
Y	expansion factor
α	temperature coefficient of resistivity, thermal diffusivity, angle of inclination of the velocity vector
β	volume coefficient of expansion, angle of inclination of the velocity vector, ratio of orifice hole to pipe diameter
γ	angle of inclination of the velocity vector, specific heat ratio
Δ	parallel beam separation
θ	yaw angle
μ	absolute viscosity
ν	kinematic viscosity
ρ	density
ρ_r	resistivity
τ	time constant
ϕ	phase angle, pitch angle
ψ	roll angel
ξ	specific weight

Subscripts

1,2	position along conduit, specific meters in same pipe
a	air, orifice thermal expansion
b	bridge, basic
c	cable, corrected, convection, collected
cca	constant-current anemometer
cta	constant-temperature anemometer
eff	effective cooling
f	fluid
F	facility in which meter is tested
g	gas, specific gravity
I	ideal
i	sample number
ℓ	gage location
m	mean, measured, mixture, manometer
M	specific flowmeter
n	arbitrary orthogonal coordinate
o	reference or stagnation conditions
p	probe, constant pressure
s	systematic, static, sensor, streamwise
t	arbitrary orthogonal coordinate, total

T	total, temperature
temp	temperature
v	constant volume
w	wall
∞	free stream conditions

VII

Thermodynamics and Heat Transfer

45

The First Law of Thermodynamics

Richard E. Sonntag
University of Michigan

45.1 System Analysis

The first law of thermodynamics is frequently called the conservation of energy (actually mass-energy), as it represents a compilation of all the energy transfers across the boundary of a thermodynamic system as the system undergoes a process. Energy transfers are of two forms, work and heat. Both work and heat are transient quantities, not possessed by a system, and both are boundary phenomena, that is, observed only in crossing a system boundary. In addition, both are path functions; that is, they are dependent on the path of the process followed by the change in state of the system. Work and heat differ in that work can always be identified as the equivalent of a force acting through a related displacement, while heat is energy transfer due to a temperature gradient or difference, from higher to lower temperature.

One common type of work is that done at a movable boundary, such as shown in Figure 45.1. The boundary-movement work equals the external force times the distance through which it acts, so that for a quasiequilibrium process in which the forces are always balanced, the work for a process between states 1 and 2 is given by the expression

$$ {}_1W_2 = \int_1^2 \delta W = \int_1^2 P\,dV \tag{45.1} $$

such that the work is equal to the area under the curve on the P–V diagram shown in Figure 45.1.

Systems in which other driving forces and their associated displacements are present result in corresponding quasiequilibrium work transfers that are the product of the force and displacement, in a manner equivalent to the compressible-substance boundary movement work of Equation 45.l. For any of these work modes, the quasiequilibrium process is an idealized model of a real process (which occurs at a finite rate because of a finite gradient in the driving force). It nevertheless is a useful model against which to compare the real process.

By convention, heat transfer to the system from its surroundings is taken as positive (heat transfer from the system therefore is negative), and work done by the system on its surroundings is positive (work done on the system is negative). For a process in which the system proceeds from initial state 1 to final state 2, the change in total energy E possessed by the system is given as

0-8493-1586-7/05/$0.00+$1.50

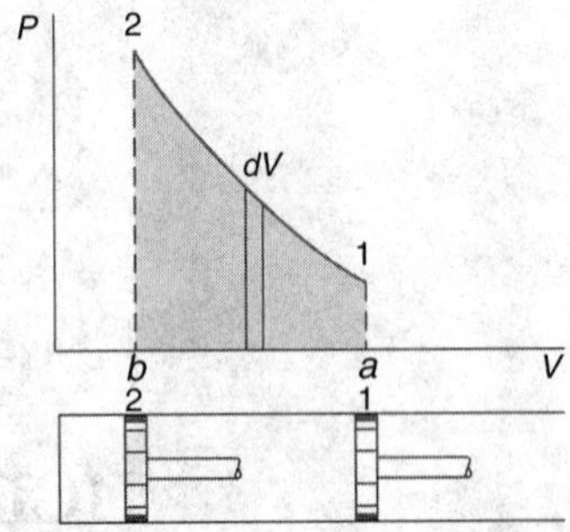

FIGURE 45.1 Boundary movement work. (*Source:* Sonntag, R. E., Borgnakke, C., and van Wylen, G. J. 2003. *Fundamentals of Thermodynamics,* 6th ed., John Wiley & Sons, Inc.)

$$E_2 - E_1 = {}_1Q_2 - {}_1W_2 \tag{45.2}$$

The total energy can be divided into the part that depends only on the thermodynamic state, the internal energy U, and the part that depends on the system's motion (kinetic energy KE) and position with respect to the chosen coordinate frame (potential energy PE), such that at any state,

$$E = U + KE + PE \tag{45.3}$$

In many applications, changes in the system kinetic energy and potential energy are negligibly small by comparison with the other energy terms in the first law.

For a constant-pressure process with boundary movement work and no changes in kinetic and potential energies, the first law can be written as

$$\begin{aligned} {}_1Q_2 &= U_2 - U_1 + P(V_2 - V_1) \\ &= (U_2 + P_2V_2) - (U_1 + P_1V_1) = H_2 - H_1 \end{aligned} \tag{45.4}$$

in which H is the property termed enthalpy.

Consider a process involving a single phase (either solid, liquid, or vapor), with possible boundary-movement work, as given by Equation 45.1. Any heat transferred to the system will then be associated with a temperature change. The specific heat is defined as the amount of heat transfer required to change a unit mass by a unit temperature change. If this process occurs at constant volume, there will be no work, and the heat transfer equals the internal energy change. If the process occurs at constant pressure, then the heat transfer, from Equation 45.4, equals the enthalpy change. Thus, there are two specific heats, which become

$$C_v = \frac{1}{m}\left(\frac{\partial U}{\partial T}\right)_v = \left(\frac{\partial u}{\partial T}\right)_v; \quad C_p = \frac{1}{m}\left(\frac{\partial H}{\partial T}\right)_p = \left(\frac{\partial h}{\partial T}\right)_p \tag{45.5}$$

The specific heat is a property that can be measured, or more precisely, can be calculated from quantities that can be measured. The specific heat can then be used to calculate internal energy or enthalpy changes. In the case of solids or liquids, the energy and enthalpy depend primarily on temperature and not very much on pressure or specific volume. For an ideal gas, a very low-density gas following the equation of state

$$P\,v = R\,T \tag{45.6}$$

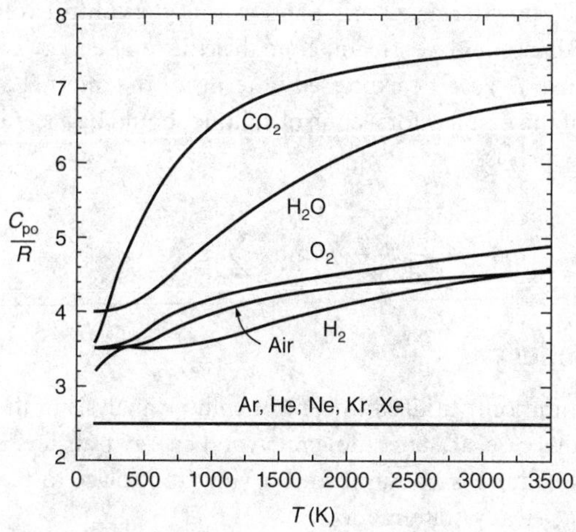

FIGURE 45.2 Ideal gas specific heats. (*Source:* Sonntag, R. E., Borgnakke, C., and van Wylen, G. J. 2003. *Fundamentals of Thermodynamics*, 6th ed., John Wiley & Sons, Inc.)

these properties depend only on temperature, and not at all on P or v. Therefore, the expressions in Equation 45.5 can be used to calculate changes in u or h as

$$u_2 - u_1 = \int_1^2 C_{v0} dT; \quad h_2 - h_1 = \int_1^2 C_{p0} dT \tag{45.7}$$

The subscript 0 is included to denote that these are the specific heats for the ideal gas model. In order to integrate these expressions, it is necessary to know the dependence of specific heat on temperature. These are commonly calculated from statistical thermodynamics and tabulated as functions of temperature. Values for several common gases are shown in Figure 45.2.

For real gases or liquids, internal energy and enthalpy dependency on pressure can be calculated using an equation of state. The changes between phases, for example, between liquid and vapor at the same temperature, can be determined using thermodynamic relations. These real properties can then all be tabulated in tables of thermodynamic properties, many of which exist in the literature.

45.2 Control Volume Analysis

In many thermodynamic applications, it is often convenient to adopt a different perspective concerning the first-law analysis. Such cases involve the analysis of a device or machine through which mass is flowing. It is then appropriate to consider a certain region in space, a control volume, and to analyze the energy being transported across its surfaces by virtue of the mass flows, in addition to the heat and work transfers. Whenever a mass δm_i flows into the control volume during the time interval δt, it is necessarily pushed into it by the mass behind it. Similarly, a mass δm_e flowing out of the control volume during δt has to push other mass out of the way. Both cases involve a local boundary movement work $Pv\,\delta m$, which must be included along with the other energy terms in the first law. For convenience, the Pv term is added to the u term for the mass flow terms, with their sum being the enthalpy. The complete first law for a control volume analysis, represented on a rate basis, is then

$$\dot{Q}_{cv} + \sum \dot{m}_i (h_i + KE_i + PE_i) = \frac{dE_{cv}}{dt} + \sum \dot{m}_e (h_e + KE_e + PE_e) + \dot{W}_{cv} \tag{45.8}$$

The summation signs on the flow terms entering and exiting the control volume are included to allow for the possibility of more than one flow stream. Note that the total energy contained inside the control volume at any instant of time, E_{cv}, can be expressed in terms of the internal energy as in Equation 45.3.

The general expression of the first law for a control volume should be accompanied by the corresponding conservation of mass, which is

$$\frac{dm_{cv}}{dt} + \sum \dot{m}_e - \sum \dot{m}_i = 0 \tag{45.9}$$

The Steady-State Model

Two model processes are commonly utilized in control volume analysis in thermodynamics. The first is the steady-state model. In this case, all states, flow rates, and energy transfers are steady with time. While the state inside the control volume is nonuniform, varying from place to place, it is everywhere steady with time. Therefore, for the steady-state model,

$$\frac{dm_{cv}}{dt} = 0; \quad \frac{dE_{cv}}{dt} = 0 \tag{45.10}$$

these terms drop from Equation 45.8 and Equation 45.9, and everything else is steady, or independent of time. The resulting expressions are very useful in describing the steady long-term operation of a machine or other flow device but of course would not describe the transient startup or shutdown of such a device. The above expressions describing the steady-state model imply that the control volume remains rigid, such that the work rate (or power) term in the first law may include shaft work or electrical work, but not boundary movement work.

There are many common examples of steady-state model applications. One is a heat exchanger, in which a flowing fluid is heated or cooled in a process that is considered to be constant pressure (in the actual case, there will be a small pressure drop due to friction of the flowing fluid at the walls of the pipe). This process may involve a single phase, gas or liquid, or it may involve a change of phase — liquid to vapor in a boiler, or vapor to liquid in a condenser. Another example of a steady-state process is a nozzle, in which a fluid is expanded in a device that is contoured such that the velocity increases as the pressure is dropping. The opposite flow process is a diffuser, in which the device is contoured such that the pressure increases as the velocity is decreasing along the flow path. Still another example is a throttle, in which the fluid flows through a restriction such that the enthalpy remains essentially constant, a conclusion reached because all the other terms in the first law are negligibly small or zero. Note that in all four of these examples of steady-state processes, the flow device includes no moving parts and no work is associated with the process. A turbine, or other flow-expansion machine, is a device built for the purpose of producing a shaft output power, this at the expense of the pressure of the fluid. The opposite device is a compressor (gas) or pump (liquid), the purpose of which is to increase the pressure of a fluid through the input of shaft work.

Several flow devices may be coupled together for a special purpose, such as the heat engine, or power plant. A particular example involving **cogeneration** is shown in Figure 45.3. High-pressure liquid water enters the steam generator, in which the water is boiled and also superheated. The vapor enters the high-pressure turbine, where it is expanded to an intermediate pressure and temperature. At this state, part of the steam is extracted and used for a specific thermal process, such as space heating. The remainder of the steam is expanded in the low-pressure turbine, producing more shaft-power output. Steam exiting the turbine is condensed to liquid, pumped to the intermediate pressure, and mixed with the condensate from the thermal process steam. All the liquid is then pumped back to the high pressure, completing the cycle and returning to the steam generator. Only a small amount of shaft power is required to pump the liquid in the two pumps in comparison to that produced in the turbine. The net difference represents a large useful power output that may be used to drive other devices, such as a generator to produce electrical power.

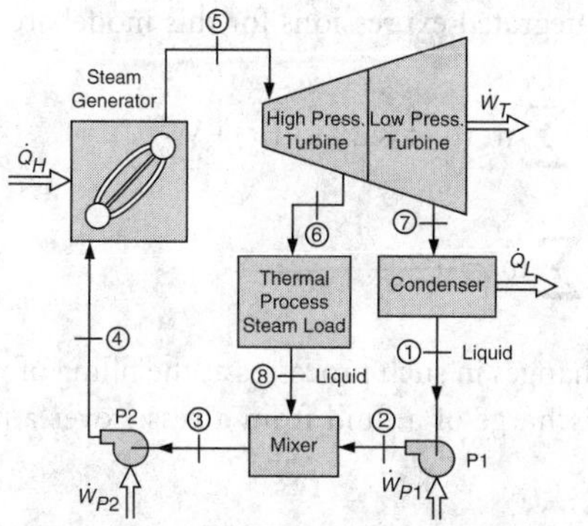

FIGURE 45.3 Cogeneration of steam and power. (*Source:* Sonntag, R. E., Borgnakke, C., and van Wylen, G. J. 2003. *Fundamentals of Thermodynamics,* 6th ed., John Wiley & Sons, Inc.)

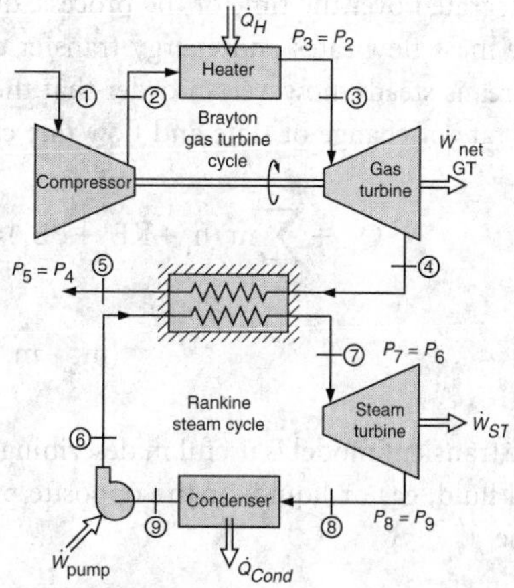

FIGURE 45.4 Combined cycle power system. (*Source:* Sonntag, R. E., Borgnakke, C., and van Wylen, G. J. 2003. *Fundamentals of Thermodynamics,* 6th ed., John Wiley & Sons, Inc.)

To improve overall thermal efficiency of a power plant, and thereby conserve energy resources, two different heat engines may be compounded together, such as in the **combined cycle** shown in Figure 45.4. In this application, the source of heat supply to the steam turbine power plant cycle is the waste heat from a higher-operating temperature gas turbine engine. In the latter, ambient air is compressed to a high pressure in the compressor, after which it enters a burner (represented in the figure as a heater) along with fuel. The combustion products enter the turbine at high temperature and are expanded to ambient pressure, producing a large power output, a portion of which is used to drive the compressor. The products exiting the gas turbine are still at a high enough temperature to serve as the heat source for the steam turbine cycle. The gas turbine cycle may be referred to as a topping cycle for the steam power plant.

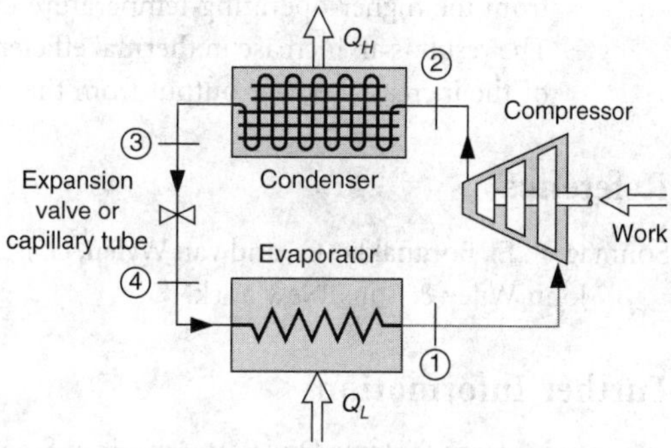

FIGURE 45.5 Refrigeration cycle. (*Source:* Sonntag, R. E., Borgnakke, C., and van Wylen, G. J. 2003. *Fundamentals of Thermodynamics,* 6th ed., John Wiley & Sons, Inc.)

Another common example of coupling several flow devices is the heat pump, or refrigerator, shown in Figure 45.5. Low-temperature vapor enters the compressor at 1 and is compressed to a high pressure and temperature, exiting at 2. This vapor is then condensed to liquid at 3, after which the liquid is throttled to low pressure and temperature. The working fluid, part liquid and part vapor at 4, now enters the evaporator, in which the remaining liquid is boiled, completing the cycle to state 1. When the reason for building this unit is to keep the cold space at a temperature below the ambient, the quantity of interest is $\dot{Q}_L$, and the machine is called a refrigerator. Likewise, when the reason for building the unit is to keep the warm space at a temperature above the ambient, the quantity of interest is $\dot{Q}_H$, and the machine is called a heat pump.

The Transient Model

The second model in common use in control volume analysis in thermodynamics concerns the analysis of unsteady-rate processes. This model is termed the transient model. Equation 45.8 and Equation 45.9

are integrated over the time of the process, during which the state inside the control volume changes, as do the mass flow rates and energy transfer quantities. It is necessary to assume that the state on each flow area is steady, however, in order that the flow terms can be integrated without detailed knowledge of the rate of change of state and flow rate change. The integrated expressions for this model are

$$Q_{cv} + \sum m_i(h_i + KE_i + PE_i) = (E_2 - E_1)_{cv} + \sum m_e(h_e + KE_e + PE_e) + W_{cv} \tag{45.11}$$

$$(m_2 - m_1)_{cv} + \sum m_e - \sum m_i = 0 \tag{45.12}$$

The transient model is useful in describing the overall changes in such processes as the filling of vessels with a fluid, gas or liquid, or the opposite process, the discharge of a fluid from a vessel over a period of time.

Defining Terms

Cogeneration — The generation of steam in a steam generator (boiler) for the dual purpose of producing shaft power output from a turbine and also for use in a specific thermal process load, such as space heating or for a particular industrial process.

Combined Cycle — Combination of heat-engine power cycles for the purpose of utilizing the waste heat from the higher-operating-temperature cycle as the supply of energy to drive the other cycle. The result is an increase in thermal efficiency as compared with the use of a single cycle because of the increased power output from the same heat source input.

Reference

Sonntag, R. E., Borgnakke, C., and van Wylen, G. J. 2003. *Fundamentals of Thermodynamics, Sixth Edition,* John Wiley & Sons, New York.

Further Information

Advanced Energy Systems Division, American Society of Mechanical Engineers, 345 E. 47th Street, New York, NY 10017.

International Journal of Energy Research, published by John Wiley & Sons, New York.

46

Second Law of Thermodynamics and Entropy

Noam Lior
University of Pennsylvania

Thermodynamics is founded on a number of axioms, laws that have not been proven in a general sense but seem to agree with all of the experimental observations of natural phenomena conducted so far. Most commonly, these axioms are formulated under the names of the first, second, third, and zeroth laws of thermodynamics, but other axiomatic definitions have been formulated and shown to be equally successful [see Callen, 1985].

The first law of thermodynamics is a statement of energy conservation, mathematically providing an energy-accounting equation useful in the analysis of **thermodynamic systems** and generally enlightening us about energy conservation and the axiom that, as far as energy is concerned, one cannot get something for nothing. It does not, however, provide guidance about the directions and limitations of the work, heat, and other energy conversion interactions in the process under consideration. For example, the assignment of an amount of heat (positive or negative) in a first law equation is not conditioned on whether that amount of heat is delivered from a low-temperature source to a high-temperature one or vice versa; heat, work, and other forms of energy are treated as having the same thermodynamic quality, and there is no restriction on how much of the heat input could be converted into work as long as overall energy conservation is maintained. The second law provides these types of guidance and thus more strongly distinguishes thermodynamics from its subset branches of physics, such as mechanics.

Several competing statements and practical corollaries of the second law have been developed over the years. They are all consistent with each other. Those that have more rigor and generality are also typically somewhat more obscure to the practitioner, whereas those that are stated in simpler and more practical terms, though still correct, do not encompass all possible processes.

The best known statement of the second law is the Kelvin–Planck one: "It is impossible to construct an engine that, operating in a **cycle**, will produce no effect other than the extraction of heat from a single **reservoir** and the performance of an equivalent amount of work." If such an engine, depicted in Figure 46.1(a), would be feasible, the produced work could then be returned to the reservoir. As a consequence of these operations, the reservoir, originally in equilibrium, is now in a new state, whereas the engine

0-8493-1586-7/05/$0.00+$1.50

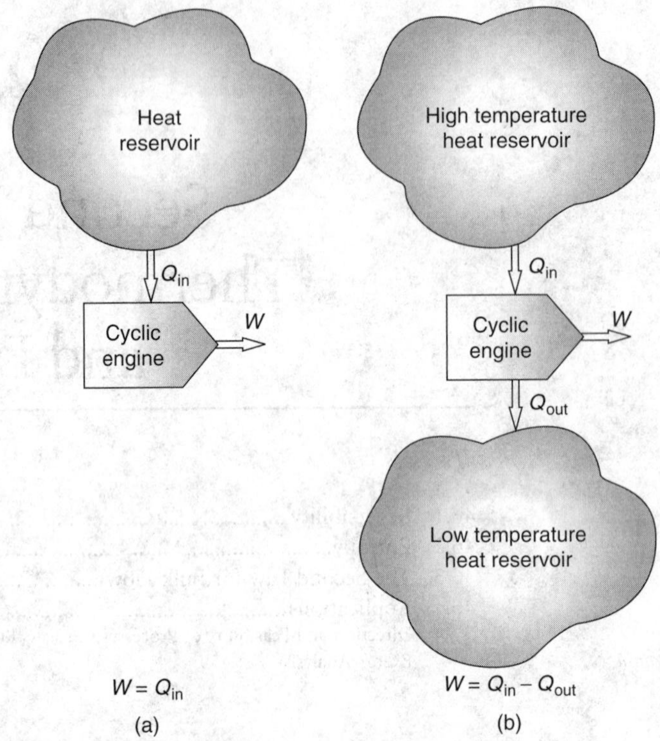

FIGURE 46.1 Comparison of power cycles (a) disallowed and (b) allowed by the second law: (a) a perpetual motion machine of the second kind (PMM2); (b) allowed power cycle.

(being cyclic) and the environment have not changed. Since a system in equilibrium (here the reservoir) can only be disequilibrated as a result of a change in its environment, the only way to satisfy the second law is by denying the possibility of a so-called *perpetual motion machine of the second kind* (PMM2) that, after having been started, produces work continuously without any net investment of energy. Rephrased, the Kelvin–Planck form of the second law thus states that a PMM2 is impossible, and, most notably, that a given quantity of heat cannot fully be converted to work: the heat-to-work conversion efficiency is always lower than 100%. Strikingly, no thermodynamic efficiency restriction exists in the opposite direction of the process: work energy can be fully converted into heat.

A work-producing cyclical engine that does not defy that second law is depicted in Figure 46.1(b). Besides making work, it has an effect of rejecting a portion of the heat gained from the heat source reservoir to a heat sink reservoir. This heat rejection causes an equivalent reduction of the amount of work produced, as dictated by the first law.

Another useful statement of the second law was made by Clausius: "It is impossible to construct a device that operates in a cycle and produces no effect other than the transfer of heat from a region of lower temperature to a region of higher temperature." If such a device, depicted in Figure 46.2(a), would be feasible, it would allow the transfer of heat from colder to warmer regions without any investment of work. Combined with the cycle of Figure 46.1(b), it would then:

1. Allow the operation of a perpetual motion machine
2. Allow the operation of refrigerators that require no work investment, and
3. Defy the empirical laws of heat transfer, which insist that heat is transferred only down the temperature gradient — from hot to cold regions — and not in the opposite direction

A cooling cycle that does not defy the second law is depicted in Figure 46.2(b). In addition to cooling (taking heat from) the lower-temperature reservoir and delivering that amount of heat to the higher-

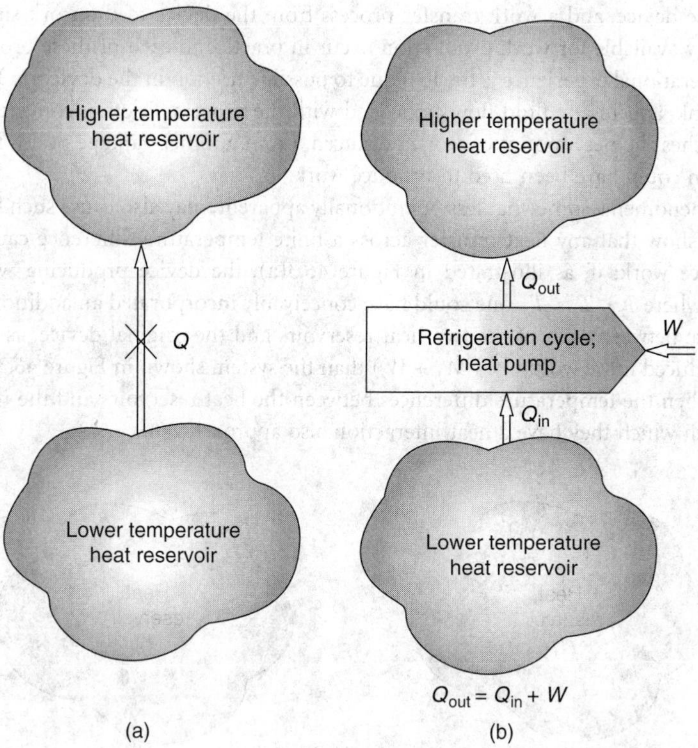

FIGURE 46.2 Comparison of conditions under which heat flow from low temperature to a higher temperature is (a) disallowed and (b) allowed by the second law: (a) heat flow from low to high temperature disallowed by the second law if the system environment does not change; (b) heat flow from low to high temperature allowed by the second law if work is supplied from the environment.

temperature reservoir, it has an effect of requiring some work from the outside, with an equivalent increase in the amount of heat delivered to the high-temperature reservoir, as dictated by the first law.

Two of the most rigorous statements of the second law are given as follows. Carathéodory proposed that "In the neighborhood of a **state** of a system there are states the system cannot reach through **adiabatic** processes" [Hatsopoulos and Keenan, 1965]. Furthermore, "A system having specified allowed states and an upper bound in volume can reach from any given state a **stable state** and leave no net effect on the **environment**" [Hatsopoulos and Keenan, 1965]. As explained already, it can be proven that these statements of the second law are fully consistent with the previous ones, in essence pronouncing axiomatically here that *stable equilibrium states exist.* A rudimentary exposition of the link between this statement and the impossibility of a PMM2 has been given in the preceding, and further clarification is available in the cited references.

46.1 Reversibility

After the disturbing second law statement that heat cannot fully be converted into work, the next logical question concerns the maximal efficiency of such energy conversion. It is obvious from the first law, and even just from operational considerations, that the most efficient heat-to-work conversion process is one that incurs minimal diversions of energy into paths that do not eventually produce work. In a generalized heat-to-work conversion process such as that in Figure 46.1(b), its components are two heat reservoirs at different temperatures, some type of device that produces work when interacting with these two heat reservoirs, and a sink for the produced work. It involves two heat transfer processes (one from the hot reservoir to the device, and one from the device to the cold reservoir), some heat-to-work conversion process inside the device, and a work transfer process from the device to the work sink. Losses in the amount of energy available for work production occur in practice in each of these processes. The most obvious from operational experience is the loss due to possible friction in the device, in the transfer from it to the work-sink, and in any fluid flow associated with the process. Friction converts work into heat and thus diminishes the net amount of work produced. From another vantage point, the force used to overcome friction could have been used to produce work.

Many other phenomena, somewhat less operationally apparent, may also cause such losses. For example, it is easy to show that any heat transfer across a finite temperature difference causes a loss in the ability to produce work: if, as illustrated in Figure 46.3(a), the device producing work W is at the temperature T_d, where $T_c < T_d < T_h$, one could have conceivably incorporated an additional heat-to-work conversion device between each of the two heat reservoirs and the original device, as shown in Figure 46.3(b), and produced more work ($W + W_h + W_c$) than the system shown in Figure 46.3(b). These losses approach zero when the temperature differences between the heat reservoirs and the respective regions of the device with which they have a heat interaction also approach zero.

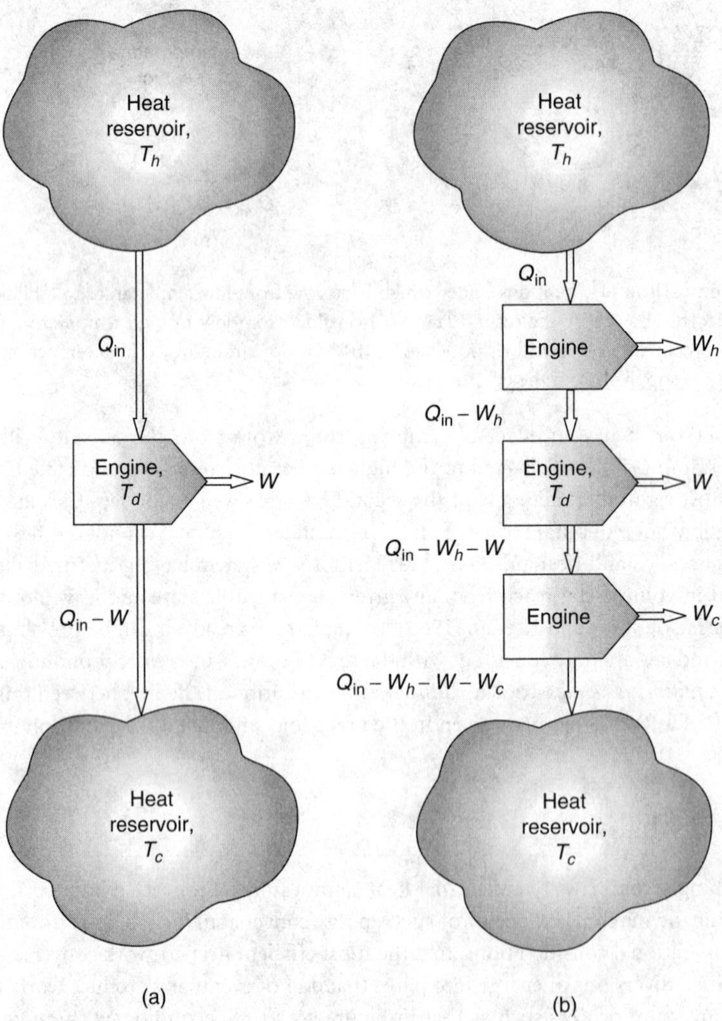

FIGURE 46.3 Power production with reservoir-engine heat transfer across finite temperature differences, $T_c < T_d < T_h$: (a) work produced is only W; (b) work produced is $W + W_h + W_c$.

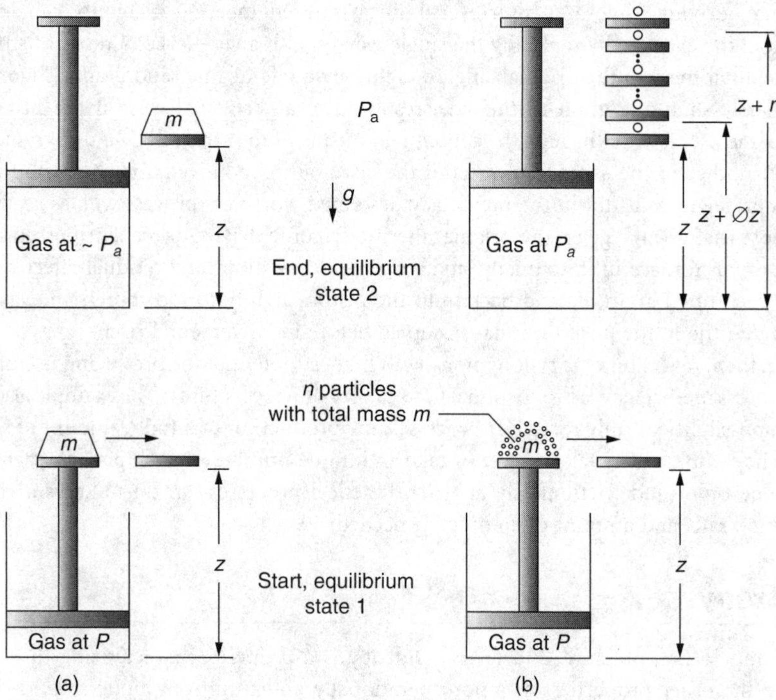

FIGURE 46.4 Illustration of the approach to reversibility: (a) abrupt expansion with no weight raise, no useful work production (highly irreversible process); (b) gradual expansion with weight raise, some useful work production (more reversible process).

Having established that temperature difference (or more strictly, temperature gradient) is a thermodynamic driving potential for work production, one can easily recognize that there are other thermodynamic potentials for work production, such as pressure gradients (piston engines and turbines, for example), concentration gradients (batteries and fuel cells), and electric potential gradients (electric motors). Analogous to the arguments presented for losses due to finite temperature differences, it is easy to recognize that similar losses occur due to phenomena such as unrestrained expansion, the mixing of two different substances, the diffusion of a single substance in a mixture across a finite concentration difference, and the passage of current across an electric potential gradient, all when unaccompanied by commensurate production of work.

It is interesting to note that the work-production losses due to all of these processes approach zero if the respective thermodynamic driving potentials approach zero. Furthermore, processes that proceed toward the new equilibrium state due to the influence of a succession of such infinitesimally small driving potentials can in the limit also be reversed in direction without any residual change in the system and its environment. A good example is a frictionless system in which some amount of gas confined in a cylinder is held at a pressure larger than the ambient pressure by means of a piston loaded down with a weight, as shown in Figure 46.4(a). To simplify the example, let us also assume that the system is originally at ambient temperature and that it is perfectly thermally insulated from the environment. Let us judge the ability of the system to do work by the height that it can lift some weight. If the weight is effortlessly slid to the side (in a direction perpendicular to gravity), the piston would pop up to the new force equilibrium state, under which the gas has expanded to a higher volume and lower pressure and temperature, but since the weight has remained at its original level the system has performed no useful work. The reversal of the process to bring the system *and the environment* to their original condition cannot be done, because work would have to be supplied for that purpose from the environment, thus changing its original state. This is an example of a process at its worst — producing no useful work and characterized by a large (finite) driving force and by irreversibility. Some engineering ingenuity can improve the situation vastly, in the extreme by replacing the single weight with a very large number of small weights having the same total mass as the original single weight, say a pile of fine sand grains [Figure 46.4(b)]. Now we effortlessly slide one grain to the side, resulting in a very small rise of the piston and the production of some useful work through the consequent raising of the sand pile *sans* one grain. Removing one grain after another in the same fashion until the last grain is removed and the piston rests at the same equilibrium height as it did in the previously described worthless process will, as seems obvious, produce the most useful work given the original thermodynamic driving force. Furthermore, since the movements of the piston are infinitesimally small, one could, without any residual effect on the environment, slide each lifted grain of sand back onto the piston slightly to recompress the gas and move the piston down a little to the position it has occupied before its incremental rise.

Generalizing, then, reversibility is synonymous with highest potential for producing useful work, and the degree of process inefficiency is proportional to its degree of irreversibility. The example also illustrates the practical impossibility of fully reversible processes: the production of a finite amount of work at this maximal efficiency would either take forever or take an infinite number of such pistons, each lifting one grain at the same time (and frictionlessly at that). Practical processes are thus always a compromise between efficiency, rate, and amount of equipment needed.

46.2 Entropy

Like volume, temperature, pressure, and energy, entropy is a property characterizing thermodynamic systems. Unlike the other properties, it is not operationally and intuitively understood by novices, a situation not made simpler by the various approaches to its fundamental definition and by its increasing cavalier metaphysical application to fields as disparate as information theory, social science, economics, religion, and philosophy. Let us begin with the fundamental, prosaic definition of entropy, S, as

$$dS = \left(\frac{\delta Q}{T}\right)_{\text{rev}} \tag{46.1}$$

showing that the differential change in entropy, dS, in a process is equal to the ratio of the differential amount of heat interaction, δQ, that takes place if the process is reversible and the absolute temperature, T, during that process. For a finite-extent process between thermodynamic states 1 and 2, Equation (46.1) can be integrated to give

$$S_2 - S_1 = \int_1^2 \left(\frac{\delta Q}{T}\right)_{\text{rev}} \tag{46.2}$$

It is noteworthy here that, since entropy is a property and thus uniquely defined at states 1 and 2, the entropy change between these two states is always the same, whether the process is reversible or irreversible.

Entropy values of various materials are available in the literature: most books on thermodynamics and chemical and mechanical engineering handbooks include values of the entropy as a function of temperature and pressure; an extensive source for gas properties particularly related to high-temperature processes are the JANAF tables [Stull and Prophet, 1971]. The entropy values are given based on a common (but somewhat arbitrary) reference state, and its units are typically kJ/kg K, Btu/lb R, or kJ/kmol K if a molar basis is used. Although not used much in engineering practice, the fundamental reference state of entropy is 0 K, where according to the third law of thermodynamics the entropy tends to 0.

If the process between states 1 and 2 is irreversible, then

$$\left(\frac{\delta Q}{T}\right)_{\text{irrev}} \neq \left(\frac{\delta Q}{T}\right)_{\text{rev}} \tag{46.3}$$

and it can be shown by using the second law that

$$S_2 - S_1 \geq \int_1^2 \left(\frac{\delta Q}{T} \right) \tag{46.4}$$

where the equality sign applies to reversible processes and the inequality to irreversible ones. Consequently, it can also be shown for isolated systems (i.e., those having no interactions with their environment) that

$$dS_{\text{isolated system}} \geq 0 \tag{46.5}$$

where again the equality sign applies to reversible processes and the inequality to irreversible ones. Since a thermodynamic system and its environment together form an isolated system by definition, Equation (46.5) also states that the sum of the entropy changes of the system and the environment is ≥ 0.

In another axiomatic approach to thermodynamics [Callen, 1985], the existence of the entropy property is introduced as an axiom and not as a derived property defined by Equation (46.1). It is basically postulated that, out of all the new equilibrium states that the system may attain at the end of a process, the one that has the maximal entropy will be realized. Equation (46.1) and other more familiar consequences of entropy are then derived as corollaries of this maximum postulate.

Equation (46.5) and the other attributes of entropy lead to several eminent mathematical, physical, and other consequences. Mathematically, the inequality Equation (46.5) indicates that many solutions may exist, which are bounded by the equality limit. In other words, out of all the new states 2 that an isolated system starting from state 1 can reach, only those at which the system entropy is larger are allowed; a unique (but practically unreachable) state 2 can be reached with no entropy change if the process is reversible. Also as a mathematical consequence, the entropy maximum principle allows the establishment of the following equations,

$$dS = 0 \quad \text{and} \quad d^2S < 0 \quad \text{at state 2} \tag{46.6}$$

which then can be used in calculating the nature of the new equilibrium state.

Among the physical consequences, the reversible isentropic process provides a unique limit on process path, which also results in maximal efficiency. Also, the fact that the entropy of real isolated systems always rises provides guidance about which new states may be reached and which may not. Through some operational arguments, but primarily by using statistical mechanics on the molecular level, entropy was shown to be a measure of disorder. This supports the observation that isolated systems undergoing any process become less orderly; mess increases unless an external agent is employed to make order, all consistent with the entropy increase law. Profoundly, entropy is thus regarded to be the scientific indicator of the direction of time — "the arrow of time," after Eddington. The inevitable increase of disorder/entropy with time has led to deeper questions about the origin of the world and indeed of time itself, about the future of the universe in which disorder continuously grows, perhaps to a state of utter disorder (i.e., death), and has raised much philosophical discourse related to these issues. Significant attempts have been made to relate entropy and the second law to almost any human endeavor, including communications, economics, politics, social science, and religion (Bazarov [1964] and Georgescu-Roegen [1971], among many), not always in a scientifically convincing manner.

46.3 The Second Law for Bulk Flow

Based on Equation (46.4), the entropy of a fixed amount of mass (which, by definition, does not give mass to its environment or receive any from it) receiving an amount δQ of heat from a heat reservoir will change as

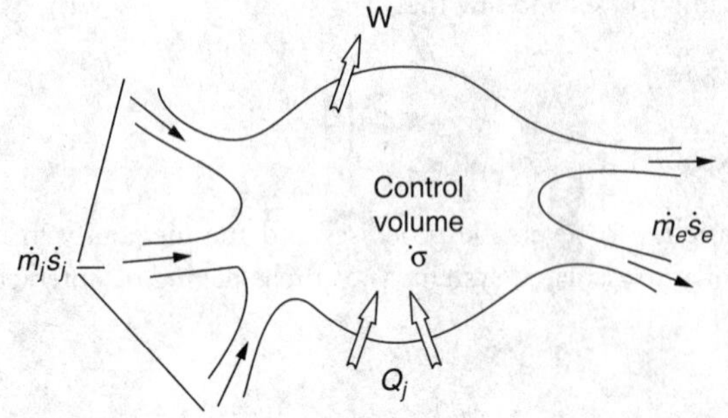

FIGURE 46.5 Entropy accounting for a control volume.

$$dS \geq \frac{\delta Q}{T} \tag{46.7}$$

The amount of entropy produced in the mass due to reversible heat transfer from the reservoir is $\delta Q/T$, and, according to the second law as expressed by Equation (46.4) and Equation (46.5), the amount of entropy produced due to the inevitable internal process irreversibilities, σ, is

$$\sigma \equiv dS - \frac{\delta Q}{T} \geq 0 \tag{46.8}$$

Addressing a transient and more general case, shown in Figure 46.5, the rate of entropy generation due to irreversibilities ($\dot{\sigma}$) in a given control volume that has (1) work, heat ($\dot{Q}$), and diffusion (mass transfer) interactions of k species with its environment, and (2) bulk molar flows ($\dot{N}$, moles/unit time) incoming (subscript i) and exiting (subscript e), can be expressed by simple entropy accounting as

$$\dot{\sigma} = \frac{dS_{cv}}{dt} - \sum_j \frac{\dot{Q}_j}{T_j} - \sum_i \left(\sum_k \dot{N}_k s_k \right)_i + \sum_e \left(\sum_k \dot{N}_k s_k \right)_e - \sum_k (\dot{N}_{k,e} - \dot{N}_{k,i}) s_{o,k} \geq 0 \tag{46.9}$$

where t is time, ds_{cv}/dt is the rate of entropy accumulation in the control volume, subscript j is the number of heat interactions with the environment, s is the specific molar entropy (entropy per mole), and subscript o refers to the conditions of this environment. Equation (46.9) expresses the fact that the entropy of this control volume changes due to heat interactions with its environment, due to the transport of entropy with entering and exiting bulk and diffusional mass flows, and due to internal irreversibilities (work interactions do not change the entropy).

46.4 Applications

Direction and Feasibility of Processes

Compliance with the second law is one of the criteria applied to examine the feasibility of proposed processes and patents. For example, we will examine such compliance for a steady-state combustion process in which it is proposed that methane preheated to 80°C be burned with 20% excess air preheated to 200°C in a leak-tight well-insulated atmospheric-pressure combustor. Since the combustor is isolated from its environment, compliance with the second law will be inspected by calculating the entropy change in the reaction

$$CH_4 + 2.4O_2 + 9.024N_2 \rightarrow CO_2 + 2H_2O + 9.024N_2 + 0.4O_2 \tag{46.10}$$

TABLE 46.1 Analysis of a Methane Combustion Reaction

	Moles n_i	y_i; p_i atm	T K	s kJ/kmol · K	s_o kJ/kmol · K	$h_o = h(T_o, p_o)$ kJ/kmol	$h(T, p)$ kJ/kmol	a_{ch} kJ/kmol	a kJ/kmol	$n_i a$ kJ/kmol CH_4
					Reactant					
CH_4	1	0.0805	353	213.5	186.3	−74,900	−72,854	830,745	824,685	824,685
O_2	2.4	0.1932	473	232.6	205.0	0	5309	−129	−3036	−7286
N_2	9.024	0.7263	473	207.8	191.5	0	5135	−102	176	1585
$\Sigma_{reactants}$	12.424	1.0000								818,984
					Product					
CO_2	1	0.0805	2241	316.2	213.7	−393,800	−286,970	13,855	90,140	90,140
O_2	0.4	0.0322	2241	273.1	205.0	0	68,492	−4568	43,639	17,456
N_2	9.024	0.7263	2241	256.1	191.5	0	64,413	−102	45,060	406,623
H_2O (v)	2	0.1610	2241	270.3	188.7	−242,000	−157,790	4138	64,031	128,062
$\Sigma_{products}$	12.424	1.0000								642,281

Note: Dead state: $p_o = 1$ atm, $T_o = 298$ K; atmospheric composition (molar fractions): $N_2 = 0.7567$, $O_2 = 0.2035$, $H_2O = 0.0303$, $CO_2 = 0.0003$.

where it may be noted that the number of moles of air was adjusted to reflect 20% excess air and that the excess oxygen and all of the nitrogen are assumed to emerge from the combustor unreacted. The entropy change can be expressed by using Equation (46.9), which for this problem is reduced to

$$\sigma = \sum_p n_p s_p - \sum_r n_r s_r \tag{46.11}$$

where the subscripts p and r refer to the reaction products and reactants, respectively, n_i is the number of moles of species i, evident from Equation (46.10), and s_i is the molar entropy of species i, to be found from the literature based on the temperature and partial pressure.

The calculation procedure is outlined in Table 46.1, and further detail can be found in several of the references, such as Howell and Buckius [1992]. The partial pressures of the participating species are easily determined from the molar composition shown in Equation (46.10). The temperature of the reactants is given, and that of the reaction products exiting the combustor, T_p, is calculated from the first law, where the enthalpies of the species, which are primarily dependent on the temperature, are obtained from reference tables or correlations. In this case it is found that $T_p = 2241$ K. Based on this information, the entropies of the species are either found in the literature or calculated from available correlations or gas state equations. The results are listed in Table 46.1, which contains, besides the entropy values, additional information that will be used in another example later.

Application of these results to Equation (46.11) gives

$$\begin{aligned} \sigma &= [1(316.2)+0.4(273.1)+9.024(256.1)+2(270.3)] \\ &\quad -[1(213.5)+2.4(232.6)+9.024(207.8)] \\ &= 630.2 \text{ kJ}/(\text{kmol} \cdot \text{K}) > 0 \end{aligned} \tag{46.12}$$

thus proving compliance of the proposed reaction with the second law and not denying its feasibility. It may be noted that such compliance is a necessary but not always sufficient condition for proving actual process feasibility; additional conditions, such as the existence of a spark in this example, may also be needed.

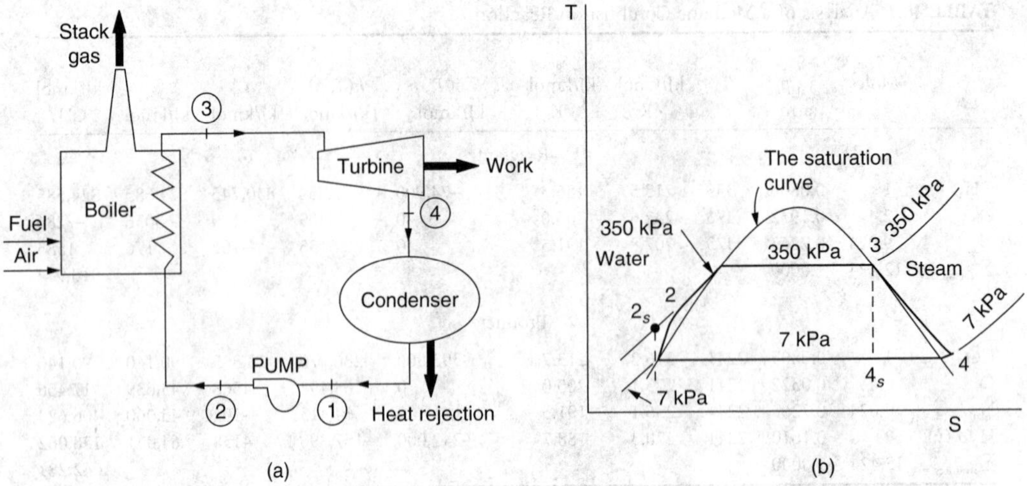

FIGURE 46.6 The Rankine cycle example: (a) the Rankine power cycle; (b) temperature-entropy diagram of the Rankine cycle.

Process Efficiency

Since reversible (isentropic for isolated systems) processes have maximal work-producing efficiency, they are often used as the basis for comparison in the definition of real process efficiencies, defined for work-producing processes (turbines, engines, etc.) as

$$\eta_{is} \equiv \frac{W_p}{W_{p,is}} \tag{46.13}$$

where η_{is} is the so-called "isentropic efficiency," subscript p refers to the work produced, and p, is to the work that would have been produced if the process was isentropic. For work-consuming processes (such as pumps, fans, etc.), the isentropic efficiency is defined as

$$\eta_{is} \equiv \frac{W_{c,is}}{W_c} \tag{46.14}$$

where subscript c refers to the work consumed in the process and c, is to the work that would have been consumed if the process was isentropic.

For example, the isentropic efficiency of the turbine in the simple Rankine cycle shown in Figure 46.6(a) and qualitatively charted on the temperature-entropy (T-S) diagram of Figure 46.6(b), in which superheated steam expands from the pressure $p_3 = 350$ kPa to the condenser pressure $p_4 = 7$ kPa, would be calculated using Equation (46.11) as follows. If the expansion from p_3 to p_4 could have been performed isentropically (i.e., reversibly in this case), the turbine would have produced the maximal amount of work, with the expansion of the steam terminating on the intersection points 4_s of the p_4 isobar and the isentrope descending vertically from point 3, with $s_{4_s} = s_3$. Having thus the values of s_{4_s} and p_4 allows the determination of the **enthalpy**, h_{4_s}, shown in Table 46.2 (only the parameters in bold in this table would be addressed in this example). Assuming that the differences between the inlet-to-exit (states 3 and 4 in Figure 46.6) elevations and kinetic energies of the steam are negligible compared to the respective enthalpy difference and that the turbine does not exchange heat with its environment, the isentropic work $W_{p,is}$ can be calculated from the first law equation for the turbine control volume, reduced to

TABLE 46.2 Data for Figure 46.6

State	p, kPa	T, K	s, kj/kg · K	h, kJ/kg	a, kJ/kg	h, kJ/kg fuel	a, kJ/kg fuel
			The Steam/Water Loop				
1	7	**312.23**	**0.5590**	**163.35**	1.36	2260.4	18.8
2_s	**350**	**312.23**	**0.5590**	**163.70**	1.71	2265.3	23.6
2	**350**	**312.38**	**0.5606**	**163.87**	1.40	2267.6	19.3
3	**350**	**782.75**	**8.2798**	**3506.00**	1042.05	48,515.7	14,419.8
2_s	7	**313.15**	**8.2798**	**2573.50**	109.55	35,611.9	1515.9
4	7	**367.25**	**8.5821**	**2676.10**	122.02	37,031.7	1688.5
			The Fuel, Air, and Stack Gas				
5	101	298.15	HHV = 50,019.00		a_{fuel} = 51,792.00	50,019.0	51,792.0
6	101	298.15	6.6999	299.03	−0.70	6158.1	−14.4
7	101	423.15	7.3388	459.81	79.40	9929.0	1714.5

Note: Dead state: p_o = 1 atm, T_o = 298 K. Atmospheric composition (molar fractions): N_2 = 0.7567, O_2 =0.2035, H_2O = 0.0303, CO_2 = 0.0003.

$$W_{p,\text{is}} = h_3 - h_{4_s} \tag{46.15}$$

In actuality the expansion is irreversible. The second law tells us that the entropy in the process will increase, with the expansion terminating on the same isobar p_4, but at a point 4 where $s_4 > s_3 = s_{4_s}$, which is indeed true in this example, as shown in Table 46.2. The work produced in this process, W_p, is again calculated from the first law:

$$W_p = h_3 - h_4 \tag{46.16}$$

Using Equation (46.13) and the enthalpy values listed in Table 46.2, the isentropic efficiency in this example is

$$\eta_{\text{is}} = \frac{h_3 - h_4}{h_3 - h_{4_s}} = \frac{35060 - 26761}{35060 - 25735} = 0.89 \tag{46.17}$$

Although not shown here, the data in Figure 46.6 include calculations assuming that the isentropic efficiency of the pump is 0.7. State 2_s in Figure 46.6(b) would have been reached if the pumping process starting from state 1 was isentropic; using Equation (46.14) and the given isentropic efficiency of the pump allows the determination of state 2 attained by the real process (note that states 2_s, 2, and 3 are here all on the same 350 kPa isobar), where s_2 is therefore greater than s_1.

Although isentropic efficiencies as defined here have a reasonable rationale and are used widely, important issue has been taken with their fundamental usefulness. Using Figure 46.6(b), it can be argued that a better definition of such efficiency (to be named effectiveness, ε) would be given if the work W_p actually produced, say in the expansion 3–4, $(W_{p,3\text{–}4})_{\text{actual}}$ [the same as the W_p used in Equation (46.14)], was compared with the work that would have been produced if the expansion from 3 to 4 was reversible $[(W_{p,3\text{–}4})_{\text{reversible}}]$,

$$\varepsilon = \frac{(W_{p,3\text{-}4})_{\text{actual}}}{(W_{p,3\text{-}4})_{\text{reversible}}} \tag{46.18}$$

rather than comparing it to work that would have been obtained in the isentropic expansion ending at a state 4_s, which does not actually exist in the real process. An example of a deficiency of the isentropic efficiency (η_{is}) compared with the effectiveness (ε) is that it does not give credit to the fact that the steam exiting the turbine at state 4 has a higher temperature than if it had exited at state 4_s, and thus has the potential to perform more work. This is especially poignant when, as usual, additional stages of reheat and expansion are present in the plant. This reasoning and further comments on effectiveness are given in the following section.

Exergy Analysis

Based on the second law statement that only a fraction of heat energy can be converted into work, a very important application of the second law is the analysis of the potential of energy to perform useful work, that is, the examination of the "quality" of energy. To lay the grounds for such analysis, let us address for simplicity a steady flow and state open-flow system such as depicted in Figure 46.5, neglecting any potential and kinetic energy effects. The amount of work produced by the system is from the first law of thermodynamics,

$$W = \sum_i h_i m_i - \sum_e m_e h_e + \sum_j Q_j \tag{46.19}$$

where h is the specific enthalpy of each incoming or exiting stream of matter m.

As explained earlier, the maximal amount of work would be produced if the process is reversible, in which case entropy generation due to process irreversibilities (σ) is zero (although entropy change of the system due to the heat interactions Q_j at the respective temperatures T_j is nonzero). Multiplying the steady-state form of Equation (46.10) by some constant temperature T (to give it energy units) and subtracting it from Equation (46.19) yields

$$W_{rev} = \left(\sum_i m_i h_i - \sum_e m_e h_e\right) - T\left(\sum_i m_i s_i - \sum_e m_e s_e\right) + \sum_j Q_j \left(1 - \frac{T}{T_j}\right) \tag{46.20}$$

As can be seen from this equation, the reversible work output in the process $i \rightarrow e$ would be further maximized if the temperature at which the heat interaction occurs, T, would be the lowest practically applicable in the considered system, say T_o, yielding an expression for the maximal work output potential of the system, as

$$W_{\max i \rightarrow e} = \sum_i m_i (h_i - T_o s_i) - \sum_e m_e (h_e - T_o s_e) + \sum_j Q_j \left(1 - \frac{T_o}{T_j}\right) \tag{46.21}$$

The term $(h - T_o s)$ appearing at the right side of this equation is thus the measure of a system's potential to perform useful work and therefore of great thermodynamics significance. Composed of thermodynamic properties at a state and of the constant T_o, it is also a thermodynamic property at that state. This term is called the flow *exergy* or flow *availability function*, *b*:

$$b \equiv h - T_o s \tag{46.22}$$

Further examination of Equation (46.22) shows that a system at state *i* would produce the maximal useful work when the new equilibrium state *e* is identical to the ambient conditions *o*, since at that state all the driving forces of the system — such as temperature, concentration, and pressure differences — are zero and the system cannot by itself produce any more useful work. The maximal work output, assuming mass conservation, is then

$$W_{\max,i\to o} = \sum_i m_i[(h_i - h_o) - T_o(s_i - s_o)] \tag{46.23}$$

The term $[(h - h_o) - T_o(s - s_o)]$ is also a property and is the measure of a system's potential to perform useful work between any given state and the so-called "dead state," at which the system can undergo no further spontaneous processes. This term is called *flow exergy* or *flow availability, a*:

$$a \equiv (h - h_o) - T_o(s - s_o) \tag{46.24}$$

Since enthalpy is the measure of the energy in flow systems, examination of Equation (46.21) through Equation (46.24) shows clearly that the portion of the energy h that cannot be converted to useful work is the product $T_o s$.

The last term on the right side of Equation (46.21) is the exergy of the heat sources Q_j, at the respective temperatures T_j, exchanging heat with the considered thermodynamic system. The form of this term is the Carnot cycle work output between T_j and the dead state temperature T_o, which indeed would produce the maximal work and is thus also the exergy of these heat sources by definition. Shaft and other mechanical power, and electric power, are pure exergy.

Turning now to real, irreversible steady state processes, their exergy accounting equation — developed using Equation (46.9), Equation (46.21), and Equation (46.24) — is

$$W = \sum_j Q_j\left(1 - \frac{T_o}{T_j}\right) + \sum_i m_i a_i - \sum_e m_e a_e - T_o\sigma \tag{46.25}$$

where the work output of the process in this control volume is, in the order of terms on the right side of the equation, produced due to (1) heat interactions Q_j at the respective temperatures T_j, with the control volume, and (2) and (3) the difference between the exergies flowing in and exiting the control volume, and is diminished by (4) the entropy generation due to process irreversibilities. This last term, $T_o\sigma$, amounts to the difference between the maximal work that could have been produced in a reversible process [Equation (46.21) and Equation (46.24)] and the amount produced in the actual irreversible process [Equation (46.25)]. It is called the *irreversibility* (*I*) or *lost work* of the process.

Beyond exergy changes due to temperature and pressure driving forces, multicomponent systems also experience exergy changes due to component mixing, phase change, and chemical reactions. It was found convenient to separate the exergy expression into its "physical" (a_{ph}) and "chemical" (a_{ch}) constituents,

$$a = a_{\text{ph}} + a_{\text{ch}} \tag{46.26}$$

where the *physical exergy* is the maximal work obtainable in a reversible physical process by a system initially at T and p and finally at the dead state T_o, p_o, and the *chemical exergy* is the maximal work obtainable from a system at dead state conditions T_o, p_o, initially at some species composition and finally at the dead state composition (such as the datum level composition of the environment). The total flow exergy, showing the thermal, mechanical, and chemical flow exergy components (segregated by the double brackets) is

$$\begin{aligned} a = {} & [[(h - h_o) - T_o(s - s_o)]]\text{(thermal)} + \left[\left[\frac{\mathbf{v}^2}{2} + gz\right]\right]\text{(mechanical)} \\ & + \left[\left[\sum_k x_k(\mu_k^\circ - \mu_{o,k})\right]\right]\text{(chemical)} \end{aligned} \tag{46.27}$$

where $\mathbf{v}$ is the flow velocity, g is the gravitational acceleration, z is the flow elevation above a zero reference level, x_k is the molar fraction of species k in the mixture, μ_k° is the **chemical potential** of species k at T_o and p_o, and $\mu_{k,o}$ is the chemical potential of species k at the system dead state defined by T_o, p_o, and the dead state composition.

The general transient exergy accounting equation that includes both physical and chemical exergy is

$$\frac{dA_{cv}}{dt}\text{(rate of exergy storage)} = \sum_j \left(1-\frac{T_o}{T_j}\right)\dot{Q}_j - \left(\dot{W}_{cv} - p_o\frac{dV_{cv}}{dt}\right) + \sum_i \dot{m}_i a_i$$

$$-\sum_e \dot{m}_e a_e \text{(rates of exergy transfer)} \tag{46.28}$$

$$-\dot{I}_{cv}\text{(rate of exergy destruction)}$$

where A_{cv} is the exergy of the control volume, $\dot{W}_{cv}$ is the work performed by the control volume, $p_o(dV_{cv}/dt)$ is the work due to the transient volume change of the control volume, a_i and a_e are the total flow exergies composed of both the physical and chemical components, and $\dot{I}_{cv}$ $(\equiv T_o\dot{\sigma})$ is the irreversibility, that is, the rate of exergy destruction. A more detailed breakdown of exergy, in differential equation form, is given in Dunbar et al. [1992].

For a *closed system* (i.e., one that does not exchange mass with its environment but can have work and heat interactions with it) at a state defined by the specific internal energy u, specific volume v, and specific entropy s, the expression for the exergy, a°, is

$$a^\circ = \left(u + \frac{\mathbf{v}^2}{2} + gz - u_o\right) + p_o(v - v_o) - T_o(s - s_o) \tag{46.29}$$

Selection of the dead state for evaluating exergy is based on the specific application considered. Thus, for example, the dead state for the analysis of a water-cooled power plant operating at a certain geographic location should consist of the cooling water temperature and the ambient atmospheric pressure and composition, all at that location. Much work has been performed in defining "universal" dead states for processes and materials [Szargut et al., 1988].

Example — Exergy Analysis of a Power Plant

To demonstrate the calculation procedure and the benefits of exergy analysis, we will perform such an analysis on the simple Rankine cycle described in Figure 46.6. The fuel used in the boiler is methane, undergoing the same combustion reaction as described in Equation (46.10). Both fuel and air enter the boiler at 25°C, 1 atm, and the combustion products exit the stack at 150°C.

Addressing the steam/water (single component) Rankine cycle loop first, the values of p, T, s, and h are already available from the example given and are listed in Table 46.2. Given the dead state conditions listed under the table, the flow exergy (a) per kg of the water and steam is calculated from Equation (46.24) and listed in the table column to the left of the vertical dividing line.

The three bottom rows of the table list the properties of the inflowing fuel and air and of the stack exhaust gas. The enthalpy of the fuel is its higher heating value (HHV), which, along with exergy, is obtained from fuel property tables [Szargut et al., 1988; Howell and Buckius, 1992; Moran and Shapiro, 1992]. The specific flow exergy values of the air and stack gas are calculated from Equation (46.28) (with negligible kinetic and potential energy components) and listed to the left of the vertical dividing line.

It is sensible to analyze the power system based on the fuel energy and exergy input, in other words, to determine how this fuel input is distributed among the different system components and pro-

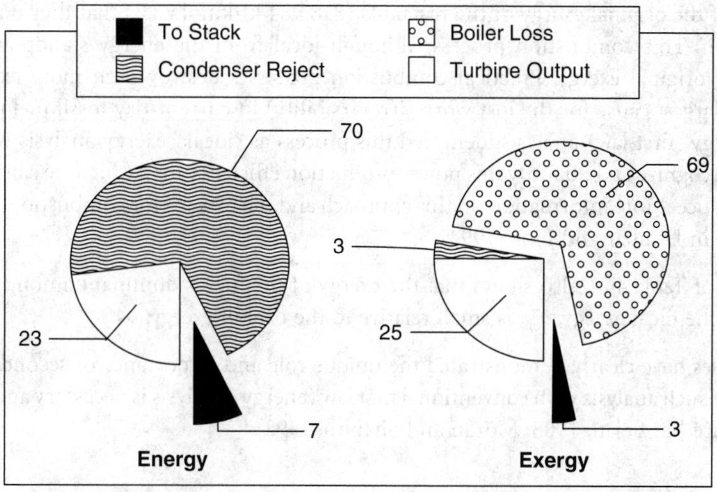

FIGURE 46.7 Energy and exergy breakdown for the Rankine power cycle example.

cesses. To that end we determine, per unit mass of fuel, the mass of cycle steam [here found to be 13.8 (kg steam)/(kg fuel) by applying the energy conservation equation to the boiler] and the mass of air and stack gas [20.6 (kg air)/(kg fuel) and 21.6 (kg exhaust gas)/(kg fuel) by applying Equation (46.10)]. These results are listed in the last two columns of the table.

Examination of the results is encouraged, and some of the most important ones, which describe the fuel energy and exergy distribution among the different system components/processes, are shown in Figure 46.7 *Energy* analysis indicates that the major energy loss, 70%, is due to the heat rejected in the condenser. Examination of the *exergy* analysis chart shows, however, that this large energy loss amounts to only 3% of the fuel exergy, and even complete elimination of the condenser heat rejection (if it were at all feasible) would increase the cycle efficiency by only three percentage points. Of course, this is because the heat rejected in the condenser is at a low temperature, only slightly elevated above that of the ambient, and thus has commensurately little potential to perform work despite its large energy. Another very significant difference is the fact that the exergy analysis identifies the major losses, 69%, to be in the boiler, due to the combustion and gas-to-steam/water heat transfer processes, whereas the energy analysis associates no loss to these processes. Finally, the exergy analysis attributes much less loss to the stack gas than the energy analysis does. Notably, the turbine output has almost the same percentage in both analyses, which is because the output is shaft power (i.e., pure exergy), and the HHV and exergy of the fuel are nearly identical. In conclusion, it is only exergy analysis that can correctly identify and evaluate the losses (irreversibilities) that diminish the ability of processes to perform useful work.

Example — Exergy Analysis of a Combustion Process

The methane combustor analyzed earlier, for which the data are given in Table 46.1, has a 100% energy efficiency because its enclosure is adiabatic and impermeable, as can also be ascertained by performing a first law energy conservation check using the data in Table 46.1. The intuitive implication of this result is that the combustor is thus perhaps "ideal" and requires no further technological improvement. It is interesting therefore to examine the effect that it has on the potential of the invested consumables (i.e., the fuel and other reactants) on producing useful work. The flow exergy a and the chemical exergy a_{ch} are calculated by using Equation (46.27), and property values from gas tables or correlations given in the literature, with the dead state (including the atmospheric composition) described at the bottom of Table 46.1. The calculation results are summarized in Table 46.1, showing, first, that the total exergy of the combustion products (642 281 kJ/mmol CH_4) is only 78.4% of the original exergy of the reactants (818 984 kJ/kmol CH_4) that they possessed prior to combustion. This combustion process, although ideal from the energy standpoint, destroyed 21.6% of the original exergy. Practical combustion processes destroy even more exergy and are typically the largest cause for the lost work (irreversibility) in current-day fossil fuel power plants. Whereas energy (first law) analysis identified this process as "ideal," exergy analysis was unique in its ability to recognize and quantify this power production efficiency loss, which we can subsequently attempt to reduce. More information on this approach and the reasons for combustion irreversibility can be found in Dunbar and Lior [1994].

Examination of Table 46.1 also shows that the exergy of the fuel is dominant among the reactants and that the chemical exergy, a_{ch}, is small relative to the overall exergy.

These examples have clearly demonstrated the unique role and importance of second law analysis. Integration of such analysis with conventional first law (energy) analysis is necessary and increasingly seen in practice [cf. Bejan, 1988; Moran and Shapiro, 1992].

Defining Terms

Adiabatic — A process or surface allowing only work interactions. Since most introductory thermodynamics texts consider only work and heat interactions, *adiabatic* is most often interpreted as a process or surface allowing no heat interactions.

Chemical potential — A potential given by $\mu_k \equiv h_k - Ts_k$, where h_k and s_k are, respectively, the enthalpy and entropy of the species k, and T is the temperature.

Cycle — A series of processes that bring a thermodynamic system back to its original state.

Enthalpy — Given by $h \equiv u + pv$, where u is the internal energy, p is the pressure, and v is the specific volume.

Environment — The environment of a thermodynamic system consists of everything outside of that system that could conceivably have some influence on it.

Equilibrium — A state in which all of the properties of a system that is not subjected to interactions do not change with time. Equilibrium states may be of various types: stable, neutral, metastable, and unstable.

Heat reservoir — A system in a stable equilibrium state such that, when subjected to finite heat interactions, its temperature remains constant. Although this is just a concept for simplifying the exposition of thermodynamics, the atmosphere, oceans, lakes, and rivers were often considered as practical examples of such a reservoir. It is noteworthy that the large and ever-increasing magnitudes of human-made heat interactions with these natural elements now invalidate their definition as "heat reservoirs" in this thermodynamic sense, because these interactions in fact change their temperature and cause increasingly adverse environmental effects.

Stable state — A system is in a stable state (or stable equilibrium state) if a finite change of state of the system cannot occur without leaving a corresponding finite change in the state of the environment. In other words a finite external influence would be needed to budge a system out of its stable state; small natural pulses and fluctuations would not suffice.

State — The state of a thermodynamic system is defined by all of the properties of the system, such as pressure, temperature, volume, composition, and so forth. A system returns to the same state when all the properties attain their original values.

Thermodynamic system — A thermodynamic system is whatever we enclose for the purpose of a study by a well-defined surface, which may be either material or imaginary and which can be isolated from everything else (i.e., from the "environment").

References

Bazarov, I. P. 1964. *Thermodynamics*, Pergamon Press, Oxford, England.

Bejan, A. 1988. *Advanced Engineering Thermodynamics,* John Wiley & Sons, New York.

Callen, H. B. 1985. *Thermodynamics and an Introduction to Thermostatistics,* 2nd ed. John Wiley & Sons, New York.

Dunbar, W. R., Lior, N., and Gaggioli, R. 1992. The component equations of energy and exergy. *ASME J. Energy Resour. Technol.* (114):75–83.

Dunbar, W. R. and Lior, N. 1994. Sources of combustion irreversibility. *Comb. Sci. Technol.* 103:41–61.

Georgescu-Roegen, N. 1971. *The Entropy Law and the Economic Process,* Harvard University Press, Cambridge, MA.

Hatsopoulos, G. N. and Keenan, J. H. 1965. *Principles of General Thermodynamics,* John Wiley & Sons, New York.

Howell, J. R. and Buckius, R. O. 1992. *Fundamentals of Engineering Thermodynamics,* McGraw-Hill, New York.

Moran, M. J. and Shapiro, H. N. 1992. *Fundamentals of Engineering Thermodynamics,* John Wiley & Sons, New York.

Stull, D. R. and Prophet, H. 1971. *JANAF Thermochemical Tables,* 2nd ed. NSRDS-NBS 37. National Bureau of Standards, Washington, DC.

Szargut, J., Morris, D. R., and Steward, F. R. 1988. *Exergy Analysis of Thermal, Chemical and Metallurgical Processes,* Hemisphere, New York.

Further Information

Anderson, E. E. *Thermodynamics,* 1994. PWS, Boston, MA. Integrates first and second law analysis.

Sonntag, R. E. and Van Wylen, G. J. *Introduction to Thermodynamics Classical and Statistical,* John Wiley & Sons, New York. Widely used textbook.

47

The Thermodynamics of Solutions

Stanley I. Sandler
University of Delaware

Hasan Orbey
Deceased

It is important to know the thermodynamic properties of mixtures to be able to predict phase behavior such as vapor–liquid, liquid–liquid, vapor–liquid–liquid, and solid–liquid equilibria necessary to design separation and purification processes, to predict when a mixture will boil or will precipitate, and to estimate the equilibrium extents of homogeneous and heterogeneous chemical reactions. Also, mixture enthalpies are needed to design heat exchange equipment and entropies for compressor calculations.

47.1 Fundamentals

There are a collection of thermodynamic properties that describe a pure fluid such as the state variables temperature T and pressure P, and the system-intensive variables, specific volume (that is, the volume per mole of substance) v, specific internal energy u, specific enthalpy h, specific entropy s, specific Helmholtz energy a, and specific Gibbs energy g. The thermodynamics of mixtures would be relatively simple if, for any specific pure fluid thermodynamic variable Θ, the corresponding mixture property Θ_{mix} at the same temperature and pressure was

$$\Theta_{mix}(T,P,x_i) = \sum_i x_i \Theta_i(T,P) \tag{47.1}$$

where the subscripts i and mix indicate the property of pure component i and the mixture, respectively, and x_i is the mole fraction of species i. However, Equation (47.1) is not generally valid for several reasons. First, for the entropy and properties that depend on the entropy (such as the Gibbs energy $g = h - Ts$ and the Helmholtz energy $a = u - Ts$) an additional term arises because at constant temperature and pressure the volume available to a molecule of species i changes in the mixing process from the pure component volume $n_i v_i$ to the total volume of the mixture $n_t v_{mix}$. For a mixture of ideal gases $v_i = v_{mix} = RT/P$, and the additional term that appears is as follows:

0-8493-1586-7/05/$0.00+$1.50

$$s_{\text{mix}}^{IG}(T,P,x_i)=\sum_i x_i s_i^{IG}(T,P)-R\sum_i x_i \ln x_i$$

$$g_{\text{mix}}^{IG}(T,P,x_i)=\sum_i x_i g_i^{IG}(T,P)+RT\sum_i x_i \ln x_i \tag{47.2}$$

$$a_{\text{mix}}^{IG}(T,P,x_i)=\sum_i x_i a_i^{IG}(T,P)+RT\sum_i x_i \ln x_i$$

The simplest liquid mixture is an ideal mixture (denoted by the superscript *IM*) for which Equation (47.1) is valid for the internal energy and volume *at all temperatures and pressures,* that is,

$$v_{\text{mix}}^{IM}(T,P,x_i)=\sum_i x_i v_i\,(T,P)$$

$$h_{\text{mix}}^{IM}(T,P,x_i)=\sum_i x_i h_i\,(T,P) \tag{47.3}$$

In this case, one can show [Sandler, 1999] that for an ideal mixture

$$s_{\text{mix}}^{IM}(T,P,x_i)=\sum_i x_i s_i\,(T,P)-R\sum_i x_i \ln x_i$$

$$g_{\text{mix}}^{IM}(T,P,x_i)=\sum_i x_i g_i\,(T,P)+RT\sum_i x_i \ln x_i \tag{47.4}$$

$$a_{\text{mix}}^{IM}(T,P,x_i)=\sum_i x_i a_i\,(T,P)+RT\sum_i x_i \ln x_i$$

Although Equation (47.4) is similar to Equation (47.2), there is an important difference. Here, the pure component and mixture properties are those of the real liquid at the temperature and pressure of the mixture, not those of ideal gases.

Equation (47.2) may not apply to real gases because they are not ideal gases. Also, Equation (47.4) does not apply to most liquid mixtures since Equation (47.3) is not, in general, valid. Another difficulty that arises is that at the temperature and pressure of interest, the pure components may not exist in the same state of aggregation as the mixture. An example of this is a mixture containing a dissolved gas or dissolved solid in a liquid.

To describe the thermodynamic behavior of a condensed phase (liquid or solid), either an excess property model or equation-of-state model may be used, depending on the mixture. For a vapor mixture, an equation of state is generally used. Both these descriptions are reviewed below.

Real Liquid Mixtures — Excess Property Description

For a liquid mixture formed by mixing pure liquids at constant temperature and pressure the following description is used:

$$\Theta_{\text{mix}}(T,P,x_i)=\Theta_{\text{mix}}^{IM}(T,P,x_i)+\Theta^{EX}(T,P,x_i) \tag{47.5}$$

Here, $\Theta^{EX}(T, P, x_i)$ is the additional change in the thermodynamic property when mixing the pure components above that on forming an ideal mixture. This excess thermodynamic property change on mixing is a function of temperature, pressure, and mixture composition. Expressions for several thermodynamic properties of real mixtures are given below:

$$
\begin{aligned}
v_{\text{mix}}(T,P,x_i) &= \sum_i x_i v_i(T,P) + v^{EX}(T,P,x_i) \\
u_{\text{mix}}(T,P,x_i) &= \sum_i x_i u_i(T,P) + u^{EX}(T,P,x_i) \\
s_{\text{mix}}(T,P,x_i) &= \sum_i x_i s_i(T,P) - R\sum_i x_i \ln x_i + s^{EX}(T,P,x_i) \\
g_{\text{mix}}(T,P,x_i) &= \sum_i x_i g_i(T,P) + RT\sum_i x_i \ln x_i + g^{EX}(T,P,x_i) \\
a_{\text{mix}}(T,P,x_i) &= \sum_i x_i a_i(T,P) + RT\sum_i x_i \ln x_i + a^{EX}(T,P,x_i)
\end{aligned} \tag{47.6}
$$

An important concept in mixture solution thermodynamics is the partial molar property $\overline{\Theta}_i(T,P,x_i)$, defined as follows:

$$
\overline{\Theta}_i(T,P,x_i) = \left(\frac{\partial n_i \Theta_{\text{mix}}}{\partial n_i}\right)_{T,P,n_j \neq i} \tag{47.7}
$$

Any solution property that is linear and homogeneous in the number of moles is related to its partial molar properties as follows [Van Ness, 1964]:

$$
\Theta_{\text{mix}}(T,P,x_i) = \sum_i x_i \overline{\Theta}_i(T,P,x_i) \tag{47.8}
$$

Also, from Equation (47.6),

$$
\overline{\Theta}_i(T,P,x_i) = \overline{\Theta}_i^{IM}(T,P,x_i) + \overline{\Theta}_i^{EX}(T,P,x_i) \tag{47.9}
$$

From this we have, as examples,

$$
\begin{aligned}
\overline{v}_i(T,P,x_i) &= v_i(T,P) + \overline{v}_i^{EX}(T,P,x_i) \\
\overline{u}_i(T,P,x_i) &= u_i(T,P) + \overline{u}_i^{EX}(T,P,x_i) \\
\overline{s}_i(T,P,x_i) &= s_i(T,P) - R\ln x_i + \overline{s}_i^{EX}(T,P,x_i) \\
\overline{g}_i(T,P,x_i) &= g_i(T,P) + RT\ln x_i + \overline{g}_i^{EX}(T,P,x_i)
\end{aligned} \tag{47.10}
$$

By definition, all excess properties must vanish in the limit of a pure component. Generally, a species excess partial molar property has its largest value when that species is at infinite dilution.

Several other thermodynamic variables of special interest are the chemical potential, μ_i,

$$
\begin{aligned}
\mu_i(T,P,x_i) &= \mu_i^o(T,P) + RT\ln x_i(T,P,x_i) + \overline{g}_i^{EX}(T,P,x_i) \\
&= \mu_i^o(T,P) + RT\ln x_i\gamma_i(T,P,x_i)
\end{aligned} \tag{47.11}
$$

the activity, a_i,

$$a_i(T,P,x_i) = \exp\left(\frac{\mu_i - \mu_i^o}{RT}\right) \tag{47.12}$$

the activity coefficient, γ_i,

$$\gamma_i(T,P,x_i) = \exp\left(\frac{\bar{g}_i^{EX}(T,P,x_i)}{RT}\right) \tag{47.13}$$

the fugacity, f_i,

$$RT\ln\left(\frac{\hat{f}_i(T,P,x_i)}{\hat{f}_i^o(T,P)}\right) = \mu_i(T,P,x_i) - \mu_i^o(T,P) \tag{47.14}$$

and the fugacity coefficient, $\hat{\phi}_i$,

$$\hat{\phi}_i(T,P,x_i) = \frac{\hat{f}_i}{x_i P} \tag{47.15}$$

In the definitions above, μ_i^o is the chemical potential of pure species i at the temperature, pressure, and state of aggregation of the mixture, and $\hat{f}_i^o$ is its fugacity in that state; this pure component state is referred to as the standard state for the component. Also, $\gamma_i(x_i \to 1) = 1$. If the pure species does not exist in the same state of aggregation as the mixture, as with a gas or solid dissolved in a liquid solvent, then other standard states are chosen. For example, if a hypothetical pure component state is chosen as the standard state with properties extrapolated based on the behavior of the component at infinite dilution, then

$$\mu_i(T,P,x_i) = \mu_i^*(T,P) + RT\ln(x_i\gamma_i^*) \tag{47.16}$$

where μ_i^* is the chemical potential in this so-called Henry's law standard state and γ_i^* is an activity coefficient defined so that $\gamma_i^*(T,P,x_i \to 0) = 1$. Another choice for the standard state of such components is the hypothetical ideal 1 molal solution, which is commonly used for electrolytes and other compounds.

The excess properties have generally been determined by experiment and then fitted to algebraic equations. These equations must satisfy the boundary condition of vanishing in the limit of a mixture going to a pure component. That is, in a binary mixture, $\Theta^{EX}(T,P,x_i)$ must go to zero as either $x_1 \to 1$ or $x_2 \to 1$. One solution model of historical interest is the strictly regular solution model, for which $u^{EX} = g^{EX} = x_1x_2\omega$ and $s^{EX} = 0$, where, from simple molecular theory, ω is the exchange energy that is the net difference on interchanging molecules between the pure species and the mixture. Another model is the athermal solution model, for which there is no excess internal energy change of mixing — that is, $u^{EX} = 0$ — but there is an excess entropy change so that $g^{EX} = -Ts^{EX}$. Most excess property models currently in use are a combination of an athermal part to account for the entropy change on mixing and a modified regular solution part to account for energy change on mixing. Several excess property models are listed in Table 47.1. There are also excess property models that are predictive (instead of merely correlative), such as the UNIFAC and ASOG models. These are described in detail elsewhere [Fredenslund et al. 1977; Sandler, 1999].

Because the electrostatic interactions between charged particles are so much stronger and longer range than between neutral molecules, electrolyte solutions are very nonideal. Consequently, specialized models, generally based on generalized Debye–Hückel theory, are used for electrolyte solutions [Sandler, 1999].

TABLE 47.1 Excess Gibbs Energy and Activity Coefficient Models

The two-constant Margules equation

$$g^{EX} = x_1 x_2 \{A + B(x_1 - x_2)\}$$

which leads to

$$RT \ln\gamma_1 = \alpha_1 x_2^2 + \beta_1 x_2^3 \text{ and } RT \ln\gamma_2 = \alpha_2 x_1^2 + \beta_2 x_1^3$$

with $\alpha_i = A + 3(-1)^{i+1} B$ and $\beta_i = 4(-1)^i B$.

The van Laar model

$$g^{EX} = x_1 x_2 \frac{2aq_1 q_2}{x_1 q_1 + x_2 q_2}$$

which leads to

$$\ln\gamma_1 = \frac{\alpha}{\left[1 + \dfrac{\alpha x_1}{\beta x_2}\right]^2} \text{ and } \ln\gamma_2 = \frac{\beta}{\left[1 + \dfrac{\beta x_2}{\alpha x_1}\right]^2}$$

with $\alpha = 2q_1 a$ and $\beta = 2q_2 a$.

The nonrandom two-liquid (NRTL) model

$$\frac{g^{EX}}{RT} = x_1 x_2 \left(\frac{\tau_{21} G_{21}}{x_1 + x_2 G_{21}} + \frac{\tau_{12} G_{12}}{x_2 + x_1 G_{12}}\right)$$

which leads to

$$\ln\gamma_1 = x_2^2 \left[\tau_{21}\left(\frac{G_{21}}{x_1 + x_2 G_{21}}\right)^2 + \frac{\tau_{12} G_{12}}{(x_2 + x_1 G_{12})^2}\right]$$

and

$$\ln\gamma_2 = x_1^2 \left[\tau_{12}\left(\frac{G_{12}}{x_1 G_{12} + x_2}\right)^2 + \frac{\tau_{21} G_{21}}{(x_1 + x_2 G_{12})^2}\right]$$

with $\ln G_{12} = -\alpha\tau_{12}$ and $\ln G_{21} = -\alpha\tau_{21}$.

The Flory-Huggins model (for polymer solutions)

$$\frac{g^{EX}}{RT} = x_1 \ln\frac{\phi_1}{x_1} + x_2 \ln\frac{\phi_2}{x_2} + \chi(x_1 + m x_2)\phi_1 \phi_2$$

which leads to

$$\ln\gamma_1 = \ln\frac{\phi_1}{x_1} + \left(1 - \frac{1}{m}\right)\phi_2 + \chi\phi_2^2 \text{ and } \ln\gamma_2 = \ln\frac{\phi_2}{x_2} + (m-1)\phi_1 + \chi\phi_1^2$$

with $\phi_1 = x_1/(x_1 + m x_2)$ and $\phi_2 = m x_2/(x_1 + m x_2)$.

Real Mixtures — Equation of State Description

Whereas the excess property description is used for the thermodynamic description of solids and liquids, a volumetric equation of state is typically used for vapor mixtures and may be used for liquid mixtures of hydrocarbons (including with light gases). The simplest volumetric equation of state is the ideal gas (which may be applicable only at low pressures):

$$P=\frac{RT}{v} \tag{47.17}$$

where R is the gas constant. The first equation of state that qualitatively described both vapors and liquids was that of van der Waals [1890]:

$$P=\frac{RT}{v-b}-\frac{a}{v^2} \tag{47.18}$$

Here b is interpreted as the hard-core volume of the molecules, and a is a parameter that represents the strength of the attractive interaction energy. Consequently, the first term results from molecular repulsions, and the second term from attractions. Many equations of state now used have this same structure, such as the Redlich–Kwong and Peng–Robinson equations:

$$P=\frac{RT}{v-b}-\frac{a(T)}{v(v+b)} \qquad P=\frac{RT}{v-b}-\frac{a(T)}{v(v+b)+b(v-b)} \tag{47.19}$$

where the a parameter has been made a function of temperature to better describe the pure component vapor pressure.

In order to apply equations of state developed for pure components to mixtures, expressions are needed that relate the mixture parameters, such as a and b above, to those of the pure fluids. This is done using mixing and combining rules. The simplest mixing rules are those of van der Waals:

$$a_{\text{mix}}=\sum_i\sum_j x_i x_j a_{ij} \quad \text{and} \quad b_{\text{mix}}=\sum_i\sum_j x_i x_j b_{ij} \tag{47.20}$$

The most commonly used combining rules are

$$b_{ij}=\frac{1}{2}(b_{ii}+b_{jj}) \quad \text{and} \quad a_{ji}=\sqrt{a_{ii}a_{jj}}\,(1-k_{ij}) \tag{47.21}$$

where a_{ii} and b_{ii} are pure component equation of state parameters and k_{ij} is an adjustable parameter. These mixing rules are applied to vapor and liquid mixtures of hydrocarbons, including light gases; more complicated rules that combine equations of state and activity coefficient models are needed for the accurate description of mixtures containing polar compounds. A recent review of equations of state and their mixing and combining rules can be found in Sandler [1994].

Once the volumetric equation of state for a mixture has been specified, the fugacity of each species in the mixture can be computed using the rigorous thermodynamic relation

$$\begin{aligned} RT\ln\hat{\phi}_i(T,P,x_i) &= \int_0^P\left(\bar{v}_i-\frac{RT}{P}\right)dP \\ &= \int_V^{\infty}\left[\left(\frac{\partial P}{\partial n_i}\right)_{T,V,n_j\neq i}-\frac{RT}{V}\right]dV-RT\ln Z \end{aligned} \tag{47.22}$$

Here, $Z = Pv/RT$ is the compressibility factor, $\bar{v}_i$ is partial molar volume of component as calculated from an equation of state, V is total volume, and n is the number of moles. The analytic expressions specific to various equations of state and their mixing rules can be found in Sandler [1999].

Solutions in the Solid Phase

Various solid solutions are encountered in engineering applications. In some cases, pure solids can exist in more than one form (for example, sulfur in rhombic and monoclinic allotropic forms, or carbon in diamond and in graphite forms), with each phase exhibiting different physical characteristics. In such cases, equilibrium may exist between the pure solid phases. Solids may also form compounds, and solutions of these compounds are common in metallurgical applications. These phenomena are beyond the scope of this chapter and the reader is referred to other sources [Gaskell, 1981; Kyle, 1992; Sandler, 1999].

47.2 Applications

The fundamental principle of equilibrium between two phases in contact is that the temperature, pressure, and chemical potentials of each species must be identical in both phases. This principle results in the following relation for vapor–liquid equilibrium [Sandler, 1999] of mixtures of condensible liquids when one of the activity coefficient models discussed earlier is used for the liquid phase:

$$x_i \gamma_i (T, P, x_i) \hat{f}_i^o (T, P, x_i) = y_i \hat{\phi}_i (T, P, y_i) P \tag{47.23}$$

To use this equation, an appropriate reference state for the liquid phase must be selected, which dictates the term $\hat{f}_i^o$, and the fugacity coefficient in the gas phase, $\hat{\phi}_i$, is evaluated from an equation of state. If the liquid phase is ideal ($\gamma_i = 1$ and $\hat{f}_i^o = P_i^{\text{vap}}$, which is the pure component vapor pressure) and the gas phase is an ideal gas mixture, the well-known Raoult's law is obtained:

$$x_i P_i^{\text{vap}} = y_i P \quad \text{and} \quad \sum_i x_i P_i^{\text{vap}} = P \tag{47.24}$$

For liquid–liquid phase separation the equilibrium relation is

$$x_i^{\text{I}} \gamma (T, x_i^{\text{I}}) = x_i^{\text{II}} \gamma (T, x_i^{\text{II}}) \tag{47.25}$$

where I and II refer to the coexisting equilibrium liquid phases. This equation has to be solved for each species, with activity coefficients obtained from the models above.

In the case where both phases can be described by an equation of state, the equation of vapor–liquid equilibrium is

$$x_i \hat{\phi}_i^L (T, P, x_i) = y_i \hat{\phi}_i^V (T, P, y_i) \tag{47.26}$$

where the fugacity coefficient of each species in each phase is calculated from Equation (47.22) and an equation of state. Note that a cubic equation of state such as Equation (47.19) may have three roots for the volume or compressibility factor at the pressure of interest and at temperatures below the critical temperature. In this case, the smallest root and the liquid compositions are used for the evaluation of the liquid phase fugacity coefficients, and the largest volume root and the vapor phase compositions are used for the calculation of the vapor phase fugacity coefficients.

In most phase equilibrium calculations, the compositions of both phases are not known in advance, so regardless of whether the activity coefficient or equation of state description is used, the solution must be found by iteration. Detailed discussions of the application of equations of state and activity coefficient models to phase equilibrium problems, and the algorithms used in their solution, can be found in Chapter 49 and in thermodynamics textbooks [e.g., Kyle, 1992; Chapter 8 of Sandler, 1999]. Calculations of equilibria of solids with solids, liquids, or vapors are important for heterogeneous chemical reactions, precipitation from solutions, and metallurgical applications. These cases are beyond the scope of this chapter and can be found elsewhere [Kyle, 1992; Gaskell, 1981; Sandler, 1999].

Defining Terms

Activity coefficient, γ_i — A term that accounts for the nonideality of a liquid mixture; related to the excess partial molar Gibbs energy.

Athermal solution — A mixture in which the excess internal energy of mixing is zero.

Chemical potential, μi — Equal to the partial molar Gibbs energy.

Equation of state — An equation relating the volume, temperature, and pressure of a pure fluid or mixture.

Excess property, $\Theta^{EX}(T, P, x_i)$ — The additional change in a thermodynamic property on forming a mixture from the pure components above that on forming an ideal mixture.

Fugacity, f_i, and fugacity coefficient, $\hat{\phi}_i$ — Terms related to the partial molar Gibbs energy or chemical potential that are computed from an equation of state.

Mixing rule — Equations to compute the parameters in a mixture equation of state from those for the pure components.

Partial molar property, $\overline{\Theta}_i\,(T, P, x_i)$ — The contribution to a thermodynamic property of a mixture made by adding a small amount on component *i*, reported on a molar basis.

Regular solution — A mixture in which the excess entropy of mixing is zero.

References

Fredenslund, A., Gmehling, J., and Rasmussen, P. 1977. *Vapor–Liquid Equilibria Using UNIFAC: A Group Contribution Method*, Elsevier, Amsterdam.

Gaskell, D. R. 1981. *Introduction to Metallurgical Thermodynamics*, 2nd ed. Hemisphere, New York.

Kyle, B. G. 1992. *Chemical and Process Thermodynamics*, 2nd ed. Prentice Hall, Englewood Cliffs, NJ.

Sandler, S. I. 1999. *Chemical and Engineering Thermodynamics*, 3rd ed. John Wiley & Sons, New York.

Sandler, S. I., ed. 1994. *Models for Thermodynamic and Phase Equilibria Calculations*, Marcel Dekker, New York.

van der Waals, J. H. 1890. Physical memoirs. (English translation by Therelfall and Adair.) *Physical Society.* 1, iii, 333.

Van Ness, H. 1964. *Classical Thermodynamics of Nonelectrolyte Solutions*, Pergamon Press, London.

Further Information

Prausnitz, J. M., Lichtenthaler, R. N., and Azevedo, E. G., 1986. *Molecular Thermodynamics of Fluid–Phase Equilibria*, 2nd ed. Prentice Hall, Englewood Cliffs, NJ.

Reid, R. C., Prausnitz, J. M., and Poling, B. E. 1987. *The Properties of Gases and Liquids*, 4th ed. McGraw-Hill, New York.

Walas, S. M. 1985. *Phase Equilibria in Chemical Engineering*, Butterworths, Boston.

48

Thermodynamics of Surfaces

William B. Krantz
University of Colorado

In writing this brief overview it was necessary to focus on those aspects of the thermodynamics of surfaces that are of considerable importance in engineering applications and that could be summarized well in this compact format. We begin by considering the basic concepts of the interface as a surface, curvature, and the generalized Gibbs phase rule. The generalized first law of thermodynamics is then stated, which forms the basis for subsequent considerations of work done on or by interfaces. The implications of surface curvature on capillary pressure, vapor pressure and solubility, and phase transition temperature are reviewed. Adsorption at interfaces is then considered with a brief introduction to surface equations of state. The first law is then used to introduce the subjects of wettability, adhesion, and contact angles. Unfortunately, space does not permit reviewing adequately other important areas involving the thermodynamics of surfaces, such as electrical aspects of surface chemistry, nucleation and growth phenomena, colloids, emulsions, foams, aerosols, and thin films. An excellent overview of these and other topics in surface science is provided in Adamson [1990].

48.1 Basic Concepts

The Interface

Two bulk phases in contact will be separated by a thin region within which the intensive thermodynamic properties change continuously from those of the one phase to those of the other. Since this region is typically only a few nanometers thick, we replace it with a surface whose thermodynamic properties are assigned to conserve the extensive thermodynamic properties of the overall system composed of the interface and the two bulk phases, which now are assumed to maintain their bulk properties up to the dividing surface. There are several conventions used to locate the dividing plane, henceforth to be

0-8493-1586-7/05/$0.00+$1.50

referred to as the *interface*; however, the interfacial intensive thermodynamic properties are independent of its location.

An interface can exist between a liquid and a gas (L/G interface), two liquid phases (L/L), a liquid and a solid (L/S), a gas and a solid (G/S), or two solid phases (S/S). Interfaces between fluid phases frequently can be assumed to be at equilibrium. However, interfaces formed with a solid phase often cannot be described by equilibrium thermodynamics, owing to slow diffusion in solids.

Curvature

The interface between two fluids takes on solidlike properties in that it can sustain a state of stress characterized by the *surface tension.* Consequently, a pressure drop can be sustained across curved interfaces, which increases with their curvature. The latter is uniquely defined at a point on the interface by the curvatures in two mutually perpendicular directions. The curvature C of an interface in a given plane is defined by

$$C \equiv \vec{n} \cdot \frac{d\vec{t}}{ds} \tag{48.1}$$

where $\vec{n}$ and $\vec{t}$ are the normal and tangential unit vectors at a point and s is the arc length along the curved interface. For a curved surface described in rectangular Cartesian coordinates by $y = y(x, z)$, Equation (48.1) implies the following for the curvatures in the yx and yz planes:

$$C_{xy} = \frac{y_{xx}}{(1+y_x^2)^{3/2}} \tag{48.2}$$

$$C_{zy} = \frac{y_{zz}}{(1+y_z^2)^{3/2}} \tag{48.3}$$

where y_c and y_{cc} denote the first and second partial derivatives of y with respect to c ($c = x$ or z). For axisymmetric bodies the two curvatures C_1 and C_2 can be represented solely in terms of y_x and y_{xx}:

$$C_1 = \frac{y_{xx}}{(1+y_x^2)^{3/2}} \tag{48.4}$$

$$C_2 = \frac{y_x}{x(1+y_x^2)^{1/2}} \tag{48.5}$$

Let us determine the curvature of a sphere of radius r defined by $x^2 + y^2 + z^2 = r^2$. Equation (48.4) and Equation (48.5) then give $C_1 = C_2 = -r^{-1}$; that is, the curvature is equal to the reciprocal of the radius of the sphere. The negative sign follows the convention that the curvature is viewed as if one is looking at the surface down the y axis. The curvature is positive when viewed from the concave side (e.g., interior of the sphere) and negative when viewed from the convex side.

The Gibbs Phase Rule

The Gibbs phase rule specifies the number of thermodynamic intensive variables f required to uniquely determine the thermodynamic state of the system. The phase rule is derived by adding up the intensive variables and then subtracting the number of independent relationships between them. For a system containing c components distributed between p phases separated by curved interfaces, the phase rule is given by

$$f = c + 1 \tag{48.6}$$

Equation (48.6) differs from the phase rule for bulk systems for which $f = c - p + 2$ because the p phases can sustain different pressures, owing to the interfacial curvature. For example, this generalized phase rule implies that a droplet of pure liquid water in equilibrium with its own vapor requires two intensive thermodynamic variables to be specified in order to determine its boiling point. Specifying the liquid and gas phase pressures, which is equivalent to specifying the curvature of the drop, is then sufficient to determine a unique boiling point.

48.2 First Law of Thermodynamics

Consider two contacting multicomponent phases α and β separated by a single interface having area A. The generalized first law of thermodynamics is then given by

$$\begin{aligned} dU &= T\,dS - P^{\alpha}dV^{\alpha} - P^{\beta}dV^{\beta} + \gamma\,dA + \sum_i \mu_i dN_i \\ &= dQ - dW + \sum_i \mu_i dN_i \end{aligned} \tag{48.7}$$

where U, S, and N_i are the total system internal energy, entropy, and moles of species i, respectively; T is the absolute temperature; μ_i is the chemical potential of i; P^p and V^p are the pressure and total volume of phase p ($p = \alpha$ or β); γ is the surface tension, an intensive thermodynamic property having units of energy per unit area (J/m^2) or force per unit length (N/m); $dQ = Td\,S$ is the heat transferred to the system; and $dW = p^{\alpha}dV^{\alpha} + P^{\beta}dV^{\beta} - \gamma dA$ is the work done by the system. Hence, changing the interfacial area can result in work being done on or by the system.

48.3 Effects of Curved Interfaces

Young and Laplace Equation

Consider the simplification of Equation (48.7) for an isolated system at equilibrium (that is, $dU = dS = dN_i = 0$). For an arbitrary displacement of the interface between α and β, $dA = (C_1 + C_2)\,dV^{\alpha}$ and $dV^{\alpha} = -dV^{\beta}$; hence,

$$\Delta P \equiv P^{\alpha} - P^{\beta} = \gamma(C_1 + C_2) \tag{48.8}$$

Equation (48.8), the Young and Laplace equation, predicts a pressure drop across a curved interface, which is referred to as a *capillary pressure effect.* For planar interfaces, $\Delta P = 0$, whereas for spherical liquid drops having a radius r, $\Delta P = 2\gamma/r$. Note the pressure is higher on the concave (positive curvature) side of the interface.

As an example of the use of Equation (48.8) consider the force required to separate two glass plates between which $V = 1$ ml of water is spread to a thickness of $H = 10$ μm, as shown in Figure 48.1. Since $C_2 << C_1$, we have

$$F = \Delta PA = \gamma C_1 \frac{V}{H} = \frac{2(72.94\cdot 10^{-3}\,\mathrm{N/m})(1\cdot 10^{-6}\ \mathrm{m}^3)}{(10\cdot 10^{-6}\ \mathrm{m})^2} = 1459\ \mathrm{N}\ (328\ \mathrm{lbf}) \tag{48.9}$$

Capillary pressure effects can be quite large and account for phenomena such as the cohesiveness of wet soils.

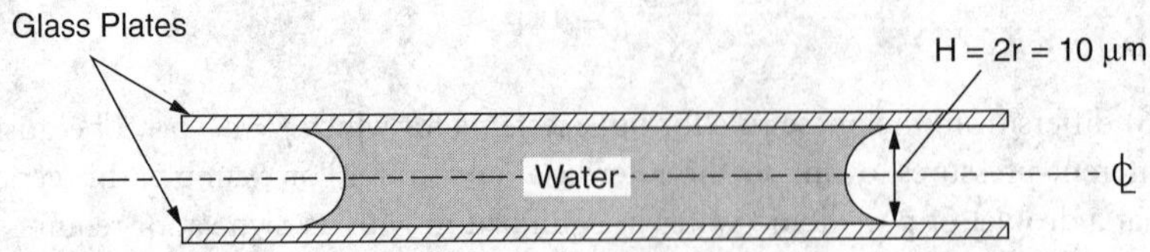

FIGURE 48.1 A wetting liquid between two parallel glass plates: $C = r^{-1} = H/2$.

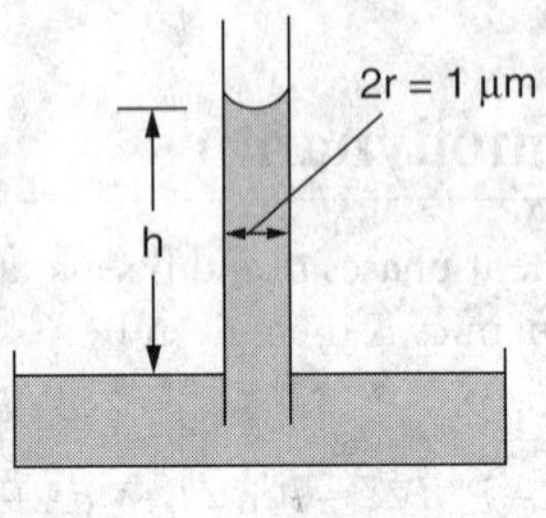

FIGURE 48.2 Capillary rise of a wetting liquid β in an ambient gas phase α: $C_1 = C_2 = r^{-1}$.

As a second example, consider the placing of a capillary tube vertically into a wetting liquid (one having a convex interface when viewed through the liquid) as shown in Figure 48.2; the liquid will rise until the pressure drop across the curved interface is equal to the hydrostatic pressure exerted by the liquid column of height h; that is,

$$\Delta P = \frac{2\gamma}{r} = \Delta\rho g h \tag{48.10}$$

where $\Delta\rho$ is the density difference between the liquid (phase β) and the ambient fluid (phase α), and g is the gravitational acceleration. For example, a 1 μm capillary can draw water to a height of 30 m (98 ft) in air. Capillary rise is involved in absorbent wicking and also is used as a method for measuring L/G and L/L surface tensions.

As a third example, consider the insertion of a capillary tube containing an insoluble gas vertically into a wetting liquid. If the volume of the gas is slowly increased, the curvature of the L/G interface will adjust to satisfy Equation (48.8) The gas-phase pressure will pass through a maximum when the bubble radius is equal to the tube radius. This is the basis of the maximum bubble pressure method for determining surface tensions.

The Kelvin Equation

The Young and Laplace equation also implies that liquids bounded by an interface having positive curvature will exert a higher vapor pressure P^v than that exerted by liquids bounded by a planar interface P_o^v; these two vapor pressures are related by the Kelvin equation:

$$P^v = P_o^v \exp\left(\frac{2\gamma \overline{V}_l}{RTr}\right) \tag{48.11}$$

where $\overline{V}_l$ is the molar volume of the liquid and R is the gas constant. A water droplet having a radius of 0.01 μm will have a $P^v = 1.11P_o^v$. The Kelvin equation can also be applied to S/L systems to predict the increased solubility of small crystals in solution. Equation (48.11) explains the formation of supersaturated vapors and Ostwald ripening whereby large crystals grow at the expense of small crystals. It also explains the falling rate period in drying during which water must be removed from progressively smaller pores, causing reduced vapor pressures.

The Gibbs–Thompson Equation

The Young and Laplace equation also implies that wetting liquids in small capillaries will have a lower freezing temperature T than the normal freezing temperature T_o; these two freezing temperatures are related by the Gibbs–Thompson equation:

$$T = T_o - \frac{\gamma T_o \overline{V}_s}{\Delta H^f}(C_1 + C_2) \tag{48.12}$$

where $\overline{V}_s$ is the molar volume of the solid phase and ΔH^f is the latent heat of fusion. Water in a capillary having a radius 0.01 μm will freeze at −5.4°C. Equation (48.12) explains why freezing in wet soils occurs over a zone rather than at a discrete plane.

48.4 Adsorption at Interfaces

The Gibbs Adsorption Equation

The interfacial concentration $\Gamma_i \equiv N_i^s / A$ —where N_i^s denotes the moles of *i* in the interface—depends on the location of the interface. It can be determined from the concentration dependence of the surface tension using the Gibbs adsorption equation, which is the interfacial analogue of the Gibbs–Duhem equation derived from the generalized first law and given by

$$d\gamma = -\sum_i \Gamma_i d\mu_i \tag{48.13}$$

Consider determining the interfacial concentration for the special case of a binary ideal solution for which the surface tension is a linear function of the bulk concentration of solute, c_2, given by $\gamma = \gamma_o - bc_2$. We will locate the interface to ensure no net adsorption of solvent 1; that is, $\Gamma_1 = 0$. This is the commonly used Gibbs convention for locating the dividing surface. Hence, Equation (48.13) implies that the interfacial concentration of solute in this convention, Γ_2^1 (where the superscript 1 denotes the Gibbs convention), is given by

$$\Gamma_2^1 = -\frac{\partial \gamma}{\partial \mu_2} = -\frac{N_2}{RT}\frac{\partial \gamma}{\partial N_2} = \frac{bc_2}{RT} = \frac{\gamma_o - \gamma}{RT} = \frac{\pi}{RT} \tag{48.14}$$

where $\pi \equiv \gamma_o - \gamma$ is referred to as the *surface pressure*, having units of force per unit length. Equation (48.14) implies that $\pi\sigma_i = RT$, where $\sigma_i \equiv 1/\Gamma_2^1$ is the area per mole of solute in the interface. This is a two-dimensional analogue to the familiar ideal gas law. Hence, by analogy one would expect a linear dependence of surface tension on bulk concentration at low surface pressures where intermolecular forces and the size of the molecules can be ignored. The two-dimensional equations of state are used to describe the $\pi - \sigma_i - T$ behavior of interfaces just as the three-dimensional analogues are used to describe the $P - V - T$ behavior of bulk phases.

Note that $b > 0 \Rightarrow \pi > 0 \Rightarrow \Gamma_2^1 > 0$; that is, surface tension decreasing with increasing solute concentration implies positive adsorption of the solute at the interface. Solutes displaying positive adsorption are said to be *surface-active* and are referred to as *surfactants.* Surfactants are used to alter the wettability of surfaces in applications such as the manufacture of cleaning detergents and tertiary oil recovery. If $b < 0 \Rightarrow \pi < 0 \Rightarrow \Gamma_2^1 < 0$, negative adsorption occurs; this is characteristic of ionizing salts in aqueous solution, which avoid the air/water interface, which has a lower dielectric constant.

If the surface pressure is eliminated from the equation of state using the prescribed equation for the surface tension dependence on bulk concentration, one obtains a relationship between the surface and bulk concentrations referred to as the *adsorption isotherm.* The Gibbs adsorption equation provides the

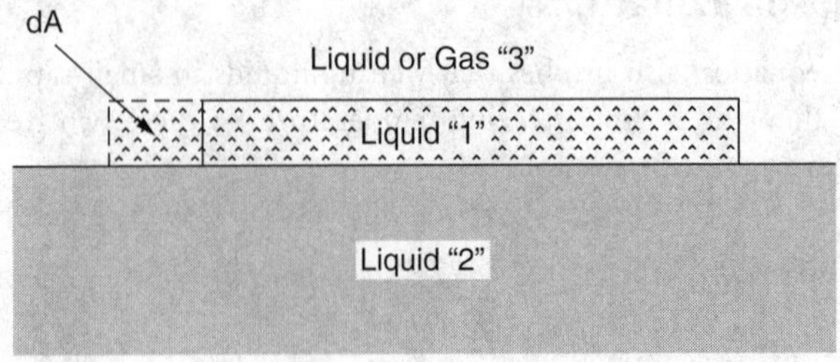

FIGURE 48.3 Spreading of liquid 1 on liquid 2 in the presence of gas or liquid 3.

link between the interfacial equation of state, the adsorption isotherm, and the surface tension dependence on bulk concentration.

48.5 Wettability and Adhesion

Wetting and Spreading

Wetting relates to the propensity of a liquid to spread over another liquid or a solid. Consider the spreading of liquid 1 on liquid 2 in the presence of gas or liquid 3, as shown in Figure 48.3, to create dA of interface between 1 and 2 and between 1 and 3 while destroying dA of interface between 2 and 3. Equation (48.7) implies that the reversible work of spreading dW_s/dA at constant T and P is given by

$$\frac{dW}{dA} = -\gamma_{12} - \gamma_{13} + \gamma_{23} \equiv S_{12} \quad (48.15)$$

where γ_{ij} denotes the surface tension between phases i and j and S_{12} is defined as the *spreading coefficient* of phase 1 on phase 2. Note that if $dW/dA = S_{12} > 0$, spreading will occur spontaneously since work can be done by the system to increase the contact area between phases 1 and 2. Consider the spreading of heptane (1) on water (2) in air (3) at 20°C, for which γ_{12} = 50.2 mN/m, γ_{13} = 20.14 mN/m, and γ_{23} = 72.94 mN/m. Hence, $dW/dA = S_{12} = 2.6$ mN/m, and we conclude that spreading will occur spontaneously. Wetting is an important consideration in developing coatings of various types, solders and welding fluxes, water-repellent fabrics, and so on.

Adhesion and Cohesion

Adhesion relates to the cohesiveness of the "bond" between two contacting phases. A measure of this property is the work of adhesion W_{ad}, the work that must be done on the system to separate the two phases in contact. Consider the separation of phase 1 from phase 2 in the presence of phase 3, as shown in Figure 48.4, to create dA of interface between 1 and 3 and between 2 and 3 while destroying dA of interface between 1 and 2. Equation (48.7) implies that the reversible work of adhesion W_{ad} at constant T and P is given by

$$W_{ad} = -\frac{dW}{dA} = \gamma_{13} + \gamma_{23} - \gamma_{12} \quad (48.16)$$

A special case of the above is the work of cohesion W_{co}, which is the work required to separate a single phase 1 in the presence of phase 3, thereby creating two interfaces between phases 1 and 3. Equation (48.16) implies that

$$W_{co} = -\frac{dW}{dA} = \gamma_{13} + \gamma_{13} = 2\gamma_{13} \quad (48.17)$$

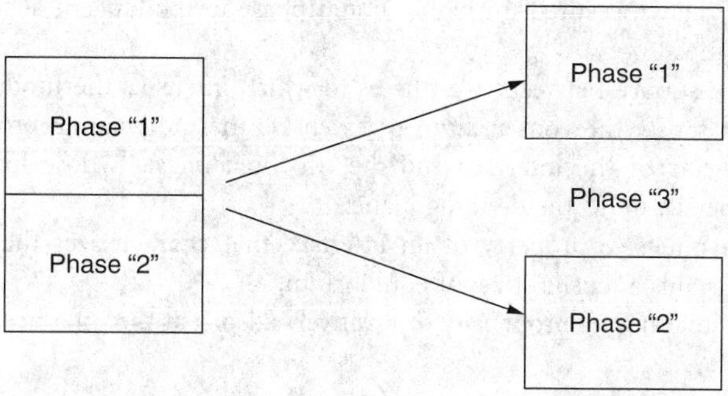

FIGURE 48.4 Work of adhesion necessary to separate phase 1 from phase 2 in the presence of phase 3.

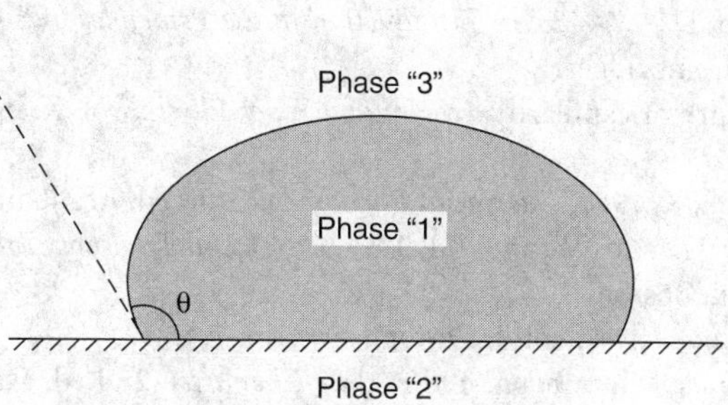

FIGURE 48.5 A drop of phase 1 resting on phase 2 in the presence of phase 3, showing the contact angle θ.

Comparison of Equation (48.15), Equation (48.16), and Equation (48.17) indicates that $S_{12} = W_{ad} - W_{co}$; that is, the spreading coefficient is equal to the difference between the work of adhesion and cohesion. Adhesion is involved in developing bonding materials, laminates, paints, printing inks, and so on.

Contact Angles

If phase 1 completely "wets" phase 2 in the presence of phase 3, it will spread over phase 2. If phase 1 does not wet phase 2, it will form a drop having a contact angle θ, where θ is measured from the interface between phases 1 and 2 through phase 1 to the tangent line at the contact line between phases 1, 2, and 3, as shown in Figure 48.5. The contact angle is an intensive thermodynamic property of the system uniquely determined by the surface tensions via Young's equation, given by

$$\gamma_{12} - \gamma_{23} + \gamma_{33}\cos\theta = 0 \tag{48.18}$$

If θ = 0, phase 1 is said to *wet* phase 2; if θ > 90°, phase 1 is said to be *nonwetting* to phase 2 in the presence of phase 3. Equation (48.16) can be combined with Equation (48.18) to obtain

$$W_{ad} = \gamma_{13}(1+\cos\theta) \tag{48.19}$$

Equation (48.19) implies that the work of adhesion of phase 1 to phase 2 can be obtained by merely measuring the interfacial tension γ_{13} and the contact angle of phase 1 on phase 2 in the presence of phase 3. This provides a simple means for determining the strength of adhesion between two materials.

Defining Terms

Capillary pressure effects — A pressure difference sustained across curved interfaces owing to surface tension.

Contact angle — The angle measured through a liquid phase to the tangent at the contact line where three phases meet.

Interface — A dividing surface between two phases to which intensive thermodynamic properties are assigned that satisfy the conservation of extensive thermodynamic properties of the overall system consisting of the interface and the two bulk phases whose intensive properties are assumed to persist up to the dividing plane.

Surface tension — An intensive property of an interface that characterizes the distinct energy state whereby it sustains a tensile stress at equilibrium.

Surfactants — Solutes that have a propensity to positively adsorb at the interface.

References

Adamson, A. W. 1990. *Physical Chemistry of Surfaces,* 5th ed. John Wiley & Sons, New York.

Aveyard, R. and Haydon, D. A. 1973. *An Introduction to the Principles of Surface Chemistry,* Cambridge University Press, Cambridge.

Buff, F. P. 1960. The theory of capillarity. In *Handbuch der Physik,* vol. X, pp. 281–304. Springer-Verlag, Berlin.

Davies, J. T. and Rideal, E. K. 1963. *Interfacial Phenomena,* 2nd ed. Academic, New York.

Edwards, D. A., Brenner, H., and Wasan, D. T. 1991. *Interfacial Transport Processes and Rheology,* Butterworth-Heinemann, Boston.

Gibbs, J. W. 1961. *Scientific Papers,* vol. 1. Dover, New York.

Hiemenz, P. C. 1986. *Principles of Colloid and Surface Chemistry,* 2nd ed. Marcel Dekker, New York.

Miller, C. A. and Neogi, P. 1985. *Interfacial Phenomena,* Surfactant Science Series, vol. 17. Marcel Dekker, New York.

Morrison, S. R. 1977. *The Chemical Physics of Surfaces,* Plenum, London.

Rosen, M. J. 1989. *Surfactants and Interfacial Phenomena,* 2nd ed. John Wiley & Sons, New York.

Further Information

Interested readers are referred to the following journals that publish research on interfacial phenomena:

Journal of Adhesion
Journal of Colloid and Interface Science
Colloids and Surfaces
Langmuir

49

Phase Equilibrium

Benjamin G. Kyle
Kansas State University

The best way to understand the rationale underlying the application of thermodynamics to phase equilibrium is to regard thermodynamics as a method or framework for processing experimentally gained information. From known properties for the system under study (information), other properties may be determined through the network of thermodynamic relationships. Thus, information about a system can be extended. The thermodynamic framework is also useful for conceptualizing and analyzing phase equilibrium. This allows extrapolation and interpolation of data and the establishment of correlations as well as evaluation and application of theories based on molecular considerations.

For the most part, the variables of interest in phase equilibrium are intensive: temperature, pressure, composition, and specific or molar properties. The phase rule determines the number of intensive variables, F, required to define a system containing π phases and C components.

$$F = C + 2 - \pi \tag{49.1}$$

The thermodynamic network also contains other intensive variables such as **fugacity** and the activity coefficient, which are necessary for computational purposes but possess no intrinsic value; they are simply useful thermodynamic artifacts. Specifically, the basis for the thermodynamic treatment of phase equilibrium is the equality of the fugacity of each component in each phase in which it exists.

49.1 Pure-Component Phase Equilibrium

Equilibrium between phases of a pure component can be visualized with the aid of Figure 49.1, which shows areas in the pressure–temperature (PT) plane labeled S, L, and V and representing conditions where, respectively, solid, liquid, and vapor exist. In accordance with the phase rule, specification of two intensive variables (here, T and P) is required to define the state of the system. Conditions where two phases are in equilibrium are represented on this diagram by lines (such as *ab*, *bc*, and *bd*) and therefore only one variable need be specified. The intersection of three lines results in a triple point, where three phases are in equilibrium and no variables may be specified. Some systems exhibit more than one triple point.

0-8493-1586-7/05/$0.00+$1.50

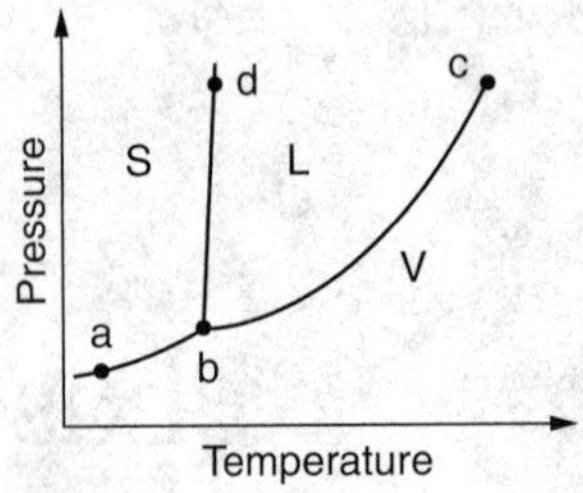

FIGURE 49.1 Pressure–temperature graph of pure-component phase equilibrium.

The Clapeyron Equation

For two phases in equilibrium, say the α and β phases, the equating of fugacities leads to the Clapeyron equation:

$$\frac{dP}{dT} = \frac{h^{\alpha} - h^{\beta}}{(v^{\alpha} - v^{\beta})T} \tag{49.2}$$

This equation relates the slope of a coexistence line on Figure 49.1 (e.g., *ab*) to differences in molar enthalpy and molar volume and can be used to determine the value of one variable when all the others are known.

The Clausius–Clapeyron Equation

When one of the phases is a vapor, the number of variables in the Clapeyron equation can be reduced; the result is the Clausius–Clapeyron equation:

$$\frac{d\ln P^0}{d(1/T)} = -\frac{\Delta h}{R} \tag{49.3}$$

If Equation (49.3) is integrated under the assumption of constant Δh, the result is

$$\ln P^0 = c - \frac{\Delta h}{RT} \tag{49.4}$$

where c is a constant of integration. Equation (49.3) and Equation (49.4) relate the vapor pressure, P^0, to temperature and the heat of vaporization or heat of sublimation, Δh, and predict that a plot of ln P^0 versus reciprocal of absolute temperature should be linear with a slope of $-\Delta h/R$. Equation (49.3) and Equation (49.4) find use in correlating, interpolating, and extrapolating vapor pressure data and in determining the heat of vaporization or sublimation from vapor pressure data.

Despite some questionable assumptions made in arriving at Equation (49.3) and Equation (49.4), the linear relationship has been found to hold over a wide range of temperature. In fact, the Antoine equation,

$$\log P^0 = A - \frac{B}{C + t} \tag{49.5}$$

which has been found to give excellent representation of vapor pressure data, can be seen to be an empiricized version of Equation (49.4). Extensive compilations of Antoine parameters (A, B, and C) are available [Boublik et al., 1973; Reid et al., 1987].

49.2 Phase Equilibrium in Mixtures

There are two approaches to the treatment of phase equilibrium in mixtures — the use of an equation of state and the use of activity coefficients. The latter is developed here; the former is delineated by Kyle [1999] and also in Chapter 47 of this text.

Vapor–Liquid Equilibrium (VLE)

Although the activity coefficient approach can be used for systems at any pressure, it is predominantly applied at low to moderate pressure. Here, consideration will be restricted to low pressure. This method is based on the ideal solution model for representing the fugacity of a component in a solution. Unlike the pure-component fugacity, which can be easily calculated or estimated, it is usually not possible to directly calculate the fugacity of a component in a solution, and a model — the ideal solution model — is therefore employed. This model is applied to both the vapor and liquid phases. It has been found to represent the vapor phase reasonably well and will be used here without correction. On the other hand, the model fits very few liquid-phase systems and needs to be corrected. The activity coefficient, γ_i, is the correction factor and is expected to depend upon temperature, pressure, and the liquid-phase composition. The effect of pressure is usually neglected and activity-coefficient equations of demonstrated efficacy are used to represent the temperature and composition dependence [Reid et al., 1987].

Employment of the equal fugacity criterion along with these models yields the basic equation for treating vapor–liquid equilibrium,

$$Py_i = P_i^0 x_i \gamma_i \tag{49.6}$$

where P is the system pressure, P_i^0 is the vapor pressure of component i, x_i is its liquid mole fraction, and y_i is its vapor mole fraction.

Determination of Activity Coefficients

Although methods for estimating activity coefficients are available (see Chapter 47), it is preferable to use experimental phase equilibrium data for their evaluation. The exception to this rule is when it is known that a system forms an ideal liquid solution and γ values are therefore unity. These systems, however, are rare and occur only when all the constituents belong to the same chemical family (e.g., paraffin hydrocarbons). Equation (49.6) is used to calculate γ_i from experimental values of x_i, y_i, and P at a given temperature where P_i^0 is known. Thus, for a binary system, γ for each component can be evaluated from a single VLE data point.

Once a set of γ_1, γ_2, and x_1 data has been determined from binary VLE data, it is customary to fit these data to an activity-coefficient equation. The Wilson equation, shown below for a binary system, is outstanding among equations of proven efficacy:

$$\ln\gamma_1 = -\ln(x_1 + x_2 G_{12}) + x_2\left(\frac{G_{12}}{x_1 + x_2 G_{12}} - \frac{G_{21}}{x_2 + x_1 G_{21}}\right) \tag{49.7}$$

$$\ln\gamma_2 = -\ln(x_2 + x_1 G_{21}) - x_1\left(\frac{G_{12}}{x_1 + x_2 G_{21}} - \frac{G_{21}}{x_2 + x_1 G_{21}}\right) \tag{49.8}$$

The G values are empirically determined parameters that are expected to show the following exponential dependence on temperature:

$$G_{12} = \frac{v_2}{v_1}\exp\left(-\frac{a_{12}}{RT}\right);\quad G_{21} = \frac{v_1}{v_2}\exp\left(-\frac{a_{21}}{RT}\right) \tag{49.9}$$

where v_1 and v_2 are the liquid molar volumes of components 1 and 2, and a_{12} and a_{21} are empirically determined parameters. Some type of parameter estimation technique is used to determine the best values of G. This can easily be done on a spreadsheet using the optimization feature to find the parameters that minimize the following objective function:

$$\sum_{i=1}^{n}(Q^{\text{exp}} - Q^{\text{cal}})_i^2 \tag{49.10}$$

where Q is defined as

$$Q = x_1 \ln\gamma_1 + x_2 \ln\gamma_2 \tag{49.11}$$

In Equation (49.10), Q^{exp} is determined from Equation (49.11) and values of γ_1 and γ_2 from experimental VLE data, while Q^{cal} is determined at the same composition from Equation (49.11) using Equation (49.7) and Equation (49.8). The difference in these Q values and, consequently, the objective function [Equation (49.10)] is a function only of the G values. Executable programs for optimizing Equation (49.10) are available [Kyle, 1999].

Sensing a certain amount of circularity, one may well question the value of this procedure that uses experimental VLE data to determine parameters in the Wilson equation that will then be used to calculate VLE by means of Equation (49.6). However, the procedure becomes more attractive when one recognizes that VLE data at one condition can be used to evaluate Wilson parameters that can then be used to calculate VLE at another condition. Or one might use the minimum amount of VLE data (one data point) to evaluate Wilson parameters and then proceed to calculate, via Equation (49.6), sufficient data to represent the system. When no experimental information is available, activity coefficients may be estimated by the UNIFAC method, which is based on the behavior of functional groups within the molecules composing the solution rather than on the molecules themselves [Reid et al., 1987]. An executable program using the UNIFAC method is available [Kyle, 1999].

Azeotropes

An azeotrope is the condition in which the vapor and liquid compositions in a system are identical. As the knowledge of the existence of azeotropes is crucial to the design of distillation columns, much effort has been expended in determining and compiling this information [Horsely, 1952, 1962]. These compilations represent a source of VLE data: values of T, P, and x_1, where $x_i = y_i$. At a specified T (or P) a value of γ can be calculated for each component via Equation (49.6), the minimum data required to determine Wilson parameters.

For an azeotrope in the 1–2 system, Equation (49.6) can be employed to obtain

$$\frac{P_1^0}{P_2^0} = \frac{\gamma_2}{\gamma_1} \tag{49.12}$$

This equation is useful in estimating the effect of temperature (which is manifested as a corresponding change in pressure) on the composition of the azeotrope. The vapor pressure ratio depends only on temperature, whereas the γ values depend on temperature and composition. If one assumes that the ratio γ_2/γ_1 is independent of temperature and uses the azeotropic data point to evaluate Wilson parameters, the right-hand side becomes a known function of x_1, and, thus, a value of x_1 can be determined from the vapor pressure ratio evaluated at any temperature.

Computational Procedures

The computations for VLE involving Equation (49.6) cannot be directly executed because all of the necessary variables cannot be specified prior to the calculation. According to the phase rule, a system of two phases possesses C degrees of freedom. If the composition of a phase is specified at the expense of $C-1$ variables, there remains only one additional variable to be specified — T or P. Because the quantities in Equation (49.6) depend on T and P, it is necessary to adopt a trial-and-error computational procedure. If, for example, T and the x_i values are specified, a value of P would be assumed and Equation (49.6) used to calculate the y_i values. If these calculated y_i values all sum to unity, the assumed value of P would be correct: if not, other values of P would be assumed until this condition is met. A similar procedure would be used for a system described in different terms.

Multicomponent Vapor–Liquid Equilibrium

In order to effectively treat multicomponent systems, it is necessary to be able to represent activity coefficients as a function of temperature and composition for use in Equation (49.6). The Wilson equation can be extended to any number of components and requires only two parameters (G_{ij} and G_{ji}) for each binary pair that can be formed from the components of the system [Kyle, 1999]. A comprehensive study [Holmes and Van Winkle, 1970] has established the ability of the Wilson equation to calculate ternary VLE with acceptable accuracy from parameters obtained from the three constituent binary systems.

Liquid–Liquid Equilibrium (LLE)

When two liquid phases are in equilibrium, the equating of fugacities results in the following expression for each component:

$$\gamma_i' x_i' = \gamma_i'' x_i'' \tag{49.13}$$

where ′ and ″ distinguish the liquid phases. Because activity coefficients are required for each phase, calculations involving LLE are computationally intensive. These calculations entail either the computation of LLE from available information concerning activity coefficients or the use of LLE to obtain information concerning activity coefficients [Kyle, 1999].

Solid–Liquid Equilibrium (SLE)

For pure solid component i in equilibrium with a liquid solution containing component i, the equating of fugacities results in

$$RT \ln x_i \gamma_i = \frac{L_{mi}(T - T_{mi})}{T_{mi}} + \Delta c_{Pi}(T_{mi} - T) + T\Delta c_{Pi} \ln \frac{T}{T_{mi}} \tag{49.14}$$

where T_{mi} is the freezing point and L_{mi} is the heat of fusion of component i. Usually, the data for the evaluation of Δc_{Pi} are unavailable and only the first right-hand term is used. Figure 49.2 shows a phase diagram for a simple binary system where the curves AE and BE are represented by Equation (49.14) with i equal to 2 and 1, respectively. The **eutectic point** E is merely the intersection of these two curves.

To apply Equation (49.14) where liquid solutions are not ideal, it will be necessary to have access to enough experimental phase equilibrium data to evaluate parameters in the Wilson, or equivalent, activity-coefficient equation. When reliable parameters are available, it has been demonstrated [Gmehling et al., 1978] that good estimates of SLE can be obtained for systems of nonelectrolytes. Alternatively, SLE data

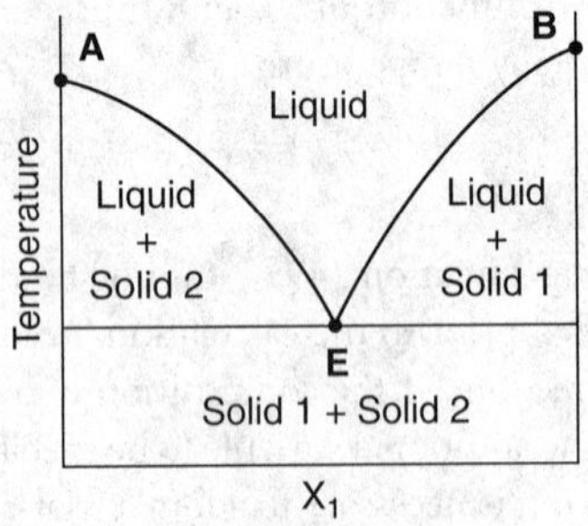

FIGURE 49.2 Phase diagram for a simple binary system.

can be used as a source of activity coefficient information for use in other types of phase equilibrium calculations involving a liquid phase.

The solid phase can be either pure or a solid solution. For a given system, solid solutions will exhibit larger deviations from the ideal solution model than will liquid solutions, and, therefore, an activity coefficient will also be required for the solid phase. Thus, calculations for this type of system will require some experimental data for the evaluation of liquid- and solid-phase activity coefficients [Kyle, 1999].

49.3 Perspective

In this work the emphasis has been placed on delineating the rationale for the application of thermodynamics to phase equilibrium rather than on amassing a set of "working equations." Such a set would be useful in dealing with routine or standard problems; however, with the exception of pure components, few such problems exist in phase equilibrium. Each problem is defined by the nature of the system, what information is desired, and what information is available; judgments concerning the suitability of data and the fitting of data may also be involved. This endeavor can be regarded as information processing in which the activity coefficient provides the means by which information (experimental phase equilibrium data) at one condition is processed in order to generate information at other conditions.

Defining Terms

Eutectic point — The existence of two solid phases and a liquid phase in equilibrium at a temperature below the melting points of the pure solid components.

Fugacity — A thermodynamic function that requires for its evaluation PVT data for a pure substance and PVT-composition data for a component in a mixture. It is often regarded as a thermodynamic pressure because it has the units of pressure and is equal to pressure for an ideal gas and equal to partial pressure for a component in an ideal gas mixture.

References

Boublik, T., Fried, V., and Hala, E. 1973. *The Vapor Pressures of Pure Substances,* Elsevier, Amsterdam.

Gmehling, J. G., Anderson, T. F., and Prausnitz, J. M. 1978. Solid–liquid equilibria using UNIFAC. *Ind. Eng. Chem. Fundam.* 17(4):269–273.

Holmes, M. J. and Van Winkle, M. 1970. Prediction of ternary vapor–liquid equilibria from binary data. *Ind. Eng. Chem.* 62(1):21–31.

Horsely, L. H. 1952. *Azeotropic Data,* Advances in Chemistry Series, No. 6. American Chemical Society, Washington, D.C.

Horsely, L. H. 1962. *Azeotropic Data — II,* Advances in Chemistry Series, No. 35. American Chemical Society, Washington, D.C.

Kyle, B. G. 1999. *Chemical and Process Thermodynamics,* 3rd ed. Prentice Hall, Englewood Cliffs, NJ.

Reid, R. C., Prausnitz, J. M., and Poling, B. E. 1987. *The Properties of Gases and Liquids*, 4th ed. McGraw-Hill, New York.

Further Information

A comprehensive treatment of phase equilibrium consistent with this work and a list of sources of phase equilibrium data is available in Kyle [1999].

A vast quantity of various types of phase equilibrium data is available in the Dortmund Data Bank via the on-line service DETHERM provided by STN International.

See Reid et al. [1987] for a detailed explication of the UNIFAC method for estimating activity coefficients and for comprehensive lists of sources of experimental data and books about phase equilibrium.

50

Thermodynamic Cycles

William J. Cook
Iowa State University

A thermodynamic cycle is a continuous series of thermodynamic processes that periodically returns the **working fluid** of the cycle to a given state. Although cycles can be executed in closed systems, the focus here is on the cycles most frequently encountered in practice: **steady-flow** cycles, cycles in which each process occurs in a steady-flow manner. Practical cycles can be classified into two groups: power-producing cycles (power cycles) and power-consuming cycles (refrigeration cycles). The working fluid typically undergoes phase changes during either a power cycle or a refrigeration cycle. Devices that operate on thermodynamic cycles are widely used in energy conversion and utilization processes since such devices operate continuously as the working fluid undergoes repeated thermodynamic cycles.

The fundamentals of cycle analysis begin with the **first law of thermodynamics.** Since each process is a steady-flow process, only the first law as it applies to steady-flow processes will be considered. For a steady-flow process occurring in a **control volume** with multiple inflows and outflows, the first law is written on a time rate basis as

$$\dot{Q} + \sum [\dot{m}(h + V^2/2 + gz)]_{\text{in}} = \dot{W} + \sum [\dot{m}(h + V^2/2 + gz)]_{\text{out}} \tag{50.1}$$

For a single-stream process between states *i* and *j*, Equation (50.1) on a unit mass basis becomes

$$_iq_j + h_i + V_i^2/2 + gz_i = {}_iw_j + h_j + V_j^2/2 + gz_j \tag{50.2}$$

where $_iq_j = {}_i\dot{Q}_j/\dot{m}$, $_iw_j = {}_i\dot{W}_j/\dot{m}$, and $\dot{m}$ is the mass rate of flow. See Sonntag et al. [2002]. In processes involved with the cycles considered here, changes in kinetic and potential energies ($V^2/2$ and gz terms, respectively) are small and are neglected. Power $\dot{W}$ is considered positive when it is transferred out of the control volume, and heat transfer rate $\dot{Q}$ is considered positive when heat transfer is to the control volume. In the figures herein that describe the transfer of power and heat energy to and from cycles, arrows indicate direction and take the place of signs. The accompanying $\dot{W}$ or $\dot{Q}$ is then an absolute quantity. Where confusion might arise, absolute value signs are used. Only power transfer and heat transfer occur across a closed boundary that encloses the complete cycle. For such a boundary,

$$\left[\sum \dot{Q} = \sum \dot{W}\right]_{\text{cycle}} \tag{50.3}$$

0-8493-1586-7/05/$0.00+$1.50

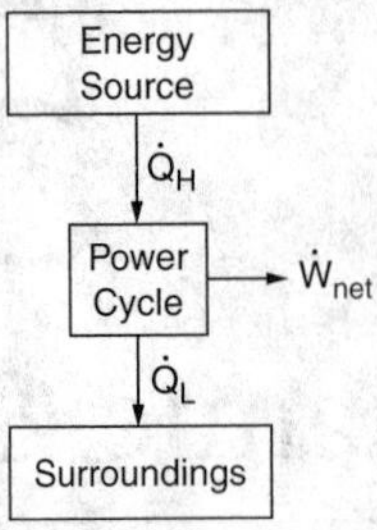

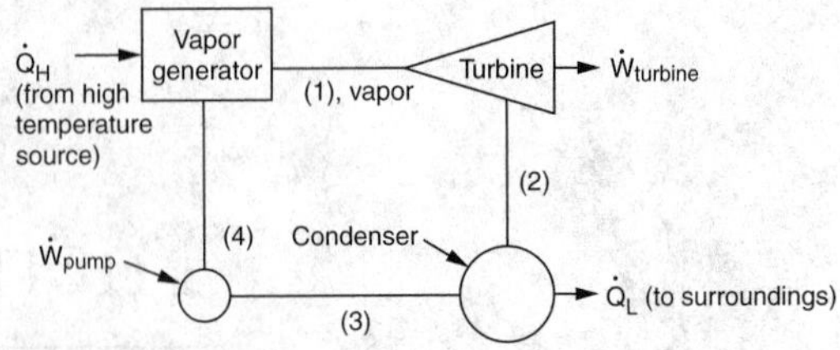

FIGURE 50.1 Descriptions of power cycles: (a) Power cycle operation. (b) The simple vapor power cycle.

50.1 Power Cycles

The purpose of a power cycle is to produce a net power output on a continuous basis from **heat energy** supplied to it from a high-temperature energy source. The device in which the power cycle is executed is sometimes referred to as a *heat engine.* Gas-turbine engines and reciprocating internal combustion engines are used widely to produce power. Strictly speaking, these engines are not classified as power cycles because their working fluids do not undergo thermodynamic cycles. Figure 50.1(a) shows a heat engine that receives heat energy at the rate $\dot{Q}_H$ from a high-temperature energy source and produces net power $\dot{W}_{net}$. As a consequence of its operation, it rejects heat energy to the lower-temperature surroundings at the rate $\dot{Q}_L$. A widely used performance parameter for a power cycle is η, the cycle thermal efficiency, defined as

$$\eta = \dot{W}_{net} / \dot{Q}_H \tag{50.4}$$

Second law considerations for thermodynamic power cycles restrict η to a value less than unity. Thus, $\dot{W}_{net}$ in Figure 50.1(a) is less than $\dot{Q}_H$. By Equation (50.3), the rate at which heat energy is rejected to the surroundings is

$$\left|\dot{Q}_L\right| = \dot{Q}_H - \dot{W}_{net} \tag{50.5}$$

It is useful to consider cycles for which the energy source and the surrounding temperatures — denoted, respectively, as T_H and T_L — are uniform. The maximum thermal efficiency any power cycle can have while operating between a source and its surroundings, each at a uniform temperature, is that for a totally reversible thermodynamic cycle (a Carnot cycle, for example) and is given by the expression

$$\eta_{max} = (T_H - T_L) / T_H \tag{50.6}$$

where the temperatures are on an absolute scale [Sonntag et al., 2002].

Figure 50.1(b) illustrates a simple vapor power cycle. Each component operates in a steady-flow manner. The vapor generator delivers high-pressure high-temperature vapor at state 1 to the turbine. The vapor flows through the turbine to the turbine exit state, state 2, and produces power $\dot{W}_{turbine}$ at the turbine output shaft. The vapor is condensed to liquid, state 3, as it passes through the condenser, which is typically cooled by a water supply at a temperature near that of the surroundings. The pump, which consumes power $\dot{W}_{pump}$, compresses the liquid from state 3 to state 4, the state at which it enters the vapor generator. Heat energy at the rate $\dot{Q}_H$ is supplied to the vapor generator from the energy source to produce vapor at state 1. Thus, the working fluid executes a cycle, in that an element of the working fluid initially at state 1 is periodically returned to that state through the series of thermodynamic processes as it flows through the various hardware components. The net power produced, $\dot{W}_{net}$, is the algebraic sum of the positive turbine power $\dot{W}_{turbine}$ and the negative pump power $\dot{W}_{pump}$.

TABLE 50.1 Properties at Cycle States for Example 50.1

State	Pressure, kPa	Temperature, °C	Quality, kg/kg	Entropy, kJ/kg K	Enthalpy, kJ/kg	Condition
1	1000	480	[a]	7.7055	3435.2	Superheated vapor
2	7	39	0.9261	7.7055	2394.4	Liquid–vapor mixture
3	7	39	0	—	163.4	Saturated liquid
4	1000	—	[a]	—	164.4	Subcooled liquid

[a] Not applicable

Example 50.1

Consider the power cycle shown in Figure 50.1(b) and let water be the working fluid. The mass flow rate $\dot{m}$ through each component is 100 kg/h, the turbine inlet pressure P_1 is 1000 kPa, and turbine inlet temperature T_1 is 480°C. The condenser pressure is 7 kPa and saturated liquid leaves the condenser. The processes through the turbine and the pump are assumed to be isentropic (adiabatic and reversible, hence constant entropy). The pressure drop in the flow direction is assumed to be negligible in both the steam generator and the condenser as well as in the connecting lines. Compute $\dot{W}_{\text{net}}$, $\dot{Q}_H$, η, and $\dot{Q}_L$.

Solution

Table 50.1 lists the properties at each state and Figure 50.2 shows the temperature (T) versus entropy (s) diagram for the cycle. Property values were obtained from *Steam Tables* by Keenan et al. [1978]. Evaluation of properties using such tables is covered in basic textbooks on engineering thermodynamics, for example, Sonntag et al. [2002]. Properties at the various states were established as follows. State 1 is in the superheat region and the values of entropy and enthalpy were obtained from the superheat table of *Steam Tables* at the noted values of P_1 and T_1. Also, since s_2 is equal to s_1,

$$s_2 = s_f + x_2(s_g - s_f) = 7.7055 = 0.5592 + x_2(8.2758 - 0.5592)$$

This yields the value for the quality x_2 as 0.9224 and allows h_2 to be calculated as

$$h_2 = h_f + x_2(h_g - h_f) = 163.4 + 0.9261(2572.5 - 163.4) = 2394.4 \text{ kJ/kg}$$

In these equations, quantities with f and g subscripts were obtained from the saturation table of *Steam Tables* at P_2. The value of enthalpy at state 4 was determined by first computing ${}_3w_4$, the work per unit mass for the process through the pump, using the expression for the reversible steady-flow work with negligible kinetic and potential energy changes [Sonntag et al., 2002]:

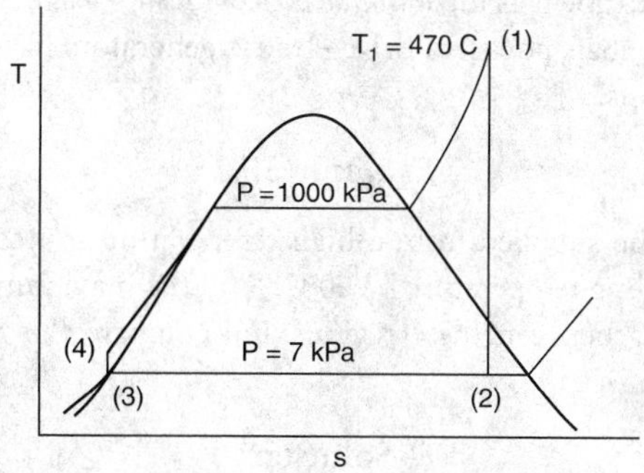

FIGURE 50.2 Temperature–entropy diagram for the steam power cycle in Example 50.1.

$$ {}_3w_4 = -\int_3^4 v\,dP = -v_3(P_4 - P_3) $$

where the specific volume v is assumed constant at v_3 since a liquid is pumped. With v_3 obtained from *Steam Tables* as v_f at P_3,

$$ {}_3w_4 = -0.001007(1000-7) = -1.00 \text{ kJ/kg} $$

Writing Equation (50.2) for the adiabatic process from state 3 to state 4,

$$ h_4 = h_3 - {}_3w_4 = 163.39 - (-1.00) = 164.4 \text{ kJ/kg} $$

Proceeding with the solution for $\dot{W}_{net}$,

$$ \begin{aligned} \dot{W}_{net} &= \dot{W}_{turbine} + \dot{W}_{pump} = \dot{m}_1 w_2 + \dot{m}_3 w_4 = \dot{m}(h_1 - h_2) + \dot{m}_3 w_4 \\ &= 100.0(3435.2 - 2394.4) + 100.0(-1.00) = 103980 \text{ kW} \end{aligned} $$

where ${}_1w_2$ was obtained by writing Equation (50.2) between states 1 and 2. Next, $\dot{Q}_H$ is determined by writing Equation (50.1) for a control volume enclosing the steam generator and noting that there is no power transmitted across its surface. Equation (50.1) reduces to

$$ \dot{Q}_H = \dot{m}h_1 - \dot{m}h_4 = 100.0(3435.2 - 164.4) = 327080 \text{ kW} $$

To find η, substitution into Equation (50.4) yields

$$ \eta = 103980/327080 = 0.318 \text{ or } 31.8\% $$

The solution for $\dot{Q}_L$ can be obtained by either of two approaches. First, by Equation (50.5),

$$ |\dot{Q}_L| = \dot{Q}_H - \dot{W}_{net} = 327080 - 103980 = 223\,100 \text{ kW} $$

The solution is also obtained by writing the first law for the process between state 2 and state 3. The result is

$$ |\dot{Q}_L| = |\dot{m}(h_2 - h_3)| = |100(163.4 - 2394.4)| = 223\,100 \text{ kW} $$

The cycle in this example is known as the *Rankine cycle with superheat.* Modified forms of this cycle are widely used to provide shaft power to drive electric generators in steam-electric power plants and other power applications.

Example 50.2

Let $\dot{Q}_H$ in Example 50.1 be supplied from a high-temperature source at a fixed temperature of 500°C and let the surrounding temperature be 20°C. Find the maximum thermal efficiency a cycle could have while operating between these regions and compare this value with η calculated in Example 50.1.

Solution

Equation (50.6) gives the expression for maximum thermal efficiency:

$$\eta_{max} = (T_H - T_L)/T_H = [(500+273)-(20+273)]/[500+273] = 0.621 \text{ or } 62.1\%$$

compared to 31.8% for Example 50.1. The maximum value of cycle thermal efficiency was not realized because of the inherent **irreversibilities** associated with heat transfer across finite temperature differences in the heat reception and heat rejection processes for the cycle.

50.2 Refrigeration Cycles

The function of a refrigeration cycle is to cause heat energy to continuously flow from a low-temperature region to a region at a higher temperature. The operation of a refrigeration cycle is illustrated in Figure 50.3(a), in which heat energy flows at the rate $\dot{Q}_L$ from the low-temperature refrigerated region, heat is rejected at the rate $\dot{Q}_H$ to the higher-temperature surroundings, and power $\dot{W}_{net}$ is required. From Equation (50.3) these are related as

$$|\dot{Q}_H| = \dot{Q}_L + |\dot{W}_{net}| \tag{50.7}$$

The performance parameter for conventional refrigeration cycles is termed *coefficient of performance* and is defined as

$$\beta = \dot{Q}_L / |\dot{W}_{net}| \tag{50.8}$$

The maximum value β can have when regions at uniform temperature T_H and T_L are considered is again derived from consideration of totally reversible cycles [Moran and Shapiro, 2000]. The expression, in terms of absolute temperatures, is

$$\beta_{max} = T_L / (T_H - T_L) \tag{50.9}$$

Figure 50.3(b) illustrates a simple vapor-compression refrigeration cycle. The compressor receives the refrigerant (working fluid) in the vapor phase at low pressure, state 1, and compresses it to state 2, where $P_2 > P_1$. Cooling at the condenser by means of a liquid or air coolant causes the vapor to condense to a liquid, state 3, after which it passes through a throttling device to the evaporator pressure. The refrigerant is a mixture of saturated liquid and saturated vapor at state 4. The liquid in the evaporator undergoes a phase change to vapor that is caused by the transfer of heat energy from the refrigerated region. The refrigerant leaves the evaporator as vapor at state 1, completing its thermodynamic cycle. The cycle illustrated in Figure 50.3(b) is the basis for practical refrigeration cycles.

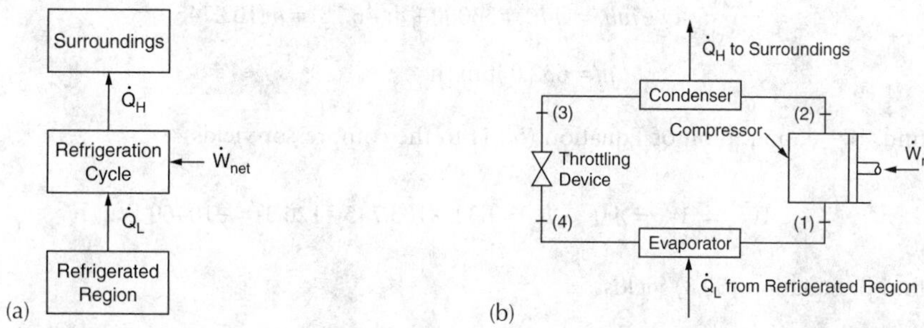

FIGURE 50.3 Descriptions of refrigeration cycles: (a) Refrigeration cycle operation. (b) The simple vapor compression refrigeration cycle.

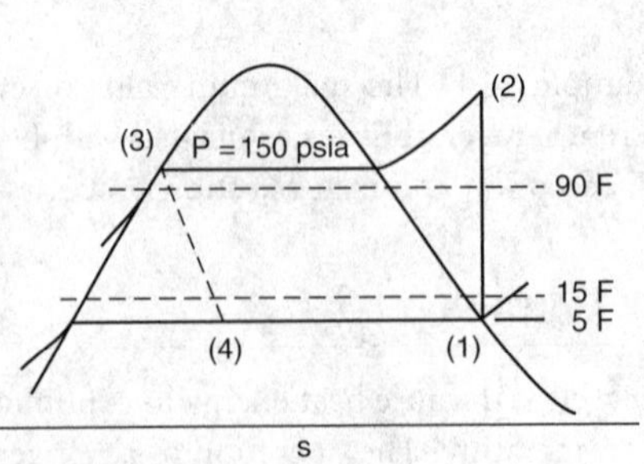

FIGURE 50.4 Temperature–entropy diagram for the refrigeration cycle in Example 50.3.

TABLE 50.2 Properties at Cycle States for Example 50.3

State	Pressure, psia	Temperature, °F	Quality, lbm/lbm	Entropy, Btu/lbm R	Enthalpy, Btu/lbm	Condition
1	23.767	5	1.00	0.22470	103.745	Saturated vapor
2	150	118.7	[a]	0.22470	120.3	Superheated vapor
3	150	105.17	0.0	0.09464	46.78	Saturated liquid
4	23.767	5	0.368	0.10210	46.78	Liquid–vapor mixture

[a] Not applicable

Example 50.3

A simple vapor compression refrigeration cycle, Figure 50.3(b), has a refrigerating capacity of three tons (36,000 Btu/h) and operates with R134a as the refrigerant. The temperature of the refrigerated region is 15°F and the surroundings are at 90°F. Saturated vapor leaves the evaporator at 5°F and is compressed isentropically by the compressor to 150 psia. The refrigerant leaves the condenser as saturated liquid at 150 psia and flows through the throttling device to the condenser, in which the temperature is uniform at 5°F. Determine $\dot{W}_{net}$, β, and the maximum coefficient of performance a refrigerator could have while operating between the refrigerated region and the surroundings.

Solution

Figure 50.4 shows the *T*-*s* diagram for the cycle and the temperatures of the two regions. Table 50.2 lists values for the various properties obtained for R134a from the *ASHRAE Handbook* [2002] at the four states. The mass rate of flow is determined by applying Equation (50.1) to the evaporator. The result is

$$\dot{Q}+\dot{m}h_1=\dot{m}h_2=36000+\dot{m}(46.78)=\dot{m}(103.745)$$

$$\dot{m}=632.0\ \text{lbm/h}$$

To find $\dot{W}_{net}$, application of Equation (50.1) to the compressor yields

$$\dot{W}_{net}={}_1\dot{W}_2=\dot{m}(h_1-h_2)=632.0(103.745-120.3)=-10460\ \text{Btu/h}$$

To find β, Equation (50.8) yields

$$\beta=36000/10460=3.44$$

The solution for maximum coefficient of performance is obtained by applying Equation (50.9) as follows:

$$\beta_{max} = [15+460]/[(90+460)-(15+460)] = 6.33$$

Irreversibilities present due to finite temperature differences associated with the heat transfer processes and the irreversibility related to the throttling process cause β to be less than β_{max}.

Defining Terms

Control volume — A region specified by a control surface through which mass flows.

First law of thermodynamics — An empirical law that in its simplest form states that energy in its various forms must be conserved.

Heat energy — Energy that is transferred across a control surface solely because of a temperature difference between the control volume and its surroundings. This form of energy transfer is frequently referred to simply as heat transfer.

Irreversibilities — Undesirable phenomena that reduce the work potential of heat energy. Such phenomena include friction, unrestrained expansions, and heat transfer across a finite temperature difference.

Steady flow — A condition that prevails in a flow process after all time transients related to the process have died out.

Working fluid — The substance that is contained within the apparatus in which the cycle is executed. The substance undergoes the series of processes that constitute the cycle.

References

ASHRAE. 2002. Refrigeration. *ASHRAE Handbook,* I-P Edition. American Society of Heating, Refrigeration and Air-Conditioning Engineers, Atlanta, GA.

Keenan, J. H., Keys, F. G., Hill, P. G., and Moore, J. G. 1978. *Steam Tables, SI Units,* John Wiley & Sons, New York.

Sonntag, R.E., Borgnakki, C., and Van Wylen, G.J. 2002. *Fundamentals of Thermodynamics,* 6th ed. John Wiley & Sons, New York.

Moran, M.J. and Shapiro, H.N., 2000. *Fundamentals of Engineering Thermodynamics,* 4th ed. John Wiley & Sons, New York.

Further Information

Proceedings of the American Power Conference, Illinois Institute of Technology, Chicago, IL. Published annually.

McQuiston, F., Parker, J.D., and Spitler, J.D., 2000. Refrigeration (chapter 15) *Heating, Ventilating, and Air Conditioning Analysis and Design,* 5th ed., John Wiley & Sons, New York

Kuehn, T.H., Ramsey, J.W, and Threlkeld, J.L., 1998. Mechanical vapor compression refrigeration cycles (Chapter 3). *Thermal Environmental Engineering,* 3rd ed. Prentice Hall, Englewood Cliffs, NJ.

51

Heat Transfer

Yildiz Bayazitoglu
Rice University
Udaya B. Sathuvalli
Rice University

There are two modes of **heat transfer** — diffusion and radiation. A *diffusion* process occurs due to the presence of a gradient (say, of temperature, density, pressure, concentration, electric potential, etc.) and *requires* a material medium. Both conduction and convection are diffusion processes. A radiative process, on the other hand, does *not* require a material medium.

This chapter will cover the basic ideas of conduction, convection, radiation, and phase change. The mode of heat transfer in solids and fluids at rest due to the exchange of the kinetic energy of the molecules or the drift of free electrons due to the application of a temperature gradient is known as *conduction.* The heat transfer that takes place due to mixing of one portion of a fluid with another via bulk movement is known as *convection.* All bodies that are at a temperature above absolute zero are thermally excited and emit electromagnetic waves. The heat transfer that occurs between two bodies by the propagation of these waves is known as *radiation. Phase change* takes place when a body changes its state (i.e., solid, liquid, or gas) either by absorption or release of thermal energy.

51.1 Conduction

Heat conduction in solids can be attributed to two basic phenomena — lattice vibrations and transport of free electrons due to a thermal gradient.

The mechanism of heat conduction in a solid can be qualitatively described as follows. The geometrical structure in which the atoms in a solid are arranged is known as a *crystal lattice.* The positions of the

0-8493-1586-7/05/$0.00+$1.50

atoms in the lattice are determined by the interatomic forces between them. When one part of a solid body is at a higher temperature than the rest of it, there is an increased amplitude of the lattice vibrations of the atoms in that region. These vibrations are eventually transmitted to the neighboring cooler parts of the solid, thus ensuring that conduction is in the direction of the temperature gradient. The mechanical oscillations of these atoms are known as *vibration modes.* In classical physics, each of these vibrations can have an *arbitrary* amount of energy. However, according to quantum mechanics, this energy is *quantized.* These quanta (in analogy with the theory of light) are called **phonons** and are the primary carriers of heat in dielectrics. The mathematical theory that studies heat conduction due to such lattice vibrations is known as the *phonon theory of conduction* (see Ziman [1960]).

On the other hand, in a metal an abundance of free electrons is present, and they are the primary carriers of heat. There is a certain amount of conduction due to phonons, but it is usually negligible. A metal can be modeled as a lattice of positively charged nuclei surrounded by closed shells of electrons and immersed in a sea of free valence electrons. The free valence electrons are the carriers of energy, and their motion in the absence of a thermal gradient is determined by the combined electric fields of the ions in the lattice. At thermal equilibrium (no net temperature gradient) there are as many electrons moving in one direction as in the opposite, and there is no net energy transport. However, when a temperature gradient is imposed, electrons that cross a given cross-section of the solid have different temperatures (different velocities) at different points because the electrons collide with the ions in the lattice and reach local thermal equilibrium. This directed migration of the electrons eventually ceases when the thermal gradient disappears and the temperature becomes uniform throughout. This model can be used to determine the heat flux in terms of the applied temperature gradient and a material property known as the *thermal conductivity.*

Fourier's Law of Heat Conduction

The important quantities in the study of heat conduction are the temperature, the heat flux, and the thermal conductivity of the material. The total amount of energy per unit time that crosses a given surface is known as the *heat flow* across the surface, Q. The energy per unit time per unit area across the surface is called the *heat flux,* **q**. Heat flux (unlike heat flow) is a vector quantity and is directed along the normal to the surface at which it is measured.

The fundamental relation between the heat flux and the temperature gradient is **Fourier's law,** which states that heat flow due to conduction in a given direction is proportional (see Figure 51.1) to the temperature gradient in that direction and the area normal to the direction of heat flow. Mathematically,

$$Q_x(x, t) = -\kappa A \frac{dT(x, t)}{dx} \tag{51.1}$$

where $Q_x(x, t)$ is the heat flow in the positive x direction through the area A, $dT(x, t)/dx$ is the temperature gradient at point x at time t, and κ is the **thermal conductivity** of the material. The minus sign on the

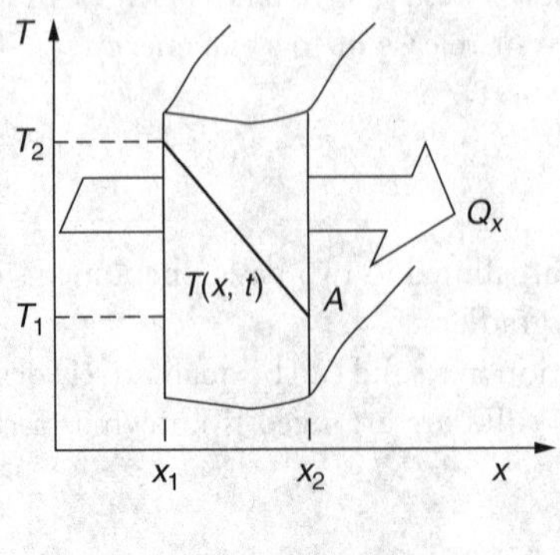

FIGURE 51.1 Flow of heat in a slab.

right-hand side of Equation (51.1) is in accordance with the second law of thermodynamics and implies that heat flow due to conduction must occur from a point at higher temperature to a point at lower temperature. Equation (51.1) can be generalized for a solid to yield

$$\mathbf{q}(\mathbf{r},\ t) = -\kappa\nabla T(\mathbf{r},\ t) \tag{51.2}$$

where $\mathbf{q}(\mathbf{r}, t)$ is the heat flux vector (i.e., heat flow per unit area, Q_x/A) and $T(\mathbf{r}, t)$ is the temperature at a point in the body whose position vector is $\mathbf{r}$ at time t. In SI units, heat flux is measured in W/m^2, temperature gradient in °C/m, and thermal conductivity in W/m°C.

When the domain thickness is on the order of the electron/phonon mean free path, the Fourier law of conduction cannot be applied.

Thermal Conductivity of Materials

Materials display a wide range of thermal conductivities. Between gases (e.g., air) and highly conducting metals (e.g., copper, silver), κ varies by a factor of 10^4. Metals have the highest thermal conductivity (~10^4 W/m°C) and gases such as hydrogen and helium are at the lower end of the range (~10^{-2} W/m°C). Materials such as oils and nonmetallic oxides have conductivities that range between 10^{-1} W/m°C and 10^2 W/m°C and are known as *thermal insulators.* Thermal conductivity is also known to vary with temperature. For some materials the variation over certain temperatures is small enough to be neglected. Generally, the thermal conductivity of metals decreases with increase in temperature. For gases (air, CO_2, H_2, etc.) and most insulators (asbestos, amorphous carbon, etc.) it increases with increase in temperature.

Generally speaking, good thermal conductors are also good electrical conductors and vice versa. The free electron theory of metals accounts for their high conductivities and shows that the thermal and electrical conductivities of metals are related by the *Weidemann–Franz law,* which states that at absolute temperature T,

$$\frac{\kappa}{\sigma_e T} = 2.23\cdot 10^{-8}\ \mathrm{W\Omega/K^2}$$

where σ_e is the electrical conductivity of the metal.

The Energy Equation

The study of heat conduction is primarily concerned with knowing the temperature distribution in a given body as a function of position and time. Consider an arbitrary body of volume V and area A, bounded by a surface S. Let the body have an internal heat source that generates heat at the rate of $g(\mathbf{r}, t)$ per unit volume (in SI units, the units for this quantity are W/m^3). Then, *conservation of energy* requires that the sum of heat fluxes that enter the body through its surface and the internal rate of heat generation in the body should equal the net rate of accumulation of energy in it. Mathematically,

$$\begin{aligned} -\oint_S \mathbf{q}(\mathbf{r},\ t)\cdot\hat{\mathbf{n}}\,dA\big|_{\text{conduction}} &= \oint_S \mathbf{q}(\mathbf{r},t)\cdot\hat{\mathbf{n}}\,dA\big|_{\text{other modes}} - \oint_V g(\mathbf{r},t)dV \\ &+ \oint_V \rho C_p \frac{\partial T(\mathbf{r},t)}{\partial T}dV \end{aligned} \tag{51.3}$$

where the left-hand side represents the net heat flux that enters the body by conduction, the first term on the right-hand side represents the corresponding term for other modes of heat transfer (i.e., convection or radiation), the second term on the right-hand side represents the total heat generated in the body, and the last term represents the accumulation of thermal energy in it. Here $\hat{n}$ is the outward drawn unit normal to the surface of the body, ρ is the density of the body, and C_p its specific heat. Equation (51.3) is known as the *integral form* of the **energy equation.** In Equation (51.3), the internal heat generation can be due to electromagnetic heating, a chemical or nuclear reaction, etc.

Using Gauss's divergence theorem, the surface integrals of the heat flux can be converted into volume integrals, and this leads to

$$-\nabla\cdot\mathbf{q}(\mathbf{r},t)\big|_{\text{conduction}}=\nabla\cdot\mathbf{q}(\mathbf{r},t)\big|_{\text{other modes}}-g(\mathbf{r},t)+\rho C_p\frac{\partial T(\mathbf{r},t)}{\partial t} \tag{51.4}$$

The above equation is known as the *distributed form* of the energy equation. When conduction is the only mode of heat transfer, Fourier's law [Equation (51.2)] can be used for a *homogeneous isotropic* medium (i.e., a medium whose thermal conductivity does not vary with position or direction) in Equation (51.4) to give

$$\nabla^2T(\mathbf{r},t)+\frac{1}{\kappa}g(\mathbf{r},t)=\frac{1}{\alpha}\frac{\partial T(\mathbf{r},t)}{\partial t} \tag{51.5}$$

where α is the **thermal diffusivity** of the medium and is defined as

$$\alpha=\kappa/(\rho C_p) \tag{51.6}$$

The thermal diffusivity of a material governs the rate of propagation of heat during transient processes. In SI units it is measured in m²/sec. When there is no internal heat generation in the body, Equation (51.5) is known as the *diffusion equation.*

In many instances the temperature in the body may be assumed to be constant over its volume and is treated only as a function of time. Then Equation (51.3) may be integrated to give

$$-Q\big|_{\text{conduction}}=Q\big|_{\text{other modes}}-G+\rho C_pV\frac{dT}{dt} \tag{51.7}$$

where T is the temperature of the body and G is the net heat generated in it. This is known as the *lumped form* of the energy equation.

One of the main problems of heat conduction is to obtain the solution of Equation (51.5) subject to appropriate boundary conditions and an initial condition. The boundary conditions for the solution of Equation (51.5) are of three kinds. The *constant temperature boundary condition* is obtained by prescribing the temperature at all points on the surface S. The *constant heat flux boundary condition* is invoked when the body is losing a known amount of heat to (or receiving heat from) the external ambient, for example, by conduction, convection, or radiation. The *mixed boundary condition* is used when a linear combination of the surface temperature and the heat flux leaving the surface is known. This situation typically occurs when a body is losing heat by convection from its surface to an ambient at a lower temperature. These boundary conditions are also known as *Dirichlet, Neumann,* and *Cauchy* boundary conditions, respectively. Once the boundary conditions and the initial condition for a given heat conduction situation are formulated, the problem reduces to a boundary value problem. The solutions to these problems are well documented in standard texts on heat conduction such as Carslaw and Jaeger [1959] and Ozisik [1993].

Limits of Fourier's Law

In certain applications that deal with transient heat flow in very small periods of time, temperatures approaching absolute zero, heat flow due to large thermal gradients, and heat flow on a nano- or microscale (such as in thin films), Fourier's law of heat conduction [Equation (51.1)] is known to be unreliable. This is because the diffusion equation predicts that the effect of a temperature gradient at a point $\mathbf{r}$ that is established at time t should be instantly felt everywhere in the medium, implying that temperature disturbances travel at an *infinite* speed. In order to account for the *finite* speed of propagation of heat, Equation (51.1) has been modified as

$$\tau\frac{\partial \mathbf{q}(\mathbf{r},t)}{\partial t}+\mathbf{q}(\mathbf{r},t)=-\kappa\nabla T(\mathbf{r},t) \tag{51.8}$$

where τ is a *relaxation time.* When this equation is used in the differential form of the energy equation, Equation (51.5), it results in a hyperbolic differential equation known as the *telegrapher's equation*:

$$\nabla^2 T(\mathbf{r},t)+\frac{1}{\kappa}\left[g(\mathbf{r},t)+\frac{\alpha}{c^2}\frac{\partial g(\mathbf{r},t)}{\partial T}\right]=\frac{1}{c^2}\frac{\partial^2 g(\mathbf{r},t)}{\partial T} \tag{51.9}$$

where c is the wave propagation speed and is given by $c=(\alpha/\tau)^{-1/2}$. The solution for Equation (51.9) indicates that heat propagates as a wave at a finite speed and is the basis of the theory of *heat waves.* For $\tau\to 0$, Equation (51.8) reduces to Fourier's law [Equation (51.2)], and when $c\to\infty$, Equation (51.9) becomes the diffusion equation, Equation (51.5).

Dimensionless Variables in Heat Conduction

If a body is subject to a mixed boundary condition given by

$$-\kappa\hat{\mathbf{n}}\cdot\nabla T(\mathbf{r},t)=h(T(\mathbf{r},t)-T_\infty)$$

on the surface S, where h is the heat transfer coefficient from the surface of the body, then Equation (51.5) can be nondimensionalized with the following variables: the *Fourier number,* Fo $=\alpha t/L^2$ (i.e., the dimensionless time variable), the dimensionless heat generation variable $[g(\mathbf{r},t)L^2]/[\kappa(T_o-T_\infty)]$, and the *Biot number,* Bi $=hL/\kappa$, where L and T_o are suitable length and temperature scales, respectively. The Fourier number denotes the ratio of the heat transferred by conduction to the heat stored in the body and is useful in solving transient problems. When the Fourier number is very large, transient terms in the solution of the diffusion equation may be neglected. The Biot number is a measure of the relative magnitudes of the heat transfer due to conduction and convection (or radiation) in the body. When the Biot number is very small (<0.1), the temperature in the body may be assumed to be a constant, and the lumped form of the energy equation may be used.

51.2 Convection

Convection takes place when a fluid moves over a solid and their temperatures are unequal. Heat transfer occurs due to actual material transport, unlike in the case of conduction. In many engineering applications the heat transfer due to convection may be calculated by using **Newton's law of cooling.** It states that if the heat transfer coefficient is h, the heat flux q due to convection from a surface at temperature T_w into a fluid at temperature T_f is given by

$$q=h(T_w-T_f) \tag{51.10}$$

There are three kinds of convection: forced, natural, and mixed. *Forced convection* takes place when the motion of the fluid that causes convection is sustained by an externally imposed pressure gradient. Forced convection typically occurs in systems such as blowers and air conditioners. Sometimes, even in the absence of external forces, pressure gradients are created due to differences in density that are caused by local heating in the fluid. This heat transfer is known as *free* or *natural* convection. *Mixed* convection, as the name implies, is the situation in which both forced and free convection are present.

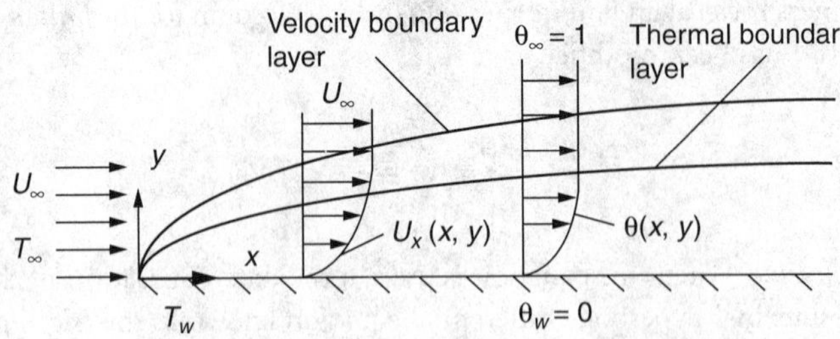

FIGURE 51.2 The thermal boundary layer.

Thermal Boundary Layer

In analogy with the momentum (or the velocity) boundary layer, a thermal boundary layer can be defined. Imagine a fluid of uniform temperature T_∞ flowing with a uniform velocity U_∞ (the free stream values) along a flat plate maintained at temperature $T_w(< T_\infty)$ (see Figure 51.2). When the fluid comes into contact with the flat plate, the layers of fluid immediately adjacent to the plate attain the temperature T_w. Deeper into the fluid (away from the plate) the temperature of the fluid increases slowly. Finally, the temperature of the fluid becomes uniform at T_∞. This suggests that there is a "temperature profile" and hence a "temperature gradient" in the layers of the fluid close to the flat plate. It is now convenient to define a dimensionless temperature $\theta(x, y)$ at each point (x, y) of the fluid as

$$\theta(x,y)=\frac{T(x,y)-T_w}{T_\infty-T_w}$$

For a point very close to the wall, for example, $y = 0$, the temperature of the fluid is the same as that of the wall and, hence, $\theta(x, y) = 0$. At points sufficiently far away from the wall, the temperature of the fluid is T_∞ and, hence, $\theta(x, y) \to 1$. For each location x along the flat plate, there is a distance y from the plate at which $\theta(x, y) = 0.99$. The locus of all such points is called the *thermal boundary layer.* For practical purposes, locations *outside* the boundary layer are thermally not affected by the plate.

Heat Transfer Coefficient

The **heat transfer coefficient** h [Equation (51.10)] is a very useful concept in the determination of the heat transfer due to convection. Consider the case shown in Figure 51.2. Very close to the wall, the fluid particles are stationary (due to the no-slip boundary condition) and the heat transfer is due to conduction in the fluid. Then the heat flux due to conduction at the wall is

$$q\big|_{\text{wall}}=-\kappa_f\left.\frac{\partial T(x,y)}{\partial y}\right|_{y=0} \tag{51.11}$$

where κ_f is the thermal conductivity of the fluid. In engineering applications the heat transfer between the fluid and the wall is related to the heat transfer coefficient using Newton's law of cooling by

$$h_x=-\kappa_f\frac{[\partial T(x,y)/\partial y]_{y=0}}{(T_\infty-T_w)} \tag{51.12}$$

The heat transfer coefficient h_x in Equation (51.12) may vary along the length of the surface shown in Figure 51.2, that is, it may be a function of x. It is therefore useful to define a *mean* or *bulk* value of the heat transfer coefficient over a finite length L as

$$h_m = \frac{1}{L}\int_0^L h_x dx \qquad (51.13)$$

In SI units h_x is measured in W/m^2°C. The heat transfer coefficient is often expressed via an important dimensionless quantity known as the *Nusselt number.* The Nusselt number is defined as

$$\mathrm{Nu}_x = \frac{h_x x}{\kappa_f} \qquad (51.14)$$

and denotes the ratio of the actual convection heat flux to the conduction heat flux that would occur through a fluid slab of thickness x. Note the similarity of the Nusselt number with the Biot number (defined earlier). As in the case of h, it is useful to define a mean or bulk Nusselt number as follows:

$$\mathrm{Nu}_m = \frac{h_m L}{\kappa_f} \qquad (51.15)$$

Similarity Parameters of Convection

Heat transfer by convection depends on the characteristics of the flow pattern of the fluid, its thermophysical properties, the geometry of the flow passage, and surface conditions. The flow patterns may be characterized as *laminar, transitional,* or *turbulent.* The thermophysical properties that determine heat transfer are the fluid density ρ, thermal conductivity κ_f, kinematic viscosity ν, and specific heat C_p. The flow may be *external* or *internal* and the flow geometry may take such forms as flow over a flat plate, a cylinder, or a sphere, flow in a channel or an enclosed space, etc. Consider the laminar flow along a flat plate in which the local heat transfer coefficient at a location x is h_x. The local heat transfer coefficient h_x is a function of x, κ_f, the local velocity U_x, ρ, ν, the volume expansion coefficient of the fluid β, a characteristic temperature difference $\Delta T(\sim[T_\infty - T_w])$, and C_p. It has been found that these variables can be grouped into a set of five nondimensional numbers in a functional relationship of the form

$$\mathrm{Nu}_x = f(\mathrm{Re}_x,\ \mathrm{Pr},\ \mathrm{Gr}_x,\ \mathrm{Ec}) \qquad (51.16)$$

where the arguments of f are the nondimensional numbers discussed as follows. The importance of Equation (51.16) is that it suggests that instead of determining h_x as a function of nine variables, it may be sought as a function of four nondimensional numbers. Further, Equation (51.16) is obtained by a method known as *similarity analysis.* Similarity analyses usually indicate the actual functional form of Equation (51.16) in advance.

Reynolds Number

The local *Reynolds number* is defined by

$$\mathrm{Re}_x = \frac{U_\infty x}{\nu} = \frac{\rho U_\infty x}{\mu} \qquad (51.17)$$

where μ is the dynamic viscosity of the fluid and is related to ν by $\nu = \mu/\rho$. It denotes the relative magnitudes of the inertial to viscous forces that govern the flow. For large values of the Reynolds number, the inertial forces dominate (low-viscosity flows), whereas the viscous forces dominate in small Reynolds number flow (creeping flows, viscous fluids).

Prandtl Number

The *Prandtl number* is defined as

$$\mathrm{Pr} = \frac{\nu}{\alpha_f} \tag{51.18}$$

and is the ratio of the kinematic viscosity and the thermal diffusivity $\alpha_f (= \kappa_f / \rho C_p)$ of the fluid. It represents the relative rates of the diffusion of the momentum and thermal boundary layers. In liquid metals, Pr << 1 and therefore the rate diffusion of thermal energy greatly exceeds that of momentum. In oils, Pr >> 1 and the converse holds.

Grashof Number

The *Grashof number* is significant in the analysis of free convection, which is discussed later. It is defined as

$$\mathrm{Gr}_x = \frac{g\beta x^3 (T_w - T_\infty)}{\nu^2} \tag{51.19}$$

where g is the acceleration due to gravity, β is the coefficient of volume expansion of the fluid, and $T_w - T_\infty$ is a characteristic temperature difference that generates free convection. This number signifies the relative importance of the buoyant and the viscous forces in the flow. When there is no natural convection (for example, in microgravity situations $g \approx 0$, or when there is not enough thermal gradient to cause appreciable convection, i.e., ΔT is very small) and there is only forced convection, Gr = 0.

Eckert Number

The *Eckert number* appears due to the inclusion of the *viscous dissipation* term in the equation for the conservation of the energy is boundary layer. The viscous dissipation term denotes the rate at which mechanical energy is being dissipated into thermal energy due to the presence of viscous forces in the fluid. The viscous dissipation is proportional to the square of the velocity gradient and the dynamic viscosity of the fluid. The Eckert number is defined as

$$\mathrm{Ec} = \frac{U_\infty^2}{C_p \Delta T} \tag{51.20}$$

and measures the kinetic energy of the flow with respect to the enthalpy difference across the thermal boundary layer.

The principal aim of convection can be said to be the determination of the Nusselt number as a function of the problem parameters that are expressed in terms of the above nondimensional numbers.

Forced Convection over Bodies (External Flows)

When there is only forced convection and the viscous dissipation is small enough (Ec ≈ 0), Equation (51.16) becomes

$$\mathrm{Nu}_x = f(\mathrm{Re}_x, \mathrm{Pr}) \tag{51.21a}$$

A more specific functional form for the above relation may be chosen as

$$\mathrm{Nu}_x = K(\mathrm{Re}_x)^a (\mathrm{Pr})^b \tag{51.21b}$$

where *K*, *a*, and *b* must be chosen by fitting curves to the experimental data or by analysis of the governing equations of convection. It is convenient to combine the Reynolds and Prandtl numbers to define a new nondimensional number, the *Peclet number*, as

$$\mathrm{Pe}_x = \mathrm{Re}_x \mathrm{Pr} = \frac{U_\infty x}{\alpha_f} \tag{51.22}$$

Actual correlations in terms of these nondimensional numbers for different flow situations are given in standard textbooks of heat transfer [Bayazitoglu and Ozisik, 1988; Kreith and Bohn, 1986].

For laminar flow, correlations of the form given by Equation (51.21) may be obtained by analytical methods. However, when the flow is turbulent, it is not easy to determine such correlations analytically and experimental methods must be used. In experiments it is often simpler to measure the drag coefficient c_x rather than the heat transfer coefficient. Furthermore, there are correlations between the heat transfer and drag coefficients. One such relation is the *Reynolds–Colburn* analogy given by

$$\mathrm{St}_x \mathrm{Pr}^{2/3} = \frac{1}{2} c_x \tag{51.23a}$$

where

$$\mathrm{St}_x = \frac{\mathrm{Nu}_x}{\mathrm{Re}_x \mathrm{Pr}} = \frac{h_x}{\rho U_\infty C_p} \tag{51.23b}$$

is the local *Stanton number* and is the nondimensional local heat transfer coefficient. The Reynolds–Colburn analogy may be used to develop expressions for local heat transfer coefficients. The relevant physical fluid properties that appear in these correlations are functions of the temperature and are usually evaluated at the **film temperature** T_f, which is defined as the mean boundary layer temperature:

$$T_f = \frac{T_w + T_\infty}{2} \tag{51.24}$$

Forced Convection in Ducts (Internal Flows)

The following concepts (in addition to the ones discussed in the previous section) are needed to understand convection due to fluid flow in ducts. When the fluid (see Figure 51.3) enters a tube whose wall is maintained at a temperature different from that of the fluid, there is heat transfer and the temperature distribution is such that, starting at the tube inlet, a thermal boundary layer develops and grows along the length of the tube until the thermal boundary layer occupies the entire width of the duct. The region from the entrance of the duct to the point where the thermal boundary layer reaches the axis of the duct is known as the *thermal entrance region*. The length of this region is called the *thermal entry length*, L_t. In this region the temperature distribution varies along the radius and the length of the duct. Beyond this region the temperature is only a function of the radial coordinate. This region is known as the *thermally developed region*. In the thermal entrance region the average temperature of the fluid at any cross-section (averaged along the radial direction) is known as the *mean fluid temperature*. If the axial velocity of the fluid inside the duct is $U(r)$, the mean temperature is given by

$$T_m(z) = \frac{\int_0^R \rho C_p U(r) T(r,z)(2\pi r)dr}{\int_0^R \rho C_p U(r)(2\pi r)dr} \tag{51.25}$$

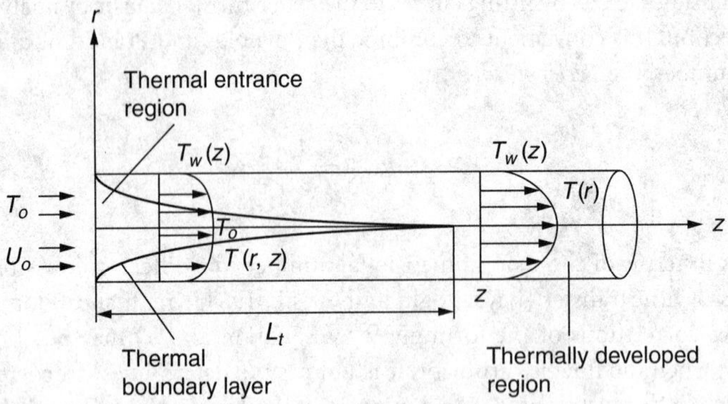

FIGURE 51.3 Thermal entrance and thermally developed regions for flow in a duct.

Based on this equation, the local heat transfer coefficient h_z is defined as

$$h_z = -\frac{\kappa_f}{T_m(z) - T_w(z)} \frac{\partial T(r,z)}{\partial r}\bigg|_{r=R,\,\text{wall}} \tag{51.26}$$

where $T_w(z)$ is the tube wall temperature at z. A quantity known as the *logarithmic mean temperature difference* (*LMTD*), $\Delta T_{\ln}$, which is useful in the determination of the total heat transfer between the fluid and wall of the duct (for example, in heat exchangers), is defined as

$$\Delta T_{\ln} = \frac{\Delta T_1 - \Delta T_2}{\ln(\Delta T_1 / \Delta T_2)} \tag{51.27}$$

where ΔT_1 and ΔT_2 are temperature differences between the fluid and the wall at the inlet and the outlet, respectively. Finally, in internal flows, the characteristic length parameter that is used to define the Nusselt number is known as the *hydraulic diameter* and is defined as $D_h = 4\, A_c/P$, where A_c is the cross-sectional area of the duct and P is its perimeter.

When the flow is turbulent, the heat transfer analysis is more complicated. There are several empirical correlations to calculate the heat transfer in ducts that carry turbulent flows. For flow in a smooth pipe of length L and diameter D, the *Dittus-Boelter* and the *Petukhov* equations are most commonly used. (See Bayazitoglu and Ozisik [1988] for the actual relations.)

Although they are incomplete, recently several analytical and experimental works have been performed to have a better understanding of heat transfer at microscale. Since the ratio of the surface area to channel cross-section is large for microchannels and steep velocity gradients exist in laminar flow, the effects of viscous heating can be significant.

Free Convection

Consider a vertical flat plate at a temperature T_w immersed in a fluid at temperature T_∞. If $T_w > T_\infty$, there is heat transfer from the plate to the fluid. The velocity and temperature profiles for this case of free convection are shown in Figure 51.4.

The Grashof number [defined in Equation (51.19)] plays an important role in free convection. In the absence of forced convection and viscous dissipation, Equation (51.16) becomes

$$\text{Nu}_x = f(\text{Gr}_x, \text{Pr}) \tag{51.28a}$$

For gases, $\text{Pr} \cong 1$ and Equation (51.28) suggests that $\text{Nu}_x = f(\text{Gr}_x)$.

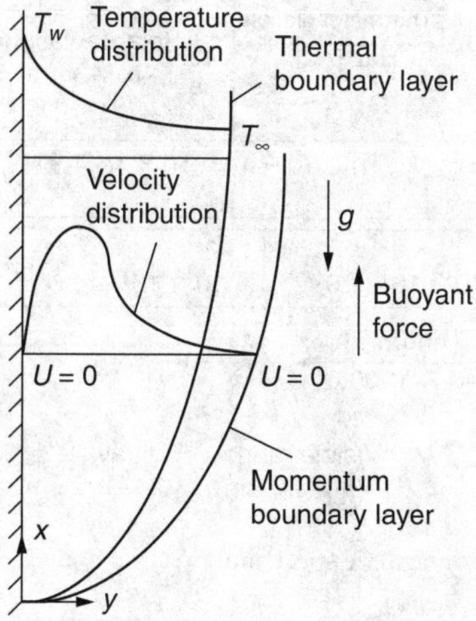

FIGURE 51.4 Free convection over a heated vertical plate.

The role of the Grashof number in free convection is similar to that of the Reynolds number in forced convection. However, free convection may take place even in situations where forced convection dominates. In such cases the relative importance of these two modes of convection is determined by the parameter Gr/Re^2. If $Gr/Re^2 << 1$, forced convection dominates and free convection may be neglected. When $Gr/Re^2 >> 1$, free convection dominates and the heat transfer correlations are of the form given by Equation (51.28a). And when $Gr/Re^2 \sim 1$, both free and forced convection must be considered. The heat transfer correlations are sometimes defined in terms of the *Rayleigh number,* Ra, given by

$$\mathrm{Ra}_x = \mathrm{Gr}_x\mathrm{Pr} = \frac{g\beta(T_w - T_\infty)x^3}{\nu\alpha} \tag{51.28b}$$

In some of the free convection correlations, the physical properties are evaluated at a mean temperature given by

$$T_m = T_w - 0.25(T_w - T_\infty) \tag{51.28c}$$

Free convection inside closed volumes involves an interesting but complicated flow phenomenon. Consider a fluid between two large horizontal plates maintained at temperatures T_h and T_c and separated by a distance *d*. If the bottom plate is at temperature T_h ($>T_c$), heat flows upward through the fluid, thus establishing a temperature profile that decreases upward. The hotter layers of fluid are at the bottom, whereas the colder (and less dense) layers are at the top. This arrangement of the fluid layers remains stationary and heat transfer is by conduction alone. This unstable state of affairs lasts as long as the buoyancy force does not exceed the viscous force. If the temperature difference (gradient) between the top and bottom plates is strong enough to overcome the viscous forces, fluid convection takes place. Theoretical and experimental investigations have shown that the fluid convection in such an enclosure occurs when the Rayleigh number reaches a critical value Ra_c, which is given in terms of the (characteristic) spacing *d* between the plates. The convective flow patterns that arise in such situations are known as *Bernard cells.*

51.3 Radiation

All bodies that are above absolute zero temperature are said to be thermally excited. When bodies are thermally excited, they emit *electromagnetic waves,* which travel at the speed of light ($c = 3 \cdot 10^8$ m/sec in

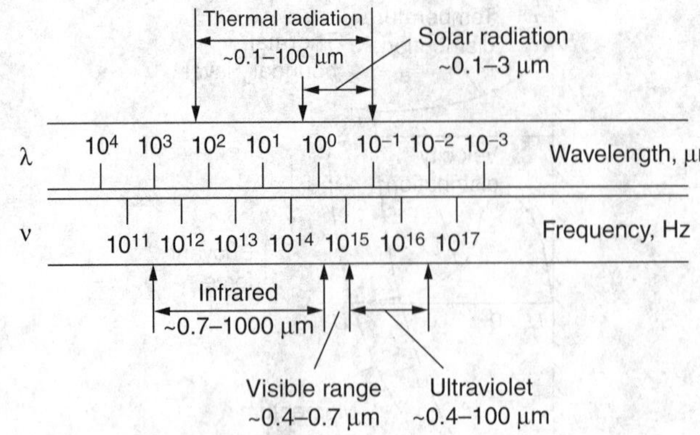

FIGURE 51.5 The range of the electromagnetic spectrum.

free space or vacuum). This radiation takes place at different wavelengths. Figure 51.5 shows the typical electromagnetic spectrum. The radiation that occurs at wavelengths between $\lambda = 0.1$ and 100 μm is called *thermal radiation*, and the ensuing heat transfer is known as *radiative heat transfer.* The wave nature of radiation implies that the wavelength λ should be associated with a frequency ν, which is given by $\nu = c/\lambda$. Bodies emit radiation at different wavelngths, and the intensity of the radiation varies with the wavelength. This is referred to as the *spectral distribution* of the radiation. Since electromagnetic waves can travel in vacuum, radiative transfer as opposed to conductive and convective transfer does *not* require a material medium.

Basic Quantities of Radiation

The radiation that is incident on the surface of a body may be absorbed, reflected, transmitted, or scattered. When all the incident radiation is absorbed by the body, the body is said to be *opaque.* If the material thickness required to absorb the radiation is very large compared to the characteristic thickness of the body, the radiation may be transmitted entirely. For example, glass is a very poor absorber of electromagnetic waves in the visible spectrum and is *transparent* to radiation in the visible spectrum. If the surface of the body is shiny and smooth, a good part of the radiation may be reflected (as with metallic surfaces). A material property can be associated with each of these phenomena: *absorptivity*, α; *reflectivity*, ρ; and *transmissivity*, τ. When these properties depend on the frequency of the incident radiation, they are said to exhibit *spectral dependence*, and when they depend on the direction of the incident radiation, they are said to exhibit *directional dependence.* When radiation travels through a relatively transparent medium that contains inhomogeneities such as very small particles of dust, it gets *scattered.* Scattering is defined as the process in which a photon collides with one or more material particles and does not lose all its energy.

When radiative properties (or quantities) depend upon the wavelength and direction of the incident radiation, they are known as *monochromatic directional* properties. When these properties are summed over all directions *above* the surface in question, they are called *hemispherical properties.* When they are summed over the entire spectrum of radiation, they are referred to as *total properties.* When summed over all the directions above the surface in question as well as the entire spectrum of radiation, they are called *total hemispherical properties.*

The *hemispherical monochromatic emissive power*, $E_\lambda(\lambda, T)$, of a body is defined as the emitted energy leaving its surface per unit time per unit area at a given wavelength λ and temperature T. It is worth emphasizing that "emitted" refers to the *original* emission from the body — that is, the radiation due to thermal excitation of the atoms in the body. The *hemispherical monochromatic radiosity*, $J_\lambda(\lambda, T)$, refers to all the radiant energy (emitted and reflected) per unit area per unit time that leaves the surface. Radiant energy that is incident on a surface from all directions is known as *hemispherical monochromatic irradi-*

ation, $G_\lambda(\lambda)$. In each of these cases the total quantity is obtained by integrating the monochromatic quantity over the entire spectrum, that is, from $\lambda = 0$ to ∞.

Radiation from a Blackbody

In the study of radiation it is useful to define an ideal surface in order to compare the properties of real surfaces. A body that absorbs *all* incident radiation regardless of its spectral distribution and directional character is known as a **blackbody.** Therefore, a blackbody is a perfect absorber. The radiation emitted by a blackbody at a given temperature is the maximum radiation that can be emitted by any body.

The emissive power $E_b\lambda(\lambda, T)$ of a blackbody at a given wavelength λ and absolute temperature T is given by **Planck's law** as

$$E_{b,\lambda}(\lambda, T) = \frac{c_1}{\lambda^5\{e^{c_2/\lambda T} - 1\}} \tag{51.29}$$

where $c_1 = 3.743 \cdot 10^8$ W-μm^4/m^2 and $c_2 = 1.4387 \cdot 10^4$ μm K. The units for c_1 and c_2 indicate that λ is measured in μm and $E_{b,\lambda}(\lambda, T)$ in W/m^2. The *total emissive power* $E_b(T)$ of the blackbody is the radiation emitted by it at all wavelengths. Mathematically,

$$E_b(T) = \int_{\lambda=0}^{\infty} E_{b,\lambda}(\lambda, T)\,d\lambda = \sigma T^4 \tag{51.30}$$

where σ is the *Stefan–Boltzmann constant*, given by

$$\sigma = \left(\frac{\pi}{c_2}\right)^4 \frac{c_1}{15} = 5.67 \cdot 10^{-8}\ \mathrm{W/m^2K^4}$$

Equation (51.30) is known as the **Stefan–Boltzmann law** of radiation. Experiments have shown that the peak values of the emission increase with T and occur at shorter wavelengths. This fact is formally given by *Wien's displacement law*, which states that

$$(\lambda T)_{max} = 2897.6\ \mu\mathrm{m\ K} \tag{51.31}$$

[This relation can also be obtained by differentiating Planck's law, Equation (51.29).] Often, in practice, it is required to know what fraction of the total emission occurs between two given wavelengths. The *blackbody radiation function*, $f_{0,\lambda}(T)$, is defined by

$$f_{0,\lambda}(T) = (\sigma T^4)^{-1} \int_{\lambda=0}^{\lambda} E_{b,\lambda}(\lambda', T)\,d\lambda' \tag{51.32}$$

and tabulated in standard textbooks. It may be used to find the emission of a blackbody between any two given wavelengths λ_1 and λ_2 by

$$f_{\lambda_1,\lambda_2}(T) = f_{0,\lambda_2}(T) - f_{0,\lambda_1}(T) \tag{51.33}$$

Intensity of Radiation

The fundamental quantity that is used to represent the amount of radiant energy transmitted in a given direction is the *intensity of radiation.* The *spectral intensity of radiation* is defined as the radiative energy

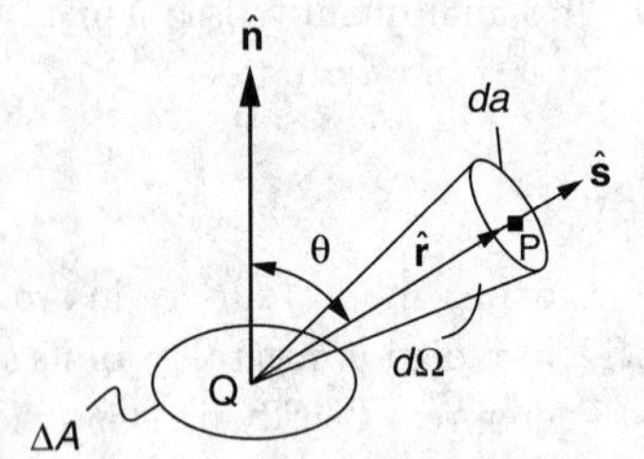

FIGURE 51.6 The definition of intensity of radiation.

per unit time per unit area normal to the direction of propagation per unit solid angle per unit wavelength. The total intensity refers to the intensity summed over all wavelengths. With reference to Figure 51.6, let ΔQ represent the energy radiated per unit time and confined to a solid angle $d\Omega$ around the solid angle $\hat{\Omega}$. The *total intensity of radiation* due to the radiating surface at Q is then defined mathematically as

$$I(\mathbf{r}, \hat{\mathbf{s}}) = \frac{dq}{d\Omega}\frac{1}{\cos\theta} = \frac{dq}{da}\frac{r^2}{\cos\theta} \tag{51.34}$$

where $q = \lim_{\Delta A \to 0}(\Delta Q / \Delta A)$ is the heat flux that is measured at the radiation source Q, and $d\Omega$ is the solid angle subtended by the area da at Q (which is equal to da/r^2). If the intensity at a point P is constant, Equation (51.34) shows that the flux intercepted by an area varies inversely with the square of its distance from the source and directly as the cosine of the angle between the normals of the radiating the intercepting surfaces. This is known as *Lambert's cosine law.* Since energy conservation implies that all the flux (emitted as well as reflected) that leaves Q must cross the hemisphere above Q,

$$q = \int_{\text{hemisphere}} I(\mathbf{r}, \hat{\mathbf{s}})\hat{\mathbf{n}} \cdot \hat{\mathbf{s}}\, d\,\Omega \tag{51.35}$$

where q is the total flux radiated by the surface (i.e., at all wavelengths). If only the intensity due to the emission from the surface is considered, the flux that leaves the hemisphere above the surface is known as the *total hemispherical emissive power* and is given by

$$E = \int_{\text{hemisphere}} I(\mathbf{r}, \hat{\mathbf{s}})\hat{\mathbf{n}} \cdot \hat{\mathbf{s}}\, d\,\Omega \tag{51.36}$$

If $I(\mathbf{r}, \hat{\mathbf{s}})$ is uniform, then the radiating surface is known as a **diffuse** emitter, and Equation (51.36) becomes

$$q = \pi I \tag{51.37}$$

The radiation emitted by a blackbody is diffuse. Therefore, the total intensity of radiation from a blackbody [due to Equation (51.30) and Equation (51.37)] is

$$I_b = \frac{\sigma T^4}{\pi} \tag{51.38}$$

Radiative Properties of Real (Nonblack) Surfaces

The important radiative properties of a real (nonblack) surface are its emissivity, absorptivity, transmissivity, and reflectivity. In the following sections we consider only the hemispherical monochromatic and total properties. For directional monochromatic quantities see Siegel and Howell [1981] or Modest [1993].

Emissivity

The *hemispherical monochromatic emissivity* of a nonblack surface at a temperature T is defined as the ratio of the hemispherical monochromatic emissive power of the surface $E_\lambda(\lambda, T)$ and the corresponding hemispherical monochromatic emissive power of a blackbody. That is,

$$\varepsilon_\lambda(\lambda, T) = \frac{E_\lambda(\lambda, T)}{E_{b,\lambda}(\lambda, T)} \tag{51.39}$$

The hemispherical total emissivity is the ratio of the corresponding hemispherical total quantities:

$$\varepsilon(T) = \frac{E(T)}{E_b(T)} = (\sigma T^4)^{-1} \int_{\lambda=0}^{\infty} \varepsilon_\lambda(\lambda, T) E_{b,\lambda}(\lambda, T)\, d\lambda \tag{51.40a}$$

A surface whose monochromatic emissivity does not depend on the wavelength is called an ideal **gray body**. For a gray body,

$$\varepsilon_\lambda(\lambda, T) = \varepsilon(T) \tag{51.40b}$$

Absorptivity

The *hemispherical monochromatic absorptivity* is defined as the fraction of the incident hemispherical monochromatic irradiation that is absorbed by the surface, that is,

$$\alpha_\lambda(\lambda, T) = \frac{G_\lambda(\lambda)|_{\text{absorbed}}}{G_\lambda(\lambda)} \tag{51.41a}$$

From the definitions of intensity and radiosity, the *hemispherical total absorptivity* of a surface may be defined as

$$\alpha(T, \text{source}) = \frac{\int_{\lambda=0}^{\infty} \alpha_\lambda(\lambda, T)[\int_{\text{hemisphere}} I_{\lambda,i}(\mathbf{r}, \lambda, \hat{\mathbf{s}}) \cos\theta_i \, \mathrm{d}\,\Omega] d\lambda}{\int_{\lambda=0}^{\infty} \int_{\text{hemisphere}} I_{\lambda,i}(\mathbf{r}, \lambda, \hat{\mathbf{s}}) \cos\theta_i \, d\Omega\, d\lambda} \tag{51.41b}$$

where $I_{\lambda,i}(\mathbf{r}, \lambda, \hat{\mathbf{s}})$ refers to the incident monochromatic intensity of radiation and θ_i is the angle between the incident radiation and the outward drawn normal to the surface. Equations (51.41) indicate that the monochromatic and the total absorptivities depend on $I_{\lambda,i}$ and, hence, on the source of radiation. Consequently, unlike the values of ε, the values of α cannot be easily tabulated.

Reflectivity

Radiation that is incident on a surface gets reflected. If the surface is smooth, the incident and reflected rays are symmetric with respect to the normal at the point of incidence, and this is known as *specular reflection.* If the surface is rough, the incident radiation is scattered in all directions. An idealized reflection law assumes that, in this situation, the intensity of the reflected radiation is constant for all angles of reflection and independent of the direction of radiation. The *hemispherical monochromatic reflectivity,* $\rho_\lambda(\lambda)$, is defined as the fraction of the reflected radiant energy that is incident on a surface. The hemispherical total reflectivity is merely

$$\rho = \int_{\lambda=0}^{\infty} \rho_\lambda(\lambda) d\lambda \tag{51.42}$$

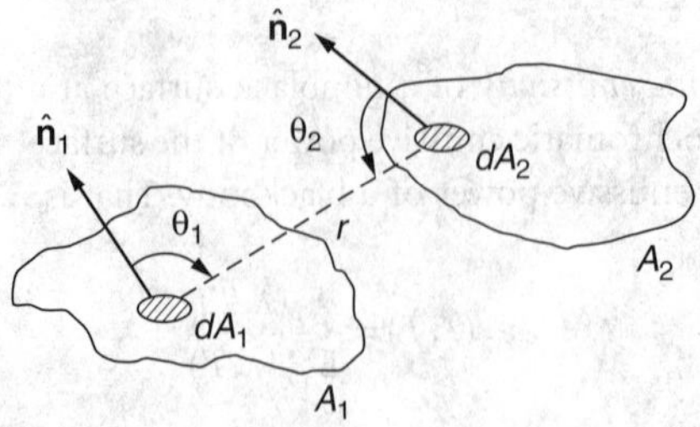

FIGURE 51.7 Determination of the shape factor.

Transmissivity

If a body is semitransparent to radiation (as glass is to solar radiation), part of the incident radiation is reflected by the surface and part of it is absorbed. In this case the sum of the absorptivity and reflectivity is less than unity, and the difference is known as *transmissivity*, τ. Therefore,

$$\alpha_\lambda + \rho_\lambda + \tau_\lambda = 1 \text{ and } \alpha + \rho + \tau = 1 \tag{51.43}$$

For an opaque body $\tau = 0$, implying that $\rho_\lambda = 1 - \alpha_\lambda$ and $\rho = 1 - \alpha$.

Kirchhoff's Law

Kirchhoff's law states that the hemispherical monochromatic emissivity $\varepsilon_\lambda(\lambda, T)$ is equal to the hemispherical monochromatic absorptivity $\alpha_\lambda(\lambda, T)$ at a given temperature T. That is,

$$\varepsilon_\lambda(\lambda, T) = \alpha_\lambda(\lambda, T) \tag{51.44}$$

However, the hemispherical total values of the emissivity and absorptivity are equal only when one of the two following conditions is met: either (1) the incident radiation of the receiving surface must have a spectral distribution that is the same as that of the emission from a blackbody at the same temeprature, or (2) the receiving surface must be an ideal gray surface — that is, $\varepsilon_\lambda(\lambda, T)$ is not a function of λ.

Shape Factors

In the exchange of radiant energy between two surfaces, it is important to find the fraction of energy that leaves one surface and strikes the other. The flux of the energy between any two given surfaces can be expressed in terms of the *shape factor* (also known as *view factor*). The shape factor is purely a function of the geometry of the two surfaces and their relative orientation. The shape factors for the two surfaces shown in Figure 51.7 are given by

$$F_{1,2} = \frac{1}{A_1} \int_{A_1} \int_{A_2} \frac{\cos\theta_1 \cos\theta_2}{\pi r^2} dA_2 dA_1 \tag{51.45a}$$

and

$$F_{2,1} = \frac{1}{A_2} \int_{A_2} \int_{A_1} \frac{\cos\theta_2 \cos\theta_1}{\pi r^2} dA_1 dA_2 \tag{51.45b}$$

If the radiation between the two surfaces is diffuse, the fraction $Q_{1,2}$ of the radiant energy that leaves surface A_1 and is intercepted by A_2 is given by

$$Q_{1,2} = A_1 J_1 F_{1,2} \tag{51.46}$$

where J_1 is the radiosity of surface A_1. In the above relations the order of the subscripts is important since $F_{1,2} \neq F_{2,1}$.

Further, shape factors possess the following properties. Any two given surfaces A_1 and A_2 satisfy the *reciprocal* property:

$$A_1 F_{1,2} = A_2 F_{2,1} \tag{51.47}$$

If a surface A_1 is divided into n parts $(A_{1_1}, A_{1_2}, \ldots, A_{1_n})$ and the surface A_2 into m parts, the following *additive property* holds:

$$A_1 F_{1,2} = \sum_n \sum_m A_{1_n} F_{1_n, 2_m} \tag{51.48}$$

If the interior of an enclosure is divided into n parts, each with a finite area A_i, $i = 1,2,\ldots, n$, then the *enclosure property* states that

$$\sum_{j=1}^{n} F_{i-j} = 1, \quad i = 1,2,\ldots,n \tag{51.49}$$

Radiative Transfer Equation

When a medium absorbs, emits, and scatters the radiant energy flowing through it, it is called a **participating medium.** An example of a participating medium is a gas with particles. Experiment shows that the intensity of radiation decays exponentially with the distance traveled in the medium. Further, if it encounters the particles in the gas, it gets scattered. Scattering takes place when one or more of diffraction, reflection, or refraction occur. Taking absorption and scattering into account, the transmissivity of a medium [also see "Radiative Properties of Real (Nonblack) Surfaces"] may be written as

$$\tau_\lambda = e^{-(\delta_\lambda + \varsigma_\lambda)s} = e^{\beta_\lambda s} \tag{51.50}$$

where δ_λ is the *monochromatic scattering coefficient*, ς_λ is the *monochromatic absorption coefficient*, and s is the thickness of the medium. β_λ is known as the *monochromatic extinction* coefficient of the medium.

In the solution of heat transfer problems that involve multiple modes of heat transfer, the energy equation must be solved. In such cases the term $\nabla \cdot q|_{\text{radiation}}$ [also see Equation (51.4)] must be calculated. Consider an absorbing, emitting, scattering medium with a monochromatic scattering coefficient δ_λ and a monochromatic absorption coefficient ς_λ. Let a beam of monochromatic radiation of intensity $I_\lambda(s, \hat{\Omega}, t)$ travel in this medium along the direction $\hat{\Omega}$. The *radiative transfer equation* (RTE) is obtained by considering the energy balance for radiation due to emission, absorption, and in-and-out scattering in a small volume of this medium. The steady divergence of the heat flux for this case is then given by

$$\nabla \cdot q|_{\text{radiation}} = 4\pi \int_0^\infty \varsigma_\lambda I_{b,\lambda}(T) d\lambda - \int_0^\infty \varsigma_\lambda [\int_{4\pi} I_\lambda(s, \hat{\Omega}) d\hat{\Omega}] d\lambda \tag{51.51}$$

where $I_{b,\lambda} = E_{b,\lambda} / \pi$ and $I_\lambda(s, \hat{\Omega})$ is obtained as a solution to the RTE.

51.4 Phase Change

Phase change occurs when a substance (solid, liquid, or gas) changes its state due to the absorption or release of energy. Below we briefly discuss melting/freezing (solid $\leftrightarrow$ liquid) and condensation/boiling (liquid $\leftrightarrow$ vapor).

Melting and Freezing

The analysis of situations in which a solid melts or a liquid freezes involves the *phase change* or *moving boundary problem.* The solutions to such problems are important in the making of ice, the freezing of food, the solidification of metals in casting, the cooling of large masses of igneous rock, the casting and welding of metals and alloys, etc. The solution of such problems is difficult because the *interface* between the solid and the liquid phases is in motion, due to the absorption or release of the latent heat of fusion or solidification. Thus, the location of the moving interface is not known beforehand and is a part of the solution to the problem. The heat transfer problem in these situations can be treated as a problem in heat conduction with moving boundaries, which requires a solution to Equation (51.6) under appropriate boundary conditions. The solution to these problems is beyond the scope of this chapter, and the interested reader is referred to works such as Ozisik [1993].

Condensation

Consider a vapor that comes into contact with a surface that is maintained at a temperature that is lower than the saturation temperature of the vapor. The resulting heat transfer from the vapor to the surface causes immediate *condensation* to occur on the surface. If the surface is cooled continuously so as to maintain it at a constant temperature, and the condensate is removed by motion due to gravity, it is covered with a thin film of condensed liquid. This process is known as *filmwise condensation.* Filmwise condensation usually occurs when the vapor is relatively free of impurities. Sometimes, when oily impurities are present on surfaces or when the surfaces are highly polished, the film of condensate breaks into droplets. This situation is known as *dropwise condensation.* In filmwise condensation the presence of the liquid film acts as a barrier for the heat transfer between the vapor and the surface. In dropwise condensation, less of a barrier exists between the vapor and the surface. As a result, heat transfer coefficients are five to ten times the corresponding values for film condensation. Since fluid motion plays a significant role in condensation and boiling, they are studied along with convective heat transfer processes. However, significant differences exist between convective heat transfer during phase change and single-phase processes. The study of condensation and boiling is useful in the design of condensers and boilers, two widely used types of heat exchangers.

Filmwise Condensation

Consider a cold vertical plate of length L maintained at a constant temperature T_w and exposed to saturated vapor at temperature T_v. Figure 51.8 shows the formation of the film of condensate on the plate. By the analysis of the boundary layer equations for the film of condensate, Nusselt obtained the heat transfer coefficient h_x at location x (with respect to Figure 51.8) for this case. Nusselt's analysis is also valid for condensation outside a long cylinder or tube if its radius is large compared to the thickness of the condensing film. The actual correlation for this case and for filmwise condensation on other surfaces can be found in standard textbooks [see, for instance, Bejan (1984)]. In condensation problems the Reynolds number of the condensate flow is given by

$$\mathrm{Re} = \frac{4\dot{m}}{\mu_f P} = \frac{4A_t h_m (T_v - T_w)}{h_{fg}\mu_f P} \tag{51.52}$$

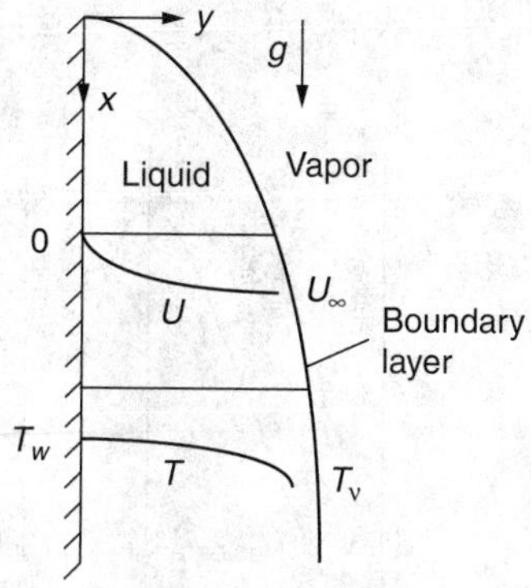

FIGURE 51.8 Filmwise condensation on a vertical surface.

where $\dot{m}$ is the mass flow rate of the condensate at the lowest part of the condensing surface. The wetted perimeter P is defined as w for a vertical plate of width w and πD for a vertical cylinder/tube with outside diameter D. A_t is the total condensing surface area, h_m is the mean heat transfer coefficient, and h_{fg} is the latent heat of condensation.

Condensers are usually designed with horizontal tubes arranged in vertical tiers. If the drainage from one tube is assumed to flow smoothly onto the tube below, then, for a vertical tier of N tubes each of diameter D,

$$\left[\mathrm{Nu}_m\right]\Big|_{\text{for } N \text{ tubes}} = \frac{1}{N^{1/4}}\left[\mathrm{Nu}_m\right]\Big|_{\text{for one tube}} \tag{51.53}$$

Dropwise Condensation

Experiments have shown that if traces of oil or other selected substances (known as **promoters**) are present either in steam or on the condensing surface, the condensate film breaks into droplets. The droplets grow, coalesce, and run off the surface, leaving a greater portion of the surface exposed to the incoming steam. Therefore, dropwise condensation is a more efficient method of heat transfer. Typical heat transfer coefficients of $5.7 \cdot 10^4$ to $50 \cdot 10^4$ W/m²°C may be obtained as opposed to heat transfer coefficients in the order of $5 \cdot 10^3$ W/m²°C for filmwise condensation. If sustained dropwise condensation can be maintained, it will result in considerable reduction in the size of condensers and their cost. Research in this area has thus been aimed at producing long-lasting dropwise condensation via the use of promoters. A satisfactory promoter must repel the condensate, stick tenaciously to the substrate, and prevent oxidation of the surface. Some of the most popular promoters are fatty acids such as oleic, stearic, and linoleic acids. In order to prevent failure of dropwise condensation as a result of oxidation, coatings of noble metals such as gold, silver, and palladium have been used in the laboratory. Such coatings have sustained over 10,000 h of continuous dropwise condensation. However, this method is too expensive for use on an industrial scale.

Another problem in dropwise condensation is the presence of a noncondensable gas. If a noncondensable gas such as air is present in the vapor even in small amounts, the heat transfer coefficient is significantly reduced. When the vapor condenses, the noncondensable gas is left at the surface of the condensate, and the incoming vapor must diffuse through this body of vapor–gas mixture before reaching the condensing surface. This diffusion process reduces the partial pressure of the condensing vapor and in turn its saturation temperature; that is, the temperature of the surface of the layer of condensate is lower than the bulk saturation temperature of the vapor. Therefore, in practical applications, there is a provision to vent the noncondensable gas that accumulates inside the condenser.

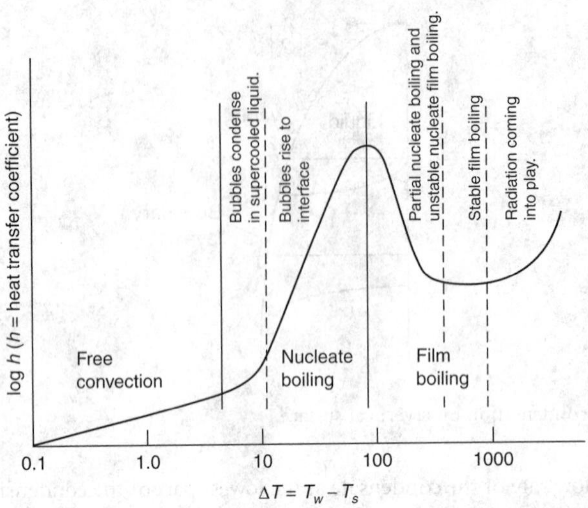

FIGURE 51.9 Principal regimes in pool boiling of water at atmospheric pressure and saturation temperature.

Pool Boiling

Boiling occurs when a liquid changes phase and becomes vapor due to the absorption of heat. When the heating surface is submerged in a quiescent pool of liquid, the boiling phenomenon is known as *pool boiling*. The boiling that might occur in a fluid that moves due to an externally imposed pressure gradient is called *forced convection boiling.*

In 1934, Nukiyama performed a systematic study of pool boiling by immersing an electric wire into a body of saturated water and heating it. He determined the heat flux and temperature from current and voltage measurements. Figure 51.9 illustrates the characteristics of pool boiling for water at atmospheric pressure. It shows the heat transfer coefficient (or the heat flux) as a function of the wire and water saturation temperatures. This curve shows that there are three distinct regimes of pool boiling: the *free-convection* regime, nucleate boiling regime, and *film boiling* regime.

Initially, heat transfer is by free convection. The temperature of the heater surface is a few degrees above the saturation temperature, and free convection is sufficient to remove heat from it. The heat transfer correlations, as expected, are of the form given by Equation (51.28).

In the **nucleate boiling** regime, bubbles are formed on the surface of the heater. There are two distinct regions in this regime. In the first, bubbles are formed at certain favored sites but are dissipated in the liquid as soon as they are detached from the surface. In the second region, the bubble generation is high enough to sustain a continuous column. Thus, large heat fluxes may be obtained in this region. In the nucleate boiling regime, the heat flux increases rapidly with increasing temperature difference until the peak heat flux is reached. The location of this peak heat flux is known as the *burnout point, departure from nucleate boiling* (DNB), or *critical heat flux.* After the peak heat flux is exceeded, an extremely large temperature difference is needed to realize the resulting heat flux, and such high temperature differences may cause "burnout" of the heating element.

As indicated in Figure 51.9, the peak heat flux occurs in the nucleation boiling regime. This maximum value must be known beforehand because of burnout considerations. That is, if the applied heat flux is greater than the peak heat flux, the transition takes place from the nucleate to the stable film boiling regime, in which (depending on the kind of fluid) boiling may occur at temperature differences well above the melting point of the heating surface.

After the peak heat flux is reached, the *unstable film boiling region* begins. No correlations are available for the heat flux in this region until the minimum point in the boiling curve is reached and the stable film boiling region starts. In the stable film boiling region the heating surface is separated from the liquid by a vapor layer across which heat must be transferred. Since vapors have low thermal conductivities, large temperature differences are needed for heat transfer in this region. Therefore, heat transfer in this regime is generally avoided when high temperatures are involved.

Defining Terms

Blackbody — A body that absorbs all incident radiation from all directions at all wavelengths without reflecting, transmitting, or scattering it.

Diffuse — The radiation from a body or a radiative property that is independent of direction.

Energy equation — The equation that makes use of the principle of conservation of energy in a given process [see Equation (51.3) through Equation (51.5)].

Film temperature — The mean boundary layer temperature [see Equation (51.24)].

Fourier's law — The law that relates the heat flow to the temperature gradient in a body [see Equation (51.2)].

Gray surface — A surface whose monochromatic emissivity is independent of the wavelength.

Heat transfer coefficient — A quantity that determines the heat flux due to convective heat transfer between a surface and a fluid that is moving over it [see Equation (51.12) and Equation (51.26)].

Newton's law of cooling — The relation that determines the heat flux due to convection between a surface and moving fluid [see Equation (2.10)].

Nucleate boiling — The stage in (pool) boiling of a fluid at which bubbles are formed at favored sites on the heating surface and detached from it.

Participating medium — A medium that absorbs, emits, and scatters the radiation that passes through it.

Phonons — A term that refers to the quantized lattice vibrations in a solid. Phonons are the energy carriers in a dielectric solid and are responsible for conduction of heat and electricity.

Planck's law — The law that relates the emissive power of a blackbody at a given wavelength to its temperature [see Equation (51.29)].

Promoters — Substances that are introduced into a pure vapor or onto a condensing surface to facilitate dropwise condensation on a surface.

Stefan–Boltzmann law — The relation that determines the heat flux due to radiation from a surface [see Equation (51.30)].

Thermal conductivity — A property that determines the ability of a substance to allow the flow of heat through it by conduction [see Equation (51.2)].

Thermal diffusivity — A material property that governs the rate of propagation of heat in transient processes of conduction [see Equation (51.6)].

References

Bayazitoglu, Y. and Ozisik, M. N. 1988. *Elements of Heat Transfer,* McGraw-Hill, New York.

Bejan, A. 1984. *Convection Heat Transfer,* John Wiley & Sons, New York.

Carslaw, H. S. and Jaeger, J. C. 1959. *Conduction of Heat in Solids,* Clarendon, Oxford.

Kreith, F. and Bohn, M. S. 1986. *Principles of Heat Transfer,* Harper and Row, New York.

Modest, M. 2003. *Radiative Heat Transfer,* Academic Press, Elsevier Science, San Diego.

Ozisik, M. N. 1993. *Heat Conduction,* 2nd ed. John Wiley & Sons, New York.

Siegel, R. and Howell, J. R. 1981. *Thermal Radiation Heat Transfer,* Hemisphere, New York.

Ziman, J. M. 1960. *Electrons and Phonons,* Oxford University Press, London.

Further Information

ASME Journal of Heat Transfer. Published quarterly by the American Society of Mechanical Engineers.

International Journal of Heat and Mass Transfer. Published monthly by Pergamon Press, Oxford; contains articles in various languages with summaries in English, French, German, or Russian.

AIAA Journal of Thermophysics and Heat Transfer. Published quarterly by the American Institute of Aeronautics and Astronautics.

52

Heat Exchangers

M. M. Ohadi
University of Maryland

A **heat exchanger** is a device that is used to transfer heat between two or more fluids that are at different temperatures. Heat exchangers are essential elements in a wide range of systems, including the human body, automobiles, computers, power plants, and comfort heating/cooling equipment. In the chemical and process industries, heat exchangers are used to concentrate, sterilize, distill, pasteurize, fractionate, crystallize, or control the fluid flow and chemical reaction rates. With the recent global promotion of energy efficiency and protection of the environment, the role of heat exchangers in efficient utilization of energy has become increasingly important, particularly for energy-intensive industries, such as electric power generation, petrochemical, air conditioning/refrigeration, cryogenics, food, and manufacturing.

The Carnot efficiency for an ideal heat engine that receives Q_H amount of heat from a high-temperature reservoir at temperature T_H and rejects Q_L amount of heat to a reservoir at temperature T_L is

$$\eta = 1 - \frac{T_L}{T_H} = 1 - \frac{Q_L}{Q_H} \tag{52.1}$$

Equation (52.1) represents the maximum possible efficiency for a heat engine operating between a low- and a high-temperature reservoir. The Carnot efficiency is a good reference against which the performance of an actual heat engine can be compared. From the Carnot efficiency, it is clear that heat exchangers play a direct role in the overall efficiency of thermal machinery and equipment. By employing a more effective heat exchanger in a thermal cycle, one can get a higher T_H and lower T_L, resulting in higher efficiencies for the cycle as a whole.

In this chapter, we will first outline the basic classifications of heat exchangers and the relevant terminology used. Next, essential features and fundamental design aspects of **shell-and-tube** and **compact heat exchangers** will be described. Recent advances in thermal performance of heat exchangers will be discussed in the last section of this chapter.

0-8493-1586-7/05/$0.00+$1.50

52.1 Heat Exchanger Types

Heat exchangers can be classified in many different ways. Shah [1981] provides a comprehensive and detailed description of the various categories, the associated terminology, and the practical applications of each type. As indicated there, heat exchangers can be broadly classified according to the six categories described below.

Transfer Processes

In this category, heat exchangers are classified into direct contact and indirect contact types. In the direct type, the heat exchanging streams (hot and cold streams) come into direct contact and exchange heat before they are separated. Such heat exchangers are more common in applications where both heat and mass transfer are present. A familiar example is evaporative cooling towers used in power plants and in comfort cooling for desert environments. In an indirect contact heat exchanger, the hot and cold fluids are separated by a solid, impervious wall representing the heat transfer surface. The conventional shell-and-tube heat exchangers fall in this category.

Number of Fluids

In many applications, the two-fluid heat exchangers are the most common type. However, in certain applications, such as cryogenics and the chemical and process industries, multifluid heat exchangers are also common.

The Degree of Surface Compactness

In this category, heat exchangers are classified according to the amount of heat transfer surface area per unit volume incorporated into the heat exchanger. This is represented by the heat transfer area-to-volume ratio ($\beta = A/V$). For example, for compact heat exchangers $\beta \approx 700$ m²/m³ or greater, for **laminar flow heat exhangers** $\beta \approx 3000$ m²/m³, and for **micro heat exchangers** $\beta \approx 10{,}000$ or greater.

Construction Features

In this category, heat exchangers are often divided into four major types: tubular, plate type, extended surface, and regenerative types. There are other types of heat exchangers with unique construction that may not fall in these four categories. However, they are not commonly used and are specific to specialized applications.

Flow Arrangements

Heat exchangers can be classified according to the manner in which the fluid flows through the tube and shell sides. The two broad types are single pass and multipass. When the fluid flows through the full length of the heat exchanger without any turns, it is considered to have made one pass. A heat exchanger is considered single pass when the hot and cold fluids make one pass in the heat exchanger. Within the single pass category, the common flow arrangements for the hot and cold fluids are parallel flow, counterflow, cross-flow, split flow, and divided flow. Multipass heat exchangers are classified according to the type of construction. The common types include extended surface, shell-and-tube, and plate heat exchangers.

Heat Transfer Regimes

Exchange of thermal energy in a heat exchanger from the hot fluid to the exchanging wall to the cold fluid can employ one or more modes of heat transfer. For example, in gas-to-gas heat exchangers, heat transfer on both the hot and cold fluids is by single phase convection. In water-cooled steam condensers, single phase convection takes place in one side and condensation in the other side. Yet, in certain cryogenic heat exchangers, condensation takes place on one side and evaporation on the other side. Basic components and essential features of the two most commonly used heat exchangers — namely, shell-and-tube and plate-fin heat exchangers — will be discussed in the following sections.

52.2 Shell-and-Tube Heat Exchangers

General Features

Among the various heat exchanger types, shell-and-tube heat exchangers are the most commonly used in a number of industries, including process, chemical, power, and refrigerating/air conditioning. In this type of heat exchanger, one fluid flows inside the tubes while the other fluid is forced through the shell side and over the tubes in a cross-flow arrangement.

The shell-and-tube heat exchangers have several main advantages that, as a whole, have contributed to their nearly universal acceptance for a wide range of applications. They can be custom designed for almost any capacity, working fluid type, operating pressure, and temperature conditions. The type of material used to construct the heat exchanger can be any of the most commonly used materials. Shell-and-tube heat exchangers yield a relatively large surface area density. Methods for cleaning and other maintenance, such as periodic replacement of gaskets and tubes, are well established and easily performed. Their broad worldwide use over the years has resulted in good methods for design and fabrication.

The size of a shell-and-tube heat exchanger can vary over a wide range from compact to supergiant configurations. Common examples of shell-and-tube heat exchangers are steam generators, condensers, evaporators, feed water heaters, oil coolers in the power and refrigeration industries, and process heat exchangers in the petroleum refining and chemical industries. Classification of shell-and-tube heat exchangers is usually according to the number of tube and shell passes. The heat exchangers schematically shown in Figure 52.1(a), Figure 52.1(b), and Figure 52.1(c) represent, respectively, one shell pass and one tube pass, two shell passes and one tube pass, and one shell and four tube passes.

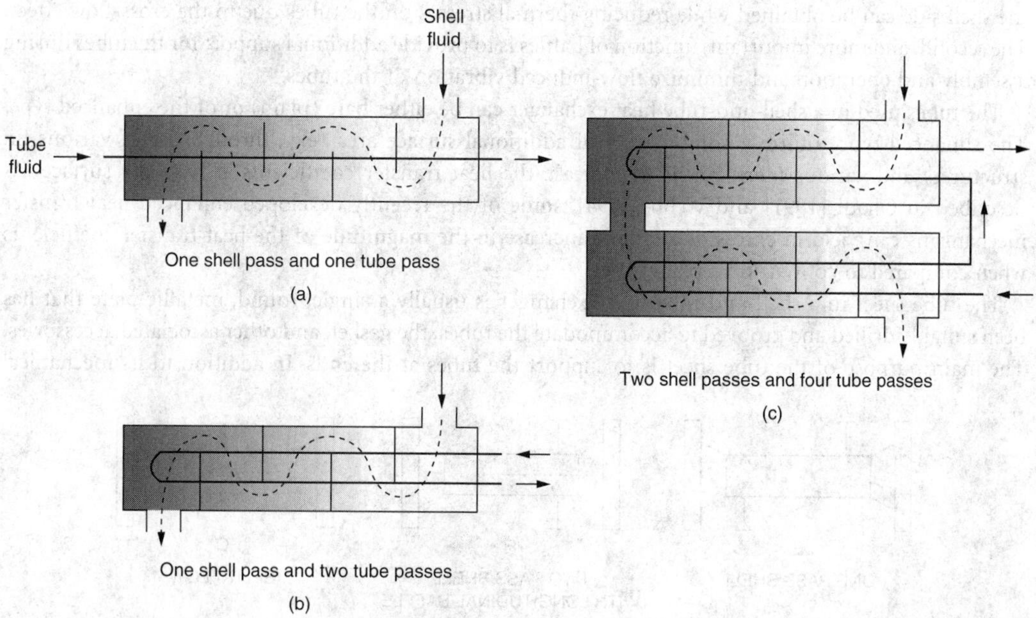

FIGURE 52.1 Examples of shell-and-tube heat exchanger configurations.

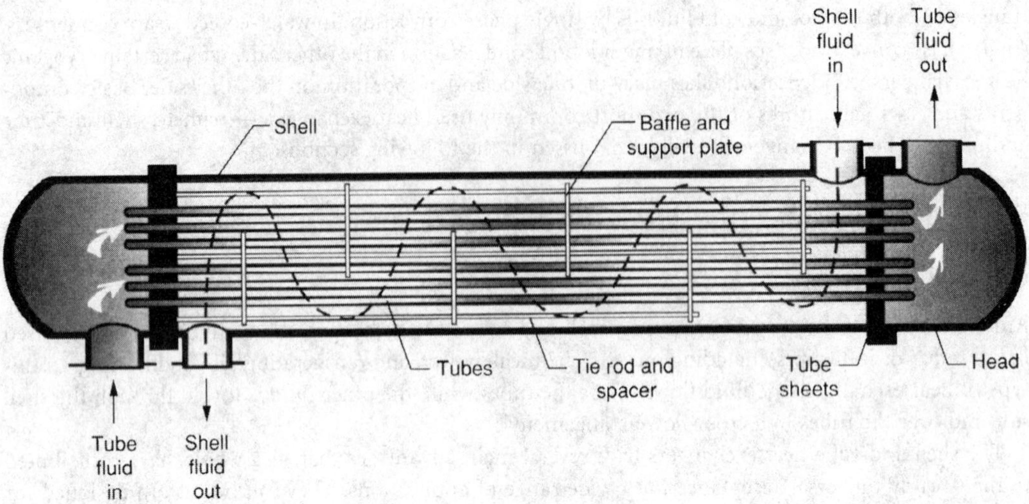

FIGURE 52.2 Major components of a shell-and-tube heat exchanger.

Major Components

Major components of a shell-and-tube heat exchanger are shown in Figure 52.2. A variety of different constructions are used in shell-and-tube heat exchangers. Major types are identifiable by a special notation developed by the Tubular Exchanger Manufacturer's Association (TEMA), in which each heat exchanger is designated by a three-letter combination. The first letter indicates the front-end head type, the second the shell type, and the third letter identifies the rear-end head type. The standard in single shell arrangement type is TEMA E, in which the entry and outlet nozzles are placed at opposite ends. Figure 52.3 shows three sample TEMA configurations.

The baffles in a shell-and-tube heat exchanger serve two tasks. First, they direct the flow in the shell side tube bundle approximately at right angles to the tubes so that higher heat transfer coefficients in the shell side can be obtained while reducing thermal stresses on the tubes due to the cross-flow effect. The second, and more important, function of baffles is to provide additional support for the tubes during assembly and operation and minimize flow-induced vibration of the tubes.

The tubes used in a shell-and-tube heat exchanger can be either bare (plain) or of the enhanced type. The enhanced types utilize a combination of additional surface area (e.g., through use of various fin structures) and various mechanisms to increase the heat transfer coefficients at the tube surface. As described in Ohadi [1991] and Webb [1994], some of the recently developed enhanced heat transfer mechanisms can yield in excess of a tenfold increase in the magnitude of the heat transfer coefficients when compared to conventional plain tubes.

The tube sheet in a shell-and-tube heat exchanger is usually a single, round, metallic plate that has been suitably drilled and grooved to accommodate the tubes, the gasket, and other associated accessories. The main purpose of the tube sheet is to support the tubes at the ends. In addition to its mechanical

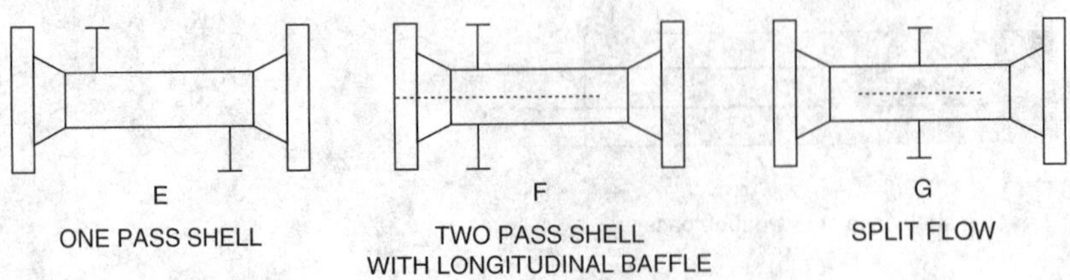

FIGURE 52.3 Sample TEMA configurations.

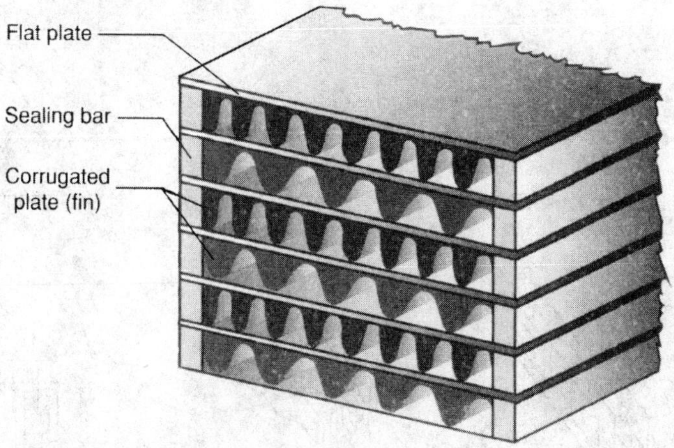

FIGURE 52.4 A basic plate-fin heat exchanger construction.

requirements, the tube sheet must be resistant to corrosion from fluids in both the tube and shell sides and must be electrochemically compatible with the tube side materials. The tube sheets are sometimes made from low-carbon steel with a thin layer of corrosion-resistant alloy.

52.3 Compact Heat Exchangers

Definition

A heat exchanger is referred to as a compact heat exchanger if it incorporates a heat transfer surface having a surface density above approximately 700 m^2/m^3 on at least one of the fluid sides, usually the gas side. Compact heat exchangers are generally of plate-fin type, tube-fin type, tube bundles with small diameter tubes, and regenerative type. In a plate-fin exchanger, corrugated fins are sandwiched between parallel plates as shown in Figure 52.4. In a tube-fin exchanger, round and rectangular tubes are most commonly used and fins are employed either on the outside, the inside, or on both the outside and the inside of the tubes, depending upon the application. Basic flow arrangements of two fluids are single pass cross-flow, counterflow, and multipass cross-counterflow.

Characteristics

Some of the subtle characteristics and uniqueness of compact heat exchangers are:

1. Usually at least one of the fluids is a gas.
2. The fluid must be clean and relatively noncorrosive.
3. Fluid pumping power is always of prime importance.
4. Operating pressure and temperatures are somewhat limited compared to shell-and-tube exchangers due to construction features.

The shape of a compact heat exchanger is usually distinguished by having a large frontal area and short flow lengths. Thus, the proper design of the header for compact heat exchangers is very important for a uniform flow distribution. A variety of surfaces are available for use in compact heat exchangers having different orders of magnitude of surface area densities. Such surfaces could introduce substantial cost, weight, or volume savings as desired by the design. Figure 52.5 shows some sample surfaces and geometries currently employed in compact heat exchangers.

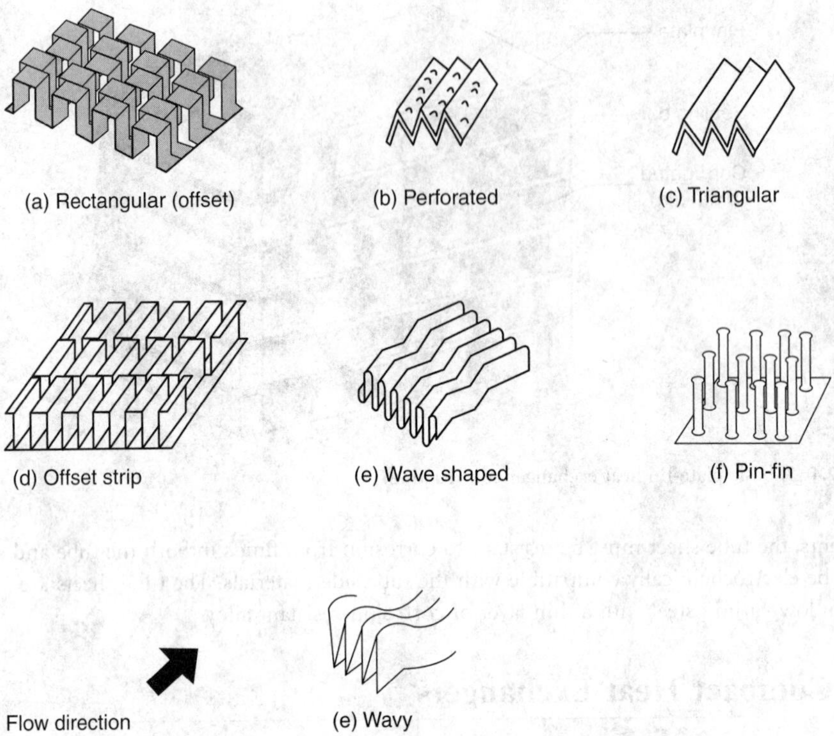

FIGURE 52.5 Examples of plate-fin surface geometries.

Applications

The major heat exchangers in cars and trucks are radiators for engine cooling, heaters for passenger compartment heating, evaporators and condensers for air conditioning, oil coolers for engine oil and transmission oil cooling, and charge air coolers and intercoolers for cooling charge air for increased mass flow rate through the engine.

A variety of compact heat exchangers are needed in space applications where the minimum weight and volume, as well as the absolute reliability and durability, are essential. Very high heat fluxes in low-weight and low-volume heat exchangers can be obtained by employing two-phase flow in offset strip fins. Such heat exchangers will have a significant impact on emerging applications, such as condensers and evaporators for thermal control of space stations, satellite instrumentation, aircraft avionics packages, thermal control of electronics packages, and thermal control of space life support and astronauts.

In commercial aircraft applications, a variety of primarily compact heat exchangers are used for cabin air conditioning and heating, avionics cooling, deicing, and oil cooling applications.

52.4 Design of Heat Exchangers

Basic Design

When designing a heat exchanger, the following requirements must be satisfied for most typical applications:

Meet the process thermal requirements within the allowable pressure drop penalty

Withstand the operational environment of the plant, including the mechanical and thermal stress conditions of the process, operational schedule, and resistance to corrosion, erosion, vibration, and fouling

Provide easy access to those components of the heat exchanger that are subject to harsh environments and require periodic repair and replacement
Should have a reasonable cost (combination of initial and maintenance costs)
Should be versatile enough to accommodate other applications within the plant, if possible

Broadly speaking, the various design tasks in a new heat exchanger can be classified into two categories: (1) the mechanical design and (2) the thermal-hydraulic design. The mechanical design deals with the proper distribution of thermal and mechanical stresses and the general integrity of the heat exchanger for the conditions specified. The thermal-hydraulic design deals with calculations of heat transfer and pressure drop coefficient and, subsequently, determination of the overall required dimensions of the heat exchanger for specified thermal operating conditions. The mechanical design aspects are beyond the scope of this article and can be found elsewhere [Hewitt et al., 1994]. An introduction to the thermal-hydraulic design criteria will be given in the following.

Thermal-Hydraulic Design

As discussed by Shah [1986], the various steps in the thermal-hydraulic design of heat exchangers can be classified into two distinct problems: the *rating problem* (also referred to as the *performance problem*) and the *sizing problem* (also referred to as the *design problem*).

The rating problem involves the determination of the overall heat transfer coefficients and pressure drop characteristics utilizing the following input data: the heat exchanger configuration, flow rates on each side, inlet temperatures and pressure data, thermophysical properties of the fluid fouling factors, and complete details on the materials and surface geometries of each side. The sizing problem utilizes the data provided in the rating problem and determines the overall (core) dimensions (core length and surface areas) for the heat exchanger. Steps involved in the rating and sizing problems are schematically shown in Figure 52.6.

The LMTD Method

It is a general practice to determine the heat transfer between two fluids in a heat exchanger using the mean quantities and the following defining equation:

$$q = FU_m A \Delta T_m \tag{52.2}$$

in which U_m and ΔT_m represent the mean heat transfer coefficient and effective temperature difference, respectively. The factor F is a function of flow arrangement and is provided graphically for various heat exchanger configurations [Shah, 1983]. The required surface area in the heat exchanger is then determined utilizing the following equation:

$$A = \frac{q}{FU_m \Delta T_m} \tag{52.3}$$

The above equation is based on several assumptions, such as constant specific heats, constant overall heat transfer coefficients in the heat exchanger, constant fluid properties, absence of heat losses to and from the surroundings, and absence of heat sources in the heat exchanger. The probable errors due to the above assumptions are regarded as the penalty for the simplicity of the method. In practice, correction factors are applied to adjust the calculations for the errors involved.

The overall heat transfer coefficient U_m is determined by taking into account the various thermal resistances involved in the heat transfer path between the hot and cold fluids:

$$U_m = \frac{1}{\sum R} = \frac{1}{(1/hA)_h + R_{fh} + R_w + R_{fc} + (1/hA)_c} \tag{52.4}$$

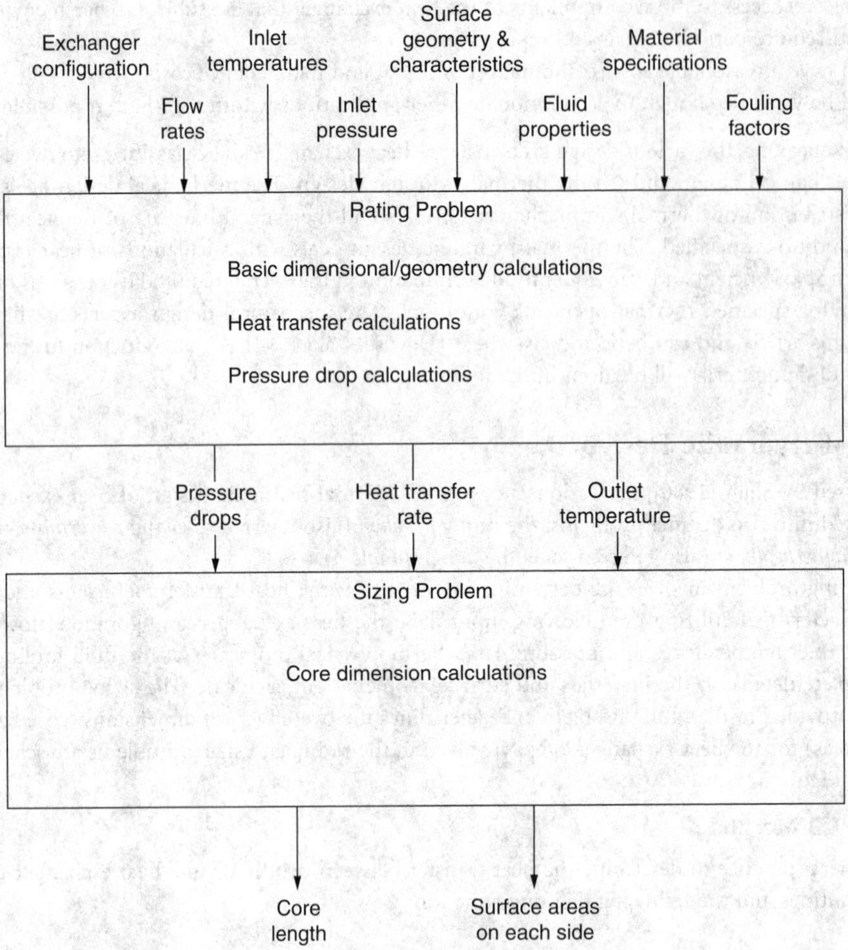

FIGURE 52.6 Schematic presentation of the heat exchanger rating and sizing problems.

in which $(1/hA)_h$ and $(1/hA)_c$, respectively, represent the convective resistance on the hot and cold sides of the heat exchanger, R_{fh} and R_{fc} represent fouling resistance on the hot and cold sides, and R_w is thermal resistance due to wall tubing. Representative values of U for common fluids are listed in Table 52.1.

To determine the mean temperature difference between the hot and cold fluids over the heat exchanger length (ΔT_m), it is a common practice to use the log mean temperature difference (LMTD), defined as

$$\Delta T_{lm} = \frac{\Delta T_i - \Delta T_o}{\ln(\Delta T_i / \Delta T_o)} \tag{52.5}$$

in which ΔT_i and ΔT_o represent the temperature difference between the hot and cold fluids at the inlet and outlet of the heat exchanger, respectively.

Effectiveness and Number of Transfer Units (NTU) Method

The LMTD method for the prediction of heat exchanger performance is useful only when the inlet and exit temperatures for the heat exchanger are known, either because they have been specified by the design or because they have been measured in a test. For situations where calculation of the inlet and outlet temperatures and flow rates are desired, use of the LMTD method requires iterative solution procedures. This can be avoided if the heat exchanger performance is expressed in terms of effectiveness and number

TABLE 52.1 Representative Values of the Overall Heat Transfer Coefficient (U)

Hot Side Fluid	Cold Side Fluid	U Btu/(hr-ft²-°F)	U W/(m²K)
Water	Water	200 to 500	800 to 2500
Water	Gas	2 to 10	10 to 50
Water	Lubricating oil	20 to 80	160 to 400
Ammonia	Water (e.g., water-cooled condenser)	150 to 500	800 to 2588
Water	Brine	100 to 200	500 to 1000
Light organics	Water	75 to 150	350 to 750
Medium organics	Water	50 to 125	240 to 610
Heavy organics	Water	5 to 75	25 to 370
Light organics	Light organics	40 to 100	200 to 500
Heavy organics	Heavy organics	10 to 40	50 to 200
Heavy organics	Light organics	30 to 60	150 to 300
Steam	Water	200 to 1000	10 to 6000
Steam	Ammonia	200 to 700	1000 to 3500
Freon 12	Water	50 to 200	300 to 1000
Steam	Heavy fuel oil	10 to 40	50 to 200
Steam	Light fuel oil	30 to 60	200 to 400
Steam	Air (air-cooled condenser)	10 to 40	50 to 200
Finned tube heat exchanger, water in tubes	Air over finned tubes	5 to 10	30 to 60
Finned tube heat exchanger, steam in tubes	Air over tubes	50 to 700	300 to 4500

of transfer units (NTU), which will be briefly discussed in the following. The heat exchanger effectiveness, ε, is defined as the ratio of actual heat transferred to the maximum heat transfer amount which can be transferred in an infinitely long counterflow heat exchanger. In an infinitely long counterflow heat exchanger, with $(mC_p)_C < (mC_p)_h$, in which subscripts c and h refer to the cold and hot streams, we can write

$$\dot{q}_{\max} = (\dot{m}C_p)_C(T_{h,\text{in}} - T_{c,\text{in}}) \tag{52.6}$$

Similarly, if $(mC_p)_h < (mC_p)_c$ then $q_{\max}$ in an infinitely long counterflow exchanger is

$$\dot{q}_{\max} = (\dot{m}C_p)_h(T_{h,\text{in}} - T_{c,\text{in}}) \tag{52.7}$$

Now if we write

$$C_{\min} = \min[(\dot{m}C_p)_h, (\dot{m}C_p)_c] \tag{52.8}$$

then the maximum heat transfer in a heat exchanger of any configuration is

$$\dot{q}_{\max} = C_{\min}(T_{h,\text{in}} - T_{c,\text{in}}) \tag{52.9}$$

From this equation, the effectiveness is expressed as

$$\varepsilon = \frac{\dot{q}}{\dot{q}_{\max}} = \frac{(\dot{m}C_p)_h(T_{h,\text{in}} - T_{h,\text{out}})}{C_{\min}(T_{h,\text{in}} - T_{c,\text{in}})} = \frac{(\dot{m}C_p)_c(T_{c,\text{out}} - T_{c,\text{in}})}{C_{\min}(T_{h,\text{in}} - T_{c,\text{in}})} \tag{52.10}$$

The number of transfer units (NTU) is defined for the hot and cold streams as

$$\mathrm{NTU}_h = \frac{AU}{(\dot{m}c_p)_h} \tag{52.11}$$

$$\mathrm{NTU}_c = \frac{AU}{(\dot{m}c_p)_c} \tag{52.12}$$

where A is the total heat exchanger area and U is the overall heat transfer coefficient. Similarly, $\mathrm{NTU}_{\min}$ corresponding to the stream having the minimum (mc_p) is defined as

$$\mathrm{NTU}_{\min} = \frac{AU}{(\dot{m}c_p)_{\min}} \tag{52.13}$$

From there it follows that

$$\varepsilon = \mathrm{NTU}_{\min}\theta \tag{52.14}$$

where θ is the ratio between the mean temperature difference ΔT_{mean} and the maximum temperature difference ΔT_{max}:

$$\theta = \frac{\Delta T_{mean}}{\Delta T_{max}} \tag{52.15}$$

The effectiveness and NTU relations for various heat exchanger configurations have been developed and are available in the literature [Mills, 1992].

52.5 Microchannel Heat Exchangers

The early work by Tuckerman and Pease (1981) demonstrated that advanced microfabrication techniques could be used in the fabrication of compact micro heat exchangers. They suggested that heat rate removal rates on the order of 1000 W/cm^2 could be possible with their original water-cooled heat exchanger. Microchannels' wide practical applications in highly specialized fields, such as bioengineering and microfabricated fluidic systems, have long been known. More recently, microchannels' wide application in the automotive air conditioning industry, fuel cells, and microelectronics has been realized. Today, microchannels have almost completely replaced circular tubes in automotive condensers and have recently become the subject of heavy research and development for cost-effective use as automotive evaporators. The advantage of the microchannel lies in its high heat transfer coefficient and significant potential in decreasing the size of heat exchangers.

Microchannels are fabricated by a variety of processes depending on the dimensions and plate material used, e.g., metals, plastics, and silicon. Conventional machining and electrical discharge machining are two typical options, while semiconductor fabrication processes are appropriate for microchannel fabrication in chip cooling applications. Using microfabrication techniques developed by the electronics industry, it is possible to manufacture three-dimensional structures with length scale as small as 0.1 μm. The prominent characters of microchannel heat sinks lie in the enhanced heat transfer expected to be realized from extremely large surface areas per unit volume (on the order of 10^5 m^2/m^3) and heat transfer coefficients (in the order of 10^2 W/cm^2-K) [Incropera, 1999]. Compared to conventional compact heat exchangers, the hydraulic diameters of microchannels are quite small, typically 1 to 2000 μm, and the fluid flow regime in microchannels in most cases is laminar.

Compared with channels of normal size, microchannels have many heat transfer advantages. Since microchannels have an increased heat transfer surface area and a large surface-to-volume ratio, they

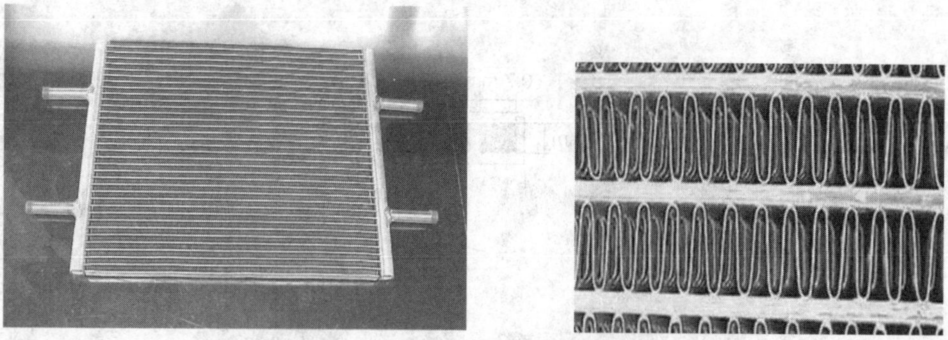

FIGURE 52.7 Schematic of refrigerant and airside flow passages in a compact automotive microchannel heat exchanger.

provide a much higher heat transfer. This feature allows heat exchangers to become compact and lightweight. In addition, microchannels can support high heat flux with small temperature gradients. However, microchannels also have weaknesses, such as large pressure drop, high cost of manufacture, dirt clogging, and flow maldistribution, especially for two-phase flows.

In recent years, major progress in compact evaporator development has been made by the automotive, aerospace, and cryogenic industries. The thermal duty and the energy efficiency increased during this period, while the space constraints became more vital. The trend was toward greater heat transfer rates per unit volume. The hot side of the evaporators in these applications was generally air, gas, or a condensing vapor. With advances in the airside fin geometry, significant improvements were achieved from increased heat transfer coefficients as well as greater surface area densities. As the airside heat transfer resistance decreased, more aggressive heat transfer designs were sought on the evaporating side, resulting in use of micro channel flow passages on the liquid side (evaporating or condensing or single-phase regimes). The major changes in recent evaporator and condenser designs for automotive and other compact heat exchanger applications involve use of individual, small-hydraulic-diameter flow passages, arranged in a multi-channel configuration on the liquid side. Figure 52.7 shows a plate-fin evaporator geometry commonly known in compact refrigerant evaporators for automotive and heat pump applications. As seen there, fins are placed in between microchannel flow slabs and the arrangement is brazed together in special ovens.

Two types of microchannel geometries are widely used in the compact heat exchanger designs. These are shown in Figure 52.8 and Figure 52.9, with typical geometric dimensions as those listed in the Table 52.2 [Zhao et al., 2001].

Despite the thin walls in microchannels, they can withstand high operating pressures. For example, a microchannel with a hydraulic diameter of 0.8 mm and a wall thickness of 0.3 mm can easily withstand operating pressures of up to 14 MPa. Another advantage of microchannels is their very large contact surface area per unit volume. This large surface area results from the definition of the hydraulic diameter, $D_h = 4A_c/p$, where A_c is the flow cross-sectional area, and p is the wetted perimeter. For a fixed total cross-sectional area, smaller D_h means larger p, which implies a larger heat transfer surface area. The high heat transfer coefficient attainable in microchannels potentially leads to significant decreases in the size of heat exchangers, thereby saving space and material. The high heat transfer coefficients are due, in part, to the increased heat transfer surface area and large surface-to-volume ratio in microchannels. Because of the high heat transfer performance, microchannels are now used routinely in most automotive condensers and have recently become the subject of study for use as automotive evaporators.

Another application area where microchannels are finding wide applicability is their use for cooling of high flux military and commercial electronics. Two widely known options exist. In one case, microchannels may be machined in a substrate or heat sink to which a chip or an array of chips is attached. In the other, they may be machined in the chip itself. A typical packaging architecture of high power laser array using a microchannel heat sink is shown in Figure 52.10 [Puchert et al., 2000].

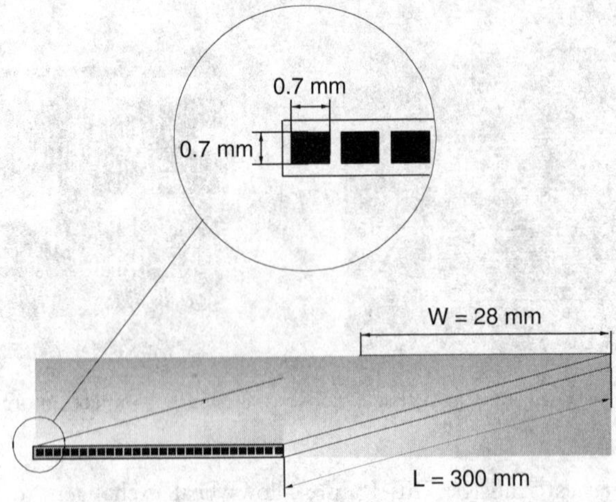

FIGURE 52.8 Typical dimensions of a rectangular microchannel for compact heat exchanger applications.

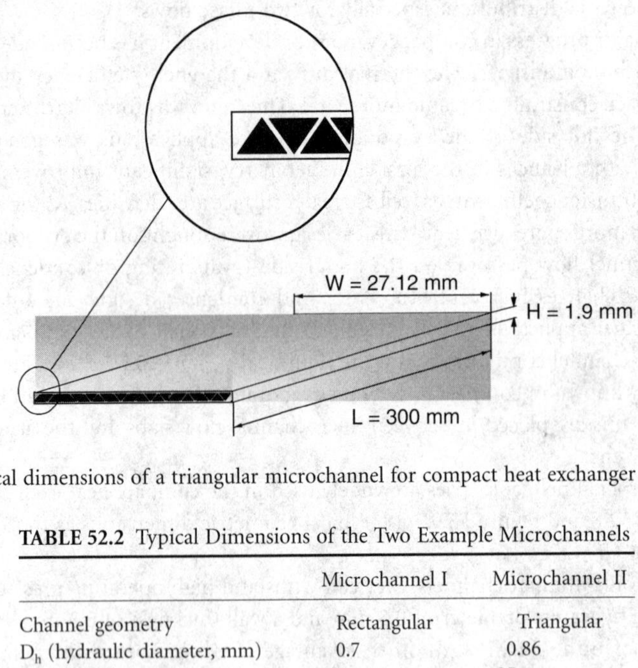

FIGURE 52.9 Typical dimensions of a triangular microchannel for compact heat exchanger applications.

TABLE 52.2 Typical Dimensions of the Two Example Microchannels

	Microchannel I	Microchannel II
Channel geometry	Rectangular	Triangular
D_h (hydraulic diameter, mm)	0.7	0.86
Number of channels	28	25
L (length, mm)	300	300
Ht (height, mm)	1.5	1.9
W (width, mm)	28	27.12
Wall thickness (mm)	0.4	0.3

The interfaces between laser array packaging materials are bonded with solders. Due to relatively low thermal conductivity of solder materials, the integrity of these solder layers is a critical factor in the thermal performance of laser diodes. Solder voids, delaminations, and other inclusions lead to sharply increased localized temperatures (hot spots) that degrade performance and can lead to the failure of laser diodes. One practical way to improve laser diode arrays is to directly etch microchannels into the silicon substrate of the laser diode arrays as shown in Figure 52.11 [Beach, 2002]. The delivery of a 41-kW-peak-power diode array module constructed using LLNL (Lawrence Livermore National Laboratory) Silicon

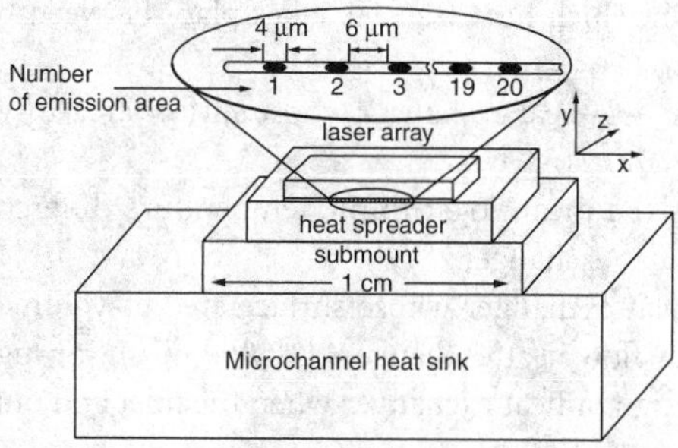

FIGURE 52.10 A schematic of a microchannel-cooled laser array on a microchannel heat sink.

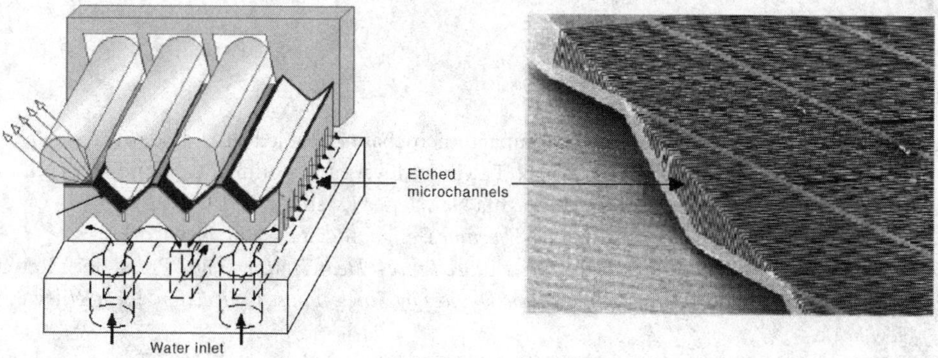

FIGURE 52.11 Sketch of a SiMMs package.

Monolithic Microchannels (SiMMs) packages was reported. It makes possible the fabrication of thousands of tiny (~30 μm-wide) cooling microchannels in very close proximity to the heat-producing laser diode bar arrays. The aggressive cooling capability of the SiMMs package enables efficient operation of laser diode arrays in a very tightly packed geometry.

Another approach to reducing the thermal resistance is to use a microchannel heat sink made of chemical-vapor-deposited diamond instead of silicon. This eliminates the impact of the thermal boundary resistance at the diamond bottom boundary and uses the high conductivity of diamond to maximize the efficiency of the microchannel-wall fins. This new configuration has the potential to yield a 75% reduction in the thermal resistance from the diode array to the water coolant compared to that for a conventional configuration based on microchannels in silicon (Goodson, 1997).

Microchannel cooling technology exhibits several advantages that provide highly efficient cooling in both earth- and space-based systems, such as higher heat transfer coefficient at the lower wall superheat temperature, eliminating contact resistance associated heat exchangers through directly fabricated-on-chip substrates, and potential application in space confined area (three-dimensional stacks of chips). However, the fluid flow and heat transfer in microchannel heat exchangers is not well understood, especially in two-phase flow regimes where flow maldistributions may be present. For very small channels, on the order of a micron or less, the heat transfer rates may not be predicted using the correlations developed for macrochannels. The design of microchannel cooling devices is still state-of-the-art, strongly depending on fabrication, channel geometry, channel surface, materials, coolant, pump, packaging, and practical system integration.

Defining Terms

Compact heat exchanger — A heat exchanger that incorporates a high surface area-to-volume density (usually 700 m^2/m^3 or higher).

Heat exchanger — A device in which heat transfer takes place between two or more fluids that are at different temperatures.

Laminar flow heat exchanger — A heat exchanger whose surface area-to-volume density is in the neighborhood of 3000 m^2/m^3.

LMTD method — Represents the mean logarithmic temperature difference between the hot and cold fluids in the entire heat exchanger.

Micro heat exchanger — A heat exchanger whose surface area-to-volume density is much higher than compact and laminar flow heat exchangers (10,000 m^2/m^3 or higher).

NTU method — Useful in design of heat exchanger when the inlet and outlet fluid temperatures are not known.

Shell-and-tube heat exchanger — A heat exchanger in which one fluid flows inside a set of tubes while the other fluid is forced through the shell and over the outside of the tubes in a cross-flow arrangement.

References

Beach, R., Freitas, B. and Rotter, M., 2002. Compact microchannel-cooled laser-diode arrays deliver over tens of kW/cm^2 onto the working surface, Lawrence Livermore National Laboratory (LLNL), Laser Science and Technology Monthly Highlights Newsletter (March).

Guyer, E. C. (Ed.) 1994. *Handbook of Applied Thermal Design,* McGraw-Hill, New York.

Hewitt, G. F., Shires, G. L., and Bott, T. R. (Eds.) 1994. *Process Heat Transfer,* CRC Press, Boca Raton, FL.

Incropera, F.R. 1999, *Liquid Cooling of Electronic Devices by Single-Phase Convection,* John Wiley & Sons, New York.

Kakac, S., Bergles, A. E., and Fernandes, E. O. (Eds.) 1988. *Two-Phase Flow Heat Exchanger: Thermal Hydraulic Fundamentals and Design,* Kluwer Academic, Dordrecht, The Netherlands.

Mills, A. F. 1992. *Heat Transfer,* Irwin, Homewood, IL.

Ohadi, M. M. 1991. Electrodynamic enhancement of single-phase and phase-change heat transfer in heat exchangers. *ASHRAE J.* 33(12):42–48.

Palen, J. W., (Ed.) 1986. *Heat Exchanger Sourcebook,* Hemisphere, New York.

Puchert, R., Bärwolff, A., Voβ, M., Menzel, U., Tomm, J.W. and Luft, J., 2000. Transient Thermal Behavior of High Power Diode Laser Arrays, *IEEE Transactions on Components, Packaging and Manufacturing Technology, Part A,* 23:95–100.

Rohsenow, W. M., Harnett, J. P., and Ganic, E. N. (Eds.) 1985. *Handbook of Heat Transfer Applications,* 2nd ed. McGraw-Hill, New York.

Taborek, J., Hewitt, G. F., and Afgan, N. (Eds.) 1983. *Heat Exchanger: Theory and Practice,* Hemisphere/McGraw-Hill, Washington, DC.

Tuckerman, D. B. and Pease, R. F. W., 1981. High performance heat sinking for VLSI, *IEEE Electronic Device Letters,* EDL-2:126–129.

Webb, R. L. 1994. *Principles of Enhanced Heat Transfer,* John Wiley & Sons, New York.

Zhao, Y., Ohadi, M. M., and Radermacher, R. 2001. Microchannel Heat Exchangers with Carbon Dioxide, Air Conditioning and Refrigeration Technology Institute (ARTI) Report No. ARTI-21CR/10020-01.

Further Information

For in-depth treatment of heat exchangers, see the following texts:

T. R. Bott, *Fouling Notebook: A Practical Guide to Minimizing Fouling in Heat Exchangers,* Association of Chemical Engineers, London (1990).

D. Chisholm, Editor, *Heat Exchanger Technology,* Elsevier Applied Science, New York (1988).

A. P. Fraas, *Heat Exchanger Design,* 2nd ed., Wiley, New York (1989).

J. P. Gupta, *Fundamentals of Heat Exchanger and Pressure Vessel Technology,* Hemisphere, Washington, D.C. (1986); reprinted as *Working with Heat Exchangers,* Hemisphere, Washington, D.C. (1990).

G. F. Hewitt, Coordinating Editor, *Hemisphere Handbook of Heat Exchanger Design*, Hemisphere, New York (1989).

S. Kakac, Editor, *Boilers, Evaporators, and Condensers,* Wiley, New York (1991).

S. Kakac, R. K. Shah, and W. Aung, Editors, *Handbook of Single-Phase Convective Heat Transfer,* Wiley, New York (1987).

S. Kakac, R. K Shah, and A. E. Bergles, Editors, *Low Reynolds Number Flow Heat Exchanger,* Hemisphere, Washington, D.C. (1983).

Y. Mori, A. E. Sheindlin, and N. H. Afgan, Editors, *High Temperature Heat Exchanger,* Hemisphere, Washington, D.C. (1986).

R. K. Shah, A. D. Kraus, and D. Metzger, Editors, *Compact Heat Exchangers — A Festschrift for A. L. London,* Hemisphere, Washington, D.C. (1990).

S. Yokell, *A Working Guide to Shell-and-Tube Heat Exchangers,* McGraw-Hill, New York (1990).

53

Industrial Combustion

Charles E. Baukal, Jr.
John Zink Co. LLC

53.1 Introduction

The subject of combustion is very broad and directly or indirectly touches nearly all aspects of our lives. The electronic devices we use are generally powered by fossil-fuel–fired power plants. The cars we drive use internal combustion engines. The planes we fly in use jet-fuel-powered turbine engines. Most of the materials we use have been made through some type of heating or melting combustion process.

Combustion is complicated by many factors. It combines heat transfer, thermodynamics, chemical kinetics, multi-phase turbulent fluid flow, and pollution formation and control, to name a few areas of physics. Therefore, the study of combustion is interdisciplinary by necessity. Space limitations allow only a cursory discussion of some of these important topics. However, many theoretical [Strehlow, 1968; Williams, 1985; Lewis and von Elbe, 1987; Bartok and Sarofim, 1991; Fristom, 1995; Glassman, 1996] and practical [Griswold, 1946; Stambuleanu, 1976; Perthuis, 1983; Keating, 1993] books have been written on the subject of combustion. Some books consider both theory and application at some length [Edwards, 1974; Barnard and Bradley, 1985; Turns, 1996; Borman and Ragland, 1998]. Some handbooks on combustion applications are also available [Segeler, 1965; Reed, 1981; Pritchard et al., 1977; Reed, 1986; IHEA, 1994]. This chapter focuses specifically on industrial combustion.

53.2 Fundamentals

Combustion Chemistry

Combustion is usually considered to be the controlled release of heat and energy from the chemical reaction between a fuel and an oxidizer. Virtually all of the combustion in industrial processes uses a hydrocarbon fuel. A generalized combustion reaction for a typical hydrocarbon fuel can be written as follows:

$$\text{fuel} + \text{oxidizer} \rightarrow CO_2 + H_2O + \text{other species} \tag{53.1}$$

The "other species" depend on a number of factors considered later, but typically include N_2, O_2, NOx, and CO.

0-8493-1586-7/05/$0.00+$1.50

Fuel Properties

The fuel has a significant influence on the combustion process. One of the most important properties is the heating value of the fuel which is used to determine how much fuel must be combusted to process the desired production rate of material that is being heated. The heating value is specified as either the higher heating value (HHV) or the lower heating value (LHV). The LHV excludes the heat of vaporization which is the energy required to convert liquid water to steam. This means that the LHV assumes all of the products of combustion are gaseous, which is generally the case for nearly all industrial combustion applications. If the combustion products were to exit the process at a temperature low enough that all of the water were converted from a gas to a liquid, then the heat of condensation would be released into the process as an additional source of energy. The HHV of a fuel includes that energy.

The fuel composition is important in determining the composition of the combustion products and the amount of oxidizer that will be needed to combust the fuel, both of which are discussed below. It is also important for determining the soot-producing tendency of the fuel.

Oxidizer Composition

The majority of those processes use air as the oxidizer. However, many of the higher-temperature processes use an oxidizer containing a higher concentration of oxygen than found in air (approximately 21% by volume) which is referred to as oxygen-enhanced combustion [Baukal, 1998]. In many cases the production rate in a heating process can be significantly increased even with only relatively small amounts of oxygen enrichment.

A common way of specifying the oxidizer composition is by calculating the O_2 mole fraction in the oxidizer which may be defined as:

$$\Omega = \frac{\text{volume flow rate of } O_2 \text{ in the oxidizer}}{\text{total volume flow rate of oxidizer}} \tag{53.2}$$

If the oxidizer is air, which contains approximately 21% O_2 by volume, $\Omega = 0.21$. If the oxidizer is pure O_2, $\Omega = 1.0$. The O_2 enrichment level is sometimes used. This refers to the incremental O_2 volume above that found in air. For example, if $\Omega = 0.35$, then the O_2 enrichment would be 14% (35% – 21% = 14%).

Mixture Ratio

A global combustion reaction using CH_4 as the fuel may be written as:

$$CH_4 + (xO_2 + yN_2) \rightarrow CO, CO_2, H_2, H_2O, N_2, NOx, O_2, \text{trace species} \tag{53.3}$$

The stoichiometry of a reaction indicates the ratio of oxygen to fuel for a given system. One method of quantifying the stoichiometry is to consider only the O_2 in the oxidizer, since the inerts in the oxidizer are generally not needed for the reaction:

$$S_1 = \frac{\text{volume flow rate of } O_2 \text{ in the oxidizer}}{\text{volume flow rate of fuel}} \tag{53.4}$$

If CH_4 is again used as an example, a global simplified stoichiometric reaction with air can be written as:

$$CH_4 + (2O_2 + 7.52N_2) \rightarrow CO_2 + 2H_2O + 7.52N_2 \tag{53.5}$$

where air is represented as $2O_2 + 7.52N_2$. In that case, $S_1 = 2/1 = 2.0$.

The most common way of defining the stoichiometry in industry in the U.S. is as follows:

$$S_2 = \frac{\text{volume flow rate of oxidizer}}{\text{volume flow rate of fuel}} \tag{53.6}$$

For Equation (53.5), this stoichiometry would be calculated as $S_2 = (2 + 7.52)/1 = 9.52$.

In the scientific community, it is common to use the equivalence ratio (ϕ) to specify the mixture ratio:

$$\phi = \frac{\text{stoichiometric volumetric ratio of oxidant:fuel}}{\text{actual volumetric ratio of oxidant:fuel}} \tag{53.7}$$

Excess O_2

Many industrial combustion processes run with approximately 2 to 3% more O_2 than is theoretically needed for perfect combustion. That is often the amount of excess O_2 that is required to minimize the emissions of unburned combustibles such as CO. This is usually due to mixing limitations between the fuel and oxidizer, especially in nonpremixed systems. Too much excess O_2 means that energy is being wasted heating excess combustion air, instead of the load. Therefore it is desirable to only use just enough excess O_2 to get low CO emissions. An example of a simplified global reaction for methane with 3% excess O_2 is the reaction below:

$$CH_4 + (2.06O_2 + 7.75N_2) \rightarrow CO_2 + 2H_2O + 0.06O_2 + 7.75N_2 \tag{8}$$

The amount of O_2 in the combustion exhaust products is often used to monitor and control the performance of combustion systems. It is usually desirable to operate at the minimum excess O_2 in the exhaust without producing significant quantities of carbon monoxide.

Combustion Properties

Combustion Products

The actual composition of the exhaust products from the combustion reaction depends on several factors including the oxidizer composition, the temperature of the gases, and the equivalence ratio. An adiabatic process means that no heat is lost during the reaction or that the reaction occurs in a perfectly insulated chamber. Note that this is not the case in an actual combustion process, where heat is lost from the flame by radiation and is only used for comparison purposes. Figure 53.1(a) and Figure 53.1(b) show the predicted major and minor species for the adiabatic equilibrium combustion of CH_4 as a function of the oxidizer composition. As N_2 is removed from the oxidant, there is an increase in the concentrations of CO, CO_2, and H_2O. For this adiabatic process, there is a significant amount of CO at higher levels of O_2 in the oxidizer.

The actual flame temperature is lower than the adiabatic equilibrium flame temperature due to imperfect combustion and radiation from the flame. The actual flame temperature is determined by how well the flame radiates its heat and how well the combustion system, including the load and the refractory walls, absorbs that radiation. A highly luminous flame generally has a lower flame temperature than a highly nonluminous flame. The actual flame temperature will also be lower when the load and the walls are more radiatively absorptive. This occurs when the load and walls are at lower temperatures and have higher radiant absorptivities. These effects are discussed in more detail in Baukal [2000].

Figure 53.2(a) and Figure 53.2(b) show the predicted major and minor species, respectively, for the equilibrium combustion of CH_4 with "air" (21% O_2, 79% N_2) as a function of the gas temperature. The highest possible temperature for the air/CH_4 reaction is the adiabatic equilibrium temperature of 3537°F (2220K). For the air/CH_4 reaction, there is very little change in the predicted gas composition as a function of temperature. Figure 53.3(a) and Figure 53.3(b) show the predicted major and minor gas species, respectively, for the adiabatic equilibrium combustion of air/CH_4 as a function of the equivalence ratio.

Figure 53.4 shows the adiabatic flame temperature as a function of the equivalence ratio for three fuels: H_2, CH_4, and C_3H_8. The peak temperature occurs at stoichiometric conditions ($\phi = 1.0$). In most real flames, the peak flame temperature often occurs at slight fuel lean conditions ($\phi < 1.0$). This is due to imperfect mixing where slightly more O_2 is needed to fully combust all of the fuel.

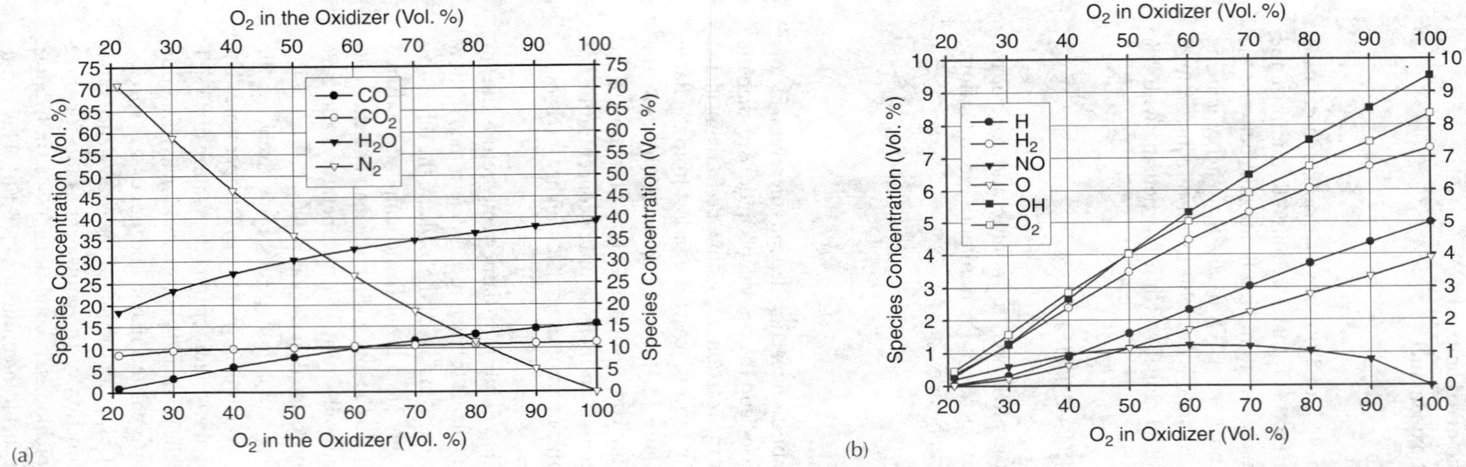

FIGURE 53.1 (a) Major and (b) minor species concentrations vs. oxidant ($O_2 + N_2$) composition, for an adiabatic equilibrium stoichiometric CH_4 flame. (*Source:* Baukal, C. E., Ed., 1998. *Oxygen-Enhanced Combustion*. CRC Press LLC, Boca Raton, FL.)

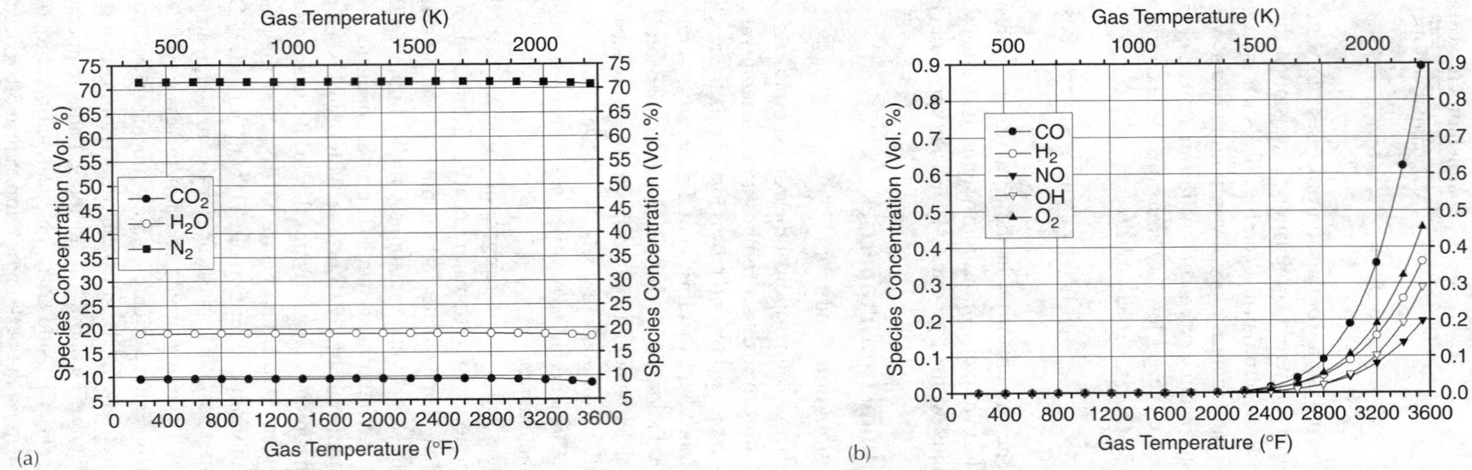

FIGURE 53.2 Equilibrium calculations for the predicted gas composition of the (a) major and (b) minor species as a function of the combustion product temperature for stoichiometric air/CH_4 flames. (*Source:* Baukal, C. E., Ed., 1998. *Oxygen-Enhanced Combustion*. CRC Press LLC, Boca Raton, FL.)

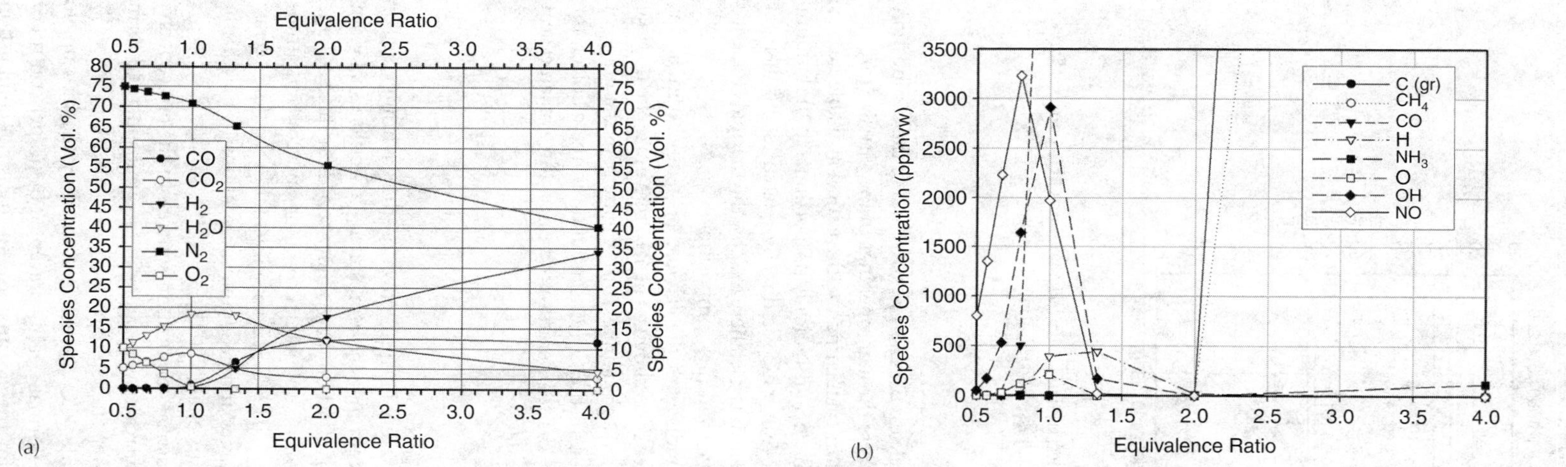

FIGURE 53.3 Adiabatic equilibrium calculations for the predicted gas composition of the (a) major and (b) minor species as a function of the equivalence ratio for air/CH_4 flames. (*Source:* Baukal, C. E. 2000. *Heat Transfer in Industrial Combustion*. CRC Press, Boca Raton, FL.)

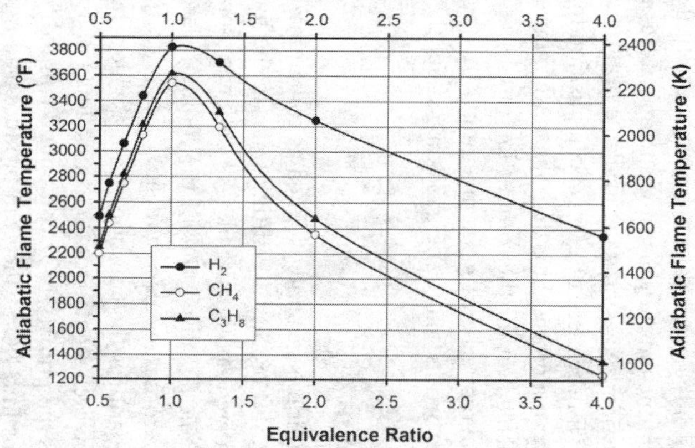

FIGURE 53.4 Adiabatic equilibrium flame temperature vs. equivalence ratio for air/H_2, air/CH_4, and air/C_3H_8 flames. (*Source:* Baukal, C. E. 2000. *Heat Transfer in Industrial Combustion*. CRC Press, Boca Raton, FL.)

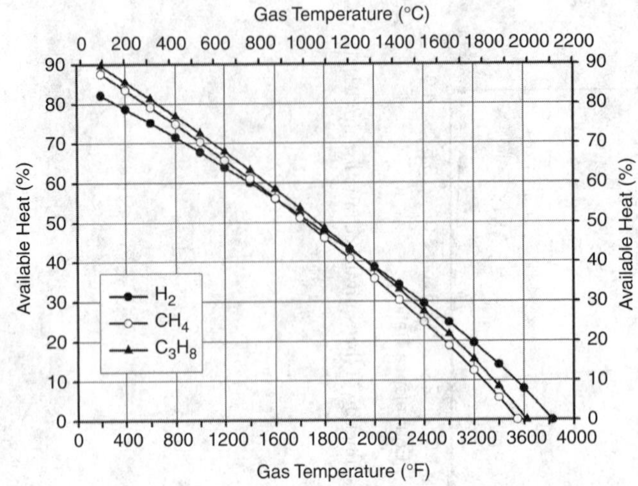

FIGURE 53.5 Available heat vs. gas temperature, for stoichiometric air/H_2, air/CH_4, and air/C_3H_8 flames. (*Source:* Baukal, C. E. 2000. *Heat Transfer in Industrial Combustion*. CRC Press, Boca Raton, FL.)

Available Heat

Available heat is defined as the gross heating value of the fuel, less the energy carried out of the combustion process by the hot exhaust gases. The heat lost from a process through openings in the furnace, through the furnace walls, or by air infiltration are not considered in calculating the theoretical available heat as those are dependent on the process. The theoretical available heat is typically proportional to the amount of energy actually absorbed by the load in an actual process, which is directly related to the thermal efficiency of the system. Therefore, the theoretical available heat is used here to show the thermal efficiency trends as functions of exhaust gas temperature and oxidizer and fuel compositions.

Figure 53.5 shows how the available heat decreases rapidly with the exhaust gas temperature and is relatively independent of the fuel composition for the three fuels shown. Then, to maximize the thermal efficiency of a process, it is desirable to minimize the exhaust gas temperature. Figure 53.6 is a graph of the available heat for the combustion of CH_4 as a function of the O_2 concentration in the oxidizer, for three different exhaust gas temperatures. As the exhaust gas temperature increases, the available heat decreases because more energy is carried out the exhaust stack.

Pollution

Air pollution is generally defined as contaminants that have a harmful effect on the environment. For example, if a process is emitting water in the form of steam into the atmosphere, it would usually not be considered a pollutant unless it was having some type of effect on the visibility at or near the discharge point. Table 53.1 lists some of the common gaseous airborne pollutants, common sources, and atmospheric removal reactions and sinks [Liu and Lipták, 1997]. The U.S. Environmental Protection Agency (EPA) quantifies the emissions from a wide range of sources and geographical locations in the U.S. to track how well regulations are controlling pollution [Elkins et al., 2001].

NO is the primary NOx compound emitted from industrial combustion sources. In the atmosphere, NO turns into NO_2, which can produce acid rain when it reacts with water in the atmosphere to form nitric acid and plays a role in ozone formation in the lower atmosphere.

A principle unburned combustible is carbon monoxide (CO), which is formed by the incomplete combustion of carbon-containing fuels. CO formation is normally easily prevented by proper control of the combustion process and is not typically a problem for the vast majority of industrial combustion

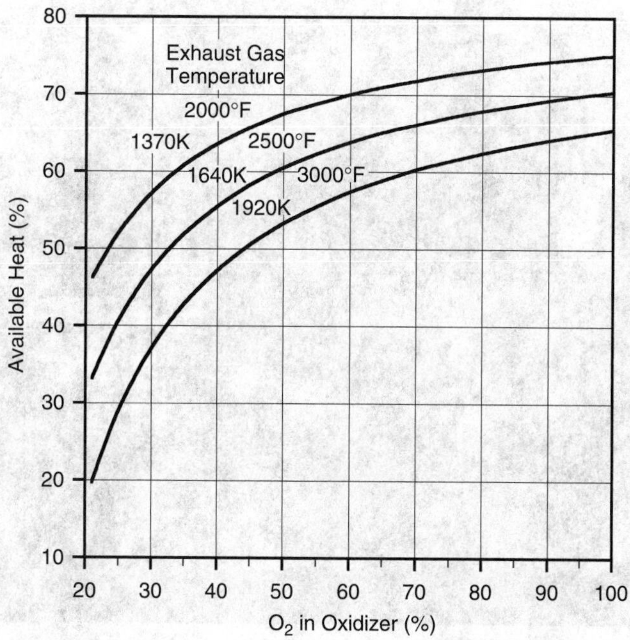

FIGURE 53.6 Available heat vs. oxidant composition, for a stoichiometric CH_4 flame, at exhaust temperatures of 2000, 2500, and 3000°F (1370, 1640, and 1920K). (*Source:* Baukal, C. E., Ed., 1998. *Oxygen-Enhanced Combustion.* CRC Press LLC, Boca Raton, FL.)

applications. Over three-fourths of the ambient CO in the atmosphere comes from automobile emissions; only a small portion comes from industrial combustion applications.

SO_2 is a pollutant emission formed in industrial combustion applications during the firing of fuels, such as heavy oil, that contain sulfur. Under typical combustion conditions, essentially all of the sulfur in the fuel is converted to SO_2. The primary environmental concern with SOx emissions is the generation of acid rain when SO_2 reacts with water in the atmosphere to form sulfuric acid.

Particulate matter is a general term used for solid and/or liquid particles emitted into the atmosphere. If the concentration of the particulate emissions is dense enough then visibility can be impaired, which

TABLE 53.1 Sources, Concentrations and Scavenging Processes of Atmospheric Pollutants

Air Pollutant	Effects
Particulates	Speeds chemical reations; obscures vision; corrodes metals; causes grime on belongings and buildings; aggravates lung illness
Sulfur oxides	Causes acute and chronic leaf injury; attacks a wide variety of trees; irritates upper respiratory tract; destroys paint pigments; erodes statuary; corrodes metals; ruins hosiery; harms textiles; disintegrates book pages and leather
Hydrocarbons (in solid and gaseous states)	May be cancer-producing (carcinogenic); retards plant growth; causes abnormal leaf and bud development
Carbon monoxide	Causes headaches, dizziness, and nausea; absorbs into blood; reduces oxygen content; impairs mental processes
Nitrogen oxides	Causes visible leaf damage; irritates eyes and nose; stunts plant growth even when not causing visible damage; creates brown haze; corrodes metals
Ozone	Discolors the upper surface of leaves of many crops, trees, and shrubs; damages and fades textiles; reduces athletic performance; hastens cracking of rubber; disturbs lung function; irritates eyes, nose, and throat; induces coughing

Source: Adapted from Liu, D. H. F. and Lipták, B. G. (eds.) 1997. *Environmental Engineers' Handbook,* 2nd ed. Lewis Publishers, Boca Raton, FL.

FIGURE 53.7 Flare on an offshore oil platform (courtesy of John Zink Co., Tulsa, OK).

is the primary environmental concern for this pollutant. Some possible sources of particulate emissions from industrial combustion include ash or char from the incomplete combustion of carbon-containing compounds and carryover of solid particles in the production process.

There are several possible sources of noise from combustion processes, including high speed flow through piping and piping components, fan noise from the rotation of the blades, and combustion roar from the combustion process, to name a few. In some cases it is relatively easy to reduce the noise by modifying the process or configuration. In other cases, some type of noise suppression, such as a muffler, may be needed.

Carbon dioxide is a product of combustion from the burning of hydrocarbon fuels. It is considered by many to be a greenhouse gas responsible for global warming. However, this is a controversial subject that does not have widespread agreement among scientists. A primary concern is the increased burning of fuels over the past century, which some believe has caused an imbalance in the ability of nature to absorb the additional CO_2 emissions.

There are a number of other pollutants that typically only apply to a limited number of industrial combustion applications. Some special pollutants are related to the use of flares, including smoke, thermal radiation, and noise. Figure 53.7 shows a flare used on an offshore oil platform, which is used to burn undesired flammable gases. Another example of a special pollutant is lead emissions, which are only present in some select applications such as lead production and waste incineration. These are normally easily controlled with the appropriate posttreatment equipment.

53.3 Combustion System Components

Six components may be important in industrial combustion processes (see Figure 53.8). One component is the burner, which combusts the fuel with an oxidizer to release heat. Another component is the load itself, which can greatly affect how the heat is transferred from the flame. In most cases, the flame and the load are located inside of a combustor, which may be a furnace, heater, or dryer, which is the third component in the system. In some cases, some type of heat recovery device may be used to increase the thermal efficiency of the overall combustion system; this is the fourth component of the system. The fifth component is the air pollution control system used to minimize the pollutants emitted from the

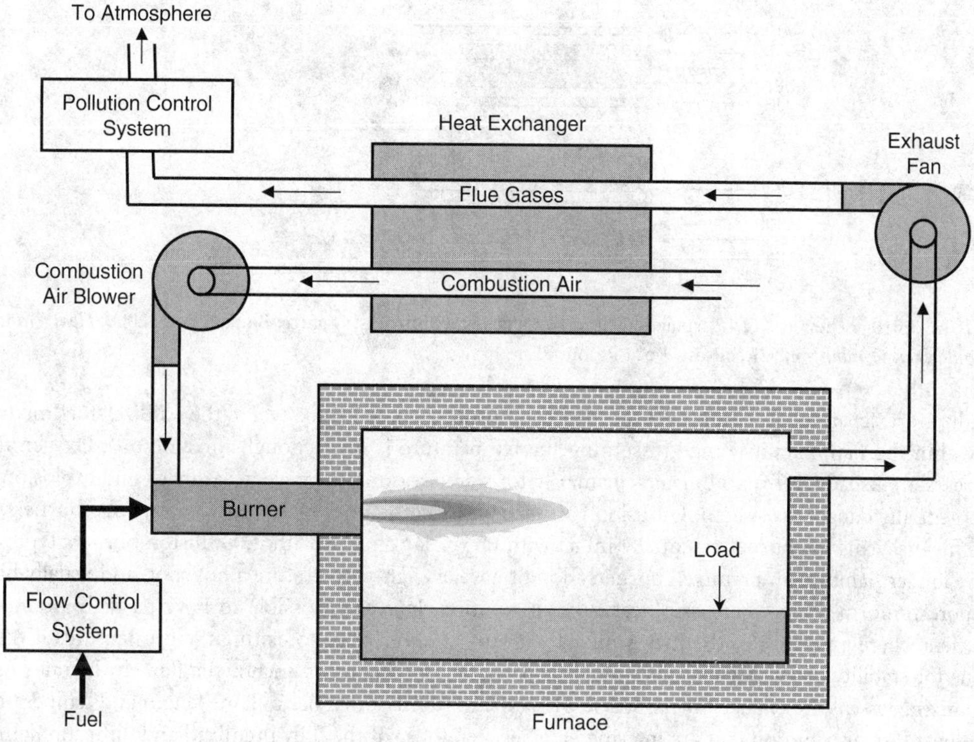

FIGURE 53.8 Schematic of the major components in a combustion system. (*Source:* Baukal, C. E., Ed., 2003. *The Handbook of Industrial Burners*, CRC Press, Boca Raton, FL.)

exhaust stack into the atmosphere. The sixth and last component is the flow control system used to meter the fuel and the oxidant to the burners. The first five are discussed next. The last component is not discussed here in the interest of space. The interested reader is referred to Gifford and Kodesh [2003] for information on combustion controls.

Burners

General Types

The first component of a typical combustion system is the burner. Burners are classified in numerous ways. One common method for classifying burners is according to how the fuel and the oxidizer are mixed. In premixed burners, shown schematically in Figure 53.9(a), the fuel and the oxidizer are completely mixed before combustion begins. Premixed burners often produce shorter and more intense flames, compared to diffusion flames. In diffusion-mixed burners, shown schematically in Figure 53.9(b), the fuel and the

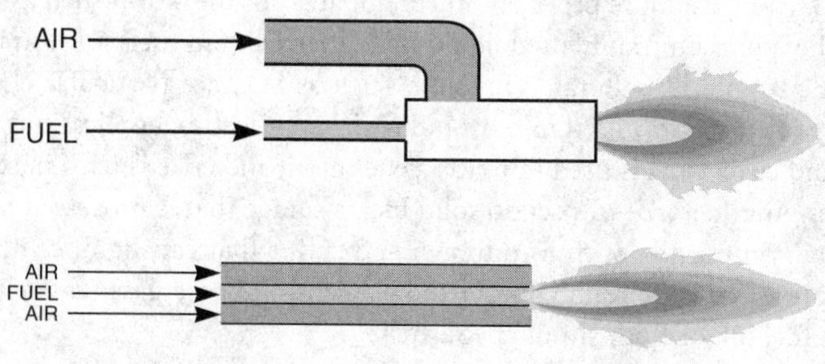

FIGURE 53.9 Schematic of (a) premixed and (b) diffusion-mixed burners. (*Source:* Baukal, C. E. 2000. *Heat Transfer in Industrial Combustion*. CRC Press, Boca Raton, FL.)

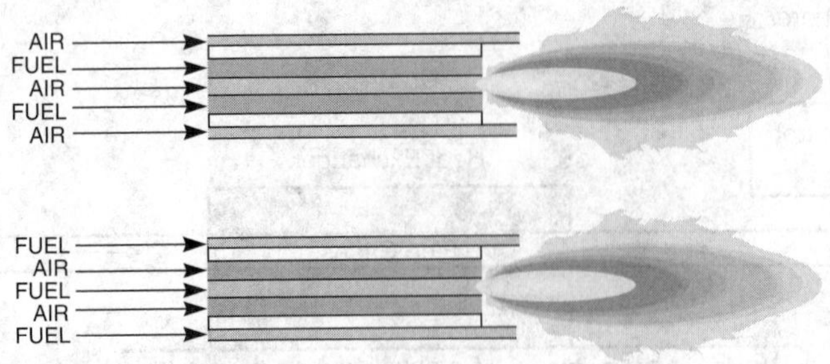

FIGURE 53.10 Schematic of (a) staged-air and (b) staged-fuel burners. (*Source:* Baukal, C. E. 2000. *Heat Transfer in Industrial Combustion*. CRC Press, Boca Raton, FL.)

oxidizer are separated and unmixed prior to combustion, which begins where the oxidizer/fuel mixture is within the flammability range (assuming the temperature is high enough for ignition). Oxygen/fuel burners are usually diffusion burners, primarily for safety reasons, to prevent flashback and explosion in a potentially dangerous system. Diffusion gas burners are sometimes referred to as "raw gas" burners, as the fuel gas exits the burner essentially intact with no oxidant mixed with it. Diffusion burners typically have longer flames than premixed burners, do not have as high-temperature a hot spot, and usually have a more uniform temperature and heat flux distribution. It is also possible to have partially premixed burners where a portion of the fuel is mixed with the oxidizer prior to exiting the burner. This is often done for stability and safety reasons where the partial premixing helps anchor the flame, while not fully premixing lessens the chance for flashback. This type of burner often has a flame length and temperature and heat flux distribution that are intermediate between those of the fully premixed and diffusion flames.

Another burner classification based on mixing is known as staging: staged air and staged fuel. Staged air and staged fuel burners are shown schematically in Figure 53.10(a) and Figure 53.10(b), respectively. Secondary and sometimes tertiary injectors in the burner are used to inject a portion of the fuel and/or the oxidizer into the flame, downstream of the root of the flame. Staging is often done to produce longer flames and reduce pollutant emissions such as NOx. These longer flames typically have a lower peak flame temperature and more uniform heat flux distribution than nonstaged flames. However, an additional challenge is that multiple longer flames may interact with each other and produce unpredictable consequences compared to single flames.

Burners may also be classified according to the fuel type. Gaseous fuel burners are the predominant type used in most of the applications considered here. In general, natural gas is the predominant gaseous fuel used because of its low cost and availability. However, a wide range of gaseous fuels are used in, for example, the chemicals industry [Baukal, 2001]. These fuels contain multiple components such as methane, hydrogen, propane, nitrogen, and carbon dioxide and are sometimes referred to as refinery fuel gases. Gaseous fuels are among the easiest to control because no vaporization is required as in liquid and solid fuels. They are also often simpler to control to minimize pollution emissions because they are more easily staged compared to liquid and solid fuels. Liquid fuel burners are used in some limited applications in the U.S. but are more prevalent in certain areas of the world such as South America. No. 2 and No. 6 oil are the most commonly used liquid fuels. Waste liquid fuels are also used in incineration processes. Fuel atomization and pollutant emissions such as SOx are particular challenges with liquid fuels. Solid fuels are not commonly used in most industrial combustion applications. The most common solid fuels are coal and coke. Coal is used in power generation, and coke is used in some primary metals production processes. Another type of pseudo solid fuel is sludge that is processed in incinerators. Solid fuels also often contain impurities such as nitrogen and sulfur that can significantly increase pollutant emissions. Some applications require the burner to be able to fire on a gaseous fuel such as natural gas, a liquid fuel such as fuel oil, or both simultaneously.

Burners and flames are sometimes classified according to the type of oxidizer used. The majority of industrial burners use air for combustion. In many of the higher-temperature heating and melting

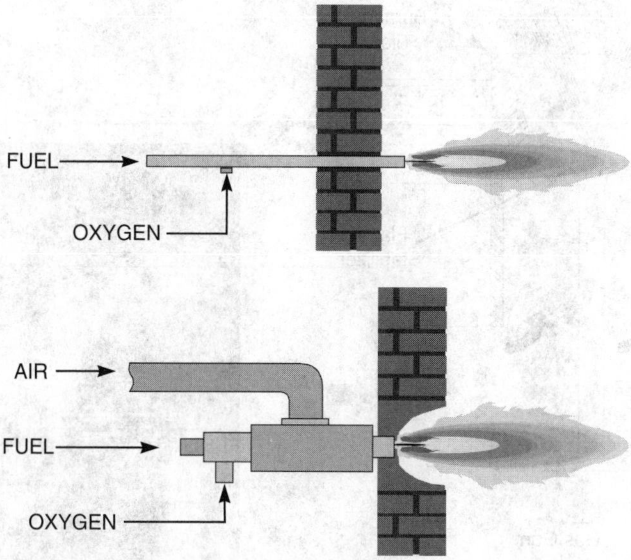

FIGURE 53.11 Schematic of an (a) oxy/fuel and (b) air-oxy/fuel burner. (*Source:* Baukal, C. E., Ed., 1998. *Oxygen-Enhanced Combustion.* CRC Press LLC, Boca Raton, FL.)

applications, such as glass production, the oxidizer is pure oxygen. In other applications, the oxidizer is a combination of air and oxygen, often referred to as oxygen-enriched air combustion. Figure 53.11(a) shows a method of using OEC commonly referred to as an oxy/fuel burner. Figure 53.11(b) shows another method of using OEC commonly referred to as an air-oxy/fuel burner.

Burners may also be classified according to how the combustion air is supplied. Most industrial burners are known as forced-draft burners. This means that the combustion air is supplied to the burner under pressure with a fan or blower. In natural-draft burners, the air used for combustion is induced into the burner by the negative draft produced in the combustor and by the motive force of the incoming fuel, which may be at a significant pressure. A schematic is shown in Figure 53.12(a). In this type of burner, the pressure drop and combustor stack height are critical in producing enough suction to induce enough combustion air into the burners. This type of burner [see photo in Figure 53.12(b)] is commonly used in the chemical and petrochemical industries in fluid heaters.

Design Factors

Many factors go into the design of a burner [Baukal, 2003]. There have been many changes in the traditional designs that have been used in burners, primarily because of the recent interest in reducing pollutant emissions. In the past, the burner designer was primarily concerned with efficiently combusting the fuel and transferring the energy to a heat load. New and increasingly stringent environmental regulations have added the requirement to consider the pollutant emissions produced by the burner. In many cases, reducing pollutant emissions and maximizing combustion efficiency are at odds with each other. For example, a well-accepted technique for reducing NOx emissions is known as staging, where the primary flame zone is deficient of either fuel or oxidizer [Reese et al., 1994]. However, radiant heat transfer from the flame may also be reduced.

In the past, the challenge for the burner designer was often to maximize the mixing between the fuel and the oxidizer to ensure complete combustion, especially if the fuel was difficult to burn, as in the case of low heating value fuels such as waste liquid fuels or process gases from chemical production. Now the burner designer must balance the mixing of the fuel and the oxidizer to maximize combustion efficiency while simultaneously minimizing all types of pollutant emissions. This is no easy task as, for example, NOx and CO emissions often go in opposite directions as shown in Figure 53.13. When CO is low, NOx may be high and vice versa.

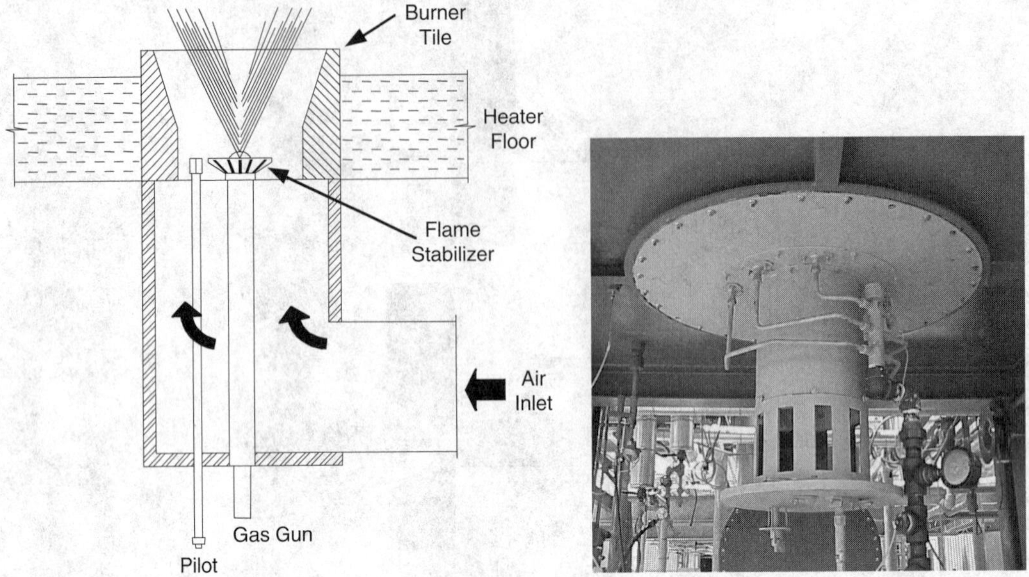

FIGURE 53.12 (a) Schematic and (b) photo of natural draft burner (courtesy of John Zink Co. LLC, Tulsa, OK).

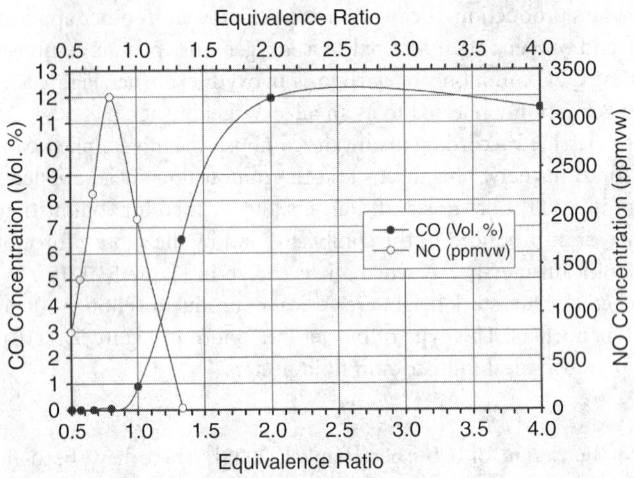

FIGURE 53.13 NOx and CO as a function of equivalence ratio. (*Source:* Baukal, C. E. 2000. *Heat Transfer in Industrial Combustion*. CRC Press, Boca Raton, FL.)

Heat Load

The second component in a combustion system is the heat load. A variety of heat loads are used in industrial applications. In petrochemical production processes, process heaters are used to heat petroleum products up to operating temperatures. The fluids are transported through the process heaters in process tubes. In some applications, heaters and burners are used to heat or dry moving substrates or webs. One common application is the use of gas-fired infrared (IR) burners to remove moisture from paper during the forming process [Longacre, 1997]. Another example of a moving substrate application is using IR burners to remove water during the production of fabrics in textile manufacturing [Smith and Baukal, 1983].

Many loads are considered opaque, which means that they absorb thermal radiation over a broad range of wavelengths. This type of load encompasses a wide range of materials including granular solids such as limestone and liquids such as molten metal. For this type of load, the heat transfers to the surface of the load and must conduct down into the material. Some loads are highly transparent, as they allow much of any incident radiant energy to penetrate below the surface. The primary example of this load type is glass, which has selective radiant transmission properties. In glass melting, the primary mode of heat transfer is by radiation. Specific radiation characteristics are preferred to enhance the heat transfer from the flame to the glass.

Combustors

The third component in a combustion system is the combustor. There are two predominant categories of combustors used in industry: process heaters and boilers for lower temperature applications (less than about 2000°F) and furnaces for higher temperature applications (greater than about 2000°F). Process heaters include, for example, ovens, heaters, reactors, and dryers. Furnaces include, for example, kilns, incinerators, and thermal crackers. Combustors are used to transform incoming charge materials by, for example, oxidation, reduction, melting, heat treating, curing, baking, and drying.

Design Considerations

A primary consideration for any combustor is the type of material that will be processed including the load handling system, which is dependent on the physical state of the material, whether it is a solid, liquid, or gas. Another factor is the transport properties of the load. For example, the solid may be granular or it might be in the form of a sheet (web). Related to that is how the solid will be fed into the combustor. A granular solid could be fed into a combustor continuously with a screw conveyor or it could be fed in with discrete charges from a front-end loader. The shape of the furnace will vary according to how the material will be transported through it. For example, limestone is fed continuously into a rotating and slightly downwardly inclined cylinder.

Combustor Classifications

Combustors can be classified in several ways, which are briefly discussed in this section. Each type has an impact on the heat transfer mechanisms in the furnace. Furnaces are often classified as to whether they are batch or continuous. In a batch furnace, the load is charged into the furnace at discrete intervals. There may be multiple load charges, depending on the application. Normally, the firing rate of the burners is reduced or turned off during the charging cycle. On some furnaces, a door may also need to be opened during charging. The heating process and heat transfer are dynamic and constantly changing as a result of the cyclical nature of the load charging. In a continuous furnace, the load is fed into and out of the combustor constantly. The feed rate may change sometimes due to conditions upstream or downstream of the combustor or due to the production needs of the plant, but the process is nearly steady state. Some furnaces are semicontinuous, where the load may be charged in a nearly continuous fashion, but the finished product may be removed from the furnace at discrete intervals. An example is an aluminum reverberatory furnace, which is charged using an automatic sidewell feed mechanism. In that process, shredded scrap is continuously added to a circulating bath of molten aluminum. When the correct alloy composition has been reached and the furnace has a full load, some or all of that load is then tapped out of the furnace.

Combustors are often classified as direct [see Figure 53.14(a)] or indirect [see Figure 53.14(b)] heating. In indirect heating, there is some type of intermediate heat transfer medium between the flames and the load that keeps the combustion products separate from the load. One example is a muffle furnace, where there is a high-temperature ceramic muffle between the flames and the load. The flames transfer their heat to the muffle, which then radiates to the load, which is usually some type of metal. The limitation of indirect heating processes is the temperature limit of the intermediate material. Although ceramic materials have fairly high temperature limits, other issues such as structural integrity over long distance spans and thermal cycling can still reduce the recommended operating temperatures. Another example of indirect heating is in process heaters where fluids are transported through metal

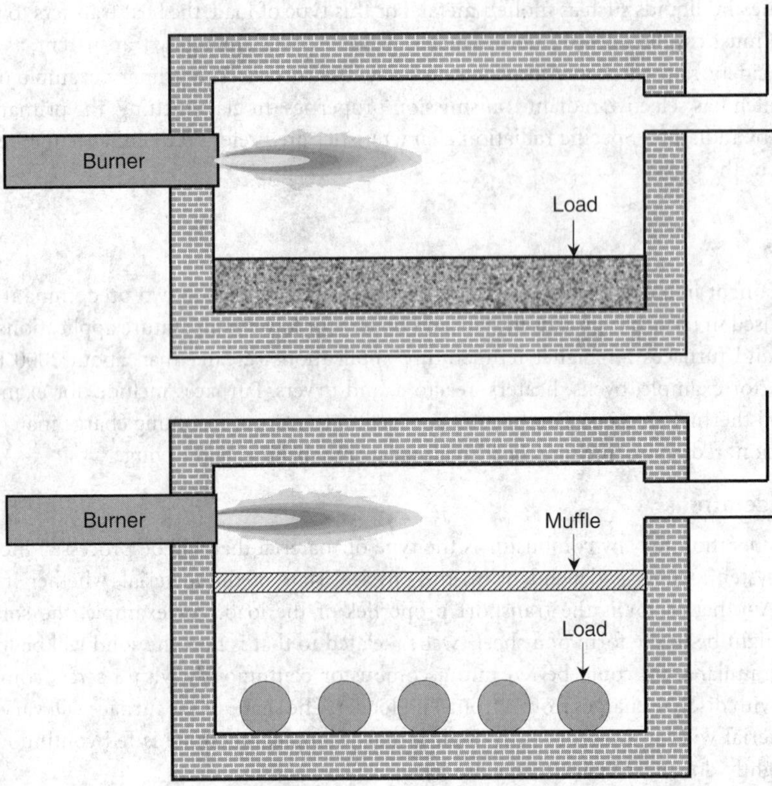

FIGURE 53.14 (a) Direct- and (b) indirect-fired process. (*Source:* Baukal, C. E., Ed., 2003. *The Handbook of Industrial Burners*, CRC Press, Boca Raton, FL.)

tubes heated by flames. Indirect heating processes often have fairly uniform heat flux distributions because the heat exchange medium tends to homogenize the energy distribution from the flames to the load. The heat transfer from the heat exchange surface to the load is often fairly simple and straightforward to compute because of the absence of chemical reactions in between. However, the heat transfer from the flames to the heat exchange surface and the subsequent thermal conduction through that surface are as complicated as if the flame was radiating directly to the load. Indirect heating may also have an advantage for reducing pollutant emissions when contact of the high temperature exhaust gases with the load could generate pollutants.

Another aspect of the geometry that is important in some applications is whether or not the furnace is moving. For example, in a rotary furnace for melting scrap aluminum, the furnace rotates to enhance mixing and heat transfer distribution. This again affects the type of analysis that would be appropriate for that system and can add some complexity to the computations.

Heat Recovery

The fourth component that may be present in a combustion system is a heat recovery unit. Heat recovery devices are often used to improve the efficiency of combustion systems. Some of these devices are incorporated into the burners, but more commonly they are another component in the combustion system, separate from the burners. These heat recovery devices incorporate some type of heat exchanger, depending on the application. The two most common types have been recuperators and regenerators, which are briefly discussed next. Reed [1987] predicts an increasing importance for heat recovery devices in industrial combustion systems for increasing heat transfer and thermal efficiencies.

Gas Recirculation

A common technique used in combustion systems is to design the burner to induce furnace gases to be drawn into the burner to dilute the flame, usually referred to as furnace gas recirculation (FGR). Even though the furnace gases are hot, they are still much cooler than the flame itself. This dilution may accomplish several purposes. One is to minimize NOx emissions by reducing the peak temperatures in the flame. Another reason to use furnace gas recirculation may be to increase the convective heating from the flame because of the added gas volume and momentum.

Recuperators

A recuperator is a low- to medium-temperature (up to about 1300°F or 700°C) continuous heat exchanger that uses the sensible energy from hot combustion products to preheat the incoming combustion air. These heat exchangers are commonly counterflow, where the highest temperatures for both the combustion products and the combustion air are at one end of the exchanger, with the coldest temperatures at the other end. Lower temperature recuperators are normally made of metal, while higher temperature recuperators may be made of ceramics. Recuperators are typically used in lower-temperature applications because of the limitations of the metals used to construct these heat exchangers.

Regenerators

A regenerator is a higher-temperature, transient heat exchanger that is used to improve the energy efficiency of high-temperature heating and melting processes, particularly in high-temperature processing industries such as glass production. In a regenerator, energy from the hot combustion products is temporarily stored in a unit constructed of firebricks. This energy is then used to heat the incoming combustion air during a given part of the firing cycle up to temperatures in excess of 2000°F (1000°C).

Regenerators are normally operated in pairs. During one part of the cycle, the hot combustion gases are flowing through one of the regenerators and heating up the refractory bricks, while the combustion air is flowing through and cooling down the refractory bricks in the second regenerator. Both the exhaust gases and the combustion air directly contact the bricks in the regenerators, although not both at the same time since each is in a different regenerator at any given time. After a sufficient amount of time (usually from 5 to 30 min), the cycle is reversed so that the cooler bricks in the second regenerator are then reheated while the hotter bricks in the first regenerator exchange their heat with the incoming combustion air. A reversing valve is used to change the flow from one gas to another in each regenerator. The burners used in these systems must be capable of not only handling the high-temperature preheated air, but also the constant thermal cycling.

Pollution Control Strategies

The fifth component of most combustion processes is the pollution control system. There are a number of issues to consider when designing a pollution control system, which may include multiple technologies to handle multiple possible pollutants or even for handling the same pollutant in some cases. Some of these issues include flammability of the gases being treated and removed, applicable construction codes, economics, corrosion, vent sizing, material handling, and chemical storage, to name a few. One of the inherent difficulties in controlling pollutant emissions is that they are sometimes inextricably linked together. For example, some of the techniques for reducing NOx emissions may increase CO emissions.

There are many factors that go into the selection of the appropriate control technologies to minimize pollution emissions from a process. Some broadly categorize these factors into: environmental, engineering, and economic [Mycock et al., 1995]. Environmental includes such things as equipment location, available space, ambient conditions, availability of utilities, and regulations. Engineering includes contaminant characteristics (physical and chemical properties, concentration, etc.), gas stream characteristics (flow rate, temperature, pressure, humidity, properties, etc.), and performance of the particular control system (power requirements, removal efficiency, temperature limitations, etc.). Economic considerations include the capital cost, ongoing operating costs, and the expected lifetime of the equipment. Spaite and

Burckle [1977] believe the two primary factors used to determine the choice of control techniques are technical feasibility and lowest cost. General pollution control strategies are discussed next with particular emphasis on those relating to the combustion process.

Pretreatment

Pretreatment basically refers to modifying the incoming feed materials to either the process or the combustion system. This would include adding or removing elements from the fuel, oxidizer, or raw materials. The fuel could be modified by cleaning out some impurities, such as removing sulfur in a pretreatment process at the plant. The fuel could also be modified by putting in an additive designed to reduce pollutant emissions. The oxidizer could be pretreated by adding high-purity oxygen to increase the overall oxygen concentration to something above the approximately 21% O_2 by volume contained in normal air. The incoming raw materials could be cleaned to remove some potential pollutant-emitting chemicals.

Process Modification

Process modification refers to either replacing the incoming feed materials to the process or the combustion system with some other alternative, or to modifying the end product in some way. Ramachandran [1997] refers to the former as raw material substitution. An example of a fuel change would be to change from a high-sulfur coal to a low-sulfur coal to reduce SOx emissions. A more radical example would be to change from fossil fuel heating to electrical heating where the emissions are moved from the plant location to the power generation facility.

Combustion Modification

This strategy involves modifying the combustion process so that fewer pollutants are generated in the first place. This is generally much less costly than removing the pollutants from the exhaust gases after they have been formed. Many combustion modification techniques have been used to reduce emissions from conventional burners, depending on the pollutant. Using NOx as an example pollutant, air staging, fuel staging, flue gas recirculation, water or steam injection, reducing air preheat temperatures, ultra-lean premix, and pulsed combustion are some of the many ways that have been used to reduce emissions. In some cases, the performance of the overall system may be reduced by these techniques. For example, injecting water into the burner or combustor does reduce NOx but it also usually reduces thermal efficiency as well.

Posttreatment

Posttreatment refers to removing pollutants from the exhaust gas stream before they are emitted into the atmosphere. This strategy may be chosen for many reasons. In some cases it may not be possible to sufficiently reduce or eliminate the formation of one or more pollutants in the combustion process. It may be more economical to remove the pollutants rather than prevent them from forming in the first place. This may occur when pollutants can be easily recycled back into the process, for example, in the case of particles carried out of the combustor. This could also occur if the changes that would need to be made to minimize pollutant formation have a negative impact on the processing of the product in the combustor. In some plants, it is more economical to duct multiple exhaust gas stacks together going to a single large posttreatment system, rather than modifying each of the individual combustion processes. It may also be necessary to use a posttreatment technique to supplement minimization strategies to economically be below regulated emission levels. Schifftner [2002] has written a good general purpose book on air pollution control equipment primarily focused on posttreatment.

Many of the posttreatment systems for removing pollutants from exhaust streams involve catalysts. There are often difficulties in using catalysts because of the high gas temperatures and dirty exhaust streams from industrial combustion processes that can foul or damage catalytic removal systems. They can be thermally deactivated by sintering, for example. However, they have been successfully used to remove a wide range of pollutants including NOx, SOx, CO, and VOCs. Heck and Farrauto [1995] have written a good general-purpose book on the use of catalysts for air pollution control.

Process Control

An important part of minimizing pollution emissions is to control all aspects of the process. This includes accurately metering the fuel and oxidizer flows and compositions to the burners, controlling the combustor itself (e.g., minimizing air infiltration into the process), and monitoring the incoming raw material feed rates into the combustor. If any of the system parameters are out of specification, then pollution emissions may be adversely affected. For example, if the fuel flow increases while the oxidizer flow rate remains the same, then carbon monoxide emissions will likely increase. If there are large air leaks into the furnace, then NOx emissions are likely to increase. If one of the incoming raw materials contains more fine particles than normal, then particulate emissions may increase.

Other

Another technique that is often important in minimizing pollutant emissions is equipment maintenance. This includes the combustor, the burners, and the control equipment. If combustors are not properly maintained, then emissions can be adversely affected. For example, if the combustor develops cracks that permit air infiltration, NOx emissions will typically increase. If broken sight ports are not replaced, then large quantities of tramp air can infiltrate into the furnace. If the refractory becomes damaged and is not replaced, hot spots on the outer shell will develop and can injure personnel who may inadvertently come in contact with these hot surfaces.

It is very important to keep burners in good operating conditions to minimize pollutant emissions. If fuel injection nozzles become plugged, the mixing patterns that produce the flame will be adversely affected, which normally would increase NOx emissions, for example. This commonly occurs in the petrochemical industry when heavy hydrocarbons are improperly fired causing the injection nozzles to clog from carbon buildup due to coking. If the oil injectors become clogged, then oil atomization can be adversely affected, which could increase particulate emissions and opacity. It is not uncommon for broken pieces of refractory to fall onto or into burners, significantly disturbing the designed burner performance. An important part of the burner that is sometimes overlooked is the burner block (also referred to as the burner tile or quarl). A cracked burner block can adversely affect the performance of the burner.

53.4 Industrial Combustion Applications

Metals Production

Metals can be classified as ferrous (iron-bearing) and nonferrous (e.g., aluminum, copper, and lead). Ferrous metal production is often high temperature because of higher metal melting points compared to nonferrous metals. Many metals production processes are done in batch, compared to most other industrial combustion processes considered here, which are typically continuous. Another unusual aspect of metal production is the very high use of recycled materials. This often lends itself to batch production because of the somewhat unknown composition of the incoming scrap materials, which may contain trace impurities that could be very detrimental to the final product if not removed. The metals are typically melted in some type of vessel and then sampled to determine the chemistry so that the appropriate chemicals can be either added or removed to achieve the desired grade of material. Another unique aspect of the metals industry is that transfer vessels are preheated prior to the introduction of molten metals into the vessel to minimize the thermal shock to the refractory. Figure 53.15 shows an example of preheating a transfer ladle.

Since metals melt at higher temperatures, higher-intensity burners are often used in these applications. This includes, for example, oxygen-enhanced combustion [Kistler and Becker, 1998; Saha and Baukal, 1998] and air preheating to increase the flame temperatures and metal melting capability. These higher-intensity burners have the potential to produce high pollutant emissions, so burner design is important to minimize these emissions.

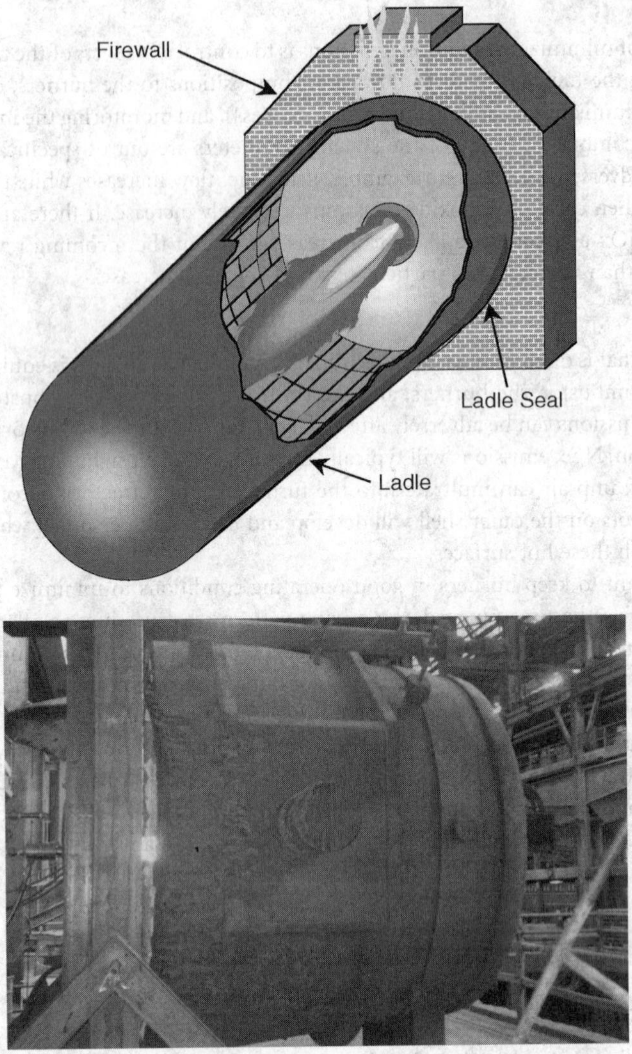

FIGURE 53.15 (a) Schematic and (b) photo of a ladle preheating process. (*Source:* Baukal, C. E., Ed., 2003. *The Handbook of Industrial Burners*, CRC Press, Boca Raton, FL.)

Another unusual aspect of metals production is that supplemental heating may be required to reheat the metals for further processing. For example, ingots may be produced in one location and then transported to another location to be made into the desired shape (e.g., wheel castings are often made by remelting aluminum ingots or sows). While this process may be economically efficient, it is energy and pollutant inefficient due to the additional heating. Burners are used in the original melting process as well as in the reheating process. This is something that has begun to attract more attention in recent years, where the entire life cycle of a product is considered rather than just its unit cost and initial energy requirements. For example, aluminum has a low life cycle cost compared to many other metals because of its high recycle ratio. While the energy consumption to make aluminum from raw ore is fairly high, remelting scrap aluminum takes only a fraction of that energy, which also means less overall pollution as well.

Minerals Production

Some common minerals processes include the production of glass, cement, bricks, refractories, and ceramics. These are typically high-temperature heating and melting applications that require a significant

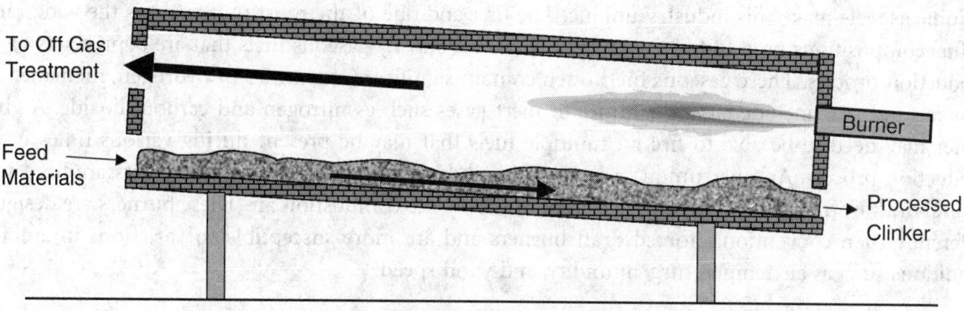

FIGURE 53.16 Schematic of a counter-rotating cement kiln. (*Source:* Baukal, C. E., Ed., 2003. *The Handbook of Industrial Burners*, CRC Press, Boca Raton, FL.)

amount of energy per unit of production. They also tend to have fairly high pollutant emissions as a result of the high temperatures and unit energy requirements. Most of the minerals applications are continuous processes, but a wide range of combustors can be used. Large glass furnaces are typically rectangularly shaped and have multiple burners. On the other hand, cement kilns are long refractory-lined rotating cylinders that are slightly inclined so that the materials flow gradually downhill (see Figure 53.16).

Chemicals Production

This is a very broad classification that encompasses many different types of production processes that have been loosely subcategorized into chemicals (organic and inorganic) and petrochemicals (organic) applications. There is some overlap in terms of the types of heating equipment used, where many of the incoming feed materials are in liquid form (e.g., crude oil) and are processed in heaters with tubes running inside them. These are generally lower-temperature applications (<2300°F or <1300°C) that incorporate heat recovery to preheat the incoming feed materials. Nearly all of the chemicals heating applications employ multiple burners but in a much more diverse configuration compared to many other industries. Burners may be fired horizontally, vertically up, vertically down, or at some angles in between depending on the specific process. Numerous configurations exist for fired process heaters (see Figure 53.17).

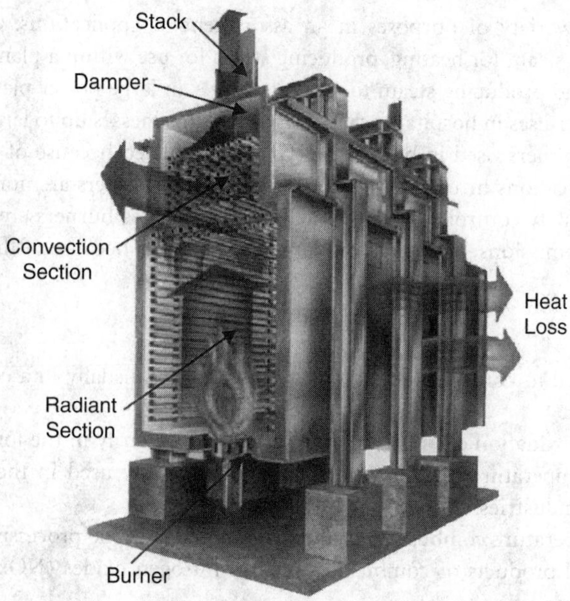

FIGURE 53.17 Schematic of a fired process heater (courtesy of John Zink Co. LLC).

Some aspects make this industry unique. The first and one of the most important is the wide range of fuel compositions used to fire the heaters. These are mostly gaseous fuels that are byproducts of the production process. These gaseous fuels often contain significant quantities of hydrogen, methane, and propane and may include large quantities of inert gases such as nitrogen and carbon dioxide. A given heater may need to be able to fire on multiple fuels that may be present during various times in the production process. Another unique aspect of this industry is that many of the heaters are fired with natural draft burners where no blower is used to supply the combustion air. These burners are designed differently than conventional forced draft burners and are more susceptible to variations in ambient conditions such as air temperature, humidity, and wind speed.

Waste Incineration

The objective of waste incineration processes is to reduce or eliminate waste products, which involves combusting those materials. Not only is the incinerator (see Figure 53.18) fired with burners, but the waste material itself is often part of the fuel that generates heat in the process. However, the waste usually has a very low heating value, hence the need for supplemental fuel. Incineration is a more complicated and dynamic process compared to most other industrial combustion processes by nature of the variability of the feed material. The waste may be very wet after a rainstorm, which may put a huge extra heat load on the incinerator. In some locations where waste materials are separated for recycling, the waste actually fed into the incinerator may have a much higher heating value compared to other incinerators where there is no separation of the waste.

A complicating factor with incinerators is that the end product, for example, the noncombustible waste, must also be disposed of, which means that one of the goals of most incineration processes is to produce minimal waste output. Because of the waste material variability, other pollutants may be generated that are not normally associated with industrial combustion processes. An example is the burning of plastics, which can produce dioxins and furans. The types of incinerators can vary greatly depending upon a variety of factors. In some cases, waste materials to be destroyed may be fed through the burners. This is particularly true of waste hydrocarbon liquids.

Industrial Boilers and Power Generation

Boilers are used for a variety of purposes in an assortment of applications. Common uses include producing hot water or steam for heating, producing steam for use within a plant such as atomizing oil for oil-fired burners, and producing steam to generate power in large power plants. Applications range from small single-burner uses in hospitals, schools, and small businesses up to large multi-burner boilers in power plants. The burners used in boilers are typically regulated because of their proliferation and widespread use in applications involving the general public. The burners are normally required to have a full complement of safety controls to ensure safe operation. These burners are often highly regulated to minimize pollutant emissions, particularly in large power plants because of the size of the source.

Defining Terms

Burners — Devices used to control combustion to generate heat, usually in a combustion chamber of some type.

Combustion — Rapid oxidation of a fuel to produce energy primarily in the form of heat.

Furnaces — Higher-temperature combustion chambers commonly used in the minerals, metals, and incineration industries.

Heaters — Lower-temperature combustion chambers often used in the processing industries.

Pollution — Unwanted products of combustion such as nitrogen oxides (NOx), sulfur oxides (SOx), and particulates.

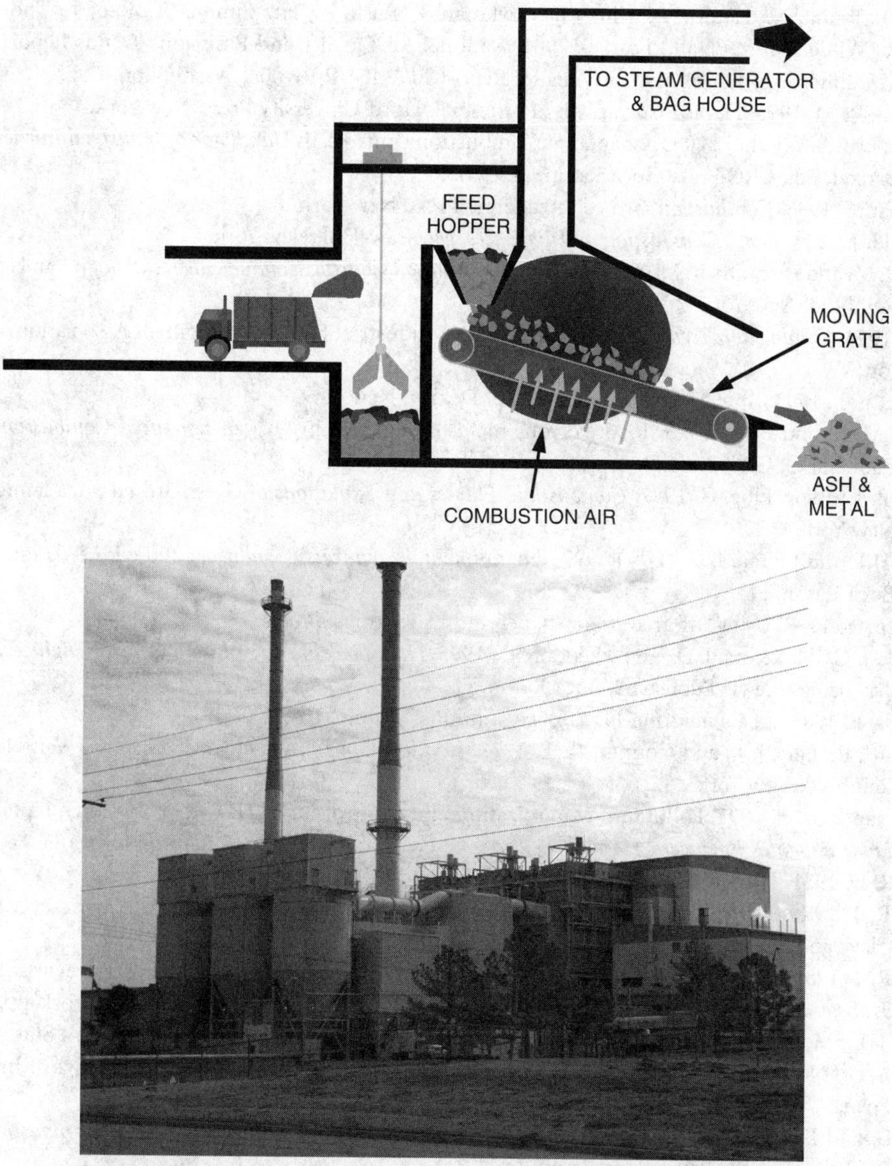

FIGURE 53.18 (a) Schematic and (b) photo of a waste incineration plant. (*Source:* Baukal, C. E. (ed.) 2003. *The Handbook of Industrial Burners*, CRC Press, Boca Raton, FL.)

References

Barnard, J. A. and Bradley, J. N. 1985. *Flame and Combustion*, 2nd ed., Chapman & Hall, London.

Bartok, W. and Sarofim, A. F. (Eds.) 1991. *Fossil Fuel Combustion*, John Wiley & Sons, New York.

Baukal, C. E. (Ed.) 1998. *Oxygen-Enhanced Combustion*. CRC Press, Boca Raton, FL.

Baukal, C. E. 2000. *Heat Transfer in Industrial Combustion*, CRC Press, Boca Raton, FL.

Baukal, C. E. (Ed.) 2001. *The John Zink Combustion Handbook*, CRC Press, Boca Raton, FL.

Baukal, C. E. (Ed.) 2003. *The Handbook of Industrial Burners*, CRC Press, Boca Raton, FL.

Borman, G. and Ragland, K. 1998. *Combustion Engineering*, McGraw-Hill, New York.

Edwards, J. B. 1974. *Combustion: The Formation and Emission of Trace Species*, Ann Arbor Science Publishers, Ann Arbor, MI.

Elkins, J., Frank, N., Hemby, J., Mintz, D., Szykman, J., Rush, A., Fitz-Simons, T., Rao, T., Thompson, R., Wildermann, E., and Lear, G. 2001. National Air Quality and Emissions Trends Report, 1999. U.S. Environmental Protection Agency, Report EPA 454/R-01-004, Washington, D.C.

Fristrom, R. M. 1995. *Flame Structure and Processes,* Oxford University Press, New York.

Gifford, J. and Kodesh, Z. 2003. Chapter 8: Combustion controls, in *Handbook of Industrial Burners,* C.E. Baukal, Ed., CRC Press, Boca Raton, FL, 2003.

Glassman, I. 1996. *Combustion,* 3rd ed., Academic Press, New York.

Griswold, J. 1946. *Fuels, Combustion and Furnaces,* McGraw-Hill, New York.

Heck, R. M. and Farrauto, R. J. 1995. *Catalytic Air Pollution Control: Commercial Technology,* Van Nostrand Reinhold, New York.

IHEA. 1994. *Combustion Technology Manual,* 5th ed., Industrial Heating Equipment Association, Arlington, VA.

Keating, E. L. 1993. *Applied Combustion,* Marcel Dekker, New York.

Kistler, M. D. and Becker, J. S. 1998. Ferrous metals, Chapter 5 in *Oxygen-Enhanced Combustion,* C. E. Baukal, Ed., CRC Press, Boca Raton, FL.

Lewis, B. and von Elbe, G. 1987. *Combustion, Flames and Explosions of Gases,* 3rd ed., Academic Press, New York.

Liu, D.H.F. and Lipták, B.G. (Eds.) 1997. *Environmental Engineers' Handbook,* 2nd ed. Lewis Publishers, Boca Raton, FL.

Longacre, S. 1997. Using infrared to dry paper and its coatings, *Process Heating,* 4:45–49.

Mycock, J. C., McKenna, J. D., and Theodore, L. 1995. *Handbook of Air Pollution Control Engineering and Technology,* Lewis Publishers, Boca Raton, FL.

Perthuis, E. 1983. *La Combustion Industrielle,* Éditions Technip, Paris.

Pritchard, R., Guy, J. J., and Connor, N. E. 1977. *Handbook of Industrial Gas Utilization,* Van Nostrand Reinhold, New York.

Ramachandran, G. 1997. Pollutants: Minimization and Control. In D. H. F. Liu and B. G. Lipták, Eds., *Environmental Engineers' Handbook,* 2nd ed. Lewis Publishers, Boca Raton, FL.

Reed, R. D. 1981. *Furnace Operations,* 3rd ed., Gulf Publishing, Houston.

Reed, R. J. 1986. *North American Combustion Handbook,* Vol. I, 3rd ed., North American Mfg. Co., Cleveland, OH.

Reed, R. J. 1987. Future Consequences of Compact, Highly Effective Heat Recovery Devices, in *Heat Transfer in Furnaces,* edited by C. Presser and D.G. Lilley, ASME HTD-Vol. 74, New York, pp. 23–28.

Reese, J. L., Moilanen, G. L., Borkowicz, R., Baukal, C., Czerniak, D., and Batten, R. 1994. State-of-the-Art of NOx Emission Control Technology, ASME paper 94-JPGC-EC-15, Proceedings of Int'l Joint Power Generation Conf., Phoenix, October 3–5.

Saha, D. and Baukal, C. E. 1998. Nonferrous metals, Chapter 6 in *Oxygen-Enhanced Combustion,* C.E. Baukal, Ed., CRC Press, Boca Raton, FL.

Schifftner, K. C. 2002. *Air Pollution Control Equipment Selection Guide,* Lewis Publishers, Boca Raton, FL.

Segeler, C. G. (Ed.) 1965. *Gas Engineers Handbook,* Industrial Press, New York.

Smith, T. M. and Baukal, C. E. 1983. Space-age refractory fibers improve gas-fired infrared generators for heat processing textile webs, *Journal of Coated Fabrics,* 12:160–173.

Spaite, P. W. and J.O. Burckle, J. O. 1977. Selection, Evaluation, and Application of Control Devices. In *Air Pollution,* Vol. 4: Engineering Control of Air Pollution, 3rd ed., A. C. Stern, Ed. Academic Press, New York.

Stambuleanu, A. 1976. *Flame Combustion Processes in Industry,* Abacus Press, Tunbridge Wells, U.K.

Strehlow, R. A. 1968. *Fundamentals of Combustion,* International Textbook Co. Scranton, PA.

Turns, S. R. 1996. *An Introduction to Combustion,* McGraw-Hill, New York.

Williams, F. A. 1985. *Combustion Theory,* Benjamin/Cummings Publishing, Menlo Park, CA.

54

Air Conditioning

Jan F. Kreider
(Second Edition)
University of Colorado and Kreider and Associates, LLC

Victor W. Goldschmidt
(First Edition)
Purdue University

Curtis J. Wahlberg
(First Edition)
Purdue University

The objective of this chapter is to introduce the reader to basic concepts in air conditioning, as well as some of its history. Details on equipment types, design, selection, installation, operation, and maintenance are not addressed. *The CRC Heating, Ventilating and Air Conditioning Handbook* [Kreider, 2001] and ASHRAE (American Society of Heating, Refrigerating and Air-Conditioning Engineers) handbooks are suggested as sources for further study and reference.

54.1 Historical Sketch

It is likely that the first air conditioning system was the result of concerns that a physician had for the comfort of ailing sailors suffering with critical fever. In the mid-1800s, Dr. John Gorrie constructed an open air-cycle refrigeration machine in order to cool two rooms in a hospital in Apalachicola, Florida, and provide his patients with some relief. Although Gorrie died frustrated, broke, and subject to criticism, he did become the first of a line of air conditioning pioneers, which also includes A. Muhl (who held the first patent for cooling residences — in this case with ether compression and expansion), Alfred R. Wolft (1859–1909; who provided comfort conditioning to more than 100 buildings, including the Waldorf Astoria, Carnegie Hall, and St. Patrick's Cathedral), and Willis Carrier (1876–1950; who not only provided the first psychrometric chart, but also set new trends in product development and marketing). Equally as fascinating as the history of air conditioning are its thermodynamics, product development, and utilization. Some of these will be addressed in the sections that follow.

54.2 Comfort

The main goal of air conditioning is to ensure human comfort in the built environment. The human body is continuously generating heat and moisture, which in turn must be removed from the environment in order to maintain constant body temperature and comfort. The transfer of heat from the body to the environment depends on the surface conditions of the body (i.e., clothing), as well as the temperature, velocity, and humidity of the surrounding air and the temperature of surrounding surfaces (affecting radiative heat transport to or from the body). Conditions for comfort are obviously influenced by an individual's age, level of activity, and clothing. In general, nominal ranges for comfort include temperature

0-8493-1586-7/05/$0.00+$1.50

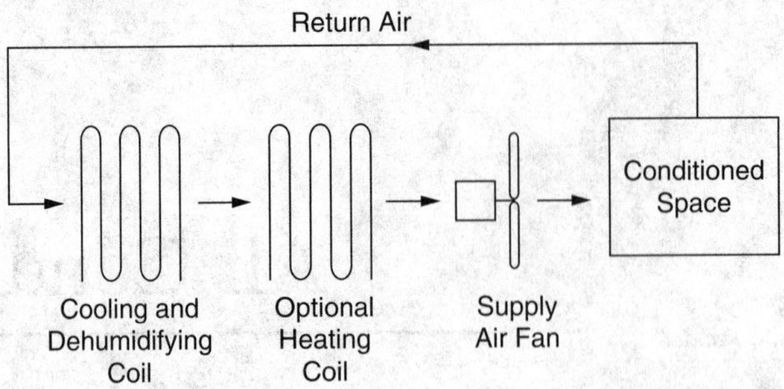

FIGURE 54.1 Typical air conditioning process.

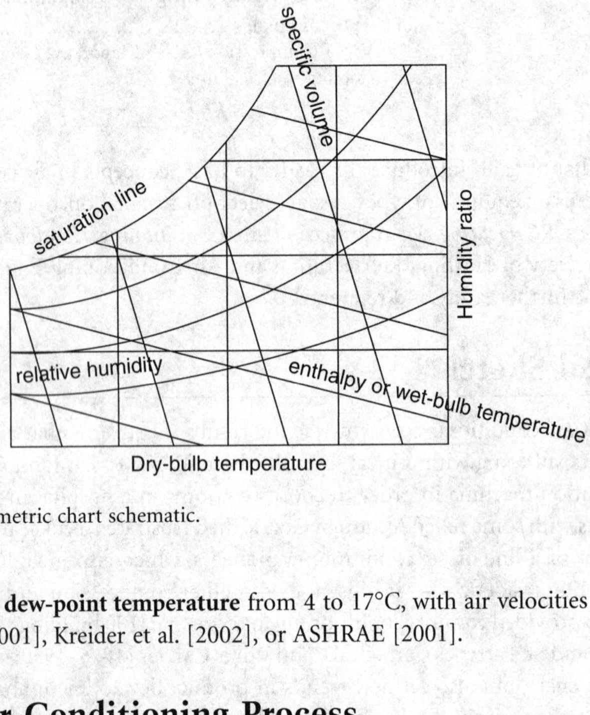

FIGURE 54.2 Psychrometric chart schematic.

from 20 to 26°C and **dew-point temperature** from 4 to 17°C, with air velocities under 0.25 m/sec. For details see Kreider [2001], Kreider et al. [2002], or ASHRAE [2001].

54.3 The Air Conditioning Process

The typical processes of air conditioning, one of which is exhibited in Figure 54.1, can best be described on a psychrometric chart. These charts present data for **relative humidity, specific volume, wet-bulb temperature**, and **enthalpy** for atmospheric air in terms of **dry-bulb temperature** and **humidity ratio**, as shown schematically in Figure 54.2. [For details see Kreider et al. (2002).]

The cooling and dehumidifying coil of Figure 54.1 brings the moist air down to its dew point, then removes moisture in the from of condensate. Air leaves the coil in conditions close to saturated and at a lower temperature than the room return air (see Figure 54.3). Latent cooling is the amount of heat removal necessary to hold the humidity ratio of room air at the comfort level. In turn, the heating section warms the building supply air stream to the dry-bulb temperature needed for comfort conditions. This common kind of system (sometimes called *reheat*) can closely control comfort conditions required. Reduced energy consumption (no need really to first cool the air and then heat it again) is achieved if the saturated cold air can be mixed with warmer return air from a conditioned space; this avoids the need for reheat (a form of energy loss).

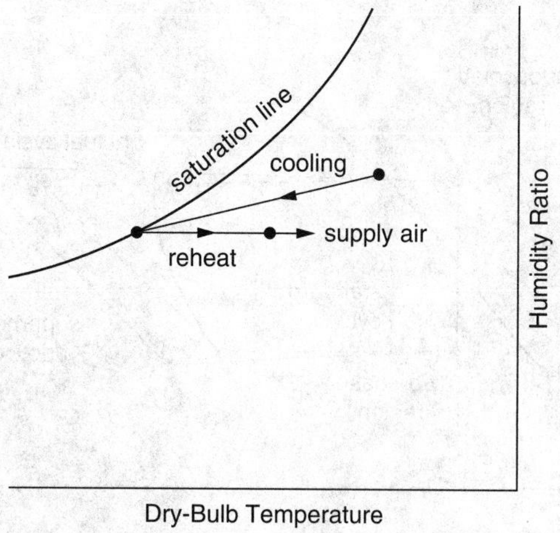

FIGURE 54.3 Cooling and dehumidification with reheat.

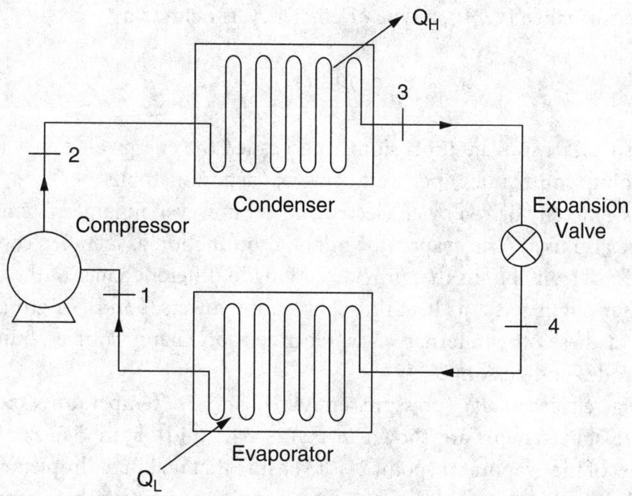

FIGURE 54.4 Vapor compression cycle components.

54.4 Typical Cycles

Vapor Compression Cycle

Figure 54.4 shows the basic components for a vapor compression cycle used as an air conditioner. Room air flows over the evaporator, thereby being cooled (sensible and latent cooling), while the two-phase refrigerant mixture within the coil boils; it is then heated slightly beyond the saturated condition into the **superheated** region (see Figure 54.5). The compressor increases the pressure of the superheated refrigerant, which is then cooled slightly below saturation temperature in the condenser where cooler outside air is blown over the coil surfaces. Through this process, the air conditioner working fluid, the refrigerant, gains thermal energy, Q_L, from the cooled indoor ambient and rejects thermal energy, Q_H, to the outdoor environment while work, W, is done on the fluid by the compressor. The line losses are generally small; hence a first-order energy balance would require that

$$W + Q_L = Q_H$$

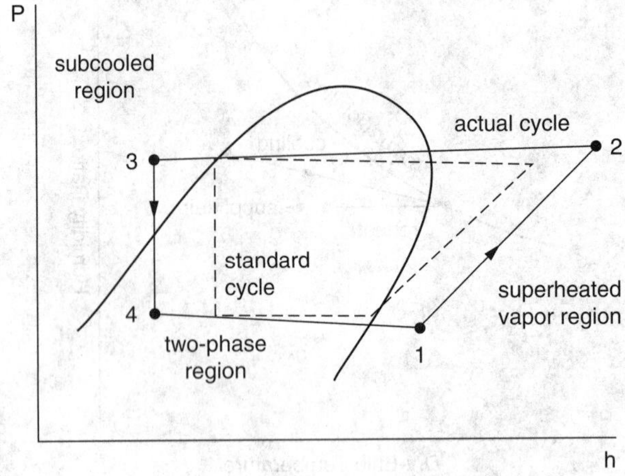

FIGURE 54.5 *Ph* diagram: vapor compression cycle.

The coefficient of performance (COP), a type of efficiency, is defined as

$$COP = Q_L / W$$

The U.S. industry also uses a similar, dimensional ratio called the energy efficiency ratio (EER), given in units of Btu/h of cooling (numerator) per watt of work (denominator).

The compressor is generally driven by an electric motor; however, in large systems gas-driven engines or steam turbines are also used. The evaporator might also function as a chiller, cooling a separate fluid in a secondary loop for terminal units that provide cooling in different zones of the building. In a similar manner, the condenser might reject its heat through cooling towers, ponds, or ground-coupled sources/sinks. The selection of these other alternatives is dependent on energy sources, building type, cost, and even local building codes and customs.

The corresponding refrigerant *Ph* (pressure, enthalpy) and *Ts* (temperature, entropy) diagrams for simple vapor compression systems are shown in Figure 54.5 and Figure 54.6, respectively. The slight superheat at the outlet of the evaporator (point 2) is to ensure that no liquid droplets enter the compressor. The **subcooled** conditions downstream of the condenser (point 3) are desirable to ensure proper response and control by the expansion valve (normally a thermostatic expansion valve or an electrostatic expansion

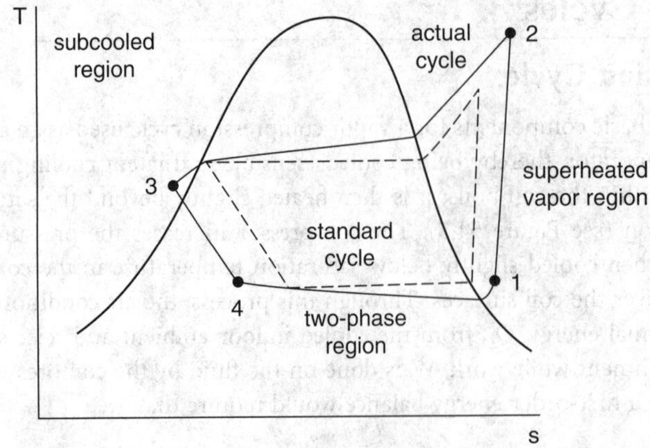

FIGURE 54.6 *Ts* diagram: vapor compression cycle.

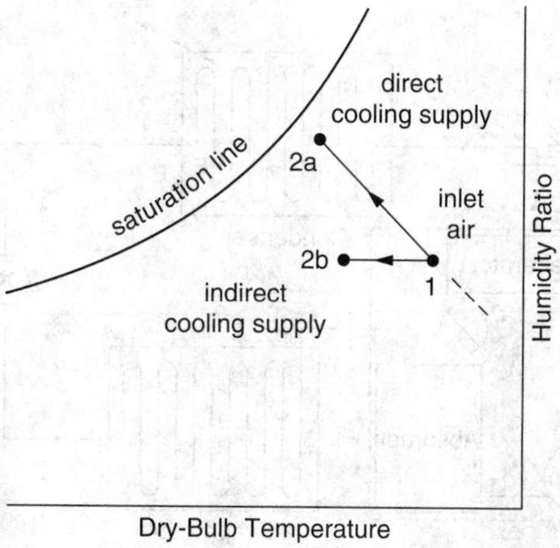

FIGURE 54.7 Evaporative cooling.

valve). These valves control superheat by adjusting the mass flow rate of refrigerant. In some cases fixed-area throttling devices might be used (such as capillary tubes or fixed-area orifices) for small systems.

Evaporative Cooling

Evaporative cooling is a cost-effective alternative for dry climates. Water, when sprayed and evaporated, will lead to a lowering of dry-bulb temperature of air supplied to a building. The cooling may be *direct* (the water is added to the air being conditioned) or *indirect.* In the case of indirect cooling, a secondary air stream (cooled by evaporation of a spray) flows through a heat exchanger, removing thermal energy from the conditioned ambient return air. The corresponding processes are shown in Figure 54.7. (For details see ASHRAE [2001] or Kreider et al. [2002].)

Absorption Cooling

Ferdinand Carre invented the adsorption system in the mid-1800s. In lieu of a compressor, as used in vapor compression systems, an absorption system uses a small pump and two chemical processes as diagrammed in Figure 54.8. First, the refrigerant leaving the evaporator (in a vapor state) is absorbed in a liquid. This operation takes place in an *absorber.* Heat is removed during this step. Second, the pressure of the solution is increased through a liquid pump. Third, the high-pressure solution is heated in order to release the high-pressure refrigerant in a vapor state and transfer it to the condenser. This third step takes place in a *generator.* The remaining solution (now with a very low concentration of refrigerant) flows through a throttling valve to the absorber. Three features of the absorption system make it sometimes preferable to a vapor-compression system. First, the compressor is exchanged for a liquid pump consuming much less power. Second, absorption cooling lends itself to the use of on site waste heat, solar energy, or similar thermal sources. Third, the working fluids in common use are not "greenhouse gases" described in the next section. Typical fluids are water–lithium bromide and water–lithium chloride, in which water is used as the refrigerant. (See Kreider et al. [2002] or Stoecker and Jones [1982].)

Refrigerants

Refrigerants were originally selected for their safety, low cost, high stability, good compatibility with materials, and excellent thermal performance (examples are R12 for refrigerators and automobile air conditioners and R22 for residential air conditioners). Data from the past two decades or more regarding ozone depletion and theoretical models explaining it placed part of the blame on chlorofluorocarbon (CFC) refrigerants such as R11 and R12. The upshot of this work is the Montreal Protocol, which required

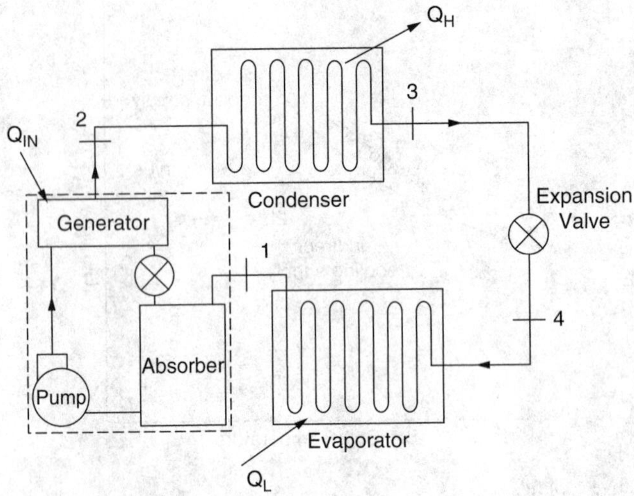

FIGURE 54.8 Absorption cooling cycle components.

an orderly phaseout of refrigerants that were thought to be involved in climate change; gases so implicated are called "greenhouse gases." Today nearly all such refrigerants are no longer in use except in well-sealed, older systems that rely on them for efficient operation. In the 1980s and 1990s, new and efficient fluids were deployed that have minimal global warming potential while achieving efficiencies very similar to those using older CFC working fluids.

Defining Terms

Dew-point temperature — Temperature at which water vapor begins to condense out of a mixture of dry air and water vapor (becomes saturated) while cooled at constant pressure from an unsaturated state.

Dry-bulb temperature — Conventional thermometer temperature.

Enthalpy — Intrinsic property of a substance determined by the summation of its internal energy and the product of its pressure and volume (denoted $h = u + Pv$).

Humidification — Process of adding moisture to an air stream (dehumidification: removal of moisture).

Humidity ratio — Mass ratio of water vapor to dry air in a mixture.

Refrigeration — Act of maintaining a low-temperature region of finite size at a selected temperature by removing heat from it.

Relative humidity — Ratio of partial pressure of water vapor in a mixture to the saturation pressure of the vapor at the same dry-bulb temperature.

Specific volume — Volume per unit mass of a substance (generally denoted v).

Subcooled — Conditions where a liquid is cooled below its saturation temperature at a given pressure.

Superheated — Conditions where a vapor is heated above its saturation temperature at a given pressure.

Wet-bulb temperature — Temperature from a conventional thermometer where the bulb is covered with a wetted wick.

References

ASHRAE. 2001. *ASHRAE Handbook Fundamentals,* ASHRAE, Atlanta, GA.

Kreider, J. F. (ed.) 2001. *Handbook of Heating, Ventilating and Air Conditioning,* CRC Press, Boca Raton, FL.

Kreider, J. F., Curtiss, P. S. and Rabl, A. 2002. *Heating and Cooling of Building,* 2nd ed., McGraw-Hill, New York.

Stoecker, W. F. and Jones, J. W. 1982. *Refrigeration and Air Conditioning,* McGraw-Hill, New York.

55

Refrigeration and Cryogenics

Randall F. Barron
Louisiana Tech University

Refrigeration involves the production and utilization of temperatures below normal ambient temperature. The general goal of conventional refrigeration is not only to achieve low temperatures but also to maintain a space or material at subambient temperatures for an extended time. Some examples of the utilization of refrigeration include:

1. Cooling of residences (air conditioning)
2. Ice making
3. Cold storage of foods
4. Condensation of volatile vapors in the petroleum and chemical industries

Cryogenics is a special field of refrigeration that involves the production and utilization of temperatures below –150°C (–240°F) or 123 K. This dividing line was chosen between conventional refrigeration and cryogenics because the refrigerants used in air conditioning, ice making, food cold storage, etc., all condense at temperatures much higher than –150°C. Typical cryogenic fluids (listed in Table 55.1) include methane, nitrogen, oxygen, argon, neon, hydrogen, and helium. All of these fluids condense at temperatures below –150°C. Some examples of the utilization of cryogenic temperatures are:

1. Production of liquefied gases for industrial and space technology uses
2. Maintaining low temperatures required for superconducting systems
3. Treatment of metals to improve the physical properties of the material
4. Destruction of defective tissue (cryosurgery)

55.1 Desiccant Cooling

Desiccant cooling has been used in connection with solar cooling systems because desiccant systems can achieve both cooling and dehumidification [Duffie and Beckman, 1991]. Lof [1955] described a closed-

0-8493-1586-7/05/$0.00+$1.50

TABLE 55.1 Properties of Cryogenic Liquids

Cryogenic Liquid	NBP[a], K	TP[b], K	Density at NBP, kg/m³	Latent Heat, kJ/kg	Specific Heat, kJ/kg-K	Viscosity, μPa-sec
Helium-3	3.19	c	58.9	8.49	4.61	1.62
Helium-4	4214	d	124.9	20.90	4.48	3.56
Hydrogen	20.27	13.9	70.8	445.6	9.68	13.2
Neon	27.09	24.54	1206.0	85.9	1.82	130
Nitrogen	77.36	63.2	807.3	50.4	2.05	158
Air	78.8	—	874	205	1.96	168
Fluorine	85.24	55.5	1507	166.3	1.54	244
Argon	87.28	83.8	1394	161.90	1.136	252
Oxygen	90.18	54.4	1141	213	1.695	190
Methane	111.7	88.7	424.1	511.5	3.461	118

[a] Normal boiling point (boiling point at 1 atm pressure).
[b] Triple point (approximately the freezing point).
[c] He-3 has no triple point.
[d] Lambda-point temperature = 2.171 K.
Source: Johnson, V. J. (Ed.) 1960. *A Compendium of the Properties of Materials at Low Temperatures, Part I.* WADD Tech. Rep. 60-56. U.S. Government Printing Office, Washington, DC.

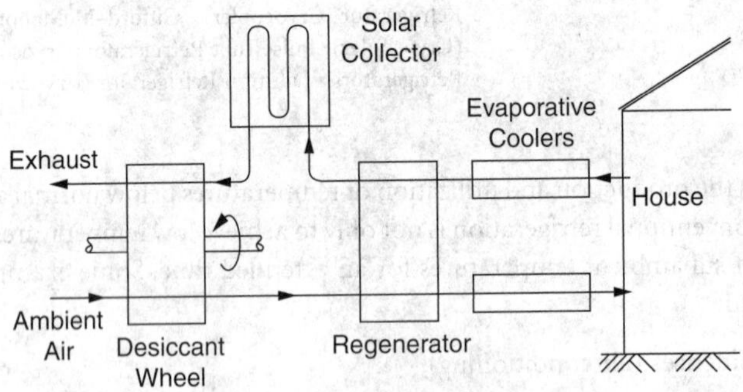

FIGURE 55.1 Solar desiccant cooler. The desiccant wheel acts to remove moisture from the ambient air entering the system.

cycle desiccant cooling system using liquid triethylene glycol as the drying agent. In this system, glycol was sprayed into an absorber to absorb moisture from the air from the building. The stream then flowed through a heat exchanger to a stripping column, where the glycol was mixed with a stream of solar-heated air. The high-temperature air removed water from the glycol, which then returned to the absorber. This system, using steam as the heating medium, has been used in hospitals and similar large installations.

An open cycle desiccant cooling system is shown in Figure 55.1 [Nelson et al., 1978]. The supply air stream is dehumidified in a desiccant regenerative heat exchanger (wheel), then cooled in a heat exchanger and an evaporative cooler before being introduced into the space to be cooled. The return-air stream from the space is evaporatively cooled before returning through the heat exchanger to a solar collector, where the air is heated. The hot air flows back through the desiccant "wheel," where moisture is removed from the desiccant before the air stream is exhausted to the atmosphere.

55.2 Heat Pumps

Heat pumps are systems that can provide either cooling or heating for a building [McQuiston and Parker, 1994]. The schematic of a typical heat pump is shown in Figure 55.2. The heat pump cycle is shown on the pressure–enthalpy plane in Figure 55.3.

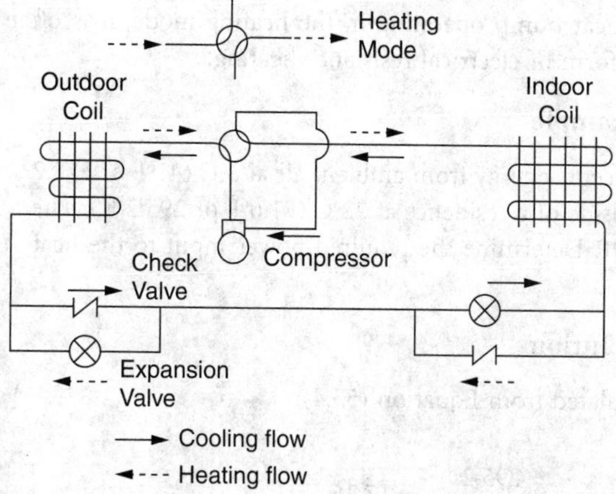

FIGURE 55.2 Heat pump. The flow directions for the cooling mode are shown as solid arrows; the flow directions for the heating mode are shown as dashed arrows.

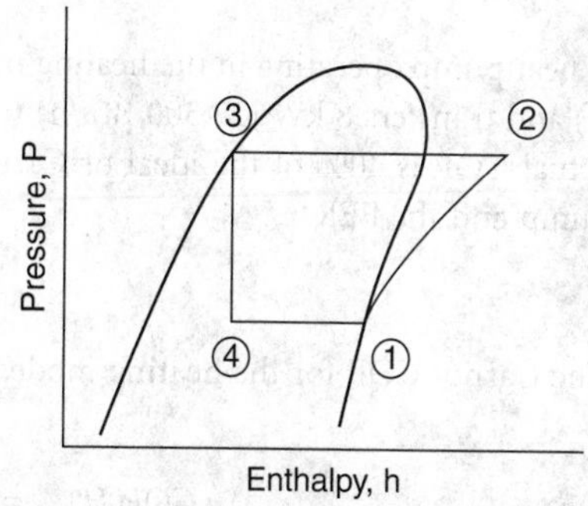

FIGURE 55.3 Pressure-enthalpy diagram for a heat pump: (1) compressor inlet; (2) compressor outlet; (3) outlet of the outdoor coil (*cooling mode*) or outlet of the indoor coil (*heating mode*); and (4) outlet of the expansion valve.

The dimensionless coefficient of performance (COP) for a heat pump operating in the cooling mode is given by:

$$COP(C) = \frac{Q_A}{-W} = \frac{h_1 - h_3}{h_2 - h_1} \tag{55.1}$$

where Q_A is the energy absorbed from the region to be cooled, W is the net work input (negative work, by convention) to the heat pump, and h is the enthalpy of the working fluid at the corresponding points in the cycle. For an ideal refrigerator (Carnot refrigerator), the COP is

$$COP(C)_{Car} = \frac{T_L}{T_H - T_L} \tag{55.2}$$

where T_H is the high temperature (absolute) in the cycle and T_L is the low temperature (absolute) in the cycle. The COP for an actual heat pump is on the order of one-third to one-half of the ideal COP [Kreider and Rabl, 1994].

For a heat pump operating in the heating mode, the COP is defined by

$$COP(H) = \frac{Q_H}{-W} = \frac{h_2 - h_3}{h_2 - h_1} \tag{55.3}$$

where Q_H is the heat added to the region being heated. For a Carnot heat pump the ideal COP is

$$COP(H)_{Car} = \frac{T_H}{T_H - T_L} \tag{55.4}$$

In the U.S., the COP for a heat pump is often expressed in units of Btu/W-h, and this ratio is called the *energy efficiency ratio* (EER), where EER = 3.412 COP

The capacity of a heat pump is strongly dependent on the outside air temperature. When the outdoor temperature falls much below 0°C (32°F) for a heat pump operating in the heating mode, it is often necessary to provide supplemental energy in the form of electrical resistance heating.

Example

A heat pump operating in the heating mode accepts energy from ambient air at 5°C (41°F or 278.2 K) and transfers 8 kW (27,300 Btu/h) to the inside of a residence at 22°C (71.6°F or 295.2 K). The actual COP is 40% of the ideal or Carnot COP. Determine the required power input to the heat pump and the EER.

Solution

The Carnot COP for the heating mode is calculated from Equation (55.4):

$$COP(H)_{Car} = \frac{T_H}{T_H - T_L} = \frac{295.2}{295.2 - 278.2} = 17.36$$

The actual COP is 40% of the ideal COP, or

$$COP(H) = (0.40)(17.36) = 6.95 = Q_H / (-W)$$

The power input to the heat pump is

$$-W = \frac{Q_H}{COP(H)} = \frac{8.00}{6.95} = 1.152\ kW$$

The energy efficiency rating is found as follows:

$$EER = 3.412\, COP(H) = (3.412)(6.95) = 23.7\ Btu/W - h$$

55.3 Cryogenics

The field of cryogenic refrigeration generally involves temperatures below –150°C (–240°F) or 123 K [Scott, 1959]. This dividing line between conventional refrigeration and cryogenics was selected because the normal boiling points of ammonia, hydrogen sulfide, and other conventional refrigerants lie above –150°C, whereas the normal boiling points of cryogens, such as liquid helium, hydrogen, oxygen, and nitrogen, all lie below –150°C.

Physical Properties of Cryogenic Liquids

The physical properties of several cryogenic liquids at the normal boiling point (NBP) are listed in Table 55.1. All of the liquids are clear, colorless, and odorless, with the exception of liquid oxygen (pale blue color) and liquid fluorine (straw-yellow color). Liquid nitrogen, liquid oxygen, liquid argon, and liquid neon are generally obtained by separating air into its constituent components in an air distillation system [Barron, 1985].

Hydrogen can exist in two distinct molecular forms: ortho-hydrogen and para-hydrogen. The equilibrium mixture at high temperatures (above room temperature) is 75% o-H_2 and 25% p-H_2, which is called *normal hydrogen*. At its NBP (20.27 K), equilibrium hydrogen is 99.8% p-H_2 and 0.2% o-H_2.

Neither isotope of helium (helium-4 or helium-3) exhibits a triple point (coexistence of the solid, liquid, and vapor phases); however, helium-4 has two liquid phases: liquid helium-I, which is a normal liquid, and helium-II, which exhibits superfluidity. Liquid helium-II exists below the lambda point (2.171 K).

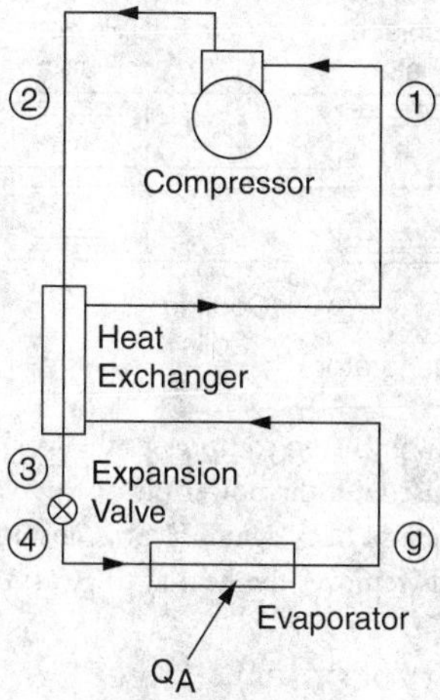

FIGURE 55.4 Joule–Thomson cryocooler.

Cryogenic Refrigeration Systems

The coefficient of performance (COP) of the thermodynamically ideal refrigerator (Carnot refrigerator) is given by

$$COP_{Car} = \frac{T_L}{T_H - T_L} \tag{55.5}$$

where T_L is the low temperature in the cycle and T_H is the high temperature in the cycle. The figure of merit (FOM) for any refrigerator or cryocooler is given by

$$FOM = \frac{COP_{act}}{COP_{Car}} \tag{55.6}$$

Joule–Thomson Refrigerator (Cryocooler)

A schematic of the Joule–Thomson cryocooler is shown in Figure 55.4. The refrigeration effect for this refrigerator is given by

$$Q_A / m = (h_1 - h_2) - (1 - \varepsilon)(h_1 - h_g) \tag{55.7}$$

where ε is the heat exchanger effectiveness, h_1 is the fluid enthalpy at temperature T_2 and pressure p_1, and h_g is the saturated vapor enthalpy at pressure p_1. Nitrogen is typically used as the working fluid for the temperature range between 65 and 115 K. Microminiature J-T cryocoolers using nitrogen–methane gas mixtures have been manufactured for operation at 65 K (–343°F) to cool electronic components [Little, 1990].

Stirling Refrigerator (Cryocooler)

The Stirling cryocooler consists of a cylinder enclosing a power piston and a displacer piston, as shown in Figure 55.5. The two chambers are connected through a regenerator (heat exchanger) that is one of the more critical components of the refrigerator [Barron, 1999]. Walker [1983] gives a detailed description of the Stirling refrigerator and its thermodynamic and mechanical performance. Ideally, the Stirling refrigerator has a FOM of unity; however, actual Stirling cryocoolers have FOM values of approximately 0.36.

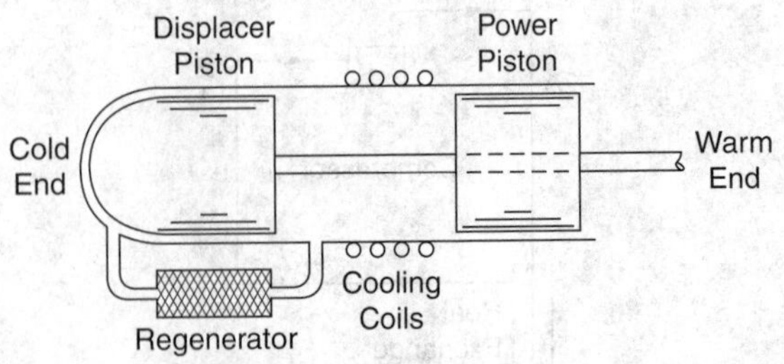

FIGURE 55.5 Stirling cryocooler. The power piston compresses the working fluid (typically, helium gas), while the displacer piston (operating 90° out of phase with the power piston) moves the working fluid from the warm space through the regenerator to the cold space and back again. Heat is absorbed from the cryogenic region at the cold end, and cooling water in the cooling coils removes the heat at the warm end.

Vuilleumier Refrigerator (Cryocooler)

The Vuilleumier (VM) cryocooler, first patented by Rudolph Vuilleumier in 1918 in the U.S., is similar to the Stirling cryocooler, except that the VM cryocooler uses a thermal compression process (Stirling engine) instead of the mechanical-drive compression process used in the Stirling system [Timmerhaus and Flynn, 1989]. A schematic of the VM refrigerator is shown in Figure 55.6. The COP for an ideal VM refrigerator is given by:

$$COP = \frac{Q_A}{Q_H} = \frac{T_L(T_H - T_o)}{T_H(T_o - T_L)} \tag{55.8}$$

where T_o is the intermediate temperature (absolute) in the cycle.

Gifford–McMahon Refrigerator (Cryocooler)

A schematic of the Gifford–McMahon (GM) cryocooler, developed by W. E. Gifford and H. O. McMahon [1960], is shown in Figure 55.7. This system has valves and seals that operate at ambient temperature, so low-temperature valve and seal problems are eliminated. The GM refrigerator is well suited for multistaging, where refrigeration is provided at more than one temperature level [Gifford and Hoffman, 1961]. For example, in systems in which thermal shields are placed around a region to be maintained at cryogenic temperatures, refrigeration may be provided for both the thermal shields (at 77 K, for example) and the low-temperature region (at 20 K, for example) by a single refrigerator.

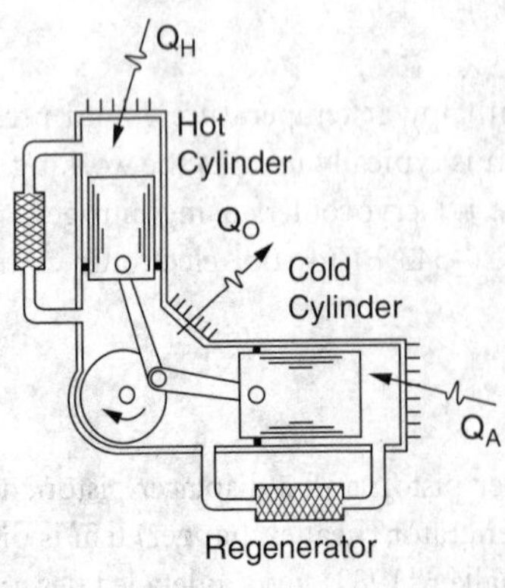

FIGURE 55.6 Vuilleumier refrigerator.

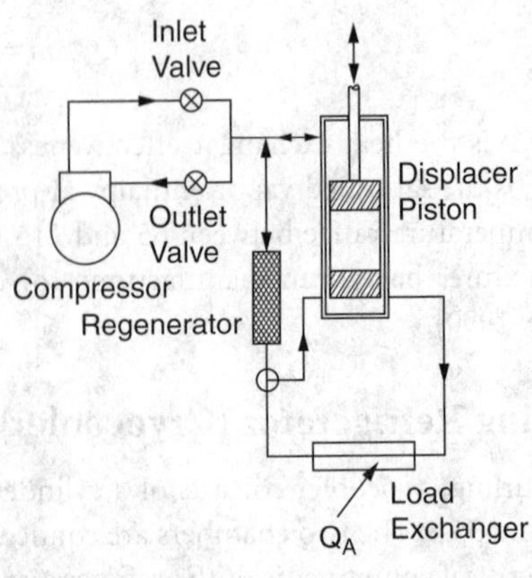

FIGURE 55.7 Gifford–McMahon refrigerator.

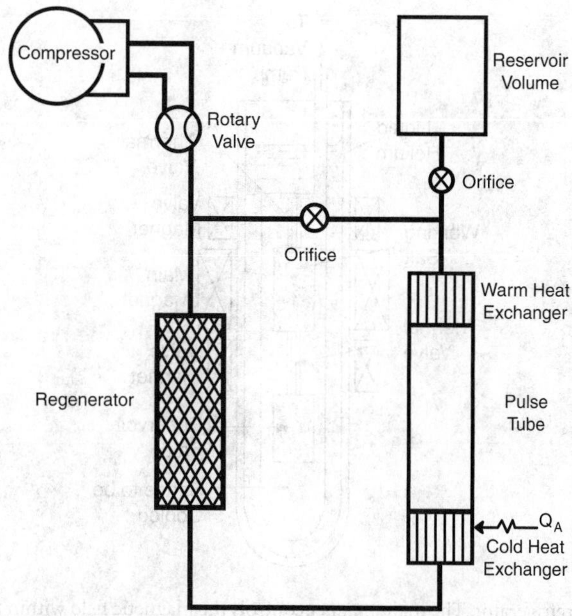

FIGURE 55.8 Pulse-tube cryocooler. The rotary valve is the source of pressure oscillation for the pulse tube. Typical switching frequencies range between 10 and 50 Hz. The orifice–reservoir volume combination is needed to control the phase angle between the mass flow and pressure in the system.

Pulse Tube Refrigerator (Cryocooler)

A schematic of the pulse tube refrigerator utilizing orifice restrictions is shown in Figure 55.8. The basic system was originally proposed by Gifford and Longsworth [1965] in 1963. The addition of the orifice impedance and buffer volume, suggested by Mikulin et al. [1984], allowed higher-frequency operation and resulted in an improvement in the refrigeration effect. Other than the compressor and reversing rotary valve, the pulse tube refrigerator has no moving parts; therefore, the reliability of the system is high. The pulse tube refrigerator has been used primarily in systems requiring cooling effects of 500 W or smaller [Weisend, 1998].

The pulse tube refrigerator consists of a compressor with a rotating valve distributor (or a source of pressure oscillation), a regenerator for heat exchange, and the pulse tube. The pulse tube is connected at the ambient temperature end to a buffer volume through an orifice, needle valve, or other suitable flow resistance element. High-pressure gas enters the regenerator, in which the gas is cooled. The cold gas leaving the regenerator enters the pulse tube at the cold end and acts as a gas piston to compress the gas already within the pulse tube. The compressed gas flows through the pulse tube to the warm end, where the gas is cooled by exchanging energy in an ambient-temperature heat exchanger. The rotary valve is turned such that the gas inlet is closed, the exhaust is opened, and the cooled gas in the pulse tube expands and experiences a decrease in temperature. The expanding gas flows back through the pulse tube, absorbs energy from the region to be refrigerated, and leaves the system through the regenerator.

The total refrigeration effect for the pulse tube refrigerator is given by [Radebaugh, 1990]:

$$Q_A = \frac{\gamma u_o A_t \Delta P \cos\phi}{2(\gamma - 1)} \tag{55.9}$$

where γ = specific heat ratio (γ = 1.67 for helium gas), u_o = maximum gas velocity in the pulse tube, A_t = cross sectional area of the pulse tube, ΔP = maximum pressure change of the gas in the pulse tube,

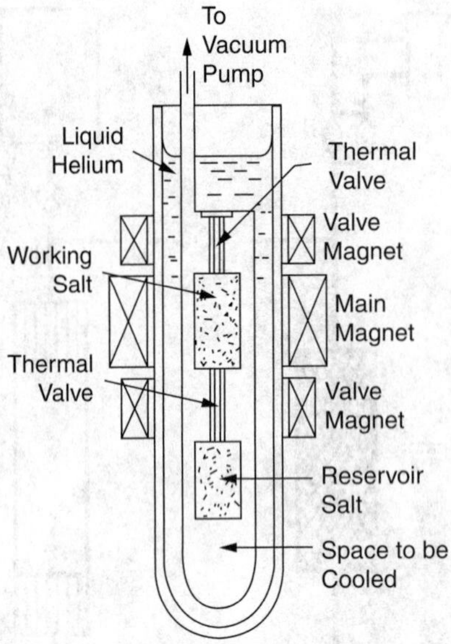

FIGURE 55.9 Magnetic refrigerator. The main magnet controls the magnetic field within the working paramagnetic salt, while the valve magnets switch the thermal valves (strips of lead) from the "off" state (superconducting) to the "on" state (normal).

and ϕ = phase angle between the gas velocity and pressure peak in the gas. The phase angle is controlled by the size of the buffer volume and the orifice impedance.

Magnetic Refrigerators

Cooling through the application of the adiabatic demagnetization process with a paramagnetic salt is one technique used to maintain temperatures below 1 K. The refrigerator consists of a paramagnetic salt such as iron ammonium alum, which is the cooling medium for the refrigerator, along with a solenoid to control the magnetic field around the working salt, as shown in Figure 55.9. Thermal valves are used to control the heat transfer to and from the working salt. Ideally, the FOM for the magnetic refrigerator would be unity; however, real magnetic refrigerators have a FOM in the range from 0.40 to 0.80.

The magnetic refrigerator has also been constructed in a "wheel" configuration, in which the magnetic field remains stationary and the paramagnetic material is moved through regions of high and low (near zero) magnetic field strength to accomplish cooling of a gas flowing through the wheel [Barclay et al., 1985].

Dilution Refrigerator (Cryocooler)

The He-3/He-4 dilution refrigerator is widely used to maintain temperatures in the range between 0.005 and 0.85 K. The basis for the operation of this refrigerator is the phase separation of mixtures of normal He-3 and superfluid He-4, discovered by K. G. Walters and W. M. Fairbank [1956], that occurs at temperatures below 0.86 K. The refrigeration effect of the dilution refrigerator is determined from

$$Q_A = n_3(h_m - h_i) \tag{55.10}$$

where n_3 is the molar flow rate of He-3, and h_m and h_i are the enthalpies of the streams leaving and entering the mixing chamber of the refrigerator, respectively. These enthalpies can be calculated from the following, for temperatures below about 0.04 K [Radebaugh, 1967].

$$h_m = (94\,J/mol - K^2)T_m^2 \tag{55.11}$$

$$h_i = (12\ J/mol - K^2)T_i^2 \tag{55.12}$$

where T_m is the temperature of the fluid leaving the mixing chamber and T_i is the temperature of the fluid entering the mixing chamber.

Defining Terms

Adiabatic demagnetization — The process of reduction of the magnetic field strength around a material while allowing negligible transfer of heat. This process results in a decrease in temperature of the material.

Closed cycle — A system in which the working fluid is recirculated.

Coefficient of performance (COP) — The ratio of the useful heat load to the required work input for a heat pump or refrigerator. The COP gives a measure of the effectiveness of the system.

Cryocooler — A refrigerator operating at cryogenic temperatures.

Cryogenics — The science and technology involving very low temperatures.

Desiccant — A material that has a high affinity for water.

Energy efficiency ratio (EER) — The coefficient of performance of a heat pump or refrigerator expressed in units of Btu/W-h.

Lambda point — The temperature at which liquid helium-4 changes from a normal fluid to a superfluid.

Normal boiling point (NBP) — The boiling point of a liquid at standard atmospheric pressure (101.325 kPa or 14.696 psia).

Open cycle — A system in which the working fluid is exhausted to the atmosphere and not recirculated.

Superconducting — The state for certain materials for which the electrical resistivity is zero and the magnetic field surrounding the material is expelled from the bulk of the material.

Superfluidity — The physical phenomenon for liquid helium-4 in which the liquid can flow with zero pressure drop through a tube heated at one end and cooled at the other end.

Triple point — The state at which all three phases (solid, liquid, and vapor) coexist in thermodynamic equilibrium.

References

Barclay, J. A., Stewart, W. F., Overton, W. C., Candler, N. J., and Harkleroad, O. D. 1985. Experimental results on a low-temperature magnetic refrigerator. *Adv. Cryogenic Eng.*, 31:743–752.

Barron, R. F. 1985. *Cryogenic Systems*, 2nd ed., Oxford University Press, New York, p. 151.

Barron, R. F. 1999. *Cryogenic Heat Transfer*, Taylor and Francis, Philadelphia, pp. 324–339.

Duffie, J. A. and Beckman, W. A. 1991. *Solar Engineering of Thermal Processes*, 2nd ed., John Wiley & Sons, New York, pp. 606–609.

Gifford, W. E. and Hoffman, T. T. 1961. A new refrigeration system for 4.2 K. *Adv. Cryogenic Eng.*, 6:82–94.

Gifford, W. E. and Longsworth, R. C. 1965. Pulse tube refrigeration progress. *Adv. Cryogenic Eng.*, 10B:69–79.

Gifford, W. E. and McMahon, H. O. 1960. A new low-temperature gas expansion cycle–Part II. In *Adv. Cryogenic Eng.*, 5:368–372.

Johnson, V. J. (Ed.) 1960. *A Compendium of the Properties of Materials at Low Temperatures, Part I.* WADD Tech. Rep. 60-56. U.S. Government Printing Office, Washington, DC.

Kreider, J. F. and Rabl, A. 1994. *Heating and Cooling of Buildings*, McGraw-Hill, New York, p. 428.

Little, W. A. 1990. Advances in Joule-Thomson cooling. *Adv. Cryogenic Eng.*, 35:1305–1314.

Lof, G. O. G. 1955. House heating and cooling with solar energy. In *Solar Energy Research*, Daniels, F. and Duffie, J. A., Eds., University of Wisconsin Press, Madison, WI, p. 33.

McQuiston, F. C. and Parker, J. D. 1994. *Heating, Ventilating, and Air Conditioning*, 4th ed., John Wiley & Sons, New York, p. 624–626.

Mikulin, E. I., Tarasov, A. A., and Shkrebyonock, M. P. 1984. Low-temperature expansion pulse tubes. *Adv. Cryogenic Eng.*, 29:629–637.

Nelson, J. S., Beckman, W. A., Mitchell, J. W., Duffie, J. A., and Close, D. J. 1978. Simulations of the performance of open cycle desiccant cooling systems. *Solar Energy*, 21:273.

Radebaugh, R. 1967. *Thermodynamic Properties of He^3-He^4 Solutions.* NBS Tech. Note 362. U.S. Government Printing Office, Washington, DC.

Radebaugh, R. 1990. A review of pulse tube refrigeration. *Adv. Cryogenic Eng.*, 35:1191–1205.

Scott, R. B. 1959. *Cryogenic Engineering,* Van Nostrand, Princeton, NJ. p. 1.

Timmerhaus, K. D. and Flynn, T. M. 1989. *Cryogenic Process Engineering,* Plenum Press, New York. pp. 156–159.

Walker, G. 1983. *Cryocoolers, Part 1,* Plenum Press, New York. pp. 280–282.

Walters, K. G. and Fairbank, W. M. 1956. Phase separation in helium-3-helium-4 solutions. *Phys. Rev.*, 103:262.

Weisend, J. G. 1998. *Handbook of Cryogenic Engineering,* Taylor and Francis, Philadelphia. pp. 314–317.

Further Information

Solar Energy Technology Handbook, W. C. Dickerson and P. N. Cheremisinoff, eds., Marcel Dekker, Inc. (1980), gives additional information on solar desiccant cooling systems.

The *ASHRAE Handbook, Refrigeration* volume and the *ASHRAE Handbook, HVAC Systems and Equipment* volume are excellent sources for heat pump design information and for examples of the application of heat pumps. These volumes are published by the American Society of Heating, Refrigerating and Air-Conditioning Engineers, 1791 Tullie Circle NE, Atlanta, GA 30329.

One of the most extensive sources of information on cryogenic systems is the series of volumes *Advances in Cryogenic Engineering,* published by Plenum Press, 233 Spring Street, New York, NY 10013 (volumes 1 through 46) and the American Institute of Physics, 2 Huntington Quadrangle, Melville, NY 11747-4502 (volumes 47 and 48). These volumes include the papers presented at the Cryogenic Engineering Conferences over the past four and one-half decades.

56

Heat Transfer to Non-Newtonian Fluids

E.F. Matthys
University of California, Santa Barbara

56.1 Introduction

Many, if not most, fluids used in industrial applications are **non-Newtonian.** Of course, the very name tells us that we lump in this category all the fluids except one type, the *Newtonian* fluids. Newtonian fluids are those that obey Newton's law relating shear stress and shear rate through a simple material property (the viscosity) dependent on basic thermodynamics variables such as temperature and pressure, but independent of flow parameters such as shear rate or time. These Newtonian fluids constitute therefore only one category of fluids: essentially the "simplest" ones. (These may, of course, nevertheless be very important. Consider water, for instance.) Conversely, non-Newtonian fluids are then defined as being all the other ones. One might therefore reasonably assume that the state of knowledge about non-Newtonian fluid heat transfer is much more extensive and sophisticated than that about Newtonian fluid heat transfer because of the size and importance of the field, but that is not the case. A major reason why this is so is that it is usually much more difficult to study non-Newtonian fluids because of their complex nature and their complex interactions with the flow field. As a result, much of the information on friction and heat transfer for non-Newtonian fluids is very empirical and consists primarily of simple correlations where the constants are determined by best-fit of experimental data.

Many engineers — when dealing with fluid mechanics or heat transfer — generally think automatically in terms of Newtonian fluids and their behavior because these are usually the main — or even the only — fluids covered in a typical undergraduate engineering education. The average engineer may therefore never had much — or any — exposure to non-Newtonian fluids before his or her first encounter with them on the job. This has important consequences as far as engineering practice, because serious mistakes can be readily made if one is not aware of the bizarre behavior of some of these fluids. Practically speaking, when the fluid is not a very simple or one-component liquid or gas such as water, oil, or air, there is a very good chance that it may indeed exhibit some non-Newtonian properties. Fluids of great importance in engineering practice, such as suspensions of particles and fibers, slurries, **polymer** solutions and melts, **surfactant** solutions, paints, foodstuffs, biofluids, soaps, inks, organic materials, adhesives, etc., may all exhibit strongly non-Newtonian behavior. The engineer should therefore be aware of the potential problems and be able to decide — on an informed basis — whether to use simpler (but perhaps leading

0-8493-1586-7/05/$0.00+$1.50

to large errors) Newtonian fluid correlations or to use the more complex information (perhaps) available on non-Newtonian fluids.

Given the breadth and complexity of the field, and space limitations, the nonspecialist reader will be best served in this author's opinion by a short chapter that focuses on giving a general idea of the basic issues and difficulties involved and on providing some useful references for deeper study. Mentioning only a few equations here may do disservice to the reader interested in this complicated field, as it is more likely to lead to inadvertent inappropriate usage than to be a timesaver. For some basic information, the reader is instead referred to some readily available handbook articles that provide numerous correlations: for example those by Irvine et al. [1987] on **purely viscous non-Newtonian fluids** and by Hartnett et al. [1998] dealing with **viscoelastic non-Newtonian fluids**. Some other references and recent work will also be mentioned hereafter. A focus of this chapter will be drag-reducing fluids because these are among the most challenging and interesting non-Newtonian fluids.

56.2 The Fluids

It should be noted first that one needs only to address the issue of *moving* fluids here, because the very concept of a non-Newtonian substance implies that the related fluid characteristics manifest themselves under flow. Accordingly, only convective heat transfer needs to be discussed here, because both conduction and radiation are normally unaffected, and require only that appropriate material properties such as thermal conductivity and emissivity be known, as in the case of Newtonian fluids. Another point to consider is that the definition of a non-Newtonian fluid covers only the *shear* viscosity of the fluid, but many non-Newtonian fluids will also show a complex *extensional* flow behavior. Some non-Newtonian fluids, perhaps the most difficult to deal with, are also viscoelastic.

The first step in trying to predict the heat transfer behavior of a non-Newtonian fluid is therefore to determine its nature and type. For simplicity, one often classifies these fluids as purely viscous vs. viscoelastic, the former lacking the elastic component of the latter. All of these may exhibit a viscosity that can either increase or decrease both with shear rate and with time, and some fluids may also exhibit a yield stress. In addition, viscoelastic fluids in general exhibit memory effects, but some may also show particularly large time-dependent effects in addition to shear rate-related variations (e.g., some surfactant solutions).

In general, the heat transfer to purely viscous fluids can be quantified by relatively simple modified Newtonian fluid correlations, and with some information on the viscosity one can then predict reasonably well the heat transfer and friction. These fluids have been studied early on, and many flow configurations have been investigated. Reviews and handbooks, even somewhat older ones, may provide most of the state-of-the-art information. For viscoelastic fluids, on the other hand, and especially in turbulent flow, much less information is available; they are still an area of active current research. Some of these fluids exhibit a fascinating property: that of reducing greatly the friction or heat transfer under turbulent flow conditions. This phenomenon is called **drag or heat transfer reduction.** For these fluids, using a Newtonian-like approach to predict heat transfer may well result in very large errors. This field is evolving rapidly; much of the current information may need to be sought in technical journals.

It is relatively easy to determine whether a fluid might be viscoelastic, even with some simple qualitative experiments such as recoil, finger-dip and fiber-pull, rod-climbing, die swell, and tubeless siphon. One could also pretty much guess that most fluids that are not viscoelastic but are nevertheless mixtures, solutions, or suspensions may likely be purely viscous non-Newtonian fluids. Beyond such simple classification, however, some data on the viscous or viscoelastic nature of the fluid may be necessary to help correlate or predict heat transfer or friction quantitatively. This information has to be acquired from some **rheology** experiments, although rough estimates can at times be found in the literature. One should be especially cautioned, however, against using viscosity models beyond their proven range of applicability, as is often done for the ubiquitous power-law model, for example. Much information, including books on rheological measurements, is readily available, and several companies produce instrumentation dedicated to such measurements.

It cannot be emphasized enough for the engineer who is not yet familiar with non-Newtonian fluids that one should be particularly wary of relying on one's experience with Newtonian fluids or one's intuition when tackling the more complex non-Newtonian fluids. The latter's behavior may indeed be very surprising. For example, a few parts per million of a polymer or surfactant in solution in water could well reduce the heat transfer by a factor of more than 10. Accordingly, the reader is strongly advised to consult some of the additional material referred to here to develop enough of an understanding of the issues to avoid such problems.

56.3 Friction

The reader is referred to another chapter in this handbook and to many articles in the literature where one can find fluid mechanics information on these fluids. An approach similar — although simpler — to that discussed here for heat transfer can usually be followed for friction. It is important to note, however, that the strong coupling between friction and heat transfer for Newtonian fluids, which enables one to predict heat transfer if one knows friction (and vice versa) for these fluids, was long believed not to hold for some non-Newtonian fluids such as drag-reducing solutions, and may be discussed as such in earlier literature. Conversely, recent studies have suggested that for these latter fluids at least, this coupling appears indeed to remain applicable. If this is indeed more generally true, a powerful new tool for the prediction of heat transfer for these fluids will have been added to the arsenal of the engineer because simpler friction experiments may then be all that is needed to predict or estimate heat transfer.

56.4 Heat Transfer

As mentioned above, one is usually reduced to using simple empirical correlations to predict the heat transfer for these fluids. In addition to the handbooks listed above, the reader is also referred to a number of reviews for additional information [Metzner, 1965; Skelland, 1967; Dimant et al., 1976; Shenoy et al., 1982; Hartnett et al., 1989; Matthys, 1991].

As in the case of Newtonian fluids, one may use the **Nusselt number** to quantify the convective heat transfer coefficient. In the case of a purely viscous non-Newtonian fluid, the Nusselt number will usually be a function of modified **Reynolds** and **Prandtl** numbers. For viscoelastic non-Newtonian fluids, it may be in addition a function of other non-dimensional numbers that may be related to the elasticity of the fluid through a relaxation time (e.g., a **Weissenberg** number) or to the extensional viscosity of the fluid or some other parameter. It is also crucial to distinguish between the several types of Reynolds numbers (and corresponding Prandtl numbers) that are used for the prediction of friction and heat transfer for these fluids. (Note that such a distinction is not always made clearly in the literature, and one should proceed cautiously.) Often, *generalized* Reynolds and Prandtl numbers are used for laminar flow, whereas *apparent* Reynolds and Prandtl numbers may be more suitable for turbulent flow. These various definitions involve the actual viscosity of the fluids in different ways. For applied engineering work, researchers often use *solvent-based* numbers, a potentially dangerous practice that leads to large errors for some fluids and should be used with great care. Indeed, even with dilute solutions, it is generally essential to measure the viscosity of the actual fluid if it is not in actuality very close to that of the solvent. This viscosity has to be measured in the laboratory and should be done as a matter of routine when dealing with non-Newtonian fluids. Indeed, if information is not available on the viscosity of the fluid, and the latter is not properly taken into account in the nondimensionalizations, significant errors may occur. Conversely, the use of proper parameters and viscosity values may, for example, reveal that a fluid at first believed to be exhibiting non-asymptotic behavior is in actuality indeed fully asymptotic [Aguilar et al., 2001]. Good discussions and definitions of these parameters can be found in some of the reviews listed above.

Since the main effects of the non-Newtonian character of the fluid on the heat transfer result from interactions between fluid and flow field through large variations in viscosity or elasticity (sometimes over several orders of magnitude), the variations in other thermophysical parameters (e.g., thermal conductivity and specific heat capacity) are often of much lesser importance. One may then often satisfactorily assume that these properties vary only with temperature in a given manner for a specific non-Newtonian fluid, or even that they remain those of the Newtonian solvent (e.g., for low-concentration solutions). Naturally, for highly concentrated suspensions of fibers or particles, for example, one has to use appropriate properties for the actual fluid, and procedures are available to predict such properties based on the concentration of the suspended material. For more complex or lesser-known fluids, one may well be forced to measure these properties directly. It is useful to note that for viscoelastic fluids, in particular, the large reductions in heat transfer seen in turbulent flow are not related to modifications of the "static" thermophysical properties, but rather to fluid/turbulence interactions, i.e., a dynamic process. This is why these fluids do not normally show such large effects on the heat transfer in the laminar regime. The distinction between laminar and turbulent flows is therefore even more crucial to make here than for Newtonian fluids. Note that the transition between the two regimes takes place under relatively similar conditions for both Newtonian and non-Newtonian fluids, and that the usual Newtonian fluid criteria for transition are often used. (In addition, some viscoelastic effects such as drag reduction may only appear after some minimum value of a flow parameter — such as the shear stress, for example — is exceeded.)

Laminar Regime — Heat transfer in the laminar regime can be fairly readily predicted for non-Newtonian fluids because of the simple nature of the flow and the absence of significant elastic effects. A good discussion on this issue for some fluids is provided in Skelland [1967]. Generally, one simply modifies the Nusselt number calculated for a Newtonian fluid with a coefficient that involves a parameter reflecting the change in velocity profile (e.g., a power-law exponent n) and possibly the aspect ratio if it is internal flow in a noncircular duct [Irvine et al., 1987; Hartnett et al., 1989]. For heat transfer in the entrance region, one has to introduce another parameter such as a Graetz number to account for the temperature profile development (as in the case of Newtonian fluids). Also as in the case of Newtonian fluids, the laminar thermal entrance region can be fairly long for high Reynolds numbers. Overall, the change in Nusselt number introduced by the non-Newtonian nature of the fluid is often moderate in the laminar regime, and if nothing else is available, the use of a Newtonian value may not be too far off in first approximation for fluids with a moderate power-law exponent, for example. For external flow, simple empirical corrections to Newtonian expressions can also be used [e.g., Irvine et al., 1987]. For free and mixed convection, relatively little information is available, and the reader is referred in particular to the review by Shenoy et al. [1982]. Specific information for polymer melts can be easily found in the processing literature. The viscosity of the latter fluids is generally very large, however, and most applications will likely involve only laminar or viscosity-dominated conditions, which simplifies the analysis considerably, in contrast to the situation for dilute solutions.

Turbulent Regime — For this regime, it becomes necessary to pay particular attention to the distinction between purely-viscous and viscoelastic fluids. Heat transfer to many purely-viscous fluids can be adequately predicted by the use of simple relations involving the Nusselt number and apparent Reynolds and Prandtl numbers or by taking advantage of the analogy between friction and heat transfer for these fluids [Metzner, 1965; Cho et al. ,1985; Matthys et al., 1987]. The thermal entrance region for these fluids is generally similar to that of Newtonian fluids and very short (e.g., about 20 to 50 diameters for flow in a tube).

For viscoelastic fluids, on the other hand, the situation is very different. Some suspensions of very long fibers may exhibit such a nature, as will polymer melts or solutions and many complex fluids. Dilute solutions of a polymer or surfactant in tube flow, in particular, are of great interest in this regard. These fluids may indeed exhibit dramatic drag and heat transfer reductions with respect to the solvent alone at the same Reynolds number. (Some good reviews for information on drag-reducing fluids can be found in Gyr and Bewersdorff [1995] and in Zakin et al. [1998].) This effect is very large, even with very small (e.g., subpercent level) traces of polymer or surfactant additives. The level of drag and heat transfer reductions at a given Reynolds number will generally increase with concentration of drag-reducing additive in the solvent. Interestingly, there is a minimum below which the drag and heat transfer can no longer be reduced, the so-called drag and heat transfer reduction asymptotes [e.g., Matthys, 1991; Aguilar et al., 2001]. One advantageous feature of this **asymptotic regime** is that the concentration no longer plays a role there in determining the friction and heat transfer, which also means that the Nusselt number is no longer a function of an additional parameter besides the Reynolds and Prandtl number. Good correlations exist for this regime, in fact surprisingly robust ones. The asymptotes were long thought to be the same for all drag-reducing fluids, but some recent work has suggested that there may well be differences between polymer and surfactant solutions, for example [Aguilar et al., 2001]. There may also be some more fundamental differences between distinct types of drag-reducing fluids [Gasljevic et al., 2001]. In the asymptotic regime, there also appears to be a strong coupling between friction and heat transfer, seen most readily if more physically meaningful nondimensional numbers are used, rather than the numbers based on simpler calculations of drag and heat transfer reductions that are commonly used [Aguilar et al., 1999]. Indeed, when working with drag-reducing fluids, one should pay particular attention to the nondimensional numbers used for the quantification of the drag and heat transfer reduction levels. For instance, the most widely used definitions of the drag and heat transfer reduction factors are based on a comparison between the solvent friction and a hypothetical fluid of zero friction. This is a convenient representation, but it has been shown instead that more physically meaningful definitions can be generated based on comparison between turbulent and laminar flows [Gasljevic and Matthys, 1999]. The use of these better parameters allows for more accurate correlations and may help explain apparently surprising results and unexpected behaviors.

In the region between the Newtonian and asymptotic limits, the friction and heat transfer are a function of the concentration (i.e., viscoelasticity) of the fluid as well as of the diameter of the pipe. At this time, there appears to be no universal technique that allows us to predict directly the heat transfer in this region based on simple material properties measurements or basic principles. The non-Newtonian (nonlinear) diameter effects, in particular, mandate that some special scaling methods be used for adequate predictions [e.g., Gasljevic et al., 1999]. Note also that for a given concentration and diameter, the level of friction and heat transfer reduction in this regime will still depend on the Reynolds number (i.e., shear stress or shear rate), and a fluid may exhibit Newtonian behavior at low Reynolds number but then progressively phase in its asymptotic behavior at high Reynolds number. In some cases there may be an onset shear stress that has to be exceeded before any drag or heat transfer reduction behavior is observed. In the nonasymptotic regime, there still appears to be coupling between friction and heat transfer, but of a more complicated nature, perhaps quantifiable by a constant nonunity turbulent Prandtl number [Aguilar et al., 1999].

Interestingly, the heat transfer is usually reduced proportionally more than the friction for drag-reducing fluids (e.g., by 95 vs. 90% for a typical asymptotic solution), but the ratio between the two remains essentially constant if proper nondimensional parameters are used [Aguilar et al., 1999]. For polymer solutions, the heat transfer and drag reductions decrease simultaneously and at the same rate as the polymer is degraded mechanically. Departures from asymptotic conditions due to dilution also take place simultaneously for friction and heat transfer. It should also be noted that the entrance lengths for these fluids are much longer than for Newtonian fluids. For polymer and surfactant solutions in tubes, for example, the thermal entrance length may easily be several hundred diameters long (compared to 20 or so for Newtonian or purely viscous fluids). Here again, more recent work suggests a good coupling between friction and heat transfer in the development region, and a substantial independence of these development effects on the velocity [Gasljevic et al., 1997]. The very long thermal entrance lengths, in particular, should be kept in mind when designing or analyzing heat exchangers for these fluids, as it is likely that the heat transfer may never reach fully developed conditions in many exchangers. Correlations exist that give the heat transfer in the entrance region as a function of distance [e.g., Matthys, 1991; Gasljevic et al., 1997]. It should also be noted that the recent evidence of strong coupling between friction and heat transfer for drag-reducing fluids suggests that it may well be possible to predict heat transfer from friction measurements, which are generally much easier to conduct. This may well save the engineer much work or aggravation when it comes to predicting heat transfer or to designing heat exchangers.

Another issue is very important when one is studying or using viscoelastic fluids: that of **degradation.** Indeed, the polymeric fluids, for example, being macromolecular in nature, may be very susceptible to mechanical degradation. This means that — when subjected to flow in tubes and especially through pumps and filters — some macromolecules will be permanently broken and that the drag and heat transfer reductions will then be lost partially — or perhaps even completely — depending on the level and duration of exposure. This can be often seen as an increase in friction or heat transfer at high Reynolds number. Degradation can also be caused by thermal or chemical processes. A consequence is that polymer solutions are not well suited for recirculating flows. Surfactant solutions, on the other hand, are much less susceptible to permanent degradation, although the drag and heat transfer reductions can be eliminated completely (but reversibly) under high shear stress (e.g., at high Reynolds number or in hydraulic components). They also possess the remarkable ability to reconstitute rapidly (depending on the temperature) after being subjected to high shear (e.g., in a centrifugal pump). In that respect they are very attractive, and even though relatively little known at this time, will undoubtedly become used more in the future. Many applications of such fluids are possible that would capitalize on their remarkable drag-reducing properties. We are studying presently, for example, their use as energy conservation agents in hydronic heating and cooling systems, a very promising application indeed. For applications involving heat exchangers, the large reductions in heat transfers exhibited by some of these fluids may introduce serious practical difficulties, however, and we have therefore developed approaches allowing us to regain and control heat transfer as needed in the heat exchangers through intentional temporary degradation of the surfactant solution in the heat exchangers. This technology has enabled us to conduct successful field tests of surfactant solutions as energy-saving thermal transport fluids for hydronic HVAC systems.

56.5 Instrumentation and Equipment

Much of the equipment encountered or needed when dealing with non-Newtonian fluids is similar to that used with Newtonian ones, although, of course, some additional equipment may be needed for the quantification of the non-Newtonian nature of the fluid. Typically, this rheological work will require the use of rheometers, perhaps over a wide range of temperature and flow conditions. Numerous such devices exist and are discussed in books on rheometry.

As far as basic instrumentation, a few words of caution are — again — in order. Pressure measurements, for instance, can be complicated by errors due to viscoelastic pressure effects in the taps. These can be minimized by paying particular attention to the shape and uniformity of the tap holes or by using differential pressure measurements whenever possible. Flow measurements may be particularly troublesome, and one should not rely automatically on flowmeters designed for water such as orifice meters or turbine meters, which can give large errors for non-Newtonian fluids unless a careful calibration has been conducted for the specific fluid and relevant flow conditions. Positive displacement devices are generally less troublesome. Ultrasonic flow meters may also be used depending on their built-in velocity profile calibration. Velocity measurements with typical LDV systems should be readily possible, but the use of hot wire anemometry would probably lead to large errors in many cases if appropriate corrections are not introduced.

For temperature measurements, thermocouples, RTDs, thermistors, and thermometers are all generally suitable under steady-state conditions. Note that the flow field may be affected, however, and also that the heat transfer may also be reduced, which may mean, for example, that a longer time might be needed before a steady-state measurement is achieved. Indeed, unsteady measurements may be more difficult. When drag-reducing flows are involved, the temperature gradients in a tube, say, may also be much larger than for Newtonian fluids at a given heat flux, because of the reduced convective heat transfer coefficients. More extensive mixing may then be necessary for bulk temperature measurements. Note also that — as discussed above — the thermal entrance length may be very long for these fluids, and that fully developed conditions may never be reached in practical situations. Some buoyancy effects appear also to be very significant for some non-Newtonian fluids and may introduce large errors in heat transfer measurements if not taken into account [Gasljevic et al., 2000].

The performance of some heat exchangers may be significantly impacted by the fluids, especially drag-reducing solutions [e.g., Gasljevic et al., 1993], and caution should be exercised there as well. The magnitude of the heat transfer reductions will of course depend greatly on the particulars of the fluids and heat exchangers. Generally speaking, tube heat exchangers will exhibit greater reductions in heat transfer than comparable plate heat exchangers, for example. Mixing devices may also perform differently with non-Newtonian fluids. Centrifugal pumps in most cases will work appropriately, even with improved efficiency at times but may also affect the fluid if excessive shear stress is applied. Pump and valve flow curves do not appear to be changed dramatically in most cases. Naturally, issues of erosion, corrosion, fouling, disposal, etc. should also be investigated for the fluid of interest.

Defining Terms

Asymptotic regime — The regime of maximum drag and heat transfer reduction that can be achieved by drag-reducing fluids. This is the point where increasing the concentration will no longer result in further decreases in drag or heat transfer, if everything else is equal.

Degradation — A modification imparted to the fluid by mechanical, thermal, or chemical effects that changes its properties (e.g., a reduction in viscosity or drag-reducing capability). Typically, mechanical degradation will be permanent in polymeric fluids but only temporary in surfactant solutions.

Heat transfer (and drag) reduction — A sometimes dramatic decrease in heat transfer (and friction) in turbulent flow due to interactions between the viscoelastic nature of the fluid and the flow field.

Non-Newtonian fluid — A fluid that does not obey Newton's law of simple proportionality between shear stress and shear rate. Its viscosity is then also a function of shear rate, time etc.

Nusselt number — A nondimensional number involving the convective heat transfer coefficient. It is a measure of the fluid capability to transfer heat to a surface.

Reynolds number — A nondimensional number involving the flow velocity and the fluid viscosity. It is a measure of the flow rate and level of turbulence in the flow.

Polymer — A material constituted of molecules made of repeating units (monomers).

Prandtl number — A nondimensional number comparing the diffusivity of momentum and heat. It reflects the relative ease of transport of momentum (friction) vs. energy at the molecular level.

Purely viscous non-Newtonian fluid — A non-Newtonian fluid that does not exhibit elasticity.

Rheology — The study of flow and deformation of matter. (Rheology is generally associated with measurements of viscosity and other material properties for *non-Newtonian* fluids.)

Surfactant — A material leading to reduced surface tension effects when added to a solvent. When in solutions, surfactants may also under some conditions form large micellar structures that may result in large viscoelastic drag and heat transfer reductions. (These are not caused by surface tension effects but rather by bulk viscoelastic effects similar to those exhibited by polymer solutions.)

Viscoelastic non-Newtonian fluid — A fluid which — in addition to variations of viscosity with shear rate — exhibits also an elastic character (i.e., has a memory, shows recoil, etc.).

Weissenberg number — A nondimensional number reflecting the viscoelasticity of a fluid through its relaxation time.

References

Aguilar G., Gasljevic, K., and Matthys, E. F. 1999. Coupling between heat and momentum transfer mechanisms for drag-reducing polymer and surfactant solutions, *Journal of Heat Transfer* 121:796–802.

Aguilar, G., Gasljevic, K., and Matthys, E. F. 2001. Asymptotes of maximum friction and heat transfer reductions for drag-reducing surfactant solutions. *International Journal of Heat and Mass Transfer*, 44:2835–2843.

Dimant, Y. and Poreh, M. 1976. Heat transfer in flows with drag reduction. *Advances in Heat Transfer*, 12:77–113.

Gasljevic, K. and Matthys, E. F. 1993. Effect of drag-reducing surfactant additives on heat exchangers. In *Developments in Non-Newtonian Flows*, ed. D. Siginer, vol. AMD-175, ASME, Washington, D.C., pp. 101–108.

Gasljevic, K. and Matthys, E. F. 1997. Experimental investigation of thermal and hydrodynamic development regions for drag-reducing surfactant solutions. *Journal of Heat Transfer* 119:80–88.

Gasljevic, K. and Matthys, E. F. 1999. Improved quantification of the drag reduction phenomenon through turbulence reduction parameters. *Journal of Non-Newtonian Fluid Mechanics*, 84:123–130.

Gasljevic K., Aguilar, G. and Matthys, E. F. 1999. An improved diameter scaling for turbulent flow of drag-reducing polymer solutions. *Journal of Non-Newtonian Fluid Mechanics* 84:131–148.

Gasljevic, K., Aguilar, G., and Matthys, E. F. 2000. Buoyancy effects on heat transfer and temperature profiles in horizontal pipe flow of drag-reducing fluids. *International Journal of Heat and Mass Transfer* 43:4267–4274.

Gasljevic K., Aguilar, G., and Matthys, E. F. 2001. On two distinct types of drag-reducing fluids, diameter scaling, and turbulent profiles. *Journal of Non-Newtonian Fluids Mechanics* 96:405–425.

Gyr, A. and Bewersdorff, H. W. 1995. *Drag Reduction of Turbulent Flows by Additives.* Fluid Mechanics and Its Applications Series, vol. 32. Kluwer, Dordrecht.

Hartnett, J. P. and Cho, Y. I. 1998. Non-Newtonian fluids. In *Handbook of Heat Transfer,* 3rd ed., W. M. Rohsenow, J. P. Hartnett, and Y. I. Cho, Eds., McGraw-Hill, New York, chap. 10.

Hartnett, J. P. and Kostic, M. 1989. Heat transfer to Newtonian and mon-Newtonian fluids in rectangular ducts. *Advances in Heat Transfer* 19:247–356.

Irvine, T. F. and Karni, J. 1987. Non-Newtonian fluid flow and heat transfer. In *Handbook of Single-Phase Convective Heat Transfer,* S. Kakac, R. K. Shah, and W. Aung, Eds., John Wiley & Sons, New York, pp. 20.1–20.57.

Matthys E. F., Ahn, H., and Sabersky, R. H. 1987. Friction and heat transfer measurements for clay suspensions with polymer additives. *Journal of Fluids Engineering* 109:307–312.

Matthys, E. F. 1991. Heat transfer, drag reduction, and fluid characterization for turbulent flow of polymer solutions: recent results and research needs. *Journal of Non-Newtonian Fluid Mechanics* 38: 313–342.

Metzner, A. B. 1965. Heat transfer in non-Newtonian fluids, *Advances in Heat Transfer* 2:357–397, Academic Press, New York.

Shenoy A. V. and Mashelkar A. R. 1982. Thermal convection in non-Newtonian fluids. *Advances in Heat Transfer, Volume 15,* Academic Press, New York.

Skelland A. H. P. 1967. *Non-Newtonian Flow and Heat Transfer,* John Wiley & Sons, New York.

Zakin J. L., Lu, B., and Bewersdorff, H. W. 1998. Surfactant drag reduction. *Reviews in Chemical Engineering* 14:253–320.

Further Information

The reader is also referred to technical journals such as *International Journal of Heat and Mass Transfer, Journal of Heat Transfer, Journal of Non-Newtonian Fluid Mechanics, Rheologica Acta, Journal of Rheology,* and many others. Symposia covering this subject are often held at conferences by ASME, AIChE, the Society of Rheology, and others. Many excellent research papers in this field have also been published by — among others — the groups led by H. W. Bewersdorff, Y. I. Cho, A. J. Ghajar, J. P. Hartnett, R. H. Sabersky, A. Steiff, H. Usui, and J. L. Zakin.

57

Heat Pipes

Jay M. Ochterbeck
Clemson University

The heat pipe is a capillary-driven two-phase (liquid–vapor) heat transfer device that transfers heat from a heat source to a heat sink. Gaugler originally conceived the basic concept of the heat pipe in 1944, but the operational characteristics of heat pipes were not widely publicized until independent developments by Trefethen in 1962 and by Grover and colleagues at the Los Alamos National Laboratory in 1964. Many types of heat pipes since have been developed and are widely used in a variety of industries. The main attractive feature is that a heat pipe is a passive heat transfer device that transfers heat over relatively long distances via the latent heat of vaporization of a working fluid. The fluid is circulated by **capillary pressure** developed in a **wick**, or porous material.

Figure 57.1 shows a schematic of a heat pipe, which may be inclined at an angle ψ relative to gravity. A heat pipe generally has three sections: an evaporator section for heat addition, an adiabatic (or transport) section, and a condenser section for heat rejection. The major components of a heat pipe are a sealed container, a wick structure, and a working fluid. The wick structure is placed on the inner surface of the heat pipe wall, is saturated with liquid working fluid, and provides the structure to develop the capillary action for liquid returning from the condenser to the evaporator section. With evaporator heat addition, the working fluid is evaporated as it absorbs an amount of heat equivalent to the latent heat of vaporization, while in the condenser section, the working fluid vapor is condensed. The mass addition to the vapor core in the evaporator and the mass rejection in the condenser result in a pressure gradient along the vapor channel that drives the vapor flow. Liquid return through the wick to the evaporator from the condenser is provided by the liquid pressure difference, which is due to the net difference in capillary pressures in the evaporator and condenser sections. Classification of heat pipes may be in terms of geometry, intended applications, or the type of working fluid utilized. Proper selection and design of the heat pipe container, working fluid, and wick structure are essential to the successful operation of a heat pipe. The **heat transfer limitations** and **effective thermal conductivity** define the operational characteristics of the heat pipe.

57.1 Heat Pipe Container, Working Fluid, and Wick Structures

The combination of the container, working fluid, and wick structure of a heat pipe determines its operational characteristics. The container and wick must be chosen to take into account chemical compatibility with the working fluid. Chemical reactions between the working fluid and the container/

0-8493-1586-7/05/$0.00+$1.50

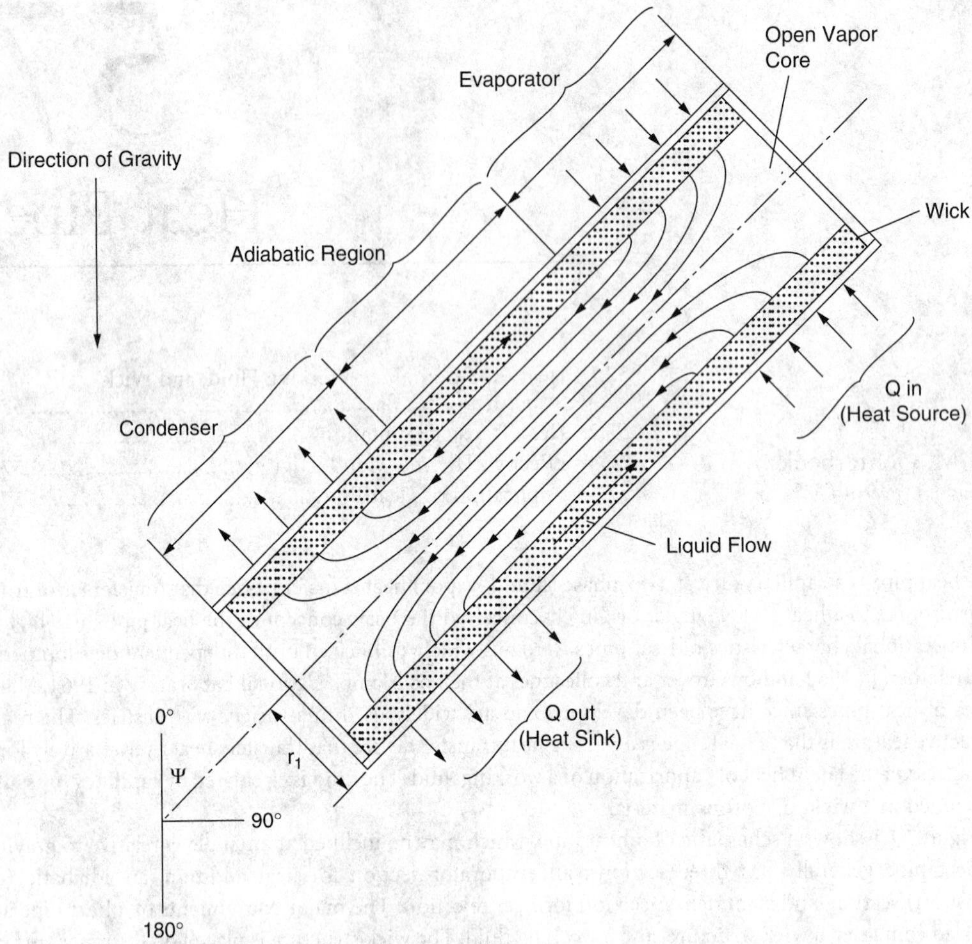

FIGURE 57.1 Schematic of a typical heat pipe.

wick (or with contamination not removed during processing) can seriously impair heat pipe performance. Noncondensible gases created during a chemical reaction will accumulate near the end of the condenser, thus decreasing the condensation surface area. This reduces the ability of the heat pipe to transfer heat to the external heat sink. The heat pipe container also must have high burst strength, low weight, and high thermal conductivity.

Using the proper working fluid for a given application is a critical element of proper heat pipe operation. The working fluid must have good thermophysical properties for the specified operational temperatures. The operational temperature range of the working fluid lies between the fluid triple point and critical point, as a two-phase saturated liquid–vapor condition must exist. Working fluids used in heat pipes range from cryogenic (e.g., nitrogen) to room temperature (e.g., water) to liquid metals (e.g., sodium). The **wettability** of the working fluid contributes to its capillary pumping and priming (filling of the wick) capability. High-surface-tension fluids are desired for heat pipe use as the surface tension relates to the necessary capillary pumping. However, the maximum heat transport of a heat pipe depends on a combination of thermophysical properties, which include the latent heat of vaporization, viscosities of the liquid and vapor, densities, and surface tension. Selections of fluids are typically based on groupings of the fluid properties referred to as a Figure of Merit found in Chi [1976] or Peterson [1994], which allow direct comparison of one fluid to another over given temperature ranges.

The wick structure provides the path for liquid return to the evaporator and provides the structure for the necessary capillary forces to be developed. Heat pipe wicks can be made from numerous porous

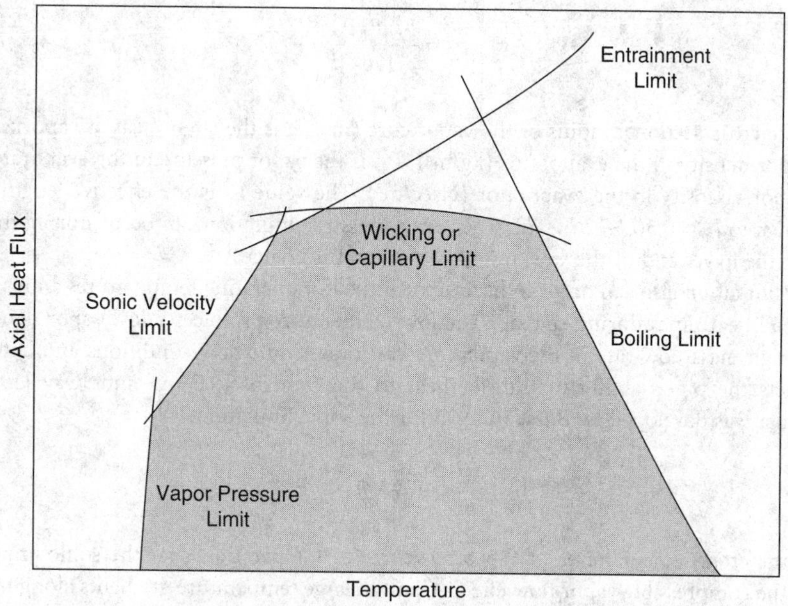

FIGURE 57.2 Heat transfer limitations in heat pipes.

materials, where some examples are wrapped screens, sintered particles, axial grooves, annular passages, and arterial. To optimize performance, more than one type of porous material can be used to balance the need for small pore sizes (high capillary pressures) with the need for high permeability (minimal flow resistance). These composite wicks can be made from variable screen mesh, screen-covered grooves, and screen tunnels with grooves. The selected pairing of the wick structure and working fluid determines the heat transfer limitations of heat pipes.

57.2 Heat Transfer Limitations

The heat transported by a heat pipe (i.e., heat transfer limitation) depends on the working fluid, the wick structure, the geometry of the heat pipe, and the heat pipe operational temperature. Figure 57.2 gives a qualitative description of the various heat transfer limitations, which include vapor pressure, sonic, entrainment, capillary, and boiling limitations. Each results from a specific, operational phenomenon within the heat pipe. The composite curve enclosing the shaded region in Figure 57.2 gives the maximum heat transfer rate transported by the heat pipe as a function of the operational temperature. The figure shows that, as the operational temperature increases, the maximum heat transfer rate of the heat pipe is limited by different physical phenomena. As long as the operational heat transfer rate falls within the shaded region, the heat pipe can function properly. It is important to note that within the heat transfer limitations, the vapor pressure and sonic limits do not represent "failure" cases — only that the heat pipe must increase the operation temperature if the heat input exceeds these values. For the capillary, boiling, and entrainment limitations, these represent failure cases as sufficient liquid flow to the evaporator is not possible and the heat pipe will cease to operate properly above these values.

The vapor pressure limitation (or viscous limitation) is reached when the required pressure drop in the vapor core reaches the same order of magnitude as the available vapor (saturation) pressure in the evaporator region. In this case, the total vapor pressure will be balanced by opposing viscous forces in the vapor channel. Thus, the total vapor pressure within the vapor region may be insufficient to sustain an increased flow. A general expression for the vapor pressure limitation is given by Dunn and Reay [1982]:

$$Q_{vp,\max} = \frac{\pi r_v^4 h_{fg} \rho_{v,e} P_{v,e}}{12\mu_{v,e} l_{\text{eff}}} \tag{57.1}$$

where r_v is the cross-sectional radius of the vapor core (m), h_{fg} is the latent heat of vaporization (J/kg), $\rho_{v,e}$ is the vapor density in the evaporator (kg/m³), $P_{v,e}$ is the vapor pressure in the evaporator (Pa), and $\mu_{v,e}$ is the vapor viscosity in the evaporator (Nsec/m²). The value l_{eff} is the effective length of the heat pipe (m), equal to $l_{\text{eff}} = 0.5(l_e + 2\, l_a + l_c)$. The vapor pressure limitation can occur during the startup of heat pipes at the lower end of the working fluid temperature range.

The sonic limitation also can occur in heat pipes at the low operating temperatures and is seen mostly in liquid metal heat pipes during startup. The low temperature produces a low vapor density. Thus, a sufficiently high mass flow rate in the vapor core can cause sonic flow conditions and choke the flow, which restricts the pipe's ability to transfer heat to the condenser. Dunn and Reay [1982] give an expression that was developed by Busse in 1973 for the sonic limitation:

$$Q_{s,\max} = 0.474 A_v h_{fg} (\rho_v P_v)^{1/2} \tag{57.2}$$

where A_v is the cross-sectional area of the vapor core (m²). Operation near the sonic limit is also not desirable as the compressible vapor flow effects result in large temperature gradients along the heat pipe, thus resulting in large deviation from the ideal, near isothermal case.

The entrainment limitation in heat pipes develops when the vapor velocity is sufficient to shear droplets of liquid from the wick surface, thus depleting the necessary liquid mass flow rate. A conservative estimate of the maximum heat transfer rate due to entrainment of liquid droplets has been given by Dunn and Reay [1976] and Faghri [1995] as

$$Q_{e,\max} = A_v h_{fg} \left[\frac{\rho_v \sigma_l}{2 r_{c,\text{ave}}} \right]^{1/2} \tag{57.3}$$

where σ_l is the surface tension (N/m) and $r_{c,\text{ave}}$ is the average capillary radius of the wick (m).

As the driving potential for the circulation of the working fluid is the capillary pressure difference, the maximum capillary pressure must be greater than the sum of all pressure losses inside the heat pipe. The pressure losses in heat pipes can be separated into the frictional pressure drops along the vapor and liquid paths and the pressure drop in the liquid as a result of body forces (e.g., gravity, centrifugal, electromagnetic). It is important to note that the inclination of the heat pipe can either be an adverse tilt (evaporator above condenser) or a favorable tilt (condenser above evaporator) such that the hydrostatic pressure either subtracts from, or adds to, the capillary pumping pressure. For many heat pipes, the frictional pressure losses in the liquid are much greater than the frictional losses in the vapor. In this case, the maximum heat transfer rate due to the capillary limitation can be expressed as [Chi, 1976]

$$Q_{c,\max} = \left[\frac{\rho_l \sigma_l h_{fg}}{\mu_l} \right] \left[\frac{A_w K}{l_{\text{eff}}} \right] \left(\frac{2}{r_{c,e}} - \left[\frac{\rho_l}{\sigma_l} \right] g L_t \cos\psi \right) \tag{57.4}$$

where K is the wick permeability (m²), A_w is the wick cross-sectional area (m²), ρ_l is the liquid density (m³), μ_l is the liquid viscosity (Nsec/m²), $r_{c,e}$ is the wick capillary radius in the evaporator (m), g is the acceleration due to gravity (9.8 m/sec²), and L_t is the total length of the pipe (m). For most practical operating conditions, this limitation is the primary limitation encountered and can be used to determine the maximum heat transfer rate in heat pipes. Chi [1976], Peterson [1994], and Faghri [1995] provide details on the full evaluation of the capillary limit when frictional vapor losses are significant.

The boiling limitation in heat pipes occurs when the degree of liquid superheat in the evaporator is large enough to cause the nucleation of vapor bubbles within the wick structure. Boiling is undesirable in heat pipes because vapor bubbles within the wick obstruct the liquid flow in the evaporator. An expression for the boiling limitation is [Chi, 1976]

$$Q_{b,\max} = \frac{2\pi L_{\text{eff}} k_{\text{eff}} T_v}{h_{fg} \rho_l \ln(r_i / r_v)} \left(\frac{2\sigma_l}{r_n} - \frac{2\sigma_l}{r_{c,e}} \right) \tag{57.5}$$

where k_{eff} is the effective thermal conductivity of the composite wick and working fluid (W/m K), T_v is the vapor saturation temperature (K), r_i is the inner container radius (m), and r_n is the nucleation radius (conservative estimate is $2.00 \cdot 10^{-6}$ m in the absence of noncondensible gas). The boiling limit differs from the other limits as it is a radial limit in the evaporator as opposed to axial transport in the case of the other limitations.

57.3 Effective Thermal Conductivity

One key attribute of the heat pipe is that it can transfer a large amount of heat while maintaining nearly isothermal conditions. The temperature difference between the external surfaces of the evaporator and the condenser can be determined from the following expression:

$$\Delta T = R_t Q \tag{57.6}$$

where R_t is the total thermal resistance (°C/W) and Q is the heat transfer rate (W). Figure 57.3 shows the thermal resistance network for a typical heat pipe and the associated thermal resistances. In most cases the total thermal resistance can be approximated by

$$R_t = R_1 + R_2 + R_3 + R_5 + R_7 + R_8 + R_9 \tag{57.7}$$

as the resistances R_{10} and R_{11} are typically several orders of magnitude greater than the other resistances, thus negating the parallel path through these resistances. The effective thermal conductivity of the heat pipe is defined as the heat transfer rate divided by the temperature difference between the heat source and heat sink,

$$k_{\text{eff}} = \frac{L_t}{R_t A_t} \tag{57.8}$$

where A_t is the overall cross-sectional area of the pipe (m²). Under normal operating conditions, the total thermal resistance is relatively small, making the temperature in the evaporator nearly equal to that in the condenser. Thus, the effective thermal conductivity of a heat pipe is very high (several orders of magnitude greater than aluminum or copper).

57.4 Application of Heat Pipes

Heat pipes have been applied to a wide variety of thermal processes and technologies. One of the first areas where heat pipes were widely used was in the aerospace industry, where heat pipes are used successfully in controlling the temperature of satellites, instruments, and space suits. Heat pipes also are applied in:

1. The electronics industry for cooling various devices (e.g., infrared sensors, parametric amplifiers)

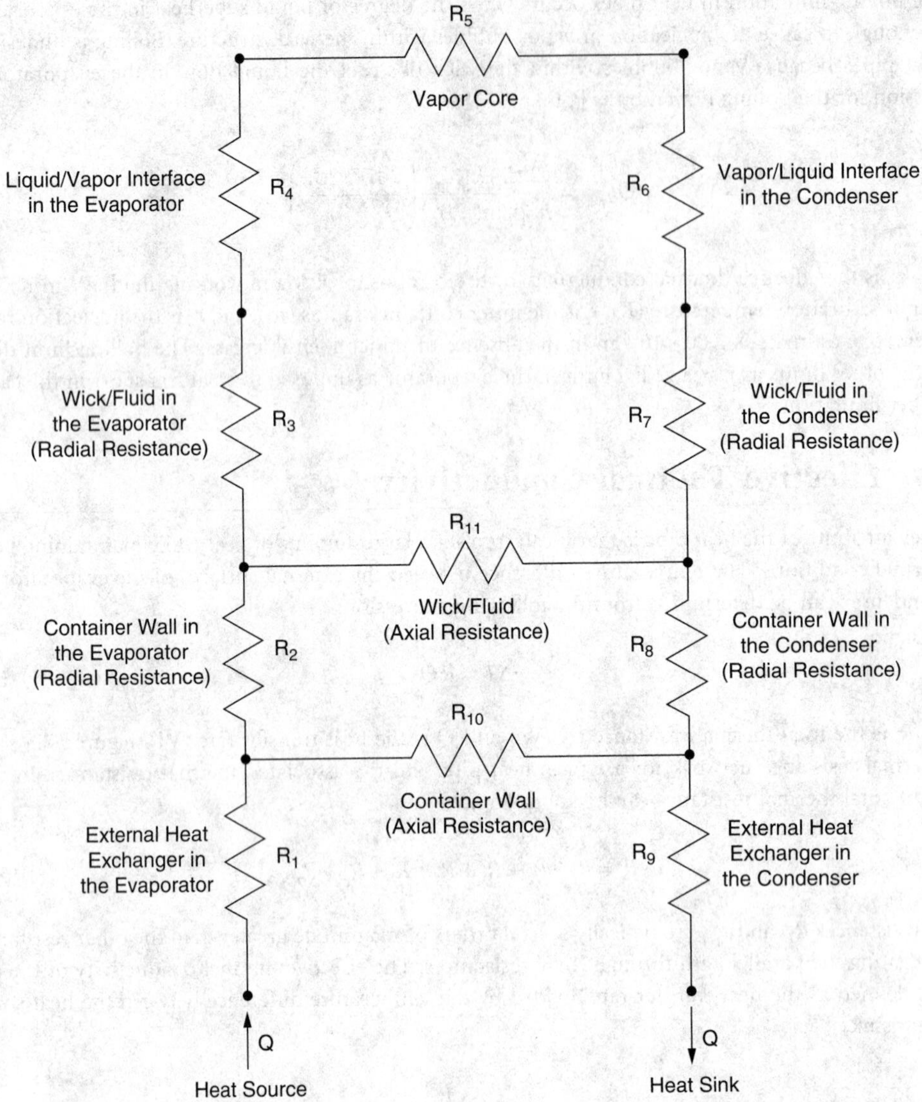

FIGURE 57.3 Thermal resistance network in a heat pipe.

2. Computers for cooling of processors
3. The medical field for surgery

Other cooling applications include:

1. Turbine blades, generators, and motors
2. Nuclear and isotope reactors
3. Heat collection from exhaust gases, and solar and geothermal energy
4. The Alaskan pipeline

In general, heat pipes have advantages over many traditional heat-exchange devices when:

1. Heat needs to be transferred nearly isothermally over relatively long distances.
2. Low weight and passive operation are essential.
3. Fast thermal-response times are required.
4. Low maintenance is mandatory.

Further advances in the heat pipe field include the development of the loop heat pipe by Maydanik and colleagues at the Russian Academy of Sciences. The loop heat pipe can transport even greater amounts of heat than standard heat pipes do as it separates the vapor passage from the liquid passage and reduces the overall size of the wick in the heat pipes. This optimizes the balance of small pores for capillary pumping and low frictional losses. Ochterbeck [2003] has further information on loop heat pipes.

Defining Terms

Capillary pressure — The pressure developed across a curved liquid–vapor interface, which is a function of the surface tension, the contact angle between the liquid and wick structure, and the pore size of the porous media.

Effective thermal conductivity — The heat transfer rate divided by the temperature difference between the evaporator and condenser outer surfaces. This relates the heat pipe to a conductivity which can be compared to conduction in a solid.

Heat transfer limitations — Limitations on the heat transfer capacity of a heat pipe which are imposed by different physical phenomena (e.g., vapor pressure, sonic, entrainment, capillary, and boiling).

Wettability — The ability of a liquid to spread over a surface. A wetting liquid spreads over a surface, whereas a nonwetting liquid forms droplets on a surface.

Wick — A porous material used to generate the capillary pressure that circulates the working fluid in a heat pipe.

References

Chi, S. W. 1976. *Heat Pipe Theory and Practice,* Hemisphere, Washington, DC.

Dunn, P. D. and Reay, D. A. 1994. *Heat Pipes,* 4th ed. Pergamon Press, Oxford, U.K.

Faghri, A., 1995. *Heat Pipe Science and Technology,* Taylor and Francis, Washington, DC.

Ochterbeck, J. M. 2003. Heat pipes. In *Handbook of Heat Transfer,* A. Bejan and A. Kraus (Eds.), John Wiley & Sons, New York.

Peterson, G., P., 1994. *An Introduction to Heat Pipes*, John Wiley & Sons, New York.

Further Information

Recent developments in heat pipe research and technology can be found in the proceedings from a number of technical conferences: (1) the International Heat Pipe Conferences, (2) the ASME IMECE, and (3) the AIAA Thermophysics Conferences. An additional book particularly strong in fundamental heat pipe theory is *The Principles of Heat Pipes* by Ivanovskii, Sorokin, and Yagodkin, available from Clarendon Press (U.K.). Further questions can be addressed to the author at jochter@clemson.edu.

[illegible]

Wettability — [illegible]

Peterson, G.P. 1994. [illegible] John Wiley & Sons, New York.

Peterson, G.P. [illegible]

Further Information

[illegible]

VIII

Separation Processes

58

Distillation

James R. Fair
University af Texas, Austin

Distillation is a method of separation that is based on the difference in composition between a boiling liquid mixture and the vapor formed from it. The composition difference is due to differing effective vapor pressures, or volatilities, of the components of the liquid mixture. When such a difference does not exist, as at an azeotropic point, separation by normal distillation is not possible. Distillation as normally practiced involves condensation of the vaporized material, usually in multiple vaporization/condensation operations. It thus differs from evaporation, which normally is applied to separate a liquid from a solid, but which can be applied to simple liquid concentration operations. In the chemical and petroleum industries, distillation represents a large fraction of plant investment, yet also is one of the largest consumers of energy among the several processing operations employed.

Distillation is the most widely used industrial method of separating liquid mixtures and is at the heart of the separation processes in many chemical and petroleum plants. The most elementary form of the method is simple **batch distillation**, in which the liquid is brought to boiling and the vapor formed is separated and condensed to form a product. In some cases, a normally gaseous mixture is condensed by using refrigeration and the resulting liquid is then amenable to distillation separation, as in the case of distillation of air. If the process is continuous with respect to feed and product flows, it is called *flash distillation.* If the feed mixture is available as an isolated batch of material, the process is a form of **batch distillation** and the compositions of the collected vapor and residual liquid are thus time dependent. The term *fractional distillation* (which may be contracted to **fractionation**) was originally applied to the collection of separate fractions of condensed vapor, each being segregated. Currently, the term is applied to distillation processes in general, where an effort is made to separate an original mixture into several components by means of distillation. When the vapors are enriched by contact with counterflowing liquid reflux, the process is called *rectification.* When fractional distillation is accomplished with a continuous feed of material and continuous removal of product fractions, the process is called **continuous distillation**. When steam is added to the vapors to reduce the partial pressures of the components to be separated, the term *steam distillation* is used.

Most distillations conducted commercially operate continuously, with a more volatile fraction recovered as **distillate** and a less volatile fraction recovered as **bottoms** (or residue). If a portion of the distillate

0-8493-1586-7/05/$0.00+$1.50

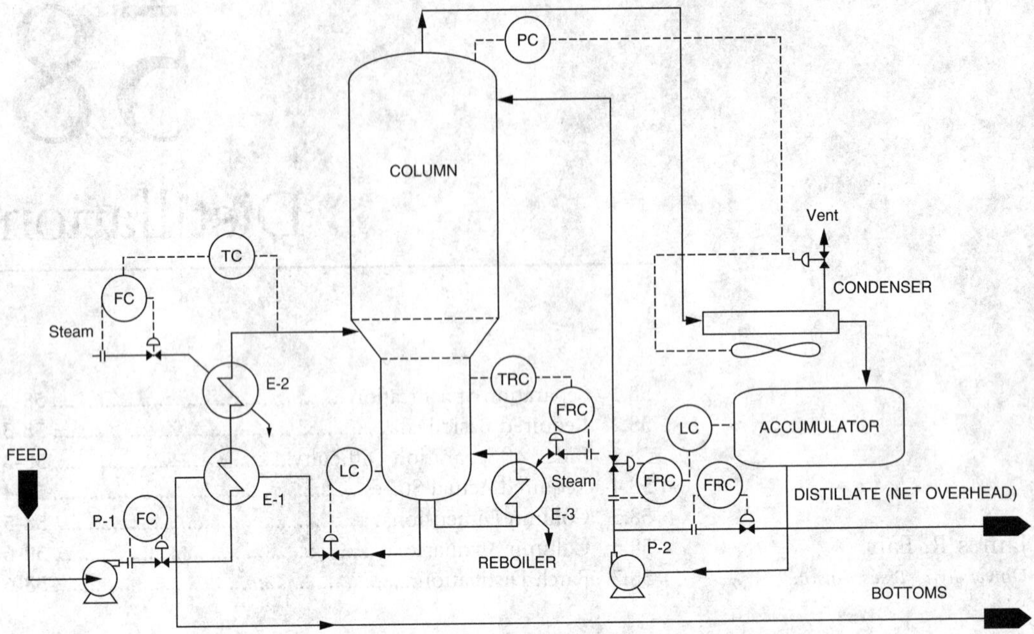

FIGURE 58.1 Flow diagram of a representative distillation system. For the case shown, the column has two diameters, commensurate with the changing flows of liquid and vapor in the column. Many columns have a single diameter. The condenser is air cooled, but water is also often used as a coolant.

is condensed and returned to the process to enrich the vapors, the liquid is called **reflux.** The apparatus in which the enrichment occurs is usually a vertical, cylindrical vessel called a still or *distillation column.* This apparatus normally contains internal devices for effecting intimate contact between rising vapor and descending liquid. The devices may be categorized as *plates* or *packings.* In some cases a batch distillation is preferred, where the **feed** mixture is placed in a stillpot and heat is added. The vaporized mixture is condensed and analyzed. Clearly, this method of operation is in a time-variant domain, unlike continuous distillation.

Distillation has been practiced in one form or another for centuries. It was of fundamental importance to the alchemists and has been in use for more than 2000 years. Because it is a process involving vaporization of a liquid, energy must be supplied if it is to function.

A representative continuous distillation system is shown in Figure 58.1. The design of such a system, or the analysis of an existing system, follows a fixed procedure. The steps in this procedure are described in the following sections, with primary emphasis on the distillation column. Such a procedure also provides an overall summary of the technology associated with distillation column design and operation.

58.1 Separation Specification

If a mixture is to be separated by distillation, it is crucial that the mixture composition be defined carefully. Serious problems can arise when some components of the mixture are disregarded or not known to be present. Next, the extent of required separation must be defined. Often there are two key **components** of the feed between which the separation specification can be related. The specification may deal with composition, recovery, or both. For example, for a benzene/toluene feed mixture, the specification may be a minimum of 95% of the benzene in the feed to be taken overhead with the distillate (the percent recovery), and the minimum purity of benzene in the distillate to be 99.5%. Simply stated, one must determine in advance the degree of separation desired.

58.2 Required Basic Data

A key parameter that is related to the ease (or difficulty) of making the separation is known as the **relative volatility.** This is a ratio of the effective vapor pressures of the components to be separated. It might be thought of as an index of the difficulty of making the separation. It is determined from the following relationship:

$$\alpha_{ij} = \frac{\gamma_i^L P_i^o}{\gamma_j^L P_j^o} \tag{58.1}$$

where components i and j are those between which the specification of separation is written (key components), γ^L is a thermodynamic parameter called the liquid phase activity coefficient, and P^o is the vapor pressure. In Equation (58.1), component i is more volatile than component j (i.e., i is more likely to be distilled overhead than is j). When liquid mixtures behave ideally (no excess energy effects when the mixture is formed), the activity coefficients have a value of 1.0. The vapor pressures of compounds are readily available from various handbooks. When the key components have a relative volatility of 1.10 or less, separation by distillation is difficult and quite expensive. For cases where α_{ij} is 1.50 or greater, the separation is relatively easy.

The determination of relative volatility involves the discipline of *solution thermodynamics*, and for distillation a sub-area is called vapor–liquid equilibrium (VLE). There are many approaches to determining VLE: direct measurement, correlation of the data of others, extrapolation of correlations beyond the area of measurement, and prediction of equilibrium data when no measurements have been made.

58.3 Index of Separation Difficulty

After the needed equilibrium data have been obtained, the next step is to determine the difficulty (or ease) of separation, using some index that is reliable and generally understood. The most used index is based on the required number of **theoretical stages.** For rigorous calculations, the stage count results from a detailed analysis of flow and composition changes throughout the column. For a simplified analysis, a basic parameter is the minimum number of theoretical stages:

$$N_{\min} = \frac{\ln\left[\left(y_i / y_j\right)^D \left(x_j / x_i\right)^B\right]}{\ln \alpha_{ij,avg}} \tag{58.2}$$

where i and j are the components between which the specified separation is to be made, y and x are mole fractions, D and B refer to distillate and bottoms. The value of N_{min} can serve as a general criterion of separation difficulty. An example material balance for a ternary separation is

Component	Feed (moles)	Distillate (moles)	Distillate (mole fraction)	Bottoms (moles)	Bottoms (mole fraction)
i	45.0	44.6	0.9911	0.4	0.0073
j	45.0	0.4	0.0089	44.6	0.8109
k	10.0	nil	nil	10.0	0.1818
	100.0	45.0	1.0000	55.0	1.0000

The key components i and j have a relative volatility of 1.40, and i is to have a minimum 99% recovery to the distillate, and a minimum purity of 99.0% in the distillate. Component k is relatively nonvolatile. The minimum theoretical stages are calculated from Equation 58.2:

$$N_{min} = \frac{\ln\left[\frac{0.9911}{0.0089} \times \frac{0.8109}{0.0073}\right]}{\ln 1.40} = 28.0 \text{ stages} \tag{58.3}$$

As a very rough rule of thumb, the number of theoretical stages at an optimum reflux rate will be about twice the minimum number of stages. In the example, about 60 total theoretical stages will be required. Equation (58.2) is suitable for making quick comparisons between the design conditions of overhead and bottoms compositions, and of relative volatility.

58.4 Required Actual Stages

The distillation column is a vertical, cylindrical vessel into which are placed contacting devices that cause the rising vapor and descending liquid to come into intimate contact. Such contact is necessary if the column is to make an efficient separation. As indicated in Figure 58.1, a return of some of the distillate product is necessary for the column to work efficiently. This flow stream is called the reflux. The ratio of this reflux flow to the flow of distillate is called the *reflux ratio.* The designer has the option of varying this ratio, and in so doing can influence the required number of theoretical stages, as shown in Figure 58.2. The asymptote for stages is N_{min} as described above. The asymptote for reflux ratio is the minimum reflux ratio R_{min}, a parameter related to the design compositions and the relative volatility. There is some optimum reflux ratio that relates to the economic conditions that exist for the design. Often the optimum reflux ratio is taken as about 1.3 times the minimum reflux ratio.

The actual number of stages required for the column is based on stage efficiency:

$$N_{act} = N_t / E_{oc} \tag{58.4}$$

where N_t, the number of theoretical stages, can be computed from

$$\frac{N_t - N_{min}}{N_t + 1} = 0.75 - 0.75\left(\frac{R - R_{min}}{R+1}\right)^{0.567} \tag{58.5}$$

In Equation 58.5, N_{min} and R_{min} are taken from a plot such as Figure 58.2 or, more likely, from rigorous calculations using modem computer programs. The reflux ratio R is a design variable, as mentioned

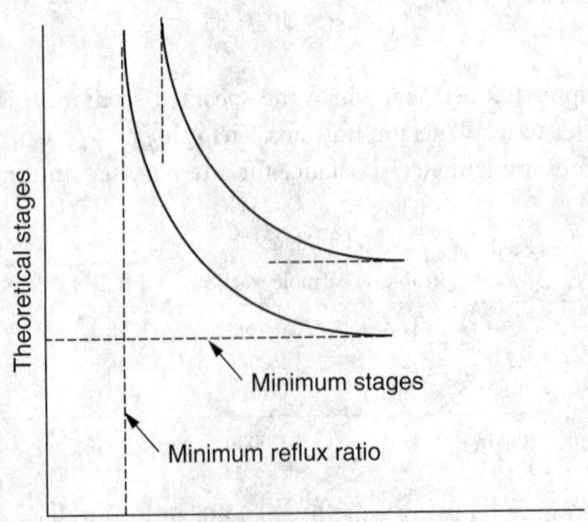

FIGURE 58.2 General relationship between stages and reflux. A given curve represents a specific separation requirement.

earlier. The value of the overall efficiency E_{oc} for use in Equation (58.4) must be obtained from experience or from predictive mathematical models. It is often in the range of 0.60 to 0.85 (60 to 85%). In the earlier example requiring about 60 theoretical stages, and for 85% efficiency, abut 70 actual stages (plates) would be used in the column.

58.5 Column Dimensions

The height of the column is a function of N_{act}, the number of actual stages required for the separation. The height of each stage is a function of the geometry of the contacting device that is used. If the device is represented by a flat, horizontal sheet metal "plate" containing perforations, then the vertical spacing of these plates will determine the required height of the column. For example, if 40 theoretical stages are needed, and the efficiency of such a plate is 80%, then 50 actual stages (plates) are required. If these plates must be spaced at 0.61 m (24 in.) intervals, then the column must be at least 30 m (98 ft) high to accommodate the vapor and liquid flows. Considering other needs for internal devices, the column might have an actual height of perhaps 35 m.

If, instead of plates, a type of internal device called *packing* is used, some equivalent dimension must be used. The required height is obtained from

$$Z = (N_t)(HETP) \tag{58.6}$$

where *HETP* is the height equivalent to a theoretical plate (stage), and its value usually is obtained from the manufacturer or vendor of the packing device. There are many different sizes and shapes of packing used commercially, and some of them allow prediction of HETP by rigorous models.

If the height of the column is determined from the needed stages, the diameter must be determined from the required flow rates of vapor and liquid in the column. These rates are based on the total amount of feed mixture to be processed, as well as the amount of reflux that is returned to the column. In turn, the device (plate or packing) selected for contacting must have the capacity to handle the required flow rates. A schematic diagram of the factors that influence column diameter is shown as Figure 58.3. The device must be able to handle the vapor flow without tending to carry the liquid upward. Likewise, the device must be able to handle the liquid flows without choking the column because of a buildup of liquid. Correlations are available for assigning numbers to the scales of Figure 58.3, and an optimum design will place the design condition well within the shaded area.

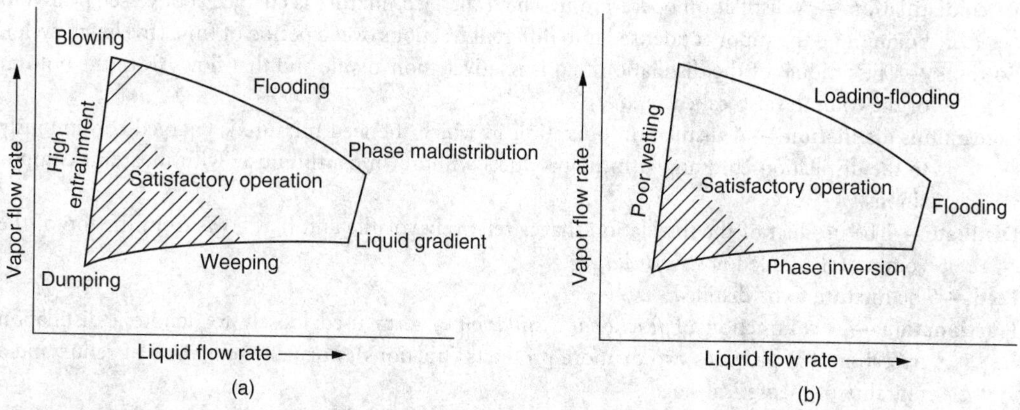

FIGURE 58.3 Generalized operating diagram for a distillation column, showing ranges of operation as a function of relative vapor and liquid flows. (a) Plate column. (b) Packed column.

58.6 Column Auxiliaries

Figure 58.1 shows that a distillation system contains more than just the column. Of the several auxiliary components, two heat exchangers are of particular importance — the reboiler and the condenser. The reboiler at the base receives the net heat input to the column (often in the form of condensing low-pressure steam) and establishes the amount of vapor to be handled by the column. Much of the heat added at the reboiler is removed in the overhead condenser, rejecting the heat directly to the atmosphere (as shown in Figure 58.1) or to a stream of cooling water. While the heat duties of the condenser and reboiler are often about the same, the overall heat balance depends also on the heat contents of the streams entering and leaving the column and on the amount of heat losses to the surroundings.

58.7 Batch Distillation

Although most commercial distillations are run continuously, batch distillation is the method of choice for certain applications. For this method, a charge, or batch, of the initial mixture is placed in a vessel, where it can be heated and distilled over a period of time. Compositions of the charge and product thus vary with time, which is not the case for continuous distillation. Particular cases where batch distillation may be preferred are:

1. Semiworks operations producing interim amounts of product in equipment that is used for multiple purposes
2. Distillations of specialty chemicals where contamination can be a problem (Batch equipment can be cleaned or sterilized between batch runs.)
3. Operations involving wide swings in feed compositions and product specifications, where batch operating conditions can be adjusted to meet the varying needs
4. Laboratory distillations where separability is being investigated without concern over the scale-up to commercial continuous operations

Batch distillations are generally more expensive than their continuous counterparts in terms of cost per unit of product. Close supervision and computer control are required, the equipment is more complex if several products are to be recovered, and total throughput is limited by the needs for changing operating modes and for recharging the system with feed material.

Defining Terms

Batch distillation — A distillation operation in which the feed mixture is charged to a vessel, heated to boiling, and the vapor condensed into different fractions over a period of time (the batch cycle).

Bottoms — The product of the distillation that is relatively nonvolatile and that flows from the bottom of the column (also called *residue*).

Continuous distillation — A distillation operation in which the feed mixture is charged continuously to the distillation column, with the products withdrawn continuously with invariant compositions.

Distillate — The product of the distillation that is relatively volatile and that flows from the top of the column (also called *net overhead*).

Feed — The mixture to be distilled.

Fractionation — A contraction of *fractional distillation,* a term used loosely to denote a distillation operation that provides two or more products (fractions), normally by means of reflux and a plurality of theoretical stages.

Reflux — Liquid that is derived from the distillate product and fed back to the top of the column for the purpose of enhancing the separability.

Relative volatility — A ratio of effective vapor pressures of the components of a mixture to be distilled. An important index of ease of separation.

Theoretical stage — A conceptual term which refers to the "ideal" situation in which the vapor leaving a defined section of a column is in thermodynamic equilibrium with the liquid leaving the same section.

References

Seader, J. D. 1997. Distillation. In *Perry's Chemical Engineers' Handbook,* 7th ed., R. H. Perry and D. W. Green, Eds., McGraw-Hill, New York. Section 13.

McCormick, J. E. and Roche, E. C. 1997. Continuous Distillation. In *Handbook of Separation Techniques for Chemical Engineers,* P. A. Schweitzer, Ed., McGraw-Hill, New York. Sections 1.1 and 1.2.

Further Information

Kister, H. Z. 1992. *Distillation-Design.* McGraw-Hill, New York.

Fair, J. R. 1987. Distillation. In *Handbook of Separation Process Technology.* R. W., Rousseau, Ed. John Wiley & Sons, New York.

Walas, S. M. 1985. *Phase Equilibria in Chemical Engineering.* Butterworths, Reading, MA.

59

Absorption and Stripping

James R. Fair
University of Texas, Austin

Absorption and stripping are two chemical process operations that normally are coupled in order to remove a minor component, the **solute**, from an incoming process gas stream and then recover that same component in a more concentrated form (see Figure 59.1). A carefully selected **solvent**, in which the solute is selectively soluble, is fed to the **absorber** (or "scrubber"), and the rich solvent is then fed to the **stripper,** where the solute is recovered. In some cases, absorbers are used separately when the **rich solvent** does not need further processing**.** Likewise, strippers can be used alone for separating a minor component from a liquid mixture. An example of the coupled system, other than the one shown in Figure 59.1, is the removal of carbon dioxide from natural gas by absorption in an aqueous amine solution, with subsequent recovery of high purity carbon dioxide by stripping at a lower pressure and a higher temperature. The gas is purified and a useful product results.

Examples of the separate systems include the absorption of ammonia in water to form ammonium hydroxide and the stripping of dissolved volatile organic compounds (VOCs) from contaminated groundwater using air as the stripping gas. A representative flow diagram of a separate stripping system is shown in Figure 59.2.

59.1 Basic Property Data

If an absorber is to be designed for efficient and economical service, it is critical to select the proper solvent. Attributes of a good solvent for coupled systems include availability, cost, stability, and volatility. Further, the solvent should be nonhazardous. But by far the most critical property is the solubility of the solute in the solvent. This is usually the first consideration. The equilibrium solubility of component i may be expressed in terms of a *Henry's Law coefficient* H_i:

$$H_i = \frac{\text{partial pressure of i in gas}}{\text{mole fraction of i in liquid}} = \frac{p_i}{x_i} \tag{59.1}$$

A related term, used both in absorption and stripping and in distillation, is the *equilibrium ratio* K_i:

0-8493-1586-7/05/$0.00+$1.50

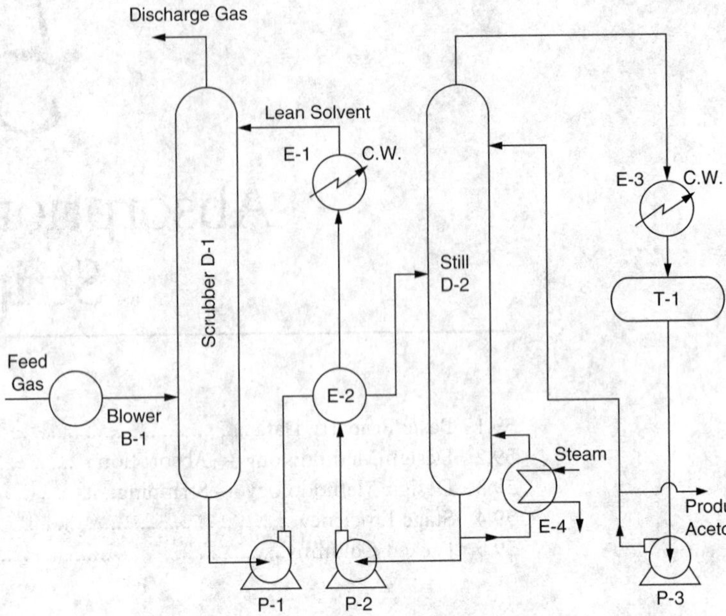

FIGURE 59.1 Coupled absorption-stripping system. Acetone is recovered from air using a water solvent. The acetone–water solution is then stripped to provide high purity acetone.

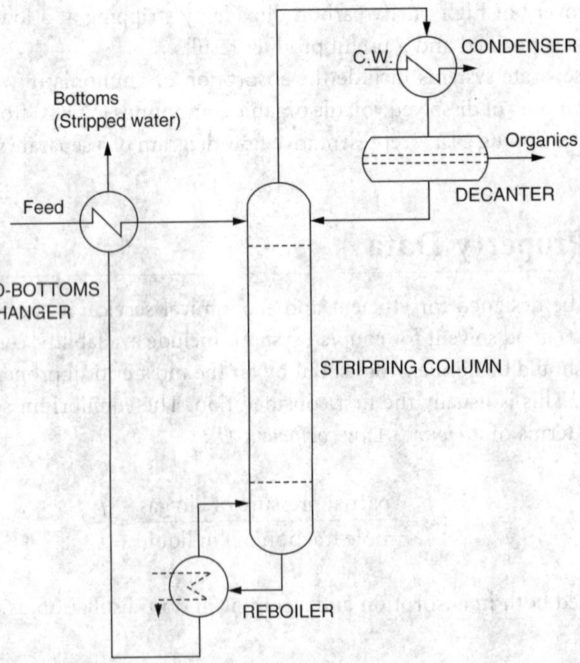

FIGURE 59.2 Flow diagram for the steam stripping of contaminants from process water. Since water is the stripped product, it can be produced *in situ* with a reboiler. Alternately, external steam can be used.

$$K_i = \frac{y_i}{x_i} \tag{59.2}$$

It is important to recognize that these terms are based on thermodynamic equilibrium. Usually, Dalton's Law is applicable, so that

$$H_i = K_i P \tag{59.3}$$

where P is the total pressure of the system. For the nonideal systems usually encountered in absorption and stripping, the K_i value can be expressed as follows:

$$K_i = \frac{\gamma_i\, p_i^{sat}}{P} \tag{59.4}$$

where γ_i is the liquid phase activity coefficient for i in the solvent, p^{sat} is the vapor pressure of i, and P is the total pressure.

Handbooks usually report solubilities as Henry's Law coefficients. However, design procedures usually employ the K value, easily obtained from H by Equation (59.3). When solubility data are not available for the conditions at hand, there is methodology for estimating values of the activity coefficient, one summary being in the book by Reid et al. Vapor pressure data are widely available.

Equation (59.1) through Equation (59.4) apply to a single solute. For dilute mixtures, equations for other solutes, if present, are considered to be independent of each other. For more concentrated mixtures, thermodynamic interactions must be determined, and effective values of K and H obtained.

59.2 Design Methodology — Absorption

The first step in design is to establish separation specifications. For example, a vent gas containing 2000 parts per million by volume (ppmv) of a pollutant must be scrubbed to a level of 200 ppmv in the exit gas. Or for a gaseous mixture which is a nonpollutant, the specification may deal with the fractional recovery (proportion absorbed) of that constituent in the rich solvent. For stripping, an example is the removal of dissolved contaminants in water to a level compatible with environmental regulations. *Fractional removal* and *limiting composition* are interchangeable in a given process, and the former will be used in the design equations given here.

In designing an absorber or stripper, one must ascertain some index of "difficulty of separation." The index to be used here is the number of required *theoretical stages.* An alternate criterion, often used for absorbers containing packing, is the required number of *transfer units.* Easy separations require few theoretical stages or transfer units. Very difficult separations may require many stages, perhaps 20 or more.

Absorption stages may be calculated from the so-called Kremser–Brown equation:

$$E_{Ai} = \frac{Y_{N+1} - Y_1}{Y_{N+1} - Y_o^*} = \frac{A_i^{N+1} - A_i}{A_i^{N+1} - 1} \tag{59.5}$$

where E_{ai} = fraction of i absorbed; Y_{N+1} = moles i per mole of solute-free gas entering; Y_1 = moles i per mole solute-free gas leaving the top of the absorber; Y_o^* = moles i per mole solute-free gas, determined as in equilibrium with the entering solvent composition; N = number of required theoretical stages; A_i = absorption factor for component $i = L/(V K_i)$; L = total moles of liquid (solvent + solute) flowing down the column; and V = total moles of vapor flowing up the column.

Note that in Equation (59.5), superscripts are exponents. If the amount absorbed is relatively low, as in most environmental applications, heat effects are minimal and the operation may be assumed to be isothermal. For this case, L and V are essentially constant.

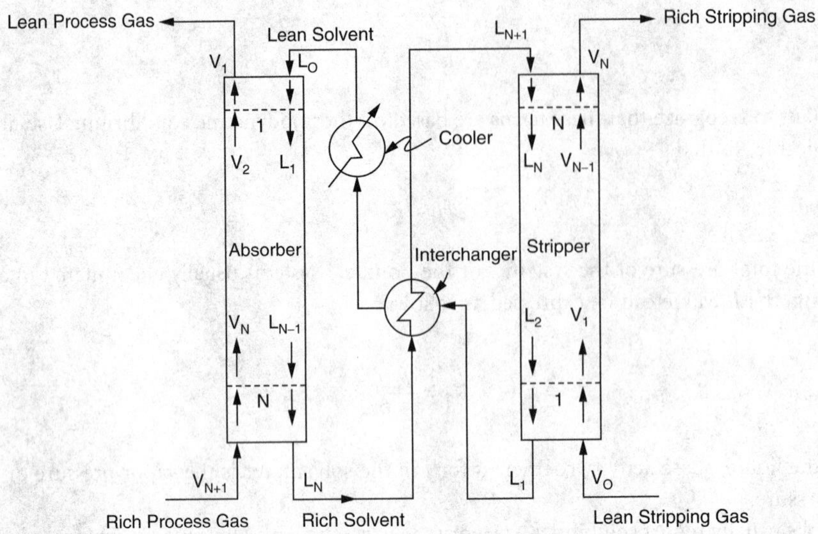

FIGURE 59.3 Nomenclature for use of the equations for absorbers or strippers. In order to maintain similarity of the design equations, for absorption the stages are numbered down from the top and for stripping the stages are numbered up from the bottom.

Equation (59.5) is a rating equation; it tells how much recovery can be obtained with a given number of stages *N*. A more useful equation is a rearrangement explicit in *N*:

$$N = \frac{\ln\left(\dfrac{E_{Ai} - A_i}{E_{Ai} - 1}\right)}{\ln A_i} - 1 \qquad (59.6)$$

The solubility effect is contained in the absorption factor A_i. If the entering solvent contains none of the solute (e.g., completely stripped out), then the term Y_o^* drops out of Equation (59.5). Note that the compositions are in terms of *mole ratios*; for very dilute mixtures, mole fractions y can be used. In Equation (59.5) and Equation (59.6), the stages are numbered down from the top. Figure 59.3 includes a guide to the nomenclature for both absorption and stripping.

59.3 Design Methodology — Stripping

For stripping, a liquid mixture is fed to the column and a stripping gas is used to remove one or more "key components" of the mixture. If the feed stream comes from a coupled absorber, then the composition is known, and the key components are likely to be those that were selectively absorbed. The stripping gas may be an external stream such as steam or air, or it may be generated by a reboiler as shown in Figure 59.1. In some respects, a reboiled stripper resembles a conventional distillation column and may even have a reflux stream.

If the stripper stages are numbered up from the bottom, Equation (59.5) and Equation (59.6) have exact counterparts:

$$E_{Si} = \frac{X_{N+1} - X_1}{X_{N+1} - X_o^*} = \frac{S_i^{N+1} - S_i}{S_i^{N+1} - 1} \qquad (59.7)$$

$$N = \frac{\ln\left(\frac{E_{Si} - S_i}{E_{Si} - 1}\right)}{\ln S_i} - 1 \tag{59.8}$$

where E_{Si} = fraction of i stripped; X_{N+1} = moles i per mole of solute-free liquid entering; X_1 = moles i per mole solute-free liquid leaving the bottom of the stripper; $X_o{}^*$ = moles i per mole solute-free liquid, determined as in equilibrium with the entering stripping gas composition; N = number of required theoretical stages; S_i = absorption factor for component $i = (V\,K_i)/L$; L = total moles of liquid (solvent + solute) flowing down the column; V = total moles of gas (stripping gas + solute) flowing up the column.

Superscripts in Equation (59.7) are exponents. If the amount stripped is relatively low, V and L are essentially constant. If the stripping gas contains no solute, X_o^* drops out of the equation.

For the methodology given here, absorption and stripping are direct counterparts. The absorption factor A is the reciprocal of the stripping factor S. A short example shows the use of Equation (59.8) [and Equation (59.6)]:

Example

A plant effluent liquid flows at 295,800 lb/h and contains 15 lb/h benzene. The benzene removal is to be 99.95%. Steam is to be the stripping gas. The stripping factor is 100. Using Equation (59.8),

$$N = \frac{\ln\left(\frac{0.9995 - 100}{0.9995 - 1}\right)}{\ln 100} - 1 = 1.65 \text{ theoretical stages}$$

Few theoretical stages are required because of the low solubility of benzene in water (high stripping factor). However, because contacting operations in stripping towers are inefficient, 20 or more actual stages may be required. This is discussed in the next subsection.

59.4 Stage Efficiency

Since gas–liquid contacting efficiency in absorbers and strippers is relatively inefficient, the required number of theoretical stages must be corrected to actual, or real, stages (plates, trays) by a stage efficiency:

$$\text{No. actual plates (trays)} = \frac{\text{No. theoretical stages}}{\text{Stage efficiency}} = \frac{N}{E_{oc}} \tag{59.9}$$

where E_{oc} is an overall column efficiency, dependent on the mass transfer characteristics of the system and the fluid mechanics of the counterflowing phases. A great deal of theory and practice is available for the estimation of E_{oc}, as described in Perry's *Chemical Engineers' Handbook.* A low solubility invariably leads to a low efficiency. For approximate work, typical values may be used:

Absorbers $E_{oc} = 0.30 = 30\%$
Strippers $E_{oc} = 0.20 = 20\%$.

The required height of the contacting zone is

$$Z = \frac{N \times TS}{E_{oc}} \tag{59.10}$$

where TS is the height spacing between actual trays.

59.5 Packed Columns

One may infer that the preceding material applies only to absorbers or strippers with trays (valve, sieve, etc.) but this is not the case. Packed columns do not contain such devices and operate in a continuum of gas and liquid flows. A more fundamental approach to estimating the required height is through the use of transfer units. There is a mathematical connection between theoretical stages and transfer units, expressed approximately as:

$$N_{og} = N\left(\frac{\ln S}{S-1}\right) \tag{59.11}$$

where N_{og} = number of overall transfer units required (gas phase basis); N = number of theoretical stages required; S = stripping factor $(V K)/L$, which apples to absorbers as well as strippers.

The objective is to determine the height of packing required for a given separation. This height is calculated by:

$$Z = (\text{HETP})(\text{N}) \tag{59.12}$$

or

$$Z = (H_{og})(N_{og}) \tag{59.12a}$$

where the height Z is expressed in the same units (ft, m) as *HETP* and H_{og}.

Values of *HETP* and H_{og} differ with the type and size of packing used. Approximate values can be obtained from vendors, but for more rigor, special mass transfer equations must be used. A discussion of the values together with the fundamental methods for calculating them may be found in Chapter 14 of Perry's *Chemical Engineers' Handbook*, as well as from many other sources.

Defining Terms

Absorption — Process for removing a minor component (or solute) from a gas stream using a liquid solvent.

Absorber (or absorption tower) — Usually a metal column containing packing materials or plates and designed to enhance gas–liquid contacting efficiency.

Rich solvent — Solvent + solute flowing from the absorber.

Solute — Component to be removed from the gas entering the absorber or from the liquid entering the stripper.

Solvent — Liquid employed for removing solute from the incoming solute-rich gas stream.

Stripper (or stripping tower) — Usually, a metal column containing packing materials or plates and designed to enhance gas–liquid contacting efficiency.

Stripping — Process for removing relatively insoluble solutes from a liquid mixture.

References

Fair, J. R. 1997. Gas absorption and gas–liquid system design. In *Perry's Chemical Engineers' Handbook*, 7th ed., R. H. Perry and D. W. Green, Eds. McGraw-Hill, New York. Section 14.

Reid, R. C., Prausnitz, J. M., and Poling, B. E. 1987. Fluid phase equilibria in multicomponent systems. In *The Properties of Gases and Liquids,* 4th ed., McGraw-Hill, New York. pp. 314–332.

Further Information

Palmer, D. A. 1987. *Handbook of Applied Thermodynamics.* CRC, Boca Raton, FL.

Sherwood, T. K., Pigford, R. L., and Wilke, C. R. 1975. *Mass Transfer.* McGraw-Hill, New York.

60

Extraction

Vincent Van Brunt
University of South Carolina

If a mixture cannot be easily separated using a direct separation such as evaporation or distillation, alternative indirect separation processes are considered. Extraction is an indirect separation that relies on the ease of separating a chemical from a solvent compared to that from its original feed.

As an example, consider the separation of acetic acid from water. Although the acid boils 18 degrees higher than water does, secondary bonding effects in the liquid phase and nonideal chemical effects in the vapor phase reduce the relative volatility of the water to acetic acid to approximately 1.1. Recovery of the acid from water requires that all the water be vaporized and be in the distillate. The low relative volatility implies that a distillation column will have a high reflux ratio and a large diameter. The combined effect is to increase both the operating and capital costs for distillation.

However, acetic acid can be recovered from dilute aqueous solutions using extraction. Several solvents are selective for acetic acid over water. A simplified extraction operation is shown in Figure 60.1. The denser aqueous *feed* flows down the extractor and countercurrently contacts the less dense *solvent* flowing up. The solvent-rich *extract* leaves the top. It contains more acid than water. The water-rich *raffinate* stream leaves the bottom with reduced acid content. The extractor is followed by a distillation column that simultaneously purifies the solvent and recovers the solute from it. Figure 60.1(a) shows a process with an ideal separation from a more volatile solvent such as ethyl or isopropyl acetate, whereas Figure 60.1(b) shows a configuration with a less volatile solvent such as diisobutyl ketone. Finally, Figure 60.1(c) shows a process configuration for acetic acid recovery with the additional column and separators needed for using a more volatile solvent. The stripping column is needed to reduce solvent loss in the raffinate.

Both the capital and the operating costs for recovering acetic acid by the extraction route are less than those for distillation. Choosing between a solvent that is more or less volatile than acetic acid depends on several factors, including relative **selectivity** for acetic acid versus water, relative capacity for acetic acid called **loading**, and relative volatility versus that of acetic acid. The purity of acid required in the product and the initial composition of acid fed to the process greatly influence solvent selection. In general, as the composition of the acid in the aqueous feed diminishes, the quantity of solvent needed for extraction will increase. This, in turn, will necessitate more solvent vaporization in the recovery step. A way around this increase in recovery cost for a more volatile solvent is to switch to a less volatile one. The acid will then become the distillate product in the recovery step, reducing the energy and operating costs for its recovery. Comparative costs are related to the entire process and not dominated by the extraction step. In fact, it usually represents only a small fraction of the overall expenses.

0-8493-1586-7/05/$0.00+$1.50

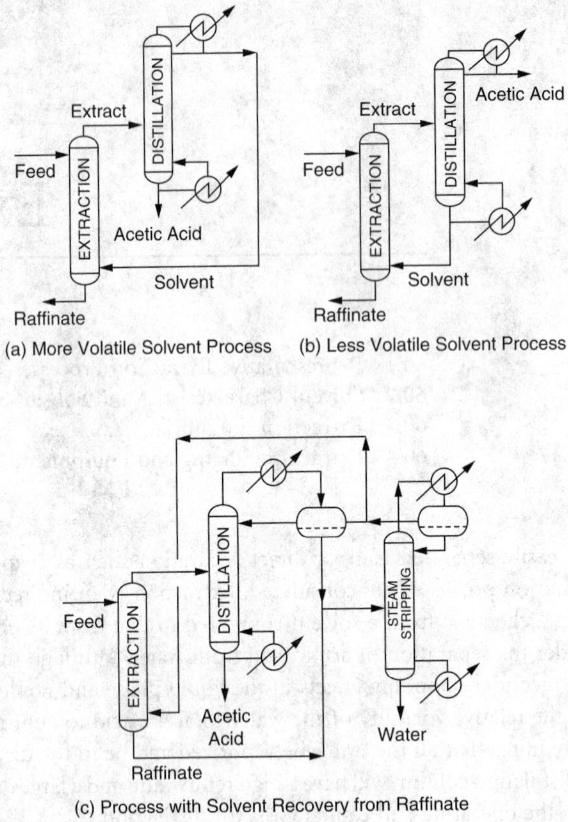

FIGURE 60.1 Acetic acid extraction processes.

The example shows that extraction needs to be considered in the context of an entire separation process. It is justified only when three factors are met. First, the solute must be easier to separate from the new solvent than the original. Second, the solvent must be able to be regenerated and recycled. And third, the solvent losses must be insignificant to the overall process. Thus, extraction is an indirect separation process that requires its evaluation and consideration with a second process step to regenerate the solvent and a third process step to recover the solute from it. In the example already given, even though the solute recovery and solvent regeneration steps were combined, still another process step is needed to reduce solvent loss in the raffinate.

60.1 Representative Extraction Processes

Extraction is used for purifying both inorganic and organic chemicals; see [Rydberg et al., 1992] and [Blumberg, 1988]. Examples of industrial uses are shown in Table 60.1 through Table 60.4. The inorganic applications are represented by purification of phosphoric acid and recovery of metals from ore leach solutions. In the latter case the metal salts are nonvolatile and usually in the presence of impurities that prevent direct recovery via, for example, electro-winning. Extraction permits the desired metal to be isolated and then recovered. The solvent for hydrometallurgical applications often relies on a specific chemistry. The examples chosen represent a sampling of the chemistries available [Sekine and Hasegawa, 1977].

The organic applications include recovering heat-sensitive chemicals such as penicillin, antibiotics, and vitamins, as well as separating close boiling mixtures. Of particular note is the separation of aromatics from aliphatics, for which there are several competing processes.

TABLE 60.1 Representative Inorganic Extraction Systems[a]

Separation	Solvent/Extractant Type	Extractant
Cobalt/nickel	Phosphoric acid	Di-2-ethyl hexyl phosphoric acid (D2EHPA)
Copper	Acid chelating — hydroxyime	2-hydroxyl 5-nonyl benzo phenone oxime (LIX65N)
Uranium	Anion exchanger — tertiary amine	n-trioctyl amine (Alamine 336)
Vanadium	Anion exchanger — quaternary ammonium salt	n-trioctylethylammonium chloride (Aliquat 336)
Zirconium/hafnium	Solvating — phosphoric acid ester	n-tributyl phosphate (TBP)
Actinides — fuel reprocessing	Solvating — phosphoric acid ester	TBP in dodecane
Phosphoric acid	Solvating — alcohols	Mixtures of butanol and pentanol
Actinides	Aqueous biphase, polyethylene glycol (PEG) rich	Crown ethers

[a] All feeds are aqueous acid solutions; all diluents are kerosene, except as noted.
Source: Lo, T. C., Baird, M. H. I., and Hanson, C. 1983. *Handbook of Solvent Extraction,* Wiley-Interscience, New York.

TABLE 60.2 Representative Organic Extraction Systems

Separation	Solvent
Aromatic/lube oil	Liquid sulfur dioxide
	Furfural
Aromatic/aliphatic	Tetrahydrothiophene-1, 1-dioxide (Sulfolane)
	Triethylene glycol–water, tetraethylene glycol–water
	N-methyl 2-pyrrolidone (MPYROL, NMP)–water
Caprolactam	Toluene, benzene
Penicillin, antibiotics	Esters such as isoamyl acetate or butyl acetate
Lipids	Methanol-water
Acetic acid	Low-molecular-weight esters such as ethyl or isopropyl acetate, n-butyl acetate
Caffeine	Supercritical carbon dioxide
Flavors and aromas	Supercritical carbon dioxide
Enzymes	Aqueous biphase, polyethylene glycol rich

Source: Lo, T. C., Baird, M. H. I., and Hanson, C. 1983. *Handbook of Solvent Extraction,* Wiley-Interscience, New York.

TABLE 60.3 Representative Dual-Solvent Extraction Systems

	Solvent 1	Solvent 2
	Inorganic	
Zinc/iron	Triisooctyl amine (TIOA)	D2EHPA
	Water strip	Sulfuric acid strip
	Organic	
Lube oil	Mix of phenol + cresol	Propane
Aromatic/aliphatic	Dimetyl sulfoxide (DMSO)–water	Paraffin

Source: Lo, T. C., Baird, M. H. I., and Hanson, C. 1983. *Handbook of Solvent Extraction,* Wiley-Interscience, New York.

Supercritical carbon dioxide is used for the recovery of essential oils, flavors, and aromas and for decaffeination [McHugh and Krukonis, 1986].

Aqueous biphase systems have a second, stable, less dense, aqueous phase formed by adding polyethylene glycol (PEG) and an inorganic salt such as ammonium sulfate. Both inorganic and organic processes are being developed that use water-soluble complexants to provide selectivity.

Figure 60.2 shows the key steps in a hydrometallurgical fractional **extraction cycle.** In this example zirconium is recovered and separated from hafnium. As in a distillation column, the feed enters between a rectifying and stripping cascade. The **extraction cascade** below the feed recovers zirconium from it.

TABLE 60.4 Representative Extraction–Reaction Systems

	Reaction
	Inorganic
Uranium/plutonium	Plutonium reduction/partitioning with hydroxylamine nitrate (HAN)
Potassium nitrate	Potassium chloride + nitric acid → potassium nitrate + hydrochloric acid
	Organic
p-cresol/m-cresol	m-cresol partitioning using NaOH to form an aqueous soluble salt
m-xylene/xylenes	Preferential extraction and isomerization of m-xylene with HF-BF_3
	Irreversible Reactions
Aromatic nitration	Use of aqueous nitric and sulfuric acids to nitrate toluene

Sources: Lo, T. C., Baird, M. H. I., and Hanson, C. 1983. *Handbook of Solvent Extraction,* Wiley-Interscience, New York, also see Van Brunt, V. and Kanel, J. S., in Kulprathipanja, S. 2002, *Reactive Separation Processes,* Taylor and Francis, New York.

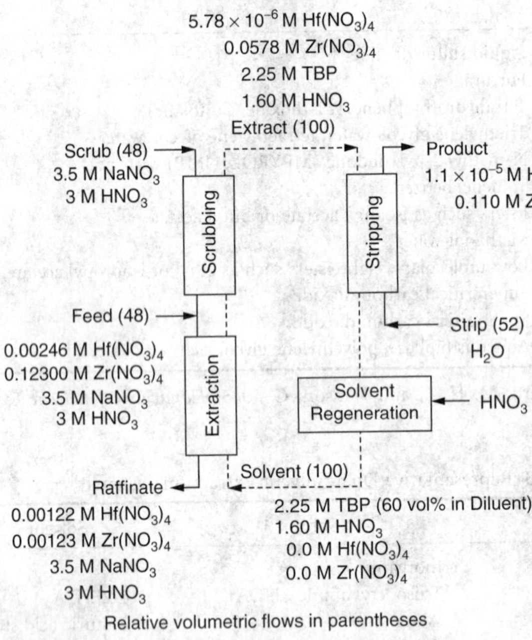

FIGURE 60.2 Zirconium–hafnium extraction cycle. (Adapted from Benedict, M. et al. 1981. *Nuclear Chemical Engineering,* 2nd ed. McGraw-Hill, New York.)

Each additional extraction stage improves the zirconium fractional recovery from the feed. Leaving the extraction cascade, the loaded solvent also contains coextracted impurities, including hafnium. As in distillation, the upper **scrubbing cascade** enhances the purity of the zirconium product by back-extracting hafnium (and other contaminants) from the solvent. Each additional scrubbing stage further decontaminates the zirconium by reducing the solvent's hafnium content.

In distillation only the reflux flow rate and its temperature can be adjusted to modify the overhead purity. In the extraction analog the *scrub* flow rate, pH, chemistry, and even temperature can all be independently adjusted to purify the zirconium. Note the presence of the nonextracted sodium nitrate in both the scrub and the feed. This **salting agent** permits an independent adjustment of the common anion, nitrate, in addition to a pH adjustment by changing the nitric acid content. The zirconium is recovered from the solvent by contacting it with an aqueous *strip* solution. Not shown are zirconium recovery from the strip solution by precipitation and the fate of the acid that is recovered and recycled.

Multiple extraction steps are used in several processes to enhance loading and to partition species extracted into the first solvent. Representative inorganic and organic mixtures are shown in Table 60.3. The zinc purification from spent electrolyte acid solution first separates zinc chloride using triisooctyl amine (TIOA). The first solvent is stripped with water before contact with the di-2-ethylhexylphosphoric acid (D2EHPA), which further purifies the zinc. Extraction of rare earths from a leach liquor is a sequential use of extraction cycles. In organic systems the second solvent may be used for solvent regeneration to lower process energy costs or to aid in solute and solvent recovery.

Both reversible and irreversible reactions are performed in the presence of two liquid phases. Extraction and reaction are used for fine separations such as isomer purification and for actinide partitioning. Isomer purification processes exploit a reversible reaction in an aqueous phase to enhance the selectivity between the isomers. Likewise, the plutonium reduction uses an irreversible reaction to effect a similar result by significantly reducing its organic phase solubility. Some representative systems are shown in Table 60.4 and several other irreversible reaction systems are tabulated in Van Brunt and Kanel.

Highly exothermic irreversible reactions such as nitrations, oxidations, and sulfonations are readily performed in liquid–liquid two-phase reactors; the second liquid phase provides a heat sink enabling reaction control. Reactive extraction needs to be considered: when reactants are immiscible, when the product undergoes an undesirable further reaction in the original reaction phase but is itself soluble in a second phase, when the products are immiscible with the reactants or catalyst, and when the product equilibrium can be enhanced by the presence of a second liquid-phase product sink. Many industrial products are produced using extraction–reaction systems; see Van Brunt and Kanel.

60.2 Solvent Characteristics and Solvent Screening

Since the solvent is present in the entire process, each of its characteristics and properties must be weighed to consider the net effect. The most significant considerations are shown in Table 60.5. The first grouping examines the solvent's relative process compatibility. A solvent or class of solvents that, even in minute quantities, reduces the value or marketability of the solute can be eliminated from consideration. Solute selectivity and loading ability must be balanced against ease of reversing and unloading it, recovering the solvent, and reducing the solvent losses.

TABLE 60.5 Desirable Solvent Characteristics

Process compatibility	High solute selectivity
	Regeneration capability
	High solute distribution coefficient
	High solute loading
	Low raffinate solubility
	Solute compatibility
Processing and equipment	Low viscosity
	Density different from the feed (>2%, preferably >5%)
	Moderate interfacial tension
	Reduced tendency to form a third phase
	Low corrosivity
Safety and environmental	Low flammability, flash point
	Low toxicity
	Low environmental impact, both low volatility and low effect or raffinate disposal costs
Overall	Low cost and availability

Adapted from Robbins, L. 1983. In *Perry's Chemical Engineers' Handbook,* 6th ed., ed. R. H. Perry and D. Green. McGraw-Hill, New York; Cusack, R. W., Fremeaux, P., and Glatz, D. 1991. A fresh look at liquid-liquid extraction. *Chem. Eng.* February.

Once a solvent has met these tests, it must pass those in the second grouping that affect processing equipment design and specification. Since extraction relies on intermittent dispersion and coalescence, desirable processing characteristics are those that enhance settling and coalescence, such as density difference with the feed and moderate interfacial tension. In general, a solvent with a lower viscosity is preferred. Likewise, the preferred solvent is less corrosive than the feed solution. Solvent chemistry can be altered to improve processing behavior.

The last grouping covers safety and environmental concerns. Safety issues such as toxicity and flammability may dominate over all others. If the raffinate or any other process stream is sent to waste treatment facilities, the solvent must be compatible with them or have low disposal costs. Finally, the preferred solvent is a readily available and low-cost commercial product.

60.3 Extraction Equilibria

Extraction equilibria affect the initial solvent screening. However, since many solvents contain more than one component, conclusions drawn from three component systems may not readily translate into the selection of a practical one. Almost always, experimental testing with the actual feed solution is needed. It is common practice to consider a solvent as having three parts, each piece contributing to desirable solvent properties. The **diluent** provides the bulk properties, such as viscosity and density. The **extractant** provides the active reversible solvent–solute interactions. Its efficiency affects both selectivity and loading. The **modifier** improves the interfacial tension and affects third-phase formation. An effective modifier will either increase the interfacial tension, thereby preventing or reducing emulsion stability or, conversely, it may be chosen to reduce the interfacial tension to enhance dispersion and interfacial area.

Organic diluents are usually petroleum fractions of varying naphthenic, aromatic, and aliphatic composition. Water may be added to aqueous soluble solvents as a diluent. Common modifiers are 2-ethylhexanol, isodecanol, tributyl phosphate and n-nonyl phenol. Ritcey and Ashbrook [1984] discusses their behavior. Extractant chemistry is discussed in Sekine and Hasegawa [1977], Alegret [1988], and Marcus and Kertes [1969].

For organic systems the thermodynamics of extraction equilibria can be used to screen potential solvents. The **distribution coefficient** is the ratio of a solute's mole fraction in the two phases. Here y refers to the organic, or less dense phase, mole fraction and x refers to the aqueous, or more dense phase, mole fraction.

$$K_i = y_i / x_i \tag{60.1}$$

For two chemical constituents the selectivity, β, can be expressed as

$$\beta_{ij} = K_i / K_j = (y_i / x_i)/(y_j / x_j) \tag{60.2}$$

Here the selectivity is written to evaluate the relative concentrations of i and j in the solvent phase, designated S, versus that in the feed phase, designated F. At equilibrium the chemical potential for each species is the same in each phase. In terms of activities this becomes

$$\gamma_i^F x_i = \gamma_i^S y_i \tag{60.3}$$

where γ is the activity coefficient. Using Equation (60.3), the distribution coefficient can be expressed as

$$K_i = \gamma_i^F / \gamma_i^S \approx \gamma_i^{\infty F} / \gamma_i^{\infty S} \tag{60.4}$$

where the ∞ indicates infinite dilution. Likewise, the selectivity becomes

TABLE 60.6 Qualitative Solvent Screening

Group	Solute	Solvent								
		1	2	3	4	5	6	7	8	9
1	Acid, aromatic OH, for example, phenol	0	–	–	–	–	0	+	+	+
2	Paraffinic OH (alcohol), water, amide or imide with active H	–	0	+	+	+	+	+	+	+
3	Ketone, aromatic nitrate, tertiary amine, pyridine, sulfone, phosphine oxide, or trialkyl phosphate	–	+	0	+	+	–	0	+	+
4	Ester, aldehyde, carbonate, nitrite or nitrate, phosphate, amide without active H; intermolecular bonding, for example, o-nitrophenol	–	+	+	0	+	–	+	+	+
5	Ether, oxide, sulfide, sulfoxide, primary, or secondary amine or imine	–	+	+	+	0	–	0	+	+
6	Multihalo-paraffin with active H	0	+	–	–	–	0	0	+	0
7	Aromatic, halogenated aromatic, olefin	+	+	0	+	0	0	0	0	0
8	Paraffin	+	+	+	+	+	+	0	0	0
9	Mono-halogenated paraffin or olefin	+	+	+	+	+	0	0	+	0

Notes: The + sign means that the solvent tends to raise the activity coefficient of the solute in a row group; the – sign means that the solvent tends to lower the activity coefficient of the solute in a row group. A 0 indicates no appreciable effect. Solvation is expected for negative group interactions. Potential solvents are those that lower the activity, that is, have minus signs.

Source: Cusack, R. W., Fremeaux, P., and Glatz, D. 1991. A fresh look at liquid-liquid extraction. *Chem. Eng.* February.

$$\beta_{ij} = \frac{\gamma_i^F/\gamma_i^S}{\gamma_j^F/\gamma_j^S} = \left(\frac{\gamma_i}{\gamma_j}\right)^F \left(\frac{\gamma_j}{\gamma_i}\right)^S \tag{60.5}$$

Since the composition of the species in the feed is known, the ratio of activity coefficients at the feed composition can be determined using one of the nonideal models, such as Van Laar, NRTL, UNIQUAC, or UNIFAC. The reader is referred to Sorensen and Arlt [1980] for more details. Examining Equation (60.5), one sees that the solvent selectivity for species i and j increases if the solvent phase activity coefficient for i is less than that for j. A preferred solvent would create solvent–solute interactions for species i, which reduces its activity coefficient below 1.0 — that is, to have negative deviations from ideality. Furthermore, the ideal solvent would preferentially reject species j and have its activity coefficient greater than 1.0 — that is, have positive deviations from ideality.

These effects can be used to identify potential solvent classes. Table 60.6 shows the expected deviations from ideality for solute–solvent pairs. For the acetic acid–water mixture the solute–solvent pairs that have negative deviations from ideality for the acid and no deviation or positive deviation for water are groups 2 through 5. Suitable solvents will probably come from those classes. In fact, solvents from those classes have all been used in practice or research. See C. J. King's chapter in Lo et al. [1983] for more details.

For the separation of benzene from n-hexane, Table 60.6 still provides some guidance. Although no negative deviations from ideality are shown, groups 3, 5, and 6 show positive deviations for the aliphatics (n-hexane) and 0 deviations for aromatics (benzene). This implies that solvents from those classes will probably be selective for benzene. The table also indicates that a solvent that is preferential for the aliphatics (n-hexane) over the aromatics (benzene) cannot be identified readily.

Another refinement for solvent screening is to use infinite dilution activity coefficients. Equation 60.5 can be approximated as

$$\beta_{ij}^{\infty} \cong \left(\frac{\gamma_i}{\gamma_j}\right)^F \left(\frac{\gamma_j^{\infty}}{\gamma_i^{\infty}}\right)^S \approx \left(\frac{\gamma_j^{\infty}}{\gamma_i^{\infty}}\right)^S \tag{60.6}$$

TABLE 60.7 Infinite Dilution Selectivity of Benzene–n-Hexane

Solvent	Table 60.6 Group	$\beta^{\infty}_{BH} \frac{\gamma^{\infty}_{n\text{-hexane}}}{\gamma^{\infty}_{benzene}}$
Sulfolane	5,5	30.5
Dimethylsulfoxide	5	22.7
Diethylene glycol	1,5	15.4
Triethylene glycol	1,5	18.3
Propylene carbonate	4	13.7
Dimethylformamide	2	12.5
n-methylpyrrolidone	3	12.5
Acetonitrile	3	9.4
Succinonitrile	3	46.8
g-butyrolacetone	3	19.5
Aniline	5	11.2
Dichloroacetic acid	1	6.1

Source: Modified from Tiegs, D. et al. 1986. *Activity Coefficients at Infinite Dilution. C1-C9*, DECHEMA Chemistry Data Series, vol. IX, part 1. DECHEMA, Frankfurt.

Here one neglects the activity coefficient ratio in the feed solvent as a given quantity and focuses attention on the ratio for the solutes to be separated in potential solvents. Further, the ratio is approximated by the ratio at infinite dilution — that is, for each of the solutes at infinite dilution in the solvent. Again, one sees that the selectivity of the solvent for i over j is *inversely* proportional to the ratio of their activity coefficients. If one uses infinite dilution activity coefficient data for common solvents from those groups suggested from Table 60.6, one obtains Table 60.7. The high values obtained for the industrial solvents are striking. Also, the difference between the two nitriles indicates an effect of carbon number. There are generally lower values for solvents having groups not identified using the qualitative screening table. Prediction of infinite dilution activity coefficients for solvent screening may also be performed using the MOSCED model and refinements to it; see Park and Carr [1987] and Hait et al. [1993].

60.4 Extraction Staging and Equipment

Procedures for extraction equilibrium-stage calculations are given in standard references, for example, Treybal [1963] and Lo et al. [1983]. For dilute systems, staging can be calculated by assuming a constant selectivity and constant solvent-to-feed ratio. An extraction factor — equivalent to a stripping factor and equal to the ratio of equilibrium to operating lines — can be used to calculate the number of equilibrium stages using the Kremser equation. For concentrated systems, neither the operating nor the equilibrium lines are of constant slope, and calculation procedures must account for the change in the solvent-to-feed ratio.

Although equilibrium-stage calculations are suitable for costing and comparing processes, detailed design *requires* an experimental evaluation of extractor performance. Because extraction uses phases that are the same order of magnitude in density, column performance is reduced by dispersion (also called *axial mixing* or *backmixing*). Dispersion will reduce extraction efficiency and can be approximated; see Lo et al. [1983] and Godfrey and Slater [1994]. However, it is extremely equipment- and chemistry-dependent.

Actual equipment sizing is tied to the physical properties of the specific extraction system being evaluated. Unlike with distillation, many different types of equipment are used. Logic diagrams for specifying extractors are given in standard texts such as Ladda and Degaleesan [1978], Lo et al. [1983], and Godfrey and Slater [1994]. Figure 60.3 presents a simplified version of the equipment decision tree [Ladda and Degaleesan, 1973]. It is strongly recommended that pilot testing be performed to determine both sizing needs and operating limits for new processes. Pilot campaigns with actual process chemicals

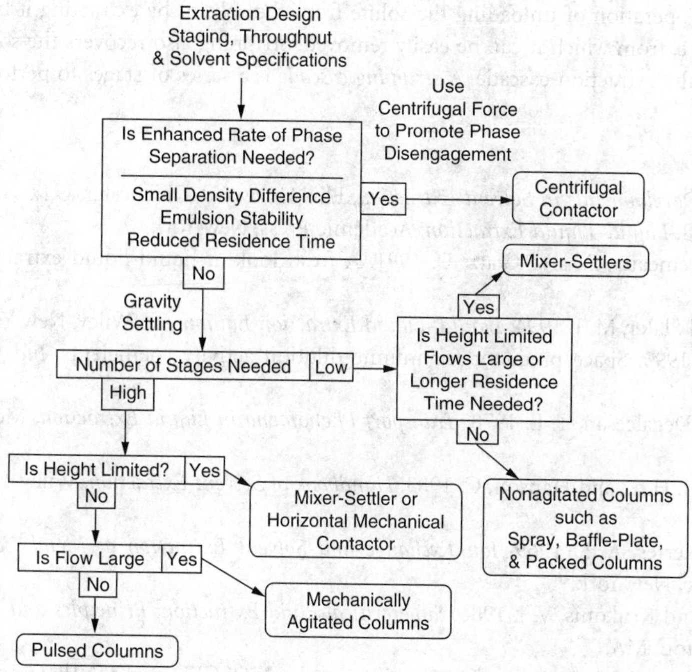

FIGURE 60.3 Equipment type selection decision tree. (Adapted from Ladda, G. S. and Degaleesan, T. E. 1978. *Transport Phenomena in Liquid Extraction*, McGraw-Hill, New York.)

will enable understanding of real processing behavior — for example, observation of interfacial solids buildup or changes in emulsion stability with degree of agitation.

Defining Terms

Diluent — The bulk constituent of a solvent, primarily used to improve solvent properties that affect processing and equipment choice.

Distribution coefficient (*K*) — The ratio of the mole fraction of a chemical species in the solvent phase divided by the mole fraction of the same chemical in the feed phase.

Extractant — The active part of a solvent that has solvating characteristics with the solute or specific solute–solvent chemistry such as chelation or ion-pair formation.

Extraction cascade — A series of stages to recover the solute from a feed solution by contacting it with a relatively immiscible solvent.

Extraction cycle — A sequence of extraction, scrubbing, and stripping cascades to recover and purify a particular solute. Cycles may themselves be staged to enhance solute purity.

Loading — The capacity of the solvent for the solute. Usually, this is expressed on a per mole solvent basis. Higher values indicate reduced solvent needs.

Modifier — A chemical added to a solvent to retard third-phase formation.

Salting agent — A nonextracted chemical, usually a salt, used to change the process chemistry. For inorganic systems the salting agent may have an anion in common with the solute but have the wrong cation valence or size to be extracted.

Scrubbing — The operation of decontaminating a loaded solvent of chemicals coextracted with the solute. Scrubbing purifies the solute-loaded solvent. A *scrubbing cascade* is a series of stages to perform scrubbing.

Selectivity (β) — The separation factor for extraction is the ratio of distribution coefficients for two solutes. The higher the value is, the greater the ability of the solvent to separate the two species.

Stripping — The operation of unloading the solute from the solvent by extracting it back into a feed-like phase, from which it can be easily removed. Stripping also recovers the solvent for recycle back to the extraction cascade. A *stripping cascade* is a series of stages to perform stripping.

References

Alegret, S. 1988. *Developments in Solvent Extraction,* Ellis Horwood, Chichester, U.K.

Blumberg, R. 1988. *Liquid–Liquid Extraction,* Academic Press, New York.

Cusack, R. W., Fremeaux, P., and Glatz, D. 1991. A fresh look at liquid-liquid extraction. *Chem Eng.* February.

Godfrey, J. C. and Slater, M. J. 1994. *Liquid–Liquid Extraction Equipment.* Wiley, New York.

Hait, M. J. et al. 1993. Space predictor for infinite dilution activity coefficients. *Ind. Eng. Chem. Res.* 32:2905–2914.

Ladda, G. S. and Degaleesan, T. E. 1978. *Transport Phenomena in Liquid Extraction,* McGraw-Hill, New York.

Lo, T. C., Baird, M. H. I., and Hanson, C. 1983. *Handbook of Solvent Extraction,* Wiley-Interscience, New York.

Marcus, Y. and Kertes, A. S. 1969. *Ion Exchange and Solvent Extraction of Metal Complexes,* Wiley-Interscience, New York.

McHugh, M. A. and Krukonis, V. J. 1986. *Supercritical Fluid Extraction, Principles and Practices,* Butterworth, Boston, MA.

Park, J. H. and Carr, P. W. 1987. Predictive ability of the MOSCED and UNIFAC activity coefficient methods. *Anal. Chem.* 59:2596–2602.

Ritcey, G. M. and Ashbrook, A. W. 1984. *Solvent Extraction, Principles and Applications to Process Metallurgy, Part I,* Elsevier, New York.

Robbins, L. Liquid–liquid extraction. In *Perry's Chemical Engineer's Handbook,* 6th ed. McGraw-Hill, New York.

Rydberg, J., Musikas, C., and Choppin, G. R. 1992. *Principles and Practices of Solvent Extraction,* M. Dekker, New York.

Schultz, W. W. and Navratil, J. D. *Science and Technology of Tributyl Phosphate* (vol. 1, 1984; vols. IIA and IIB, 1987; vol. III, 1990; vol. IV, 1991). CRC Press, Boca Raton, FL.

Sekine, T. and Hasegawa, Y. 1977. *Solvent Extraction Chemistry, Fundamentals and Applications,* M. Dekker, New York.

Sorensen, J. M. and Arlt, W. *Liquid–Liquid Equilibrium: Chemistry Data Series, Volume V* (part 1, Binary Systems, 1979; part 2, Ternary Systems, 1980; part 3, Ternary and Quaternary Systems, 1980; supplement 1, 1987, by Macedo, E. A. and Rasmussen, P.). DECHEMA, Frankfurt.

Tiegs, D., 1986. *Activity Coefficients at Infinite Dilution, C1-C9,* DECHEMA Chemistry Data Series, vol. IX, part 1. DECHEMA, Frankfurt.

Treybal, R. E. 1963. *Liquid Extraction,* 2nd ed. McGraw-Hill, New York.

Van Brunt, V. and Kanel, J. S. 2002. Extraction with reaction: In *Reactive Separation Processes* S. Kulprathipanja, Ed. Taylor and Francis, New York, Chapter 3.

Wisniak, J. and Tamir, A. *Liquid–Liquid Equilibrium and Extraction: A Literature Source Book* (vol. 1, 1980; vol. 2, 1980; supplement 1, 1987). Elsevier, New York.

Further Information

The first general reference for extraction chemistry, equipment, and its operation is the *Handbook of Solvent Extraction.* Equipment is further described in *Transport Phenomena in Liquid Extraction* and in *Liquid–Liquid Extraction Equipment.*

The European Federation of Chemical Engineering has established three systems for evaluation of extraction operations. These guidelines are available in Misek, T. 1984. *Recommended Systems for Liquid–Liquid Extraction, Institute of Chemical Engineers.* Rugby, UK.

Sources of reliable extraction data include all the volumes of Sorenson and Arlt [1979, 1980, 1980, 1987] as well as all the volumes of Wisniak and Tamir [1980, 1980, 1987]. The text by Francis, A. W., 1963, *Liquid–Liquid Equilibriums,* Wiley-Interscience, New York also contains references to reliable data.

The International Solvent Extraction Conference (ISEC) is held every three years with bound proceedings available. The conference presents the latest information about extraction chemistry and equipment. See listings under *Proceedings of ISEC.* Each set of proceedings has been published by a different professional organization. The latest advances in extraction chemistry and processing may be found in the proceedings from ISEC '99 in Barcelona, Spain, *Solvent Extraction for the 21st Century,* M. Cox, M. Hidalgo, and M. Valiente, Eds., SCI, London 2001 and ISEC 2002 in Capetown. South Africa, *Proceedings of ISEC 2002,* Chris van Rensburg Publications Ltd, Melville, South Africa 2002. Each of these proceedings is published in two volumes.

Many extraction processes and refinements of them are discussed in the journals *Solvent Extraction and Ion Exchange* and *Separation Science and Technology.* The series *Ion Exchange and Solvent Extraction* contains tutorial articles about specific topics.

The science of extraction has been greatly influenced by nuclear applications, including nuclear reprocessing. Many references are available, including chapters in Long, J. T. 1978. *Engineering for Nuclear Fuel Reprocessing.* American Nuclear Society, La Grange Park, La Grange, IL; Benedict, M., Pigford, T. H., and Levi, H. L. 1981. *Nuclear Chemical Engineering,* 2nd ed. McGraw-Hill, New York; and all four volumes of Schultz and Navratil [1984, 1987, 1990, 1991].

61

Adsorption

Shivaji Sircar
Air Products and Chemicals, Inc.

Adsorption is a surface phenomenon. When a pure fluid (gas or liquid) is contacted with a solid surface (adsorbent), fluid–solid intermolecular forces of attraction cause some of the fluid molecules (adsorbates) to be concentrated at the surface. This creates a denser region of fluid molecules, which extends several molecular diameters near the surface (adsorbed phase). For a multicomponent fluid mixture, certain components of the mixture are preferentially concentrated (selectively adsorbed) at the surface due to differences in the fluid–solid forces of attraction between the components. This creation of an adsorbed phase having a composition different from that of the bulk fluid phase forms the basis of separation by adsorption technology.

Adsorption is a thermodynamically spontaneous process. Energy is released (exothermic) during the process. The reverse process by which the adsorbed molecules are removed from the surface to the bulk fluid phase is called **desorption.** Energy must be supplied to the adsorbed phase (endothermic) for the desorption process. Both adsorption and desorption form vital steps in a practical separation process in which the adsorbent is repeatedly used. This concept of regenerative ad(de)sorption is key to the practical use of this technology. It has found numerous commercial applications in chemical, petrochemical, biochemical, and environmental industries for separation and purification of fluid mixtures, as listed in Table 61.1.

61.1 Adsorbent Materials

A key element in the development of adsorption technology has been the availability of a large spectrum of micro- and mesoporous adsorbents. They have large specific surface areas (500 to 1500 m^2/g), varying pore structures, and surface properties (polar and nonpolar) that are responsible for selective adsorption of specific components of a fluid mixture. These include activated carbons, zeolites, aluminas, silica gels, polymeric adsorbents, and ion-exchange resins. Adsorbents may be energetically homogeneous, containing adsorption sites of identical adsorption energy, or heterogeneous, containing a distribution of sites of varying adsorption energies.

0-8493-1586-7/05/$0.00+$1.50

TABLE 61.1 Key Commercial Applications of Adsorption Technology

Gas Separation	Liquid Separation	Environmental Separation	Bioseparation
Gas drying	Liquid drying	Municipal and industrial waste treatment	Recovery of antibiotics
Trace impurity removal	Trace impurity removal	Ground and surface water treatment	Purification and recovery of enzymes
Air separation	Olefin-paraffin separation	VOC removal	Purification of proteins
Carbon dioxide-methane separation	Xylene, cresol, cymene isomer separation		Removal of microorganisms
Carbon monoxide-hydrogen separation	Fructose and glucose separation		Recovery of vitamins
Hydrogen and carbon dioxide recovery from SMR off gas	Breaking azeotropes		
Production of ammonia-synthesis gas			
Normal-iso paraffin separation			
Ozone enrichment			
Solvent vapor recovery			

61.2 Adsorption Equilibria

Adsorption equilibria determine the thermodynamic limits of specific amounts of adsorption (moles/g) of a pure fluid or the components of a mixture under a given set of conditions [pressure (P), temperature (T), and mole fraction (y)] of the bulk fluid phase. A convenient way to represent adsorption equilibria is in terms of adsorption isotherms in which the specific amount adsorbed of a pure gas (n_i^o) or that of component $i(n_i)$ from a multicomponent gas mixture is expressed as a function of P (pure gas) or as functions of P and y_i (fluid mixtures) at constant T. The values n_i^o and n_i decrease with increasing T for a given P and y_i. Adsorption isotherms can have many different shapes, but most microporous adsorbents exhibit the shape (type I) shown in Figure 61.1 (pure gas) and Figure 61.2 (binary gas).

The simplest thermodynamic system for adsorption from liquids consists of a binary liquid mixture. The adsorption isotherm in this case is measured in terms of Gibbs surface excess (moles/g) of component $i(n_i^e)$:

$$n_i^e = n_i - y_i \sum_i n_i; \quad \sum_i n_i^e = 0, \quad i = 1,2; \quad n_i^e = 0 \text{ for pure liquid} \tag{61.1}$$

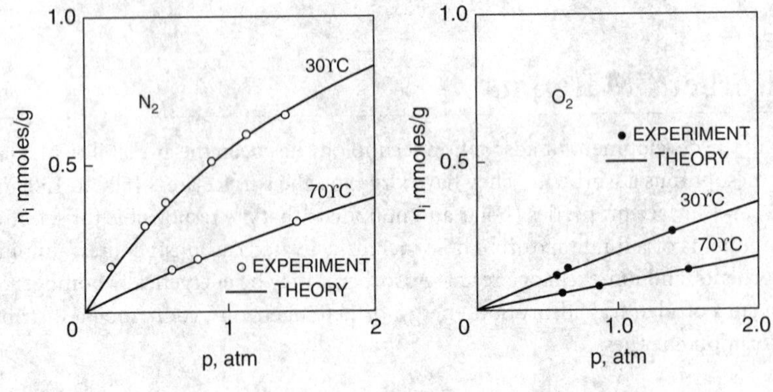

FIGURE 61.1 Langmuirian pure gas adsorption isotherms on Na-mordenite. (*Source:* Kumar, R. and Sircar, S. 1986. *Chem. Eng. Sci.* 41:2215–2223.)

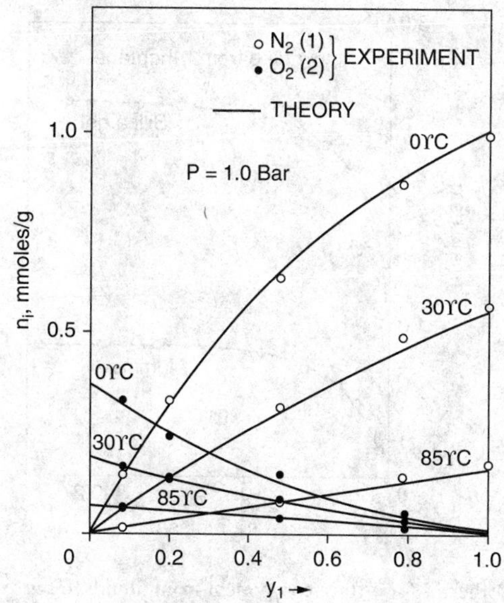

FIGURE 61.2 Langmuirian binary gas adsorption isotherms on Na-mordenite. (*Source:* Kumar, R. and Sircar, S. 1986. *Chem. Eng. Sci.* 41:2215–2223.)

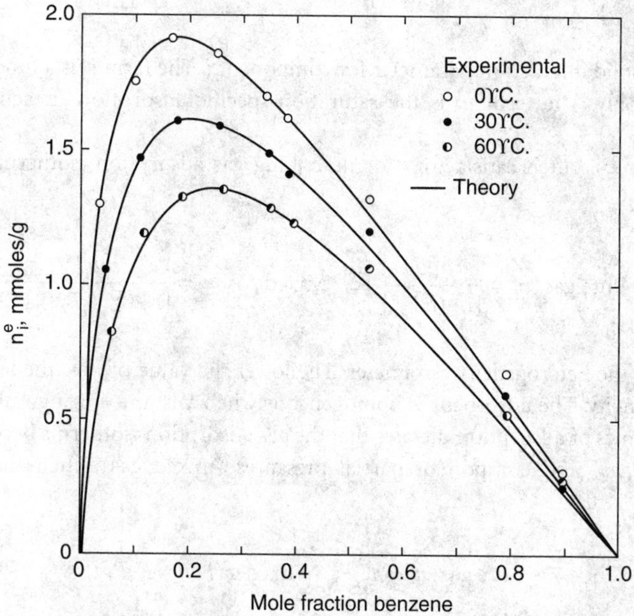

FIGURE 61.3 Benzene–cyclohexane liquid phase surface excess isotherms on silica gel. (*Source:* Sircar, S., Novosad, J., and Myers, A. L. 1972. *Ind. Eng. Chem. Fundam.* 11:249–254.)

The liquid phase surface excess is a function of y_i and T. Figure 61.3 shows an example of bulk binary liquid phase isotherm. For adsorption of a very selective trace component ($y_i \ll 1$), n_i^e is approximately equal to n_i. The isotherm in that case (Figure 61.4) is type I in shape.

The simplest model to describe pure and multicomponent gas adsorption isotherms on a homogeneous adsorbent is the Langmuir equation,

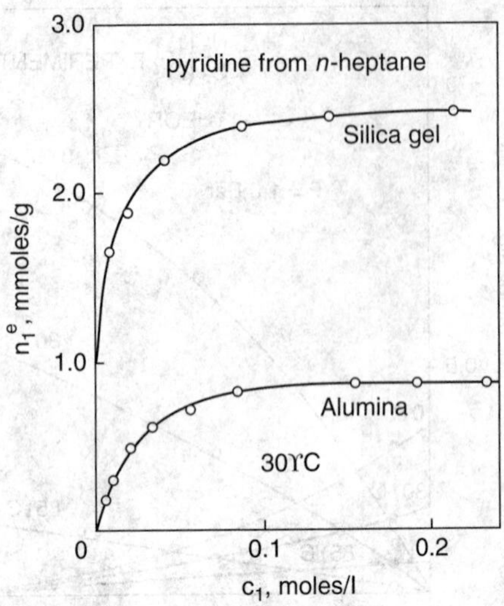

FIGURE 61.4 Surface excess isotherms for a trace component from liquids. (*Source:* Rahman, M. A. and Ghosh, A. K. 1980. *J. Colloid Interface Sci.* 77:50–52.)

$$\text{Pure gas } n_i^o = \frac{mb_iP}{1+b_iP}; \quad \text{Mixed gas } n_i = \frac{mb_iPy_i}{1+\Sigma_i b_iPy_i} \tag{61.2}$$

where b_i is the gas–solid interaction parameter for component *i*. The term b_i is a function of temperature [Equation (61.7)] only. The term *m* is the saturation-specific adsorption capacity (moles/g) for the components.

The Toth equation is often satisfactory for describing gas adsorption isotherms on heterogeneous adsorbents:

$$\text{Pure gas } n_i^o = \frac{mb_iP}{[1+(b_iP)^k]^{1/k}}; \quad \text{Mixed gas } n_i = \frac{mb_iPy_i}{[1+(\Sigma_i b_iPy_i)^k]^{1/k}} \tag{61.3}$$

The term k (≤ 1) is the heterogeneity parameter. The lower the value of *k* is, the larger is the degree of adsorbent heterogeneity. The adsorbent is homogeneous when *k* is unity (Langmuir model).

Statistical mechanics of adsorption dictates that the gas adsorption isotherms become linear functions of pressure (pure gas) and component partial pressures (mixed gas) when the total gas pressure approaches zero ($P \to 0$):

$$\text{Pure gas } n_i^o = K_iP; \quad \text{Mixed gas } n_i = K_iPy_i \tag{61.4}$$

K_i is called the Henry's law constant for pure gas *i*. It is a function of temperature only [Equation (61.7)].

The simplest model isotherm for adsorption of an ideal binary liquid mixture of equal adsorbate sizes on a homogeneous adsorbent is given by

$$n_1^e = \frac{my_1y_2(S-1)}{Sy_1+y_2} = -n_2^e \tag{61.5}$$

where *S* is the selectivity of adsorption of component 1 over component 2.

61.3 Heat of Adsorption

The thermodynamic variable of practical use that describes the differential heat evolved (consumed) during the ad(de)sorption process due to a differential change in the adsorbate loading (n_i^o or n_i) of a pure gas (q_i^o) or the components of a gas mixture (q_i) is called the isosteric heat of adsorption (kcal/mole). It is given by

$$\text{Pure gas } q_i^o = RT^2\left[\frac{\partial \ln P}{\partial T}\right]_{n_i^o}; \quad \text{Mixed gas } q_i = RT^2\left[\frac{\partial \ln Py_i}{\partial T}\right]_{n_i}, \quad i = 1, 2, \ldots \tag{61.6}$$

It follows from Equation (61.2), Equation (61.3), Equation (61.4), and Equation (61.6) that

$$K_i = K_i^* \exp[q_i^* / RT]; \quad b_i = b_i^* \exp[q_i^* / RT] \tag{61.7}$$

where q_i^* is the isosteric heat of adsorption of pure gas *i* at the limit $P \rightarrow 0$ (Henry's law region), and K_i^* and b_i^* are constants.

The terms q_i^o and q_i are constants and equal to q_i^* for any adsorbate loading for a homogeneous adsorbent. They are functions of loadings for a heterogeneous adsorbent. Thus, for the Toth model,

$$q_i^0 = q_i^* + \left(\frac{RT^2}{k}\right)\left(\frac{d \ln k}{dT}\right) F_i(\theta_i^0); \quad q_i = q_i^* + \left(\frac{RT^2}{k}\right)\left(\frac{d \ln k}{dT}\right) F(\theta) \tag{61.8}$$

$$F(\theta_i^o) = \frac{[1-(\theta_i^o)^k]\ln[1-(\theta_i^o)^k] + (\theta_i^o)^k \ln(\theta_i^o)^k}{[1-(\theta_i^o)^k]} \tag{61.9}$$

where (θ_i^o) is fractional coverage ($= n_i^o / m$) by pure gas *i*. The term $\theta (= \Sigma\, \theta_i)$ is the total fractional coverage ($= \Sigma\, n_i/m$) by all adsorbates for the mixture. $F(\theta)$ has the same mathematical form as Equation (61.9) except that θ_i^o is replaced by θ.

Equation (61.8) shows that q_i^o or q_i decreases with increasing θ_i^o or θ_i for a heterogeneous adsorbent. The higher energy sites of the adsorbent are predominantly filled at lower adsorbate loadings, and the lower energy sites are progressively filled at higher coverages.

Figure 61.5 shows examples of isosteric heats of adsorption of pure gases on heterogeneous adsorbents. Figure 61.6 shows the variation of isosteric heats of the components of a binary gas mixture with coverage (or gas composition) at a constant total gas pressure according to the Toth model.

An integral heat of adsorption called *heat of immersion* (kcal/mole) can be measured by contacting a clean adsorbent with a pure liquid (ΔH_i^o) or a liquid mixture (ΔH). The typical variation of (ΔH) as a function of bulk liquid phase concentration is shown in Figure 61.7. For an ideal binary liquid system [Equation (61.5)], ΔH is given by

$$\Delta H = \frac{Sy_i \Delta H_1^o + y_2 \Delta H_2^0}{Sy_1 + y_2} \tag{61.10}$$

61.4 Thermodynamic Selectivity of Adsorption

Most practical adsorptive separations are based on thermodynamic selectivity. The selectivity S_{ij} ($= n_i y_j / n_j y_i$) of adsorption of component *i* over component *j* from a mixture determines the maximum achievable separation between the components under equilibrium conditions. Component *i* is selectively adsorbed

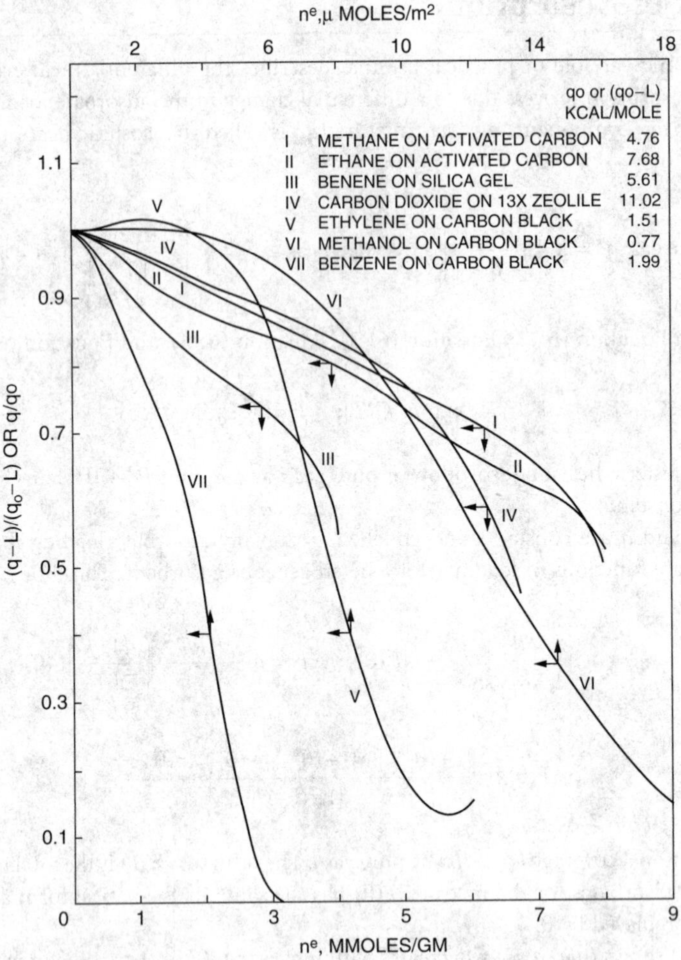

FIGURE 61.5 Isosteric heats of adsorption of pure gases on heterogeneous adsorbents. (*Source:* Sircar, S. and Gupta, R. 1981. *Am. Inst. Chem. Eng. J.* 27:806–812.)

over component j when $S_{ij} > 1$. S_{ij} can approach infinity if component j is excluded from entering the pores of the adsorbent (molecular sieving).

The selectivity of adsorption (S_{ij}^*) in the Henry's law region $(P \to 0)$ is given by

$$S_{ij}^* = \frac{K_i}{K_j} = \left(\frac{K_i^*}{K_j^*}\right) \exp\left[\frac{q_i^* - q_j^*}{RT}\right] \tag{61.11}$$

S_{ij} values can be strong functions of adsorbate loadings of the components when the adsorbates have different molecular sizes and when the adsorbent is heterogeneous, as shown in Figure 61.8.

For adsorption of an ideal binary liquid system [Equation (61.5)],

$$S = \exp[\phi_2^o - \phi_1^o]/m\,RT \tag{61.12}$$

where ϕ_i^o is the surface potential for adsorption of pure liquid i. S for liquid mixtures can be strong functions of bulk phase concentrations when the adsorbates have different sizes or when the adsorbent is heterogeneous.

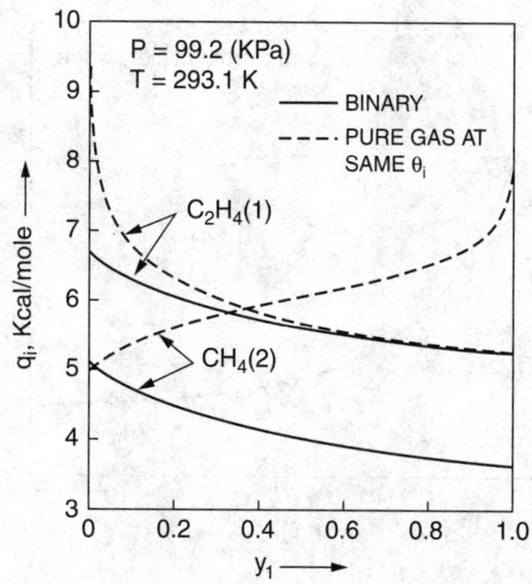

FIGURE 61.6 Isosteric heats of adsorption of binary gas on heterogeneous adsorbents. (*Source:* Sircar, S. 1991a. *Langmuir* 7:3065–3069.)

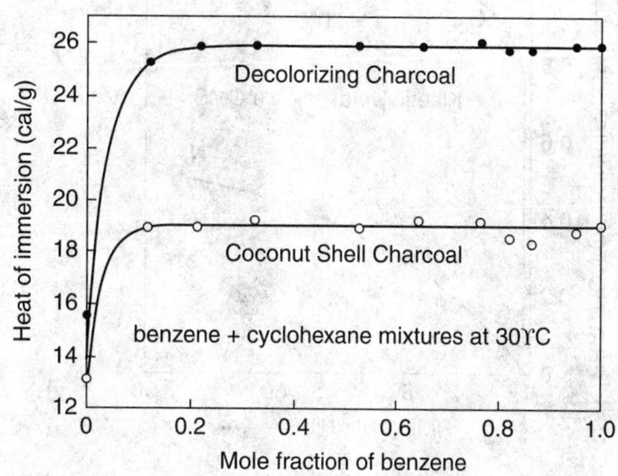

FIGURE 61.7 Heats of immersion of binary liquid mixtures. (*Source:* Wright, E. H. M. 1967. *Trans. Faraday Soc.* 63:3026–3038.)

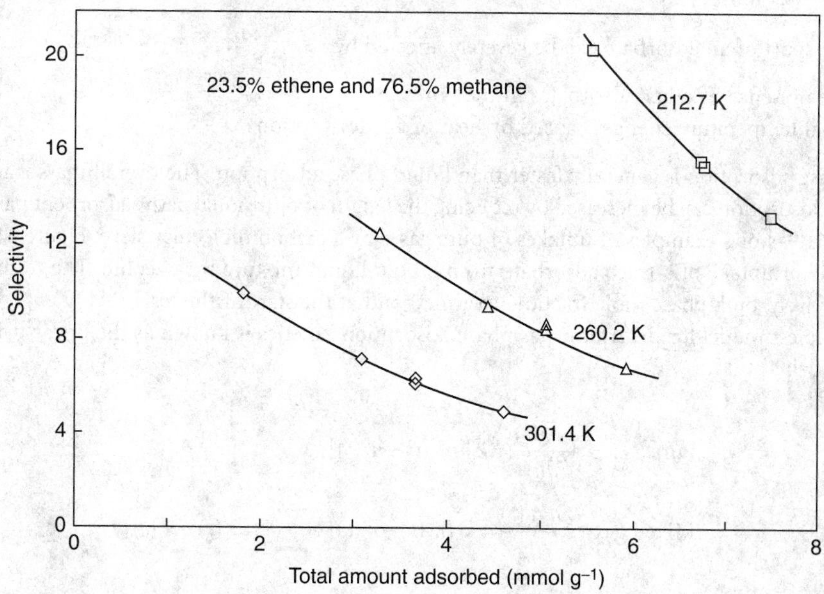

FIGURE 61.8 Binary selectivity of adsorption on activated carbon. (*Source:* Sircar, S. and Myers, A. L. 1985. *Adsorp. Sci. Technol.* 2:69–87.)

61.5 Adsorption Kinetics

The actual physisorption process is very fast (milliseconds to reach equilibrium). However, a finite amount of time may be required for an adsorbate molecule to travel from the bulk fluid phase to the adsorption site in a microporous adsorbent. This rate process is generally referred to as *adsorption kinetics.* Adsorbate mass transfer resistance may be caused by:

1. Fluid film outside the adsorbent particle (for mixture adsorption)
2. Anisotropic skin at the particle surface
3. Internal macro- and microporous diffusional resistances (pore and surface diffusion)

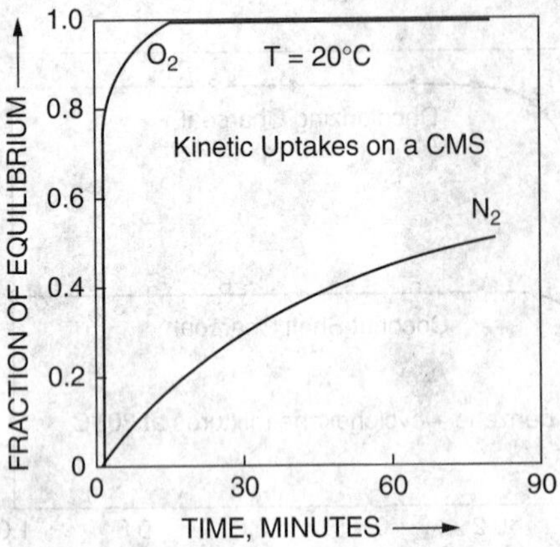

FIGURE 61.9 Pure gas adsorption kinetics on carbon molecular sieve. (*Source:* Sircar, S. 1994. In *Separation Technology: The Next Ten Years*, ed. J. Garside, Institute of Chemical Engineers, Rugby, Warwickshire, U.K. p. 49.)

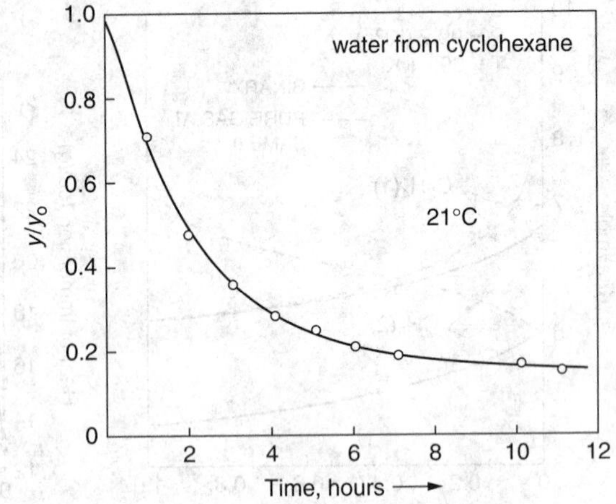

FIGURE 61.10 Adsorption kinetics of a trace liquid adsorbate on 3A zeolite. (*Source:* Sircar, S. and Myers, A. L. 1986. *Sep. Sci. Technol.* 21:535–562.)

The transport of an adsorbate can be severely affected by:

1. The presence of other adsorbates in the pores
2. Local temperature changes caused by heat of ad(de)sorption

Gas phase adsorption is generally faster than liquid phase adsorption. The overall mass transfer coefficient for adsorption can be increased by reducing the length of diffusional path (adsorbent particle size).

Figure 61.9 shows examples of uptakes of pure gases by a carbon molecular sieve. Figure 61.10 shows an example of uptake of a trace adsorbate from a bulk liquid mixture by a zeolite. The terms y and y_o are, respectively, bulk phase mole fractions at time t and at the start of the test.

The simplest model for describing gas phase adsorption kinetics is known as the linear driving force (LDF) model:

$$\text{Pure gas} \quad \frac{d\,n_i^o(t)}{dt} = k_i^o[n_i^{o*}(t) - n_i^o(t)];$$
$$\text{Mixed gas} \quad \frac{d\,n_i(t)}{dt} = k_{ii}[n_i^*(t) - n_i(t)] + \sum_j k_{ij}[n_j^*(t) - n_j(t)] \tag{61.13}$$

The terms $n_i^o(t)$ and n_i are, respectively, the transient adsorbate loading of pure component i, and that for component i from a mixture at time t. The terms $n_i^*(t)$ and $n_i^o\,(t)$ are, respectively, the corresponding equilibrium adsorbate loadings of component i at the instantaneous bulk phase conditions. The term k_i^o (sec^{-1}) is the overall mass transfer coefficient for pure component i. The terms k_{ii} and k_{jj} are the overall straight and cross (between component i and j) mass transfer coefficients for component i in the mixture. The temperature and adsorbate loading dependence of the mass transfer coefficients is determined by the governing transport mechanism. The LDF model can also describe the kinetics of adsorption of liquid mixtures by rewriting Equation (61.13) in terms of transient surface excess $n_i^e(t)$ and equilibrium surface excess n_i^{e*} of component i.

An experimental gas adsorption kinetics process is generally nonisothermal, whereas liquid phase adsorption kinetics can be measured isothermally due to high heat capacity of the liquid.

Separation of the components of a fluid mixture can also be achieved by a kinetic selectivity when certain components of the mixture are adsorbed at a much faster rate than the others, even though there

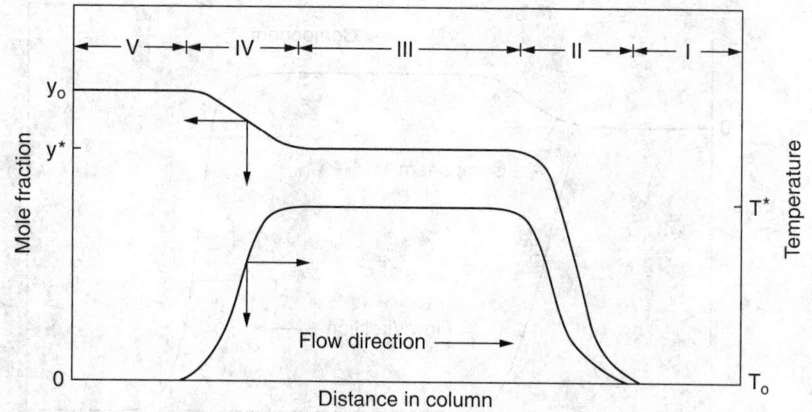

FIGURE 61.11 Type I column dynamics. (*Source:* Sircar, S. and Myers, A. L. 1985. *Adsorp. Sci. Technol.* 2:69–87.)

is no thermodynamic selectivity of adsorption between the components. Figure 61.9 is an example of such a case.

61.6 Adsorption Column Dynamics

Practical separation and purification of fluid mixtures are carried out in packed columns of adsorbent materials. The dynamics of the ad(de)sorption process in columns is determined by the adsorption equilibria, heats, and kinetics and by the modes of operation of the process. The simplest case of the adsorption dynamics is to flow a binary gas mixture consisting of a single adsorbate (mole fraction y^o) and an inert carrier gas through a packed column that has previously been saturated with the pure inert gas at the pressure (P^o) and temperature (T^o) of the feed gas. Two types of behavior may be observed:

Type I. Two pairs of mass and heat transfer zones are formed in the column, as shown in Figure 61.11. The column ahead (section I) of the front zones (section II) remains saturated with the carrier gas at initial conditions. The column (section III) between the front and rear zones (section IV) is equilibrated with a gas mixture of mole fraction y^* ($< y^o$) at temperature T^* ($>T^o$). The column (section V) behind the rear zones is equilibrated with feed gas mixture at feed conditions.

Type II. A pure heat transfer zone (section II) is formed, followed by a pair of mass and heat transfer zones (section IV), as shown in Figure 61.12. The adsorbate is absent in sections I to III in this case. The column behind (section V) the rear zones remains equilibrated with feed gas at feed conditions.

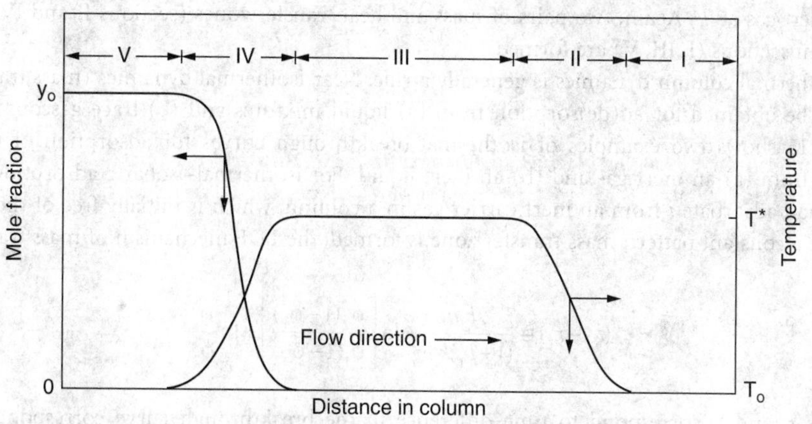

FIGURE 61.12 Type II column dynamics. (*Source:* Sircar, S. and Myers, A. L. 1985. *Adsorp. Sci. Technol.* 2:69–87.)

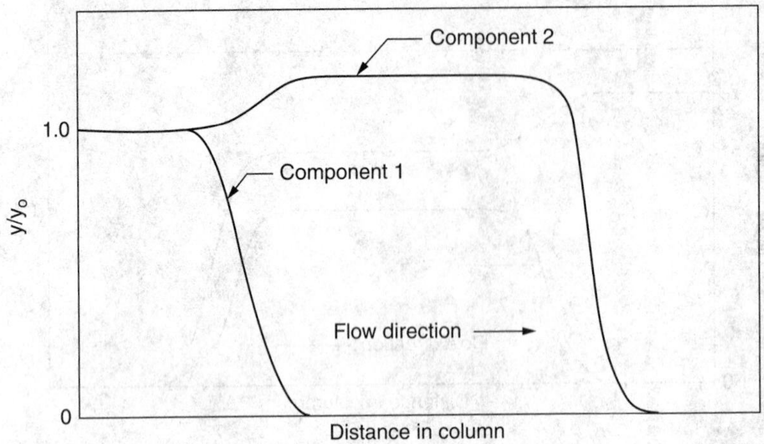

FIGURE 61.13 Rollover effect for binary adsorbate system. (*Source:* Sircar, S. and Myers, A. L. 1985. *Adsorp. Sci. Technol.* 2:69–87.)

Approximate criteria for formation of these two types of systems are as follows:

$$\text{Type I: } n^o / y^o < C_s / C_g; \quad \text{Type II: } n^o / y^o > C_s / C_g \tag{61.14}$$

The term n^o is adsorbate loading at feed conditions. C_s and C_g are heat capacities of the adsorbent and the feed gas. The zones move through the column as more feed gas is introduced.

Adsorbates with type I adsorption equilibria generally yield constant pattern (unchanging shape and size) front zones (type I dynamics) and rear zones (type II dynamics) in a long column. The zones can be proportionate pattern (expanding in size with time) when the adsorption isotherms are linear. The zones eventually leave the column, and the measured adsorbate concentration–time and temperature–time profiles (breakthrough curves) are mirror images of these profiles within the column. Multiple transfer zones and equilibrium sections are formed in systems with multicomponent adsorbates. They are often characterized by rollover effects in which the more strongly adsorbed species (component 1) displaces the weaker species (component 2) as the zones propagate. This effect is illustrated in Figure 61.13 for isothermal adsorption of a binary adsorbate from inert carrier gas.

Well-defined mass and heat transfer zones and equilibrium sections can also be formed in the column during the desorption process. Figure 61.14 is an example for isobaric desorption of a single adsorbate from a column saturated with the adsorbate at mole fraction y^o and temperature T^o that is being purged with an inert gas at T^o. Again, two pairs of mass and heat transfer zones (sections II and IV) and three equilibrium sections (I, III, V) are formed.

Nonisothermal column dynamics is generally a rule. Near-isothermal dynamics (mass transfer zones only) can be obtained for ad(de)sorption from (a) liquid mixtures and (b) trace gaseous adsorbates. Figure 61.15 shows two examples of isothermal breakthrough curves for adsorption of trace single adsorbate from (a) an inert gas and (b) an inert liquid. For isothermal–isobaric adsorption of a trace Langmuirian adsorbate i from an inert carrier gas in a column, which is initially free of the adsorbate, and where a constant pattern mass transfer zone is formed, the LDF mechanism of mass transfer yields

$$(t_2 - t_1) \cong \frac{b_1 m}{(1+b_1)k_i^o n_i^o} \ln\left[\frac{\phi_2(1-\phi_1)}{\phi_1(1-\phi_2)}\right] + \ln\frac{\phi_1}{\phi_2} \tag{61.15}$$

The terms t_1 and t_2 correspond to time difference in the breakthrough curve corresponding to two arbitrary composition levels ϕ_1 and ϕ_2 [$\phi = y_i(t) / y_i^o$].

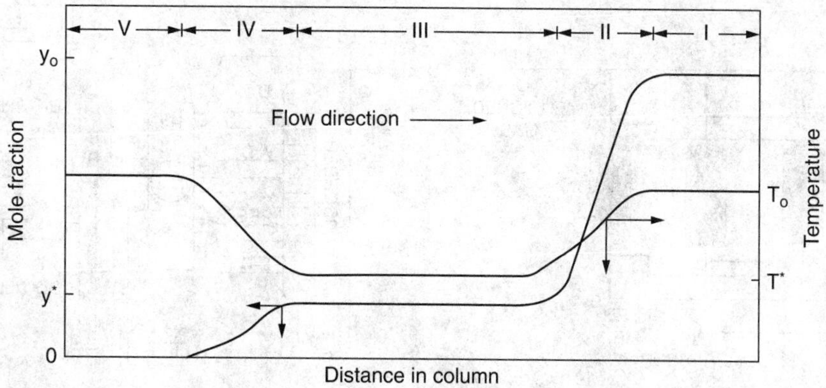

FIGURE 61.14 Column dynamics for desorption by purge. (*Source:* Sircar, S. and Myers, A. L. 1985. *Adsorp. Sci. Technol.* 2:69–87.

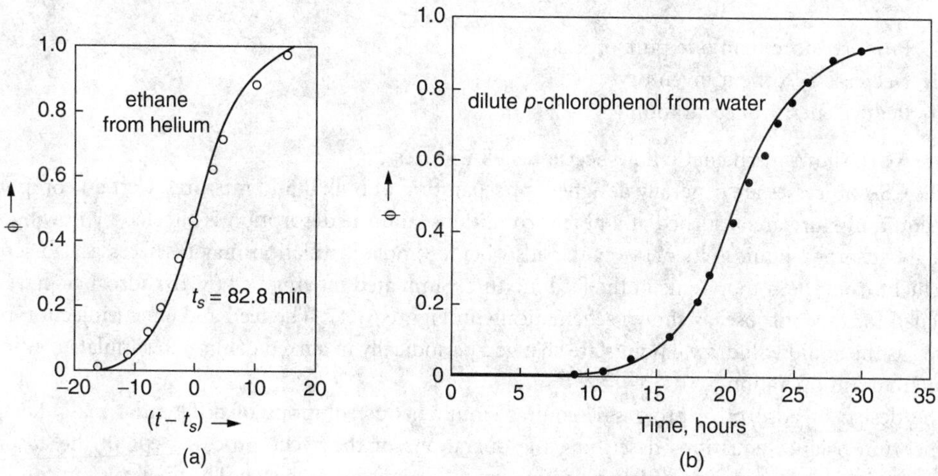

FIGURE 61.15 Isothermal breakthrough profiles for a single-trace adsorbate. [*Sources:* (a) Garg, D. R. and Ruthven, D. M. 1974. *Chem. Eng. Sci.* 29:571–581. (b) Mathews, A. P. 1984. In *Proceedings of First International Conference on Fundamentals of Adsorption,* ed. A. L. Myers, Engineering Foundation, New York. p. 345.]

61.7 Adsorptive Separation Processes and Design

Three generic adsorptive process schemes have been developed to serve most of the applications described by Table 61.1. These are (a) **thermal swing adsorption** (TSA), (b) **pressure swing adsorption** (PSA), and (c) **concentration swing adsorption** (CSA).

TSA is by far the most frequently used industrial application of this technology. It is used for removal of trace impurities as well as for drying gases and liquids. The adsorption is carried out at near ambient temperature and the desorption is achieved by directly heating the adsorbent using a part of the cleaned fluid or steam. The adsorbent is then cooled and reused.

PSA is primarily used for bulk gas separations and for gas drying. These processes consist of a series of sequential cyclic steps. The adsorption step is carried out at higher partial pressures of the adsorbates and the desorption is achieved by lowering their partial pressures in the column by (a) decreasing the total pressure of the column and (b) flowing an adsorbate free gas through the column. Many complementary process steps are also used in modern PSA processes in order to:

1. Increase the recovery and purity of desired products from a multicomponent feed gas mixture

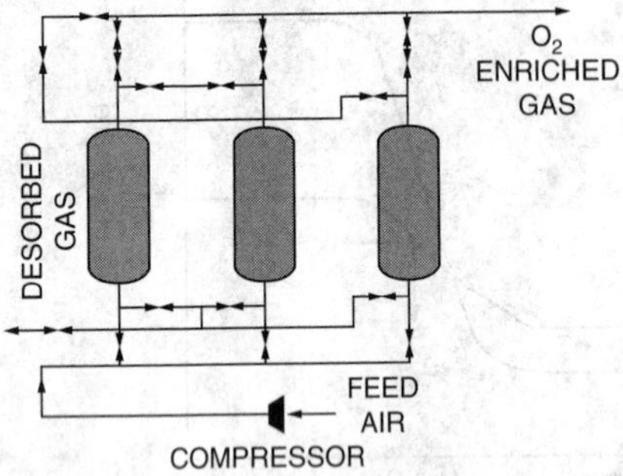

FIGURE 61.16 Schematic flowsheet of a PSA process. (*Source:* Sircar, S. 1989. In *Adsorption Science and Technology,* ed. A. I. Rodrigues et al. NATO ASI Series E158. Kluwer Academic, Dordrecht, The Netherlands. p. 285.)

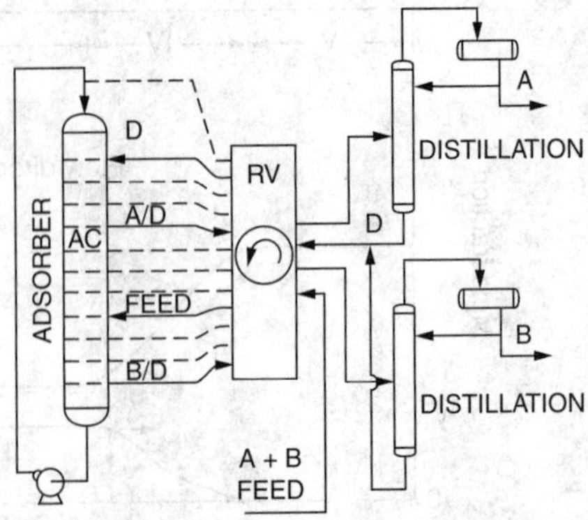

FIGURE 61.17 Schematic flowsheet of an SMB process. (*Source:* Keller, G. E., Anderson, R. A., and Yon, C. M. 1987. In *Handbook of Separation Process Technology,* ed. R. W. Rousseau, John Wiley & Sons, New York. p. 644.)

2. Produce more than one pure product
3. Decrease adsorbent inventory
4. Reduce energy of separation

Figure 61.16 shows a schematic flowsheet for a PSA process.

The CSA processes are generally designed for separation of bulk liquid mixtures. Certain components of a liquid mixture are adsorbed at ambient conditions, and the desorption is effected by flowing a less strongly adsorbed liquid (eluent) over the adsorbent. Simple distillation may be necessary to separate the eluent from the components of the feed mixture. Simulated moving bed (SMB) adsorbers have been designed for this purpose, as shown schematically in Figure 61.17. The feed and eluent injection points as well as the liquid withdrawal points are changed periodically in a fixed column to simulate continuous countercurrent operation.

The design of adsorptive processes requires simultaneous solutions of differential mass, heat, and momentum balance equations describing the operations of the cyclic process steps in the adsorbent column using the appropriate initial and boundary conditions for each step. The final column conditions at the end of a step become the initial conditions for the next step. Numerical integration of the equations is often necessary in order to reach a cyclic steady state solution. Multicomponent adsorption equilibria, heats, and kinetics from the key input variables for the solution. Bench- and pilot-scale process performance data are generally needed to confirm design calculations.

Defining Terms

Adsorbent — A material used for carrying out the adsorption process.

Adsorption — The surface phenomenon by which the molecules of a bulk fluid phase are attracted by a solid surface in contact with the fluid.

Adsorption equilibria — The thermodynamic property describing the extent of adsorption of a fluid species by a solid surface.

Adsorption kinetics — The measure of travel time of an adsorbate molecule from bulk fluid phase to the adsorption site.

Column dynamics — Defines behavior of mass and heat transfer zone movements within an adsorption column during the ad(de)sorption process.

Concentration swing adsorption — An adsorptive process in which desorption is effected by changing the fluid phase concentration of the adsorbate.

Desorption — The process of removing the adsorbed molecules from the solid surface to the bulk fluid phase.

Heat of adsorption — The measure of thermal energy released during the exothermic adsorption process.

Pressure swing adsorption — An adsorptive process in which desorption is effected by lowering the partial pressure of the adsorbate.

Selectivity of adsorption — The measure of extent of separation of a component of a fluid mixture by adsorption process.

Thermal swing adsorption — An adsorptive process in which desorption is effected by heating the adsorbent.

References

Basmadjian, D. 1997. *The Little Adsorption Book,* CRC Press, Boca Raton, FL.

Broughton, D. B. and Gembicki, S. A. 1984. Adsorptive separations by simulated moving bed technology. In *Fundamentals of Adsorption,* ed. A. L. Myers and G. Belfort, Engineering Foundation, New York. p. 115.

Crittenden, B. and Thomas, W. J. 1998. *Adsorption Technology and Design,* Butterworth-Heinemann, Oxford.

Do, D. D. 1998. *Adsorption Analysis: Equilibria and Kinetics,* Imperial College Press, London.

Garg, D. R. and Ruthven, D. M. 1974. The performance of molecular sieve adsorption columns: system with micropore diffusion control. *Chem. Eng. Sci.* 29:571–581.

Keller, G. E., Anderson, R. A., and Yon, C. M. 1987. Adsorption. In *Handbook of Separation Process Technology,* ed. R. W. Rousseau, John Wiley & Sons, New York. p. 644.

Kumar, R. and Sircar, S. 1986. Skin resistance for adsorbate mass transfer into extruded adsorbent pellets. *Chem. Eng. Sci.* 41:2215–2223.

Mathews, A. P. 1984. Dynamics of adsorption in a fixed bed of polydisperse particles. In *Proceedings of First International Conference on Fundamentals of Adsorption,* ed. A. L. Myers, Engineering Foundation, New York. p. 345.

Rahman, M. A. and Ghosh, A. K. 1980. Determination of specific surface area of ferric oxide, alumina and silica gel powders. *J. Colloid Interface Sci.* 77:50–52.

Rouquerol, F., Rouquerol, J., and Sing, K.S.W. 1999. *Adsorption by Powders and Porous Solids,* Academic Press, London.

Ruthven, D. M. 1984. *Principles of Adsorption and Adsorption Processes,* John Wiley & Sons, New York.

Ruthven, D. M., Farouq, S., and Knaebel, K. S. 1994. *Pressure Swing Adsorption,* VCH, New York.

Sircar, S. 1989. Pressure swing adsorption technology. In *Adsorption Science and Technology,* ed. A. I. Rodrigues et al. NATO ASI Series E158. Kluwer Academic, Dordrecht, The Netherlands. p. 285.

Sircar, S. 1991a. Isosteric heats of multicomponent gas adsorption on heterogeneous absorbents. *Langmuir* 7:3065–3069.

Sircar, S. 1991b. Pressure swing adsorption — research needs by industry. In *Proceedings of Third International Conference on Fundamentals of Adsorption,* ed. A. Mersmann, Engineering Foundation, New York. p. 815.

Sircar, S. 1993. Novel applications of adsorption technology. In *Proceedings of Fourth International Conference on Fundamentals of Adsorption,* ed. M. Suzuki, Kodansha, Tokyo. p. 3.

Sircar, S. 1994. Adsorption technology: A versatile separation tool. In *Separation Technology: The Next Ten Years,* ed. J. Garside, Institute of Chemical Engineers, Rugby, Warwickshire, U.K. p. 49.

Sircar, S. and Gupta, R. 1981. A semi empirical adsorption equation for single component gas solid equilibrium. *Am. Inst. Chem. Eng. J.* 27:806–812.

Sircar, S. and Myers, A. L. 1985. Gas adsorption operations: equilibrium, kinetics, column dynamics and design. *Adsorp. Sci. Technol.* 2:69–87.

Sircar, S. and Myers, A. L. 1986. Liquid adsorption operations: equilibrium, kinetics, column dynamics, and applications. *Sep. Sci. Technol.* 21:535–562.

Sircar, S., Novosad, J., and Myers, A. L. 1972. Adsorption from liquid mixtures on solids: thermodynamics of excess properties and their temperature coefficients. *Ind. Eng. Chem. Fundam.* 11:249–254.

Wright, E. H. M. 1967. Thermodynamic correlation for adsorption from non-ideal solutions at the solid-solution interface. *Trans. Faraday Soc.* 63:3026–3038.

Yang, R. T. 1987. *Gas Separation by Adsorption Processes,* Butterworths, London.

Young, D. M. and Crowell, A. D. 1962. *Physical Adsorption of Gases,* Butterworths, London.

Further Information

Barrer, R. M. 1978. *Zeolites and Clay Minerals as Sorbents and Molecular Sieves,* Academic Press, New York.

Breck, D. W. 1974. *Zeolite Molecular Sieves,* Wiley-Interscience, New York.

Capelle, A. and deVooys, F. (Eds.) 1983. *Activated Carbon — A Fascinating Material,* Norit N. V., Amersfoort, The Netherlands.

Gregg, S. J. and Sing, K. S. W. 1982. *Adsorption Surface Area and Porosity,* Academic Press, London.

Karger, J. and Ruthven, D. M. 1992. *Diffusion in Zeolites,* Wiley-Interscience, New York.

Rousseau, R. W. (Ed.) 1981. *Handbook of Separation Technology,* John Wiley & Sons, New York.

Suzuki, M. 1990. *Adsorption Engineering,* Kodansha, Tokyo.

Wankat, P. C. 1986. *Large Scale Adsorption and Chromatography,* CRC Press, Boca Raton, FL.

62

Crystallization and Evaporation

Richard C. Bennett
Swenson Process Equipment, Inc.

A **crystal** is a solid bounded by plane surfaces. Crystallization is important as an industrial process because a large number of commodity chemicals, pharmaceuticals, and specialty chemicals are marketed in the form of crystals. The wide use of crystallization is due to the highly purified and attractive form in which the compounds can be obtained from relatively impure solutions by means of a single processing step. Crystallization can be performed at high or low temperatures, and it generally requires much less energy for separation of pure materials than other commonly used methods of purification do. While crystallization may be carried on from vapor or a melt, the most common industrial method is from a solution.

A solution is made up of a liquid (solvent) — most commonly water — and one or more dissolved species that are solid in their pure form (solute). The amount of solute present in solution may be expressed in several different units of concentration. For engineering calculations, expressing the solubility in mass units is the most useful. The solubility of a material is the maximum amount of solute that can be dissolved in a solvent at a particular temperature. Solubility varies with temperature and, with most substances, the amount of solute dissolved increases with increasing temperature.

For crystallization to occur, a solution must be supersaturated. **Supersaturation** means that, at a given temperature, the actual solute concentration exceeds the concentration under equilibrium or saturated conditions. A supersaturated solution is metastable, and all crystallization occurs in the metastable region. A crystal suspended in saturated solution will not grow. Supersaturation may be expressed as the ratio between the actual concentration and the concentration at saturation [Equation (62.1)] or as the difference in concentration between the solution and the saturated solution at the same temperature [Equation (62.2)].

$$S = C/C_s \tag{62.1}$$

$$\Delta C = C - C_s \tag{62.2}$$

0-8493-1586-7/05/$0.00+$1.50

where C is the concentration (g/100 g of solution), and C_s is the concentration (g/100 g of solution) at saturation. This difference in concentration may also be referenced to the solubility diagram and expressed as degrees (°C) of supersaturation.

Nucleation is the birth of a new crystal within a supersaturated solution. Crystal growth is the layer-by-layer addition of solute to an existing crystal. Both of these phenomena are caused by supersaturation. Nucleation is a relatively rapid phenomenon that can occur in a matter of seconds. Growth is a layer-by-layer process on the surface of an existing crystal and takes considerably more time. The ratio of nucleation to growth controls the size distribution of the crystal product obtained. Generating a high level of supersaturation spontaneously leads to both nucleation and growth. The competition between these two processes determines the character of the product produced.

62.1 Methods of Creating Supersaturation

Supersaturation may be created by cooling a solution of normal solubility into the metastable zone. Typically, the amount of supersaturation that can be created in this way without causing spontaneous nucleation is in the range of 1 to 2°C. Evaporation of solvent at a constant temperature also produces supersaturation by reducing the amount of solvent available to hold the solute. The reaction of two or more chemical species, which causes the formation of a less soluble species in the solvent, can also produce supersaturation. Finally, the addition of a miscible nonsolvent in which the solute is not soluble to a solvent will cause a decrease in the solubility of the solute in the solution. This technique is most often used in pharmaceutical operations involving the addition of alcohol or similar solvents to the primary solvent (water).

62.2 Reasons for the Use of Crystallization

Crystallization is important as an industrial process because a large number of materials can be marketed in the form of crystals that have good handling properties. Typically, crystalline materials can be separated from relatively impure solutions in a single processing step. In terms of energy requirements, the energy required for crystallization is typically much less than for separation by distillation or other means. In addition, crystallization can often be performed at relatively low temperatures on a scale that involves quantities from a few pounds up to thousands of tons per day.

62.3 Solubility Relations

Equilibrium relations for crystallization systems are expressed in the form of solubility data, which are plotted as phase diagrams or solubility curves. The starting point in designing any crystallization process is knowledge of the solubility curve, which is ordinarily plotted in terms of mass units as a function of temperature. An example is given in Figure 62.1 for the solubility of magnesium sulfate in water as a function of temperature. At any concentration and temperature, the information on the diagram allows one to predict the mixture of solids and solution that exists. Note that, in the case of magnesium sulfate, a number of different hydrates can exist in addition to the solution itself, or ice plus the solution. The line that forms a boundary between the solution area and the various crystal hydrate areas is a solubility curve.

Starting from point (1) at 50°C and cooling to 30°C at point (2) is a path that crosses the solubility line. During the cooling process, crossing the line in this manner indicates that the solution has become supersaturated for the concentration in question. If the supersaturation is within the metastable range — which is approximately 1°C — then growth can occur on existing crystals, but no substantial amount of nucleation will occur. If the cooling proceeds further, the system can become unstably supersaturated, and spontaneous nucleation takes place. If spontaneous nucleation takes place, very small crystals or nuclei form, and they will grow as long as the solution remains supersaturated.

As growth takes place, the concentration drops in the direction of point (3), and, as it approaches the solubility line, growth ceases because the driving force approaches zero. Organic and inorganic materials

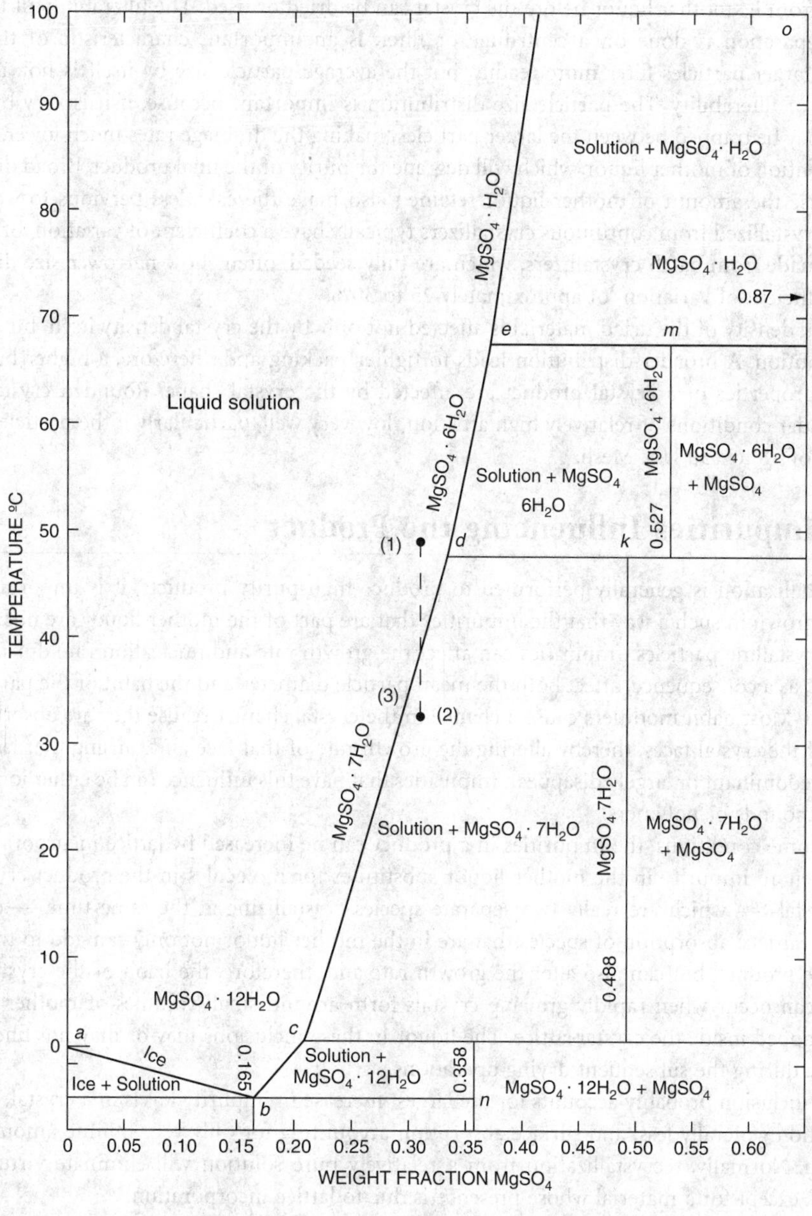

FIGURE 62.1 Weight fraction of $MgSO_4$ versus temperature. (*Source:* Courtesy of Swenson Process Equipment Inc.)

have similar solubility curves and they vary in concentration and temperature for each compound. Some materials have no hydrates and others exhibit a wide range of hydrates similar to those shown in Figure 62.1. Solubility information on most compounds is available from the literature in publications, such as the *International Critical Tables* [Campbell and Smith, 1951] and *Lang's Solubility of Inorganic and Organic Compounds* [Linke, 1958], and in various software packages that are becoming available.

62.4 Product Characteristics

The shape and size of a crystal are determined by its internal structure as well as external conditions that occur during its growth cycle. These external conditions include growth rate, the solvent system, the level of agitation, and the effect of impurities that may be present. Crystalline material is almost always separated from its mother liquor before the crystal can be dried or used. The filterability of the crystals, whether separation is done on a centrifuge or filter, is an important characteristic of the product. Generally, larger particles filter more readily, but the average particle size by itself is not an unfailing indication of filterability. The particle size distribution is important because, if it is very broad, small particles may be trapped between the larger particles, making the drainage rates much lower. This could lead to retention of mother liquor, which will degrade the purity of the final product. Broad distributions that increase the amount of mother liquor retained also make the cake less pervious to wash liquids. Products crystallized from continuous crystallizers typically have a coefficient of variation of 45 to 50%. Products made from batch crystallizers, which are fully seeded, often show narrower size distributions with a coefficient of variation of approximately 25 to 30%.

The bulk density of the dried material is affected not only by the crystal density itself, but also by the size distribution. A broader distribution leads to tighter packing and, therefore, a higher bulk density. The flow properties of a crystal product are affected by the crystal shape. Rounded crystals that are formed under conditions of relatively high attrition flow very well, particularly if the particles are in the size range of –8 to +30 U.S. Mesh.

62.5 Impurities Influencing the Product

Since crystallization is generally performed to produce high-purity products, it is important that the crystal be grown in such a way that the impurities that are part of the mother liquor are not carried out with the crystalline particles. Impurities can affect the growth rate and nucleation rate during crystallization and, as a consequence, affect both the mean particle diameter and the habit of the particles being crystallized. Most habit modifiers cause a change in the crystal shape because they are absorbed on one or more of the crystal faces, thereby altering the growth rate of that face and causing that face to either become predominant or largely disappear. Impurities that have this influence can be either ionic, surface-active compounds or polymers.

Under some conditions, the impurities in a product can be increased by lattice incorporation, which occurs when an impurity in the mother liquor substitutes for molecules in the product crystal lattice. Mixed crystals — which are really two separate species crystallizing at the same time — can also be produced. Surface absorption of species that are in the mother liquor not only can add to the impurity level of the product, but can also alter the growth rate and, therefore, the habit of the crystals. Solvent inclusion can occur when rapidly growing crystals form around small volumes of mother liquor that become trapped inside the crystal lattice. The liquor in these inclusions may or may not find its way to the surface during the subsequent drying operations.

Solvent inclusion probably accounts for the largest increase in impurity levels in a crystal, with lattice incorporation generally less, and surface absorption accounting for only very minute amounts of contamination. Normally, recrystallization from a relatively pure solution will eliminate virtually all the impurities, except for a material whose presence is due to lattice incorporation.

62.6 Kinds of Crystallization Processes

Crystallization can be carried on in either a batch or continuous manner, irrespective of whether evaporation, cooling, or solvent change is the method of creating supersaturation. Batch processes are almost always used for small capacities and have useful application for large capacities when a very narrow particle size distribution is required, such as with sugar, or when materials (e.g., pharmaceuticals) that require very accurate inventory control are being handled.

A continuous crystallization process normally must operate around the clock because the retention times typically used in crystallizers range from about 1 to 6 h. As such, it takes at least four to six retention times for the crystallizer to come to equilibrium, which means there may be off-spec product when the system is started up. To minimize this, the unit should be kept running steadily as long as possible. The cost of at least three operators per day and the instrumentation required to continuously control the process represent a substantially greater investment than what is required for batch processing. This disadvantage can only be overcome by utilizing that labor and investment at relatively high production rates.

62.7 Calculation of Yield in a Crystallization Process

In order to calculate the yield in a crystallization process, it is necessary that the concentration of feed, mother liquor, and any change in solvent inventory (evaporation) be known. In most crystallization processes, the supersaturation in the residual mother liquor is relatively small and can be ignored when calculating the yield. With some materials, such as sugar, a substantial amount of supersaturation can exist, and under such circumstances the exact concentration of the solute in the final mother liquor must be known in order to make a yield calculation. The product crystal may be hydrated, depending on the compound and temperature at which the final crystal is separated from the mother liquor.

Shown below is a formula method for calculating the yield of a hydrated crystal from a feed solution [Myerson, 1993].

$$P = R\frac{100W_o - S(H_o - E)}{100 - S(R-1)} \tag{62.3}$$

where

P = weight of product

$R = \dfrac{\text{mole weight of hydrate crystal}}{\text{mole weight of anhydrous crystal}}$

S = solubility at the mother liquor (final) temperature in units/100 units of solvent

W_o = weight of anhydrous solute in feed

H_o = weight of solvent in feed

E = evaporation

62.8 Mathematical Models of Continuous Crystallization

Randolph and Larsen [1988] developed a method of modeling continuous crystallizers in which the growth rate is independent of size and the slurry is uniformly mixed. Such crystallizers are often referred to as the mixed-suspension mixed-product removal (MSMPR) type. For operation under steady conditions, the population density of an MSMPR crystallizer (FC and DTB types shown in Figure 62.2 and Figure 62.3. respectively) is

$$n = n^o e^{-L/GT} \tag{62.4}$$

where

n = population density, number/mm

G = growth rate, mm/h

T = retention time, h

L = characteristic length, mm

n^o = nuclei population density (i.e., intercept at size $L = O$)

A plot of the ln n versus L will be a straight line if the system is operating under the conditions assumed above. The nucleation rate and the mean particle size (by weight) are

$$B^o = Gn^o \tag{62.5}$$

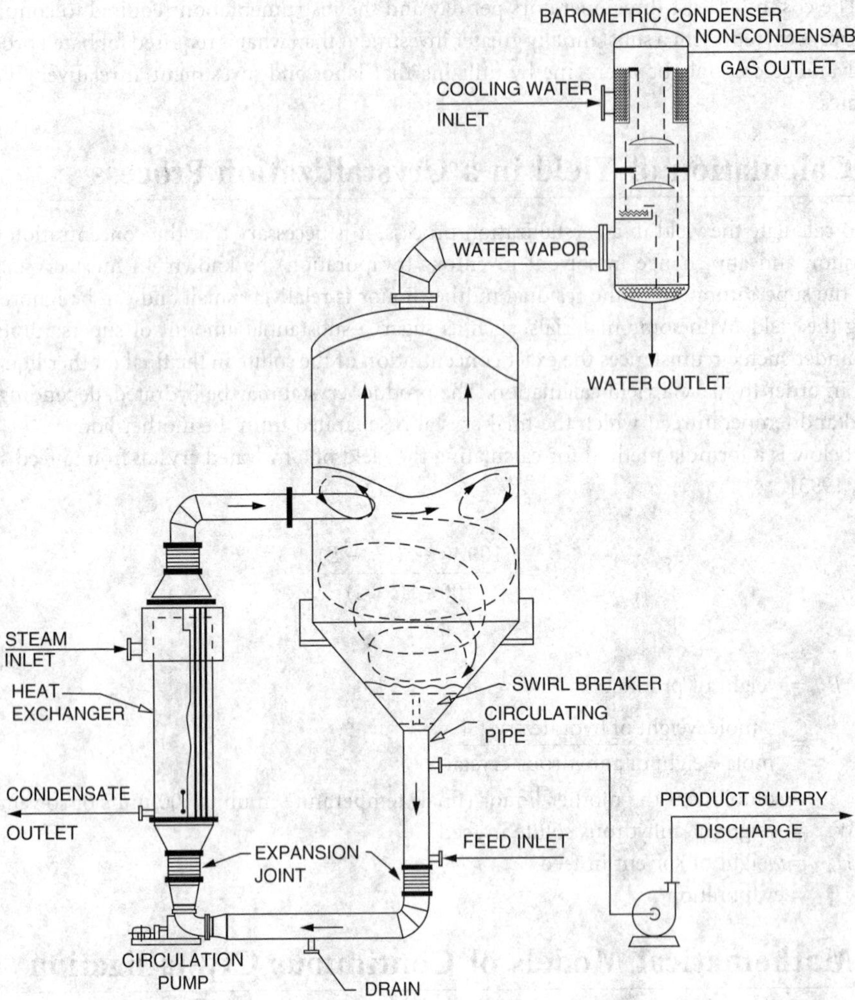

FIGURE 62.2 Swenson forced-circulation crystallizer. (*Source:* Courtesy of Swenson Process Equipment Inc.)

$$L_a = 3.67GT \tag{62.6}$$

where

L_a = average particle by weight
B^o = nucleation rate, number/cc-sec

It is possible to calculate the particle size distribution by weight if the assumptions above are valid and if the plot of ln n versus L is a straight line. The weight fraction up to any size L is

$$W_x = 1 - e^{-x}\left(1 + x + \frac{x^2}{2} + \frac{x^3}{6}\right) \tag{62.7}$$

where

x = L/GT
W_x = cumulative weight fraction up to size L

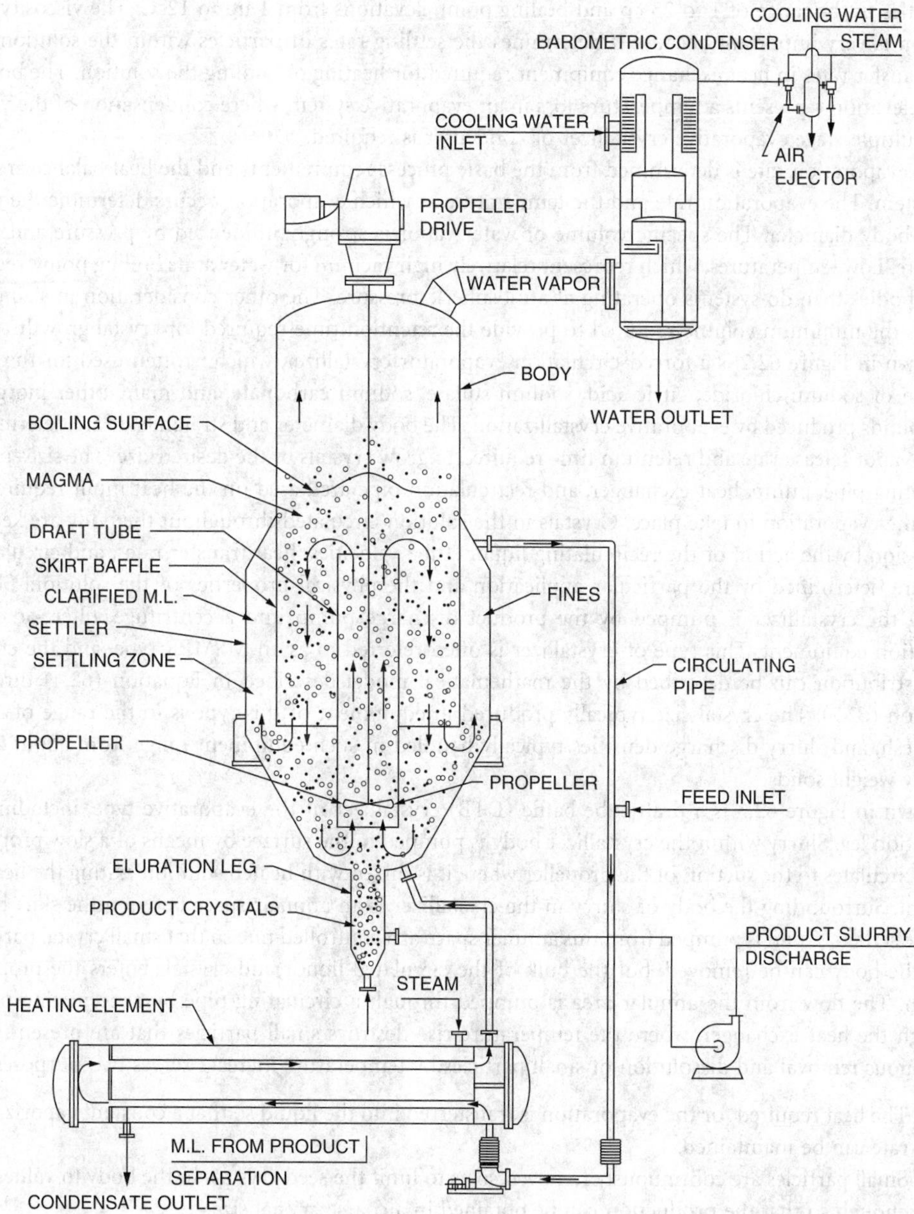

FIGURE 62.3 Swenson draft-tube baffle crystallizer. (*Source:* Courtesy of Swenson Process Equipment Inc.)

In solving Equation (62.4) and Equation (62.5), it must be remembered that the growth rate and the nucleation rate must be measured under the same conditions. In evaluating performance of crystallization equipment, it is necessary to know the heat balance, material balance, and population balance of the particles being used as seed (when used), as well as the product population balance.

62.9 Equipment Designs

While many solvent systems are possible, most large-scale industrial crystallizers crystallize solutes from water. Organic solvents are sometimes encountered in the petroleum industry, and alcohol solutions or mixtures of alcohol and water are found in pharmaceutical applications. Typically, water solutions have viscosities in the range of 1 to 25 cp and boiling point elevations from 1 up to 12°C. The viscosity of a solution is very important because it determines the settling rates of particles within the solution and heat transfer rates in heat exchange equipment required for heating or cooling the solution. The boiling point elevation represents a temperature loss in an evaporative system where condensation of the vapor in a multiple-stage evaporative crystallizer or condenser is required.

The evaporation rate is determined from the basic process requirements and the heat balance around the system. The evaporation rate and the temperature at which evaporation occurs determine the minimum body diameter. The specific volume of water vapor is strongly influenced by pressure and temperature. Low temperatures, which represent relatively high vacuum for water at its boiling point, require larger bodies than do systems operating at atmospheric pressure. The other consideration in sizing the body is the minimum volume required to provide the retention time required for crystal growth.

Shown in Figure 62.2 is a forced-circulation evaporator-crystallizer, which is often used for the production of sodium chloride, citric acid, sodium sulfate, sodium carbonate, and many other inorganic compounds produced by evaporative crystallization. The body diameter and straight side are determined by the vapor release rate and retention time required to grow crystals of the desired size. The sizes of the circulating pipe, pump, heat exchanger, and recirculation pipe are based on the heat input required to cause the evaporation to take place. Crystals in the solution circulated throughout the body are kept in suspension by the action of the recirculating liquor. Tube velocities, heat transfer rates, and circulation rates are determined by the particular application and the physical properties of the solution. Slurry leaving the crystallizer is pumped by the product discharge pump into a centrifuge, filter, or other separation equipment. This type of crystallizer is often referred to as an MSMPR type, and the crystal size distribution can be described by the mathematical model described in Equation (62.4) through Equation (62.7). The crystal size typically produced in equipment of this type is in the range of 30 to 100 Mesh, and slurry discharge densities typically handled in such equipment range from about 20 to 40% by weight solids.

Shown in Figure 62.3 is a draft-tube baffle (DTB) crystallizer of the evaporative type, including an elutriation leg. Slurry within the crystallizer body is pumped to the surface by means of a slow propeller and recirculates to the suction of the propeller where it is mixed with heated solution exiting the heating element. Surrounding the body of slurry in the crystallizer is an annular space between the skirt baffle and the settler. Liquid is pumped from this annular space at a controlled rate so that small crystal particles from the body can be removed, but the bulk of the circulated liquor and crystals enters the propeller suction. The flow from the annular area is pumped through a circulating pipe by a circulating pump through the heat exchanger, where the temperature rise destroys small particles that are present. This continuous removal and dissolution of small particles by temperature increase serves two purposes:

1. The heat required for the evaporation is transferred into the liquid so that a constant vaporization rate can be maintained.
2. Small particles are continuously removed so as to limit the seed crystals in the body to values low enough so that the production can be obtained in a coarse crystal size.

When the crystals become too large to be circulated by the propeller, they settle into the elutriation leg, where they are washed by a countercurrent stream of mother liquor pumped from behind the baffle. Crystals leaving the leg are therefore classified generally at a heavier slurry density than would be true if they were pumped from the body itself. This combination of removal of unwanted fines for destruction and classification of the particle size being discharged from the crystallizer encourages the growth of larger particles than would be obtained in a crystallizer such as the forced circulation type in Figure 62.2. Typically, the DTB crystallizer is used for products in the range of 8 to 20 Mesh with materials such as ammonium sulfate and potassium chloride.

Shown in Figure 62.4 is a surface-cooled crystallizer, which is frequently used at temperatures close to ambient or below. Slurry leaving the body is pumped through a heat exchanger and returns to the body through a vertical inlet. Surrounding the circulating slurry is a baffle that permits removal of unwanted fine crystals or provides for the removal of clarified mother liquor to increase the slurry density

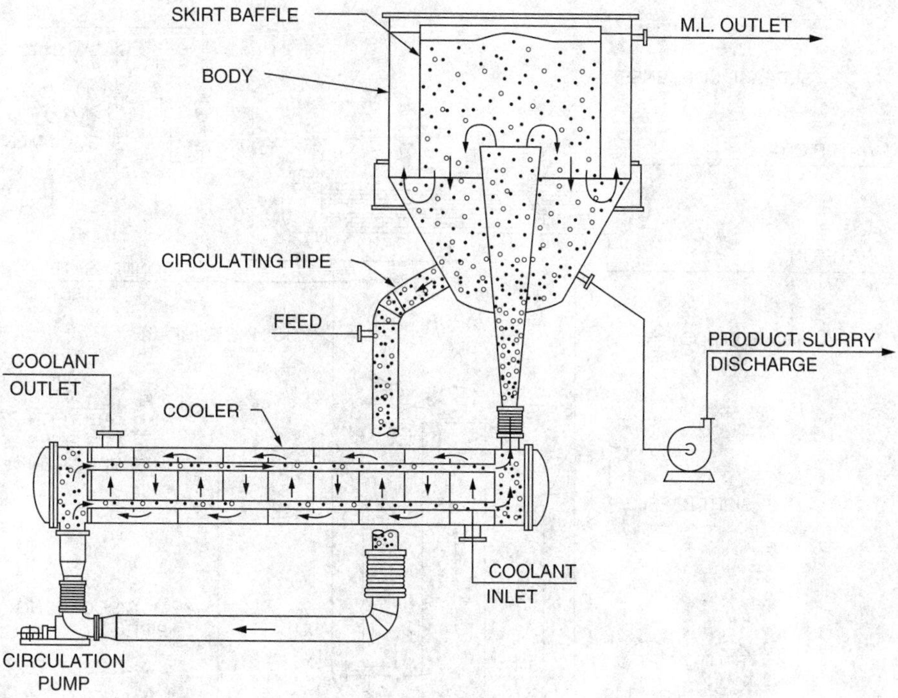

FIGURE 62.4 Swenson surface-cooled crystallizer. (*Source:* Courtesy of Swenson Process Equipment Inc.)

within the crystallizer body. Slurry pumped through the tubes of the cooler is chilled by a coolant that is circulated outside the tubes. The temperature difference between the coolant and the slurry flowing through the tubes must be limited to approximately 3 to 8°C. The temperature drop of the slurry passing through the tubes is normally about 0.5°C. These very low values are required in order to minimize the growth of solids on the tubes. Crystallizers of this type produce a product that ranges between 20 and 150 Mesh in size. Common applications are for the production of copper sulfate pentahydrate, sodium chlorate, sodium carbonate decahydrate, and sodium sulfate decahydrate.

Shown in Figure 62.5 is a reaction-type DTB crystallizer. This unit, while in many respects similar to the DTB crystallizer shown in Figure 62.3, has the important difference that no heat exchanger is required to supply the heat required for evaporation. The heat of reaction of the reactants injected into the crystallizer body supplies this heat. Typically, this type of equipment is used for the production of ammonium sulfate, where sulfuric acid and gaseous ammonia are mixed in the draft tube of the crystallizer so as to produce supersaturation with respect to ammonium sulfate. The heat of reaction is removed by vaporizing water, which can be recirculated to the crystallizer and used for the destruction of fines. Whenever a chemical reaction causes a precipitation of crystalline product, this type of equipment is worth considering because the conditions used in crystallization are compatible with low temperature rises and good heat removal required in reactors. By combining the reactor and crystallizer, there is better control of the particle size with an obvious decrease in equipment costs.

62.10 Evaporation

When a solution is boiled (evaporated) at constant pressure, the total pressure above the solution represents the sum of the partial pressures of the liquids that are boiling. If only water is present, then the pressure above the solution at any temperature corresponds to water at its boiling point at that pressure. If there is more than one component present and that component has a vapor pressure at the temperature of the liquid, then the total pressure represents the vapor pressure of water plus the vapor

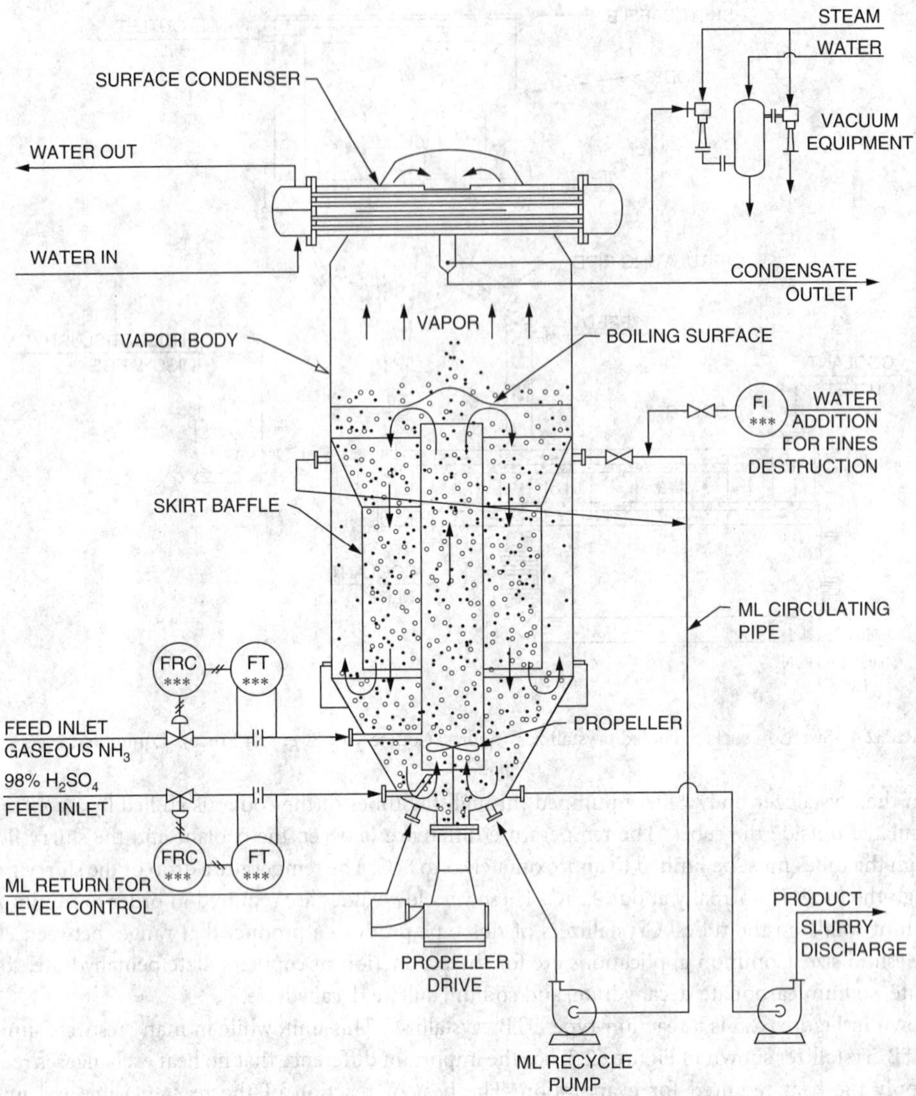

FIGURE 62.5 Swenson reaction-type DTB crystallizer. (*Source:* Courtesy of Swenson Process Equipment Inc.)

pressure of the other component. Vapor leaving such a system, therefore, represents a mixture of solvents in the ratio of their partial pressures. In a sense, an evaporator is a single plate distillation column. In most applications, the vapor pressure of the solute is negligible and only water is removed during boiling, which can be condensed in the form of a pure solution. However, when volatile compounds are present (e.g., H_3BO_3, HNO_3), some of the volatile material will appear in the overhead vapor.

Since the heat required to vaporize water is approximately 556 cal/kg (1000 Btu/lb), it is important to reduce the amount of energy required as much as possible so as to improve the economics of the process. For this reason, multiple-effect evaporators were developed in the middle of the 19th century and continue today as an important means for achieving good economy during evaporation or crystallization. A multiple-effect falling-film evaporator consisting of three vessels and a condenser is shown in Figure 62.6. In this type of equipment, the vapor boiled from the first effect (the vessel where the steam enters) is conducted to the heat exchanger of the second effect, where it acts as the heating medium. Vapor boiled in the second effect is conducted to the third effect, where it again acts as the heating medium. Vapor leaving the third effect, in this case, is condensed in a condenser utilizing ambient-

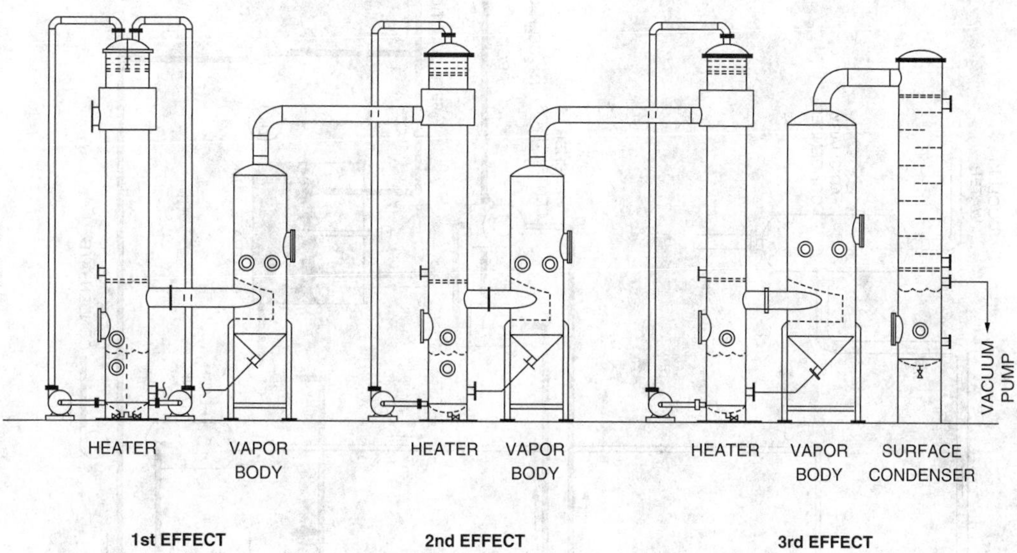

FIGURE 62.6 Swenson triple-effect falling-film evaporator. (*Source:* Courtesy of Swenson Process Equipment Inc.)

temperature water. The flow of feed solution to the evaporator can be either forward, backward, or parallel. In a forward feed evaporator, the feed enters the first effect, then passes to the second effect, and is ultimately removed from the third effect as concentrated liquor. With this type of flowsheet, heat exchange means must be employed to minimize the sensible heat required for the liquid fed to the first effect. In a backward feed evaporator, this is not normally done.

An alternative means for reducing energy consumption during evaporation is shown in the recompression evaporative crystallizer in Figure 62.7. The technique can be employed on both evaporation and crystallization equipment. In this case, a single vessel is employed, and the vapor boiled out of the solvent is compressed by a centrifugal compressor and used as the heating medium in the heat exchanger. The compressed vapor has a higher pressure and a higher condensing temperature so that there is a change in temperature between the vapor being condensed in the heater and the liquid being heated in the heat exchanger. In utilizing this technique, it must be remembered that the boiling point elevation decreases the pressure of the vapor above the liquid at any given temperature and, thereby, represents a pressure barrier that must be overcome by the compressor. The efficiency of this process varies greatly with the boiling point elevation. As a practical matter, such techniques are limited to those liquids which have boiling point elevations of less than about 13°C. Typically, such compressors are driven at constant speed by an electric motor. The turndown ratio on a constant speed compressor is about 40%. A variable-speed drive would give a greater range of evaporative capacity.

During the last 100 years, a wide variety of evaporator types has evolved, each offering advantages for certain specific applications. The forced-circulation crystallizer shown in Figure 62.2 is utilized for many applications where no crystallization occurs, but the liquids being handled are viscous, and use of the circulation system is needed to promote heat transfer. A number of evaporator types have been developed that require no external circulating system. For the most part, these rely upon thermo-syphon effects to promote movement of liquid through the tubes as an aid to heat transfer. The calandria evaporator (or Roberts type) shown in Figure 62.8 is a design that has been widely used since the 19th century for both crystallization and evaporation applications. It relies on natural circulation in relatively short tubes (1 to 2 m) to maintain heat transfer rates; a relatively large amount of recirculation occurs through the tubes. Since there is no recirculation pump or piping, this type of equipment is relatively simple to operate and requires a minimum of instrumentation. The volume of liquid retained in this vessel is much larger than in some of the rising or falling film designs and, therefore, in dealing with heat-sensitive materials where concentration must proceed at relatively short retention times, the calandria would be

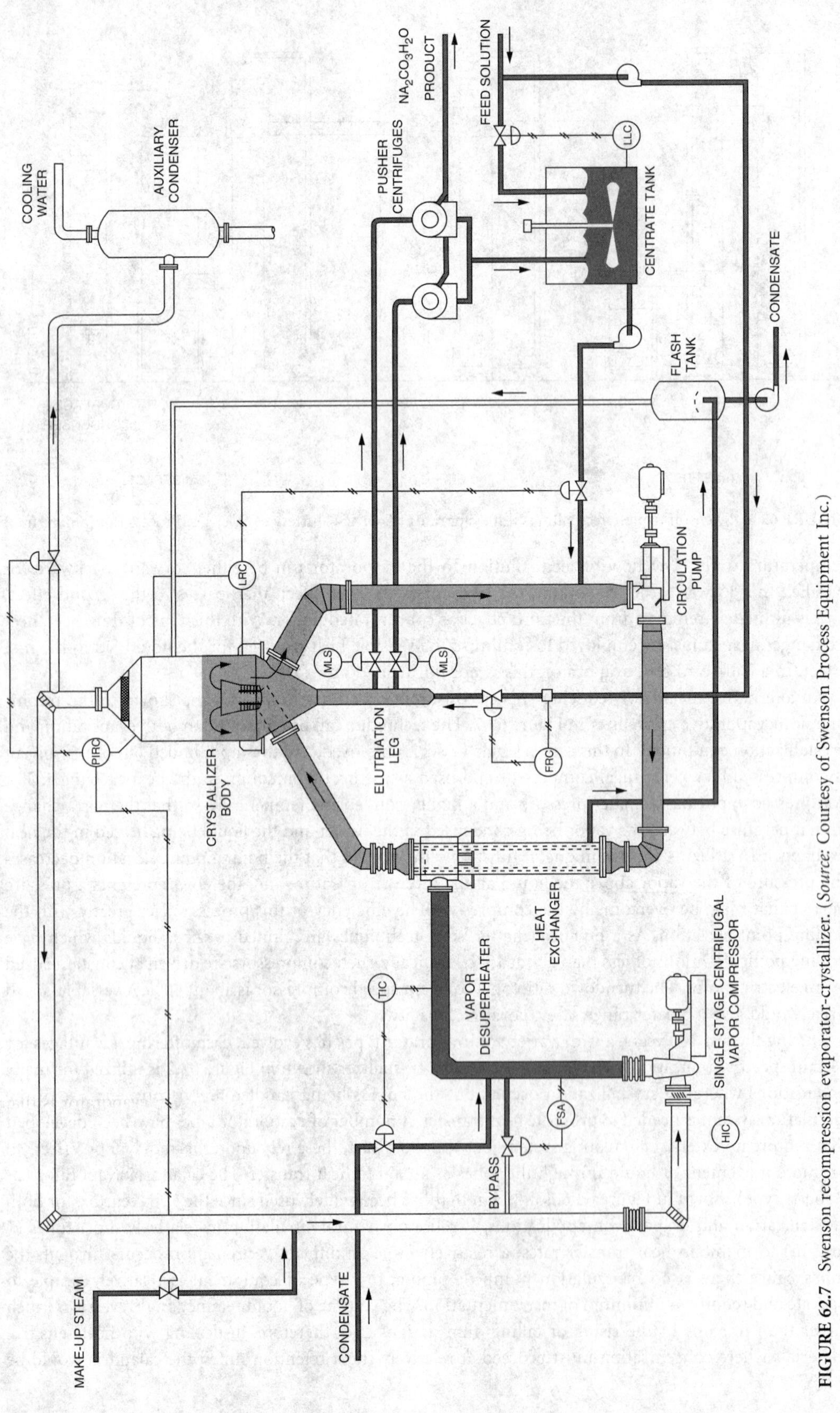

FIGURE 62.7 Swenson recompression evaporator-crystallizer. (*Source:* Courtesy of Swenson Process Equipment Inc.)

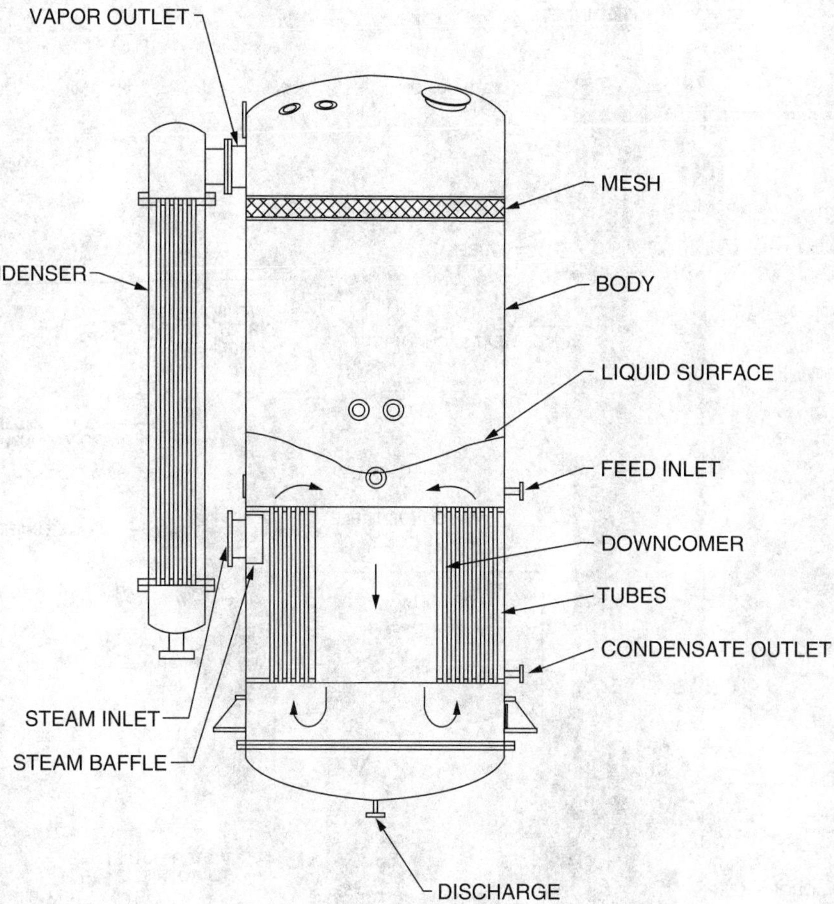

FIGURE 62.8 Swenson calandria evaporator. (*Source:* Courtesy of Swenson Process Equipment Inc.)

a poor choice. In many situations, however, especially where some crystallization may occur, this evaporator may be operated successfully in a semi-batch or continuous manner.

The falling-film evaporator shown in Figure 62.9 is similar to the rising-film evaporator, except that there must be sufficient liquid at all times entering the heater at the feed inlet to wet the inside surface of the tubes in the heat exchanger bundle. With insufficient circulation, solute material can dry on the tubes and cause a serious reduction in heat transfer. Many falling-film evaporators operate with a recirculating pump between the concentrated liquor outlet and the feed inlet to be certain that the recirculation rate is adequate to maintain a film on the tubes at all times. If this is done, the system can operate stably through a wide range of capacities and achieve very high rates of heat transfer, often 50 to 100% more than are obtained in a rising-film evaporator. The other advantage of the falling film evaporator is that it can operate with very low temperature differences between the steam and the liquid since there is no hydrostatic pressure drop of consequence within the tubes to prevent boiling at the inlet end of the heat exchanger. As a result, this type of design has found wide application as a recompression evaporator.

Even though evaporators are typically used where no precipitation of solids occurs, there is often a trace of precipitation in the form of scaling components that coat the inside of the tubes over a relatively long period of time. This scaling is analogous to that which occurs in boilers and many other types of heat transfer equipment. Typically it is due to either a small amount of precipitation or, because of the composition of the materials being concentrated, some inverted solubility components. Such scaling may often be reduced by a technique known as "sludge recirculation." This is commonly done in cooling tower blowdown evaporation and in the evaporation of salt brines where scaling components are present.

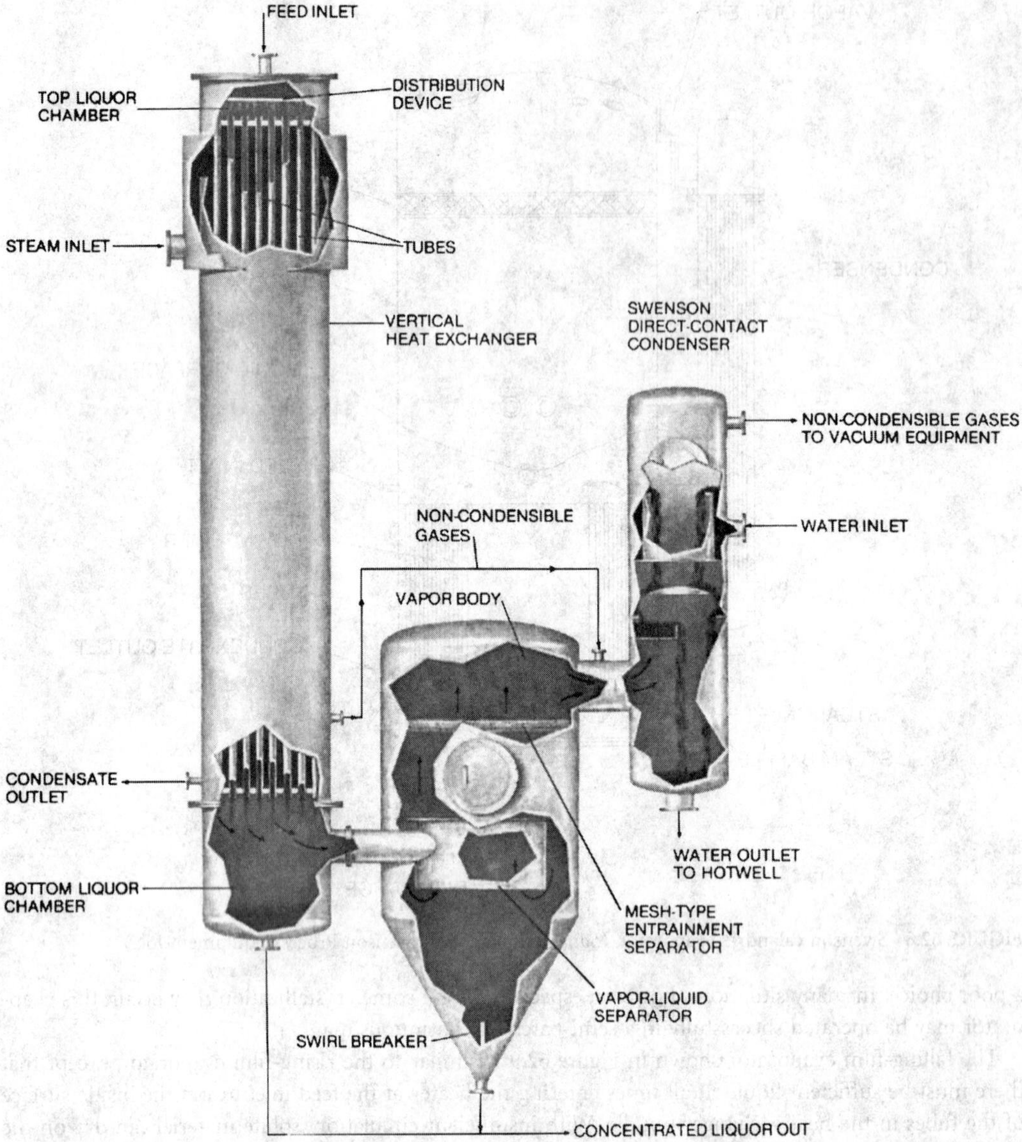

FIGURE 62.9 Swenson falling-film evaporator. (*Source:* Courtesy of Swenson Process Equipment Inc.)

In these cases, the evaporator flowsheet is designed in such a way that a thickened slurry of the scaling component can be recirculated from the discharge of the evaporator back to the feed side. By maintaining an artificial slurry density of the scaling component, which is higher than the natural slurry density, it is often possible to reduce the growth of scale which occurs on heat transfer surfaces.

Defining Terms

Crystal — A solid bounded by plane surfaces that has an internal order with atoms or molecules in a fixed lattice arrangement.

Crystallizer — An apparatus for causing the crystallization of solutes from solvents by means of changes in heat or solvent inventory.

Evaporator — An apparatus for causing water or other solvents to be removed from a solution in order to increase the concentration of the solution.

Nucleation — The birth of a new crystal within a supersaturated solution.

Recompression — A process for collecting the vapor boiled from the solution in an evaporator or crystallizer and compressing to a higher pressure, where it can be used as the heating medium for said evaporator or crystallizer.

Supersaturation — A metastable condition in a solution that permits nucleation and growth of crystals to occur.

References

Campbell, A. N. and Smith, N. O. 1951. *Phase Rule,* 9th ed. Dover, Mineola, NY.

Linke, W. F. (Ed.) 1958. *Solubilities, Inorganic and Metal-Organic Compounds,* Van Nostrand, New York.

Myerson, A. (Ed.) 1993. *Handbook of Industrial Crystallization,* Butterworth, Boston. p. 104.

Randolph, A. D. and Larson, M. A. 1988. *Theory of Particulate Processes,* 2nd ed., Academic Press, Boston. p. 84.

Washburn, E. W. (Ed.) 1926. *International Critical Tables,* McGraw-Hill, New York.

63

Membrane Separation

Theodore T. Moore
(Second Edition)
Georgia Institute of Technology

Shilpa Damle
(Second Edition)
Georgia Institute of Technology

David Wallace
(Second Edition)
Georgia Institute of Technology

William J. Koros
(First Edition)
Georgia Institute of Technology

In its simplest form, a membrane is either a porous or dense (i.e., nonporous) material that is used to separate mixtures of gases or liquids. Membranes used for separations such as microfiltration and ultrafiltration are porous and rely upon hydrodynamic sieving of particles from suspending fluids, and are covered separately in other chapters under their respective headings. Unlike such hydrodynamic sieving devices, the separation methods covered in this chapter — dialysis, reverse osmosis, and gas separation membranes — operate at a molecular scale (3 to 5 Å) to achieve selective passage of one or more components in a feed stream. Figure 63.1 illustrates idealized flow schemes of these processes and the materials that they separate. Penetrant partitioning into the membrane and thermally activated diffusion within the membrane cooperate to yield a *solution–diffusion* process defined as *permeation.* The two primary measures of membrane performance are the membrane *permeability* (i.e., productivity) of the desired penetrant and the membrane *selectivity* (i.e., separation effectiveness) between penetrants. Industrially relevant membranes are generally **asymmetric** and typically "packaged" in one of three basic configurations: flat sheet, tubular, and hollow fiber (Figure 63.2). Implementation of these materials in actual processing environments introduces challenges that could affect membrane stability and performance if not addressed. These considerations as well as a brief section on the future directions of membrane technology are covered in the present chapter.

63.1 Dialysis

Dialysis is a primarily concentration-driven membrane separation, in contrast to most other membrane processes, which are predominantly pressure driven. Dialysis occurs through a semipermeable membrane to separate components based primarily on their different diffusivities through the membrane. Dialysis processes are most effective for systems having solutes that differ significantly in molecular size from the solvent. Although the concentration-driven nature of dialysis limits the flux, it allows application in systems where the species to be separated are sensitive to high shear rates or high temperatures, particularly bioseparations.

0-8493-1586-7/05/$0.00+$1.50

Process	Concept	Materials Passed	Driving Force	Material Retained
Dialysis	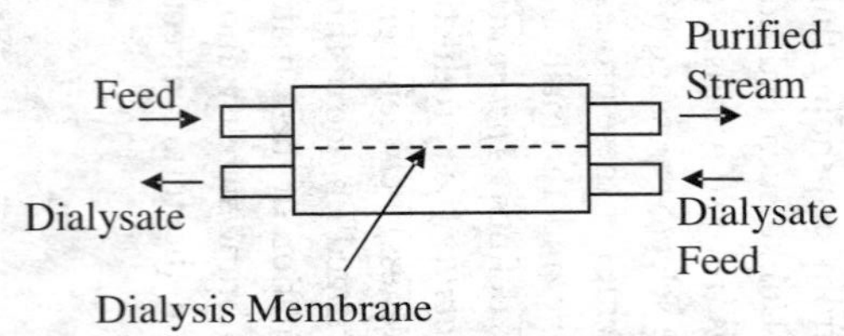	Ions and Low-Molecular Weight Organics	Concentration Difference	Dissolved and Suspended Material with Molecular Weight > 1000
Reverse Osmosis	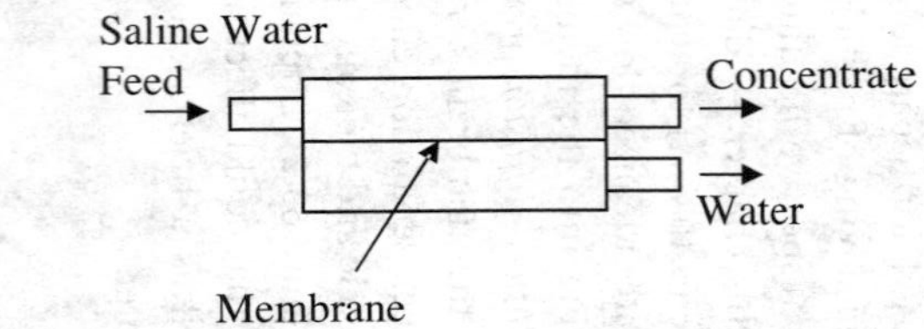	Water	Pressure Difference, Typically 100-1500 psi	Virtually All Suspended and Dissolved Material
Gas and Vapor Separations	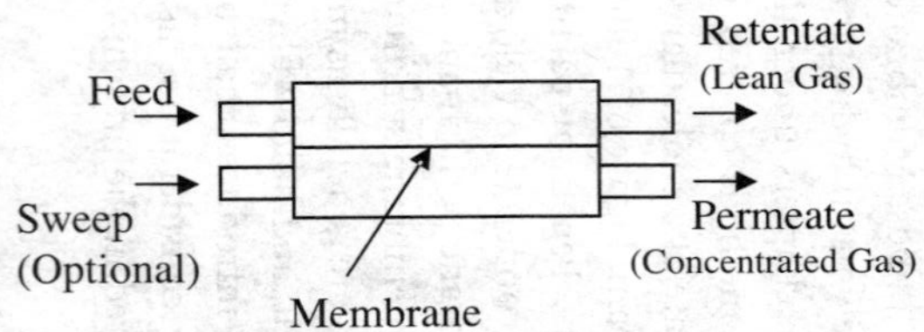	Gases and vapors	Pressure Difference, Typically 15-1500 psi	Membrane-Impermeable Gases and Vapors
Vapor-Liquid Separations	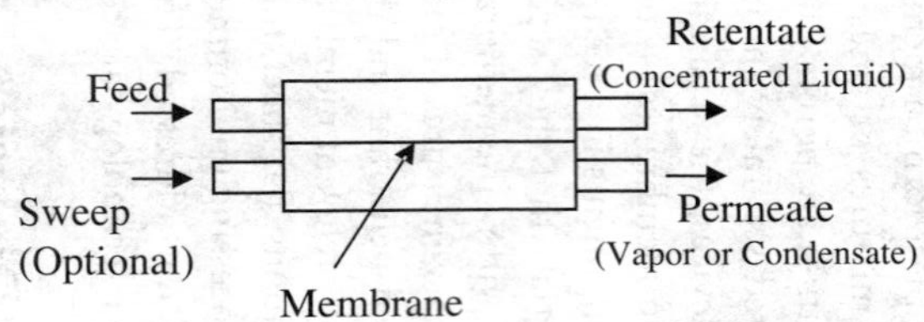	Vapors	Pressure Difference and Vapor-Liquid Equilibrium	Membrane-Impermeable Liquids and Vapors

FIGURE 63.1 Idealized flow schemes and materials separated.

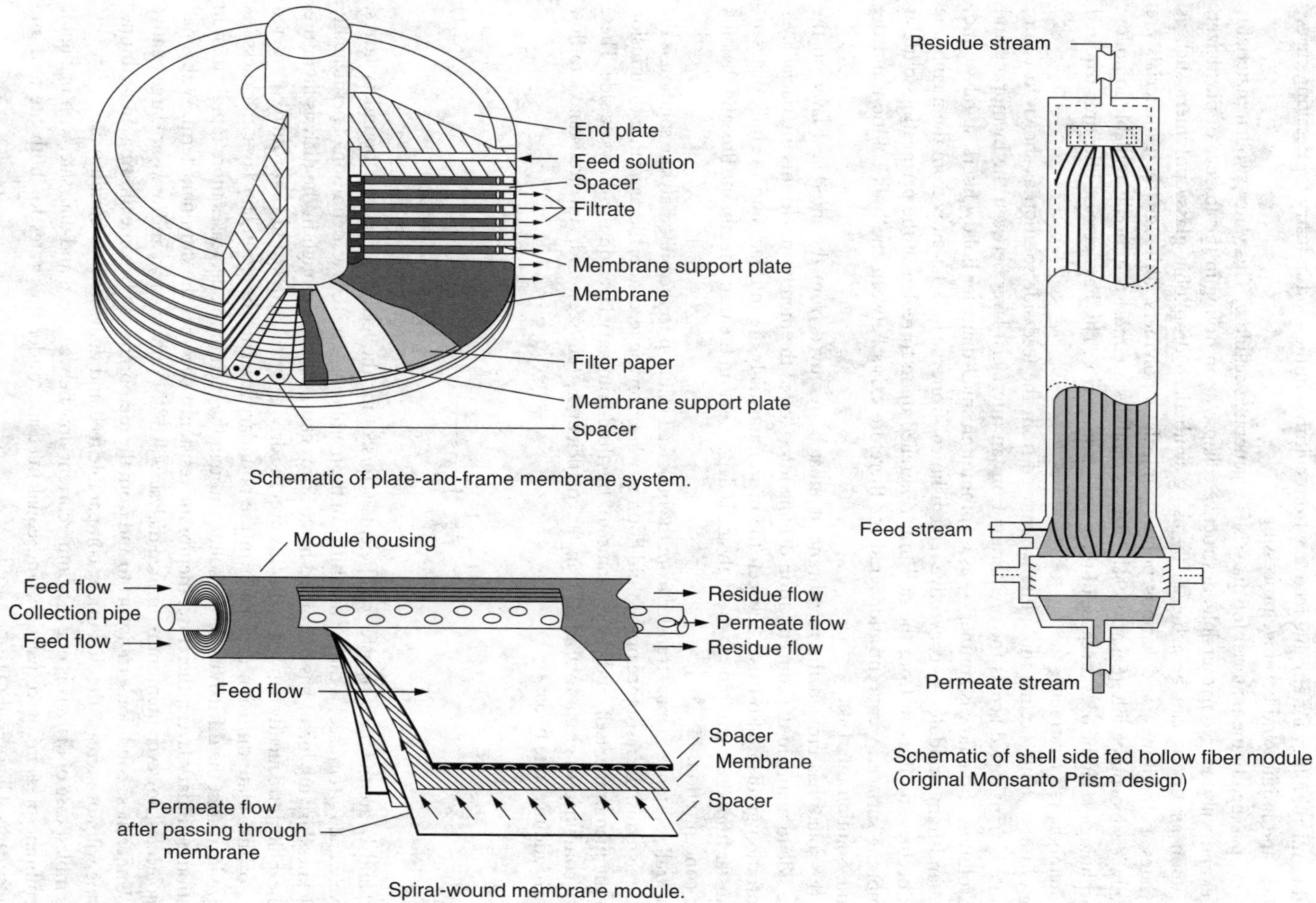

FIGURE 63.2 Configurations of membrane modules. (*Source:* Koros, W. J. and Fleming, G. K. 1993. *J. Membrane Sci.*, 83:1. With permission.)

The most prominent use of dialysis membranes is in the treatment of end-stage renal disease, or kidney failure. In this process, low- and intermediate-molecular-weight waste products are removed from the blood of a patient. Hemodialysis represents the largest market for membrane materials, although growth in this market is limited compared to other membrane applications. To a much smaller extent, dialysis is applied in microbiology for enzyme recovery from cultures, in the food industry for desalting cheese whey solids, and in juice and beverage processing.

The principal membrane material for dialysis is regenerated cellulose. This water-swollen material is a hydrogel and will undergo irreversible, destructive collapse upon drying in the absence of **plasticizer.** Stability upon air drying can be achieved by using synthetic hydrophobic **glassy polymers** such as methacrylates, polysulfones, and various polyacrylonitrile copolymers, although these materials are intrinsically less biocompatible [Sakai, 1994]. Dialysis modules are most commonly available in hollow fiber morphologies (Figure 63.2) because of their high surface areas. This has decreased the time required for each hemodialysis treatment.

As dialysis of a solute occurs, the solute is removed from the feed side solution as shown in Figure 63.1. Solute is transported across the membrane from an interfacial layer between the membrane and the bulk solution. If the flow is laminar — as is common in small-diameter hollow fibers, spiral-wound, and plate-and-frame modules — convective replenishment from the bulk far away from the membrane does not occur, and diffusion from the bulk must be relied upon to replenish the permeating solute at the membrane surface. A concentration gradient therefore exists between the bulk solution and the membrane–solution interface.

After the solute has crossed the membrane, it must be removed from the interfacial layer on the **dialysate** side to maintain the concentration driving force across the membrane. In this interfacial layer a concentration gradient also exists between the membrane–dialysate interface and the bulk dialysate. Mass transfer from the membrane to the dialysate is dependent on the transmembrane flux and the bulk dialysate concentration.

The overall resistance to mass transfer from the bulk feed solution to the bulk dialysate solution is the sum of three terms: the resistance of the fluid boundary layer on the feed side, the resistance to diffusion through the membrane itself, and the resistance of the fluid boundary layer on the dialysate side. These terms are combined to give a total mass transfer coefficient, k_T (cm/sec), and, since resistances to flow are linearly additive, one finds:

$$\frac{1}{k_T} = \frac{1}{k_F} + \frac{\ell}{P_M} + \frac{1}{k_D} \tag{63.1}$$

where k_F and k_D are the mass transfer coefficients (cm/sec) for the feed and dialysate boundary layers, respectively, and P_M (cm^2/sec) is the **permeability** of the membrane. The IUPAC definition of permeability (flux·thickness/driving force) has been used here [Koros et al., 1996], although in the dialysis literature, the permeability is frequently defined as the flux over the driving force. Both k_F and k_D are often considered the ratio of the solute diffusion coefficient in the boundary layer to the boundary layer thickness. As solute size increases, the **diffusivity** of the solute through the membrane decreases more rapidly than in the solution boundary layers; therefore, membrane resistance becomes the dominant term. With small solute molecules, however, membrane-limited dialysis will occur only when k_F and/or k_D are large relative to P_M/ℓ. Under these conditions, transport to and from the membrane can become limiting, and a thinner membrane will not increase the flux under comparable feed and dialysate flow conditions.

The simplest case of dialysis is one of countercurrent flow between feed and dialysate, with negligible transmembrane pressure or convection, and equal inlet and outlet flow rates for both the feed and dialysate streams. The overall mass balance is

$$\dot{m} = Q_F(C_{Fi} - C_{Fo}) = Q_D(C_{Do} - C_{Di}) \tag{63.2}$$

where $\dot{m}$ is the mass flow rate of the solute across the membrane from feed to dialysate streams (g/sec), Q is the volumetric flow rate (cm³/sec), F and D refer to feed and dialysate solutions, respectively, and C_i and C_o are the solute concentrations in and out of the dialyzer, respectively (g/cm³). In terms of an overall mass transfer coefficient, k_T, the mass flow rate of solute is

$$m = k_T A \frac{(C_{Fo} - C_{Di}) - (C_{Fi} - C_{Do})}{\ln[(C_{Fo} - C_{Di})/(C_{Fi} - C_{Do})]} \tag{63.3}$$

where A is the membrane area (cm²). This is analogous to heat transfer in a shell-and-tube heat exchanger. Performance of the dialyzer can be expressed as a *dialysance*, D^*, which is defined as the ratio of the mass flow rate and the inlet concentration difference as follows (cm³/sec):

$$D^* = \frac{\dot{m}}{C_{Fi} - C_{Di}} \tag{63.4}$$

A more detailed discussion of the preceding theoretical considerations is given in the *Membrane Handbook* [Ho and Sirkar, 1992]. The handbook also discusses expressions for other flow configurations including cases where convective transport and ultrafiltration of the feed occur.

Donnan dialysis and *electrodialysis* use charged membranes to more efficiently separate ions. Donnan dialysis is the separation of ionic components in a feed based on their ability to diffuse through an ion exchange membrane. Anion exchange membranes with fixed positive charges exclude positive ions except for protons because of their small size but allow diffusion of anions. Similarly, in the absence of an applied electrical field, positively charged ions can only diffuse through ion exchange membranes with fixed negative charges. Materials with negative fixed charges (e.g., Nafion®; Du Pont, Wilmington, DE) are readily available; however, high-efficiency cationic membranes are still under development [Ho and Sirkar, 1992]. Donnan dialysis has found applications in the recovery of acids from metals-processing waste streams.

Another enhanced dialysis method is **electrodialysis.** In electrodialysis, alternating cation- and anion-exchange membranes are aligned between a cathode and an anode. Feed solution passes between each pair of membranes. Positively charged ions flow toward the cathode but can only traverse cation-exchange membranes. Although most are swept away, they tend to accumulate at anion-exchange membranes. Negatively charged ions flow toward the anode, accumulating at cation-exchange membranes. In this manner, alternating streams become enriched and depleted of ions. The accumulation of ions on oppositely charged membranes makes scale formation a major concern (see Section 63.6). This has been overcome by periodically reversing the electrical field to remove the accumulated ions from their respective membranes and then purging the system before introducing fresh feed. Electrodialysis is most economical with total dissolved salt concentrations of less than 1% [Lee and Koros, 2002], and can be used for desalination or protein desalting.

Once again, a detailed theoretical treatment can be found in Ho and Sirkar [1992], but the principal equations are given here. The maximum ionic flux is determined by the **limiting current density**, which occurs when the ion concentration at either the anion-exchange or cation-exchange membrane becomes zero due to a concentration drop in the boundary layer between the membrane and the bulk feed. The limiting current density, i_{LIM} (A/cm³), is given by:

$$i_{LIM} = \frac{C_B z_+ F k}{t_+ - t'_+} \tag{63.5}$$

where C_B (mol/cm³) is the concentration of the bulk feed, z_+ is the charge of the positively charged ion, F is Faraday's constant, k is the mass transfer coefficient (cm/sec), and t_+ and t_+' are the transport numbers (i.e., fraction of total current carried by the positively charged ions) in the membrane and solution, respectively. The permselectivity, α_+, is given by:

$$\alpha_+ = \frac{t_+ - t'_+}{1 - t'_+} \tag{63.6}$$

Another important factor governing electrodialysis is the **current utilization**. Clearly, a high current utilization is desired for the most economical operation; however, actual current utilization will be less than 100% due to imperfect membrane selectivity, osmotic and ion-bound water transport, and current passing through the stack manifold [Ho and Sirkar, 1992]. The actual required current, I (A), is then given by the following equation:

$$I = \frac{z_+ F Q_F \Delta C_+}{\xi} \tag{63.7}$$

where Q_F (cm³/sec) is the volumetric flow rate of the feed and ΔC_+ (mol/cm³) is the concentration difference between the feed and purified solution. Equation (63.5) through Equation (63.7) are written in terms of the cation, but they can also be written for the anion.

63.2 Reverse Osmosis

Reverse osmosis (RO) uses a pressure driving force to overcome osmotic pressure and force the solvent, usually water, through the membrane, thereby separating it from the solute. Closely related to reverse osmosis is **nanofiltration**, also called "loose" RO, which often uses an ion exchange membrane to allow for lower pressure drops, while retaining larger solutes with good rejection of polyvalent salts. Ideal reverse osmosis membranes should display high water flux; high salt retention; hydrolytic, chemical, and biological stability; and minimal change in properties with time.

The economics are most favorable for low solute concentrations since high concentrations require large energy input to overcome higher osmotic pressures. By far, the largest application of reverse osmosis and nanofiltration membranes is in water and wastewater purification. Specific applications include desalination of brackish water and seawater, purification of municipal wastewater, removal of organic contaminants from process streams, treatment of coal wash water, concentration of waste from metal processing, and recovery of dyes from textile wash water [Lee and Koros, 2002]. However, the membranes also find applications in the paper industry for the concentration and partial fractionation of lignosulfonates in wood pulping and in the food industry for the concentration of whey, milk, and maple syrup.

Initially, cellulose acetate was the most common RO membrane material, but polyamides and polyaramides have become more popular recently because of their chemical and biological stability [Khedr, 2003]. These membranes are fabricated as flat sheet composites with porous support membranes or as monolithic hollow fibers to provide good stability and extremely high flux.

It is common for RO feeds to be pretreated before being pumped through the membrane module(s) to reduce fouling (see Section 63.6) or eliminate chemically or biologically reactive species. The feed pressure must be balanced so that the exit pressure is greater than the osmotic pressure, but not so high as to cause membrane failure. This requirement sometimes dictates the use of booster pumps in the later stages of the process. The typical required driving pressure (operating pressure – osmotic pressure) of 100 to 1500 psi requires a high-pressure pump, which is the dominant energy consumer. A discussion of the economics of reverse osmosis is given by Malek et al. [1996]. Current work focuses on reducing the operating pressure necessary to accomplish the separation. The development of nanofiltration has aided this goal, but nanofiltration cannot efficiently reject monovalent ions.

Reverse osmosis is a solution–diffusion membrane separation process that relies upon an applied pressure to overcome the osmotic pressure of the feed solution. At low solute concentrations, the osmotic pressure, π (kPa), of the solution at temperature T (K) with a solute molar concentration C_s (mol/cm³) is given by:

$$\pi = C_s RT \tag{63.8}$$

where R is the ideal gas constant. The concentration term must include dissociation effects of the solute since π is a colligative property of the solution (e.g., in NaCl, C_S refers to the concentration of Na^+ and Cl^- ions since they exist as dissociated species in water). Some solute passes through the membrane in response to the concentration gradient that develops. Fixed charges within the membrane, as in nanofiltration, can help suppress uptake of ionic solutes by *Donnan exclusion* (see Donnan dialysis in Section 63.1) and minimize the undesirable solute flux from the feed to the permeate side.

The molar flux, N (mol/cm^2 · sec), of desired solvent, typically water, through a reverse osmosis membrane of thickness ℓ (cm) is proportional to the difference between the applied pressure difference, Δp (kPa), and the osmotic pressure difference, $\Delta\pi$ (kPa), across the membrane:

$$N_w = \frac{C_w D_w v_w}{RT}\left(\frac{\Delta p - \Delta\pi}{\ell}\right) \tag{63.9}$$

where C_w is the concentration of the solvent in the membrane (mol/cm^3), D_w is the diffusivity of the solvent in the membrane (cm^2/sec), and v_w is the partial molar volume of the solvent (cm^3/mol). The solute flux is given as:

$$N_s = \frac{D_s S_s \Delta C_s}{\ell} \tag{63.10}$$

where D_S is the diffusivity of the solute in the membrane and S_s is a dimensionless partition coefficient for the solute in the membrane material. From Equation (63.9) and Equation (63.10) it is apparent that the solvent flux increases as the pressure applied to the feed is raised, whereas the solute flux does not increase. Thus, by increasing the pressure, rejection of the solute may be 99% or higher.

63.3 Gas and Vapor Separations

Gas separation membranes are starting to compete well with more mature separation technologies, such as pressure swing adsorption and cryogenic distillation, because they are simple in construction, modular in nature, and economical. Development of membrane materials for separation purposes depends on the targeted application. In the most general sense, a membrane material should be chemically and physically stable under anticipated operating conditions, have the productivity and selectivity dictated by process volume and economics, and be conveniently fabricated into modular membrane form. Specifically, materials may be tailored to carry out separations based on size and shape or chemical properties of the penetrants. Because they offer a broad range of functional and structural conformations and are economical (~\$2.0/ft^2), polymers are the most frequently used membrane materials and, therefore, the focus of the following discussion. Polymers cannot, however, easily be used in extreme temperature and harsh chemical conditions. For extreme operating conditions, specialty membranes comprised of ceramics, metals, and carbons are also available.

In an idealized sense, gas separation membranes act as molecular-scale filters separating a feed mixture of A and B into a **permeate** of pure A and nonpermeate (**retentate**) of pure B. Figure 63.3 shows that gas separations by membranes can be performed using four types of transport mechanisms: Knudsen diffusion, solution–diffusion, molecular sieving, and selective surface flow. Transport through dense polymeric membranes is best described using the solution–diffusion mechanism. Transport through these materials occurs due to the thermally activated motion of polymer chain segments that create penetrant-scale transient gaps in the matrix, allowing diffusion from the upstream to the downstream side of the membrane.

The driving force for gas separation is the transmembrane partial pressure typically achieved by pressurization of the feed gas. The permeability of the polymer for component i, P_i, is defined as:

Knudsen Flow	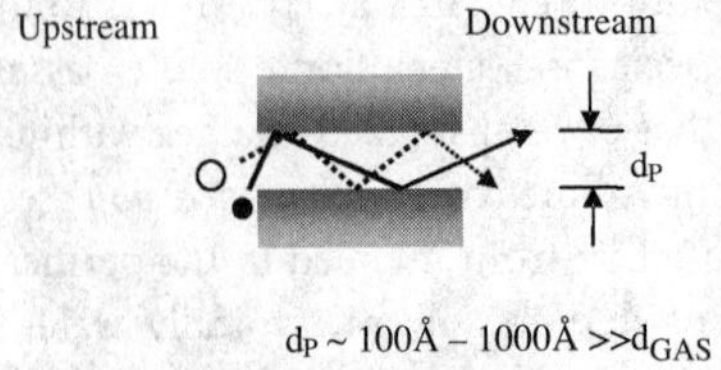	Separation is based on the inverse square root ratio of the molecular weights of A & B. Selectivity is very limited.
Solution – Diffusion	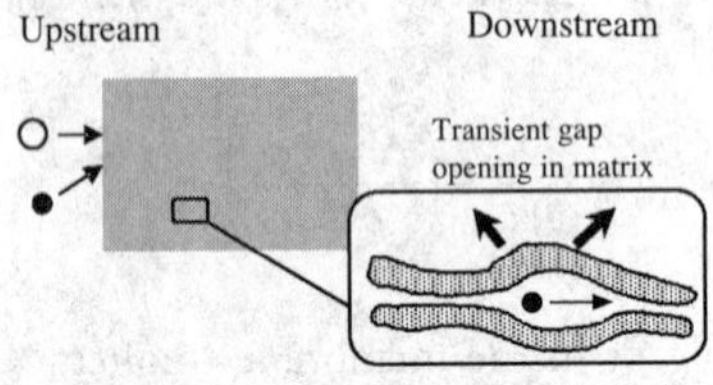	Separation is based on both solubility and diffusive considerations. In polymers, transient gaps allow diffusive "jumps" by the penetrants. Permeability and selectivity are limited.
Molecular Sieving	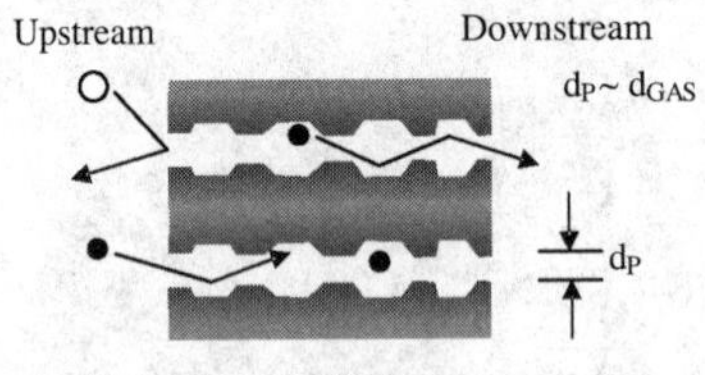	Separation in rigid ultramicroporous materials is based primarily on the much higher diffusion coefficients of the *smaller* penetrant, but sorption may be important for similarly sized molecules. Selectivity can be infinite.
Selective Surface Flow	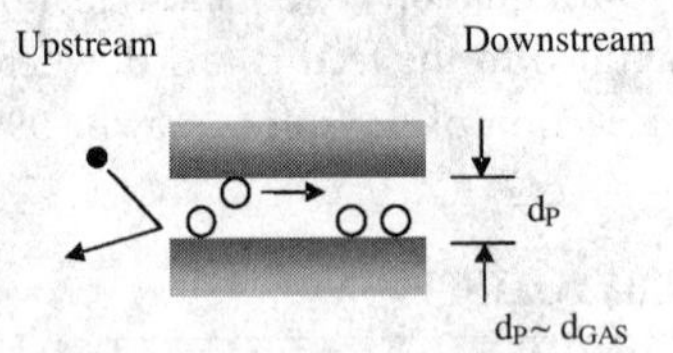	Separation in rigid microporous materials is based on the much higher sorption affinity of the more *condensable* penetrant, which is usually the larger penetrant. Selectivity can be very high.

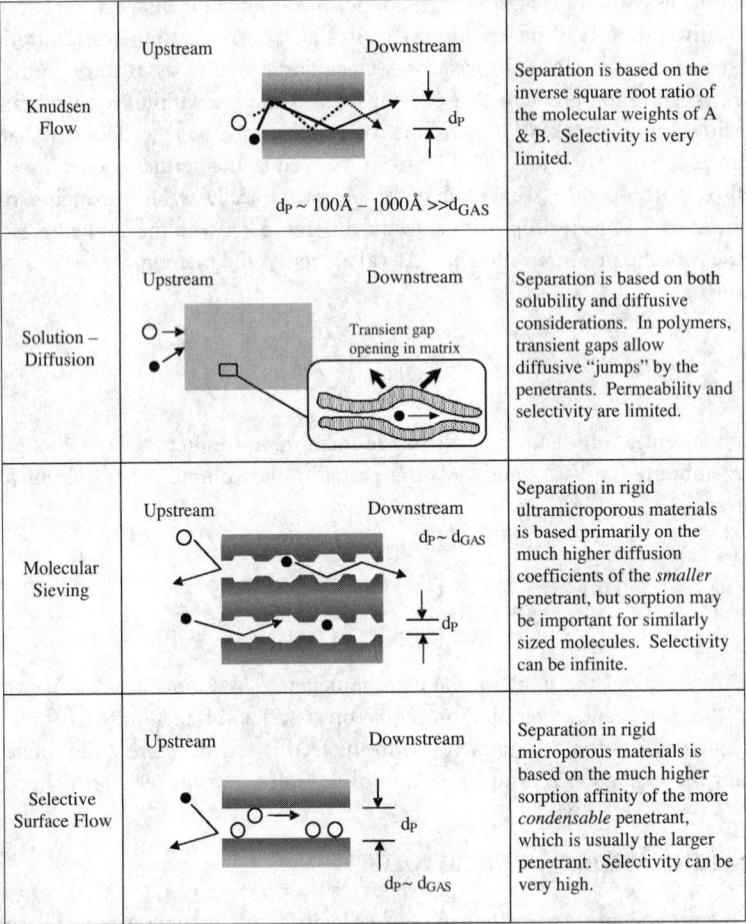

FIGURE 63.3 Transport mechanisms for gas separations.

$$P_i = \frac{N_i}{(\Delta p_i / \ell)} \tag{63.11}$$

where N_i is the gas flux [cm³ (STP)/(cm² · sec)], Δp_i is the partial pressure difference between the upstream and downstream faces of the membrane (cm Hg), and ℓ is the thickness of the selective layer of the membrane (cm). Polymer permeabilities are often expressed in Barrers:

$$1 \text{ Barrer} = \frac{10^{-10} \text{ cm}^3 \text{ (STP)} \cdot \text{cm}}{\text{cm}^2 \cdot \text{cmHg} \cdot \text{s}} \tag{63.12a}$$

Closely related is the measure of productivity, known as *permeance*, through an asymmetric membrane. The permeance, P/ℓ, is simply a pressure normalized flux, and is expressed in Gas Permeation Units, or GPUs:

$$1 \text{ GPU} = \frac{10^{-6} \text{ cm}^3 \text{ (STP)}}{\text{cm}^2 \cdot \text{cmHg} \cdot \text{s}} \tag{63.12b}$$

The permeation of penetrants occurs due to a solution–diffusion mechanism, so the permeability coefficient is expressed as a product of two terms:

$$P_i = D_i S_i \tag{63.13}$$

where D_i is the average diffusion coefficient of the penetrant across the membrane (cm^2/sec) and S_i is the effective solubility coefficient of the penetrant in the membrane [cm^3 (STP)/($cm^3 \cdot$ cm Hg)]. The solubility coefficient is defined as:

$$S_i = C_i / p_i \tag{63.14}$$

where C_i is the concentration of the penetrant i in the polymer [cm^3 (STP)/cm^3] at equilibrium when the external partial pressure is p_i (cm Hg).

The ideal separation factor, $\alpha_{A/B}^*$, is based on the individual permeabilities of two gases A and B and is given as:

$$\alpha_{A/B}^* = \frac{P_A}{P_B} \tag{63.15}$$

The ideal separation factor provides a measure of the intrinsic **permselectivity** of a membrane material for mixtures of A and B. In the absence of strong polymer-penetrant interactions, $\alpha_{A/B}^*$ in mixed feed situations can typically be approximated to within 10 to 15% using the more easily measured ratio of permeabilities of pure components A and B. If Equation (63.13) is substituted into Equation (63.15), the ideal separation factor can be separated into two parts;

$$\alpha_{A/B}^* = \left[\frac{D_A}{D_B}\right]\left[\frac{S_A}{S_B}\right] \tag{63.16}$$

where D_A/D_B is the diffusivity selectivity and S_A/S_B is the solubility selectivity. The diffusivity selectivity is determined by the ability of the polymer to discriminate between the penetrants based on their sizes and shapes and is governed by intrasegmental motions and intersegmental packing. The sorption selectivity is thermodynamic in nature and depends on the condensability of the components, polymer/penetrant interactions, and free volume within the polymer.

The actual separation factor between two components A and B is defined in terms of permeate (Y_A and Y_B) and feed (X_A and X_B) mole fractions.

$$SF = \frac{(Y_A / Y_B)}{(X_A / X_B)} \tag{63.17}$$

Equation (63.17) can be written in terms of D and S but also includes partial pressure terms.

$$SF = \left[\frac{D_A}{D_B}\right]\left[\frac{S_A}{S_B}\right]\left[\frac{\Delta p_A / p_{A\,feed}}{\Delta p_B / p_{B\,feed}}\right] \tag{63.18}$$

From Equation (63.18) it is apparent that the actual separation factor may be less than the ideal selectivity. Also apparent is that when the downstream pressure is nearly zero, Equation (63.18) simplifies to Equation (63.16).

The above equations represent the most common methods of evaluating membrane performance for ideal gas separations. For nonideal gases under high pressures, fugacity driving forces and possible bulk flow phenomena should be addressed [Kamaruddin and Koros, 1997]. In practical membrane applications, the retentate and permeate pressures will be nonzero, and the driving force varies along the surface of the membrane. Under these realistic conditions, even more complex analysis is needed to account for the variations of driving force terms used in Equation (63.18) [Koros and Fleming, 1993].

Applications of membrane-based gas separation technology tend to fall into four major categories: hydrogen separation from a wide variety of slower-permeating **supercritical** components such as CO, CH_4, and N_2; acid gas (CO_2 and H_2S) separation from natural gas; oxygen or nitrogen enrichment of airl; and vapor/gas separations, including gas drying and organic vapor recovery.

The order of the various types of gas–gas applications given in this list provides a qualitative ranking of the relative ease of performing these three types of separations. The fourth application, involving the removal of vapors from fixed gases, is generally carried out with the use of rubbery polymers, whereas the first three applications rely upon glassy materials.

The first large-scale applications of membranes for gas separation were for hydrogen recovery. The most well-known recovery processes include H_2 recovery from ammonia purge gas, the separation of H_2 from CO to optimize the reactant stoichiometry for methanol production, and the separation of H_2 from fuel gas [Lee and Koros, 2002]. The extraordinarily small molecular size of H_2 makes it highly permeable through membranes and easily collected as a permeate product compared to many larger-sized gases such as N_2, CH_4, and CO.

Acid gases such as carbon dioxide and hydrogen sulfide can be found in large quantities in natural gas streams. These components are responsible for the "sour" nature of gases from various sources, and their removal from the fuel improves the heating value of the gas and also helps reduce corrosion of pipelines and transmission equipment [Lee and Koros, 2002]. The relatively high solubilities of CO_2 and H_2S in membranes at low partial pressures, coupled with the low diffusivity and solubility of the bulky supercritical methane molecule, give high membrane productivity for these penetrants with good selectivity over CH_4.

The products of air separation, oxygen and nitrogen of various purities, can be utilized in many applications. Combustion efficiency and biochemical processes benefit from oxygen-enriched air containing 30–40% O_2. Nitrogen at 90–99% purity provides an inert atmosphere useful in many applications, including blanketing fuel storage tanks and pipelines to minimize fire hazards, reducing oxidation during annealing, and retarding the spoilage of foods during transport and storage. This is intrinsically the most difficult of the three types of gas–gas separations because both the diffusivities and solubilities of this gas pair in most membranes are similar. This leads to similar permeabilities for the two components and thus modest selectivities, rendering high-purity separation difficult. Nevertheless, well-engineered commercial systems do exist for producing nitrogen-enriched air above 99% purity.

Gas drying and organic vapor recovery from air or nitrogen are both important separations on which membrane technology has a significant impact. Increasing environmental awareness and the rising costs of energy and chemicals have resulted in efforts aimed at using membrane processes as a means of recovering those resources. The separation of vapors from gases is easily accomplished in rubbery polymers because of the condensable vapor's high solubility selectivity coupled with the near unity diffusivity selectivity. The resulting overall selectivity is quite high. Silicone rubber is a common choice for such applications since it offers high transport rates. This type of separation requires great care to minimize excessive partial pressure buildup of the vapor in the permeate. In many cases a sweep gas or partial vacuum is utilized to reduce the permeate partial pressure buildup at the downstream face of the membrane.

63.4 Vapor–Liquid Separations

Pervaporation is a membrane-mediated evaporation process for separating liquid mixtures. The rate of evaporation of each liquid is governed by the permeability of the component through the separating material. A closely related process utilizes a saturated vapor feed. By either using a vacuum or providing a sweep stream on the downstream side of the membrane, the permeate is collected and either condensed as a liquid or continuously withdrawn. The final composition of permeate is dictated by the composition of the feed stream, which establishes the inherent driving force for transport, and by the permselectivity of the membrane. The separation factor for the pervaporation process, β_{pervap}, is the product of the evaporation separation factor, β_{evap}, and the membrane separation factor, β_{mem}, as shown below:

$$\beta_{pervap} = \beta_{evap} \cdot \beta_{mem} \tag{63.19}$$

At low permeate pressures, the membrane separation factor, $\beta_{mem,}$ is simply equal to the ratio of permeability of the individual components. This is completely analogous to the previous discussion of ideal selectivity in gas separations (Equation 63.15). As with gas separations, material development is highly dependent on both the characteristics of the mixtures to be separated, process temperatures, and process pressures.

Opportunities exist for pervaporation technology where distillation techniques are cumbersome. This is especially true when the mixtures to be separated have close boiling points and their composition cannot be changed by conventional distillation unless the thermodynamic equilibrium is shifted by adding an additional component. The dominant use of pervaporation technology is in the separation of ethanol from water, though production scale separations of isopropanol and water have also been attained [Lee and Koros, 2002; Kujawski, 1999; Staudt-Bickel and Lichtenthaler, 1994].

63.5 Practical Implementation of Membrane Systems

Most current membranes are made with an asymmetric morphology. In order to increase flux without sacrificing mechanical integrity, asymmetric membranes are composed of a thin separating layer supported by a thicker, porous layer (Figure 63.4). Several methods can be employed to achieve this asymmetric morphology. Integrally skinned or "simple" membrane structures are constructed from a variety of polymers, including cellulosics, polyimides, polyamides, and other soluble heterochain polymers such that a single membrane polymer is used for both the support and thin selective layers. Polysulfone is easily formed into porous structures but is difficult to prepare with the truly pore-free skin that is necessary for gas separations. This obstacle is generally overcome using a *caulking* procedure developed by Henis and Tripodi of Monsanto [Henis and Tripodi, 1980]. Another approach involves the generation of a *composite* membrane comprising a nominally unskinned porous support with subsequent attachment of a separating layer. Asymmetric and composite structures permit the preparation of membranes with thin separating skin thicknesses of 1000 Å or less. If the resistance to transport across the porous support layer exceeds roughly 10% of the resistance of the dense selective layer, the efficacy of the membrane is severely limited. This effect places a premium on the ability to generate low-resistance open-celled porous supports.

Once the microscopic morphology of the membrane material itself has been defined (asymmetric, composite, caulked, etc...), there are several choices for membrane module design, including plate-and-frame, spiral-wound, and hollow fiber, as shown in Figure 63.2. Each of the three options has its own advantages and shortcomings.

Plate-and-frame modules are the simplest design, comprised of flat membrane sheets set in stackable frames. Additional membrane area is achieved by increasing the size of the individual membrane sheets, adding additional plates, or by using pleated sheets. The chief advantages are simplicity of design, production, and operation (high flow rates and easy cleaning). Plate-and-frame modules are limited in their surface area-to-volume ratio. This makes them the design of choice for separations that foul easily (see Section 63.6) and do not require high surface areas.

Spiral-wound modules consist of flat membrane sheets layered with spacers and then wound around a central tube. The spacers provide channels for the flow of feed and permeate. Spiral-wound modules retain many of the positive attributes of plate-and-frame modules. Although slightly more difficult to manufacture, operation and cleaning are still relatively simple, and surface area per volume is significantly increased.

The highest surface area-to-volume ratio is achieved in hollow fiber membranes. With areas over an order of magnitude greater than spiral-wound modules of the same volume, hollow fibers dominate markets that require large surface areas where particulate fouling is not a major issue. Area/volume ratio

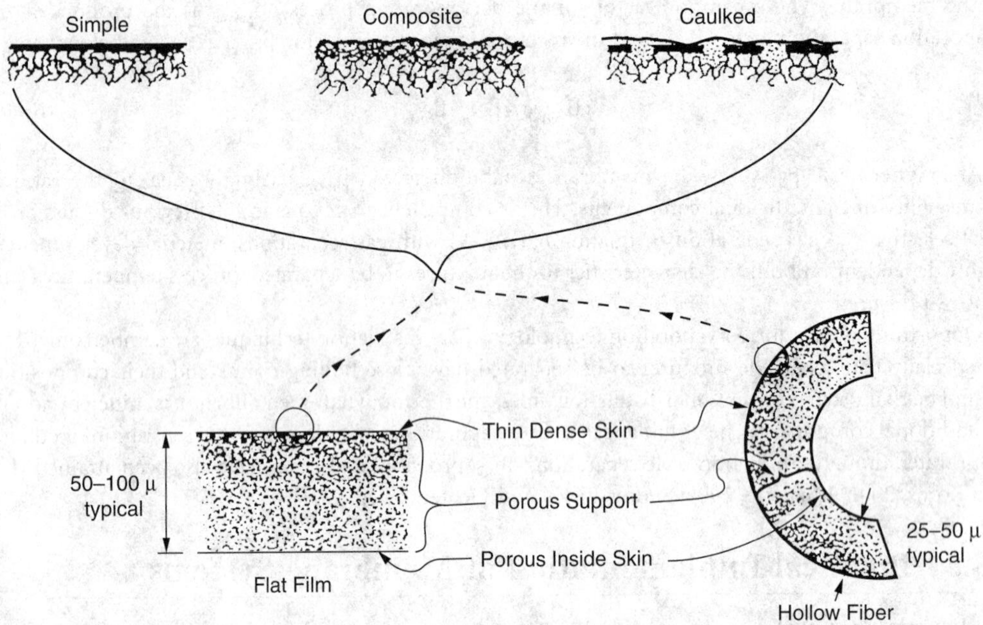

FIGURE 63.4 Asymmetric membrane structure. (*Source:* Koros, W.J. 1988. In *Encyclopedia of Chemical Processing and Design*, vol. 29, McKenna, J. J. and Cunningham, W. A., Eds., Marcel Dekker, New York, p. 301. With permission.)

is inversely proportional to the diameter of the hollow fibers. Smaller fibers allow for more area in a confined space but are more difficult to clean and are more susceptible to fouling.

Another critical factor in module design and operation is the driving force for permeation. Separations rely on a difference in the activity of a species between the upstream and downstream sides of a membrane to drive permeation. It is important to establish and maintain this driving force for effective separations to occur. Either the upstream or downstream activity can be influenced in a number of ways. One of the simplest options, pressure, is found in several applications, especially gas separations and reverse osmosis. By increasing the upstream pressure and/or pulling partial vacuum on the downstream, permeation rates are increased. Another common technique found in many separations is the use of a downstream sweep. In hemodialysis, for example, clean dialysate flowing on the downstream side of the membrane sweeps toxins away, maintaining a near-zero activity and constant driving force for permeation.

Other design considerations include flow patterns, temperatures, pressures, and integration of the membrane unit into the overall process. Numerous articles have been published on design considerations for a variety of applications, including air separation [Ho and Sirkar, 1992], natural gas purification [Tabe-Mohammadi, 1999], petrochemical gas/vapor separations [Baker et al., 1998], and desalination [Matsura, 2001]. These authors and others [Bhide and Stern, 1991a and b; Prasad et al., 1994] also consider the economics of membrane separations for various applications and compare them to competitive technologies. During the design phase, it is also helpful to have a predictive model to assist in sizing the system. One example is the model published by Kovvali et al. [1994] for gas separation with hollow fibers, and numerous models exist for other flow patterns and module designs.

63.6 Practical Membrane Challenges

Membrane research often leads to materials that are ideal candidates for separations when tested in the lab. If the specific feed conditions (e.g., contaminants and minor components) are not carefully matched to the membrane material and configuration, field use can result in below-optimum performance. Membranes will often perform well initially, but over time their performance will deteriorate, usually in the form of reduced selectivity. Replacement costs must be factored into operating costs when considering membrane economics. Membrane modules are designed to be easily replaced, but if performance can not be maintained for at least 6 to 12 months, replacement costs may become prohibitive. Several factors can contribute to this performance decline, including physical aging, feed fluctuations, and mechanical instability. However, the two main challenges are fouling and chemical stability.

Fouling (e.g., scale formation, microbial growth) occurs when the surface of a membrane becomes contaminated, limiting the productivity of the membrane. Contaminants in a feed stream that are too large to permeate, but are unable to diffuse away or be swept along with the retentate due to interactions with the membrane or insufficient feed flow, will often collect on the upstream side of a membrane. Over time, a layer can build up that significantly increases the resistance of the membrane, reducing productivity. Because this additional resistive layer is not selective, selectivity frequently decreases as well. In order to deal with these issues, systems that are prone to fouling must be designed to accommodate regular cleaning. Cleaning options include physical scrubbing, back-flushing, and chemical cleaning.

Concentration polarization is another deleterious effect that can precede fouling. In this case, a decline in productivity is not caused by the additional permanent resistance of a physical boundary layer, but is due to a boundary layer that must be crossed before permeation through the membrane can occur. Thus, even if the bulk feed composition remains constant, the lower concentration at the membrane interface results in decreased productivity. This effect can often be reduced by increasing the feed flow rate to more effectively shear and thin the boundary layer. Intelligent initial module design can also minimize concentration polarization.

The chemical stability of membranes must also be considered, particularly for those that involve sensitive solution–diffusion transport mechanisms. For example, carbon dioxide and larger hydrocarbons and aromatics will often interact with polymer membranes to cause *plasticization*. This occurs when the penetrant's effect on the polymer matrix is an increase in polymer chain mobility, causing permeability to rise and selectivity to fall. Many of the more aggressive gas separation applications encounter this challenge, including natural gas sweetening and treatment of streams containing aggressive hydrocarbon components.

Pretreatment is one common method for preventing performance decline due to either fouling or plasticization. A coarse filter placed before a solution–diffusion membrane can be effective in removing the particulates that might cause fouling. One major chemical contaminant is water, which could be removed from a gas stream via a coalescing filter before it adversely affects a membrane. Within the membrane itself, modifications can also be made to improve stability. Biofouling in aqueous feeds, for example, can be reduced by modifying the surface of a membrane to make it inhospitable for the organisms that would otherwise accumulate and cause fouling. Several material options are currently under investigation for the reduction of plasticization effects in polymeric membranes. One of the most promising approaches is the use of *crosslinked* membrane materials. By crosslinking the polymer matrix, penetrant-induced chain mobility is reduced, and polymeric membranes are able to withstand much more aggressive feeds without losing selectivity over time.

63.7 Future Directions of Membrane Technology

Mixed Matrix Approach

Though membrane technology has made great strides in increasing productivity and selectivity in large-scale applications such as O_2/N_2 separations, simply maintaining this performance is not sufficient for capturing new markets. To capture the increased selectivity of inorganic zeolites and carbon molecular sieves, while also maintaining the superior processability of polymers, heterogeneous membrane structures are being considered. In these *mixed matrix* materials, high selectivity inserts, such as zeolites or carbon molecular sieves, are embedded in a polymer matrix producing a material that has selectivities well beyond state of the art polymer matrices [Mahajan and Koros, 2000]. The insert may operate either via a molecular sieving mechanism or a selective surface flow mechanism. Molecular sieves favor smaller molecules, while selective surface flow inserts tend to favor larger, more condensable molecules. Thus, the mixed matrix approach can ideally be used to separate both large and small molecules from feed streams.

Fuel Cell Technology

Fuel cells are electrochemical devices that directly convert available chemical free energy in a fuel into electrical energy. The fuel cell assembly consists of an ion conducting film sandwiched between two platinum-based electrodes. Hydrogen fuel is supplied to the anode while an oxidant is applied to the cathode. Hydrogen dissociates at the anode to yield electrons and protons. These protons migrate through a proton exchange membrane, such as Nafion, while the electrons travel to the cathode through an external circuit. The protons and electrons react with oxygen at the cathode to produce water and heat. The driving force for the reaction manifests itself in the voltage that drives the electrons through the external circuit [Singh, 1999].

Fuel cells have a large advantage over traditional mobile energy generation because of the inherent efficiency (twice as high as internal combustion engines) and negligible emissions. There are still a number of technical issues to overcome prior to large-scale commercialization, including developing the means of supplying the fuel safely and lowering the cost of the membrane. However, once these hurdles are overcome, fuel cell technology has the potential for a large economic and environmental impact [Lee and Koros, 2002].

Controlled Release Materials

A large concern in the pharmaceutical sector is delivering drugs effectively at a constant dosage over an extended period for the required therapeutic effect. Membrane materials developed for these controlled release applications are an effective approach for safe and timely drug delivery. Delivery can be either passive or active. In passive controlled release, a pharmaceutical agent can be dissolved in excess of its solubility limit in a membrane-surrounded reservoir. The excess agent maintains saturation conditions and provides a constant permeation driving force to supply solute to the external sink at a constant rate until the concentration drops below saturation. The release rate of the agent depends on its solubility and diffusivity in the membrane, and can be tuned by tailoring the properties of the membrane. In active controlled release, the membrane is able to respond and alter its permeability to a pharmaceutical agent as the body environment changes. An example is an active controlled release membrane that responds to changes in glucose levels in the body by changing its permeability to insulin. This approach combines biosensing and control functions in a single membrane device.

63.8 Conclusions

In their simplest form, membranes can separate mixtures based on size exclusion principles, and in their most complex, they can act as sophisticated biosensing agents that alter permeation properties with the changing environment. They have a broad spectrum of potential and established uses, and when optimized, are simple, modular, and economical devices that have successfully been implemented in industrial processing environments [Hagg, 1998]. This chapter has focused on the basic principles that are core to the development, implementation, and stabilization of membranes for use in reverse osmosis, dialysis, gas/vapor, and vapor/liquid separations.

References

Baker, R. W., Wijmans, J. G. and Kaschemekat, J. H. 1998. The design of membrane vapor-gas separation systems, *J. Membr. Sci.*, 151:55.

Bhide, B. D. and Stern, S. A. 1991a. A new evaluation of membrane processes for the oxygen enrichment of air. I. Identification of optimum operating conditions and process configuration, *J. Membr. Sci.*, 62:13.

Bhide, B. D. and Stern, S. A. 1991b. A new evaluation of membrane processes for the oxygen enrichment of air. II. Effects of economic parameters and membrane properties, *J. Membr. Sci.*, 62:37.

Hagg, M.-B. 1998. Membranes in chemical processing. A review of applications and novel developments, *Sep. Purification Meth.*, 27:51.

Henis, J. M. S. and Tripodi, M. K. 1980. Multicomponent Membranes for Gas Separations, U.S. Patent 4230463, Monsanto Co., 1980.

Ho, W. W. and Sirkar, K. K. 1992. *Membrane Handbook,* Chapman and Hall, New York.

Kamaruddin, H. D. and Koros, W. J. 1997. Some observations about the application of Fick's first law for membrane separation of multicomponent mixtures, *J. Membr. Sci.* 135:147.

Khedr, M. G. 2003. Development of reverse osmosis desalination membranes composition and configuration: future prospects, *Desalination,* 153:295.

Koros, W.J. and Fleming, G.K. 1993. Membrane-based gas separation, *J. Membr. Sci.,* 83:1.

Koros, W. J., Ma, Y. H. and Shimidzu, T. 1996. Terminology for membranes and membrane processes, *J. Membr. Sci.,* 120:149.

Kovvali, A. S., Vemury, S. and Admassu, W., 1994. Modeling of multicomponent countercurrent gas permeators, *Ind. Eng. Chem. Res.,* 33:896.

Kujawski, W. 1999. Separation of liquid mixtures by pervaporation, *Environ. Prot. Eng.,* 25:49.

Lee, E. K. and Koros, W. J. 2002. Membranes, synthetic, applications, in *Encyclopedia of Polymer Science and Technology,* vol. 9, Academic Press, New York, p. 279.

Mahajan, R. and Koros, W. J. 2000. Factors controlling successful formation of mixed-matrix gas separation materials, *Ind. Eng. Chem. Res.,* 39:2692.

Malek, A., Hawlader, M. N. A. and Ho, J. C. 1996. Design and economics of RO seawater desalination, *Desalination,* 105:245.

Matsuura, T. 2001. Progress in membrane science and technology for seawater desalination — a review, *Desalination,* 134:47.

Prasad, R., Shaner, R. L. and Doshi, K. J. 1994. Comparison of membranes with other gas separation technologies, in *Polymic Gas Separation Membranes,* Paul, D. R. and Yampol'skii, Y. P., Eds., CRC Press, Boca Raton, FL, p. 513.

Sakai, K. 1994. Determination of pore size and pore size distribution 2. Dialysis membranes, *J. Membr. Sci.,* 96:91.

Singh, R. 1999. Will developing countries spur fuel cell surge? *Chem. Eng. Prog.,* 85:59.

Staudt-Bickel, C. and Lichtenthaler, R. N. 1994. Pervaporation — thermodynamic properties and selection of membrane polymers, *Vysokomolekulyarnye Soedineniya, Seriya A i Seriya B,* 36:1924.

Tabe-Mohammadi, A. 1999. A review of the applications of membrane separation technology in natural gas treatment, *Separation Sci. Tech.,* 34:2095.

64

Solid–Liquid Separation

Shiao-Hung Chiang
University of Pittsburgh

Solid–liquid separation plays a key role in nearly all manufacturing industries, including chemical, mineral, paper, electronics, food, beverage, pharmaceutical, and biochemical industries, as well as in energy production, pollution abatement, and environmental control. It also serves to fulfill vital needs of our daily life, since we must have cartridge oil/fuel filters for operating an automobile, a paper filter for the coffee machine, a sand filter bed for the municipal water treatment plant, and so on. In fact, modern society cannot function properly without the benefit of the solid–liquid separation.

Technically, solid–liquid separation involves the removal and collection of a discrete phase of matter (particles) existing in a dispersed or colloidal state in suspension. This separation is most often performed in the presence of a complex medium structure in which physical, physicochemical and/or electrokinetic forces interact. Their analysis requires combined knowledge of fluid mechanics, particle dynamics, solution chemistry, and surface/interface sciences.

Although the industrial equipment classified as solid–liquid separation devices are too numerous to be cited individually, it is generally accepted that these may be grouped into six categories of unit operations: (1) **screening**, (2) **sedimentation**, (3) **centrifugation**, (4) **hydrocycloning**, (5) **flotation**, and (6) **filtration**.

64.1 Unit Operations in Solid–Liquid Separation

A description of each unit operation in solid–liquid separation is presented in this section with the exception of filtration. Liquid filtration, one of the most commonly used industrial operations, is discussed separately in subsequent sections to illustrate the fundamental concept and design considerations for solid–liquid separation.

Screening

Screening is the simplest mechanical operation to separate solid particles based on their sizes. When solids are placed on a screen, particles smaller than the screen opening pass through while the larger particles are retained by the screen. In this manner, feed solids can be separated into two different parts,

0-8493-1586-7/05/$0.00+$1.50

namely, the undersized and the oversized portions. Often two or more screens of graded openings are used in series to separate a material into different size fractions. In many instances, screening is used as an analytical tool to determine particle size distribution in a sample of solid material. The particle size distribution can be used as a basis for equipment selection in solid–liquid separation (see Section 64.2 below). In addition to size separation, an important usage of screening is to perform mechanical dewatering (often combined with washing) of solid materials [Svarovsky, 1985]. For industrial applications, screens are made of various metals in the forms of wire mesh and slotted or perforated plates. The openings of standard screens range from 10 cm (4 in.) down to as small as a few micrometers. In screening operations, mechanical vibration and shaking are often applied to the screen surface to enhance the effectiveness of separation [Perry and Green, 1997].

Sedimentation

Sedimentation is a unit operation designed to separate suspended solid from a liquid stream by particle settling under the influence of a body force, most commonly gravity. From an operation standpoint, gravity sedimentation can be divided into two basic types: clarification and thickening. The objective of clarification is to remove small quantities of suspended particulates from the liquid stream to produce a clarified effluent or overflow stream. On the other hand, thickening is to concentrate dilute suspensions for their subsequent treatment in filters or centrifuges.

The settling behavior of suspended particulates in a gravitational field is affected by three factors: the particle size, the solid concentration, and the aggregation status of particles. In a dilute suspension, the settling solid behaves as individual particles and the process is regarded as *particulate or free settling regime.* Most clarifier operations fall into this regime. As the solid concentration increases, the suspended particles have more chances to approach each other closely and to form aggregates. Once the concentration reaches a level at which the suspended particles settle as a mass, the corresponding sedimentation is known as *hindered* or *zone settling.* In this regime, the settling behavior is related more to the solid concentration than to the particle size. As the solid concentration increases further, a settled bed of sediment mass is compressed by the overburden of sediment on top of it. Liquid is expressed from the lower sediment layers and flows upward through the sediment. This regime is termed *compression regime.* Sedimentation with the addition of chemical flocculant usually falls into this regime. A feed suspension in a thickener (or clarifier) can be operated in any regime. Therefore, the design of sedimentation equipment must consider all three regimes.

Figure 64.1 shows a schematic diagram of a thickener that exhibits three distinct zones: a clean liquid (or clarification) zone at the top, a compression zone at the bottom, and a transition zone in between. The thickener consists of several basic components: a tank to contain the slurry, a feed well for feed supply (with or without flocculant), a rotating rake mechanism, an underflow solids-withdrawal and an overflow launder. In addition, an underflow recirculation system (not shown in the diagram) is often used. The physical size of a conventional thickener can vary from a few meters to more than 100 meters in diameter. For the operation of large vessels, careful consideration must be given to the design of the supporting structure for the rotating rake mechanism and the control scheme for liquid levels and flow rates. Detailed descriptions of major components and instrumentations used for different types of thickeners can be found in the literature [Perry and Green, 1997; Schweitzer, 1997].

Centrifugation

Centrifuges are equipment that employs centrifugal force for the effective separation of solid–liquid suspension. The centrifugal force used in such equipment ranges up to 10,000 times the gravitational acceleration. Liquid–solid separation centrifuges can be broadly divided into two types: **sedimentation centrifuges** and **filtering centrifuges.**

Due to its much stronger force field, **sedimentation centrifuges** can be used to separate very fine particles as well as emulsions, which might normally be stable in a gravitational field. These centrifuges

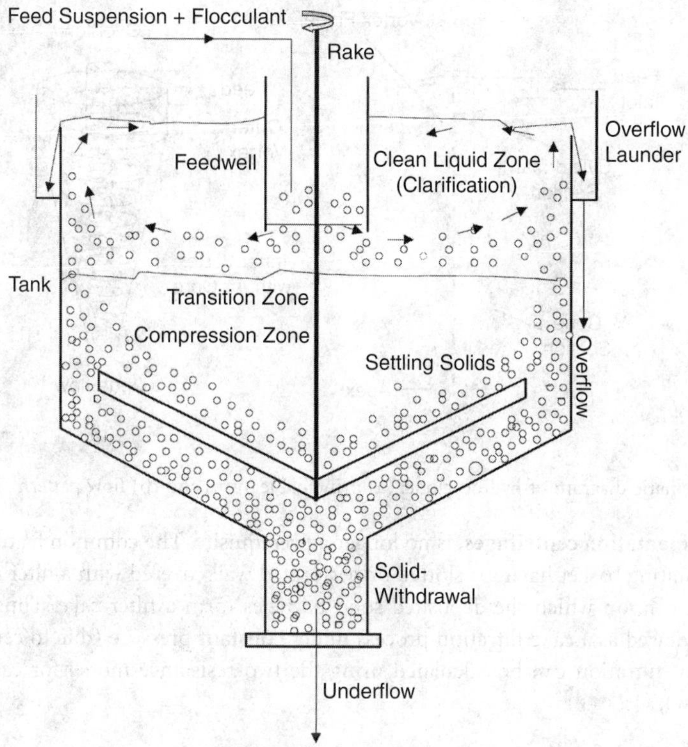

FIGURE 64.1 Schematic diagram of thickener operation.

are not usually sensitive to feed solid concentration because the liquid does not have to move through the solids or a medium. In order for a particle of a given size to be removed from the liquid, a sufficient time should be allowed for the particle to settle and reach the wall of the separator bowl. For example, in a simple tubular centrifuge, as shown in Figure 64.2, the bowl consists of a vertical tube with a large height to diameter ratio, which rotates at a high speed about its vertical axis [McCabe et al., 2001]. The feed point is at the bottom and the liquid discharge is at the top. The incoming suspension starts to rotate with the bowl, and its angular velocity will soon become identical with that of the bowl. There is therefore no tangential flow in the bowl. The rotating liquid moves upward through the bowl at a constant velocity, carrying solid particles with it. In the meantime, under the influence of high centrifugal forces the solid particles begin to settle toward the wall. The total settling time is limited by the residence time of the liquid in the bowl. At the end of this time if the particle does not reach the wall, it leaves the centrifuge with the liquid. Only those particles that reach the wall within the residence time are removed from the liquid.

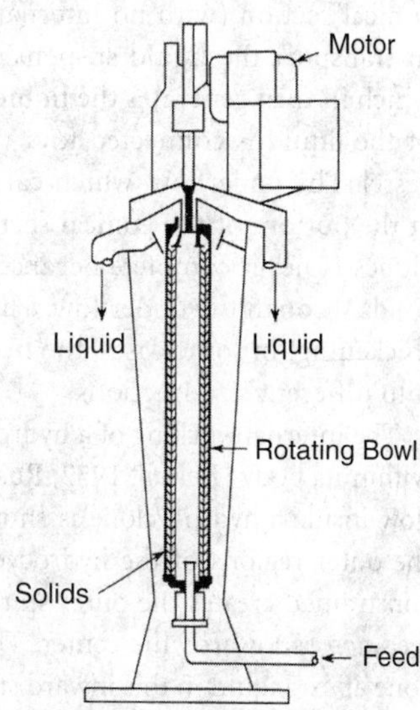

FIGURE 64.2 Tubular centrifuge. (*Source:* McCabe, W. L., Smith, J. C., and Harriott, P. 2001. *Unit Operations of Chemical Engineering,* 6th ed., Figure 29.36, p. 1049. McGraw-Hill, New York.)

Filtering centrifuges separate solid particles and liquid from a solid–liquid suspension by employing pressure resulting from the centrifugal action to force the liquor through the filter medium, leaving the solid particles behind [Zeitsch, 1990]. The density difference between the solids and the liquid, which governs the sepa-

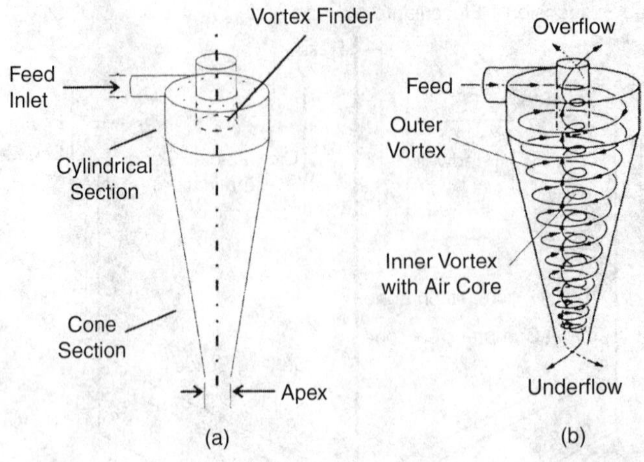

FIGURE 64.3 Schematic diagram of hydrocyclone: (a) principal features and (b) flow pattern.

ration in the **sedimentation centrifuges,** is no longer a prerequisite. The common feature of all filtering centrifuges is a rotating basket having a slotted or perforated wall covered with a filter medium, such as canvas or metal cloth, on which the deposited solid particles form a filter cake. Thus, the centrifugal filtration can be viewed as a cake filtration process under constant pressure (due to centrifugal acceleration). The rate of filtration can be calculated using the two-resistance model for cake filtration (see Section 64.4 below).

Hydrocycloning

The hydrocyclone is another device using centrifugal force to separate solids from liquid based on differences in density and particle size. A typical hydrocyclone consists of a cylindrical section and a conical section (with no internal rotating parts) as shown in Figure 64.3. An external pump is used to transport the liquid suspension to the hydrocyclone through a tangential inlet at high velocity, which in turn generates the liquid rotation and the necessary centrifugal force. The outlet for the bulk of the liquid is connected to a vortex finder located on the axis of upper cylindrical section of the vessel. The underflow, which carries most of the solids, leaves through an adjustable opening (apex) at the bottom of the conical section. It should be noted that the solid–liquid separation in hydrocyclones is never complete because there is always a significant amount of liquid discharging with the solids through the underflow. This feature limits the applications of hydrocyclone to clarification and thickening. In some cases, the hydrocyclone is also used as a classifier to separate suspended particles into different size fractions.

The internal working of a hydrocyclone is best described in terms of a double spiral liquid flow pattern within its body [Kelsall, 1952; Rushton et al., 2000; and Svarovsky, 1985]. A schematic view of the spiral flow inside a hydrocyclone is shown in Figure 64.3(b). Liquid on entry commences downward flow in the outer regions of the hydrocyclone body. This combined with the rotational motion to which it is constrained creates the outer spiral. At the same time, some of the downward-moving liquid begins to feed across towards the center. The amount of inward motion of liquid increases as it approaches the cone apex. Liquid in this inward stream ultimately reverses its direction and flows upwards to the cyclone overflow outlet via the vortex finder. The reversal applies only to the vertical component of velocity, and the inner spiral rotates in the same circular direction as the outer one. Wall friction causing "obstruction" of tangential velocity results in a nontangential motion. Consequently, a strong axially directed current occurs near the wall, which carries solid particles to the apex opening and out of the hydrocyclone. Thus, it achieves the desired solid–liquid separation.

Flotation

The use of bubbles to float fine particles in a liquid is commonly known as *flotation process.* Such a process consists of attaching gas bubbles to the suspended solid particles to alter their apparent density for selective levitation of particles to be separated. The flotation operation involves not only the adhesion of small particles to gas bubbles, but also the collection of the gassy particles in the form of froth. Thus, the bubble flotation is also named as *froth flotation.*

The flotation process is fundamentally different from other mechanical separation techniques in that flotation is a surface property-driven process, which depends upon complex phenomena occurring at the interface of solid particles and gas bubbles [Fuerstenau et al., 1985; Jaycock and Parfitt, 1981; and Zettlemoyer, 1969]. In a bubble-particle attachment process, the tendency of the particle to replace its solid–liquid interface by the solid–vapor interface is termed hydrophobicity or floatability. If a surface is completely wetted by water, it would be denoted as high-energy surface (i.e., hydrophilic). Most metals and minerals exhibit high-energy surfaces. On the other hand, hydrocarbon surfaces are low energetic (i.e., hydrophobic). The particles with a low free energy have a high floatability.

The contact between particles and gas bubbles in a suspension is considered as a two-step process: (1) the collision between the particle and the bubble and (2) the attachment of the particle onto the gas bubble. Each step can be modeled as a stochastic event. Thus, the overall probability of particle collection by gas bubble is defined as the product of the probability of particle–bubble collision and the probability of adhesion after the collision. The collision probability depends mainly on the hydrodynamic characteristics of the flotation cell while the adhesion probability is related to the hydrophobicity of the particle.

Traditionally, flotation is carried out in an open cell equipped with a gas-inducing agitator (turbine or impeller). As shown in Figure 64.4, air is induced through the air passage in the agitator shaft by suction. The rotational motion of the agitator disperses air bubbles into the suspension. These bubbles attach to suspended particles to form *aggregates.* The particle–bubble aggregates float upward to the froth layer, which is mechanically skimmed off or flows over a weir into the discharge launder as a froth product. The nonfloatable particles are withdrawn from the bottom of the cell as tailings. A more recent development in flotation is the use of bubble column as a flotation device [Finch et al., 1995]. In Figure 64.5, it shows that the space in a flotation column can be divided into two parts: the collection zone and the froth zone. The feed enters the column via a feed port at the middle and flows downward to the base of the column. The gas bubbles are generated either by an internal sparger near the bottom of the column or an external gas bubble generator. To minimize the effect of unexpected particle entrainment, a wash water device is added near the top of the column just below the overflow weir for cleaning the froth. The operating performance of a flotation column is generally superior to that of open cell flotation.

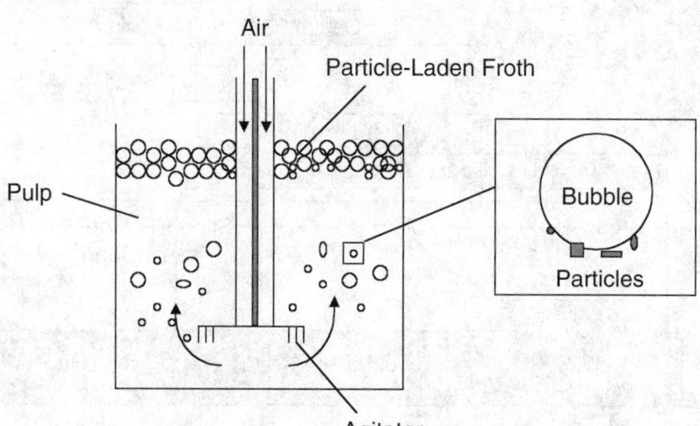

FIGURE 64.4 Conventional open-cell flotation.

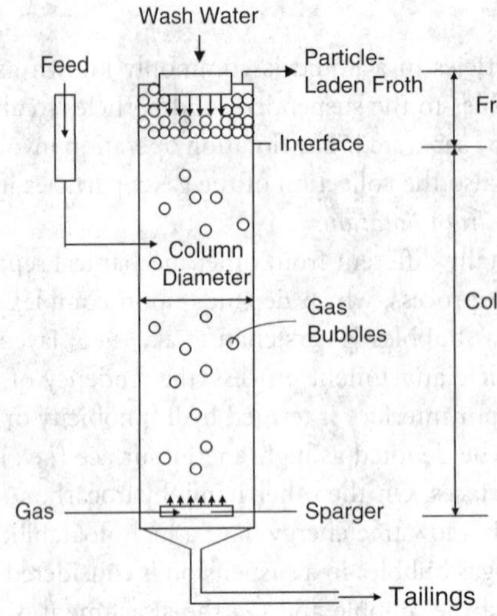

FIGURE 64.5 Column flotation.

64.2 Equipment

The most important criterion for the selection of equipment for a given application of solid–liquid separation is the particle size of the system. Figure 64.6 shows the general range of applicability of major types of equipment in terms of the particle size and representative materials involved. Of course, this representation is an oversimplification of the selection process, as many other factors are not considered. For example, the solid concentration in the feed mixture (suspension) can influence the choice of equipment type. In general, deep-bed filtration is best for treating dilute slurry with solid concentration less than 1%, whereas cake filtration is the method of choice for slurries having solid concentration much greater than 1%.

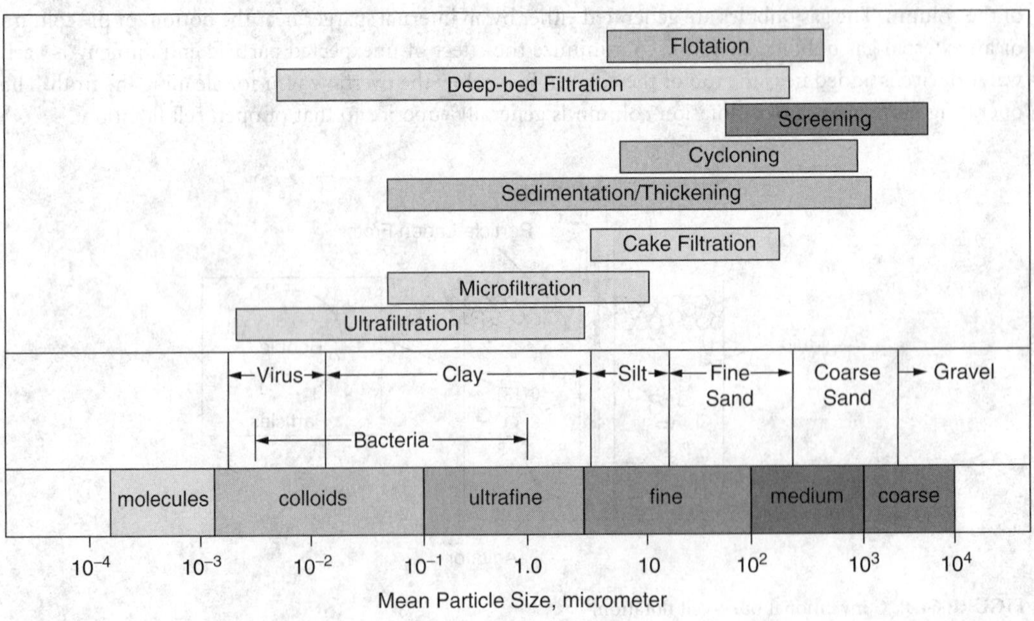

FIGURE 64.6 Equipment selection for solid–liquid separation based on particle size.

TABLE 64.1 Filtration Equipment

Discontinuous Filters	Semicontinuous Filters	Continuous Filters
Plate and frame filter press	Rotary pan filter	Drum filter
Leaf filter	Semicontinuous belt filter	Rotary disk filter
Tray filter	Automatic filter press	Vacuum belt filter
	Electrical precipitator	Rotary disk cross-flow filter
		Rotating cylinder cross-flow filter

It should also be pointed out that the various filtration processes (deep-bed filtration, cake filtration, microfiltration, and ultrafiltration) cover nearly the entire range of particle size. Therefore, the term *filtration* is often used as a synonym to represent the field of *solid–liquid separation.* The most commonly used filtration equipment is given in Table 64.1.

A detailed procedure for equipment selection for a given requirement in solid–liquid separation can be found in *Perry's Chemical Engineers' Handbook* [Perry and Green, 1997].

64.3 Fundamental Concept

There are two general types of operations for separating solid particulate matter from a liquid phase. In the first type, the separation is accomplished by moving the particles through a constrained liquid phase. The particle movement is induced by a body force, such as gravity or centrifugal acceleration. For example, in sedimentation, the solid particles settle due to a difference in density between solid and liquid under the influence of gravity. In centrifugation and hydrocycloning, the separation is effected by centrifugal acceleration.

In the second type of operation, exemplified by the filtration process, the separation is accomplished by contacting the solid–liquid suspension with a porous medium (see Figure 64.7). The porous medium acts as a semipermeable barrier that allows the liquid to flow through its capillary channels and retains the solid particles on its surfaces. Depending on the mechanism for arrest and accumulation of particles, this type of separation can be further divided into two classes [Perry and Green, 1997]: deep-bed filtration and cake filtration.

Deep-bed filtration is also known by terms such as *blocking filtration, surface filtration,* and *clarification* [see Figure 64.7(b)]. This type of filtration is preferred when the solid content of the suspension is less than 1%. In such an operation, a deep bed of packing material (e.g., sand, diatomite, or synthetic fibers) is used to capture the fine solid particles from a dilute suspension. The particles to be removed are several orders of magnitude smaller than the size of the packing material, and they will penetrate a considerable depth into the bed before being captured. The particles can be captured by several mechanisms [Tien, 1989]:

1. The direct-sieving action at the constrictions in the pore structure
2. Gravity settling
3. Brownian diffusion
4. Interception at the solid–liquid interfaces
5. Impingement
6. Attachment due to electrokinetic forces

Cake filtration is the most commonly used industrial process for separating fine particles from a solid–liquid suspension. In cake filtration, the filtered particles are stopped by the surface of a filter medium (a porous barrier) and then piled upon one another to form a cake of increasing thickness [see Figure 64.7(a)]. This cake of solid particles forms the "true" filtering medium. In the case of liquid filtration, a filter cake with filtrate (the liquid) trapped in the void spaces among the particles is obtained at the end of the operation. In many instances where the recovery of the solids is the ultimate objective, it is necessary that the liquid content in the cake be as low as possible. In order to reduce the liquid

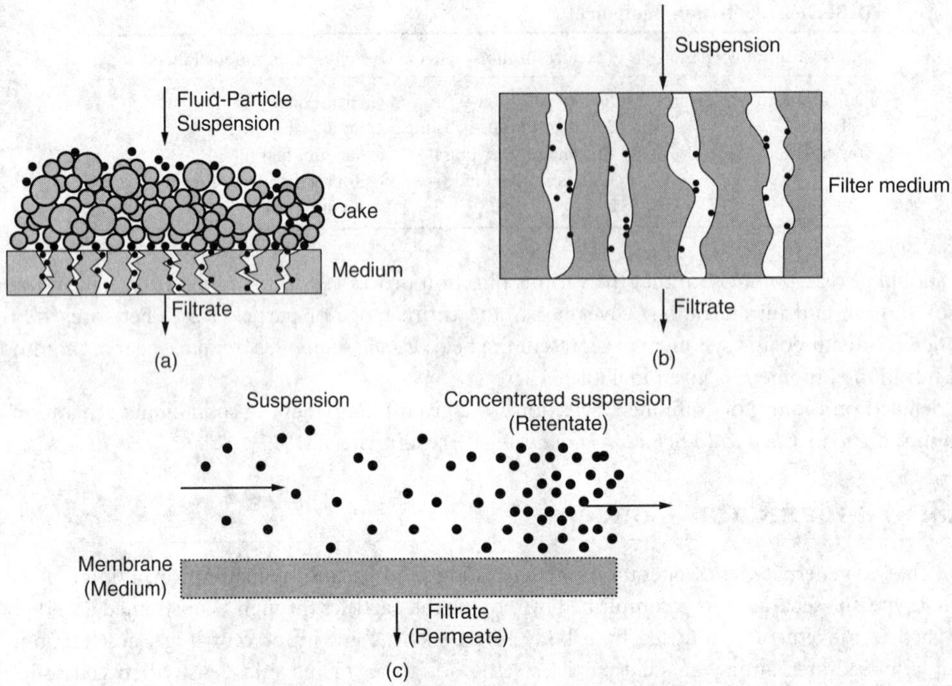

FIGURE 64.7 Mechanisms of filtration: (a) cake filtration, (b) deep-bed filtration, (c) cross-flow filtration (ultrafiltration). (*Source:* McCabe, W. L., Smith, J. C., and Harriott, P. 2001. *Unit Operations of Chemical Engineering,* 6th ed., Figure 29.3, p. 992. McGraw-Hill, New York.)

content, the cake is subjected to desaturating forces. These forces can be mechanical, hydrodynamic, electrical, or acoustic in nature [Muralidhara, 1989].

When the mean particle size is less than a few micrometers, the conventional cake filtration operation becomes ineffective, primarily due to the formation of high-resistance filter cake. To overcome this obstacle, cross-flow filtration (often coupled with ultrafiltration) is used to limit the cake growth. In the cross-flow configuration (e.g., in continuous ultrafiltration), the solid–liquid suspension flows tangentially to the filter medium rather than perpendicularly to the medium as in conventional filtration. The shear forces of the flow in the boundary layer adjacent to the surface of the medium continuously remove a part of the cake and thus prevent the accumulation of solid particles on the medium surface. In this manner, the rate of filtration can be maintained at a high level to ensure a cost-effective operation.

64.4 Design Principles

Cake Filtration

In the design of a cake filtration process, the pressure drop, Δp, the surface area of the cake, A, and the filtration time, t, are important parameters to be determined. As the filtration proceeds, particles retained on the filter medium form a filter cake (see Figure 64.8). For an incompressible cake the pressure drop, Δp, across the filter cake and filter medium can be expressed as:

$$\Delta p = p_a - p_b = \left(\frac{\alpha m_c}{A} + R_m \right) \mu u \tag{64.1}$$

where μ is the viscosity of the filtrate, u is the velocity of the filtrate, m_c is the total mass of solids in the cake, R_m is the filter-medium resistance, and α is defined as the specific cake resistance. The specific cake

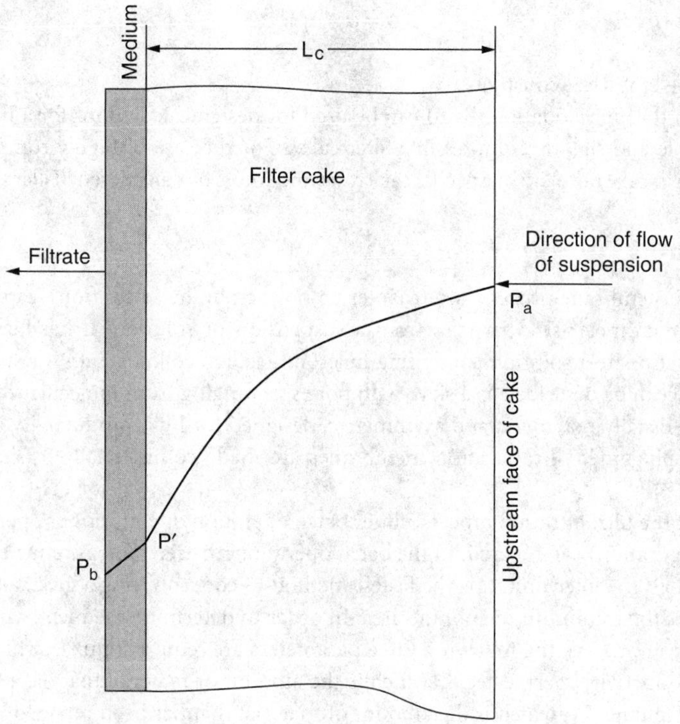

FIGURE 64.8 Pressure gradient in filter medium and cake.

resistance depends on particle size, shape, and distribution. It is also a function of porosity of filter cake and pressure drop. For incompressible cakes, α is independent of the pressure drop and the position in the filter cake.

For data analysis, Equation (64.1) is usually rewritten as follows: [McCabe et al., 2001]

$$\frac{dt}{dV} = \frac{\mu}{A(-\Delta p)}\left[\frac{\alpha c V}{A} + R_m\right] \tag{64.2}$$

where t is the filtration time, V is the volume of filtrate, and c is the mass of solid per unit filtrate volume.

In order to use this equation for design of a cake filtration operation, the specific cake resistance and filter-medium resistance must first be determined by performing experimental tests. Equation (64.2) can be further integrated under constant pressure to give:

$$\frac{t}{V} = \left(\frac{K_c}{2}\right)V + \frac{1}{q_o} \tag{64.3}$$

where

$$K_c = \frac{\mu c \alpha}{A^2 \Delta p} \quad \text{and} \quad \frac{1}{q_o} = \frac{\mu R_m}{A \Delta p} \tag{64.4}$$

A plot of t/V against V yields a straight line with a slope of $(K_c/2)$ and an intercept of $(1/q_o)$. Using Equation (64.4), the values of α and R_m can be calculated. For slightly compressible cake, α becomes a function of pressure drop and can be represented by the following correlation:

$$\alpha = \alpha_0 (\Delta p)^a \tag{64.5}$$

where α_0 and a are empirical constants.

Equation (64.2) through Equation (64.5) can be used for design calculations for a filtration operation with incompressible and slightly compressible filter cakes. For the case of highly compressible cakes the effect of variations in cake porosity on specific cake resistance must be considered [Tiller and Shirato, 1964].

Ultrafiltration

Ultrafiltration is a membrane process (see Chapter 63 on membrane separation) capable of separating or collecting submicrometer-size particles from a suspension or solution. It has been widely used to concentrate or fractionate a solution containing macromolecules, colloids, salts, or sugars. The ultrafiltration membrane can be described as a sieve with pore size ranging from molecular dimension to a few micrometers. It is usually polymeric and asymmetric, designed for high productivity (permeation flux) and resistance to plugging. Ultrafiltration membranes are made commercially in sheet, capillary, and tubular forms.

In the design of the ultrafiltration process, either batch operation or continuous operation (employing a cross-flow configuration) can be used. In the batch operation, the retentate is returned to the feed tank for recycling through the filter unit. It is the fastest method of concentrating a given amount of material and it also requires the minimum membrane area. In order to determine the membrane surface area, A, for the ultrafiltration process, the following three parameters are required: flux, J, which is a measure of the membrane productivity; permeate, V_p, which is the amount of material that has passed through the membrane; and retentate, V_R, which is the amount of material that has been retained by the membrane. During the batch ultrafiltration operation, flux decreases because of an increase in concentration in the recycled stream. Furthermore, the phenomenon of concentration polarization tends to cause a higher concentration at the membrane surface than that in the bulk. Therefore, an average flux should be used in the design. The average flux, J_{av}, can be estimated by the following equation:

$$J_{\text{av}} = J_f + 0.33(J_i - J_f) \tag{64.6}$$

where J_f is the final flux at the highest concentration and J_i is the initial flux. The material balance gives

$$V_f = V_r + V_p \tag{64.7}$$

where V_f, V_r, and V_p are volume of feed, retentate, and permeate, respectively.

The membrane area can be expressed as

$$A = (V_f - V_r)/J_{\text{av}} \tag{64.8}$$

Equation (64.6) to Equation (64.8) can be used to estimate the membrane surface area required for a given ultrafiltration operation [Cheryan, 1986].

64.5 Economics

The cost for a given solid–liquid separation process varies widely. For example, the cost for purchasing industrial filtration equipment can vary from several hundred dollars to over ten thousand dollars per square meter of filter area. Such a large variation in cost is due to a wide variety of individual features and materials of construction required by specific applications. A good source of information on the cost of common industrial filtration and other solid–liquid separation equipment can be found in *Perry's Handbook* [Perry and Green, 1997].

Defining Terms

Cake filtration — The separation of particles is effected by contacting the solid–liquid suspension with a porous filter medium (made of cloth, synthetic fibers, or metals). The filter medium allows the liquid to flow through its pores while it retains the particles on its surface to form a cake. As filtration proceeds, the cake of solid particles grows in thickness and becomes the "true" filtering medium.

Centrifugation — Centrifugation is a separation process based on the centrifugal force either to hold the material in it or to let the material pass through it. Separation is achieved due to the difference in density.

Hydrocycloning — Hydrocycloning is a centrifugal separation process. The feed is introduced tangentially into the cylindrical portion of a cyclone, causing it to flow in a tight conical vortex. The bulk of the liquid leaves upward through a pipe located at the center of the vortex. Solid particles are thrown to the wall and discharged with a small portion of the liquid through the bottom apex of the cyclone.

Deep-bed filtration — In this type of filtration a deep bed of packing materials, such as sand, diatomite, or synthetic fibers, is used as the filter medium. The particles are captured within the packed bed while the liquid passes through it.

Flotation — Flotation is a gravity separation process based either on the use of a dense medium in which the desired particles will float or on the attachment of gas bubbles to particles, which are then carried to the liquid surface to be separated.

Membrane filtration — In membrane filtration a thin permeable film of inert polymeric material is used as the filter medium. The pore size of the membrane ranges from molecular dimension to a few micrometers. It is widely used to collect or fractionate macromolecules or colloidal suspensions. It is also applied to beverage filtration and preparation of ultrapure water.

Screening — Screening is an operation by which particles are introduced onto a screen of a given aperture size to separate particles of different sizes.

Specific cake resistance — Specific cake resistance is the resistance of a filter cake having unit weight of dry solids per unit area of filtration surface.

Thickening/sedimentation — Thickening/sedimentation is a gravity-settling process that removes the maximum quantity of liquid from a slurry and leaves a sludge for further processing.

Ultrafiltration — Ultrafiltration is a special type of membrane filtration. It is used for concentration and purification of macromolecular solutes and colloids in which the solution is caused to flow under pressure parallel to a membrane surface (in a cross-flow configuration). Solutes (or submicrometer particles) are rejected at the semipermeable membrane while the solvents and small solute molecules pass through the membrane.

References

Cheryan, M., 1986. *Ultrafiltration Handbook.* Technomic, Lancaster, PA.

Finch, J. A., Uribe-Salas, A. and Xu, M. 1995. Column flotation, In: Flotation Science and Engineering, Matis, K. A., Ed. Marcel Dekker, New York. pp. 291–330.

Fuerstenau, M. C., Miller, J. D., and Kuhn, M. C. 1985. *Chemistry of Flotation,* Society of Mining Engineers, New York, p. 2.

Jaycock, M. J. and Parfitt, G. D. 1981. *Chemistry of Interfaces,* Ellis Horwood Limited, New York.

Kelsall, D. F. 1952. A study of the motion of solid particles in a hydraulic cyclone. *Trans. Inst. Chem. Eng.* 30:87–104.

McCabe, W. L., Smith, J. C., and Harriott, P. 2001. *Unit Operations of Chemical Engineering,* 6th ed. McGraw-Hill, New York. pp. 986–1056.

Muralidhara, H. S. (Ed.) 1989. *Solid/Liquid Separation.* Battelle Press, Columbus, OH.

Perry, R. H. and Green, D. W. (Ed.) 1997. *Perry's Chemical Engineers' Handbook,* 7th ed., McGraw-Hill, New York. Chapters 18 and 22.

Rushton, A., Ward, A. S., and Holdich, R. G. 2000. *Solid–Liquid Filtration and Separation Technology,* 2nd ed., Wiley-VCH, Weinheim, Germany. Chapters 2 and 3.

Svarovsky, L. 1985. *Solid–Liquid Separation Processes and Technology,* Elsevier, Amsterdam, pp. 68–71.

Schweitzer, P. A. 1997. *Handbook of Separation Techniques for Chemical Engineers,* 3rd ed., McGraw-Hill, New York. pp. 4.140–4.156.

Tien, C. 1989. *Granular Filtration of Aerosols and Hydrosols.* Butterworths, Stoneham, MA.

Tiller, F. M. and Shirato, M. 1964. The role of porosity in filtration: VI. new definition of filtration resistance. *AlChE J.* 10(1):61–67.

Zeitsch, K. 1990. Centrifugal filtration. In Svarovsky, L. Ed. *Solid–Liquid Separation,* 3rd ed., Butterworths, London. pp. 476–532.

Zettlemoyer A. C. 1969. Hydrophobic surfaces. In Fowkes, F. M. Ed. *Hydrophobic Surfaces,* Academic Press, New York. pp. 1–27.

Further Information

An excellent in-depth discussion on the theory and practice of solid–liquid separation is presented in *Solid–Liquid Separation,* 3rd ed., by Ladislav Svarovsky, Butterworths, London, 1990.

The proceedings of the annual American Filtration and Separation Society meeting and the World Filtration Congress document new developments in all aspects of solid–liquid separation.

Four major journals cover the field of solid–liquid separation:

Solid–Liquid Separation Journal. Published by the American Filtration and Separation Society, Houston, TX.

Particulate Science and Technology: An International Journal. Published by Taylor and Francis, Washington, DC.

Transactions of Filtration Society. Published by the Filtration Society, Leics. LE67 8PP, UK.

Separations Technology. Published by Butterworth-Heinemann, Stoneham, MA.

65

Other Separation Processes

William C. Corder
CONSOL, Inc.

Simon P. Hanson
CONSOL, Inc.

For the purposes of this publication, "other separation processes" will be confined to sublimation, diffusional separations, adsorptive bubble separation, dielectrophoresis, and electrodialysis.

Sublimation is the transformation of a substance from the solid into the vapor state without formation of an intermediate liquid phase. Desublimation is the reverse. **Sublimation processes** may be used to separate solids not easily purified by more common techniques.

Diffusional separation processes may be employed when the substances to be separated are quite similar (e.g., the separation of isotopes). Gaseous diffusion, thermal diffusion, pressure diffusion, and mass diffusion fit this category. Gaseous diffusion provides separation by selectively permitting constituents of a gas mixture to flow through extremely small holes in a barrier or membrane. Molecules of species in the mixture for which the mean free path is smaller than the holes will pass through (i.e., separate), whereas the bulk flow of the gas will not. Thermal diffusion depends on a temperature gradient for separation and applies to liquids and gases.

Use of pressure diffusion for separation of gaseous mixtures has been largely confined to the laboratory, although considerable work has been done on the use of the gas centrifuge for separating isotopes of uranium. The separation method depends on the imposition of a pressure gradient. Mass diffusion involves the separation of a gas mixture by diffusion into a third component, or **sweep gas.** The sweep gas establishes a partial pressure gradient by introducing it as a gas through a porous wall and then letting it condense as a liquid at another region of the process. The sweep gas sweeps the less diffusible component along with it, thus separating this component of the mixture.

Adsorptive bubble separation utilizes the selective attachment of materials onto the surfaces of gas bubbles passing through a solution or, commonly, a suspension [Lemlich, 1966]. The bubbles rise to create a foam or froth that is swept away, allowing the collapse of the bubbles and recovery of the adsorbed material. Additives commonly are included in the suspension to depress the adsorption of some species and enhance the attachment of others. Probably the greatest use of adsorptive bubble separation is in the minerals-processing industries.

Dielectrophoresis (DEP), related electrophoresis, and electrodialysis all depend on electric gradients to effect a separation. Electrophoresis involves the motion of charged particles in a uniform electric field, whereas DEP applies to neutral, polarizable matter in a nonuniform electric field. DEP works best with

0-8493-1586-7/05/$0.00+$1.50

larger-than-molecular-sized particles. Electrodialysis is the removal of electrolytes from a solution through an ion-selective membrane by means of an applied electric potential gradient. The primary application of electrodialysis has been the desalination of seawater and brackish water. Most commercial membranes were developed for this purpose.

65.1 Sublimation

Sublimation has been used most often for separation of a volatile component from other components that are essentially nonvolatile. There has been some interest in separating mixtures of volatile components by sublimation [Gillot and Goldberger, 1969].

The vapor pressure of the subliming components must be greater than their partial pressures in the gas phase in contact with the solid. Usually, the solid must be heated and the gaseous environment in contact with the solid must be controlled. Vacuum operation and the use of nonreactive gas or **entrainer** are means of controlling the gaseous environment. The use of vacuum operation has been the most common commercial approach. For pure substances or mechanical mixtures containing only one volatile component, no theoretical limit to the purity of the product obtained in a sublimation process exists.

In a sublimation process comprised of a sublimer (to volatize the sublimable components) and a condenser (to recover them), the loss per pass of entrainer gas through the system for a system consisting of two sublimable components is

$$\text{Percent loss} = \frac{r(P_{AC}+P_{BC})/(P_{AS}+P_{BS})}{(1+r)-[(P_{AC}+P_{BC}-\Delta P)/(P_{AS}+P_{BS})]}\times 100 \tag{65.1}$$

where r is the ratio of moles of inert gas (either unavoidable, as in a vacuum operation, or intentional, as with an entrainer) to the moles of solids sublimed (i.e., $r = P_1/(P_{AS} + P_{BS}) = (P - P_{AS} - P_{BS})/(P_{AS} + P_{BS})$. P_A and P_B are the vapor pressures of components A and B, subscripts S and C refer to the sublimer and condenser, P_1 is the partial pressure of inert gas, ΔP is the total pressure drop between sublimer and condenser, and P is the total pressure in the sublimer.

To calculate the **yield per pass,** a material balance must be made on the sublimer and condenser using the calculated loss per pass. For vacuum sublimation where recycling of gas is not possible, the yield of condensed solids is simply the percent loss calculated by Equation (65.1) subtracted from 100.

65.2 Diffusional Separations

Gaseous diffusion requires many stages to result in a clean separation. It is probably economically feasible only for large-scale separation of heavy isotopes (e.g., uranium). The flow rate of the gases to be separated across the gaseous diffusion barrier is described by the equations for Knudsen and Poiseuille flow, the combination of which provides

$$N_T = \frac{a}{\sqrt{M}}(P_F - P_B) + \frac{b}{\mu}\left(P_F^2 - P_B^2\right) \tag{65.2}$$

where N_T is the molar flow rate of gas per unit area of barrier, a and b are functions of temperature and barrier properties, M is the molecular weight of the gas, P_F and P_B are the high and low side pressures on the barrier, and μ is the gas viscosity. A comprehensive treatment of the design equations for gaseous diffusion is provided by Pratt [1967].

As with gaseous diffusion, thermal diffusion requires many stages to provide a clean separation. Thus, its use has been confined to isotope separations. The preferred equipment for thermal diffusion evolved into the thermal diffusion or thermogravitational column. In the column, the fluid mixture has a horizontal temperature gradient imposed on it. Thermal convection currents create a countercurrent flow effect providing a large number of individual separation stages in a single piece of equipment. The fundamental equations for separation by thermal diffusion are heavily dependent on the physical properties of the mixture, the properties of the separating column, and the temperatures imposed.

Gas centrifuges and separation nozzles are the tools used to achieve separation by pressure diffusion. In both, a high pressure gradient is created to segregate the lighter components from the heavier ones. The gas centrifuge is the more highly developed. High energy requirements for operation have confined the use of the technology to the separation of mixtures that are very hard to separate (e.g., isotopes). The maximum separative work a gas centrifuge can accomplish per unit time [Olander, 1972] is

$$\Delta U = \frac{\pi ZCD}{2} \frac{(\Delta MV^2)^2}{(2RT)} \tag{65.3}$$

where ΔU is the separation in moles/units time, Z is the length of the centrifuge, C is the molar density, D is the diffusivity of the gas, ΔM is the mass difference between the gases to be separated, V is the peripheral velocity of the centrifuge, T is the absolute temperature, and R is the gas constant.

Mass diffusion can be conducted in continuous countercurrent flow columns. The sweep gas moves in a closed cycle between liquid and vapor and (returns to) liquid. The less diffusible component is enriched at the bottom of the column, and the more diffusible one is enriched at the top of the column. A treatment of the theory is provided in Pratt [1967].

65.3 Adsorptive Bubble Separation

Adsorptive bubble separation techniques are most often employed for the removal of small amounts of either liquid or solids from large amounts of liquid. In its most common use (i.e., **flotation**), fine particulates are removed from liquid. Adsorptive bubble separations can be conducted in staged cells or in columns, the former being the more common type of equipment used. The mechanisms influencing the effectiveness of the separation are adsorption, bubble size and formation, foam overflow and drainage, foam coalescence, and foam breaking. Perry [1984] provides equations applicable to each mechanism. The techniques, in addition to recovery of minerals and coal, have industrial applications in the areas of pollution control, papermaking, food processing, and the removal of organic materials from water.

65.4 Dielectrophoresis

Separation by DEP is dependent on the fact that, in a nonuniform electric field, a net force will act on even neutral particles in a fluid. Particles will behave differently depending on their polarizability. The net force will direct different particles to regions of varying field strength. A series of equations describing the technique is presented in Pohl [1978]. Application of DEP is reasonably widespread and includes electrofiltration to remove particles from a fluid medium, orientation and separation of biological materials (e.g., cells), and imaging processes.

65.5 Electrodialysis

The ion-selective membrane used in electrodialysis is key. Ion-selective membranes are identified as anionic or cationic depending on the ion permitted passage through the membrane. Their electrochemical behavior is characterized by a conductivity, a transference number, and a transport number. The conductivity is simply the amount of current that passes through the membrane per unit of imposed electric potential gradient. The transference number is the amount of any substance transported through the membrane per unit of current. The transport number is the fraction of the current carried by a particular ionic species. The conductivity, transference number, and transport number are similarly defined for the solution in contact with the membrane.

If the transport number for an ionic species in solution differs from that inside a membrane, separation will occur when an electric current passes through the membrane. The effectiveness of an ion-selective membrane is quantified by the *permselectivity*, as defined [Winger et al., 1957] for anionic and cationic membranes, respectively, by the equations

$$\Psi_a = \frac{t_a^m - t_a^s}{1 - t_a^s} \tag{65.4}$$

$$\Psi_c = \frac{t_c^m - t_c^s}{1 - t_c^s} \tag{65.5}$$

where Ψ is the permselectivity, t is the transport number, subscripts a and c designate anion or cation, and superscripts m and s designate membrane or solution. At low solute concentrations, the transport number for anions in an anionic membrane, or cations in a cationic membrane, approaches unity. As the concentration of solute increases, membrane conductivity increases, but the transport number decreases. This is due to compression of the diffuse charge double layer within the membrane's pores such that passage of excluded electrolyte species cannot be prevented.

Electrodialysis is a selective transport process that directly uses electrical energy to effect separation of charged species from a solvent. The theoretical thermodynamic minimum energy to effect a separation [Spiegler, 1958] is given by

$$U = 2RT(C_i - C_{\text{od}})\left[\frac{\log\left(\dfrac{C_i}{C_{\text{oc}}}\right)}{\left(\dfrac{C_i}{C_{\text{oc}}} - 1\right)} - \frac{\log\left(\dfrac{C_i}{C_{\text{od}}}\right)}{\left(\dfrac{C_i}{C_{\text{od}}} - 1\right)}\right] \tag{65.6}$$

where R is the ideal gas constant, T is the absolute temperature, C is the concentration in equivalents, and the subscripts i, oc, and od indicate the inlet, concentrate outlet, and dilute outlet streams. The direct use of electric energy is most effective at low concentrations since energy is expended only on removing the contaminant. The actual energy consumed is usually 10 to 20 times the theoretical.

Electrodialysis is competitive with other separation processes for low concentration feeds or for situations in which the specificity of electrodialysis makes it uniquely suitable.

Defining Terms

Entrainer — Nonreactive, gaseous diluent used to assist in the sublimation process.

Flotation — The process of removing fine particulate matter from aqueous solution by attachment of the particulate to air bubbles.

Sublimation process — Either one or both of a combination of sublimation and desublimation steps in equipment designed for that purpose.

Sweep gas — The vapor introduced in mass diffusion separations which moves through the process and assists in the separation.

Yield per pass — The amount of the desirable constituent recovered per pass of the entrainer gas through the sublimation process, usually expressed as a percent of the total.

References

Gillot, J. and Goldberger, W. M. 1969. *Chem. Eng. Prog. Symp. Ser.* 65(91):36–42.

Lemlich, R. (Ed.) 1972. *Adsorptive Bubble Separation Techniques,* Academic Press, New York.
Olander, D. R. 1972. Technical basis of the gas centrifuge. *Adv. Nucl. Sci. Technol.* 6:105–174.
Perry. 1984. *Perry's Chemical Engineers Handbook.* McGraw-Hill, New York.
Pohl, H. A. 1972. *Dielectrophoresis: The Behavior of Matter in Non-Uniform Electric Fields,* Cambridge University Press, New York.
Pratt, H. R. C. 1967. *Countercurrent Separation Processes,* Elsevier, Amsterdam.
Spiegler, K. S. 1958. Transport processes in ionic membranes. *Trans. Faraday Soc.* 54:1408.
Winger, A. G., Bodamer, G. W., and Kunin, R. 1957. Some electrochemical properties of new synthetic ion exchange membranes. *J. Electrochem. Soc.* 100:178.

Further Information

Wilson, J. R. (Ed.). 1960. *Demineralization by Electrodialysis.* Butterworths Scientific Publications, London.
Fuerstenau, D. W. (Ed.) 1962. *Froth Flotation.* AIME, New York.

IX

Fuels and Energy Conversion

66

Fuels

Safwat M. A. Moustafa
California Polytechnic University

Materials that possess chemical energy are known as *fuels.* The general categories of fuels are fossil fuels, nuclear fuels, and renewable energy sources. Fossil fuels release their chemical energy during combustion. Nuclear fuels are those that release their chemical energy by nuclear reaction.

There are three general classes of fossil fuels — coal, oil, and natural gas. Other fuels, such as shale oil, tar-sand, and fossil-fuel derivatives, are commonly lumped under one of the three main fossil fuel categories.

Fossil fuels were produced from fossilized carbohydrate compounds with the chemical formula $C_x(H_2O)_y$. These compounds were produced by living plants in the photosynthesis process. After the plants died, the carbohydrates were converted by pressure and heat, in the absence of oxygen, into hydrocarbon compounds with a general chemical formula of C_xH_x. Although hydrocarbon compounds are composed of only carbon and hydrogen, in some molecules the same number of hydrogen atoms can be arranged in various structures to produce compounds that are strikingly different in chemical and physical properties.

66.1 Coal

Coal is the most abundant fossil fuel. It is thought to be fossilized vegetation. At least 20 ft of compacted vegetation is necessary to produce a 1-ft seam of coal. The compacted vegetation — in the absence of air and in the presence of high pressure and temperature — is converted into peat (a very low-grade fuel), then brown coal, then lignite, then subbituminous coal, then **bituminous coal**, and finally anthracite coal. As the aging process progresses, the coal becomes harder, the hydrogen and oxygen content decrease, the moisture content usually decreases, and the carbon content increases. Coal is normally found in the earth's crust. The average seam thickness in the U.S. is approximately 1.65 m.

The American Society for Testing Materials (ASTM) has classified coal into four major classes (Table 66.1) according to the length of its aging process — the oldest being anthracitic coal, followed by bituminous coal, subbituminous coal, and lignitic coal.

0-8493-1586-7/05/$0.00+$1.50

TABLE 66.1 ASTM Classification of Coals (ASTM D388)

Class	Group	Fixed Carbon Limit% (dry mineral matter-free)		Volatile Matter Limit% (dry mineral matter-free)		Calorific Value Limit, Btu/lb (moist mineral matter-free)		Agglomerating Character
		= or > than	< than	> than	= or < than	= or > than	< than	
	a. Mataanthracite	98	—	—	2	—	—	
1. Anthracitic	b. Anthracite	92	98	2	8	—	—	Nonagglomerating
	c. Semianthracite	86	92	8	14	—	—	
	a. Low-volatile bituminous coal	78	86	14	22	—	—	
	b. Medium-volatile bituminous coal	69	78	22	31	—	—	
2. Bituminous	c. High-volatile A bituminous coal	—	69	31	—	14,000	—	Common agglomerating
	d. High-volatile B bituminous coal	—	—	—	—	13,000	14,000	
	e. High-volatile C bituminous coal	—	—	—	—	10,500	13,000	Agglomerating
	a. Subbituminous A coal	—	—	—	—	10,500	11,500	Nonagglomerating
3. Subbituminous	b. Subbituminous B coal	—	—	—	—	9,500	10,500	
	c. Subbituminous C coal	—	—	—	—	8,300	9,500	
4. Lignitic	a. Lignite A	—	—	—	—	6,300	8,300	
	b. Lignite B	—	—	—	—	—	6,300	

Source: ASTM, Standards on Gaseous Fuels, Coal and Coke.

Selection of a coal for a particular application involves consideration of its specific chemical and physical properties, as well as general factors involving storage, handling, or pulverizing. Factors related to furnace design include volume, grate area, and the amount of radiant heating surface.

Coal Class

The higher-rank coals are classified according to fixed carbon on a dry basis, whereas the lower-rank coals are classified according to heating value (Btu) on a moist basis.

Coal Analysis

The two basic coal analyses are the proximate analysis and the ultimate analysis. In any coal seam there are two components that can show significant variation throughout the seam — the moisture and the ash. The ash fraction varies because ash is essentially the inorganic matter deposited with the organic material during the compaction process. The moisture content of coal varies significantly, depending on the exposure to groundwater before mining and during transportation and the storage before the coal is burned.

The proximate analysis gives the mass fraction of fixed carbon (FC), volatile matter (VM), and ash of the coal. The analysis can be made by weighing, heating, and burning small samples of coal. A powdered coal sample is carefully weighed and then heated to 110°C (230°F) for 20 min. The sample is then weighed again; the mass loss divided by the original mass gives the mass fraction of moisture (M) in the sample. The remaining sample is then heated to 954°C (1750°F) in a closed container for 7 min, after which the sample is weighed. The resulting mass divided by the original mass is equal to the mass fraction of volatile matter in the sample.

The sample is then heated to 732°C (1350°F) in an open crucible until it is completely burned. The residue is then weighed; the final weight divided by the original weight is the ash fraction (A). The mass fraction of fixed carbon is obtained by subtracting the moisture, volatile matter, and ash fraction from unity. In addition to the FC, the VM, the M, and the A, most proximate analyses list separately the sulfur mass fraction (S) and the higher heating value (HHV) of the coal.

The ultimate coal analysis is a laboratory analysis that lists the mass fraction of carbon (C), hydrogen (H_2), oxygen (O_2), sulfur (S) and nitrogen (N_2) in the coal along with the higher heating value (HHV). It also lists the moisture, M, and ash, A. This analysis is required to determine the combustion-air requirements for a given combustion system and the size of the draft system for the furnace.

The Heating Value (Btu)

The heating value represents the amount of chemical energy in a pound mass of coal (Btu/1bm) on a moist basis. There are two heating values — a higher or gross heating value (HHV) and a lower or net heating value (LHV). The difference between these two values is essentially the latent heat of vaporization of the water vapor present in the exhaust products when the fuel is burned in dry air. In actual combustion systems, this includes the water present in the fuel as burned (the moisture) and the water produced from the combustion of hydrogen, but it does not include any moisture that is introduced by the combustion of air.

Agglomerating Character

Coals are considered agglomerating if, in a test to determine the amount of volatile matter, they produce either a coherent button that will support a 500 g weight without pulverizing or a button that shows swelling or cell structure.

Fixed Carbon (FC) Limit

Two formulas used for calculating the fixed carbon limit, on a mineral matter-free (mm-free) basis, are the Parr formula,

$$\text{Dry, mm-free FC} = \frac{\text{FC} - 0.15S}{100 - (1.08A + 0.55S)} \times 100 \tag{66.1}$$

and the approximate formula,

$$\text{Dry, mm-free FC} = \frac{\text{FC}}{10 - (M + 1.1A + 0.1S)} \times 100 \tag{66.2}$$

Coal Heating Value

Two formulas are used for calculating the coal heating value: the Parr formula,

$$\text{Moist, mm-free Btu} = \frac{\text{Btu} - 50S}{100 - (1.08A + 0.55S)} \times 100 \tag{66.3}$$

and the approximate formula,

$$\text{Moist, mm-free Btu} = \frac{\text{Btu}}{100 - (1.1A + 0.1S)} \times 100 \tag{66.4}$$

Percent Volatile Matter (VM)

The formula used for calculating coal VM value is as follows:

$$\text{Dry, mm-free VM} = 100 - \text{dry, mm-free FC} \tag{66.5}$$

66.2 Oil

Oil Formation

Petroleum or crude oil is thought to be partially decomposed marine life. It is normally found in large domes of porous rock. Crude oils are normally ranked in three categories, depending on the type of residue left after the lighter fractions have been distilled from the crude. Under this system the petroleum is classified as paraffin-based crude, asphalt-based crude, or mixed-based crude.

Oil Composition

Crude oil is composed of many organic compounds, yet the ultimate analysis of all crude oils is fairly constant. The carbon mass fraction ranges from 84 to 87%, and the hydrogen mass fraction from 11 to 16%. The sum of the oxygen and nitrogen mass fractions ranges from 0 to 7%, and the sulfur mass fraction ranges from 0 to 4%.

Crude Oil Refining

Crude oils are more valuable when defined as petroleum products. Distillation separates the crude oil into fractions equivalent in the boiling range to **gasoline, kerosene**, gas oil, lubricating oil, and a residual. Thermal or **catalytic cracking** is used to convert kerosene, gas oil, or residual to gasoline, lower-boiling

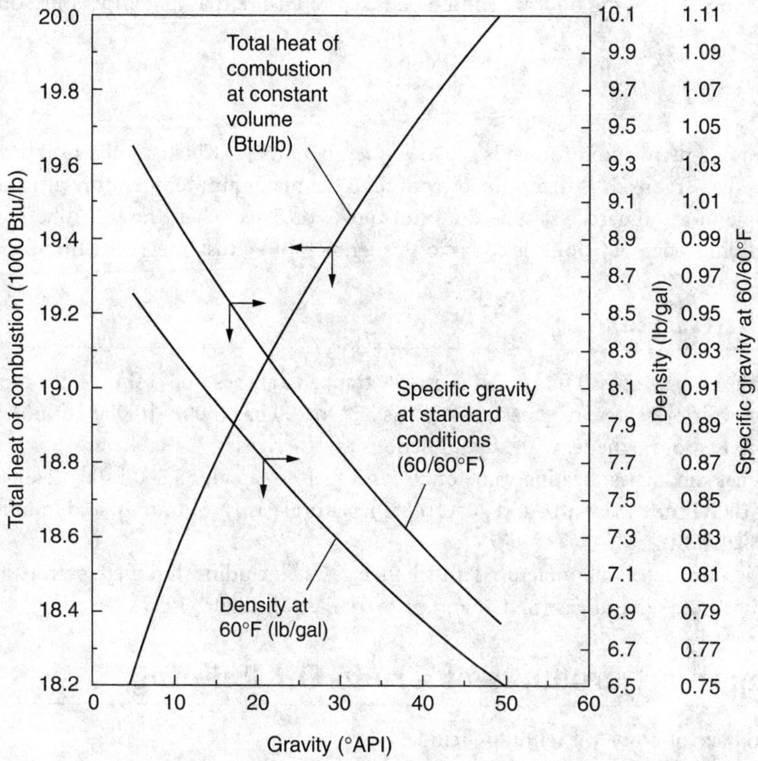

FIGURE 66.1 Properties of petroleum derivatives. (*Source:* Babcock & Wilcox Company. 1972. *Steam — Its Generation and Use,* 38th ed. Babcock & Wilcox Company, New York. With permission.)

fractions, and a residual coke. The major finished products are usually blends of a number of stocks, plus additives.

Properties of Petroleum Derivatives

The main variables of petroleum derivatives are the heating value, the specific gravity (s), the flash point, and the pour point. The heating value (usually the higher heating value) is reported in units of either kilojoules per kilogram (or Btu per pound mass) or kilojoules per liter (or Btu per gallon).

The heating value of crude oil and petroleum products is shown as a function of the specific gravity in Figure 66.1. The heating value (on unit mass basis) of petroleum derivatives increases as the specific gravity of the product decreases.

The specific gravity of any liquid is the density of the liquid divided by the density of water at 60°F (15.6°C). The specific gravity of crude oil is usually between 0.80 and 0.97. The corresponding °API (American Petroleum Institute) gravity is 45 to 15 degrees. The relationship between the specific gravity and the °API is as follows:

$$°\text{API} = \frac{141.5}{\text{Specific gravity at } 60/60\,°\text{F}} - 131.5 \tag{66.6}$$

Flash Point

The flash point of a liquid fuel is the minimum fluid temperature at which the vapor coming from the fluid surface will just ignite. At a slightly higher temperature, called the *fire point*, the vapors will support combustion. When storing the oil, care should be taken to ensure that its temperature does not exceed its flash point.

Pour Point

The pour point of a petroleum product is the lowest temperature at which an oil or oil product will flow under standard conditions. It is determined by finding the maximum temperature at which the surface of an oil sample in a standard test tube does not move for 5 sec when the test tube is rotated to the horizontal position. The pour point is equal to this temperature plus 5 degrees Fahrenheit.

66.3 Natural Gas

Natural gas is the only true fossil fuel gas. It is usually trapped in limestone casing on the top of petroleum reservoirs. Reservoir pressures may reach as high as 350 to 700 bar (5000–10,000 1b/in.^{2}). Natural gas is primarily composed of methane, with small fractions of other gases.

Natural gas has the highest heating value of all fossil fuels — about 55,800 kJ/kg (24,000 Btu/lbm), or 37,000 kJ/m^3 (1000 Btu/ft^3) at 1 atm and 20°C (68°F). Natural gas is commonly sold in units of "therms" (1 therm = 100,000 Btu).

There are a number of manufactured fossil fuel gases, including liquified petroleum gas (LPG), synthetic or substitute natural gas (SNG), and primary flash distillate (PFD).

66.4 Important Products of Crude Oil Refining

Important products of crude oil refining include:

1. *Natural gas.* Natural gas is composed mainly of CH_2 and some N_2, depending on the source. It is separated from crude oil by natural sources. Its principal use is as a fuel gas.
2. *Liquified petroleum gas (LPG).* LPG is composed of propane and butane. It is usually stripped from "wet" natural gas or from a crude oil cracking operation. LPG is widely used for industrial and domestic applications where natural gas is not available.
3. *Primary flash distillate (PFD).* PFD is composed of propane and butane dissolved in the gasoline–kerosene range of liquids. It is produced by preliminary distillation of crude oil.
4. *Gasoline.* Gasoline has a boiling range of 300 to 450 K (80 to 350°F). It is produced by primary distillation and reforming processes to improve its performance. Its principal use is spark-ignition internal combustion engines.
5. *Kerosene.* Kerosene is a paraffinic hydrocarbon with a boiling range of 410 to 575 K (280 to 575°F). It is produced by distillation and cracking of crude oil. Its principal uses are in agricultural tractors, lighting, heating, and aviation gas turbines.
6. *Gas oil.* Gas oil is a saturated hydrocarbon with a boiling range of 450 to 620 K (350 to 660°F). It is composed of saturated hydrocarbons produced by distillation and hydrodesulfurization of crude oil. Its principal uses are in diesel fuel, heating, and furnaces and as a feed to cracking units.
7. *Diesel fuel.* **Diesel fuel** is a saturated hydrocarbon with a boiling range of 450 to 650 K (350 to 680°F). It is composed of saturated hydrocarbons produced by distillation cracking of crude oil. Its principal uses are in diesel engines and furnace heating. The ASTM grades for diesel fuel are as follows:
   Grade 1-D. A volatile distillate fuel. Suitable for engines in service requiring frequent speed and load changes.
   Grade 2-D. A distillate fuel oil of lower volatility. Suitable for engines in industrial and heavy mobile service.
   Grade 4-D. A fuel oil for low- and medium-speed engines.
8. *Fuel oil.* Fuel oil has a boiling range of 500 to 700 K (440 to 800°F). It is composed of residue of primary distillation, blended with distillates. Its primary use is in large-scale industrial heating.

9. *Lubricating oil.* There are three types of lubricating oils — mainly aromatic, mainly aliphatic, and mixed — which are manufactured by vacuum distillation of primary distillation residue-solvent extraction.
10. *Wax.* Wax is composed of paraffins produced by chilling residue of vacuum distillation. It is used primarily in food, candles, and petroleum jelly.
11. *Bitumen.* Bitumen is produced from residue of vacuum distillation or by oxidation of residue from primary distillation. It is used for road surfacing and waterproofing.

Nomenclature

A	Percentage of ash
API	American Petroleum Institute
Btu	Coal heating value, Btu per pound on moist basis
C	Mass fraction of carbon
FC	Percentage of fixed carbon
H_2	Hydrogen
HHV	Higher heating value
M	Percentage of moisture
N_2	Nitrogen
O_2	Oxygen
S	Percentage of sulfur
s	Specific gravity
VM	Percentage of volatile matter

Defining Terms

Bituminous coal — Soft coal containing large amounts of carbon. It has a luminous flame and a great deal of smoke.

Catalytic cracking — A refinery process that converts a high-boiling fraction of petroleum (gas oil) to gasoline, olefin feed for alkylation, distillate, fuel oil, and fuel gas by use of a catalyst and heat.

Diesel fuel — Fuel for diesel engines obtained from the distillation of crude oil. Its quality is measured by cetane number.

Gasoline — Light petroleum product obtained by refining crude oil with or without additives, blended to form a fuel suitable for use in spark-ignition engines. Its quality is measured by octane rating. The four classes are leaded regular (87–90 octane), unleaded regular (85–88 octane), mid-grade unleaded (88–90 octane), and premium (greater than 90 octane).

Kerosene — Colorless low-sulfur oil products used in space heaters, cook stoves, and water heaters.

Octane — A rating scale used to grade gasoline according to its antiknock properties. Also, any of several isometric liquid paraffin hydrocarbons, C_8H_{18}. Normal octane is a colorless liquid found in petroleum boiling at 124.6°C.

References

Bobcock and Wilcox Company. 1972. *Steam — Its Generation and Use,* 38th ed. Babcock & Wilcox Company, New York.

Francis, W. 1965. *Fuels and Fuel Technology,* Pergamon Press, New York.

Harker, J. H. and Allen, D. A. 1972. *Fuel Science,* Oliver and Boyd, Edinburgh, U.K.

Johnson, A. J. and Auth, G. A. 1951. *Fuel and Combustion Handbook,* McGraw-Hill, New York.

Popovich, M. and Hering, C. 1959. *Fuels and Lubricants,* John Wiley & Sons, New York.

Further Information

American Gas Association
American Petroleum Institute
National Petroleum Council
U.S. Department of the Interior publications *International Petroleum Annual* and *U.S. Petroleum and Natural Gas Resources*

67

Solar Electric Systems[1]

Thomas R. Mancini
Sandia National Laboratories

Roger Messenger
Florida Atlantic Univeristy

Jerry Ventre
Florida Solar Energy Center

67.1 Solar Thermal Electric Systems

Thomas R. Mancini

Solar thermal power systems, which are also referred to as concentrating solar power systems, use the heat generated from the concentration and absorption of solar energy to drive heat engine/generators and, thereby, produce electrical power. Three generic solar thermal systems, trough, power tower, and dish-engine systems, are used in this capacity. Trough systems use linear parabolic concentrators to focus sunlight along the focal lines of the collectors. In a power tower system, a field of two-axis tracking mirrors, called heliostats, reflects the solar energy onto a receiver that is mounted on top of a centrally located tower. Dish-engine systems, the third type of solar thermal system, continuously track the sun, providing concentrated sunlight to a thermal receiver and heat engine/generator located at the focus of the dish.

Trough Systems

Of the three solar thermal technologies, trough-electric systems are the most mature with 354 MW installed in the Mojave Desert of Southern California. Trough systems produce about 75 suns concentration and operate at temperatures of up to 400°C at annual efficiencies of about 12%. These systems use linear-parabolic concentrators to focus the sunlight on a glass-encapsulated tube that runs along the focal line of the collector, shown in Figure 67.1. Troughs are usually oriented with their long axis north–south, tracking the sun from east to west, to have the highest collection efficiencies. The oil working fluid is heated as it circulates through the receiver tubes and before passing through a steam-generator heat exchanger. In the heat exchanger, water boils producing the steam that is used to drive a conventional Rankine-cycle turbine generator. The optimal size for a trough-electric system is thought to be about 200 megawatts, limited mainly by the size of the collector field.

A major challenge facing these plants has been to reduce the operating and maintenance costs, which represent about a quarter of the cost of the electricity they produce. The newer plants are designed to operate at 10 to 14% annual efficiency and to produce electricity for \$.08 to \$.14 a kilowatt-hour, depending on interest rates and tax incentives with maintenance costs estimated to be about \$.02 a

[1]The Concentrating Solar Power activities presented herein are funded by the U.S. Department of Energy through Sandia National Laboratories, a multi-program laboratory operated by Sandia Corporation, a Lockheed Martin Company, for the U.S. Department of Energy under Contract DE-AC04-94-AL85000.

0-8493-1586-7/05/\$0.00+\$1.50

FIGURE 67.1 Solar collector field at a SEGS plant located at Kramer Junction, CA. (Photograph courtesy of Sandia National Laboratories.)

kilowatt hour. New systems are currently being designed for use in the Southwest U.S. that will utilize ongoing research and development addressing issues such as improved receiver tubes, advanced working/storage fluids, and molten-salt storage for trough plants.

Power Towers

In the Spring of 2003, the first commercial power towers are being designed for installation in Spain in response to a very attractive solar incentive program. Power towers, shown in Figure 67.2, are also known

FIGURE 67.2 Warming the receiver at Solar 2 in Dagget, CA. The "wings" one either side of the receiver result from atmospheric scattering of the sunlight reflected from the heliostat field. (Photograph courtesy of Sandia National Laboratories.)

as central receivers and, while close, they are not as commercially mature technologically as trough systems. In a power tower, two-axis tracking mirrors called heliostats reflect solar energy onto a thermal receiver that is located at the top of a tower in the center of the heliostat field. To maintain the sun's image on the centrally located receiver, each heliostat must track a position in the sky that is midway between the receiver and the sun. Power towers have been designed for different working fluids, including water/steam, sodium nitrate salts, and air. Each working fluid brings an associated set of design and operational issues; the two active tower design concepts in Europe and the U.S. utilize air and molten-nitrate salt as the working fluids. Both system designs incorporate thermal storage to increase the operating time of the plant, thereby allowing the solar energy to be collected when the sun shines, stored, and used to produce power when the sun is not shining. This feature, which allows power to be dispatched when needed, increases the "value" of the electricity generated with solar energy.

In the U. S., the choice of a molten-nitrate salt as the working fluid provides for high-temperature, low-pressure operation and thermal storage in the hot salt. The molten-salt approach uses cold salt (290°C) that is pumped from a cold storage tank, through the receiver where it is heated with 800 suns to 560°C. The hot salt is delivered to a hot salt storage tank. Electrical power is produced when hot salt is pumped from the hot tank and through a steam generator; the steam is then used to power a conventional Rankine turbine/generator. The salt, which has cooled, is returned to the cold tank. The optimal size for power towers is in the 100- to 300-megawatt range. Studies have estimated that power towers could operate at annual efficiencies of 15 to 18% and produce electrical power at a cost of $.06 to $.11 per kilowatt-hour. Advanced systems are focusing on reducing the cost of heliostats and on operational issues associated with plant operation and thermal storage.

Dish–Stirling Systems

Dish–Stirling systems track the sun and focus solar energy into a cavity receiver where it is absorbed and transferred to a heat engine/generator. Figure 67.3 is a picture of a Dish–Stirling system. Although a Brayton engine has been tested on a dish and some companies are considering adapting microturbine technology to dish engine systems, kinematic Stirling engines are currently being used in all four Dish–Stirling systems under development today. Stirling engines are preferred for these systems because of their high efficiencies (thermal-to-mechanical efficiencies in excess of 40% have been reported), high power density (40 to 70 kW/liter for solar engines), and their potential for long-term, low-maintenance

FIGURE 67.3 Advanced dish development system (ADDS) on test at Sandia National Laboratories' National Solar Thermal Test Facility. (Photograph courtesy of Sandia National Laboratories)

operation. Dish–Stirling systems have demonstrated the highest efficiency of any large solar power technology, producing more than 3000 suns concentration, operating at temperatures in excess of 750°C, and at annual efficiencies of 23% solar-to-electric conversion. Dish–Stirling systems are modular, i.e., each system is a self-contained power generator, allowing their assembly into plants ranging in size from a few kilowatts to tens of megawatts. The near-term markets identified by the developers of these systems include remote power, water pumping, grid-connected power in developing countries, and end-of-line power conditioning applications.

The major issues being addressed by system developers are reliability, due to high-temperature, high-heat flux conditions, system availability, and cost. Current systems range in size from 10 to 25 kW, primarily because of the available engines. These systems are the least developed of the three concentrating solar power systems. Research and development is focusing on reliability improvement and solar concentrator cost reductions.

67.2 Photovoltaic Power Systems

Roger Messenger and Jerry Ventre

In 2002, worldwide shipments of photovoltaic (PV) modules passed the 500-megawatt mark as PV module shipments continued to grow at an annual rate near 25% [Maycock, 2002]. Another important milestone in the PV industry occurred in 2001. Of the 395 MW of PV installed during 2001, the capacity of grid-connected PV systems installed (204 MW) surpassed the capacity of stand-alone systems installed (191 MW). Improvements in all phases of deployment of PV power systems have contributed to the rapid acceptance of this clean electric power-producing technology.

The photovoltaic effect occurs when a photodiode is operated in the presence of incident light when connected to a load. Figure 67.4a shows the I-V characteristic of the photodiode. Note that the characteristic passes through the first, third, and fourth quadrants of the I-V plane. In the first and third quadrants, the device dissipates power supplied to the device from an external source. But in the fourth quadrant, the device generates power, since current leaves the positive terminal of the device. Photovoltaic cells are characterized by an open circuit voltage and a short circuit current, as shown in Figure 67.4b. Note that in Figure 67.4b, the current axis is inverted such that a positive current represents current leaving the device positive terminal. The short circuit current is directly proportional to the incident light intensity, while the open circuit voltage is proportional to the logarithm of the incident light intensity. Figure 67.4b shows that the cell also has a single point of operation where its output power is maximum. This point, labeled P_{max}, occurs at current, I_{mp} and voltage, V_{mp}. Since cells are relatively expensive, good system designs generally ensure that the cells operate as close as possible to their maximum power points.

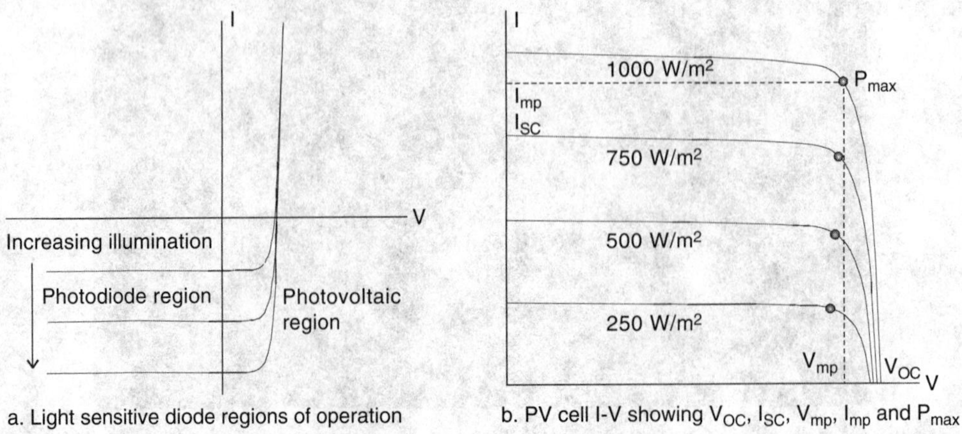

FIGURE 67.4 I-V characteristics of light sensitive diode.

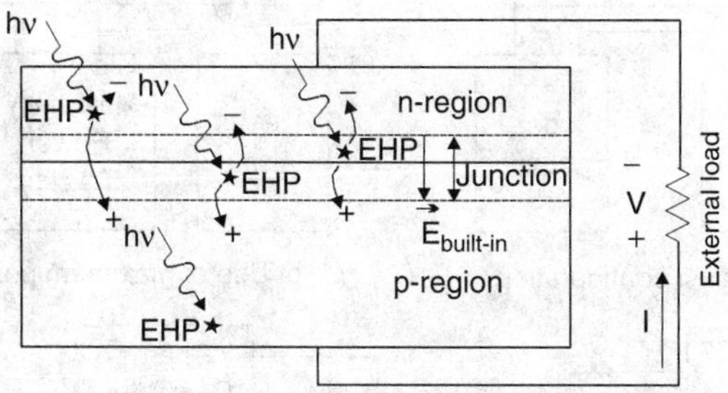

FIGURE 67.5 The illuminted pn junction showing the creation of electron-hole pairs.

The power generation mechanism that applies to conventional PV cells is shown in Figure 67.5. Photovoltaic power generation requires the presence of a pn junction. The pn junction is formed from a material, such as crystalline silicon, that has holes as the dominant charge carriers on the p-side and electrons as dominant carriers on the n-side. In simple terms, incident photons with sufficient energy interact with crystal lattice atoms in the PV cell to generate electron-hole pairs (EHP). If the EHP is generated sufficiently close to the pn junction, the negatively charged electron will be swept to the n-side of the junction and the positively charged hole will be swept to the p-side of the junction by the built-in electric field of the junction. The separation of the photon-generated charges creates a voltage across the cell, and if an electrical load is connected to the cell, the photon-generated current will flow through the load as shown in Figure 67.5.

Individual PV cells, with cell areas ranging from small to approximately 225 cm^2, generally produce only a few watts or less at about half a volt. Thus, to produce large amounts of power, cells must be connected in series–parallel configurations. Such a configuration is called a module. A typical crystalline silicon module will contain 36 cells and be capable of producing about 100 watts when operated at approximately 17 V and a current of approximately 6 A. Presently, the largest available module is a 300-watt unit that measures approximately 4 × 6 ft (see http://www.asepv.com/aseprod.html for more information). Although higher conversion efficiencies have been obtained, typical conversion efficiencies of commercially available crystalline silicon PV modules are in the 14% range, which means the modules are capable of generating approximately 140 W/m^2 of module surface area under standard test conditions. However, under some operating conditions where the module temperatures may reach 60°C, the module output power may be degraded by as much as 15%.

If the power of a module is still inadequate for the needs of the system, additional modules may be connected in series or in parallel until the desired power is obtained.

Stand-Alone PV Systems

Figure 67.6 shows a hierarchy of stand-alone PV systems. Stand-alone PV systems may be as simple as a module connected to an electrical load. The next level of complexity is to use a maximum power tracker or a linear current booster between module and load as a power matching device to ensure that the module delivers maximum power to the load at all times. These systems are common for pumping water and are the equivalent of a dc-dc matching transformer that matches source resistance to load resistance.

The next level of complexity involves the use of storage batteries to allow for the use of the PV-generated electricity when the sun is not shining. These systems also generally include an electronic controller to prevent battery overcharge and another controller to prevent overdischarge of the batteries. Many charge controllers perform both functions.

If loads are ac, an inverter can be incorporated into the system to convert the dc from the PV array to ac. A wide range of inverter designs are available with a wide range of output waveforms ranging from square waves to relatively well-approximated sine waves with minimal harmonic distortion. Most good inverters are capable of operating at conversion efficiencies greater than 90% over most of their output power range.

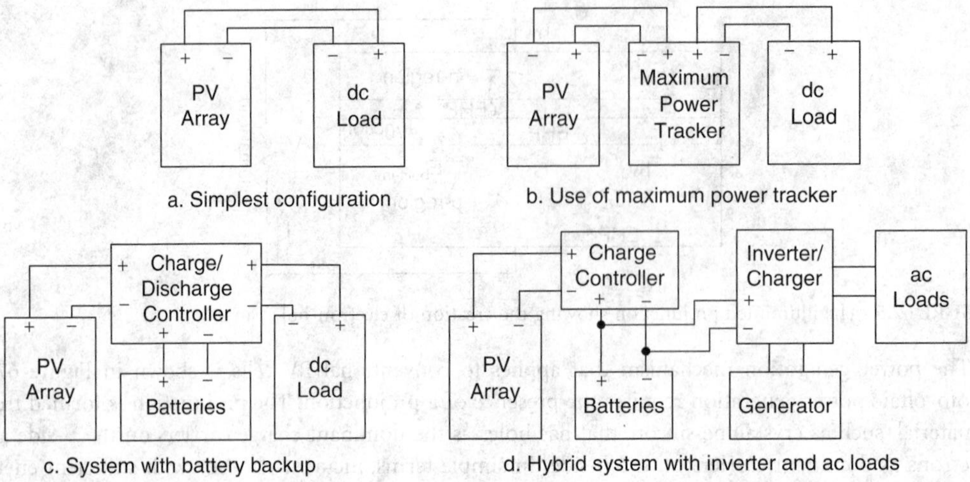

FIGURE 67.6 Four examples of stand-alone PV systems.

Available sunlight is measured in kWh/m²/day. However, since peak solar power, or "peak sun" is defined as 1 kW/m², the term peak sun hours (psh) is often used as an equivalent expression for available solar energy (i.e., 1 psh = 1 kWh/m²/day). A number of tabulations of psh are available [Sandia National Laboratories, 1995; see also www.nrel.gov and http://solstice.crest.org/renewables/solrad/]. In some regions, there is a large difference in seasonal available sunlight. In these regions with large seasonal differences in psh, it often makes sense to back up the solar electrical production with a fossil-fueled or wind generator. When more than one source of electricity is incorporated in a stand-alone system, the system is called a hybrid system. Elegant inverter designs incorporate circuitry for starting a generator if the PV system battery voltage drops too low.

Grid-Connected PV Systems

Grid-connected, or utility interactive, PV systems connect the output of the PV inverter directly to the utility grid and are thus capable of supplying power to the utility grid. An inverter operating in the utility interactive mode must be capable of disconnecting from the utility in the event of a utility failure. Elegant control mechanisms have been developed that enable the inverter to detect either an out-of-frequency-range or an out-of-voltage-range utility event. In essence, the output of the inverter monitors the utility voltage and delivers power to the grid as a current source rather than as a voltage source. Since the inverters are microprocessor controlled, they are capable of monitoring the grid in intervals separated by less than a millisecond, so it is possible for the inverter to keep the phase of the injected current very close to the phase of the utility voltage. Most inverter operating power factors are close to unity.

The simplest utility-interactive PV system connects the PV array to the inverter and connects the inverter output to the utility grid. It is also possible to incorporate battery backup into a utility-interactive PV system so that in the event of loss of the grid, the inverter will be able to power emergency loads. The emergency loads are connected to a separate emergency distribution panel so the inverter can still disconnect from the utility if the utility fails but then instantaneously switch from current source mode to voltage source mode to supply the emergency loads. Figure 67.7 shows the two types of utility interactive systems.

Defining Terms

Grid-connected PV systems — Photovoltaic systems that interact with the utility grid, using the utility grid as a destination for excess electricity generated by the PV system or as a source of additional electricity needed to supply loads that exceed the PV system capacity.

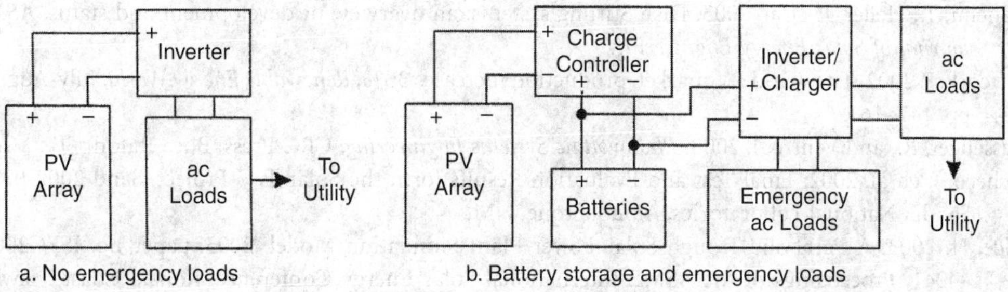

FIGURE 67.7 Utility interactive PV systems.

Heliostats — A system of mirrors that are controlled to reflect sunlight onto a power tower.

Photovoltaic cells — Cells made of various materials that are capable of converting light directly into electricity.

Photovoltaic systems — Systems that incorporate series and parallel combinations of photovoltaic cells along with power conditioning equipment to produce high levels of dc or ac power.

Power tower — A central receiving station upon which the sun is reflected by a system of mirrors to produce solar thermal electricity.

Solar thermal electric — The process of using the sun to heat a material that can be used to generate steam to drive a turbine to generate electricity.

Stand-alone PV systems — Photovoltaic systems that operate independently of a utility grid.

Trough electric systems — A system of parabolic troughs in which a liquid is passed through the focal point for superheating to produce solar thermal electricity.

References

Cohen, G. E., Kearney, D. W., and Kolb, G. J. 1999. Final Report on the Operation and Maintenance Improvement Program for Concentrating Solar Power Plants, Sandia National Laboratories Report, SAND 99-1290, June.

Davenport, R., Mayette, J., and Forristall, R. 2001. The Salt River Project SunDish Dish/Stirling System, ASME Solar Forum 2001, Washington, D.C., April 21–25.

Diver, R., and Andraka, C. 2003. Integration of the Advanced Dish Development System, paper no. ISEC2003-44238, Proceedings of the ASME International Solar Energy Conference, Kohala Coast, Hawaii, March 15–18.

Diver, R. B., Andraka, C. E., Rawlinson, K. S., Goldberg, V., and Thomas, G. 2001. The Advanced Dish Development System Project, Proceedings of Solar Forum 2001: Solar Energy: The Power to Choose, Washington, D.C. April 21–25.

Diver, R., Andraka, C., Rawlinson, K., Moss, T., Goldberg, V., and Thomas, G. 2003. Status of the Advanced Dish Development System Project, paper no. ISEC2003-44237, Proceedings of the ASME International Solar Energy Conference, Kohala Coast, Hawaii, March 15–18.

Falcone, P. 1986. A Handbook for Solar Central Receiver Design, Sand 86-8009, Sandia National Laboratories, Albuquerque, NM.

Haeger, M., Keller, L., Monterreal, R., and Valverde, A. 1994. PHOEBUS Technology Program Solar Air Receiver (TSA): Experimental Set-up for TSA at the CESA Test Facility of the Plataforma Solar de Almeria (PSA), Solar Engineering 1994, proceedings of ASME/JSME/JSES International Solar Energy Conference, ISBN 0-7918-1192-1. pp. 643–647.

Heller, P., Baumüller, A., and Schiel, W. 2000. EuroDish — The Next Milestone to Decrease the Costs of Dish/Stirling System towards Competitiveness, 10th International Symposium on Solar Thermal Concentrating Technologies, Sydney.

Lotker, M. 1991. Barriers to Commercialization of Large-Scale Solar Electricity: Lessons Learned form the LUZ Experience. Sandia National Laboratories, Albuquerque, NM, SAND91-7014.

Mancini, T., Heller, P. et al. 2003. Dish Stirling systems: an overview of development and status, *ASME Journal of Solar Energy Engineering.*

Maycock, P. 2002. The world PV market, production increases 36%, *Renewable Energy World,* July-August. pp. 147–161.

Messenger, R., and Ventre, J. 2000. *Photovoltaic Systems Engineering,* CRC Press, Boca Raton, FL.

Pacheco, J. et. al. 2002. Final Test and Evaluation Results form the Solar Two Project, Sand 2002-0120, Sandia National Laboratories, Albuquerque, NM.

Price, H. 2003. A Parabolic Trough Solar Power Plant Simulation Model. 2003. paper no. ISEC2003-44241, Proceedings of the ASME International Solar Energy Conference, Kohala Coast, Hawaii, March 15–18.

Price, H., and Kearney, D. 2003. Reducing the Cost of Energy from Parabolic Trough Solar Power Plants, paper no. ISEC2003-44069, Proceedings of the ASME International Solar Energy Conference, Hawaii, March 15–18.

Reilly, H., Ang G. Kolb, 2001. An Evaluation of Molten-Salt Power Towers Including Results of the Solar Two Project, Sand 2001-3674, Sandia National Laboratories, Albuquerque, NM, November.

Sandia National Laboratories. 1995. *Stand-Alone Photovoltaic Systems: A Handbook of Recommended Design Practices,* Sandia National Laboratories, Albuquerque, NM.

Stone, K., Leingang, E., Liden, R., Ellis, E., Sattar, T., Mancini, T. R., and Nelving, H. 2001. SES/Boeing Dish Stirling System Operation, Solar Energy: The Power to Choose, ASME/ASES/AIA/ASHRAE/SEIA, Washington, DC, April 21–25.

Stone, K., Rodriguez, G., Paisley, J., Nguyen, J. P., Mancini, T. R., and Nelving, H. 2001. Performance of the SES/Boeing Dish Stirling System, Solar Energy: The Power to Choose, ASME/ASES/AIA/ASHRAE/SEIA, Washington, D.C., April 21–25.

Tamme, R., Laing, D. and Steinmann, W. 2003. Advanced Thermal Energy Storage Technology for Parabolic Trough, paper no. ISEC2003-44033, Proceedings of the ASME International Solar Energy Conference, Kohala Coast, Hawaii, March 15–18.

Tyner, C., Kolb, G., Geyer, M., and Romero, M . Concentrating Solar Power in 2001: An IEA/SolarPACES. SolarPACES Task I: Electric Power Systems, Solar Paces Task 1, Electric Power Systems.

68

Internal Combustion Engines

Alan A. Kornhauser
Virginia Polytechnic Institute and State University

An internal combustion (i.c.) engine is a heat engine in which the thermal energy comes from a chemical reaction within the working fluid. In external combustion engines, such as steam engines, heat is transferred to the working fluid through a solid wall and rejected to the environment through another solid wall. In i.c. engines, heat is released by a chemical reaction in the working fluid and rejected by exhausting the working fluid to the environment.

Internal combustion engines have two intrinsic advantages over other engine types:

1. They require no heat exchangers (except for auxiliary cooling). Thus, weight, volume, cost, and complexity are reduced.
2. They require no high-temperature heat transfer through walls. Thus, the maximum temperature of the working fluid can exceed maximum allowable wall material temperature.

They also have some intrinsic disadvantages:

1. Practically, working fluids are limited to air and products of combustion.
2. Nonfuel heat sources (waste heat, solar, nuclear) cannot be used.
3. There is little flexibility in combustion conditions because they are largely set by engine requirements. This can make low-emissions combustion hard to attain.

The advantages far outweigh the disadvantages. I.c. engines comprise more individual units and more rated power than all other types of heat engines combined.

0-8493-1586-7/05/$0.00+$1.50

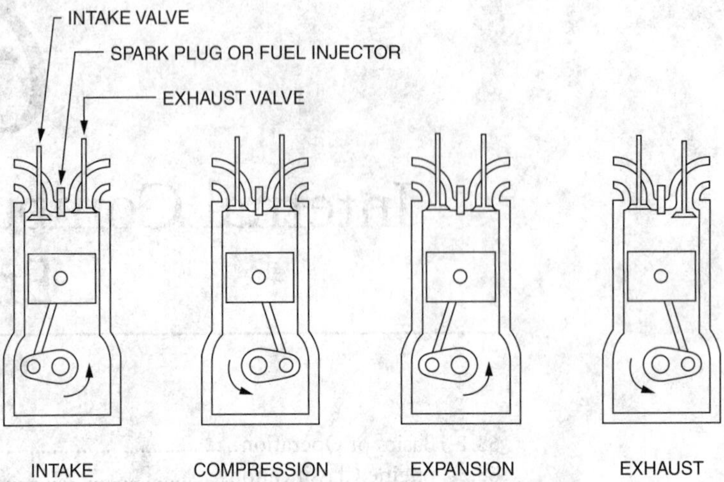

FIGURE 68.1 Operating cycle for a four-stroke i.c. engine.

According to the definition given above, i.c. engines include reciprocating types, rotary (Wankel) types, and gas turbines. In customary usage, however, the title "internal combustion" is used only for the first two of these three types. A more proper designation might be "positive displacement internal combustion" engines. These are the engines described in this chapter.

68.1 Basics of Operation

The basic operation of an i.c. engine is shown in Figure 68.1. The typical engine cycle is divided into four steps:

1. *Intake.* Engine working volume increases. Intake valve opens to admit air or air/fuel mixture into the working volume.
2. *Compression.* Engine working volume decreases. Valves are closed, and the air or mixture is compressed. Work is done on the working fluid.
3. *Combustion* and *expansion.* Air/fuel mixture burns and releases chemical energy. If fuel was not admitted previously, it is injected at this point. Pressure and temperature inside the working volume increase dramatically. Working volume increases, and work (much greater than that of compression) is done by the working fluid.
4. *Exhaust.* Engine working volume decreases. Exhaust valve opens to expel combustion products from the working volume.

The engine shown is a four-stroke reciprocating type; details would vary for two-stroke or rotary engines.

68.2 Engine Classifications

I.c. engines can be classified in various ways. Some important classifications are:

Spark ignition/compression ignition. In **spark ignition** (s.i., gasoline, petrol, or Otto) engines, the fuel is either mixed with the air prior to the intake stroke or shortly after inlet valve closure. An electric spark ignites the mixture. In **compression ignition** (c.i., oil, or diesel) engines, the fuel is injected after the compression process. The high temperature of the compressed gas causes ignition.

Four-stroke/two-stroke. In **four-stroke** engines, the working cycle is as shown in Figure 68.1. A complete four-stroke cycle takes two crankshaft revolutions, with each stage (intake, compression, expansion, exhaust) comprising about 180°. A complete two-stroke cycle takes only one crankshaft revolution. In a **two-stroke,** engine intake and exhaust strokes are eliminated: gas exchange occurs

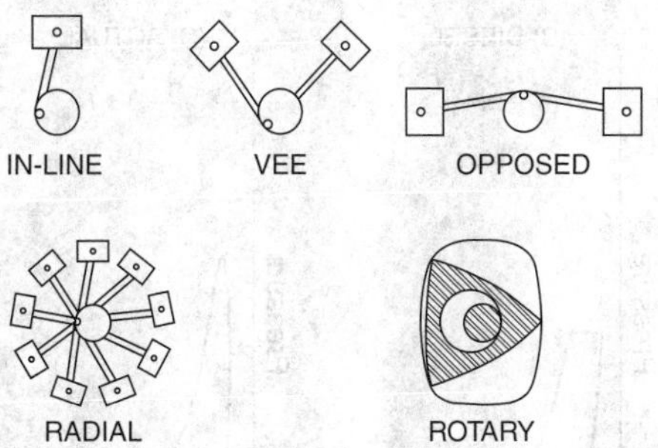

FIGURE 68.2 Engine arrangements.

when the piston is near bottom center position between the expansion and compression strokes. Because the piston does not provide pumping action, some external device is required to ensure that fresh air or mixture replaces the combustion products.

Mechanical layout. Various mechanical layouts are shown in Figure 68.2. Reciprocating i.c. engines use multiple piston-cylinder arrangements driving a single crankshaft. The total number of cylinders per engine ranges from one to 20 or more, with one, four, six, and eight the most common. The cylinders can be arranged in line, in a vee, radially, or horizontally opposed. Rotary i.c. engines use an approximately triangular rotor which revolves eccentrically in a lobed stator. The spaces between the rotor and the stator go through essentially the same processes shown in Figure 68.1. A single rotor-stator pair is thus equivalent to three cylinders. Additional rotor-stator pairs can be stacked on a single shaft to form larger engines.

Intake system. In **naturally aspirated** engines, the pumping action of the piston face draws air into the cylinder. In crankcase **scavenged** engines, the pumping action of the back side of the piston in the crankcase forces air into the cylinder. In **supercharged** engines, a compressor, typically driven off the crankshaft, forces air into the cylinder. In **turbocharged** engines, the compressor is driven by a turbine that recovers work from the exhaust gas.

Besides the major classifications above, engines can be classified by valve number and design (two, three, or four valves per cylinder; rocker arm or overhead cam; cross-, loop-, or uniflow-scavenged), by fuel addition method (carbureted, fuel injected), by combustion chamber shape (tee, ell, flat, wedge, hemisphere, bowl-in-piston), and by cylinder wall cooling method (air, water).

68.3 Spark Ignition Engines

Idealized and Actual Cycles

The spark ignition (s.i) engine can be idealized as an **Otto** cycle using an ideal gas with constant specific heat [Figure 68.3(a)]. The Otto cycle consists of isentropic compression, constant volume heating (simulating combustion), isentropic expansion, and constant volume cooling (simulating intake and exhaust). The **thermal efficiency** of an Otto cycle (Figure 68.4) is $\eta_t = 1 - r_v^{1-\gamma}$, where r_v is the **compression ratio** and γ is the gas specific heat ratio.

The actual engine "cycle" [Figure 68.3(c)] differs from the Otto cycle:

1. In that heat transfer occurs during compression and expansion
2. In that combustion takes place gradually during compression and expansion rather than instantaneously
3. In the presence of intake and exhaust processes
4. In the variation in gas composition and gas specific heat

For a given r_v, the efficiency of a typical s.i. engine is considerably lower than that of the ideal cycle (Figure 68.4), and actual engine r_v is limited by combustion **knock.**

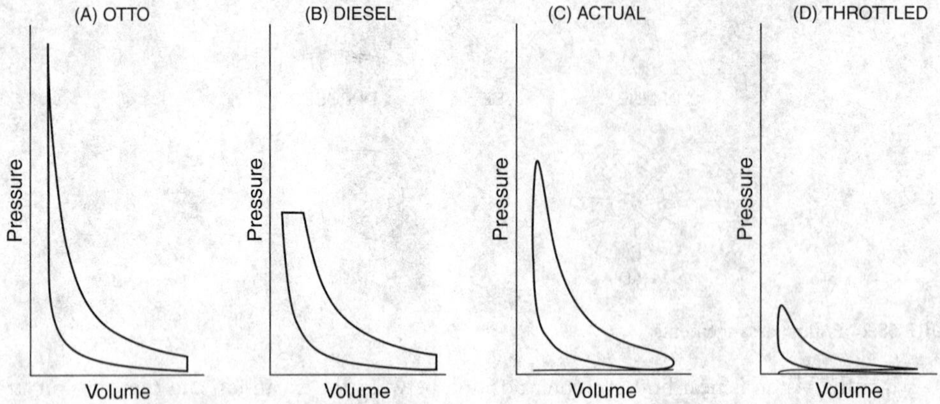

FIGURE 68.3 Pressure-volume diagrams for ideal and actual engine cycles.

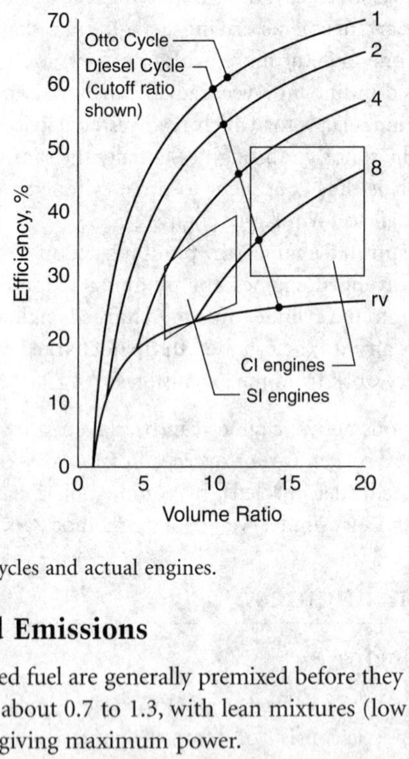

FIGURE 68.4 Efficiency of ideal cycles and actual engines.

Combustion, Fuels, and Emissions

In an s.i. engine, air and vaporized fuel are generally premixed before they enter the cylinder. **Equivalence ratio** (ϕ) generally ranges from about 0.7 to 1.3, with lean mixtures (low ϕ) giving maximum efficiency and the rich mixtures (high ϕ) giving maximum power.

The mixture is heated by compression, but not enough to cause autoignition. Combustion is initiated by an electric spark. If the engine is operating properly, a turbulent flame front travels smoothly and rapidly across the cylinder space. It takes a 15 to 25° crank angle for the first 10% of the mixture to burn, while the next 85% is burned within an additional 35 to 60° [Figure 68.5(a)]. The low numbers of these ranges correspond to $\phi \approx 1$, high turbulence combustion chambers, and low engine speeds. The high numbers correspond to rich or lean mixtures, low turbulence combustion chambers, and high engine speeds. To time the heat release optimally, the spark is typically discharged 5 to 40° before top center, with this advance automatically varied according to engine speed and load. Because higher engine speeds result in increased turbulence and, thus, in increased flame speeds, the total crank angle for combustion increases only slightly as engine speed changes.

Under some operating conditions, the flame does not burn smoothly. In these cases the mixture ahead of the flame front is heated by compression and autoignites before the flame arrives. The resulting deto-

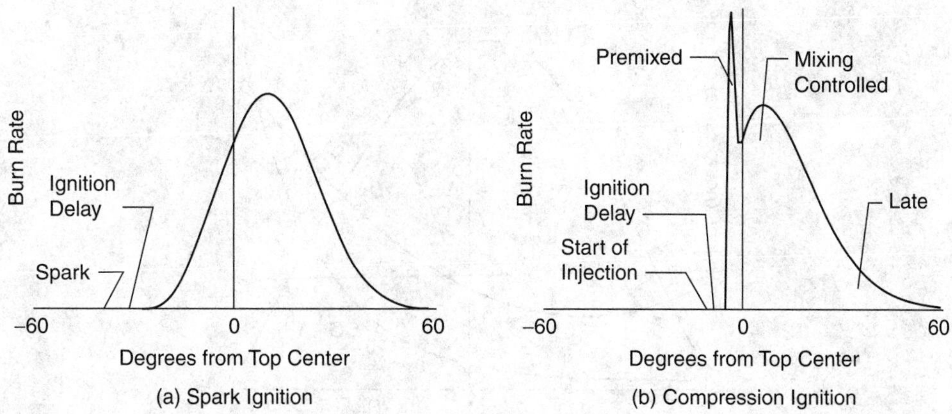

FIGURE 68.5 Heat release rates for s.i. and c.i. engines.

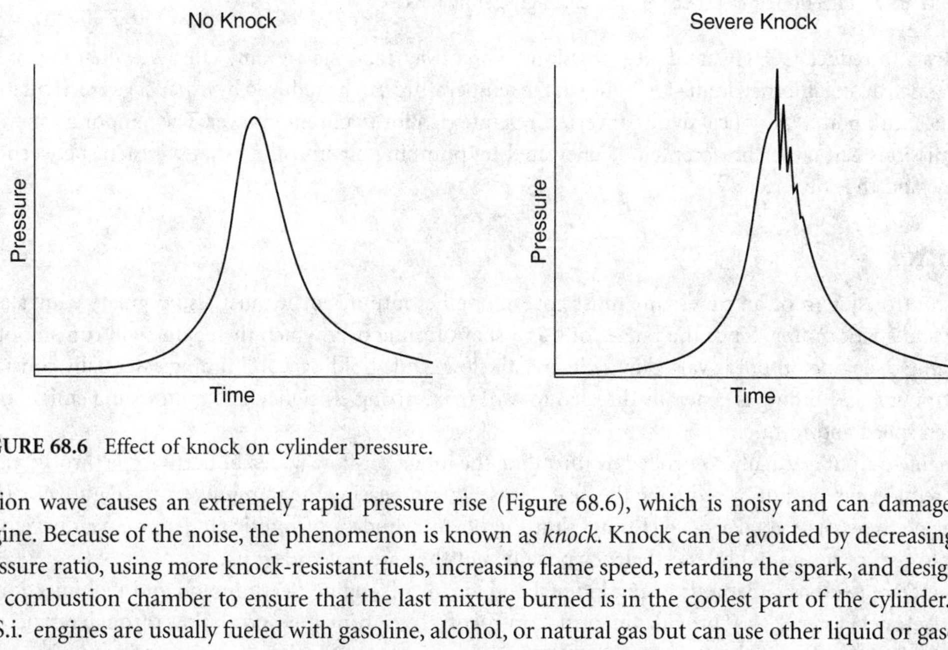

FIGURE 68.6 Effect of knock on cylinder pressure.

nation wave causes an extremely rapid pressure rise (Figure 68.6), which is noisy and can damage the engine. Because of the noise, the phenomenon is known as *knock*. Knock can be avoided by decreasing the pressure ratio, using more knock-resistant fuels, increasing flame speed, retarding the spark, and designing the combustion chamber to ensure that the last mixture burned is in the coolest part of the cylinder.

S.i. engines are usually fueled with gasoline, alcohol, or natural gas but can use other liquid or gaseous fuels. It is important that any s.i. fuel be resistant to autoignition. This resistance is expressed in terms of the **octane number** of the fuel, based on an empirical scale on which iso-octane has been assigned a rating of 100 and *n*-heptane a rating of zero. Typical gasolines have octane numbers in the 85 to 105 range; these octane numbers are usually obtained with the aid of additives. Liquid s.i. engine fuels must be adequately volatile to evaporate fully prior to ignition but not so volatile as to cause problems with storage and transfer.

Besides carbon dioxide and water, the combustion process in s.i. engines produces several pollutants (Figure 68.7): carbon monoxide (CO), unburned hydrocarbons (HCs), and nitric oxide (NO). Large amounts of CO are formed as an equilibrium product in rich mixtures, while smaller amounts remain in the products of lean mixtures due to chemical kinetic effects. HCs are left over from the combustion of rich mixtures and from flame quenching at walls and crevices in lean mixtures. NO is formed from air at high temperatures, and the chemical kinetics allow it to remain as the burned gas cools.

Most contemporary engines meet emissions standards by using **catalytic converters** in their exhaust systems. Some engines use **two-way catalysts** to reduce CO and HC emissions, while relying on lean operation to reduce flame temperature and give acceptable NO emissions. Other engines use **three-way**

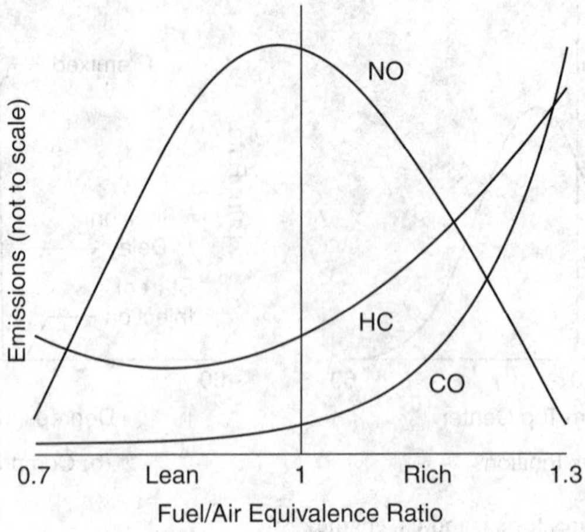

FIGURE 68.7 Effect of equivalence ratio on s.i. engine emissions.

catalysts to reduce CO, HC, and NO emissions. Three-way catalysts are only effective when the engine operates near stoichiometric air–fuel ratio; flame temperature can be reduced by exhaust gas recirculation. Engines with both types of catalytic converters operate rich for maximum power. The temporary increase in emissions can either be accepted or alleviated by pumping air into the exhaust system between the engine and the converter.

Control

The control system of an s.i. engine must govern engine output, but it must also regulate equivalence ratio and spark timing. Since the ranges of ϕ and spark timing over which the engine will run smoothly are limited, engine output is varied by reducing air flow while holding ϕ and timing essentially constant. Control of ϕ and timing is generally directed toward maximizing efficiency and minimizing emissions at a given speed and torque.

Engine output is usually controlled by throttling the intake air flow with a butterfly-type throttle valve. This reduces net output by reducing the heat release and increasing the pumping work [Figure 68.3(d)]. Other methods (late intake valve closing, shutting down cylinders of multicylinder engines) have been tried to reduce output with less efficiency penalty, but they are not widely used.

There are two basic methods of mixing fuel and air for s.i. engines: carburetion and fuel injection. A **carburetor** [Figure 68.8(a)] provides intrinsic control of ϕ by putting fuel and air flow through restrictions with the same differential pressure. The intrinsic control is imperfect because air is compressible while liquid fuels are not. Various corrective methods are used to provide near-constant lean ϕ over most of the air flow range, with enrichment to $\phi > 1$ for starting and maximum power operation. The manifold between the carburetor and the cylinder(s) must be arranged so that fuel evaporates fully and is evenly distributed among the cylinders. Due to the difficulty in obtaining low emissions levels, no contemporary U.S. production automobiles use carburetors.

A **fuel injector** [Figure 68.8(b)] injects a spray of fuel into the air stream. In throttle body injection, a single injector serves for multiple cylinders; in port injection (more common), each cylinder has its own injector. Port injectors are timed to spray fuel while the inlet valve is closed, to allow evaporation time. They are typically controlled by digital electronics. The volume of injected fuel is controlled in response to various measurements, including speed, inlet manifold vacuum, and exhaust oxygen concentration. The mixture is kept lean (two-way catalyst) or stoichiometric (three-way catalyst), except for starting and maximum power. Figure 68.8 shows carburetors and fuel injectors used for liquid fuels; the arrangements for gaseous fuels are similar.

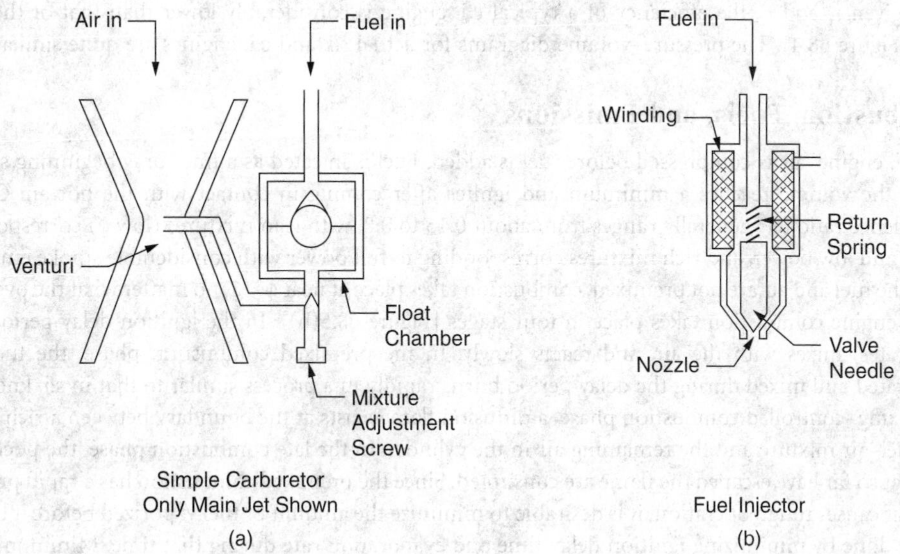

FIGURE 68.8 S.i. engine fuel addition devices.

The high voltage (10 to 25 kV) for the ignition spark is provided either by interrupting current through a choke or discharging a capacitor. The spark advance is typically regulated in response to engine speed and manifold vacuum; high speeds and high vacuums require more advance. The switching required for spark generation and control can be done either mechanically or electronically. Some electronically controlled engines incorporate vibrational knock sensors to retard the spark if required. For older designs, control is almost entirely mechanical, with the necessary adjustments to the fuel addition and ignition systems made through pressure-driven diaphragms, centrifugal speed sensors, and linkages. On newer designs, control is mainly through electronic sensors, digital electronics, and solenoid actuators.

Advantages

Relative to compression ignition engines, s.i. engines have higher mass and volume power density, lower first cost, greater fuel availability (for automotive use), and wider speed range. Emissions are lower with use of a catalytic converter. The advantages of s.i. engines become more pronounced for smaller sizes.

68.4 Compression Ignition Engines

Idealized and Actual Cycles

The compression ignition (c.i.) engine can be idealized as a diesel cycle using an ideal gas with constant specific heat [Figure 68.3(b)]. The **diesel** cycle consists of isentropic compression, constant pressure heating (simulating combustion), isentropic expansion, and constant volume cooling (simulating intake and exhaust). The thermal efficiency of a diesel cycle (Figure 68.4) is $\eta_t = 1 - r_v^{1-\gamma}(r_c^{\gamma} - 1)/(r_c - 1)/\gamma$. The **cutoff ratio,** r_c, idealizes the volume ratio over the fuel addition period.

The actual engine cycle [Figure 68.3(c)] differs from the diesel cycle:

1. In that heat transfer occurs during compression and expansion
2. In that combustion takes place at varying rather than constant pressure
3. In that combustion continues after the end of fuel addition
4. In the presence of intake and exhaust processes
5. In the variation in gas composition and gas specific heat

For a given r_v and r_c, the efficiency of a typical c.i. engine is considerably lower than that of the ideal cycle (Figure 68.4). The pressure–volume diagrams for actual s.i. and c.i. engines are quite similar.

Combustion, Fuels, and Emissions

In a c.i. engine, air is compressed before fuel is added. Fuel is injected as a fine spray beginning slightly before the volume reaches a minimum and ignites after coming in contact with the hot air. Overall equivalence ratio (ϕ) generally ranges from about 0.15 to 0.8, with lean mixtures (low ϕ) corresponding to idle and low power, and rich mixtures corresponding to full power with considerable smoke emission. Since the fuel and air are not premixed, combustion takes place at near $\phi = 1$, no matter what the overall ϕ.

C.i. engine combustion takes place in four stages [Figure 68.5(b)]. In the ignition delay period, fuel evaporates, mixes with the air, and reacts slowly. In the premixed combustion phase, the fuel that evaporated and mixed during the delay period burns rapidly in a process similar to that in s.i. knock. In the mixing–controlled combustion phase, a diffusion flame exists at the boundary between a rich atomized fuel–air mixture and the remaining air in the cylinder. In the late combustion phase, the pockets of fuel that so far have escaped the flame are consumed. Since the premixed combustion has a rapid pressure rise that causes rough operation, it is desirable to minimize the amount of fuel vaporized before it begins. This is done by minimizing ignition delay time and evaporation rate during that time. Minimum delay is obtained by injecting at the optimum time (10 to 15° before top center), with high cylinder wall temperature, high compression ratio, and high cetane number fuel. Indirect injection (see below) gives little premixed combustion.

The combustion process in c.i. engines does not speed up with increased turbulence as much as the process in s.i. engines does. For large, low-speed engines, combustion is adequate when the fuel is injected directly into the center of a relatively quiescent combustion chamber. For medium-size, medium-speed engines, the combustion chamber must be designed for increased turbulence in order for combustion to take place in the time available. In small, high-speed engines, combustion is initiated in a small, hot, highly turbulent prechamber. The fuel and burned gases from the prechamber then expand into the main combustion chamber and combine with the remaining air. Engines with a single chamber are known as **direct injection** (d.i.) engines, while those with a prechamber are known as **indirect injection** (i.d.i.) engines (Figure 68.9).

C.i. engines are fueled with petroleum oils consisting of longer-chain molecules than those in gasolines. Depending on the engine design, oils ranging from crude to kerosene can be used. It is important that any c.i. fuel have adequate autoignition properties. The ignition quality is expressed in terms of the **cetane number** of the fuel, based on an empirical scale on which *n*-hexadecane (cetane) has been assigned a rating of 100 and heptamethylnonane (isocetane) a rating of 15. Typical c.i. fuels have cetane numbers in the 30 to 60 range. For heavy, low-cost c.i. fuels, high pour point can be a problem. Some of these fuels must be heated before they can be pumped.

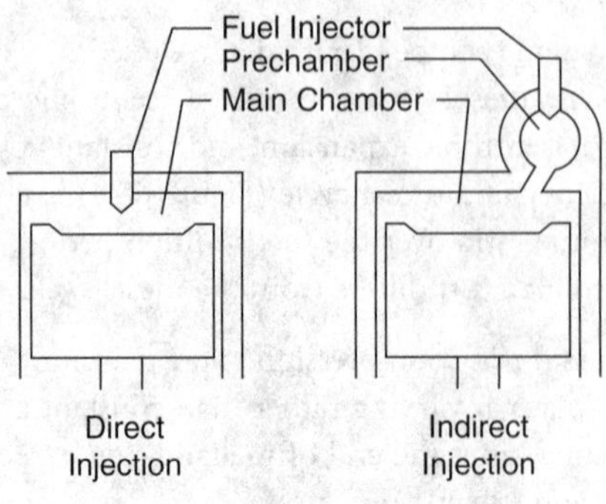

FIGURE 68.9 C.i. engine combustion chamber types.

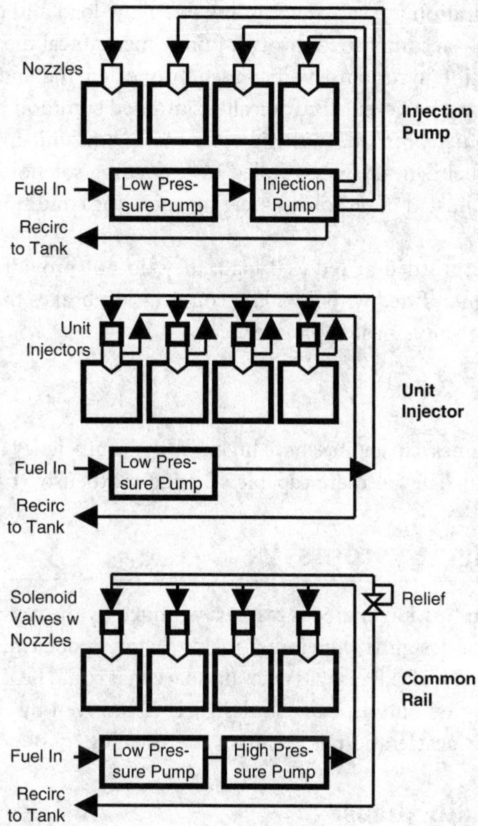

FIGURE 68.10 C.i. fuel systems.

Besides carbon dioxide and water, the combustion process in c.i. engines produces several pollutants, the most important of which are soot (carbon plus hydrocarbons), nitric oxide (NO), and nitrogen dioxide (NO_2). Carbonaceous soot is formed by fuel pyrolysis in rich regions near the flame front. Hydrocarbons then adsorb onto the soot particles during expansion and exhaust. Soot emissions are highest at high loads. NO and NO_2 are formed from air at high temperatures, and the chemical kinetics allow them to remain as the burned gases cool. Carbon monoxide and gaseous hydrocarbon emissions from diesel engines are relatively small.

Control

C.i. engines are not throttled, but are controlled by regulating the amount of fuel injected and the injection timing. Because of the required high injection pressures, large diesel engine fuel injectors are mechanically rather than electrically driven; systems for small engines may use electric pumps. Three types of systems are used: injection pump, unit injector, and common-rail (Figure 68.10). In an **injection pump** system, a central pump timed to the camshaft delivers fuel to nozzles located at each cylinder. The pump typically has individual barrels for each cylinder, but a single barrel with a fuel distributor is also used. In a **unit injector** system, there are a pump and nozzle on each cylinder, driven by a shaft running over all the cylinder heads. In a **common rail** system, a central pump supplies high-pressure fuel to a common header, and each cylinder has a solenoid-operated valve and nozzle. Common rail systems are suitable only for low injection pressures (small engines); injection pump systems are suitable for medium injection pressure (medium engines); unit injector systems are suitable for high injection pressures (large engines).

The injection start and duration are varied according to engine load and operating conditions. In the past, the control was generally accomplished through purely mechanical means and consisted mainly of increasing the injection duration in response to increased torque demand and advancing injection timing at higher speeds. In recent years, however, electronically controlled common rail systems, injector pumps, and unit injectors have become common. In the injector pump and unit-injector systems, the power is supplied mechanically, but fuel delivery is controlled by unloading solenoids. Electronic control allows fuel delivery to be carefully adjusted in response to engine operating conditions and is useful in achieving low emissions.

Since c.i. engines are not throttled at reduced load, they do not provide engine braking. For heavy vehicle use, c.i. engines are often fitted with auxiliary compression brakes that increase engine pumping work by opening the exhaust valves near top center.

Advantages

Relative to spark ignition engines, c.i. engines have higher thermal efficiency at full load and much higher thermal efficiency at low load. They also are capable of using inexpensive fuels such as heavy fuel oil.

68.5 Gas Exchange Systems

The torque of an internal combustion engine is primarily limited by the mass of air that can be captured in the cylinder. Intake system design is therefore a major factor in determining the torque for a given engine displacement. Since residual exhaust gas takes up space that could be used for fresh charge, exhaust system design also affects engine output. Exhaust design is also driven by the need to muffle the noise generated by the sudden flow acceleration at exhaust valve opening.

Four-Stroke Intake and Exhaust

Air flow into a naturally aspirated four-stroke engine is optimized by reducing charge temperature, by reducing flow friction in the intake system, by reducing residual exhaust gas, and by tuning and extended valve opening.

It has been found experimentally that engine air flow and torque are inversely proportional to the square root of the stagnation temperature of the air entering the cylinder. In s.i. engines, torque is increased by the cooling effect of fuel evaporation. This effect is much larger with alcohol fuels, which are therefore used in many racing cars. Torque is adversely affected by heat transfer to the intake air from the hot cylinder and intake manifold. In carbureted s.i. engines, some heat transfer is necessary to prevent fuel from puddling in the intake manifold.

According to both experiment and theory, engine torque is proportional to the pressure of the air entering the cylinder. This pressure is increased by minimizing the intake pressure drop. Intake valves are thus made as large as possible. High performance engines utilize two or three intake valves per cylinder to maximize flow area. Intake piping is normally designed for minimum pressure drop. However, in carbureted s.i. engines, the intake manifold is often designed for optimal fuel evaporation and distribution rather than for minimum flow friction. Intake air cleaners are designed for minimum flow resistance consistent with adequate dirt removal.

Cylinder pressure just prior to exhaust valve opening is much higher than atmospheric, but pressure falls rapidly when the exhaust valve opens. Back pressure due to exhaust system pressure drop increases the concentration of burned gas in the charge and thus reduces torque. The effect of exhaust system pressure drop is less than that of intake system pressure drop, but two exhaust valves per cylinder are often used to minimize pressure drop. Exhaust system pressure drop is usually increased by the use of a muffler to reduce exhaust noise.

Good intake and exhaust system design makes use of the dynamic effects of gas acceleration and deceleration. Most engine designs incorporate open periods well over 180° of crank angle: intake valves open 5 to 30° before top center and close 45 to 75° after bottom center; exhaust valves open 40 to 70° before bottom center and close 15 to 35° after top center. (The longer valve open times correspond to high performance engines.) Valves are thus open when piston motion is in the opposite direction from the desired gas flow. At high engine speeds, the correct flow direction is maintained by pressure differences and by gas inertial effects. At low engine speeds, an extended valve open period is detrimental to performance. Some engines incorporate variable valve timing to obtain optimum performance over a range of speeds.

The extended valve open period is generally used in combination with intake and exhaust **tuning.** Intake systems are often acoustically tuned as organ-pipe resonators, Helmholtz resonators, or more complex resonating systems. By tuning the intake to three, four, or five times the cycle frequency, pressure at the intake valve can be increased during the critical periods near valve opening and closing. Such tuning is usually limited to diesel, port fuel injected, or one carburetor per cylinder engines because the design of intake manifolds for other carbureted engines and throttle-body injected engines is dominated by the need for good fuel distribution. The tuning penalizes performance at some speeds away from the design speed. Branched exhaust systems are tuned so that, at design speed, expansion waves reflected from the junctions arrive at the exhaust valve when it is near closing. Individual cylinder exhaust systems are tuned as organ-pipe resonators. In either case, performance away from the design speed is penalized.

Two-Stroke Scavenging

In a two-stroke engine, intake and exhaust take place simultaneously, and some means of air pumping is needed for gas exchange (Figure 68.11). Small s.i. engines are generally crankcase scavenged — the bottom face of the piston is used to pump the air and oil is generally added to the fuel to lubricate the crank bearings. Larger engines use either rotary superchargers or turbochargers. These allow more freedom in crankcase lubrication. The cylinder and piston are arranged to maximize inflow of fresh charge and outflow of exhaust while minimizing their mixing. Cross-scavenging [Figure 68.12(a)] and loop-scavenging [Figure 68.12(b)] require only cylinder wall ports; uniflow scavenging [Figure 68.12(c)] requires poppet valves as well.

Scavenging spark ignition engines involves a trade-off between residual gas left in the cylinder and air–fuel mixture lost out the exhaust. In compression ignition engines, only air is lost through the exhaust.

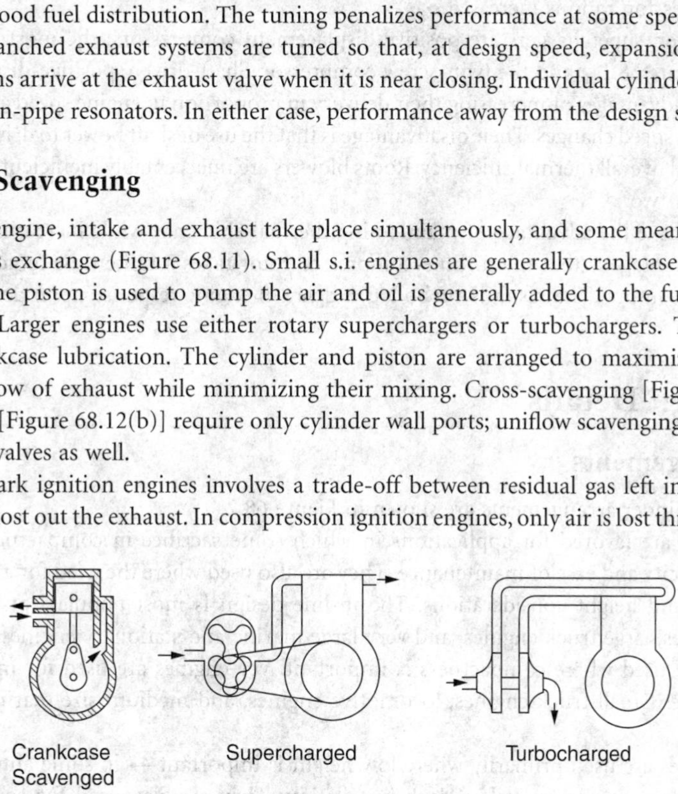

FIGURE 68.11 Scavenging and supercharging systems.

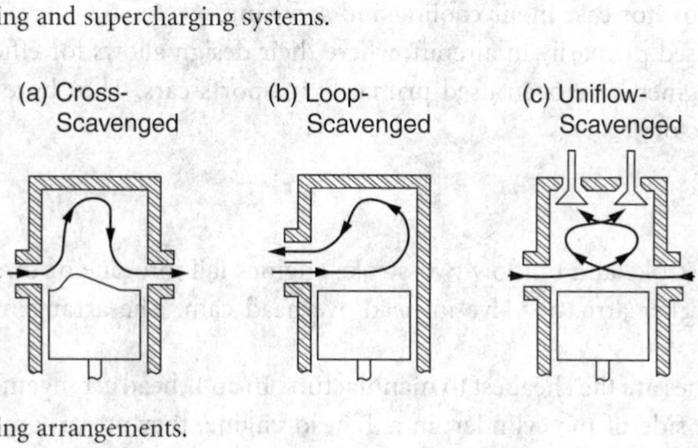

FIGURE 68.12 Scavenging arrangements.

Two-stroke s.i. engines are thus used mainly where low weight and first cost are of primary importance, while two-stroke c.i. engines can be built to be suitable for any service. In recent years some two-stroke s.i. engines have featured fuel injection into the cylinder after the exhaust port closes, eliminating loss of fuel out the exhaust.

Supercharging and Turbocharging

The output of a given i.c. engine can be increased by providing an auxiliary air compressor, or supercharger, to increase the pressure and, thus, the density of the air entering the cylinder intake. Although it actually applies to all types of auxiliary air compression systems, the term *supercharger* is generally used to describe systems driven by the engine output shaft. Air compression systems powered by an exhaust gas-driven turbine are known as *turbochargers.* When supercharging or turbocharging is added to a naturally aspirated engine, the engine is usually modified to reduce its compression ratio. However, the overall compression ratio is increased.

Most shaft-driven superchargers are positive displacement compressors; the most common are the Roots blower (Figure 68.11) and the twin-screw compressor. Shaft-driven, positive-displacement superchargers have the advantages of increasing their delivery in proportion to engine speed and of responding almost instantly to speed changes. Their disadvantage is that the use of shaft power to drive the compressor results in decreased overall thermal efficiency. Roots blowers are unacceptably inefficient at pressure ratios greater than about two.

Turbochargers have the advantage of maintaining thermal efficiency while increasing engine output. Their main disadvantage is that the turbine-compressor rotor takes a finite time to accelerate upon an increase in engine output. The result is **turbo lag**, a delay in engine response upon demand for a rapid increase in power.

68.6 Design Details

Engine Arrangements

Various engine cylinder arrangements are shown in Figure 68.2.

In-line engines are favored for applications in which some sacrifice in compactness is justified by mechanical simplicity and ease of maintenance. They are also used where the need for a narrow footprint overrides length and height considerations. The in-line design is most popular for small utility and automobile engines, large truck engines, and very large marine and stationary engines.

Vee engines are used where compactness is important. Vee engines are used for medium and large automobile engines, small truck engines, locomotive engines, and medium-size marine and stationary engines.

Opposed engines are used primarily where low height is important — in some automobiles and for small marine engines meant for below-deck installation. They are also used for some small aircraft engines, where they allow for ease in air cooling and servicing.

Radial engines are used primarily in aircraft, where their design allows for efficient air cooling.

Rotary (Wankel) engines have been used primarily in sports cars. They have not captured a major share of any market sector.

Valve Gear

Poppet valves on four-stroke and uniflow two-stroke engines fall into one of three categories: valve-in-block, valve-in-head/rocker arm, or valve-in-head/overhead cam. The arrangements are illustrated in Figure 68.13.

Valve-in-block engines are the cheapest to manufacture. In an L-head arrangement, intake and exhaust valves are on the same side of the cylinder; in a T-head engine, they are on opposite sides. The valves are directly driven by a camshaft located in the block, gear, or chain driven at half the crankshaft speed.

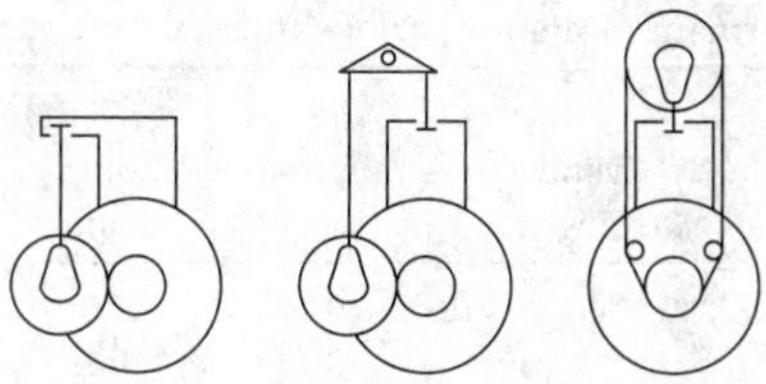

FIGURE 68.13 Valve arrangements.

The performance of these engines suffers from the elongated shape of the combustion chamber, and the designs are currently used only for inexpensive utility engines.

Valve-in-head/**rocker arm** engines have the valves installed in the cylinder head while maintaining a camshaft in the block. The design allows compact combustion chambers, but control of valve motion suffers from the slack in the long mechanical drive train. A majority of production automobile engines have this type of valve drive.

Valve-in-head/**overhead cam** engines have valves in the cylinder head directly driven by a camshaft running over all the heads. The design allows both compact combustion chamber design and accurate control of valve motion, but is more expensive to manufacture and more difficult to maintain than rocker arm designs. In recent years, overhead cam designs have become increasingly common in high-performance automobile engines.

Various arrangements for variable valve timing are now marketed, but none has achieved a large market share. Most engines have fixed valve timing.

Lubrication

The bearings of most i.c. engines are plain or grooved journal bearings. In the crankshaft bearings, a hydrodynamic film is maintained by rotation; in the piston pin bearings, the maintenance of a film depends on the oscillating nature of the load. For the connecting rod bearings the two effects are combined.

The most critical lubrication areas in an i.c. engine are at the piston rings, which are required to seal the high-pressure gas in the cylinder and prevent excess oil from entering the cylinder. Typical designs have two compression rings to seal the gases and an oil control ring to wipe oil from the cylinder wall. Piston rings ride on a hydrodynamic film at midstroke, but are in a boundary lubrication regime near top and bottom center. Lubrication is aided by good ring (alloy cast iron) and cylinder wall (cast iron or chrome-plated steel) materials.

Low-cost four-stroke engines are splash lubricated by running the crankshaft partly in an oil pan, but most engines have force-feed lubrication systems that deliver filtered oil to the bearings, the cylinder walls below the piston, and the valve train. Smaller engines depend on convection from the oil pan to cool the oil, while others have auxiliary oil coolers.

On small two-stroke engines with crankcase scavenging, lubrication is provided by mixing a small percentage of oil into the fuel entering the crankcase. On these engines rolling-contact bearings are often used on crankshafts, piston pins, and connecting rods.

Cooling

Most large i.c. engines are liquid cooled, and most small engines are air cooled. About a third of the energy input to a typical engine is dissipated through the cooling system. Liquid-cooled engines use either water or an aqueous ethylene glycol solution as coolant. When the glycol is used, it gives lower freezing and higher boiling points, but also increases the viscosity of the coolant. Although some natural-convection cooling systems have been built, most engines have the coolant pumped through numerous passages in the cylinder walls and heads and then into a heat exchanger where the heat is transferred to the environment. Small marine engines are typically cooled directly with water from the environment.

TABLE 68.1 Design and Performance Data for Various Internal Combustion Engines

Application and Type	Cylinders/ Arrangement	Displ. (1)	Comp. Ratio	Rated Power (kW)	Rated Speed (rpm)	Mass (kg)
Utility, two-stroke, s.i., c.s.	1	0.10	9.0	8.9	9000	5.0
Marine, two-stroke, s.i., c.s.	1	0.13	10.5	7.5	8000	6.6
Utility, four-stroke, s.i., n.a.	1	0.17	6.2	2.5	3600	13.9
Motorcycle, two-stroke, s.i., c.s.	2/in-line	0.30	7.1	19.4	7000	—
Utility, 4–stroke, s.i., n.a.	1	0.45	8.7	11.9	3600	38.2
Motorcycle, four-stroke, s.i., n.a.	2/in-line	0.89	10.6	61	6800	—
Automobile, four-stroke, s.i., n.a.	4/in-line	2.2	9.0	73	5200	—
Automobile, four-stroke, s.i., t.c.	4/in-line	2.2	8.2	106	5600	—
Automobile, four-stroke, c.i., n.a.	4/in-line	2.3	23	53	4500	—
Automobile, four-stroke, c.i., t.c.	4/in-line	2.3	21	66	4150	—
Aircraft, four-stroke, s.i., n.a.	4/opposed	2.8	6.3	48	2300	76
Automobile, four-stroke, c.i., n.a.	8/vee	5.0	8.4	100	3400	—
Truck/bus, two-stroke, c.i., t.c.	8/vee	9.5	17	280	2100	1100
Truck/bus, four-stroke, c.i., t.c.	6/in-line	10	16.3	201	1900	890
Aircraft, four-stroke, c.i., s.c.	9/radial	30	7.21	1140	2800	670
Locomotive, two-stroke, c.i., t.c.	16/vee	172	16	2800	950	16,700
Locomotive, four-stroke, c.i., t.c.	16/vee	239	12.2	3400	1000	25,000
Large marine, two-stroke, c.i., t.c.	12/in-line	14,500	—	36,000	87	$1.62 \cdot 10^6$

Note: c.s.: crankcase scavenged; n.a.: naturally aspirated; t.c.: turbocharged.

Source: Based on Taylor, C. F. 1985. *The Internal Combustion Engine in Theory and Practice,* 2nd ed. MIT Press, Cambridge, MA.

Air-cooled engines have finned external surfaces on their pistons and heads to improve heat transfer and fans to circulate air over the engine. The larger passages needed for air require that the cylinders be more widely spaced than for liquid-cooled engines. While most air-cooled engines are small, many large aircraft engines have been air cooled.

68.7 Design and Performance Data for Typical Engines

Design and performance data for various engines are given in Table 68.1.

Defining Terms

Carburetor — Controls fuel–air mixture by flowing air and fuel across restrictions with the same differential pressure.

Catalytic converter — Uses catalyst to speed up chemical reactions, normally slow, which destroy pollutants.

Cetane number — Empirical number quantifying ignition properties of c.i. engine fuels.

Common rail — Fuel injection with solenoids controlling flow from a single high-pressure header.

Compression ignition engine — Fuel and air compressed separately, ignited by high air compression temperatures.

Compression ratio — Ratio of maximum working volume to minimum working volume.

Cutoff ratio — Fraction of expansion stroke during which heat is added in diesel cycle.

Diesel cycle — Thermodynamic idealization of compression ignition engine.

Direct injection c.i. engine — Fuel is injected directly into the main combustion chamber.

Equivalence ratio — Fuel/air ratio relative to fuel/air ratio for stoichiometric combustion.

Four-stroke engine — One power stroke per cylinder per two revolutions.

Fuel injector — Controls fuel–air mixture by metering fuel in proportion to measured or predicted air flow.

Indirect injection c.i. engine — Fuel is injected into a prechamber connected to the main combustion chamber.

Injection pump — Delivers metered high-pressure fuel to all fuel injector nozzles of a c.i. engine.

Knock — Spark ignition engine phenomenon in which the fuel–air mixture detonates instead of burning smoothly.

Naturally aspirated engine — Piston face pumping action alone draws in air.

Octane number — Empirical number quantifying antiknock properties of s.i. fuels.

Otto cycle — Thermodynamic idealization of spark ignition engine.

Overhead cam engine — Valves are in head, driven by camshaft running over top of head.

Rocker arm engine — Valves are in head, driven from camshaft in block of push rods and rocker arms.

Scavenging — Intake/exhaust process in two-stroke engines.

Spark ignition engine — Fuel and air compressed together, ignited by electric spark.

Supercharged engine — Shaft-driven air compressor forces air into cylinder.

Thermal efficiency — Engine work divided by heat input or lower heating value of fuel used.

Three-way catalyst — Catalyst that reduces CO, HC, and NO emissions. Engine must run stoichiometric.

Tuning — Designing intake and exhaust so that flow is acoustically reinforced at design speed.

Two-stroke engine — One power stroke per cylinder per revolution.

Two-way catalyst — Catalyst that reduces CO and HC emissions. Engine may run lean or stoichiometric.

Turbo lag — The delay in response of a turbocharged engine upon demand for rapid power increase.

Turbocharged engine — Air forced into cylinder by compressor driven by exhaust gas turbine.

Unit injector — Combination pump and nozzle which delivers metered fuel to a single c.i. engine cylinder.

Valve-in-block engine — Low-cost design in which valves are driven directly by a camshaft in the cylinder block.

References

Benson, R. S. and Whitehouse, N. D. 1989. *Internal Combustion Engines,* Pergamon, New York.

Cummins, L. C., Jr. 1989. *Internal Fire,* rev. ed. Society of Automotive Engineers, Warrendale, PA.

Heywood, J. B. 1988. *Internal Combustion Engine Fundamentals,* McGraw-Hill, New York.

Obert, E. F. 1968. *Internal Combustion Engines and Air Pollution,* 3rd ed. Harper Collins, New York.

Taylor, C. F. 1985. *The Internal Combustion Engine in Theory and Practice,* 2nd ed. MIT Press, Cambridge, MA.

Further Information

Internal Combustion Engines and Air Pollution by Edward F. Obert and *The Internal Combustion Engine in Theory and Practice* by Charles Fayette Taylor are comprehensive and highly readable texts on i.c. engines. Although they are somewhat dated, they are still invaluable sources of information.

Internal Combustion Engine Fundamentals by John B. Heywood is an up-to-date and comprehensive text, but is less accessible to those with no previous i.c. engine background than the texts above.

Internal Fire by Lyle C. Cummins, Jr., is a fascinating history of the i.c. engine.

The Society of Automotive Engineers publishes *SAE Transactions* and a wide variety of books and papers on internal combustion engines. For more information contact: SAE, 4000 Commonwealth Drive, Warrendale, PA, 15096, USA. Phone (412) 776–4841. Internet http://www.sae.org.

69

Gas Turbines

Lee S. Langston
University of Connecticut

George Opdyke, Jr.
Dykewood Enterprises

In the history of energy conversion, the gas turbine is a relatively new prime mover. The first practical gas turbine used to generate electricity ran at Neuchatel, Switzerland, in 1939 and was developed by the Brown Boveri firm. The first gas turbine-powered airplane flight also took place in 1939, in Germany, using the gas turbine developed by Hans P. von Ohain. In England the 1930s invention and development of the aircraft gas turbine by Frank Whittle resulted in a similar British flight in 1941.

The name *gas turbine* is somewhat misleading, for it implies a simple turbine that uses gas as a working fluid. Actually, a gas turbine (as shown schematically in Figure 69.1) has a *compressor* to draw in and compress gas (usually air), a *combustor* (or burner) to add fuel to heat the compressed gas, and a *turbine* to extract power from the hot gas flow. The gas turbine is an internal combustion (IC) engine employing a continuous combustion process, as distinct from the intermittent combustion occurring in a diesel or Otto cycle IC engine.

Because the 1939 origin of the gas turbine lies both in the electric power field and in aviation, there has been a profusion of "other names" for the gas turbine. For land and marine applications it is generally called a *gas turbine*, but also a *combustion turbine*, a *turboshaft engine*, and sometimes a *gas turbine engine*. For aviation applications it is usually called a *jet engine*, and various other names (depending on the particular aviation configuration or application) such as *jet turbine engine, turbojet, turbofan, fanjet*, and *turboprop* or *prop jet* (if it is used to drive a propeller). The compressor–combustor–turbine part of the gas turbine (Figure 69.1) is commonly termed the *gas generator*.

69.1 Gas Turbine Usage

In an aircraft gas turbine, all of the turbine power is used to drive the compressor (which may also have an associated fan or propeller). The gas flow leaving the turbine is then accelerated to the atmosphere in an exhaust nozzle [Figure 69.1(a)] to provide *thrust* or *propulsion power*. Gas turbine or jet engine thrust power is the mass flow momentum increase from engine inlet to exit, multiplied by the flight velocity.

A typical jet engine is shown in Figure 69.2. Such engines can range form about 100 pounds thrust (1bt)(445 N) to as high as 100,000 lbt, (445,000 N), with dry weights ranging from about 30 lb (134 N) to 20,000 lb (89 000 N). The jet engine of Figure 69.2 is a *turbofan* engine, with a larger-diameter compressor-mounted fan. Thrust is generated both by air passing through the fan (bypass air) and through the gas generator itself. With a large frontal area, the turbofan generates peak thrust at low (takeoff) speeds, making it most suitable for commercial aircraft. A *turbojet* does not have a fan and

0-8493-1586-7/05/$0.00+$1.50

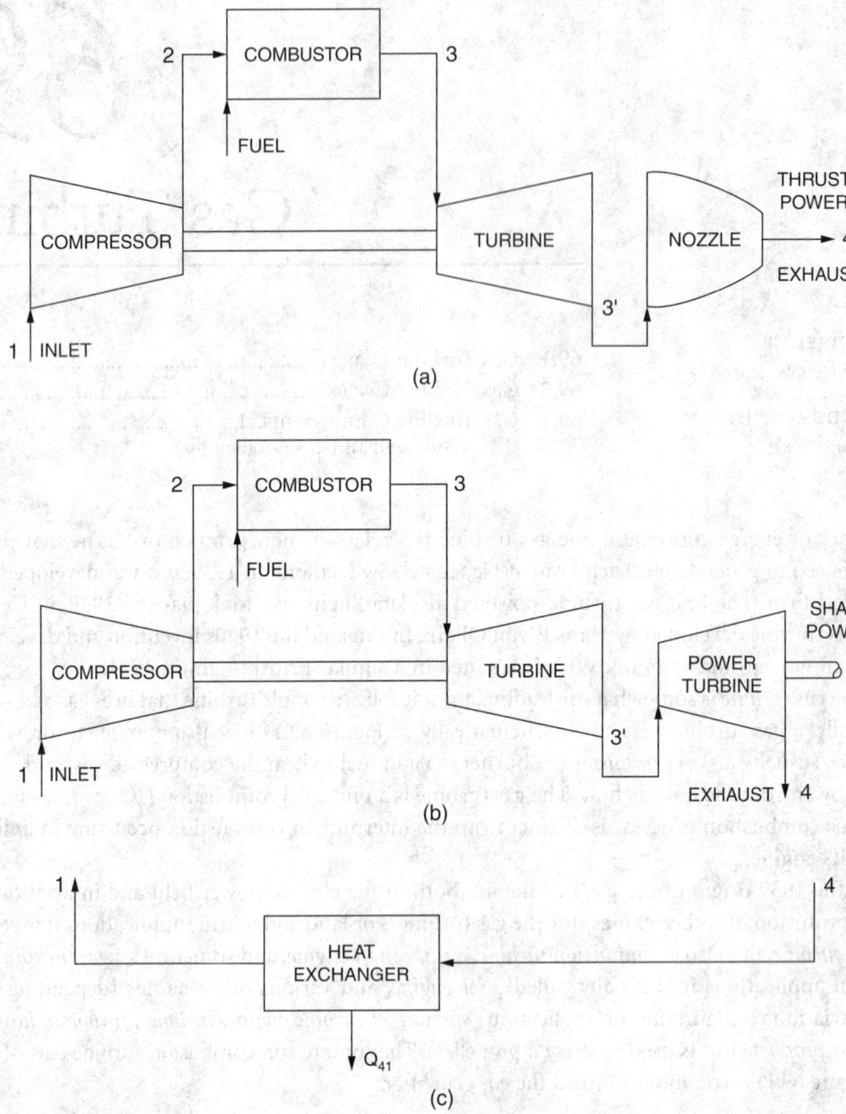

FIGURE 69.1 Gas turbine schematics. (a) Jet engine, a gas turbine (open cycle) used to produce thrust power. (b) Gas turbine (open cycle) used to produce shaft power. (c) Heat exchanger added to (b) yields a closed-cycle gas turbine (combustor becomes another heat exchanger since closed-cycle working fluid cannot sustain combustion).

generates all of its thrust from air that passes through the gas generator. Turbojets have smaller frontal areas and generate peak thrusts at high speeds, making them most suitable for fighter aircraft.

In nonaviation gas turbines, only part of the turbine power is used to drive the compressor. The remainder, the "useful power," is used as output *shaft power* to turn an energy conversion device [Figure 69.1(b)] such as an electrical generator or a ship's propeller. [The second "useful power" turbine shown in Figure 69.1 (b) need not be separate from the first turbine.]

A typical land-based gas turbine is shown in Figure 69.3. Such units can range in power output from 0.05 to as high as 240 MW. The unit shown in Figure 69.3 is also called an *industrial* or *frame* machine. Lighter-weight gas turbines derived from jet engines (such as in Figure 69.2) are called *aeroderivative* gas turbines and are most frequently used to drive natural gas line compressors and ships and to provide peaking and intermittent power for electrical utility applications.

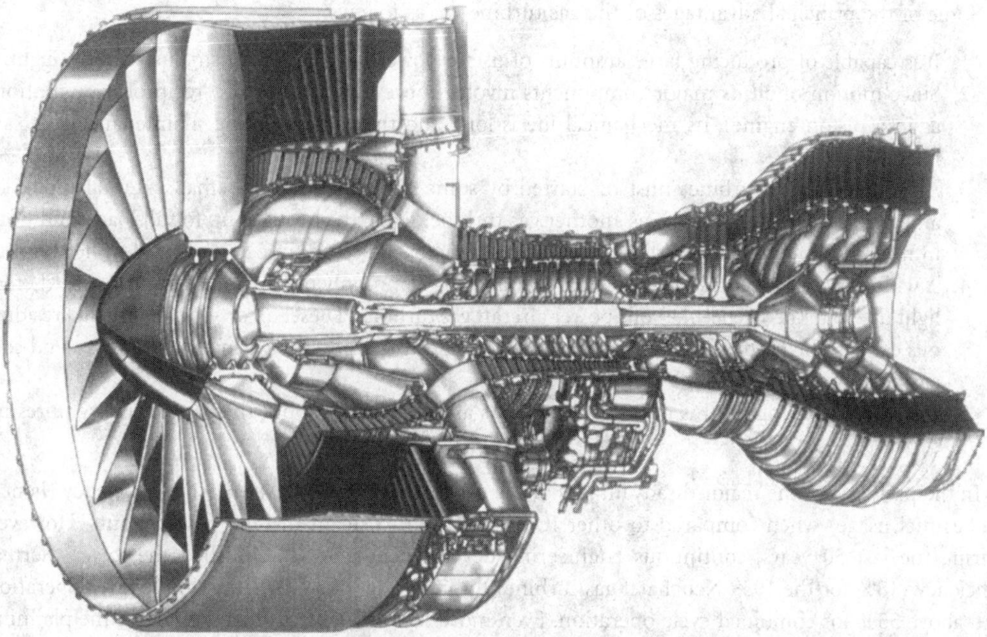

FIGURE 69.2 A modern jet engine used to power Boeing 777 aircraft. This is a Pratt and Whitney PW4084 turbofan, which can produce 84,000 pounds (374 kN) of thrust (about 63 MW at 168 m/sec). It has a 112 in. (2.85 m) diameter front-mounted fan, a flange-to-flange length of 192 inches (4.87 m), and a mass of about 15,000 pounds (6804 kg). (Courtesy of Pratt and Whitney.)

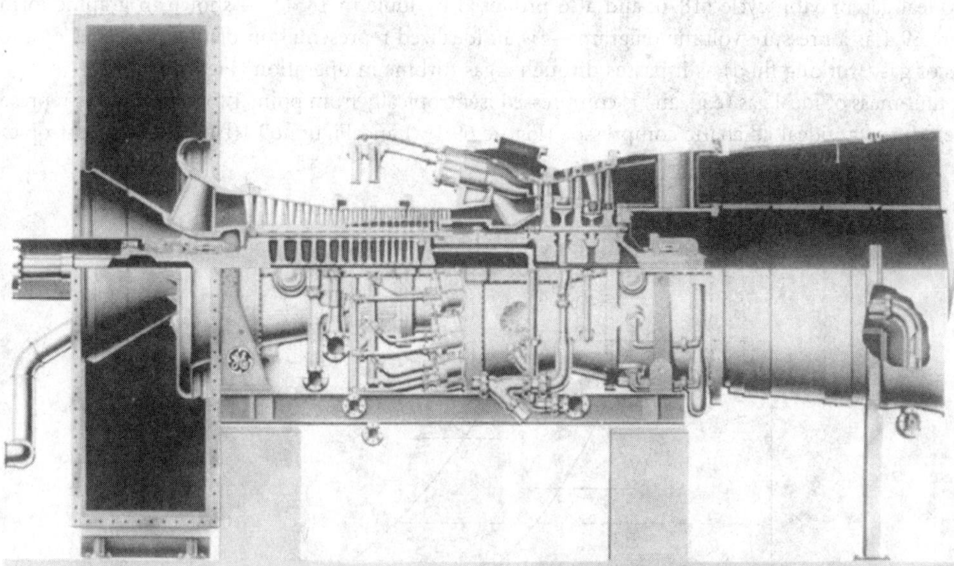

FIGURE 69.3 A modern land-based gas turbine used for electrical power production and for mechanical drives. This General Electric MS7001FA gas turbine is rated at 168 MW and is about 44 ft (13.4 m) in length and weighs approximately 377,000 lb (171,000 kg). Similar units have been applied to mechanical drive applications up to 108,200 hp. (Courtesy of General Electric.)

Some of the principal advantages of the gas turbine are as follows:

1. It is capable of producing large amounts of useful power for a relatively small size and weight.
2. Since motion of all its major components involves pure rotation (i.e., no reciprocating motions as in a piston engine), its mechanical life is long and the corresponding maintenance costs are relatively low.
3. Although the gas turbine must be started by some external means (a small external motor or another gas flow source, such as another gas turbine), it can be brought up to full load conditions in minutes, in contrast to a steam turbine plant, whose start-up time is measured in hours.
4. A wide variety of fuels can be utilized. Natural gas is commonly used in land gas turbines, whereas light distillate (kerosene-like) oils power aircraft gas turbines. Diesel oil or specially treated residual oils can also be used, as well as combustible gases derived from blast furnaces, refineries, and coal gasification.
5. The usual working fluid is atmospheric air. As a basic power supply, the gas turbine requires no coolant (e.g., water).

In the past, one of the major disadvantages of the gas turbine was its lower thermal efficiency (hence, higher fuel usage) when compared to other IC engines and to steam turbine power plants. However, during the last 50 years, continuous engineering development work has pushed the lower thermal efficiency (18% for the 1939 Neuchatel gas turbine) to present levels of 40% for simple cycle operation and above 55% for combined cycle operation. Even more fuel-efficient gas turbines are in the planning stages, with simple cycle efficiencies predicted as high as 45 to 47% and combined cycles in the 60% range. These projected values are significantly higher than values for other prime movers, such as steam power plants.

69.2 Gas Turbine Cycles

The ideal gas Brayton cycle (1876, and also proposed by Joule in 1851) — shown in graphic form in Figure 69.4 as a pressure-volume diagram — is an idealized representation of the properties of a fixed mass of gas (working fluid) as it passes through a gas turbine in operation (Figure 69.1).

A unit mass of ideal gas (e.g., air) is compressed isentropically from point 1 to point 2. This represents the effects of an ideal adiabatic compressor [Figure 69.1(a) and Figure 69.1(b)] and any isentropic gas

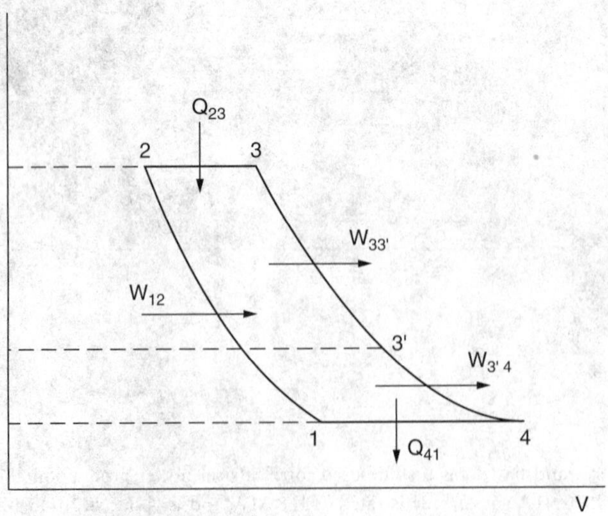

FIGURE 69.4 Brayton cycle pressure (*P*) versus volume (*V*) diagram for a unit mass of working fluid (e.g., air), showing work (*W*) and heat (*Q*) inputs and outputs.

flow deceleration for the case of an aviation gas turbine in flight [Figure 69.1(a)]. The ideal work necessary to cause the compression is represented by the area between the pressure axis and the isentropic curve 1–2.

The unit gas mass is then heated at constant pressure from 2 to 3 in Figure 69.4 by the exchange of heat input, Q_{23}. This isobaric process is the idealized representation of heat addition caused by the combustion of injected fuel into the combustor in Figure 69.1 (a) and Figure 69.1(b). The mass flow rate of the fuel is very much lower than that of the working fluid (roughly 1:50), so the combustion products can be neglected as a first approximation.

The unit gas mass is then isentropically expanded (lower pressure and temperature and higher volume) from 3 to 4 in Figure 69.4, where $P_4 = P_1$. In Figure 69.1 this represents flow through the turbine (to point 3′) and then flow through the exit nozzle in the case of the jet engine [Figure 69.1(a)], or flow through the power turbine in Figure 69.1(b) (point 4). In Figure 69.4 the area between the pressure axis and the isentropic curve 3–3′ represents the ideal work, $W_{33'}$, derived from the turbine. This work has to be equal to the ideal compressor work W_{12} (the area bounded by curve 1–2). The ideal "useful work," $W_{3'4}$, in Figure 69.4 (area bounded by the isentropic curve 3′–4) is that which is available to cause output *shaft power* [Figure 69.1(b)] or *thrust power* [Figure 69.1(a)].

The Brayton cycle is completed in Figure 69.4 by a constant pressure process in which the volume of the unit gas mass is decreased (temperature decrease) as heat Q_{41} is rejected. Most gas turbines operate in an *open cycle* mode, in which, for instance, air is taken in from the atmosphere (point 1 in Figure 69.1 and Figure 69.4) and discharged back into the atmosphere (point 4), with exiting working fluid mixing to reject Q_{41}. In a *closed cycle* gas turbine facility the working fluid is continuously recycled by ducting the exit flow (point 4) through a heat exchanger [shown schematically in Figure 69.1(c)] to reject heat Q_{41} at (ideally) constant pressure and back to the compressor inlet (point 1). Because of its confined, fixed mass working fluid, the closed cycle gas turbine is *not* an internal combustion engine, so the combustor shown in Figure 69.1 is replaced with an input heat exchanger supplied by an external source of heat. The latter can take the form of a nuclear reactor, the fluidized bed of a coal combustion process, or some other heat source.

The ideal Brayton cycle thermal efficiency, η_B, can be shown [Bathie, 1984] to be

$$\eta_B = 1 - \frac{T_4}{T_3} = 1 - \frac{1}{r^{(k-1)/k}} \tag{69.1}$$

where the absolute temperatures T_3 and T_4 correspond to the points 3 and 4 in Figure 69.4, $k = c_p/c_v$ is the ratio of specific heats for the ideal gas, and r is the compressor pressure ratio,

$$r = \frac{P_2}{P_1} \tag{69.2}$$

Thus, the Brayton cycle thermal efficiency increases with both the turbine inlet temperature, T_3, and compressor pressure ratio, r. Modern gas turbines have pressure ratios as high as 30:1 and turbine inlet temperatures approaching 3000°F (1649°C).

The effect of real (nonisotropic) compressor and turbine performance can be easily seen in the expression for the net useful work $W_{\text{net}} = W_{3'4}$ [Huang, 1988],

$$W_{\text{net}} = \eta_T c_P T_3 \left[1 - \frac{1}{r^{(k-1)/k}}\right] - \frac{c_P T_1}{\eta_C}[r^{(k-1)/k} - 1] \tag{69.3}$$

where c_P is the specific heat at constant pressure, and η_T and η_C are the thermal (adiabatic) efficiencies of the turbine and compressor, respectively. This expression shows that for large W_{net} one must have high values for η_T, η_C, r, and T_3. For modern gas turbines, η_T can be as high as 0.92 to 0.94 and η_C can reach 0.88.

A gas turbine that is configured and operated to closely follow the ideal Brayton cycle (Figure 69.4) is called a *simple cycle* gas turbine. Most aircraft gas turbines operate in a simple cycle configuration since attention must be paid to engine weight and frontal area.

However, in land or marine applications, additional equipment can be added to the simple cycle gas turbine, leading to increases in thermal efficiency and/or the net work output of a unit. Three such modifications are regeneration, intercooling, and reheating.

Regeneration involves the installation of a heat exchanger through which the turbine exhaust gases [point 4 in Figure 69.1(b) and Figure 69.4] pass. The compressor exit flow [point 2 in Figure 69.1(b) and Figure 69.4] is then heated in the exhaust gas heat exchanger before the flow enters the combustor. If the regenerator is well designed — that is, if the heat exchanger effectiveness is high and the pressure drops are small — the cycle efficiency will be increased over the simple cycle value. However, the relatively high cost of such a regenerator must also be taken into account.

Intercooling also involves use of a heat exchanger. An intercooler is a heat exchanger that cools compressor gas during the compression process. For instance, if the compressor consists of a high- and a low-pressure unit, the intercooler could be mounted between them to cool the flow and decrease the work necessary for compression in the high-pressure compressor. The cooling fluid could be atmospheric air or water (e.g., sea water in the case of a marine gas turbine). It can be shown that net work output of a given gas turbine is increased with a well-designed intercooler. Recent studies of gas turbines equipped with extensive turbine convective and film cooling show that intercooling can also allow increases in thermal efficiency by providing cooling fluid at lower temperatures, thereby allowing increased turbine inlet temperatures [T_3 in Equation (69.1)].

Reheating occurs in the turbine and is a way to increase turbine work without changing compressor work or exceeding the material temperature limits in the turbine. If a gas turbine has a high-pressure and a low-pressure turbine, a reheater (usually another combustor) can be used to "reheat" the flow between the two turbines. Reheat in a jet engine is accomplished by adding an afterburner at the turbine exhaust, thereby increasing thrust, with a greatly increased fuel consumption rate.

A *combined cycle* gas turbine power plant, frequently identified by the abbreviation *CCGT*, is essentially an electrical power plant in which a gas turbine provides useful work (W_B) to drive an electrical generator. The gas turbine exhaust energy (Q_{BR}) is then used to produce steam in a heat exchanger (called a *heat recovery steam generator*) to supply a steam turbine (see Chapter 74) whose useful work output (W_R) provides the means to generate more electricity. If the steam is used for heat (e.g., heating buildings), the unit would be called a *cogeneration plant* (see Chapter 75). The sketch in Figure 69.5 is a simplified thermodynamic representation of a CCGT and shows it to be two heat engines (Brayton and Rankine) coupled in series. The "upper" engine is the gas turbine (represented as a Brayton cycle heat engine), whose energy input is Q_{IN}. It rejects heat (Q_{BR}) as the input energy to the "lower" engine (the steam turbine, represented as a Rankine cycle heat engine). The Rankine heat engine then rejects unavailable energy (heat) as Q_{OUT} by means of a steam condenser. The combined thermal efficiency (η_{CC}) can be derived fairly simply [Huang, 1988] and is given as

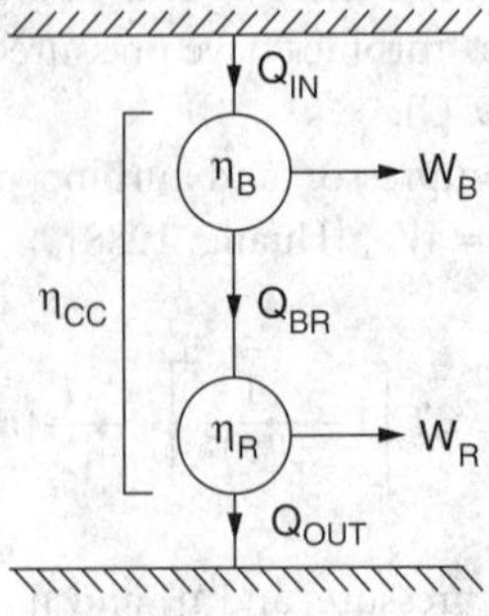

FIGURE 69.5 Schematic of combined cycle (CC), Brayton (B), and Rankine (R) heat engines, showing work (W) and heat (Q) inputs and outputs. (Courtesy Global Gas Turbine News.)

$$\eta_{CC} = \eta_B + \eta_R - \eta_B \eta_R \tag{69.4}$$

where $\eta_B = W_B/Q_{IN}$ and $\eta_R = W_R/Q_{BR}$ are the thermal efficiencies of the Brayton and Rankine cycles, respectively.

Taking $\eta_B = 40\%$ (a good value for modern gas turbines) and $\eta_R = 30\%$ (a reasonable value at typical CCGT conditions), the sum minus the product in Equation (69.4) yields $\eta_{CC} = 58\%$, a value of combined cycle efficiency greater than either of the individual efficiencies.

This remarkable equation gives insight into why CCGTs are so efficient. The calculated value of η_{CC} represents an upper bound on an actual CCGT, since Equation (69.4) does not account for efficiencies associated with transferring Q_{BR} (duct losses, irreversibilities due to heat transfer, etc.) Actual efficiency values as high as 52 to 54% have been attained with CCGT units during the last few years, thus coming close to values given by Equation (69.4).

69.3 Gas Turbine Components

A greater understanding of the gas turbine and its operation can be gained by considering its three major components (Figure 69.1, Figure 69.2, and Figure 69.3): the compressor, the combustor, and the turbine. The features and characteristics of turbines and compressors will be touched on here only briefly. A longer treatment of the combustor will be given, since environmental considerations make knowledge of its operation and design important to a wider audience.

Compressors and Turbines

The turbine and compressor components are mated by a shaft, since the former powers the latter. A *single-shaft* gas turbine has but one shaft connecting the compressor and turbine components. A *twin-spool* gas turbine has two concentric shafts, a longer one connecting a low-pressure compressor to a low-pressure turbine (the low spool), which rotates inside a shorter, larger-diameter shaft. The latter connects the high-pressure turbine with the higher-pressure compressor (the high spool), which rotates at higher speeds than the low spool. A *triple-spool* engine would have a third, intermediate-pressure compressor-turbine spool.

Gas turbine compressors can be centrifugal, axial, or a combination of both. Centrifugal compressors (radial outflow) are robust, generally cost less, and are limited to pressure ratio of 6 or 7 to 1. They are found in early gas turbines or in modern, smaller gas turbines.

The more efficient, higher-capacity axial flow compressor is used on most gas turbines (e.g., Figure 69.2 and Figure 69.3). An axial compressor is made up of a number of stages, each stage consisting of a row of rotating blades (airfoils) and a row of stationary blades (called *stators*) configured so the gas flow is compressed (adverse or unfavorable pressure gradient) as it passes through each stage. It has been said that compressor operation can founder upon a metaphoric rock, and that rock is called *stall.* Care must be taken in compressor operation and design to avoid the conditions that lead to blade stall, or flow separation. The collective behavior of blade separation can lead to compressor stall or surge, which manifests itself as an instability of gas flow through the entire gas turbine.

Turbines are generally easier to design and operate than compressors, since the flow is expanding in an overall favorable pressure gradient. Axial flow turbines (Figure 69.2 and Figure 69.3) will require fewer stages than an axial compressor for the same pressure change magnitude. There are some smaller gas turbines that utilize centrifugal turbines (radial inflow) but most utilize axial turbines (e.g., Figure 69.3).

Turbine design and manufacture are complicated by the need to ensure turbine component life in the hot gas flow. The problem of ensuring durability is especially critical in the first turbine stage, where temperatures are highest. Special materials and elaborate cooling schemes must be used to allow metal alloy turbine airfoils that melt at 1800 to 1900°F (982 to 1038°C) to survive in gas flows with temperatures as high as $T_3 = 3000$°F (1649°C).

Combustors

A successful combustor design must satisfy many requirements and has been a challenge from the earliest gas turbines of Whittle and von Ohain. The relative importance of each requirement varies with the application of the gas turbine, and, of course, some requirements are conflicting, requiring design compromises to be made. The basic design requirements can be classified as follows:

1. High combustion efficiency at all operating conditions.
2. Low levels of unburned hydrocarbons and carbon monoxide, low oxides of nitrogen at high power, and no visible smoke.
3. Low pressure drop. Three to four percent is common.
4. Combustion stability limits must be wider than all transient operating conditions.
5. Consistently reliable ignition must be attained at very low ambient temperatures, and at high altitudes (for aircraft).
6. Smooth combustion, with no pulsations or rough burning.
7. A small exit temperature variation for good turbine life requirements.
8. Useful life (thousands of hours), particularly for industrial use.
9. Multifuel use. Characteristically, natural gas and diesel fuel are used for industrial applications and kerosene for aircraft.
10. Length and diameter compatible with engine envelope.
11. Designed for minimum cost, repair, and maintenance.
12. Minimum weight (for aircraft applications).

A combustor consists of at least three basic parts: a casing, a flame tube, and a fuel injection system. The casing must withstand the cycle pressures and may be a part of the structure of the gas turbine. It encloses a relatively thin-walled flame tube, within which combustion takes place, and a fuel injection system. Sometimes the diffusing passage between the compressor and the combustor is considered a part of the combustor assembly as well.

The flame tube can be either tubular or annular. In early engines the pressure casings were tubular and enclosed the flame tubes. This configuration is called a *can combustor.* One or more can be used on an engine. If an annular casing encloses can-type flame tubes, this is called a *can-annular* or *cannular* combustor. An annular combustor consists of both an annular casing and an annular flame tube and is a common design used in aircraft gas turbines. Industrial engines often use can or cannular combustors because of their relative ease of removal.

Fuel injectors can be pressure atomizing, air atomizing, or air blast types. The specific styles vary widely and are often designed and manufactured by specialist suppliers. For industrial engines an injector may be required to supply atomized liquid fuel, natural gas, and steam or water (for nitric oxide reduction).

Compared to other prime movers, gas turbines are considered to produce very low levels of combustion pollution. The gas turbine emissions of major concern are unburned hydrocarbons, carbon monoxide, oxides of nitrogen (NO_X), and smoke. Although the contribution of jet aircraft to atmospheric pollution is less than 1% [Koff, 1994], jet aircraft emissions injected directly into the upper troposphere have doubled between the latitudes of 40 and 60 degrees north, increasing ozone by about 20%. In the stratosphere, where supersonic aircraft fly, NO_x will deplete ozone. Both effects are harmful, so NO_x reduction in gas turbine operation is a challenge for the twenty-first century [Schumann, 1993].

Defining Terms

Aeroderivative — An aviation propulsion gas turbine (jet engine) used in a nonaviation application (e.g., an electric power plant).

Brayton cycle — Ideal gas cycle that is approximated by a gas turbine simple cycle (see Figure 69.4).

Combined cycle — The combination of a gas turbine power plant and a steam turbine power plant. The gas turbine exhaust is used as heat input to generate steam.
Gas generator — The compressor, combustor, and turbine components of a gas turbine.
Jet engine — An aviation gas turbine used for aircraft propulsion.
Rankine cycle — Ideal gas cycle that is approximated by a simple steam turbine power plant cycle.

References

Bathie, W. W. 1984. *Fundamentals of Gas Turbines,* John Wiley & Sons, New York.
Huang, F. F. 1988. *Engineering Thermodynamics,* 2nd ed. Macmillan, New York.
Koff, B. L. 1994. Aircraft gas turbine emissions challenge. *J. Eng. Gas Turbines Power* 116:474–477.
Schumann, U. 1993. On the effect of emissions from aircraft engines on the state of the atmosphere. *Fuels and Combustion Technology for Advanced Aircraft Engines.* AGARD Conference Proceedings 536, Sept.

Further Information

Cohen, H., Rogers, G. F. C., and Saravanamuttoo, H. I. H. 1987. *Gas Turbine Theory,* 3rd ed. Longman Scientific and Technical, Essex, England.
Horlock, J. H. 1993. *Combined Power Plants,* Pergamon Press, Oxford.
Kerrebrock, J. L. 1980. *Aircraft Engines and Gas Turbines,* MIT Press, Cambridge, MA.
Lefebvre, A. H. 1983. *Gas Turbine Combustion,* McGraw-Hill, New York.
Northern Research and Engineering Corporation. 1964. *The Design and Performance Analysis of Gas Turbine Combustion Chambers* and *The Design and Development of Gas Turbine Combustors* 1980. (They are popularly called the “Orange Books.”)
Transactions of the ASME, Published quarterly by the American Society of Mechanical Engineers, New York. Two ASME gas turbine journals are *Journal of Turbomachinery* and *Journal of Engineering for Gas Turbines and Power.*

70

Nuclear Power Systems

David M. Woodall
(First Edition)
University of Idaho

Scott W. Heaberlin
(Second Edition)
Pacific Northwest National Laboratory

Nuclear power systems use the controlled release of **nuclear energy** as a heat source for the generation of electricity or for direct thermal heat. The advantages of nuclear power include extremely high power density, large energy release per unit mass of fuel, and the production of energy without the emission of greenhouse gases. The disadvantages include the relatively high capital cost, the production of radioactive materials requiring careful handling and disposal, and the operation of components in a severe radiation environment.

70.1 Nuclear Power Applications

There are both terrestrial and space applications of nuclear power systems: commercial nuclear power production of electricity; nuclear propulsion of ocean-going vessels; and high-reliability electricity and heat for remote locations, including energy sources for space satellites, space science probes, and space propulsion systems.

Terrestrial Nuclear Power

Commercial Nuclear Electric Power

The commercial application of nuclear energy followed closely on the military release of nuclear energy in the nuclear weapons developed during World War II. The first commercial nuclear power plant was completed in the late 1950s, and there are presently (in 2002) 103 operating nuclear power plants in the U.S., generating 20% of the national electricity supply. Worldwide there are 436 operating nuclear power plants, with 13 countries obtaining more than one-third of their electricity from nuclear power, including France, Belgium, Sweden, Ukraine, and South Korea. The principal types of commercial nuclear power systems in use include the **pressurized water reactor (PWR)** and the **boiling water reactor (BWR).** These two reactor types comprise over 80% of the current operating power reactors. A third type of power reactor is the pressurized heavy water reactor (PHWR), typified by the Canadian-designed CANDU reactor series. Reactor designs being considered for future applications include the **liquid metal fast**

0-8493-1586-7/05/$0.00+$1.50

breeder reactor (LMFBR) and **high-temperature gas cooled reactor (HTGR).** The former Soviet block countries operate two Russian designs, a PWR-like reactor called the VVER and a graphite-moderated, water-cooled thermal reactor, the **pressure-tube graphite reactor (PTGR),** also called the RBMK. The RBMK is the reactor type that experienced the Chernobyl Unit 4 reactor accident.

The design and operation of commercial nuclear power systems require the use of complex nuclear and thermal-hydraulic design codes incorporating basic engineering principles and engineering correlations. While scoping analysis principles are outlined in the following sections, the serious practitioner is encouraged to read the reference materials and the current technical literature.

Nuclear Propulsion

The U.S. and Russian navies have a major commitment to nuclear propulsion for their surface and submarine fleets. Nuclear power enables a vessel to travel long distances at high speed without refueling and supports an extended submerged operation, as no oxygen is consumed for combustion. The U.S. has 75 submarines and nine aircraft carriers that are nuclear powered. The Russian fleet is similarly equipped with nuclear propulsion systems. The pressurized water reactor (PWR) commercial reactor type was demonstrated by the naval nuclear propulsion program. The storage of spent reactor cores as high-level waste (HLW) is a technical issue for naval operations, especially for the Russian navy, which has a large inventory of obsolete and surplus nuclear vessels but limited infrastructure to deal with radioactive materials. International programs are being developed to assist the Russian Federation in responsible stewardship of these materials.

Space Nuclear Power

Power sources for space operations are limited by the high cost of launching a system to orbit. A high premium is placed on system reliability, due to the inaccessibility of systems in use and the high cost of alternative power systems. The options for space power include the conversion of solar energy to electricity. Solar electricity has two principal disadvantages: the requirement for battery backup power during periods of unavailability of sunlight and the decreased solar radiation for deep space missions outward from the sun. Nuclear space power systems do not suffer from either of these limitations but have been limited by public acceptance of the launch of nuclear materials, due primarily to the perception of the environmental risk of launch accident–initiated dispersal of radioactive materials. However, recognizing that nuclear energy is an enabling technology for deep space exploration, NASA has requested $950 million over 5 years beginning in FY2003 to develop both radioisotope power systems and fission reactors for propulsion and planetary power systems.

Radioisotope Systems

The nuclear power system that has found the most application in U.S. space missions is the radioisotope thermoelectric generator (RTG). Electricity is produced by a thermoelectric converter, powered by the thermal energy from the decay of a long-lived alpha-emitting isotope of plutonium, ${}_{94}Pu^{238}$, with a half-life of 88 years. Several planetary exploration missions have been powered by RTGs. The two Voyager spacecraft launched in 1977, now over 10 billion kilometers from Earth, continue to send data due to their long-lived RTG power sources.

Space Reactors

Small nuclear reactors were used in the former Soviet Union's space power program. Reactors powering thermoelectric energy conversion systems were routinely used in earth observation satellites. The U.S. space program launched and operated in space only one nuclear reactor, SNAP-10A, during the early space program. Fission reactors have been proposed to provide power for Mars-based habitats including the production of oxygen separated from the Martian atmosphere to fuel the return to Earth. Fission reactor system concepts for propulsion include **nuclear thermal propulsion (NTP),** which uses a nuclear reactor to heat an exhaust propellant, and **nuclear electric propulsion (NEP),** which uses nuclear energy to produce electricity, which then is used to power an ion drive propulsion system. Either concept represents a near order of magnitude improvement in drive energy per launch mass over conventional chemical rockets.

70.2 Nuclear Power Fundamentals

Fission

The basic source of energy in nuclear power is the fissioning of the nucleus of an isotope of uranium or plutonium. The energy released per fission depends on the particular isotope, and is approximately 3.2×10^{-11} J (joules) for the thermal neutron fission of $_{92}U^{235}$, the most common nuclear fuel. Thus approximately 3.1×10^{10} fissions/sec are required to produce 1 W of thermal energy. Typical commercial nuclear power systems produce 1000 MW of electrical power at a thermal efficiency of about 33%. In such a commercial power system 1 kg of nuclear fuel will produce about 240,000 kilowatt hours of energy, whereas 80,000 kg of coal would be required to produce the same amount of energy.

The fission process is initiated by the absorption of a neutron into the nucleus of a heavy isotope. The energy released from the fission of a nucleus is distributed among various products of the process, including nuclear fragments, neutrons, and gamma rays. An average fission releases two or three excess neutrons. These excess neutrons are available to initiate further fissions; hence, nuclear fission in a suitable assemblage of materials can be a self-sustaining process. Control of the power of a nuclear power system depends on the control of the neutron population, typically through the use of neutron absorbers in movable structures called **control rods.**

In most fissions, approximately 80% of the energy released is carried by two large nuclear fragments, or fission fragments. The fission fragments have very short ranges in the fuel material and come to rest releasing thermal energy and capturing electrons to become fission product atoms. Fission products are typically radioactive, having a spectrum of radioactive decay products and half-lives. Fission products are neutron rich, and a small fraction decay by neutron emission, adding to the neutron population in the reactor. Such fission products are termed *fission product precursors,* and their radioactive decay influences the time-dependent behavior of the overall neutron population in the reactor. These **delayed neutrons** play an essential role in the control of the reactor power.

The thermal energy deposited in nuclear fuel, coolant, and structural material is dependent on the type of radiation carrying the energy. Neutrons are absorbed in stable nuclei, causing the **transmutation** to a radioactive element. Such a capture event releases an energetic photon, a capture gamma, which deposits its energy in the surrounding material. Subsequent radioactive decay of these radionuclides releases additional radiation energy in the form of decay gamma and decay beta radiation. All of the energy released in beta decay is not available to be locally deposited, due to the extremely long range of the antineutrinos, which share the energy available with the beta radiation. Table 70.1 delineates the distribution of energy from a typical fission, including information on the heat produced in the reactor from each source. The convenient unit of energy used is the MeV (mega-electron-volt), where 1 MeV is equal to 1.6×10^{-13}J.

TABLE 70.1 Energy Distribution among Products of Fission

Energy Source	Energy (MeV)	Heat Produced
Fission fragments	168	168
Fast neutrons	5	5
Prompt gammas	7	7
Decay gammas	7	7
Capture gammas	—	5
Decay betas	20	8
Total	207	200

Radioactivity

Elements that undergo radioactive decay are called radioisotopes. Such isotopes hold excess energy in their nuclei and are therefore unstable. They move toward stability by the process of radioactive decay. A radioisotope can be characterized by the energetic radiation that is emitted in the decay process and by the probability of decay per unit time, the decay constant, λ. The products from a radioactive decay include the daughter isotope and one or more radiation emissions. The three most common radiation emissions are alpha, beta, and gamma radiation. Alpha radiation consists of an energetic helium nucleus, beta radiation is an energetic electron, and gamma radiation is a photon of electromagnetic radiation released from nuclear energy level transitions. The activity of a radioactive substance, A, is the number of decays per second and the number of radiations emitted per second, assuming a single radiation per decay:

$$A = \lambda N = \lambda N_0 e^{-\lambda t} \quad (70.1)$$

where t is the time in seconds and N_0 is the initial number of atoms present at $t = 0$. The unit of activity is the Bequerel (Bq), with 1 Bq corresponding to 1 decay per second (dps). The non-SI unit in common use for activity is the curie (Ci), with $1\text{Ci} = 3.7 \times 10^{10}$ dps. The half-life of an isotope is the time for half of the atoms of a given sample to decay, $T_{1/2}$. The decay constant and half-life are related through the following formula:

$$T_{1/2} = \ln 2 / \lambda \quad (70.2)$$

Criticality

The assemblage of nuclear fuel and other material into a device which will support sustained fission is called a **nuclear reactor.** The central region of the reactor is involved in fission energy release, containing the nuclear fuel and related structural and energy transfer materials, and is called the **reactor core.** The typical nuclear fuel for a commercial power reactor is an oxide of uranium that has been enriched in the fuel isotope, ${}_{92}U^{235}$. While a number of fuel forms are in use, the predominant form is compressed and sintered pellets of the ceramic oxide UO_2. These pellets are stacked in cladding tubes of **zircaloy,** forming fuel pins, which are held in rectangular arrays of fuel elements by structural guides. The fuel elements are arranged in a rectangular array by core guides to form the reactor core. Coolant passes axially along the fuel elements and through the core.

The neutrons born in fission are typically high-energy neutrons, with an average energy of about 3×10^{-13} J. However, the likelihood of producing fission is higher for low-energy neutrons, those in collisional (thermal) equilibrium with the reactor. Such thermal neutrons have energies of about 4×10^{-21} J. Neutrons in the reactor lose energy through elastic and inelastic collisions with light nuclei, called moderator materials. Collisions with light nuclei are more efficient in reducing neutron energy; therefore, typical materials used to moderate the energy of the neutrons are hydrogen in the form of water, deuterium in water, and carbon in graphite.

Reactor Kinetics

Neutrons are born in the fission process at high energy. Such **fast neutrons** travel through the core, scattering from nuclei and losing energy until they are absorbed, leak from the reactor core, or are thermalized. Thermal neutrons diffuse through the reactor core, gaining or losing energy in thermal equilibrium with the nuclei of core materials until they are absorbed or leak from the system. Those neutrons absorbed in nuclear fuel may cause fission, leading to additional fast neutrons. The behavior of neutrons in the reactor can be described in terms of a neutron life cycle, with the ratio of neutrons in one generation to those in the previous generation given by the multiplication constant k. The time variation of the reactor power is a function of the multiplication constant, with the reactor condition

TABLE 70.2 Critical Constant versus Reactor Power Behavior

Condition of Reactor	Multiplication Constant	Reactor Power
Critical	k = 1	Steady
Supercritical	k > 1	Increasing
Subcritical	k < 1	Decreasing

termed subcritical, critical, or supercritical, depending on the value of k for the present configuration, as shown in Table 70.2.

The value of k is dependent on geometric and material properties in the reactor and can be varied by introducing or withdrawing absorbing materials or control rods. Six factors make up the average value of k for the reactor:

$$k = P_{fnl} P_{tnl} \varepsilon \eta f p \tag{70.3}$$

The first two factors are geometric, relating to the leakage of neutrons out of the core of the reactor to be lost by absorption in surrounding material. The fast nonleakage probability, P_{fnl}, is the probability that a fast neutron born in fission does not leak from the reactor before it is thermalized. The thermal nonleakage probability, P_{tnl}, is the probability that a thermal neutron does not leak from the reactor prior to being absorbed. These two terms are a function of the geometry of the core and can be approximated as follows:

$$P_{fnl} P_{tnl} = 1(1 + M^2 B^2) \tag{70.4}$$

where M^2 is the neutron migration length and B^2 is the geometric buckling. The neutron migration length is

$$M^2 = L^2 + \tau \tag{70.5}$$

τ is the Fermi age, the mean square distance that a fast neutron travels in the reactor prior to becoming thermalized. L^2 is the mean square distance that a thermal neutron diffuses through the reactor prior to being absorbed. τ can be calculated from scattering theory, but this is beyond the scope of this treatment. L^2 is given in terms of the thermal neutron diffusion coefficient and the **mean free path** of thermal neutrons to absorption, λ_a, which are determined from the properties of the moderator, fuel, and structural material in the reactor core.

$$L^2 = \sqrt{D\lambda_a} \tag{70.6}$$

The geometric buckling is determined by the physical size and shape of the reactor. For a finite length cylinder (the common shape for power reactors), geometric buckling is given as:

$$B^2 = (2.405/R)^2 + (\pi/H)^2 \tag{70.7}$$

where R = radius and H = height.

The next four factors are based on the average properties of the materials that make up the reactor core and are independent of geometry:

ε is the fast fission factor, which is the number of neutrons produced in fission from all neutron energies divided by the number of neutrons produced by fissions from thermal neutrons. For a typical thermal reactor this parameter ranges from 1.0 to 1.05.

η is the number of fast neutrons produced per thermal neutron absorbed in the fuel. Thus η is a function of the mean free path of thermal neutrons to fission, the absorption mean free path, and the number of neutrons produced in an average thermal fission, ν. For a typical PWR, ν has a value of about 2.5.

$$\eta = \nu(\lambda_a^F / \lambda_f) \tag{70.8}$$

f is the thermal utilization, which is the fraction of absorbed thermal neutrons that are absorbed in the fuel material; thus it is a ratio including the mean free path for absorption in fuel, λ_a^F, and the mean free path including all absorptions, λ_a^T:

$$f = \lambda_a^T / \lambda_a^F \tag{70.9}$$

p is the resonance escape probability, the likelihood that a neutron makes the transition from fast to thermal energy without being captured by a nuclear resonance, which leads to loss of the neutron without producing fission. While generally speaking the probability of neutron absorption increases with decreasing neutron energy, there exist specific neutron energies, or resonance energies, that align with the properties of heavy nuclei and have very high probability of neutron absorption. The resonance escape probability depends substantially on the heterogeneous geometry of the reactor, because neutrons leaving the fuel can thermalize in a moderator, thereby skipping over the resonance energies prior to reentering the fuel by diffusion. **Neutron transport theory** must be used to compute p.

In practice, the reactivity, ρ, is often used instead of the multiplication constant, where

$$\rho = (k-1)/k \tag{70.10}$$

Changes in core configuration or material properties can be characterized as reactivity insertions, where ρ equal to zero corresponds to steady power operation, a positive reactivity insertion (withdrawing an absorber control rod, for example) leads to a power increase, and a negative reactivity insertion (inserting an absorber control rod) leads to a power decrease.

Intelligent reactor design will seek to establish self-stabilizing dynamic behavior of the reactor. This is accomplished through the careful consideration of reactivity coefficients during design. Reactivity coefficients are measures of feedback characteristics of the design. Two important reactivity coefficients are the moderator coefficient and the fuel Doppler coefficient.

The moderator coefficient measures the feedback effect on reactor reactivity with a change in the density of the moderator. If the reactor is designed with slightly less than the optimum amount of moderator, then any decrease in moderator density, either through an increase in temperature or a loss of moderator, will result in a decrease in reactivity and hence a decrease in reactor power. This results in a self-stabilizing condition since an increase in temperature will decrease reactor power, hence restabilizing temperature.

The fuel Doppler coefficient addresses the property of resonance capture within the nuclear fuel. As temperature increases, the range of energies over which any given resonance is effective increases. If the reactor is designed properly this can be a very strong and very rapid self-stabilizing effect protecting the fuel against even very quick events.

The value of these reactivity coefficients is that their effects are inherent in the design and require no active controls or human interaction.

Reactor Thermal Hydraulics

The transfer of thermal energy through a coolant to an energy conversion system is more complex for a nuclear power system than for a combustion-fueled power system. Most of the fission energy is deposited in the nuclear fuel, typically a high-temperature ceramic with poor conductivity. Transfer of heat to the coolant

TABLE 70.3 Annual Radiation Exposures

Occupational limit	50 mSv
General public limit	1.0 mSv
Average background exposure from nonnuclear power sources [natural and medical radiation (dental x-rays and medical treatment)]	3.5 mSv

occurs through conduction in the fuel pellet; conduction, radiation, and convection in the gap between the pellet and the cladding; conduction through the cladding; and convection to the coolant, flowing in channels along the fuel pins. Material properties of the fuel change over time with fuel **burnup,** due to fission product entrainment, radiation damage, and thermal cycling, which leads to pellet swelling and cracking.

The balance of plant for a nuclear power plant is similar to that of a conventional power plant; for additional details refer to Chapter 71, on power plants, and Chapter 74, on turbines. However, the presence of coolant-borne radioisotopes may place additional constraints on the operation of steam generators and primary coolant pumps in a nuclear system. Thermal energy management systems are the major source of operational and maintenance costs for a commercial nuclear power system.

Radiation Protection

Analysis of the radiation hazard from nuclear energy systems must include the effect of such systems on materials, biological systems, and the environment. The radiation environment of the fission process requires shielding of radiation-sensitive systems from the radiation produced during reactor operation, principally neutrons and gammas. The fission products produced and the **transuranic** (TRU) isotopes require environmental isolation and long-term storage.

Radiation protection concepts include methods of dealing with both external and internal sources of radiation. National and international standards on radiation protection are set by the National and International Commissions on Radiation Protection (NCRP and ICRP). Standards are enforced through guidelines and procedures promulgated by federal agencies, such as the U.S. Nuclear Regulatory Commission (NRC) and the U.S. Environmental Protection Agency (EPA). The general public and the radiation worker are protected from exposure to radiation far below levels known to cause harm.

Radiation damage occurs due to the deposition of energy in materials, which leads to ionization and atomic displacement damage. The damage to biological systems is a function of the type of radiation and is due to the production of oxidizing radicals following radiation's ionizing effect. Each source of radiation is characterized by its relative biological effectiveness (*RBE*), which is the damage of a given amount of the particular radiation relative to the damage of the same energy deposited by 100 keV x-rays. The unit of absorbed dose is the gray, where 1 gray (Gy) = 1 J (absorbed energy)/kg (mass of material). The unit of radiation exposure is the sievert (Sv), where

$$\text{Dose (Sv)} = \text{Absorbed dose (Gy)} \times RBE \qquad (70.11)$$

The radiation exposure principle that is applied in all occupational exposures is "as low as reasonably achievable" (ALARA). Thus the actual average occupational exposure and public exposure due to nuclear power operations is much less than allowed (see Table 70.3).

Radiation protection from external sources of radiation relies on three principles: time, distance, and shielding. The exposure to radiation can be minimized by (a) minimizing the time of the exposure to the source; (b) maximizing the distance from the source; (c) interposing absorbing materials, or shielding materials, between the source and the individual; or (d) some appropriate combination of these three strategies. Exposure to radiation from internal sources is controlled by minimizing the ingestion or inhalation of such sources. Regulations exist on the allowable maximum permissible concentration (MPC) of each radioisotope in air and water, based on daily consumption and keeping lifetime exposures below acceptable limits.

Nuclear Fuel Cycle

The nuclear fuel cycle encompasses the following processes from the exploration for uranium ore to the final disposition of the nuclear waste in a geological repository:

Exploration: Uranium and thorium are naturally occurring elements that exist in the earth's crust. Geological evaluation of the potential of a region is followed by monitoring for the low levels of natural radiation produced by the ores.

Mining: Both open-pit and underground mining is done for uranium ore (pitchblende). Major suppliers of uranium ore include Canada, Australia, Niger, Namibia, Uzbekistan, Russia, Kazakhstan, the U.S., and South Africa (listed in order of 2000 production). Canada supplied nearly a third of the uranium mined in 2000, and Australia and Canada combined accounted for more than half.

Milling: Uranium is separated from the ore by physical and chemical means and converted to U_3O_8 (yellowcake).

Enrichment: Natural uranium contains 0.711% (by weight) fuel isotope ${}_{92}U^{235}$. The remainder (>99%) is ${}_{92}U^{238}$, which can be used to breed fuel but is not itself usable as fuel in a thermal reactor. Some process must be used to increase the ${}_{92}U^{235}$ concentration in order to make the fuel usable in light water reactors. Enrichment methods include gaseous diffusion of UF_6 gas, centrifugal enrichment, and laser enrichment. In the U.S., gaseous diffusion plants originally built for the nuclear weapons program are employed for commercial enrichment. In Europe and Russia, gaseous diffusion has been largely replaced with more energy-efficient centrifuge systems. Laser enrichment remains under development.

Fabrication: The enriched UF_6 must be converted to the ceramic form (UO_2) and pressed into pellets, which are sintered and encased in a cladding material. Fuel pins are bound together into fuel assemblies.

Use in reactor: Fuel assemblies are placed in the core of the reactor and over a period of reactor operation, fission energy is extracted. A typical reactor operation time is 18 months, with fuel allowing extensions to 24 months being developed. Fuel management often dictates installing a third of the core as unburned fuel assemblies and discharging a third of the core as spent fuel at each refueling operation. When initially charged to the reactor, the fuel contains about 4% ${}_{92}U^{235}$ and 96% ${}_{92}U^{238}$. When the nuclear fuel is discharged it contains about 1% ${}_{92}U^{235}$, 1% plutonium created from neutron absorptions in ${}_{92}U^{238}$, 95% ${}_{92}U^{238}$, and 3% fission products.

Interim storage: The spent fuel removed from the core must be cooled in water-filled storage pits for 6 months to a year, in order for the residual heat and radioactivity to decay. Following use in a reactor, the spent fuel elements are depleted in the useful nuclear fuel isotopes and contain fission products and transuranic elements. While they are no longer useful for production of electricity, special handling is required due to the residual radioactivity. The decay of fission products continues to provide an internal heat source for the fuel element, which requires cooling during an interim storage period. The heat source is a function of the duration of reactor operation, but following steady power operation the time dependence has the following variation:

$$P(t) = 0.066P_0[t^{-1/5} - (t + t_0)^{-1/5}] \tag{70.12}$$

where P_0 is the power level during operation for time t_0, and t is the time in seconds after shutdown. Table 70.4 shows the relative heat generation of spent fuel assuming a 1-year operating period. Note that while the heat generation does decrease markedly, heat generation does continue for an extended period and adequate cooling must be provided.

With the significant delays in long-term disposal facilities, the period of interim storage has been extended longer than initially planned. In some reactor facilities the space in water-filled storage pools has been exceeded. In these cases, older fuel with reduced heat generation has been successfully placed in dry cask storage. Dry casks are large steel and/or concrete vessels that can be adequately cooled by passive convection to ambient air.

Reprocessing: As can be seen from the description of discharged fuel above, the spent fuel still contains significant usable material with only 3% of it actually being waste. Recovery and reuse of this material

TABLE 70.4 Decay Heat of Spent Nuclear Fuel after One Year of Operation

Time after Shutdown	Fraction of Operating Power (%)
1 minute	2.7
1 hour	1.1
1 day	0.47
1 month	0.14
1 year	0.027
10 years	0.0025

can be achieved through reprocessing, where the spent fuel is dissolved and the waste products chemically separated from usable material. However, this process produces separated plutonium, which generates weapons proliferation concerns and the expense of the process is currently challenged by relatively inexpensive new nuclear fuel. The United States has foregone reprocessing because of the proliferation concern and intends to directly dispose of spent fuel without recycling. European and Japanese nuclear programs do use reprocessing to recycle nuclear fuel.

Waste disposal: High-level radioactive waste storage includes both the storage of spent fuel elements in a geological repository and the storage of vitrified waste from the reprocessing stream in such a repository. A geologic repository would place high level waste far underground.

Transportation: The transporting of nuclear materials poses special hazards, due to the potential radiological hazard of the material being shipped. Specially designed and extremely robust shipping casks have been developed with government funding that are able to withstand a highway accident including potential vehicle fire without the release of radionuclides to the environment.

Nuclear safety: Handling of nuclear materials within the nuclear fuel cycle requires special consideration of criticality safety, in order to keep these materials from accidentally coming into a critical configuration outside the confined environment of the reactor core.

Nuclear material safeguards: Nuclear materials, in particular those materials which are key to the development of a nuclear weapon, such as separated plutonium, must be protected from diversion into the hands of terrorists. Special safeguard procedures have been developed to control special nuclear material (SNM). Special nuclear material includes plutonium and highly enriched uranium. However, uranium fuel used in common commercial reactors has enrichment far below that required for weapon uses.

Reactor Types

The most common commercial power reactor type is the **pressurized water reactor (PWR).** In this design, water is used as both a coolant and a moderator. Water is circulated through the reactor core picking up the heat from the nuclear fission. The heated water is routed to steam generators which are large tube and shell heat exchanges. The shell side water is held at a reduced pressure relative to the primary circuit water and is allowed to boil. The generated steam is then routed to a relatively conventional Rankine cycle electrical generating power plant.

The second most common commercial power reactor is the **boiling water reactor (BWR).** This design is similar to the PWR using water as a coolant and moderator; however, it differs in that it operates at a lower pressure allowing water to boil directly in the primary vessel. This avoids the need for steam generators, but since the primary water will have some entrained radioactive material in the form of activated corrosion products, the steam plant will be contaminated, complicating maintenance. Between them, PWR and BWR reactors account for over 80% of the current commercial nuclear power plants.

The next most common design is the **pressurized heavy water reactor (PHWR).** In this design the moderator and coolant is heavy water, that is water comprised of oxygen and deuterium, the first heavy isotope of hydrogen. Although a less efficient moderator than hydrogen, deuterium has the advantage of a much smaller probability of neutron absorption. This allows a critical reactor to be fueled with natural uranium and avoids the expensive process of isotopic enrichment for the nuclear fuel. This cost advantage

TABLE 70.5 Common Commercial Nuclear Power Systems

Reactor Type	Output (thermal/electric)	Fuel Form	Coolant/Moderator
Pressurized water reactor (PWR)	3400 MWth/1150 MWe	UO_2 pellets, 2 to 4% enriched uranium	H_2O/H_2O
Boiling water reactor (BWR)	3579 MWth/1178 MWe	UO_2 pellets, 2 to 4% enriched uranium	H_2O/H_2O
Pressure tube heavy water reactor (CANDU)	2180 MWth/648 MWe	UO_2 pellets, natural uranium	D_2O/D_2O
Pressure tube graphite reactor (PTGR)	3200 MWth/950 MWe	UO_2 pellets, 1.8 to 2.4% enriched uranium	H_2O/graphite
Liquid metal fast breeder reactor (LMFBR)	3000 MWth/1200 MWe	Mixed UO_2 and PuO_2 pellets, 10 to 20% plutonium	Liquid sodium/none (fast reactor)

is offset by the need to isotopically separate deuterium from normal water. Canada is the major producer of PHWRs with its CANDU reactor series. India has produced an indigenous PHWR, albeit based on the CANDU design. PHWRs make up just under 8% of the world's commercial nuclear power plants.

The nuclear power systems in common use throughout the world are listed in Table 70.5, along with information on the fuel form, coolant type, moderator, and typical size.

Reactor Operations, Licensing, and Regulation

The licensing of nuclear power systems in the U.S. is the responsibility of the NRC. State public utility commissions are involved in general oversight of electric power generation by electric utilities, including those with nuclear power plants. Industry associations are involved in the development and promulgation of best industry practices. Two such organizations are the Institute for Nuclear Power Operations (INPO) and the Nuclear Energy Institute (NEI). Similar organizations exist to support international nuclear power operations, including the International Atomic Energy Agency (IAEA), a branch of the United Nations, and the World Association of Nuclear Operators (WANO), an international nuclear power utility industry association.

Radioactive Waste Management

The radioactive wastes generated by nuclear power systems must be handled responsibly, minimizing the radiation exposure to employees and the general public, while providing long-term environmental isolation. Radioactive wastes are characterized as low-level wastes (LLW) or high-level wastes (HLW).

LLW consists of materials contaminated at low levels with radioisotopes but requiring no shielding during processing or handling. LLW is generated not only by nuclear power plant operations but also from medical procedures, university research activities, and industrial processes that use radioisotopes. LLW is judged to have minimal environmental impact, and the preferred method of disposal is shallow land burial in approved facilities. LLW is handled by individual states and state compacts, under authority granted by the U.S. Congress in the Low-Level Radioactive Waste Policy Act of 1980. In 1995, U.S. commercial power reactors each produced an average of about 100 cubic meters of LLW. With the high cost of LLW disposal, efforts to reduce this volume have been undertaken.

HLW includes spent nuclear fuel and the residue of nuclear fuel reprocessing. There is presently (2002) no commercial reprocessing of spent nuclear fuel in the U.S., based on U.S. nonproliferation policy. However, there is a substantial amount of HLW in the U.S. that was generated by the nuclear weapons production program. This is currently stored at U.S. DOE sites, and the required facilities for processing prior to long-term storage are being constructed. European countries and Japan have active programs for the reprocessing of spent reactor fuel. In the U.S., under the Nuclear Waste Policy Act of 1982, HLW is scheduled for long-term geological repository storage, with the U.S. Department of Energy assigned to manage both storage and disposal programs. A national geological repository is being developed at

Yucca Mountain, Nevada. The Department of Energy has formally recommended the site, and in July 2002, the U.S. Senate approved the plan. The NRC now has the disposal plan under review. It is not expected to be ready for permanent storage of HLW until 2010. In the interim, electric utilities are using on-site wet-well or dry-cask storage of their spent fuel.

70.3 Economics of Nuclear Power Systems

Existing commercial power plants are currently very competitive with other forms of electrical generation. The majority of the cost of nuclear electrical production is amortization of the capital cost of the plant; fuel and O&M account for a small fraction of the cost of power. Nuclear fuel and O&M costs are roughly equivalent to coal-fired plants and significantly less than natural gas-fired plants with their high fuel cost. The current generation of U.S. nuclear units has largely amortized their original capital cost and is now producing electricity at relatively low cost.

Capacity factors have steadily improved, from 76% in 1996 to over 89% in 2001. Useful fuel life has been extended with the period between refueling outages going from 12 months to 18 and promised further extensions to 24 months. Refueling outages have been dramatically shortened and forced outages significantly reduced. The number of nuclear operating utilities has been reduced as a set of highly efficient nuclear-dominated operators has bought up nuclear generating assets. The result is a smaller number of better operated and more economically viable nuclear operators.

New construction within the U.S. is still limited by the high capital cost and uncertainties in the regulatory environment. New plant concepts exist to address construction cost and schedule, and the NRC has developed a streamlined licensing process; however, neither the new concepts nor NRC process have yet been put to the test.

Alternative Sources of Electrical Power

The primary competing sources of electric power are fossil fuel-fired generation, that is, coal and natural gas. While coal retains approximately a 50% share of electrical generation, essentially all of the near-term future expansion in the U.S. is slated to be natural gas. The extremely low capital cost and short construction times of natural gas gives that source a great advantage over its competitors. Natural gas also produces less environmental pollution than does coal. It produces significantly less sulfur and nitrous oxides and, due to its greater carbon efficiency, less greenhouse gases than coal.

The public is becoming more aware of the greenhouse gas produced by fossil fuels and concerned about the resulting global warming. While there has been some recognition that nuclear generation of electricity avoids all greenhouse gas emissions, this has not resulted in a large increase in public support for nuclear energy. Misunderstandings of the magnitude and true hazard of HLW and of power plant safety still limit public support. Lack of a demonstration of economical new construction in the U.S. limits investments in new commercial power reactors in this country. However, according to the World Nuclear Association, worldwide 27 new commercial power reactors are under construction (as of September 2002) and an additional 31 are being planned.

Next-Generation Commercial Systems

In conjunction with the Electrical Power Research Institute and three major U.S. reactor vendors, the U.S. Department of Energy developed three advanced light water reactor designs. These were the ABB Combustion Engineering's System 80+ PWR design, Westinghouse's AP600, and General Electric's Advanced Boiling Water Reactor (ABWR) design. All three were reviewed by the NRC and given final design approval. None of these systems has been built in the U.S. However, two ABWRs have been constructed and are operating in Japan. These plants were complete in 52 months. The Koreans have two PWRs under construction that are largely based on the system 80+ design, and construction is expected to take only 48 months.

These advanced light water reactors are driving down the capital cost of nuclear units. The AP600 designers are attempting to bring the total cost of power of their concept down to 3.2¢/kilowatt hour. That price would be instantly competitive in the U.S. market.

Beyond light water reactors, considerable attention has been given to pebble-bed modular reactors. In this concept the fuel is uranium carbide microspheres encased in silicon carbon shells and compressed into tennis ball-sized spheres. The fuel spheres or pebbles are held in a vessel with helium gas forced through as a coolant. The hot helium is then used to drive a conventional Brayton cycle gas turbine. The benefits of the concept are extremely high thermal efficiency and an essentially indestructible fuel form. Even in the event of a loss of all forced cooling, the fuel will retain its fission products. Other graphite-moderated, gas-cooled configurations have been proposed, also offering high thermal efficiency and extremely robust nuclear fuel.

While decreased relative to prior years, research continues on breeder reactor concepts. These concepts have been challenged by high cost and the proliferation concerns of plutonium production. However, their strong appeal is the utilization of a majority of the 95% of the energy content currently abandoned in a once-through reactor fuel cycle.

Defining Terms

Boiling water reactor (BWR) — Allows the liquid coolant to boil in the core, and the primary loop includes steam separators and dryers and the generation of steam within the primary vessel.

Burnup — The quantity of energy released, in megawatt-days per metric ton of fuel; also the change in quantity of nuclear fuel due to fissioning.

CANDU — A reactor using natural uranium with deuterated water for both cooling and moderation, the preferred reactor of the Canadian nuclear power program. No fuel enrichment is required, but deuterium must be separated from water.

Control rods — Devices including elements that strongly absorb neutrons, used to control the neutron population in a nuclear reactor.

Delayed neutrons — Neutrons born from the decay of neutron-rich fission products. The emission of these neutrons is controlled by the radioactive decay of their parent fission product; hence, they are delayed from the fission event.

Fast neutrons — Neutrons born in fission, typically with energies well above the thermal range.

Liquid metal fast breeder reactor (LMFBR) — A reactor that operates on a fast neutron spectrum and breeds additional nuclear fuel from nonfuel ${}_{92}U^{238}$. Sodium metal is the liquid coolant, and a secondary sodium coolant loop transfers energy through a heat exchanger to a tertiary loop of water, which generates steam for a turbine.

Mean free path — The average distance that a neutron of a given energy travels through matter prior to an interaction of the type specified (i.e., scattering or absorption).

Moderator — A material that scatters neutrons rather than absorbing them, typically a light element that causes neutrons to lose energy efficiently during scattering.

Nuclear energy — The release of energy from fissioning the nucleus, as opposed to fossil energy, which releases energy from chemical combustion.

Nuclear reactor — An assemblage of materials designed for the controlled release of nuclear energy.

Neutron transport theory — The theory that includes the detailed spatial and time behavior of neutrons traveling through matter; transport analysis requires detailed understanding of the nuclear properties of materials.

Pressure-tube graphite reactor (PTGR) — A reactor in which fuel assemblies are contained in water-filled tubes that provide flowing cooling. Those tubes penetrate a graphite block, with the graphite providing neutron moderation.

Pressurized water reactor (PWR) — The PWR has a pressurized primary loop that transfers heat through a heat exchanger to a secondary water loop that includes the steam generator. Heat transfer in the core occurs in the subcooled phase.

Reactor core — The central region of a nuclear reactor, where fission energy is released.

Thermal neutrons — Neutrons that have kinetic energies comparable to the thermal energy of surrounding materials and hence are in collisional equilibrium with those materials.

Transmutation — The change of a nucleus to a different isotope through a nuclear reaction such as the absorption of a neutron and the subsequent emission of a photon (capture gamma).

Transuranic — Isotopes higher in atomic number than uranium; they are created by nuclear transmutations from neutron captures.

Zircaloy — An alloy of zirconium that is used for cladding of nuclear fuel. Zirconium has a very small probability of absorbing neutrons, so it makes a good cladding material.

References

Knief, R. A. 1992. *Nuclear Engineering: Theory and Technology of Commercial Nuclear Power,* 2nd ed. Hemisphere, New York.

Lamarsh, J. R. 1983. *Introduction to Nuclear Engineering,* 2nd ed. Addison-Wesley, Reading, MA.

Murray, R. L. 1994. *Nuclear Energy,* 4th ed. Pergamon Press, Oxford.

Todreas, N. E. and Kazimi, M. S. 1990. *Nuclear Systems I, Thermal Hydraulic Fundamentals,* Hemisphere, New York.

Further Information

Research and development related to nuclear power system technologies is published in *Transactions of the ANS,* published by the American Nuclear Society (ANS), La Grange Park, IL, and in *IEEE Transactions on Nuclear Science,* published by the IEEE Nuclear and Plasma Sciences Society, Institutes of Electrical and Electronics Engineers, New York.

The National Council of Examiners for Engineering and Surveying (NCEES) sets examinations for professional engineering registration in the nuclear engineering area. The major work behaviors for those working in the nuclear power systems areas are outlined in *Study Guide for Professional Registration of Nuclear Engineers,* published by the ANS.

Further details on the history and present status of space nuclear technologies can be obtained from *Space Nuclear Power,* by Joseph A. Angelo, Jr. and David Buden or from the *Space Nuclear Power Systems* conference proceedings (1984–present), edited by Mohamed S. El-Genk and Mark D. Hoover, all published by Orbit Book Co., Malabar, FL.

71
Power Plants

Mohammed M. El-Wakil
University of Wisconsin

Power plants convert a primary source of energy to electrical energy. The primary sources are:

1. *Fossil fuels,* such as coal, petroleum, and gas.
2. *Nuclear fuels,* such as uranium, plutonium, and thorium in fission and deuterium and tritium in fusion.
3. *Renewable energy,* such as solar, wind geothermal, hydro, and energy from the oceans. The latter could be due to tides, waves, or the difference in temperature between surface and bottom, called *ocean-temperature energy conversion* (OTEC).

Systems that convert these primary sources to electricity are in turn generally classified as follows:

1. The **Rankine cycle,** primarily using water and steam as a working fluid, but also other fluids, such as ammonia, a hydrocarbon, a freon, and so on. It is widely used as the conversion system for fossil and nuclear fuels, solar energy, geothermal energy, and OTEC.
2. The **Brayton cycle,** using, as a working fluid, hot air–fossil fuel combustion products or a gas, such as helium, that is heated by nuclear fuel.
3. The combined cycle, a combination of Rankine and Brayton cycles in series.
4. Wind or water turbines, using wind, hydropower, ocean tides, or ocean waves.
5. Direct energy devices, which convert some primary sources to electricity directly (without a working fluid), such as photovoltaic cells for solar energy and fuel cells for some gaseous fossil fuels.

In the mid-1990s, U.S. power plants generated more than 550,000 megawatts. About 20% of this capacity was generated by fission nuclear fuels using the Rankine cycle. A smaller fraction was generated by hydropower, and a meager amount by other renewable sources. The largest portion used fossil fuels

0-8493-1586-7/05/$0.00+$1.50

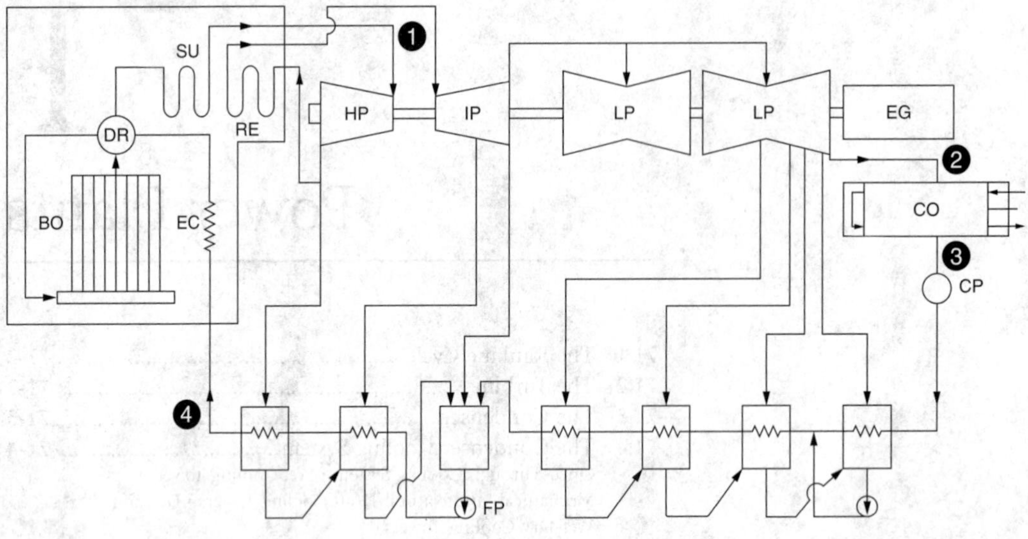

FIGURE 71.1 A flow diagram of a fossil-fuel Rankine-cycle power plant with one closed feedwater heater with drains pumped forward, five closed feedwater heaters with drains cascaded backward, and one open feedwater heater. HP = high-pressure turbine. IP = intermediate-pressure turbine. LP = low-pressure turbine. EG = electric generator. CO = condenser. CP = condensate pump. FP = feedwater pump. EC = economizer. DR = steam drum. BO = boiler. SU = superheaters. RE = reheaters.

and the Rankine cycle. Nuclear power plants and some of the renewables are described elsewhere in this chapter. The following section describes fossil-Rankine-type power plants.

71.1 The Rankine Cycle

Rankine is a versatile cycle that can use a wide variety of heat sources. In its most common form, it uses water and steam as a working fluid. It can be built to generate large quantities of electric power, exceeding 1000 megawatts in a single power plant. It has the highest conversion efficiency (ratio of electrical energy generated to heat energy added) of all large practical conversion systems.

Figure 71.1 shows a flow diagram of a Rankine cycle. High-pressure, superheated steam is admitted to a **steam turbine** at 1, commonly at 170 bar (about 2500 psia) and 540°C (about 1000°F), though new developments call for higher values with pressures in the supercritical range, above 221 bar (3208 psia). Steam expands through the turbine to 2, becoming a two-phase mixture of steam and water, usually 80% steam by mass, where the pressure and temperature are typically 0.07 bar and 40°C (about 1 psia and 104°F) but vary according to the available cooling conditions in the **condenser.**

The turbine exhaust at 2 is cooled in a condenser at constant pressure and temperature, condensing to a saturated liquid at 3. The condensate is then pumped through a feedwater system, where it is heated in stages by a series of **feedwater heaters,** to 4, where the temperature is just below the boiling temperature at the maximum pressure in the cycle. Heat is then added to feedwater in a **steam generator,** converting it to steam at 1.

71.2 The Turbine

The energy imparted by the steam from 1 to 2 is converted to mechanical work by the turbine, which in turn drives an electric generator to produce electricity, according to the following formula:

$$W_T = \sum (m \cdot \Delta h) \times \eta_T \qquad (71.1)$$

where W_T = turbine mechanical power, kW or British thermal units (Btu)/h; m = mass flow rate of steam through each turbine section, kg/h or lb/h; Δh = enthalpy drop of steam through each turbine section, kJ/kg or Btu/lb; and η_T = overall turbine efficiency = ratio of turbine shaft power to power imparted by the steam to the turbine, and

$$W_G = W_T \times \eta_G \tag{71.2}$$

where W_g = electrical generator power and η_G = electrical generator efficiency.

Modern power plant turbines are made of multiple sections, usually in *tandem* (on one axis). The first section, a high-pressure turbine, made largely of **impulse blading,** receives inlet steam and exhausts to a reheater in the steam generator. The reheated steam, at about 20% of the pressure and about the same temperature as at 1, enters an intermediate-pressure turbine made of **reaction blading,** from which it leaves in two or three parallel paths to two or three low-pressure turbines, *double-flow* and also made of reaction blading. Steam enters each in the center and exhausts at both ends, resulting in four or six paths to the condenser. This configuration divides up the large volume of the low-pressure steam — and therefore the height and speed of the turbine blades — and eliminates axial thrust on the turbine shaft. Chapter 74 of this text describes steam turbines in greater detail.

71.3 The Condenser

The process of condensation is necessary if net power is to be generated by a power plant. (If the turbine exhaust were to be pumped back directly to the steam generator, the pumping power would be greater than the electrical power output, resulting in net negative power. Also, the second law of thermodynamics stipulates that not all heat added to a thermodynamic cycle can be converted to work; hence, some heat must be rejected.) The condenser is where heat is rejected.

To increase the cycle efficiency, the rejected heat must be minimized so that a higher percentage of the heat added is converted to work. This is done by operating the condenser at the lowest temperature, and hence the lowest pressure, possible by using the lowest-temperature coolant available, usually water from a nearby large supply, such as a river, lake, or ocean. Most power plants are situated near such bodies of water. When cooling water goes through a condenser, its temperature rises before it is readmitted to its source. To minimize this heating and its undesirable effect on the environment and to conserve water, cooling towers may be used. The heat rejected to the environment by the condenser Q_R is given by

$$Q_R = m_c(h_2 - h_3) \tag{71.3}$$

where m_c = steam mass flow rate to condenser = mass flow rate of turbine inlet steam at 1 minus steam bled from the turbine for feedwater heating, as discussed later.

The most common condenser is the *surface condenser,* Figure 71.2. It is a shell-and-tube heat exchanger, composed of a steel shell with water boxes on each side connected by water tubes. Cooling water from the coolest part of the source is cleaned of debris by an intake mechanism and pumped by large circulating pumps to one of the water boxes, from which it goes through the tubes, exiting at the other box, and back to its source, at such a location to avoid reentry of the heated water to the condenser. Such a condenser is of the *one-pass* kind. A *two-pass* condenser is one in which one box is divided into two compartments. The incoming water enters half the tubes from one compartment, reverses direction in the second box, and returns through the other half of the tubes to the second compartment of the first box. One-pass condensers require twice the quantity of cooling water as two-pass condensers but result in lower condenser pressures and higher power plant efficiencies and are used where there are ample supplies of water. Surface condensers are large in size, often exceeding 100,000 m^2 (more than a million square feet) of tube surface area, and 15 to 30 m (50 to 100 ft) tube lengths in large power plants.

Another type, called the *direct-contact* or *open* condenser, is used in special applications, such as with geothermal power plants, with OTEC, and when dry cooling towers (below) are used. A direct-contact

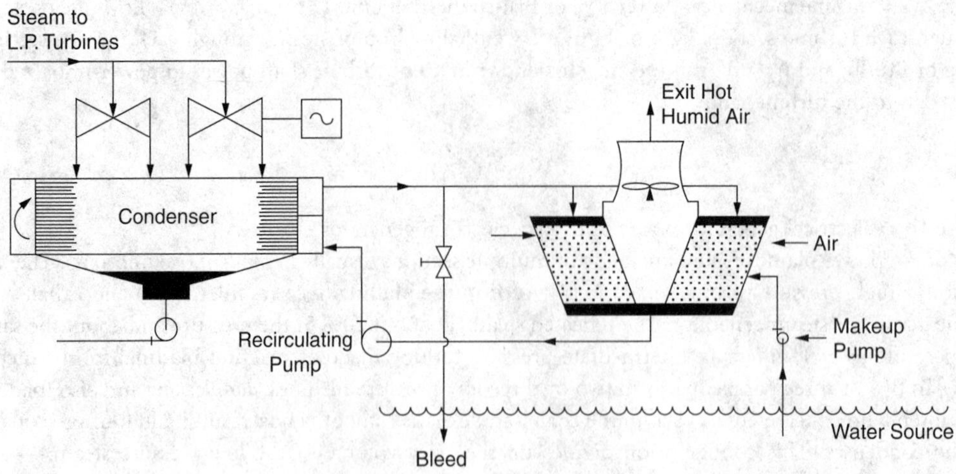

FIGURE 71.2 A flow diagram of a power plant cooling system, with a two-pass surface condenser and a wet, mechanical-induced-draft, cross-flow cooling tower.

condenser is further classified as a *spray* condenser, a *barometric* condenser, or a *jet* condenser. The latter two are not widely used.

A spray direct-contact condenser is one in which demineralized cooling water is mixed with the turbine exhaust via spray nozzles. The mixture becomes a saturated liquid condensate. A fraction of it equal to the turbine flow goes to the cycle; the balance is cooled in a dry cooling tower and then recirculated to the condenser spray nozzles. The ratio of cooling water to turbine flow is large, about 20 to 25. In geothermal plants the fraction equal to the turbine flow may be returned to the ground. In OTEC it is returned to the ocean.

71.4 The Condenser Cooling System

A condenser cooling system may be open (or *once-through*) or partially closed, using **cooling towers.** The latter are classified into *wet natural-draft cooling towers, wet mechanical-draft cooling towers,* and *dry cooling towers.*

Once-Through Cooling System

Here, cooling water is taken from the source (usually at a depth where it is sufficiently cool), passed through the condenser, then returned to the source at a point that ensures against short circuiting of the warmer water back to the condenser, such as downstream of the intake. Once-through systems are the most efficient means of cooling a condenser but require large quantities of water and discharge warm water back to the source. Environmental regulations often prohibit the use of once-through systems, in which case cooling towers are used.

Wet Cooling Towers

In a wet cooling tower, the warm condenser water is essentially cooled by direct contact with atmospheric air. It is sprayed over a lattice of slats or bars, called a *fill* or *packing,* which increases its surface-to-volume ratio, as in Figure 71.2. Atmospheric air passes by the water in a *cross-flow* or *counterflow* manner. The water is cooled by exchanging heat with the cooler air and, more importantly, by partial evaporation into the heated and, hence, lower-relative-humidity air.

Because of evaporative losses, wet towers do not eliminate the need for water, but they appreciably reduce it, since these losses are a fraction of the total water flow. An additional loss is due to *drift,* in which unevaporated water drops escape with the air. *Drift eliminators* are added to reduce this loss. Evaporative losses depend upon the climatic conditions and could be as high as 1.5% of total water flow. Drift could be as high as 2.5% of the evaporative losses. An additional loss is *blowdown* or *bleed.* Warm water in the tower contains suspended solids and is fully aerated. Chemical additives are used to inhibit microbiological growth and scales. Thus, a certain percentage of the circulating cooling water is bled to maintain low concentrations of these contaminants. The bleed, nearly as high as the evaporative loss if high purity is to be maintained, is often returned to the source after treatment to minimize pollution. All losses must be compensated for by *makeup*; power plants using wet cooling towers are also sited near bodies of water. Other problems of wet towers are icing and fogging due to exiting saturated air in cold weather.

Mechanical- and Natural-Draft Cooling Towers

Atmospheric air flows through cooling towers either mechanically or naturally. In the former, it is moved by one or more fans. Because of distribution problems, leaks, and possible recirculation of the hot humid exit air, most mechanical towers move the air by *induced-draft* fans. These are placed at the top and suck the hot air through the tower. Such towers are usually multicell with several fans placed in stacks atop a bank of towers, the number depending upon the size of the power plant. The fans are usually multibladed (made of aluminum, steel, or fiberglass), are driven at low speeds by electric motors through reduction gearing, and could be as large as 10 m (33 ft) in diameter. Mechanical-draft cooling towers consume power and are relatively noisy.

In the natural-draft cooling tower, air flows by a natural driving force, F_D, caused by the density differences between the cool air outside and the warm air inside the tower, given by

$$F_D = (\rho_0 - \rho_i)Hg \tag{71.4}$$

where ρ_o and ρ_i = average densities of air outside and inside the tower; H = height of the tower; and g = the gravitational acceleration.

Because the difference between the densities is small, H is large — about 130 m (430 ft). The towers are imposing structures that are visible from afar and are costly to build, but consume no power. The water distribution system and fill are placed at the bottom, and most of the tower height is open space of circular cross-section. The vertical profile is hyperbolic, which offers good resistance to wind pressures. Natural-draft cooling towers are usually made of reinforced concrete and sit on stilts and are mostly of the counterflow type.

A compromise between mechanical and natural draft towers is called the *hybrid* or *fan-assisted hyperbolic* cooling tower. A number of forced draft fans surround the bottom to augment the natural driving force of a shorter hyperbolic tower. The hybrid consumes less power than a mechanical and is smaller and less costly than a natural tower.

Dry and Wet-Dry Cooling Towers

Dry cooling towers are used when a power plant is sited far from adequate sources of water, near coal mines or other abundant fuel to reduce transportation costs, near large power consumers to reduce electrical transmission costs, or in arid areas. They are essentially closed-type heat exchangers in which warm condenser water passes through a large number of finned tubes cooled by atmospheric air, and no water is lost due to evaporation, drift, and so on. They are usually easier to maintain than wet towers and do not suffer from fogging or icing.

Dry cooling towers, however, are not as effective, Lacking evaporative cooling, they have lower heat transfer capabilities, resulting in large heat exchanger surfaces and land areas. This also results in higher condenser water temperatures and hence higher back pressures on the turbine than with wet towers, resulting in lower power plant efficiencies. The problem is aggravated further during periods of high atmospheric air temperatures.

Dry cooling towers may be mechanical or natural draft. They also may be designed to operate in *direct* or *indirect* modes. In the direct mode, the turbine exhaust steam passes through large finned tubes that are cooled by the atmospheric air. Indirect dry towers, more common, use a conventional surface condenser with an intermediate coolant, such as water, or a two-phase fluid, such as ammonia. The latter, under development, improves heat transfer and results in lower penalties on the cycle efficiency.

Wet-dry cooling towers are combinations of the above. Warm condenser water enters a dry section of the tower — reducing its temperature partially — then goes on to a wet section. Parallel air flows to each section combine to a common exit. This reduces fogging and evaporative losses, but at the expense of more complexity and cost.

71.5 The Feedwater System

The condensate at 3 (in Figure 71.1) is returned to the cycle to be converted to steam for reentry to the turbine at 1. Called the *feedwater,* it is pumped by condensate and feedwater pumps — to overcome flow pressure losses in the feedwater system and the steam generator — and enters the turbine at the desired pressure. The feedwater is heated successively to a temperature close to the saturated temperature at the steam generator pressure. This process, called *regeneration* or *feedwater heating,* results in marked improvement in cycle efficiency and is used in all modern Rankine cycle power plants, both fossil and nuclear.

Regeneration is done in stages in feedwater heaters, which are of two types: (1) *closed* or *surface* type and (2) *open* or *direct-contact* type. The former are further classified into *drains cascaded backwards* and *drains pumped forward.* All types use steam bled from the turbine at pressures and temperatures chosen to match the temperatures of the feedwater in each feedwater heater. The amount of steam bled from the turbine is a small fraction of the total turbine flow because it essentially exchanges its latent heat of vaporization with sensible heat of the single-phase feedwater.

Closed feedwater heaters are shell-and-tube heat exchangers where the feedwater flows inside tubes and the bled steam condenses over them. Thus, they are much like condensers but are smaller and operate at higher pressures and temperatures. The steam that condenses is returned to the cycle. It is either cascaded backwards — that is, throttled to the next lower-pressure feedwater heater — or pumped forward into the feedwater line. The cascade type is most common (see Figure 71.1).

Open feedwater heaters, on the other hand, mix the bled steam with the feedwater, resulting in saturated liquid. The mix is then pumped by a feedwater pump to the next higher-pressure feedwater heater. Most power plants use one open-type feedwater heater, which doubles as a means to rid the system of air and other noncondensable gases; this type is often referred to as a *deaerating* or *DA* heater. It is usually placed near the middle of the feedwater system, where the temperature is most conductive to deaeration.

The mass flow rate of the bled steam to the feedwater heaters is obtained from energy balances on each heater [El-Wakil, 1984]. This determines the mass flow rate in each turbine section [which is necessary to evaluate the turbine work [Equation (71.1)] and the heat rejected by the condenser [Equation (71.3)].

71.6 The Steam Generator

A modern fossil-fuel power plant steam generator is a complex system. Combustion gases pass successively through the boiler, superheaters, reheaters, economizer, and air preheater and finally leave through a stack.

The Fuel System

Fuel is burned in a furnace with excess air (more than stoichiometric or chemically correct). This combustion air is forced through the system from the atmosphere by a **forced-draft fan,** resulting in combustion gases at about 1650°C (3000°F). At steam generator exit the air (now called *flue gases*) is drawn out by an *induced-draft fan* at about 135 to 175°C (275 to 350°F) into the stack. This seemingly high temperature represents an energy loss to the system but is necessary to prevent condensation of water vapor in the gases, which would combine with other combustion products to form acids and to facilitate flue gas dispersion into the atmosphere.

Pulverized Coal Firing

Furnaces have undergone much evolution. With coal, the old mechanical stokers have given way to **pulverized coal** firing in most modern systems. To pulverize, *run-of-the-mill* (as shipped from the mine) coal, averaging about 20 cm (8 in.) in size, is reduced to below 2 cm (0.75 in.) by *crushers*, which are of several types, including *rings, Bradford breakers,* and *hammer mills.* The crushed coal is then dried by air at 345°C (650°F) or more, obtained from the air preheater. The coal is then ground by pulverizers, which are usually classified according to speed. A common one is the medium-speed (75 to 225 rpm) *ball-and-race pulverizer,* which grinds the coal between two surfaces. One surface consists of steel balls that roll on top of the other surface, similar to a large ball bearing. Hot air then carries the powdery coal in suspension to a *classifier,* which returns any escaping large particles back to the grinders.

Pulverized coal is classified as 80% passing a #200 mesh screen (0.074 mm openings) and 99.99% through a #50 mesh screen (0.297 mm). It is fed to the furnace burners via a set of controls that also regulate primary (combustion) air to suit load demands. Large steam generators have more than one pulverizer system, each feeding a number of burners for a wide range of load control. Burners may be designed to burn pulverized coal only or to be multifuel, capable of burning pulverized coal, oil, or gas.

Cyclone Furnaces

A **cyclone furnace** burns crushed coal (about 95% passing a #4 mesh screen, about 5 mm). It is widely used to burn poorer grades of coal that contain high percentages of ash and volatile matter. Primary air, about 20% of the total combustion air, and the rest, secondary and tertiary air, enter the burner successively and tangentially, imparting a centrifugal motion to the coal. This good mixing results in high rates of heat release and high combustion temperatures that melt most of the ash into a molten slag. This drains to a tank at the bottom of the cyclone where it gets solidified, broken, and removed. Ash removal materially reduces erosion and fouling of steam generator surfaces and reduces the size of particulate matter-removal equipment such as electrostatic precipitators and bag houses. The disadvantages of cyclone firing are high power requirements and, because of the high temperatures, the production of more pollutants, such as oxides of nitrogen, NO_x.

Fluidized-Bed Combustion

Another type of furnace uses **fluidized-bed combustion.** Crushed coal particles, 6 to 20 mm (0.25 to 0.75 in.) in size, are injected into a bed above a bottom grid. Air from a plenum below flows upwards at high velocity so that the drag forces on the particles are at least equal to their weight, and the particles become free or fluidized with a swirling motion that improves combustion efficiency. Combustion occurs at lower temperatures than in a cyclone, reducing NO_x formation. About 90% of the sulfur dioxide that results from sulfur in the coal is largely removed by the addition of limestone (mostly calcium carbonate, $CaCO_3$, plus some magnesium carbonate, $MgCO_3$) that reacts with SO_2 and some O_2 from the air to form calcium sulfate, $CaSO_4$, and CO_2. The former is a disposable dry waste. Technical problems, such as the handling of the calcium sulfate, are under active study.

The Boiler

The boiler is that part of the steam generator that converts saturated water or low-quality steam from the economizer to saturated steam. Early boilers included fire-tube, scotch marine, straight-tube, and Stirling boilers. The most recent are water-tube–water-wall boilers. Water from the economizer enters a steam drum, then flows down insulated down-comers, situated outside the furnace to a header. The latter feeds vertical closely spaced water tubes that line the furnace walls. The water in the tubes receives heat from the combustion gases and boils to a two-phase mixture. The density difference between the downcomer and the tubes causes a driving force that circulates the mixture up the tubes and into the drum. The tubes also cool the furnace walls. There are several water-wall designs. A now-common one is the *membrane* design, in which 2.75- to 3-in. tubes on 3.75- to 4-in. centers are connected by welded membranes that act as fins to increase the heat-transfer surface as well as form a pressure-tight wall protecting the furnace walls. The steam drum now contains a two-phase bubbling mixture, from which dry steam is separated by gravity and mechanically with baffles, screens, and centrifugal separators.

Superheaters and Reheaters

Dry-saturated steam from the boiler enters a primary and then a secondary superheater in series, which convert it to superheated steam. The superheaters are made of 2- to 3-in. diameter U-tube bundles made of special high-strength alloy steels of good strength and corrosion resistance, suitable for high-temperature operation. The bundles are usually hung from the top, and called *pendant tubes,* or from the side, and called *horizontal tubes.* Another type, supported from the bottom and called the *inverted tube,* is not widely used.

Superheated steam enters the high-pressure turbine and exhausts from it to return to the steam generator, where it is reheated to about the same turbine inlet temperature in a set of reheaters, downstream of and similar in design to the superheaters. The reheated steam enters the intermediate-pressure and then the low-pressure turbines, as explained above.

Superheaters and reheaters may be of the *radiant* or *convection* types. The former, in view of the luminous combustion flames in the furnace, receives heat primarily by radiation. This heat transfer mode causes the exit steam temperature to decrease with increasing load (steam flow). The latter receives heat by convection, the main form of heat transfer in superheaters and reheaters, which causes the exit steam temperature to increase with load. To obtain fairly constant steam temperatures, *attemperators* are placed between the primary and secondary sections of superheaters and reheaters. In its most common form, an attemperator maintains the desired temperatures by spraying regulated amounts of lower-temperature water from the economizer or boiler directly into the steam.

The Economizer

The flue gases leave the reheaters at 370 to 540°C (700 to 1000°F). Rather than reject their energy to the atmosphere, with a consequent loss of plant efficiency, flue gases now heat the feedwater leaving the last (highest pressure) feedwater heater to the inlet temperature of the steam generator. This is done in the *economizer.* At high loads the economizer exit may be low-quality water–steam mixture. Economizers are usually made of tubes, 1.75 to 2.75 in. in diameter, arranged in vertical sections between headers and placed on 1.75 to 2 in. spacings. They may be plain-surfaced, finned, or studded to increase heat transfer. Smaller spacings and studs are usually used with clean ash-free burning fuels, such as natural gas.

Air Preheater

Flue gases leave the economizer at 315 to 425°C (600 to 800°F). They are now used in an *air preheater* to heat the atmospheric air, leaving the forced-draft fan to about 260 to 345°C (500 to 650°F) before admitting it to the furnace, thus reducing total fuel requirements and increasing plant efficiency. Air preheaters may be recuperative or regenerative. *Recuperative* preheaters are commonly counterflow shell-and-tube heat exchangers in which the hot flue gases flow inside and the air outside vertical tubes, 1.5 to 4 in. in diameter. A hopper is placed at bottom to collect soot from inside the tubes. *Regenerative* preheaters use an intermediate medium. The most common, called *ljungstrom,* is rotary and is driven by an electric motor at 1 to 3 rpm through reduction gearing. The rotor has 12 to 24 sectors that are filled with a heat-absorbing material such as corrugated steel sheeting. About half the sectors are exposed to and are heated by the hot flue gases moving out of the system at any one instant; as the sectors rotate, they become exposed to and heat the air that is moving in the opposite direction (into the system)

Environmental Systems

Besides cyclone and fluidized-bed combustion, there are other systems that reduce the impact of power generation on the environment. Flue gas desulfurization systems, also called **scrubbers,** use aqueous slurries of lime–limestone to absorb SO_2. **Electrostatic precipitators** remove particulate matter from the flue gases. Here, wire-to-discharge electrodes carry a 40 to 50 kV current and are centrally located between grounded plates or collection electrodes. The resulting current charges the soot particles, which migrate to the plates, where they are periodically removed. Fabric filters or **baghouses** also remove particulate matter. They are made of a large number of vertical hollow cylindrical elements — 5 to 15 in. in diameter and up to 40 ft high, made of various porous fabrics (wool, nylon, glass fibers, etc.) — through which the flue gases flow and get cleaned in the manner of a household vacuum cleaner. The elements are also periodically cleaned.

71.7 Cycle and Plant Efficiencies and Heat Rates

The heat added to the power plant, Q_A, and the cycle, Q_C, are given by the following:

$$Q_A = m_f \times \text{HHV} \tag{71.5a}$$

$$Q_C = Q_A \times \eta_{sg} = m_1(h_1 - h_4) \tag{71.5b}$$

where m_f = the mass flow rate of fuel to the furnace, in kg/h or 1b/h, and HHV = higher heating value of fuel, in kJ/kg or Btu/lb.

The plant efficiency, η_p, and cycle efficiency, η_C, are given by the following:

$$\eta_P = W_G / Q_A \tag{71.6a}$$

$$\eta_C = W_T / Q_C \tag{71.6b}$$

The value η_p, given above, is often referred to as the plant *gross efficiency.* Since some of the generator power, W_G, is used within the plant to power various equipment, such as fans, pumps, pulverizers, lighting, and so on, *a net efficiency* is often used, in which W_G is reduced by this auxiliary power. Another parameter that gives a measure of the economy of operation of the power plant is called the **heat rate,** HR. It is given by the ratio of the heat added in Btu/h to the plant power in kW, which may be gross or net. For example,

$$\text{Net plant HR, Btu/kWh} = (Q_A\text{, Btu/h})/(W_G - \text{auxiliary power, kW}) \tag{71.7}$$

The lower the value of HR is, the better. A benchmark net HR is 10,000, equivalent to a net plant efficiency of about 34%. It could be as high as 14,000 for older plants and as low as 8500 for modern plants.

Defining Terms

Baghouse — Removes particulate matter from the flue gases by porous fabric filters.

Brayton cycle — A cycle in which a gas (most commonly air) is compressed, heated, and expanded in a gas turbine to produce mechanical work.

Condenser — A heat exchanger in which the exhaust vapor (steam) of the turbine in a Rankine cycle is condensed to liquid, usually by cooling water from an outside source, for return back to the steam generator.

Cooling tower — A heat exchanger in which the condenser cooling water is in turn cooled by atmospheric air and returned back to the condenser.

Cyclone furnace — A furnace in which crushed coal is well mixed with turbulent air, resulting in good heat release and high combustion temperatures that melt the coal ash content into removable molten slag, thus reducing furnace size and the fly ash content of the flue gases and eliminating the cost of coal pulverization.

Electrostatic precipitator — A system that removes particulate matter from the flue gases by using one electrode at high voltage to electrically charge the particles, which migrate to the other grounded electrode, where they are periodically removed.

Feedwater heaters — Heat exchangers that successively heat the feedwater before entering the steam generator using steam that is bled from the turbine.

Fluidized-bed furnace — A furnace in which crushed coal is floated by upward air — resulting in a swirl motion that improves combustion efficiency, which in turn gives lower combustion temperatures and reduced NO_x in the flue gases — and in which limestone is added to convert much of the sulfur in the coal to a disposable dry waste.

Forced-draft fan — The fan that forces atmospheric air into the steam generator to be heated first by an air preheater and then by combustion in the furnace.

Heat rate — The rate of heat added to a power plant in Btu/h to produce one kW of power.

Impulse blades — Blades in the high-pressure end of a steam turbine and usually symmetrical in shape that ideally convert kinetic energy of the steam leaving a nozzle into mechanical work.

Once-through cooling — The exhaust vapor from the turbine of a Rankine cycle is condensed by cool water obtained from an available supply such as a river, lake, or the ocean, and then returned to that same supply.

Pulverized coal — A powdery coal that is prepared from crushed and dried coal and then ground, often between steel balls and a race.

Rankine cycle — A closed cycle that converts the energy of a high-pressure and high-temperature vapor produced in a steam generator (most commonly steam) into mechanical work via a turbine, condenser, and feedwater system.

Reaction blades — Blades downstream of impulse blades in a steam turbine and having an airfoil shape that convert some of both kinetic and enthalpy energies of incoming steam to mechanical work.

Scrubbers — A desulfurization system that uses aqueous slurries of lime–limestone to absorb SO_2 in the flue gases.

Steam generator — A large complex system that transfers the heat of combustion of the fuel to the feedwater, converting it to steam that drives the turbine. The steam is usually superheated at subcritical or supercritical pressures (critical pressure = 3208 psia or 221 bar). A modern steam generator is composed of economizer, boiler, superheater, reheater, and air preheater.

Steam turbine — A machine that converts steam energy into the rotary mechanical energy that drives the electric generator. It is usually composed of multiple sections that have impulse blades at the high-pressure end, followed by reaction blades.

References

El-Wakil, M. M. 1984. *Powerplant Technology,* McGraw-Hill, New York.

Singer, J. G. (Ed.) 1991. *Combustion, Fossil Power,* Combustion Engineering, Windsor, CT.

Stultz, S. C., and Kitto, J. B. (Eds.) 1992. *Steam, Its Generation and Use,* Babcock & Wilcox, Barberton, OH.

Further Information

Proceedings of the American Power Conference. American Power Conference, Illinois Institute of Technology, Chicago, IL 60616.

ASME publications (ASME, 345 E. 47th Street, New York, NY 10017):

ASME Boiler and Pressure Vessel Code

Mechanical Engineering

Journal of Energy Resources Technology
Journal of Gas Turbines and Power
Journal of Turbomachinery
Combustion and Flame. The Journal of the Combustion Institute, published monthly by Elsevier Science, 655 Avenue of the Americas, New York, NY 10010.
Department of Energy, Office of Public Information, Washington, D.C. 20585.
EPRI Journal. The Electric Power Research Institute, P.O. Box 10412, Palo Alto, CA 94303.
Energy, International Journal. Elsevier Science Ltd., Bampfylde Street, Exeter EX1 2AH, England.
Construction Standards for Surface Type Condensers for Ejector Service. The Heat Exchange Institute, Cleveland, OH.
Standards for Closed Feedwater Heaters. The Heat Exchange Institute, Cleveland, OH.
Power. McGraw-Hill, P.O. Box 521, Hightstown, NJ 08520.
Power Engineering. 1421 Sheridan Road, Tulsa, OK 74112.

72

Wind Energy

Kyle K. Wetzel

Wetzel Engineering, Inc.

Wind turbines are mechanical devices that convert the kinetic energy of the wind into useful shaft power. Figure 72.1 shows an example of a modern wind farm employing large electric-generating wind turbines, the 112-MW Gray County Wind Farm built in 2001 in Kansas.

72.1 Power in the Wind

Wind turbines extract energy from the air. The power available in the wind is in the form of kinetic energy. The power, P, contained in a wind of speed V and air density ρ passing through an area A normal to the wind equals the product of the rate of mass flow of air through that area and the kinetic energy per unit of mass of the wind, resulting in the equation

$$P = (\rho A V) \cdot \left(\frac{1}{2} V^2 \right) = \frac{1}{2} \rho A V^3 \tag{72.1}$$

where P is generally measured in units of watts, ρ in kilograms per cubic meter, A in square meters, and V in meters per second. Equation (72.1) reflects the fundamental and critically important result that the power from a wind turbine is proportional to the area swept by the rotor, the air density, and the cube of the wind speed.

Wind power potential at a given geographic site is based on the long-term average wind speed. Two different classification systems exist, as summarized in Table 72.1 and Table 72.2. The system shown in Table 72.1 is widely used in the United States and serves as the basis for the wind resource atlas of the U.S. shown in Figure 72.2 [Elliott et al., 1986]. This classification is based upon the mean wind speed as measured at a height of 10 m above the ground.

0-8493-1586-7/05/$0.00+$1.50

FIGURE 72.1 Gray County Wind Farm near Montezuma, Kansas. (Photo by K. Wetzel.)

TABLE 72.1 Classes of Wind Power Density, NREL System

Wind Power Class	10 m (33 ft) Wind Power W/m²	10 m Speed m/sec	10 m Speed mph	50 m (164 ft) Wind Power W/m²	50 m Speed m/sec	50 m Speed mph
1	0	0.0	0.0	0	0.0	0.0
2	100	4.4	9.8	200	5.6	12.5
3	150	5.1	11.5	300	6.4	14.3
3	200	5.6	12.5	400	7.0	15.7
4	250	6.0	13.4	500	7.5	16.8
5	300	6.4	14.3	600	8.0	17.9
6	400	7.0	15.7	800	8.8	19.7
7	1000	9.4	21.1	2000	11.9	26.6

Source: Elliott, D. L. 1986. *Wind Energy Resource Atlas of the United States,* DOE/CH 10093-4. Solar Technical Information Program, National Renewable Energy Laboratory, Golden, CO. p. 12.

TABLE 72.2 IEC Classes for Design of Wind Turbines

Wind Turbine Generator System Class	I	II	III	IV	S
Average hub height wind speed, V_{ave} (m/sec)	10	8.5	7.5	6.0	Specified by the wind turbine manufacturer
Turbulence intensity, I_{15}, Category A	18%				
Turbulence variation, a, Category A	2				
Turbulence intensity, Category B	16%				
Turbulence variation, a, Category B	3				

Source: IEC Standard 61400-1: Wind turbine generator systems — Part 1: Safety requirements, Edition 2. International Electrotechnical Commission, Geneva.

Class 6 sites are rare but presently most economical for development of wind power, while Class 5 sites are frequently developed. Class 4 sites are presently marginal, but areas shaded as Class 4 in Figure 72.2 may contain local sites with Class 5 or even Class 6 resource. In recent years, researchers have begun combining the data contained in Figure 72.2 with models of atmospheric physics and geographical information system (GIS) databases of local topography to create high-resolution maps of local wind resource. Some maps have resolution as fine as a quarter mile. The National Wind Technology Center at the National Renewable Energy Laboratory in Golden, Colorado, has sponsored much of this research and maintains a library of these maps. Interactive versions of some of the maps are available online at

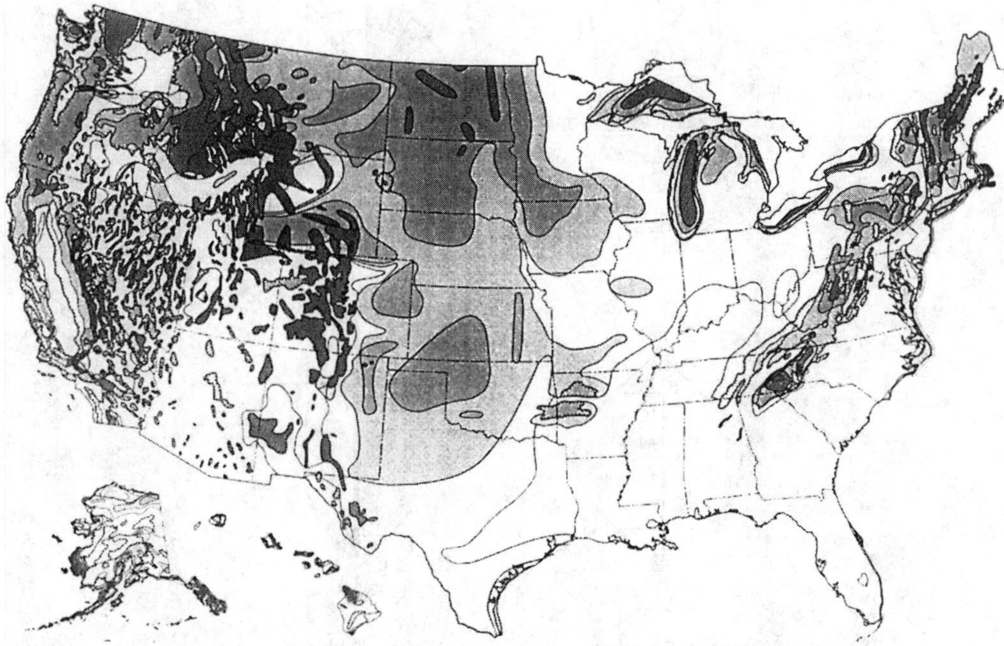

FIGURE 72.2 U.S. wind resource map. (*Source:* Elliott, D. L. 1986. *Wind Energy Resource Atlas of the United States*, DOE/CH 10093-4. Solar Technical Information Program, National Renewable Energy Laboratory, Golden, CO. p. 12.)

the NWTC website (www.nrel.gov/wind). While such maps are useful for identifying potential sites, wind power developers routinely install meteorological towers with anemometers and record wind speed data for at least a year before proceeding with a project.

The data shown in Table 72.1 for 50-m height is based upon an assumption that the wind speed aloft follows the commonly employed one-seventh power law where the wind speed, V, at some height, h, above ground follows the trend $V(h) \sim h^{1/7}$. Although it is frequently assumed that this relationship applies to flat grassy terrain, vertical gradients in wind speed can vary significantly from the one-seventh power law. Zero gradients and even negative gradients have been observed, particularly in hilly or mountainous terrain. Therefore, meteorological towers are generally instrumented with anemometers at several heights, typically the height of the bottom, middle, and top of the rotor of the wind turbine a developer plans to use.

Table 72.2 summarizes the classification defined by the International Electrotechnical Commission [IEC, 1999], which is based upon the average wind speed at the height of the wind turbine rotor hub (i.e., the axis of rotation). Turbulence Classes A and B are also defined based upon the average turbulence at the hub height. The IEC Classifications are used exclusively for turbine design certification purposes as described in Section 72.8. Most large-scale turbines are presently designed for and installed in IEC Class II and III sites.

72.2 Types of Wind Turbines

Wind turbine configurations are most fundamentally classified as drag devices or lift devices depending upon whether they derive their motion from aerodynamic drag or aerodynamic lift.

The simplest form of a wind turbine is the Savonius type (illustrated in Figure 72.3), which relies upon differential drag to provide its driving force. This type has found some limited application in water-pumping activities where high starting torque is of great benefit.

Modern commercial wind turbines designed to generate electricity utilize aerodynamic lift to provide their driving force. The rotational rate of a drag device is limited because the tangential velocity of the

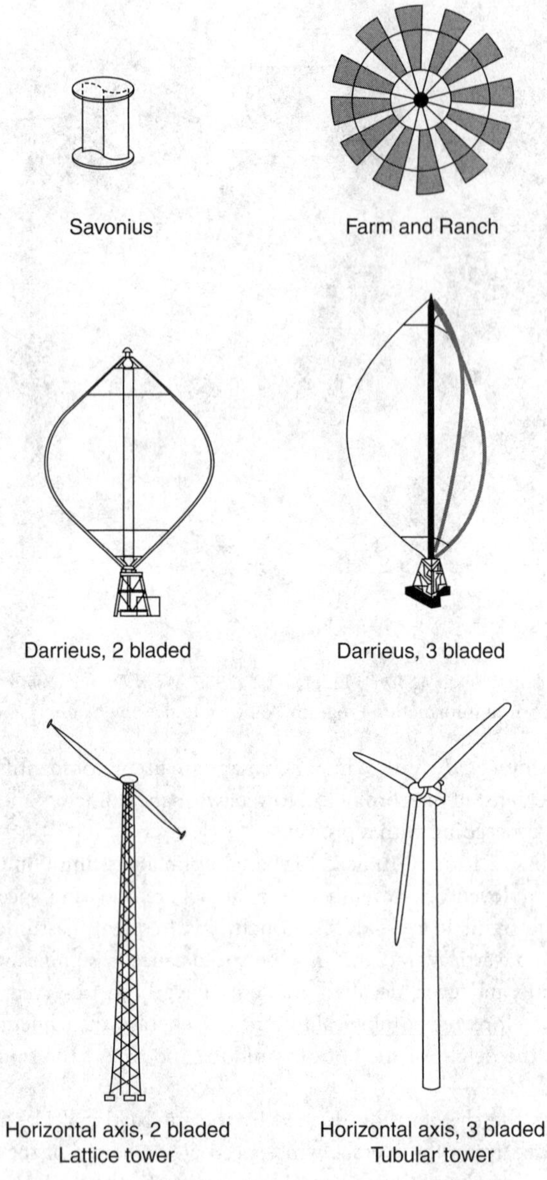

FIGURE 72.3 Typical wind turbine configurations.

device cannot exceed the speed of the wind. This means that the aerodynamic forces acting on a drag device are limited to those produced by the wind alone. A lifting device can rotate much faster, meaning that the velocity relative to the rotor can be much higher than the wind speed. A lifting rotor can utilize the vector sum of the rotational speed and the wind speed, yielding a power magnification on the order of a factor of 100 [Rohatgi and Nelson, 1994].

There are two broad categories of wind turbines: **horizontal axis (HAWT)** and **vertical axis (VAWT).** Both types typically have two or three blades, although HAWTs have been built with single blades and with more than three blades. Augmentation devices have been proposed for both types, but the additional energy extraction has not proved sufficient to justify the increased cost and complexity.

VAWTs can receive wind from any direction and thus require no aligning devices. They also benefit from having their generators at ground level, which simplifies servicing. However, severe alternating

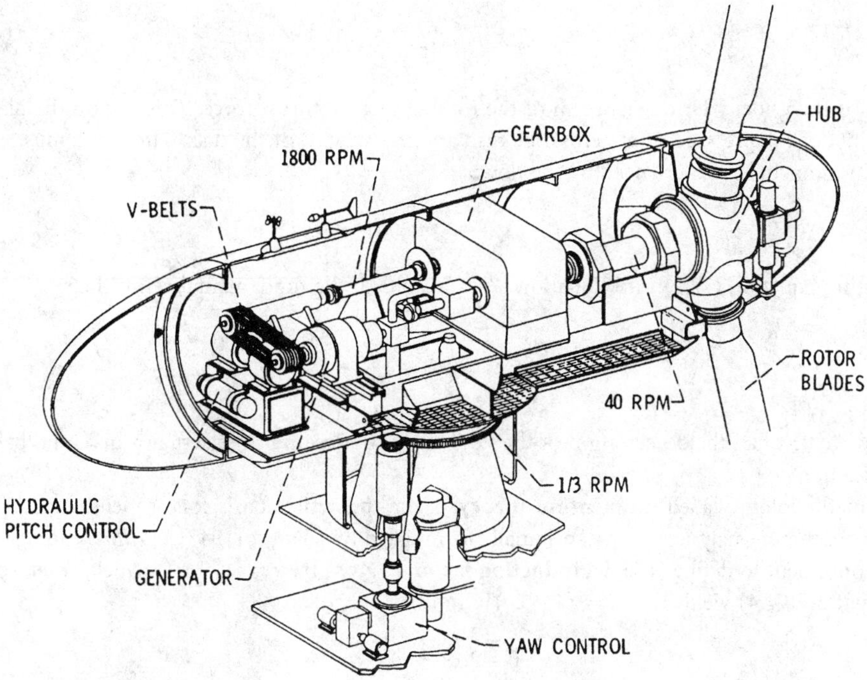

FIGURE 72.4 Cut-away view of a typical horizontal axis wind turbine. (*Source:* Eldridge, F. R. 1975. *Wind Machines,* National Science Foundation, Grant Number AFR-75-12937. Energy Research and Development Administration, Washington, DC. p. 30.)

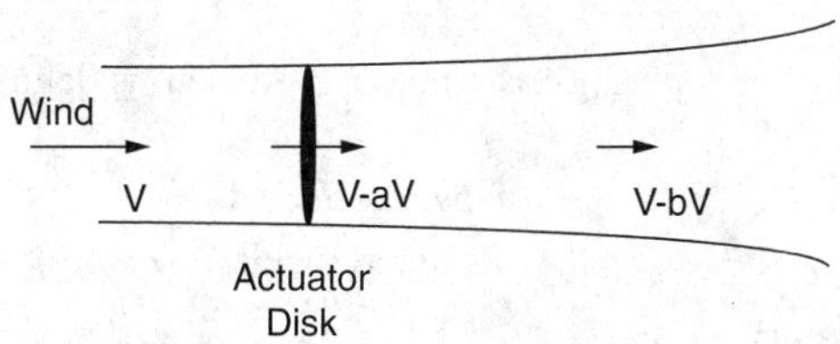

FIGURE 72.5 Actuator disk.

stresses are induced by the windstream on the blades as they rotate about the vertical shaft, and VAWTs cannot exploit the higher winds aloft from which HAWTs on tall towers benefit. The large, utility-scale wind turbine industry has abandoned VAWTs during the past decade in favor of HAWTS. Figure 72.4 illustrates the major components of a modern HAWT.

72.3 Power from a Wind Turbine: Axial Momentum Theory

By extracting energy from the wind, a wind turbine reduces the velocity of the air passing through its rotor. Figure 72.5 illustrates an actuator disk placed in an airstream without a duct or shroud. In this simplified model, the air flow experiences a change in pressure as it passes through the disk. If one assumes that the air flow is incompressible and inviscid, then it follows from Bernoulli's Law that the total pressure, p_t, remains constant upstream of the disk as the flow decelerates from its freestream velocity, V, to the velocity at the rotor disk, u. Then the flow experiences a discontinuous drop in total and static pressure, Δp, as it passes through the disk. Again the total pressure remains constant as the flow continues to decelerate further downstream, returning the static pressure far downstream of the rotor to ambient. The pressure drop across the rotor is given by

$$\Delta p = p_{t,upstream} - p_{t,downstream} = \frac{1}{2}\rho V^2 - \frac{1}{2}\rho u_1^2 \tag{72.2}$$

where u_l is the velocity far downstream of the rotor. The axial thrust force, T, acting on the disk equals the product of the pressure drop across the disk and the area, A, of the disk. This force must equal the change in momentum experienced by the flow:

$$T = \Delta pA = (\rho uA)\cdot(V - u_l) \tag{72.3}$$

Substituting Equation (72.2) into Equation (72.3) results in the fundamental result that

$$u = \frac{1}{2}(V + u_1) \tag{72.4}$$

or, in words, that half the deceleration of the flow occurs upstream of the actuator disk, and half occurs downstream.

This methodology, called **momentum theory,** is the theoretical basis for propeller, helicopter, and wind turbine rotor analysis, and was originally formulated by Glauert [1947].

It is convenient to define an **axial induction factor**, a, given by $u = (1 - a)V$, which, when combined with Equation (72.4) yields that $u_l = (1 - 2a)V$, and that

$$T = 2\rho AV^2 a(1 - a) \tag{72.5}$$

The power, P, extracted from the wind by the turbine equals the product of the thrust force, T, and the air velocity at the disk, u, resulting in

$$P = 2\rho AV^3 a(1 - a)^2 \tag{72.6}$$

To calculate the efficiency of energy capture, a power coefficient is defined as

$$C_p = \frac{\text{Power extracted}}{\text{Power available}} \tag{72.7}$$

Substituting Equation (72.1) for the denominator and Equation (72.6) for the numerator in Equation (72.7) yields

$$C_p = 4a(1 - a)^2 \tag{72.8}$$

Since mass flow through the disk and downstream must be maintained for energy extraction, it is impossible to extract all of the energy available in the airstream by bringing the freestream velocity to zero. There is a limit to the available energy capture, which is reflected as a maximum value of C_{pmax} = 16/27 = 0.593 called the **Betz limit.** It can be calculated by taking the first derivative of Equation (72.8) with respect to a, setting it equal to zero, and solving for a = 1/3. The maximum power extraction is limited to 59.3% of the power available in the wind stream and occurs when $u = \frac{2}{3}V$ and $u_l = \frac{1}{3}V$, that is, the freestream wind velocity is slowed to two-thirds of its original value at the rotor disk.

An axial force coefficient can be defined by normalizing the thrust by the dynamic pressure

$$C_T = \frac{T}{0.5\rho AV^2} = 4a(1 - a) \tag{72.9}$$

Finding the maximum for this equation shows that C_{Tmax} = 1 occurs when a = 0.5.

Momentum theory is constrained to values of the induced flow factor, a, less than 0.5, since values of a greater than 0.5 imply flow reversal downstream of the actuator disk, in which case momentum theory

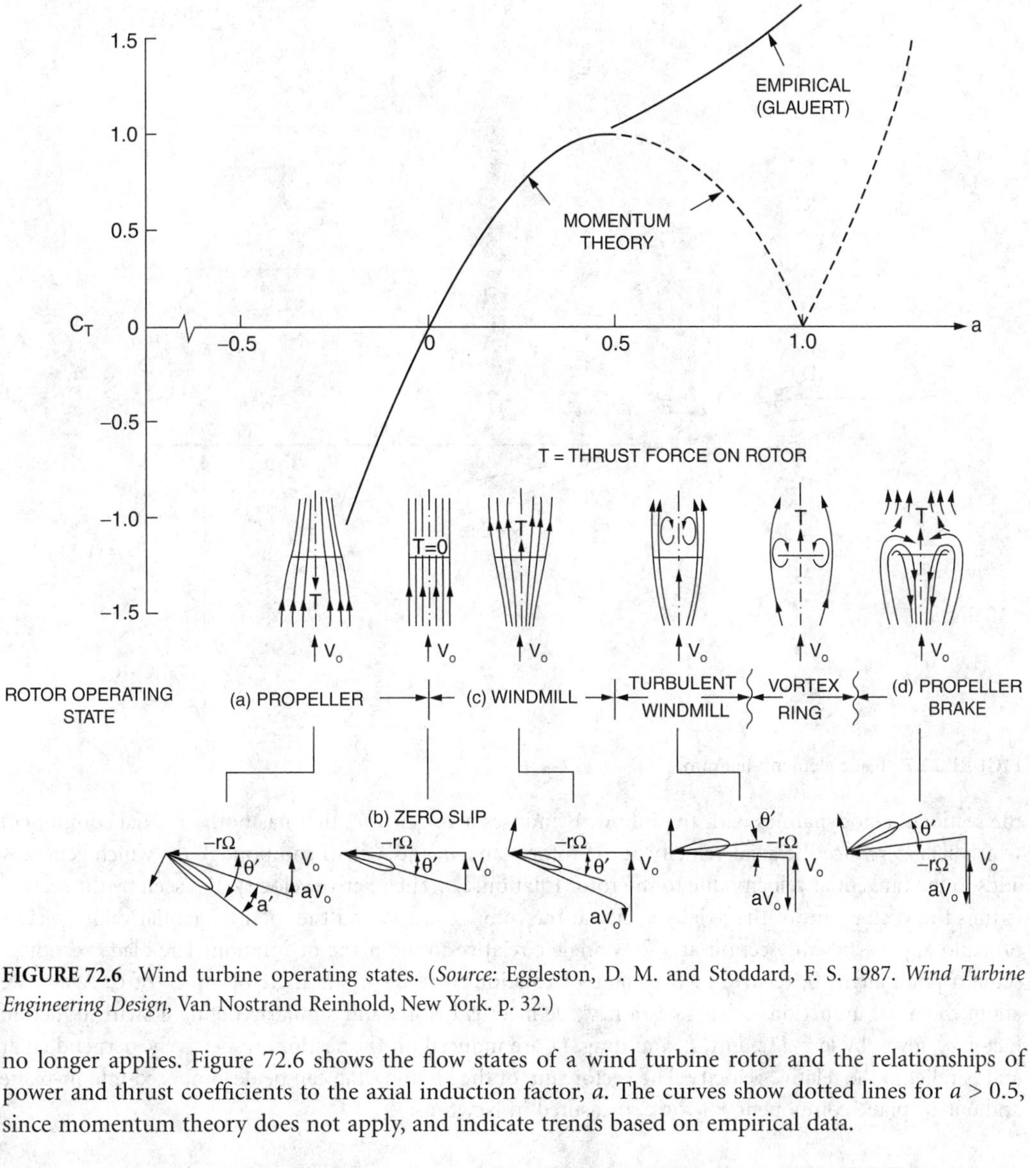

FIGURE 72.6 Wind turbine operating states. (*Source:* Eggleston, D. M. and Stoddard, F. S. 1987. *Wind Turbine Engineering Design*, Van Nostrand Reinhold, New York. p. 32.)

no longer applies. Figure 72.6 shows the flow states of a wind turbine rotor and the relationships of power and thrust coefficients to the axial induction factor, *a*. The curves show dotted lines for $a > 0.5$, since momentum theory does not apply, and indicate trends based on empirical data.

72.4 Power from a Wind Turbine: Blade Element Momentum (BEM) Theory

The Axial Momentum Theory presented above does not account for the swirl induced by the rotor, the aerodynamic drag acting on the rotor blades, or the radial variation in the flow through a rotor. **Blade Element Momentum (BEM) Theory** is often used to account for these effects.

The BEM method assumes first that the airstream flowing in to an actuator disk as shown in Figure 72.5 can be treated as a series of concentric and independent annular stream tubes and that mass and momentum do not flow across the boundaries of these imaginary stream tubes. The axial momentum theory presented in the previous section can be applied independently to the portion of the total flow in each such annulus. In that sense each radial section of a rotor blade on a horizontal axis wind turbine interacts with the freestream flow in one of these distinct annular stream tubes.

Figure 72.7 shows the airfoil cross-section of one such radial station, *r*, on a rotating wind turbine rotor blade, where it is assumed that the inflow is axial and that the blades are rigid. The interaction of

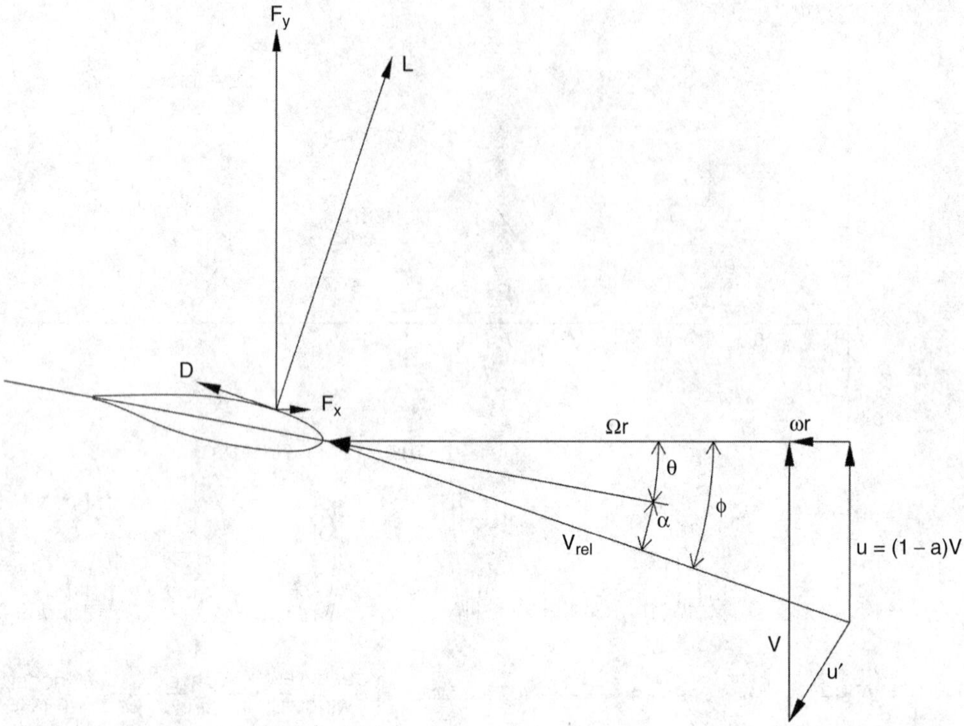

FIGURE 72.7 Blade element diagram.

the sectional aerodynamics with the inflow, V, induces a velocity, u', that has both an axial component which acts to reduce the axial velocity at the rotor plane, u, and a swirl component, ωr, which generally adds to the tangential velocity due to the rotor rotation, Ωr. The relative velocity, V_{rel}, seen by the section is thus the vector sum of the axial velocity at the rotor plane, u, and the total tangential velocity, $(\Omega + \omega)r$, and approaches the section at a flow angle ϕ relative to the plane of rotation. The blade section is set at a pitch angle, θ, relative to the plane of rotation to result in an angle of attack of α. Again, we adopt the axial induction factor notation, a, defined as before, and similarly define a swirl induction factor, a', given by $\omega = a'\Omega$. Lift, L, and drag, D, are induced on the section, respectively, perpendicular and parallel to the relative velocity. The vector sum of the lift and drag can be decomposed into in-plane and out-of-plane components of force, measured in Newtons:

$$dF_x = \frac{1}{2}\rho c \cdot [V^2(1-a)^2 + \Omega^2 r^2 (1+a')^2] \cdot \{C_L \sin\phi - C_D \cos\phi\} dr \tag{72.10}$$

$$dF_y = \frac{1}{2}\rho c \cdot [V^2(1-a)^2 + \Omega^2 r^2 (1+a')^2] \cdot \{C_L \cos\phi + C_D \sin\phi\} dr \tag{72.11}$$

where standard definitions of lift and drag coefficient, C_L and C_D, from airfoil theory are employed, c is the sectional chord length in meters, and ρ is the air density in kilograms per cubic meter. The in-plane force, dF_x, is the component that induces torque about the axis of rotation, while the out-of-plane force, dF_y, induces out-of-plane deflection of the rotor blades.

Figure 72.7 and Equation (72.10) and Equation (72.11) illustrate several important points. First, increasing the shaft speed, Ω, will significantly increase the force up to the point where the inflow angle, ϕ, becomes so small that the term $\{C_L \sin\phi - C_D \cos\phi\}$ in Equation (72.10) starts to decline faster than the $(\Omega r)^2$ term is increasing. This generally limits optimal rotor blade tip speeds to less than ten times the freestream wind speed, while most modern wind turbine rotor blades are optimized for a **Tip Speed Ratio**, $X = \Omega R/V$, of between five and eight. Second, for relatively high tip speed ratios, mid-span and outboard regions of the blade will experience significantly higher out-of-plane forces than in-plane loading.

It can be shown [Glauert, 1947] that Equation (72.4) and Equation (72.5) still apply to each annular stream tube, and conservation of momentum dictates that the sum of the out-of-plane forces acting on the same radial section of all of the B blades on a rotor must equal the change in momentum of the stream tube passing through those sections:

$$dF_y = 2\rho V^2 a(1 - a)\cdot(2\pi r)dr \tag{72.12}$$

It can also be shown [Glauert, 1947] that the change in angular momentum of the flow far downstream of the rotor is twice that at the rotor. Balancing the in-plane force with the change in angular momentum results in:

$$dQ = dF_x r = (2\omega r)\cdot\rho u\cdot(2\pi r)rdr = (2a'\Omega r)\cdot\rho u\cdot(2\pi r)rdr \tag{72.13}$$

Equation (72.10) through Equation (72.13) can be solved for the axial and swirl induction factors

$$a = \frac{\sigma C_y}{4\sin^2\phi + \sigma C_y} \tag{72.14}$$

$$a' = \frac{\sigma C_x}{4\sin\phi\cos\phi - \sigma C_x} \tag{72.15}$$

where σ is the sectional blade solidity, given by $\sigma = Bc/2\pi r$ and C_x and C_y are, respectively, the coefficients of in-plane and out-of-plane loading given by $C_x = \{C_L\sin\phi - C_D\cos\phi\}$ and $C_y = \{C_L\cos\phi + C_D\sin\phi\}$. Because of the dependence of ϕ on a and a', Equation (72.14) and Equation (72.15) must be solved iteratively. During the design process the twist angle and chord length of each blade section are adjusted so as to maximize the edgewise force, dF_x, at each station for the design tip speed ratio and pitch setting. For off-design tip speed ratios and pitch settings, the same equations can be used to determine the performance.

Once dF_x is known at all radial stations, the total mechanical power produced by B blades can be calculated by integrating the torque produced by this force from the blade root to the blade tip and multiplying by the shaft speed:

$$P = B\Omega \int_{r_{root}}^{r_{tip}} r\,dF_x \tag{72.16}$$

Rotor power curves are typically plotted against the tip speed ratio, X. These curves are unique for a given rotor with fixed pitch and can be performance-tailored by modifying the airfoil section characteristics, pitch angle of the blades, and the blade twist over the span of the rotor. Sample power coefficient–tip speed ratio curves for various rotors are shown in Figure 72.8. One will note that high-speed rotors with low solidity have higher conversion efficiency, although they generally produce less torque than high-solidity rotors, especially at startup.

Because of electrical considerations mentioned in the next section, wind turbines are often operated at a fixed shaft speed. One shortcoming of fixed-speed operation is that as the wind speed varies the rotor blades are frequently not operating at optimal tip speed ratio, and the power coefficient will be less than the design value, meaning that energy capture is reduced. Transient loads are also generally higher on fixed-speed turbines since the rotor cannot accelerate or decelerate in response to wind gusts. If the shaft speed can be varied with the wind speed to maintain constant tip speed ratio, then the energy

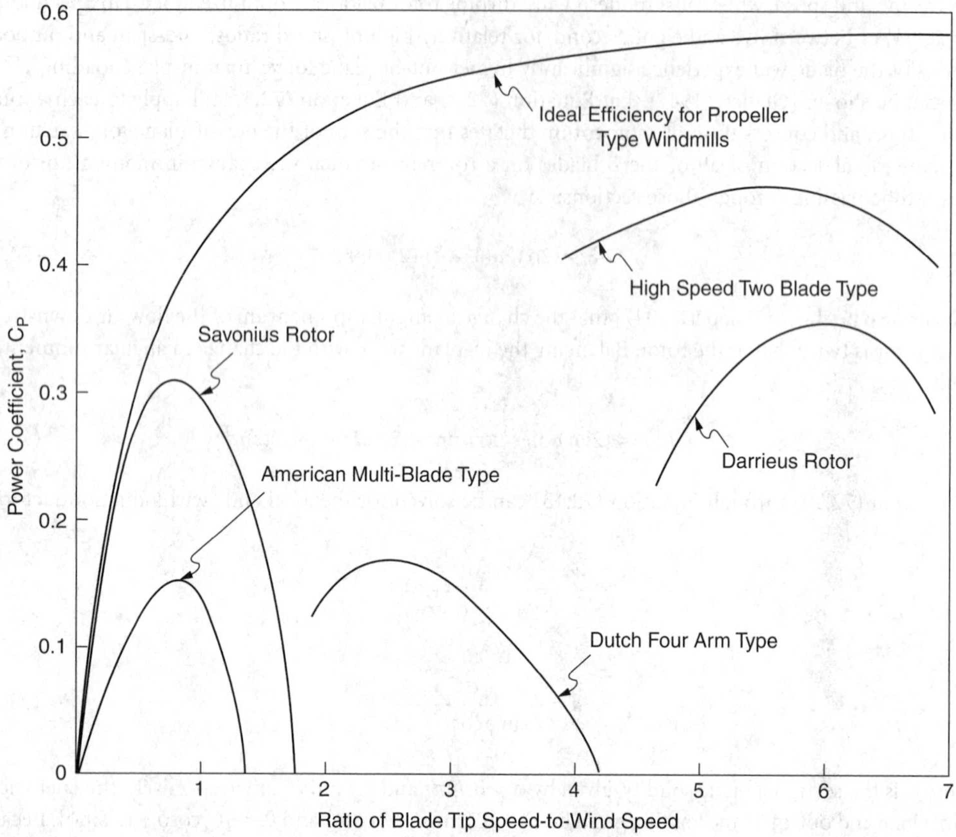

FIGURE 72.8 Typical performance curves for various wind turbine configurations. (*Source:* Eldridge, F. R. 1975. *Wind Machines,* National Science Foundation, Grant Number AER-75-12937. Energy Research and Development Administration, Washington. DC. p. 55.)

capture is maximized and transient loads can be absorbed better. These two types of turbines are generally referred to simply as **fixed speed** or **variable speed** wind turbines.

Steady-state BEM algorithms for wind turbine rotor design and analysis were incorporated into a code called PROPSH [Tangler, 1983], which in that form and in its current form of WT_Perf [Buhl, 2000] have been widely used for designing wind turbine rotors.

72.5 Electromechanical Considerations

The power calculated by the BEM analysis is the mechanical power produced by the rotor. The electrical power produced by a wind turbine is also a product of the efficiency of the mechanical and electrical systems. Figure 72.4 illustrates the construction of a typical horizontal-axis wind turbine.

Wind turbine drive train arrangements are generally divided into two classes: direct drive generators and geared systems. Most small wind turbines (10 kW and smaller) dispense with the gearbox and directly connect the aerodynamic rotor to the generator. Nearly all of the commercially available machines presently employ permanent magnet alternators. Since wind turbines both small and large have similar tip speeds, small turbines exhibit much higher shaft speeds on the order of hundreds of revolutions per minute, at which PM alternators can efficiently operate. These turbines are generally variable speed, generating a variable frequency current which is rectified and either used for battery charging or other DC applications or is reinverted at line frequency (50 or 60 Hz) for applications where the system is connected to the grid.

Large, utility-scale wind turbines rated on the order of 1 MW and larger typically exhibit low-speed shaft speeds of under 25 rpm, well below the shaft speeds typical of electric generators. Therefore, direct drive configurations in this size are rare but do exist.

In most large wind turbines, however, the power from the rotor is transmitted to the generator via a speed-increasing gearbox. These gear units usually have two or three stages producing gear ratios of between 1:20 and 1:100. Mechanical efficiencies of wind turbine gearboxes are typically on the order of 95% near rated.

Nearly all utility-scale wind turbine manufacturers employ induction generators, mostly squirrel cage designs, which are basically the same as AC induction motors. This is in contrast to most of the power industry, which employs synchronous generators. The primary reason that wind turbines have traditionally employed induction generators is that they are relatively easy to connect to an AC grid despite the constant fluctuations in wind speed and hence shaft torque. With no load applied, an induction motor or generator connected to an electrical grid with alternating current will turn at its synchronous speed, N_s, which is given in rpm by the formula

$$N_s = \frac{120 f_{AC}}{n_p} \text{ [rpm]} \tag{72.17}$$

where f_{AC} is the AC line frequency (i.e., 50 or 60 Hz depending upon where in the world one is located) and n_p is the number of poles in the generator. For example, a six-pole generator operating in the U.S. has a synchronous speed, N_s, of 1,200 rpm. If mechanical power is put into the machine as in the generator mode, then the shaft speed, N, will increase above the synchronous speed. The generator will continue to generate at line frequency. The slip, s, is the fraction of the synchronous speed below it at which the machine operates, given by $s = (N_s - N)/N_s$, where s is generally negative for generator mode according to electrical industry convention. The power increases as the slip increases, but most machines are designed to reach rated power at small values of slip typically on the order of 1 to 3%. This means that such turbines operate at a relatively fixed speed. Induction generators typically have efficiencies of 94 to 96% at rated.

The problem with variable speed operation using either a squirrel cage induction generator or a conventional synchronous generator (either wound rotor or permanent magnet rotor) is that the generator produces current with a variable frequency that cannot be directly connected to the grid. The current must be rectified and reinverted using expensive power electronics. Such power conversion also adds an additional source of electrical losses on the order of 2%.

An alternative electrical configuration that allows for a wider range of shaft speed employs the use of a doubly fed induction generator, in which the rotor and stator are wound separately. Only the stator is connected directly to the AC grid. The generator rotor is allowed to generate a variable frequency AC current, which is rectified and reinverted at the line frequency using power electronics and is then fed to the grid. Values of slip can range ±30% both subsynchronous and supersynchronous. In the subsynchronous mode, power is actually fed to the rotor from the line, while in the supersynchronous mode, power is fed from the rotor to the line. The stator always feeds power to the grid. At full supersynchronous slip of 30%, approximately 24% of the total power output of the generator is produced by the rotor, and the frequency converter power electronics must be sized to accommodate this power instead of 100% power conversion necessary with a variable speed squirrel cage electric machine. The industry trend in recent years has been towards wider use of variable speed operation.

A typical wind turbine power curve reflecting both aerodynamic power coefficient and electromechanical efficiencies is shown in Figure 72.9. This illustrates an important feature of wind power: a wind turbine will not always produce power at its nameplate rated capacity if the wind is not blowing strongly enough to produce that much power from the size of rotor installed on the turbine. The wind speed at which the turbine reaches its rated power is referred to as the **Rated Wind Speed**, below which the turbine will produce less than rated power. The rated wind speed of a turbine can be reduced by increasing the

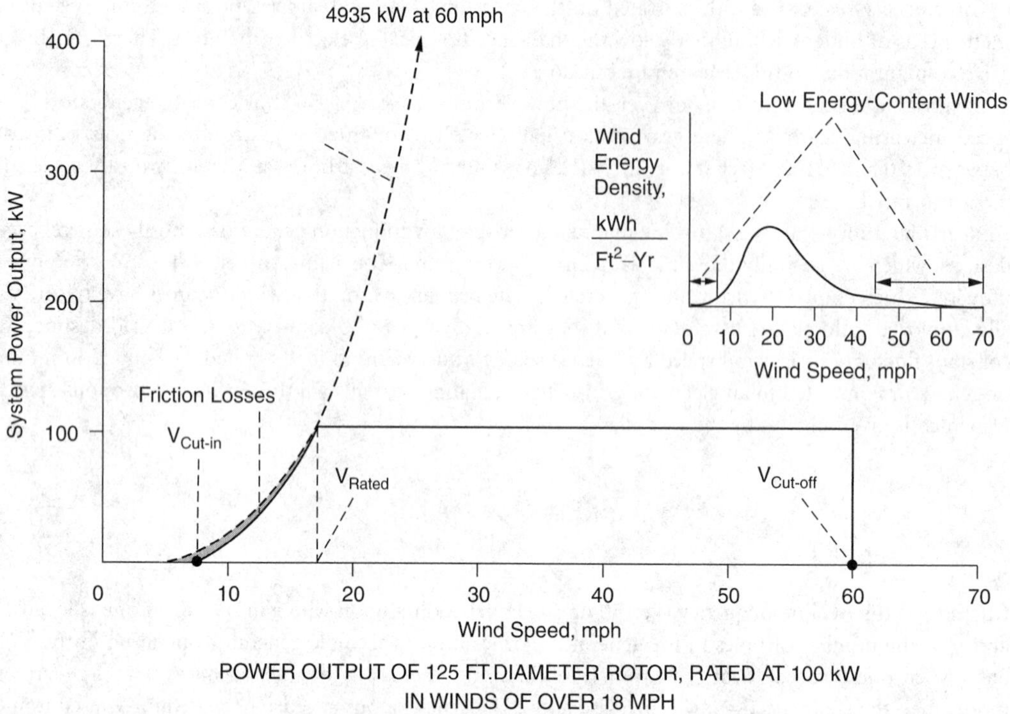

FIGURE 72.9 Wind turbine power curve. (*Source:* Eldridge, F. R. 1975. *Wind Machines,* National Science Foundation, Grant Number AER-75-12937. Energy Research and Development Administration, Washington, DC. p. 56.)

size of the rotor, but this will also increase the loads for which the structural and mechanical systems must be designed. For large, utility-scale wind turbines designed for IEC Class II, the rotors are typically sized so that the power loading is approximately 400 watts per square meter swept by the rotor. For example, a typical 1.5-MW turbine will feature a 70-m diameter rotor. Turbines designed for IEC Class III will typically employ rotors with 10% larger diameters.

The power output of a wind turbine can also generally be improved at most sites by installing it on a taller tower to take advantage of vertical gradients in wind speed. This also increases the cost of the tower, however, and so some optimal height generally exists. The trend in the industry has been that tower heights are typically 100 to 130% of the rotor diameter.

72.6 Power Regulation and Control

Several power regulation and control issues must be addressed in a modern HAWT. When the wind speed increases to a value at which the generator is producing rated power, some control action must occur so that the generator does not exceed its rated capacity and overheat. Typical methods of power regulation at **rated wind speed** are **pitch regulation**, **stall regulation**, and **yaw control** or **furling**. Pitch regulation is accomplished by providing rotating bearings at the blade root and actively changing the blade pitch angle relative to the wind, thus regulating power. Stall control is accomplished by designing the rotor so that aerodynamic stall is reached at rated wind speed and the rotor power is limited by airfoil stall. Yaw control or furling turns the entire rotor out of the wind either vertically or horizontally at rated wind speed and regulates power by reducing the rotor area exposed to the wind. The latter approach is generally only employed on small wind turbines.

In addition to power regulation, loss of generator load and overspeed protection is required in case the load on the rotor is lost during operation. Loss of load will allow the rotor speed and thrust to rise rapidly and will result in turbine destruction in moderate winds if not brought under control rapidly.

Pitch control rotors usually have a fail-safe pitch change to the feather position in an emergency shutdown, whereas stall control turbines use rotor tip brakes, a large mechanical brake that activates upon overspeed, or a combination of the two. Yaw control turbines simply yaw out of the wind to control overspeed.

Additional controls regulate turbine cut-in when the wind increases above starting wind speed, and cut-out in very high winds or in case of excessive vibration or other problem. Finally, most large HAWTs employ yaw drives to maintain alignment with the wind.

72.7 Energy Capture and Cost of Energy

Once a turbine power curve is established, it can be used with wind frequency information in order to estimate the annual energy production available at a given site. Wind speeds fluctuate continuously but have been shown to generally follow a Weibull frequency distribution. One special type of Weibull distribution is the Rayleigh distribution, which is often used for energy capture estimation. The Rayleigh distribution is based only on average wind speed at the hub height where the turbine is to be located and can be calculated using the following equation:

$$f(V) = \frac{\pi}{2}\frac{V}{V_{ave}^2}\exp\left[-\frac{\pi}{4}\left(\frac{V}{V_{ave}}\right)^2\right] \tag{72.18}$$

where V is the center of a wind speed bin of width 1 m/sec, ft/sec, and so on in consistent units, and V_{ave} is the average wind speed, also in consistent units.

Assuming a Rayleigh distribution at hub height, annual energy can be estimated by first adjusting the power curve for air density variation due to site elevation, if required, and then using the Rayleigh frequency distribution and power curve to determine the number of hours per year the turbine will operate at each wind speed interval and summing the number of operating hours times the power level for each interval over all wind speeds. This results in an estimation of the **Gross Annual Energy Production (GAEP).** The **Net Annual Energy Production** (**NAEP**) is determined by adjusting GAEP for turbine availability and subtracting losses such as line losses in the electrical collection system. Availability of 95 to 97% is common.

Once NAEP and the initial capital cost of the wind turbine system are known, the cost of energy (COE) can be calculated. The COE from a wind turbine is calculated by adding the life cycle annual cost of the initial capital investment (including turbine, tower, construction and development costs, etc.), annualized equipment replacement costs (referred to as Levelized Parts Replacement or LPR), and annual operation and maintenance (O&M) costs and dividing this total annualized cost by the net annual energy production. Units are usually expressed in dollars or cents per kWh. The equation is as follows:

$$\mathrm{COE} = \frac{\mathrm{ACC} + \mathrm{LPR} + \mathrm{O\&M}}{\mathrm{NAEP}} \tag{72.19}$$

where ACC is the annualized capital cost, which is a function of the initial capital investment and the cost of financing the project.

The COE from windpower has declined significantly during the past 20 years, from roughly $0.40 per kWh in 1980 to approximately $0.035 per kWh today at sites where wind farms are typically being developed in the U.S.

72.8 Design and Certification of Wind Turbines

The design of wind turbines has undergone a significant transformation during the past decade, with the development by the International Electrotechnical Commission of Standard 61400 for the design of Wind Turbine Generator Systems. The 61400 standard includes the following Parts:

- IEC 61400-1 (1999) Part 1: Safety requirements (Edition 2)
- IEC 61400-2 (1996) Part 2: Safety of small wind turbines
- IEC 61400-11 (2002) Part 11: Acoustic noise measurement techniques
- IEC 61400-12 (1998) Part 12: Wind turbine power performance testing
- IEC 61400-13 (2001) Part 13: Measurement of mechanical loads
- IEC 61400-21 (2001) Part 21: Measurement and assessment of power quality characteristics of grid connected wind turbines
- IEC 61400-23 (2001) Part 23: Full-scale structural testing of rotor blades
- IEC 61400-24 (2002) Part 24: Lightning protection

A second edition of Part 2 has been developed and should be published during 2004.

These standards provide guidelines defining the operating conditions for which wind turbines must be designed and tested. Manufacturers of turbines work with certification agencies to ensure that the standards have been followed in the development of new turbine designs, and Type Certificates are issued to manufacturers certifying that the turbine as designed is suitable for operation in one of the IEC Classes defined in Table 72.2. They also issue Site Suitability Certificates certifying that a given turbine design is suitable for use at a particular site where meteorological data have been collected and against which the turbine design has been verified.

The IEC 61400-1 standard prescribes the Normal Wind Conditions, including the wind speed frequency distribution, the normal vertical wind speed profile (NWP), and the Normal Turbulence Model (NTM). The wind speed probability distribution is presumed to follow a Rayleigh distribution as shown in Equation (72.18) for the purposes of design load calculations. During site suitability analyses, however, actual meteorological data may exist from which the actual wind speed frequency distribution may be calculated. IEC 61400-1 also prescribes a vertical wind profile exponent of $\alpha = 0.2$, where

$$V(z) = V_{hub}(z/z_{hub})^{\alpha} \tag{72.20}$$

where z is the height above ground, z_{hub} is the height of the rotor hub, $V(z)$ is the wind speed at height z, and V_{hub} is the wind speed at hub height. In the NTM the standard deviation, σ_l, of longitudinal wind speed is prescribed to vary with wind speed as follows:

$$\sigma_l = I_{15}(15 + aV_{hub})/(a + 1) \tag{72.21}$$

where I_{15} and a are given in Table 72.2 for turbulence classes A and B and V_{hub} is in units of m/sec.

IEC 61400-1 also defines a number of extreme wind conditions, including Extreme Wind Speed (EWS), Extreme Operating Gust (EOG), Extreme Direction Change (EDC), Extreme Coherent Gust (ECG), Extreme Coherent Gust with Direction Change (ECD), and Extreme Wind Shear (EWS) for both horizontal and vertical shear. Most of the extreme wind conditions are defined for both 1-year and 50-year frequencies of occurrence. The Standard also prescribes a series of Design Load Conditions (DLCs) for which a turbine must be designed, including several different DLCs for each of (1) power production, (2) power production with a system fault, (3) startup, (4) normal shutdown, (5) emergency shutdown, (6) parked or idling, (7) parked with fault conditions, and (8) transport, assembly, maintenance, and repair. Partial safety factors for loads, material properties, and consequences of failure are defined. Requirements for fatigue analysis are also defined.

72.9 Dynamic and Structural Analysis of Wind Turbines

The Blade Element Momentum (BEM) method discussed in Section 72.4 assumes the inflow is normal to the rotor plane, spatially uniform, and steady. In reality, the wind approaching a rotor typically is highly turbulent. It varies spatially across the face of the rotor, includes significant horizontal and vertical components in the rotor plane, and varies temporally at fairly high frequencies. In the process of engineering wind turbine designs, a number of computer codes can be employed that will generate models of the inflow. IECWind can be used to generate relatively simple models of the various IEC extreme wind conditions. These models prescribe a time history of the hub height wind speed as well as corresponding time histories of the inflow yaw angle, vertical component of the inflow, and the horizontal and vertical shear. SNLWIND-3D [Veers, 1988; Kelley, 1993] can be used to generate time histories of full field turbulence. Both codes are available from the National Renewable Energy Laboratory.

BEM methods can still be used to analyze the loads on a blade in a turbulent flow in a sense that is discrete both spatially and temporally. A number of computer codes exist that perform this analysis. One of the most commonly used in the U.S. is AeroDyn [Laino and Hansen, 2002], which is available through NREL. The blade is modeled aerodynamically as a series of elements, and at any moment in time the aerodynamic loads acting on each element on each blade must satisfy the BEM equations presented above. AeroDyn and similar codes use the inflow models produced by codes such as SNLWIND-3D or IECWind as input. They are also coupled to separate codes that can model the dynamic response of the wind turbine system to the loads induced by the wind. A number of sophisticated dynamic simulators exist. These codes generally model the tower, drive train, and rotor blades with a series of elements, joints, and constraints which approximate the mass, inertial, stiffness, and mechanical characteristics of the various components. The models also include simulations of the electrical and control characteristics of the system.

When a simulation is run, at any one time step codes such as AeroDyn receive information from the dynamic simulator about the position and motion of every element on each rotor blade and also receive information from the wind inflow file about the inflow characteristics. AeroDyn then calculates the aerodynamic loads on each element of each blade, taking into account all effects that influence the wind velocity at the blade element, including the inflow and the rotation of the blade, as well as the rates of out-of-plane deflection and torsional deflection of the blade. The calculated loads are then provided as input back to the dynamic simulator, which integrates the response of the turbine system forward to the next time step. Time steps are typically very small, on the order of 1% of the period of the rotor revolution or less. In this manner, one can conduct a time-stepping simulation of the operation of a wind turbine during either an extreme wind condition or during turbulent operation.

Such simulations can be used to generate time histories of the turbine dynamics, including critical information such as the maximum deflection of the rotor blades, as well as time histories of key loads on the blades, hub, tower top, tower base, and elsewhere. They can also be used to generate simulations of the power performance of the turbine during operation in turbulent conditions. Simulations are typically run for all of the DLCs required by IEC as well as multiple simulations of operation in turbulent conditions over a range of mean wind speeds at which the turbine will operate.

The extreme loads and time histories of transient loads are used as input to models of the turbine structure. Finite Element Analysis methods are applied to all the major components in a wind turbine, especially the blades, hub, shaft, main frame, and tower. The structural integrity of these components is checked against both extreme loads and the fatigue induced by transient loads.

The dynamic and structural simulations serve as the basis for finalizing wind turbine designs before prototypes are built and tested. Type certification is granted based upon the results of these analyses as well as testing conducted in accordance with IEC Standard 61400, Parts 11, 12, 13, 21, and 23, as discussed in Section 72.8.

72.10 Applications and Development of Wind Power

In excess of 100,000 small electric-generating wind turbine systems of less than 20 kW capacity are in use worldwide. Most small turbines are installed as single units or sometimes as a group of a small number of units. Many small wind turbines — particularly those in the 3 to 20 kW rating — are connected to the power grid, offsetting energy that would otherwise be supplied by the utility. Many other small turbines are installed in stand-alone applications, such as the 10-kW turbine shown in Figure 72.10, supplying power to remote locations where the cost of utility connection is uneconomic. Sometimes coupled with diesel generators or solar electric panels in hybrid configurations, they are capable of

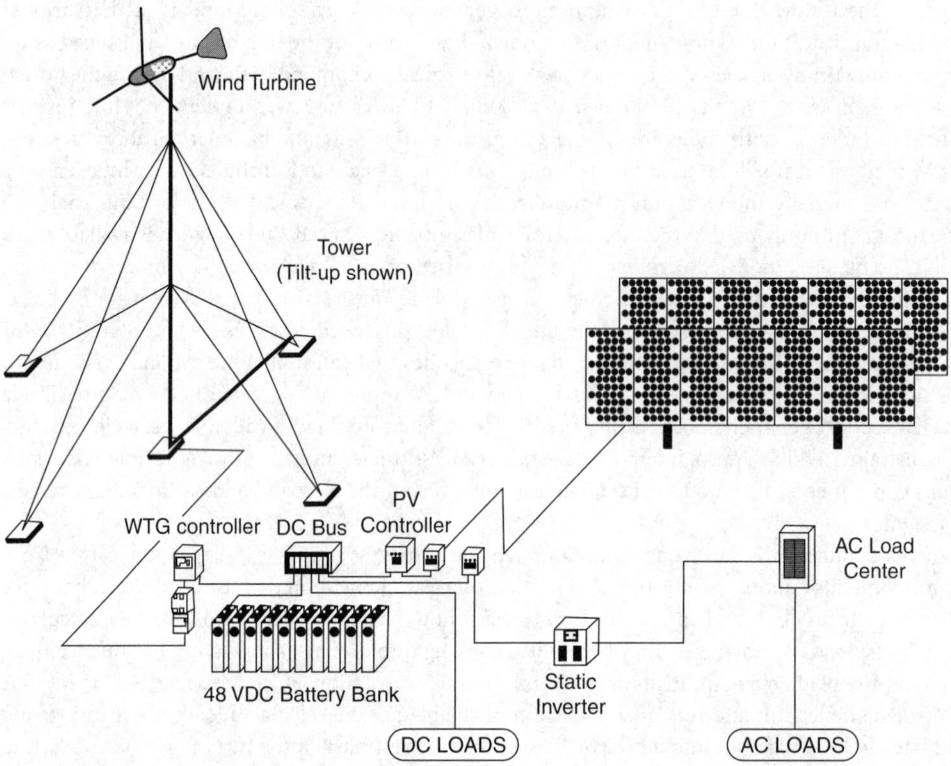

FIGURE 72.10 Example of a stand-alone hybrid wind system. (*Source:* Bergey, M. 1989. *An Overview of Wind Power for Remote Site Telecommunications Facilities,* Renewable Energy Power Supplies for Telecommunications Conference, British Wind Energy Association, London. p. 11. With permission.)

providing reliable power for basic necessities such as water pumping, lighting, refrigeration of medical supplies, or power for remote telecommunication facilities.

The vast majority of the world's wind power capacity is installed in the form of large, utility-scale wind turbines. Worldwide installed wind power capacity has grown at an annual rate of 30% in recent years, topping 31,000 MW at the end of 2002, as shown in Figure 72.11. Development has been greatest in Europe — led by Germany, Spain, and Denmark — the United States, and India (Figure 72.12). In the United States, much of the recent development has occurred in the form of large wind farms (Figure 72.1) consisting of dozens or hundreds of large wind turbines and with total capacities of hundreds of megawatts. In Europe many turbines are installed singly or in small numbers, while the most recent trend is the development of off-shore wind farms utilizing multi-megawatt turbines. At the time of this writing, off-shore wind farms are being built with turbines rated as large as 3.6 MW each. In Europe, some off-shore sites are not too deep to sink foundations, and off-shore sites offer the advantage of consistent winds, generally lower turbulence levels, and elimination of land use issues. Interest is also growing in off-shore applications on the Eastern seaboard of the U.S.

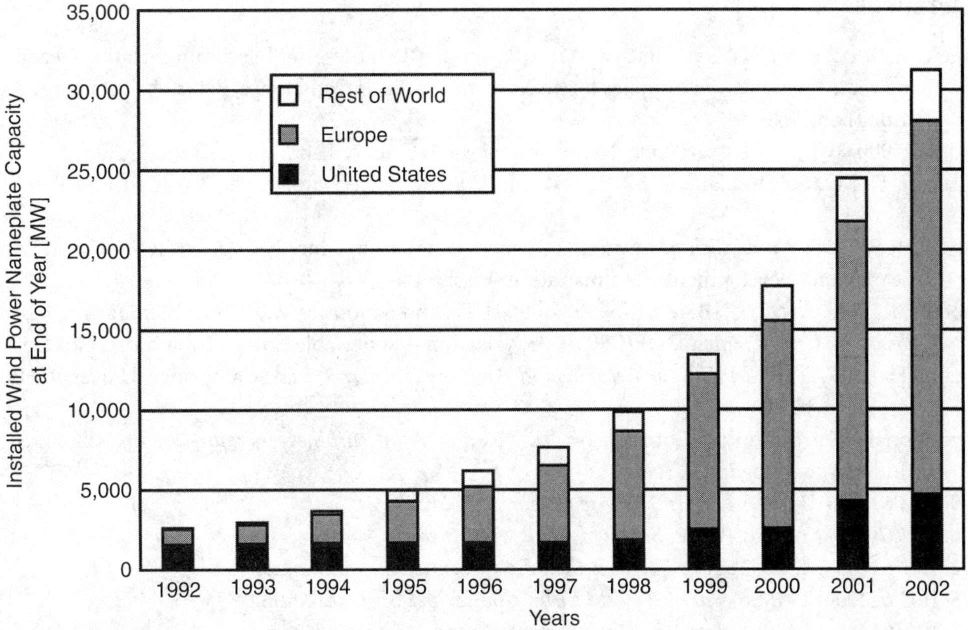

FIGURE 72.11 Growth in worldwide installed wind power capacity (*Source:* "Windicator" column published quarterly in *Windpower Monthly.)*

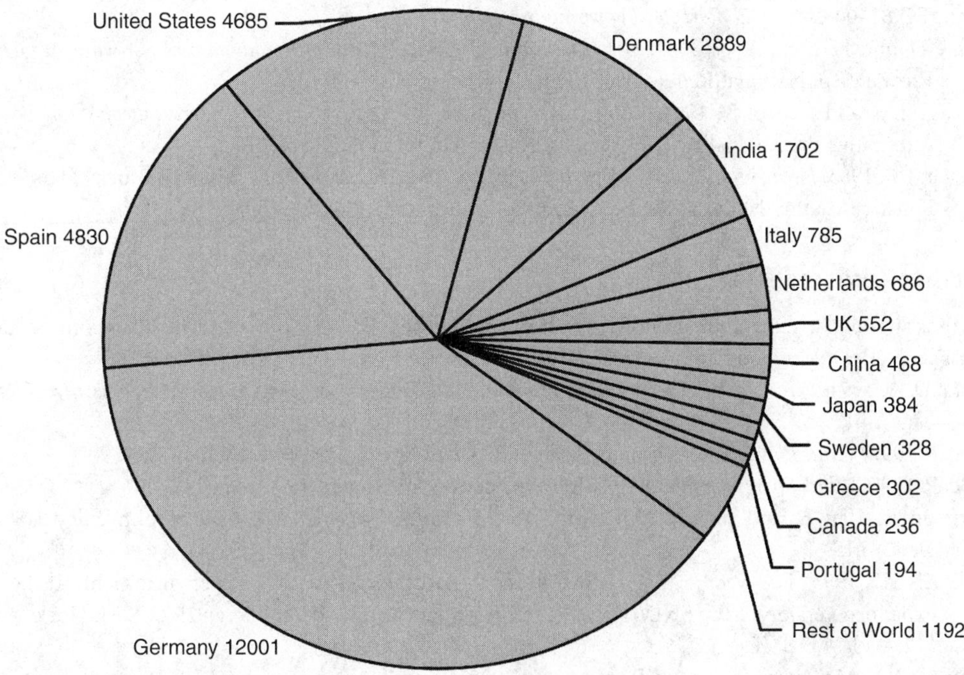

FIGURE 72.12 Distribution of worldwide windpower capacity in megawatts as of December 31, 2002. (*Source:* "Windicator" column published quarterly in *Windpower Monthly.*)

References

Bergey, M. L. S. 1989. An Overview of Wind Power for Remote Site Telecommunications Facilities. Renewable Energy Power Supplies for Telecommunications Conference, British Wind Energy Association, London.

Buhl, M. 2000. *WT_Perf User's Guide.* National Renewable Energy Laboratory, Golden, CO.

Eggleston, D. M. and Stoddard, F. S. 1987. *Wind Turbine Engineering Design,* Van Nostrand Reinhold, New York.

Eldridge, F. R. 1975. *Wind Machines,* National Science Foundation, Grant Number AER-75-12937. Energy Research and Development Administration, Washington, DC.

Elliott, D. L., Holladay, C. G., Barchet, W. R. Foote, H. D., and Sandusky, W. F. 1986. *Wind Energy Resource Atlas of the United States,* DOE/CH 10093-4. National Renewable Energy Laboratory, Golden, CO.

Glauert, H. 1947. *The Elements of Aerofoil and Airscrew Theory,* 2nd ed. Cambridge University Press, Cambridge, U.K.

International Electrotechnical Commission. *IEC 61400: Wind turbine generator systems.* IEC, Geneva, including the following Parts:

1999. IEC 61400-1 (1999) *Part 1: Safety requirements (Edition 2)*

1996. IEC 61400-2 (1996) *Part 2: Safety of small wind turbines*

2002. IEC 61400-11 (2002) *Part 11: Acoustic noise measurement techniques*

1998. IEC 61400-12 (1998) *Part 12: Wind turbine power performance testing*

2001. IEC 61400-13 (2001) *Part 13: Measurement of mechanical loads*

2001. IEC 61400-21 (2001) *Part 21: Measurement and assessment of power quality characteristics of grid connected wind turbines*

2001. IEC 61400-23 (2001) *Part 23: Full-scale structural testing of rotor blades*

2002. IEC 61400-24 (2002) *Part 24: Lightning protection*

Laino, D. and Hansen, A. C. 2002. *User's Guide to the Wind Turbine Aerodynamics Software AeroDyn, Version 12.5,* National Renewable Energy Laboratory, Golden, CO.

Rohatgi, J. S. and Nelson, V. 1994. *Wind Characteristics: An Analysis for the Generation of Wind Power,* Alternative Energy Institute, West Texas A & M University, Canyon, TX.

Tangler, J. L. 1983. *Horizontal Axis Wind Turbine Performance Prediction Code PROPSH,* Rocky Flats Wind Research Center, National Renewable Energy Laboratory, Golden, CO.

Further Information

For those wishing to expand their knowledge of this topic, the following sources are highly recommended:

Freris, L. L. 1990. *Wind Energy Conversion Systems,* Prentice Hall, Hemel Hempstead, U.K.

Spera, D. A. 1994. *Wind Turbine Technology: Fundamental Concepts of Wind Turbine Engineering,* ASME, New York.

Gipe, P. 1993. *Wind Power for Home and Business,* Chelsea Green, Post Mills, VT.

Gipe, P. 1995. *Wind Energy Comes of Age,* John Wiley & Sons, New York.

Harrison, R., Hau, E., and Snel, H. 2001. *Large Wind Turbines: Design and Economics,* John Wiley & Sons, New York.

Kelley, N. D. 1993, Full vector (3-D) simulation in natural and wind farm environments using an expanded version of the SNLWIND (Veers) turbulence code, *Wind Energy 1993,* S.M. Hock (ed.), SED-Vol.14, ASME.

Manwell, J. F., McGowan, J. G., and Rogers, A. L. 2002. *Wind Energy Explained,* John Wiley & Sons, New York.

Righter, R. 1996. *Wind Energy in America: A History,* University of Oklahoma Press, Norman, OK.

Veers, P.S. 1988. *Three-Dimensional Wind Simulation, SAND 88-0512,* Sandia National Laboratory, Albuquerque, NM.

Windpower Monthly News Magazine, Redding, CA.

73

Hydraulic Turbines

Roger E.A. Arndt
St. Anthony Falls Laboratory
University of Minnesota

A hydraulic turbine is a mechanical device that converts the potential energy associated with a difference in water elevation (**head**) into useful work. Modern hydraulic turbines are the result of many years of gradual development. Economic incentives have resulted in the development of very large units (exceeding 800 megawatts in capacity) with efficiencies that are sometimes in excess of 95%.

The emphasis on the design and manufacture of very large turbines is shifting to the production of smaller units, especially in developed nations, where much of the potential for developing large base-load plants has been realized. At the same time, the escalation in the cost of energy has made many smaller sites economically feasible and has greatly expanded the market for smaller turbines. The increased value of energy also justifies the cost of refurbishment and increasing the capacity of older facilities. Thus, a new market area is developing for updating older turbines with modern replacement runners having higher efficiency and greater capacity.

73.1 General Description

Typical Hydropower Installation

As shown schematically in Figure 73.1, the hydraulic components of a hydropower installation consist of an intake, penstock, guide vanes or distributor, turbine, and draft tube. Trash racks are commonly provided to prevent ingestion of debris into the turbine. Intakes usually require some type of shape transition to match the passageway to the turbine and also incorporate a gate or some other means of stopping the flow in case of an emergency or turbine maintenance. Some types of turbines are set in an open flume; others are attached to a closed-conduit penstock.

Turbine Classification

There are two types of turbines, denoted as impulse and reaction. In an *impulse turbine* the available head is converted to kinetic energy before entering the runner; the power available is extracted from the

0-8493-1586-7/05/$0.00+$1.50

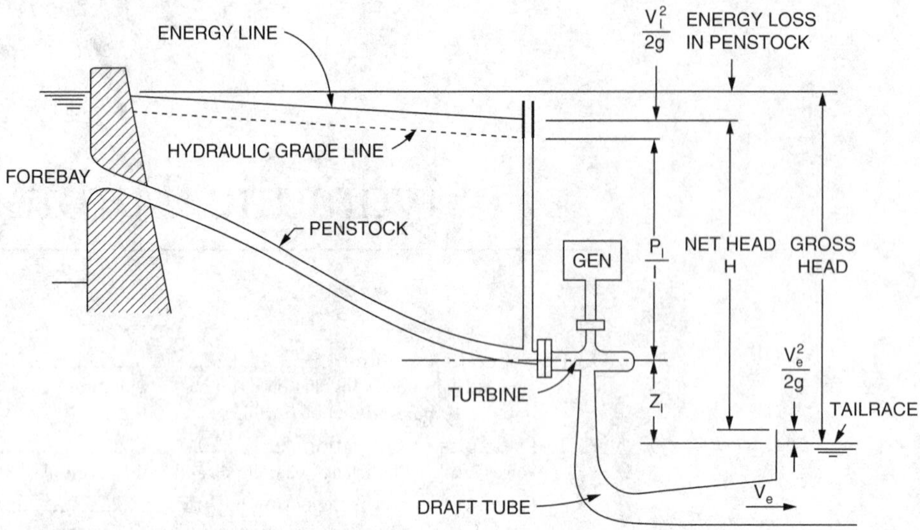

FIGURE 73.1 Schematic of a hydropower installation.

flow at approximately atmospheric pressure. In a *reaction turbine* the runner is completely submerged and both the pressure and the velocity decrease from inlet to outlet. The velocity head in the inlet to the turbine runner is typically less than 50% of the total head available.

Impulse Turbines

Modern impulse units are generally of the Pelton type and are restricted to relatively high head applications (Figure 73.2). One or more jets of water impinge on a wheel containing many curved buckets. The jet stream is directed inwardly, sideways, and outwardly, thereby producing a force on the bucket, which in turn results in a torque on the shaft. All kinetic energy leaving the runner is "lost." A **draft tube** is generally not used since the runner operates under approximately atmospheric pressure and the head represented by the elevation of the unit above tailwater cannot be utilized. (In principle, a draft tube

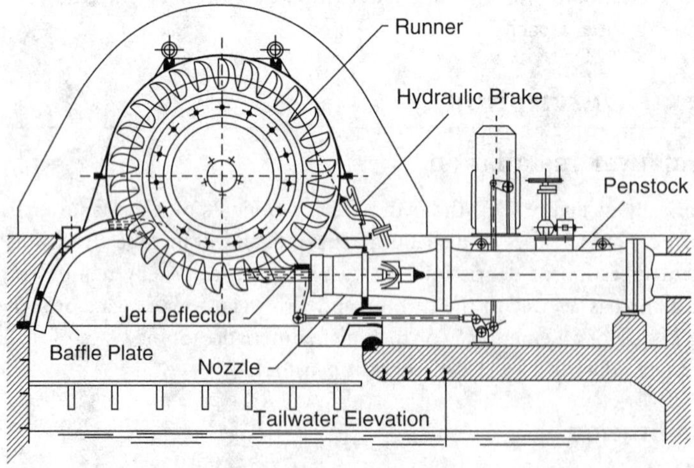

FIGURE 73.2 Cross-section of a single-wheel, single-jet Pelton turbine. This is the third highest head Pelton turbine in the world, $H = 1447$ m, $n = 500$ rpm, $P = 35.2$ MW, $N_s \sim 0.038$. (Courtesy of Vevey Charmilles Engineering Works. Adapted from Raabe, J. 1985. *Hydro Power: The Design, Use, and Function of Hydromechanical, Hydraulic, and Electrical Equipment*, VDI Verlag, Dusseldorf, Germany.)

could be used, which requires the runner to operate in air under reduced pressure. Attempts at operating an impulse turbine with a draft tube have not met with much success.) Since this is a high-head device, this loss in available head is relatively unimportant. As will be shown later, the Pelton wheel is a low–specific speed device. Specific speed can be increased by the addition of extra nozzles, the specific speed increasing by the square root of the number of nozzles. Specific speed can also be increased by a change in the manner of inflow and outflow. Special designs such as the Turgo or crossflow turbines are examples of relatively high specific speed impulse units [Arndt, 1991].

Most Pelton wheels are mounted on a horizontal axis, although newer vertical-axis units have been developed. Because of physical constraints on orderly outflow from the unit, the number of nozzles is generally limited to six or fewer. Whereas **wicket gates** control the power of a reaction turbine, the power of the Pelton wheel is controlled by varying the nozzle discharge by means of an automatically adjusted needle, as illustrated in Figure 73.2. Jet deflectors or auxiliary nozzles are provided for emergency unloading of the wheel. Additional power can be obtained by connecting two wheels to a single generator or by using multiple nozzles. Since the needle valve can throttle the flow while maintaining essentially constant jet velocity, the relative velocities at entrance and exit remain unchanged, producing nearly constant efficiency over a wide range of power output.

Reaction Turbines

Reaction turbines are classified according to the variation in flow direction through the runner. In radial- and mixed-flow runners, the flow exits at a radius different from the radius at the inlet. If the flow enters the runner with only radial and tangential components, it is a radial-flow machine. The flow enters a mixed-flow runner with both radial and axial components. Francis turbines are of the radial- and mixed-flow type, depending on the design specific speed. A Francis turbine is illustrated in Figure 73.3.

Axial-flow propeller turbines are generally either of the fixed-blade or Kaplan (adjustable-blade) variety. The "classical" propeller turbine, illustrated in Figure 73.4, is a vertical-axis machine with a scroll case and a radial wicket gate configuration that is very similar to the flow inlet for a Francis turbine. The flow enters radially inward and makes a right-angle turn before entering the runner in an axial direction. The Kaplan turbine has both adjustable runner blades and adjustable wicket gates. The control system is designed so that the variation in blade angle is coupled with the wicket gate setting in a manner that achieves best overall efficiency over a wide range of flow rates.

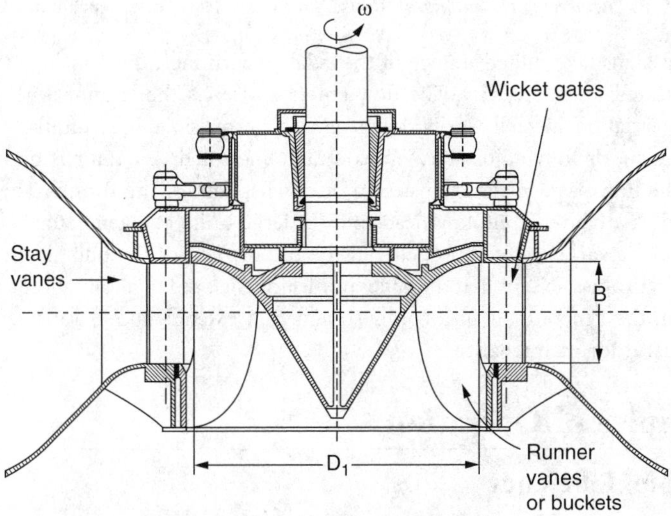

FIGURE 73.3 Francis turbine, N_s ~ 0.66. (Adapted from Daily, J. W. 1950. Hydraulic machinery. In *Engineering Hydraulics*, H. Rouse, Ed., Wiley, New York. Reprinted with permission.)

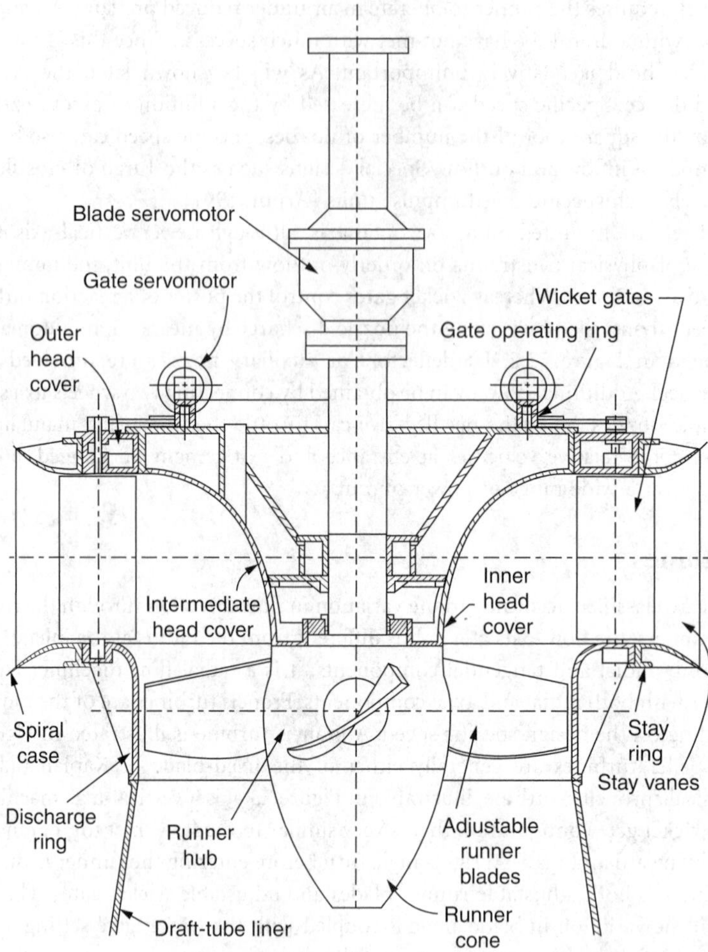

FIGURE 73.4 Smith–Kaplan axial-flow turbine with adjustable-pitch runner blades, $N_S \sim 2.0$. (From Daily, J. W. 1950. Hydraulic machinery. In *Engineering Hydraulics*, H. Rouse, Ed., Wiley, New York. Reprinted with permission.)

Some modern designs take full advantage of the axial-flow runner; these include the tube, bulb, and Straflo types illustrated in Figure 73.5. The flow enters and exits the turbine with minor changes in direction. A wide variation in civil works design is also permissible. The tubular type can be fixed-propeller, semi-Kaplan, or fully adjustable. An externally mounted generator is driven by a shaft that extends through the flow passage either upstream or downstream of the runner. The bulb turbine was originally designed as a high-output, low-head unit. In large units, the generator is housed within the bulb and is driven by a variable-pitch propeller at the trailing end of the bulb. Pit turbines are similar in principle to bulb turbines, except that the generator is not enclosed in a fully submerged compartment (the bulb). Instead, the generator is in a compartment that extends above water level. This improves access to the generator for maintenance.

73.2 Principles of Operation

Power Available, Efficiency

The power that can be developed by a turbine is a function of both the head and flow available:

$$P = \eta \rho g Q H \tag{73.1}$$

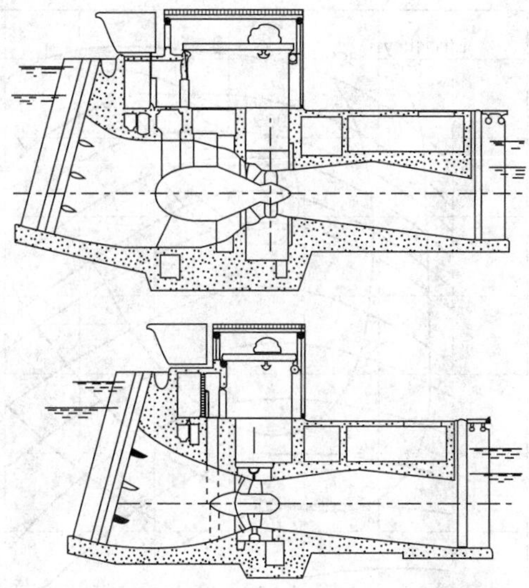

FIGURE 73.5 Comparison between bulb (upper) and Straflo (lower) turbines. (Courtesy U.S. Dept. of Energy.)

where η is the turbine efficiency, ρ is the density of water (kg/m^3), g is the acceleration due to gravity (m/sec^2), Q is the flow rate (m^3/sec), and H is the net head in meters. *Net head* is defined as the difference between the *total head* at the inlet and the tailrace, as illustrated in Figure 73.1. Various definitions of net head are used in practice, which depend on the value of the exit velocity head, $V_e^2/2g$, that is used in the calculation. The International Electrotechnical Test Code uses the velocity head at the draft tube exit.

The efficiency depends on the actual head and flow utilized by the turbine runner, flow losses in the draft tube, and the frictional resistance of mechanical components.

Similitude and Scaling Formulas

Under a given head, a turbine can operate at various combinations of speed and flow depending on the inlet settings. For reaction turbines the flow into the turbine is controlled by the wicket gate angle, α. The nozzle opening in impulse units typically controls the flow. Turbine performance can be described in terms of nondimensional variables,

$$\psi = \frac{2gH}{\omega^2 D^2} \tag{73.2}$$

$$\phi = \frac{Q}{\sqrt{2gH}D^2} \tag{73.3}$$

where ω is the rotational speed of the turbine in radians per second and D is the diameter of the turbine.

The hydraulic efficiency of the runner alone is given by

$$\eta_h = \frac{\phi}{\sqrt{\psi}}\left(C_1 \cos\alpha_1 - C_2 \cos\alpha_2\right) \tag{73.4}$$

where C_1 and C_2 are constants that depend on the specific turbine configuration, and α_1 and α_2 are the inlet and outlet angles that the absolute velocity vectors make with the tangential direction. The value of cos α_2 is approximately zero at peak efficiency. The terms ϕ, ψ_2, α_1, and α_2 are interrelated. Using model test data, isocontours of efficiency can be mapped in the $\phi\psi$ plane. This is typically referred to as a *hill diagram*, as shown in Figure 73.6.

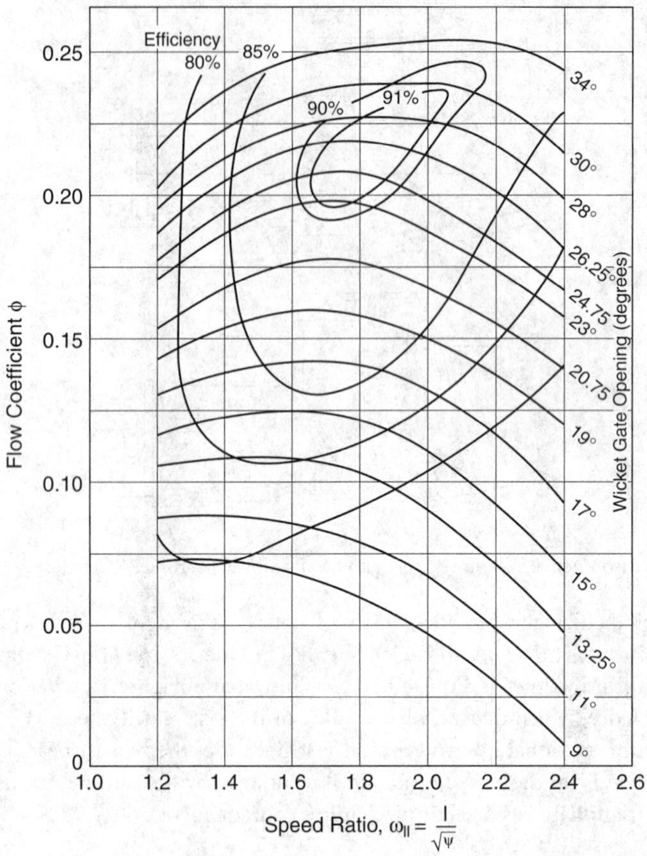

FIGURE 73.6 Typical hill diagram. Information of this type is obtained in a turbine test stand (See Figure 73.9). (Adapted from Wahl, T. L. 1994. Draft tube surging times two: The twin vortex problem. *Hydro Rev.* 13(1):60–69, 1994. With permission.)

The **specific speed** is defined as

$$N_s = \frac{\omega\sqrt{Q}}{(2gH)^{3/4}} = \sqrt{\frac{\phi}{\psi}} \tag{73.5}$$

A given specific speed describes a specific combination of operating conditions that ensures similar flow patterns and the same efficiency in geometrically similar machines regardless of the size and rotational speed of the machine. It is customary to define the design specific speed in terms of the value at the design head and flow where peak efficiency occurs. The value of specific speed so defined permits a classification of different turbine types.

The specific speed defined herein is dimensionless. Many other forms of specific speed exist that are dimensional and have distinct numerical values depending on the system of units used [Arndt, 1991]. (The literature also contains two other minor variations of the dimensionless form. One differs by a factor of $1/\pi^{1/2}$ and the other by $2^{3/4}$.) The similarity arguments used to arrive at the concept of specific speed indicate that a given machine of diameter D operating under a head H will discharge a flow Q and produce a torque T and power P at a rotational speed given by

$$Q = \phi D^2\sqrt{2gH} \tag{73.6}$$

$$T = T_{11}\rho D^3 2gH \tag{73.7}$$

$$P = P_{11}\rho D^2 (2gH)^{3/2} \tag{73.8}$$

$$\omega = \frac{2u_1}{D} = \omega_{11}\frac{\sqrt{2gH}}{D}, \quad \left[\omega_{11} = \frac{1}{\sqrt{\psi}}\right] \tag{73.9}$$

with

$$P_{11} = T_{11}\omega_{11} \tag{73.10}$$

where T_{11}, P_{11}, and ω_{11} are also nondimensional. (The reader is cautioned that many texts, especially in the American literature, contain dimensional forms of T_{11}, P_{11}, and ω_{11}.) In theory, these coefficients are fixed for a machine operating at a fixed value of specific speed, independent of the size of the machine. Equation (73.6) through Equation (73.10) can be used to predict the performance of a large machine using the measured characteristics of a smaller machine or model.

73.3 Factors Involved in Selecting a Turbine

Performance Characteristics

Impulse and reaction turbines are the two basic types of turbines. They tend to operate at peak efficiency over different ranges of specific speed. This is due to geometric and operational differences.

Impulse Turbines

Of the head available at the nozzle inlet, a small portion is lost to friction in the nozzle and to friction on the buckets. The rest is available to drive the wheel. The actual utilization of this head depends on the velocity head of the flow leaving the turbine and the setting above tailwater. Optimum conditions, corresponding to maximum utilization of the head available, dictate that the flow leaves at essentially zero velocity. Under ideal conditions this occurs when the peripheral speed of the wheel is one half the jet velocity. In practice, optimum power occurs at a speed coefficient, ω_{11}, somewhat less than 1.0. This is illustrated in Figure 73.7. Since the maximum efficiency occurs at fixed speed for fixed H, V_j must remain constant under varying flow conditions. Thus the flow rate Q is regulated with an adjustable nozzle. However, maximum efficiency occurs at slightly lower values of ω_{11} under partial power settings. Present nozzle technology is such that the discharge can be regulated over a wide range at high efficiency.

A given head and penstock configuration establishes the optimum jet velocity and diameter. The size of the wheel determines the speed of the machine. The design specific speed is approximately

$$N_S = 0.77\frac{d_j}{D} \text{ (Pelton turbines)} \tag{73.11}$$

Practical values of d_j/D for Pelton wheels to ensure good efficiency are in the range 0.04 to 0.1, corresponding to N_s values in the range 0.03 to 0.08. Higher specific speeds are possible with multiple nozzle designs. The increase is proportional to the square root of the number of nozzles. In considering an impulse unit, one must remember that efficiency is based on net head; the net head for an impulse unit is generally less than the net head for a reaction turbine at the same gross head because of the lack of a draft tube.

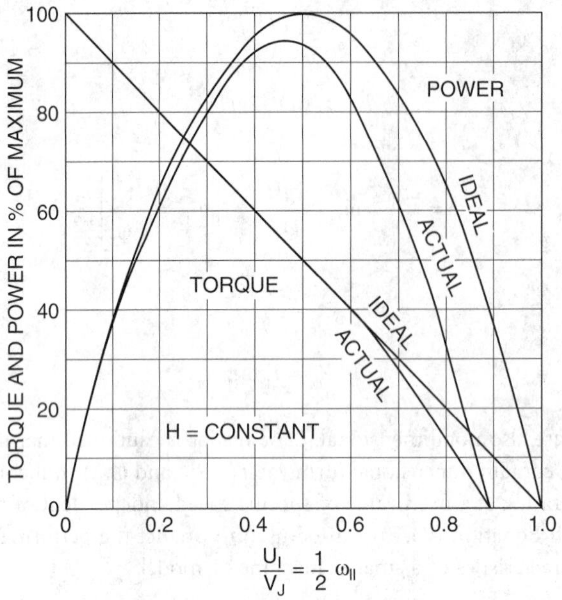

FIGURE 73.7 Ideal and actual variable-speed performance for an impulse turbine. (Adapted from Daily, J. W. 1950. Hydraulic machinery. In *Engineering Hydraulics*, H. Rouse, Ed., Wiley, New York. With permission.)

Reaction Turbines

The main difference between impulse units and reaction turbines is that a pressure drop takes place in the rotating passages of the reaction turbine. This implies that the entire flow passage from the turbine inlet to the discharge at the tailwater must be completely filled. A major factor in the overall design of modern reaction turbines is the draft tube. It is usually desirable to reduce the overall equipment and civil construction costs by using high–specific speed runners. Under these circumstances the draft tube is extremely critical for both flow stability and efficiency. (This should be kept in mind when retrofitting on older, low-specific speed turbine with a new runner of higher capacity.) At higher specific speed a substantial percentage of the available total energy is in the form of kinetic energy leaving the runner. To recover this efficiently, considerable emphasis should be placed on the draft tube design.

The practical specific speed range for reaction turbines is much broader than for impulse wheels. This is due to the wider range of variables that control the basic operation of the turbine. The pivoted guide vanes allow for control of the magnitude and direction of the inlet flow. Because there is a fixed relationship between blade angle, inlet velocity, and peripheral speed for shock-free entry, this requirement cannot be completely satisfied at partial flow without the ability to vary blade angle. This is the distinction between the efficiency of fixed-propeller and Francis types at partial loads and the fully adjustable Kaplan design.

In Equation (73.4), optimum hydraulic efficiency of the runner would occur when α_2 is equal to 90°. However, the overall efficiency of the turbine is dependent on the optimum performance of the draft tube as well, which occurs with a little swirl in the flow. Thus, the best overall efficiency occurs with $\alpha_2 \approx 75°$ for high–specific speed turbines.

The determination of optimum specific speed in a reaction turbine is more complicated than for an impulse unit since there are more variables. For a radial-flow machine, an approximate expression is

$$N_S = 1.64\left[C_V \sin\alpha_1 \frac{B}{D_1}\right]^{1/2} \omega_{11} \text{ (Francis turbines)} \tag{73.12}$$

where C_v is the fraction of net head that is in the form of inlet velocity head and B is the height of the inlet flow passage (see Figure 73.3). N_s for Francis units is normally found to be in the range 0.3 to 2.5.

Standardized axial-flow machines are available in the smaller size range. These units are made up of standard components, such as shafts and blades. For such cases,

$$N_S \sim \frac{\sqrt{\tan\beta}}{n_B^{3/4}} \quad \text{(Propeller turbines)} \tag{73.13}$$

where β is the blade pitch angle and n_B is the number of blades. The advantage of controllable pitch is also obvious from this formula; the best specific speed is simply a function of pitch angle.

It should be further noted that ω_{11} is approximately constant for Francis units and N_S is proportional to $(B/D_1)^{1/2}$. It can also be shown that velocity component based on the peripheral speed at the throat, ω_{11e}, is proportional to N_S. In the case of axial-flow machinery, ω_{11} is also proportional to N_S. For minimum cost, peripheral speed should be as high as possible — consistent with cavitation-free performance. Under these circumstances N_S would vary inversely with the square root of head (H is given in meters):

$$N_S = \frac{C}{\sqrt{H}} \tag{73.14}$$

where the range of C is 8 to 11 for fixed-propeller units and Kaplan units and 6 to 9 for Francis units.

Performance Comparison

The physical characteristics of various runner configurations are summarized in Figure 73.8. It is obvious that the configuration changes with speed and head. Impulse turbines are efficient over a relatively narrow

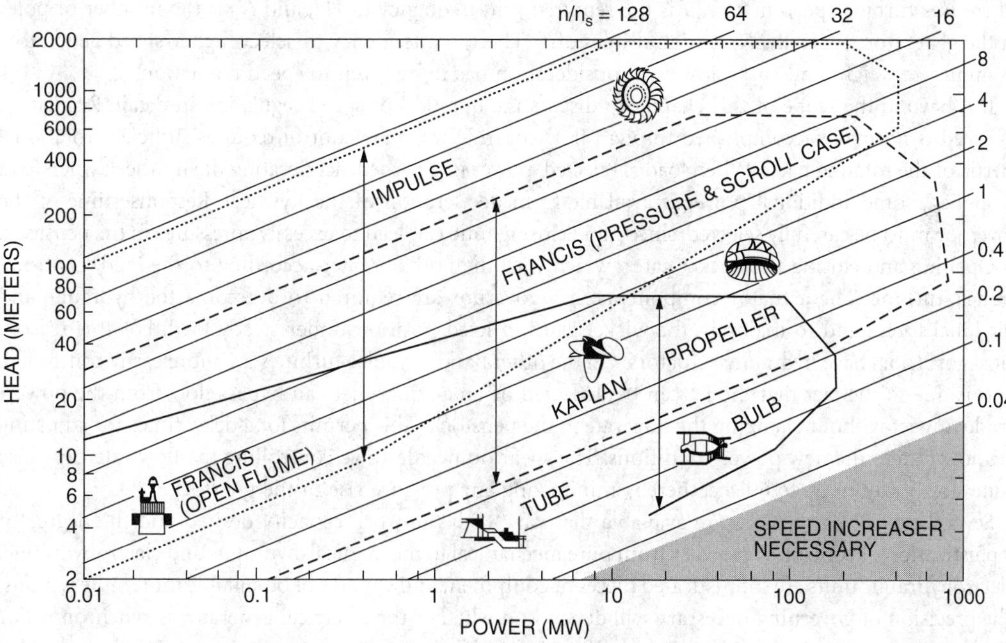

FIGURE 73.8 Application chart for various turbine types (n/n_S is the ratio of turbine speed in rpm, n, to specific speed defined in the metric system, $n_S = nP^{1/2}/H^{5/4}$, with P in kilowatts). (From Arndt, R. E. A. 1991. Hydraulic turbines. In *Hydropower Engineering Handbook,* J. S. Gulliver and R. E. A. Arndt, Eds., pp. 4.1–4.67. McGraw-Hill, New York. With permission.)

range of specific speed, whereas Francis and propeller turbines have a wider useful range. An important consideration is whether or not a turbine is required to operate over a wide range of load. Pelton wheels tend to operate efficiently over a wide range of power loading because of their nozzle design. In the case of reaction machines that have fixed geometry, such as Francis and propeller turbines, efficiency can vary widely with load. However, Kaplan and Deriaz [an adjustable-blade mixed-flow turbine (see Arndt, 1991)] turbines can maintain high efficiency over a wide range of operating conditions. The decision of whether to select a simple configuration with a relatively "peaky" efficiency curve or incur the added expense of installing a more complex machine with a broad efficiency curve will depend on the expected operation of the plant and other economic factors.

Note in Figure 73.8 that there is an overlap in the range of application of various types of equipment. This means that either type of unit can be designed for good efficiency in this range, but other factors, such as generator speed and cavitation, may dictate the final selection.

Speed Regulation

The speed regulation of a turbine is an important and complicated problem. The magnitude of the problem varies with size, type of machine and installation, type of electrical load, and whether the plant is tied into an electrical grid. It should also be kept in mind that runaway or no-load speed can be higher than the design speed by factors as high as 2.6. This is an important design consideration for all rotating parts, including the generator.

The speed of a turbine has to be controlled to a value that matches the generator characteristics and the grid frequency:

$$n = \frac{120f}{N_P} \tag{73.15}$$

where n is turbine speed in rpm, f is the required grid frequency in Hz, and N_P is the number of poles in the generator. Typically, N_p is in multiples of 4. There is a tendency to select higher-speed generators to minimize weight and cost. However, consideration has to be given to speed regulation.

It is beyond the scope of this chapter to discuss the question of speed regulation in detail. Regulation of speed is normally accomplished through flow control. Adequate control requires sufficient rotational inertia of the rotating parts. When load is rejected, power is absorbed, accelerating the flywheel; when load is applied, some additional power is available from deceleration of the flywheel. Response time of the governor must be carefully selected, since rapid closing time can lead to excessive pressures in the penstock.

Opening and closing the wicket gates, which vary the flow of water according to the load, control a Francis turbine. The actuator components of a governor are required to overcome the hydraulic and frictional forces and to maintain the wicket gates in fixed position under steady load. For this reason, most governors have hydraulic actuators. On the other hand, impulse turbines are more easily controlled. This is due to the fact that the jet can be deflected or an auxiliary jet can bypass flow from the power-producing jet without changing the flow rate in the penstock. This permits long delay times for adjusting the flow rate to the new power conditions. The spear on needle valve controlling the flow rate can close quite slowly, say, in 30 to 60 sec, thereby minimizing any pressure rise in the penstock.

Several types of governors are available that vary with the work capacity desired and the degree of sophistication of control. These vary from pure mechanical to mechanical-hydraulic and electrohydraulic. Electrohydraulic units are sophisticated pieces of equipment and would not be suitable for remote regions. The precision of governing necessary will depend on whether the electrical generator is synchronous or asynchronous (induction type). There are advantages to the induction type of generator. It is less complex and therefore less expensive but typically has slightly lower efficiency. Its frequency is controlled by the frequency of the grid it feeds into, thereby eliminating the need for an expensive conventional governor. It cannot operate independently but can only feed into a network and does so with lagging power factor, which may or may not be a disadvantage, depending on the nature of the load. Long transmission lines, for example, have a high capacitance, and, in this case, the lagging power factor may be an advantage.

Speed regulation is a function of the flywheel effect of the rotating components and the inertia of the water column of the system. The start-up time of the rotating system is given by

$$t_s = \frac{I\omega^2}{P} \tag{73.16}$$

where I = moment of inertia of the generator and turbine, kg · m^2 [Bureau of Reclamation, 1966].

The start-up time of the water column is given by

$$t_p = \frac{\sum LV}{gH} \tag{73.17}$$

where L = the length of water column and V = the velocity in each component of the water column.

For good speed regulation, it is desirable to keep $t_s/t_p > 4$. Lower values can also be used, although special precautions are necessary in the control equipment. It can readily be seen that higher ratios of t_s/t_p can be obtained by increasing I or decreasing t_p. Increasing I implies a larger generator, which also results in higher costs. The start-up time of the water column can be reduced by reducing the length of the flow system, by using lower velocities, or by addition of surge tanks, which essentially reduce the effective length of the conduit. A detailed analysis should be made for each installation, since, for a given length, head, and discharge, the flow area must be increased to reduce t_p, which leads to associated higher construction costs.

Cavitation and Turbine Setting

Another factor that must be considered prior to equipment selection is the evaluation of the turbine with respect to tailwater elevations. Hydraulic turbines are subject to pitting, loss of efficiency, and unstable operation due to cavitation [Arndt, 1981, 1991; Arndt et al., 2000]. For a given head, a smaller, lower-cost, high-speed runner must be set lower (i.e., closer to tailwater or even below tailwater) than a larger, higher-cost, low-speed turbine runner. Also, atmospheric pressure or plant elevation above sea level is a factor, as are tailwater elevation variations and operating requirements. This is a complex subject that can only be accurately resolved by model tests. Every runner design will have different cavitation characteristics. Therefore, the anticipated turbine location or setting with respect to tailwater elevations is an important consideration in turbine selection.

Cavitation is not normally a problem with impulse wheels. However, by the very nature of their operation, cavitation is an important factor in reaction turbine installations. The susceptibility for cavitation to occur is a function of the installation and the turbine design. This can be expressed conveniently in terms of Thoma's sigma, defined as

$$\sigma_T = \frac{H_a - H_v - z}{H} \tag{73.18}$$

where H_a is the atmospheric pressure head, H is the vapor pressure head (generally negligible), and z is the elevation of a turbine reference plane above the tailwater (see Figure 73.1). Draft tube losses and the exit velocity head have been neglected.

The term σ_T must be above a certain value to avoid cavitation problems. The critical value of σ_T is a function of specific speed [Arndt, 1991]. The Bureau of Reclamation [1966] suggests that cavitation problems can be avoided when

$$\sigma_T > 0.26 N_s^{1.64} \tag{73.19}$$

Equation (73.19) does not guarantee total elimination of cavitation, only that cavitation is within acceptable limits. Cavitation can be totally avoided only if the value of σ_T at an installation is much greater than the limiting value given in Equation (73.19). The value of σ_T for a given installation is known as the plant sigma, σ_P. Equation (73.19) should only be considered a guide in selecting σ_P, which is normally determined by a model test in the manufacturer's laboratory. For a turbine operating under a given head, the only variable controlling σ_P is the turbine setting z. The required value of σ_P then controls the allowable setting above tailwater:

$$z_{\text{allow}} = H_a - H_v - \sigma_P H \tag{73.20}$$

It must be borne in mind that H_a varies with elevation. As a rule of thumb, H_a decreases from the sea-level value of 10.3 m by 1.1 m for every 1000 m above sea level.

73.4 Performance Evaluation

Model Tests

Model testing is an important element in the design and development phases of turbine manufacture. Manufacturers own most laboratories equipped with model turbine test stands. Major hydro projects have traditionally had proof-of-performance tests in model scale as part of the contract (at either an independent laboratory or the manufacturer's laboratory). In addition, it has been shown that competitive model testing at an independent laboratory can lead to large savings at a major project because of improved efficiency. Recently, turbine design procedures have been dramatically improved through the use of sophisticated numerical analysis of the flow characteristics. These analysis techniques, linked with design programs, provide the turbine designer with powerful tools for achieving highly efficient turbine designs. In spite of this progress, computational methods require fine-tuning with model tests. In addition, model testing is necessary for determining performance over a range of operating conditions and for determining quasi-transitory characteristics. Model testing can also be used to eliminate or mitigate problems associated with vibration, cavitation, hydraulic thrust, and pressure pulsation [Fisher and Beyer, 1985].

A typical turbine test loop is shown in Figure 73.9. All test loops perform basically the same function. A model turbine is driven by high-pressure water from a head tank and discharges into a tail tank, as shown. The flow is recirculated by a pump, usually positioned well below the elevation of the model to ensure cavitation-free performance of the pump while performing cavitation testing with the turbine model. One important advantage of a recirculating turbine test loop is that cavitation testing can be done over a wide range of cavitation indices at constant head and flow.

The extrapolation of model test data to prototype values has been a subject of considerable debate for many years. Equation 73.6 through Equation 73.10 can be used to predict prototype values of flow, speed, power, etc., from model tests. Unfortunately, there are many factors that lead to scale effects, i.e., the prototype efficiency and model efficiency are not identical at a fixed value of specific speed. The cited scale-up formulae are based on inviscid flow. There are several sources of energy loss, which lead to an efficiency that is less than ideal. All of these losses follow different scaling laws and, in principle, perfect similitude can only be achieved by testing the prototype. There have been several attempts at rationalizing the process of scaling up model test data. The International Electrotechnical Test Code and various ASME publications outline in detail the differences in efficiency between model and prototype. It should also be pointed out that other losses such as in the draft tube and "shock losses" at the runner inlet might not be independent of Reynolds number.

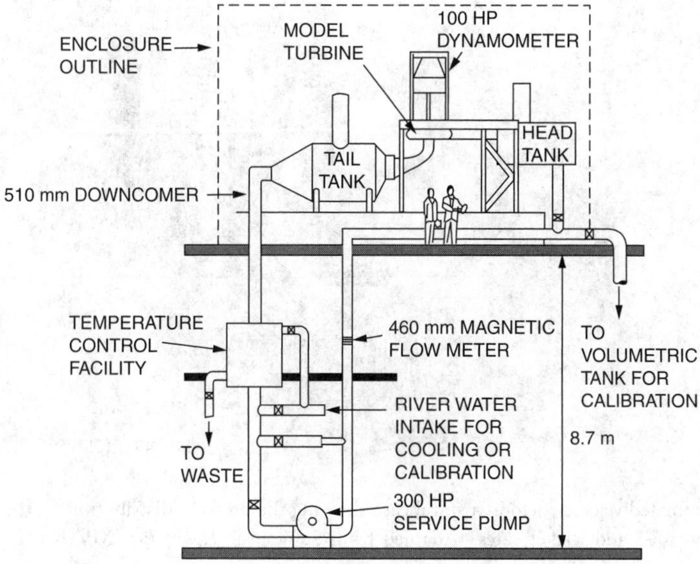

FIGURE 73.9 Schematic of the SAFL Independent Turbine Test Facility. (Courtesy of the St. Anthony Falls Laboratory, University of Minnesota.)

Numerical Simulation[1]

Until very recently, the analysis of turbines and other components of hydropower facilities was largely dependent on approximate models such as the Euler Equation and Reynolds Averaged Navier–Stokes (RANS) models because the complete Navier–Stokes equations were considered to be too difficult to be solved for hydropower components. (See Section 33.4) The Euler equation model has been applied with reasonable success for turbine runner simulation, but energy losses and the efficiency could not be calculated. RANS models have been applied for the spiral case and the draft tube simulations with limited success. Goede et al. [1991] contains a good summary of experiences with the application of these computational methods to various hydropower components. Very recently, reliable commercial codes have become available and are being used with increasing frequency. Many of the current commercial codes rely on variants of RANS models although rapid progress is being made in adapting the large eddy simulation method (LES).

LES is a step forward in the application of CFD [Song et al., 1995]. This technique is able to more accurately capture the effects of turbulent flow in a turbine than previous techniques. At the present time, it requires a supercomputer to achieve sufficient resolution and good accuracy for final design purposes. Parallel processing with desktop computers shows promise and can presently be used for relatively simple geometry or for preliminary evaluation purposes. However, progress in its application is very rapid and it is anticipated that an entire computation may be carried out on a high-end desktop computer in the near future.

The components that require simulation include the spiral case, wicket gates, the runner, and the draft tube. Often the spiral case including stay vanes and wicket gates is modeled as a unit. This is necessary because the stay vanes and wicket gates are so close to each other that their mutual interactions cannot be ignored, and each stay vane may be of slightly different shape and orientation and cannot be modeled separately. A typical spiral case contains more than 20 stay vanes, and an equal number of wicket gates, requiring extensive computational resources for a complete simulation. A sample calculation is shown in Figure 73.10. In this example, the calculated energy loss through this device is 2.62% of the net available

[1]This section prepared by Professor Charles Song of the Saint Anthony Falls Laboratory, University of Minnesota.

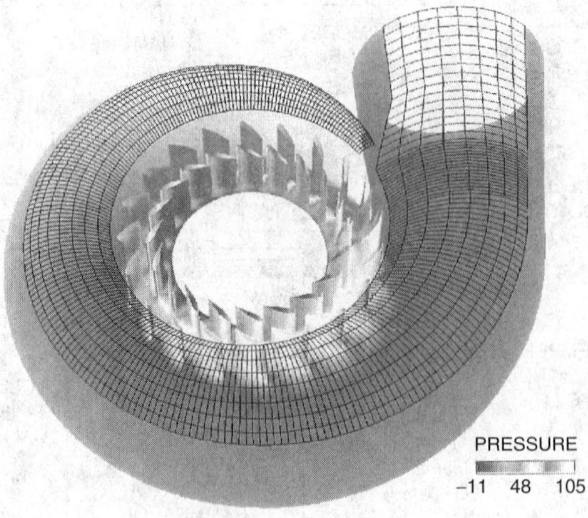

FIGURE 73.10 Simulated velocity field in a spiral case showing the pressure distribution on the boundaries of the spiral case, the stay vanes and wicket gates. (Adapted from Song et al. *Hydro Rev.* XIV, No. 4, pp. 104–111. With permission.)

energy for this particular case. This is significant and justifies additional computational effort to minimize the losses.

The runner is the most extensively studied component of a turbine. Since all the blades in a runner are of the same geometric shape, only one or two flow passage models are commonly used for runner simulation. A complete model is required if vibration or cavitation due to nonsymmetrical modes of interactions between blades and vortices is to be studied. An important application of computer simulation is in the design of runners for units used for pumped storage. **Pumped-storage** schemes are becoming very popular for smoothing out the difference between energy demand and supply. Special care is required in the design because a runner must be designed to act efficiently both as a turbine and a pump. Because of viscous effects, a runner optimized under the turbine mode may have poor efficiency in the pump mode. Flow in the pumping mode can be unstable and more difficult to calculate. An LES-based analysis greatly facilitates the optimum design of this type of runner. Figure 73.11 is an example of the calculated flow in the pump mode. A small flow separation near the entrance can be observed. This kind of information is very useful to determine how the blade geometry can be modified to improve the performance.

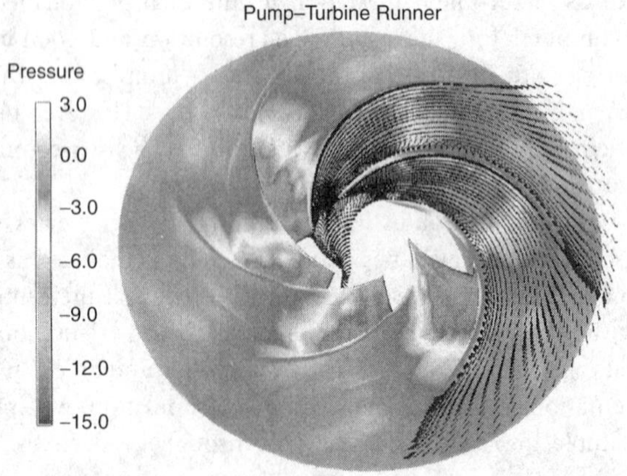

FIGURE 73.11 Simulation of the velocity and pressure distribution in a pump turbine runner operating in the pumping mode. (Adapted from Song et al. *Hydro Rev.* XIV, No. 4, pp. 104–111. With permission.)

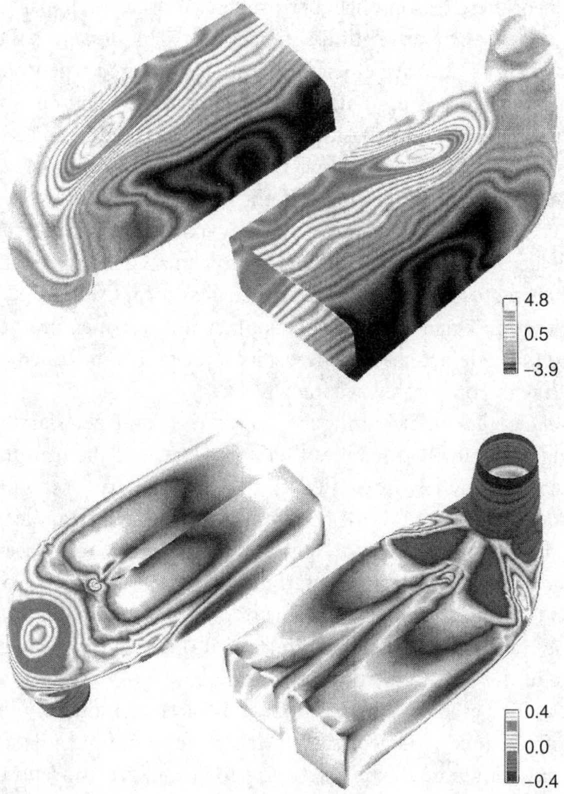

FIGURE 73.12 Simulation showing a comparison of the flow in a draft tube with and without a dividing wall. The pressure patterns on the walls indicate a very asymmetrical flow pattern without a divider wall (Figure 73.12a). With a wall in place a very uniform flow pattern is evident (Figure 73.12b). (Adapted from Song et al. *Hydro Rev.* XIV, No. 4, pp. 104–111. With permission.)

Draft tube design is an important factor in the efficiency and stability of operation of a turbine. Although a typical draft tube geometry is somewhat simpler than that of a runner, it takes much more advanced computational techniques to accurately simulate its performance. This is because the diffuser-like flow produces secondary currents, three-dimensional vortex shedding, and horseshoe vortices that are all very important contributors to energy loss. Current RANS models are ineffective for this flow. At the present time, only the LES model can fulfill the requirements for draft tube simulation. Figure 73.12 illustrates the complexity of the problem. The instantaneous pressure distribution on the walls of an elbow-type draft tube with a divider wall is compared with the same draft tube with the divider wall removed. A dramatic change in the flow pattern and pressure distribution occurs when the divider wall is removed. By removing the wall, the draft tube becomes a diffuser of large angle with very unstable flow. Clearly, the divider wall stabilizes the flow and reduces the energy loss due to vortex shedding. These types of simulations are invaluable in evaluating draft tube performance. This is underscored by the fact that many projects involve refurbishing existing units. Typically only the runner is replaced, usually with increased design flow. On many occasions, the existing draft tube is unable to operate efficiently at higher flow rates, canceling out any improvements that a new runner can provide.

Field Tests

Model tests and numerical simulations are only valid when geometric similitude is adhered to, i.e., there is no guarantee that the prototype machine is an accurate reproduction of the design. In addition, approach flow conditions, intake head losses, the effect of operating other adjacent units, etc., are not simulated in model tests. For these reasons, field performance tests are often performed. There are several different types of field tests, which serve different purposes. The *absolute* efficiency is measured for acceptance or performance tests. *Relative* efficiency is often measured when operating information or fine-tuning of turbine performance is desired. Field tests are also carried out for commissioning a site and for various problem-solving activities. Basic procedures are covered by codes of the American Society of Mechanical Engineering and the International Electrotechnical Commission. The major difference between an "absolute" and a relative or index test is in the measurement of flow rate. Net head is evaluated in the same manner for each procedure. There are a variety of methods for measuring flow that are code-accepted. These include the pressure-time technique, tracer methods (salt velocity, dye dilution), area-velocity (Pitot tubes or current meters), volumetric (usually on captive pumped storage sites), Venturi meters, and weirs. The thermodynamic method is actually a direct measure of efficiency. Flow is not measured. In addition to the code-accepted methods, it has been demonstrated that acoustic meters can measure flow in the field with comparable accuracy.

The *pressure-time technique* relies on measuring the change in pressure necessary to decelerate a given mass of fluid in a closed conduit. The method requires the measurement of the piezometric head at two cross-sections spaced a distance L apart. A downstream valve or gate is necessary for this procedure. This technique requires load rejection for each test point and the need to estimate or measure any leakage. An adequate length of conduit is required and the conduit geometry must be accurately measured [Hecker and Nystrom, 1987].

The *salt velocity method* is based on measuring the transit time, between two sensors, of an injected cloud of concentrated salt solution. Given the volume of the conduit between sensors, the flow rate may be calculated from the average transit time. Electrodes that measure the change in conductivity of the liquid detect the passage of the salt cloud at a given location.

The *dye-dilution method* is based on conservation of a tracer continuously injected into the flow. A sufficient length for complete mixing is necessary for accurate results. The data required are the initial concentration and injection flow rate of the tracer and the measured concentration of the fully mixed tracer at a downstream location. The method is quite simple, but care is necessary to achieve precise results.

In principle, *area-velocity measurements* are also quite simple. Either Pitot tubes or propeller-type current meters are used to measure point velocities that are integrated over the flow cross-section. The method is applicable to either closed conduits or open channels. A relatively uniform velocity distribution is necessary for accurate results. A single unit can be traversed across the conduit or a fixed or movable array of instruments can be used to reduce the time for data collection.

The *thermodynamic method* is a direct indication of turbine efficiency. Flow rate is not measured. In its simplest form, the method assumes adiabatic conditions, i.e., no heat transfer from the flow to its surroundings. Under these conditions, that portion of the available energy not utilized in the machine to produce useful work results in increased internal energy of the fluid, which is sensed as an increase in temperature.

Acoustic flow meters have been developed which produce results with a precision equal to or greater than the code-accepted methods. Flow velocity is determined by comparing acoustic travel times for paths diagonally upstream and downstream between pairs of transducers. The speed of sound is assumed constant. The difference in travel time is related to the component of flow velocity along the acoustic path (increased travel time upstream, decreased travel time downstream). An extensive evaluation and comparison of this method has been reported [Sullivan, 1983].

Index tests circumvent the problem of accurate flow measurement by measuring relative flow detected by the differential pressure between two points in the water passages leading to the runner. Often the differential pressure is measured with Winter–Kennedy taps which are positioned at the inner and outer radii of the spiral case of a turbine. Calibration of properly placed Winter–Kennedy taps shows that flow rate is very closely proportional to the square root of the pressure difference. Index testing is useful for calibration of relative power output versus gate opening and for optimizing the various combinations of gate opening and blade setting in Kaplan units. The use of index testing to optimize cam settings in Kaplan turbines has resulted in substantial increases in weighted efficiency (i.e., a flatter efficiency curve over the full range of operation).

Defining Terms

Draft tube — The outlet conduit from a turbine that normally acts as a diffuser. This is normally considered an integral part of the unit.

Forebay — The hydraulic structure used to withdraw water from a reservoir or river. This can be positioned a considerable distance upstream from the turbine inlet.

Head — The specific energy per unit weight of water. *Gross head* is the difference in water surface elevation between the forebay and tailrace. *Net head* is the difference between *total head* (the sum of velocity head, $V^2/2g$, pressure head, p/ρ_g, and elevation head, z) at the inlet and outlet of a turbine. Some European texts use specific energy per unit mass, for example, specific kinetic energy is $V^2/2$.

Pumped storage — A scheme in which water is pumped to an upper reservoir during off-peak hours and used to generate electricity during peak hours.

Runner — The rotating component of a turbine in which energy conversion takes place.

Specific speed — A universal number for a given machine design.

Spiral case — The inlet to a reaction turbine.

Surge tank — A hydraulic structure used to diminish overpressures in high-head facilities due to water hammer resulting from the sudden stoppage of a turbine.

Wicket gates — Pivoted, streamlined guide vanes that control the flow of water to the turbine.

References

Arndt, R. E. A. 1981. Cavitation in fluid machinery and hydraulic structures. *Annu. Rev. Fluid Mech.* 13:273–328.

Arndt, R. E. A. 1991. Hydraulic turbines. In *Hydropower Engineering Handbook*, ed. J. S. Gulliver and R. E. A. Arndt, pp. 4.1–4.67. McGraw-Hill, New York.

Arndt, R. E. A., Keller, A., and Kjeldsen, M. 2000 Unsteady operation due to cavitation. *Proceedings 20th IAHR Symposium on Hydraulic Machinery and Systems,* Charlotte, NC, August.

American Society of Mechanical Engineers, 1992. *Power Test Code 18.*

Bureau of Reclamation. 1966. *Selecting Hydraulic Reaction Turbines,* Engineering Monograph No. 20.

Daily, J. W. 1950. Hydraulic machinery. In *Engineering Hydraulics,* ed. H. Rouse. Wiley, New York.

Fisher, R. K. and Beyer, J. R. 1985. The value of model testing for hydraulic turbines. *Proc. Am. Power Conf.,* ASME vol. 47, 1122–1128.

Goede, E., Cuenod, R., Grunder, R., and Pestalozzi, J. 1991. A new computer method to optimize turbine design and runner replacement, *Hydro Review,* Vol. X, No. 1.

Hecker, G. E. and Nystrom, J. B. 1991. Which flow measurement technique is best? *Hydro Rev.* 6(3), June

IEC. 1991. *International Code for the Field Acceptance Tests of Hydraulic Turbines,* Publication 41. International Electrotechnical Commission.

Raabe, J. 1985. *Hydro Power: The Design, Use, and Function of Hydromechanical, Hydraulic, and Electrical Equipment,* VDI Verlag, Dusseldorf, Germany.

Song, C. C. S., Chen, X., He, J., Chen, C., and Zhou, F., 1995. Using computational tools for hydraulic design of hydropower plants. *Hydro Rev.* XIV, No. 4, pp. 104–111, July.

Sullivan, C. W. 1983. Acoustic flow measurement systems: economical, accurate, but not code accepted. *Hydro Rev.* 6(4), August.

Wahl, T. L. 1994. Draft tube surging times two: The twin vortex problem. *Hydro Rev.* 13(1):60–69.

Further Information

J. Fluids Eng. Published quarterly by the ASME.

ASME Symposia Proc. on Fluid Machinery and Cavitation. Published by the Fluids Eng. Div.

Hydro Rev. Published eight times per year by HCI Publications, Kansas City, MO.

Moody, L. F. and Zowski, T. 1992. Hydraulic machinery. In *Handbook of Applied Hydraulics*, ed. C. V. Davis and K. E. Sorenson. McGraw-Hill, New York.

Waterpower and Dam Construction. Published monthly by Reed Business Publishing, Surrey, UK.

74

Steam Turbines and Generators

Otakar Jonas
Jonas, Inc.

A **steam turbine** is a rotary engine that uses superheated or saturated steam produced by a steam generator (boiler) to convert the thermal energy into work (mechanical energy). Steam turbines are used for driving electric generators as "prime movers" or driving mechanical equipment such as compressors, fans, and pumps. Turbines have been in use for over 100 years, and no new technology currently available can replace them for high-energy-output applications. Steam turbine generators produce 80% of the 750,000 megawatts needed in the U.S. and the 3000 gigawatts worldwide. This chapter focuses on the basics of turbine and electric generator design, working principles, and descriptions of key components. Steam turbine problems and steam chemistry and corrosion are also briefly discussed. There is a large volume of information dealing with all aspects of turbine and electric generator design, operation, and maintenance that is beyond the intent of this chapter. Additional information is in the "References" and "Further Information" sections.

74.1 Design of Steam Turbines [1–13]

Development of modern steam turbines started in the 1800s. In Sweden, Gustav Laval introduced the first useful turbine in 1889. A turbine with multiple blade stages closer to today's turbines was designed in 1883 in England by Parsons. Steam turbines are the simplest and most efficient engines for converting large amounts of heat energy into mechanical work. They use superheated or saturated steam generated

0-8493-1586-7/05/$0.00+$1.50

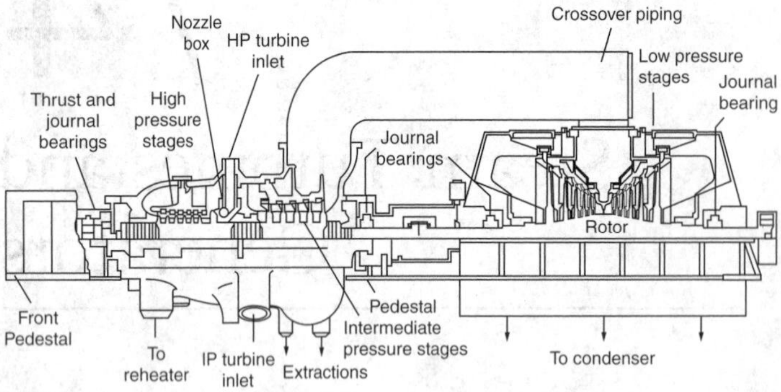

FIGURE 74.1 Typical tandem compound, single reheat condensing turbine. (Courtesy of EPRI.)

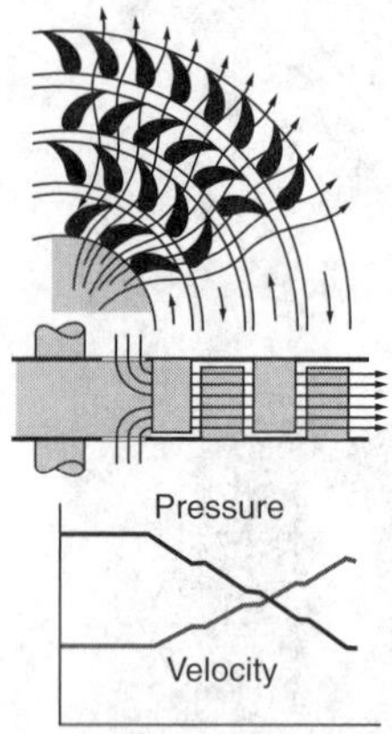

FIGURE 74.2 Radial flow double-rotation reaction turbine has only moving blades. (Courtesy of *Power* Magazine.)

in one of many types of steam generators [8–12]. As the steam expands, it acquires high velocity and exerts force on the turbine blades. The steam conditions range from a few psig saturated to 4500 psig, superheated to 1112°F and steam flows up to 15 million pounds per hour (nuclear). Turbines range in size from a few kilowatts (kW) or horsepower (hp) for one-stage units to 1500 MW for multiple-stage multiple-component units comprising high-pressure, intermediate-pressure, and up to three low-pressure double-flow turbines. Mechanical drive turbines range from single and double stage turbines to over 20 stages, depending on horsepower. The common size range for these machines is 5 hp to over 100,000 hp. Most larger modern turbines are multiple-stage axial flow horizontal units. Figure 74.1 shows a typical utility tandem-compound reheat turbine with a combined high pressure (HP) and intermediate pressure (IP) turbine and a two-flow low pressure (LP) turbine. There are other, infrequently used, designs of smaller turbines such as the radial flow designs (see Figure 74.2).

In Figure 74.1, steam enters from the main steam lines (pipes) through stop and control valves into the HP section. The first (control) stage is spaced somewhat apart from subsequent stages to allow for stabilization of the flow. After passing through the HP turbine, cold reheat piping carries the steam to the reheater and returns it in the hot reheat piping to the integrated HP and IP cylinder to pass through the IP turbine section. The flow exits the IP turbine through the IP exhaust hood and then passes through crossover piping to the LP turbine and exits to the condenser through the LP exhaust. The typical modern steam turbine has a number of extraction points throughout all sections for the steam to the feedwater heaters.

During its expansion through the LP turbine blades, the steam crosses the saturation line. The region where condensation begins, termed the "Phase Transition Zone" (PTZ) or "Wilson Line," is the location where many of the corrosion damage mechanisms occur [14, 15]. In single reheat turbines at full load, this zone is usually at the last minus 1 stage, which is also in the transonic flow region where, at the sonic velocity (Mach = 1), sonic shock waves can be a source of blade excitation and cyclic stresses causing fatigue or corrosion fatigue [15–18].

Steam turbines can also be classified by their configuration. A steam turbine generator for electrical power generation can be classified as a tandem compound or cross compound unit. A tandem compound unit operates with all of the turbine sections rotating on a common shaft connected to one generator. A cross compound unit has the turbine sections on two separate shafts: the HP and IP turbines are on a single shaft connected to a single generator, whereas the LP turbine is on a separate shaft connected to a separate generator. Virtually all new steam turbine installations are the tandem compound type. Many **combined cycles** also have the gas and steam turbine and electric generator on one shaft (connected by couplings). The number of exhaust flows to the condenser is a function of the number of low-pressure turbine sections. Utility turbine generators can have two, four, or six exhausts to the condenser. Large utility turbogenerators can be over 200 feet long and weigh over 2,000 tons.

Steam turbines are used in the following cycles [2, 4, 5, 8–11]:

- Fossil fuel drum and once-through (supercritical and subcritical) boiler cycles
- Combined gas turbine and steam turbine with HRSG (heat recovery steam generator)
- Nuclear PWR (pressurized water reactor), BWR (boiling water reactor), other
- Waste heat boiler
- Trash, baggage, black liquor, etc. in drum boiler cycles
- Geothermal
- Solar
- Ocean thermal energy (contemplated)

Most **steam turbine generators** for nonnuclear power generation rotate at the constant "speed" of 3600 rpm when they drive 60-Hz electric generators (U.S., Canada, etc.) and 3000 rpm for 50 Hz generators (Europe, etc.). Most nuclear turbines operate at 1800 rpm (60 Hz) and 1500 rpm (50 Hz) to accommodate higher steam flows using longer blades; since the blades are longer, the lower speed is necessary to reduce blade stresses. Steam flows over 12 million lb/h are needed for the large nuclear turbines because the steam inlet pressures and temperatures are low. The power output from a steam turbine is controlled by varying the control valve position on the steam chest, thereby admitting more or less steam to the turbine. The increased torque on the steam turbine generator shaft resulting from the increased steam flow increases the power (MW) output.

Mechanical drive turbines generally operate over a variable speed range with maximum speed over 20,000 rpm. They are useful in driving equipment that frequently operates at lower loads. As the load on a pump or fan is reduced, the turbine can slow down and reduce the capacity of the pump or fan. As the demand increases, the turbine increases speed as required to produce the flow. This infinite type of speed adjustment can result in large power savings compared to the constant speed operation of driven equipment.

The engineering disciplines used in steam turbine design include:

- Thermodynamics and flow (cycle, blade path, controls)
- Mechanical design (stresses, vibration, strength at low and high temperatures, creep)
- Electrical design (generator, controls and monitoring)
- Materials (strength, fatigue, fracture mechanics)
- Corrosion and water and steam chemistry

The modern design tools include:

- Finite element stress, temperature, and vibration analysis
- Computational flow dynamics (viscous, nonviscous, with condensation, with steam chemistry)
- Life prediction and fracture mechanics methods
- Prototype testing (performance, efficiency, vibration, etc.)
- Material mechanical fracture, fatigue, creep, erosion, and corrosion properties

Steam Turbine Developments

After the hundred years of history, major developments are still taking place in the design, operation, and maintenance of steam turbines. The driving forces for these developments are higher efficiency, increased reliability, and lower maintenance costs (including longer inspection intervals). The main long-term steam cycle and turbine developments are governed by the availability of better materials, which can operate for the **30-year turbine design life** at higher temperatures and are more resistant to fatigue, corrosion, and fracture. The developments of the last decades include:

TABLE 74.1 Steam Turbine Materials

Component	Material
Rotor	CrMoV, NiCrMoV low-alloy steel forging
Discs	NiCrMoV, CrMoV, NiCrMo low-alloy steel forging, 12Cr weld repair
Shell and piping	Carbon steel, low-alloy steels
Blades and shrouds	12Cr stainless steels, 15-5PH, 17-4 PH, Ti6-4, PH13-8Mo
Erosion shields	Stellite Type 6B — weld deposited or soldered, same as blade, hardened blades
Stationary blades	304 stainless steel and other stainless steels
Expansion bellows	AISI Types 321 or 304 stainless steels, Inconel 600
Bearings	High tin babbit cast on bronze
Valves	Body — low-alloy steel, stems — stainless steel, seats — stellite
Bolts	High temp. — heat-resistant alloys, low temp. — medium-strength steels

- Turbines for advanced steam conditions for fossil fuel cycles (up to 4500 psi pressure and 1112°F superheat) — over 6% efficiency gain
- Turbines for cycling and peaking duty (can be shut down once a day)
- Longer last-stage blades and hollow stationary blades with moisture extraction slots leading to better efficiency
- Better blade path flow design (three-dimensional viscous flow) and curved and twisted stationary "banana" blades leading to better efficiency and cyclic stresses
- Better overall design allowing over 10-year inspection intervals
- Better control of steam purity preventing deposits which lead to corrosion and loss of efficiency and MW
- Turbine bypass for faster startups
- Retractable internal packings to prevent rubbing during startups
- Tilt pad bearings for better alignment
- Better rotor and blade materials
- Superconducting electric generators (under development)
- Magnetic bearings (under development)

Bearings

Turbine and driven equipment bearings range from pressure-lubricated journal type for large turbines through ring-oiled journal bearings for small units and ball bearings for very small turbines. The oil-lubricated journal bearings rely on maintaining an oil film between the bearing babbit and the rotor shaft. Typically, each separate turbine section (HP, IP, and LP) has two radial bearings, and the whole turbine set has one axial bearing because the steam pressure differential across turbine stages creates a net thrust along the shaft.

Steam Turbine Materials

Typical materials used for various turbine components are listed in Table 74.1. Requirements for these materials include resistance to creep and high temperature oxidation for the high pressure and intermediate pressure turbines and resistance to fatigue, corrosion fatigue, stress corrosion, pitting, and brittle fracture for all components [3, 7, 11, 16–18].

74.2 Working Principles [1–6]

The main steam turbine working principle [i.e., conversion of heat and pressure energy (Equation 74.1) to kinetic energy of steam and work] was demonstrated over 2200 years ago by a Greek named Hero — see Figure 74.3. In his reaction turbine, the steam generated in a heated container (boiler) was shooting

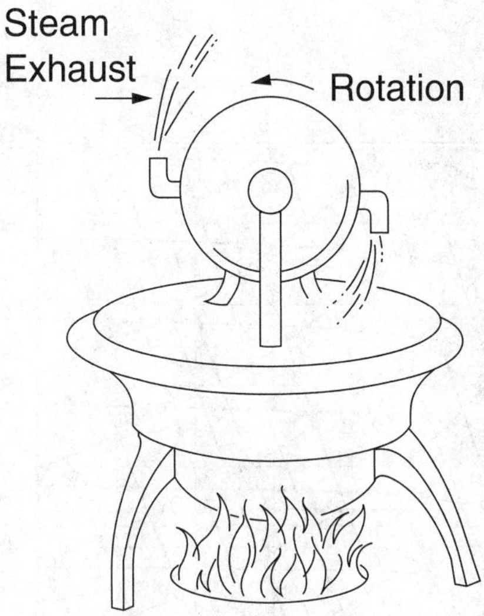

FIGURE 74.3 Hero's reaction turbine.

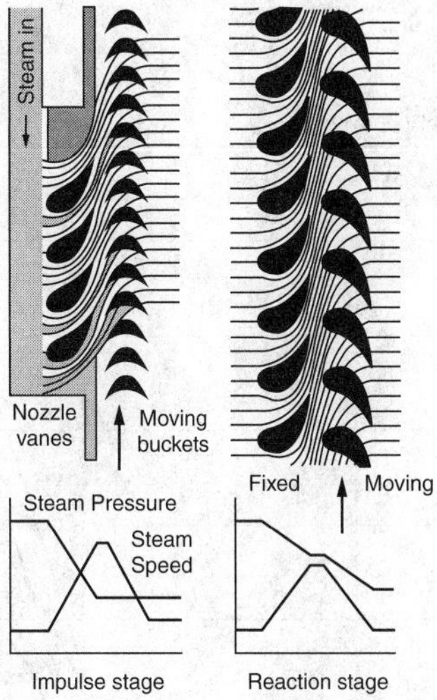

FIGURE 74.4 Steam flow through impulse and reaction blading. (Courtesy of *Power* Magazine.)

out through two nozzles, spinning the whole device. In modern turbines, after multiple nozzles (stationary blades), there are rotating blades or buckets attached to a rotor.

Impulse vs. Reaction Blades

There are two fundamental types of steam turbine rotating blades from the perspective of how energy is transferred from the steam to the turbine shaft. The energy is transferred by the steam expanding through a series of stationary and rotating blade sections (stages). These blade sections can be of the **impulse** and **reaction** types and for longer blades, the two types are often combined.

In an **impulse** turbine blade stage, the steam is expanded through stationary nozzles to impact the rotating blades. The energy to rotate the turbine comes from the force of the steam impacting on the buckets.

In a **reaction stage,** the stationary nozzles and rotating blades are of similar shape, similar to an airplane wing profile. The steam expands, increases in velocity, and loses pressure as it passes through the blade sections. Pressure exists on the concave side of the profile and suction on the convex side. The resulting force generated by the velocity turns the rotor. Figure 74.4 illustrates the principles of impulse and reaction blading. A blade stage is a stationary plus rotating row of blades.

74.3 Thermodynamics and Efficiency [1–12]

The steam conditions in turbines and in the whole steam cycle are measured and calculated according to established practices. Steam tables and thermodynamic diagrams are used to derive related steam and water properties such as pressure, temperature, energy, and % moisture (quality) [2–4, 6, 12, 13]. For steam turbines, a Mollier (entropy–enthalpy) diagram is often used (see Figure 74.5 and Figure 74.6). The steam expansion lines in these diagrams represent the average blade or flow path conditions. The local conditions along the longer blades and elsewhere are different. The steam cycle conditions are usually shown in a **heat balance diagram** (see Figure 74.7 for a fossil fuel drum boiler cycle).

The **amount of power** that can be generated by a steam turbine is a function of the initial steam pressure and temperature, steam flow, exhaust pressure, and efficiency of the machine. Additionally, the bearings, electrical generator, and other losses need to be considered, since these losses reduce the power output.

The heat content of a gas (steam) is equal to C_vT and is called intrinsic energy. The gas can also do work when its pressure is higher than some reference pressure. For steam turbines, the reference pressure

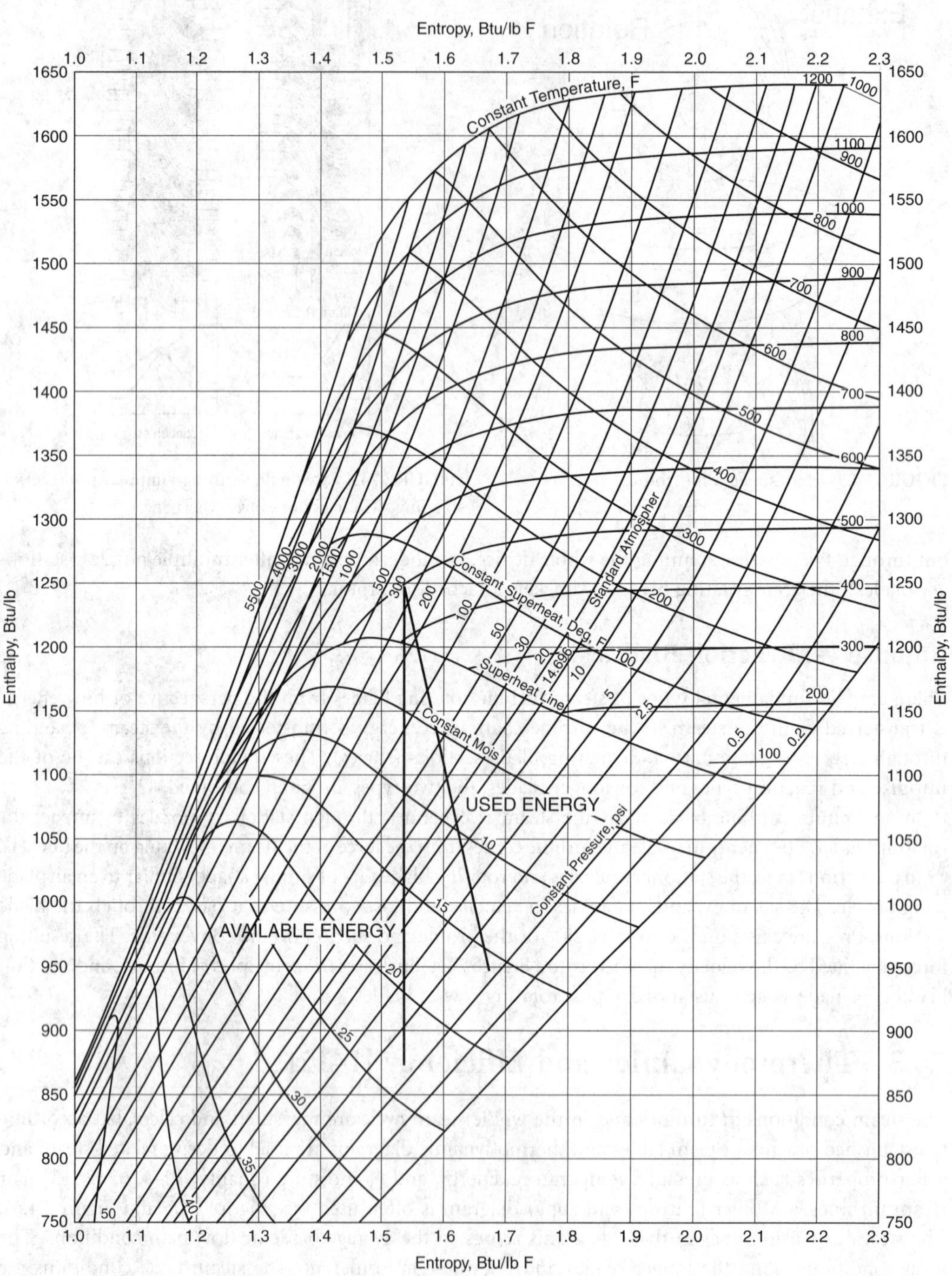

FIGURE 74.5 Mollier (entropy–enthalpy) chart with a turbine steam expansion line (used energy) and isentropic expansion line (available energy).

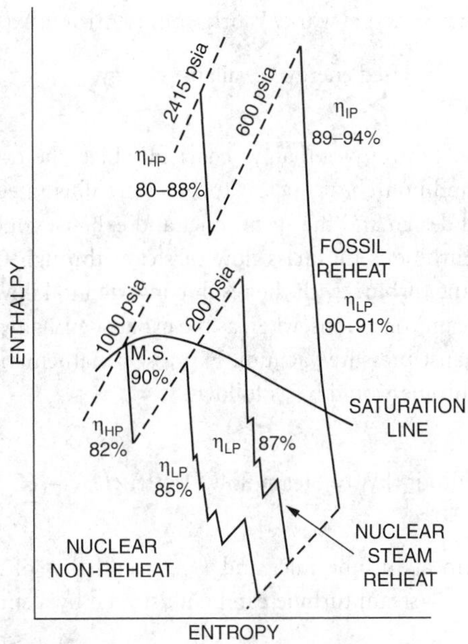

FIGURE 74.6 Mollier diagram with typical utility turbine expansion lines and efficiencies.

is usually the condenser pressure. This energy is referred to as the work term, Pυ. The sum of the intrinsic energy and the work energy is called enthalpy, H, which in equation form is:

$$H = \left[C_\upsilon T + \frac{P\upsilon}{J}\right](\text{BTU/lbm}) \tag{74.1}$$

Both the intrinsic energy, $C_\upsilon T$, and the work term, Pυ, are determined by measuring P and T of the gas at rest. The total energy of a moving fluid such as steam flowing through a turbine has an additional kinetic energy term, $V^2/2gJ$. Thus, the total energy is:

$$\text{Total Energy} = \left[C_\upsilon T + \frac{P\upsilon}{J} + \frac{V^2}{2gJ}\right](\text{BTU/lbm}) \tag{74.2}$$

where P = pressure (lb/ft^2) absolute; υ = specific volume (ft^3/lbm); C_υ = specific heat at constant volume (BTU/lbm/°F); T = absolute temperature ($T_{measured}$ + 459.6°); J = mechanical equivalent of heat (778.26 ft-lbm/BTU); g = acceleration of gravity (32 ft/sec^2); and V = velocity (ft/sec).

In a steam turbine system, the mass and energy are preserved, and in the turbine itself, most of the steam enthalpy is converted to the kinetic energy of the moving steam by the blades and is then used to generate mechanical work.

The **efficiency** of a steam turbine is defined as the actual work produced divided by the (theoretical) work produced by an isentropic (constant entropy) expansion. An isentropic process is an idealized process that represents the amount of available energy (enthalpy). The second law of thermodynamics, however, states that the conversion of this thermal energy to useful work cannot be 100% efficient. In practical use, it will be less than 100% because of the second law and because of flow, condensation, mechanical, and electrical losses in the turbine generator itself.

Expressed mathematically (see Figure 74.5), the efficiency of a steam turbine is as follows:

$$\begin{aligned}\text{Efficiency} &= \text{Actual work / Work from isentropic expansion}\\ &= \text{Used energy / Available energy}\end{aligned} \tag{74.3}$$

Figure 74.6 shows a Mollier (entropy–enthalpy) chart in which the expansion of steam in a utility reheat turbine and in reheat and nonreheat nuclear turbines are illustrated. The efficiency is a function of the machine type, size, and design and the steam inlet and exhaust conditions.

Power output of a steam turbine is the mass flow of steam through the turbine multiplied by the difference in **enthalpy** across the turbine, with the result converted to kilowatts or horsepower. The inlet enthalpy is known from the steam conditions, whereas the exhaust enthalpy is a function of the efficiency of the expansion and the exhaust pressure/vacuum. Expressed mathematically, the power output for a simple nonextraction steam turbine would be as follows:

$$\text{Power output (kW)} = \text{Steam flow (1b/h)} \times \left(H_{\text{in}} - H_{\text{out}}\right)/3413 \tag{74.4}$$

where H_{in} = enthalpy of steam at turbine inlet and H_{out} = enthalpy of steam at turbine outlet. The calculation of power output for a steam turbine can be illustrated by a simple example.

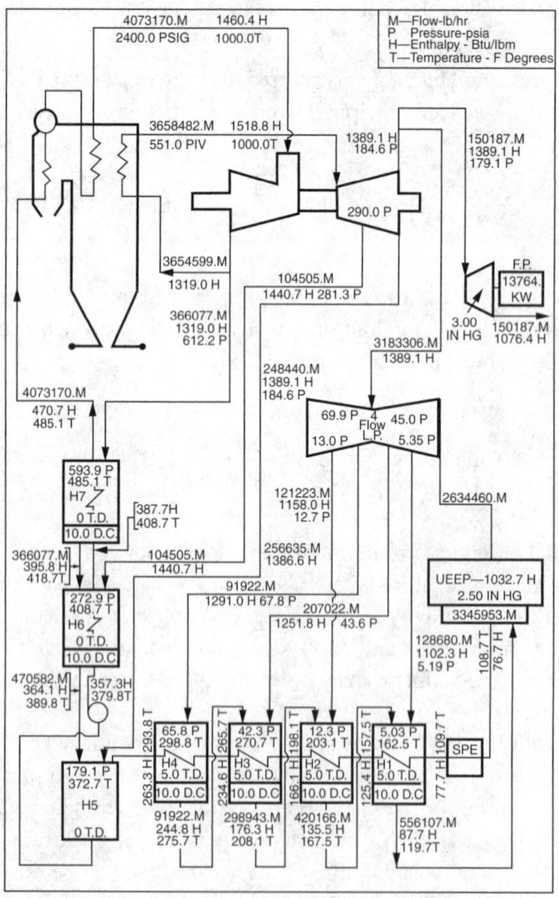

FIGURE 74.7 Heat balance diagram for a utility fossil fuel drum boiler cycle. (Courtesy of ASME.)

Example

An industrial, condensing, nonextraction steam turbine with a throttle pressure of 400 psig and 500 F (Figure 74.5) and a steam flow of 100,000 1b/h exhausts to a condenser at 3.0″ Hga. The manufacturer states that the efficiency of this machine is 80.0%. Determine the power output of this machine.

Solution

From steam tables and Figure 74.5:

Enthalpy of steam at 400 psig/500 F = 1243.2 Btu/lb (H1)
Entropy of steam at 400 psig/500 F = 1.5225 Btu/lb-°F (S1)
Enthalpy of steam at S1 and 3.0" HgA = 868.26 Btu/lb (H2S)
Exhaust enthalpy = [(H2S − H1) × 0.80] + H1 = 943.23 Btu/lb (H2)
Power output = [100,000 lb/h × (1243.2 − 943.23)]/3413 = 8789 kW

where 3413 is in Btu/kWh.

In this example, the enthalpy of the supply steam could be determined from pressure and temperature measurements. The exhaust enthalpy, however, would not be known. The determination of enthalpy in a two-phase region (where steam and moisture coexist) by pressure and temperature measurements will only yield estimates of the actual enthalpy since the actual % moisture in the steam is not known. The exhaust enthalpy, however, could be estimated by back-calculating, since the power output would be measured during testing.

The above simplified example is based on a single-stage turbine with no steam extraction or induction. All of the steam entering the turbine produces power and then leaves the turbine exhaust. In practice, steam can be extracted from a turbine for the purposes of exporting steam to a process or improving cycle efficiency by heating boiler feed water.

Frequently, turbine performance is expressed as *turbine heat rate.* The heat rate is defined as the number of Btu that must be added to the working fluid (the steam) to generate one kWh of electrical power (Btu/kWh). The heat rate is a function not only of the process conditions but also of the cycle design. In a power station, the heat rate is affected by the number of feedwater heaters in the cycle, in addition to steam pressure, temperature, exhaust pressure, steam flow, the type of unit (reheat versus nonreheat), control valve position, and cycle losses.

Turbine heat rate is mathematically defined as

$$\text{Heat input to working fluid} / \text{kW output} \tag{74.5}$$

Heat rate can also be expressed in terms of efficiency, as follows:

$$\text{Heat rate} = 3413/(\text{efficiency}) \tag{74.6}$$

where efficiency is expressed as a decimal.

Two important but frequently misunderstood parameters are the **expansion line end point (ELEP)** and the **used energy end point (UEEP).** The expansion line end point represents the turbine exhaust enthalpy that would exist if there was no exhaust loss at the turbine exit. However, in practice, the expansion of the steam from the low-pressure turbine into the condenser results in a loss that is a function of the velocity through the turbine exhaust opening (exhaust hood). In the calculation of power output, the UEEP should be used (not the ELEP) because the exhaust loss does not contribute to the generation of power.

Steam cycle efficiency is much lower than the turbine efficiency, ranging from about 20 to 46% (for condensing cycles). The main loss is the exhaust loss where heat is removed by the condenser cooling water [12].

$$\text{Exhaust loss} = \text{UEEP} - \text{ELEP (Btu/lb)} \tag{74.7}$$

74.4 Controls and Instrumentation

Control of turbine operation includes speed (rpm) and load control for turbogenerators operating at constant speed and speed and extraction and exhaust pressure control for industrial turbines where steam is used in various processes. There are other controls of the auxiliary systems and the supervisory instrumentation monitoring the turbine conditions. The industry trend is for more automatic control; today, some large power plants are fully automatic, having only one operator for the whole plant during normal operation.

Small-turbine needs in controls and instruments represent a minimum: speed governor, overspeed trip, throttle-pressure gage, throttle thermometer, exhaust-pressure gage, and tachometer. Where the turbine controls or affects exhaust-header pressure, its speed governor includes an exhaust-pressure regulator to adjust the governor control.

The list of requisite equipment grows with turbine capacity and rising steam conditions. As an example, the controls needed by some reheat turbines for automatic operation are shown in Figure 74.8. Shaft speed is one of the most important variables in running a turbine. Governing systems control speed in turbines large and small.

Speed governors are used to measure shaft speed and adjust the control/governor-valve openings to pass the needed steam flow to keep speed within governor's regulation range. To hold exact speed, speed-changer spring or bellows tension must be adjusted manually or by an automatic frequency-measuring device. When generators work in parallel, the speed changers are used to divide the total load between the turbines on the line. Governors can use a flyweight (see Figure 74.9) or hydraulic pump connected to the turbine shaft to sense its rotation.

A *load-limiting meter* overrides the speed governor to hold maximum load at any point the operator wants. The limiter controls steam flow during startup by setting governor valves as needed.

Stop and Control Valves

Valves control the flow of steam through a turbine. **Stop valves** are provided on the turbine to admit steam during normal operation or to shut off the steam very quickly in the event of an emergency. This function is often performed by a throttle-stop (also stop-throttle) valve, which controls steam flow to all inlet nozzles during turbine warmup and is fully open during load operation. These valves are normally fully opened or closed. The **control or governing valves** are located in the steam chest at the turbine

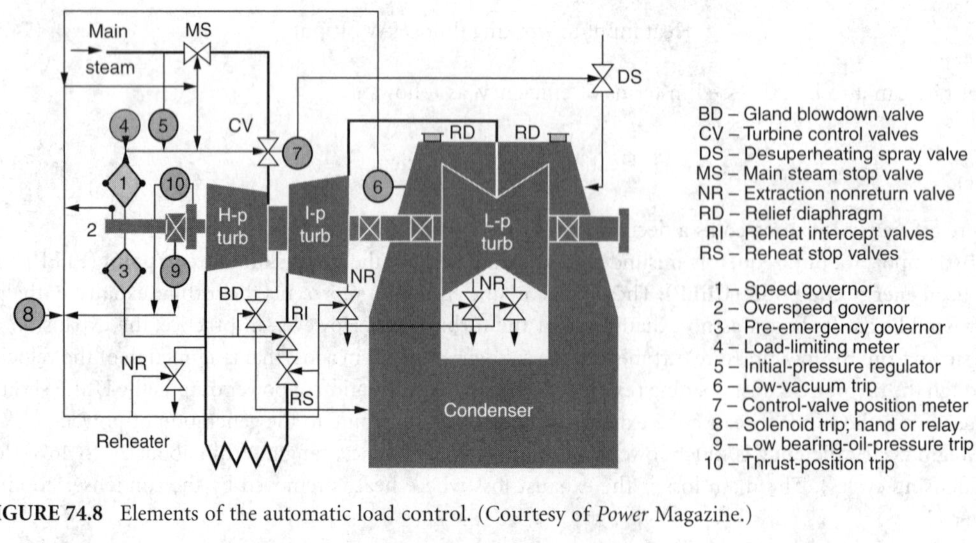

FIGURE 74.8 Elements of the automatic load control. (Courtesy of *Power* Magazine.)

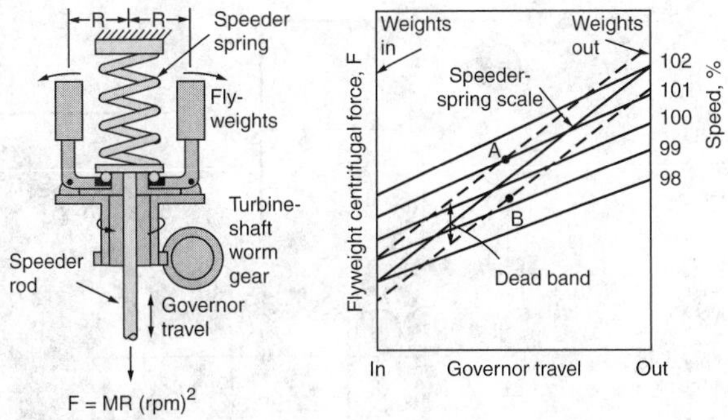

FIGURE 74.9 Flyweight governor balances the force of springs and rotating weights and moves speeder rod which is connected to a control valve. (Courtesy of *Power* Magazine.)

inlet and control the flow of steam to the turbine. On larger units, multiple control valves are used to provide better efficiency over the load range. Figure 74.8 shows locations of the valves at the turbine steam inlet. The turbine control valves can be operated in either a **full arc admission** or **partial arc admission mode.** In full arc admission, all of the control valves are opened simultaneously, with the stop valve used to control flow. In this mode of operation, there is more even heating of the turbine rotor and casing. In partial arc admission, one control valve is opened at a time (sequentially). The turbine is not as evenly heated; but the efficiency of the turbine is better, since valve-throttling losses are reduced. The valves should be periodically exercised to ensure their reliable operation and prevent turbine overspeed and possible destruction.

Other Valves

For control of the steam extractions, **extraction valves** internal to the turbine or on the extraction piping are used. In large turbine piping, **reheat intercept valves** may be used to prevent turbine overspeed in case of turbine trip.

Supervisory Instruments

A supervisory instrument system monitors several operating variables on a steam turbine. It uses sensing elements or detectors, mounted on the turbine, which transmit electrical or fiber optic signals to remote recording and/or indicating instruments. Measured variables generally include turbine-shaft vibration, shaft eccentricity, shell expansion, differential expansion, control-valve position, turbine speed, and various turbine-metal temperatures. An example of the supervisory instrumentation for a utility turbine is given in Figure 74.10.

74.5 Electric Generators [21–25]

The electric generator (see Figure 74.11) is a rotating machine that converts the mechanical work of the turbine into electrical power. This conversion is accomplished by inducing an electromotive force by the relative motion between a conductor and a magnetic field. The stationary part of a generator is called a *stator* or armature. The moving part is called a *rotor* or field. In order to create the magnetic field, an **exciter,** which is a separate rotating machine on the same shaft as the generator, provides the DC electricity to magnetize the rotor. The frequency of the AC voltage generated by a synchronous generator is proportional to the speed of the rotor in revolutions per unit time. A two-pole synchronous generator for a nonnuclear-type power plant must revolve at 3600 rpm to generate a 60-Hz voltage, as the following equation illustrates:

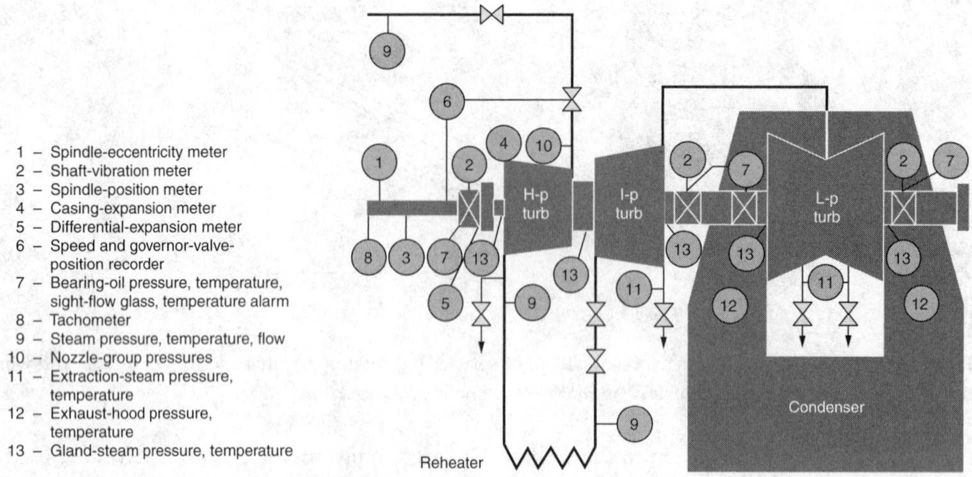

FIGURE 74.10 Typical supervisory instrumentation for a utility turbine. (Courtesy of *Power* Magazine.)

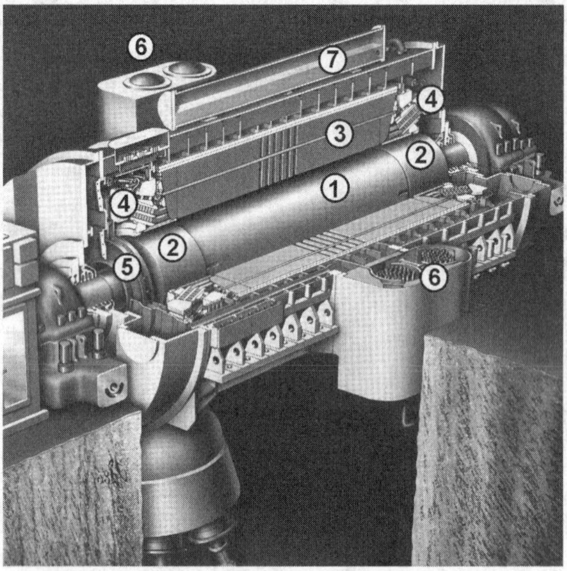

FIGURE 74.11 Schematic of an electric generator: 1, rotor; 2, retaining rings; 3, stator core; 4, stator bars; 5, fan; 6, hydrogen coolers; 7, head tank for cooling water. (Courtesy of Alstom.)

$$f = (P/2) \times (N/60) \tag{74.8}$$

where f = frequency in Hz, P = the number of magnetic poles, and N = the speed of the rotor in rpm. A nuclear turbine operating at 1800 rpm would require a four-pole generator to generate 60 Hz. Standard voltages range from 550 to 20,000 volts.

Generator Cooling

Generators must be cooled to remove the heat produced by the windings.

Stator-Winding Cooling — Liquid flow through individual strands of the stator bar has been the method used to cool stators on large generators since the mid-1950s. Low-conductivity demineralized water, the coolant used today, is supplied to the stator winding by an external pumping and cooling system. This closed system is designed to operate as an independent subloop in the turbine/generator control system. The generator is protected from loss of coolant flow and high coolant temperature by a load runback circuit, or a time-delayed trip circuit, depending on the load-following capability of the steam-supply system.

Rotor Cooling — For years, all large turbine/generators have been designed to operate in a hydrogen atmosphere, to capitalize on the improved cooling capability of hydrogen compared to air. A gas-control system performs the dual function of supplying hydrogen to the generator to maintain proper coolant operating pressure and carbon dioxide during hydrogen purging and filling operations. This system consists of regulators and controls for supplying the hydrogen and carbon dioxide and a control cabinet with the instruments necessary for monitoring generator gas pressure, purity, and temperature. Shaft oil seals and associated control equipment prevent hydrogen from leaking out of the generators. Some smaller turbine generators use air instead of hydrogen for rotor cooling.

74.6 Turbine Generator Auxiliaries

Many auxiliary systems that are required for operation are provided with a steam turbine generator. Some of the major auxiliary systems are briefly discussed below.

Steam Seals

At locations where the steam turbine shaft penetrates the casing(s), a steam seal system is used to prevent steam from leaking out of the seals which are above atmospheric pressure and air leaking into the LP turbine seals which are below atmospheric pressure. For medium and large turbines, **labyrinth seals** are used. For smaller turbines, shaft sealing can be accomplished by **graphite ring seals.** Labyrinth seals are noncontact seals between the turbine casing and rotor where ridges or strips are used to reduce steam leakage by multiple pressure drops through the seal.

A steam seal system uses steam leak-off from the high and intermediate seals during operation to seal the low-pressure seals. When the turbine is on-line, the machine is said to be "self-sealing." When the turbine is being brought on-line, a separate source of steam is used to seal the turbine prior to establishing a condenser vacuum. Figure 74.12 shows the typical flow pattern for a labyrinth steam seal system on a unit startup and during normal operation.

Internal seals that force most of the steam to flow through the blade path are also noncontact seals using similar principles as the labyrinth seals. Some use removable seal strips which are refurbished during turbine overhauls.

Turning Gear

During periods when the turbine is shut down and prior to startup, the hot rotor must be rotated to prevent bowing of the rotor shaft. This action is taken to prevent excessive vibration on startup and possible bearing damage and is usually accomplished with a motor-driven device operating at 3 to 5 rpm.

Lube Oil System

The turbine generator must be provided with an oil lubrication system for the bearings. The lubrication system consists of a number of shaft-driven and electric motor-driven pumps, filters, oil coolers, and an oil reservoir. The shaft-driven pump is used when the turbine is above 90% of rated speed. Below this speed, AC or DC motor-driven pumps are used. The high-pressure control oil system, which operates the control and stop valves and turbine governor, is supplied from the lube oil system. Bearing oil temperature is maintained at ~120°F at the inlet to the bearings. The bearings are supported in pedestals (metal boxes), from which the oil drains back into the oil reservoir.

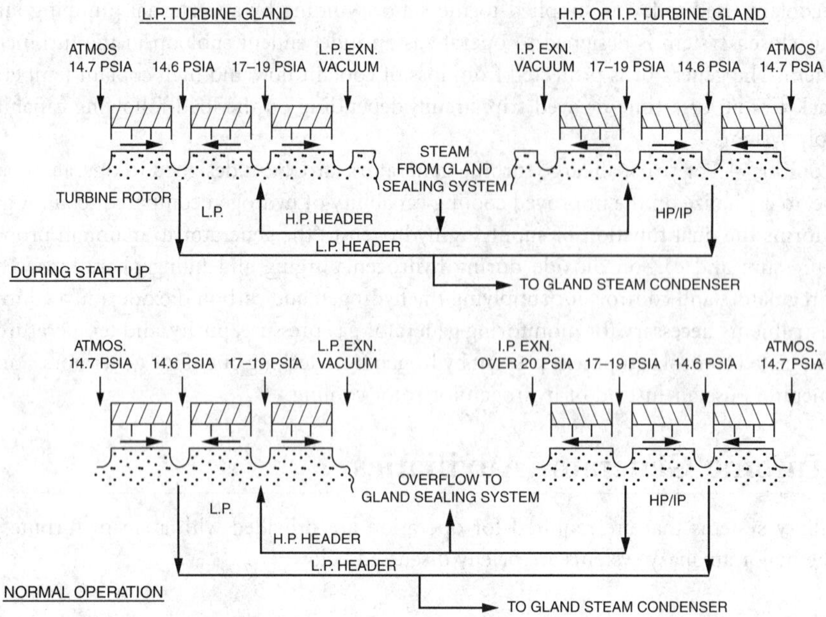

FIGURE 74.12 Typical flow pattern for a labyrinth steam seal system on a unit startup and during normal operation.

Hydraulic Fluid System

In modern turbines, the electro-hydraulic control system (EHC) uses hydraulic fluid for valve actuators. The EHC fluid system powers the hydraulic actuator at each steam valve. A fire resistant phosphate-ester fluid is usually used to minimize the fire hazard in case of a leak. The hydraulic fluid is also used in the emergency overspeed trip system.

Water Induction Protection

It is possible to induct water into the turbine from cold piping during a unit startup, an extraction line, the boiler, or a desuperheater. Water induction or **water hammer** incidents are more likely to occur during unit trips or load changes. The dense water, in comparison to the steam, can cause considerable damage to the blades, and the cold water in contact with hot turbine metal can cause cracking of metal or rubbing between moving and stationary parts due to differential expansion.

The ASME has developed a standard that should be followed when designing systems associated with a steam turbine. ASME TDP-1 provides design details on how to prevent water induction. Although the standard applies mainly to steam turbine generators, the guidelines are also applicable to mechanical drive turbines.

74.7 Steam Turbine Problems [1–7, 11, 15–19]

Turbines and their attached equipment are typically designed for a 30-year service, but many steam generating units in the United States are over 40 years old. The maintenance period for turbine disassembly, inspection, and major maintenance has now been extended up to 10 years, and up to 20-year warranties, during which there should not be any major damage, are being offered. Problems occur by wear, corrosion, erosion, cavitation, fatigue, and creep at high temperatures due to normal operation and operating errors. The main impact occurs when the turbine component damage results in an unexpected "forced outages." About 90% of the cost of forced outages is the cost of lost production and replacement power; it can be up to $1 million/day. Table 74.2 lists the major problems.

TABLE 74.2 Turbine Corrosion, Erosion, and Deposition Problems

Problem	Damage Mechanisms[a]	Root Causes[b]	Inspection and Detection[c]	Possible Safety Issues	Cost per Event[d] ($ millions)
Disc/blade attachment cracking	SCC, cf. P	D, CH	UI, V, MP	Turbine wreck	1 to 20
Disc bore cracking	SCC, P	D, M	UI	Turbine wreck	2 to 20
Blade airfoil cracking	CF, P, WE	D, CH	V, VSA	Penetration of casing	0.1 to 2
Blade root cracking	CF, SCC, P	D, CH	V, UI	Penetration of casing	0.2 to 2
Blade airfoil and valve erosion	SPE	D, O of boiler	CM, V	None	0.5 to 3
Blade airfoil damage	FOD	Cleanliness, M	V, CM	None	0.1
Stationary blade cracking	CF, P	D, O	V, DP	None	Low
HP/IP turbine rotor cracking	LCF, CR	D, O, H, M	UI, MP	Turbine wreck	2 to 30
Rotor cracking	CF, F, P	D, O	MP, VSA	Turbine wreck	2 to 10
High temperature steam Pipe cracking	Creep, LCF, graphitization	D, O	UI, MP, MET	Failure	1 to 20
LP Casing weld cracking	SCC	D, M	V, DP, UI, MP	None	Low
Casing and extraction piping	FAC	D, CH	V, UI	Steam leak	0.1 to 1
Cross-over pipe, expansion bellows	SCC	D, CH	V, UI, leak	Steam leak	0.1 to 2
Bearing wear	Wear, F	Dirty oil, O	Vibration, V, temperature	None	0.1 to 2
Thrust bearing wear	Wear	CH — deposits on blades	V, rotor position	Turbine wreck	Up to 10
Loss of MW and efficiency	Deposits, P, FOD, WE, SPE	CH, D, O, A	V, performance monitoring	None	Up to 2/year
Destructive overspeed	Fracture	O, CH	Not Applicable	Turbine wreck	Up to 200
Turbine rubbing	Wear	O, CH	Vibration	None	0.2 to 5

[a] cf., corrosion fatigue; F, fatigue; SCC, stress corrosion cracking; P, pitting; FAC, flow-accelerated corrosion; LCF, low cycle fatigue; LCCF, low cycle corrosion fatigue; WE, water droplet erosion; FOD, foreign object damage; SPE, solid particle erosion; CR, creep.

[b] D, design and material selection; CH, chemistry; O, operation; A, age; M, manufacturing and maintenance; CM, condition monitoring; MET, metallography.

[c] V, visual; UI, ultrasonic inspection; MP, magnetic particle; EC, eddy current; DP, dye penetrant; VSA, vibration signature analysis.

[d] Lost production and repairs per one event. The cost of lost production is typically much higher than the loss from repairs with a ratio of up to 10:1.

74.8 Steam Chemistry and Turbine Corrosion [11, 12, 15–19]

Chemical composition of turbine steam is important for prevention of turbine deposits, which could lower turbine efficiency and MW generating capacity and cause corrosion. Maintaining **steam purity** is often neglected. It is one of the main functions of boiler design and operation and steam cycle water purification to produce steam of the desired purity, which is determined by the turbine steam conditions. For superheated steam turbines, the concentrations of steam impurities, such as salts, hydroxides, silica, and metal oxides, should be in the low part per billion, ppb range (1 ppb is 10^{-9} by weight). For turbines using saturated and wet steam, the concentration can be higher.

When the concentration of a specific impurity in superheated steam exceeds its solubility, the impurity precipitates and deposits — having generally harmful effects. The solubility of steam impurities decreases as the steam expands and is the lowest at the lowest pressure turbine stages before condensation occurs. For a utility reheat turbine, solubility of NaCl and NaOH at this point is about 5 ppb and solubility of silica about 20 ppb. Salts, acids, and hydroxides, which concentrate by deposition on steam turbine surfaces, are hygroscopic, particularly near the saturation line, and form concentrated aqueous solutions that can be very corrosive. In the lower-pressure turbines, NaCl can concentrate from low ppb in

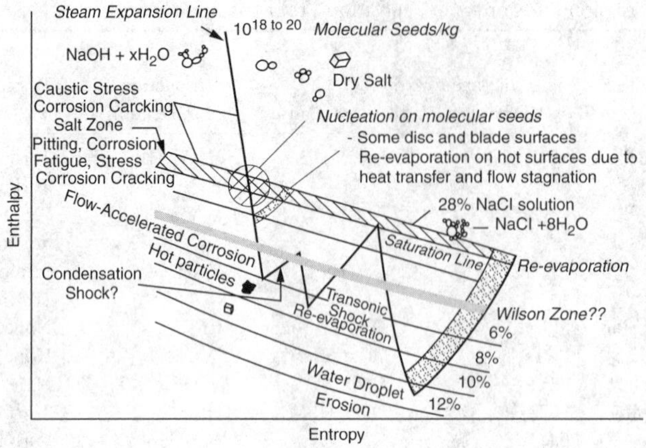

FIGURE 74.13 Mollier diagram with the LP steam turbine expansion line and regions of steam impurity concentration and corrosion.

superheated steam up to 28% solution in the so-called "salt zone." Behavior of typical steam impurities in steam turbines is illustrated in the Mollier diagram shown in Figure 74.13.

Defining Terms

Blade Stage — The combination of one stationary and one rotating row of blades.

Buckets — Turbine blades.

Condensing turbine — Any turbine with an exhaust below atmospheric pressure. Turbine exhaust steam contains moisture droplets.

ELEP — Expansion line end point; enthalpy of exhaust steam if the exhaust loss at the turbine exit is neglected.

Enthalpy — A measure of the internally stored energy; the ability of a substance to provide useful work from energy, expressed in Btu/lb-°F.

Entropy — A measure of the ability of steam to give up heat to produce work. As the entropy of steam increases, the ability of the steam to do work decreases.

Feedwater heater — A heat exchanger that uses extraction steam to heat feedwater and condensate in a power station.

Governor — Regulates the turbine speed.

Heat balance diagram — Diagram that shows the temperature, pressure, enthalpy, and flow of all streams in the cycle.

Impulse blades — Blades that use the force of the steam impacting on the blades or buckets to rotate the turbine.

Mechanical drive turbine — A steam turbine connected to a pump, fan, or a similar device to provide motive power.

Mollier diagram — Entropy–enthalpy diagram, usually used for steam turbines because it conveniently shows the steam expansion lines and all parameters needed for efficiency and blade stage evaluations.

Noncondensing or back-pressure turbine — Any turbine with an exhaust above atmospheric pressure.

Reaction blade — Blades that use the force generated by the velocity in the stages to rotate the turbine.

Steam chemistry — The concentration of contaminants or additives in steam and resulting chemical characteristics of deposits and water droplets [11, 12, 14–19].

Steam purity — The concentration of contaminants or additives in steam.

Steam quality — Percent of saturated steam by weight in a steam/water mixture. Sometimes imporperly used for chemical purity and water and steam chemistry.

Steam tables — Properties of steam (pressure, temperature, volume, enthalpy, entropy, etc.) shown in tabular or electronic form [13].
Steam turbine — A rotary engine that converts thermal energy to useful mechanical work by the impulse or reaction of steam.
Superheated, saturated, and wet steam — Terms describing the amount of liquid water in the steam.
Turbine-generator — A steam turbine connected to an electrical generator to produce electric power. Also, turbogenerator.
Turbine rotor — Shaft to which the rotating blades are attached.
UEEP — Used energy end point; enthalpy of steam at the turbine exhaust.
Water induction — The admission of water into a steam turbine.

References

1. Stodola, A. 1945. *Steam and Gas Turbines,* Peter Smith, New York.
2. Scegljajev, A. V. 1983. *Parni Turbiny (Steam Turbines),* SNTL, Prague.
3. Sanders, W. 2002. *Turbine Steam Path Damage and Maintenance (Volumes I and II),* Pennwell, Tulsa, OK.
4. Cotton, K. 1993. *Evaluating* and *Improving Steam Turbine Performance,* Cotton Fact, Rexford, NY.
5. *Main Turbine Performance Upgrade Guideline,* EPRI, Palo Alto, CA: January 1997. TR-106230.
6. Salisbury, J. 1974. *Steam Turbines and Their Cycles,* Kreiger, Huntington, NY.
7. *Turbine Steam Path Damage: Theory and Practice,* EPRI. Palo Alto, CA: June 1998. AP-108943.
8. El-Wakil, E., 1984. *Power Plant Technology,* McGraw-Hill, New York.
9. Stultz, S. and Kitto, J. eds. 1992. *Steam: Its Generation and Use,* Babcock & Wilcox, Barberton, OH.
10. Singer, J. 1981. *Combustion Fossil Power Systems,* Combustion Engineering, Inc. Windsor, CT.
11. ASME. 1989. *The ASME Handbook on Water Technology for Thermal Power Systems,* ASME, New York.
12. O. Jonas. 1983. Steam. *Kirk-Othmer Encyclopedia of Chemical Technology, Vol. 21, 3rd Edition.* John Wiley & Sons, New York.
13. ASME. 2000. *ASME International Steam Tables for Industrial Use,* ASME, New York.
14. Moisture Nucleation in Steam Turbines. EPRI, Palo Alto, CA: October 1997. TR-108942.
15. Jonas, O. Steam Turbine Efficiency and Corrosion: Effects of Surface Finish, Deposits, and Moisture. EPRI, Palo Alto, CA: October 2001. Report 1003997.
16. Jonas, O. 1985. Steam Turbine Corrosion. *Materials Performance,* 24(2):9–18.
17. *Turbine Steam, Chemistry and Corrosion: Experimental Turbine Tests.* EPRI. Palo Alto, CA: June 1997. TR-108185.
18. Jonas, O. and Dooley, B. 1997. Major turbine problems related to steam chemistry: R&D, root causes, and solutions. *Proceedings: Fifth International Conference on Cycle Chemistry in Fossil Plants.* EPRI, Palo Alto, CA. TR-108459.
19. Jonas, O. 1994. On-Line Diagnosis of Turbine Deposits and First Condensate, 55th Annu. Int. Water Conf., Pittsburgh, PA, 1994.
20. Review of Corrosion Resistant Coatings for Steam Turbine Components. EPRI. Palo Alto, CA: November 1981. CS-2124.
21. Kaiser, J. 1991. *Electrical Power: Motors, Controls, Generators, Transformers,* Goodheart-Willcox, Tinley Park, IL.
22. Belove, C. 1986. *Handbook of Modern Electronics and Electrical Engineering,* John Wiley & Sons, Inc. New York.
23. Generator Cooling System Operating Guidelines: Cooling System Maintenance and Performance Guidelines During Start-Up, Operation, and Shutdown Prevention of Flow Restrictions in Generator Stator Water Cooling Circuits. EPRI. Palo Alto, CA. December 2001. TR-1004004.
24. Svoboda, R. and Jonas, O. 2003. Electric Generators. *Low Temperature Corrosion Problems in Fossil Power Plants: State of Knowledge Report.* EPRI. Palo Alto, CA.

25. Scarlin, R. et al. 1984. Environment induced cracking of generator rotor retaining rings, In *Corrosion in Power Generating Equipment*, Speidel, M. and Atrens, A., Eds., Plenum Press, New York.

Further Information

Elliott, T. C. 1989. *Standard Handbook of Powerplant Engineering,* McGraw-Hill, New York.

Potter, P. J. 1959. *Power Plant Theory And Design,* John Wiley & Sons, New York.

ASME (American Society of Mechanical Engineers) — provides standards and codes for design and testing, organizes conference and research

ASME Performance Test Codes: PTC6, Steam Turbines (1996), PTC20.1, Speed and Load Governing Systems for Steam Turbine-Generator Units (1988), PTC20.3, Pressure Control Systems Used on Steam Turbine-Generator Units (1986), PTC20.2, Overspeed Trip Systems for Steam Turbine Generator Units, (1986), PTC19.11, Water and Steam in the Power Cycle (1974).

Electric Power Research Institute (EPRI): Numerous reports on corrosion, design, maintenance, etc.

ASTM (American Society for Testing and Materials) — publishes material and testing standards, organizes conferences.

ASM International (American Society of Materials) — publishes magazines and handbooks, organizes conferences.

NACE International (National Association of Corrosion Engineers) — publishes magazines and handbooks, organizes conferences.

IAPWS (International Association for Properties of Water and Steam) — compiles thermodynamic and transport properties of water and steam (steam tables) and properties of chemical solutions relevant to steam generation, organizes periodic conferences, recommends research.

pc-GAR (Generating Availability Report, Statistical Data 1982-2000). North American Electric Reliability Council (NERC). 2002. Updated annually.

75

Cogeneration: Combined Heat and Power Systems

M. Krarti
University of Colorado

This chapter provides a general description of the basic concepts of combined heat and power generation systems, their benefits, and their applications. First, an overview of the U.S. power industry with some historical background is presented. Then, a review of the fundamental theory of power cycles and cogeneration systems is outlined, followed by a description of commercially available cogeneration systems and their components. In addition, some guidelines and examples are presented to carry out feasibility analysis of cogeneration systems. Finally, a cogeneration case study is presented.

75.1 Introduction

Combined heat and power (CHP) and cogeneration are terms used interchangeably to denote the simultaneous generation of power (electricity) and usable thermal energy (heat) in a single, integrated system. A CHP plant derives its efficiency, and hence lower costs, by recovering and utilizing the heat produced as a byproduct of the electricity generation process, which would normally be wasted to the environment. The overall fuel efficiency of typical CHP installations can be in the range of 70 to 90%, compared with 35 to 50% for conventional electricity generation. Overall, CHP achieves a 35% reduction in primary energy usage compared with remote power stations and heat-only boilers. Moreover, CHP avoids transmission and distribution losses since it generally supplies electricity close to the generation site.

0-8493-1586-7/05/$0.00+$1.50

While the CHP concept is not new, it has been applied to a wide range of commercial buildings only recently. Indeed, until the 1980s, cogeneration systems were used only in large industrial or institutional facilities with high electricity demand (typically over 1000 kW). After the energy crisis of 1973, during which fuel and electricity prices had increased significantly (by a factor of five), the U.S. government passed in 1978 the National Energy Act (NEA), which includes the Public Regulatory Policies Act (PURPA). The PURPA regulations have forced utilities to purchase electricity and to provide supplementary or backup power to any qualified cogeneration facilities. The Energy Policy Act of 1992 has increased even more the appeal of cogeneration systems by opening up transmission line access and retail wheeling.

In addition to the favorable regulations, the development of energy-efficient cogeneration systems and small preengineered packaged cogeneration units has provided the needed incentives to encourage the implementation of systems capable of generating electricity and heat for commercial, institutional, and even residential applications. Currently, cogeneration systems are available over a wide range of sizes from less than 50 kW (micro-systems) to over 100 MW. Moreover, advances in controls have provided better procedures to operate and integrate the various components of cogeneration systems (including prime movers, electrical generators, and heat recovery systems). More recently, energy service companies (ESCOs) have increased awareness and interest of the public in onsite generation technologies.

In addition to their better overall energy efficiency compared to conventional utility plants, cogeneration systems offer the following benefits:

- Cleaner power generation sources with reduced No_x and carbon emissions
- Increased reliability and quality of electrical power since customers are less vulnerable to blackouts from utility power lines
- Added national energy security with diversified sources and locations of power generation
- Avoidance of transmission and distribution costs since no new lines would be needed if distributed cogeneration plants are available

To evaluate the feasibility of a cogeneration system, several technical and economic aspects as well as regulatory issues should be considered. Although this chapter provides some discussion of the main regulatory considerations and financial options for cogeneration in the U.S., the main focus of the chapter is to provide technical information and engineering principles that are necessary to understand the design and the operation of cogeneration systems. First, an overview of past, current, and future status of the electricity generation industry in the U.S. is presented.

75.2 Electricity Generation in the U.S.

Energy is essential to modern industrial societies. The availability of adequate and reliable energy supplies is required to maintain economic growth and to improve living standards. The major energy sources include fossil fuels (petroleum, natural gas, and coal), hydropower, and nuclear energy. Table 75.1 illustrates the progression of energy consumption by region throughout the world. As expected, the industrialized countries (including North America and Western Europe) consumed more than 50% of the total energy used throughout the world during 1997. The U.S. alone, with less than 5% of the world's population, used about one-fifth of the world total energy consumption in 1997. Associated with the energy consumption are environmental and health impacts that have been only properly investigated in the last decade. In particular, the burning of fossil fuels has significantly increased levels of carbon emissions as indicated in Table 75.1. The carbon emissions are believed to have major effects on changing the global climate by increasing global temperatures. To reduce the environmental impacts of producing electricity, several federal and state governments in the U.S. and abroad have set regulations and environmental restrictions to foster new alternatives for power generation including CHP plants and distributed generation systems. Additional incentives have been proposed for reducing energy consumption through energy efficiency and demand-side management programs.

To better understand the potential for combined heat and power systems, a brief review of the U.S. electrical power generation is provided below.

TABLE 75.1 Energy Consumption and Carbon Emissions by Region

Region	Energy Consumption [Quadrillion Btu]			Carbon Emissions (Million Metric Tons)		
	1990	1997	2010[a]	1990	1997	2010[a]
Industrialized countries	183	204	240	2850	3039	3563
Eastern Europe	76	53	63	1337	878	1151
Developing countries:						
Africa	9	11	16	180	214	292
Asia	51	75	126	1067	1522	2479
Central/South America	14	18	30	174	225	399
Middle East	13	18	26	229	297	552
Total	87	122	198	1649	2258	4930
Total (World)	346	379	500	5836	6175	8146

[a] Projections.

Source: OECD, 1999, Economic Statistics by the Organization for Economic Cooperation and Development, http://www.ocde.org.

TABLE 75.2 Annual U.S. Electric Energy Generated by Utilities by Primary Energy Sources (in Billion kWh)

Primary Energy Source	1972	1982	1992	1998	2001
Coal	771	1192	1576	1873	1904
Natural gas	376	305	264	531	613
Petroleum products	274	147	89	129	126
Nuclear power	54	283	619	674	769
Renewable energy	274	314	254	400	297
Total	1749	2241	2797	3212	3719

Source: EIA, 2003, *Annual Energy Review,* Department of Energy, Energy Information Administration, http://www.doe.eia.gov.

TABLE 75.3 Annual U.S. Electric Energy Sold by Utilities by Sector (in Billion kWh)

End-Use Sector	1972	1982	1992	1998	2001
Residential	539	730	936	1130	1201
Commercial	359	526	761	979	1085
Industrial	641	745	973	1051	994

Source: EIA, 2003, *Annual Energy Review,* Department of Energy, Energy Information Administration, http://www.doe.eia.gov.

In the U.S., electricity is generated from either power plants fueled from primary energy sources (i.e., coal, natural gas, or fuel oil) or from nuclear power plants or renewable energy sources (such as hydroelectric, geothermal, biomass, wind, photovoltaic, and solar thermal sources). Coal is the fuel of choice for most existing U.S. electrical power plants as shown in Table 75.2. However, gas-fired power plants are expected to be more common in the future due partly to environmental concerns but also due the availability of more efficient and reliable combustion turbines.

The quantity of electricity sold by U.S. utilities has increased steadily for all end-use sectors as indicated by the data summarized in Table 75.3. The increase in electricity consumption could be even higher without the various energy conservation programs implemented by the federal or state governments and utilities. For instance, it is estimated that the demand-side-management programs provided by utilities saved about 35 billion kWh in electrical energy use during 1992 and over 56 billion kWh in 1997 (EIA, 2003).

The prices of electricity for all end-use sectors have actually decreased since 1982 after a recovery period from the 1973 energy crisis, as illustrated in Table 75.4. As expected, industrial customers enjoyed

TABLE 75.4 Average Retail Prices of Electric Energy Sold by U.S. Utilities by Sector in 1992 cents per kWh

End-Use Sector	1972	1982	1992	1998	2000
Residential	7.2	9.8	8.2	8.3	8.2
Commercial	6.9	9.8	7.7	7.4	7.2
Industrial	3.6	7.1	4.8	4.5	4.5

Source: EIA, 2003, *Annual Energy Review*, Department of Energy, Energy Information Administration, http://www.doe.eia.gov.

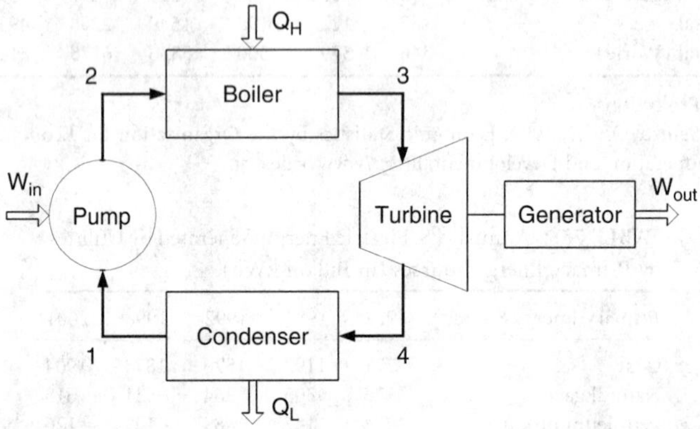

FIGURE 75.1 A rankine cycle used for power generation.

the lowest electricity prices over the years. Meanwhile, the cost of electricity for residential customers remained the highest.

75.3 Basics of Power Generation Cycles

Two types of heat engines are commonly used to generate electricity for both power-only plants and combined heat and power systems:

- **External combustion** engines use **water** as a working fluid. Steam power plants, operating as rankine cycles, are examples of external combustion systems. A typical rankine cycle includes a boiler, one or several turbines, a condenser, and a pump. Each of these components should be analyzed to determine the overall performance of the rankine cycle. One of the main advantages of a rankine cycle power plant is its capability of utilizing several types of fuel including coal, chemical wastes, heavy and light oils, biomass, and even natural gas. Figure 75.1 illustrates common components for a rankine cycle used to generate electricity.
- **Internal combustion engines** use **air** (actually a mixture of air, fuel, and combustion products) as a working fluid. Air-standard assumptions or cold-air-standard assumptions are generally used to analyze the internal combustion engines. Otto cycle (used for gasoline reciprocating engines), Diesel cycle (used for diesel reciprocating engines), and Brayton cycle (used for gas-turbines) are common internal combustion engines. The thermal efficiency of common internal combustion engines is summarized in Table 75.5 as a function of the compression ratio, r, the cutoff ratio, r_c, and the pressure ratio, r_p [Cengel and Turner, 2001].

The thermal performance of all heat engines is expressed in terms of the first law thermal efficiency, η_{th}, as the ratio of net work produced, W_{net}, to the total heat input, Q_H:

TABLE 75.5 Basic Processes and Thermal Efficiency of Common Internal Combustion Cycles

Cycle	Basic Processes of the Simplified Cycle	Thermal Efficiency
Otto for spark-ignition (SI) engines (closed cycle)	**1-2:** isentropic compression; **2-3:** constant volume heat addition; **3-4:** isentropic expansion; **4-1:** constant volume heat rejection	$\eta_{th,Otto} = 1 - \frac{1}{r^{k-1}}$
Diesel for compression-ignition (CI) engines (closed cycle)	**1-2:** isentropic compression; **2-3:** constant pressure heat addition; **3-4:** isentropic expansion; **4-1:** constant volume heat rejection	$\eta_{th,Diesel} = 1 - \frac{1}{r^{k-1}} \left[\frac{r_c^k - 1}{k(r_c - 1)} \right]$
Brayton for gas-turbines (open cycle)	**1-2:** isentropic compression; **2-3:** constant pressure heat addition; **3-4:** isentropic expansion; **4-1:** constant pressure heat rejection	$\eta_{th,Diesel} = 1 - \frac{1}{r_p^{(k-1)/k}}$

$$\eta_{th} = \frac{W_{net}}{Q_H} = \frac{W_{out} - W_{in}}{Q_H} \tag{75.1}$$

where W_{out} is the power output from the turbine-generator system and W_{in} is the power required to operate the cycle as illustrated in Figure 75.1 for a rankine cycle.

To better understand how to improve the thermal performance of power cycles, it is a common practice to consider the Carnot cycle with only reversible processes. Indeed, the Carnot cycle is the most efficient power cycle that can be theoretically executed between a heat source, T_H, and a heat sink, T_L. Carnot showed that the maximum possible thermal efficiency for a reversible power cycle operating between temperatures T_H and T_L is:

$$\eta_{th,Carnot} = 1 - \frac{T_L}{T_H} \tag{75.2}$$

Figure 75.2 illustrates the T-S (Temperature–Entropy) diagram of a Carnot power cycle operating between temperatures T_H and T_L. Equation (75.2) indicates that increasing the temperature T_H and lowering the temperature T_L are two options to increase the thermal efficiency of a power cycle. For a Rankine cycle power plant, T_L is typically 15°C (59°F) if the condenser is cooled by a nearby lake or river, and T_H is the steam temperature produced by the boiler. The maximum steam temperature achieved by the current boiler technology is about 620°C (1150°F). New materials such as ceramics offer promising applications to build boilers that can withstand higher steam pressure and temperatures. Thus, the highest Carnot efficiency that can be achieved today for vapor power cycles is about 68%. Actual efficiencies for utility power-only plants range from 30 to 45%.

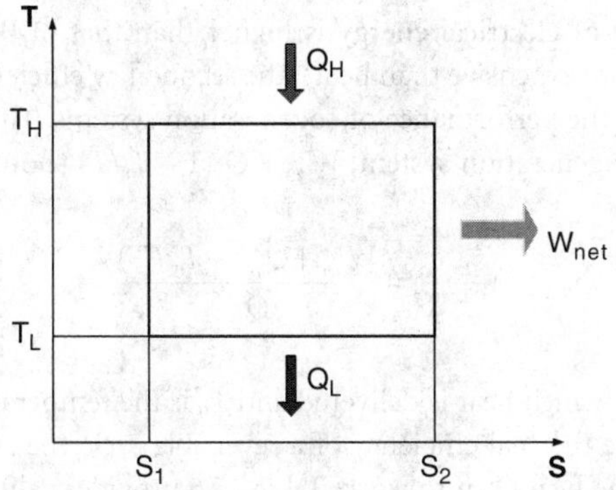

FIGURE 75.2 A schematic T-S diagram for a Carnot power cycle.

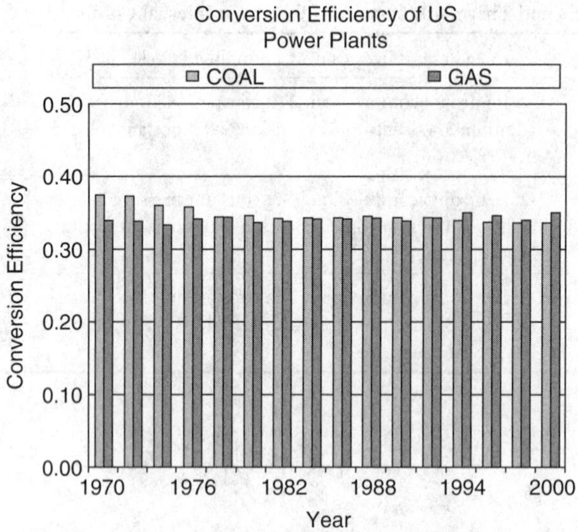

FIGURE 75.3 Annual average efficiencies for power-only utilities in the U.S since 1970.

Figure 75.3 illustrates the historical variation of the conversion efficiency achieved by U.S. electrical utility power plants from 1970 to 2000. Since 1970, the efficiency has remained almost constant at about 34% for both coal-fired and natural gas-fired power plants. These low efficiencies are inherent to the rankine cycle with the need to dispose of heat in the condenser as well as in the boiler exhausts. Indeed, a rankine cycle has to reject (thus waste) heat Q_L to the environment as indicated in Figure 75.1.

If a portion of the heat rejection, Q_L, can be recovered and used through heat recovery systems, then the overall efficiency of the plant can be significantly increased. A combined heat and power (or cogeneration) plant provides a single and integrated system to generate electrical power, W_{net}, and useful thermal energy, Q_u ($Q_u < Q_L$). While the concept of recovering heat from a power cycle (i.e., combined heat and power generation) has been known and available since the 1800s, cogeneration has regained popularity only recently due to the incentives provided by the Public Utilities Regulatory Policies Act (PURPA) of 1978.

The first law thermal efficiency of a cogeneration plant that uses an energy input Q_H to produce a net electrical power, W_{net}, and a useful thermal energy, Q_u, can be defined as:

$$\eta^{I}_{th,cogen} = \frac{W_{net} + Q_u}{Q_H} \tag{75.3}$$

However, since the quality of electrical energy is higher than that of thermal energy (since work is usually two or three times more expensive than heat), the second law efficiency, using the exergy concept, would be a better measure of the performance of cogeneration systems. Thus, instead of using $W_{net} + Q_u$ to quantify the output of a cogeneration system, $W_{net} + Q_u(1 - T_0/T)$ should be considered:

$$\eta^{II}_{th,cogen} = \frac{W_{net} + Q_u(1 - T_0/T)}{Q_H} \tag{75.4}$$

where T is the temperature at which heat is delivered and T_0 is the temperature of the environment. The term $(1 - T_0/T)$ represents the thermal efficiency of a reversible cycle (i.e., Carnot cycle), as provided by Equation (75.2), to convert the heat Q_u into work. Table 75.5 provides values for the term $(1 - T_0/T)$ for selected heat delivery temperatures, T, when the environment is set at $T_0 = 20°C$ (68°F)

TABLE 75.6 Values for $(1 - T_0/T)$ for Various Heat Delivery Temperatures with $T_0 = 20°C$ (68°F)

Delivery Temperature, °C (°F)	100°C (212°F)	200°C (392°F)	300°C (572°F)	400°C (752°F)	500°C (932°F)	600°C (212°F)
$(1 - T_0/T)$	0.214	0.381	0.489	0.565	0.621	0.664

It should be noted that typical cogeneration systems deliver thermal energy at temperatures in the range of 300°C (572°F) and 400°C (752°F), thus a value of 0.5 can be used for the term $(1 - T_0/T)$ to estimate the second law thermal efficiency of most cogeneration systems as expressed by Equation (75.4). As discussed later in this chapter, the PURPA efficiency is actually based on Equation (75.4) using a value of 0.5 for the term $(1 - T_0/T)$.

75.4 History of Cogeneration

Systems for combined heat and power generation have been in existence since the 1880s throughout the U.S. and Europe. Indeed, several industrial facilities generated their own electricity and steam using coal-fired boilers and steam-turbine generators. It is estimated that CHP systems produced up to 58% of the total electricity generated in the U.S.

However, in the mid-1900s, large central electrical power plants have been built with reliable utility grids. The electricity from these plants was produced at relatively low cost. As a consequence, industrial facilities started to purchase electricity from the power plants and thus reduced gradually their reliance on on-site generation. The contribution of CHP systems to the total U.S. electric power generation represented 15% in 1950 and only 4% in 1974 [EIA, 2000]. Large electric utility holding companies have been formed and expanded since the early 1900s. In 1920, the majority of the U.S. power industry was controlled by few privately owned electric power holding companies. These holding companies abused their power and charged consumers higher prices for electricity.

To reduce the monopoly of the privately owned electric utility holding companies, the federal government intervened by passing the Public Utility Holding Company Act (PUHCA) during 1935. Under the provisions of the PUHCA, the electric utility holding companies became regulated by the Securities and Exchange Commission. In a further effort to reduce electricity prices, government-owned hydroelectric power facilities were built including the Hoover Dam in 1936. The Bonneville Project Act of 1937 provided the federal government with means to oversee transmission and marketing of power produced from hydroelectric facilities. By 1941, power produced from publicly owned facilities represented 12% of the total utility generation.

Until the early 1970s, utilities were able to meet the ever-increasing electrical power demands at decreasing prices due to the economies of scale, technological advances, and declining fuel costs. However, a series of events that occurred during the 1970s has significantly impacted the U.S. electric power industry. These events include the oil embargo in 1973–1974 and the passage of the Clean Air Act in 1970 as well as the Energy Supply and Environmental Coordination Act in 1974. As a result of these events, prices of electrical power increased dramatically in the 1980s. Between 1973 and 1985, the prices of fuels and electricity increased by a factor of five.

As a reaction to the oil embargo, the federal government passed in 1978 the National Energy Act (NEA) in order to reduce U.S. dependence on foreign oil, develop alternative energy sources, and encourage energy conservation. The NEA comprises five different statutes:

- Public Utility Policy Act (PURPA)
- Energy Tax Act
- National Energy Conservation Policy Act
- Powerplant and Industrial Fuel Use Act
- Natural Gas Policy Act

PURPA has the most significant impact on the electric power industry and on the development of cogeneration in the U.S. Indeed, PURPA allowed nonutility facilities that meet certain ownership and efficiency criteria to sell electric power to utility companies. Specifically, PURPA sets the following legal obligations for the electric utility companies toward cogenerators defined as Qualified Facilities (QFs):

- Utility companies have to purchase cogenerated energy and capacity from QFs.
- Utility companies have to sell energy and capacity to QFs.
- Utility companies have to provide to QFs supplementary power, backup power, maintenance power, and interruptible power.
- Utility companies have to provide access to transmission grid to "wheel" to other electric utility companies.

The term "wheeling" refers to the process by which utilities can buy or sell electricity to or from other utilities in order to meet peak demands or shed excess generation. The wheeling process enables utilities to spend less on peaking power plants, thereby lowering capital expenditures required to meet high periods of electric power demands.

Additional legislation passed in the 1990s, including the Clean Air Act Amendments (CAAA) of 1990 and the Energy Policy Act (EPACT) of 1992, has provided new opportunities for cogeneration. In particular, EPACT provides incentives for nonutility generators to enter the wholesale market for electrical power by exempting them from the PUHCA constraints. The law creates a new category of power producers known as Exempt Wholesale Generators (EWGs). These EWGs are different from PURPA QFs in that they are not required to meet PURPA's cogeneration criteria. In addition, utilities are not required to purchase electric power from EWGs.

Table 75.7 summarizes the major federal legislation that affected the U.S. power industry since the early 1930s.

Because of the passage of PURPA, multiple megawatt CHP projects have been developed and built especially at large industrial facilities including pulp and paper, steel, chemical, and refining plants. Recent advances in reciprocating engines and microcombustion turbines have made CHP more cost-effective for small applications, such as fast-food restaurants, as well as commercial buildings.

It is estimated that in 2000, CHP accounted for about 7.5% of electricity generation capacity and almost 9% of electricity generated in the U.S. [EIA, 2003]. Both the Department of Energy (DOE) and the Environmental Protection Agency (EPA) have recently set a goal to double U.S. CHP capacity between 1999 and 2010, from 46 to 92 GW. If this goal is achieved, CHP will represent about 14% of U.S. electric generating capacity. Figure 75.4 illustrates the progression of the actual and projected CHP contribution in generating U.S. electrical power. Unlike some European countries, the share of CHP in the total U.S. electricity generation remains rather small as indicated in Table 75.8. In Denmark, for instance, CHP generates over 40% of the total national electric power.

75.5 Components of Cogeneration Systems

Types of Cogeneration Systems

There are several types of cogeneration systems that are commercially available. In general, three categories of cogeneration systems can be considered:

1. *Conventional Cogeneration Systems:* These systems consist of large cogeneration units (more than 1000 kW) and require a thorough design process to select the size of all equipment and components (i.e., prime movers, electrical generators, and heat recovery systems).
2. *Packaged Cogeneration Systems:* These systems are small (below 1000 kW) and are easy to design and install since they are preengineered and preassembled units.
3. *Distributed Generation Technologies:* Some cogeneration systems can use a number of technologies that have been recently developed to produce both electricity and heat, including fuel cells.

TABLE 75.7 Major Federal Legislation Affecting the U.S Electric Power Industry

Legislation	Year	Main Scope
Public Utility Holding Company Act (PUHCA)	1935	To reduce utility industry abuses by giving the Securities and Exchange Commission the authority to oversee holding companies
Bonneville Project Act	1937	To create the Bonneville Power Administration (BPA) to oversee the transmission and marketing of power produced in Northwest dams
Clean Air Act	1970	To establish programs to reduce emissions
Energy Supply and Environmental Coordination Act (ESECA)	1974	To allow the federal government to prohibit electric utilities from burning natural gas or petroleum products
Public Utility Regulatory Policies Act (PURPA)	1978	To promote conservation of electric energy by allowing nonutility generators and qualified cogenerators to sell power to utilities
Energy Tax Act (ETA)	1978	To allow tax credits for investment in cogeneration equipment and renewable technologies; these incentives were curtailed in the mid-1980s
National Energy Conservation Policy Act	1978	To require utilities to develop residential energy conservation plans in order to reduce growth in electricity demand
Powerplant and Industrial Fuel Use Act	1978	To replace ESECA of 1974 and extend federal government prohibition on the use of natural gas and petroleum in new power plants
Pacific Northwest Electric Power Planning and Conservation (PNEPPC) Act	1980	To create PNEPPC Council in order to coordinate the conservation and resource acquisition of the BPA
Electric Consumers Protection Act (ECPA)	1986	To set new environmental criteria in licensing hydroelectric power plants; to reduce significantly PURPA benefits for new hydroelectric projects; and to increase the enforcing powers of FERC
Clean Air Act Amendments (CAAA)	1990	To establish a new emissions-reduction program; generators of electricity are made responsible for large portion of the sulfur dioxide and nitrogen oxide reductions
Energy Policy Act (EPACT)	1992	To create a new category of electricity producers and to authorize FERC to open up the national electricity transmission system to wholesale suppliers

Source: EIA, 2000, *The Changing Structure of the Electric Power Industry 2000: An Update*, Report from Energy Information Administration, Washington, DC.

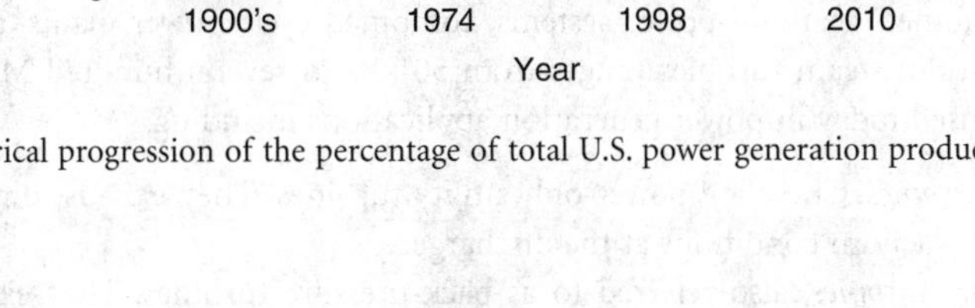

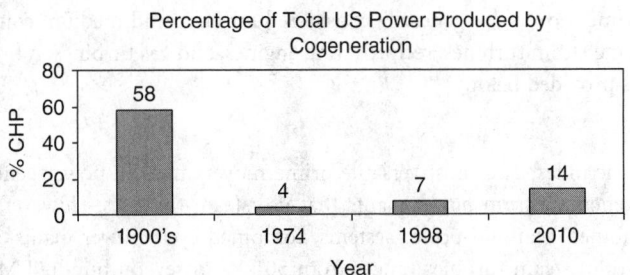

FIGURE 75.4 Historical progression of the percentage of total U.S. power generation produced by CHP.

TABLE 75.8 Share of CHP in Total National Power Generation for Selected Countries

Country	Denmark	Netherlands	Germany	Italy	U.S.	U.K.	France
Contribution of CHP (%)	40	30	14	12	7	5	3

Source: EIA, 2000, *The Changing Structure of the Electric Power Industry 2000: An Update*, Report from Energy Information Administration, Washington, DC.

Conventional Cogeneration Systems

A typical cogeneration plant consists of several pieces of equipment to produce electricity and heat (in the form of either steam or hot water). The quantity and the type of equipment in a cogeneration plant depend on the size of the system and the procedure used to generate electricity and heat. Generally, a conventional cogeneration system includes the following components:

1. A prime mover: this is the most important equipment in a cogeneration system. It is typically a turbine that generates mechanical power using a primary source of fuel. Three turbine types are commonly used in cogeneration plants: turbines operated by steam generated from boilers, gas turbines fueled by natural gas or light petroleum products, and internal combustion engines fueled by natural gas or distillate fuel oils.
2. A generator: a device that converts the mechanical power to electrical energy.
3. A heat recovery system: a set of heat exchangers that can recover heat from exhaust or engine cooling and convert it into a useful form (i.e., steam or hot water).

To operate a cogeneration plant, a robust control system is needed to ensure that all the individual pieces of equipment provide the expected performance. Two basic operation cycles are used to generate electricity and heat: either a bottoming cycle or a topping cycle.

Bottoming Cycle — In this cycle, the generation of heat is given the priority to supply process heating to the facility. Thermal energy is produced directly from fuel combustion (in the prime mover). Heat is then recovered and fed to the generator to produce electricity as illustrated in Figure 75.5(a). Industrial plants characterized by high-temperature heat requirements (such as steel, aluminum, glass, and paper industries) typically use bottoming cycle cogeneration systems.

Topping Cycle — Unlike the bottoming cycle, the generation of electricity takes precedence over the production of heat as indicated in Figure 75.5(b). The waste heat is then recovered and converted to either steam or hot water. Most existing cogeneration systems are based on topping cycles. A hybrid of a topping cycle commonly used by several industrial facilities and even by electrical utilities is the combined cycle as depicted in Figure 75.5(c). In this cycle, a gas turbine is typically used to produce electricity. The exhaust gas is then fed to a heat recovery steam generator to generate more electricity using a steam turbine. For a cogeneration plant, a small portion of steam can be converted into a useful form of thermal energy.

Three types of prime movers are generally considered for large and medium conventional cogenerations systems including steam turbines, reciprocating engines, and gas turbines. A brief overview of each prime mover type is provided below.

Steam Turbines

Steam turbines are the oldest and most versatile prime movers used in power generation. In the U.S., most electricity is generated from power plants that use steam turbines. However, steam turbines are also utilized in combined heat and power systems, combined cycle power plants, and district heating systems. The capacity of steam turbines ranges from 50 kW to several hundred MW. Several types of steam turbines are used today in power generation applications including:

- *Condensing turbines.* These are power-only utility turbines. They exhaust directly to condensers that maintain vacuum conditions at the discharge.
- *Noncondensing turbines,* also referred to as back-pressure turbines. They exhaust steam to the facility mains at conditions close to the process heat requirements.
- *Extraction turbines.* They have openings in their casing for extraction of a portion of the steam at some intermediate pressure before condensing the remaining of the steam.

The electrical generating efficiency of steam turbines varies from 37% for large electric utility plants to 10% for small plants that produce electricity as a byproduct of steam generation. The common applications of steam turbines for combined heat and power systems involve industrial processes where

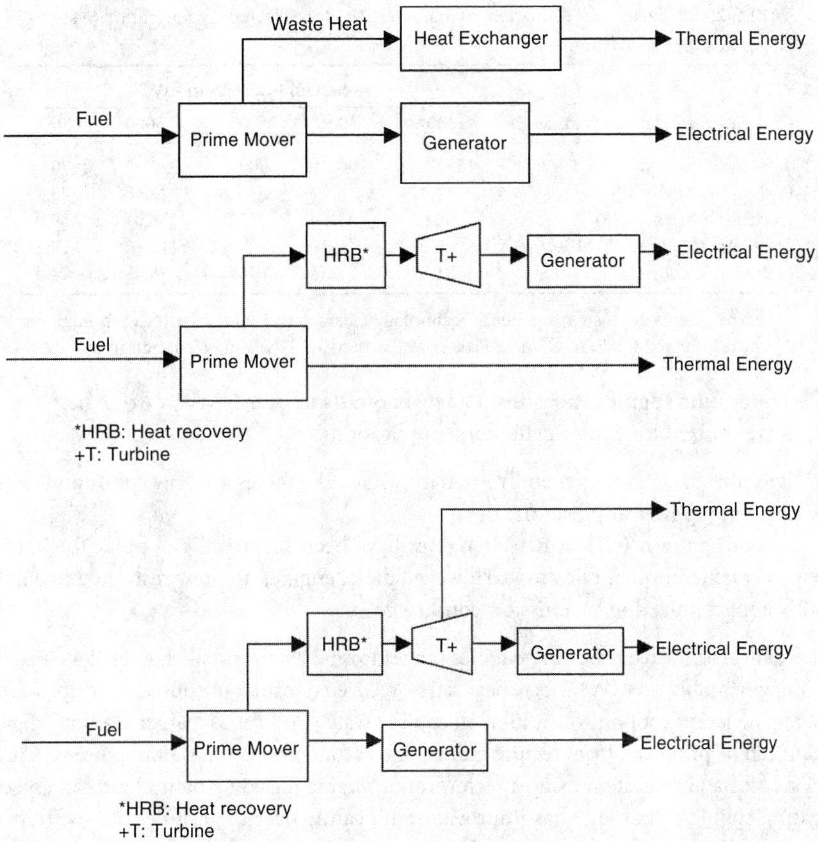

FIGURE 75.5 (a) Topping cycle cogeneration system; (b) Bottoming cycle cogeneration system; (c) Combined cycle cogeneration system. (*Source:* Krarti, M. 2000. *Energy Audit of Building Systems: An Engineering Approach*, CRC Press, Boca Raton, FL.)

TABLE 75.9 Cost and Performance Characteristics for Selected Cogeneration Systems Using Steam Turbines

	System Capacity in kW		
Cost and Performance Parameters	500	3000	15,000
Total installed cost ($/kW)	918.0	385.0	349.0
Fuel input (MMBtu/hr)	26.7	147.4	549.0
Electric efficiency (%)	6.4	6.9	9.3
Steam to process (MMBtu/hr)	19.6	107.0	386.6
Overall efficiency (%)	79.6	79.5	79.7

Source: Energy Nexus Group, 2002. Technology Characterization: Steam Turbines, A report prepared for EPA, Climate Protection Partnership Division, Washington, DC.

solid or waste fuels are readily available. In the U.S., it is estimated that more than 580 industrial and institutional facilities use steam engines to produce about 19,000 MW of electric capacity. Table 75.9 lists typical cost and performance parameters for selected cogeneration systems using steam turbines.

Reciprocating Engines

Reciprocating internal combustion engines are widely used technology to generate power for several applications including institutional and industrial facilities and combined heat and power systems. The

TABLE 75.10 Cost and Performance Characteristics for Selected Cogeneration Systems Using Reciprocating Engines

	System Capacity in kW				
Cost and Performance Parameters	100	300	800	3000	5000
Total installed cost (2001 $/kW)	1515	1200	1000	920	920
Fuel input (MMBtu/hr)	1.11	3.29	8.20	28.48	43.79
Electrical efficiency (%)	30.6	31.1	33.3	36.0	39.0
Total heat recovered (MMBtu/hr)	0.57	1.51	3.50	11.12	15.28
Overall efficiency (%)	81.0	77.0	76.0	75.0	74.0

Source: Energy Nexus Group, 2002. Technology Characterization: Reciprocating Engines, A report prepared for EPA, Climate Protection Partnership Division, Washington, DC.

capacity of reciprocating engines varies from a few kilowatts to over 5 MW. Two basic types of reciprocating engines are utilized for power generation applications:

- Spark ignition (SI) engines generally use natural gas. SI engines are now commonly used for duty-cycle stationary power applications.
- Compression ignition (CI) or diesel engines have been historically popular for both small and large power generation applications. However, diesel engines are now restricted to emergency and standby applications due to emission concerns.

The electric efficiency of reciprocating engines ranges from 28% for small engines (less than 100 kW) to 40% for large engines (above 3 MW). Waste heat can be recovered from four sources in reciprocating engines: exhaust gas, engine jacket cooling water, lube oil cooling water, and turbocharger cooling. High- and low-pressure steam can be produced from reciprocating engines for combined heat and power applications. The overall efficiency for a CHP system using a reciprocating engine fueled by natural gas can exceed 70%.

Reciprocating engine technology has improved significantly over the last decades with increased fuel efficiency, reduced emissions, improved reliability, and low first cost. The use of reciprocating engines for combined heat and power generation applications is expected to continue to grow in the next decade. Currently, it is estimated that 1055 CHP systems operating in the U.S. use reciprocating engines with an overall power capacity of 800 MW. Table 75.10 provides typical cost and performance parameters for commercially available reciprocating engines suitable for cogeneration applications.

Gas Turbines

Gas turbines provide one of the cleanest means for electric power generation with very low emissions of carbon dioxide (CO_2) and oxides of nitrogen (NO_x). The available capacity of gas turbines ranges from 500 kW to 250 MW. Gas turbines are well suited for CHP applications because high-pressure steam (as high as 1200 psig) can be generated from their high-temperature exhaust using heat recovery steam generators (HRSGs). It is estimated that over 575 industrial and institutional facilities in the U.S. use gas turbines to generate power and heat with a total capacity of 40,000 MW [Energy Nexus Group, 2002c].

Table 75.11 lists typical cost and performance characteristics for selected commercially available gas turbine CHP systems. As indicated in Table 75.11, both the electrical efficiency and the overall CHP efficiency increase with the size of the gas turbine.

Packaged Cogeneration Systems

For cogeneration facilities requiring small systems ranging from less than 50 kW to about 1 MW, preengineered and factory-assembled cogeneration units are currently available with reduced construction, installation, and operation costs. In addition, small packaged systems with capacities ranging from 4 to 25 kW have been developed and can be installed in a short period of time with little interruption of service. Almost all packaged cogeneration systems are equipped with advanced controls to improve the reliability and the energy efficiency of the units.

TABLE 75.11 Cost and Performance Characteristics for Selected Cogeneration Systems Using Gas Turbines

	System Capacity in kW				
Cost and Performance Parameters	1000	5000	10,000	25,000	40,000
Total installed cost (2000 $/kW)	1780	1010	970	860	785
Fuel input (MMBtu/hr)	15.6	62.9	117.7	248.6	368.8
Electrical efficiency (%)	21.9	27.1	29.0	34.3	37.0
Steam output (MMBtu/hr)	7.1	26.6	49.6	89.8	128.5
Overall efficiency (%)	68.0	69.0	71.0	73.0	74.0

Source: Energy Nexus Group, 2002. Technology Characterization: Gas Turbines, A report prepared for EPA, Climate Protection Partnership Division, Washington, DC.

TABLE 75.12 Cost and Performance Characteristics for Selected Cogeneration Systems Using Microturbines

	System Capacity in kW			
Cost and Performance Parameters	30	70	100	350
Total installed cost (2000 $/kW)	2516	2031	1561	1339
Fuel input (MMBtu/hr)	0.437	0.948	1.264	4.118
Electrical efficiency (%)	23.4	25.2	27.0	29.0
Heat output (MMBtu/hr)	0.218	0.369	0.555	1.987
Overall efficiency (%)	73.0	64.0	71.0	77.0

Source: Energy Nexus Group, 2002. Technology Characterization: Fuel Cells, A report prepared for EPA, Climate Protection Partnership Division, Washington, DC.

Packaged cogeneration units are often sold as turn-key installations. In particular, manufacturers are generally responsible for the testing and installation of the complete cogeneration systems. The development of packaged systems has enlarged the appeal of cogeneration to a wide range of facilities including office buildings, restaurants, homes, and multifamily complexes. In addition, packaged cogeneration systems are now typically more cost-effective than conventional systems for small and medium hospitals, schools, and hotels. However, each facility has to be thoroughly evaluated to determine the economic feasibility of any packaged cogeneration system.

Packaged cogeneration systems typically use reciprocating engines. Recently, microturbines have been developed; they have been commercially available since 2000. Microturbines are small electricity generators that operate at very high speeds (over 60,000 rpm). They are available in sizes ranging from 30 to 350 kW. They have been promoted as ideal generators for distributed generation applications including cogeneration systems due to their connection flexibility (they can be stacked in parallel), their reliability, and their low emissions. Table 75.12 summarizes typical cost and performance characteristics for selected commercially available microturbines suitable for CHP applications.

Distributed Generation Technologies

Distributed generation is a relatively recent approach proposed to produce electricity using small and modular generators. The small generators with capacities in the range of 1 kW to 10 MW can be assembled and relocated in strategic locations (typically near customer sites) to improve power quality, reliability, and flexibility in order to to meet a wide range of customer and distribution system needs. Some technologies have emerged in the last decade that allow generation of electricity with reduced waste, cost, and environmental impact, which may make the future of distribution generation promising especially in a competitive deregulated market. Among these technologies are renewable energy sources (wind and solar) and fuel cells, as well as microturbines, combustion turbines, gas engines, and diesel engines. Perhaps the recent developments in fuel cells represent the best opportunity for distributed generation

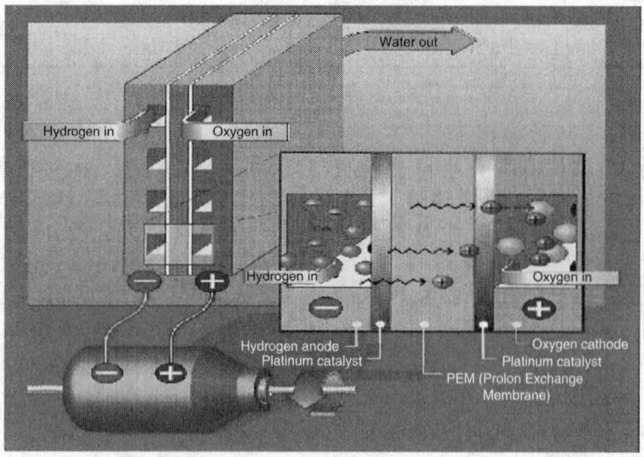

FIGURE 75.6 A basic operation of a fuel cell.

and CHP applications. It is expected that fuel cells will play a significant part in the twenty-first century electricity market.

The principle of the fuel cell was first demonstrated over 150 years ago. In its simplest form, the fuel cell is constructed similar to a battery with two electrodes in an electrolyte medium, which serves to carry electrons released at one electrode (anode) to the other electrode (cathode). Typical fuel cells use hydrogen (derived from hydrocarbons) and oxygen (from air) to produce electrical power with other byproducts (such as water, carbon dioxide, and heat). High efficiencies (up to 73%) can be achieved using fuel cells. Figure 75.6 illustrates the operation of a typical fuel cell.

Table 75.13 summarizes characteristics for various types of fuel cells that are commercially available or under development. Each fuel cell type is characterized by its electrolyte, fuel (source of hydrogen), oxidant (source of oxygen), and operating temperature range. CHP systems using fuel cells are available in sizes ranging from few to thousands of kW. Table 75.14 lists some cost and performance parameters for selected commercially available CHP systems using fuel cell technology.

TABLE 75.13 Types and Characteristics of Available Fuel Cell Technologies

Fuel Cell Name	Electrolyte	Efficiency (%)	Fuel	Oxidant	Operating Temperatures (°C)
PAFC	Phosphoric acid	35 to 45	Pure hydrogen	Clear air (without CO_2)	200
AFC	Alkaline	32 to 40	Pure hydrogen	Pure oxygen and water	60 to 120
SPFC	Solid polymer	25 to 35	Pure hydrogen	Pure oxygen	60 to 100
MCFC	Molten carbonate	40 to 50	Hydrocarbons	Air and oxygen	650
SOFC	Solid oxide	45 to 55	Any fuel	Air	900 to 1000

TABLE 75.14 Cost and Performance Parameters for Selected CHP Using Fuel Cells

	System Capacity in kW			
Cost and Performance Parameters	10	100	200	2000
Fuel cell type	PEM	SOFC	PAFC	MCFC
Total installed cost (2002 $/kW)	5500	3500	4500	2800
Fuel input (MMBtu/hr)	0.10	0.80	1.90	14.80
Electrical efficiency (%)	30.0	45.0	36.0	46.0
Heat output (MMBtu/hr)	0.04	0.19	0.74	3.56
Overall efficiency (%)	68.0	70.0	75.0	70.0

Source: Energy Nexus Group, 2002d).

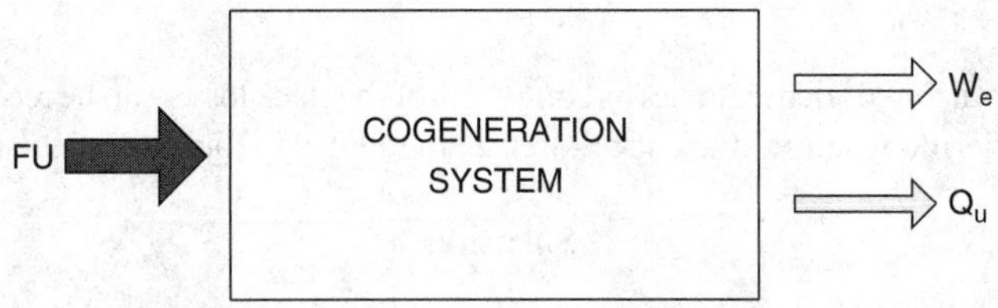

FIGURE 75.7 Input/output energy flows for a cogeneration system.

75.6 Evaluation of Cogeneration Systems

To evaluate the technical and economic feasibility of a cogeneration system for a given facility, it is important to collect accurate data about the facility and its energy consumption. In particular, information about current and projected future energy consumption and costs need to be available. For a detailed evaluation analysis, hourly electrical and thermal energy data are required. However, monthly and even yearly energy data can be sufficient for a preliminary feasibility analysis of cogeneration systems.

In this section, basic considerations involved in performing a feasibility analysis of cogeneration systems are discussed.

Efficiency of Cogeneration Systems

To account for the fact that a typical cogeneration system produces both electrical power, W_e, and useful thermal energy, Q_u, from fuel energy, *FU*, as illustrated in Figure 75.7, the overall efficiency, $\eta_{overall}$, of the cogeneration system is defined using the first law energy efficiency [refer to Equation (75.3)] as follows:

$$\eta_{overall} = \frac{W_e + Q_u}{FU} \tag{75.5}$$

It should be noted that for a cogeneration facility to meet the criteria specified by the PURPA, it has to comply with certain efficiency standards. These standards use a "PURPA efficiency," η_{PURPA}, which is based on the second law efficiency defined by Equation (75.4). Using the energy flows of Figure 75.7, the PURPA efficiency is defined as follows:

$$\eta_{PURPA} = \frac{W_e + Q_u/2}{FU} \tag{75.6}$$

PURPA states that to be a Qualified Facility (QF) for cogeneration, the efficiency, η_{PURPA}, has to be:

1. At least 45% for cogeneration facilities with a useful thermal energy fraction that is larger than 5%.
2. At least 42.5% for cogeneration facilities with a useful thermal fraction that is larger than 15%.

The main purpose of the PURPA efficiency is to ensure that a sufficient amount of thermal energy is produced so that the cogeneration facility is more efficient than the electric utility.

Example 75.1 illustrates how to estimate overall thermodynamic and PURPA efficiencies for conventional cogeneration systems [Krarti, 2000].

Example 75.1

Consider a 20-MW cogeneration power plant in a campus complex. An energy balance analysis indicates the following energy fluxes for the power plant:

- Electricity generation 33%
- Condenser losses 30%
- Stack losses 30%

- Radiation losses 7%

It is estimated that all the condenser losses but only 12% of the stack losses can be recovered. Determine both the overall thermodynamic efficiency as well as the PURPA efficiency of the power plant.

Solution

First, the recovered thermal energy is determined:

$$E_t = \text{condenser losses} + \text{part of the stack losses}$$

This thermal energy output can be expressed in terms of the fuel use (FU) of the power plant:

$$E_t = 30\% * \text{FU} + 12\% * [30\%*\text{FU}] = 0.34*\text{FU}$$

The energy flow for the campus power plant is summarized in the diagram below:

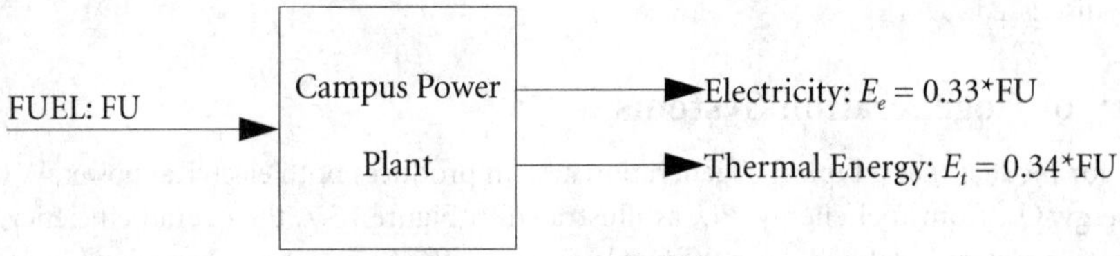

Thus, the overall thermodynamic efficiency of the power plant can be easily determined using Equation (75.5):

$$\eta_{overall} = \frac{E_e + E_t}{FU} = \frac{0.33 * FU + 0.34 * FU}{FU} = 0.67$$

The PURPA efficiency can be calculated using Equation (75.6),

$$\eta_{PURPA} = \frac{E_e + E_t / 2}{FU} = \frac{0.33 * FU + 0.34 * FU / 2}{FU} = 0.50$$

Therefore, the campus power plant meets the PURPA criteria ($\eta_{PURPA} > 45\%$) and is thus a qualified cogeneration facility.

Feasibility Analysis Procedure

To determine if a cogeneration system is cost-effective, feasibility analysis should first be performed. A further evaluation with more detailed energy analysis tools may be warranted to determine the optimal design specifications of the cogeneration system.

A cogeneration feasibility analysis generally includes an evaluation of the facility electrical and thermal energy use, an assessment of client requirements, and a good understanding of the increasingly complex relationships between power purchase, fuel purchase, power sales, and the strategic benefits of combined heat and power generation. Specific tasks for a feasibility analysis study may include:

- Evaluate electricity and heat demand, profiles, and cost, including site shutdowns/holidays
- Determine possible system configurations and permutations to be considered
- Obtain information on the plant operation and needs for steam/hot water use on site

- Determine operation requirements for equipment such waste incinerators, air compressors, and refrigeration systems
- Assess the availability of a heat sink where that surplus heat can be disposed of, such as a river or a lake

Ideally, the cogeneration system can be sized to match exactly both the electrical and thermal loads. Unfortunately, an exact match is almost never possible. Therefore, the cogeneration system has to be designed to meet specific load requirements such as the base-load thermal demand, base-load electrical demand, peak thermal demand, or peak electrical demand. The main features of each design scenario are briefly described below:

Base-load cogeneration systems produce only a portion of the facility's electrical and thermal requirements. Thus, production of supplemental thermal energy (using a boiler for instance) and purchase of additional electrical energy are generally required. Base-load cogeneration systems are suitable for facilities characterized by variable thermal and electrical loads but not willing or able to sell electrical power.

Thermal-tracking cogeneration systems are those systems that produce all the thermal energy required by a facility. In case the generated electrical energy exceeds the electrical demand, the facility has to sell power to the utility. In case the generated electrical energy is lower than the electrical demand, additional power has to be purchased from the utility. Thermal-tracking cogeneration systems are increasingly becoming attractive to small buildings that have to pay higher utility rates than large industrial and commercial facilities do.

Electricity-tracking cogeneration systems are designed to match electrical loads. Any supplemental energy requirements are produced through boilers. These systems are typically suitable for large industrial facilities with fairly high and constant electrical loads and lower but variable thermal loads.

Peak-shaving cogeneration systems: In the case where the cost associated with peak electrical demand is high, it may be cost effective to design cogeneration systems specifically for peak shaving even though these systems may operate only few hours (less than 1000 hours per year).

Simplified calculation procedures as well as computer tools can be utilized to design cogeneration plants. Krarti [2000] provides a simplified analysis method to determine the optimum size for cogeneration systems. Simulation tools such as COGEN [Argonne National Laboratory, 1981; Fisher and Schmidt, 1983], CELCAP [Lee, 1988], and DOE 2 [LBL, 1984] can also be used for designing and assessing the cost-effectiveness of cogeneration systems. Example 75.2 illustrates a simplified calculation procedure to determine the cost-effectiveness of installing a cogeneration system for a hospital building [Krarti, 2000].

Example 75.2

Consider a 60-kW cogeneration system that produces electricity and hot water with the following efficiencies: (a) 26% for the electricity generation and (b) 83% for the combined heat and electricity generation. Determine the annual savings of operating the cogeneration system compared to a conventional system that consists of purchasing electricity at a rate of $0.08/kWh and producing heat from a boiler with 70% efficiency. The cost of fuel is $5/MMBtu. The maintenance cost of the cogeneration system is estimated at $1.00 per hour of operation. Assume that all the generated thermal energy and electricity are utilized during 6500 h/year. *Determine the payback period of the cogeneration system if the installation cost is $2,500/kW.*

Solution

First, the cost of operating the cogeneration system is compared to that of the conventional system on an hourly basis:

1. *Cogeneration System:* For each hour, 60 kW of electricity is generated (at an efficiency of 26%) with fuel requirements of 0.787 MMBtu [=60 kW*0.003413 MMBtu/kW/0.26]. In the same time,

a thermal energy of 0.449 MMBtu [=0.787 MMBtu *(0.83 – 0.26)] is obtained. The hourly flow of energy for the cogeneration system is summarized in the diagram below:

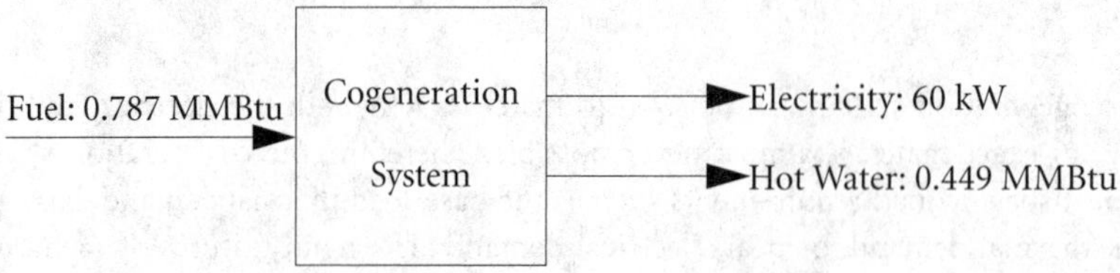

Thus, the cost of operating the cogeneration on an hourly basis can be estimated as follows:
Fuel Cost: 0.787 MMBtu/h * $5/MMBtu = $3.93/h
Maintenance Cost: $1.00/h
Total Cost: $4.93/h

2. *Conventional System:* For this system, the 60 kW electricity is directly purchased from the utility, while the 0.449 MMBtu of hot water is generated using a boiler with an efficiency of 0.65. Thus the costs associated with utilizing a conventional system are as follows:
Electricity Cost: 60 kWh/h * 0.08 = $4.80/h
Fuel Cost (Boiler): (0.449 MMBtu/h)/0.65 * $ 5/MMBtu $3.45/h
Total Cost: $8.25/h

Therefore, the annual savings associated with using the cogeneration system are:

$$\Delta Cost = (\$8.25/hr - \$4.93/hr) * 6500 hr/yr = \$21{,}580/yr$$

Thus, the simple payback period for the cogeneration system:

$$SPB = \frac{\$2500/kW * 60kW}{\$21{,}580} = 7.0 years$$

A Life Cycle Cost analysis may be required to determine if the investment in the cogeneration system is really warranted.

Financing Options

To finance a cogeneration system, several options are generally available. Selecting the most favorable financial arrangement is critical to the success of a cogeneration project. A number of factors affect the selection of the most appropriate financial arrangement for a given cogeneration project. These factors include ownership arrangements, risk tolerance, tax laws, credit markets, and cogeneration regulations. In the U.S., the most common financial approaches for cogeneration facilities include the following [Krarti, 2000]:

1. *Conventional ownership and operation:* In this financing structure, the owner of the cogeneration facility funds either totally or partially the project from internal sources. In the case of partial funding, the owner can borrow the remaining funds from a conventional lending institution. Operation and maintenance of the cogeneration system can be performed by an external contractor.
2. *Joint venture partnership:* This structure is an alternative to the conventional ownership and operation with a shared financing and ownership with a second partner such as an electric utility. Indeed, PURPA regulations provide the option for an electric utility to own up to 50% of a cogeneration facility. The joint venture financing structure reduces the risks for both partners but may increase the complexity of the various contracts between all the involved parties including

the owner and its partner, gas provider, electric utility, lending institution, and possible operation and maintenance contractor.

3. *Leasing:* In this financing option, a company builds the cogeneration facility with a leasing agreement from the owner to use part or all of the thermal and electrical energy output of the cogeneration plant. The construction of the cogeneration system by the lessor (i.e., the builder of the facility) can be financed through funds from lenders and/or investors. The owner is generally heavily involved in the construction phase of the cogeneration facility.
4. *Third-party ownership:* This financing structure is similar to that described for the leasing case. However, in third-party ownership, the owner is not involved in either financing or construction of the cogeneration facility. Instead, a third party or a lessee develops the project and arranges for gas/fuel supply, electrical power and heat sales, and operation and maintenance agreements. The finances can be arranged by a lessor through funds from investors and/or lenders.
5. *Guaranteed savings contracts:* In this financing option, a developer first builds and maintains the cogeneration facility. Then, the developer enters in a guaranteed savings contract with the energy consumer (the owner). This contract is typically made for a period ranging from 5 to 10 years with a guaranteed fixed savings per year. This type of financial structure is common for small cogeneration systems (i.e., packaged units) since it shifts all the financing and operation risks from the owner to the facility developer (i.e., the guaranteed savings contractor).

75.7 Case Study: Cogeneration System at the University of Colorado

The Power House at the University of Colorado was constructed in 1909. Coal was the original fuel source for the plant and was delivered to the site on railroad cars. The plant was converted to clean-burning natural gas in 1950, with the addition of No. 6 fuel oil stored in underground tanks on-site beginning in the early 1960s as backup fuel. Due to increasing demands in electricity and steam use, the university retrofitted the Power House in 1992 to build a cogeneration plant. The specifications of the components for the cogeneration systems are summarized in Table 75.15. Figure 75.8 provides a top view of the cogeneration plant.

The operation of the University of Colorado (CU) cogeneration system is outlined in the schematics provided in Figure 75.9. The primary systems of the CU cogeneration plant include two combustion gas turbines, two heat recovery steam generators (HRSGs), high-pressure steam boilers, and low-pressure

TABLE 75.15 Equipment Used for the Cogeneration Plant at the University of Colorado

Utility	Quantity	Equipment	Description
Electric power	2	Mitsubishi industrial gas turbine sets	16 MW each; each includes a dual-fuel Mitsubishi Heavy Industries MF-111AB gas turbine driving a RENK single reduction gearbox coupled to a Brush two pole synchronous generator; operates with a shaft speed of 9645 rpm
Electric power	1	Dresser rand steam turbine set	1 MW induction generator double-ended low- and high-pressure steam
Steam	2	Zurn (HRSG) heat recovery steam generators	Supplemental fire capability—Davis duct burners 80,000 lb/h, maximum steam output 300 psig
Steam	1	Erie City 1966 boiler	Front-fired 150,000 lb/h, maximum steam output 130 psig
Steam	1	Combustion Engineering 1957 boiler	Tangentially fired 115,000 lb/hr maximum steam output 130 psig
Chilled water	3	Steam absorption chillers	1 Trane 1470 tons cooling; 2 York 900 tons cooling; each utilizes 10 psig steam lithium bromide

Source: CU, 2002. Utilities: Generation and Distribution at the University of Colorado, http://www.colorado.edu/utilities.

FIGURE 75.8 Outside view of the Power House where the cogeneration plant is located at the University of Colorado.

absorption chillers. The two gas turbines are each capable of producing 16 MW of power while providing sufficient hot exhaust gas to generate 80,000 lb/h of 300 psig steam in each heat recovery steam generator (HRSG). The steam turbine generator consists of dual topping turbines driving a common generator. The turbine reduces the incoming steam pressure by expanding it through the turbines which, in turn, drive the generator. The exhaust steam is then exported to either the 130-psig or the 10-psig steam header. The 130-psig steam is delivered to various buildings in the campus for heating purposes. The 10-psig steam is utilized to operate three absorption chillers that deliver chiller water to the campus. The high pressure steam boilers deliver 300 psig steam to the gas turbines for NOx control and steam injection for power augmentation [CU, 2002].

The CU cogeneration plant was designed to deliver over 30 MW of electrical power. Absorption chillers with over 3000-ton capacity were added to increase the thermal load on the plant especially during summer months. A large portion of the cogenerated electricity is generally sold to the local utility. As stipulated by PURPA, the plant can purchase electricity from the same utility in case of maintenance or

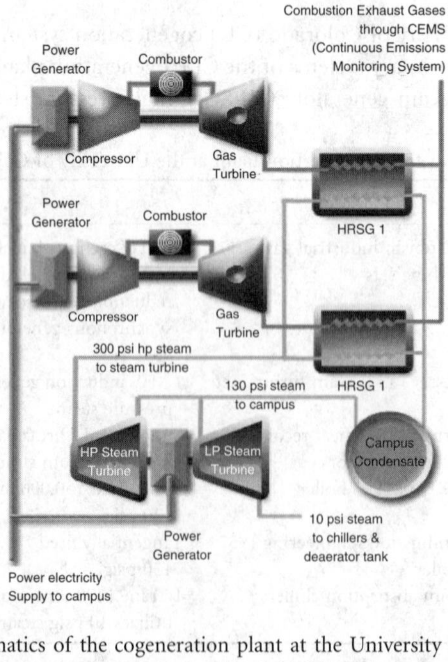

FIGURE 75.9 Operation schematics of the cogeneration plant at the University of Colorado. (Source: CU. 2002. Utilities: Generation and Distribution at the University of Colorado, http://www.colorado.edu/utilities.)

TABLE 75.16 Monthly Fuel Used and Electricity/Steam Produced by the CU Cogeneration Plant During 2002

		CU Cogeneration Performance 2002			
		Electricity			130 psig
	Nat. Gas	Used	Sold	Total	Steam
Month	MMBtu	kWh	kWh	kWh	1000 lb
Jan	195,300	9,803,800	5,952,000	15,755,800	84,932
Feb	161,000	9,332,820	5,376,000	14,708,820	79,452
Mar	141,600	10,011,670	5,952,000	15,963,670	84,932
Apr	141,600	10,055,890	5,760,000	15,815,890	82,192
May	150,350	9,776,410	5,952,000	15,728,410	84,932
Jun	141,000	9,899,130	5,760,000	15,659,130	82,192
Jul	181,757	10,752,740	5,952,000	16,704,740	84,932
Aug	201,802	10,604,980	5,952,000	16,556,980	84,932
Sep	180,480	10,492,980	5,760,000	16,252,980	82,192
Oct	150,474	10,389,270	5,952,000	16,341,270	84,932
Nov	166,230	10,553,990	5,760,000	16,313,990	82,192
Dec	154,039	9,441,120	5,952,000	15,393,120	84,932
Year	1,965,632	121,114,800	70,080,000	191,194,800	1,000,000

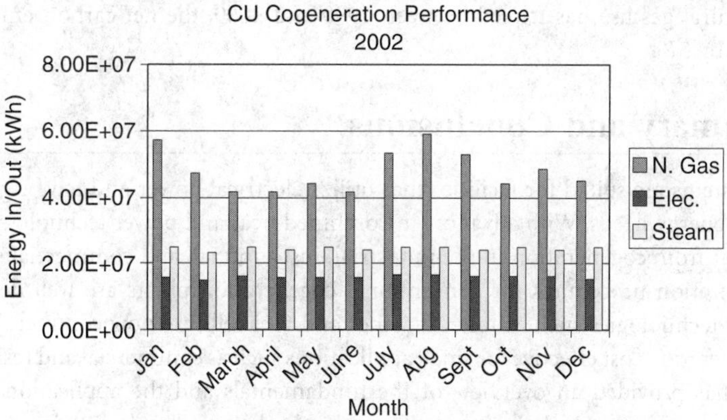

FIGURE 75.10 Monthly energy input and output of the CU cogeneration during 2002.

emergency periods. A typical monthly performance of the CO cogeneration plant is outlined in Table 75.16 and Figure 75.10.

As indicated in Table 75.16, the University used about 63% of the electricity produced by the cogeneration plant while the remaining 37% was sold to the local utility. In 1992, the cogeneration sold more than 50% of its electric power. Due to the ever-increasing electricity demands at the University (associated with new buildings and additional research laboratories), the cogeneration plant may not be able to meet the University load in the near future. The University is currently investigating plans to add a new cogeneration plant.

From the data provided in Table 75.16 and Figure 75.10, the electrical efficiency and the overall efficiency of the CU cogeneration plant during 2002 are estimated to be 33 and 80%, respectively. The PURPA efficiency [refer to Equation (75.5)] of the plant is over 56%, well above the requirements set by PURPA.

Over the last two decades, electricity and steam consumption have increased substantially at the University of Colorado with a growth rate of about 5%. Table 75.17 illustrates the progression of energy use and carbon emissions at the University of Colorado between 1990 (before the construction of the cogeneration plant) and 2000 (after the construction of the cogeneration plant). As indicated in Table 75.17, the cogeneration plant has substantially decreased the carbon output per unit of energy consumed.

TABLE 75.17 Comparison of Annual Energy Use and Carbon Emissions in 1990 and 2000 at the University of Colorado

	Year 1990		Year 2000	
	Energy Purchased or Sold	Carbon Emissions (tons)	Energy Purchased or Sold	Carbon Emissions (tons)
	Natural Gas Purchases			
Central Plant/Cogeneration	634,159 MMBtu	10,115	1,936,341 MMBtu	30,885
Individual Buildings	116,500 MMBtu	1,849	632,419 MMBtu	10,080
	Electricity Purchases			
Central Plant/Cogeneration	66,024,000 kWh	18,476	39,937,364 kWh	10,637
Individual Buildings	20,782,510 kWh	5,816	9,970,008 kWh	2,656
	Electricity Sales			
Cogeneration Plant	0 kWh	—	74,893,631 kWk	(19,947)
Total Natural Gas Purchased	750,659 MMBtu	11,994	2,568,760 MMBtu	40,965
Net Electricity Purchased/Sold	86,806,510	24,292	(24,986,259 kWh)	(6,655)
Net Carbon Emissions	—	36,296	—	34,310

Even though natural gas use has tripled between 1990 and 2000, the net carbon emissions have been actually reduced by 5%.

75.8 Summary and Conclusions

Cogeneration systems are suited for facilities that utilize electrical power and have significant and continuous thermal energy needs. With advances in combined heat and power technology, several facilities are able to benefit from cogeneration in the industrial, institutional, and commercial sectors. The components and operation procedures for conventional cogeneration plants are well developed and well established. New technologies such as microturbines and fuel cells are now commercially available and are making cogeneration cost effective for small applications such as commercial and residential buildings.

This chapter has provided an overview of the fundamentals and the applications of cogeneration systems. In particular this chapter has presented historical background, technical design issues, regulatory considerations, and financial options for cogeneration systems. Moreover, the chapter included one case study to illustrate the operation of a typical cogeneration system in an institutional facility.

In the future, cogeneration is expected to become more attractive for small buildings, especially with new developments in fuel cell technologies and microprocessor-based control systems. These developments will make small cogeneration systems cost effective, reliable, and efficient for even nontraditional cogeneration applications such as residential buildings. With market conditions and trends favoring distributed generation and with involvement of ESCOs, combined heat and power technology is expected to have a promising future in the U.S. over the next few decades.

References

Argonne National Laboratory, 1981. OASIS CODE Application to Proposed Argonne National Cogenertain Plant. ANL/CNVS-TM-67, Argonne, IL.

Cengel, Y. A. and Turner R. H. 2001. *Fundamentals of Thermal-Fluid Sciences*, McGraw-Hill, New York.

CU. 2002. Utilities: Generation and Distribution at the University of Colorado, http://www.colorado.edu/utilities.

EIA. 2000. *The Changing Structure of the Electric Power Industry 2000: An Update,* Report from Energy Information Administration, Washington, DC.

EIA. 2003. *Annual Energy Review,* Department of Energy, Energy Information Administration, http://www.doe.eia.gov.

Energy Nexus Group. 2002a. Technology Characterization: Steam Turbines, A report prepared for EPA, Climate Protection Partnership Division, Washington, DC.

Energy Nexus Group. 2002b. Technology Characterization: Reciprocating Engines, A report prepared for EPA, Climate Protection Partnership Division, Washington, DC.

Energy Nexus Group. 2002c. Technology Characterization: Gas Turbines, A report prepared for EPA, Climate Protection Partnership Division, Washington, DC.

Energy Nexus Group. 2002d. Technology Characterization: Fuel Cells, A report prepared for EPA, Climate Protection Partnership Division, Washington, DC.

Fisher, D. B. and Schmidt, P. S. 1983. Analysis of In-Plant Cogeneration with a Microcomputer, Proceedings of the 1983 Industrial Energy Technology Conference, Houston, TX.

Krarti, M. 2000. *Energy Audit of Building Systems: An Engineering Approach,* CRC Press, Boca Raton, FL.

Lawrence Berkeley Laboratory. 1993. DOE-2 Supplement Version 2.1E, Report prepared for the Department of Energy by the LBL, Berkeley, CA.

Lee, T. Y. 1988. Cogeneration system selection using the Navy's CELCAP code, *Energy Engineering,* 85:2–24.

OECD. 1999. Economic Statistics by the Organization for Economic Cooperation and Development, http://www.ocde.org.

76

Electric Machines

Iqbal Husain
The University of Akron

A significant percentage of the world's energy is associated with electromechanical energy conversion processes for both efficient energy transportation and mechanical work done at the receiving end. The ease of transmitting energy with exceptional control over long distances in electrical form is the primary motivation for converting the world's available energy from chemical, mechanical, or nuclear form into electrical form. Again, from a user's perspective, the electrical energy needs to be converted to mechanical energy for useful work. Electric machines accomplish the task of converting electrical energy to mechanical form and vice versa. The term "motor" is used for the electric machine when energy is converted from electrical to mechanical and the term "generator" is used when power flow is in the opposite direction with the machine converting mechanical energy into electrical energy. The term braking or regeneration is used to describe the generating mode when electric machines convert residual mechanical or magnetic energy into electrical energy. The goal of this chapter is to introduce the different types of machines available for electromechanical energy conversion.

The torque in electric machines is produced utilizing one of two basic principles of electromagnetic theory:

1. The Lorentz force principle, where torque is produced by the mutual interaction of two orthogonal magnetomotive forces (mmf)
2. The reluctance principle, where the rotor produces torque while moving towards the minimum reluctance position in a varying reluctance path.

0-8493-1586-7/05/$0.00+$1.50

Direct current (DC) and alternating current (AC) machines, including permanent magnet machines work on the first principle, while reluctance machines work on the latter principle.

76.1 DC Machines

DC machines have two sets of windings, one in the rotor and the other in the stator, which establish the two fluxes needed to produce the torque. The orthogonality of the two magnetomotive forces, which is essential for maximum torque production, is maintained by a set of mechanical components called commutators and brushes. The winding in the rotor is called the armature winding, while the winding in the stationary part of the machine is called the field windings. Both the armature and the field windings are supplied with DC currents. The armature windings carry the bulk of the current, while the field windings carry a small field excitation current. The armature and the field currents in the respective windings establish the armature and field mmf. The magnitude of the mmf is the product of the number of turns in the windings and the current.

The advantage of a DC machine is the ease of control due to linearity, which makes it amenable to independent torque and flux control. The manufacturing technology is also well established. The disadvantage of DC machines is the brush wear that leads to high maintenance. The machines have a low power-to-weight ratio.

DC Generator

In a DC generator, a prime mover supplies power to rotate the shaft in the direction of applied shaft torque T_S, and voltage is induced in the armature windings. The induced or generated voltage can deliver power to an electrical load. The electromagnetic torque T_e is in a direction opposite to the rotational direction, as shown in Figure 76.1.

The armature equivalent circuit of a DC machine consists of the armature winding resistance R_A, the self-inductance of aramture winding L_{AA} and the back-emf E_A. Figure 76.1 shows the armature circuit for a DC generator with armature current i_A flowing out of the circuit. In the figure, V_A is the armature voltage, ω_m is the shaft speed, and ϕ is the armature linking flux (primarily from field current). Applying KVL around the armature circuit, the voltage balance equation is

$$V_A = -R_A i_A - L_{AA}\frac{di_A}{dt} + E_A \tag{76.1}$$

where

$$\begin{aligned} E_A &= K\phi\omega_m \\ T_e &= K\phi i_A \end{aligned} \tag{76.2}$$

K is a machine constant that depends on the machine construction, number of windings, and core material properties. The field equivalent circuit of the DC machine is shown in Figure 76.2. The field

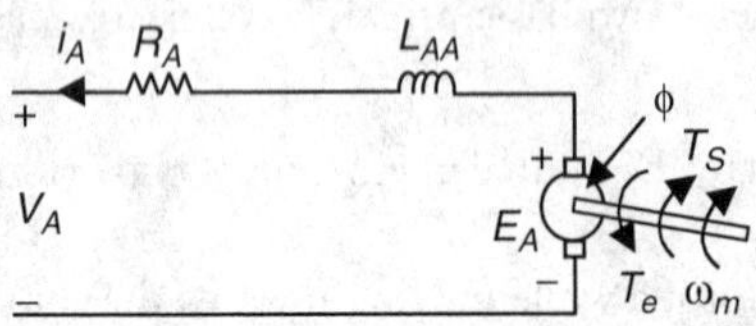

FIGURE 76.1 Armature equivalent circuit of a DC generator. (From Husain, I. 2003. *Electric and Hybrid Vehicles: Design Fundamentals,* CRC Press, Boca Raton, FL.)

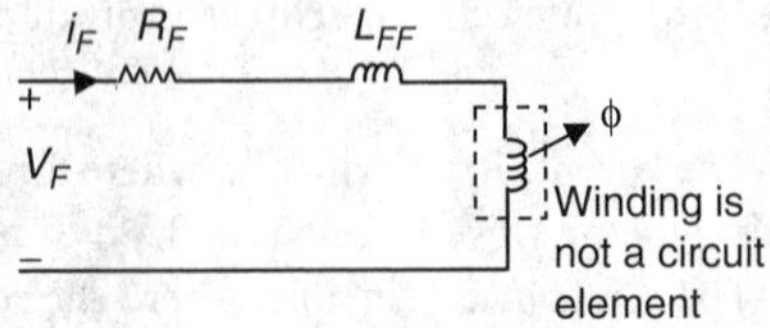

FIGURE 76.2 DC machine field equivalent circuit. (From Husain, I. 2003. *Electric and Hybrid Vehicles: Design Fundamentals,* CRC Press, Boca Raton, FL.)

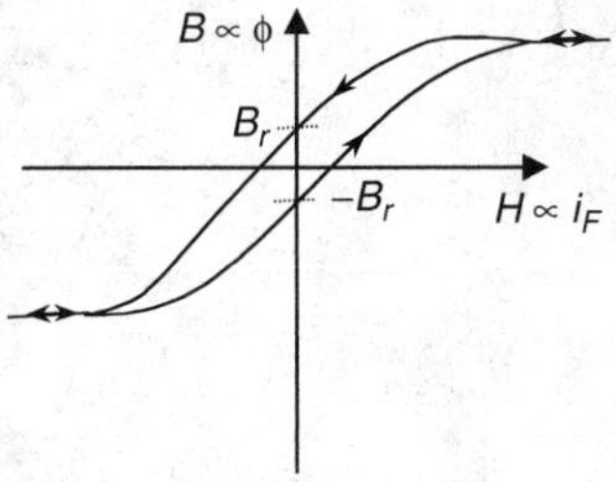

FIGURE 76.3 Typical machine magnetization characteristics. (From Husain, I. 2003. *Electric and Hybrid Vehicles: Design Fundamentals,* CRC Press, Boca Raton, FL.)

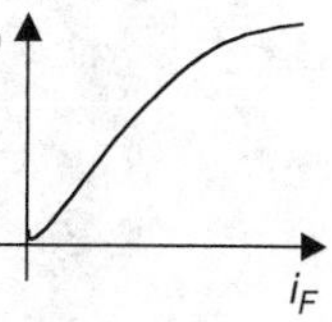

FIGURE 76.4 Magnetization characteristics of DC machines. (From Husain, I. 2003. *Electric and Hybrid Vehicles: Design Fundamentals,* CRC Press, Boca Raton, FL.)

circuit consists of the field winding resistance R_F and the self inductance of the field winding L_{FF}. V_F is the voltage applied to the field. The field circuit equation is

$$V_F = R_F i_F + L_{FF}\frac{di_F}{dt}$$

The transient response in the field circuit is much faster than the armature circuit. The field voltage is also typically not adjusted frequently, and for all practical purposes, a simple resistor fed from a DC source characterizes the electrical unit of the field circuit. The field current establishes the mutual flux or field flux, which is responsible for torque production in the machine. The field flux is a nonlinear function of field current and can be described by $\phi = f(i_F)$.The electromagnetic properties of the machine core materials are defined by the relationship

$$B = \mu H$$

where B is the magnetic flux density in Tesla or weber/m^2, H is the magnetic field intensity in Ampere-turn/m, and μ is the permeability of the material. The permeability in turn is given by $\mu = \mu_0 \mu_r$ where $\mu_0 = 4\pi \times 10^{-7}$. *H/m* is the permeability of free space and μ_r is the relative permeability. The relative permeability of air is 1. The *B-H* relationship of magnetic materials is nonlinear and is difficult to describe by a mathematical function. The properties of core materials are often described graphically in terms of the *B-H* characteristics as shown in Figure 76.3. The nonlinearity in the characteristics is due to the saturation of flux for higher currents and the hysteresis effects. The hysteresis causes magnetic flux density *B* to be a multivalued function that depends on the direction of magnetization. The magnetic effect that remains in the core after the complete removal of magnetization force is known as the residual magnetism (denoted by B_r in Figure 76.3). The direction of the residual flux, as mentioned previously, depends on the direction of field current change. The *B-H* characteristics can also be interpreted as the ϕ-i_F characteristics, since *B* is proportional to ϕ and *H* is proportional to i_F for a given motor. The energy required to cause change in the magnetic orientations is wasted in the core material and is referred to as hysteresis loss. The area of the hysteresis loop in the magnetization characteristics is proportional to the hysteresis loss.

For most applications, it is sufficient to show the magnetic properties of core materials through a single-valued yet nonlinear function, which is known as the DC magnetization curve. The magnetization curve of a DC machine is typically shown as a curve of open-circuit induced voltage (which is essentially the back-emf E_A) versus field current i_F at a particular speed. Therefore, the shape of this characteristic, shown in Figure 76.4, is similar to that of the magnetic characteristics of the core material.

DC Motor

The principles of electromagnetic operation of a DC motor are the same as that of a DC generator, but the power and energy flow is in the opposite direction. The armature equivalent circuit is the same as

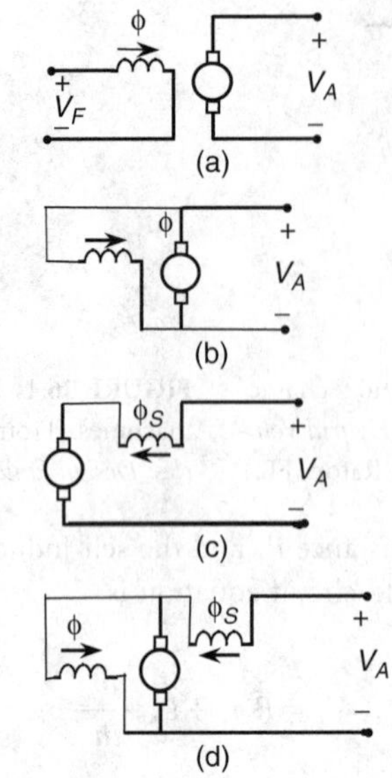

FIGURE 76.5 Different types of DC machines: (a) separately excited, (b) shunt, (c) series, and (d) compound.

in Figure 76.1 except that the current flows into the windings when the armature is connected to a DC voltage supply. In the case of a DC motor, the voltage balance equation is

$$V_A = R_A i_A + L_{AA}\frac{di_A}{dt} + E_A \tag{76.3}$$

The back-emf E_A and torque T_e expressions are the same as those in Equation (76.2).

Types of DC Machines

Depending on the number of supply sources and the type of connection between the armature and field windings, there can be several types of DC machines (Figure 76.5) [Sen, 1997]. When the armature and field windings are connected to independent DC voltages, then it is known as a separately excited DC machine. The separately excited DC machine offers the maximum flexibility of torque and speed control through independent control of the armature and field currents. The DC shunt machine has the similar parallel configuration of armature and field windings, but the same DC source supplies both the windings. In another type of DC machine, known as the series DC machine, the armature and the series windings are connected in series and the machine is connected to a single terminal DC voltage. Since the two windings carry the same current, the field is wound with a few turns of heavy gauge wires to deliver the same mmf or ampere-turns as in the separately excited machine. A machine may have both a series winding and a shunt winding, in which case the machines are called compound machines. The series field may be assisting the shunt field or they may be opposing the shunt field, giving rise to cumulatively compound or differentially compound DC machines.

In the case of DC generators, the terminal voltage-current characteristics are important for performance evaluation. The terminal characteristics of several types of DC generators are shown in Figure 76.6. The speed–torque characteristics of DC motors are of interest for performance evaluation. The speed–torque relationship of a DC motor can be derived from Equation (76.2) and Equation (76.3) and is given by

$$\omega_m = \frac{V_A}{K\phi} - \frac{R_A}{(K\phi)^2}T_e \tag{76.4}$$

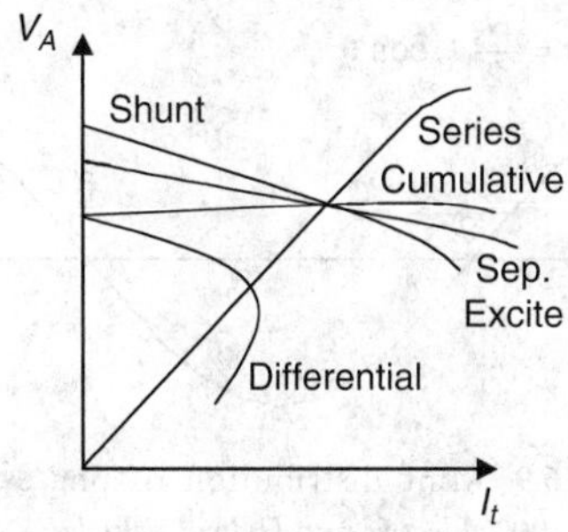

FIGURE 76.6 Terminal voltage-current characteristics of DC generators.

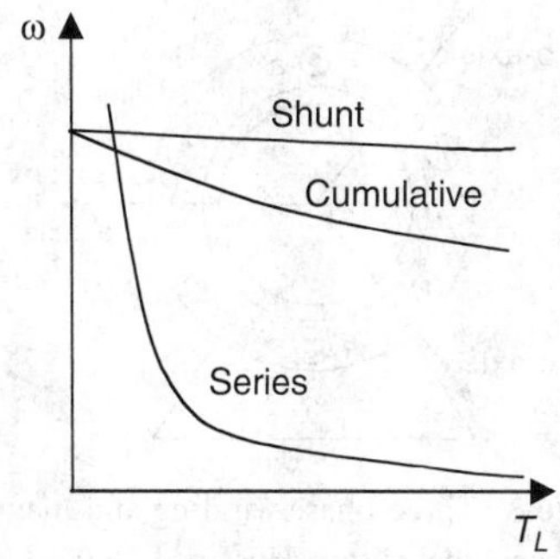

FIGURE 76.7 Speed–torque characteristics of DC motors.

The speed–torque characteristics of three types of DC motors are shown in Figure 76.7. The greatest advantage of the series machine is the very high starting torque that helps achieve rapid acceleration. However, the control flexibility is lost due to the series connection of armature and field windings.

76.2 Induction Machines

AC Machines

The primary difference between AC machines and DC machines is that the armature circuit of the former is located in the stationary piece of the structure. The major advantage of this arrangement is the elimination of the commutator and brushes of the DC machines. The machines are fed from AC sources and can be of single phase or multiple phase type. Single phase AC machines are used for low-power appliance applications, while higher power machines are always of three-phase configuration. The second mmf required for torque production in AC machines (equivalent to the field mmf of DC machines) comes from the rotor circuit. The method through which the rotor mmf is established differentiates the different types of AC machines. Broadly, the AC machines can be classified into two categories, synchronous machines and asynchronous machines. In synchronous machines, the rotor always rotates at synchronous speed. The rotor mmf is established by using either a permanent magnet or an electromagnet created by feeding DC currents in a rotor coil. The latter type synchronous machines are typically the large machines used in electric power generating systems. In the asynchronous-type AC machine, the rotor rotates at a speed that is different but close to the synchronous speed. These machines are known as induction machines, which in the more common configurations are fed only from the stator. The stator windings common to all AC machines will be presented first before discussing induction machines.

Sinusoidal Stator Windings

The three-phase stator windings of AC machines are distributed spatially around the stator circumference to produce a sinusoidal radial magnetic field along the stator circumference. The windings for the three phases are 120° shifted with respect to each other as shown in Figure 76.8. In practice, the phase windings are not concentrated in one location as shown but distributed in slots to produce a sinusoidal mmf.

The current i_a flowing through the phase-*a* stator windings will result in a sinusoidal phase-*a* mmf as shown in Figure 76.9. The sinusoidal conductor-density distribution in phase-*a* winding can theoretically be represented as [Krasue and Wasynchuk, 1986]

$$n_s(\theta) = \frac{N_s}{2}\sin\theta, \qquad 0 \le \theta \le \pi \tag{76.5}$$

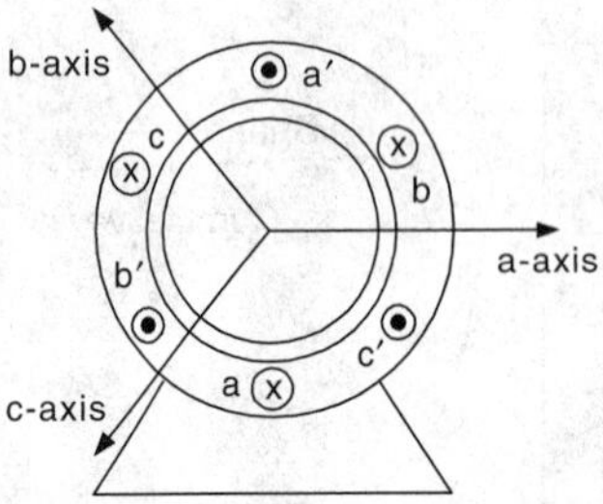

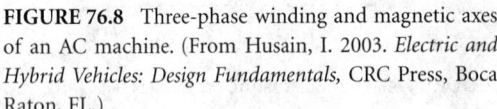

FIGURE 76.8 Three-phase winding and magnetic axes of an AC machine. (From Husain, I. 2003. *Electric and Hybrid Vehicles: Design Fundamentals,* CRC Press, Boca Raton, FL.)

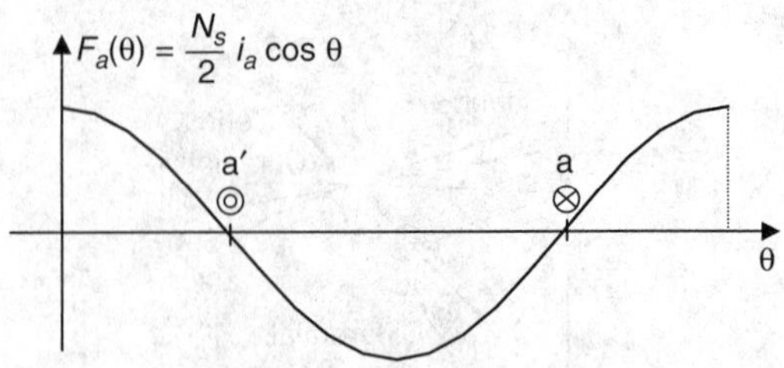

FIGURE 76.9 Mmf distribution of phase a. (From Husain, I. 2003. *Electric and Hybrid Vehicles: Design Fundamentals,* CRC Press, Boca Raton, FL.)

The resulting phase-*a* stator mmf when current flows through the phase-*a* windings, derived using Ampere's Law is

$$F_a(\theta) = \frac{N_s}{2} i_a \cos\theta \qquad (76.6)$$

The mmf, flux intensity and field intensity are all 90 degrees phase shifted in space with respect to the winding distribution. The angle θ is measured in the counterclockwise direction with respect to the phase-*a* magnetic axis. The field distribution in Figure 76.9 is for positive current. Irrespective of the direction of current, the peak of the mmf (positive or negative) will always appear along the phase-*a* magnetic axis, which is the characteristic of mmf produced by a single phase winding.

The resulting mmf due to currents in phases *b* and *c* can be expressed as

$$F_b(\theta) = \frac{N_s}{2} i_b \cos(\theta - \frac{2\pi}{3}) \quad \text{and} \quad F_c(\theta) = \frac{N_s}{2} i_c \cos(\theta + \frac{2\pi}{3}) \qquad (76.7)$$

Number of Poles

The two equivalent phase-*a* conductors in Figure 76.8 represent two poles of the machines. Electric machines are designed with multiple pairs of poles for efficient utilization of the stator and rotor magnetic core material. In multiple pole pair machines, the electrical and magnetic variables (such as induced voltages, mmf, and flux denisity) complete more cycles during one mechanical revolution of the motor. The electrical and mechanical angles of revolution and the corresponding speeds are related by

$$\theta_e = \frac{P}{2}\theta_m \quad \text{and} \quad \omega_e = \frac{P}{2}\omega_m \qquad (76.8)$$

where *P* is the number of poles. The phase *a* mmf of a *P* pole machine is mathematically represented as

$$F_a(\theta_e) = \frac{N_s}{P} i_a \cos(\theta_e) \qquad (76.9)$$

Resultant Mmf in a Balanced System

Let us consider only the stator circuit and assume that all the current that is flowing through the stator winding is the magnetizing current required to establish the stator mmf. The 3-phase currents have the same magnitude and frequency, but are 120 degrees shifted in time with respect to each other. The currents in the time domain can be expressed as

$$
\begin{aligned}
i_a(t) &= \hat{I}_M \cos\omega t \\
i_b(t) &= \hat{I}_M \cos(\omega t - 120^0) \\
i_c(t) &= \hat{I}_M \cos(\omega t - 240^0).
\end{aligned}
\tag{76.10}
$$

For the above three-phase variables, the space vector method can be used where a compact expression is derived from the individual phase variables as follows

$$
\vec{i}_M(t) = i_a(t) + i_b(t)\angle 120 + i_c(t)\angle 240 \tag{76.11}
$$

The resulting space vector for the balanced set of currents is

$$
\vec{i}_M(t) = \frac{3}{2}\hat{I}_M \angle \omega t \tag{76.12}
$$

The *space vector representation* is a convenient method of expressing the equivalent resultant effect of the sinusoidally space distributed electrical and magnetic variables in AC machines. The space vectors provide a very useful and compact form of representing the machine equations, which not only simplifies the representation of three-phase variables but also facilitates the transformation between three-phase and two-phase variables. The space vectors are similar but more complex than the phasors, since they represent time variation as well as space variation. The space vectors, just like any other vectors, have a magnitude and an angle, but the magnitude can be time varying.

The resultant stator mmf space vector is

$$
\vec{F}_{ms}(t) = \frac{N_S}{2}\vec{i}_M(t) = \frac{3}{2}\frac{N_S}{2}\hat{I}_M \angle \omega t = \hat{F}_{ms} \angle \omega t \tag{76.13}
$$

The result shows that the stator mmf has a constant peak amplitude $\hat{F}_{ms}$ (since N_S and $\hat{I}_M$ are constants) that rotates around the stator circumference at a constant speed equal to the angular speed of the applied stator voltages. This speed is known as the synchronous speed. Unlike the single-phase stator mmf, the peak of the stator mmf resulting in the three-phase AC machine is rotating synchronously along the stator circumference with the peak always located at $\theta = \omega t$. The mmf wave is a sinusoidal function of the space angle θ. The wave has a constant amplitude and a space-angle ωt, which is a linear function of time. The angle ωt provides rotation of the entire wave around the air-gap at a constant angular velocity ω. The three-phase stator mmf is known as the rotating mmf, which can be equivalently viewed as a permanent magnet rotating around the stator circumference at a constant speed.

Mutual Inductance L_m and Induced Stator Voltage

In an ideal situation, the equivalent electrical circuit for the stator windings with no rotor existing consists of the applied stator voltage source and a set of winding that is represented by an inductance known as magnetizing or mutual inductance. The practical circuit extends on this ideal circuit by adding the stator winding resistance and the stator leakage inductance in series with the magnetizing inductance. The magnetizing inductance for the three-phase AC machine including the effects of mutual coupling among the three phases can be shown to be [Novotny and Lipo, 1996]

$$
L_m = \frac{3}{2}\left[\frac{\pi \mu_0 r l}{l_g}\left(\frac{N_S}{2}\right)^2\right] \tag{76.14}
$$

where r is the radius to the air gap, l is the rotor axial length and l_g is the air gap length. The voltage induced in the stator windings due to the magnetizing current flowing through L_m in space vector form is

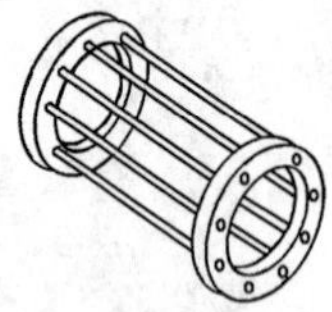

FIGURE 76.10 The squirrel cage of an induction motor. (From Husain, I. 2003. *Electric and Hybrid Vehicles: Design Fundamentals,* CRC Press, Boca Raton, FL.)

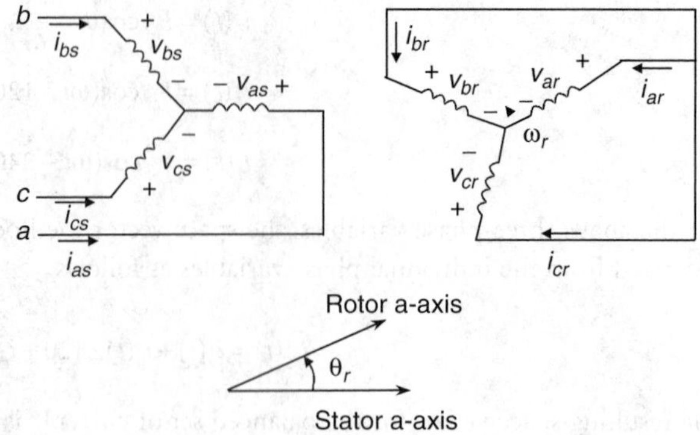

FIGURE 76.11 Stator and rotor electric circuit and magnetic axes of a three-phase induction machine. (From Husain, I. 2003. *Electric and Hybrid Vehicles: Design Fundamentals,* CRC Press, Boca Raton, FL.)

$$\vec{e}_{ms}(t) = j\omega L_m \vec{i}_M(t) \tag{76.15}$$

Types of Induction Machines

The low-cost, rugged construction and almost maintenance-free operation of three-phase induction machines have made them the workhorse of the industry in numerous motor drive applications. The three-phase stator windings of an induction machine are fed from a balanced three-phase supply, which creates a rotating magnetomotive force in the stator field.

The two types of induction machines are the squirrel cage machine and the wound rotor machine. Squirrel cage induction motors are widely used because of their rugged construction and low cost. The rotor, usually made of stacked laminations, has copper or aluminum rotor bars molded around the periphery in the axial direction. The bars are short-circuited at the ends through electrically conducting end rings forming the shape of a squirrel cage. The squirrel cage of an induction motor is shown in Figure 76.10. The rotor winding terminals of wound rotor induction machines are brought outside with the help of slip rings for external connections, which are used for speed control. Squirrel cage induction motors are of greater interest, and hence, will be discussed further.

The electrical equivalent circuit of a three-phase induction machine, along with the direction of phase-*a* stator and rotor magnetic axes, is shown in Figure 76.11. When a balanced set of voltages is applied to the stator windings, a magnetic field is established, which rotates at synchronous speed as described earlier in this chapter. By Faraday's Law, ($e = Blv$), as long as the rotor rotates at a speed other than the synchronous speed (even at zero rotor speed), the rotor conductor is cutting the stator magnetic field and there is a rate of change of flux in the rotor circuit, which will induce a voltage in the rotor bars. This is also analogous to transformer action, where a time-varying AC flux established by the primary winding induces voltage in the secondary set of windings. The induced voltage will cause rotor currents to flow in the rotor circuit, since the rotor windings or bars are short-circuited in the induction machine. The rotor induced voltages and the current have a sinusoidal space distribution, since these are created by the sinusoidally varying (space sinuoids) stator magnetic field. The resultant effect of the rotor bar currents is to produce a sinusoidally distributed rotor mmf acting on the air gap. The interaction of the stator and rotor mmfs produces the electromagnetic torque.

The difference between the rotor speed and the stator synchronous speed is the speed by which the rotor is slipping from the stator magnetic field, and is known as the slip speed

$$\omega_{slip} = \omega_e - \omega_m \tag{76.16}$$

where ω_e is the synchronous speed and ω_m is the motor or rotor speed. The slip speed expressed as a fraction of the synchronous speed is known as the slip

$$s = \frac{\omega_e - \omega_m}{\omega_e} \tag{76.17}$$

The rotor bar voltages, current and magnetic field are of the slip speed or slip frequency with respect to the rotor. The slip frequency is given by

$$f_{slip} = \frac{\omega_e - \omega_m}{2\pi} = sf \text{ , where } f = \frac{\omega_e}{2\pi}. \tag{76.18}$$

From the stator perspective, the rotor voltages, currents and rotor mmf all have the synchronous frequency, since the rotor speed of ω_m is superimposed on the rotor variables' speed of ω_{slip}.

Per-phase Equivalent Circuit

The steady-state analysis of induction motors is often carried out using the per-phase equivalent circuit. A single phase equivalent circuit is used for the three-phase induction machine assuming a balanced set as shown in Figure 76.12. The per-phase equivalent circuit consists of the stator loop and the rotor loop with the magnetic circuit parameters in the middle. For the stator and rotor electrical parameters, the circuit includes the stator winding resistance and leakage reactance and the rotor winding resistance and leakage reactance. A slip dependent equivalent resistance represents the mechanical power delivered at the shaft due to the energy conversion in the air gap coupled electromagnetic circuit. The electrical input power supplied at the stator terminals converts to magnetic power and crosses the air gap. The air gap power P_{ag} is converted to mechanical power delivered at the shaft after overcoming the losses in the rotor circuit. Note that the voltages and currents described here in relation to the per-phase equivalent circuit are phasors and not space vectors.

Although the per-phase equivalent circuit is not enough to develop controllers that demand good dynamic performance, the circuit provides the basic understanding of induction machines. The vast majority of applications of induction motors are for adjustable speed drives where controllers designed for good steady-state performance are adequate. The circuit does allow the analysis of a number of steady-state performance features.

The power and torque relations are

$$P_{ag} = \text{Air gap power} = 3|I_r'|^2 \frac{R_r'}{s}$$

$$P_{dev} = \text{Developed mechanical power} = 3|I_r'|^2 \frac{(1-s)R_r'}{s} = (1-s)P_{ag} = T_e\omega_m \tag{76.19}$$

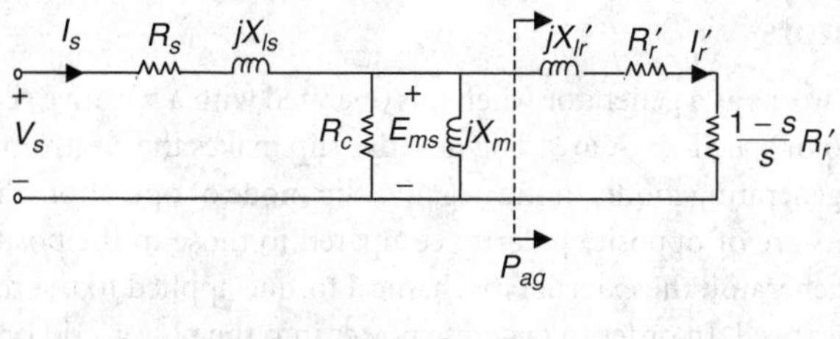

FIGURE 76.12 Steady-state per-phase equivalent circuit of an induction motor. (From Husain, I. 2003. *Electric and Hybrid Vehicles: Design Fundamentals,* CRC Press, Boca Raton, FL.)

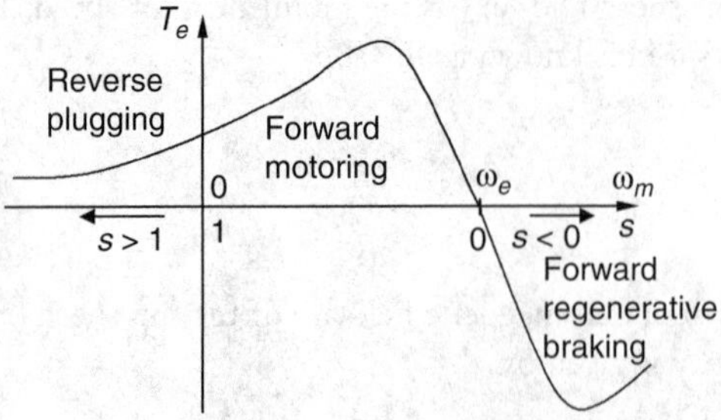

FIGURE 76.13 Steady-state torque speed characteristics of an induction motor. (From Husain, I. 2003. *Electric and Hybrid Vehicles: Design Fundamentals,* CRC Press, Boca Raton, FL.)

The electromagnetic torque with the assumption that $(R_c \| \dot{d}x_m) >> (R_s + \dot{d}X_{ls})$ is given by

$$T_e = 3|I_r'|^2 \frac{(1-s)R_r'}{s\omega_m} = \frac{3R_r'}{s\omega_s} \frac{V_s^2}{(R_{ls} + R_r'/s)^2 + (X_{ls} + X_{lr}')^2} \tag{76.20}$$

The steady-state torque-speed characteristics of the machine are as shown in Figure 76.13. The torque produced by the motor depends on the slip and the stator currents among other variables. The induction motor starting torque, while depending on the design, is lower than the peak torque achievable from the motor. The motor is always operated in the linear region of the torque-speed curve to avoid the higher losses associated with the high slip operation.

The value of the rotor circuit resistance determines the speed at which the maximum torque will occur. In general, the starting torque is low and the maximum torque occurs close to the synchronous speed when the slip is small. The motor draws a large current during line starting from a fixed AC source, which gradually subsides as the motor reaches the steady-state speed. If the load requires a high starting torque, the motor will accelerate slowly. This will make a large current flow for a longer time, thereby creating a heating problem.

The speed of an induction motor can be controlled by varying either the stator terminal voltage or stator frequency. Changing the terminal voltage changes the torque output of the machine as is evident from Equation (76.20). Note that changing the applied voltage does not change the slip for maximum torque. The speed control through changing the applied frequency is based on the frequency and synchronous speed relation

$$\omega_e = \frac{4\pi f}{p}$$

Changing f changes ω_e. The adjustable speed drives consist of a variable voltage, variable frequency power electronics drive that adjusts the supply voltage and frequency of the induction motor.

Induction Generators

The induction machine works as a generator when it is operated with a negative slip, i.e., the synchronous speed is less than the motor speed $\omega_e < \omega_m$. The negative slip makes the electromagnetic torque negative during regeneration or generating mode. In the negative slip mode of operation, the voltages and currents induced in the rotor bars are of opposite polarity compared to those in the positive slip mode.

When operated as a generator, the external mechanical torque applied to the rotor drives the machine beyond the synchronous speed. In order to generate power into the power grid or any electrical network, the machine requires reactive power for its excitation. The excitation power can be provided by external capacitors connected to the generator terminal, and no separate AC supply is needed. In the case of a grid-connected generator, the reactive power is supplied from the infinite bus. The advantage of induction generators is the absence of a separate field circuit and flexibility in speed. These advantages make the induction generator attractive for wind power generating systems.

Regenerative Braking

In the regenerative braking mode, the kinetic energy of the motor drive system is processed by the electric machine and returned to the energy source. From the machine perspective, this is no different than operating the machine in the generator mode. The electric machine converts the mechanical power available from the system kinetic energy and converts it to electrical energy. The electromagnetic torque acts on the rotor to oppose the rotor rotation, thereby decelerating the rotor. Induction motor drives are often used for four-quadrant drives, meaning that the electric motor is controlled by the drive to deliver positive or negative torque at positive or negative speed. Traction drives are examples of such applications.

dq Modeling

dq modeling relates to the transformation of three-phase variables in the *abc* coordinate system into an equivalent two-phase coordinate system that has an arbitrary speed in a given reference frame [Krasue and Wasynchuk, 1986; Lyon, 1954]. In the *dq* coordinate system, the *d*-axis is along the direct magnetic axis of the resultant mmf, while the *q*-axis is in quadrature to the direct axis. The *dq* modeling of AC machines enables the development of an electric machine controller, which operates in the inner loop with respect to the outer loop of the system level controller. The *dq* modeling analysis provides the necessary transformation equations required to implement the inner loop controller. Furthermore, the electromagnetic torque expression in terms of machine variables (current, flux linkage, etc.) in the *dq* model is often used to estimate the torque for closed loop control.

The space vector approach is retained in the *dq* reference frame, since these vectors help express the complex three-phase equations of AC machines in a compact form and also provide a simple relation for transformation between *abc* and *dq* reference frames. Several choices exist to define the relation between the *abc*-axes variables and the *dq*-axes variables. One choice is to take the *dq*-variables to be 2/3 times the projection of *a*,*b* and *c* variables on the *d* and *q* axes. The 2/3rd factor gives the same space vector magnitude as the peak value of the individual phase time-phasor variables. Another possible choice is to take the dq-variables to be $\sqrt{2/3}$ times the projection of *a*, *b* and *c* variables on the *d* and *q* axes. The $\sqrt{2/3}$ ratio between the *dq*-variables and the *abc*-variables conserves power without any multiplying factor in the *dq* and *abc* reference frames, and hence, is known as the power invariant transformation. The transformation matrices between the *abc* and *dq* variables for a multiplying factor of 2/3 are

$$\begin{bmatrix} f_d(t) \\ f_q(t) \\ 0 \end{bmatrix} = T_{abc->dq} \begin{bmatrix} f_a(t) \\ f_b(t) \\ f_c(t) \end{bmatrix}, \text{ where } T_{abc->dq} = \frac{2}{3} \begin{bmatrix} \cos(\theta) & \cos(\theta - 2\pi/3) & \cos(\theta + 2\pi/3) \\ \sin(\theta) & \sin(\theta - 2\pi/3) & \sin(\theta + 2\pi/3) \\ 0.5 & 0.5 & 0.5 \end{bmatrix} \tag{76.21}$$

$$T_{dq->abc} = \begin{bmatrix} \cos(\theta) & \sin(\theta) & 1 \\ \cos(\theta - 2\pi/3) & \sin(\theta - 2\pi/3) & 1 \\ \cos(\theta + 2\pi/3) & \sin(\theta + 2\pi/3) & 1 \end{bmatrix} \tag{76.22}$$

f is used to represent voltage, current, or flux linkages. These transformations are known as Park's transformations [Park, 1929].

The projections from both the *abc* frame and the *dq* frame to form the same space vector for a three-phase AC machine current variable are shown in Figure 76.14. The orientation of the *dq*-axes with respect to the *abc*-axes is at an arbitrary angle θ. The *dq*-reference frame can be stationary with respect to the stator or rotating at an arbitrary speed, such as at rotor speed or at synchronous speed. Again, when a

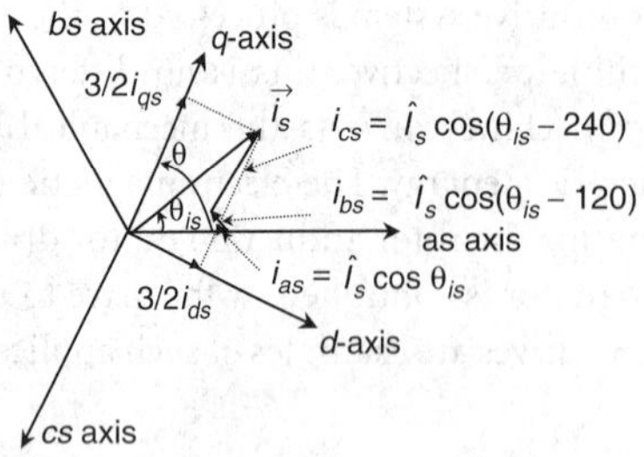

FIGURE 76.14 Transformation from three-phase variables to dq-axes variables. (From Husain, I. 2003. *Electric and Hybrid Vehicles: Design Fundamentals,* CRC Press, Boca Raton, FL.)

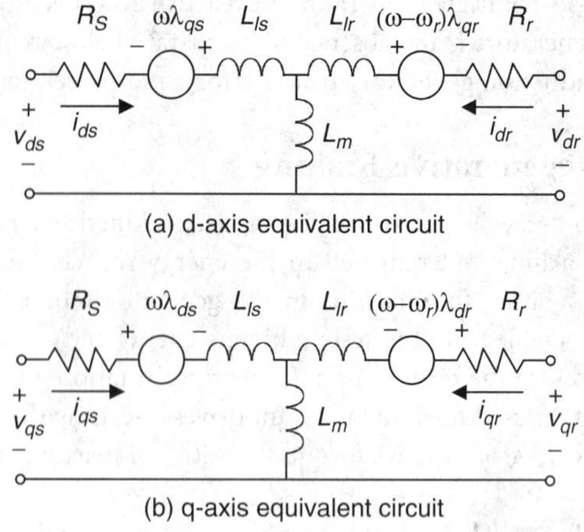

FIGURE 76.15 d and q-axes circuits of the induction machine. (From Husain, I. 2003. *Electric and Hybrid Vehicles: Design Fundamentals,* CRC Press, Boca Raton, FL.)

stationary dq reference frame is used, the dq-axes may be at any arbitrary angle with respect to our chosen reference phase a-axis.

The three speeds of dq windings commonly used for transformation are $0, \omega_m$, and ω_e. The zero speed is known as the stationary reference frame where, typically, the stationary d-axis is aligned with the phase-a axis of the stator. The d and q axes variables oscillate at the synchronous frequency in the balanced sinusoidal steady state. When ω_e is chosen as the speed of the reference dq frame, all the associated variables in the stator and in the rotor dq windings appear as DC variables in the balanced sinusoidal steady state. For an arbitrary speed of reference dq windings, the angle of transformation is

$$\theta = \int_0^t \omega(\xi)d\xi + \theta_0 \tag{76.23}$$

The dq-equivalent circuit model for the induction machine in circuit schematic form is shown in Figure 76.15.

Power and Electromagnetic Torque

The power into the induction machine is the product of the phase voltage and phase currents given as

$$P_{in} = (v_{as}i_{as} + v_{bs}i_{bs} + v_{cs}i_{cs}) + (v_{ar}i_{ar} + v_{br}i_{br} + v_{cr}i_{cr}) \tag{76.24}$$

The input power expression in terms of dq components in scalar form is

$$P_{in} = \frac{3}{2}(v_{ds}i_{ds} + v_{qs}i_{qs} + v_{dr}i_{dr} + v_{qr}i_{qr}) \tag{76.25}$$

The multiplying factor 3/2 is due to our choice of the factor 2/3 for the ratio of dq and abc variables.

The electromagnetic torque for a P-pole machine is

$$T_e = \frac{3}{2}\frac{P}{2}\mathrm{Im}\left[\vec{i}_{qds}\vec{\lambda}_{qds}^{\,*}\right] = \frac{3}{2}\frac{P}{2}(\lambda_{ds}i_{qs} - \lambda_{qs}i_{ds}) \tag{76.26}$$

Several alternative forms of the electromagnetic torque can be derived using the stator and rotor flux linkage expressions [Novotny and Lipo, 1996]. The torque expressions in terms of *dq* variables are used in the vector control of induction motor drives. Vector control implementations are accomplished in one of the several available choices of reference frames, such as rotor flux oriented reference frame, stator flux oriented reference frame, or air gap flux oriented reference frame. The *abc* variables at the input of the controller are converted to *dq* variables in the chosen reference frame at the input of the controller. The control computations take place in terms of *dq* variables, and the generated command outputs are again converted back to *abc* variables. The inverter controller executes the commands to establish the desired currents or voltages in the drive system.

76.3 Synchronous Machines

The synchronous machine is an AC machine with a separate field circuit to supply the magnetizing flux [Sen, 1997; Krasue and Wasynchuk, 1986; Novotny and Lipo, 1996]. The stator has three-phase windings sinusoidally distributed around the stator circumference similar to induction machine windings. DC current is fed to the rotor field windings through slip rings and brushes. The flux produced by the separate field circuit generates an induced voltage in the armature, the peak value of which can be approximated at low levels of flux by the relationship

$$E_{af} = \omega_e L_m I_f$$

where E_{af} is the induced voltage, ω_e is the angular electrical frequency, L_m is the mutual coupling inductance and I_f is the field current. At larger field currents, the mutual inductance changes due to saturation. This causes the relationship between the field current and the induced voltage to be nonlinear.

Let us consider synchronous generator operation first. When the generator supplies a load connected to the terminal, stator currents flowing through the armature or stator windings produce a synchronously rotating field at the same angular speed as the rotor magnetic field. The stator magnetic field induces a second voltage in the stator windings. The stator winding voltage is also known as the armature reaction voltage, since it is directly proportional to the amount of stator current flowing. The three-phase synchronous machine is conveniently analyzed using the single phase equivalent circuit representing the steady state characteristics. Let I_a be the phasor representation of the stator current in the single phase equivalent circuit. The armature reaction voltage drop can be expressed as

$$E_{ar} = -jX_{ar} I_a$$

where X_{ar} is an equivalent armature reaction impedance. The leakage reactance (X_{al}) caused by fluxes linking the stator conductors only is added to X_{ar} to form what is commonly referred to as synchronous reactance $X_S = X_{ar} + X_{al}$. The per-phase equivalent circuit of a synchronous generator can be drawn as shown in Figure 76.16. In the figure, V_t is the terminal voltage of the synchronous generator, R_a is the per-phase stator winding resistance, θ is the power factor angle at the terminal, and δ is the power angle between the terminal and the internal voltage. The generated internal voltage obtained by applying Kirchoff's voltage law is

$$E_a \angle \delta = V_t \angle 0 + I_a \angle \theta * (R_a + jX_S) \tag{76.27}$$

The power and torque characteristics of a synchronous machine are analyzed neglecting the stator phase winding resistance R_a. The synchronous generators used in power systems are normally connected to the infinite (fixed voltage) bus of a power grid. The maximum power that the machine can deliver depends on the maximum mechanical torque that the prime mover can apply without loss of synchro-

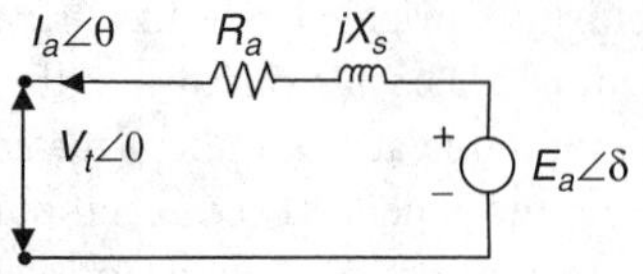

FIGURE 76.16 Single-phase armature equivalent circuit of a synchronous generator.

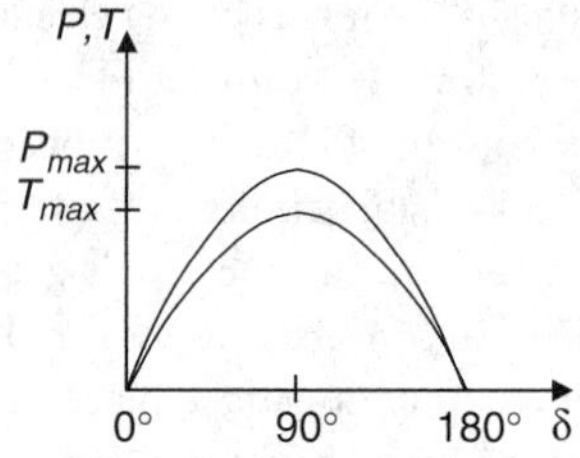

FIGURE 76.17 Power and torque angle characteristics of three-phase synchronous generators.

nism. The real and reactive three-phase powers for the synchronous generator can be derived from the equivalent circuit of Figure 76.16 as follows:

$$P = 3\frac{E_a V_t}{X_S}\sin\delta \tag{76.28}$$

$$Q = 3\frac{V_t}{X_S}\left(E_a\cos\delta - V_t\right) \tag{76.29}$$

The torque applied by the prime mover is given by

$$T_e = 3\frac{E_a V_t}{\omega_e X_S}\sin\delta \tag{76.30}$$

The power and torque angle characteristics are shown in Figure 76.17. The maximum power is obtained when the power angle is 90 degrees, which is also the static stability limit. The machine will lose synchronism if $\delta > 90°$. The torque is also maximum at $\delta > 90°$. The maximum torque is known as pull-out torque. In most cases, the angle δ is small (15° or less) and cos δ is approximately 1.0. Thus, E_a cos $\delta - V_t \approx E_a - V_t$. Note that, if $E_a > V_t$, $Q > 0$ (overexcited generator operating with leading power factor) and if $E_a < V_t$, $Q < 0$ (underexcited generator operating with lagging power factor).

In the case of synchronous motor, the current flows into the machine (opposite to the direction shown in Figure 76.16), and the terminal voltage equation becomes

$$V_t\angle 0 = E_a\angle\delta + I_a\angle\theta * (R_a + jX_S)$$

The real power of the synchronous motor can be expressed by Equation (76.28) or alternatively by $P = 3V_t I_a\cos\theta$.

The round rotor synchronous machines discussed so far have uniform air-gap, and the armature reaction mmf produces the same flux irrespective of the rotor position. Low-speed, multipolar machines have saliency in the rotor, which causes the reactance measured at the stator terminal to vary as a function of the rotor position. The magnetic reluctance is low along the pole axis or *d*-axis and high between the poles which is known as quadrature or *q*-axis. The armature reaction mmf will produce more flux if it is acting along the *d*-axis and less flux if it is acting along the *q*-axis.

76.4 Permanent Magnet Machines

The machines that use magnets to produce air-gap magnetic flux instead of field coils as in DC machines or magnetizing component of stator current as in induction machines are the permanent magnet or PM

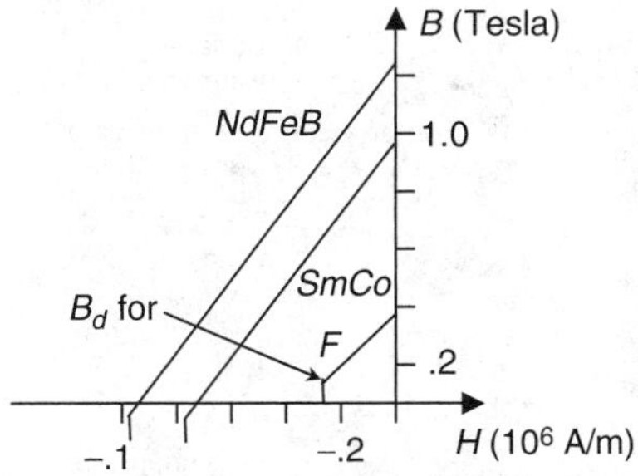

FIGURE 76.18 Characteristics of commonly used permanent magnets. (From Husain, I. 2003. *Electric and Hybrid Vehicles: Design Fundamentals,* CRC Press, Boca Raton, FL.)

machines [Miller, 1989]. This configuration eliminates the rotor copper loss as well as the need for maintenance of the field exciting circuit. PM machines can be broadly classified into PM synchronous machines and trapezoidal or squarewave machines.

The permanent magnets provide a loss-free excitation in a compact way without complications of connections to external stationary electric circuits. This is especially true for smaller machines, since there is always an excitation penalty associated with providing the rotor field through electrical circuits. For smaller machines, the mmf required is small and the resistive effects often become comparable and dominating resulting in lower efficiency.

The permanent magnets are a source of mmf much like a constant current source with relative permeability μ_r just greater than air, i.e., $\mu_r \approx 1.05$ to 1.07. The PM characteristics are displayed in the second quadrant of the *B-H* plot as shown in Figure 76.18, conforming with the fact that these are sources of mmf. The common type of magnets used in PM machines are the ferrites, samarium cobalt (SmCo) and neodmium-iron-boron (NdFeB). The magnets remain permanent as long as the operating point is within the linear region of its *B-H* characteristics. However, if the flux density is reduced beyond the knee-point of the characteristics (B_d), some magnetism will be lost permanently. On removal of the demagnetizing field greater than the limit, the new characteristics will be another straight line parallel to, but lower than, the original.

Permanent magnets machines are not only expensive, but also sensitive to temperature and load conditions, which constitutes the major drawback of PM machines. Most of the PM machines are found in small- to medium-power applications, although such machines are used in some high-power applications.

PM Synchronous Motors

The PMSM has a stator with a set of three-phase sinusoidally distributed copper windings similar to the windings described earlier for AC machines. A balanced set of applied three-phase voltages forces a balanced set of sinusoidal currents in the three-phase stator windings, which in turn establishes the constant amplitude rotating mmf in the air gap. The stator currents are regulated using rotor position feedback so that the applied current frequency is always in synchronism with the rotor. The permanent magnets in the rotor are appropriately shaped, and their magnetization directions are controlled such that the rotor flux linkage created is sinusoidal. The electromagnetic torque is produced at the shaft by the interaction of these two stator and rotor magnetic fields.

The PMSM has a high efficiency and a cooling system that is easier to design. The use of rare earth magnet materials increases the flux density in the air gap and accordingly increases the motor power density and torque-to-inertia ratio. In high-performance motion control systems that require servo-type operation, the PMSM can provide fast response, high power density, and high efficiency. The PMSM, similar to induction and DC machines, is fed from a power electronic inverter for operation of the system. Smooth torque output is maintained in these machines by shaping the motor currents, which necessitates a high-resolution position sensor and current sensors. A flux-weakening operation that enables a constant

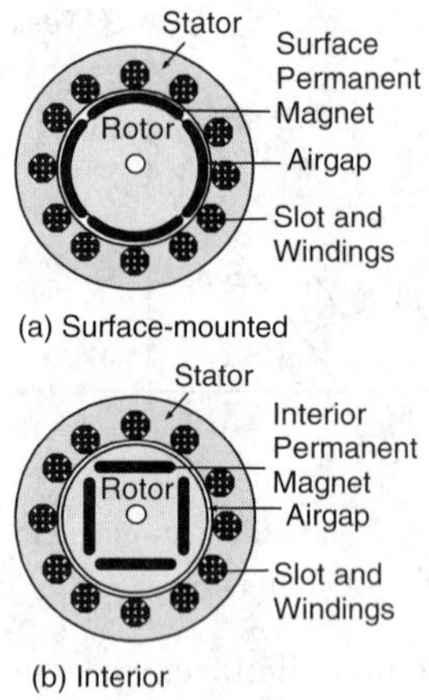

FIGURE 76.19 Permanent magnet machine: (a) surface-mounted and (b) interior. (From Husain, I. 2003. *Electric and Hybrid Vehicles: Design Fundamentals,* CRC Press, Boca Raton, FL.)

power mode of operation is possible in PMSM by applying a stator flux in opposition to the rotor magnet flux. The motor high speed limit depends on the motor parameters, its current rating, the back-emf waveform, and the maximum output voltage of the inverter.

PMSMs are classified according to the position and shape of the permanent magnets in the rotors. The three common arrangements of the rotors are surface mounted, inset, and interior or buried. The surface mounted and interior PM machine configurations are shown in Figure 76.19. The direct and quadrature-axes inductances of surface-mounted PMSM are approximately equal, since permeability of the path that the flux crosses between the stator and the rotor is equal all around the stator circumference. The space needed to mount the magnets increases the radial distance of the effective air gap, making the self-inductance relatively smaller in PMSMs. The interior PMSM has its magnets buried inside the rotor. The manufacturing process is complicated and expensive for the interior PMSM. The quadrature axis inductance L_q in the interior PMSMs can be much larger than the direct axis inductance L_d. The larger differences in the d and q-axes inductances make the interior PM more suitable for flux weakening operation delivering a wider constant power region compared to the other type of PMSMs. The extended constant power range capability is extremely important for applications requiring wide speed range operation. Because of the unequal reluctance paths in the direct and quadrature axes, a reluctance torque exists in buried and inset PMSMs.

The modeling of a PMSM can be accomplished either in the stationary reference frame or in the synchronous reference frame. In the synchronous reference frame, the dq reference frame is locked to the rotor frame. The d-axis is aligned with the magnet flux direction, while the q-axis lags the d-axis by 90 degrees of space angle. The stator applied voltage is balanced by the stator winding resistance drop and the induced voltage in the winding. The dq model of a PMSM in the rotor reference frame is

$$\begin{aligned} v_q &= R_s i_q + \frac{d}{dt}\lambda_q + \omega_r \lambda_d, \quad \text{with } \lambda_q = L_q i_q \\ v_d &= R_s i_d + \frac{d}{dt}\lambda_d - \omega_r \lambda_q, \quad \text{with } \lambda_d = L_d i_d + \lambda_f \end{aligned} \tag{76.31}$$

where i_d, i_q are the dq axis stator currents, v_d, v_q are the dq axis stator voltages, R_s is the stator phase resistance, $L_d = L_{ls} + L_{md}$ and $L_q = L_{ls} + L_{mq}$ are the dq axis phase inductances, λ_f is the amplitude of the flux linkage established by the PM, λ_d, λ_q are dq axis flux linkages and ω_r is the rotor speed in electrical radians/sec. L_{ls} is the leakage inductance. The electromagnetic torque for a P pole machine is

$$T_e = \frac{3}{2}\frac{P}{2}\left[\lambda_f i_q + (L_d - L_q) i_d i_q\right] \quad (76.32)$$

The rotor position information gives the position of d and q-axes. The control objective is to regulate the voltages v_d and v_q or the currents i_d and i_q by controlling the firing angles of the inverter switches. The rotor position is given by

$$\theta_r = \int_0^t \omega_r(\xi)d\xi + \theta_r(0)$$

PM Brushless DC Motors

Permanent magnet AC machines with trapezoidal back-emf waveforms are known as PM brushless DC machines or electronically commutated machines (ECMs). PM brushless DC motors are widely used in a range of applications starting from computer drives to sophisticated medical equipment. The reason behind the popularity of these machines is the simplicity of control. Only six discrete rotor positions per electrical revolution are needed in a three-phase machine to synchronize the phase currents with the phase back-emfs for effective torque production. A set of three Hall sensors mounted on the stator facing a magnet wheel fixed to the rotor and placed 120 degrees apart can easily give this position information. This eliminates the need for a high-resolution encoder or position sensor required in PM synchronous machines, but the penalty paid for position sensor simplification is in the performance. Vector control is not possible in PM brushless DC machines because of the trapezoidal shape of back-emfs.

The three-phase back-emf waveforms and the ideal phase currents of a PM brushless DC motor are shown in Figure 76.20. Square wave phase currents are supplied such that they are synchronized with the back-emf peak of the respective phase. The three stator windings for the three phases are identical with 120 degrees (electrical) phase displacement among them. Therefore, the stator winding resistances and the self-inductance of each of the three phases can be assumed to be identical. Let R_s = stator phase winding resistance, $L_{aa} = L_{bb} = L_{cc} = L$ = stator phase self inductance, and $L_{ab} = L_{ac} = L_{bc} = M$ = stator mutual inductance.

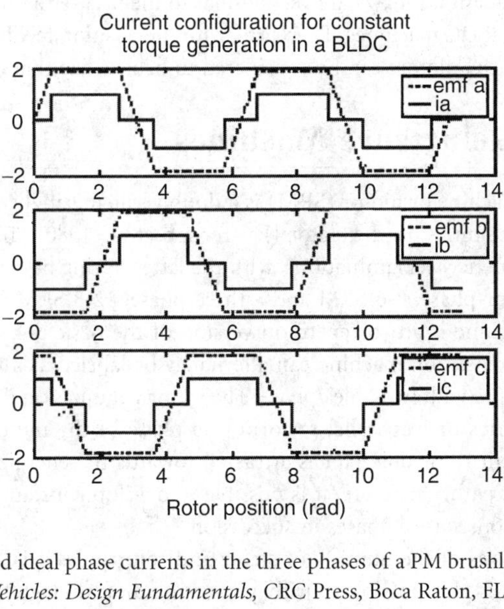

FIGURE 76.20 Back-emf and ideal phase currents in the three phases of a PM brushless DC motor. (From Husain, I. 2003. *Electric and Hybrid Vehicles: Design Fundamentals*, CRC Press, Boca Raton, FL.)

The voltage balance equation for the three-phase PM brushless DC motor is

$$\begin{bmatrix} v_a \\ v_b \\ v_c \end{bmatrix} = R \cdot \begin{bmatrix} i_a \\ i_b \\ i_c \end{bmatrix} + \begin{bmatrix} L & M & M \\ M & L & M \\ M & M & L \end{bmatrix} \cdot p \cdot \begin{bmatrix} i_a \\ i_b \\ i_c \end{bmatrix} + \begin{bmatrix} e_a \\ e_b \\ e_c \end{bmatrix} \tag{76.33}$$

where p is the operator d/dt, and e_a, e_b, and e_c are the back-emfs in the three phases.

The electrical power transferred to the rotor is equal to the mechanical power $T_e\omega_r$ available at the shaft. Using this equality, the electromagnetic torque for the PM brushless DC motor is

$$T_e = \frac{e_a \cdot i_a + e_b \cdot i_b + e_c \cdot i_c}{\omega_r} \tag{76.34}$$

For the control strategy where only two phase-currents are active at one time, the torque expression for equal currents in two phases simplifies to

$$T_e = \frac{2 \cdot e_{\max} \cdot I}{\omega_r} \tag{76.35}$$

Since the currents are controlled to synchronize with the maximum back-emf only, e_{max} has been used in Equation (76.35) instead of e as a function of time or rotor position. For a peak phase flux of $\phi_{\max}$, e_{max} can be shown to be

$$e_{\max} = K \cdot \phi_{\max} \cdot \omega_r \tag{76.36}$$

where K is a machine constant. The torque expression of Equation (76.34) can also be written as

$$T_e = K \cdot \phi_{\max} \cdot I \tag{76.37}$$

Equation (73.36) and Equation (76.37) are very similar to the back-emf and torque equations [Equation (76.2)] for conventional DC machines. Therefore, with the position feedback–based inverter control strategy, the PM brushless DC motors can be considered to behave like that of a DC machine.

76.5 Switched Reluctance Machines

The switched reluctance machine or motor (SRM) is a doubly salient, singly excited reluctance machine with independent phase windings on the stator [Lawrenson et al., 1980; Miller, 1993]. The stator and the rotor are made of magnetic steel laminations, with the latter having no windings or magnets. Cross-sectional diagrams of a four-phase, 8-6 SRM and a three-phase, 12-8 SRM are shown in Figure 76.21. The three-phase, 12-8 machine is a two-repetition version of the basic 6-4 structure within the single stator geometry. The two-repetition machine can alternately be labeled as a four-poles/phase machine, compared to the 6-4 structure with two poles/phase. The stator windings on diametrically opposite poles are connected either in series or in parallel to form one phase of the motor. When a stator phase is energized, the most adjacent rotor pole-pair is attracted towards the energized stator to minimize the reluctance of the magnetic path. Therefore, it is possible to develop constant torque in either direction of rotation by energizing consecutive phases in succession.

Several other combinations of the number of stator and rotor poles exist, such as 10-4, 12-8, etc. A 4-2 or a 2-2 configuration is also possible, but they have the disadvantage that if the stator and rotor poles are aligned exactly, then it would be impossible to develop a starting torque. Configurations with

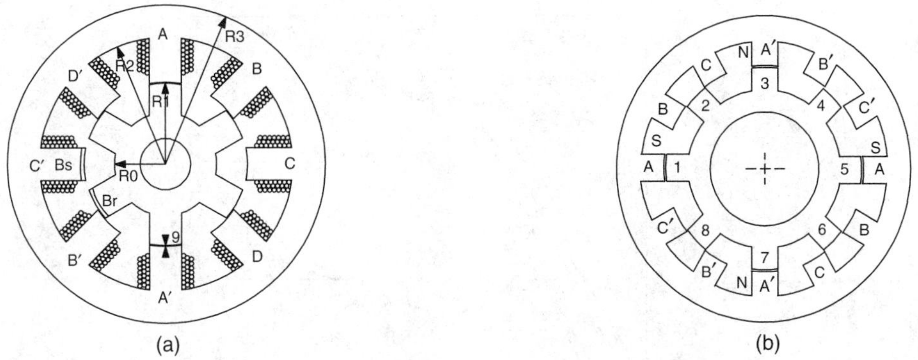

FIGURE 76.21 Cross-sections of three-phase SR machines: (a) four-phase 8/6 structure; (b) 12/8, two-repetition structure. (From Husain, I. 2002. Switched reluctance machines, in *Power Electronics Handbook*, Skvarenina, T. L., Ed., CRC Press, Boca Raton, FL.)

higher numbers of stator/rotor pole combinations have less torque ripple and do not have the problem of starting torque.

Advantages and Drawbacks

The SRM possesses a few unique features that make it a strong competitor to existing AC and DC motors in various adjustable speed drive and servo applications. The major advantage of an SRM is the simple and low-cost machine construction due to the absence of rotor winding and permanent magnets. The fault tolerance capability in SRM because of the phase independence and natural protection against shoot-through faults in the converter is also an extremely attractive feature. In SRM, the bulk of the losses appears in the stator, which is relatively easier to cool. The maximum permissible rotor temperature is higher, since there is no permanent magnet. The rotor inertia is low, enabling extremely high speeds with a wide constant power region operation.

Among the disadvantages of SRM are the higher torque ripple and higher acoustic noise compared to other machines. The absence of permanent magnets imposes the burden of excitation on the stator windings and converter, which increases the converter kVA requirement. However, the maximum speed at constant power is not limited by the fixed magnet flux as in the PM machine, and hence, an extended constant power region of operation is possible in SRMs. The control can be simpler than the field-oriented control of induction machines, although for torque ripple minimization, significant computations may be required for an SRM drive.

Basic Principle of Operation

The general equation governing the flow of stator current in one phase of an SRM can be written as

$$V_{ph} = i_{ph}R_s + \frac{d\lambda_{ph}}{dt} = i_{ph}R_s + \frac{\partial\lambda_{ph}}{\partial i_{ph}}\frac{di_{ph}}{dt} + \frac{\partial\lambda_{ph}}{\partial\theta}\frac{d\theta}{dt} \tag{76.39}$$

where V_{ph} is the DC bus voltage, i_{ph} is the instantaneous phase current, R_s is the winding resistance and λ_{ph} is the flux linking the coil. The SRM is always driven into saturation to maximize the utilization of the magnetic circuit; hence, the flux-linkage λ_{ph} is a nonlinear function of stator current and rotor position

$$\lambda_{ph} = \lambda_{ph}(i_{ph}, \theta)$$

The electromagnetic profile of an SRM is defined by the $\lambda - i - \theta$ characteristics shown in Figure 76.22.

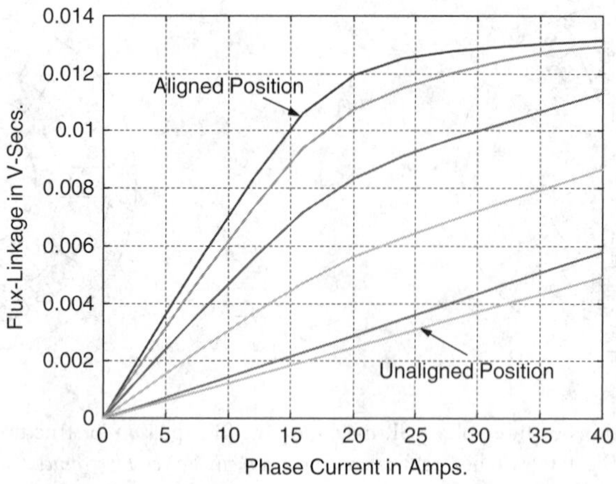

FIGURE 76.22 Flux-angle-current characteristics of a four-phase SRM. (From Husain, I. 2002. Switched reluctance machines, in *Power Electronics Handbook*, Skvarenina, T. L., Ed., CRC Press, Boca Raton, FL.)

The last term in Equation (76.38) is the "back-emf" or "motional-emf" and has the same effect on SRM as the back-emf has on DC motors or electronically commutated motors. However, the back-emf in SRM is generated in a different way from the DC machines or ECMs where it is caused by a rotating magnetic field. In an SRM, there is no rotor field and back-emf depends on the instantaneous rate of change of phase flux-linkage.

Assuming magnetic linearity [where $\lambda_{ph} = L_{ph}(\theta)i_{ph}$] for simplifying the analysis, the instantaneous input power can be expressed as

$$P_{in} = V_{ph}i_{ph} = i_{ph}^{2}R + \frac{d}{dt}\left(\frac{1}{2}L_{ph}i_{ph}^{2}\right) + \frac{1}{2}i_{ph}^{2}\frac{dL_{ph}}{d\theta}\omega \tag{76.39}$$

The first term represents the stator winding loss, the second term denotes the rate of change of magnetic stored energy, and the third term is the mechanical output power. The rate of change of magnetic stored energy always exceeds the electromechanical energy conversion term. The supplied energy is used most effectively when the current is maintained constant during the positive $dL_{ph}/d\theta$ slope. The magnetic stored energy is not necessarily lost but can be retrieved by the electrical source if an appropriate converter topology is used. In the case of a linear SRM, the energy conversion effectiveness can be at most 50%. The drawback of lower effectiveness is the increase in converter volt–amp rating for a given power conversion of the SRM. The division of input energy increases in favor of energy conversion if the motor operates under magnetic saturation, which is the primary reason for operating the SRM always under saturation.

Torque Production

The general expression for instantaneous phase torque for a device operating under the reluctance principle is

$$T_{ph}(\theta, i_{ph}) = \left.\frac{\partial W'(\theta, i_{ph})}{\partial \theta}\right|_{i=constant} \quad \text{with co-energy} \quad W' = \int_0^i \lambda_{ph}(\theta, i_{ph})di \tag{76.40}$$

Obviously, the instantaneous torque is not constant. The total instantaneous torque of the machine is given by the sum of the individual phase torques

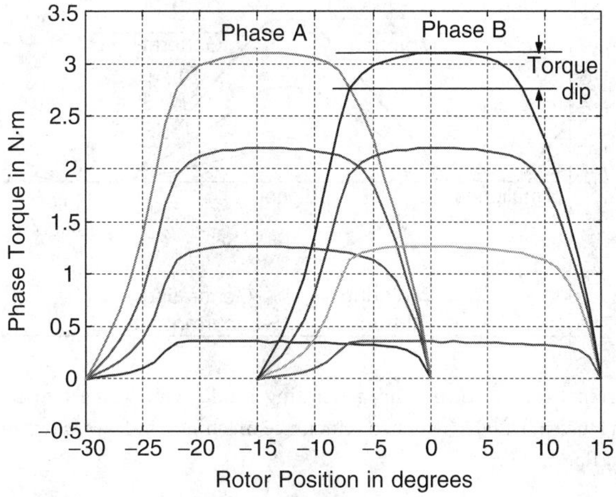

FIGURE 76.23 Torque-angle-current characteristics of a four-phase SRM for four constant current levels. (From Husain, I. 2002. Switched reluctance machines, in *Power Electronics Handbook,* Skvarenina, T. L., Ed., CRC Press, Boca Raton, FL.)

$$T_{inst}(\theta,i)=\sum_{phases}T_{ph}(\theta,i_{ph}). \tag{76.41}$$

The SRM electromechanical properties are defined by the static $T-i-\theta$ characteristics of a phase, an example of which is shown in Figure 76.23.

When magnetic saturation can be neglected, the instantaneous phase torque expression becomes

$$T_{ph}(\theta,i)=\frac{1}{2}i_{ph}^{2}\frac{dL_{ph}(\theta)}{d\theta} \tag{76.42}$$

The linear torque expression also follows from the energy conversion term (last term) in Equation (76.39).

The phase current needs to be synchronized with the rotor position for effective torque production. For positive or motoring torque, the phase current is switched such that the rotor moves from the unaligned position towards the aligned position. The linear SRM model is very insightful in understanding these situations. Equation (76.42) clearly shows that for motoring torque, the phase current must coincide with the rising inductance region. On the other hand, the phase current must coincide with the decreasing inductance region for braking or generating torque. The phase currents for motoring and generating modes of operation are shown in Figure 76.24 with respect to the phase inductance profiles.

The power electronic circuits for SRM drives are quite different from those of AC motor drives. The torque developed in an SRM is independent of the direction of current flow. Therefore, unipolar converters are sufficient to serve as the power converter circuit for the SRM, unlike the situation with induction motors or synchronous motors, which require bidirectional currents. This unique feature of the SRM, together with the fact that the stator phases are electrically isolated from one another, generates a wide variety of power circuit configurations. The type of converter required for a particular SRM drive is intimately related to motor construction and the number of phases. The choice also depends on the specific application. The most flexible and the most versatile four-quadrant SRM converter is the bridge converter shown in Figure 76.25, which requires two switches and two diodes per phase.

Appropriate positioning of the phase excitation pulses relative to the rotor position is the key in obtaining effective performance out of an SRM drive system. Therefore, a rotor position transducer is essential to provide the position feedback signal to the controller. The turn-on time, the total conduction period, and the magnitude of the phase current determine torque, efficiency, and other performance parameters.

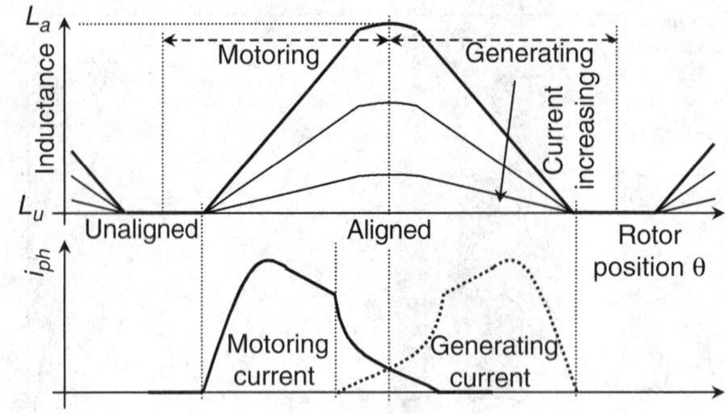

FIGURE 76.24 Phase currents for motoring and generating modes with respect to rotor position and idealized inductance profiles. (From Husain, I. 2002. Switched reluctance machines, in *Power Electronics Handbook*, Skvarenina, T. L., Ed., CRC Press, Boca Raton, FL.)

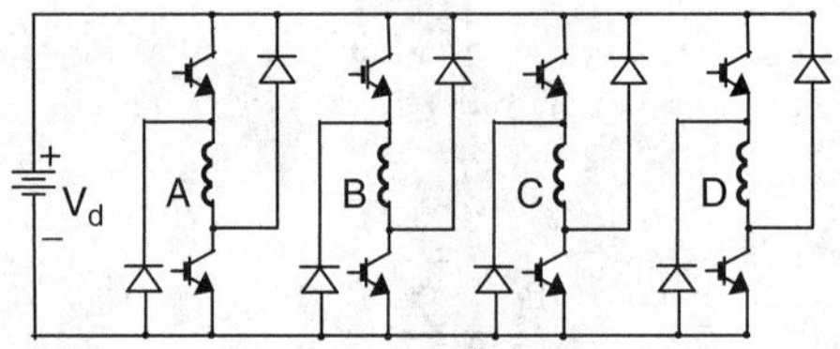

FIGURE 76.25 Classic bridge SRM power converter. (From Husain, I. 2003. *Electric and Hybrid Vehicles: Design Fundamentals*, CRC Press, Boca Raton, FL.)

76.6 Synchronous Reluctance Machine

The variable reluctance type synchronous machine is a viable alternative for variable speed drives for applications where rated torque at low speed is required [Lipo et al., 1992]. For such applications, the low-speed rotor losses of the induction machine make it a suboptimal choice. The synchronous reluctance machine (abbreviated as SynRM) rotor is constructed with saliency to employ the principle of reluctance torque production for electromechanical energy conversion. The machine retains many of the advantages of a switched reluctance motor, but at the same time overcomes some of its disadvantages. For example, the SynRM can be operated with unipolar currents in the same manner as in a SRM. Alternatively, the stator can be wound conventionally as in an induction or synchronous machine to produce the sinusoidal uniformly rotating air gap mmf to overcome the acoustic noise and torque ripple problems of SRM. The three-phase machine is fed from a inverter with a balanced set of three-phase AC voltages.

The SynRM are charactecterized by the absence of rotor windings or currents and special salient pole rotor configurations. The modern axially laminated rotor [Figure 76.26(a)] has high L_d/L_q ratio providing high torque density (Nm/kg), power factor, and efficiency that are comparable or superior to those of induction motors. The conventionally laminated anisotropic rotor [Figure 76.26(b)] has $L_d/L_q < 3$ leading to lower-power-factor (0.45 to 0.5) operation. The major difficulty with the axially laminated SynRM is the high manufacturing cost of the rotor, which could be eliminated or minimized only through mass production. The absence of rotor currents and ideally zero rotor core losses (in reality there are some losses due to harmonic flux currents) eliminates the need for cooling the rotor, which enables the motor to operate at very low speeds and high torque provided the stator is adequately cooled.

The *d-q* axes transformation theory can be used for the analysis of the SynRM. The torque produced by the machine can be expressed as

$$T_e = \frac{3}{2}\cdot\frac{P}{2}\cdot(\lambda_{ds} i_{qs} - \lambda_{qs} i_{ds}) \tag{76.43}$$

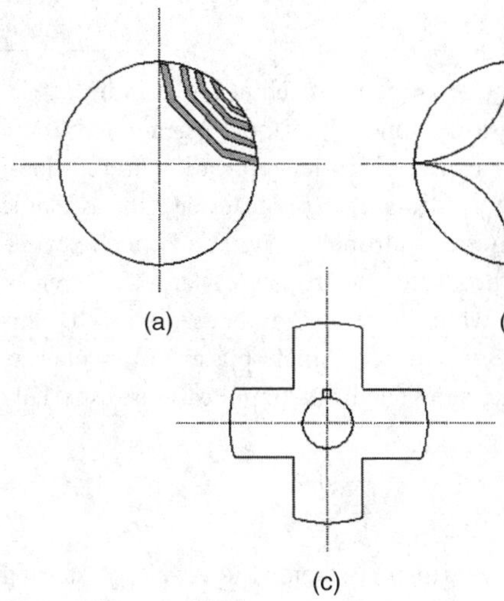

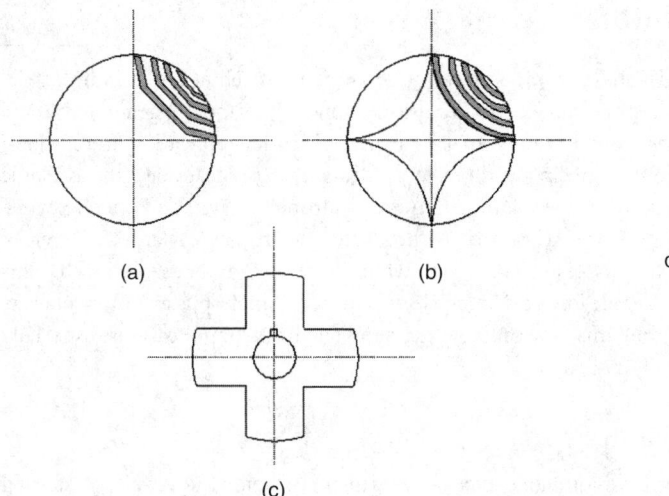

FIGURE 76.26 SynRM rotors: (a) rotor with multiple flux barrier, (b) modern axially laminated, and (c) conventional. (From Boldea, I. and Nasar, S. A. 1999. *Electric Drives,* CRC Press, Boca Raton, FL.)

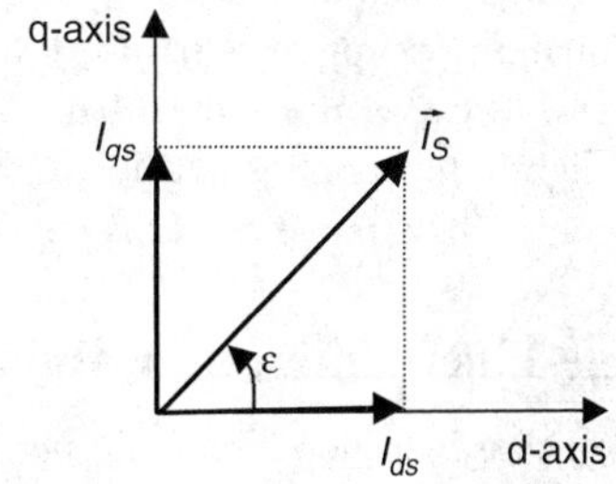

FIGURE 76.27 Mmf angle ε and *dq* currents in a synchronous reluctance machine.

In normal steady-state conditions when the rotor currents are zero, the *d* and *q*-axes flux linkages are given by $\lambda_{ds} = L_{ds} i_{ds}$ and $\lambda_{qs} = L_{qs} i_{qs}$. L_{ds} and L_{qs} are the *d* and *q* axes inductances with $L_{ds} \neq L_{qs}$. In general, L_{ds} and L_{qs} both have a leakage component and a magnetizing component, i.e.,

$$\begin{aligned} L_{ds} &= L_{ls} + L_{md} \\ L_{qs} &= L_{ls} + L_{mq} \end{aligned} \tag{76.44}$$

Therefore, the torque expression in terms of the magnetizing component is

$$T_e = \frac{3}{2} \cdot \frac{P}{2} \cdot (L_{md} - L_{mq}) i_{qs} i_{ds} \tag{76.45}$$

Expressing the *d* and *q* axes currents in terms of the stator current amplitude and the mmf angle ε (see Figure 76.27)

$$\begin{aligned} i_{ds} &= I_S \cos\varepsilon \\ i_{qs} &= I_S \sin\varepsilon \end{aligned} \tag{76.46}$$

The electromagnetic torque in terms of the stator current amplitude and mmf angle ε is

$$T_r = \frac{3}{2} \cdot \frac{P}{2} \cdot (L_{md} - L_{mq}) I_S^{2} \frac{\sin 2\varepsilon}{2} \tag{76.47}$$

Clearly, the torque is maximized when ε = 45°. For loads less than and up to rated conditions, the angle ε is maintained at 45°, while the current amplitude is varied, minimizing the I^2R loss as well as the converter loss.

76.7 Small Electric Motors

Motors for small loads are often single-phase as opposed to the three-phase machines used in industrial applications [Sen, 1997]. Although the three-phase motors use the majority of power generated, there are a large number of smaller machines used to move small loads in residential application and in servo applications. Small electric motors are available in a wide variety of designs in order to meet the available power supply and cost requirements. Some of these motors require electronics to start and operate, while others can run directly from a single-phase AC source without any electronics. A large number of fractional power applications use PM brushless DC motors, which have already been discussed. This section will cover some of the other available low-cost small electric motors. Single-phase motors operate with slightly higher slips and losses, but this is acceptable in many small load applications, especially when only single-phase power is readily available.

Single-Phase Induction Machines

Single-phase induction motors do not have a rotating flux wave, which is essential to develop a starting torque. The machines are built with an auxiliary mechanism to provide a second current that is out of phase with the main winding so that the interaction of the windings produces a rotating mmf to start the motor.

One method of providing starting torque to single-phase induction machines is to place a capacitor in series with the auxiliary winding of the motor. These motors are known as capacitor motors. The mmf of the starting current in the auxiliary winding can be adjusted to be equal to the current in the main winding, and the phase angle of the auxiliary winding current can be made to lead the main phase winding current by 90° by proper selection of the capacitor size. The 90° phase difference in the two currents results in a rotating mmf, which is required for the starting torque. A centrifugal switch in series with the capacitor circuit disconnects the capacitor once the motor starts running (Figure 76.28). Alternately, the capacitor can be kept permanently connected in the circuit, eliminating the need for the centrifugal switch. These permanent split capacitor motors tend to run more smoothly with higher efficiency and power factor. However, the starting torque is lower, since the capacitor is sized to balance the currents during operation with normal loads. In a more complex two-capacitor motor, known as capacitor-start, capacitor-run motor, a larger capacitor value is used for starting to ensure a high starting torque. Once the motor attains sufficient speed, the centrifugal switch opens, lowering the capacitor value to balance currents for normal loads.

Split-phase induction motors use a main winding and an auxiliary winding without a capacitor with their axes displaced by 90° in space. The auxiliary winding has a large *R/X* ratio, which is accomplished by using smaller wire-gauge for the winding. The centrifugal switch in series with the auxiliary winding opens at some set speed leaving only the main winding for operation. Split-phase motors have a moderate starting torque with very low starting current. The direction of rotation of any split phase motor or capacitor-start motor can be reversed by switching the connections of the auxiliary winding.

Another method of providing the starting torque for single-phase induction machines is to employ a separate short-circuited coil (called a shading coil) on one side of the main winding coil (Figure 76.29).

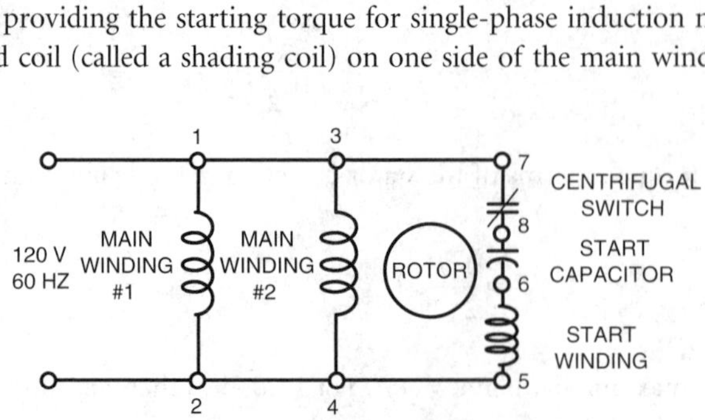

FIGURE 76.28 Connections for single-phase capacitor start motors.

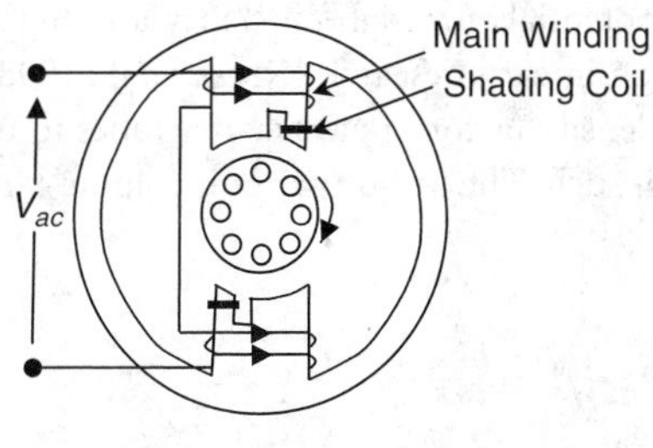

FIGURE 76.29 Shaded-pole induction motor.

These motors are known as shaded pole motors. The shading coils induce a current that resists a change of flux, therefore effectively delaying the flux buildup in the part of the pole with the coil. This creates a small rotating flux that can generate enough starting torque to start a motor that is lightly loaded. The direction of rotation of these motors cannot be reversed. The torque and rotation are in the direction from the unshaded to the shaded portion of the pole face. These are also less efficient and have much higher slip.

Universal Motors

Universal motors are essentially DC motors with series windings. The motors can be operated from either a DC or an AC source. Since the torque in a DC motor is proportional to both the armature and the field currents, connecting the windings in series ensures that the polarity of both the armature and field windings reverses simultaneously producing unidirectional torque.

Universal motors can operate at up to 20,000 rpm and are widely used for vacuum cleaners, portable drills, food mixers, and fans. These motors tend to be noisy when supplied from an AC source with high torque ripples and are also not very efficient. A triac connected in series with the supply can be used to regulate the supply voltage, thereby enabling the use of these machines for low-cost variable-speed applications.

Stepper Motors

The stepper motor is an electromagnetic actuator used for positioning of the rotor without any feedback. The motor converts digital pulse inputs designating a certain degree of rotation to an actual rotation of the rotor shaft. The machines have zero steady-state error and high torque density. The stepper motors develop a holding torque rather than a rotating torque when one phase is activated, thereby accurately retaining the position with load. Typical applications of stepper motors are printers, plotters, machine tools, and robotics.

Stepper motors are designed on the concept of either the variable (switched) reluctance motor or the PM synchronous motor. In the variable reluctance-type stepper motor, a large number of teeth is used in the rotor to create saliency. When one phase is aligned, the other phases are unaligned, similar to the switched reluctance machine. Permanent magnet stepper motors use permanent magnets in the rotor. There also exist hybrid stepper motors, where the rotor has an axial permanent magnet in the middle and ferromagnetic teeth at the outer sections.

76.8 Transverse Flux Machines

Conventional motor drive systems typically use a high-speed motor with a speed reduction gearbox. The drawbacks of gearing mechanisms are additional space, elasticity, backlash, and efficiency. A direct-drive system can eliminate the gearing mechanism in applications where the tip speed and other speed-related effects are not important. The direct-drive systems make sense functionally as well as economically only when a high-torque density machine is available. The concept of transverse flux (TF) machines with permanent magnets (PM) emerged about 12 years ago as a result of the search for optimized magnet paths in electric machines. The potentials for high efficiency and high torque density in TF machines have been demonstrated to be high in early research [Kruse et al., 1998].

Let us first discuss the torque density factor of electric machines to understand how TF machines are designed to achieve a high value of such. The torque per unit volume can be generally expressed as [Beaty and Kirtley, 1998]

$$\frac{T}{V_a} = \frac{\tau}{2}\frac{r}{h_s + t_b}\frac{l}{l + 2\Delta l} \tag{76.48}$$

where τ is the average shear stress producing the electromagnetic interaction, r is the rotor radius, l is the axial length of the machine, h_s is the radial length extension to accommodate stator windings, t_b is the radial length extension for the back iron and Δl is the axial length extension for the end windings. It can be concluded from Equation (76.48) that torque density increases directly with rotor radius, and large pole numbers reduce the impact of both t_b and h_s. The value of large pole numbers hits a point of diminishing return when $t_b < h_s$.

High-torque machines must be designed with short pole-pitch that has the advantage of reducing the required end-turn length and back-iron depth, especially for low-speed machines. However, short pole-pitch machines require relatively large magnetizing mmf, since every pole must be excited. Furthermore, low-speed machines are typically inefficient, because conduction losses are normally associated with torque, while power is torque times speed.

The TF machine uses a topology with a torroidal armature coil, where the current flows parallel to the direction of rotation. The winding arrangement gives an extremely short current path in the machine, thereby reducing conduction losses. The high specific torque in these machines is obtained by increasing the current loading through simply increasing the pole pairs without increasing the armature ampere-turns. The decoupled structure of flux-paths and armature coils enables such a design. The stator core is essentially salient with a large number of poles.

In PM TF machines, the magnets provide for the relatively large mmf requirements in these machines. In direct drive motors, the amount of magnet material is also proportionately high due to the high torque, which works against PM TF machines for economic reasons in the face of high magnet costs. The basic operating principle is the same for all PM TF machine variants. A stator phase winding having a ring or circular form produces a homopolar mmf distribution in the air-gap. This mmf is modulated by a pattern of stator poles or teeth to interact with a heteropolar pattern of PMs placed on the rotor. The basic structure of a PM TFM is shown in Figure 76.30. The number of stator poles is usually half of the number of rotor PMs, but much greater than for conventional machines with heteropolar topology. The machine thus obtained is capable of producing power densities up to three times greater than conventional machines. The PM TF machine must have two or more phases to produce continuous rotation and to

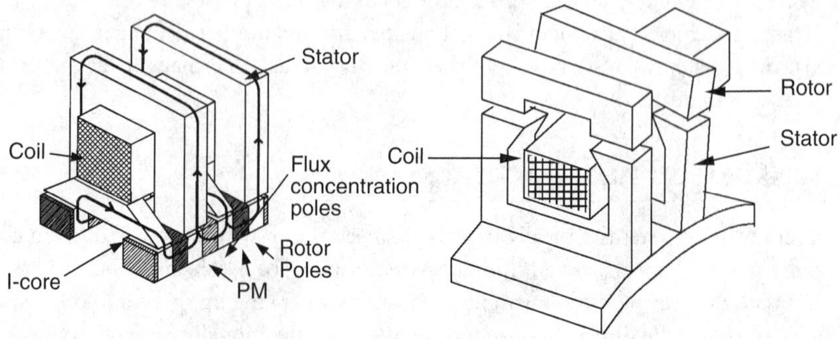

FIGURE 76.30 Transverse flux machines: (a) PM type, (b) reluctance type. (Figure 76.30(a) is from Henneberger, G. and Viorel, I. A. 2001. *Variable Reluctance Electric Machines*, Shaker Verlag, Aachen, Germany.)

avoid the starting difficulties. In contrast to conventional machines, there is no common rotating field in the three-phase structure of the TF machine, only three independent alternating fields, which are electrically shifted by 120°. The necessary mechanical shift of 120° for the prototype is obtained in the rotor by shifting the magnets. The PM TF machine stator windings must be supplied from an electronic power converter, which is controlled with respect to the rotor position. The control of the TF machines with PMs on the rotor is similar to that of the conventional synchronous machines.

The rotor in TF machines can also be made of salient poles instead of using permanent magnets [Kruse et al., 1998]. This type of machine is known as the reluctance TFM and is shown in Figure 76.31. The reluctance TF machine has quite the same features as a switched reluctance machine, the main difference being the homopolar type ring winding placed on the stator. The reluctance TF machine has the same number of poles on both stator and rotor. The phases are placed axially as is usual for TF machines. A reluctance TF machine can have three or more than three independent phases, which are fed independently as in a switched reluctance machine.

The TF machine has an outstanding torque-to-volume ratio compared to all other types of machines but has a lower power factor due to its homopolar feature. The constructional difficulty is the other main drawback of these machines.

References

Beaty, H. W. and Kirtley, J. L. 1998. *Electric Machines Handbook,* McGraw-Hill, New York.

Henneberger, G. and Viorel, I. A. 2001. *Variable Reluctance Electrical Machines,* Shaker-Verlag, Aachen, Germany.

Krasue, P. C. and Wasynchuk, O. 1986. *Analysis of Electric Machinery,* McGraw Hill, New York.

Kruse, R., Pfaff, G., and Pfeiffer, C. 1998. Transverse flux reluctance motor for direct servodrive applications, IEEE-IAS Annual Meeting, St. Louis, pp. 655–662.

Lawrenson, P. J., Stephenson, J. M., Blenkinsop, P. T., Corda, J. and Fulton, N. N. 1980. Variable-speed switched reluctance motors, *IEE Proc.,* Pt. B, 127:253–265.

Lipo, T. A., Vagati, A., Malesani, L., and Fukao, T. 1992. Synchronous Reluctance Motors and Drives — A New Alternative, Tutorial Presentation, IEEE-IAS Annual Meeting, Houston, TX.

Lyon, W. V. 1954. *Transient Analysis of Alternating Current Machinery,* John Wiley & Sons, Inc., New York.

Miller, T. J. E. 1989. *Brushless Permanent Magnet and Switched Reluctance Motor Drives,* Oxford University Press, Oxford.

Miller, T. J. E. 1993. *Switched Reluctance Motors and their Control,* Magna Physics Publishing, Hillsboro, OH.

Novotny, D. W. and Lipo, T. A. 1996. *Vector Control and Dynamics of AC Drives,* Oxford University Press, New York.

Park, R. H. 1929. Two-reaction theory of synchronous machines — generalized method of analysis — Part I, *AIEE Transactions,* 48:716–727.

Sen, P. C. 1997. *Principles of Electric Machines and Power Electronics,* John Wiley & Sons, Inc., New York.

77

Fuel Cells

Gregor Hoogers
University of Applied Sciences Trier

As early as 1838/1839 Friedrich Wilhelm Schönbein and William Grove discovered the basic operating principle of fuel cells by reversing water electrolysis to generate electricity from hydrogen and oxygen [Bossel, 2000]. This has not changed since:

A fuel cell is an electrochemical device that continuously converts chemical energy into electric energy — and heat — for as long as fuel and oxidant are supplied.

Fuel cells therefore bear similarities both to batteries, with which they share the electrochemical nature of the power generation process, and with engines which, unlike batteries, will work continuously consuming a fuel of some sort. When hydrogen is used as fuel, they generate only power and pure water: they are therefore called zero-emission engines. Thermodynamically, the most striking difference is that thermal engines are limited by the well-known Carnot efficiency, while fuel cells are not. This is an advantage at low-temperature operation — if one considers, for example, the speculative Carnot efficiency of a biological system working, instead, as a heat engine. At high temperatures, the Carnot efficiency theoretically surpasses the electrochemical process efficiency. Therefore, one has to look in detail at the actual efficiencies achieved in working thermal and fuel cell systems where, in both cases, losses rather than thermodynamics dominate the actual performance.

Fuel cells are currently being developed for three main markets: automotive propulsion, electric power generation, and portable systems. Each main application is dominated by specific systems requirements and even different types of fuel cells, while the underlying operating principle remains the same.

77.1 Fundamentals

Operating Principle

The underlying principle of fuel cell operation is the same for all types that have emerged over the past 160 years: a so-called redox reaction is carried out in two half-reactions located at two electrodes separated by an electrolyte. The purpose of the electrolyte is electronic separation of and ionic connection between the two electrodes, which will assume different electrochemical potentials. The advantage of carrying out

0-8493-1586-7/05/$0.00+$1.50

the reaction in two parts is direct electrochemical conversion of chemical energy into electric energy by employing the resulting potential difference between the two electrodes, anode and cathode. The simplest and most relevant reaction in this context is the formation of water from the elements,

$$H_2 + \int O_2 = H_2O \quad (77.1)$$

Rather than carrying out reaction (77.1) as a gas phase reaction, i.e., thermally after supplying activation energy (ignition) or, more gently, by passing the two reactants over an oxidizing catalyst such as platinum and generating merely heat, the electrochemical reaction requires the half-reactions

$$H_2 = 2H^+ + 2e^- \quad (77.2)$$

and

$$\int O_2 + 2H^+ + 2e^- = H_2O \quad (77.3)$$

to take place at an anode (77.2) and cathode (77.3), respectively. Therefore, the two electrodes assume their electrochemical potentials at — theoretically — 0 and 1.23 V, respectively, and an electronic current will flow when the electrodes are connected through an external circuit. The anode reaction is the hydrogen oxidation reaction (HOR); the cathode reaction is the oxygen reduction reaction (ORR). In this specific case, the electrolyte is an acidic medium which only allows the passage of protons, H^+, from anode to cathode. Water is formed at the cathode.

Fuel Cell Types

The different fuel cells differ from each other by and are named after the choice of electrolyte. The electrolyte will also determine the nature of the ionic charge carrier and whether it flows from anode to cathode or cathode to anode. Electrode reactions and electrolytes are listed in Table 77.1.

Grove's fuel cell operating in dilute sulfuric acid electrolyte, but also the phosphoric acid fuel cell (PAFC, operating at around 200°C) and the proton exchange membrane fuel cell (PEMFC or PEFC, operating at 80°C, also called solid polymer fuel cell, SPFC) are examples for cells with acidic, proton-conducting electrolytes as in reactions (77.2) and (77.3).

In contrast, when using an alkaline electrolyte, the primary charge carrier is the OH^- ion, which flows in the opposite direction, so water is formed at the anode. The overall reaction (77.1) does not change. The resulting fuel cell is called an alkaline fuel cell or AFC and leads a very successful niche existence in supplying electric power to space craft such as Apollo and the Shuttle.

TABLE 77.1 Fuel Cell Types, Electrolytes, and Electrode Reactions

Fuel Cell Type	Electrolyte	Charge Carrier	Anode Reaction	Cathode Reaction	Operating Temperature
Alkaline FC (AFC)	KOH	OH^-	$H_2 + 2OH^- = 2H_2O + 2e^-$	$1/2\ O_2 + H_2O + 2e^- = 2OH^-$	60–120°C
Proton exchange membrane FC (PEMFC, SPFC)	Solid polymer (such as Nafion)	H^+	$H_2 = 2H^+ + 2e^-$	$1/2\ O_2 + 2H^+ + 2e^- = H_2O$	50–100°C
Phosphoric acid FC (PAFC)	Phosphoric acid	H^+	$H_2 = 2H^+ + 2e^-$	$1/2\ O_2 + 2H^+ + 2e^- = H_2O$	~220°C
Molten carbonate FC (MCFC)	Lithium and potassium carbonate	CO_3^{2-}	$H_2 + CO_3^{2-} = H_2O + CO_2 + 2e^-$	$1/2\ O_2 + CO_2 + 2e^- = CO_3^{2-}$	~650°C
Solid oxide FC (SOFC)	Solid oxide electrolyte (yttria stabilized zirconia, YSZ)	O^{2-}	$H_2 + O^{2-} = H_2O + 2e^-$	$1/2\ O_2 + 2e^- = O^{2-}$	~1000°C

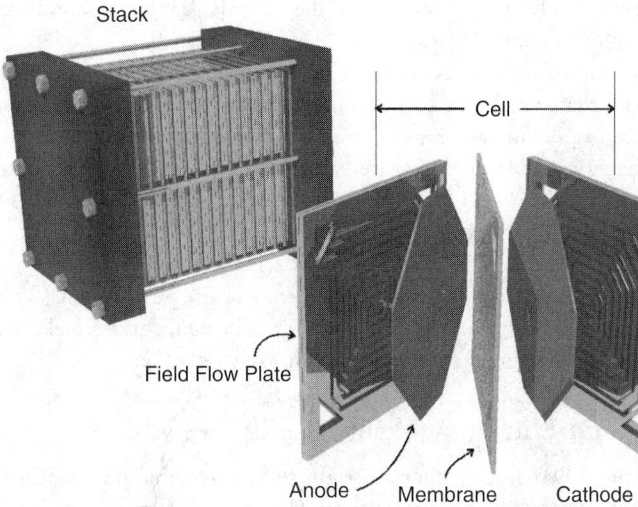

FIGURE 77.1 Fuel cell stack made up of flow field plates (or bipolar plates) and MEAs (shown in the insert).

Phosphoric acid fuel cells (PAFCs) use molten H_3PO_4 as an electrolyte. The PAFC has been mainly developed for the medium-scale power generation market, and 200 kW demonstration units have now clocked up many thousands of hours of operation. However, in comparison with the two low-temperature fuel cells, alkaline and proton exchange membrane fuel cells (AFCs, PEMFCs), PAFCs achieve only moderate current densities.

The proton exchange membrane fuel cell, PEMFC, takes its name from the special plastic membrane that it uses as its electrolyte. Robust cation exchange membranes were originally developed for the chlor-alkali industry by Du Pont and have proven instrumental in combining all the key parts of a fuel cell, anode and cathode electrodes and the electrolyte, in a very compact unit. This membrane electrode assembly (MEA), not thicker than a few hundred microns, is the heart of a PEMFC and, when supplied with fuel and air, generates electric power at cell voltages up to 1 V and power densities of up to about $1 Wcm^{-2}$.

The membrane relies on the presence of liquid water to be able to conduct protons effectively, and this limits the temperature up to which a PEMFC can be operated. Even when operated under pressure, operating temperatures are limited to below 100°C. Therefore, to achieve good performance, effective electrocatalyst technology is required. The catalysts form thin (several microns to several tens of microns) gas-porous electrode layers on either side of the membrane. Ionic contact with the membrane is often enhanced by coating the electrode layers using a liquid form of the membrane ionomer.

The MEA is typically located between a pair of current collector plates with machined flow fields for distributing fuel and oxidant to anode and cathode, respectively (compare Figure 77.1). A water jacket for cooling may be inserted at the back of each reactant flow field followed by a metallic current collector plate. The cell can also contain a humidification section for the reactant gases, which helps to keep the membrane electrolyte in a hydrated, proton-conduction form. The technology is given a more thorough discussion in the following chapter.

Another type of fuel cell is already built into most cars today: the lambda sensor measuring the oxygen concentration in the exhaust of four-stroke spark ignition engines is based on high temperature oxygen ion, O^{2-}, conductors. Typically, this is yttria stabilized zirconia or YSZ, which is also used as solid electrolyte in high temperature (up to 1100°C or 1850°F) solid oxide fuel cells (SOFC). In fact, a lambda sensor works rather well as a high-temperature fuel cell when a hydrogen or a hydrocarbon is used instead of exhaust gas. Table 77.1 shows the electrode reactions.

The second high-temperature fuel cell uses molten carbonate salts at 650°C (1200°F) as electrolyte which conducts carbonate ions, CO_3^{2-}. It is unique that in this type of fuel cell, the ions are formed not from the reactants but from CO_2 that is injected into the cathode gas stream, and usually recycled from the anode exhaust — compare Table 77.1.

The two high-temperature fuel cells, solid oxide and molten carbonate (SOFC and MCFC), have mainly been considered for large scale (MW) stationary power generation. In these systems, the electrolytes consist of anionic transport materials, as O^{2-} and CO_3^- are the charge carriers. These two fuel cells have two major advantages over low-temperature types. They can achieve high electric efficiencies — prototypes have achieved over 45% — with over 60% currently targeted. This makes them particularly attractive for fuel-efficient stationary power generation.

The high operating temperatures also allow direct, internal processing of fuels such as natural gas. This reduces the system complexity compared with low-temperature power plants, which require hydrogen generation in an additional process step. The fact that high-temperature fuel cells cannot easily be turned off is acceptable in the stationary sector, but most likely only there.

Thermodynamics and Efficiency of Fuel Cells

Thermodynamics [Atkins, 1994] teaches that the right thermodynamic potential to use for processes with nonmechanical work is ΔG, the Gibbs free energy. The Gibbs free energy determines the maximum of the nonmechanical work that can be expected from such a reaction, for example, a biological or electrochemical reaction. Another useful quantity in this context is the standard potential of an electrochemical reaction, E_0. E_0 is related to the Gibbs free energy of the reaction by

$$\Delta G = nFE_0 \tag{77.4}$$

where n is the number of charges in the reaction, and $F = 96485$ Cmol^{-1} is the Faraday constant. ΔG essentially determines the upper limit of the electric work coming out of a fuel cell. Both Gibbs free energies and standard reaction potentials are tabulated. So are standard reaction enthalpies, ΔH.

In order to derive the highest possible cell voltage and the cell efficiency, let us consider reaction (77.1). The reaction Gibbs free energy is determined in the usual way from the Gibbs free energies of formation, ΔG_f, of the product side minus ΔG_f on the reactant side.

$$\Delta G = \Delta G_f(H_2O) - \Delta G_f\,(H_2) - 1/2\,\Delta G_f\,(O_2) = \Delta G_f\,(H_2O) = -237.13 \text{ kJmol}^{-1} \tag{77.5}$$

(The Gibbs free energies of elements in their standard states are zero.)

From (77.4) we can now calculate the maximum cell voltage for a cell based on reaction (77.1):

$$E_0 = \Delta G(H_2O)/nF = 237.13 \text{ kJmol}^{-1}/(2 \bullet 96485 \text{ C mol}^{-1}) = 1.23 \text{ V} \tag{77.6}$$

Here, n = 2 because reaction (77.3) shows that two electrons are exchanged for the formation of one mol of water.

The higher heating value (HHV) of the same reaction is simply ΔH, the reaction enthalpy for reaction (77.1) with liquid product water. In analogy with reaction (77.5), we can calculate the reaction enthalpy from the tabulated enthalpies of formation. Again:

$$\Delta H = \Delta H_f\,(H_2O, l) = -285.83 \text{ kJmol}^{-1} \tag{77.7}$$

Comparison with (77.5) shows that an electrochemical cell based on reaction (77.1) can at best achieve an electric efficiency of

$$\eta_{el}^{max} = \Delta G/\Delta H = 0.83 = 83\%$$

based on the HHV of the fuel.

Measuring the actual electric efficiency of a fuel cell is extremely simple: A measurement of the operating cell voltage E is sufficient:

$$\eta_{el} = E/E_0 \bullet \Delta G/\Delta H = (nF/\Delta H) \bullet E = 0.68\ \mathrm{V}^{-1}\ E \tag{77.8}$$

For example, a hydrogen/air fuel cell achieving a cell voltage of 0.7 V under operation converts 0.68 V^{-1} • 0.7 V = 0.47 = 47% of the chemical energy supplied into electric energy.

Effects of Pressure and Temperature

Everything that has been said so far was based on standard values for pressure and temperature, i.e., T = 298 K and P = 1 bar. In practice, fuel cells usually operate at elevated temperatures and, often, above ambient pressure.

Generally, for a chemical reaction of the type

$$a\mathrm{A} + b\mathrm{B} \quad m\mathrm{M} + n\mathrm{N} \tag{77.9}$$

the effect of the partial pressures (strictly speaking: activities) of the reactants and products on the change in Gibbs free energy is given as

$$\Delta G = \Delta G^\circ + RT\ln\frac{P_\mathrm{M}^m P_\mathrm{N}^n}{P_\mathrm{A}^a P_\mathrm{B}^b} \tag{77.10}$$

When converted to potentials, using Equation 77.4, this turns into the well-known Nernst equation. It is clear from (77.1), (77.9), and (77.10) that an increase in either $P(H_2)$ or $P(O_2)$ will lead to a higher cell potential.

The temperature dependence is given by the temperature dependence of the Gibbs free energy, $\Delta G(T)$. Since $\Delta G = \Delta H - T\Delta S$, and ΔH depends only weakly on T, the temperature derivative of ΔG is given by:

$$dG/dT = -\Delta S \text{ or, using (77.4)}$$

$$dE_0/dT = -\Delta S/nF \tag{77.11}$$

System Efficiency

The *system efficiency* in this case will, of course, be lower due to the consumption of the power system itself and due to wastage of hydrogen fuel or air. The reason for this is often system related, i.e., a certain amount of hydrogen is passing through the cell unused in order to remove continuously water or contaminants from the anode. The hydrogen leaving the cell is either recycled or burned. The hydrogen excess is quantified by two terms used by different authors, stoichiometry and the fuel utilization. A 50% excess of hydrogen feed is expressed in a fuel stoichiometry of 1.5 or in a fuel utilization of 1/1.5 = 2/3. An air surplus is also used routinely. This will affect the system efficiency through the energy lost on compressor power.

Another consideration is linked to the method of providing hydrogen. When hydrogen is made from another chemical that is more apt for storage (for example, on board a vehicle) or for supply from the natural gas grid, the fuel processor or *reformer* will also have a certain efficiency.

Finally, the DC power generated by a fuel cell is often converted into AC, either for an electric motor or for supplying mains current. The alternator, again, affects the overall system efficiency by a load-dependent loss factor.

Kinetics of Fuel Cell Reactions

As may be expected, the treatment of the kinetics of fuel cell reactions is harder than the thermodynamics but nonetheless much simpler than the details of the energy conversion in thermal engines.

A full treatment of the electrochemical kinetics (Butler–Volmer Equation) is beyond the scope of this chapter and can be found in the general electrochemical [Hamann, 1998] or dedicated fuel cell literature

[Hoogers, 2003]. It is a striking result of the kinetic work on fuel cells that anode reaction kinetics are far superior to cathode kinetics and that the anode potential — with the cell operating on clean hydrogen — is close to the reversible hydrogen potential at 0 V. The cell performance is therefore dominated by the cathode potential, i.e., $E_{cell} = E_c$.

Essentially, the kinetic description of electrode kinetics predicts an exponential dependence of the cell current on the cell voltage. For practical purposes, it is often more desirable to express the cell potential as a function of current drawn from the cell, i.e.,

$$E_c = E_r - b\ log_{10}(i/i_o) - i\ r \tag{77.12}$$

where E_r = reversible potential for the cell; i_o = exchange current density for oxygen reduction; b = Tafel slope for oxygen reduction; r = (area specific) ohmic resistance; and i = current density, in which ohmic losses have been included by adding the term ($-ir$).

This is the so called *Tafel equation* (with addition of the *ir*-dependence). The *Tafel slope b* is determined by the nature of the electrochemical process. b can be expressed as

$$b = RT/\beta F \tag{77.13}$$

where $R = 8.314\ Jmol^{-1}K^{-1}$ denotes the universal gas constant, F is again the Faraday constant, T is the temperature (in K), and β is the transfer coefficient, a parameter related to the symmetry of the chemical transition state, usually taken to be 0.5, when no first principle information is available. For the oxygen reduction reaction in practical proton exchange membrane fuel cells, b is usually experimentally determined with values between 40 and 80 mV.

The main factor controlling the activation overpotential and hence the cell potential, $E_{cell} = E_c$, is the (apparent) exchange current density i_0. Equation. (77.12) demonstrates that, due to the logarithm, a tenfold increase in i_0 leads to an increase in cell potential at the given current by one unit of b, or typically 60 mV. It is important to dwell on this point. While the reversible potential E_r is given by thermodynamics and the Tafel slope b is dictated by the chemical reaction (and the temperature), the value for i_0 depends entirely on reaction kinetics. Ultimately, it depends on the skill of the MEA and electrocatalyst producer to increase this figure.

For doing so there are, in principle, the following possible approaches:

- The magnitude of i_0 can be increased (within limits) by adding more electrocatalyst to the cathode. As today's electrocatalysts (used in low temperature fuel cells) contain platinum, there are economic reasons why MEA makers do not just put more platinum inside their products.
- Clearly, there have been many attempts to do away with platinum as the leading cathode catalyst for low-temperature fuel cells altogether. Unfortunately, to date, there appears to be no convincing alternative to platinum or related noble metals. This is not merely due to the lack of catalytic activity of other catalyst systems but often is a result of insufficient chemical stability of the materials considered.
- A logical and very successful approach is the more effective use of platinum in fuel cell electrodes. A technique borrowed from gas phase catalysis is the use of supported catalysts with small, highly dispersed platinum particles. Of course, electrocatalysts have to use electrically conducting substrate materials, usually specialized carbons.
- Of course, it is not sufficient to improve the surface area of the catalyst employed; there has to be good electrochemical contact between the membrane and the catalyst layer. *In-situ* measurement of the effective platinum surface area (EPSA) is a critical test for the quality of an electrode structure. The EPSA may be measured by electronic methods or, more commonly, by carbon monoxide adsorption and subsequent electrooxidation to carbon dioxide with charge measurement. For more information see Chapter 6 in [Hoogers, 2003]. Impedance measurements under practical loads can give valuable information on the catalyst utilization under operating conditions.

77.2 Fueling and Fuel Cell Systems

Not all fuel cells require these two reactants, but all fuel cells will function well with hydrogen and oxygen or air. In stationary power systems, the most convenient fuel is usually natural gas from the national grid. This has to be converted into the right fuel for the fuel cell — see below. In automotive systems, but also for portable power fuel, storage is an issue.

Hydrogen Storage

Pressure Cylinders

The need for lighter gas storage has led to the development of lightweight composite rather than steel cylinders. Carbon-wrapped aluminum cylinders can store hydrogen at pressures of up to 55 MPa (550 bar/8000 PSI). In most countries, gas cylinders are typically filled up to a maximum of 24.8 or 30 MPa (248 bar/3600 PSI and 300 bar/4350 PSI, respectively). At the higher pressure, a modern composite tank reaches a hydrogen mass fraction of approximately 3%, i.e., only 3% of the weight of the full cylinder consists of hydrogen. In a further development, so called "conformable" tanks have been produced in order to give a better space filling than packed cylinders.

General Motors' current compressed hydrogen gas storage systems typically hold 2.1 kg of hydrogen in a 140 liter/65 kg tank at 350 bar, which is good for 170 km (106 miles). The target here is a 230 liter/110 kg tank that would hold 7 kg of hydrogen at 700 bar, giving the same range as liquid hydrogen (see below), 700 km (438 miles) (H&FC, 2001).

Californian Quantum Technologies WorldWide has demonstrated a composite hydrogen pressure storage tank with a nominal operating pressure of almost 700 bar (10,000 PSI) giving an 80% capacity increase over tanks operating at 350 bar. The new tank underwent a hydrostatic burst test at which it failed under 1620 bar (23,500 psi). This test was done along the lines given in the draft regulations by the European Integrated Hydrogen Project (EIHP). The tank has an in-tank regulator that provides a gas supply under no more than 10 bar (150 PSI).

Liquid Hydrogen

It is unfortunate that the critical temperature of hydrogen, i.e., the temperature below which the gas can be liquefied, is at 33 K. Storage in cryogenic tanks at the boiling point of hydrogen, 20.39 K (–252.76°C/–422.97°F) at 1 atm (981 hPa), allows higher storage densities at the expense of the energy required for the liquefaction process. Lowering the temperature of hydrogen to its boiling point at 20.39 K (–252.76°C/–422.97°F) at atmospheric pressure requires approximately 39.1 kJ/g or 79 kJ mol^{-1}. To put this figure into perspective, this energy amounts to over a quarter of the higher heating value (286 kJ mol^{-1}) of hydrogen.

Another problem with cryogenic storage is hydrogen boil-off. Despite good thermal insulation, the heat influx into the cryogenic tank is continuously compensated by boiling off quantities of the liquid (heat of evaporation). In cryogenic storage systems on-board cars, the boil-off rate is estimated by most developers at approximately 1% per day, which results in further efficiency losses.

Cryogenic tanks consist of a multi-layered aluminum foil insulation. A typical tank stores 120 l of cryogenic hydrogen or 8.5 kg, which corresponds to an extremely low (liquid) density of 0.071 kg dm^{-3}. The empty tank has a volume of approximately 200 l and weighs 51.5 kg [Larminie, 2000]. This corresponds to a hydrogen mass fraction of 14.2%.

In General Motors' HydroGen1, 5 kg of hydrogen are stored in a 130 l/50 kg tank that gives the vehicle a 400 km (250 mile) drive range. The future target is a 150-l tank that is lighter yet, holding 7 kg for a range of 700 km (438 miles), as well as reduced boil-off time via an additional liquefied/dried air cooling shield developed by Linde.

The actual handling of cryogenic hydrogen poses a problem to the filling station, requiring special procedures such as fully automated, robotic filling stations.

Metal Hydrides

Most elements form ionic, metallic, covalent, or polymeric hydrides or mixtures thereof [Greenwood, 1984]. Ionic and metallic types are of particular interest, as they allow reversible storage of hydrogen [Sandrock, 1994].

The formation of the hydride is an exothermal process. Important parameters in this context are the enthalpy of formation of the hydride, which may range between several kJ and several hundred kJ per mol of hydrogen stored, and the resulting temperature and pressure to release the hydrogen from the hydride. In order to adjust these figures to technically acceptable levels, intermetallic compounds have been developed. Depending on the hydride used, the mass fraction of hydrogen ranges between 1.4 and 7.7% of total mass. Many hydrides actually store more hydrogen by volume than liquid hydrogen does. However, the storage process itself is exothermal and may amount to 25 to 45% of the hydrogen HHV (for full details, see Chapter 5 in [Hoogers, 2003]). Clearly, for large storage capacities such as the typical automotive tank, this is entirely unacceptable. Yet, for small, hydrogen-based portable systems, metal hydrides may form a convenient method of energy storage if cartridges are made universally available for purchase and recycling.

Sodium Borohydride

Sodium borohydride, $NaBH_4$, has recently received much attention through the work of Millenium Cell and DaimlerChrysler. Millennium Cell has patented a process that releases hydrogen from an aqueous solution of sodium borohydride, $NaBH_4$, in an exothermal reaction. (Sodium borohydride is usually made from borax using diborane, a highly reactive, highly toxic gas.) Hydrogen is only produced when the liquid fuel is in direct contact with a catalyst. The only other reaction product, sodium metaborate (analogous to borax), is water soluble and environmentally benign. The 35 wt% solution (35 wt% $NaBH_4$, 3 wt% NaOH, 62 wt% H_2O) will store 7.7 wt% of hydrogen or 77 g/921 standard liters of hydrogen in one liter of solution.

DaimlerChrysler has presented a fuel cell vehicle, the *Natrium*, that incorporates a sodium borohydride tank of about the size of a regular gas tank, which can power the concept vehicle about 300 miles.

There are numerous questions regarding production, energy efficiency, infrastructure, and stability of the solutions (Millenium quotes a half-life of 450 days, equivalent to 0.15% decomposition per day — this would be far less than the liquid hydrogen boil-off). The technique may well have its merits for special applications, particularly if an environmentally benign production process can be developed.

Fuel Reforming

Hydrogen is currently produced in large quantities for mainly two applications. Roughly 50% of the world hydrogen production is consumed for the hydro-formulation of oil in refineries producing mainly automotive fuels. Approximately 40% is produced for subsequent reaction with nitrogen to ammonia, the only industrial process known to bind atmospheric nitrogen. Ammonia is used in a number of applications, with fertilizer production playing a key role.

Storage of some hydrocarbon-derived liquid fuel followed by hydrogen generation on-board vehicles has been one option for chemical storage of hydrogen. So far, practical prototype cars have only been produced with methanol reformers (DaimlerChrysler). In contrast, the work on gasoline reformers by GM/Toyota/Exxon on the one hand and HydrogenSource (Shell Hydrogen and UTC Fuel Cells) on the other has not yet led to working fuel cell prototype cars.

Fortunately, there is no dispute about natural gas reforming for stationary power generation. An exhaustive discussion of the catalysis involved in fuel processing is presented in [Trimm, 2001].

The main techniques are briefly described in the following.

Steam Reforming (SR)

Steam reforming, SR, of methanol is given by the following chemical reaction equation:

$$CH_3OH + H_2O \rightarrow CO_2 + 3\ H_2 \qquad \Delta H = 49\ kJmol^{-1} \tag{77.14}$$

Methanol and water are evaporated and react in a catalytic reactor to carbon dioxide and hydrogen, the desired product. Methanol steam reforming is currently performed at temperatures between 200 and 300°C (390 and 570°F) over copper catalysts supported by zinc oxide [Emonts, 1998]. One mole of methanol reacts to three moles of dihydrogen. This means that an extra mole of hydrogen originates from the added water.

In practice, reaction (77.14) is only one of a whole series, and the raw reformer output consists of hydrogen, carbon dioxide, and carbon monoxide. Carbon monoxide is converted to carbon dioxide and more hydrogen in a high temperature shift, HTS, stage followed by a low temperature shift, LTS, stage. In both stages, the ***water–gas shift reaction***

$$CO + H_2O \rightarrow CO_2 + H_2 \qquad \Delta H = -41 kJmol^{-1} \tag{77.15}$$

takes place.

As water–gas shift is an exothermal reaction, if too much heat is generated, it will eventually drive the reaction towards the reactant side (Le Chatelier's principle). Therefore, multiple stages with interstage cooling are used in practice. The best catalyst for the HTS reaction is a mixture of iron and chromium oxides (Fe_3O_4 and Cr_2O_3) with good activity between 400 and 550°C (750 and 1020°F). LTS uses copper catalysts similar to and under similar operating conditions to those used in methanol steam reforming (77.14).

Steam reforming of methane from natural gas is the standard way of producing hydrogen on an industrial scale. It is therefore of general importance to a hydrogen economy. In addition, smaller-scale methane steam reformers have been developed to provide hydrogen for stationary power systems based on low-temperature fuel cells, PEMFC and PAFC.

The methane steam reforming reaction is described by

$$CH_4 + H_2O \rightarrow CO + 3\ H_2 \qquad \Delta H = 206\ kJmol^{-1} \tag{77.16}$$

It is again followed by the shift reactions (77.15).

Methane steam reforming is usually catalyzed by nickel [Ridler, 1996] at temperatures between 750 and 1000°C (1380 and 1830°F), with excess steam to prevent carbon deposition ("coking") on the nickel catalyst [Trimm, 2001].

Partial Oxidation (POX)

The second important reaction for generating hydrogen on an industrial scale is partial oxidation. It is generally employed with heavier hydrocarbons or when there are special preferences because certain reactants (for example, pure oxygen) are available within a plant.

It can be seen as oxidation with less than the stoichiometric amount of oxygen for full oxidation to the stable end products, carbon dioxide and water. For example, for methane

$$CH_4 + \int O_2 \rightarrow CO + 2H_2 \qquad \Delta H = -36\ kJmol^{-1} \tag{77.17a}$$

and/or

$$CH_4 + O_2 \rightarrow CO_2 + 2H_2 \qquad \Delta H = -319\ kJmol^{-1} \tag{77.17b}$$

While the methanol reformers used in fuel cell vehicles presented by DaimlerChrysler are based on steam reforming, Epyx (a subsidary of Arthur D. Little, now part of Nuvera) and Shell (now incorporated into HydrogenSource) are developing partial oxidation reactors for processing gasoline.

Autothermal Reforming (ATR)

Attempts have been made to combine the advantages of both concepts, steam reforming and partial oxidation. Ideally, the exothermal reaction (77.17) would be used for startup and for providing heat to the endothermal process (77.16) during steady-state operation. The reactions can either be run in separate reactors that are in good thermal contact or in a single catalytic reactor.

Comparison of Reforming Technologies

So, when does one use which technique for reforming? The first consideration is the ease by which the chosen fuel can be reformed using the respective method. Generally speaking, methanol is most readily reformed at low temperatures and can be treated well in any type of reformer. Methane and, similarly, LPG (liquefied petroleum gas) require much higher temperatures but again can be processed by any of the methods discussed. With higher hydrocarbons, the current standard fuels used in the automotive sector, one usually resorts to POX reactors.

Table 77.2 compares the theoretical (thermodynamic) efficiencies for proton exchange membrane fuel cells operating directly (DMFC) or indirectly on methanol (using one of the reformers discussed). The efficiencies currently achieved are labeled $\eta_{el}(tech)$*-typ*, while the technical limits are estimated by the author at $\eta_{el}(tech)$*-max*.

Table 77.3 and Table 77.4 show gas compositions from methanol reformers calculated as upper limit (by the author) and determined experimentally, respectively. Table 77.3 also includes information on the feed mix, i.e., X, the molar percentage, and wt%, the weight percentage of methanol in the methanol/water feed. A disadvantage of the POX reactors is the presence of nitrogen (from partial oxidation with ambient air) in the reformer output. This leads to lower hydrogen concentrations.

TABLE 77.2 Thermodynamics of PEM Fuel Cell Systems Operating on Methanol

	ΔH [kJ/mol]	ΔG [kJ/mol]	η_{el} (theor.)	η_{el} (tech) – typ	η_{el} (tech) – max
DMFC		–702,35	0,97	0,27	0,40
SR + H_2 – PEM	130,98	–711,39	0,83	0,32	0,54
POX + H_2 – PEM	–154,85	–474,26	0,65	0,25	0,42
ATR + H_2 – PEM	0	–602,73	0,83	0,32	0,54

Note: The DMFC is a fuel cell that can electro-oxidize methanol directly while the other systems are based on hydrogen fuel cells with reformer: Steam-reformer (SR), partial oxidation reformer (POX), and autothermal (ATR) reformer. The efficiencies have been determined by thermodynamics (theo.) or estimated by the author (typ./max.).

TABLE 77.3 Stoichiometric Input Feed and Output Gas Compositions for Different Concepts of Methanol Reforming

	X (MeOH)	wt% MeOH	H_2	CO_2	N_2
DMFC	0,5	64			
SR + H_2 – PEM	0,5	64	75%	25%	0%
POX + H_2 – PEM	1,0	100	41%	20%	39%
ATR + H_2 – PEM	0,65	77	58%	23%	20%

TABLE 77.4 Experimental Gas Compositions Obtained as Reformer Outputs

	H_2	CO_2	N_2	CO
SR [PASE 00]	67%	22%	–	
POX [PASE 00]	45%	20%	22%	
ATR [GOLU 98]	55%	22%	21%	2%

Steam reforming gives the highest hydrogen concentration. At the same time, a system relying entirely on steam reforming operates best under steady-state conditions because it does not lend itself to rapid dynamic response. This also applies to startup. Partial oxidation, in contrast, offers compactness, fast startup, and rapid dynamic response, while producing lower concentrations of hydrogen. In addition to differences in product stoichiometries between SR and POX reformers, the output of a POX reformer is necessarily further diluted by nitrogen. Nitrogen is introduced into the system from air, which is usually the only economical source of oxygen, and carried through as an inert. Autothermal reforming offers a compromise.

Yet, the fuel processor cannot be seen on its own. Steam reforming is highly endothermic. Heat is usually supplied to the reactor, for example, by burning extra fuel. In a fuel cell system, (catalytic) oxidation of excess hydrogen exiting from the anode provides a convenient way of generating the required thermal energy. In stationary power generation, it is worth considering that the PAFC fuel cell stack operates at a high enough temperature to allow generating steam and feeding it to the fuel processor. Steam reforming may be appropriate here whereas autothermal reforming could be considered in a PEMFC system, which has only low-grade heat available.

Fuel efficiency also deserves careful attention. Though always important, the cost of fuel is the most important factor in stationary power generation (on a par with plant availability). Hence, the method offering the highest overall hydrogen output from the chosen fuel, usually natural gas, is selected. Steam reforming delivers the highest hydrogen concentrations. Therefore, the fuel cell stack efficiency at the higher hydrogen content may offset the higher fuel demand for steam generation. This is probably the reason why steam reforming is currently also the preferred method for reforming natural gas in stationary power plants based on PEM fuel cells.

In automotive applications, the dynamic behavior of the reformer system may control the whole drive train, depending on whether backup batteries, super-capacitors or other techniques are used for providing peak power. A POX reformer offers the required dynamic behavior and fast startup and is likely to be the best choice with higher hydrocarbons. For other fuels, in particular, methanol, an ATR should work best. Yet, the reformers used by DaimlerChrysler in their NECAR 3 and 5 vehicles are steam reformers. This perhaps surprising choice can be reconciled when one considers that during startup, additional air is supplied to the reformer system to do a certain degree of partial oxidation. During steady-state operation, the reformer operates solely as SR with heat supplied from excess hydrogen. Clearly, it is not always possible to draw clear borderlines between different types of reformers.

Striking advantages of liquid fuels are their high energy storage densities and their ease of transport and handling. Liquid fuel tanks are readily available, and their weight and volume are essentially dominated by the fuel itself. LPG is widely applied in transportation in some countries, and storage is comparably straightforward since LPG is readily liquefied under moderate pressures (several bar).

CO Removal/Pd-Membrane Technology

As was noted in the introduction to this chapter, different fuel cells put different demands on gas purity. CO removal is of particular concern to the operation of the PEMFC, less so with fuel cells operating at higher temperatures.

After reforming and water gas shift, the CO concentration in the reformer gas is usually reduced to 1 to 2%. A PEMFC will require further cleanup down to levels in the lower ppm range. Another reason for having further CO cleanup stages is the risk of CO spikes, which may result from rapid load changes of the reformer system as can be expected from automotive applications.

There are a number of ways to clean the raw reformer gas of CO. Alternatively, ultra-pure hydrogen void of any contaminant can be produced using Pd membrane technology.

We will discuss these methods in turn.

Preferential Oxidation

Oxidative removal of CO is one option often applied in fuel cell systems working with hydrocarbon reformers. Unfortunately, this increases the system complexity because well-measured concentrations of air have to be added to the fuel stream.

The reaction

$$CO + \int O_2 \rightarrow CO_2 \qquad \Delta H = -260 \text{ kJmol}^{-1} \tag{77.18}$$

works surprisingly well despite the presence of CO_2 and H_2 in the fuel gas. This is due to the choice of catalyst, which is typically a noble metal such as platinum, ruthenium, or rhodium supported on alumina. Gold catalysts supported on reducible metal oxides have also shown some benefit, particularly at temperatures below 100°C [Plzak, 1999]. CO bonds very strongly to noble metal surfaces at low to moderate temperatures. So, the addition reaction (77.18) takes place on the catalytic surface, in preference to the undesirable direct catalytic oxidation of hydrogen. Therefore, this technique is referred to as preferential oxidation, or PROX. The selectivity of the process has been defined as the ratio of oxygen consumed for oxidizing CO divided by the total consumption of oxygen.

The term selective oxidation, or SELOX, is also used. But this should better be reserved for the case where CO removal takes place *within* the fuel cell.

An elegant way of making the fuel cell more carbon monoxide-tolerant is the development of CO-tolerant anode catalysts and electrodes [Cooper, 1997]. A standard technique is the use of alloys of platinum and ruthenium. Another, rather crude, way of overcoming anode poisoning by CO is the direct oxidation of CO by air in the anode itself [Gottesfeld, 1988]. One may see this as an internal form of the preferential oxidation discussed above. In order to discriminate the terminology, this method is often referred to as selective oxidation or SELOX. The air for oxidizing CO is "bled" into the fuel gas stream at concentrations of around 1%. Therefore, this technique has been termed *air bleed.* It is a widely accepted way of operating fuel cells on reformer gases.

Bauman et al. have also shown that anode performance after degradation due to CO "spikes," which are likely to appear in a reformer-based fuel cell system upon rapid load changes, recovers much more rapidly when an air bleed is applied [Bauman, 1999].

Pd-Membranes

In some industries, such as the semiconductor industry, there is a demand for ultra-pure hydrogen. Since purchase of higher-grade gases multiplies the cost, hydrogen is either generated on site or low-grade hydrogen is further purified.

A well-established method for hydrogen purification (and only applicable to hydrogen) is permeation through palladium membranes [McCabe, 1997], and plug-and-play hydrogen purifiers are commercially available. Palladium only allows hydrogen to permeate and retains any other gas components, such as nitrogen, carbon dioxide, carbon monoxide, and any trace impurities, on the upstream side. As carbon monoxide adsorbs strongly onto the noble metal, concentrations in the lower percent range may hamper hydrogen permeation through the membrane, unless membrane operating temperatures high enough to oxidize carbon monoxide (in the presence of some added air) to carbon dioxide are employed. In a practical test, operating temperatures in excess of 350°C and operating pressures above 20 hPa (20 bar or 290 PSI) had to be used [Emonts, 1998].

For economic reasons, thin film membranes consisting of palladium/silver layers deposited on a ceramic support are being developed. Thin film membranes allow reducing the amount of palladium employed and improve the permeation rate. Silver serves to stabilize the desired metallic phase of palladium under the operation conditions. Yet, thermal cycling and hydrogen embrittlement pose potential risks to the integrity of membranes no thicker than a few microns [Emonts, 1998].

The main problems with palladium membranes in automotive systems appear to be the required high pressure differential, which takes its toll from overall systems efficiency, the cost of the noble metal, and/or membrane lifetime. Other applications, in particular compact power generators, may benefit from reduced system complexity.

A number of companies, including Misubishi Heavy Industries [Kuroda, 1996] and IdaTech [Edlund, 2000], have developed reformers based on Pd membranes. Here, the reforming process takes place inside a membrane tube or in close contact with the Pd membrane. IdaTech achieved CO and CO_2

levels of less than 1 ppm with a SR operating inside the actual cleanup membrane unit using a variety of fuels.

Methanation

Methanation,

$$CO + 3H_2 \rightarrow CH_4 + H_2O \qquad \Delta H = -206 \text{ kJmol}^{-1} \tag{77.19}$$

is another option for removing CO in the presence of large quantities of H_2.

Unfortunately, due to current lack of selective catalysts, the methanation of CO_2 is usually also catalyzed. Therefore, for cleaning up small concentrations of CO in the presence of large concentrations of CO_2, methanation is not possible. IdaTech uses methanation in conjunction with Pd-membranes as second cleanup stage. This is possible in this particular instance because the membrane removes both CO and CO_2 down to ppm levels.

Other System Components

Fuel Cell Stack

Figure 77.1 demonstrates how MEAs are supplied with reactant gases and put together to form a fuel cell stack. The gas supply is a compromise between the flat design necessary for reducing ohmic losses and sufficient access of reactants. Therefore, so-called flow field plates are employed to feed hydrogen to the anodes and air/oxygen to the cathodes present in a fuel cell stack. Other stack components include cooling elements, current collector plates for attaching power cables, end plates, and, possibly, humidifiers. End plates give the fuel cell stack mechanical stability and enable sealing of the different components by compression. A number of designs have been presented. Ballard Power Systems has used both threaded rods running along the whole length of the stack and metal bands tied around the central section of the stack for compression.

Bipolar Plates

Flow field plates in early fuel cell designs — and still in use in the laboratory — were usually made of graphite into which flow channels were conveniently machined. These plates have high electronic and good thermal conductivity and are stable in the chemical environment inside a fuel cell. Raw bulk graphite is made in a high-temperature sintering process that takes several weeks and leads to shape distortions and the introduction of some porosity in the plates. Hence, making flow field plates is a lengthy and labor-intensive process, involving sawing blocks of raw material into slabs of the required thickness, vacuum-impregnating the blocks or the cut slabs with some resin filler for gas-tightness, and grinding and polishing to the desired surface finish. Only then can the gas flow fields be machined into the blank plates by a standard milling and engraving process. The material is easily machined but abrasive. Flow field plates made in this way are usually several millimeters (1 mm = 0.04 inch) thick, mainly to give them mechanical strength and allow the engraving of flow channels. This approach allows the greatest possible flexibility with respect to designing and optimizing the flow field.

When building stacks, flow fields can be machined on either side of the flow field plate such that it forms the cathode plate on one and the anode plate on the other side. Therefore, the term *bipolar plate* in often used in this context. The reactant gases are then passed through sections of the plates and essentially the whole fuel cell stack — compare Figure 77.1.

Graphite-Based Materials

The choice of materials for producing bipolar plates in commercial fuel cell stacks is not only dictated by performance considerations but also cost. Present blank graphite plates cost between US$20 and US$50 apiece in small quantities, i.e., up to US$1000/m^2, or perhaps over US$100 per kW assuming one plate per MEA plus cooling plates at an MEA power density of 1 Wcm^{-2}. Again, automotive cost targets are well beyond reach, even ignoring additional machining and tooling time.

This dilemma has sparked off several alternative approaches. Ballard Power Systems has developed plates based on (laminated) graphite foil, which can be cut, molded, or carved in relief in order to generate a flow field pattern. This may open up a route to low-cost volume production of bipolar plates. Potential concerns are perhaps the uncertain cost and the availability of the graphite sheet material in large volumes.

Another cost-effective volume production technique is injection or compression molding. Difficulties with molded plates lie in finding the right composition of the material, which is usually a composite of graphite powder in a polymer matrix. While good electronic conductivity requires a high graphite fill, this hampers the flow and hence the moldability of the composite. Thermal stability and resistance towards chemical attack of the polymers limit the choice of materials.

Plug Power has patented flow field plates that consist of conducting parts framed by nonconducting material, which may form part of the flow field systems.

Metallic Bipolar Plates

Metals are very good electronic and thermal conductors and exhibit excellent mechanical properties. Undesired properties include their limited corrosion resistance and the difficulty and cost of machining.

The metals contained in the plates bear the risk of leaching in the harsh electrochemical environment inside a fuel cell stack; leached metal may form damaging deposits on the electrocatalyst layers or could be ion-exchanged into the membrane or the ionomer, thereby decreasing the conductivity. Corrosion is believed to be more serious at the anode, probably due to weakening of the protective oxide layer in the hydrogen atmosphere.

Several grades of stainless steel (310, 316, 904L) have been reported to survive the highly corroding environment inside a fuel cell stack for 3000 h without significant degradation [Davies, 2000] by forming a protective passivation layer.

Clearly, the formation of oxide layers reduces the conductivity of the materials employed. Therefore, coatings have been applied in some cases. In the simplest case, this may be a thin layer of gold or titanium. Titanium nitride layers are another possibility and have been applied to lightweight plates made of aluminum or titanium cores with corrosion-resistant spacer layers. Whether these approaches are commercially viable depends on the balance between materials and processing costs.

Meanwhile, mechanical machining of flow fields into solid stainless steel plates is difficult. A number of companies such as Microponents (Birmingham, U.K.) and PEM (Germany) are attempting to achieve volume production of flow field plates by employing chemical etching techniques. Yet, etching is a slow process and generates slurries containing heavy metals, and it is hence of limited use for mass production.

Another solution to the problem of creating a (serpentine) flow field is using well-known metal stamping techniques. To date, there appear to be no published data on flow fields successfully produced in this fashion.

Humidifiers and Cooling Plates

A fuel cell stack may contain other components. The most prominent ones are cooling plates or other devices and techniques for removing reaction heat and, possibly, humidifiers.

Cooling is vital to maintain the required point of operation for a given fuel cell stack. This may either be an isothermal condition, or perhaps a temperature gradient may be deliberately superimposed in order to help water removal. Relatively simple calculations show that for very high power densities such as those attained in automotive stacks, liquid cooling is mandatory. This is traditionally done by introducing dedicated cooling plates into the stack, through which water is circulated.

In less demanding applications, such as portable systems, where the system has to be reduced to a bare minimum of components, air cooling is sometimes applied. In the simplest case, the cathode flow fields are open to ambient, and reactant air is supplied by a fan, at the same time providing cooling. No long-term performance data have been reported for this type of air-cooled stack.

A second function sometimes integrated into the stack is reactant humidification. It is currently unknown whether innovative membrane concepts will ever allow unhumidified operation in high-performance fuel cell stacks. In most automotive stacks to date, both fuel gas and air are most likely humidified because maximum power is required, which is only achieved with the lowest possible membrane resistance.

The literature knows several types of humidifiers, bubblers, membrane, or fiber-bundle humidifiers and water evaporators. The simplest humidifier is the well-known "bubbler," essentially the wash bottle design with gas directly passing through the liquid. Clearly, this approach allows only poor control of humidification, is less suited within a complex fuel cell system, and may cause a potential safety hazard due to the direct contact of the fluids. Another approach is using a membrane humidifier. A semipermeable membrane separates a compartment filled with water from a compartment with the reactant gas. Ideally, the gas is conducted along the membrane and continually increases its humidity up to or close to saturation as it passes from the gas inlet to the gas outlet. Some concepts combine humidification and cooling [Vitale, 2000].

This concludes the list of the most important functional components of a fuel cell stack. The fuel cell system contains a large number of other components for fuel generation, pumping, compression, etc., which are usually just summarized under the term balance-of-plant, BOP.

Direct Methanol Fuel Cells

No doubt one of the most elegant solutions would be to make fuel cells operate on a liquid fuel. This is particularly so for the transportation and portable sectors. The direct methanol fuel cell (DMFC), a liquid- or vapor-fed PEM fuel cell operating on a methanol/water mix and air, therefore deserves careful consideration. The main technological challenges are the formulation of better anode catalysts, which lower the anode overpotentials (currently several hundred millivolts at practical current densities), and the improvement of membranes and cathode catalysts in order to overcome cathode poisoning and fuel losses by migration of methanol from anode to cathode. Current prototype DMFCs generate up to 0.2 Wcm^{-2} (based on the MEA area) of electric power, but not yet under practical operating conditions or with acceptable platinum loadings.

Therefore, there is currently little hope that DMFCs will ever be able to power a commercially viable car. The strength of the DMFC lies in the inherent simplicity of the entire system and the fact that the liquid fuel holds considerably more energy than do conventional electrochemical storage devices such as primary and secondary batteries. Therefore, DMFCs are likely to find a range of applications in supplying electric power to portable or grid-independent systems with small power consumption but long, service-free operation.

Applications

It now looks as though fuel cells are eventually coming into widespread commercial use. Before starting a fuel cell development project, one should consider the benefits to be expected in the respective application.

Table 77.5 lists a range of applications for each type of fuel cell.

Transportation

In the transportation sector, fuel cells are probably the most serious contenders as competitors to internal combustion engines. They are highly efficient as they are electrochemical rather than thermal engines. Hence they can help to reduce the consumption of primary energy and the emission of CO_2.

What makes them most attractive for transport applications, though, is the fact that they emit zero or ultra-low emissions. And this is what mainly inspired automotive companies and other fuel cell developers in the 1980s and 1990s to start developing fuel cell-powered cars and buses. Leading developers realized that although the introduction of the three-way catalytic converter had been a milestone, keeping up the pace in cleaning up car emissions further was going to be very tough indeed. Legislation such as California's Zero Emission Mandate has come in and initially, only battery-powered vehicles were seen as a solution to the problem of building zero-emission vehicles. Meanwhile, it has turned out that the storage capacity of batteries is unacceptable for practical use because customers ask for the same drive range they are used to from internal combustion engines (ICEs). In addition, the battery solution is

TABLE 77.5 Currently Developed Types of Fuel Cells and Their Characteristics and Applications

Fuel Cell Type	Electrolyte	Charge Carrier	Operating Temperature	Fuel	Electric Efficiency (System)	Power Range/ Application
Alkaline FC (AFC)	KOH	OH	60–120°C	Pure H_2	35–55%	<5 kW, niche markets (military, space)
Proton exchange membrane FC (PEMFC)	Solid polymer (such as Nafion)	H^+	50–100°C	Pure H_2 (tolerates CO_2)	35–45%	Automotive, CHP (5–250 kW), portable
Phosphoric acid FC (PAFC)	Phosphoric acid	H^+	~220°C	Pure H_2 (tolerates CO_2, approx. 1% CO)	40%	CHP (200 kW)
Molten carbonate FC (MCFC)	Lithium and potassium carbonate	CO_3^{2-}	~650°C	H_2, CO, CH_4, other hydrocarbons tolerates CO_2	>50%	200 kW-MW range, CHP and standalone
Solid oxide FC (SOFC)	Solid oxide electrolyte (yttria, zirconia)	O^{2-}	~1000°C	H_2, CO, CH_4, other hydrocarbons tolerates CO_2	>50%	2 kW-MW range, CHP and standalone

unsatisfactory for yet another reason: with battery-powered cars the point where air pollution takes place is only shifted back to the electric power plant providing the electricity for charging. This is the point where fuel cells were first seen as the only viable technical solution to the problem of car-related pollution.

It has now become clear that buses will make the fastest entry into the market because the hydrogen storage problem has been solved with roof-top pressure tanks. Also, fueling is not an issue due to the fleet nature of buses. For individual passenger cars, the future remains unclear. Most developers have now moved away from on-board reforming and promote direct hydrogen storage. Yet, this will require a new fuel infrastructure, which is not easy to establish.

Another automotive application is auxiliary power units (APUs). APUs are designed not to drive the main power train but to supply electric power to all devices on-board conventional internal combustion engines, even when the main engine is not operating. A typical example for a consumer requiring large amounts of power (several kW) is an air-conditioning system. Currently, due to the difficulties with reforming gasoline, high-temperature fuel cells are considered a good option.

Stationary Power

Cost targets were first seen as an opportunity: The reasoning went that, when fuel cells were going to meet automotive cost targets, other applications, including stationary power, would benefit from this development, and a cheap multipurpose power source would become available.

Meanwhile, stationary power generation, in addition to buses, is viewed as the leading market for fuel cell technology. The reduction of CO_2 emissions is an important argument for the use of fuel cells in small stationary power systems, particularly in combined heat and power generation. In fact, fuel cells are currently the only practical engines for micro-CHP systems in the domestic environment at less than, say, 5 kW of electric power output. The higher capital investment for a CHP system would be offset against savings in domestic energy supplies and — in more remote locations — against power distribution cost and complexity. It is important to note that due to the use of CHP, the electrical efficiency is less critical than in other applications. Therefore, in principle all fuel cell types are applicable. In particular, PEMFC systems and SOFC systems are currently being developed in this market segment. PEMFC systems offer the advantage that they can be easily turned off when not required. SOFC systems, with their higher operating temperature, are usually operated continuously. They offer more high-grade heat and provide simpler technology for turning the main feed — natural gas, sometimes propane — into electric power than the more poison-sensitive PEMFC technology.

In the 50 to 500 kW range there will be competition with spark or compression ignition engines modified to run on natural gas. So far, several hundred 200-kW phosphoric acid fuel cell plants manufactured by ONSI (IFC) have been installed worldwide. Yet, in this larger power range, high-temperature fuel cells offer distinct advantages. They allow simpler gas pretreatment than low-temperature fuel cells do and achieve higher electrical efficiencies. This is important when one considers that a standard motor CHP unit at several hundred kilowatts of electric power output will generally achieve close to 40% electrical efficiency — at a fraction of the cost of a fuel cell. High-temperature fuel cells also offer process heat in the form of steam rather than just hot water. Both MCFC and SOFC are currently being developed for industrial CHP systems.

Portable Power

The portable market is less well defined, but a potential for quiet fuel cell power generation is seen in the sub-1kW portable range. The term "portable fuel cells" often includes grid-independent applications such as camping, yachting, and traffic monitoring. Fuels considered vary from one application to another. And fuels are not the only aspect that varies. Different fuel cells may be needed for each subsector in the portable market.

It is currently not clear whether DMFC systems will ever be able to compete with lithium ion batteries in mass markets such as cellular phones or portable computers. But in applications where size does not matter and long, unattended service at low power levels is required, DMFCs are the ideal power source.

There may also be room for hydrogen-fueled PEMFC systems where higher peak power demand is needed for short durations. Storage of hydrogen and the implementation of a supply infrastructure remain problems to be solved.

At more than several hundred watts of electric power, reformer-based systems remain the only option to date. This is the range of the APU (see Transportation, above), and it is likely that similar systems will provide power to yachts and motor homes, mountaineering huts, etc. They may be fueled by propane or perhaps some liquid fuel — compare the following section.

Choice of Fuel

Before even deciding which fuel cell type to use, one needs to identify the best fuel base and the allowable size and complexity of the resulting fuel cell system.

Of the three key applications for fuel cells — automotive propulsion, stationary power generation and portable power — the automotive case is most readily dealt with. For generating the propulsion power of a car, bus, or truck, only PEMFCs are currently being considered for their superior volumetric power density with operation on hydrogen. Leading developers originally favored hydrogen generation from methanol on board and, more recently, either hydrogen generation from reformulated gasoline on board or hydrogen storage on board. As hydrogen generation from gasoline is technically difficult, questionable for energy efficiency reasons, and methanol as a fuel base has been demonstrated but is unwanted by most developers, hydrogen storage on board appears to emerge as the overall compromise. Moreover, automotive propulsion is still at least 10 years away from widespread market introduction; it is a highly challenging, specialist job that is well underway at the leading developers' laboratories — and it should be left there.

The other extreme is portable power, where again the PEMFC is the only fuel cell seriously considered. In this case, there are two options for fuel, hydrogen or direct methanol, i.e., without prior conversion to hydrogen. The latter type of PEMFC is usually referred to as DMFC. For systems larger than, say, 1 kW that operate continuously, direct methanol is not an option because of the expense of the fuel cell. Direct hydrogen is not an option because of the lack of storage capacity. Therefore, one may consider converting a chemical such as LPG, propane, butane, methanol, gasoline, or diesel into hydrogen. Such larger systems will be considered among the stationary power systems.

Stationary power generation, at the first glance, is the least clear application in terms of the technology base: four types of fuel cell are currently being developed for stationary power generation.

In contrast, the fuel base is rather clear in this particular case: for residential and small combined heat and power generation, natural gas is the usual source of energy, and fuel cell systems have to adapt to this fuel. In areas without a gas — and electricity — grid, LPG may be an option and is being actively pursued by a number of developers.

Table 77.6 summarizes the most likely applications and fueling options for all types of fuel cells.

Defining Terms

AFC — Alkaline fuel cell.

APU — Auxiliary power unit — supplying electric power to cars independent of the main engine.

Bipolar plate — Metal, graphite or composite plate separating two adjacent cells in planar PEMFC, AFC, PAFC, and MCFC stacks (possibly also planar SOFC). One side consequently acts as positive electrode (cathode), while the opposite side acts as negative electrode (anode) to the following cell. The two sides contain channels to distribute the reactants, usually hydrogen-rich gas (anode) and air (cathode).

CHP — Combined heat and power generation.

DMFC — Direct methanol fuel cell. Liquid or vaporized methanol/water fueled version of PEMFC.

Humidifier — Unit required for providing high water vapor concentration in inlet gas streams of PEMFC.

MCFC — Molten carbonate fuel cell.

MEA — Membrane electrode assembly — central electrochemical part of a PEMFC. Consists of anode, cathode and membrane electrolyte.

PAFC — Phosphoric acid fuel cell.

PEMFC — Proton exchange membrane fuel cell — also called SPFC.

Reformer — Gas or liquid processing unit providing hydrogen-rich gas to run fuel cells off readily available hydrocarbon fuels such as methane, methanol, propane, gasoline, or diesel.

SPFC — Solid polymer fuel cell — see PEMFC.

SOFC — Solid oxide fuel cell.

Stack — Serial combination (pile) of alternating planar cells and bipolar plates in order to increase the overall voltage.

TABLE 77.6 Fuel Cell Applications and Likely Fueling Options

	Automotive		Portable		Stationary		
Fuel Cell	Main Drive Train	Auxiliary Power	Battery Replacement Long Operation, Low Power	Battery Replacement Short Operation, High Power	Residential Remote Power or Residential Heat and Power (few kW range)	Small-Scale Heat and Power (100 kW to several MW)	Large Central Power Station (many MW)
PEMFC	50 to 100 kW hydrogen powered (liquid or compressed) other fuels less likely	Less likely	0.1 to 500 W direct methanol powered = DMFC	100 W to few kW hydrogen (compressed or metal hydride)	1 to 10 kW natural gas or LPG powered (possibly methanol or fuel oil)	100 kW to 10 MW natural gas powered less likely	
MCFC		Less likely			Less likely	100 kW to 10 MW natural gas powered	Perhaps, in combination with gas turbine coal gas fired
SOFC		Several kW gasoline or diesel powered			1 to 10 kW natural gas or LPG powered	100 kW to 10 MW natural gas powered	Perhaps, in combination with gas turbine coal gas fired
PAFC						100 kW to 10 MW natural gas powered less likely	

References

Atkins, P. W. 2000. *Physical Chemistry,* Oxford University Press, Oxford.

Bauman, J. W., Zawodzinski, T.A., Jr., and Gottesfeld, S. 1999. In S. Gottesfeld and T. F. Fuller (Eds.), Proceedings of the Second International Symposium on Proton Conducting Membrane Fuel Cells II, Electrochem. Soc, Pennington, NJ.

Bossel, U. 2000. *The Birth of the Fuel Cell 1835–1845,* European Fuel Cell Forum, Oberrohrdorf, Switzerland.

Cooper, S. J., Gunner, A. G., Hoogers, G., and Thompsett, D. 1997. Reformate tolerance in proton exchange membrane fuel cells: electrocatalyst solutions. In *New Materials for Fuel Cells and Modern Battery Systems II,* O. Savadogo and P. R. Roberge (Eds .).

Davies, D. P., Adcock, P. L., Turpin, M., and Rowen, S. J. 2000. *J . Power Sources* 86:237.

Edlund, D. 2000. A versatile, low-cost, and compact fuel processor for low-temperature fuel cells, Fuel Cells Bulletin No. 14.

Emonts, B., Bøgild Hansen, J., Lœgsgaard Jørgensen, S., Höhlein, B., and Peters, R. 1998. Compact methanol reformer test for fuel-cell powered light-duty vehicles, *J. Power Sources* 71:288.

Gottesfeld, S. and Pafford, J. 1988. A new approach to the problem of carbon monoxide poisoning in fuel cells operating at low temperatures, *J. Electrochem Soc.* 135:2651.

Greenwood, N. N. and Earnshaw, A. 1984. *Chemistry of the Elements*, Pergamon Press, Oxford.

Hamann, C., Hamnett, A., and Vielstich, W. 1998, *Electrochemistry.* Wiley-VCH ,

H&FC. 2001. *Hydrogen & Fuel Cell Letter,* June.

Kuroda, K., Kobayashi, K., O'Uchida, N., Ohta, Y., and Shirasaki, Y. 1996. Study on performance of hydrogen production from city gas equipped with palladium membranes, *Mitsubishi Juko Giho* 33(5).

McCabe, R. W. and Mitchell, P. J. 1987. *J. Catal.* 103:419.

Plzak, V., Rohland, B., and Jörissen, L. 1999. Preparation and screening of Au/MeO catalysts for the preferential oxidation of CO in H2-containing gases, Poster, 50th ISE Meeting, Pavia.

Ridler, D. E. and Twigg, M. V. 1996. Steam reforming, In M. V. Twigg (Ed.) *Catalyst Handbook*, Manson Publishing, London. p. 225.

Sandrock, G. 1994. Intermetallic hydrides: history and applications, in P. D. Bennett and T. Sakai (Eds.), Proceedings of the Symposium on Hydrogen and Metal Hydride Batteries.

Trimm, D. L. and Önsan, Z. 2001. On-board fuel conversion for hydrogen-fuel-cell-driven vehicles, *Catalysis Reviews* 43:31–84.

Vitale, N. G. and Jones, D. O. (Plug Power Inc). 2000. U.S. Patent US6,066,408.

Further Reading

Hoogers, G. (Ed.) 2003. *Fuel Cell Technology Handbook,* CRC Press, Boca Raton, FL.

Larminie, J. and Dicks, A. 2000. *Fuel Cell Systems Explained,* Wiley-VCH .

Vielstich, W., Lamm, A., and Gasteiger, H. (Eds.) 2003. *Handbook of Fuel Cells — Fundamentals, Technology, Applications* (4 Volumes). Wiley-VCH .

X

Kinetics and Reaction Engineering

78

Reaction Kinetics

K. H. Lin

Oak Ridge National Laboratory

This chapter presents a brief overview of reaction kinetics primarily for engineers who are not directly involved in the investigation of reaction kinetics or in the design of chemical reactors. For a comprehensive treatise on reaction kinetics, the reader should consult the references at the end of the chapter.

In contrast to the static and equilibrium concept of thermodynamics, reaction kinetics is concerned with dynamics of chemical changes. Thus, reaction kinetics is the science that investigates the rate of such chemical changes as influenced by various process parameters and attempts to understand the mechanism of the chemical changes. For any reaction system to change from an initial state to a final state, it must overcome an energy barrier. The presence of such an energy barrier is commonly manifested in the observed relationship between the reaction rate and temperature, which will be discussed in some detail in the sections that follow.

78.1 Fundamentals

Basic Terms and Equations

One of the key terms in reaction kinetics is the **rate of reaction**, r_A, the general definition of which is given by

$$r_A = \frac{1}{y}\frac{dN_A}{dt} \tag{78.1}$$

which expresses the rate r_A as the amount of a chemical component of interest being converted or produced per unit time per unit quantity of a reference variable y [e.g., volume of reacting mixture (V) or of reactor (V_R), mass (W), surface area (S), etc.]. It delineates the time dependence of chemical changes involving component A as a derivative. In homogeneous fluid reactions, y is normally represented by V or V_R, whereas the mass (W) or the surface area (S) of the solid reactant may be taken as y in heterogeneous solid–fluid reactions. Here, N_A refers to the amount of component A, and t is time. By convention, r_A is negative when A is a reactant and positive when A is a product. Molal units are commonly used as the

0-8493-1586-7/05/$0.00+$1.50

amount of N_A, but other units such as mass, radioactivity, pressure, and optical property are also used. When V remains constant (as in a liquid-phase batch reactor), Equation (78.1) is simplified to

$$r_A = dC_A/dt \tag{78.2}$$

where C_A represents the concentration of component A.

Rate Constant and Elementary Reactions

In general, the rate of reaction in terms of component i is a function of the concentration of all components participating in the reaction, C; the temperature, T; the pressure, P; and other parameters, m:

$$r_i = \text{function}\left(C, T, P, m\right) \tag{78.3}$$

Thus, the rate expression for a simple irreversible reaction in terms of the reacting components may assume the following form:

$$-r_A = k(C_A)^{n_1}(C_B)^{n_2}\cdots(C_i)^{n_i} \tag{78.4}$$

The proportionality constant, k, is the **rate constant** that is markedly influenced by the temperature and may be subject to the influence of pressure, pH, kinetic isotopes, and so on, and the presence of catalysts. The exponents $n_1, n_2, \ldots, n_i$ are *orders of reaction* with respect to individual reacting components $A, B, \ldots, i$. The *overall order of reaction* refers to the sum of $n_1, n_2, \ldots, n_i$, which does not have to be an integer and may be determined empirically.

The units and value of rate constant k vary with the units of C, the specific component that k refers to, and the reaction order. The effect of temperature on k was first described by Arrhenius through the following equation:

$$k = Ae^{-\frac{E}{RT}} \tag{78.5}$$

where A, termed the *frequency factor*, has the same units as k, E is **activation energy**, and R is the gas law constant. E was considered by Arrhenius as the amount of energy that a reacting system must have in excess of the average initial energy level of reactants to enable the reaction to proceed.

A *simple* or *elementary* reaction is one in which the order of reaction is identical to the *molecularity* (the number of molecules actually participating in the reaction). Under these circumstances, the chemical stoichiometric equation represents the true reaction mechanism, and the rate equation may therefore be derived directly from the stoichiometric equation. Thus, for an elementary reaction $n_1\,A + n_2\,B = n_3\,D$, the rate equation in terms of disappearance of A would be $-r_A = k\,(C_A)^{n_1}(C_B)^{n_2}$. The values of n_1 and n_2 in this equation are positive integers.

Complex (or Multiple) Reactions

A reaction that proceeds by a mechanism involving more than a single reaction path or step is termed a *complex reaction*. Unlike elementary reactions, the mechanisms of complex reactions differ considerably from their stoichiometric equations. Most industrially important reactions are complex reactions, the mechanisms of which can often be determined by assuming that the overall reaction consists of several elementary reaction steps. The resulting overall rate expression is then compared with the experimental data, and the procedure is repeated until a desired degree of agreement is obtained. Each of the elementary reaction steps may proceed *reversibly, concurrently*, or *consecutively*.

A *reversible reaction* is one in which conversion of reactants to products is incomplete at equilibrium because of an increasing influence of the reverse reaction as the forward reaction approaches equilibrium. For a reversible reaction of the type

$$A + B \underset{k_r}{\overset{kf}{\Leftrightarrow}} D + E$$

the net forward rate of reaction in terms of disappearance of A is

$$-r_A = k_f C_A C_B - k_r C_D C_E \tag{78.6}$$

which assumes both forward and reverse reactions to be elementary.

A simple example of *consecutive reaction* is illustrated by

$$A \overset{k_1}{\rightarrow} B \overset{k_2}{\rightarrow} D \tag{78.7}$$

Again, assuming an elementary reaction for each reaction step, the following rate equations result:

$$-r_A = k_1 C_A \tag{78.8}$$

$$r_D = k_2 C_B \tag{78.9}$$

$$r_B = k_1 C_A - k_2 C_B \tag{78.10}$$

Parallel or *simultaneous reactions* are those involving one or more reactants undergoing reactions of more than one scheme, as in

$$A \overset{k_1}{\rightarrow} B \qquad A \overset{k_2}{\rightarrow} D$$

The rate equation may assume the following forms:

$$-r_A = k_1 (C_A)^a + k_2 (C_A)^b \tag{78.11}$$

$$r_B = k_1 (C_A)^a \tag{78.12}$$

$$r_D = k_2 (C_A)^b \tag{78.13}$$

Under a constant-volume condition, $r_B = dC_B/dt$ and $r_D = dC_D/dt$. Therefore, the relative rate of formation of B and D is derived from Equation (78.12) and Equation (78.13) as

$$dC_B / dC_D = (k_1 / k_2)(C_A)^{a-b} \tag{78.14}$$

The ratio dC_B/dC_D is termed the *point selectivity,* which is the ratio of the *rate* of formation of product B to the *rate* for product D. The *overall* (or *integrated*) *selectivity* is obtained by integration of this ratio, and it represents the ratio of the overall *amount* of product B to that of product D. Equation (78.14) implies that the relative rate of formation of B is proportional to C_A when $a > b$, whereas it is inversely proportional to C_A when $a < b$.

In a chemical process involving complex reactions that consist of several reaction steps, one or more steps may represent major factors in governing the overall rate of reaction. Such reaction steps are termed the **rate-controlling steps.** The rate-controlling reaction steps are observed in homogeneous complex reactions and heterogeneous reactions.

Uncatalyzed Heterogeneous Reactions

Heterogeneous reactions involve more than one phase (e.g., gas–liquid, gas–solid, liquid–solid, and gas–liquid–solid) and are generally more complicated than homogeneous reactions are due to interaction between physical and chemical processes; that is, reactants in one phase have to be transported (physical process) to the other phase, containing other reactants where the reactions take place.

In a gas–solid reaction, for example, the reaction may proceed in several steps, as follows:

1. Reactants in the gas phase diffuse to the gas–solid interface.
2. When there is a layer of solid product and/or inert material at the interface (e.g., ash), the reactants from the gas phase would have to diffuse through this layer before they can reach the unreacted solid core containing other reactants.
3. The chemical reaction takes place between the reactants from the gas phase and those in the unreacted solid core.
4. The reaction products diffuse within the solid phase and/or diffuse out of the solid phase into the bulk of gas phase.

The step that controls the overall reaction rate will be determined by the nature of the phases and specific reactions involved and by process conditions. Thus, the overall reaction rate is subject to the influence of parameters that affect both the physical and chemical processes, including (a) patterns of phase contact, (b) the reactor geometry, (c) fluid dynamic factors (e.g., velocity and degree of turbulence), (d) interfacial surface area, (e) mass transfer factors, (f) chemical kinetics of reactions involved, and (g) process parameters (e.g., temperature and pressure). Some of these parameters may interact with one another. For example, in a reaction involving two distinct fluid phases (e.g., gas–liquid or liquid–liquid), parameters (d) and (e) would be affected by parameter (c).

The overall reaction rate expression of a heterogeneous reaction is fairly complex, since it considers all of these parameters. Further, the form of rate equation varies with the type of heterogeneous reaction system and with the nature of the controlling step. Some examples of the industrially significant uncatalyzed heterogeneous reactions are given in Table 78.1.

Homogeneous and Heterogeneous Catalytic Reactions

A catalytic reaction is a chemical reaction, the rate of which is modified in the presence of a catalyst. A **catalyst** is a substance that may or may not change chemically during the reaction and is regenerated at the end of the reaction. The catalytic reaction proceeds appreciably faster than does an uncatalyzed reaction, presumably because an intermediate compound, formed between the catalyst and some reactants, reacts with other reactants by a mechanism that requires a lower activation energy to form desired products. In *homogeneous catalysis* the catalyst forms a homogeneous phase with the reaction mixture. In *heterogeneous catalysis,* however, the catalyst is present in a phase different from that of the reaction mixture.

Homogeneous Catalysis

Most homogeneous catalysis takes place in the liquid phase. Perhaps the most widely studied type of liquid-phase catalysis is the acid-base catalysis that exerts influence on the rates of many important organic reactions, including (a) esterification of alcohols, (b) hydrolysis of esters, and (c) inversion of sugars. One of the industrially important gas-phase catalytic reactions is the oxidation of SO_2 to SO_3 in the production of sulfuric acid, catalyzed by nitric oxide in the lead chamber.

TABLE 78.1 Examples of Uncatalyzed Heterogeneous Reactions

Gas–Liquid Reactions	Liquid–Solid Reactions
Production of ammonium nitrate by reaction between ammonia gas and nitric acid Hydrogenation of vegetable oil with hydrogen gas Production of nitric acid by absorption of nitric oxide in water	Reaction of aqueous sulfuric acid with phosphate rock Ion exchange process Recovery of uranium by leaching of uranium ores with sulfuric acid
Gas–Solid Reactions	**Solid–Solid Reactions**
Gasification of coal Production of hydrogen gas by reaction of steam with iron Production of volatile uranium chloride by reaction of uranium oxide with chlorine gas	Production of calcium carbide by reaction of carbon with lime Production of Portland cement by reaction of limestone with clay Production of glass by melting a mixture of calcium carbonate, sodium carbonate, and silica
Liquid–Liquid reactions	**Gas–Liquid–Solid Reaction**
Aqueous sulfuric acid treatment of petroleum liquid Nitration of organic solvents with aqueous nitric acid Production of soaps by reaction of aqueous alkalies and fatty acids	Liquefaction of coal by reaction of hydrogen with coaloil slurry

Solid-Catalyzed Reaction

This reaction, the most common type of heterogeneous catalysis, finds extensive applications in many important industrial processes that produce inorganic and organic chemicals. Well-known examples of such chemicals include HNO_3, HCl, ammonia, aniline, butadiene, ethanol, formaldehyde, methanol, organic polymers, and petrochemicals. The generally accepted, simplified mechanism of solid-catalyzed fluid (gas or liquid) phase reactions is outlined as follows:

1. Reactants diffuse from the main body of the fluid phase to the exterior surface of catalyst pellets and subsequently into catalyst pores.
2. Reactants are adsorbed onto both the catalyst exterior and pore surfaces.
3. Products are formed from interaction of the reactants on the surfaces (catalyst exterior and pore).
4. Products thus formed are desorbed (or released) from the surfaces; those formed on the pore surface are released to the fluid phase within the pores and then diffuse out of pores to the exterior surface of the catalyst pellet.
5. Products then diffuse from the exterior surfaces into the bulk of the fluid phase.

The relative importance of these steps in influencing the overall reaction rate depends upon a variety of factors, among which are thermal factors, fluid dynamic factors, properties of the catalyst, and diffusion characteristics of the reactants and products. Besides the process steps described earlier, there are various deactivation processes that cause loss of catalytic efficiency, such as fouling and poisoning.

78.2 Analysis of Kinetic Data

Data Acquisition

Because chemical reactions involved in most industrial processes are complex, development of the database for the design of the chemical reactor facility can be quite time consuming. Accordingly, selection of experimental methods and equipment for acquisition of kinetic data is crucial in determining the development cost as well as the accuracy and reliability of the data obtained. In essence, the selection process is concerned with both the methods and the equipment to conduct the reactions and to monitor the progress of the reactions. The type of equipment to be used is generally determined by the experimental method for data acquisition. Acquisition of the kinetic data for homogeneous reactions is frequently carried out using a batch reactor because of its relatively simple design and versatility. In heterogeneous reactions, a flow reactor is often utilized for the data acquisition. For detailed discussion on various methods and equipment for obtaining the experimental kinetic data, the reader is referred to the references at the end of the chapter.

Evaluation of Reaction Mechanism

Understanding of the reaction mechanism is important in the selection and design of an industrial reactor for a specific reaction. Full elucidation of the mechanism, however, is not always possible, in which case derivation of the rate equation may have to resort to the empirical method (see the following subsection). No simple standardized method is available for evaluation of the reaction mechanism. Nevertheless, a trial-and-error method is often used based on the experimental kinetic data acquired — including analysis of the reaction mixture to determine the distribution of the residual reactants, intermediates, and final products — following these steps:

1. Assume a simple elementary reaction mechanism and a corresponding stoichiometry and derive a rate equation from the assumed mechanism.
2. Evaluate the experimental data based on the proposed reaction mechanism and the corresponding rate equation using the integral method (described in the section to follow) first since it is relatively easy to use.
3. If the experimental data do not agree with the proposed mechanism, possibly suggesting a nonelementary reaction, propose a new mechanism that consists of several elementary reaction steps with formation of intermediate compounds.
4. Develop rate equations for individual elementary reaction steps and combine the individual rate equations to represent the overall rate expression.
5. If the experimental kinetic data do not fit into the rate equation developed above, assume an alternate mechanism, and repeat step 4. Continue this process until a desired degree of agreement is reached between the experimental data and the rate expression.
6. Evaluation of the reaction mechanism may also be accomplished using the differential method (see the next subsection), especially for complicated reactions. The method of approach is similar to that using the integral method, but it requires more accurate and extensive experimental data.

An example of a nonelementary complex reaction consisting of several elementary reaction steps is the formation of hydrogen bromide from hydrogen and bromine:

$Br_2 \leftrightarrow Br^{\cdot} + Br^{\cdot}$	Initiation and termination
$Br^{\cdot} + H_2 \rightarrow HBr + H^{\cdot}$	Propagation
$H^{\cdot} + Br_2 \rightarrow HBr + Br^{\cdot}$	Propagation
$H^{\cdot} + HBr \rightarrow Br^{\cdot} + H_2$	Propagation
$Br_2 \rightarrow H_2 \rightarrow 2HBr$	Overall stoichiometric reaction

Based on the above reaction mechanism, the following rate equation has been derived:

$$r_{HBr} = \frac{k_1 C_{H_2} C_{Br_2}^{0.5}}{1 + k_2 (C_{HBr}/C_{Br_2})} \tag{78.15}$$

Development of Rate Equation

For reactions with simple mechanisms under isothermal conditions, either the integral method or the differential method (discussed in a later section) may be used in the derivation of rate equations. It is assumed that the data representing the extent of an isothermal reaction are available in terms of the time variation of a selected component *A*.

Integral Method

An elementary reaction mechanism is first assumed; for example, a second-order isothermal homogeneous reaction under *constant-volume* conditions, $A + B \rightarrow C + D$, results in the rate equation of the form

$$-r_A = \frac{dC_A}{dt} = kC_A C_B \tag{78.16}$$

and the integrated rate equation becomes

$$kt = \frac{1}{C_{B_0} - C_{A_0}} \ln \frac{C_{A_0}(C_A + C_{B_0} - C_{A_0})}{C_A C_{B_0}} \tag{78.17}$$

where C_{A_0} and C_{B_0} represent the initial concentrations of reactants *A* and *B*, respectively. One way to confirm the proposed reaction mechanism is to compute values of *k* at various values of C_A and *t*. If the values of *k* remain nearly constant, the proposed mechanism is accepted. Otherwise, another mechanism is assumed, and the process is repeated until a desired degree of agreement is reached.

When the volume of the reaction mixture varies with the extent of reaction, the rate equation becomes more complicated. In this case, if it is assumed that the volume varies linearly with the extent of reaction, derivation of the rate equation could be simplified. Using the fractional conversion x_A to replace C_A as the variable, the volume *V* is expressed as

$$V = V_0(1 + f_A x_A) \tag{78.18}$$

Here, V_0 represents the initial volume of the reaction mixture and f_A is the fractional change in *V* between no conversion and complete conversion with respect to reactant *A*, as defined by

$$f_A = (V_{x_A=1} - V_{x_A=0}) / V_{x_A=0} \tag{78.19}$$

Thus, for a *variable-volume* reaction, the reaction rate defined by Equation (78.1) assumes the following form,

$$-r_A = -\frac{1}{V}\frac{dN_A}{dt} = -\frac{1}{V_0(1 + f_A x_A)}\frac{d}{dt} N_{A_0}(1 - x_A) \tag{78.20}$$

which simplifies to

$$-r_A = \frac{C_{A_0}}{1 + f_A x_A}\frac{dx_A}{dt} \tag{78.21}$$

and the integrated rate equation is

$$t = C_{A_0}\int_0^{x_A} \frac{dx_A}{(1 + f_A x_A)(-r_A)} \tag{78.22}$$

where $-r_A$ stands for the rate expression for the assumed reaction mechanism to be evaluated. For example, for a first-order homogeneous reaction,

$$-r_A = kC_A = kC_{A_0}\frac{1-x_A}{1+f_A x_A} \tag{78.23}$$

Introducing this expression for $-r_A$ into Equation (78.22),

$$Kt = -\ln(1 - x_A) \tag{78.24}$$

The steps to be taken to evaluate the assumed reaction mechanism for the variable-volume case are identical to those for the constant-volume case.

Differential Method

This approach is based on the direct application of the differential rate equation in the analysis of the experimental kinetic data to determine the reaction mechanism. The method requires more accurate and extensive experimental data than the integral method. The basic principle of the method is illustrated by an example of assumed *n*th-order reaction under the isothermal and constant-volume condition:

$$aA + bB = dD + eE$$

With initially equimolal concentrations of *A* and *B* (i.e., $C_A = C_B$), the rate equation may take the following form,

$$-r_A = kC_A^a C_B^b = kC_A^{a+b} = kC_A^n \tag{78.25}$$

which is rearranged into

$$\log(-r_A) = \log(-dC_A/dt) = \log k + n\log C_A \tag{78.26}$$

The assumption of the *n*th-order reaction mechanism is confirmed if a log-log plot of (dC_A/dt) versus C_A results in a straight line.

The next step is to evaluate the values of the rate constant *k* and the overall order of reaction *n* from the plot. With the known values of *k* and *n*, the reaction orders with respect to *A* and *B* can be determined by the method that follows. Rearranging Equation (78.25),

$$-r_A = kC_A^a C_B^{n-a} = kC_B^n (C_A/C_B)^a \tag{78.27}$$

On further rearrangement,

$$\log\left(-\frac{r_A}{kC_B^n}\right) = a\log(C_A/C_B) \tag{78.28}$$

A log-log plot of Equation (78.28) yields the reaction order *a* with respect to *A*, whereas the reaction order *b* is obtained as the difference between *n* and *a*.

Empirical Method

This method, which finds uses when the reaction mechanism appears to be complex, is often based on a mathematical approach using the curve-fitting procedure. The method involves a trial-and-error technique to fit the experimental data to a relatively simple form of the empirical equation, including (a) linear form, $y = a + bx$, (b) semilogarithmic form, $y = ae^{bx}$, (c) logarithmic form, $y = c + ax^n$, and so forth. The initial step usually consists of plotting the experimental data on graph papers of different coordinates that may produce a straight line. Upon selection of a proper form of the empirical equation, the constants in the empirical equation are determined either by the graphic means or by the analytical technique using the method of averages or the method of least squares. Computer software is available for performing the curve-fitting procedure.

Determination of Rate Constant and Arrhenius Parameters

The rate constant k can be obtained by using either a differential form [e.g., Equation (78.16)] or an integrated form [e.g., Equation (78.17)] of the rate equation. It is an average of values calculated at various experimental kinetic data points (i.e., reactant concentrations at various reaction times).

Arrhenius parameters consist of the activation energy E and the frequency factor A. The value of E may be calculated from the rate constants at two distinct but adjacent temperatures, T_1 and T_2, as follows:

$$k_1 = Ae^{-E/RT_1}; \quad k_2 = Ae^{-E/RT_2}$$

Combining the above two equations,

$$E = R\frac{\ln(k_2/k_1)}{1/T_1 - 1/T_2} \tag{78.29}$$

The value of A is calculated from one of the Arrhenius equations shown above.

Defining Terms

Activation energy — A parameter associated with the Arrhenius equation and considered by Arrhenius as the energy in excess of the average energy level of reactants required to enable the reaction to proceed.

Catalyst — A substance that accelerates the reaction, presumably by making available a reaction path that requires a lower activation energy. The catalyst may or may not change chemically during the reaction and is regenerated at the end of the reaction.

Rate constant — A proportionality constant in the rate equation. The rate constant is markedly influenced by temperature and, to a lesser degree, by pressure and the presence of catalysts. The units and value of the rate constant depend on the specific chemical component to which it refers, the units for concentration (or other quantity) of the component, and the reaction order.

Rate-controlling step — The slow steps that tend to control the overall rate of reaction. In a complex reaction consisting of several chemical reaction steps (and physical process steps in heterogeneous reaction), the reaction rate (and physical process rate) of one or more steps may be much slower than other steps.

Rate equation — A functional expression describing the relationship between the rate of reaction and the amounts (e.g., concentrations) of selected chemical components participating in the reaction at any time under isothermal condition.

Rate of reaction — The amount of a chemical component of concern being converted or produced per unit time per unit quantity of a reference variable. Examples of the reference variable include the volume of reacting mixture, the reactor volume, the mass of solid (solid–fluid reaction), and the surface area of solid.

References

Carberry, J. J. 1976. *Chemical and Catalytic Reaction Engineering,* McGraw-Hill, New York.

Connors, K. A. 1990. *Chemical Kinetics — The Study of Reaction Rates in Solution,* VCH, New York.

Katakis, D. and Gordon, G. 1987. *Mechanisms of Inorganic Reactions,* John Wiley & Sons, New York.

Lin, K. H. 1984. Reaction kinetics, reactor design (section 4). In *Perry's Chemical Engineers' Handbook*, 6th ed., ed. R. H. Perry and D. W. Green, McGraw-Hill, New York. pp. 4.1–4.52.

Moore, J. W. and Pearson, R. G. 1981. *Kinetics and Mechanism*, 3rd ed. John Wiley & Sons, New York.

Further Information

The following professional journals provide good sources for examples of basic and applied reaction kinetic studies on specific reactions of industrial importance:

AIChE Journal. Published monthly by the American Institute of Chemical Engineers, New York.

Chem. Eng. Sci. Published semimonthly by Elsevier Science, Oxford, U.K.

Ind. Eng. Chem. Res. Published monthly by the American Chemical Society, Washington, D.C.

J. Am. Chem. Soc. Published biweekly by the American Chemical Society, Washington, D.C.

J. Catal. Published monthly by Academic Press, Orlando, FL.

J. Chem. Soc. — Faraday Trans. Published semimonthly by The Royal Society of Chemistry, Cambridge, U.K.

Trans. Inst. Chem. Eng. (London) — Chem. Eng. and Design. Published bimonthly by the Institute of Chemical Engineers, Basinstoke, U.K.

79

Chemical Reaction Engineering

H. Scott Fogler
University of Michigan

Chemical reaction engineering (CRE) sets chemical engineers apart from other engineers. Students and professionals can easily learn the elements of CRE because it has a very logical structure. The six basic pillars that hold up what you might call the "temple" of chemical reaction engineering are shown in Figure 79.1.

The pillar structure shown in Figure 79.1 allows one to develop a few basic concepts and then to arrange the parameters (equations) associated with each concept in a variety of ways. Without such a structure, one is faced with the possibility of choosing or perhaps memorizing the correct equation from a multitude of equations that can arise for a variety of reactions, reactors, and sets of conditions. This chapter shall focus on five types of chemical reactors commonly used in industry: **batch, semibatch, CSTR, plug flow,** and **packed bed reactors.** Table 79.1 describes each of these reactors.

By using an algorithm to formulate CRE problems, we can formulate and solve CRE problems in a very logical manner. Step 1 in the CRE algorithm is to begin by choosing the mole balance for one of the five types of reactors shown. In step 2 we choose the rate law and in step 3 we specify whether the reaction is gas or liquid phase. Finally, in step 4 we combine steps 1, 2, and 3 and obtain an analytical solution or solve the equations using an *ordinary differential equation (ODE) solver* [Sacham and Cutlip, 1988].

79.1 The Algorithm

We now address each of the individual steps in the algorithm to design isothermal reactors: (1) mole balances, (2) rate laws, (3) stoichiometry, and (4) combine.

Mole Balances

The general mole balance equation (GBE) on species j in a system volume V is:

0-8493-1586-7/05/$0.00+$1.50

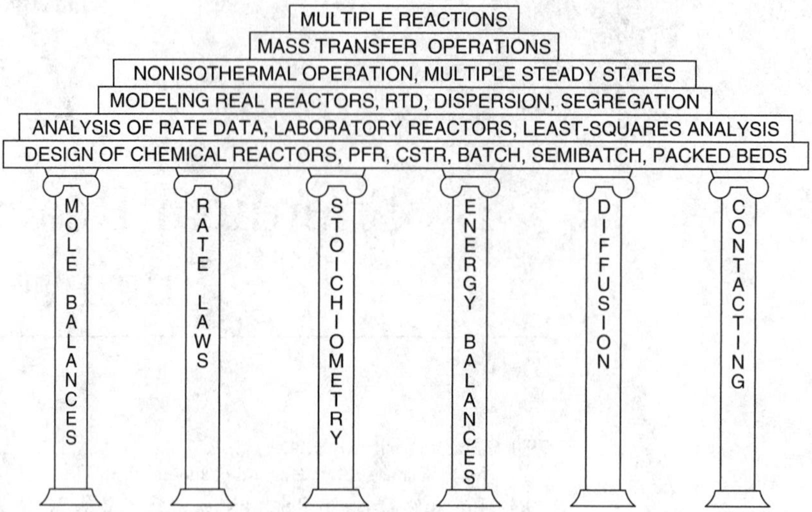

FIGURE 79.1 Pillars of the temple of chemical reaction engineering. (*Source:* Fogler, H. S. 1992. *The Elements of Chemical Reaction Engineering,* 2nd ed. Prentice Hall, Englewood Cliffs, NJ.)

$$\begin{bmatrix}\text{Molar flow}\\ \text{rate}\\ \text{IN}\end{bmatrix} - \begin{bmatrix}\text{Molar flow}\\ \text{rate}\\ \text{OUT}\end{bmatrix} + \begin{bmatrix}\text{Molar rate}\\ \text{of}\\ \text{GENERATION}\end{bmatrix} = \begin{bmatrix}\text{Molar rate}\\ \text{of}\\ \text{ACCUMULATION}\end{bmatrix}$$

$$F_{j_0} - F_1 + \int_0^V r_j dV = \frac{dN_j}{dt} \tag{79.1}$$

We now make use of the definition of conversion, X, with respect to the limiting reactant, which we shall call species A,

Batch	Flow
$X = (N_{A0} - N_A)/N_{A0}$	$X = (F_{A0} - F_A)/F_{A0}$

and apply the GME to each of the following reactors: batch, continuous stirred tank reactors (CSTR), plug flow reactor (PER), and packed bed reactor (PBR). The CSTR, PER, and PBR are all operated as steady state (i.e., $dN_j/dt = 0$) and it is assumed that the PBR and PFR are in plug flow (no radial gradients or dispersion) and that the contents of the CSTR are well mixed. There is no in-flow or out-flow ($F_{j0} = F_j = 0$) in the batch reactor. When these conditions and the definition of conversion are applied to the general mole balance, the design equations [(Equation (79.2) to Equation (79.8)] in Table 79.2 result.

In order to evaluate the design equations given in Table 79.2, we must determine the form of the rate of formation, r_A. We do this with the aid of a rate law.

Rate Laws

The power law model is one of the most commonly used forms for the rate law. It expresses the rate of reaction as a function of the concentrations of the species involved in the reaction.

For the irreversible reaction in which A is the limiting reactant,

$$A + \frac{b}{a}B \rightarrow \frac{c}{a}C + dD \tag{79.9}$$

TABLE 79.1 Comparison of Five Types of Chemical Reactors

Type of Reactor	Characteristics	Usage	Advantages	Disadvantages
Batch	Reactor is charged (filled) via two holes in the top of the tank; while reaction is carried out, nothing else is put in or taken out until reaction is done; tank easily cooled or heated by jacket	Small-scale production Intermediate or one-shot productions Pharmaceuticals Fermentations	High conversion per unit volume for one pass Same reactor can be used to produce one product one time and a different product the next	High operating cost (labor) Product quality more variable than with continuous operation
Semibatch	Either one reactant is charged and the other is fed continuously (at small concentrations) or else one of the products can be removed continuously (to avoid side reactions)	Small-scale production Competing reactions	Good selectivity; feed can be controlled so as to minimize side runs.	High operating labor cost Product quality more variable than with continuous operation
Continuously stirred tank reactor (CSTR)	Run at steady state with continuous flow of reactants and products; the feed assumes a uniform composition throughout the reactor, exit stream has the same composition as in the tank	When agitation is required Series configuration for different concentration streams	Continuous operation Good temperature control Good control Simplicity of construction Low operating (labor) cost	Lowest conversion per unit volume Bypassing and channeling possible with poor agitation
Plug flow reactor (PFR)	Arranged as one long reactor or many short reactors in a tube bank; no radial variation in reaction rate (concentration); concentration charges with length down the reactor	Large-scale production Homogeneous reactions Heterogeneous reactions Continuous production High temperature	Highest conversion per unit volume Low operating labor cost Continuous operation Good heat transfer	Undesired thermal gradients may exist Poor temperature control Shutdown, cleaning may be expensive
Tubular packed bed reactor (PBR)	Tubular reactor that is packed with solid catalyst particles	Used primarily in heterogeneous gas phase reactions with a catalyst	Highest conversion per unit mass of catalyst Low operating cost Continuous operation	Undesired thermal gradients may exist Poor temperature control Channeling may occur

TABLE 79.2 Design Equations

Reactor	Differential	Algebraic	Integral
Batch	$N_{A0}\frac{dX}{dt} = -r_A V$ (79.2)		$t = N_{a0}\int_0^x \frac{dX}{-r_A V}$ (79.3)
CSTR		$t = N_{a0}\int_0^x \frac{dX}{-r_A V}$ (79.3)	
PFR	$F_{A0}\frac{dX}{dV} = -r_A$ (79.5)		$V = F_{A0}\int_0^x \frac{dX}{-r_A}$ (79.6)
PBR	$F_{A0}\frac{dX}{dW} = -r_A'$ (79.7)		$W = F_{A0}\int_0^x \frac{dX}{-r_A'}$ (79.8)

the rate law is

$$-r_A = kC_A^{\alpha}C_B^{\beta} \tag{79.10}$$

We say the reaction is α order in A, β order in B, and overall order = $\alpha + \beta$. For example, if the reaction $A + B \rightarrow C + D$ is said to be second order in A and first order in B and overall third order, then the rate law is

$$-r_A = kC_A^2 C_B \tag{79.11}$$

The temperature dependence of specific reaction rate, k, is given by the Arrhenius equation,

$$k = Ae^{-E/RT} \tag{79.12}$$

where A is the frequency factor and E the activation energy. Taking the natural log of both sides of Equation (79.12),

$$\ln k = \ln A - \frac{E}{R}\left(\frac{1}{T}\right) \tag{79.13}$$

we see the slope of a plot of ln k versus $(1/T)$ will be a straight line equal to $(-E/R)$.

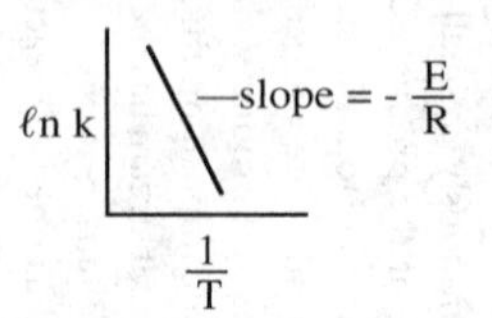

The specific reaction rate at temperature T is commonly written in terms of the specific reaction rate, k_1, at a reference temperature T_1 and the activation energy E. That is,

$$k = k_1(T_1)\exp\left[\frac{E}{R}\left(\frac{1}{T_1} - \frac{1}{T}\right)\right] \tag{79.14}$$

Example

The following reaction is carried out in a constant volume batch reactor:

$$A \rightarrow \text{Products}$$

Determine the appropriate linearized concentration-time plots for zero-, first-, and second-order reactions.

Solution

Use the algorithm to determine the concentration of A as a function of time.

Mole balance:

$$\frac{dN_A}{dt} = r_A V \tag{79.15}$$

Rate law:

$$-r_A = kC_A^{\alpha} \tag{79.16}$$

Stoichiometry:

$$V = V_O \quad \therefore \quad C_A = N_A / V_0 \tag{79.17}$$

Combine:

$$-\frac{dC_A}{dt} = kC_A^{\alpha} \tag{79.18}$$

Solving Equation (79.18) for a first-, second-, and third-order rate law, we can arrive at the following linearized concentration-time plots.

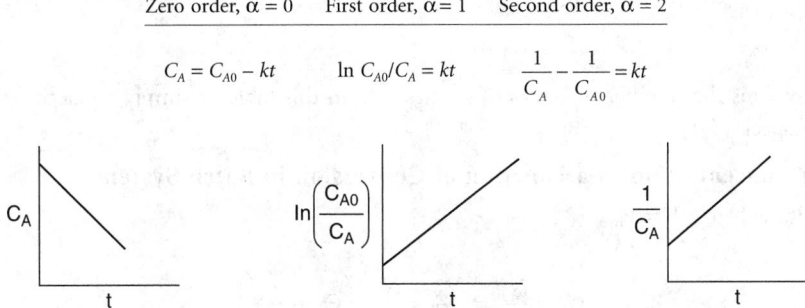

Zero order, $\alpha = 0$	First order, $\alpha = 1$	Second order, $\alpha = 2$
$C_A = C_{A0} - kt$	$\ln C_{A0}/C_A = kt$	$\frac{1}{C_A} - \frac{1}{C_{A0}} = kt$

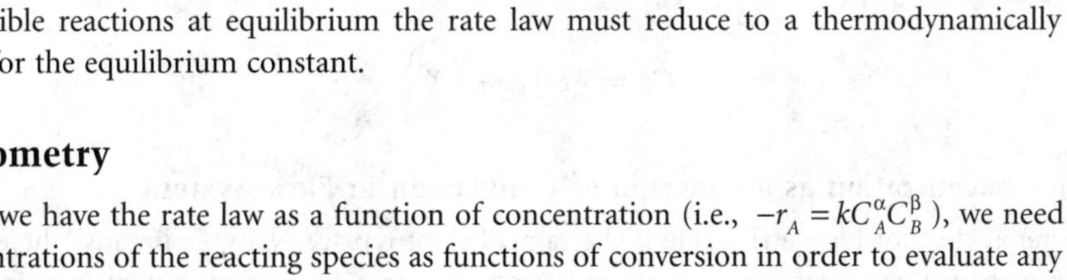

For reversible reactions at equilibrium the rate law must reduce to a thermodynamically consistent equation for the equilibrium constant.

Stoichiometry

Now that we have the rate law as a function of concentration (i.e., $-r_A = kC_A^{\alpha}C_B^{\beta}$), we need to express the concentrations of the reacting species as functions of conversion in order to evaluate any one of the reactor design equations.

Concentration

We start by defining concentration for a flow system and a batch system. For a flow system,

$$C_i = \frac{F_i}{\upsilon} \tag{79.19}$$

where υ is the volumetric flow rate. For a batch system,

$$C_i = \frac{N_i}{V} \tag{79.20}$$

The next step is to express N_i and F_i as a function of conversion using a stoichiometric table.

The Stoichiometry Table

Using our definition of conversion we can construct the following stoichiometric table.

Stoichiometry $A + \frac{b}{a}B \rightarrow \frac{c}{a}C + \frac{d}{a}D$

Batch Systems

Species	Symbol	Initial	Change	Remaining
A	A	N_{A0}	$-NA_0X$	$N_A = N_{A0}(1-X)$
B	B	$N_{A0}\Theta_B$	$-\frac{b}{a}N_{A0}X$	$N_B = N_{A0}\left(\Theta_B - \frac{b}{a}X\right)$
C	C	$N_{A0}\Theta_C$	$+\frac{c}{a}N_{A0}X$	$N_C = N_{A0}\left(\Theta_C + \frac{c}{a}X\right)$
D	D	$N_{A0}\Theta_D$	$+\frac{d}{a}N_{A0}X$	$N_C = N_{A0}\left(\Theta_D + \frac{d}{a}X\right)$
Inert	I	$N_{A0}\Theta_I$	—	$N_I = N_{A0}\Theta_I$
		N_{T0}		$N_T = N_{T0} + \delta N_{A0}X$

where $\delta = \frac{d}{a} + \frac{c}{a} - \frac{b}{a} - 1, \varepsilon = y_{A0}\delta$, and $\Theta_i = \frac{N_{i0}}{N_{A0}} = \frac{y_{i0}}{y_{A0}} = \frac{C_{i0}}{C_{A0}}$

For flow systems the number of moles of species i, N_i in this table, is simply replaced by the molar flow rates of species i, F_i.

Expressing Concentration as a Function of Conversion in Batch System

Constant volume batch: $V = V_0$.

$$C_B = \frac{N_B}{V} = \frac{N_B}{V_0} = \frac{N_{A0}}{V_0}\left(\Theta_B - \frac{b}{a}X\right)$$

$$C_B = C_{A0}\left(\Theta_B - \frac{b}{a}X\right) \tag{79.21}$$

Expressing Concentration as a Function of Conversion in Flow System

For flow systems, the stoichiometric table is the same, except replace N_i by F_i. Because there is hardly ever a volume change with reaction, the concentration of A in a *liquid* flow system is as follows. For liquid systems,

$$C_A = \frac{F_A}{\upsilon_0} = \frac{F_{A0}}{\upsilon_0}(1-X) = C_{A0}(1-X) \tag{79.22}$$

For gas systems,

$$C_A = \frac{F_A}{\upsilon} \tag{79.23}$$

In ideal gas systems the gas volumetric flow rate, υ, can change with conversion, temperature, and pressure according to the following equation:

$$\upsilon = \upsilon_0 \left(\frac{F_T}{F_{T0}}\right)\frac{P_0}{P}\frac{T}{T_0} \tag{79.24}$$

Taking the ratio of F_T/F_{T0} and then using the stoichiometric table, we arrive at the following equation for the volumetric flow rate at any point in the reactor.

$$\upsilon = \upsilon_O(1+\varepsilon X)\frac{P_0}{P}\frac{T}{T_0} \tag{79.25}$$

Substituting this result and Equation (79.22) into Equation (79.23) gives

$$C_A = \frac{C_{A0}(1-X)}{(1+\varepsilon X)}\frac{P}{P_0}\frac{T_0}{T} \tag{79.26}$$

We now will apply the algorithm described earlier to a specific situation. Suppose we have, as shown in Figure 79.2, mole balances for three reactors, three rate laws, and the equations for concentrations for both liquid and gas phases. In Figure 79.3, the algorithm is used to formulate the equation to calculate the PFR reactor volume for a first-order gas-phase reaction. The pathway to arrive at this equation is shown by the ovals connected to the dark lines through the algorithm. The dashed lines and the boxes represent other pathways for other solutions. For the reactor and reaction specified, we follow these steps:

1. Choose the *mole balance* on species A for a PFR.
2. Choose the *rate law* for an irreversible first-order reaction.
3. Choose the equation for the concentration of A in the gas phase (*stoichiometry*).
4. Finally, *combine* to calculate the volume necessary to achieve a given conversion or calculate the conversion that can be achieved in a specified reaction volume.

For the case of isothermal operation with no pressure drop, we were able to obtain an analytical solution, given by equation (A) in Figure 79.2, which gives reactor volume necessary to achieve a conversion X for a gas phase reaction carried out isothermally in a PFR. However, in the majority of situations, analytical solutions to the ordinary differential equations appearing in the combine step are not possible. By using this structure, one should be able to solve reactor engineering problems through reasoning rather than memorization of numerous equations together with the various restrictions and conditions under which each equation applies (i.e., whether there is a change in the total number of moles, etc.). In perhaps no other area of engineering is mere formula plugging more hazardous; the number of physical situations that can arise appears infinite, and the chances of a simple formula being sufficient for the adequate design of a real reactor are vanishingly small.

Example

The elementary gas phase reaction

$$2A + B \rightarrow C$$

$$-r_A = k_A C_A^2 C_B$$

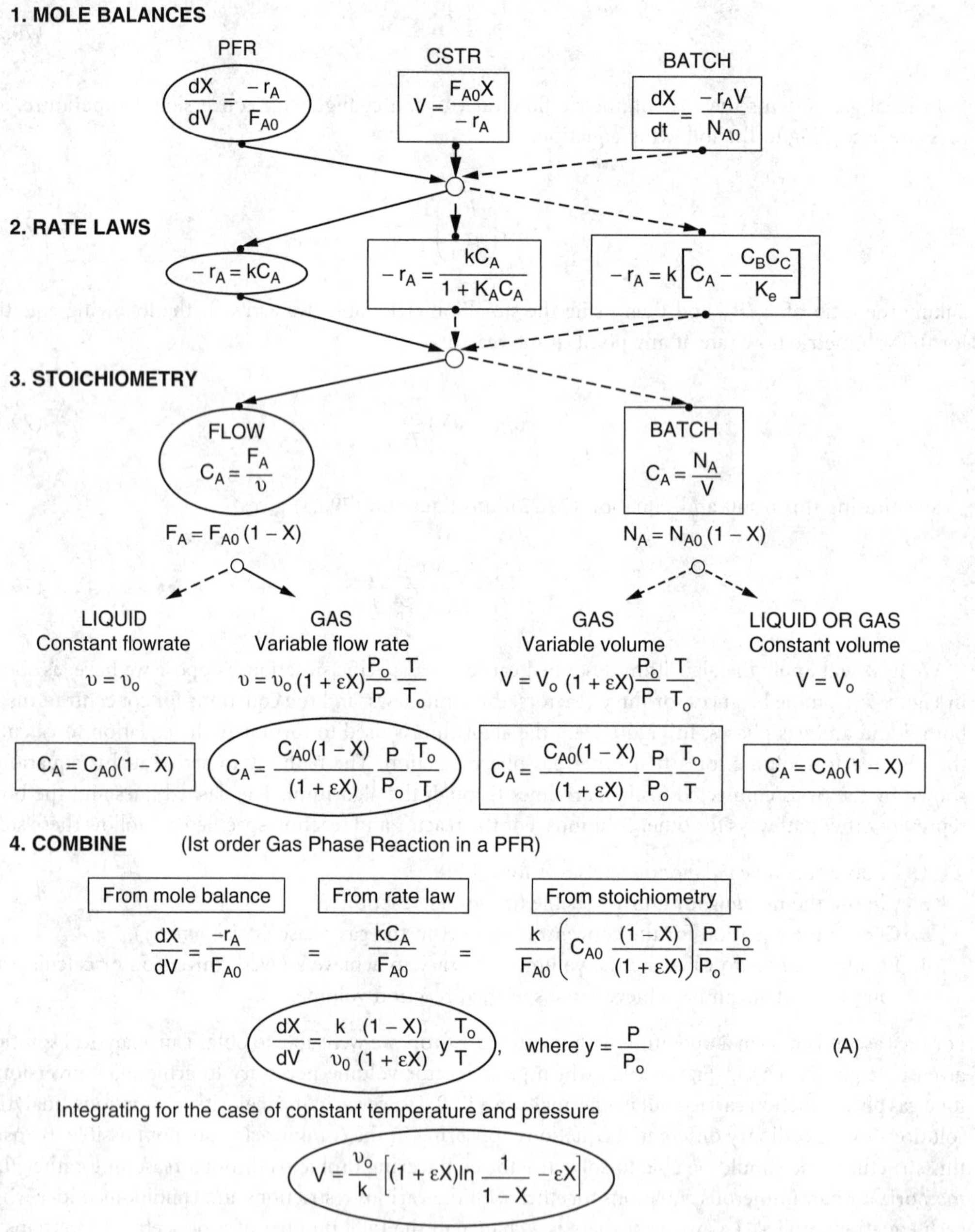

1. MOLE BALANCES
PFR
$\frac{dX}{dV} = \frac{-r_A}{F_{A0}}$
CSTR
$V = \frac{F_{A0}X}{-r_A}$
BATCH
$\frac{dX}{dt} = \frac{-r_A V}{N_{A0}}$
2. RATE LAWS
$-r_A = kC_A$
$-r_A = \frac{kC_A}{1 + K_A C_A}$
$-r_A = k\left[C_A - \frac{C_B C_C}{K_e}\right]$
3. STOICHIOMETRY
FLOW
$C_A = \frac{F_A}{\upsilon}$
BATCH
$C_A = \frac{N_A}{V}$
$F_A = F_{A0}(1 - X)$
$N_A = N_{A0}(1 - X)$
LIQUID
Constant flowrate
$\upsilon = \upsilon_0$
GAS
Variable flow rate
$\upsilon = \upsilon_0 (1 + \varepsilon X)\frac{P_0}{P}\frac{T}{T_0}$
GAS
Variable volume
$V = V_0 (1 + \varepsilon X)\frac{P_0}{P}\frac{T}{T_0}$
LIQUID OR GAS
Constant volume
$V = V_0$
$C_A = C_{A0}(1 - X)$
$C_A = \frac{C_{A0}(1 - X)}{(1 + \varepsilon X)}\frac{P}{P_0}\frac{T_0}{T}$
$C_A = \frac{C_{A0}(1 - X)}{(1 + \varepsilon X)}\frac{P}{P_0}\frac{T_0}{T}$
$C_A = C_{A0}(1 - X)$
4. COMBINE
(Ist order Gas Phase Reaction in a PFR)
From mole balance
From rate law
From stoichiometry
$\frac{dX}{dV} = \frac{-r_A}{F_{A0}} = \frac{kC_A}{F_{A0}} = \frac{k}{F_{A0}}\left(C_{A0}\frac{(1 - X)}{(1 + \varepsilon X)}\right)\frac{P}{P_0}\frac{T_0}{T}$
$\frac{dX}{dV} = \frac{k}{\upsilon_0}\frac{(1 - X)}{(1 + \varepsilon X)} y \frac{T_0}{T}$, where $y = \frac{P}{P_0}$ (A)
Integrating for the case of constant temperature and pressure
$V = \frac{\upsilon_0}{k}\left[(1 + \varepsilon X)\ln\frac{1}{1 - X} - \varepsilon X\right]$

FIGURE 79.2 Algorithm for isothermal reactors. (*Source:* Fogler, H. S. 1996. *Elements of Chemical Reaction Engineering.* 3rd ed. Prentice Hall, Englewood Cliffs, NJ.)

is carried out at constant T (500 K) and P (16.4 atm) with $k_A = 10$ dm^6/mol^2·sec. Determine the CSTR reactor volume necessary to achieve 90% conversion when the feed is 50% mole A and 50% B.

Solution

The feed is equal molar in A and B; therefore, A is the limiting reactor and taken as our basis of calculation:

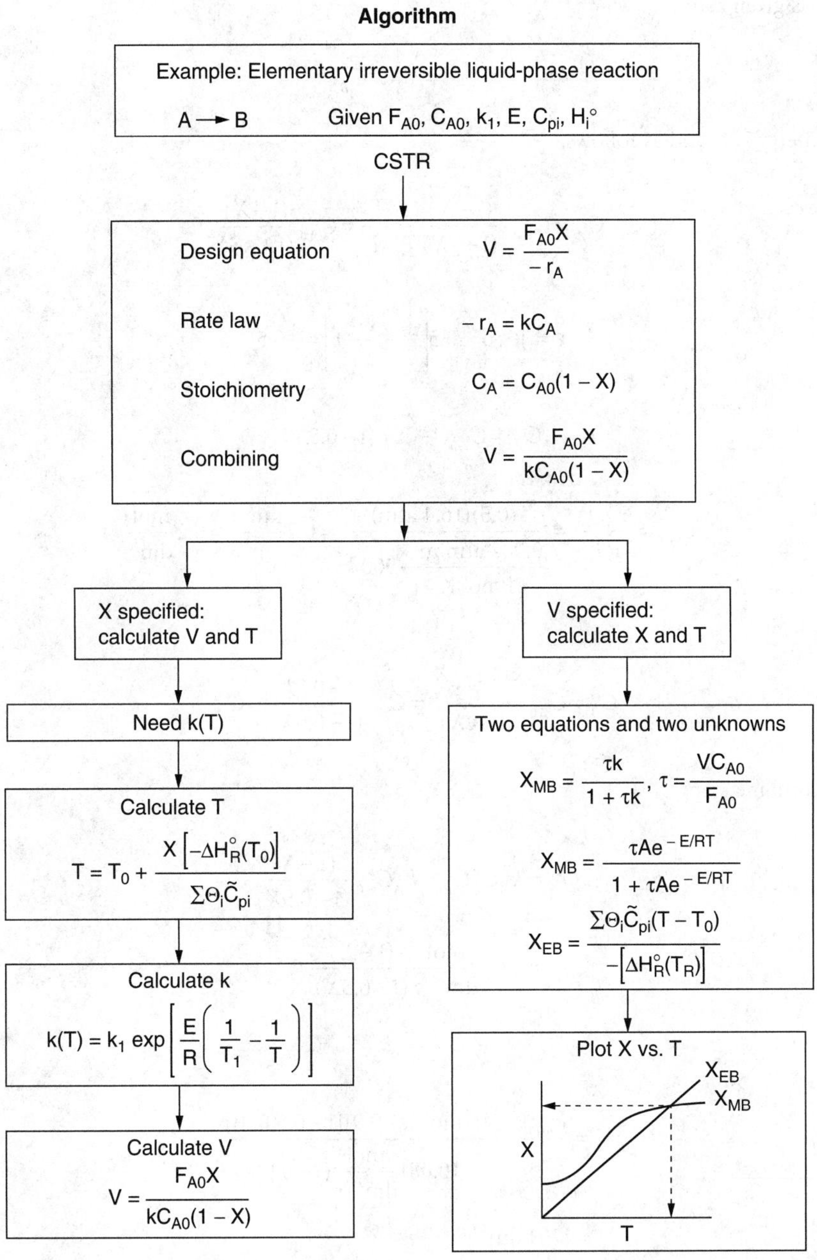

FIGURE 79.3 Algorithm for nonisothermal CSTR design. (*Source:* Fogler, H. S. 1992. *The Elements of Chemical Reaction Engineering,* 2nd ed. Prentice Hall, Englewood Cliffs, NJ.)

$$A + \frac{B}{2} \rightarrow \frac{C}{2}$$

Mole balance is given as:

$$V = \frac{F_{A0}X}{-r_A} \tag{79.27}$$

Rate law is given as:

$$-r_A = k_A C_A^2 C_B \tag{79.28}$$

Stoichiometry is found as follows:

$$C_A = C_{A0} \cdot \frac{(1-X)}{(1+\varepsilon X)} \frac{P}{P_0} \frac{T_0}{T} = C_{A0} \frac{(1-X)}{(1+\varepsilon X)}$$

$$\varepsilon = y_{A0}\delta = 0.5\left[\frac{1}{2} - \frac{1}{2} - 1\right] = -0.5 \tag{79.29}$$

$$C_A = C_{A0}(1-X)/(1-0.5X)$$

$$C_{A0} = \frac{y_{A0}P_0}{RT_0} = \frac{(0.5)(16.4\ \text{atm})}{\frac{0.082\ \text{atm m}^3}{\text{kmol K}} \cdot 500\ k} = 0.2\frac{\text{kmol}}{\text{m}^3} = 0.2\frac{\text{mol}}{\text{dm}^3} \tag{79.30}$$

$$C_B = C_{A0} \frac{\Theta_B - \frac{1}{2}X}{(1+\varepsilon X)} = C_{A0} \frac{(1-0.5X)}{(1-0.5X)} = C_{A0} \tag{79.31}$$

For the combine step,

$$\begin{aligned} -r_A &= k_A C_A^2 C_B = k_A C_{A0}^3 \frac{(1-X)^2}{(1-0.5X)^2} \\ &= 0.08 \frac{\text{mol}}{\text{dm}^3 \cdot \text{s}} \frac{(1-X)^2}{(1-0.5X)^2} \end{aligned} \tag{79.32}$$

For a CSTR,

$$\begin{aligned} V &= \frac{F_{A0}X}{-r_A} = \frac{(5\ \text{mol/s})(0.9)[1-0.5(0.9)]^2}{(0.08)\frac{\text{mol}}{\text{dm}^3 \cdot \text{s}}(1-0.9)^2} \\ &= 1701\ \text{dm}^3 \end{aligned} \tag{79.33}$$

79.2 Pressure Drop in Reactors

If pressure drop is not accounted for in gas phase reactions, significant underdesign of the reactor size can result. This variation is handled in the stoichiometry step, in which concentration is expressed as a function of conversion, temperature, and total pressure. The change in total pressure is given by the Ergun equation [Fogler, 1992]:

$$\frac{dP}{dL} = -\frac{G(1-\phi)}{\rho g_C D_p \phi^3}\left[\frac{150(1-\phi)\mu}{D_p} + 1.75\ G\right] \tag{79.34}$$

For isothermal operation the density is (assuming ideal gas)

$$\rho = \frac{\rho_0}{(1+\varepsilon X)} \frac{P}{P_0} \tag{79.35}$$

The catalyst weight, *W*, and length down the reactor, *L*, are related by the equation

$$W = LA_c(1-\Phi)\rho_{\text{cat}}$$

Substituting back in the Ergun equation,

$$\frac{dP}{dW} = -\frac{\alpha_P}{2} \frac{(1+\varepsilon X)}{\frac{1}{P_0}\left(\frac{P}{P_0}\right)} \tag{79.36}$$

where

$$\alpha_p = \frac{\frac{G(1-\Phi)}{\rho_0 g_c D_p \Phi^3}\left[\frac{150(1-\Phi)\mu}{D_P} + 1.75\,G\right]}{A_c(1-\Phi)\rho_{\text{cat}} P_0}$$

We now need to solve this differential equation to obtain the pressure as a function of the weight of catalyst the gas has passed over. We can obtain an analytical solution of $\varepsilon = 0$. Otherwise, we must solve the equation numerically and simultaneously with the mole balance. For an analytical solution,

$$\frac{d(P/P_0)^2}{dW} = -\alpha_p(1+\varepsilon X) \tag{79.37}$$

For POLYMATH solution, letting $y = P/P_0$,

$$\frac{dy}{dW} = -\frac{\alpha_p(1+\varepsilon X)}{2y} \tag{79.38}$$

Example

To understand the effect pressure drop has on gas phase reaction in a packed bed, we analyze the reaction $A \rightarrow B$ carried out in a packed bed reactor (PBR). Mole balance (PBR) is

$$F_{A0}\frac{dX}{dW} = -r_A'$$

Wherever pressure drop occurs in a PBR, we must use the differential from of the mole balance to separate variables. Pressure drop only affects C_A, C_B, and so on, as well as $-r_A'$.

Rate law is second order in *A* and irreversible, according to the formula $-r_A' = kC_A^2$.

Stoichiometry is given by

$$C_A = C_{A0}\frac{(1-X)}{(1+\varepsilon X)}\frac{P}{P_0}\frac{T_0}{T}$$

For $\varepsilon = 0$ and isothermal operation,

$$\frac{P}{P_0} = (1-\alpha_p W)^{1/2}$$

$$C_A = C_{A0}(1-X)\frac{P}{P_0} = C_{A0}(1-X)(1-\alpha_p W)^{1/2}$$

combining,

$$F_{A0}\frac{dX}{dW} = -r_A' = kC_{A0}^2(1-X)^2(1-\alpha_p W)$$

By integrating with limits $W = 0$, $X = 0$ we obtain the desired relationship between conversion and catalyst weight.

$$\frac{X}{1-X} = \frac{kC_{A0}^2}{F_{A0}}\left[W - \frac{\alpha_p W^2}{2}\right]$$

79.3 Multiple Reactions

There are three basic types of multiple reactions: series, parallel, and independent. In *parallel reactions* (also called *competing reactions*) the reactant is consumed by two different reactions to form different products:

$$A \xrightarrow{k_1} B$$
$$A \xrightarrow{k_2} C$$

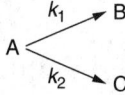

In *series reactions*, also called *consecutive reactions*, the reactant forms an intermediate product, which reacts further to form another product:

$$A \xrightarrow{k_1} B \xrightarrow{k_2} C$$

Multiple reactions involve a combination of both series and parallel reactions, such as

$$A + B \rightarrow C + D$$
$$A + C \rightarrow E$$

Independent reactions are of the type

$$A \rightarrow B$$
$$C \rightarrow D + E$$

and occur in feed stocks containing many reactants. The cracking of crude oil to form gasoline is an example of an independent reaction.

To describe selectivity and yield we consider the following competing reactions.

$$A + B \xrightarrow{k_D} D \quad \text{(desired)}$$

$$A + B \xrightarrow{k_U} U \quad \text{(undesired)}$$

The rate laws are

$$r_D = k_D C_A^{\alpha_1} C_B^{\beta_1} \tag{79.39}$$

$$r_U = k_U C_A^{\alpha_1} C_B^{\beta_2} \tag{79.40}$$

We want the rate of D, r_D, to be high with respect to the rate of formation U, r_U. Taking the ratio of these rates, we obtain a rate *selectivity parameter, S*, which is to be maximized:

$$S_{DU} = \frac{r_D}{r_U} \tag{79.41}$$

Substituting Equation (79.39) and Equation (79.40) into Equation (79.41) and letting $a = \alpha_1 - \alpha_2$ and $b = \beta_2 - \beta_1$, where a and b are both positive numbers, we have

$$S_{DU} = \frac{r_D}{r_U} = \frac{k_1 C_A^a}{k_2 C_B^b}$$

To make S_{DU} as large as possible, we want to make the concentration of A high and the concentration of B low. To achieve this result, use the following:

A semibatch reactor in which B is fed slowly into a large amount of A
A tubular reactor with side streams of B continually fed to the reactor
A series of small CSTRs with A fed only to the first reactor and B fed to each reactor

Another definition of selectivity used in the current literature is given in terms of the flow rates leaving the reactor:

$$\tilde{S}_{DU} = \text{Selectivity} = \frac{F_D}{F_U} = \frac{\text{Exit molar flow rate of desired product}}{\text{Exit molar flow rate of undesired product}} \tag{79.42}$$

For a batch reactor, the selectivity is given in terms of the number of moles of D and U at the end of the reaction time:

$$\tilde{S}_{DU} = \frac{N_D}{N_U} \tag{79.43}$$

One also finds that the reaction yield, like the selectivity, has two definitions: one based on the ratio of reaction rates and one based on the ratio of molar flow rates. In the first case the yield at a point can be defined as the ratio of the reaction rate of a given product to the reaction rate of the key reactant A [Carbery, 1967]:

$$Y_D = \frac{r_D}{-r_A} \tag{79.44}$$

In the case of reaction yield based on molar flow rates, the yield is defined as the ratio of moles of product formed at the end of the reaction to the number of moles of the key reactant, *A*, that have been consumed. For a batch system,

$$\tilde{Y}_D = \frac{N_D}{N_{A0} - N_A} \tag{79.45}$$

For a flow system,

$$\tilde{Y}_D = \frac{F_D}{F_{A0} - F_A} \tag{79.46}$$

Because of the various definitions for selectivity and yield, when reading literature dealing with multiple reactions, check carefully to ascertain the definition intended by the author.

79.4 Heat Effects

For nonisothermal reaction in CRE we must choose which form of the energy balance to use (e.g., PFR, CSTR) and which terms to eliminate (e.g., $Q = 0$ for adiabatic operation). The structure introduced to study these reactors builds on the isothermal algorithm by introducing the Arrhenius equation, $k = Ae^{-E/RT}$ in the **rate law** step, which results in one equation with two unknowns, *X* and *T*, when we finish with the combine step. For example, using again the PFR mole balance and conditions in Figure 79.2 [Equation (A)], we have, for constant pressure,

$$\frac{dX}{dV} = \frac{k(1-X)}{\upsilon_0(1+\varepsilon X)}\frac{T_0}{T} \tag{79.47}$$

$$\frac{dX}{dV} = \frac{Ae^{-E/RT}(1-X)}{\upsilon_0(1+\varepsilon X)}\left(\frac{T_0}{T}\right) \tag{79.48}$$

We can now see the necessity of performing an energy balance on the reactor to obtain a second equation relating *X* and *T*. We will use energy balance to relate *X* and *T*.

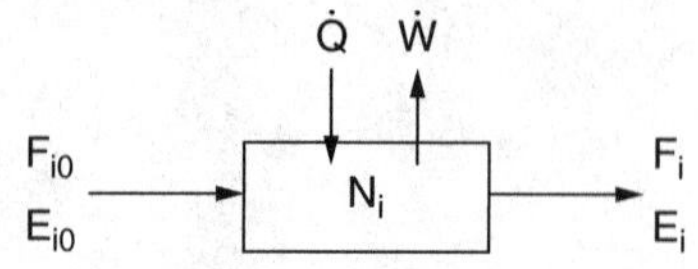

$$\begin{bmatrix} \text{Rate of} \\ \text{accumulation} \\ \text{of energy} \\ \text{within the} \\ \text{system} \end{bmatrix} = \begin{bmatrix} \text{Rate of flow} \\ \text{of heat to} \\ \text{the system} \\ \text{from the} \\ \text{surroundin gs} \end{bmatrix} - \begin{bmatrix} \text{Rate of work} \\ \text{done by} \\ \text{the system} \\ \text{on the} \\ \text{surroundin gs} \end{bmatrix}$$

$$+ \begin{bmatrix} \text{Rate of energy} \\ \text{added to the} \\ \text{system by mass} \\ \text{flow into the} \\ \text{system} \end{bmatrix} - \begin{bmatrix} \text{Rate of} \\ \text{energy leaving} \\ \text{system by mass} \\ \text{flow out of} \\ \text{the system} \end{bmatrix}$$

$$\frac{dE}{dt} = \dot{Q} - \dot{W} + \sum F_{\text{in}} E_{\text{in}} - \sum F_{\text{out}} E_{\text{out}}$$

Neglecting potential and kinetic energy, the energy E_i is just the internal energy of species i, substituting for W in terms of the flow work and the shaft work, W_S, and using the definition of enthalpy gives for steady-state operation

$$\dot{Q} - \dot{W}_s + \sum F_{i0} H_{i0} - \sum F_i H_i = 0 \tag{79.49}$$

For adiabatic operation, no work done on the system and constant heat capacity and $\Delta C_P = 0$, the energy balance reduces to

$$T = T_0 + \frac{(-\Delta H_R)X}{\sum \theta_i C_{pi} + \Delta C_P X} \tag{76.50}$$

We now use this relationship to solve adiabatic reactor design problems.

The procedure for nonisothermal reactor design can be illustrated by considering the first-order irreversible liquid-phase reaction $A \rightarrow B$. The CSTR design equation is

$$V = \frac{F_{A0} X}{-r_A}$$

Rate law is found by

$$-r_A = kC_A \tag{79.51}$$

with the Arrhenius equation:

$$k = Ae^{-E/RT}$$

Stoichiometry for the liquid phase (i.e., $\upsilon = \upsilon_0$) is given by

$$C_A = C_{A0}(1 - X)$$

Combining yields

$$V = \frac{\upsilon_0}{Ae^{-E/RT}}\left(\frac{X}{1-X}\right) \tag{79.52}$$

Continuing from this point requires two distinct cases. For the first case, the variables X, υ_0, C_{A0}, and F_{i0} are specified and the reactor volume, V, must be determined. The procedure is as follows:

1. Evaluate Equation (79.50) to find the temperature, T, for the conditions specified.
2. Calculate k from the Arrhenius equation.
3. Calculate the reactor volume, V, from Equation (79.52).

For the second case, the variables υ_0, C_{A0}, V, and F_{i0} are specified and the exit temperature, T, and conversion, X, are unknown quantities. The procedure is as follows:

1. Solve the energy balance for X as a function of T. If adiabatic, Equation (79.50) becomes

$$X_{EB} = \frac{\sum \Theta_i \tilde{C}_{pi}(T - T_0)}{-[\Delta H_R^\circ(T_R)]} \tag{79.53}$$

2. Solve the mole balance [Equation (79.52)] for X as a function of T.

$$X_{MB} = \frac{\tau A e^{-E/RT}}{1 + \tau A e^{-E/RT}} \tag{79.53}$$

where $\tau = V/\upsilon_0$

3. Plot the previous two steps on the same graph to determine the intersection. At this point the values of X and T satisfy both the energy balance and mole balance. As an alternative, one may equate the equations for X from the previous two steps and solve numerically.

An energy balance on a PFR with heat exchange yields the second equation we need relating our independent variables X and T:

$$\frac{dT}{dV} = \frac{[UA_c(T_a - T) + (r_A)(\Delta H_R)]}{F_{A0} C_{P_A}} \tag{79.54}$$

The differential equation describing the change of temperature with volume (i.e., distance) down the reactor,

$$\frac{dT}{dV} = g(X, T) \tag{79.55}$$

must be coupled with the mole balance, Equation (79.5),

$$\frac{dX}{dV} = \frac{-r_A}{F_{A0}} f(X, T) \tag{79.56}$$

and solved simultaneously. A variety of numerical integration schemes and ODE solvers can be used to solve these two equations simultaneously.

79.5 Summary

By arranging chemical reaction engineering in a structure analogous to a French menu, we can study a multitude of reaction systems with very little effort. This structure is extremely compatible with a number of user-friendly ordinary differential equation (ODE) solvers. Using **ODE solvers** such as POLYMATH, the student is able to focus on exploring reaction engineering problems rather than crunching numbers. Thus, the teacher is able to assign problems that are more open ended and give students practice at developing their creativity. Practicing creativity is extremely important, not only in CRE, but in every course in the curriculum if students are to compete in the world arena and succeed in solving the relevant problems that they will be faced with in the future.

Nomenclature

A	frequency factor (appropriate units)
A_c	cross-sectional area, m^2
C_i	concentration of species i ($i = A, B, C, D$), mol/dm^3
C_{P_i}	heat capacity of species i, J/g/K
D_P	particle diameter, m
E	activation energy, J/mol
F_i	entering molar flow rate of species i, mol/sec
G	superficial gas velocity g/m^2/sec
g_c	conversion factor
k	specific reaction rate (constant), appropriate units
K_A	adsorption equilibrium constant (dm^3/mol)
K_e	equilibrium constant, appropriate units
L	length down the reactor, m
N_i	number of moles of species i, mol
P	pressure, kPa
r_i	rate of formation of species i per unit volume, mol/sec/dm^3
r_i'	rate of formation of species i per unit mass of catalyst, mol/sec/g
R	ideal gas constant, J/mol/K
t	time, sec
T	temperature, K
U	overall heat transfer coefficient, J/(dm^{3}sec K)
V	volume, dm^3
W	catalyst weight, g
X	conversion
y	pressure drop parameter (P/P_o)
y_A	mole fraction of A
a	ambient temperature
A	refers to species A
cat	catalyst density kg/m^3
EB	energy balance
MB	mole balance
T	total number of moles
0	entering or initial condition
a	reaction order
α_p	pressure drop parameter, g^{-1}
β	reaction order
ΔH_R	heat of reaction, J/mole A
δ	change in the total number of moles per mole of A reacted
ε	bolume change parameter $= y_{A0}\delta$
ϕ	porosity
μ	viscosity, cp
ρ	density, g/dm^3
υ	volumetric flow rate, dm^3/sec
Θ_i	N_i/N_{A0}

Defining Terms

Batch reactor — A closed vessel (tank) in which there is no flow in or out of the vessel during the time the reaction is taking place.

Continuous stirred tank reactor (CSTR) — A reactor in which the reactant and products flow continuous into and out of (respectively) the tank. A reactor where the contents are well mixed.

ODE solver — A user-friendly software package that solves ordinary differential equations, for example, Mathematica, POLYMATH, Matlab.

Packed bed reactor — Usually a tubular reactor packed with solid catalyst pellets.

Plug flow reactor — Usually a tubular reactor used for gas phase reactions in which it is assumed there are no radial gradients in temperature or concentration as well as no dispersion of reactants.

Semibatch reactor — A reactor (vessel) in which one of the reactants is placed in the reactor and a second reactant is slowly added to the reactor.

References

Carbery, J. J. 1967. Applied kinetics and chemical reaction engineering, In *Chemical Engineering Education*, ed. R. L. Gorring and V. W. Weekman, American Chemical Society, Washington, DC. p. 89.

Fogler, H. S. 1992. *The Elements of Chemical Reaction Engineering,* 2nd ed. Prentice Hall, Englewood Cliffs, NJ.

Shacham, M. and Cutlip, M. B. 1988. Applications of a microcomputer computation package in chemical engineering. *Chemical Engineering Education* 12(1):18.

Further Information

Professional Organizations

The American Institute of Chemical Engineers (three national meetings per year), 345 E. 47th St., New York, NY 10017. Phone (212) 705-7322.

The American Chemical Society (several national meetings each year), 1155 16th St., Washington, D.C. 20036. Phone (202) 872-4600.

Special Meetings and Conferences

The Engineering Foundation Conferences on Chemical Reaction Engineering, 345 E. 47th St., New York, NY 10017. Phone (212) 705-7835.

International Symposia on Chemical Reaction Engineering (even years), sponsored by American Institute of Chemical Engineers, American Chemical Society, Canadian Society for Chemical Engineering, and the European Federation of Chemical Engineering.

Professional Journals

AIChE Journal. Published monthly by the American Institute of Chemical Engineers, New York.

Chem. Eng. Sci. Published semimonthly by Elsevier Science, Oxford, U.K.

80

The Scaleup of Chemical Reaction Systems from Laboratory to Plant

J. B. Cropley
Union Carbide Corporate Fellow (Retired)

Scaleup is one of those overworked terms that has come to mean almost anything and everything, depending on who is using it. In this article we will use it very little, and then only to indicate the generic process of commercializing new chemical technology. The alternative to scaleup is rational design, which utilizes mathematical relationships and computer simulation to develop the best design for the reactor. The mathematical relationships describe both the reaction kinetics and the attributes of the reactor and its associated auxiliary equipment.

Geometric scaleup was practiced routinely in the chemical industry as a design protocol until a few decades ago, but, today, rational design — based on laboratory data and correlations — has largely replaced it for most types of industrial chemical reaction systems. To understand why this is so, it is necessary to note how the chemical industry has changed over the years.

Forty or 50 years ago, merely producing a chemical on an industrial scale was usually sufficient to ensure a profit for the manufacturer. Chemical processes were labor intensive, but capital and energy were cheap and the selling price of a pound of finished product was typically several times the raw material cost. Furthermore, the environment had not yet been discovered either by industry or by the public at large.

0-8493-1586-7/05/$0.00+$1.50

It was really unnecessary to design reactors rationally for most kinds of processes in that era, because any questions about productive capacity could be addressed simply by making the reactor larger, raw material selectivities usually were not economically critical, and the large quantities of energy that were expended in complex distillation trains to remove byproducts and impurities were both practicable and inexpensive.

In contrast, the petrochemical industry today utilizes chemical reactions that produce the desired products much more directly and cleanly, with as little waste as possible. Raw material cost is frequently the largest component of the final product cost, and the crude product must not contain unexpected byproducts that the refining system cannot remove adequately and efficiently. The failure to meet tight product specifications can produce chemicals that either cannot be sold at all without expensive reprocessing or can be sold only for little more than their value as fuel.

In any case, both today's marketplace and concerns for the environment demand that chemical reaction systems produce no more than extremely small amounts of waste or off-specification product per pound of refined salable product. Reactors must be accurately designed and operated because today's chemistry frequently is strongly dependent upon carrying out just the desired amount of reaction in order to avoid the production of unwanted byproducts by over-reaction. The energy efficiencies of refining systems strongly depend upon their receiving crude product of uniform and predictable composition, because product specifications are usually tight and must be met at minimum cost. Simply put, today's chemical reaction systems must operate as intended.

80.1 General Considerations in the Rational Design of Chemical Reactors

Reaction Kinetics Models and Reactor Models

The rational design of any type of reactor involves the marriage of one or more reaction kinetics models and a reactor model. It is important to recognize exactly what these two kinds of models describe:

- *Reaction kinetics models* describe the response of reaction rates to the reaction environment — that is, to temperature and the concentrations of virtually everything in the system — reactants, products, byproducts, catalysts, and contaminants. For design purposes, it is necessary and sufficient that the kinetic model reflect the reaction stoichiometry accurately and that it predict reaction rates accurately. It is not important that it reflect the actual reaction mechanism.
- *Reactor models* describe how the reaction environment is shaped by the geometry of the reactor, by physical processes such as fluid dynamics and heat and mass transport, and by process variables and conditions such as mean reactor residence time and residence time distribution, flow rate, pressure, and temperature.

These distinctions are subtle but important. It is sufficient to remember that kinetics models contain *only* temperature and concentration terms, whereas reactor models may contain these as well as everything else that influences the conduct of the reaction.

Kinetics of a Simple Hypothetical System of Reactions

Real systems will have their own individual structures and characteristics and will reflect the particular stoichiometry of the reaction system at hand. In this chapter, we will use a general, somewhat simplified set of reactions and reactor equations for illustration. Consider the general group of n_r reactions presented below, in which chemical species A and B react to produce several products P_k according to the following scheme:

$$\alpha_{A1}A+\alpha_{B1}B \xrightarrow{r_1} \sum_{1}^{n_c}(\alpha_{k1}P_k)$$

$$\alpha_{A2}A+\alpha_{B2}B \xrightarrow{r_2} \sum_{1}^{n_c}(\alpha_{k2}P_k)$$

$$\vdots$$

$$\alpha_{Aj}A+\alpha_{Bj}B \xrightarrow{r_j} \sum_{1}^{n_c}(\alpha_{kj}P_k)$$

$$\vdots$$

$$\alpha_{An_r}A+\alpha_{Bn_r}B \xrightarrow{r_{n_r}} \sum_{1}^{n_c}(\alpha_{kn_r}P_k)$$

Kinetic models for the rational design of reaction systems will usually be of one of two basic mathematical forms, each given in moles/volume/time and describing the response of the reaction rates r_j to temperature and concentration. For exponential models,

$$r_j = K_{oj}e^{-\frac{E_{a_j}}{RT}}C_A^{a_j}C_B^{b_j}C_{P_1}^{P_{1_j}}\ldots C_{P_n}^{p_{n_j}} \tag{80.1}$$

For hyperbolic models,

$$r_j = \frac{K_{o_j}e^{-\frac{E_{a_j}}{RT}}C_AC_B}{1+K_{A_j}C_A+K_{B_j}C_B+\sum_{1}^{n_c}(K_{P_{kj}}C_{P_k})} \tag{80.2}$$

Each of these two types of kinetic models has its own preferred uses. Their development will be discussed in a later section of this chapter. For now, assume that either of these types is to be used in a combined kinetics and reactor model to predict reaction rates, as described in the following section.

Combined Kinetics and Reactor Models: The General Continuity Equation

Virtually all combined reaction kinetics and reactor models utilize some form of the general continuity equation for flow and reaction in an element of volume of some kind of reactor, such as that shown in Figure 80.1. In its simplest form, this equation states that the quantity of each component k that enters the volume element of the reactor must either leave, react, or accumulate:

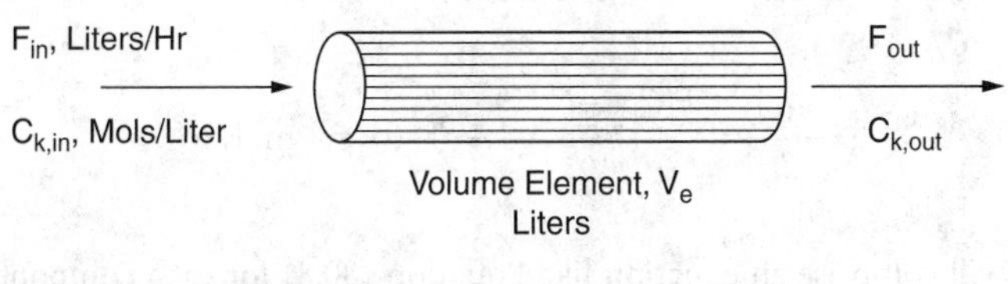

FIGURE 80.1 Flow and concentration in a volume element of a chemical reactor.

$$F_{e_{in}} C_{k_{in}} = F_{e_{out}} C_{k_{out}} - V_e \sum_1^{n_r} (\alpha_{kj} r_j) + \frac{d(V_e C_{k_e})}{dt} \text{ mol/time} \tag{80.3}$$

Different types of idealized reactors can be represented by this equation, simply by noting what terms are not appropriate and eliminating them. Thus, a simple *batch reactor* has no flow in or out, and therefore terms containing F drop out. So, for a simple batch reactor,

$$\frac{d(V_e C_{k_e})}{dt} = \sum_1^{n_r} (\alpha_{kj} r_j) V_e \tag{80.4}$$

which may be further simplified by canceling the V_e terms as well if the reaction volume is constant. Note that the volume element in this case is simply the entire filled volume of the reactor, because the concentration is assumed to be uniform throughout.

A steady-state *continuous stirred tank reactor (CSTR)* may have the same flow rate in and out, and the entire contents of the reactor comprise the volume element, as in the batch reactor. The time-derivative term is absent because the reactor is at steady state. The concentrations of all species in the reactor are the same as in the outlet. Thus we have, for the CSTR,

$$FC_{k_{in}} = FC_{k_{out}} - V \sum_1^{n_r} (\alpha_{kj} r_j) \tag{80.5}$$

which can be rearranged to

$$\sum_1^{n_r} \alpha_{kj} r_j = \frac{F(C_{k_{out}} - C_{k_{in}})}{V} \text{ mol/volume/time} \tag{80.6}$$

Weight of catalyst, W_c, replaces reactor volume, V, for a catalytic reaction:

$$\sum_1^{n_r} \alpha_{kj} r_j = \frac{F(C_{k_{out}} - C_{k_{in}})}{W_c} \text{ mol/wt. catalyst/time} \tag{80.7}$$

This simple relationship makes the CSTR the preferred type of reactor for many kinds of kinetics studies, in which the net rates of formation or disappearance [that is, $\sum_1^{n_r} (\alpha_{kj} r_j)$] for each individual component can be observed directly.

For design purposes, it is more convenient to integrate the unsteady-state form of the general continuity equation for the CSTR until the steady-state concentrations are attained. (The model will behave very much as the real reactor would in this respect.) This procedure avoids the need to use constrained nonlinear estimation to predict the reactor outlet concentrations. Again, assuming that flows in and out of the reactor are the same and that volume is constant,

$$\frac{d(C_k)}{dt} = \frac{F(C_{k_{in}} - C_{k_{out}})}{V} + \sum_1^{n_r} (\alpha_{kj} r_j) \text{ mol/h} \tag{80.8}$$

The reactor model will comprise an equation like Equation (80.8) for each component k in the CSTR. Another advantage of the unsteady-state model is that stoichiometry is automatically preserved without the need for any constraints. It may be used readily to simulate a system comprising a large number of components and reactions in a multistage system of CSTRs, which is otherwise mathematically intractable. And, of course, it may be used to study the dynamic behavior of the system as well. Therefore, it is the preferred design approach for multistage CSTR systems.

In the *ideal steady-state plug flow reactor*, all elements of fluid that enter the reactor together travel down its length and exit together, having thus stayed in the reactor for identical lengths of time. The volume element will be only a differential slice of the reactor cross-section, denoted by *dV.* There will be no accumulation term, since the reactor is at steady state. The differential concentration difference across the differential volume element will be denoted by *dC.* Thus, Equation (80.3) is once again applicable and simplifies to

$$F_e dC_k = \sum_1^{n_r} (\alpha_{kj} r_j) dV_e \tag{80.9}$$

If F changes as the reaction proceeds (as with many gas-phase reactions), then this can be accommodated down the length of the plug flow reactor by modifying the above equation to

$$d(F_e C_k) = \sum_1^{n_r} (\alpha_{kj} r_j) dV_e$$

whence

$$\begin{aligned} F_e \frac{dC_k}{dV_e} + C_k \frac{dF_e}{dV_e} &= \sum_1^{n_r} (\alpha_{kj} r_j) dV_e \\ \frac{dC_k}{dV_e} &= \frac{\sum_1^{n_r} (\alpha_{kj} r_j)}{F_e} - \frac{C_k}{F_e} \frac{dF_e}{dV_e} \end{aligned} \tag{80.10}$$

Equation (80.10) describes the rate of change of concentration C of species k with volume down a plug flow reactor as a function of the reaction rates, concentration, and flow. It also reflects the change in molar flow as the reaction proceeds.

A complete *isothermal* plug flow reactor model can readily be constructed using as many Equations (80.10) as there are components k, and as many kinetic models as there are reactions j. For *nonisothermal* reactors, differential equations that describe the temperature changes down the length of the reactor can be constructed in an analogous fashion, using the molar heat generation for each reaction j and its corresponding reaction rate r_j and heat transfer terms appropriate to the reactor type and geometry. For a multitube plug flow reactor with coolant on the outside of the tubes, the equation for reaction temperature (in degrees/volume) is as follows:

$$\frac{dT_r}{dV_e} = \frac{\sum_1^{n_r} (r_j \Delta H_j) - \frac{4U}{D_t}(T_r - T_c)}{F_e \rho_p c_p} \tag{80.11}$$

For coolant temperature,

$$\frac{dT_c}{dV_e} = \frac{\frac{4U}{D_t}(T_r - T_c)(\text{Mode})}{F_c \rho_c c_c} \tag{80.12}$$

Combined model equations like these are simplified in the sense that they do not account for departures from ideality because of nonideal mixing patterns in the case of stirred reactors, radial and axial diffusion effects in the case of tubular catalytic reactors, or the very specialized phenomena in fluidized beds and fixed-bed multiphase reactors. Yet they are surprisingly applicable in many industrial applications — and will certainly be preferred to geometric scaleup in almost all cases.

80.2 Protocol for the Rational Design of Chemical Reactors

Given the distinctions between kinetics models and reactor models, as well as the characteristics of combined reaction and reactor models, we can now establish a general protocol for rational design that will apply to many types of chemical reactors in industrial situations. The protocol comprises several steps, each of which is discussed in the following sections.

Step 1: Select the Type of Reactor for the Commercial Process

First, select one or more potentially useful types of reactors for the commercial process. In many instances, the preferred reactor type will be known from past experience with the same or similar reactions. Even so, the scaleup characteristics and requirements for two important types of reaction — batch reactions and solid-catalyzed reactions — merit special attention here. The reader is referred to the large open literature for additional information. Texts by Froment and Bischoff [1979] and Levenspiel [1972] are classic and are particularly recommended.

Scaleup of Laboratory Batch Reactions

Plant-Size Batch Reactors — Many reactions are conveniently studied in laboratory batch reactors, but batch reactors often are not preferred for full-scale operations, particularly if the reaction is rapid or if the planned plant will be quite large. Plant-size batch reactors are costly to operate, simply because they must be shut down, emptied, and recharged after a fairly short time — typically after only a few hours. This means that each reactor produces chemicals only on a part-time basis. As a consequence, batch reactors are usually preferred only for fairly high-priced specialty chemicals such as pharmaceuticals that are produced at fairly low volumes — say, under 50 million pounds per year — so that the required number and size of batch reactors are reasonable.

Ideal Plug Flow Reactors (PFRs) — PFRs have the same residence-time distribution as ideal batch reactors — that is, each element of feed is exposed to reaction conditions for exactly the same length of time. They are particularly useful for high-volume, low-priced commodity chemicals, for which the laboratory batch reactor is preferred. If the reaction time is short — say, under an hour — it may be practicable to use some type of large-scale plug flow reactor (PFR) for the plant. A baffled column is a common example. More often than not, however, batch times are at least several hours, and a plug flow continuous reactor would be too large and too costly for full-scale plant use. However, PFRs are routinely used in industry for solid-catalyzed reactions, as discussed later. The difference here is that the catalyst dramatically accelerates the reaction so that large-scale plug flow reactors are quite practicable.

Single and Multistage CSTRs — The single-stage continuous stirred tank reactor (CSTR) is relatively inexpensive and provides good temperature control, but it has a broad residence-time distribution. This means that some of the feed may be under reaction for a very short time and some may be in the reactor for an extended time. Also, concentrations of reactants and products throughout a CSTR will be the same as their exit concentrations, so that the reactions are conducted at minimum reactant concentration and maximum product concentration. This will mean a relatively slow reaction rate with maximum exposure of products to further reaction, which may lead to relatively low production rates and relatively high by-product formation rates.

Residence-Time Distribution and the Effects of Staging — To overcome some of the disadvantages of both the PFR and CSTR reactors for reactions that are ideally carried out by batch in the laboratory, several CSTRs are frequently connected in series. Their residence-time distributions will be intermediate

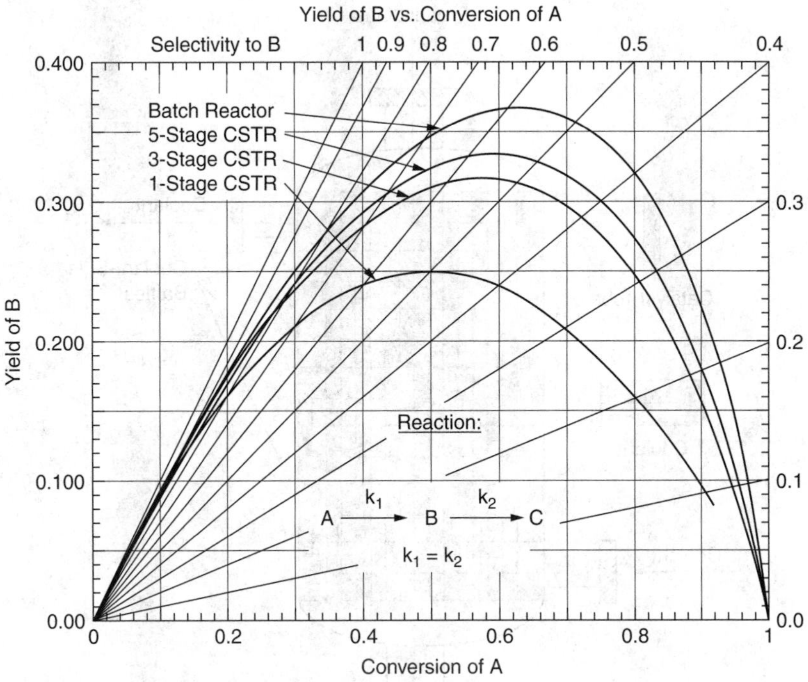

FIGURE 80.2 Effects of residence-time distribution.

between the very narrow distributions of the batch reactor or PFR and the very broad distributions of the CSTR. Such a multistage reactor may be a good compromise, but it will not operate identically to the laboratory batch reactor.

The effect of staging on a chemical reaction may be inferred from Figure 80.2, in which a simple sequential reaction of *A* going to *B* (desired) and then to *C* (undesired) is assumed. The figure shows that the batch reactor has the highest conversion of *A* and the highest yield of *B*, and that the CSTR has the lowest. The three- and five-stage CSTRs are intermediate between the batch and CSTR reactors. In practice, multistage systems of two to five CSTRs are common. Froment and Bischoff [1979], Levenspiel [1972], and others have written extensively on residence-time distribution and its impact on product yields and selectivities. These concepts are important to the successful commercialization of batchwise laboratory reaction technology.

Scaleup of Solid-Catalyzed Reactions

Solid catalysts are widely used in the chemical industry in several types of packed-bed reactors, slurry reactors, and fluidized bed reactors. Commercial processes frequently utilize solid-catalyzed gas-phase reactions, but it is not uncommon for them to use both gas and liquid streams, which greatly complicates the hydrodynamics. Reactors for solid-catalyzed reactions are all subject to scaleup problems because of differences in mass and heat transport between large-scale equipment and laboratory equipment.

Packed-bed reactors are probably used more than any other type for solid-catalyzed reactions. They tend to have residence-time distributions that closely approach plug flow, especially for plant-scale single-phase systems in which the bed length is at least 5 to 10 meters. Their narrow residence-time distribution makes them preferred from the standpoint of being able to control product distribution in systems of sequential reactions. Equation (80.10) through Equation (80.12) are appropriate here.

Shell-and-tube packed-bed reactors have excellent heat removal characteristics, particularly if the tube diameters are fairly small — say, 1 to 1.5 in. A schematic of a typical plant-scale shell-and-tube reactor is shown in Figure 80.3. Although relatively expensive, such reactors can usually be designed rationally with confidence.

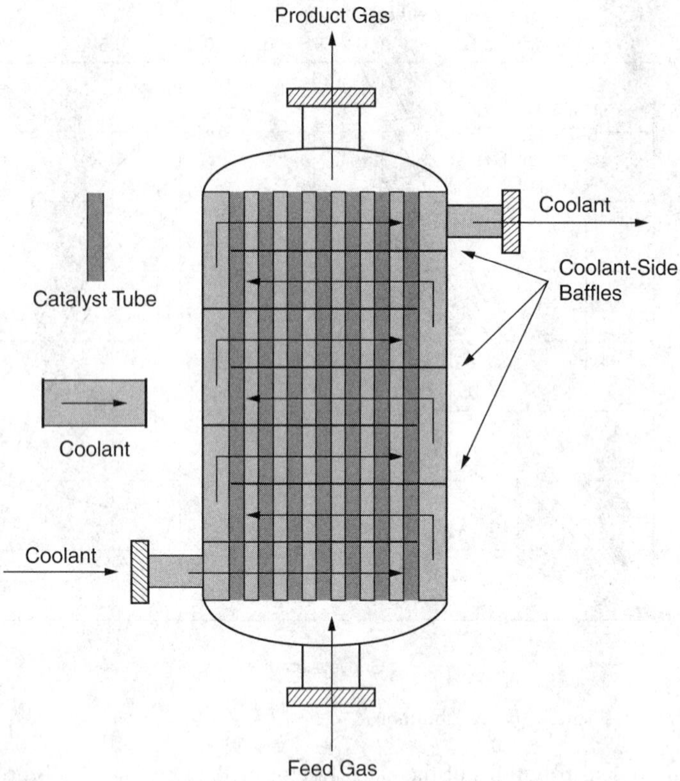

FIGURE 80.3 Conceptual shell-and-tube fixed-bed reactor. (*Source:* Reproduced with permission of the American Institute of Chemical Engineers from Cropley, J. B. 1990. *Chemical Engineering Progress* 86(2):32–39. Copyright 1990 AIChE. All rights reserved.)

Adiabatic packed-bed reactors are attractive because of their relatively low cost. (At their simplest, they can be little more than an empty tank filled with catalyst pellets.) However, they suffer from the effects of an uncontrolled reaction temperature and tend to drift towards the maximum possible attainable temperature, which usually means that the limiting reactant is completely consumed or that equilibrium has been reached. This is a desirable property for some reactions (e.g., hydrogenations), but usually the chemical selectivity suffers as a result. Unwanted byproducts may result, even in hydrogenation systems. The adiabatic reactor usually is not recommended where temperature control is important.

Multiphase packed-bed reactors (e.g., trickle beds) are widely used, but the transport of reactants and products between the flowing gas and liquid and the solid catalyst is a major uncertainty. A lot of work remains to be done before the rational design of these reactors can be undertaken with confidence. Instead, back-and-forth experimentation and mathematical analysis of the hydrodynamics is necessary. Suffice it to note here that the transport characteristics of multiphase reactors are different in each of the several hydrodynamic regimes that may be experienced (e.g., trickle flow, bubbling flow, pulsing flow, slug flow, mist flow). A priori calculations using relationships from the open literature can often predict whether reaction kinetics or mass transfer will limit the reactor's performance, and the preliminary design sometimes can be developed rationally thereafter. Pilot-scale experimental verification of the performance of the final design in the *same hydrodynamic region* as the plant reactor is nonetheless essential to avoid surprises. There is an abundant literature on this type of reactor, and the excellent text by Ramachandran and Chaudhari [1983] is a good place to obtain an introduction to this complex technology.

Fluidized-bed reactors are widely used in fluidized catalytic cracking, the manufacture of polyethylene and polypropylene, the manufacture of silicones from silicon, and some other commercial reactions. The fluid and solid dynamics of these systems are extremely complex, and available models are best described as learning models, rather than predictive models for reaction system design. Scaleup is typically done incrementally in a series of pilot-plant reactors. Despite the well-known advantages of these systems for some purposes (excellent temperature control, absence of diffusional and transport restrictions), the decision to use fluidized systems for new applications must not be taken lightly. Although the solid phase may usually be assumed to be well mixed, the residence-time distribution of the fluid phase is complex and largely unpredictable. It follows that the performance of a scaled-up reactor is also largely unpredictable if the fluid-phase residence-time distribution is important.

It is worth noting that, with few exceptions, all successful commercial fluidized bed reaction systems involve a solid phase that is always in some kind of rapid transition. Fluidized catalytic-cracking catalyst becomes coked and inactive in only 3 to 5 sec and is continuously removed from the fluidized riser reactor and regenerated by burning in dilute oxygen. The solid phase is in fact the product polyolefin in fluidized systems such as Union Carbide's Unipol™ polyolefins process. Dimethyldichloro silane (an intermediate in the manufacture of silicone oils and other products) is manufactured by reacting silicon metal and methyl chloride in a fluidized-bed reactor. Here, the silicon reactant is in the form of a fine metal powder that is continually consumed by the reaction and that comprises the solid phase in the reactor. An exception to the above observation is the Badger process for the manufacture of phthalic anhydride in a fluidized-bed reactor, in which the solid is a catalyst that is not undergoing any rapid change. It is carried out in a fluidized bed because the temperature control is important to the process.

The technology of fluidization has been studied in depth since before World War II and continues to be studied still because of its importance in those industries that depend upon it. AIChE has published several volumes on fluidization in its symposium series, and more appear periodically. Some excellent texts exist on fluidization phenomena; that by Kunii and Levenspiel [1969] is considered a classic.

Slurry reactors are attractive for many gas-liquid systems that are catalyzed by solid catalysts, but their design is nearly as complicated as for fluidized-bed systems. There is an extensive open literature on this type of reactor. The Air Products Company has published a number of reports on work done under contract for the U.S. Department of Energy and has a large pilot plant at LaPorte, Texas, for the manufacture of methanol and higher alcohols from CO and hydrogen. These reports and the text on multiphase reactors by Ramachandran and Chaudhari [1983] are recommended.

Step 2: Design the Laboratory to Generate Reaction Kinetics Data

The second step in the protocol for rational design is to design the laboratory reaction system specifically to generate kinetics rate data. By now it should be clear that the kinetic model is the link between the laboratory operations and the large-scale plant design. The experimental reactor for kinetics studies will ordinarily not look at all like the final plant reactor. Rather, it will be designed to obtain the kinetic data necessary for kinetic model development. Three primary types of laboratory reactor are suitable for design-quality kinetics studies:

- The batch reactor
- The long-tube reactor with sample points along its length
- The continuous stirred tank reactor, or CSTR

The following paragraphs discuss each of these types, with the emphasis upon what kinds of reactions are suitable for each and any special considerations that may apply.

Laboratory batch reactors are found in virtually limitless variety in most industrial chemical laboratories. For kinetics studies it is imperative that chemical analyses be obtained at several times during the reaction period. This sometimes poses problems if the reaction starts before the desired reaction temperature has been reached or if the reaction proceeds so quickly that multiple samples are not practical. Likewise, the loss of reacting volume due to sampling may be important, particularly if a solid catalyst is involved. Both of these problems are best handled during mathematical analysis of the data by forcing the simulated time–temperature and time-reacting volume profiles to be the same as the observed ones.

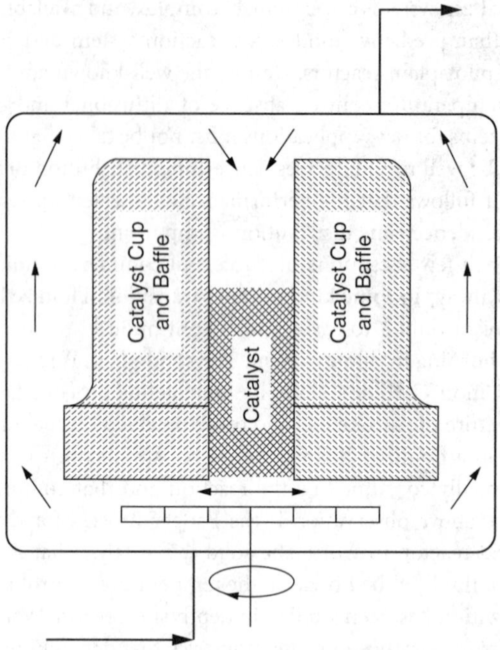

FIGURE 80.4 The Berty reactor for experimental catalyst testing and kinetics.(*Source:* Reproduced with permission of the American Institute of Chemical Engineers from Cropley, J. B. 1990. *Chemical Engineering Progress* 86(2):32–39.)

Batch laboratory reactors are suitable for either uncatalyzed or catalyzed liquid-phase or gas–liquid reactions. In special situations they may be useful for gas-phase reactions as well. They are especially useful for slurry-catalyzed reactions, but they can also be used with pelleted catalysts, provided that provision is made to retain the pellets in a basket or container through which the fluid passes. Both the Berty reactor (see Figure 80.4 and [Berty, 1974]) and the Carberry rotating basket reactor [Levenspiel, 1972] have been widely used for batchwise kinetic studies of catalytic reactions, although their best use is probably in continuous gas-phase kinetics studies.

Long-tube reactors for kinetics studies will typically be of a length to promote both plug flow characteristics and good mass and heat transfer so that these physical processes do not mask the chemical reaction rates. Tube length per se is not critical, although it is highly desirable that the length-to-diameter ratio be at least 100:1 to avoid the effects of axial mixing and departure from plug flow behavior. Tube diameter is not particularly important but should be small enough to ensure good heat transfer. If solid-catalyst pellets are to be used, the tube diameter should be no more than four or five pellet diameters to avoid radial temperature gradients. It is important that provision be made for samples to be taken at multiple points down the tube length. Reactors such as these are particularly well suited for kinetics studies using plant-scale catalyst pellets and for the study of the kinetics of sequential byproduct formation, in which the desired product reacts to form unwanted byproducts.

Perhaps the most difficult requirement from the standpoint of experimental reactor design is to provide for multiple sampling points down the length of the tube. It is relatively easy to make such a reactor using lengths of stainless steel tubing connected by tubing tees, which can then be used simultaneously as sample taps and for the insertion of thermocouples. Tubing diameter will typically be between 0.25 and 0.50 in., and the length will be 5 to 6 feet. Sample taps at 12-in. intervals will provide good composition and temperature profiles. Excellent temperature control can be attained in such a reactor if it is immersed in a thermostated heat transfer fluid or fluidized sand-bath heater.

Continuous stirred tank reactors (CSTRs) are ideally suited for kinetics studies for many types of reactions, owing largely to the ease with which reaction rates can be measured directly. Kinetics models are readily developed from CSTR data, as discussed by Cropley [1978]. The Berty and Carberry reactors cited previously are especially useful for continuous studies of catalytic kinetics, using real catalyst pellets. An abundant literature on their use exists; Cropley [1990] describes an overall strategy for their use in catalytic reactor design.

Step 3: Use Statistically Valid Experimental Programs and Models

The third element of the protocol is to utilize statistically valid experimental programs and data analysis for kinetic model development. Usually, kinetic data should be generated from statistically designed experimental programs, such as the factorial or central composite design, for which an abundant literature is available. The writings of Hendrix [1979] are recommended. Cropley [1978] describes the heuristic development of both exponential and hyperbolic kinetic models from a statistically designed data set. Table I from that article is an example of a central composite statistical design for the study of the kinetics of a fictitious catalytic reaction (the oxidation of Dammitol to Valualdehyde). The experimental reactor was assumed to be a CSTR like the Berty reactor, which permitted kinetic rates to be observed directly, as in Equation (80.7). The synthetic data in that table were developed from the following "true" model, with the incorporation of 20% normally distributed random error [reproduced with permission of the American Chemical Society from (Cropley, 1978)].

$$r_{\text{Val}} = \frac{4.67(10^{11})e^{-20000/RT}(P_{O_2})^{0.5}(P_{\text{Dam}})^{1.0}}{1+5.52(10^{-4})e^{5000/RT}(P_{O_2})^{0.5}+7.64(10^{-4})e^{5000/RT}(P_{\text{Val}})^2}\ \text{Gmols/Kgcat/H} \tag{80.13}$$

If a CSTR is used as the experimental reactor, the experimental design should include the independent control of any product species that might influence the reaction rates to avoid systematic bias in the kinetic parameters, as discussed by Cropley [1987].

In the heuristic study cited above, log-linear multiple regression was used for the development of exponential models, and the Nelder–Mead nonlinear search algorithm [Nelder and Mead, 1965] was used for estimation of the nonlinear parameters in the hyperbolic models. (Although Nelder–Mead tends to converge more slowly than some other algorithms, it is robust, stable, and reliable for nonlinear estimation and optimization studies.)

Step 4: Develop Computer Programs for Reactor Simulation and Design

The fourth element of the protocol is to develop computer programs for reactor simulation and design. The typical reaction system design model will comprise several dozen differential equations and a number of nonlinear constraints as well. Consider very carefully just how complex the final models should be. The determination of the true economic optimum may require that the reactor be simulated iteratively many thousands of times, and the model at this point should be no more complex than necessary in order to minimize computer time. Therefore, simulation models for reactor design and optimization should in most cases be based primarily upon the kind of simple idealized reactors represented by Equation (80.3) through Equation (80.12), as discussed earlier. Even so, simulation models can become unwieldy. For example, a five-stage CSTR model for a system of four chemical components and temperature will comprise 25 nonlinear differential equations similar to Equation (80.8). Any of a number of numerical integration techniques may be used for the solution of the differential equations; the fourth-order Runge–Kutta algorithm is perhaps the most common.

Step 5: Develop the Economically Optimum Reactor Design

The fifth protocol for rational design is to develop economically optimum reactor designs. The potentially useful reactor types should be compared to one another at their individual economic optimum design and operating conditions. Arrive at these for each type by varying the simulated reactor size and geometry

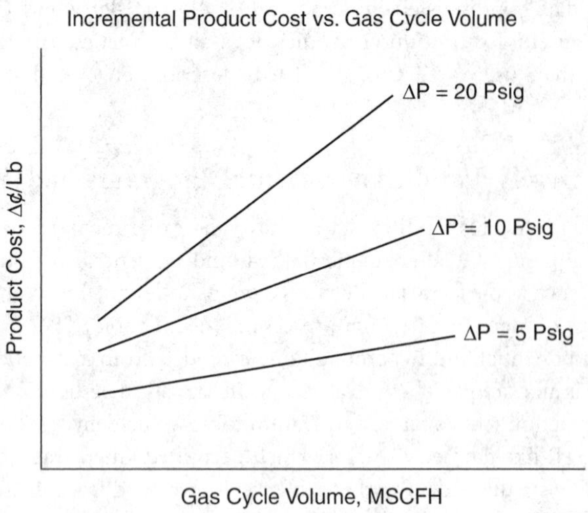

FIGURE 80.5 Elements of product cost. (*Source:* Reproduced with permission of the American Institute of Chemical Engineers from Cropley, J. B. 1990. *Chemical Engineering Progress* 86(2):32–39. Copyright 1990 AIChE. All rights reserved.)

(e.g., diameter, length and number of tubes, or agitator horsepower and vessel diameter and height) and operating conditions (like pressure, flow rate, reactant concentration, coolant temperature, and so on). A suitable optimization procedure, such as the Nelder–Mead algorithm cited previously [Nelder and Mead, 1965], should be used.

The Objective Function for Optimization

The design process will include the computation of objective economic or reactor performance criteria for the comparison of optimized design alternatives. This objective function must be reasonably accurate, yet simple enough to be evaluated easily and quickly for each iteration. It is convenient and quite accurate to use simple linear relationships for optimization criteria like that in Figure 80.5, which illustrates the dependence of incremental product cost on cycle gas flow rate and catalyst bed pressure drop for a hypothetical process. Several similar relationships can comprise an easily used objective function for optimization. Relationships like this can be developed easily from detailed economic analyses of a small number of base case designs. Their development was discussed by Cropley [1990].

Explicit and Implicit Constraints

The optimization package will normally utilize both explicit and implicit constraints to relate the reactor to the rest of the process. In a typical design, for example, reactor inlet pressure and reactor gas flow rate might be *explicitly* constrained to reasonable ranges. Likewise, the maximum reactor temperature might be *implicitly* constrained not to exceed a stipulated maximum. Constraints like these are at the very heart of the optimization process. At the same time, their formulation is both esoteric and beyond the scope of this article, and they are not extensively treated in the open literature. It is recommended that an optimization or numerical analysis specialist be consulted in their development.

Step 6: Validate the Design in a Pilot Plant Reactor

The final step in the protocol for rational design is to validate the overall optimum design by operation of a pilot reactor system. This is the *only time* that scaleup, per se, will be considered in the design. Here again, this may not necessarily involve a pilot reactor that is geometrically similar to the final plant design, although it may be in some cases. It will be important to design the pilot reactor to be able to confirm the predicted reaction yields and selectivities under design conditions. It is vital that all important recycle streams be incorporated into the pilot plant system — which will then be, to the extent possible, a scaled-down version of the integrated commercial process. Typically, recycle streams include previously unreacted raw materials, byproducts that can be reverted to useful product, or potential pollutants that can simply be destroyed by further reaction in the reactor. But trace components and minor species can build up in recycle systems to many times their single-pass concentrations, and they may have adverse effects on catalyst life, equilibrium conversion, crude product quality, and so on. It is important that the effects of recycled species be incorporated into the kinetics model for operational monitoring and control. Unanticipated effects of recycle streams probably account for a significant fraction of scaleup problems in commercial systems.

Finally, please note that this chapter has not discussed the *size* of the pilot plant reactor — size per se really is not an issue. What *is* important is that it function in such a way as to test the rational design before it is built. More discussion of process optimization and validation can be found in Cropley [1990].

Nomenclature

Uppercase Symbols

A,B	Reactants
C_k	Concentration of *k*th chemical component, in mols/volume units
D_t	Tube diameter, in linear units
Dam	Fictitious component Dammitol
F_e, F_c	Flow rate of process fluid (*e*) or of coolant (*c*) through the volume element, in volume/time units
ΔH_j	Heat of reaction of *j*th reaction in heat/mol units
K_{0_j}, E_{a_j}	Arrhenius kinetic parameters for reaction *j*
$K_{A_j}, K_{B_j}, K_{P_{k_j}}$	Kinetic parameters associated with reactants *A* and *B* and products P_k for reaction *j*
Mode	(−1) if coolant flows countercurrent to process stream flow (+1) if flows are cocurrent (0) if coolant is isothermal (infinite flow, boiling, etc.)
P_k	Reaction products
R	Gas constant, typically 1.987 cal/Gmol/K for kinetics models
T_r, T_c	Reaction and coolant temperature, respectively
U	Overall heat transfer coefficient, in heat/area/time/temperature units
Val	Fictitious component Valualdehyde
V_e	An element of volume of any reactor
W_c	Weight of catalyst in the reactor, in weight units

Lowercase and Greek Symbols

α_{kj}	Stoichiometric coefficient for species *k* in reaction *j*
c_c	Heat capacity of coolant stream, in heat/mass/temperature units
c_p	Heat capacity of flowing process stream, in heat/mass/temperature units
n_c	Total number of products *k* in the kinetic model
n_r	Number of reactions *j* in the kinetic model
n_t	Number of tubes in a tubular reactor
ρ_p, ρ_c	Density of process stream (*p*) or coolant (*c*), mass/volume units
r	Kinetic reaction rate, mols/volume/time (or mols/wt. catalyst/time for catalytic reactions)
t	Time, in units consistent with *F*, *r*, *U*

Defining Terms

Chemical reaction — The chemical transformation of one or more reactant species into one or more chemical products, usually with the evolution or absorption of heat.

Chemical reactor — The vessel in which a chemical reaction is conducted.
Continuous stirred tank reactor — A well-mixed continuous reactor, characterized by a broad residence time distribution.
Conversion — The fraction of a feed component that undergoes chemical transformation in the reactor.
Plug flow reactor — A type of reactor in which all entering elements of fluid have the same residence time. The residence time distribution is thus extremely narrow.
Residence time — The amount of time that a reactive mixture spends in a chemical reactor.
Residence-time distribution (RTD) — The spread of residence times that different elements entering a reactor spend in it. RTD is one of the distinguishing characteristics of different reactor types.
Selectivity — The fraction of all of a feed component that is converted to form a specified product.
Yield — The fraction of all of a feed component entering a reactor that is converted to a specified product. Yield = (Selectivity) (Conversion).

References

Berty, J. M. 1974. Reactor for vapor-phase catalytic studies. *Chem. Eng. Prog.* 70:578–584.

Cropley, J. B. 1978. Heuristic approach to complex kinetics. In *Chemical Reaction Engineering — Houston,* D. Luss and V. Weekman, eds. ACS Symposium Series 65. Paper 24.

Cropley, J. B. 1987. Systematic errors in recycle reactor kinetics studies. *Chemical Engineering Progress.* 83(2):46–50.

Cropley, J. B. 1990. Development of optimal fixed bed catalytic reaction systems. *Chemical Engineering Progress.* 86(2):32–39.

Froment, G. F. and Bischoff, K. B. 1979. *Chemical Reactor Analysis and Design.* John Wiley & Sons, New York.

Hendrix, C. D. 1979. What every technologist should know about experimental design. *Chemtech.* 9(3):167–174.

Kunii, D. and Levenspiel, O. 1969. *Fluidization Engineering.* John Wiley & Sons, New York.

Levenspiel, O. 1972. *Chemical Reaction Engineering,* 2nd ed. John Wiley & Sons, New York.

Nelder, J. A. and Mead, J. R. 1965. A simplex method for function minimization. *Computer Journal.* 7:308–313.

Ramachandran, P. A. and Chaudhari, R. V. 1983. *Three-Phase Catalytic Reactors.* Gordon and Breach Science, New York.

Further Information

Professional Organizations

The American Institute of Chemical Engineers (three national meetings per year). 345 E. 47th St., New York, NY 10017. Phone (212) 705-7322.

The American Chemical Society (several national/regional meetings each year). 1155 16th St., N. W., Washington, DC, 20036. Phone (202) 872-4600.

Special Meetings and Conferences

The Engineering Foundation Conferences on Chemical Reaction Engineering. 345 East 47th Street, New York, NY 10017. Phone (212)705-7835.

International Symposia on Chemical Reaction Engineering (even years). Sponsored by American Institute of Chemical Engineers, American Chemical Society, Canadian Society for Chemical Engineering, and the European Federation of Chemical Engineering.

XI

Geotechnical

81

Soil Mechanics

Braja M. Das
California State University, Sacramento

Soil mechanics is the branch of science that deals with the study of the physical properties of soil and the behavior of soil masses while being subjected to various types of forces. Soils engineering is the application of the principles of soil mechanics to practical problems.

81.1 Weight–Volume Relationship

The three phases of a soil sample are solid, water, and air, as shown in Figure 81.1. Thus, a given soil sample of weight W can be expressed as

$$W = W_s + W_w + W_a \tag{81.1}$$

where W_s is the weight of the soil solids, W_w is the weight of the water, and W_a is the weight of air. Assuming that $W_a \approx 0$,

$$W = W_s + W_w \tag{81.2}$$

As shown in Figure 81.1, the total volume of the soil sample is V. The volumes occupied by solid, water, and air are, respectively, V_s, V_w, and V_a. Thus, the volume of void, V_v, is

$$V_v = V_w + V_a \tag{81.3}$$

Common weight and volume relationships are given in Table 81.1.

81.2 Hydraulic Conductivity

In 1856, Darcy published a relationship for the discharge velocity of water through saturated soil (usually referred to as Darcy's law), according to which

$$v = ki \tag{81.4}$$

where v is the discharge velocity, i is the **hydraulic gradient**, and k is the *coefficient of permeability or hydraulic conductivity.* The unit of hydraulic conductivity is LT^{-1}. See Table 81.2.

0-8493-1586-7/05/$0.00+$1.50

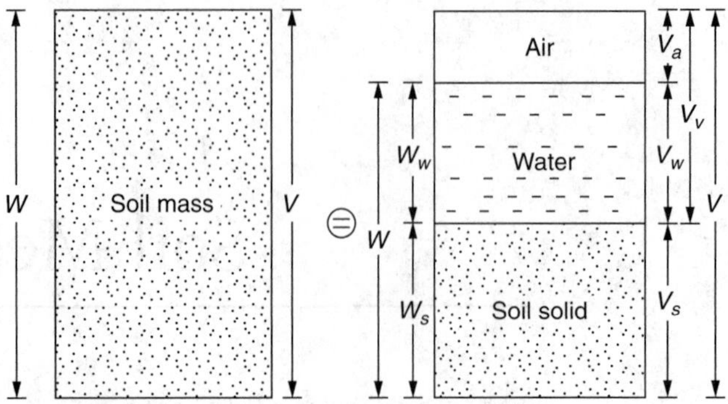

FIGURE 81.1 Weight–volume relationship.

TABLE 81.1 Weight–Volume Relationships

Volume Relationships

Void ratio, $e = \dfrac{V_v}{V_s} = \dfrac{n}{1-n}$

Porosity, $n = \dfrac{V_v}{V} = \dfrac{e}{1+e}$

Degree of saturation, $S = \dfrac{V_w}{V_v} = \dfrac{wG_s}{e}$

Weight Relationships

Moisture content, $w = \dfrac{W_w}{W_s}$

Moist unit weight, $\gamma = \dfrac{W}{V}$

$$\gamma = \frac{(1+w)G_s\gamma_w}{1+e}$$

$$\gamma = \frac{(G_s + Se)\gamma_w}{1+e}$$

$$\gamma = G_s\gamma_w(1-n)(1+w)$$

Dry unit weight, $\gamma_d = \dfrac{\gamma}{1+w}$

$$\gamma_d = \frac{G_s\gamma_w}{1+e}$$

$$\gamma_d = G_s\gamma_w(1-n)$$

$$\gamma_d = \frac{G_s\gamma_w}{1+wG_s/S}$$

Saturated unit weight, $\gamma_{sat} = \dfrac{(G_s+e)\gamma_w}{1+e}$

$$\gamma_{sat} = [(1-n)G_s + n]\gamma_w$$

$$\gamma_{sat} = \gamma_d + n\gamma_w$$

Note: G_s = Specific gravity of soil solids; γ_w = Unit weight of water (62.4 lb/ft^3; 9.81 kN/m^3)

TABLE 81.2 Typical Range of Hydraulic Conductivity

Soil Type	k (cm/sec)	Relative Hydraulic Conductivity
Clean gravel	10^2 to 10^0	High
Coarse sand	10^0 to 10^{-2}	High to medium
Fine sand	10^{-2} to10^{-3}	Medium
Silt	10^{-3} to 10^{-5}	Low
Clay	Less than 10^{-6}	Very low

One of the most well-known relationships for hydraulic conductivity is the Kozeny–Carman equation, which is of the form

$$k = \frac{1}{\eta} \frac{1}{s_p t^2 s_s^2} \frac{n^3}{(1-n)^2} \tag{81.5}$$

where η is viscosity, s_p is pore shape factor, t is tortuosity, s_s is specific surface per unit volume, and n is porosity. Table 81.2 gives the general magnitude of the hydraulic conductivity.

81.3 Effective Stress

In saturated soil the **total stress** in a given soil mass can be divided into two parts — a part that is carried by water present in continuous void spaces, which is called *pore water pressure,* and the remainder that is carried by the soil solids at their points of contact, which is called the *effective stress.* Or

$$\sigma = \sigma' + u \tag{81.6}$$

where σ is total stress, σ' is effective stress, and u is pore water pressure.

In partially saturated soil, water in the void space is not continuous, and the soil is a three-phase system — that is, solid, pore water, and pore air. In that case,

$$\sigma = \sigma' + u_a - \chi(u_a - u_w) \tag{81.7}$$

where u_a is pore air pressure and u_w is pore water pressure.

81.4 Consolidation

Consolidation is the *time-dependent* volume change of saturated clayey soil due to the expulsion of water occupying the void spaces. When a load is applied to a saturated compressible soil mass, the increase in stress is initially carried by the water in the void spaces (that is, increase in pore water pressure) due to its relative incompressibility. With time, the water is squeezed out, and the stress increase is gradually transferred to effective stress. The effective stress increase results in consolidation settlement of the clayey soil layer(s).

A clay is said to be *normally consolidated* when the present *effective* overburden pressure (p_o) is the maximum pressure to which the soil has been subjected in the past. When the present effective overburden pressure of a clay is less than that which it experienced in the past, it is referred to as an *overconsolidated* clay. The maximum past effective overburden pressure is called the *preconsolidation pressure* (p_c).

The *primary consolidation* settlement (S) of a saturated clay layer of thickness H (Figure 81.2) due to a load application can be calculated by the following relationships. In Figure 81.2, for the clay layer, p_o is the effective overburden pressure before the load application, Δp is the increase in stress at the middle

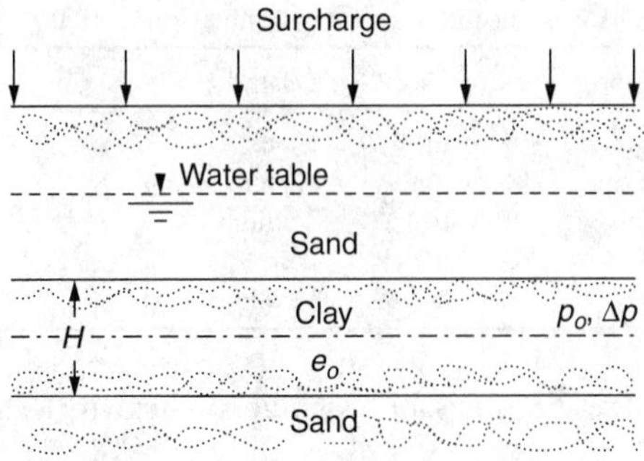

FIGURE 81.2 Consolidation settlement of a clay layer.

of the clay layer due to the load application, and e_o is the initial void ratio. For normally consolidated clay ($p_o = p_c$),

$$S = \frac{C_c H}{1+e_o} \log\left(\frac{p_o + \Delta p}{p_o}\right) \tag{81.8}$$

For overconsolidated clay with $p_o + \Delta p \leq p_c$,

$$S = \frac{C_s H}{1+e_o} \log\left(\frac{p_o + \Delta p}{p_o}\right) \tag{81.9}$$

For overconsolidated clay with $p_o < p_c < p_o + \Delta p$,

$$S = \frac{C_s H}{1+e_o} \log\left(\frac{p_c}{p_o}\right) + \frac{C_c H}{1+e_o} \log\left(\frac{p_o + \Delta p}{p_c}\right) \tag{81.10}$$

where C_c is the compression index and C_s is the swelling index. The magnitudes of C_c and C_s can be obtained from laboratory consolidation tests. In the absence of the laboratory results the following empirical approximations can be used. From Skempton [1944],

$$C_c = 0.009(\mathrm{LL} - 10) \tag{81.11}$$

where *LL* is liquid limit, in percent. From Nishida [1956],

$$C_c = 1.15(e_0 - 0.27) \tag{81.12}$$

And, finally, from Rendon-Herrero [1980],

$$C_c = 0.156 e_o + 0.0107 \tag{81.13}$$

$$C_s \approx 0.1 - 0.2 C_c \tag{81.14}$$

Time Rate of Consolidation

The average degree of consolidation (U) of a clay layer due to an applied load can be defined as

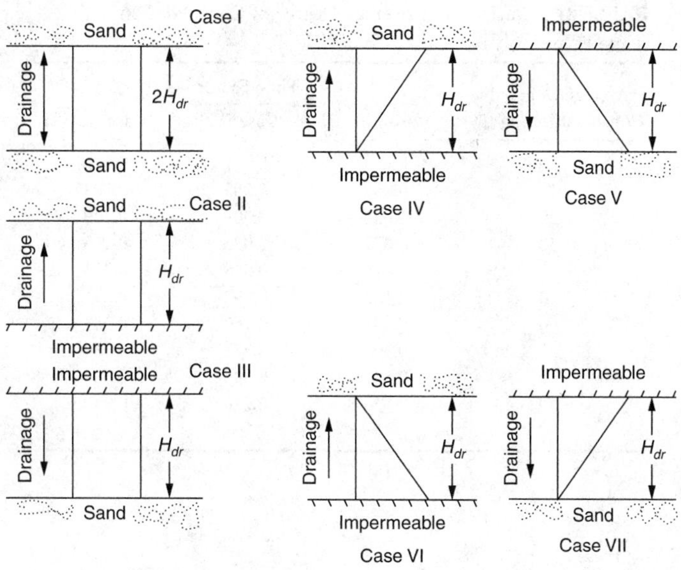

FIGURE 81.3 Definition of H_{dr}.

$$U = \frac{S_t}{S} \tag{81.15}$$

where S_t is primary consolidation settlement after time t of load application, and S is the ultimate consolidation settlement. The degree of consolidation is a function of time factor T_v, which is a nondimensional quantity. Or

$$U = f(T_v) \tag{81.16}$$

$$T_v = \frac{C_v t}{H_{\mathrm{dr}}^2} \tag{81.17}$$

where C_v is the **coefficient of consolidation**, t is the time after load application, and H_{dr} is the length of the smallest drainage path. Figure 81.3 shows the definitions of H_{dr} and the initial excess pore water pressure (u_o) distribution for seven possible cases. The variations of U with T_v for these cases are shown in Table 81.3.

81.5 Shear Strength

Shear strength of a soil mass is the internal resistance per unit area that the soil mass can offer to resist failure and sliding along any plane inside it. For most soil mechanics problems, it is sufficient to approximate the shear strength by the *Mohr–Coulombe failure criteria*, or

$$s = c + \sigma' \tan\phi \tag{81.18}$$

where c is the cohesion, σ' is the effective normal stress, and ϕ is the drained friction angle. For normally consolidated clays and sands, $c \approx 0$.

Figure 81.4 shows an approximate correlation for ϕ with porosity and **relative density** for coarse-grained soils. Table 81.4 gives typical values of ϕ for sand and silt.

TABLE 81.3 Variation of Average Degree of Consolidation with Time Factor

Average Degree of Consolidation, U (%)	Time Factor, T_v		
	Case I, II, III	Case IV, V	Case VI, VII
0	0	0	0
10	0.008	0.003	0.0047
20	0.031	0.009	0.100
30	0.071	0.025	0.158
40	0.126	0.048	0.221
50	0.197	0.092	0.294
60	0.287	0.160	0.383
70	0.403	0.271	0.500
80	0.567	0.440	0.665
90	0.848	0.720	0.940
100	∞	∞	∞

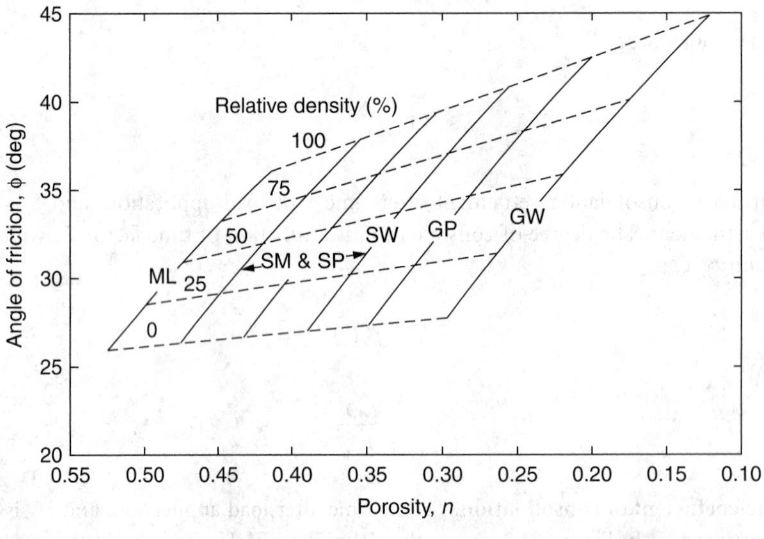

FIGURE 81.4 Approximate correlation for ϕ of coarse-grained soil with porosity and relative density. Note: GW = well-graded gravel, GP = poorly graded gravel, SW = well-graded sand, SP = poorly graded sand, SM = silt sand, ML = silt with low plasticity. (After Department of the Navy. 1971. *Soil Mechanics, Foundations, and Earth Structures — NAVFAC DM-7,* U. S. Government Printing Office, Washington, DC.)

TABLE 81.4 Typical Values of ϕ for Sand and Silt

Soil	ϕ (deg)
Sand	
Loose	28–35
Medium	30–40
Dense	35–45
Silt	25–35

Defining Terms

Coefficient of consolidation — The coefficient of consolidation is defined by the relationship

$$C_v = \frac{k}{\gamma_w \left(\frac{\Delta e}{\Delta p} \frac{1}{1+e_o} \right)}$$

where Δe is the change in void ratio due to a pressure increase in Δp.

Hydraulic gradient — The ratio of the loss of head to the length of flow over which the loss occurred.

Relative density — Relative density is defined as $(e_{max} - e)/(e_{max} - e_{min})$, where e_{max} and e_{min} are, respectively, the maximum and minimum possible void ratios for a soil, and e is the *in situ* void ratio.

Total stress — The total stress at a given elevation is the force per unit gross cross-sectional area due to soil solids, water, and surcharge.

References

Darcy, H. 1856. *Les Fontaines Publiques de la Ville de Dijon,* Dalmont, Paris.

Department of the Navy. 1971. *Soil Mechanics, Foundations, and Earth Structures — NAVFAC DM-7,* U. S. Government Printing Office, Washington, DC.

Nishida, Y. 1956. A brief note on compression index of soils. *J. Soil Mech. Found. Div., ASCE.* 82:1027-1–1027-14.

Rendon-Herrero, O. 1980. Universal compression index equation. *J. Geotech. Engr. Div., ASCE.* 106:1179–1200.

Skempton, A. W. 1944. Notes on the compressibility of clays. *J. Geol. Soc. Lond.* 100:119–135.

Further Information

Das, B. M. 1994. *Principles of Geotechnical Engineering,* 3rd ed. PWS, Boston, MA.

XII

Transportation

82

Transportation Planning

Michael D. Meyer
Georgia Institute of Technology

Transportation planning is undertaken for a variety of reasons. With the provision of much of the world's transportation infrastructure the responsibility of governments, transportation planning is undertaken primarily to support public officials in their choice of most cost-effective investments. Because transportation investment has a strong influence on how a community evolves, transportation planning must necessarily consider a variety of factors when assessing the cost effectiveness of alternative investment options. For example, transportation investment can strongly influence land use patterns, the attractiveness of different parts of a region for economic development, the equitable distribution of mobility benefits among different population groups, and the environmental consequences of both the construction and operation of transportation facilities. Transportation planning must therefore be forward-looking, as well as give attention to current problems in the transportation system.

82.1 Basic Framework of Transportation Planning

The basic framework for transportation planning that could be applied at any scale of application is shown in Figure 82.1. The steps shown in this framework are discussed in the following sections.

Define a Vision

The transportation system can impact society in a variety of ways — providing mobility and accessibility, promoting economic development, contributing to quality of life, as well as negatively affecting the natural environment. The first step in transportation planning thus usually consists of defining what it is that the nation, state or region desires in terms of its future characteristics.

Identify Goals and Objectives

Once a desired vision is articulated, *goals* can be identified that relate the vision to the ultimate achievement of a transportation plan. *Objectives* are more specific statements that indicate the means by which these goals will be achieved. Goals and objectives not only provide overall direction to the transportation planning process, but they also help define the criteria, known as *measures of effectiveness*, that are used later in the process for evaluating alternative courses of action.

0-8493-1586-7/05/$0.00+$1.50

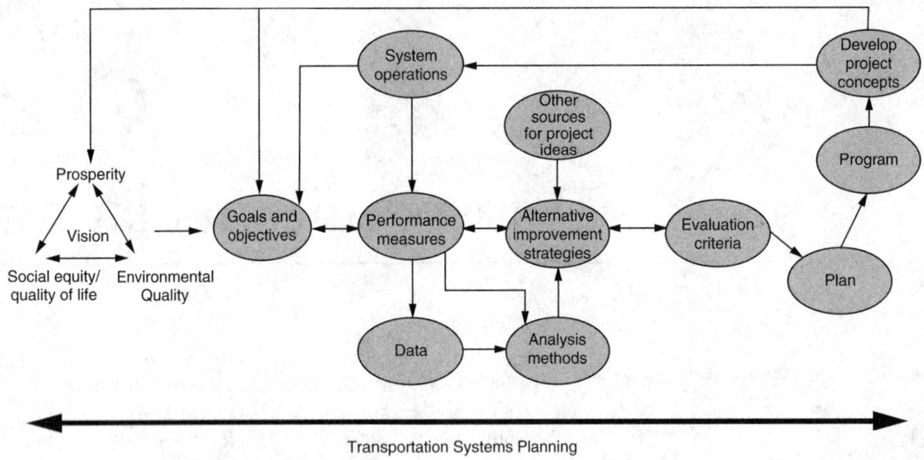

FIGURE 82.1 Transportation planning process.

Identify Performance Measures

An important aspect of a continuing transportation planning process is the monitoring of system performance. This monitoring systematically identifies areas where improvements might occur, and, in addition, helps transportation officials assess the effectiveness of previously implemented actions. Performance measures can focus explicitly on transportation system operations, e.g., the level of freeway delay during the morning peak travel hours, or on other issues of importance to transportation officials, e.g., the level of transportation-related air pollutants emitted during specified periods of time.

Collect Data

Given that transportation investment is usually aimed at upgrading the *physical condition* of a facility (e.g., repaving a road or building a new bridge) or at improving its *performance* (e.g., providing new person-carrying capacity by setting aside highway lanes for multi-occupant vehicles or by building a new road), engineers are continually collecting data on the many different components of the transportation system. The base condition or performance of all the different facilities or services that make up a transportation system is called an *inventory.*

Forecasting future demand for transportation requires engineers and planners to characterize the current and likely future states of the factors that influence this demand. Thus, for example, the type of data that is collected includes such things as current land use and socioeconomic characteristics of the traveling population. Current land use is readily attained through land use inventories. The methods of estimating future land use range from trends analysis to large-scale land use models that predict household and employment sites decades into the future. Important socioeconomic characteristics include level of household income, number of members of the household, number of autos in the household, number of children, age of the head of household, and highest level of education achieved. Each of these factors has been shown through research to influence the amount and type of travel associated with a typical household.

Use Analysis Tools to Identify System Deficiencies or Opportunities

The analysis tools and methods used to identify transportation deficiencies and improvement opportunities can vary widely. In some cases, computer-based transportation network models are used to estimate future traffic volumes and transit ridership, with the results then compared to existing system capacity to handle such volumes. This comparison relies on one of the more popular performance measures used in transportation planning today, the volume-to-capacity (V/C) ratio. However, given the many different goals and objectives that can characterize a transportation planning process, a wide variety of measures are often used for determining system deficiencies. Other types of analysis tools include time–distance diagrams, queuing models, fluid-flow approximation methods, macro- and micro-simulation models, and mathematical programming techniques.

Develop and Analyze Alternatives

Various types of strategies can result from the planning process:

1. Improving the *physical infrastructure* of the transportation system — for example, adding new highway lanes or extending an existing subway line
2. Improving *system operations* — for example, coordinating traffic signals, improving traffic flow through improved geometric design of intersections, or making transit operations more efficient through schedule coordination
3. Reducing *travel demand* so that the transportation system can handle peak loads more effectively — for example, flexible working hours, increasing average vehicle occupancy through such measures as carpools or transit use, or raising the "price" of travel through the use of tolls

In the past 10 years, the application of advanced transportation technologies to the operation of the transportation system, known as *intelligent transportation systems* (ITS), has become an important type of strategy in many cities. Thus, it is not uncommon for major cities to now have a centralized traffic management center, with a regional surveillance and traveler communication system that permits transportation system managers to communicate to travelers the best times for travel and which routes are least congested.

Evaluate Alternatives

Evaluation brings together all of the information gathered on individual alternatives/plans and provides a systematic framework to compare the relative worth of each. This evaluation process most often relies on the various measures of effectiveness that link to the goals and objectives defined at the beginning of the process. Different types of evaluation methods include use of benefit/cost ratios, cost-effectiveness indices, goals matrix analysis, and subjective assessment of the merits of individual alternatives.

Develop Transportation Plan

One of the most important products of the transportation planning process is the *transportation plan.* The plan outlines the many different strategies and projects that are necessary to meet the challenges and opportunities facing a state or region. In the U.S., federal law requires that every state and every metropolitan area over 50,000 population have a transportation plan. The state department of transportation (DOT) is responsible for preparing the state transportation plan; an agency called the metropolitan planning organization (MPO) is responsible for preparing the metropolitan transportation plan.

Implement Plan

Another major product of the transportation planning process is a strategy for implementing all of the actions identified in the plan. In the U.S., federal law requires each state and every metropolitan area over 50,000 population to produce a *transportation improvement program* that lists the projects that will be implemented over the next 3 to 5 years, identifies which agency is responsible for each project, and describes the source of project funding.

The implemented projects will affect the performance of the transportation system. Through a continuing monitoring process, linked directly to important performance measures, the performance of individual projects or of the entire transportation system can be fed back into the planning process as a means of identifying new problems.

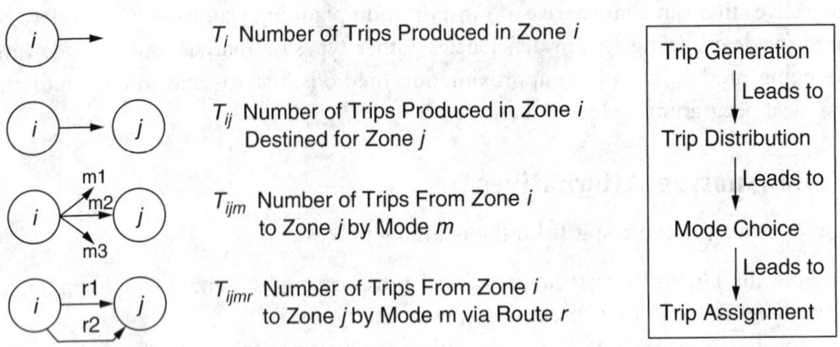

FIGURE 82.2 Transportation modeling framework.

82.2 Transportation Modeling

The level of transportation analysis can vary according to the level of complexity and scale of application of potential solution strategies. Thus, for example, the consideration of a new subway system would necessarily have to be examined from a metropolitan level, while the transportation impacts of a new development site would likely be analyzed at a subregional level. In most cases, however, the modeling process consists of four major steps — trip generation, trip distribution, mode split, and trip assignment. Even though recent models combine some of these steps together during the analysis process, the concept of the "trip" consisting of these four stages still holds. Each study area (whether a nation, state, metropolitan area, or community) is divided into *zones* of homogeneous characteristics (e.g., similar household incomes) that can then be used as the basic foundation for estimating trips from or attracted to that zone. Most planning studies define these zones to be similar to those used in other data collection activities (e.g., the U.S. census tracts) so that data useful to the transportation study collected by others can be readily linked to the transportation analysis. The transportation system is represented in models as a network of links and nodes. Links represent line-haul facilities, such as roads or transit lines, and nodes represent points of connection, such as an intersection or transit terminal. Given the complex nature of transportation systems, the typical transportation network consists of links representing only highly used facilities or other facilities that are critical to the overall performance of the transportation system.

The steps in a typical modeling exercise are shown in Figure 82.2. Basic to this approach is the concept of *derived demand*. Derived demand means that a trip is taken to accomplish some activity at a destination, and that the trip itself is simply a means of reaching this activity. There is no intrinsic value of the trip itself. Thus, modeling trip-making requires linking travel behavior to the characteristics of the trip-maker and to the activities at the origin and destination ends of the trip that will influence the way the trips are made.

Trip generation is the process of analytically deriving the number of trips that will be generated from a location or zone based on socioeconomic characteristics of the household, or in the case of freight movement, the zonal economic characteristics. Trip generation also includes predicting the number of trips that will be attracted to each zone in the study area.

$$\text{Number of trips produced in a zone} = f\,(\text{Population socio-economic characteristics, land use, transportation mode availability})$$

$$\text{Number of trips attracted to a zone} = f\,(\text{Attractiveness of the zone})$$

Two approaches are often used to estimate the number of trips generated. The first uses trip rate models that are based on trip-making behavior as compared to important variables. For example, see Table 82.1. The other approach is to use regression models that are estimated either from survey data

TABLE 82.1 Cross-Classification Analysis, Trips per Day, by Household Size and Income

	Number of People in Households		
	1	2	3+
Low income	2.4	3.3	4.5
Medium income	3.5	3.8	4.8
High income	3.9	4.2	5.4

collected throughout the study area or from some other data source, such as the U.S. Census. The following regression equations illustrate this approach.

Zone Trip Productions: $T_i = 184.2 + 120.6\ (\text{Workers}_i) + 34.5\ (\text{Autos}_i)$

Household Trip Productions: $T_{ih} = 0.64 + 2.3 + (\text{Employee}_i) + 1.5\ (\text{HHAuto}_i)$

Zonal Attractions: $T_j = 54.2 + 0.23\ (\text{Office}_j) + 0.43\ (\text{Retail}_j)$

where T_i = total number of trips generated in zone *i*; T_{ih} = total trips generated per household in zone *i*; T_j = total trips attracted to zone *j*; Workers_i = number of workers in zone *i*; Autos_i = number of autos in zone *i*; Employee_i = number of employees per household in zone *i*; HHAuto_i = number of autos per household in zone *i*; Office_j = number of office employees in zone *j*; and Retail_j = number of retail employees in zone *j*.

Trip distribution is the process of estimating the number of trips that travel from each zone to every other zone in the study area. The results of the trip distribution process is a matrix called the *trip table,* which shows the number of trips traveling between each origin–destination (O-D) pair for the time period being examined. A common method for distributing trips in a zonal system is the gravity model, which is of the following form:

$$T_{ij} = P_i \times \frac{A_i \times F_{ij} \times K_{ij}}{\sum (A_j \times F_{ij} \times K_{ij})}$$

where T_{ij} = total trips originating in zone *i* and destined to zone *j*; P_i = number of trips produced in zone *i*; A_j = level of attractiveness of zone *j* (e.g., number of retail employees); F_{ij} = friction or impedance factor between zones *i* and *j* (a value usually a function of travel time); and K_{ij} = socioeconomic adjustment factors for trips between zones *i* and *j* (a value that represents variables that influence trip making not accounted for by other variables).

Mode choice is the process of estimating the percentage of travelers who will use one mode of transportation versus the others available for a given trip. The basic approach in making this estimation is that each mode has associated with it some empirically known characteristics that, when combined with characteristics of the traveler in a mathematical equation, can define that mode's *utility.* Variables such as travel time, travel cost, modal reliability, and so on are often incorporated into a mode's *utility function,* along with socioeconomic characteristics of the traveler. Freight models use a similar concept in estimating commodity flows by mode. One of the most familiar forms of mode choice models, based on the concept of consumer choice, is the logit model, which predicts mode shares based on the following equation:

$$P_{ik} = \frac{e^{U}_{k}}{\sum e^{U}_{m}} \quad \text{for all modes } n$$

where P_{ik} = probability of individual *i* choosing mode *k*; U_k = utility of mode *k*; U_m = utility of mode *m*; *n* = number of modes available for trip.

The utility of each mode is often represented as a linear function of those variables found to influence an individual's choice of mode. For example, a utility function for the automobile mode might be of the form,

$$U_a = 6.3 - 0.21\ (X_1) - 0.43\ (X_2) - 0.005\ (X_3)$$

where U_a = utility of automobile; X_1 = access and egress time when automobile is chosen; X_2 = line-haul travel time; and X_3 = cost of travel.

The utility functions of other modes available for a specific trip would be similarly specified. The respective probabilities would then be multiplied by the total number of trips between an origin and destination to obtain the number of trips made by mode.

Trip assignment is the process of estimating the trip paths through a transportation network based on a trip table (which is produced in trip distribution). The basic concept found in all trip assignment methods is that travelers choose modes that will minimize travel time, that is, they will choose the shortest path through a network (once again, the assumption of derived demand influencing the analysis approach). Link performance functions that relate travel time to the number of vehicles or riders on that link are used to iteratively update estimated link travel times so that minimum path travel times reflect the effect of congestion (see Figure 82.3). A portion of the total O-D travel demand is assigned to the network, with travel times then updated based on the link performance function, given the volume on each link. An additional portion of the O-D travel is next assigned given the updated travel times, still following the minimum travel time path through the network. This process continues until all estimated trips have been assigned to a link path in the network. Stochastic assignment is also used in many planning studies. This assignment recognizes that, in certain cases, some subset of trip routes will have associated with them some characteristics that attract specific types of travelers, even if the travel time is longer. A probabilistic approach takes these characteristics into account.

In order to develop more behaviorally based travel models, researchers in recent years have focused on the fact that travel arises out of the need to participate in out-of-home activities (work, shopping, school, etc.). This directly leads to the conclusion that what one should study in the first instance is not travel per se, but rather the participation in the *activities* that ultimately generate travel. This approach has been referred to as "activity-based modeling." Figure 82.4 shows the difference in the traditional approach toward modeling and the activity-based approach. Many activity-based models are being implemented within a *micro-simulation* framework, within which the behavior of each individual is dynamically simulated over time.

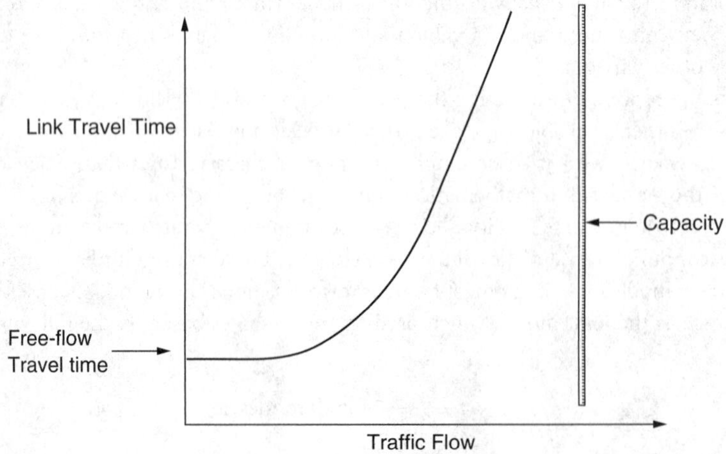

FIGURE 82.3 Link performance function.

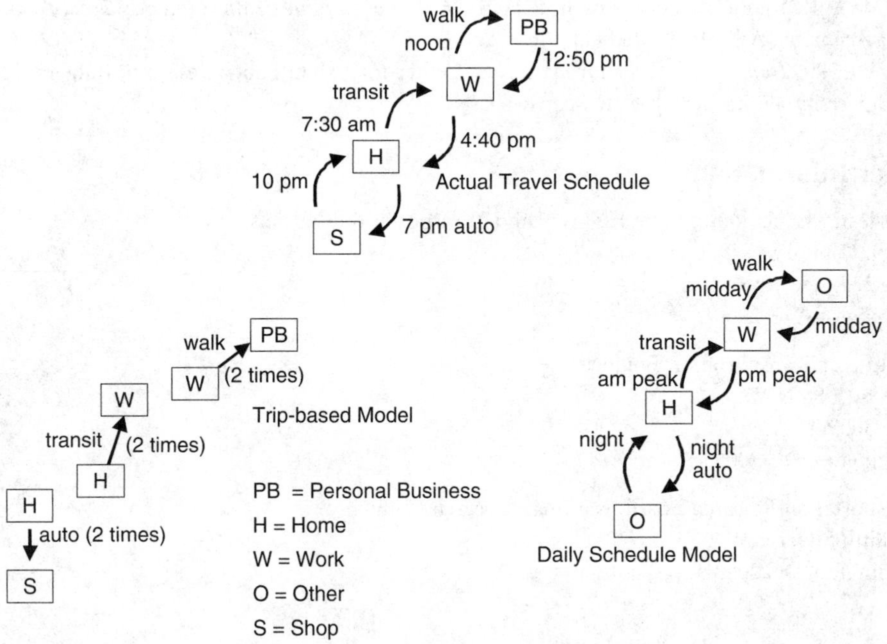

FIGURE 82.4 Difference between trip-based modeling and activity-based modeling.

Defining Terms

Demand management — Reducing the demand for travel during specific time periods by shifting trips to other times, diverting trips to other routes or modes, or reducing the need for trip-making to begin with.

Derived demand — An assumption that travelers make a trip to accomplish some objective at the destination and that the trip itself is simply a means of reaching that activity.

Intelligent transportation systems — Application of surveillance, communication, and control technologies to the management of the transportation system, and in some cases, to the control of individual vehicles.

Transportation network — A transportation system is represented in models as a network of links and nodes. Links represent line-haul facilities, such as roads and transit lines, and nodes represent points of connection.

Utility function — A mathematical formulation that assigns a numerical value to the attractiveness of individual modes of transportation based primarily on that mode's characteristics.

Zonal system — Each study area (whether nation, state, metropolitan region, or community) is divided into zones of homogeneous characteristics that can then be used as the basic foundation for estimating trips from, or attracted to, that zone.

References

Goulias, K. (Ed). 2003. *Transportation Systems Planning,* CRC Press, Boca Raton, FL.

Grava, S. 2003. *Urban Transportation Systems,* McGraw-Hill, New York.

Hall, R. (Ed). 2003. *Handbook of Transportation Science,* 2nd ed. Kluwer, Boston.

Institute of Transportation Engineers. 1997. *Trip Generation Handbook,* 6th ed. ITE, Washington, DC.

Meyer, M. and Miller, E. 2001. *Urban Transportation Planning: A Decision-Oriented Approach,* 2nd ed. McGraw-Hill, New York.

Ortuzar, J. and Willumsen. L.G. 1994. *Modelling Transport,* 2nd ed. John Wiley & Sons, New York.

Taylor, M. A. P., Young, W. and Bonsall, P. W. 1996. *Understanding Traffic Systems: Data, Analysis and Presentation*, Ashgate, Brookfield, VT.

Vuchic, V. 1999. *Transportation for Livable Cities*, Center for Urban Policy Research, Rutgers, The State University of New Jersey, New Brunswick, NJ.

Further Information

American Association of State Highway and Transportation Officials
444 N. Capitol St. NW
Suite 225
Washington, DC 20001

Institute of Transportation Engineers
1099 14th St. NW
Suite 300W
Washington, DC 20005

Transportation Research Board, National Research Council
500 Fifth Street, NW
Washington, DC 20001

83

Design of Transportation Facilities

John Leonard II
Georgia State Road and Tollway Authority

Michael D. Meyer
Georgia Institute of Technology

The efficient movement of people and goods requires transportation systems and facilities that are designed to provide sufficient capacity for the demands they face in as safe a manner as possible. In addition, in most modern societies, the design of transportation facilities must explicitly minimize harm to the natural and human-made environment while providing for mitigation measures that relate to those impacts that are unavoidable. In many ways the critical challenge to today's designers of transportation projects is successfully designing a facility that minimally harms the environment.

The design of a transportation facility almost always takes place within the context of a much broader **project development process.** This process can vary in complexity with the type of project under design and with the scale of implementation. The importance of the project development process to the designer is that it:

- Establishes the key characteristics of the project that must be considered in the design
- Indicates the time frame that will be followed for project design
- Establishes which agencies and groups will be involved in the process and when this involvement will likely occur
- Links the specific elements of the project design with other tasks that must be accomplished for the project to be constructed

0-8493-1586-7/05/$0.00+$1.50

- Satisfies legal requirements for a design process that is open for public review and comment
- Indicates the specific products that must be produced by the designers to complete the project design process

In most cases the project development process consists of a well-defined set of tasks that must be accomplished before the next task can occur. These tasks include both technical activities and public involvement efforts that are necessary for successful project development.

83.1 Components of the Project Development Process

Identify Project Need

A project need can be identified through a formal planning process or from a variety of other sources, including suggestions from elected officials, agency managers, transportation system users, and citizens. Important in this early portion of project development is an indication of what type of improvement is likely to be initiated. For example, a project could relate to one or more of the following types of improvement strategies:

- *New construction.* A transportation facility constructed at a new location
- *Major reconstruction.* Addition of new capacity or significant changes to the existing design of a facility, but usually occurring within the area where the current facility is located
- *Rehabilitation/restoration.* Improvements to a facility usually as it is currently designed and focusing on improving the physical condition of the facility or making minor improvements to enhance safety
- *Resurfacing.* Providing new pavement surface to a transportation facility that prolongs its useful life
- *Spot improvements.* Correction of a problem or hazard at an isolated or specific location

Establish Project Limits and Context

One of the very first steps in the design process is to define the boundaries or limits of the project. This implies establishing how far the project will extend beyond the area being targeted for improvement and the necessary steps to ensure smooth connections to the existing transportation system. Project boundaries also have important influence on the amount of right-of-way that might have to be purchased by an agency to construct a project.

Establish Environmental Impact Requirements

The design of a project will most likely be influenced by environmental laws or regulations that require design compliance with environmental mandates. These mandates could relate to such things as wetland protection, preservation of historic properties, use of public park lands, maintaining or enhancing water quality, preserving navigable waterways, protecting fish and wildlife, reducing air pollutants and noise levels, and protecting archaeological resources. One of the first steps in project development is to determine whether the likely project impacts are significant enough to require a detailed environmental study.

Develop Strategy for Interagency Coordination and Public Involvement

Depending on the complexity and potential impact of a project, the project designer could spend a great deal of time interacting with agencies having some role in or jurisdictional control over areas directly related to the project. These agencies could have jurisdiction by law (e.g., wetlands) or have special expertise that is important to project design (e.g., historic preservation). In addition to interagency coordination, transportation project development is often subject to requirements for public outreach and/or public hearings. An important aspect of recent project development efforts is to develop very early in the process a consensus among involved agencies on what environmental impacts will have to be carefully studied and on the definition of the project purpose and need.

Initiate Project Design and Preliminary Engineering

Topographic data of the study area and forecasted vehicular volumes expected to use the facility in the design year are used as input into the preliminary design of the horizontal and vertical alignment of the facility, that is, the physical space the facility will occupy once finished. This preliminary engineering step also includes the preparation of initial right-of-way (ROW) plans, which indicate the amount of land that must be available to construct the facility. Preliminary engineering is a critical step for environmental analysis in that it provides the first detailed examination of the scope and extent of potential environmental impacts.

Project Engineering

Once preliminary engineering has provided the basic engineering information for the project, the more detailed project design begins. This entails specific layouts of horizontal and vertical geometry, soils/subsurface examination and design, design of utility location, drainage design, more detailed ROW plans, and initial construction drawings. Concurrent with this design process, the environmental process continues with updated information on project changes that might cause additional environmental harm, the initiation of any permitting process that might be needed to construct the project (e.g., environmental agency permission to affect wetlands), and public hearings/meetings to keep the public involved with project development.

Final Engineering

The final engineering step is the culmination of the design process, which completes the previous design plans to the greatest level of detail. This step includes finalizing ROW plans, cost estimates, construction plans, utility relocation plans, and any agreements with other agencies or jurisdictions that might be necessary to complete the project. Environmental permits are received and final project review for environmental impacts is completed.

Context-Sensitive Design

One of the important characteristics of transportation facility design is the potentially negative impact that new facilities could have on the surrounding community and natural environment. Engineers and planners have begun to consider such impacts earlier in the project development process so that the context within which a facility is constructed is incorporated into the design itself. This process is called **context-sensitive design.**

83.2 Basic Concepts of Project Design

Human Factors

Human factors have a great deal of influence on the design of transportation facilities in such things as width of facility, length and location of access/egress points, vehicle braking distance, location of information/guidance aids such as signs, and geometric characteristics of the facility's alignment. The driver-vehicle-roadway interface is shown in Figure 83.1.

Vehicle or User Performance Factors

The dynamics of vehicle motion play an important role in determining effective and safe design. The key vehicle characteristics that relate to facility design criteria include:

- *Vehicle size.* Influences vertical and horizontal clearances, turning radii, alignment width, and width of vehicle storage berths.
- *Vehicle weight.* Influences strength of material needed to support vehicle operations.

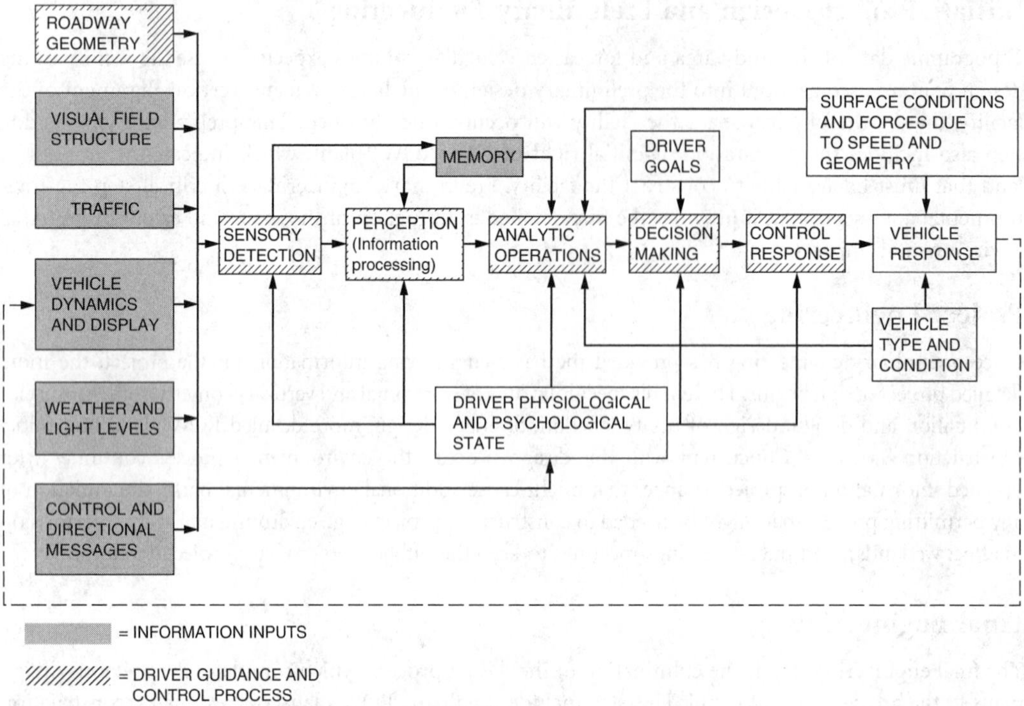

FIGURE 83.1 Driver-vehicle-roadway interface.

- *Vehicle or user performance.* Influences specifications for horizontal and vertical geometry, braking distances, operational performance and needed capacity to allow passing and successful maneuvering (e.g., assumed walking speed of pedestrians crossing a road that dictates how long a traffic signal must remain red).

Classification Schemes

The transportation system serves many functions, ranging from providing access to specific locations to providing high-speed, high-capacity movement over longer distances. Classification schemes are used to represent these various roles and influence the design criteria that are associated with the facilities in each classification category. A common **functional classification** scheme for highways is shown in Figure 83.2.

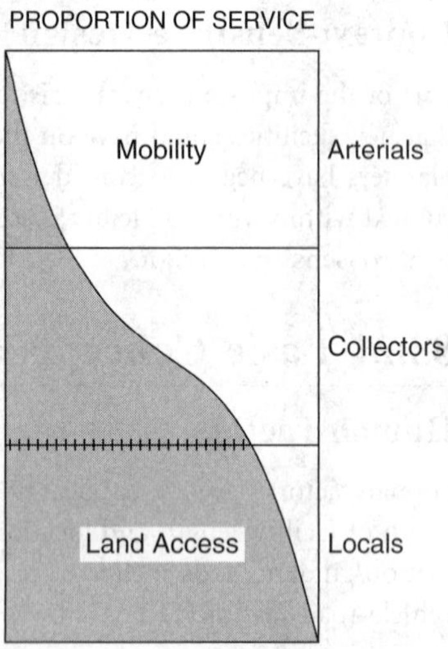

FIGURE 83.2 Relationship of functionally classified systems in relation to traffic mobility and land access. (*Source:* American Association of State Highway and Transportation Officials. 2001. *A Policy on the Geometric Design of Highways and Streets,* AASHTO, Washington, DC. Figure 1 through Figure 5.)

Capacity and Level of Service

Every design usually begins with some estimation of the demand for the transportation facility that will likely occur if the facility is built. The key design question then becomes, what facility capacity (e.g., number of road lanes, runways, transit lines, or vehicle departures) is necessary if a certain level of performance is desired? These different levels of performance are referred to as **level of service (LOS).** Level of service is a critical element in establishing important design factors (see Figure 83.3).

FREEWAYS

Level of Service	Maximum Density (pc/mi/ln)	Minimum Speed (mph)	Max Service Flow Rate (pcphpl)	Maximum *v/c* Ratio
		Free-Flow Speed = 70 mph		
A	10.0	70.0	700	0.318/0.304
B	16.0	70.0	1120	0.509/0.487
C	24.0	68.5	1644	0.747/0.715
D	32.0	63.0	2015	0.916/0.876
E	36.7/39.7	60.0/58.0	2200/2300	1.000
F	var	var	var	var
		Free-Flow Speed = 65 mph		
A	10.0	65.0	650	0.295/0.283
B	16.0	65.0	1040	0.473/0.452
C	24.0	64.5	1548	0.704/0.673
D	32.0	61.0	1952	0.887/0.849
E	39.3/43.4	56.0/53.0	2200/2300	1.000
F	var	var	var	var
		Free-Flow Speed = 60 mph		
A	10.0	60.0	600	0.272/0.261
B	16.0	60.0	960	0.436/0.417
C	24.0	60.0	1440	0.655/0.626
D	32.0	57.0	1824	0.829/0.793
E	41.5/46.0	53.0/50.0	2200/2300	1.000
F	var	var	var	var
		Free-Flow Speed = 55 mph		
A	10.0	55.0	550	0.250/0.239
B	16.0	55.0	880	0.400/0.383
C	24.0	55.0	1320	0.600/0.574
D	32.0	54.8	1760	0.800/0.765
E	44.0/47.9	50.0/48.0	2200/2300	1.000
F	var	var	var	var

Note: In table entries with split values, the first value is for four-lane freeways, and the second is for six- and eight-lane freeways.

PEDESTRIAN WALKWAYS

		Expected flows and Speeds		
Level of Service	Space (sq ft/ped)	Ave. Speed, *S* (ft/min)	Flow Rate, *v* (ped/min/ft)	Vol/Cap Ratio, *v/c*
A	≥130	≥260	≤2	≤0.08
B	≥40	≥250	≤7	≤0.28
C	≥24	≥240	≤10	≤0.40
D	≥15	≥225	≤15	≤0.60
E	≥6	≥150	≤25	≤1.000
F	<6	<150	—Variable—	

*Average conditions for 15 min.

FIGURE 83.3 Example level of service characteristics (*Source:* Transportation Research Board, 2000. *Highway Capacity Manual,* National Academy Press, Washington, DC.) *Continued.*

SIGNALIZED INTERSECTIONS

Level of Service	Stopped Delay per Vehicle (sec)
A	≤5.0
B	5.1 to 15.0
C	15.1 to 25.0
D	25.1 to 40.0
E	40.1 to 60.0
F	>60.0

ARTERIAL ROADS

Arterial Class	I	II	III
Range of Free Flow Speeds (mph)	45 to 35	35 to 30	35 to 25
Typical Free Flow Speed (mph)	40 mph	33 mph	27 mph
Level of Service	**Average Travel Speed (mph)**		
A	≥35	≥30	≥25
B	≥28	≥24	≥19
C	≥22	≥18	≥13
D	≥17	≥14	≥ 9
E	≥13	≥10	≥ 7
F	<13	<10	< 7

FIGURE 83.3 *Continued.*

Design Standards

Design standards dictate minimum or maximum values of project characteristics that are associated with a particular facility type. Design standards usually result from extensive study of the relationship between various facility characteristics, vehicle performance, and the safe handling of the vehicles by human operators. Design standards often vary by the "design speed" of the facility (and thus the importance of the facility classification) and by the "design vehicle." Design standards are often the basis for developing typical cross-sections (see Figures 83.4 and 83.5).

83.3 Intermodal Transportation Terminals or Transfer Facilities

Terminals or transfer facilities are locations where users of the transportation system change from one mode of travel to another. The effective design of such facilities is a critical element of successful transportation system performance, given the potential bottlenecks they represent if not designed appropriately. The design of terminals and transfer facilities must pay special attention to the needs of the users of the facility, in that they serve to establish the effective capacity of the facility, for example:

1. Internal pedestrian movement facilities and areas (stairs, ramps, escalators, elevators, corridors, etc.)
2. Line-haul transit access area (entry control, fare collection, loading, and unloading)
3. Components that facilitate movements between access modes and the station (ramps or electric doors)

RECOMMENDED ROADWAY SECTION WIDTHS

Functional Class	U/R	Number of Lanes	Travel Lane	Shoulder	
				Right	Left[1]
Freeway	Urban	4–8	12	10	4[2]
	Rural	4–8	12	10	4[2]
Arterial	Urban	Multilane with median	12	10	4
	Urban	Multilane without median	11–12	8–10*	N/A
	Rural	2 lane	12	See Table 5.2	N/A
	Rural	Multilane with median	12	8–10	4

WIDTH OF USABLE SHOULDER—EACH SIDE OF TRAVEL WAY
RURAL TWO-LANE ARTERIAL

	Design Traffic Volume				
	Current ADT Under 400	Current ADT Over 400	DHV 100–200	DHV 200–400	DHV Over 400
All design speeds	4 ft	6 ft	6 ft	8 ft	10 ft

RECOMMENDED WIDTH OF TRAVEL WAY AND GRADED SHOULDER
RURAL COLLECTOR

	Design Traffic Volume				
Design Speed (mph)	Current ADT Under 400	Cuurent ADT Over 400	DHV 100–200	DHV 200–400	DHV Over 400
30	20 ft	20 ft	20 ft	22 ft	24 ft
40	20 ft	22 ft	22 ft	22 ft	24 ft
50	20 ft	22 ft	22 ft	24 ft	24 ft
60	22 ft	22 ft	22 ft	24 ft	24 ft
	Graded Shoulder (Each Side)*				
All speeds	2 ft	4 ft	6 ft	8 ft	8 ft

*If right-of-way permits

FIGURE 83.4 Example design criteria. (*Source:* Massachusetts Department of Public Works. 1988. *Highway Design Manual,* Boston, MA.)

4. Communications (public address systems and signage)
5. Special provisions for disabled patrons (elevators and ramps)

The criteria that could relate to the design of such a facility include threshold values for pedestrian level of service, delay at access points, connectivity from one area of the facility to another, and low-cost maintenance. For the vehicle side of such terminals, special consideration must be given to the performance of the design vehicle (e.g., turning radii of buses or semitrailer trucks) and the vehicle storage requirements (e.g., the number, size, and orientation of loading/unloading berths).

83.4 Advanced Technology Projects

One of the characteristics of transportation system development in recent years has been the increased application of advanced (usually electronic) technologies to improve system performance. Known as **intelligent transportation systems (ITS),** the following steps apply in their design.

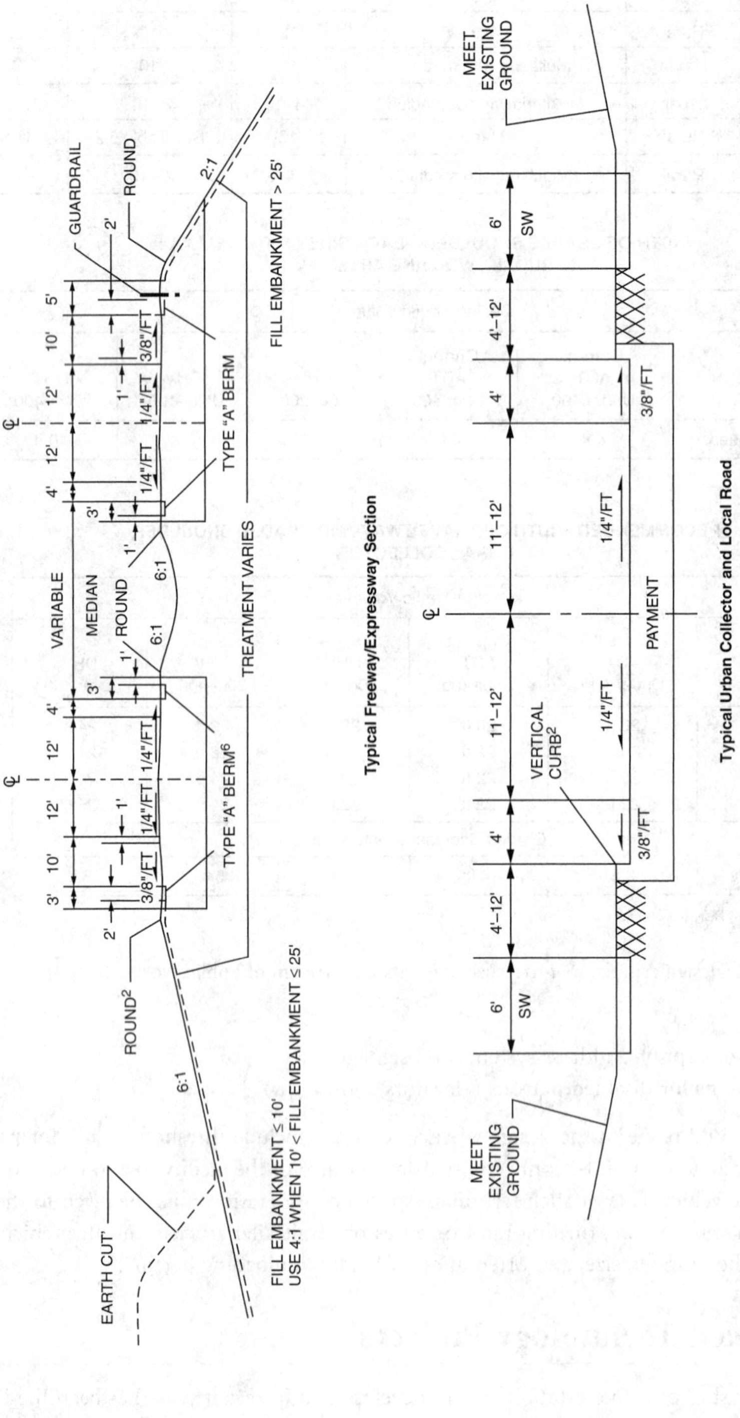

FIGURE 83.5 Example cross-sections.

Define Problems and Needs

Typical problems or needs might relate to congestion, excessively high accident rates, or improving current system capabilities and levels of service.

Define System

A system definition should include a mission statement, listing of physical components (e.g., roads, travelers, buses, rolling stock, existing rail lines, control centers, and communication links), and the physical relationship between those components.

Define Users

System users is a rather broad description of all individuals, organizations, and other systems that might interact or have a stake in the fully implemented transportation system under study.

Establish Institutional Framework and Partnerships

Various organizations possess differing missions, priorities, and policies — sometimes in conflict. Strong emphasis on coalition building during the early stages of project planning and engineering can help diffuse potential project-stopping disagreements later in the process.

Develop User Service Plan

The development of a user service plan consists of the following steps: (1) establish user services, (2) identify technology areas, and (3) map user services to technology areas. User services might include:

- Traveler information services
- Freight and fleet management services
- Emergency vehicle management services
- Traffic management services
- Public transport services

Available technologies fall within one of the following functional areas: (1) surveillance, (2) communications, (3) traveler interface, (4) control strategies, (5) navigation/guidance, (6) data processing, and (7) in-vehicle sensors.

Define System Architecture

A logical architecture consists of a *block diagram* identifying the major systems and subsystems, the participating agencies, and users. Through the use of arrows, the flow of information between these elements is identified. A logical architecture also shows the allocation of responsibilities throughout the transportation system.

Evaluate Alternative Technologies

Some of the factors to be considered in this evaluation include (1) cost, (2) performance, (3) reliability, (4) compatibility, (5) environmental impacts, and (6) compliance to standards.

Defining Terms

Context-sensitive design — A design process in which the community and environmental context is considered very early in the project development process. Mitigation and avoidance of significant impacts are proposed at this stage.

Design standard — Physical characteristics of a proposed facility that are professionally accepted and often based on safety considerations.

Functional classification — Classifying a transportation facility based on the function it serves in the transportation system. Such classification becomes important in that design standards are often directly related to the functional classification of a facility.

Intelligent transportation systems — Use of information and surveillance technologies to monitor and control the operation of the transportation system. Many of these systems are targeted at specific travel markets.

Level of service — An assessment of the performance of a transportation facility based on measurable physical characteristics (e.g., vehicular speed, average delay, density, flow rate). Level of service is usually subjectively defined as ranging from level of service A (good performance) to level of service F (bad or heavily congested performance).

Project development process — The steps that are followed to take a project from initial concept to final engineering. This process includes not only the detailed engineering associated with a project design but also the interaction with the general public and with agencies having jurisdiction over some aspect of project design.

References

American Association of State Highway and Transportation Officials. 2001. *A Policy on the Geometric Design of Highways and Streets,* AASHTO, Washington, DC.

Federal Highway Administration, 1997. *Flexibility in Highway Design,* Report FHWA-PD-97-062, FHWA, Washington, DC.

Transportation Research Board, 1999. Transit Capacity and Quality of Service Manual, Transit Cooperative Research Web Document 6, National Academy Press, Washington, D.C., www.gulliver.trb.org/publications/tcrp/tcrp_webdoc_6-a.pdf.

Transportation Research Board. 2000. *Highway Capacity Manual,* National Academy Press, Washington, D.C.

Further Information

Transportation Research Board, National Academy of Sciences, 2101 Constitution Ave., N.W., Washington, D.C. 20418

American Association of State Highway and Transportation Officials, 444 N. Capitol St., N.W., Suite 225, Washington, D.C. 20001

Institute of Transportation Engineers, 525 School St., S.W., Suite 410, Washington, D.C. 20024

Federal Highway Administration, 400 7th St. S.W., Washington, D.C. 20590

84

Operations and Environmental Impacts

Michael D. Meyer
(Second Edition)
Georgia Institute of Technology

Paul W. Shuldiner
(First Edition)
University of Massachusetts, Amherst

Kenneth B. Black
(First Edition)
University of Massachusetts, Amherst

The safe and efficient movement of vehicles and people are the two most important goals of highway design. This is accomplished through careful geometric design of the layout and physical features of the roads, and through operational procedures and devices designed to guide, advise, and regulate vehicle operators in their use of the road. Proper design is based on a thorough understanding of the complex interactions among driver, vehicle, and roadway characteristics. For example, the design of a highway passing lane reflects the performance characteristics of the vehicle, the reaction time of the driver, and the ability of the road itself to provide sufficient distance that allows safe vehicle passing.

The following sections present the fundamental equations of vehicle flow and engineering principles upon which good highway design is based.

84.1 Fundamental Equations

Flow, q, is given by the following formula:

$$q = \frac{n \times 3600}{T} \quad \text{vehicles per hour} \tag{84.1}$$

where n = number of vehicles passing a point in T seconds.

Density, k, is given by

$$k = \frac{n}{L} \times 5280 \quad \text{vehicles per mile} \tag{84.2}$$

where n = number of vehicles occupying a length of road, L, and L is in feet, or by

0-8493-1586-7/05/$0.00+$1.50

$$k = \frac{n}{L} \times 1000 \quad \text{vehicles per kilometer} \tag{84.3}$$

where n = number of vehicles occupying a length of road, L, and L is in meters.

Average speed, $\bar{u}$, is given by

$$\bar{u}_t = \frac{1}{n}\sum_{i=1}^{n} u_i \tag{84.4}$$

where u_i = speed of the ith vehicle, termed the *time mean speed*, and by

$$\bar{u}_s = \frac{n \cdot s}{\sum_{i=1}^{n} t_i} \tag{84.5}$$

where t_i = time for the ith vehicle to traverse distance s, termed the *space mean speed*. The values $\bar{u}_s$ and $\bar{u}_t$ usually differ very slightly from one another unless there are wide variations among the speeds of individual vehicles.

84.2 Flow, Speed, and Density Relationships

The fundamental equation relating flow, density, and speed can be formulated as follows:

$$q = k \times \bar{u}_s \tag{84.6}$$

For simplicity, it is often assumed that average speed, u, decreases linearly as density, k, increases. Given this assumption and the dimensional identity expressed in Equation (84.6), it follows that flow, density, and speed are related as shown in Figure 84.1. It may be seen from Figure 84.1 that

1. When density on the highway is zero, there are no vehicles on the highway and the flow is zero.
2. As density goes to its maximum, flow goes to zero as vehicles line up — essentially bumper to bumper.

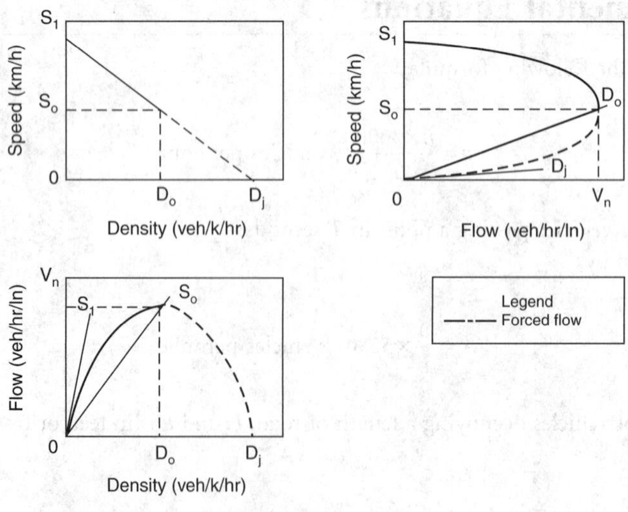

FIGURE 84.1 Relationships among flow, density, and speed.

3. As density increases, flow also increases — up to the point where $k = kj/2$ — and then decreases as density increases.
4. As flow increases from zero to a maximum (the capacity of the roadway), speed also decreases. Once the flow approaches capacity, any impediment to the steady movement of traffic will reduce both flow and speed abruptly, with a concomitant rapid increase in density.

84.3 Level of Service (LOS)

Level of service is a measure of how freely a given transportation system is operating. It is affected primarily by traffic **volume**, but other factors such as proportion of heavy vehicles, roadway geometrics, and incidents also affect the level of service. LOS is designated by letters A through F. LOS A is characterized as *free flow* and is applied to conditions of very low traffic volume in which there is little or no restriction to traffic flow. LOS F is characterized as *forced flow* with congested, stop-and-go movement. LOS C involves some restriction to free movement of traffic and is the level of service commonly used as the target for road design. The maximum flow along a facility occurs at LOS E.

Different elements of the road system emphasize different LOS measures. For example, the following measures are the key determinants in the LOS for the identified element of the road network.

Urban street:	Average travel speed for through vehicles
Freeway:	Average speed, freedom to maneuver, and proximity of other vehicles
Freeway ramp:	Vehicle density
Signalized intersection:	Average traffic control delay per vehicle
Two-lane highways:	Percent time spent following other vehicles, and average speed
Pedestrian walkways:	Space per pedestrian, flow rate, and average speed

The calculation of road level of service varies by type of road. For example, the *Highway Capacity Manual*, the manual that defines how level of service should be calculated, describes different methodologies for the following transportation facilities: urban streets, signalized intersections, unsignalized intersections, pedestrian walkways, bicycle facilities, two-lane highways, multi-lane highways, freeway facilities, basic freeway segments, freeway weaving sections, ramps, interchange ramp terminals, and transit services.

As an example, the LOS for a freeway segment is calculated with the following two equations:

$$\mathrm{FFS} = \mathrm{BFFS} - f_{LW} - f_{LC} - f_N - f_{ID} \tag{84.7}$$

where FFS = free flow speed (mi/h); BFFS = base free flow speed (70 mi/h for urban and 75 mi/h for rural); f_{LW} = adjustment for lane width; f_{LC} = adjustment for right-shoulder lateral clearance; f_N = adjustment for number of lanes; and f_{ID} = adjustment for interchange density.

$$D = \frac{V}{S \times (\mathrm{PHF} \times N \times f_{HV} \times f_P)} \tag{84.8}$$

where D = density (pc/mi/ln); S = average passenger car speed (mi/h); V = hourly volume (veh/h); PHF = peak hour factor (variation in traffic low within hour); N = number of lanes; f_{HV} = heavy vehicle adjustment factor; and f_P = driver population factor.

The results of these two equations for the circumstances of a specific freeway segment are then entered into a table that defines a corresponding level of service.

84.4 Highway Capacity

The **capacity** of a highway, that is the maximum flow of vehicles or passengers, depends on many geometric and traffic characteristics. The capacity of a road occurs at LOS E. Equation 84.7 and Equation

84.8 indicate that the most important of these factors include number of lanes, width of lanes, lateral clearance to obstructions, percentage of the traffic stream made up of trucks and buses, characteristics of the driver population and percent grade of the roadway (incorporated into the heavy vehicle adjustment factor). Lane widths will vary according to the type of road; however, for high-capacity, high-speed roads such as freeways, the lane width should be at least 12 ft. Obstructions should be a minimum of 6 ft from the edge of pavement. Vertical grades of 3% or more that are one-half mile long or longer will reduce the capacity of a highway, especially where commercial vehicles are present. For freeways, maximum lane capacity under ideal conditions is generally assumed to be 2400 vehicles/hr per lane. Ideal conditions, however, are seldom achieved.

84.5 Intersection Capacity

The *Highway Capacity Manual* provides a detailed method for estimating the capacity of a signalized intersection. This method examines the geometric, traffic and signalization conditions for each lane approach to the intersection. Similar to Equation 84.7 and Equation 84.8, the following equation is used to determine the saturation flow rate for each group of lanes entering an intersection.

$$S = s_0 \times N \times f_W \times f_{HV} \times f_g \times f_p \times f_{bb} \times f_a \times f_{LU} \times f_{LT} \times f_{RT} \times f_{Lpb} \times f_{Rpb} \tag{84.9}$$

where S = saturation flow rate for each group of lanes (veh/h); s_0 = base saturation flow rate per lane (pc/h/ln); N = number of lanes in lane group; f_W = adjustment for lane width; f_{HV} = adjustment for heavy vehicles (trucks and buses); f_g = adjustment for approach grade; f_p = adjustment for existence of parking lane and parking activity adjacent to lane group; f_{bb} = adjustment for blocking effect of local buses that stop within intersection area; f_a = adjustment for area type; f_{LU} = adjustment for lane utilization; f_{LT} = adjustment for left turns in lane group; f_{RT} = adjustment for right turns in lane group; f_{Lpb} = pedestrian–bicycle adjustment factor for left-turn movements; and f_{Rpb} = pedestrian–bicycle adjustment factor for right-turn movements.

The result of Equation 84.9 is then multiplied by the amount of green time per total signal cycle time for the lane group being analyzed to obtain the capacity of the lane group. As shown, Equation 84.9 illustrates the basic concept of first starting with an "ideal" capacity or saturation flow and then modifying this value with adjustments that reflect physical characteristics of the existing facility and area.

84.6 Traffic Control Devices

Traffic signs, signals, and pavement markings (**traffic control devices**) are used to regulate, warn, and guide the users of the road. The effectiveness of such devices is greatly enhanced through the uniformity of their design — including shape, color, and message — and through their use and placement on the roadway. Uniformity is ensured by adherence to the standards promulgated in the *Manual of Uniform Traffic Control Devices* [U.S. DOT and FHWA, 2000] or an equivalent manual adopted by each state highway agency. As stated in the **Manual**, to be effective, a traffic control device should meet five basic requirements: fulfill a need, command attention, convey a simple and clear meaning, command respect from road users, and give adequate time for proper response.

One of the ways that traffic control devices gain respect from road users is by using them in situations where such devices are clearly warranted. Experience over many decades and engineering judgement have become the basis for "warrants," guidelines that are used by engineers to justify the use of particular traffic control devices. Two examples of how this principle is used by engineers are provided below for STOP signs and traffic signals.

STOP signs should be used only where one or more of the following conditions or warrants hold:

1. At an intersection of a less important road with a main road where application of the normal right-of-way rule would not be expected to provide reasonably safe operation

2. Where a street enters a through highway or street
3. At an unsignalized intersection in a signalized area
4. At other intersections where a combination of high speed, restricted view, and crash records indicate a need for control by a STOP sign

Traffic control signals should not be installed unless at least one of several signal warrants is met (the *Manual* defines eight warrants). Eleven signal warrants are used to account for such factors as traffic volume, crash experience, school crossings, and pedestrian volumes. In addition to a review of these warrants, an engineering study should be conducted to show that the installation of such a signal would improve the overall safety and operation of the intersection.

Consideration should be given to other, less restrictive measures before these warrants are applied.

84.7 Coordinated System Operations

The efficient operation of a road network requires more than simply placing traffic control devices at warranted locations. A regional road network for a modern metropolitan area consists of many different road types, whose jurisdiction is shared by numerous different organizations. One of the key challenges facing many metropolitan areas today is coordinating the many different responsibilities for operating a regional road network. This can be accomplished through ad hoc working groups focusing on specific operations strategies (e.g., coordinating traffic signals across jurisdictions) or by the creation of a regional entity with a legislative mandate to operate all transportation services in a region (much more common in countries other than the U.S.). In the U.S., some of the strategies used to coordinate system operations include:

1. Standardized strategies to be implemented in response to expected incidents or events
2. Shared databases and information retrieval
3. Corridor teams that focus on network operations
4. Regional programs, e.g., incident management efforts or traffic signal coordination, which target key determinants of system performance
5. Regional traffic management centers that collect data on system performance and provide information and guidance to road network users
6. Formal information and technological systems architecture that specifies protocols and processes for computer-supported approaches to system operations

With the advent of more powerful computers and improved sensors, it seems likely that future developments in coordinated system operations will to a large extent rely on the application of advanced technology.

84.8 Environmental Impacts

One of the most important characteristics of transportation system performance is the corresponding impact on the natural and human environment. This impact occurs not only during the construction of transportation facilities themselves but also from the resulting operations. For example, automobile travel in many urban areas is a primary source of many air pollutants. In terms of energy consumption, total auto travel accounts for approximately 25% of U.S. oil consumption, with about two-thirds of this consumption occurring in urban areas. Table 84.1 shows the environmental impacts that are of concern to transportation officials. Of those shown, the impacts that are directly linked to the operations of a facility or transportation system include air quality, noise, vibration, stormwater, energy consumption, and ecological effects.

It is often difficult to determine the best operations strategy for minimizing these impacts. For example, the air quality impacts of motor vehicles are a function of the number, type, and condition of vehicles on the highway and of the speed and efficiency with which these vehicles operate. Carbon monoxide, hydrocarbons, and oxides of nitrogen are the primary pollutants emitted by vehicular traffic. The level

TABLE 84.1 Transportation System Impacts of Concern to Transportation Officials

Natural System Impacts
Terrestrial ecology (habitats and animals)
Aquatic ecology (habitats and animals)
Physical Impacts
Air quality
Noise
Vibration
Water quality
Hazardous wastes
Stormwater
Energy consumption
Erosion and sedimentation
Farmland conversion
Social and Cultural Impacts
Historic and archaeological
Displacement of people
Community cohesion
Resource consumption
Land use
Aesthetics
Infrastructure effects
Accessibility of facilities, services, and jobs
Environmental justice
Employment, income, and business activity

of emission of each pollutant is complicated and difficult to predict. In general, the emission of carbon monoxide and hydrocarbons is minimized at speeds between 35 and 45 mi/h, whereas the emission of oxides of nitrogen increases over this speed range. Therefore, there is no ideal speed at which the level of emission of all three pollutants will be minimized. Nevertheless, smoothly flowing traffic will tend to produce less pollution, and, therefore, whatever can be done to reduce delay and congestion — especially when combined with a reduction in total vehicle miles traveled — will result in an improvement in air quality.

The reader is referred to the following site for information on the many environmental impacts that are of concern to transportation officials: http://www.fhwa.dot.gov/environment/index.htm

Defining Terms

Capacity — A maximum flow of vehicles or persons at a specific location, given the characteristics of the facility.

Density (*k*) — Number of vehicles or persons occupying a given length of a facility at an instant in time; usually expressed as veh/mi or veh/km for vehicles and persons/ft^2 or persons/m^2 for individuals.

Flow (*q*) — Time rate of flow of facility users; usually expressed as veh/h or passenger/h.

Level of service (LOS) — The quality of movement experienced by the users of a facility in a traffic stream; usually expressed in terms of the freedom of movement of facility users, ranging from little or no restriction (LOS A) to stopped flow (LOS F).

Speed (*u*) — Average speed of vehicles in a traffic stream; usually expressed as miles per hour (mph) or kilometers per hour (kph).

Traffic control devices — Signs, signals, and pavement markings placed on streets or highways to guide, inform, or regulate the movement of vehicles on those facilities.

References

Garber, N. and L. Hoel. 1997. *Traffic and Highway Engineering,* 2nd ed., PWS Publishing, Boston.

Institute of Transportation Engineers. 2001. *Traffic Control Devices Handbook,* Washington, DC.

Khisty, C.J. and B. K. Lall. 2003. *Transportation Engineering, An Introduction,* 3rd ed., Prentice Hall, New York.

Meyer, M. and E. Miller. 2001. *Urban Transportation Planning: A Decision-Oriented Approach,* 2nd ed., McGraw-Hill, New York.

TRB. 2000. *Highway Capacity Manual,* Transportation Research Board, Washington, DC.

U.S. DOT and FHWA. 2001. *Manual on Uniform Traffic Control Devices,* U.S. Department of Transportation, Federal Highway Administration, Washington, DC.

Further Information

American Society of Civil Engineers
1801 Alexander Bell Drive
Reston, VA 20191
(800) 548-2723

Federal Highway Administration
U.S. Department of Transportation
400 7th Street, S.W.
Washington D.C. 20590
http://www.ops.fhwa.dot.gov/traffic/

Institute of Transportation Engineers
1099 14th Street, NW
Suite 300 West
Washington, D.C. 20005-3438
202-289-0222
Fax 202-289-7722

Transportation Research Board
Keck Center of the National Academies
500 Fifth Street, NW
Washington, D.C. 20001
202-334-2934
Fax 202-334-2003

85

Transportation Systems

Paul Schonfeld
University of Maryland

The various forms of transportation that have been developed over time are called **modes.** The classification of modes may be very broad (e.g., highway transportation or air transportation) or more restrictive (e.g., chartered helicopter service). The major distinctions among transportation modes that help to classify them include:

1. Medium (e.g., air, space, surface, underground, water, underwater)
2. Users (e.g., passengers vs. cargo, general-purpose vs. special trips or commodities, common vs. private carrier)
3. Service type (scheduled vs. **demand responsive**, fixed vs. variable route, nonstop vs. express or local, mass vs. personal)
4. Right-of-way type (exclusive, semiexclusive, shared)
5. Technology:
   A. Propulsion (e.g., electric motors, diesel engines, gas turbines, linear induction motors, powered cables)
   B. Energy sources (e.g., petroleum fuels, natural gas, electric batteries, electric power from conducting cables)
   C. Support (e.g., aerodynamic lift, flotation on water, steel wheels on two steel rails, monorails, air cushions, magnetic levitation, suspension from cables)
   D. Flow type (e.g., discrete vehicles vs. continuous flow, as in pipelines, conveyor belts and escalators)
   E. Local control (e.g., lateral control by steering wheels, wheel flanges on railroad vehicles, rudders, longitudinal control by humans or automatic devices)
   F. Network guidance and control systems (with various degrees of automation and optimization)

A mode may be defined by its combination of such features. The number of conceivable combinations greatly exceeds the number of modes that have been actually tried, which, in turn, exceeds the number of successful modes. Success may be limited to relatively narrow markets and applications (e.g., for helicopters or aerial cablecars) or may be quite general. Thus, automobiles are successful in a very broad range of applications and have become the basis for distinct transportation modes such as taxis, carpools, or ambulances.

0-8493-1586-7/05/$0.00+$1.50

The relative success of various transportation modes depends on available technology and socioeconomic conditions at any particular time, as well as on geographic factors. As technology or socioeconomic conditions change, new transportation modes appear, develop, and may later decline as more effective competitors appear. For many centuries water transportation was considerably cheaper than overland transportation. Access to waterways was quite influential in the location of economic activities and cities. Access to good transportation is still very important to industries and communities. Technological developments have so drastically improved the relative effectiveness of air transportation that within a short period (approximately 1950 to 1965), aircraft almost totally replaced ships for transporting passengers across oceans. It is also notable that as economic prosperity grows, personal transportation tends to shift from the walking mode to bicycles, motorcycles, and then automobiles. Geography can significantly affect the relative attractiveness of transportation modes. Thus, natural waterways are highly valuable where they exist. Hilly terrain decreases the economic competitiveness of artificial waterways or conventional railroads while favoring highway modes. In very mountainous terrain, even highways may become uncompetitive compared to alternatives such as helicopters, pipelines, and aerial cablecars.

The relative shares of U.S. intercity passenger and freight traffic are shown in Table 85.1. The table shows the relative growth since 1929 of airlines, private automobiles, and trucks and the relative decline of railroad traffic.

85.1 Transportation System Components

The major components of transportation systems are:

1. Links
2. Terminals
3. Vehicles
4. Control systems

Certain "continuous-flow" transportation systems, such as pipelines, conveyor belts, and escalators, have no discrete vehicles and, in effect, combine the vehicles with the link.

Transportation systems may be developed into extensive networks. The networks may have a hierarchical structure. Thus, highway networks may include freeways, arterials, collector streets, local streets, and driveways. Links and networks may be shared by several transportation modes (e.g., cars, buses, trucks, taxis, bicycles, and pedestrians on local streets). Exclusive lanes may be provided for particular modes (e.g., pedestrians or bicycles) or groups of modes (e.g., buses and carpools).

Transportation terminals provide interfaces among modes or among vehicles of the same mode. They may range from marked bus stops or truck loading zones on local streets to huge airports or ports.

85.2 Evaluation Measures

Transportation systems are evaluated in terms of their effects on their suppliers, users, and environment. Both their costs and benefits may be classified into supplier, user, and external components. Private transportation companies normally seek to maximize their profits (i.e., total revenues minus total supplier costs). Publicly owned transportation agencies should normally maximize net benefits to their jurisdictions, possibly subject to financial constraints.

From the supplier's perspective, the major indicators of performance include measures of **capacity** (maximum throughput), speed, **utilization rate** (i.e., fraction of time in use), **load factor** (i.e., fraction of maximum payload actually used), energy efficiency (e.g., Btu per ton-mile or per passenger mile), and labor productivity (e.g., worker hours per passenger mile or per ton-mile). Measures of environmental impact (e.g., noise decibels or parts of pollutant per million) are also increasingly relevant. To users, price and service quality measures, including travel time, wait time, access time, reliability, safety, security, comfort (ride quality, roominess), simplicity of use, and privacy are relevant in selecting modes, routes, travel times, and suppliers.

TABLE 85.1 Volume of U.S. Intercity Freight and Passenger Traffic

	Millions of Revenue Freight Ton-Miles and Percentage of Total												
Year	Railroads	%	Trucks	%	Great Lakes	%	Rivers and Canals	%	Oil Pipelines	%	Air	%	Total
1929	454,800	74.9	19,689	3.2	97,322	16.0	8,661	1.4	26,900	4.4	3	0.0	607,375
1939	338,850	62.3	52,821	9.7	76,312	14.0	19,937	3.7	55,602	10.2	12	0.0	543,534
1944	746,912	68.6	58,264	5.4	118,769	10.9	31,386	2.9	132,864	12.2	71	0.0	1,088,266
1950	596,940	56.2	172,860	16.3	111,687	10.5	51,657	4.9	129,175	12.2	318	0.0	1,062,637
1960	579,130	44.1	285,483	21.7	99,468	7.6	120,785	9.2	228,626	17.4	778	0.1	1,314,270
1970	771,168	39.8	412,000	21.3	114,475	5.9	204,085	10.5	431,000	22.3	3,295	0.2	1,936,023
1980	932,000	37.5	555,000	22.3	96,000	3.9	311,000	12.5	588,000	23.6	4,840	0.2	2,486,840
1986	889,000	35.5	634,000	25.3	68,000	2.7	325,000	13.0	578,000	23.1	7,340	0.3	2,501,340
1987	968,000	36.3	668,000	25.1	78,000	2.9	358,000	13.4	585,000	22.0	8,720	0.3	2,665,720

	Millions of Revenue Passenger-Miles and Percentage of Total (Except Private)											
Year	Railroads	%	Buses	%	Air Carrier	%	Inland Waterways	%	Total (except private)	Private Automobiles	Private Airplanes	Total (including private)
1929	33,965	77.1	6,800	15.4	—	—	3,300	7.5	44,065	175,000	—	219,065
1939	23,669	67.7	9,100	26.0	683	2.0	1,486	4.3	34,938	275,000	—	309,938
1944	97,705	75.7	26,920	20.9	2,177	1.7	2,187	1.7	128,989	181,000	1	309,990
1950	32,481	47.2	26,436	38.4	8,773	12.7	1,190	1.7	68,880	438,293	1,299	508,472
1960	21,574	28.6	19,327	25.7	31,730	42.1	2,688	3.6	75,319	706,079	2,228	783,626
1970	10,903	7.3	25,300	16.9	109,499	73.1	4,000	2.7	149,702	1,026,000	9,101	1,184,803
1980	11,000	4.5	27,400	11.3	204,400	84.2	NA	—	242,800	1,300,400	14,700	1,557,900
1986	11,800	3.4	23,700	6.9	307,900	89.7	NA	—	343,400	1,450,100	12,400	1,805,900
1987	12,300	3.4	22,800	6.2	329,100	90.4	NA	—	364,200	1,494,900	12,400	1,871,500

Note: Railroads includes all classes, including electric railways, Amtrak and Auto-Train.
Source: Transportation Policy Associates and Railroad facts, 1988 Edition, Association of American Railroads.

85.3 Air Transportation

Air transportation is relatively recent, having become practical for transporting mail and passengers in the early 1920s. Until the 1970s, its growth was paced primarily by technological developments in propulsion, aerodynamics, materials, structures, and control systems. These developments improved its speed, load capacity, energy efficiency, labor productivity, reliability, and safety, to the point where it now dominates long-distance mass transportation of passengers overland and practically monopolizes it over oceans. Airliners have put ocean passenger liners out of business because they are much faster and also, remarkably, more fuel efficient and labor efficient when transporting passengers. However, despite this fast growth, the cargo share of air transportation is still small.

For cargoes that are perishable, high in value, or urgently needed, air transportation is preferred over long distances. For other cargo types, ships, trucks, and railroads provide more economic alternatives. In the 1990s, the nearest competitors to air cargo were containerships over oceans and trucks over land. For passengers, air transportation competes with private cars, trains, and intercity buses over land, with practically no competitors over oceans. The growth of air transportation has been restricted to some extent by the availability of adequate airports, by environmental concerns (especially noise), by security concerns, and by the fear of flying of some passengers.

There are approximately 10,000 commercial jet airliners in the world, of which the largest (as of 2003) are Boeing B-747 types, of approximately 800,000 lb gross takeoff weight, with a capacity of 550 passengers. The economic cruising speed of these and smaller "conventional" (i.e., **subsonic**) airliners has stayed at around 560 mph since the late 1950s. A few supersonic transports (SSTs) capable of cruising at approximately 1300 mph were built in the 1970s (the Anglo-French Concorde and the Soviet Tu-144) but, due to high capital and fuel costs, were unprofitable to operate.

The distance that an aircraft can fly depends on its payload, according to the following equation:

$$R=\frac{V}{c'}\left(\frac{L}{D}\right)\ln(W_{\mathrm{TO}}/W_L) \tag{85.1}$$

where R = range (mi); c' = specific fuel consumption (lb fuel/lb thrust × h); (L/D) = lift-to-drag ratio (dimensionless); W_{TO} = aircraft takeoff weight (lb) = $W_L + W_F$; W_L = aircraft landing weight (lb) = $W_E + W_R + P$; W_E = aircraft empty weight (lb); W_R = reserve fuel weight (lb); W_F = consumed fuel weight (lb); and P = payload (lb).

This equation assumes that the difference between the takeoff weight and landing weight is the fuel consumed. For example, suppose that for a Boeing B-747 the maximum payload carried (based on internal fuselage volume and structural limits) is 260,000 lb, maximum W_{TO} is 800,000 lb, W_{R} = 15,000 lb, W_E = 370,000, L/D = 17, V = 580 mph, and c' = 0.65 lb/lb thrust × h. The resulting weight ratio $[W_{\mathrm{TO}}/W_L = 800/(370+15+260)]$ is 1.24 and the range R is 3267 mi. Payloads below 260,000 allow higher ranges.

Most airline companies fly scheduled routes, although charter services are common. U.S. airlines are largely free to fly whatever routes (i.e., origin–destination pairs) they prefer in the U.S. In much of the rest of the world, authority to serve particular routes is regulated or negotiated by international agreements. The major components of airline costs are direct operating costs (e.g., aircraft depreciation or rentals, aircrews, fuel, and aircraft maintenance) and indirect operating costs (e.g., reservations, advertising and other marketing costs, in-flight service, ground processing of passengers and bags, and administrations).

The efficiency and competitiveness of airline service is heavily dependent on efficient operational planning. Airline scheduling is a complex problem in which demand at various times and places, route authority, aircraft availability and maintenance schedules, crew availability and flying restrictions, availability of airport gates and other facilities, and various other factors must all be considered. Airline management problems are discussed in [Wells, 1984].

Airports range from small unmarked grass strips to major facilities requiring many thousands of acres and billions of dollars. Strictly speaking, an airport consists of an airfield (or "airside") and terminal (or "landside"). Airports are designed to accommodate specified traffic loads carried by aircraft up to a "design aircraft," which is the most demanding aircraft to be accommodated. The design aircraft might determine such features as runway lengths, pavement strengths, or terminal gate dimensions at an airport. Detailed guidelines for most aspects of airport design (e.g., runway lengths and other airfield dimensions, pavement characteristics, drainage requirements, allowable noise and other environmental impacts, allowable obstruction heights, lighting, markings, and signing) are specified by the U.S. Federal Aviation Administration (FAA) in a series of circulars.

Airport master plans are prepared to guide airport growth, usually in stages, toward ultimate development. These master plans:

1. Specify the airport's requirements
2. Indicate a site if a new airport is considered
3. Provide detailed plans for airport layout, land use around the airport, terminal areas, and access facilities
4. Provide financial plans, including economic and financial feasibility analysis

Major new airports tend to be very expensive and very difficult to locate. Desirable airport sites must be reasonably close to the urban areas they serve yet far enough away to ensure affordable land and acceptable noise impacts. Many other factors — including airspace interference with other airports, obstructions (e.g., hills, buildings), topography, soil, winds, visibility, and utilities — must be reconciled. Hence, few major new airports are being built, and most airport engineering and planning work in the U.S. is devoted to improving existing airports. Governments sometimes develop multiairport system plans for entire regions or countries.

National agencies (such as the FAA in the U.S.) are responsible for traffic control and airspace management. Experienced traffic controllers, computers, and specialized sensors and communication systems are required for this function. Increasingly sophisticated equipment has been developed to maintain safe operations even for crowded airspace and poor visibility conditions. For the future we can expect increasing automation in air traffic control, relying on precise aircraft location with global positioning satellite (GPS) systems and fully automated landings. Improvements in the precision and reliability of control systems are increasing (slowly) the capacity of individual runways as well as reducing the required separation among parallel runways, thus allowing capacity increases in restricted airport sites.

85.4 Railroad Transportation

The main advantages of railroad technology are low frictional resistance and automatic lateral guidance. The low friction reduces energy and power requirements but limits braking and hill-climbing abilities. The lateral guidance provided by wheel flanges allows railroad vehicles to be grouped into very long trains, yielding economies of scale and, with adequate control systems, high capacities per track. The potential energy efficiency and labor productivity of railroads is considerably higher than for highway modes, but is not necessarily realized due to regulations, managerial decisions, demand characteristics, or terrain.

The main competitors of railroads include automobiles, aircraft, and buses for passenger transportation, and trucks, ships, and pipelines for freight transportation. To take advantage of their scale economies, railroad operators usually seek to combine many shipments into large trains. Service frequency is thus necessarily reduced. Moreover, to concentrate many shipments, rail cars may be frequently re-sorted into different trains, rather than moving directly from origin to destination, which results in long periods spent waiting in classification yards, long delivery times, and poor vehicle utilization. An alternative operational concept relying on direct nonstop "unit trains" is feasible only when demand is sufficiently large between an origin–destination pair.

Substantial traffic is required to cover the relatively high fixed costs of railroad track. Moreover, U.S. railroads, which are privately owned, must pay property taxes on their tracks, unlike their highway competitors. By 1920, highway developments had rendered low-traffic railroad branch lines noncompetitive in the U.S. Abandonment of such lines has greatly reduced the U.S. railroad network, even though the process was retarded by political regulation.

The alignment of railroad track is based on a compromise between initial costs and operating costs. The latter are reduced by a more straight and level alignment, which requires more expensive earthwork, bridges, or tunnels. Hay [1982] provides design guidelines for railroads.

In general, trains are especially sensitive to gradients. Thus, compared to highways, railroad tracks are more likely to go around rather than over terrain obstacles, which increases the **circuity factors** for railroad transportation.

The resistance for railroad vehicles may be computed using the Davis equation [Hay, 1982]:

$$r = 1.3 + 29/w + bV + CAV^2/wn + 20G + 0.8D \tag{85.2}$$

where G= gradient (%); D = degree of curvature; r = unit resistance (lb of force per ton of vehicle weight); w = weight (tons per axle of car or locomotive); n = number of axles; b = coefficient of flange friction, swaying, and concussion (0.045 for freight cars and motor cars in trains, 0.03 for locomotives and passenger cars, and 0.09 for single-rail cars); C = drag coefficient of air [0.0025 for locomotives (0.0017 for streamlined locomotives) and single- or head-end-rail cars, 0.0005 for freight cars, and 0.00034 for trailing passenger cars, including rapid transit]; A = cross-sectional area of locomotives and cars (usually 105 to 120 ft² for locomotives, 85 to 90 ft² for freight cars, 110 to 120 ft² for multiple-unit and passenger cars, and 70 to 110 ft² for single- or head-end-rail cars); and V = speed (mph).

The coefficients shown for this equation reflect relatively old railroad technology and can be significantly reduced for modern equipment [Hay, 1982]. The equation provides the unit resistance in pounds of force per ton of vehicle weight. The total resistance of a railroad vehicle (in lb) is

$$R_v = rwn \tag{85.3}$$

The total resistance of a train R is the sum of resistances for individual cars and locomotives. The rated horsepower (hp) required for a train is:

$$\text{hp} = \frac{RV}{375\eta} \tag{85.4}$$

where η = transmission efficiency (typically about 0.83 for a diesel electric locomotive).

The hourly fuel consumption for a train may be computed by multiplying hp by a specific fuel consumption rate (approximately 0.32 lb/hp × h for a diesel electric locomotive).

Diesel electric locomotives with powers up to 5,000 hp haul most trains in the U.S. Electric locomotion is widespread in other countries, especially those with low petroleum reserves. It is especially competitive on high-traffic routes (needed to amortize electrification costs) and for high-speed passenger trains. Steam engines have almost disappeared.

The main types of freight rail cars are box cars, flat cars (often used to carry truck trailers or intermodal containers), open-top gondola cars, and tank cars. Passenger trains may include restaurant cars and sleeping cars. Rail cars have tended toward increasing specialization for different commodities carried, a trend that reduces opportunities for back hauls. Recently, many "double-stack" container cars have been built to carry two tiers of containers. Such cars require a vertical clearance of nearly 20 ft, as well as reduced superelevation (banking) on horizontal curves. In the U.S., standard freight rail cars with gross weights up to 315,000 lb are used.

High-speed passenger trains have been developed intensively in Japan, France, Great Britain, Italy, Germany, and Sweden. Some of the most advanced French and Japanese trains (in 2003) have cruising speeds of 186 mph and double-deck cars. At such high speeds, trains can climb long, steep grades (e.g.,

3.5%) without slowing down much. Construction costs in hilly terrain can thus be significantly reduced. Even higher speeds are being tested in experimental railroad and magnetic levitation (MAGLEV) trains.

85.5 Highway Transportation

Highways provide very flexible and ubiquitous transportation for people and freight. A great variety of transportation modes, including automobiles, buses, trucks, motorcycles, bicycles, pedestrians, animal-drawn vehicles, taxis, and carpools, can share the same roads. From unpaved roads to multilane freeways, roads can vary enormously in their cost and performance. Some highway vehicles may even travel off the roads in some circumstances. The vehicles also range widely in cost and performance, and at their lower range (e.g., bicycles) are affordable for private use even in poor societies.

Flexibility, ubiquity, and affordability account for the great success of highway modes. Personal vehicles from bicycles to private automobiles offer their users great freedom and access to many economic and social opportunities. Trucks increase the freedom and opportunities available to farmers and small business. Motor vehicles are so desirable and affordable that in the U.S. the number of registered cars and trucks approximates the number of people of driving age. Other developed countries are approaching the same state despite strenuous efforts to discourage motor vehicle use.

The use of motor vehicles brings significant problems and costs. These include:

1. Road capacity and congestion. Motor vehicles require considerable road space, which is scarce in urban areas and costly elsewhere. Shortage of road capacity results in severe congestion and delays.
2. Parking availability and cost.
3. Fuel consumption. Motor vehicles consume vast amounts of petroleum fuels. Most countries have to import such fuels and are vulnerable to price increases and supply interruptions.
4. Safety. The numbers of people killed and injured and the property damages in motor vehicle accidents are very significant.
5. Air quality. Motor vehicles are major contributors to air pollution.
6. Regional development patterns. Many planners consider the low-density "sprawl" resulting from motor vehicle dominance to be inefficient and inferior to the more concentrated development produced by mass transportation and railroads.

In the U.S., trucks have steadily increased their share of the freight transportation market, mostly at the expense of railroads, as shown in Table 85.1. They can usually provide more flexible, direct, and responsive service than railroads do, but at higher unit cost. They are intermediate between rail and air transportation in both cost and service quality. With one driver required per truck, the labor productivity is much lower than for railroads, and there are strong economic incentives to maximize the load capacity for each driver. Hence, the tendency has been to increase the number, dimensions, and weights allowed for trailers in truck-trailer combinations, which requires increased vertical clearances (e.g., bridge over-passes), geometric standards for roads, and pavement costs.

Various aspects of highway flow characteristics, design standards, and safety problems were presented in Chapter 82 through Chapter 84. The main reference for highway design is the AASHTO manual [AASHTO, 1990]. For capacity, the main reference is the Transportation Research Board *Highway Capacity Manual* [TRB, 1985]. Extensive software packages have been developed for planning, capacity analysis, geometric design, and traffic control.

Currently (2003), major research and development efforts are being devoted to exploiting advances in information technology to improve highway operations. The Intelligent Transportation Systems (ITS) program of the U.S. Department of Transportation includes, among other activities, an Advanced Traffic Management System (ATMS) program to greatly improve the control of vehicles through congested road networks, an Advanced Travelers Information System (ATIS) program to guide users through networks, and, most ambitiously, an Automated Highway System (AHS) program to replace human driving with hardware. Such automation, when it becomes feasible and safe, has the potential to drastically improve lane capacity at high speeds by greatly reducing spacing between vehicles. Other potential benefits include reduced labor costs for trucks, buses, and taxis; higher and steadier speeds; improved routings through networks; remote self-parking vehicles; and use of vehicles by nondrivers such as children and handicapped persons. However, substantial technological, economical, and political problems will have to be surmounted.

Very elaborate traffic control and management systems are currently being developed to minimize traffic delays in transportation networks. These often use information from numerous distributed sensors to forecast, simulate, and optimize traffic flows. Since the demand for road trips may greatly exceed the available road capacity, transportation system management policies, such as pricing schemes, lane use priorities, parking controls, land use controls, and incentives for public transportation use, may be helpful in shifting some of the demand to less congested travel routes, modes, or times of day. As telecommunications improve over time, they may also replace some trips currently made by highways or other modes.

As the infrastructure for highways and other transportation modes matures, maintaining existing facilities without excessive costs or service disruptions becomes at least as challenging as building new facilities. Hence, considerable efforts are being devoted to materials research, to maintenance planning and scheduling, and to the development of efficient inspection, maintenance, and repair techniques.

85.6 Water Transportation

Water transportation may be classified into (1) marine transportation across seas and (2) inland waterway transportation; their characteristics differ very significantly. Inland waterways consist mostly of rivers, which may be substantially altered to aid transportation. Lakes and artificial canals may also be part of inland waterways. Rivers in their natural states are often too shallow, too fast, or too variable in their flows. All these problems may be alleviated by impounding water behind dams at various intervals. (This also helps generate electric power.) Boats can climb or descend across dams by using **locks** or other elevating systems [Hochstein, 1981]. In the U.S. inland waterway network, there are well over 100 major lock structures, with chambers up to 1200 ft long and 110 ft wide. Such chambers allow up to 18 large barges (35 × 195 ft) to be raised or lowered simultaneously.

In typical inland waterway operations, large diesel-powered "towboats" (which actually push barges) handle a rigidly tied group of barges (a "tow"). Tows with up to 48 barges (35 × 195 ft, or about 1300 tons/barge) are operated on the lower Mississippi, where there are no locks or dams. On other rivers, where locks are encountered at frequent intervals, tow sizes are adjusted to fit through locks. The locks constitute significant bottlenecks in the network, restricting capacity and causing significant delays.

Table 85.1 indicates that the waterway share of U.S. freight transportation has increased substantially in recent years. This is largely attributable to extensive improvements to the inland waterway system undertaken by the responsible agency, the U.S. Army Corps of Engineers.

The main advantage of both inland waterway and marine transportation is low cost. The main disadvantage is relatively low speed. Provided that sufficiently deep water is available, ships and barges can be built in much larger sizes than ground vehicles. Ship costs increase less than proportionally with ship size, for ship construction, crew, and fuel. Energy efficiency is very good at low speeds [e.g., 10 to 20 knots (nautical mi/h)]. However, at higher speeds the wave resistance of a conventional ship increases with the fourth power of speed (V^4). Hence, the fuel consumption increases with V^4 and the power required increases with V^5. Therefore, conventional-displacement ships rarely exceed 30 knots in commercial operation. Higher practical speed may be obtained by lifting ships out of the water on hydrofoils or air cushions. However, such unconventional marine vehicles have relatively high costs and limited markets at this time. Over time, ships have increased in size and specialization. Crude oil tankers of up to 550,000 tons (of payload) have been built. Tankers carrying fluids are less restricted in size than other ships because they can pump their cargo from deep water offshore without entering harbors. Bulk carriers (e.g., for coal, minerals, or grains) have also been built in sizes exceeding 300,000 tons. They may also be loaded through conveyor belts built over long pier structures to reach deep water. General cargo ships and containerships are practically always handled at shoreline berths and require much storage space nearby.

The use of intermodal containers has revolutionized the transportation of many cargoes. Such containers greatly reduce the time and cost required to load and unload ships. Up to 8000 standard 20-ft containers ($20 \times 8 \times 8$ ft) can be carried at a speed of about 25 knots on the largest recently built (2003) containerships.

Port facilities for ships should provide shelter from waves and sufficiently deep water, including approach channels to the ports. In addition, ports should provide adequate terminal facilities, including loading and unloading equipment, storage capacity, and suitable connections to other transportation networks. Ports often compete strenuously with other ports and strive to have facilities that are at least equal to those of competitors. Since ports generate substantial employment and economic activities, they often receive financial and other support from governments.

Geography limits the availability of inland waterways and the directness of ship paths across oceans. Major expensive canals (e.g., Suez, Panama, Kiel, Welland) have been built to provide shortcuts in shipping routes. These canals may be so valuable that ship dimensions are sometimes compromised (i.e., reduced) to fit through these canals. In some parts of the world (e.g., Baltic, North Sea, most U.S. coasts) the waters are too shallow for the largest ships in existence. Less efficient, smaller ships must be used there. The dredging of deeper access channels and ports can increase the allowable ship size, if the costs and environmental impacts are considered acceptable.

85.7 Public Transportation

Public transportation is the term for ground passenger transportation modes available to the general public. It connotes public availability rather than ownership. "Conventional" public transportation modes have fixed routes and fixed schedules and include most bus and rail transit services. "Unconventional" modes (also labeled "paratransit") include taxis, carpools and van pools, rented cars, dial-a-ride services, and subscription services.

The main purposes of public transportation services, especially conventional mass transportation services in developed countries, are to provide mobility for persons without automobiles (e.g., children, poor, nondrivers); to improve the efficiency of transportation in urban areas; to reduce congestion effects, pollution, accidents, and other negative impacts of automobiles; and to foster preferred urban development patterns (e.g., strong downtowns and concentrated rather than sprawled development).

Conventional services (i.e., bus and rail transit networks) are quite sensitive to demand density. Higher densities support higher service frequencies and higher network densities, which decrease user wait times and access times, respectively. Compared to automobile users, bus or rail transit users must spend extra time in access to and from stations and in waiting at stations (including transfer stations). Direct routes are much less likely to be available, and one or more transfers (with possible reliability problems) may be required. Thus, mass transit services tend to be slower than automobiles unless exclusive rights-of-way (e.g., bus lanes, rail tunnels) can favor them. Such exclusive rights-of-way can be quite expensive if placed on elevated structures or in tunnels. Even when unhindered by traffic, average speeds may be limited by frequent stops and allowable acceleration limits for standing passengers. Prices usually favor mass transit, especially if parking for automobiles is scarce and expensive.

The capacity of a transit route can be expressed as:

$$C = F\,L\,P \tag{85.5}$$

where C = one-way capacity (passengers/hour) past a certain point; F = service frequency (e.g., trains/hour); L = train length (cars/train); and P = passenger capacity of cars (spaces/car).

For rail transit lines where high capacity is needed in peak periods, C can reach 100,000 passengers/hour (i.e., 40 trains/hour $\times$ 10 cars/train $\times$ 250 passenger spaces/car). There are few places in the world where such capacities are required. For a bus line the train length L would usually be 1.0. If no on-line stops are allowed, an exclusive bus lane also has a large capacity (e.g., 1000 buses/hour $\times$ 90 passenger spaces/bus), but such demand levels for bus lanes have not been observed.

The average wait time of passengers on a rail or bus line depends on the headway, which is the interval between successive buses or trains. This can be approximated by:

$$\overline{W} = \overline{H}/2 + \text{var}(H)/2\overline{H} \tag{85.6}$$

where $\overline{W}$ = average wait time (e.g., minutes); $\overline{H}$ = average headway (e.g., minutes); var(H) = variance of headway (e.g., minutes^2).

It should be noted that the headway is the inverse of the service frequency.

The minimum number of vehicles N required to serve a route (before factoring reserve vehicles or those undergoing maintenance) is:

$$N = RFL \tag{85.7}$$

where R = vehicle round trip time on route (e.g., hours).

The effectiveness of a public transportation system depends on many factors, including demand distribution and density, network configuration, routing and scheduling of vehicles, fleet management, personnel management, pricing policies, and service reliability. Demand and economic viability of services also depend on automobile ownership and on how good and uncongested the road system is for automobile users.

Engineers can choose from a great variety of options for propulsion, support, guidance and control, vehicle configurations, facility designs, construction methods, and operating concepts. New information and control technology can significantly improve public transportation systems. It will probably foster increased automation and a trend toward more personalized (i.e., taxi-like) service rather than mass transportation.

Defining Terms

Capacity — The maximum flow rate that can be expected on a transportation facility. "Practical" capacity is sometimes limited by "acceptable" delay levels, utilization rates, and load factors.

Circuity factor — Ratio of actual distance on network to shortest airline distance.

Delay — Increase in service time due to congestion or service interruptions.

Demand-responsive — A mode whose schedule or route is adjusted in the short term as demand varies, such as taxis, charter airlines, and "TRAMP" ships.

Load factor — Fraction of available space or weight-carrying capability that is used.

Lock — A structure with gates at both ends, which is used to lift or lower ships or other vessels.

Mode — A distinct from of transportation.

Subsonic — Flying below the speed of sound (Mach 1), which is approximately 700 mph at cruising altitudes of approximately 33,000 ft.

Utilization rate — Fraction of time that a vehicle, facility, or equipment unit is in productive use.

References

AASHTO (American Society of State Highway and Transportation Officials). 1990. *A Policy on Geometric Design of Highways and Streets,* AASHTO, Washington, DC.

Brun, E. 1981. *Port Engineering,* Gulf Publishing Co., Houston.

Hay, W. W. 1985. *Railroad Engineering,* John Wiley & Sons, New York.

Hochstein, A. 1981. *Waterways Science and Technology,* Final Report DACW 72-79-C-0003. U.S. Army Corps of Engineers, August.

Homburger, W. S. 1985. *Transportation and Traffic Engineering Handbook,* Prentice Hall, Englewood Cliffs, NJ.

Horonjeff, R. and McKelvey, F. 1994. *Planning and Design of Airports,* McGraw-Hill, New York.

Morlok, E. K. 1976. *Introduction to Transportation Engineering and Planning,* McGraw-Hill, New York.
TRB (Transportation Research Board). 1985. *Highway Capacity Manual,* Special Report 209. TRB, Washington, DC.
Vuchic, V. 1981. *Urban Public Transportation,* Prentice Hall, Englewood Cliffs, NJ.
Wells, A. T. 1984. *Air Transportation,* Wadsworth Publishing Co., Belmont, CA.
Wright, P. H. and Paquette, R. J. 1987. *Highway Engineering,* John Wiley & Sons, New York.

Further Information

The ITE Handbook [Homburger, 1992] and Morlok [1978] cover most transportation modes. Horonjeff and McKelvey [1994], Hay [1982], Wright and Paquette [1987], Brun [1981], and Vuchic [1981] are more specialized textbooks covering airports, railroads, highways, ports, and urban public transportation systems, respectively. Periodicals such as *Aviation Week & Space Technology, Railway Age, Motor Ship,* and *Mass Transit* cover recent developments in their subject areas.

86

Intelligent Transportation Systems*

Yorgos J. Stephanedes
Institute of Computer Science Foundation for Research and Technology — Hellos, Greece

*Note: This document contains very substantial portions of text, largely unchanged, from References 1, 2, 32, 38, 41, and 46.

0-8493-1586-7/05/$0.00+$1.50

86.1 Introduction

Surface transportation systems in the United States today face a number of significant challenges. Congestion and safety continue to present serious problems in spite of the nation's superb roadway systems. Congestion imposes an exorbitant cost on productivity, costing the nation an estimated $40 billion per year. Vehicle crashes cause another $150 billion burden to the economy and result in the loss of 40,000 lives annually. Inefficient surface transportation, whether in privately owned vehicles, commercial motor carriers, or public transit vehicles, constitutes a burden on the nation's quality of life through wasted energy, increased emissions, and serious threats to public safety [41]. In addition, it directly impacts national economic growth and competitiveness.

Over the last two decades, demand for mobility has continued to increase, but the available capacity of the roadway system is nearly exhausted. Vehicle travel has increased 70%, while road capacity has increased only slightly more than 1% [47].

Except for fine-tuning and relatively modest additions, the road system cannot be expanded in many areas. The only means left for increasing available travel capacity is to use the available capacity more effectively, e.g., redirect traffic to avoid congestion, provide assistance to drivers and other travelers on planning and following optimal routes, increase the reliability of and access to public transportation, and refocus safety efforts on accident avoidance rather than merely minimizing the consequences of accidents [1].

Responding to this need, Intelligent Transportation Systems (ITS) is the integrated application of well-established technologies in advanced information processing and communications, sensing, control, electronics, and computer hardware and software to improve surface transportation performance, both in the vehicle and on the highway [1,2,32].

This simple definition underlies what has been a substantial change in surface transportation in the United States and around the world. Development of ITS was motivated by the increased difficulty — social, political, and economic — of expanding transportation capacity through conventional infrastructure building. ITS represents an effort to harness the capabilities of advanced technologies to improve transportation on many levels. ITS is intended to reduce congestion, enhance safety, mitigate the environmental impacts of transportation systems, enhance energy performance, and improve productivity [32].

Intelligent Transportation Systems, formerly Intelligent Vehicle–Highway Systems (IVHS), offers technology-based solutions to the compelling challenges confronting the nation's surface transportation systems while concurrently establishing the basis for dealing with future demands through a strategic, intermodal view of transportation. ITS applications offer proven and emerging technologies in fields such as data processing, communications, control, navigation, electronics, and the supporting hardware and software systems capable of addressing transportation challenges. Although ITS technology applications alone cannot completely satisfy growing transportation needs, they provide the means to revise current approaches to problem solving, and they improve the efficiency and effectiveness of existing systems. When deployed and integrated effectively, ITS technologies will enable the surface transportation system to operate as multimodal, multijurisdictional entities providing meaningful benefits, including more efficient use of infrastructure and energy resources, complemented by measurable improvements in safety, mobility, productivity, and accessibility [41].

Some of the effects that ITS could have in transportation operations, safety, and productivity are described next.

86.2 Role of ITS in Tomorrow's Transportation Systems

Operations

The essence of ITS as it relates to transportation operations is the improved ability to manage transportation services as a result of the availability of accurate, real-time information and to greatly enhance control of traffic flow and individual vehicles. With ITS, decisions that individuals make as to time, mode, and route choices can be influenced by information that currently is not available when it is needed or is incomplete, inconvenient, or inaccurate. For example, ITS technology would enable operators to detect incidents more quickly; to provide information immediately to the public on where the incident is located, its severity, its effect on traffic flow, and its expected duration; to change traffic controls to accommodate changes in flow brought about by the incident; and to provide suggestions on better routes and information on alternative means of transportation [2].

The availability of this information would also enable the development of new transportation control strategies. For example, to obtain recommended routing information, drivers will have to specify their origins and destinations. Knowledge of origin and destination information in real time will enable the development of traffic assignment models that will be able to anticipate when and where congestion will occur (origins and destinations can also be estimated in real time). Control strategies that integrate the operation of freeway ramp metering systems, driver information systems, and arterial traffic signal control systems, and that meter flow into bottleneck areas, can be developed to improve traffic control. Eventually, perhaps toward the third decade of the 21st century, totally automated facilities may be built on which vehicles would be controlled by electronics in the highway [2].

Safety

Whereas many safety measures developed over the years have been aimed at reducing the consequences of accidents (such as vehicle crashworthiness and forgiving roadside features), many ITS functions are directed toward the *prevention* of accidents. A premise of the European PROMETHEUS program, for example, was that 50% of all rear-end collisions and accidents at crossroads and 30% of head-on collisions could be prevented if the driver was given another half-second of advance warning and reacted correctly. Over 90% of these accidents could be avoided if drivers took the appropriate countermeasures 1 second earlier. ITS technologies that involve sensing and vehicle-to-vehicle communications are initially designed to automatically warn the driver, providing enough lead time for him or her to take evasive actions. The technologies may also assume some of the control functions that are now totally the responsibility of drivers, compensating for some of their limitations and enabling them to operate their vehicles closer together but safer. Even before these crash-avoidance technologies become available to the public, ITS holds promise for improving safety by providing smoother traffic flow. For example, driver information systems provide warnings on incident blockages ahead, and this may soften the shock wave that propagates as a result of sudden and abrupt decelerations caused by unanticipated slowdowns [2].

Potential safety dangers must, however, also be acknowledged. A key issue involves driver distraction and information overload from the various warning and display devices in the vehicle. Other issues include dangers resulting from system unreliability (for example, a warning or driver-aid system that fails to operate) and the incentive for risky driving that ITS technologies may provide. These are important research issues that must be addressed before such systems are widely implemented [2].

Productivity

The availability of accurate, real-time information will be especially useful to operators of vehicle fleets, including transit, high-occupancy vehicle (HOV), emergency, fire, and police services, as well as truck fleets. Operators are able to know where their vehicles are and may receive an estimate on how long a trip can be expected to take; thus, they will be able to advise on best routes to take and will be able to manage their fleets better [2].

There is great potential for productivity improvements in the area of regulation of commercial vehicles. Automating and coordinating regulatory requirements through application of ITS technologies can, for example, reduce delays incurred at truck weigh stations, reduce labor costs to the regulators, and minimize the frustration and costs of red tape to long-distance commercial vehicle operators. There is also potential to improve coordination among freight transportation modes; as an example, if the maritime and trucking industries used the same electronic container identifiers, as has been the case in certain limited applications, freight-handling efficiencies would be greatly improved [2].

86.3 ITS Categories

At its early stages, by general agreement, IVHS (the precursor of ITS) had been subdivided into six interlocking system areas, three focused on technology and three on applications [1]:

Technology oriented:
- Advanced Traffic Management Systems (ATMS)
- Advanced Traveler Information Systems (ATIS)
- Advanced Vehicle Control Systems (AVCS)

Applications oriented:
- Advanced Public Transportation Systems (APTS)
- Commercial Vehicle Operations (CVO)
- Advanced Rural Transportation Systems (ARTS)

Advanced Traffic Management Systems

ATMS addresses technologies to monitor, control, and manage traffic on streets and highways. ATMS technologies include [1]:

- Traffic management centers (TMCs) in major metropolitan areas to gather and report traffic information, and to control traffic movement to enhance mobility and reduce congestion through ramp, signal, and lane management; vehicle route diversion; etc.
- Sensing instrumentation along the highway system, which consists of several types of sensors, including magnetic loops and machine vision systems, that provide current information on traffic flow to the TMC
- Variable message signs that provide current information on traffic conditions to highway users and suggest alternate routes
- Priority control systems to provide safe travel for emergency vehicles when needed
- Programmable, directional traffic signal control systems
- Automated dispatch of tow, service, and emergency vehicles to accident sites

Advanced traffic management systems have six primary characteristics differentiating them from the typical traffic management system of today [2]. In particular, ATMS:

- Work in real time.
- Respond to changes in traffic flow. In fact, an ATMS will be one step ahead, predicting where congestion will occur based on collected origin–destination information.
- Include areawide surveillance and detection systems.
- Integrate management of various functions, including transportation information, demand management, freeway ramp metering, and arterial signal control.
- Imply collaborative action on the part of the transportation management agencies and jurisdictions involved.
- Include rapid-response incident management strategies.

To implement ATMS, real-time traffic monitoring and data management capabilities are being developed, including advanced detection technology, such as image processing systems, automatic vehicle location and identification techniques, and the use of vehicles as probes. New traffic models are being created, including real-time dynamic traffic assignment models, real-time traffic simulation models, and corridor optimization techniques. The applicability of artificial intelligence and expert systems techniques is assessed, and applications such as rapid incident detection, congestion anticipation, and control strategy selection are being developed and tested [2].

Advanced Traveler Information Systems

ATIS address technologies to assist travelers with planning, perception, analysis, and decision making to improve the convenience and efficiency of travel. In the automobile, ATIS technologies include [1]:

- Onboard displays of maps and roadway signs (in-vehicle signing)
- Onboard navigation and route guidance systems
- Systems to interpret digital traffic information broadcasts
- Onboard traffic hazard warning systems (e.g., icy road warnings)

Outside the vehicle, ATIS technologies include [1]:

- Trip planning services
- Public transit route and schedule information available online at home, office, kiosks, and transit stops.

Advanced traveler information systems provide drivers with information about congestion and alternate routes, navigation and location, and roadway conditions through audio and visual means in the vehicle. This information can include incident location, location of fog or ice on the roadway, alternate routes, recommended speeds, and lane restrictions. ATIS provide information that assist in trip planning at home, at work, and by operators of vehicle fleets. ATIS also provide information on motorist services such as restaurants, tourist attractions, and the nearest service stations and truck and rest stops (this has been called the yellow pages function.) ATIS can include onboard displays that replicate warning or navigational roadside signs when they may be obscured during inclement weather or when the message should be changed, as when speed limits should be lowered on approaches to congested freeway segments or fog areas. An automatic Mayday feature may also be incorporated, which would provide the capability to automatically summon emergency assistance and provide vehicle location [2].

A substantial effort is required to define the communications technology, architecture, and interface standards that will enable two-way, real-time communication between vehicles and a management center. Possibilities include radio data communications, cellular systems, roadside beacons used in conjunction with infrared or microwave transmissions or low-powered radio signals, and satellite communications. Software methods to fuse the information collected at the management center and format it for effective use by various parties must also be developed. These parties include commuters, tourists, other trip makers, and commercial vehicle operators, both before they make a trip and en route; operators of transportation management systems; and police, fire, and emergency response services [2].

A number of critical human factors issues must also be investigated. These include identifying the critical pieces of information and the best way of conveying them to different individuals. The human factors issues also include a critical examination of in-vehicle display methods [2].

Advanced Vehicle Control Systems

AVCS address technologies to enhance the control of vehicles by facilitating and augmenting driver performance and, ultimately, relieving the driver of some tasks, through electronic, mechanical, and communications devices in the vehicle and on the roadway. AVCS technologies include [1]:

- Adaptive cruise control, which slows a cruise-controlled vehicle if it gets too close to a preceding vehicle
- Vision enhancement systems, which aid driver visibility in the dark or in adverse weather
- Lane departure warning systems, which help drivers avoid run-off-the-road crashes
- Automatic collision avoidance systems, i.e., automatic braking upon obstacle detection
- Automated Highway Systems (AHS), automatically controlling vehicles in special highway lanes to increase highway capacity and safety

Whereas the other categories of ITS primarily serve to make traveling more efficient by providing more timely and accurate information about transportation, AVCS serve to greatly improve safety and potentially make dramatic improvements in highway capacity by providing information about changing conditions in the immediate environment of the vehicle, sounding warnings, and assuming partial or total control of the vehicle [2].

Early implementation of AVCS technologies may include a number of systems to aid with the driving task. These include hazard warning systems that sound an alarm or actuate a light when a vehicle moves dangerously close to an object, such as when backing up or when moving into the path of another vehicle when changing lanes. Infrared imaging systems may also be implemented that enhance driver visibility at night. AVCS technologies also include adaptive cruise control and lane-keeping systems that automatically adjust vehicle speed and position within a lane through, for example, radar systems that detect the position and speed of a lead vehicle, or possibly through electronic transmitters in the pavement that detect the position of vehicles within the lane and send messages to a computer in the vehicle that has responsibility for partial control functions. As technology advances, lanes of traffic may be set aside exclusively for automated operation, known as platooning highway systems. These automated facilities have the potential to greatly increase highway capacity while at the same time providing for safer operation [2].

Much research, development work, and testing are needed before such systems can be built and implemented, and much of it is taking place today. Perhaps the most important issues, though, relate to the role of humans in the system — that is, public acceptability and how it is likely to affect system effectiveness. Other human factors issues include driver reaction to partial or full control — whether it will cause them to lose alertness or drive more erratically. Another important area is AVCS reliability and the threat of liability [2].

The vision for the AHS program is to create a fully automated system that evolves from today's roads, beginning in selected corridors and routes; provides fully automated, "hands-off" operation at better levels of performance than there are today, in terms of safety, efficiency, and comfort; and allows equipped vehicles to operate in both urban and rural areas and on highways that are instrumented and not instrumented [7].

Although full deployment of an AHS is certainly a long-term goal, pursuit of this goal is extremely important. A new level of benefits could be realized with the complete automation of certain facilities. By eliminating human error, an automated highway could provide a nearly accident-free driving environment. In addition, the precise, automated control of vehicles on an automated vehicle–highway system could result in an increase of two to three times the capacity of present-day facilities while encouraging the use of more environmentally benign propulsion methods. Initial AHS deployments might be on heavily traveled urban or interstate highway segments, and the automated lanes might be comparable to the HOV vehicle lanes on today's highways. If successful, the AHS could evolve into a major advance of the nation's heavily traveled roadways or the interstate highway system [7].

Advanced Public Transportation Systems

APTS addresses applications of ITS technologies to enhance the effectiveness, availability, attractiveness, and economics of public transportation. APTS strives to improve performance of the public transportation system at the unit level (vehicle and operator) and at the system level (overall coordination of facilities and provision of better information to users). APTS technologies include [1]:

- Fleet monitoring and dispatch management
- Onboard displays for operators and passengers
- Real-time displays at bus stops
- Intelligent fare collection (e.g., using smart cards)
- Ride-share and HOV information systems

Applications of ITS technologies could lead to substantial improvements in bus and paratransit operations in urban and rural areas. Dynamic routing and scheduling could be accomplished through onboard devices, communications with a fleet management center, and public access to a transportation information system containing information on routes, schedules, and fares. Automated fare collection systems could also be developed that would enable extremely flexible and dynamic fare structures and relieve drivers of fare collection duties [2].

Commercial Vehicle Operations

CVO addresses applications of ITS technologies to commercial roadway vehicles (trucks, commercial fleets, and intercity buses). Many CVO technologies, especially for interstate trucking, relate to the automated, no-stop-needed handling of the routine administrative tasks that have traditionally required stops and waiting in long lines: toll collection, road-use calculation, permit acquisition, vehicle weighing, etc. Such automation can save time, reduce air pollution (most, and the worst, emissions are produced during acceleration and deceleration), and increase the reliability of record keeping and fee collection. CVO technologies include [1]:

- Automatic vehicle identification (AVI)
- Weigh-in motion (WIM)
- Automatic vehicle classification (AVC)
- Electronic placarding or bill of lading
- Automatic vehicle location (AVL)
- Two-way communications between fleet operator and vehicles
- Automatic clearance sensing (ACS)

The application of ITS technologies holds great promise for improving the productivity, safety, and regulation of all commercial vehicle operations, including large trucks, local delivery vans, buses, taxis, and emergency vehicles. Faster dispatching, efficient routing, and more timely pickups and deliveries are possible, and this will have a direct effect on the quality and competitiveness of businesses and industries at both the national and international levels. ITS technologies can reduce the time spent at weigh stations, improve hazardous material tracking, reduce labor costs to administer government truck regulations, and minimize costs to commercial vehicle operators [2].

ITS technologies manifest themselves in numerous ways in commercial vehicle operations. For example, for long-distance freight operations, onboard computers not only will monitor the other systems of the vehicle but also can function to analyze driver fatigue and provide communications between the vehicle and external sources and recipients of information. Applications include automatic processing of truck regulations (for example, commercial driver license information, safety inspection data, and fuel tax and registration data), avoiding the need to prepare redundant paperwork and leading to "transparent borders"; provision of real-time traffic information through advanced traveler information systems; proof of satisfaction of truck weight laws using weigh-in motion scales, classification devices, and automated vehicle identification transponders; and two-way communication with fleet dispatchers using automatic vehicle location and tracking and in-vehicle text and map displays. Regulatory agencies would be able to take advantage of computerized record systems and target their weighing operations and safety inspections at those trucks that are most likely to be in violation [2].

Advanced Rural Transportation Systems

ARTS address applications of ITS technologies to rural needs, such as vehicle location, emergency signaling, and traveler information. The issues involved in implementing ITS in rural areas are significantly different from those in urban areas, even when services are similar. Rural conditions include low population density, fewer roads, low amount of congestion, sparse or unconventional street addresses, etc. Different technologies and communications techniques are needed in rural ITS to deal with those conditions. Safety is a major issue in ARTS; over half of all accidents occur on rural roads. ARTS technologies include [1]:

- Route guidance
- Two-way communications

- Automatic vehicle location
- Automatic emergency signaling
- Incident detection
- Roadway edge detection

Application of ARTS technologies can address the needs of rural motorists who require assistance either because they are not familiar with the area in which they travel (tourists) or because they face extreme conditions such as weather, public works, and special events. Provision of emergency services is particularly important in rural areas. A state study [28] has determined that notification of spot hazardous conditions and collision avoidance at nonsignalized intersections are highly important issues that ARTS could address in the short term; long-term issues include construction zone assistance, transit applications, inclement weather trip avoidance and assistance, tourist en route information and traffic control, and in-vehicle Mayday devices.

86.4 ITS Restructuring and Progress

With the enactment of the Intermodal Surface Transportation Efficiency Act (ISTEA) in 1991, Congress set a new course for transportation by mandating increased efficiency and safety on the existing highway and transit infrastructure through increased emphasis on intermodalism — the seamless integration of multiple modes of transportation. In response to ISTEA, the U.S. Department of Transportation (DOT) initiated a multifaceted ITS program involving research and field operational testing of promising ITS applications. With the passage of the Transportation Equity Act for the 21st Century (TEA-21) in June 1998, Congress reaffirmed the U.S. DOT's role in continuing the development of ITS technologies and launching the transition nationwide and the integrated deployment of ITS applications to foster the management of multiple transportation resources as unified systems delivering increased efficiency, safety, and customer satisfaction. We have thus witnessed the restructuring of the ITS program from the program areas established during the ISTEA era into the new organization reflecting congressional direction in TEA-21, which emphasizes deployment and integration of ITS. The advent of TEA-21 catalyzed a restructuring of ITS program activities into intelligent infrastructure categories and the Intelligent Vehicle Initiative (IVI) [41].

The program reorientation reflects the evolution of emphasis to deployment whose output is infrastructure or vehicles. Metropolitan ITS infrastructure inherits the research in Advanced Traffic Management Systems, Advanced Public Transportation Systems, and Advanced Traveler Information Systems. The Rural ITS infrastructure encompasses the activities of the Advanced Rural Transportation Systems (ARTS) program, including the application of technologies under development for metropolitan and commercial vehicle infrastructure that are adaptable to rural community needs. The commercial vehicle ITS infrastructure continues to build on the research endeavors of the Commercial Vehicle Operations program and is heavily focused on the deployment of Commercial Vehicle Information Systems and Networks (CVISN). The Intelligent Vehicle Initiative integrates the work accomplished in various facets of intelligent vehicle research and development to include the Advanced Vehicle Control and Safety Systems (AVCSS) program and the Automated Highway Systems [41].

The enabling research and technology program area continues to provide crosscutting support to each of the four functional components constituting the program's foundation. Figure 86.1 provides a crosswalk depicting the dynamics of the realignment.

The restructured ITS program places emphasis in two major areas: deploying and integrating intelligent infrastructure, and testing and evaluating intelligent vehicles. Intelligent infrastructure and intelligent vehicles, working together, will provide the combinations of communications, control, and information management capabilities needed to improve mobility, safety, and traveler decision making in all modes of travel. Intelligent infrastructure comprises the family of technologies that enable the effective operation of ITS services in metropolitan areas, in rural and statewide settings, and in commercial vehicle applications. Intelligent vehicle technologies foster improvements in safety and mobility of vehicles. The

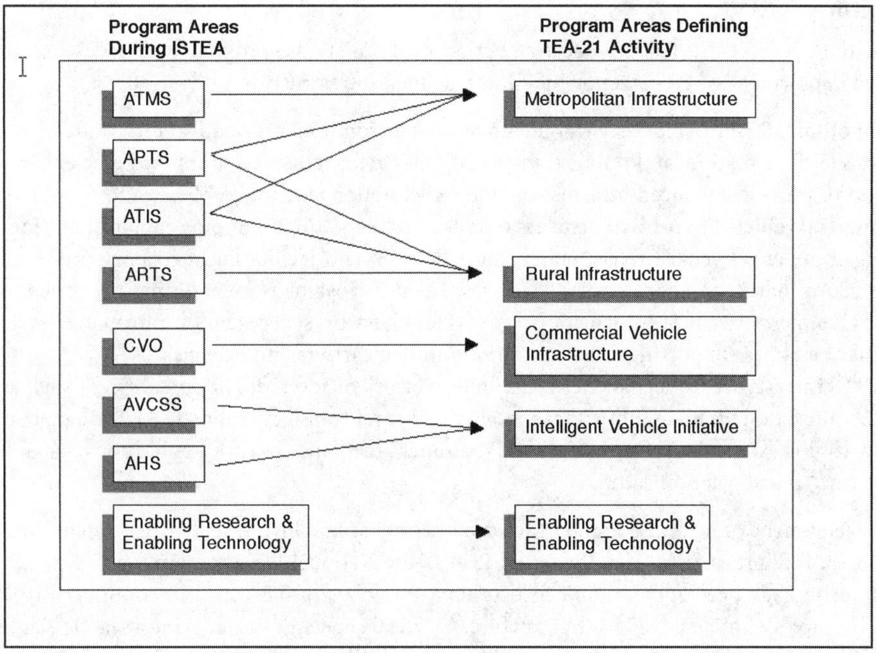

FIGURE 86.1 ITS program reorientation. (From ITS JPO, Department of Transportation's Intelligent Transportation Systems (ITS) Projects Book, U.S. DOT, FHWA Operations Core Business Unit, FTA Office of Mobility Innovation, National Highway Traffic Safety Administration, Washington, D.C., 2001.)

Intelligent Vehicle Initiative embraces four classes of vehicles: light vehicles (ranging from passenger automobiles and vans to light trucks), transit vehicles (buses), commercial vehicles (trucks and interstate buses), and specialty vehicles (emergency response, enforcement, and maintenance vehicles).

Within this restructuring, intelligent infrastructure and intelligent vehicle program development objectives are pursued through four program areas: metropolitan ITS infrastructure, rural ITS infrastructure, commercial vehicle ITS infrastructure, and the Intelligent Vehicle Initiative, as described below, which includes light vehicles, transit vehicles, trucks, and emergency and specialty vehicles.

The metropolitan ITS infrastructure program area is focused on deployment and integration of technologies in that setting. The rural ITS infrastructure program area emphasizes deployment of high-potential technologies in rural environments. Commercial vehicle ITS infrastructure program objectives are directed at safety and administrative regulation of interstate trucking. Intelligent vehicle program objectives are centered on in-vehicle safety systems for all classes of vehicles in all geographic environments.

There are no specific ITS applications that hold the potential for addressing all of the current or projected transportation system needs. The potential for success lies in developing a national transportation system incorporating integrated and interoperable ITS services. The ITS program envisions a gradual and growing interaction between infrastructure and vehicles to produce increased benefits in mobility and traveler safety.

The documents guiding ITS program direction are evolving. The U.S. DOT's goals, key activities, and milestones for fiscal years (FY) 1999 through 2003 are documented in the National Intelligent Transportation Systems Program 5-year horizon plan [38]. This plan was followed by a 10-year program plan [46] that presents the next generation research agenda for ITS. These two documents, coupled with the Intelligent Transportation Society of America's national deployment strategy, satisfy congressional direction in TEA-21 to update the National ITS Program Plan published in 1995 and address ITS deployment and research challenges for stakeholders at all levels of government and the private sector. Within the restructured framework, the ITS program is focused on activities impacting both near-term and long-term horizons.

Near Term

Through the end of FY 2003, the effective period of TEA-21, the program will focus on facilitating integrated deployment of ITS components in the defined infrastructure categories.

Metropolitan ITS infrastructure will integrate various components of advanced traffic management, traveler information, and public transportation systems to achieve improved efficiency and safety and to provide enhanced information and travel options for the public.

Commercial vehicle ITS infrastructure is oriented on integrating technology applications for improving commercial vehicle safety, enhancing efficiency, and facilitating regulatory processes for the trucking industry and government agencies. The principal instrument of this component is known as Commercial Vehicle Information Systems and Networks, a system of information systems that link the nodes supporting communications among carriers and agencies.

Rural ITS infrastructure is characterized by a framework of seven development tracks such as surface transportation weather and winter mobility and rural transit mobility. ITS technologies are demonstrating exceptional effectiveness and customer acceptance in such applications that are tailored to rural transportation settings.

The development of a robust market fueled by private sector investment is dependent on a critical mass of basic ITS infrastructure. In the era of ISTEA, the National ITS Program focused principally on research, technology development, and field testing; the focus of TEA-21 will continue this legacy by building on successes to deploy ITS infrastructure. A critical challenge in achieving a seamless, intermodal transportation system is ensuring interoperability through the use of an open, nonproprietary architecture and the adoption of ITS standards.

Long Term

The long-term focus will be directed at supporting research, development, and testing of advanced technologies demonstrating potential for deployment in the 5- to 20-year horizon. The in-vehicle component of this effort will be consolidated into a single Intelligent Vehicle Initiative centered heavily on applying driver assistance and control intervention systems to reduce vehicle crashes. A companion effort seeks to integrate driving assistance and motorist information functions to facilitate information processing, decision making, and more effective vehicle operation.

A summary of ITS projects, tests, and studies initiated through September 2000 that have been partially or totally financed from federal ITS funds can be found in Reference 41.

86.5 What We Have Learned [32]

Now, with the National ITS Program more than a decade old in the United States, it certainly is timely, appropriate, and necessary to ask: Have we succeeded in deploying ITS? Has that deployment had a positive effect on surface transportation? What have we learned from these ITS deployments that can guide us in the future?

A recent study by the Federal Highway Administration (FHWA) [32] has addressed these questions. While many possible definitions exist, success here was tied to effectiveness — that is, whether an ITS application addresses major societal goals such as enhanced safety and improved quality of life — and to deployment of each particular ITS technology or application. Implicit in this metric for success is the belief in the test of the marketplace and the ITS community's ability to select those technologies and applications it sees as cost-effective and beneficial.

How It Was Done

The ITS Joint Program Office (JPO) of the FHWA of the U.S. DOT funds the development of several databases that are used to judge various ITS technologies and applications. These databases include the following:

- Metropolitan ITS Deployment Tracking Database, maintained by the Oak Ridge National Laboratory
- Commercial Vehicle Information Systems Network Deployment Tracking Database, maintained by the John A. Volpe National Transportation Systems Center (Volpe Center)
- 1998 Survey of Transit Agencies, conducted by the Volpe Center
- ITS cost database, maintained by Mitretek Systems

Deployment levels for various technologies were defined as follows:

- Deployed in fewer than 10% of the possible sites = limited deployment
- Deployed in 10 to 30% of the possible sites = moderate deployment
- Deployed in more than 30% of the possible sites = widespread deployment

Deployment levels are based on the actual presence of particular technologies, not future plans to deploy, even if funding for the deployment has already been secured. However, simply identifying an ITS technology or application as unsuccessful (i.e., not adequately deployed) is not a sufficient base for understanding how to subsequently advance in that area. The study, therefore, included the reason for the lack of success, choosing among three fundamental causes for a technology or application not being deployed:

1. The technology simply did not function effectively in a real-world environment.
2. While the technology or application worked in a technical sense, it was too costly, meaning any one of the following:
   A. It was simply too expensive to deploy compared with the potential benefits that accrued from its deployment.
   B. The absolute costs of acquisition, operations, and maintenance were considered too large by the deploying organization.
   C. The technology used was not suitable for a particular application.
3. Institutional barriers prevented the effective deployment of the technology or application.

Any of these reasons for lack of success could potentially be overcome in the future. Technologies can be enhanced; prices of various technologies can and do fall, often dramatically; and institutional barriers, while often tenacious, can be overcome with careful work over the long term. Further, a particular technology or application may not have had time to develop a "following" in the marketplace, given development and deployment cycles. Therefore, each technology was characterized as one of the following: successful, unsuccessful, holds promise, or jury is still out. To be sure, the deployment level does not necessarily relate directly to success. For example, a technology that is only moderately deployed could be considered successful because it serves as an appropriate technological solution, though only for a small segment of the market.

The areas included within the scope of this study are [32] freeway, incident, and emergency management, and electronic toll collection (ETC); arterial management; traveler information systems; Advanced Public Transportation Systems; Commercial Vehicle Operations (CVO); crosscutting technical issues; and crosscutting institutional issues.

Freeway, Incident, and Emergency Management, and Electronic Toll Collection

This area includes a number of different, albeit related, technologies. Various technologies, including transportation management centers, ramp metering, dynamic message signs, roadside infrastructure, and dynamic lane and speed control, form the basis of these applications. ETC is one of the fundamental and earliest-deployed ITS technologies. It is also the most common example of the electronic linkage between vehicle and infrastructure that characterizes ITS. Freeways (i.e., limited-access highways) represent a major and early ITS application area. Incident management on those facilities is of primary importance in reducing nonrecurrent congestion. Emergency management predates ITS as a concept but is enhanced by the addition of ITS technologies.

Although a number of systems have seen widespread deployment, much more can be accomplished. Institutional issues preventing truly integrated services are a major barrier. An important technical advancement would be to upgrade such systems to be predictive (in the sense of predicting when congestion will occur in the future as a function of current traffic patterns and expectations about the future) as opposed to the responsive systems currently in place. There is a need to institutionalize operation budgets for these kinds of systems as well as a need to attract high-quality technical staff for deployment and operations support.

Arterial Management

Arterials are high-capacity roadways controlled by traffic signals, with access via cross-streets and often abutting driveways. Arterial management predates ITS, with early deployments going back to the 1960s; it is a useful ITS application with current deployment. However, adaptive control strategies, which make real-time adjustments to traffic signals based on sensing conditions (e.g., queues), at arterials are not in widespread use. While some argue that such control strategies have potential for substantial benefits, only a handful are deployed nationally, of which four are federally funded field operational tests. The reasons for this deployment lag include cost issues and concerns that algorithms for adaptive traffic control simply do not perform well. In particular, when traffic volumes are heavy, the state-of-the-art algorithms appear to break down (although vendors claim otherwise). Also, system complexity drives the need for additional training.

Widespread deployment has not yet occurred for traveler information systems for arterials, even though studies suggest safety and delay reduction benefits. The hope is that with the addition of cellular phones, or cellular phone geolocation for traffic probes, and implementation of a national three-digit traveler information number (511), more deployment will occur. Integration of various traffic management technologies with arterial management is an important next step. Integration of arterial management with emergency vehicle management, transit management, and freeway management would represent important and useful advances.

Traveler Information Systems

Traveler information is one of the core concepts of ITS. Among the items valued by consumers are high-quality information, easy and timely accessibility to that information, a high-quality user interface, and low prices, preferably free. Consumer demand for traveler information is a function of the amount of congestion on the regional transportation network, the overall characteristics of that network, what is provided on the supply side in terms of information quality and user interface, characteristics of individual trips, and driver and transit user characteristics.

Examples abound of various kinds of traveler information systems, with extensive deployment of various kinds of systems. While people value high-quality traveler information in the conceptual sense, they are not necessarily willing to pay for it. After all, free information — although often of lower quality — is universal (e.g., radio helicopter reports). So, whether traveler information systems can be a viable stand-alone commercial enterprise is likewise unclear. More likely, transportation information will be offered as part of some other package of information services. The Internet is likely to be a major basis of traveler information delivery in the future.

The analysis of traveler information systems brings home the fact that ITS operates within the environment of people's expectations for information. In particular, timeliness and quality of information are on a continually increasing slope in many non-ITS applications, with people's expectations heightened by the Internet and related concepts. Traveler information providers, whether in the public or private sector, need to be conscious of operating in the context of these changed expectations. Further, the effective integration of traveler information with network management, or transportation management systems, of which freeway and arterial management are examples, is currently virtually nonexistent. Both network management and traveler information systems would benefit by more substantial integration, as would the ultimate customers — travelers and freight carriers — of these systems.

Advanced Public Transportation Systems

That transit has difficulty attracting market share is a well-established fact. Reasons include the following: land-use patterns incompatible with transit use; lack of high-quality service, with travel times too long and unreliable; lack of comfort; security concerns; and incompatibility with the way people currently travel (for example, transit is often not suited for trip chaining). The hypothesis is that ITS transit technologies — including automatic vehicle location, passenger information systems, traffic signal priority, and electronic fare payment — can help ameliorate these difficulties, improving transit productivity, quality of service, and real-time information for transit users.

Using ITS to upgrade transit clearly has potential. However, deployment has, for the most part, been modest, stymied by a number of constraints: lack of funding to purchase ITS equipment, difficulties in integrating ITS technologies into conventional transit operations, and lack of human resources needed to support and deploy such technologies. Optimistically, there will be a steady but slow increase in the use of ITS technologies for transit management as people with ITS expertise join transit agencies. However, training is needed, and inertia must be overcome in deploying these technologies in a chronically capital-poor industry. Integrating transit services with other ITS services is potentially a major intermodal benefit of ITS transit deployments; it is hoped that this integration, including highway and transit, multiprovider services and intermodal transfers, will be feasible in the near term. Still, the question remains: How can we use ITS to fundamentally change transit operations and services? The transit industry needs a boost, and it can be vital in providing transportation services, especially in urban areas, and in supporting environmentally related programs. Can ITS be the mechanism by which the industry reinvents itself? The jury is certainly still out on that question.

Commercial Vehicle Operations

This review is limited to the public sector side of CVO systems (i.e., it does not include fleet management) as states fulfill their obligation to ensure safety and enforce other regulations related to truck operations on their highways. These systems fall under the CVISN rubric and deal with roadway operations, including safety information exchange and electronic screening, as well as back-office applications such as electronic credentialing.

While CVISN is experiencing some deployment successes, much remains to be done. Participation by carriers is voluntary in most programs, and requiring use of transponders by truckers may be difficult. Certainly these facts make universal deployment challenging. Also important as a barrier to deployment is consistency among states, particularly contiguous ones. Recognizing trucking as a regional or even national business, the interface between the trucking industry and the various states needs to be consistent for widespread deployment to occur. While each state has its own requirements for such systems, driven by its operating environment, states must work toward providing interstate interoperability. Expanded public–public partnerships are needed among states and between the federal government and states.

Some public and private sector tensions occur in the CVISN program, as well. A good example is how truckers like the technologies that support weigh station bypass, whereby they are not required to stop at a weigh station if they have been previously checked. In such systems, the information is passed down the line from an adjoining station or even another state. At the same time, truckers are concerned about equity in tax collection and the privacy of their origin–destination data, because of competitive issues. Ironically, the same underlying CVISN system drives both applications. Public–private partnerships need to be developed in this application for public and private benefits to be effectively captured.

Crosscutting Technical and Programmatic Issues

Advanced technology is at the heart of ITS, so it is helpful to consider technical issues that affect ITS functions and applications. Technical issues include how one deals with rapidly changing technologies and how this aspect relates to the need for standards. Rapid obsolescence is a problem. All in all, technology issues are not a substantial barrier to ITS deployment. Most technologies perform; the question is, are they priced within the budget of deploying organizations, and are those prices consistent with the benefits that can be achieved? Two core technologies are those used for surveillance and communication.

Surveillance technologies have experienced some successes in cellular phone use for incident reports and in video use for incident verification, but the jury is still out on cellular phone geolocation for traffic probes. The lack of traffic flow sensors in many areas and on some roadway types continues to inhibit the growth of traveler information and improved transportation management systems.

Communications technologies have experienced some success with the Internet for pretrip traveler information and credentials administration in CVO. Emerging technologies include wireless Internet and automated information exchange. The growth rate of these technologies is high. In particular, the number of Americans having access to the Internet is growing rapidly, portending increased use of ITS applications [48].

Crosscutting Institutional Issues

Institutional issues are the key barrier to ITS deployment. The ten most prominent issues are awareness and perception of ITS, long-range operations and management, regional deployment, human resources, partnering, ownership and use of resources, procurement, intellectual property, privacy, and liability. Awareness and public and political appreciation of ITS as a system that can help deal with real and meaningful issues (e.g., safety and quality of life) are central to deployment success. Building a regional perspective to deployment using public–public and public–private partnerships is important. Recognizing that one must plan for sustained funding for operations in the long term is critical. Dealing with procurement questions is an important institutional concern, and public sector agencies are not accustomed to procuring high-technology components where intellectual property is at issue.

Fundamentally, ITS deployment requires a cultural change in transportation deployment organizations that have traditionally focused on providing conventional infrastructure. No silver bullet exists for achieving this cultural change; rather, it is a continuing, ongoing, arduous process and one that must be undertaken if ITS is to be successfully deployed.

Conclusions

A useful typology for assessing the above seven areas is along the three major dimensions commonly used to characterize transportation issues: technology, systems, and institutions [35]. Technology includes infrastructure, vehicles, and hardware and software that provide transportation functionality. Systems are one step removed from the immediacy of technology and deal with how holistic sets of components perform. An example is transportation networks. Institutions refer to organizations and interorganizational relationships that provide the basis for developing and deploying transportation programs.

Technology

Four technologies are central to most ITS applications:

1. Sensing: typically the position and velocity of vehicles on the infrastructure
2. Communicating: from vehicle to vehicle, between vehicle and infrastructure, and between infrastructure and centralized transportation operations and management centers
3. Computing: processing of the large amounts of data collected and communicated during transportation operations
4. Algorithms: typically computerized methods for dynamically operating transportation systems

One overarching conclusion is that the quality of technology is not a major barrier to the deployment of ITS. Off-the-shelf technology exists, in most cases, to support ITS functionality. An area where important questions about technology quality still remain is algorithms. For example, questions have been raised about the efficacy of software to perform adaptive traffic signal control. Also, the quality of collected information may be a technical issue in some applications.

Issues do remain on the technology side. In some cases, technology may simply be considered too costly for deployment, operations, and maintenance, particularly by public agencies that see ITS costs as not commensurate with the benefits to be gained by their deployment. In other cases, the technology may be too complex to be operated by current agency staff. Also, in some cases, technology falters because it is not easy to use, either by operators or transportation customers. Nonintuitive kiosks and displays for operators that are less than enlightening are two examples of the need to focus more on user interface in providing ITS technologies.

Systems

The most important need at the ITS systems level is integration of ITS components. While exceptions can certainly be found, many ITS deployments are stand-alone applications (e.g., ETC). It is often cost-effective in the short run to deploy an individual application without worrying about all the interfaces and platforms required for an integrated system. In their zeal to make ITS operational, people often have opted for stand-alone applications — not necessarily an unreasonable approach for the first generation of ITS deployment. However, for ITS to take the next steps forward, it will be important, for reasons of both efficiency and effectiveness, to think in terms of system integration. For example, the integration of services for arterials, freeways, and public transit should be on the agenda for the next generation of ITS deployments. Further integration of services, such as incident management, emergency management, traveler information, and intermodal services, must be accomplished. While this integration certainly adds complexity, it is also expected to provide economies of scale in system deployment and improvements in overall system effectiveness, resulting in better service for freight and traveling customers.

Another aspect of system integration is interoperability — ensuring that ITS components can function together. Possibly the best example of this function is interoperability of hardware and software in vehicles and on the infrastructure (e.g., ETC devices). The electronic linkage of vehicles and infrastructure must be designed using system architecture principles and open standards to achieve interoperability. It is quite reasonable for the public to ask whether their transponders will work with ETC systems across the country or even regionally. Unfortunately, the answer most often is no. Additionally, while it is important to make this technology operate properly on a broad geographic scale, it should also work for public transportation and parking applications. Systems that need to work at a national scale, such as CVO, must provide interoperability among components. No doubt, institutional barriers to interoperability exist (e.g., different perspectives among political jurisdictions), and these barriers inhibit widespread deployment.

Another important example of needed integration is between Advanced Transportation Management Systems and Advanced Traveler Information Systems. The former provides for operations of networks, the latter for traveler information, pretrip and in-vehicle, to individual transportation customers. For the most part, these two technologies, while conceptually interlinked, have developed independently. Currently, there are limited evaluative data on the technical, institutional, and societal issues related to integrating ATMS and ATIS, whereby ATMS, which collect and process a variety of network status data and estimates of future demand patterns, provide travelers (via ATIS services) with dynamic route guidance. This integration, together with ATMS-derived effective operating strategies for the network — which account for customer response to ATIS-provided advice, can lead to better network performance and better individual routes.

Institutions

The integration of public and private sector perspectives on ITS, as well as the integration of various levels of public sector organizations, is central to advancing the ITS agenda. The major barriers to ITS deployment are institutional in nature. This conclusion should come as no surprise to observers of the ITS scene; the very definition of ITS speaks of applying "well-established technologies," so technological breakthroughs are not needed for ITS deployment. But looking at transportation from an intermodal, systemic point of view requires a shift in institutional focus that is not easy to achieve. Dealing with intra- and interjurisdictional questions, budgetary frameworks, and regional-level perspectives on transportation systems; shifting institutional foci to operations rather than construction and maintenance; and training, retaining, and compensating qualified staff are all institutional barriers to widespread deployment of ITS technologies. Thinking through how to overcome various institutional barriers to ITS is the single most important activity we can undertake to enhance ITS deployment and develop successful implementations.

Operations

Recent years have brought an increasing emphasis on transportation operations, as opposed to construction and maintenance of infrastructure, as a primary focus. ITS is at the heart of this initiative, dealing as it does with technology-enhanced operations of complex transportation systems. The ITS community has argued that this focus on operations through advanced technology is the cost-effective way to go, given the extraordinary social, political, and economic costs of conventional infrastructure, particularly in urban areas. Through ITS, it is argued, one can avoid the high up-front costs of conventional infrastructure by investing more modestly in electronic infrastructure, then focusing attention on effectively operating that infrastructure and the transportation network at large.

While ITS can provide less expensive solutions, they are not free. There are up-front infrastructure costs and additional spending on operating and maintaining hardware and software. Training staff to support operations requires resources. Spending for ITS is of a different nature than spending for conventional infrastructure, with less up front and more in the out years. Therefore, planning for operations requires a long-term perspective by transportation agencies and the political sector. For that reason, it is important to institutionalize operations within transportation agencies. Stable budgets need to be provided for operations and cannot be the subject of year-to-year fluctuation and negotiation, which is how maintenance has traditionally been, if system effectiveness and efficiency are to be maintained. Human resource needs must be considered as well.

To justify ITS capital costs as well as continuing costs, it is helpful to consider life cycle costing in the evaluation of such programs. The costs and benefits that accrue over the long term are the important metric for such projects. But organizations need to recognize that a lack of follow-through will cause those out-year benefits to disappear as nonmaintained ITS infrastructure deteriorates and algorithms for traffic management are not recalibrated.

Mainstreaming

The term *mainstreaming* is used in different ways in the ITS setting. Some argue that mainstreaming means integrating ITS components into conventional projects. A good example is the Central Artery/Ted Williams Tunnel project in Boston, which includes important ITS elements as well as conventional infrastructure. Another is the Woodrow Wilson Bridge on I-95, connecting Maryland and Virginia, currently undergoing a major redesign, which includes both conventional infrastructure and ITS technologies and applications. This approach has the advantage of serving as an opportunity for ITS deployment within construction or major reconstruction activities. Typically, the ITS component is a modest fraction of total project cost. Even so, ITS technologies and applications can sometimes come under close political scrutiny well beyond their financial impact on the project. For example, on the Woodrow Wilson Bridge, the decommitting of various ITS elements is being considered [49].

Another definition of ITS mainstreaming suggests that ITS projects not be protected by special funds sealed for ITS applications but that ITS should compete for funding with all other transportation projects. The advantage of this method is that ITS would compete for a much larger pool of money; the disadvantage is that ITS, in the current environment, might not compete particularly successfully for that larger pool. Those charged with spending public funds for transportation infrastructure have traditionally spent virtually all their money on conventional projects. Convincing these decision makers that funds are better spent on ITS applications may be difficult.

This issue is clearly linked to human resource development. Professionals cannot be expected to select ITS unless they are knowledgeable about it, so education of the professional cadre is an essential precondition for success of mainstreaming — by either definition. Of course, the National ITS Program must also be prepared to demonstrate that the benefits of ITS deployments are consistent with the costs incurred. Protected ITS funds — funds that can be spent only on ITS applications — may be a good transition strategy as professional education continues and ITS benefits become clearer, but in the longer run, there are advantages to ITS being mainstreamed.

Human Resources

An important barrier to success in the deployment of new technologies and applications embodied in ITS is a lack of people to support such systems. The ITS environment requires skilled specialists representing new technologies. It also needs broad generalists with policy and management skills who can integrate advanced thinking about transportation services based on new technologies [35]. The ITS community has recognized these needs, and various organizations have established substantial programs for human resource development. FHWA's Professional Capacity Building program is a premier example but not the only one. Universities have also developed relevant programs that, along with graduate transportation programs undergoing substantial ITS-related changes around the country, can provide a steady stream of talented and newly skilled people for the industry. However, we must emphasize that institutional changes in transportation organizations are needed if these people are to be used effectively and retained, as people with high-technology skills can often demand much higher salaries than are provided by public sector transportation organizations. Cultural change, along with appropriate rewards for operations staff, for example, will be necessary in organizations where the culture strongly favors conventional infrastructure construction and maintenance.

The need for political champions for ITS has long been understood in the ITS community. Here, though, we emphasize the need at all levels of implementing organizations for people with the ability to effectively deploy ITS. The political realities may require public sector organizations to "contract in" staff to perform some of the high-technology functions inherent in ITS, as opposed to permanently hiring such individuals. Also, "contracting out" — having private sector organizations handle various ITS functions on behalf of the public sector — is another option. In the short run, these options may form useful strategies. In the long run, developing technical and policy skills directly in the public agency has important advantages for strategic ITS decision making.

The Positioning of ITS

Almost from its earliest days, ITS has unfortunately been subject to overexpectations and overselling. Advocates have often promoted the benefits of ITS technologies and applications and have minimized the difficulties in system integration during deployment. Often ITS has been seen by the public and politicians as a solution looking for a problem. Overtly pushing ITS can be counterproductive. Rather, ITS needs to be put to work in solving problems that the public and agencies feel truly exist.

Safety and quality of life are the two most critical areas that ITS can address. Characterizing ITS benefits along those dimensions when talking to the public or potential deploying agencies is a good strategy. The media can also help to get the story out about ITS [37].

Operator versus Customer Perspective

Information is at the heart of ITS. The provision of information to operators to help them optimize vehicle flows on complex systems is one component. The flow of information to customers (drivers, transit users, etc.) so they can make effective choices about mode, route choice, etc. is another component. There is a great deal of overlap in these two information sets, yet sharing information between operators and customers is often problematic. Operators are usually public sector organizations. From their perspective, the needs of individual travelers should be subordinate to the need to make the overall network perform effectively. On the other hand, private sector information providers often create and deliver more tailored information focusing on the needs of particular travelers rather than overall system optimization.

It is not surprising that the agendas of the public sector agencies operating the infrastructure and those of the information-provider private sector companies differ. Nonetheless, it seems clear that the ultimate customer — the traveler — would benefit from a more effective integration of these two perspectives. This issue is both a technical and an institutional one and is an important example of the need for service integration.

Regional Opportunities

From a technological and functional point of view, ITS provides, for the first time, an opportunity to manage transportation at the scale of the metropolitan-based region. Along with state or even multistate geographic areas, metropolitan-based regions — the basic geographic unit for economic competition and growth [34] and for environmental issues — can now be effectively managed from a transport point of view through ITS. While a few regions in Europe and the United States have made progress, ITS technologies generally have not been translated into a regionally scaled capability. The institutional barriers are, of course, immense, but the prize from a regional viability perspective is immense, as well. Thinking through the organizational changes that will allow subregional units some autonomy, but at the same time allow system management at the regional scale, is an ITS issue of the first order [33]. Indeed, this approach could lead to new paradigms for strategic planning on a regional scale, supported by the information and organizational infrastructure developed in the context of ITS.

The strategic vision for ITS is as the integrator of transportation, communications, and intermodalism on a regional scale [35,37]. Multistate regions with traffic coordination over very large geographic areas, as in the mountain states, is an important ITS application. Corridors such as I-95, monitored by the I-95 Corridor Coalition and stretching from Maine to Virginia, represent an ITS opportunity, as well.

Surface Transportation as a Market

Surface transportation needs to be thought of as a market with customers with ever-rising and individual expectations. Modern markets provide choices. People demand choices in level of service and often are willing to pay for superior service quality; surface transportation customers will increasingly demand this service differentiation, as well. While a market framework is not without controversy in publicly provided services, surface transportation operators can no longer think in terms of "one size fits all." High-occupancy toll (HOT) lanes, where people driving a single-occupancy vehicle are permitted to use an HOV lane if they pay a toll, are an early example of this market concept in highway transportation. HOT lanes are enabled by ITS technologies. Other market opportunities building on ITS will doubtless emerge, as well.

Assessment summaries of technologies are in Tables 86.1 through 86.9 in the Appendix.

86.6 Benefits of ITS

It is interesting to reflect on how technology has influenced transportation in the United States. The steam engine improved travel by boat and railroad, resulting in coast-to-coast systems. The internal combustion engine freed the vehicle from a fixed guideway or waterway and encouraged the construction of farm-to-market roads as well as enabled travel by air. ITS will move another step forward by providing traveler information and by operating traffic management and control systems. This quantum leap will have a major impact on today's lifestyle [6].

Attempting to quantify the benefits of widely deployed ITS technologies at the birth of what was then IVHS was similar to what planners of the U.S. interstate highway system tried to do in the 1950s. It was impossible to anticipate all of the ways that applications of ITS technology may affect society, just as planners of the interstate highway system could not have anticipated all of its effects on American society. Recognizing the importance of the issue, however, Mobility 2000, an ad hoc coalition of industry, university, and federal, state, and local government participants, whose work led to the establishment of ITS America, addressed the potential benefits of applying ITS technology in the United States. Numerous benefits were predicted for urban and rural areas and for targeted groups, such as elderly and disadvantaged travelers. Positive benefits were also found in regard to the environment [2].

ITS represents a wide collection of applications, from advanced signal control systems to ramp meters to collision warning systems. In order to apply ITS technologies most effectively, it is important to know which technologies are most effectively addressing the issues of congestion and safety. Some technologies provide more cost-effective benefits than others, and as technology evolves, the choices to deployers change. Often, several technologies are combined in a single integrated system, providing synergistic benefits that exceed the benefits of any single technology. It is important to know which technologies and technology combinations provide the greatest benefits, so that transportation investments can be applied most effectively to meet the growing transportation demands of our expanding economy [40].

Since 1994, the U.S. DOT's ITS Joint Program Office has been actively collecting information on the impacts that ITS and related projects have on the operation and management of the nation's surface transportation system. The evaluation of ITS is an ongoing process. Significant knowledge is available for many ITS services, but gaps in knowledge also exist [39]. In general, all ITS services have shown some positive benefit, and negative impacts are usually outweighed by other positive results. For example, higher speeds and improved traffic flow result in increases in nitrous oxides, while other measures that indicate increased emissions, such as fuel consumption, travel time, and delay, are reduced. Because of the nature of the data, it is often difficult to compare data from one ITS project to another. This is because of the differences in context or conditions between different ITS implementations. Thus, statistical analysis of the data is not done across data points. In several cases, ranges of reported impacts are presented and general trends can be discussed. These cases include traffic signal systems, automated enforcement, ramp metering, and incident management [39].

Most of the data collected to date are concentrated within metropolitan areas. The heaviest concentrations of such data are in arterial management systems, freeway management, incident management, transit management, and regional multimodal traveler information. Most of the available data on traffic signal control systems are from adaptive traffic control. For freeway management, most data are concentrated around benefits related to ramp metering. There are also recent studies on the benefits of ITS at highway–rail intersections.

There has been an increase in the implementation and evaluation of rural ITS. Several state and national parks are now examining and implementing improved tourism and travel information systems, and several rural areas are implementing public travel services. Many states are examining the benefits of incorporating ITS, specifically weather information, into the operation and maintenance of facilities and equipment. Many of the data reported for rural ITS are concentrated in the areas of crash prevention and security. A significant amount of information is available for road weather management activities, including winter weather-related maintenance, pavement condition monitoring, and dissemination of road weather information.

ITS for Commercial Vehicle Operations (ITS/CVO) continues to provide benefits to both carriers and state agencies. ITS/CVO program areas usually report benefits data from directly measurable effects. Therefore, it might be expected that these data are accurate and only a few data points would be necessary to convince carriers, states, and local authorities of the possible benefits of implementing these systems. To date, most of the data collected for ITS/CVO are for cost, travel time, and delay savings for carrier operations.

ITS program areas and user services associated with driver assistance and specific vehicle classes are still being developed and planned. As market penetrations increase and improved systems are developed, there will be ample opportunity to measure and report data based on actual measurements [39].

Taxonomy and Measures of Effectiveness

To track the progress toward meeting ITS program goals, the JPO has identified and established a set of measures of effectiveness. These measures are termed "A Few Good Measures" and are used as a standard in the reporting of much of the ITS benefits data currently available. Data collected are not limited to these measures; additional measures are also reported when available. The few good measures are:

- Safety: usually measured by impacts on crashes, injuries, fatalities
- Delay: usually measured in units of time
- Cost: measured in monetary amount
- Effective capacity: measured in throughput or traffic volumes
- Customer satisfaction: usually results from user surveys
- Energy and environment: usually measured in fuel consumption and emissions

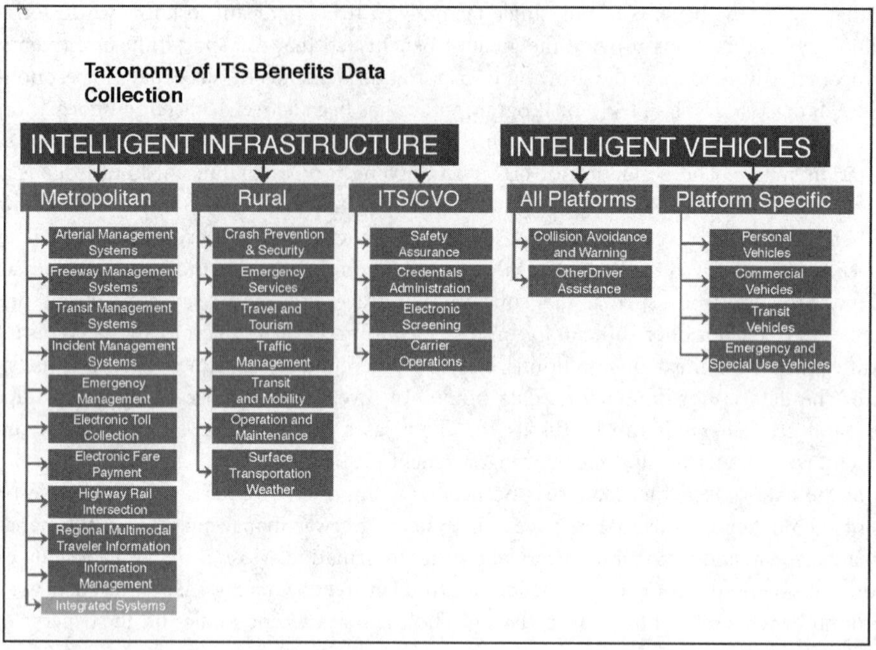

FIGURE 86.2 ITS benefits taxonomy. (Mitretek Systems, *Taxonomy for Classification of ITS Benefits*, Department of Transportation ITS JPO, Washington, D.C., June 2000.)

The benefits database desk reference (Figure 86.3) provides a brief summary of the metropolitan data available in the online database. The desk reference is updated regularly and is also available at the database web site. It is based on a taxonomy developed [42] for the classification of benefits (Figure 86.2).

A cost database is also available and can be found through the ITS web page of the U.S. DOT [44], with more details in Reference 45. The ITS unit cost database consists of cost estimates for a set of ITS elements. These cost estimates are categorized as capital, and operating and maintenance (O&M) costs (also known as nonrecurring and recurring costs, respectively). These costs are presented in a range to capture the lows and highs of the cost elements from the different data sources identified on the cost data sources page. The cost data are useful in developing project cost estimates during the planning process. However, the user is encouraged to find local and regional data sources and current vendor data to perform a more detailed cost estimate. The set of ITS elements is based primarily on the unit cost elements in the National ITS Architecture Cost Analysis and the ITS Deployment Analysis System (IDAS) equipment list. IDAS is software developed by the Federal Highway Administration [44] that can be used in planning for ITS deployment. IDAS can estimate relative costs and benefits for more than 60 types of ITS investments. Practitioners will find a number of useful features that enhance ITS planning.

One of the powerful aspects of ITS is the capability of components to share information and resources with other components. This integration of individual components allows the formation of a unified regional traffic control and management system. To better describe the flow of information between components, a number of integration links have been developed for the metropolitan ITS infrastructure. These links represent both inter- and intracomponent sharing of information. Each of the links has been assigned a number and an origin or destination path from one component to another. For example, metropolitan integration link number 29 is from transit management to incident management and represents the ability of transit agencies to notify incident management agencies of incident location, severity, and type. Figure 86.4 depicts the links in metropolitan integration, and definitions of the links can be found in Table 86.10 in the Appendix [43].

For a more complete understanding of these components, integration, and how they can be interpreted, refer to the following documents: "Tracking the Deployment of Integrated Metropolitan Intelligent Transportation Systems Infrastructure in the USA: FY 1997 Results," Document 5883, September 1998;

Metropolitan Benefits By Program Area

Program Area/Benefit Measure		Summary
Arterial Management Systems	Safety Improvements	Automated enforcement of traffic signals has reduced red-light violations 20–75%.
	Delay Savings	Adaptive signal control has reduced traffic delay 14–44%. Transit signal priority has reduced bus journey times by 7%.
	Throughput	
	Customer Satisfaction	In Michigan, 72% of surveyed drivers felt "better off" after signal control improvements.
	Cost Savings	Transit signal priority on a Toronto Transit Line allowed same level-of-service with less rolling stock.
	Environmental	Improvements to traffic signal control have reduced fuel consumption 2–13%.
	Other	Between 1969 and 1976, traffic signal preemption systems in St. Paul, MN, reduced emergency vehicle accidents by 71%.
Freeway Management Systems	Safety Improvements	Ramp Metering has shown a 15–50% reduction in crashes.
	Delay Savings	In Minneapolis-St. Paul, MN, ramp metering has reduced freeway travel time 22%, for an annual savings of 25,121 vehicle-hours.
	Throughput	Ramp metering has increased throughput 13–16%.
	Customer Satisfaction	After the Twin Cities ramp meter shutdown test, 69% of travelers supported modified continued operations.
	Cost Savings	The GA Navigator (integrated system) supported incident delay reductions, for an annual savings of $44.6 million.
	Environmental	
	Other	Ramp metering has shown an 8–60% increase in freeway speeds.
Transit Management Systems	Safety Improvements	In Denver, AVL systems with silent alarms have supported a 33% reduction in bus passenger assaults.
	Delay Savings	CAD/AVL has improved on-time bus performance 9–23%.
	Throughput	
	Customer Satisfaction	In Denver, installation of CAD/AVL decreased customer complaints by 26%.
	Cost Savings	In San Jose, AVL has reduced paratransit expense from $4.88 to $3.72 per passenger.
	Environmental	
	Other	More efficient bus utilization has resulted in a 4–9% reduction in fleet size.
Incident Management Systems	Safety Improvements	In San Antonio, integrated VMS and incident management systems decreased accidents by 2.8%.
	Delay Savings	Incident management in city and regional areas has saved 0.95–15.6 million vehicle-hours of delay per year.
	Throughput	
	Customer Satisfaction	Customers have been very satisfied with service patrols (hundreds of letters).
	Cost Savings	Cost savings have ranged from 1–45 million dollars per year, depending on coverage area size.
	Environmental	Models of the Maryland CHART system have shown fuel savings of 5.8 million gallons per year.
	Other	The I-95 TIMS system in PA has decreased highway incidents 40% and cut closure time 55%.
Emergency Management Systems	Safety Improvements	In Palm Beach, GPS/AVL systems have reduced police response times by 20%.
	Delay Savings	
	Throughput	
	Customer Satisfaction	95% of drivers equipped with PushMe Mayday system felt more secure.
	Cost Savings	
	Environmental	
	Other	

FIGURE 86.3 Metropolitan benefits.

and "Measuring ITS Deployment and Integration," Document 4372, January 1999. Both documents are available on the FHWA electronic document library [44].

Figure 86.4 illustrates the numbered links that represent the flow of information between metropolitan ITS components. Much of the data collected regarding integration illustrates benefits to delay and travel time savings or cost savings. A few evaluation studies are currently planned or in progress that may

Metropolitan Benefits By Program Area

Program Area	Benefit Measure	Summary
Electronic Toll Collection	Safety Improvements	Driver uncertainty about congestion contributed to a 48% increase in accidents at E-PASS toll stations in Florida.*
	Delay Savings	The New Jersey Turnpike Authority (NJTA) E-Zpass system has reduced vehicle delay by 85%.
	Throughput	Tappan Zee Bridge: Manual lane 400–450 vehicles/hour (vph), ETC lane 1000 vph.
	Customer Satisfaction	
	Cost Savings	ETC has reportedly reduced roadway maintenance and repair costs by 14%
	Environmental	NJTA models indicate E-Zpass saves 1.2 mil gallons of fuel per yr, 0.35 tons of VOC per day, and 0.056 tons NOx per day.
	Other	20% of travelers on two bridges in Lee County, FL, adjusted their departure times as a result of value pricing at electronic tolls.
Electronic Fare Payment	Safety Improvements	
	Delay Savings	
	Throughput	
	Customer Satisfaction	Europe has enjoyed a 71–87% user acceptance of smart cards for transit/city coordinated services.
	Cost Savings	The Metro Card System saved New York approximately $70 million per year.
	Environmental	
	Other	
Highway Rail Intersections	Safety Improvements	In San Antonio, VMS with railroad crossing delay information decreased crashes by 8.7%.
	Delay Savings	
	Throughput	
	Customer Satisfaction	School bus drivers felt in-vehicle warning devices enhanced awareness of crossings.
	Cost Savings	
	Environmental	Automated horn warning systems have reduced adjacent noise impact areas by 97%.
	Other	
Regional Multimodal Traveler Information	Safety Improvements	IDAS models show the ARTIMIS traveler information system has reduced fatalities 3.2% in Cincinnati and Northern Kentucky.
	Delay Savings	A model of SW Tokyo shows an 80% decrease in delay if 15% of vehicles shift their departure time by 20 min.
	Throughput	
	Customer Satisfaction	38% of TravTek users found in-vehicle navigation systems useful when travelling in unfamiliar areas.
	Cost Savings	
	Environmental	EPA-model estimates of SmarTraveler impacts in Boston show 1.5% less NOx and 25% less VOC emissions.
	Other	Models of Seattle show freeway-ATIS is 2x more effective in reducing delay if integrated with arterial ATIS.

Source: http://www.benefitcost.its.dot.gov *Database also includes negative impacts of ITS Date: 12/31/2001

FIGURE 86.3 Metropolitan benefits (*Continued*).

include results for several integration links. Few data have been reported for components that use information collected using arterial management (links 1 to 4). It is expected that the primary benefit for these integration links would be delay and travel time savings. The sharing of information between arterial management and freeway management (links 2 and 11), which can be used to change ramp-metering rates and traffic signal times, may yield significant advantages [43].

86.7 Five-Year Plan [38]

With the passage of the Transportation Equity Act for the 21st Century in 1998, Congress reaffirmed the role of the U.S. DOT in advancing the development and integrated deployment of ITS technologies. A five-year plan was developed presenting the U.S. DOT's goals, key activities, and milestones for the National ITS Program for FY 1999 through 2003.

Metropolitan Benefits By Measure

Benefit Measure/Program Area		Summary
Safety Improvements	Arterial Management	Automated enforcement of traffic signal has reduced red-light violations 20–75%.
	Freeway Management	Ramp metering has shown a 15–50% reduction in crashes.
	Transit Management	In Denver, AVL systems with silent alarms have supported a 33% reduction in bus passenger assaults.
	Incident Management	In San Antonio, integrated VMS and incident management systems decreased accidents by 2.8%.
	Emergency Management	In Palm Beach, GPS/AVL systems have reduced police response times by 20%.
	Electronic Toll Collection	Driver uncertainty about congestion contributed to a 48% increase in accidents at E-PASS toll stations in Florida.*
	Electronic Fare Payment	
	Highway Rail Intersection	In San Antonio, VMS with railroad crossing delay information decreased crashes by 8.7%.
	Regional Traveler Info.	IDAS models show the ARTIMIS traveler information system has reduced fatalities 3.2% in Cincinnati and Northern KY.
Delay Savings	Arterial Management	Adaptive signal control has reduced traffic delay 14–44%. Transit signal priority has reduced bus journey times by 7%.
	Freeway Management	In Minneapolis-St. Paul, MN ramp metering has reduced freeway travel time 22%, for an annual savings of 25,121 vehicle-hours.
	Transit Management	CAD/AVL has improved on-time bus performance 9–23%.
	Incident Management	
	Emergency Management	
	Electronic Toll Collection	The New Jersey Turnpike Authority (NJTA) E-Zpass system has reduced vehicle delay by 85%.
	Electronic Fare Payment	
	Highway Rail Intersection	
	Regional Traveler Info.	A model of SW Tokyo shows an 80% decrease in delay if 15% of vehicles shift their departure time by 20 min.
Throughput	Arterial Management	
	Freeway Management	Ramp metering has increased throughput 13–16%.
	Transit Management	
	Incident Management	
	Emergency Management	
	Electronic Toll Collection	Tappan Zee Bridge: Manual lane 400–450 vehicles/hour (vph), ETC lane 1000 vph.
	Electronic Fare Payment	
	Highway Rail Intersection	
	Regional Traveler Info.	
Customer Satisfaction	Arterial Management	In Michigan, 72% of surveyed drivers felt "better off" after signal control improvements.
	Freeway Management	After the Twin Cities ramp meter shutdown test, 69% of travelers supported modified continued operations.
	Transit Management	In Denver, installation of CAD/AVL decreased customer complaints by 26%.
	Incident Management	Customers have been very satisfied with service patrols (hundreds of letters).
	Emergency Management	95% of drivers equipped with PushMe Mayday system felt more secure.
	Electronic Toll Collection	
	Electronic Fare Payment	Europe has enjoyed a 71–87% user acceptance of smart cards for transit/city coordinated services.
	Highway Rail Intersection	School bus drivers felt in-vehicle warning devices enhanced awareness of crossings.
	Regional Traveler Info.	38% of TravTek users found in-vehicle navigation systems useful when travelling in unfamiliar areas.

FIGURE 86.3 Metropolitan benefits (*Continued*).

Transition from Research to Deployment

Under ISTEA, the National ITS Program focused primarily on research, technology development, and field testing that advanced the state of technology, demonstrated substantial public benefits, and fostered new models of institutional cooperation. The program began to lay the foundation for an

Metropolitan Benefits By Measure

Benefit Measure/Program Area		Summary
Customer Satisfaction	Arterial Management	In Michigan, 72% of surveyed drivers felt "better off" after signal control improvements.
	Freeway Management	After the Twin Cities ramp meter shutdown test, 69% of travelers supported modified continued operations.
	Transit Management	In Denver, installation of CAD/AVL decreased customer complaints by 26%.
	Incident Management	Customers have been very satisfied with service patrols (hundreds of letters).
	Emergency Management	95% of drivers equipped with PushMe Mayday system felt more secure.
	Electronic Toll Collection	
	Electronic Fare Payment	Europe has enjoyed a 71–87% user acceptance of smart cards for transit/ city coordinated services.
	Highway Rail Intersection	School bus drivers felt in-vehicle warning devices enhanced awareness of crossings.
	Regional Traveler Info.	38% of TravTrek users found in-vehicle navigation systems useful when travelling in unfamiliar areas.
Cost Savings	Arterial Management	Transit signal priority on a Toronto Transit Line allowed same level-of-service with less rolling stock.
	Freeway Management	The GA Navigator (integrated system) supported incident delay reductions, for an annual savings of $44.6 million.
	Transit Management	In San Jose, AVL has reduced paratransit expense from $4.88 to $3.72 per passenger.
	Incident Management	Cost savings have ranged from 1–45 million dollars per year, depending on coverage area size.
	Emergency Management	
	Electronic Toll Collection	ETC has reportedly reduced roadway maintenance and repair costs by 14%.
	Electronic Fare Payment	The Metro Card System saved New York approximately $70 million per year.
	Highway Rail Intersection	
	Regional Traveler Info.	
Environmental	Arterial Management	Improvements to traffic signal control have reduced fuel consumption 2–13%.
	Freeway Management	
	Transit Management	
	Incident Management	Models of the Maryland CHART system have shown fuel savings of 5.8 million gallons per year.
	Emergency Management	
	Electronic Toll Collection	NJTA models indicate E-Zpass saves 1.2 mil gallons of fuel per yr, 0.35 tons of VOC per day, and 0.056 tons NOx per day.
	Electronic Fare Payment	
	Highway Rail Intersection	Automated horn warning systems have reduced adjacent noise impact areas by 97%.
	Regional Traveler Info.	EPA-model estimates of SmarTraveler impacts in Boston show 1.5% less NOx and 25% less VOC emissions.
Other	Arterial Management	Between 1969 and 1976, traffic signal preemption systems in St. Paul, MN, reduced emergency vehicle accidents by 71%.
	Freeway Management	Ramp metering has shown an 8–60% increase in freeway speeds.
	Transit Management	More efficient bus utilization has resulted in a 4–9% reduction in fleet size.
	Incident Management	The I-95 TIMS system in PA has decreased highway incidents 40% and cut closure time 55%.
	Emergency Management	
	Electronic Toll Collection	20% of travelers on two bridges in Lee County, FL, adjusted their departure times as a result of value pricing at electronic tolls.
	Electronic Fare Payment	
	Highway Rail Intersection	
	Regional Traveler Info.	Models of Seattle show freeway-ATIS is 2x more effective in reducing delay if integrated with arterial ATIS.

Source: http://www.benefitcost.its.dot.gov * Database also includes negative impacts of ITS Date: 12/31/2001

FIGURE 86.3 Metropolitan benefits (*Continued*).

information and communications infrastructure that would enable the nation to realize the vision set forth by Congress — to manage multiple transportation facilities as unified systems for greater efficiency, safety, customer service, and quality of life. TEA-21 continues the legacy of ISTEA by building on the success of research and development to date. Today, many ITS technologies are available, and

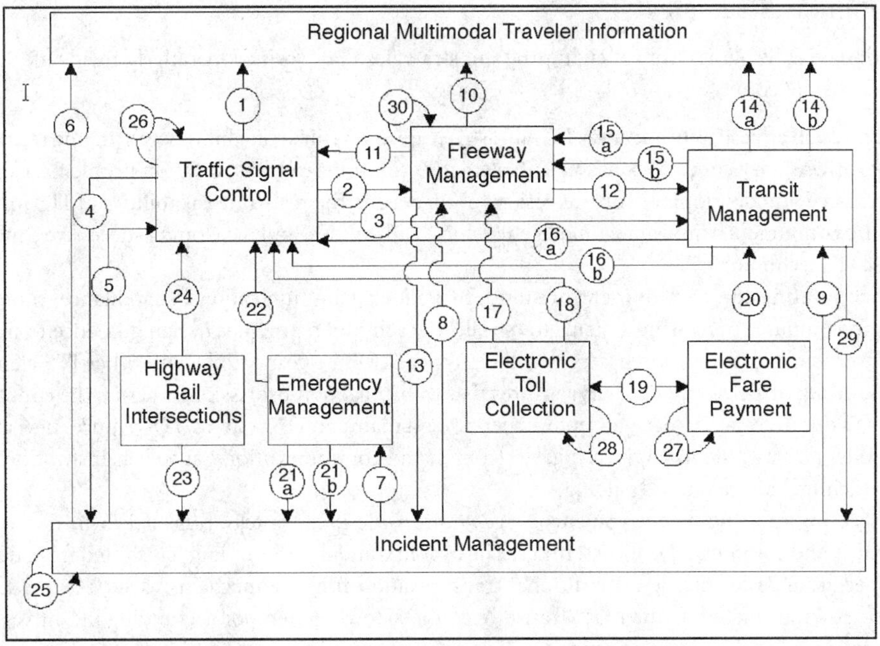

FIGURE 86.4 Metropolitan integration links. (From Mitretek Systems, ITS Benefits: Data Needs Update 2000, prepared in connection with ITS Benefits Data Needs Workshop, August 2000.)

the National ITS Program will shift its emphasis accordingly to the deployment of proven ITS technologies in an integrated fashion, while continuing to advance ITS capabilities through further research.

Intelligent Infrastructure and Intelligent Vehicles

The National ITS Program focuses on two main objectives: deployment of intelligent infrastructure and testing and evaluation of intelligent vehicle technologies. Intelligent infrastructure and intelligent vehicles provide the information and control needed to better manage surface transportation facilities (highways, roads, transit, and rail), to improve the safety of vehicles operating on those facilities, and to help users of all modes make better decisions about travel. Intelligent infrastructure is the necessary network of technologies — a communications and information backbone — that supports and unites key ITS services for metropolitan, rural, statewide, and commercial vehicle application. Intelligent vehicle technologies improve safety and enhance mobility of the vehicles that operate on our roadways. Such technologies apply to four classes of vehicles: light vehicles (passenger cars, vans, and light trucks), transit vehicles (buses), commercial vehicles (trucks and interstate buses), and specialty vehicles (emergency response, enforcement, and highway maintenance vehicles). ITS products and services must be seamlessly integrated and interoperable. Over time, intelligent vehicles will increasingly interact with intelligent infrastructure to yield even greater gains in mobility and traveler safety.

Intelligent infrastructure and intelligent vehicle objectives are addressed through four program areas: metropolitan ITS infrastructure, rural ITS infrastructure, commercial vehicle ITS infrastructure, and intelligent vehicles. Each program area targets a specific environment in which ITS capabilities are used. The metropolitan and rural ITS infrastructure program areas address network-based technologies deployed in those two settings. The commercial vehicle ITS infrastructure program area focuses on the integrated technologies needed specifically for safety and administrative regulation of interstate trucking. Finally, the intelligent vehicles program area targets in-vehicle safety systems for all users and geographic settings.

ITS Program Strategies

The National ITS Program utilizes eight program strategies that work cooperatively to advance the state of ITS across the country:

1. **Conducting research:** advances ITS infrastructure and vehicle capabilities by bringing technologies from visionary concepts to viable and attractive solutions to transportation problems. Continued research and development are necessary for increasing the real-time capabilities of ITS infrastructure components, improving intelligent vehicle capabilities, and developing successive generations of ITS technologies.
2. **Accelerating the development of standards:** allows communications, surveillance, monitoring, and computer processing systems to "speak" to each other, provides design guidance to manufacturers, and reassures purchasers that their systems will be compatible with other ITS elements.
3. **Building professional capacity:** ensures that transportation professionals across the country have the skills necessary to design, deploy, operate, and maintain ITS systems. ITS requires new technical skills, such as systems engineering, electronics, and communications, as well as institutional skills, which include coalition building.
4. **Creating funding incentives:** encourages more widespread integration of ITS in metropolitan, rural, and commercial vehicle settings. The overall trend in ITS funding has shifted from dedicated special funds to the use of traditional transportation funding mechanisms such as the Highway Trust Fund (including the Mass Transit Account). However, temporary funding incentives are still necessary to foster integration and national interoperability and to accelerate deployment.
5. **Providing guidance and technical assistance:** aids implementers seeking to deploy integrated ITS. The U.S. DOT provides specialized technical support through its federal field staff, through the publication of guidance documents on best practices for ITS deployment, and with the Peer-to-Peer Network, a resource composed of professionals from the private and public sectors who are on call to provide short-term, no-cost technical assistance to transportation colleagues across the country.
6. **Ensuring consistency with the National ITS Architecture and standards:** helps in planning for ITS integration, reducing development time and cost, and laying the groundwork for a seamless national ITS network. The U.S. DOT is working with stakeholders to develop federal policy on consistency and is actively training stakeholders on this issue.
7. **Evaluating the program:** essential for understanding the value and effectiveness of ITS activities and for measuring progress toward deployment goals. Tracking and evaluation are consistent with the spirit of the Government Performance and Results Act and allow for the continual refinement of the National ITS Program.
8. **Showcasing benefits:** communicates positive results realized through the use of ITS technologies to multiple decision makers. By understanding the benefits of ITS, decision makers can compare ITS to other transportation options when addressing local transportation issues. Showcasing benefits also encourages integration of ITS systems. For example, deployment sites demonstrate successful interjurisdictional working relationships and interagency coordination. By learning about the deployment sites, decision makers better understand the operation and management planning that is necessary to achieve integration in their areas.

Program Area Goals, Key Activities, Milestones

The eight ITS program strategies are used to meet specific goals in each of the four ITS program areas: metropolitan, rural, and commercial vehicle infrastructure, and intelligent vehicles.

Metropolitan ITS Infrastructure Program Area

The metropolitan ITS program has demonstrated proven technologies for metropolitan application. Model deployments in metropolitan settings have been successful and will continue to be showcases for other areas. However, while individual systems are being purchased and installed around the country, sites are just beginning to integrate systems across jurisdictions and modes. Since integration has been limited, communities are not yet reaping the full benefits of ITS.

The metropolitan component of the National ITS Program is focused on meeting the goals for integrated deployment laid out by former Secretary of Transportation Federico Peña in 1996 and reiterated by former Secretary of Transportation Rodney Slater. Known as Operation TimeSaver, the U.S. DOT's objective is to facilitate integrated deployment of basic ITS services in 75 metropolitan areas by 2006. At present, 36 sites are considered to have some elements of integration, and additional sites show the clear beginnings of integrated systems.

Goal

By 2003, the metropolitan program aims to have 64 sites achieve the Operation TimeSaver goal for integrated deployment.

Key Activities and Milestones

The Operation TimeSaver goal will be met using all eight ITS program strategies as follows:

Conducting research: Traffic management and transit management research will be advanced under TEA-21. New models for traffic management, such as the ITS Deployment Analysis System, have been developed that more accurately represent the impact of ITS, allowing transportation planners and designers to compare ITS with other transportation options more effectively. This model was made available to the ITS planning community in 1999. A more sophisticated planning model — the Transportation Analysis and Simulation System (TRANSIMS) — is under development and will be available for initial use in 2002. In transit management, research will focus on the application of integrated transit systems through operational tests in areas such as fleet management, electronic fare payments, and traveler information. From 1999 to 2003, this research will be guided by the Federal Transit Administration's (FTA) 5-year research and technology plan.

Accelerating the development of standards: Development of standards, such as the National Transportation Communications for ITS Protocol (NTCIP) and the Transit Communications Interface Profiles (TCIP), will help facilitate integration in metropolitan areas. NTCIP will allow traffic management and operations personnel to better control, manage, and monitor virtually all the devices used on the roadway. TCIP will allow data to be shared among transit departments and other operating entities, such as emergency response services and regional traffic management centers. Because these standards are fundamental to metropolitan transportation operations, a training course on each standard is necessary. The course on NTCIP is already available, on a request basis, through the Institute of Traffic Engineers, and it will be available to stakeholders through 2003. A course on TCIP will be developed, with delivery expected to occur from 2001 to 2003.

Building professional capacity: Training courses will continue to be offered on all aspects of metropolitan ITS deployment. Courses will be updated as new information is made available.

Creating funding incentives: TEA-21 funding incentives are being offered to metropolitan public sector applicants to support technical integration and jurisdictional coordination of ITS infrastructure. Funding will be offered to both highway and transit projects, and the U.S. DOT will work with Congress and the funding recipients to ensure that both the spirit and intent of TEA-21 funding criteria are met. The U.S. DOT will allocate funding incentives annually based on programmatic goals and the criteria defined in TEA-21. Funding available for integration was set by TEA-21 at \$75 million in 1999, \$83 million in 2000, \$83 million in 2001, \$85 million in 2002, and \$85 million in 2003. A maximum of 90% of this funding is available to metropolitan areas.

Providing guidance and technical assistance: Special guidance and technical assistance are being offered to transportation officials through federal field staff expertise, guidance documents, and the Peer-to-Peer Network to assist in the planning, design, operation, and maintenance of metropolitan ITS. In addition, federal field staff will work with their state and local partners to develop "ITS service plans" that outline local technical guidance needs and plans for delivery. Development of ITS service plans began with a focus on the top 78 metropolitan areas and has expanded over time to include statewide concerns that typically involve rural ITS applications. The U.S. DOT started with 62 service plans being implemented (49 from the top 78 metropolitan areas and 13 statewide plans) in FY 2000. U.S. DOT expects to expand over time and to include activities in other metropolitan areas beyond the top 78.

Ensuring consistency with the National ITS Architecture and standards: The National ITS Architecture and standards will be instrumental in catalyzing integrated ITS deployment across the country, enabling areas to meet local needs while reducing development costs and risks, facilitating future expansion capability, and fostering interoperability. Interim policy guidance on consistency with the National ITS Architecture and standards was issued in 1999. The interim guidance was implemented until release of the final policy; the final policy will be implemented through 2003 and beyond.

Evaluating the program: Program evaluations track levels of deployment and integration in the 75 metropolitan areas. Evaluations are being used to demonstrate ITS benefits and to measure progress toward the Operation TimeSaver goal. From 1999 through 2003, U.S. DOT will conduct annual tracking surveys, assemble the data received, and report findings. Using this information, the metropolitan program will be refined as appropriate.

Showcasing benefits: The four metropolitan model deployment sites funded under ISTEA — Phoenix, Arizona; Seattle, Washington; San Antonio, Texas; and the New York–New Jersey–Connecticut metropolitan area — will continue to showcase the benefits of metropolitan ITS technologies under TEA-21. These sites have brought together public and private sector partners to integrate existing infrastructure with new traveler information systems. In addition, results of deployment evaluations will be incorporated into publications to disseminate benefits information.

Through these eight strategies, the metropolitan ITS program will continue to pursue the deployment of integrated, intelligent transportation systems — including advanced traffic management, traveler information, and public transit systems — that will improve urban transportation management in the 75 largest urban areas. At the same time, there are 340 major metropolitan areas nationwide that could benefit from advanced technologies, and the U.S. DOT's field staff is working actively with all interested communities.

Rural and Statewide ITS Infrastructure Program Area

Information technologies are currently being applied in rural settings to help improve the safety and mobility of rural travelers. However, rural and statewide ITS applications are not yet as well defined as metropolitan and commercial vehicle applications. Under TEA-21, the rural program will focus primarily on research and field operational testing to further develop rural infrastructure components. Through these tests, the U.S. DOT will identify solutions that reduce the public sector costs of providing, operating, and maintaining rural ITS infrastructure. Lessons learned from the metropolitan program will be leveraged to the maximum possible extent, as will rural program resources. For example, the U.S. DOT will cooperate with other organizations and other federal departments involved in the mobility of people (such as Health and Human Services) in order to develop innovative ITS-supported services such as mobility management. Systems such as multiagency mobility management, automatic collision notification, tourist information, and weather information will be the primary focus in the early years of the program.

Goal

By 2003 the rural ITS program aims to have demonstrated in ten locations a statewide information network that is multijurisdictional and multimodal within a state and able to share data across state lines.

Key Activities and Milestones

To reach this goal, the rural ITS program will focus primarily on conducting research through operational tests. The other seven program strategies will be used to a more limited extent.

Conducting research: Seven areas have been identified for further research: surface transportation weather and winter mobility, emergency services, statewide and regional traveler information infrastructure, rural crash prevention, rural transit mobility, rural traffic management, and highway operations and maintenance. While activities are expected in all seven areas, the U.S. DOT has worked with stakeholders to categorize and prioritize the list. Initial efforts will focus on multiagency mobility management services, weather information, emergency services, and regional traveler information. Operational tests are currently under way for all four services, and additional rounds of tests will be conducted through 2003.

Accelerating the development of standards: The U.S. DOT is just beginning to identify what standards may be necessary for rural-specific ITS applications. Standards are identified by assessing user needs, defining rural ITS infrastructure, and modifying the National ITS Architecture. The rural ITS program is actively seeking stakeholder participation in this process, and modifications to the National ITS Architecture are being made as rural ITS applications are defined. Once the National ITS Architecture is revised, ITS standards requirements can be identified. U.S. DOT defined unique rural user services in 1999 and proceeded to develop them.

Building professional capacity: Professional capacity building for rural practitioners involves modifying existing ITS courses to reflect the needs of rural ITS users and exploring distance-learning opportunities. Practitioners with rural expertise will help tailor existing courses to a rural audience. Initiatives are under way to overcome barriers of limited time and travel funding experienced most acutely by rural partners. U.S. DOT is exploring methods to deliver training through satellite broadcast, CD-ROM, and the Internet. As with other parts of the National ITS Program, the U.S. DOT has planned to transfer course delivery to the National Highway Institute and National Transit Institute in 2001.

Creating funding incentives: TEA-21 funding incentives are being offered to rural public sector applicants to support deployment of individual project components and the integration of existing ITS components. Funding will be offered to both highway and transit projects, and the U.S. DOT will work with Congress and the funding recipients to ensure that both the spirit and intent of TEA-21 funding criteria are met. U.S. DOT will allocate funding annually based on programmatic goals and TEA-21 criteria. A minimum of 10% of available funding will be used in rural areas.

Providing guidance and technical assistance: The U.S. DOT will provide guidance and technical assistance primarily by disseminating the results of rural field operational tests to stakeholders. Materials, such as lessons learned and simple solutions compendia, technical toolboxes, and catalogs of available systems, will be compiled and packaged for stakeholders in an Advanced Rural Transportation Systems toolbox. Assistance also will be available to rural stakeholders through federal field staff and the Peer-to-Peer Network.

Ensuring consistency with the National ITS Architecture and standards: Interim policy guidance on consistency with the National ITS Architecture and standards was issued in October 1998. The U.S. DOT expected to issue a final policy on consistency in FY 2001. This policy will be instrumental in catalyzing integrated ITS deployment across the country. In rural areas, stakeholders will be engaged in the policy development process to work through consistency issues at the statewide planning level. The interim guidance is being implemented until release of the final policy; the final policy will be implemented through 2003 and beyond.

Evaluating the program: Rural ITS infrastructure components must be defined before they can be tracked, so program evaluation activities are just beginning. Once the components are defined, quantifiable indicators will be identified as they have been for metropolitan applications.

Showcasing benefits: In these early stages of the rural ITS program, benefits of rural ITS applications are being showcased through field operational tests. These tests are not of the same scale as the metropolitan model deployments, but they still provide rural stakeholders the opportunity to see rural ITS technologies in operation and the benefits to rural America. Tests include automatic collision notification, traveler information, weather information technologies, and traveler information in a national park setting.

Commercial Vehicle ITS Infrastructure Program Area

The commercial vehicle ITS infrastructure program focuses on increasing safety for commercial drivers and vehicles while improving operating efficiencies for government agencies and motor carriers. At the center of the program is the deployment of Commercial Vehicle Information Systems and Networks, which link existing information systems to enable the electronic exchange of information. The initial implementation of CVISN, known as Level 1, addresses safety information exchange, credentials administration, and electronic screening; it is being prototyped in two states and piloted in eight states nationwide. The U.S. DOT expected CVISN Level 1 capabilities to be achieved in Maryland, Virginia, Washington, Kentucky, California, Colorado, Connecticut, Michigan, Minnesota, and Oregon by 2000 or 2001, depending on the availability of discretionary deployment incentive funding from Congress.

Goal

The U.S. DOT has set a goal of having 26 to 30 states deploy CVISN Level 1 capabilities by 2003. Achievement of this goal will depend on the extent to which funds authorized for CVISN in TEA-21 are appropriated for that use.

Key Activities and Milestones

All eight ITS program strategies are being used to meet the commercial vehicle program area goal as follows:

Conducting research: Research efforts will continue the development, testing, and implementation of technologies necessary to support commercial vehicle safety enforcement and compliance goals. Under TEA-21, FHWA and the Federal Motor Carrier Safety Administration (FMCSA) will undertake coordinated activities intended to reduce or eliminate transportation-related incidents and the resulting deaths, injuries, and property damage. These activities include demonstrating cost-effective technologies for achieving improvement in motor carrier enforcement, compliance, and safety while keeping up with the latest technological advances. The U.S. DOT will define CVISN Level 2 capabilities and expects to demonstrate prototype technologies in two or three states from 2000 through 2003.

Accelerating the development of standards: The U.S. DOT will continue to update and maintain the CVISN architecture to ensure consistency and interoperability, to include lessons learned from deployments, and to keep current with changing technology. In addition, two standards — electronic data interchange (EDI) and dedicated short-range communication (DSRC) — are essential to the demonstration of CVISN. EDI supports safety information and credential information exchange and has been approved. The U.S. DOT has completed a DSRC standard at 5.9 GHz, which is necessary for vehicle-to-roadside exchange of information. The U.S. DOT's emphasis has shifted to developing guidelines for compatibility and certification testing of the DSRC standard. Ultimately, independent testing organizations will be responsible for certification testing of DSRC.

Building professional capacity: Professional capacity building is critical to states, vendors, and FHWA and FMCSA project managers in order to implement CVISN. In addition to the current suite of commercial vehicle ITS awareness and deployment courses, training and technical assistance will be available to states in the areas of interoperability testing for conformance with the National ITS Architecture, systems integration issues and lessons learned, and commercial vehicle ITS project monitoring and maintenance.

Creating funding incentives: TEA-21 authorized $184 million over 6 years to deploy CVISN in a majority of states. Funding was allocated based on programmatic goals and TEA-21 criteria as follows: $27.2 million in 1999, $30.2 million in 2000, $32.2 million in 2001, $33.5 million in 2002, and $35.5 million in 2003. The funding will assist prototype and pilot states, as well as other interested states, in reaching CVISN Level 1 capabilities.

Providing guidance and technical assistance: The U.S. DOT has developed an integrated strategy to support states through the deployment of CVISN. From 1999 through 2003, U.S. DOT will continue to provide support to states through tool kits, guides, the Peer-to-Peer Network, and outreach.

Ensuring consistency with the National ITS Architecture and standards: The interim guidance issued in October 1998 and the final policy expected in 2001 apply equally to commercial vehicle ITS applications. At the heart of CVISN is the need for interoperability among federal, state, carrier, and other commercial vehicle systems and networks that allow the exchange of data. The development of a policy to ensure consistency with the National ITS Architecture and approved standards supports this interoperability. Federal field staff will implement the policy through 2003 and beyond.

Evaluating the program: Deployment tracking surveys will be conducted for all 50 states at 2-year intervals from 1999 to 2003. In addition, field operational tests will be completed and results will be incorporated into ITS costs and benefits databases.

Showcasing benefits: All eight pilot states serve as model deployments to showcase the benefits of CVISN. Benefits information were collected from the sites and incorporated into brochures and materials for distribution to stakeholders in 1999 and 2000. CVISN technologies were also showcased across the country in 1999 and 2000 with the commercial vehicle technology truck, a traveling classroom that contains commercial vehicle technologies and provides an interactive learning environment for stakeholders.

Intelligent Vehicle Initiative Program Area

Under ISTEA, U.S. DOT research in crash avoidance, in-vehicle information systems, and Automated Highway Systems pointed to new safety approaches and promising solutions that could significantly reduce motor vehicle crashes. Preliminary estimates by the National Highway Traffic Safety Administration showed that rear-end, lane change, and roadway departure crash avoidance systems have the potential, collectively, to reduce motor vehicle crashes by one sixth, or about 1.2 million crashes annually. Such systems may take the form of warning drivers, recommending control actions, and introducing temporary or partial automated control of the vehicle in hazardous situations. These integrated technologies can be linked to in-vehicle driver displays that adhere to well-founded human factors requirements. The U.S. DOT has harnessed these efforts into one program, the Intelligent Vehicle Initiative (IVI).

Goal

IVI is focused on working with industry to advance the commercial availability of intelligent vehicle technologies and to ensure the safety of these systems within the vehicles.

Key Activities and Milestones

This program is solely a research effort; therefore, only the conducting research program strategy applies.

Conducting research: Intelligent vehicle research aims to identify in-vehicle technologies to counter a series of problems that are major causes of vehicle crashes. To help speed the development of solutions, IVI has been organized into manageable tasks by dividing the spectrum of problems into eight problem areas and segmenting vehicle types into four vehicle platforms. Each problem area will be studied in the platform(s) where new technologies are most needed and can be readily adopted. Currently, the IVI program is moving forward through pilot research and testing projects within each platform. Projects range from defining safety needs for specialty vehicles to widespread initial trial deployment of automatic collision notification systems for light vehicles. In general, the light and commercial vehicle platforms are further along in the process because they benefited from prior research. However, the transit and specialty vehicle platforms will advance rapidly by adapting research conducted in the other platforms. In addition to the core in-vehicle technologies, the IVI program will also begin to explore possible vehicle infrastructure cooperative technologies as well as ways to help improve the ability of drivers to receive and process more information in the vehicle.

Eight IVI problem areas:

- Rear-end collision avoidance
- Lane change and merge collision avoidance
- Road departure collision avoidance

Intersection collision avoidance
Vision enhancement
Vehicle stability
Safety-impacting services
Driver condition warning

Four IVI platforms:
Light vehicles
Commercial vehicles
Transit vehicles
Specialty vehicles

To accomplish programmatic objectives, IVI is undertaking public and private partnerships with the motor vehicle industry and infrastructure providers. For transit, key partnerships with fleet operators will also be necessary, as transit vehicle designs are influenced not only by the vehicle manufacturers but also by transit agencies. The U.S. DOT will use multiple platforms to allow the program to focus initial research on the classes of vehicles where new technology will be adopted most quickly. Other vehicle types can then be equipped with the proven technology. The U.S. DOT will also study linkages with intelligent infrastructure, multiple systems integration, generations of vehicles with increased capabilities, and human factors. Finally, peer review will be used to help keep the goals and objectives of the program on target. Under TEA-21, the intelligent vehicle program was to form a public or private partnership to mutually govern and conduct enabling research for intelligent vehicles, engage the Transportation Research Board for a multiyear peer review, and complete initial operational tests on all platforms by 2001.

Additional Areas Covered in the Plan

In addition to program area goals and activities, this report also covers the National ITS Architecture and ITS standards, emerging program activities, and an update of ITS user services.

The National ITS Architecture and ITS Standards

The full benefits of ITS cannot be realized unless systems are integrated, rather than deployed as individual components. At the urging of public and private sector stakeholders, the U.S. DOT is facilitating system integration and technical interoperability through the development of the National ITS Architecture and ITS standards. The National ITS Architecture is a framework that defines the functions performed by ITS components and the ways in which components can be integrated into a single system. It can be used to help agencies plan and design both projects and deployment approaches that meet near-term needs while keeping options open for eventual system expansion and integration. The U.S. DOT will ensure that the National ITS Architecture responds to changing needs of the National ITS Program and the ITS industry by keeping the architecture up-to-date and relevant as new ITS applications emerge.

Since the inception of the ITS Program under ISTEA, stakeholders have recognized that ITS standards are necessary to achieve technical interoperability. Without technical standards, state and local governments, as well as consumers, risk buying products that do not necessarily work together or consistently in different parts of the country. The U.S. DOT is facilitating the creation of technical standards to minimize public sector risk in procuring these products. The overall goal of the ITS standards program is to have a comprehensive set of ITS standards developed and routinely used as states and localities deploy integrated, intermodal systems.

Over the past several years, the U.S. DOT has funded standards development organizations in conjunction with industry volunteer support to accelerate the traditional standards development process. Under TEA-21, the U.S. DOT expects that all ITS standards identified in the baseline National ITS Architecture will be developed and that the ITS standards program will increasingly focus on implementation. A first step in this direction will include the testing of approved ITS standards under realistic transportation conditions. Additionally, the U.S. DOT worked with the ITS user community in FY 1999 to identify critical ITS standards. In a report to Congress, 17 standards were identified as critical for national interoperability or as foundation standards for the development of other critical standards. Development of critical standards will be actively monitored, and provisional standards may be established.

Emerging Program Activities

As the National ITS Program evolves and transportation opportunities arise, it becomes apparent that new areas can benefit from ITS. Five such areas will be addressed under TEA-21: intermodal freight, ITS data archiving, rail transit, pedestrian and bicycle safety, and accessibility.

The goal of the emerging **intermodal freight** program is to facilitate goods movement around congested areas, across multiple modes, and with international trading partners to the north and south. The application of advanced information and communications technologies to the intermodal system offers opportunities to strengthen the links between the separate modal systems that currently operate as competitors. Under TEA-21, the intermodal freight program will conduct field operational tests to identify benefits and opportunities for ITS applications for border and corridor safety clearance applications and for intermodal freight applications that enhance operational efficiency. By 2001, the U.S. DOT expected to have enough information to develop an intermodal freight ITS to be added to the National ITS Architecture.

ITS data archiving addresses the collection, storage, and distribution of ITS data for transportation planning, administration, policy, operations, safety analyses, and research. The recently approved archived data user service, the 31st user service in the National ITS Architecture, addresses this new area and was integrated into the architecture in early FY 2000.

Rail transit is an important transit mode that historically has used advanced technologies in its operations. However, little attention has focused on how rail can benefit from system integration and ITS information. The U.S. DOT aims to address rail transit as a part of identifying integrated transit systems across agencies, modes, or regions.

Efforts toward **pedestrian and bicycle safety** focus on creating more pedestrian-friendly intersections through the use of adaptive crosswalk signals, inclusion of pedestrian and bicycle flows in traffic management models, and the promotion of in-vehicle technologies to detect and avert impending vehicle–pedestrian collisions.

Accessibility: Improvement can be made in this area, especially for rural Americans, with better information coordination and dispatching for ride sharing, paratransit, and other public transit efforts. Moreover, efforts will be aimed at improving mobility and safety for two user groups underserved by current pedestrian crossings: the elderly and the disabled. The U.S. DOT will support ITS solutions that meet the needs of these Americans.

Update of ITS User Services

The National ITS Program focuses on the development and deployment of a collection of interrelated user services. These are areas in which stakeholders have identified potential benefits from advanced technologies that improve surface transportation operations. The user services have guided the development of the National ITS Architecture and ITS standards, as well as research and development of ITS systems. In 1993, the ITS America National Program Plan introduced a set of ITS user services and subservices. When the 1995 National ITS Program Plan was published, 29 user services were identified. However, in keeping with the evolving nature of the National ITS Architecture, two new services have been identified: highway–rail intersection and the archived data user service.

ITS solutions for highway–rail intersections aim to avoid collisions between trains and vehicles at highway–rail grade crossings. Examples of intersection control technologies include advisories and alarms to train crews, roadside variable message signs, in-vehicle motorist advisories, warnings, automatic vehicle stopping, improved grade crossing gates and equipment, and automated collision notification.

Archived data services require ITS-related systems to have the capability to receive, collect, and archive ITS-generated operational data for historical purposes and for secondary users. ITS technologies generate massive amounts of operational data. These data offer great promise for application in areas such as transportation administration, policy, safety, planning, operations, safety analyses, and research. Intelligent transportation systems have the potential to provide data needed for planning performance monitoring, program assessment, policy evaluation, and other transportation activities useful to many modes and for intermodal applications. Below are listed all 31 ITS user services, including the two new ones. The user services have been grouped together in seven areas: travel and traffic management, public transportation management, electronic payment, Commercial Vehicle Operations, emergency management, advanced vehicle safety systems, and information management.

ITS user services are defined not along lines of common technologies but rather by how they meet the safety, mobility, comfort, and other needs of transportation users and providers. They represent essential, but not exclusive, ITS products and services.

Travel and traffic management:
- Pretrip travel information
- En route driver information
- Route guidance
- Ride matching and reservation
- Traveler services information
- Traffic control
- Incident management
- Travel demand management
- Emissions testing and mitigation
- Highway–rail intersection*

Public transportation management:
- Public transportation management
- En route transit information
- Personalized public transit
- Public travel security

Electronic payment:
- Electronic payment services

Commercial Vehicle Operations:
- Commercial vehicle electronic clearance
- Automated roadside safety inspection
- Onboard safety monitoring
- Commercial vehicle administrative processes
- Hazardous material incident response
- Commercial fleet management

Emergency management:
- Emergency notification and personal security
- Emergency vehicle management

Advanced vehicle safety systems:
- Longitudinal collision avoidance
- Lateral collision avoidance
- Intersection collision avoidance
- Vision enhancement for crash avoidance
- Safety readiness
- Precrash restraint deployment
- Automated vehicle operation

Information management:
- Archived data user service*

---

*User service added since the first edition of the National ITS Program Plan in 1995.

86.8 "The National Intelligent Transportation Systems Program Plan: A Ten-Year Vision" [46]

The Goals

"The National Intelligent Transportation Systems Program Plan: A Ten-Year Vision" sets forth the next-generation research agenda for ITS. It identifies benefits areas and associated goals against which change and progress can be measured. These goals provide the guideposts for fully realizing the opportunities that ITS technology can provide in enhancing the operation of the nation's transportation systems, in improving the quality of life for all citizens, and in increasing user satisfaction, whether for business or personal travel. The goals include:

Safety: reduce annual transportation-related fatalities by 15% overall by 2011, saving 5000 to 7000 lives per year

Security: have a transportation system that is well protected against attacks and responds effectively to natural and manmade threats and disasters, enabling the continued movement of people and goods, even in times of crisis

Efficiency and economy: save at least $20 billion per year by enhancing throughput and capacity with better information, better system management, and the containment of congestion by providing for the efficient end-to-end movement of people and goods, including quick, seamless intermodal transitions

Mobility and access: have universally available information that supports seamless, end-to-end travel choices for all users of the transportation system

Energy and environement: save a minimum of 1 billion gallons of gasoline each year and reduce emissions at least in proportion to this fuel saving

This plan develops a series of programmatic and enabling themes to describe the opportunities, benefits, and challenges of the transportation system of the future and activities required to realize this system.

Programmatic Theme 1

A new, bold transportation vision is needed to set the directions and mold the institutions for the next 50 years. This new, bold vision is based on information management and availability, on connectivity, and on system management and optimization — in short, the creation of an Integrated Network of Transportation Information.

Programmatic Theme 2

Transportation-related safety is clearly more than safe driving. However, in recent years, motor vehicle crashes have resulted in more than 40,000 fatalities and more than 3 million injuries each year. Driver error remains the leading cause of crashes, cited in more than 80% of police crash reports. In-vehicle systems, infrastructure improvements, and cooperative vehicle infrastructure systems can help drivers avoid hazardous mistakes by minimizing distraction, helping in degraded driving conditions, and providing warnings or control in imminent crash situations.

Programmatic Theme 3

Getting emergency response teams as quickly as possible to the scene of a crash or other injury-producing incident is critical to saving lives and returning roadways to normal, unimpeded operation. ITS technologies, coupled with computer-aided dispatch, wireless communications, records management systems, private call centers, and websites, can be used to achieve these objectives.

Programmatic Theme 4

Advanced transportation management involves using advanced technology to intelligently and adaptively manage the flow of goods and people through the physical infrastructure.

Enabling themes set the stage and lay the groundwork for the application of technology to surface transportation.

Enabling Theme 1

A culture of transportation systems management and operations will be created over the next 10 years to focus increasingly on safety, security, customer service, and systems performance. The demands of both the external and internal environments are generating changes in the culture of both service providers and users.

Enabling Theme 2

ITS and the information management and communications capabilities that it brings will support a new level of cooperative operations among multiple agencies, across boundaries and travel modes. An increase in the level of investment in ITS by the public sector will improve the cost–benefit balance of the transportation network as a whole.

Enabling Theme 3

Traditional business–government partnerships need to be redefined to enhance private sector opportunities in the commercial marketplace. Government needs to help accelerate deployment by adopting and encouraging the adoption by others of appropriate ITS products and services.

Enabling Theme 4

While the new information opportunities that ITS creates are clearly valuable — in many cases essential, the sheer volume of information also creates potential problems, e.g., overload, distraction, and confusion. ITS designers must consider what the vehicle operator is capable of doing while operating a vehicle safely. User-centered design is a fundamental concept within human factors engineering and is a proven method of promoting effective, successful, and safe design.

The Stakeholders

More than a dozen major stakeholders are identified and called on to contribute to the realization of this plan. Most of these stakeholders fall into one of three macrolevel groups: the public sector, the private sector, and universities.

86.9 Case Study: Incident Management

Although congestion is recognized as a problem for commuters and motor carriers alike, information on the scope and cost of congestion is limited. A 1984 staff study by the FHWA found that freeway congestion in the nation's 37 largest metropolitan areas was responsible for 2 billion vehicle hours of delay at a cost of $16 billion [10,11]. By 2005, those figures could rise to as high as 8 billion vehicle hours and $88 billion annually. Most of the cost of congestion is borne by large cities. A dozen large urban areas account for more than 80% of freeway congestion cost. New York, Los Angeles, San Francisco, and Houston have the highest congestion costs, about $2 billion each in current dollars; Detroit, Chicago, Boston, Dallas, and Seattle, about $1 billion each; and Atlanta, Washington, D.C., and Minneapolis, about $500 million each. The patterns of past growth and the trends for the immediate future all point toward the conclusion that congestion will continue to be a significant metropolitan and national issue. Without attention, congestion will sap the productivity and competitiveness of our economy, contribute to air pollution, and degrade the quality of life in our metropolitan areas [10].

The term *recurring problem* is used to describe congestion when it routinely occurs at certain locations and during specific time periods. The term *nonrecurring problem* is used to describe congestion when it is due to random events such as accidents or, more generally, incidents [18].

Recurring Congestion

The most common cause of recurring congestion is excessive demand, the basic overloading of a facility that results in traffic stream turbulence. For instance, under ideal conditions, the capacity of a freeway is approximately 2000 to 2200 passenger cars per lane per hour. When the travel demand exceeds this number, an operational bottleneck will develop. An example is congestion associated with nonmetered freeway ramp access. If the combined volume of a freeway entrance ramp and the main freeway lanes creates a demand that exceeds the capacity of a section of freeway downstream from the ramp entrance, congestion will develop on the main lanes of the freeway, which will result in queuing upstream of the bottleneck. The time and location of this type of congestion can be predicted [18].

Another cause of recurring congestion is the reduced capacity created by a geometric deficiency, such as a lane drop, difficult weaving section, or narrow cross section. The capacity of these isolated sections, called geometric bottlenecks, is lower than that of adjacent sections along the highway. When the demand upstream of the bottleneck exceeds the capacity of the bottleneck, congestion develops and queuing occurs on the upstream lanes. As above, the resulting congestion can also be predicted [18].

Nonrecurring Congestion: Incidents

Delays and hazards caused by random events constitute another serious highway congestion problem. Referred to as temporary hazards or incidents, they can vary substantially in character. Included in this category is any unusual event that causes congestion and delay [18]. According to FHWA estimates, incidents account for 60% of the vehicle hours lost to congestion. Of the incidents that are recorded by police and highway departments, the vast majority, 80%, are vehicle disablements — cars and trucks that have run out of gas, have a flat tire, or have been abandoned by their drivers. Of these, 80% wind up on the shoulder of the highway for an average of 15 to 30 minutes. During off-peak periods when traffic volumes are low, these disabled vehicles have little or no impact on traffic flow. However, when traffic volumes are high, the presence of a stalled car or a driver changing a flat tire in the breakdown lane can slow traffic in the adjacent traffic lane, causing 100 to 200 vehicle hours of delay to other motorists [10].

An incident that blocks one lane of three on a freeway reduces capacity in that sense of travel by 50% and even has a substantial impact on the opposing sense of travel because of rubbernecking [12]. If traffic flow approaching the incident is high (near capacity), the resulting backup can grow at a rate of about 8.5 miles per hour — that is, after 1 hour, the backup will be 8.5 miles long [12,13]. Traffic also backs up on ramps and adjacent surface streets, affecting traffic that does not even intend to use the freeway. Observations in Los Angeles indicate that in off-peak travel periods, each minute of incident duration results in 4 or 5 minutes of additional delay. In peak periods, the ratio is much greater [12,14].

Accidents account for only 10% of reported incidents. Most are the result of minor collisions, such as sideswipes and slow-speed rear-end collisions [10]. Forty percent of accidents block one or two lanes of traffic. These often involve injuries or spills. Each such incident typically lasts 45 to 90 minutes, causing 1200 to 2500 vehicle hours of delay [10,15]. It is estimated that major accidents make up 5 to 15% of all accidents and cause 2500 to 5000 vehicle hours of delay per incident [10,16]. Very few of these major incidents, typically those involving hazardous materials, last 10 to 12 hours and cause 30,000 to 40,000 vehicle hours of delay. These incidents are rare, but their impacts can be catastrophic and trigger gridlock [10]. To be sure, these statistics are location specific and may differ across areas in the United States.

Incident Management

Incident congestion can be minimized by detecting and clearing incidents as quickly as possible and diverting traffic before vehicles are caught up in the incident queue. Most major incidents are detected within 5 to 15 minutes; however, minor incidents may go unreported for 30 minutes or more [10]. Traffic information for incident detection is typically collected from loop detectors and includes occupancy and volume averaged at 20- to 60-second intervals, usually across all lanes. Detector spacing along the freeway is a half-mile on the average. Certain systems in the United States and Canada (e.g., California I-880 and Ontario's Queen Elizabeth Way) also use paired detectors to collect speed data [20].

During an incident the queue continues to build until the incident is cleared and traffic flow is restored. The vehicle hours of delay that accrue to motorists are represented in the exhibit by the area that lies between the normal flow rate and the lower incident flow rates. If the normal flow of traffic into the incident site is reduced by diverting traffic to alternate routes, the vehicle hours of delay are minimized (shaded area). If normal traffic flow is not diverted, additional vehicle hours of delay (hatched area) are accrued [10]. The time saved by an incident management program depends on how well the stages of an incident are managed.

Effective incident detection requires consideration of all major false alarm sources. In particular, traffic flow presents a number of inhomogeneities, hard to distinguish from those driven by incidents. Events producing traffic disturbances include bottlenecks, traffic pulses, compression waves, random traffic fluctuations, and incidents. Sensor failure, also treated as an event, is related only to the measurement component of detection systems. The major characteristics of each event are described below [20].

Incidents

Incidents are unexpected events that block part of the roadway and reduce capacity. Incidents create two traffic regimes, congested flow upstream (high occupancies) and uncongested downstream (low occupancies). Two shock waves are generated and propagate upstream and downstream, each accompanying its respective regime. The congested-region boundary propagates upstream at approximately 16 kilometers per hour (10 miles per hour), where the value depends on incident characteristics, freeway geometry, and traffic level. Downstream of the incident, the cleared region boundary propagates downstream at a speed that can reach 80 kilometers per hour (50 miles per hour) [50].

The evolution and propagation of each incident is governed by several factors, the most important of which are incident type, number of lanes closed, traffic conditions prior to incident, and incident location relative to entrance and exit ramps, lane drops or additions, sharp turns, grade, and sensor stations. Other, less important factors that are harder to model include pavement condition, traffic composition, and driver characteristics.

Incident patterns vary depending on the nature of the incident and prevailing traffic conditions [50]. The most distinctive pattern occurs when the reduced capacity from incident blockage falls below oncoming traffic volume so that a queue develops upstream. This pattern, which is clearest when traffic is flowing freely prior to the incident, is typical when one or more moving lanes are blocked following severe accidents. The second pattern type occurs when the prevailing traffic condition is freely moving but the impact of the incident is not severe. This may result, e.g., from lane blockage that still yields reduced capacity higher than the volume of incoming traffic. This situation may lead to missed detection, especially if the incident is not located near a detector. The third type characterizes incidents that do not create considerable flow discontinuity, as when a car stalls on the shoulder. These incidents usually do not create observable traffic shock waves and have limited or no noticeable impact on traffic operations. The fourth type of incident occurs in heavy traffic when a freeway segment is already congested. The incident generally leads to clearance downstream, but a distinguishable traffic pattern develops only after several minutes, except in a very severe blockage. This type of incident is often observed in secondary accidents at the congested region upstream of an incident in progress.

Bottlenecks

Bottlenecks are formed where the freeway cross section changes, e.g., in lane drops or additions. While incidents have only a temporary effect on occupancies, bottlenecks generally result in longer-lasting spatial density or occupancy discrepancies.

Traffic Pulses

Traffic pulses are created by platoons of cars moving downstream. Such disturbances may be caused by a large entrance ramp volume; for instance, a sporting event letting out. The observed pattern is an increase in occupancy in the upstream station followed by a similar increase in the downstream station. When ramp metering is present, traffic pulses are rarely observed.

Compression Waves

Compression waves occur in heavy, congested traffic, usually following a small disturbance, and are associated with severe slow-down, speed-up vehicle speed cycles. Waves are typically manifested by a sudden, large increase in occupancy that propagates through the traffic stream in a direction counter to the traffic flow. Compression waves result in significantly high station occupancies of the same magnitude as that in incident patterns.

Random Fluctuations

Random fluctuations are often observed in the traffic stream as short-duration peaks of traffic occupancy. These fluctuations, although usually not high in magnitude, may form an incident pattern or obscure real incident patterns.

Detection System Failures

Detection system failures may be observed in several forms, but a particular form often results in a specific pattern. This pattern is observed with isolated high-magnitude impulses in the 30-second volume and occupancy measurements, appearing simultaneously in several stations. These values are considered outliers or impulsive data noise [20].

Formulation of Incident Detection Problem

Incident detection can be viewed as part of a statistical decision framework in which traffic observations are used to select the true hypothesis from a pair, i.e., incident or no incident. Such a decision is associated with a level of risk and cost. The cost of a missed detection is expressed in terms of increased delays, and the cost of a false alarm is expressed in terms of incident management resources dispatched to the incident location. The objective of incident detection is to minimize the overall cost.

To formulate the incident detection problem in a simple incident versus no-incident environment, we observe the detector output that has a random character and seek to determine which of two possible causes, incident or normal traffic, produced it. The possible causes are assigned to a hypothesis, i.e., incident H_1 versus no-incident (normal traffic) H_0. Traffic information is collected in real time and processed through a detection test, in which a decision is made based on specific criteria. Traffic information, such as occupancy, represents the observation space. We can assume that the observation space corresponds to a set of N observations denoted by the observation vector **r**. Following a suitable decision rule, the total observation space Z is divided into two subspaces, Z_1 and Z_0. If observation **r** falls within Z_1, the decision is d_1; otherwise the decision is d_0.

To discuss suitable decision rules, we first observe that each time the detection test is performed four alternatives exist, depending on the true hypothesis H_i and the actual decision d_i, i = 0 or 1:

1. H_0 true; choose H_0 (correct "no incident" decision)
2. H_0 true; choose H_1 (false alarm)
3. H_1 true; choose H_1 (correct "incident" decision)
4. H_1 true; choose H_0 (missed incident)

The first and third alternatives correspond to correct choices; the second and fourth correspond to errors.

The Bayes' minimum error decision rule is based on the assumption that the two hypotheses are governed by probability assignments, known as a priori probabilities P_0 and P_1. These probabilities represent the observer's information about the sources (incident or no incident) before the experiment (testing) is conducted. Further, costs C_{00}, C_{10}, C_{11}, and C_{01} are assigned to the four alternatives. The first subscript indicates the chosen hypothesis and the second the true hypothesis. The costs associated with a wrong decision, C_{10}, C_{01}, are dominant. Each time the detection test is performed, the minimum error rule considers the risk (cost) and attempts to minimize the average risk. The risk function is written,

$$R = C_{00}P_0P(d_0/H_0) + C_{01}P_1P(d_0/H_1) + C_{10}P_0P(d_1/H_0) + C_{11}P_1P(d_1/H_1) \tag{86.1}$$

that is,

$$R = \sum\sum C_{ij}\,P_j\,P(d_i/H_j) \qquad i, j = 0, 1$$

where the conditional probabilities $P(d_i/H_j)$ result from integrating $p(\mathbf{r}/H_j)$, the conditional probability to observe the vector $\mathbf{r}$ over Z_i, the observation subspace in which the decision is d_i. In particular, the probability of detection is

$$P(d_1/H_1) = P_D = \int_{Z_1} p(\mathbf{r}/H_1)\,dr$$

and the probability of false alarm is

$$P(d_1/H_0) = P_F = \int_{Z_1} p(\mathbf{r}/H_0)\,dr$$

Minimizing the average risk yields the *likelihood ratio* test:

$$\Lambda(\mathbf{r}) = \frac{p(\mathbf{r}/H_1)}{p(\mathbf{r}/H_0)} \mathop{\gtrless}_{H_0}^{H_1} \frac{(c_{10} - c_{00})\,P_0}{(c_{01} - c_{11})\,P_1}$$

where the second part in the inequality represents the test threshold, and the conditional and a priori probabilities can be estimated through time observations of incident and incident-free data. However, obtaining an optimal threshold requires realistic assignment of costs to each alternative. This is further impeded by the fact that incidents (or false alarms) are not alike in frequency, impact, and consequences. Therefore, an optimal threshold cannot practically be established. Previous attempts to use the Bayes' decision rule employed a simplified risk function to overcome the cost assignment issue and reduce the calibration effort. For instance, Levin and Krause [51] obtained a suboptimal threshold by maximizing the expression

$$R = P(d_0/H_0) + P(d_1/H_1)$$

using the relative spatial occupancy difference between adjacent stations as the observation parameter.

An alternative procedure to Bayes' rule, applicable when assigning realistic costs or a priori probabilities is not feasible, is the Neyman–Pearson (NP) criterion. The NP criterion views the solution of the optimization of the risk function in Equation (86.1) as a constrained maximization problem. This is necessitated by the fact that minimizing P_F and maximizing P_D are conflicting objectives. Therefore, one must be fixed while the other is optimized:

Constrain $P_F \leq \alpha$ and design a test to maximize P_D under this constraint.

Similarly to the minimum error criterion, the NP test results in a likelihood ratio test:

$$\Lambda(\mathbf{r}) \geq \lambda$$

where the threshold λ is a function of P_F only. Decreasing λ is equivalent to increasing Z_1, the region where the decision is d_1 (incident). Thus, both P_F and P_D increase as λ decreases. The Neyman–Pearson *lemma* [52] implies that the maximum P_D occurs at $P_F = \alpha$.

The lemma holds since P_D is a nondecreasing function of P_F. In practical terms, an NP procedure implies that after an incident test has been designed, it is applied to a data set initially employing a high (restrictive) threshold, which results in low P_F. The threshold is incrementally reduced until P_F increases to the upper tolerable limit α. The corresponding P_D represents the detection success of the test at false alarm α. An NP procedure seems more applicable to incident detection than a minimum error procedure for two reasons. First, the only requirement is the constraint on P_F, which can easily be assessed by traffic engineers to a tolerable limit. Second, an NP procedure does not require separate threshold calibration since no optimal threshold, in the Bayesian sense, is sought. Instead, thresholds result from the desirable P_F.

The decision process is facilitated by the likelihood ratio $\Lambda(\mathbf{r})$. In signal detection practice, $\Lambda(\mathbf{r})$ is replaced by a *sufficient statistic* $l(\mathbf{r})$, which is simpler than the $\Lambda(\mathbf{r})$ function of the data. The values of the sufficient statistic are then compared to appropriate thresholds to decide which hypothesis is true. In incident detection applications, however, the tests of an algorithm are designed empirically so that they only approximately can be considered as sufficient statistics.

Need for All Incident Management Stages to Perform

Classical incident management strategies at the incident management components of detection, verification, response, and traffic management are aimed at minimizing the negative effects of nonrecurrent congestion that are due to incidents. The basic idea is that fast clearance of the incident scene can help to alleviate the incident-related congestion. Early and reliable detection and verification of the incident, together with integrated motorway and nonmotorway traffic management strategies, are important contributions that improve the efficiency of the incident response, i.e., the actions taken once an incident has occurred. However, it would still be better if the incident had been avoided in the first place. A first requirement, then, is that one can recognize conditions in which an incident is more likely to occur. The component of incident probability estimation should be developed and added to the incident management suite for this purpose.

Automatic incident detection (AID) involves two major elements: a traffic detection system that provides the traffic information necessary for detection and an incident detection algorithm that interprets the information and ascertains the presence or absence of a capacity-reducing incident.

Most AID algorithms have been developed based on loop detector data. Detection has typically been based on models that determine the expected traffic state under normal traffic conditions and during incidents. Comparative (or pattern recognition) algorithms establish predetermined incident patterns in traffic measurements and attempt to identify these patterns by comparing detector output against preselected thresholds. One of these involves separating the flow-occupancy diagram into areas corresponding to different states of traffic conditions (e.g., congested and not congested) and detecting incidents after observing short-term changes of the traffic state. These algorithms operate on a detector output of 30- to 60-second occupancy and volume data.

Time series and statistical algorithms employ simple statistical indicators or time series models to describe normal traffic conditions and detect incidents when measurements deviate significantly from the model output. A third class includes algorithms that involve macroscopic traffic flow modeling to describe the evolution of traffic variables; the diversity of incident patterns requires development of a large number of pattern-specific models, and this has limited the potential of these algorithms for practical applications. Other methods include detection of stationary or slow-moving vehicles, filtering to reduce the undesired effects of traffic disturbances, application of fuzzy sets, transform analysis, and neural networks to take advantage of learning processes.

Recent work addressed the vehicle reidentification problem, lexicographic optimization, and derivation of section-related measures of traffic system performance using current inductive loops that provide vehicle waveforms. Another promising recent work performs real-time detection and characterization of motorway incidents using a three-step process, i.e., symptom identification of anomalous changes in traffic characteristics, signal processing for stochastic estimation of incident-related lane traffic characteristics, and pattern recognition.

In Europe, algorithms tested with data from loops are of the comparative type (e.g., HERMES I; German I, II, and IV; and Dutch MCSS), time series type (GERDIEN), or the type employing filtering (HERMES II). They use typical aggregate data (speed, volume, and occupancy) and aim to detect congestion and slow-moving or stopped vehicles. Other AID techniques extract traffic data from radar, such as the Millimetric Radar System (MMW) and German III. Using machine vision, AID systems serve as loop detector emulators (CCATS VIP and IRB), qualitative traffic state detectors (IMPACTS), or vehicle tracking detectors (TRISTAR and CCIDS).

Despite substantial research, algorithm implementation has been hampered by limited performance reliability, substantial implementation needs, and strong data requirements. Several problems require the attention of developers:

False alarm rate (FAR): The high number of false alarms has discouraged traffic engineers from integrating these algorithms in automated traffic operations. Algorithm alarms typically trigger the operator's attention; the operator verifies the validity of the alarm using closed-circuit TV cameras and decides on the appropriate incident response. In most cases, incident response is initiated only after an incident has been reported by the police or motorists.

Calibration: The need for algorithm calibration has not been extensively assessed, and lack of adequate calibration often leads to significantly deteriorated algorithm performance. Calibration by optimization of a set of different algorithms on the same field data set is the most reliable way for comparative evaluation across algorithms.

Evaluation: The major method adopted for comparatively evaluating the performance of AID algorithms is that of the operating characteristic curves. Performance tests have shown the following [53]:

United States

- Time series algorithms performed worse than comparative ones.
- DELOS, an algorithm based on filtering, produced 50% fewer false alarms than comparative algorithms, e.g., California type.
- The time series algorithm by Persaud et al. produced a good detection rate but too many false alarms to be practical.

France

- Comparative algorithms produced at least 30% fewer false alarms than single-variable time series algorithms (Standard Normal Deviate, Double Exponential, and ARIMA) at all detection levels.
- DELOS performed better than the time series algorithm developed by Persaud et al.

IN-RESPONSE Project in Europe

- DELOS and Algorithm 8 were evaluated against machine vision methods, and the results showed each to have its strengths under given conditions.

Canada

- The time series algorithm by Persaud et al. produced fewer false alarms than the California algorithm and was adopted as its replacement.

Transferability: Some understanding of algorithm transferability potential has been achieved, mainly in the IN-RESPONSE, HERMES, MARGOT-LLAMD, and EUROCOR projects and in a comparative evaluation in Minnesota and California (analysis of 213 incidents over 1660 hours, 24 hours a day) [54].

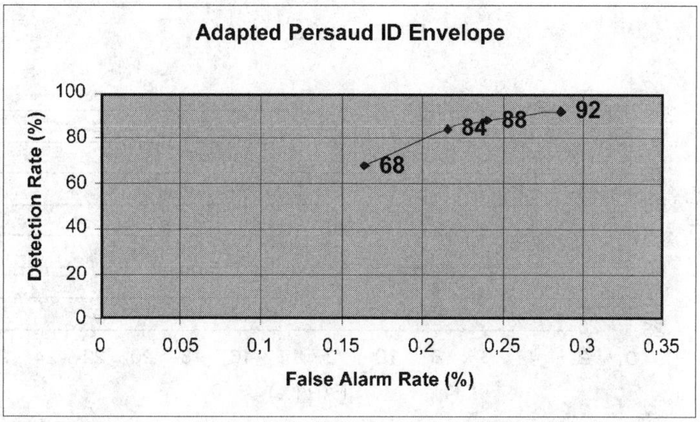

FIGURE 86.5 Performance envelope. (Used with permission from J. Barcelo and L. Montero.)

Traffic management objectives: While most U.S. efforts seek to remove the incident and achieve smooth traffic flow, work in Europe focuses on warning drivers of congestion even if no incident has occurred and on assisting stopped vehicles. Work in rural areas has focused on achieving AID with sparse instrumentation. The latter two objectives can often best be addressed by AID systems that are based on machine vision. Such systems have achieved performance equivalent to that of loop detectors. However, the additional advantage of the new systems is that they can detect incidents that do not influence traffic substantially or that cannot be detected by loop-based systems but are still a risk to the motorist.

Addressing the need for determining improved performance of incident detection methods under varying conditions, a recent project, PRIME, tested incident detection algorithms that have not been extensively tested in Europe, and more advanced sensing hardware. The project addressed all incident management components, i.e., estimation of incident probability, incident detection, incident verification, and integrated incident response strategies. Recent results from the project indicate that the **incident detection** component has satisfied the specifications in terms of detection rate and false alarm rate. For instance, application of the modified Persaud algorithm in Barcelona resulted in the performance envelope shown in Figure 86.5 [55].

The real-time **estimation of incident probability** is sparsely documented.

From IN-RESPONSE [56], it was concluded that the incident probability estimation model was a promising way of linking real-time 1-minute traffic and weather data to static data on road geometry for estimating incident probability. The technique could not be properly evaluated because of the shortage of incident data and the inaccuracy of the time stamps.

The same lack of incident data was reported in Reference 57. The authors presented several empirical methods for analyzing incident data. A key issue raised was whether an accident was responsible for the measured variability in traffic conditions or whether these conditions were caused by the accident.

The problem of data availability was not reported in Reference 58, which used a model similar to that in IN-RESPONSE (binary logit) to establish relationships between incident likelihood and explanatory variables such as weather and traffic conditions.

A method of overcoming this shortage of incident data is to simulate incident situations [59]. The cellular automaton-based microscopic simulation model TRANSIMS was used to estimate the probability of accidents. The model used a relationship between the probability of an incident and a safety criterion (braking power).

PRIME developed EIP hierarchical logit, logit, and fuzzy models with online data and simulation. The performance was good in terms of estimation rate but had several false alarms; see the performance envelope shown in Figure 86.6 with data from Barcelona.

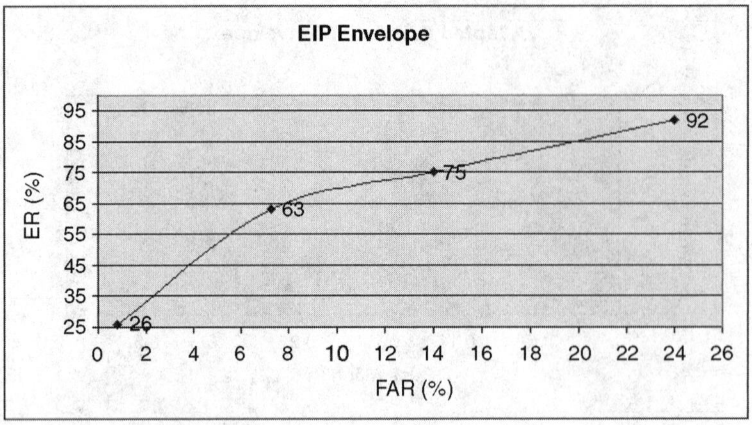

FIGURE 86.6 Performance envelope. (Used with permission from J. Barcelo and L. Montero.)

Incident verification (IV) aims to accumulate evidence and information about possible detected incidents and use this additional information to drop false alarms, merge repeated alarms, and provide complete incident reports in case of real incidents.

Most countries with an operational traffic management system are using one or more incident verification methods, primarily CCTV and patrol vehicles. Realizing the potential of using cellular telephones as an incident management tool, many highway agencies have formed partnerships with cellular telephone carriers to implement programs that encourage drivers to report randomly occurring motorway incidents.

However, information obtained from cellular phones varies in the detail and quality, and the incident may be reported after considerable time has elapsed. Therefore, the feasibility of motorway surveillance systems utilizing cellular phones needs to be carefully evaluated. A survey of 42 traffic management centers in the United States found that 75% use cellular detection. However, in most states, such as Texas, video cameras are deployed for verification. Weaknesses of cellular phones include a very low rate of detecting small incidents, the highest rate of false alarms, and limited information on the incident severity. Also, cellular phone messages need further verification and cannot tell when the incident is cleared. Incidents reported by cellular phones show greater incident duration by 14 minutes on the average than similar incidents reported by the CHP/MSP. This extra delay is due to the incident verification process by dispatching an officer.

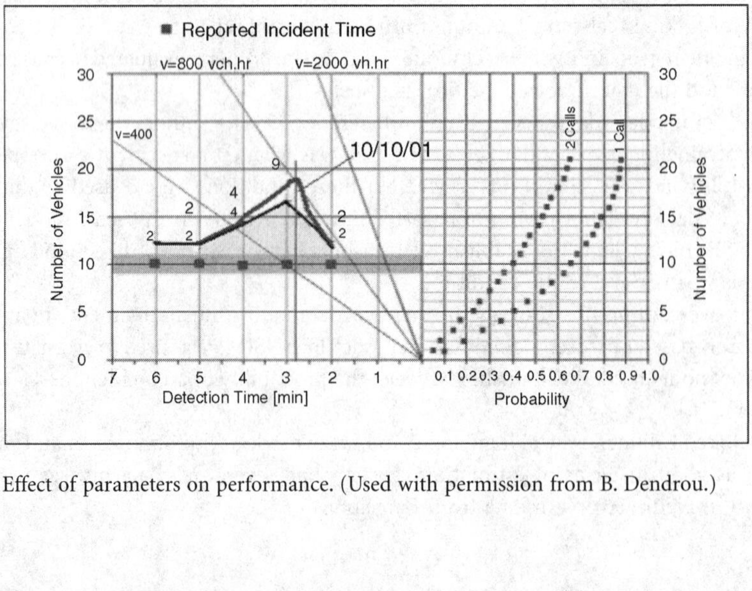

FIGURE 86.7 Effect of parameters on performance. (Used with permission from B. Dendrou.)

Cellular phone false alarms fall into two categories: (1) reporting incorrect or incomplete information regarding the location of the incident and its severity and (2) erroneous calls, including fake or prank calls. On the other hand, wireless phone users can report incidents that traditional methods cannot capture, such as debris, flooding, or wandering animals.

Incident management requirements for incident verification within advanced transport telematics (ATT) systems cannot rely solely on cellular phones. Cellular phone reports may contribute significantly to the incident detection in combination with other sources and may be used in the verification of incidents, including those detected by other methods. This would require proper fusion of cellular phone data with information from other sources and use of appropriate technologies, such as video surveillance.

When all sites use the same automated and reliable procedure, consistent information about the incidents will be collected. Information that could be retrieved to verify an incident in this way can include one or more incident attributes, e.g.:

- Location (road number, travel direction, kilometer point, and lane)
- Type and severity (injuries, fire, trapped injuries, or hazardous goods)
- Identity of source
- Type of assistance needed (mechanical, police, or emergency)
- Certainty

Results with data from Athens, Greece indicate the effect of parameters on performance, i.e., the effect of the number of calls required, mobile phone penetration, traffic volume, etc.; see Figure 86.7 [55].

Additional performance improvements in PRIME were attained with the development of specialized hardware and software, such as panoramic camera and accompanying algorithms that transform the original image acquired by a panoramic camera (Figure 86.8a) to a bird's-eye view that can be further processed (Figure 86.8b) and, through a homography-based transformation, transform the original image of a machine vision camera (Figure 86.8c) to top-down view (Figure 86.8d) for easier calculations and error reduction [55].

FIGURE 86.8 (a) Original image acquired by panoramic camera. (b) Bird's-eye transformation of (a). (Courtesy of Computer Vision and Robotics Lab, Institute of Computer Science, FORTH, Heraklion, Crete, Greece.)

References

1. TRB Committee on Intelligent Vehicle Highway Systems, Primer on Intelligent Vehicle Highway Systems, Transportation Research Circular 412, Transportation Research Board, National Research Council, Washington, D.C., August 1993.
2. Euler, G.W., Intelligent vehicle/highway systems: definitions and applications, *ITE J.*, 60, 17, 1990.
3. Minnesota ITS Planning Group, GuideStar: Guiding the Future of Minnesota's Highways, Minnesota Department of Transportation and University of Minnesota, Minneapolis, 1990.
4. Beaubien, R.F., Deployment of intelligent vehicle-highway systems, *ITE J.*, 63, 15, 1993.
5. Constantino, J., ITS America two years later, *ITE J.*, 63, 19, 1993.
6. Kraft, W.H., ITS and the transportation profession, *ITE J.*, 63, 23, 1993.
7. Carlson, E.D., Federal actions to deliver the ITS program, *ITE J.*, 63, 26, 1993.
8. Davis, G., Private communication, University of Minnesota, Minneapolis, October 1993.
9. Inside ITS, FHWA Awards Contracts for ITS Architecture: Intelligent Vehicle/Highway Systems Update, Vol. 3, Waters Information Services, Inc., September 1993, p. 1.
10. Grenzeback, L.R. and Woodle, C.E., The true costs of highway congestion, *ITE J.*, 62, 16, 1992.
11. Lindley, J.A., Quantification of Urban Freeway Congestion and Analysis of Remedial Measures, Report RD-871052, Federal Highway Administration Staff, October 1986; Urban freeway congestion: quantification of the problem and effectiveness of potential solutions, *ITE J.*, 57, 27, 1987.
12. Judycki, D.C. and Robinson, J.R., Managing traffic during nonrecurring congestion, *ITE J.*, 62, 21, 1992.
13. Morales, J.M., Analytical procedures for estimating freeway traffic congestion, *Public Roads*, 50, 55, 1986.
14. Judycki, D.C. and Robinson, J.R., Freeway incident management, in 1988 Compendium of Technical Papers, Institute of Transportation Engineers, Washington, D.C., 1988, p. 161.
15. JHK & Associates, Estimate Based on Delay Evaluation Procedures in Alternative Surveillance Concepts and Methods for Freeway Incident Management, Vol. 2: Planning and Trade-Off Analyses for Low-Cost Alternatives, Federal Highway Administration, Washington, D.C., March 1978; A Freeway Management Handbook, Vol. 2: Planning and Design, FHWA, Washington, D.C., May 1983; modifications recommended by California Department of Transportation.
16. Recker, W.W., An Analysis of the Characteristics and Congestion Impacts of Truck-Involved Freeway Accidents, Institute of Transportation Studies, Irvine, CA, December 1988.
17. Robinson, J.R. and McDade, J.D., Incident Management Programs in the United States, Office of Traffic Operations, Federal Highway Administration, Washington, D.C., October 1990.
18. Capelle, D.G., Freeway Traffic Management, Final Report, NCHRP Project 20-3D, Transportation Research Board, National Research Council, Washington, D.C., 1979.
19. Stephanedes, Y.J. and Chassiakos, A.P., Freeway incident detection through filtering, *Transp. Res. C*, 1C, 219, 1993.
20. Chassiakos, A.P. and Stephanedes, Y.J., Smoothing algorithms for incident detection, *Transp. Res. Rec.*, 1394, 8, 1993.
21. Ritchie, S.G. and Stack, R., A Real-Time Expert System for Freeway Incident Management in Orange County, California, paper presented at Fifth International Conference on Computing in Civil Engineering, ASCE, New York, 1993.
22. Stephanedes, Y.J. and Liu, X., Artificial Neural Networks for Freeway Incident Detection, paper presented at 73rd Annual Meeting of the Transportation Research Board, Washington, D.C., January 1994.
23. Transportation Research Board, ITS-IDEA, program announcement, Washington, D.C., April 1993.
24. French, R.L., Intelligent vehicle/highway systems in action, *ITE J.*, 60, 23, 1990.
25. Rowe, E., ITS: making it work, pulling it all together, *ITE J.*, 63, 45, 1993.
26. Stephanedes, Y.J. and Chang, K., Optimal control of freeway corridors, *J. Transp. Eng. ASCE*, 119, 50, 1993.

27. Stephanedes, Y.J. and Kwon, E., Adaptive demand-diversion prediction for integrated control of freeway corridors, *Transp. Res. C*, 1C, 1993.
28. Castle Rock Consultants, Rural ITS Scoping Study, Minnesota GuideStar, St. Paul, April 1994.
29. Stephanedes, Y.J., Implementation of On-Line Zone Control Strategies for Optimal Ramp Metering in the Minneapolis Ring Road, Road Traffic Monitoring and Control, 391, The Institution of Electrical Engineers, London, 1994.
30. Executive Office of the President of the United States, A Vision of Change for America, U.S. Government Printing Office, Washington, D.C., February 17, 1993.
31. Federal Highway Administration, ITS Staffing and Educational Needs, Final Report, DTFH61-92-C-00145, U.S. Department of Transportation, Washington, D.C., 1993.
32. Federal Highway Administration, What Have We Learned about Intelligent Transportation Systems? U.S. Department of Transportation, Washington, D.C., 2000.
33. Hardy, C., Are we all federalists now? in *Beyond Certainty: The Changing Worlds of Organizations*, Harvard Business School Press, Boston, MA, 1996.
34. Porter, M., *On Competition*, Harvard Business School Press, Boston, MA, 1998.
35. Sussman, J.M., *Introduction to Transportation Systems*, Artech House, Boston, MA, 2000.
36. Sussman, J.M., It happened in Boston, thoughts on ITS column, *ITS Quarterly*, Washington, D.C., Spring 2000.
37. Sussman, J.M., ITS deployment and the "competitive region," thoughts on ITS column, *ITS Quarterly*, Washington, D.C., Spring 1996.
38. ITS Joint Program Office, Five-Year Horizon National Intelligent Transportation Systems Program Plan, U.S. Department of Transportation, Washington, D.C., August 2000.
39. Mitretek Systems, Intelligent Transportation Systems Benefits: 2001 Update, FHWA-OP-01-024, Department of Transportation Intelligent Transportation Systems Joint Program Office, Washington, D.C., June 2001.
40. Bureau of Transportation Statistics, The Changing Face of Transportation, BTS00-007, U.S. Department of Transportation, Washington, D.C., 2000.
41. Intelligent Transportation Systems Joint Program Office, Department of Transportation's Intelligent Transportation Systems (ITS) Projects Book, U.S. Department of Transportation, Federal Highway Administration Operations Core Business Unit, Federal Transit Administration Office of Mobility Innovation, National Highway Traffic Safety Administration, Office of Associate Administrator for Research and Development, Federal Motor Carrier Safety Administration Office of Research and Technology, Washington, D.C.
42. Mitretek Systems, *Taxonomy for Classification of ITS Benefits*, Department of Transportation Intelligent Transportation Systems Joint Program Office, Washington, D.C., June 2000.
43. Mitretek Systems, ITS Benefits: Data Needs Update 2000, prepared in connection with the July 12 ITS Benefits Data Needs Workshop, August 2000.
44. Intelligent Transportation Systems websites, www.its.dot.gov/welcome.htm; www.its.dot.gov/eval/itsbenefits.htm.
45. Mitretek Systems, Working Paper National Costs of the Metropolitan ITS Infrastructure: Update to the FHWA 1995 Report, 2nd revision, FHWA-OP-01-147, Department of Transportation Intelligent Transportation Systems Joint Program Office, Washington, D.C., July 2001.
46. Intelligent Transportation Society of America, Delivering the Future of Transportation: The National Intelligent Transportation Systems Program Plan: A Ten-Year Vision, Federal Highway Administration, Washington, D.C., January 2002.
47. Johnson, C.M., The Future of ITS, keynote presentation at Intelligent Transportation Society of America, Eleventh Annual Meeting, Miami Beach, FL, June 4, 2001.
48. A Survey of Government on Internet: The Next Revolution, *The Economist*, June 14, 2000.
49. *The Washington Post*, June 29, 2000, p. A-15.
50. Payne, H.J. and Tignor, S.C., Freeway incident detection algorithms based on decision trees with states, *Transp. Res. Rec.*, 682, 30, 1978.

51. Levin, M. and Krause, G.M., Incident detection: a Bayesian approach, *Transp. Res. Rec.*, 682, 52, 1978.
52. Van Trees, H., *Detection, Estimation, and Modulation Theory Part I*, John Wiley & Sons, New York, 1968.
53. Stephanedes, Y.J. and McDonald, M., Improved Methods for Incident and Traffic Management, paper presented at IST 2000, Torino, Italy, 2000.
54. Stephanedes, Y.J. and Hourdakis, J., Transferability of freeway incident detection algorithms, *Transp. Res. Rec.*, 1554, 184, 1996.
55. PRIME Project, Draft Deliverable D61, Brussels, 2002.
56. IN-RESPONSE, Final Report, Brussels, 1999.
57. Hughes & Council, 1999.
58. Madanat and Liu.
59. Ree et al., 2000.
60. Sussman, J.M., Educating the "new transportation professional," *ITS Quarterly*, Washington, D.C., Summer 1995.

Further Information

Anderson, I., Graham, A.W., and Whyte, D.G., FEDICS: First Year Feedback, paper presented at 8th International Conference on Road Traffic Monitoring and Control, IEE, London, 1996, p. 28.

Atkins, W.K., Driver Reactions to Variable Message Traffic Signs in London, Stage 2 Report, report to Department of Transport by W.S. Atkins Planning Limited, 1995.

Cellular Telecommunications Industry Association, Semi-Annual Industry Survey, Washington, D.C., April, 1998; www.wow-com.com.

Chassiakos, A.P. and Stephanedes, Y.J., Smoothing algorithms for incident detection, *Transp. Res. Rec.*, 1394, 8, 1993.

Christenson, R.C., Evaluation of Cellular Call-In Programs for Incident Detection and Verification, Department of Civil Engineering, Texas A&M University, College Station, 1995.

Cohen, S., Comparative Assessment of Conventional and New Incident Detection Algorithms, paper presented at 7th International Conference on Road Traffic Monitoring and Control, IEE, London, 1995, p. 156.

CORD, Incident Detection Review, Incident Detection Task Force, DRIVE, Brussels, 1994.

Dörge, L., Vithen, C., and Lund-Sørensen, P., Results and Effects of VMS Control in Aalborg, paper presented at 8th International Conference on Road Traffic Monitoring and Control, IEE, London, 1996, p. 150.

Durand-Raucher, Y. and Santucci, J.C., Socio-Economic Benefits of the Paris Region Policy, Balanced between Traffic Management and Information, paper presented at 2nd World Congress on Intelligent Transport Systems, Yokohama, Japan, 1995, p. 1883.

Fall Creek Consultants, Wireless Location Services, E911, 1998; http://www.comm-nav.com.

Haj-Salem, H. et al., Field Trial Results of VMS Travel Display on the Corridor Périphérique of Paris, paper presented at 4th International Conference on Application of Advanced Technologies in Transportation Engineering, Capri, 1995, p. 171.

Hart, P.D., Research Associates, Inc., Attitudes Toward Wireless Telephones, Washington, D.C., March 1996, p. 2.

Hobbs, A. et al., The Use of VMS for Strategic Network Management: The PLEIADES Experience, paper presented at First World Congress of Application of Transport Telematics and Intelligent Vehicle-Highway Systems, Paris, 1994, p. 1237.

McDonald, M. and Richards, A., Urban Incident Management Using Integrated Control and Information Systems, paper presented at 8th International Conference on Road Traffic Monitoring and Control, IEE, London, 1996, p. 188.

McDonald, M., Richards, A., and Shinakis, E.G., Managing an Urban Network through Control and Information, paper presented at VNIS Conference in conjunction with Pacific Rim TransTech Conference, Seattle, 1995a, p. 516.

McDonald, M., Richards, A., and Shinakis, E.G., Integrated Urban Transport Management in Southampton, paper presented at 2nd World Congress on Intelligent Transport Systems, Yokohama, Japan, 1995b.

McLean, C.H., Cellular Phones: A Key Traffic Management Component, in ITE 1991 Compendium of Technical Papers, Institute of Transportation Engineers, Washington, D.C., 1991.

Motyka, V. and James, B., Concrete Application of Road Informatics Strategies: VMS in Ile-de-France: Detailed Quantitative Evaluation and First Glimpse of Socio-Economic Benefits, paper presented at First World Congress of Application of Transport Telematics and Intelligent Vehicle-Highway Systems, Paris, 1994, p. 1364.

Mussa, R.N. and Upchurch, J.E., Simulator Evaluation of Incident Detection Using Wireless Communications, submitted to 79th Annual Meeting of the Transportation Research Board, Washington, D.C., January 9–13, 2000.

Persaud, B.N., Hall, F.L., and Hall, L.M., Congestion identification aspects of the McMaster incident detection algorithm, *Transp. Res. Rec.*, 1287, 151, 1990.

Richards, A., Lyons, G., and McDonald, M., Network Routing Effects of Variable Message Signs, paper presented at Third World Congress of Application of Transport Telematics and Intelligent Vehicle-Highway Systems, Orlando, 1996.

Ritchie, S.G. and Baher, A., Development, Testing and Evaluation of Advanced Techniques for Freeway Incident Detection, California PATH Working Paper UCB-ITS-PWP-97-22, University of California, Irvine, July 1997.

Sheu, J. and Ritchie, S.G., A Sequential Detection Approach for Real-Time Freeway Incident Detection and Characterisation, submitted to Transportation Research Board, Washington, D.C., 2000.

Shinakis, E.G., Richards, A., and McDonald, M., The Use of VMS in Integrated Urban Traffic Management, paper presented at International Conference on Application of New Technology to Transport Systems, Melbourne, Australia, 1995, p. 195.

Skabardonis, A., Chavala, T.C., and Rydzewski, D., The I-880 Field Experiment: Effectiveness of Incident Detection Using Cellular Phones, paper presented at 77th Annual Meeting of the Transportation Research Board, Washington, D.C., January 1998.

Stephanedes, Y.J. and Liu, X., Artificial neural networks for freeway incident detection, *Transp. Res. Rec.*, 1494, 91, 1995.

Sun, C. et al., Use of Vehicle Signature Analysis and Lexicographic Optimisation for Vehicle Reidentification on Freeways, submitted to Transportation Research Board, Washington, D.C., 2000.

Swann, J. et al., Results of Practical Applications of Variable Message Signs (VMS): A64/A1 Accident Reduction Scheme and Forth Estuary Driver Information and Control System (FEDICS), paper presented at Seminar G of 23rd PTRC European Transport Forum, University of Warwick, Coventry, U.K., 1995, p. 149.

Tarry, S. and Graham, A., The role of evaluation in ATT development: 4. Evaluation of ATT systems, *Traffic Eng. Control*, 12, 688, 1995.

Tavana, H., Mahmassani, H.S., and Haas, C.C., Effectiveness of Wireless Phones in Incident Detection: A Probabilistic Analysis, submitted to 78th Annual Meeting of the Transportation Research Board, Washington, D.C., January 1999.

True Position, Inc., Time Difference of Arrival Technology for Locating Narrowband Cellular Signals, http://www.trueposition.com/tdoa.htm, 1997.

Van Eeden, P.G.M.A. et al., Dynamic Route Information in the Netherlands, Effects and Research, paper presented at 8th International Conference on Road Traffic Monitoring and Control, IEE, London, 1996, p. 145.

Walters, C.H., Wiles, P.B., and Cooner, S.A., Incident Detection Primarily by Cellular Phones: An Evaluation of a System for Dallas, Texas, submitted to 78th Annual Meeting of the Transportation Research Board, Washington, D.C., January 1999.

Wratten, B.M. and Higgins, R., Queensland's Smart Road: ITS Applications for the Pacific Motorway, *Traffic Technology International*, April/May 1999.

Appendix

TABLE 86.1 Incident Management Assessment Summary

Technology	Deployment Level	Limiting Factors	Comments
Service patrols	Widespread deployment	Cost, staffing	Successful
Common communication frequencies	Limited deployment[a]	Cost, institutional issues	Successful
Automated incident detection algorithms	Medium deployment[a]	Technical performance	Mixed
Cellular communication for incident detection	Widespread deployment	Availability, institutional issues	Jury is still out
Motorist call boxes	Limited deployment[a]	Being replaced by cell phone use	Successful
CCTV (ground, airborne, high magnification)	Widespread deployment	Cost	Successful
Cellular geolocation (old generation)	Operational testing[a]	Accuracy	Unsuccessful
Cellular geolocation (emerging generation)	Operational testing[a]	Availability, institutional issues	Jury is still out
Regional incident management programs	Limited deployment[a]	Institutional isues	Holds promise

[a] Quantitative deployment tracking data not available. Deployment level determined by expert judgment.

Source: Federal Highway Administration, What Have We Learned about Intelligent Transportation Systems? U.S. Department of Transportation, Washington, D.C., 2000.

TABLE 86.2 Freeway Management Assessment Summary

Technology/System	Deployment Level	Limiting Factors	Comments
Transportation management centers (may incorporate multiple technologies)[a]	Widespread deployment[b]	Implementation cost, staffing	Successful
Portable transportation management centers (may incorporate multiple technologies)	Limited deployment[b]	Implementation cost, staffing	Successful
Road closure and restriction systems (may incorporate multiple technologies)	Limited deployment[b]	Institutional issues	Successful
Vehicle detection systems (may incorporate multiple technologies)	Widespread deployment	Cost, maintenance	Mixed — depends on technology
Vehicles as probes (may incorporate multiple technologies)	Limited deployment	Cost, integration	Jury is still out
Ramp metering (includes multiple technologies)	Medium deployment	Politics, user appearance	Successful
Dynamic message signs (includes multiple technologies)	Widespread deployment	Cost, changing technology	Mixed — due to operations quality
Highway advisory radio (includes multiple technologies)	Medium deployment	Staffing	Mixed — due to operations quality
Dynamic lane control	Medium deployment	Not in MUTCD for main lanes[c]	Successful — especially on bridges and in tunnels
Dynamic speed control/variable speed limit	Technical testing[b]	Not in MUTCD; may require local legislation to be enforceable	Holds promise
Downhill speed warning and rollover warning systems	Limited deployment[b]	Cost	Successful

[a] A transportation management center may control several of the systems listed in the table and will possibly utilize additional technologies, such as video display systems, local area networks, flow monitoring algorithms, geographic information systems, graphic user interfaces, and database management systems.

[b] Quantitative deployment tracking data not available. Deployment level determined by expert judgment.

[c] Main lanes are freeway lanes that are not tunnels or bridges.

Source: Federal Highway Administration, What Have We Learned about Intelligent Transportation Systems? U.S. Department of Transportation, Washington, D.C., 2000.

TABLE 86.3 Emergency Management Assessment Summary

Technology	Deployment Level	Limiting Factors	Comments
GPS/differential GPS on emergency management fleets	Widespread deployment	Cost	Successful
Mayday systems	Widespread deployment[a]	Cost, vehicle choice	Successful
Mayday processing centers/customer service centers	Widespread deployment[a]	Cost	Successful
Public safety answering points	Widespread deployment[a]	Cost, staffing	Successful
CDPD communication	Limited deployment[a]	Availability	Jury is still out
Onboard display	Widespread deployment	Cost, user acceptance	Successful
Preemption infrared signal system	Widespread deployment	Institutional issues, lack of standards	Successful
Computer-aided dispatch	Widespread deployment	Cost, support staffing	Successful
Automatic vehicle location	Widespread deployment	Cost	Successful
Networked systems among agencies	Limited deployment[a]	Institutional issues, integration cost	Holds promise

[a] Quantitative deployment tracking data not available. Deployment level determined by expert judgment.

Source: Federal Highway Administration, What Have We Learned about Intelligent Transportation Systems? U.S. Department of Transportation, Washington, D.C., 2000.

TABLE 86.4 Electronic Toll Collection Assessment Summary

Technology	Deployment Level	Limiting Factors	Comments
Dedicated short-range communication	Widespread deployment	Need for standard	Successful
Smart cards	Limited deployment	Commercial and user acceptance; need for standard	Successful
Transponders	Widespread deployment	Privacy	Successful
Antennas	Widespread deployment	Technical performance	Successful
License plate recognition	Limited deployment[a]	Technical performance	Jury is still out

[a] Quantitative deployment tracking data not available. Deployment level determined by expert judgment.

Source: Federal Highway Administration, What Have We Learned about Intelligent Transportation Systems? U.S. Department of Transportation, Washington, D.C., 2000.

TABLE 86.5 Arterial Management Asssessment Summary

Technology	Deployment Level	Limiting Factors	Comments
Adaptive control strategies	Limited deployment	Cost, technology, perceived lack of benefits	Jury is still out — has shown benefits in some cases; cost still a prohibitive factor; some doubt among practitioners on its effectiveness
Arterial information for ATIS	Moderate deployment	Limited deployment of appropriate surveillance; difficulty in accurately describing arterial congestion	Holds promise — new surveillance technology likely to increase the quality and quantity of arterial information
Automated red-light-running enforcement	Moderate deployment[a]	Controversial; some concerns about privacy, legality	Successful — but must be deployed with sensitivity and education
Automated speed enforcement on arterial streets	Limited deployment[a]	Controversial; some concerns about privacy, legality	Jury is still out — public acceptance lacking; very controversial
Integration of time-of-day and fixed-time signal control across jurisdictions	Widespread deployment	Institutional issues still exist in many areas	Successful — encouraged by spread of closed-loop signal systems and improved communications
Integration of real-time or adaptive control strategies across jurisdictions (including special events)	Limited deployment	Limited deployment of adaptive control strategies; numerous institutional barriers	Holds promise — technology is becoming more available; institutional barriers falling
Integration with freeway (integrated management)	Limited deployment	Institutional issues exist; lack of standards between systems preventing integration	Holds promise — benefits have been realized from integrated freeway arterial corridors
Integration with emergency (signal preemption)	Widespread deployment	None	Successful
Integration with transit (signal priority)	See Chapter 5, "What Have We Learned about Advanced Public Transportation Systems?"	See Chapter 5, "What Have We Learned about Advanced Public Transportation Systems?"	See Chapter 5, "What Have We Learned about Advanced Public Transportation Systems?"

[a] Quantitative deployment tracking data not available. Deployment level determined by expert judgment.

Source: Federal Highway Administration, What Have We Learned about Intelligent Transportation Systems? U.S. Department of Transportation, Washington, D.C., 2000.

TABLE 86.6 ATIS Assessment Summary

ATIS Service	Deployment Level	Limiting Factors	Comments
Real-time traffic information on the Internet	Widespread deployment	While deployment is widespread, customer satisfaction with the services seems related to local traffic conditions and website information quality	Mixed — the characteristics of the websites vary, depending on the availability and quality of the user interface and underlying traffic data
Real-time transit status information on the Internet	Limited deployment	Transit authorities have limited funds for ATIS investments and little data that establish a relationship between ridership and ATIS	Holds promise — where the service is available, reports suggest that there is high customer satisfaction with the service
Static transit system information on the Internet	Widespread deployment	N/A	Successful
Real-time traffic information on cable television	Limited deployment	Limited by information quality and production costs, although one service provider has developed a way to automate production	Successful — as evaluated in a highly congested metropolitan area where consumers value the easy, low-tech access to traffic information
Real-time transit status information at terminals and major bus stops	Limited deployment	Cost	Successful — where evaluated in greater Seattle
Dynamic message signs	Widespread deployment	Positive driver response is a function of sign placement, content, and accuracy	Successful — drivers really appreciate accurate en route information
In-vehicle navigation systems (no traffic information)	Limited deployment[a]	Purchase cost	Holds promise — as prices fall, more drivers will purchase the systems
In-vehicle dynamic route guidance (navigation with real-time traffic information)	No commercial deployment; the San Antonio MMDI installed prototype systems in public agency vehicles[a]	Irregular coverage and data quality, combined with conflicting industry geocode standards, have kept this product from the market	Holds promise — manufacturers are poised to provide this service once issues are resolved
Fee-based traffic and transit information services on palm-type computers	Unknown deployment	Service providers make this service available through their websites; actual subscription levels are unknown	Jury is still out — requires larger numbers of subscribers becoming acclimated to mobile information services

Note: N/A = not applicable.

[a] Quantitative deployment tracking data not available. Deployment level determined by expert judgment.

Source: Federal Highway Administration, What Have We Learned about Intelligent Transportation Systems? U.S. Department of Transportation, Washington, D.C., 2000.

TABLE 86.7 APTS Assessment Summary

Technology	Deployment Level	Limiting Factors	Comments
Automatic vehicle location	Moderate deployment	Cost, fleet size, service type, staff technological competence	Successful — use continues to grow; new systems principally use GPS technology, but usually augmented by dead reckoning
Operations software	Widespread deployment	N/A	Successful
Fully automated dispatching for demand response	Research and development[a]	Still in research and development stage	Jury is still out
Mobile data terminals	Moderate deployment[a]	Most frequently deployed with automatic vehicle location systems	Successful — reduces radio frequency requirements
Silent alarm/covert microphone	Moderate deployment[a]	Most frequently deployed with automatic vehicle location systems	Successful — improves security of transit operations
Surveillance cameras	Limited deployment[a]	Cost	Holds promise — enhances onboard security; deters vandalism
Automated passenger counters	Limited deployment	Cost	Holds promise — provides better data for operations, scheduling, planning, and recruiting at lower cost
Pretrip passenger information	Widespread deployment	N/A	Successful — improves customer satisfaction
En route and in-vehicle passenger information	Limited deployment	Cost, lack of evidence of ridership increases	Jury is still out
Vehicle diagnostics	Limited deployment	Cost, lack of data on benefits	Jury is still out
Traffic signal priority	Limited deployment	Institutional issues, concerns about impacts on traffic flows	Holds promise — reduces transit trip times; may reduce required fleet size
Electronic fare payment	Limited deployment	Cost	Holds promise — increases customer convenience

Note: N/A = not applicable.

[a] Quantitative deployment tracking data not available. Deployment level determined by expert judgment.

Source: Federal Highway Administration, What Have We Learned about Intelligent Transportation Systems? U.S. Department of Transportation, Washington, D.C., 2000.

TABLE 86.8 CVISN Assessment Summary

Technology	Deployment Level	Limiting Factors	Comments
		Safety Information Exchange	
Laptop computers with Aspen or equivalent	Widespread deployment	N/A	Successful
Wireless connections to SAFER at roadside	Moderate deployment	Technical challenges with communications among systems	Holds promise — for identifying frequent violators of safety laws
CVIEW or equivalent	Limited deployment	Connections to legacy state system	Jury is still out — being tested in three or four states
		Electronic Screening	
One or more sites equipped with DSRC	Widespread deployment (number of states); limited deployment (number of carriers)	Interoperability	Holds promise — deployment trend is positive
		Electronic Credentialing	
End-to-end IRP and IFTA processing	Limited deployment	Challenges and costs of connecting legacy systems	Holds promise — potential for significant cost savings to states and carriers
Connection to IRP and IFTA clearinghouses	Limited deployment	Institutional issues	Jury is still out — cost savings can be realized only with widespread deployment

TABLE 86.9 Crosscutting Technical Issues Assessment Summary

Technology	Deployment Level	Limiting Factors	Comments
	Sensor and Surveillance Technologies		
Cell phones for incident reporting	Widespread deployment[a,b]	N/A	Successful
Cell phones for emergency notification	Limited deployment[a,b]	Relatively new; mostly sold in new vehicles; takes long time to reach 30% of vehicle fleet	Successful — number of equipped vehicles growing rapidly
GPS for position, determination, automatic vehicle location	Moderate deployment in fleets (transit, trucking, emergency vehicles) [c]	N/A	Successful — use continuing to grow[c]
Video surveillance	Widespread deployment	N/A	Successful
DSRC (toll tags) for travel time data	Limited deployment	Mostly used only in areas with electronic toll collection; requires power and communications to readers	Successful — holds promise
Direct link between Mayday systems and public safety answering points	Limited deployment[b]	Still in research and test phase; significant institutional policy and technical issues	Jury is still out — no known deployments
Cellular geolocation for traffic probes	Limited deployment	New technologies just beginning field trials	Jury is still out — older technology unsuccessful
	Communications Technologies		
Loop detectors	Widespread deployment	N/A	Successful
Alternatives to loop detectors	Widespread deployment	Initial cost, familiariy	Holds promise — video widespread; others limited; many cities use only for a few locations
Real-time, in-vehicle traffic information	Limited deployment[a,b]	Cost, commercial viability	Jury is still out
LIDAR for measuring automotive emissions	Limited deployment[b]	Minnesota test was unsuccessful; technology didn't work well enough	Unsuccessful — no known deployment
Internet for traveler information	Widespread deployment	N/A	Successful — free services Jury is still out — on commerical viability
High-speed Internet	Limited deployment[b]	Slow rollout; availability limited	Holds promise
Fully automated Internet-based exchange	Limited deployment[b]	New technology	Holds promise
DSRC	Widespread deployment	N/A	Successful — current use mostly limited to electronic toll collection
DSRC at 5.9 GHz	Limited deployment[b]	Frequency just recently approved for use; standards in development	Jury is still out — no known deployment in the U.S., but used in other countries at 5.8 GHz
Fiber optics for wire line communications	Widespread deployment	N/A	Successful
Digital subscriber line	Limited deployment	New technology; first applied to ITS in 1999	Holds promise — several deployments; many more locations considering
220-MHz radio channels for ITS	Limited deployment	ITS is too small a market to support unique communications systems	Unsuccessful — only known use during Atlanta test during the 1996 Olympic Games

TABLE 86.9 Crosscutting Technical Issues Assessment Summary (*Continued*)

Technology	Deployment Level	Limiting Factors	Comments
High-speed FM subcarrier for ITS	Limited deployment[a,b]	Low demand to date for in-vehicle real-time data	Jury is still out — multiple conflicting "standards" and proprietary approaches; competition from other wireless technologies
CDPD for traveler information	Limited deployment[a,b]	Lack of real-time information to send; limited use of CDPD by consumers	Unsuccessful — CDPD will soon be overtaken by other wireless data technologies
Wireless Internet	Limited deployment[a,b]	New technology	Jury is still out — on ITS uses; general use predicted to grow rapidly
Local area wireless	Limited deployment	New technology	Jury is still out
Low-power FM	Limited deployment[b]	Just legalized by FCC; first licenses not yet granted	Jury is still out — brand new; no deployments yet
High-speed fixed wireless	Limited deployment[b]	New technology	Jury is still out
	Analysis Tools		
Models incorporating operations into transportation planning	Limited deployment[b]	Emerging technology; cost and institutional issues may become factors for some approaches	Jury is still out — IDAS available; PRUEVIIN methodology demostrated; TRANSIMS in development

Note: N/A = not applicable; FCC = Federal Communications Commission.

[a] Quantitative deployment tracking data are not available. Deployment level was determined by expert judgment.

[b] For in-vehicle consumer systems, deployment levels are based on the percent of users or vehicle fleet, not number of cities available. For example, real-time in-vehicle traffic is available in more than two dozen cities, but the percentage of drivers subscribing to it is small.

[c] For AVL using GPS in transit, the moderate-level assessment is based on the percent of transit agencies using the technology according to a 1998 survey of 525 transit agencies conducted by the John A. Volpe National Transportation Systems Center. This measure was used for consistency with the transit section of this report. If the 78 major metropolitan areas are used as a measure, then the deployment level is "widespread," as 24 of 78 cities use GPS-based AVL.

Source: Federal Highway Administration, What Have We Learned about Intelligent Transportation Systems? U.S. Department of Transportation, Washington, D.C., 2000.

TABLE 86.10 Definitions of the Metropolitan Integration Links

Definitions of the metropolitan integration links represent both inter- and intracomponent sharing of information. Each of the links has been assigned a number and an origin or destination path from one component to another. The definitions used are from the most recent version of the draft report titled "Tracking the Deployment of the Integrated Metropolitan Intelligent Transportation Systems Infrastructure in the USA: FY 1999 Results," prepared by the Oak Ridge National Laboratory and Science Applications International Corporation for the U.S. Department of Transportation's ITS Joint Program Office, dated March 2000.

Link 1: Arterial management to regional multimodal traveler information: Arterial travel time, speed, and condition information is displayed by regional multimodal traveler information media.

Link 2: Arterial management to freeway management: Freeway management center monitors arterial travel times, speeds, and conditions using data provided from arterial management to adjust ramp meter timing, lane control, or HAR in response to changes in real-time conditions on a parallel arterial.

Link 3: Arterial management to transit management: Transit management adjusts transit routes and schedules in response to arterial travel time, speed, and condition information collected as part of arterial management.

Link 4: Arterial management to incident management: Incident management monitors real-time arterial travel times, speeds, and conditions using data provided from arterial management to detect arterial incidents and manage incident response activities.

Link 5: Incident management to arterial management: Arterial management monitors incident severity, location, and type information collected by incident management to adjust traffic signal timing or provide information to travelers in response to incident management activities.

Link 6: Incident management to regional multimodal traveler information: Incident location, severity, and type information is displayed by regional multimodal traveler information media.

Link 7: Incident management to emergency management: Incident severity, location, and type data collected as part of incident management are used to notify emergency management for incident response.

Link 8: Incident management to freeway management: Incident severity, location, and type data collected by incident management are monitored by freeway management for the purpose of adjusting ramp meter timing, lane control, or HAR messages in response to freeway or arterial incidents.

Link 9: Incident management to transit management: Transit management adjusts transit routes and schedules in response to incident severity, location, and type data collected as part of incident management.

Link 10: Freeway management to regional multimodal traveler information: Freeway travel time, speed, and condition information is displayed by regional multimodal traveler information.

Link 11: Freeway management to arterial management: Freeway travel time, speed, and condition data collected by freeway management are used by arterial management to adjust arterial traffic signal timing or arterial VMS messages in response to changing freeway conditions.

Link 12: Freeway management to transit management: Transit management adjusts transit routes and schedules in response to freeway travel time, speed, and condition information collected as part of freeway management.

Link 13: Freeway management to incident management: Incident management monitors freeway travel time, speed, and condition data collected by freeway management to detect incidents or manage incident response.

Link 14a: Transit management to regional multimodal traveler information: Transit routes, schedules, and fare information is displayed on regional multimodal traveler information media.

Link 14b: Transit management to regional multimodal traveler information: Transit schedule adherence information is displayed on regional multimodal traveler information media.

Link 15a: Transit management to freeway management: Freeway ramp meters are adjusted in response to receipt of transit vehicle priority signal.

Link 15b: Transit management to freeway management: Transit vehicles equipped as probes are monitored by freeway management to determine freeway travel speeds or travel times.

Link 16a: Transit management to arterial management: Traffic signals are adjusted in response to receipt of transit vehicle priority signal.

Link 16b: Transit management to arterial management: Transit vehicles equipped as probes are monitored by arterial management to determine arterial speeds or travel times.

Link 17: Electronic toll collection to freeway management: Vehicles equipped with electronic toll collection tags are used as probes and monitored by freeway management to determine freeway travel speeds or travel times.

Link 18: Electronic toll collection to arterial management: Vehicles equipped with electronic toll collection tags are used as probes and monitored by arterial management to determine arterial travel speeds or travel times.

Link 19: Electronic toll collection to electronic fare payment: Transit operators accept electronic toll collection–issued tags to pay for transit fares.

Link 20: Electronic fare payment to transit management: Ridership details collected as part of electronic fare payment are used in transit service planning by transit management.

Link 21a: Emergency management to incident management: Incident management is notified of incident location, severity, and type by emergency management to identify incidents on freeways or arterials.

TABLE 86.10 Definitions of the Metropolitan Integration Links (*Continued*)

Link 21b: Emergency management to incident management: Incident management is notified of incident clearance activities by emergency management to manage incident response on freeways or arterials.

Link 22: Emergency management to arterial management: Emergency management vehicles are equipped with traffic signal priority capability.

Link 23: Highway–rail intersection to incident management: Incident management is notified of crossing blockages by highway–rail intersection to manage incident response.

Link 24: Highway–rail intersection to arterial management: Highway–rail intersection and arterial management are interconnected for the purpose of adjusting traffic signal timing in response to train crossing.

Link 25: Incident management intracomponent: Agencies participating in formal working agreements or incident management plans coordinate incident detection, verification, and response.

Link 26: Arterial management intracomponent: Agencies operating traffic signals along common corridors share information and possible control of traffic signals to maintain progression on arterial routes.

Link 27: Electronic fare payment intracomponent: Operators of different public transit services share common electronic fare payment media.

Link 28: Electronic toll collection intracomponent: Electronic toll collection agencies share a common toll tag for the purpose of facilitating "seamless" toll transactions.

Link 29: Transit management to incident management: Transit agencies notify incident management agencies of incident locations, severity, and type.

Link 30: Freeway management intracomponent: Agencies operating freeways within the same region share freeway travel time, speed, and condition data.

Source: Mitretek Systems, ITS Benefits: Data Needs Update 2000, prepared in connection with ITS Benefits Data Needs Workshop, August 2000.

XIII

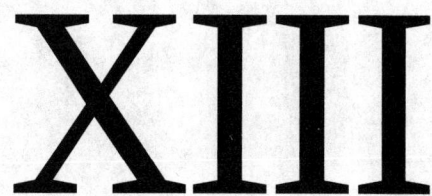

Coastal and Ocean Engineering

87

Shallow Water and Deep Water Engineering

John B. Herbich
Texas A & M University Consulting & Research Services, Inc.

Ocean engineering is a relatively new branch of engineering. The need for this new specialty was recognized in the 1960s. Several universities, including Texas A&M, MIT, Florida Atlantic, the U.S. Coast Guard Academy, and the U.S. Naval Academy, have established undergraduate degree programs in ocean engineering. Several universities have also developed programs at the graduate level specializing in ocean engineering.

Ocean and coastal engineering covers many topics, generally divided between shallow water (coastal engineering) and deep water (ocean engineering), shown in Figure 87.1 and Figure 87.2.

87.1 Wave Phenomena

Wave phenomena are of great importance in coastal and ocean engineering. Waves determine the composition and geometry of beaches. Since waves interact with human-made shore structures or offshore structures, safe design of these structures depends to a large extent on the selected wave characteristics. The structural stability criteria are often stated in terms of extreme environmental conditions (wave heights, periods, water levels, astronomical tides, storm surges, tsunamis, and winds). Waves in the ocean constantly change and are irregular in shape, particularly when under the influence of wind; such waves are called *seas*. When waves are no longer under the influence of wind and are out of the generating area, they are referred to as *swells*. Many wave theories have been developed, including the Airy, cnoidal, solitary, stream function, Stokian, and so forth. Figure 87.3 describes the regions of validity for various wave theories. Cnoidal and stream function theories apply principally to shallow

0-8493-1586-7/05/$0.00+$1.50

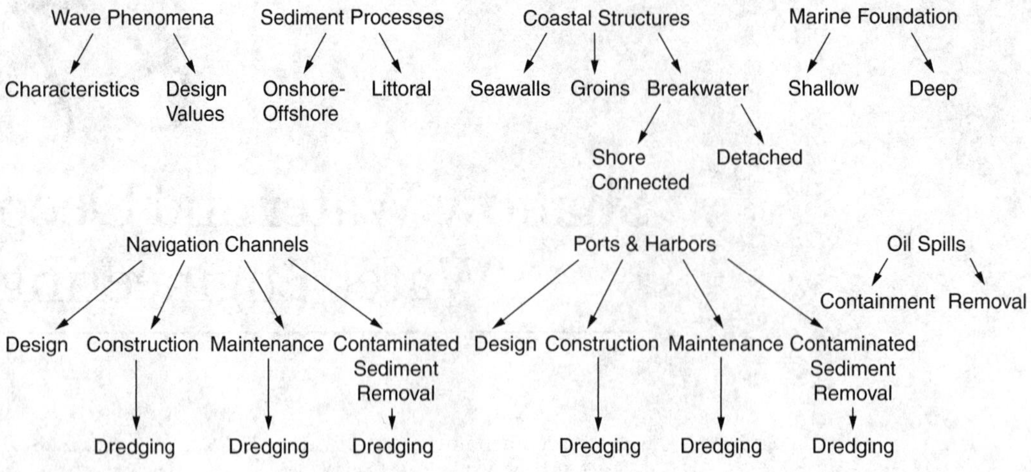

FIGURE 87.1 Coastal engineering (shallow water).

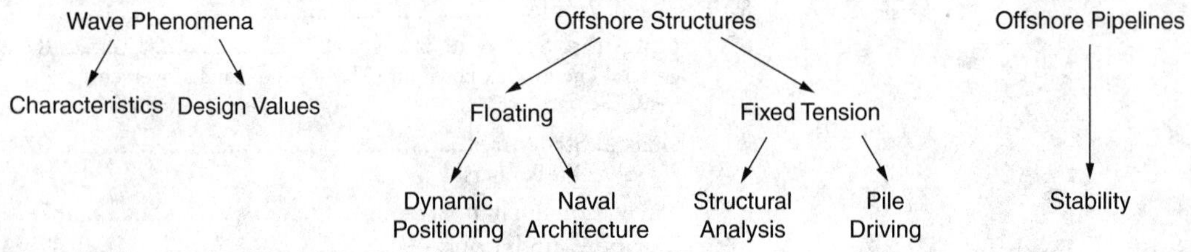

FIGURE 87.2 Ocean engineering (deep water).

and transitional water, whereas Airy and Stokian theories apply to transitional and deep water (Airy applies to low amplitude waves).

Airy (Low Amplitude)

Wavelength is given by the following equations.

Shallow water
$$L = T\sqrt{gh} = CT \tag{87.1}$$

Transitional water
$$L = \frac{gT^2}{2\pi}\tanh\left(\frac{2\pi h}{L}\right) \tag{87.2}$$

Deep water
$$L_o = \frac{gT^2}{2\pi} = C_o T \tag{87.3}$$

where T = wave period; g = acceleration due to gravity; h = water depth; and C = wave celerity. Subscript o denotes deep water conditions.

Cnoidal (Shallow Water, Long Waves)

The theory originally developed by Boussinesq [1877] has been studied and presented in more usable form by several researchers. Wavelength is given by

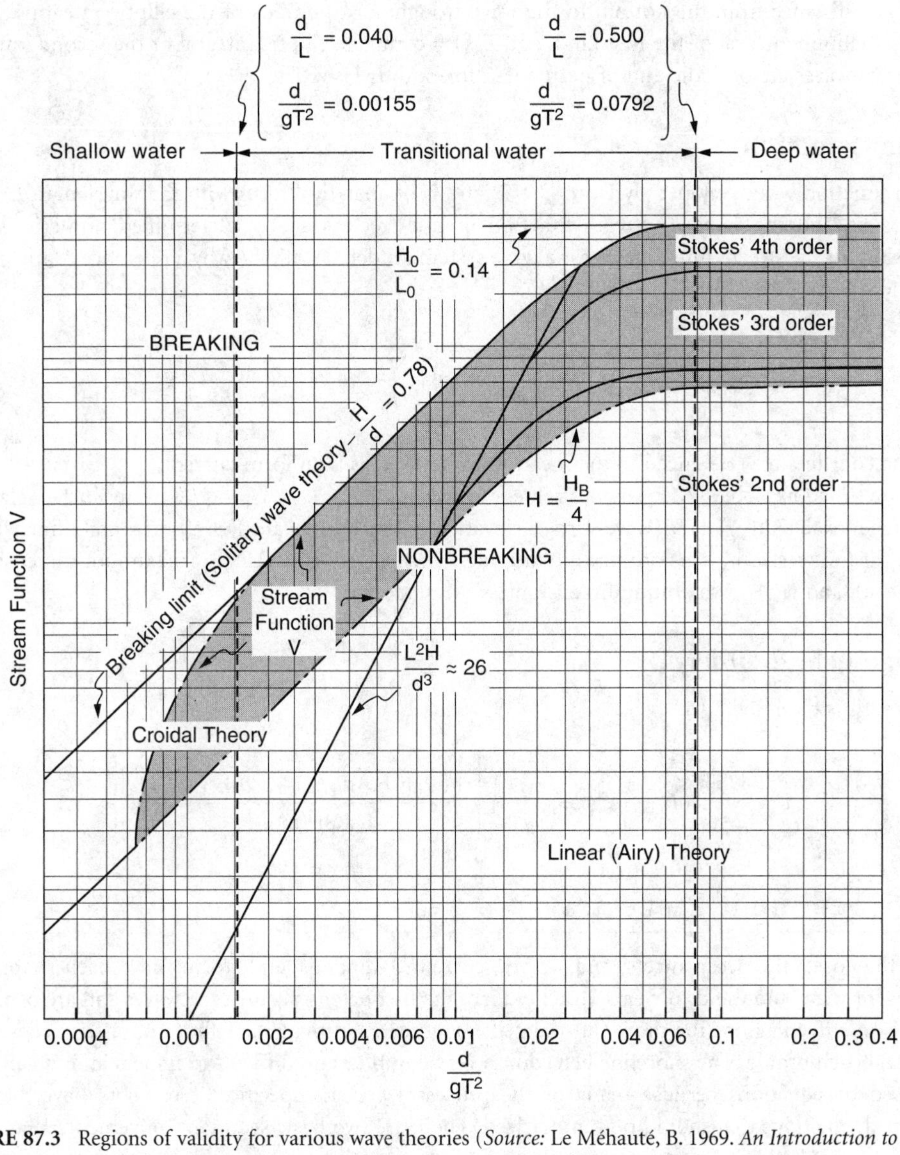

FIGURE 87.3 Regions of validity for various wave theories (*Source:* Le Méhauté, B. 1969. *An Introduction to Hydrodynamics and Water Waves,* Report No. ERL 118-POL3-1&2. U.S. Department of Commerce, Environmental Science Services Administration, Washington, DC.)

$$L = \sqrt{\frac{16d^3}{3H} kK(k)} \tag{87.4}$$

and wave period by

$$T\sqrt{\frac{g}{h}} = \sqrt{\frac{16y_t}{3H}} \frac{h}{y_t} \left[\frac{kK(k)}{1 + \frac{H}{y_t k^2}\left(\frac{1}{2} - \frac{E(k)}{K(k)}\right)} \right] \tag{87.5}$$

where y_t = distance from the bottom to the wave trough; k = modulus of the elliptic integrals; $K(k)$ = complete elliptic integral of the first kind; and $E(k)$ = complete elliptic integral of the second kind.

Cnoidal waves are periodic and of permanent form; thus $L = CT$.

Stream Function

Stream function was developed by Dean [1977] and is of analytical form with the wavelength L, coefficients $X(n)$, and the value of stream function on the free surface ψ_η determined numerically. The expression for the stream function, ψ, for a wave system rendered stationary by a reference frame moving with the speed of the wave, C, is

$$\psi = \left(\frac{L}{T} - U\right) z + \sum_{n=1}^{NN} X(n) \sinh\left[\frac{2\pi n}{L}(h+z)\right] \cos\left(\frac{2\pi n x}{L}\right) \tag{87.6}$$

with the coordinate z referenced to the mean water level; U is a uniform current.

Stream function (Table 87.1) provides values of wavelength $L' = L/L_o$, $\eta'_c = \eta_c/H$ (water surface elevation above mean water), $\eta'_t = \eta_t/H$ (wave surface elevation below mean water), u'_c (horizontal dimensionless velocity at the crest), w'_m (maximum dimensionless vertical velocity), $(F'_D)_m$ (maximum dimensionless drag force), and $(F'_I)_m$ (maximum dimensionless inertia force).

Stokian (Third Order)

Wavelength is given by

$$L = \frac{gT^2}{2\pi} \tanh\left(\frac{2\pi h}{L}\right) \left\{1 + \left(\frac{\pi H}{L}\right)^2 \left[\frac{5 + 2\cosh(4\pi h/L) + 2\cosh^2(4\pi h/L}{8\sinh^4(2\pi d/L)}\right]\right\} \tag{87.7}$$

87.2 Sediment Processes

Along the coasts the ocean meets land. Waves, currents, tsunamis, and storms have been shaping the beaches for many thousands of years. Beaches form the first defense against the waves and are constantly moving on, off, and along the shore (littoral drift). Figure 87.4 provides a definition for terms describing a typical beach profile. The shoreline behavior is very complex and difficult to understand; it cannot be expressed by equations because many of the processes are site specific. Researchers have, however, developed equations that should be summarized. There are two basic sediment movements:

1. On- and offshore
2. Parallel to the shore and at an angle to the shore.

87.3 Beach Profile

Information on beach profiles is essential in designing structural modifications (such as seawalls, revetments, and breakwaters, both connected and detached, pipeline crossings, and beach replenishment. Bruun [1954] indicated that many beach profiles (Figure 87.5) can be represented by

$$h(x) = Ax^{2/3}$$

where h is the water depth at a distance x offshore, and A is a dimensional scale parameter.

Dean [1977] showed that H_b/wT is an important parameter distinguishing **barred** profiles from nonbarred profiles (where H_b is breaking wave height, w is fall velocity of sediment in water, and T is wave period). This parameter is consistent with the following beach profiles in nature:

TABLE 87.1 Selected Summary of Tabulated Dimensionless Stream Function Quantities

Case	h/L_o	H/L_0	L'	η'_c	η'_t	u'_c	w'^*_m	$\theta(w'_m)^*$	$(F'_D)_m$	$(F'_I)^*_m$	$\theta(F'_I)^*_m$	p'_{Dc} (Bottom)
1-A	0.002	0.00039	0.120	0.910	−0.090	49.68	13.31	10°	2574.0	815.6	10°	1.57
1-B	0.002	0.00078	0.128	0.938	−0.062	47.32	15.57	10°	2774.6	1027.0	10°	1.45
1-C	0.002	0.00117	0.137	0.951	−0.049	43.64	14.98	10°	2861.0	1043.5	10°	1.35
1-D	0.002	0.00156	0.146	0.959	−0.041	40.02	13.63	10°	2985.6	1001.7	10°	1.29
2-A	0.005	0.00097	0.187	0.857	−0.143	29.82	8.70	20°	907.0	327.1	20°	1.46
2-B	0.005	0.00195	0.199	0.904	−0.096	29.08	9.29	10°	1007.9	407.1	10°	1.36
2-C	0.005	0.00293	0.211	0.927	−0.073	26.71	9.85	10°	1060.7	465.7	10°	1.23
2-D	0.005	0.00388	0.223	0.944	−0.056	23.98	9.47	10°	1128.4	465.2	10°	1.11
3-A	0.01	0.00195	0.260	0.799	−0.201	19.83	6.22	30°	390.3	162.1	30°	1.34
3-B	0.01	0.00389	0.276	0.865	−0.135	19.87	7.34	20°	457.3	209.0	20°	1.28
3-C	0.01	0.00582	0.292	0.898	−0.102	18.47	6.98	20°	494.7	225.6	10°	1.16
3-D	0.01	0.00775	0.308	0.922	−0.078	16.46	6.22	10°	535.4	242.4	10°	1.04
4-A	0.02	0.00390	0.359	0.722	−0.278	12.82	4.50	30°	156.3	82.2	30°	1.18
4-B	0.02	0.00777	0.380	0.810	−0.190	13.35	5.38	30°	197.6	103.4	20°	1.16
4-C	0.02	0.01168	0.401	0.858	−0.142	12.58	5.29	20°	222.9	116.1	20°	1.06
4-D	0.02	0.01555	0.422	0.889	−0.111	11.29	4.99	20°	242.4	113.5	20°	0.97
5-A	0.05	0.00975	0.541	0.623	−0.377	7.20	3.44	50°	44.3	37.6	50°	0.93
5-B	0.05	0.01951	0.566	0.716	−0.284	7.66	3.69	50°	59.1	38.5	50°	0.94
5-C	0.05	0.02916	0.597	0.784	−0.216	7.41	3.63	30°	72.0	47.1	30°	0.88
5-D	0.05	0.03900	0.627	0.839	−0.161	6.47	3.16	30°	85.5	45.1	20°	0.76
6-A	0.10	0.0183	0.718	0.571	−0.429	4.88	3.16	75°	17.12	22.62	75°	0.73
6-B	0.10	0.0366	0.744	0.642	−0.358	5.09	3.07	50°	22.37	23.67	50°	0.73
6-C	0.10	0.0549	0.783	0.713	−0.287	5.00	2.98	50°	28.79	23.64	30°	0.70
6-D	0.10	0.0730	0.824	0.782	−0.218	4.43	2.44	50°	36.48	22.43	30°	0.62
7-A	0.20	0.0313	0.899	0.544	−0.456	3.63	3.05	75°	6.69	13.86	75°	0.46
7-B	0.20	0.0625	0.931	0.593	−0.407	3.64	2.93	75°	8.60	13.61	75°	0.47
7-C	0.20	0.0938	0.981	0.653	−0.347	3.54	2.49	50°	11.31	13.31	50°	0.47
7-D	0.20	0.1245	1.035	0.724	−0.276	3.16	2.14	50°	15.16	11.68	50°	0.44
8-A	0.50	0.0420	1.013	0.534	−0.466	3.11	2.99	75°	2.09	6.20	75°	0.090
8-B	0.50	0.0840	1.059	0.570	−0.430	3.01	2.85	75°	2.71	6.21	75°	0.101
8-C	0.50	0.1260	1.125	0.611	−0.389	2.86	2.62	75°	3.53	5.96	75°	0.116
8-D	0.50	0.1681	1.194	0.677	−0.323	2.57	1.94	50°	4.96	5.36	50°	0.120
9-A	1.00	0.0427	1.017	0.534	−0.466	3.09	2.99	75°	1.025	3.116	75°	0.004
9-B	1.00	0.0852	1.065	0.569	−0.431	2.98	2.85	75°	1.329	3.126	75°	0.005
9-C	1.00	0.1280	1.133	0.609	−0.391	2.83	2.62	75°	1.720	3.011	75°	0.008
9-D	1.00	0.1697	1.211	0.661	−0.339	2.60	1.99	75°	2.303	2.836	50°	0.009
10-A	2.00	0.0426	1.018	0.533	−0.467	3.09	2.99	75°	0.513	1.558	75°	−0.001
10-B	2.00	0.0852	1.065	0.569	−0.431	2.98	2.85	75°	0.664	1.563	75°	0.000
10-C	2.00	0.1275	1.134	0.608	−0.392	2.83	2.63	75°	0.860	1.510	75°	−0.001
10-D	2.00	0.1704	1.222	0.657	−0.343	2.62	2.04	75°	1.137	1.479	50°	0.0000

Notes: (1) Except where obvious or noted otherwise, dimensionless quantities are presented for mean water elevation. (2) The maximum dimensionless drag and inertial forces apply for a piling extending through the entire water column. (3) Subscripts *m, c,* and *t* denote "maximum," "crest," and "trough," respectively.

Source: Dean, R. G. 1991. Beach profiles. In *Handbook of Coastal and Ocean Engineering, Volume 2,* ed. J. B. Herbich. Gulf, Houston.

$$\text{Milder slope profiles}\begin{cases}\text{High waves}\\ \text{Short periods}\\ \text{Small sediment diameter}\end{cases}$$

$$\text{Steeper profiles}\begin{cases}\text{Low waves}\\ \text{Long periods}\\ \text{Large sediment diameter}\end{cases}$$

$$\text{When } \frac{H_b}{wT} > 0.85 \text{, one can expect } \textbf{bar} \text{ formation.} \tag{87.8a}$$

$$\text{When } \frac{H_b}{wT} < 0.85 \text{, a monotonic profile can be expected.} \tag{87.8b}$$

Later, on the basis of large laboratory data, Kriebel et al. [1986] found the value of 2.3 rather than 0.85 in Equation (87.8a) and Equation (87.8b).

87.4 Longshore Sediment Transport

The longshore transport (Q) is the volumetric rate of sand movement parallel to the shoreline. Much longshore transport occurs in the surf zone and is caused by the approach of waves at an angle to the shoreline.

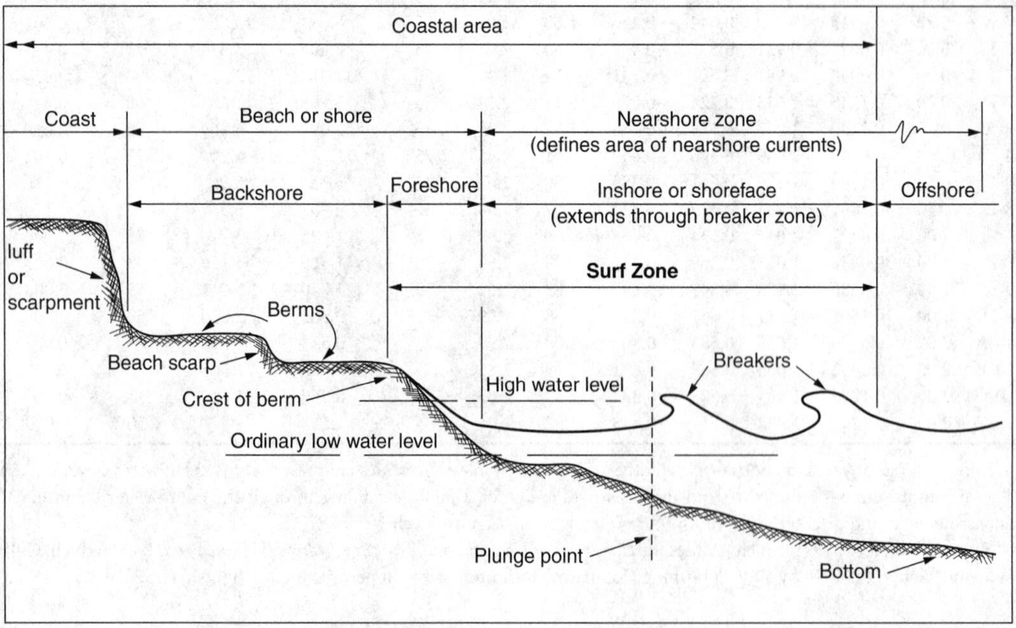

FIGURE 87.4 Visual definition of terms describing a typical beach profile. (*Source:* Department of the Army. 1987. *Shore Protection Manual,* vols. I and II. Department of the Army, Corps of Engineers, Coastal Engineering Research Center, Waterways Experiment Station, Vicksburg, MS.)

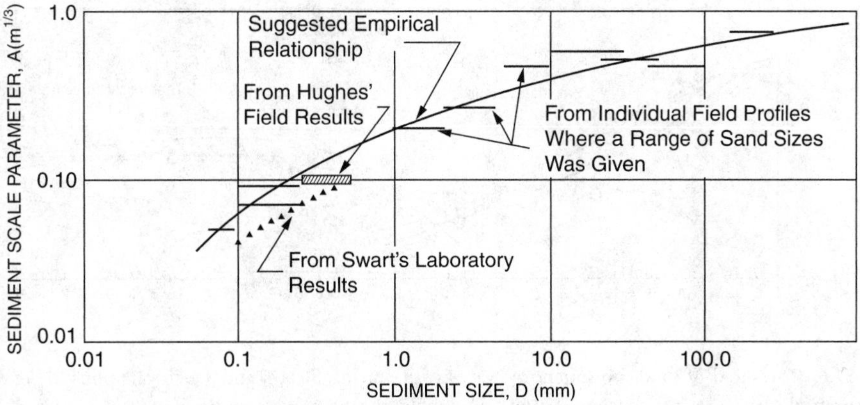

FIGURE 87.5 Beach profile scale factor, A, versus sediment diameter, D, in relationship $h = Ax^{2/3}$. (*Source:* Dean, R. G. 1991. Beach profiles. In *Handbook of Coastal and Ocean Engineering, Volume 2,* ed. J. B. Herbich. Gulf, Houston. Copyright 1990 by Gulf Publishing Company, Houston. Used with permission. All rights reserved.)

Longshore transport rate (Q, given in unit volume per second) is assumed to depend upon the longshore component of wave energy flux, P_{ls} (Department of the Army, 1984):

$$Q = \frac{K}{(\rho_s - \rho)g\alpha} P_{ls} \tag{87.9}$$

where K = dimensionless empirical coefficient (based on field measurements) = 0.39; ρ_s = density of sand; ρ = density of water; g = acceleration due to gravity; and a = ratio of the volume of solids to total volume, accounting for sand porosity = 0.6.

General Energy Flux Equation

The energy flux per unit length of wave crest or, equivalently, the rate at which wave energy is transmitted across a plane of unit width perpendicular to the direction of wave advance, is

$$P = ECg \tag{87.10}$$

where E is wave energy density and C_g is wave group speed. The wave energy density is calculated by

$$E = \frac{\rho g H^2}{8} \tag{87.11}$$

where ρ is mass density of water, g is acceleration of gravity, and H is wave height.

If the wave crests make an angle α with the shoreline, the energy flux in the direction of wave advance per unit length of beach is

$$P\cos\alpha = \frac{\rho g H^2}{8} C_g \cos\alpha \tag{87.12}$$

The longshore component of wave energy flux is

$$P_l = P\cos\alpha \sin\alpha = \frac{\rho g H^2}{8} C_g \cos\alpha \sin\alpha \tag{87.13}$$

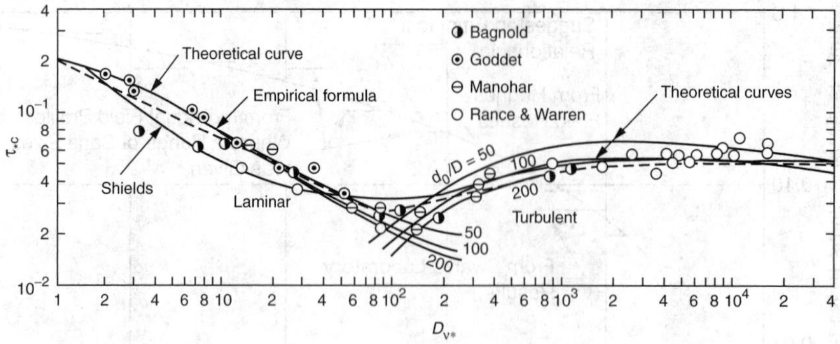

FIGURE 87.6 Threshold of sand movement by waves with Shields, Sleath, and Tsuchiya empirical curves, as well as the theoretical curve. (*Source:* Tsuchiya, Y. 1991. Threshold of sand movement. In *Handbook of Coastal and Ocean Engineering, Volume 2,* ed. J. B. Herbich. Gulf, Houston. Copyright 1990 by Gulf Publishing Company, Houston. Used with permission. All rights reserved.)

or

$$P_l = \frac{\rho g}{16} H^2 C_g \sin 2\alpha \tag{87.14}$$

Threshold of Sand Movement by Waves

The threshold of sand movement by wave action has been investigated by a number of researchers [e.g., Tsuchiya, 1991]. Figure 87.6 shows the modified Shields diagram, where $\tau_{*c} = 1/\varepsilon\psi_\iota(D_{\nu*})$, and $\psi_\iota(D_{\nu*})$ is a function of sediment-fluid number only, plotted as a function of $D_{\nu*}$.

The empirical formula shown by dashed lines is as follows:

$$\begin{aligned} \tau_{*c} &= 0.20 && \text{for } D_{\nu*} \le 1 \\ &= 0.20 D_{\nu*}^{-2/3} && \text{for } 1 \le D_{\nu*} \le 20 \\ &= 0.010 D_{\nu*}^{1/3} && \text{for } 20 \le D_{\nu*} \le 125 \\ &= 0.050 && \text{for } 125 \le D_{\nu*} \end{aligned} \tag{87.15}$$

87.5 Coastal Structures

Wave forces act on coastal and offshore structures; the forces may be classified as due to non-breaking, breaking, and broken waves. Fixed coastal structures include:

1. Wall-type structures such as **seawalls,** bulkheads, revetments, and certain types of breakwaters
2. Pile-supported structures such as piers and offshore platforms
3. Rubble structures such as breakwaters, **groins,** and revetments

Seawalls

Forces due to nonbreaking waves may be calculated using Sainflou or Miche–Rundgren formulas. Employing the Miche–Rundgren formula, the pressure distribution is

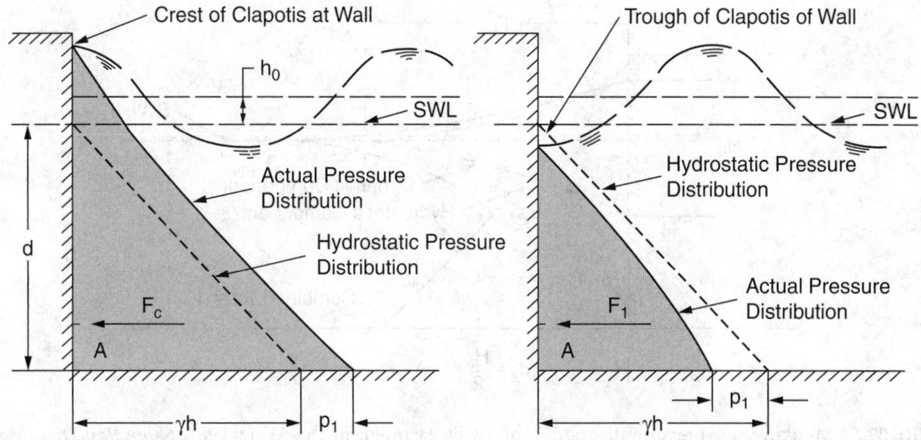

FIGURE 87.7 Pressure distributions for nonbreaking waves. (*Source:* Department of the Army. 1987. *Shore Protection Manual,* vols. I and II. Department of the Army, Corps of Engineers, Coastal Engineering Research Center, Waterways Experiment Station, Vicksburg, MS.)

$$p_1 = \left(\frac{1+\chi}{2}\right)\frac{\gamma H_i}{\cosh(2\pi h / L)} \tag{87.16}$$

where χ = wave reflection coefficient; γ = unit weight of water; H_i = incident wave height; h = water depth; and L = wavelength.

Figure 87.7 shows the pressure distribution at a vertical wall at the crest and trough of a clapotis.

Forces due to breaking waves may be estimated by Minikin and Goda methods. The Minikin method described by the Department of the Army [1984] estimates the maximum pressure (assumed to act on the SWL) to be:

$$p_m = 101\gamma \frac{H_b}{L_D}\frac{d_s}{D}(D+d_s) \tag{87.17}$$

where p_m is the maximum dynamic pressure, H_b is the breaker height, d_s is the depth at the toe of the wall, D is the depth one wavelength in front of the wall, and L_D is the wavelength in water depth D. The distribution of dynamic pressure is shown in Figure 87.8. The pressure decreases parabolically from p_m at the WL to zero at a distance of $H_b/2$ above and below the SWL. The force represented by the area under the dynamic pressure distribution is

$$R_m = \frac{p_m H_b}{3} \tag{87.18}$$

Goda's method [1985] assumes a trapezoidal pressure distribution (Figure 87.9). The pressure extends to a point measured from SWL at a distance given by η^*:

$$\eta^* = 0.75(1 + \cos\beta)H_{\max} \tag{87.19}$$

in which β denotes the angle between the direction of wave approach and a line normal to the breakwater.

The wave pressure at the wall is given by

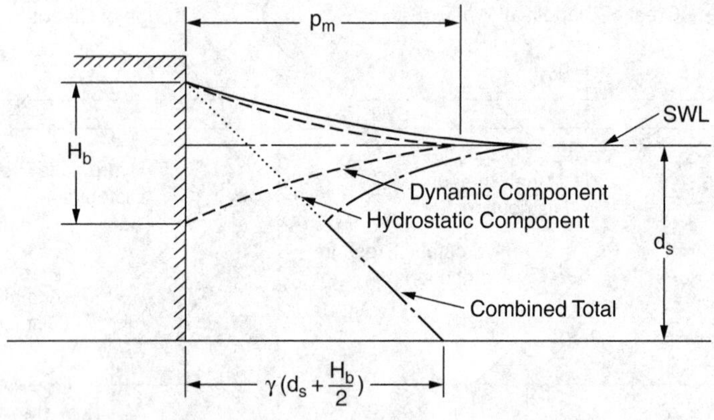

FIGURE 87.8 Minikin wave pressure diagram. (*Source:* Department of the Army. 1987. *Shore Protection Manual,* vols. I and II. Department of the Army, Corps of Engineers, Coastal Engineering Research Center, Waterways Experiment Station, Vicksburg, MS.)

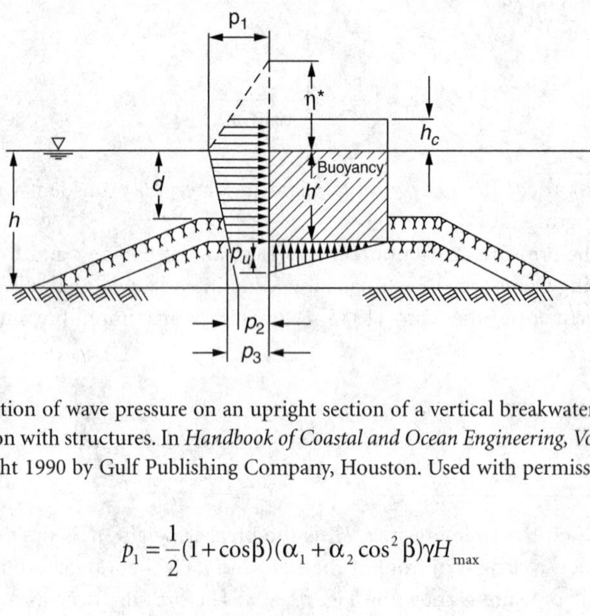

FIGURE 87.9 Distribution of wave pressure on an upright section of a vertical breakwater. (*Source:* Goda, Y. 1990. Random wave interaction with structures. In *Handbook of Coastal and Ocean Engineering, Volume 1,* ed. J. B. Herbich. Gulf, Houston. Copyright 1990 by Gulf Publishing Company, Houston. Used with permission. All rights reserved.)

$$p_1 = \frac{1}{2}(1+\cos\beta)(\alpha_1 + \alpha_2 \cos^2\beta)\gamma H_{max} \tag{87.20}$$

$$p_2 = \frac{p_1}{\cosh(2\pi h / L)} \tag{87.21}$$

$$p_3 = \alpha_3 p_1 \tag{87.22}$$

in which

$$\alpha_1 = 0.6 + 0.5\left[\frac{4\pi h / L}{\sinh(4\pi h / L)}\right]^2 \tag{87.23}$$

$$\alpha_2 = \min\left[\frac{h_b - d}{3h_b}\left(\frac{H_{max}}{d}\right)^2, \frac{2d}{H_{max}}\right] \tag{87.24}$$

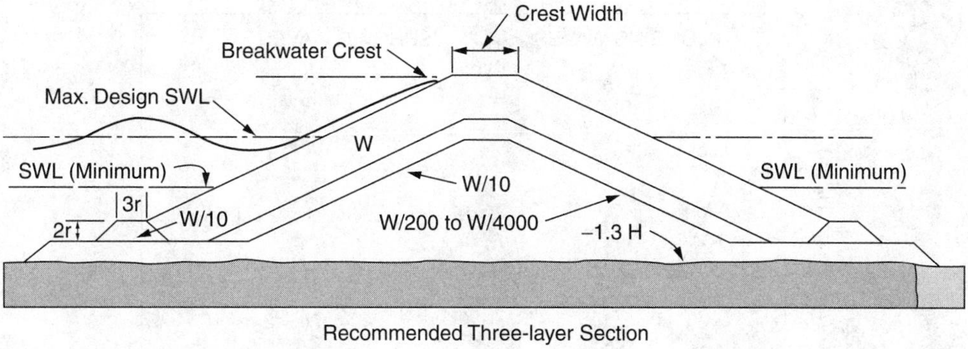

FIGURE 87.10 Rubble-mound section for wave exposure on both sides with moderate overtopping conditions. (*Source:* Department of the Army. 1987. *Shore Protection Manual,* vols. I and II. Department of the Army, Corps of Engineers, Coastal Engineering Research Center, Waterways Experiment Station, Vicksburg, MS.)

$$\alpha_3 = 1 - \frac{h'}{h}\left[1 - \frac{1}{\cosh(2\pi h / L)}\right] \tag{87.25}$$

Breakwaters

Rubble-mound breakwaters are the oldest form of breakwaters, dating back to Roman times. The rubble mound is protected by larger rocks or artificial concrete units. This protective layer is usually referred to as **armor** or cover layer.

$$W = \frac{\gamma_r H^3}{K_D (S_r - 1)^3 \cot\theta} \tag{87.26}$$

where W = weight in newtons or pounds of an individual armor unit in the primary cover layer; γ_r = unit weight (saturated surface dry) of armor unit in N/m^3 or lb/ft^3; S_r = specific gravity of armor unit, relative to the water at the structure ($S_r = w_r/w_\omega$); γ_ω = unit weight of water: freshwater = 9800 N/m^3 (62.4 lb/ft^3); seawater = 10,047 N/m^3 (64.0 lb/ft^3); θ = angle of structure slope measured from horizontal in degrees; and K_D = stability coefficient that varies primarily with the shape of the armor units, roughness of the armor unit surface, sharpness of edges, and degree of interlocking obtained in placement.

Figure 87.10 presents the recommended three-layer section of a rubble-mound breakwater. Note that underlayer units are given in terms of *W*, the weight of armor units.

Automated coastal engineering system (ACES) describes the computer programs available for the design of breakwaters using Hudson and related equations.

Van der Meer [1987] developed stability formulas for plunging (breaking) waves and for surging (nonbreaking) waves. For plunging waves,

$$H_s / \Delta D_{n50} * \sqrt{\xi_z} = 6.2 P^{0.18} (S / \sqrt{N^{0.2}}) \tag{87.27}$$

For surging waves,

$$H_s / \Delta D_{n50} = 1.0 P^{-0.13} (S / \sqrt{N^{0.2}}) \sqrt{\cot\alpha}\, \xi_z^P \tag{87.28}$$

where

H_s = **significant wave** height at the toe of the structure

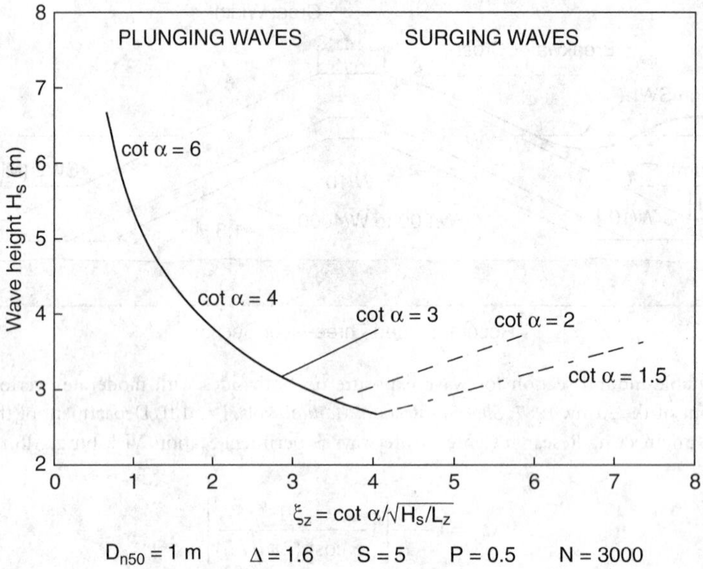

FIGURE 87.11 Influence of slope angle. (*Source:* Van der Meer, J. W. 1990. Rubble mounds — Recent modifications. In *Handbook of Coastal and Ocean Engineering, Volume 1,* ed. J. B. Herbich. Gulf, Houston, TX. Copyright 1990 by Gulf Publishing Company, Houston, TX. Used with permission. All rights reserved.)

ξ_z = surf similarity parameter, $\xi_z \dfrac{\tan\alpha}{\sqrt{2\pi H_s / gT_z^2}}$

T_z = zero up-crossing wave period

α = slope angle

Δ = relative mass density of the stone, $\Delta = \rho_a/(\rho - 1)$

ρ_a = mass density of the stone

ρ = mass density of water

D_{n50} = nominal diameter of the stone, $D_{n50} = (W_{50}/\rho_a)^{1/3}$

W_{50} = 50% value (median) of the mass distribution curve

P = permeability coefficient of the structure

S = damage level, $S = A/D^2_{n50}$

A = erosion area in a cross-section

N = number of waves (storm duration)

Influence of breakwater slope angle is depicted in Figure 87.11.

87.6 Navigational Channels

The development of very large commercial craft (VLCC) and ultralarge commercial craft (ULCC) forced many government planners and port managers to evaluate existing channels. Navigational channels allow large vessels to reach harbors. Of paramount design consideration is the safety of vessels in a channel, particularly when passing [Herbich, 1992].

Vessel behavior in channels is a function of bottom suction, bank suction, interference of passing ships, waves, winds, and currents. Most major maritime countries have criteria regarding the depth and width of channels. The international commission ICORELS (sponsored by the Permanent International Association of Navigation Congresses — PIANC) recommends that general criteria for gross underkeel clearances can be given for drawing up preliminary plans:

TABLE 87.2 General Criteria for Channel Widths

	Minimum Channel Width in Percent of Beam			
	Vessel Controllability			
Location	Very Good	Good	Poor	Channels with Yawing Forces
Maneuvering lane, straight channel	160	180	200	Judgment[a]
Bend, 26° turn	325	370	415	Judgment[a]
Bend, 40° turn	385	440	490	Judgment[a]
Ship clearance	80	80	80	100 but not less than 100 ft
Bank clearance	60	60 plus	60 plus	150

[a] Judgment will have to be based on local conditions at each project.
Source: U.S. Army Corps of Engineers. 1983. *Engineering Manual: Hydraulic Design of Deep Draft Navigation Projects,* EM 1110-2-1613. U.S. Army Corps of Engineers, Washington, DC.

- *Open sea area.* When exposed to strong and long stern or quarter swells where speed may be high, the gross underkeel clearance should be about 20% of the maximum draft of the large ships to be received.
- *Waiting area.* When exposed to strong or long swells, the gross underkeel clearance should be about 15% of the draft.
- *Channel.* For sections exposed to long swells, the gross underkeel clearance should be about 15% of the draft.

The *Engineering Manual* [U.S. Army Corps of Engineers, 1983] provides guidance for the layout and design of deep-draft navigation channels. Table 87.2 provides the general criteria for channel widths.

87.7 Marine Foundations

Design of marine foundations is an integral part of any design of marine structures. The design criteria require a thorough understanding of marine geology; geotechnical properties of sediments at a given location; and wind, wave, currents, tides, and surges during maximum storm conditions. In the arctic areas information on fast ice and pack ice is required for the design of offshore structures (on artificial islands) and offshore pipelines.

A number of soil engineering parameters are required, as shown in Table 87.3. Many of the properties may be obtained employing standard geotechnical methods. Geotechnical surveys and mapping of seabed characteristics have reached a high degree of sophistication. High-resolution geophysical surveys determine water depth, seafloor imagery, and vertical profiles. Bottom-mapping systems include multibeam bathymetry, sea beam, side-scan sonars, and subbottom profilers (including shallow, medium, and deep penetration types).

The geotechnical investigation is designed to include sediment stratigraphy; sediment types; and sediment properties, including density, strength, and deformational characteristics. Deployment systems employed for sampling *in situ* include self-contained units, drilling rigs, and submersibles. (Figure 87.12 shows the deployment systems.)

There are many *in situ* testing devices; these include the vane shear test, cone penetrometer test, pressure meter, shear vane velocity tools, temperature probes, natural gamma logger, and so forth [Young, 1991].

87.8 Oil Spills

The best method of controlling oil pollution is to prevent oil spills in the first place. This may include such techniques as rapid removal of oil from stricken tankers, continuous monitoring of oil wells, killing wild wells at sea, and containing oil spills under the water surface. Spilled oil, being lighter than water, floats on the water surface and spreads laterally. As oil is spilled, several regimes are generally assumed:

TABLE 87.3 Soil Engineering Parameters Normally Required for Categories of Geotechnical Engineering Applications

				Strength Properties				Common Properties			
				Clay		Sand		Clay		Sand	
Application	Soil Classification	Grain Size	Atterberg Limits	S_u, S_t	$\bar{c}$, ϕ'	ϕ'	ϕ or S_u	C_v, k	C_c	C_c	Subbottom Depth of Survey
Shallow foundation	Yes	Yes	Yes	Yes	Yes	Yes	Yes	Yes	Yes	Yes	1.5 to 2 × foundation width
Deadweight anchors	Yes	No	No	Yes	Yes	Yes	No	No	No	No	1.5 to 2 × anchor width
Deep pile foundations	Yes	Yes	Yes	Yes	Yes	Yes	No	Yes	Yes	No	1 to 1.5 × pile group width, below individual pile tips
Pile anchors	Yes	Yes	Yes	Yes	Yes	Yes	No	No	No	No	To depth of pile anchor
Direct-embedment anchors	Yes	Yes	No	Yes	Yes	Yes	Yes	Yes	No	No	To expected penetration of anchor, maximum 33 to 50 ft clay; 13 to 33 ft sand
Drag anchors	Yes	Yes	No	Yes	No	No	No	No	No	No	33 to 50 ft clay; 10 to $16\frac{1}{2}$ ft sand for large anchors
Penetration	Yes	Yes	No	Yes	No	Yes	Yes	No	No	No	33 to 50 ft clay; 13 to 33 ft sand
Breakout	Yes	Yes	Yes	Yes	Yes	Yes	Yes	No	No	No	1 × object width plus embedment depth
Scour	Yes	Yes	No	Yes	No	No	No	No	No	No	3.3 to $16\frac{1}{2}$ ft; related to object size and water motion
Slope stability	Yes	Yes	Yes	Yes	Yes	Yes	No	No	No	No	33 to 100 ft; more on rare occasions

Note: Su = udrained shear strength; S_t = sensitivity; $\bar{c}$ = drained cohesion intercept; ϕ' = drained friction angle; ϕ = undrained friction angle for sands rapidly sheared; C_v = coefficient of consolidation; k = permeability; C_c = compression index.

Source: Marine Board, National Research Council. 1989. *Our Seabed Frontier — Challenges and Choices,* National Academy Press, Washington, DC.

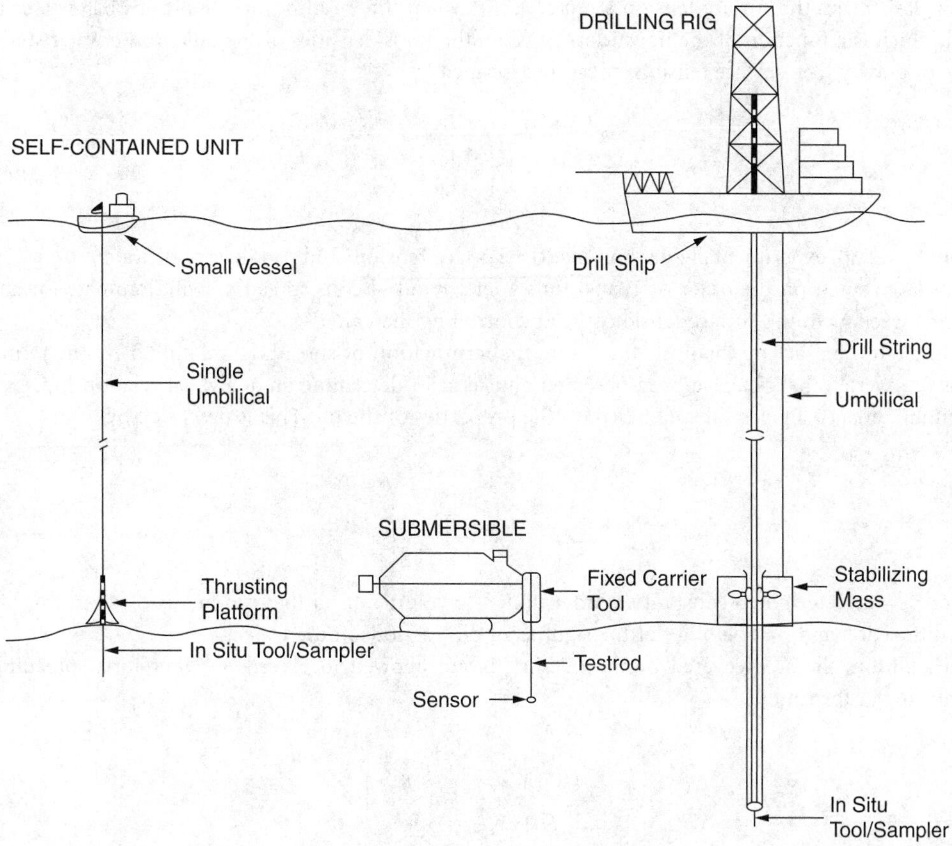

FIGURE 87.12 Deployment systems used for sampling, in situ, and experimental testings. (*Source:* Marine Board, National Research Council. 1989. *Our Seabed Frontier — Challenges and Choices,* National Academy Press, Washington, DC.)

gravity-inertial, gravity-viscous, and surface tension. In the early stage, generally less than 1 h, the gravity-inertial regime, or inertial spread, dominates and is described by

$$R = k_4(\Delta g L t^2)^{1/4} \tag{87.29}$$

where R = radius of the oil slick; k_4 = nondimensional coefficient experimentally determined to be 1.14; Δ = the ratio of the absolute difference between the densities of sea water and the oil to that of seawater; g = force of gravity; L = original volume of oil spilled; and t = time.

When the oil film thickness becomes equal to the viscous layer in the water, a transition occurs from the gravity-inertial regime to the gravity-viscous regime. This viscous spreading is described by

$$\text{Radius of oil slock} = R = k_5\left(\frac{\Delta g L^2 t^{3/2}}{\nu^{1/2}}\right)^{1/6} \tag{87.30}$$

where k_5 is the nondimensional coefficient determined to be about 1.45, ν is the kinematic viscosity of water, Δ is the ratio of the difference between density of seawater and oil, L is the original volume of spilled oil, and t is the time.

The last phase, the surface tension regime, occurs when the oil film thickness drops below a critical level, which is a function of the net surface tension, the mass densities of the oil and the water, and the force of gravity. The surface tension spread is described by

$$R = k_6 \left(\frac{\sigma^2 t^3}{\rho^2 \nu} \right)^{1/4} \tag{87.31}$$

where $k_6 = 2.30$, experimentally determined; σ = surface tension; and ρ = density of water.

For large spills, on the order of 10,000 tons, inertial and viscous spreading will dominate for about the first week, with the surface tension spread controlling thereafter.

Although the exact mechanisms that cause the termination of spreading are unknown, the terminal areas of several oil slicks have been observed and used to determine an analytical relationship for the maximum area of a given oil spill based on the properties of the oil. This is described by

$$A_T = K_a \left(\frac{\sigma^2 V^6}{\rho^2 \nu D^3 s^6} \right)^{1/8} \tag{87.32}$$

where K_a = undetermined constant or order unit; V = volume of oil that can be dissolved in this layer; D = diffusivity; and s = solubility of the significant oil fractions in the water.

In addition, the area covered by the oil slick is not allowed to exceed A_T; therefore, spreading is terminated at the time

$$t = \left(\frac{V\rho}{s\sigma} \right)^{1/2} \left(\frac{\nu}{D} \right)^{1/4} \left(\frac{K_a}{\pi k_6^2} \right)^{2/3} \tag{87.33}$$

Oil may be set up by wind and current against a barrier; any containment device must take the setup estimates into account. There are a number of containment devices (barriers) that prevent oil from spreading. Most mechanical-type oil containment barriers fail in wave heights greater than 2 ft, when the wave steepness ratio is greater than 0.08, and in currents normal to the barrier greater than about 0.7 knots.

Oil may also be removed from the water surface by skimming devices. Most mechanical skimming devices have only been able to work in waves less than 2 to 3 ft in height, in moderate currents.

87.9 Offshore Structures

Many types of offshore structures have been developed since 1947, when the first steel structure was installed in 18 feet of water. Since that time over 4100 template-platforms have been constructed on the U.S. continental shelf in water depths less than 600 feet (Figure 87.13).

Deep-water marine structures include gravity platforms, fixed platforms, guyed tower, tension-leg platform, and a buoyant compliant tower (Figure 87.14).

Wave forces on certain types of offshore platforms are computed by the Morrison equation, which is written as the sum of two individual forces, inertia and drag. The equation may be written as

$$f(t) = C_M \rho \frac{\pi}{4} D^2 \dot{u}(t) + \frac{1}{2} C_D \rho D |u(t)| u(t) \tag{87.34}$$

The force, f, as a function of time, t, is written as a function of the horizontal water particle velocity, $u(t)$, and the horizontal water particle acceleration, $\dot{u}(t)$, at the axis of the cylinder, and is dependent on

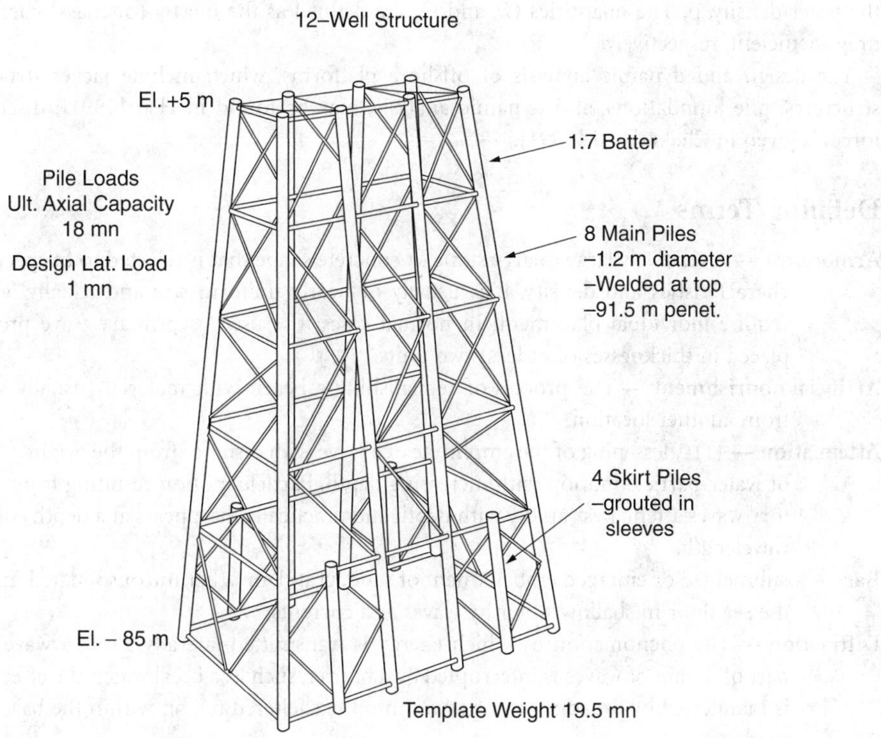

FIGURE 87.13 Template-type pile foundation structure. (*Source:* Young, A. G. 1991. Marine foundation studies. In *Handbook of Coastal and Ocean Engineering, Volume 2,* ed. J. B. Herbich. Gulf, Houston. Copyright 1990 by Gulf Publishing Company, Houston. Used with permission. All rights reserved.)

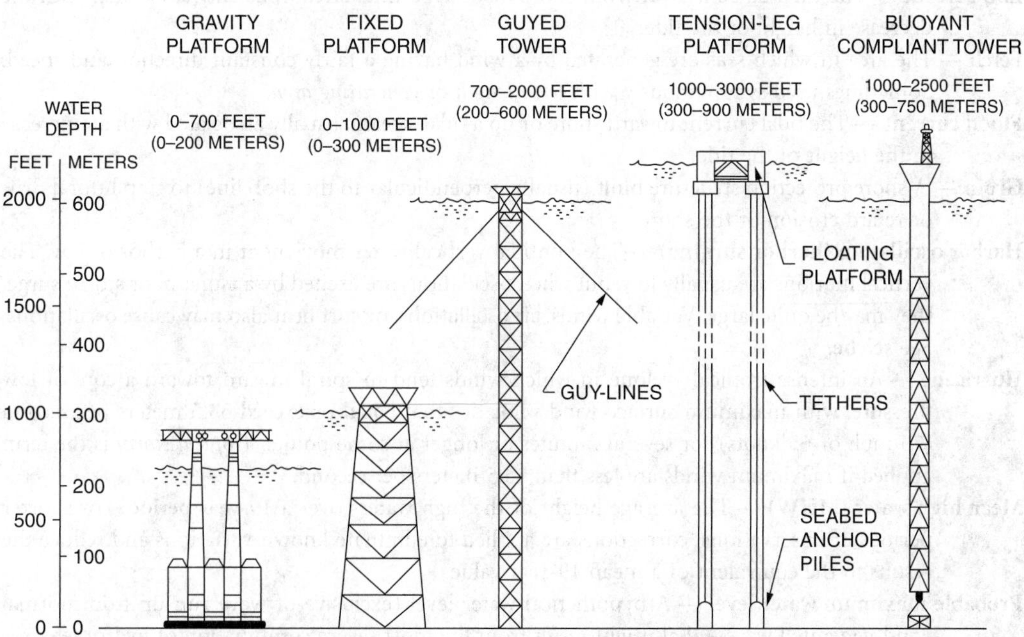

FIGURE 87.14 Range of water depths for various types of deep-water marine structures. (*Source:* Marine Board, National Research Council. 1989. *Our Seabed Frontier — Challenges and Choices,* National Academy Press, Washington, DC.)

the water density, ρ. The quantities C_M and C_D are defined as the inertia (or mass) coefficient and the drag coefficient, respectively.

The design and dynamic analysis of offshore platforms, which include jacket structures, topside structures, pile foundations, and dynamic analysis, may be found in Hsu [1991]; discussion of wave forces is given in Chakrabarti [1991].

Defining Terms

Armor unit — A relatively large quarry stone or concrete shape that is selected to fit specified geometric characteristics and density. It is usually of nearly uniform size and usually large enough to require individual placement. In normal cases it is used as primary wave protection and is placed in thicknesses of at least two units.

Artificial nourishment — The process of replenishing a beach with material (usually sand) obtained from another location.

Attenuation — (1) A lessening of the amplitude of a wave with distance from the origin. (2) The decrease of water-particle motion with increasing depth. Particle motion resulting from surface oscillatory waves attenuates rapidly with depth and practically disappears at a depth equal to a surface wavelength.

Bar — A submerged or emerged embankment of sand, gravel, or other unconsolidated material built on the sea floor in shallow water by waves and currents.

Diffraction — The phenomenon by which energy is transmitted laterally along a wave crest. When a part of a train of waves is interrupted by a barrier, such as a breakwater, the effect of diffraction is manifested by propagation of waves into the sheltered region within the barrier's geometric shadow.

Dunes — (1) Ridges or mounds of loose, wind-blown material, usually sand. (2) Bed forms smaller than bars but larger than ripples that are out of phase with any water-surface gravity waves associated with them.

Ebb current — The tidal current away from shore or down a tidal stream, usually associated with the decrease in height of the tide.

Fetch — The area in which seas are generated by a wind having a fairly constant direction and speed. Sometimes used synonymously with *fetch length* or *generating area.*

Flood current — The tidal current toward shore or up a tidal stream, usually associated with an increase in the height of the tide.

Groin — A shore protection structure built (usually perpendicular to the shoreline) to trap littoral drift or retard erosion of the shore.

Harbor oscillation (harbor surging) — The nontidal vertical water movement in a harbor or bay. The vertical motions are usually low, but when oscillations are excited by a tsunami or storm surge, they may be quite large. Variable winds, air oscillations, or surf beat also may cause oscillations. See **seiche.**

Hurricane — An intense tropical cyclone in which winds tend to spiral inward toward a core of low pressure, with maximum surface wind velocities that equal or exceed 33.5 meters per second (75 mph or 65 knots) for several minutes or longer at some points. *Tropical storm* is the term applied if maximum winds are less than 33.5 meters per second.

Mean high water (MHW) — The average height of the high waters over a 19-year period. For shorter periods of observations, corrections are applied to eliminate known variations and reduce the results to the equivalent of a mean 19-year value.

Probable maximum water level — A hypothetical water level (exclusive of wave run-up from normal wind-generated waves) that might result from the most severe combination of hydrometeorological, geoseismic, and other geophysical factors and that is considered reasonably possible in the region involved, with each of these factors considered as affecting the locality in a maximum manner. This level represents the physical response of a body of water to maximum applied phenomena such as hurricanes, moving squall lines, other cyclonic meteorological events, tsunamis, and astronomical tide, combined with maximum probable ambient hydrological conditions such as wave setup, rainfall, runoff, and river flow. It is a water level with virtually no risk of being exceeded.

Refraction — (1) The process by which the direction of a wave moving in shallow water at an angle to the contours is changed. The part of the wave advancing in shallower water moves more slowly than that part still advancing in deeper water, causing the wave crest to bend toward alignment with the underwater contours. (2) The bending of wave crests by currents.

Scour — Removal of underwater material by waves and currents, especially at the base or toe of a shore structure.

Seawall — A structure separating land and water areas, primarily designed to prevent erosion and other damage due to wave action.

Seiche — (1) A standing wave oscillation of an enclosed water body that continues, pendulum fashion, after the cessation of the originating force, which may have been either seismic or atmospheric. (2) An oscillation of a fluid body in response to a disturbing force having the same frequency as the natural frequency of the fluid system. Tides are now considered to be seiches induced primarily by the periodic forces caused by the sun and moon.

Significant wave — A statistical term relating to the one-third highest waves of a given wave group and defined by the average of their heights and periods. The composition of the higher waves depends upon the extent to which the lower wave are considered.

Wave spectrum — In ocean wave studies, a graph, table, or mathematical equation showing the distribution of wave energy as a function of wave frequency. The spectrum may be based on observations or theoretical considerations. Several forms of graphical display are widely used.

References

Boussinesq, J. 1877. Essai sur la theorie des eaux courantes, *Mem. divers Savants a L'Academie des Science,* No. 32:56.

Bruun, P. 1954. *Coast Erosion and the Development of Beach Profiles,* Tech. Memo. No. 44, 1954. Beach Erosion Board, U.S. Army Corps of Engineers.

Chakrabarti, S. K. 1991. Wave forces on offshore structures. In *Handbook of Coastal and Ocean Engineering, Volume 2,* ed. J. B. Herbich. Gulf Publishing Co., Houston.

Dean, R. G. 1977. *Equilibrium Beach Profiles: U.S. Atlantic and Gulf Coasts,* Ocean Engineering T.R. No. 12. Department of Civil Engineering, University of Delaware, Newark, DE.

Dean, R. G. 1990. Stream function wave theory and applications. In *Handbook of Coastal and Ocean Engineering, Volume 1,* ed. J. B. Herbich. Gulf Publishing Co., Houston.

Dean, R. G. 1991. Beach profiles. In *Handbook of Coastal and Ocean Engineering, Volume 2,* ed. J. B. Herbich. Gulf Publishing Co., Houston.

Department of the Army. 1987. *Shore Protection Manual,* vols. I and II. Department of the Army, Corps of Engineers, Coastal Engineering Research Center, Waterways Experiment Station, Vicksburg, MS.

Department of the Army. 1992. *Automated Coastal Engineering System,* Department of the Army, Corps of Engineers, Coastal Engineering Research Center, Waterways Experiment Station, Vicksburg, MS.

Goda, Y. 1985. *Random Seas and Design of Maritime Structures,* Tokyo University Press, Tokyo,

Goda, Y. 1990. Random wave interaction with structures. In *Handbook of Coastal and Ocean Engineering, Volume 1,* ed. J. B. Herbich. Gulf Publishing Co., Houston.

Herbich, J. B. (Ed.) 1990 (vol. 1), 1991 (vol. 2), 1992 (vol. 3). *Handbook of Coastal and Ocean Engineering,* Gulf Publishing Co., Houston.

Hsu, T. H. 1991. Design and dynamic analysis of offshore platforms. In *Handbook of Coastal and Ocean Engineering, Volume 2,* ed. J. B. Herbich. Gulf Publishing Co., Houston.

Kriebel, D. L., Dally, W. R., and Dean, R. G. 1986. *Undistorted Froude Number for Surf Zone Sediment Transport,* Proc. 20th Coastal Engineering Conference, ASCE. pp. 1296–1310.

Le Méhauté, B. 1969. *An Introduction to Hydrodynamics and Water Waves,* Report No. ERL 118-POL3-1&2. U.S. Department of Commerce, Environmental Science Services Administration, Washington, DC.

Tsuchiya, Y. 1991. Threshold of sand movement. In *Handbook of Coastal and Ocean Engineering, Volume 2,* ed. J. B. Herbich. Gulf Publishing Co., Houston.

U.S. Army Corps of Engineers. 1983. *Engineering Manual: Hydraulic Design of Deep Draft Navigation Projects,* EM 1110-2-1613. U.S. Army Corps of Engineers, Washington, DC.

Van der Meer, J. W. 1987. Stability of breakwater armor layers — Design formula. *J. Coastal Engin.* 11(3):219–239.

Van der Meer, J. W. 1990. Rubble mounds — Recent modifications. In *Handbook of Coastal and Ocean Engineering, Volume 1,* ed. J. B. Herbich. Gulf Publishing Co., Houston.

Young, A. G. 1991. Marine foundation studies. In *Handbook of Coastal and Ocean Engineering, Volume 2,* ed. J. B. Herbich. Gulf Publishing Co., Houston, TX.

Further Information

ASCE Journal of Waterway, Port, Coastal and Ocean Engineering: Published bimonthly by the American Society of Civil Engineers. Reports advances in coastal and ocean engineering.

ASCE specialty conference proceedings: Published by the American Society of Civil Engineers. Report advances in coastal and ocean engineering.

PIANC Bulletin: Published quarterly by the Permanent International Association of Navigation Congresses, Brussels, Belgium. Reports case studies.

Coastal Engineering Research Center (Technical reports, contract reports, miscellaneous papers): Published by the Army Corps of Engineers, Waterways Experiment Station, Vicksburg, MS.

Sea Technology: Published monthly by Compass Publications, Inc., Arlington, VA.

IEEE proceedings of ocean conferences: Published by the Institute of Electrical and Electronics Engineers. Report advances in ocean engineering.

Offshore Technology Conference Preprints: Published by the Offshore Technology Conference, Dallas, TX. Report annually on topics in ocean engineering.

Marine Board, National Research Council reports: Published by the National Academy Press, Washington, DC.

American Gas Association project reports: Published by the American Gas Association, Arlington, VA.

American Petroleum Institute standards: Published by the American Petroleum Institute, Dallas.

Marine Technology Society conference proceedings: Published by the Marine Technology Society, Houston.

World Dredging, Mining & Construction: Published monthly by Wodcon Association, Irvine, CA.

Terra et Aqua: Published by the International Association of Dredging Companies, The Hague, the Netherlands.

Center for Dredging Studies abstracts: Published by the Center for Dredging Studies, Texas A&M University, College Station, TX.

Komar, P. D. 1983. *Handbook of Coastal Processes and Erosion,* CRC Press, Boca Raton, FL. A series of papers on coastal processes, beach erosion, and replenishment.

Bruun, P. 1989–90. *Port Engineering,* vols. 1 and 2, 4th ed. Gulf, Houston. A comprehensive treatment on port and harbor design.

International Dredging Review: Bimonthly, Fort Collins, CO.

Technical Standards for Port and Harbour Facilities in Japan, 1980: Published by the Overseas Coastal Area Development Institute of Japan, 3-2-4 Kasumigaseki, Chiyoda-ku, Tokyo, Japan.

Herbich, J. B., Schiller, R. E., Jr., Watanabe, R. K., and Dunlap, W. A. 1987. *Seafloor Scour.* Marcel Dekker, New York. Design guidelines for ocean-founded structures.

Grace, R. A. 1978. *Marine Outfalls Systems,* Prentice Hall, Englewood Cliffs, NJ. A comprehensive treatment of marine outfalls.

Herbich, J. B. 1981. *Offshore Pipelines Design Elements,* Marcel Dekker, New York. Information relating to design of offshore pipelines.

Herbich, J. B. 1992. *Handbook of Dredging Engineering,* McGraw-Hill, New York. A comprehensive treatise on the subject of dredging engineering.

XIV

Environmental System and Management

88

Drinking Water Treatment

Appiah Amirtharajah
Georgia Institute of Technology

S. Casey Jones
Georgia Institute of Technology

The goal of drinking water treatment is to provide a water supply that is both safe and pleasing to consume. To meet this goal, conventional water treatment plants utilize the physicochemical processes of coagulation, sedimentation, filtration, and disinfection. The following sections will review the important aspects of water quality, drinking water regulations, and the main processes in conventional water treatment.

88.1 Water Quality

Microbial Contamination

Microorganisms present in water supplies can cause immediate and serious health problems. Infections by bacteria, viruses, and protozoa usually cause gastrointestinal distress; however, some, such as the bacterium *Vibrio cholerae,* can result in death. The protozoa *Giardia lamblia* and *Cryptosporidium* form chlorine-resistant cysts, and just a few cysts can cause disease. Beyond these, a vast number of pathogenic organisms exist, and water suppliers cannot feasibly monitor for all of them. Therefore, they monitor for **indicator organisms** instead. The total coliform group of bacteria is the most common indicator. Unfortunately, some pathogens (e.g., viruses and protozoa) are more resistant to conventional water treatment processes than are total coliforms.

Chemical Contamination

Inorganic Contaminants

Toxic metals and other inorganic compounds contaminate water supplies from both human-made and natural sources. Nitrates, common in groundwaters, cause methemoglobinemia or "blue-baby syndrome" in infants. Fluoride, added by many water suppliers in small doses to prevent tooth decay, causes a

0-8493-1586-7/05/$0.00+$1.50

weakening of the bones called *skeletal fluorosis* at concentrations above 4 mg/L. Radon, a naturally occurring radionuclide, may cause lung cancer from long-term exposures in the air after being released from water.

Organic Contaminants

Most organic contaminants are either volatile organic chemicals (VOCs) or synthetic organic chemicals (SOCs). Dissolved VOCs transfer to the gas phase when exposed to the atmosphere. They are typically found in groundwaters that have been contaminated by leaks from industrial storage facilities. Examples include trichloroethylene (TCE) and tetrachloroethylene (PCE), both probable carcinogens. SOCs are more soluble in water and include pesticides and pollutants from leaking underground gasoline storage tanks, such as benzene and toluene. The health effects of SOCs range from central nervous system damage to cancer.

Aesthetic Aspects of Water Quality

Color and Turbidity

Inorganic metals such as iron and organic compounds such as natural organic matter (NOM) cause color. In addition to being aesthetically undesirable, color in the form of NOM is a precursor to the formation of **disinfection by-products (DBPs)**, which may cause cancer. Turbidity is the cloudiness of a water and is determined by measuring the amount of light scattered by suspended particles in water. The unit of turbidity is the nephelometric turbidity unit (NTU). Although not a direct threat to health, turbidity decreases the efficiency of disinfection, and particles that cause turbidity can transport harmful chemicals through a treatment plant.

Taste and Odor

Zinc, copper, iron, and manganese can be detected by taste at concentrations of 1 mg/L. Hydrogen sulfide, a common contaminant in groundwaters, is detectable at concentrations of 100 ng/L. Many tastes and odors in surface waters result from biological activity of filamentous bacteria and blue-green algae. These organisms produce geosmin and methylisoborneol (MIB), which cause an earthy or musty smell. Both are detected at concentrations of 10 ng/L [Tate and Arnold, 1990].

Alkalinity

Alkalinity is a measure of the buffering capacity of a water. Alkalinity determines the magnitude of pH changes during coagulation and affects the solubility of calcium carbonate in the distribution system. In natural waters the carbonate system dominates alkalinity. In such systems, bicarbonate (HCO_3^-), carbonate (CO_3^{2-}), and hydroxide (OH^-) ions are the major species of alkalinity.

Temperature and pH

Temperature and pH affect coagulation, disinfection, and corrosion control. Equilibrium constants and reaction rates vary with temperature. The hydrogen ion concentration, measured as pH, is an important chemical species in these processes. Furthermore, the density and viscosity of water vary with temperature; thus, it is an important variable in the design of mixing, flocculation, sedimentation, and filtration process units.

88.2 Drinking Water Regulations

In the U.S., the Environmental Protection Agency (EPA) sets standards to regulate drinking water quality. Typically, the EPA establishes a maximum contaminant level goal (MCLG) and a maximum contaminant level (MCL) for each contaminant. An MCLG is the level at which no adverse health effect occurs. An MCL is set as close to the MCLG as is economically and technically feasible. A primary MCL is a legally enforceable standard based on a potential health risk. A secondary MCL is a nonenforceable standard based on a potential adverse aesthetic effect. Table 88.1 lists some MCLs of important drinking water contaminants.

TABLE 88.1 Some Drinking Water Standards

Contaminant	MCL[a]	Sources
Arsenic	0.05	Geological, pesticide residues; industrial waste and smelter operations
Asbestos	7 mfl[b]	Natural mineral deposits; also in asbestos/cement pipe
Benzene	0.005	Fuel (leaking tanks); solvent commonly used in manufacture of industrial chemicals, pharmaceuticals, pesticides, paints, and plastics
Cadmium	0.005	Natural mineral deposits; metal finishing; corrosion product plumbing
Chromium	0.1	Natural mineral deposits; metal finishing; textile, tanning, and leather industries
Mercury	0.002	Industrial/chemical manufacturing; fungicide; natural mineral deposits
Nitrate (as N)	10	Fertilizers, feedlots, sewage; naturally in soil, mineral deposits
Polychlorinated biphenyls (PCBs)	0.0005	Electric transformers, plasticizers; banned in 1979
Radium 226/228	5 pCi/L[c]	Radioactive waste; geological/natural
Tetrachloroethylene (PCE)	0.005	Dry cleaning; industrial solvent
Toluene	1	Chemical manufacturing; gasoline additive; industrial solvent
Trichloroethylene (TCE)	0.005	Waste from disposals of dry cleaning materials and manufacturing of pesticides, paints, waxes, and varnishes; paint stripper; metal degreaser

[a] Maximum contaminant level, in milligrams per liter unless otherwise noted.
[b] Million fibers per liter, with fiber length greater than 10 μm.
[c] Picocurie (pCi) is the quantity of radioactive material producing 2.22 nuclear transformations per minute.
Source: U.S. Environmental Protection Agency, 1991. *Fact Sheet: National Primary Drinking Water Standards.*

In 1974, the U.S. Congress enacted the Safe Drinking Water Act (SDWA), which was the first law to cover all public drinking water utilities in the U.S. In 1986, Congress amended the SDWA, and, as a result, in 1989 the EPA promulgated the Surface Water Treatment Rule (SWTR). The SWTR established standards for all water treatment plants that use surface water as a source. Subsequently, the EPA also modified the standards for total coliforms, lead, and copper.

Total Coliform Rule

The total coliform rule set the MCLG for total coliforms at zero. The MCL is currently based on a presence–absence test. Utilities must monitor the distribution system for total coliforms taking a specified number of samples based on the population served by the utility. No more than 5.0% of the samples taken per month should be positive for total coliforms. Small systems taking fewer than 40 samples per month can have no more than one positive sample.

Surface Water Treatment Rule

Turbidity

Filtered water turbidity must never exceed 5 NTU and should meet the MCL in 95% of the samples taken either continuously or every 4 h. The MCL is either 0.5 or 1.0 NTU depending on the type of filter used. For conventional and direct filtration the MCL is 0.5 NTU.

Disinfectant Residual

The disinfectant concentration or residual entering the distribution system cannot be less than 0.2 mg/L for more than 4 h. A disinfectant residual must be detectable in 95% of the samples taken monthly at consumers' taps for 2 consecutive months.

Treatment Techniques

Instead of setting an MCL for the pathogens *Giardia lamblia* and viruses, the U.S. EPA specified a treatment technique, that is, disinfection and possibly filtration. To ensure the required disinfection, utilities must determine *CT* values, where *C* is the concentration of disinfectant and *T* is the contact time between disinfectant and water. The *CT* value required for *Giardia lamblia* corresponds to 99.9% (3 log) removal or inactivation, and for viruses it corresponds to 99.99% (4 log) removal or inactivation. Filtration must be practiced unless the utility demonstrates that adequate treatment occurs without it and that the water source is free from significant contamination. Conventional filtration plants receive a *CT* credit of 2.5 log removal for *Giardia lamblia* and 2.0 log removal of viruses.

Lead and Copper Rule

The EPA also specified a treatment technique for lead and copper. However, in this case, the treatment technique need not be implemented unless an action level is exceeded. Action levels of 0.015 mg/L for lead and 1.3 mg/L for copper are set at the consumer's tap. If these action levels are exceeded during an extensive sampling program, the utility must perform a corrosion control study to determine the best technique to minimize the corrosivity of the water entering the distribution system.

Future Regulations

At present, no regulation exists regarding the protozoa *Cryptosporidium*, in part because of a lack of technical information about the organism. The EPA is in the process of collecting this information. Regulations are also expected in the near future for DBPs. The EPA is expected to lower the MCLs for total trihalomethanes (THMs) from 0.10 to 0.08 mg/L and to set MCLs for total haloacetic acids at 0.04 mg/L. In addition, removals of total organic carbon, a precursor to DBP formation, are to be specified under an Enhanced Surface Water Treatment Rule.

88.3 Water Treatment Processes

Figure 88.1 shows the treatment processes in a conventional surface water treatment plant. After being withdrawn from a source (lake or river), raw water is a suspension of small, stable colloidal particles whose motions are governed by molecular diffusion. In coagulation these particles are destabilized by the addition of a coagulant during rapid mixing. Flocculation promotes the collisions of these unstable particles to produce larger particles called *flocs*. In sedimentation, these flocs settle under the force of gravity. The particles that do not settle are removed during filtration. A disinfectant such as chlorine is then added, and, after a certain amount of contact time, the treated water is distributed to consumers. Direct filtration plants omit the sedimentation and occasionally the flocculation processes. These plants are suitable for raw waters with low to moderate turbidities and low color. The following sections describe

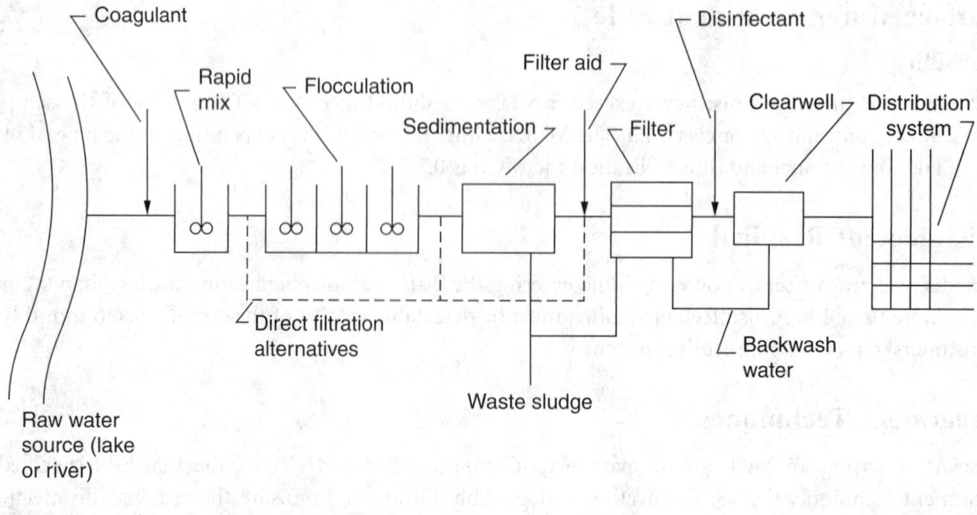

FIGURE 88.1 Schematic of a conventional water treatment plant.

the underlying theory and design of each of the major processes: coagulation, sedimentation, filtration, and disinfection.

Coagulation

In coagulation, small particles combine into larger particles. Coagulation consists of three separate and sequential processes: coagulant formation, particle destabilization, and interparticle collisions. The first two steps occur during rapid mixing, whereas the third occurs during flocculation. In natural waters, particles (from 10 nm to 100 μm in size) are stable, because they have a negative surface charge.

Mechanisms of Destabilization

The possible mechanisms of particle destabilization are double layer compression, polymer bridging, **charge neutralization**, and **sweep coagulation.** In water treatment the last two mechanisms predominate; however, when organic polymers are used as coagulants, polymer bridging can occur. In charge neutralization the positively charged coagulant, either the hydrolysis species of a metal salt (alum or ferric chloride) or polyelectrolytes, adsorbs onto the surface of the negatively charged particles. As a result, the particles have no net surface charge and are effectively destabilized. In sweep coagulation a metal salt is added in concentrations sufficiently high to cause the precipitation of a metal hydroxide (e.g., aluminum hydroxide). The particles are enmeshed in the precipitate, and it "sweeps" the particles out of the water as it forms and settles.

Solution Chemistry

With metal salt coagulants, the mechanism of coagulation is determined by the coagulant dose and the pH of the equilibrated solution. The most common coagulant is alum $[Al_2(SO_4)_3 \cdot 14.3\ H_2O]$. The alum coagulation diagram, shown in Figure 88.2, indicates the regions where each mechanism dominates. A

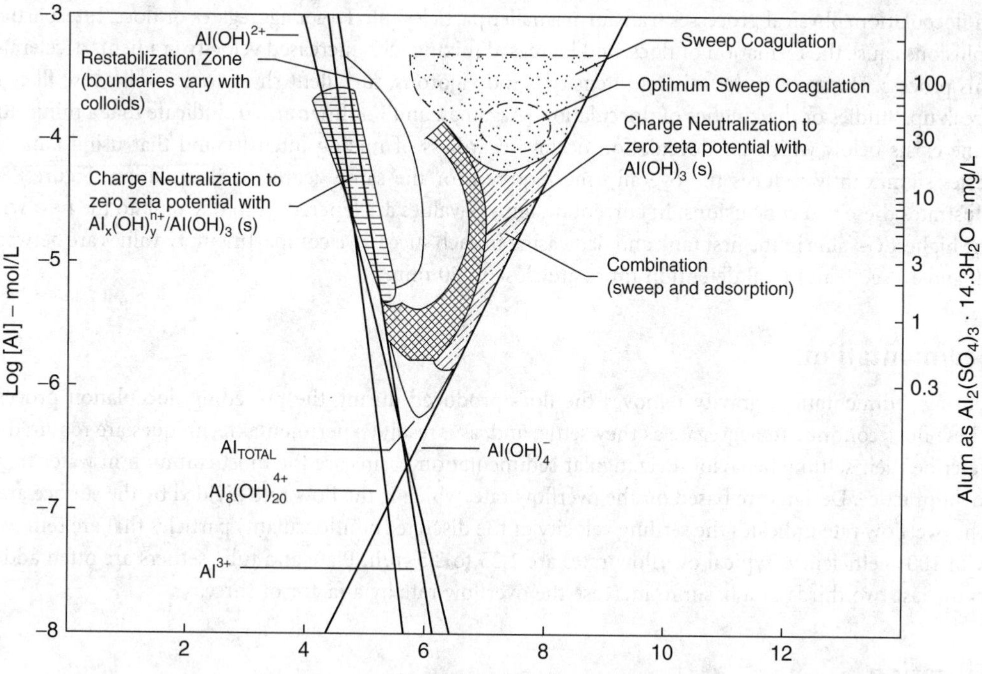

FIGURE 88.2 The alum coagulation diagram that defines the mechanism of coagulation based on pH and alum dose. (*Source:* Amirtharajah, A. and Mills, K. M. 1982. Rapid mix design for mechanisms of alum coagulation. *J. AWWA.* 74(4):210–216.)

similar diagram exists for ferric chloride [Amirtharajah and O'Melia, 1990]. Important considerations in using the alum coagulation diagram are that the boundaries of the restabilization zone vary with the surface area of the raw water particles and that significant concentrations of NOM rather than turbidity can control the alum dose required for effective treatment.

Rapid Mix Design

At a fundamental level the rapid-mixing unit provides encounters between molecules and particles in the source water and the coagulant species. These encounters are controlled by the hydrodynamic parameters and geometry of the mixer, molecular properties of the source water, and the kinetics of the coagulation reactions. Research indicates that coagulation by sweep coagulation is insensitive to mixing intensity. Although its applicability is questionable on theoretical grounds, the ***G*-value** is widely used to represent mixing intensity in both rapid mix and flocculation units. The *G*-value is computed as

$$G = \sqrt{\frac{P}{\mu V}} = \sqrt{\frac{\varepsilon}{\nu}} \tag{88.1}$$

where G is the velocity gradient (sec^{-1}), P is the *net* power input to the water (W), μ is the dynamic viscosity of water ($\text{N} \cdot \text{sec/m}^2$), V is the mixing volume (m^3), ε is the rate of energy dissipation per mass of fluid (W/kg), and ν is the kinematic viscosity (m^2/sec) Mixing time, t, is an important design parameter, and it can vary from less than 1 sec in some in-line mixers to over a minute in back-mix reactors. In general, short times (<1 sec) are desired for the charge neutralization mechanism and longer times (10 to 30 sec) for sweep coagulation.

Flocculator Design

In flocculation, physical processes transform smaller particles into larger aggregates or flocs. Interparticle collisions cause the formation of flocs, and increased mixing with increased velocity gradients accelerates this process. However, if the mixing intensity is too vigorous, turbulent shear forces will cause flocs to break up. Studies of the kinetics of flocculation [Argaman and Kaufman, 1970] indicate that a minimum time exists below which no flocculation occurs regardless of mixing intensity and that using tanks in series significantly reduces the overall time required for the same degree of flocculation. Figure 88.3 illustrates these two conclusions. In current designs, *G*-values are tapered from one tank to the next with the highest *G*-value in the first tank and decreasing in each successive compartment. *G*-values are between 60 and 10 sec^{-1}, and total detention times are close to 20 min.

Sedimentation

During sedimentation, gravity removes the flocs produced during the preceding flocculation process. These flocs continue to aggregate as they settle, and, as a result, experimental techniques are required to describe their settling behavior. Rectangular sedimentation basins are the most common in water treatment practice. Designs are based on the overflow rate, which is the flow rate divided by the surface area. The overflow rate indicates the settling velocity of the discrete (nonflocculant) particles that are removed with 100% efficiency. Typical overflow rates are 1.25 to 2.5 m/h. Plate and tube settlers are often added to the last two thirds of a basin to increase the overflow rate by a factor of three.

Filtration

In the U.S., the most common filters are dual-media filters, in which water flows by gravity through a porous bed of two layers of granular media. The top layer is anthracite coal, and the bottom layer is sand. Filters are operated until one of two criteria is exceeded — the effluent turbidity standard or the allowable

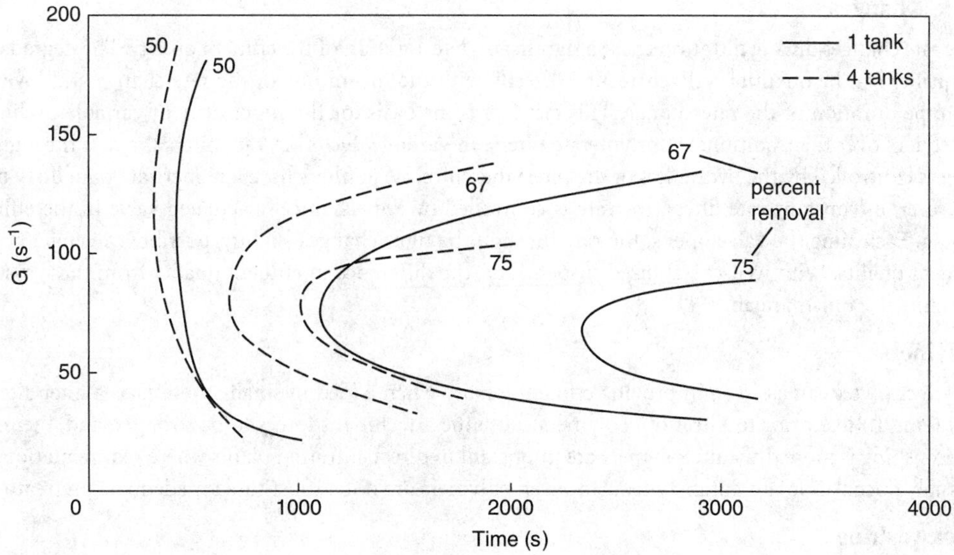

FIGURE 88.3 A graph illustrating the benefit of tanks in series for flocculation. (*Source:* Argaman, Y. and Kaufman, W. J. 1970. Turbulence and flocculation. *ASCE J. Sanitary Engineering Div.* 96(SA2):223–241.)

head loss through the filter. The filters are cleaned by backwashing to remove the particles that have been collected on the filter media.

The removal of particles in a dual-media filter occurs within the pores of the filter and is mediated by transport mechanisms that carry small particles across fluid streamlines to distances close to the filter grains (also called *collectors*). When particles are very close to the collectors, short-range surface forces cause the collector to capture the particle. The dominant transport mechanisms in water filtration are diffusion and sedimentation. Diffusion is transport resulting from random Brownian motion by bombardment of the particle by molecules of water. Diffusion is increasingly important for particles less than 1 μm in size. Sedimentation is due to the force of gravity and the associated settling velocity of the particle, which causes it to cross streamlines and reach the collector. This mechanism becomes increasingly important for particles greater than 1 μm in size (for a size range of 5 to 25 μm). The combination of these two mechanisms results in a minimum net transport efficiency for a size of approximately 1 μm. It is interesting to extrapolate this result to two important microbial contaminants. Cysts of *Giardia lamblia,* with dimensions of 10 to 15 μm, are probably removed by the sedimentation mechanism, whereas *Cryptosporidium,* with a dimension close to 3 to 5 μm, is probably close to the minimum net transport efficiency. Unfortunately, a theory of filtration that is sufficiently general and predictive does not yet exist. Therefore, designers must rely on empirical evidence from pilot-scale tests for guidance.

Chemical Pretreatment

Evidence clearly shows that chemical pretreatment for particle destabilization is the most important variable affecting filtration efficiency. Plant-scale studies have shown that filtration rates between 5 and 15 m/h can treat water equally well given adequate chemical pretreatment.

Initial Degradation and Filter Ripening

Just after backwashing, filters typically have poor effluent quality, and the quality improves during the course of the run. The improvement with time is called *filter ripening*. Studies have shown that greater than 90% of the particles passing through a well-operated filter do so during the initial stages of filtration. If the duration of filter ripening is short, then the initial filter effluent can be wasted until the effluent turbidity reaches a desired level. Particles removed during a filter run function as collectors themselves, resulting in filter ripening.

Rate Changes

Any rate changes during filtration cause a significant deterioration of the effluent quality. The degradation in quality can be quantitatively correlated directly with the magnitude of the rate change and inversely with the duration of the rate change. This fact forms the basis for the superiority of variable declining-rate filters over the traditional constant-rate filters. In variable declining-rate filters the rate through the filter is controlled by the hydraulics of the filter; therefore, clean filters have a higher rate than dirty ones. However, in constant-rate filters, the rate is controlled by a mechanically actuated valve in the effluent piping. Each time the valve opens, the rate through the filter changes slightly, possibly causing a poorer effluent quality. With well-destabilized suspensions, the difference in effluent quality from these systems is minimal [Amirtharajah, 1988].

Polymers

Polyelectrolytes can assist in improving effluent quality when added in small amounts as a filter aid (0.1 to 1.0 mg/L) just prior to filtration. Polymers cause the attachment forces to be stronger, and, therefore, backwashing is more difficult. Polymers are important in direct-filtration plants where sedimentation and possibly flocculation are not included. However, polymers are not a substitute for adequate pretreatment.

Backwashing

In the U.S., filters have traditionally been backwashed by fluidizing the filter media for a specific period of time. However, very few particle contacts occur between fluidized particles. Hence, particulate fluidization with water alone is an intrinsically weak process. Air scour with subfluidization water wash causes abrasions between particles throughout the depth of the bed. Surface wash causes collisions at the top of the bed. Both processes are effective auxiliaries for cleaning. When polymers are used, air scour or surface wash is necessary.

Disinfection

A variety of disinfectants are available in water treatment, including chlorine, chloramines, chlorine dioxide, and ozone. In the U.S., however, chlorine is the most common disinfectant. Chlorine gas is added to water to form hypochlorous acid (HOCl). At pHs between 6 and 9, HOCl dissociates to form the hypochlorite ion (OCl^-) and hydrogen ion (H^+). HOCl has the greatest disinfection power. The extent of disinfection in a water treatment plant is determined by computing *CT* values as mentioned in Section 88.2. The *CT* value required varies with chlorine concentration, pH, and temperature.

Although increasing the *CT* value may provide a large factor of safety against microbial contamination, disinfection causes the formation of disinfection byproducts (DBPs), which are suspected carcinogens. DBPs result from reactions between disinfectants and NOM, which is ubiquitous in natural waters. The most common DBPs from chlorine are the THMs: chloroform, bromodichloromethane, dibromochloromethane, and bromoform. Other technologies such as membranes and adsorption will be used in the future, together with the traditional water treatment processes, to reduce the threat from both microorganisms and DBPs.

Defining Terms

Alkalinity — The ability of a water to resist changes in pH. Includes the following species: carbonate, bicarbonate, and hydroxide.

Charge neutralization — A mechanism of coagulation in which positively charged coagulants adsorb to the surface of negatively charged colloids. The resulting colloids, with no net surface charge, are effectively destabilized.

Disinfection byproducts (DBPs) — Byproducts that result from the reactions of disinfectants such as chlorine with the natural organic matter that is present in all natural waters. DBPs such as trihalomethanes are suspected carcinogens.

***G*-value** — A measure of mixing intensity in turbulent mixing vessels. Used in the design of both rapid mix and flocculation units.

Indicator organisms — Easily detectable organisms that act as a surrogate for waterborne pathogens. Although not pathogenic themselves, indicators are present in the same environs as pathogens but in larger concentrations.

Sweep coagulation — A mechanism of coagulation in which particles are enmeshed in a precipitate of metal hydroxide, such as aluminum hydroxide.

References

Amirtharajah, A. and Mills, K. M. 1982. Rapid mix design for mechanisms of alum coagulation. *J. AWWA*. 74(4):210–216.

Amirtharajah, A. 1988. Some theoretical and conceptual views of filtration. *J. AWWA*. 80(12):36–46.

Amirtharajah, A. and O'Melia, C. R. 1990. Coagulation processes: Destabilization, mixing, and flocculation. In *Water Quality and Treatment,* 4th ed., ed. F. W. Pontius, pp. 269–365. McGraw-Hill, New York.

Argaman, Y. and Kaufman, W. J. 1970. Turbulence and flocculation. *ASCE J. Sanit. Eng. Div.* 96(SA2):223–241.

Tate, C. H. and Arnold, K. F. 1990. Health and aesthetic aspects of water quality. In *Water Quality and Treatment,* 4th ed., ed. F. W. Pontius, McGraw-Hill, New York. pp. 63–156.

U.S. Environmental Protection Agency. 1991. *Fact Sheet: National Primary Drinking Water Standards,* U.S. E.P.A., Washington, DC.

Further Information

Design Textbooks

Kawamura, S. 1991. *Integrated Design of Water Treatment Facilities,* John Wiley & Sons, New York.

James Montgomery Consulting Engineers. 1988. *Water Treatment Principles and Design,* John Wiley & Sons, New York.

Introductory Textbooks

Davis, M. L. and Cornwell, D. A. 1991. *Introduction to Environmental Engineering,* 2nd ed. McGraw-Hill, New York.

Peavy, H. S., Rowe, D. R., and Tchobanoglous, G. 1988. *Environmental Engineering,* McGraw-Hill, New York.

Specialized Textbooks

Amirtharajah, A., Clark, M. M., and Trussell, R. R. (Eds.) 1991. *Mixing in Coagulation and Flocculation,* American Water Works Association Research Foundation, Denver, CO.

Stumm, W. and Morgan, J. J. 1981. *Aquatic Chemistry,* 2nd ed. John Wiley & Sons, New York.

Snoeyink, V. L. and Jenkins, D. 1980. *Water Chemistry,* John Wiley & Sons, New York.

Journals

Journal of the American Water Works Association
Water Research
ASCE Journal of Environmental Engineering
Environmental Science and Technology

89

Air Pollution

F. Chris Alley
Clemson University (Emeritus)

C. David Cooper
University of Central Florida

Air pollution is defined as contamination of the atmosphere with solid particles, liquid mists, or gaseous compounds in concentrations that can harm people, plants, or animals, or reduce environmental quality. Particulate matter (PM) constitutes a major class of air pollution. Ambient (outdoor) air quality standards exist for particulate matter less than 10 μm (PM-10), and many industries have emission limits on total PM. Particles have various shapes, different chemical and physical properties, and a wide range of sizes — from less than 0.01 to over 100 μm in **aerodynamic diameter.** The major gaseous pollutants that are emitted into the air include sulfur oxides, nitrogen oxides, carbon monoxide, and volatile organic compounds (such as petroleum fuels and organic chemicals). Major sources of air pollution include combustion processes (especially fossil fuel power plants), motor vehicles, and the materials processing and petrochemical industries.

89.1 Control of Particulate Matter

Control of PM involves separation and removal of the PM from a flowing stream of air. A control device is chosen based on the size and properties (density, corrosivity, reactivity) of the particles, the chemical and physical characteristics of the gas (flow rate, temperature, pressure, humidity, chemical composition), and the collection efficiency, E, desired. The calculation of E is based on the fraction of mass removed:

$$E = (M_i - M_e)/M_i \tag{89.1}$$

where E is the collection efficiency (fraction), M is the mass flow rate of the pollutant (g/sec), and i and e are subscripts indicating the input or exit stream. The efficiency of most devices is a strong function of particle size, decreasing rapidly as particle size decreases.

A key operating parameter of most control devices is the pressure drop (ΔP). In general, to increase efficiency of a device, more energy must be expended, which often shows up as ΔP through the system. But moving the air through that ΔP can account for a major portion of the operating cost. The fan power required is given by

$$W = Q\Delta P/\eta \tag{89.2}$$

where W is the power (kW), Q is the volumetric flow rate (m^3/sec), ΔP is the pressure drop (kPa), and η = fan/motor efficiency.

0-8493-1586-7/05/$0.00+$1.50

TABLE 89.1 Ranges of Values for Key ESP Parameters

Parameter	Range of Values
Drift velocity w_e	1.0 to 10 m/min
Channel width D	15 to 40 cm
Specific collection area (plate area/gas flow)	0.25 to 2.1 $m^2/(m^3/min)$
Gas velocity u	1.2 to 2.5 m/sec (70 to 150 m/min)
Aspect ratio R (duct length/height)	0.5 to 1.5 (not less than 1.0 for $\eta >$ 99%)
Corona power ratio P_c/Q (corona power/gas flow)	1.75 to 17.5 $W/(m^3/min)$
Corona current ratio I_c/A (corona current/plate area)	50 to 750 $\mu A/m^2$
Power density versus resistivity	
Ash Resistivity, ohm-cm	Power Density, W/m^2
10^4 to 10^7	43
10^7 to 10^8	32
10^9 to 10^{10}	27
10^{11}	22
10^{12}	16
10^{13}	10.8
Plate area per electrical set A_s	460 to 7400 m^2
Number of electrical sections N_s	
a. In the direction of gas flow	2 to 8
b. Total	1 to 10 bus sections/(1000 m^3 min)

Source: Cooper, C. D. and Alley, F. C. 1994. *Air Pollution Control — A Design Approach,* 2nd ed. Waveland Press, Prospect Heights, IL. With permission.

Cyclones are moderate-efficiency mechanical separators that depend on centrifugal force to separate PM from the air stream. These precleaners are typically much less expensive than the more efficient devices discussed herein: electrostatic precipitators, fabric filters, and wet scrubbers [Cooper and Alley, 1994].

Electrostatic Precipitators

An electrostatic precipitator (ESP) applies electrical force to separate PM from the gas stream. A high voltage drop is established between the electrodes (many sets of large plates in parallel, with vertical wires in between). The particles being carried by the gas flowing between the electrodes acquire a charge, and then are attracted to an oppositely charged plate, while the cleaned gas flows through the device. The plates are cleaned by rapping periodically, and the dust is collected in hoppers in the bottom of the device.

The classic Deutsch equation (first published in 1922) is used for preliminary design and performance evaluation:

$$E = 1 - \exp(-wA/Q) \tag{89.3}$$

where w is the drift velocity, A is the collection area, and Q is the gas volumetric flow rate, all in a consistent set of units.

In design, this equation is used to estimate the total plate area needed, and thus the size and cost of the ESP. The drift velocity, w, is a key parameter, and is a function of many variables (including the particle's diameter and resistivity, the electrical field strength, and the gas viscosity). Ranges of values of some operating and design parameters are given in Table 89.1.

The resistivity of the dust is a measure of its resistance to electrical conduction. Resistivity is very important to the design and operation of an ESP because

1. For different dusts, it can range over many orders of magnitude.

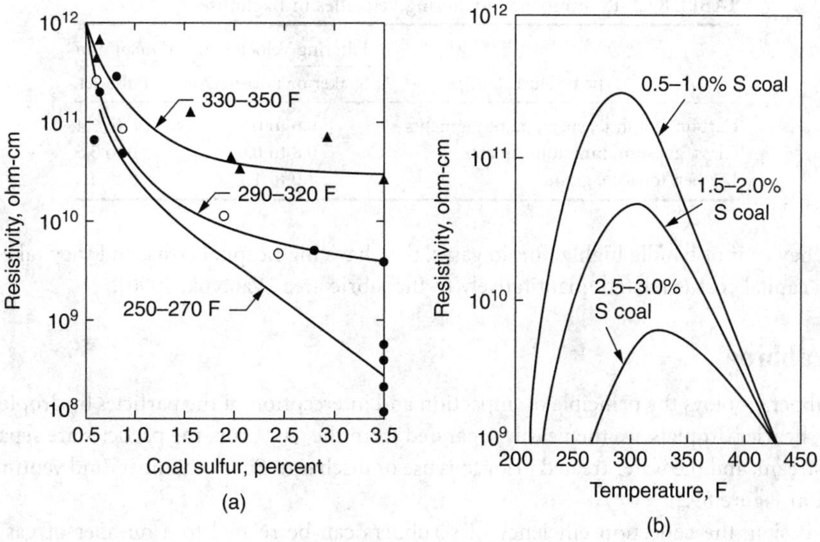

FIGURE 89.1 Variations of resistivity of coal fly ash. (*Source:* Cooper, C. D. and Alley, F. C. 1994. *Air Pollution Control — A Design Approach,* 2nd ed. Waveland Press, Prospect Heights, IL. With permission.)

2. For a given ESP, an increase in dust resistivity can decrease the drift velocity markedly, reducing the collection efficiency.

Resistivity is also quite dependent on operating conditions. For example, for coal fly ash, the resistivity is a strong function of temperature and sulfur content of the coal (see Figure 89.1).

ESPs can be designed for efficiencies above 99% on many types of dry dusts or wet fumes, and can handle large volumes of air with very low ΔP (less than 1 in. H_2O). However, they tend to have high capital costs and take up a lot of space. The capital cost is a quantitative function of the plate area [Vatavuk, 1990].

Fabric Filtration

A fabric filter (or baghouse) operates on the age-old principle of filtering air through a cloth to remove dust. The air passes through the cloth, leaving the dust behind and providing a clean air stream. The dust builds up as a loosely packed mat on the cloth until removed by shaking or blowing; the dust is then collected in hoppers. There are three main types of baghouses (classified by cleaning method): shaker, reverse-air, and pulse-jet. The first two types have parallel compartments; each compartment is taken off-line, cleaned, and returned to service sequentially while the filtering continues in the other compartments. The pulse-jet baghouse has only one compartment, and the bags are cleaned in sequence by blasts of high-pressure air while filtration occurs.

The efficiency of a fabric filter system is extremely high and is almost independent of particle size because of the mat of dust that builds up on the cloth. However, as the thickness of the dust mat increases, so does the pressure drop. The baghouse ΔP can be related to key operating variables through the filter drag model:

$$\Delta P/V = K_1 + K_2 W \tag{89.4}$$

where V is the filtering velocity (m/min), W is the fabric area dust density (g/m^2), and K_1 and K_2 are empirical constants. V is defined simply as the volumetric gas flow rate divided by the on-line fabric area. The filtering velocity is a key design parameter: some values are shown in Table 89.2.

Baghouses often achieve efficiencies above 99.9% on many types of dry dusts, are fairly flexible to operating changes, and can handle large volumes of air with reasonable pressure drops (6 to 12 in. H_2O).

TABLE 89.2 Recommended Filtering Velocities in Baghouses

	Filtering Velocity, m^3/m^2 or m/min	
Type of Dust	Shaker or Reverse-Air	Pulse-Jet
Carbon, graphite, sugar, paint pigments	0.6 to 0.7	1.6 to 2.4
Glass, gypsum, limestone, quartz	0.8 to 0.9	3.0 to 3.5
Leather, tobacco, grains	1.0 to 1.1	3.8 to 4.2

However, they cannot handle highly humid gases, they have high capital costs, and they take up a lot of space. The capital cost is related quantitatively to the fabric area [Vatavuk, 1990].

Wet Scrubbing

A wet scrubber employs the principle of impaction and interception of the particles by droplets of water. The larger, heavier droplets are then easily separated from the gas. Later, the particles are separated from the water stream, and the water treated prior to reuse or discharge. Spray chambers and venturi scrubbers are shown in Figure 89.2.

During design, the collection efficiency of scrubbers can be related to a number of gas and liquid parameters through specific equations [Cooper and Alley, 1994)]. Some of the most important variables include the liquid-to-gas ratio, the particle diameter, and, in a venturi, the gas velocity (which is directly related to pressure drop). In high-efficiency venturis, ΔP can exceed 60 in. H_2O. Extrapolating the performance of existing scrubbers can be done successfully using the contacting power approach. Contacting power is the energy expended per unit volume of gas treated and is related to collection efficiency through the following equation:

$$E = 1 - \exp[a(P_T)^b] \tag{89.5}$$

where P_T is the contacting power, kW/1000 m^3/h, and a and b are empirical constants.

Scrubbers have been used on a wide variety of dry or wet PM with efficiencies as high as 99%, while removing some soluble gases as well. Successful use of wet scrubbers also requires (1) good humidification of the gases prior to entering the scrubber and (2) good separation of the water mist before exhausting the scrubbed gases. The disadvantages include potential for corrosion and the production of a liquid effluent that must be further treated. Capital costs of scrubbers have been related to the gas scrubbing capacity [Peters and Timmerhaus, 1991].

89.2 Control of Gaseous Pollutants

The most widely used processes for gaseous pollution control are absorption, adsorption, and incineration. The selection of the most practical process for a specific control application is based primarily on the chemical and physical properties of the pollutant to be removed. Some of the properties include vapor pressure, chemical reactivity, toxicity, solubility, flammability, and corrosiveness.

The following sections describe the design basics for each process and some typical applications. This section will refer to gases and vapors where gases are defined as substances far removed from the liquid state and vapors are substances existing near condensation conditions.

Absorption Processes

Absorption refers to the transfer of a material from a gas or vapor mixture to a contacting liquid, which in most cases is either water or an aqueous alkaline or acidic solution. The contacting process is typically carried out in a cylindrical tower partially filled with an inert packing to provide a large wetted surface to contact the incoming gas stream. The process is shown schematically in Figure 89.3, where V and L designate the flow rates in mol/h of the vapor and liquid streams. The concentrations of the liquid streams

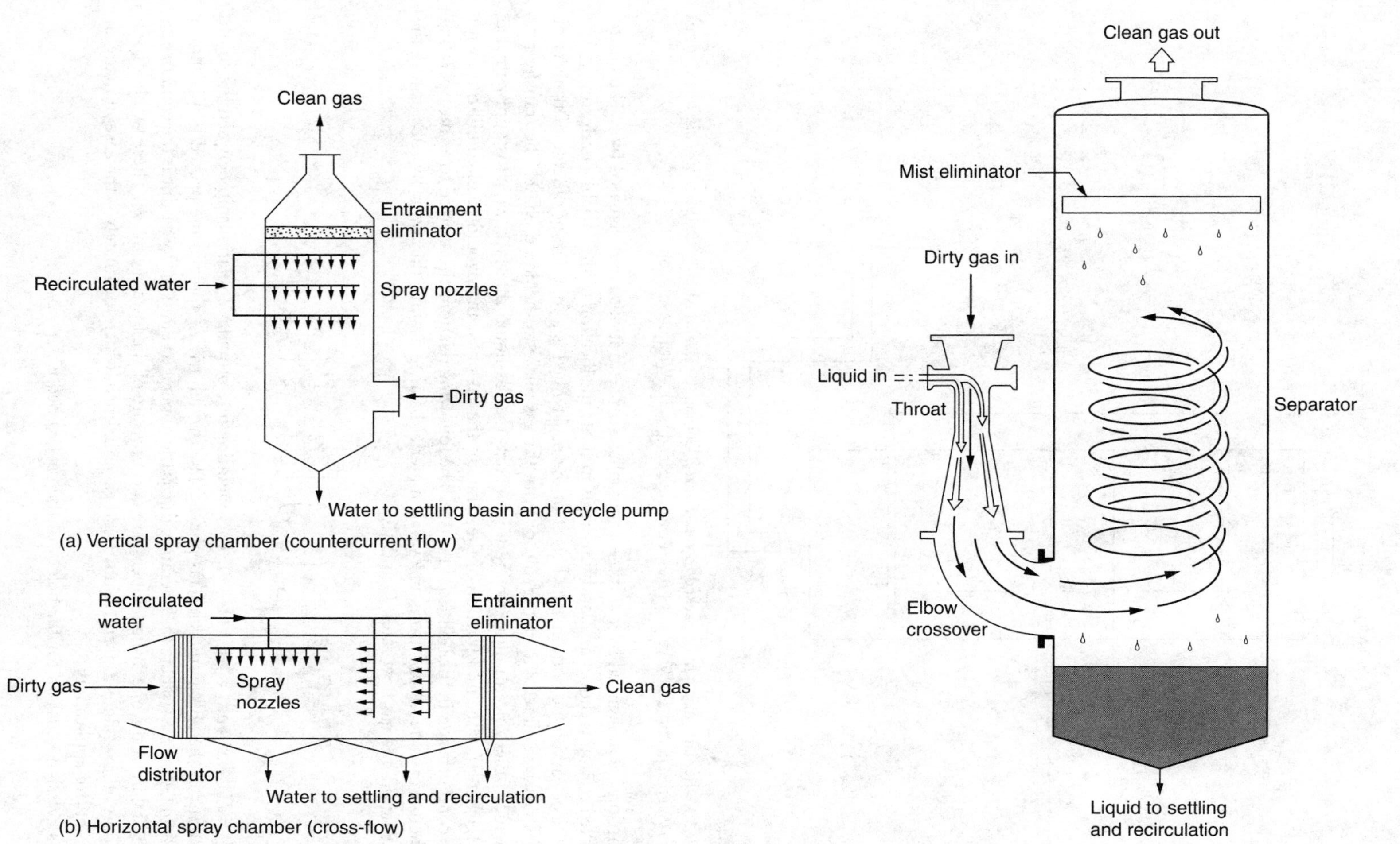

FIGURE 89.2 Common types of wet scrubbers. (a), (b) Spray chambers. (*Source:* Cooper, C. D. and Alley, F. C. 1994. *Air Pollution Control — A Design Approach,* 2nd ed. Waveland Press, Prospect Heights, IL. With permission.) (c) venturi scrubber with cyclone separator.

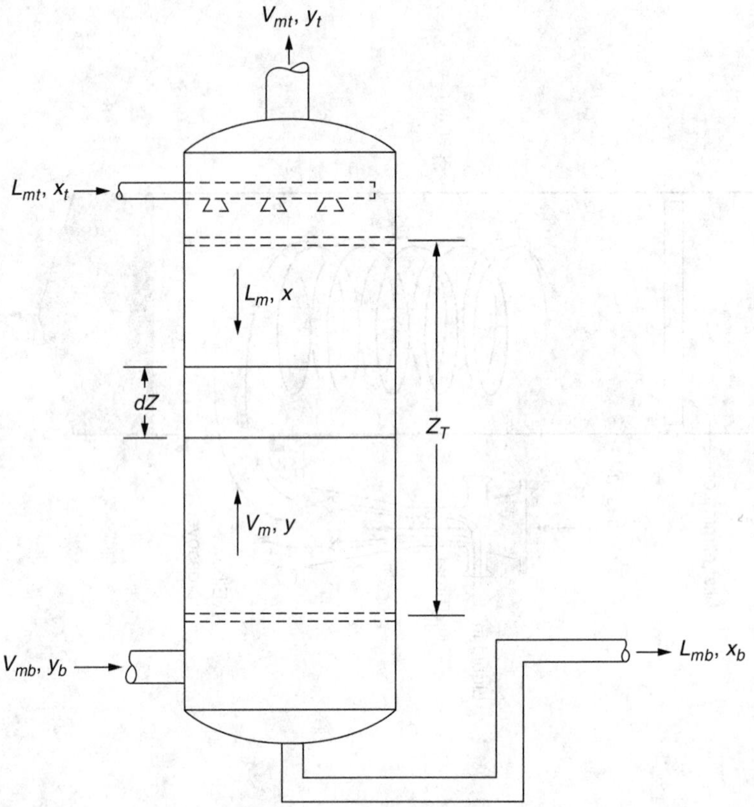

FIGURE 89.3 Schematic diagram of packed tower. (*Source:* Cooper, C. D. and Alley, F. C. 1994. *Air Pollution Control — A Design Approach,* 2nd ed. Waveland Press, Prospect Heights, IL. With permission.)

in mol fraction are designated by x and the vapor phase concentrations are designated by y. The height of the packed section is shown as Z_t and subscripts indicate the location of the variable, t referring to the top of the tower and b the bottom. The height of the packed bed is a function of the rate of transfer of material (which we will refer to as the solute) from the gas phase to the liquid phase. The transfer rate is dependent on the concentration driving force across the gas-liquid interface, which is approximated by $(y - y^*)$, where y^* is the mol fraction of the solute in the gas phase that would be in equilibrium with the solute concentration in the bulk liquid phase. The following relationship referred to as the operating line for the tower is developed from an overall material balance around the tower by substituting the solute-free liquid and gas flow rates, L' and V'.

$$\frac{y}{1-y} = \frac{L'_m}{V'_m}\left(\frac{x}{1-x}\right) + \left[\frac{y_b}{1-y_b} - \frac{L'_m}{V'_m}\left(\frac{x_b}{1-x_b}\right)\right] \tag{89.6}$$

where $L'_m = L(1-x)$ and $V'_m = V(1-y)$. The operating line provides a relationship between the solute concentrations in the bulk liquid and gas phases. The overall driving force in the packed bed is shown graphically in Figure 89.4, where the operating line is the upper curve and the lower line is the equilibrium line for the solute at the operating temperature and pressure of the tower (y^* vs. x). These two lines will be linear only at low solute concentrations (lean gas mixtures), which fortunately is the case in many air pollution control applications.

The basic design equation for the height of the packed bed is

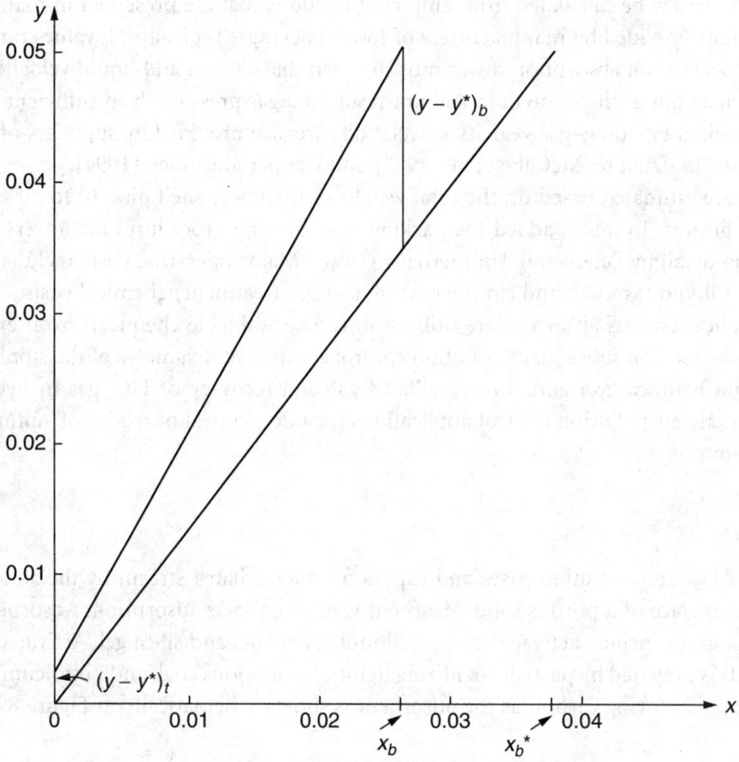

FIGURE 89.4 Operating the equilibrium lines. (*Source:* Cooper, C. D. and Alley, F. C. 1994. *Air Pollution Control — A Design Approach*, 2nd ed. Waveland Press, Prospect Heights, IL. With permission.)

$$\left(\frac{K_y a}{G_{my}}\right) Z_t = \int_{y_t}^{y_b} \frac{dy}{(1-y)(y-y^*)} \tag{89.7}$$

The reciprocal of the term in parentheses, the overall mass transfer rate divided by the molar gas phase flux, is referred to as H_y, the height of a transfer unit based on the overall gas phase driving force. The value of the integral may then be thought of as the number of transfer units, N_{ty}. Hence, Equation (89.7) may also be written as

$$Z_t = (N_{ty})(H_y) \tag{89.8}$$

For the case of linear operating and equilibrium lines, the integral can be replaced with

$$\int_{y_t}^{y_b} \frac{dy}{(1-y)(y-y^*)} = \frac{y_b - y_t}{\Delta y_{\text{LM}}} \tag{89.9}$$

where

$$\Delta y_{\text{LM}} = (y - y^*)_{\text{LM}} = \frac{(y_b - y_b^*) - (y_t - y_t^*)}{\ln\left[\dfrac{(y_b - y_b^*)}{(y_t - y_t^*)}\right]}$$

The value of H_y may be calculated from empirical relations that are presented in many texts or from experimental plots provided by manufacturers of tower packings. Typically, H_y values range from 1.0 to 3.0 ft. The diameter of an absorption tower must be such that the gas and liquid velocities will provide good phase contact but at the same time will not result in a gas pressure drop sufficient to cause liquid hold-up or flooding. Flooding–gas velocity correlations are also provided by suppliers of tower packing and are discussed in detail by McCabe et al. [1985] and Cooper and Alley [1994].

Tower costs are estimated based on the total weight of the tower shell plus 10 to 20% additional for manholes and fittings. To this is added the packing cost. Costing procedures for towers and auxiliaries are presented in detail by Peters and Timmerhaus [1991]. Major operating costs include blower power, spent absorbent liquid disposal, and (in the case of acid gas treatment) chemical costs.

In many applications gas absorbers are utilized to recover valuable chemicals from exhaust streams, which reduces the load on subsequent pollution control equipment. Examples of this application include recovery of light hydrocarbon ends from wellhead gas and recovery of HCl gas in hydrochloric acid production. Strictly air pollution control applications include removal of oxides of sulfur and nitrogen, chlorine, and ammonia.

Adsorption

The removal of low-concentration gases and vapors from an exhaust stream by the adherence of these materials to the surface of a porous solid adsorbent is referred to as adsorption. Adsorbents used in air pollution applications include activated carbon, alumina, bauxite, and silica gel. Activated carbon (most frequently used) is prepared by partially oxidizing lignite, bituminous coal, and petroleum coke. A typical fixed-bed system employing carbon as the adsorbent is shown schematically in Figure 89.5.

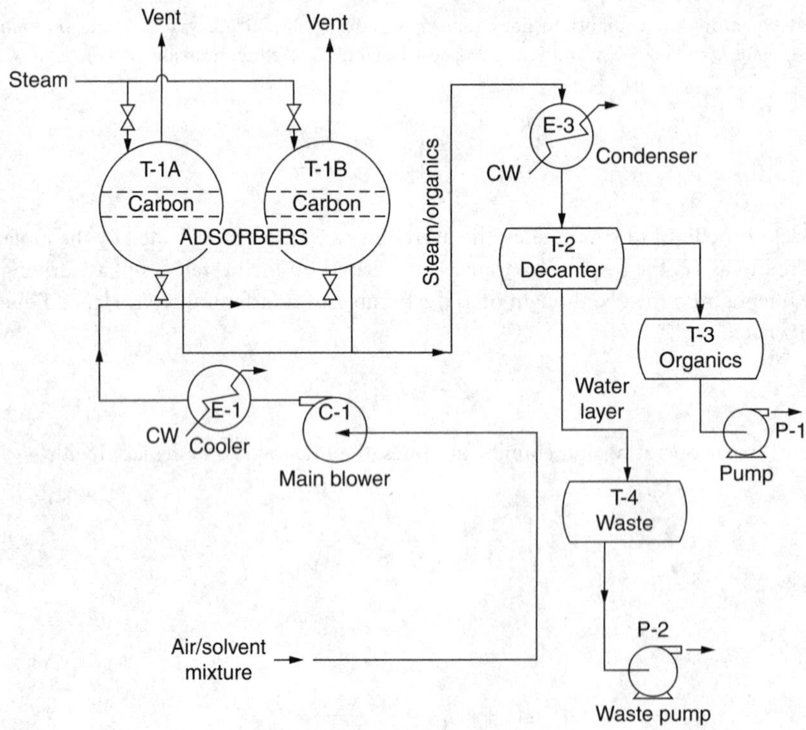

FIGURE 89.5 Flow sheet for a fixed-bed system. (*Source:* Cooper, C. D. and Alley, F. C. 1994. *Air Pollution Control — A Design Approach,* 2nd ed. Waveland Press, Prospect Heights, IL. With permission.)

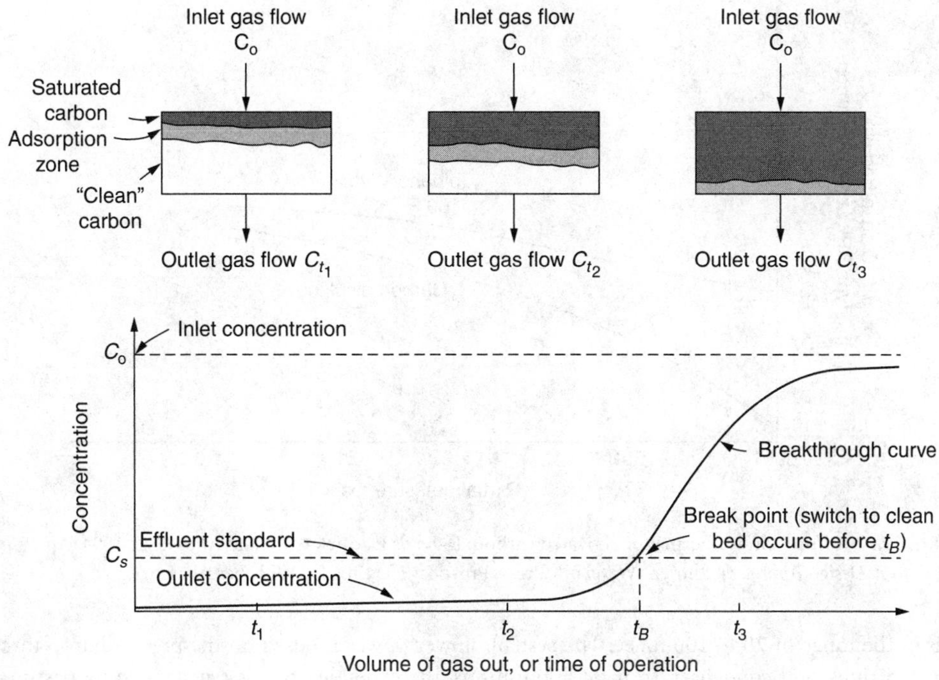

FIGURE 89.6 The adsorption wave and breakthough curve. (*Source:* Cooper, C. D. and Alley, F. C. 1994. *Air Pollution Control — A Design Approach,* 2nd ed. Waveland Press, Prospect Heights, IL. With permission.)

This system incorporates two horizontal cylindrical vessels, each containing a bed of granular activated carbon supported on a heavy screen. An air-solvent or air-pollutant mixture from a plant source flows through a cooler and then into one carbon bed while the other bed is being regenerated. A monitor-controller located in the vent stack detects a preset maximum concentration (breakthrough) and switches the incoming exhaust stream to the other bed, which has completed a regeneration cycle. The bed undergoing regeneration is contacted with saturated steam, which drives off the adsorbed material and carries it to a condenser, where it is condensed along with the steam. If the adsorbed material is insoluble in water, the condensed mixture goes to a decanter and is separated for final disposal.

When a material is adsorbed it may be weakly bonded to the solid (physical adsorption) or it may react chemically with the solid (chemisorption). Physical adsorption permits economical regeneration of the adsorbent and is practical in air pollution control applications such as solvent recovery and the removal of low-concentration odorous and toxic materials. The design of fixed-bed physical adsorption systems is described below.

The actual adsorption onto a carbon bed occurs in a concentration zone called the adsorption zone, as shown in Figure 89.6. The zone is bounded by saturated carbon upstream and clean carbon downstream. Unlike the steady state conditions in an absorption tower, adsorption is an unsteady state process in which the mass transfer driving force goes to zero as the carbon bed becomes saturated with respect to the partial pressure of the adsorbate (material adsorbed) in the gas stream.

The quantity of adsorbent in each bed is based on the saturation capacity for the adsorbate at the operating conditions of the adsorber. The operating adsorbent capacity is normally supplied by the manufacturer of the adsorbent or estimated by using 30% of the saturation capacity, as shown on the capacity curve (isotherm) illustrated in Figure 89.7.

For a two-bed system each bed must contain sufficient adsorbent to adsorb all solvent or pollutant in the incoming gas stream while the other bed is being regenerated, usually a period of 0.5 to 1.0 h. The cross-sectional area of each bed is determined by assuming that the superficial velocity of the gas in the

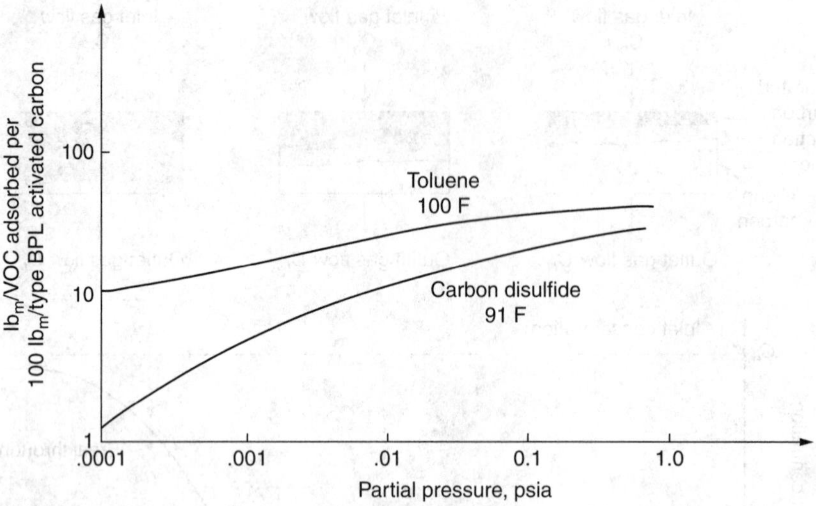

FIGURE 89.7 Adsorption isotherms for activated carbon. (*Source:* Cooper, C. D. and Alley, F. C. 1994. *Air Pollution Control — A Design Approach,* 2nd ed. Waveland Press, Prospect Heights, IL. With permission.)

bed is in the range of 70 to 100 ft/sec. The system blower power is based on the pressure drop through the bed, fittings, and condenser. In most applications, the expended bed is regenerated by passing low pressure-saturated steam through it at a rate of 0.3 to 0.5 pounds of steam per pound of carbon. The size of the condenser is based on 70 to 80% of this heat load. Cooper and Alley [1994] present examples of typical fixed-bed adsorber design.

Vatavuk [1990] reported that the purchased equipment cost (PEC) of packaged adsorber systems containing up to 14,000 pounds of carbon can be estimated by

$$\text{PEC} = 129(M_C)^{0.848} \tag{89.10}$$

where PEC = purchase cost, 1988 dollars and M_C = mass of carbon, pounds.

Typical grades of carbon utilized in adsorbers cost from 2 to 3 dollars per pound and have a normal service life of 5 to 7 years. The major utility costs are for steam, blower power, and cooling water. Average blower horsepower requirement is 6 to 8 hp/1000 SCFM of gas throughput.

Carbon adsorption has been shown to be applicable for the control of many types of gaseous pollutants, including volatile organic compounds (VOC), chlorinated solvents, and odorous materials.

Incineration Processes

Incineration of polluted exhaust streams offers an effective but costly means of controlling a wide variety of contaminants, including VOCs, which constitute a major fraction of the total annual emissions in the U.S. and play an important role in the **photochemical smog** process. A pollution control incinerator, also referred to as an afterburner or thermal oxidizer, is shown schematically in Figure 89.8. The incoming polluted air is injected into the flame mixing chamber, heated, and held in the reaction chamber until the desired pollutant destruction efficiency is obtained.

The process design of an incinerator requires specification of an operating temperature and a residence time in the reaction chamber. Typical operating condition ranges to provide the required "three Ts" of combustion are temperature (1000 to 1500°F), residence time (0.3 to 0.5 sec), and turbulence (mixing velocity 20 to 40 ft/sec). Burner fuel requirements are estimated from a steady state enthalpy balance, as shown in Equation (89.11):

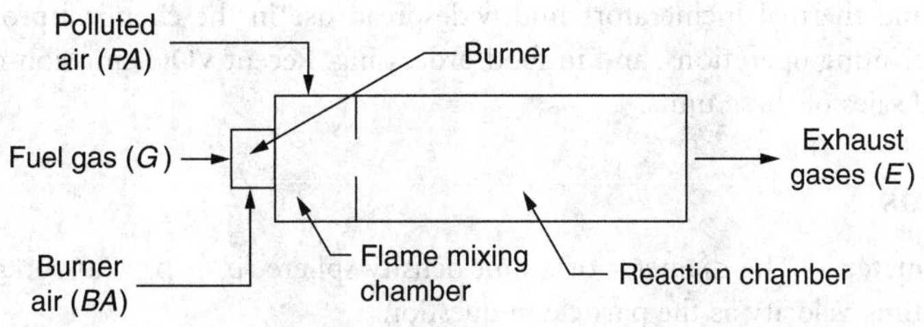

FIGURE 89.8 Schematic diagram of a vapor incinerator. (*Source:* Cooper, C. D. and Alley, F. C. 1994. *Air Pollution Control — A Design Approach,* 2nd ed. Waveland Press, Prospect Heights, IL. With permission.)

$$0 = \dot{M}_{PA}h_{PA} + \dot{M}_G h_G + \dot{M}_{BA}h_{BA} - \dot{M}_E h_E + \dot{M}_G(-\Delta H_c)_G$$
$$+\Sigma \dot{M}_{\text{VOC}_i}(-\Delta H_c)_{\text{VOC}_i} X_i - q_L \tag{89.11}$$

where $\dot{M}$ = mass flow rate, kg/min or lbm/min; h = specific enthalpy, kJ/kg or Btu/lbm; $-\Delta H_c$ = net heat of combustion (lower heating value), kJ/kg or Btu/lbm; X_i = fractional conversion of VOC_i; and q_L = rate of heat loss from the incinerator, kJ/min or Btu/min.

In most designs, the heat generated by combustion of the pollutant is ignored, but at concentrations in the 1000 ppm range, this heat may be roughly 10% of the burner heat requirement. Insurance regulations limit the combustible pollutant concentration entering the incinerator to 25% of the **lower explosion limit** (LEL).

To reduce the required operating temperature in the reaction chamber, the combustion mixture may be passed over a catalyst bed, which greatly increases the pollutant oxidation rate. Typically, the catalyst is either palladium or platinum, although Cr, Mn, Co, and Ni are used. The catalyst is deposited on an alumina support or wire screen, which is placed just downstream of the burner mix chamber. The total amount of catalyst surface area required is in the range of 0.2 to 0.5 ft^2 per SCFM of waste gas. Typical operating temperature for catalytic incinerators is in the range of 600 to 900°F.

Combustion kinetics are extremely complicated and often incinerator design is based on empirical data or the results of pilot scale tests. The application of simplified kinetic reaction models to incinerator design is described by Cooper and Alley [1994].

The choice between a thermal or catalytic incinerator is in most cases a trade-off between capital and operating cost. The installed cost of a catalytic unit will normally be 40 to 70% higher than that of a thermal unit, but the catalytic unit will require 30 to 50% less fuel for the same waste gas throughput. Offsetting this saving, however, is the service life of the catalyst, which is usually 4 to 6 years. Fuel costs for both units may be lowered by incorporating heat recovery equipment in the overall design. The cost of a packaged thermal unit handling up to 30,000 SCFM with facilities for 50% heat recovery may be estimated with the following equation [Vatavuk, 1990]:

$$P = \$4920Q^{0.389} \tag{89.12}$$

where P = manufacturer's f.o.b. price, 1988 dollars and Q = waste gas flow rate, SCFM.

The cost of catalytic units (not including the catalyst) equipped for 50% heat recovery may be found from [Vatavuk, 1990]

$$P = \exp(11.7 + 0.0354Q) \tag{89.13}$$

where P = manufacturer's f.o.b. price, 1988 dollars and Q = waste gas flow rate, thousands of SCFM.

Catalyst costs in 1988 were \$3000 per cubic foot for precious metals and \$600 per cubic foot for common metals.

Both catalytic and thermal incinerators find widespread use in the chemical process industries, in coating and film printing operations, and in food processing. Recent VOC emission regulations should promote increased sales of these units.

Defining Terms

Aerodynamic diameter — The diameter of a unit density sphere ($\rho_p = \rho_w = 1000$ kg/m^3) that has the same settling velocity as the particle in question.

Contacting power — The quantity of energy dissipated per unit volume of gas treated.

Lower explosion limit — The lowest concentration of a flammable gas, vapor, or solid in air that will provide flame propagation throughout the mixture.

Photochemical smog — An air pollution condition resulting from a series of photochemical reactions involving various volatile organic compounds and oxides of nitrogen in the lower atmosphere.

References

Cooper, C. D. and Alley, F. C. 1994. *Air Pollution Control — A Design Approach,* 2nd ed. Waveland Press, Prospect Heights, IL.

McCabe, W. L., Smith, J. C., and Harriott, P. 1985. *Unit Operations of Chemical Engineering,* 4th ed. McGraw-Hill, New York.

Peters, M. S. and Timmerhaus, K. D. 1991. *Plant Design and Economics for Chemical Engineers,* 4th ed. McGraw-Hill, New York.

Vatavuk, W. M. 1990. *Estimating Costs of Air Pollution Control,* Lewis, Chelsea, MI.

Further Information

Journal of the International Air and Waste Management Association. Published monthly by A&WMA, Pittsburgh, PA.

Air Pollution Engineering Manual, Anthony J. Buonicore and Wayne T. Davis, eds., Van Nostrand Reinhold, New York, 1993.

90

Wastewater Treatment and Current Trends

Frank R. Spellman
Norfolk, Virginia

Figure 90.1 shows a basic schematic of a wastewater treatment process providing primary and secondary treatment using the **Activated Sludge Process.** This is the model, the prototype, of the system in common use today. In secondary treatment (which provides BOD removal beyond what is achievable by simple sedimentation), there are actually three commonly used approaches: **trickling filter**, **activated sludge**, and **oxidation ponds**. Our focus, for illustrative and instructive purposes, is mainly on the activated sludge process. The purpose of Figure 90.1 is to allow the reader to follow the treatment process step-by-step as it is presented (as it is actually configured in the real world) and to assist understanding of how all the various unit processes sequentially follow and tie into each other.

90.1 Wastewater Sources, Classification, and Characteristics

Wastewater treatment is designed to use (to mimic) the natural purification processes (the self-purification processes of streams and rivers) to the maximum level possible. It completes the process in a controlled environment rather than over many miles of stream or river. We can say that a wastewater treatment plant is a stream (the self-purification process) in a box (the treatment plant). The treatment plant is also designed to remove other contaminants, which are not normally subjected to natural processes, as well as treating the solids, which are generated throughout the treatment unit steps. A typical wastewater treatment plant is designed to achieve many different purposes:

- Protect public health
- Protect public water supplies
- Protect aquatic life

0-8493-1586-7/05/$0.00+$1.50

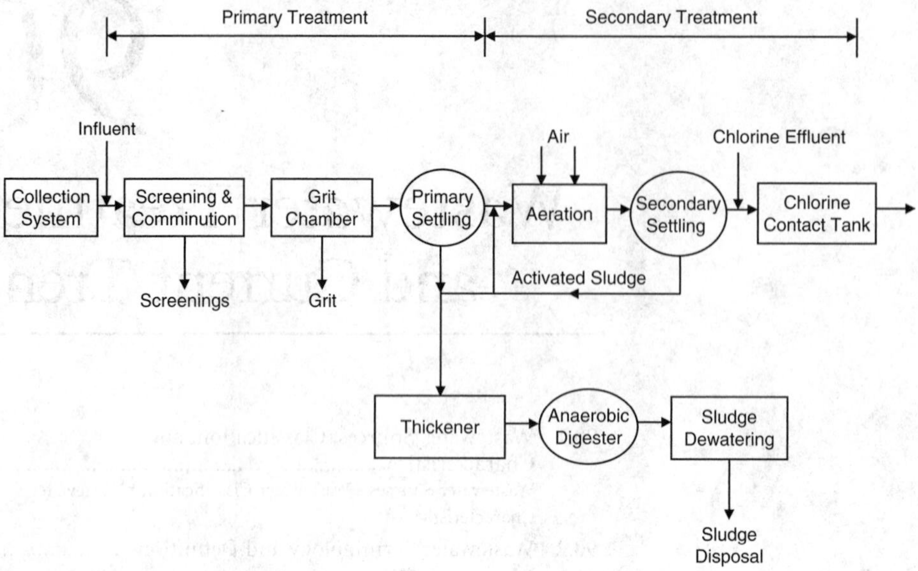

FIGURE 90.1 Schematic of an example wastewater treatment process providing primary and secondary treatment using activated sludge process.

- Preserve the best uses of the waters
- Protect adjacent lands

Wastewater Sources

The principal **sources** of domestic wastewater in a community are the residential areas and commercial districts. Other important sources include institutional and recreational facilities and stormwater (runoff) and groundwater (infiltration). Each source produces wastewater with specific characteristics.

Wastewater is generated by five major sources:

1. *Human and animal wastes* — contain the solid and liquid discharges of humans and animals and are considered by many to be the most dangerous from a human health viewpoint. The primary health hazard is present because of the millions of bacteria, viruses, and other microorganisms (some of which may be pathogenic) present in the wastestream.
2. *Household wastes* — are wastes, other than human and animal wastes, discharged from the home. Household wastes usually contain paper, household cleaners, detergents, trash, garbage, and other substances the homeowner discharges into the sewer system.
3. *Industrial wastes* — includes industry-specific materials that can be discharged from industrial processes into the collection system. Typically contains chemicals, dyes, acids, alkalis, grit, detergents, and highly toxic materials.
4. *Stormwater runoff* — Many collection systems are designed to carry both the wastes of the community and stormwater runoff. In this type of system when a storm event occurs, the wastestream can contain large amounts of sand, gravel, and other grit as well as excessive amounts of water.
5. *Groundwater infiltration* — Groundwater will enter older collection systems through cracks or improperly sealed pipe joints. Not only can this add large amounts of water to wastewater flows but also additional grit.

Wastewater Classification

Wastewater can be classified according to the sources of flow: domestic, sanitary, industrial, combined, and stormwater.

1. *Domestic (sewage) wastewater* — mainly contains human and animal wastes, household wastes, small amounts of groundwater infiltration, and small amounts of industrial wastes.
2. *Sanitary wastewater* — consists of domestic wastes and significant amounts of industrial wastes. In many cases, the industrial wastes can be treated without special precautions. However, in some cases the industrial wastes will require special precautions or a pretreatment program to ensure the wastes do not cause compliance problems for the wastewater treatment plant.
3. *Industrial wastewater* — industrial wastes only. Often an industry will determine that it is safer and more economical to treat its waste independent of domestic waste.
4. *Combined wastewater* — is the combination of sanitary wastewater and stormwater runoff. All the wastewater and stormwater of the community is transported through one system to the treatment plant.
5. *Stormwater* — a separate collection system (no sanitary waste) that carries stormwater runoff including street debris, road salt, and grit.

Wastewater Characteristics

Wastewater is characterized in terms of its physical, chemical, and biological agents. The **physical characteristics** of wastewater are based on color, odor, temperature, and flow.

The **chemical characteristics** include organic matter, the measurement of organic matter, inorganic matter, and gases. For the sake of simplicity, we specifically describe chemical characteristics in terms of alkalinity, biochemical oxygen demand (BOD), chemical oxygen demand (COD), dissolved gases, nitrogen compounds, pH, phosphorus, solids (organic, inorganic, suspended, and dissolved solids), and water.

When it enters the treatment plant, wastewater can be characterized in **biological terms** because influent typically contains millions of microorganisms, including algae, bacteria, protozoa, and viruses. The majority of these organisms are nonpathogenic; however, several pathogenic organisms may also be present (these may include the organisms responsible for typhoid, tetanus, hepatitis, dysentery, gastroenteritis, and other diseases).

90.2 Wastewater Terminology and Definitions

Wastewater treatment, like many other technical fields, has its own unique terms with their own meaning. Even though some of these terms are unique to wastewater treatment, many are also common to other professions. Wastewater treatment is a combination of engineering, biology, mathematics, hydrology, chemistry, physics, and other disciplines; many of the terms used in these disciplines are also used in wastewater treatment. However, terms such as *sludge and biosolids; unit processes; systems;* and *primary, secondary treatment* frequently appear in the literature, and their usage is not always consistent.

Activated sludge is the unit process where solids are formed when microorganisms are used to treat wastewater. **Biosolids,** according to *Merriam-Webster's Collegiate Dictionary, Tenth Edition* (1998): biosolid *n* (1977) — "solid organic matter recovered from a sewage treatment process and used especially as fertilizer — usually used in plural." At present, the term biosolids is used in many places (and even in describing activated "biosolids") to replace the standard term sludge. The author (along with others in the field) views the term sludge as an ugly four-letter word that is inappropriate to use in describing biosolids. Biosolids is a product that can be reused; it has some value. Because biosolids has some value, it should not be classified as a "waste" product, and when beneficial reuse of biosolids is practiced, it is not.

As mentioned, wastewater treatment is a series of steps. Each of the steps can be accomplished using one of more treatment processes or types of equipment. The major categories of treatment steps are a series of **unit processes** (conversion processes). The entire process is made of individual unit processes tied together — from influent end to effluent end — in trainlike fashion. Note that the term unit process should not be confused with the term **unit operations,** which have been used to describe contaminant removal by physical forces (screening, sedimentation, and filtration).

The term **reactor** refers to a vessel or containment structure. These vessels are more commonly referred to as unit process structures such as clarifiers, contact basins, aeration basins, digesters, and others.

Preliminary treatment removes materials that could damage plant equipment or would occupy treatment capacity without being treated. **Primary treatment** removes settleable and floatable solids (may not be present in all treatment plants). **Secondary treatment** removes BOD_5 (Biological Oxygen Demand) and dissolved and colloidal suspended organic matter by biological action; organics are converted to stable solids, carbon dioxide, and more organisms. **Advanced waste treatment** (sometimes referred to as tertiary treatment when the objective is nutrient removal only) uses physical, chemical and biological processes to remove additional BOD_5, solids, and nutrients.

90.3 Wastewater Treatment

Wastewater collection systems collect and convey wastewater to the treatment plant. The complexity of the system depends on the size of the community and the type of system selected. Methods of collection and conveyance of wastewater include gravity systems, force main systems, vacuum systems, and combinations of all three types of systems.

The initial stage in the wastewater treatment process (following collection and influent pumping) is **preliminary treatment.** Raw influent entering the treatment plant may contain many kinds of materials (trash). The purpose of preliminary treatment is to protect plant equipment by removing these materials which could cause clogs, jams, or excessive wear to plant machinery. In addition, the removal of various materials at the beginning of the treatment process saves valuable space downstream within the treatment plant. Preliminary treatment may include many different processes, each designed to remove a specific type of material, which is a potential problem for the treatment process. Processes include wastewater collections — influent pumping, screening, shredding, grit removal, flow measurement, preaeration, chemical addition, and flow equalization; the major processes are shown in Figure 90.1.

The purpose of **primary treatment** (primary sedimentation or primary clarification) is to remove settleable organic and flotable solids. Normally, each primary clarification unit can be expected to remove 90 to 95% settleable solids, 40 to 60% total suspended solids, and 25 to 35% BOD_5.

Sedimentation may be used throughout the plant to remove settleable and floatable solids. It is used in primary treatment, secondary treatment, and advanced wastewater treatment processes.

Upon completion of screening, degritting, and settling in sedimentation basins, large debris, grit, and many settleable materials have been removed from the wastestream. What is left is referred to as **primary effluent.** Usually cloudy and frequently gray in color, primary effluent still contains large amounts of dissolved food and other chemicals (nutrients).

The main purpose of **secondary treatment** is to provide biochemical oxygen demand (BOD) removal beyond what is achievable by primary treatment. Essentially, secondary treatment refers to those treatment processes that use biological processes to convert dissolved, suspended, and colloidal organic wastes to more stable solids that can either be removed by settling or discharged to the environment without causing harm. There are three commonly used approaches, all of which take advantage of the ability of microorganisms to convert organic wastes (via biological treatment), into stabilized, low-energy compounds. Two of these approaches, the *trickling filter* [and/or its variation, the *rotating biological contactor (RBC)*] and the *activated sludge* process, sequentially follow normal primary treatment. The third, *ponds* (oxidation ponds or lagoons), however, can provide equivalent results without preliminary treatment.

Most secondary treatment processes decompose solids aerobically, producing carbon dioxide, stable solids, and more organisms. Since solids are produced, all of the biological processes must include some form of solids removal (settling tank, filter, etc.).

Secondary treatment processes can be separated into two large categories: fixed film systems and suspended growth systems.

Fixed film systems are processes, which use a biological growth (biomass or slime), which is attached to some form of media. Wastewater passes over or around the media and the slime. When the wastewater and slime are in contact, the organisms remove and oxidize the organic solids. The media may be stone, redwood, synthetic materials or any other substance that is durable (capable of withstanding weather conditions for many years), provides a large area for slime growth while providing open space for ventilation, and is not toxic to the organisms in the biomass. Fixed film devices include trickling filters and rotating biological contactors (RBCs).

Trickling filters have been used to treat wastewater since the 1890s. It was found that if settled wastewater was passed over rock surfaces, slime grew on the rocks and the water became cleaner (i.e., a fixed film biological treatment method designed to remove BOD_5 and suspended). Today we still use this principle but, in many installations, instead of rocks we use plastic media. In most wastewater treatment systems, the *trickling filter* follows primary treatment and includes a secondary settling tank or clarifier. Trickling filters are widely used for the treatment of domestic and industrial wastes.

The **Rotating Biological Contactor (RBC)** is a biological treatment system and is a variation of the trickling filter. Still relying on microorganisms that grow on the surface of a medium, the RBC also is a **fixed film** biological treatment device. An RBC consists of a series of closely spaced (mounted side by side), circular, plastic (synthetic) disks, that are typically about 3.5 m in diameter and attached to a rotating horizontal shaft. Approximately 40% of each disk is submersed in a tank containing the wastewater to be treated. As the RBC rotates, the attached biomass film (slime) that grows on the surface of the disk moves into and out of the wastewater. While submerged in the wastewater, the microorganisms (slime) absorb organics; while they are rotated out of the wastewater, they are supplied with needed oxygen for aerobic decomposition. As the slime reenters the wastewater, excess solids and waste products are stripped off the media as **sloughings.** These sloughings are transported with the wastewater flow to a settling tank for removal.

Suspended growth systems are processes that use a biological growth, which is mixed with the wastewater. Typical suspended growth systems consist of various modifications of the **activated sludge process.**

As shown in Figure 90.1, the **activated sludge process** follows primary settling. The basic components of an activated sludge sewage treatment system include an aeration tank and a secondary settling basin, or clarifier. Primary effluent is mixed with settled solids recycled from the secondary clarifier and is then introduced into the aeration tank. Compressed air is injected continuously into the mixture through porous diffusers located at the bottom of the tank, usually along one side.

In operation, wastewater is fed continuously into an aerated tank, where the microorganisms metabolize and biologically flocculate the organics. Microorganisms (activated sludge) are settled from the aerated *mixed liquor* under quiescent conditions in the final clarifier and are returned to the aeration tank. Left uncontrolled, the number of organisms would eventually become too great; therefore, some must periodically be removed (wasted). A portion of the concentrated solids from the bottom of the settling tank must be removed from the process (waste activated sludge or WAS). Clear supernatant from the final settling tank is the plant effluent.

Like potable water, wastewater effluent is **disinfected** to protect public health in general. In wastewater treatment, this is particularly important when the secondary effluent is discharged into a body of water used for swimming and/or for a downstream water supply.

Wastewater treatment can be accomplished using **ponds**. Ponds are relatively easy to build and to manage, they accommodate large fluctuations in flow, and they can also provide treatment that approaches conventional systems (producing a highly purified effluent) at much lower cost. The actual degree of treatment provided depends on the type and number of ponds used. Ponds can be used as the sole type of treatment or they can be used in conjunction with other forms of wastewater treatment; that is, other treatment processes followed by a pond or a pond followed by other treatment processes.

90.4 Advanced Wastewater Treatment

Biological nitrification is the first basic step of *biological nitrification–denitrification.* In nitrification, the secondary effluent is introduced into another aeration tank, trickling filter, or biodisc. Because most of the carbonaceous BOD has already been removed, the microorganisms that drive this advanced step are the nitrifying bacteria *Nitrosomonas* and *Nitrobacter.* In nitrification, the ammonia nitrogen is converted to nitrate nitrogen, producing a *nitrified effluent.* At this point, the nitrogen has not actually been removed, only converted to a form that is not toxic to aquatic life and that does not cause an additional oxygen demand.

The nitrification process can be limited (performance affected) by alkalinity (requires 7.3 parts alkalinity to 1.0 part ammonia nitrogen); pH; dissolved oxygen availability; toxicity (ammonia or other toxic materials); and process mean cell residence time (sludge retention time). As a general rule, biological nitrification is more effective and achieves higher levels of removal during the warmer times of the year.

Biological denitrifcation removes nitrogen from wastewater. When bacteria come in contact with a nitrified element in the absence of oxygen, they reduce the nitrates to nitrogen gas, which escapes the wastewater. The denitrification process can be accomplished in either an anoxic activated sludge system (suspended growth) or in a column system (fixed growth). The denitrification process can remove up to 85% or more of nitrogen. After effective biological treatment, little oxygen-demanding material is left in the wastewater when it reaches the denitrification process.

The denitrification reaction will only occur if an oxygen demand source exists when no dissolved oxygen is present in the wastewater. An oxygen demand source is usually added to reduce the nitrates quickly. The most common demand source added is soluble BOD or methanol. Approximately 3 mg/l of methanol is added for every 1 mg/l of nitrate–nitrogen. Suspended growth denitrification reactors are mixed mechanically but only enough to keep the biomass from settling without adding unwanted oxygen.

Submerged filters of different types of media may also be used to provide denitrification. A fine media downflow filter is sometimes used to provide both denitrification and effluent filtration. A fluidized sand bed where wastewater flows upward through a media of sand or activated carbon at a rate to fluidize the bed may also be used. Denitrification bacteria grow on the media.

Land application is application of secondary effluent onto a land surface. This procedure can provide an effective alternative to the expensive and complicated advanced treatment methods. A high-quality polished effluent (i.e., with low effluent levels of TSS, BOD, phosphorus, and nitrogen compounds as well as refractory organics) can be obtained by the natural processes that occur as the effluent flows over the vegetated ground surface and percolates through the soil.

Limitations are involved with land application of wastewater effluent. For example, the process needs large land areas. Soil type and climate are also critical factors in controlling the design and feasibility of a land treatment process.

Recent experience has shown that **biological nutrient removal (BNR)** systems are reliable and effective in removing nitrogen and phosphorus. The process is based upon the principle that, under specific conditions, microorganisms will remove more phosphorus and nitrogen than is required for biological activity. Several patented processes are available for this purpose. Performance depends on the biological activity and the process employed.

90.5 Solids (Sludge/Biosolids)

The wastewater treatment unit processes described to this point remove solids and BOD from the waste-stream before the liquid effluent is discharged to its receiving waters. What remains to be disposed of is a mixture of solids and wastes, called *process residuals* — more commonly referred to as *sludge* or *biosolids.*

The most costly and complex aspect of wastewater treatment can be the collection, processing, and disposal of sludge. This is the case because the quantity of sludge produced may be as high as 2% of the original volume of wastewater, depending somewhat on the treatment process being used.

Because sludge can be as much as 97% water content, and because the cost of disposal will be related to the volume of sludge being processed, one of the primary purposes or goals (along with **stabilizing** it so it is no longer objectionable or environmentally damaging) of sludge treatment is to separate as much of the water from the solids as possible. Sludge treatment methods may be designed to accomplish both of these purposes.

TABLE 90.1 Typical Water Content of Sludges

Water Treatment Process	% Moisture of Sludge Generated	lb Water/lb Sludge Solids
Primary sedimentation	95	19.0
Trickling filter		
Humus — low rate	93	13.3
Humus — high rate	97	32.3
Activated sludge	99	99.0

Source: USEPA's *Operational Manual: Sludge Handling and Conditioning*, EPA-430/9-78-002, 1978.

When we speak of **sludge** or **biosolids**, we are speaking of the same substance or material; each is defined as the suspended solids removed from wastewater during sedimentation, then concentrated for further treatment and disposal or reuse. The difference between the terms **sludge** and **biosolids** is determined by their ultimate end use.

Sludge forms initially as a 3 to 7% suspension of solids, and with each person typically generating about 4 gallons of sludge per week, the total quantity generated each day, week, month, and year is significant. Because of the volume and nature of the material, sludge management is a major factor in the design and operation of all water pollution control plants.

The composition and characteristics of sewage sludge vary widely and can change considerably with time. Notwithstanding these facts, the basic components of wastewater sludge remain the same. The only variations occur in quantity of the various components as the type of sludge and the process from which it originated changes.

The main component of all sludges is *water*. Prior to treatment, most sludge contains 95 to 99+% water (see Table 90.1). This high water content makes sludge handling and processing extremely costly in terms of both money and time. Sludge handling may represent up to 40% of the capital cost and 50% of the operation cost of a treatment plant. As a result, the importance of optimum design for handling and disposal of sludge cannot be overemphasized. The water content of the sludge is present in a number of different forms. Some forms can be removed by several sludge treatment processes, thus allowing the same flexibility in choosing the optimum sludge treatment and disposal method.

Sludge treatment processes can be classified into a number of major categories: thickening, digestion (or stabilization), de-watering, incineration, and land application. Each of these categories is further subdivided according to the specific processes that are used to accomplish sludge treatment.

The solids content of primary, activated, trickling-filter, or even mixed sludge (i.e., primary plus activated sludge) varies considerably, depending on the characteristics of the sludge. Note that the sludge removal and pumping facilities and the method of operation also affect the solids content. **Sludge thickening** (or *concentration*) is a unit process used to increase the solids content of the sludge by removing a portion of the liquid fraction. By increasing the solids content, more economical treatment of the sludge can be effected. Sludge thickening processes include gravity thickeners, flotation thickeners, and solids concentrators.

Equipment used for **aerobic digestion** consists of an aeration tank (digester), which is similar in design to the aeration tank used for the activated sludge process. Either diffused or mechanical aeration equipment is necessary to maintain the aerobic conditions in the tank. Solids and supernatant removal equipment is also required.

In operation, process residuals (sludge/biosolids) are added to the digester and aerated to maintain a dissolved oxygen (D.O.) concentration of 1.0 mg/L. Aeration also ensures that the tank contents are well mixed. Generally, aeration continues for approximately 20 days retention time. Periodically, aeration is stopped and the solids are allowed to settle. Sludge and the clear liquid supernatant are withdrawn as needed to provide more room in the digester. When no additional volume is available, mixing is stopped for 12 to 24 h before solids are withdrawn for disposal. Process control testing should include alkalinity, pH, % solids, and % volatile solids for influent sludge, supernatant, digested sludge, and digester contents.

Anaerobic digestion is the traditional method of sludge stabilization. It involves using bacteria that thrive in the absence of oxygen and is slower than aerobic digestion, but has the advantage that only a small percentage of the wastes is converted into new bacterial cells. Instead, most of the organics are converted into carbon dioxide and methane gas.

The purpose of composting sludge is to stabilize the organic matter, reduce volume, and to eliminate pathogenic organisms. In a **composting operation,** dewatered solids are usually mixed with a bulking agent (i.e., hardwood chips) and stored until biological stabilization occurs. The composting mixture is ventilated during storage to provide sufficient oxygen for oxidation and to prevent odors. After the solids are stabilized, they are separated from the bulking agent. The composted solids are then stored for curing and applied to farmlands or used for other beneficial uses.

In the **lime stabilization** process, residuals are mixed with lime to achieve a pH of 12.0. This pH is maintained for at least 2 h. The treated solids can then be dewatered for disposal or directly land applied.

Thermal treatment (or wet air oxidation) subjects sludge to high temperature and pressure in a closed reactor vessel. The high temperature and pressure rupture the cell walls of any microorganisms present in the solids and cause chemical oxidation of the organic matter. This process substantially improves dewatering and reduces the volume of material for disposal. It also produces a very high-strength waste, which must be returned to the wastewater treatment system for further treatment.

Chlorine oxidation also occurs in a closed vessel. In this process chlorine (100 to 1000 mg/L) is mixed with a recycled solids flow. The recycled flow and process residual flow are mixed in the reactor. The solids and water are separated after leaving the reactor vessel. The water is returned to the wastewater treatment system, and the treated solids are dewatered for disposal. The main advantage of chlorine oxidation is that it can be operated intermittently. The main disadvantage is production of extremely low pH and high chlorine content in the supernatant.

Digested sludge removed from a digester is still mostly liquid. **Sludge dewatering** is used to reduce volume by removing the water to permit easy handling and economical reuse or disposal. Dewatering processes include sand drying beds, vacuum filters, centrifuges, filter presses (belt and plate), and incineration.

Drying beds have been used successfully for years to dewater sludge. Composed of a sand bed (consisting of a gravel base, underdrains and 8 to 12 inches of filter-grade sand), drying beds include an inlet pipe, splash pad containment walls, and a system to return filtrate (water) for treatment. In some cases, the sand beds are covered to provide drying solids protection from the elements.

Rotary vacuum filters have also been used for many years to dewater sludge. The vacuum filter includes filter media (belt, cloth, or metal coils), media support (drum), vacuum system, chemical feed equipment, and conveyor belt(s) to transport the dewatered solids.

Pressure filtration differs from vacuum filtration in that the liquid is forced through the filter media by a positive pressure instead of a vacuum. Several types of presses are available, but the most commonly used types are plate and frame presses and belt presses.

Centrifuges of various types have been used in dewatering operations for at least 30 years and appear to be gaining in popularity.

The purpose of **land application of biosolids** is to dispose of the treated biosolids in an environmentally sound manner by recycling nutrients and soil conditioners. In order to be land applied, wastewater biosolids must comply with state and federal biosolids management/disposal regulations. Biosolids must not contain materials that are dangerous to human health (i.e., toxicity, pathogenic organisms, etc.) or dangerous to the environment (i.e., toxicity, pesticides, heavy metals, etc.). Treated biosolids are land applied by either direct injection or application and plowing in (incorporation).

90.6 Future of Wastewater Treatment

Earth was originally allotted a finite amount of water — we have no more or no less than that original allotment today. Thus, it logically follows that, in order to sustain life, as we know it, we must do everything we can to preserve and protect our water supply. Moreover, we also must purify and reuse the water we presently waste (i.e., wastewater).

The Paradigm Shift

Historically, the purpose of wastewater treatment has been to protect the health and well being of our communities. The purpose of wastewater treatment processes has not changed. However, primarily because of new regulations that include (1) protecting against protozoan and virus contamination; (2) implementation of the multiple barrier approach to microbial control; and (3) regulations for trihalomethanes (THMs) and disinfection byproducts (DBPs), the paradigm has shifted.

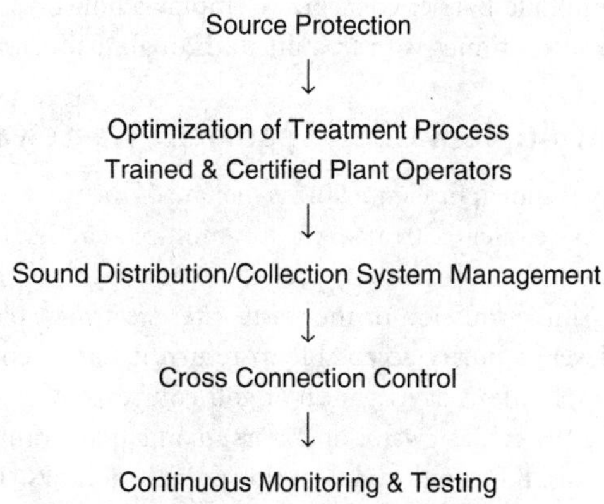

FIGURE 90.2 Multiple barrier approach.

A **paradigm shift** is defined as a major change in the way things are thought about, especially scientifically. Once a problem can no longer be solved in the existing paradigm, new laws and theories emerge and form a new paradigm, overthrowing the old if it is accepted. Simply, a paradigm shift represents "a profound change in the thoughts, perceptions, and values that form a particular vision of reality" [Capra, 1982]. For our purposes, we use the term paradigm shift to mean a change in the way things are understood and done.

Multiple-Barrier Concept

Drinking water standards are regulations that the U.S. Environmental Protection Agency (USEPA) sets to control the level of contaminants in the nation's drinking water. These standards are part of the Safe Drinking Water Act's "**multiple barrier approach**" to drinking water protection. Note that the multiple barrier approach can be applied to protecting potable water as well as to wastewater treatment. First, we will examine how this concept applies to our potable water supply (see Figure 90.2).

The multiple barrier approach includes the following elements.

1. **Assessing and protecting drinking water sources** — means doing everything possible to prevent microbes and other contaminants from entering water supplies. Minimizing human and animal activity around our watersheds is one part of this barrier.
2. **Optimizing treatment processes** — provides a second barrier. This usually means filtering and disinfecting the water. It also means making sure that the people who are responsible for our water are properly trained/certified and knowledgeable of the public health issues involved.
3. **Ensuring the integrity of distribution systems** — this consists of maintaining the quality of water as it moves through the system on its way to the customer's tap.
4. **Effecting correct cross-connection control procedures** — this is a critical fourth element in the barrier approach. It is critical because the greatest potential hazard in water distribution systems is associated with cross-connections to nonpotable waters. There are many connections between potable and nonpotable systems — every drain in a hospital constitutes such a connection — but cross-connections are those through which backflow can occur [Angele, 1974].
5. **Continuous monitoring and testing of the water before it reaches the tap** — is a critical element in the barrier approach. It should include having specific procedures to follow should potable water ever fail to meet quality standards.

With the involvement of USEPA, local governments, drinking water utilities, and citizens, these multiple barriers ensure that the tap water in the United States and territories is safe to drink. Simply, in the multiple-barrier concept, we employ a holistic approach to water management that begins at the source and continues with treatment, through disinfection and distribution.

Multiple-Barrier Approach: Wastewater Operations

Not shown in Figure 90.1 is the fate of the used water; that is, what happens to the wastewater produced? Wastewater is treated via the multiple-barrier treatment train, which consists of the combination of pollution prevention, pretreatment programs, and unit processes used in the treatment plant. The primary mission of the wastewater treatment plant (and the operator) is to treat the wastestream to a level of purity acceptable to return it to the environment or for immediate reuse (i.e., reuse in such applications as irrigation of golf courses, etc.).

Water/wastewater operators maintain a continuous urban water cycle, a "hidden function," on a daily basis. B.D. Jones sums up this point as follows [Jones, 1980]:

> Delivering services is the primary function of municipal government. It occupies the vast bulk of the time and effort of most city employees, is the source of most contacts that citizens have with local governments, occasionally becomes the subject of heated controversy, and is often surrounded by myth and misinformation. Yet, service delivery remains the "hidden function" of local government.

Wastewater service professionals treat the urban wastestream to remove pollutants before discharging the effluent into the environment. Water and wastewater treatment services are the urban circulatory system [Cox, 1939; AJPH, 1930]. In addition, like the human circulatory system, the urban circulatory system is less than effective if flow is not maintained as per design.

Maintaining flow is what wastewater treatment is all about. Seems easy enough, water has been flowing literally for eons — no problem there. However, this is not to say that wastewater operations are not without problems and/or challenges. The dawn of the twenty-first century brought with it, for many of us, aspirations of good things ahead in the constant struggle to provide quality food and water for humanity. However, the only way in which we can hope to accomplish this is to stay on the cutting edge of technology and to face all challenges head on. Some of these "challenges" are addressed in the following section.

Management Problems Facing Wastewater Operations

Problems come and go, shifting from century to century, decade to decade, year to year, and also from site to site. They range from the problems caused by natural forces (storms, earthquakes, fires, floods, and droughts) to those caused by social forces, currently including terrorism.

In general, five areas are of concern to many wastewater managers:

1. Complying with regulations, and coping with new and changing regulations
2. Maintaining infrastructure
3. Privatization and/or reengineering
4. Benchmarking
5. Upgrading security

1. Compliance with New, Changing, and Existing Regulations[1]

Adapting the workforce to the challenges of meeting changing regulations and standards for wastewater treatment is a major concern. In regards to wastewater collection and treatment, the National Pollution Discharge Elimination System program (NPDES) issues permits that control wastewater treatment plant discharges. "Meeting permit" is always a concern for wastewater treatment managers because the effluent discharged into water bodies affects those downstream of the release point. Individual point source dischargers must use the best available technology (BAT) to control the levels of pollution in the effluent

[1]This section adapted from Drinan, J.E., *Water & Wastewater Treatment: A Guide for the Nonengineering Professional.* Lancaster, PA: Technomic Publishing Company, Lancaster, PA, 2001. pp. 2–3.

they discharge into streams. As systems age and best available technology changes, meeting permit with existing equipment and unit processes becomes increasingly more difficult.

2. Maintaining Infrastructure

During the 1950s and 1960s, the U.S. government encouraged the prevention of pollution by providing funds for the construction of municipal wastewater treatment plants, water-pollution research, and technical training and assistance. New processes were developed to treat sewage, analyze wastewater, and evaluate the effects of pollution on the environment. In spite of these efforts, however, expanding population with its corresponding industrial and economic growth caused the pollution and corresponding implications to increase.

In response to the need to make a coordinated effort to protect the environment, the National Environmental Policy Act (NEPA) was signed into law on January 1, 1970. In December of that year, the USEPA was created to bring under one agency all of the pollution-control programs related to air, water, and solid wastes. In 1972, the Water Pollution Control Act Amendments expanded the role of the federal government in water pollution control and significantly increased federal funding for construction of wastewater treatment plants.

Many of the wastewater treatment plants in operation today are the result of federal grants made over the years. For example, because of the 1977 Clean Water Act Amendment to the Federal Water Pollution Control Act of 1972 and the 1987 Clean Water Act reauthorization bill, funding for wastewater treatment plants was provided.

Some large sanitation districts, with their multiple plant operations, and even a larger number of single plant operations in smaller communities still in operation today are a result of these early environmental laws and federal grants.

Many of these locally or federally funded treatment plants are aging and becoming less efficient. Complicating the problems associated with natural aging and associated wear on equipment is the increasing pressure on older systems to meet demands of increased population and urban growth. Facilities built in the 1960s and 1970s are now 30 to 40 years old, and not only are they showing signs of wear and tear, they simply were not designed to handle the level of growth (increase in flow) that has occurred in many municipalities.

Regulations often necessitate a need to upgrade. By matching funds or providing federal money to cover some of the costs, municipalities can take advantage of a window of opportunity to improve their facilities at a lower direct cost to the community. Federal dollars, of course, do not come without strings attached; they are to be spent on specific projects in specific areas. On the other hand, many times new regulatory requirements are put in place without the financial assistance needed to implement them. When this occurs, either the local community ignores the new requirements (i.e., until caught and forced to comply) or faces the situation and implements local tax hikes to pay the cost of compliance.

An example of how a change in regulations can force the issue is demonstrated by the demands made by the Occupational Safety and Health Administration (OSHA) and USEPA in their Process Safety Management (PSM)/Risk Management Planning (RMP) regulations. These regulations put the use of elemental chlorine (and other listed hazardous materials) under scrutiny. Moreover, because of these regulations, plant managers throughout the country are forced to choose which side of a double-edged sword cuts their way the most. One edge calls for full compliance with the regulations (analogous to stuffing the regulation through the eye of a needle). The other edge calls for substitution, that is, replacing toxic elemental chlorine with a nonlisted hazardous chemical (e.g., hypochlorite) or a physical (ultraviolet irradiation, UV) disinfectant — either way, a very costly undertaking.

3. Privatization and/or Re-engineering[1]

As mentioned, wastewater treatment operations are undergoing a paradigm shift. We explained that this paradigm shift focuses on the holistic approach to treating water. The shift is, however, more inclusive.

---

[1]Adapted from Johnson, R. and Moore, A., Policy Brief 17 Opening the Floodgates: Why Water Privatization will Continue. Reason Public Institute. January 2002, pp. 1-3, [www.rppi.org/pbrief 17]. Accessed May 14, 2002.

It also includes "thinking outside the box." That is, in order to remain efficient and therefore competitive in the real world of operations, wastewater facilities have either bought into the new paradigm shift or have been forcibly "shifted" to doing other things (often these "other" things have little to do with wastewater operations).

Experience has shown that few words conjure up more fear among municipal plant managers (and other employees) than "privatization" or "reengineering." **Privatization** means allowing private enterprise to compete with government in providing public services, such as water and wastewater operations. **Existing management,** on the other hand, can accomplish reengineering internally, or it can be used (and usually is) during the privatization process. Reengineering is the systematic transformation of an existing system into a new form to realize quality improvements in operation, system capability, functionality, performance, or evolvability at a lower cost, better schedule, or lower risk to the customer.

Many on-site managers consider privatization and/or reengineering schemes threatening. In the worst case scenario, a private contractor could bid the existing staff out of their jobs. What might happen in the best case? Privatization and/or re-engineering are often a very real threat that forces on-site managers into workforce cuts, improving efficiency and cutting costs. At the same time, on-site managers must still ensure that the community receives safe drinking water and that the facility meets standards and permits, with fewer workers — without causing injury or accident to workers, and/or damage to the facility and/or the environment.

Local officials may take a hard look at privatization and/or re-engineering for a number of reasons:

- **Decaying infrastructures.** As mentioned, many wastewater operations include infrastructure that dates back to the early 1900s. The most recent systems were built with federal funds during the 1970s, and even these now need upgrading or replacing.
- **Mandates.** The federal government has reduced its contributions to local wastewater systems over the past 30 years, while at the same time imposing stricter water quality and effluent standards under the Clean Water Act and Safe Drinking Water Act. Moreover, as previously mentioned, new unfunded mandated safety regulations, such as OSHA's Process Safety Management and USEPA's Risk Management Planning, are expensive to implement using local sources of revenues or state revolving loan funds.
- **Hidden function.** Earlier we stated that much of the work of wastewater treatment is a "hidden function." Because of this lack of visibility, it is often difficult for local officials to commit to making the necessary investments in local wastewater systems. Simply, local politicians lack the political will; wastewater pipes and/or interceptors are not visible and not perceived as immediately critical for adequate funding to fund operations. Thus, it is easier for elected officials to ignore utility needs in favor of expenditures of more visible services, such as police and fire. Additionally, raising sewage rates to cover operations and maintenance is not always effected because it is an unpopular move for elected officials, meaning that sewer rates often do not adequately cover the actual cost of providing services in many municipalities.

In many locations throughout the United States, expenditures for wastewater services are the largest facing local governments today. (Note: This is certainly the case for those municipalities struggling to implement the latest stormwater requirements). Thus, this area presents a great opportunity for cost savings. Through privatization, private wastewater companies can take advantage of advanced technology, more flexible management practices, and streamlined procurement and construction practices to lower costs and make the critical improvements more quickly.

4. Benchmarking

Primarily out of self-preservation (to retain their lucrative positions), many utility directors work against the trend to privatize water, wastewater, and other public operations. Usually the real work to prevent privatization is delegated to the individual managers in charge of each specific operation. Moreover, working against privatization by these "local" managers is also in their own self-interest and in the interest of their workers, whose jobs may be at stake.

Start ⟶ Plan ⟶ Research ⟶ Observe ⟶ Analyze ⟶ Adapt

FIGURE 90.3 Benchmarking process.

The question is, of course, how does one go about preventing his or her wastewater operation from being privatized? The answer is rather straightforward and clear: Efficiency must be improved at reduced cost. In the real world, this is easier said than done — but is not impossible. For example, for those facilities under Total Quality Management (TQM), the process can be much easier. The advantage TQM offers the plant manager is the variety of tools to help plan, develop, and implement wastewater efficiency measures. These tools include self-assessments, statistical process control, International Organization for Standards (ISO) 9000 and 14,000, process analysis, quality circle, and benchmarking (see Figure 90.3).

Benchmarking is a process for rigorously measuring your performance vs. "best-in-class" operations, and using the analysis to meet and exceed the best in class.

What benchmarking is:

1. Benchmarking vs. best practices gives wastewater managers a way to evaluate their operations overall.
   A. How effective
   B. How cost effective
2. Benchmarking shows plant managers both how well their operations stack up and how well those operations are implemented.
3. Benchmarking is an objective-setting process.
4. Benchmarking is a new way of doing business.
5. Benchmarking forces an external view to ensure correctness of objective setting.
6. Benchmarking forces internal alignment to achieve plant goals.
7. Benchmarking promotes teamwork by directing attention to those practices necessary to remain competitive.

Potential results of benchmarking:

1. Benchmarking may indicate direction of required change rather than specific metrics.
2. Costs must be reduced.
3. Customer satisfaction must be increased.
4. Return on assets must be increased.
5. Maintenance must be improved.
6. Operational practices must be improved.
7. Best practices are translated into operational units of measure.

Targets:

1. Consideration of available resources converts benchmark findings to targets.
2. A target represents what can realistically be accomplished in given time frame.
3. Can show progress toward benchmark practices and metrics.
4. Quantification of precise targets should be based on achieving benchmark.

When forming a benchmarking team, the goal should be to provide a benchmark that evaluates and compares privatized and reengineered wastewater treatment operations to the operation in order to be more efficient and remain competitive and make continual improvements. It is important to point out that benchmarking is more than simply setting a performance reference or comparison; it is a way to facilitate learning for continual improvement. The key to the learning process is looking outside one's own plant to other plants that have discovered better ways of achieving improved performance.

As shown in Figure 90.3, the benchmarking process consists of five steps.

Step 1: **Planning.** Managers must select a process (or processes) to be benchmarked. A benchmarking team should be formed. The process of benchmarking must be thoroughly understood and documented. The performance measure for the process should be established (i.e., cost, time, and quality).

Step 2: **Research.** Information on the "best-in-class" performer must be determined through research. The information can be derived from the industry's network, industry experts, industry and trade associations, publications, public information, and other award-winning operations.

Step 3: **Observation.** The observation step is a study of the benchmarking subject's performance level, processes, and practices that have achieved those levels, and other enabling factors.

Step 4: **Analysis.** In this phase, comparisons in performance levels among facilities are determined. The root causes for the performance gaps are studied. To make accurate and appropriate comparisons, the comparison data must be sorted, controlled for quality, and normalized.

Step 5: **Adaptation.** This phase is putting what is learned throughout the benchmarking process into action. The findings of the benchmarking study must be communicated to gain acceptance, functional goals must be established, and a plan must be developed. Progress should be monitored and, as required, corrections in the process made.

The bottom line on privatization — Privatization is becoming of greater concern. Governance boards see privatization as a potential way to shift liability and responsibility from the municipality's shoulders, with the attractive bonus of cutting costs. Wastewater facilities face constant pressure to work more efficiently, more cost effectively, with fewer workers, to produce a higher quality product; that is, all functions must be value added. Privatization is increasing, and many municipalities are seriously considering outsourcing parts or all of their operations to contractors.

5. Upgrading Security

You may say Homeland Security is a Y2K problem that doesn't end January 1 of any given year.

— Governor Tom Ridge [Henry, 2002]

One consequence of the events of September 11 is USEPA's directive to establish a Water Protection Task Force to ensure that activities to protect and secure water supply/wastewater treatment infrastructure are comprehensive and carried out expeditiously. Another consequence is a heightened concern among citizens in the United States over the security of their critical water/wastewater infrastructure. The nation's water/wastewater infrastructure, consisting of several thousand publicly owned water/wastewater treatment plants, more than 100,000 pumping stations, hundreds of thousands of miles of water distribution and sanitary sewers, and another 200,000 miles of storm sewers, is one of America's most valuable resources, with treatment and distribution/collection systems valued at more than $2.5 trillion. Wastewater treatment operations taken alone include the sanitary and storm sewers forming an extensive network that runs near or beneath key buildings and roads and is contiguous to many communication and transportation networks. Significant damage to the nation's wastewater facilities or collection systems would result in loss of life; catastrophic environmental damage to rivers, lakes, and wetlands; contamination of drinking water supplies; long-term public health impacts, destruction of fish and shellfish production; and disruption of commerce, the economy, and our normal way of life.

Governor Tom Ridge points out the security role for the public professional (including water/wastewater professionals):

> Americans should find comfort in knowing that millions of their fellow citizens are working every day to ensure our security at every level — federal, state, county, municipal. These are dedicated professionals who are good at what they do. I've seen it up close, as Governor of Pennsylvania....But there may be gaps in the system. The job of the Office of Homeland Security will be to identify those gaps and work to close them [Henry, 2002].

It is to shore up the "gaps in the system" that has driven many water/wastewater facilities to increase security. Moreover, USEPA in its *Water Protection Task Force Alert #IV: What Wastewater Utilities Can do*

Now to Guard Against Terrorist and Security Threats (October 24, 2001) made several recommendations to increase security and reduce threats from terrorism. These recommendations include:

1. Guarding against unplanned physical intrusion
    - Lock all doors and set alarms at your office, pumping stations, treatment plants, and vaults, and make it a rule that doors are locked and alarms are set.
    - Limit access to facilities and control access to pumping stations and chemical and fuel storage areas, giving close scrutiny to visitors and contractors.
    - Post guards at treatment plants, and post "Employees Only" signs in restricted areas.
    - Control access to storm sewers.
    - Secure hatches, metering vaults, manholes, and other access points to the sanitary collection system.
    - Increase lighting in parking lots, treatment bays, and other areas with limited staffing.
    - Control access to computer networks and control systems, and change the passwords frequently.
    - Do not leave keys in equipment or vehicles at any time.
2. Making security a priority for employees
    - Conduct background security checks on employees at hiring and periodically thereafter.
    - Develop a security program with written plans and train employees frequently.
    - Ensure all employees are aware of communications protocols with relevant law enforcement, public health, environmental protection, and emergency response organizations.
    - Ensure that employees are fully aware of the importance of vigilance and the seriousness of breaches in security, and make note of unaccompanied strangers on the site and immediately notify designated security officers or local law enforcement agencies.
    - Consider varying the timing of operational procedures if possible so if someone is watching the pattern changes.
    - Upon the dismissal of an employee, change passcodes and make sure keys and access cards are returned.
    - Provide customer service staff with training and checklists of how to handle a threat if it is called in.
3. Coordinating actions for effective emergency response
    - Review existing emergency response plans, and ensure they are current and relevant.
    - Make sure employees have necessary training in emergency operating procedures.
    - Develop clear protocols and chains-of-command for reporting and responding to threats along with relevant emergency, law enforcement, environmental and public health officials, as well as consumers and the media. Practice the emergency protocols regularly.
    - Ensure key utility personnel (both on and off duty) have access to crucial telephone numbers and contact information at all times. Keep the call list up to date.
    - Develop close relationships with local law enforcement agencies, and make sure they know where critical assets are located. Request they add your facilities to their routine rounds.
    - Work with local industries to ensure that their pretreatment facilities are secure.
    - Report to county or state health officials any illness among the employees that might be associated with wastewater contamination.
    - Report criminal threats, suspicious behavior, or attacks on wastewater utilities immediately to law enforcement officials and the relevant field office of the Federal Bureau of Investigation.
4. Investing in security and infrastructure improvements
    - Assess the vulnerability of collection/distribution system, major pumping stations, water/wastewater treatment plants, chemical and fuel storage areas, outfall pipes, and other key infrastructure elements.
    - Assess the vulnerability of the storm water collection system. Determine where large pipes run near or beneath government buildings, banks, commercial districts, or industrial facilities or are contiguous with major communication and transportation networks.

- Move as quickly as possible with the most obvious and cost-effective physical improvements, such as perimeter fences, security lighting, tamper-proofing manhole covers and valve boxes, etc.
- Improve computer system and remote operational security.
- Use local citizen watches.
- Seek financing for more expensive and comprehensive system improvements.

The bottom line on security — Again, when it comes to the security of our nation and even of water/wastewater treatment facilities, few have summed it up better than Governor Ridge.

> Now, obviously, the further removed we get from September 11, I think the natural tendency is to let down our guard. Unfortunately, we cannot do that. The government will continue to do everything we can to find and stop those who seek to harm us. And I believe we owe it to the American people to remind them that they must be vigilant, as well [Henry, 2002].

Technical Management vs. Professional Management

Wastewater treatment management is directed toward providing treatment of incoming raw influent (no matter what the quantity), at the right time, to meet regulatory requirements, and at the right price to meet various requirements.

The techniques of management are manifold both in water resource management and wastewater treatment operations. In water treatment operations, for example, management techniques may include "storage to detain surplus water available at one time of the year for use later, transportation facilities to move water from one place to another, manipulation of the pricing structure for water to reduce demand, use of changes in legal systems to make better use of the supplies available, introduction of techniques to make more water available through watershed management, cloud seeding, desalination of saline or brackish water, or area-wide educational programs to teach conservation or reuse of water" [Mather, 1984]. Many of the management techniques employed in water treatment operations are also employed in wastewater treatment. In addition, wastewater treatment operations employ management techniques that may include upgrading present systems for nutrient removal, reuse of process residuals in an earth-friendly manner, and area-wide educational programs to teach proper domestic and industrial waste disposal practices.

The manager of a wastewater treatment plant, in regards to expertise, must be a well-rounded, highly skilled individual. No one questions the need for incorporation of these highly trained practitioners in the profession — well-versed in the disciplines of sanitary engineering, biology, chemistry, hydrology, environmental science, safety principles, accounting, auditing, technical aspects, and operations. Based on personal experience, however, engineers, biologists, chemists, scientists, and others with no formal management training are often hindered (limited) in their ability to solve the complex management problems currently facing both industries.

There are those who will view this opinion with some disdain. However, in the current environment where privatization, the need to upgrade security, and other pressing concerns are present, skilled "management professionals" are needed to manage and mitigate these problems.

References

A national movement for cleaner cities, *AJPH* March 1930. 20:296–297.

Angele, F. J., Sr. 1974. *Cross Connections and Backflow Protection*, 2nd ed. American Water Association, Denver.

Capra, F. 1982. *Turning Point: Science, Society and the Rising Culture*, Simon and Schuster, New York. p. 30.

Cox, G. W., 1939. Sanitary services of municipalities, *Texas Muncipalities* 26 (August):218.

Henry, K. 2002. New face of security. *Government Security*, pp. 30–37.

Jones, B. D. 1980. *Service Delivery in the City: Citizen Demand and Bureaucratic Rules*, Longman, New York. p. 2.

Mather, J. R. 1984. *Water Resources: Distribution, Use, and Management*, John Wiley & Sons, New York. p. 384.

91

Solid Wastes

Ross E. McKinney
Environmental Consultant

Solid wastes are primarily a problem for urban and industrial areas. Although solid wastes have been a problem for urban centers since the beginning of time, it was the rapid growth of the industrial society that generated the current problems. As the standard of living increased, the consumption of goods produced solid wastes that created environmental pollution when they were discarded without proper treatment. Recently, the industrialized countries have begun to examine solid wastes and to develop methods for reducing their production as well as processing them for return to the environment without creating pollution. Although progress has been made, more is required.

91.1 Regulations

The primary solid waste regulations are at the local government level. One of the primary functions of local government is to protect the health of its citizens. Solid waste collection and processing are primarily a public health problem that can only be managed at the local level. Most communities have developed ordinances to handle solid waste storage, collection, and processing.

Local government derives its authority from state government. Since solid wastes are public health problems, many states have placed their solid waste regulations under state departments of health. In recent years, a number of states have moved control of solid wastes to departments of environmental pollution control or departments of natural resources. The authority for all solid waste regulations resides in each state legislature, with the development of detailed regulations assigned to the appropriate state agency.

In 1965, the federal government became involved in solid waste legislation when Congress passed the first federal solid waste legislation as an addendum to the 1965 Federal Air Pollution legislation. The primary purpose of this federal legislation was to reduce the number of open burning dumps that were creating air pollution and to determine the status of state solid waste programs. The Resource Recovery Act was passed by Congress in 1970 to assist in the development of new systems of processing solid wastes and recovering materials for reuse. This was followed by the Resource Conservation and Recovery Act of 1976 (RCRA). The most important components of this legislation were Subtitle C, which dealt with hazardous solid wastes, and Subtitle D, which dealt with sanitary landfills. Problems with buried hazardous wastes resulted in Congress's passing the Comprehensive Environmental Response, Compensation and Liability Act of 1980 (CERCLA), also known as the Superfund Act. RCRA was amended by Congress

0-8493-1586-7/05/$0.00+$1.50

in 1980 and in 1984. The 1984 amendments were designated the Hazardous and Solid Waste Amendments. In 1986 the Superfund Amendments and Reauthorization Act (SARA) passed Congress. Congress has had recent difficulty passing new legislation, but it has allowed existing legislation to continue, permitting the federal Environmental Protection Agency (EPA) to develop regulations covering the areas of concern. As a net result, it is essential to keep informed on the continuous stream of regulations published in the *Federal Register.*

Overall, federal regulations provide national policies in solid wastes, whereas the state regulations are designed to implement the federal regulations. However, it must be recognized that ultimate control of solid waste operations is at the local level. It is at the local level that federal and state policies and regulations are converted to action. For the most part, regulations are being developed in response to public perception of environmental problems, and the two are not always in a logical sequence.

91.2 Characteristics

One of the more interesting aspects of solid waste management is the simple fact that very little real data exist on current municipal solid waste characteristics. It is too difficult to collect detailed data on solid waste characteristics for even the largest cities. A few detailed studies have been made when the economic considerations for specific solid waste-processing systems were high enough to justify determination of solid waste characteristics. Most engineers use the data generated by the U.S. EPA [EPA, 2002] as the basis for municipal solid waste characteristics. The EPA data represent approximate characteristics based on a material flow methodology developed at Midwest Research Institute in 1969. The EPA composition data are a reasonable estimate on a national basis. It should be recognized that solid waste characteristics vary with the section of the country, the seasons of the year, and economic conditions.

The EPA Office of Solid Waste recognized that solid wastes are composed of many different materials, making composition analyses difficult if a common set of criteria are not used. As a result, a broad classification of nine categories was used. Although several subcategories have been added for specific projects, the nine categories have gained widespread acceptance. Table 91.1 gives the percentage of total solid wastes for the nine categories as estimated by the EPA for the year 2000 [EPA, 2002]. It is estimated that 75% of municipal solid waste is combustible and 25% is noncombustible. The combustibility characteristics of solid wastes are important when solid wastes are to be processed by incineration, whereas the general solid waste characteristics are used for evaluating solid waste recycling and solid waste reduction projects. It should be noted that the EPA solid waste characteristics are based on an "as collected" basis, which includes about 20% moisture. Variations in moisture content are important when evaluating solid waste characteristics for incineration. Care should be taken before using general solid waste characteristics in solid waste treatment facilities design.

TABLE 91.1 Characteristics of Solid Wastes Generated in 2000

Type of Waste	Percentage of Total Solid Waste
Paper and paperboard	37.4
Glass	5.5
Metals	7.8
Plastics	10.7
Rubber, leather, and textiles	6.7
Wood	5.5
Food wastes	11.2
Yard wastes	12.0
Other wastes	3.2

Source: Environmental Protection Agency. 2002. *Municipal Solid Waste in the United States: 2000 Facts and Figures,* EPA/530-R-02-0001. p. 6.

91.3 Generation

Data on solid waste generation are a little better than data on solid waste characteristics, since more communities have gathered data on the weight of solid wastes handled than on solid waste analyses. However, it should be recognized that weight data may not be much better than characterization data. Accurate weight data require checking the scales at regular intervals to ensure accuracy. It is also important to weigh trucks both before and after unloading to determine the real weight of the solid wastes. Use of a truck tare weight to minimize the number of weight measurements will often result in high weight values. The key to proper evaluation of generation data is regular collection and weighing of solid wastes from the same areas of the community.

Generation data are generally reported in pounds per person per day. With a uniform generation rate, it is easy to determine the total solid waste production for various sizes of communities. The EPA data on solid waste generation [EPA, 2002] indicated a change from 2.7 lb/person/day in 1960 to 4.5 lb/person/day in 2000. It has been found that solid waste generation is primarily a function of economic conditions. Municipal solid waste generation decreases during poor economic periods and increases with good economic growth. Concern over the generation of solid wastes has resulted in changes in the packaging of materials. Smaller packages and lighter packaging materials have contributed to the decrease in the weight of packaging wastes. Increased advertising has generated more pages of newspapers and magazines that quickly become outdated. Careful evaluation of social conditions indicates that there is a limit to how much solid waste a person can actually generate on a daily basis. A study for the Institute for Local Self-Reliance [Platt et al., 1990] found that residential solid waste generation for 15 cities ranged from 1.9 to 5.5 lb/person/day with a median of 3.3 lb/person/day. Unfortunately, the economics of some solid waste projects depend upon a reasonable estimation of the solid waste generation rate. The ease of collecting solid waste generation data should allow engineers to obtain sufficient data on municipal solid waste generation for project design.

91.4 Collection

Solid wastes must be collected where they are generated. This means that a system must be developed to go to every building in the community at some time frequency. Currently, residential solid wastes can be conveniently collected once weekly. A few communities employ twice-weekly collection. Commercial buildings may require more frequent collection, depending on the rate of generation and the size of the storage containers. Restaurants often require daily solid waste collection.

Every solid waste generator must have a suitable storage container to hold solid wastes between collections. Residences tend to use individual containers, each having a maximum capacity of 30 gallons, to minimize the weight in any single container. Recently, large plastic containers with wheels have been used in more communities. The householder is able to wheel the solid wastes from the garage to the curb for pickup without difficulty. Special mechanisms on the collection trucks allow the solid waste collector to transfer the solid wastes from the wheeled container to the collection truck with a minimum of effort. Apartment buildings, office buildings, and commercial establishments produce larger quantities of solid wastes than residences and require larger solid waste containers. Storage containers range in size from 1 to 10 yd^3. Small storage containers can be dumped into rear-loading collection trucks or side-loading trucks, whereas large storage containers require front-loading collection trucks.

Solid wastes weigh between 200 and 250 lb/yd^3 in loose containers. Collection trucks use hydraulic compaction to maximize the quantity of solid wastes that can be collected before dumping. Although it is possible to compact solid wastes to 600 to 700 lb/yd^3, most collection trucks average 500 lb/yd^3 after compaction. Rear-loading compaction trucks range in size from 9 to 32 yd^3, with 20 yd^3 trucks being the most commonly used for residential collection. Side-loading compaction trucks range from a small, 6 yd^3, to a large, 35 yd^3. Front-loading compaction trucks range from 22 to 42 yd^3, with 30 and 35 yd^3 trucks being most widely used for commercial routes. Roll-off units are used by industrial and special commercial accounts having large quantities of stable solid wastes. Roll-off units range in size from 10 to 30 yd^3 and may be attached to stationary compaction units to increase the capacity of the roll-off unit. When the roll-off unit is full, the collection truck pulls the roll-off unit onto the truck frame for transport to the processing site. The collection trucks all have special unloading mechanisms to remove the compacted solid wastes at the desired processing site.

Collection crew size depends upon the specific collection pattern of each community. One-person crews are used for some side-loading residential collections, roll-off collections, and some front-loading commercial collections. Two-person crews are used in which one person is the driver and the other person is the collector. The two-person crews are quite common in areas where collection can only be made on one side of the street at a time. Three-person crews are most common in residential collection from both sides of the street. Four-person crews have been used in large communities. The size of collection crews and collection trucks depends on the size of the collection routes and the time allotted for collection activities. Time–motion studies are very important for evaluating crew collection efficiency, whereas good operation and maintenance records are important for evaluating the drivers and determining equipment replacement.

91.5 Transfer and Transport

Transport from the collection route to the solid waste processing site is an important parameter affecting the cost of operations. It is part of the rest time for collectors. If the transport distance is too long, valuable collection time can be lost. As a result, transfer stations have been constructed in large cities to permit smaller collection trucks to unload and return to collection routes. The solid wastes in the transfer station are placed into larger transfer trucks, 50 to 70 yd^3 capacity, for transport to the final processing site. The large transfer trucks normally have only a single driver, reducing labor costs. Transfer stations are economical only in very large cities and require special design for efficient traffic flow. Small transfer stations where individuals bring their solid wastes in their own automobiles or trucks have been used in rural areas.

Recently, large cities have examined transporting solid wastes by rail for final processing. Rail transport is very expensive and has been used only in crowded areas where several communities are contiguous and no processing site can be found in any of the communities. Barge transport has also been used to move large quantities of solid wastes more economically. Transport of solid wastes from one area to other distant areas has created social and political problems that have limited long-distance transport of unprocessed solid wastes.

91.6 Processing and Resource Recovery (Recycling)

Solid waste processing has moved from incineration and open dumps to burial in sanitary landfills to partial recycling. Incineration has been the primary processing system for solid wastes in large cities over the past 100 years. The British pioneered incineration with energy recovery at the end of the last century, but it has been in only the last two decades that many energy recovery incineration systems have been constructed in the U.S. Poor designs and poor operations limited the use of energy recovery incineration even though it was quite successful in Europe. Incineration has application only in the very crowded areas of the country where heat energy can be easily used by adjacent industrial operations. It is important to understand that incineration simply converts combustible solid wastes to gaseous wastes that must be discharged to the atmosphere for final disposal and the noncombustible solid wastes to ash that must be returned to the land. Because of the potential for air pollution, the EPA has required extensive air pollution control equipment for solid waste incinerators, increasing the capital and operating costs. The U.S. Supreme Court has indicated that municipal solid waste ash must pass the hazardous waste criteria or be treated as hazardous waste. The high cost of incineration has limited its use to highly populated areas that produce large quantities of solid wastes on a continuous basis. Even in these areas, social and political pressures have limited the use of incineration for processing solid wastes. New York City has used incineration of municipal solid wastes as extensively as any large city in the United States but has begun phasing out all of its incinerators without replacing them. New York City is planning to use modern waste-to-energy incinerators outside of the city limits.

Sanitary landfills are the engineered burial of solid wastes. Fundamentally, sanitary landfills are sound processing systems for municipal solid wastes. The two most common forms of sanitary landfills are trench landfills and area landfills. Both methods employ heavy equipment to compact the solid wastes to at least 1000 lb/yd^3. Unfortunately, inadequate designs and improper operations have combined to create a poor public image for sanitary landfills. The major problems with sanitary landfills have been (1) the failure to apply adequate soil cover on the compacted solid waste at the end of each day, allowing wind to blow the solid wastes over the landfill surface and odors to escape, and (2) allowing water to enter the landfill, creating leachate and methane gas. Leachate passing through the sanitary landfill has contaminated groundwater as well as surface waters in some locations. Methane gas formation has produced fires in a few landfills, creating concern in the media over the safety of sanitary landfills. The simplicity of sanitary landfills has made them the most popular form of solid waste processing with local government. The failure of local government to operate sanitary landfills properly has led the EPA to develop more complex regulations under Subtitle D of RCRA. Leachate and gas collection systems — as well as water barriers and sampling systems to prove that leachate and gas are not leaving the landfill and creating environmental pollution — will be required in the future. The cost of sanitary landfill processing of solid wastes will increase significantly as the EPA develops more stringent operational requirements. The number of operational sanitary landfills in the United States has decreased from 7924 in 1988 to 1967 in 2000.

Reuse and recycling have long been important parts of the solid waste processing system. Most countries with limited economic bases do not have serious solid waste problems. These countries use and reuse all their resources as a necessity. As countries grow economically, their production of solid wastes increases. In 1970, the United States demonstrated that municipal solid waste recycling was technically feasible. Unfortunately, recycling of solid wastes was not as cost effective as using raw materials and did not gain much acceptance. Environmental groups brought pressure on various industrial groups to begin recycling. The aluminum can industry developed the best recycling system and survived the challenge from environmentalists. The plastic industry recognized the dangers that adverse media exposure posed and began to develop methods for plastic recycling. Glass was easily recycled, but was too heavy to ship very far and was made from an abundant material widely available in nature. Office paper was the easiest paper product to recycle. Corrugated cardboard was also recyclable. Newspapers were abundant but had to be used at a lower product level or de-inked for fiber recovery. The problem with lower product formation was that ultimately the material had to be processed back into the environment. De-inking was expensive and produced water pollution with concentrated solids that had to be processed before being returned to the environment.

The EPA established a Municipal Solid Waste Task Force in 1988 to develop a national agenda for solid wastes [EPA, 1989]. With the help of various environmental organizations they published *The Solid Waste Dilemma: An Agenda for Action.* They called for a national goal of a reduction of 25% in solid wastes by source reduction/recycling by 1992. It was recognized that 17% of the solid wastes was composed of yard wastes that could be removed and composted for reuse. This fact meant that only 8% more solid waste reduction was necessary in the form of metal cans, glass bottles, and newspapers. A number of state legislatures quickly accepted the 25% reduction goal and eliminated the discharge of yard wastes to sanitary landfills. Local communities had to eliminate collection of yard wastes or establish separate collections and composting programs. A few states legislated mandatory recycling. Unfortunately, mandating recycling created a sudden increase in recyclables that overwhelmed the markets, driving the prices down and increasing the costs. This meant that the excess solid wastes were buried in sanitary landfills rather than being recycled. It was quickly learned that even the commodity of solid waste was governed by supply and demand economics. The basic problem was excessive social and political pressures for an instant solution to problems that required time and patience to resolve. During the past decade most communities have established composting projects to remove yard wastes from the municipal solid waste stream. Some communities have established special collections for paper, glass, aluminum and steel cans, and specific plastics. The EPA [2002] estimated that 30.1% of the municipal solid wastes produced in 2000 were "recovered and recycled or composted." Part of the problem is measuring the solid waste reduction. The EPA developed methodology to assist everyone in using the same parameters to measure solid waste production and reduction. The total solid waste stream is the sum of the solid wastes sent to the disposal site and the solid wastes recycled. The fraction of the solid wastes recycled is simply the solid waste recycled divided by the total solid wastes produced. Since the recycled wastes are often processed by private handlers, care must be taken to verify that the recycled solid waste data are correct.

91.7 Final Disposal

There is no doubt that solid waste reduction, reuse, and recycling are the best methods for processing solid wastes. The major issue is the same today as it was in 1970 — economics. Society has not learned how to reduce the production of solid wastes while maintaining an economically viable operation. Solving the problem of solid waste involves determining what materials should not be produced and how to efficiently employ people in meaningful endeavors. Reuse and recycling are more attractive since work is required to reprocess the waste materials into useful products. It should be recognized that all materials come from the earth and most materials are still available in solid wastes. Recycling is the ultimate solid waste process, with wastes being converted back to raw materials for manufacture to new products to benefit all people. The strength of democratic societies lies in their ability to adapt to current conditions with a minimum of wasted effort. Until complete recycling systems are developed that can economically process solid wastes, storage will continue to be the final disposal of solid wastes. Storage may take many forms, including burial below the ground surface or above the ground surface, as in a sanitary landfill. When society needs the buried materials, sanitary landfills will become mines for future generations. Few people recognize that the future of current societies depends on their ability to process solid wastes back into the environment without creating serious pollution problems. The treatment of polluted air and polluted water also results in the production of solid wastes that must be returned to the environment. Survival does not depend on the heads of state or various legislative bodies, but on the action of a few individuals at the local level. Those dedicated individuals who collect and process all the solid wastes that society produces provide the environment that makes life worth living and allow the rest of society to do the things it finds most enjoyable. The future is bright because the solid waste profession has met all the challenges to date.

Defining Terms

Area landfill — A landfill constructed by placing the solid waste against a natural hill, compacting the solid waste to 1000 lb/yd^3, and covering with the minimum of 2 ft soil for final cover. The next cell is constructed by placing the solid waste against the previous cell and raising the land surface to the top of the landfill.

Ash — Residual inorganic material remaining after incineration.

Incineration — Controlled burning of mixed solid wastes at a temperature of 1800°F for a sufficient time to oxidize 99% of the organic matter in the solid waste.

Solid wastes — All solid materials that have no significant economic value to the owner and are discarded as wastes.

Trench landfill — A landfill constructed by digging a trench 10 to 15 ft deep and 20 to 30 ft wide, filling it with solid waste, compacting the solid waste to 1000 lb/yd^3, and covering with a minimum of 2 ft soil for final cover.

Waste recycling — The reuse of solid waste materials after separation and reprocessing to new products.

Waste reduction — The reduction in the production of solid wastes, primarily by reducing the production and consumption of goods by society.

Waste reuse — The direct reuse (without major processing) of solid wastes by members of society.

References

Environmental Protection Agency. 1989. *The Solid Waste Dilemma: An Agenda for Action,* EPA/530-SW-89-019, EPA, Washington, DC.

Environmental Protection Agency. 2002. *Municipal Solid Waste in the United States: 2000 Facts and Figures,* EPA/530-R-02-001, EPA, Washington, DC.

Platt, B., Doherty, C., Broughton, A. C., and Morris, D. 1990. *Beyond 40 Percent: Record-Setting Recycling and Composting Programs,* Institute for Local Reliance, Washington, DC.

Walsh, D. C. 2002. The evolution of refuse incineration, *Environ. Sci. Technol.* 36:316A–322A.

Further Information

Current federal regulations can be obtained from the Office of Solid Wastes, U.S. Environmental Protection Agency, Washington, D.C.

EPA reports are available through the National Technical Information Service (NTIS).

The Institute for Local Reliance is located at 2425 18th Street N.W., Washington, D.C. 20009. It is a nonprofit research and educational organization, providing technical information to local governments and interested citizens. *Environmental Science & Technology* is published monthly by the American Chemical Society, 1155 16th Street, N.W., Washington, D.C. 20036.

92

Hazardous Waste Management

Harold M. Cota
California Polytechnic State University

David Wallenstein
Woodward–Clyde Consultants

Hazardous waste management is a broad and evolving field. Applicable state and federal regulations comprising over 60,000 pages are continually being updated. Many of these regulations overlap and are subject to differences in interpretation that often lead to court rulings. Regulations, economic pressures, and public perception are forcing companies to rapidly change the way they manufacture products in order to minimize hazardous waste generation.

Over 200 million tons of solid hazardous waste are generated annually in the U.S. Huge quantities of hazardous waste deposited in landfills, ponds, fields, and other locations require removal or *in situ* treatment. Common hazardous wastes include solvents, acids, bases, heavy metals, pesticides, plating, and heat-treating wastes. Six major effects of improper hazardous waste management are groundwater contamination, contamination of surface runoff, air pollution, fire and explosion, and adverse health effects via direct contact or via the food chain.

This chapter provides a general overview of federal regulations governing hazardous waste management, as well as a brief review of the types of hazardous waste, waste minimization, and treatment and

0-8493-1586-7/05/$0.00+$1.50

disposal technologies. Four types of hazardous waste will be discussed here: chemical waste, radioactive waste, infectious waste, and mixed waste.

92.1 Regulatory Overview

Federal, state, and local governments regulate hazardous waste management. Federal regulatory agencies sometimes delegate authority to the state agencies provided that the state programs are at least as strict. Only federal regulations are discussed here. With a few exceptions, the Environmental Protection Agency (EPA) governs most federal hazardous waste management activities. The Department of Transportation (DOT) governs the transportation of all hazardous wastes except for high level nuclear waste.

The Nuclear Regulatory Commission (NRC) governs nuclear wastes from the private sector and coregulates mixed wastes with the Department of Energy (DOE). Other agencies, such as the Occupational Safety and Health Administration (OSHA) and the Bureau of Land Reclamation, have responsibilities within the hazardous waste management process. Important legislation is summarized in the following sections.

RCRA

Congress passed the Resource Conservation and Recovery Act (RCRA) in 1976 as an amendment to the Solid Waste Act of 1965. RCRA is composed of 10 parts covering all aspects of hazardous waste management with the exception of nuclear waste. It governs generation, storage, transportation, treatment, recycling, and disposal. The generator of hazardous waste is responsible for the waste through each of these steps and beyond. EPA was authorized by RCRA to establish and implement regulations governing hazardous waste management. This program involved:

1. Defining hazardous waste
2. Tracking the transportation of the waste
3. Establishing construction and operation standards for treatment, storage, and disposal facilities (TSDFs)
4. Establishing guidelines for state hazardous waste management programs

The Medical Waste Tracking Act (MWTA) of 1988 expanded RCRA to include regulating the disposal and treatment of medical waste.

CERCLA

The Comprehensive Environmental Response, Compensation and Liability Act (CERCLA) was passed in 1980 to address the cleanup of toxic waste sites. CERCLA provides for liability, compensation, cleanup, and emergency response to existing and closed hazardous waste sites. The heart of CERCLA is the Superfund, which started in 1980 as a $1.6 billion fund. The Superfund Amendments and Reauthorization Act (SARA) of 1986 revised and broadened CERCLA, creating a $8.5 billion fund to be collected through special taxes. Congress asserted that regardless of who was responsible for causing the contamination, the sites needed remediation in order to protect human health and the environment.

EPA involvement begins with the identification of a potential hazardous waste site. The EPA has developed the Emergency and Remedial Response Information System (CERCLIS) to document all of the sites in the U.S. that may be candidates for remedial action. As of 1993 the list was approaching 30,000 sites. In order to be eligible for cleanup under CERCLA, a site must first be placed on the National Priorities List (NPL). The NPL identifies the hazardous waste sites of immediate concern and ranks them according to their relative risk to public health and the environment. As of August 1990, 1187 sites were on the NPL list identified as "Superfund sites." The EPA addresses the cleanup of each site on a case-by-case basis. Part of EPA's process involves the identification of the potentially responsible parties (PRPs) for:

1. State or federal remedial action
2. Recovery of necessary response cost incurred
3. Damages to natural resources

The PRPs may include past owners or operators, generators of the hazardous substances that are at the site, or the transporters who brought the waste to the site. EPA tries to get the PRPs to initiate cleanup or to pay for the remediation, but if the PRPs cannot, the EPA uses money from the Superfund for the cleanup and tries to recover the monies later. No amount of care can guarantee a person handling hazardous substances protection from liability. The liability cannot be transferred to another party.

Under CERCLA, the EPA can respond to incidents involving hazardous substances in three ways:

1. Immediate removal and emergency response to prevent significant harm (completed within 6 months)
2. Planned removal, which is an expedited response to a situation that is not necessarily an emergency
3. Remedial response, which requires additional time and money intended to achieve a site solution that is a permanent remedy

TSCA

The Toxic Substance Control Act (TSCA) was passed in 1976 to prevent environmentally unsound chemicals such as DDT and PCBs from being released to the environment. TSCA gives the EPA 90 days to investigate the potential deleterious effects of new chemicals on the environment before allowing them to be manufactured. If a chemical poses a significant threat, the EPA may impose special restrictions, from warning labels to an outright ban. The EPA can also require testing of existing chemicals if there is insufficient information to form a risk assessment.

HSWA

Congress passed the Hazardous and Solid Waste Amendments (HSWA) in 1984. These amendments significantly broadened the scope and effectiveness of RCRA. A major theme of the HSWA is the protection of groundwater through the following measures:

- Hazardous waste landfill design requirements
- Requirements for small-quantity generators
- Requirements for underground storage tanks
- Requirements for landfilling of municipal wastes
- Restrictions on future land disposal
- Treatment requirements for waste deposited in landfills

Provisions of the HSWA ban the land disposal of specified types of hazardous wastes unless the wastes are first treated to levels that are protective of human health and the environment. The regulations specifically prohibit the disposal of bulk (noncontainerized) liquid wastes in landfills and surface impoundments. It prohibits disposal of containerized liquids unless no alternative is available. The amendments set out a series of deadlines for ending land disposal of various groups of hazardous wastes.

There are three exceptions to the land ban:

1. The waste meets EPA and state treatment standards.
2. A petitioner demonstrates to the EPA and/or state that the waste in question will not migrate from the disposal unit for as long as the wastes remain hazardous.
3. EPA has issued a variance extending the deadline.

In addition, the Occupational Health and Safety Act of 1970, the Hazardous Materials Transportation Act (HMTA) of 1975, portions of the 1970 Clean Air Act and as amended in 1990, the Clean Water Act (1972), and the Safe Drinking Water Act (1974) also regulate certain aspects of hazardous waste management.

Code of Federal Regulations

Once Congress passes legislation, federal agencies prepare and codify regulations that are published annually in the Code of Federal Regulations (CFR). The CFR has 50 titles. Title 10 contains the Atomic Energy Act and related NRC and DOE regulations, Title 29 the OSHA regulations, Title 40 the EPA regulations, and Title 49 the DOT regulations. The federal government publishes the daily amendments and potential changes to the annual CFR and court interpretations in the *Federal Register (FR).* Several private companies sell computerized versions of the CFR as well as state regulations, including recent updates and legal interpretations. Users can access these databases via modem, floppy disks, or CD-ROM format. The computer databases allow for smaller storage space requirements, provide quicker access to specific regulations, and provide various advanced searching methods.

92.2 Definition of Hazardous Waste

Hazardous waste has been defined by RCRA as a "solid waste, or combination of solid wastes, which because of its quantity, concentration, or physical, chemical, or infectious characteristics may (1) cause, or significantly contribute to, an increase in mortality or an increase in serious irreversible or incapacitating reversible illness or, (2) pose a substantial present or potential hazard to human health or the environment when improperly treated, stored, transported or disposed of or otherwise managed" [EPA, 1990]. Other regulations may define hazardous waste differently, but RCRA sets the standard. The term *solid waste* includes solids and certain liquids or gases.

As shown in Figure 92.1, the RCRA definition has a number of conditions and exceptions that are a result of overlap with other regulations and vestiges of earlier definitions of hazardous waste. Most of these exempted wastes are either regulated by other federal legislation or are regulated on the state and/or local level. The RCRA definition of solid waste does not include radioactive waste because this waste is governed by the Atomic Energy Act. The flowchart illustrates the general philosophy behind defining and classifying hazardous waste. We advise the reader to consult the CFR prior to classifying a waste. As shown on the chart, RCRA classifies a waste as hazardous if the CFR lists it as a hazardous waste or if it has hazardous characteristics.

Listed Wastes

EPA has defined and listed certain wastes as hazardous. A mixture of a listed hazardous waste and a nonhazardous solid waste is a hazardous waste. The "mixture rule" implies that any concentration of listed waste in a material makes it a hazardous waste. In recent years the EPA has allowed a more flexible interpretation that classifies certain wastes based on characteristic hazardous components and concentration. Furthermore, one may be able to de-list a listed waste by following procedures in 40 CFR 260. RCRA regulates unrinsed containers, inner linings, and any residue or soil contamination from hazardous waste as hazardous waste.

Nonlisted Wastes

If the waste is neither listed nor specifically exempted, the generator must make a judgment based on the hazardous characteristics of the waste. In some cases the generator can apply process knowledge of the waste or past experience to classify the waste as hazardous; otherwise, the waste must be tested at an approved laboratory. If the waste has any of the following four characteristics, it is hazardous:

1. Reactivity (coded D003)
   - Normally unstable
   - Ignites on contact with water
   - Releases toxic gases in contact with water

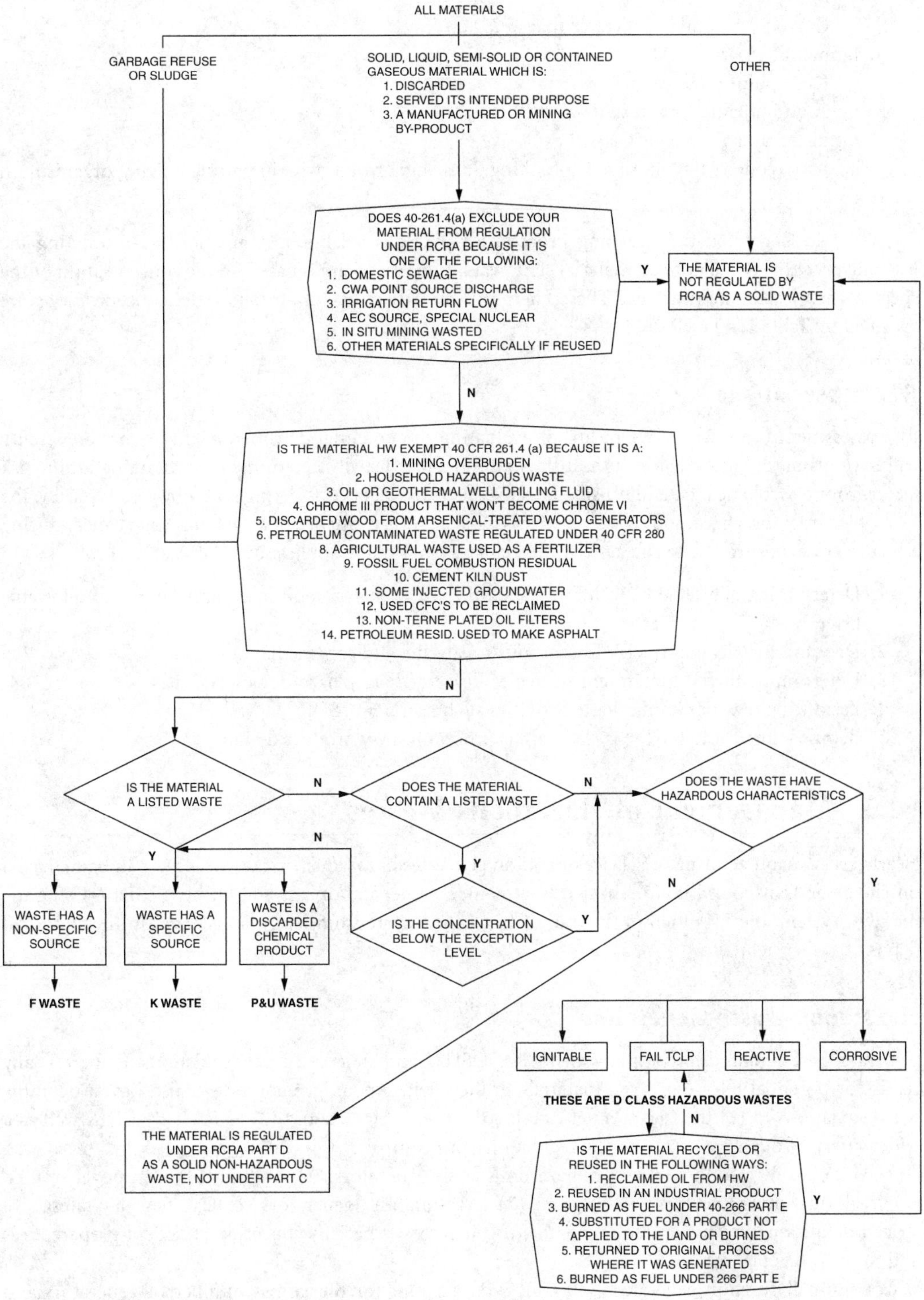

FIGURE 92.1 Definitions of hazardous wastes.

2. Corrosivity (coded D002)
   - Aqueous pH < 2 or > 12.5
   - Corrodes SAE steel at greater than 0.25 in./year
3. Ignitability (coded D003)
   - Flash point < 60°C
   - Solid ignitable compressed gas oxidizer
   - Listed as a Class A or Class B explosive
4. Toxicity (coded D###); capable of killing, injuring, or otherwise impairing a living organism on contact

The Toxic Characteristics Leaching Procedure (TCLP) is a laboratory method for estimating the leaching potential of a landfilled material. The evaluation of toxicity is based on the concentration of the specific chemical in the leachate. These chemicals and their maximum allowable concentrations are specified by Table 1, 40 CFR 261.

Risk Assessment

Risk assessment is a quantitative means of evaluating the potential detrimental effects that an activity will have primarily on human health. Risk is a probability usually determined by obtaining a unit risk factor from toxicological data and using effective exposure data considering all pathways. Typically, the allowable target for a hazardous waste activity is a risk of $1 \cdot 10^{-6}$ chance of a person getting cancer during a lifetime of exposure due to the activity. Risk analysis has many applications, for example:

1. Determining allowable concentrations in the TCLP analysis and in concentrations of allowable listed wastes in solid wastes
2. Determining cleanup levels of hazardous waste sites under remediation
3. Determining the best treatment strategies for remediating hazardous waste sites
4. Permitting new hazardous waste activities such as a new TSDF
5. Being factored into the life cycle analysis process in new product design

92.3 Management of Hazardous Wastes

Nearly every manufacturing or service operation generates hazardous waste. This waste can be managed on-site or be transported to a permitted treatment, storage, and/or disposal facility (TSDF). Using the manifest system, the EPA follows the trail of hazardous waste from its initial generation through final deposit or treatment.

Hazardous Waste Generators

Generators are required to comply with RCRA, OSHA, and state and local regulations. If there is any release or potential release of toxic materials to the environment, generators will also have to comply with regulations in (1) the Clean Water Act, (2) the Clean Air Act, and (3) SARA Title III, as well as a wide variety of other industry-specific regulations and controls.

The EPA has defined three classes of hazardous waste generators: (1) conditionally exempt generators, (2) small-quantity generators (SQGs), and (3) large-quantity generators (LQGs). Several states have stricter definitions and tighter requirements for generators, especially concerning accident preparedness and contingency plans.

Conditionally exempt generators (40 CFR 261) account for the largest number of generators and produce the least quantity of hazardous waste — less than 100 kg/month of hazardous waste or less than 1 kg/month of acutely hazardous waste — or store less than 1000 kg of waste on site. Although RCRA exempts them from most of the extensive permitting requirements, it still requires them to (1) identify their hazardous waste, and (2) treat hazardous waste on-site or transport it to a permitted TSDF or recycling facility.

SQGs generate less than 1000 kg/month or less than 1 kg/month of acutely hazardous waste or store less than 6000 kg of waste on site. RCRA allows SQGs to store wastes up to 180 days or 270 days if the nearest TSDF is more than 200 miles away. SQGs must obtain a 12-digit EPA identification number, follow proper waste transportation and storage requirements, and perform proper record keeping and reporting, including tracking of manifests.

LQGs are those that generate more than 1000 kg/month, or 1 kg/month of acutely toxic waste, or store more than 6000 kg of waste on site. RCRA requires these generators to comply with all requirements in 40 CFR 262–265 and related regulations outside RCRA. LQGs do not require a transfer storage and disposal facility (TSDF) permit if they (1) store only their wastes on-site (do not accept wastes from outside the facility), (2) store all wastes in a designated area, (3) store the wastes in proper containers properly marked as hazardous waste, and (4) comply with the requirements for owners or operators in subparts C (preparedness and prevention) and D (Contingency Plan and Emergency Procedures) in 40 CFR 265.

Storage of Hazardous Waste

Proper storage of hazardous waste provides for appropriate separation of wastes, separation of noncompatibles, easy access to the waste, reasonable protection from vandalism, and appropriate secondary containment. These practices result in the reduction of the incidence and severity of accidents, simplified compliance, reduced/eliminated fines, and more economical uses of space and management time. Hazardous wastes are stored in a variety of approved containers including drums, overpack containers, tanks, and impoundments.

Underground Storage Tanks

There are approximately 1.5 to 2 million underground storage tanks (USTs) in the U.S. USTs are regulated by the EPA through a mixture of statutory and regulatory provisions or by the states to which the authority has been delegated. In many instances, inconsistent federal, state, and local laws have resulted in conflicts regarding the designation of "responsible parties." RCRA Subtitle I regulates tanks that contain a regulated Superfund substance or a liquid petroleum product with at least 10% of its volume underground (including piping connected to the tank). The statute requires tank owners to notify the state that they possess such tanks and directs the EPA to establish detection, prevention, release-detection requirements, and performance standards. Subtitle I does not cover approximately two-thirds of tanks, because either (1) other regulations cover these tanks, or (2) they do not pose a threat to human health and the environment. These tanks include

- Tanks holding 110 gallons or less
- Farm and residential tanks holding 1100 gallons or less of motor fuel used for noncommercial purposes
- Tanks storing heating oil burned on the premises where it is stored
- Tanks on or above the floor of underground areas such as basements or tunnels
- Septic tanks and systems for collecting stormwater and wastewater
- Flow through process tanks
- Emergency spill and overflow tanks
- Surface impoundments and pits
- Tanks storing hazardous waste (RCRA Subtitle C)

Approximately 200,000 USTs are leaking or have leaked at one time. The four major causes of releases are corrosion, faulty installation, piping failure, and overfilling. The majority of older tanks are bare steel and, therefore, galvanic corrosion is the most common fault. Federal regulations issued September 13, 1988 require that new UST systems be properly installed according to industry codes, equipped with spill and overfill prevention devices, protected from corrosion, and equipped with leak detection devices. Tanks no longer in use must be decommissioned either by removal or by filling with an inert material.

Transportation of Hazardous Waste

The DOT regulates all aspects of the transportation, packaging, labeling, marking, and placarding of hazardous waste under the Hazardous Materials Tracing Act of 1975 (HMTA), 49 CFR 172–179. The HMTA was reauthorized in 1993 and governs all hazardous material shipments.

The DOT has five divisions that govern the transportation of hazardous waste/materials:

1. Coast Guard
2. Federal Aviation Administration
3. Federal Highway Administration
4. Federal Railroad Administration
5. Research and Special Programs Administration

Under RCRA, the EPA is directed to establish certain standards for transporters of hazardous materials and to coordinate transportation regulatory activities with the DOT. The transportation of hazardous waste may also be regulated on the local level. State and local governments may require shippers and carriers to provide information about the types of materials they handle, origins and destinations of shipments, routes followed, miles covered in a given year, proof of insurance coverage, vehicle inspection dates, and drivers employed. Some states require hazardous waste registration, special training and certification, prenotifications of shipments, and periodic summaries of activities. Local authorities may also restrict the routes that hazardous waste transporters take.

Generator Requirements

The generator is required to ship wastes in the proper containers and accurately mark and label them (40 CFR 262). In addition, the generator is responsible for providing the transporter with the proper placards per 49 CFR 172 F. The placards alert the first responders to a spill of hazardous materials. The placards are DOT approved and include the hazard symbol, ID number, and UN hazard class number. The placards are placed on the ends and sides of motor vehicles, rail cars, and freight containers. Each container must be labeled with a hazardous waste label, which includes the generator's name and address, the proper shipping name, DOT hazard class, EPA hazard class, and waste type. Other appropriate labels are required, such as "flammable" or "corrosive." Containers should be labeled so that the contents of a package can be identified if it is separated from its shipping papers.

Waste Manifests

The uniform hazardous waste manifest provides the mechanism for "cradle-to-grave" tracking of hazardous wastes. All hazardous waste shipments, except those from conditionally exempt SQGs, must be accompanied by this form. States with EPA authorization can print their own version, and they may have additional requirements. Part 262 D sets forth generator reporting and record-keeping requirements; hazardous waste import and export requirements are covered in subparts E and F, respectively. When shipping interstate, the shipper should use the receiving state's manifest. A copy of the manifest is kept by the generator, the transporter, and the owner of the receiving facility, and one copy is sent to the state. A second copy is returned to the generator upon arrival at the TSDF within 45 days of shipment. The manifest must be kept for 3 years or longer. SQGs are not required to submit copies of their manifests to the state, but LQGs are. The generator must keep a log of the manifest numbers used.

The manifest requires information about the generator, waste type, DOT shipping description (49 CFR 173.2), waste quantity alternate acceptance facility, spill cleanup precautions, TSDF authorization numbers, billing information, listing of hazardous materials and non-RCRA-regulated wastes, location of generator, hazardous waste numbers, physical state of waste, hazard code(s), and intended treatment or disposal method in code. It also requires the generator and transporter to certify that the wastes are in proper condition for transportation.

92.4 Hazardous Waste Treatment

> Treatment means any method, technique, or process, designed to change the physical, chemical, or biological character or composition of any hazardous waste so as to neutralize such waste, or so as to recover energy or material resources from the waste, or so as to render such waste non-hazardous, or less hazardous; safer to transport, store, or dispose of; or amenable for recovery, amenable for storage, or reduced in volume. (40 CFR 260)

A facility that performs the above is required to obtain a RCRA Part B permit or operate under interim status until a permit is obtained. There are certain exceptions to these requirements, as specified in 40 CFR 265.

Treatment Options

Table 92.1 summarizes many of the hazardous waste treatment technologies available today. Many of these technologies have been adapted from other manufacturing fields, such as chemical, mining, plating, water and wastewater treatment, and concrete manufacture.

The four basic classifications of treatment technologies used here are physical, biological, chemical, and thermal. Physical treatments such as air stripping, sedimentation, and adsorption separate the hazardous constituent from the waste, resulting in a reduced volume of hazardous waste. Other physical treatments such as compaction and shredding reduce the size of the waste but not the total mass. Biological treatments such as bioreactors and composting result in the biological degradation of the hazardous constituents in a waste stream. Chemical treatments such as fixation, sterilization, and chemical oxidation destroy or reduce the toxicity or mobility of the contaminant by breaking and/or making chemical bonds in the hazardous constituents of the waste. Thermal treatments such as incineration, microwave treatment, and vitrification destroy the contaminants or reduce their mobility through intense heat.

The cost information provided in Table 92.1 is intended to serve as a general guideline only. The costs are from published project costs, vendor estimates, and EPA compiled reports, and have been estimated in terms of 1994 dollars. The cost of any hazardous waste treatment method is highly dependent on the specific contaminant, concentration, quantity of waste to be treated, treatment goals, regulatory fees, and restrictions, as well as a myriad of other performance goals.

Selecting a Suitable Waste Treatment Facility

The generator is ultimately responsible for the waste if the transporter or TSDF fails to handle the waste properly. It is therefore very important to exert adequate effort in investigating a TSDF. The following are some items to check out prior to shipping wastes anywhere.

Facilities need a RCRA Part B permit or interim status permit. One should obtain copies of all permits: state, local, and air and water emissions and check to see that the application or permits cover the appropriate waste types and processes. In addition, check for

- Notices of violations and court actions
- Plant security, alarms, and emergency response
- Safety practices and written procedures
- Spill containment and groundwater monitoring
- Laboratory certification
- Personnel qualifications and experience
- Housekeeping and waste-handling practices
- Financial stability — ability to pay for closure and postclosure monitoring, insurance; ability to compensate injured parties and indemnify clients against loss

TSDFs are usually very specific in the type of waste that they can accept. Generators may rely on the services of a waste broker to arrange for treatment or disposal. The broker may also arrange vendor

TABLE 92.1 Summary of Hazardous Waste Treatment Technologies

Technology	Contaminant	Type	Requirements	Products	Cost[a]
Adsorption	Organic contaminants	G, WW, GW, OW	Adsorbents such as activated carbon added to water and filter out, a cost highly dependent on concentration	Treated streamaste carbon; carbon can absorb up to 55 organics	Carbon is $2 to 3/lb
Aerobic bioreactor	Readily degradable toxic organics	GW, OW	Tank with air injection and biological population	Inerts, biomass, and reduced BOD waters	200/ton
Air stripping	Petroleum, hydrocarbons, chlorinated solvents	W	Vacuum air stream	Gaseous or condensed contaminants and water	Variable
Anaerobic bioreactors	More recalcitrant and higher-concentration organics	W, WW	Anaerobic tank and biological population and possibly methane injection	Inerts, biomass, acids, and reduced-BOD water	200/ton
Biological land treatment, composting, heaping	Petroleum and wood-treating wastes	S, SOL, SL	Open land and mixing equipment	Lesser-contaminated materials	50/ton
Chelating	Heavy metals and radionuclides	S	Tank or *in situ* and chelating agents	Treated soil and liquid wastes	N/A
Chemical oxidation	Wide range of compounds include chlorinated HCs, cyanides, mercaptans, and phenols	SS, L, LW	Chemical oxidants include ozone, chlorine, hydrogen peroxide, and UV radiation	Inerts and intermediate products of oxidation	200/ton
Compaction, grinding, shredding, baling, section, acid digestion	Biological and/or radioactive hazard	SOL	Appropriate equipment	Reduced volume of original waste but not reduced hazard or mass	On-site
Encapsulation or microencapsulation	Toxic leachable compounds and radioisotopes	L, OW, S, SED	Organophilic polymers and binders	Encapsulated landfillable or incinerable waste	100 to 200/ton
Evaporation/distillation	Wide variety of aqueous or solvent-based contaminants including radionuclides	L, WW	Evaporation ponds or distillation equipment	Still bottoms, purified liquid	10 to 100/ton
Gas/vapor sterilization	Biological hazard	SOL	Formaldehyde or ethylene oxide in vessel, ground solids	Disinfected wastes: May still have hazardous components	On-site
Incineration: thermal oxidizers of many types	Wide variety of organics, infectious waste, and nuclear wastes	G, S, SL, SED, SOL	Controlled high-temperature burner, fuel, air and emissions control	Inerts, ash, trace air pollutants	500 to 1000/ton
Ion exchange	Some metals, cyanide, and other cations/anions including radioisotopes	W, WW	Appropriate resins	Treated water, contaminated resin	20/ton
Irradiation	Biological hazard	SOL	Ultraviolet radiation, dry environment	Disinfected waste: May still have hazardous components	On-site
Magnetic separation	Metals, radionuclides, and nitrates	WW, GAS	Magnetic core, selective absorbers, organic polymers	Treated water to PPM level	N/A, high capital
Membrane processes: electrodialysis reverse osmosis, and ultrafiltration	Extreme toxics not removed by other processes; works on solutes	W	Specialized membranes and high-pressure pumps or direct electric current	Contaminated brine water and clean water	20/ton
Microwave treatment	Biological hazard	SOL	Microwave radiation and moisture used to heat waste, shredded	Disinfected waste: May still have hazardous components	On-site

Neutralization	Contaminants with pH <2 or >12.5	L, WW, S	Mixing equipment and lime, soda ash, or free acids	Large volume or neutral material	100 to 200/ton
Physical treatments including screening, filtering, centrifugation	Pretreatment to remove large nonhazardous compounds including radionuclides	L, WW, S	Tanks, loading and unloading equipment	Clean or treated water requiring further treatment and contaminated sludge	50/ton
Physical treatments including precipitation flocculation/coagulation, gas floatation, and clarification	Dissolved and suspended solids and some solutes usually inorganic; may include radionuclides	GW, WW	Settling tanks and flocculants such as alum; polymers and sludge removal equipment may require air supply	Clean or treated water requiring further treatment and concentrated contaminated sludge	150/ton
Pyrolysis/molten salt baths	Wide variety of toxic organics	L	Controlled nonoxidizing burner	Cracked and thermally broken down organics suitable for fuel, further treatment, or disposal	N/A
Slurry bioreactors	Heavy and recalcitrant organics	S, SL	Tanks or lagoons, mixers and surfactants	Biomass, inerts, intermediates, and wet treated soil	200/ton
Soil vapor extraction	Petroleum, hydrocarbons, chlorinated solvents	S	Vapor extraction equipment	Gaseous contaminants and clean soil	50 to 100/ton
Solidification	Heavy metals, some organics and leachable liquids, and radioisotopes	S, SL, L	Binders, asphalts, hydraulic cements, ureaformaldehyde	Better structural materials, nonleaching suitable for landfilling or leaving in place	30 to 80/ton
Stabilization/fixation	Heavy metals, acid sludges, some organics	S, SOL	Pozzolanic reagents and lime silicates	Nonleaching landfillable materials or remediated sites	30 to 80
Steam sterilization/autoclaving	Small-size solid biological hazard	SOL	High-temperature steam and oven	Disinfected waste: May still have hazardous components	On-site
Steam stripping	Volatile and semi-volatile compounds	W	High-pressure steam and vacuuextraction	High-concentration stream and clear water	Variable
Stream stripping, *ex situ* and *in situ*	Volatile organics	S	High-pressure stream and vacuum extraction	Hot water with contaminants	32AZ80 200/cubic yard
Supercritical fluid extraction	Extractable organics	S, SL, W	High-pressure and -temperature fluids	Recovered organics and oxidized organics	160/ton
Thermal desorption	Petroleum, hydrocarbons, chlorinated solvents	S, SL	Thermal desorption ovens, natural gas or electric heaters, air pollution controls	Gaseous contaminants and sterilized soil	50 to 100/ton
Thermal inactivation	Biological hazard	SOL, L	High-temperature oven or vessel	Disinfected wastes: May still have hazardous components	On-site
Vitrification	Nonspecific, asbestos, and radioactive waste	S, SED, SOL	High-temperature and/or high-electric current air emissions control	Glassified wastes suitable for disposal or leave in place	N/A

Note: G = gaseous waste; GW = contaminated groundwater; HC = hydrocarbons; L = liquid wastes; OW = oily water; S = contaminated soil; SED = sediments; SL = process sludges; SOL = waste solids; W = contaminated water; WW = industrial waste water.

[a] In 1994 U.S. dollars

contracts, make arrangements for waste acceptance, and hire a transporter. The generator is ultimately responsible for the decisions the broker makes, and it is imperative that these selections be reviewed by a responsible member of the company.

Waste Disposal

The land disposal of hazardous waste is a highly discouraged practice in the U.S. However, it is inevitable that some wastes be disposed of because not enough treatment capacity or complete treatment technologies are available for all the wastes generated. Table 92.2 summarizes the various disposal technologies currently available.

Remediation

Cleaning up hazardous waste sites involves the implementation of the CERCLA regulations. The cleanup process involves investigation, feasibility studies, risk assessment, record of decision, design of remediation systems, construction and implementation, and long-term monitoring. This topic is covered in more detail in another chapter.

92.5 Infectious Waste Management

Infectious wastes are biological wastes potentially contaminated with disease-producing organisms. They can pose a hazard to personnel who handle their disposal as well as to the general public. The EPA regulates medical waste management under RCRA Subtitle J, promulgated in the Medical Waste Tracking Act (MWTA) of 1988. RCRA defines regulated medical waste as "any solid waste generated in the diagnosis, treatment, or immunization of human beings or animals, in research pertaining thereto, or in production or testing of biologicals" (40 CFR 259). The MWTA directs the EPA to establish a demonstration program for managing and tracking medical waste. Much of the regulation is controlled by state and local government and varies widely from state to state. The EPA has seven recommended infectious waste categories:

1. Isolation wastes (wastes from patients with communicable diseases)
2. Cultures and stocks of infectious agents and associated biologicals
3. Human blood and blood products
4. Pathological wastes
5. Contaminated sharps
6. Contaminated animal carcasses, body parts, and bedding
7. Unused sharps

Storage of infectious wastes should consider the integrity of the packaging, storage temperature, the duration of storage, and the location and design of the storage area. In selecting the appropriate packaging, the waste type and handling and treatment methods should be taken into consideration. Protection from sharps is another important factor. The biohazard symbol must be prominent.

The EPA now requires that medical waste be sterilized prior to disposal to kill all pathogens in the wastes. Not all states enforce this yet for landfills. Incineration and steam sterilization are the most frequently used treatments; however, other processes are effective. There are approximately 6700 on-site incinerators at hospitals throughout the country, and approximately 60% of waste is treated on-site. If no longer hazardous after treatment, the material may be disposed of at a sanitary landfill or in the sewer if locally approved.

Biological indicators, which consist of a resistant strain of bacteria, test the effectiveness of treatment technologies. The indicators are placed in the waste prior to treatment, and their destruction indicates the success of the treatment. Table 92.2 includes several infectious waste treatment options.

TABLE 92.2 Summary of Hazardous Waste Disposal Options

Disposal	Description	Waste Type	Advantages	Limitations
Above-grade vaults	Engineered large steel and concrete structure above natural grade	LLW, segregated	Easy access to waste	Vulnerable to intrusion and erosion
Augured holes	Shallow drilled holes above water table	Class C low level	Tested by DOE, good worker protection	Difficult to remove wastes, small capacity
Below-grade vaults	Engineered structure below grade designed to be compatible with soil	LLW, segregated	Less vulnerable to intrusion and erosion	Waste less accessible
Carlsbad, NM, waste isolation pilot plant	DOE project designed to hold 6,000,000 cubic feet in salt formation	Transuranic defense wastes	Remote, already in pilot-plant stage	Controversial
Deep space	Canisters placed in indefinite orbit or into sun	HLW	Studied by NASA and DOE	
Deep well injection	Acid waste pumped to great depths into fractured rock	HLW and transuranic below critical level	Waste far removed from environment	Waste nonrecoverable
Earth-mounded concrete bunkers	Combined above- and below-grade disposal of structurally strong waste containers	LLW, HLW	Lower cost	Vulnerable to intrusion and erosion
Hanford mixed waste trenched	427,000 cubic feet trenches for storage of mixed wastes from cleanup of Hanford site	Mixed wastes	First and only mixed waste disposal	
Ice sheet	Canisters dropped through ice meltdown	Self-heating HLW	Remote locations, heat removal	Conceptual only, potential ice movement
Improved shallow land disposal	Waste placed in deeper trenches with impermeable cap	LLW	Required since 1982, not allowed in some states	More costly and complex than older technology
Injection wells classes I, IV, V	Injection into porous rock formations not communicating with class I groundwater	HW, petroleum wastes, sewage, LLW	Widely practiced, secure	Potential drinking water contamination, subject to land ban
Landfill	Waste placed in subsurface cells with liners, cover, leachate, and groundwater collection monitoring and venting	Hazardous and medical wastes pretreated or not banned	Simply proven	Heavily taxed and discouraged, ultimate liability, land ban
Mined cavities	Hollowed-out regions in mineral deposits	LLW	No intruder accessibility	Hard to access
Modular concrete containers	Use of an overpack or "mini vault" to encase drums	LLW	Waste can be easily removed	
Ocean disposal	Concrete encased in drums at ocean bottom	LLW	Cheap, arguably low risk	Now banned, some drums have leaked

TABLE 92.2 Summary of Hazardous Waste Disposal Options (*Continued*)

Disposal	Description	Waste Type	Advantages	Limitations
Ocean plate disposal	Canisters of waste between continental shelf and moving plates through holes drilled	HLW and transuranic	Geologic stability	Conceptual, not good for some isotopes
Salt mines	Liquid wastes placed into hollowed mines	Nonspecific	Cheaper, inaccessible	Nonrecoverable
Shale grout injection	High-pressure water cracks shale and then waste in cement, slurry is pumped into formation	HLW, trans	Waste relatively immobile, far away from surface	Waste nonrecoverable
Shallow land disposal	Waste placed in unlimited earthen trenches up to 30 feet deep	LLW	Cheap and easy	Now banned
Surface impoundment	Placement of wastes into impoundments open to atmosphere	Liquid hazardous waste, uranium mine tailings	Reduces volume, low technology	Essentially short-term storage, historically problematic
Yucca Mountain facility	Underground repository for waste in canisters will open approximately 2010	Nuclear spent fuel, defense wastes	Retrievable waste, very remote, highly regulated	Very expensive, capacity limited to 70,000 tons

Note: HLW = high-level radioactive waste; HW = hazardous waste; LLW = low-level radioactive waste. Class A waste = LLW not requiring stabilization, short half-life; Class B waste = LLW requiring stabilization but will decay within 100 years; Class C waste = LLW requiring stabilization and shilding during handling, longer half-life.

92.6 Radioactive Waste Management

Sources

Radioactive wastes come from a wide variety of sources, including nuclear power plants, government reactors, private industry, defense, and medical activities. Nuclear power waste sources include mine tailings, enrichment, fuel fabrication, reactor operation, spent nuclear fuel, and decommissioning of reactors. Industrial sources include production processes, research and manufacturing of radiopharmaceutical compounds, smoke detectors, and watch dials. Defense wastes include waste from manufacture of nuclear weapons and operation of nuclear-powered vessels. The volume of defense waste generated is 10 times that of commercial waste. Medical sources include wastes from medical diagnosis, treatment, and research.

There are three classifications of nuclear waste: high level (HLW), transuranic, and low level (LLW). HLWs contain materials that were exposed to high-level nuclear reactions or contain spent fuel from nuclear reactions. High-level wastes currently exist at federal government facilities such as the Hanford Reservation Center in Washington, the Idaho National Engineering Laboratory, the former Western New York Nuclear Service Center in West Valley, New York, and Savannah River. Transuranic wastes are compounds having more protons than uranium, for example, plutonium. These wastes are primarily produced by the reprocessing of spent fuels and by processing weapons-grade material. LLW can be produced from any action involving radioactive material. The largest source is from mine tailings of uranium and thorium mines. Other sources mainly consist of contaminated clothing, equipment, structural material, cleanup materials, and a large amount of potentially contaminated material — such as packing and paper — and activated material.

Government Involvement

The Atomic Energy Act (AEA) of 1954 was the first legislation to include controlling radioactive waste management. As of 1986, 18 federal laws governed radioactive waste management programs. These laws are enforced by the DOE, NRC, EPA, DOT, the Department of the Interior, various smaller agencies, and state-mandated programs. Congress determined LLW to be a state issue. The Nuclear Waste Policy Act of 1982 established a framework for developing and maintaining a system for HLW management and disposal in the U.S. As part of this act the U.S. Treasury established a nuclear waste fund to receive payments for disposal fees. The DOE, through the Office of Civilian Radioactive Waste Management (OCRWM), develops geological repositories in compliance with 10 CFR 60.

The NRC reviews nuclear safety, siting, construction, and operation of repositories for the commercial sector. It also inspects the activities of the OCRWM. The NRC is responsible for regulating the receipt, possession, and the use and transfer of radioactive materials to protect public health and safety and the common defense and security of the U.S.

CFR Title 10 contains the NRC regulations. Part 20 provides the basic standards for protection from radiation. Other technical requirements are spelled out in 29 CFR 1926; OSHA "Safety and Health Requirements for Construction"; the ASME code for pressure and piping; the American Concrete Institute "Code Requirements for Nuclear Safety Related Concrete Structures" (ACI 349–80); and relevant parts of the Uniform Building Code, the Uniform Mechanical Code, and the National Electric Code.

The EPA is responsible for developing environmental standards to protect the health and safety of the public and the environment from potential hazards in the management and disposal of nuclear wastes.

Health Concerns with Radiation Exposure

Radiation exposure may cause severe acute health effects and chronic carcinogenic, teratogenic, or mutagenic effects. The principle of *as low as reasonably achievable* (ALARA) is the basis for the management of nuclear waste originally established in the Atomic Energy Act of 1954. The hazardous nature of a waste stream is measured by a number of factors:

1. Activity level expressed as nanocuries/gram (nCi/g) of waste
2. The type of radioactive emission: alpha, beta, gamma
3. The half-life of the waste isotopes
4. The relative physical stability of the waste
5. The mobility, leachability, or in some cases volatility of the waste

Activity is measured in units of curies. One curie (Ci) is equivalent to the disintegration rate of one gram of radium or $3.7 \cdot 10^{-10}$ disintegrations/sec. A rad is the dose of ionizing radiation that produces energy absorption of 100 erg/g in any medium. The *relative biological effectiveness* (RBE) is used as a measure of the specific biological damage caused by different types of radiation. It is used to calculate the rem (roentgen-equivalent, man) dose unit, which is determined by multiplying the absorbed dose in rad by the RBE.

The EPA sets standards for exposure to radioactive materials in the environment: "the combined annual dose equivalent to any member of the public in the general environment resulting from: (1) Discharges of radioactive material and direct and indirect radiation from such management and storage and (2) all operations covered by part 190 [the uranium fuel cycle] shall not exceed 25 millirems to the whole body, 75 millirems to the thyroid, and 25 millirems to any other critical organ" (40 CFR 191). Waste management facilities must meet these criteria with "reasonable expectation" for 1000 years and provide a "reasonable expectation" that releases will be limited for 10,000 years (40 CFR 191.16).

Disposal/Storage

Nuclear wastes cannot be detoxified or the hazards reduced like those of other waste through traditional treatment methods. The relative hazards of nuclear waste decrease over time such that proper storage of wastes over time can render the wastes nonhazardous. In many cases the disposal of nuclear waste is a storage problem. The treatment that is performed on the wastes usually involves either concentrating the wastes to reduce volume or stabilizing the physical form of the wastes to reduce the mobility and simplify handling of the waste. Special precautions must be taken to avoid exceeding the "criticality" of the fissile material waste that could cause a runaway nuclear reaction.

Table 92.1 and Table 92.2 summarize applicable treatment and disposal technologies for radioactive wastes. HLW and transuranic waste present a health risk.

Disposal facilities for HLW and transuranic wastes must be on federally owned land and are regulated solely by the Atomic Energy Act regulations in 10 CFR 60. The design concept for the disposal of HLW and transuranic wastes is to place the appropriate canisters in a deep geologic depository. The DOE has collected $10 billion since 1988 in taxes to cover the cost of building storage facilities. Construction is scheduled to begin in the year 2010; however, the DOE has not yet cited an interim storage permit. This design concept includes a temporary storage facility known as a *monitored retrievable storage* (MRS). As listed in Table 92.2, numerous other technologies have been suggested; however, there is no current plan of further development of these alternative technologies.

Management of LLW

The states have responsibility to provide for the safe management of LLW generated within their borders. States may license the LLW facility themselves or the NRC licenses them. In addition to radiological protection, the states must regulate other aspects of the management of the facilities. Certain requirements for LLW facilities are provided in 10 CFR 61.

As of June 1994, only two sites were open for LLW disposal: Barnwell County, South Carolina, and Hanford, Washington. These two sites, however, accept waste from a limited number of locations. Other disposal sites have submitted license applications, but the federal government under environmental pressure has been slow to approve them. The lack of permanent disposal facilities for LLW has caused the thousands of generators throughout the country to store the wastes at their own sites and may cause those generators without adequate storage capacity to shut down.

States have the right to regulate radioactive wastes that are exempt from NRC enforcement. The NRC exempts wastes that:

1. Cause < 10 millirem exposure/year/person
2. Cause < 1 millirem of exposure to a large group/year
3. Guarantee that no person receives > 100 millirem/year from all activities at the licensed facility

Transportation

The DOT primarily governs transportation of radioactive materials as specified in 49 CFR 173–178. These regulations specify performance criteria for packaging, package activity limits, driver-training requirements, and routing restrictions. The NRC regulates the receipt, possession, use, and transfer of byproducts and sources of special nuclear materials.

The NRC also sets standards for the design and performance of packages used to transport high-level radioactive materials and conducts inspection of its licenses. Other NRC regulations require the advance notification to the states of certain shipments and provide for physical security measures.

92.7 Mixed Wastes

Mixed waste (MW) is both hazardous and radioactive. Since mixed waste is considered hazardous under RCRA and radioactive under the Atomic Energy Act, both the NRC and the EPA work to address the management of the wastes. It is generated by pharmaceutical and biomedical research laboratories, universities and research laboratories, nuclear reactors, commercial facilities, analytical laboratories, DOE/DOD facilities, new TSDFs, and old disposal sites.

DOE estimates that it generated 22,000 cubic meters of mixed wastes in 1990 and has an inventory of 107,000 cubic meters of MW from past operations. It also estimates that they will generate 600,000 cubic meters remediating their sites. In contrast, commercial mixed waste is generated at a rate of 140,000 cubic feet per year; however, 70% of commercial mixed waste is incinerated and some materials are pretreated, thereby further reducing the volume.

The generation and characterization of mixed wastes will likely increase as cleanup of CERCLA sites proceeds. Although land disposal restrictions require treatment of mixed wastes, there is very little treatment or disposal capacity for mixed wastes in the U.S. A new mixed waste facility is scheduled to be opened in Hanford, Washington (see Table 92.2); however, it is only designed for mixed waste from on-site remediation activities. Consequently, most mixed waste is stored either on-site or at designated facilities. Storing of mixed wastes can be problematic because RCRA may not allow storing of hazardous waste under a facilities permit. Therefore, the EPA enacted a policy to relax enforcement of storage regulations of certain mixed waste generators. Most generators are not equipped to store mixed wastes, which can pose a health and safety problem for workers.

92.8 Corrective Action

Corrective action refers to the investigation and remediation of releases of hazardous constituents from facilities that have or are seeking hazardous waste facility permits or are operating under interim status. RCRA Section 3004(u) requires that any permits issued after 8 November 1984 require corrective action for all releases of hazardous waste or hazardous constituents from any solid waste management unit at the facility. Corrective action beyond the facility boundary must be performed where necessary to protect human health and the environment, as part of ongoing permit conditions.

Under RCRA, the term *release* is broader than as defined under CERCLA. The RCRA classification includes releases that are exempt under CERCLA and covered by the Clean Air Act and NPDES permits. The EPA broadly defines the term *release* to include a release of any amount to soil, surface water, groundwater, or air of any hazardous waste or hazardous material listed in Appendix VIII of 40 CFR 261.

Once there has been a release, the corrective action process begins in five steps:

1. An RCRA facility assessment is made, which is a desktop review of all information available, a site inspection, and optional sampling.
2. A corrective action order is given as specified in permit conditions.
3. An RCRA facility investigation is performed to characterize the extent of contamination resulting from the spill, including risk assessment and contaminant movement modeling. This investigation includes feasibility studies of treatment technologies.
4. Interim measures are taken to address releases of immediate concern.
5. A corrective measure study sets forth the proposed remediation alternatives. This study is required if the contamination found exceeds an "action level."

Cleanup levels required under corrective action are discussed in 40 CFR 254. These levels are subject to interpretation based on health risk assessments.

92.9 Waste Minimization

Waste minimization entails modifying a process in order to reduce the volume or the toxicity of the waste generated prior to any treatment, storage, or disposal. It includes source reduction and environmentally sound recycling; equipment and process changes; product reformulation; raw material substitution; and improvements in inventory control and housekeeping, maintenance, and training.

Waste generators can achieve significant savings by reducing raw material consumption, waste disposal, or treatment fees and by taking advantage of tax credits given for waste minimization efforts. Federal law explicitly states that waste reduction is the preferred antipollution method. The 1984 amendments to RCRA promulgated a new federal policy that, wherever feasible, the generation of hazardous waste is to be reduced or eliminated as expediently as possible. The EPA ranks management options as follows:

1. Waste reduction
2. Waste separation and concentration
3. Waste exchange and recycling off-site
4. Energy/material recovery
5. Incineration/treatment
6. Secure land disposal

The EPA published guidance on hazardous waste minimization in the 1993 publication *Hazardous Waste Minimization: Interim Final Guidance for Generators* (EPA/530-F-93-009). In this document, the EPA lists six basic elements that generators should incorporate into their waste minimization plan:

1. Top management support
2. Characterization of waste generation/management costs
3. Periodic waste minimization assessments
4. Cost allocation
5. Encouragement of technology transfer
6. Program implementation and evaluation

Life Cycle Design

Environmental groups have criticized waste minimization in the past for being too narrowly focused and sometimes resulting in process modifications that only shift pollutants from one medium to another. The solution to this problem is *life cycle design* (LCD). LCD is a method of product and process design that incorporates the minimization of waste production throughout every step of design and every step of the life of the product. This "cradle-to-grave" design method considers raw material extraction, product manufacturing, use, and final residual disposal. LCD establishes a more coherent means of integrating environmental requirements with more traditional concerns in product development, such as performance, cost, and cultural and legal requirements.

Life cycle analysis (LCA) is one of the more promising systematic approaches for identifying and evaluating opportunities to improve the environmental performance of industrial activity. It is a useful tool for evaluating the environmental consequences of a product. LCA can be used in conjunction with LCD or be employed separately. By focusing on source reduction as well as reuse, recovery, and treatment, LCA gives a more accurate portrayal of the environmental impacts and true costs of a project. The EPA has put together a life cycle design manual as a guide for conducting and interpreting life cycle inventories.

92.10 Right-to-Know Laws

SARA Title III and the OSHA Hazard Communication Standard of 1985 established a regulatory program that requires disclosure of information to workers and the general public about the potential dangers of hazardous chemicals. These standards also require development of emergency response plans for chemical emergencies. Congress enacted these laws in response to the more than 2000 deaths caused by the release of a toxic chemical in Bhopal, India in 1984. The Clean Air Act amendments of 1990 placed additional responsibilities on the EPA and OSHA to enact legislation requiring further sharing of risk information and process hazard analysis. States and municipalities have passed additional right-to-know laws in response to specific companies' failures to disclose information about their process chemicals. Title III also requires the reporting of annual releases (42 USC 9061) and that facilities that release extremely hazardous chemicals over threshold amounts must immediately notify the community emergency coordinator and the state commission. The EPA now requires monitoring for 320 chemicals. A proposed new rule could double the requirements to 633 monitored chemicals.

The majority of state right-to-know laws address both community and employee access to information about workplace hazards. The requirements of right-to-know laws most relevant to hazardous material planning and emergency response include (1) providing public access to information on hazardous materials present, (2) conducting inventories or surveys, (3) establishing record-keeping and exposure-reporting systems, and (4) complying with container-labeling regulations for workplaces.

OSHA now requires chemical manufacturers and importers to prepare *material safety data sheets* (MSDSs) for all hazardous materials produced or used. The MSDSs must be available to a state agency, local fire chief, and the public as part of their community right-to-know programs.

The format of the MSDS form, also known as OSHA Form 20, is up to the provider. It must contain at least nine elements, as follows:

1. Manufacturer's name, address, and telephone number; chemical name, synonyms, and formula
2. Hazardous ingredients, approximate concentration, and threshold limit value (TLV)
3. Physical data of product
4. Fire and explosion hazard data for product
5. Health hazard data
6. Reactivity data
7. Spill or leak procedures
8. Special protective information
9. Special precautions

92.11 Computer Usage in Hazardous Waste Management

The use of computers has become increasingly important in hazardous waste management. Computers provide updates to current regulations, prepare reports and manifests, perform complex modeling, perform risk assessments and LCAs, serve as design tools, track waste disposal, manage inventory, and provide various other convenience services. Approximately 1500 software programs are available for pollution management activities. Several periodicals and trade magazines review the available software and provide annual summaries of the packages offered.

Regulatory agencies will often produce their own software, which industry uses to do reporting. These software and databases are either on-line or available by computer disk. These systems include databases, training and procedure review packages, pollution-modeling programs, and emission calculation programs. One purpose for the software is to standardize calculations. Several regulatory agencies accept quarterly reports and release report information on computer disk. Computers provide direct access to environmental databases and provide on-line reports to the EPA and other regulatory bodies. Databases are available either on floppy disks, CD-ROM, or through modem access. These databases provide chemical and toxicological data, current environmental regulations and interpretations, and abstracts of published research articles.

Defining Terms

Code of Federal Regulations (CFR) — Contains detailed information on federal regulations including those dealing with the generation, treatment, reuse, storage, and disposal of hazardous waste.

Hazardous wastes — Wastes that may cause or contribute to mortality or may pose a threat to human health. More detailed definitions are given in regulations such as RCRA.

Mixed wastes — Wastes that contain both hazardous waste and radioactive waste.

Resource Conservation and Recovery Act (RCRA) — Initially passed by Congress in 1976, the original act and its amendments cover aspects of hazardous waste management.

Risk assessment — Involves determining potential health effects due to a specific chemical and potential human exposure. This leads to an estimation of the risk of developing an effect such as cancer due to a lifelong exposure.

References

Berlin, R. E. and Stanton, C. C. 1992. *Radioactive Waste Management,* John Wiley & Sons, New York.

Bureau of National Affairs, various handbooks and CD ROM databases, Washington, DC. (800) 372-1033.

Environmental Protection Agency. 1992. *Risk Assessment Guidance for Superfund, Vol I — Human Health Evaluation Manual,* EPA/540/1-89/002. U.S. EPA, Washington, DC.

Environmental Protection Agency. 1990. *RCRA Orientation Manual, 1990 Edition,* EPA/530-SW-90–036. U.S. EPA, Washington, DC.

Keolein, G. A. and Menerey, D. 1994. Sustainable development by design-review of life cycle design and related approaches. *J. Air Waste Management Assoc.* 44:645.

LaGrega, M. D. and Buckingham, P. L. 1994. *Hazardous Waste Management,* McGraw-Hill, New York.

Landrum, V. J. 1991. *Medical Waste Management and Disposal,* Noyes Data Corporation, Park Ridge, NJ.

Lindgren, G. F. 1983. *Guide to Managing Industrial Hazardous Waste,* Butterworth, Woburn, MA.

Phifer, R. W. 1988. *Handbook of Hazardous Waste Management for Small Generators,* Lewis, Chelsea, MI.

Plater, Z. J. B. 1992. *Environmental Law and Policy,* West, St. Paul, MN.

Wentz, C. A. 1992. *Hazardous Waste Management,* McGraw-Hill, New York.

Further Information

Professional Organizations

Air and Waste Management Association
American Academy of Environmental Engineers
American Chemical Society
American Institute of Chemical Engineering
American Society of Civil Engineers
Water Environment Federation

Journals

Journal of Air and Waste Management Association
Environmental Science and Technology
Environmental Progress
Journal of Water Environment Federation
Journal of American Industrial Hygiene Association
Journal of Environmental Engineering
Journal of Environmental Permitting
Journal of Environmental Regulation
Environmental Impact Assessment Review
Pollution Prevention
Pollution Engineering
Chemical Engineering
National Environmental Journal
The Generator's Journal
Hazardous Materials
Waste Business Magazine
Hazmat World

Conferences

Hazmat Conferences
Hazmacon

93

Soil Remediation

Ronald C. Sims
Utah State University

J. Karl C. Nieman
Utah State University

93.1 Regulations

Federal and analogous state programs that address soil remediation include the Comprehensive Environmental Response, Compensation, and Liability Act (CERCLA, or Superfund) and the Resource Conservation and Recovery Act (RCRA). CERCLA, enacted in 1980, was the first comprehensive federal law addressing releases of hazardous substances into the environment. The primary goal of CERCLA was to establish a mechanism to respond to releases of hazardous substances at abandoned or uncontrolled hazardous waste sites that posed a threat to human health and the environment. The Superfund Amendments and Reauthorization Act of 1986 (SARA), Section 121, Cleanup Standards, stipulates that the U.S. Environmental Protection Agency (U.S. EPA) is required to select remedial actions involving treatment that "permanently and significantly reduce the volume, toxicity, or mobility of hazardous substances, pollutants, and contaminants." Remedial actions must also meet all applicable or relevant and appropriate requirements (ARARs). Applicable requirements refer to standards, requirements, criteria, or limitations that specifically address a hazardous substance or waste-contaminated soil. Relevant and appropriate requirements refer to those cleanup standards that address problems or situations sufficiently similar to those encountered at the site so that their use is suitable. Under CERCLA, a revision of the National Contingency Plan (NCP) in 1985 addressed the need for rapid characterization of site risk and evaluation of remedial actions using a remedial investigation and feasibility study prior to selection of an action. For contaminated soil, the RI/FS process involves characterization of the extent of soil contamination at a site, determination of goals for cleanup, and evaluation of alternatives for reaching those goals. The

0-8493-1586-7/05/$0.00+$1.50

remedial design (RD) occurs after the selection of the remedial action(s). Federal regulatory agencies require the potentially responsible parties (PRPs) to evaluate and remediate contaminated soil, as the Superfund Law provided for both a response and an enforcement mechanism with financial resources originating from the PRPs or from the Superfund trust fund created by a tax on oil and chemical companies [Sims, 1990; Sellers, 1999; Shah, 2000].

The Resource Conservation and Recovery Act (RCRA) of 1976 addresses the treatment, storage, and disposal (TSD) of hazardous wastes. The Hazardous and Solid Waste Amendments (HSWA) were enacted in 1984 and included the land disposal restrictions (LDRs), or land ban. LDRs prohibit placement of hazardous waste on or in the soil for treatment without prior treatment [Sellers, 1999]. RCRA Corrective Action (RCRA-CA) represents the U.S. EPA's program to address the investigation and remediation of contamination at or from hazardous waste treatment, storage, or disposal facility. RCRA-CA involves the conduct of an RCRA facility assessment (RFA), and investigation (RFI), followed by a corrective measures study (CMS) that is followed by corrective measures implementation (CMI). A focus on results, rather than process, was the message of the May 1, 1996 Advance Notice of Proposed Rulemaking (ANPR) for the RCRA Corrective Action program. The results-based approach promulgated includes inviting innovative technical approaches for soil and site characterization and treatment. The U.S. EPA believes that there are approximately 6400 facilities subject to RCRA-CA [U.S. EPA, 2000].

A third program addressing contaminated soil, among other media, was initiated by the U.S. EPA in 1995 and is titled the Brownfield Economic Redevelopment Plan [Sellers, 1999; Shah, 2000]. Brownfields are parcels of land that have some contamination and are not developed because of perceived or real environmental contamination. The Brownfield Plan was developed to address concerns with regard to the multiple agency regulations, liability issues, and economic disincentives [Shah, 2000]. Reuse and revitalization of previously contaminated land for economic development is a high priority for the U.S. EPA. Both CERCLA and RCRA-CA sites can become involved in the Brownfields initiative, where appropriate [U.S. EPA, 2000]. As part of this initiative, 27,000 sites have been removed from the Superfund list and the U.S. EPA will not pursue Superfund action at these sites [Shah, 2000].

Therefore contaminated soil characterization and remediation may be a part of the CERCLA, RCRA-CA, and/or Brownfield regulatory processes, and the following technologies are applicable to all three regulatory programs. The following sections identify and describe current soil remediation technologies and processes for different classes of chemical contaminants and for different site conditions. Cost information is also included where available.

93.2 Mechanical/Chemical Treatments

Mechanical and chemical soil treatments rely on energy-intensive processes to physically remove, contain, or destroy soil contaminants. They include soil vapor extraction (SVE), solidification/stabilization, incineration, thermal desorption, chemical oxidation/reduction, containment, soil flushing, soil washing, and chemical extraction. In general, they are more costly than biological treatments, but usually have the advantage of shorter treatment times. Unless otherwise noted, the following technology descriptions are based on excerpts from the Federal Remediation Technologies Roundtable (FRTR) Remediation Technologies Screening Matrix, 4th edition [FRTR, 2002]

93.3 Soil Vapor Extraction

Soil vapor extraction (SVE) is an *in situ* unsaturated (vadose) zone soil remediation technology in which a vacuum is applied to the soil to induce the controlled flow of air and remove volatile and some semivolatile contaminants from the soil. The gas leaving the soil may be treated to recover or destroy the contaminants, depending on local and state air discharge regulations. Vertical extraction vents are typically used at depths of 1.5 meters (5 feet) or greater and have been successfully applied as deep as 91 meters (300 feet). Horizontal extraction vents (installed in trenches or horizontal borings) can be used as warranted by contaminant zone geometry, drill rig access, or other site-specific factors. Geomembrane covers are often placed over the soil surface to prevent short-circuiting and to increase the radius of influence of the wells.

SVE is applicable to sites that have high soil permeability and are contaminated with volatile organic compounds (**VOCs**). The vapor pressure of the contaminants should be greater than about 0.5 torr for effective performance. SVE is not appropriate for semivolatile compounds such as **PCBs** and many pesticides. Drier soils are easier to clean than soils with high moisture content, but the process has been applied to partially saturated soils. The duration of operation and maintenance for *in situ* SVE is typically medium to long term.

93.4 Solidification/Stabilization

Solidification/stabilization (S/S) reduces the mobility of hazardous substances and contaminants in the environment through both physical and chemical means. Unlike other remedial technologies, S/S seeks to trap or immobilize contaminants within their "host" medium (i.e., the soil, sand, and/or building materials that contain them) instead of removing them through chemical or physical treatment. Leachability testing is typically performed to measure the immobilization of contaminants. S/S techniques can be used alone or combined with other treatment and disposal methods to yield a product or material suitable for land disposal or, in other cases, that can be applied to beneficial use.

Solidification/stabilization techniques include auger/caisson systems and injector head systems that apply S/S agents to soils to trap or immobilize contaminants. Implementation of this technology is highly dependent on the physical properties of soil. Another S/S technology is vitrification that can be applied either *in situ* or *ex situ*. Vitrification uses an electric current to melt soil or other earthen materials at extremely high temperatures (1600 to 2000°C or 2900 to 3650°F) and thereby immobilize most inorganics and destroy organic pollutants by pyrolysis. Radionuclides and heavy metals are retained within the molten soil. The timeframe for *in situ* S/S is short to medium term, while the *in situ* vitrification (ISV) process is typically short term.

93.5 Incineration

Incineration involves the use of high temperatures, 870 to 1200°C (1600 to 2200°F), to volatilize and combust (in the presence of oxygen) halogenated and other refractory organics in hazardous wastes. Often, auxiliary fuels are employed to initiate and sustain combustion. The destruction and removal efficiency (DRE) for properly operated incinerators exceeds the 99.99% requirement for hazardous waste and can be operated to meet the 99.9999% requirement for PCBs and dioxins. Off gases and combustion residuals generally require treatment.

Types of incinerators include circulating bed combustors (CBC), fluidized beds, infrared combustion, and rotary kilns. Incineration, primarily off-site, has been selected or used as the remedial action at more than 150 Superfund sites. Incineration is subject to a series of technology-specific regulations, including the following federal requirements: CAA (air emissions), TSCA (PCB treatment and disposal), RCRA (hazardous waste generation, treatment, storage, and disposal), NPDES (discharge to surface waters), and NCA (noise). The duration of incineration technology ranges from short to long term.

93.6 Thermal Desorption

Thermal desorption is a physical separation process and is not designed to destroy organics. Excavated wastes are heated to volatilize water and organic contaminants. A carrier gas or vacuum system transports volatilized water and organics to the gas treatment system. The bed temperatures and residence times designed into these systems will volatilize selected contaminants but will typically not oxidize them. Contaminated soils are excavated and then transported to a thermal desorption facility or processed on-site in a mobile thermal desorption unit.

Two common thermal desorption designs are the rotary dryer and thermal screw. Rotary dryers are horizontal cylinders that can be indirect or direct fired. The dryer is normally inclined and rotated. For the thermal screw units, screw conveyors or hollow augers are used to transport the medium through an enclosed trough. All thermal desorption systems require treatment of the off-gas to remove particulates and contaminants.

Based on the operating temperature of the desorber, thermal desorption processes can be categorized into two groups: high temperature thermal desorption (HTTD, 320 to 560°C [600 to 1000°F]) and low temperature thermal desorption (LTTD, 90 to 320°C [200 to 600°F]). HTTD is frequently used in combination with incineration, solidification/stabilization, or dechlorination, depending upon site-specific conditions. The technology has proven it can produce a final contaminant concentration level below 5 mg/kg for the target contaminants identified. LTTD is a full-scale technology that has been proven successful for remediating petroleum hydrocarbon contamination in all types of soil. Contaminant destruction efficiencies in the afterburners of these units are greater than 95%. The same equipment could probably meet stricter requirements with minor modifications, if necessary. Decontaminated soil retains its physical properties. Unless heated to the higher end of the LTTD temperature range, organic components in the soil are not damaged, which enables treated soil to retain the ability to support future biological activity.

93.7 Chemical Oxidation/Reduction

Oxidation chemically converts hazardous contaminants to nonhazardous or less toxic compounds that are more stable, less mobile, and/or inert. The chemical oxidants most commonly employed to date include peroxide, ozone, and permanganate (traditional oxidants including hypochlorites, chlorine, and chlorine dioxide may also be utilized). Oxidants have been able to cause the rapid and complete chemical destruction of many toxic organic chemicals; other organics are amenable to partial degradation as an aid to subsequent bioremediation. In general, the oxidants have been capable of achieving high treatment efficiencies (e.g., >90%) for unsaturated aliphatic (e.g., trichloroethylene [TCE]) and aromatic compounds (e.g., benzene), with very fast reaction rates (90% destruction in minutes). Matching the oxidant and *in situ* delivery system to the contaminants of concern and the site conditions is the key to successful implementation and achieving performance goals.

Oxidation using liquid hydrogen peroxide (H_2O_2) in the presence of native or supplemental ferrous iron (Fe^{+2}) produces Fenton's Reagent, which yields free hydroxyl radicals. These strong, nonspecific oxidants can rapidly degrade a variety of organic compounds. Fenton's Reagent oxidation is most effective under very acidic pH (e.g., pH 2 to 4) and becomes ineffective under moderate to strongly alkaline conditions. The reactions are extremely rapid and follow second-order kinetics.

Ozone gas can oxidize contaminants directly or through the formation of hydroxyl radicals. Like peroxide, ozone reactions are most effective in systems with acidic pH. The oxidation reaction proceeds with extremely fast, pseudo-first-order kinetics. Due to ozone's high reactivity and instability, O_3 is produced on-site, and it requires closely spaced delivery points (e.g., air sparging wells). *In situ* decomposition of the ozone can lead to beneficial oxygenation and biostimulation.

The reaction stoichiometry of permanganate (typically provided as liquid or solid $KMnO_4$, but also available in Na, Ca, or Mg salts) in natural systems is complex. Due to its multiple valence states and mineral forms, Mn can participate in numerous reactions. The reactions proceed at a somewhat slower rate than the previous two reactions, according to second-order kinetics. Depending on pH, the reaction can include destruction by direct electron transfer or free radical advanced oxidation — permanganate reactions are effective over a pH range of 3.5 to 12.

93.8 Landfilling and Containment

Excavation and off-site disposal is a well proven and readily implementable technology. Prior to 1984, excavation and off-site disposal was the most common method for cleaning up hazardous waste sites.

CERCLA includes a statutory preference for treatment of contaminants, and excavation and off-site disposal is now less acceptable than in the past. The disposal of hazardous wastes is governed by RCRA (40 CFR Parts 261-265), and the U.S. Department of Transportation (DOT) regulates the transport of hazardous materials (49 CFR Parts 172–179, 49 CFR Part 1387, and DOT-E 8876). Some pretreatment of the contaminated media usually is required in order to meet land disposal restrictions.

On-site containment involves installation of impermeable barriers around the contaminated region. Examples of barriers are trenches backfilled with a bentonite slurry (slurry walls) or lined with a synthetic geomembrane. The top of the contaminated site is usually capped with a clay layer. In some cases, the contaminated soil must be excavated and the empty hole lined with a barrier material before the contaminated soil is backfilled into the hole.

93.9 Soil Flushing

In situ soil flushing is the extraction of contaminants from the soil with water or other suitable aqueous solutions. Soil flushing is accomplished by passing the extraction fluid through in-place soils using an injection or infiltration process. Extraction fluids must be recovered from the underlying aquifer and, when possible, they are recycled. Types of extraction fluids include solvents (e.g., ethanol) and surfactants. Solvent addition is known as cosolvent flushing. Soil flushing can be applied to soils to dissolve either the source of contamination or the contaminant plume emanating from it. The flushing mixture is normally injected upgradient of the contaminated area, and the dissolved contaminants are extracted downgradient and treated above ground. Treatment of the extractant may be required to separate or destroy the contaminants prior to reinjection or release to a publicly owned treatment works. Also, residual flushing additives in the soil may be a concern and should be evaluated on a site-specific basis. The duration of soil flushing process is generally short to medium term.

93.10 Soil Washing

Soil washing is similar to soil flushing but takes place in a controlled reactor after the contaminated soil has been excavated and often involves particle size separation. This *ex situ* soil separation process, mostly based on mineral processing techniques, is widely used in Northern Europe and America for the treatment of contaminated soil. The process removes contaminants from soils by dissolving or suspending them in the wash solution (which can be sustained by chemical manipulation of pH for a period of time) or by concentrating them into a smaller volume of soil through particle size separation, gravity separation, and attrition scrubbing (similar to those techniques used in sand and gravel operations).

The concept of reducing soil contamination through the use of particle size separation is based on the finding that most organic and inorganic contaminants tend to bind, either chemically or physically, to clay, silt, and organic soil particles. The silt and clay, in turn, are attached to sand and gravel particles by physical processes, primarily compaction and adhesion. Washing processes that separate the fine (small) clay and silt particles from the coarser sand and gravel soil particles effectively separate and concentrate the contaminants into a smaller volume of soil that can be further treated or disposed of. Complex mixtures of contaminants in the soil (such as a mixture of metals, nonvolatile organics, and SVOC s) and heterogeneous contaminant compositions throughout the soil mixture make it difficult to formulate a single suitable washing solution. For these cases, sequential washing, using different wash formulations and/or different soil to wash fluid ratios, may be required. The duration of soil washing is typically short to medium term.

93.11 Chemical Extraction

Chemical extraction does not destroy wastes but is a means of separating hazardous contaminants from soils, sludges, and sediments, thereby reducing the volume of the hazardous waste that must be treated.

The technology uses an extracting chemical and differs from soil washing, which generally uses water or water with wash-improving additives. Two basic types of chemical extraction are acid extraction (for heavy metals) and solvent extraction (for organic contaminants). Physical separation steps are often used before chemical extraction to grade the soil into coarse and fine fractions, with the assumption that the fines contain most of the contamination. Physical separation can also enhance the kinetics of extraction by separating out particulate heavy metals, if these are present in the soil. The duration of operations and maintenance for chemical extraction is medium term.

93.12 Biological Treatments

As the ability of indigenous microorganisms to degrade many contaminants has been explored, biological treatments as engineering solutions at contaminated sites have become more prevalent. Many of these technologies rely on minimal site manipulation and in many cases may be less expensive than more intensive treatment processes. Most biological processes are considered to be long-term technologies, and extended analytical and administrative costs must be assessed.

93.13 Monitored Natural Attenuation

Natural subsurface processes such as dilution, volatilization, biodegradation, adsorption, and chemical reactions with subsurface materials are allowed to reduce contaminant concentrations to acceptable levels. Biological processes are often the primary mechanisms involved with organic contaminants. The term "monitored natural attenuation" (MNA) has been adopted to indicate that a remedial action is taking place at the site and that the solution is not merely perceived as a "no action" decision.

Consideration of the MNA option usually requires modeling and evaluation of contaminant degradation rates and pathways and predicting contaminant concentration in treated soils and at down gradient receptor points, especially when aqueous contaminant plumes are still expanding/migrating. The primary objective of site modeling is to demonstrate that natural degradation processes will reduce contaminant concentrations below regulatory standards or risk-based levels before potential exposure pathways are completed. In addition, long-term monitoring must be conducted throughout the process to confirm that degradation is proceeding at rates consistent with meeting cleanup objectives.

Compared with other remediation technologies, natural attenuation has the following advantages:

- Less generation or transfer of remediation wastes
- Less intrusive as few surface structures are required
- May be applied to all or part of a given site, depending on site conditions and cleanup objectives
- May be used in conjunction with, or as a follow-up to, other (active) remedial measures
- Overall cost likely lower than with active remediation

93.14 Bioremediation

Enhanced bioremediation is a process in which microorganisms (e.g., fungi, bacteria, and other microbes) degrade (metabolize) organic contaminants found in soil and/or groundwater, converting them to innocuous end products. Most often, these microorganisms are indigenous to the contaminated soils, but inoculation of acclimated microorganisms is also practiced. Nutrients, oxygen, or other amendments may be used to enhance bioremediation and contaminant desorption from subsurface materials.

The bioremediation of contaminated soils often involves the delivery of oxygen through tillage (landfarming) or venting (bioventing, see Section 93.15 below) to provide aerobic microorganisms an electron acceptor for the oxidation of organic contaminants. Excavated soils can also be placed into "biopiles" with oxygen distribution systems or composted with amendments that provide improved soil structure or act as substrates for fungal growth. Anaerobic degradation processes can also be encouraged through the addition of easily degradable carbon sources that cause subsurface oxygen depletion and force alternate electron acceptors (e.g., nitrate and sulfate) to be utilized.

Although soil microorganisms with the ability to degrade many petroleum hydrocarbons have been shown to be ubiquitous, each site must be evaluated on an individual basis. Contaminant toxicity or extremely low contaminant concentrations (or bioavailability) may slow the bioremediation process. As with other technologies, laboratory- or field-scale feasibility studies are recommended to determine applicability. Enhanced bioremediation may be classified as a medium- to long-term technology.

93.15 Bioventing

Bioventing is a well-studied *in situ* method of bioremediaiton that stimulates the natural biodegradation of any aerobically degradable compounds in soil by providing oxygen to existing soil microorganisms. In contrast to soil vapor extraction, bioventing uses low air flow rates to provide only enough oxygen to sustain microbial activity. Oxygen is most commonly supplied through direct air injection into residual contamination in soil. In addition to degradation of adsorbed fuel residuals, volatile compounds are biodegraded as vapors move slowly through biologically active soil.

The U.S. Air Force (USAF) has produced a technical memorandum that summarizes the results of bioventing treatability studies of fuels conducted at 145 USAF sites [AFCEE, 1996]. The memorandum discusses overall study results and presents cost and performance data and lessons learned. Regulatory acceptance of this technology has been obtained in 30 states and in all 10 EPA regions, and the use of this technology in the private sector is growing rapidly following USAF leadership. Bioventing is a medium- to long-term technology. Cleanup ranges from a few months to several years.

93.16 Phytoremediation

Phytoremediation is a process that uses plants to remove, transfer, stabilize, and destroy contaminants in soil and sediment. The mechanisms of phytoremediation include enhanced rhizosphere biodegradation, phytoextraction (also called phyt-accumulation), phytodegradation, and phytostabilization. Enhanced rhizosphere biodegradation takes place in the soil immediately surrounding plant roots. Natural substances released by plant roots supply nutrients to microorganisms, which enhances their biological activities. Plant roots also loosen the soil and then die, leaving paths for transport of water and aeration. This process tends to pull water to the surface zone and dry the lower saturated zones. Phytoaccumulation is the uptake of contaminants by plant roots and the translocation/accumulation (phytoextraction) of contaminants into plant shoots and leaves.

Phytodegradation is the metabolism of contaminants within plant tissues. Plants produce enzymes, such as dehalogenase and oxygenase, that help catalyze degradation. Phytostabilization is the phenomenon of production of chemical compounds by plants to immobilize contaminants at the interface of roots and soil.

The most commonly used flora in phytoremediation projects are poplar trees, primarily because the trees are fast-growing and can survive in a broad range of climates. In addition, poplar trees can draw large amounts of water (relative to other plant species) as it passes through soil or directly from an aquifer. This may draw greater amounts of dissolved pollutants from contaminated media and reduce the amount of water that may pass through soil or an aquifer, thereby reducing the amount of contaminant flushed though or out of the soil or aquifer.

Phytoremediation may be applicable for the remediation of metals, pesticides, solvents, explosives, crude oil, PAH s, and landfill leachates. Some plant species have the ability to store metals in their roots. They can be transplanted to sites to filter metals from wastewater. As the roots become saturated with metal contaminants, they can be harvested. Hyper-accumulator plants may be able to remove and store significant amounts of metallic contaminants. Phytoremediation is considered a long-term treatment technology.

Defining Terms

Ex situ — A process conducted on excavated soil.

In situ — A process conducted in place, without excavation of soil.

PCBs — Polychlorinated biphenyls, dielectric compounds previously used in high-voltage electrical equipment such as transformers and capacitors.

Remedial action — Specific steps taken to decontaminate a site.

Surfactants — Surface active chemicals, typically found in soaps, that assist solubilization of nonpolar organic materials in water.

SVOCs — Semivolatile organic compounds, which slightly volatilize at room temperature.

VOCs — Volatile organic compounds, which readily volatilize at room temperature.

TABLE 93.1 Acronyms Used in Soil Remediation

ARAR	Applicable or relevant and appropriate requirements
CERCLA	Comprehensive Environmental Response Compensation and Liability Act
CMI	Corrective Measures Implementation (RCRA-CA)
CMS	Corrective Measures Study (RCRA-CA)
HRS	Hazardous ranking system (EPA's system for ranking the priority of cleanup sites)
LDR	Land Disposal Restrictions (HSWA 1984)
MCL	Maximum contaminant level, enforceable standards under the Safe Drinking Water Act
NCP	National Contingency Plan
NPL	National Priority List, a list of sites eligible for federal cleanup
PRPs	Potentially responsible parties, who may be held liable for cleanup costs pursuant to CERCLA
RA	Remedial action
RCRA	Resource Conservation and Recovery Act
RD	Remedial design
RFA	Remedial Facility Assessment (RCRA-CA)
RFI	Remedial Facility Investigation (RCRA-CA)
RI/FS	Remedial investigation/feasibility study
ROD	Record of decisions
SARA	Superfund Amendment Reauthorization Act
SITE	Superfund Innovative Technology Evaluation
TCLP	Toxicity characteristics leaching procedure
TSCA	Toxic Substance Control Act

TABLE 93.2 Application and Costs for Soil Treatment Technologies

Technology	Number of vendors[a]	Number of sites[a]	Number of NPL sites [b]	Cost range[c] ($/cu. yard)	Notes
Soil vapor extraction	175	290	152	$2 to $1040; avg. $140	
Solidification/ stabilization	88	215	70	$40 to $60[d] (shallow); $150 to $250 [d] (deep)	Costs given for *in situ* soil mixing/ auger technique
Incineration	62	152	150	$240 $2100[e]; avg. $820	Costs given for on-site incineration (off-site average: $790/yd^3)
Natural attenuation	Not listed		146	Variable	Costs vary widely depending on the scope of the monitoring plan involved
Thermal desorption	57	131	49	$50 to $840[e]; avg. $330	
Bioremediation	58	90	69	$10 to $1220; avg. $270	
Chemical oxid./ reduction	32	79	8	$150 to $500[d]	
Bioventing	30	66	12	$1 to $330; avg. $35	Also applied at 145 U.S. Air Force sites
Landfilling and containment	Not listed		42	$230 without stabilization; $360 with stabilization	NPL site number based on search for "containment cell" or "off-site disposal"; costs based on RCRA Haz. Waste Landfill costs
Soil flushing	25	40	14	$25 to $250[d]	Cost depends highly on type of surfactants or cosolvents used
Soil washing	20	32	15	$250[d]	
Phytoremediation	9	23	10	$20 to $40[d]	Estimated cost for lead-contaminated soil
Chemical extraction	13	18	12	$150 to $600[d,e]	

[a] Numbers reported on the U.S. EPA Remediation and Characterization Innovative Technologies website (www.eparea-chit.gov)

[b] Superfund National Priorities List (NPL) Fact Sheets CD, May 2001 (order at www.epa.gov). Numbers reported are based on the number of NPL fact sheets that discuss the given technology.

[c] Unless otherwise noted, costs are from the U.S. EPA Remediation Technology Cost Compendium — Year 2000. EPA-542-R-01-009. Available at www.frtr.gov.

[d] Federal Remediation Technologies Roundtable (FRTR) (2002). Remediation Technologies Screening Matrix, 4th Edition, April 2002. available at: http://www.frtr.gov/matrix2/top_page.html go to http://www.frtr.gov/matrix2/section3/table3_2.html for the screening matrix table.

[e] Costs converted to $/cu. yard based on bulk density of 1.5 ton/cu. yard.

TABLE 93.3 Applicable Contaminant Classes for Soil Treatment Technologies[a]

Technology	Nonhalogenated VOCs	Halogenated VOCs	Nonhalogenated SVOCs	Halogenated SVOCs	Fuels	Inorganics	Radionulides	Explosives
Soil vapor extraction	•	•			•			
Solidification/stabilization			•	•		•	•	
Incineration	•	•	•	•	•			•
Natural attenuation[a]	•	•	•	•	•	•		•
Thermal desorption	•	•	•	•	•			•
Bioremediation	•	•	•		•			•
Chemical oxidation/reduction	•	•		•		•		•
Bioventing	•		•		•			
Landfilling and containment	•	•	•	•	•	•		•
Soil flushing	•	•	•	•	•	•		
Soil washing	•	•	•	•	•	•		
Phytoremediation	•	•	•		•	•		
Chemical extraction	•	•	•	•	•	•	•	

Note: Selection based on a rating of "better" or "average" for the technology-contaminant combination given in the Remediation Technologies Screening Matrix.

[a] While not listed as a treatment technology by FRTR (see source note below), natural attenuation has potential applicability to most contaminant classes with the exception of radioactive wastes.

Source: Federal Remediation Technologies Roundtable (FRTR). 2002. *Remediation Technologies Screening Matrix, 4th Edition,* April 2002.

TABLE 93.4 Appropriate Technologies for Potentially Limiting Site or Contaminant Characteristics (General Guidelines for Technology Consideration)

Technology	Saturated Soils	Extreme pH	Deep soils	Saline Soils	Clay Soils	Hydrophobic Contaminants	Contaminant Toxicity
Soil vapor extraction		•	•	•			•
Solidification/stabilization		•		•	•	•	•
Incineration		•		•	•	•	•
Natural attenuation	•		•		•	•	
Thermal desorption		•		•	•	•	•
Bioremediation	•		•			•	
Chemical oxid./reduction	•	•	•	•		•	•
Bioventing			•			•	
Landfilling and containment		•		•	•	•	•
Soil flushing	•	•	•	•			•
Soil washing		•		•			•
Phytoremediation	•					•	
Chemical extraction		•		•	•		•

References

AFCEE. 1996. *Bioventing Performance and Cost Results from Multiple Air Force Test Sites,* Air Force Center for Environmental Excellence, Tecnology Transfer Division. Brooks AFB, Texas. Available at: http://www.afcee.brooks.af.mil/er/ert/download/Biov02.pdf

Federal Remediation Technologies Roundtable (FRTR). 2002. *Remediation Technologies Screening Matrix, 4th Edition,* April 2002, available at: http://www.frtr.gov/matrix2/top_page.html;go to http://www.frtr.gov/matrix2/section3/table3_2.html for the screening matrix table

Sellers, K. 1999. *Fundamentals of Hazardous Waste Site Remediation,* Lewis Publishers, Boca Raton, FL.

Shah, K. L. 2000. *Basics of Solid And Hazardous Waste Management Technology,* Prentice Hall, Columbus, OH.

Sims, R. C. 1993. Soil remediation techniques at uncontrolled hazardous waste sites: a critical review. *J. Air Waste Manage. Assoc.* 49(5): 703–732.

U.S. EPA. 2000. *RCRA Corrective Action Workshop/Handbook On Results-Based Project Management,* U.S. EPA Corrective Action Programs Branch Office of Solid Waste. July 11-13, Region 1.

94

Urban Storm Water Design and Management

James F. Thompson
J.F. Thompson, Inc.

Philip B. Bedient
Rice University

The application of the principles of hydrology and hydraulics in urban storm water management systems from a design and flood control standpoint is addressed within this chapter. The basic design approaches employed to develop such storm water systems are presented, including the incorporation of urban hydrology, floodplain hydraulics, and the design of urban storm drainage facilities such as inlets, storm sewers, and open channels. Also included within this discussion are the elements of flood control and mitigation, as well as sediment and pollutant control amenities to these urban systems.

Particular storm water design and management methods are presented, which include conventional and advanced analysis procedures using industry-standard computer models as well as next-generation technologies. The focus of this chapter is on bringing the theories of urban hydrology and hydraulics into real applications as are utilized to analyze and solve complex urban watershed problems.

94.1 Storm Water Design and Management Overview

Storm water management systems are commonly considered to be those manmade features such as storm sewers and open channels designed and constructed for the purposes of facilitating storm water runoff occurring from particular storm events. In actuality, the overall system is inherently complex as other features are also included that affect the performance of the system, such as streets, inlets, inlet leads, and special features such as detention basins and pollutant control amenities. In aggregate, these features form the basis of a complex urban storm water management system, as is commonly implemented today.

The basic analysis and design process is generally the same when first approaching a particular storm water project. These basic steps involve:

1. Define the drainage areas serving the project area.
2. Determine the hydrologic response and runoff from these drainage areas given a particular design or actual storm event.
3. Design the storm water conveyance conduits.

0-8493-1586-7/05/$0.00+$1.50

4. Analyze the effects of increased storm water quantity and decreased quality if any.
5. Mitigate any project negative effects.

While the above steps are indeed a simplification to the detailed process, this basic overview process is fundamental.

In almost all cases, developed urban communities already have some form of design guidelines or criteria. These may be established by the local community agency such as the city, or in many cases, these local guidelines are deferred to other agency criteria such as those established by various State departments of transportation or even various federal agencies. The point is that when approaching an urban storm water analysis or design project, one should remember that usually, some established guideline or procedure already exists. As urban environments become more complex in terms of development, thus restricting the construction of certain storm water features, project analyses and designs become inherently more complicated. This in turn requires even more complex approaches to solve commonly encountered urban storm water and flood control needs.

As a final overview note, urban flooding is as historic as the concept of urbanization itself. No matter what manmade features are designed and constructed, flooding cannot be eliminated with absolute certainty, as there will always be a potential storm event occurring over a project area that is of greater intensity than what was planned for. As engineers, we can only move, store, and treat storm water — we cannot control the quantity handed to us.

94.2 Design Procedures

A typical urban storm water system consists of streets constructed with curbs, gutters, inlets, and roadside ditches; underground storm sewers; and open outfall channels such as streams, bayous, and rivers receiving runoff from these features. When designing a system, the first basic step is the determination of the contributing drainage areas to the system such that the runoff from these areas may be determined. This entails the development of a drainage area map based upon the topography of the project area. Figure 94.1 illustrates a typical urban setting upon which a drainage area map has been developed to define the contributing drainage areas to storm sewer system junctions.

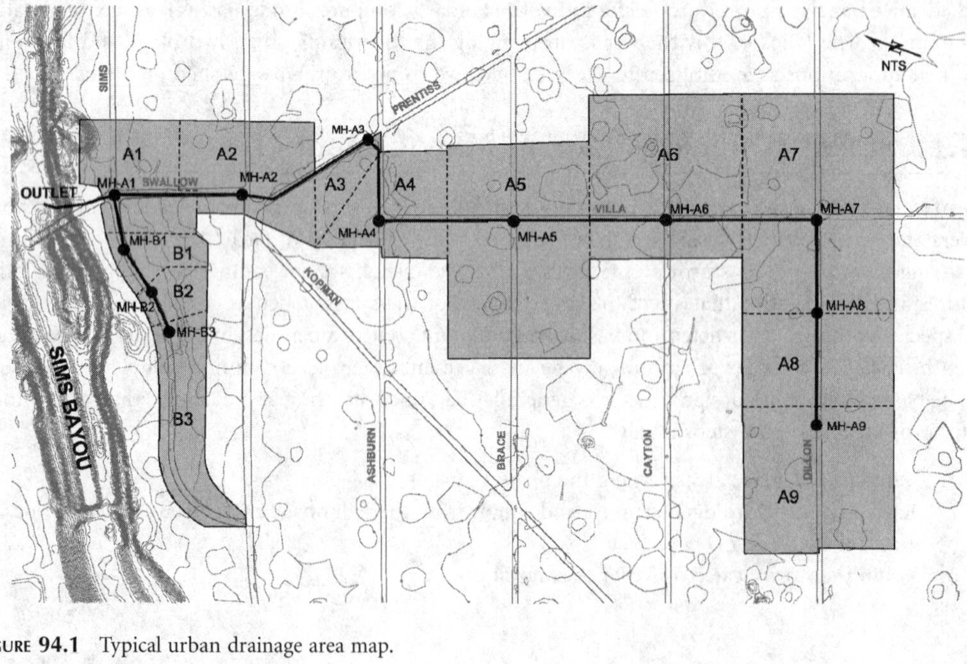

FIGURE 94.1 Typical urban drainage area map.

TABLE 94.1 Design Events and Frequencies

Design Event Frequency	Percent Probability of Occurrence in any Given Year
2	50
5	20
10	10
25	4
50	2
100	1
500	0.2

More explicit drainage area maps are often prepared delineating the individual drainage areas based upon inlet placement, but as is commonly simplified, the drainage area map illustrated is intended for the design of the storm sewer system. Inlet design and placement is critical and is discussed later in this chapter.

The development of a proper drainage area map is a fundamental building block of any urban storm water system. Existing topographic maps and other map sources illustrating relative topographic contours may be used to develop the map. More advanced methods may be used such as aerial photography, which is in turn used to develop a **digital elevation model** (DEM). Further developments in remote sensing techniques are being applied today such as **light detection and ranging** (LIDAR), which is used to produce very accurate three-dimensional digital maps for use in the development of drainage area boundaries [American Society for Photogrammetry and Remote Sensing, 2003]. Regardless of the source, it is imperative that the drainage area map reflect the true pattern of the surface flow of rainfall runoff. Once these patterns and contributing drainage areas are known, the quantity of storm water runoff may be determined.

Prior to the determination of runoff quantities within a project area, it must first be established what is the particular design event for the project at hand. In most cases, urban storm drainage systems are designed for a particular **synthetic design event.** These synthetic events are usually based upon some compilation of weather data such as Technical Paper No. 40 (TP-40), which is a Rainfall Frequency Atlas of the United States [National Weather Service, 2003]. Design storm events are typically associated with a particular return frequency as shown in Table 94.1.

The most commonly used method for determining peak runoff rates for a particular urban project area is the Rational Method using the formula

$$Q = (CA) \cdot i \tag{94.1}$$

where Q = peak flow (cfs), C = runoff coefficient, i = rainfall intensity (in/hr), and A = drainage area (acres).

The Rational Method is simplistic in that it relies on the basic physical characteristics of a drainage area (i.e., size and runoff coefficient) multiplied by a rainfall intensity as based upon the design event being considered. The runoff coefficient, C, varies dramatically based upon the developed state. While many reference manuals have complete listings of runoff coefficients, Table 94.2 lists some commonly utilized runoff coefficients.

TABLE 94.2 Common Runoff Coefficients

Land Use	Runoff Coefficient
Residential	0.35 to 0.55
Multifamily	0.65 to 0.80
Business	0.70 to 0.90
Industrial	0.60 to 0.85
Unimproved	0.10 to 0.30

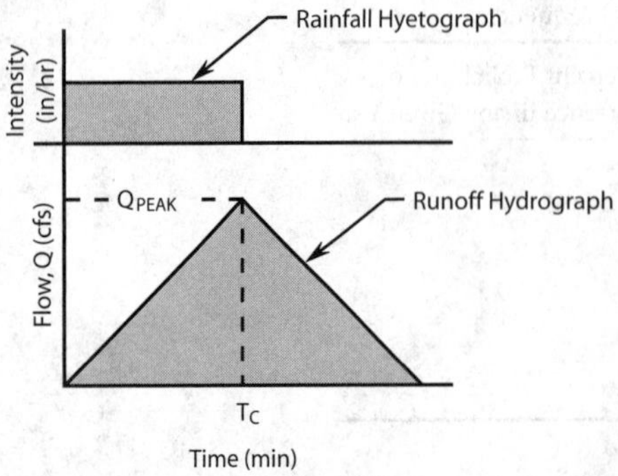

FIGURE 94.2 Rational formula hyetograph and runoff hydrograph.

The rainfall intensity is a function of the return frequency or design event and the time-of-concentration (*Tc*). *Tc* is normally thought of as the time in minutes it takes for water to travel from the furthest hydraulically remote part of the drainage area to the point of consideration or design — in other words, how many minutes it takes for a water particle to travel the longest distance within a drainage area being considered. The basic fundamental assumption of *Tc* in the Rational Method is that the time-to-peak, *Tp*, is equal to *Tc* and that the runoff hydrograph for a project area has a simplified triangular relationship as shown in Figure 94.2, resulting from a constant rainfall intensity over the drainage area for a time period equal to *Tc*. Using sources relating total rainfall depth to a rainfall duration period (i.e., 6, 12, and 24 hr) such as TP-40, intensity-duration-frequency (IDF) curves may be developed which relate *Tc*, the design event or frequency, and the runoff intensity measured in inches per hour. Most major municipalities and state entities have such IDF curves readily available for use. Furthermore, these same entities typically specify the design event to be utilized on a given project. Typical IDF curves are illustrated in Figure 94.3.

Numerous reference materials deal with the above application of the Rational Method [Urbonas and Roesner, 1993; American Concrete Pipe Association, 2000; Mays, 2001]. Also, it should be noted that there are also several other methods for determining runoff rates for given drainage areas. A partial list of these other methods is included in Table 94.3.

The Rational Method is generally considered applicable to drainage areas less than 200 acres. When dealing with contributing drainage areas greater than 200 acres, other methods as tabulated in Table 94.3 may prove more applicable.

The actual design of urban storm sewers and open channels, or conduits, is a process of sizing the conveyance conduit to facilitate the estimated peak flow rate at that point in the system considering the upstream drainage area(s). In using the Rational Method, as the drainage areas are summed moving downstream along the new or proposed storm drainage conduit, *Tc* is naturally increased. This provides

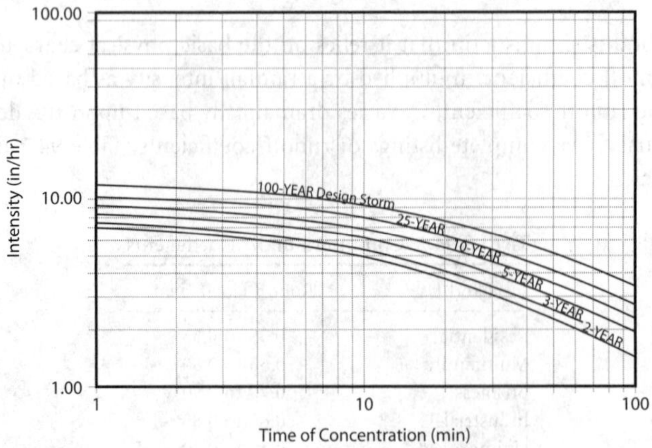

FIGURE 94.3 IDF curves for Houston, Texas.

TABLE 94.3 Various Runoff Determination Methods

Method	Description
Rational	Simplest and most commonly used method for urban storm drain systems; usually limited to watersheds of less than 200 acres
Area/runoff (Regional Regression Equations)	Usually used to estimate peak discharges for larger watersheds in a specific geographical region at an ungauged site
TR-20	NRCS hydrology program for small- to large-ranging systems to estimate peak discharge and runoff hydrographs
HEC-HMS	Used in large complex drainage systems where storage; multiple tributaries; and hydraulic structures, such as dams, weirs, etc. come into play; and rainfall and runoff vary over both time and space

some consideration for the dampening or attenuation of the storm flood wave as it propagates downstream through the conduits. To size the conduits themselves, it is common to use Manning's Equation as

$$Q = \frac{1.49}{n} AR^{\frac{2}{3}} S^{\frac{1}{2}} \tag{94.2}$$

where

Q = flow (cfs)
n = Manning's roughness coefficient = 0.011 to 0.013 for concrete and 0.020 to 0.035 for turf lined ditches
A = cross-sectional area (ft^2)
R = A/P, the hydraulic radius (ft)
P = wetted perimenter (ft)
S = conduit slope (ft/ft)

In parallel with the sizing of the conduits, the relative position of the resultant water surface profile should be examined. This is achieved by the calculation of the **hydraulic grade line** (HGL), starting at the most downstream point in the system, typically the storm drainage system outlet. The resultant HGL represents the locus of elevations to which the storm water would rise if open to atmospheric pressure, such as within piezometer tubes, and can be used to evaluate the adequacy of the design and identify areas of potential surface flooding. The difference in elevation of the water surface in successive conduit reaches represents the friction and other minor losses in the conduits assuming gradually varied flow conditions. If the conduit has a slope equal to the friction slope, then the HGL would be parallel to the conduit. Considering Equation 94.2, the slope, S, is equal to the friction loss, h_l, divided by the length of conduit, L, ignoring minor losses. Solving for h_l, Equation 94.3 represents Manning's Equation as utilized for determining the friction head loss of a conduit reach.

$$h_l = L\left(\frac{Qn}{1.49AR^{\frac{2}{3}}}\right)^2 \tag{94.3}$$

where h_l = head loss in conduit reach (ft), L = length of conduit reach (ft), and other terms as previously stated.

It is also important for a designer to understand the relationship of the HGL to the **energy grade line** (EGL). The EGL represents the total energy head, or the HGL plus the velocity head, at any given point along a conduit computed by Equation 94.4.

$$EGL = HGL + \frac{V^2}{2g} \tag{94.4}$$

where V = velocity of flow (ft/sec) and g = gravitational constant (32.2 ft/sec^2).

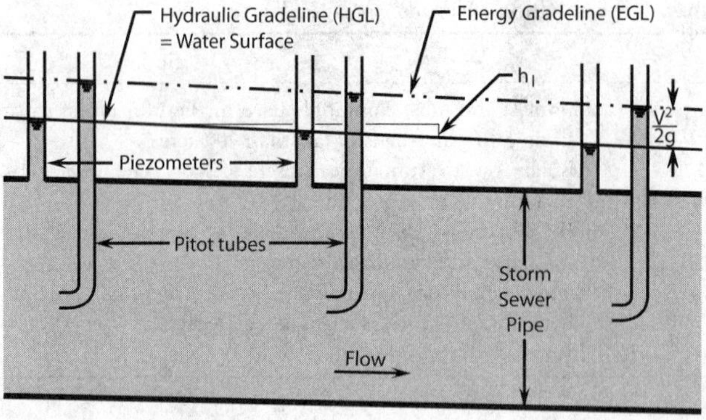

FIGURE 94.4 HGL and EGL representation.

In a physical sense, the EGL would be represented by a series of pitot tubes (small tubes with their end openings pointed upstream) intercepting the kinetic energy of the flow. Figure 94.4 represents an illustration of the relationship between the HGL, the EGL, and a conduit (in this case a storm sewer pipe). Notice that if the starting water surface elevation of the downstream outlet of a conduit is below the soffit of the conduit itself, then the conduit may be flowing partially full. Conversely, if the outlet is fully submerged, then the conduit will be under pressure, at least at the outlet, assuming that the conduit is a pipe. Open channels behave similarly as closed conduits except that there are no bounding physical conditions that would cause an open channel to become pressurized as does typically occur in a closed conduit such as a pipe. Further reference on the behavior of open and closed conduit systems is encouraged to fully understand the behavior of such systems [Brater and King, 1976; Chow, 1959].

The relationship of the HGL to the surrounding environment is critical in the design of a storm drainage system. In a common urban environment, storm drainage systems usually exist as roadside ditches, roadside ditches in combination with an underground storm sewer, or a depressed curb-and-gutter street with an underground storm sewer. When two physical drainage facilities exist in conjunction with each other such as with a depressed curb-and-gutter street and an underground storm sewer system, the storm sewer system acts as the primary drainage mechanism while the street acts as a secondary drainage mechanism (i.e., storm water flows within the street as an open channel). This relationship demonstrates the importance of the relative position of the primary drainage system HGL. For example, if a calculated HGL elevation is shown to be above the natural ground elevation along a project route, then surface flooding, even beyond the street, is indicated. The design of the inlet systems to provide connectivity between the primary and secondary drainage mechanisms is also extremely important, as discussed later within this chapter.

Figure 94.5 illustrates a typical storm sewer system design spreadsheet. Such a spreadsheet may be expanded to include other applicable data such as critical natural ground elevations for comparison to the HGL and other pertinent information. Once the fundamentals of the above equations are understood, the actual design development is rather straightforward. As can be seen from reference back to Figure 94.1, the drainage area map provides the fundamental basis of the design accuracy. The design proceeds from upstream to downstream, summing the combined $C \times A$ values from the Rational Method, thus providing a weighted $C \times A$ value to account for different C values along the project route, and realizing the changing Tc, the various intensity values are determined from IDF curves (or intensity formulas if they are available for the project locale). The flow rates, Q, are then calculated for each design reach and the conduit is sized accordingly using Equation 94.2.

After the conduits are sized, it is common to then calculate the HGL or water surface profile of the system. In closed conduit systems, this entails the computation of the friction losses through the storm sewers and adding those losses, on a reach-by-reach basis, to the starting water surface elevation at the outlet; hence, the importance of defining a proper starting water surface elevation at the outlet. Considering the design event utilized, the water surface elevation at the storm drainage system outlet must have

Storm Sewer Design Spreadsheet

$Q = (CA) \cdot i \qquad Q = \frac{1.49}{n} AR^{2/3} S^{1/2}$

$Q = VA \qquad h = L\left(\frac{Qn}{1.49AR^{2/3}}\right)^2$

Project: Garden Villas
Designer: MSK
Date: April 5, 2003

Design Storm Event: 2 Year
HGL Starting Elev. at Outlet: 20.00

Min. Pipe Size: 18-in.
Min. Tc: 10 min.

1	2	3	4	5	6	6	7	8	9	10	11	12	13	14	15	16	17	18	19	20	21	22	23	24	25
From	To	Drainage Area				Total D.A.	Sum			Design Flow	Pipe Information									Flow			Actual	Headloss	Upstream
Manhole	Manhole	No.	(Acre)	C	C x A	(Acre)	C x A	Tc	I	Q	Elevation (ft)		Dia/Hgt	Length	Slope	n	D	D	Dia.	Capacity	Area	Velocity	Velocity	h_l	MH HGL
								(min)	(in/hr)	(cfs)	u/s	d/s	(in)	(ft)	%		calc	calc(in)	(in)	(cfs)			(fps)	(ft)	(ft)
Line A																									
A9	A8	A9	7.44	0.45	3.35	7.44	3.35	29	3.14	10.51	31.70	31.19	24	425	0.22%	0.013	1.99	23.92	24	10.61	3.14	3.38	3.35	0.92	24.37
A8	A7	A8	4.77	0.45	2.15	12.21	5.49	31	3.07	16.85	30.96	29.45	24	355	0.18%	0.013	2.47	29.65	30	17.40	4.91	3.55	3.44	0.60	23.45
A7	A6	A7	11.36	0.45	5.11	23.57	10.61	32	2.97	31.47	28.80	27.60	48	587	0.10%	0.013	3.49	41.83	42	31.82	9.62	3.31	3.27	0.57	22.86
A6	A5	A6	8.48	0.45	3.82	32.05	14.42	33	2.92	42.10	27.53	26.66	54	595	0.12%	0.013	3.76	45.09	48	49.76	12.56	3.96	3.35	0.51	22.28
A5	A4	A5	10.42	0.45	4.69	42.47	19.11	34	2.87	54.92	26.45	26.44	60	485	0.10%	0.013	4.30	51.55	54	62.19	15.90	3.91	3.46	0.38	21.77
A4	A3	A4	3.48	0.45	1.57	45.95	20.68	35	2.86	59.16	26.43	25.86	60	270	0.13%	0.013	4.21	50.46	54	70.91	15.90	4.46	3.72	0.24	21.39
A3	A2	A3	1.09	0.45	0.49	47.04	21.17	35	2.86	60.48	25.56	25.40	60	580	0.10%	0.013	4.45	53.45	54	62.19	15.90	3.91	3.80	0.55	21.15
A2	A1	A2	4.31	0.45	1.94	51.35	23.11	35	2.84	65.70	25.39	24.51	60	530	0.10%	0.013	4.59	55.13	60	82.37	19.63	4.20	3.35	0.34	20.60
Line B																									
B3	B2	B3	2.50	0.65	1.63	2.50	1.63	27	3.29	5.35	28.07	26.44	18	166	0.26%	0.013	1.50	17.99	18	5.36	1.77	3.03	3.03	0.43	21.82
B2	B1	B2	0.87	0.45	0.39	3.37	2.02	27	3.25	6.56	26.39	25.56	18	193	0.39%	0.013	1.50	18.00	18	6.56	1.77	3.71	3.71	0.75	21.39
B1	A1	B1	1.35	0.7	0.95	4.72	2.96	28	3.20	9.49	25.54	24.81	24	215	0.20%	0.013	1.95	23.43	24	10.12	3.14	3.22	3.02	0.38	20.64
Line A																									
A1	Outlet	A1	5.42	0.45	2.44	61.49	28.51	36	2.81	80.21	22.18	21.94	72	270	0.10%	0.013	4.95	59.42	60	82.37	19.63	4.20	4.09	0.26	20.26

FIGURE 94.5 Typical storm sewer system design spreadsheet.

some relation and may be used as provided by other studies and analyses or as determined from approximations, if acceptable, using normal depth calculations within the outlet receiving channel. In many instances, the design criterion allows for simplified assumptions pertaining to starting water surface elevations at the outlet for the purposes of computing the HGL along a designed storm sewer system. This may be the soffit of the last reach of pipe at the outlet or even perhaps some arbitrary elevation, as prescribed by a governing flood control agency. In any case, the HGL is computed upstream throughout the storm sewer system to detect instances where significant head loss may exist. Increasing the size of the conduit increases the wetted perimeter of the conduit, thereby reducing the friction loss and the slope of the HGL. In cases where certain HGL criteria must be met, a trial-and-error type of approach is often exercised to finalize a design that satisfies the design intent.

Manhole and junction losses as well as bend losses may be considered in the development of the HGL profile [American Iron and Steel Institute, 1980]. In some cases, these losses are considered minor and insignificant; yet, in other cases, especially where relative velocities are rather high, the losses may need to be considered and added to the conduit friction losses at the junctions. Equations 94.5 and 94.6 represent the equations for the calculation of manhole/junction losses and bend losses, respectively. Notice that the primary influencing parameter in these equations is the velocity term. When the velocity is rather low, these losses are indeed very minor. In general, it is desirable to keep the minimum velocities within conduits above 3 ft/sec. Maximum velocities should not exceed 8 ft/sec without due consideration of the use of energy dissipation features to control the effects of high velocities such as erosion. Again, local design guidelines generally dictate this controlling criterion.

$$h_j = K_j\left(\frac{V^2}{2g}\right) \tag{94.5}$$

where h_j = manhole/junction loss without a curved deflector (ft), K_j = loss coefficient for manhole/junction = 0.05 for straight-through (no bend at junction) = 0.37 for 45° angle of storm sewers at junction = 1.33 for 90° angle of storm sewers at junction, and other terms as previously stated.

$$h_b = K_b\left(\frac{V^2}{2g}\right) \tag{94.6}$$

where h_b = bend loss (ft); K_b = loss coefficient for bend, 0.19 for 15°; 0.47 for 45°; 0.70 for 90°, and other terms as previously stated.

Equally important as the design of the conduits is the design of inlets and other storm water capture facilities. After all, the conduit system may be sized adequately for a certain design event, but if the storm water cannot enter the conduits at the rate designed for, then the hydraulic connectivity of the system is compromised. This is especially true of depressed curb-and-gutter street sections using conventional on-grade and sag inlets. Inlets work hand-in-hand with the surface streets in urbanized storm sewer systems. As previously mentioned, the storm sewers act as the primary drainage system, while the surface streets act as the secondary system. Those storm water flows not facilitated by the storm sewers are conveyed overland by the street system. Inlets provide the vital function of hydraulic connectivity between the primary and secondary systems. Much study has been done on the subject of inlet behavior, and these studies are still ongoing and evolving. The inlet capacity is designed based upon allowable street ponding widths, street gutter flow rates, street physical features such as slope and friction value, and the inlet physical features. It is highly recommended that inlet computations be precisely considered using established guidelines such as those developed by the Hydraulic Engineering Circular No. 22 (HEC 22) as published by the Federal Highway Administration (FHWA) [Federal Highway Administration, 2001].

Inlets act as weirs until submerged, and then they transition to act as orifices. In curb opening inlets, these two states transition at depths between 1.0 and 1.4 times the inlet opening height. In these cases, the inlet behavior should be analyzed using both the weir and orifice equations, and the lesser of these two capacities computed should be used in the design. Test studies have shown that inlet physical features such as inlet throat depressions and grate types greatly affect the actual performance of an inlet. For these reasons, it is recommended to refer to detailed guidelines as previously mentioned or check with local governing agencies that likely have published allowable inlet capacities for those types of inlets commonly utilized in the project area. Absent of these detailed procedures, inlet behavior may be approximated using weir and orifice equations as depicted in Equations 94.7 and 94.8; yet, due consideration must be given to the other parameters that may limit inlet capacities such as inlets on steep grades which induce the likelihood of substantial inlet bypass flow.

$$Q = C_w (L + 1.8W) h^{1.5} \tag{94.7}$$

where Q = weir condition inlet flow (cfs), C_w = weir coefficient = 3.087 (English) and 1.25 (metric), L = length of curb inlet opening (ft), h = head at inlet opening (ft), and W = lateral width of inlet throat depression (ft).

$$Q = C_o d_o L \sqrt{2gh} \tag{94.8}$$

where Q = orifice condition inlet flow (cfs), C_o = orifice coefficient = 0.67, L = length of curb opening (ft), d_o = depth of curb opening (ft), g = acceleration due to gravity (32.2 ft/sec^2), and h = effective head at the center of the orifice throat.

The importance of the behavior of the primary and secondary drainage systems, working collectively, cannot be overemphasized. When surface flooding does occur, ponded sections within roadways and other areas act as storage junctions. These junctions detain storm water flows until the primary drainage system can facilitate that stored volume. In cases where there is insufficient storage volume on the surface for a particular storm event, then the water surface elevation (WSEL) of the flooded secondary system, commonly a street, increases to a point in which structural flooding of adjacent homes and buildings may occur. Furthermore, storm water typically flows overland between the surface storage junctions. This is a very common occurrence in urban drainage systems where overland flow is conveyed between surface storage junctions that may exist as street intersections, for example. In these cases, the relative position of the overland flow routes in a project area must be fully understood as to limit the potential for flooded WSELs building up to the point at which structural flooding of homes and businesses occurs. Such surface conduits (i.e., the streets) may be considered using weir equations or even calculated as conventional open channels. Mother Nature developed flow patterns along our surface topography long before we imposed urbanized development upon our lands. These natural flow patterns must be recognized and respected, as there is always another natural storm that may or eventually will occur above and beyond that which was planned for in the design development of our urban drainage systems.

94.3 Routing and Flood Wave Propagation

The previous discussion pertaining to the design of urban storm water systems is based upon standard design practices using conventional steady state analyses based upon the simplified form of the continuity equation as

$$Q = V_1 A_1 = V_2 A_2 \tag{94.9}$$

where Q = flow (cfs), V = velocity (ft/sec), and A = cross-sectional area (ft^2).

This approach has proven very applicable to most standard urban systems; yet, in reality, there is a recognized change in storage relative to time throughout the system. The recognition of the storage and time elements becomes increasingly important when dealing with large storm drainage systems or when considering the issues of adverse hydraulic impacts and pollutant transport, both discussed later in this chapter. The fundamentals of these applications are based upon a more dynamic approach to the continuity equation whereby

$$Q_{in} - Q_{out} = \frac{\Delta S}{\Delta t} \tag{94.10}$$

where Q_{in} = inflow (cfs), Q_{out} = outflow (cfs), ΔS = change in storage (ft^3), and Δt = change in time (sec).

Given the importance of the behavior of storm drainage systems in our developed environments, the basic design procedures previously described must be coupled with the understanding of the dynamic system as it will actually exist. This includes the principles of flood wave propagation in storm sewers and open channels as well as the routing of flows through these systems. From a basic design standpoint, these analyses can be inherently complex, as the numerical process of calculating changes in storage in conjunction with elements of changes in momentum becomes very taxing for simplified design procedures. Invariably, computer models are used to account for these complex procedures, and it is important to understand the fundamentals of these basic routing concepts.

There are two basic categories of routing techniques: hydrologic and hydraulic. Hydrologic routing is more simplified and involves the basic application of the continuity equation, whereby inflow and outflow are balanced with the change in storage relative to time. Hydraulic routing is more complex and, if applied correctly, can provide more accurate results. The basic premise of full hydraulic routing involves the simultaneous computation of the continuity equation and the momentum equation for unsteady nonuniform flow solutions. This method of hydraulic routing is based upon the Saint Venant equations. For simplification purposes, the full Saint Venant equations may be shortened to account for a simplified continuity and uniform flow relationship commonly termed **kinematic wave routing** [Bedient and Huber, 2002]. Other routing methods exist as well, including the modified Puls, Muskingum, and Muskingum–Cunge methods. Table 94.4 illustrates the various forms of the momentum equation, and Table 94.5 lists various routing methods commonly applied.

TABLE 94.4 Forms of the Momentum Equation

Type of Flow	Momentum Equation
Kinematic wave (steady uniform)	$S_f = S_o$
Diffusion (noninertia) model	$S_f = S_o - \partial y/\partial x$
Steady nonuniform	$S_f = S_o - \partial y/\partial x - (v/g)\partial v/\partial x$
Unsteady nonuniform	$S_f = S_o - \partial y/\partial x - (v/g)\partial v/\partial x - (1/g)\partial v/\partial t$

TABLE 94.5 Commonly Applied Routing Methods

Title	Type
Unsteady flow using St. Venant equations	Hydraulic method
Diffusion approximation of St. Venant equations (Muskingum–Cunge method)	Hydraulic method
Kinematic-wave approximation of St. Venant equations	Hydraulic method
Modified Puls	Hydrologic method
Muskingum	Hydrologic method
Combination of steady-state backwater computations and Modified Puls	Hydrologic method

Kinematic waves assume that the gravity force of water is approximately balanced by the conduit bed friction, and these waves appear as uniform, unsteady flow with the water surfaces and conduit bed parallel to each other and to the energy grade line. Dynamic waves appear as gradually varied, unsteady flow with water surface profiles and the conduit bed not being parallel to each other. These two scenarios are best illustrated in Figure 94.6. In most cases dealing with open channel floodplain hydraulics, steady or gradually varied flow exists, making kinematic wave and diffusion approximations of the Saint Venant equations applicable. In urban closed conduit systems subject to rapid changes in flow and storage, unsteady flow analyses using the full Saint Venant equations provide a more exact solution given suitable and adequate input data.

In dealing with flood wave propagation and routing within our urban drainage systems, we are interested in such characteristics as wave timing, wave peaks, relative peak attenuation, and flood storage and discharge. These are some of the basic design variables that are of interest when confronted with issues above and beyond the standard base design as provided by simplified procedures. These other design issues typically include the identification and mitigation of adverse hydraulic impacts caused by a new storm drainage project or with issues pertaining to pollutant and sedimentation transport. In these cases, it is common to utilize computer models for computing the performance of a designed system.

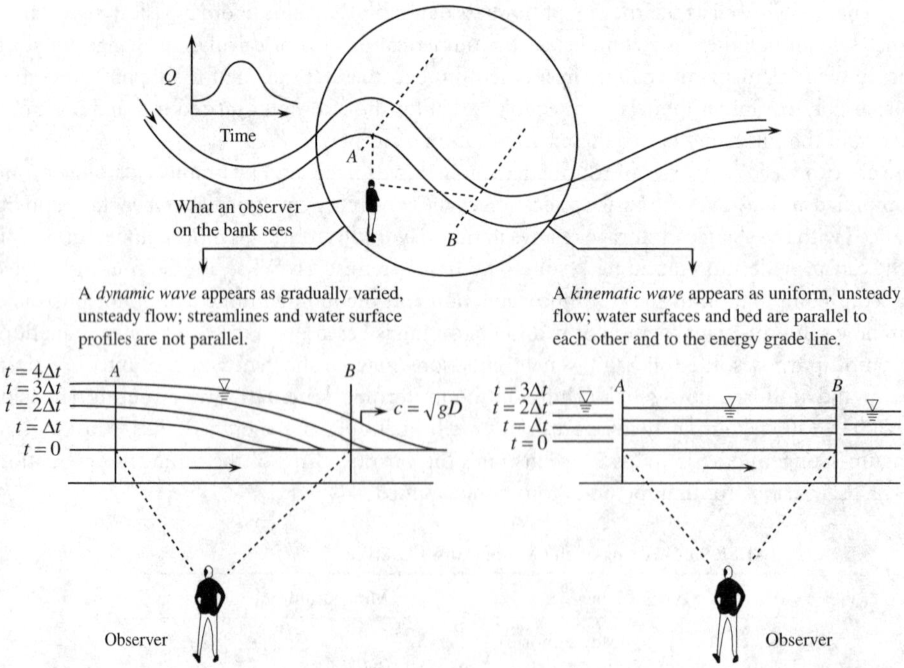

FIGURE 94.6 Visualization of dynamic and kinematic flood waves. (From Bedient, P. B. and Huber, W. C. 2002. *Kinematic Wave Routing, Hydrology and Floodplain Analysis*, 3rd ed., 2002, chap. 4.6. With permission.)

TABLE 94.6 Commonly Applied Computer Models

Model	Author	Description
HEC-HMS	HEC	Hydrologic modeling system (replace HEC-1)
HEC-RAS	HEC	River analysis system (replace HEC-2)
TR-20	NRCS	Hydrologic simulation model
SWMM	Huber and Dickinson	Storm water management model

The modeling tools that can be used in the design of urban storm water systems are varied. From simple spreadsheet functions to complex computer models, all types of modeling tools can prove very valuable to the successful development of a project. For open channel hydrology and hydraulics, the U.S. National Flood Insurance Program (NFIP), in conjunction with the Federal Emergency Management Agency (FEMA), utilizes the U.S. Army Corps of Engineers' HEC-HMS and RAS programs, which supercede the HEC-1 and 2 programs [Department of the Army, 2003]. For complex urban storm drainage systems, there exist various proprietary versions of the U.S. Environmental Protection Agency's Storm Water Management Model (SWMM) [U.S. Environmental Protection Agency, 2003], which facilitate the full design and analysis of closed and open conduit systems using fully dynamic conditions. In addition, several other public and proprietary software models exist with varying degrees of complexity that can aid in the design and analysis of urban drainage systems. Table 94.6 lists several of the more commonly used models and their application.

94.4 Hydraulic Impact Determination and Mitigation

The development of a given urban storm water management project within an existing watershed can consist of not only the construction of new facilities on raw land but also the alteration of existing impervious levels and drainage characteristics in previously developed areas. As such, new developments, channel modifications, roadway reconstructions, and other common civil works projects may lead to the creation of undesirable **hydraulic impacts.** The identification, quantification, and mitigation of such impacts to receiving outlet channels or conduits and adjoining properties have become a primary focus in a typical project development life cycle.

The term "impact" is used extensively in hydraulic engineering as related to a particular project's development. By strict definition, all changes to the behavior of existing storm water flows as routed from or through a project area could be deemed an impact. These impacts may be negative or positive. The alteration of an existing watershed's drainage characteristics, as caused by an urban storm water project, often leads to increased flow rates to the receiving outlet or downstream collection system. In many cases, these increased flow rates equate to rises in open channel water surface elevations or the surcharging of existing storm sewer systems, which in turn often directly relate to increased levels of surface flooding.

While the elimination or significant reduction of the increase of surface flooding is desired, not all increases in flow rates from a given project area negatively impact receiving outlet channels or closed conduit systems. This would be the case if an outlet system had sufficient capacity to facilitate these increased flows without any effect on or alteration to levels of surface flooding. Provided that no onsite, upstream, or downstream flooding conditions are created or worsened as caused by the increased discharges and any resultant rise in water surface elevations, it can be argued that no negative impacts have occurred. If an outlet system has no excess hydraulic capacity, and any increase in flows would result in surface flooding, possibly causing physical flood damage, then obviously an increase in flow rate and corresponding increase in water surface elevation would be considered a negative impact.

Considering this discussion on various baseline parameters for the identification of negative hydraulic impacts, the term *impact* actually implies *adverse impact.* An adverse impact would exist from a project development standpoint if the resultant increase in existing runoff flow rates caused by the alteration to the watershed characteristics directly results in increased surface flooding, erosion, or other such physically damaging phenomena. After a storm water management project is designed using either advanced or conventional means, the system should be analyzed to ascertain if any adverse hydraulic impacts do indeed exist as a result of the new or proposed improvements. If these impacts do exist, then mitigation measures should be undertaken. Typically, the determination and mitigation of hydraulic impacts usually entails a rather complex analysis utilizing some of the computer modeling tools described in the previous section. Simplified approaches are available in urbanized areas where storage coefficients or other such watershed parameters have been established to facilitate the development of storm water management improvements. This may be the case, for example, if detention basin requirements are simplified to per-acre-of-detention-needed versus the size of a given project area. As always, design engineers are encouraged to check local governing entities for such established criteria.

In most urbanized settings, even when some simplified guidelines have been established, hydraulic impacts must be analyzed to insure that no adverse effects will exist from the project's development. The basic steps in performing an impact analysis build upon the design procedure outlined in Section 94.2 as follows:

1. Perform the design of the storm water management improvements as based upon a particular design event or other such established criteria.
2. Model the existing conditions using computed contributing drainage areas and peak discharges.
3. Change the drainage area characteristics, if needed, for the proposed conditions.
4. Model the new or proposed conditions by incorporating all of the designed storm water facilities (ditches, storm sewers, streets, inlets, etc.) into the proposed model.
5. Compare the existing model results to those of the new or proposed model.
6. Identify adverse hydraulic impacts.
7. Determine and analyze mitigation methods (detention, in-line restriction, etc.).
8. Refine/change the base design within the new or proposed model to reflect the incorporated mitigation measures.
9. Remodel the new or proposed conditions to demonstrate the mitigation results.

The steps are rather straightforward and can vary given certain requirements of reviewing entities; however, the fundamental approach is simply to analyze the existing conditions, compare those hydraulic conditions to the new or proposed conditions that would result due to project improvements, and mitigate any impacts as required.

One of the more obvious means of identifying adverse hydraulic impacts from a storm water management project is the comparison of the existing condition outlet discharge hydrograph to that of the new or proposed condition outlet discharge hydrograph. The base design, as previously described, is usually performed using some sort of synthetic design reoccurrence interval (i.e., a 2-, 5-, 10-, 25-, or 50-year frequency event). Hydraulic impacts are commonly analyzed for not only the lesser storm frequency events but more importantly for the extreme storm frequency events such as the 100-year frequency event. This is to ensure that the project design performs adequately in such extreme events without inducing adverse hydraulic impacts, which would typically result in increased structural flooding in or near the project area.

One of the most common forms of mitigation employed is the utilization of a detention basin within the project design. There are numerous detailed references on detention design [Stahre and Urbonas, 1990], but the basic fundamental sizing of a detention basin is based on the comparison of the existing to proposed discharge hydrograph at the outlet whereby the needed storage volume is determined from the area under the proposed to the existing hydrograph as illustrated in Figure 94.7.

Figure 94.8 illustrates an actual overlay of an urban project area's existing condition discharge hydrograph at the outlet to the receiving river with that of the new or proposed condition discharge hydrograph. Noticeably, there is an increase in peak of the proposed hydrograph. This is not necessarily an indication of the induction of adverse hydraulic impacts in the receiving river, stream, or channel. By modeling the river flood wave in existing versus proposed conditions (using HEC-HMS and RAS for example) and taking into consideration the alteration of the inflow hydrograph for the project area, it could be documented that the increase in peak discharge from the project area occurs well ahead of the

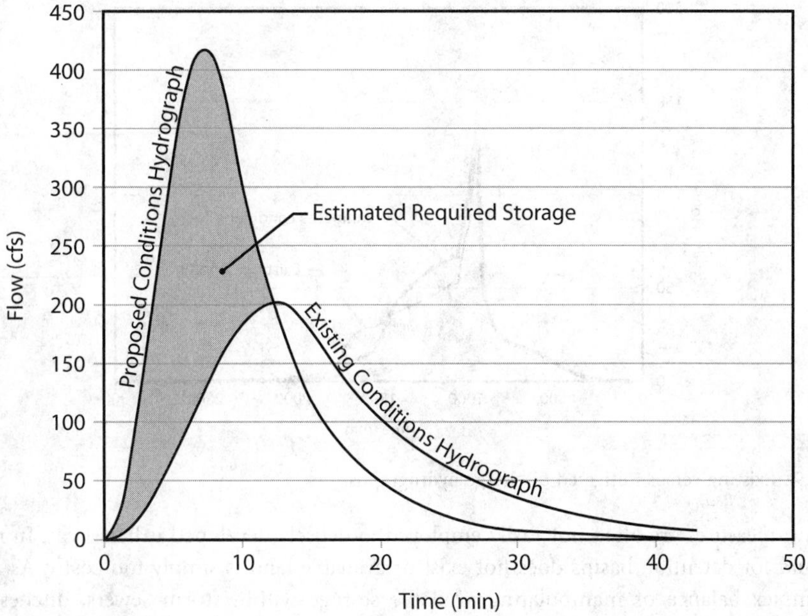

FIGURE 94.7 Detention basin sizing.

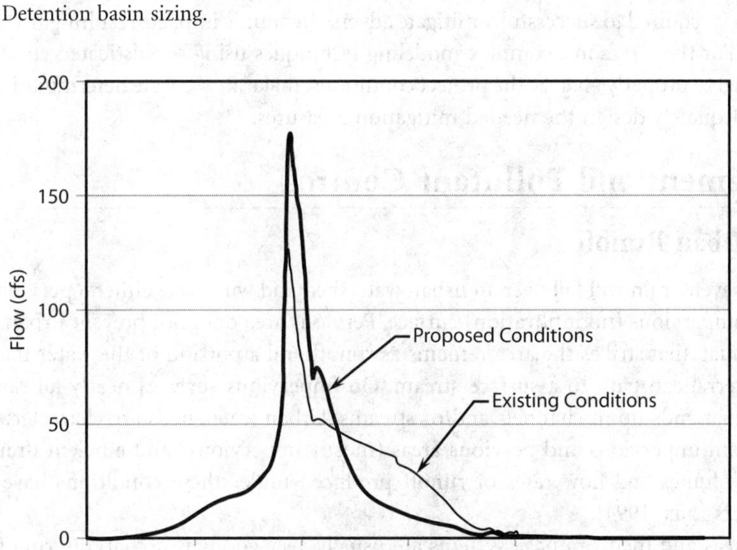

FIGURE 94.8 Existing versus proposed discharge hydrographs.

propagating river flood wave in such a manner as to not result in any increased water surface elevations along the river. In this case, it was noticed that the peak discharge of the proposed conditions model at the outlet coincided fairly closely to the peak in the river flood wave, and thus, a rise in water surface elevation along the river was noticed for several river miles downstream of the project area. To address this adverse hydraulic impact, a detention basin was utilized within the project area, and after several iterations of sizing the basin, the peak of the proposed condition discharge hydrograph was reduced as seen in Figure 94.9 to mimic the existing conditions peak, and the proposed condition hydrograph was then input in the river hydrologic and hydraulic models and routed downstream to ensure that no rises in water surface elevations were caused by the project improvements.

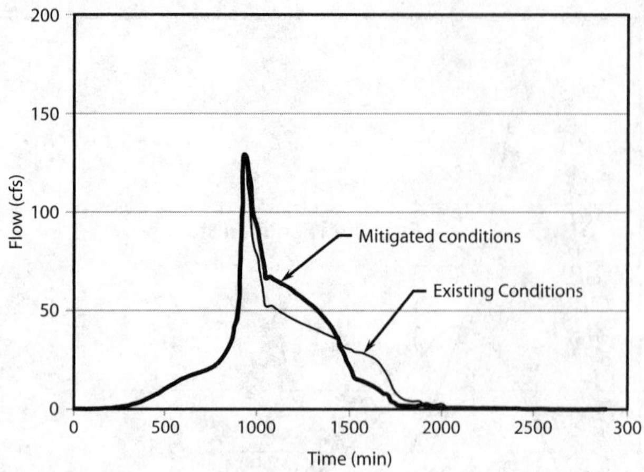

FIGURE 94.9 Existing versus mitigated discharge hydrographs.

Mitigation measures are often not easily employed in densely developed urban areas. In many cases, available land for detention basins does not exist or available land is simply too costly. As is often the case, a complex balance of manipulating available storage within storm sewers, ditches, and even depressed curb-and-gutter streets in conjunction with adjusting the routing of flows within the storm drainage system is required to successfully mitigate adverse hydraulic impacts resulting from the proposed improvements. For these reasons, complex modeling techniques using sophisticated computer software are often required to properly analyze the project conditions, make an accurate determination of hydraulic impacts, and adequately design the needed mitigation measures.

94.5 Sediment and Pollutant Control

Effects of Urban Runoff

During a storm event, rain will fall over an urban watershed and will strike either a pervious (infiltration allowed) or an impervious (no infiltration) surface. Pervious area does not prevent urban runoff, rather some of the rainfall that strikes the area remains as runoff and a portion of the water that is infiltrated will take a subterranean path to a surface stream. On impervious surfaces nearly all rainfall becomes surface runoff that ends up in channels and/or streams. Urban watersheds are characterized by surface water runoff from impervious and pervious areas (mostly impervious) and efficient drainage systems. The increased volumes and flow rates of runoff produced under these conditions have a number of deleterious effects [Nix, 1994].

Developed areas and their drainage systems are usually very good at discharging runoff — so much so that they can cause flooding at downstream locations. On the other hand, older or inadequately designed drainage systems can themselves be overwhelmed by runoff from increased urbanization. The result is flooding at the "headwater" of the drainage system. The increased runoff accompanying urban development increases the bed load, or sediment-carrying, capacity of a stream [Whipple et al., 1983]. This increase of sediment in the stream is the result of stream bed and stream bank erosion. Increased runoff does not only affect the integrity of the stream itself, but the delicate ecosystem surrounding the stream. The runoff changes the habitats equilibrium and the thermal loads associated with runoff can also disrupt the ecosystem.

As rainfall moves vertically through the atmosphere, it may wash out air pollutants and carry them to the surface. Upon reaching the ground, rainfall will dislodge some particles and dissolve other materials. The freed and dissolved contaminants will either be carried to drainage systems by surface runoff, or infiltrate the soil at pervious areas that may threaten subterranean aquifers. The pollutants carried from the watershed surface come from a number of sources, as listed in Table 94.7.

TABLE 94.7 Sources of Storm Water Pollutants

Transportation (tire and brake wear and exhaust dust)
Industrial activities (petroleum, chemical, energy, and manufacturing)
Decaying vegetation (leaf litter and lawn waste)
Soil erosion (especially from construction sites)
Animal wastes
Fertilizer/pesticide application on lawns and gardens
Deicing agents
General litter

TABLE 94.8 Opportunities for Urban Runoff Pollution

Leaking sanitary sewers
Direct connections of sanitary sewer to storm sewers
Poorly operating septic systems
Illegal disposals of oils, paints, etc. to storm sewer systems
Accidental spills
Leaking underground storage tanks
Leachate from landfills
Leakage from hazardous waste sites

Pollutants may also be contributed by the watershed's drainage system. Such a system may consist of natural or artificial channels as well as sewerage. A system that is designed exclusively for stormwater flows is known as a **separate sewer system** (most modern systems), and a system that is designed to carry both stormwater and sanitary sewer flows is known as a **combined sewer system** (seen throughout the older systems of the northeastern U.S.). The contamination of the urban water runoff can be enhanced by erosion of natural channels and sediment resuspension in artificial channels and sewer systems. Other opportunities for urban runoff are listed in Table 94.8. Figure 94.10 shows possible sources of urban runoff contamination in a typical urban area as well as the inlets and outlets of a typical drainage system.

The pollutants are carried in runoff at varying concentrations. Table 94.9 presents concentration ranges for some specific pollutants in urban runoff as compared to other wastewater streams. Urban stormwater is roughly characterized by a heavy suspended solid load and relatively low **BOD** and nutrient levels when compared to untreated sanitary sewage. Combined sewage, being a combination of sanitary sewage and stormwater runoff, has characteristics somewhere between the two. Table 94.10 summarizes the impacts associated with the various kinds of pollutants and Table 94.11 shows the EMC s for all the nationwide urban runoff program [Nix, 1994].

Urban Runoff Transport

A point source is a regulatory term meaning a source that is discharged through a pipe at a known position, whereas a nonpoint source originates from flow distributed over the land surface. To simplify matters, urban runoff is going to be considered a nonpoint source because it is a diffuse source of pollution and the flow follows the temporal and spatial characteristics of rainfall over a large area. If the concentration of the contaminant during a storm were constant, the mass versus time (loadograph) would match the hydrograph. What usually takes place is that the concentration versus time (pollutograph) exhibits a high concentration at the beginning of a storm; this is known as the first flush. The first flush is a phenomenon that is due to high rainfall intensities at the beginning of the storm creating higher runoff and larger sediment wash-off (due to buildup prior to runoff). The first flush can be readily seen in sewers where the solids were deposited during dry weather and washed away during the first storm event; it is less evident in highly urbanized areas where factors such as street cleaning, traffic, and wind remove the sediments just as fast as they are deposited. On pervious surfaces, buildup plays a lesser role, and entrainment of water quality constituents in runoff is due more to erosion and solution mechanisms. Constituents may be adsorbed onto particular matter and thus be subject to transport as solids [Huber, 1992].

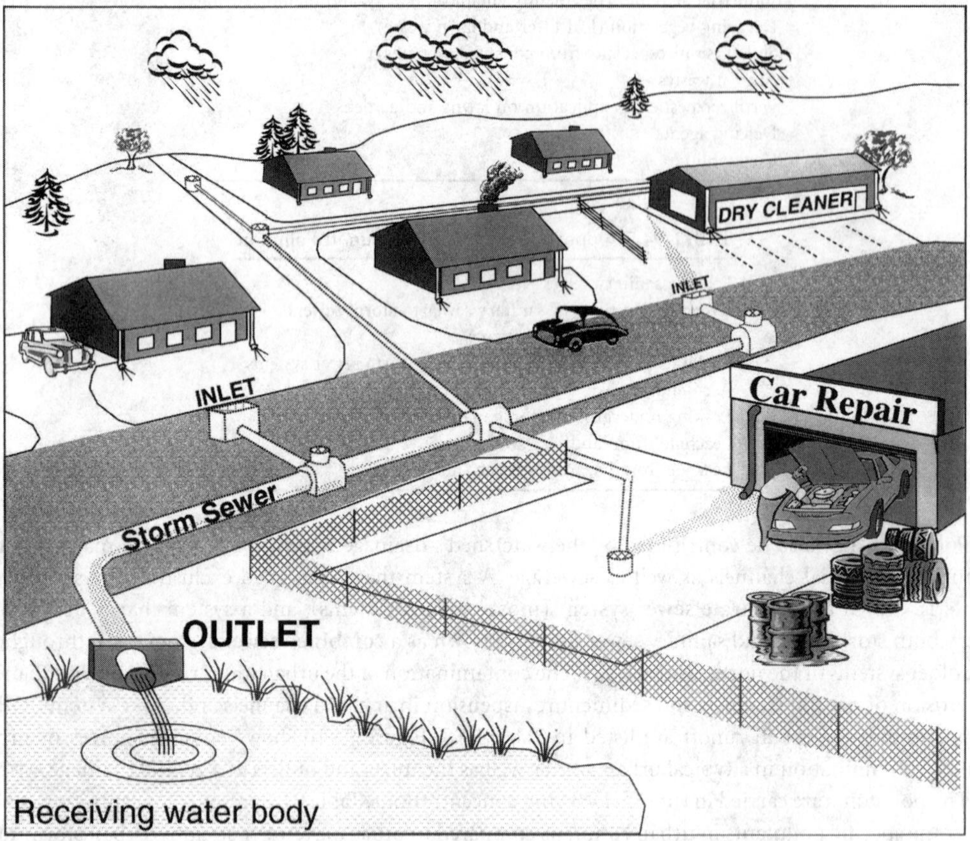

FIGURE 94.10 Typical urban stormwater system.

TABLE 94.9 Characteristics of Urban Wastewater Streams

Wastewater Type	BOD_5 (mg/L)	SS (mg/L)	Total N (mg/L)	Total P (mg/L)	Lead (mg/L)	Total Coliforms (MPN/100 mL)
Stormwater	10 to 250[c]	3 to 11,000	3 to 10	0.2 to 1.7	0.03 to 3.1	10^3 to 10^8
CSOs[a]	60 to 200[c]	100 to 1000	3 to 24	1 to 11	0.4	10^5 to 10^7
Untreated sewage	160[c]	235	35	10	—	10^7 to 10^9
POTW[b] effluent	20	20	30	10	—	10^4 to 10^6

[a] Combined sewer overflows.
[b] Publicly owned treatment works.
[c] Single numbers are means. Others represent a range.
Source: Ellis, J.B. 1986. Pollution aspects of urban runoff, in *Urban Runoff Pollution*, Torno, H., J. Marsalek, and M. Desbordes, Eds., NATO ASI Series, Series G: Ecological Sciences, Vol. 10, Springer-Verlag, Berlin. p. 1.

TABLE 94.10 Summary of Urban Runoff Pollution Effects

Sediment (Sand, Silt, Clay in Colloidal Suspension)

Suspended sediment decreases recreational values, reduces fishery habitat, reduces sunlight penetration, thereby impairing photosynthesis of aquatic plants, adds to the mechanical wear of water supply pumps and distribution systems, and adds treatment costs for water supplies.

Nutrients (Phosphorus, Nitrogen)

Nutrient enrichment of surface waters may cause excessive algae and aquatic plant growth, choking open waters consuming oxygen (mainly through plant die-off).

Oxygen-Demanding Organics (Human, Animal Excreta; Decaying Plant, Animal Matter; Discarded Litter, Food Wastes)

Organic materials (natural or synthetic) may enter surface waters dissolved or suspended in runoff; natural decomposition of these materials may deplete dissolved oxygen supplies in the surface waters.

Toxic Substances (Heavy Metals, Pesticides, Oil, and Other Petroleum Products)

The principal concerns about toxic substances in surface water are their entry into the food chain, bioaccumulation, toxic effects on fish, wildlife, and microorganisms, and habitat degradation of public water supply sources; the ground water impacts are primarily related to water supply sources.

Pathogens (Bacteria, Viruses)

The principal concerns about pathogens are the survival and transmission of such organisms and their impacts on drinking water supplies, contact recreation waters, and fish and wildlife or domestic animals.

Source: New York State Department of Environmental Conservation. 1992. *Reducing the Impacts of Stormwater Runoff from New Development*, New York State Department of Environmental Conservation, Albany, NY.

TABLE 94.11 Median Event Mean Concentrations (EMCs) for all EPA Nationwide Urban Runoff Program (NURP) Sites by Land Use Categories

		Residential		Mixed		Commercial		Open/Nonurban	
Constituent	Units	Median	CV	Median	CV	Median	CV	Median	CV
5-day biochemical oxygen demand	mg/L	10.0	0.41	7.8	0.52	9.3	0.31	—	—
Chemical oxygen demand	mg/L	73	0.55	65	0.58	57	0.39	40	0.78
Total suspended solids	mg/L	101	0.96	67	1.14	69	0.85	70	2.92
Total lead	μg/L	144	0.75	114	1.35	104	0.68	30	1.52
Total copper	μg/L	33	0.99	27	1.32	29	0.81	—	—
Total zinc	μg/L	135	0.84	154	0.78	226	1.07	195	0.66
TKN	μg/L	1900	0.73	1288	0.50	1179	0.43	965	1.00
NO_2- N + NO_3-N	μg/L	736	0.83	558	0.67	572	0.48	543	0.91
Total phosphorus	μg/L	383	0.69	263	0.75	201	0.67	121	1.66
Soluble phosphorus	μg/L	143	0.46	56	0.75	80	0.71	26	2.11

Note: Mixed land use consists primarily of low- and medium-density residential and commercial land uses.
Source: Environmental Protection Agency. 1986. *Results of the Nationwide Urban Runoff Program*, vol. I, Final Report, NTIS PB84-185552, Washington, DC.

The **event mean concentration** (EMC) is the total storm load (mass) divided by the total runoff volume. Mathematically,

$$EMC = C = \frac{M}{V} = \frac{\int C(t)Q(t)dt}{\int Q(t)dt} \tag{94.11}$$

where $C(t)$ = time-variable concentration, $Q(t)$ = time-variable flow, M = pollutant mass, and V = runoff volume.

TABLE 94.12 Water Quality Characteristics of Typical Urban Runoff

Constituent	Units	Event-to-Event Variability in EMCs (Coefficient of Variation)	Site Median EMC For Median Urban Site	Site Median EMC For 90th Percentile Urban Site
BOD_5	mg/L	0.5 to 1.0	9	15
COD	mg/L	0.5 to 1.0	65	140
TSS	mg/L	1 to 2	100	300
Total lead	µg/L	0.5 to 1.0	144	350
Total copper	µg/L	0.5 to 1.0	34	93
Total zinc	µg/L	0.5 to 1.0	160	500
TKN	µg/L	0.5 to 1.0	1500	3300
NO_{2}_N + NO_3_N	µg/L	0.5 to 1.0	680	1750
Total phosphorus	µg/L	0.5 to 1.0	330	700
Soluble phosphorus	µg/L	0.5 to 1.0	120	210

Source: Environmental Protection Agency. 1986. *Results of the Nationwide Urban Runoff Program*, vol. I, Final Report, NTIS PB84-185552, Washington DC.

It is clear that the EMC results from a flow-weighted average, not simply a time average of concentration. Thus, when the EMC is multiplied by the runoff volume, an estimate of the loading and receiving water is provided. The instantaneous concentration during a storm can be higher or lower than the EMC, but the use of EMC as an event characterization replaces the actual time variation of C versus t in a storm with a pulse of constant concentration having equal mass and duration as the actual event. This ensures that mass loadings from storms will be correctly represented [Huber, 1992; Mustard et al., 1987].

Just as instantaneous concentrations vary within a storm, event mean concentrations vary from storm to storm and from site to site as well. The median or fiftieth percentile EMC at a site, estimated from a time series, is called the site median EMC. When site median EMCs from different locations are aggregated, their variability can be quantified by their median and coefficient of variation (CV = standard deviation divided by the mean) to achieve an overall description of the runoff characteristics of a constituent across various sites. An indication of this variability is shown for combined sewer overflows in Table 94.12.

When dealing with the quality of urban runoff, it is no longer necessary to design for large infrequent storms, rather design storms for quality control are generally storms less than the 1-year storm. The reason why these smaller storms are more important is because they are more frequent (causing more frequent flushes of contaminants) and also because capturing the 2-month storm (six overflows per year) in a basin designed to do so will treat about 90% of the runoff events. The runoff water that enters basins designed for smaller storm events will receive primary treatment as the flow passes through the basin. Along with designing controls for the smaller storm, there are four other basic factors involved in the control of urban runoff quality: preventing the deposition of pollutants, source control of pollutants, minimizing directly connected impervious areas, and using the treatment train concept [Urbonas and Roesner, 1992].

Preventing the deposition of pollutants is the best and theoretically the easiest way to prevent the contamination of urban stormwater systems, but it takes awareness and cooperation of the citizens to properly work. Examples of problems associated with the deposition of pollutants are draining motor oil, antifreeze, gasoline, pesticides, herbicides, paints, and solvents onto an empty lot or down storm sewers. It takes management practices as well as public awareness for this control procedure to work properly.

Source controls of pollutants involve the prevention of contaminants coming into contact with rainfall and runoff. These controls involve the covering of chemical storage areas, diking around chemical unloading and potential spill areas, minimizing the use of deicing chemicals, proper use and handling of herbicides and pesticides, and disconnecting illicit wastewater connections (wastewater sewers from buildings, drains from carwashes, and chemical storage) to storm sewers. Being aware of source pollutant potentials can decrease the amount of contaminants entering water runoff.

Minimizing directly connected impervious area means reducing the amount of impermeable area that drains directly into a gutter, ditch, or pipe. This is effective because it allows a longer traveling time of the stormwater, which creates an opportunity of maximum infiltration where the rainfall lands. Grassed areas not only allow permeable landscape, but also remove pollutants from the rainfall [Urbonas and Roesner, 1992].

The treatment methods used for the treatment train can be grouped into two categories: infiltration practices and detention. Infiltration practices can be further divided into three classes: swales and filter strips, percolation trenches, and infiltration basins. Swales and filter strips use vegetation such as grasses to minimize the amount of directly connected impervious area, which also reduces the runoff velocity and allows the water an extended amount of time to infiltrate the soil. A swale is a shallow trench seen along streets and highways, and a filter strip is a strip of land across that stormwater, from an impervious area, sheet flows before entering drainage. The vegetation acts as a filter, removing approximately 80% of the contaminants during runoff [Whalen and Callum, 1988].

Percolation trenches are devices used in areas of water runoff ranging in size up to 10 acres. There are two main types of percolation trenches, the open surface type and the underground type. The purpose of a percolation trench is to be able to store water runoff for 48 to 72 h, allowing infiltration through the bottom and sidewalls of the trench. The infiltration trench is a long dug-out trench that is lined with a filter cloth or granular filter media to prevent the adjacent soils from entering the trench; it is filled with stone aggregate to allow the storage of water in the pores of the media. The trench bottom must be 4 ft above the seasonal groundwater high and bedrock.

An infiltration basin is either constructed in a natural depression in the landscape or by an embankment that captures rainwater runoff and allows infiltration of the water into the soil in a maximum of 72 h. The basin must be constructed out of soil that is very permeable and the bottom must be at least 4 ft above the seasonal groundwater high or bedrock. Vegetation adds aesthetic beauty and also breaks up the surface soil, which keeps the surface area from getting plugged [Urbonas and Roesner, 1992].

The aforementioned water quality practices are good for very permeable soils that drain small areas, but if the area is impermeable or large, then the controls above will require too much land and be impractical. As a result, a detention pond is needed. Detention ponds can be classified either as an extended detention basin (dry detention basin) or a retention pond (wet detention pond). An extended detention basin completely empties between storm events, while a retention pond has a permanent pool of water and stores the storm water above its permanent water surface. Retention ponds are much more effective in removing pollutants and this is especially true for nutrients. Retention ponds remove two to three times more phosphorus and about two times more total nitrogen than extended detention basins do. Extended detention basins work by allowing the settling of pollutants, but since detention basins allow water to steadily pass, soluble pollutants are not effectively removed. Their efficiency is good if the detention time is greater than 24 h, but if the time is under 12 h, the efficiency is very poor. Since retention

TABLE 94.13 Potential Pollutant Removal Rates, in Percent, of Various Treatment Practices

Type of Practice	Total Suspended Solids	Total P	Total N	Zinc	Lead	BOD	Bacteria
Porous pavement	85 to 95	65	75 to 85	98	80	80	N/A
Infiltration	0 to 99	0 to 75	0 to 70	0 to 99	0 to 99	0 to 90	75 to 98
Percolation trench	99	65 to 75	60 to 70	95 to 99	N/A	90	98
Retention ponds	91	0 to 79	0 to 80	0 to 71	9 to 95	0 to 69	N/A
Extended detention	50 to 70	10 to 20	10 to 20	30 to 60	75 to 90	—	50 to 90
Wetland	41	9 to 58	21	56	73	18	N/A
Sand filters	60 to 80	60 to 80	0	10 to 80	60 to 80	60 to 80	N/A

Source: Urbonas, B. R. (ed.) 1990. BMP Practices Assessment for the Development of Colorado's Stormwater Management Program, final report of the Assessment Subcommittee of Colorado's Stormwater Task Force to Colorado Water Quality Control Division, Denver.

ponds have a constant pool of water, they remove pollutants by settling and dissolved pollutants biochemically by phytoplankton growth. The pond must be designed so that the permanent pool is deep enough to allow minimal sunlight penetration at the bottom reducing weed growth, but not too deep or the bottom becomes anoxic. A depth of 3.5 to 12 ft appears to work best. In areas of dry climate, a retention pond is not feasible because such ponds cannot retain a permanent pool. A water budget must be performed on the area to make sure that evapotranspiration and exfiltration do not cause too great of a loss in the retention pond. Table 94.13 shows the potential percent removal of the different treatment methods [Stahre and Urbonas, 1990].

Defining Terms

BOD — The amount of oxygen that would be consumed if all the organics in volume of water were oxidized by bacteria and protozoa.

Combined sewer system — A system of conduits that carries storm runoff and domestic sewage together.

Digital elevation model (DEM) — An array of numbers representing the regularly spaced distribution of geographic elevations for a specific area horizontally referenced to either a UTM or geographic coordinate system in a digital form.

Energy grade line (EGL) — A line connecting the set of points along a conduit representing the total energy of the flow within the conduit referenced to a common elevation datum.

Event mean concentration (EMC) — The total storm load (mass) divided by the total runoff volume.

Hydraulic impact — The negative effect of developed project increased runoff rates, which directly results in increased flooding, erosion, or other such physically damaging phenomena.

Hydraulic grade line (HGL) — A line connecting the set of points along a conduit representing the elevation head and pressure head of the flow within the conduit referenced to a common elevation datum.

Kinematic wave routing — A simplied form of wave routing whereby the water surfaces, bed slope, and energy grade lines along a conduit are parallel.

Light detection and ranging (LIDAR) — An active sensory system that uses laser light to measure distances between an airborne platform and points on the ground, including trees, buildings, etc., to collect and generate densely spaced and highly accurate elevation data.

Separate sewer system — A system of conduits that carry storm runoff and domestic sewage separately.

Synthetic design event — A design storm whose properties such as total rainfall, intensity, duration, and distribution are empirically and statistically derived from the analysis of historical storm data for some defined geographic area.

References

American Concrete Pipe Association. 2000. *Concrete Pipe Design Manual,* American Concrete Pipe Association, Irving, TX.

American Iron and Steel Institute. 1980. *Modern Sewer Design,* 1st ed. American Iron and Steel Institute, Falls Church, VA. 1980.

American Society for Photogrammetry & Remote Sensing. 2003. The Imaging and Geospatial Information Society, ASPRS Online. 15 Apr. 2003 http://www.asprs.org.

Bedient, P. B. and Huber, W. C. 2002. Kinematic wave routing, *Hydrology and Floodplain Analysis,* 3rd ed., Prentice Hall, Inc., Upper Saddle River, NJ. Chapter 4.6.

Brater, E. F. and King, H. W. 1976. *Handbook of Hydraulics,* 6th ed., McGraw-Hill Book Company, New York.

Chow, V. T. 1959. *Open-Channel Hydraulics,* McGraw-Hill, Inc., New York.

Department of the Army, Corps of Engineers. 2003. U.S. Army Corps of Engineers' HEC-HMS and RAS program. 15 April. http://www.hec.usace.army.mil/default.html.

Federal Highway Administration, 2001. Hydraulic Engineering Circular No. 22 (HEC 22); *Urban Drainage Design Manual,* Chapter 4.4.

Huber, W. C. 1992. Contaminant transport in surface water, in *Handbook of Hydrology,* Maidment, D. R., Ed., McGraw-Hill, New York. pp. 14.1–14.50.

Mays, L. W., 2001. *Water Resources Engineering,* John Wiley & Sons, New York.

Mustard, M. H., Driver, N. E., Chyr, J., and Hansen, B. G. 1987. U.S. Geological Survey Urban Stormwater Data Base of Constituent Storm Loads; Characteristics of Rainfall, Runoff, and Antecedent Conditions; and Basin Characteristics, Water Resources Investigations Report 87-4036, U.S. Geological Survey, Denver.

National Weather Service. 2003. National Weather Service Souther Region Headquarters, 22 Apr. National Oceanic and Atmospheric Administration (NOAA); National Climatic Data Center (NCDC); Rainfall Frequency Atlas of the U.S., 15 Apr. 2003 http://www.srh.noaa.gov/lub/wx/precip_fred/precip_index.htm.

Nix, S. J. 1994. *Urban Stormwater Modeling and Simulation,* Lewis Publishers, Boca Raton, FL.

Stahre, P., and Urbonas, B. 1990. *Stormwater Detention for Drainage, Water Quality, and CSO Management,* Prentice Hall, Englewood Cliffs, NJ.

Urbonas, B. R. and Roesner, L. A. 1993. Hydrologic design for urban drainage and flood control, in *Handbook of Hydrology,* Maidment, D. R., Ed., McGraw-Hill, 1992, pp. 28.1–28.52.

U.S. Environmental Protection Agency. 2003. Home page. 15 Apr. U.S. Environmental Protection Agency's Storm Water Management Model (SWMM), 1 May 2003 http://www.epa.gov/ednnrmrl/swmm/.

Whalen, P. J. and Callum, M. G. 1988. An assessment of urban land use/stormwater runoff quality relationships and treatment efficiencies of selected stromwater management systems, South Florida Water Management District, Technical Publications 88-9.

Whipple, W., Grigg, S., Grizzard, T., Randall, C. W., Shubinski, R. P. and Tucker, L. S. 1983. *Stormwater Management in Urbanizing Areas,* Prentice Hall, Englewood Cliffs, NJ.

XV

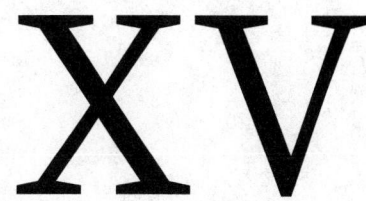

Water Resources Engineering

95 Hydraulics

Barbara Hauser

Bay de Noc Community College

Hydraulics deals with the principles that govern the behavior of liquids at rest and in motion. This is the study of the mechanics of water and its control by man. Hydraulics deals with pressurized systems and open channel flow, and includes principles of pressure and force, energy theorem, **flow** calculations and measurement, friction losses, pumps, and pumping applications.

95.1 Flow Characteristics

Laminar flow occurs at extremely low **velocity;** water molecules move in straight parallel lines called *laminae,* which slide upon each other as they travel, evidenced in groundwater flow; friction losses are minimal. *Turbulent flow,* normal pipe flow, occurs because of roughness encountered on the inner conduit walls. Outer layers of water are thrown into the inner layers; movement in different directions and at different velocities generates turbulence. *Steady-state flow* occurs if at any one point flow and velocity are unchanging. Hydraulic calculations almost always assume steady state flow. *Uniform flow* occurs when the magnitude and direction of velocity do not change from point to point.

95.2 Equation of Continuity

At a given flow, water velocity is dependent upon the cross-sectional area of the conduit. This statement expresses the most basic hydraulic equation, the equation of continuity:

$$Q = AV \tag{95.1}$$

where Q is the flow in cfs, A is the cross-sectional area in ft^2, and V is the velocity in ft/sec.

0-8493-1586-7/05/$0.00+$1.50

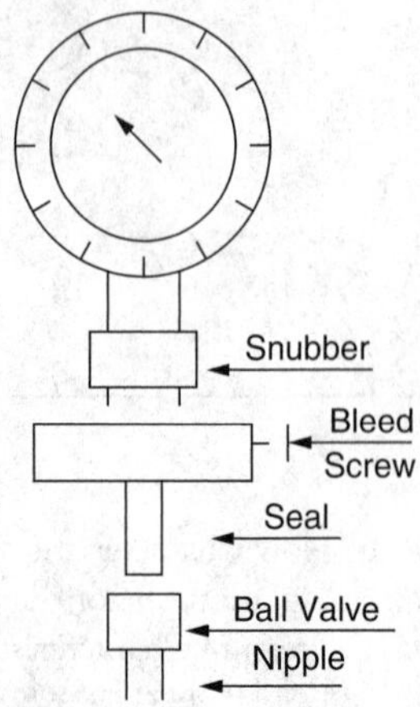

FIGURE 95.1 Pressure gage.

95.3 Pressure Characteristics

Unit Pressure

Water **pressure** is due to its weight (62.4 lb/ft^3) and the depth of water above the point of measurement (equating pressure to depth, 0.433 psi/ft water depth, or 2.31 ft water depth/psi). Water pressure does not affect liquid density, and *Pascal's law* states that the pressure at any one point in a static liquid is exerted with equal intensity in all directions.

Pressure Measurement

The *piezometer* is an open-ended vertical tube inserted at the point of pressure measurement and is impractical for measuring anything but the smallest pressures. The *manometer* is a modification of the piezometer with internal liquid fill of a higher specific gravity, bent into a U shape for easy reading. When connected to two separate sources of pressure, the unity yields a differential reading.

The *pressure gage,* called a *Bourdon tube,* may read feet, psi, or inches of mercury. It provides a direct, easy-to-read, static connection to the source of pressure (Figure 95.1).

The earth's atmosphere, about 200 miles deep, exerts a pressure of 14.7 psi or the equivalent of 34 ft water pressure upon the surface of the earth at sea level, the standard reference point for all pressure measurements. *Gage pressure* (psig) is the pressure read on a gage above or below atmospheric. *Absolute pressure* (psia) includes atmospheric pressure in the reading and is employed when calculating pumping suction **heads.**

Total Pressure

Force is registered in pounds and is calculated as follows:

$$\begin{gathered} \text{Force} = \text{Pressure} \times \text{Area} \\ \text{lb} = \text{lb}/\text{ft}^2 \times \text{ft}^2 \end{gathered} \tag{95.2}$$

Components of Pressure

Pressure head (PH) is due to the depth of the water and is measured as feet of water or registered on a pressure gage. *Velocity head* (VH) is the distance the water can move due to velocity energy; it does not register on a pressure gage, but can be captured by a pitot gage (described later) and is calculated as follows:

$$\text{VH} = \frac{V^2(\text{ft/s})^2}{2g(\text{ft/s}^2)} \tag{95.3}$$

Elevation head (Z) is pressure due to elevation above the point of reference, measured as feet of water or registered on a pressure gage.

Bernoulli's Theorem

In a fluid system employing steady-state flow, the theoretical total energy is the same at every point in the path of flow, and the energies are composed of pressure head, velocity head, and elevation head (Z). Expressed in terms of actual energy change between two points in a dynamic system where a pressure decrease or head loss (HL) occurs,

$$\text{PH}_1 + \text{VH}_1 + Z_1 = \text{PH}_2 + \text{VH}_2 + Z_2 + \text{HL} \tag{95.4}$$

95.4 Effects of Pressure — Dynamic Systems

Water hammer (hydraulic shock) is the momentary increase in pressure that occurs in a moving water system when there is a sudden change in direction or velocity of the water. **Suppressors** are installed where water hammer is encountered frequently in order to minimize shock and protect piping and appurtenances. *Surge,* a less severe form of hammer, is a slow-motion mass oscillation of water caused by internal pressure fluctuations in the system; it can be controlled by surge suppressors or spring-loaded pressure relief valves. *Thrust,* caused by an imbalance of pressures, is the force that water exerts on a pipeline as it turns a corner. Its intensity is directly proportional to water **momentum** and acts perpendicular to the outside corner of the pipe, affecting bends, tees, reducers, and dead ends, pushing the coupling away from both sections of pipeline. To calculate total pounds of thrust,

$$\text{Thrust} = 2TA \times \sin\frac{1}{2}\theta \tag{95.5}$$

where T is the test pressure of system in psf, plus 100 psi for hammer, and A is the cross-sectional area of fitting.

For pipes using push-on or **mechanical joints,** thrust restraint is desired. *Thrust blocks* are concrete blocks cast in place onto the pipe and around the outside corner of the turn. Block-bearing face must be large enough so that its pressure does not exceed soil-bearing strength (variable, <1000 to 10,000 lb/ft^2, depending on soil type) (Figure 95.2).

To calculate bearing face area of the thrust block:

$$\text{Area} = \frac{\text{Total thrust (lb)}}{\text{Bearing strength of soil (lb/ft}^2\text{)}} \tag{95.6}$$

In locations where it is difficult to use thrust blocks, *restrained joint pipe* is an alternative. Extra locking rings stabilize the joint under thrust conditions and transfer the load from the pipe directly to the surrounding soil.

95.5 Pressure Loss

Major head loss occurs because of friction dropping pressure along the conduit length. *Minor head loss* is caused by extra turbulence at bends, fittings, and diameter changes in the pipeline. In open channel systems, slope equates to the amount of pipe incline, as feet of drop per foot of pipe length; it is designed

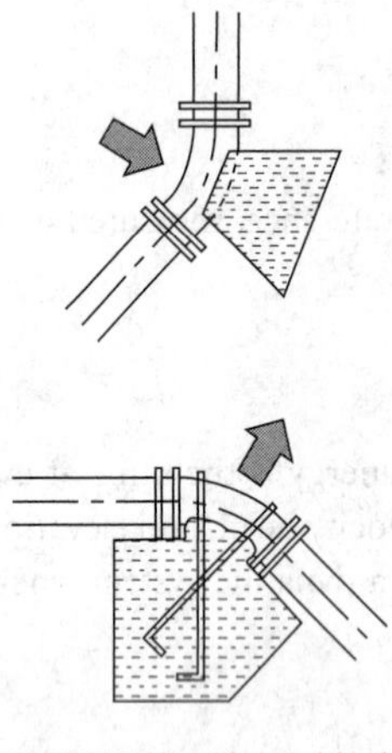

FIGURE 95.2 Thrust blocks.

TABLE 95.1 Hazen–Williams Roughness Coefficient (*C*) Value

Type of Material	*C* Value
Asbestos cement	140
Brass	140
Brick sewer	100
Cast iron: 10 years old	110
20 years old	90
Ductile iron (cement lined)	140
Concrete: Smooth	140
Rough	110
Copper	140
Fire hose (rubber lined)	135
Galvanized iron	120
Glass	140
Lead	130
Masonry conduit	130
Plastic	150
Steel: Coal-tar enamel lined	150
Unlined	140
Riveted	110
Vitrified	120

to be just enough to overcome friction losses so that velocity will remain constant. In closed conduit systems under pressure, the pipe is taken as horizontal unless otherwise indicated, and slope relates directly to the loss of pressure per foot of pipe:

$$\text{Slope} = \frac{\text{Head loss}}{\text{Length}} \tag{95.7}$$

Head Loss — Physical Components

Interior pipe roughness is dependent upon pipe material and increases with corrosion and age, designated by the *C* factor, the roughness coefficient (Table 95.1).

Length, velocity head, and diameter (inversely) also affect pressure loss. A widely used formula for flow, velocity, or head loss calculation in a closed pipe system, derived from these physical components, was developed by Hazen and Williams:

$$Q = 0.435 \times C \times d^{2.63} \times s^{0.54} \tag{95.8}$$

For field use, a **nomograph** has become popular (Figure 95.3).

Variations of this formula are easily recognizable, and with any of them approximate values can be measured or calculated for flow, velocity, slope, or length. Accuracy is limited, however, by the roughness coefficient *C*, which can only be estimated based on pipe type and age and a knowledge of water quality.

Compound Pipe Systems

When pipes are laid in series, flow is continuous through the system, and head losses in the component segments are additive. Pipes laid in parallel split flow among the components; head loss in each is identical and is the same as the total head loss. Indirect solutions may be obtained by the equivalent pipe method, creating a single pipe with head loss equivalent to that in the compound set.

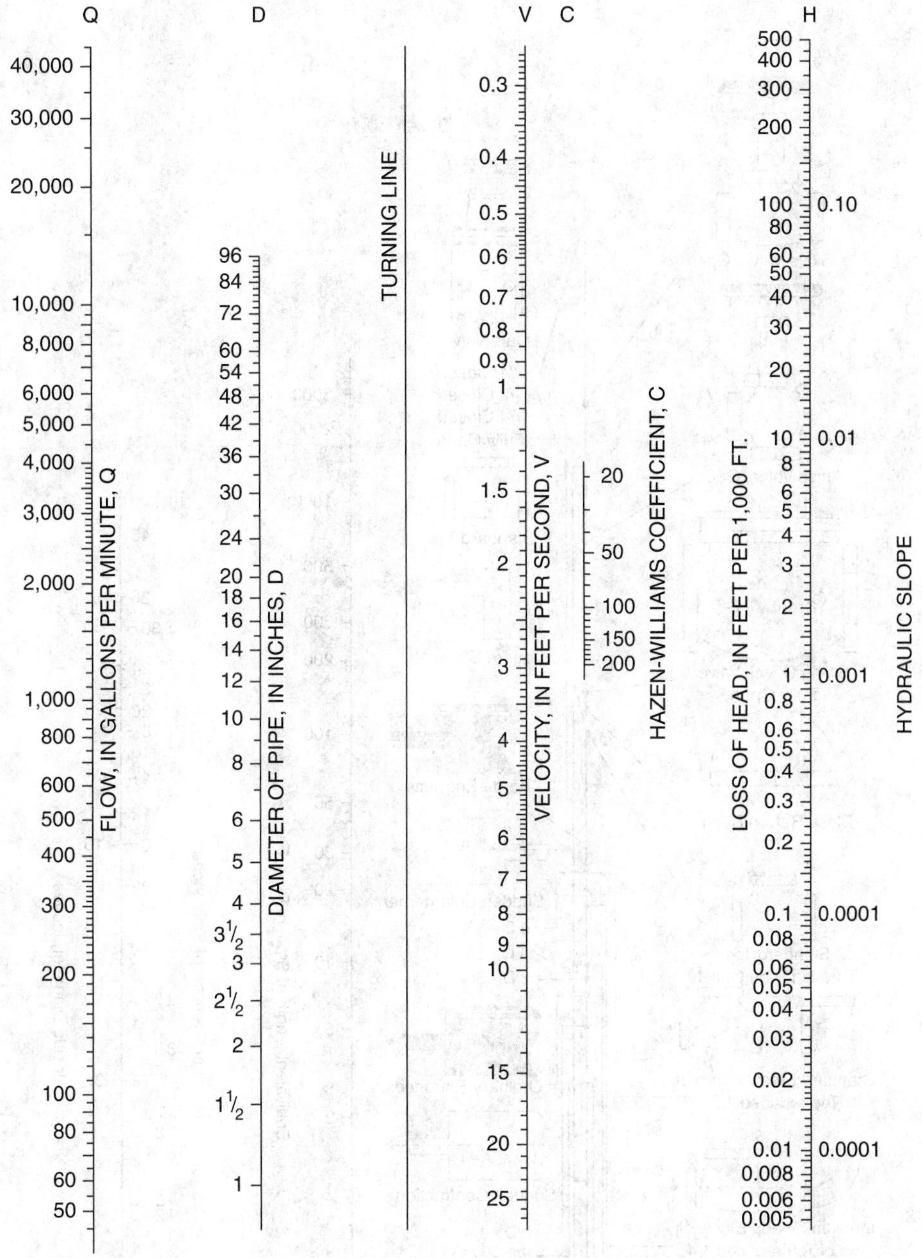

FIGURE 95.3 Hazen–Williams alignment chart.

Minor Head Loss

For determination of minor head loss, a standard nomograph is often used, with which each fitting is converted to an equivalent length of straight pipe of the same diameter. Head loss calculations using the new length of pipe will include both major and minor losses (Figure 95.4).

A few minor losses may be expressed in terms of velocity head ($V^2/2g$). Useful in calculations where Bernoulli's theorem is in use and a velocity head value is readily at hand, this method directly converts to head loss (ordinary exit: HL = 1VH; ordinary entrance: HL = 0.5VH; Borda entrance: HL = 1VH).

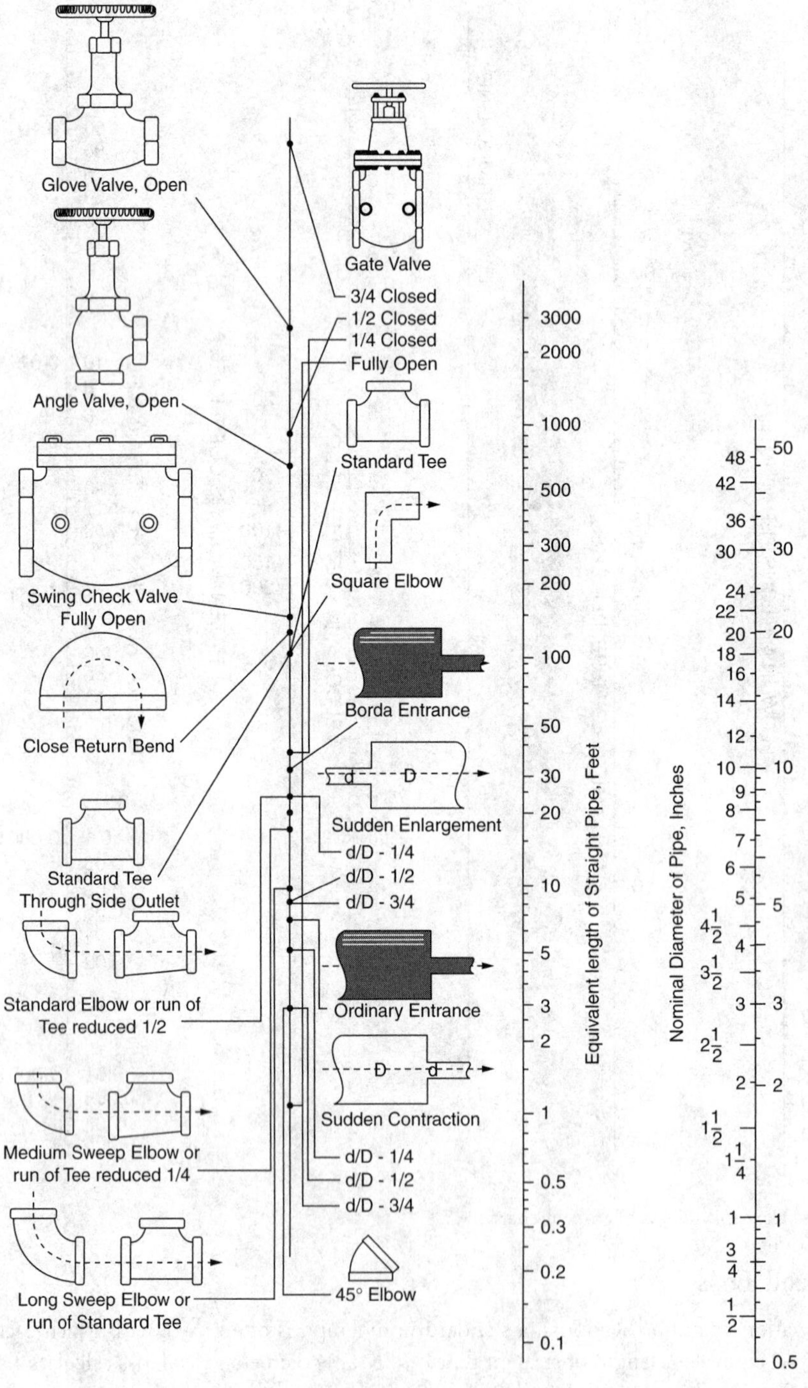

FIGURE 95.4 Resistance of valves and fittings to flow of fluids.

FIGURE 95.5 Hydraulic jump.

95.6 Open Channel Flow

In conduits where the water has a free surface exposed to atmospheric pressure, velocity head is the energy driving flow. The channel must be physically sloped enough to overcome friction losses so that velocity is maintained. In a properly sloped channel, at steady state flow, the water surface is parallel to the channel bottom and the hydraulic grade line follows. Energy loss is negated by the slope of the channel, and water depth remains constant throughout. In an open channel with a horizontal bottom, the water encounters friction and decreases in velocity, "piling up" behind; it produces the pressure head needed as a greater depth at the upstream end, and the slope of the water surface registers its progressive loss downstream. If the channel is sloped more than is necessary to overcome friction losses, velocity will increase and water depth will decrease; the steep slope creates extra velocity head. A stream bed that slopes sharply, then levels off, will carry water at a shallow depth where the slope is steep and velocity is high. Downstream, the water will be deeper, and the velocity slower (Figure 95.5).

In locations where waters of two different velocities meet, a short section of deeper water occurs; water level rises at the point of velocity change before the surface evens out again; there is extra turbulence at this point, and the water traps air and expands. Called *hydraulic jump*, the phenomenon occurs dramatically at dams and flumes, less so as shoreline waves or those created in a ship's wake, or as ripples that form when a stone is thrown into water.

Manning's formula is widely used for calculation of velocity or flow in open channels:

$$V = \frac{1.486}{n} \times R^{0.66} \times s^{0.5} \tag{95.9}$$

Manning's formula employs a roughness coefficient, n, which is specialized for materials of which open channels are constructed (Table 95.2). In this equation, diameter has been replaced by hydraulic radius (R), making the formula flexible enough to adapt to all cross-sectional areas. R is a measure of the efficiency with which the conduit transmits water and is determined by dividing the cross-sectional area of the water by the wetted perimeter.

$$R = \frac{\text{Wetted area}}{\text{Wetted perimeter}} \tag{95.10}$$

For pipes less than full, wetted area and wetted perimeter are difficult to ascertain, and a hydraulic elements curve is an indirect but accurate method of obtaining the desired value based on its percentage of the full pipe value (Figure 95.6).

TABLE 95.2 Manning Roughness Coefficient (n) Value

Type of Material	n Value
Pipe	
Cast iron: Coated	0.012 to 0.014
Uncoated	0.013 to 0.015
Wrought iron: Galvanized	0.015 to 0.017
Black	0.012 to 0.015
Steel: Riveted	0.015 to 0.017
Corrugated	0.021 to 0.026
Wood stave	0.012 to 0.013
Concrete	0.012 to 0.017
Vitrified	0.013 to 0.015
Clay, drainage tile	0.012 to 0.014
Lined Channels	
Metal: Smooth semicircular	0.011 to 0.015
Corrugated	0.023 to 0.025
Wood: Planed	0.010 to 0.015
Unplaned	0.011 to 0.015
Cement lined	0.010 to 0.013
Concrete	0.014 to 0.016
Cement rubble	0.017 to 0.030
Unlined Channels	
Earth: Uniform	0.017 to 0.025
Winding	0.023 to 0.030
Stony	0.025 to 0.040
Rock: Uniform	0.025 to 0.035
Jagged	0.035 to 0.045

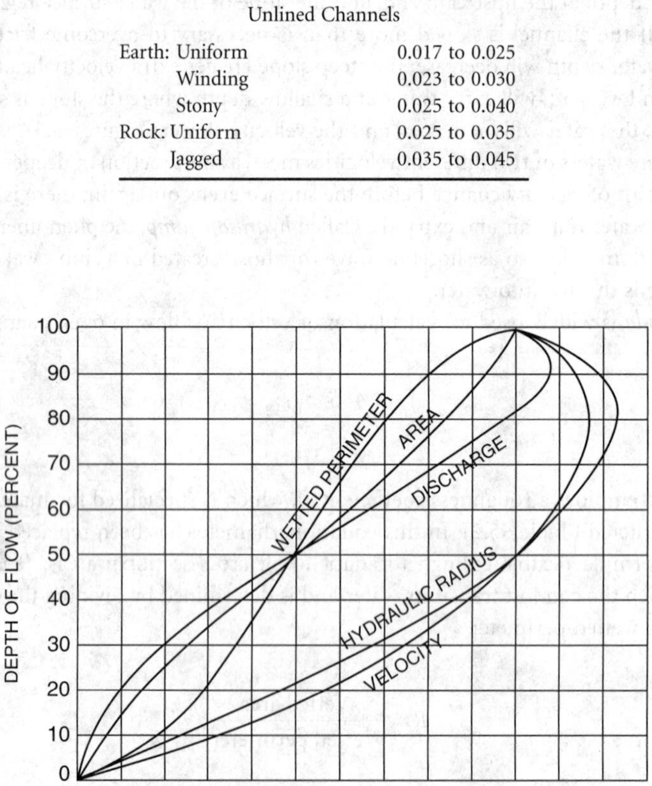

FIGURE 95.6 Hydraulic elements curve.

95.7 Flow Measurement

Orifice Meter

An orifice meter is a flat steel plate with a precisely sized small-diameter hole at the center that is installed between flanges in a pipeline; the pressure differential created across the orifice is measured by gages upstream and at the orifice discharge. Derived from Bernoulli's formula, calculation is based upon change in velocity head passing through the orifice:

$$Q = C_d A \sqrt{(\text{PH}_1 - \text{PH}_2) \times 2g} \tag{95.11}$$

where C_d = coefficient of discharge (0.6–0.9); A = area of the orifice (ft²); PH_1 = pressure head upstream; and PH_2 = pressure head at discharge.

Venturi Meter

The Venturi meter is the most accurate and widely used closed conduit pressure differential meter — a constricted tube, with converging section, throat, and longer diverging outlet section. Gages are placed just upstream from the convergence and at the throat; flow is smooth through the unit, and head loss is minimal (C_d = 0.98).

$$Q = C_d A \sqrt{\frac{(\text{PH}_1 - \text{PH}_2)}{1 - (d_2/d_1)^4} \times 2g} \tag{95.12}$$

Pitot Gage

The Pitot gage is a flowmeter that relies on a direct measurement of velocity head. A pressure gage with a double sensing unit (or a differential manometer) is installed into the pipeline. One end is bent backwards into the flow to capture velocity head as well as pressure head. The differential reading is velocity head, which can be converted to flow.

Magnetic Flowmeter

A magnetic flowmeter consists of a set of magnetic coils that surround the pipe, creating an electromagnetic field; an opposed pair of electrodes mounted at right angles registers the induced voltage (converted to a current signal), which is directly proportional to the velocity of the water passing through the unit.

Ultrasonic Meter

The transmissive type of ultrasonic meter sends ultrasonic beams through the pipe from opposite transmitter/receivers mounted at a diagonal to the flow stream; beam differential is directly proportional to water velocity. The reflective type sends a single sonic beam into the water from a transmitter mounted on the pipe; the beam bounces off solids in the water and is picked up at a different frequency. The magnitude of frequency change is directly proportional to water velocity.

A variation sends a beam to the water surface from an overhead transmitter and relates the return time to water depth, which is convertible to flow.

Positive Displacement Meter

Positive displacement meters are service meters suitable for residential customers; the unit consists of a measuring chamber enclosing a disk or piston; with each pulse a magnetic contact is made to a register that totalizes flow.

Turbine Meter

Turbine meters consist of a measuring chamber with a rotor that turns in response to the velocity of the water. Large customer flows (hotels, industries) are recorded with turbine or the more efficient "turbo" meters.

Compound Meter

Compound meters are installed when accurate reading at both high and low flows is required for customer billing; this device consists of a turbine meter on the main line and a positive displacement meter on the bypass.

Weir

The weir is the least costly open-channel flow meter, a flat plate over which the flow passes. For a rectangular weir,

$$Q = 3.33 \times L \times h^{1.5} \tag{95.13}$$

where L is the width of the weir and h is the head on the weir. For a 90-degree V-notch weir,

$$Q = 2.5 \times h^{2.5} \tag{95.14}$$

Parshall Flume

The Parshall flume is widely used for wastewater and irrigation water flow measurement due to low head loss; this device has self-cleansing capacity and the ability to operate accurately over a significant flow range; it consists of an inlet, a downward-inclining throat, and a diverging outlet. Depth measurement is taken from a stilling well at the inlet. The following formula applies to throat widths of 1 to 8 ft and a medium range of flows:

$$Q = 4W \times H_a^{1.52} \times W^{0.026} \tag{95.15}$$

where W is the width of the throat and H_a is the depth in the stilling well upstream.

Parshall flume formulas are based on low flows; at higher flows, a hydraulic jump forms at the outlet, which may submerge the throat, restricting flow. Formulas will not yield a true flow value; a stilling well at the throat bottom measures downstream depth H_b, and with percent submergence, the correction graph (Figure 95.7) can be referred to.

95.8 Centrifugal Pump

Centrifugal pumps are mechanical devices that convert other forms of energy to hydraulic energy; these pumps create the pressure needed for flow to occur. Pumping technology was limited to positive displacement and screw devices until the nineteenth century, when the centrifugal pump was developed. This device is a small, efficient unit employing a rapidly revolving impeller; high water velocity is developed, which is then converted to pressure upon exit. The pressure developed, the horsepower required, and the resulting efficiency vary with the discharge. These values are diagrammed graphically for each pump on a pump characteristic curve (Figure 95.8), which has been developed from the formula

$$\frac{\text{gpm} \times \text{TDH}}{3960} = \frac{\text{WHP}}{\text{Pump efficiency}} = \text{BHP} \tag{95.16}$$

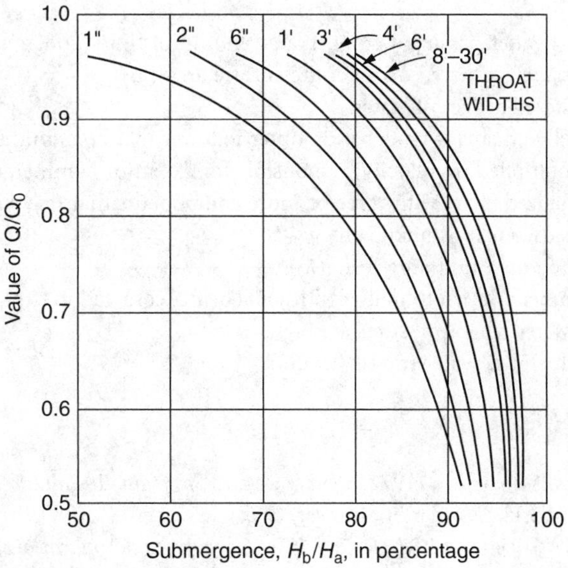

FIGURE 95.7 Corrections for Parshall flume.

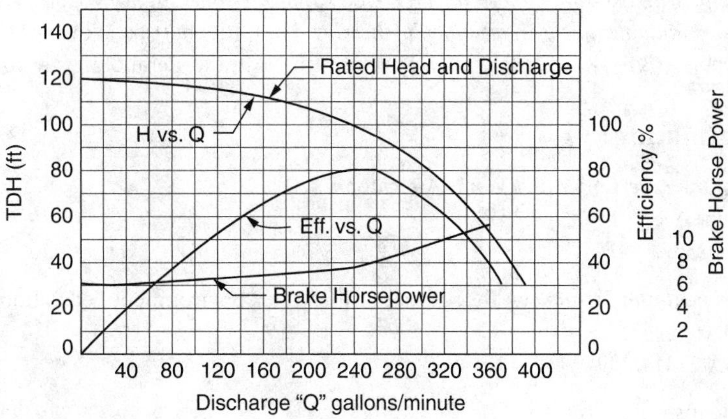

FIGURE 95.8 Pump characteristic curve.

where TDH is the total dynamic head (the work the pump must do in overcoming lift and losses to move the water), WHP is the water horsepower (the power needed to move the water), and BPH is the brake horsepower (the power that must be available to the pump).

Each pump is built to operate at its *design point,* the head and flow at which it achieves maximum efficiency. The point at which it does operate is dependent upon the characteristics of the system — the arrangement of the pipes and appurtenances through which the flow must pass. A centrifugal pump will perform according to its characteristic curve. Pump characteristics, however, may be changed by adjusting the pumping speed (change rpm or impeller size), and oak tree curves are designed to demonstrate characteristics using various sizes of impellers.

Pumps installed in series increase the pumping head, as in booster pumping (multiple pumps) or high-pressure or deep well pumps (multiple impellers). Pumps arranged in parallel increase the flow; the head remains that of one pump working.

For further information on incompressible fluids, refer to Chapter 35.

Defining Terms

Flow — The quantity of water passing a point in a given unit of time (gpd, gpm, cfs).
Force — The total pressure registered on an entire surface area (lb).
Head — Pressure, registered as feet of water.
Mechanical joint — Bell and spigot type; has an outer follower that bolts to the flanged end.
Momentum — Mass multiplied by velocity; responsible for extent of hammer, thrust, surge.
Nomograph — A specialized chart with three or more components that may yield an answer with one or more of its dimensions unknown.
Pressure — Force exerted on a unit area (psf, psi).
Suppressor — Air chamber or open container with small orifice connection to pipeline; allows temporary exit of some water when pressures are high.
Velocity — Speed at which the water travels (ft/sec).

References

American Water Works Association. 1972. *Water Meters: Selection, Installation, Testing, Maintenance,* American Water Works Association, Denver, CO.
American Water Works Association. 1980. *Basic Science Concepts and Applications,* American Water Works Association, Denver, CO.
French, R. 1985. *Open Channel Hydraulics,* McGraw-Hill, New York.
Hauser, B. 1995. *Practical Hydraulics Handbook,* Lewis, Chelsea, MI.
Kanen, J. 1986. *Applied Hydraulics for Technology,* CBS College Publishing, New York.
Prasuhn, A. 1987. *Fundamentals of Hydraulic Engineering,* Holt, Rinehart & Winston, New York.
Walski, T. 1984. *Analysis of Water Distribution Systems,* Van Nostrand Reinhold, New York.

Further Information

The Journal of the American Water Works Association
6666 W. Quincy Ave.
Denver, CO 80235

Water Environment and Technology, the journal of the Water Environment Federation
601 Wythe St.
Alexandria, VA 22314-1994

The Hydraulic Institute
712 Lakewood Center N
14600 Detroit Ave.
Cleveland, OH 44107

WATERNET, AWWA's database of the water and wastewater industries.
See AWWA's Computer Search Service, (303)347-6170.

96 Hydrology

Vijay P. Singh
Louisiana State University

Hydrology can be defined as the science that deals with occurrence, movement, distribution, and storage of water with respect to both its quantity and quality over and below the land surface in space, time, and frequency domains. Water quantity encompasses the physical aspects, and water quality, the chemical and biological aspects. One might sum up hydrology as the study of water in all aspects at macro and higher scales.

The study of hydrology originated in the design of hydraulic works. This historical underpinning continues to dominate the scope and the range of hydrologic investigations. It is therefore no surprise that most often civil engineering is the home of hydrology. With the changing environmental landscape, however, hydrology is beginning to establish its niche in the study of earth, environment, and ecology, and there are signs of hydrology becoming a geophysical science in its own right.

96.1 Classification of Hydrology

It is instructive to peruse the various classifications of hydrology [Singh, 1993, 1997a]. By definition, hydrology can be classified as physical hydrology, chemical hydrology, or biological hydrology; as water quantity hydrology or water quality hydrology; as surface-water hydrology or subsurface hydrology. Depending on the type of **watershed** for which the study of water is undertaken, it can be classified as agricultural hydrology, forest hydrology, urban hydrology, mountainous hydrology, desert hydrology,

0-8493-1586-7/05/$0.00+$1.50

wetlands hydrology, or coastal hydrology. Considering the form of water or where water occurs predominantly, this study can be classified as snow hydrology, ice and glacier hydrology, atmospheric hydrology, or lake hydrology. Depending on the particular emphasis on land phase or channel phase, it can be classified as watershed hydrology or river hydrology. Hydrology is also classified based on the tools employed for investigation of hydrologic systems. Parametric hydrology, theoretical hydrology, mathematical hydrology, statistical hydrology, probabilistic hydrology, stochastic hydrology, systems hydrology, and digital hydrology form this classification. The various classifications of hydrology are useful in that they point to its scope and the range of techniques and scales employed in its study.

96.2 Hydrologic Cycle

Water originates in the atmosphere when water vapors are transformed into droplets forming precipitation that falls on the land surface. Part of this precipitation, which eventually returns to the atmosphere through evaporation, is intercepted during its fall by vegetative canopy, buildings, and so on. Another part, which may fall on water surfaces such as streams, lakes, ponds, and seas, may either run off, evaporate, or get stored and finally evaporate. The remainder fills in the depressions on the ground, meets the infiltrative demand of the soil, and runs off the ground to form stream flow. The infiltrated water percolates down and recharges the groundwater and may eventually become stream flow. The final destination of all streams is the ocean, so stream flows finally reach the ocean. Part of the oceanic water returns to the atmosphere through evaporation. Part of the infiltrated water as well as of the surface flow returns to the atmosphere through evapotranspiration. Thus the cyclic movement of water from the atmosphere through precipitation to the land, through stream flow to the ocean, and through evapotranspiration back to the atmosphere is designated the **hydrologic cycle.** This movement, of course, follows devious paths. The study of the hydrologic cycle can then be defined as hydrology. A sketch of the hydrologic cycle [Ackermann et al., 1955], as shown in Figure 96.1, is meaningful in that it brings out the complexity as well as the challenge encountered in the study of water.

96.3 Laws of Science

The laws that govern the movement of water and the constituents it carries with it over and below the ground are the laws of conservation of mass, momentum, and energy. These laws are supplemented by some flux laws. The conservation of mass is expressed as a continuity equation, and that of momentum as an equation of motion. Depending on the type of flow, these equations are expressed in a variety of forms, as will be clear from the following discussion.

Surface Flow

For simplicity, only the one-dimensional form of the governing equations using a control volume is given here. The continuity equation can be expressed as

$$\frac{\partial A}{\partial t}+\frac{\partial Q}{\partial x}=q(x,t)-i(x,t)-e(x,t) \tag{96.1}$$

the momentum equation as

$$\frac{\partial u}{\partial t}+u\frac{\partial u}{\partial x}+g\frac{\partial h}{\partial x}=g(S_0-S_f)-\frac{(q-i)(u-v)}{A} \tag{96.2}$$

and the energy equation as

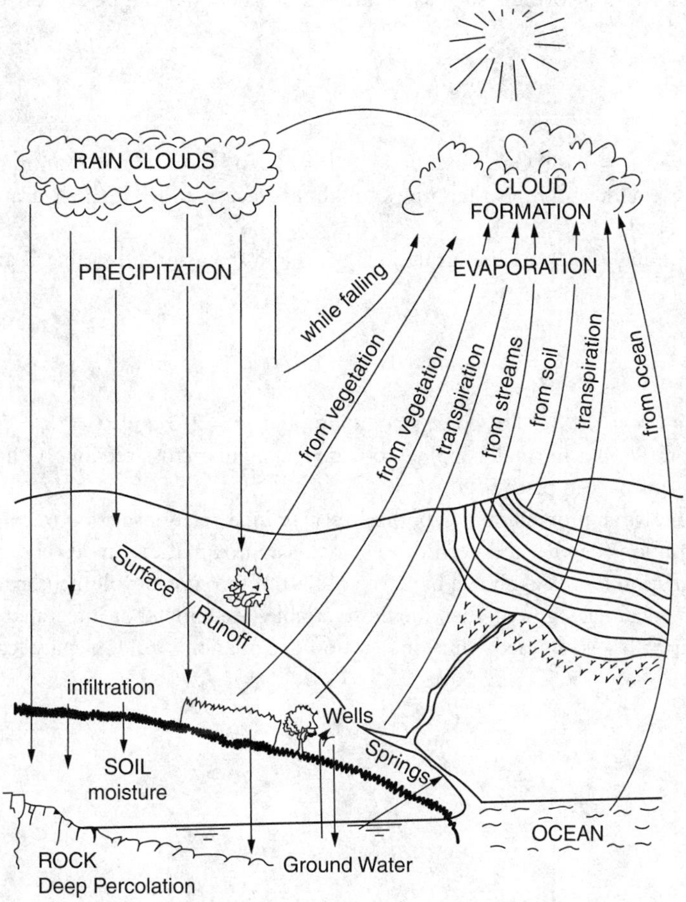

FIGURE 96.1 A schematic of hydrologic cycle. (From Ackermann, W. C., Colman, E. A., and Ogrosky, H. O. 1955. In *U.S. Department of Agriculture Yearbook 1955*, USDA, Washington, DC. pp. 41–51.)

$$\frac{\partial u}{\partial t}+u\frac{\partial u}{\partial x}+g\frac{\partial h}{\partial x}=g(S_0-S_f)+\frac{u-v(v/u)}{2A}q \tag{96.3}$$

where A is flow cross-sectional area, Q is discharge (volumetric rate = $u \cdot A$), u is average flow velocity, h is depth of flow, S_0 is bed slope, S_f is frictional slope, q is lateral inflow per unit length of flow, i is infiltration per unit length, e is evaporation rate and other abstractions per unit length, v is velocity of lateral inflow in the longitudinal direction, x is distance in the longitudinal direction, and t is time. Except for the term expressing the influence of lateral inflow or outflow, Equation (96.2) and Equation (96.3) are equivalent.

Equation (96.1), in conjunction with either Equation (96.2) or Equation (96.3), can be employed to model **surface flows** on plains and/or in channels. Two popular approximations of Equation (96.2) are the diffusion-wave and kinematic-wave approximations [Lighthill and Whitham, 1955; Dooge, 1973; Singh, 1996], which can be expressed, respectively, as follows:

$$\frac{\partial h}{\partial x}=S_0-S_f \tag{96.4}$$

$$S_0=S_f \tag{96.5}$$

With use of a uniform flow formula such as Manning's or Chezy's, S_f can be expressed as

$$S_f = \beta \frac{u^2}{R^a} \tag{96.6}$$

where $\beta = 1/C^2$ and $a = 1$ for Chezy's equation; $\beta = n_m^2$ and $a = 4/3$ for Manning's equation; C is Chezy's roughness coefficient, n_m is Manning's roughness factor, and R is the hydraulic radius (= A/P, P = wetted perimeter).

Substitution of Equation (96.6) into Equation (96.5) and the assumption that R and h are uniquely related leads to

$$u = \alpha h^m, \quad m > 0 \quad \text{or} \quad Q = \alpha h^n, \quad n = m + 1 \tag{96.7}$$

where $m = 0.5$ and $\alpha = C(S_0)^{0.5}$ for Chezy's equation, and $m = 2/3$ and $\alpha = (S_0)^{0.5}/n_m$ for Manning's equation. Equation (96.7) is normally applied to wide rectangular cross-sections. Otherwise, Q can also be expressed in terms of A in place of h.

The kinematic-wave approximation hypothesizes a unique relationship between the flux (average velocity), concentration (depth), and position. Thus, this approximation can also be expressed in forms different from Equation (96.7), as shown by Beven [1979]. If the control volume is extended to the scale of a watershed or a channel segment, then the flow variables are lumped or integrated in space and only their temporal variability is retained. Thus, integration of Equation (96.1) in space leads to

$$\frac{dS}{dt} = Q - I(t) - f(t) - E(t) \tag{96.8}$$

where

$$S = \int_{x_1}^{x_2} A\,dx, \quad Q = Q(x_2, t), \quad I = Q(x_1, t) + \int_{x_1}^{x_2} q\,dx,$$

$$f = \int_{x_1}^{x_2} i\,dx, \quad E = \int_{x_1}^{x_2} e\,dx$$

where S is storage or volume and Q is discharge as volumetric rate. Equation (96.8) is a volume balance or water budget equation with two unknowns, S and Q. Its solution requires another equation relating S to Q, I, and/or other variables. A very general relation between S and I and Q is [Singh, 1988]:

$$S = \sum_{j=0}^{M} a(Q, I) \frac{d^j Q}{dt^j} + \sum_{i=0}^{N} b(Q, I) \frac{d^i I}{dt^i} \tag{96.9}$$

where a and b are coefficients, and M and N are some integers. A special case, involving one of the most frequently used relations in hydrology, is $S = S(Q)$:

$$S = KQ, \quad S = kQ^\beta \tag{96.10}$$

where K is the storage parameter (lag time for $\beta = 1$), and k and β are parameters.

Since Equation (96.8) is derived from Equation (96.1), Equation (96.10) can be derived from the momentum equation. As an example, consider Equation (96.7) with $n = 1$. By multiplying both sides by $\Delta x = x_2 - x_1$ and recalling that $S = \Delta x \cdot h \cdot 1$ and Q is volumetric flow rate, Equation (96.10) results immediately.

Unsaturated Flow

In an unsaturated porous medium, part of the pore space is occupied by air, so the degree of saturation is to be taken into account in dealing with unsaturated flow. The moisture content θ in the medium (volume of water per unit volume of porous medium) is a function of the capillary pressure $\psi < 0$, and likewise is the medium's hydraulic conductivity $K(\psi)$. The basic governing equations for **unsaturated flow** are the continuity equation and a flux law given by Darcy's equation in lieu of the momentum equation. This flux law can also be derived from energy conservation considerations. The three-dimensional continuity equation, under the assumption of incompressible water, can be written as

$$\frac{\partial q_x}{\partial x}+\frac{\partial q_y}{\partial y}+\frac{\partial q_z}{\partial z}=-\frac{\partial \theta}{\partial t} \tag{96.11}$$

and Darcy's equation as

$$q_s=-K_s(\psi)\frac{\partial h}{\partial s}, \quad s=x,y,z; \quad \vec{q}=\{q_x,q_y,q_z\} \tag{96.12}$$

where h is the hydraulic head and q_s is the flux in the s direction. Substituting Equation (96.12) into Equation (96.11) and recalling that $h = \psi + z$, one gets

$$\frac{\partial}{\partial x}\left(K_x(\psi)\frac{\partial \psi}{\partial x}\right)+\frac{\partial}{\partial y}\left(K_y(\psi)\frac{\partial \psi}{\partial y}\right)+\frac{\partial}{\partial z}\left(K_z(\psi)\frac{\partial \psi}{\partial z}+K_z(\psi)\right)=C(\psi)\frac{\partial \psi}{\partial t}, \qquad C(\psi)=\frac{\partial \theta}{\partial \psi} \tag{96.13}$$

where $C(\psi)$ is the specific moisture capacity. This is the well-known Richards equation [Richards, 1931]. Based on simplifications of porous media properties (anisotropy and heterogeneity) and the nature of flow, a number of simpler versions can be derived [Singh, 1997b]. On the other hand, if the control volume is extended to a soil element, then spatially lumped equations can be derived. For example, Equation (96.11) can be integrated over space and expressed in the form of a water balance equation as

$$\frac{dS(t)}{dt}=f_s(t)-f(t) \tag{96.14a}$$

where $S(t)$ is the potential water storage space in the soil element, $f_s(t)$ is the seepage rate from the element, and $f(t)$ is the infiltration rate. If the initial storage space available in the element is S_0, then the amount of water storage at any time t is

$$W(t)=S_0-S(t)=\int_0^t [f(t)-f_s(t)\,dt \tag{96.14b}$$

which is an integral expression of continuity. Another relation in lieu of Equation (96.12) can be expressed [Singh and Yu, 1990] as

$$f(t)=f_s(t)+\frac{a[S(t)]^m}{[S_0-S(t)]^n} \tag{96.15}$$

where a, m, and n are positive real constants.

Saturated Flow

The governing equations for **saturated flow** are the continuity equation and the flux law specified by Darcy's equation. A three-dimensional form of continuity equation for incompressible flow is

$$\frac{\partial q_x}{\partial x}+\frac{\partial q_y}{\partial y}+\frac{\partial q_z}{\partial z}=-S_s\frac{\partial h}{\partial t} \tag{96.16}$$

where S_s is the specific storage for confined formations, or specific yield divided by the saturated thickness for unconfined formations. Darcy's equation can be written as

$$q_s=-K_s\frac{\partial h}{\partial s},\quad s=x,y,z;\quad \vec{q}=\{q_x,q_y,q_z\} \tag{96.17}$$

where K_s = the saturated hydraulic conductivity in the s direction. Substitution of Equation (96.17) into Equation (96.16) gives the general flow equation, which specializes — depending on the simplifications of porous media properties and the nature of flow — into a number of equations, such as the Laplace equation, the diffusion equation, the Theis equation, the Poisson equation, and the Boussinesq equation [Singh, 1997b].

Sediment Transport

The governing equations for transport of suspended sediment by convection and turbulent diffusion under gravity are the conservation of mass for sediment and the shallow-water equations of momentum and mass conservation for sediment-laden water. The latter two equations are Equation (96.1) and Equation (96.2). The three-dimensional form of sediment mass conservation can be expressed as

$$\frac{\partial C}{\partial t}+u\frac{\partial C}{\partial x}+v\frac{\partial C}{\partial y}+w\frac{\partial C}{\partial z}=w_s\frac{\partial C}{\partial z}+\frac{\partial}{\partial x}\left(\varepsilon_x\frac{\partial C}{\partial x}\right)+\frac{\partial}{\partial y}\left(\varepsilon_y\frac{\partial C}{\partial y}\right)+\frac{\partial}{\partial Z}\left(\varepsilon_z\frac{\partial C}{\partial z}\right) \tag{96.18}$$

where C is the concentration of sediment by volume; u, v, and w are velocity components in the x, y, and z directions; w_s is the particle fall velocity; and ε_s is the turbulent diffusion coefficient for sediment particle in the s direction ($s = x, y, z$). A number of simplifying assumptions are often made regarding the flow, which gives rise to simplifying sediment transport models [Yang, 1997]. For example, the flow is frequently assumed one-dimensional, and ε_s is considered constant or independent of the direction.

Solute Transport

The governing equations for solute transport are the conservation of solute mass and flux laws and the shallow water equations of conservation of mass and momentum of flow containing solute. For expressing the solute mass conservation, advection, diffusion, and dispersion fluxes; adsorption and desorption; and loss and gain of solute have to be expressed. If the medium is unsaturated with moisture content θ, then the solute-mass conservation can be expressed as

$$\frac{\partial}{\partial x}\left(\theta D_h\frac{\partial C}{\partial x}-qC\right)-\frac{\partial}{\partial t}(\theta C+\rho_s S)=\mu_w\theta C+\mu_s\rho_s S-\gamma_w\theta-\gamma_s\rho_s \tag{96.19}$$

where C is the solute concentration, q is Darcy's flux of water, ρ_s is the porous media bulk density, μ_w is the rate constant for first-order decay in the liquid phase, μ_s is the rate constant for first-order decay in the solid phase, γ_w is the rate constant for zero-order production in the liquid phase, γ_s is the rate constant for zero-order production in the solid phase, S is the adsorbed concentration, and D_h is the coefficient of hydrodynamic dispersion. Depending on the nature of solute and flow, a number of models can be derived from simplification of Equation (96.19) [Singh, 1997b].

Microbial Transport

The fate of microorganisms in the subsurface environment depends on their survival and retention by soil particles. The governing equations for transport and retention of microorganisms are obtained from the conservation of microbial particles in porous media [Kommalapati et al., 1991–1992]. The first governing equation is for the deposited particles, which can be written as

$$\frac{\partial(\rho\sigma)}{\partial t} = K_c(\theta C) - k_d(\rho\sigma) - R_{dd} + R_{gd} \tag{96.20}$$

where R_{dd} and R_{gd} are decay and growth terms of the deposited microbes, ρ is density of microbial particles, σ is volume of deposited bacteria per unit volume of bulk soil, C is concentration of suspended microbial particles per unit volume of flowing suspension, and θ is effective porosity.

The second governing equation is the mass balance equation, including growth and decay terms:

$$\frac{\partial(\rho\sigma)}{\partial t} + \frac{\partial(\theta C)}{\partial t} = -\nabla[-\theta D\nabla C + \theta C(v_f + v_g + v_m)] + [\theta C + \rho\sigma](\mu - b) \tag{96.21}$$

where D is the coefficient of hydrodynamic dispersion, v_f is superficial longitudinal velocity of flow, v_g is settling velocity, v_m is migration velocity, μ is a specific growth term, and b is the specific decay rate.

The third governing equation is derived from the mass conservation for the organic matter:

$$\frac{\partial(\rho_s S_F)}{\partial t} + \frac{\partial(\theta C_F)}{\partial t} = -\nabla(-D_e\theta\nabla C_F) + \theta v_f C_F) - \frac{\mu}{Y}(\theta C + \rho\sigma) \tag{96.22}$$

where ρ_s is the bulk mass density of dry soil, S_F is the mass of adsorbed substrate per unit mass of soil particles, C_f is the mass of the substrate per unit volume, D_e is the effective diffusivity coefficient, and Y is the true yield coefficient.

Depending upon the nature of flow, properties of porous media, and characteristics of microorganisms, a number of simplifications of Equation (96.20) through Equation (96.22) have been made, which then form the basis of simpler models.

Initial and Boundary Conditions

In hydrology, all three types of partial differential equations are involved. For example, Equation (96.1) and Equation (96.2) are nonlinear hyperbolic; Equation (96.11) and Equation (96.12) are nonlinear parabolic; and Equation (96.16), with unsteady state term dropped, and Equation (96.17) are nonlinear elliptic. Initial and boundary conditions are needed to obtain a unique solution to a given problem. As an example, for surface flow the usual initial condition is one of dry surface or zero flow, the upstream boundary condition is one of zero discharge, and the downstream boundary condition is specified based on the type of flow or the existence of a control. For subcritical flow in a channel, it may be given by a known control; for supercritical flow another upstream boundary is specified. In a similar manner, conditions are to be specified for unsaturated flow, saturated flow, sediment transport, solute transport, or biological transport.

96.4 Approaches to Hydrologic Problems

Hydrologic systems are analyzed in one or more of three domains: (1) time, (2) space, and (3) frequency. They are also analyzed at different space–time scales. Most of the approaches developed in hydrology can be classified on the basis of domains and scales. Let us consider a hydrologic variable Y at any location (x, y, z) as a function of time. Then one can write

$$Y(t) = \overline{Y}(t) + \varepsilon(t) \tag{96.23}$$

where $\overline{Y}(t)$ represents the mean value of Y and ε the fluctuations around the mean. If Y is entirely deterministic, then ε vanishes and any approach for determination of Y is deterministic. If Y is entirely probabilistic, then Y is completely specified through modeling of ε and the approach is entirely probabilistic. If Y is part deterministic and part stochastic, the approach employed for determination of Y is mixed. Determination of $\overline{Y}$ constitutes the subject matter of deterministic or mathematical hydrology, and that of ε the subject matter of statistical hydrology. All the deterministic approaches are either empirical, systems based, or physically based. Empirical approaches are based on data, systems approaches are based on the volume balance equation and a type of storage-discharge relation, and physically based approaches are based on the continuity equation together with an equation of motion or energy.

Methods for description of $\varepsilon(t)$ are statistical, probabilistic, or stochastic. Statistical methods are mostly empirical and yield certain moment characteristics or descriptors such as mean, variance, skewness, and so on. Probabilistic approaches employ certain axioms, conceptually or physically based, and proceed to derive the probabilistic structure of ε. The stochastic approaches, on the other hand, can be likened to systems approaches and strive to describe the entire time series of ε without, however, necessarily deriving its probabilistic structure. Time series analysis is an example of a stochastic approach where co-variance structure, correlogram, and spectral analysis are frequently employed. These approaches constitute the subject matter of statistical hydrology. Frequently, the terms *statistical, probabilistic*, and *stochastic* are used interchangeably in hydrology. In this chapter the term *statistical* will be used to mean all three types.

96.5 Tools for Hydrologic Analyses

With continuing evolution of hydrology and its expanding role in environmental studies, a greater range of scientific, mathematical, and statistical tools are becoming increasingly important. In addition to physical, chemical, and biological training, hydrologic analyses require a good level of training in mathematics (at the level of partial differential equations and finite element method), statistics (at the level of stochastic processes, time series analysis, reliability analysis, spectral analysis, and multivariate analysis), and operations research (at the level of dynamic programming). Furthermore, to take full advantage of this knowledge, hydrologists of today have to be conversant with GIS, database management, computer graphics and imaging, computer languages, and word processing. Methods of laboratory and field experimentation will receive increasing attention in the years to come. Hydrologic concepts and theories will have to be based more and more on what actually transpires in the field.

96.6 Components of Hydrology Cycle — Deterministic Hydrology

Precipitation

Precipitation forms input to hydrologic systems. It greatly varies in space and time and its space–time structure is highly random, especially at small time scales. The spatial and temporal variability of precipitation is sampled by a network of gages. These must be optimally located. An optimum number of rain gages N corresponding to an assigned percentage error ε in estimation of mean areal rainfall can be obtained as

$$N = \left(\frac{C_v}{\varepsilon}\right)^2, \quad C_v = \frac{100 S_p}{\overline{P}} \tag{96.24}$$

where C_v is the coefficient of variation of the rainfall values of the gages, S_p is the standard deviation of rainfall values, and $\overline{P}$ is the mean of rainfall values.

Precipitation measurements are often inconsistent and incomplete. Methods for checking inconsistency are either graphical or statistical [Buishand, 1982, 1984]. One of the popular methods for correcting it is the double mass curve [Singh, 1989]. Statistical methods include the von Neumann ratio test, cumulative deviations, likelihood ratio test, Bayesian tests, and run test. A multitude of methods exist for filling in the missing values, including the normal ratio method, arithmetic average, isohyetal method, the inverse distance method (IDM), ratio method, and linear programming [Kruizinga and Yperlaan, 1978; Tung, 1983]. The IDM is quite popular and is based on the assumption that the dependence between any two gages is directly proportional to the inverse of some power (between 1.5 and 2.0) of the distance between them. Thus, the missing precipitation value at a gage x for a given time interval can be computed as

$$P(x) = \sum_{i=1}^{m} w_i P_i, \quad w_i = \frac{1/(D_i^a)}{\sum_{i=1}^{m} 1/(D_i^a)} \tag{96.25a}$$

where P_i is the precipitation at the ith gage, D_i is the distance between the gages x and i, m is the number of gages used in filling (usually between three and five), and w_i is the weight assigned to the ith gage.

For hydrologic modeling, mean areal precipitation $\overline{P}$ is often needed and is obtained in a variety of ways [Singh and Birsoy, 1975; Singh and Chowdhury, 1986], including unweighted mean, grouped area-aspect weighted mean, Thiessen polygons, individual area-altitude weighted mean, triangular area weighted mean, Myers method, isohyetal method, trend surface analysis, reciprocal distance-squared method, two-axis method, double Fourier series, modified polygon method, finite element method, analysis of variance, and kriging. These methods can be expressed as

$$\overline{P} = \sum_{i=1}^{N} a_i P_i \tag{96.25b}$$

where P_i is the precipitation value at the ith gage, N is the number of gages, and a_i is the weight assigned to the ith gage. The various methods differ in computation of a_i values. For example, $a_i = 1/N$ for the arithmetic average, and $a_i = A_i/A$ for the Thiessen polygon method, where A_i is the area of the polygon surrounding the ith gage and A is the watershed area. For monthly and yearly values, all of these methods yield comparable results, but for short time intervals (e.g., hourly), isohyetal-type methods may be preferable.

An estimate of the mean areal rainfall is prone to random and systematic errors. Random errors are caused by storm characteristics, rain gage density, and the representativeness of gages, whereas systematic errors are due to improper siting, poor exposure, and change in observer and in gage. Statistical methods for estimating the error in $\overline{P}$ have been reported by Zawadzki [1973] and Bras and Rodriguez-Iturbe [1976].

With improved satellite and radar technology, it is now possible to obtain the spatial field of precipitation for a given event. Such a description can be directly input to distributed hydrologic models [Singh and Frevert, 2002a, b]. To address complex water resources, environmental and ecological problems, distributed models are being increasingly utilized these days.

Evaporation and Transpiration

Evaporation is an energy exchange process through which water is transformed to vapor, and transpiration is the process by which plants transpire water. These are the only processes by which water is returned to the atmosphere to sustain the hydrologic cycle. Evaporation from a water body differs from that from land only if the latter has limited water that may be insufficient to satisfy the evaporative demand. The evaporation occurring from water bodies is referred to as *potential evaporation* (PE). If the soil has limited water and evaporative demand is not fully satisfied, then the evaporation occurs at a rate less than the potential and is called *actual evaporation* (AE).

Evaporation from a water body depends upon atmospheric conditions such as temperature, pressure, humidity, radiation, sunshine, and wind velocity. As a result, a number of methods are available for estimating evaporation, some based on temperature, some on radiation, some on humidity, and some on combinations of the controlling factors [Jensen et al., 1990; Jones, 1992]. Perhaps the best known is the Penman–Monteith method [Penman, 1948; Monteith, 1981]. This method combines the mass-transfer and energy balance approaches, and can be expressed as

$$E = \frac{\Delta}{\Delta + \gamma}(R_n + G) + \frac{\gamma}{\Delta + \gamma}E_a \tag{96.26}$$

where Δ is the slope of the saturation vapor pressure curve for water, γ is the psychrometric constant, R_n is the net radiation, G is the sensible heat flux, and E_a is the evaporation due to water vapor saturation deficit at some height. Monteith [1981] incorporated aerodynamic and canopy resistance by modifying γ. Allen [1985] presents an account of several variants of the Penman method.

In addition to limiting soil moisture, evaporation from croplands is also affected by crop characteristics. In this case, evaporation from the soil and transpiration are combined to form *evapotranspiration* (ET). This phenomenon is also commonly referred to as *consumptive use.* One of the popular methods for its determination is the Jensen–Haise method [Jensen and Haise, 1963], expressed as

$$\text{ET} = C_T(T - T_x)R_s \tag{96.27}$$

where C_T is the temperature constant = 0.014 and T_x is the intercept axis = 26.5 for T in °F; and C_T = 0.025 and $T_x = -3$ for T in °C.

The Blaney–Criddle method [Blaney and Criddle, 1962] is also quite popular for computing ET:

$$\text{ET} = KF = \sum kf, \quad f = \frac{TP}{100} \tag{96.28}$$

where ET is the consumptive use in inches of water during the growing season, K is the seasonal consumptive use coefficient for a crop, F is the sum of monthly consumptive use factors, T is the mean monthly air temperature in °F, P is the mean monthly percentage of daytime hours, and k is the monthly consumptive use coefficient.

Extraction of water by plants is limited by the availability of soil moisture. Holmes and Robertson [1959] showed that the ratio of AE to PE varies with the drying of soil and that the nature of this variation depends upon the type of soil and vegetation as well as drying rate. Thus, PE is modulated to account for soil moisture stress for determining AE [Singh and Dickinson, 1975].

Infiltration and Soil Moisture

The process by which water enters into the soil at its surface its called *infiltration.* The subsequent downward movement of water is referred to as *percolation.* If water availability is not the limitation, then water will infiltrate the soil at the maximum rate, called *infiltration capacity,* f_p. Under water ponding, the soil's infiltration capacity declines exponentially in time. If availability of water is limited, then the infiltration rate, f, is less than the capacity rate. A number of factors affecting infiltration include soil characteristics, land use, vegetative cover, and rainfall characteristics [Smith, 2002].

A number of models for computing f_p are available. The physically based models are based on Equation (96.11) and Equation (96.12), the conceptual models on Equation (96.14a) and Equation (96.15), and empirical models on data [Smith, 2002]. Singh and Yu [1990] showed that a number of empirical models could be derived using the systems-theoretic framework based on Equation (96.14a) and Equation (96.15) and that these are actually conceptual models. Examples of some well-known conceptual models are the Philip two-term model [Philip, 1969],

$$F = At + st^{0.5} \tag{96.29}$$

the Green–Ampt model [Green and Ampt, 1911],

$$Kt = F - \eta S \ln\left(\frac{\eta S + F}{\eta S}\right) \tag{96.30}$$

and the Horton model [Horton, 1940],

$$F = f_c t + \frac{1}{k}(f_0 - f_c)[1 - \exp(-kt)] \tag{96.31}$$

where F is cumulative infiltration, t is time, A is the coefficient $\simeq$ saturated hydraulic conductivity, s is sorptivity, η is wettable porosity, S is the capillary section at the setting front, K is hydraulic conductivity, f_0 is the initial infiltration rate, f_c is the steady infiltration rate, and k is a constant depending upon the soil type and initial condition. By coupling these models with a rainfall event, the actual infiltration can be determined for that event.

The downward movement of water permits determination of soil moisture. This moisture is evapotranspired, so the status of soil moisture is predicted with the use of an appropriate evapotranspiration model.

Surface Runoff

Surface runoff originates on the land surface and includes both overland flow and channel flow. Two fundamental problems in surface runoff are its time distribution for a specified rainfall event and the amount of surface runoff (also called *yield*) generated from it. A great deal of attention has historically been directed at these two problems, principally because of their ubiquitous application in the design of hydraulic works.

Surface Runoff Hydrograph

The physically based models of the hydrograph are based on Equation (96.1) and Equation (96.2) or Equation (96.4) and Equation (96.5). The most popular models for overland flow are based on the kinematic-wave approximation and those for channel flow on diffusion-wave approximation [Singh, 1990, 1996]. For realistic field conditions, equations are solved numerically [Liggett and Woolhiser, 1967].

The conceptual models most commonly employed in hydraulic design are based on Equation (96.8) and Equation (96.9). The fundamental assumptions made in these models are that rainfall and infiltration are combined, forming excess rainfall, and that the watershed may be represented fictitiously through a network of reservoirs and/or channels, or through a geomorphologic network of average channels and cumulative overland areas [Singh, 1988]. If the watershed is assumed linear, then it is sufficient to compute the *instantaneous unit hydrograph* (IUH) for a delta-function excess rainfall, which, for a watershed represented by a linear reservoir [Dooge, 1959], is

$$h(t) = \frac{1}{k}\exp(-t/k) \tag{96.32}$$

where k is the reservoir lag time. For any excess rainfall the surface runoff hydrograph is then given by the convolution integral:

$$Q(t) = \int_0^t I(\tau)h(t-\tau)\,d\tau \tag{96.33}$$

If the watershed is represented by a series of n equal reservoirs each with lag time k, as done by Nash [1957], then

$$h(t) = \frac{1}{K\Gamma(n)}\left(\frac{t}{k}\right)^n \exp(-t/k) \tag{96.34}$$

where $\Gamma\ (n)$ is the gamma function, with n as its argument. Convolution of this IUH with an excess rainfall yields the runoff hydrograph.

Surface Runoff Volume

Determination of runoff volume for a given event is rather complicated, for the determination of the exact amount of infiltration and other abstractions in a watershed has been elusive. One of the popular methods for estimating storm runoff from agricultural areas is the SCS–curve number (SCS–CN) method [Soil Conservation Service, 1971; Mishra and Singh, 2003]. This method is based on two main hypotheses:

1. The ratio of the actual amount of runoff to the potential amount of runoff is equal to the ratio of the actual amount of infiltration to the potential infiltration.
2. An initial amount of abstraction must be satisfied before commencement of any runoff, and that consists of interception, surface storage, and infiltration.

The SCS-CN method can be written as

$$\frac{V_p - V_r - V_Q}{V_R} = \frac{V_Q}{V_p - V_r} \tag{96.35a}$$

where V_R is maximum possible retention, V_r is initial abstraction, V_p is amount of rainfall, and V_Q is amount of runoff. Note that $V_r = aV_R, a \simeq 0.1$ to 0.2. The term V_R depends upon the characteristics of soil cover complex and antecedent soil moisture conditions. SCS expresses V_R as

$$V_r = \frac{1000}{C_N} - 10 \tag{96.35b}$$

where C_N is the curve number on a scale of 10 to 100. For computation of daily, weekly, monthly, or yearly runoff, water balance models are employed. Such models are also employed for computation of storm runoff. Reviews of such models have been presented by Sorooshian [1983] and Renard et al. [1982], [Singh and Frevert, 2002a, b] among others.

Sediment Transport and Yield

Sediment has been characterized as the greatest carrier of pollutants. Sediment yield generated by a storm or on a daily, monthly, or yearly basis has been modeled in a number of ways [Singh, 1989]. By far the best known yield model for small watersheds is the *universal soil loss equation* (USLE) [Wischmeier and Smith, 1978], which can be written as

$$A = RKLSCP \tag{96.36}$$

where A is the soil loss per unit area, in the units of K and for the period of R; R is the rainfall–runoff factor; K is the erodibility factor; L is the slope-length factor; S is the slope steepness factor; C is the cover and management factor; and P is the support practice factor. The values of these factors have been extensively tabulated for a wide range of soil–vegetation–land use conditions.

The conceptual models of sediment discharge are derived based on the unit hydrograph concept [Singh, 1989; Singh et al., 1982; Williams, 1978; Sharma and Dickinson, 1980]. Consequently, a unit sediment graph for a watershed is defined as the graph of sediment discharge for a given duration that accumulates to 1 ton. The graph is generated by an effective sediment erosion intensity distributed uniformly in time and in space of the watershed. Singh [1989] has presented a comprehensive discussion of conceptual sediment yield models.

A large volume of literature on sediment transport exists [Vanoni, 1975]. For comprehensive sediment transport models, the entire transport continuum, comprising detachment, deposition, degradation, suspension, and bed-load transport, has to be taken into account. Many of these processes can be modeled using the kinematic-wave theory [Singh, 1997].

Solute Transport

The sources of water pollution can be distinguished as environmental, domestic, industrial, and agricultural. Depending upon the nature of the solute, the transport process in a hydrologic environment may entail advection, diffusion, dispersion, adsorption, desorption, decay of contaminants, chemical reactions, solubilization, precipitation, volatilization, particulate transport, and miscibility. The mechanisms for transport are advection, dispersion, diffusion, solute–solid interaction, chemical reactions, and decay phenomena. The process by which solute is transported by the bulk motion of flowing water is called *advection.* When a solute moves, it tends to spread out from its advection path. This spreading, called *hydrodynamic dispersion,* comprises mechanical dispersion and molecular diffusion. Mechanical dispersion is caused entirely by mixing during the motion of the fluid, which is impacted by the geometry of the conduit, initial and boundary conditions and the sources and sinks. In molecular diffusion, molecular constituents move under the influence of their thermal kinetic energy in the direction of their concentration gradient. Many solutes react with soil through the process of adsorption. This reaction results in partitioning of the solute into the mobile solution phase and the immobile soil surface phase. The reverse of adsorption is the process of dislodgement of chemicals from soil, called *desorption.*

The fundamental law governing the solute transport is the law of conservation of solute mass, which for one-dimensional flow through unsaturated porous media is given by Equation (96.19). Based on the consideration of sources and sinks, four groups of models can be identified [Singh, 1997b]:

1. Models with no production and no decay terms ($\gamma = \mu = 0$)
2. Models with zero production only ($\gamma \neq 0, \mu = 0$),
3. Models with first-order decay only ($\mu \neq 0, \gamma = 0$)
4. Models with zero-order production and first-order decay

If transport mechanisms constitute the basis of model classification, then they can be distinguished as kinematic-wave models and advection–dispersion models.

Microbial Transport

Contaminated soils pose one of the most serious threats to surface and groundwater quality. *In situ* bioremediation is being touted as a viable technology to remedy this problem. Fundamental to development of this technology is the modeling of microbial transport. When microbes are injected into the subsurface environment to augment degradation, one of the problems faced is the limited capacity to transport and disperse bacteria by the soil through the zone of soil contamination [Jackson et al., 1994]. The retention of bacteria by the soil matrix restricts transport of bacteria and is controlled by straining and adsorption. The models for bacterial transport are derived from Equation (96.20) through Equation (96.22). Kommalapati et al. (1991–1992) numerically solved the first two equations by assuming the absence of substrate for biological growth. In a similar manner, simplifications of these equations lead to a variety of simple models.

Models of Hydrologic Cycle

The models of hydrologic cycle are also the stream flow simulation models or the watershed hydrology models. These models are either event based or continuous-time. A large number of simulation models of both types are available. Singh [1989] has provided a short review of these models. A large number of these models are described in Singh (1995), and Singh and Frevert (2002 a, b).

96.7 Statistical Hydrology

Statistical analyses in hydrology have been developed along three lines, as mentioned previously. The following sections present short discussions of each.

Empirical Analyses

These analyses may involve determination of moments, regression and correlation, or entropy for given hydrologic data. If x_i, $i = 1, 2,\ldots, N$ represents a set of observations, then the mean $\bar{x}$, the variance S_x^2, and the skewness G are given, respectively, as

$$\bar{x} = \frac{1}{N}\sum_{i=1}^{N} x_i; \quad S_1^2 = \frac{1}{N}\sum_{i=1}^{N}(x_i - \bar{x})^2;$$
$$G = \frac{1}{(n-1)(n-2)}\sum_{i=1}^{N}(x_i - \bar{x})^3 \tag{96.37}$$

The correlation coefficient r between two data sets of data $X = \{x_i, x_2,\ldots, x_N\}$ and $Y = \{y_1, y_2, y_3,\ldots, y_N\}$ is

$$r = \frac{\text{cov }(x, y)}{[\text{var }(x)\text{ var }(y)]^{1/2}} = \frac{\Sigma_{i=1}^{N}(x_i - \bar{x})(y_i - \bar{y})}{NS_xS_y} = \frac{\Sigma_{i=1}^{N} x_i y_i - N\overline{xy}}{NS_xS_y} \tag{96.38}$$

where cov (x, y) is the covariance of X and Y, var (•) is the variance of (•), S_x is the standard deviation of X, and S_y is the standard deviation of Y. The coefficient of determination is the square of r and gives the amount of variability explained by the relationship between Y and X.

The regression analysis permits establishment of a relationship between the dependent variable Y and the set of independent variables X_i, $i = 1, 2, \ldots, M$ as

$$Y = f(X_i, i = 2,\ldots,M;\ a_i,\ \ i = 0, 1, 2,\ldots,M) \tag{96.39}$$

where a_i values are parameters. The parameters appearing in Equation (96.39) are estimated using the least squares method.

Another empirical analysis frequently employed in hydrology involves finding a frequency distribution that best fits the given set of data [Rao and Hamed, 2000]. This can be done in three ways: (1) graphically, (2) testing the fit of an assumed distribution, and (3) applying the principle of maximum entropy subject to some information about the hydrologic system the data represent [Singh and Fiorentino, 1992]. A number of distributions have been used in hydrology, and entropy has been used to derive a number of these distributions [Singh, 1998]. Other examples of empirical analyses are pattern recognition-based analyses for in-filling of missing records, stream-flow forecasting, and so on [Unny, 1982; Panu, 1992].

Phenomenological Analyses

These analyses make certain postulates about the behavior of a random variable and employ them to derive the probability distribution of the random variable. Examples of such analyses are the extreme-value analysis based on the random number of random variables [Todorovic, 1982], the point process theory of rainfall [Waymire et al., 1984], and the storage theory of reservoirs [Phatarford, 1976]. As an example, probabilistic models of floods employ such assumptions as independence, stationarity, and Markov property that concern the stochastic structure of the family of the random variables involved. A random number of random variables (RNRV) model considers flood exceedances above a threshold as a sequence of independent random variables, and the number of such exceedances within the selected time interval, say a year, also as a random variable. The choice of probability distributions or stochastic processes for these random variables leads to a number of RNRV-based flood models [Todorovic, 1982].

Stochastic Analyses

Stochastic analyses involve constructing the entire time series of a variable without explicitly knowing the probabilistic structure of the variable. Techniques based on time series analysis, spectral analysis, and so on belong to this class. As an example, consider a random variable X that is observed in a sequential manner as $X_1, X_2, \ldots$, where the subscript may represent intervals of time, distance, and so forth. When the interval is time, this sequence is a time series and is stochastic. The set of random variables $X_1, X_2, \ldots$, associated with its underlying probability distribution is a stochastic process. A time series model has a certain mathematical form and a set of parameters. A comprehensive discussion of time series models in hydrology is given by Salas et al. [1980].

96.8 Hydrologic Design

The type of project determines the need for a particular type of hydrologic information. Hydrologic design of many water resources projects is based on either peak discharge or the complete discharge hydrograph. The selection of a design flood for water resources projects such as dams and spillways involves selecting safety criteria and estimating the flood that meets these criteria. Depending upon the size of a water resources project, three types of design floods are recognized: (1) probable maximum flood (PMF), (2) standard project flood (SPF), and (3) frequency-based flood (FBF). Design of large dams is often based on the concepts of probable maximum precipitation (PMP) and probable maximum flood (PMF). These are either determined deterministically or statistically. Urban drainage is designed using design hydrographs and an acceptable level of risk. Reservoirs are designed using either the storage theory or flow duration curves. Methods of estimating a design flood can be distinguished, depending upon the data requirements, as rainfall-runoff methods, frequency-based methods, and risk-based methods.

Defining Terms

Hydrologic cycle — Denotes the endless movement of water among atmospheric, earth, and oceanic systems through the processes of precipitation, evaporation, infiltration, and stream flow.

Saturated flow — Includes the flow of water occurring in saturated geologic formations.

Surface flow — Includes flow of water occurring over the land.

Unsaturated flow — Includes the flow of water occurring in the vadose zone.

Watershed — Denotes the drainage area draining into an outlet; there is no flow across this area's boundaries. The magnitude of this area varies with the position of the outlet.

References

Ackermann, W. C., Colman, E. A., and Ogrosky, H. O. 1955. From ocean to sky to land to ocean. In *U.S. Department of Agriculture Yearbook 1955,* USDA, Washington, DC. pp. 41–51.

Allen, R. G. 1985. A Penman formula for all seasons. *Journal of Irrigation and Drainage Engineering, ASCE.* 112:348–368.

Beven, K. 1979. On the generalized kinematic routing method. *Water Resources Research.* 15(5): 1238–1242.

Blaney, H. F. and Criddle, W. D. 1962. *Determining Consumptive Use and Irrigation Water Requirements.* Tech. Bull. 1275. USDA, Washington, DC.

Bras, R. L. and Rodriguez-Iturbe, I. 1976. Evaluation of mean square error involved in approximating the areal average of a rainfall event by a discrete summation. *Water Resources Research.* 12:181–183.

Buishand, T. A. 1982. Some methods for testing the homogeneity of rainfall records. *Journal of Hydrology.* 58:11–27.

Buishand, T. A. 1984. Testing for detecting a shift in the mean of hydrological time series. *Journal of Hydrology.* 73:51–69.

Dooge, J. C. I. 1959. A general theory of the unit hydrograph. *Journal of Geophysical Research.* 64:241–256.

Dooge, J. C. I. 1973. *Linear Theory of Hydrologic Systems.* Tech. Bull. 1468. USDA, Agricultural Research Service, Washington, DC.

Green, W. H. and Ampt, C. A. 1911. Studies on soil physics. 1. Flow of air and water through soils. *Journal of Agricultural Sciences.* 4:1–24.

Holmes, R. M. and Robertson, G. W. 1959. A modulated soil moisture budget. *Monthly Weather Review.* 87(3):1–5.

Horton, R. E. 1940. An approach toward a physical interpretation of infiltration capacity. *Soil Science Society of America Proceedings.* 5:399–417.

Jackson, A., Roy, D., and Breitenbeck, G. 1994. Transport of a bacterial suspension through a soil matrix using water and an anionic surfactant. *Water Research.* 28:943–949.

Jensen, M. E., Burman, R. D., and Allen, R. G. (Eds.) 1990. *Evaporation and Irrigation Water Requirements.* ASCE Manuals and Reports on Engineering Practice No. 70. ASCE, New York.

Jensen, M. E. and Haise, H. R. 1963. Estimating evapotranspiration from solar radiation. *Journal of Irrigation and Drainage Division, ASCE.* 89:15–41.

Jones, F. E. 1996. *Evaporation of Water with Emphasis on Applications and Measurements,* Lewis Publishers, Inc., Chelsea, MI.

Kommalapati, R. R., Wang, G. T., Roy, D., and Adrian, D. D. 1991–1992. Transport and retention of microorganisms in porous media: comparison of numerical techniques and parameter estimation. *Journal of Environmental Systems.* 21:121–142.

Kruizinga, S. and Yperlaan, G. J. 1978. Spatial interpolation of daily totals of rainfall. *Journal of Hydrology.* 36:65–73.

Liggett, J. A. and Woolhiser, D. A. 1967. Finite-difference solutions of the shallow water equations. *Journal of the Engineering Mechanics Division, ASCE.* 93(EMZ):39–71.

Lighthill, M. J. and Whitham, G. B. 1955. On kinematic waves: 1. Flood movement in long rivers. *Proceedings of the Royal Society of London.* Series A. Vol. 229, pp. 281–316.

Mishra, S. K. and Sing, V. P. 2003. *Soil Conservation Service Curve Number Methodology.* Kluwer Academic Publishers, Boston.

Monteith, J. L. 1981. Evaporation and environment. *Symposium of Society for Experimental Biology.* 19:205–235.

Nash, J. E. 1957. The form of the instantaneous unit hydrograph. *IAHS.* 45(3):114–121.

Panu, U. S. 1996. Application of some entropic measures in hydrologic data infilling procedures. In *Entropy and Energy Dissipation in Water Resources,* ed. V. P. Singh and M. Fiorentino, Kluwer Academic, Dordrecht, The Netherlands. pp. 175–196.

Penman, H. L. 1948. Natural evaporation from open water, bare soil and grass. *Proceedings of the Royal Society of London.* Series A. 193:120–145.

Phatarford, R. M. 1976. Some aspects of stochastic reservoir theory. *Journal of Hydrology.* 30:199–217.

Philip, J. R. 1969. Theory of infiltration. In *Advances in Hydroscience, Vol. 5,* ed. V. T. Chow, Academic Press, New York. pp. 215–296.

Ponce, V. M. 1986. Diffusion wave modeling of catchment dynamics. *Journal of Hydraulic Engineering.* 112:716–727.

Rao, A. R. and Hamed, K. H., 2000. *Flood Frequency Analysis,* CRC Press, Boca Raton, FL.

Renard, K. G., Rawls, W. J., and Fogel, M. M. 1982. Currently available models. In *Hydrologic Modeling of Agricultural Watershed,* ed. C. T. Haan, pp. 507–522. ASAE Monograph No. 5. American Society of Agricultural Engineers, St. Joseph, MI.

Richards, L. A. 1931. Capillary conduction of liquids through porous mediums. *Physics.* 1:318–333.

Salas, J. D., Delleur, J. W., Yevjevich, V., and Lane, W. L. 1980. *Applied Modeling of Hydrologic Time Series,* Water Resources Publications, Littleton, CO.

Sharma, T. C. and Dickinson, W. T. 1980. System model of daily sediment yield. *Water Resources Research.* 16:501–506.

Singh, V. P. 1988. *Hydrologic Systems: Vol. 1. Rainfall-Runoff Modeling,* Prentice Hall, Englewood Cliffs, NJ.

Singh, V. P. 1989. *Hydrologic Systems: Vol. 2. Watershed Modeling,* Prentice Hall, Englewood Cliffs, NJ.

Singh, V. P. 1990. Hydraulic considerations for water resources modeling. *V. U. B. Hydrologie 17,* 280 pp. Vrije Universiteit Brussel, Brussels, Belgium.

Singh, V. P. 1993. *Elementary Hydrology,* Prentice Hall, Englewood Cliffs, NJ.

Singh, V. P. (ed.) 1995. *Computer Models of Watershed Hydrology,* Water Resources Publications, Littleton, Co.

Singh, V. P. 1996. *Kinematic Wave Modeling in Water Resources: Surface Water Hydrology,* John Wiley & Sons, New York.

Singh, V. P. 1997a. Hydrology: Perspectives and issues. In *Proceedings of International Symposium on Emerging Trends in Hydrology,* Vol. 1, University of Roorkee, Roorkee, India.

Singh, V. P. 1997b. *Kinematic Wave Modeling in Water Resources: Environmental Hydrology,* John Wiley & Sons, New York.

Singh, V. P. 1998. *Entropy-Based Parameter in Hydrology,* Kluwer Academic Publishers, Boston.

Singh, V.P., Baniukiewicz, A., and Chen, V. J. 1982. An instantaneous unit sediment graph study for small upland watersheds. In *Modeling Components of Hydrologic Cycle,* ed. V. P. Singh, Water Resources Publications, Littleton, CO. pp. 534–554.

Singh, V. P. and Birsoy, Y. K. 1975. Comparison of the methods of estimating mean areal rainfall. *Nordic Hydrology.* 6:222–241.

Singh, V. P. and Chowdhury, P. K. 1986. Comparing some methods of estimating mean areal rainfall. *Water Resources Bulletin.* 22:275–282.

Singh, V. P. and Dickinson, W. T. 1975. An analytical method to determine daily soil moisture. *Proceedings of the Second World Congress on Water Resources.* New Delhi, India. IV:355–365.

Singh, V. P. and Fiorentino, M. (Eds.) 1996. *Entropy and Energy Dissipation in Water Resources,* Kluwer Academic, Dordrecht, The Netherlands.

Singh, V. P. and Frevert, D. K. (eds.) 2002a. *Mathematical Models of Large Watershed Hydrology,* Water Resources Publications, Highlands Ranch, CO.

Singh, V. P. and Frevert, D. K. (eds.) 2002b. *Mathematical Models of Small Watershed Hydrology and Applications,* Water Resources Publications, Highlands Ranch, CO.

Singh, V. P. and Yu, F. X. 1990. Derivation of infiltration equation using systems approach. *Journal of Irrigation and Drainage Engineering.* 116:837–858.

Smith, R. E. 2002. Infiltration Theory for Hydrologic Applications, Water Resources Monograph 15, American Geophysical Union, Washington, DC.

Soil Conservation Service. 1971. Hydrology, In *SCS National Engineering Handbook* (section 4). USDA, Washington, DC.

Sorooshian, S. 1983. Surface water hydrology: On line estimation. *Reviews of Geophysics and Space Physics.* 21:706–721.

Todorovic, P. 1982. Stochastic modeling of floods. In *Rainfall–Runoff Relationship,* ed. V. P. Singh. Water Resources, Littleton, CO. pp. 597–650.

Tung, Y. K. 1983. Point rainfall estimation for a mountainous region. *Journal of Hydraulic Engineering, ASCE.* 109:1386–1393.

Unny, T. E. 1982. Pattern analysis for hydrologic modeling. In *Statistical Analysis of Rainfall and Runoff,* ed. V. P. Singh, Water Resources Publications, Littleton, CO. pp. 349–387.

Vanoni, V. (Ed.) 1975. *Sedimentation Engineering.* ASCE Manual and Reports on Engineering Practice No. 54. ASCE, New York.

Waymire, E., Gupta, V. K., and Rodriguez-Iturbe, I. 1984. A spectral theory of rainfall intensity at the meso-β scale. *Water Resources Research.* 20:1454–1465.

Williams, J. R. 1978. A sediment graph model based on an instantaneous unit sediment graph. *Water Resources Research.* 14:659–664.

Wischmeier, W. H. and Smith, D. D. 1978. *Predicting Rainfall Erosion Losses — A Guide to Conservation Planning.* Agricultural Handbook No. 537. Science and Education Administration, USDA, Washington, DC.

Yang, C. T. 1996. *Sediment Transport: Theory and Practice,* McGraw-Hill Book Company, New York.

Zawadzki, I. I. 1973. Errors and fluctuations of raingage estimates of areal rainfall. *Journal of Hydrology,* 18:243–255.

Further Information

Advances in Water Resources. Published quarterly by Elsevier Science Publishers.

Hydrological Sciences Journal. Published quarterly by International Association of Hydrological Sciences.

Journal of Hydrology. Published monthly by Elsevier Science Publishers.

Maidment, D. R. (Ed.) 1996. *Handbook of Hydrology.* McGraw-Hill, New York.

Water Resources Research. Published monthly by American Geophysical Union.

97

Sedimentation

Everett V. Richardson
Ayres Associates

Sedimentation studies in water resources engineering require knowledge of and the ability to determine the yield of sediment from the land surface, the transport of sediment by streams, erosion and erosion control, and reservoir sedimentation. These studies require an understanding of **fluvial geomorphology**, sediment properties, hydraulics of open channel flow, river and sediment transport mechanics, and **bed forms** in alluvial channels.

97.1 Fluvial Geomorphology

Fluvial geomorphology is the science dealing with the profiles and planforms of streams and rivers. Rivers are dynamic — always changing their shape, position, and other morphological characteristics — with long- and short-term changes in water or sediment discharge, climate, and tectonic and humanity's activities. A local change in a river or stream, such as construction of a dam or bridge, causes modifications of a river both up- and downstream. With a dam, changes in water and sediment discharge can cause degradation downstream, and backwater can cause aggradation upstream. The fluvial geomorphology of a stream determines its mode of sediment transport and its response to changes in climate, tectonic activity, and man's activity, as well as changes in sediment and water discharge.

Geomorphic factors affecting stream morphology include stream size, flow habit (ephemeral, perennial but flashy, or perennial), bed and bank material, valley setting, flood plains, natural levies, apparent incision, channel boundaries, vegetation, stream planform, variability of width, and development of bars [Vanoni, 1975; Schumm, 1977; Simons and Senturk, 1992; and Richardson et al., 2001]. Stream planform is broadly classified as **meandering, straight, braided,** or **anabranching.** These broad classifications have been subclassified by geomorphologists and engineers, but for sedimentation studies these broad classifications normally will be sufficient.

Lane [1957, Richardson et al., 2001], studying the relationship between **slope,** S, **discharge,** Q, and **channel planform,** observed that, when $SQ^{1/4} < 0.0017$, a sand bed stream had a meandering planform. Also, when $SQ^{1/4} > 0.01$, a stream had a braided planform. A stream with a value between the two could have either a meandering or a braided planform. A meandering stream whose slope increased such that the value of $SQ^{1/4}$ is larger than 0.0017 might become a braided stream. Conversely, a braided stream that

0-8493-1586-7/05/$0.00+$1.50

has its discharge or slope decreased such that the value of $SQ^{1/4}$ is smaller than 0.01 might change to a meandering planform.

Lane [1955; Schumm, 1972] also proposed the following qualitative relationship to predict the response of a stream or river to changes in **water discharge,** Q, **slope,** S, **sediment discharge,** Q_s, and **median diameter** of the bed material, D_{50}:

$$QS \sim Q_s D_{50} \tag{97.1}$$

This relationship shows that, with a decrease in Q and no changes in the qualities on the right, the slope increases. The increase in slope results from deposition of Q_s. Changes in other qualities would induce a response to keep a balance between the left and right sides of the relationship. Richardson et al. [2001] give examples of the uses of this relationship and also expand on the effect of additional variables on the response of streams to change. Leopold and Maddock, Mackin, and others [Schumm, 1972, 1977] give additional insight into the response of streams to change and the importance of fluvial geomorphology in sedimentation.

97.2 Sediment Properties

Sediment properties of value in sedimentation are the physical size, fall velocity, and density of the particles and bulk properties of the bed and bank material and sediment deposits. Sediments are composed of clay (0.0002 to 0.004 mm), silt (0.004 to 0.062 mm), sand (0.062 to 0.2 mm), gravel (2.0 to 64 mm), cobble (64 to 250 mm), or boulder (250 to 4000 mm) material. Bulk properties are described by the size frequency distributions, specific weight, and porosity. Size distributions are determined by sieving, visual accumulation tube analysis, pebble count, and pipette methods. Size distributions are usually expressed as a percentage finer than a given size in the distribution. Also of major importance is the viscosity of suspensions with large concentrations of silts and clays. Sediment properties and methods to determine them are discussed in detail by Brown [1950], Vanoni [1975], Simons and Senturk [1992], and Richardson et al. [2001].

97.3 Beginning of Motion

Knowledge of the point at which fluid forces are large enough to move sediment particles is important in sediment transport, erosion, and design of riprap. Shields [Brown, 1950; Vanoni, 1975; Buffington, 1999; and Richardson, et al., 2001] experimentally determined a relationship at beginning of motion between the ratio of the **critical shear stress** τ_c to move a particle and its **submerged weight** expressed as $(S_s - 1)\gamma D$ and the shear velocity, particle size Reynolds number $[(gRS)^{0.5} D/v]$, where S_s is the specific gravity of the particle, γ is the unit weight of water, g is the acceleration of gravity, R is the hydraulic radius, D is the particle size, and v is the kinematic viscosity. Lane, Fortier and Scobey, Keown, and others [Brown, 1950; Vanoni, 1975; and Richardson et al., 2001] give values for the critical shear or critical velocity for the beginning of motion of silts, clays, sand, and coarser particles. Equations for determining **shear stress** on a boundary are given in the above references. Richardson et al. [2001] and Richardson and Davis [2001] give the following equation for the critical velocity at the beginning of motion of a particle:

$$V_C = [K_s(S-1)D]^{1/2}\, y^{1/6} / n \tag{97.2}$$

where V_c = critical velocity above which bed material of size D and smaller will be transported (ft/sec or m/sec), D = particle size (ft or m), K_s = Sheilds parameter for beginning of motion, S_s = specific gravity of bed material particles, y = depth of flow (ft or m) and n = Manning n. Taking S_s equal to 2.65 — a typical value for sand — $K_s = 0.039$ and $n = 0.041\ D^{1/6}$, Equation (97.2) in metric units reduces to

$$V_C = 6.19 y^{1/6} D^{1/3} \quad (97.3)$$

97.4 Sediment Yield

Determining the amount of erosion (*sediment yield*) from the land has great significance in water resources engineering. Erosion is classified into overland processes and stream processes. Overland processes include sheet wash, rilling, and gullying. Stream processes include bed and bank erosion and will be described later.

The *universal soil loss equation* (USLE), developed by the U.S. Department of Agriculture, is the most widely used regression equation for predicting sediment yield from overland flow [Vanoni, 1975; Simons and Senturk, 1992]. The five major factors used to determine the average annual soil loss A, in U.S. tons per acre, are the rainfall factor R, in inches per hour; soil erodibility factor k, in tons per acre per unit of rainfall factor R; topography factors S for slope (land gradient) and L for length of slope; cropping and management factor C; and erosion control practices factor P.

97.5 Bed Forms

Flow in sand and medium gravel alluvial channels (D_{50} from 0.062 to 16 mm) is divided into lower and upper flow regime, separated by a transition zone on the basis of the **bed form, resistance to flow**, and **bed material discharge.** In the lower flow regime the bed forms in natural channels are dunes, with large resistance to flow (Manning n ranges from 0.02 to 0.04) and low bed material discharge (concentrations ranging from 200 to 2000 ppm by weight). In the upper flow regime the bed forms are plane beds or antidunes, with low resistance to flow (Manning n ranging from 0.012 to 0.020) and large bed material discharge (concentrations of 1000 ppm and larger). In the transition between the two regimes the bed form, resistance to flow, and bed material discharge range between characteristics of the two. The bed form in an alluvial stream depends on the discharge, slope, depth of flow, size of bed material, and viscosity of the fluid [Simons and Richardson, 1963; Vanoni, 1975; Richardson et al., 2001].

Alluvial streams with steep slopes may flow in the upper flow regime all the time, whereas streams with lower slopes change from lower to upper flow regimes depending upon the discharge, bed material size, and fluid viscosity. This tendency results in streams that at low flow are in the lower flow regime and at high flow are in the upper flow regimes. These streams can have a dune bed form in the summer and washed-out dunes, plane beds, or antidunes in the fall and winter. An example is the Missouri River along the border between Nebraska and Iowa. In the summer, when the water temperature is high (70 to 80°F), the bed form is in the lower flow regime and the Manning n is 0.020; in the fall, water temperatures are much lower (35 to 65°F) and the bed form is in the transition and Manning n is 0.015 with, consequently, a lower depth and higher velocity for the same discharge (U.S. Army Corps of Engineers, 1969). These changes in flow regime with discharge and temperature produces discontinuous or shifting rating curves for many sand channel streams and affects the determination of discharge, depths of flow, velocity, and bed material discharge in these streams.

97.6 Sediment Transport

The quantity of sediment transported by a stream consists of the **bed material discharge** and the **fine material discharge** (**washload discharge**) from the watershed and banks. The bed material discharge is determined by the flow variables and fluid and sediment properties and is, with some degree of accuracy, subject to calculation. The fine sediment discharge depends on availability of fine sediment, is not transported at the capacity of the stream to transport it, is not functionally related to measurable hydraulic variables, and must be measured.

Bed material and fine material are transported in suspension by the turbulence of the stream (**suspended sediment discharge**) or by rolling along in contact with the bed (**contact sediment discharge**). Laursen [Richardson et al., 2001, Richardson and Davis, 2001] determined that, when the ratio of the shear velocity [$(gRS)^{0.5}$] to the fall velocity (ω) of the bed material is less than 0.5, the bed material discharge is mostly contact bed material discharge; ratios between 0.5 and 2.0 suggest some suspended bed material discharge, and those larger than 2.0 suggest mostly suspended bed material discharge. When the bed material discharge is mostly in contact with the bed, the Meyer–Peter–Muller equation is normally used to determine bed material discharge. Suspended bed material discharge is measured or calculated by integrating the velocity and sediment concentration at a point through the vertical and across the stream. Suspended fine sediment discharge is measured by integrating the velocity and sediment concentration at a point through the vertical and across the stream using depth integrating samplers. To calculate suspended bed material discharge, an equation for the distribution of sediment particles in the vertical (developed by Rouse) is combined with a velocity distribution equation and integrated through the depth. Various investigators then combined the suspended bed material discharge with the contact bed material discharge to develop an equation to determine the total bed material discharge. To determine the total sediment discharge of a stream or river, the fine sediment discharge, which has to be measured, is added to the measured to calculated bed material discharge.

The Meyer–Peter–Muller, Einstein, modified Einstein (developed by Colby and Hembree), and other equations for calculating total bed material discharge are given by Brown [1950], Vanoni [1975], Richardson et al. [2001], and Simons and Senturk [1992].

Colby [1964] developed a very useful graphical method of determining total bed material discharge. The method is given in Figure 97.1 and Figure 97.2 and Equation (97.4) and Equation (97.5):

$$q_T = [1 + (K_1 K_2 - 1)K_3]q_n \tag{97.4}$$

$$Q_s = W q_T \tag{97.5}$$

where q_T is total bed material discharge per unit width (U.S. tons/day/ft), the K_s are correction coefficients determined from Figure 97.2, q_n is the discharge of bed material determined from Figure 97.1, and Q_s (U.S. tons/day) is the total bed material discharge for a channel of width W (ft).

The uncorrected total bed material discharge (q_n) is determined from Figure 97.1 for a given velocity, median diameter of the bed material, and two depths that bracket the desired depth by interpolating on a logarithmic graph of depth versus q_n to obtain the bed material discharge per unit width. The corrected bed material discharge per unit width (q_T) is determined using Equation (97.4) and Figure 97.2. The total bed material discharge is then obtained using Equation (97.5).

The modified Einstein method determines the **total bed material discharge** and **total sediment discharge** from measurements of the suspended sediment discharge to determine the **unmeasured sediment discharge** and, thus, is probably the most accurate of the methods used to calculate the total bed material discharge and the total sediment discharge of a stream. The **unmeasured sediment discharge** is composed of the contact bed material discharge and the suspended sediment discharge in the unmeasured zone. Suspended sediment samplers, except in special cases, only measure within a set distance to the bed (0.2 to 0.4 ft, depending on the sampler). Suspended sediment samplers that are used in very turbulent streams or in specially constructed turbulence flumes (all the bed material is in suspension, no contact discharge) or that traverse the total depth as at the nape of a weir measure the total sediment discharge of a stream. In fact, this method was used to obtain data for the development of the modified Einstein method and other equations. Suspended sediment samplers and measurement methods are described by Vanoni [1975], Simons and Senturk [1992], and Richardson et al. [2001]. Mathematical computer models have been developed to determine total bed material discharge using currently available equations. Models are described by Richardson et al. [2001] and Simons and Senturk [1992].

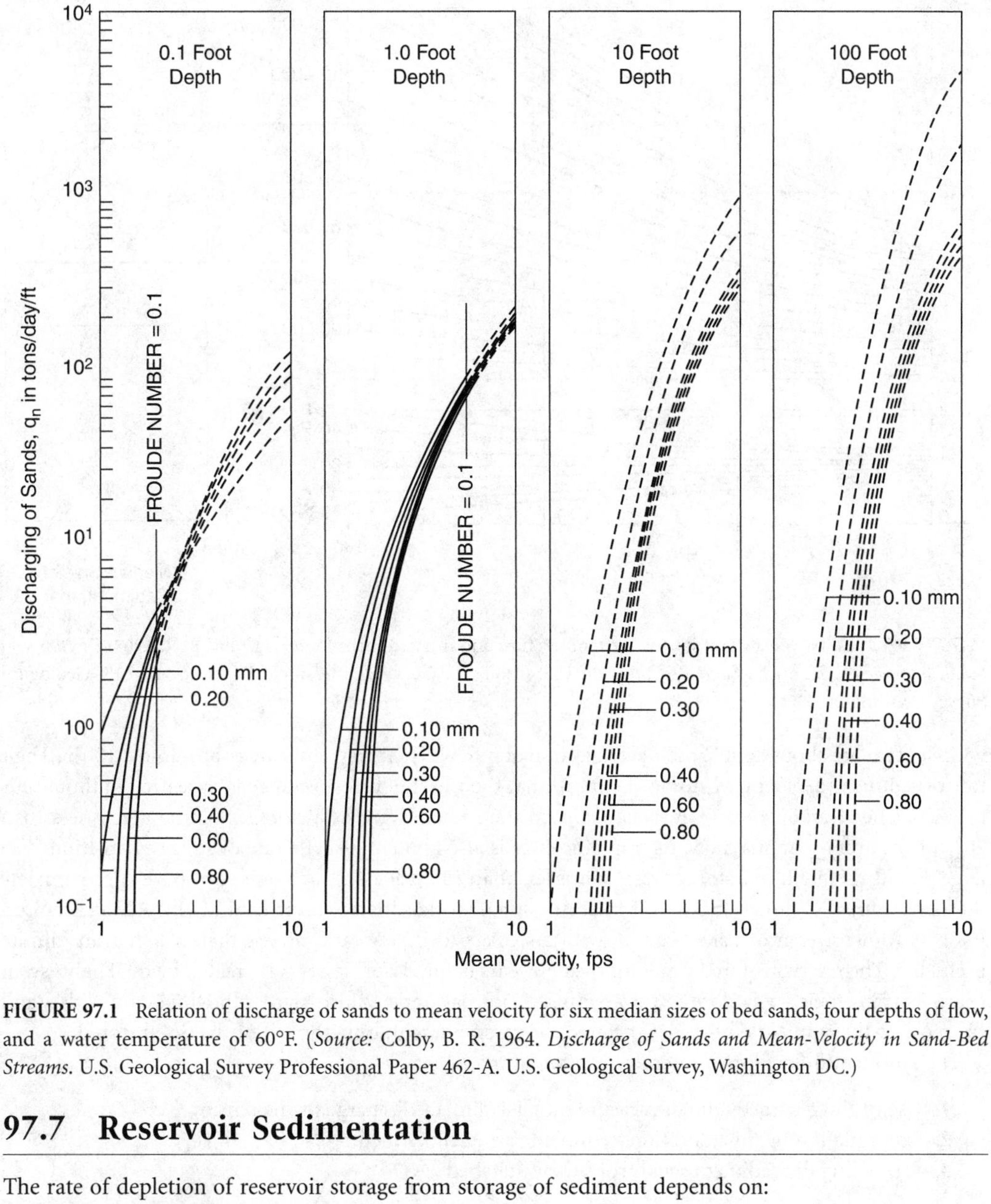

FIGURE 97.1 Relation of discharge of sands to mean velocity for six median sizes of bed sands, four depths of flow, and a water temperature of 60°F. (*Source:* Colby, B. R. 1964. *Discharge of Sands and Mean-Velocity in Sand-Bed Streams.* U.S. Geological Survey Professional Paper 462-A. U.S. Geological Survey, Washington DC.)

97.7 Reservoir Sedimentation

The rate of depletion of reservoir storage from storage of sediment depends on:

1. The volume of the reservoir in relation to inflow
2. Sediment inflow
3. Reservoir trap efficiencies
4. The specific weight (density) of the sediment deposits

The distribution of the sediment in the reservoir can sometimes be important. Sedimentation damages a reservoir over time if it decreases storage volume to such an extent that it no longer can serve its design function.

Reservoirs with large storage volume in relation to average annual inflow of water have a lower rate of storage loss due to sedimentation than reservoirs that do not — even with the same trap efficiency [Richardson, 1996]. Lake Mead on the Colorado River and Lake Nasser on the Nile River have storage

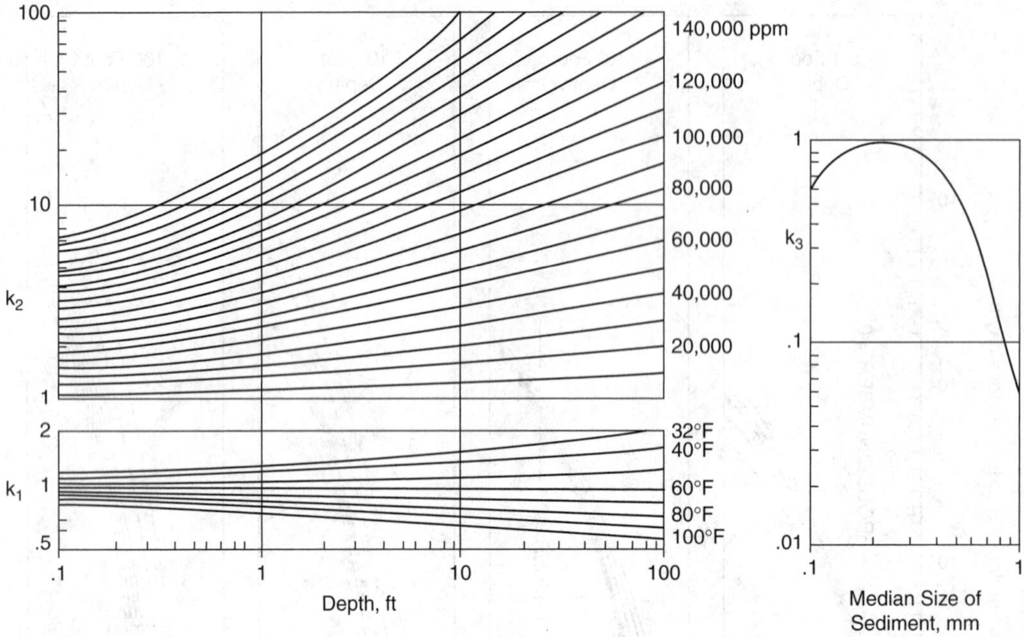

FIGURE 97.2 Colby's correction curves for temperature and fine sediment (*Source:* Colby, B. R. 1964. *Discharge of Sands and Mean-Velocity in Sand-Bed Streams.* U.S. Geological Survey Professional Paper 462-A. U.S. Geological Survey Washington, DC.)

volumes that are about double their average annual inflow. Their trap efficiencies are almost 100%. Their rate of sediment depletion of storage is so low that their useful life expectancy is measured in thousands of years. The reservoir behind Tarbella Dam on the Indus River has a storage volume that is less than 20% of the average annual flow. Its trap efficiency is also about 100%. Its rate of storage depletion is so large that its useful life as a storage reservoir is less than 100 years. All three reservoirs have approximately the same annual sediment inflow. Of interest is the fact that the reservoir behind Old Aswan Dam on the Nile (downstream of Lake Nasser), with a storage-to-inflow ratio of less than 3%, has an infinite useful life. The reservoir stored water for over 50 years before Lake Nasser was created by the High Aswan Dam. Its trap efficiency is close to zero because (1) it has under-sluices across the total dam width, and (2) most of the annual flood passes through the reservoir, with only the tail of the flood stored.

The sediment inflow into a reservoir can be estimated by

1. Use of the recorded annual measured total sediment discharge of the stream
2. Computation by flow duration–sediment rating curve method
3. Estimating the sediment yield from the watershed

The use of recorded annual measured total sediment discharge of a stream is limited to those sites that have a historical record based on frequent sampling to establish a reliable estimate of the sediment inflow into the reservoir. Normally, the record of sediment inflow is less than the stream flow record and is only suspended sediment discharge. However, with studies, the suspended sediment record can be adjusted to obtain the total sediment discharge using methods described in the previous section.

The flow duration–sediment rating curve method, which extends the available total sediment discharge record to the historical stream flow record, is the most desirable. In this method, a sediment rating curve is made by relating the daily sediment discharge (normally expressed in tons per day) to the daily discharge. There will be a large scatter in the data, some of which may be seasonal, but with study a reliable set of curves can be developed. A flow duration curve is also prepared for the entire stream flow record. From these two sets of relationships the average annual sediment discharge is determined [Vanoni,

1975; Simons and Senturk, 1992; Albertson et al., 1996]. The sediment rating curve may be developed using measured suspended sediment discharge, measured total sediment discharge (suspended sediments plus unmeasured sediment discharge), or a measured suspended sediment discharge corrected for the unmeasured sediment discharge.

The sediment yield from a watershed is estimated using the universal soil loss equation given earlier. This value is then used with the appropriate trap efficiency to determine sedimentation rates in a reservoir. This method is often used for small reservoirs.

Trap efficiency is the measure of the percentage of the sediment inflow retained (trapped) in the reservoir. It depends on the velocity of flow through the reservoir as well as sediment size. The velocity of flow depends on the size of the reservoir, type of dam, and operating procedures. Reservoirs that are formed from large embankments (soil, concrete, or rock), with large storage volumes and over-year storage, and outlets that normally discharge less flow than the incoming floods have large trap efficiencies (close to 100%). These reservoirs may have some sediment removed by density currents or large velocities when the reservoir is low, but these amounts are ignored in consideration of the other approximations and uncertainties in the determination of the sediment inflow. Small reservoirs with large velocities or reservoirs with under-sluices that can pass large inflows will have low trap efficiencies. The Old Aswan Dam described earlier is an example of the latter. Operation of the dam to have a low pool with high velocities part of each year will decrease trap efficiency.

To determine trap efficiency, Churchill — from a study of TVA reservoirs — developed a relationship between the percentage of incoming silt passing through the reservoir and the ratio of period of retention divided by mean velocity; Brune developed a relationship between percentage of sediment trapped and a ratio of reservoir capacity divided by the mean annual inflow [Vanoni, 1975; Simons and Senturk, 1992]. In using these relations, engineering judgement must be used to determine trap efficiency for a particular reservoir.

The specific weight of the sediment deposited in a reservoir is needed to convert to volume the estimate of sediment deposited in the reservoir, which is normally given in terms of weight. The specific weight of the sediment deposits in the reservoir increases with time as they consolidate. Coarse sediments (sand and gravels) will consolidate faster than finer silts and clays and will reach their ultimate weight faster. Also, sediments consolidate faster if they are not always submerged. Vanoni [1975] and Simons and Senturk [1992] present methods for estimating the specific weight developed by Lane and Koelzer, which takes into account the type of sediment and degree of submergence.

The distribution of sediment deposits in reservoirs and/or movement of sediments through reservoirs depends on the size composition of the sediment flowing into a reservoir and the management of the outflow. Coarse sediments are deposited in the upper reaches of the reservoir and finer sediments farther down. If the water storage in the reservoir is managed so as to have a low water level part of the year, then the coarser materials are moved farther down into or even through the reservoir. Coarse sediment deposits at the upper end of the reservoir can increase the backwater effects upstream. Vanoni [1975], Simons and Senturk [1992] and Albertson et al. [1996] discuss the location of sediment deposits and present methods to evaluate the location.

Sediment deposits in reservoirs are measured using sonic sounders and surveying techniques to monitor the loss of storage with time. Vanoni [1975] and Albertson et al. [1999] describes methods for conducting these surveys.

Defining Terms

Anabranch — A stream whose flow is divided at normal or lower discharges by large, relatively permanent islands. The channels are more permanent and more widely and distinctly separated than those of a braided stream.

Bed form — A relief feature on the bed of a stream, such as dunes, plane bed, or antidunes. Also called *bed configuration.*

Bed material — Material found in the bed of a stream. May be transported in contact with the bed or in suspension.

Bed material discharge — The part of the total sediment discharge of a stream that is composed of grain sizes found in the bed and is equal to the transport capability of the flow.

Braided stream — A stream whose flow is divided at normal and low flow into several channels by bars, sandbars, or islands. The bars and islands change with time, sometimes with each runoff event. A braided stream may have the aspect of a single large channel at large flows.

Contact sediment discharge — Sediment that is transported in a stream by rolling, sliding, or skipping along in contact with the bed. Also called *bed load* or *contact load.*

Critical shear stress — The minimum amount of shear (force) exerted by the flow on a particle or group of particles that is required to initiate particle motion.

Discharge — Time rate of the movement of a quantity of water or sediment passing a given cross section of a stream or river.

Fine sediment discharge (washload discharge) — That part of the total sediment discharge of a stream that is not found in appreciable quantities in the stream bed. Normally, the fine sediment discharge in a sand bed stream is composed of particles finer than sand (0.062 mm). In coarse gravel, cobble, or boulder bed stream, silts, clays and sand could be fine sediment or washload discharge.

Flow duration curve — A graph indicating the percentage of time a given discharge is exceeded.

Fluvial — Related to stream or rivers.

Geomorphology — The branch of physiography and geology that deals with the general configuration (form) of the earth's surface and the changes that take place as the result of the forces of nature.

Hydraulic radius — The cross-sectional area of a stream, divided by its wetted perimeter. Equals the depth of flow when the width is larger than 10 times depth.

Meandering stream — A stream with sinuous S-shaped flow pattern.

Median diameter — The particle diameter at which 50% of a sample's particles are coarser and 50% are finer (D_{50}).

Sediment yield — The total sediment outflow from a unit of land (field, watershed, or drainage area) at a point of reference and per unit of time.

Shear stress, tractive force — The force or drag on the channel boundaries that is caused by the flowing water. For uniform flow, average shear stress is equal to the unit weight of water times the hydraulic radius times the slope. Usually expressed as force per unit area.

Shear velocity — The square root of the shear stress divided by the mass density of water, in units of velocity.

Slope — Fall per unit length of the channel bottom, water surface, or energy grade line.

Suspended sediment — Sediment particles that are suspended in the flow by the turbulence of the stream.

Suspended sediment discharge — The quantity of suspended sediment passing through a stream cross-section per unit of time.

Total bed material discharge — The sum of the suspended sediment bed material discharge (suspended load) and the contact sediment discharge (bed-load).

Total sediment discharge — The sum of the suspended sediment discharge and the contact sediment discharge, the sum of the bed material discharge and the washload discharge, or the sum of the measured sediment discharge and the unmeasured sediment discharge.

Unmeasured sediment discharge — The sediment discharge that is not measured by suspended sediment samplers. It consists of the suspended sediment discharge in the unsampled zone and the contact sediment discharge. Suspended sediment samplers normally do not measure to the bed of a stream.

References

Albertson, M. L., Molinas, A., and Hotchkiss, R., eds. 1996. Reservoir Sedimentation. Vol. 102, Proc. Inter. Conference on Reservoir Sedimentation, Civil Engineering Department, Colorado State University, Ft. Collins, CO.

Brown, C. B. 1950. Sediment transportation. In *Engineering Hydraulics,* ed. H. Rouse, John Wiley & Sons, New York. pp. 769–858.

Buffington, J. M. 1999. The legend of A. F. Schields. *ASCE Jour. of Hydr. Eng.* 125(4), 376–387.

Colby, B. R. 1964. Discharge of Sands and Mean-Velocity in Sand-Bed Streams, U.S. Geological Survey Professional Paper 462-A. U.S. Geological Survey, Reston, VA.

Lane, E. W. 1955. The importance of fluvial morphology in hydraulic engineering. *Pro. ASCE.* 81(745), 745-1 to 745-17.

Lane, E. W. 1957. A Study of the Shape of Channels Formed by Natural Streams Flowing in Erodible Material. Missouri River Div. Sed. Series No. 9, U. S. Army Corps of Engineers, Omaha, NE.

Richardson, E. V. 1996. Estimating Reservoir Life, Proc. Inter. Conf. on Reservoir Sedimentation, Vol. 2, Civil Engineering Department, Colorado State University, Fort Collins, CO, p. 777.

Richardson, E. V. and Davis, S. R. 2001. *Evaluating Scour at Bridges.* HEC 18, Pub. No. FHWA-NHI-01-001. FHWA, Washington, DC.

Richardson, E. V., Simons, D. B., and Lagasse, P. F. 2001. *River Engineering for Highway Encroachments, Highways in the River Environment.* HDS 6, Pub. No. FHWA-NHI-01-004. FHWA, Washington, DC.

Schumm, S. A., ed.. 1972. *River Morphology.* Benchmark Papers in Geology. Dowden, Huchinson and Ross, Stroudsburg, PA.

Schumm, S. A. 1977. *The Fluvial System.* John Wiley & Sons, New York.

Simons, D. B. and Richardson, E. V., 1963. Forms of bed roughness in alluvial channels. *ASCE Trans.* 128, 284–323.

Simons, D. B. and Senturk, F. 1992. *Sediment Transport Technology,* Water Resources Publications, Littleton, CO.

U.S. Corps of Engineers. 1969. *Missouri River Channel Regime Studies.* MRD Sed. Series No. 13.B. U.S. Corps of Engineers, Omaha, NE.

Vanoni, V. A. (Ed.) 1975. *Sedimentation Engineering,* ASCE Manual No. 54, ASCE, New York.

Further Information

The *ASCE Journal of Hydraulic Engineering, Transactions* and *Hydraulic Engineering* report advances in sedimentation. For subscription information contact: ASCE, Reston, VA., Web site www.pubs.asce.org

The U.S. Geological Survey publishes data on the sediment discharge of streams in the U.S., techniques for measuring sediment discharge, and recent advances in the science of sedimentation. Catalogs of their publications are available from USGS Map Distribution, Box 25286, MS 306, Federal Center, Denver, CO 80225, Web site www.water.usgs.gov.

The U.S. Army Corps of Engineers publishes reports on their research in sedimentation. These reports focus on studies of a specific river problem as well as the general subject of sedimentation. The library at the Waterways Experiment Station, Vicksburg, MS 39180, maintains copies of most Corps publications.

American Geophysical Union (AGU) is an excellent source for additional information on sedimentation. They can be contacted at AGU, 2000 Florida Ave. NW, Washington, DC. 20009, Web site www.agu.org.

The Water Resources Bulletin of AWRA contains papers on sedimentation. Their address is AWRA, 5410 Grosvenor Lane, Suite 220, Bethesda, MD 20814-2192.

Agencies of the federal government such as ARS and SCS (U.S. Department of Agriculture), FEMA and USBR (U.S. Department of the Interior) have extensive information on sedimentation of both specific and general nature. Access to their publications can be made through their local offices or offices in Washington, DC.

XVI

Linear Systems and Models

98

Transfer Functions and Laplace Transforms

C. Nelson Dorny
University of Pennsylvania

We perceive a system primarily through its behavior. Therefore, our mental image of a system usually includes representative response **signals.** The *step response,* the behavior when we suddenly turn on the system, is such a system-characterizing signal. We should view the step response as a description of the system. The *impulse response* is another description of the system. For a system represented by linear differential equations, the unit-step response is the integral of the unit-impulse response.

Let us represent time differentiation (d/dt) by the *time-derivative operator, p.* Then we can denote the time derivative of a signal y by py, its second derivative by p^2y, its integral with respect to time by $(1/p)y$, and so on. This *operator notation* simplifies the expressions for differential equations. We shall use the expression *system equations* to mean a set of differential equations that determines fully the behaviors of the dependent variables that appear in those equations. We can reduce a *linear* set of system equations to a single **input-output system equation** by eliminating all but one dependent variable from the set. The *transfer function* associated with that dependent variable is a mathematical expression that contains all the essential information embodied in the system differential equation.

The Laplace transformation converts signals (functions of time) to functions of a *complex-frequency variable,* $s = \sigma + j\omega$. A one-to-one correspondence exists between a signal and its Laplace transform. We can retrieve the time function by inverse transformation. Laplace transformation produces images that have some properties that are more convenient than those of the original signals. In particular, time differentiating a signal corresponds to multiplying its Laplace transform by the complex-frequency variable s. Hence, the transformation converts linear constant-coefficient differential equations to linear algebraic equations. Such simplifications of time-domain operations make Laplace transformation useful.

The Laplace transformation also converts the impulse response of a system variable to the transfer function for that variable. As a consequence, we can view the differential equation that represents a linear system as an expression of the response of that system to an impulsive input.

98.1 Transfer Functions

The node displacements x_1 and x_2 and the compressive forces f_1 and f_2 within the branches of the lumped model of Figure 98.1 are related to each other by the spring equations, the damper equation, and the balance of forces at node 2. The spring equation is $f_2 = k(x_1 - x_2)$. The equation for the damper is $f_1 =$

0-8493-1586-7/05/$0.00+$1.50

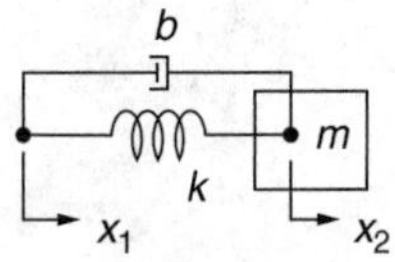

FIGURE 98.1 The lumped model of a mechanical system.

$b(px_1 - px_2)$. The balance of forces requires that $f_1 + f_2 = mp^2x_2$. These equations describe fully the behavior of the system if the spring and mass are unenergized. (If the mass were moving and/or the spring were compressed, we would have to express separately their initial energy states to describe fully the future relations among the variables.)

Eliminating f_1 and f_2 from the equations to obtain the operational equation:

$$(mp^2 + bp + k)x_2 = (bp + k)x_1 \tag{98.1}$$

This differential equation describes fully the **zero-state** relation between x_1 and x_2. Rearrange Equation (98.1) to form the ratio

$$\frac{x_2}{x_1} = \frac{bp + k}{mp^2 + bp + k} \tag{98.2}$$

We call Equation (98.2) the *transfer function* from x_1 to x_2. The transfer function focuses attention on the mathematical operations that characterize the behavioral relationships rather than on the particular natures of the variables. [Note that the transfer function from v_1 to v_2 where $v_1 = px_1$ and $v_2 = px_2$ is the same as the transfer function given by Equation (98.2).]

In general, suppose that y_1 and y_2 are two variables related (in operator notation) by the linear differential equation

$$y_2 = G(p)y_1 \tag{98.3}$$

We formally define the *transfer function* from y_1 to y_2 by

$$G(p) = \left.\frac{y_2}{y_1}\right|_{ZS} \tag{98.4}$$

where the notation ZS means **zero state.** If y_1 is an independent variable, then $G(p)$ is the *input-output transfer function* for the variable y_2 and accounts fully for its behavior owing to the **input signal** y_1. We can determine from that transfer function the behavior of the system for any source waveform and any initial state.

98.2 The Laplace Transformation

The *one-sided Laplace transformation,* L, is an integral operator that converts a signal $f(t)$ to a complex-valued function $F(s)$ in the following fashion:

$$L[f(t)] \equiv F(s) \triangleq \int_{0^-}^{\infty} f(t)e^{-st}\,dt \tag{98.5}$$

We refer to the transformed function $F(s)$ as the *Laplace transform* of the signal $f(t)$. Picture the lower limit 0^- of the integral as a *specific* instant prior to but infinitesimally close to $t = 0$. It is customary to use a lowercase symbol (f) to represent a signal waveform and an uppercase symbol (F) to represent its Laplace transform. [Although we speak here of time signals, there is nothing in Equation (98.5) that requires $f(t)$ to be a function of time. The transformation can be applied to functions of any quantity t.]

We shall use the Laplace transformation to transform the signals of **time-invariant** linear systems. The behavior of such a system for $t \geq 0$ depends only on the input signal for $t \geq 0$ and on the prior **state** of the output variable (at $t = 0^-$.

The process of finding the time function $f(t)$ that corresponds to a particular Laplace transform $F(s)$ is called *inverse Laplace transformation*, and is denoted by L^{-1}. We also call $f(t)$ the *inverse Laplace transform* of $F(s)$. Since the one-sided Laplace transformation ignores $t < 0^-$, $F(s)$ contains no information about $f(t)$ for $t < 0^-$. Therefore, inverse Laplace transformation cannot reconstruct $f(t)$ for $t < 0^-$. We shall treat all signals as if they are defined only for $t \geq 0^-$. Then there is a one-to-one relation between $f(t)$ and $F(s)$.

To illustrate the Laplace transformation, we find the Laplace transform of the decaying exponential, $f(t) = e^{-\alpha t}$, $t \geq 0^-$. The transform is

$$F(s) = \int_{0^-}^{\infty} e^{-\alpha t} e^{-st}\, dt = \left. \frac{e^{-(s+\alpha)t}}{-(s+\alpha)} \right|_{0^-}^{\infty}$$
$$= \left. \frac{e^{-(\sigma+\alpha)t} e^{-j\omega t}}{-(s+\alpha)} \right|_{0^-}^{\infty} = \frac{1}{s+\alpha} \quad \text{for } \mathrm{Re}[s] > -\alpha \tag{98.6}$$

We must require $\sigma > -\alpha$, where σ is the real part of s, in order that the real-exponent factor converges to zero at the upper limit. (The magnitude of the complex-exponent factor remains 1 for all t.) Therefore, the Laplace transform of the decaying exponential is defined only for $\mathrm{Re}[s] > -\alpha$. This restriction on the domain of F in the complex s plane is comparable to the restriction $t \geq 0^-$ on the domain of f.

The significant features of the complex-frequency function $1/(s + \alpha)$ are the existence of a single pole and the location of that pole, $s = -\alpha$ rad/sec. [The pole defines the left boundary of that region of the complex s plane over which the transform $1/(s + \alpha)$ is defined.] The significant features of the corresponding time function are the fact of decay and the rate of decay, with the exponent $-\alpha$ rad/sec. There are clear parallels between the features of $f(t)$ and $F(s)$. We should think of the whole complex-valued function F as representing the whole time waveform f.

As a second transformation example, let $f(t) = \delta(t)$, the unit impulse, essentially a unit-area pulse of very short duration. It acts at $t = 0$ — barely within the lower limit of the Laplace integral. It has value zero at $t = 0^-$. (Because we use 0^- as the lower limit of the defining integral, it does not matter whether the impulse straddles $t = 0$ or begins to rise at $t = 0$.) The impulse is nonzero only for $t \approx 0$, where $e^{-st} \approx 1$.Therefore, the Laplace transform is

$$\Delta(s) = \int_{0^-}^{\infty} \delta(t) e^{-st}\, dt \approx \int_{0^-}^{\infty} \delta(t)\,(1)\, dt = 1 \tag{98.7}$$

It is not necessary to derive the Laplace transform for each signal that we use in the study of systems. Table 98.1 gives the transforms for some signal waveforms that are common in dynamic systems.

98.3 Transform Properties

A number of useful properties of the Laplace transformation L are summarized in Table 98.2. According to the derivative property, the multiplier s acts precisely like the time-derivative operator, but in the domain of Laplace-transformed signals. When we Laplace transform the equation for an energy-storage element such as a mass or a spring, the derivative property automatically incorporates the prior energy

Table 98.1 Laplace Transform Pairs

$f(t) = L^{-1}[F(s)],\ t \geq 0^-$	$F(s) = L[f(t)]$
1. Unit impulse $\delta(t)$	1
2. Unit step $u_s(t)$	$\frac{1}{s}$
3. $t^n,\ n = 1, 2, \ldots$	$\frac{n!}{s^{n+1}}$
4. $e^{-\alpha t}$	$\frac{1}{s+\alpha}$
5. $t^n e^{-\alpha t},\ n = 1, 2, \ldots$	$\frac{n!}{(s+\alpha)^{n+1}}$
6. $\sin(\omega_0 t)$	$\frac{\omega_0}{s^2+\omega_0^2}$
7. $\cos(\omega_0 t)$	$\frac{s}{s^2+\omega_0^2}$
8. $e^{-\alpha t}\sin(\omega_d t)$	$\frac{\omega_d}{(s+\alpha)^2+\omega_d^2}$
9. $e^{-\alpha t}\cos(\omega_d t)$	$\frac{s+\alpha}{(s+\alpha)^2+\omega_d^2}$

Source: Dorny, C. N. 1993. *Understanding Dynamic Systems,* Prentice Hall, Englewood Cliffs, NJ. p. 412. With permission.

state of the element — essentially the value of the variable at $t = 0^-$. When we Laplace transform the input-output system equation for a particular system variable, the derivative property automatically incorporates the whole prior system state. As a consequence, we can find the solution to the system equation without having to determine the initial conditions (at $t = 0^+$) — a considerable simplification of the solution process.

Since $F(s)$ contains all information about $f(t)$ for $t \geq 0^-$, it is possible to find some features of the signal $f(t)$ from the transform $F(s)$ without performing an inverse Laplace transformation. Properties 7 through 9 of Table 98.2 provide three of these features — namely, the initial value ($t \to 0^+$), the final value ($t \to \infty$), and the area under the waveform. The remaining properties in the table show the effect on the transform of various changes in the signal waveform.

The usual approach to finding inverse transforms is to use a table of transform pairs. That table might be stored in a software package such as CC, MATLAB, MAPLE, and so on. Table 98.1 demonstrates that transforms of typical system signals are ratios of polynomials in s. A ratio of polynomials can be decomposed into a sum of *simple* polynomial fractions — a process referred to as *partial fraction expansion.* Hence, the inversion process can be accomplished by a computer program that incorporates a brief table of transforms.

98.4 Transformation and Solution of a System Equation

Suppose that an independent external source applies a specific velocity pattern $v_1(t)$ to node 1 of Figure 98.1. To obtain the input-output system equation that relates the velocity v_2 of node 2 to the input signal v_1, multiply Equation (98.2) by p and substitute v_1 for px_1 and v_2 for px_2. The result is

$$(mp^2 + bp + k)v_2 = (bp + k)v_1 \tag{98.8}$$

TABLE 98.2 Properties of the Laplace Transformation, L

1. Magnification	$L\,[af(t)] = a\mathbf{F}(s)$
2. Addition	$L\,[f_1(t) + f_2(t)] = \mathbf{F}_1(s) + \mathbf{F}_2(s)$
3. Derivative	$L\,[\dot{f}(t)] = \mathbf{sF}(s) - f(0^-)$
4. Derivatives	$L[\ddot{f}(t)] = s^2F(s) - sf(0^-) - \dot{f}(0^-)$
5. Integral	$L\left[\int_o^t f(t)dt\right] = \frac{F(s)}{s}$
6. Convolution	$L\left[\int_o^b f_1(\lambda) f_2(t-\lambda)\right] = F_1(s)F_2(s)$
7. Initial value	$f(0^+) = \lim_{t\to 0^+} f(t) = \lim_{s\to\infty} s\mathbf{F}(s)$
8. Final value	$f(\infty) = \lim_{t\to\infty} f(t) = \lim_{s\to 0} s\mathbf{F}(s)$ if finite
9. Definite integral	$\int_0^\infty f(t)dt = \lim_{s\to 0} \mathbf{F}(s)$ if finite
10. Exponential decay	$L[e^{-at}f(t)] = \mathbf{F}(s+\alpha)$
11. Delay	$L[f(t-t_0)u_s(t-t_0)] = e^{-t_0 s}\,F(s)$ For $t_0 \geq 0$
12. Time multiplication	$L[tf(t)] = -\frac{dF(s)}{ds}$
13. Time division	$L\left[\frac{f(t)}{t}\right] = \int_s^\infty F(s)ds$
14. Time scaling	$L[f(at)] = \frac{F(s/a)}{a}$

Source: Dorny, C. N. 1993. *Understanding Dynamic Systems,* Prentice Hall, Englewood Cliffs, NJ. p. 413. With permission.

The two sides of Equation (98.8) are identical functions of time. Therefore, the Laplace transforms of the two sides of Equation (98.8) are equal. Since the Laplace transformation is linear (properties 1 and 2 of Table 98.2), and since the coefficients of the differential equation are constants, the Laplace transform can be applied separately to the individual terms of each side. The result is

$$m[s^2V_2(s) - sv_2(0^-) - \dot{v}_2(0^-)] + b[sV_2(s) - v_2(0^-)] + kV_2(s) = b[sV_1(s) - v_1(0^-)] + kV_1(s) \tag{98.9}$$

where the derivative properties of the Laplace transformation (properties 3 and 4 of Table 98.2) introduce the prior values $v_1\,(0^-)$, $v_2\,(0^-)$, and $\dot{v}_2(0^-)$ into the equation. According to Equation (98.9), to fully determine the transform $V_2(s)$ of the behavior $v_2(t)$, we must specify these prior values and also $V_1(s)$. It can be shown that specifying the three prior values is equivalent to specifying the energy states of the spring and mass.

Let us assume that the independent source applies the constant velocity $v_1(t) = v_c$ beginning at $t = 0$. The corresponding transform, by item 3 of Table 98.1 and property 1 of Table 98.2, is $V_1(s)\ v_c/s$. Substitute the transform $V_1(s)$ into Equation (98.9) and solve for

$$V_2(s) = \frac{(bs+k)v_c + ms\dot{v}_2(0^-) + bs[v_2(0^-) - v_1(0^-)] + ms^2v_2(0^-)}{s(ms^2+bs+k)} \tag{98.10}$$

We could find the output signal waveform $v_2(t)$ as a function of the model parameters m, k, b, the source-signal parameter v_c, and the prior state information $v_1(0^-)$, $v_2(0^-)$, and $\dot{v}_2(0^-)$, but the expression for the solution would be messy. Instead, we complete the solution process for specific numbers: $m = 2$

kg, $b = 4$ N · sec/m, $k =$ N/m, $\dot{v}_2(0^-) = 0$ m/sec², $v_1(0^-) = 0$ m/sec, $v_2(0^-) = -1$ m/sec, and $v_c = 1$ m/sec. The partial-fraction expansion of the transform and the inverse transform, both obtained by a commercial computer program, are

$$V_2(s) = \frac{1}{s} - \frac{2s+2}{(s+1)^2 + 2^2} \tag{98.11}$$

$$v_2(t) = 1 - 2e^{-t}\cos(2t), \quad \text{for} \quad t \geq 0 \tag{98.12}$$

We can take Laplace transforms of the system equations at any stage in their development. We can even write the equations directly in terms of transformed variables if we wish. The process of eliminating variables can be carried out as well in one notation as in another. For example, the operator $G(p)$ in Equation (98.3) represents a ratio of polynomials in the time-derivative operator p. Therefore, Laplace transforming the differential Equation (98.3) introduces the prior values of various derivatives of y_1 and y_2. If the prior values of all these derivatives are zero, then the Laplace-transformed equation is

$$Y_2(s) = G(s)Y_1(s) \tag{98.13}$$

where the operator p in Equation (98.3) is replaced by the complex-frequency variable s in Equation (98.13). It is appropriate, therefore, to define the transfer function directly in terms of Laplace-transformed signals:

$$G(s) = \left.\frac{Y_2(s)}{Y_1(s)}\right|_{PV=0} \tag{98.14}$$

where $Y_1(s)$ and $Y_2(s)$ are the Laplace transforms of the signals $y_1(t)$ and $y_2(t)$, and the notation PV = 0 means that the prior values (at $t = 0^-$) of $y_1(t)$ and $y_2(t)$ and the various derivatives mentioned above in connection with Equation (98.13) are set to zero. The *frequency domain* definition [Equation (98.14)] is equivalent to the time *domain* definition [Equation (98.4)].

Suppose that the input signal $y_1(t)$ is the unit impulse $\delta(t)$. Then the response signal $y_2(t)$ is the unit-impulse response of the system. Since the Laplace transform of the unit impulse is $Y_1(s) = \Delta(s) = 1$ by entry 2 of Table 98.1, Equation (98.13) shows that the Laplace transform $Y_2(s)$ of the unit-impulse response is identical to the zero-state transfer function (expressed in the transform domain).

The transfer function for a linear system has two interpretations. Both interpretations characterize the system. In the frequency domain, the transfer function $G(s)$ is the multiplier that produces the response — by multiplying the source-signal transform, as in Equation (98.13). In the time domain, we use a representative response signal — the impulse response — to characterize the system. The transfer function $G(s)$ is the Laplace transform of that characteristic response.

Defining Terms

Input — An independent variable.

Input-output system equation — A differential equation that describes the behavior of a single dependent variable as a function of time. The dependent variable is viewed as the system output. The independent variable(s) are the inputs.

Output — A dependent variable.

Signal — An observable variable; a quantity that reveals the behavior of a system.

State — The state of an nth-order linear system corresponds to the values of a dependent variable and its first $n - 1$ time derivatives.

Time invariant — A system that can be represented by differential equations with constant coefficients.
Zero state — A condition in which no energy is stored or in which all variables have the value zero.

References

Franklin, G. F., Powell, J. D., and Emami-Naeini, A. 1998. *Feedback Control of Dynamic Systems,* 3rd ed. Addison Wesley, Reading, MA.
Kuo, B. C. 1991. *Automatic Control Systems,* 6th ed. Prentice Hall, Englewood Cliffs, NJ.
Nise, N. S. 1992. *Control Systems Engineering,* Benjamin Cummings, Redwood City, CA.

Further Information

A thorough mathematical treatment of Laplace transforms is presented in *Advanced Engineering Mathematics,* by C. Ray Wylie and Louis C. Barrett. *Understanding Dynamic Systems,* by C. Nelson Dorny, applies transfer functions and related concepts in a variety of contexts. The following journals publish papers that use transfer functions and Laplace transforms:

IEEE Transactions on Automatic Control. Published monthly by the Institute of Electrical and Electronics Engineers.
IEEE Transactions on Systems, Man, and Cybernetics. Published bimonthly.
Journal of Dynamic Systems, Measurement, and Control. Published quarterly by the American Society of Mechanical Engineers.

99

Block Diagrams

Taan ElAli
Wilberforce University

99.1 Introduction

Block diagrams are representations of physical systems using blocks. They are among the many representations such as transfer function representation, impulse response representation, difference equation representation, and state-space representation. Each of these representations, including the block diagram representation, gives a complete description of the system under consideration. Some of them are better than the others, and that depends on the system itself. Individual blocks can be put together to represent the physical system in block diagram form. Individual blocks can be the basic blocks or they can be subsystems. The discussion we will undertake will be applicable to discrete or analogue linear time-invariant systems. We will use analogue systems in the discussion in the Laplace domain with s being the Laplace domain variable. We will start by examining block diagrams that are built from the very basic components, and then we will look at block diagrams that are built from subsystems. We will also look at methods of reducing block diagrams with the main goal of obtaining the transfer function represen-

0-8493-1586-7/05/$0.00+$1.50

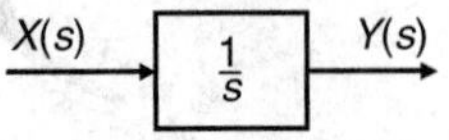

FIGURE 99.1 The ideal integrator block.

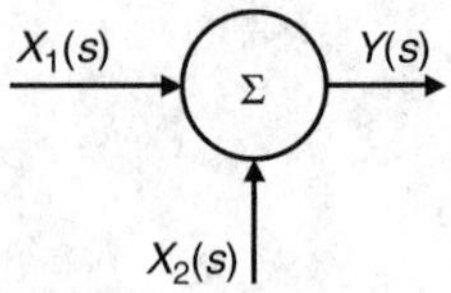

FIGURE 99.2 The adder block.

tations. We will also see in the coming discussion that block diagram representations are not unique. We can have many different block diagram representations, but the transfer function relating the inputs to the outputs is unique. Among the many block diagram representations, we are mainly concerned with about five. The first three are the canonical controllable, the canonical observable, and the diagonal or the Jordan forms. These three representations will be presented in block diagrams using the basic building blocks. The other two are the series and parallel forms and they will be built using subsystems. The first three representations are very important in the design of feedback and state estimators in the state-space domain. These three representations are also important in the process of building actual prototype circuits using operational amplifier. The series and the parallel representations are important in the process of building analogue or digital filters using first- and second-order transfer functions.

Every dynamical linear time-invariant system can be put in the transfer function representation form. Thus our starting point in building block diagrams will be the transfer function representation. Also we will deal with only causal and realizable systems. This requires that in the transfer function representation, the degree of the numerator is less than the degree of the denominator. Thus the transfer function is in the strictly proper form.

99.2 Basic Block Diagram Components

The Ideal Integrator

The integrator block diagram is seen in Figure 99.1.

The output in this case is

$$Y(s)=\frac{1}{s}X(s) \tag{99.1}$$

The Adder

The adder block diagram is seen in Figure 99.2.

The output in this case is

$$Y(s)=X_1(s)+X_2(s) \tag{99.2}$$

The Subtractor

The subtractor block diagram is seen in Figure 99.3.

The output in this case is

$$Y(s)=X_1(s)-X_2(s) \tag{99.3}$$

The Multiplier

The multiplier block diagram is seen in Figure 99.4.

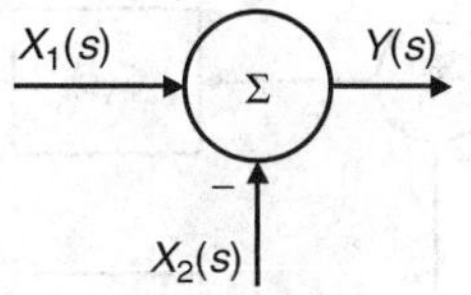

FIGURE 99.3 The subtractor block.

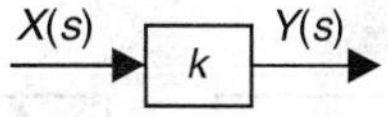

FIGURE 99.4 The multiplier block.

The output in this case is

$$Y(s) = kX(s) \tag{99.4}$$

99.3 Block Diagrams as Interconnected Subsystems

We will look at four important configurations. These are the general transfer function representation, the series representation, the parallel representation, and the basic feedback representation.

The General Transfer Function Representation

A block diagram representation for a system with $x(t)$ as the input, $y(t)$ as the output, and $h(t)$ as the impulse response can be represented as shown in Figure 99.5. Mathematically we write

$$Y(s) = X(s)H(s) \tag{99.5}$$

The Parallel Representation

If we have two subsystems connected in parallel as shown in Figure 99.6, then the output is written as

$$Y(s) = X(s)[H_1(s) + H_2(s)] \tag{99.6}$$

If we have more subsystems connected the procedure is the same; you add all the individual transfer functions for the subsystems and multiply by the input $X(s)$ to get $Y(s)$. You can find the equivalent transfer function for the whole system as

$$H(s) = \frac{Y(s)}{X(s)} = H_1(s) + H_2(s) + H_3(s) + \ldots \tag{99.7}$$

The Series Representation

If we have two subsystems connected in series as in Figure 99.7, then mathematically we write the output as

$$Y(s) = X(s)H_1(s)H_2(s) \tag{99.8}$$

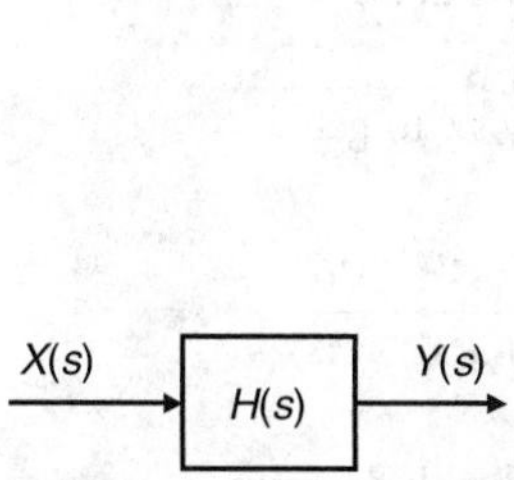

FIGURE 99.5 The general transfer function representation.

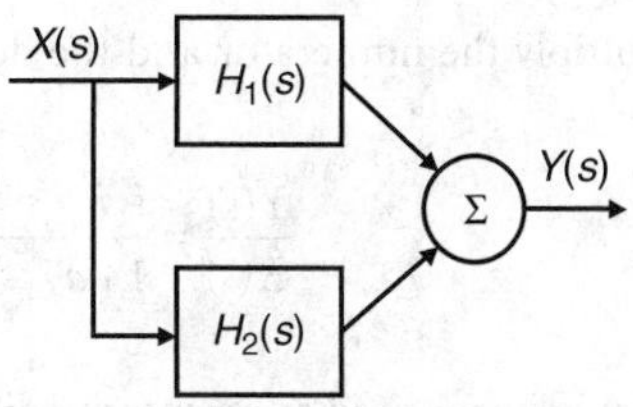

FIGURE 99.6 The parallel representation.

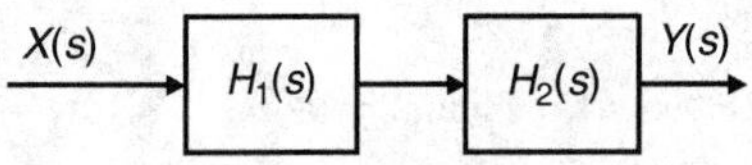

FIGURE 99.7 The series representation.

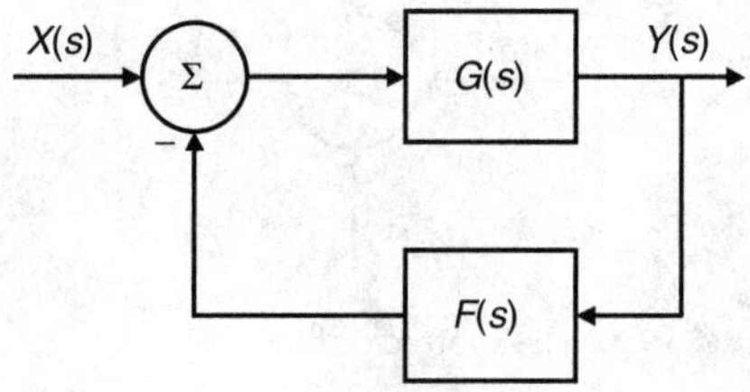

FIGURE 99.8 Basic feedback block diagram.

If we have more subsystems connected, the procedure is the same; you multiply all the individual transfer functions for the subsystems first and then multiply the result by the input $X(s)$ to get $Y(s)$. You can find the equivalent transfer function for the whole system as

$$H(s) = \frac{Y(s)}{X(s)} = H_1(s)H_2(s)H_3(s)... \quad (99.9)$$

The Basic Feedback Representation

A very basic and fundamental feedback block diagram representation is shown in Figure 99.8.

From Figure 99.8 we have

$$E(s) = X(s) - F(s)Y(s) \quad (99.10)$$

and the output is

$$Y(s) = G(s)E(s) \quad (99.11)$$

If we substitute Equation (99.10) in Equation (99.11) we have

$$H(s) = \frac{G(s)}{1+G(s(F(s)} \quad (99.13)$$

99.4 The Controllable Canonical Form (CCF) Block Diagrams with Basic Blocks

Consider the transfer function representation

$$\frac{Y(s)}{X(s)} = \frac{b_{n-1}s^{n-1} + b_{n-2}s^{n-2} + b_{n-3}s^{n-3} + ... + b_0}{s^n + a_{n-1}s^{n-1} + a_{n-2}s^{n-2} + a_{n-3}s^{n-3} + ... + a_0} \quad (99.14)$$

Let us multiply the numerator and the denominator in (99.14) by s^{-n} to get

$$\frac{Y(s)}{X(s)} = \frac{b_{n-1}s^{-1} + b_{n-2}s^{-2} + b_{n-3}s^{-3} + ... + b_0 s^{-n}}{1 + a_{n-1}s^{-1} + a_{n-2}s^{-2} + a_{n-3}s^{-3} + ... + a_0 s^{-n}} \frac{V(s)}{V(s)} \quad (99.15)$$

where $V(s)$ is introduced as an intermediate signal in the desired block diagram.

From (99.15) we can write two separate equations

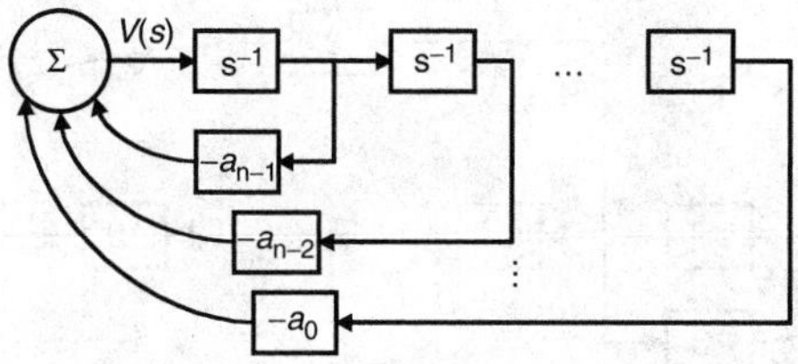

FIGURE 99.9 Initial block for CCF diagram.

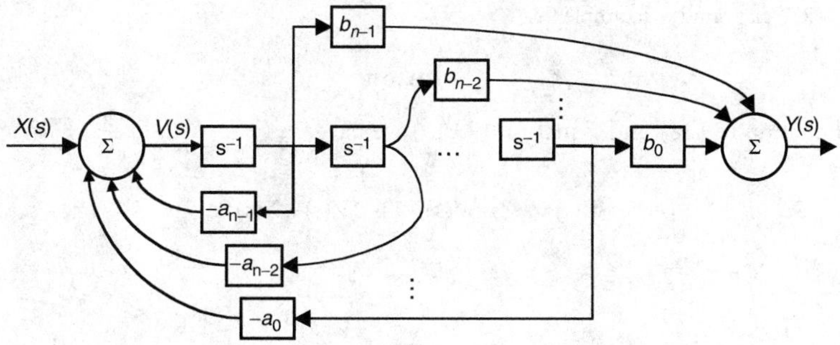

FIGURE 99.10 Final block for CCF diagram.

$$Y(s)=[b_{n-1}s^{-1}+b_{n-2}s^{-2}+b_{n-3}s^{-3}+\ldots+b_0s^{-n}]V(s) \tag{99.16}$$

and

$$X(s)=[1+a_{n-1}s^{-1}+a_{n-2}s^{-2}+a_{n-3}s^{-3}+\ldots+a_0s^{-n}]V(s) \tag{99.17}$$

We can solve for the intermediate variable $V(s)$ from Equation (99.17) to get

$$V(s)=-a_{n-1}s^{-1}V(s)-a_{n-2}s^{-2}V(s)-a_{n-3}s^{-3}V(s)-\ldots-a_0s^{-n}V(s)+X(s) \tag{99.18}$$

We can also solve for the output from Equation (99.16) to get

$$Y(s)=b_{n-1}s^{-1}V(s)+b_{n-2}s^{-2}V(s)+b_{n-3}s^{-3}V(s)+\ldots+b_0s^{-n}V(s) \tag{99.19}$$

From Equation (99.18) we can draw the partial block diagram as shown in Figure 99.9.

Note than we have n basic integrators in the diagram, where n is the order of the system or the number of states.

Next we can add to the diagram in Figure 99.9 the equivalent of Equation (99.19) to get the final block diagram in the CCF as shown in Figure 99.10.

Example 99.1

Consider the system

$$\frac{Y(s)}{X(s)}=\frac{2s+10}{s^2+3s+11}$$

Draw the CCF block diagram.

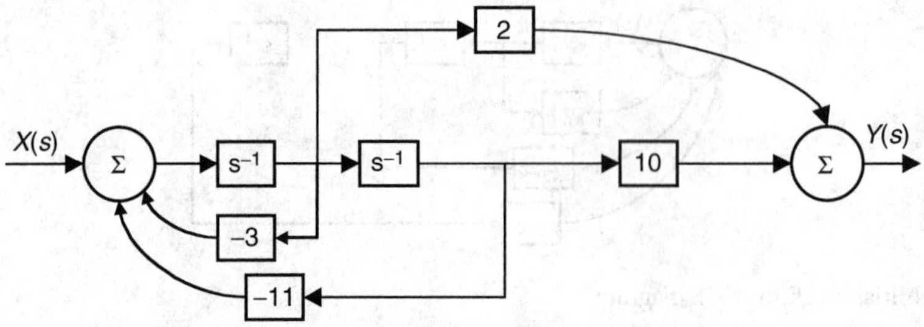

FIGURE 99.11 CCF diagram for Example 99.1.

Solution

Similar to Equation (99.18) and Equation (99.19) we have

$$V(s)=-3s^{-1}V(s)-11s^{-2}V(s)+X(s)$$

and

$$Y(s)=2s^{-1}V(s)+10s^{-2}V(s)$$

The block diagram in the CCF is shown in Figure 99.11. The number of integrators is also *n*, which is the order of the system or the number of states.

99.5 The Observable Canonical Form (OCF) Block Diagrams with Basic Blocks

Consider Equation (99.15) again.

$$\frac{Y(s)}{X(s)}=\frac{b_{n-1}s^{-1}+b_{n-2}s^{-2}+b_{n-3}s^{-3}+\ldots+b_0s^{-n}}{1+a_{n-1}s^{-1}+a_{n-2}s^{-2}+a_{n-3}s^{-3}+\ldots+a_0s^{-n}}\frac{V(s)}{V(s)} \tag{99.20}$$

Let us cross multiply to get

$$Y(s)[1+a_{n-1}s^{-1}+a_{n-2}s^{-2}+a_{n-3}s^{-3}+\ldots+a_0s^{-n}]=[b_{n-1}s^{-1}+b_{n-2}s^{-2}+b_{n-3}s^{-3}+\ldots+b_0s^{-n}]X(s) \tag{99.21}$$

We can solve for the output $Y(s)$ from Equation (99.21) and write

$$Y(s)=[-a_{n-1}s^{-1}-a_{n-2}s^{-2}-a_{n-3}s^{-3}-\ldots-a_0s^{-n}]Y(s)+[b_{n-1}s^{-1}+b_{n-2}s^{-2}+b_{n-3}s^{-3}+\ldots+b_0s^{-n}]X(s) \tag{99.22}$$

From Equation (99.22) we can draw the block diagram in Figure 99.12.

Example 99.2

Consider the same system as in Example 99.1.

$$\frac{Y(s)}{X(s)}=\frac{2s+10}{s^2+3s+11}$$

Draw the OCF block diagram.

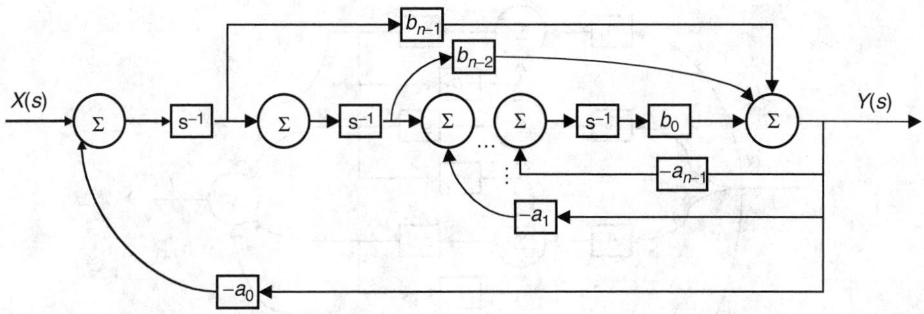

FIGURE 99.12 Final block for OCF diagram.

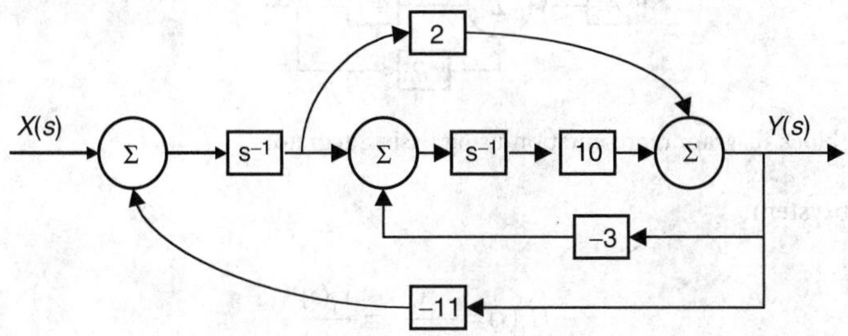

FIGURE 99.13 OCF diagram for Example 99.2

Solution

Similar to Equation (99.22), we have

$$Y(s) = [-3s^{-1} - 11s^{-2}]Y(s) + [2s^{-1} + 10s^{-2}]X(s)$$

and the block diagram is shown in Figure 9.13.

99.6 The Diagonal Form (DF) Block Diagrams with Basic Blocks

Distinct Roots Case

Consider the same transfer function we considered before

$$\frac{Y(s)}{X(s)} = \frac{b_{n-1}s^{n-1} + b_{n-2}s^{n-2} + b_{n-3}s^{n-3} + \ldots + b_0}{s^n + a_{n-1}s^{n-1} + a_{n-2}s^{n-2} + a_{n-3}s^{n-3} + \ldots + a_0} \tag{99.23}$$

The transfer function in Equation (99.23) can be written in partial fraction form as

$$\frac{Y(s)}{X(s)} = \frac{A_1}{s+p_1} + \frac{A_2}{s+p_2} + \frac{A_3}{s+p_3} + \ldots + \frac{A_n}{s+p_n} \tag{99.24}$$

The output in this case can be written as the sum of all outputs obtained from each individual subsystem such as the subsystem

$$\frac{A_1}{s+p_1}$$

with $X(s)$ being the input to all subsystems.

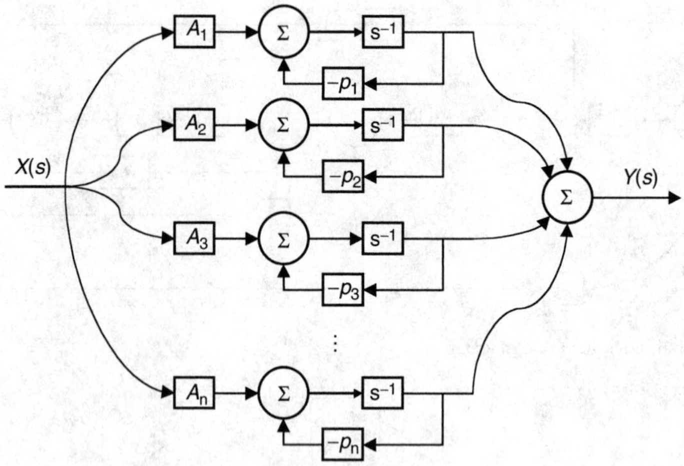

FIGURE 99.14 DF block diagram representation using basic integrators.

For the first subsystem

$$H_1(s) = \frac{A_1}{s+p_1} = \frac{Y_1(s)}{X(s)}$$

we have

$$Y_1(s)[s+p_1] = A_1X(s) \tag{99.25}$$

If we take the inverse Laplace transform on Equation (99.25) we get

$$y'_1(t) = -p_1y_1(t) + A_1x(t) \tag{99.26}$$

For the second subsystem we also get

$$y'_2(t) = -p_2y_2(t) + A_2x(t) \tag{99.27}$$

We can get similar results for the second subsystem, the third subsystem, and so on.

For the nth subsystem we have

$$y'_n(t) = -p_ny_n(t) + A_nx(t) \tag{99.28}$$

and the block diagram represented in what is called the diagonal form is shown in Figure 99.14.

Repeated Roots Case

This is best dealt with using an example. Consider the following transfer function representation in partial fraction form.

$$\frac{Y(s)}{X(s)} = \frac{1}{(s+p_1)^2(s+p_1)} = \frac{A_1}{(s+p_1)^2} + \frac{A_2}{s+p_1} + \frac{A_3}{s+p_2} \tag{99.29}$$

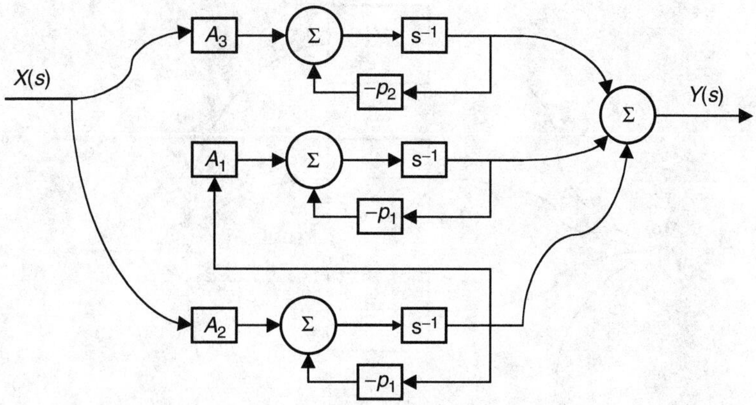

FIGURE 99.15 JF block diagram representation using basic integrators.

We will have three parallel subsystems, but it seems that we need four basic integrators to realize this system in the block diagram form. If fact, we can realize this system with three integrators where the integrator for

$$\frac{A_2}{s+p_1}$$

will be shared by

$$\frac{A_1}{(s+p_1)^2}$$

The block diagram is shown in Figure 99.15.

When we have repeated roots we call the representation in Figure 99.15 the Jordan form (JF) block diagram representation.

99.7 The Parallel Block Diagrams with Subsystems

Parallel block diagrams can arise frequently in implementing analogue or digital filters. We already have fixed designs for first- and second-order analogue or digital filters. If we are given a transfer function of order higher than two, we can simplify the transfer function and write it as the sum of first- or/and second-order filters. We will use second-order filters in the implementation when we have complex poles since it is difficult to realize a single-pole subsystem if the pole is complex. For real systems, complex poles come as complex conjugates, and thus the subsystem is second order with real coefficients.

Distinct Roots Case

Consider the same transfer function we considered before

$$\frac{Y(s)}{X(s)}=\frac{b_{n-1}s^{n-1}+b_{n-2}s^{n-2}+b_{n-3}s^{n-3}+\ldots+b_0}{s^n+a_{n-1}s^{n-1}+a_{n-2}s^{n-2}+a_{n-3}s^{n-3}+\ldots+a_0} \tag{99.30}$$

The transfer function in Equation (99.30) can be written in partial fraction form as

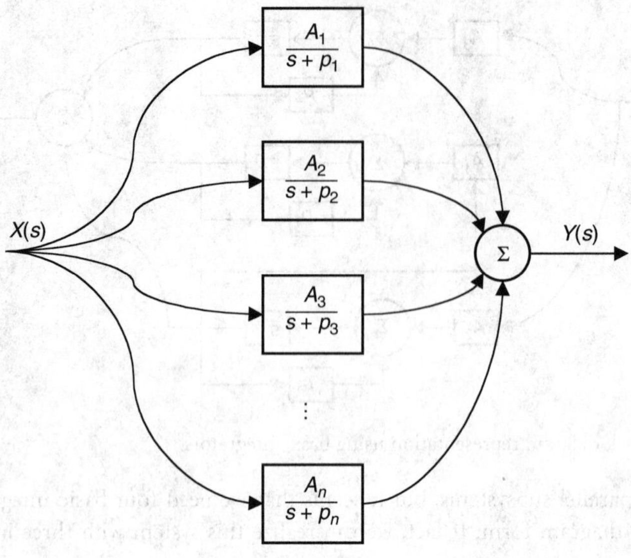

FIGURE 99.16 DF block diagram representation using subsystems.

$$\frac{Y(s)}{X(s)} = \frac{A_1}{s+p_1} + \frac{A_2}{s+p_2} + \frac{A_3}{s+p_3} + \ldots + \frac{A_n}{s+p_n} \tag{99.31}$$

The output in this case can be written as the sum of all outputs obtained from each individual subsystem such as the subsystem

$$\frac{A_1}{s+p_1}$$

with $X(s)$ being the input to all subsystems. The output in this case is

$$Y(s) = X(s)\frac{A_1}{s+p_1} + X(s)\frac{A_2}{s+p_2} + X(s)\frac{A_3}{s+p_3} + \ldots + X(s)\frac{A_n}{s+p_n}$$

The block diagram represented in what is called the diagonal form is shown in Figure 99.16.

Repeated Roots Case

This is best dealt with using an example. Consider the following transfer function representation in partial fraction form.

$$\frac{Y(s)}{X(s)} = \frac{1}{(s+p_1)^2(s+p_1)} = \frac{A_1}{(s+p_1)^2} + \frac{A_2}{s+p_1} + \frac{A_3}{s+p_2} \tag{99.32}$$

We will have three simple single-pole parallel subsystems, but it seems that we need four to realize this system in the block diagram form. In fact, we can realize this system with three simple single-pole subsystems where the output for subsystem

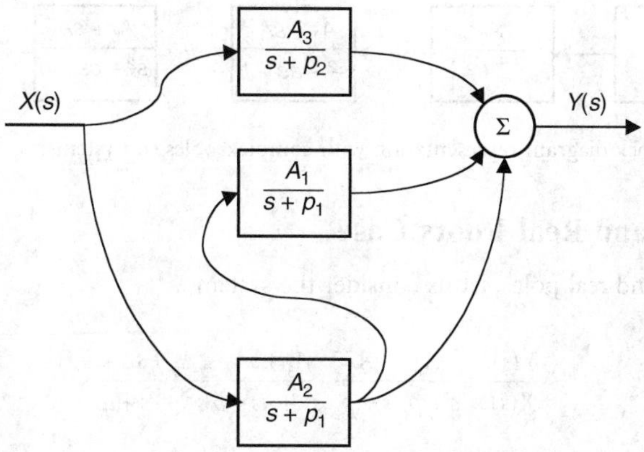

FIGURE 99.17 JF block diagram representation using subsystems.

$$\frac{A_2}{s+p_1}$$

will be the input to subsystem

$$\frac{A_1}{(s+p_1)^2}$$

The block diagram is shown in Figure 99.17.

When we have repeated roots, we call the representation in Figure 99.17 the Jordan form (JF) block diagram representation.

99.8 The Series Block Diagrams with Subsystems

Series or cascade block diagrams can arise frequently in implementing analogue or digital filters. We already have fixed designs for first- and second-order analogue or digital filters. If we are given a transfer function of order higher than two, we can simplify the transfer function and write it as the product of first- or/and second-order filters. We will use second-order filters in the implementation when we have complex poles since it is difficult to realize a single-pole subsystem if the pole is complex. For real systems, complex poles come as complex conjugates, and thus the subsystem is second order with real coefficients.

Distinct Real Roots Case

Consider the transfer function

$$\frac{Y(s)}{X(s)} = \frac{A_1}{s+p_1}\frac{A_2}{s+p_2}\frac{A_3}{s+p_3}\cdots\frac{A_n}{s+p_n} \tag{99.33}$$

in the series or the cascade form. The block diagram is shown in Figure 99.18.

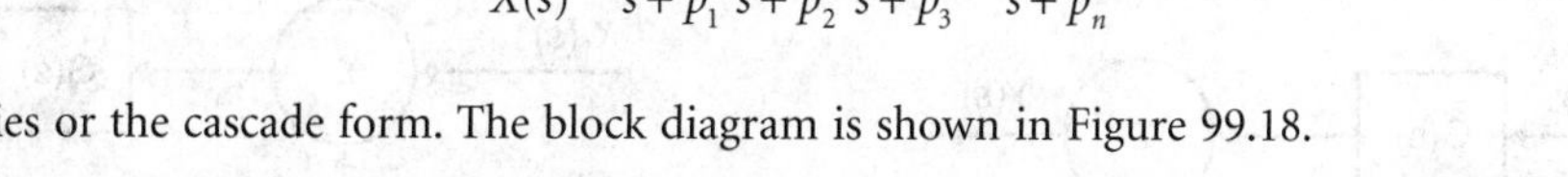

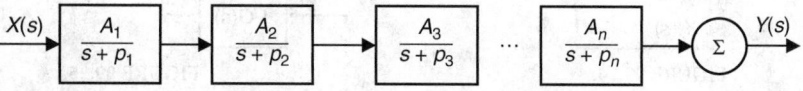

FIGURE 99.18 Series block diagram representation using real single-pole subsystems.

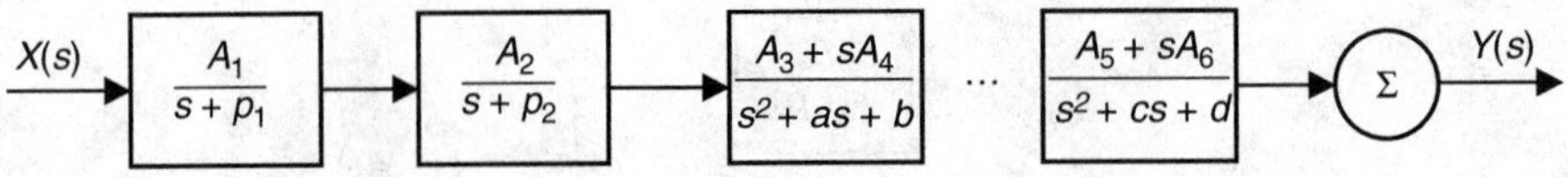

FIGURE 99.19 Series block diagram representation with complex-poles subsystems.

Mixed Complex and Real Roots Case

With mixed complex and real poles, let us consider the system

$$\frac{Y(s)}{X(s)} = \frac{A_1}{s+p_1}\frac{A_2}{s+p_2}\frac{A_3+sA_4}{s^2+as+b}\frac{A_5+sA_6}{s^2+cs+d}$$

where we have two simple real single-pole subsystems and two complex two-pole subsystems. All parameters in these subsystems are real, so they can easily be implemented. This case is shown in Figure 99.19.

99.9 Block Diagram Reduction Rules

Using the Reduction Rules

We list in Table 99.1 some techniques in reducing block diagrams. Other techniques were presented at the beginning of the chapter under series, parallel, and basic feedback block diagrams in Figure 99.6, Figure 99.7, and Figure 99.8.

TABLE 99.1 Equivalent Block Diagrams

Original Block	Equivalent Block
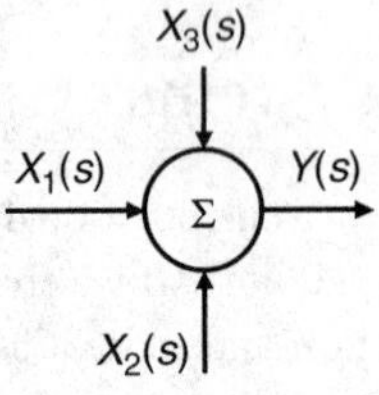 **FIGURE 99.20.**	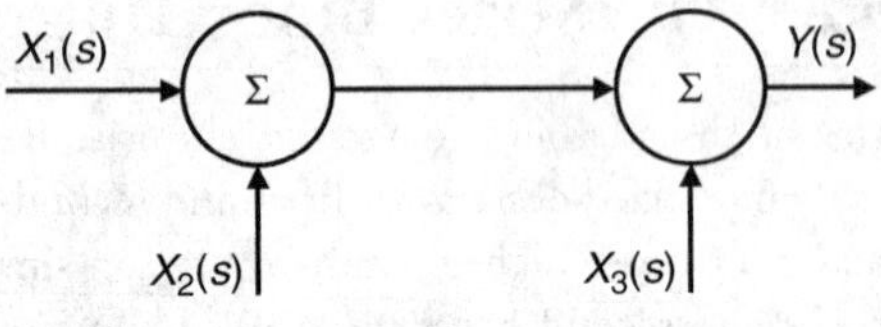 **FIGURE 99.21.**
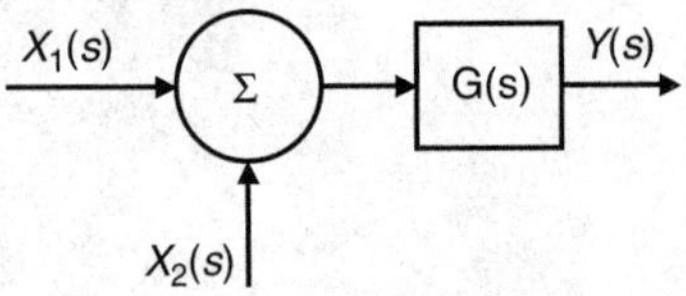 **FIGURE 99.22.**	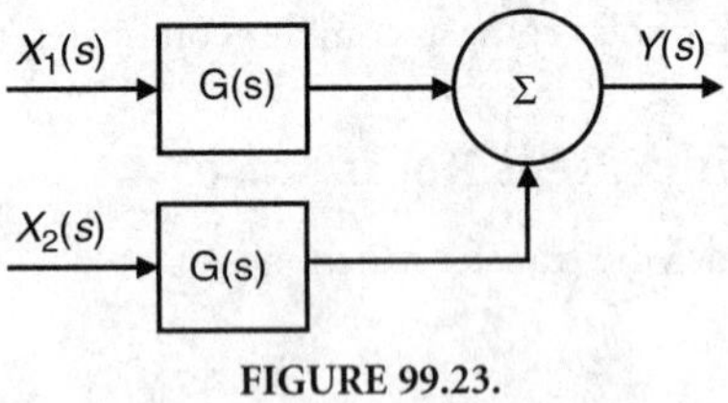 **FIGURE 99.23.**
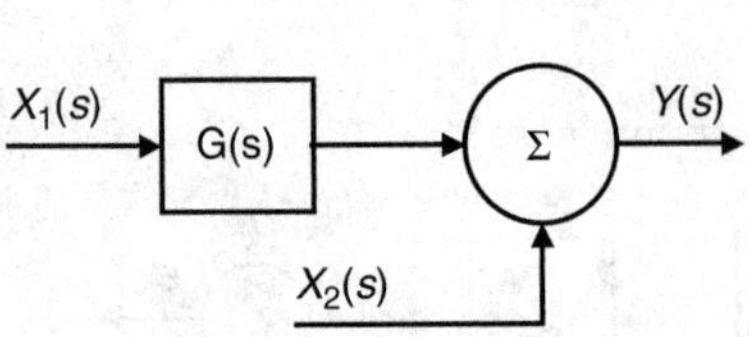 **FIGURE 99.24.**	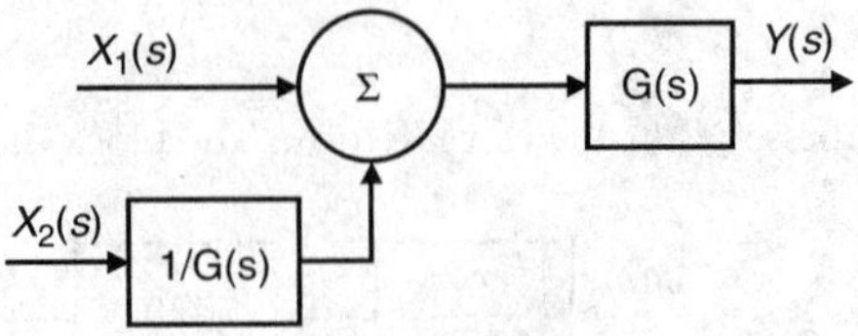 **FIGURE 99.25.**

Using Mason's Rule

In using Mason's rule we will apply the following formula that relates any input to any output in a block diagram. The formula is

$$\frac{y}{x} = \frac{\sum_i p_i \Delta_i}{\Delta} \tag{99.34}$$

where x is any of the single inputs and y is any of the single outputs in the block diagram. The p_i is gain of the forward path that connects x to y. We can have i of these forward paths. The denominator in Equation 99.34 is given by

$$\begin{aligned}\Delta = 1 &- (\text{sum of all loop gains in the diagram}) + (\text{sum of products of all 2-nontouching loops})\\ &- (\text{sum of products of all 3-nontouching loops}) + \ldots\end{aligned} \tag{99.35}$$

The Δ_i is the same as Δ but with setting the gains that the pi path touches to zero.

99.10 Block Diagram Reduction Examples

In the following, we will give some examples where we will try to simplify the block diagrams and get the system in its transfer function form or its state-space representation form. If the system is given in state-space form with the matrices A, B, C, and D as

$$\mathbf{v}'(t) = A\mathbf{v}(t) + B\mathbf{x}(t)$$

$$y(t) = C\mathbf{v}(t) + D\mathbf{x}(t) \tag{99.36}$$

then the transfer function representation is

$$H(s) = C(sI - A)^{-1}B + D \tag{99.37}$$

If the system in block diagram is given as an interconnection of simple integrators, constant gains, and summers, we will use state-variables in simplifying the diagram. Then we will use Equation 99.37 to get the transfer function. If the block diagram is given in terms of interconnected subsystems, then it is easier to use the block diagram reduction rules given in Table 99.1 or Mason's formula to simplify the diagram and get the transfer function.

From Block Diagrams with Basic Block Components to Transfer Functions

Example 99.3

Consider the system in Figure 99.26. What is the transfer function representation?

Solution

We assign the output of every integrator as one state variable. Thus the input of the integrator is the derivative of the state variable. Then from the block diagram we have

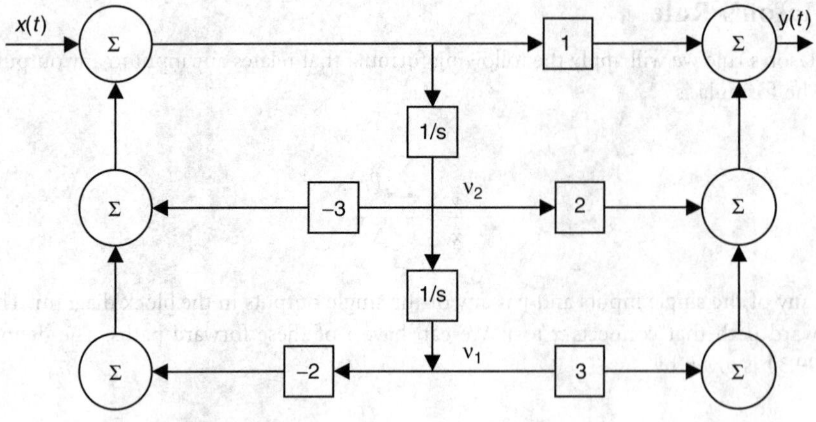

FIGURE 99.26 Block diagram for Example 99.3.

$$v_1'(t) = v_2(t)$$

$$v_2'(t) = x(t) - 2v_1(t) - 3v_2(t)$$

$$y(t) = x(t) - 2v_1(t) - 3v_2(t) + 2v_2(t) + 3v_1(t) = x(t) + v_1(t) - v_2(t)$$

Thus the state-space system is

$$\mathbf{v}'(t) = \begin{pmatrix} 0 & 1 \\ -2 & -3 \end{pmatrix}\mathbf{v}(t) + \begin{pmatrix} 0 \\ 1 \end{pmatrix}\mathbf{x}(t)$$

$$y(t) = \begin{pmatrix} 1 & -1 \end{pmatrix}\mathbf{v}(t) + (1)\mathbf{x}(t)$$

The transfer function representation can be obtained using Equation 99.37 or using Matlab as in the script

```
A=[0 1; -2 -3];
B=[0;1];
C=[1 -1];
D=[1];
[num, den]=ss2tf(A,B,C,D)
```

The output of the script is

```
num = 1.0000 2.0000 3.0000
den = 1 3 2
```

The transfer function representation is then

$$H(s) = \frac{s^2 + 2s + 3}{s^2 + 3s + 2}$$

Example 99.4

Consider the system in Figure 99.27. What is the transfer function representation?

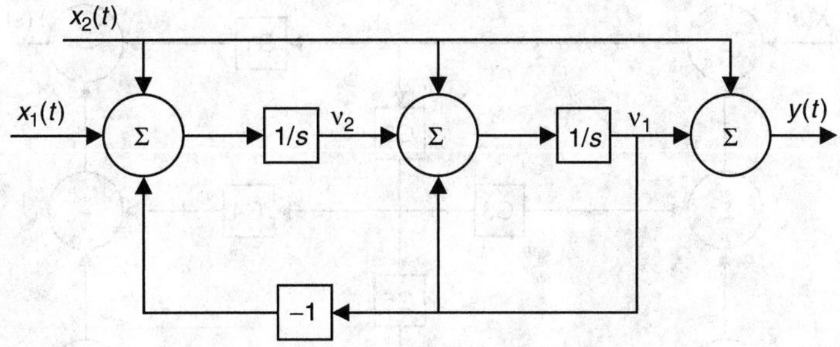

FIGURE 99.27 Block diagram for Example 99.4.

Solution

Again, we assign the output of every integrator as one state variable. Thus the input of the integrator is the derivative of the state variable. Then from the block diagram we have

$$v_1'(t) = x_2(t) + v_2(t) + v_1(t)$$

$$v_2'(t) = x_2(t) + x_1(t) - v_1(t)$$

$$y(t) = x_2(t) + v_1(t)$$

Thus the state-space system is

$$\mathbf{v}'(t) = \begin{pmatrix} 1 & 1 \\ -1 & 0 \end{pmatrix} \mathbf{v}(t) + \begin{pmatrix} 0 & 1 \\ 1 & 1 \end{pmatrix} \mathbf{x}(t)$$

$$y(t) = \begin{pmatrix} 1 & 0 \end{pmatrix} \mathbf{v}(t) + \begin{pmatrix} 0 & 1 \end{pmatrix} \mathbf{x}(t)$$

The transfer function representation can be obtained using Equation 99.37 or using Matlab as in the script

```
A=[1 1; -1 0];
B=[0 1; 1 1];
C=[1 0];
D=[0 1];
[num, den]=ss2tf(A,B,C,D,1)%1 for the first input x1(t)
```

The output of the script is

```
num = 0 0 1
den = 1.0000 -1.0000 1.0000
```

The transfer function representation is then

$$H_1(s) = \frac{Y(s)}{X_1(s)} = \frac{1}{s^2 - s + 1}$$

For the second transfer function we use the script

```
A=[1 1; -1 0];
B=[0 1; 1 1];
C=[1 0];
D=[0 1];
```

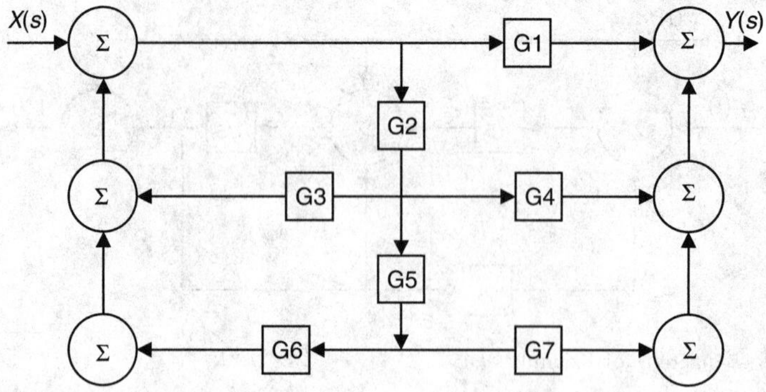

FIGURE 99.28 Block diagram for Example 99.5.

```
[num, den]=ss2tf(A,B,C,D,2)%1 for the second input x2(t)to get
num = 1 0 2
den = 1.0000 -1.0000 1.0000
```

with

$$H_2(s)=\frac{Y(s)}{X_2(s)}=\frac{s^2+2}{s^2-s+1}$$

From Block Diagrams with Interconnected Subsystems to Transfer Functions

Example 99.5

Consider the block diagram in Figure 99.28. Use the Mason's formula to find the transfer function representation.

Solution

From the figure we can see that we have only three forward paths from the only input $x(t)$ to the only output $y(t)$. They are

$$p_1 = \mathrm{G1} \qquad p_2 = \mathrm{G2G4} \qquad p_3 = \mathrm{G2G5G7}$$

We also have two loops in the figure with the gains G2G3 and G2G5G6, and they are touching loops. In this case we have

$$\Delta = 1 - \mathrm{G2G3} - \mathrm{G2G5G6}$$

All the forward paths touch the loops in the figure. Thus

$$p_1\Delta_1 = \mathrm{G1}(1)$$

$$p_2\Delta_2 = \mathrm{G2G4}(1)$$

$$p_3\Delta_3 = \mathrm{G2G5G7}(1)$$

and the transfer function is

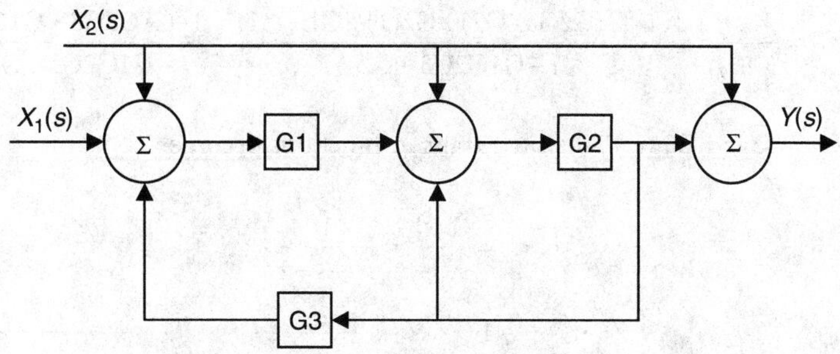

FIGURE 99.29 Block diagram for Example 99.6.

$$H(s) = \frac{G1 + G2G4 + G2G5G7}{1 - G2G3 - G2G5G6}$$

If G1= 1, G2 = 1/s, G3 = –3, G4 = 2, G5 = 1/s, G6 = –2, and G7 = 3, then the transfer function becomes

$$H(s) = \frac{s^2 + 2s + 3}{s^2 + 3s + 2}$$

Example 99.6

Consider the block diagram in Figure 99.29. Use Mason's formula to find the transfer function representation.

Solution

From the figure we can see that we have two touching loops with the gains G1G2G3 and G2. Thus we have

$$\Delta = 1 - G1G2G3 - G2$$

From the input $x_1(t)$ to the output $y(t)$, we have one forward path that touches all the loops in the figure. Thus we have p_1 = G1G2 and $\Delta_1 = 1$. The transfer function is then

$$H_1(s) = \frac{Y(s)}{X_1(s)} = \frac{G1G2}{1 - G1G2G3 - G2}$$

From the input $x_2(t)$ to the output $y(t)$ we have three forward paths. They are with the gains 1, G2, and G1G2. The forward path with the unity gain does not touch any of the loops in the figure. Thus we have

$$p_1 = 1 \qquad p_2 = G2 \qquad p_3 = G1G2$$

and

$$\Delta_1 = 1 - G1G2G3 - G2 \qquad \Delta_2 = 1 \qquad \Delta_3 = 1$$

The second transfer function is

$$H_1(s) = \frac{Y(s)}{X_2(s)} = \frac{1(1\text{-G1G2G3-G2}) + \text{G2}(1) + \text{G1G2}(1)}{1 - \text{G1G2G3} - \text{G2}} = \frac{1\text{-G1G2G3+G1G2}}{1 - \text{G1G2G3} - \text{G2}}$$

If G1= 1/s, G2 = 1/s, and G3 = -1 then the transfer functions become

$$H_1(s) = \frac{Y(s)}{X_1(s)} = \frac{1}{s^2 - s + 1}$$

and

$$H_2(s) = \frac{Y(s)}{X_2(s)} = \frac{s^2 + 2}{s^2 - s + 1}$$

Example 99.7

Consider the block diagram in Figure 99.30. Find the transfer function using the block diagram simplification rules.

Solution

In Figure 99.30 we have three loops, two inner loops and one outer. We can simplify the two inner loops (they are simple feedback loops) to get the intermediate simplified block in Figure 99.31.

The inner summer is of no importance. Now we still have a single simple feedback loop to simplify. The final block diagram is shown in Figure 99.32.

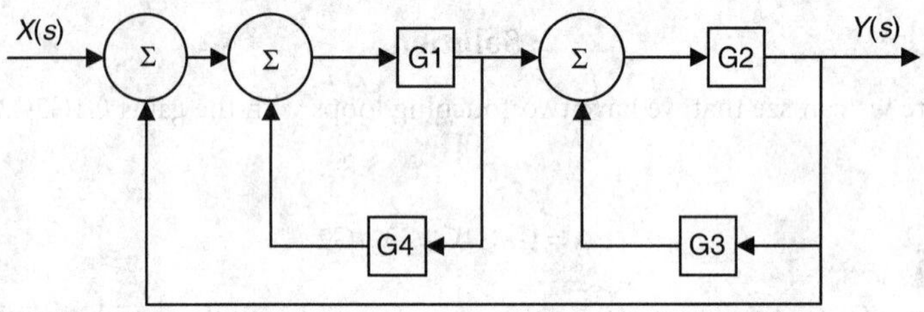

FIGURE 99.30 Block diagram for Example 99.7.

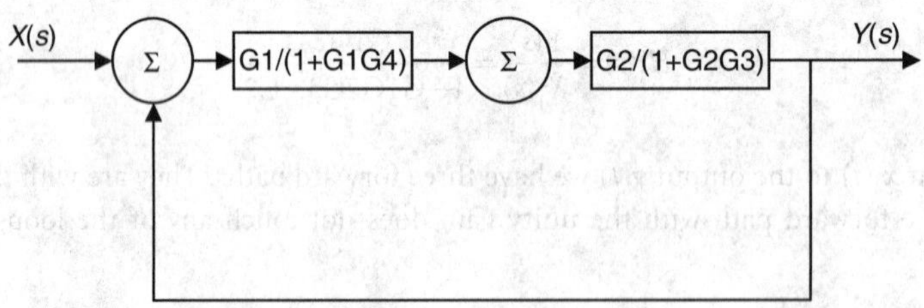

FIGURE 99.31 Block diagram for Example 99.7.

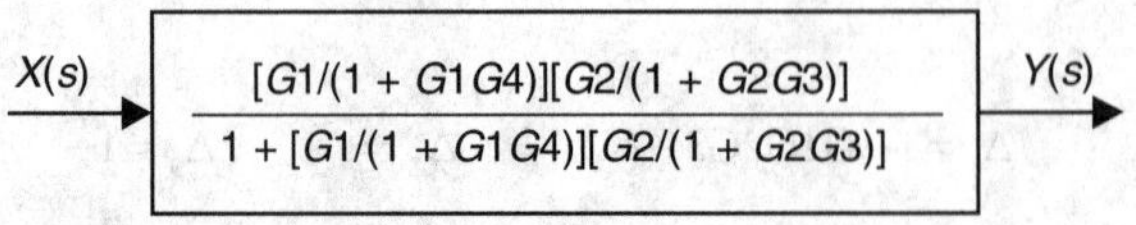

FIGURE 99.32 Final block diagram for Example 99.7.

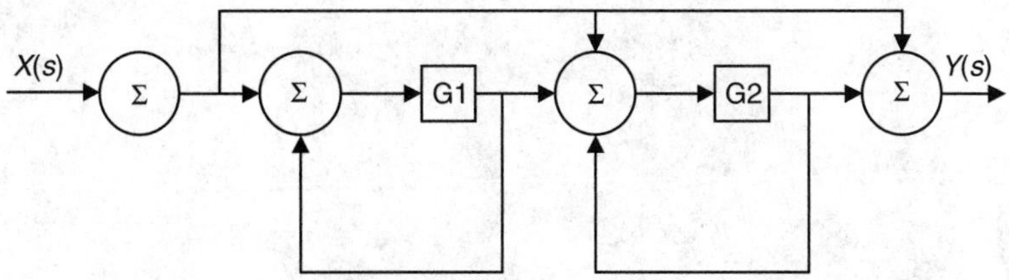

FIGURE 99.33 Block diagram for Example 99.8.

Example 99.8

Consider the block diagram in Figure 99.33. Find the transfer function using the block diagram simplification rules.

Solution

We will start with the inner loops one at a time to get the intermediate diagram in Figure 99.34.

Now it looks like we have a parallel connection at the forward parallel paths with gains unity and G1/(1 + G1). This section can be simplified, giving the diagram in Figure 99.35.

At this point we have a series connection between [1 + G1/(1 + g1)] and G2/(1 + G2). We can combine this series connection to get the product G = [1 + G1/(1 + g1)] [G2/(1 + G2)]. This product is now in parallel with the forward gain of unity. So we will have 1 + G as the result of combining G with the unity gain. The three summers in the diagram serve no purpose and we can eliminate them. The final block diagram is shown in Figure 99.36.

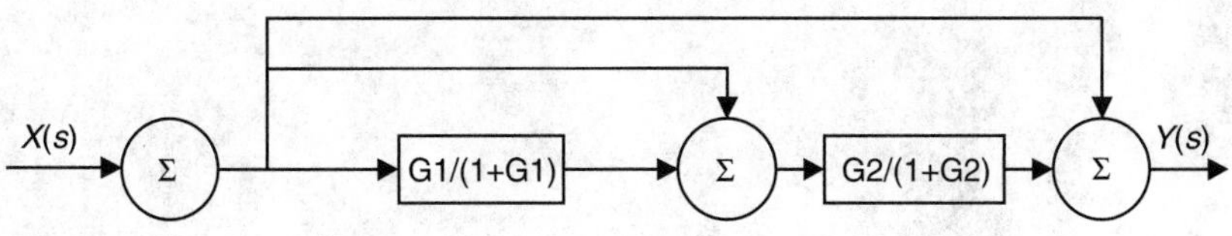

FIGURE 99.34 Block diagram for Example 99.8.

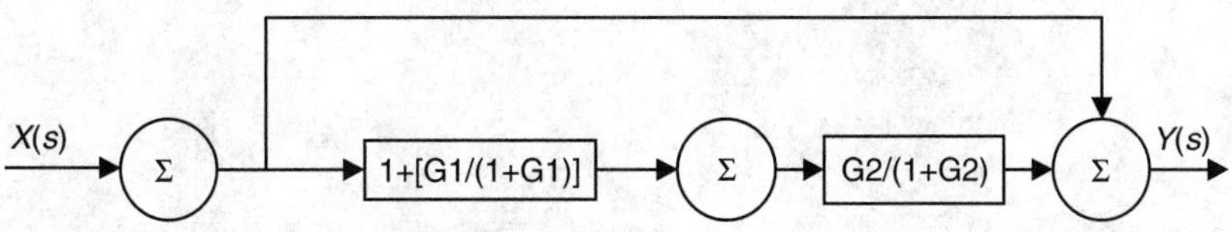

FIGURE 99.35 Block diagram for Example 99.8.

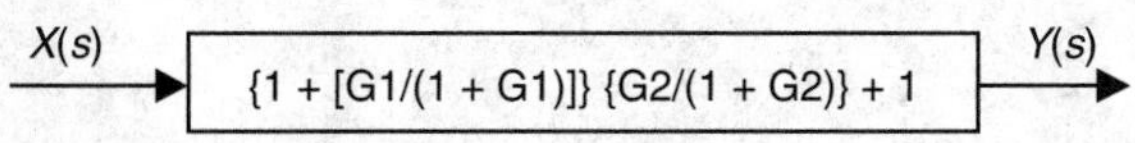

FIGURE 99.36 Final block diagram for Example 99.8.

100 Signal Flow Analysis

Partha P. Banerjee
University of Dayton

Signal flow graphs are a viable alternative to block diagrammatic representation of a system. What makes signal flow graphs attractive is that certain features from graph theory can be applied to the simplification and the synthesis of complex systems.

100.1 Introduction

The relationship between the input and output of a certain system can be represented in terms of a *block diagram*. The block diagram represents the operator that operates on the input to produce the output, and can be represented either in the time domain or in the Laplace domain for a time-dependent input and output. The relationship between the input and the output in the Laplace domain is called the *transfer function* of the system. In this case, the input is the independent variable and the output is the dependent variable. Sometimes, when there are intermediate dependent variables, the relationships between each other as well as the input(s) and output(s) can also be represented by block diagrams. Alternatively, instead of block diagrams, the dependent variables and the inputs can be denoted as *nodes*, and connections or *paths* between the nodes can denote the mathematical operator linking the two variables or nodes. This is used to draw what is called a *signal flow graph*. A simple diagram representing the similarities between a block diagram and a signal flow graph is shown in Figure 100.1.

100.2 Signal Flow Graphs for Feedback Systems

In many systems, there is feedback (positive or negative) from the output to the input. Negative feedback makes a system more stable, while positive feedback causes a system to become unstable and is the principle behind the operation of oscillators. Feedback is depicted in a block diagram through a feedback transfer function $G(s)$ between the output and the input, as shown in Figure 100.2. Note that in this case, $U(s)$ is the input, and the output $X_2(s)$ is fed back to the input through $G(s)$. The input $X_1(s)$ to $H(s)$ can be expressed as

$$X_1(s) = U(s) - X_2(s)G(s) \tag{100.1}$$

0-8493-1586-7/05/$0.00+$1.50

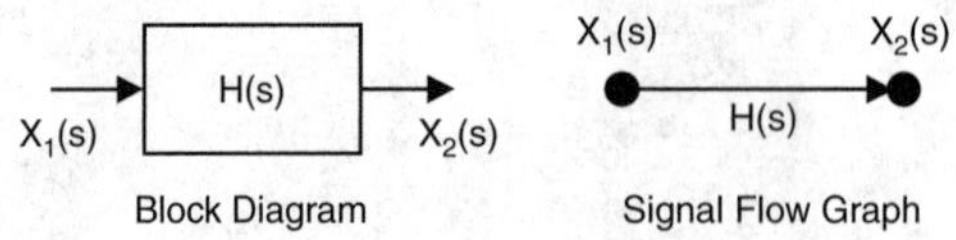

FIGURE 100.1 Connection between a block diagram and a signal flow graph.

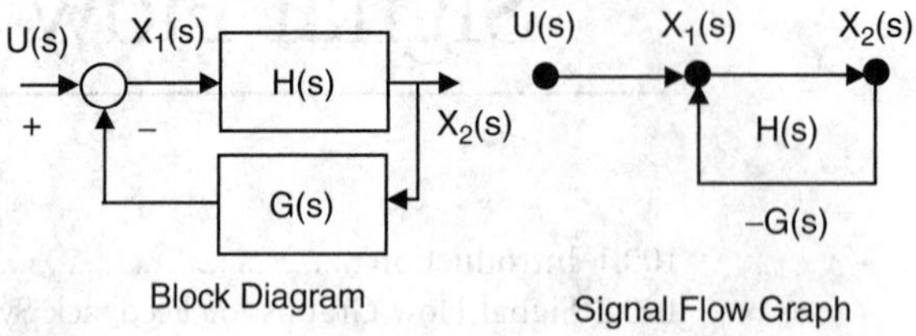

FIGURE 100.2 Connection between a block diagram and a signal flow graph for a system with negative feedback.

where

$$X_2(s) = X_1(s)H(s) \tag{100.2}$$

The same is depicted in the signal flow graph drawn on the right. Note that upon manipulating Equation (100.1) and Equation (100.2), a direct relationship can be found between the input $U(s)$ and the output $X_2(s)$ as

$$X_2(s) = \frac{H(s)}{1+G(s)H(s)}U(s) = H_{eq}U(s) \tag{100.3}$$

In other words, the feedback system represented by the block diagram in Figure 100.2 can be reduced to a block diagram similar to Figure 100.1, where the input is now $U(s)$ and the transfer function relating the output $X_2(s)$ to the input is now $H_{eq}(s)$, defined in Equation (100.3). This is shown in Figure 100.3. The equivalent signal flow graph also reduces to a form similar to Figure 100.1, with $X_1(s)$ replaced by $U(s)$ and $H(s)$ replaced by $H_{eq}(s)$, as shown in Figure 100.3. This also suggests that in a signal flow graph, it may be possible to reduce the number of nodes through a systematic node elimination procedure. This is facilitated through using Mason's theorem for reduction of systems, to be described below.

In general, signal flow graphs may be more complicated and comprise nodes, paths, and *loops*. An example of a *feedback loop* appears in the signal flow graph of Figure 100.2; however, *self-loops* are possible as well. An example of a more complicated signal flow graph involving many loops is shown in Figure 100.4. This signal flow graph corresponds to the set of equations

$$\begin{aligned} X_1(s) &= U(s) - X_2(s)G_1(s), \\ X_2(s) &= X_1(s)H_1(s) + X_2(s)G_2(s), \\ Y(s) &= X_2(s)H_2(s). \end{aligned} \tag{100.4}$$

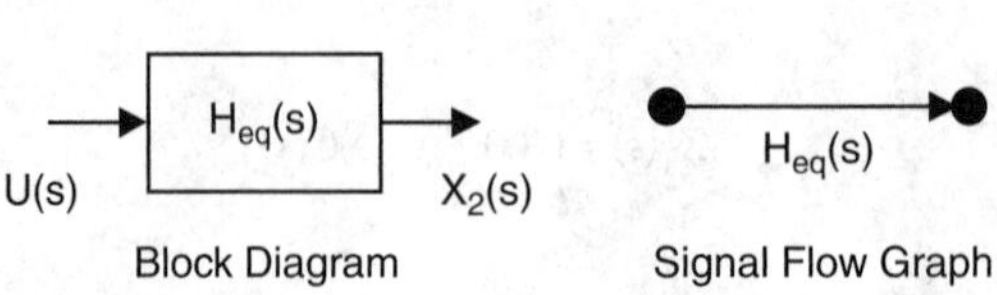

FIGURE 100.3 Reduced block diagram and corresponding signal flow graph from Figure 100.2.

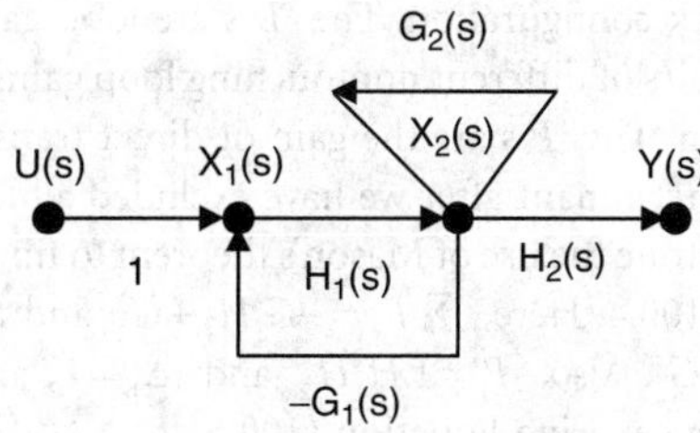

FIGURE 100.4 A more complicated signal flow graph corresponding to the system of equations showing feedback loops including self-loops.

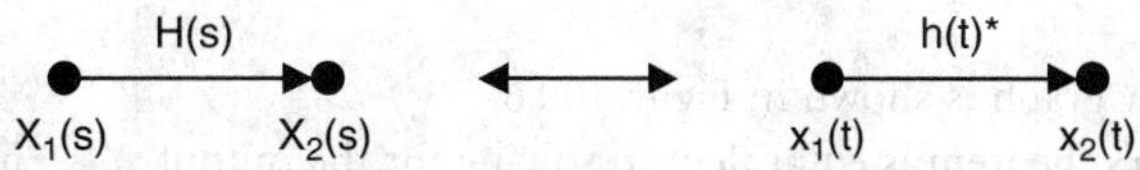

FIGURE 100.5 Equivalence between signal flow graphs in Laplace and time domains.

One can readily go from the set of equations to the signal flow graph and vice versa.

In passing, we would like to point out that the second of the relations in Equation (100.4) above can be rewritten in the form

$$X_2(s) = X_1(s)\frac{H_1(s)}{1-G_2(s)}$$

This implies that self-loops can be eliminated by dividing all incoming path gains by $1-G_i(s)$, where $G_i(s)$ is the self-loop gain for the node X_i. Other types of simplification of signal flow graphs, such as *node elimination,* are discussed later. We would like to remind readers that signal flow graphs can be drawn for signals depicted in either the Laplace domain or the time domain. The time domain equivalent of the signal flow graph in Figure 100.4 would involve the same nodes and loops, except that the nodes would be depicted as $u(t), x_1(t), x_2(t), y(t)$, which are the inverse Laplace transforms of $U(s), X_1(s), X_2(s), Y(s)$, respectively, and the loops would correspond to operators in the time domain such as $h_1(t), h_2(t), g_1(t), g_2(t)$, which are the inverse Laplace transforms of $H_1(s), H_2(s), G_1(s), G_2(s)$, respectively. Figure 100.5 shows the equivalence between the signal flow graphs in the Laplace and time domains. It should be noted that multiplication in the Laplace domain corresponds to convolution in time, denoted as a * in Figure 100.5.

For instance, if $H_1(s) = s$, then $h_1(t) = d[\delta(t)]/dt$, and it can be shown from the properties of convolution that $h_1(t) * x_1(t) = [d[\delta(t)]/dt] * x_1(t) = dx_1(t)/dt$, so that the operator $h_1(t)* = d/dt$. If $H_1(s) = c$, a constant, then $h_1(t)* = c$, which is the same multiplicative constant.

100.3 Reduction of Signal Flow Graphs

We now enunciate *Mason's theorem* for reduction of a signal flow graph. It states that the equivalent transfer function from input $U(s)$ to output $Y(s)$ can be written as

$$H_{eq}(s) = Y(s)/U(s) = \sum_i P_i\Delta_i/\Delta \tag{100.5}$$

where

$$\Delta = 1 - \sum L_j + \sum{}' L_k L_l - \sum{}' L_m L_n L_o + \ldots \tag{100.6}$$

is the determinant of the feedback configuration. The L_i s are loop gains; ΣL_j is the sum of all loop gains; $\Sigma' L_k L_l$ is the sum of all pairs of different nontouching loop gains, etc. Two loops are nontouching if they have no nodes in common. The P_i s are the gain of direct transmittances from the input to the output. Also Δ_i is the system determinant after we have excluded all loops that touch the P_i path.

As an example, we will demonstrate the use of Mason's theorem to find the equivalent transfer function for the system shown in Figure 100.4. Here, $\Sigma L_j = -G_1H_1 + G_2$, and all higher order sums in (100.6) are zero, so that $\Delta = 1 + G_1H_1 - G_2$. Also, $P_1 = 1.H_1H_2$ and $\Delta_1 = 1$, and there is only one direct path from the input to the output. Hence, using Equation (100.5),

$$H_{eq}(s) = Y(s)/U(s) = \frac{H_1H_2}{1 - G_2 + G_1H_1} \tag{100.7}$$

The reduced signal flow graph is shown in Figure100.6.

In retrospect, Mason's theorem is equivalent to solving for the output $Y(s)$ in terms of the (known) input $U(s)$ from a set of linear algebraic equations of the form $\underline{\underline{A}}\underline{X} = \underline{B}$. According to Cramer's rule, the solution for the j-th component of the vector $\underline{X}$ is $Xj = |\underline{\underline{Dj}}| / |\underline{\underline{A}}|$, where the matrix D_j has $\underline{B}$ as its j-th column, and the corresponding columns of $Y(s)$ as its other columns. Upon applying this to the example depicted in Figure 100.4, we see that the dependent variables X_1, X_2, Y can be solved by first rewriting (100.4) in the form of a vector-matrix equation of the type $\underline{\underline{A}}\underline{X} = \underline{B}$ as

$$\begin{bmatrix} 1 & G_1 & 0 \\ H_1 & 1-G_2 & 0 \\ 0 & H_2 & 1 \end{bmatrix} \begin{bmatrix} X_1 \\ X_2 \\ Y \end{bmatrix} = \begin{bmatrix} U \\ 0 \\ 0 \end{bmatrix} \tag{100.8}$$

the solution for Y, using Cramer's rule is

$$Y = \begin{vmatrix} 1 & G_1 & U \\ H_1 & 1-G_2 & 0 \\ 0 & H_2 & 0 \end{vmatrix} \Bigg/ \begin{vmatrix} 1 & G_1 & 0 \\ H_1 & 1-G_2 & 0 \\ 0 & H_2 & 1 \end{vmatrix} = \frac{H_1H_2}{(1-G_2) + H_1G_1} U, \tag{100.9}$$

which yields the result for the equivalent transfer function identical to Equation (100.7) above.

The reduction of the signal flow graph shown above using Mason's theorem can also be achieved through a repeated *node elimination* process. The rules of elimination are as follows. Assume that we would like to eliminate the node X_2. First, we need to eliminate the self-loop around X_2. The self-loop of gain G_2 is eliminated by dividing all incoming transmittances by $1 - G_2$. This makes the transmittance from X_1 to X_2 equal to $H_1/(1-G_2)$. Now, in the reduced signal flow diagram, all nodes except X_2 are drawn, and all original branches not entering or leaving X_2 are inserted. Finally, we add branches representing every possible path (in the signal flow diagram without self-loops) through X_3. For instance, we can go from X_1 to X_1 through $-H_1G_1/(1-G_2)$, and X_1 to Y through $H_2G_1/(1-G_2)$. The reduced signal flow graph is shown in Figure 100.7.

A similar procedure can be used to eliminate the node X_1. As before, it entails first removing the self-node at X_1, which makes the transmittance from U to X_1 equal to $(1-G_2)/(1-G_2+H_1G_1)$. Upon now eliminating X_1, the transmittance from U to Y is $[(1 - G_2)/(1 - G_2 + H_1G_1)] \times [H_2G_1/(1 - G_2)]$

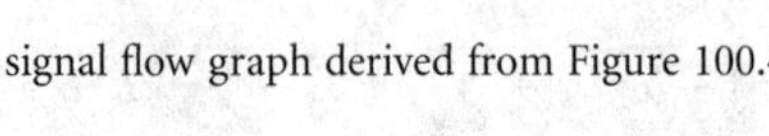

FIGURE 100.6 Equivalent reduced signal flow graph derived from Figure 100.4.

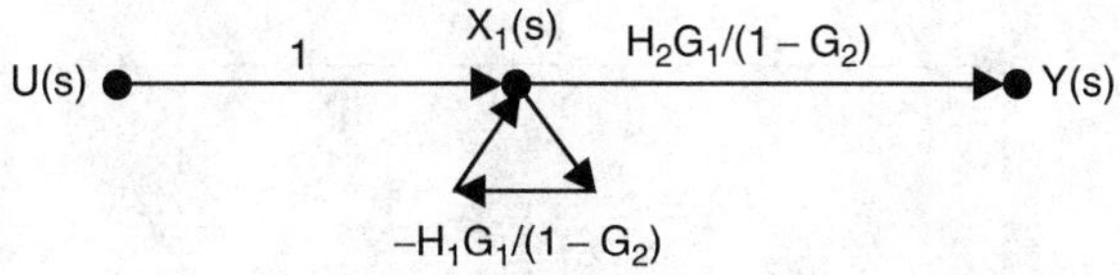

FIGURE 100.7 Reduction of the signal flow graph in Figure 100.4 through elimination of node X_2.

$= H_2G_1/(1 - G_2 + H_1G_1)$, which is identical to the equivalent transfer function $H_{eq}(s)$ in Equation (100.7) derived using Mason's theorem. The resulting signal flow graph is as shown in Figure 100.6.

100.4 Realization of Transfer Functions

Thus far, we have discussed procedures for simplifying signal flow graphs to derive the transfer function of the system. We will now turn our attention to synthesizing signal flow graphs given a particular transfer function. Using the $H_{eq}(s)$ in Equation (100.7) as an illustration, assume that

$$H_1(s) = s + a_1,\ H_2(s) = a_2,\ \ G_1(s) = s + b_1,\ G_2(s) = s + b_2\,. \tag{100.10}$$

Then

$$H_{eq}(s) = \frac{Y(s)}{U(s)} = \frac{a_2 s + a_1 a_2}{s^2 + (a_1 + b_1 - 1)s + (a_1 b_1 - b_2 + 1)}\,. \tag{100.11}$$

As is often the case, the degree of the polynomial in the denominator is equal to or greater than the degree of the polynomial in the numerator. Then the degree of the polynomial in the denominator is defined as the *order* of the system, and is equal to the number of *states* of the system. We can therefore define two state variables $X_1(s), X_2(s)$ for the system, related through $X_1(s) = X_2(s)/s$, or equivalently, $X_2(s) = sX_1(s)$. In the time domain this implies that $\chi_2(t) = d\chi_1(t)/dt$, where $\chi_1(t), \chi_2(t)$ are the inverse Laplace transforms of $X_1(s), X_2(s)$, respectively. Conversely, $\chi_1(t)$ is the integral of $\chi_1(t)$.

For convenience, Equation (100.11) is reexpressed in the form

$$H_{eq}(s) = \frac{Y(s)}{U(s)} = \frac{a_2/s + a_1 a_2/s^2}{1 + (a_1 + b_1 - 1)/s + (a_1 b_1 - b_2 + 1)/s^2}\,. \tag{100.12}$$

Since this is a second-order system, one needs two integrators. The integrator outputs are called $X_1(s), X_2(s)$, and the integrator inputs are called $X_1'(s), X_2'(s)$, respectively, as shown in Figure 100.8.

The second step is the realization of the denominator in Equation (100.12). Since Mason's theorem states that all loops that touch have a $\Delta = 1 - \sum L_j$, it is convenient to construct loops incorporating feedback which have a node in common, viz., $X_1(s)$, and with feedback loop gains equal to $-(a_1 + b_1 - 1)/s$ and $-(a_1 b_1 - b_2 + 1)/s^2$, as illustrated by the dashed lines in Figure 100.9. Finally, to construct the numerator, we ensure that all direct paths also pass through one node, viz., $X_1(s)$. The path gains are a_2/s and $a_1 a_2/s^2$, as illustrated by the dotted lines in Figure 100.9.

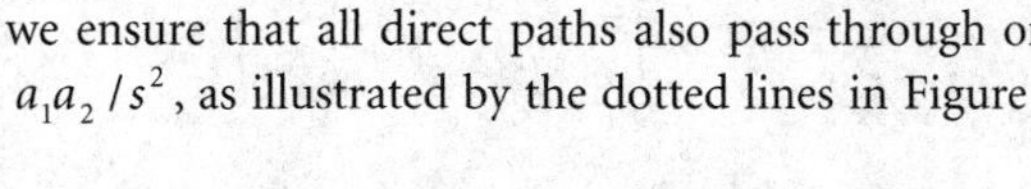

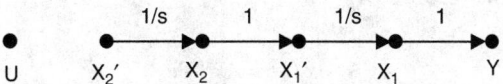

FIGURE 100.8 First step in the realization of the transfer function in Equation (100.12).

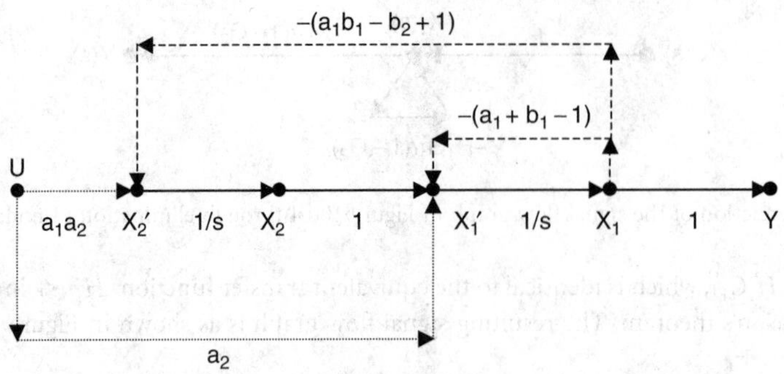

FIGURE 100.9 Signal flow graph for the transfer function in Equation (100.12).

The signal flow graph shown in Figure 100.9, also called *Type I* realization, is not unique. *Type II* is an alternate realization that assumes that all feedback loops and parallel paths go through $X_2'(s)$ rather than $X_1(s)$. Finally, *Type III* involves a realization that is based on first decomposing Equation (100.12) into partial fractions in the form

$$H_{eq}(s) = \sum a_i/(s+b_i),\ b_i \in \Re \tag{100.13}$$

by first factorizing the denominator. This yields a realization of the transfer function in terms of parallel loops. In cases where the denominator has complex roots, it can be decomposed into partial fractions involving sums of terms as in Equation (100.11) and Equation (100.13). Details can be found in Truxal (1972).

100.5 Boundary Conditions and Signal Flow Graphs

Signal flow graphs can be suitably adapted to incorporate initial conditions imposed on a certain state. For instance, assume that in the time domain, states $x_1(t), x_2(t)$ are related through the set of coupled differential equations as

$$\begin{aligned} dx_1(t)/dt &= x_2, \\ dx_2(t)/dt &= -a_2x_1 - a_1x_2 \end{aligned} \tag{100.14}$$

where a_1, a_2 are constants. Equation (100.14) is the state variable formulation of a second order ODE of the form $d^2x/dt^2 + a_1dx/dt + a_2x = 0$. With the definitions $x_1 = x, x_2 = dx_1/dt$, this ODE can be rewritten as Equation (100.14). Note also that Equation (100.14) can be recast in the form

$$\frac{d}{dt}\begin{pmatrix} x_1 \\ x_2 \end{pmatrix} = \begin{bmatrix} 0 & 1 \\ -a_2 & -a_1 \end{bmatrix}\begin{pmatrix} x_1 \\ x_2 \end{pmatrix}, \tag{100.15}$$

which is a special case of the vector ODE

$$d\underline{x}/dt = \underline{\underline{A}}\underline{x} + \underline{\underline{B}}u \tag{100.16a}$$

Together with the output equation

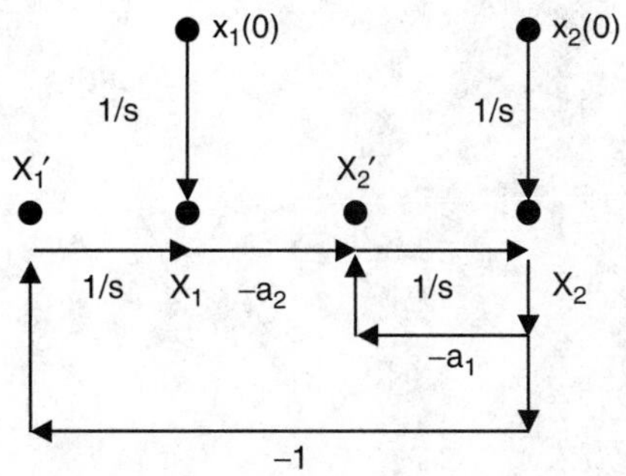

FIGURE 100.10 Signal flow diagram for realization of the system modeled by Equation (100.15).

$$\underline{y} = \underline{\underline{C}}\underline{x} + \underline{\underline{D}}\underline{u} \tag{100.16b}$$

one can describe the behavior of the entire linear system.

Upon Laplace transforming Equation (100.14), we get

$$\begin{aligned} sX_1(s) + X_2(s) &= x_1(0), \\ a_2X_1(s) + (s + a_1)X_2(s) &= x_2(0). \end{aligned} \tag{100.17}$$

Similar to the way the signal flow graph from the transfer function in Equation (100.12) was realized, we can draw the signal flow diagram for Equation (100.17), as shown in Figure 100.10.

100.6 Conclusion

We have summarized the salient points of signal flow graphs, their reduction, and their synthesis. As seen from the discussions above, they are an analogue to block diagrams in the analysis of linear systems. In some cases, signal flow graphs can give valuable information about the *controllability* and *observability* of linear systems as well. Loosely speaking, a state is controllable if it can be changed by an appropriate set of inputs. A state is observable if the output(s) depend on the particular state. However, formal tests for controllability and observability can be made on the basis of the matrices $\underline{\underline{A}}, \underline{\underline{B}}, \underline{\underline{C}}, \underline{\underline{D}}$ defined in Equation (100.16) above. This is outside the scope of this chapter.

References

Kuo, B.C. and Golnaraghi, F. 2002. *Automatic Control Systems,* Wiley, New York.

Truxal, J.G. 1972. *Introductory Systems Engineering,* McGraw-Hill, New York.

101

Linear State–Space Models

Boyd D. Schimel
Washington State University

Walter J. Grantham
Washington State University

101.1 State–Space Models

The general **state–space** model that we will consider for a continuous-time dynamical system consists of a system of n_x first-order differential equations of the form

$$\begin{aligned} \frac{dx_1}{dt} &= f_1(x_1,\ldots,x_{n_x},u_1,\ldots,u_{n_u}) \\ &\vdots \\ \frac{dx_{n_x}}{dt} &= f_{n_x}(x_1,\ldots,x_{n_x},u_1,\ldots,u_{n_u}) \end{aligned} \tag{101.1}$$

where t denotes time; $\mathbf{x}(t) = [x_1 \ldots x_{n_x}]^{\mathrm{T}}$ is an n_x-dimensional **state vector; $\mathbf{u} = [u_1 \ldots u_{n_u}]^{\mathrm{T}}$ is** an n_u-dimensional control or **input vector**; and $[\cdot]^{\mathrm{T}}$ denotes the transpose. Loosely speaking, the dimension of the state vector equals the number of initial conditions required to determine the motion, given the input — say $\mathbf{u}(t)$ or $\mathbf{u}(\mathbf{x}, t)$ — and the model of Equation (101.1). As an example, the motion of a point mass m satisfying Newton's equations may be described by

$$m\frac{d^2\mathbf{y}}{dt^2} = \mathbf{F}$$

where $\mathbf{y} = [y_1\ y_2\ y_3]^{\mathrm{T}}$ is the position vector and $\mathbf{F} = [F_1\ F_2\ F_3]^{\mathrm{T}}$ is the applied force. This second-order system can be converted to first-order state–space form by, for example, taking the state vector as $\mathbf{x} = [y_1\ y_2\ y_3\ dy_1/dt\ dy_2/dt\ dy_3/dt]^{\mathrm{T}}$ and the input vector as $\mathbf{u} = (1/m)\mathbf{F}$, yielding

0-8493-1586-7/05/$0.00+$1.50

$$\dot{x}_1 = x_4$$
$$\dot{x}_2 = x_5$$
$$\dot{x}_3 = x_6$$
$$\dot{x}_4 = u_1$$
$$\dot{x}_5 = u_2$$
$$\dot{x}_6 = u_3$$

where $(\dot{\cdot}) = d()/dt$. This state–space formulation makes it clear that the required initial conditions include not only the initial position $\mathbf{y}(t)$ but also the initial velocity $\dot{\mathbf{y}}(0)$. Writing dynamic system models as first-order differential equations also allows numerical analysts, for example, to focus on one type of numerical integration procedure, rather than having one scheme for first-order systems, another for second-order systems, and various other schemes for nth-order systems, since all systems of higher-order differential equations can be converted to equivalent systems of first-order differential equations, as illustrated in the preceding equations.

It should be noted that Equation (101.1) does not explicitly cover systems governed by partial differential equations, such as the vibration of a drumhead, in which the motion of the object is distributed over space as well as time. Nor does Equation (101.1) cover time-delay systems, in which the motion depends not only on the current state $\mathbf{x}(t)$ and current input $\mathbf{u}(t)$, but also on a history of past conditions, such as the state $\mathbf{x}(t - \tau)$ at some time τ seconds in the past. Both of these situations could be (and typically are) converted to state–space models of the form in Equation (101.1), by discretizing the spatial region for partial differential equations or discretizing the time-delay interval for time delay systems. In the limit as the discretization intervals become small the dimension of the state vector would become infinite.

In addition to the state and input vectors, there are typically various output or measurement quantities of interest, which are related to the state and input vectors by algebraic **output equations** of the general form

$$\begin{aligned} y_1 &= g_1(x_1,\ldots,x_{n_x}, u_1,\ldots,u_{n_u}) \\ \vdots \quad & \quad \vdots \\ y_{n_y} &= g_{n_y}(x_1,\ldots,x_{n_x}, u_1\ldots,u_{n_u}) \end{aligned} \tag{101.2}$$

where $\mathbf{y} = [y_1 \ldots y_{n_y}]^{\mathrm{T}}$ is an n_y-dimensional **output vector.** For our point-mass example, taking the position vector as the output would correspond to the output equations

$$y_1 = x_1$$
$$y_2 = x_2$$
$$y_3 = x_3$$

In vector form the state equations [Equations (101.1)] and the output equations [Equations (101.2)] can be written as

$$\dot{\mathbf{x}} = \mathbf{f}(\mathbf{x}, \mathbf{u}) \tag{101.3}$$

$$\mathbf{y} = \mathbf{g}(\mathbf{x}, \mathbf{u}) \tag{101.4}$$

where $(\dot{\cdot}) = d()/dt$, $\mathbf{f} = [f_1 \ldots f_{n_x}]^{\mathrm{T}}$, and $\mathbf{g} = [g_1 \ldots g_{n_y}]^{\mathrm{T}}$. Systems in which time t appears explicitly in the right-hand sides of Equation (101.3) or Equation (101.4) can be handled, for example, either by including such terms in the input vector **u** or by treating t as a state variable, say $x_{n_x} = t$ with the differential equation $\dot{t} = 1$.

101.2 Linearization

For a system of the form

$$\dot{\mathbf{X}} = \mathbf{f}(\mathbf{X}, \mathbf{U}) \tag{101.5}$$

with output equations

$$\mathbf{Y} = \mathbf{g}(\mathbf{X}, \mathbf{U}) \tag{101.6}$$

let $\mathbf{U}(t) = \overline{\mathbf{U}}(t)$ be a reference input. For a given initial state $\overline{\mathbf{X}}(0)$ let $\overline{\mathbf{X}}(t)$ be the corresponding solution to Equation (101.5) generated by $\overline{\mathbf{U}}(t)$ and let $\overline{\mathbf{Y}}(t)$ be the resulting output, given by Equation (101.6). Let $\mathbf{x}(t)$, $\mathbf{y}(t)$, and $\mathbf{u}(t)$ denote small deviations from the reference state, output, and input, respectively. Substituting $\mathbf{X}(t) = \overline{\mathbf{X}}(t) + \mathbf{x}(t)$, $\mathbf{Y}(t) = \overline{\mathbf{Y}}(t) + \mathbf{y}(t)$, and $\mathbf{U}(t) + \overline{\mathbf{U}}(t) + \mathbf{u}(t)$ into Equation (101.5) and Equation (101.6), expanding the results using Taylor's theorem, and retaining only the first-order (linear) terms yields the time-varying **linear state equations,**

$$\dot{\mathbf{x}} = \mathbf{A}(t)\mathbf{x} + \mathbf{B}(t)\mathbf{u} \tag{101.7}$$

and the **linear output equations**

$$\mathbf{y} = \mathbf{C}(t)\mathbf{x} + \mathbf{D}(t)\mathbf{u} \tag{101.8}$$

where $\mathbf{A}(t) = [a_{ij}(t)] = [\partial f_i / \partial X_j]$, i = row, j = column is an $n_x \times n_x$ matrix; $\mathbf{B}(t) = [\partial f_i / \partial U_j]$ is an $n_x \times n_u$ matrix; $\mathbf{C}(t) = [\partial g_i / \partial X_j]$ is an $n_y \times n_x$ matrix; $\mathbf{D}(t) = [\partial g_i / \partial U_j]$ is an $n_y \times n_u$ matrix; and all matrices are evaluated along the reference conditions $\overline{\mathbf{X}}(t)$, $\overline{\mathbf{U}}(t)$.

The most common occurrence of linear systems arises when the reference input is a constant $\overline{\mathbf{U}}(t) = \overline{\mathbf{U}}$ and the reference state is an equilibrium (i.e., constant) state $\overline{\mathbf{X}}(t) = \overline{\mathbf{X}}$. Then the reference output $\overline{\mathbf{Y}}(t) = \overline{\mathbf{Y}}$ is also constant. In this case the matrices in the linear model are also constant. This yields a constant-coefficient **multiple-input, multiple-output** (**MIMO**) linear state–space model, with linear state equations

$$\dot{\mathbf{x}} = \mathbf{A}\mathbf{x} + \mathbf{B}\mathbf{u} \tag{101.9}$$

and the linear output equations

$$\mathbf{y} = \mathbf{C}\mathbf{x} + \mathbf{D}\mathbf{u} \tag{101.10}$$

Henceforth we will be concerned only with constant-coefficient linear state–space models of the form given in Equation (101.9) and Equation (101.10).

For the special case of a single input u and a single output y ($n_u = n_y = 1$) we have a **single-input, single-output (SISO)** state–space system,

$$\dot{\mathbf{x}} = \mathbf{A}\mathbf{x} + \mathbf{B}u \tag{101.11}$$

$$y = \mathbf{C}\mathbf{x} + Du \tag{101.12}$$

where u and y are scalar variables, D is a scalar constant, $\mathbf{B} = [b_1 \ldots b_{n_x}]^{\mathrm{T}}$, and $\mathbf{C} = [c_1 \ldots c_{n_x}]$.

101.3 Linear System Representations

The representation of a linear state–space system is generally not unique. For example, a change in the coordinate system for **x** will change the matrices in the state–space representation. In addition, there are certain state–space representations and other types of representations that may be more convenient for various types of analyses.

State–Space Systems

There are several special representations for a state–space system that occur frequently in control systems analysis:

Decoupled Form

The simplest representation of a state–space system occurs when the state equations are decoupled, that is, when the **A** matrix is in the diagonalized form,

$$\mathbf{A} = \text{diag}\,[\lambda_1,\ldots,\lambda_{n_x}] = \begin{bmatrix} \lambda_1 & & \mathbf{0} \\ & \ddots & \\ \mathbf{0} & & \lambda_{n_x} \end{bmatrix} \tag{101.13}$$

so that the evolution of each state variable depends only on itself and the inputs, but not on the other state variables.

Block Diagonal Form

For a particular state–space system it may not be possible to completely diagonalize the **A** matrix so that each state variable is decoupled from the others. In this case a generalization of the diagonal structure occurs where the **A** matrix is in block diagonal form,

$$\mathbf{A} = \text{diag}\,[\mathbf{A}_1,\ldots,\mathbf{A}_n] \tag{101.14}$$

in which the $\mathbf{A}_i$ are square matrices on the diagonal of **A**.

Companion Form

For a special class of SISO state–space systems, a representation exists that can lead to a description of the system in terms of a single higher-order differential equation. A single-input, single-output state–space system is said to be in companion form if the **A** matrix is a companion matrix,

$$\mathbf{A} = \begin{bmatrix} 0 & 1 & 0 & & 0 \\ 0 & 0 & 1 & \cdots & 0 \\ & & & \ddots & \\ 0 & 0 & 0 & \cdots & 1 \\ -a_1 & -a_2 & -a_3 & \cdots & -a_{n_x} \end{bmatrix} \tag{101.15}$$

and the scalar control input enters only in the $\dot{x}_{n_x}$ equation, with the column matrix **B** of the form $\mathbf{B} = [0\ 0\ldots1]^T$. As we will see, a companion form representation has a particularly simple relationship to a representation involving a single n_x-order differential equation.

Input-Output Systems

Equations (101.11) and (101.12) are the general representation of a constant-coefficient SISO system. Another common representation for a special class of SISO systems is referred to as an **input-output (IO)** representation. Using the notation $y^{(n)} = d^n y/dt^n$, we call an SISO system an IO system if the system can be represented by a single n_x-order differential equation of the form

$$y^{(n_x)} + p_{n_x-1}y^{(n_x-1)} + \cdots + p_1\dot{y} + p_0 y = q_0 u + q_1\dot{u} + \cdots + q_{n_x}u^{(n_x)} \tag{101.16}$$

Note that the left-hand side of Equation (101.16) is expressed in terms of the output $y(t)$ and the right-hand side is expressed in terms of the input $u(t)$. In order for the output to depend on the input, at least one of the right-hand coefficients must be nonzero.

There is a close relationship between the state–space representation of an SISO system in companion form and the IO representation of the same system, provided that certain conditions are satisfied. For example, if the output is just the first state component — that is, the **C** matrix is of the form $\mathbf{C} = [1\ 0\ldots 0]$ and $D = 0$, with $\mathbf{B} = [0\ldots 0\ 1]^{\mathrm{T}}$ — then from Equation (101.11) and Equation (101.15) the last state equation can be written as

$$\dot{x}_{n_x} + a_{n_x}x_{n_x} + \cdots + a_2x_2 + a_1x_1 = u \tag{101.17}$$

Since $x_1 = y$, it follows from the other state equations of Equation (101.11) and Equation (101.15), $\dot{x}_i = x_{i+1}$ for $i = 1,\ldots, n_x - 1$, that

$$\begin{aligned} x_2 &= \dot{x}_1 = \dot{y} \\ x_3 &= \dot{x}_2 = \ddot{y} \\ &\vdots \\ x_{n_x} &= \dot{x}_{n_x-1} = y^{(n_x-1)} \end{aligned}$$

Thus Equation (101.17) becomes

$$y^{(n_x)} + a_{n_x}y^{(n_x-1)} + \cdots + a_2\dot{y} + a_1 y = u$$

which is in the IO format.

If the output is not just the first state component, then it may not be possible to convert an SISO system in companion form to an equivalent IO representation. In the next section we will discuss a general requirement that, when satisfied, will guarantee that a state–space SISO system (whether in companion form or not) can be converted to an equivalent IO representation.

101.4 Transforming System Representations

For the remainder of this chapter we will be concerned with transforming between various types of representations of constant-coefficient linear systems. It turns out that two fundamental concepts, controllability and observability, govern whether or not certain equivalent representations can be achieved.

Controllability and Observability

Controllability is concerned with whether a control input $\mathbf{u}(t)$ exists that will transfer the state $\mathbf{x}(t)$ from a given initial point $\mathbf{x}(0)$ to a specified final state $\mathbf{x}(t_f)$ in some finite time interval $0 \le t \le t_f$. A linear constant-coefficient system of the form in Equation (101.9) is completely controllable (from any initial state to any final state) if and only if the Kalman controllability criterion, rank $[\mathbf{P}] = n_x$, is satisfied, where the $n_x \times n_x n_u$ matrix,

$$\mathbf{P} = [\mathbf{B}, \mathbf{AB}, \mathbf{A}^2\mathbf{B}, \ldots, \mathbf{A}^{n_x - 1}\mathbf{B}] \tag{101.18}$$

is called the *controllability matrix.* Here, rank $[\mathbf{P}]$ is the maximum number of linearly independent rows (or columns) in $\mathbf{P}$ and is equal to the size of the largest nonzero square determinant obtained by deleting various rows and/or columns of $\mathbf{P}$. For a single-input system, $\mathbf{P}$ is square and the controllability criterion requires $|\mathbf{P}| \neq 0$, where $|\mathbf{P}|$ denotes the determinant of $\mathbf{P}$.

Observability is concerned with the problem of determining the state based on the measured outputs. In particular, a linear state–space system is observable if it is possible to uniquely determine the initial state $\mathbf{x}(0)$ given the output and input histories $\mathbf{y}(t)$ and $\mathbf{u}(t)$ over a finite time interval $0 \leq t \leq t_f$. A linear constant-coefficient system of the form in Equation (101.9) and Equation (101.10) is completely observable (over any nonzero time interval) if and only if the Kalman observability criterion, rank $[\mathbf{Q}] = n_x$, is satisfied, where the $n_x n_y \times n_x$ matrix,

$$\mathbf{Q} = \begin{bmatrix} \mathbf{C} \\ \mathbf{CA} \\ \mathbf{CA}^2 \\ \vdots \\ \mathbf{CA}^{n_x - 1} \end{bmatrix} \tag{101.19}$$

is called the *observability matrix.* For a single-output system, $\mathbf{Q}$ is square and the observability criterion requires $|\mathbf{Q}| \neq 0$.

State–Space Transformations

As indicated previously, the state–space representation of a linear system is not unique. Let $\mathbf{M}$ be any constant nonsingular $n_x \times n_x$ matrix with inverse $\mathbf{M}^{-1}$. Then the coordinate transformation $\mathbf{z} = \mathbf{M}^{-1}\mathbf{x}$ transforms the system given in Equation (101.9) and Equation (101.10) into

$$\dot{\mathbf{z}} = \hat{\mathbf{A}}\mathbf{z} + \hat{\mathbf{B}}\mathbf{u} \tag{101.20}$$

$$\mathbf{y} = \hat{\mathbf{C}}\mathbf{z} + \mathbf{D}\mathbf{u} \tag{101.21}$$

where

$$\hat{\mathbf{A}} = \mathbf{M}^{-1}\mathbf{AM}, \qquad \hat{\mathbf{B}} = \mathbf{M}^{-1}\mathbf{B}, \qquad \hat{\mathbf{C}} = \mathbf{CM} \tag{101.22}$$

Diagonalization

For $i = 1,\ldots n_x$ let λ_i denote the scalar **eigenvalues** of $\mathbf{A}$, with corresponding **eigenvectors** $\xi_i \neq \mathbf{0}$. That is,

$$\mathbf{A}\xi_i = \lambda_i \xi_i \tag{101.23}$$

where the eigenvalues satisfy the n_x-order polynomial characteristic equation

$$0 = |\lambda \mathbf{I} - \mathbf{A}| = \mathcal{P}(\lambda) = \lambda^{n_x} + p_{n_x - 1}\lambda^{n_x - 1} + \cdots + p_1\lambda + p_0 \tag{101.24}$$

Suppose that the n_x eigenvectors are linearly independent. A sufficient condition for this, from linear algebra, is that the eigenvalues be distinct $(\lambda_i \neq \lambda_j \text{ for } i \neq j)$. Then the eigenvector matrix (also called the *modal matrix*),

$$\mathbf{M} = [\xi_1 \ldots \xi_{n_x}] \tag{101.25}$$

whose columns are the eigenvectors of **A**, has an inverse. Using the eigenvector matrix and the coordinate transformation $\mathbf{z} = \mathbf{M}^{-1}\mathbf{x}$ applied to the system given in Equation (101.9) and Equation (101.10) yields the transformed system of Equation (101.20) and Equation (101.21), in which $\hat{\mathbf{A}} = \mathbf{M}^{-1}\mathbf{AM}$ is a diagonal matrix, with eigenvalues on the main diagonal in the same order as the eigenvector columns of **M**. That is, the state equations become decoupled in the new state variables.

In component form the decoupled state equations are

$$\dot{z}_i = \lambda_i z_i + \hat{\mathbf{b}}_i^{\mathrm{T}}\mathbf{u}, \quad i = 1, \ldots, n_x \tag{101.26}$$

and the output equations can be written as

$$\mathbf{y} = z_1\hat{\mathbf{c}}_1 + \cdots + z_{n_x}\hat{\mathbf{c}}_{n_x} + \mathbf{Du} \tag{101.27}$$

where $\hat{\mathbf{b}}_i^{\mathrm{T}}$ is the *i*th row of $\hat{\mathbf{B}} = \mathbf{M}^{-1}\mathbf{B}$ and $\hat{\mathbf{c}}_i$ is the *i*th column of $\hat{\mathbf{C}} = \mathbf{CM}$.

In terms of the decoupled system equations, we have direct alternate tests for controllability and observability. The system of Equation (101.9) and Equation (101.10) is controllable if and only if $\hat{\mathbf{B}}$ contains no zero row, so that **u** affects each eigenstate z_i. Similarly, the system of Equation (101.9) and Equation (101.10) is observable if and only if $\hat{\mathbf{C}}$ contains no zero column, so that **y** reflects each eigenstate z_i.

Block Diagonalization

If the n_x eigenvectors of **A** are not linearly independent, then it is not possible to transform **A** to a diagonal matrix. That is, we cannot find a coordinate system in which each state variable is decoupled from the other state variables. However, we can always find a coordinate transformation matrix **M** so that $\mathbf{z} = \mathbf{M}^{-1}\mathbf{x}$ transforms **A** to an $n_x \times n_x$ block diagonal *Jordan matrix*,

$$\mathbf{J} = \text{diag}\,[\mathbf{J}_1, \mathbf{J}_2, \ldots, \mathbf{J}_n] \tag{101.28}$$

where n is the number of linearly independent eigenvectors and each Jordan block $\mathbf{J}_i$, associated with an eigenvalue λ_i **A**, is either a 1×1 matrix $[\lambda_i]$ for distinct eigenvalues or, for repeated eigenvalues, a square submatrix in upper triangular form with λ_i on the diagonal, ones above and adjacent to the diagonal, and zeroes elsewhere:

$$\mathbf{J}_i = \begin{bmatrix} \lambda_i & 1 & & & \\ & \lambda_i & 1 & 0 & \\ & & \lambda_i & \ddots & \\ & 0 & & \ddots & 1 \\ & & & & \lambda_i \end{bmatrix} \tag{101.29}$$

The resulting transformed system, of the form

$$\dot{\mathbf{z}}_i = \mathbf{J}_i\mathbf{z}_i + \mathbf{B}_i\mathbf{u} \tag{101.30}$$

will be composed of block-decoupled subsystems, whose state variables are decoupled from those of the other subsystems.

The development of the matrix **M**, so that $\mathbf{z} = \mathbf{M}^{-1}\mathbf{x}$ transforms **A** to Jordan block diagonal form, is essentially the same as in the case of complete diagonalization. The difference is that in order for the matrix **M** to have an inverse, any linearly dependent eigenvectors of **A** are replaced by "generalized eigenvectors" to form an **M** matrix with n_x linearly independent columns. For a repeated eigenvalue λ with eigenvector $\boldsymbol{\xi}$, a *generalized eigenvector* $\hat{\boldsymbol{\xi}}$ is a nonzero solution to the equation

$$[\boldsymbol{\lambda}_i \mathbf{I} - \mathbf{A}]\hat{\boldsymbol{\xi}} = -\boldsymbol{\xi} \tag{101.31}$$

This equation can be used repeatedly, if necessary, with the right-hand side being either an eigenvector or a previously generated generalized eigenvector.

Let $\lambda_1,\ldots, \lambda_{n_x}$ and $\xi_1,\ldots, \xi_{n_x}$ be the eigenvalues and the associated linearly independent eigenvectors or generalized eigenvectors. Order them such that each generalized eigenvector $\boldsymbol{\xi}_i$ is generated from $\boldsymbol{\xi}_{i-1}$ by

$$[\boldsymbol{\lambda}_i \mathbf{I} - \mathbf{A}]\boldsymbol{\xi}_i = -\boldsymbol{\xi}_{i-1} \tag{101.32}$$

whereas each eigenvector $\boldsymbol{\xi}_1$ satisfies

$$[\boldsymbol{\lambda}_i \mathbf{I} - \mathbf{A}]\boldsymbol{\xi}_i = \mathbf{0} \tag{101.33}$$

Then the transformation $\mathbf{z} = \mathbf{M}^{-1}\mathbf{x}$, with $\mathbf{M} = [\boldsymbol{\xi}_1 \ldots \xi_{n_x}]$, converts **A** to a Jordan block diagonal form $\mathbf{J} = \mathbf{M}^{-1}\mathbf{AM}$, since $\mathbf{MJ} = \mathbf{AM}$. For more details on block diagonalization see Brogan [1982, p. 143].

Instead of block diagonalization, one can employ a matrix perturbation technique and only consider the case where the eigenvalues of **A** are distinct, so that the eigenvectors are linearly independent and complete diagonalization is possible. This technique is based on the fact that distinct eigenvalues can always be achieved by making a small perturbation ε in the elements of **A** that cause repeated eigenvalues [Luenberger, 1979, p. 149]. After any ensuing analysis has been performed — for example, for the solution $\mathbf{x}(t, \varepsilon)$ — the results can be examined in the limit as $\varepsilon \to 0$.

Companion Form Systems

An SISO state–space system such as in Equation (101.11) and Equation (101.12) can be transformed to a unique companion form if, and only if, the system is controllable. By way of construction, we note from Equation (101.11) and Equation (101.15) that in companion form the state variables z_i all follow from z_1 in a cascade: $z_{i+1} = \dot{z}_i$, $i = 1,\ldots, n_x - 1$. Thus we can construct the transformation matrix by finding an n_x-dimensional vector $\boldsymbol{\rho}$ such that, by choosing $z_1 = \boldsymbol{\rho}^T\mathbf{x}$, repeated differentiation yields a system in companion form. In particular, we choose

$$\begin{aligned}
z_1 &= \boldsymbol{\rho}^T\mathbf{x} && \\
z_2 = \dot{z}_1 &= \boldsymbol{\rho}^T\mathbf{A}\mathbf{x} && \text{with } \boldsymbol{\rho}^T\mathbf{B} = \mathbf{0} \\
z_3 = \dot{z}_2 &= \boldsymbol{\rho}^T\mathbf{A}^2\mathbf{x} && \text{with } \boldsymbol{\rho}^T\mathbf{AB} = \mathbf{0} \\
&\vdots \\
z_{n_x} = \dot{z}_{n_x-1} &= \boldsymbol{\rho}^T\mathbf{A}^{n_x-1}\mathbf{x} && \text{with } \boldsymbol{\rho}^T\mathbf{A}^{n_x-2}\mathbf{B} = \mathbf{0} \\
\dot{z}_{n_x} &= \boldsymbol{\rho}^T\mathbf{A}^{n_x}\mathbf{x} + \mathbf{u} && \text{with } \boldsymbol{\rho}^T\mathbf{A}^{n_x-1}\mathbf{B} = \mathbf{1}
\end{aligned} \tag{101.34}$$

From the left-hand sides of Equation (101.34) and $\mathbf{z} = \mathbf{M}^{-1}\mathbf{x}$, we have

$$\mathbf{M}^{-1} = \begin{bmatrix} \boldsymbol{\rho}^{\mathrm{T}} \\ \boldsymbol{\rho}^{\mathrm{T}}\mathbf{A} \\ \boldsymbol{\rho}^{\mathrm{T}}\mathbf{A}^2 \\ \vdots \\ \boldsymbol{\rho}^{\mathrm{T}}\mathbf{A}^{n_x-1} \end{bmatrix} \tag{101.35}$$

From the right-hand sides of Equation (101.34), $\boldsymbol{\rho}$ is the solution to

$$\boldsymbol{\rho}^{\mathrm{T}}[\mathbf{B}, \mathbf{AB}, \mathbf{A}^2\mathbf{B}, \ldots, \mathbf{A}^{n_x-1}\mathbf{B}] = [\mathbf{0} \quad \mathbf{0} \ldots \mathbf{1}]$$

For SISO systems, the controllability matrix $\mathbf{P}$ is square and the controllability condition ensures the existence of $\mathbf{P}^{-1}$. Thus

$$\boldsymbol{\rho}^{\mathrm{T}} = [0\ 0 \ldots 1]\mathbf{P}^{-1} \tag{101.36}$$

Hence $\boldsymbol{\rho}^{\mathrm{T}}$ is the last row of $\mathbf{P}^{-1}$ and we construct $\mathbf{M}^{-1}$ as shown in Equation (101.35). The matrix $\mathbf{M}^{-1}$ has an inverse, since the right-hand sides of Equation (101.34) imply that the rows in Equation (101.35) are linearly independent [Luenberger, 1979, p. 292].

Input-Output Systems

We can transform a state–space SISO system to a unique equivalent n_x-order IO form if, and only if, the system is observable. To perform the transformation we differentiate the output $y(t)$ n_x times. Using the state equation [Equation (101.11)] to substitute for $\dot{\mathbf{x}}$ at each step yields the following system of equations:

$$\begin{aligned}
y &= \mathbf{Cx} + Du \\
\dot{y} &= \mathbf{CAx} + D\dot{u} + \mathbf{CB}u \\
\ddot{y} &= \mathbf{CA}^2\mathbf{x} + D\ddot{u} + \mathbf{C}\{\mathbf{B}\dot{u} + \mathbf{AB}u\} \\
y^{(3)} &= \mathbf{CA}^3\mathbf{x} + Du^{(3)} + \mathbf{C}\{\mathbf{B}\ddot{u} + \mathbf{AB}\dot{u} + \mathbf{A}^2\mathbf{B}u\} \\
\vdots \quad & \quad \vdots \\
y^{(n_x-1)} &= \mathbf{CA}^{n_x-1}\mathbf{x} + Du^{(n_x-1)} + \mathbf{C}\{\mathbf{B}u^{(n_x-2)} + \mathbf{AB}u^{(n_x-3)} + \cdots + \mathbf{A}^{n_x-2}\mathbf{B}u\} \\
y^{(n_x)} &= \mathbf{CA}^{n_x}\mathbf{x} + Du^{(n_x)} + \mathbf{C}\{\mathbf{B}u^{(n_x-1)} + \mathbf{AB}u^{(n_x-2)} + \cdots + \mathbf{A}^{n_x-1}\mathbf{B}u\}
\end{aligned} \tag{101.37}$$

The first n_x of these equations can be solved for $\mathbf{x}$ in terms of y, u, and their derivatives. For SISO systems the observability matrix $\mathbf{Q}$ is square and the observability condition ensures the existence of an inverse. Thus $\mathbf{x}$ will be a unique function of y, u, and their derivatives. Substituting this result into the $y^{(n_x)}$ equation in Equation (101.37) and collecting terms yields an IO system of the form in Equation (101.16).

State Equations from the Transfer Matrix

For a multiple-input, multiple-output state–space system of the form in Equation (101.9) and Equation (101.10), taking the Laplace transform of both equations and setting the initial condition terms to zero yields

$$\mathbf{Y}(s) = \mathbf{G}(s)\mathbf{U}(s) \tag{101.38}$$

where $\mathbf{Y}(s) = \mathscr{L}[\mathbf{y}(t)]$ and $\mathbf{U}(s) = \mathscr{L}[\mathbf{u}(t)]$ are the Laplace transforms of $\mathbf{y}(t)$ and $\mathbf{u}(t)$, respectively, and

$$\mathbf{G}(s) = \mathbf{C}[s\mathbf{I} - \mathbf{A}]^{-1}\mathbf{B} + \mathbf{D} \tag{101.39}$$

is the $n_y \times n_u$ transfer matrix. Suppose that the transfer matrix is known, perhaps from experimental results, and is given by

$$\mathbf{G}(s) = \frac{1}{\mathcal{P}(s)} = \begin{bmatrix} \mathcal{Q}_{11}(s) & \dots & \mathcal{Q}_{1n_u}(s) \\ \vdots & \ddots & \vdots \\ \mathcal{Q}_{n_y 1}(s) & \dots & \mathcal{Q}_{n_y n_u}(s) \end{bmatrix} \tag{101.40}$$

where each $\mathcal{P}_{ij}(s)$ is a polynomial of order less than or equal to the order n_x of the characteristic polynomial $\mathcal{P}(s) = |s\mathbf{I} - \mathbf{A}|$. We wish to construct a state–space model from the transfer matrix model.

We will first obtain a decoupled model in which the **A** matrix is diagonal. Then a suitable coordinate transformation is applied to yield a state–space representation having specified properties, such as a desired set of eigenvectors or an **A** matrix that is in companion form.

We compute the eigenvalues from the n_x-order characteristic equation $\mathcal{P}(\lambda) = 0$. Assuming that these eigenvalues are distinct, we define the diagonalized state matrix as $\hat{\mathbf{A}} = \text{diag}\ [\lambda_1, \dots, \lambda_{n_x}]$. The next step is to determine an $n_x \times n_u$ matrix $\hat{\mathbf{B}}$, an $n_y \times n_x$ matrix $\hat{\mathbf{C}}$, and an $n_y \times n_u$ matrix **D** such that the following conditions hold:

1. The transfer matrix corresponds to a state–space system, that is, $\mathbf{G}(s) = \hat{\mathbf{C}}[s\mathbf{I} - \hat{\mathbf{A}}]^{-1}\hat{\mathbf{B}} + \mathbf{D}$.
2. $\hat{\mathbf{B}}$ has no zero rows (controllability is satisfied).
3. $\hat{\mathbf{C}}$ has no zero columns (observability is satisfied).

Usually, the system of equations that result from equating like powers of s in condition 1 involves fewer equations than unknowns in $\hat{\mathbf{B}}$, $\hat{\mathbf{C}}$, and **D**, so there is some degree of freedom in choosing the elements of $\hat{\mathbf{B}}$, $\hat{\mathbf{C}}$, and **D**. The resulting decoupled state–space system is

$$\begin{aligned} \dot{\mathbf{z}} &= \hat{\mathbf{A}}\mathbf{z} + \hat{\mathbf{B}}\mathbf{u} \\ \mathbf{y} &= \hat{\mathbf{C}}\mathbf{z} + \mathbf{D}\mathbf{u} \end{aligned} \tag{101.41}$$

From the diagonalized representation, which may have complex-valued matrices, we can change to a final set of state variables $\mathbf{x} = \mathbf{M}\mathbf{z}$ by choosing a set of desired linearly independent eigenvectors as the columns of the transformation matrix $\mathbf{M} = [\boldsymbol{\xi}_1 \dots \boldsymbol{\xi}_{n_x}]$. This transformation yields

$$\begin{aligned} \dot{\mathbf{x}} &= \mathbf{A}\mathbf{z} + \mathbf{B}\mathbf{u} \\ \mathbf{y} &= \mathbf{C}\mathbf{x} + \mathbf{D}\mathbf{u} \end{aligned} \tag{101.42}$$

where $\mathbf{A} = \mathbf{M}\hat{\mathbf{A}}\mathbf{M}^{-1}$, $\mathbf{B} = \mathbf{M}\hat{\mathbf{B}}$, and $\mathbf{C} = \hat{\mathbf{C}}\mathbf{M}^{-1}$. The matrix **A** will have eigenvalues λ_i and the chosen eigenvectors. For real-valued final matrices, complex conjugate eigenvectors should be chosen for any corresponding conjugate eigenvalues.

As an example, consider the IO system,

$$\ddot{y} + 3\dot{y} + 2y = u_1 + 3u_2 + \dot{u}_2$$

that has the transfer matrix representation

$$Y(s) = \begin{bmatrix} \dfrac{1}{s^2 + 3s + 2} & \dfrac{s+3}{s^2 + 3s + 2} \end{bmatrix} \begin{bmatrix} U_1(s) \\ U_2(s) \end{bmatrix} = \mathbf{G}(s)\mathbf{U}(s)$$

The characteristic equation $\lambda^2 + 3\lambda + 2 = 0$ yields the eigenvalues $\lambda_1 = -1, \lambda_2 = -2$. Thus the diagonalized state equations will have the $\hat{\mathbf{A}}$ matrix given by

$$\hat{\mathbf{A}} = \begin{bmatrix} -1 & 0 \\ 0 & -2 \end{bmatrix}$$

The condition $\mathbf{G}(s) = \hat{\mathbf{C}}\,[s\mathbf{I} - \hat{\mathbf{A}}]^{-1}\,\hat{\mathbf{B}} + \mathbf{D}$ yields

$$\frac{[1 \;\; s+3]}{s^2+3s+2} = [\hat{c}_1 \;\; \hat{c}_2] \begin{bmatrix} \dfrac{1}{s+1} & 0 \\ 0 & \dfrac{1}{s+2} \end{bmatrix} \begin{bmatrix} \hat{b}_{11} & \hat{b}_{12} \\ \hat{b}_{21} & \hat{b}_{22} \end{bmatrix} + [d_1 \;\; d_2]$$

Thus the elements of $\hat{\mathbf{B}}$, $\hat{\mathbf{C}}$, and $\mathbf{D}$ must satisfy

$$1 = d_1 s^2 + \left(\hat{c}_1\hat{b}_{11} + \hat{c}_2\hat{b}_{21} + 3d_1\right)s + \left(2\hat{c}_1\hat{b}_{11} + \hat{c}_2\hat{b}_{21} + 2d_1\right)$$

$$s+3 = d_2 s^2 + \left(\hat{c}_1\hat{b}_{12} + \hat{c}_2\hat{b}_{22} + 3d_2\right)s + \left(2\hat{c}_1\hat{b}_{12} + \hat{c}_2\hat{b}_{22} + 2d_2\right)$$

Equating like powers of s and solving the resulting equations yields

$$d_1 = d_2 = 0, \quad \hat{c}_1\hat{b}_{11} = -1, \quad \hat{c}_2\hat{b}_{21} = -1, \quad \hat{c}_1\hat{b}_{12} = 2, \quad \hat{c}_2\hat{b}_{22} = -1$$

The observability condition in 3 requires that $\hat{c}_1 \neq 0$ and $\hat{c}_2 \neq 0$. Therefore, these results can be solved for the $\hat{b}_{ij}$ in terms of $\hat{c}_1$ and $\hat{c}_2$. The resulting $\hat{b}_{ij}$ are all nonzero, so the controllability condition in 2 is also satisfied. For this example, we choose $\hat{c}_1 = \hat{c}_2 = 1$, yielding the decoupled system of Equation (101.41) with

$$\hat{\mathbf{A}} = \begin{bmatrix} -1 & 0 \\ 0 & -2 \end{bmatrix}, \quad \hat{\mathbf{B}} = \begin{bmatrix} -1 & 2 \\ -1 & -1 \end{bmatrix}, \quad \hat{\mathbf{C}} = [1 \;\; 1], \quad \mathbf{D} = [0 \;\; 0]$$

Since the eigenvalues were all real, the diagonalized state–space model is real, so further transformation is not required. However, to complete the example, suppose we want to change from this diagonal form to a state–space system having a specified set of eigenvectors, such as $\xi_1 = [1 \; -1]^{\mathrm{T}}$ and $\xi_2 = [1 \; -2]^{\mathrm{T}}$. As a final step, we employ a coordinate transformation $\mathbf{x} = \mathbf{Mz}$, where $\mathbf{M} = [\xi_1 \; \xi_2]$. This yields a state–space system as in Equation (101.42), with

$$\mathbf{A} = \begin{bmatrix} 0 & 1 \\ -2 & -3 \end{bmatrix}, \quad \mathbf{B} = \begin{bmatrix} 0 & 1 \\ 1 & 0 \end{bmatrix}, \quad \mathbf{C} = [1 \;\; 0], \quad \mathbf{D} = [0 \;\; 0]$$

Note that our choice of eigenvectors happened to produce an **A** matrix in companion form, with the same eigenvalues as $\hat{\mathbf{A}}$, but with different eigenvectors. If our objective had been to directly produce an **A** matrix in companion form, we could have chosen $\xi_1 = [1 \; \alpha]^{\mathrm{T}}$ and $\xi_2 = [1 \; \beta]^{\mathrm{T}}$ as variables, computed $\mathbf{A} = \mathbf{M}\hat{\mathbf{A}}\mathbf{M}^{-1} = [a_{ij}]$, and then solved for the parameters α and β to satisfy the two companion matrix conditions $a_{11} = 0$ and $a_{12} = 1$.

Defining Terms

Cause and effect — In this context for the IO formulation, a condition that requires that the order of any derivative in the input be less than or equal to the highest order of derivative in the output.

Controllability — The ability to drive a system from an arbitrary initial state to an arbitrary final state in finite time.

Eigenvalue — Any scalar λ satisfying the equation $\det[\lambda \mathbf{I} - \mathbf{A}] = 0$, where $\mathbf{A}$ is an $n_x \times n_x$ matrix and $\mathbf{I}$ is the identity matrix; alternately, a scalar λ satisfying the equation $\mathbf{A}\boldsymbol{\xi} = \lambda\boldsymbol{\xi}$, where $\boldsymbol{\xi}$ is an n_x-dimensional eigenvector.

Eigenvector — Any nonzero n_x-dimensional vector $\boldsymbol{\xi}$ satisfying the equation $\mathbf{A}\boldsymbol{\xi} = \boldsymbol{\lambda\xi}$, where $\mathbf{A}$ is an $n_x \times n_x$ dimensional matrix and λ is an eigenvalue.

Input-output (IO) model — A representation of a dynamic system in terms of a single n_x-order differential equation relating the output $y(t)$ to the input $u(t)$.

Input vector — An n_u-dimensional column vector consisting of the variable quantities, other than state variables, that affect the evolution of the state of a dynamic system.

Linear output equations — A set of algebraic equations defining the output variables in terms of linear combinations of the state and input variables.

Linear state equations — A set of first-order differential equations that model the behavior of a physical system in terms of a linear combination of the state and input variables.

Multiple-input, multiple-output (MIMO) model — A state–space model with an input vector, $\mathbf{u}$, of dimension n_u greater than one and output vector, $\mathbf{y}$, of dimension n_y greater than one.

Observability — The ability to determine a system's initial state given the output and input histories over a finite time interval.

Output equations — A set of algebraic equations defining the output variables as functions of the state variables and the input variables.

Output vector — An n_y-dimensional column vector whose elements model the measurements of a physical system.

Single-input, single-output (SISO) model — A state–space model with a single scalar input, u, and single scalar output, y.

State equations — A set of first-order differential equations that model the behavior of a physical system.

State space — A geometric space with dimension n_x equal to the number of state variables. Any possible state of a dynamic model can be represented as a point in state space.

State variables — The smallest set of time-differentiated variables whose initial conditions, along with the inputs and the model, allow complete prediction of the behavior of a dynamic system. It is possible to define more than one set of state variables for any particular model. However, the number of state variables is a unique quantity for a system.

State vector — An n_x-dimensional column vector consisting of the state variables of a model.

References

Brogan, W. L. 1982. *Modern Control Theory,* Prentice Hall, Englewood Cliffs, NJ.

Luenberger, D. G. 1979. *Introduction to Dynamic Systems,* John Wiley & Sons, New York.

Further Information

The following two texts present examples and in-depth information on state–space models:

Friedland, B. 1986. *Control System Design: An Introduction to State–Space Methods,* McGraw-Hill, New York.

Grantham, W. J. and Vincent, T. L. 1993. *Modern Control Systems Analysis and Design,* John Wiley & Sons, New York.

102

Frequency Response

Paul Neudorfer
Seattle University
Pierre Gehlen
Seattle University

Frequency response in stable, linear systems is defined as "the frequency-dependent relation in both gain and phase difference between steady-state sinusoidal inputs and the resultant steady-state sinusoidal outputs" [IEEE, 1988]. The frequency response characteristics of a system can be found analytically from its transfer function. They are also commonly measured in laboratory or field tests. A single-input/single-output linear time-invariant system is shown in Figure 102.1.

For systems with no time delay, the transfer function $H(s)$ is in the form of a ratio of polynomials in the complex frequency s,

$$H(s) = K\frac{N(s)}{D(s)}$$

where K is a frequency-independent constant. For a system in the sinusoidal steady state, s is replaced by the sinusoidal frequency $j\omega$ ($j = \sqrt{-1}$) and the system function becomes

$$H(j\omega) = K\frac{N(j\omega)}{D(j\omega)} = |H(j\omega)|e^{j\angle H(j\omega)}$$

$H(j\omega)$ is a complex quantity. Its magnitude $|H(j\omega)|$ and its angle or argument $\angle H(j\omega)$ relate, respectively, the amplitudes and phase angles of sinusoidal steady-state input and output signals. Referring to Figure 102.1, if the input and output signals are

$$x(t) = X\cos(\omega t + \theta_x)$$

$$y(t) = Y\cos(\omega t + \theta_y)$$

then the output's magnitude Y and phase angle θ_y are related to those of the input by the two equations

$$Y = |H(j\omega)|X$$

$$\theta_y = \angle H(j\omega) + \theta_x$$

0-8493-1586-7/05/$0.00+$1.50

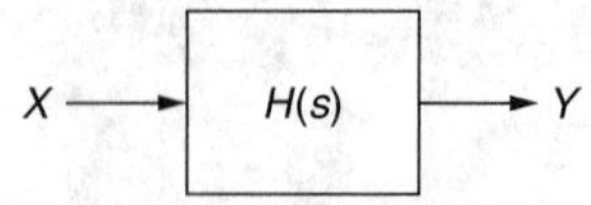

FIGURE 102.1 A single-input/single-output linear system.

The phrase *frequency response characteristic* usually implies a complete description of a system's sinusoidal steady-state behavior as a function of frequency. Because $H(j\omega)$ is complex, frequency response characteristics cannot be graphically displayed as a single curve plotted with respect to frequency. Instead, the magnitude and angle of $H(j\omega)$ can be separately plotted as functions of frequency. It is often advantageous to plot frequency response curves on other-than-linearly scaled Cartesian coordinates. **Bode diagrams** (developed in the 1930s by H.W. Bode of Bell Labs) use a logarithmic scale for frequency and a decibel measure for magnitude. In **Nyquist plots** (from Harry Nyquist, also of Bell Labs), $H(j\omega)$ is displayed in Argand (polar) diagram form on the complex number plane, with $\mathrm{Re}[H(j\omega)]$ on the horizontal axis and $\mathrm{Im}[H(j\omega)]$ on the vertical. Frequency is a parameter of such curves. It is sometimes numerically identified at selected points of the curve and sometimes omitted. The **Nichols chart** (developed by N. B. Nichols) graphs magnitude versus phase for the system function, frequency again being a parameter of the curve.

Frequency response techniques are most obviously applicable to topics such as communications and electrical filters in which the frequency response behaviors of systems are central to an understanding of their operations. It is, however, in the area of control systems that frequency response techniques are most fully developed as analytical and design tools. The Nichols chart, for instance, is used exclusively in the analysis and design of classical feedback control systems. Although frequency response concepts are perhaps most often associated with electrical engineering, they are also commonly used in other branches of engineering and the natural sciences. Two nonelectrical examples are given in Figure 102.2 and Figure 102.3, which show a frequency response test of a scale model of the suspension system of a truck and the modulation transfer function (MTF) of a photographic film. MTFs provide an accurate description of the quality of photographic systems and have largely replaced more subjective parameters, such as resolving power, sharpness, and acutance. Note that frequency response concepts are equally applicable to problems involving temporal (Figure 102.2) and spatial (Figure 102.3) frequencies.

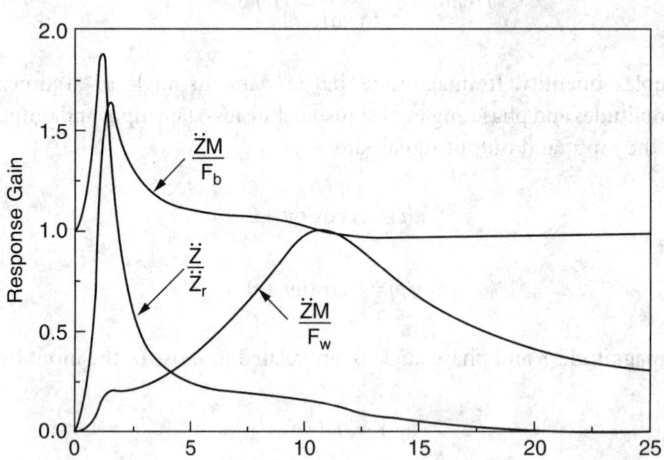

FIGURE 102.2 Frequency response test of a truck cab and suspension. (*Source:* Gillespie, T.B. 1992. *Fundamentals of Vehicle Dynamics*, Society of Automotive Engineers, Warrendale, PA. Courtesy of the Society of Automotive Engineers, Inc.)

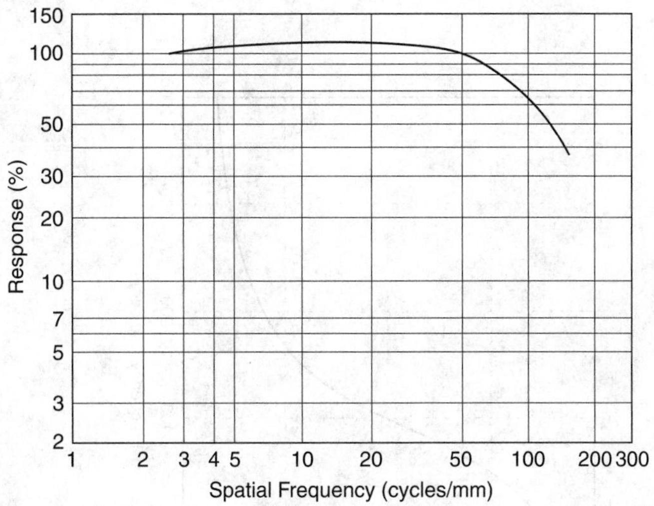

FIGURE 102.3 Modulation transfer function of photographic film. (*Source:* Williams, J. B. 1991. *Image Clarity: High Resolution Photography,* Focal, Boston. Courtesy of Eastman Kodak.)

102.1 Frequency Response Plotting

Frequency response plots are prepared by computing the magnitude and angle of $H(j\omega)$.

Linear Plots

In linear plots $|H(j\omega)|$ and $\angle H(j\omega)$ are shown in separate diagrams as functions of frequency (either f or ω). Cartesian coordinates are used, and all scales are linear.

Example 102.1

Consider the transfer function

$$H(s)=\frac{160000}{s^2+220s+160000}$$

The complex frequency variable s is replaced by the sinusoidal frequency $j\omega$ and the magnitude and angle are found.

$$H(j\omega)=\frac{160000}{(j\omega)^2+220(j\omega)+160000}$$

$$|H(j\omega)|=\frac{160000}{\sqrt{(160000-\omega^2)^2+(220\omega)^2}}$$

$$\angle H(j\omega)=-\tan^{-1}\frac{220\omega}{160000-\omega^2}$$

The plots of magnitude and angle are shown in Figure 102.4. Linear plots are most useful when the frequency range of interest is small, as is commonly the case for mechanical systems. Such plots give a straightforward representation of system response.

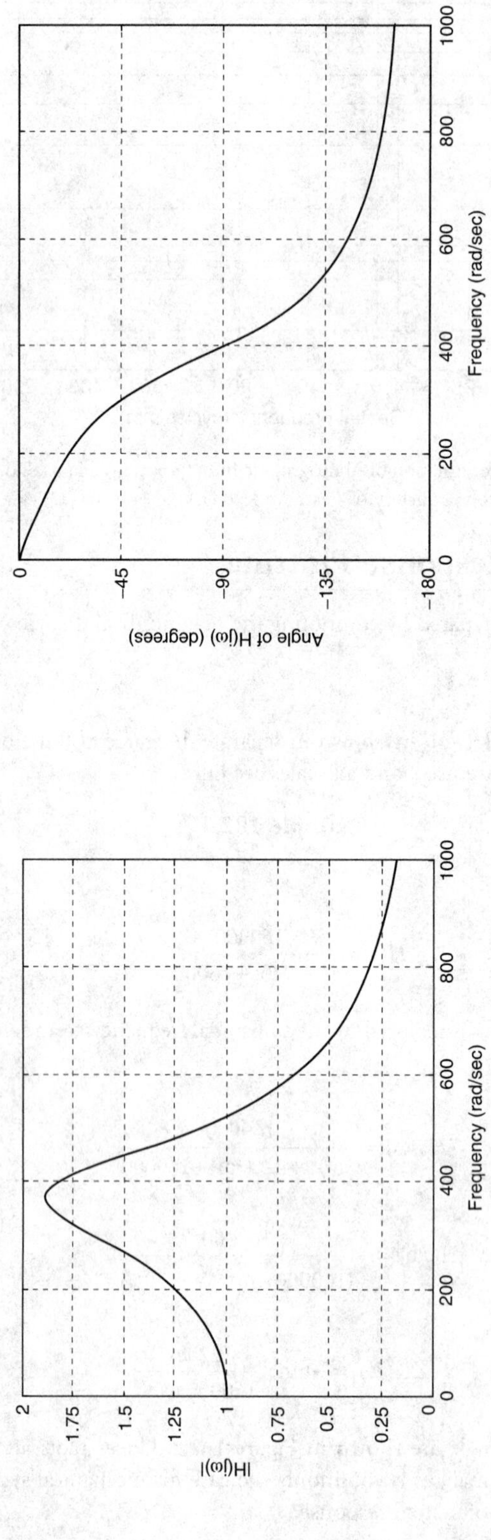

FIGURE 102.4 Linear frequency response curves of $H(j\omega)$.

The peaking of the magnitude curve near $\omega = 400$ rad/s in Figure 102.4 is a reflection of the phenomenon of **resonance.** Resonance may exist in lightly damped second- or higher-order systems. Examples of resonance are common in both the natural and manmade worlds. Organ pipes, for example, are designed to resonate at desired pitches. Resonances can have undesirable effects, such as when the steering mechanism of a vehicle with misaligned front wheels shimmies at certain operating speeds. Two well-known examples of the destructive effects of resonance are the Tacoma Narrows suspension bridge, which underwent catastrophic failure during a moderate windstorm in 1940, and the 1981 collapse of an elevated walkway in the lobby of a Kansas City hotel under the load of a crowd of dancers. All mechanical systems resonate at certain critical frequencies. The careful design engineer makes sure that vibrations that might affect the system's structural integrity are well outside the service range.

Bode Diagrams

A Bode diagram consists of plots of the gain and angle of a transfer function, each with respect to logarithmically scaled frequency axes. In addition, the gain of the transfer function is scaled in **decibels (dB)** according to the definition

$$H_{dB} = 20\log_{10}\left|H(j\omega)\right|$$

Bode diagrams have the advantage of clearly identifying system features even if they occur over wide ranges of frequency and dynamic response. Before constructing a Bode diagram, the transfer function is normalized so that each pole or zero term (except those at $s = 0$) has a DC gain of one. For instance,

$$H(s) = K\frac{s+\omega_z}{s(s+\omega_p)} = \frac{K\omega_z}{\omega_p}\frac{s/\omega_z+1}{s(s/\omega_p+1)} = K'\frac{s\tau_z+1}{s(s\tau_p+1)}$$

It is common to draw Bode diagrams directly from $H(s)$ without making the formal substitution $s = j\omega$.

When drawn by hand, Bode magnitude and angle curves are developed by adding the individual contributions of the factored terms of the transfer function's numerator and denominator. In general, these factored terms may include (1) a constant K, (2) a simple s term corresponding to either a zero or a pole at the origin, (3) a term such as $(s\tau + 1)$ corresponding to a real-valued (nonzero) pole or zero, and (4) a quadratic term with a possible standard form of $[(s/\omega_n)^2 + 2\zeta(s/\omega_n) + 1]$ corresponding to a pair of complex conjugate poles or zeros and for which $0 < \zeta < 1$. With the exception of quadratic terms having small ζ (**damping ratio**), the Bode magnitude and angle curves for all such expressions can be reasonably approximated by a series of straight line segments. Detailed procedures for drawing Bode diagrams from these basic forms are described in many references. Examples are given in Figure 102.5, which shows straight-line approximations for both a numerator-factored term $(s/\omega_z + 1)$ and a denominator-factored term, $1/(s/\omega_p + 1)$. The approximations are shown as dotted lines. The exact curves are shown as solid lines.

Note in Figure 102.5 that both decibel magnitude and angle are plotted semilogarithmically. The frequency axis is logarithmically scaled so that every tenfold, or **decade,** change in frequency occurs over an equal distance. The magnitude axis is given in decibels. Positive decibel magnitudes correspond to amplifications between input and output that are greater than one (output amplitude larger than input). Negative decibel gains correspond to attenuation between input and output (output amplitude smaller than input). Notice in Figure 102.5 that at low and high frequencies the straight-line approximations are virtually identical to the exact curves. The straight-line approximations differ most greatly from the exact curves at points where the approximations change slope. In magnitude curves these are called **breakpoints.** At breakpoints the straight-line approximation and the exact curve differ by 3dB for each pole or zero.

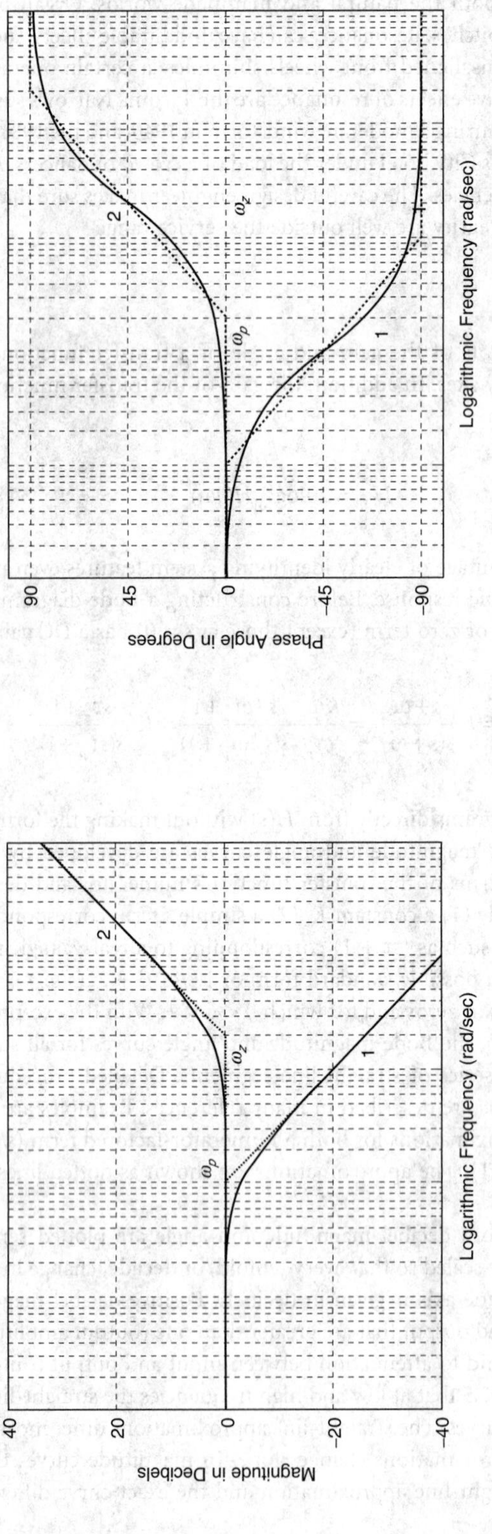

FIGURE 102.5 Bode curves for (1) a pole at $s = -\omega_p$ and (2) a zero at $s = -\omega_z$.

Bode diagrams are easily constructed by hand because, with the exception of lightly damped quadratic terms, each contribution can be reasonably approximated with straight lines. Also, the overall frequency response curve is found by adding the individual contributions. Today, many commercially available mathematical analysis software packages have built-in utilities for creating Bode diagrams. These have rendered the plotting of Bode diagrams by hand nearly obsolete. Still, there is benefit in understanding the traditional methods because they give insight into the meanings of Bode diagram features. Breakpoints, for instance, can be related to the locations of poles and/or zeros and slopes can be related to the numbers of poles and/or zeros. Two examples of Bode diagram construction follow.

Example 102.2

$$A(s) = \frac{10^4 s}{s^2 + 1100s + 10^5} = \frac{10^4 s}{(s+100)(s+1000)} = 10^{-1}\frac{s}{(s/100+1)(s/1000+1)}$$

In Figure 102.6, the individual contributions of the four factored terms of $H(s)$ are shown as dotted lines. The overall straight-line approximations for gain and angle are shown with dashed lines. The exact curves are plotted with solid lines.

Example 102.3

$$G(s) = \frac{1000(s+500)}{s^2 + 70s + 1000} = \frac{50(s/500+1)}{(s/100)^2 + 2(.35)(s/100) + 1}$$

Note that, for the quadratic term in the denominator, the damping ratio is 0.35, an indication of resonance. For small damping ratios the straight-line approximations of Bode magnitude and phase plots can vary significantly from the exact curves. For improved accuracy the approximations would have to be adjusted near the frequency of $\omega = 100$ rad/sec. This is not a consideration when a computer is used to generate a Bode diagram. Figure 102.7 shows the exact gain and angle frequency response curves for $G(s)$.

102.2 A Comparison of Methods

This chapter ends with the frequency response of a simple system plotted in three different ways.

Example 102.4

$$T(s) = \frac{10^7}{(s+100)(s+200)(s+300)}$$

Figure 102.8 shows linear frequency response curves for $T(s)$. Corresponding Bode and Nyquist diagrams are shown in Figure 102.9 and Figure 102.10, respectively. The information contained in the three sets of diagrams is the same.

Defining Terms

Bode diagram — A frequency response plot of 20-log gain and phase angle on a log-frequency base.

Breakpoint — A point of abrupt change in slope in the straight-line approximation of a Bode magnitude curve.

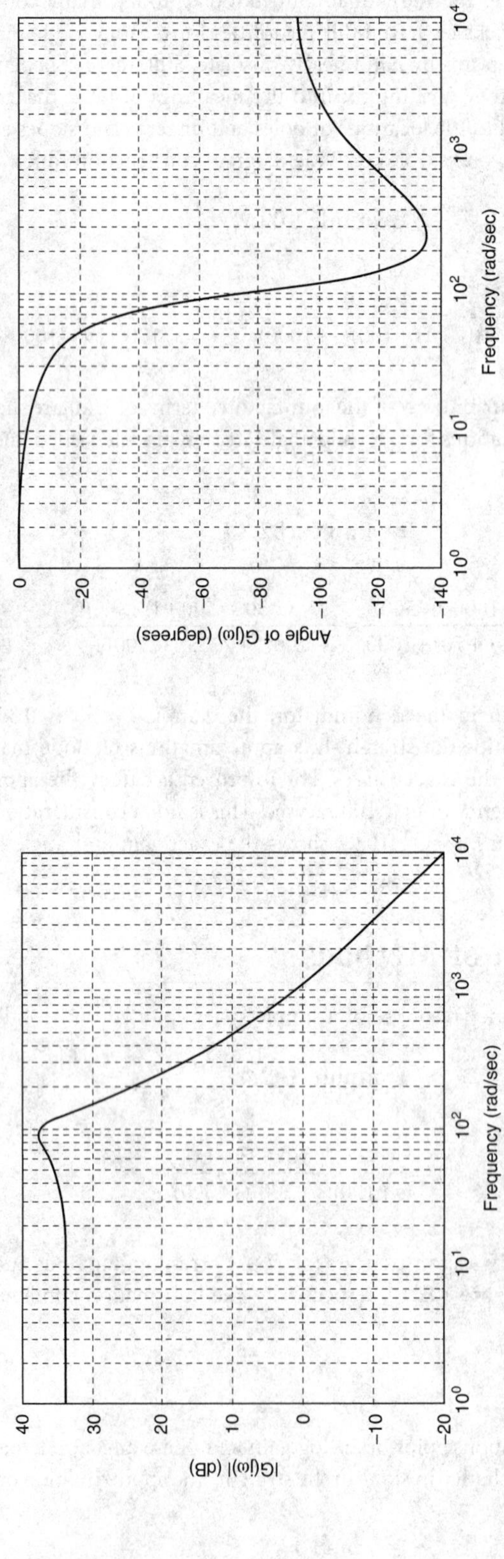

FIGURE 102.6 Bode diagram curves for $G(j\omega)$.

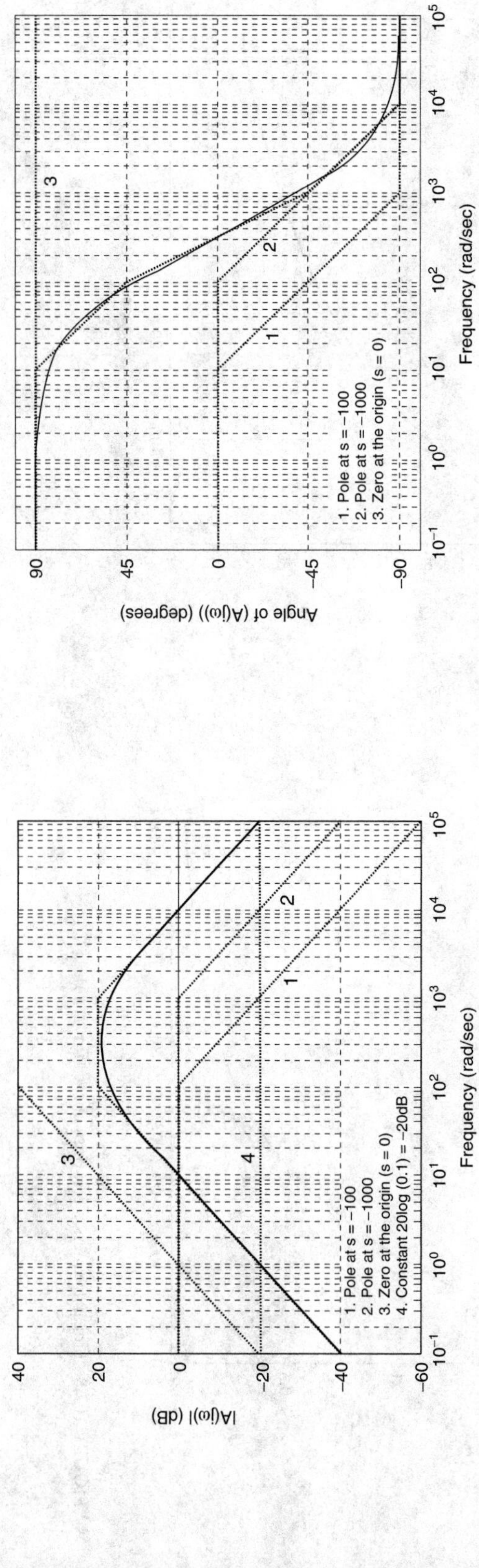

FIGURE 102.7 Bode diagram curves for $A(j\omega)$.

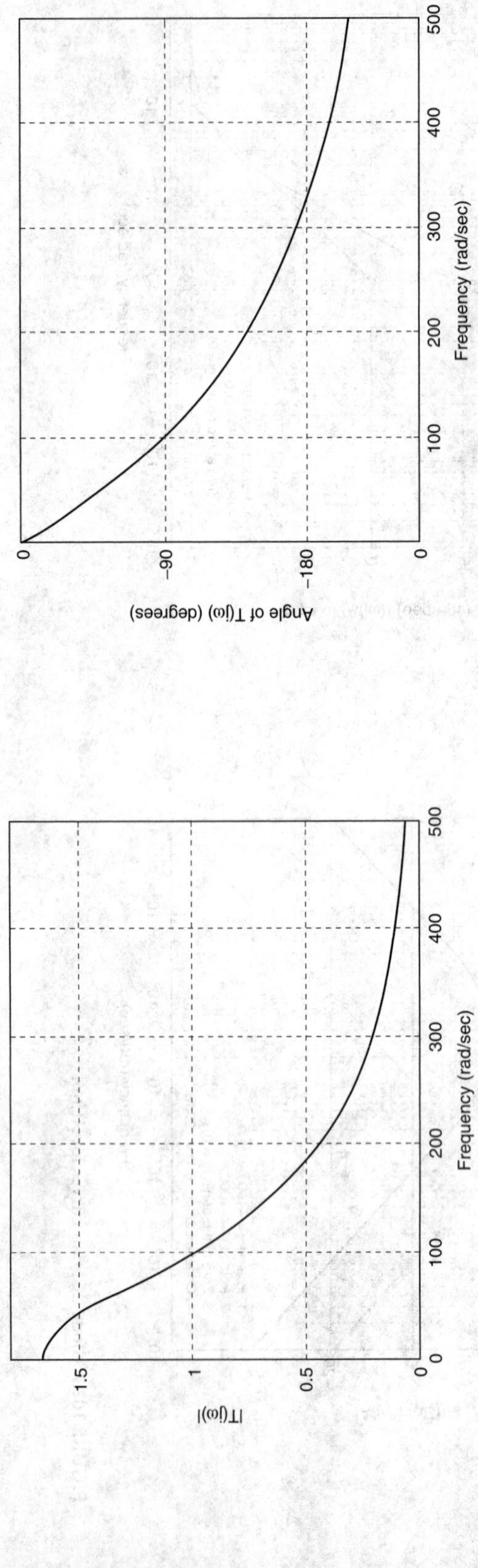

FIGURE 102.8 Linear frequency response curves for $T(j\omega)$.

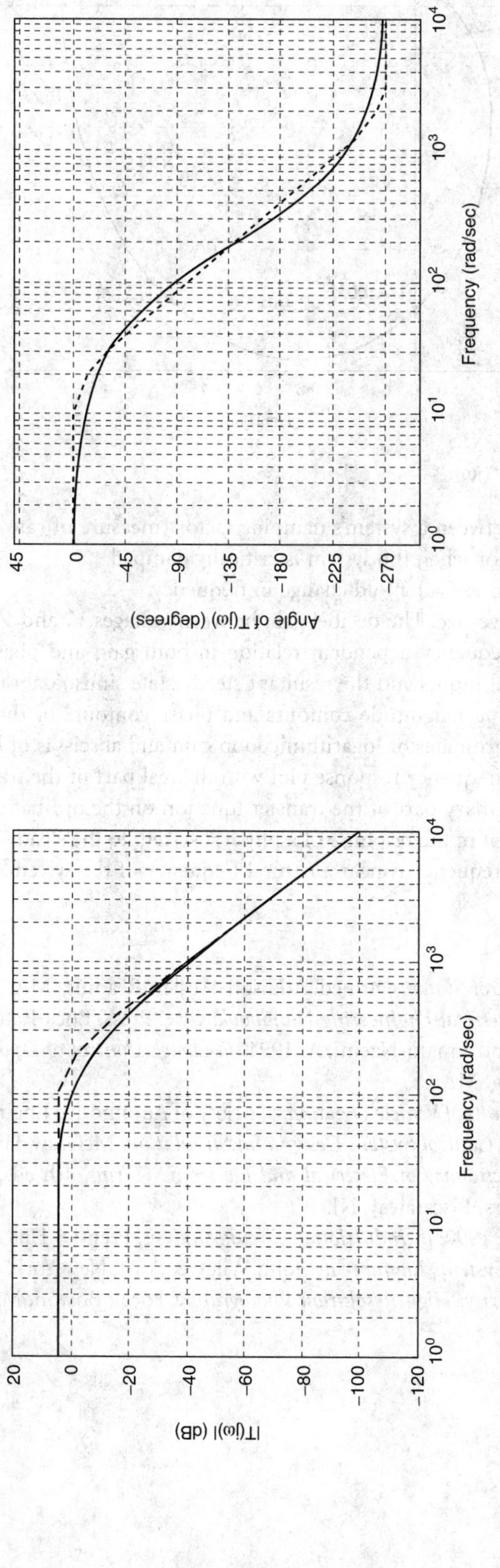

FIGURE 102.9 Bode diagram curves for $T(j\omega)$.

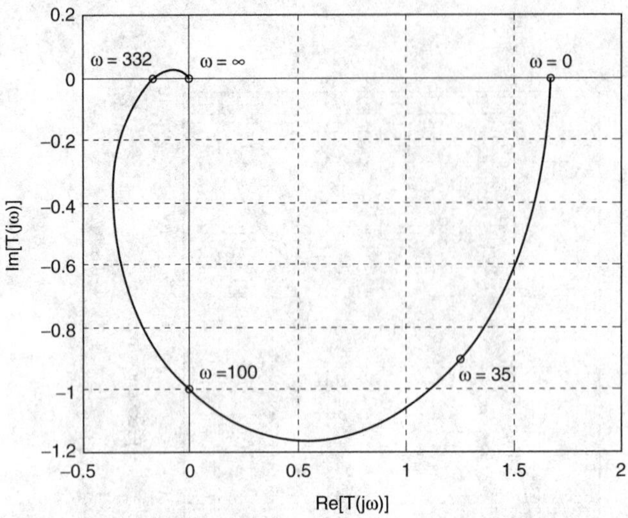

FIGURE 102.10 Nyquist plot of $T(j\omega)$.

Damping ratio — The ratio between a system's damping factor (measure of rate of decay of response) and the damping factor when the system is critically damped.

Decade — Power of ten. In context, a tenfold change in frequency.

Decibel — A measure of relative size. The decibel gain between voltages V_1 and V_2 is 20 $\log_{10}(V_1/V_2)$.

Frequency response — The frequency-dependent relation in both gain and phase difference between steady-state sinusoidal inputs and the resultant steady state sinusoidal outputs.

Nichols chart — A plot showing magnitude contours and phase contours of the closed-loop transfer function referred to ordinates of logarithmic loop gain and abscissas of loop phase.

Nyquist plot — A parametric frequency response plot with the real part of the transfer function on the abscissa and the imaginary part of the transfer function on the ordinate.

Resonance — The enhancement of the response of a physical system to the steady-state sinusoidal input when the excitation frequency is near a natural frequency of the system.

References

Dorf, R. C. 1986. *Modern Control Systems,* 4th ed. Addison-Wesley, Reading, MA.

Dorf, R. C. (Ed.) 1993. *The Electrical Engineering Handbook,* CRC Press, Boca Raton, FL.

Franklin, G. F., Powell, J. D., and Emani-Naeini, A. 1994. *Feedback Control of Dynamic Systems,* 3rd ed. Addison-Wesley, Reading, MA.

Gillespie, T. B. 1992. *Fundamentals of Vehicle Dynamics,* Society of Automotive Engineers, Warrendale, PA.

Golten, J. and Verwer, A. 1991. *Control System Design and Simulation,* McGraw-Hill, New York.

IEEE. 1988. *IEEE Standard Dictionary of Electrical and Electronics Terms,* 4th ed. Institute of Electrical and Electronics Engineers, Piscataway, NJ.

Neudorfer, P. O. and Hassul, M. 1990. *Introduction to Circuit Analysis,* Prentice Hall, Englewood Cliffs, NJ.

Palm, W. J., III, 1986. *Control Systems Engineering,* John Wiley & Sons, New York.

Williams, J. B. 1991. *Image Clarity: High Resolution Photography,* Focal, Boston, MA.

103

Convolution Integral

Rodger E. Ziemer

University of Colorado, Colorado Springs

The mathematical operation of **convolution** is defined in the first sections of this chapter, both for **continuous-time** and **discrete-time signals**. Properties of convolution are given and illustrated. Several applications are then enumerated and illustrated. Further discussions and illustrations of convolution are given in Ziemer et al. [1998] and Phillips and Parr [1999]. A good reference on engineering mathematics — including many of the system analysis tools arising in linear systems where the convolution integral arises — is Kreyszig [1998].

103.1 Fundamentals

Continuous-Time Convolution

Let $x(t)$ and $h(t)$ be two continuous-time signals defined for $-\infty < t < \infty$. Their **continuous-time convolution** is defined as the integral operation

$$y(t) = \int_{-\infty}^{\infty} x(\tau)h(t-\tau)\, d\tau \tag{103.1}$$

With the change of variables $\lambda = t - \tau$, Equation (103.1) can be written in the equivalent form

$$y(t) = \int_{-\infty}^{\infty} x(t-\lambda)h(\lambda)\, d\lambda \tag{103.2}$$

Inspection of Equation (103.1) reveals that the convolution operation is composed of four steps with respect to the variable of integration:

1. **Shifting** of $h(\tau)$, represented by replacing τ by $\tau - t$
2. Reversal with respect to the independent variable, also known as **folding**, which gives the signal $h(t-\tau)$

0-8493-1586-7/05/$0.00+$1.50

3. Multiplication by the second signal to produce the integrand $x(\tau)h(t-\tau)$
4. Integration of the product with respect to τ for all values of the delay variable t, which is the independent variable of the new signal $y(t)$

Example 103.1

Consider the convolution of the following two signals:

$$x(t)=\begin{cases}\exp(-\alpha t), & t\geq 0, \quad \alpha>0\\ 0, & t<0\end{cases} \tag{103.3}$$

$$h(t)=\begin{cases}\exp(-\beta t), & t\geq 0, \quad \alpha\neq\beta>0\\ 0, & t<0\end{cases} \tag{103.4}$$

The convolution of these two signals, using Equation (103.1), is given by

$$\begin{aligned} y(t) &= \int_0^t \exp(-\alpha\tau)\exp[-\beta(t-\tau)]d\tau \\ &= \exp(-\beta t)\int_0^t \exp[-(\alpha-\beta)\tau]d\tau \\ &= \frac{\exp(-\beta t)-\exp(-\alpha t)}{\alpha-\beta}, t\geq 0 \end{aligned} \tag{103.5}$$

where the lower limit on the integral is 0, by virtue of $x(\tau) = 0$ for $\tau < 0$, and the upper limit is t, by virtue of $h(t-\tau) = 0$ for $t-\tau < 0$ or $\tau > t$. Figure 103.1(a–d) illustrates the factors in the integrand for various values of t; the final result of the full convolution is shown in Figure 103.2 for $0 \leq t \leq 6$.

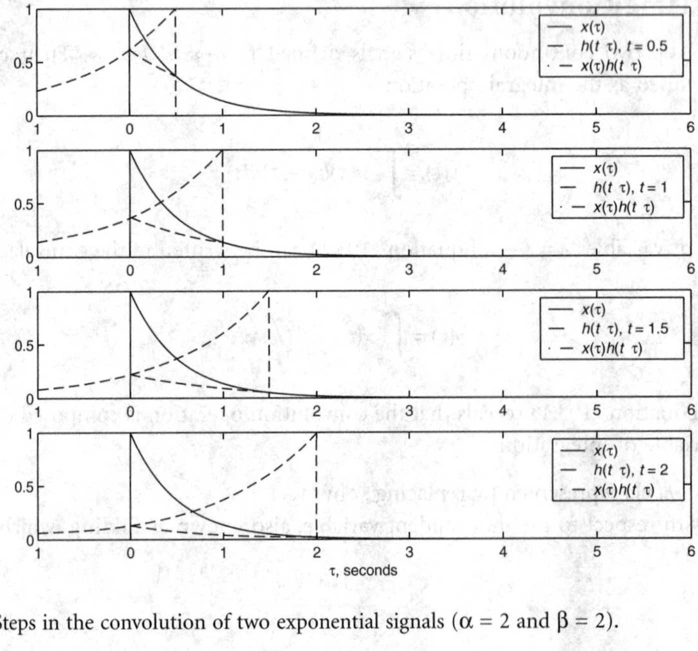

FIGURE 103.1 Steps in the convolution of two exponential signals ($\alpha = 2$ and $\beta = 2$).

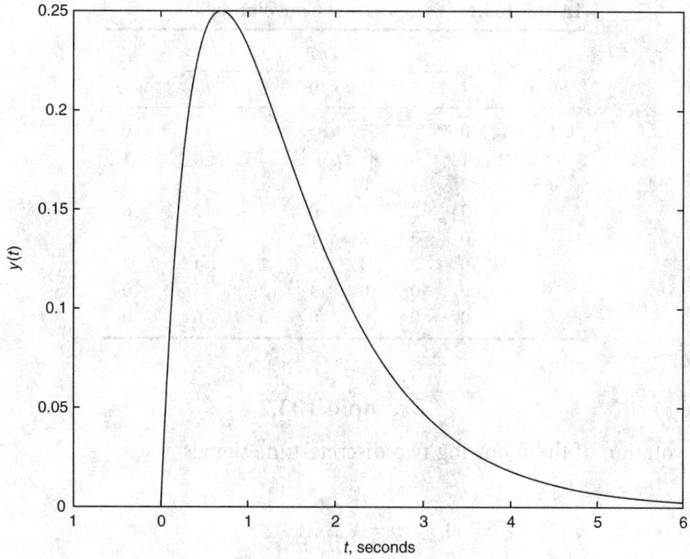

FIGURE 103.2 Final result for the convolution of two exponential signals ($\alpha = 2$ and $\beta = 2$).

Convolution is often used to find the output of a **linear time-invariant system** to an input signal. (The linearity property means that superposition holds; that is, an arbitrary linear combination of two inputs applied to a linear system results in the same linear combination of outputs due to the inputs applied separately. *Time invariant* means that if a given input results in a certain output, the output due to the input delayed is the original output delayed by the same amount as the input.) When used in this context, Equation (103.1) or Equation (103.2) involves the following functions or **signals:** the input signal $x(t)$, the output signal $y(t)$, and the **impulse response,** $h(t)$, which is the response of the system to a unit impulse function applied at time zero. The latter completely characterizes the system. A system is **causal** if its output does not anticipate the application of an input, which is manifested in the impulse response being zero for $t < 0$. The system is **stable** if every bounded input results in a bounded output, which is manifested by the impulse response being **absolutely integrable.**

Discrete-Time Convolution

Given two discrete-time signals $x(n)$ and $h(n)$ defined for $-\infty < n < \infty$, their **discrete-time convolution** (sum) is

$$y(n) = \sum_{k=-\infty}^{\infty} x(k)h(n-k) \tag{103.6}$$

With the change of variables $j = n - k$, this can be written as

$$y(n) = \sum_{j=-\infty}^{\infty} x(n-j)h(j) \tag{103.7}$$

Discrete-time convolution involves the same operations as continuous-time convolution except that integration is replaced by summation.

TABLE 103.1 Discrete-Time Convolution

	$x(n)$						
$h(-n)$	1	1	1	0	0	$n =$	$y(n)$
0 3 2 1	0					−1	0
0 3 2	1	0				0	1
0 3	2	1	0			1	3
0	3	2	1	0		2	6
	0	3	2	1	0	3	5
	0	0	3	2	1	4	3
	0	0	0	3	2	5	0
	0	0	0	0	3	6	0

Example 103.2

Consider the convolution of the following two discrete-time signals:

$$x(n) = \begin{cases} 1, & n = 0, 1, 2 \\ 0, & n < 0 \quad \text{and} \quad n > 2 \end{cases} \tag{103.8}$$

$$h(n) = \begin{cases} 0, & n \leq 0 \quad \text{and} \quad n > 3 \\ 1, & n = 1 \\ 2, & n = 2 \\ 3, & n = 3 \end{cases} \tag{103.9}$$

The discrete-time convolution can be carried out as shown in Table 103.1. Note that the duration of $x(n)$ is 3, that of $h(n)$ is 3, and that of $y(n)$ is $3 + 3 - 1 = 5$.

103.2 Properties of the Convolution Operation

Several convenient **properties of convolution** can be proved. These are listed in Table 103.2. Hints at proving these properties are given in the right-hand column.

103.3 Applications of the Convolution Integral

Filtering

From property 7 of Table 103.2, it follows that

$$\begin{aligned} |Y(f)| &= |H(f)||X(f)| \\ \angle Y(f) &= \angle H(f) \angle X(f) \end{aligned} \tag{103.10}$$

where $f = \omega/2\pi$ is frequency in hertz. The first equation shows that the magnitude of the spectral content of $x(t)$ is enhanced or attenuated by the magnitude of $H(f)$, which is known as the **amplitude response** of the system. Similarly, the phase components of the spectrum of $x(t)$ are shifted by the argument of $H(f)$, which is known as the **phase response** of the system. Standard design techniques can be used to design filters to modify the amplitude and phase components of $x(t)$ as desired. Similar statements can be applied to a discrete-time system where $H(e^{j2\pi fT})$ plays a role similar to $H(f)$.

TABLE 103.2 Properties of the Convolution Integral

Number	Property	Comments on Proof
1	$x(t) * h(t) = h(t) * x(t)$ or $x(n) * h(n) = h(n) * x(n)$ (commutative)	Compare Equation (103.1) and Equation (103.2) or Equation (103.6) and Equation (103.7).
2	$x(t) * [\alpha h(t)] = \alpha[x(t) * h(t)]$ or $x(n) * [\alpha h(n)] = \alpha[x(n) * h(n)]$ α constant	Convolution is an integral from which the constant α can be taken outside.
3	$x(t) * [y(t) * z(t)] = [x(t) * y(t)] * z(t)$ or $x(n) * [y(n) * z(n)] = [x(n) * y(n)] * z(n)$ (associative)	Write the two convolution operations as integrals and reverse the orders of integration.
4	$x(t) * [y(t) + z(t)] = x(t) * y(t) + x(t) * z(t)$ or $x(n) * [y(n) + z(n)] = x(n) * y(n) + x(n) * z(n)$ (distributive)	Separate the integral involving the convolution of $x(t)$ with the sum of the other two signals into the sum of two convolution integrals.
5	If the duration of $x(t)$ is T_1 and the duration of $y(t)$ is T_2 then the duration of their convolution is $T_1 + T_2$. For discrete-time sequences, the length of the convolution is $N_1 + N_2 - 1$, where $x(n)$ has length N_1 and $y(n)$ has length N_2.	Evident by sketching time-limited versions of the integrand or summand factors and considering the overlap.
6	$x(t) * \delta(t - \tau) = x(t - \tau)$, τ constant $x(n) * \delta(n - n_0) = x(n - n_0)$, n_0 constant	Use the sifting property of the delta function inside convolution integral, or the sifting property of the unit pulse function inside the convolution sum.
7	$\mathcal{F}[x(t) * h(t)] = \mathcal{F}[x(t)]\,\mathcal{F}[h(t)]$ $\mathcal{L}[x(t) * h(t)] = \mathcal{L}[x(t)]\,\mathcal{L}[h(t)]$ $\mathcal{Z}[x(n) * h(n)] = \mathcal{Z}[x(n)]\,\mathcal{Z}[h(n)]$ $\mathcal{F}(\cdot)$ = Fourier transform $\mathcal{L}(\cdot)$ = Laplace transform $\mathcal{Z}(\cdot)$ = z transform	Write the Fourier or Laplace transform integrals and interchange the order of transforming and convolution. For the z transform, interchange the order of the z transform and convolution sums.
8	$x(t) * y(-t) = \int_{-\infty}^{\infty} x(\lambda) y(t+\lambda) d\lambda$ = correlation of $x(t)$ and $y(t)$	Write out the convolution integral with the argument of the second signal negated. A similar expression holds for discrete-time sequences.
9	$x(t) * e^{j\omega t} = H(j\omega) e^{j\omega t}$ where $H(j\omega) = \int_{-\infty}^{\infty} h(t) e^{j\omega t}\, dt$ $x(n) * e^{jn\omega T} = H(e^{jn\omega T}) e^{jnwT}$ where $H(e^{j\omega T}) = \sum_{k=-\infty}^{\infty} h(k) e^{jk\omega T}$	Do a direct substitution into the convolution integral, factor out the term exp $(j\omega t)$, and use the given definition of $H(j\omega)$. A similar derivation holds for discrete-time sequences.

Spectral Analysis

The approach for determining the presence of spectral components in a signal is similar to filtering, where now the amplitude response function of the filter is designed to pass the desired spectral components of the signal. In the case of random signals, spectral analysis requires the averaging of the Fourier transforms of several segments of the random signal in order to decrease the variance of the estimates at specific frequencies [Oppenheim and Schafer, 1999].

Correlation or Matched Filtering

An important problem in signal detection involves maximizing the peak signal-to-rms-noise ratio at the output of a filter at a certain time. The filter that does this is called the **matched filter** for the particular signal under consideration. Its impulse response is the time reverse of the signal to which it is matched, or, for a signal $s(t)$, the matched filter impulse response is

$$h_m(t) = s(t_0 - t) \tag{103.11}$$

where t_0 is the time at which the output peak signal-to-rms-noise ratio is a maximum. The constant t_0 is often chosen to make the impulse response of the matched filer causal. In this case **causality** can be taken to mean that the impulse response is zero for $t < 0$. If $x(t) = 0$ outside of the interval $[0, T]$, Equation (103.1) can be used to write the output of a matched filter to signal alone at its input as

$$y(t) = \int_0^T s(\tau)s(t_0 - t + \tau)d\tau \tag{103.12}$$

If $t_0 = T$ and the output is taken at $t = T$, Equation (103.12) becomes the energy in the signal:

$$y(T) = \int_0^T s(\tau)s(\tau)\, d\tau = \int_0^T s^2(\tau)\, d\tau = E \tag{103.13}$$

The peak output signal squared-to-mean-square noise ratio can be shown to be

$$\mathrm{SNR}_{\mathrm{out}} = \frac{2E}{N_0} \tag{103.14}$$

where N_0 is the single-sided power spectral density of the white input noise.

103.4 Two-Dimensional Convolution

Convolution of two-dimensional signals, such as images, is important for several reasons, including pattern recognition and image compression. This topic will be considered briefly here in terms of discrete-spatial-variable signals. Consider two-dimensional signals defined as $x(m, n)$ and $h(m, n)$. Their **two-dimensional convolution** is defined as

$$y(m, n) = \sum_{j=-\infty}^{\infty} \sum_{k=-\infty}^{\infty} x(j, k)(h(m-j, n-k) \tag{103.15}$$

A double change of summation variables permits this to also be written as

$$y(m, n) = \sum_{j=-\infty}^{\infty} \sum_{k=-\infty}^{\infty} x(m-j, n-k)h(j, k) \tag{103.16}$$

Properties similar to those shown for the one-dimensional case can also be given.

Example 103.3

The shift property given in Table 103.2 (generalized to two dimensions) and superposition will be used to convolve the two-dimensional signals given in the table that follows (assumed zero outside of its definition array), with

$$h(m, n) = \delta(m-1, n-1) + \delta(m-2, n-1) \tag{103.17}$$

Let $x(m, n)$ be given by

$$x(m,n) = \begin{bmatrix} 2 & 2 & 1 & 0 & 0 \\ 2 & 1 & 0 & 0 & 0 \\ 1 & 0 & 0 & 0 & 0 \\ 0 & 0 & 0 & 0 & 0 \\ 0 & 0 & 0 & 0 & 0 \end{bmatrix}$$

The two-dimensional convolution of these two signals is given by

$$y(m,\ n) = x(m-1,\ n-1) + x(m-2,\ n-1) \tag{103.18}$$

The result of the two-dimensional convolution is given in the table that follows, for $y(m, n)$:

$$x(m,n) = \begin{bmatrix} 0 & 0 & 0 & 0 & 0 \\ 0 & 2 & 2 & 1 & 0 \\ 0 & 2 & 1 & 0 & 0 \\ 0 & 1 & 0 & 0 & 0 \\ 0 & 0 & 0 & 0 & 0 \end{bmatrix} + \begin{bmatrix} 0 & 0 & 0 & 0 & 0 \\ 0 & 0 & 0 & 0 & 0 \\ 0 & 2 & 2 & 1 & 0 \\ 0 & 2 & 1 & 0 & 0 \\ 0 & 1 & 0 & 0 & 0 \end{bmatrix} = \begin{bmatrix} 0 & 0 & 0 & 0 & 0 \\ 0 & 2 & 2 & 1 & 0 \\ 0 & 4 & 3 & 1 & 0 \\ 0 & 3 & 1 & 0 & 0 \\ 0 & 1 & 0 & 0 & 0 \end{bmatrix}$$

103.5 Time-Varying System Analysis

The form of the convolution integral given in Equation (103.1), when applied to finding the output of a system to a given input, assumes that the system is time-invariant — that is, its properties do not change with time. In this context $x(t)$ is viewed as the input to the system, $y(t)$ its output, and $h(t)$ the response of the system to a delta function applied at time zero. The generalization of Equation (103.1) to time-varying (i.e., *not* time-invariant) systems is

$$y(t) = \int_{-\infty}^{\infty} x(\tau)h(t,\tau)\, d\tau \tag{103.19}$$

where now $h(t, \tau)$ is the response of the system at time t to a delta function applied at time τ.

Defining Terms

Amplitude response — The amount of attenuation or amplification given to a steady state sinusoidal input by a filter or time-invariant linear system. This is the magnitude of the Fourier transform of the impulse response of the system at the frequency of the sinusoidal input.

Causal — A property of a system stating that the system does not respond to a given input before that input is applied. Also, for a linear time-invariant system, **causality** implies that its impulse response is zero for $t < 0$.

Continuous-time signal (system) — A signal for which the independent variable — quite often time — takes on a continuum of values. When applied to a system, continuous time refers to the fact that the system processes continuous-time signals.

Convolution — The process of taking two signals, reversing one in time and shifting it, then multiplying the signals point by point and integrating the product (**continuous-time convolution**). The result of the convolution is then a signal whose independent variable is the shift used in the operation. When done for all shifts, the resulting convolution then produces the output signal for all values of its independent variable, usually time. When convolving discrete-time signals, the integration of continuous-time convolution is replaced by summation and referred to as the **convolution sum** or **discrete-time convolution.**

Discrete-time signal (system) — A signal for which the independent variable takes on a discrete set of values. When applied to a system, discrete time refers to the fact that the system processes discrete-time signals.

Folding — The process of reversing one of the signals in the convolution integral or sum.

Impulse response — The response of an LTI system (see below) to a unit impulse function applied at time zero. For a system that is linear but not time invariant, the impulse response, $h(t, \tau)$, is the response of the system at time t to a unit impulse function applied at time τ.

Linear time-invariant (LTI) system — A system for which superposition holds and for which the output is invariant to time shifts of the input.

Matched filter — An LTI system whose impulse response is the time reverse of the signal to which it is matched.

Phase response — The amount of phase shift given to a steady state sinusoidal input by a filter or linear time-invariant system. This is the argument of the Fourier transform of the impulse response of the system at the frequency of the sinusoidal input.

Properties of convolution — Properties of the convolution operation given in Table 103.2 result from the form of the defining integral (continuous-time case) or sum (discrete-time case). They can be used to simplify application of the convolution integral or sum.

Signal — Usually a function of time, but also may be a function of spatial and time variables. Signals can be classified in many different ways. Two classifications used here are continuous time and discrete time.

Shifting — An operation used in evaluating the convolution integral or sum. It consists of time-shifting the folded signal.

Stable — A system is stable if every bounded input results in a bounded output. For an LTI system, stability means that the impulse response is **absolutely integrable.**

Superposition — A term that can refer to the superposition integral or sum, which is identical to the convolution integral or sum except that it refers to the response of an LTI system or can apply to the property of superposition that defines a linear system.

Time-invariant system — A system whose properties do not change with time. When expressed in terms of the impulse response of a time-invariant linear system, this means that the impulse response is a function of a single variable, because the unit impulse function that is applied to give the impulse response can be applied at time zero.

Two-dimensional convolution — The process of convolution applied to signals dependent on two variables, usually spatial.

References

Kreyszig, E. 1998. *Advanced Engineering Mathematics,* 8th ed. John Wiley & Sons, New York.

Oppenheim, A. V. and Schafer, R. W. 1999. *Discrete-Time Signal Processing,* 2nd ed., Prentice Hall, Upper Saddle River, NJ.

Phillips, C. L. and Parr, J. M. 1999. *Signals, Systems, and Transforms,* 2nd ed. Prentice Hall, Upper Saddle River, NJ.

Rioul, O. and Vetterli, M. 1991. Wavelets in signal processing. *IEEE Signal Processing Magazine.* 8:14–38, October.

Ziemer, R. E., Tranter, W. H., and Fannin, D. R. 1998. *Signals and Systems: Continuous and Discrete,* 4th ed. Prentice Hall, Upper Saddle River, NJ.

Further Information

The convolution integral finds application in linear system analysis and signal processing. In the book by Ziemer, Tranter, and Fannin [1998], Chapter 2 is devoted to the convolution integral and its application to continuous-time linear system analysis, whereas Chapter 8 includes material on the discrete-time convolution sum.

The *IEEE Signal Processing Magazine* has tutorial articles on various aspects of signal processing, whereas the *IEEE Transactions on Signal Processing* include research papers on all aspects of signal processing. One notable and timely application of convolution is to wavelet transform theory, which has many potential applications including signal and image compression. The April 1992 issue of the *IEEE Signal Processing Magazine* contains a tutorial article on wavelet transforms.

104

Stability Analysis

Raymond T. Stefani
California State University, Long Beach

The output response of a linear system has two important components: the zero-state response caused by the input and the zero-input response caused by the initial conditions. It is desirable that each of those responses be well behaved. The speed of an automobile, for example, must provide a comfortable ride for the passengers. The term *stability* refers to how those two response components behave. The following sections define types of stability, relationships between stability types, and methods of determining the type of stability for a given system. Several response examples are provided. Some practical design considerations are given.

104.1 Response Components

Figure 104.1 shows two important response components for a linear system. The **zero-state** component occurs when the initial condition vector x_0 is zero and the input r is nonzero. Conversely, the **zero-input** response occurs when the input r is zero but some part of the initial condition vector x_0 is nonzero. Then, when both inputs are nonzero, the response contains both zero-input and zero-state responses.

Table 104.1 shows the Laplace transform and resulting time response where the zero-state and zero-input responses are defined using classical mathematics (based on transfer functions) and state variable mathematics (based on system matrices). In the time domain the state variable-based description is

$$dx/dt = Ax + Br \tag{104.1}$$

$$y = Cx \tag{104.2}$$

The classical transfer function can be related to the (A, B, C) system by

$$H(s) = Y(s)/R(s) = C[sI - A]^{-1}B \tag{104.3}$$

An important type of zero-state response is the **impulse response** in which the input r is a unit impulse function $\delta(t)$ whose Laplace transform $R(s)$ equals 1. Notice in Table 104.1 the close relationship between the form of the zero-state impulse response $C\phi(t)B$ and the zero-input response $C\phi(t)x_0$. Thus stability definitions based on the impulse response bear a close relationship to stability definitions based on the zero-input response.

0-8493-1586-7/05/$0.00+$1.50

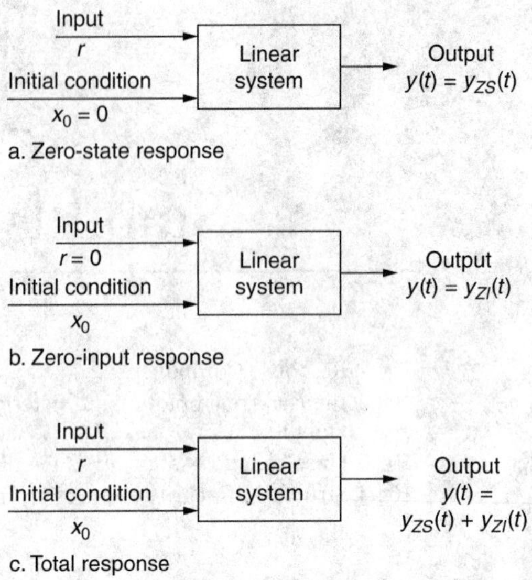

FIGURE 104.1 Zero-state and zero-input components.

TABLE 104.1 Classical and State Variable Forms for Zero-State and Zero-Input Responses

	Laplace	Time
	Zero State	
Classical (general)	$H(s)R(s) = \frac{N(s)}{D(s)}R(s)$	$\int_0^t h(\tau)r(t-\tau)\,d\tau$
Classical (impulse)	$H(s)$	$h(t)$
State variable (general)	$C\phi(s)BR(s)$	$\int_0^t \phi(\tau)Br(t-\tau)\,d\tau$
State variable (impulse)	$C\phi(s)B$	$C\phi(t)B$
	Zero Input	
Classical	$H_1(s) = \frac{N_1(s,x_0)}{D(s)}$	$h_1(t)$
State variable	$C\phi(s)x_0$	$C\phi(t)x_0$

Note: Transfer function $= H(s) = C\phi(s)B$; resolvant matrix $= H(s) = \phi(s) = [sI - A]^{-1}$; characteristic polynomial $= D(s) = |sI - A|$; initial state vector $= x_0$.

104.2 Internal (Asymptotic) and External (BIBO) Stability

Stability definitions relate either to the zero-input response (**internal stability**) or to the zero-state response (**external stability**). The latter is called *external* because performance depends on the external input as it influences the external output (internal performance is ignored). Table 104.2 contains two commonly used stability definitions: **asymptotic stability** relates to zero-input (internal) stability, whereas the **bounded-input, bounded-output (BIBO)** stability relates to zero-state (external) stability.

Asymptotic stability implies that the zero-input response tends to zero for all initial conditions. A necessary and sufficient condition for asymptotic stability is that all the eigenvalues A are in the left half plane (LHP). BIBO stability implies that all bounded inputs will cause the output to be bounded. A

TABLE 104.2 Stability Definitions for Linear Systems

Type of Stability	Requirement	Necessary and Sufficient Condition
Asymptotic (internal) (zero input)	$C\phi(t)x_0 \to 0$ as $t \to \infty$ for all x_0	All eigenvalues of A (closed-loop poles) are in the LHP
BIBO (external) (zero state)	If $\lvert r(t)\rvert \le M_1 < \infty$ then $\lvert y(t)\rvert \le M_2 < \infty$	$\int_0^\infty \lvert h(t)\rvert dt \le M_2 < \infty$ (all poles of $H(s)$ are in the LHP)

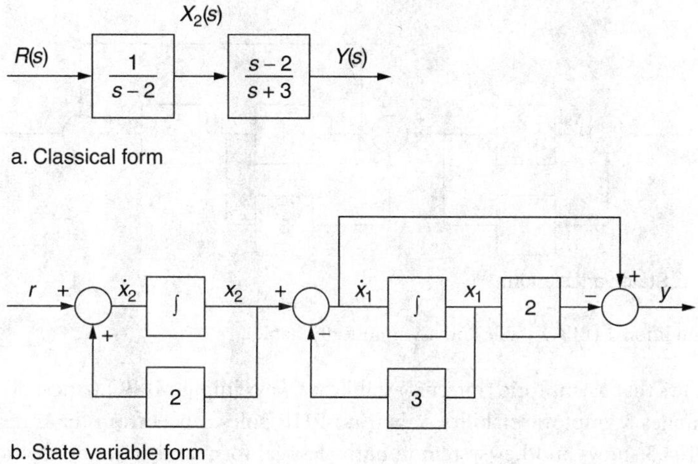

FIGURE 104.2 Condition 1 (BIBO stable but asymptotically unstable).

necessary and sufficient condition for BIBO stability is that the integral of the magnitude of the impulse response is bounded, which occurs if all the poles of $H(s)$ are in the LHP. These two stability definitions will result in different conclusions when there are right half plane (RHP) poles that are cancelled by RHP zeros of $H(s)$, obscuring internal responses that may not be bounded.

Figure 104.2 shows a system in both classical form and state variable form. From either form of the second-order system, notice that the transfer function $H(s) = 1/(s + 3)$ is actually first order, because an RHP pole at –2 is cancelled by an RHP zero at the same location. The impulse response of $y(t)$ is e^{-3t} (whose integral is bounded), causing the system to be BIBO (externally) stable. However, for an impulse input, $X_2(s)$ is $1/(s - 2)$ and $x_2(t)$ is e^{2t}, which is not bounded. An example would be an aircraft with a malfunctioning component that (fortunately) does not influence output response, perhaps velocity, under the current flight condition, which does not activate that component. Should the flight condition change, causing the zero to become –1, for example, the component is activated, causing $H(s)$ to become $(s - 1)/[(s - 2)(s + 3)]$. Unfortunately, the aircraft velocity would be unbounded, possibly causing the aircraft to crash unless the component is disconnected quickly and control is reestablished.

That potentially undesirable effect is easily revealed by the state variable description

$$A = \begin{bmatrix} -3 & 1 \\ 0 & 2 \end{bmatrix}, \quad B = \begin{bmatrix} 0 \\ 1 \end{bmatrix}, \quad C = \begin{bmatrix} -5 & 1 \end{bmatrix} \tag{104.4}$$

Although $H(s) = C[sI - A]^{-1} B = 1/(s + 3)$, passing the test for BIBO stability, the eigenvalues of A are –3 and 2, failing the test for asymptotic stability.

TABLE 104.3 Equivalence of Stability Types

Equivalence	Requirement
Asymptotic → BIBO	Always
BIBO → asymptotic	No cancellation of RHP poles in $H(s)$

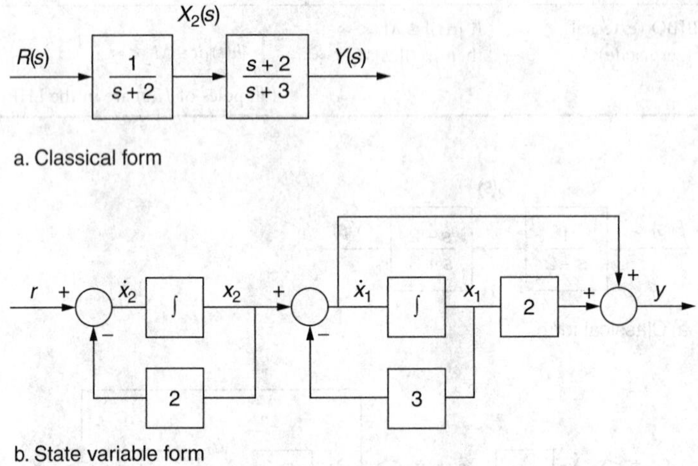

FIGURE 104.3 Condition 2 (BIBO stable and asymptotically stable).

Table 104.3 shows that asymptotic (internal) stability always implies BIBO (external) stability but that BIBO stability implies asymptotic stability only if no RHP poles cancel from the transfer function. For example, Figure 104.3 shows another system in both classical form and state variable form. From either form, notice that $H(s) = 1/(s + 3)$ (no RHP poles cancel this time, just one LHP pole), so the impulse response of $y(t)$ is e^{-3t} (whose integral is bounded), causing the system to be BIBO (externally) stable. Now, for an impulse input, $X_2(s)$ is $1/(s + 2)$ and $x_2(t)$ is e^{-2t}, which is also bounded. From the state variable description,

$$A = \begin{bmatrix} -3 & 1 \\ 0 & -2 \end{bmatrix}, \quad B = \begin{bmatrix} 0 \\ 1 \end{bmatrix}, \quad C = \begin{bmatrix} -1 & 1 \end{bmatrix} \tag{104.5}$$

As before, $H(s) = C[sI - A]^{-1} B = 1/(s + 3)$, passing the test for BIBO stability; but now the eigenvalues of A are -3 and -2, also passing the test for asymptotic stability.

Figure 104.4 shows three stable $H(s)$ functions and the corresponding impulse responses $h(t)$. Respectively, $H(s)$ has one real LHP pole, complex conjugate LHP poles, and double real LHP poles. A bounded input produces a bounded output in each case.

104.3 Unstable and Marginally Stable Responses

A system that does not pass the test for asymptotic stability and/or BIBO stability may be classified as **unstable** or **marginally stable** as regards the appropriate type of stability, as in Table 104.4. As before, BIBO and asymptotic stability may differ if zeros at the same place cancel certain poles of $H(s)$. An unstable system has a BIBO response that is unbounded for all bounded inputs and a zero-input response that diverges for at least one initial condition. A system is unstable in the BIBO sense if there is at least one RHP pole of $H(s)$ and/or there are repeated poles along the imaginary axis (IA). A system is asymptotically unstable if there is at least one RHP eigenvalue of A and/or there are repeated eignevalues

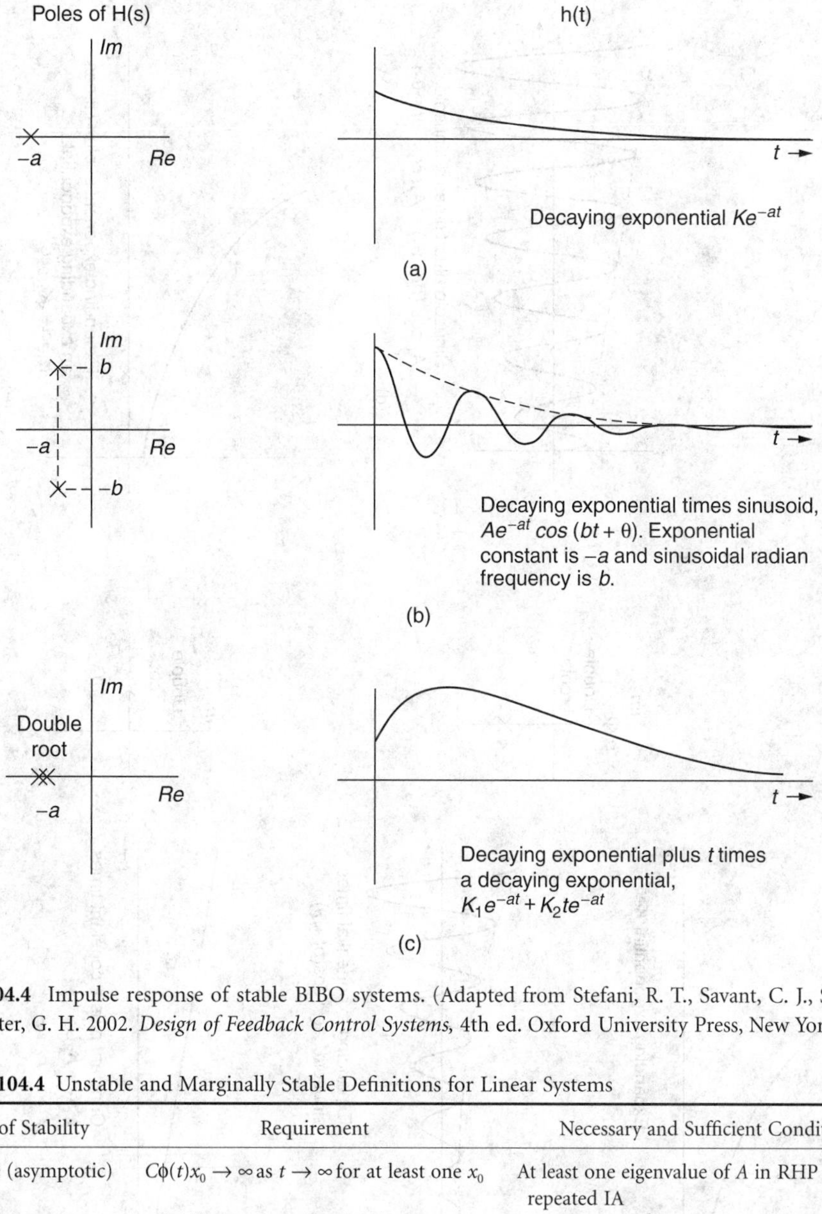

FIGURE 104.4 Impulse response of stable BIBO systems. (Adapted from Stefani, R. T., Savant, C. J., Shahian, B., and Hostetter, G. H. 2002. *Design of Feedback Control Systems,* 4th ed. Oxford University Press, New York.)

TABLE 104.4 Unstable and Marginally Stable Definitions for Linear Systems

Type of Stability	Requirement	Necessary and Sufficient Condition
Unstable (asymptotic)	$C\phi(t)x_0 \to \infty$ as $t \to \infty$ for at least one x_0	At least one eigenvalue of A in RHP and/or repeated IA
Unstable (BIBO)	If $\lvert r(t)\rvert \le M_1 < \infty$ then $\lvert y(t)\rvert \to \infty$	At least one pole of $H(s)$ in RHP and/or repeated IA
Marginally stable (asymptotic)	$C\phi(t)\, x_0 \to C_0$ $0 < \lvert C_0\rvert < \infty$ as $t \to \infty$ for all $_{x0}$	One eigenvalue of A at 0 and/or nonrepeated complex conjugated IA; the rest are LHP
Marginally stable (BIBO)	For some but not all $\lvert r(t)\rvert \le M_1 < \infty$ then $\lvert y(t)\rvert \le M_2 < \infty$	One pole of $H(s)$ at 0 and/or nonrepeated complex conjugate IA; the rest are LHP

along the imaginary axis (IA). Figure 104.5 shows five examples of unstable (BIBO) $H(s)$ functions and the corresponding impulse responses $h(t)$. Respectively, there is one real RHP pole, one pair of complex conjugate RHP poles, double poles at the origin, double complex conjugate poles along the IA and double RHP real poles. Clearly, the impulse response is unbounded in each case. Notice in Figure 104.5(b) and Figure 104.5(d) that although the response goes to zero periodically, the response passes through zero instantaneously and reaches a peak that increases with each cycle.

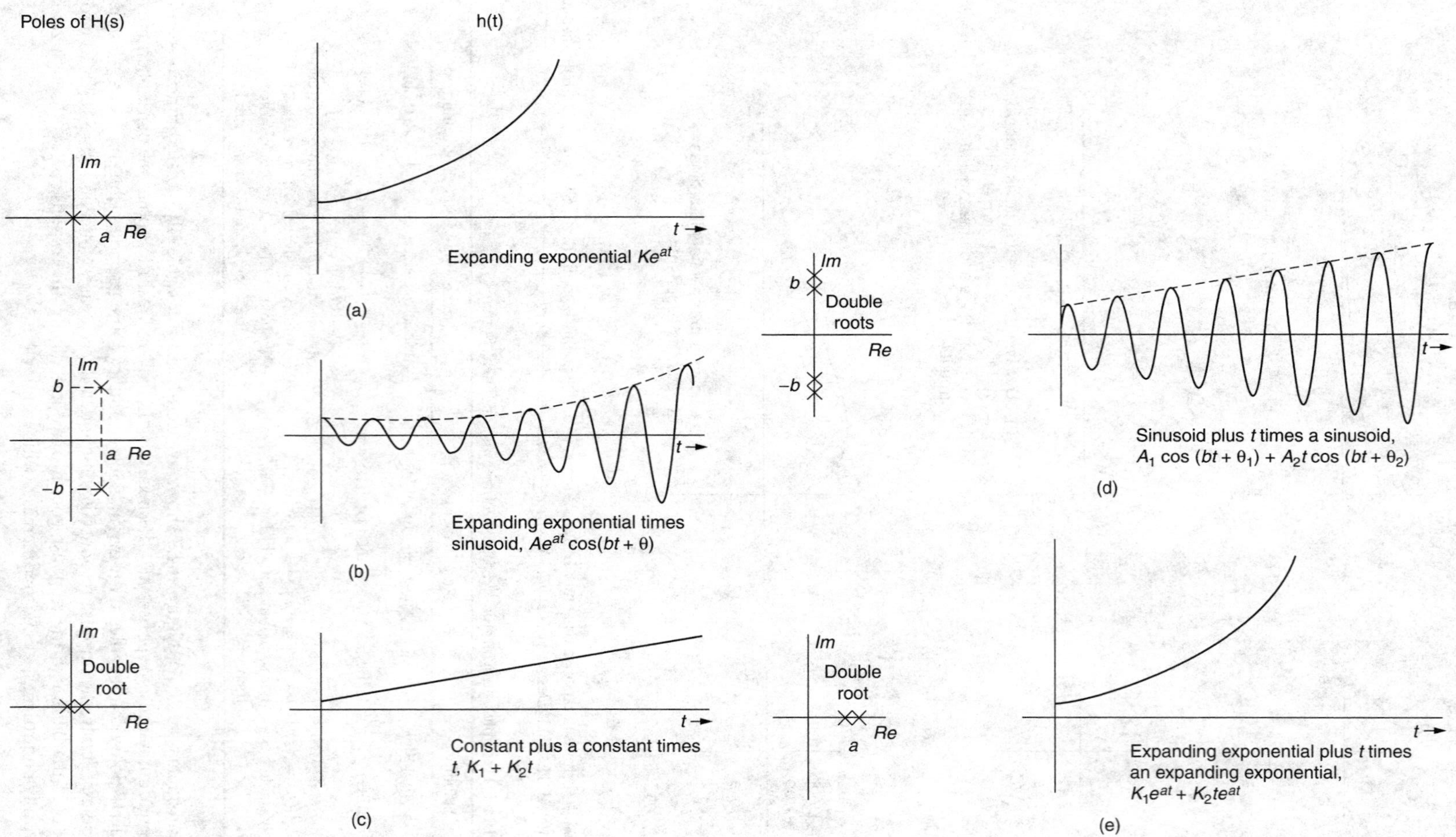

FIGURE 104.5 Impulse response of unstable BIBO systems. (Adapted from Stefani, R. T., Savant, C. J., Shahian, B., and Hostetter, G. H. 2002. *Design of Feedback Control Systems*, 4th ed. Oxford University Press, New York.)

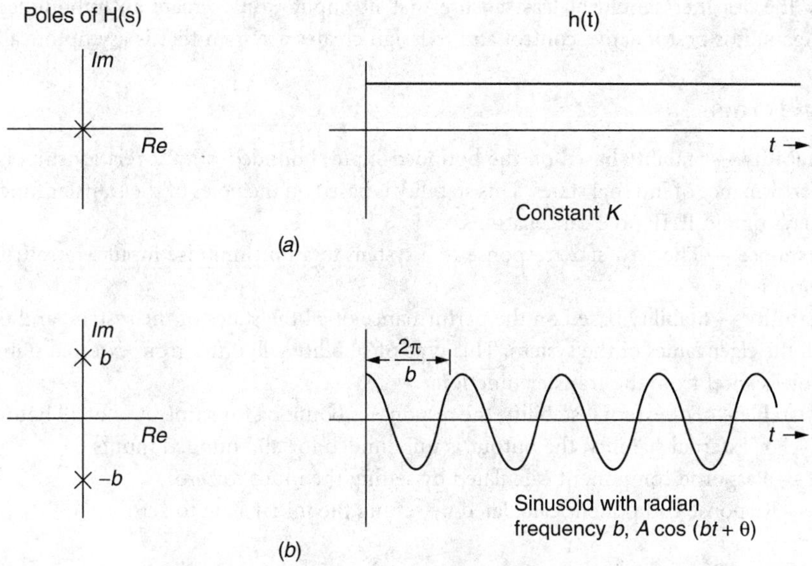

FIGURE 104.6 Impulse response of marginally stable BIBO systems. (Adapted from Stefani, R. T., Savant, C. J., Shahian, B., and Hostetter, G. H. 2002. *Design of Feedback Control Systems,* 4th ed. Oxford University Press, New York.)

A marginally stable system has a BIBO response that is bounded for some inputs but unbounded for others and a zero-input response that is bounded (going neither to zero nor infinity) for all initial conditions. A system is marginally stable for closed loop poles (BIBO) and eigenvalues (asymptotic) that are IA and nonrepeated. Figure 104.6 shows two marginally stable (BIBO) $H(s)$ functions and the corresponding impulse responses $h(t)$. For Figure 104.6(a), $H(s) = 1/s$, so the zero-state impulse response is a constant (which is bounded). However, if the input is a step, the zero-state response becomes $1/s^2$ and the time response for this case is the same as the impulse response in Fig. 104.5(c), which is obviously unbounded. Similarly, the example of Figure 104.6(b) has a bounded sinusoidal impulse response for IA poles of $H(s)$, but if the bounded $r(t)$ is also sinusoidal and the IA poles of $R(s)$ are at the same place as for $H(s)$, then the multiple IA poles of $Y(s)$ create a time response like that of Figure 104.5(d), which is unbounded.

104.4 Structural Integrity and Design Implications

Assuming a bounded input, a stable BIBO system always has a bounded output response. That bounded response can sometimes have a very large and undesirable value when multiple (LHP) poles are present [as in Figure 104.4(c)]. In such cases, even though the output should tend toward zero thereafter, assuming the system operates in a linear region, structural integrity can be compromised, followed by nonlinear behavior. For example, a bridge over Tacoma Narrows collapsed in 1939 because winds created a lift/drag force with an $R(s)$ having poles near the bridge's $H(s)$ poles. Similarly, there are many examples of earthquakes providing a shaking motion having an $R(s)$ with poles near some poles of a structure's $H(s)$. The oscillating vertical motion, with an upper containing envelope similar to Figure 104.4(c), can exceed the structural tolerance and the ensuing failure can have tragic consequences. The design solution is to reinforce the structure, moving the poles of $H(s)$ away from those of $R(s)$, thus maintaining the response within structural limits. An engineer must examine more than just stability when ensuring that a design results in a satisfactory time response.

In summary, it should concern any designer when RHP eigenvalues occur for A. Active control and redesign should then be applied to move the RHP eigenvalues into the LHP, creating a system that is both BIBO and asymptotically stable. If the system is marginally stable from either BIBO or asymptotic viewpoints, the designer should at least ensure that no input would create an unbounded zero-state response. Again, it is best if active control and redesign creates a system that is asymptotically stable.

Defining Terms

External stability — Stability based on the bounded-input, bounded-output relationship, ignoring the performance of internal states. This stability is based on the poles of the transfer function, which may contain RHP pole cancellations.

Impulse response — The zero-state response of a system to a unit impulse input whose Laplace transform is 1.

Internal stability — Stability based on the performance of all the states of the system, and therefore on all the eigenvalues of the system. This type of stability will differ from external stability if RHP poles cancel from the transfer function.

Marginally stable — For external stability, the response is bounded for some but not all bounded inputs.

Unstable — For external stability, the output is unbounded for all bounded inputs.

Zero input — Response component calculated by setting the input to zero.

Zero state — Response component calculated by setting the initial state to zero.

References

Kailath, T. *Linear Systems* 1980. Prentice Hall, Englewood Cliffs, NJ.

Kuo, B. C. 1987. *Automatic Control Systems,* 5th ed. Prentice Hall, Englewood Cliffs, NJ.

Stefani, R. T., Shahian, B., Savant, C. J., and Hostetter, G. H. 2002. *Design of Feedback Control Systems,* 4th ed. Oxford University Press, New York.

Further Information

IEEE Transactions on Automatic Control is a rather theoretically oriented journal including aspects of stability related to a wide spectrum of linear, nonlinear, and adaptive control systems.

Control Systems is a magazine published by IEEE with a more basic level of presentation than the *IEEE Transactions on Automatic Control.* Applications are made to the stability of various practical systems.

IEEE Transactions on Systems, Man, and Cybernetics is a journal providing coverage of stability as it relates to a variety of systems with human interface or models that emulate the response of humans.

Automatica and the *International Journal of Control* cover stability concepts for an international audience.

105

z Transform and Digital Systems

Rolf Johansson
Lund University

A digital system (or discrete-time system or sampled-data system) is a device such as a digital controller or a digital filter or, more generally, a system intended for digital computer implementation and usually with some periodic interaction with the environment and with a supporting methodology for analysis and design. Of particular importance for modeling and analysis are recurrent algorithms — for example, difference equations in input-output data — and the *z* transform is important for the solution of such problems.

The ***z* transform** is being used in the analysis of linear time-invariant systems and discrete-time signals — for example, for digital control or filtering — and may be compared to the Laplace transform as used in the analysis of continuous-time signals and systems, a useful property being that the convolution of two time-domain signals is equivalent to multiplication of their corresponding *z* transforms. The *z* transform is important as a means to characterize a linear time-invariant system in terms of its pole-zero locations, its transfer function and Bode diagram, and its response to a large variety of signals. In addition, it provides important relationships between temporal and spectral properties of signals. The *z* transform generally appears in the analysis of difference equations as used in many branches of engineering and applied mathematics.

105.1 The *z* Transform

The *z* transform of the sequence $\{x_k\}_{-\infty}^{+\infty}$ is defined as the generating function

$$X(z) = \mathcal{Z}\{x\} = \sum_{k=-\infty}^{\infty} x_k z^{-k} \tag{105.1}$$

where the variable *z* has the essential interpretation of a forward shift operator so that

0-8493-1586-7/05/$0.00+$1.50

$$\mathcal{L}\{x_{k+1} = z\mathcal{L}\{x_k\} = zX(z) \tag{105.2}$$

The z transform is an infinite power series in the complex variable z^{-1} where $\{x_k\}$ constitutes a sequence of coefficients. As the z transform is an infinite power series, it exists only for those values of z for which this series converges and the *region of convergence* of $X(z)$ is the set of z for which $X(z)$ takes on a finite value. A sufficient condition for existence of the z transform is convergence of the power series

$$\sum_{k=-\infty}^{\infty} |x_k| \cdot |z^{-k}| < \infty \tag{105.3}$$

The region of convergence for a finite-duration signal is the entire z plane except $i = 0$ and $z = \infty$. For a one-sided infinite-duration sequence $\{x_k\}_{k=0}^{\infty}$, a number r can usually be found so that the power series converges for $|z| > r$. Then, the *inverse z transform* can be derived as

$$x_k = \frac{1}{2\pi i} \oint X(z) z^{k-1}\, dz \tag{105.4}$$

where the contour of integration encloses all singularities of $X(z)$. In practice it is standard procedure to use tabulated results; some standard z transform pairs are to be found in Table 105.1.

TABLE 105.1 Properties of the z Transform

z transform	$\mathcal{L}(\{f_k\}) = F(z)$
Convolution	$\mathcal{L}(\{f_k * g_k\}) = \mathcal{L}(\{f_k\}) \cdot \mathcal{L}(\{g_k\})$ $\mathcal{L}(\{f_k \cdot g_k\}) = \mathcal{L}(\{f_k\}) * \mathcal{L}(\{g_k\})$
Forward shift	$\mathcal{L}(\{f_{k+1}\}) = z\mathcal{L}(\{f_k\}) = zF(z)$
Backward shift	$\mathcal{L}(\{f_{k-1}\}) = z^{-1}\mathcal{L}(\{f_k\}) = z^{-1}F(z)$
Linearity	$\mathcal{L}(\{af_k + bg_k\}) = a\mathcal{L}(\{f_k\}) + b\mathcal{L}(\{g_k\})$
Multiplication	$\mathcal{L}(\{a^k f_k\}) = F(a^{-1}z)$
Final value	$\lim_{k\to\infty} f_k = \lim_{z\to 1}(1 - z^{-1})F(z)$
Initial value	$f_0 = \lim_{z\to\infty} F(z)$

	Time Domain		z Transform
Impulse	$\delta_k = \begin{cases} 1, & k = 0 \\ 0, & k \neq 0 \end{cases}$	$\leftrightarrow$	$\mathcal{L}\{\delta_k\} = 1,\ z \in C$
Step function	$\sigma_k = \begin{cases} 0, & k < 0 \\ 1, & k \geq 0 \end{cases}$	$\leftrightarrow$	$\mathcal{L}\{\sigma_k\} = \frac{z}{z-1},\ \lvert z\rvert > 1$
Ramp function	$x_k = k \cdot \sigma_k$	$\leftrightarrow$	$X(z) = \frac{z}{(z-1)^2},\ \lvert z\rvert > 1$
Exponential	$x_k = a^k \cdot \sigma_k$	$\leftrightarrow$	$X(z) = \frac{z}{z-a},\ \lvert z\rvert > \lvert a\rvert$
Sinusoid	$x_k = \sin \omega k \cdot \sigma_k$	$\leftrightarrow$	$X(z) = \frac{z \sin\omega}{z^2 - 2z\cos\omega + 1},\ \lvert z\rvert > 1$

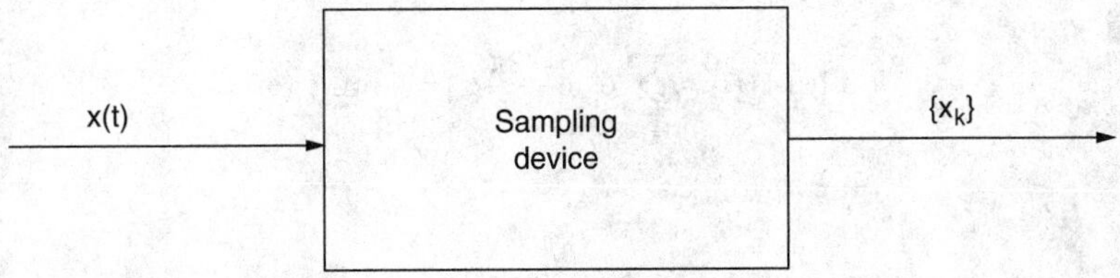

FIGURE 105.1 A continuous-time signal $x(t)$ and a sampling device that produces a sample sequence $\{x_k\}$.

105.2 Digital Systems and Discretized Data

Periodic sampling of signals and subsequent computation or storing of the results requires the computer to schedule sampling and to handle the resulting sequences of numbers. A measured variable $x(t)$ may be available only as periodic observations of $x(t)$ as sampled with a time interval T (the sampling period). The sample-sequence can be represented as

$$\{x_k\}_{-\infty}^{\infty}; \quad x_k = x(kT) \quad \text{for} \quad k = \ldots, -1, 0, 1, 2, \ldots \tag{105.5}$$

and it is important to ascertain that the sample sequence adequately represents the original variable $x(t)$; see Figure 105.1. For ideal sampling it is required that the duration of each sampling be very short, and the sampled function may be represented by a sequence of infinitely short impulses $\delta(t)$ (the Dirac impulse). Let the sampled function of time be expressed thus:

$$x_\Delta(t) = x(t) \cdot T \sum_{k=-\infty}^{\infty} \delta(t - kT) = x(t) \cdot ш_T(t) \tag{105.6}$$

where

$$ш_T(t) \triangleq T \sum_{k=-\infty}^{\infty} \delta(t - kT) \tag{105.7}$$

and where the sampling period T is multiplied to ensure that the averages over a sampling period of the original variable x and the sampled signal x_Δ, respectively, are of the same magnitude. A direct application of the discretized variable $x_\Delta(t)$ in Equation (105.6) verifies that the spectrum of x_Δ is related to the z transform $X(z)$ as

$$X_\Delta(i\omega) = \mathcal{F}\{x(t)\, ш(t)\} = T \sum_{k=-\infty}^{\infty} x_k \exp(-i\omega kT) = TX(e^{i\omega T}) \tag{105.8}$$

Obviously, the original variable $x(t)$ and the sampled data are not identical, and thus it is necessary to consider the distortive effects of discretization. Consider the spectrum of the sampled signal $x_\Delta(t)$ obtained as the Fourier transform

$$X_\Delta(i\omega) = \mathcal{F}\{x_\Delta(t)\} = \mathcal{F}\{x(t)\} * \mathcal{F}\{ш_T(t)\} \tag{105.9}$$

where

$$\mathcal{F}\{\text{Ш}_T(t)\} = \sum_{k=-\infty}^{\infty} \delta\left(\omega - \frac{2\pi}{T}k\right) = \frac{T}{2\pi}\,\text{Ш}_{2\pi/t}(\omega) \tag{105.10}$$

so that

$$X_\Delta(i\omega) = \mathcal{F}\{x(t)\} * \mathcal{F}\{\text{Ш}_T(t)\} = \sum_{k=-\infty}^{\infty} X\left[i\left(\omega - \frac{2\pi}{T}k\right)\right] \tag{105.11}$$

Thus, the Fourier transform X_Δ of the sampled variable has a periodic extension of the original spectrum $X(i\omega)$ along the frequency axis with a period equal to the sampling frequency $\omega_s = 2\pi/T$. There is an important result based on this observation known as the *Shannon sampling theorem,* which states that the continuous-time variable $x(t)$ may be reconstructed from the samples $\{x_k\}_{-\infty}^{+\infty}$ if and only if the sampling frequency is at least twice that of the highest frequency for which $X(i\omega)$ is nonzero. The original variable $x(t)$ may thus be recovered as

$$x(t) = \sum_{k=-\infty}^{\infty} x_k \frac{\sin\frac{\pi}{T}(t-kT)}{\frac{\pi}{T}(t-kT)} \tag{105.12}$$

The formula given in Equation (105.12) is called *Shannon interpolation,* which is often quoted though it is valid only for infinitely long data sequences and though it would require a noncausal filter to reconstruct the continuous-time signal $x(t)$ in real-time operation. The frequency $\omega_n = \omega_s/2 = \pi/T$ is called the *Nyquist frequency* and indicates the upper limit of distortion-free sampling. A nonzero spectrum beyond this limit leads to interference between the sampling frequency and the sampled signal (*aliasing*); see Figure 105.2.

The Discrete Fourier Transform

Consider a finite length sequence $\{x_k\}_{k=0}^{N-\infty}$ that is zero outside the interval $0 \le k \le N-1$. Evaluation of the z transform $X(z)$ at N equally spaced points on the unit circle $z = \exp(i\omega_k T) = \exp[i\,(2\pi/NT)kT]$ for $k = 0, 1,\ldots, N-1$ defines the *discrete Fourier transform* (DFT) of a signal x with a sampling period h and N measurements

$$X_k = \text{DFT}\{x(kT)\} = \sum_{l=0}^{N-1} x_l \exp(-i\omega_k lT) = X(e^{i\omega_k T}) \tag{105.13}$$

Notice that the discrete Fourier transform $\{X_k\}_{k=0}^{N-1}$ is only defined at the discrete frequency points

$$\omega_k = \frac{2\pi}{NT}k, \quad \text{for} \quad k = 0, 1,\ldots, N-1 \tag{105.14}$$

In fact, the discrete Fourier transform adapts the Fourier transform and the z transform to the practical requirements of finite measurements. Similar properties hold for the discrete Laplace transform with $z = \exp(sT)$, where s is the Laplace transform variable.

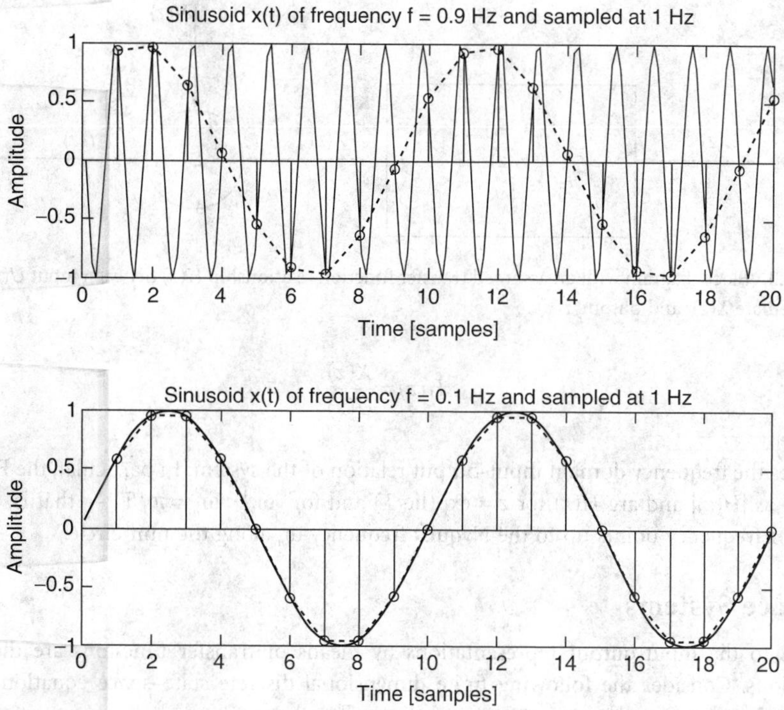

FIGURE 105.2 Illustration of aliasing appearing during sampling of a sinusoid $x(t) = \sin 2\pi \cdot 0.9\, t$ at the insufficient sampling frequency 1 Hz (sampling period $T = 1$) (*upper graph*). The sampled signal exhibits aliasing with its major component similar to a signal $x(t) = \sin 2\pi \cdot 0.1\, t$ sampled with the same rate (*lower graph*).

105.3 The Transfer Function

Consider the following discrete-time linear system with input sequence $\{u_k\}$ (stimulus) and output sequence $\{y_k\}$ (response). The dependency of the output of a linear system is characterized by the convolution-type equation and its z transform,

$$y_k = \sum_{m=0}^{\infty} h_m u_{k-m} + v_k = \sum_{m=-\infty}^{k} h_{k-m} u_m + v_k, \qquad k = \ldots, -1, 0, 1, 2, \ldots \tag{105.15}$$

$$Y(z) = H(z)U(z) + V(z)$$

where the sequence $\{v_k\}$ represents some external input of errors and disturbances and with $Y(z) = \mathrm{L}\{y\}$, $U(z) = \mathrm{L}\{u\}$, $V(z) = \mathrm{L}\{v\}$ as output and inputs. The *weighting function* $h(kT) = \{h_k\}_{k=0}^{\infty}$, which is zero for negative k and for reasons of causality is sometimes called *pulse response* of the digital system (compare *impulse response* of continuous-time systems). The pulse response and its z transform, the *pulse transfer function*,

$$H(z) = \mathcal{L}\{h(kT)\} = \sum_{k=0}^{\infty} h_k z^{-k} \tag{105.16}$$

determine the system's response to an input $U(z)$; see Figure 105.3. The pulse transfer function $H(z)$ is obtained as the ratio

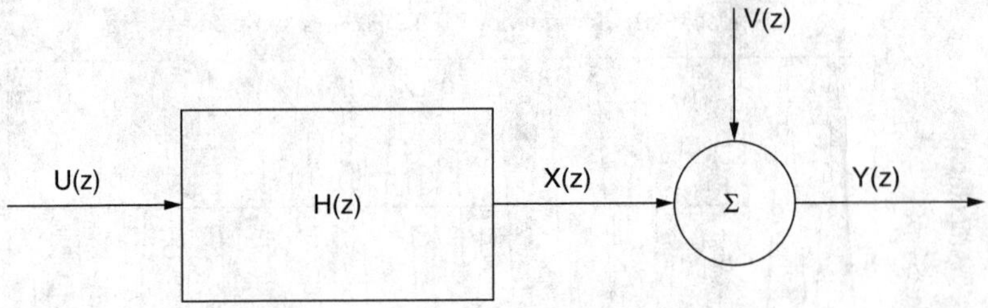

FIGURE 105.3 Block diagram with an assumed transfer function relationship $H(z)$ between input $U(z)$, disturbance $V(z)$, intermediate $X(z)$, and output $Y(z)$.

$$H(z) = \frac{X(z)}{U(z)} \tag{105.17}$$

and provides the frequency domain input-output relation of the system. In particular, the Bode diagram is evaluated as $|H(z)|$ and arg $H(z)$ for $z = \exp(i\omega_k T)$ and for $|\omega_k| < \omega_n = \pi / T$ — that is, when $H(z)$ is evaluated for frequency points up to the Nyquist frequency ω_n along the unit circle.

State–Space Systems

Alternatives to the input-output representations by means of transfer functions are the state–space representations. Consider the following finite dimensional discrete state–space equation with a state vector $x_k \in \mathbb{R}^n$, input $u_k \in \mathbb{R}^p$, and observations $y_n \in \mathbb{R}^m$.

$$\begin{cases} x_{k+1} = \Phi x_k + \Gamma u_k \\ y_k = Cx_k + Du_k \end{cases}, \quad k = 0,1\ldots \tag{105.18}$$

with the pulse transfer function

$$H(z) = C(zI - \Phi)^{-1}\Gamma + D \tag{105.19}$$

and the output variable

$$Y(z) = C\sum_{k=0}^{\infty} \Phi^k z^{-k} x_0 + H(z)U(z) \tag{105.20}$$

where possible effects of initial conditions x_0 appear as the first term. Notice that the initial conditions x_0 can be viewed as the net effects of the input in the time interval $(-\infty, 0)$.

105.4 Digital Systems Described by Difference Equations (ARMAX Models)

An important class of nonstationary stochastic processes is one in which some deterministic response to an external input and a stationary stochastic process are superimposed. This is relevant, for instance, when the external input cannot be effectively described by some probabilistic distribution. A discrete-time model can be formulated in the form of a difference equation with an external input $\{u_k\}$ that is usually considered to be known:

$$y_k = -a_1 y_{k-1} - \cdots - a_n y_{k-n} + b_1 u_{k-1} + \cdots + b_n u_{k-n} + w_k + c_1 w_{k-1} + \cdots + c_n w_{k-n} \tag{105.21}$$

Application of the z transform permits formulation of Equation (105.21) as

$$A(z^{-1})Y(z) = B(z^{-1})U(z) + C(z^{-1})W(z) \tag{105.22}$$

where

$$\begin{aligned} A(z^{-1}) &= 1 + a_1 z^{-1} + \cdots + a_n z^{-n} \\ B(z^{-1}) &= 1 + b_1 z^{-1} + \cdots + b_n z^{-n} \\ C(z^{-1}) &= 1 + c_1 z^{-1} + \cdots + c_n z^{-n} \end{aligned} \tag{105.23}$$

Stochastic models including the A polynomial, according to Equation (105.22) and Equation (105.23), are known as **autoregressive models** (AR), and models including the C polynomial are known as **moving-average** models (MA), whereas the B polynomial determines the effects of the external input (X). Notice that the term *moving average* is here somewhat misleading, as there is no restriction that the coefficients should add to 1 or that the coefficients are nonnegative. An alternative description is *finite impulse response* or *all-zero filter.*

Thus, the full model of Equation (105.22) is an **autoregressive moving average model** with external input (ARMAX) and its pulse transfer function $H(z) = B(z^{-1})/A(z^{-1})$ is stable if and only if the *poles* — that is, the complex numbers $z_1,\ldots,z_n$ solving the equation $A(z^{-1}) = 0$ — are strictly inside the unit circle, that is, $|z_i| < 1$. The *zeros* of the system — that is, the complex numbers $z_1,\ldots,z_n$ solving the equation $B(z^{-1}) = 0$ — may take on any value without any instability arising, although it is preferable to obtain zeros located strictly inside the unit circle, that is, $|z_i| < 1$ (*minimum-phase zeros*). By linearity, $\{y_k\}$ can be separated into one purely deterministic process $\{x_k\}$ and one purely stochastic process $\{v_k\}$:

$$\begin{cases} A(z^{-1})X(z) = B(z^{-1})U(z) \\ A(z^{-1})V(z) = C(z^{-1})W(z) \end{cases} \quad \text{and} \quad \begin{cases} y_k = x_k + v_k \\ Y(z) = X(z) + V(z) \end{cases} \tag{105.24}$$

The type of decomposition [Equation (105.24)] that separates the deterministic and stochastic processes is known as the *Wold decomposition.*

105.5 Prediction and Reconstruction

Consider the problem of predicting the output d steps ahead when the output $\{y_k\}$ is generated by the ARMA model,

$$A(z^{-1})Y(z) = C(z^{-1})W(z) \tag{105.25}$$

which is driven by a zero-mean white noise $\{w_k\}$ with covariance $\mathcal{E}\{w_i\, w_j\} = \sigma_w^2 \delta_{ij}$. In other words, assuming that observations $\{y_k\}$ are available up to the present time, how should the output d steps ahead be predicted optimally? Assume that the polynomials $A(z^{-1})$ and $C(z^{-1})$ are mutually prime with no zeros for $|z| \geq 1$. Let the C polynomial be expanded according to the *Diophantine equation,*

$$C(z^{-1}) = A(z^{-1})F(z^{-1}) + z^{-d}\, G(z^{-1}) \tag{105.26}$$

which is solved by the two polynomials

$$\begin{aligned} F(z^{-1}) &= 1 + f_1 z^{-1} + \cdots + f_{n_f} z^{-n_F}, & n_F &= d - 1 \\ G(z^{-1}) &= g_0 + g_1 z^{-1} + \cdots + g_{n_g} z^{-n_G}, & n_G &= \max(n_A - 1, n_C - d) \end{aligned} \tag{105.27}$$

Interpretation of z^{-1} as a *backward shift operator* and application of Equation (105.25) and Equation (105.26) permit the formulation

$$y_{k+d} = F(z^{-1})w_{k+d} + \frac{G(z^{-1})}{C(z^{-1})}y_k \tag{105.28}$$

Let us, by $\hat{y}_{k+d|k}$, denote linear *d*-step predictors of y_{k+d} based upon the measured information available at time *k*. As the zero-mean term $F(z^{-1})\,y_{k+d}$ of Equation (105.28) is unpredictable at time *k*, it is natural to suggest the following *d*-step predictor

$$\hat{y}_{k+d|k} = \frac{G(z^{-1})}{C(z^{-1})}y_k \tag{105.29}$$

The prediction error satisfies

$$\begin{aligned}\varepsilon_{k+d} &= (\hat{y}_{k+d|k} - y_{k+d}) \\ &= \frac{G(z^{-1})}{C(z^{-1})}y_k - \frac{A(z^{-1})F(z^{-1}) + z^{-d}G(z^{-1})}{C(z^{-1})}y_{k+d} \\ &= -F(z^{-1})w_{k+d}\end{aligned} \tag{105.30}$$

Let $\mathscr{E}\{\cdot|\mathscr{F}_k\}$ denote the *conditional mathematical expectation* relative to the measured information available at time *k*. The conditional mathematical expectation and the covariance of the *d*-step prediction relative to available information at time *k* is

$$\begin{aligned}\mathscr{E}\{\hat{y}_{k+d|k} - y_{k+d}|\mathscr{F}_k\} &= \mathscr{E}\{-F(z^{-1})w_{k+d}|\mathscr{F}_k\} = 0 \\ \mathscr{E}\{(\hat{y}_{k+d|k} - y_{k+d})^2|\mathscr{F}_k\} &= \mathscr{E}\{[-F(z^{-1})w_{k+d}]^2|\mathscr{F}_k\} \\ &= \mathscr{E}\{(w_{k+d} + f_1 w_{k+d-1} + \cdots + f_{d-1}w_{k+1})^2|\mathscr{F}_k\} \\ &= (1 + f_1^2 + \cdots + f_{n_F}^2)\sigma_w^2 = 0\end{aligned} \tag{105.31}$$

If follows that the predictor of Equation (105.29) is unbiased and that the prediction error only depends on future, unpredictable noise components. It is straightforward to show that the predictor of Equation (105.29) achieves the lower bound of Equation (105.31) and that the predictor of Equation (105.29) is optimal in the sense that the prediction error variance is minimized.

Example 105.1 — An Optimal Predictor for a First-Order Model

Consider for the first-order ARMA model

$$y_{k+1} = -a_1 y_k + w_{k+1} + c_1 w_k \tag{105.32}$$

The variance of a one-step-ahead predictor $\mathscr{E}\{(\hat{y}_k$ is

$$\begin{aligned}\mathscr{E}\{(\hat{y}_{k+1|k} - y_{k+1})^2|F_k\} &= \mathscr{E}\{(\hat{y}_{k+1|k} + a_1 y_k - c_1 w_k)^2|\mathscr{F}_k\} + \mathscr{E}\{w_{k+1}^2|\mathscr{F}_k\} \\ &= \mathscr{E}\{(\hat{y}_{k+1|k} + a_1 y_k - c_1 w_k)^2|\mathscr{F}_k\} + \sigma_w^2 \geq \sigma_w^2\end{aligned} \tag{105.33}$$

The optimal predictor satisfying the lower bound in Equation (105.33) is obtained from Equation (105.33) as

$$\hat{y}^{o}_{k+1|k} = -a_1 y_k + c_1 w_k \tag{105.34}$$

which, unfortunately, is not realizable as it stands because w_k is not available to measurement. Therefore, the noise sequence $\{w_k\}$ has to be substituted by some function of the observed variable $\{y_k\}$, and Equation (105.25) suggests that $W(z)$ be chosen by the filtering relationship $C^{-1}(z^{-1})A(z^{-1})Y(z)$ which, in turn, gives Equation (105.29). Accordingly, a linear predictor is

$$\hat{y}_{k+1|k} = \frac{G(z^{-1})}{C(z^{-1})} y_k = \frac{c_1 - a_1}{1 + c_1 z^{-1}} y_k \tag{105.35}$$

105.6 The Kalman Filter

Consider the linear state–space model

$$x_{k+1} = \Phi x_k + v_k, \quad x_k \in \mathbb{R}^n$$

$$y_k = C x_k + w_k, \quad y_k \in \mathbb{R}^m \tag{105.36}$$

where $\{v_k\}$ and $\{w_k\}$ are assumed R to be independent zero-mean white-noise processes with covariances Σ_v and Σ_w, respectively. It is assumed that $\{y_k\}$ but not $\{x_k\}$ is available to measurement and that it is desirable to predict $\{x_k\}$ from measurements of $\{y_k\}$.

Introduce the state predictor,

$$\hat{x}_{k+1|k} = \Phi \hat{x}_{k|k-1} - K_k(\hat{y}_k - y_k), \quad \hat{x}_{k|k-1} \in \mathbb{R}^n$$

$$\hat{y}_k = C\hat{x}_{k|k-1}, \quad y_k \in \mathbb{R}^m \tag{105.37}$$

The predictor of Equation (105.37) has the same dynamics matrix Φ as the state–space model of Equation (105.36) and, in addition, there is a correction term $K_k(\hat{y}_k - y_k)$ with a factor K_k to be chosen. The prediction error is

$$\tilde{x}_{k+1|k} = \hat{x}_{k+1|k} - x_{k+1} \tag{105.38}$$

The prediction-error dynamics is

$$\tilde{x}_{k+1} = (\Phi - K_k C)\tilde{x}_k + v_k - K_k w_k \tag{105.39}$$

The mean prediction error is governed by the recursive equation

$$\mathscr{E}\{\tilde{x}_{k+1}\} = (\Phi - K_k C)\mathscr{E}\{\tilde{x}_k\} \tag{105.40}$$

and the mean square error of the prediction error is governed by

$$\begin{aligned}\mathscr{E}\{\tilde{x}_{k+1}\tilde{x}^T_{k+1}\} &= \mathscr{E}\{[(\Phi - K_k C)\tilde{x}_k + v_k - K_k w_k][(\Phi - K_k C)\tilde{x}_k + v_k - K_k w_k]^T\} \\ &= (\Phi - K_k C)\mathscr{E}\{\tilde{x}_k \tilde{x}^T_k\}(\Phi - K_k C)^T + \Sigma_v + K_k \Sigma_w K_k\end{aligned} \tag{105.41}$$

If we denote

$$P_k = \mathscr{E}\{\tilde{x}_k \tilde{x}_k^T\}$$
$$Q_k = \Sigma_w + CP_kC^T \tag{105.42}$$

then Equation (105.41) is simplified to

$$P_{k+1} = \Phi P_k \Phi^T - K_k C P_k \Phi - \Phi^T P_k C^T K_k^T + \Sigma_v + K_k Q_k K_k^T \tag{105.43}$$

By completing squares of terms containing K_k we find

$$P_{k+1} = \Phi P_k \Phi^T + \Sigma_v - \Phi P_k C^T Q_k^{-1} C P_k \Phi^T + (K_k - \Phi P_k C^T Q_k^{-1}) Q_k (K_k - \Phi P_k C^T Q_k^{-1})^T \tag{105.44}$$

where only the last term depends on K_k. Minimization of P_{k+1} can be done by choosing K_k such that the positive semidefinite K_k-dependent term in Equation (105.44) disappears. Thus P_{k+1} achieves its lower bound for

$$K_k = \Phi P_k C^T (\Sigma_w + CP_kC^T)^{-1} \tag{105.45}$$

and the *Kalman filter* (or *Kalman–Bucy filter* for continuous-time systems) takes the form

$$\begin{aligned}
\hat{x}_{k+1|k} &= \Phi \hat{x}_{k|k-1} - K_k(\hat{y}_k - y_k) \\
\hat{y}_k &= C\hat{x}_{k|k-1} \\
K_k &= \Phi P_k C^T (\Sigma_w + CP_kC^T)^{-1} \\
P_{k+1} &= \Phi P_k \Phi^T + \Sigma_v - \Phi P_k C^T (\Sigma_w + CP_kC^T)^{-1} C P_k \Phi^T
\end{aligned} \tag{105.46}$$

which is the optimal predictor in the sense that the mean square error [Equation (105.41)] is minimized in each step.

Example 105.2 — Kalman Filter for a First-Order System

Consider the state–space model

$$\begin{aligned}
x_{k+1} &= 0.95 x_k + v_k \\
y_k &= x_k + w_k
\end{aligned} \tag{105.47}$$

where $\{v_k\}$ and $\{w_k\}$ are zero-mean white-noise processes with covariances $\mathscr{E}\{v_k^2\} = 1$ and $\mathscr{E}\{w_k^2\} = 1$, respectively.

The Kalman filter takes on the form

$$\begin{aligned}
\hat{x}_{k+1|k} &= 0.95 \hat{x}_{k|k-1} - K_k(\hat{x}_{k|k-1} - y_k) \\
K_k &= \frac{0.95 P_k}{1 + P_k} \\
P_{k+1} &= 0.95^2 P_k + 1 - \frac{0.95^2 P_k^2}{1 + P_k}
\end{aligned} \tag{105.48}$$

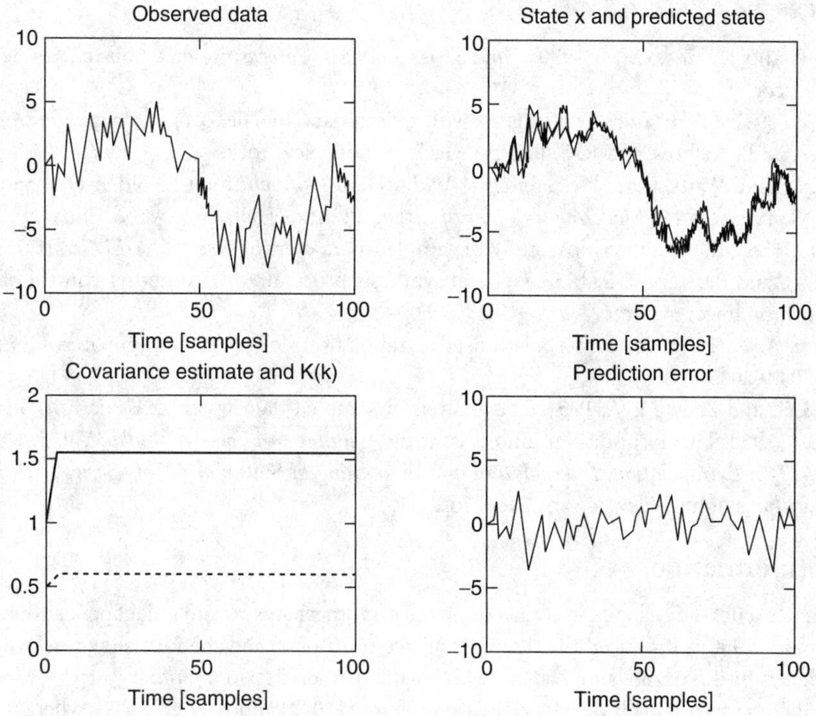

FIGURE 105.4 Kalman filter applied to one-step-ahead prediction of x_{k+1} in Equation (105.47). The observed variable $\{y_k\}$, the state $\{x_k\}$, and the predicted state $\{\hat{x}_k\}$, the estimated variance $\{P_k\}$ and $\{K_k\}$, and the prediction error $\{\tilde{x}_k\}$ are shown in a 100-step realization of the stochastic process. (*Source:* Johansson, R. 1993. *System Modeling and Identification,* Prentice Hall, Englewood Cliffs, NJ.)

The result of one such realization is shown in Figure 105.4.

Defining Terms

Autoregressive model (AR) — An autoregressive time series of order n is defined via $y_k = -\sum_{m=1}^{n} a_m y_{k-m} + w_k$. The sequence $\{w_k\}$ is usually assumed to consist of zero-mean identically distributed stochastic variables w_k.

Autoregressive moving average model (ARMA). An autoregressive moving average time series of order n is defined via $y_k = -\sum_{m=1}^{n} a_m y_{k-m} + \sum_{m=0}^{n} c_m w_{k-m}$. The sequence $\{w_k\}$ is usually assumed to consist of zero-mean identically distributed stochastic variables w_k.

Discrete Laplace transform. The discrete Laplace transform is counterpart to the Laplace transform with application to discrete signals and systems. The discrete Laplace transform is obtained from the z transform by means of the substitution $z = \exp(sT)$, where T is the sampling period.

Moving average process (MA). A moving average time series of order n is defined via $y_k = \sum_{m=0}^{n} c_m w_{k-m}$. The sequence $\{w_k\}$ is usually assumed to consist of zero-mean identically distributed stochastic variables w_k.

Rational model — AR, MA, ARMA, and ARMAX are commonly referred to as rational models.

Time series — A sequence of random variable $\{y_k\}$, where k belongs to the set of positive and negative integers.

***z* transform** — A generating function applied to sequences of data and evaluated as a function of the complex variable z with interpretation of frequency.

References

Box, G. E. P. and Jenkins, G. M. 1970. *Time Series Analysis: Forecasting and Control,* Holden-Day, San Francisco.

Hurewicz, W. 1947. Filters and servo systems with pulsed data. In *Theory of Servomechanisms,* ed. H. M. James, N. B. Nichols, and R. S. Philips. McGraw-Hill, New York.

Jenkins, G. M. and Watts, D. G. 1968. *Spectral Analysis and Its Applications,* Holden-Day, San Francisco.

Johansson, R. 1993. *System Modeling and Identification,* Prentice Hall, Englewood Cliffs, NJ.

Jury, E. I. 1956. Synthesis and critical study of sampled-data control systems. *AIEE Trans.* 75:141–151.

Kalman, R. E. and Bertram, J. E. 1958. General synthesis procedure for computer control of single and multi-loop linear systems. *Trans. AIEE.* 77:602–609.

Kolmogorov, A. N. 1939. Sur l'interpolation et extrapolation des suites stationnaires. *C. R. Acad. Sci.* 208:2043–2045.

Ragazzini, J. R. and Zadeh, L. A. 1952. The analysis of sampled-data systems. *AIEE Trans.* 71: 225–234.

Tsypkin, Y. Z. 1950. Theory of discontinuous control. *Avtomatika i Telemekhanika.* Vol. 5.

Wiener, N. 1949. *Extrapolation, Interpolation and Smoothing of Stationary Time Series with Engineering Applications.* John Wiley & Sons, New York.

Further Information

Early theoretical efforts developed in connection with servomechanisms and radar applications [Hurewicz, 1947]. Tsypkin [1950] introduced the discrete Laplace transform, and the formal z transform definition was introduced by Ragazzini and Zadeh [1952] with further developments by Jury [1956]. Much of prediction theory was originally developed by Kolmogorov [1939] and Wiener [1949], whereas state–space methods were forwarded by Kalman and Bertram [1958]. Pioneering textbooks on time-series analysis and spectrum analysis are provided by Box and Jenkins [1970] and Jenkins and Watts [1968].

Detailed accounts of time-series analysis and the z transform and their application to signal processing are to be found in:

- Hayes, M. H. 1996. Statistical Digital Signal Processing and Modeling. John Wiley & Sons, New York.
- Oppenheim, A. V. and Schafer, R. W. 1999. *Discrete-Time Signal Processing,* (2nd ed.). Prentice Hall, Upper Saddle River, NJ.
- Proakis, J. G. and Manolakis, D. G. 1995. *Digital Signal Processing: Principles, Algorithms and Applications,* (3rd Ed.). Prentice Hall Maxwell, Upper Saddle River, NJ.

Theory of time-series analysis and its application to discrete-time control is to be found in:

- Åström, K. J. and Wittenmark, B. 1996. *Computer-Controlled Systems,* (3rd ed.). Prentice Hall, Upper Saddle River, NJ.

Theory of time-series analysis and methodology for determination and validation of discrete-time models and other aspects of system identification are to be found in:

- Johansson, R. 1993. *System Modeling and Identification,* Prentice Hall, Englewood Cliffs, NJ.

Good sources to monitor current research are:

- *IEEE Transactions on Automatic Control*
- *IEEE Transactions on Signal Processing*

An easy-to-read survey article for signal processing applications is:

- Cadzow, J. A. 1990. Signal processing via least-squares error modeling. *IEEE ASSP Magazine.* 7:12–31, October.

XVII

Circuits

106

Passive Components

Henry Domingos
Clarkson University

Passive components such as resistors, capacitors, and inductors can be an integral part of a more complex unit such as an operational amplifier, digital integrated circuit, or microwave circuit. However, this chapter deals with discrete components purchased as individual parts. Their importance lies in the fact that virtually every piece of electronic equipment incorporates a multitude of discrete passive components, and sale of these components parallels the fortunes of the electronics industry as a whole.

Although passive components have been in use for hundreds of years, one should not assume they represent a static industry: There is a continuing evolution in the design and fabrication of these parts to make them smaller and cheaper with better performance and better reliability. In fact, the useful life of modern components exceeds by far that of the electronic equipment in which they are used. Failure is more often the result of misapplication. The continuous improvement in quality, price, performance, and general usefulness contributes to greater convenience and functionality for consumers at a lower and lower price.

106.1 Resistors

Resistance is a property of a device that relates the current through the device with the voltage or electric potential across it. It is a dissipative property in that energy is converted from electrical energy to heat. The resistance of a device is directly related to the **resistivity** of the material of which it is composed. The resistivity is a relative property that varies from that of good conductors such as aluminum or copper (resistivity $\approx 10^{-6}$ ohm-cm) to semiconductors such as silicon ($\approx 10^{-3} - 10^{+3}$ ohm-cm) to that of good insulators such as alumina ($\approx 10^{14}$ ohm-cm). The resistance of a device is determined by its resistivity and its geometry. Figure 106.1 and Equation (106.1) show the relationship for objects of uniform resistivity and cross-sectional area.

$$R = \rho \frac{L}{A} \tag{106.1}$$

In Equation (106.1), R is the resistance in ohms, ρ is the resistivity in ohm-cm, L is the length in cm, and A is the cross-sectional area in cm^2. For example, #12 AWG copper wire has a diameter of 0.0808 in. and a resistivity of $1.724\ 1 \cdot 10^{-6}$ ohm-cm. Using Equation (106.1) and converting units as appropriate, the resistance of 1000 feet of wire is 1.588 ohms.

0-8493-1586-7/05/$0.00+$1.50

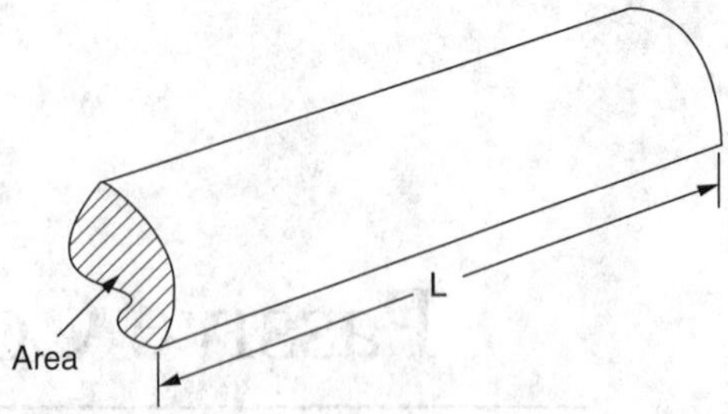

FIGURE 106.1 The resistance of a conductor with uniform resistivity and cross-section is given by Equation (106.1).

When resistive films are deposited on substrates or impurities are implanted in layers in silicon devices, the resistivity is not uniform and/or the thickness of the layer may not be known accurately. It is convenient then to define a **sheet resistance** that takes into account the average resistivity and the thickness so that the resistance is given by

$$R = R_s \frac{L}{W} \tag{106.2}$$

where R_s is the sheet resistance in ohms per square (the "square" is dimensionless), L is the length of the film, and W its width. Suppose, for example, a resistor ink is silk-screened onto a ceramic substrate and fired so that its sheet resistance is 100 ohms per square. If the width of the resistor is 10 mil (a mil is one one-thousandth of an inch), the length required to make a 1500-ohm resistor is, from Equation (106.2), $L = 150$ mil.

Using the concepts of resistivity and sheet resistance, a discrete resistor can be fabricated in a variety of ways. The common types are carbon composition resistors, wire-wound resistors, and film resistors. Carbon composition resistors were at one time the most popular type. They are low priced, high in reliability, and available in a wide range of resistance values. The drawbacks are poor long-term stability and tolerance of only 5% or larger. The resistor is fabricated from a silica-loaded resin with resistivity controlled by the addition of carbon particles. Axial leads are inserted and the outer shell, of the same material as the core, minus the carbon granules, insulates the assembly.

Wire-wound resistors are power wire wounds or precision wire wounds. In either case the resistance wire is either a nickel-chromium alloy with a resistivity of $1.33 \cdot 10^{-4}$ ohm-cm for high resistance values, or a copper-nickel alloy with a resistivity of $5 \cdot 10^{-5}$ ohm-cm for lower resistances. Power wire wounds are intended to operate at higher power dissipation and are composed of single layers of bare wire with substrates and packages of high-temperature materials. Precision resistors are stable, highly accurate components. The wire is multilayered, and therefore must be insulated, and all materials can be low temperature. The main disadvantages of wire-wound resistors are the high cost and low operating frequency.

Film-type resistors are made from various resistive materials deposited on an insulating substrate. Film thicknesses range from 0.005 micrometers deposited by evaporation for precision film resistors to as much as 100 micrometers deposited as a resistive ink. Sheet resistances range from 10 to 10,000 ohms per square. The final resistance value can be adjusted by cutting a spiral path through the resistive film until the desired value is reached. The resistors are stable, accurate, and low cost and are the most popular types today. Resistor networks, chip resistors, and film resistors deposited directly on a hybrid substrate provide a wide selection of styles.

The important specifications for a resistor are the resistance value, the tolerance, and the power rating. For carbon composition resistors the resistance value and tolerance are indicated by the familiar color code in Table 106.1. Values range from 1 ohm to 100 megohms. Power rating ranges from 1/20 to 2 W. The power rating is somewhat misleading since the temperature rise is the crucial variable, not the power dissipation. A 2 W resistor cannot dissipate 2 W if it is conformably coated and situated where cooling

TABLE 106.1 Resistor Color Code

Color	First Band,[a] Significant Figure	Second Band, Significant Figure	Third Band, Multiplier	Fourth Band,[b] Tolerance (%)	Fifth Band,[b] Failure Rate (%/1000 h)
Black	0	0	1	—	—
Brown	1	1	10	—	1
Red	2	2	10^2	—	0.1
Orange	3	3	10^3	—	0.01
Yellow	4	4	10^4	—	0.001
Green	5	5	10^5	—	—
Blue	6	6	10^6	—	—
Violet	7	7	10^7	—	—
Gray	8	8	10^8	—	—
White	9	9	10^9	—	—
Silver	—	—	0.01	10	—
Gold	—	—	0.1	5	—
None	—	—	—	20	—

[a] The first band is the one closest to one end of the resistor. A first band wider than the others indicates a wire-wound resistor.
[b] Certain MIL parts.

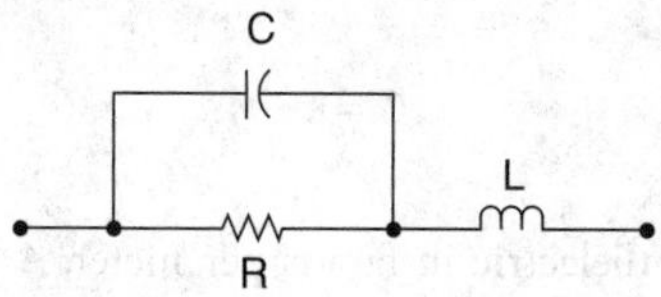

FIGURE 106.2 At high frequencies, the effects of parasitic capacitance and inductance of a resistor must be accounted for. This shows one possible equivalent representation.

air is restricted. On the other hand, if a suitable heat sink is provided and forced air cooling is available, the rating could be higher than specified. Under pulse conditions, the instantaneous power can be orders of magnitude larger than its nominal rating.

No resistance is ideal, and over wide frequency ranges the parasitic inductance and capacitance limit the usefulness. Figure 106.2 shows a common equivalent circuit for a resistor. The series inductance can be large for wire-wound resistors and limits their usefulness to frequencies less than about 50 to 100 kHz. For other resistors the leads constitute the main source of inductance, estimated to be of the order of 20 nH per inch. In general, low values of resistance and thin films tend to have lower inductance. Capacitance tends to dominate at high frequencies and with higher values of resistance. Figure 106.3 illustrates typical variation of impedance of a resistor with frequency.

Other resistor characteristics include temperature coefficient, voltage coefficient, stability, and noise. For a discussion of these the reader is referred to Dorf [1993] and Harper [1977].

106.2 Capacitors

Capacitance is a property that exists between two conductors separated by a **dielectric**. With equal but opposite charges on the two conductors a potential difference can be measured between them, and the capacitance is defined by the equation

$$C = \frac{Q}{V} \tag{106.3}$$

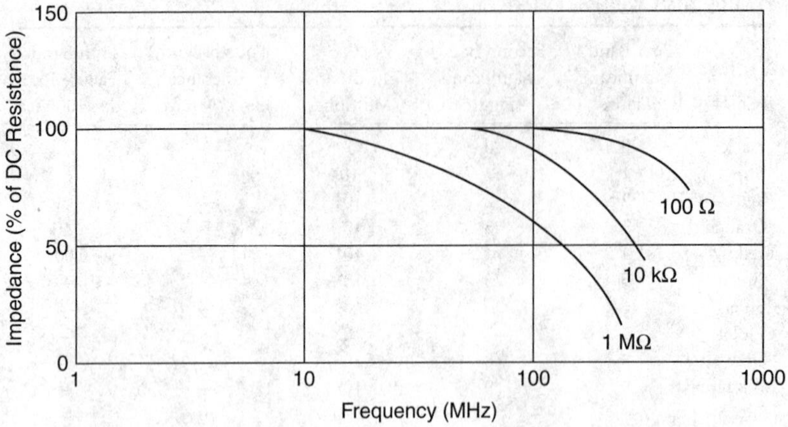

FIGURE 106.3 Variation of total impedance of a resistor with frequency for a typical film resistor.

C is the capacitance in farads, Q is the magnitude of the charge on either plate in coulombs, and V is the voltage between the plates in volts. The capacitance in the ideal case of two plane-parallel plates is given by

$$C = \varepsilon \frac{A}{d} \tag{106.4}$$

where ε is the **permittivity** of the dielectric in farads per meter, A is the area of each plate in meters squared, and d is the plate separation in meters. ε is equal to $\varepsilon_r\ \varepsilon_0$, where ε_r is the relative permittivity (sometimes referred to as the *dielectric constant*, k) = 1 for vacuum, ≈ 1 for air, and ranging from 2 to 10 for common dielectrics. ε_0 is the permittivity of free space, equal to $8.854 \cdot 10^{-12}$ farads per meter. All **capacitors** use appropriate values of ε, A, and d in Equation (106.4) to achieve the wide range of values available in commercial capacitors.

The constitutive relationship between voltage and current in a capacitor is

$$v(t) = v_0 + \frac{1}{C}\int_0^t i\,dt \tag{106.5}$$

where $v(t)$ is the instantaneous voltage across the capacitor, v_0 is the initial voltage on the capacitor at $t = 0$, and i is the instantaneous current through the capacitor. By differentiating Equation (106.5),

$$i(t) = C\frac{dv}{dt} \tag{106.6}$$

Unlike a resistor, an ideal capacitor does not dissipate energy. Instead, it accumulates energy during the charging process and releases energy to the electrical circuit as it discharges. The energy stored on a capacitor is given by

$$W = \frac{1}{2}CV^2 = \frac{1}{2}\frac{Q^2}{V} \tag{106.7}$$

where W is in joules.

There are two main classes of capacitors — electrolytic and electrostatic. Electrolytic capacitors include aluminum and tantalum types, used in applications where large capacitance values are needed. Electrostatic capacitors include plastic, ceramic disk, ceramic chip, mica, and glass. Aluminum electrolytics are constructed from high-purity aluminum foils that are chemically etched to increase the surface area, then anodized to form the dielectric. The thickness of this anodized layer determines the voltage rating of the capacitor. If only one foil is anodized, the capacitor is a polarized unit, and the instantaneous voltage cannot be allowed to reverse polarity. Porous paper separates the two foils and is saturated with a liquid electrolyte; therefore, the unit must be sealed, and leakage is a common failure mode. During extended periods of storage, the anodized layer may partially dissolve, requiring the unit to be reformed before rated voltage can be applied. Aluminum electrolytics exhibit a large series inductance, which limits the useful range of frequencies to about 20 kHz. They also have a large leakage current. Nevertheless, because of low cost and very large values of capacitance (up to 1 F), they are a popular choice for filtering applications.

Tantalum electrolytics are available in foil, wet slug, and solid varieties. All types require a tantalum electrode to be anodized, but the means used to make electrical contact to the plates differs in each case. The overall characteristics are superior to those of aluminum, but the cost is greater. Because of their small physical size and large capacitance, they are used for the same applications as aluminum electrolytics.

The common type of plastic capacitor is the metalized film type, made by vacuum deposition of aluminum a fraction of a micron thick directly onto a plastic dielectric film. Common film types include polyester, polycarbonate, polypropylene, and polysulfone. Although the relative permittivity is low, the films can be made extremely thin, resulting in a large capacitance per unit volume. This type of capacitor exhibits an interesting self-healing feature. If breakdown should happen to occur — for instance, at a defect in the film or during a momentary electrical overstress — the metalization around the breakdown site will act like a fuse and vaporize, clearing the breakdown and restoring operation to an equal or higher voltage capability.

Plastic film capacitors have good electrical characteristics and are capable of operation at higher voltages and much higher frequencies than are electrolytic capacitors. They have found wide application in filtering, bypassing, coupling, and noise suppression.

Ceramic capacitors employ ferroelectric dielectrics — commonly, barium titanate — that have an extremely high relative permittivity, 200 to 100,000. For disk capacitors the appropriate combination of powders is mixed, pressed into disks, and sintered at high temperatures. Electrodes are silk-screened on, leads are attached, and a protective coating applied. By varying the area, thickness (limited, to maintain mechanical strength), and especially the dielectric formulation, wide ranges of capacitance value are achieved. Very high voltage capability is also an option that can be achieved by increasing the thickness.

Ceramic chip capacitors are made from a slurry containing the dielectric powder and cast onto stainless-steel or plastic belts. After drying, electrodes are printed on the sheets, which are then arranged in a stack and fired. Electrodes are attached and the unit is encapsulated. By adjusting the number of plates and the plate area, a wide range of values can be obtained. The extremely small size makes this type especially useful in hybrid integrated circuits and on printed circuit boards.

Mica and glass capacitors are used whenever high quality and excellent stability with respect to temperature and aging are required, and where high frequency operation is required. They are used in tuning circuits and for coupling where high performance and reliability are essential. Mica capacitors are made from sheets of mica that are alternately stacked with foils. Alternate foils are extended and folded over the ends. After leads are attached, the unit is encapsulated. Glass capacitors are made from glass ribbons that are stacked alternately with foils. Leads are attached, and the entire assembly is sealed in glass at high temperature and pressure.

Capacitors used in power system applications employ oil-impregnated paper as a dielectric. Because of the very high voltage ratings, these capacitors are shipped and stored with a shorting bar across the terminals as a safety precaution. Even though a capacitor has been discharged after high voltage operation or testing, it is possible for charge and voltage to reaccumulate by a phenomenon known as *dielectric absorption.*

Figure 106.4, Figure 106.5, and Figure 106.6 show ranges of capacitance values, voltage ratings, and frequency ranges for the different capacitors described in this section.

All capacitors dissipate a small amount of power due to resistive losses in the conductors, leakage current across the dielectric, and losses in the dielectric under AC operation. If the capacitor is represented

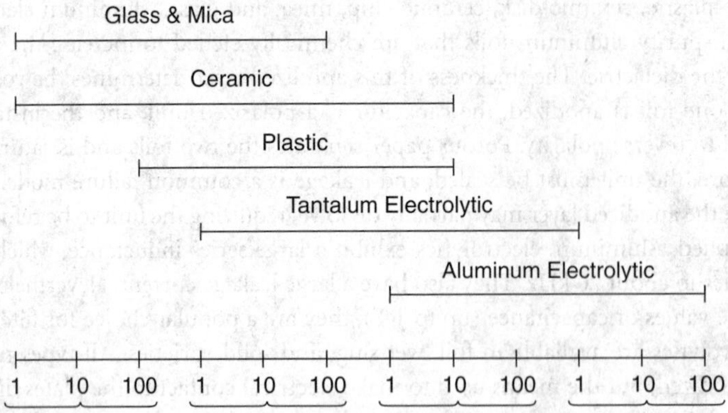

FIGURE 106.4 Range of values for various types of capacitors.

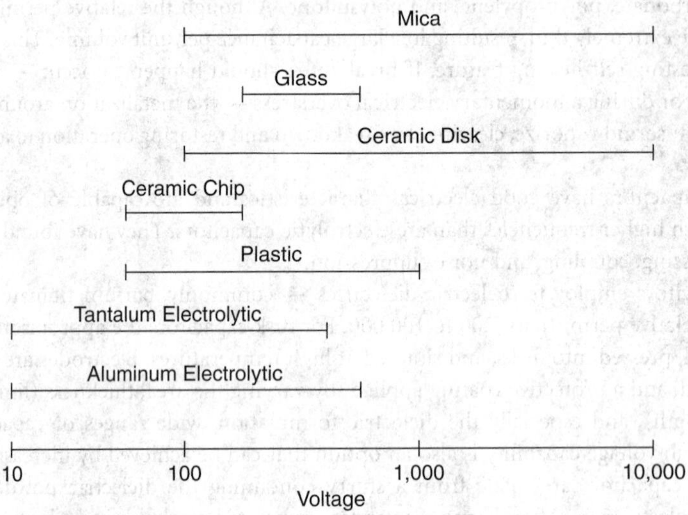

FIGURE 106.5 Voltage ratings of various capacitors.

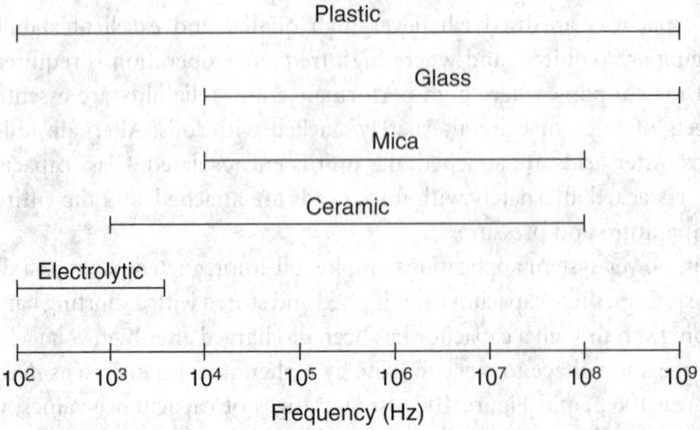

FIGURE 106.6 Frequency range of various capacitors.

by its capacitance value in series with a resistor to simulate these losses, the resistance is called the **equivalent series resistance (ESR).** Capacitance bridges often measure the **dissipation factor (DF),** which is given by

$$DF = 2\pi f C_s R_s \tag{106.8}$$

where C_s is the series capacitance, f is the frequency in hertz, and R_s is the ESR. The reciprocal of the dissipation factor is the **quality factor** (Q) of the capacitor.

106.3 Inductors

Inductance is the property of a device that relates a magnetic field to the current that produces it. In Figure 106.7, the current I flowing in a counterclockwise direction in the coil produces, by the right-hand rule, a magnetic flux. The inductance of the coil is defined by

$$L = \frac{\Phi}{I} \tag{106.9}$$

where L is the inductance in henrys, Φ is the flux in webers, and I is the current in amperes.

In an **inductor** the instantaneous voltage and current are related by the equation

$$v(t) = L\frac{di}{dt} \tag{106.10}$$

where $v(t)$ is the instantaneous voltage in volts and di/dt is the rate of change of current with respect to time in amperes per second. Alternatively,

$$i(t) = i_0 + \frac{1}{L}\int_0^t v\,dt \tag{106.11}$$

where i_0 is the initial current through the inductor at $t = 0$.

The energy stored in an inductor is

$$W = \frac{1}{2}I^2 L \tag{106.12}$$

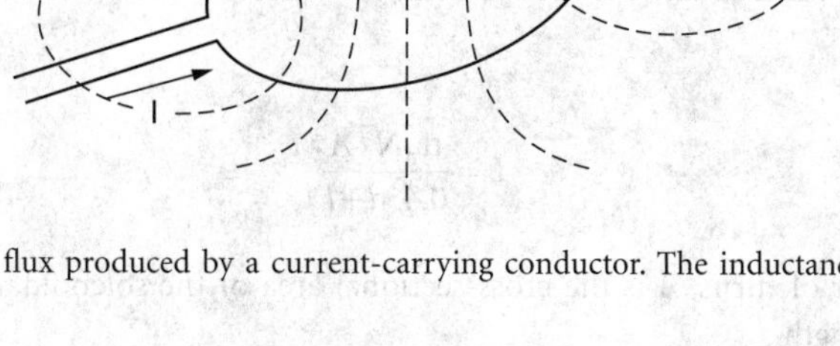

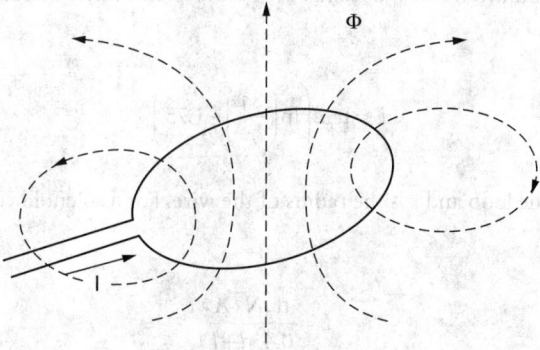

FIGURE 106.7 Magnetic flux produced by a current-carrying conductor. The inductance is defined by Equation (106.9).

Since the finite resistance of the wire in an inductor causes some energy loss, the quality factor, or Q, of an inductor — relating energy stored and energy dissipated — is an important quantity. It is given by

$$Q = \frac{2\pi f L}{R} \tag{106.13}$$

where f is the frequency in hertz and R is the series resistance of the inductor.

In general, it is not possible to find an exact analytical expression for the inductance of a configuration. A number of empirical formulas are given as follows, and some general principles can be stated.

1. The inductance is proportional to the size of the coil. Thus, larger loops have larger inductance.
2. Inductance is proportional to the number of turns squared. In Figure 106.7, if there are two turns, the flux is doubled; if the flux is changing, then twice the induced voltage will appear in each turn, for a total of four times the voltage in a one-turn loop.
3. The inductance is proportional to the permeability of the surrounding medium. Most materials have a permeability μ_0 very close to that of free space; $\mu_0 = 4 \cdot 10^{-7}$ henrys per meter. Magnetic materials have a permeability ranging from about 100 to 100,000 times μ_0. They are used to concentrate and intensify the flux in transformers, motors, relays, and inductors.

Approximate formulas for the inductance of several configurations can be given. The internal inductance per unit length of a round solid wire is

$$L = \frac{\mu_0}{8\pi} \tag{106.14}$$

The inductance per unit length of two parallel wires carrying current in opposite directions is

$$L = 0.4 \cdot 10^{-6} \ln\left(\frac{2D}{d}\right) \tag{106.15}$$

where D is the center-to-center spacing of the wires and d is the diameter of the wire. For a coaxial cable the inductance per unit length is

$$L = \frac{\mu_0}{2\pi} \ln\left(\frac{r_b}{r_a}\right) \tag{106.16}$$

where the subscripts b and a refer to the radius of the outer and inner conductors, respectively. For a circular loop of wire,

$$L = \mu_0 a \left(\ln\left(\frac{8a}{r}\right) - 1.75 \right) \tag{106.17}$$

where a is the radius of the loop and r is the radius of the wire. For a solenoid consisting of a single layer of turns,

$$L = \frac{\mu_0 N^2 A}{0.45d + l} \tag{106.18}$$

where N is the number of turns, A is the cross-sectional area of the solenoid, d is the diameter of the solenoid, and l is its length.

In many cases the objective is to minimize the parasitic inductance of a layout. To achieve this, one should minimize the area of current-carrying loops and use rectangular or flat conductors.

Defining Terms

Capacitance — A property of an arrangement of two conductors equal to the charge on a conductor divided by the voltage between them.

Capacitor — An electronic component that accumulates charge and energy in an electric circuit.

Dielectric — An insulating material, for example, between the plates of a capacitor.

Dissipation factor (DF) — The ratio of the energy dissipated per cycle to two times the maximum energy stored at a given frequency. The lower the DF, the more nearly ideal the capacitor.

Equivalent series resistance (ESR) — The resistance in series with a capacitor that accounts for energy loss in a nonideal capacitor.

Inductance — The property of a conductor arrangement that relates the magnetic flux to the current in the conductor.

Inductor — A passive component that stores energy in its magnetic field whenever it is conducting a current.

Permittivity — The property of a material that relates the electric field intensity to the electric flux density. The capacitance between two conductors is directly proportional to the permittivity of the dielectric between them.

Quality factor (Q) — A value equal to two times the ratio of the maximum stored energy to the energy dissipated per cycle at a given frequency in a capacitor or inductor. The larger the value of Q, the more nearly ideal the component. For a capacitor, Q is the inverse of DF.

Resistivity — The property of a material that relates current flow to electric field. The current density is equal to the electric field divided by the resistivity.

Sheet resistance — Sheet resistance (or sheet resistivity) is the average resistivity of a layer of material divided by its thickness. If the layer has the same length as width, the resistance between opposite edges of the "square" is numerically equal to the sheet resistance, regardless of the size of the square.

References

Dorf, R. C. 1993. *The Electrical Engineering Handbook*, CRC Press, Boca Raton, FL.

Harper, C. C. 1977. *Handbook of Components for Electronics*, McGraw-Hill, New York.

Meeldijk, V. 1995. *Electronic Components: Selection and Application Guidelines*, John Wiley & Sons, New York.

Further Information

The *IEEE Transactions on Components, Packaging, and Manufacturing Technology* has a number of technical articles on components, and the CPMT Society sponsors the Electronic Components and Technology Conference. The transactions and the conference proceedings are the main sources for the latest developments in components.

The Capacitor and Resistor Technology Symposium (CARTS) is devoted entirely to the technology of electronic components. For further information on CARTS, contact: Components Technology Institute, Inc., 904 Bob Wallace Ave., Suite 117, Huntsville, AL 35801.

107

RL, RC, and RLC Circuits

Michael D. Ciletti
University of Colorado, Colorado Springs

Circuits that contain only resistive elements can be described by a set of algebraic equations that are obtained by systematically applying Kirchhoff's current and voltage laws to the circuit. When a circuit contains energy storage elements — that is, inductors and capacitors — Kirchhoff's laws are still valid, but their application leads to a differential equation (DE) model of the circuit instead of an algebraic model. A DE model can be solved by classical DE methods, by time-domain convolution, and by Laplace transform methods.

107.1 RL Circuits

A series RL circuit can be analyzed by applying Kirchhoff's voltage law to the single loop that contains the elements of the circuit. For example, writing Kirchhoff's voltage law for the circuit in Figure 107.1 leads to the following first-order DE stating that the voltage drop across the inductor and the resistor must equal the voltage drop across the voltage source:

$$L\frac{di}{dt}+i(t)R=v_g(t) \tag{107.1}$$

This DE belongs to a family of linear, constant-coefficient, ordinary differential equations. Various approaches can be taken to find the solution for $i(t)$, the current through the elements of the circuit. The first is to solve the DE by classical (time-domain) methods; the second is to solve the equation by using Laplace transform theory. A third approach, waveform convolution, will not be considered here (see Ziemer et al. [1993].)

The complete time-domain solution to the DE is formed as the sum of two parts, called the **natural solution** and the **particular solution:**

$$i(t)=i_N(t)+i_P(t) \tag{107.2}$$

0-8493-1586-7/05/$0.00+$1.50

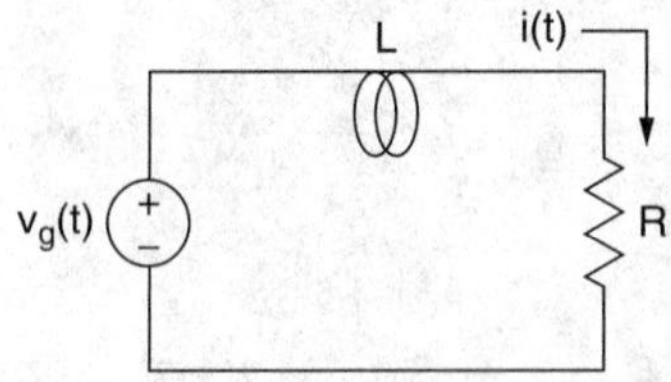

FIGURE 107.1 Series RL circuit.

The natural solution, $i_N(t)$, is obtained by solving the (homogeneous) DE with the sources (or forcing functions) turned off. The circuit in this example has the following homogeneous DE:

$$L\frac{di}{dt}+i(t)R=0 \tag{107.3}$$

The solution to this equation is the exponential form: $i_N(t) = Ke^{st}$, where K is an arbitrary constant. To verify this solution, substitute Ke^{st} into Equation (107.3) and form

$$LKse^{st}+RKe^{st}=0 \tag{107.4}$$

Rearranging gives:

$$[SL+R]Ke^{st}=0 \tag{107.5}$$

The factor Ke^{st} can be canceled because the factor e^{st} is nonzero, and the constant K must be nonzero for the solution to be nontrivial. This reduction leads to the characteristic equation

$$sL+R=0 \tag{107.6}$$

The **characteristic equation** of this circuit specifies the value of s for which $i_N(t) = Ke^{st}$ solves Equation (107.3). By inspection, $s = -R/L$ solves Equation (107.6), and $i_N(t) = Ke^{-t/\tau}$ satisfies Equation (107.3), and $\tau = L/R$ is called the *time constant* of the circuit. The value $s = -R/L$ is called the **natural frequency** of the circuit.

In general, higher-order circuits (those with higher-order differential equation models and corresponding higher-order algebraic characteristic equations) will have several natural frequencies, some of which may have a complex value — that is, $s = \sigma + j\omega$ — where σ is an exponential damping factor and ω is the undamped frequency of oscillation. The natural frequencies of a circuit play a key role in governing the dynamic response of the circuit to an input by determining the form and the duration of the transient waveform of the response [Ciletti, 1988].

The particular solution of this circuit's DE model is a function $i_P(t)$ that satisfies Equation (107.1) for a given source $v_g(t)$. For example, the constant input signal $v_g(t) = V$ has the particular solution $i_P(t) = V/R$. [This can be verified by substituting this expression into Equation (107.1).] The complete solution to the differential equation when $v_g(t) = V$ is given by

$$i(t)=Ke^{(-R/L)t}+V/R=Ke^{-t/\tau}+V/R \tag{107.7}$$

The sources that excite physical circuits are usually modeled as being turned off before being applied at some specific time, say t_0, and the objective is to find a solution to its DE model for $t \geq t_0$. This leads to consideration of boundary conditions that constrain the behavior of the circuit and determine the unknown parameters in the solution of its DE model. The boundary conditions of the DE model of a physical circuit are determined by the energy that is stored in the circuit when the source is initially

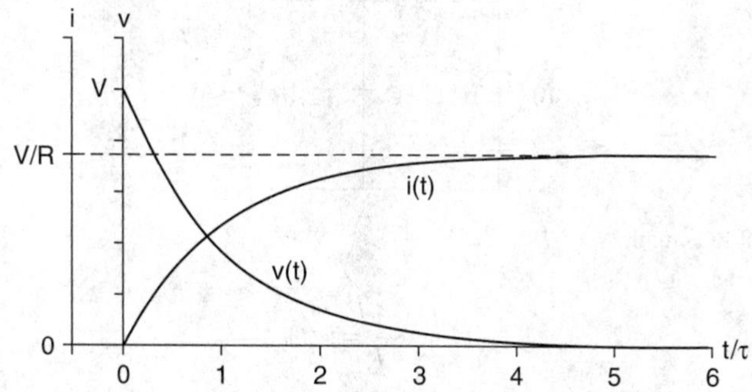

FIGURE 107.2 Waveforms of $v(t)$ and $i(t)$ in the series RL circuit.

applied. For example, the inductor current in Figure 107.1 could have the constraint given as $i(t_0) = i_0$, where i_0 is a constant. If a circuit is modeled by a constant-coefficient DE, the time of application can be taken to be $t_0 = 0$ without any loss in generality. (Note: The physical conditions that created i_0 are not of concern.)

The value of the parameter K in Equation (107.7) is specified by applying the given boundary condition to the waveform of Equation (107.7), as follows:

$$i(0^+) = i_0 = K + V/R \tag{107.8}$$

and so $K = i_0 - V/R$.

The response of the circuit to the applied input signal is the complete solution of the DE for $t \geq 0$ evaluated with coefficients that conform to the boundary conditions. In the RL circuit example, the complete response to the applied step input is

$$i(t) = [i_0 - V/R]e^{-t/\tau} + V/R, \qquad t \geq 0 \tag{107.9}$$

The waveforms of $i(t)$ and $v(t)$, the inductor voltage, are shown in Figure 107.2 for the case in which the circuit is initially at rest, that is, $i_0 = 0$. The time axis in the plot has been normalized by τ.

The physical effect of the inductor in this circuit is to provide inertia to a change in current in the series path that contains the inductor. An inductor has the physical property that its current is a continuous variable whenever the voltage applied across the terminals of the inductor has a bounded waveform (that is, no impulses). The current flow through the inductor is controlled by the voltage that is across its terminals. This voltage causes the accumulation of magnetic flux, which ultimately determines the current in the circuit. The initial voltage applied across the inductor is given by $v(0^+) = V - i(0^+)R$. If the inductor is initially relaxed — that is, $i(0^-) = 0$ — the continuity property of the inductor current dictates that $i(0^+) = i(0^-) = 0$. So $v(0^+) = V$. All of the applied voltage initially appears across the inductor. When this voltage is applied for an interval of time, a magnetic flux accumulates, and current is established through the inductor. Mathematically, the integration of this voltage causes the current in the circuit. The current waveform in Figure 107.2 exhibits exponential growth from its initial value of 0 to its final (steady state) value of V/R. The inductor voltage decays from its initial value to its steady state value of 0, and the inductor appears to be a "short circuit" to the steady state DC current.

107.2 RC Circuits

A capacitor acts like a reservoir of charge, thereby preventing rapid changes in the voltage across its terminals. A capacitor has the physical property that its voltage must be a continuous variable when its current is bounded. The DE model of the parallel RC circuit in Figure 107.3 is described according to Kirchhoff's current law by

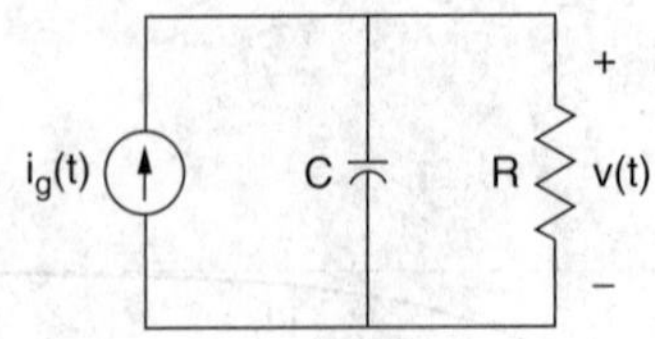

FIGURE 107.3 Parallel RC circuit.

$$C\frac{dv}{dt}+\frac{v}{R}=i_g(t) \tag{107.10}$$

Taking the one-sided Laplace transform [Ziemer, 1993] of this differential equation gives

$$sCV(s)-Cv(0^-)+\frac{V(s)}{R}=I_g(s) \tag{107.11}$$

where $V(s)$ and $I_g(s)$ denote the Laplace transforms of the related time-domain variables. (Note that the Laplace transform of the derivative of a variable explicitly incorporates the initial condition of the variable into the model of the circuit's behavior.) Rearranging this algebraic equation gives the following:

$$V(s)=\frac{I_g(s)}{sC+1/R}+\frac{Cv(0)}{sC+1/R} \tag{107.12}$$

$$V(s)=I_g(s)H(s)+\frac{v(0)}{s+1/(RC)} \tag{107.13}$$

where the s-domain function

$$H(s)=\frac{1}{sC+(1/R)}=\frac{1/C}{s+(1/RC)} \tag{107.14}$$

is called the input/output **transfer function** of the circuit. The expression in Equation (107.13) illustrates an important property of linear circuits: the response of a linear RLC circuit variable is the superposition of the effect of the applied source and the effect of the initial energy stored in the circuit's capacitors and inductors. Another important fact is that the roots of the denominator polynomial of $H(s)$ are the natural frequencies of the circuit (assuming no cancellation between numerator and denominator factors).

If a circuit is initially relaxed — that is, all capacitors and inductors are deenergized [set $v(0^-) = 0$ in Equation (107.13)], then the transfer function defines the ratio of the Laplace transform of the circuit's response to the Laplace transform of its stimulus. Alternatively, the transfer function and the Laplace transform of the input signal determine the Laplace transform of the output signal according to the simple product

$$V(s) = I_g(s)H(s) \tag{107. 15}$$

The transfer function of a circuit can define a voltage ratio, a current ratio, a voltage-to-current ratio (impedance) or a current-to-voltage ratio (admittance). In this circuit, $H(s)$ relates the output (response of the capacitor voltage) to the input (i.e., the applied current source). Thus, $H(s)$ is actually a generalized impedance function. If a circuit is initially relaxed its transfer function contains all of the information necessary to determine the response of the circuit to any given input signal.

Suppose that the circuit has an initial capacitor voltage and that the applied source in Equation (107.13) is given by $i_g(t) = Iu(t)$, a step of height I. The Laplace transform [Ciletti, 1988] of the step waveform is given by $I_g(s) = I/s$, so the Laplace transform of $v(t)$, denoted by $V(s)$, is given by

$$V(s) = \frac{I/(sC)}{s+1/(RC)} + \frac{v(0^-)}{s+1/(RC)} \tag{107.16}$$

This expression for $V(s)$ can be expanded algebraically into partial fractions as

$$V(s) = \frac{IR}{s} - \frac{IR}{s+1/(RC)} + \frac{v(0^-)}{s+1/(RC)} \tag{107.17}$$

Associating the s-domain Laplace transform factor $1/(s + a)$ with the time-domain function $e^{-at}u(t)$ and taking the inverse Laplace transform of the individual terms of the expansion gives

$$v(t) = IR + [v(0^-) - IR]e^{-t/(RC)}, \qquad t \geq 0 \tag{107.18}$$

When the source is applied to this circuit, the capacitor charges from its initial voltage to the steady state voltage given by $v(\infty) = IR$. The response of $v(t)$ is shown in Figure 107.4 with $\tau = RC$ and $v_0 = v(0^-)$. The capacitor voltage follows an exponential transition from its initial value to its steady state value at a rate determined by its time constant, τ. A similar analysis would show that the response of the capacitor current is given by

$$i(t) = [I - v(0^-)/R]e^{-t/RC}, \qquad t \geq 0 \tag{107.19}$$

Capacitors behave like short circuits to sudden changes in current. The initial capacitor voltage determines the initial resistor current by Ohm's law: $v(0^+)/R$. Any initial current supplied by the source in excess of this amount will pass through the capacitor as though it were a short circuit. As the capacitor builds voltage, the resistor draws an increasing amount of current and ultimately conducts all of the current supplied by the constant source. In steady state the capacitor looks like an open circuit to the constant source — $i(\infty) = 0$ — and it conducts no current.

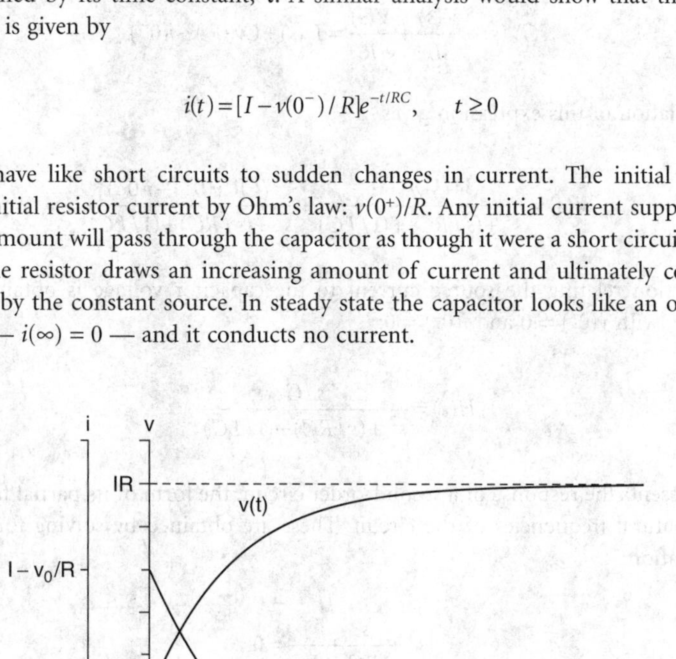

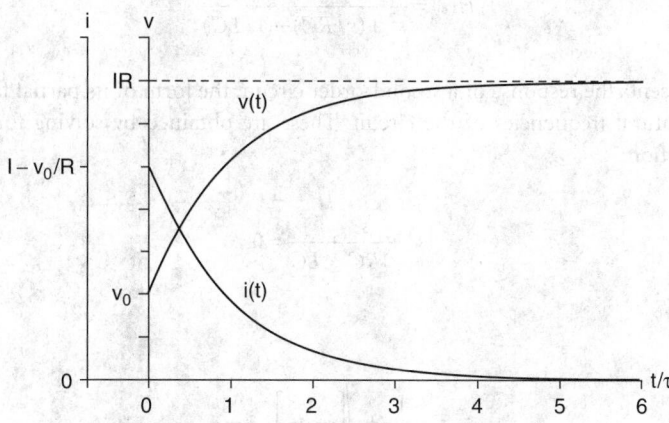

FIGURE 107.4 Step response of an initially relaxed RC circuit.

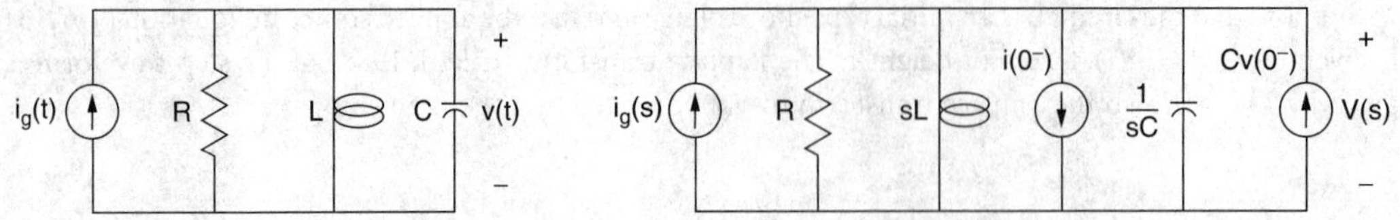

FIGURE 107.5 Parallel RLC circuit.

107.3 RLC Circuits

Circuits that contain inductors and capacitors exhibit dynamical effects that combine the inductor's inertia to sudden changes in current with the capacitor's inertia to sudden changes in voltage. The topological arrangement of the components in a given circuit determines the behavior that results from the interaction of the currents and voltages associated with the individual circuit elements.

The parallel RLC circuit in Figure 107.5(a) has an *s*-domain counterpart [see Figure 107.5(b)], sometimes referred to as a *transformed circuit,* that is obtained by replacing each time-domain variable by its Laplace transform, and each physical component by a Laplace transform model of the component's voltage-current relationship. Here, for example, the physical capacitor is replaced by a model that accounts for the impedance of the capacitor and the capacitor's initial voltage. The additional sources account for the possibly nonzero initial values of capacitor voltage and inductor current. Algebraic expressions of Kirchhoff's laws are written from the Laplace model of the circuit. Applying Kirchhoff's current law to the circuit of Figure 107.5 gives

$$sCV(s)+\frac{V(s)}{sL}+\frac{V(s)}{R}=I_g(s)+Cv(0^-)-i(0^-) \tag{107.20}$$

Algebraic manipulation of this expression gives

$$V(s)=\frac{(s/C)I_g(s)}{s^2+(s/RC)+(1/LC)}+\frac{s/C[Cv(0^-)-i(0^-)]}{s^2+(s/RC)+(1/LC)} \tag{107.21}$$

The transfer function relating the source current to the capacitor voltage is obtained directly from Equation (107.21), with $v(0^-)=0$ and $i(0^-)=0$:

$$H(s)=\frac{s/C}{s^2+(s/RC)+(1/LC)} \tag{107.22}$$

Because $V(s)$ represents the response of a second-order circuit, the form of its partial fraction expansion depends on the natural frequencies of the circuit. These are obtained by solving for the roots of the characteristic equation:

$$s^2+\frac{s}{RC}+\frac{1}{LC}=0 \tag{107.23}$$

The roots are

$$s_1=\frac{-1}{2RC}+\sqrt{\left[\frac{1}{2RC}\right]^2-\frac{1}{LC}} \tag{107.24}$$

$$s_2 = \frac{-1}{2RC} - \sqrt{\left[\frac{1}{2RC}\right]^2 - \frac{1}{LC}} \tag{107.25}$$

To illustrate the possibilities of the second-order response, we let $i_g(t) = u(t)$, and $I_g(s) = 1/s$.

Case 1: Overdamped Circuit

If both roots of the second-order characteristic equation are real valued the circuit is said to be *overdamped.* The physical significance of this term is that the circuit response to a step input exhibits exponential decay and does not oscillate. The form of the step response of the circuit's capacitor voltage is

$$v(t) = K_1 e^{s_1 t} + K_2 e^{s_2 t} \tag{107.26}$$

and K_1 and K_2 are chosen to satisfy the initial conditions imposed by $v(0^-)$ and $i(0^-)$.

Case 2: Critically Damped Circuit

When the two roots of a second-order characteristic equation are identical, the circuit is said to be *critically damped.* The circuit in this example is critically damped when $s_1 = s_2 = -1/(2RC)$. In this case, Equation (107.21) becomes

$$V(s) = \frac{I/C}{[s+(1/2RC)]^2} + \frac{s}{C}\frac{[Cv(0^-) - i(0^-)]}{[s+(1/2RC)]^2} \tag{107.27}$$

The partial fraction expansion of this expression is

$$V(s) = \frac{1/C}{[s+(1/2RC)]^2} + \frac{1}{C}\frac{[Cv(0^-) - i(0^-)]}{s+(1/2RC)} - \frac{1}{2RC^2}\frac{[Cv(0^-) - i(0^-)]}{[s+(1/2RC)]^2} \tag{107.28}$$

Taking the inverse Laplace transform of $V(s)$ gives

$$v(t) = \frac{1}{C} t e^{-t/(2RC)} + \frac{1}{C}[Cv(0^-) - i(0^-)]e^{-t/(2RC)} - \frac{1}{2RC^2}[Cv(0^-) - i(0^-)]te^{-t/(2RC)} \tag{107.29}$$

The behavior of the circuit in this case is called *critically damped* because a reduction in the amount of damping in the circuit would cause the circuit response to oscillate.

Case 3: Underdamped Circuit

The component values in this case are such that the roots of the characteristic equation are a complex conjugate pair of numbers. This leads to a response that is oscillatory, having a damped frequency of oscillation, ω_d, given by

$$\omega_d = \sqrt{\frac{1}{LC} + \left[\frac{1}{2RC}\right]^2} \tag{107.30}$$

and a damping factor, α, given by

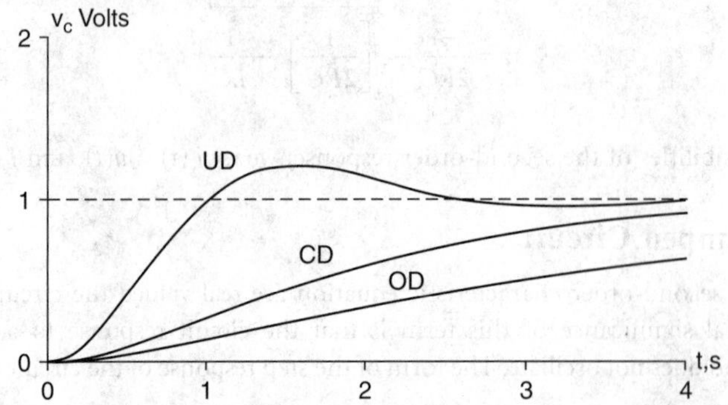

FIGURE 107.6 Overdamped, critically damped, and underdamped responses of a parallel RLC circuit with a step input.

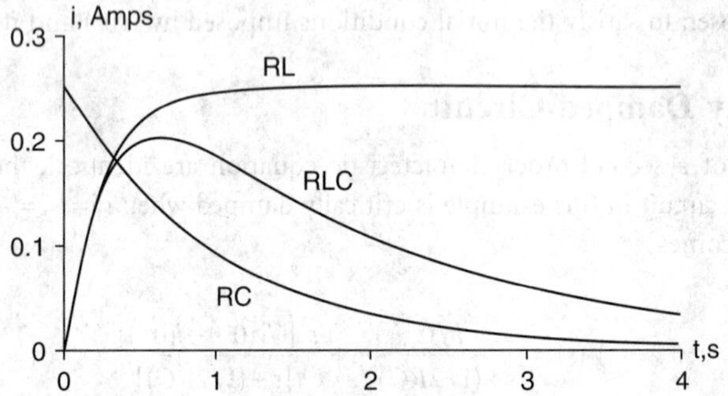

FIGURE 107.7 A comparison of current in series RL, RC, and RLC circuits.

$$\alpha = \frac{1}{2RC} \tag{107.31}$$

The form of the response of the capacitor voltage to a unit step input current source is

$$v(t) = 2|K|e^{-\alpha t}\sin(\omega_d t + \varphi) \tag{107.32}$$

where the parameters K (a nonnegative constant) and φ are chosen to satisfy initial conditions.

A comparison of overdamped, critically damped, and underdamped responses of the capacitor voltage is given in Figure 107.6 for a unit step input, with the circuit initially relaxed. Additional insight into the behavior of RL, RC, and RLC circuits can be obtained by examining the response of the current in the series RL, RC, and RLC circuits for corresponding component values. For example, Figure 107.7 shows the waveforms of the step response of $i(t)$ for the following component values: $R = 4\ \Omega$, $L = 1$ H, and $C = 1$ F. The RL circuit has a time constant of 0.25 sec, the RC circuit has a time constant of 2.00 sec, and the overdamped RLC circuit has time constants of 0.29 and 1.69 sec. The inductor in the RL circuit blocks the initial flow of current, the capacitor in the RC circuit blocks the steady-state flow of current, and the RLC circuit exhibits a combination of both behaviors. Note that the time constants of the RLC circuit are bounded by those of the RL and RC circuits.

RLC Circuit — Frequency Response

A circuit's transfer function defines a relationship between the frequency-domain (spectral) characteristics of any input signal and the frequency-domain characteristics of its corresponding output signal. In many engineering applications, circuits are used to shape the spectral characteristics of a signal. For

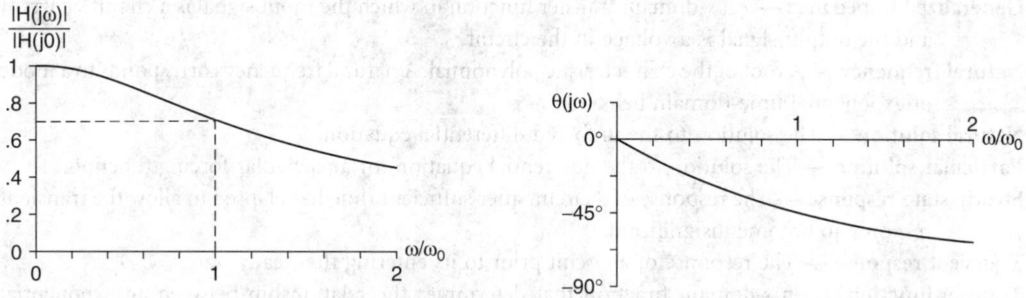

FIGURE 107.8 Magnitude and phase response of a parallel RC circuit.

example, in a communications application the frequency response could be chosen to eliminate noise from a signal. The frequency response of a circuit consists of the graphs of $|H(j\omega)|$ and $\theta(j\omega)$, the magnitude response and the phase response, respectively, of the transfer function of a given circuit voltage or current. The term $|H(j\omega)|$ determines the ratio of the amplitude of the *sinusoidal steady-state response* of the circuit to a sinusoidal input, and $\theta(j\omega)$ determines the phase shift (manifest as a time-axis translation) between the input and the output waveforms. Values of $|H(j\omega)|$ and $\theta(j\omega)$ are obtained by taking the magnitude and angle, respectively, of the complex value $H(j\omega)$.

The magnitude and phase responses play an important role in filter theory, where for distortionless transmission (i.e., the filter output waveform is a scaled and delayed copy of the input waveform) it is necessary that $|H(j\omega)| = K$, a constant (flat response), and $\theta(j\omega)$ be linear over the pass band of a signal that is to be passed through a filter.

In the parallel RC circuit of Figure 107.3, the transfer function relating the capacitor voltage to the current source has

$$H(j\omega) = \frac{R}{1 + j\omega RC} \tag{107.33}$$

The arithmetic of complex numbers gives the following:

$$|H(j\omega)| = \frac{R}{\sqrt{1 + \omega^2 R^2 C^2}} \tag{107.34}$$

$$\theta(j\omega) = \angle H(j\omega) = -\tan^{-1}(\omega RC) \tag{107.35}$$

The graphs of $|H(j\omega)|$ and $\theta(j\omega)$ are shown in Figure 107.8. The capacitor voltage in the parallel RC circuit has a "low pass" filter response, meaning that it will pass low frequency sinusoidal signals without significant attenuation provided that $\omega < \omega_0[\omega_0 = 1/RC)]$, the cutoff frequency of the filter. The approximate linearity of the phase response within the passband is also evident.

A filter's component values determine its cutoff frequency. An important design problem is to determine the values of the components so that a specified cutoff frequency is realized by the circuit. Here, increasing the size of the capacitor lowers the cutoff frequency, or, alternatively, reduces the bandwidth of the filter. Other typical filter responses that can be formed by RL, RC, and RLC circuits (with and without op amps) are high-pass, band-pass, and band-stop filters.

Defining Terms

Characteristic equation — The equation obtained by setting the denominator polynomial of a transfer function equal to zero. The equation defines the natural frequencies of a circuit.

Generalized impedance — An *s*-domain transfer function in which the input signal is a circuit's current and the output signal is a voltage in the circuit.

Natural frequency — A root of the characteristic polynomial. A natural frequency corresponds to a mode of exponential time-domain behavior.

Natural solution — The solution to the unforced differential equation.

Particular solution — The solution to the differential equation for a particular forcing function.

Steady-state response — The response of a circuit after sufficient time has elapsed to allow the transient response to become insignificant.

Transient response — The response of a circuit prior to its entering the steady state.

Transfer function — An *s*-domain function that determines the relationship between an exponential forcing function and the particular solution of a circuit's differential equation model. It also describes a relationship between the *s*-domain spectral description of a circuit's input signal and the spectral description of its output signal.

References

Ciletti, M. D., 1988. *Introduction to Circuit Analysis and Design*, Holt, Rinehart and Winston, New York.

Ziemer, R. E., Tranter, W. H., and Fannin, D. R. 1993. *Signals and Systems: Continuous and Discrete*, Macmillan, New York.

Further Information

For further information on the basic concepts of RL, RC, and RLC circuits, see *Circuits, Devices, and Systems* by R. J. Smith and R. C. Dorf. For a treatment of convolution methods, Fourier transforms, and Laplace transforms, see *Signals and Systems: Continuous and Discrete* by R. E. Ziemer et al. For a treatment of RL, RC, and RLC circuits with op amps, see *Introduction to Circuit Analysis and Design* by Ciletti.

108

Node Equations and Mesh Equations

James A. Svoboda
Clarkson University

Node equations and mesh equations are sets of simultaneous equations that are used to analyze electric circuits. Engineers have been writing and solving node equations and mesh equations for a long time. For example, procedures for formulating node equations and mesh equations are found in textbooks from the late 1950s and 1960s [Seshu and Balabanian, 1959; Seshu and Reed, 1961; Desoer and Kuh, 1969]. This longevity is likely due to two facts. First, node equations and mesh equations are easy to write. Second, node equations and mesh equations are both relatively small sets of simultaneous equations.

Electric circuits are interconnections of electrical devices, for example, resistors, independent and dependent sources, capacitors, inductors, op amps, and so on. Circuit behavior depends both on how the devices work and on how the devices are connected together. **Constitutive equations** describe how the devices in the circuit work. Ohm's law is an example of a constitutive equation. The Kirchhoff's law equations describe how the devices are connected together to form the circuit. Both the node equations and the mesh equations efficiently organize the information provided by the constitutive equations and the Kirchhoff's law equations.

In order to better appreciate the advantages of using node and mesh equations, an example will be performed without them. The constitutive equations and Kirchhoff's law equations comprise a set of simultaneous equations. In this first example, illustrated in Figure 108.1, these equations are used to analyze an electric circuit. This example also illustrates the use of MathCAD to solve simultaneous equations. The availability of such a convenient method of solving equations influences the way that node equations and mesh equations are formulated.

The circuit shown in Figure 108.1 consists of six devices. Two variables — a branch current and a voltage — are associated with each branch. There are 12 variables associated with this small circuit. The constitutive equations describe each of the six devices:

$$\begin{aligned} &v_1 = 3i_1, \quad v_2 = 4i_2, \quad v_3 = 5i_3, \\ &v_4 = 12, \quad i_5 = 2i_2, \quad v_6 = 5v_1 \end{aligned} \tag{108.1}$$

Kirchhoff's laws provide six more equations:

$$\begin{aligned} &i_1 + i_2 + i_4 = 0, \quad -i_2 + i_3 + i_5 = 0, \quad -i_1 - i_3 + i_6 = 0, \\ &v_1 - v_3 - v_2 = 0, \quad v_2 + v_5 - v_4 = 0, \quad v_3 + v_6 - v_5 = 0 \end{aligned} \tag{108.2}$$

0-8493-1586-7/05/$0.00+$1.50

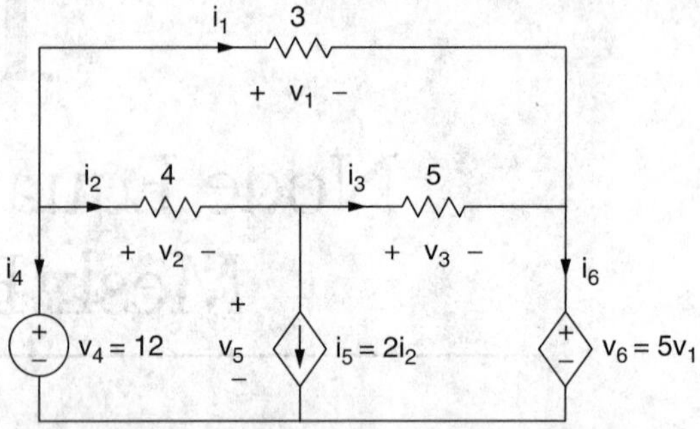

↓branch.MCD↓ 0 79 auto

Solving a Circuit using Branch Equations

Guess the values of the branch voltages and currents. Any guess will do. A good guess is not required. In this example, V4 is known to be 12. Zero is a suitable guess for the other branch voltages and currents.

I1 : = 0	V1 : = 0	I2 : = 0	V2 : = 0
I3 : = 0	V3 : = 0	I4 : = 0	V4 : = 12
I5 : = 0	V5 : = 0	I6 : = 0	V6 : = 0

Enter the equations describing the branches of the network. The word "Given" marks the beginning of these equations.

Given

Enter the Kirchhoff's Current Law (KCL) equations. (Use <ALT> = for the wiggly equal signs.)

I1 + I2 + 14 ≈ 0   −I2 + I3 + 15 ≈ 0   −I1 − I3 + 16 ≈ 0

Enter the Kirchhoff's Voltage Law (KVL) equations.

V1 − V3 − V2 ≈ 0   V2 + V5 − V4 ≈ 0   V3 + V6 − V5 ≈ 0

Enter the Constitutive Equations.

V1 ≈ 3 · I1	V2 ≈ 4 · I2	V3 ≈ 5 · I3	V4 ≈ I2
I5 ≈ 2 · I2	V6 ≈ 5 · V1		

Ask MathCAD to solve these equations. The word "Find" marks the end of the equations.

$$\text{Find (I1, I2, I3, I4, I5, I6, V1, V2, V3, V4, V5, V6)} = \begin{bmatrix} 0.667 \\ -2 \\ 2 \\ 1.333 \\ -4 \\ 2.667 \\ 2 \\ -8 \\ 10 \\ 12 \\ 20 \\ 10 \end{bmatrix}$$

FIGURE 108.1 Example illustrating the use of MathCAD to solve simultaneous equations.

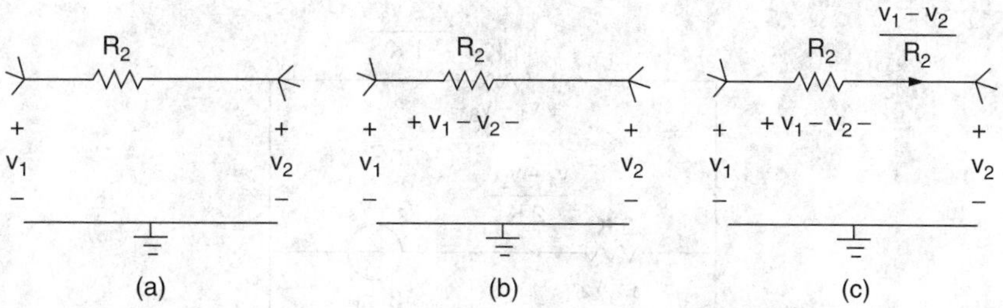

FIGURE 108.2 Expressing branch voltages and currents as functions of node voltages.

Although it is not practical to solve 12 equations in 12 unknowns by hand, these equations can easily be solved using a personal computer with appropriate software. Figure 108.1 includes a screen dump that illustrates the use of MathCAD [Wieder, 1992] to solve these equations.

Writing all of these equations is tediou,s and the process becomes more tedious as the size of the circuit increases. It would be convenient to use a smaller set of simultaneous equations, hence the interest in node equations and mesh equations. The size of the set of simultaneous equations is very important when the equations must be solved by hand. Most contemporary treatments of node equations and mesh equations [Dorf, 1993; Irwin, 1993; Nilsson, 1993] describe procedures that result in as small a set of equations as possible. The availability of the personal computer with appropriate software reduces, but does not eliminate, the importance of the size of the set of equations.

108.1 Node Equations

This section describes a procedure for obtaining node equations to represent a connected circuit. This procedure is based on the observation that an independent set of equations can be obtained by applying Kirchhoff's current law (KCL) at all of the nodes of the circuit except for one node [Seshu and Balabanian, 1959; Seshu and Reed, 1961; Desoer and Kuh, 1969]. The node at which KCL is not applied is called the **reference node.**

The voltage at any node, with respect to the reference node, is called the **node voltage** at that node. Figure 108.2 shows that the current in a resistor can be expressed in terms of the voltages at the nodes of the resistor. This is accomplished in three steps. First, in Figure 108.2(a), the node voltages corresponding to the nodes of the resistor are labeled. Next, in Figure 108.2(b), the branch voltage is expressed in terms of the node voltages (notice the polarities of the voltages). Finally, in Figure 108.2(c), Ohm's law is used to express the branch current in terms of the node voltages (notice the polarities of the branch voltage and current).

Consider writing simultaneous equations to represent the circuit shown in Figure 108.3. When the reference node is selected as shown in Figure 108.3, the node voltages are v_1, v_2, and v_3. The resistor currents have been expressed in terms of the node voltages using the technique described in Figure 108.2. Since all of the branch currents have been labeled, a set of simultaneous equations can now be obtained by applying KCL at all of the nodes except for the reference node.

$$\begin{aligned} .001 &= \frac{v_1 - v_2}{2000} + \frac{v_1 - v_3}{1000} \\ .002 &= .003 + \frac{v_1 - v_2}{2000} \\ \frac{v_1 - v_3}{1000} &= .003 + \frac{v_3}{3000} \end{aligned} \tag{108.3}$$

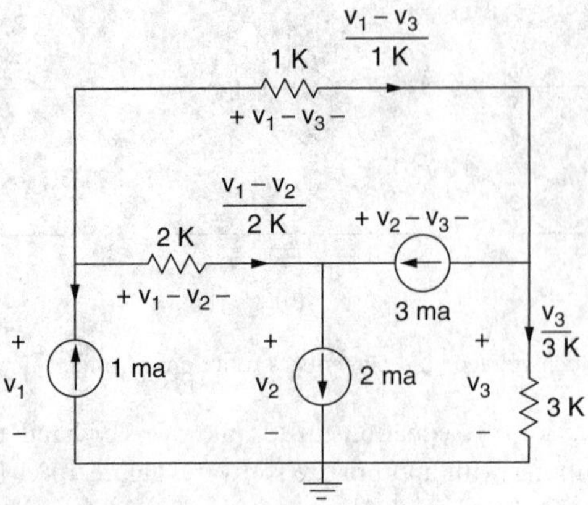

FIGURE 108.3 A circuit consisting of resistors and current sources.

The independent variables in these equations are node voltages, and so this set of simultaneous equations is called the *node equations.* Representing the branch voltages and branch currents as functions of the node voltages made it possible to represent this circuit by a smaller set of variables, the node voltages.

Consider again the circuit shown in Figure 108.1. This circuit contains two voltage sources (one independent and one dependent). In general, there is no easy way to express the current in a voltage source as a function of the node voltages. Since the size of the set of simultaneous equations is not of critical importance, it is appropriate to add the currents of the voltage sources to the list of independent variables. The circuit has been redrawn in Figure 108.4. A reference node has been selected and labeled. The independent variables, that is, the node voltages and currents in the voltage sources, have been labeled.

The voltage across each voltage source can be expressed in two ways. The source voltage is given by the constitutive equation. The branch voltage is the difference of the node voltages at the nodes of the voltage source. These expressions must be equivalent so the branch voltage can be equated to the source voltage. In Figure 108.4 this means that

$$v_4 = 12 \text{ and } v_6 = 5(v_4 - v_6) \tag{108.4}$$

The circuit in Figure 108.4 contains two dependent sources. Dependent sources will not be a problem if the controlling variables of the dependent sources are first expressed as functions of the independent variables of the circuit. In Figure 108.4 this means that i_2 and v_1 must be expressed as functions of v_4, v_5, v_6, i_4, and i_6.

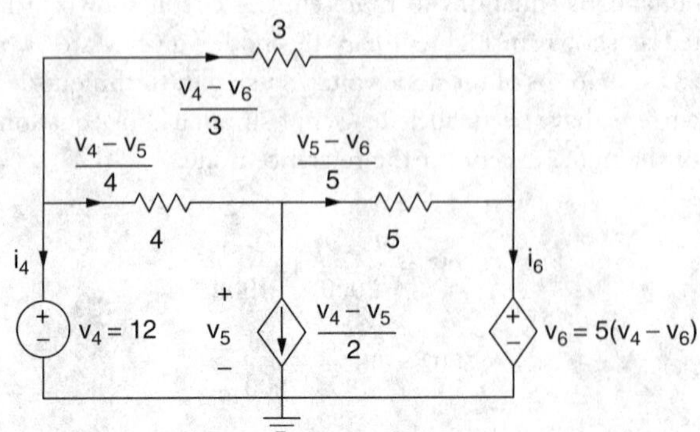

FIGURE 108.4 Writing node equations for the circuit from Figure 108.1.

$$i_2 = \frac{v_4 - v_5}{4} \quad \text{and} \quad v_1 = v_4 - v_6 \tag{108.5}$$

Next, the controlled variables of the dependent source can be expressed as functions of the independent variables of the circuit:

$$i_5 = 2\left(\frac{v_4 - v_5}{4}\right) = \frac{v_4 - v_5}{2} \quad \text{and} \quad v_6 = 5(v_4 - v_6) \tag{108.6}$$

Now all of the branch currents have been labeled. Apply KCL at all of the nodes except the reference node to get

$$\begin{aligned} \frac{v_4 - v_6}{3} + \frac{v_4 - v_5}{4} + i_4 &= 0 \\ -\frac{v_4 - v_5}{4} + \frac{v_4 - v_5}{2} + \frac{v_5 - v_6}{5} &= 0 \\ -\frac{v_5 - v_6}{5} - \frac{v_4 - v_6}{3} + i_6 &= 0 \end{aligned} \tag{108.7}$$

Equation (108.4) and Equation (108.7) comprise the node equations representing this circuit.

In summary, the following procedure is used to write node equations:

1. Choose and label the reference node. Label the independent variables: node voltages and voltage source currents.
2. Express branch currents as functions of the independent variables and the input variables.
3. Express the branch voltage of each voltage source as the difference of the node voltages at its nodes. Equate the branch voltage to the voltage source voltage.
4. Apply KCL at each node except for the reference node.

108.2 Mesh Equations

Mesh equations are formulated by applying Kirchhoff's voltage law (KVL) to each **mesh** of a circuit. A **mesh current** is associated with each mesh. Figure 108.5 shows how to express the branch voltage and branch current of a resistor as functions of the mesh currents.

The procedure for formulation of mesh equations is analogous to the procedure for formulating node equations:

1. Label the independent variables: mesh currents and current source voltages.
2. Express branch voltages as functions of the independent variables and the input variables.

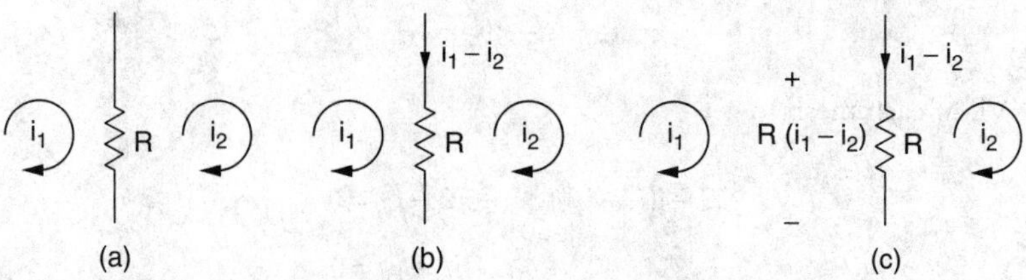

FIGURE 108.5 Expressing branch voltages and currents as functions of mesh currents.

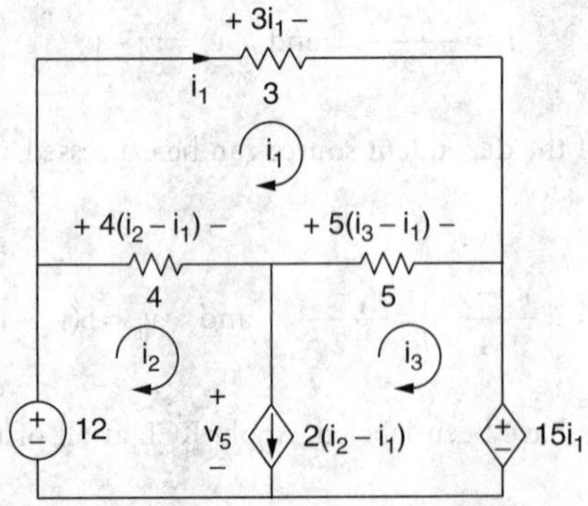

↓mesh_eqn.MCD↓ 23 77

Solving a Circuit using Mesh Equations

Guess the values of the mesh currents and current source voltage. Any guess will do. A good guess is not required.

i1 := 0 i2 := 0 i3 := 0 v5 := 0

Given

branch current = i2 − i3 ≈ 2 · (i2 − i1) = current source current

Apply KVL to each mesh.

−4 · (i2 − i1) + 3 · i1 − 5 · (i3 − i1) ≈ 0 −12 + 4 · (i2 − i1) + v5 ≈ 0

−v5 + 5 · (i3 − i1) + 15 · i1 ≈ 0

$$\text{Find (i1, i2, i3, v5)} = \begin{bmatrix} 0.667 \\ -1.333 \\ 2.667 \\ 20 \end{bmatrix}$$

FIGURE 108.6 Writing mesh equations for the circuit from Figure 108.1.

3. Express the branch current of each current source as the difference of the mesh currents of its meshes. Equate the branch current to the current source current.
4. Apply KVL to each mesh.

As an example, consider again the circuit shown in Figure 108.1. This circuit is redrawn in Figure 108.6 to show the mesh currents i_1, i_2, and i_3. The branch voltage of the current source, v_5, is also labeled, because it will be added to the list of independent variables. In Figure 108.6, the branch voltages have been expressed as functions of the independent variables i_1, i_2, i_3, and v_5.

The current in the dependent current source can be expressed in two ways. This source current is given by the constitutive equation. This branch current is also the difference of two mesh currents. These expressions must be equal so

$$i_2 - i_3 = 2(i_2 - i_1) \tag{108.8}$$

Next, apply KVL to each mesh to get

$$\begin{aligned} -4(i_2 - i_1) + 3i_1 - 5(i_3 - i_1) &= 0 \\ -12 + 4(i_2 - i_1) + v_5 &= 0 \\ -v_5 + 5(i_3 - i_1) + 15i_1 &= 0 \end{aligned} \tag{108.9}$$

Equation (108.8) and Equation (108.9) comprise the mesh equations. Figure 108.6 illustrates the use of MathCAD to solve the circuit using mesh equations.

Defining Terms

Constitutive equations — Equations that describe the relationship between the branch current and branch voltage of an electric device. Ohm's law is an example of a constitutive equation.

Mesh — A loop that does not contain any branch in its interior. Only planar circuits have meshes. Redrawing a planar circuit can change the meshes [Seshu and Reed, 1961; Desoer and Kuh, 1969].

Mesh current — A current associated with a mesh. This current circulates around the mesh.

Node voltage — The voltage at any node, with respect to the reference node.

Reference node — An independent set of equations can be obtained by applying KCL at all of the nodes of the circuit except for one node. The node at which KCL is not applied is called the *reference node.* Any node of the circuit can be selected to be the reference node. In electronic circuits, where one node of the network is the ground node of the power supplies, the reference node is almost always selected to be the ground node.

References

Desoer, C. A. and Kuh, E. S. 1969. *Basic Circuit Theory,* McGraw-Hill, New York.
Dorf, R. C. 1993. *Introduction to Electric Circuits,* John Wiley & Sons, New York.
Irwin, J. D. 1993. *Basic Engineering Circuit Analysis,* Macmillan, New York.
Nilsson, J. W. 1993. *Electric Circuits,* Addison-Wesley, Reading, MA.
Seshu, S. and Balabanian, N. 1959. *Linear Network Analysis,* John Wiley & Sons, New York.
Seshu, S. and Reed, M. B. 1961. *Linear Graphs and Electric Networks,* Addison-Wesley, Reading, MA.
Wieder, S. 1992. *Introduction to MathCAD for Scientists and Engineers,* McGraw-Hill, New York.

Further Information

Computer Methods for Circuit Analysis and Design by Jiri Vlach and Kishore Singhal describes procedures for formulating circuit equations that are well suited to computer-aided analysis.

109

Sinusoidal Excitation and Phasors

Muhammad H. Rashid
Purdue University

A sinusoidal forcing function known as a **sinusoid** is one of the most important excitations. In electrical engineering, the carrier signals for communications are sinusoids, and the sinusoid is also the dominant signal in the power industry. Sinusoids abound in nature, as, for example, in the motion of a pendulum, in the bouncing of a ball, and in the vibrations of strings and membranes.

Because of the importance of sinusoids, the output response of a circuit due to a sinusoidal input signal is an important criterion in determining the performance of the circuit. In this section, we will define a sinusoidal function, represent it in a phasor form, and then illustrate the techniques for determining the sinusoidal responses.

109.1 Sinusoidal Source

A sinusoidal source (independent or dependent) produces a signal that varies sinusoidally with time. In electrical engineering, it is normally a voltage or a current. A sinusoid can be expressed as a sine function or a cosine function. There is no clear-cut choice for the use of either function. However, the sine function is more commonly used. Using a sine function, the instantaneous voltage, which is shown in Figure 109.1, can be represented by

$$v(t) = V_m \sin \omega t \tag{109.1}$$

where V_m is the maximum amplitude, and ω is the *angular* or *radian frequency* in rad/sec.

A sinusoid is a *periodic* function defined generally by the property

$$v(t + T) = v(t) \tag{109.2}$$

where T is the period, the time for one complete cycle. That is, the function goes through one complete cycle every T seconds and is then repeated. The period T is related to the angular frequency ω by

$$T = \frac{2\pi}{\omega} \tag{109.3}$$

0-8493-1586-7/05/$0.00+$1.50

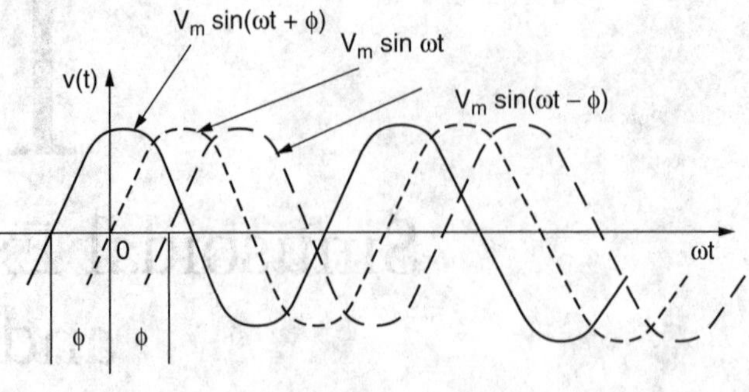

FIGURE 109.1 Three sinusoids with different phase angles.

In 1 sec, the function goes through $1/T$ cycles, or periods. The *frequency* is then

$$f = \frac{1}{T} = \frac{\omega}{2\pi} \tag{109.4}$$

The number of cycles per second, or *hertz* (abbreviated Hz), named for the German physicist Heinrich R. Hertz (1857–1894), is the standard unit of frequency. The frequency and the angular frequency are related by

$$\omega = 2\pi f \tag{109.5}$$

So far, we have assumed that a sinusoid starts at $t = 0$. However, it could be phase shifted by an angle ϕ and has the more general expression given by

$$v(t) = V_m \sin(\omega t + \phi) \tag{109.6}$$

where ϕ is the *phase angle*, or simply *phase.* Since ωt is in radians, ϕ should also be expressed in radians. However, it is often convenient to specify ϕ in degrees. That is,

$$v(t) = V_m \sin\left(\omega t + \frac{\pi}{3}\right) \tag{109.7}$$

or

$$v(t) = V_m \sin(\omega t + 60^\circ) \tag{109.8}$$

is acceptable, although Equation (109.8) is mathematically inconsistent. While computing the value of $\sin(\omega t + \phi)$, one should use the same units (radians or degrees) for both ωt and ϕ. If ϕ has a positive value, the sinusoid is said to have a *leading phase angle.* If ϕ has a negative value, it is said to have a *lagging phase angle.*

109.2 Phasor

A **phasor** is a complex number that represents the amplitude and phase angle of a sinusoidal function. It transforms a sinusoidal function from time domain to the complex number domain. The phasor concept, which is generally credited to electrical engineer Charles Proteus Steinmetz (1865–1923), is based on the Euler's exponential function of the trigonometric function

$$e^{j\theta} = \cos\theta + j\sin\theta \tag{109.9}$$

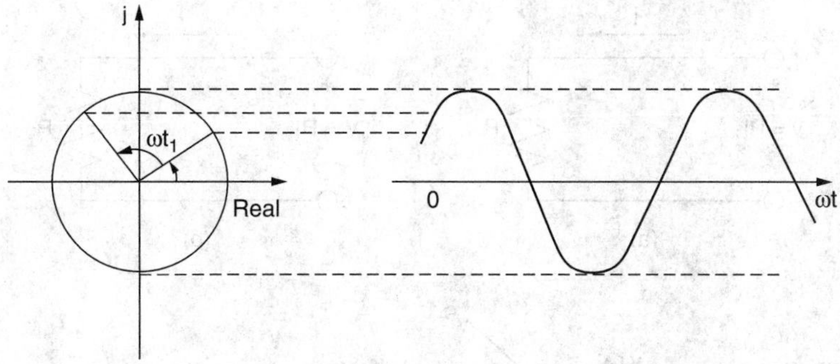

FIGURE 109.2 Rotating phasor.

Thus, a cosine function may be regarded as the real part of the exponential function and a sine function as the imaginary part of the exponential function. That is,

$$\cos\theta = \mathrm{Re}(e^{j\theta}) \tag{109.10}$$

$$\sin\theta = \mathrm{Im}(e^{j\theta}) \tag{109.11}$$

Therefore, we can write the sinusoidal voltage function of Equation (109.6) as

$$v(t) = V_m \sin(\omega t + \phi) \tag{109.12}$$

$$= V_m \,\mathrm{Im}\{e^{j(\omega t+\phi)}\} = \mathrm{Im}\{(V_m e^{j\phi})e^{j\omega t}\} \tag{109.13}$$

which indicates that the coefficient of the exponential $e^{j\omega t}$ is a complex number. It is the *phasor representation*, or *phasor transform*, of the sinusoidal function and is represented by a boldface letter. Thus, the phasor **V** becomes

$$\mathbf{V} = V_m e^{j\phi} \text{ (exponential form)} \tag{109.14}$$

$$= V_m \cos\theta + jV_m \sin\theta \text{ (polar form)} \tag{109.15}$$

Since phasors are used extensively in analysis of electrical engineering circuits, the exponential function $e^{j\phi}$ is abbreviated in a shorthand notation for the sake of simplicity as

$$e^{j\phi} = 1\angle\phi \tag{109.16}$$

A graphical relationship between a phasor and sinusoid is shown in Figure 109.2. A unit phasor may be regarded as a unit vector having an initial phase displacement of ϕ and rotating in the counterclockwise direction at an angular speed of ω.

109.3 Passive Circuit Elements in Phasors Domain

If we want to find the response of an electrical circuit due to a sinusoidal forcing function, first we need to establish the relationship between the phasor voltage across and the phasor current through passive elements such as the resistor, inductor, and capacitor.

Resistor

Let us assume that a sinusoidal current of

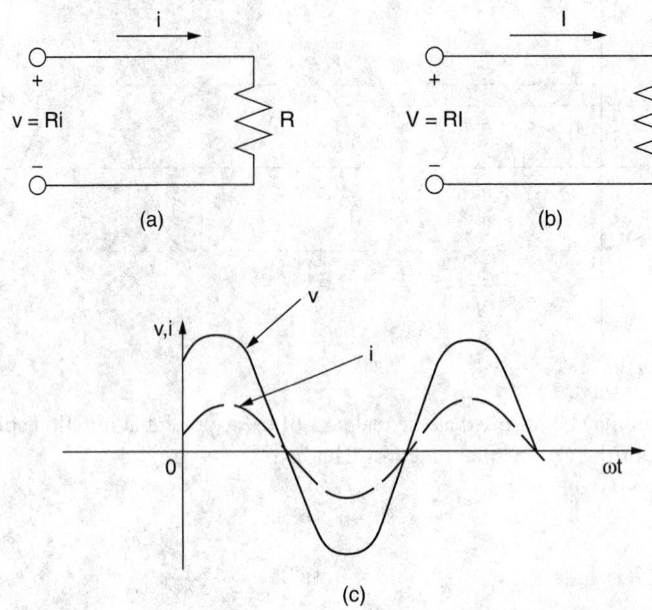

FIGURE 109.3 Voltage-current relationships of a resistor.

$$i(t) = I_m \sin(\omega t + \theta) \tag{109.17}$$

where I_m is the maximum amplitude of the current in amperes, and θ is the phase angle of the current.

Using Ohm's law for the resistor in Figure 109.3(a), the instantaneous voltage across the terminals of the resistor is related to its instantaneous current by

$$v(t) = Ri(t) = R\{I_m \sin(\omega t + \theta)\} = RI_m \sin(\omega t + \theta) \tag{109.18}$$

which can be represented as a phasor by

$$\mathbf{V} = RI_m e^{j\theta} = RI_m \angle\theta \tag{109.19}$$

Since $I_m\angle\theta = \mathbf{I}$ is the phasor representation of the current, we can write

$$\mathbf{V} = R\mathbf{I} \tag{109.20}$$

That is, the phasor voltage across the terminals of a resistor is the resistor times the phasor current. Figure 109.3(b) shows the **V** and **I** relationship of a resistor. A resistor has no phase shift between the voltage and its current. The current is said to be in time phase with the voltage, as shown in Figure 109.3(c).

Inductor

Assuming a sinusoidal current of $i = I_m \sin(\omega t + \theta)$, the voltage across an inductor shown in Figure 109.4(a) is related to its current by

$$v = L\frac{di}{dt} = \omega L I_m \cos(\omega t + \theta) \tag{109.21}$$

Using the trigonometric identity of $\cos A = \sin(A + \pi/2)$, Equation (109.21) can be rewritten as

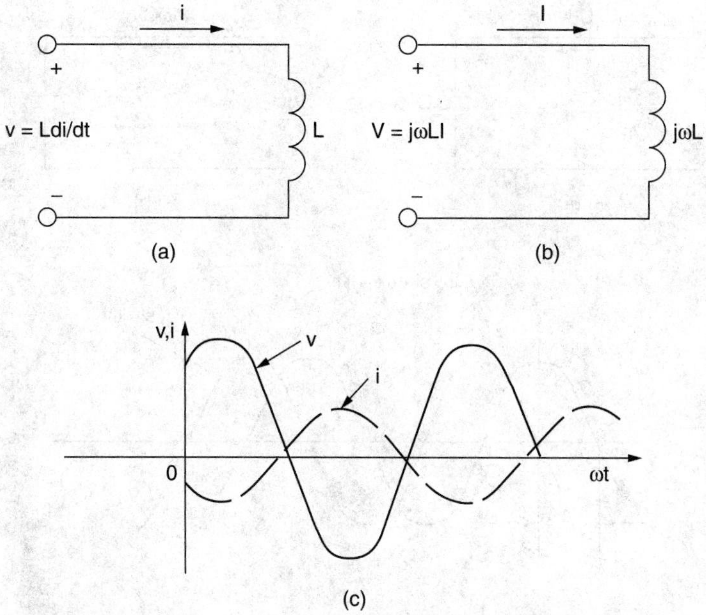

FIGURE 109.4 Voltage-current relationships of an inductor.

$$v = \omega L I_m \sin\left(\omega t + \theta + \frac{\pi}{2}\right) \tag{109.22}$$

The phasor representation of the voltage given by Equation (109.22) is

$$\begin{aligned} \mathbf{V} &= \omega L I_m e^{j(\theta+\pi/2)} = \omega L I_m \angle(\theta + \pi/2) \\ &= \omega L I_m e^{j\theta} e^{j\pi/2} \\ &= \omega L I_m e^{j\theta}(\cos \pi/2 + j \sin \pi/2) \\ &= j\omega L I_m e^{j\theta} = j\omega L \mathbf{I} \end{aligned} \tag{109.23}$$

Thus, the phasor voltage across the terminals of an inductor equals $j\omega L$ times the phasor current. Since the operator j gives a phase shift of +90°, then $j\mathbf{I} = jI\angle\theta = I_m\angle(\theta + \pi/2)$. That is, the current lags behind the voltage by 90°, or the voltage leads the current by 90°. The voltage and current relationship in the phasor domain is shown in Figure 109.4(b) and in the time domain in Figure 109.4(c).

Capacitor

Assuming a sinusoidal voltage of $v = V_m \sin(\omega t + \phi)$, the current through a capacitor shown in Figure 109.5(a) is related to its voltage by

$$i = C\frac{dv}{dt} = j\omega C V_m \cos(\omega t + \phi) \tag{109.24}$$

Using the trigonometric identity, Equation (109.24) can be rewritten as

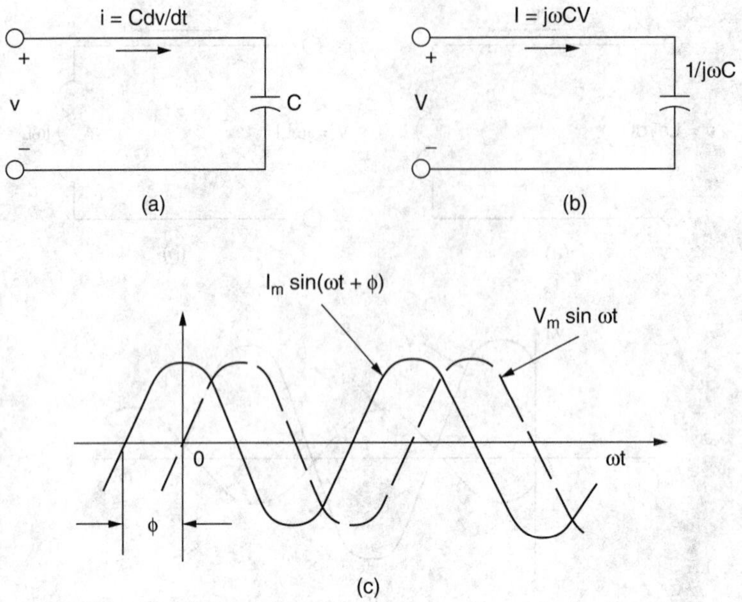

FIGURE 109.5 Voltage-current relationships of a capacitor.

$$i = \omega C V_m \sin\left(\omega t + \phi + \frac{\pi}{2}\right) \tag{109.25}$$

The phasor representation of the current given by Equation (109.25) is

$$\mathbf{I} = j\omega C\mathbf{V} \tag{109.26}$$

Thus, the phasor current through a capacitor equals $j\omega C$ times the phasor voltage. That is, the voltage lags behind the current by 90°, or the current leads the voltage by 90°. The voltage and current relationship in the phasor domain is shown in Figure 109.5(b) and in the time domain in Figure 109.5(c).

Sinusoidal Responses

We can conclude from the previous discussions that the phasor relationship between the voltage and the current of an element takes the general form of

$$\mathbf{V} = \mathbf{ZI} \tag{109.27}$$

where **Z** is the **impedance** of the circuit element. That is, the impedance of a resistor is R, the impedance of an inductor is $j\omega L$, and the impedance of a capacitor is $1/j\omega C$. Thus, for a circuit having L and/or C, the impedance **Z** will be a complex number with a real part R and an imaginary part X such that

$$\mathbf{Z} = R + jX = Z\angle\theta \tag{109.28}$$

where

$$Z = [R^2 + X^2]^{1/2} \tag{109.29}$$

and

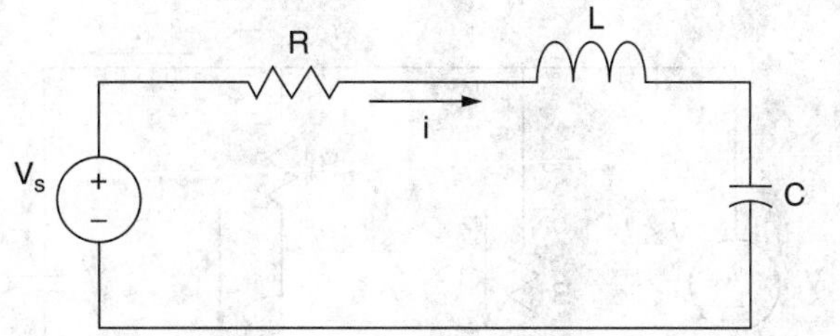

FIGURE 109.6 A series *RLC* circuit.

$$\theta = \tan^{-1}(X/R) \tag{109.30}$$

If the Kirchhoff's voltage law for a set of n sinusoidal voltages in the time domain is given by

$$v_1 + v_2 + v_3 + v_4 + \cdots + v_n = 0 \tag{109.31}$$

then the equivalent statement in the phasor domain can be written as

$$\mathbf{V}_1 + \mathbf{V}_2 + \mathbf{V}_3 + \mathbf{V}_4 + \cdots + \mathbf{V}_n = 0 \tag{109.32}$$

Similarly, if the Kirchhoff's current law for a set of n sinusoidal currents in the time domain is given by

$$i_1 + i_2 + i_3 + i_4 + \cdots + i_n = 0 \tag{109.33}$$

then the equivalent statement in the phasor domain can be written as

$$\mathbf{I}_1 + \mathbf{I}_2 + \mathbf{I}_3 + \mathbf{I}_4 + \cdots + \mathbf{I}_n = 0 \tag{109.34}$$

Let us consider the *RLC* circuit shown in Figure 109.6. We will use the phasor concept in finding its current in response to an input source voltage of $v_s = V_m \sin(\omega t + \phi)$. We use the phasor notation

$$\mathbf{V}_s = V_m \angle \phi \tag{109.35}$$

Applying the Kirchhoff's voltage law, we get

$$\mathbf{V}_s = (R + j\omega L + 1/j\omega C)\mathbf{I} = \mathbf{ZI} \tag{109.36}$$

where $\mathbf{Z}$ is the total impedance of the series loop formed by R, L, and C. That is,

$$\begin{aligned} \mathbf{Z} &= R + j(\omega L - 1/\omega C) = [R^2 + (\omega L - 1/\omega C)^2]^{1/2} \angle \theta \\ &= Z \angle \theta \end{aligned} \tag{109.37}$$

where Z is the impedance magnitude and is given by

$$Z = [R^2 + (\omega L - 1/\omega C)^2]^{1/2} \tag{109.38}$$

and θ is the impedance angle given by

$$\theta = \tan^{-1}\left(\frac{\omega L - 1/\omega C}{R}\right) \tag{109.39}$$

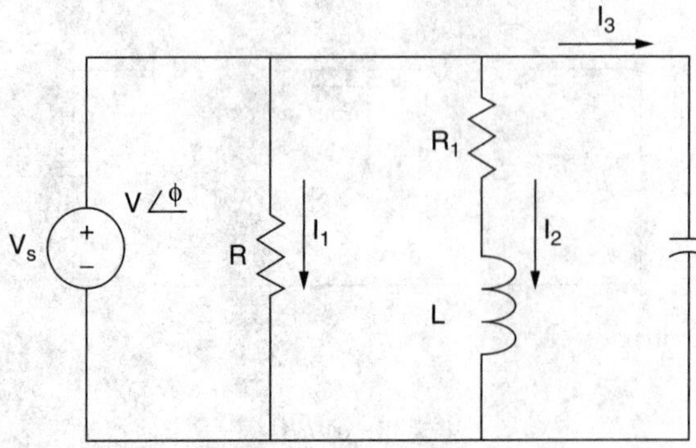

FIGURE 109.7 A parallel *RLC* circuit.

Dividing the phasor voltage by the phasor impedance gives the phasor current as

$$\mathbf{I} = \frac{\mathbf{V}_s}{\mathbf{Z}} = \frac{V_s\angle\phi}{Z\angle\theta} = \frac{V_s\angle\phi-\theta}{Z} \tag{109.40}$$

which indicates that the current lags the input voltage by an angle of θ. Converting to the time domain, the current is given by

$$i = \frac{V_m}{\sqrt{R^2+(\omega L-1/\omega C)^2}}\sin(\omega t+\phi-\theta) \tag{109.41}$$

Let us consider the parallel *RLC* circuit shown in Figure 109.7. Applying the Kirchhoff's current law at the node *a*, we get

$$\mathbf{I} = \mathbf{I}_1 + \mathbf{I}_2 + \mathbf{I}_3 \tag{109.42}$$

Since the three impedances are $\mathbf{Z}_R = R$, $\mathbf{Z}_1 = R_1 + j\omega L$, and $\mathbf{Z}_C = 1/j\omega C$, we can substitute for **I** by using Ohm's law. That is,

$$\begin{aligned}\mathbf{I} &= \frac{V_s\angle\phi}{R} + \frac{V_s\angle\phi}{R_1+j\omega L} + \frac{V_s\angle\phi}{1/j\omega C}\\ \mathbf{I} &= V_s\angle\phi\left[\frac{1}{R} + \frac{1}{\sqrt{R_1^2+(\omega L)^2}}\angle\theta_1 + j\omega C\right]\end{aligned} \tag{109.43}$$

where θ_1 is the impedance angle of R_1 and L, and it is given by

$$\theta_1 = \tan^{-1}\frac{\omega L}{R_1} \tag{109.44}$$

Equation (109.42) can be written as

$$\mathbf{I} = V_s\angle\phi Y\angle{-\theta} = V_s Y\angle(\phi-\theta) \tag{109.45}$$

where $Y\angle{-\theta}$ is the equivalent admittance of R, R_1, and L such that it is related to the equivalent impedance **Z** by

$$Y\angle -\theta = \frac{1}{Z\angle\theta} = \frac{1}{R} + \frac{1}{R_1 + j\omega L} + j(\omega C) \tag{109.46}$$

Equation (109.43) can be written in the time domain as

$$i(t) = V_m Y \sin(\omega t + \phi - \theta) \tag{109.47}$$

where θ is the impedance angle of $(R_1 + j\omega L)$, which is in parallel with R and $1/j\omega C$. In converting from the phasor domain to the time domain, we have assumed that the maximum amplitude of a sinusoid is the magnitude of its phasor representation. However, in practice, the magnitude of a sinusoid is normally quoted in its root-mean-square (rms) value, which is defined by

$$V_{\text{rms}} = \sqrt{\frac{1}{T}\int V_m^2 \cos^2(\omega t + \phi)dt} \tag{109.48}$$

which, after completing the integration and simplification, becomes

$$V_{\text{rms}} = \frac{V_m}{\sqrt{2}} \tag{109.49}$$

Thus, $v = V_m \sin(\omega t + \phi) = \sqrt{2}V_{\text{rms}} \sin(\omega t + \phi)$, and it will be represented in a phasor form by $V_{\text{rms}}\angle\phi$. For example, $170\sin(\omega t + \phi) \equiv (170/\sqrt{2})\angle\phi = 120\angle\phi$.

Defining Terms

Impedance — An impedance is a measure of the opposition to a current flow due to a sinusoidal voltage source. It is a complex number and has a real and an imaginary part.

Phasor — A phasor is the vector representation of a sinusoid in the complex domain. It represents the amplitude and the phase angle of a sinusoid.

Sinusoid — A sinusoid is a sinusoidal time-dependent periodic forcing function which has a maximum amplitude and can also have a phase delay.

References

Balabania, N. 1994. *Electric Circuits,* McGraw-Hill, New York.

Dorf, R. C. 1993. *Introduction to Electric Circuits,* John Wiley & Sons, New York.

Hayt, W. H., Jr., and Kemmerley, J. E. 1993. *Engineering Circuit Analysis,* McGraw-Hill, New York.

Irwin, D. J. 1989. *Basic Engineering Circuit Analysis,* Macmillan, New York.

Jackson, H. W. 1986. *Introduction to Electric Circuits,* Prentice Hall, Englewood Cliffs, NJ.

Johnson, D. E., Hilbert, J. L., and Johnson, J. R. 1989. *Basic Electric Circuit Analysis,* Prentice Hall, Englewood Cliffs, NJ.

Nilson, J. W. 1993. *Electric Circuits,* Addison-Wesley, Reading, MA.

Further Information

A general review of linear circuits is presented in *Linear Circuits* by M. E. Van Valkenburh, Prentice Hall (Chapter 9 through Chapter 11).

The applications of phasors in determining frequency responses are presented in *Basic Network Theory* by J. Vlach, Van Nostrand Reinhold (Chapter 8).

Many examples of computer-aided simulations by the Industry Standard Circuit Simulator SPICE are given in *SPICE for Circuits and Electronics Using PSpice* by M. H. Rashid, Prentice Hall.

The monthly journal *IEEE Transactions on Circuits and Systems — Fundamental Theory and Applications* reports advances in the techniques for analyzing electrical and electronic circuits. For subscription information contact IEEE Service Center, 445 Hoes Lane, P.O. Box 1331, Piscataway, NJ 08855-1331. Phone (800) 678-IEEE.

110

Three-Phase Circuits

Norman Balabanian
University of Florida, Gainesville

A very important use of electricity is the driving of industrial equipment, such as electric motors, in the AC steady state. Suppose that the instantaneous AC voltage and current of such a load is given by

$$\begin{aligned} v(t) &= \sqrt{2}|V|\cos(\omega t + \alpha) \\ i(t) &= \sqrt{2}|I|\cos(\omega t + \beta) \end{aligned} \tag{110.1}$$

Then the power to the load at any instant of time is

$$p(t) = |V||I|[\cos(\alpha - \beta) + \cos(2\omega t + \alpha + \beta)] \tag{110.2}$$

The instantaneous power has a constant term and a sinusoidal term at twice the frequency. The quantity in brackets fluctuates between a minimum value of $\cos(\alpha - \beta) - 1$ and a maximum value of $\cos(\alpha - \beta) + 1$. This fluctuation of power delivered to the load has a great disadvantage when the load is an electric motor. A motor operates by receiving electric power and transmitting mechanical (rotational) power at its shaft. If the electric power is delivered to the motor in spurts, the motor is likely to vibrate. For satisfactory operation in such a case, a physically larger motor, with a larger shaft and flywheel, will be needed to provide more inertia for smoothing out the fluctuations than would be the case if the delivered power were constant.

This problem is overcome in practice by the use of what is called a *three-phase* system. This chapter will provide a brief discussion of three-phase power systems.

Consider the circuit in Figure 110.1. This is an interconnection of three AC sources to three loads connected in such a way that each source/load combination shares the return connection from O to N. The three sources can be viewed collectively as a single source, and the three loads — which are assumed to be identical — can be viewed collectively as a single load. Each of the individual sources and loads is referred to as one *phase* of the three-phase system.

0-8493-1586-7/05/$0.00+$1.50

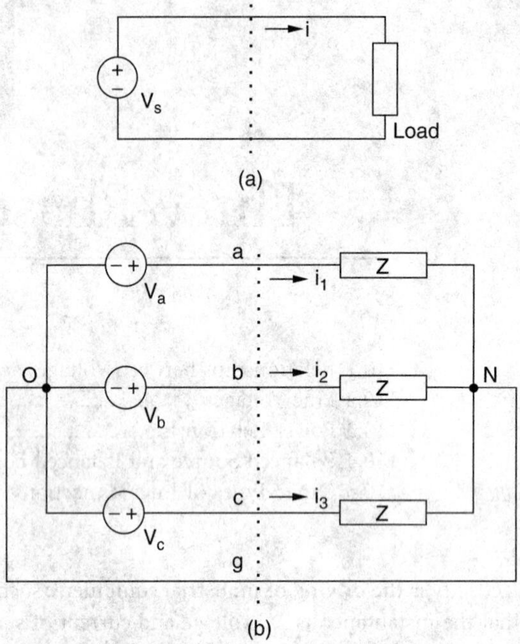

FIGURE 110.1 Flow of power from source to load.

110.1 Relationships between Voltages and Currents

The three sources are assumed to have the same frequency; for this reason, they are said to be *synchronized*. It is also assumed that the three phase voltages have the same rms values and that the phase difference between each pair of voltages is ±120° (2π/3 rad). Thus, the voltages can be written:

$$\begin{aligned} v_a &= \sqrt{2}|V|\cos(\omega t + \alpha_1) \quad \leftrightarrow \quad V_a = |V|e^{j0^0} \\ v_b &= \sqrt{2}|V|\cos(\omega t + \alpha_2) \quad \leftrightarrow \quad V_b = |V|e^{-j120^0} \\ v_c &= \sqrt{2}|V|\cos(\omega t + \alpha_3) \quad \leftrightarrow \quad V_c = |V|e^{j120^0} \end{aligned} \tag{110.3}$$

The **phasors** representing the sinusoids have also been shown. For convenience the angle v_a has been chosen as the reference for angles; v_b *lags* v_a by 120° and v_c *leads* v_a by 120°.[1]

There are two options for choosing the sequence of the phases. Once the particular phase that is to be the reference for angles is chosen and named, "a," there are two possible sequences for the other two: either "abc" or "acb." This fact is hardly earthshaking; all it means is that the leading and lagging angles can be interchanged. Obviously, nothing fundamental is different in the second sequence. Hence, the discussion that follows is limited to the abc *phase sequence.*

Because the loads are identical, the rms values of the three currents shown in Figure 110.1 will also be the same, and the phase difference between each pair of them will be ±120°. Thus, the currents can be written as

[1]Observe that the principal value of the angle lying between ±180° is used. One could add 360° to the negative angle and use the value 240° instead.

$$
\begin{aligned}
i_1 &= \sqrt{2}|I|\cos(\omega t+\beta_1) &\leftrightarrow\quad I_1 &= |I|e^{j\beta_1} \\
i_2 &= \sqrt{2}|I|\cos(\omega t+\beta_2) &\leftrightarrow\quad I_2 &= |I|e^{j(\beta_1-120^0)} \\
i_3 &= \sqrt{2}|I|\cos(\omega t+\beta_3) &\leftrightarrow\quad I_3 &= |I|e^{j(\beta_1+120^0)}
\end{aligned}
\tag{110.4}
$$

Perhaps a better form of visualizing the voltages and currents is a graphical one. Phasor diagrams for the voltages and the currents are shown separately in Figure 110.2. The value of angle β_1 will depend on the load. Something significant is clear from these diagrams. First, V_2 and V_3 are each the other's conjugate. So if they are added, the imaginary parts cancel and the sum will be real, as illustrated by the construction in the voltage diagram. Furthermore, the construction shows this sum to be negative and equal in magnitude of V_1. Hence, *the sum of the three voltages is zero.* The same is true of the sum of the three currents, as can be established graphically by a similar construction. The same results can be confirmed analytically by converting the phasor voltages and currents into rectangular form.

By Kirchhoff's current law applied at node N in Figure 110.1, we find that the current in the return line is the sum of the three currents in Equation (110.4). But since this sum was found to be zero, *the return line carries no current.* Hence, it can be removed entirely without affecting the operation of the system. The resulting circuit is redrawn in Figure 110.3. It can be called a *three-wire* three-phase system. Because of its geometrical form, this connection of both the sources and the loads is said to be a **wye** (Y) **connection,** even though *t* is an upside-down Y.

Notice that the circuit in Figure 110.3 is planar, with no lines crossing any other lines. That simplicity has been achieved at a price. Notice how the sequence (abc) of sources has been laid out geometrically.

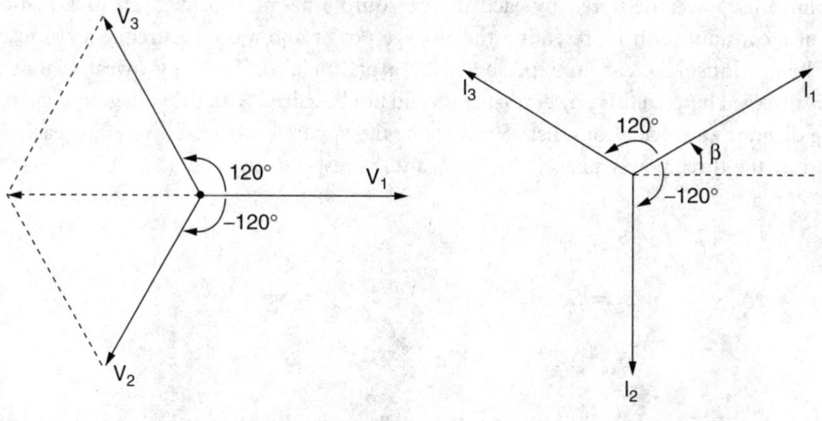

FIGURE 110.2 Voltage and current phasor diagrams.

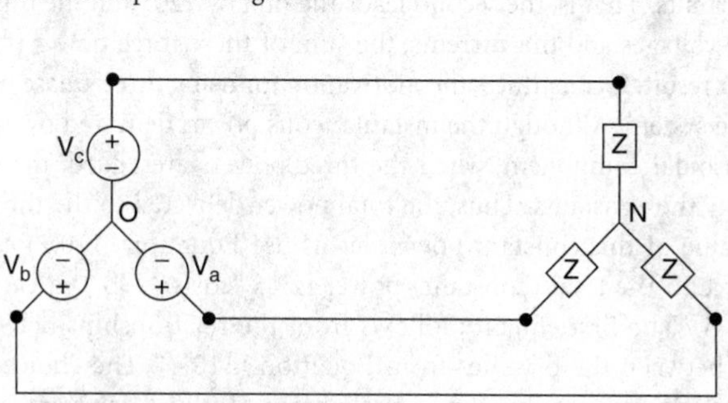

FIGURE 110.3 Wye-connected three-phase system.

Clearly, with the connections shown, the sequence of the loads is not the same as that of the sources. Having the same sequence would require interchanging the connections of the b and c sources with the bottom two loads. Doing that would result in one branch crossing another. However, nothing fundamental would change with either connection, assuming equal loads.

110.2 Line Voltages

In the three-wire three-phase system in Figure 110.3, the neutral point O is not accessible, so phase voltages cannot be measured. The voltages that *are* available for measurement are the *line-to-line* or simply the *line* voltages: V_{ab}, V_{bc}, and V_{ca}. By Kirchhoff's voltage law,

$$\begin{aligned} V_{ab} &= V_a - V_b = |V| - |V|e^{-j120^0} = \sqrt{3}|V|e^{j30^0} \\ V_{bc} &= V_b - V_c = |V|e^{-j120^0} - |V|e^{j120^0} = \sqrt{3}|V|e^{-j90^0} \\ V_{ca} &= V_c - V_a = |V|e^{j120^0} - |V| = \sqrt{3}|V|e^{j150^0} \end{aligned} \tag{110.5}$$

The interesting result is that all the line-voltage magnitudes are equal at $\sqrt{3}$ times the phase-voltage magnitude. Thus, a 220 V line voltage corresponds to a phase voltage of 127 V. The line-voltage angles have the same mutual relationships as the phase-voltage angles; they are separated by ±120°.

110.3 Power Relationship

The instantaneous power delivered by each of the sources has the form given in Equation (110.2), consisting of a constant term representing the average power and a double-frequency sinusoidal term. The latter, being sinusoidal, can be represented by a phasor also. The only caveat is that a different frequency is involved here, so this power phasor should not be mixed with the voltage and current phasors in the same diagram or calculations. Let $|S| = |V||I|$ be the apparent power delivered by each of the three sources, and let the three power phasors be S_a, S_b, and S_c, respectively. Then,

$$\begin{aligned} S_a &= |S|e^{j(\alpha_1+\beta_1)} = |S|e^{j\beta_1} \\ S_b &= |S|e^{j(\alpha_2+\beta_2)} = |S|e^{j(-120^0+\beta_1-120^0)} = |S|e^{j(\beta_1+120^0)} \\ S_c &= |S|e^{j(\alpha_3+\beta_3)} = |S|e^{j(120^0+\beta_1+120^9)} = |S|e^{j(\beta_1-120^0)} \end{aligned} \tag{110.6}$$

It is evident that the phase relationships between these three phasors are the same as the ones between the voltages and the currents. That is, the second leads the first by 120° and the third lags the first by 120°. Hence, just as with the voltages and the currents, the sum of these three power phasors will also be zero. This is a very significant result. It constitutes the motivation for using three-phase power over the pulsating power of a single-phase system. Although the instantaneous power delivered by each load has a constant component and a sinusoidal component, when the three powers are added, the sinusoidal components add to zero, leaving only the constants. Thus, the total power delivered to the three loads is constant.

To determine the value of this constant power, let us use Equation (110.2) as a model. The contribution of the kth source to the total (constant) power is $|S|\cos(\alpha_k - \beta_k)$. It can be easily verified that $\alpha_k - \beta_k = \alpha_1 - \beta_1 = -\beta_1$. The first equality follows from the relationships between the α values from Equation (110.3) and between the β values from Equation (110.4). The choice of $\alpha_1 = 0$ leads to the last equality. Hence, each phase contributes an equal amount to the total average power. If P is the total average power, then

$$P = P_a + P_b + P_c = 3P_a = 3|V||I|\cos(\alpha_1 - \beta_1) \qquad (110.7)$$

Although the angle α_1 has been set equal to zero, it is shown in this equation for the sake of generality.

A similar result can be obtained for the reactive power. The reactive power of the kth phase is $|S|\sin(\alpha_k - \beta_k) = |S|\sin(\alpha_1 - \beta_1)$. If Q is the total reactive power, then

$$Q = 3|S|\sin(\alpha_1 - \beta_1)$$

110.4 Balanced Source and Balanced Load

What has just been described is a *balanced* three-phase three-wire power system. The three sources in practice are not three independent sources but consist of three different parts of the same generator. The same is true of the loads.[1] What has been described is ideal in a number of ways. First, the circuit can be *unbalanced* — for example, by the loads being somewhat unequal. Second, since the real devices whose ideal model is a voltage source are coils of wire, each source should be accompanied by a branch consisting of the coil inductance and resistance. Third, since the power station (or the distribution transformer at some intermediate point) may be at some distance from the load, the parameters of the physical line carrying the power (the line inductance and resistance) must also be inserted in series between the source and the load.

The analysis of this chapter does not apply to an unbalanced system. An entirely new analytical technique is required to do full justice to such a system.[2] An understanding of balanced circuits is a prerequisite to tackling the unbalanced case.

The last two of the conditions that make the circuit less than ideal (winding and line impedances) introduce algebraic complications but change nothing fundamental in the preceding theory. If these two conditions are taken into account, the appropriate circuit takes the form shown in Figure 110.4. Here the internal impedance of a source (the winding impedance labeled Z_w) and the line impedance Z_l connecting that source to its load are both connected in series with the corresponding load. Thus, instead of the impedance in each phase being Z, it is $Z + Z_w + Z_l$. Hence, the rms value of each current is

$$|I| = \frac{|V|}{|Z + Z_w + Z_l|} \qquad (110.8)$$

instead of $|V|/|Z|$. All other previous results remain unchanged — namely, that the sum of the phase currents adds to zero and that the sum of the phase powers is a constant. The detailed calculations just become a little more complicated.

[1]An AC power generator consists of (a) a *rotor* that is rotated by a *prime mover* (say a turbine) and produces a magnetic field that also rotates, and (b) a *stator* on which is wound one or more coils of wire. In three-phase systems the number of coils is three. The rotating magnetic field induces a voltage in each of the coils. The frequency of the induced voltage depends on the number of magnetic poles created on the rotor and the speed of rotation. These are fixed so as to "synchronize" with the 60 Hz frequency of the power system. The 120° leading and lagging phase relationships between these voltages are obtained by distributing the conductors of the coils around the circumference of the stator so that they are separated geometrically by 120°. Thus, the three sources described in the text are in reality a single physical device, a single generator. Similarly, the three loads might be the three windings on a three-phase motor, again a single physical device. Or they might be the windings of a three-phase transformer.

[2]The technique for analyzing unbalanced circuits utilizes what are called *symmetrical components.*

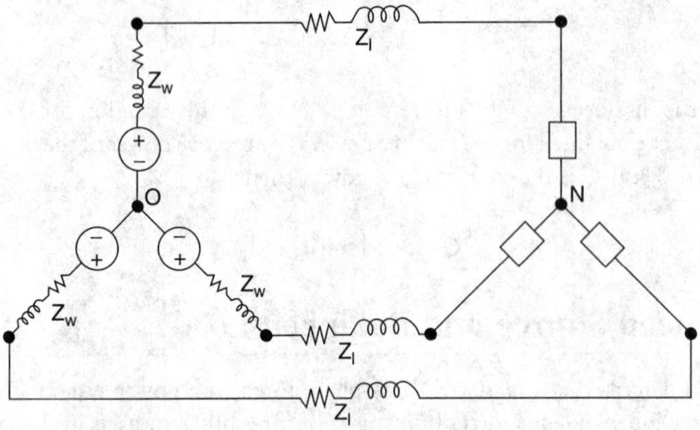

FIGURE 110.4 Three-phase circuit with nonzero winding and line impedances.

110.5 Other Types of Interconnections

All of the preceding development was based on both the sources and the loads being connected in a wye connection. Although the upside-down Y structure looks geometrically a little different from an upside-down tee circuit, electrically, the two are exactly the same. The wye is not, however, the only possible way to connect the phases of a three-phase system. Another possibility, the **delta connection,** so named because it looks like the Greek letter **Δ**, is shown in Figure 110.5. (In this figure the boxes can represent either sources or impedances.)

By proper choice of the branch parameters, a tee can be made equivalent to a pi (∏) at the terminals. We note that the delta is just an upside-down pi. As a pi, the junction between A and B is usually extended as the common terminal of a two-port.

If the structures in Figure 110.5 are to be equivalent, the line voltages V_{ab}, V_{bc}, and V_{ca} should be the same in both circuits. Similarly, the currents into the terminals should be the same in both. Note that, in the delta connection, the phase voltages are not evident; the only voltages available are the line voltages. Thus the voltages in the delta are the line voltages given in Equation (110.3). In the wye, the phase currents are also the currents in the lines. For the delta, however, the line currents are the difference between two phase currents, as noted in Figure 110.5. For the line currents, a set of equations similar to Equation (110.5) can be written in terms of the phase currents. Since the same 120° difference of angle exists between the phase currents as between the phase voltages, we would expect that the result for currents would be similar to the result for voltages in Equation (110.5) — namely, that the line-current magnitudes in a delta connection would be $\sqrt{3}$ times the phase-current magnitudes.

In a three-phase circuit the sources, the loads, or both can be replaced by a delta equivalent; four different three-phase circuits can therefore be imagined: wye-wye, wye-delta, delta-wye, and delta-delta. There are no fundamental differences in analyzing each of these four circuits.

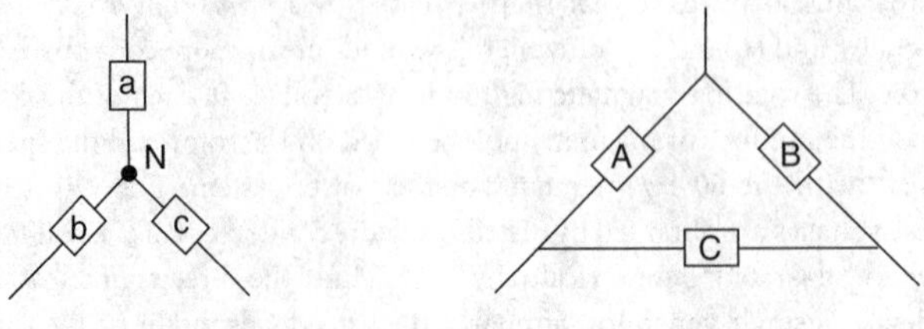

FIGURE 110.5 Wye connection and delta connection.

Example

A balanced, 120 V, three-wire, three-phase transmission system in a wye-wye connection is represented by the circuit in Figure 110.4. Assume that the winding impedances are negligible but that the line impedances are given by $Z_l = 0.1 + j0.2\ \Omega$. Each load impedance is $Z = 20 + j5\ \Omega$. The following quantities are to be determined: (a) the line current magnitude; (b) the magnitude of the voltage across each load; (c) the average power, reactive power, and apparent power delivered to the load by each phase; (d) the average power, reactive power, and apparent power delivered by each source; and (e) the fraction of the power delivered by the system that is lost in the lines.

Solution

The solution is completely straightforward. First, the line current is found by dividing the phase voltage by the sum of the load and line impedances; the load voltage follows from the product of the load impedance by the line current. Thus,

$$|I| = \frac{120}{\sqrt{(20+0.1)^2 + (5+0.2)^2}} = 5.78\ \text{A}$$

$$|V_L| = |I||Z| = 5.78\sqrt{20^2 + 5^2} = 119.16\ \text{V}$$

The power calculations then follow:

$$|S_L| = |V_L||I| = 119.16(5.78) = 688.7\ \text{VA or}$$

$$|S_L| = |I|^2|Z_L| = 5.78^2\sqrt{20^2 + 5^2} = 688.7\ \text{VA}$$

$$P_L = R_L|I|^2 = 20(5.78)^2 = 668.2\ \text{W}$$

$$Q_L = X_L|I|^2 = 5(5.78)^2 = 167.0\ \text{VAR}$$

$$= \sqrt{|S_L|^2 - P_L^2} = \sqrt{688.7^2 - 668.2^2} = 166.8\ \text{VAR}$$

Perhaps the best way to find the power delivered by the sources is to determine the power lost in the line and then add this to the load power. Carrying out this approach leads to the following result:

$$P_l = 0.1|I|^2 = 3.34\ \text{W} \qquad P_s = 3.34 + 668.2 = 671.5\ \text{W}$$

$$Q_l = 0.2|I|^2 = 6.68\ \text{VAR} \qquad Q_s = 6.68 + 167.0 = 173.7\ \text{VAR}$$

Finally, the fraction of the source power that is lost in the line is 3.34/671.5 = 0.005 or 0.5%.

Defining Terms

Delta connection — The sources or loads in a three-phase system connected end-to-end, forming a closed path, like the Greek letter Δ.

Phasor — A complex number representing a sinusoid; its magnitude and angle are the rms value and the phase of the sinusoid, respectively.

Wye connection — The three sources or loads in a three-phasor system connected to have one common point, like the letter Y.

References

del Toro, V. 1992. *Electric Power Systems,* Prentice Hall, Englewood Cliffs, NJ.
Gungor, B. R. 1988. *Power Systems,* Harcourt Brace Jovanovich, San Diego, CA.
Peebles, P. Z. and Giuma, T. A. 1991. *Principles of Electrical Engineering,* McGraw-Hill, New York.

111

Filters (Passive)

Albert J. Rosa
University of Denver

A **filter** is a frequency-sensitive two-port circuit that transmits with or without amplification signals in a band of frequencies and rejects (or attenuates) signals in other bands. The electric filter was invented during the First World War by two engineers working independently of each other — the American engineer G. A. Campbell and the German engineer K. W. Wagner. O. Zobel followed in the 1920s. These devices were developed to serve the growing areas of telephone and radio communication. Today, filters are found in all types of electrical and electronic applications from power to communications. Filters can be both active and passive. In this section we will confine our discussion to those filters that employ no active devices for their operation. The main advantage of passive filters over active ones is that they require no power (other than the signal) to operate. The disadvantage is that they often employ inductors that are bulky and expensive.

111.1 Fundamentals

The basis for filter analysis involves the determination of a filter circuit's sinusoidal steady state response from its transfer function $T(j\omega)$. Some references use $H(j\omega)$ for the transfer function. The filter's transfer function $T(j\omega)$ is a complex function and can be represented through its gain $|T(j\omega)|$ and phase $\angle T(j\omega)$ characteristics. The gain and phase responses show how the filter alters the amplitude and phase of the input signal to produce the output response. These two characteristics describe the *frequency response* of the circuit since they depend on the frequency of the input sinusoid. The signal-processing performance of devices, circuits, and systems is often specified in terms of their frequency response. The gain and phase functions can be expressed mathematically or graphically as *frequency-response* plots. Figure 111.1 shows examples of gain and phase responses versus frequency, ω.

The terminology used to describe the frequency response of circuits and systems is often based on the form of the gain plot. For example, at high frequencies the gain in Figure 111.1 falls off so that output signals in this frequency range are reduced in amplitude. The range of frequencies over which the output is significantly attenuated is called the *stopband.* At low frequencies the gain is essentially constant and there is relatively little attenuation. The frequency range over which there is little attenuation is called a *passband.* The frequency associated with the boundary between a passband and an adjacent stopband is called the *cutoff frequency* ($\omega_C = 2\pi f_C$). In general, the transition from the passband to the stopband, called the *transition band*, is relatively gradual, so the precise location of the cutoff frequency is a matter of definition. The most widely used approach defines the cutoff frequency as the frequency at which the gain has decreased by a factor of $1/\sqrt{2} = 0.707$ from its maximum value in the passband.

0-8493-1586-7/05/$0.00+$1.50

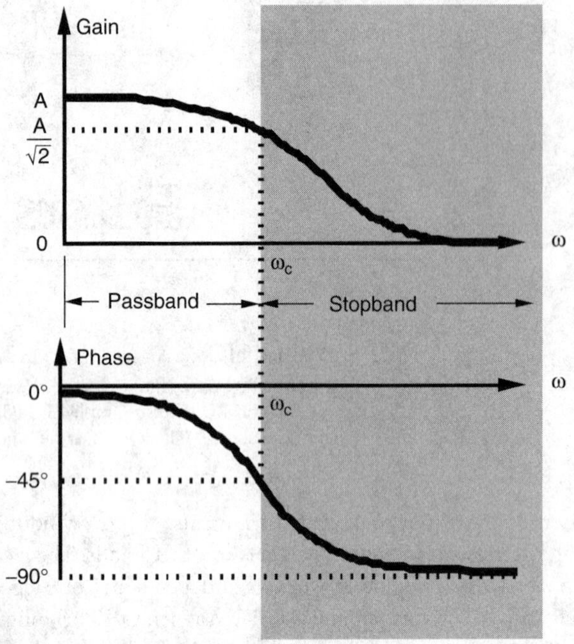

FIGURE 111.1 Low-pass filter characteristics showing passband, stopband, and the cutoff frequency, ω_C.

This particular definition is based on the fact that the power delivered to a resistor by a sinusoidal current or voltage waveform is proportional to the square of its amplitude. At a cutoff frequency the gain is reduced by a factor of $1/\sqrt{2}$ and the square of the output amplitude, and thusly also its power, is reduced by a factor of one half. For this reason the cutoff frequency is also called the *half-power frequency.*

There are four prototypical filters. These are *low pass* (LP), *high pass* (HP), *band pass* (BP), and *bandstop* (BS). Figure 111.2 shows how the amplitude of an input signal consisting of three separate equal-amplitude frequencies is altered by each of the four-prototypical filter responses. The low-pass filter passes frequencies below its cutoff frequency ω_C, called its *passband,* and attenuates the frequencies above the cutoff, called its *stopband.* The high-pass filter passes frequencies above the cutoff frequency ω_C and attenuates those below. The band-pass filter passes those frequencies that lie between two cutoff frequencies, ω_{C1} and ω_{C2}, its passband, and attenuates those frequencies that lie outside the passband. Finally, the bandstop filter attenuates those frequencies that lie in its reject or stopband, between ω_{C1} and ω_{C2}, and passes all others.

The *bandwidth* of a gain characteristic is defined as the frequency range spanned by its passband. For the band-pass case in Figure 111.2, the bandwidth is the difference in the two cutoff frequencies.

$$BW = \omega_{C2} - \omega_{C1} \tag{111.1}$$

This equation applies to the low-pass response with the lower cutoff frequency ω_{C1} set to zero. In other words, the bandwidth of a low-pass circuit is equal to its cutoff frequency ($BW = \omega_C$). The bandwidth of a high-pass characteristic is infinite since the upper cutoff frequency ω_{C1} is infinity. For the bandstop case, Equation (111.1) defines the bandwidth of the stopband rather than the passbands.

Frequency-response plots are usually made using logarithmic scales for the frequency variable because the frequency ranges of interest often span several orders of magnitude. A logarithmic frequency scale compresses the data range and highlights important features in the gain and phase responses. The use of a logarithmic frequency scale involves some special terminology. A frequency range whose end points have a 2:1 ratio is called an *octave* and one with 10:1 ratio is called a *decade.* Straight-line approximations

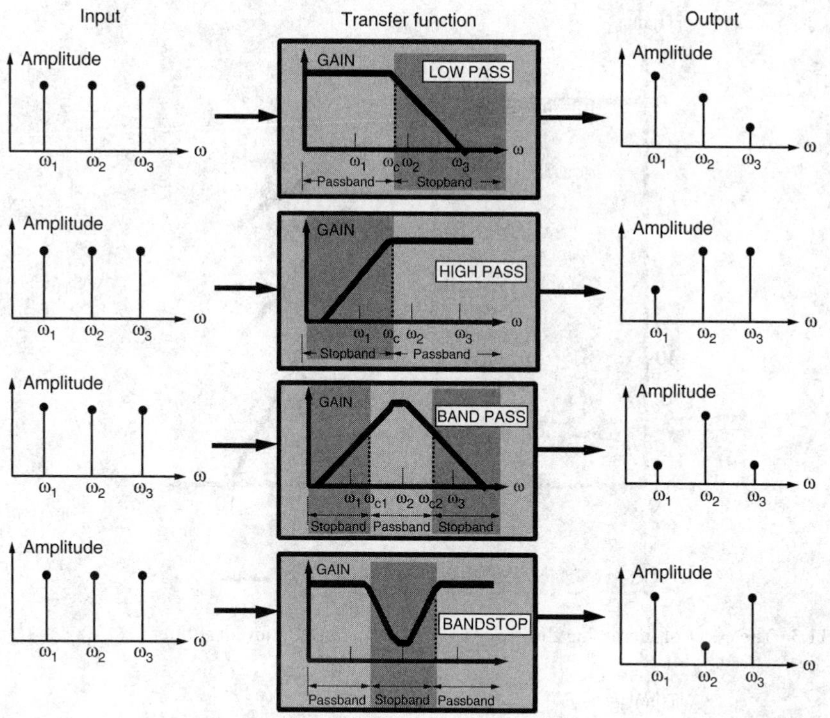

FIGURE 111.2 Four prototype filters and their effects on an input signal consisting of three frequencies.

of these plots, called Bode (Bow-dee) plots, are often used to describe the general behavior of the devices, circuits, or systems.

In frequency-response plots the gain $|T(j\omega)|$ is often expressed in *decibels* (dB), defined as

$$|T(j\omega)|_{dB} = 20 \log_{10}|T(j\omega)| \tag{111.2}$$

Gains expressed in decibels can be either positive, negative, or zero. A gain of zero dB means that $|T(j\omega)| = 1$ — that is, the input and output amplitudes are equal. A positive dB gain means the output amplitude exceeds the input since $|T(j\omega)| > 1$, whereas a negative dB gain means the output amplitude is smaller than the input since $|T(j\omega)| < 1$. A cutoff frequency usually occurs when the gain is reduced from its maximum passband value by a factor $1/\sqrt{2}$ or 3 dB.

Figure 111.3 shows the asymptotic gain characteristics of ideal and real low-pass filters. The gain of the *ideal filter* is unity (0 dB) throughout the passband and zero ($-\infty$ dB) in the stopband. It also has an infinitely narrow transition band. The asymptotic gain responses of real low-pass filters show that we can only approximate the ideal response. As the order of the filter or number of poles n increases, the approximation improves since the asymptotic slope or "rolloff" in the stopband is $-20 \times n$ dB/decade. On the other hand, adding poles requires additional stages in a cascade realization, so there is a trade-off between (1) filter complexity and cost and (2) how closely the filter gain approximates the ideal response.

Figure 111.4 shows how low-pass filter requirements are often specified. To meet the specification, the gain response must lie within the unshaded region in the figure, as illustrated by the two responses shown in Figure 111.4. The parameter T_{max} is the *passband gain.* In the passband the gain must be within 3 dB of T_{max} and must equal $T_{max}/\sqrt{2}$ at the cutoff frequency ω_C. In the stopband the gain must decrease and remain below a gain of T_{min} for all $\omega \geq \omega_{min}$. A low-pass filter design requirement is usually defined by specifying values for these four parameters. The parameters T_{max} and ω_C define the passband response, whereas T_{min} and ω_{min} specify how rapidly the stopband response must decrease.

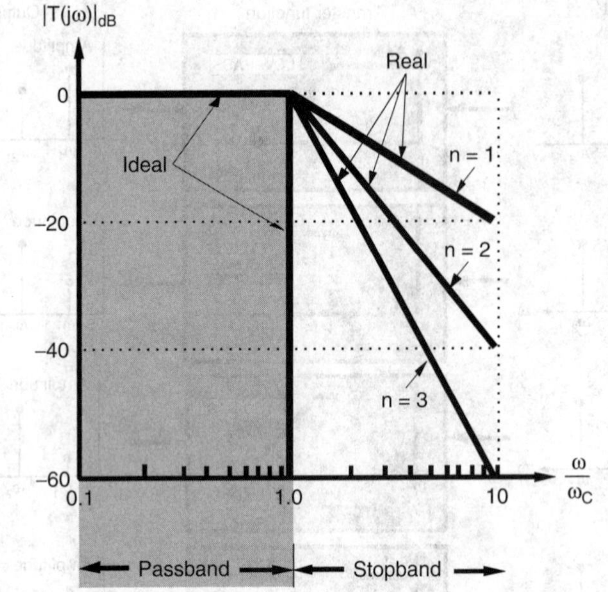

FIGURE 111.3 The effect of increasing the order *n* of a filter relative to an ideal filter.

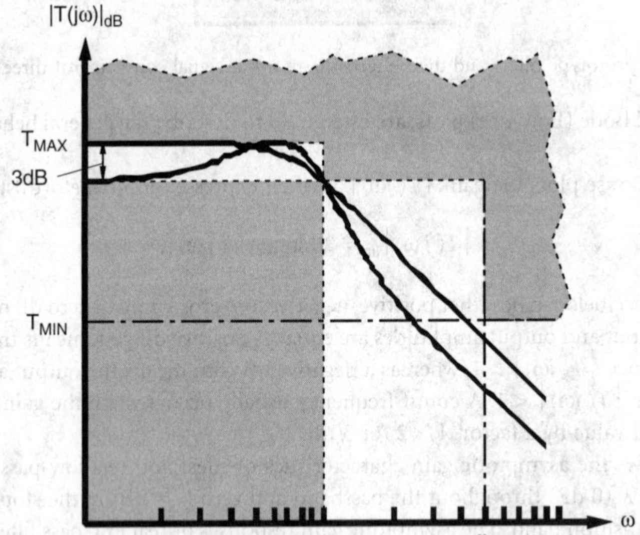

FIGURE 111.4 Parameters for specifying low-pass filter requirements.

111.2 Applications

Simple RL and RC Filters

A first-order LP filter has the following transfer function:

$$T(s) = \frac{K}{s+\alpha} \tag{111.3}$$

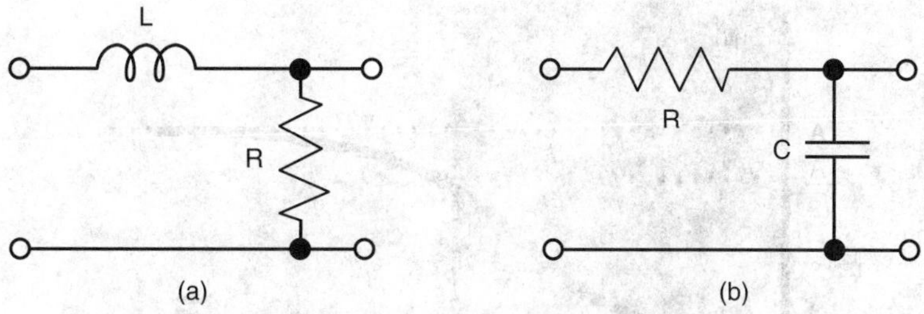

FIGURE 111.5 Single-pole LP filter realizations: (a) RL, (b) RC.

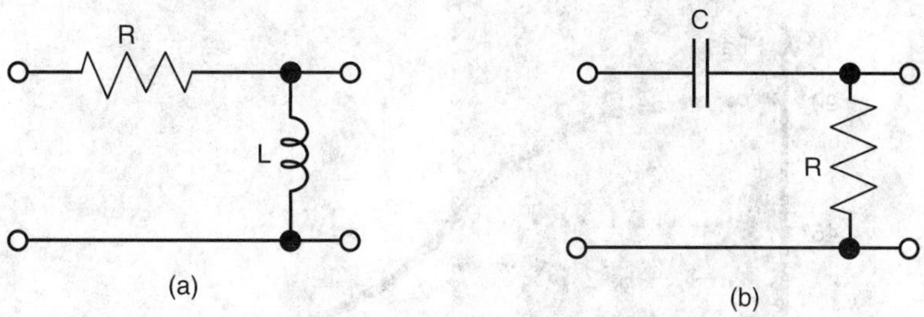

FIGURE 111.6 Single-pole HP filter realizations: (a) RL, (b) RC.

where for a passive filter $K \le \alpha$ and $\alpha = \omega_C$. This transfer function can be realized in several ways including using either of the two circuits shown in Figure 111.5.

For sinusoidal response the respective transfer functions are

$$T(j\omega)_{RL} = \frac{R/L}{j\omega + (R/L)}; \qquad T(j\omega)_{RC} = \frac{1/RC}{j\omega + (1/RC)} \tag{111.4}$$

For these filters the passband gain is equal to one and the cutoff frequency is determined by *R/L* for the RL filter and 1/*RC* for the RC filter. The gain $|T(j\omega)|$ and phase $\angle T(j\omega)$ plots of these circuits are shown back in Figure 111.1.

A first-order HP filter is given by the following transfer function:

$$T(s) = \frac{Ks}{s + \alpha} \tag{111.5}$$

where, for a passive filter, $K \le 1$ and α is the cutoff frequency. This transfer function can also be realized in several ways including using either of the two circuits shown in Figure 111.6. For sinusoidal response the respective transfer functions are

$$T(j\omega)_{RL} = \frac{j\omega}{j\omega + (R/L)}; \qquad T(j\omega)_{RC} = \frac{j\omega}{j\omega + (1/RC)} \tag{111.6}$$

For the LP filters the passband gain is one and the cutoff frequency is determined by *R/L* for the RL filter and $1/RC$ for the RC filter. The gain $|T(j\omega)|$ and phase $\angle T(j\omega)$ plots of these circuits are shown in Figure 111.7.

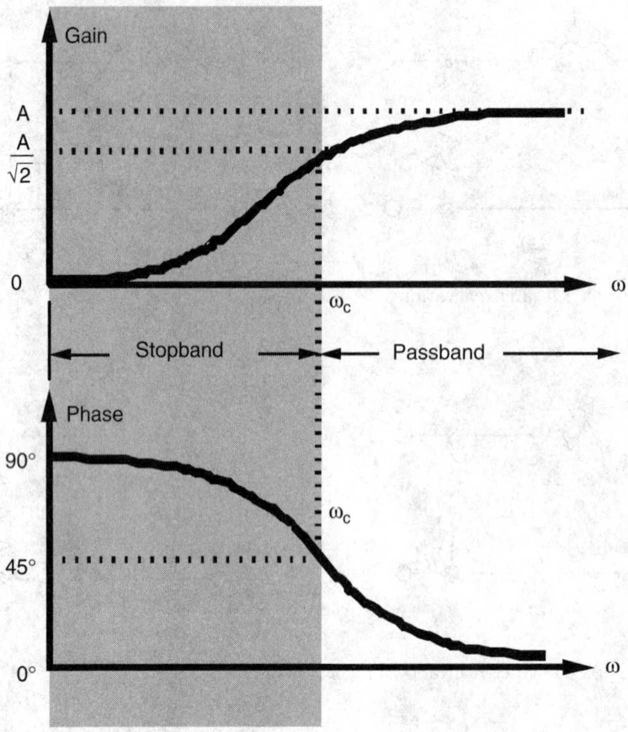

FIGURE 111.7 High-pass filter characteristics showing passband, stopband, and the cutoff frequency, ω_c.

Simple RLC Filters

Simple second-order LP, HP, or BP filters can be made using series or parallel RLC circuits. Series or parallel RLC circuits can be connected to produce the following transfer functions:

$$T(j\omega)_{LP} = \frac{K}{-\omega^2 + 2\zeta\omega_0 j\omega + \omega_0^2}$$

$$T(j\omega)_{HP} = \frac{-K\omega^2}{-\omega^2 + 2\zeta\omega_0 j\omega + \omega_0^2} \tag{111.7}$$

$$T(j\omega)_{BP} = \frac{Kj\omega}{-\omega^2 + 2\zeta\omega_0 j\omega + \omega_0^2}$$

where for a series RLC circuit $\omega_0 = \sqrt{LC}$ and $\zeta = \frac{R}{2}\sqrt{\frac{C}{L}}$. ω_0 is called the undamped natural frequency and is related to the cutoff frequency in the HP and LP case and is the center frequency in the BP case. ζ is called the damping ratio and determines the nature of the roots of the equation that translates to how quickly a transition is made from the passband to the stopband. ζ in the BP case helps define the bandwidth of the circuit, that is, $Bw = 2\zeta\omega_0$. Figure 111.8 shows how a series RLC circuit can be connected to achieve the transfer functions given in Equation 111.7. The gain $|T(j\omega)|$ and phase $\angle T(j\omega)$ plots of these circuits are shown in Figure 111.9 through Figure 111.11.

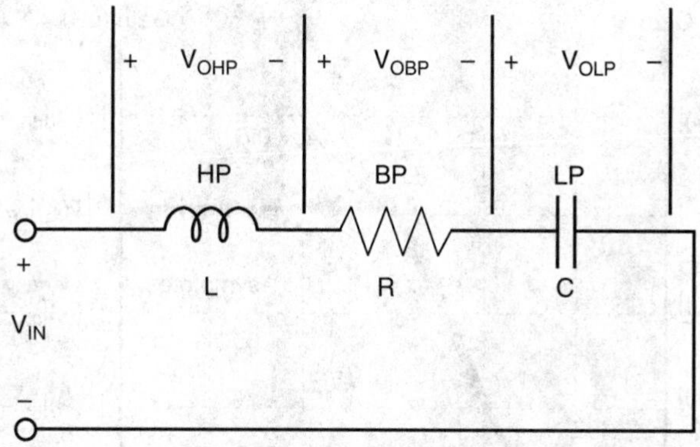

FIGURE 111.8 RLC circuit connections to achieve LP, HP, or BP responses.

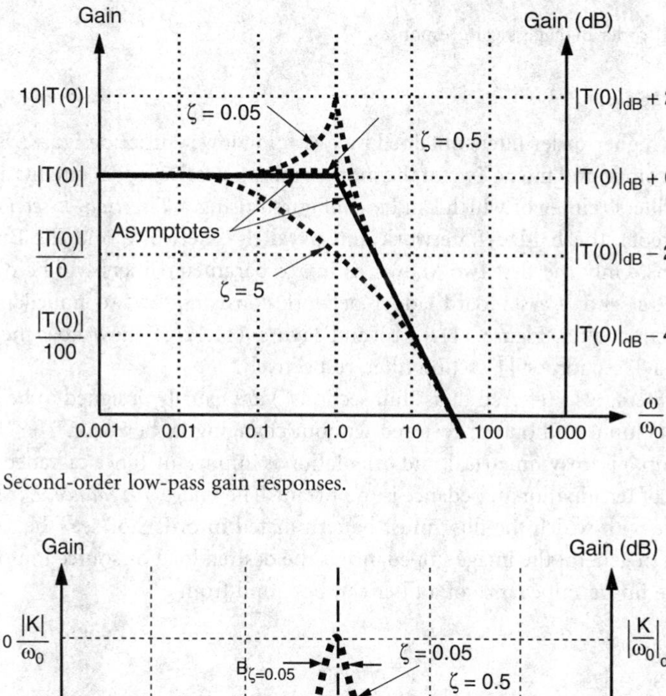

FIGURE 111.9 Second-order low-pass gain responses.

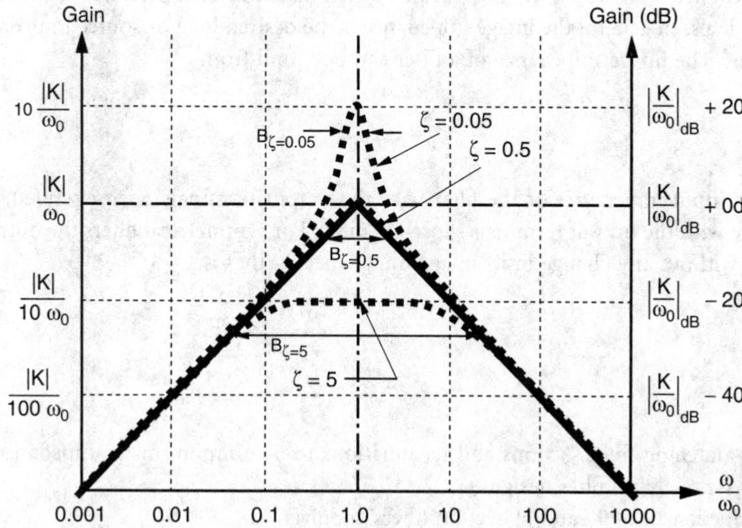

FIGURE 111.10 Second-order band-pass gain responses.

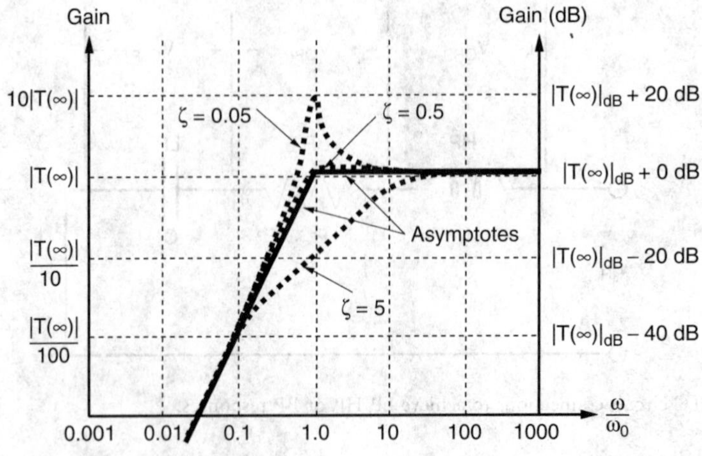

FIGURE 111.11 Second-order high-pass gain responses.

Compound Filters

Compound filters are higher-order filters obtained by cascading lower-order designs. *Ladder circuits* are an important class of compound filters. Two of the more common passive ladder circuits are the *constant-k* and the *m-derived* filters (either of which can be configured using a *T-section*, *π-section*, or *L-section*, or combinations thereof), the **bridge-T network** and **parallel-T network,** and the **Butterworth** and **Chebyshev** realizations. Only the first two known as image-parameter filters will be discussed in this section. Figure 111.12(a) shows a standard ladder network consisting of two impedances, Z_1 and Z_2, organized as an L-section filter. Figure 111.12(b) and Figure 111.12(c) show how the circuit can be redrawn to represent a T-section or Π-section filter, respectively.

T- and Π-section filters (also referred to as "full sections") are usually designed to be symmetrical so that either can have its input and output reversed without changing its behavior. The "L-section" (also known as a "half section") is unsymmetrical, and orientation is important. Since cascaded sections "load" each other, the choice of termination impedance is important. The *image impedance*, Z_i, of a symmetrical filter is the impedance with which the filter must be terminated in order to "see" the same impedance at its input terminals. In general the image impedance is the desired load or source impedance to which the filter matches. The image impedance of a filter can be found from

$$Z_i = \sqrt{Z_{1O} Z_{1S}} \tag{111.8}$$

where Z_{1O} is the input impedance of the filter with the output terminals open circuited, and Z_{1S} is its input impedance with the output terminals short-circuited. For symmetrical filters the output and input can be reversed without any change in its image impedance — that is,

$$Z_{1i} = \sqrt{Z_{1O} Z_{1S}} \quad \text{and} \quad Z_{2i} = \sqrt{Z_{2O} Z_{2S}}$$
$$Z_{1i} = Z_{2i} = Z_i \tag{111.9}$$

The concept of matching filter sections and terminations to a common image impedance permits the development of symmetrical filter designs.

The image impedances of T- and Π-section filters are given as

$$Z_{iT} = \sqrt{Z_{1O} Z_{1S}} = \sqrt{\frac{1}{4} Z_1^2 + Z_1 Z_2}$$

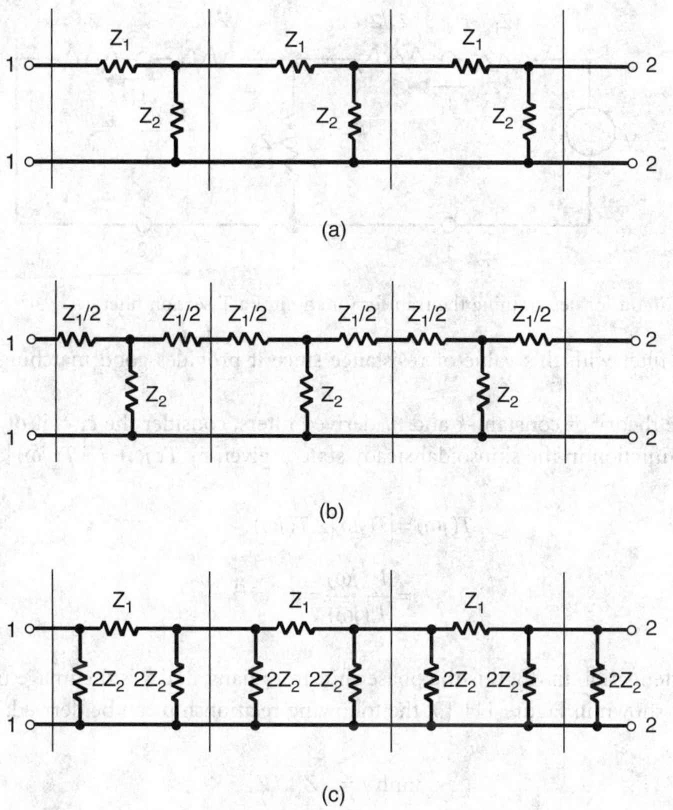

FIGURE 111.12 Ladder networks: (a) standard L-section, (b) T-section, (c) Π-section.

and

$$Z_{i\Pi} = \sqrt{Z_{1O}Z_{1S}} = \frac{Z_1Z_2}{\sqrt{(1/4)Z_1^2 + Z_1Z_2}} \tag{111.10}$$

These expressions also describe the input (or output) impedance when the filter's output (or input) is terminated with appropriate image impedance, i.e., Z or Z_T.

The image impedance of an L-section filter, being unsymmetrical, depends on the terminal pair being calculated. For the L-section shown in Figure 111.12(a), image impedances are

$$Z_{1iL} = \sqrt{\frac{1}{4}Z_1^2 + Z_1Z_2} = Z_{iT} \quad \text{and} \quad Z_{2iL} = \frac{Z_1Z_2}{\sqrt{(1/4)Z_1^2 + Z_1Z_2}} = Z_{i\Pi} \tag{111.11}$$

These equations show that the image impedance of an L-section at its input is equal to the image impedance of a T-section, whereas the image impedance of an L-section at its output is equal to the image impedance of a Π-section. This relationship is important in achieving an optimum termination when cascading L-sections with T- and/or Π-sections to form a composite filter.

Since Z_1 and Z_2 vary significantly with frequency, the image impedances of T- and Π-sections will also change. This condition does not present any particular problem in combining any number of equivalent filter sections together, since their impedances va4ry equally at all frequencies. But this does make it difficult to terminate these filters exactly, causing a limitation of these types of filters. However, there is a frequency within the filter's passband where the image impedance becomes purely resistive. It is useful

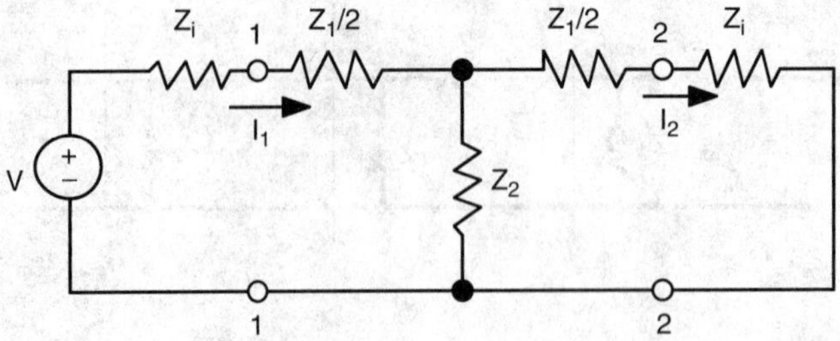

FIGURE 111.13 Circuit for determining the transfer function of a T-section filter.

to terminate the filter with this value of resistance since it provides good matching over much of the filter's passband.

To develop the theory of constant-*k* and *m*-derived filters, consider the circuit of Figure 111.13. The current transfer function in the sinusoidal steady state is given by $T(j\omega) = |T(j\omega)|\angle T(j\omega) = I_2/I_1$:

$$\begin{aligned} T(j\omega) &= |T(j\omega)|\angle T(j\omega) \\ &= \frac{I_2(j\omega)}{I_1(j\omega)} = e^{-\alpha}e^{-j\beta} = e^{-\gamma} \end{aligned} \tag{111.12}$$

where α is the attenuation in dB, β is the phase shift in radians, and γ is the image transfer function.

For the circuit shown in Figure 111.13, the following relationship can be derived:

$$\tanh\gamma = \sqrt{Z_{1S}/Z_{1O}} \tag{111.13}$$

This relationship and those in Equation (111.12) will be used to develop the constant-*k* and *m*-derived filters.

Constant-*k* Filters

During the 1920s O. Zobel developed an important class of symmetrical filters called *constant-k filters* with the conditions that Z_1 and Z_2 are purely reactive — that is, $\pm X(j\omega)$ and

$$Z_1Z_2 = k^2 = R^2 \tag{111.14}$$

In modern references an R replaces the k. Note that the units of k are ohms. The advantage of this type of filter is that the image impedance in the passband is a pure resistance, whereas in the stopband it is purely reactive. Hence if the termination is a pure resistance and equal to the image impedance, all the power will be transferred to the load since the filter itself is purely reactive. Unfortunately, the value of the image impedance varies significantly with frequency, and any termination used will result in a mismatch except at one frequency.

In LC constant-*k* filters, Z_1 and Z_2 have opposite signs, so that $Z_1Z_2 = \pm jX_1 \mp jX_2 = +X_1X_2 = R^2$. The image impedances become

$$Z_{iT} = R\sqrt{1-(-Z_1/4Z_2)} \quad \text{and} \quad Z_{i\Pi} = \frac{R}{\sqrt{1-(-Z_1/4Z_2)}} \tag{111.15}$$

Therefore, in the stopband and passband, we have the following relations for standard T- or Π-sections, where n represents the number of identical sections:

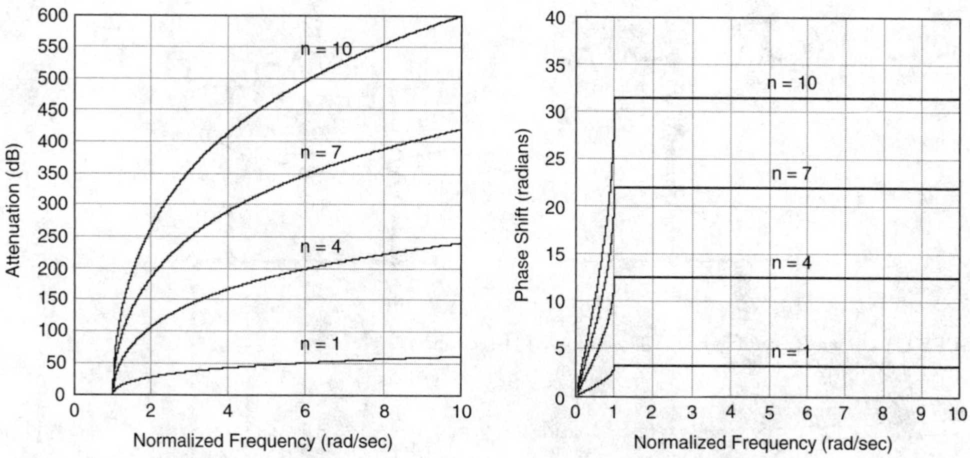

FIGURE 111.14 Normalized plots of attenuation and phase angle for various numbers of sections *n*.

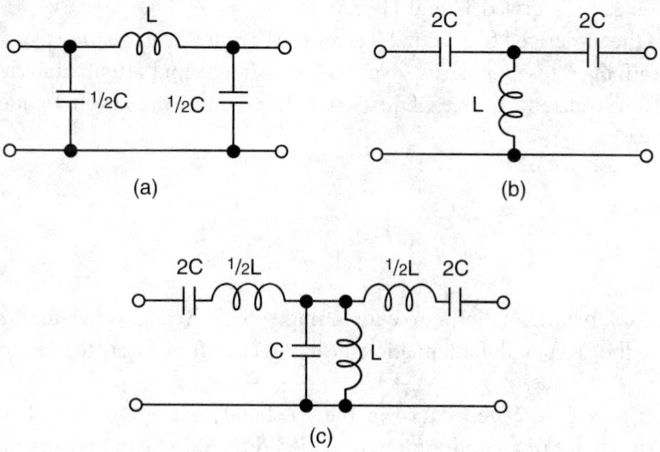

FIGURE 111.15 Typical sections: (a) LP Π-section, (b) HP T-section, (c) BP T-section.

Stopband	Passband	
$\alpha = 2n\cosh^{-1}\sqrt{-Z_1/4Z_2}$	$\alpha = 0$	(111.16)
$\beta = \pm n\pi, \pm 3n\pi, \ldots$	$\beta = 2n\sin^{-1}\sqrt{-Z_1/4Z_2}$	

The ultimate roll-off of constant-*k* filters is 20 dB per decade per reactive element or 60 dB per decade for the T- or Π-section, 40 dB per decade per L-section. Figure 111.14 shows normalized plots of α and β versus $\sqrt{-Z_1/4Z_2}$. These curves are generalized and apply to low-pass, high-pass, band-pass, or band-reject filters. Figure 111.15 shows examples of a typical LP Π-section, an HP T-section, and a BP T-section.

m-Derived Filters

The need to develop a filter section that could provide high attenuation in the stopband near the cutoff frequency prompted the development of the *m*-derived filter. O. Zobel developed a class of filters that had the same image impedance as the constant-*k* but had a higher attenuation near the cutoff frequency. The impedances in the *m*-derived filter were related to those in the constant-*k* as

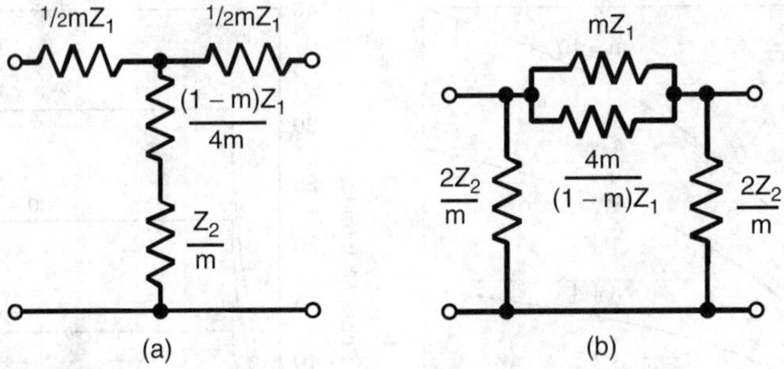

FIGURE 111.16 *m*-derived filters: (a) T-section, (b) Π-section.

$$Z_{1m} = mZ_{1k} \quad \text{and} \quad Z_{2m} = \left(\frac{1-m^2}{4m}\right)Z_{1k} + \frac{1}{m}Z_{2k} \tag{111.17}$$

where m is a positive constant ≤ 1. If $m = 1$ then the impedances reduce to those of the constant-k. Figure 111.16 shows generalized m-derived T- and Π-sections.

The advantage of the m-derived filter is that it gives rise to infinite attenuation at a selectable frequency, ω_∞, just beyond cutoff, ω_C. This singularity gives rise to a more rapid attenuation in the stopband than can be obtained using constant-k filters. Equation 111.18 relates the cutoff frequency to the infinite attenuation frequency.

$$m = \sqrt{1 - \frac{\omega_C^2}{\omega_\infty^2}} \tag{111.18}$$

Figure 111.17 shows the attenuation curve for a single m-derived LP stage for two values of m. The smaller m becomes, the steeper the attenuation near the cutoff, but also the lesser the attenuation at higher frequencies.

Constant-k filters have image impedance in the passband that is always real but that varies with frequency, making the choice of an optimum termination difficult. The impedance of an m-derived filter also varies, but how it varies depends on m. Figure 111.18 shows how the impedance varies with frequency (both normalized) and m. In most applications, m is chosen to be 0.6, keeping the image impedance nearly constant over about 80% of the passband.

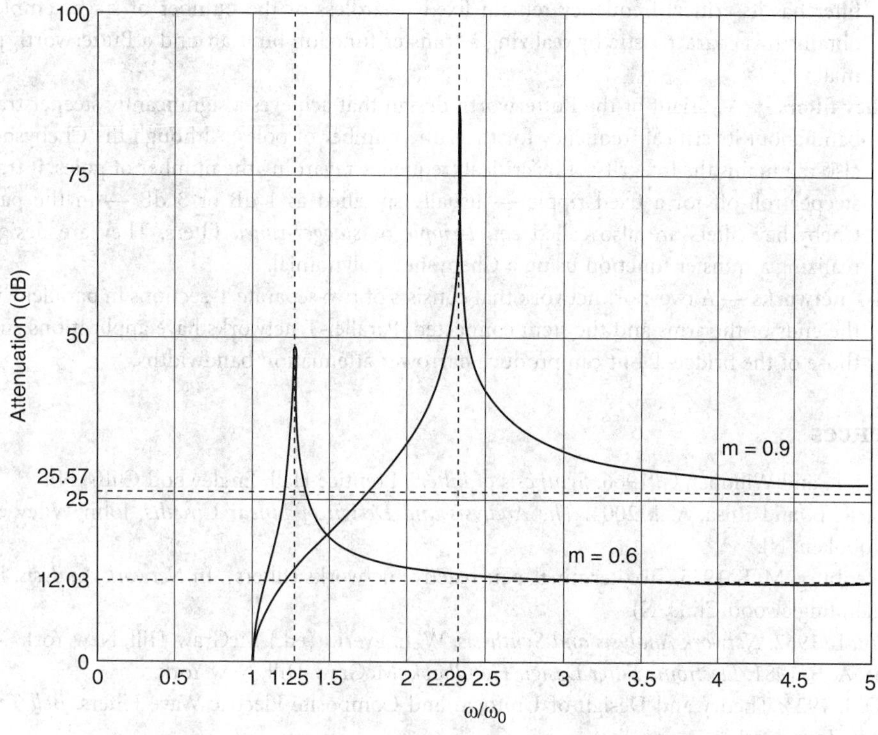

FIGURE 111.17 Attenuation curves for a single-stage filter with $m = 0.6$ and $m = 0.9$.

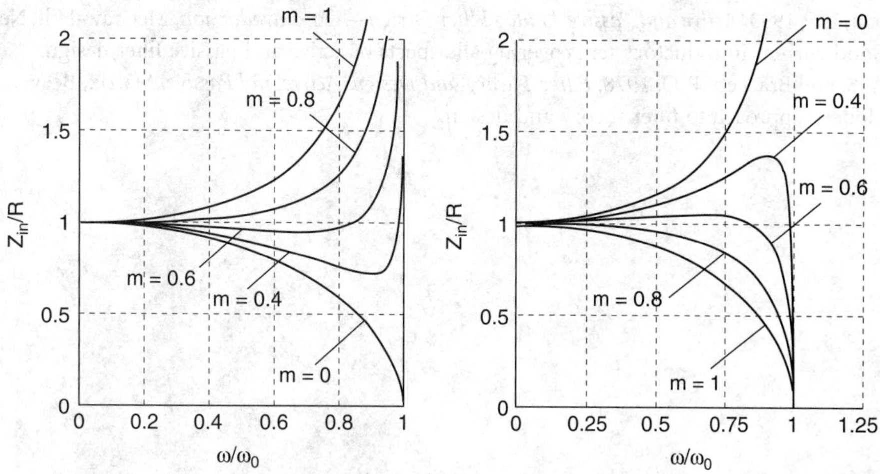

FIGURE 111.18 Z_{iT}/R and $Z_{i\Pi}/R$ versus normalized frequency for various values of m.

Defining Terms

Bridge-T network — A two-port network that consists of a basic T-section and another element connected so as to "bridge across" the two arms. Such networks find applications as band rejection filters, calibration bridges, and feedback networks.

Butterworth filters — Ladder networks that enjoy a unique passband frequency response characteristic that remains very constant until near the cutoff, hence the designation "maximally flat." This filter has its critical frequency remain fixed regardless of the number of stages employed. It obtains this characteristic by realizing a transfer function built around a Butterworth polynomial.

Chebyshev filters — A variant of the Butterworth design that achieves a significantly steeper transition band about its critical frequency for the same number of poles. Although the Chebyshev filter also maintains the integrity of its critical frequency regarding the number of poles, it trades the steeper roll-off for a fixed ripple — usually specified as 1 dB or 3 dB — in the passband. Chebyshev filters are also called *equal-ripple* or *stagger-tuned* filters. They are designed by realizing a transfer function using a Chebyshev polynomial.

Parallel-T networks — A two-port network that consists of two separate T-sections in parallel with only the ends of the arms and the stem connected. Parallel-T networks have applications similar to those of the bridge-T but can produce narrower attenuation bandwidths.

References

Herrero, J. L. and Willoner, G. 1966. *Synthesis of Filters,* Prentice Hall, Englewood Cliffs, NJ.

Thomas, R. E. and Rosa, A. J. 2004. *The Analysis and Design of Linear Circuits*, John Wiley & Sons, Hoboken, NJ.

Van Valkenburg, M. E. 1955. Two-terminal-pair reactive networks (filters). In *Network Analysis,* Prentice Hall, Englewood Cliffs, NJ.

Weinberg, L. 1962. *Network Analysis and Synthesis,* W. L. Everitt (ed.) McGraw-Hill, New York.

Williams, A. B. 1981. *Electronic Filter Design Handbook,* McGraw-Hill, New York.

Zobel, O. J. 1923. Theory and Design of Uniform and Composite Electric Wave Filters. *Bell Telephone Syst. Tech. J.* 2:1.

Further Information

Huelsman, L. P. 1993. *Active and Passive Analog Filter Design — An Introduction,* McGraw-Hill, New York. Good current introductory text covering all aspects of active and passive filter design.

Sedra, A. S. and Brackett, P. O. 1978. *Filter Theory and Design: Active and Passive,* Matrix, Beaverton, OR. Modern approach to filter theory and design.

112

Power Distribution

Robert Broadwater
Virginia Polytechnic Institute and State University

Albert Sargent
Entergy Corporation

Murat Dilek
Electrical Distribution Design, Inc.

The function of power distribution is to deliver to consumers economic, reliable, and safe electrical energy in a manner that conforms to regulatory standards. Power distribution systems receive electric energy from high-voltage transmission systems and deliver energy to consumer service-entrance equipment. Systems typically supply alternating current at voltage levels ranging from 120 V to 46 kV.

Figure 112.1 illustrates aspects of a distribution system. Energy is delivered to the distribution substation (shown inside the dashed line) by three-phase transmission lines. A transformer in the substation steps the voltage down to the distribution primary system voltage — in this case, 12.47 kV. Primary distribution lines leave the substation carrying energy to consumers. The substation contains a breaker that may be opened to disconnect the substation from the primary distribution lines. If the breaker is opened, outside the substation there is normally an open supervisory switch that may be closed in order to provide an alternate source of power for the customers normally served by the substation. The substation also contains a capacitor bank used for either voltage or power factor control.

Four types of customers, along with representative distribution equipment, are shown in Figure 112.1. A set of loads requiring high reliability of service is shown being fed from an underground three-phase secondary network cable grid. A single fault does not result in an interruption to this set of loads. A residential customer is shown being supplied from a two-wire, one-phase overhead lateral. Commercial and industrial customers are shown being supplied from the three-phase, four-wire, overhead primary feeder. At the industrial site, a capacitor bank is used to control the power factor. Except for the industrial customer, all customers shown have 240/120 V service. The industrial customer has 480Y/277 V service.

For typical electric utilities in the U.S., investment in distribution ranges from 35 to 60% of total capital investment.

112.1 Equipment

Figure 112.1 illustrates a typical arrangement of some of the most common equipment. Equipment may be placed into the general categories of transmission, protection, and control.

Arresters protect distribution system equipment from transient over-voltages due to lightning or switching operations. In over-voltage situations the arrester provides a low-resistance path to ground for currents to follow.

0-8493-1586-7/05/$0.00+$1.50

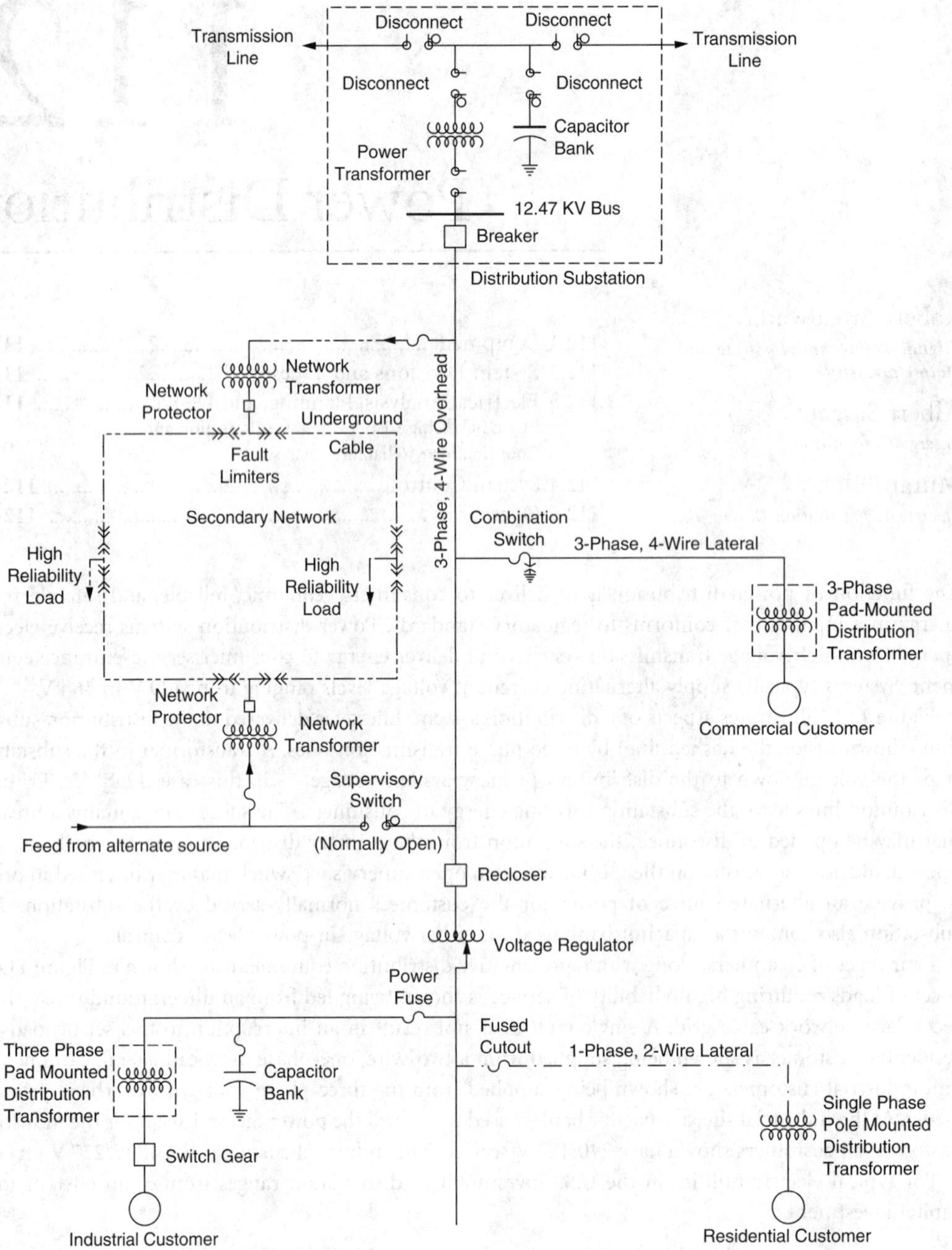

FIGURE 112.1 Distribution system schematic.

Capacitor banks are energy storage devices primarily used to control voltage and power factor. System losses are reduced by the application of capacitors.

Conductors are used to transmit energy and may be either bare or insulated. Bare conductors have better thermal properties and are generally used in overhead construction where contact is unlikely. Insulated cables are used in underground/conduit construction and in overhead applications where minimum right-of-way is available. Concentric neutral and tape-shielded cables provide both a phase conductor and a return path conductor in one unit.

Distribution lines are made up of conductors and are classified according to primary voltage, the number of phases, number of conductors, and return path. The three-phase, four-wire, multi-grounded system is the most common primary system, where one conductor is installed for each of the three phases and the fourth conductor is a neutral that provides a return current path. Multi-grounded means that the neutral is grounded at many points, so that the earth provides a parallel path to the neutral for return current. Three-phase, three-wire primary systems, or delta-connected systems, are rarely used because faults therein are more difficult to detect. A lateral is a branch of the system that is shorter in length, more lightly loaded, or has a smaller conductor size than the primary feeder.

Distribution transformers step the voltage down from the primary circuit value to the customer utilization level, thus controlling voltage magnitude. Sizes range from 1.5 to 2500 kVA. Distribution transformers are installed on poles, ground-level pads, or in underground vaults. A specification of 7200/12,470Y V for the high-voltage winding of a single-phase transformer means the transformer may be connected in a line-to-neutral "wye" connection for a system with a line-to-line voltage of 12,470 V or in a line-to-line "delta" connection for a system with a line-to-line voltage of 7200 V. A specification of 240/120 V for the low-voltage winding means the transformer provides a three-wire connection with 120 V mid-tap voltage and 240 V full-winding voltage. A specification of 480Y/277 V for the low voltage winding means the winding is permanently wye-connected with a fully insulated neutral avilable for a three-phase, four-wire service to deliver 480 V line-to-line and 277 V line-to-neutral.

Distribution substations consist of one or more step-down power transformers configured with switch gear, protective devices, and voltage regulation equipment for the purpose of supplying, controlling, switching, and protecting the primary feeder circuits. The voltage is stepped down for safety and flexibility of handling in congested consumer areas. Over-current protective devices open and interrupt current flow in order to protect people and equipment from fault current. Switches are used for control to interrupt or redirect power flow. Switches may be operated manually, automatically with PLC control, or remotely with supervisory control. Switches are usually rated to interrupt load current and may be either pad or pole mounted.

Power transformers are used to control and change voltage level. Power transformers equipped with tap-changing mechanisms can control secondary voltage over a typical range of plus or minus 10%.

Voltage regulators are autotransformers with tap-changing mechanisms that may be used throughout the system for voltage control. If the voltage at a remote point is to be controlled, then the regulator can be equipped with a line drop compensator that may be set to regulate the voltage at the remote point based upon local voltage and current measurements. Modern microprocessor-based controls enable regulators and line capacitors to work together to provide optimal voltage regulation.

112.2 System Divisions and Types

Distribution transformers separate the primary system from the secondary. Primary circuits transmit energy from the distribution substation to customer distribution transformers. Three-phase distribution lines that originate at the substation are referred to as primary feeders or primary circuits. Primary feeders are illustrated in Figure 112.1. Secondary circuits transmit energy from the distribution transformer to the customer's service entrance. Line-to-line voltages range from 208 to 600 V.

Radial distribution systems provide a single path of power flow from the substation to each individual customer. This is the least costly system to build and operate and thus the most widely used.

Primary networks contain at least one loop that generally may receive power from two distinct sources. This design results in better continuity of service. A primary network is more expensive than the radial system design because more protective devices, switches, and conductors are required.

Secondary networks are normally underground cable grids providing multiple paths of power flow to each customer. A secondary network generally covers a number of blocks in a downtown area. Power is supplied to the network at a number of points via network units, consisting of a network transformer in series with a network protector. A network protector is a circuit breaker connected between the secondary winding of the network transformer and the secondary network itself. When the network is operating properly, energy flows into the network. The network protector opens when reverse energy flow is detected, such as may be caused by a fault in the primary system.

112.3 Electrical Analysis, Planning, and Design

The distribution system is planned, designed, constructed, and operated based on the results of electrical analysis. Generally, computer-aided analysis is used.

Line impedances are needed by most analysis applications. Distribution lines are electrically unbalanced due to loads, unequal distances between phases, dissimilar phase conductors, and single-phase or two-phase laterals. Currents flow in return paths due to the imbalance in the system. Three-phase, four-wire, multigrounded lines have two return paths — the neutral conductor and earth. Three-phase, multigrounded concentric neutral cable systems have four return paths. The most accurate modeling of distribution system impedance is based upon Carson's equations. With this approach a 5 × 5 impedance matrix is derived for a system with two return paths, and a 7 × 7 impedance matrix is derived for a system with four return paths. For analysis, these matrices are reduced to 3 × 3 matrices that relate phase voltage drops (i.e., ΔV_A, ΔV_B, ΔVc) to phase currents (i.e., I_A, I_B, I_C) as indicated by

$$\begin{bmatrix} \Delta V_A \\ \Delta V_B \\ \Delta V_C \end{bmatrix} = \begin{bmatrix} Z_{AA} & Z_{AB} & Z_{AC} \\ Z_{BA} & Z_{BB} & Z_{BC} \\ Z_{CA} & Z_{CB} & Z_{CC} \end{bmatrix} \begin{bmatrix} I_A \\ I_B \\ I_C \end{bmatrix}$$

Load analysis forms the foundation of system analysis. Load characteristics are time varying and depend on many parameters, including connected consumer types and weather conditions. The load demand for a given customer or group of customers is the load averaged over an interval of time, say 15 min. The peak demand is the largest of all demands. The peak demand is of particular interest since it represents the load that the system must be designed to serve. Diversity relates to multiple loads having different time patterns of energy use. Due to diversity, the peak demand of a group of loads is less than the sum of the peak demands of the individual loads. For a group of loads,

$$\text{Diversity factor} = \frac{\text{Sum of individual load peaks}}{\text{Group peak}}$$

Loads may be modeled as either lumped parameter or distributed. Lumped parameter load models include constant power, constant impedance, constant current, voltage-dependent, and combinations thereof. Generally, equivalent lumped parameter load models are used to model distributed loads. Consider the line section of length L shown in Figure 112.2(a), with a uniformly distributed load current that varies along the length of the line as given by

$$i(x) = \frac{I_2 - I_1}{L} x + I_1$$

The total load current drawn by the line section is thus

$$I_L = I_2 - I_1$$

An equivalent lumped parameter model for the uniformly distributed current load is shown in Figure 112.2(b).

Metered load values are used for analysis when available. Otherwise, estimated loads are calculated using kWHr-to-Peak-KW conversion factors, daily load shapes, and diversity factor curves based on

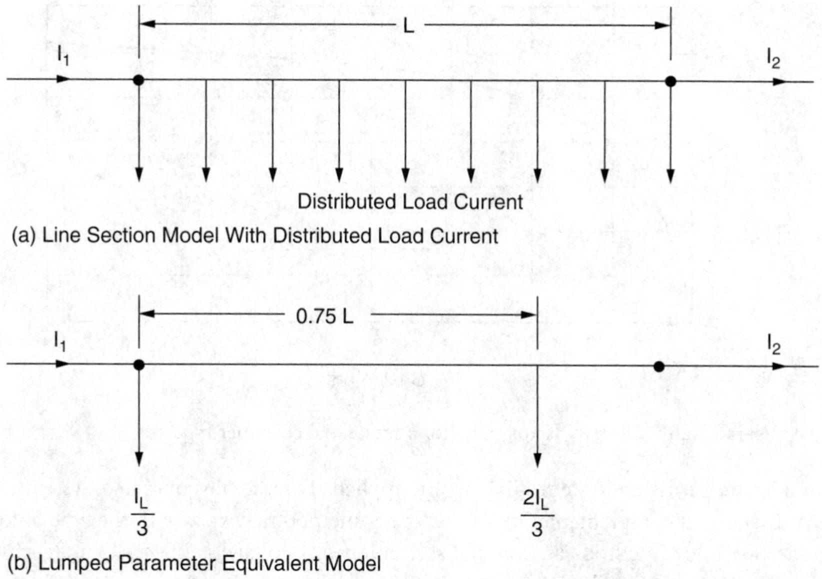

FIGURE 112.2 (a) Line section model with distributed load current; (b) lumped parameter equivalent model.

customer class. These estimation parameters are derived from load research data gathered from a set of randomly selected test customers. A representative diversified load curve for a residential customer type is shown in Figure 112.3, and a representative diversity factor curve for a residential customer type is shown in Figure 112.4.

Load forecasting is concerned with determining load magnitudes during future years from customer growth projections. Short-range forecasts generally have time horizons of approximately five years, whereas long-range forecasts project to around 20 years.

Power flow analysis determines system voltages, currents, and power flows. Power flow results are checked to ensure that voltages fall within allowable limits, that equipment overloads do not exist, and that phase imbalances are within acceptable limits. For primary and secondary networks, power flow

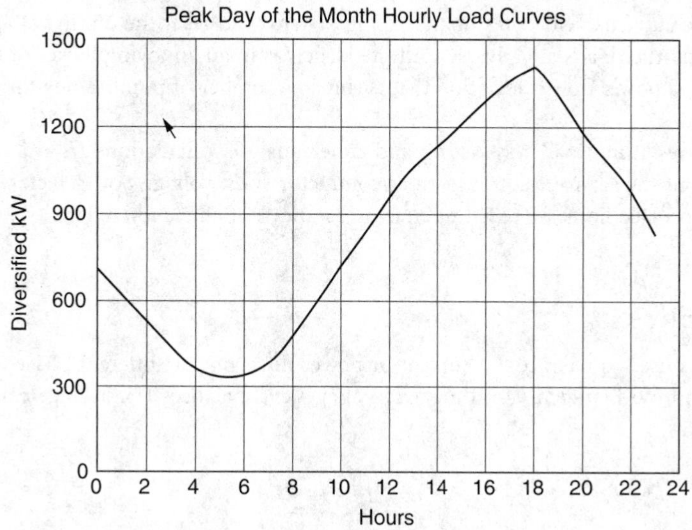

FIGURE 112.3 Representative diversified load curve for a residential customer type.

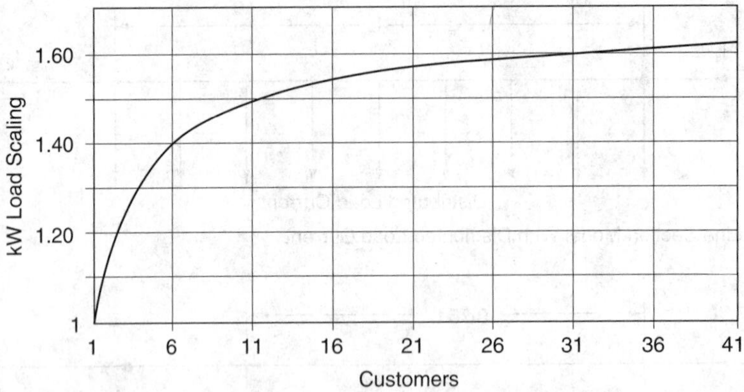

FIGURE 112.4 Representative diversity factor curve for a residential customer type.

methods used in transmission system analysis are applied. For radially operated systems, the ladder method is used. The actual implementation of the ladder method may vary with the type of load models used. All ladder load flow methods assume the substation bus voltage is known. An algorithm for the ladder method consists of the following five steps:

1. **Step 1.** Assume a value for all node voltages throughout the circuit. Generally, assumed voltages are set equal to the substation voltage.
2. **Step 2.** At each load in the circuit, calculate the current from the known load value and assumed voltage.
3. **Step 3.** Starting at the ending nodes, sum load currents to obtain line section current estimates, performing summation until the substation is reached.
4. **Step 4.** Having estimates of all line section currents, start at the substation and calculate line section voltage drops and new estimates of node voltages.
5. **Step 5.** Compare new node voltages with estimates of previous iteration values. The algorithm has converged if the change in voltage is sufficiently small. If the algorithm has not converged, return to Step 2.

Dynamic load analysis includes such studies as motor-starting studies. Rapid changes in large loads can result in large currents, with a resultant drop in system voltage. If the dip in voltage is too large or too frequent, then other loads are adversely affected, such as in an annoying flicker of lights. This study generally employs a power flow calculation that is run at a number of points along the dynamic characteristic of the load.

Planning involves using load forecasting and other analysis calculations to evaluate voltage level, substation locations, feeder routes, transformer/conductor sizes, voltage/power factor control, and restoration operations. Decisions are based upon considerations of efficiency, reliability, peak demand, and life cycle cost.

Phase Balancing

Phase balancing is used to balance the current or power flows on the different phases of a line section. This results in improved efficiency and primary voltage level balance. The average current in the three phases is defined as

$$I_{avg} = \frac{I_A + I_B + I_C}{3}$$

The maximum deviation from I_{avg} is given by

$$\Delta I_{dev} = \text{maximum of}\left\{\left|I_{avg} - I_A\right|, \left|I_{avg} - I_B\right|, \left|I_{avg} - I_C\right|\right\}$$

Phase imbalance is defined as

$$\text{Phase imbalance} = \frac{\Delta I_{dev}}{I_{avg}}$$

Fault Analysis

Fault analysis provides the basis for protection system design. Thus, in the model used to calculate fault currents, load currents are neglected. Sources of fault current are the substation bus, distributed resource generators, and large synchronous motors on the feeder or neighboring feeders. A variety of fault conditions are considered at each line section, including three-phase-to-ground, single-phase-to-ground, and separate phases contacting one another. In performing the calculations, both bolted (i.e., zero-impedance) faults and faults with impedance in the fault path are considered. Of interest are the maximum and minimum phase and return path fault currents, as well as the fault types that result in these currents.

At a multi-phase grounded node in a linear distribution system, post-fault voltages are related to pre-fault voltages and fault currents as given by

$$\mathbf{V}_f = \mathbf{V}_0 - \mathbf{Z}_{th}\mathbf{I} \tag{112.1}$$

where $\mathbf{V}_f$ denotes phase voltages (voltages between phase A and ground, B and ground, and C and ground) of the node during the fault, $\mathbf{V}_0$ is the array of phase voltages before the fault occurs, $\mathbf{I}$ is the array of fault currents that will flow out of the phases of the node during the fault, and $\mathbf{Z}_{th}$ (a 3×3 matrix) represents the phase Thevenin matrix looking into the node.

Once $\mathbf{Z}_{th}$ and pre-fault voltages at the node are available, $\mathbf{V}_f$ can be written in terms of $\mathbf{I}$ depending upon conditions imposed by the fault. Then Equation (112.1) can be solved for $\mathbf{I}$. The pre-fault system model represents the system behavior before the fault occurs. On the pre-fault model, a power flow calculation may be used to obtain the voltages $\mathbf{V}_0$.

The post-fault model represents the system behavior during the fault. For fault calculations, the circuit model used is modified in several ways from the pre-fault model. Usually, load currents are neglected in the post-fault model, and instead superposition is used to add load currents obtained from the pre-fault power flow analysis to fault currents. For the fault calculations, the circuit model is assumed to be linear. Other changes for the post-fault circuit analysis include neglecting slow-acting control devices (such as substation transformer tap changers) that do not have time to react during the time of the fault; and inserting appropriate Thevenin equivalent source impedances, representing the Thevenin impedance seen by the distribution substation looking back into the transmission or subtransmission system. Using the post-fault circuit model assumptions, a power flow calculation using constant current injections at the fault point may be used to perform the fault calculations. This is the approach described here.

The calculation of $\mathbf{Z}_{th}$ may be performed by inserting a small test load sequentially at every individual phase of the faulted node. Prior to any test load insertion, the phase voltages of the node are obtained from a power flow solution. Let these voltages be $\mathbf{V}_i$. After inserting a test load, the power flow is used again to obtain the current flowing into the test load and the voltages at all phases of the node. Ratios of changes in $\mathbf{V}_i$ to the current drawn by the test load constitute the columns of $\mathbf{Z}_{th}$. For instance, if the test load is inserted between phase A and ground at grounded node N, the results calculated are the first column of $\mathbf{Z}_{th}$.

To elaborate, refer to Figure 112.5 where a grounded node N is considered. Here, a general node at which any phase may exist is assumed. The power flow is run on the post-fault system model, and phase-to-neutral voltages at node N are obtained as V_{an}, V_{bn}, and V_{cn} for the phases A, B, and C, respectively

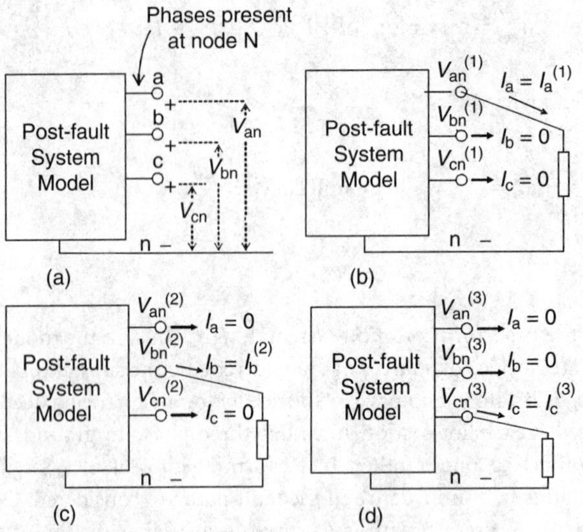

FIGURE 112.5 Constructing $\mathbf{Z}_{th}$ at a grounded node N. (a) Voltages before inserting any test load. (b) A test load being inserted between phase A and ground. (c) A test load being inserted between phase B and ground. (d) A test load being inserted between phase C and ground.

[Figure 112.5(a)]. The neutral n is regarded to be the same as ground. A test load will be inserted between A and n, B and n, and C and n sequentially [Figure 112.5(b) through Figure 112.5(d)]. During each load insertion, line currents and phase-to-neutral voltages may be obtained from a power flow calculation.

The elements z_{ij} of $\mathbf{Z}_{th}$ may be determined in the following manner:

$$z_{11} = \frac{V_{an} - V_{an}^{(1)}}{I_a^{(1)}}, \quad z_{21} = \frac{V_{bn} - V_{bn}^{(1)}}{I_a^{(1)}}, \quad z_{31} = \frac{V_{cn} - V_{cn}^{(1)}}{I_a^{(1)}}$$

$$z_{12} = \frac{V_{an} - V_{an}^{(2)}}{I_b^{(2)}}, \quad z_{22} = \frac{V_{bn} - V_{bn}^{(2)}}{I_b^{(2)}}, \quad z_{32} = \frac{V_{cn} - V_{cn}^{(2)}}{I_b^{(2)}}$$

$$z_{13} = \frac{V_{an} - V_{an}^{(3)}}{I_c^{(3)}}, \quad z_{23} = \frac{V_{bn} - V_{bn}^{(3)}}{I_c^{(3)}}, \quad z_{33} = \frac{V_{cn} - V_{cn}^{(3)}}{I_c^{(3)}}$$

$\mathbf{Z}_{th}$ represents the relationship between the voltage changes and the current changes at N. Suppose phase-to-neutral voltages at N before the fault are V_{an}^i, V_{bn}^i, and V_{cn}^i. Assume that a fault occurs at N and causes currents I_a, I_b and I_c to flow out of phases A, B, and C, respectively, resulting in phase-to-neutral voltages V_{an}^f, V_{bn}^f, and V_{cn}^f.

Then voltage changes at N are related to the currents drawn, via $\mathbf{Z}_{th}$ as

$$\begin{bmatrix} \Delta V_{an} \\ \Delta V_{bn} \\ \Delta V_{cn} \end{bmatrix} = \begin{bmatrix} z_{11} & z_{12} & z_{13} \\ z_{21} & z_{22} & z_{23} \\ z_{31} & z_{32} & z_{33} \end{bmatrix} \begin{bmatrix} I_a \\ I_b \\ I_c \end{bmatrix} \tag{112.2}$$

where $\Delta V_{kn} = V_{kn}^i - V_{kn}^f$ for k = a,b,c.

Equation (112.2) denotes a general case. Suppose node N is a double-phase location having phases A and B but no phase C. Then, all the elements in the third row and third column of $\mathbf{Z}_{th}$ are zero.

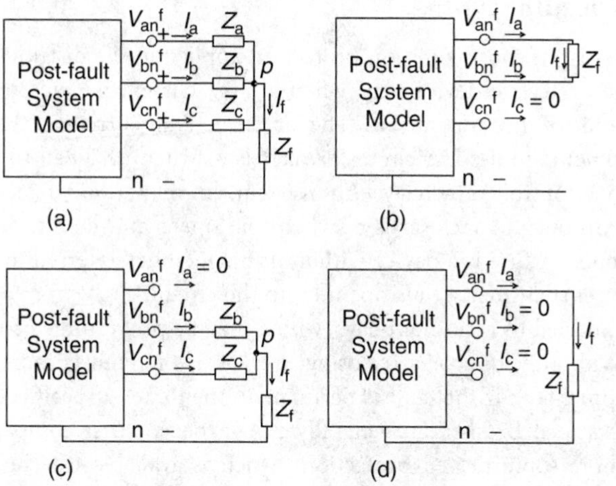

FIGURE 112.6 Various faults at a grounded node N. (a) Three-phase fault. (b) Phase-to-phase fault. (c) Double-phase-to-ground fault. (d) Single-phase-to-ground fault.

Various fault cases at N are shown in Figure 112.6. A general case of a three-phase-to-ground fault at N is represented in Figure 112.6(a). Here, each phase has its own fault impedance (Z_a, Z_b, and Z_c for phases A, B, and C, respectively) to the common point p. Z_f is the impedance between p and n. Consider solving for a three-phase-to-ground fault. Let V^i_{kn} and V^f_{kn} denote pre-fault and post-fault phase-to-ground voltages of phase k, respectively. Then the boundary conditions are:

$$I_f = I_a + I_b + I_c, \tag{112.3}$$

$$\Delta V_{an} = V^i_{an} - V^f_{an} = V^i_{an} - \left(I_a Z_a + I_f Z_f\right), \tag{112.4}$$

$$\Delta V_{bn} = V^i_{bn} - V^f_{bn} = V^i_{bn} - \left(I_b Z_b + I_f Z_f\right), \tag{112.5}$$

$$\Delta V_{cn} = V^i_{cn} - V^f_{cn} = V^i_{cn} - \left(I_c Z_c + I_f Z_f\right). \tag{112.6}$$

First using Equation (112.3) in Equation (112.4), Equation (112.5), and Equation (112.6), then substituting Equation (112.4), Equation (112.5), and Equation (112.6) into Equation (112.2) gives the fault currents as

$$\begin{bmatrix} I_a \\ I_b \\ I_c \end{bmatrix} = \begin{bmatrix} z_{11} + Z_a + Z_f & z_{12} + Z_f & z_{13} + Z_f \\ z_{21} + Z_f & z_{22} + Z_b + Z_f & z_{23} + Z_f \\ z_{31} + Z_f & z_{32} + Z_f & z_{33} + Z_c + Z_f \end{bmatrix}^{-1} \begin{bmatrix} V^i_{an} \\ V^i_{bn} \\ V^i_{cn} \end{bmatrix}$$

Any fault event imposes a set of boundary conditions. Initial voltages (pre-fault voltages) can be readily calculated from the power flow. The final voltages are expressed under the boundary conditions in terms of the fault currents and fault impedances. Then, Equation (112.2) is solved for fault currents. Using this approach, fault currents for cases b through d shown in Figure 112.6 may be evaluated. For an ungrounded node, the phase-to-phase voltages instead of phase-to-neutral voltages are employed.

Protection and Coordination

Over-current protection is the most common protection applied to the distribution system. With over-current protection, the protective device trips when a large current is detected. The time to trip is a function of the magnitude of the fault current. The larger the fault current is, the quicker the operation. Various types of equipment are used. A circuit breaker is a switch designed to interrupt fault current, the operation of which is controlled by relays. An over-current relay, upon detecting fault current, sends a signal to the breaker to open. A recloser is a switch that opens and then recloses a number of times before finally locking open. A fuse is a device with a fusible member, referred to as a fuse link, which in the presence of an over current melts, thus opening up the circuit.

Breakers may be connected to reclosing relays, which may be programmed for a number of opening and reclosing cycles. With a recloser or a reclosing breaker, if the fault is momentary, then the power interruption is also momentary. If the fault is permanent, then after a specified number of attempts at reclosing the device locks open. Breakers are generally more expensive than comparable reclosers. Breakers are used to provide more sophisticated protection, which is available via choice of relays. Fuses are generally used in the protection of laterals.

Protective equipment sizing and other characteristics are determined from the results of fault analysis. Moving away from the substation in a radial circuit, both load current and available fault current decrease. Protective devices are selected based on this current grading. Protective devices are also selected to have different trip-delay times for the same fault current. With this time grading, protective devices are coordinated to work together such that the device closest to a permanent fault clears the fault. Thus reclosers can be coordinated to protect load-side fuses from damage due to momentary faults.

Reliability Analysis

Reliability analysis involves determining indices that relate to continuity of service to the customer. Reliability is a function of tree conditions, lightning incidence, equipment failure rates, equipment repair times, and circuit design. The reliability of a circuit generally varies from point to point due to protection system design, placement of switches, and availability of alternative feeds. Many indices are used in evaluating system reliability. Common ones include system average interruption frequency index (SAIFI), system average interruption duration index (SAIDI), customer average interruption frequency index (CAIFI), and customer average interruption duration index (CAIDI) as defined by

$$\text{SAIFI} = \frac{\text{Total number of customer interruptions}}{\text{Total number of customers served}}$$

$$\text{SAIDI} = \frac{\text{Sum of customer interruption durations}}{\text{Total number of customers}}$$

$$\text{CAIFI} = \frac{\text{Total number of customer interruptions}}{\text{Total number of customers affected}}$$

$$\text{CAIDI} = \frac{\text{Sum of customer interruption durations}}{\text{Total number of customers affected}}$$

112.4 System Control

Voltage control is required for proper operation of customer equipment. For instance, in the U.S., "voltage range A" for single-phase residential users specifies that the voltage may vary at the service entrance from 114/228 V to 126/252 V. Regulators, tap-changing under load transformers, and switched capacitor banks are used in voltage control.

Power factor control is used to improve system efficiency. Due to the typical load being inductive, power factor control is generally achieved with fixed and/or switched capacitor banks.

Power flow control is achieved with switching operations. Such switching operations are referred to as system reconfiguration. Reconfiguration may be used to balance a load among interconnected distribution substations. Such switching operations reduce losses while maintaining proper system voltage.

Load control may be achieved with voltage control and also by remotely operated switches that disconnect load from the system. Generally, load characteristics are such that if the voltage magnitude is reduced, then the power drawn by the load will decrease for some period of time. Load control with remotely operated switches is also referred to as load management.

Effective system control is essential to provide adequate power quality. Power quality may be defined as the absence of service interruptions, voltage dips and sags, and voltage spikes and surges. Proper control of system voltage is more critical now than ever before because many microprocessor-based controls and adjustable speed drives have voltage tolerances less than 10%.

112.5 Operations

The operations function includes system maintenance, construction, and service restoration. Maintenance, such as trimming trees to prevent contact with overhead lines, is important to ensure a safe and reliable system. Interruptions may be classified as momentary or permanent. A momentary interruption is one that disappears very quickly — for instance, a recloser operation due to a fault from a tree limb briefly touching an overhead conductor. Power restoration operations are required to repair damage caused by permanent interruptions.

While damaged equipment is being repaired, power restoration operations often involve reconfiguration in order to restore power to interrupted areas. With reconfiguration, power flow calculations may be required to ensure that equipment overloads are not created from the switching operations.

Defining Terms

Current return path — The path that current follows from the load back to the distribution substation. This path may consist of either a conductor (referred to as the neutral) or earth, or the parallel combination of a neutral conductor and the earth.

Fault — A conductor or equipment failure or unintended contact between conductors or between conductors and grounded objects. If not interrupted quickly, fault current can severely damage conductors and equipment.

Phase — Relates to the relative angular displacement of the three sinusoidally varying voltages produced by the three windings of a generator. For instance, if phase A voltage is $120\angle\ 0^\circ$ V, phase B voltage $120\angle\ -120^\circ$ V, and phase C voltage $120\angle\ 120^\circ$ V, the phase rotation is referred to as ABC. Sections of the system corresponding to the phase rotation of the voltage carried are commonly referred to as phase *A*, *B*, or *C*.

Tap-changing mechanism — A control device that varies the voltage transformation ratio between the primary and secondary sides of a transformer. The taps may only be changed by discrete amounts, say 0.625%.

References

Broadwater, R. P., Shaalan, H. E., Oka, A., and Lee, R. E. 1993. Distribution system reliability and restoration analysis. *Electric Power Sys. Res. J.* 29(2):203–211.

Carson, J. R. 1926. Wave propagation in overhead wires with ground return. *Bell System Tech. J.* 5:40–47.

Engel, M. V., Greene, E. R., and Willis, H. L. (Eds.) 1992. IEEE Tutorial Course: Power Distribution Planning. Course Text 92 EHO 361-6-PWR IEEE Service Center, Piscataway, NJ.

Kersting, W. H. and Mendive, D. L. 1976. An Application of Ladder Network Theory to the Solution of Three-Phase Radial Load Flow Problems. IEEE Winter Meeting, New York.

Further Information

Redmon, J. R. 1988. IEEE Tutorial Course on Distribution Automation. Course Text 88 EH0 280-8-PWR IEEE Service Center, Piscataway, NJ.

Electric Utility Engineers, Westinghouse Electric Corporation. 1950. *Electrical Transmission and Distribution Reference Book,* Westinghouse Electric Corporation, Pittbsurgh, PA.

Kersting, W. H. 2002. *Distribution System Modeling and Analysis,* CRC Press, Boca Raton, FL.

Lakervi, E. and Holmes, E. J. 1989. *Electricity Distribution Network Design,* Peter Peregrinus, London.

Pansini, A. J. 1992. *Electrical Distribution Engineering,* Fairmont Press, Liburn, GA.

113

Grounding, Shielding, and Filtering

William G. Duff
(First Edition)
Computer Sciences Corporation

Arindam Maitra
(Second Edition)
EPRI PEAC Corporation

Kermit Phipps
(Second Edition)
EPRI PEAC Corporation

Anish Gaikwad
(Second Edition)
EPRI PEAC Corporation

Many electromagnetic compatibility problems are caused by low immunity to emissions and poor facility and data wiring and grounding which further affect equipment performance. Besides, radiated and conducted emissions generated by a piece of equipment may also affect that equipment's own performance. To avoid problems, electromagnetic interference (EMI) control measures must be incorporated into the initial circuit design. Grounding, shielding, and filtering are some of the important factors that must be considered during the initial design of electronic circuits.

The purpose of this chapter is to give engineers and facility engineers who are unfamiliar with EMI-related problems insights and information necessary to improve equipment compatibility. The goal here is not to duplicate information currently available, but rather to collect information in a single location and then supplement it to provide adequate information and procedures to applications personnel in effectively limiting the spurious emissions given off by electronics and ensuring that electronic equipment is not adversely affected by such emissions.

0-8493-1586-7/05/$0.00+$1.50

113.1 Analyzing and Solving Problems Associated with Electromagnetic Interference

What Is Electromagnetic Interference?

Electromagnetic interference (EMI) is any natural or manmade electrical or electromagnetic energy that results in unintentional and undesirable equipment responses. Electromagnetic energy travels in the form of emissions, either conducted or radiated.

Conducted emissions are generated inside electrical or electronic equipment and may be transmitted outward through the equipment's data input or output lines, its control leads, or its power conductors. Conducted emissions may cause an EMI problem between equipment that generates useful emissions and other equipment with low immunity to those same emissions.

Radiated emissions are radio-frequency electromagnetic energy that travels through the air. Radiated emissions are also generated by electrical or electronic equipment and may be emitted from power and data cables that are poorly shielded or unshielded, leaky equipment apertures, equipment housings that are inadequately shielded, or equipment antennae that may or may not be operating normally.

Current trends in the electronics industry (such as increases in the quantity of electronic equipment, reliance on electronic devices in critical applications, higher clock frequencies of computing devices, higher power levels, lower sensitivities, increased packaging densities, use of plastics, etc.) will tend to create more EMI problems. Whether conducted or radiated, emissions include three properties: amplitude, frequency, and waveform. EMI can occur in equipment with low immunity to emissions when any or all of these properties vary from normal, for example, emissions that are too high in amplitude, are too low or too high in frequency, or whose waveforms are distorted. EMI can also occur when these properties are within normal operating parameters, usually resulting from equipment's low immunity to emissions. Examples of intentional and unintentional conducted and radiated emissions are illustrated in Table 113.1.

Causes of EMI

EMI is generally common-mode (CM) noise, which is induced onto a signal with respect to a reference ground. The noise is coupled to ground from the power cables through the capacitance between the power cable and ground. Figure 113.1 demonstrates this principle.

The capacitance between the cable and ground increases as the length of cable increases. Therefore, short lengths of cable have a low risk of common-mode noise. As the length of cable increases, the risk of common-mode noise increases and the need for EMI solutions rises.

As shown in Figure 113.2, the common-mode ground current $I_{ao} = C_{l\text{-}g}\, dv/dt$. This characteristic of common-mode current makes the adjustable-speed drive a prime source of common-mode noise because of its abrupt voltage transitions on the drive output terminal. The conducted noise will be created as the individual pulses on the drive output couple with the ground conductor. Some common symptoms of EMI-related problems are:

- Unexplained drive trips with no correlation with voltage disturbances.
- Malfunctions of barcode/vision systems, ultrasonic sensors, and weighing and temperature sensors.
- Intermittent data errors in drive-control interfaces such as encoder feedback, I/O, and 0-10-V analog out.
- Interference with TV, AM radio, and radio-controlled devices.

Radiated emissions from many types of electronic equipment, including ASDs, lighting systems, broadcast communication equipment, and medical equipment, have been shown to cause electromagnetic interference with other types of sensitive electronic equipment. Figure 113.1 shows how conducted and radiated emissions propagated through the electromagnetic environment may interfere with sensitive electronic medical equipment that is microprocessor based.

TABLE 113.1 Examples of Intentional and Unintentional Conducted and Radiated Emissions

Examples of Conducted Emission Waveform or Electromagnetic Disturbance	Description
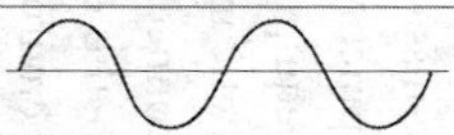 Nominal 60 Hz Line Voltage	No electrical noise, or conducted emissions, are present on ideal noise-free nominal voltage.
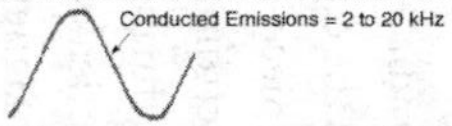 Nominal Voltage with Continuous Conducted Emissions	Power-line carriers, which are often continuous emissions superimposed upon nominal voltage, carry useful information to other devices on the building wiring, such as automatic clock systems.
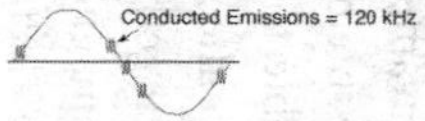 Nominal Voltage with Pulsed Conducted Emissions	Power-line carriers may also be pulsed, or short bursts of conducted emissions to operate equipment, such as energy management systems, transmitters and receivers on the building wiring system.
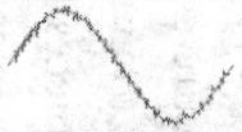 Nominal Voltage with Continuous Electrical Noise	Switching electronic equipment, which may contain a switch-mode power supply, and rotating electrical machinery, may generate continuous unintentional electrical noise, or conducted emissions.
 Nominal Voltage with Pulsed Electrical Noise	The turning on and off of electrical loads and electronic equipment may generate unintentional conducted emissions.

Radiated Emission Waveform or Electromagnetic Disturbance	Description
 Amplitude Modulated (AM) Radiated Emissions	AM radio broadcast towers and some high-frequency medical equipment generate useful AM radiated emissions ranging from 535 kHz to 1605 kHz
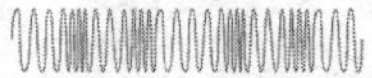 Frequency Modulated (FM) Radiated Emissions Broadcast Transmission)	FM radio broadcast towers generate useful FM radiated emissions ranging from 88 MHz to 108 MHz
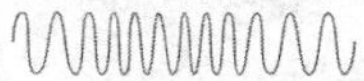 Frequency Modulated (FM) Radiated Emissions (Wireless Communications Devices)	Land-mobile transmitter towers and analog wireless communications devices generate useful FM radiated emissions ranging from 1605 kHz to 459 MHz.
 Television Broadcast Radiated Emissions	Television transmissions generate useful radiated emissions from 54 MHz to 890 MHz.
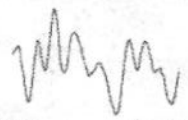 Analog Cellular Telephone Radiated Emissions	Analog cellular telephones generate a continuous type of useful radiated emissions ranging from 824 MHz to 893 MHz.
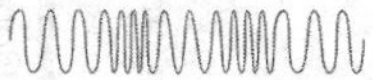 Telemetry Signal Radiated Emissions with Undesired Radiated Interference	Radiated emissions from telemetry transmitters may become distorted by undesired emissions from nearby sources of other radiated emissions.
 Normal High-Energy Radiated Emissions from High-Frequency Medical Equipment	High-frequency medical equipment such as electrosurgical units (ESU) generate useful radiated emissions that are used to perform medical procedures during surgery.
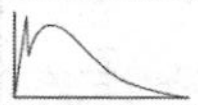 Electrostatic Discharges (ESDs) Generate Radiated Emissions	ESD events may generate unwanted radiated emissions from the electromagnetic fields that collapse around the discharge of current from one surface to another.

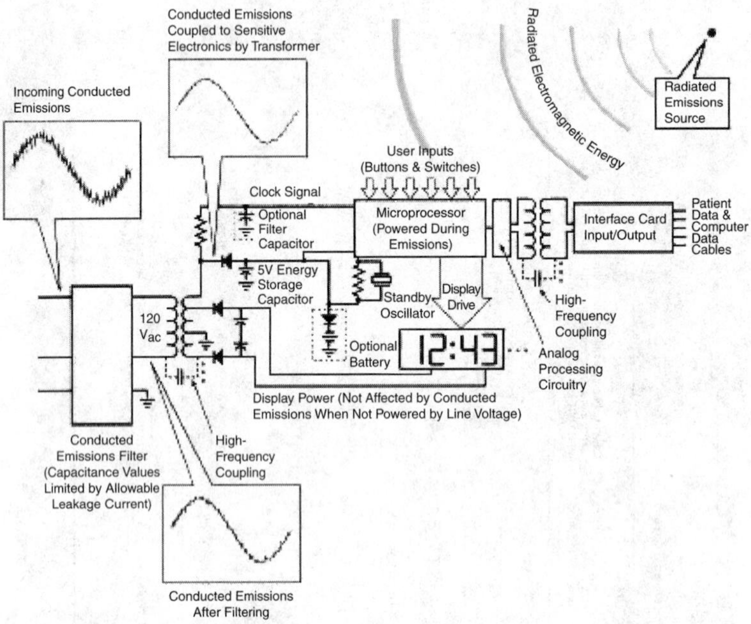

FIGURE 113.1 Capacitive-coupling from phase-to-ground conductor.

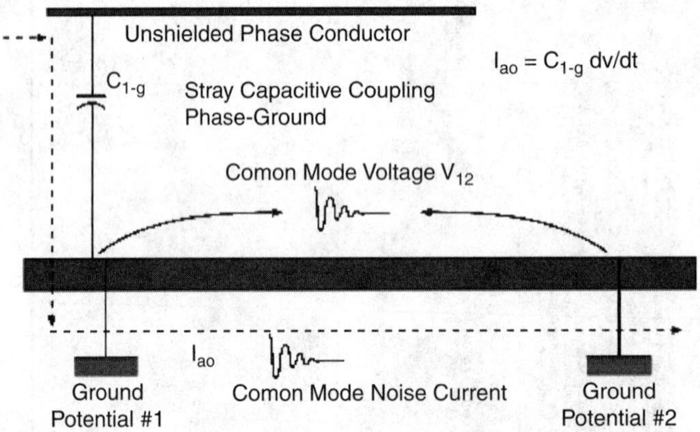

FIGURE 113.2 Examples of conducted and radiated emissions interfering with equipment.

Ungrounded shields or floating shields can act like an antenna to unwanted radiated emissions. In the case of a shielded cable, a cable connector also helps to connect the cable shield to the equipment enclosure electromagnetically. Connectors are used to mechanically secure a cable to a piece of equipment. The pieces of a connector must fit together properly and securely to ensure the electromagnetic integrity of the connector. Particular attention must be paid to how the pieces fit together during connector installation. Improperly fitted or loose connector joints can cause electromagnetic leaks in the connector, which can allow unwanted radiated electromagnetic energy from the electromagnetic environment to penetrate the cable system.

Solutions to EMI Problems

Methods to mitigate EMI in an industrial facility could include proper grounding and shielding of sensitive equipment, attenuating emissions at the source, capturing and returning emissions to the source, etc.

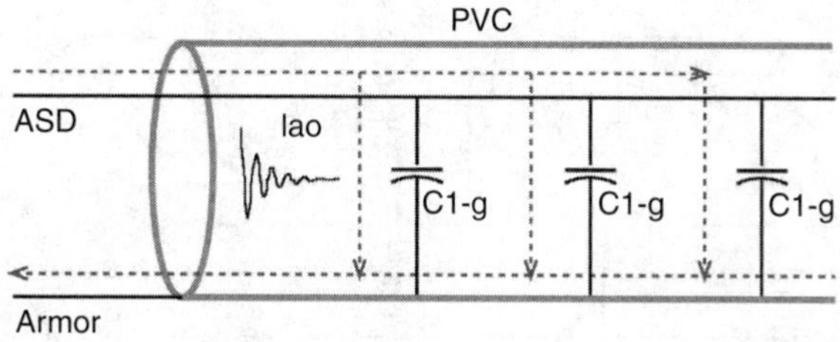

FIGURE 113.3 Use of a shielded/armor cable to reduce EMI.

Proper Grounding and Shielding of Sensitive Equipment

The practice of using unshielded phase conductors in a cable tray from an ASD to a motor could introduce common-mode noise into the system. The use of a shielded-armor power cable from a drive to a motor will provide a path for the common-mode noise to return to the source. Figure 113.3 demonstrates this concept.

Signal shields reduce external coupling but may introduce EMI if the shield is connected to a noisy ground potential. The standard practice is to ground the shield at the source side of the cable. If the standard practice does not eliminate the EMI, it becomes necessary to do whatever it takes to fix the problem, including grounding on both ends, grounding on the other end, or not grounding at all.

The path of common-mode emissions can be diverted from sensitive equipment by separating control and signal cables from high-voltage wires. It is also best if the power conductors include a ground wire and are placed in a conduit. The conduit should be bonded to an ASD cabinet, the motor junction box, and the ground wire should be connected to the cabinet ground bus and motor ground stud. The ground wire and conduit setup parasitic capacitive paths within the conduit and couple high dV/dt pulses and return them back to the source of emissions.

Often it is necessary to isolate the conduit coupling at the point of the motor to prevent the coupled emissions from traveling on the outer surface of the conduit. Insulated motor pairs and ground are recommended to prevent inadvertent grounding of the conduit where new ground loops may be established to radiate the noise. In this practice, it is critical to carry the safety ground within the conduit and ensure proper bonding at the motor ground stud to ensure NEC compliance. This practice is usually necessary where interference levels may be in the low frequency band, for example, 10 to 100 kHz.

Another often-recommended and important practice is to separate control and signal cables from power cables in cable trays. The practice of placing covers on a signal cable tray will further reduce the noise coupled to the signal cables from the power cables.

Attenuating Emissions at the Source

The best way to eliminate system emissions is to attenuate emissions at the source. The use of a common-mode choke (CMC) is one way to achieve this. A CMC is an inductor with phases A, B, and C conductors wound in the same direction through a common magnetic core. It provides high impedance and high inductance to any line-to-ground capacitive current emissions. Unlike a line reactor/inductor, a CMC does not affect the power-line circuit. This device is available from drive vendors. Figure 113.4 shows an example of a CMC application.

Capturing and Returning Emissions to the Source

Another method to reduce EMI is to capture emissions and return them to the source. This can be accomplished with an EMI filter. Figure 113.5 demonstrates the use of an EMI filter. This figure shows that the CM current I_{ao} will collect in the ground conductors and return to the drive through the EMI filter. The filter contains a large common-mode core inductance and individual phase capacitors that limit the high frequency ground return current to low levels in the main AC supply.

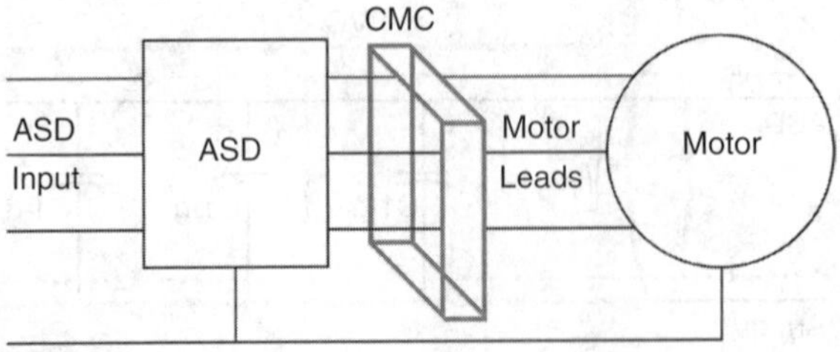

FIGURE 113.4 Application of a common-mode choke.

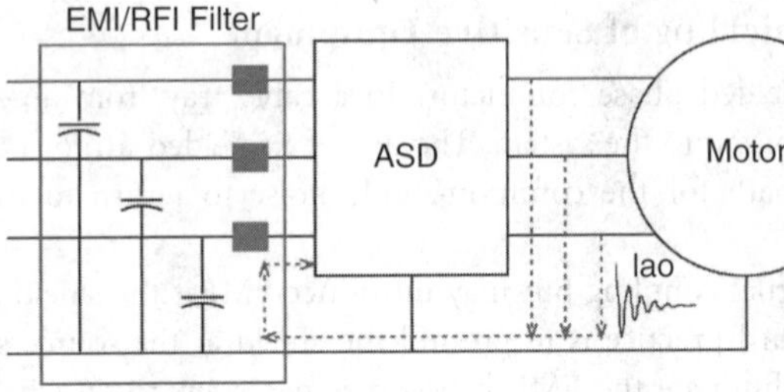

FIGURE 113.5 Application of EMI filter.

113.2 Grounding

With electrical power systems, there are at least two groups or classes of grounding:

1. System grounding
2. Equipment grounding

System grounding pertains to the manner in which a circuit conductor of a system is intentionally connected to earth, or to some conducting body that is effectively connected to earth or serves in place of earth. Equipment grounding is the process of bonding together, with equipment-grounding conductors, all conductive enclosures for conductors and equipment within each circuit. Article 250-50 of the NEC and other important standards require that at each building or structure served, metal underground water pipes, supplemental electrodes, metal frames of the building, and concrete-encased electrodes (rebar) be bonded together to form the grounding electrode system. This grounding practice establishes a zero-voltage reference for an electrical power distribution system. Fundamental design principles for system and equipment grounding and the basic factors that influence the selection of the type of system/equipment grounding are extensively covered in IEEE Color books, including IEEE Std. 141-1993, IEEE Std. 142-1991, IEEE Std. 446-1995, IEEE Std. 1100-1999, NFPA 70-1999 National Electric Code (NEC), and Federal Information Processing Standards (FIPS) 1994.

Note that the basic objectives for grounding circuits, cables, equipment, and systems are to prevent a shock hazard; to protect circuits and equipments; and to reduce EMI due to electromagnetic field, common ground impedance, or other forms of interference coupling. However, grounding is one of the least understood and most significant factors in many EMI problems. Most equipment manufacturers will provide details on grounding of their equipment and may often violate NEC and other important standards. The EMI part of the problem is emphasized in the subsequent sections.

Characteristics of Ground Conductors

Ideally, a ground conductor should provide a zero-impedance path to all signals for which it serves as a reference. If this were the situation, signal currents from different circuits would return to their respective sources without creating unwanted coupling between circuits. Many interference problems occur because

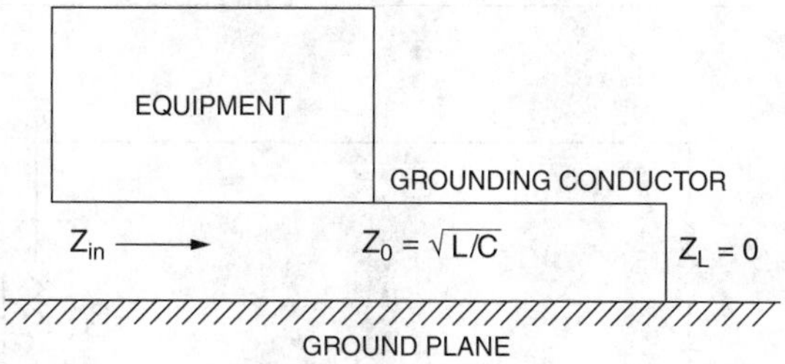

FIGURE 113.6 Idealized equipment grounding.

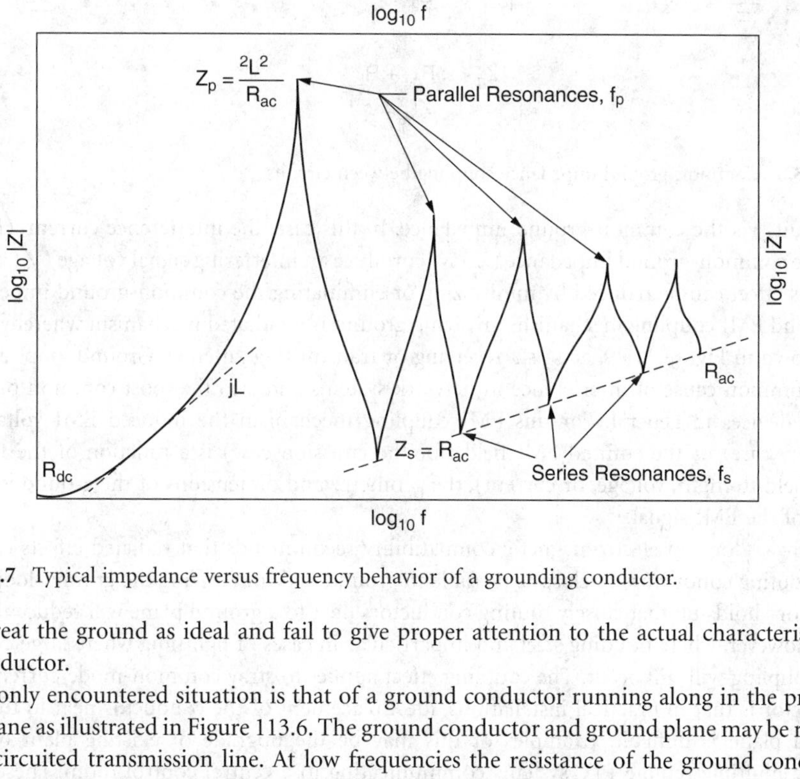

FIGURE 113.7 Typical impedance versus frequency behavior of a grounding conductor.

designers treat the ground as ideal and fail to give proper attention to the actual characteristics of the ground conductor.

A commonly encountered situation is that of a ground conductor running along in the proximity of a ground plane as illustrated in Figure 113.6. The ground conductor and ground plane may be represented as a short-circuited transmission line. At low frequencies the resistance of the ground conductor will predominate. At higher frequencies the series inductance and the shunt capacitance to ground will become significant, and the ground conductor will exhibit alternating parallel and series resonance as illustrated in Figure 113.7. To provide a low impedance to ground, it is necessary to keep the length of the grounding conductor short relative to wavelength (i.e., less than 1/20 of the wavelength).

Ground voltage equalization of voltage differences between parts of an automated data processing (ADP) grounding system is accomplished in parts when the equipment-grounding conductors are connected to the grounding point of a single power supply. However, if the equipment grounding conductors are long, it is difficult to achieve a constant potential throughout the grounding system, particularly for high frequency noise. Supplemental conductors, low-inductance ground plates, and grounding and bonding of raised floor pedestals may be necessary. Detailed discussions and standard practices and procedures are extensively covered in IEEE Color books, including IEEE Std. 142-1991 and IEEE Std. 1100-1999.

Ground-Related EMI Coupling

Ground-related EMI involves one of two basic coupling mechanisms. The first mechanism results from circuits sharing the ground with other circuits. Figure 113.8 illustrates EMI coupling between culprit and

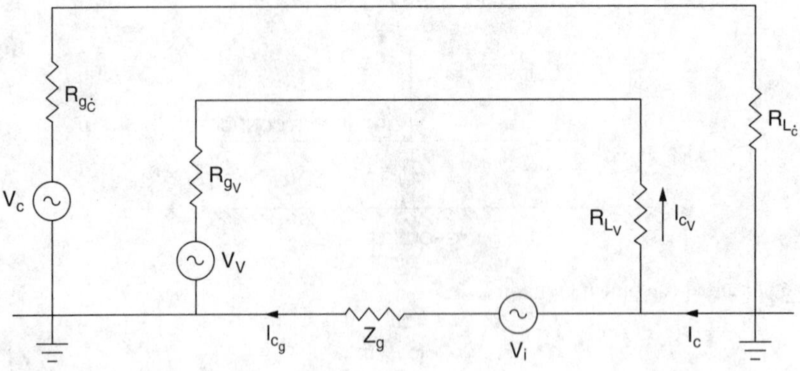

$$Z_g << R_{g_V} + R_{L_V}$$
$$I_{c_V} << I_{c_g} \;\&\; I_{c_g} - \sim I_c$$

FIGURE 113.8 Common-ground impedance coupling between circuits.

victim circuits via the common-ground impedance. In this case, the interference current (I_{c_g}) flowing through the common-ground impedance (Z_g) will produce an interfering signal voltage (V_i) in the victim circuit. This effect can be reduced by minimizing or eliminating the common-ground impedance.

The second EMI coupling mechanism involving ground is a radiated mechanism whereby the ground loop, as shown in Figure 113.9, acts as a receiving or transmitting antenna. Ground loops are probably the most common cause of interference in network systems and also the most common problem with multi-port devices in general. For this EMI coupling mechanism the induced EMI voltage (for the susceptibility case) or the emitted EMI field (for the emission case) is a function of the EMI driving function (field strength, voltage, or current), the geometry and dimensions of the ground loop, and the frequency of the EMI signal.

Common wisdom on electromagnetic compatibility recommends that radiated effects can be minimized by routing conductors as close as possible to ground and minimizing the ground-loop area. Even though theory holds up that closely routing conductors next to a ground plane will reduce the effects of coupling, however, where building steel and in particular in cases of lightning where large currents may flow, the coupling will still occur. The coupling effect applies to stray common-mode currents also. The primary factor is that in practical installations, ideal placement of the conductor next to the surface of the ground plane is difficult. Examples of this may be the upgrade of existing plant operation to incorporate multiple remote PLC systems communicating to a central control room. These cable runs are seldom able to achieve the ideal installation.

Figure 113.10 uses linear scales to emphasize the very rapid rise of induced voltage near a conductor and the small additional gain after a few meters. However, if "close to" were interpreted by an installer as a few centimeters, all the expected benefits from the "close" installation would be lost — in other words, it is not very effective to attempt minimizing induced voltages by casual routing of unshielded

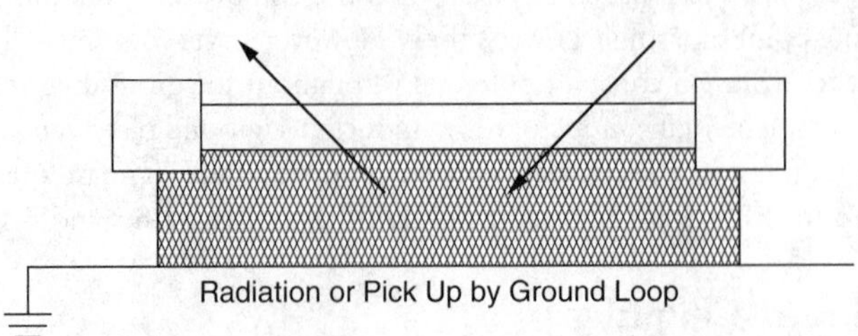

FIGURE 113.9 Common-mode radiations into and from ground loops.

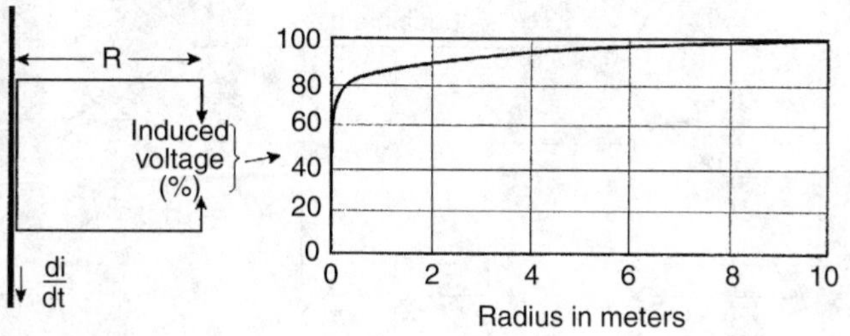

FIGURE 113.10 Voltage induced by the field into the loop.

cables "near" the ground planes. From Figure 113.10 it can be clearly seen that when a cable is just a few centimeters above the ground plane, the cable will still have significant amounts of induced noise.

Grounding Configurations

Typical electronic equipment may have a number of different types of functional signals as shown in Figure 113.11. To mitigate interference due to common-ground impedance coupling, as many separate grounds as possible should be used.

The grounding scheme for a collection of circuits within equipment can assume any one of several configurations. Each of these configurations tends to be optimum under certain conditions and may contribute to EMI problems under other conditions. In general, the ground configurations are a floating ground, a single-point ground, a multiple-point ground, or some hybrid combination. The determination to use single-point grounding or multiple-point grounding typically depends on the frequency range of interest. Analog circuits with signal frequencies up to 300 kHz may be candidates for single-point grounding. Digital circuits with signal frequencies in the MHz range should utilize multiple-point grounding.

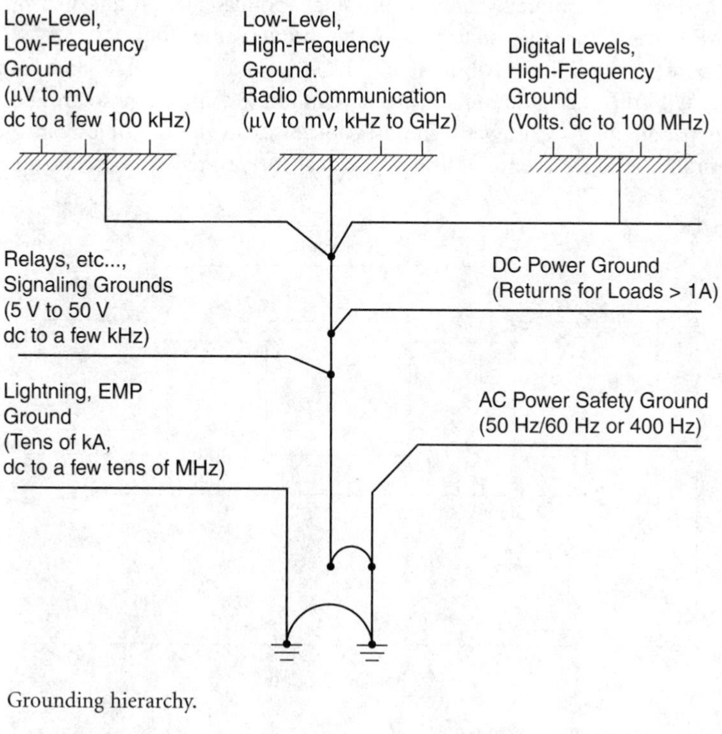

FIGURE 113.11 Grounding hierarchy.

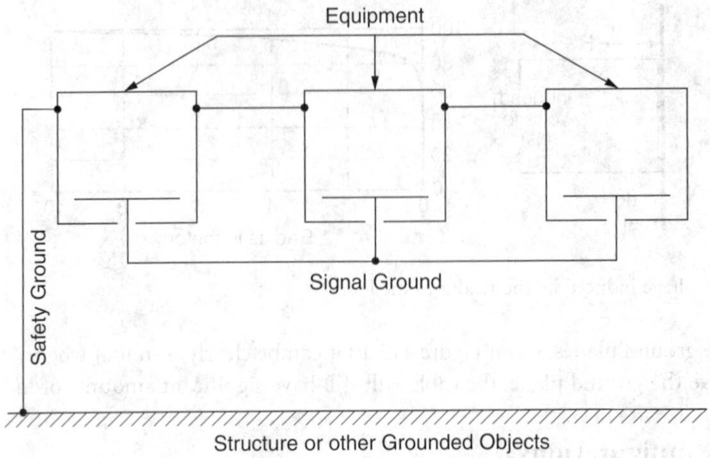

FIGURE 113.12 Floating signal ground.

A floating ground configuration is illustrated in Figure 113.12. The signal ground is electrically isolated from the equipment ground and other conductive objects. Hence, equipment noise currents present in the equipment and power ground will not be conductively coupled to the signal circuits.

The effectiveness of floating ground configurations depends upon their true isolation from other nearby conductors; that is, to be effective, floating ground systems must really float. It is often difficult to achieve and maintain an effective floating system. A floating ground configuration is most practical if only a few circuits are involved and power is supplied from either batteries or DC-to-DC converters.

A single-point ground configuration is illustrated in Figure 113.13. An important advantage of the single-point configuration is that it helps control conductively coupled interference. As illustrated in Figure 113.13, EMI currents or voltages in the equipment ground are not conductively coupled into the signal circuits via the signal ground. Therefore, the single-point signal ground network minimizes the effects of any EMI currents that may be flowing in the equipment ground.

The multiple-point ground illustrated in Figure 113.14 is the third configuration frequently used for signal grounding. With this configuration, circuits have multiple connections to ground. Thus, in equipment, numerous parallel paths exist between any two points in the multiple-point ground network. Multipoint grounding is more economical and practical for printed circuits and integrated circuits.

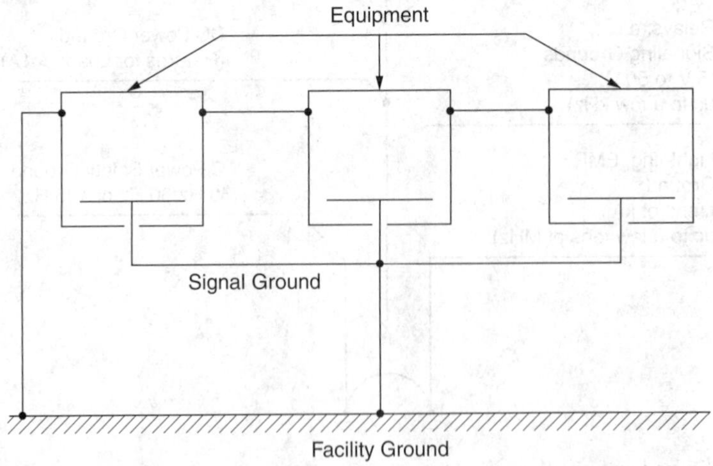

FIGURE 113.13 Single-point signal ground.

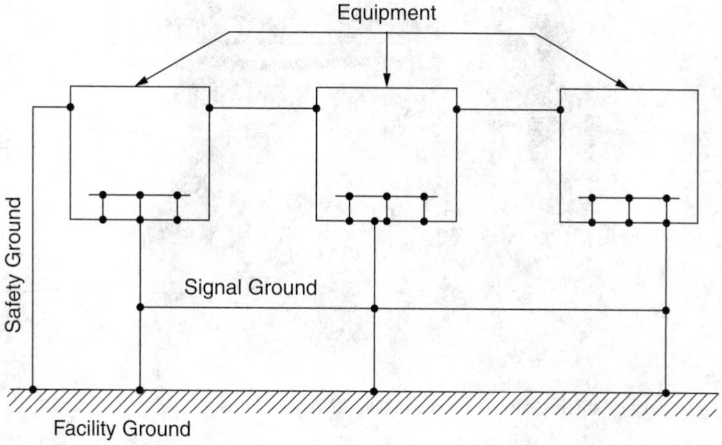

FIGURE 113.14 Multiple-point ground configuration.

Interconnection of these components through wafer risers, mother boards, and so forth should use a hybrid grounding approach in which single-point grounding is used to avoid low-frequency ground loops and/or common-ground impedance coupling; multipoint grounding is used otherwise.

Summary of Grounding Considerations

It should be noted that both the conducted and radiated EMI coupling mechanisms identified earlier involve a "ground loop." It is important to recognize that ground loop EMI problems can exist without a physical connection to ground. In particular, at RF frequencies, capacitance-to-ground can create a ground loop condition even though circuits or equipments are floated with respect to ground.

A properly designed ground configuration is one of the most important engineering elements in protecting against the effects of EMI. The ground configuration should provide effective isolation between power, digital, high-level analog, and low-level analog signals. In designing the ground it is essential to consider the circuit, signal characteristics, equipment, cost, maintenance, and so forth. In general, either floating or single-point grounding is optimum for low-frequency situations, and multiple-point grounding is optimum for high-frequency situations. In many practical applications, a hybrid ground approach is employed to achieve the single-point configuration for low frequencies and the multiple-point configuration for high frequencies.

113.3 Shielding

Shielding is one of the most effective methods for controlling radiated EMI effects at the component, circuit, equipment, subsystem, and system levels. Good EMC design reduces EMI signals to a level where they do not cause problems within the equipment or with any other equipment.

Shielding effectiveness is a measure of how well a shield blocks radiated emissions. Reducing the level of radiated emissions incident upon the equipment requires selecting the proper shielding material (see Figure 113.15), and correct installation and maintenance to ensure the integrity of the shield.

Radiated Electromagnetic Waves

The performance of shields is a function of the characteristics of the incident electromagnetic fields. All electromagnetic waves consist of two oscillating fields that operate at right angles to each other. One of these fields is the electric (E) field, whose strength is measured in volts per meter. Perpendicular to the E field is the magnetic (H) field, whose strength is measured in amps per meter. H fields usually are generated by high current, low voltage, low impedance circuits. In contrast, E fields are produced by

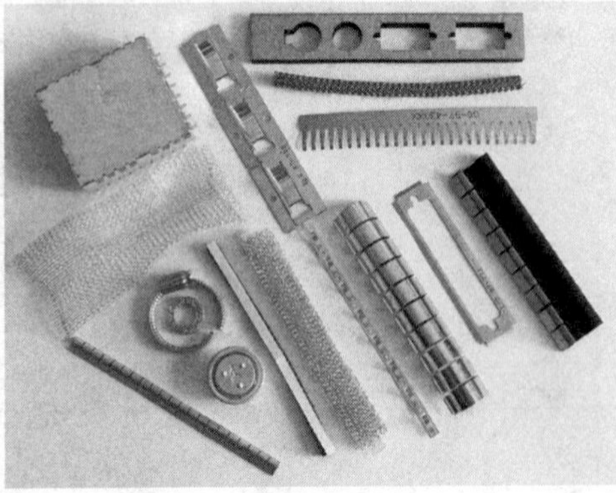

FIGURE 113.15 A variety of shields can be used to solve EMI problems.

devices that have high voltage, but relatively low current and high impedance. The ratio of E field to H field is called the wave impedance. The H and E fields vary in relative magnitude according to the distance of the wave from the generating source and the nature of the source itself.

The material selected for the shield depends upon characteristics of the source of the radiated emissions and the reflective and absorbent properties of the shielding material (see Figure 113.16). These include the impedance of the electromagnetic fields of the source creating the emissions, which depends on the distance from the emissions source to the malfunctioning equipment. In most cases, this distance (referred to as the "far-field") is such that the emissions are primarily due to electric fields resulting in field impedance that is constant and can be easily approximated. If the equipment is relatively close to the source of emissions, the impedance will be lower where the equipment is referred to as being in the "near-field," and where the magnetic field component must be considered. Before a shielding material can be selected, one must determine whether the affected equipment is in the "near-" or "far-field."

In the "far-field," a moderately good conductor may be used to shield against radiated emissions from electric fields. For emissions above about 1 MHz, the conductivity of the material, which is partially based on its thickness, does not have a significant effect on the material's ability to block electric fields. The primary dependence is on conductivity, which is the material's ability to carry electrons from the field and thus, provide a low impedance path to ground for the radiated emissions. However, at distances closer to the source of the emissions, more attention must be placed on selecting the shielding material.

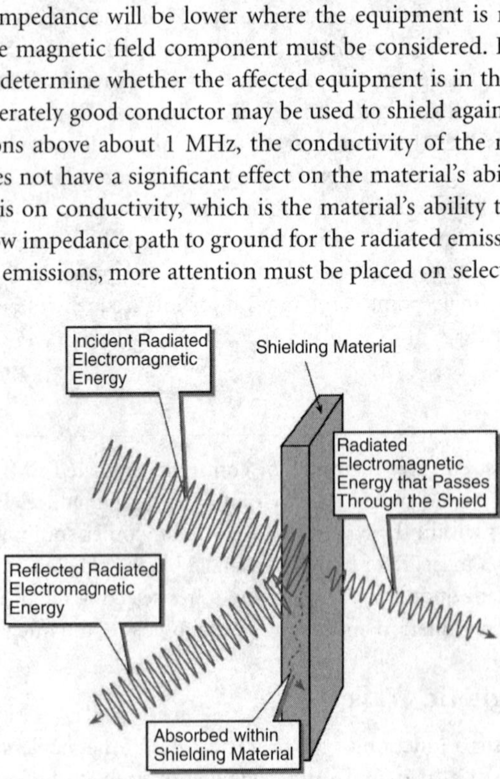

FIGURE 113.16 Diagram of a shielding material absorbing and reflecting emission.

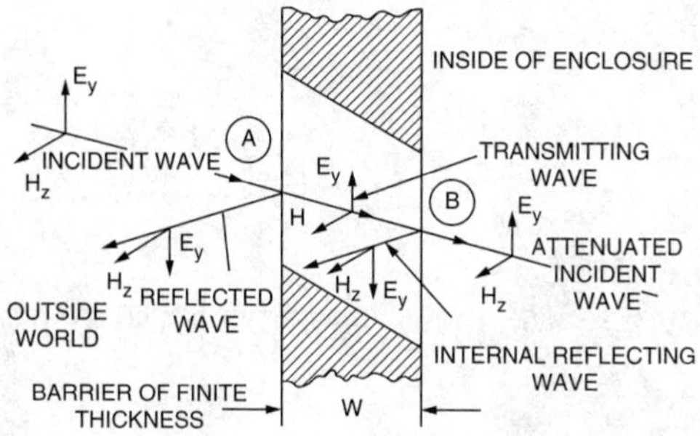

FIGURE 113.17 Shielding of plane waves.

Unlike the mitigation of electric fields in the "far-field," which only requires the shield to be reflective, equipment in the "near-field," which may be in an area containing both electric and magnetic-field components, requires a material that will reflect and absorb electromagnetic energy. Therefore, shielding considerations in the near-field region of an EMI source may be significantly different from shielding considerations in the far-field region.

Shielding Theory

If a metallic barrier is placed in the path of an electromagnetic field as illustrated in Figure 113.17, only a portion of the electromagnetic field may be transmitted through the barrier. Several effects may occur when the incident wave encounters the barrier. First, a portion of the incident wave may be reflected by the barrier. Second, the portion of the incident wave that is not reflected will penetrate the barrier interface and may experience absorption loss while traversing the barrier. Third, additional reflection may occur at the second barrier interface, where the electromagnetic field exits the barrier. Usually, this second reflection is insignificant relative to the other effects that occur, and it may be neglected.

The shielding effectiveness of the barrier may be defined in terms of the ratio of the impinging field intensity to the exiting field intensity. For high-impedance electromagnetic fields or plane waves, the shielding effectiveness is given by

$$\mathrm{SE}_{\mathrm{dB}} = 20\log\left(\frac{E_1}{E_2}\right) \tag{113.1}$$

where E_1 is the impinging field intensity in volts per meter and E_2 is the exiting field intensity in volts per meter. For low-impedance magnetic fields, the shielding effectiveness is defined in terms of the ratio of the magnetic field strengths.

The total shielding effectiveness of a barrier results from the combined effects of reflection loss and absorption loss. Thus, the shielding effectiveness, S, in dB is given by

$$S_{\mathrm{dB}} = R_{\mathrm{dB}} + A_{\mathrm{dB}} + B_{\mathrm{dB}} \tag{113.2}$$

where R_{dB} is the reflection loss, A_{dB} is the absorption loss, and B_{dB} is the internal reflection loss. Characteristics of the reflection and absorption loss are discussed in the following sections.

Reflection Loss

When an electromagnetic wave encounters a barrier, a portion of the wave may be reflected. The reflection occurs as a result of a mismatch between the wave impedance and the barrier impedance. The resulting reflection loss, R, is given by

$$R_{dB} = 20\log_{10}\frac{(K+1)^2}{4K}, \qquad K = \frac{Z_w}{Z_b}$$
$$= 20\log_{10}\left(\frac{Z_w}{4Z_b}\right), \qquad K \geq 10 \tag{113.3}$$

where Z_w is the wave impedance = E/H, and Z_b is the barrier impedance.

Absorption Loss

When an electromagnetic wave encounters a barrier, a portion of the wave penetrates the barrier. As the wave traverses the barrier, the wave may be reduced as a result of the absorption loss that occurs in the barrier. This absorption loss, A, is independent of the wave impedance and may be expressed as follows:

$$A_{dB} = 8.68t/\delta = 131t\sqrt{f_{MHz}\mu_r\sigma_r} \tag{113.4}$$

where t is the thickness in mm, f_{MHz} is the frequency in MHz, μ_r is the permeability relative to copper, and σ_r is the conductivity relative to copper.

Shielding Material Characterization

The ability of a material to absorb radiated emissions depends on the frequency of the source, and the conductivity, permeability, and thickness of the shielding material. An increase in any of these variables will increase absorption. However, the variables of concern when selecting a material to absorb emissions in the "near-field" are the ones that vary widely among available materials. These are permeability, which is the extent to which a material can be magnetized, and thickness. Thus, to mitigate against low-frequency magnetic fields in the "near-field" such as those originating from power distribution equipment such as transformers from some medical equipment utilizing 60-hertz power and from electrical appliances, a material of the proper thickness such as iron, which can be magnetized and provide a path for the magnetic field emissions, is required. A screened or solid shield made of copper or other material to attenuate electric fields, some type of steel composite or other material to attenuate magnetic fields, or a combination of both may be installed around equipment or inside the walls of the room where equipment is used.

The total shielding effectiveness resulting from the combined effects of reflection and absorption loss are plotted in Figure 113.18 for copper and iron materials having thicknesses of 0.025 mm and 0.8 mm, having electric and magnetic fields and plane-wave sources, and having source-to-barrier distances of 2.54 cm and 1 m.

As shown in Figure 113.18, good shielding efficiency for plane waves or electric (high-impedance) fields is obtained by using materials of high conductivity, such as copper and aluminum. However, low-frequency magnetic fields are more difficult to shield because both the reflection and absorption loss of nonmagnetic materials, such as aluminum, may be insignificant. Consequently, to shield against low-frequency magnetic fields, it may be necessary to use magnetic materials.

Conductive Coatings

Conductive coatings applied to nonconductive materials such as plastics will provide some degree of EMI shielding. The principal techniques for metalizing plastic are the following:

- Conductive paints
- Plating (electrolytic)
- Electroless plating

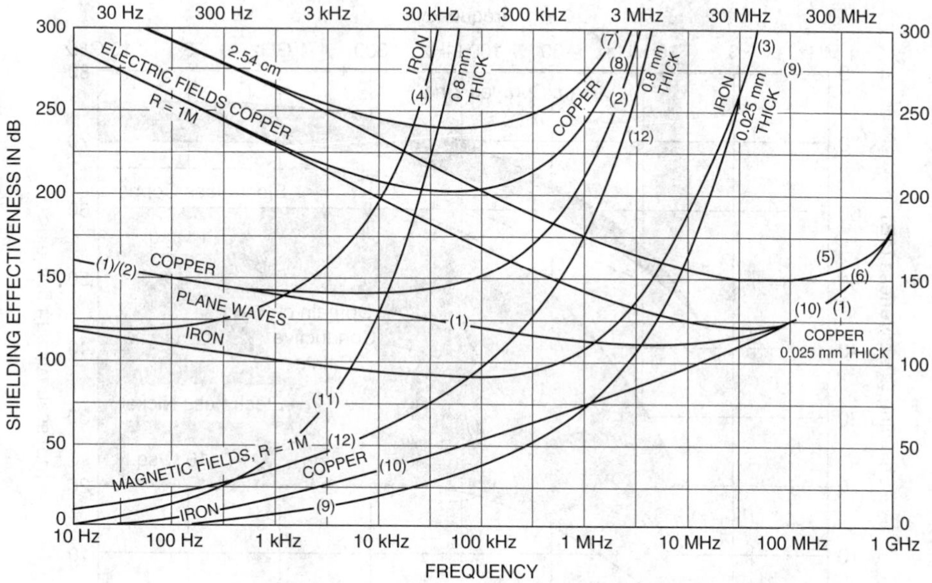

FIGURE 113.18 Total shielding effectiveness.

- Flame spray
- Arc spray
- Ion (plasma torch) spray
- Vacuum deposition

Because the typical conductive coatings provide only a thin film of conductive material, the shielding results from reflection losses that are determined by the ratio of the wave impedance to the conductive barrier impedance. The surface resistance (in ohms per square) will determine shielding effectiveness. Figure 113.19 shows comparative data for shielding effectiveness for various conductive coatings. The most severe situation (i.e., a low-impedance magnetic field source) has been assumed.

Aperture Leakages

Various shielding materials are capable of providing a high degree of shielding effectiveness under somewhat idealized conditions. However, when these materials are used to construct a shielded housing, the resulting enclosure will typically have holes and seams that may severely compromise the overall shielding effectiveness.

Figure 113.20 shows a rectangular aperture in a metal (or metalized) panel. A vertically polarized incident electric field will induce currents in the surface of the conductive panel. If the aperture dimensions are much less than a half wavelength of the incident electric field, the aperture leakage will be small. On the other hand, as the aperture dimensions approach a half wavelength, the aperture leakage will be significant. An aperture with dimensions equal to or greater than a half wavelength will provide almost no shielding (i.e., the incident field will propagate through the aperture with very little loss). In general, the shielding effectiveness of a conductive panel with an aperture may be approximated by the following equation:

$$SE_{dB} \cong 100 - 20\log L_{mm} \times F_{MHz} + 20\log\left(1 + \ln\frac{L}{S}\right) \tag{113.5}$$

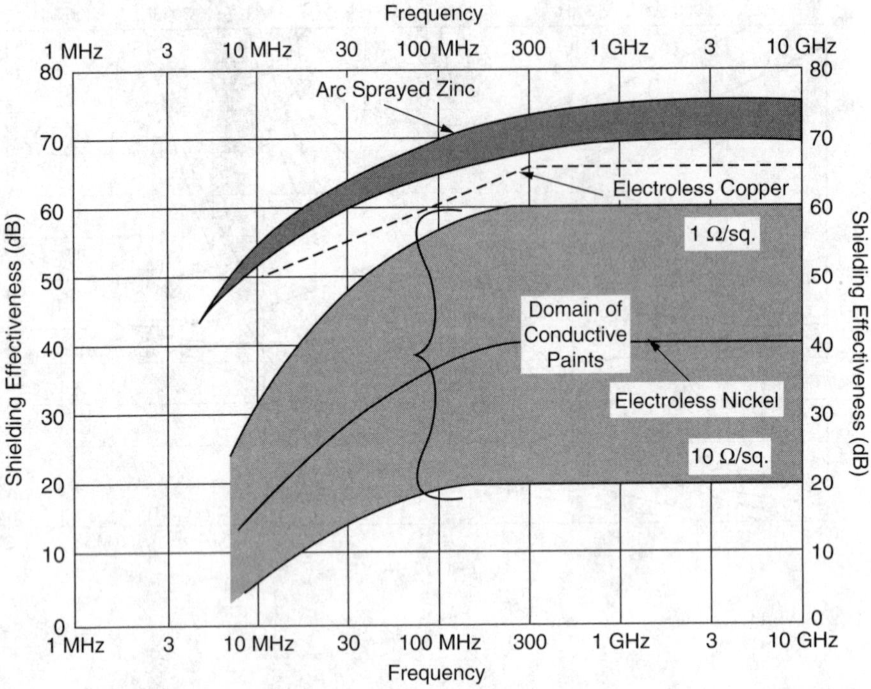

FIGURE 113.19 Shielding of conductive coatings. (By standard 30 cm distance test. Near field attenuation is given against the H field.) For paints, thickness is typically 2 mil = 0.05 mm.

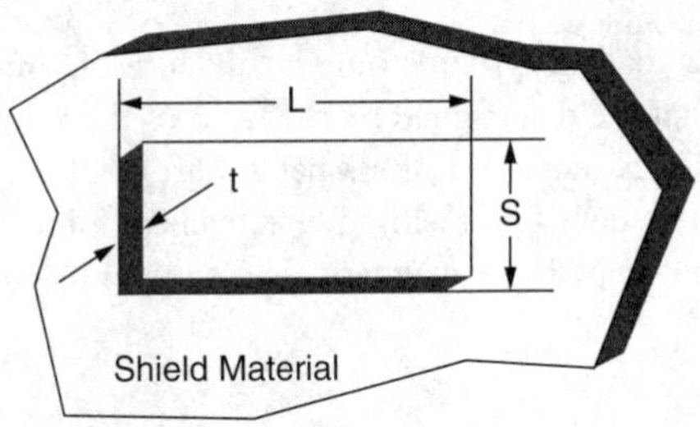

FIGURE 113.20 Slot and aperture leakage.

To maintain shielding integrity for an equipment enclosure, it may be necessary to provide EMI protection for the apertures.

Summary of Shielding Considerations

Shielding can provide an effective means of controlling radiated EMI effects. To ensure that shielding effectiveness requirements are met, it is necessary to:

- Select a material that is capable of providing the required shielding
- Minimize the size of openings to control aperture leakages
- Subdivide large openings into a number of smaller ones
- Protect leaky apertures (e.g., cover with wire screen)
- Use EMI gaskets on leaky seams
- Filter conductors at points where they enter or exit a shielded compartment

113.4 Using Shielded Isolation Transformers to Improve Equipment Compatibility

Some EMI problems can be solved with appropriate power-conditioning devices that contain EMI filtering and shielding. For EMI problem-solving purposes, a shielded isolation transformer can be used to reduce emissions common to both power and neutral conductors induced by "ground loops" or multiple-current paths in the grounding system, for which in many cases a source cannot be identified. Subsequent sections provide an overview of this technology and their effectiveness to reduce EMI problems. More realistic methods of measurements for shielding and attenuation performance of isolation transformers are also discussed.

Shielded Isolation Transformers

Shielding isolation transformers are not true "power conditioning" devices but rather low frequency "noise rejectors" that provide good common-mode noise isolation. The main characteristic that distinguishes an isolation transformer (IT) from a general-purpose dry-type transformer (GPT) is the introduction of a grounded electrostatic shield made from an aluminum or copper sheet, called a Faraday shield, between the primary and secondary windings (as shown in Figure 113.21). The electrostatic shield is wrapped and insulated on its edges to prevent a short circuit in the loop around the core. As shown in Figure 113.21, the shield is then connected to ground, which establishes two new capacitances that link the primary and secondary windings to the static shield, as shown in Figure 113.21.

By grounding the shield, the original distributed capacitances, C_P and C_S, are now referenced to ground. Inter-winding capacitance, marked as C_W in the figure, is the primary path for noise. The grounded electrostatic shield effectively reduces C_W by a factor of 10 to 100 and increases the total capacitance between line and neutral on the primary and secondary. The reputed reduction of high-frequency common-mode noise resulting from the electrostatic shield ranges from 20 to 60 dB, depending on the material used to make the shield. If an isolation transformer is required to reduce high-level conducted emissions, some are available with a "super-isolation" type electrostatic shield, which further decouples the primary and secondary windings.

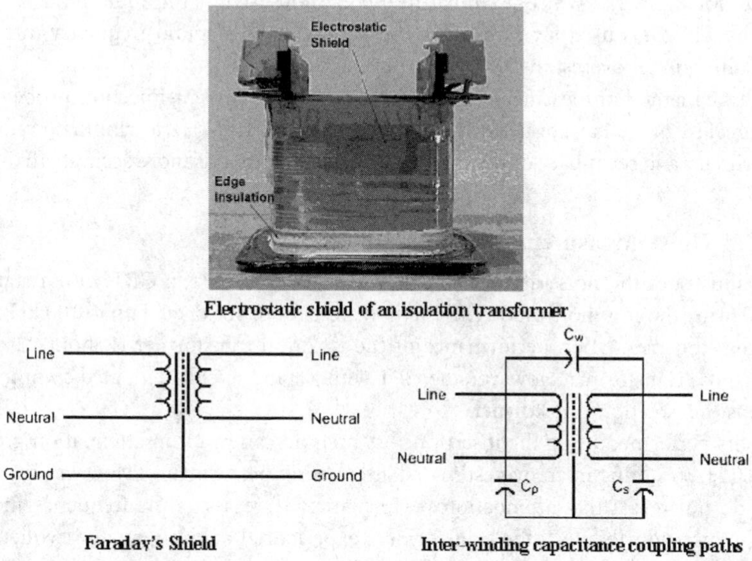

FIGURE 113.21 Primary to secondary EMI coupling in a shielded isolation transformer.

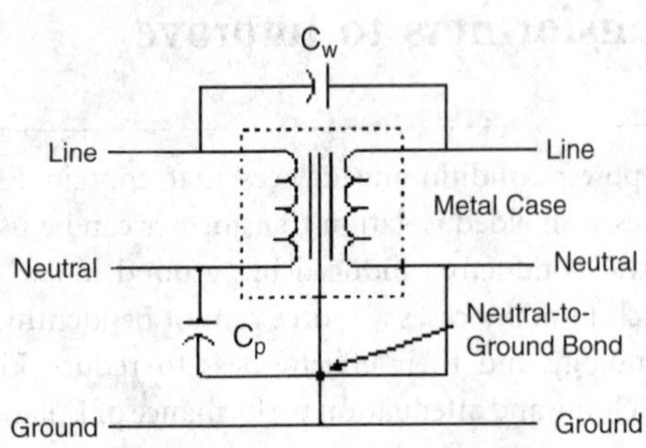

FIGURE 113.22 Capacitance-coupling paths for properly bonded GPT (model shown here has a static shield).

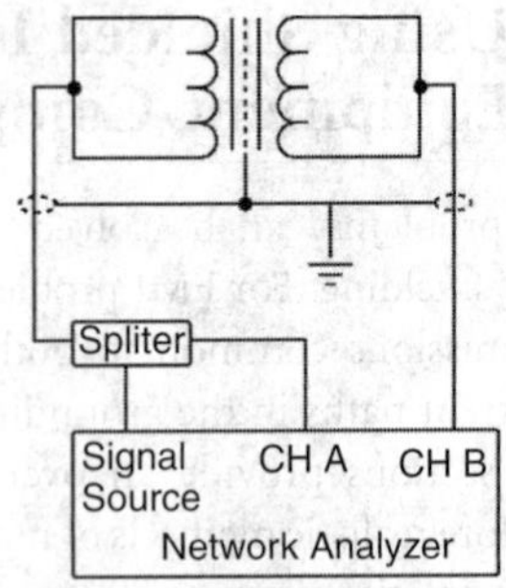

FIGURE 113.23 Test setup for CM noise measurement using insertion-loss method (IT shown in the test circuit).

It is noteworthy to mention here that although ITs are known for superior CM isolation, GPTs may provide enough noise attenuation without the expensive electrostatic-shield design. In most cases, simply establishing a neutral-to-ground bond on the secondary of a transformer will reduce common-mode noise caused by the capacitances between the transformer windings. Figure 113.22 shows the capacitance-coupling paths resulting from a neutral-to-ground bond on the secondary of a GPT (compare to Figure 113.21).

Common-Mode Noise Measurements (Single Phase)

Presently, there is no recognized standard for measuring the noise-attenuation performance of transformers. The transformer-manufacturing industry has adopted methods similar to those described in the test procedure in MIL-Std-T27E, *Transformers and Inductors (Audio Power and High Power Pulses)*, which was designed to measure the performance of a electrostatic shield used in an IT. The transformer industry may use a no-power test and select a bypass capacitor placed on the output to achieve and describe better attenuation performance during the test. The audio industry, however, uses balanced networks on the input and output to perform the common-mode attenuation measurements. In both cases, measurements of greater than 100 dB attenuation can be made, resulting in an overestimated attenuation performance. Another method similar to MIL-Std-T27E is known as MIL-Std-220A, *Method of Insertion Loss Measurement,* which is similar to the method used by the audio industry. This method, shown in Figure 113.23, is used for measuring the performance of radio-frequency interference (RFI) filters, also resulting in an overestimated attenuation performance.

A more realistic method of measuring noise attenuation is the current-injection-probe (CIP) method, shown in Figure 113.24. Tests were performed from 10 kHz to 100 kHz to eliminate radiated coupling between the primary and secondary of the transformer. This method is more accurate than the insertion-loss method.

Results of CM Noise Measurements (Single Phase)

Figure 113.25 illustrates the noise-attenuation performance of a GPT, a GPT with a static shield, and low-leakage IT using the common-mode CIP method. Tests were performed up to 100 kHz. These results demonstrate that the attenuation performance of the isolation transformer is about 22 dB better than the general-purpose transformer. However, the GPT with a static shield attenuated common-mode noise nearly as well as the isolation transformer.

Measurements performed using the insertion-loss method can result in attenuation greater than 100 dB at 50 and 60 Hz, resulting in an overestimated attenuation performance. However, most transformer specifications do not reveal that attenuation performance decreases as the frequency increases. When tested using the insertion-loss test, the transformers experienced an even greater frequency-dependent attenuation performance. One can see that attenuation greater than 100 dB decreases quickly as the frequency increases. Figure 113.26 shows measurements from a MIL-Std-220A insertion-loss test for a GPT, a shielded GPT, and a medical-grade low-leakage IT.

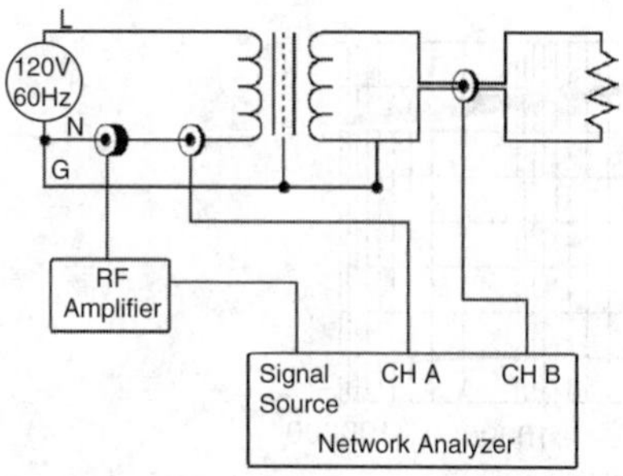

FIGURE 113.24 Test setup for common-mode noise measurement using the CIP method (IT shown in the test circuit).

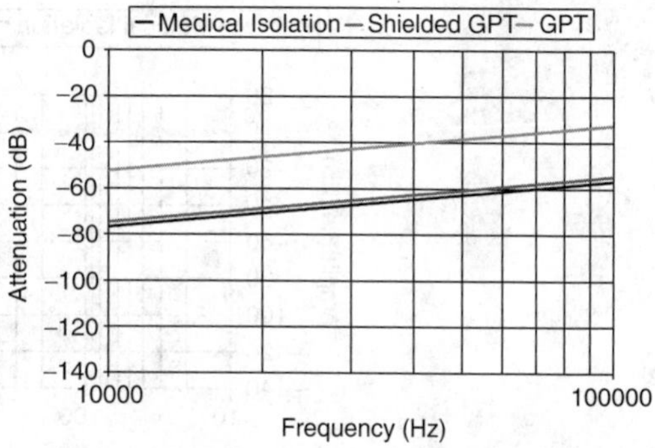

FIGURE 113.25 Attenuation of common-mode noise for three different transformers using the CIP method.

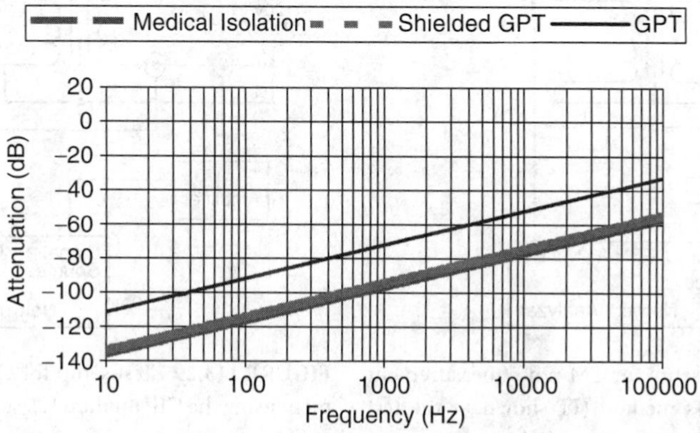

FIGURE 113.26 Attenuation of common-mode noise for three different transformers using the insertion-loss method.

The difference in performance between the CIP and insertion-loss test methods occurs for two reasons. The primary reason is that not enough energy is present to excite the core in the insertion-loss method. The difference between the impedance of the source and the load, which also affects the performance but to a lesser degree, is another difference between the test methods.

Insufficient excitation of the core will cause the mutual flux to be nonuniformly distributed within the core. This results in poor inductive coupling between the windings, which yields an overestimate of attenuation performance. The inter-winding and core-to-winding capacitances continue to provide capacitive coupling between the windings, which provides for some attenuation. However, once the mutual flux in the core has been uniformly established — as it is in the CIP method — then coupling between the windings at higher frequencies improves. This results in a magnetic path, which allows the transfer of the higher-frequency components and a lower (and more realistic) attenuation performance.

Transverse-Mode Noise Measurements (Single Phase)

Transverse-mode (TM) noise, also known as across-the-line or line-to-line noise, is less likely to upset end-use equipment than common-mode noise is. TM noise is measured from line to line or line to neutral. This noise passes through an IT without appreciable attenuation and is not affected by the electrostatic shielding of the transformer for frequencies up to 100 kHz (see Figure 113.27). In most cases, end-use equipment is usually designed to handle this type of noise.

Standard electronic power supplies used in end-use equipment are designed to filter conducted emissions, to meet limits defined in EMI conducted-emissions standards that are promulgated, for example,

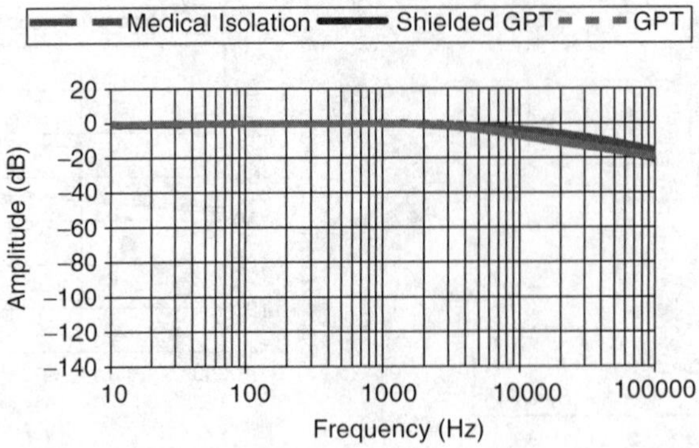

FIGURE 113.27 Illustration of a typical noise attenuation.

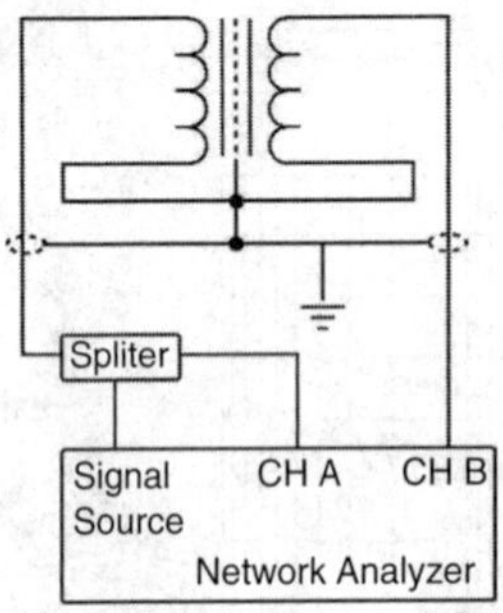

FIGURE 113.28 Test setup for TM noise measurement using the insertion-loss method (IT shown in the test circuit).

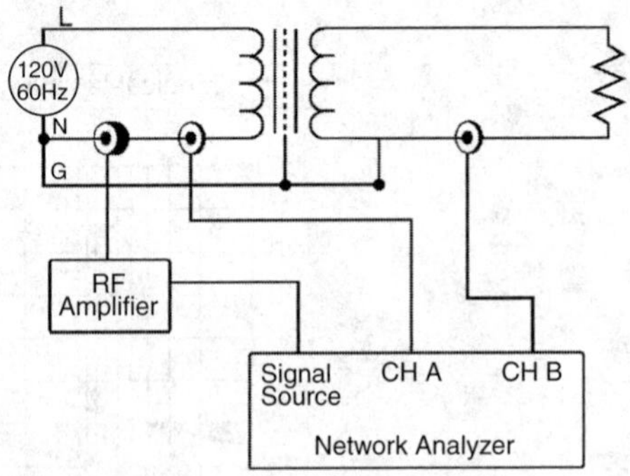

FIGURE 113.29 Test setup for CM/TM noise measurement using the CIP method (IT shown in the test circuit).

by the U.S. Federal Communication Commission (FCC) and the Electrotechnical Commission (IEC). Normally, the basic construction or design of the transformer will reduce TM noise somewhat at frequencies above 1 MHz.

Figure 113.28 illustrates the transverse-mode insertion-loss test setup used to measure TM noise coupling. Figure 113.29 shows common/transverse-mode CIP test setup.

Results of Transverse-Mode Noise Measurements (Single Phase)

Test results indicate little difference in the performance of shielded and nonshielded transformers when the transformer is grounded correctly, as shown in Figure 113.30 and Figure 113.31. One can see here almost zero attenuation in this mode up to about 100 kHz. These test results provide additional evidence and support for the statement, "Grounding the secondary alone in most cases will reduce common-mode noise." A transformer by design will attenuate most high-frequency noise components because of the simple reactive drop associated with inductors. However, as shown in Figure 113.27, the significant noise attenuation (greater than 20 dB, times 10 reduction) for most transformers does not begin until about 100 kHz using the MIL-Std-220A measurement technique.

Three-Phase Isolation Transformer Shielding and Attenuation Performance

Just as in the case of single-phase isolation transformers, traditional evaluation techniques for shielding performance for three-phase isolation transformers are not well documented or standardized. Most manufacturers have adopted the same methods for balanced and unbalanced insertion loss tests to determine attenuation. As may be seen in Figure 113.32, it is very apparent no significant gain in signal rejection occurs at lower frequencies. By using single or double shields, there is relatively small difference

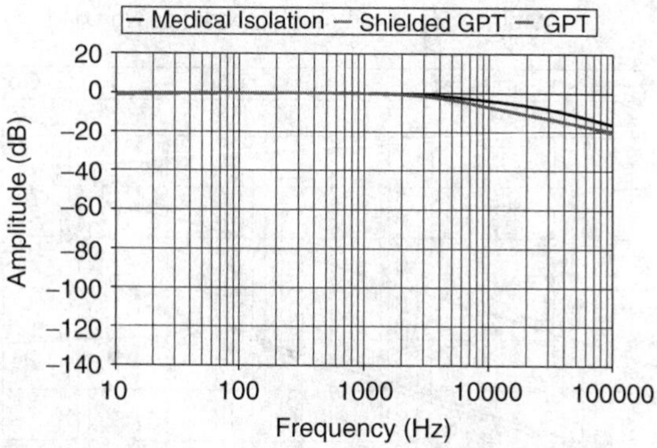

FIGURE 113.30 Attenuation of transverse-mode noise for three different transformers using the insertion-loss method.

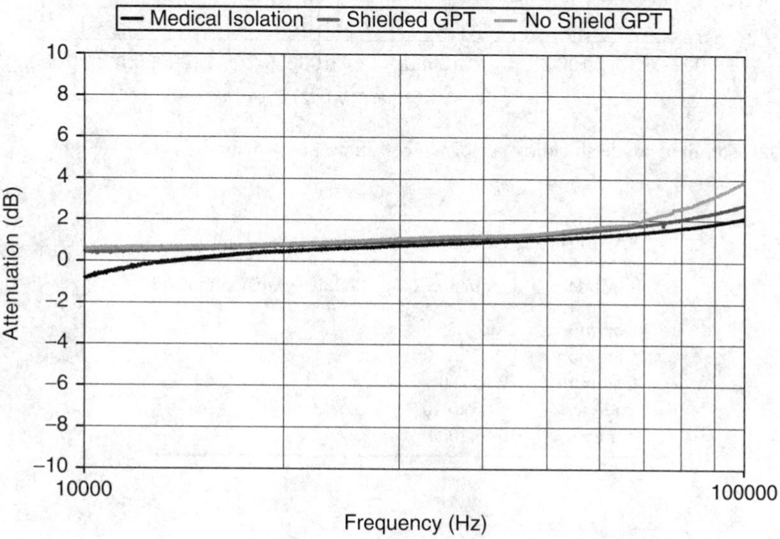

FIGURE 113.31 Attenuation of transverse-mode noise for three different transformers using the CIP method.

in increased performance. Hence, choosing one or more shields may only be based on a performance-driven specification, where the winning method may only differ by less than 0.01%, for example, 1.4:1 (3 dB) when compared to the overall shielding ratios of 10,000:1 (60 dB) and 10,000:1 (80 dB).

From the experience of measuring actual attenuation performance of power line filters, the method of measurement by using RF current injection was chosen and adapted to gain an understanding of performance for three-phase isolation transformers under-load in a laboratory environment. These data may not be competitive in the market against MIL-Std-220A curves, but the yield data may be used by a customer who wishes to know a relative performance ability. These data may now be used to promote a new competitive environment and to ensure that the customer understands the actual attenuation performance versus shielding effectiveness of certain materials and construction techniques. An example of a specification is shown in Table 113.2 below.

For transformers of normal construction such as standard E-cores, it is already apparent that this is not a realistic specification even in nonloaded test systems and particularly in the traverse mode, where standard transformers will function up to several kHz with very little loss. Standard loss of such a transformer is designed in the normal construction of a four to one reduction from the delta to the Y side and has a factor of −12dB in the operating region of the transformer when converting from 480 to 120V, as seen in Figure 113.33.

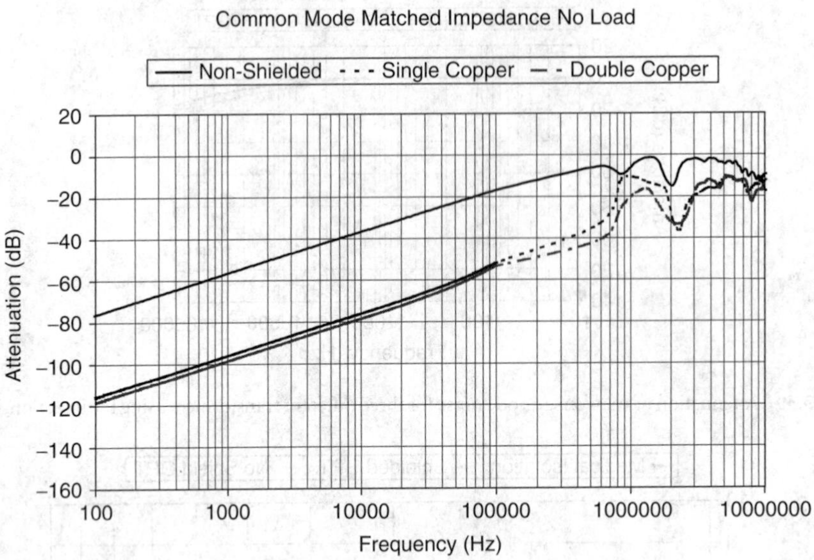

FIGURE 113.32 Common-mode shielding performances under no load conditions.

TABLE 113.2 Performance Specifications

Mode	Frequency Range (kHz)	Attenuation (dB)
Common	0 to 1.5	−120
Common	1.5 to 10	−90
Common	10 to 100	−65
Transverse	1.5 to 10	−52
Transverse	10 to 1000	−30

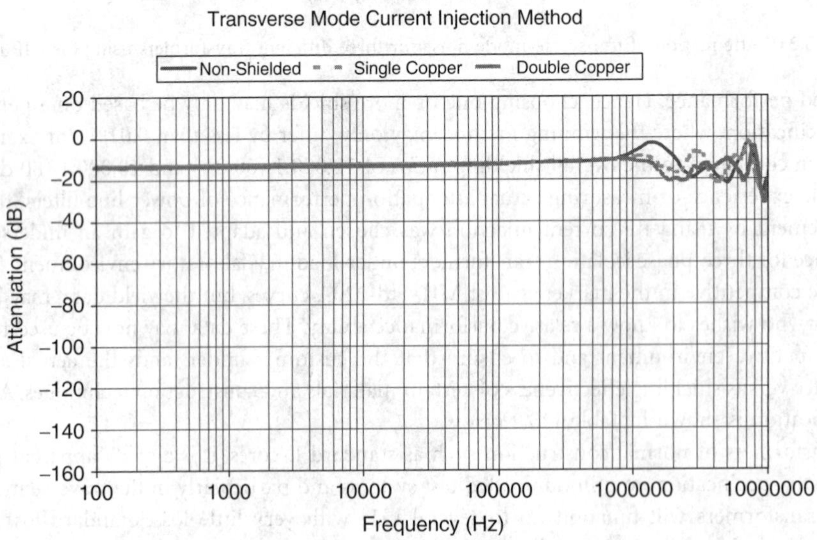

FIGURE 113.33 Transverse mode three-phase isolation under load shielding performance.

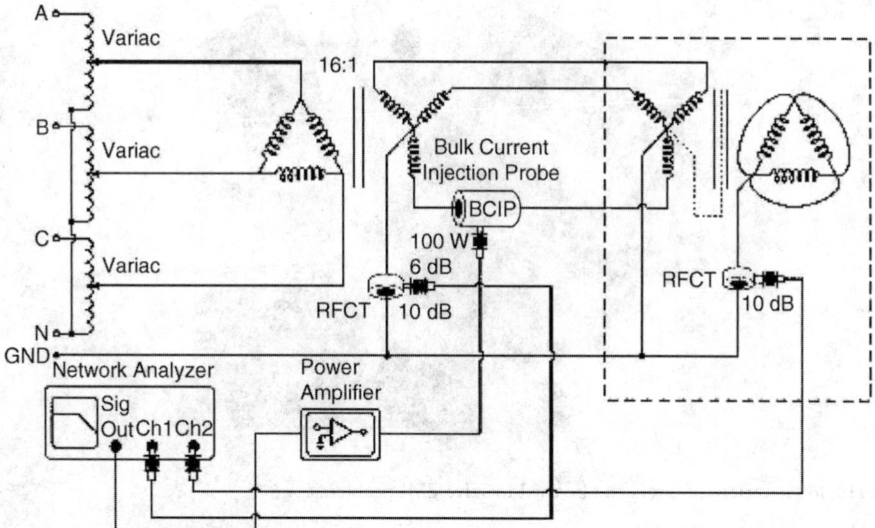

FIGURE 113.34 Common mode current injection.

When testing large three-phase transformers, it is difficult to supply large currents. Therefore, elaborate setups must be established in order to load the transformer for delta or wye configuration testing. Figure 113.34 represents the setup requirements, where the delta side is shorted and then the Y side is loaded by a 16:1 transformer to limit the power frequency current during the test. By energizing the core, we have established that the magnetic path that must be considered in attenuation performance of the transformer; one cannot just depend upon the traditional shielding effectiveness measurement.

One can see that the transformer is more than capable of transferring the signal well past 100 kHz before achieving any significant loss or gain in the MHz region. The most practical method of achieving the specifications shown in Table 113.2 is the use of three-phase filters constructed within the transformer. Some manufacturers may add such filtering inside the case of the transformer. Also, a special class of transformer known as CVTs may exhibit more than satisfactory attenuation results.

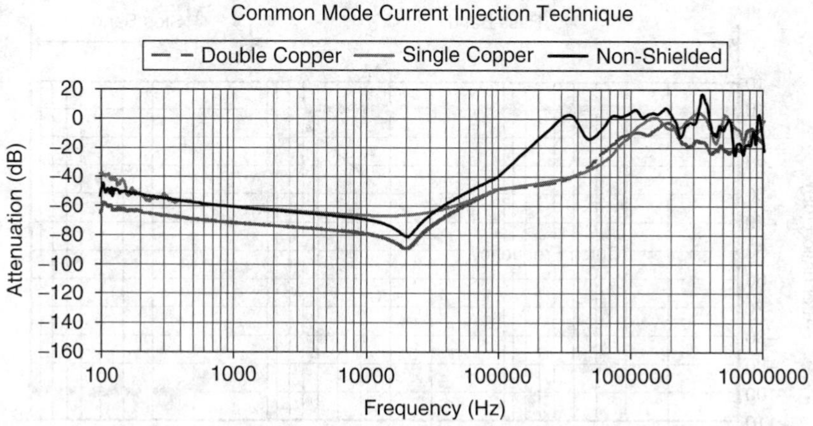

FIGURE 113.35 Common mode three-phase isolation transformer shielding performance under load.

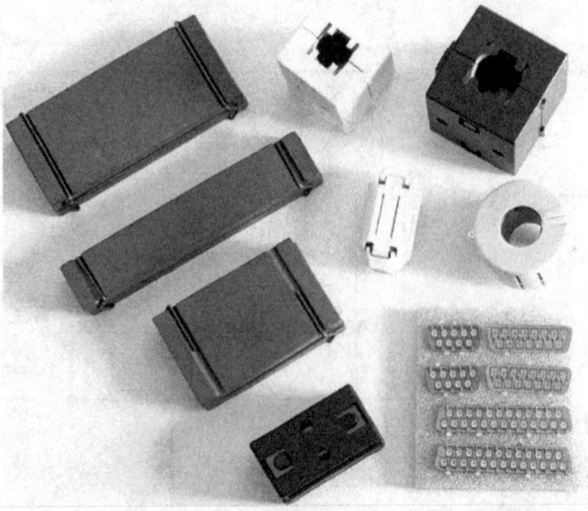

FIGURE 113.36 A variety of filters can be used to solve EMI problems.

113.5 Using Filtering Technologies to Improve Equipment Compatibility

Many commercial and industrial electrical environments contain equipment that is sensitive to radio-frequency (RF) interference. Theft detectors, digital sensors, and medical-telemetry systems have all been known to malfunction in the presence of RF interference. To prevent RF signals from entering a facility via the power lines, a variety of filters (as shown in Figure 113.36), installed between sensitive loads and the power source, can be used to block unwanted emissions from sensitive electronic equipment. A screened or solid shield made of copper to attenuate electric fields or some type of steel composite to attenuate magnetic fields may be installed around equipment or in the room where equipment is used. If properly designed and installed, it can protect the equipment from electromagnetic interference. In some situations, both filters and a shield may be needed at an installation site.

Line filters prevent electrical noise from traveling into sensitive electronic equipment from the building wiring system by blocking higher-frequency noise that has been shown to cause EMI problems but passing 60 Hz power. They also eliminate noise generated by the protected equipment from feeding back into the building wiring system. As shown in Figure 113.37, a power-line filter passes frequencies in its pass

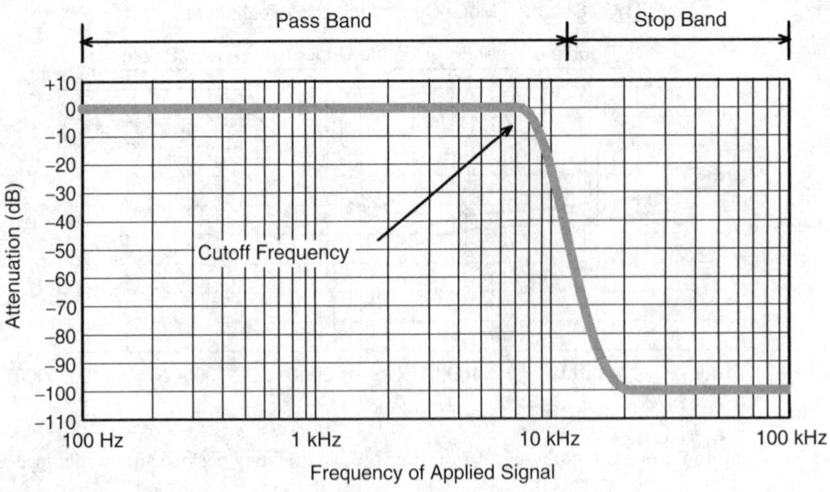

FIGURE 113.37 Attenuation plot for an ideal power-line filter.

band (including the frequency of the power source, either 60 or 50 Hz) but blocks frequencies in its stop band (including the frequencies of RF interference).

The attenuation plot in Figure 113.37 is entirely conceptual; the *in situ* performance of a power-line filter will certainly fail to meet this ideal and may even fail to meet the performance predicted by the manufacturer's specifications. Therefore, power-line filters are routinely tested to determine their actual performance in the field. One of the parameters of filter performance is *insertion loss,* which is the difference between the level of an interference signal without and with a filter in the circuit.

Harmonic filters trap harmonic currents, which are low-frequency conducted emissions. They range in size from small units for plug-connected loads to high-power units that are hard-wired to electrical distribution panels. These filters are selected based on the predominant harmonic currents to be blocked.

The subsequent sections provide an overview of the current standard being used to measure the insertion loss of a power-line filter. The existing standard has inherent shortcomings and needs substantial modifications. IEEE Standard for Method of Tests for Filter Performance, P1560 Working Group, is an effort towards improving the testing methods of measuring insertion loss. An overview of the testing method recommended by P1560 Working Group is also provided.

An Overview of the Current Standard

The main standard currently used to measure the insertion loss of a power-line filter is MIL-STD-220, *Method of Insertion Loss Measurement.* This standard is used to measure the attenuation performance of a filter that is connected to a 50-ohm power source and a 50-ohm load, which is called "matched impedance." The standard was originally developed in 1952 to measure the insertion loss of filters used to mitigate radio interference in mobile communication systems. Since 1952, MIL-STD-220 has been widely used to create general performance specifications for all types of filters. The problem arises from the application of MIL-STD-220 to determine the insertion loss of a power-line filter used with an unmatched source and load. In the real world, power line filters must operate under a wide range of impedances (see Schlicke 1976), as well as a wide range of load types.

Unfortunately, MIL-STD-220A has become the industry norm, largely because of the lack of a viable alternative. The problems associated with the test methods defined in MIL-STD-220A have been known since it was first used. In fact, the standard itself warns users about its limitations. Nevertheless, this standard has been referenced in other standards and applied in thousands of applications throughout the military and commercial industries to characterize power-line filters.

Although MIL-STD-220A was revised in 2000 to MIL-STD-220B (some describe this revision as a facelift), its test methods are virtually unchanged. MIL-STD-220B describes itself as a quality-control standard. The standard itself emphasizes that its test methods are not intended to predict actual performance of filters that are used in mismatched-impedance situations. Yet, people continue to use this standard to predict the behavior of filters in the field.

Many other standards have also been developed since the creation of MIL-STD-220A. Two examples of standards used to measure power-line filters are SAE ARP4244 – 1998, *Recommended Insertion Loss Test Methods for EMI Power Line Filters,* and C.I.S.P.R. 17 1981, *Methods of Measurement of the Suppression Characteristics of Passive Radio Interference Filters and Suppression Components.* The authors of these standards have attempted to correct the problems associated with using MIL-STD-220A. At best, these standards address the issues pertinent to the requirements for a particular organization or application but do not directly address the root of the problem — test methods incorrectly defined for actual applications with mismatched-impedance conditions. The next section describes the typical test setup used for MIL-STD-220B standard.

Limitations of MIL-STD-220B

Frequency Range

If testing adheres to the 100-kHz minimum frequency of MIL-STD-220B, then some significant filter behaviors may be missed. For example, Figure 113.38 shows the effect of inductor saturation and

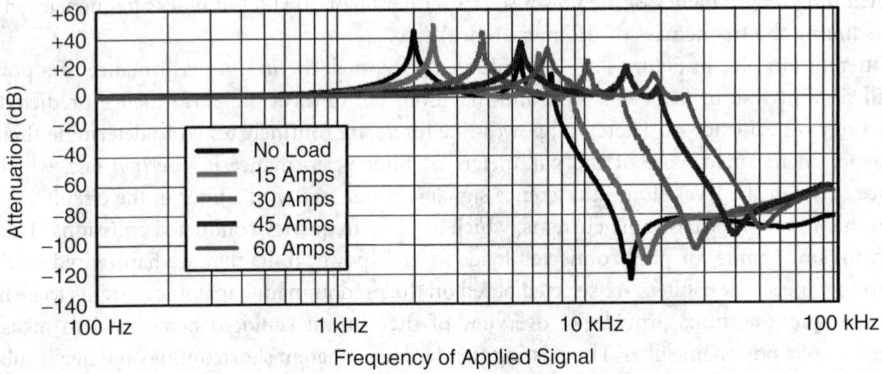

FIGURE 113.38 Effect of inductor saturation under DC current loading.

resonance on filter attenuation for frequencies below 100 kHz. Resonance may develop when a filter is loaded. The peaks in the attenuation indicate this. Also, note that the cutoff frequency shifts to the right as the current through the filter increases as a result of inductor saturation. Testing at a minimum of 100 kHz would obscure this behavior.

Inductor Saturation

A buffer network, according to MIL-STD-220B, is used to isolate the test equipment to prevent short-circuiting the direct current (DC) source under loaded conditions. Not only is the frequency range of a typical buffer network insufficient for accurate predictions of field performance, a buffer network poses yet another problem in measurement accuracy. As shown in Figure 113.39, a buffer network, when placed in a test circuit, adds additional LC elements to the already complicated network of the filter under test. This is an extreme analogy to Heisenburg's Uncertainty Principle, wherein the introduction of a measuring device introduces an uncertainty in the measured value. With the addition of these LC elements, it is difficult to ascertain the effects that the additional elements will have on the overall filter-attenuation measurement being performed, unless one compensates for the insertion loss of the buffer network for the filter parameters.

Inaccuracy of 220B Measurements

Matched-impedance testing under no-load conditions should not be used to predict the performance of filters used in an unmatched-impedance environment. Unmatched impedance causes an unbalancing of the filter network, which can result in less-than-expected attenuation of the stop band or, in some cases, random resonance or even an amplification of the pass band.

Another problem with determining the performance of a filter under loaded conditions using the MIL-STD-220B method is that it is best suited for evaluating inductor saturation and traditionally uses

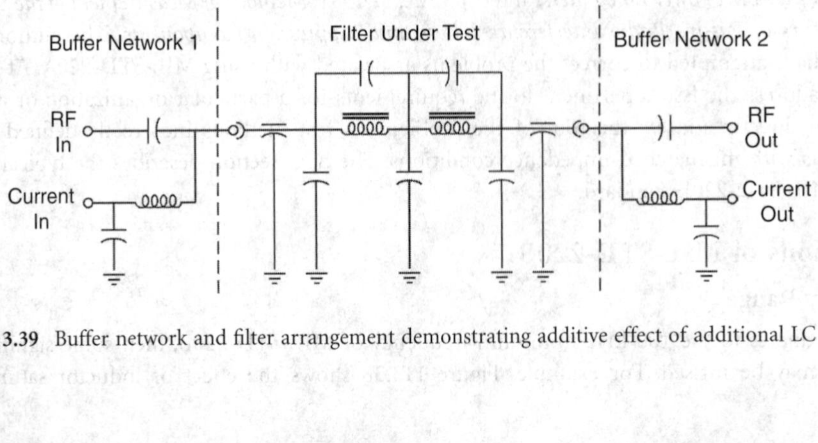

FIGURE 113.39 Buffer network and filter arrangement demonstrating additive effect of additional LC networks.

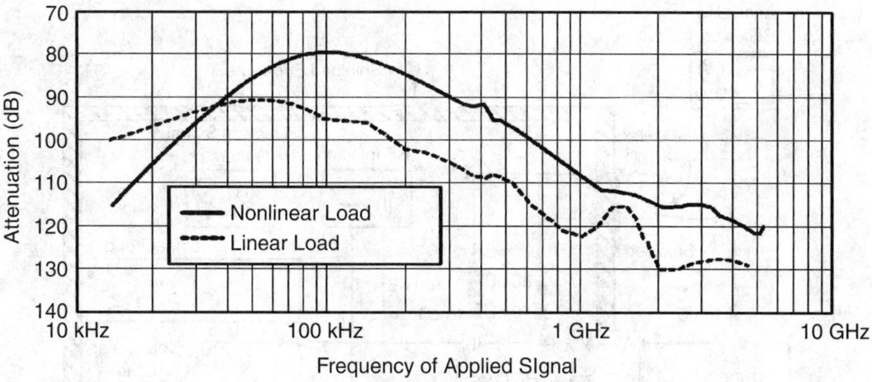

FIGURE 113.40 Effect of inductor saturation under DC current load using strictly RF current injection.

only DC current to do so. When the inductor becomes saturated, it loses most of its inductive properties. This effect varies according to core construction and the core material chosen. The loss of inductance due to saturation alters the attenuation properties of the filter at frequencies below 100 kHz. As shown in Figure 113.38, the cutoff frequency increases as the core approaches saturation. In addition, the gains from 2 to 7 kHz vary as a function of filter load. These filter characteristics may not be detected using MIL-STD-220B.

Although the application of current through buffer networks seems like a reasonable way to predict the attenuation performance of a powered filter, it still does not account for realistic source and load impedances. More often than not, when a filter is tested with a load, the load is usually resistive. Fifty years ago, using a purely resistive load in a filter test may have been an acceptable method because most building wiring systems powered linear (passive) loads, such as motors. Today, however, silicon technologies permeate the electrical environments where power-line filters are used. For example, motors are rarely connected directly to the power source. Instead, they are controlled by motor drives, which are power-electronic, nonlinear loads.

Millions of end-use devices now use nonlinear power supplies. According to Briggs (1994), problems with applying power-line filters to these nonlinear loads are often not fully realized until the filter has failed, either in expected performance or in a catastrophic event, such as the failure of capacitors inside the filter. Figure 113.40 illustrates the dynamic difference of filter performance under linear and nonlinear loading using a method called "RF current-injection," which can be found in MIL-STD-461. This standard is used to determine the susceptibility of equipment to conducted RF interference by injecting RF currents into the cables and harnesses of the equipment.

IEEE P1560: A Step Forward

To address the shortcomings of MIL-STD-220B, IEEE P1560 Working Group is working towards a new standard. The main objectives of the working group are:

- To develop alternative methods for characterizing the performance of power-line filters that account for real-world situations.
- To assemble the accepted and proven test methods into a single standard for the frequency range of 100 Hz to 10 GHz.

The latest draft, IEEE P1560 D/1.3 (March 5, 2003) is now available for review. The draft specifies no-load and loaded test methods, which are extended down to 100 Hz, as opposed to MIL-STD-220B, which extends down to only 100 kHz for the loaded test and 150 kHz for the loaded method using standardized buffer networks. The draft discourages the use of buffer networks and line-impedance stabilization networks (LISNs) for measurements below 100 kHz and proposes techniques using RF

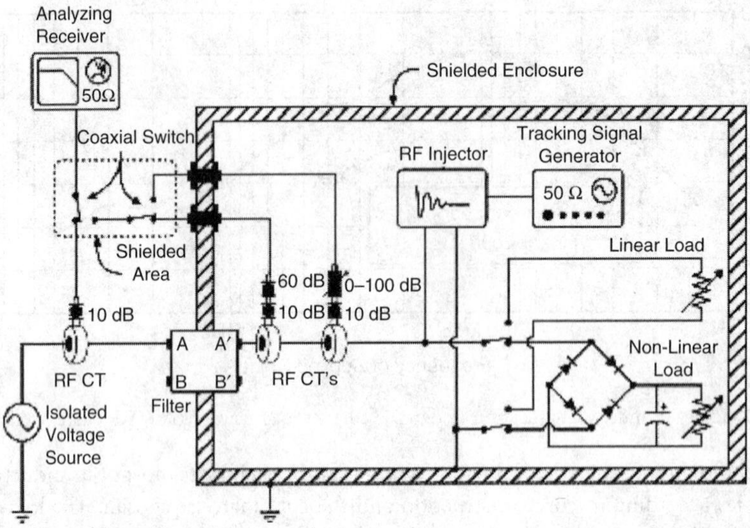

FIGURE 113.41 Typical common mode test setup: 100 Hz to 100 kHz.

current injection probes to be utilized from 100 Hz to 100 kHz. The test setup for common mode tests is shown in Figure 113.41.

Defining Terms

Ground — Any reference conductor that is used for a common return.

Near-field/far-field transition distance — For electrically small radiators (i.e., dimensions << wavelength), the near-field/far-field transition occurs at a distance equal to approximately one sixth of a wavelength from the radiating source.

Plane wave — Far-field electromagnetic wave with an impedance of 377 ohms in air.

Reference — Some object whose potential (often 0 volts with respect to earth or a power supply) is the one to which analog and logic circuits, equipments, and systems can be related or benchmarked

Return — The low (reference) voltage side of a wire pair (e.g., neutral), outer jacket of a coax, or conductor providing a path for intentional current to get back to the source.

Wavelength — The distance corresponding to a period for the electromagnetic wave spatial variation. Wavelength (meters) = 300/frequency (MHz).

References

Denny, H. W. 1983. *Grounding for the Control of EMI,* Interference Control Technologies, Gainesville, VA.

Duff, W. G. 1989. *Grounding for the Control of EMI,* EMC EXPO 89 Symposium Record. Interference Control Technologies, Gainesville, VA.

Duff, W. G. 1991. *Electromagnetic Shielding,* EMC EXPO 91 Symposium Record. Interference Control Technologies, Gainesville, VA.

Mardiguian, M. 1988. *Grounding and Bonding, Volume 2 — A Handbook Series on Electromagnetic Interference and Compatibility,* Interference Control Technologies, Gainesville, VA, 1988

White, D. R. J. and Mardiguian, M. *Electromagnetic Shielding, Volume 3 — A Handbook Series on Electromagnetic Interference and Compatibility,* Interference Control Technologies, Gainesville, VA.

FIPS Pub 94-1983, Guideline on Electrical Power for ADP Installations.

IEEE Std. 141-1993, *IEEE Recommended Practice for Electric Power Distribution for Industrial Plants (IEEE Red Book).*

IEEE Std. 142-1991, *IEEE Recommended Practice for Grounding of Industrial and Commercial Power Systems (IEEE Green Book).*

IEEE Std. 446-1995, *IEEE Recommended Practice for Emergency and Standby Power System for Industrial and Commercial Applications (IEEE Orange Book).*

IEEE Std. 518-1982, *IEEE Guide for the Installation of Electrical Equipment to Minimize Electrical Noise Inputs to Controllers from External Sources.*

NFPA 70-1999, *National Electrical Code.*

NFPA 75-1999, *Protection of Electronic Computer/Data Processing Equipment.*

IEEE Std. 1100-1999, *IEEE Recommended Practice for Powering and Grounding Sensitive Electronic Equipment (The Emerald Book).*

Phipps, K. O., Keebler P. E., and Connatser B. R. 2002. *Improving the Way We Measure Insertion Loss,* Interference Technology Engineers' Master, Robar Industries, Inc . PA.

Phipps, K. O., Keebler P. E., and Connatser B. R. *Isolation Transformers — Are They Worth It?,* Interference Technology Engineers' Master, Robar Industries, Inc. PA.

Ozenbaugh, R. L. 1996. *EMI Filter Design,* Marcel Dekker, New York.

IEEE P1560/D1.3-2003, *IEEE Standard for Methods of Measurements of Radio Frequency Interference Filter Suppression Capability in the Range of 100Hz to 40GHz.*

EPRI BR-112113. 1998. *Electromagnetic Compatibility for Healthcare,* Palo Alto, CA.

Martzloff, F. D., Mansoor, A., Phipps, K. O., and Grady, W. M. 1995. *Surging The Upside-Down House: Measurements and Modeling Results,* PQA, New York.

Key, T. S. and Martzloff, F. D. 1994. Surging the Upside-Down House: Looking into Upsetting Reference Voltages, Proc. Third International Conference on Power Quality — End-Use Applications and Perspectives, Amsterdam, The Netherlands, PQA, Amsterdam.

Further Information

IEEE Transactions on EMC. Published quarterly by the Institute of Electrical and Electronic Engineers.

IEEE International EMC Symposium Records. Published annually by the Institute of Electrical and Electronic Engineers

ANSI/IEEE Std. C63.13-1991, *Guide on the Application and Evaluation of EMI Power-Line Filters for Commercial Use.*

UL1283, *Standard for Electromagnetic Interference Filters,* August 24, 1993

C.I.S.P.R-17, 1981, *Methods of Measurements of the Suppression Characteristics of Passive Radio Interference Filters and Suppression Components.*

MIL-STD-220B:2000, *Method of Insertion Loss Measurement.*

MIL-STD-461E:1999, *Requirements for the Control of Electromagnetic Interference Characteristics of Subsystems and Equipment.*

SAE ARP 4244:1996, *Recommended Insertion Loss Test Methods for EMI Power Line Filters.*

114
Electromagnetics

M.N.O. Sadiku
Prairie View A&M University

C.M. Akujuobi
Prairie View A&M University

Electromagnetics (EM) is that branch of physics or electrical engineering in which the interactions between charges at rest and in motion are examined. It entails the analysis, synthesis, physical interpretation, and application of electric and magnetic fields. EM theory is indispensable to the design and operation of many practical devices. EM principles find applications in diverse areas such as microwaves, antennas, electric machines, bioelectromagnetics, plasmas, fiber optics, satellite communication, electromagnetic compatibility, radar meteorology, and remote sensing.

The foundational basis of electromagnetic theory was established in 1873 by James Clerk Maxwell, who for the first time unified observations of Michael Faraday, Karl Friedrich Gauss, and Andre-Marie Ampere. In this chapter, we present Maxwell's equations in time-domain and frequency-domain and the constitutive equations. SI units are assumed throughout.

114.1 Maxwell's Equations

At the macroscopic level, electric and magnetic phenomena are described by Maxwell's equation. Maxwell's work was based on a large body of theoretical and empirical works of others such as Gauss, Ampere, and Faraday. In differential form, Maxwell's equations are given by

$$\nabla x E = -\frac{\partial B}{\partial t} \quad \text{(Faraday's law)} \tag{114.1}$$

$$\nabla x H = J + \frac{\partial D}{\partial t} \quad \text{(Ampere's law)} \tag{114.2}$$

$$\nabla \bullet D = \rho_v \quad \text{(Gauss's law)} \tag{114.3}$$

$$\nabla \bullet B = 0 \quad \text{(Gauss's law)} \tag{114.4}$$

where **E**, **H**, **D**, **B**, **J**, and ρ_v are real functions of space and time. **E** is electric field intensity (volts/meter), **H** is the magnetic field intensity (amperes/meter), **D** is the electric flux density (coulombs/square meter),

0-8493-1586-7/05/$0.00+$1.50

B is the magnetic flux density (webers/square meter), **J** is the conduction current density (amperes/square meter), and ρ_v is the electric field charge density (coulombs/cubic meter). The differential form of Maxwell's equations as in Equation (114.1) to Equation (114.4) is the most widely used representation to solve boundary-value electromagnetic problems. Although the equations are simply a set of coupled, partial differential equations, they are sufficient to understand the full range of electromagnetic phenomena.

Maxwell's equations contain an important relationship between charge and current. Taking the divergence of both sides of Equation (114.2) and applying the vector identity $\nabla\bullet(\nabla xA)=0$ yields

$$\nabla(\nabla xH)=\nabla J+\nabla\frac{\partial D}{\partial t}=0 \tag{114.5}$$

Introducing (114.3) leads to

$$\nabla J+\frac{\partial\rho_v}{\partial t}=0 \tag{114.6}$$

which is the continuity equation expressing the conservation of charge and current.

Another equation associated with Maxwell's equation is the Lorentz force law, which relates the electromagnetic fields to measurable forces. If a charge q moves in an electric field **E** and magnetic field **B** at a velocity **v**, the force on the charge is given by Lorentz force law:

$$F=q(E+vxB) \tag{114.7}$$

When the charge is not discrete, Lorentz force law becomes

$$F=\rho_v E+JxB \tag{114.8}$$

The integral form of Maxwell's equations can be obtained from the differential form by applying divergence and Stokes's theorems to Equation (114.1) to Equation (114.4). They are derived as

$$\oint_L E\cdot dl=-\frac{\partial}{\partial t}\oiint_S B\cdot dS \text{ (Faraday's law)} \tag{114.9}$$

$$\oint_L H\cdot dl=\oiint_S J\cdot dS+\frac{\partial}{\partial t}\oiint_S D\cdot dS \text{ (Ampere's law)} \tag{114.10}$$

$$J\oiint_S D\cdot dS=\oiint_V \rho_v dv \text{ (Gauss's law)} \tag{114.11}$$

$$\oiint_S B\cdot dS=0 \text{ (Gauss's law)} \tag{114.12}$$

The integral form of Maxwell's equations depicts the underlying physical laws, whereas the differential form is used more frequently in solving problems.

114.2 Constitutive Relations

Maxwell's equations contain more variables than equations. To solve an EM problem using Maxwell's equations therefore requires some additional set of relations between the field quantities **E, H,** and **D, B.**

Thus, we need the constitutive relations to supplement Maxwell's equation. For an isotropic medium, the constitutive relations are

$$D = \varepsilon E \tag{114.13}$$

$$B = \mu H \tag{114.14}$$

$$J = \sigma E \tag{114.15}$$

where ε, μ, and σ are, respectively, the permittivity, the permeability, and the conductivity of the medium. In electromagnetics, a homogeneous and isotropic region with no sources in it is called *free space.* For free space,

$$\varepsilon_o = 8.85x10^{-12} \text{ (F/m)} \tag{114.16}$$

$$\mu_o = 4\pi x10^{-7} \text{ (H/m)} \tag{114.17}$$

$$\sigma = 0 \text{ (S/m)} \tag{114.18}$$

Equation (114.11) to Equation (114.13) are regarded as the *constitutive relations*, while ε, μ, and σ are known as the *constitutive parameters.* The constitutive parameters are often used to characterize the properties of a material. For example, a dielectric material is said to be linear if $D = \varepsilon E$ applies and if ε does not change with the applied **E** field; it is homogeneous if ε does not change from point to point; and it is isotropic if ε does not change with direction. A material is said to be dispersive if its constitutive parameters are functions of frequency.

For anisotropic media, the constitutive relations become

$$\overline{D} = \overline{\varepsilon} \cdot \overline{E} \tag{114.19}$$

$$\overline{B} = \overline{\mu} \cdot \overline{H} \tag{114.20}$$

where $\overline{\varepsilon}$ and $\overline{\mu}$ are permittivity and permeability tensors, respectively. In this case, the vector field **D** is no longer parallel to **E** and the vector field **B** is no longer parallel to **H.** For example, in the Cartesian coordinate system, Equation (114.19) becomes

$$\begin{bmatrix} D_x \\ D_y \\ D_x \end{bmatrix} = \begin{bmatrix} \varepsilon_{xx} & \varepsilon_{xy} & \varepsilon_{xz} \\ \varepsilon_{yx} & \varepsilon_{yy} & \varepsilon_{yz} \\ \varepsilon_{zx} & \varepsilon_{zy} & \varepsilon_{zz} \end{bmatrix} \begin{bmatrix} E_x \\ E_y \\ E_x \end{bmatrix} \tag{114.21}$$

For crystals,

$$\overline{\varepsilon} = \begin{bmatrix} \varepsilon_x & 0 & 0 \\ 0 & \varepsilon_y & 0 \\ 0 & 0 & \varepsilon_z \end{bmatrix} \tag{114.22}$$

indicating that crystals are described by symmetric permittivity tensor.

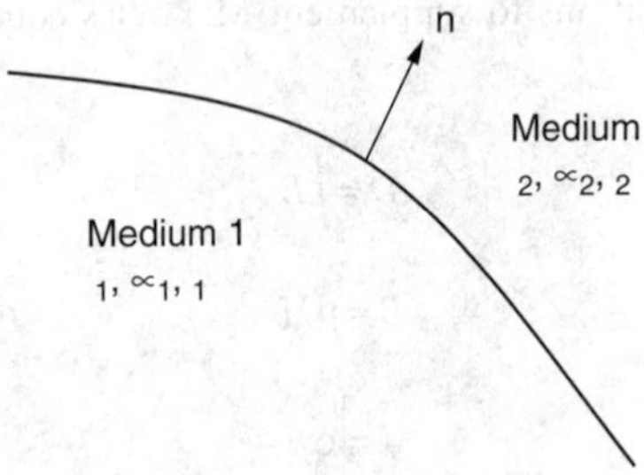

FIGURE 114.1 Boundary condition at the interface between two media.

114.3 Boundary Conditions

Along the interface where the media involved exhibit discontinuity in electrical properties or sources exist along the interface, the field vectors themselves are discontinuous, and their behavior across the interface is governed by the boundary conditions. In fact, the solution of Maxwell's equation is not complete until the boundary conditions are imposed. However, Maxwell's equations in integral form can be used to obtain the boundary conditions at the interface between two media. A sufficient set of boundary conditions is

$$nX(H_2 - H_1) = J_s \tag{114.23}$$

$$(E_2 - E_1)xn = M_s \tag{114.24}$$

where **n** is unit normal directed from medium 1 (with $\varepsilon_1, \mu_1, \sigma_1$) to medium 2 (with $\varepsilon_2, \mu_2, \sigma_2$) as shown in Figure 114.1. J_s and M_s are, respectively, the electric and magnetic surface currents. Equation (114.23) and Equation (114.24) can be written in terms of the tangential components of the fields as

$$H_{2t} = H_{1t} + J_s \tag{114.25}$$

$$E_{2t} = E_{1t} + M_s \tag{114.26}$$

If one medium is a perfect conductor, the boundary conditions in Equation (114.25) and Equation (114.26) become

$$H_t = J_s \tag{114.27}$$

$$E_t = 0 \tag{114.28}$$

A boundary on which Equation (114.28) is valid is known as electric wall since the tangential components of E are "shorted out" and must vanish at the surface of the conductor. Dual to this is the magnetic wall, which is a boundary on which H vanishes. For this case,

$$H_t = 0 \tag{114.29}$$

$$E_t = M_s \tag{114.30}$$

Finally, when dealing with problems that have one or more infinite boundaries, it is required that at an infinite distance from the source, the fields must be vanishingly small.

114.4 Power and Energy

Energy can be transported from one point to another by means of EM waves. (This is the case in wireless communication systems.) The rate of such energy transportation can be obtained from Maxwell's equations. Dotting both sides of Equation (114.2) with **E** and replacing **J** by σ**E** leads to

$$E(\nabla xH) = \sigma E^2 + E\varepsilon \frac{\partial E}{\partial t} \tag{114.31}$$

Using the vector identity

$$\nabla(AxB) = B(\nabla xA) - A(\nabla xB) \tag{114.32}$$

gives

$$H(\nabla xE) + \nabla(HxE) = \sigma E^2 + \frac{1}{2}\varepsilon \frac{\partial E^2}{\partial t} \tag{114.33}$$

From Equation (114.1) and Equation (114.14),

$$H(\nabla XE) = H\left(-\mu \frac{\partial H}{\partial t}\right) = -\frac{\mu}{2}\frac{\partial H^2}{\partial t} \tag{114.34}$$

Substituting Equation (114.34) into Equation (114.33),

$$\nabla(ExH) = -\frac{1}{2}\varepsilon \frac{\partial E^2}{\partial t} - \frac{1}{2}\mu \frac{\partial H^2}{\partial t} - \sigma E^2 \tag{114.35}$$

Taking the volume integral of both sides and applying the divergence theorem to the left-hand side yields

$$\underbrace{\oint_S (E \times H)\cdot dS}_{\text{Total power leaving the volume}} = \underbrace{-\frac{\partial}{\partial t}\int_v \left[\frac{1}{2}\varepsilon E^2 + \frac{1}{2}\mu H^2\right] dv}_{\text{Rate of decrease in energy stored in EM fields}} - \underbrace{\int_v \sigma E^2 dv}_{\text{Ohmic power loss}} \tag{114.36}$$

where volume v is bounded by the closed surface S. This expresses a conservation of power or Poynting's theorem. The first term on the right-hand side is interpreted as the rate of decrease in energy stored in the electric and magnetic fields. The second term is the power dissipated due to a conducting medium ($\sigma \neq 0$). The quantity $\mathbf{E} \times \mathbf{H}$ on the left-hand side of Equation (114.36) is known as the Poynting vector **P**

$$\mathbf{P} = \mathbf{E} \infty \mathbf{H} \tag{114.37}$$

P is the flow of power flux per unit area. It represents the power flux density and its unit is watts/m^2.

114.5 Vector and Scalar Potentials

It is sometimes convenient in electromagnetics to introduce a vector magnetic potential from which both electric and magnetic fields may be derived. This reduces the number of equations to be solved. We define this potential (one of several possible potentials) in terms of the magnetic flux density:

$$\nabla XA = B \tag{114.38}$$

But $\nabla x E = -\partial B / \partial t$ so that

$$\nabla x \left(E + \frac{\partial A}{\partial t} \right) = 0 \tag{114.39}$$

Since the curl of the gradient of any scalar field is zero, the bracket term in Equation (114.39) can be replaced by the gradient of a scalar potential V.

$$E + \frac{\partial A}{\partial t} = -\nabla V$$

or

$$E = -\nabla V - \frac{\partial A}{\partial t} \tag{114.40}$$

From Equation (114.38) and Equation (114.40), we can determine the vector fields **B** and **E** provided the potentials **A** and V are known. By taking the divergence of Equation (114.40) and introducing Equation (114.3), we have

$$\nabla E = \frac{\rho_v}{\varepsilon} = -\nabla^2 V - \frac{\partial}{\partial t}(\nabla A)$$

or

$$\nabla^2 V + \frac{\partial}{\partial t}(\nabla A) = -\frac{\rho_v}{\varepsilon} \tag{114.41}$$

Taking the curl of Equation (114.38) and incorporating Equation (114.2) and Equation (114.40) results in

$$\nabla x \nabla x A = \mu J - \mu\varepsilon \nabla \left(\frac{\partial V}{\partial t} \right) - \mu\varepsilon \frac{\partial^2 A}{\partial t^2} \tag{114.42}$$

Applying the vector identity

$$\nabla x \nabla x A = \nabla(\nabla A) - \nabla^2 A \tag{114.43}$$

leads to

$$\nabla^2 A - \nabla(\nabla A) = -\mu J + \mu\varepsilon \nabla \left(\frac{\partial V}{\partial t} \right) + \mu\varepsilon \frac{\partial^2 A}{\partial t^2} \tag{114.44}$$

A vector field is uniquely defined when its curl and divergence are specified. The curl of **A** has been specified in Equation (114.38). We may choose the divergence of **A** as

$$\nabla A = -\mu\varepsilon \frac{\partial V}{\partial t} \tag{114.45}$$

This choice relates **A** and V, and it is called the *Lorentz condition.* By imposing this condition, Equation (114.41) and Equation (114.44) become

$$\nabla^2 V - \mu\varepsilon \frac{\partial^2 V}{\partial t^2} = -\frac{\rho_v}{\varepsilon} \tag{114.46}$$

and

$$\nabla^2 A - \mu\varepsilon \frac{\partial^2 A}{\partial t^2} = -\mu J \tag{114.47}$$

which are wave equations. Once Equation (114.46) and Equation (114.47) are solved, the fields **E** and **B** are obtained from Equation (114.38) and Equation (114.40). The magnetic vector potential **A** does not have the physical significance of the scalar potential V and is solely a mathematical convenience in determining EM fields.

114.6 Time-Harmonic Fields

In many practical applications involving EM fields, the time variations are of sinusoidal form and are referred to as time-harmonic. Also, it is expedient to first understand the behavior of monochromatic waves (which have only a single frequency) and from them construct any other waveform using Fourier analysis. For time-harmonic fields, the time variation can be represented by $e^{j\omega t}$. Hence, for time-harmonic field **A**,

$$A(x, y, z, t) = \mathrm{Re}\left[A_s(x, y, z)e^{j\omega t}\right] \tag{114.48}$$

where $\mathbf{A}_s$ is the phasor form of **A** and is only a function of position in space (x,y,z). **A** in Equation (114.48) stands for EM fields **E, H, B**, or **D**. We notice from Equation (114.48) that the differential and integral operations respectively yield

$$\frac{\partial A}{\partial t} \rightarrow j\omega A_s \tag{114.49}$$

$$\int A \partial t \rightarrow \frac{A_s}{j\omega} \tag{114.50}$$

Thus, for time-harmonic fields, Maxwell's equations can be written in simpler forms, as shown in Table 114.1. Also, Maxwell's equations are easier to solve for time-harmonic fields. For time-harmonic fields,

TABLE 114.1 Time-Harmonic Maxwell's Equations

Differential Form	Integral Form
$\nabla D_s = \rho_{vs}$	$\oint D_s \cdot dS = \int \rho_{vs} dv$
$\nabla B_s = 0$	$\oint B_s \cdot dS = 0$
$\nabla x E_s = -j\omega B_s$	$\oint E_s \cdot dl = -j\omega \int B_s \cdot dS$
$\nabla x H_s = J_s + j\omega D_s$	$\oint H_s \cdot dl = \int (J_s + j\omega D_s) \cdot dS$

the boundary conditions are not independent of each other. If, for example, the tangential components of the E and H fields satisfy the boundary conditions, the normal components of the same fields automatically satisfy their appropriate boundary conditions.

We can also show that for time-harmonic fields, the time-average Poynting vector is

$$P_{ave} = \frac{1}{T}\int_0^T ExH\partial t = \frac{1}{2}\mathrm{Re}\left[E_s x H_s^*\right] \tag{114.51}$$

where the asterisk (*) denotes complex conjugate.

114.7 Wave Equations

Consider a linear, isotropic, homogeneous, lossless dielectric medium ($\sigma = 0$) that is charge free ($\rho_v = 0$). Assuming and suppressing the time factor $e^{j\omega t}$, Maxwell's equations (see Table 114.1) become

$$\nabla E_s = 0 \tag{114.52}$$

$$\nabla H_s = 0 \tag{114.53}$$

$$\nabla x E_s = -j\omega\mu H_s \tag{114.54}$$

$$\nabla x H_s = j\omega\varepsilon E_s \tag{114.55}$$

Taking the curl of both sides of (114.54) gives

$$\nabla x \nabla x E_s = -j\omega\mu\nabla x H_s \tag{114.56}$$

Applying the vector identity in Equation (114.43) to the left-hand side of Equation (114.56) and invoking Equation (114.52) and Equation (114.55), we get

$$\nabla^2 E_s + \omega^2\mu\varepsilon E_s = 0 \tag{114.57}$$

Using separation of variables, we find that the solution to Equation (114.57) is plane waves. The simplest solution for the electric field $\mathbf{E}_s$ is

$$E_s = E_o\cos(kz - \omega t)a_x \tag{114.58}$$

where $k = \omega\sqrt{\mu\varepsilon}$ and ω is the angular frequency of the wave. The EM wave is polarized along the x-direction, propagates in the +z-direction, travels at the speed

$$u = \frac{1}{\sqrt{\mu\varepsilon}} \tag{114.59}$$

and has a wave impedance

$$\eta = \sqrt{\frac{\mu}{\varepsilon}} \tag{114.60}$$

The wave repeats itself at a regular interval. The frequency f is defined as

$$f = \frac{\omega}{2\pi} \tag{114.61}$$

The EM waves obey all other usual wave phenomena of reflection and refraction when incident on interfaces between two different media.

References

Balanis, C. A. 1989. *Advanced Engineering Electromagnetics,* John Wiley & Sons, New York. pp. 1–32.

Buris, N. E. 2001. Maxwell's equations, in M. Golio (ed.), *The RF and Microwave Handbook,* CRC Press, Boca Raton, FL. pp. 9.1–9.12.

Kong, J. A. 1997. Electromagnetic fields, in R. C. Dorf (ed.), *The Electrical Engineering Handbook,* CRC Press, Boca Raton, FL. pp. 889–897.

Ishimaru, A. 1991. *Electromagnetic Wave Propagation, Radiation, and Scattering.* Prentice Hall, Englewood Cliffs, NJ. pp. 5–29.

Sadiku, M. N. O. 2001. *Elements of Electromagnetics.* Oxford University Press, New York. pp. 1–2, 384–392.

Staelin, D. H., Morgenthaler, A. W. and Kong, J. A. 1994. *Electromagnetic Waves,* Prentice Hall, Englewood Cliffs, NJ. pp. 1–14.

XVIII

Electronics

115

Operational Amplifiers

Paul J. Hurst
University of California, Davis

The operational amplifier (op amp) is one of the most versatile building blocks for analog circuit design. The earliest op amps were constructed of discrete devices — first vacuum tubes and later transistors. Today, an op amp is constructed of numerous transistors along with a few resistors and capacitors, all fabricated on a single piece of silicon. Parts are sold with one, two, or four op amps in a small integrated circuit package.

Op amps are popular because they are small, versatile, easy to use, and inexpensive. Despite their low cost, modern op amps offer superb performance. An op amp provides high voltage gain, high input impedance, and low output impedance. Many op amps are commercially available, and they differ in specifications such as input noise, bandwidth, gain, offset voltage, output swing, and supply voltage.

Op amps are used as high-gain amplifiers in negative feedback circuits. An advantage of such circuits is that design and analysis is relatively simple, with the overall gain depending only on external passive components that provide the feedback around the op amp. In the following sections, analysis and design of op-amp feedback circuits are covered.

115.1 The Ideal Op Amp

The schematic symbol for an op amp is shown in Figure 115.1(a). The op amp has two inputs, v_+ and v_-, and a single output. The voltages v_+, v_-, and v_{OUT} are measured with respect to ground. The op amp amplifies the voltage difference $v_+ - v_-$ to produce the output voltage v_{OUT}. A simple model for the op amp that is valid when the op amp is operating in its linear high-gain region is shown in Figure 115.1(b). The model consists of an input impedance Z_i, output impedance Z_o, and voltage gain a. Here, the voltage difference $v_+ - v_-$ is called v_E. If Z_i, Z_o, and a are known, this model can be used to analyze an op-amp circuit if the op amp is biased in the linear region, which is usually the case when the op amp is in a negative feedback loop. A typical op amp has large Z_i, small Z_o, and large a. As a result, the op-amp model is often further simplified by setting $Z_i = \infty$ and $Z_o = 0$, as shown in Figure 115.1(c).

For an example of an op-amp feedback circuit, consider the inverting gain amplifier shown in Figure 115.2. Using the model of Figure 115.1(c) in Figure 115.2, the gain can be found by summing currents at the v_- input of the op amp. Since the current into the op amp is zero (due to $Z_i = \infty$), $i_1 = i_2$; therefore,

$$\frac{v_{IN} - (-v_E)}{R_1} = i_1 = i_2 = \frac{-v_E - v_{OUT}}{R_2} \tag{115.1}$$

0-8493-1586-7/05/$0.00+$1.50

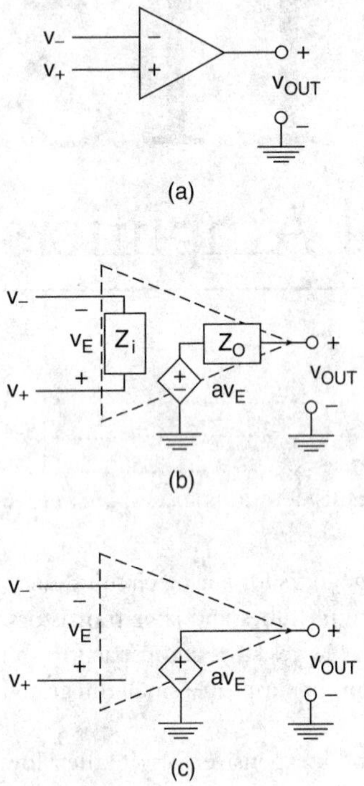

FIGURE 115.1 (a) The schematic symbol for an op amp. (b) A simple three-element model for the op amp. (c) The model in (b) simplified further by setting $Z_i = \infty$ and $Z_o = 0$.

Using the relationship $v_E = v_{OUT}/a$, Equation (115.1) yields

$$\frac{v_{OUT}}{v_{IN}} = -\frac{R_2}{R_1}\frac{1}{1+(R_1+R_2)/aR_1} \tag{115.2}$$

Assuming that the op-amp gain is very large ($a \to \infty$), Equation (115.2) becomes

$$\frac{v_{OUT}}{v_{IN}} = -\frac{R_2}{R_1} \tag{115.3}$$

The gain depends on the ratio of resistors and is independent of the op-amp parameters.

The above analysis can be made easier by assuming that the op amp is ideal ($Z_i = \infty$, $Z_o = 0$, and $a = \infty$) before beginning the analysis. This **ideal op-amp model** greatly simplifies analysis and gives results that are surprisingly accurate. The assumption $Z_i = \infty$ ensures that the currents flowing into the op-amp input terminals are zero. With $Z_o = 0$, the controlled source directly controls the output, giving $v_{OUT} = av_E$. The final assumption $a = \infty$ leads to $v_E = v_{OUT}/a = 0$ if v_{OUT} is bounded, which is typically true in negative feedback circuits. The condition $v_E = 0$ is referred to as a *virtual short circuit* because the op-amp input voltages are equal ($v_+ = v_-$), even though these inputs are not actually connected together. Negative feedback forces v_E to be equal to zero.

To demonstrate the advantage of the ideal op-amp model, consider the circuit in Figure 115.3. If the op amp is ideal, $v_+ = v_-$, and therefore $v_- = v_{IN}$ as a result of the virtual short circuit. Since no current flows into the op-amp input, the currents i_3 flowing through resistor R_3 and i_4 through R_4 must be equal:

$$\frac{0 - v_{IN}}{R_3} = i_3 = i_4 = \frac{v_{IN} - v_{OUT}}{R_4} \tag{115.4}$$

The resulting gain is positive, or noninverting, and is given by

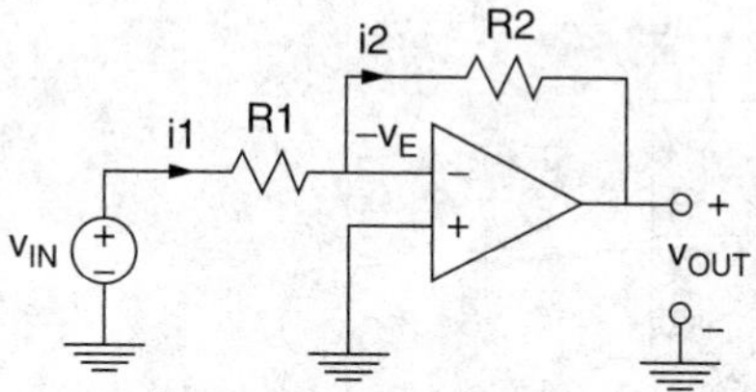

FIGURE 115.2 An inverting gain amplifier.

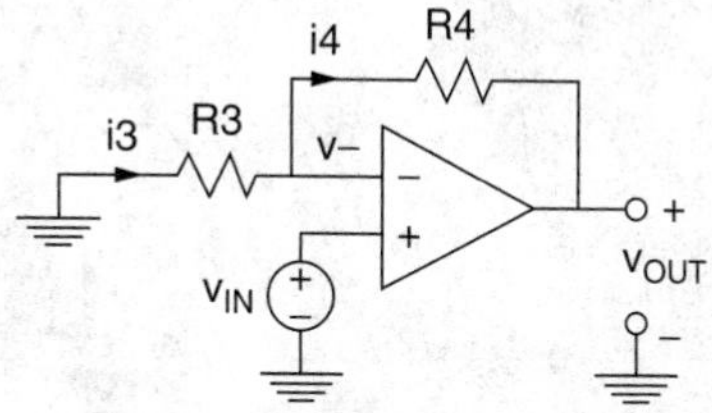

FIGURE 115.3 A noninverting gain amplifier.

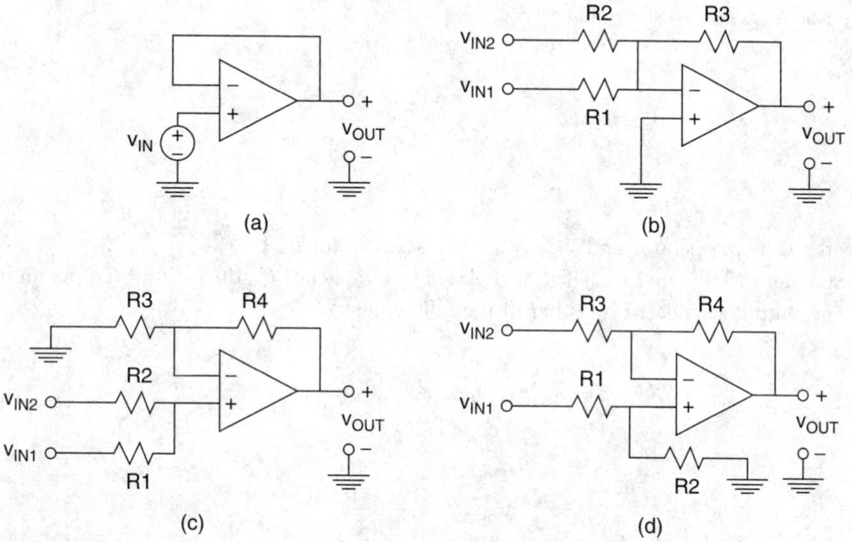

FIGURE 115.4 (a) A voltage buffer, $v_{OUT} = v_{IN}$. (b) A two-input inverting gain amplifier, $v_{OUT} = -(R_3/R_1)v_{IN1} - (R_3/R_2)v_{IN2}$. (c) A two-input noninverting stage, $v_{OUT} = \{[R_2/(R_1 + R_2)]v_{IN1} + [R_1/(R_1 + R_2)]v_{IN2}\}(R_3 + R_4)/R_3$; (d) a differencing amplifier, $v_{OUT} = -(R_4/R_33)v_{IN2} + [R_2/(R_1 + R_2)][(R_3 + R_4)/(R_3)]v_{IN1}$.

$$\frac{v_{OUT}}{v_{IN}} = 1 + \frac{R_4}{R_3} \tag{115.5}$$

This example demonstrates how the ideal op-amp model can be used to quickly analyze op-amp feedback circuits.

A number of interesting op-amp feedback circuits are shown in Figure 115.4. The corresponding expressions for v_{OUT} are given in the caption for Figure 115.4, assuming an ideal op amp. The simplest circuit is the voltage buffer of Figure 115.4(a), which has a voltage gain of 1. The circuits in Figure 115.4(b–d) have multiple inputs. Figure 115.4(b) is an inverting, summing amplifier if $R_1 = R_2$. Figure 115.4(c) is a noninverting, summing amplifier if $R_1 = R_2$. If $R_1/R_2 = R_3/R_4$ in Figure 115.4(d), this circuit is a differencing amplifier with

$$v_{OUT} = \frac{R_4}{R_3}(v_{IN1} - v_{IN2}) \tag{115.6}$$

Op amps can be used to construct filters for signal conditioning. If R_1 in Figure 115.2 is replaced by a capacitor C_1, the circuit becomes a differentiator and its input and output are related by either of the following:

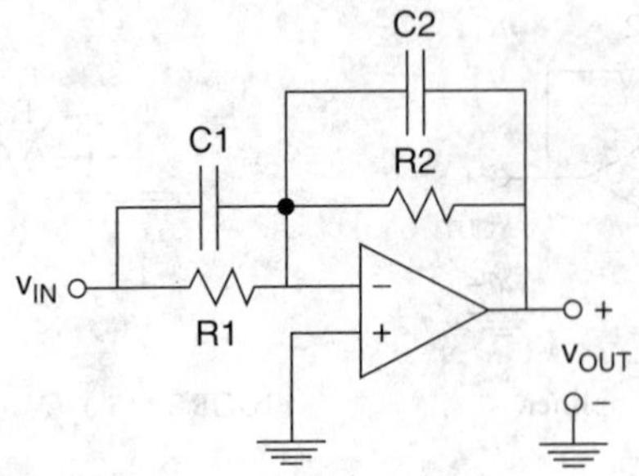

FIGURE 115.5 A one-pole, one-zero filter.

$$v_{OUT}(t) = -R_2 C_1 \frac{dv_{IN}(t)}{dt} \tag{115.7a}$$

$$V_{OUT}(s) = -sR_2 C_1 V_{IN}(s) \tag{115.7b}$$

where s is the Laplace operator and $V(s)$ is the Laplace transform of $v(t)$.

Again, starting with Figure 115.2, if R_2 is replaced by a capacitor C_2, the circuit becomes an integrator. Its input and output are related by either of the following:

$$v_{OUT}(t) = -\frac{1}{R_1 C_2} \int_0^t v_{IN}(\tau)d\tau \tag{115.8a}$$

$$V_{OUT}(s) = -\frac{V_{IN}(s)}{sR_1 C_2} \tag{115.8b}$$

Here, v_{OUT} $(t = 0) = 0$ is assumed. Since there is no DC feedback from the op-amp output to its input through capacitor C_2, an integrator will only work properly when used in an application that provides DC feedback around the integrator for biasing. Integrators are often used inside feedback loops to construct filters or control loops.

Figure 115.5 is a one-pole, one-zero filter with transfer function

$$\frac{V_{OUT}(s)}{V_{IN}(s)} = -\frac{R_2}{R_1} \frac{1 + sR_1 C_1}{1 + sR_2 C_2} \tag{115.9}$$

Higher-order filters can be constructed using op amps [Jung, 1986].

115.2 Feedback Circuit Analysis

The goal of every op-amp feedback circuit is an input/output relationship that is independent of the op amp itself. An exact formula for the closed-loop gain A of an op-amp feedback circuit (e.g., Figure 115.2) is given by Rosenstark [1986]:

$$A = \frac{v_{OUT}}{v_{IN}} = A_\infty \frac{\text{RR}}{1 + \text{RR}} + \frac{d}{1 + \text{RR}} \tag{115.10}$$

Here, A_∞ is the gain when the op-amp gain $a = \infty$. The term $d = v_{OUT}/v_{IN}|_{a=0}$ accounts for feed-forward directly from input to output; typically, d is zero or close to zero. RR is the return ratio for the controlled

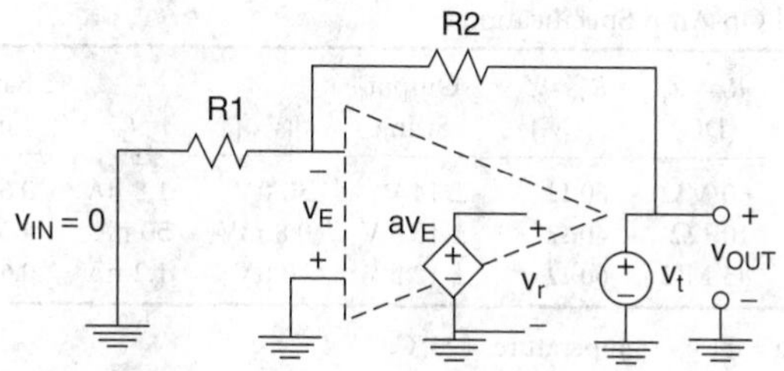

FIGURE 115.6 Figure 115.2 modified to allow calculation of the return ratio.

source *a*. From Equation (115.10), it can be seen that gain *A* approaches the ideal value A_∞ as RR approaches infinity. The return ratio for the controlled source *a* can be found by (1) setting all independent sources to zero, (2) breaking the connection between the controlled source *a* and the rest of the circuit, (3) driving the circuit at the break with a voltage source with value v_t, and (4) finding the resulting voltage v_r across the dependent source. Then $\mathrm{RR} = -v_r/v_t$. For negative feedback, RR is positive. Ideally, RR is large so that the gain *A* is close to A_∞.

For example, consider Figure 115.2. Using the op-amp model of Figure 115.1(c), the return ratio for the dependent source *a* can be found using Figure 115.6, and the result is

$$\mathrm{RR} = -\frac{v_r}{v_t} = a\frac{R_1}{R_1 + R_2} \tag{115.11}$$

With this model, *d* is zero, since setting *a* equal to 0 forces v_{OUT} equal to 0. Therefore, Equation (115.10) simplifies to

$$A = \frac{v_{\mathrm{OUT}}}{v_{\mathrm{IN}}} = A_\infty \frac{\mathrm{RR}}{1+\mathrm{RR}} \tag{115.12}$$

The RR in Equation (115.11) can be used in Equation (115.10) to determine *A* at DC or as a function of frequency if $a(s)$ is known. Information on $a(s = j\omega)$ is usually provided graphically on a data sheet.

Since op-amp circuits use feedback, stability is a concern. In circuits with passive feedback and load elements, the terms A_∞, RR, and *d* in Equation (115.10) are stable (all poles are in the left half of the *s* plane). Therefore, stability of the closed-loop transfer function $A(s)$ is determined by the location of the zeroes of $[1 + \mathrm{RR}(s)]$, which are poles of $A(s)$. The location of the zeroes of $[1 + \mathrm{RR}(s)]$ can be determined by examining the phase and gain margins of RR [Rosenstark, 1986]. The phase margin is 180° – [the phase of RR $(s = j\omega_U)$], where ω_U is the frequency where the magnitude of RR is unity. The **gain margin** is $-20 \log_{10} |\mathrm{RR}(s = j\omega_{180})|$, where ω_{180} is the frequency where the phase of RR is –180°. The phase and gain margins are positive for a stable circuit. Using an accurate frequency-dependent model for the op amp, or using the frequency response plot of $a(s = j\omega)$ on the data sheet, the gain and phase margins can be found. Roughly, $A(s = j\omega) = A_\infty$ for frequencies below ω_U, and $A(s = j\omega)$ will deviate from A_∞ at frequencies near and above ω_U.

Most op amps are designed to be unity-gain stable — that is, they are stable when connected as a buffer [see Figure 115.4(a)]. In that configuration, $\mathrm{RR}(s) = a(s)Z_i/(Z_i + Z_o) \approx a(s)$ (the approximation follows from $|Z_o| \ll |Z_i|$).

115.3 Input and Output Impedances

The input impedance of an op-amp circuit can be found by applying a test voltage source across the input port and measuring the current that flows into the port. The applied voltage divided by the resulting current is the input impedance. (When computing the output impedance, the input source(s) must be

TABLE 115.1 Typical Op-Amp Specifications

Part #	a at DC	$R_i = Z_i$ (DC)	$R_0 = Z_o$ (DC)	Output Swing	$\lvert V_{OS}\rvert$	I_B	Bandwidth for $A_\infty = 1$	Slew Rate
OP-07	$5 \cdot 10^5$	60 MΩ	60 Ω	± 14 V	30 μV	1.8 nA	0.6 MHz	0.3 V/μs
LF411	$2 \cdot 10^5$	10^{12} Ω	40 Ω	± 13.5 V	0.8 mV	50 pA	4 MHz	15 V/μs
OP-177F	$1.2 \cdot 10^7$	45 MΩ	60 Ω	± 12.5 V	10 μV	1.2 nA	0.6 MHz	0.3 V/μs

Note: Supply voltage = ±15 V; temperature = 25°C.

set to zero and then the same procedure is carried out on the output port.) Using the model in Figure 115.1(b), the input impedance for the voltage buffer in Figure 115.4(a) can be found by following these steps, and the result is

$$Z_{\text{in}}(\text{with feedback}) = (Z_i + Z_o)\cdot\left(1 + a\frac{Z_i}{Z_o + Z_i}\right) \approx aZ_i \tag{115.13}$$

where the approximation is valid since $|Z_o| \ll |Z_i|$ and $a \gg 1$. The negative feedback increases the input impedance from Z_i to aZ_i. For the OP-07 op amp (see Table 115.1), $Z_i = 60$ MΩ and $a = 500\,000$ at DC, giving a remarkably large Z_{in} (with feedback) = $3 \cdot 10^{13}$ Ω at DC. Calculation of the output impedance in Figure 115.4(a) gives

$$Z_{\text{out}}(\text{with feedback}) = \frac{Z_i \parallel Z_o}{1 + a(Z_i)/(Z_o + Z_i)} \approx \frac{Z_o}{a} \tag{115.14}$$

where $x \parallel y = xy/(x + y)$, and $Z_i \parallel Z_o \approx Z_o$ because $|Z_o| \ll |Z_i|$. The feedback reduces the output impedance. Again, using values for the OP-07, Z_{out} (with feedback) = 0.1mΩ at DC.

In some cases, the ideal op-amp model can be used to quickly determine the input impedance. For example, the input resistance in Figure 115.2 is approximately R_1 because feedback forces v_- to be close to 0. The output impedance of an ideal op-amp circuit is zero because $Z_o = 0$. The actual value can be found by using the model in Figure 115.1(b) and carrying out the computation described earlier.

115.4 Practical Limitations and Considerations

The ideal op-amp model is easy to use, but practical limitations are not included in this simple model. Specifications for a few popular op amps are given in Table 115.1 [Analog Devices, 1992; National Semiconductor, 1993]. The LF411 uses field-effect input transistors to give extremely high-input resistance and low-input bias current. The input bias current I_B is the DC current that flows into the op-amp inputs; this is bias current required by the input transistors in the op amp. (This current is not included in the simple models of Figure 115.1, and it cannot be computed using Z_i.)

The output voltage swing over which the linear model is valid is limited by the power supply voltage(s) and the op-amp architecture. The peak output current is limited by internal bias currents and/or protection circuitry. The output slew rate, which is the maximum slope of the output signal before distortion occurs, is limited by op-amp internal currents and capacitances. The common-mode input voltage range is the range of the common-mode (or average) input voltage, which is $(v_+ + v_-)/2$, for which the linear op-amp model is valid. In Figure 115.2, it is important that ground be within this range since $(v_+ + v_-)/2 \approx 0$ in this circuit. In Figure 115.3, v_{IN} must stay in the common-mode input range since $(v_+ + v_-)/2 \approx v_{\text{IN}}$.

The **input offset voltage** V_{OS} is the DC op-amp input voltage that causes v_{OUT} to equal 0 V DC. The offset voltage varies from part to part; it is dependent on imbalances within the op amp. The input offset voltage can be included in the op-amp model, as shown in Figure 115.7. When this model is used in Figure 115.2, for example, the output voltage depends on the input voltage and the offset voltage:

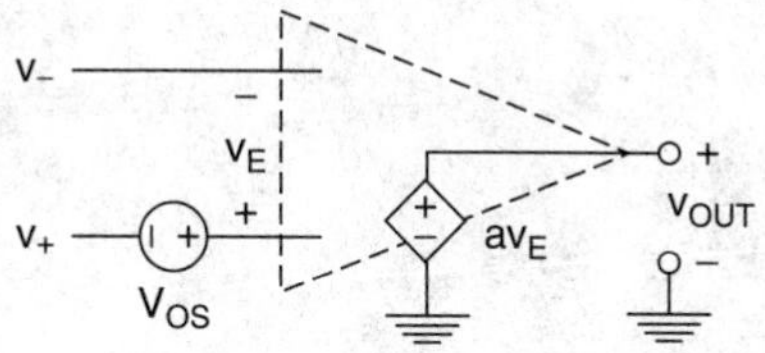

FIGURE 115.7 The op-amp model of Figure 115.1(c) with the addition of the input offset voltage V_{OS}.

$$v_{OUT} = -\frac{R_2}{R_1} v_{IN} + \left(1 + \frac{R_2}{R_1}\right) V_{OS} \tag{115.15}$$

For high gain ($R_2 \gg R_2$), the amplification of V_{OS} can cause problems, in which case a capacitor can be added in series with R_1. This eliminates the amplification of V_{OS}, but it also causes the DC gain from v_{IN} to v_{OUT} to be zero.

These parameters and many more are specified on the data sheet and are incorporated into an op-amp macro model, which is an interconnection of circuit elements that model the linear and nonlinear behavior of the op amp. Most companies provide macro models for their op amps that can be used in a circuit simulation program such as SPICE.

Defining Terms

Ideal op-amp model — Has $Z_i = \infty$, $Z_o = 0$, and $a = \infty$, which leads to $v_E = v_+ - v_- = 0$ (the virtual-short-circuit condition).

Input offset voltage — The op-amp DC input voltage, $v_+ - v_-$, that causes v_{OUT} to equal 0 V DC.

Phase margin — 180° – [the phase of RR($s = j\omega_U$)], where ω_U is the frequency where the magnitude of RR is unity (i.e., $|$ RR($s = j\omega_U$)$| = 1$). Phase margin is measured in degrees.

Gain margin — Is $-20 \log_{10} |$ RR($s = j\omega_{180}$)$|$, where ω_{180} is the frequency where the phase of RR is −180° (i.e., phase of RR($s = j\omega_{180}$) = −180°). Gain margin is measured in decibels. Since RR is some factor times $a(s)$ [e.g., see Equation (115.11)], plots of the magnitude and phase of RR versus frequency can be generated from plots of $a(s = j\omega)$. The gain and phase margins can be found from magnitude and phase plots of RR($s = j\omega$).

SPICE — A computer program that can simulate circuits with linear and nonlinear elements. Many versions of SPICE are available.

References

Analog Devices Inc., 1992. *Amplifier Reference Manual.* ADI, Norwood, MA.
Frederiksen, T. M. 1984. *Intuitive IC Op Amps.* C.M.C., Milpitas, CA.
Jung, W. G. 1986. *IC Op-Amp Cookbook.* Howard W. Sams, Indianapolis, IN.
National Semiconductor Corp. 1993. *Operational Amplifiers Databook.* Santa Clara, CA.
Rosenstark, S. 1986. *Feedback Amplifier Principles.* Macmillan, New York.

Further Information

Two books that include many useful op-amp feedback circuits and valuable practical information are the *IC Op-Amp Cookbook* by Walter G. Jung and *Intuitive IC Op Amps* by Thomas M. Frederiksen. Other good sources of practical information are the op-amp data and application books that are available from op-amp manufacturers. Stability and return ratio are covered in *Feedback Amplifier Principles* by Sol Rosenstark and in *Theory of Linear Active Networks* by E. S. Kuh and R. A. Rohrer.

116

Active RC Filters

Michael A. Soderstrand
University of California, Davis, and Oklahoma State University

Active RC filters arose out of the need for filters that are compatible with modern integrated circuit (IC) technology. Attempts to implement passive RLC filters in IC technology have largely failed due to difficulties in implementing inductors using IC technology. Circuit theorists, however, have shown that adding an active device (e.g., an operational amplifier) to resistors and capacitors makes it possible to implement any filter function that can be implemented using passive components (i.e., passive R, L, C filters). Since operational amplifiers (op amps), resistors, and capacitors are all compatible with modern IC technology, active RC filters are very attractive.

While the primary motivation for active RC filters comes from the desire to implement filter circuits in IC technology, there are additional advantages over passive RLC filters:

- Op amps, resistors, and capacitors perform much more closely to their ideal characteristics than inductors, which typically exhibit large parasitic resistances and significant nonlinearities.
- Passive RLC filters are not able to take advantage of the component tracking and very accurate matching of component ratios available in IC technologies. (Note: Filter characteristics depend on ratios of element values rather than the absolute value of the components.)
- Passive RLC filters cannot realize power gain, only attenuation.

In their infancy, active filters often suffered from a tendency to oscillate. Early work in active filters eliminated these problems but led to filters with high sensitivity to component values. Finally, in the 1970s, active filters emerged that clearly outperformed passive filters in cost, power consumption, ease of tuning, sensitivity, and flexibility. While IC technology allows for active filters that take full advantage of the new active RC filter design techniques, even active RC filters designed with discrete components will perform as well as, if not better than, their passive counterparts.

0-8493-1586-7/05/$0.00+$1.50

116.1 History of Active Filters

The first practical circuits for active RC filters emerged during World War II and were documented much later in a classic paper by Sallen and Key [1955]. With the advent of IC technology in the early 1960s, active filter research flourished with the emphasis on obtaining stable filters [Mitra, 1971]. During the early 1970s, research focused on reducing the sensitivity of active filters to their component values [Schaumann et al., 1976]. By the end of the 1970s, active filters had been developed that were better than passive filters in virtually every aspect [Bowron and Stephenson, 1979; Ghausi and Laker, 1981; Schaumann et al., 1990; Tsividis and Voorman, 1993]. Unfortunately, most active filter handbooks and books that concentrate on the practical design of active filters have used the old Sallen and Key approach to active filter design. In this chapter we briefly discuss op-amp-based active RC filter design techniques. (Note: In VLSI applications transconductance-based active filters offer an excellent alternative, which we not discuss [Tsividis and Voorman, 1993].) Then we introduce a table-based design procedure for one of the modern op-amp techniques that yields active RC filters as good as or better than passive RLC filters.

116.2 Active Filter Design Techniques

In this section we define three categories of active RC filter design using very different approaches. Within each category, we briefly describe several of the specific methods available.

Cascaded Biquads

The simplest and most popular active RC filter design techniques are based on a cascade of second-order (biquadratic or biquad) sections. The numerator and denominator of the filter function $H(s)$ (expressed in terms of the complex frequency variable s) are factored into second-order factors (plus one first-order, in the case of an odd-order filter). Second-order transfer functions are then formed by combining appropriate numerator terms with each second-order denominator term such that the product of the second-order transfer functions is equal to the original $H(s)$. (Note: In the case of odd-order filters, there will also be one first-order section.) Active RC biquad circuits are then used to implement each second-order section. In the case of odd-order filters, a separate first-order RC section may be added or the real pole can be combined with one of the second-order sections to form a third-order section. The different cascade methods are defined based on which biquad circuits are used and whether the ordering of the biquads is considered as part of the design procedure.

The advantages of the cascade design lie in the ease of design and the fact that the biquads are isolated and thus can be separately tuned during the manufacturing process. The disadvantage of the cascade design also derives from the isolated sections, as it allows only for feedback around individual second-order stages without overall feedback around the entire filter structure. Thus the cascade of second-order sections is essentially operated open loop.

Cascade of Sallen and Key Filters

This approach is the basis for most active RC filter handbooks and many practical texts on active RC filter design. The resulting filters have the advantage of being canonical (i.e., having the fewest possible components — two resistors, two capacitors, and one op amp per second-order stage). The performance of these filters is generally acceptable but is not as good as passive RLC filters. The primary disadvantage of this approach lies in the fact that it is not possible to use matched op amps to compensate for the nonideal properties of the operational amplifier, since the sections only use one op amp. A complete description of the design of a cascade of Sallen and Key filters can be found in Williams [1975]. Williams also describes a slightly improved design using state-variable-based second-order sections. However, one should use optimum biquads rather than state-variable biquads since the optimum biquads have the same number or fewer components and outperform the state-variable filters.

Cascade of Optimum Biquads

Optimum biquads require at least two active elements in order to make use of matched op amps to reduce the effects of the nonideal properties of the active elements. Optimum two-op-amp biquad circuits are based on the gyrator circuits introduced by Antoniou [1967] and modified by Hamilton and Sedra [1971]. The design of these circuits is covered later in this chapter.

Ordered Cascade of Optimum Biquads

In order to obtain the best performance possible using the cascade technique, it is not sufficient to use optimum biquads. The second-order sections must also be ordered in an optimum fashion. In this chapter we show how to design using the optimum biquads, but we do not go into the details of ordering the biquads. For those who need the ultimate in performance, an excellent discussion of ordering is provided in chapter 10 of Sedra and Brackett [1978].

Simulation of LC Ladders

This approach was originally designed simply to take advantage of the vast reservoir of knowledge on passive LC ladder filter circuit design in the design of active filters. The simplest approach uses a gyrator [Antoniou, 1967] and a capacitor to simulate an inductor and replaces the inductors in the passive RLC ladder with this simulation (chapter 11 of Sedra and Brackett [1978]). More sophisticated techniques simulate the voltage and current equations of the passive RLC ladder (chapter 12 of Sedra and Brackett [1978]) or the scattering parameters [Haritantis et al., 1976]. The latter techniques have resulted in the best filter circuits ever constructed.

At first it may seem that simulating passive RLC filters would at best yield filters equal to, but not better than, passive filters. However, the poor quality of passive inductors coupled with the ability of active RC filters to be constructed in IC technology, where very accurate ratios of components can be obtained, gives a huge advantage to active RC filters compared with passive RLC filters. Even for filters realized with discrete components (i.e., not using IC technology), the advantage of not using an inductor and the use of dual matched op amps yields circuits superior to those available with passive RLC filters.

Component Simulation of LC Ladders

The primary attraction of this technique is the ease of design. You simply replace the inductors in a passive RLC filter obtained from any passive RLC design handbook [Hansell, 1969; Zverev, 1967] with active RC simulated inductors. Unfortunately, the floating inductors (inductors which do not have one terminal grounded) required for most passive RLC filters are one of the poorer-performing active RC components. However, modifications of this approach using *frequency-dependent negative resistors* (FDNRs) proposed by Bruton et al. [1972] have largely eliminated these problems. For details on the component simulation of LC ladders, see chapter 11 of Sedra and Brackett [1978].

Operational Simulation of LC Ladders

In this approach active RC components are used as an *analog computer* to simulate the differential equations of Kirchhoff's current and voltage laws at each node in a passive RLC circuit. This approach has resulted in some of the best filter circuits ever designed. For details, see chapter 12 of Sedra and Brackett [1978] (see also [Ghausi and Laker, 1981; Schaumann et al., 1990]).

Wave Analog Filters

This approach is similar in concept to the operational simulation of RLC ladders except that scattering matrices of the passive RLC filter are simulated rather than Kirchhoff's current and voltage laws. This approach has become very popular in Europe, yielding filters every bit as good as those designed using the operational simulation of RLC ladders. For details see either Schaumann et al. [1990] or Haritantis et al. [1976].

Coupled Biquads

The *coupled biquad* approach uses the same optimum biquads that we use in the cascade approach later in this chapter, but provides feedback around the biquad sections. The advantage of this approach is that this additional feedback makes it possible to obtain filter performance equivalent to the filters realized by simulating RLC ladders, but using the modular, easily tuned biquad sections. However, the coupling of the biquads makes it much more difficult to tune the structure than in the case of cascaded biquads. For an excellent discussion of the coupled biquad design approach, see chapter 5 of Ghausi and Laker [1981] (see also chapter 10 of Sedra and Brackett [1978] and chapter 5 of Schaumann et al. [1990]).

Current Feedback Active Filters

All of the filters we have considered to this point were based upon the use of voltage feedback operational amplifiers. Recently a number of authors have praised the virtues of current-feedback operational amplifiers in active filter circuits [Becvar et al., 2000; Chiu et al., 1996; Chong and Smith, 1986; Roberts and Sedra, 1989; Toumazou and Lidgey, 1990]. As with most technologies, current-feedback active filters provide some improvements over voltage feedback active filters, but also show some disadvantages. National Semiconductor provides a series of application notes that pretty much set the record straight [Buck, 1992; Brandenberg, 1998; National Semiconductor, 1993; Potson, 1988]. In particular the article by Brandenberg [1998] provides a very good comparison of the two technologies. She concludes that for most applications the two technologies are equivalent. However, current-feedback does offer independence between the gain and bandwidth that results in faster slew rates and lower distortion at the expense of restrictions on the value of the feedback resistors when compared to voltage-feedback circuits. On the other hand, voltage feedback offers lower noise, better DC performance, and freedom of selection of feedback resistance that is not available in current-feedback circuits. In this chapter we shall only discuss voltage-feedback circuits, but it should be noted that most of the voltage-feedback circuits have analogies in current-feedback and therefore it is easy to substitute current-feedback into the circuits discussed in this chapter [Brandenberg, 1998].

116.3 Filter Specifications and Approximations

Low-pass filters are specified by four parameters: (1) α_p, the passband ripple; (2) ω_p, the passband frequency; (3) α_s, the stopband attenuation; and (4) ω_s, the stopband frequency. Figure 116.1 illustrates these parameters. A filter meeting these specifications must have $|H(s)|$ lie between 0 dB and α_p from DC to frequency ω_p, and below α_s for frequencies above ω_s. For passive filters and odd-order active filters, α_p is always negative. For active even-order filters, α_p is positive and these filters exhibit power gain. Note that the gain for all active filter designs is normalized to 0 dB at DC (normalization is different for even-order passive filters). In practice, active RC filters are capable of providing an overall gain K, which multiplies the system function $H(s)$ with the effect of scaling the vertical axis in Figure 116.1.

The specifications of Figure 116.1 are only half the story, however. In addition to the magnitude characteristics, filters have group delay characteristics related to the phase response of the filter. Figure 116.2 shows a typical group delay specification where (1) τ_0 is the nominal group delay, (2) $\Delta\tau$ is the passband group delay tolerance, and (3) ω_g is the frequency at which the group delay has decayed to 50% of τ_0. Unfortunately, magnitude characteristics cannot be specified independent of group delay characteristics. Thus designers must make a choice between which set of specifications they wish to implement and must settle for whatever they get with the other specifications. Filter approximations are specific techniques for achieving one or the other set of specifications. Each approximation offers a different compromise between meeting the magnitude specification and meeting the group delay specification.

Filters can be classified into two broad classes: (1) all-pole filters and (2) filters with finite zeros. Figure 116.3 shows the basic asymptotic response for all-pole filters in the left column and the basic asymptotic response for filters with finite zeros in the right column. In each column the response of the three basic

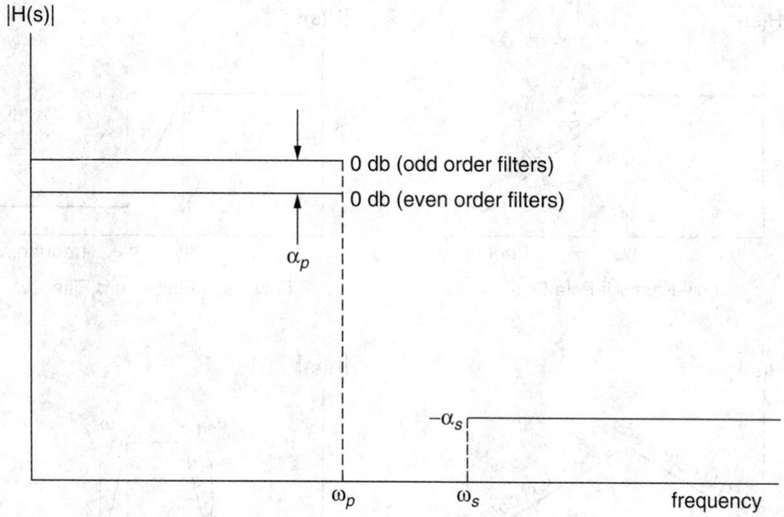

FIGURE 116.1 Magnitude specifications for a low-pass filter.

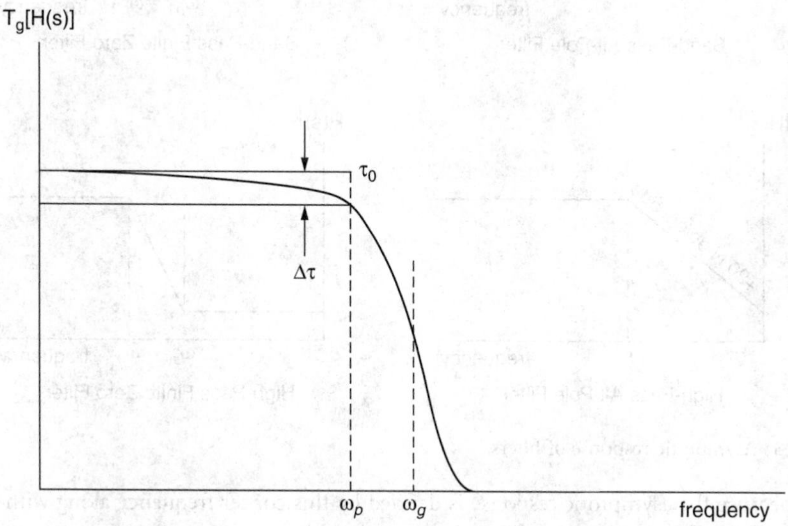

FIGURE 116.2 Delay specifications for a low-pass filter.

filters, low-pass, high-pass, and band-pass, are shown. A fourth filter type, called a notch filter or band-elimination filter, is not shown because this type of filter exhibits an asymptotic response that is 0 dB at all frequencies, with a notch (area of high attenuation) around the corner frequency ωc. Most filter design procedures concentrate on designing low-pass filters and use standard transformations to convert these low-pass filters to high-pass, band-pass, or notch filters. (A good explanation of how to do this can be found in chapter 6 of Sedra and Brackett [1978].)

All passive RLC and active RC filters, regardless of the filter approximation or the method of implementation, have the same asymptotes for the some-order filter (*order* refers to the number of effective reactive elements — inductors and capacitors in a passive RLC filter, and capacitors in an active RC filter). The various filter approximations that we are about to introduce primarily affect how the filter performs in the transition band between frequencies ω_p and ω_s. In general, the wider the transition band, the better the group delay performance, but at the expense of significant deviation from the asymptotic response in the transition band. Note that the corner frequency ωc is an important parameter for all-

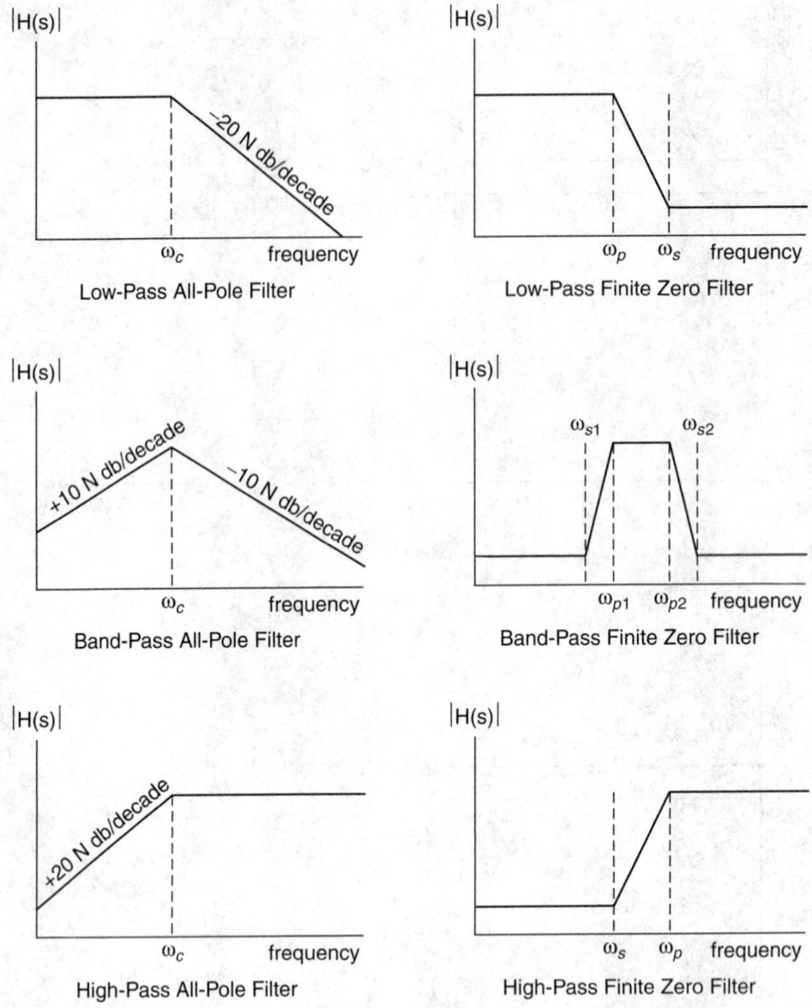

FIGURE 116.3 Asymptotic response of filters.

pole filters because the asymptotic response is defined by this corner frequency along with the order of the filter N and the filter gain K.

Bessel–Thompson Filters

The focus of the Bessel–Thompson filters is on meeting the group delay specification. The group delay is fixed at τ_0 for DC, and as many derivatives of the group delay function as possible are set to zero at DC. This yields what is called a *maximally flat* delay response. Mathematically it can be shown that this will obtain the closest possible match to the ideal "brick wall" delay response without any delay in excess of the prescribed group delay τ_0. These filters were originally developed by Thompson and make use of Bessel functions — hence the name Bessel–Thompson filters. Bessel–Thompson filters exhibit the low-pass all-pole asymptotic response of Figure 116.3.

Although Bessel–Thompson filters offer the best possible group delay response, they do this at the expense of very poor magnitude response in the transition band. Furthermore, for a given order of filter, the transition band is much larger than for the other filter approximations. Other group delay approximations such as the equal-ripple group delay approximation provide only slightly better magnitude performance.

It is often very useful to calculate the required order of a Bessel–Thompson filter from the specifications for that filter. Since Bessel–Thompson filters are primarily dependent on the group delay specifications, the exact formula does not involve the standard magnitude specifications. However, the following approximate formula estimates the order of a Bessel–Thompson filter:

$$N \approx \frac{\alpha_s}{20\log_{10}(\omega_s/\omega_c)} \tag{116.1}$$

with αs in dB and ωc and ωs in radians (or hertz).

Butterworth Filters

Butterworth filters provide a maximally flat passband response for the magnitude of the transfer function at the expense of some peaking in the group delay response around the passband frequency ω_p. In most cases the large improvement in magnitude response compared with Bessel–Thompson filters more than compensates for the peaking of the group delay response. Butterworth filters exhibit the low-pass all-pole asymptotic response of Figure 116.3 with exactly 3 dB of attenuation at the corner frequency ω_c. This is less droop (i.e., attenuation at the corner frequency) than Bessel–Thompson filters, but more than the other filter approximations we discuss.

It is often very useful to calculate the required order of a Butterworth filter from the specifications. The following formula gives this relationship:

$$N \geq \frac{\log_{10}(K_\alpha)}{\log_{10}(K_\omega)} \tag{116.2}$$

where

$$K_\alpha = \sqrt{\frac{10^{\alpha_s/10}-1}{10^{\alpha_p/10}-1}} \tag{116.3}$$

$$K_\omega = \omega_s/\omega_p \tag{116.4}$$

with α_p and α_s in dB and ω_p and ω_s in radians (or hertz).

Chebyshev Filters

Chebyshev filters provide an equal-ripple response in the passband (i.e., the magnitude of $H(s)$ oscillates or ripples between zero and α_p in the passband). Chebyshev filters are all-pole filters that can be shown to provide the narrowest transition band for a given filter order N of any of the all-pole filters. This narrow transition band, however, comes at the expense of significant and often unacceptable peaking and oscillation in the group delay response.

It is often very useful to calculate the required order of a Chebyshev filter from the specifications for that filter. The following formula gives this relationship:

$$N \geq \frac{\cosh^{-1}(K_\alpha)}{\cosh^{-1}(K_\omega)} \tag{116.5}$$

or

$$N \geq \frac{\log_{10}\left(K_\alpha + \sqrt{K_\alpha^2 - 1}\right)}{\log_{10}\left(K_\omega + \sqrt{K_\omega^2 - 1}\right)} \qquad (116.6)$$

where K_α and K_ω are as defined in Equations (116.3) and (116.4), respectively.

Inverse Chebyshev Filters

Inverse Chebyshev filters provide the same maximally flat magnitude response of Butterworth filters, including the slight, but usually acceptable, peaking in the group delay response, with a transition band exactly the same as for the Chebyshev filters. However, the inverse Chebyshev filters exhibit the low-pass finite-zero asymptotic response of Figure 116.3 with equal ripple in the stopband.

It is often very useful to calculate the required order of an inverse Chebyshev filter from the specifications. Since the inverse Chebyshev filter has the same order as the Chebyshev filter, Equations (116.5) or (116.6) can also be used to calculate the order of inverse Chebyshev filters.

Elliptical–Cauer Filters

The elliptical–Cauer filters exhibit the low-pass finite-zero asymptotic response of Figure 116.3 with equal ripple in both the passband (like Chebyshev filters) and the stopband (like inverse Chebyshev filters). While the group delay of these filters is poor, like that of the Chebyshev filters, elliptical–Cauer filters can be shown to have the narrowest transition band of any of the filter approximations.

It is often very useful to calculate the required order of an elliptical–Cauer filter. The following formula gives this relationship:

$$N \geq \frac{CEI(1/K_\omega)CEI\left(\sqrt{1 - 1/K_\alpha^2}\right)}{CEI(1/K_\alpha)CEI\left(\sqrt{1 - 1/K_\omega^2}\right)} \qquad (116.7)$$

where *CEI* is the *complete elliptic integral* function. (The *CEI* function is denoted by K or q in most textbooks, but we have chosen *CEI* so as not to confuse it with other filter parameters.) Although the *CEI* function is not typically provided on calculators or in computer mathematics libraries, computer programs are readily available for calculating the *CEI* function. (For example, MATLAB has the algorithm *ellipord* for calculating the order of both digital and analog elliptic filters. Also, Daniels [1974] contains a computer algorithm for calculating *CEI*, and Lindquist [1977] contains an approximate formula for N. K_α and K_ω are as defined in Equations (116.3) and (116.4), respectively.

116.4 Filter Design

Figure 116.4 represents a biquad circuit that can be used for low-pass all-pole filters, and Figure 116.5 shows a biquad that can be used for low-pass finite-zero filters. Both biquads exhibit optimum performance with regard to both passive and active sensitivities [Sedra and Brackett, 1978]. The element values for the circuit of Figure 116.4 are given by

$$R = Q \quad C = 1/\omega_0 \qquad (116.8)$$

where ω_0 and Q are the normalized pole frequency and Q of the second-order filter section. Practical element values are obtained by impedance scaling to practical impedance levels (i.e., multiply each resistor by the desired impedance level and divide each capacitor by the desired impedance level) and frequency scaling to the desired ω_p (i.e., divide each capacitor by the desired frequency in radians per second or by

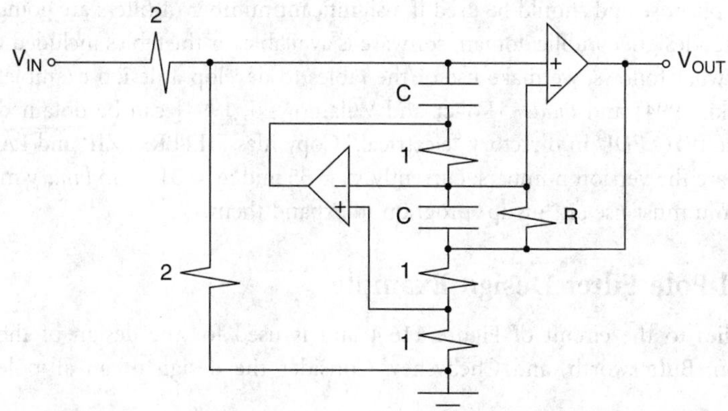

FIGURE 116.4 Biquad circuit for all-pole low-pass filters.

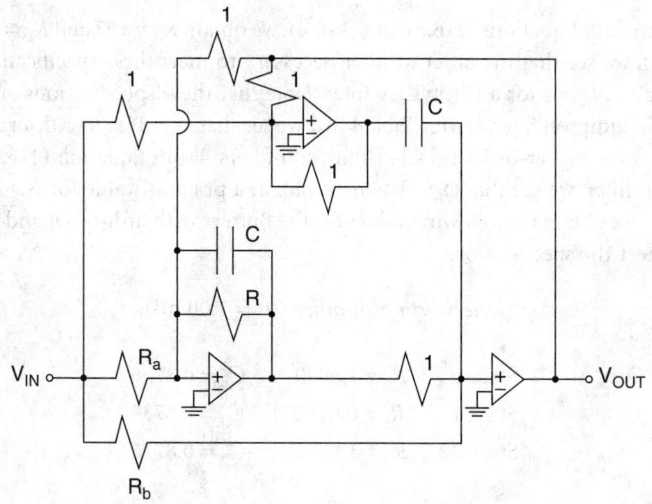

FIGURE 116.5 Biquad circuit for finite-zero low-pass filters.

2π times the desired frequency in hertz). Unlike most filter design tables, the tables with this chapter of the handbook are normalized to $\omega_p = 1$ rather than $\omega_c = 1$. This makes the tables somewhat longer but simplifies the use of the tables.

Figure 116.5 is used for the filters with finite zeros. The element values for the circuit of Figure 116.5 are given by

$$R_a = \frac{Q^2\omega_n^2}{Q^2\omega_n^2 + \omega_0^2} \qquad R_b = \frac{Q\omega_n^2}{\omega_0^2} \qquad R_c = \frac{\omega_n^2}{\omega_0^2} \tag{116.9}$$

$$R = Q \qquad C = 1/\omega_0$$

where ω_0 and Q are the normalized resonant frequency and Q for the second-order filter section and ω_n is the normalized frequency of the finite zero. As for the all-pole filter section, our tables have been normalized to $\omega_p = 1$. Practical resistor and capacitor values are obtained by impedance scaling to the desired impedance level and frequency scaling to the desired ω_p.

In order to determine what the resonant frequencies, Q's, and zero frequencies are for various types of filters, one may either use a computer program or a table. Many commercial computer programs are available for this purpose and should be used if a significant number of filters are going to be designed. For the casual filter designer, public domain software is available, or the tables included with this chapter may be used. In what follows, we make use of the tables to develop a design example. (The programs *filter* [Wilamowski, 1994] and *ladder* [Koller and Wilamowski, 1994] can be obtained by anonymous FTP to *PLAINS.UWYO.EDU* in directory "electrical." Copy files FILTERxx.ZIP and LADDERyy.ZIP — where xx and yy are the version numbers; currently xx = 35 and yy = 31 — in *binary* mode, as these are in "zip" format. You must use an "unzip" program to expand them.)

Low-Pass All-Pole Filter Design Example

Table 116.1 applies to the circuit of Figure 116.4 and is used for the design of the all-pole filters: Bessel–Thompson, Butterworth, and Chebyshev. Consider the design of an all-pole filter with the specifications

$$\alpha_p = 1\text{ dB}, \quad \omega_p = 1\text{ kHz}, \quad \alpha_s = 40\text{ dB}, \quad \text{and} \quad \omega_s = 3\text{ kHz}$$

Plugging these values into Equations (116.3) and (116.4), we obtain $K_\alpha = 197$ and $K_\omega = 3$. Using Equations (116.2) and (116.5), we see that the order of filter necessary to meet these specifications is $N_B = 5$ for a Butterworth filter and $N_C = 4$ for a Chebyshev filter. Note that these specifications are too restrictive to be met by a Bessel–Thompson filter. (From Table 116.1 we see that $\omega_c \approx 3$ for a sixth-order Bessel–Thompson filter and $\omega_c > 3$ for higher-order Bessel–Thompson filters. From Equation (116.1) for the order of a Bessel–Thompson filter, we see that $\omega_s = 3 > \omega_c$ to obtain a practical value for N.)

From Table 116.1 we obtain the following values for the Butterworth fifth-order and Chebyshev fourth-order filters that meet the specifications

Butterworth fifth-order (α_p = 1.00 dB):

Stage 0	$R_0 = 1.0000$	$C_0 = 0.8736$
Stage 1	$R_1 = 0.6180$	$C_1 = 0.8736$
Stage 2	$R_2 = 1.6180$	$C_2 = 0.8736$

Chebyshev fourth-order (α_p = 1.00 dB):

Stage 1	$R_1 = 0.7845$	$C_1 = 1.8919$
Stage 2	$R_2 = 3.5590$	$C_2 = 1.0068$

To obtain practical values, we must denormalize by impedance scaling to a practical impedance level (we have chosen 10 kΩ) and frequency scaling to ω_p = 1 kHz. Thus we multiply all resistors by 10^4 and divide all capacitors by 10^4 to impedance-scale, and we then divide all capacitors by $2\pi \times 10^3$ to frequency-scale. For active filter implementation, we recommend *metal film* resistors and *polystyrene* capacitors. *Mica* capacitors are slightly superior to polystyrene capacitors but have a more restricted range of values. *Polycarbonate* capacitors have a larger range of values but a poorer "retrace" property. *Mylar* capacitors provide large capacitance values in a small package, but at the expense of a much poorer temperature coefficient. The practical values for the resistors and the capacitors of our example after scaling are as follows:

Butterworth fifth-order (α_p = 1.00 dB):

Stage 0	$R_0 = 10.0$ kΩ	$C_0 = 13.9$ nF
Stage 1	$R_1 = 6.18$ kΩ	$C_1 = 13.9$ nF
Stage 2	$R_2 = 16.2$ kΩ	$C_2 = 13.9$ nF

TABLE 116.1 Design Table for Low-Pass All-Pole Filters

			Bessel–Thompson				Butterworth				Chebyshev			
N	α_p	S	R	C	α_c	ω_c	R	C	α_c	ω_c	R	C	α_c	ω_c
2	0.10	1	0.5774	0.1511	−4.77	6.620	0.7071	0.3914	−3.00	2.555	0.7674	0.5505	−2.30	1.817
	0.25	1		0.2367		4.225		0.4948		2.021	0.8093	0.6878	−1.84	1.454
	0.50	1		0.3323		3.009		0.5910		1.692	0.8637	0.8121	−1.27	1.231
	1.00	1		0.4619		2.165		0.7133		1.402	0.9565	0.9524	−0.39	1.050
	2.00	1		0.6432		1.555		0.8745		1.143	1.1286	1.1023	+1.05	0.907
	3.00	1		0.7842		1.275		1.0000		1.000	1.3047	1.1885	+2.31	0.841
3	0.10	0	1.0000	0.1459	−6.25	7.295	1.0000	0.5344	−3.00	1.871	1.0000	1.0316	−0.83	1.178
		1	0.6910	0.1333			1.0000	0.5344			1.3409	0.7693		
	0.25	0	1.0000	0.2299		4.634	1.0000	0.6243		1.602	1.0000	1.3034	−0.32	1.016
		1	0.6910	0.2098			1.0000	0.6243			1.5080	0.8643		
	0.50	0	1.0000	0.3229		3.293	1.0000	0.7045		1.419	1.0000	1.5963	−0.03	0.902
		1	0.6910	0.2950			1.0000	0.7045			1.7062	0.9356		
	1.00	0	1.0000	0.4545		2.352	1.0000	0.7986		1.252	1.0000	2.0236	−0.18	0.788
		1	0.6910	0.4125			1.0000	0.7986			2.0177	1.0029		
	2.00	0	1.0000	0.6264		1.696	1.0000	0.9148		1.093	1.0000	2.7107	−1.25	0.690
		1	0.6910	0.5723			1.0000	0.9148			2.5516	1.0623		
	3.00	0	1.0000	0.7578		1.409	1.0000	1.0000		1.000	1.0000	3.3487	−2.51	0.632
		1	0.6910	0.6906			1.0000	1.0000			3.0677	1.0916		
4	0.10	1	0.5219	0.1740	−8.13	4.666	0.5412	0.6250	−3.00	1.600	0.6188	1.2670	+0.09	0.956
		2	0.8055	0.2625			1.3066	0.6250			2.1829	0.8671		
	0.25	1	0.5219	0.2512		3.232	0.5412	0.7024		1.424	0.6572	1.4828	+0.16	0.858
		2	0.8055	0.3790			1.3066	0.7024			2.5361	0.9277		
	2.00	1	0.5219	0.5025		1.616	0.5412	0.9352		1.069	0.9294	2.1244	+0.04	0.675
		2	0.8055	0.7581			1.3066	0.9352			4.5939	1.0377		
	3.00	1	0.5219	0.5706		1.513	0.5412	1.0000		1.000	1.0765	2.2589	+0.19	0.649
		2	0.8055	0.8608			1.3066	1.0000			5.5789	1.0523		
5	0.10	0	1.0000	0.1248	−8.80	8.682	1.0000	0.6866	−3.00	1.457	1.0000	1.8556	−0.09	0.841
		1	0.5635	0.1205			0.6180	0.6866			0.9145	1.2540		
		2	0.9165	0.1068			1.6180	0.6866			3.2820	0.9148		
	0.25	0	1.0000	0.1970		5.495	1.0000	0.7538		1.327	1.0000	2.2886	−0.22	0.765
		1	0.5635	0.1902			0.6180	0.7538			1.0359	1.3654		
		2	0.9165	0.1686			1.6180	0.7538			3.8757	0.9554		
	0.50	0	1.0000	0.2775		3.887	1.0000	0.8103		1.234	1.0000	2.7600	−0.24	0.716
		1	0.5635	0.2678			0.6180	0.8103			1.1778	1.4483		
		2	0.9165	0.2375			1.6180	0.8103			4.5450	0.9826		
	1.00	0	1.0000	0.3914		2.755	1.0000	0.8736		1.145	1.0000	3.4543	−0.25	0.662
		1	0.5635	0.3778			0.6180	0.8736			1.3988	1.5262		
		2	0.9165	0.3349			1.6180	0.8736			5.5564	1.0059		
	2.00	0	1.0000	0.5482		1.945	1.0000	0.9478		1.055	1.0000	4.5807	−0.07	0.611
		1	0.5635	0.5292			0.6180	0.9478			1.7751	1.5949		
		2	0.9165	0.4692			1.6180	0.9478			7.2323	1.0248		
	3.00	0	1.0000	0.6645		1.614	1.0000	1.0000		1.000	1.0000	5.6329	−0.02	0.578
		1	0.5635	0.6415			0.6180	1.0000			2.1375	1.6286		
		2	0.9165	0.5687			1.6180	1.0000			8.8178	1.0336		
6	−0.10	1	0.5103	0.1159	−10.1	9.289	0.5176	0.7310	−3.00	1.368	0.5995	1.9486	+0.06	0.775
		2	0.6112	0.1101			0.7071	0.7310			1.3316	1.1983		
		3	1.0233	0.0976			1.9319	0.7310			4.6329	0.9410		
	−0.25	1	0.5103	0.1831		5.875	0.5176	0.7902		1.266	0.6370	2.2519	+0.25	0.713
		2	0.6112	0.1738			0.7071	0.7902			1.5557	1.2597		
		3	1.0233	0.1542			1.9319	0.7902			5.5204	0.9698		
	−0.50	1	0.5103	0.2583		4.150	0.5176	0.8392		1.192	0.6836	2.5238	+0.47	0.679
		2	0.6112	0.2453			0.7071	0.8392			1.8104	1.3019		
		3	1.0233	0.2175			1.9319	0.8392			6.5128	0.9887		

TABLE 116.1 Design Table for Low-Pass All-Pole Filters (*Continued*)

			Bessel–Thompson				Butterworth				Chebyshev			
N	α_p	S	R	C	α_c	ω_c	R	C	α_c	ω_c	R	C	α_c	ω_c
	−1.00	1	0.5103	0.3647		2.938	0.5176	0.8935		1.119	0.7609	2.8317	+0.75	0.645
		2	0.6112	0.3463			0.7071	0.8935			2.1980	1.3390		
		3	1.0233	0.3071			1.9319	0.8935			8.0037	1.0047		
	−2.00	1	0.5103	0.5110		2.104	0.5176	0.9563		1.046	0.9016	3.1634	+0.99	0.614
		2	0.6112	0.4852			0.7071	0.9563			2.8443	1.3698		
		3	1.0233	0.4303			1.9319	0.9563			10.4616	1.0175		
	−3.00	1	0.5103	0.6232		1.725	0.5176	1.0000		1.000	1.0443	3.3557	+1.05	0.600
		2	0.6112	0.5917			0.7071	1.0000			3.4581	1.3843		
		3	1.0233	0.5247			1.9319	1.0000			12.7801	1.0234		
7	−0.10	0	1.0000	0.1100	−11.3	9.875	1.0000	0.7645	−3.009	1.308	1.0000	2.6541	−0.27	0.726
		1	0.5324	0.1079			0.5550	0.7645			0.8464	1.7402		
		2	0.6608	0.1017			0.8019	0.7645			1.8472	1.1522		
		3	1.1263	0.0904			2.2470	0.7645			6.2332	0.9568		
	−0.25	0	1.0000	0.1738		6.254	1.0000	0.8172		1.224	1.0000	3.2510	−0.19	0.679
		1	0.5324	0.1705			0.5550	0.8172			0.9596	1.8802		
		2	0.6608	0.1606			0.8019	0.8172			2.1904	1.1902		
		3	1.1263	0.1428			2.2470	0.8172			7.4678	0.9782		
	−0.50	0	1.0000	0.2454		4.428	1.0000	0.8605		1.162	1.0000	3.9037	−0.48	0.646
		1	0.5324	0.2409			0.5550	0.8605			1.0916	1.9847		
		2	0.6608	0.2268			0.8019	0.8605			2.5755	1.2155		
		3	1.1263	0.2017			2.2470	0.8605			8.8418	0.9920		
	−1.00	0	1.0000	0.3461		3.140	1.0000	0.9080		1.101	1.0000	4.8682	−0.49	0.612
		1	0.5324	0.3397			0.5550	0.9080			1.2969	2.0831		
		2	0.6608	0.3199			0.8019	0.9080			3.1559	1.2371		
		3	1.1263	0.2845			2.2470	0.9080			10.8967	1.0037		
	−2.00	0	1.0000	0.4868		2.232	1.0000	0.9624		1.039	1.0000	6.4375	−1.76	0.578
		1	0.5324	0.4778			0.5550	0.9624			1.6464	2.1699		
		2	0.6608	0.4500			0.8019	0.9624			4.1151	1.2545		
		3	1.1263	0.4001			2.2470	0.9624			14.2801	1.0129		
	−3.00	0	1.0000	0.5928		1.834	1.0000	1.0000		1.000	1.0000	7.9061	−2.34	0.557
		1	0.5324	0.5817			0.5550	1.0000			1.9829	2.2127		
		2	0.6608	0.5479			0.8019	1.0000			5.0214	1.2626		
		3	1.1263	0.4872			2.2470	1.0000			17.4645	1.0172		

Chebyshev fourth-order (α_p = 1.00 dB):

Stage 1	$R_1 = 7.85\ \text{k}\Omega$	$C_1 = 30.1\ \text{nF}$
Stage 2	$R_2 = 35.6\ \text{k}\Omega$	$C_2 = 16.0\ \text{nF}$

Note: All resistors labeled 1 Ω in Figure 116.4 become 10 kΩ, and the two resistors labeled 2 Ω become 20 kΩ. These values are well suited for metal film resistors and polystyrene capacitors.

Low-Pass Finite-Zero Filter Design Example

Table 116.2 applies to the circuit of Figure 116.5 and is used for the design of the finite-zero filters: inverse Chebyshev and elliptical–Cauer. Consider the design of a finite-zero filter with the same specifications as in the first example:

$$\alpha_p = 1\text{dB}, \quad \omega_p = 1\ \text{kHz}, \quad \alpha_s = 40\ \text{dB}, \quad \text{and} \quad \omega_s = 3\ \text{kHz}$$

Plugging these values into Equations (116.3) and (116.4), we obtain $K_\alpha = 197$ and $K_\omega = 3$. Using Equations (116.6) and (116.7), respectively, we see that the order of filter necessary to meet these specifications is

$N_{IC} = 4$ for an inverse Chebyshev filter and $N_{EC} = 3$ for an elliptical–Cauer filter. From Table 116.2 we obtain the following values for the inverse Chebyshev fourth-order and elliptical–Cauer third-order filters that meet the specifications:

Inverse Chebyshev fourth-order ($\alpha_s = 40$ dB and $\alpha_p = 1.00$ dB):

Stage 1	$Q_1 = 0.5540$	$\omega_1 = 1.3074$	$\omega_{n1} = 2.5312$
Stage 2	$Q_2 = 1.4780$	$\omega_2 = 1.1832$	$\omega_{n2} = 6.1109$

Elliptical–Cauer third-order ($\alpha_s = 40$ dB and $\alpha_p = 1.00$ dB):

Stage 0		$\omega_0 = 0.5237$	
Stage 1	$Q_1 = 2.2060$	$\omega_1 = 1.0027$	$\omega_{n1} = 2.7584$

With Table 116.2, an extra step is required in order to find the normalized element values for the circuit of Figure 116.5. We must use Equation (116.9) to find the element values for each stage from the values found in the table for $Q(Q_1,Q_2,\ldots)$, $\omega_0(\omega_1, \omega_2,\ldots)$, and $\omega_n(\omega_{n1}, \omega_{n2},\ldots)$. For odd-order filters, we must also add a first-order RC low-pass filter to the cascade with normalized values $R = 1.0$ and $C = \omega_0$. To obtain practical values, we then denormalize by impedance scaling to a practical impedance level (we have chosen 10 kΩ) and frequency scaling to $\omega_p = 1$ kHz. Thus we multiply all resistors by 10^4 and divide all capacitors by 10^4 to impedance-scale, and we then divide all capacitors by $2\pi \times 10^3$ to frequency-scale. The practical values for the resistors and the capacitors become:

Inverse Chebyshev fourth-order ($\alpha_p = 1.00$ dB):

Stage 1	$R_a = 5.35$ kΩ	$R_b = 20.8$ kΩ	$R_c = 37.5$ kΩ
	$R = 5.54$ kΩ	$C = 12.2$ nF	
Stage 2	$R_a = 9.83$ kΩ	$R_b = 394$ kΩ	$R_c = 267$ kΩ
	$R = 14.8$ kΩ	$C = 13.5$ nF	

Elliptical–Cauer third-order ($\alpha_p = 1.00$ dB)

Stage 0	$R_0 = 10.0$ kΩ	$C_0 = 30.4$ nF	
Stage 1	$R_a = 9.74$ kΩ	$R_b = 167$ kΩ	$R_c = 75.7$ kΩ
	$R = 22.1$ kΩ	$C = 15.9$ nF	

Note: All resistors labeled 1 Ω in Figure 116.5 become 10 kΩ.

Defining Terms

Active RC filter — A filter circuit that uses an active device (FET, transistor, op amp, transconductance amp, etc.) with resistors and capacitors without the need for inductors. Active RC filters are particularly suited for integrated circuit applications.

Bessel–Thompson filter — A mathematical approximation to an ideal filter in which the group delay of the transfer function in the frequency domain is maximally flat. Bessel–Thompson filters are optimum in the sense that they offer the least droop without overshoot in the group delay.

Biquad — A filter circuit that implements a second-order filter function. High-order filters are often built from a cascade of second-order blocks. Each of these second-order blocks is referred to as a biquad because in the *s* domain these filters are characterized by a quadratic function of *s* in both the numerator and denominator.

TABLE 116.2 Design Table for Low-Pass Finite-Zero Filters

Specifications				Inverse Chebyshev				Cauer (Elliptical)			
α_s	α_p	N	S	Q	ω_0	ω_n	ω_s	Q	ω_0	ω_n	ω_s
40 dB	0.10 dB	2	1	0.7107	2.5616	25.6162	18.11336	0.7719	1.8218	18.1134	12.81790
		3	0		1.9437		5.51714		1.0100		3.51952
			1	1.0455	1.8591	6.3706		1.4269	1.2936	4.0429	
		4	1	0.5540	1.7282	3.3461	3.09139	0.6420	0.8643	2.0662	1.92976
			2	1.4780	1.5640	8.0782		2.6744	1.1362	4.7252	
		5	0		1.7469		2.22010		0.6706		1.41762
			1	0.6811	1.5870	2.3344		1.0852	0.8940	1.4691	
			2	2.0218	1.3997	3.7771		5.0104	1.0690	2.1727	
		6	1	0.5346	1.7437	1.8689	1.80515	0.6349	0.6933	1.2276	1.20454
			2	0.8653	1.4716	2.5529		1.9787	0.9366	1.5033	
			3	2.6828	1.2968	6.9768		9.3816	1.0360	3.6245	
	0.25 dB	2	1	0.7107	2.0293	20.2927	14.34906	0.8144	1.4557	14.3491	10.15870
		3	0		1.6686		4.73634		0.8031		3.03512
			1	1.0455	1.5960	5.4691		1.6166	1.1552	3.4800	
		4	1	0.5540	1.5487	2.9986	2.77030	0.6847	0.7472	1.8693	1.75051
			2	1.4750	1.4016	7.2392		3.1856	1.0700	4.2159	
		5	0			1.6087	2.04445			0.5554	1.33131
			1	0.6811	1.4515	2.1497		1.2523	0.8351	1.3761	
			2	2.0218	1.2889	3.4782		6.2260	1.0344	2.0009	
		6	1	0.5346	1.6360	1.7534	1.69357	0.6783	0.6162	1.1779	1.15808
			2	0.8653	1.3806	2.3951		2.4101	0.9079	1.4204	
			3	2.6828	1.2166	6.5436		12.1365	1.0174	1.3636	
	0.50 dB	2	1	0.7107	1.6949	16.9489	11.98464	0.8699	1.2335	11.9847	8.48923
		3	0		1.4839		4.21200		0.6591		2.71147
			1	1.0455	1.4193	4.8636		1.8439	1.0706	3.1031	
		4	1	0.5540	1.4245	2.7579	2.54801	0.7366	0.6669	1.7568	1.62843
			2	1.4780	1.2891	6.6583		3.7584	1.0303	3.9200	
		5	0		1.5115		1.92086		0.4700		1.27264
			1	0.6811	1.3731	2.0197		1.4499	0.7995	1.3126	
			2	2.0218	1.2110	3.2680		7.6747	1.0144	1.8800	
		6	1	0.5346	1.5594	1.6713	1.61432	0.7318	0.5635	1.1443	1.12698
			2	0.8653	1.3160	2.2830		2.9188	0.8925	1.3623	
			3	2.6828	1.1597	6.2374		15.4850	1.0071	3.1569	
	1.00 dB	2	1	0.7107	1.4054	14.0540	9.93763	0.9644	1.0526	9.9377	7.04485
		3	0		1.3143		3.73075		0.5237		2.41619
			1	1.0455	1.2571	4.3079		2.2060	1.0027	2.7584	
		4	1	0.5540	1.3074	2.5312	2.33855	0.8255	0.6015	1.6096	1.51549
			2	1.4780	1.1832	6.1109		4.7456	0.9993	3.5253	
		5	0		1.4186		1.80280		0.3854		1.21869
			1	0.6811	1.2887	1.8956		1.7634	0.7727	1.2538	
			2	2.0218	1.1366	3.0671		10.0100	0.9994	1.7643	
		6	1	0.5346	1.4857	1.5923	1.53798	0.8200	0.5174	1.1138	1.09888
			2	0.8653	1.2538	2.1751		3.7287	0.8831	1.3070	
			3	2.6828	1.1049	5.9425		21.0133	0.9997	2.9628	
	2.00 dB	2	1	0.7107	1.1478	11.4782	8.11633	1.1402	0.9107	8.1164	5.76106
		3	0		1.1540		3.27555		0.3953		2.13924
			1	1.0455	1.1037	3.7823		2.8401	0.9513	2.4338	
		4	1	0.5540	1.1936	2.3109	2.13499	0.9861	0.5469	1.4902	1.40843
			2	1.4780	1.0802	5.5790		6.4348	0.9771	3.1959	
		5	0		1.3270		1.68641		0.3008		1.16811
			1	0.6811	1.2055	1.7732		2.3131	0.7567	1.982	
			2	2.0218	1.0632	2.8691		14.2439	0.9895	1.6501	
		6	1	0.5346	1.4124	1.5137	1.46212	0.9809	0.4812	1.0856	1.07317
			2	0.8653	1.1920	2.0678		5.1692	0.8808	1.2529	
			3	2.6828	1.0504	5.6493		31.3724	0.9952	2.7640	

TABLE 116.2 Design Table for Low-Pass Finite-Zero Filters (*Continued*)

Specifications				Inverse Chebyshev				Cauer (Elliptical)			
α_s	α_p	N	S	Q	ω_0	ω_n	ω_s	Q	ω_0	ω_n	ω_s
60 dB	0.10 dB	2	1	0.7075	2.5599	80.9518	57.24160	0.7678	1.8206	57.2417	40.4791
		3	0		1.8865		11.80927		0.9780		7.45970
			1	1.0095	1.8687	13.6362		1.3588	1.2986	8.6040	
		4	1	0.5449	1.6396	5.8410	5.39637	0.6255	0.8121	3.5549	3.25974
			2	1.3577	1.5886	14.1015		2.3212	1.1480	8.4379	
		5	0		1.5664		3.40515		0.5883		2.04438
			1	0.6402	1.5100	3.5804		0.9746	0.8373	2.1363	
			2	1.7657	1.4333	5.7933		3.8628	1.0830	3.3302	
		6	1	0.5241	1.5299	2.6201	2.53074	0.6138	0.5928	1.5903	1.54869
			2	0.7705	1.4215	3.5791		1.5802	0.8868	2.0567	
			3	2.2442	1.3334	9.7783		6.3316	1.0491	5.3017	
	0.25 dB	2	1	0.7075	2.0271	64.1023	45.32720	0.8098	1.4541	45.3272	32.05510
		3	0		0.6157		10.11407		0.7748		6.39522
			1	1.0095	1.6004	11.6788		1.5305	1.1567	7.3732	
		4	1	0.5449	1.4623	5.2094	4.81286	0.6654	0.6971	3.1456	2.91918
			2	1.3577	1.4168	12.5767		2.7212	1.0755	7.4297	
		5	0		1.4332		3.11565		0.4814		1.88699
			1	0.6402	1.3816	3.2760		1.1118	0.7747	1.9694	
			2	1.7657	1.3114	5.3007		4.6482	1.0416	3.0475	
		6	1	0.5241	1.4250	2.4406	2.35734	0.6538	0.5203	1.4996	1.46216
			2	0.7705	1.3241	3.3339		1.8792	0.8521	1.9229	
			3	2.2442	1.2420	9.1083		7.8068	1.0242	4.9008	
	0.50 dB	2	1	0.7075	1.6923	53.5142	37.84029	0.8643	1.2316	37.8403	26.7618
		3	0		1.4334		8.97286		0.6336		5.67935
			1	1.0095	1.4199	10.3610		1.7346	1.0693	6.5451	
		4	1	0.5449	1.3390	4.7699	4.40683	0.7147	0.6195	2.8889	2.68325
			2	1.3577	1.2973	11.5157		3.1795	1.0311	6.7941	
		5	0		1.3388		2.91045		0.4028		1.77664
			1	0.6402	1.2906	3.0603		1.2727	0.7354	1.8523	
			2	1.7657	1.2250	4.9516		5.5513	1.0166	2.8471	
		6	1	0.5241	1.3499	2.3119	2.23304	0.7032	0.4704	1.4357	1.40138
			2	0.7705	1.2543	3.1581		2.2239	0.8317	1.8275	
			3	2.2442	1.1765	8.6280		9.5116	1.0080	4.6250	
	1.00 dB	2	1	0.7075	1.4022	44.3421	31.35464	0.9573	1.0503	31.3548	22.17680
		3	0		1.2656		7.92242		0.5004		5.02121
			1	1.0095	1.2536	9.1480		2.0563	0.9984	5.7834	
		4	1	0.5449	1.2221	4.3535	4.02211	0.7966	0.5513	2.6465	2.46079
			2	1.3577	1.1840	10.5103		3.8880	0.9954	6.1909	
		5	0		1.2479		2.71285		0.3255		1.67162
			1	0.6402	1.2030	2.8525		1.5249	0.7037	1.7406	
			2	1.7657	1.1419	4.6154		6.9480	0.9968	2.6541	
		6	1	0.5241	1.2768	2.1867	2.11216	0.7848	0.4261	1.3748	1.34354
			2	0.7705	1.1864	2.9871		2.7576	0.8168	1.7351	
			3	2.2442	1.1128	8.1609		12.1690	0.9980	4.3467	
	2.00 dB	2	1	0.7075	1.1439	36.1740	25.57890	1.1298	0.9076	25.5790	18.09400
		3	0		1.1062		6.92494		0.3745		4.39728
			1	1.0095	1.0958	7.9962		2.6103	0.9436	5.0609	
		4	1	0.5449	1.1076	3.9459	3.64547	0.9461	0.4945	2.4102	2.24440
			2	1.3577	1.0732	9.5261		5.0926	0.9684	0.9637	
		5	0		1.1574		2.51614		0.2493		1.56860
			1	0.6402	1.1158	2.6457		1.9585	0.6807	1.6307	
			2	1.7657	1.0591	4.2808		9.3369	0.9823	2.4619	
		6	1	0.5241	1.2033	2.0609	1.99060	0.9334	0.3894	1.3150	1.28693
			2	0.7705	1.1181	2.8152		3.6694	0.8081	1.6428	
			3	2.2442	1.0488	7.6913		16.7738	0.9900	4.0639	

Butterworth filter — A mathematical approximation to an ideal filter in which the magnitude of the transfer function in the frequency domain is maximally flat. Butterworth filters are optimum in the sense that they provide the least droop without overshoot in the magnitude response.

Cascade — A circuit topology in which a series of simple circuits (often biquads) are connected with the output of the first to the input of the second, the output of the second to the input of the third, and so forth. Using this configuration, a complex filter can be constructed from a "cascade" of simpler circuits.

Chebyshev filter — A mathematical approximation to an ideal filter in which the magnitude of the transfer function has a series of ripples in the passband that are of equal amplitude. Chebyshev filters are optimum in the sense that they offer the sharpest transition band for a given filter order for an all-pole filter.

Elliptical–Cauer filter — A mathematical approximation of an ideal filter in which the magnitude of the transfer function has a series of ripples in both the passband and stopband that are of equal amplitude. Elliptical–Cauer filters are optimum in the sense that they provide the sharpest possible transition band for a given filter order.

Filter circuit — An electronic circuit that passes some frequencies and rejects others so as to separate signals according to their frequency content. Common filter type are low pass, band pass, and high pass, which refer to whether they pass low, middle, or high frequencies, respectively.

Inverse Chebyshev filters — A mathematical approximation to an ideal filter exhibiting a Butterworth magnitude response in the passband and an elliptical–Cauer magnitude response in the stop-band.

LC ladder filter circuits — A planar circuit topology resembling a ladder in which the rungs of the ladder (referred to as the *parallel circuit elements*) consist of inductors or capacitors, one rail of the ladder is the ground plane, and the other rail (referred to as the *series circuit elements*) consists of a series connection of inductors and capacitors.

Passband — The frequencies for which a filter passes the signal from the input to the output without significant attenuation.

Passive RLC filter — A filter circuit that uses only passive devices, resistors, capacitors, inductors, and mutual inductors.

Pole frequency (ω_0) — A parameter of a biquad filter used along with Q to define the location of the complex second-order pole pair in the s domain. The pole frequency ω_0 is the distance of the second-order pole pair from the origin in the s plane. This frequency is approximately equal to the cutoff frequency for low-pass and high-pass filters and to the resonant frequency for band-pass filters (see Q).

Q — A parameter of a biquad circuit used along with ω_0 to define the location of the complex second-order pole pair in the s plane. The real part of the pole pair is located in the s plane at $\omega_0/2Q$. The two poles are at a distance ω_0 from the origin in the s plane.

Stopband — The frequencies for which a filter provides significant attenuation for signals between the input and output.

Transition band — The frequencies between a passband and stopband for which there are no filter specifications that must be met. Transition bands are necessary to allow for practical filter circuits, but it is desirable to keep transition bands as narrow as possible (i.e., a sharp transition band).

References

Antoniou, A. 1967. Gyrator using operational amplifiers. *Electron. Lett.*, 3:350–352.

Becvar, D. et al. 2000. Novel universal active block: a universal current conveyor, *Proc. ISCAS 2000*, pp. 471–474.

Bowron, P. and Stephenson, F. W. 1979. *Active Filters for Communications and Instrumentation*. McGraw-Hill, New York.

Brandenburg, D. 1998. Current vs. voltage feedback amplifiers, National Semiconductor Application Note OA-30, January.

Bruton, L. T., Pederson, R. T., and Treleaven, D. H. 1972. Low-frequency compensation of FDNR low-pass filters. *Proc. IEEE*, 60:444–445.

Buck, A. 1992. Current-feedback myths debunked, National Semiconductor Application Note OA-30, July.

Chiu, W. et al. 1996. CMOS differential difference current conveyors and their applications, *IEE Proc. Circuits Devices Sys.*, 143:91–96.

Chong, C.P. and Smith, K.C. 1986. Biquadratic filter sections employing a single current conveyor," *Electron. Lett.*, 22:1162–1164.

Daniels, R. W. 1974. *Approximation Methods for Electronic Filter Design.* McGraw-Hill, New York.

Ghausi, M. S. and Laker, K. R. 1981. *Modern Filter Design.* Prentice Hall, Englewood Cliffs, NJ.

Hamilton, T. A. and Sedra, A. S. 1971. A novel application of a gyrator-type circuit. In Proceedings, Fifth IEEE Asilomar Conference on Circuits and Systems, Pacific Grove, CA, pp. 343–348.

Hansell, G. E. 1969. *Filter Design and Evaluation.* Van Nostrand Reinhold, New York.

Haritantis, I., Constantinides, A. G., and Deliyannis, T. 1976. Wave active filters. *Proc. IEE (England)*, 123:676–682.

Koller, R. D. and Wilamowski, B. M. 1994. "Ladder" Filter Design Program. Electrical Engineering Department, University of Wyoming, Laramie.

Lindquist, C. S. 1977. *Active Network Design with Signal Filtering Applications.* Steward & Sons, Long Beach, CA.

Mitra, S. K. 1971. *Active Inductorless Filters.* IEEE Press, New York.

National Semiconductor. 1993. A tutorial on applying op-amps to RF applications, National Semiconductor Applications Note OA-11, September.

Potson, D. 1988. Current feedback op-amp applications circuit guide, National Semiconductor Applications Note OA-07, May.

Roberts, G. W. and Sedra, A. S. 1989. All current-mode frequency selective circuits. *Electron. Lett.*, 25:759–761.

Sallen, R. P. and Key, E. L. 1955. A practical method of designing RC active filters. *IRE Trans. Circuits Syst.*, CT-2:74–85.

Schaumann, R., Ghausi, M. S., and Laker, K. R. 1990. *Design of Analog Filters: Passive, Active-RC and Switched Capacitors.* Prentice-Hall, Englewood Cliffs, NJ.

Schaumann, R., Soderstrand, M. A., and Laker, K. 1976. *Modern Active Filter Design.* IEEE Press, New York.

Sedra, A. S. and Brackett, P. O. 1978. *Filter Theory and Design: Active and Passive.* Matrix Publishers, Champaign, IL.

Tiliute, D. E. 1998. A SPICE model for high-performance current conveyor. In: Proceedings of SIMSIS'98, University of Galati, pp. 205–207.

Toumazou, C. and Lidgey, J. 1990. Universal current-mode analogue amplifiers. In: Toumazou, C. and Ligdey, F. J., eds., *Analogue IC Design: The Current Mode Approach*, pp. 127–128.

Tsividis, Y. P. and Voorman, J. O. 1993. *Integrated Continuous-Time Filters: Principles, Design, and Applications.* IEEE Press, New York.

Wilamowski, B. M. 1994. "Filter" Filter Design Program. Electrical Engineering Department, University of Wyoming, Laramie, WY.

Williams, A. B. 1975. *Active Filter Design.* Artech House, Dedham, MA.

Zverev, A. I. 1967. *Handbook of Filter Synthesis.* John Wiley & Sons, New York.

Further Information

The following are recommended for further information on active filter design:

Schaumann, R., Ghausi, M. S., and Laker, K. R. 1990. *Design of Analog Filters: Passive, Active-RC and Switched Capacitors.* Prentice-Hall, Englewood Cliffs, NJ.

Sedra, A. S. and Brackett, P. O. 1978. *Filter Theory and Design: Active and Passive.* Matrix Publishers, Champaign, IL.

Tsividis, Y. P. and Voorman, J. O. 1993. *Integrated Continuous-Time Filters: Principles, Design, and Applications.* IEEE Press, New York.

117

Diodes and Transistors

Sidney Soclof
California State University, Los Angeles

Transistors form the basis of all modern electronic devices and systems, including the integrated circuits used in systems ranging from radio and TVs to computers. Transistors are solid-state electron devices made out of a category of materials called *semiconductors.* The most widely used semiconductor for transistors, by far, is silicon, although gallium arsenide, which is a compound semiconductor, is used for some very-high-speed applications. We start off with a very brief discussion of semiconductors. Next is a short discussion of PN junctions and diodes, followed by a section on the three major types of transistors — the bipolar junction transistor (BJT), the junction field-effect transistor (JEET), and the metal-oxide silicon field-effect transistor (MOSFET).

117.1 Semiconductors

Semiconductors are a category of materials with an electrical conductivity that is intermediate between that of the good conductors and the insulators. The good conductors, which are all metals, have electrical resistivities down in the range of $10^{-6}\ \Omega \cdot$ cm. The insulators have electrical resistivities that are up in the range of 10^{6} to as much as about $10^{12}\ \Omega \cdot$ cm. Semiconductors have resistivities that are generally in the range of 10^{-4} to $10^{4}\ \Omega \cdot$ cm. The resistivity of a semiconductor is strongly influenced by impurities, called *dopants*, which are purposely added to the material to change the electronic characteristics.

We will first consider the case of the pure, or intrinsic, semiconductor. As a result of the thermal energy present in the material, electrons can break loose from covalent bonds and become free electrons able to move through the solid and contribute to the electrical conductivity. The covalent bonds left behind have an electron vacancy called a *hole.* Electrons from neighboring covalent bonds can easily move into an adjacent bond with an electron vacancy, or hole, and thus the hole can move from one covalent bond to an adjacent bond. As this process continues, we can say that the hole is moving through the material. These holes act as if they have a positive charge equal in magnitude to the electron charge, and they can also contribute to the electrical conductivity. Thus, in a semiconductor there are two types of mobile electrical charge carriers that can contribute to the electrical conductivity: the free electrons and the holes. Since the electrons and holes are generated in equal members and recombine in equal numbers, the free electron and hole populations are equal.

0-8493-1586-7/05/$0.00+$1.50

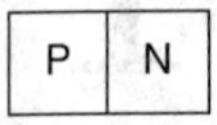

FIGURE 117.1 PN junction.

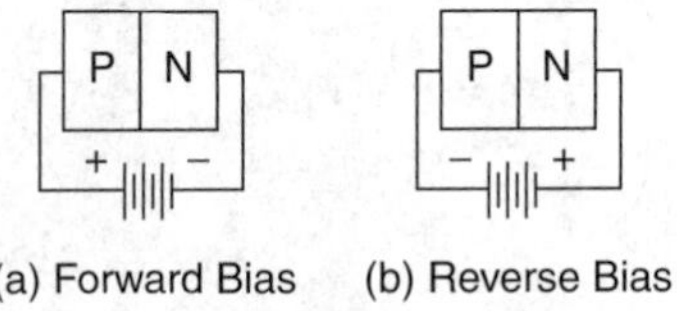

FIGURE 117.2 Diode.

In the extrinsic or doped semiconductor, impurities are purposely added to modify the electronic characteristics. In the case of silicon, every silicon atom shares its four valence electrons with each of its four nearest neighbors in covalent bonds. If an impurity or dopant atom with a valency of five, such as phosphorus, is substituted for silicon, four of the five valence electrons of the dopant atom will be held in covalent bonds. The extra, or fifth, electron will not be in a covalent bond and is loosely held. At room temperature, almost all of these extra electrons will have broken loose from their parent atoms and become free electrons. These pentavalent dopants thus donate free electrons to the semiconductor and are called *donors*. These donated electrons upset the balance between the electron and hole populations, so there are now more electrons than holes. This is now called an *N-type semiconductor*, in which the electrons are the majority carriers and holes are the minority carriers. In an N-type semiconductor the free electron concentration is generally many orders of magnitude larger than the hole concentration.

If an impurity or dopant atom with a valency of three, such as boron, is substituted for silicon, three of the four valence electrons of the dopant atom will be held in covalent bonds. One of the covalent bonds will be missing an electron. An electron from a neighboring silicon-to-silicon covalent bond, however, can easily jump into this electron vacancy, thereby creating a vacancy, or hole, in the silicon-to-silicon covalent bond. These trivalent dopants thus accept free electrons, thereby generating holes, and are called *acceptors*. These additional holes upset the balance between the electron and hole populations, so there are now more holes than electrons. This is now called a *P-type semiconductor*, in which the holes are the majority carriers and the electrons are the minority carriers. In a P-type semiconductor the hole concentration is generally many orders of magnitude larger than the electron concentration.

Figure 117.1 shows a single crystal chip of silicon that is doped with acceptors to make it a P-type on one side and doped with donors to make it N-type on the other side. The transition between the two sides is called the *PN junction*. As a result of the concentration difference of the free electrons and holes there will be an initial flow of these charge carriers across the junction, which will result in the N-type side attaining a net positive charge with respect to the P-type side. This results in the formation of an electric potential "hill" or barrier at the junction. Under equilibrium conditions the height of this potential hill, called the *contact potential*, is such that the flow of the majority carrier holes from the P-type side up the hill to the N-type side is reduced to the extent that it becomes equal to the flow of the minority carrier holes from the N-type side down the hill to the P-type side. Similarly, the flow of the majority carrier free electrons from the N-type side is reduced to the extent that it becomes equal to the flow of the minority carrier electrons from the P-type side. Thus, the net current flow across the junction under equilibrium conditions is zero.

In Figure 117.2 the silicon chip is connected as a diode, or two-terminal electrode device. The situation in which a bias voltage is applied is shown. In Figure 117.2(a) the bias voltage is a forward bias, which produces a reduction in the height of the potential hill at the junction. This allows for a large increase in the flow of electrons and holes across the junction. As the forward bias voltage increases, the forward current will increase at approximately an exponential rate and can become very large.

In Figure 117.2(b) the bias voltage is a reverse bias, which produces an increase in the height of the potential hill at the junction. This essentially chokes off the flow of (1) electrons from the N-type side to the P-type side, and (2) holes from the P-type side to the N-type side. The only thing left is the very small trickle of electrons from the P-type side and holes from the N-type side. Thus, the reverse current of the diode will be very small.

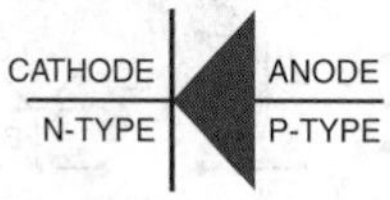

FIGURE 117.3 Diode symbol.

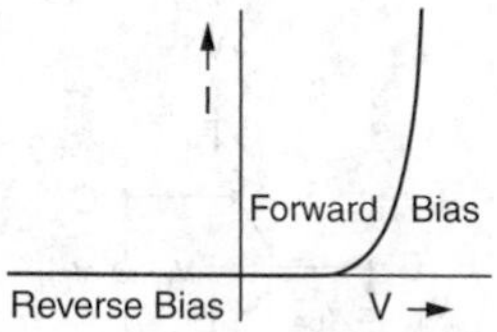

FIGURE 117.4 Current versus voltage curve.

In Figure 117.3 the circuit schematic symbol for the diode is shown, and in Figure 117.4 a graph of the current versus voltage curve for the diode is presented. The P-type side of the diode is called the *anode*, and the N-type side is the cathode of the diode. The forward current of diodes can be very large — in the case of large power diodes, up into the range of 10 to 100 A. The reverse current is generally very small, often down in the low nanoampere or even picoampere range.

The diode is basically a one-way voltage-controlled current switch. It allows current to flow in the forward direction when a forward bias voltage is applied, but when a reverse bias is applied the current flow becomes extremely small. Diodes are used extensively in electronic circuits. Applications include rectifiers to convert AC to DC, wave shaping circuits, peak detectors, DC level shifting circuits, and signal transmission gates. Diodes are also used for the demodulation of amplitude-modulated (AM) signals.

117.2 Bipolar Junction Transistors

A basic diagram of the bipolar junction transistor, or BJT, is shown in Figure 117.5. Whereas the diode has one PN junction, the BJT has two PN junctions. The three regions of the BJT are the emitter, base, and collector. The middle, or base, region is very thin, generally less than 1 micrometer wide. This middle electrode, or base, can be considered as the control electrode that controls the current flow through the device between emitter and collector. A small voltage applied to the base (i.e., between base and emitter) can produce a large change in the current flow through the BJT.

BJTs are often used for the amplification of electrical signals. In these applications the emitter-base PN junction is turned "on" (forward biased) and the collector-base PN junction is "off" (reverse biased). For the NPN BJT shown in Figure 117.5, the emitter will emit electrons into the base region. Since the P-type base region is so very thin, most of these electrons will survive the trip across the base and reach the collector-base junction. When the electrons reach the collector-base junction they will "roll downhill" into the collector and thus be collected by the collector to become the collector current, I_C. The emitter and collector currents will be approximately equal, so $I_C \cong I_E$. There will be a small base current, I_B, resulting from the emission of holes from the base across the emitter-base junction into the emitter. There will also be a small component of the base current due to the recombination of electrons and holes in the base. The ratio of collector current to base current, given by the parameter β or h_{FE}, is $\beta = I_C/I_B$, which will be very large, generally up in the range of 50 to 300 for most BJTs.

In Figure 117.6(a) the circuit schematic symbol for the NPN transistor is shown, and in Figure 117.6(b) the corresponding symbol for the PNP transistor is given. The basic operation of the PNP transistor is similar to that of the NPN, except for a reversal of the polarity of the algebraic signs of all DC currents and voltages.

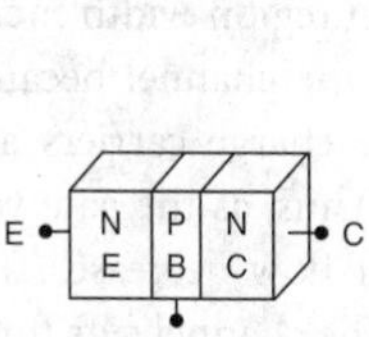

FIGURE 117.5 Bipolar junction transistor.

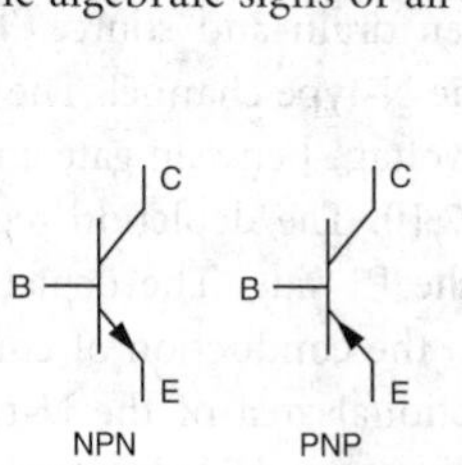

FIGURE 117.6 BJT schematic symbols.

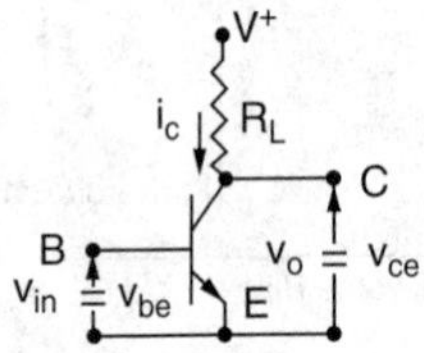

FIGURE 117.7 BJT amplifier.

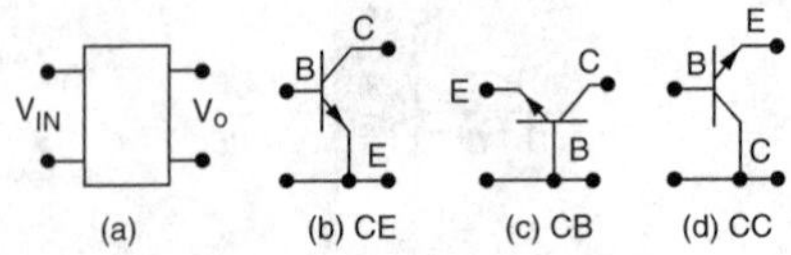

FIGURE 117.8 The BJT as a two-port device.

In Figure 117.7 the operation of a BJT as an amplifier is shown. When the BJT is operated as an amplifier the emitter-base PN junction is turned "on" (forward biased) and the collector-base PN junction is "off" (reverse biased). An AC input voltage applied between base and emitter, $v_{in} = v_{be}$, can produce an AC component, i_c, of the collector current. Since i_c flows through a load resistor, R_L, an AC voltage, $v_o = v_{ce} = -i_c \cdot R_L$, will be produced at the collector. The AC small-signal voltage gain is $A_V = v_o/v_{in} = v_{ce}/v_{be}$.

The collector current, I_C, of a BJT when operated as an amplifier is related to the base-to-emitter voltage, V_{BE}, by the exponential relationship $I_C = I_{CO} \cdot \exp(V_{BE}/V_T)$, where I_{CO} is a constant and $V_T =$ thermal voltage = 25 mV. The rate of change of I_C with respect to V_{BE} is given by the transfer conductance, $g_m = dI_C/dV_{BE} = I_C/V_T$. If the net load driven by the collector of the transistor is R_L, the AC small-signal voltage gain is $A_V = v_{ce}/v_{be} - g_m \cdot R_L$. The negative sign indicates that the output voltage will be an amplified but inverted replica of the input signal. If, for example, the transistor is biased at a DC collector current level of $I_C = 1$ mA and drives a net load of $R_L = 10\ \text{k}\Omega$, then $g_m = I_C/V_T = 1\ \text{mA}/25\ \text{mV} = 40$ mS, and $A_V = v_c/v_{be} = -g_m \cdot R_L = -40\ \text{mS} \cdot 10\ \text{k}\Omega = -400$. Thus, we see that the voltage gain of a single BJT amplifier stage can be very large, often up in the range of 100 or more.

The BJT is a three-electrode or *triode* electron device. When connected in a circuit it is usually operated as a two-port, or two-terminal pair device, as shown in Figure 117.8. Therefore, one of the three electrodes of the BJT must be common to both the input and output ports. Thus, there are three basic BJT configurations: common-emitter (CE), common-base (CB), and common-collector (CC). The most often-used configuration, especially for amplifiers, is the common-emitter (CE), although the other two configurations are used in some applications.

The BJT is often used as a switching device, especially in digital circuits, and in high-power applications. When used as a switching device, the transistor is switched between the *cutoff region*, in which both junctions are off, and the *saturation region*, in which both junctions are on. In the cutoff region the collector current is reduced to a very small value, down in the law nanoampere range, so the transistor looks essentially like an open circuit. In the saturation region the voltage drop between collector and emitter becomes very small, usually less than 0.1 volts, and the transistor looks like a very small resistance.

117.3 Junction Field-Effect Transistors

A *junction field-effect transistor* (JFET) is a type of transistor in which the current flow through the device between the drain and source electrodes is controlled by the voltage applied to the gate electrode. A simple physical model of the JFET is shown in Figure 117.9. In this JFET an N-type conducting channel exists between drain and source. The gate is a heavily doped P-type region (designated as P^+) that surrounds the N-type channel. The gate-to-channel PN junction is normally kept reverse biased. As the reverse bias voltage between gate and channel increases, the depletion region width increases, as shown in Figure 117.10. The depletion region extends mostly into the N-type channel because of the heavy doping on the P^+ side. The depletion region is depleted of mobile charge carriers and thus cannot contribute to the conduction of current between drain and source. Thus, as the gate voltage increases, the cross-sectional area of the N-type channel available for current flow decreases. This reduces the current flow between drain and source. As the gate voltage increases, the channel gets further constricted and the current flow gets smaller. Finally when the depletion regions meet in the middle of the channel, as shown in Figure 117.11, the channel is pinched off in its entirety, all of the way between the source

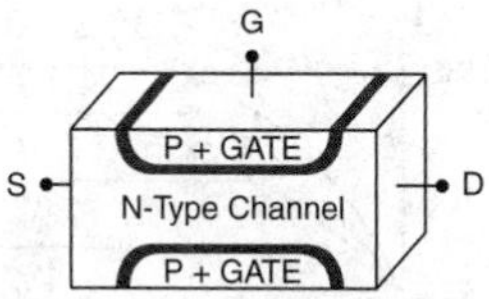

FIGURE 117.9 JFET model.

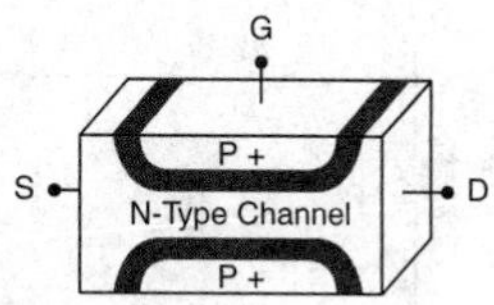

FIGURE 117.10 JFET with increased gate voltage.

and the drain. At this point the current flow between drain and source is reduced to essentially zero. This voltage is called the *pinch-off voltage*, V_P. The pinch-off voltage is also represented as *VGS* (off), as being the gate-to-source voltage that turns the drain-to-source current, I_{DS}, off.

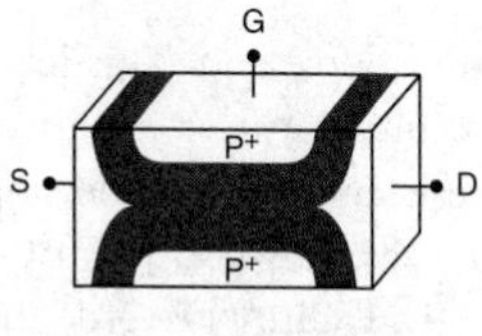

FIGURE 117.11 JFET with pinched-off channel.

We have been considering here an N-channel JFET. The complementary device is the P-channel JFET, which has a heavily doped N-type (N^+) gate region surrounding a P-type channel. The operation of a P-channel JFET is the same as for an N-channel device, except the algebraic signs of all DC voltages and currents are reversed.

We have been considering the case for V_{DS} small compared to the pinch-off voltage, such that the channel is essentially uniform from drain to source, as shown in Figure 117.12(a). Now let us see what happens as V_{DS} increases. As an example, let us consider an N-channel JFET with a pinch-off voltage of $V_P = -4$ V. We will see what happens for the case of $V_{GS} = 0$ as V_{DS} increases. In Figure 117.12(a) the situation is shown for the case of $V_{DS} = 0$ in which the JFET is fully on and there is a uniform channel from source to drain. This is a point A on the I_{DS} versus V_{DS} curve of Figure 117.13. The drain-to-source conductance is at its maximum value of g_{ds} (on), and the drain-to-source resistance is correspondingly at its minimum value of r_{ds} (on). Now let us consider the case of $V_{DS} = +1$ V, as shown in Figure 117.12(b). The gate-to-channel bias voltage at the source end is still $V_{GS} = 0$. The gate-to-channel bias voltage at the drain end is $V_{GD} = V_{GS} - V_{DS} = -1$ V, so the depletion region will be wider at the drain end of the channel than at the source end. The channel will thus be narrower at the drain end than at the source end and this will result in a decrease in the channel conductance, g_{ds}, and correspondingly, an increase in the channel resistance, r_{ds}. So the slope of I_{DS} versus V_{DS} curve that corresponds to the channel conductance will be smaller at $V_{DS} = 1$ V than it was at $V_{DS} = 0$, as shown at point B on the I_{DS} versus V_{DS} curve of Figure 117.13.

In Figure 117.12(c) the situation for $V_{DS} = +2$ V is shown. The gate-to-channel bias voltage at the source end is still $V_{GS} = 0$, but the gate-to-channel bias voltage at the drain end is now $V_{GD} = V_{GS} - V_{DS} = -2$V, so the depletion region will now be substantially wider at the drain end of the channel than at

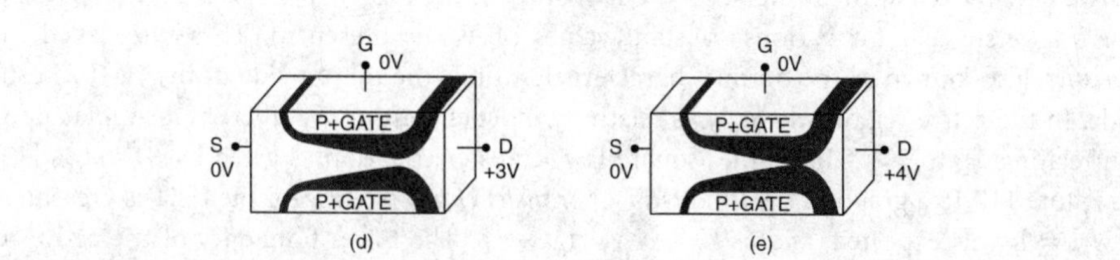

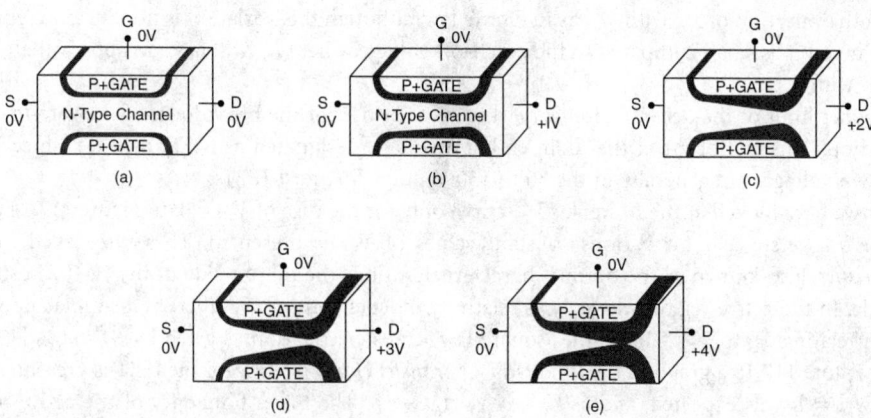

FIGURE 117.12 N-type channel.

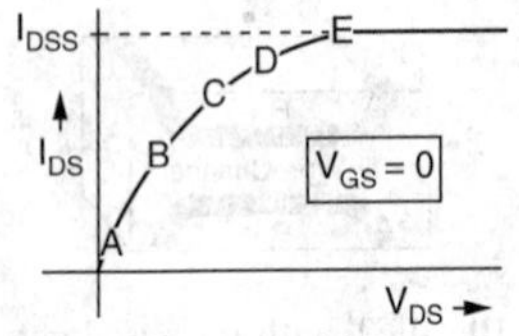

FIGURE 117.13 I_{DS} versus V_{DS} curve.

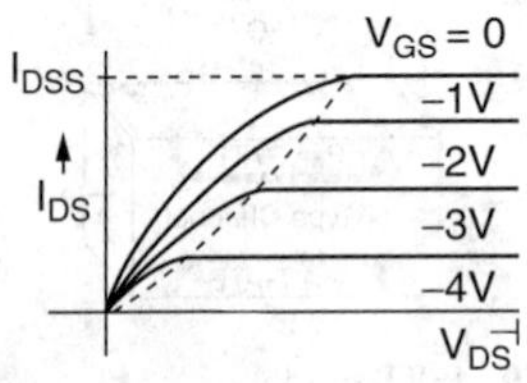

FIGURE 117.14 JFET drain characteristics.

the source end. This leads to a further constriction of the channel at the drain end and this will again result in a decrease in the channel conductance, g_{ds}, and correspondingly, as increase in the channel resistance, r_{ds}. So, the slope of the I_{DS} versus V_{DS} curve will be smaller at V_{DS} = 2 V than it was at V_{DS} = 1 V, as shown at point C on the I_{DS} versus V_{DS} curve of Figure 117.13.

In Figure 117.12(d) the situation for V_{DS} = +3 V is shown, and this corresponds to point D on the I_{DS} versus V_{DS} curve of Figure 117.13.

When $V_{DS} = +4$ V, the gate-to-channel bias voltage will be $V_{GD} = V_{GS} - V_{DS} = 0 - 4\text{ V} = -4\text{ V} = V_P$. As a result the channel is now pinched off at the drain end, but it is still wide open at the source end since $V_{GS} = 0$, as shown in Figure 117.12(e). It is very important to note that the channel is pinched off just for a very short distance at the drain end, so that the drain-to-source current, I_{DS}, can still continue to flow. This is not at all the same situation as for the case of $V_{GS} = V_P$, wherein the channel is pinched off in its entirety, all of the way from source to drain. When this happens, it is like having a big block of insulator the entire distance between source and drain, and I_{DS} is reduced to essentially zero. The situation for $V_{DS} = +4\text{ V} = -V_P$ is shown at point E on the I_{DS} versus V_{DS} curve of Figure 117.13.

For $V_{DS} > +4$V, the current essentially saturates and does not increase much with further increases in V_{DS}. As V_{DS} increases above +4 V, the pinched-off region at the drain end of the channel gets wider, which increases r_{ds}. This increase in r_{ds} essentially counterbalances the increase in V_{DS} such that I_D does not increase much. This region of the I_{DS} versus V_{DS} curve, in which the channel is pinched off at the drain end, is called the *active region* and is also known as the *saturated region.* It is called the active region because when the JFET is to be used as an amplifier it should be biased and operated in this region. The saturated value of drain current up in the active region for the case of $V_{GS} = 0$ is called I_{DSS}. Since there is not really a true saturation of current in the active region, I_{DSS} is usually specified at some value of V_{DS}. For most JFETs, the values of I_{DSS} fall in the range of 1 to 30 mA. In the current specification, I_{DSS}, the third subscript "S" refers to I_{DS} under the condition of the gate *shorted* to the source.

The region below the active region where $V_{DS} < +4\text{ V} = -V_p$ has several names. It is called the *nonsaturated region*, the *triode region*, and the *ohmic region*. The term *triode region* apparently originates from the similarity of the shape of the curves to that of the vacuum tube triode. The term *ohmic region* is due to the variation of I_{DS} with V_{DS} as in Ohm's law, although this variation is nonlinear except for the region where V_{DS} is small compared to the pinch-off voltage, where I_{DS} will have an approximately linear variation with V_{DS}.

The upper limit of the active region is marked by the onset of the breakdown of the gate-to-channel PN junction. This will occur at the drain end at a voltage designated as BV_{DG}, or BV_{DS}, since $V_{GS} = 0$. Breakdown voltages are generally in the 30 to 150 V range for most JFETs.

So far we have looked at the I_{DS} versus V_{DS} curve only for the case of $V_{GS} = 0$. In Figure 117.14 a family of curves I_{DS} versus V_{DS} for various constant values of V_{GS} is presented. These are called the *drain characteristics*, also known as the output characteristics, since the output side of the JFET is usually the drain side. In the active region where I_{DS} is relatively independent of V_{DS}, there is a simple approximate equation relating I_{DS} to V_{GS}. This is the "square law" transfer equation as given by $I_{DS} = I_{DSS}\,[1 - (V_{GS}/V_p)]^2$. In Figure 117.15 a graph of the I_{DS} versus V_{GS} *transfer characteristics* for the JFET is presented. When $V_{GS} = 0$, $I_{DS} = I_{DSS}$ as expected, and as $V_{GS} \to V_P$, $I_{DS} \to 0$. The lower boundary of the active region is controlled by the condition that the channel be pinched off at the drain end. To meet this condition the basic requirement is that the gate-to-channel bias voltage at the drain end of the channel, V_{GD}, be greater than the pinch-off voltage V_P. For the example under consideration with $V_P = -4$ V, this means that $V_{GD} = V_{GS} - V_{DS}$ be more negative than -4 V. Therefore, $V_{DS} - V_{GS} \geq +4$ V. Thus, for $V_{GS} = 0$, the active region will begin at $V_{DS} = +4$ V. When $V_{GS} = -1$V, the active region will begin at $V_{DS} = +3$ V, for now $V_{GD} = -4$ V. When $V_{GS} = -2$ V, the active region begins at $V_{DS} = +2$ V, and when $V_{GS} = -3$ V, the active region begins at $V_{DS} = +1$ V. The dotted line in Figure 117.14 marks the boundary between the nonsaturated and active regions.

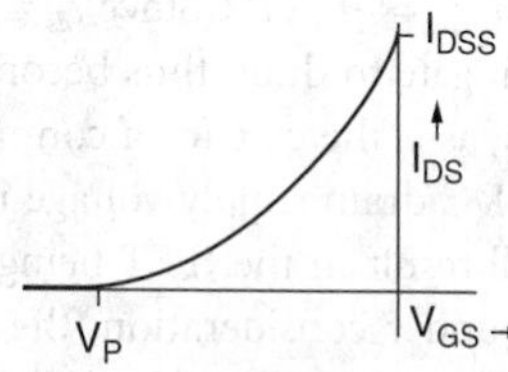

FIGURE 117.15 JFET transfer characteristics.

The upper boundary of the active region is marked by the onset of the avalanche breakdown of the gate-to-channel PN junction. When $V_{GS} = 0$, this breakdown occurs at $V_{DS} = BV_{DS} = BV_{DG}$. Since $V_{DG} = V_{DS} - V_{GS}$ and breakdown occurs when $V_{DG} = BV_{DG}$, as V_{GS} increases the breakdown voltages decrease as given by $BV_{DG} = BV_{DS} - V_{GS}$. Thus, $BV_{DS} = BV_{DG} + V_{GS}$. For example, if the gate-to-channel breakdown voltage is 50 V, the V_{DS} breakdown voltage will start off at 50 V when $V_{GS} = 0$ but decreases to 46 V when $V_{GS} = -4$ V.

In the nonsaturated region I_{DS} is a function of both V_{GS} and I_{DS}, and in the lower portion of the nonsaturated region where V_{DS} is small compared to V_P, I_{DS} becomes an approximately linear function of V_{DS}. This linear portion of the nonsaturated is called the *voltage-variable resistance* (VVR) region, for in this region the JFET acts like a linear resistance element between source and drain. The resistance is variable in that it is controlled by the gate voltage.

JFET as an Amplifier — Small-Signal AC Voltage Gain

Let us consider the common-source amplifier of Figure 117.16. The input AC signal is applied between gate and source, and the output AC voltage between is taken between drain and source. Thus, the source electrode of this triode device is common to input the output, hence the designation of this JFET configuration as a common-source (CS) amplifier.

A good choice of the DC operating point or quiescent point ("Q-point") for an amplifier is in the middle of the active region at $I_{DS} = I_{DSS}/2$. This allows for the maximum symmetrical drain current swing, from the quiescent level of $I_{DSQ} = I_{DSS}/2$, down to a minimum of $I_{DS} \cong 0$, and up to maximum of $I_{DS} = I_{DSS}$. This choice for the Q-point is also a good one from the standpoint of allowing for an adequate safety margin for the location of the actual Q-point due to the inevitable variations in device and component characteristics and values. This safety margin should keep the Q-point well away from the extreme limits of the active region, and thus ensure operation of the JFET in the active region under most conditions. If $I_{DSS} = +10$ mA, then a good choice for the Q-point would thus be around +5 mA. The AC component of the drain current, i_{ds}, is related to the AC component of the gate voltage, v_{gs}, by $i_{ds} = g_m \cdot v_{gs}$, where g_m is the dynamic transfer conductance and is given by $g_m = 2\sqrt{I_{DS} \cdot I_{DSS}}/(-V_P)$. If $V_P = -4$ V, then $g_m = \sqrt{5\text{ mA} \cdot 10\text{ mA}}/4\text{ V} = 3.54$ mA/V $= 3.54$ mS. If a small AC signal voltage, v_{gs}, is superimposed on the quiescent DS gate bias voltage $V_{GSQ} = V_{GG}$, only a small segment of the transfer characteristic adjacent to the Q-point will be traversed, as shown in Figure 117.17. This small segment will be close to a straight line, and as a result the AC drain current, i_{ds}, will have a waveform close to that of the AC voltage applied to the gate. The ratio of i_{ds} to v_{gs} will be the slope of the transfer curve as given by $i_{ds}/v_{gs} \cong dI_{DS}/dV_{GS} = g_m$. Thus, $i_{ds} \cong g_m \cdot v_{gs}$. If the net load driven by the drain of the JFET is the drain

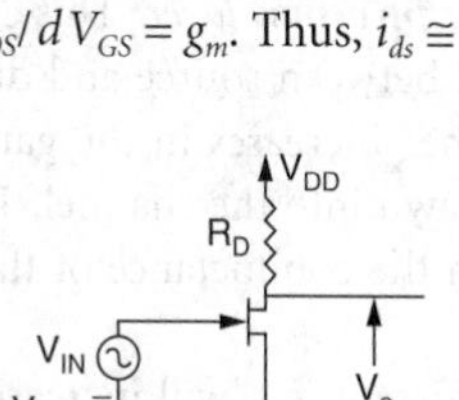

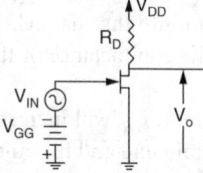

FIGURE 117.16 Common source amplifier.

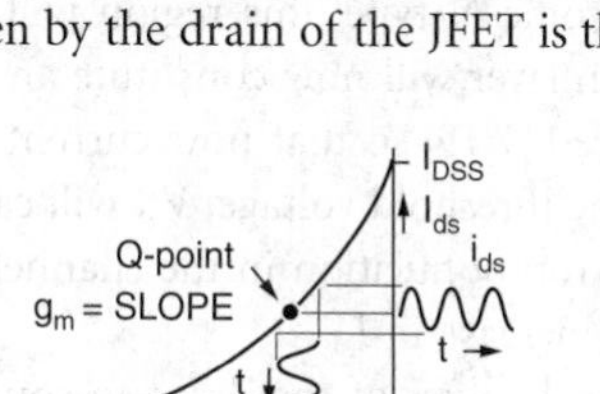

FIGURE 117.17 JFET transfer characteristic.

load resistor, R_D, as shown in Figure 117.16, then the AC drain current i_{ds} will produce an AC drain voltage of $v_{ds} = -i_{ds} \cdot R_D$. Since $i_{ds} = g_m \cdot v_{gs}$, this becomes $v_{ds} = -g_m \cdot v_{GS} \cdot R_D$. The AC small-signal voltage gain from gate to drain thus becomes $A_V = v_O/v_{IN} = v_{ds}/v_{gs} = -g_m \cdot R_D$. The negative sign indicates signal inversion, as is the case for a common-source amplifier.

If the DC drain supply voltage is V_{DD} = +20V, a quiescent drain-to-source voltage of $V_{DSQ} = V_{DD}/2$ = +10V will result in the JFET being biased in the middle of the active region. Since I_{DSQ} = 5 mA in the example under consideration, the voltage drop across the drain load resistor, R_D, is 10 V. Thus, R_D = 10 V/5 mA = 2 kΩ. The AC small-signal voltage gain, A_V, thus becomes $A_V = -g_m \cdot R_D$ = –3.54 mS · 2 kΩ = –7.07. Note that the voltage gain is relatively modest compared to the much larger voltage gains that can be obtained in a bipolar-junction transistor (BJT) common-emitter amplifier. This is due to the lower transfer conductance of both JFETs and MOSFETs compared to BJTs. For a BJT the transfer conductance is given by $g_m = I_C/V_T$, where I_C is the quiescent collector current and $V_T = kT/q \cong$ 25 mV is the "thermal voltage." At I_C = 5 mA, g_m = 5 mA/25 mV = 200 mS for the BJT, as compared to only 3.5 mS for the JFET in this example. With a net load of 2 kΩ, the BJT voltage gain will be –400 as compared to the JFET voltage gain of only 7.1.

Thus, FETs have the disadvantage of a much lower transfer conductance and therefore lower voltage gain than BJTs operating under similar quiescent current levels, but they do have the major advantage of a much higher input impedance and a much lower input current. In the case of a JFET the input signal is applied to the reverse-biased gate-to-channel PN junction and thus sees very high impedance. In the case of a common-emitter BJT amplifier the input signal is applied to the forward-biased base-emitter junction and the input impedance is given approximately by $r_{IN} = r_{BE} \cong 1.5 \cdot \beta \cdot V_T/I_C$. If I_C = 5 mA and β = 200, for example, then $r_{IN} \cong$ 1500 Ω. This moderate input resistance value of 1.5 kΩ is certainly no problem if the signal source resistance is less than around 100 Ω. However, if the source resistance is above 1 kΩ, then there will be a substantial signal loss in the coupling of the signal from the signal source to the base of the transistor. If the source resistance is in the range of above 100 kΩ, and certainly if it is above 1 MΩ, there will be severe signal attenuation due to the BJT input impedance, and a FET amplifier will probably offer a greater overall voltage gain. Indeed, when high impedance signal sources are encountered, a multistage amplifier with a FET input stage, followed by cascaded BJT stages is often used.

117.4 Metal-Oxide Silicon Field-Effect Transistors

A metal-oxide silicon field-effect transistor (MOSFET) is similar to a JFET, in that it is a type of transistor in which the current flow through the device between the drain and source electrodes is controlled by the voltage applied to the gate electrode. A simple physical model of the MOSFET is shown in Figure 117.18. The gate electrode is electrically insulated from the rest of the device by a thin layer of silicon dioxide (SiO_2). In the absence of any gate voltage there is no conducting channel between source and drain, so the device is off and $I_{DS} \cong 0$. If now a positive voltage is applied to the gate, electrons will be drawn into the silicon surface region immediately underneath the gate oxide. If the gate voltage is above the threshold voltage, V_T, there will be enough electrons drawn into this silicon surface region to make the electron population greater than the hole population. This surface region under the oxide will thus become N-type; this region will be called an *N-type surface inversion layer*. This N-type surface inversion layer will now constitute an N-type conducting channel between source and drain, as shown in Figure 117.19, so that now current can flow and $I_{DS} > 0$. Further increases in the gate voltage, V_{GS}, above the threshold voltage, V_T, will cause more electrons to be drawn into the channel. The increase in the electron population in the channel will result in an increase in the conductance of the channel and thus an increase in I_{DS}.

At small values of the drain-to-source voltage, V_{DS}, the drain current, I_{DS}, will increase linearly with V_{DS}, but as V_{DS} increases the channel will become constricted at the drain end and the curve of I_{DS} versus V_{DS} will start to bend over. Finally, if V_{DS} is large enough such that $V_{GS} - V_{DS} < V_T$, the channel becomes pinched off at the drain end and the curve of I_{DS} versus V_{DS} will become almost horizontal, as shown in

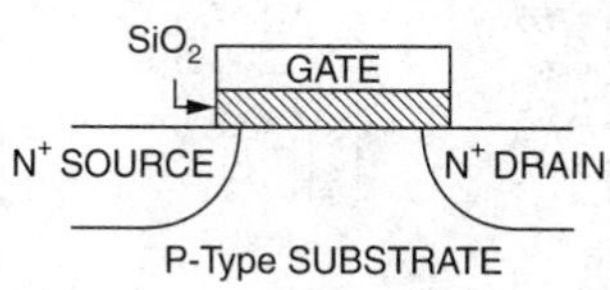

FIGURE 117.18 MOSFET physical model.

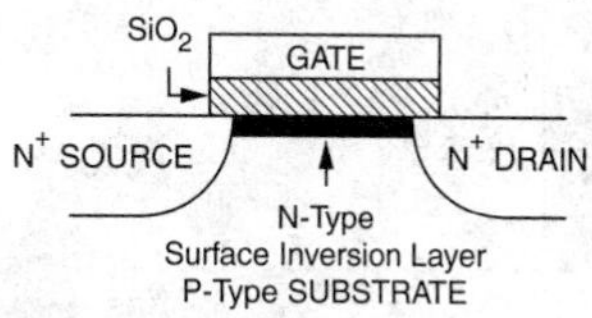

FIGURE 117.19 $V_{GS} > V_T$.

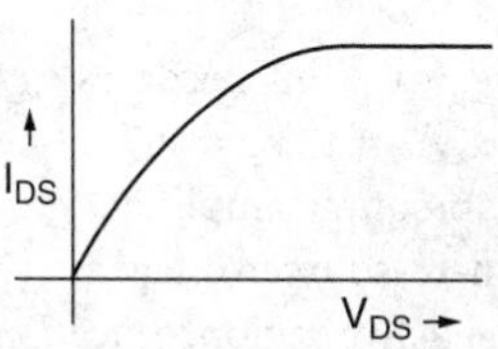

FIGURE 117.20 I_{DS} versus V_{DS} curve.

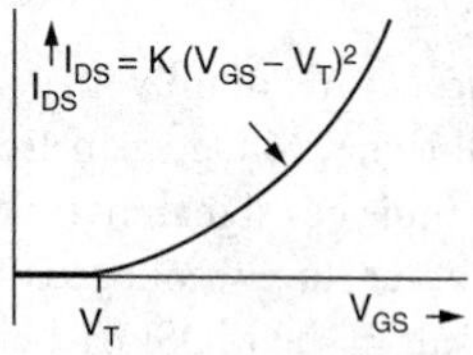

FIGURE 117.21 Transfer characteristic.

Figure 117.20. This region where I_{DS} becomes relatively independent of V_{DS} is called the *active region.* When the MOSFET is used as an amplifier it should be operated in the active region.

In the active region I_{DS} is relatively independent of V_{DS} but is a strong function V_{GS}. The transfer relationship between I_{DS} and V_{GS} in the active region is given approximately by the square law equation $I_{DS} = K(V_{GS} - V_T)^2$, where K is a constant. In Figure 117.21, a graph of the I_{DS} versus V_{GS} transfer characteristics of an N-channel MOSFET is shown.

In the active region, under small-signal conditions, the AC component of the drain current is related to the AC component of the gate voltage by $i_{ds} = g_m \cdot v_{gs}$, where g_m is the *dynamic transfer conductance.* We have that $g_m = dI_{DS}/dV_{GS} = 2K(V_{GS} - V_T)$. Since $K = I_{DS}/(V_{GS} - V_T)^2$, this can be expressed as $g_m = 2I_{DS}/(V_{GS} - V_T)$. Since $(V_{GS} - V_T) = \sqrt{I_{DS}/K}$, this can also be rewritten as $g_m = 2\sqrt{K \cdot I_{DS}}$.

MOSFET as an Amplifier — Small-Signal AC Voltage Gain

In Figure 117.22 some MOSFET symbols are shown for both N-channel and P-channel devices. Let us consider the MOSFET common-source amplifier circuit of Figure 117.23. The input AC signal is applied between the gate and the source as v_{gs}, and the output AC voltage is taken between the drain and the source as v_{ds}. Thus, the source electrode of this triode device is common to input and output — hence the designation of this MOSFET configuration as common-source (CS) amplifier.

Let us assume a quiescent current level of $I_{DSQ} = 10$ mA and $V_{GS} - V_T = 1$V. The AC component of the drain current, i_{ds}, is related to the AC component of the gate voltage, v_{gs}, by $i_{ds} = g_m \cdot v_{gs}$, where g_m is the *dynamic transfer conductance*, and is given by $g_m = 2K(V_{GS} - V_T) = 2I_{DS}/(V_{GS} - V_T) = 2 \cdot 10\text{ mA}/1\text{ V} = 20$ mS. If a small AC signal voltage, v_{gs}, is superimposed on the DC gate bias voltage V_{GS}, only a small segment of the transfer characteristic adjacent to the Q-point will be traversed, as shown in Figure 117.24. This small segment will be close to a straight line, and as a result the AC drain current, i_{ds}, will have a waveform close to that of the AC voltage applied to the gate. The ratio of i_{ds} to v_{gs} will be the slope of the transfer curve, is given by $i_{ds}/v_{gs} \cong dI_{DS}/dV_{GS} = g_m$. Thus, $i_{ds} \cong g_m \cdot v_{gs}$. If the net load driven by the drain of the MOSFET is the drain load resistor, R_D, as shown in Figure 117.23, then the AC drain current

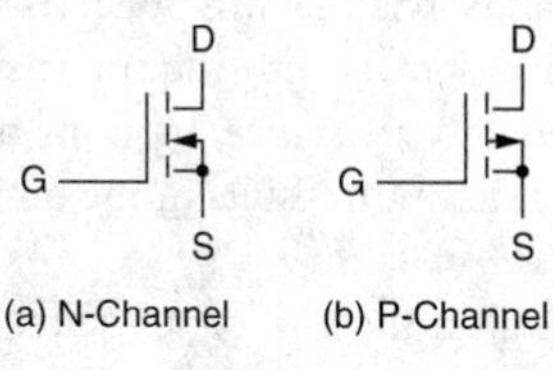

FIGURE 117.22 MOSFET symbols.

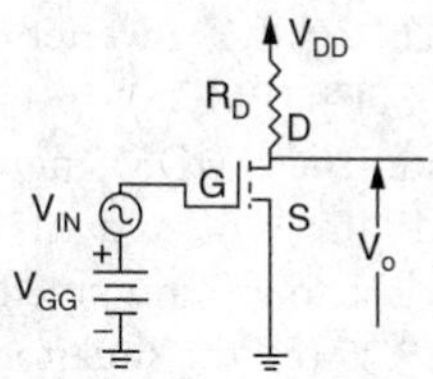

FIGURE 117.23 Common-source amplifier.

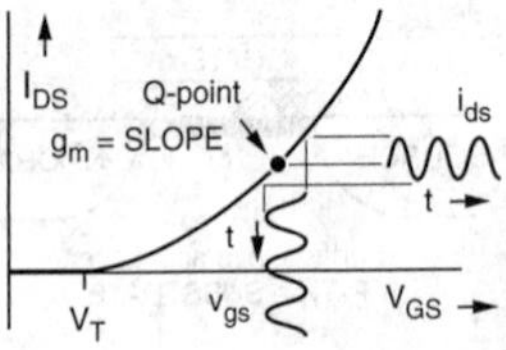

FIGURE 117.24 Transfer characteristic.

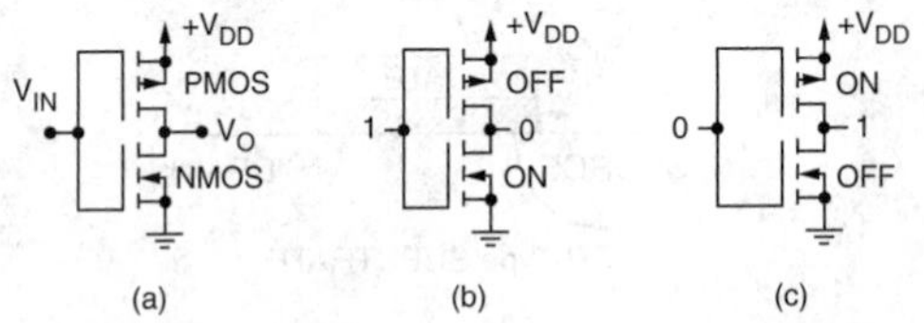

FIGURE 117.25 CMOS circuit.

i_{ds} will produce an AC drain voltage of $v_{ds} = -i_{ds} \cdot R_D$. Since $i_{ds} = g_m \cdot v_{gs}$, this becomes $v_{ds} = -g_m v_{gs} \cdot R_D$. The AC small-signal voltage gain from gate to drain thus becomes $A_V = v_o/v_{in} = v_{ds}/v_{gs} = -g_m \cdot R_D$. The negative sign indicates signal inversion, as is the case for a common-source amplifier.

If the DC drain supply voltage is $V_{DD} = +20$ V, a quiescent drain-to-source voltage of $V_{DSQ} = V_{DD}/2 = +10$ V will result in the MOSFET being biased in the middle of the active region. Since $I_{DSQ} = 10$ mA in the example under consideration and the voltage drop across the drain load resistor (R_D) is 10 V, we obtain $R_D = 10$ V/10 mA = 1 kΩ. The AC small-signal voltage gain, A_V, thus becomes $A_V = -g_m \cdot R_D = -20$ mS · 1 kΩ = −20. Note that the voltage gain is relatively modest, as compared to the much larger voltage gains that can be obtained in a bipolar-junction transistor (BJT) common-emitter amplifier. This is due to the lower transfer conductance of both JFETs and MOSFETs compared to BJTs. For a BJT the transfer conductance is given by $g_m = I_C/V_T$, where I_C is the quiescent collector current and $V_T = kT/q \cong 25$ mV is the thermal voltage. At $I_C = 10$ mA, $g_m = 10$ mA/25 mV = 400 mS, compared to only 20 mS for the MOSFET in this example. With a net load of 1 kΩ, the BJT voltage gain will be −400, compared to the MOSFET voltage gain of only −20.

Thus, FETs do have the disadvantage of a much lower transfer conductance and therefore lower voltage gain than BJTs operating under similar quiescent current levels, but they do have the major advantage of a much higher input impedance and a much lower input current. In the case of a MOSFET the input signal is applied to the gate, which is insulated from the rest of the device by the thin SiO_2 layer, and thus sees very high impedance. In the case of a common-emitter BJT amplifier, the input signal is applied to the forward-biased base-emitter junction and the input impedance is given approximately by $r_{IN} = r_{BE} \cong 1.5 \cdot \beta \cdot V_T/I_C$, where V_T is the thermal voltage of about 25 mV. If $I_C = 5$ mA and $\beta = 200$, for example, then $r_{IN} \cong 1500$ Ω. This moderate input resistance value of 1.5 kΩ is certainly no problem if the signal source resistance is less than around 100 Ω. However, if the source resistance is above 1 kΩ, there will be a substantial signal loss in the coupling of the signal from the signal source to the base of the transistor. If the source resistance is in the range of above 100 kΩ, and certainly if it is above 1 MΩ, there will be severe signal attenuation due to the BJT input impedance, and an FET amplifier will probably offer a greater overall voltage gain. Indeed, when high-impedance signal sources are encountered, a multistage amplifier with an FET input stage followed by cascaded BJT stages is often used.

MOSFETs for Digital Circuits

MOSFETs are used very extensively for digital applications. For high-density integrated circuits, MOSFETs offer great advantages over BJTs and JFETs from the standpoint of a much smaller size and lower power consumption.

A very important MOSFET configuration is the *complementary symmetry MOSFET*, or CMOS circuit. In Figure 117.25, a CMOS inverter circuit is shown, comprising an N-channel MOSFET (NMOS) and a P-channel MOSFET (PMOS). In Figure 117.25(b), the situation is shown with the input signal in the high or "1" state. The NMOS is now on and exhibits a moderately low resistance, typically on the order of 100 Ω. The PMOS is off and acts as a very high resistance. This situation results in the output voltage, V_o going low, close to ground potential (0 V).

In Figure 117.25 (c), the situation is shown with the input signal in the low, or "0" state. The NMOS is off and acts as a very high resistance, and the PMOS is on and exhibits a moderately low resistance, typically on the order of 100 Ω. This situation results in the output voltage, V_o, being pulled up high (the "1" state), close to the V_{DD} supply.

We note that under both input conditions one of the transistors will be off, and as a result the current flow through the two transistors will be extremely small. Thus, the power dissipation will also be very small, and, indeed, the only significant amount of power dissipation in the CMOS pair occurs during the short switching interval when both transistors are simultaneously on.

117.5 New Trends

The need for high-speed transistors for various communications applications has led to various developments. These include decreases in device dimensions for reduced capacitance, and for MOSFETs reduced channel lengths down to the range of less than 100 nm. Also, there is the use of III-V compound semiconductors, notably gallium arsenide, in which the substantially higher electron mobility as compared to silicon offers the possibility of a substantially higher speed. The addition of a small amount of germanium to silicon also produces substantially higher electron mobility. Heterostructure transistors using SiGe as the base material can also result in substantially higher speed devices. The GaAs and SiGe transistors do, however, have the disadvantage of higher processing costs as compared to silicon transistors.

Defining Terms

Active region — The region of transistor operation in which the output current is relatively independent of the output voltage. For the BJT, this corresponds to the condition that the emitter base junction is on, and the collector base junction is off. For the FETs, this corresponds to the condition that the channel is on or open at the source end and pinched off at the drain end.

Acceptors — Impurity atoms that, when added to a semiconductor, contribute holes. In the case of silicon, acceptors are atoms from the third column of the periodic table, such as boron.

Anode — The P-type side of a diode.

Cathode — The N-type side of a diode.

CMOS — The complementary-symmetry MOSFET configuration composed of an N-channel MOSFET and a P-channel MOSFET, operated such that when one transistor is on, the other is off.

Contact potential — The internal voltage that exists across a PN junction under thermal equilibrium conditions, when no external bias voltage is applied.

Donors — Impurity atoms that, when added to a semiconductor, contribute free electrons. In the case of silicon, donors are atoms from the fifth column of the periodic table, such as phosphorus, arsenic, and antimony.

Dopants — Impurity atoms that are added to a semiconductor to modify electrical conduction characteristics.

Doped semiconductor — A semiconductor that has had impurity atoms added to modify the electrical conduction characteristics.

Extrinsic semiconductor — A semiconductor that has been doped with impurities to modify the electrical conduction characteristics.

Forward bias — A bias voltage applied to the PN junction of a diode or transistor that makes the P-type side positive with respect to the N-type side.

Forward current — The large current flow in a diode that results from the application of a forward-bias voltage.

Hole — An electron vacancy in a covalent bond between two atoms in a semiconductor. Holes are mobile charge carriers with an effective charge that is opposite to the charge on an electron.

Intrinsic semiconductor — A semiconductor with a degree of purity such that the electrical characteristics are not significantly affected.

Majority carriers — In a semiconductor, the type of charge carrier with the larger population. For example, in an N-type semiconductor, electrons are the majority carriers.

Minority carriers — In a semiconductor, the type of charge carrier with the smaller population. For example, in an N-type semiconductor, holes are the minority carriers.

N-type semiconductor — A semiconductor that has been doped with donor impurities to produce the condition that the population of free electrons is greater than the population of holes.

Ohmic, nonsaturated, or triode region — These three terms all refer to the region of FET operation in which a conducting channel exists all of the way between source and drain. In this region the drain current varies with both the gate voltage and the drain voltage.

Output characteristics — The family of curves of output current versus output voltage. For the BJT, output characteristics are curves of collector current versus collector voltage for various constant values of base current or voltage and are also called the *collector characteristics*. For FETs, these will be curves of drain current versus drain voltage for various constant values of gate voltage and are also called the *drain characteristics*.

P-type semiconductor — A semiconductor that has been doped with acceptor impurities to produce the condition that the population of holes is greater than the population of free electrons.

Pinch-off voltage, V_P — The voltage that, when applied across the gate-to-channel PN junction, will cause the conducting channel between drain and source to become pinched off. This is also represented as V_{GS} (off).

Reverse bias — A bias voltage applied to the PN junction of a diode or transistor that makes the P-type side negative with respect to the N-type side.

Reverse current — The small current flow in a diode that results from the application of a reverse bias voltage.

Thermal voltage — The quantity kT/q where k is Boltzmann's constant, T is absolute temperature, and q is electron charge. The thermal voltage has units of volts and is a function only of temperature, being approximately 25 mV at room temperature.

Threshold voltage — The voltage required to produce a conducting channel between source and drain in a MOSFET.

Transfer conductance — The AC or dynamic parameter of a device that is the ratio of the AC output current to the AC input voltage. The transfer conductance is also called the *mutual transconductance* and is usually designated by the symbol g_m.

Transfer equation — The equation that relates the output current (collector or drain current) to the input voltage (base-to-emitter or gate-to-source voltage).

Triode — The three-terminal electron device, such as a bipolar junction transistor or a field-effect transistor.

References

Bogart, T. F., Beasley, J. S., and Rico, G. 2001. *Electronic Devices and Circuits*, 5th ed. Prentice Hall, Englewood Cliffs, NJ.

Boylestad, R. and Nashelsky, L. 2000. *Electronic Devices and Circuit Theory*, 8th ed. Prentice Hall, Englewood Cliffs, NJ.

Cathey, J. 2002. *Electronic Devices and Circuits*. McGraw-Hill, New York.

Dailey, D. 2001. *Electronic Devices and Circuits: Discrete and Integrated*. Prentice Hall, Englewood Cliffs, NJ.

Fleeman, S. 2003. *Electronic Devices*, 7th ed. Prentice Hall, Englewood Cliffs, NJ.

Floyd, T. L. 2002. *Electronic Devices*, 6th ed. Prentice Hall, Englewood Cliffs, NJ.

Hassul, M. and Zimmerman, D. E. 1996. *Electronic Devices and Circuits*. Prentice Hall, Englewood Cliffs, NJ.

Mauro, R. 1989. *Engineering Electronics*. Prentice Hall, Englewood Cliffs, NJ.

Millman, J. and Grabel, A. 1987. *Microelectronics*, 2nd ed. McGraw-Hill, New York.

Mitchell, F. H., Jr. and Mitchell, F. H., Sr. 1992. *Introduction to Electronics Design*, 2nd ed. Prentice Hall, Englewood Cliffs, NJ.

Muller, R. S., Kamins, T. I., and Chan, M. 2002. *Device Electronics for Integrated Circuits*, 3rd ed. John Wiley & Sons, New York.

Paynter, R. T. 1996. *Introductory Electronic Devices and Circuits*, 4th ed. Prentice Hall, Englewood Cliffs, NJ.

Roden, M. S. 2001. *Electronic Design*, 4th ed. Discovery Press, Los Angeles.

Sedra, A. S. and Smith, K. C. 1998. *Microelectronics Circuits*, 4th ed. Oxford University Press, Oxford.

Streetman, B. G. and Banerjee, S. 1999. *Solid State Electronic Devices*, 5th ed. Prentice Hall, Englewood Cliffs, NJ.

Williams, Gerald E. 1996. *Analog Electronics: Devices, Circuits and Techniques.* Delmar Learning, Clifton Park, NY.

118

Analog Integrated Circuits

Sidney Soclof

California State University, Los Angeles

An integrated circuit is an electronic device in which there is more than one circuit component in the same package. Most integrated circuits contain many transistors, together with diodes, resistors, and capacitors. Integrated circuits may contain tens, hundreds, or even many thousands of transistors. Indeed, some integrated circuits for computer and image-sensing applications may have millions of transistors or diodes on a single silicon chip.

A monolithic integrated circuit is one in which all of the components are contained on a single-crystal chip of silicon. This silicon chip typically measures from 1 × 1 × 0.25 mm thick for the smallest integrated circuits to 10 × 10 × 0.5 mm for the larger integrated circuits.

A hybrid integrated circuit has more than one chip in the package. The chips may be monolithic integrated circuits and separate or "discrete" devices such as transistor or diode chips. There can also be discrete passive components such as capacitor and resistor chips. The chips are usually mounted on an insulating ceramic substrate, usually alumina (Al_2O_3), and are interconnected by a thin-film or thick-film conductor pattern that has been deposited on the ceramic substrate. Thin-film patterns are deposited by vacuum evaporation techniques and are usually about 1 micrometer in thickness, whereas thick-film patterns are pastes that are printed on the substrate through a screen and usually range in thickness from 10 to 30 micrometers.

Integrated circuits can be classified according to function, the two principal categories being *digital* and *analog* (also called *linear*) integrated circuits. A digital integrated circuit is one in which all of the transistors operate in the switching mode, being either off (in the cut-off mode of operation) or on (in

0-8493-1586-7/05/$0.00+$1.50

the saturation mode) to represent the high and low (1 and 0) digital logic levels. The transistors during the switching transient pass very rapidly through the active region. Virtually all digital integrated circuits are of the monolithic type and are composed almost entirely of transistors, usually of the MOSFET type. Some digital integrated circuits contain more than one million transistors on a single silicon chip.

Analog or linear integrated circuits operate on signal voltages and currents that are in analog or continuous form. The transistors operate mostly in the active (or linear) mode of operation. There are many different types of analog integrated circuits, such as operational amplifiers, voltage comparators, audio power amplifiers, voltage regulators, voltage references, video (wide-bandwidth) amplifiers, radio-frequency amplifiers, modulators and demodulators for AM and FM, logarithmic converters, function generators, voltage-controlled oscillators, phase-locked loops, digital-to-analog and analog-to-digital converters, and other devices. The majority of analog integrated circuits are of the monolithic type, although there are many hybrid integrated circuits of importance.

In addition to the two basic functional categories of analog and digital integrated circuits, there are many integrated circuits that have both analog and digital circuitry in the same integrated circuit package, or even on the same chip. Some of the analog integrated circuits mentioned previously, such as the digital-to-analog and analog-to-digital converters, contain both types of circuitry.

Almost all integrated circuits are fabricated from silicon. The principal exceptions are some very high-speed digital integrated circuits that use gallium arsenide (GaAs) to take advantage of the very high electron mobility in that material. Integrated circuits are fabricated using the same basic processes as for other semiconductor devices. In integrated circuits the vast majority of devices are transistors and diodes, with relatively few passive components such as resistors and capacitors. In many cases no capacitors at all are used, and in some cases there are no resistors either. The active devices contained in an integrated circuit are transistors, including bipolar junction transistors (BJTs), junction field-effect transistors (JFETs), and metal-oxide silicon field-effect transistors (MOSFETs). Digital integrated circuits are made up predominantly of MOSFETs, with generally very few other types of components. Some analog integrated circuits use mostly, or even exclusively, BJTs, although many integrated circuits use a mixture of BJTs and field-effect transistors (either JFETs or MOSFETs), and some are even exclusively FETs, with no BJTs at all.

The sections that follow give very brief descriptions of some of the analog integrated circuits.

118.1 Operational Amplifiers

An operational amplifier is an integrated circuit that produces an output voltage, V_O, that is an amplified replica of the difference between two input voltages, as given by the equation $V_O = A_{OL}(V_1 - V_2)$, where A_{OL} is called the *open-loop gain*. The basic symbol for the operational amplifier is shown in Figure 118.1. Most operational amplifiers are of the monolithic type, and there are hundreds of different types of operational amplifiers available from dozens of different manufacturers.

The operational amplifier was one of the first types of analog integrated circuits developed. The term *operational amplifier* comes from one of the earliest uses of this type of circuit — in analog computers dating back to the early and middle 1960s. Operational amplifiers were used in conjunction with other circuit components, principally resistors and capacitors, to perform various mathematical operations,

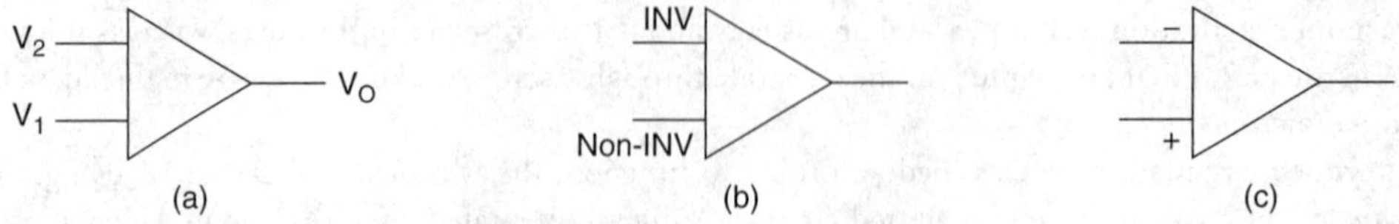

FIGURE 118.1 Operational amplifier symbols: (a) basic operational amplifier symbol; (b) symbol with input polarities indicated explicitly; (c) symbol with input polarities indicated explicitly. (*Source:* Soclof, S. 1991. *Design and Applications of Analog Integrated Circuits.* Prentice Hall, Englewood Cliffs, NJ.)

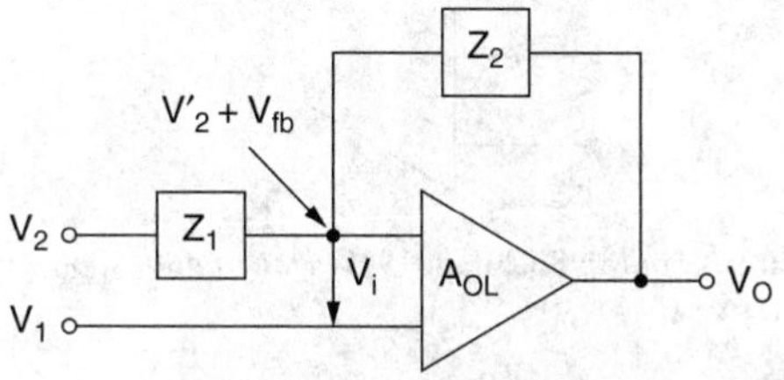

FIGURE 118.2 Closed-loop (negative feedback) operational amplifier system. (*Source:* Soclof, S. 1991. *Design and Applications of Analog Integrated Circuits.* Prentice Hall, Englewood Cliffs, NJ.)

such as addition, subtraction, multiplication, integration, and differentiation — hence the name "operational amplifier." The range of applications of operational amplifiers has vastly expanded since these early beginnings; operational amplifiers are now used to perform a multitude of tasks through the entire field of electronics.

Operational amplifiers are usually used in a feedback, or closed-loop, configuration, as shown in Figure 118.2. Under the assumption of a large open-loop gain, A_{OL}, the output voltage is given by

$$V_O = V_1[1 + (Z_2/Z_2)] - V_2(Z_2/Z_1).$$

The following is a list of some important applications of operational amplifiers:

Difference amplifier. This produces an output voltage proportional to the difference of two input voltages.

Summing amplifier. This produces an output voltage that is a weighted summation of a number of input voltages.

Current-to-voltage converter. This produces an output voltage that is proportional to an input current.

Voltage-to-current converter. This produces an output current that is proportional to an input voltage but is independent of the load being driven.

Active filters. This is a very broad category of operational amplifier circuit that can be configured as low-pass, high-pass, band-pass, or band-stop filters.

Precision rectifiers and clipping circuits. This is a broad category of wave-shaping circuits that can be used to clip off or remove various portions of a waveform.

Peak detectors. This produces an output voltage proportional to the positive or negative peak value of an input voltage.

Logarithmic converters. This produces an output voltage proportional to the logarithm of an input voltage.

Exponential or antilogarithmic converters. This produces an output voltage that is an exponential function of an input voltage.

Current integrator or charge amplifier. This produces an output voltage proportional to net flow of charge in a circuit.

Voltage regulators. This produces an output voltage that is regulated to remain relatively constant with respect to changes in the input or supply voltage and with respect to changes in the output or load current.

Constant current sources. This produces an output current that is regulated to remain relatively constant with respect to changes in the input or supply voltage and with respect to changes in the output or load impedance or voltage.

Amplifiers with electronic gain control. These are amplifiers in which the gain can be controlled over a wide range by the application of an external voltage.

Function generators. These are circuits that can be used to generate various types of waveforms, including square waves and triangular waves.

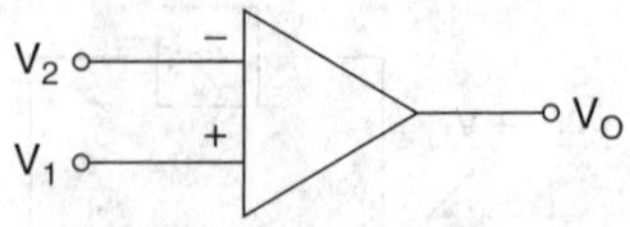

FIGURE 118.3 Voltage comparator. (*Source:* Soclof, S. 1991. *Design and Applications of Analog Integrated Circuits.* Prentice Hall, Englewood Cliffs, NJ.)

Clamping circuits. These circuits produce an output voltage that has the same AC waveform as the input single, but the DC level is shifted by an amount controlled by a fixed reference voltage.

Analog signal multiplexer. This circuit combines several input single for transmission over a signal communications link by means of *time-domain multiplexing.*

Sample-and-hold circuit. The input signal is sampled over a short period of time, and the sampled value is then held at that value until the next sample is taken.

Analog multiplier. This produces an output voltage proportional to the product of two input voltages.

118.2 Voltage Comparators

A voltage comparator is an integrated circuit as shown in Figure 118.3 that is used to compare two input voltages and produce an output voltage that is in the high (or "1") stage if $V_1 > V_2$ and in the low (or "0") stage if $V_1 < V_2$. It is essentially a one-bit analog-to-digital converter.

In many respects, voltage comparators are similar to operational amplifiers, and, indeed, operational amplifiers can be used as voltage comparators. A voltage comparator is, however, designed specifically to be operated under open-loop conditions, basically as a switching device. An operational amplifier, on the other hand, is almost always used in a closed-loop configuration and is usually operated as a linear amplifier.

Being designed to be used in a closed-loop configuration, the frequency response characteristics of an operational amplifier are generally designed to ensure an adequate measure of stability against an oscillatory type of response. This results in a sacrifice being made in the bandwidth, rise time, and slewing rate of the device. In contrast, since a voltage comparator operates as an open-loop device, no sacrifices have to be made in the frequency response characteristics, so a very fast response time can be obtained.

An operational amplifier is designed to produce a zero output voltage when the difference between the two input signals is zero. A voltage comparator, in contrast, operates between two fixed output voltage levels, so the output voltage is either in the high or low states.

The output voltage of an operational amplifier will saturate at levels that are generally about 1 or 2 V away from the positive and negative power supply voltage levels. The voltage comparator output is often designed to provide some degree of flexibility in fixing the high- and low-state output voltage levels and for ease in interfacing with digital logic circuits.

There are many applications of voltage comparators. These include pulse generators, square-wave and triangular-wave generators, pulse-width modulators, level and zero-crossing detectors, pulse regenerators, line receivers, limit comparators, voltage-controlled oscillators, analog-to-digital converters, and time-delay generators.

118.3 Voltage Regulators

A voltage regulator is an electronic device that supplies a constant voltage to a circuit or load. The output voltage of the voltage regulator is regulated by the internal circuitry of the device to be relatively independent of the current drawn by the load, the supply or line voltage, and the ambient temperature. A voltage regulator may be part of some larger electronic circuit but is often a separate unit or module, usually in the form of an integrated circuit. A voltage regulator, as shown in Figure 118.4, is composed of three basic parts:

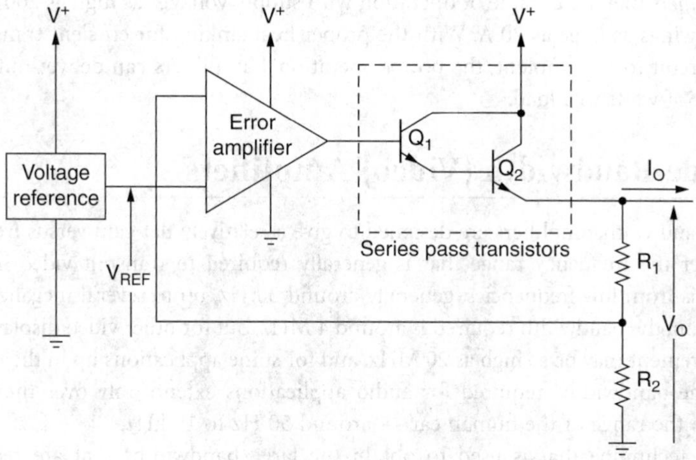

FIGURE 118.4 Voltage regulator: basic block diagram. (*Source:* Soclof, S. 1991. *Design and Applications of Analog Integrated Circuits.* Prentice Hall, Englewood Cliffs, NJ.)

A voltage reference circuit that produces a reference voltage that is independent of the temperature and supply voltage

An amplifier to compare the reference voltage with the fraction of the output that is fed back from the voltage regulator output to the inverting input terminal of the amplifier

A series-pass transistor or combination of transistors to provide an adequate level of output current to the load being driven

Voltage regulators usually include protection circuitry such as current limiting and thermal limiting to protect the integrated circuit against overheating and possible damage.

An important type of voltage regulator is the *switching-mode regulator,* in which the series-pass transistors are not on continuously but, rather, are rapidly switched from being completely on to completely off. The output voltage level is controlled by the fraction of time that the series-pass transistors are on (i.e., the duty cycle). Switching-mode regulators are characterized by having very high efficiencies, often above 90%.

118.4 Power Amplifiers

Although most integrated circuit amplifiers can deliver only small amounts of power to a load, generally well under 1 watt, there are integrated circuits that are capable of supplying much larger amounts of power, up in the range of several watts, or even several tens of watts. There are a variety of integrated circuit audio power amplifiers available that are used in the range of frequencies up to about 10 or 20 kHz for amplification of audio signal for delivery to loudspeakers. These integrated circuit audio power amplifiers also are used for other applications, such as relay drivers and motor controllers.

There are also available a variety of power operational amplifiers. The maximum current available from most operational amplifiers is generally in the range of about 20 to 25 mA. The maximum supply voltage rating is usually around 36 V with a single supply or +18 V and −18 V when a split supply is used. For an operational amplifier with a 36 V total supply voltage, the maximum peak-to-peak output voltage swing available will be around 30 V. With a maximum output current rating of 20 mA, the maximum AC power that can be delivered to a load is $P_L = 30\text{ V} \times 20\text{ mA}/4 = 150\text{ mW}$. Although this level of output power is satisfactory for many applications, a considerably larger power output is required for some applications.

Larger AC power outputs from operational amplifiers can be obtained by adding external *current boost* power transistors that are driven by the operational amplifier to the circuit. There are also available *power operational amplifiers* that are capable of operation with supply voltages as high as 200 V, and with peak output current swings as large as 20 A. With the proper heat sinking for efficient transfer of heat from the integrated circuit to the ambient, the power operational amplifiers can deliver output power up in the range of tens of watts to a load.

118.5 Wide-Bandwidth (Video) Amplifiers

Video, or wide-bandwidth, amplifiers are designed to give a relatively flat gain versus frequency response characteristic over the frequency range that is generally required to transmit video information. This frequency range is from low frequencies, generally around 30 Hz, up to several megahertz. For standard television reception, the bandwidth required is around 4 MHz, but for other video display applications the bandwidth requirement may be as high as 20 MHz, and for some applications up in the range of 50 MHz.

In contrast, the bandwidths required for audio applications extend only over the frequency range corresponding to the range of the human ear — around 50 Hz to 15 kHz.

The principal technique that is used to obtain the large bandwidths that are required for video amplifiers is the trading off of reduced gain in each amplifier stage for increased bandwidth. This tradeoff is accomplished by the use of reduced load resistance for the various gain stages of the amplifier and by the use of negative feedback. In many video amplifiers both techniques are employed. Adding additional gain stages can compensate for the reduction in the gain of the individual stages.

Included in the category of video amplifiers are the very-wide-bandwidth operational amplifiers. Most operational amplifiers are limited to a bandwidth of around 1 to 10 MHz, but there are wide-bandwidth operational amplifiers available that can be used up in the range of 100 to 200 MHz.

118.6 Modulators, Demodulators, and Phase Detectors

This is a category of integrated circuit that can be used to produce amplitude-modulated (AM) and frequency-modulated (FM) signals. These same integrated circuits can also be used for the demodulation, or detection, of AM and FM signals. In the AM case these integrated circuits can be used to generate and to demodulate *double-sideband/suppressed carrier* (DSB/SC) and *single-sideband/suppressed carrier* (SSB/SC) signals. Another application of this type of integrated circuit is as a *phase detector*, in which an output voltage proportional to the phase difference of two input signals is produced.

118.7 Voltage-Controlled Oscillators

A voltage-controlled oscillator (VCO) is an oscillator circuit in which the frequency of oscillation can be controlled by an externally applied voltage. VCOs are generally designed to operate over a wide frequency range, often with a frequency ratio of 100:1. One important feature that is often required for VCOs is a linear relationship between the oscillation frequency and the control voltage. Many VCOs have a maximum frequency of operation of around 1 MHz, but there are some emitter-coupled VCOs that can operate up to 50 MHz.

118.8 Waveform Generators

A waveform generator is an integrated circuit that generates the following three types of voltage waveforms: square waves, triangular waves, and sinusoidal waves. The square and triangular waves can be generated by the same type of circuits as used for VCOs.

For the generation of a sinusoidal waveform a feedback amplifier using an LC tuned circuit or an RC phase shift network in the feedback loop can be used. These feedback oscillators can produce a very low-distortion since wave, but it is difficult to modulate the oscillation frequency over a very wide range by means of a control voltage. The VCO, on the other hand, is capable of a frequency sweep ratio as large

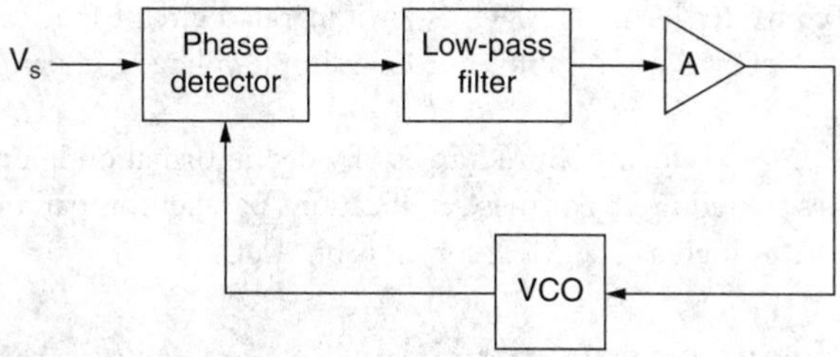

FIGURE 118.5 Phase-locked loop. (*Source:* Soclof, S. 1991. *Design and Applications of Analog Integrated Circuits.* Prentice Hall, Englewood Cliffs, NJ.)

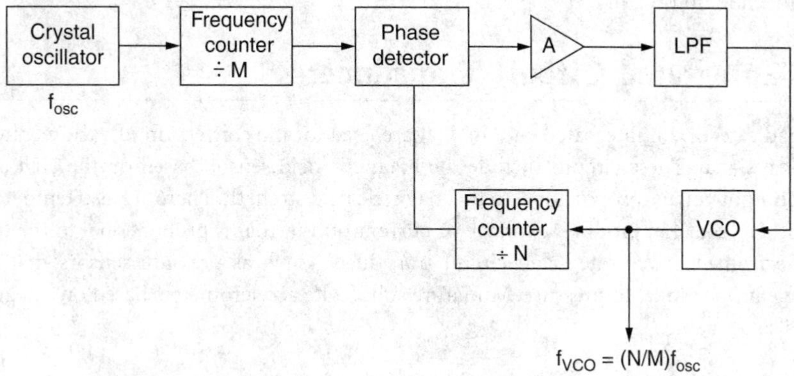

FIGURE 118.6 PLL frequency synthesizer. (*Source:* Soclof, S. 1991. *Design and Applications of Analog Integrated Circuits.* Prentice Hall, Englewood Cliffs, NJ.)

as 100:1, with very good linearity between the frequency and the control voltage. A sinusoidal waveform can be obtained from a VCO by using a wave-shaping network to convert the triangular wave output to a sine wave.

118.9 Phase-Locked Loops

A phase-locked loop (PLL) is a feedback loop comprising a phase detector, low-pass filter, and a voltage-controlled oscillator (VCO) as shown in Figure 118.5. When the PLL has locked in on a signal, the frequency of the VCO will exactly follow the signal frequency.

The PLL can be used as an FM demodulator or detector. In this case the VCO control voltage is proportional to the frequency deviation of the FM signal and represents the demodulated output voltage. Another closely related application is the demodulation of frequency-shift keying (FSK) signals, a process that is similar to FM except that the signal frequency is shifted between just two values.

On important application of PLLs is in frequency synthesis, in which a precise series of frequencies is produced, all derived from a stable crystal-controlled oscillator. In Figure 118.6, a PLL frequency synthesizer circuit is shown.

118.10 Digital-to-Analog and Analog-to-Digital Converters

A digital-to-analog converter (D/A or DAC) is an integrated circuit that converts a digital input signal to an analog output voltage (or current) that is proportional to the digital signal. DACs vary in resolution from 4 to 16 bits.

An analog-to-digital converter (A/D or ADC) is an integrated circuit that converts an analog input signal to a digital output. ADCs vary in resolution from a simple voltage comparator used as a 1-bit ADC to 16-bit ADCs.

For some applications, such as storing entire frames of video information in one-thirtieth of a second, very high conversion rates are required. For these applications parallel comparator (or "flash") ADCs are used with conversion rates as high as 500 MHz for an 8-bit ADC.

118.11 Radio-Frequency Amplifiers

There are radio-frequency (R-F) integrated circuits that use tuned circuits and operate as band-pass amplifiers. These are used in communications circuits, such as AM and FM radio, and in television for signal amplification and mixing.

118.12 Integrated Circuit Transducers

This is a broad category of integrated circuits that are used for the conversion of various physical inputs to an electrical signal. These circuits include the magnetic field sensor, based on the Hall effect, which produces an output voltage proportional to the magnetic field strength. There are also temperature sensor integrated circuits that can produce a voltage or current output that is proportional to the temperature. There are electromechanical integrated circuit transducers such as pressure sensors that produce an output voltage proportional to pressure. Miniature solid-state accelerometers based on integrated circuit sensors are also available.

118.13 Optoelectronic Devices

A very important category of integrated circuit transducers is that of the optoelectronic devices. These devices range from photodiode or phototransistor-amplifier modules to image sensors containing in the range of one million individual photodiode image-sensing elements. These image-sensing integrated circuit chips also include additional circuitry to properly transfer out (line by line, in a serial output) the information from the two-dimensional array of the image sensor.

There are also integrated circuits used with light emitting diodes (LEDs) and laser diodes for the generation of light pulses for optoelectronic communications systems, including fiber optic systems.

In the case of a fiber optic communications system, a small diameter glass fiber is used as a conduit or wave guide to guide a beam of a light from transmitter to receiver. Optoelectronic integrated circuits are used at the transmitting end with LEDs or laser diodes for the generation and emission of the optical signal. Optoelectronic integrated circuits are used at the receiving end with photodiodes or phototransistors for the detection, amplification, and processing of the received signal.

118.14 Recent Trends

The first integrated circuits were based on bipolar junction transistors. Present day digital integrated circuits are almost entirely composed of MOSFET devices, especially in the form of the CMOS configuration. There has also been the trend in analog integrated circuits to the use of more MOSFET devices, especially in the CMOS configuration. Analog integrated circuits can incorporate both bipolar and MOSFET devices on the same chip and are often referred to as BiCMOS circuits.

Advances in the processing technology of analog integrated circuits have led to circuits offering higher speeds, greater power outputs, and greater precision. Also mixed-signal integrated circuits with a combination of analog devices and digital devices have become more important, offering the possibility of having an entire electronic system on a chip.

Defining Terms

Hybrid integrated circuit — An electronic circuit package that contains more than one chip. These chips can be a mixture of monolithic ICs, diodes, transistors, capacitors, and resistors.

Integrated circuit — An electronic circuit package that contains more than one circuit element.

Monolithic integrated circuit — A single-crystal chip of a semiconductor, generally silicon, that contains a complete electronic circuit.

References

Franco, S. 2002. *Design with Operational Amplifiers and Analog Integrated Circuits*, 3rd ed. McGraw-Hill, New York.

Gray, P. R. and Meyer, R. G. 2001. *Analysis and Design of Analog Integrated Circuits*, 3rd ed. John Wiley & Sons, New York.

Irvine, R. G. 1987. *Operational Amplifiers — Characteristics and Applications.* Prentice Hall, Englewood Cliffs, NJ.

Kennedy, E. J. 1988. *Operational Amplifier Circuits.* Holt, Rinehart and Winston, New York.

Laker, K. and Sansen, W. 1994. *Design of Analog Integrated Circuits and Systems.* McGraw-Hill, New York.

McMenamin, J. M. 1985. *Linear Integrated Circuits: Operation and Applications.* Prentice Hall, Englewood Cliffs, NJ.

Michael, J. J. 1993. *Applications and Design with Analog Integrated Circuits*, 2nd ed. Prentice Hall, Englewood Cliffs, NJ.

Razavi, B. 2001. *Design of Analog CMOS Integrated Circuits.* McGraw-Hill, New York.

Rutenbar, R. A. 2002. *Computer-Aided Design of Analog Integrated Circuits and Systems.* John Wiley & Sons, New York.

Sedra, A. S. and Smith, K. C. 1998. *Microelectronic Circuits*, 4th ed. Oxford University Press, New York.

Seippel, R. G. 1983. *Operational Amplifiers.* Prentice Hall, Englewood Cliffs, NJ.

Serdijn, W. 1995. *Low-Voltage Low-Power Analog Integrated Circuits.* Kluwer, Dordrecht.

Soclof, S. 1991. *Design and Applications of Analog Integrated Circuits.* Prentice Hall, Englewood Cliffs, NJ.

119

Optoelectronic Devices

Sidney Soclof
California State University, Los Angeles

The subject of optoelectronics deals with the interaction of light and electronic devices. There are three basic categories of optoelectronic devices: (1) light sensing devices, such as photocells, photodiodes, and phototransistors, in which the incoming light produces an electrical output, (2) light emitting devices, such as light-emitting diodes and laser diodes in which an electrical input produce a light output, (3) power conversion devices, such as a solar cell, in which a light input, usually sunlight, is converted to electrical power.

119.1 Photoconductive Cells

The simplest type of optoelectronic device is the photoconductive cell that is based on the phenomenon of photoconductivity. Figure 119.1 presents a piece of semiconductor. Contacts are made to the two ends and a battery or DC power supply is connected. In response to the applied voltage V, a current I will flow. According to Ohm's Law, $I = V/R = V \cdot G$, where G is the conductance of the semiconductor material. The conductance G is related to the *conductivity*, σ, by $G = \sigma A/L$, where L = length and A = area. Now there will be light incident of this material. If the photons entering the semiconductor have energy, E_{PHOTON}, greater than the energy gap, E_G, then they will be absorbed. The photon energy will be transferred

0-8493-1586-7/05/$0.00+$1.50

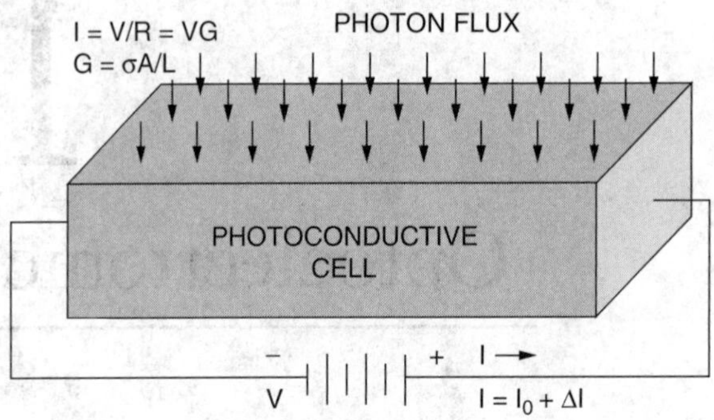

FIGURE 119.1 Photoconductive cell.

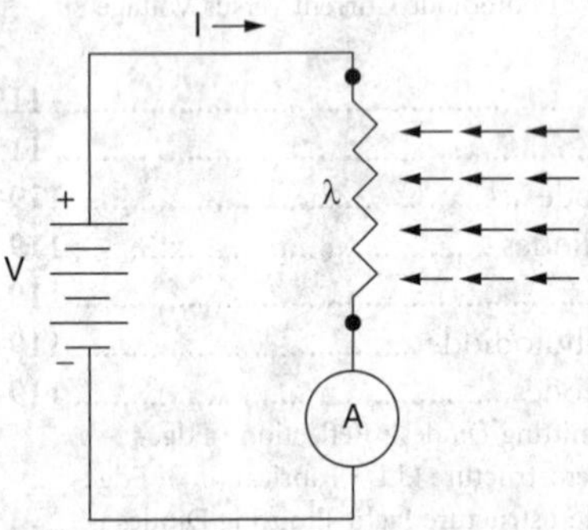

FIGURE 119.2 Simple light meter circuit.

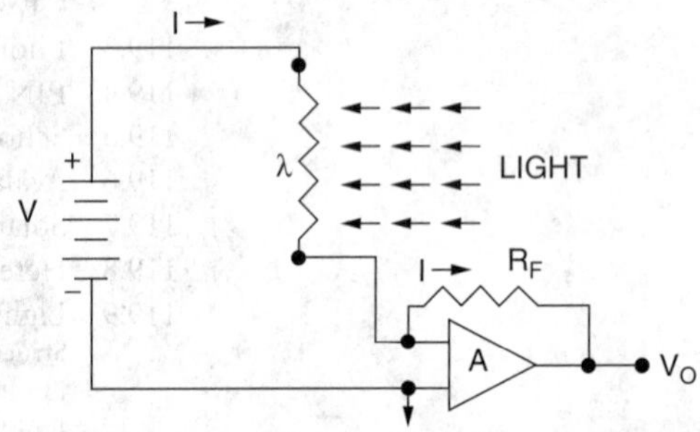

FIGURE 119.3 Light-sensing circuit using an operational amplifier.

to electrons in covalent bonds, allowing them to break loose, becoming free electrons, and leaving holes behind. This is the photogeneration process.

As a result of the increase in the free electron and hole population the conductivity will increase to become $\sigma = \sigma_o + \Delta\sigma$, where σ_o is the conductivity under dark conditions, and $\Delta\sigma$ is the increase in σ due to the photogeneration process. As a result of the $\Delta\sigma$, there will be an increase in the conductance, ΔG, and correspondingly an increase in the current flow, ΔI. This change in the conductivity due to the photogeneration process is the basis of the photoconductive cell. The photoconductive cell is also known as a photocell or photoresistor.

Figure 119.2 shows a very simple circuit using a *photocell.* The light shining on the photocell reduces its resistance, and the resulting current flow can be measured by a current meter or DMM. A resistor can be used to convert the current to a voltage, which can then be amplified by the amplifier. Another circuit shown in Figure 119.3 uses this *current-to-voltage* operational amplifier circuit. Current I will flow through the operational amplifier feedback resistor, R_F. If we assume that the operational amplifier has $I_{BIAS} << I$ and $A_{OL} >> 1$, then $V_O = -I \cdot R_F$, where $I = V \cdot G_{PHOTOCELL}$.

The most commonly used semiconductors for photoconductive cells are II–VI compound semiconductors such as cadmium sulfide (CdS), cadmium selenide (CdSe), and zinc sulfide (ZnS). The maximum spectral response of these materials is in the visible region of the spectrum, and they are often used for light meters for photographic purposes. They are also used for such applications as automatically turning on lights. A major disadvantage of these photoconductive cells is their very slow response times of milliseconds up to seconds. Consequently, they are not viable for high-speed applications such as optical communications. For long-wavelength detectors in the infrared region III-V, semiconductors with very small energy gaps such as InAs and InSb are used. For detection out into the far infrared region, detectors operating at cryogenic temperatures are used, such as germanium operating at liquid nitrogen (77°K) or liquid helium (4.2°K) temperatures.

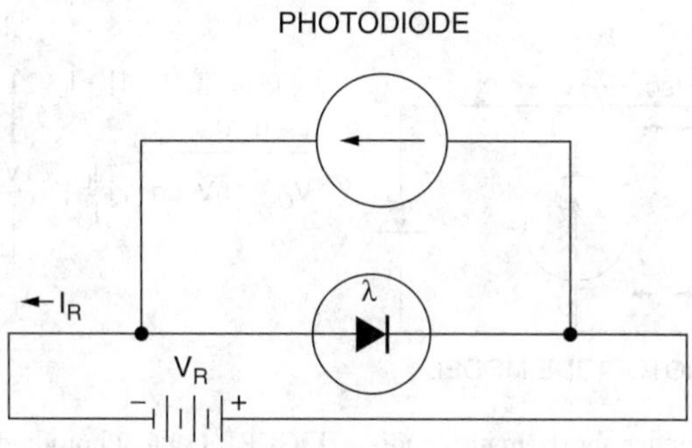

FIGURE 119.4 Circuit representation of a photodiode.

119.2 Photodiodes

In this section, we discuss what happens when light is incident on a PN junction diode. Photons penetrate into the semiconductor. If the photon energy, E_{PHOTON}, is greater than the energy gap, E_G, then there will be photon absorption and the photogeneration of electron–hole pairs. As the photons penetrate into the semiconductor, the photon flux will diminish due to the absorption of photons and the resulting photogeneration process. These are the *photogenerated minority carriers*. The minority carrier population on both sides is normally very small, but can be drastically increased as a result of the photogeneration process.

The photogenerated minority carrier electrons on the P-type side will diffuse to the junction, roll down the hill and be collected by the N-type side. Every electron collected by the N-type side will result in an electron being expelled from the N-type side into the metal wire and then flowing around the circuit.

The photogenerated minority carrier holes on the N-type side will diffuse to the junction, roll down the hill, and be collected by the P-type side. Every hole collected by the P-type side will result in an electron being drawn into the P-type side from the metal wire. This results in an electron flowing around the circuit. Note that the flow of the PMCs on both sides of the junction results in an external current flow in the same direction as the dark reverse current. The illumination of a photodiode can indeed result in a very large increase in the reverse current, I_R, by many orders of magnitude.

A photodiode is basically the same as an ordinary PN junction diode, except that it is placed in a package with a transparent window so that light can get in. It is often of much larger surface area than an ordinary diode to increase the collection of photons. A photodiode can be represented as a constant current source, I_{PHOTO}, in parallel with an ordinary PN junction diode as shown in Figure 119.4. The current I_L represents the photocurrent resulting from the photogeneration process. Under dark conditions, $I_{PHOTO} = 0$, and we have just the characteristics of an ordinary diode. However, under illuminated conditions, $I_{PHOTO} > 0$, the total reverse current, I_R, is now the dark reverse current plus I_{PHOTO}. The photocurrent, I_{PHOTO}, is directly proportional to the photon flux. That is the basic operation of the photodiode. Under forward bias conditions, a very large forward current will flow due to the majority carriers climbing the potential hill at the junction. Under dark reverse bias conditions, the current flow will be very small due to the very small density of minority carriers. However, upon illumination, the reverse current flow can be increased by many orders of magnitude. Next we will look at the equation for current as a function of voltage for the photodiode, and then look at curves of current versus voltage.

The diode equation is now modified by inclusion of I_L, giving $I = I_O[\exp(V/nV_T) - 1] - I_L$. Under short-circuit conditions, as shown in Figure 119.5, $V = 0$, so $I = I_{SC} = -I_L$; thus in these conditions, the photodiode will produce a current flow as a result of the photogeneration process. This current will be directly proportional to the incident photo flux, ϕ, or light intensity, so $I_{SC} = -I_L \propto \phi$.

Now consider open-circuit conditions with $I = 0$, so that $V = V_{OC}$ = *open-circuit voltage* as shown in Figure 119.6. Since $I = 0$, the photocurrent I_L must produce a forward current flow through the PN junction equal to I_L, and the diode equation can be written as $I_O[\exp(V/nV_T) - 1] - I_L = 0$. Solving for

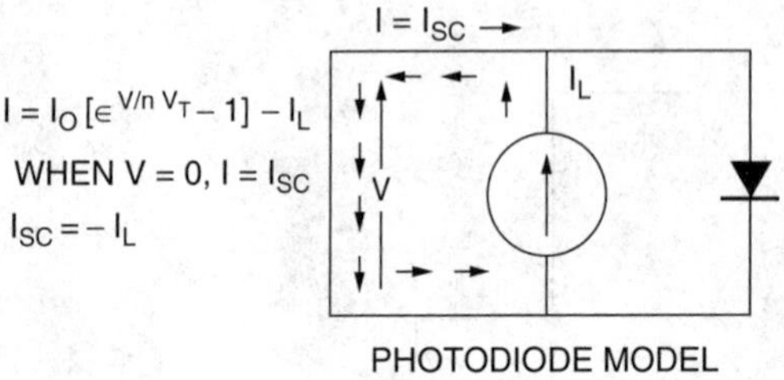

FIGURE 119.5 Photodiode under short-circuit conditions.

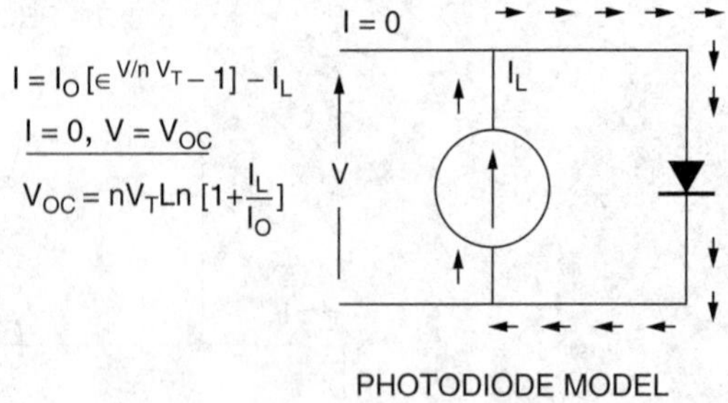

FIGURE 119.6 Photodiode under open-circuit conditions.

the exponential term gives exp $(V/nV_T) = 1 + I_L/I_O$, so that solving for the *open-circuit voltage* gives $V_{OC} = nV_T \cdot \ln[1 + (I_L/I_O)]$.

Since I_O is generally extremely small, often down in the femtoampere (10E-15 A) range, we usually will have $I_L >> I_O$. Thus, $V_{OC} \cong n \cdot \text{Ln}\,(I_L/I_O)$, that is, a logarithmic function of the illumination level. We see that the short-circuit current, I_{SC}, is a linear function of the light level, whereas the open-circuit voltage, V_{OC}, is a logarithmic function of the light level.

Current Responsivity

In this section, we look at current responsivity. Consider the case of an ideal photodetector such as an "ideal photodiode" as illustrated in Figure 119.7. Photon flux density, ϕ photons/sec-cm^2, is incident on the photodetector. The photons enter the semiconductor and are absorbed, producing photogenerated free electrons and holes. This results in a photocurrent, I_{PHOTO}, flowing around the circuit. For an ideal photodetector every incident photon results in the flow of one electron around the circuit.

Let the active area of the device be A cm^2. A photon flux of ϕ photons/sec-cm^2 will result in an electron flow of $\phi \cdot A$ electrons/sec around the circuit. This electron flow will correspond to a current flow of $I = q \cdot$ (electron flow) so $I = q \cdot \phi \cdot A$. The current responsivity is the ratio of the photocurrent, I, to the optical power, P, incident on the device active area. Thus current responsivity = I/P, where P = photon flux times the energy per photon. The energy per photon is $E_{PHOTON} = hc/\lambda$, so the incident optical power is $P = \phi \cdot A \cdot (hc/\lambda)$. We get this for the current responsivity. The ϕ and A terms in the numerator and denominator can be canceled out, resulting in this expression. The wavelength λ can be moved up to the numerator giving this equation. Since hc/q = 1.239 V-µm, the expression for the current responsivity can be written like this. We see that for the "ideal photodetector," the current responsivity is directly proportional to the wavelength.

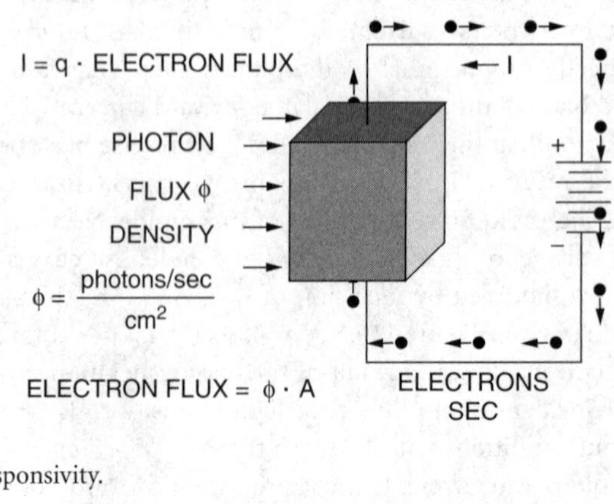

FIGURE 119.7 Current responsivity.

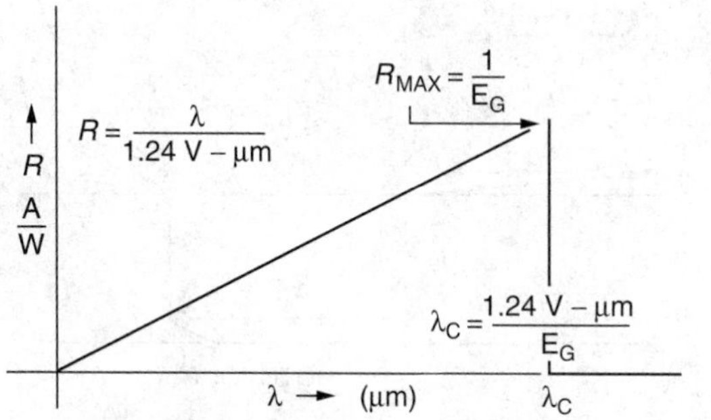

FIGURE 119.8 Current responsivity versus wavelength for ideal photodiode.

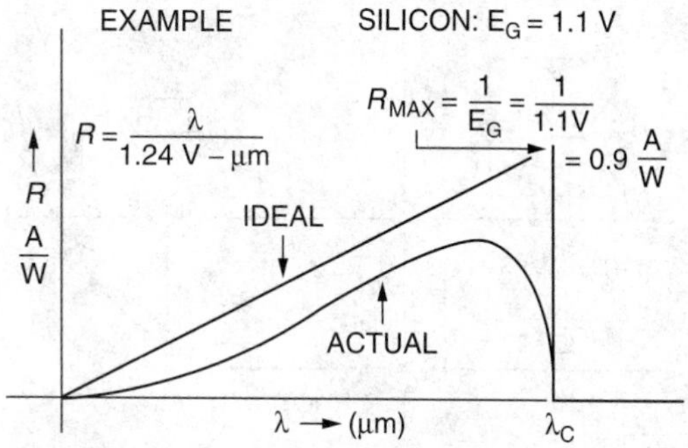

FIGURE 119.9 Current responsivity versus wavelength for real photodiode.

We will now look at a graph of current responsivity versus wavelength, λ, as shown in Figure 119.8. We note that if λ is expressed in units of μm, the current responsivity will have units of reciprocal volts (1/V). Since watts = volts · amperes, the units of 1/volts = amperes/watts = A/W. For the "ideal photodetector," a straight line like this is obtained.

For any real semiconductor photodetector, there will be no photon absorption when $E_{PHOTON} < E_G$. This corresponds to $\lambda > \lambda_C$, the *critical wavelength*. Beyond $\lambda = \lambda_C$, there will be no photon absorption and no photogeneration will be possible, and thus the current responsivity will be zero. For the ideal photodetector, the maximum current responsivity will be right at $\lambda = \lambda_C$.

At $\lambda = \lambda_C$, $E_{PHOTON} = E_G = 1.24$ V-μm/λ_C, so that $\lambda_C = 1.24$ V-μm/E_G. Thus, λ_C/1.24 V-μm = 1/E_G, so the maximum current responsivity = 1/E_G. For the case of silicon, the energy gap is E_G = 1.1 V. Thus the maximum current responsivity = 1/E_G = 1/(1.1 V), which gives a value of 0.9/V = 0.9 A/W for the maximum current responsivity.

The typical *spectral response* curve for an actual silicon photodiode is shown in Figure 119.9. Due to various loss factors, the current responsivity of an actual photodiode will be substantially less than that of the ideal device. The maximum current responsivity for a silicon photodiode is typically around 0.5 A/W. The loss factors include the reflection of some of the incident photons at the photodiode surface, incomplete absorption of photons, especially at the longer wavelengths, and the recombination of the photogenerated minority carriers.

Photodiode Current versus Voltage Curves

The focus of this section is the current versus voltage characteristics of photodiodes. Figure 119.10 presents the current and voltage axes and reference directions of voltage and current. The current versus

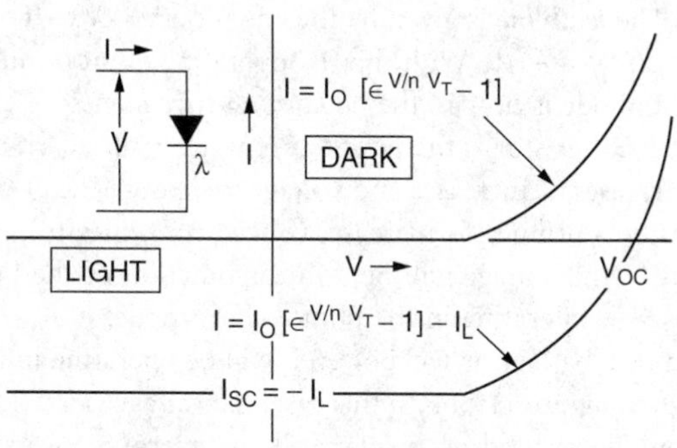

FIGURE 119.10 Photodiode current versus voltage curves.

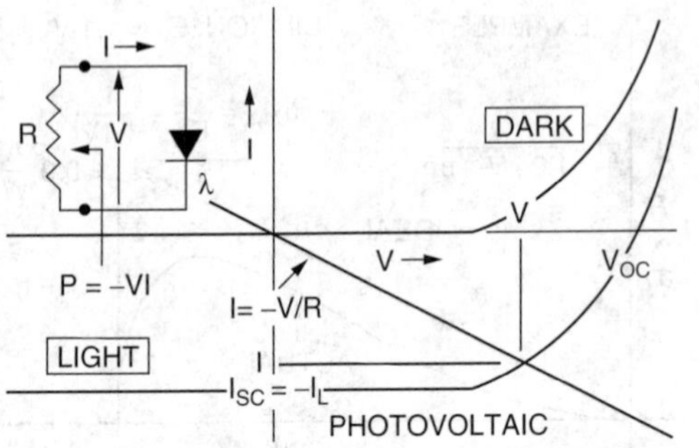

FIGURE 119.11 Photodiode current versus voltage curves with load line.

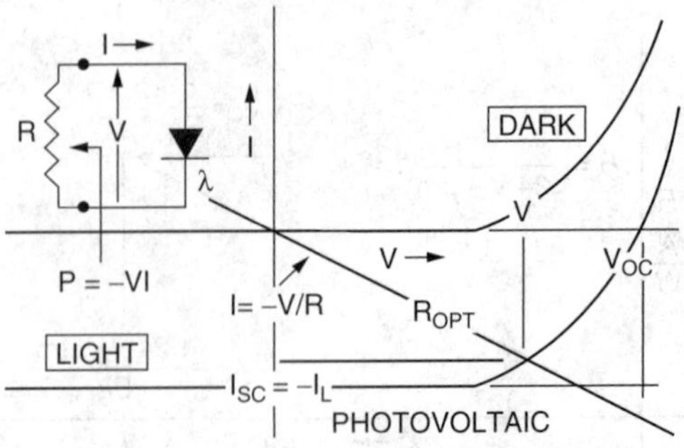

FIGURE 119.12 Photovoltaic mode peak power point and power output.

voltage curve under dark conditions is the same as for an ordinary diode. Under illuminated conditions the curve shifts down by an amount equal to the photocurrent, I_L. Under short-circuit conditions, $V = 0$. The current when $V = 0$ is $I(V = 0) = I_{SC} = -I_L$. Under open-circuit conditions, $I = 0$ and $V(I = 0) = V_{OC} = n \cdot V_T \cdot \text{Ln}[\,(I_L/I_O) + 1]$. In Figure 119.11, a load resistor, R, is connected across the photodiode. The *load line* is given by $I = -V/R$, and thus has a *slope* $= -1/R$. The photodiode is now operating in the photovoltaic mode, because it is generating a voltage that drives the current around the loop. The intersection of the *load line* with the current versus voltage curve is the *operating point*. The power delivered to the load is $P = -V \cdot I$. Since $V > 0$ and $I < 0$, $P > 0$, so in the *photovoltaic mode* with the current versus voltage curve in the fourth quadrant, the photodiode converts light energy to electrical energy. This is photovoltaic energy conversion, and is the basis of the solar cell, which is a large area photodiode designed for the conversion of sunlight to electricity.

The area of this rectangle is equal to the power delivered to the load. If a smaller value for R is used, the load line slope will be larger and since the current is approximately equal to the short-circuit current, I_{SC}, but if the voltage is smaller, the power output will be smaller. If a larger value for R is used, the load line slope will be smaller and since voltage is approximately equal to the open-circuit voltage, V_{OC}, but if the current is smaller, the power output will be smaller.

The optimum load resistance, R_{OPT}, is one that results in the maximum output power. The load line for R_{OPT} intersects the current versus voltage curve at the peak power point. The value of R_{OPT} will change as the illumination level changes.

The "ideal" current versus voltage curve for a solar cell would have a rectangular shape as shown in Figure 119.12. The output power would then be $P = V_{OC} \cdot I_{SC}$. The actual output power when $R = R_{OPT}$ will be $P = V \cdot I$. The *curve factor* $= (V \cdot I)/(V_{OC} \cdot I_{SC})$, and is typically around 0.6. The photodiode or solar cell is now operating in the fourth quadrant in the *photovoltaic mode*.

The photodiode can be biased for operation in the third quadrant by a battery or DC power supply as shown in Figure 119.13. The *load line* is given by the equation $V = V_B - I \cdot R$. Thus, $V = V_B$ when $I = 0$, and the slope is given by $dI/dV = -1/R$. With the DC operating point or quiescent point (Q-point) in the third quadrant, the photodiode is now in the *photoconductive mode*.

The current through the load resistor is the photocurrent, I_L. Under dark conditions, $I_L = 0$ and V = VB. As the light intensity increases, I_L increases and voltage increases. The Q-point moves along the *load line*, and as the light intensity continues to increase, voltage continues to increase. Since I_L is a linear function of the light intensity, the voltage will be a linear function of the light intensity. When VB is removed, the photodiode is now operating in the photovoltaic mode.

If the diode is not biased by a battery or DC power, it will be operating in the fourth quadrant in the photovoltaic mode as shown in Figure 119.14. In this case, the voltage is not a linear function of the light intensity. When the photodiode is used as a solar cell for energy conversion, it is operated in the photovoltaic mode. However, when the photodiode is used for light sensing and optical communications, it is usually best to operate it in the photoconductive mode. This is because of the linear voltage versus

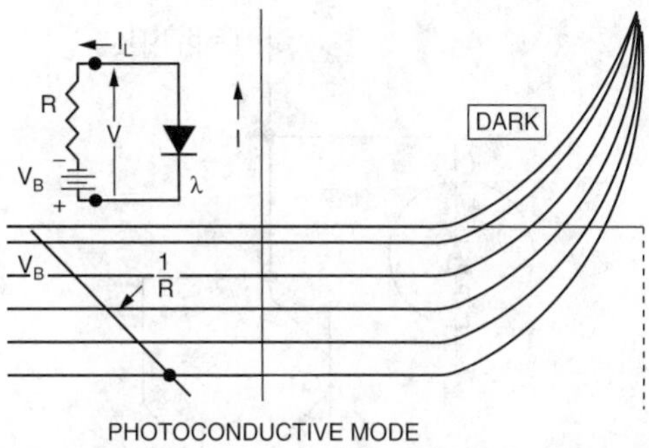

FIGURE 119.13 Photodiode current versus voltage curves — photoconductive mode.

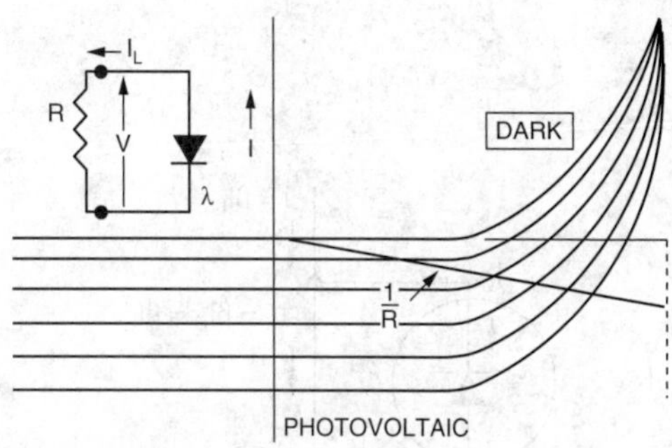

FIGURE 119.14 Photodiode current versus voltage curves — photovoltaic mode.

light intensity transfer relationship. Also, the reverse bias voltage across the PN junction reduces the junction capacitance, so that a faster response time and a wider bandwidth is available.

119.3 Phototransistors

Now we will look at a *phototransistor*, which is an integrated combination of a photodiode and a transistor as shown in Figure 119.15. This combination of the photodiode and transistor constitutes a phototransistor. The photodiode generates a photocurrent, I_L. If $V_{CE} > +0.2$ V, the transistor will be in the active region, and $I_C = \beta \cdot I_B$. Since $I_B = I_L$, $I_C = \beta \cdot I_B = \beta \cdot I_L$, as shown in Figure 119.16.

The current flowing through the phototransistor is $I = I_C + I_L = \beta \cdot I_L + I_L$.

This can be written as $I = (\beta + 1)\, I_L$. Since $\beta >> 1$, this can be expressed as $I \cong \beta \cdot I_L$, as shown in Figure 119.17. The current produced by a phototransistor is thus greater than that produced by a photodiode by a factor of the transistor current gain, β, and since $\beta \sim 100 - 300$ for phototransistors, this can represent a very large gain indeed. Whereas a photodiode may have responsivity of about 0.5 mA/mW at 0.9 μm, a phototransistor may have responsivity up in the 50 to 150 mA/mW range.

Now let us consider a *photodarlington*, which is an integrated combination of a photodiode and a Darlington transistor, as shown in Figure 119.18. The photocurrent I_L is multiplied by the net current gain of the Darlington pair, which is approximately $\beta_1 \cdot \beta_2$. Since $\beta \sim 100$, this current gain will be of the order of 10,000. Thus, current responsivity values of the order of 5 mA/mW are possible.

Regarding phototransistor structure, starting with an N-type substrate, there is a P-type (boron) diffusion into the substrate. Then there is an N^+ diffusion producing the result shown in Figure 119.19.

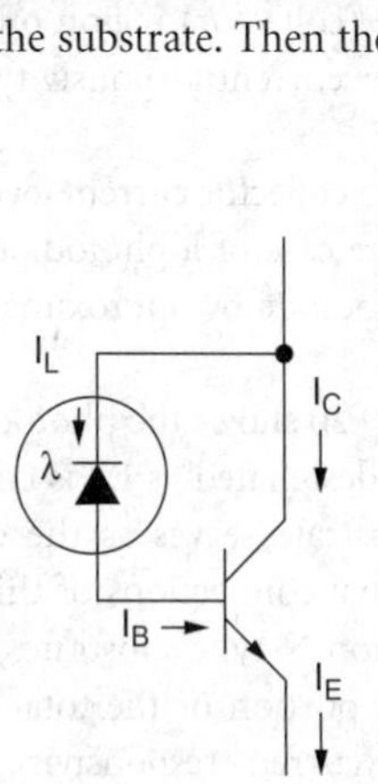

FIGURE 119.15 Phototransistor.

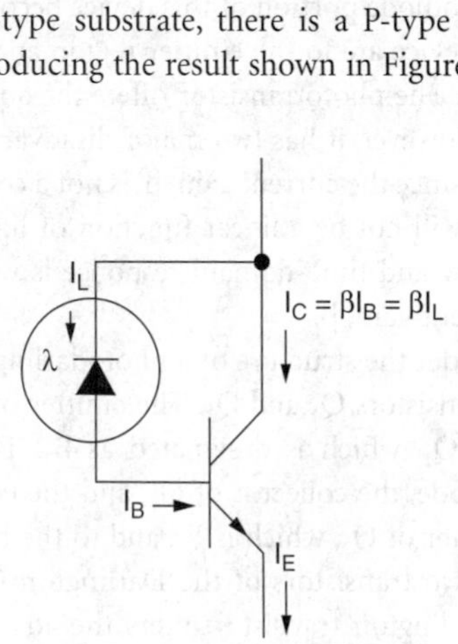

FIGURE 119.16 Phototransistor — internal current gain.

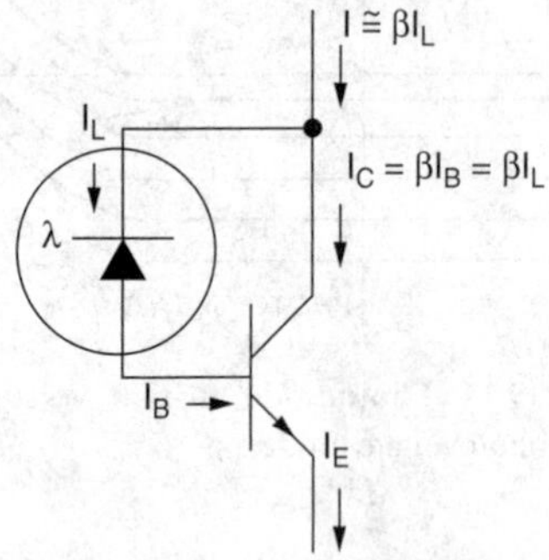

FIGURE 119.17 Phototransistor — current gain.

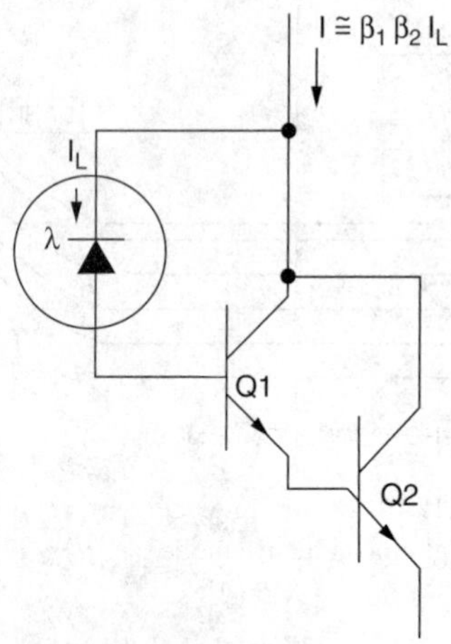

FIGURE 119.18 Photodarlington.

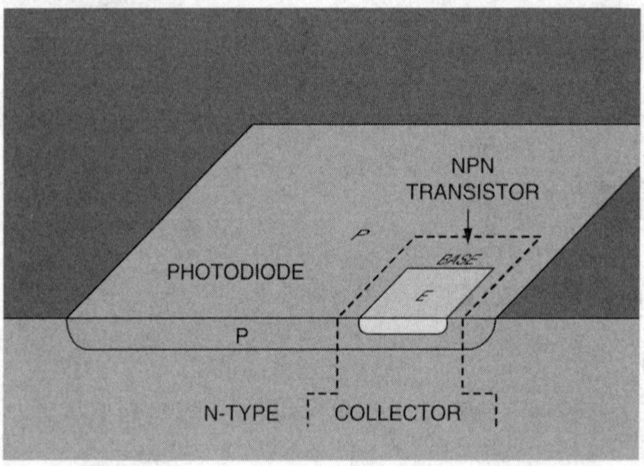

FIGURE 119.19 Phototransistor structure.

This is an integrated combination of a photodiode and a transistor. The photocurrent produced by the photodiode portion of this device becomes the base current of the transistor. The two contacts made to this device are to the emitter region and to the N-type substrate (collector) region on the bottom of the chip. The phototransistor offers the advantage of a much higher current responsivity than a photodiode. However, it has two major disadvantages.

First, since the current gain, β, is not a constant, but does vary with collector current level, the resulting current will not be a linear function of light intensity, as it is in the case of a photodiode. Second, the frequency and time-domain response is worse than that of a photodiode by approximately a factor of the current gain, β.

Consider the structure of a photodarlington transistor. Figure 119.20 shows the photodarlington with NPN transistors Q_1 and Q_2. The emitter of transistor Q_1, which is designated as E_1, is connected to the base of Q_2, which is designated as B_2. The common N-type substrate serves as the cathode of the photodiode, the collector of Q_1, and the collector of Q_2. The external connections of this device are to the emitter of Q_2, which is E_2, and to the bottom side of the common N-type substrate.

The two transistors of the Darlington pair occupy only a small portion of the total chip area. The photodarlington transistor offers the advantage of a much higher current responsivity than that of a photodiode. However, as in the case of the phototransistor, it has two major disadvantages. First, since both β_1 and β_2 are not constants, but do vary with collector current level, the resulting current will not

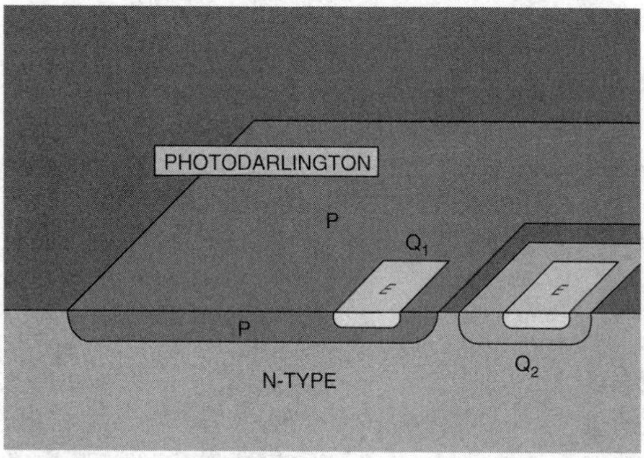

FIGURE 119.20 Photodarlington structure.

be a linear function of light intensity, as it is in the case of a photodiode. Second, the frequency and time-domain response is much worse than that of a photodiode by approximately a factor of the net current gain, $\beta_1 \cdot \beta_2$.

119.4 PIN Photodiode

First, we will look at the structure of an "ordinary" photodiode. Openings are etched in the oxide layer as shown in Figure 119.21. A donor dopant such as phosphorus is deposited into the silicon surface. The donors are then diffused into the silicon at a high temperature (~1000 to 1200°C). The openings in the oxide layer are reoxidized during this process producing the result shown in Figure 119.22. Openings are etched in the oxide again. A metal film is deposited on the wafer surface and then etched to form this contact area.

The PIN photodiode is designed for very high-speed operations. In this case, a heavily doped P+ substrate is used, upon which a lightly doped P-type layer is deposited, as shown in Figure 119.23. This lightly doped P-type layer is called an "epitaxial layer" since it is a crystallographic continuation of the substrate. The thickness of the epitaxial layer is designated as t_{EPI}, as shown in Figure 119.24, and is generally in the range of 10 μm to 50 μm.

A reverse bias voltage, V_R, is applied across the PN junction causing an increase in the width of the depletion region. With a large enough reverse bias voltage, the depletion region can extend all of the way across the lightly doped epitaxial layer as is Figure 119.25. Now the mobile charge carriers are rapidly swept out of this region due to the strong electric field. This region thus acts as an "insulating region," hence the "I" in the name of the PIN diode. The width of this "fully depleted" region is $W = t_{EPI} - x_J$. The device capacitance under these "fully depleted" conditions is at a minimum given by $CMIN = \varepsilon \cdot A_J/W$, as shown in Figure 119.26. With a thick epitaxial layer, this capacitance can be made very small, and the small capacitance leads to a faster response time and a wider bandwidth.

Now let us consider what happens with photons incident on the PIN photodiode. The photons penetrate into the semiconductor and are absorbed. This results in the photogeneration of minority carriers as shown in Figure 119.27. The photogenerated minority carriers (electrons) in the epitaxial layer will flow toward the PN junction under the influence of the strong electric field in this region as in Figure 119.28. These photogenerated minority carriers will travel at a relatively high speed toward the PN junction due to this electric field, so not many will be lost due to recombination. As a result, the collection efficiency will be high. In addition, the transit time will be very small, which is another reason for this being a very fast device. Thus, the PIN photodiode will have a very fast response time due to the small junction capacitance and the short transit time of the photogenerated minority carriers. It will also have an improved response at longer wavelengths due to the improved collection efficiency.

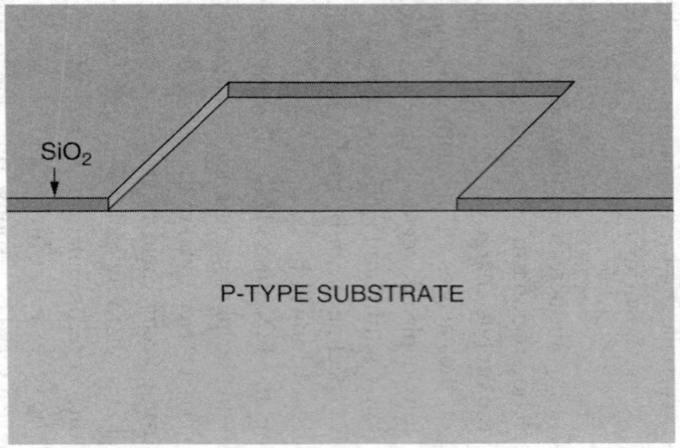

FIGURE 119.21 Photodiode structure.

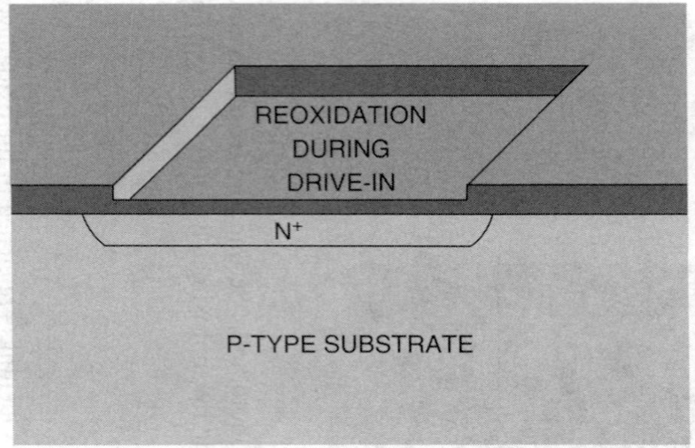

FIGURE 119.22 Photodiode structure.

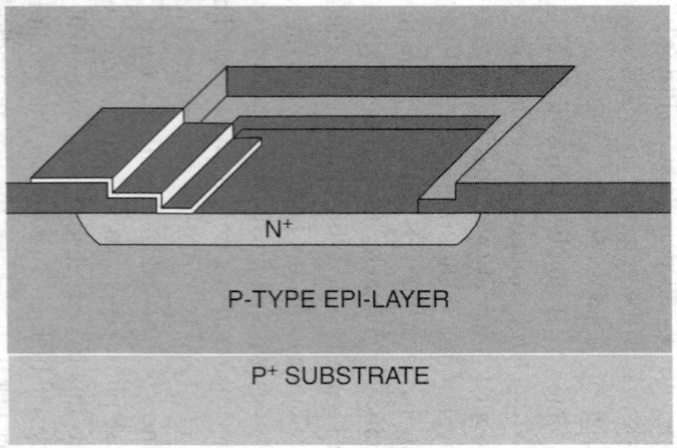

FIGURE 119.23 PIN photodiode.

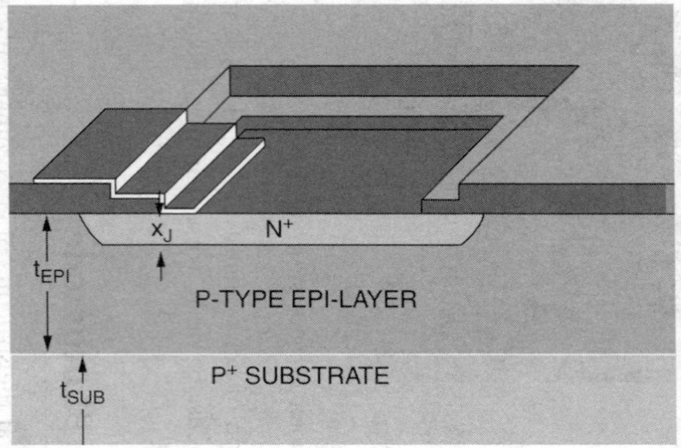

FIGURE 119.24 PIN photodiode epitaxial layer thickness.

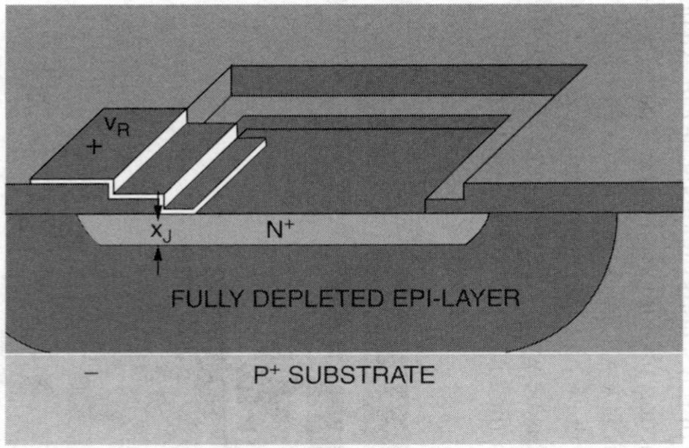

FIGURE 119.25 PIN photodiode — fully depleted epitaxial layer.

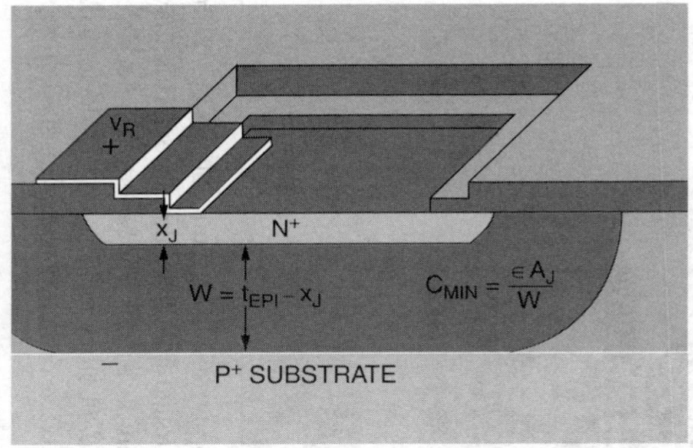

FIGURE 119.26 PIN photodiode — capacitance.

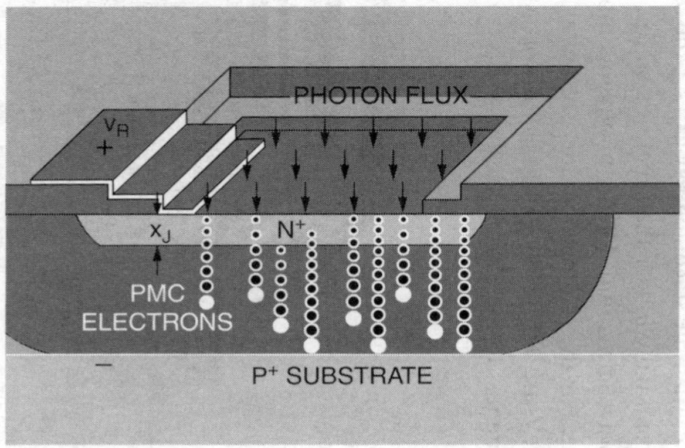

FIGURE 119.27 PIN photodiode — photon absorption and photogeneration.

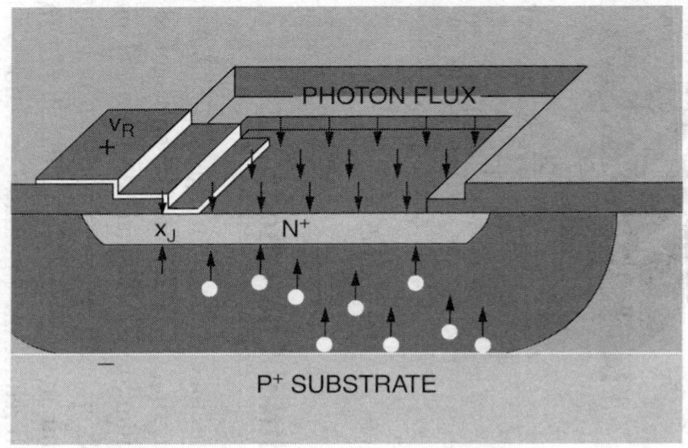

FIGURE 119.28 PIN photodiode — collection of photo-generated minority carriers.

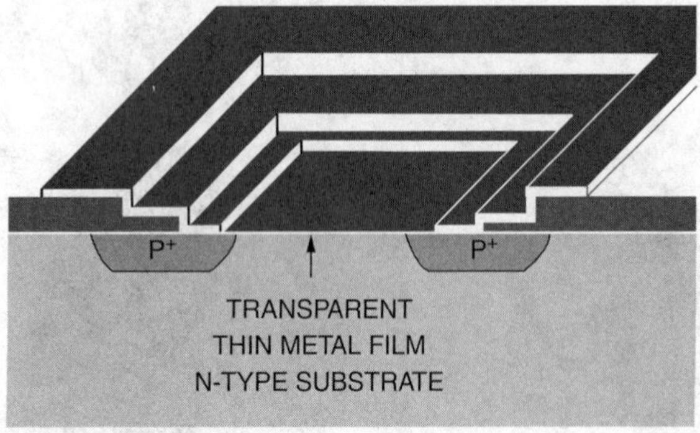

FIGURE 119.29 Shottky-barrier photodiode.

119.5 Schottky Photodiode

The Schottky photodiode is shown in Figure 119.29. A very thin film of metal, about 150 angstroms in thickness, is thin enough to allow most of the light to pass through and enter the silicon. This thin transparent metal film forms the metal side of a metal-silicon Schottky barrier. The Schottky photodiode often uses an N/N^+ epitaxial structure as shown in Figure 119.30. Photons incident on this device will pass through the thin metal film and enter the silicon. An antireflective coating is often deposited on top of the metal film to reduce the amount of reflection.

The absorption of photons results in the photogeneration of free electrons and holes. The photogenerated holes then diffuse toward the metal-semiconductor Schottky barrier. As the photogenerated holes enter the metal, they immediately recombine with electrons. Other electrons flow into the metal layer to replace those lost to recombination with the holes, and this constitutes the photocurrent. Another process that can occur is the *photoemission* of electrons from the metal into the semiconductor. For this to happen the electrons must be given enough energy to surmount the potential barrier of height ϕ_B at the metal-silicon interface.

Since ϕ_B is smaller than the silicon energy gap E_G, the response of the Schottky photodiode can extend to wavelengths substantially longer than that of a silicon PN junction photodiode. For example, if ϕ_B = 0.65 eV, the long wavelength limit for the photoemission process is λ = 1.24 eV-μm/ϕ_B = 1.24 eV-μm/0.65 eV = 1.91 μm. For the photogeneration process in silicon, the long wavelength limit is λ_C = 1.24 eV-μm/E_G = 1.24 eV-μm/1.1eV = 1.13 μm. Thus, for wavelengths between 1.13 μm and 1.91 μm, the response will be due just to the photoemission of electrons from the metal to the semiconductor.

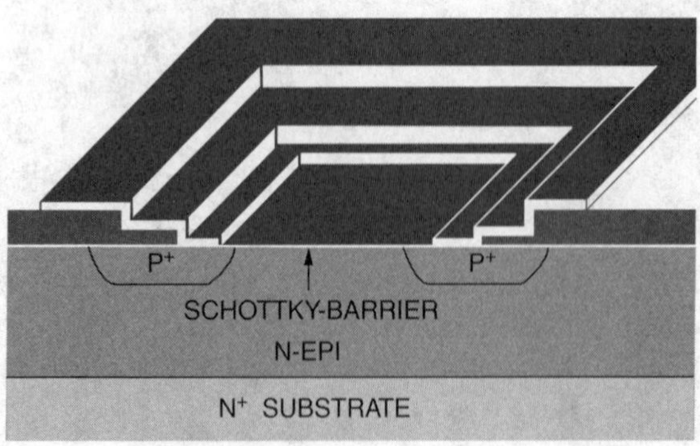

FIGURE 119.30 Shottky-barrier photodiode with epitaxial layer.

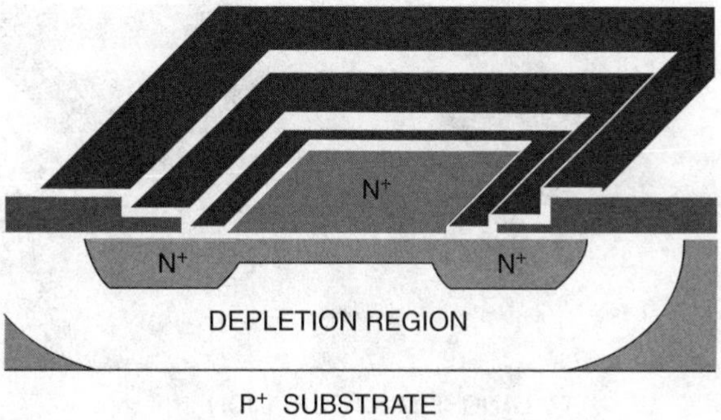

FIGURE 119.31 Avalanche photodiode.

119.6 Avalanche Photodiodes

Now we examine an *avalanche photodiode* or *APD*. We start off with a P-type epitaxial layer on a heavily doped P+ substrate; this deep N^+ guard ring is then diffused, followed by the diffusion of the much shallower N^+ region. The finished device with the cathode contact metallization is shown in Figure 119.31.

The APD is usually operated under a relatively large reverse bias voltage in the 30 to 50 V range, such that the epitaxial layer is fully depleted, as shown in Figure 119.32. There is a strong electric field in this depletion region. Upon illumination with photons of energy greater than the energy gap E_G, there will be the photogeneration of free electrons and holes in the depletion region. These photogenerated charge carriers will then be accelerated by the strong electric field in this region. If the electric field is large enough, the photogenerated charge carriers will produce additional free electrons and holes by the impact ionization process.

In the absence of the N^+ guard ring there would be a strong intensification of the electric field around the periphery of the PN junction where there is a high degree of curvature. This is shown in Figure 119.33. As a result, most of the avalanche multiplication would occur in this small region rather than over the much larger flat area of the junction. The purpose of the more deeply diffused N^+ guard ring is to reduce the electric field intensification in this region, and thus prevent the avalanche multiplication occurring here. Now the avalanche multiplication occurs over the larger flat area of the PN junction.

The junction area of the APD is generally kept small to reduce the probability of crystallographic defects that could result in the premature breakdown of the junction at a reduced voltage. This would cause the breakdown voltage to be below the value required for a uniform avalanche multiplication over the entire flat area of the junction. The junction diameter is typically in the 10 to 100 mm range, and this also helps to reduce the junction capacitance for high-speed performance. The avalanche photodiode is thus a

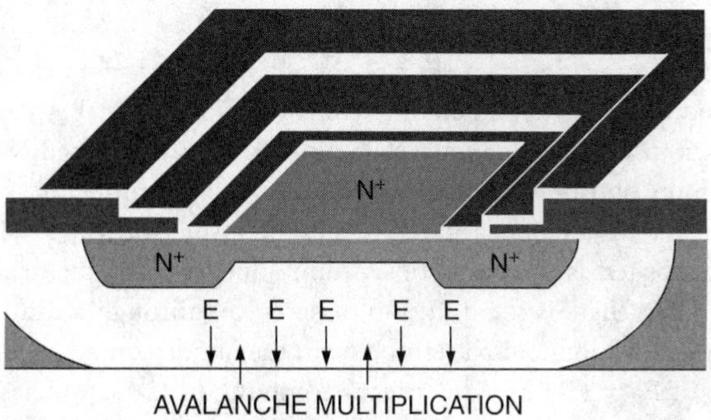

FIGURE 119.32 Avalanche photodiode — carrier multiplication.

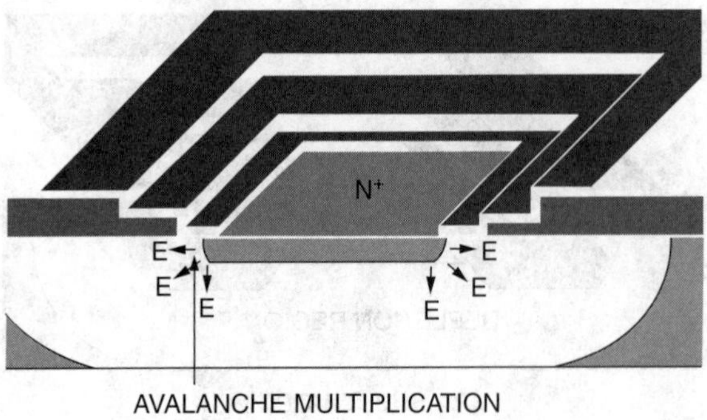

FIGURE 119.33 Avalanche photodiode — electric field intensification.

photodiode with a built-in gain, which is the avalanche multiplication factor, M, which can be in excess of 100. The principal advantages of the avalanche photodiode are high gain and speed, with gain-bandwidth products in excess of 100 GHz, and it is especially useful for fiber optic communication systems.

The principal disadvantage of the avalanche photodiode is the high noise generated as a result of the random nature of the avalanche multiplication process.

For a high gain, the avalanche photodiode must be biased at a voltage that is close to the junction breakdown voltage. As a result, a very stable, temperature-compensated biasing source is required, generally up in the 30 to 50V range.

119.7 Solar Cells

Solar cells are photodiodes made specifically to convert solar energy to electrical energy. The process is called *photovoltaic energy conversion*. Conversion efficiencies for silicon solar cells are typically in the 15% to 20% range, and GaAs cells can offer efficiencies close to 25%.

The construction of typical silicon solar cell is shown in Figure 119.34. Phosphorus diffusion is used to produce a thin N^+ diffused layer on a P-type substrate. The N^+ diffused layer is very thin, only about 1 μm in thickness. Then metal grid lines and a contact strip to collect the current are deposited as shown in Figure 119.35.

The top surface metallization occupies only about 5% of the top surface area, so that most of the silicon is exposed to the light. The bottom surface is completely covered by metal. There will be a substantial amount of incident light reflected from the silicon surface. To minimize this light reflection, a thin antireflective coating is deposited on the solar cell. The solar cell has an area in the range of 1 cm^2 up to 10 cm^2.

119.8 Heterostructure Photodiode

A heterostructure photodiode uses more than one semiconductor material. An example of a heterostructure photodiode is shown in Figure 119.36. We start off with a heavily doped N-type indium phosphide (N^+ InP) substrate. Indium phosphide is a III-V compound semiconductor. A more lightly doped N-type indium phosphide (N InP) epitaxial layer is then deposited on the N^+ InP substrate. The next epitaxial layer that is deposited is a ternary III-V compound semiconductor, N-type indium gallium arsenide, $In_{0.53}Ga_{0.47}As$. Then there is the diffusion of acceptors through a diffusion mask to produce a P+ region of InGaAs. Metallization and patterning form the anode contact of this photodiode. The InP layer has an energy gap of $E_G = 1.35$ eV. The ternary compound $In_{.53}Ga_{.47}As$ has a much smaller energy gap of $E_G = 0.75$ eV. The bottom contact metallization is deposited and patterned to have an opening or window for light to enter the device, as shown in Figure 119.37. The dimensions of these diagrams are not to scale, and in actuality the N^+ InP substrate is much thicker than the InP and InGaAs epitaxial layers.

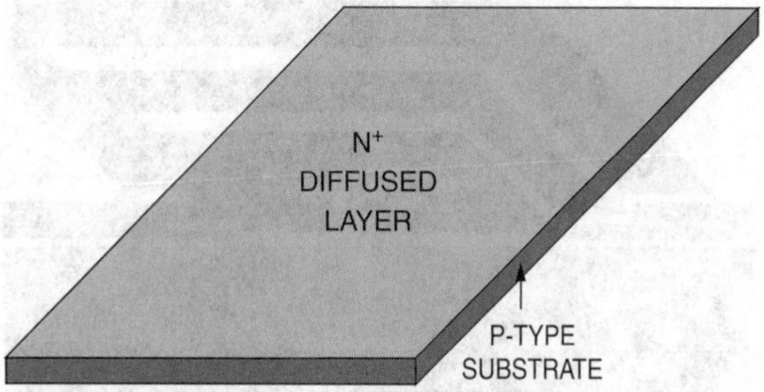

FIGURE 119.34 Solar cell.

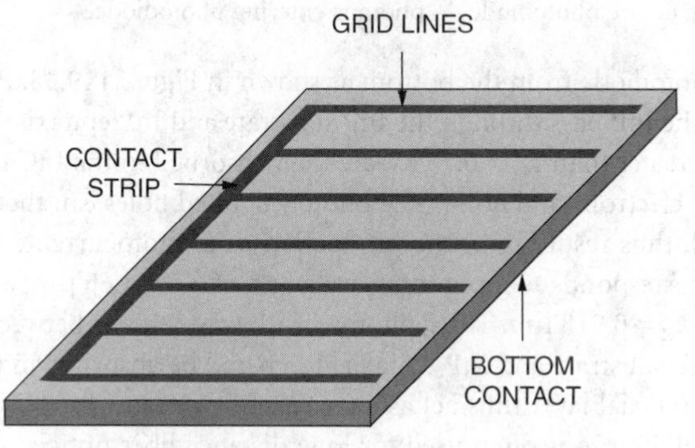

FIGURE 119.35 Solar cell — grid lines.

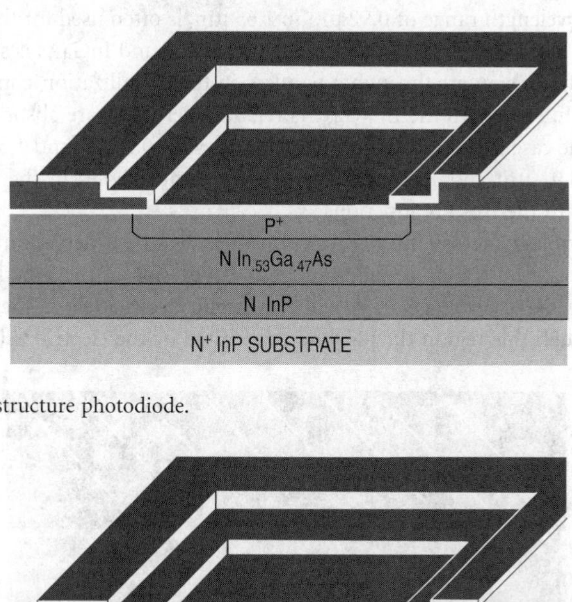

FIGURE 119.36 Heterostructure photodiode.

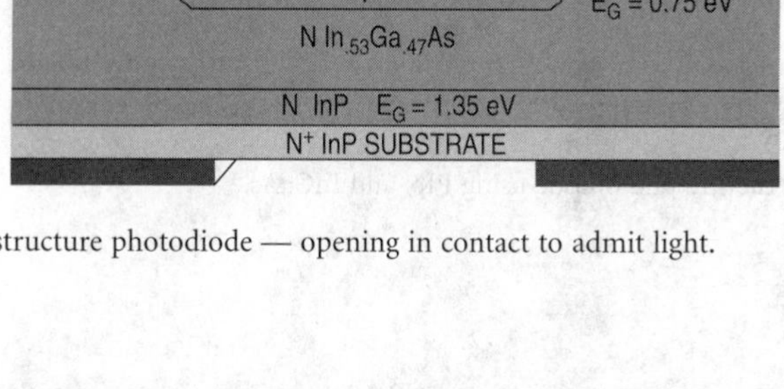

FIGURE 119.37 Heterostructure photodiode — opening in contact to admit light.

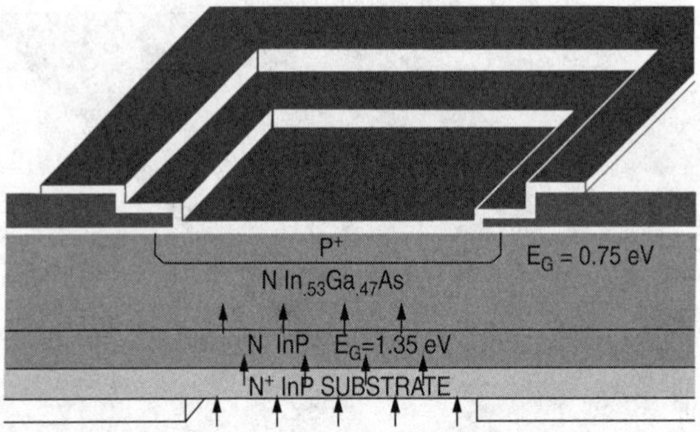

FIGURE 119.38 Heterostructure photodiode — photons entering photodiode.

Photons enter the photodiode from the bottom as shown in Figure 119.38. Photons with energies less than the energy gap of the InP pass through the InP substrate and InP epitaxial layer to the InGaAs layer. Photons with energies greater than $E_G = 0.75$ eV are then absorbed in the InGaAs layer, and result in the photogeneration of free electrons and holes. The photogenerated holes can then diffuse to the P+ anode region and be collected, thus resulting in the production of a photocurrent. The $In_{.53}Ga_{.47}As$, arsenide energy gap of 0.75 eV corresponds to a photon wavelength of $\lambda_C = 1.65$ μm, and the InP energy gap of 1.35 eV corresponds to $\lambda_C = 0.918$ μm. Thus, photons with wavelengths between 0.918 μm and 1.65 μm will pass through the InP substrate and InP epitaxial layer, and be absorbed in the $In_{.53}Ga_{.47}As$ layer. The InP substrate and InP epitaxial layer thus act as a transparent window for this wavelength range.

A wavelength of $\lambda = 1.55$ mm is often used for long distance fiber optic systems, because the optical absorption of the glass fibers is a minimum in this wavelength range. Thus, this type of heterostructure photodiode, with its wavelength range of 0.92 μm to 1.65 μm, is often used for these fiber optic systems.

Another example of heterostructure photodiode that uses InP and InGaAs is shown in Figure 119.39. In this case also, the photons enters through a window in the metallization contact area. The photons then pass through the InP layers to the $In_{.53}Ga_{.47}As$ region where they are absorbed as shown in Figure 119.40. This again is the case for photons with energies between 0.75 eV and 1.35 eV, corresponding to wavelengths between 0.92 μm and 1.65 μm. The photogenerated holes in the $In_{.53}Ga_{.47}As$ absorption region will diffuse toward the InP PN junction.

With a reverse bias voltage across the PN junction there will be a depletion region adjacent to the junction, extending mostly on the more lightly doped N-type side of the junction as shown in Figure 119.41. With a large reverse bias voltage, there will be a strong electric field in the depletion region; thus, as the holes travel through this region they will be accelerated by the electric field.

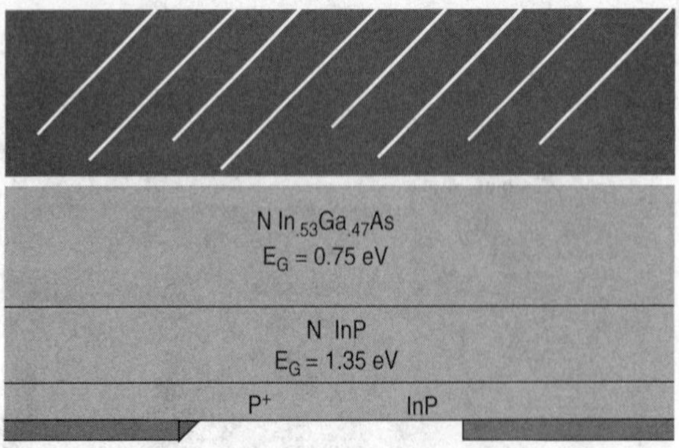

FIGURE 119.39 Heterostructure photodiode using PIN and InGaAs.

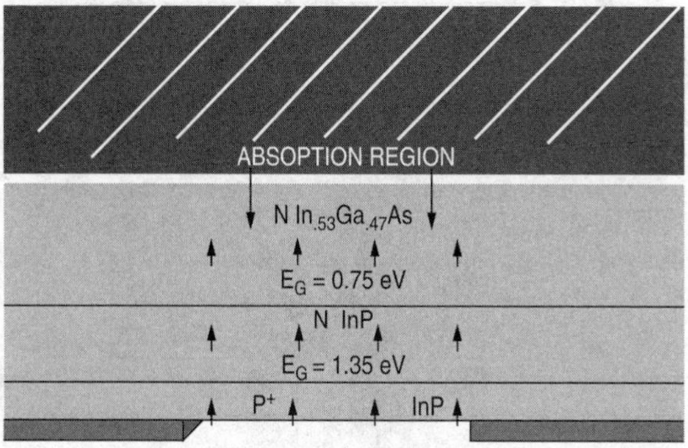

FIGURE 119.40 Heterostructure photodiode — absorption region.

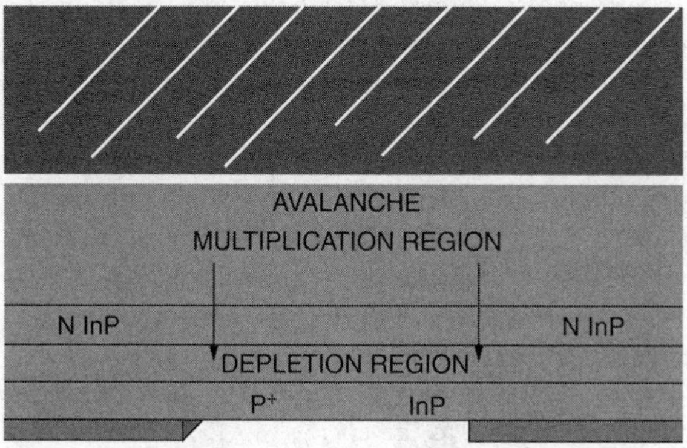

FIGURE 119.41 Heterostructure photodiode — avalanche multiplication.

With a large enough electric field, an avalanche multiplication process occurs, with additional holes and free electrons being produced by the impact ionization process. As a result of the avalanche multiplication process, the collection of electrons and holes will increase along with a corresponding increase in the photocurrent. This is, therefore, another example of an avalanche photodiode.

119.9 Light-Emitting Diodes

A light-emitting diode (LED) is to a great extent the opposite of a photodiode. A photodiode can be used to produce an electric current as a result of light shining on it, whereas an LED produces light as a result of current passing through it. The light emitted by an LED increases with increasing forward current, although the relationship between light intensity and current is not a linear one. LEDs and laser diodes are used in a wide variety of optoelectronic systems. In particular, laser diodes are widely used as the light source for high-speed fiber optic communication systems.

Figure 119.42 shows a PN junction. A forward bias voltage, V_F, makes the P-type side or anode positive with respect to the N-type side or cathode. The forward current, I_F, flows from the anode (P-type side) to the cathode (N-type side). Electrons are emitted from the cathode into the anode. As the electrons flow into the P-type side, they recombine with holes. As a result, the electron density decreases exponentially with distance as given by $n(x) = n(0) \cdot \exp(-x/L_N)$, where L_N = minority carrier (electron) diffusion length.

Holes are emitted from the anode into the cathode. As the holes flow into the N-type side, they recombine with electrons. The hole density decreases exponentially with distance as given by $p(x) = p(0) \cdot \exp(-x/$

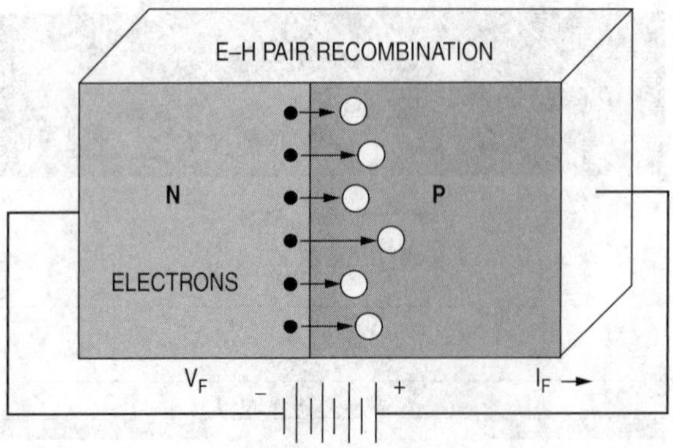

FIGURE 119.42 Light-emitting diode — electron–hole pair recombination.

L_P), where L_P = minority carrier (hole) diffusion length. Electrons and holes flow across the junction to the opposite sides, where they become *minority carriers* and recombine with the *majority carriers*.

Consider the case of a PN junction where the N-type side is very heavily doped (N^+) as compared to the P-type side. In this case, most of the diode current flow is due to the flow of electrons across the junction from N^+ to P. The electrons recombine with holes in the P-type side, mostly within a short distance from the junction. Electron–hole recombination occurs in this region.

In the case of silicon and germanium, almost all of the recombination events result in the generation of heat. In the III-V compound semiconductors, such as GaAs, most of the recombination events result in the generation of photons as illustrated in Figure 119.43. This is called *radiative recombination*. The photon energy is related to the photon wavelength by $E_{PHOTON} = hc/\lambda$ where h = Planck's constant and c = velocity of light. Solving for the wavelength gives $\lambda = hc/E_{PHOTON}$. The energy evolved in an electron-hole pair recombination is usually close to the energy gap E_G, so $\lambda = hc/E_G$. If the energy gap E_G is expressed with units of electron-volts, E_G (in joules) = $q \cdot E_G$, where q is the magnitude of the electron–charge. We thus have $\lambda = hc/q \cdot E_G$, where E_G is given in units of electron volts; hc/q = 1.239 eV-μm, so λ = 1.239 eV-μm/E_G. For GaAs, E_G = 1.43V, so λ = 0.9 μm, which is in the near-infrared region. The photon emission occurs randomly through this region near the junction, and the photons travel out in all directions. A *laser diode* is a specially constructed LED in which the photons are all emitted in phase and travel in the same direction. The result is a very narrow collimated beam of light. The photons all have close to the same wavelength, so this is approximately a monochromatic light source. For this to happen in a laser diode, the forward current IF must be greater than some threshold value, so $I_F > I_{TH}$. If $I_F < I_{TH}$, then the laser diode will act like an ordinary LED.

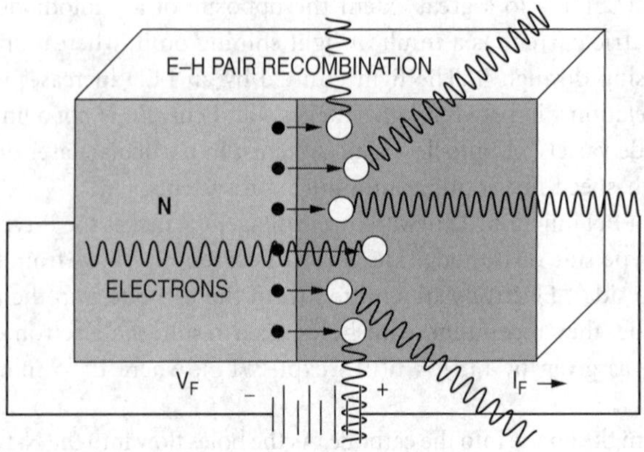

FIGURE 119.43 Light-emitting diode — photon emission.

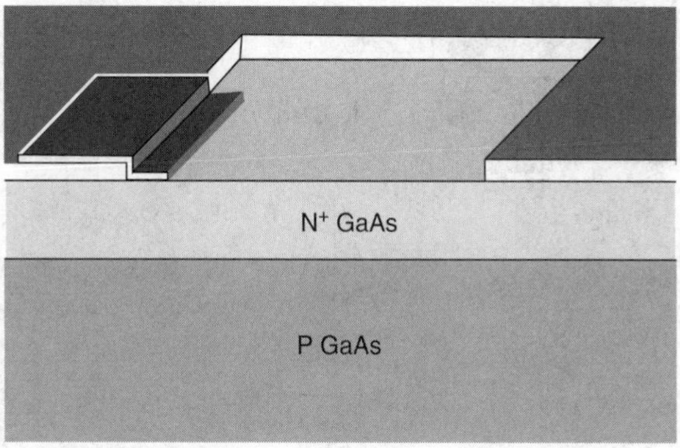

FIGURE 119.44 Structure of light-emitting diode.

Structure of Light-Emitting Diodes

In this section, we focus on the structure and operation of light-emitting diodes. We will start off with a gallium arsenide (GaAs) substrate as shown in Figure 119.44. Next, a thin heavily-doped N^+ epitaxial layer of GaAs is deposited. Then there is the deposition of an SiO_2 layer and a photolithographic step to open windows in the SiO_2 layer. Metallization is then deposited and patterned for the cathode contact of the LED. The anode contact is made to the bottom of the P-type GaAs substrate. When the PN junction is forward biased, most of the current flow is due to the emission of electrons from the N^+ region into the P-type substrate. After traveling a short distance the minority carrier electrons in the P-type substrate recombine with the majority carrier holes. The average distance the electrons travel before recombination is very short, on the order of just 1 μm. Electron–hole pair recombination occurs close to the junction.

Reflection

In GaAs almost all of the recombination events are radiative, resulting in the emission of photons. The photons travel off in all directions, and only a few of the photons get out. Many photons are not traveling in the right direction to get out. Many photons are reflected at the GaAs surface. GaAs has a rather high index of refraction, $n_2 = 3.66$, so a substantial amount of the incident photon flux, ϕ_i, gets reflected. The reflection factor at the interface between two optical media for normal incidence is $R = \phi_r/\phi_i = (n_2 - n_1)^2/(n_2 + n_1)^2$. For a GaAs ($n_2 = 3.66$)/air($n_1 = 1.0$) interface, we have $R = (3.66 - 1)^2/(3.66 + 1)^2$. This evaluates to $R = 0.33$, so one third of the incident photon flux is reflected at the GaAs/air interface. Thus, the transmission factor at the GaAs/air interface for normal incidence is $T = 1 - R = \phi_t/\phi_i = 0.67$.

As the angle of incidence, θ_i, increases, the reflection factor increases. For $\theta_i > \theta_c$, the reflection factor becomes unity, and we have *total internal reflection.* The angle θ_c is the critical angle for total internal reflection, and is given by the condition that $\sin\theta_c = n_1/n_2$. For the GaAs/air interface we have $\sin\theta_c = 1/3.66$, so $\theta_c = 15.9°$. Thus, only those photons incident at the surface within a cone of half-angle $\theta_c = 15.9°$ stand any chance of being emitted as shown in Figure 119.45. This represents only 2% of the total photon flux generated in the device, so we see that this is a severe problem. For the most favorable case of normal incidence the transmission factor is $T = 0.67$. We see that overall only about 1.3% of the internally-generated photons can get out.

Another problem is the reabsorption of photons within the GaAs itself, especially since most of the photon generation takes place on the P-type side of the P-N junction. The photon reabsorption problem can be decreased by using a P-type epitaxial layer on an N^+ substrate as shown in Figure 119.46. Now we have the emission of electrons upward from the N^+ GaAs substrate into the thin P-type GaAs epitaxial layer. There the injected minority carrier electrons can recombine with the holes. We have the subsequent emission of photons. One problem with this is that many of the electrons will travel all of the way across the thin P-type layer and be trapped by crystallographic defects at the surface called *surface states.* The

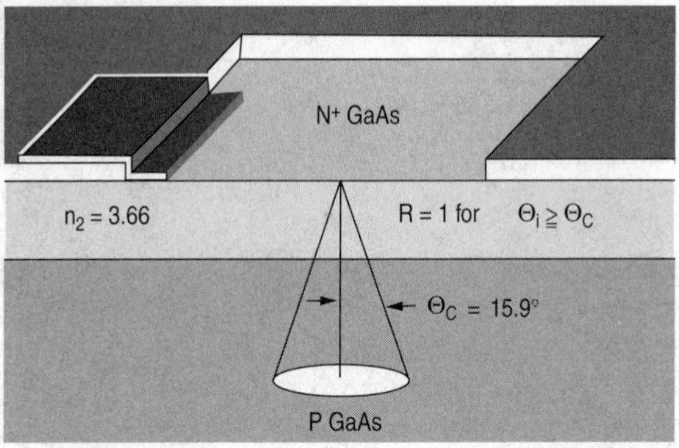

FIGURE 119.45 Light-emitting diode — effect of critical angle on light emission.

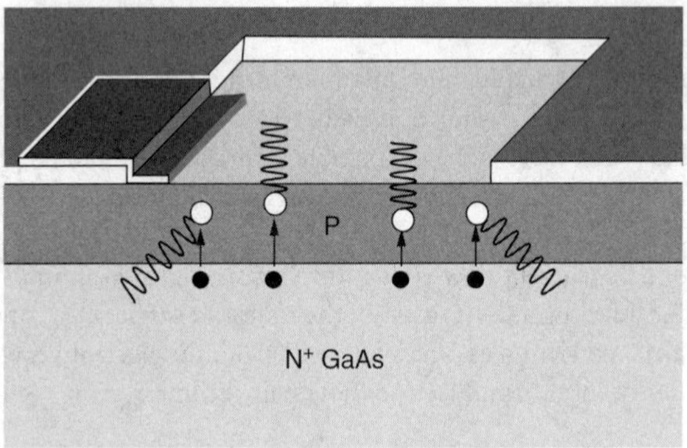

FIGURE 119.46 Light-emitting diode — photon emission.

trapped electrons will eventually recombine with holes, but with a nonradiative recombination process involving the evolution of heat rather than photons.

Now we will look at some heterojunction LEDs that help to solve some of these problems. Here in Figure 119.47 is a diagram of a heterostructure using GaAs and the ternary III-V compound semiconductor GaAlAs. There is a thin layer of GaAs sandwiched between two GaAlAs layers. The GaAs has an energy gap of E_G = 1.5 eV, and GaAlAs has E_G = 3.0 eV. Under forward bias conditions electrons from the N^+ GaAlAs, and holes from the P-type GaAlAs flow into the GaAs layer. As a result of the differences of the energy gaps, the injected electrons become mostly confined to the GaAs layer as shown in Figure 119.48. Electron–hole pair recombination occurs in this GaAs carrier confinement region. There is the

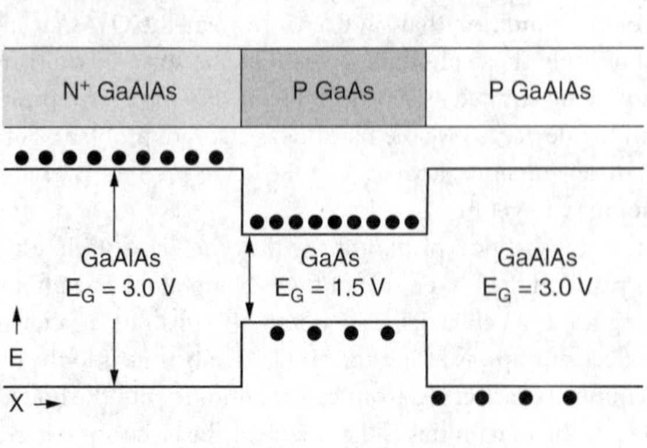

FIGURE 119.47 Heterostructure light-emitting diode — carrier confinement.

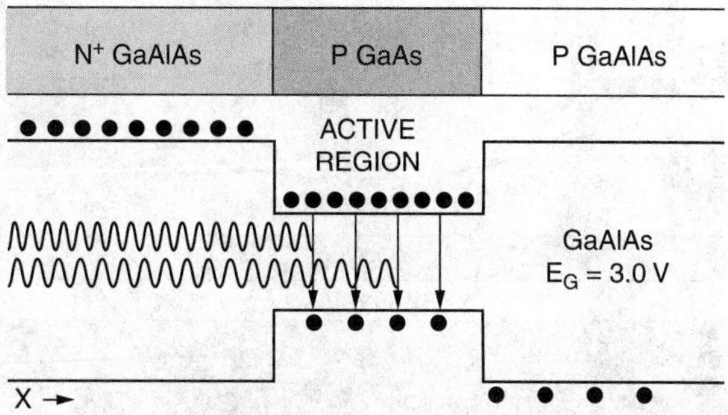

FIGURE 119.48 Heterostructure light-emitting diode — photon emission from active region.

subsequent emission of photons. Since the photon energy corresponds to the energy gap of the GaAs layer with E_G = 1.5 eV, there is no photon reabsorption occurring in the GaAlAs layers with E_G = 3.0 eV. Thus, the GaAlAs layers serve as transparent windows for the photon emission from the GaAs carrier confinement active layer.

Consider now the fabrication of a heterostructure LED, shown in Figure 119.49. We start with a heavily doped P^+ GaAs substrate of about 200 μm in thickness. A thin P-type GaAlAs epitaxial layer is then deposited. This is followed by the deposition of the thin P-type GaAs active layer. We then have the deposition of a heavily doped N^+ GaAlAs layer. The top metal contacts are then deposited and patterned, and a bottom contact to the substrate is also made. Under forward bias conditions there is the emission of electrons from the N^+ GaAlAs into the GaAs active layer as shown in Figure 119.50. There is also the emission of holes from the P-type GaAlAs into the GaAs active layer. In the GaAs active layer there will the recombination of electrons and holes, and the emission of the photons.

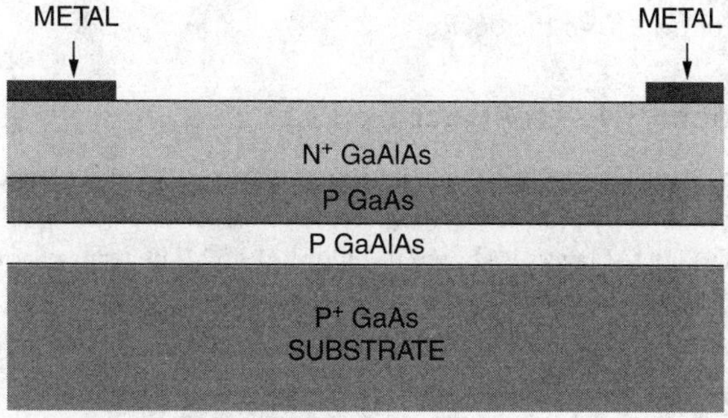

FIGURE 119.49 Structure of heterostructure light-emitting diode.

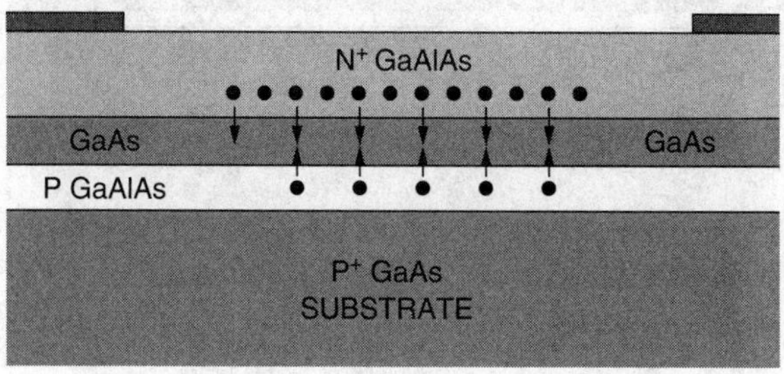

FIGURE 119.50 Heterostructure light-emitting diode — recombination in active layer.

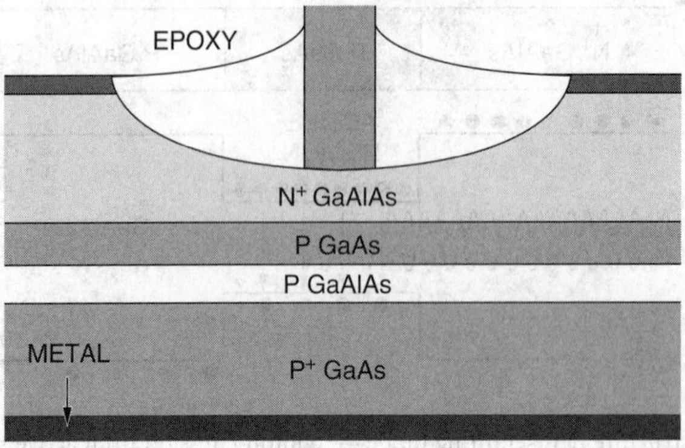

FIGURE 119.51 Coupling of light-emitting diode to optical fiber.

In many cases, the LED is used as the light source in a fiber optic system. For more efficient coupling of the light to a small diameter fiber, a cavity can be etched into the N$^+$ GaAlAs top layer as shown in Figure 119.51. The optical fiber is then butted up against the LED surface. The cavity is then refilled with an epoxy to hold the fiber into place. The bottom metal contact can also serve to reflect some of the photons back upward toward the fiber.

The close spacing of the end of the fiber to the active region and the higher refractive index of the fiber greatly improves the optical coupling. For example, with a fiber with an index of refraction of $n_1 = 1.5$ and GaAlAs with $n_1 \cong 3.0$, we now have a critical angle of $\theta_c \cong 30°$, as compared to 16° for the GaAs/air interface. Now the fraction of photons incident on the surface within a cone of half-angle $\theta_c \cong 30°$ is about 7%, as compared to 2% for the GaAs/air interface.

To further improve the efficiency of the surface-emitting LED, the bottom or anode contact can be limited to just a narrow strip. A layer of SiO_2 is deposited on the bottom surface, and a narrow opening is etched. Then the metal anode contact is deposited, and the LED mounted on a metal heat sink for more efficient transfer of heat from the device.

Edge-Emitting Double Heterostructure LED

We have been examining the surface-emitting LED. Now we turn to another basic type of heterostructure LED, the edge-emitting double heterostructure LED. A heterostructure LED with a thin GaAs active layer sandwiched between two GaAlAs layers is shown in Figure 119.52. The ternary compound semiconductor

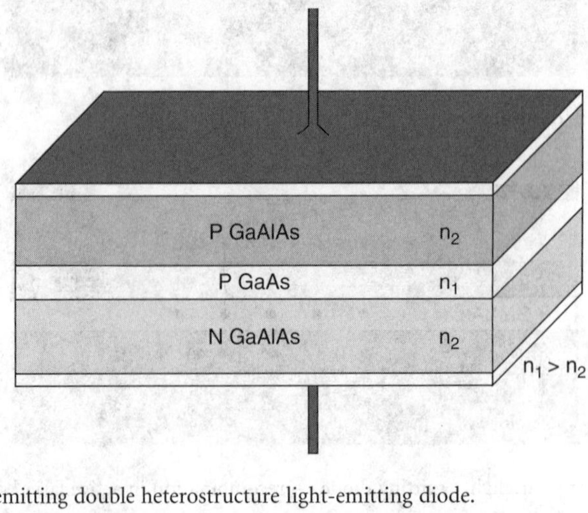

FIGURE 119.52 Edge-emitting double heterostructure light-emitting diode.

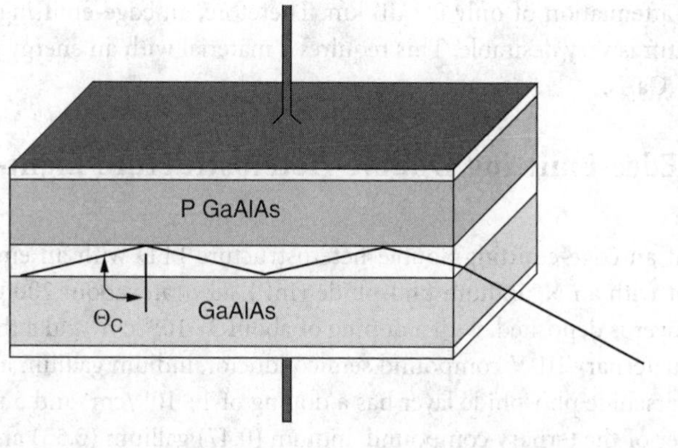

FIGURE 119.53 Light emission from edge-emitting double-heterostructure light-emitting diode.

GaAlAs has an energy gap that is larger than that of GaAs, and an index of refraction that is substantially smaller than that of GaAs. Photons that are generated in the GaAs active layer that have an angle of incidence at the GaAs/GaAlAs interface greater than the critical angle, θ_c will undergo total internal reflection. These photons can then propagate through the thin active layer. The photons are then emitted out the edge, as shown in Figure 119.53, so this is called an edge-emitting LED.

To further improve the efficiency of this device, the top metal contact can be confined to a narrow strip about 50 μm wide. The current flow will then be confined to a small region as shown here, which will improve the efficiency of the LED. The confinement of the LED current to a small region will also be especially important for LASER diodes

The structure of an edge-emitting heterostructure LED is shown in Figure 119.54. Starting off with a GaAs substrate, about 200 μm thick, a heavily-doped N^+ GaAs layer with a doping of about $1 \cdot 10^{18}/cm^3$ is deposited. Next comes an N-type layer of $Ga_{0.3}Al_{0.7}As$ with a doping of about $1 \cdot 10^{18}/cm^3$ and a thickness of about 1 μm. This is followed by the deposition of the P-type GaAs active layer with a doping in the $1 \cdot 10^{18}/cm^3$ range, and a thickness of about 0.1 to 0.2 μm. On top of the GaAs active layer is a P-type $Ga_{0.3}Al_{0.7}As$ layer with a thickness of about 1 μm and a doping of about $1 \cdot 10^{18}/cm^3$. The topmost layer is a heavily-doped P^+ GaAs layer, with a doping up in the $1 \cdot 10^{19}/cm^3$ range, and a thickness of about 0.1 μm. A layer of SiO_2 is then deposited. The SiO_2 layer is etched to produce an opening in the form of a narrow strip. The top metal anode contact is then deposited, and a bottom cathode metal contact is also deposited to complete the LED structure.

This edge-emitting LED emits at wavelengths near $\lambda = 1.24$ eV-μm/E_G, where $E_G = 1.43$ eV is the energy gap of the GaAs active layer. This gives an emission wavelength near $\lambda = 0.9$ μm. Good-quality glass optical fibers exhibit absorption minima at 1.3 μm, with an attenuation of about 0.5 dB/km, and

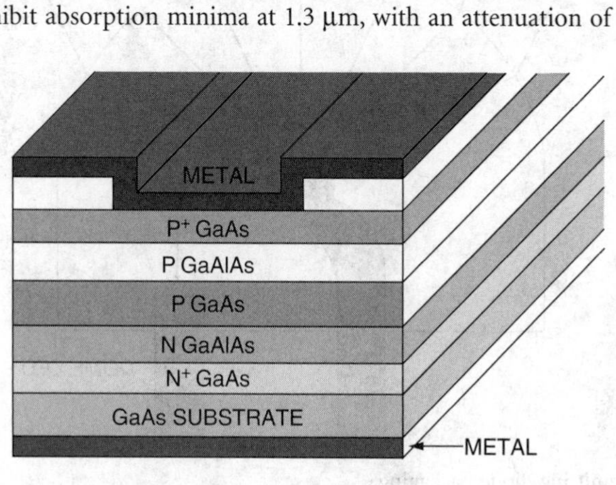

FIGURE 119.54 Structure of edge-emitting double-heterostructure light-emitting diode.

at 1.55 μm with an attenuation of only 0.2 dB/km. Therefore, an edge-emitting LED with an emission wavelength of 1.55 μm is very desirable. This requires a material with an energy gap that is substantially smaller than that of GaAs.

Fabrication of Edge-Emitting Double-Heterostructure Light-Emitting Diodes

We will now look at an edge-emitting double-heterostructure LED with an emission wavelength near 1.55 μm. We start of with an N^+ indium phosphide (InP) substrate, about 200 μm in thickness. An N^+ indium phosphide layer is deposited, with a doping of about $3 \cdot 10^{18}/cm^3$ and a thickness of 0.2 μm. This is followed by the quaternary III-V compound semiconductor, indium gallium arsenide phosphide. The N^+ indium gallium arsenide phosphide layer has a doping of $1 \cdot 10^{18}/cm^3$ and a thickness of 1 μm. Next comes the active layer of the ternary compound, indium (0.47) gallium (0.53) arsenide with a doping in the $1 \cdot 10^{18}/cm^3$ range, and a thickness of 0.1 to 0.2 μm. On top of the active layer is a P+ indium gallium arsenide phosphide layer with a doping of $10^{18}/cm^3$ and a thickness of 1 μm.

The top low-resistance contact layer is heavily doped P^+ indium phosphide, with a doping up in the $10^{19}/cm^3$ range, and a thickness of 0.1 μm. Next is the deposition of an SiO_2 insulating layer, and then the etching of this layer to produce an opening in the form of a narrow strip. Then comes the top surface anode contact metallization, followed by the bottom surface cathode contact metallization. The indium (0.47) gallium (0.53) arsenide active layer has an energy gap of $E_G = 0.8$ eV. The corresponding emission wavelength is $\lambda = 1.55$ μm. This emission wavelength corresponds to the wavelength of minimum attenuation of the optical fiber.

Packaging of Light-Emitting Diodes

Now we focus on the packaging of a typical surface-emitting dome LED, as shown in Figure 119.55. We start off with the metal leads for the diode anode and cathode contacts. The LED chip is bonded to a flat area on one of the metal contact leads. A small-diameter top contact wire is then connected between the LED chip and the other contact lead. Next is the injection molding of a transparent plastic encapsulation. This plastic encapsulation serves to protect the device, and at the same time allows the light to get out. The shape of the plastic encapsulation is such that there is a focusing action that narrows the beam width of the emitted light and thus increases the directivity of the LED.

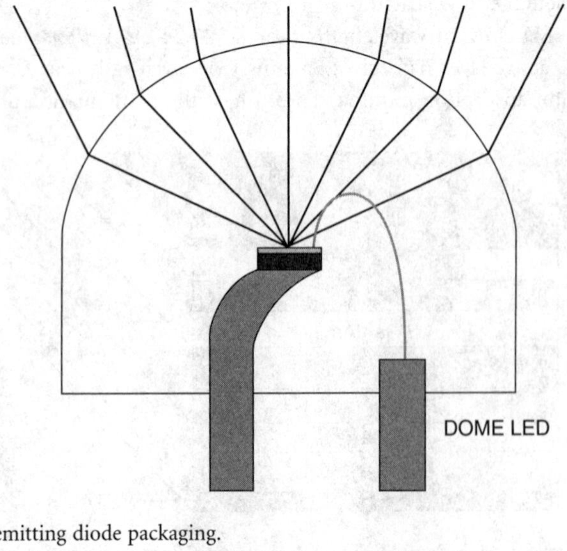

FIGURE 119.55 Light-emitting diode packaging.

The index of refraction of the plastic of about 1.6 increases the critical angle for total internal reflection, θ_c, in the semiconductor from about 16° to about 26°. This increase in the critical angle from 16° to 26° will produce an increase in the emitted photon flux by a factor of about three.

119.10 Semiconductor Diode Lasers

In the recombination process in a semiconductor, electrons recombine with holes and give off energy E_G. Most of the recombination events in the III-V compound semiconductors are radiative recombinations, with photons being emitted. In an LED these recombination events occur spontaneously, and the emitted photons travel off in all directions.

In the *stimulated emission* process a photon produced by a recombination event stimulates the recombination of another free electron with a hole. This results in the emission of a second photon. The two photons are exactly at the same frequency and phase, and travel in the same direction. These photons then stimulate the recombination of another free electron and hole. This results in the emission of another photon. These photons will be at exactly the same frequency and phase, and travel in the same direction. This process of the stimulated emission of radiation can then continue through the active region of the device. Indeed, the word *laser* is an acronym for *light amplification by the stimulated emission of radiation.* The resulting emission is of photons that are at the same frequency, and hence this is monochromatic radiation. The photons are all in phase with each other, so this is coherent radiation. The photons are all traveling in the same direction resulting in a collimated beam of light. In actuality, both spontaneous and stimulated emission processes occur.

In an ordinary LED, the spontaneous emission process is by far the dominant process, so the emitted light is not monochromatic, but does have a spread of wavelengths. It is also not coherent and not collimated. In order to make the stimulated emission process the dominant one, the concentration of free electrons and holes in the active region must be sufficiently large. For this to happen, the forward current must be greater than some minimum threshold value, I_{TH}, for lasing action. In addition, the photons produced by this process must selectively be made to pass many times through the active volume of the device.

The stimulated emission process results in an increase in the photon flux, so this can be viewed as a light amplification process. There will also photon absorption taking place, and in order to have a net photon gain or amplification, the stimulated emission rate must exceed the photon absorption rate. For this to happen there must be a very high population density of free electrons and holes in the active region of the device. In order for this process to continue, there must be a feedback mechanism to feed photons back into the active volume. To accomplish this second objective, the laser diode is made in the form of an optical cavity with two parallel reflecting surfaces or mirrors. An electromagnetic wave that is traveling in a direction perpendicular to the end mirrors will be reflected upon reaching the end mirror. The reflected wave will now travel in the opposite direction. Upon reaching the other end there will be another reflection, and this process continues. The net electric and magnetic field distribution in this cavity will be the vector sum of all of the waves that are bouncing back and forth between the two mirrors. Let the spacing between the two reflecting surfaces be L. If L is an integer multiple of $\lambda/2$ such that $n \cdot (\lambda/2) = L$, then there will be a standing wave pattern in the cavity, and the waves will reinforce each other as shown in Figure 119.56. Wavelengths that do not satisfy the condition that $n \cdot (\lambda/2) = L$ will suffer destructive interference and will rapidly decay to zero. At the wavelengths that satisfy the condition that $n \cdot (\lambda/2) = L$, this is a resonant cavity. In the microwave region this is called a cavity resonator, and is somewhat analogous to an L-C resonant circuit. In the optical part of the spectrum, it is called a Fabry-Perot resonator.

Photons that satisfy the wavelength condition and are traveling perpendicular to the reflecting surfaces will bounce back and forth between the two reflecting surfaces. These photons will stimulate the emission of other photons and there will be a build-up of this stimulated emission process. In order to allow the laser light to get out, one of the reflecting surfaces is actually only partially reflecting as shown in

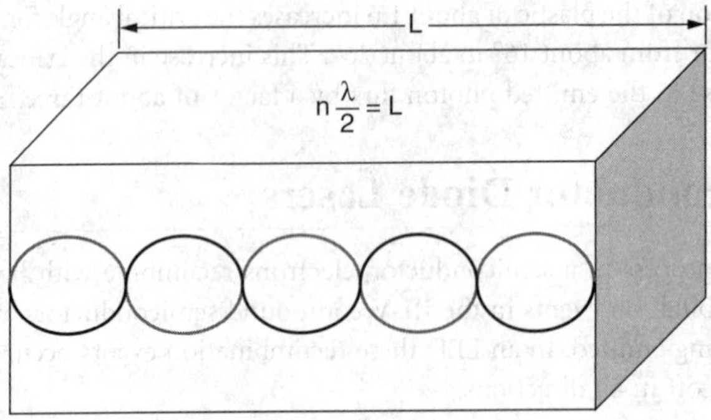

FIGURE 119.56 Standing wave pattern in an optical cavity.

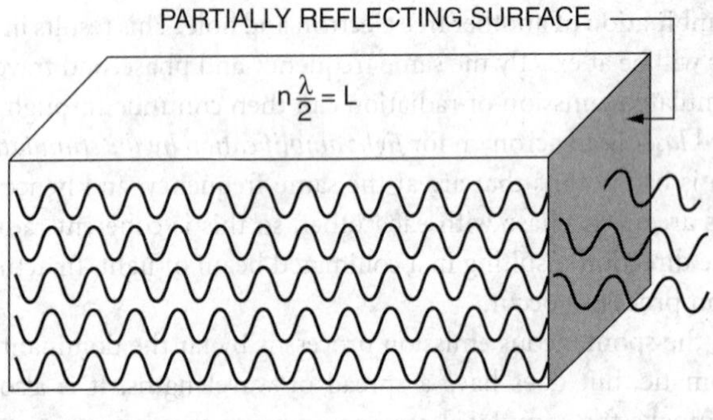

FIGURE 119.57 Light emission from laser diode.

Figure 119.57. As a result, the laser beam can escape from the partially reflecting-partially transmitting surface. Note that the laser light will actually not be truly monochromatic, but will consist of a number of discrete wavelength or modes that satisfy the condition that $n \cdot (\lambda/2) = L$, where n is an integer. However, since the photon energies must be close to the energy gap E_G, not all of the possible cavity modes will actually be produced. Only those modes that correspond to photon energies close to E_G will be produced.

The emission spectrum of an LED is shown in Figure 119.58. The peak wavelength is $\lambda_{peak} \cong hc/(qE_G) = 1.24$ eV-μm/E_G, and is in the 0.5- to 1.5-μm range for most LEDs. The radiation is in a narrow band of wavelengths, typically about 50 nm wide, centered about the peak wavelength, λ_{peak}. When lasing action occurs only those modes will appear that correspond to wavelengths within the LED emission spectrum will thus be present. As the forward current increases further above the threshold for lasing action, there will be a further narrowing of the emission spectrum. With further increases in the forward current the spectrum will be even narrower. Finally, with a large enough value of the forward current, all of the

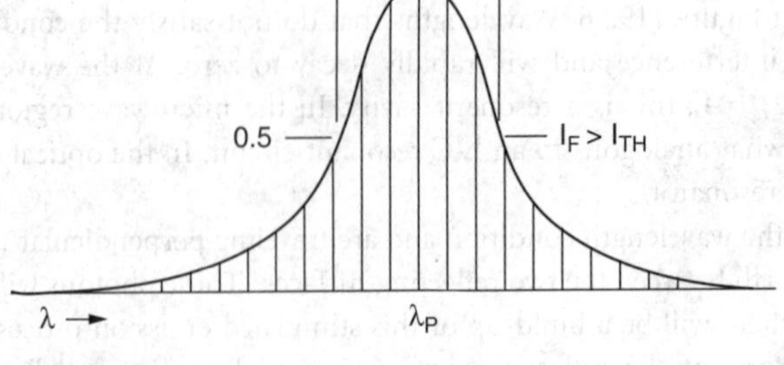

FIGURE 119.58 Emission spectrum.

energy will be concentrated in just a few modes or spectral lines, and an approximately monochromatic output will be obtained.

Typical laser diodes are represented by GaAlAs with λ_{peak} in the 750 to 850 nm range and power outputs up to 5 watts. Quaternary compounds such as InGaAlP have λ_{peak} = 670 nm and power outputs up to 5 mW. Another example is InGaAsP with λ_{peak} in the 1300- to 1550-nm range and power outputs up to 100 mW.

InGaAsP lasers are especially useful for long-range fiber optic applications. This is a result of the emission wavelength of 1550 nm = 1.55 μm being in the range of minimum attenuation for the fiber. In addition, the power output of up to 100 mW for InGaAsP is another favorable factor for long-range fiber optic applications.

Some specially constructed laser diodes can be made to operate as just a single-mode laser, and this is especially useful for long-range wide-bandwidth fiber optic systems.

Edge-Emitting Double-Heterostructure Laser

Now we consider the fabrication of an edge-emitting double-heterostructure laser diode. A P-type GaAs active layer is sandwiched between two GaAlAs layers, as shown in Figure 119.59. The GaAs has a smaller energy gap than GaAlAs, and also has a larger index of refraction so $n_1 > n_2$. Light rays that are incident on the GaAs/GaAlAs interface with an angle of incidence θ_i greater than the critical angle for total internal reflection will undergo total reflection at the interface, as in Figure 119.60. The two opposite faces of the

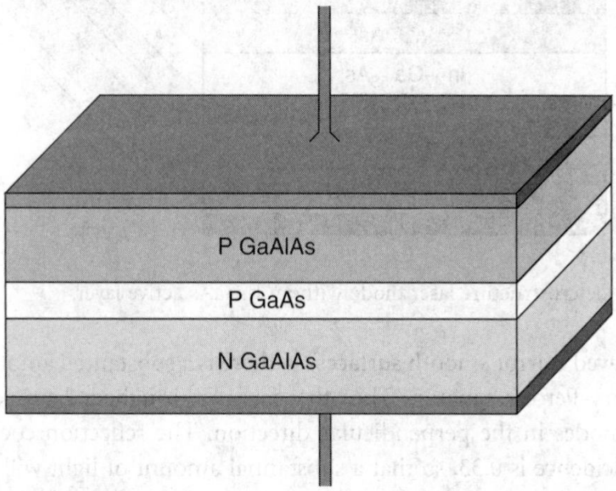

FIGURE 119.59 Laser diode structure.

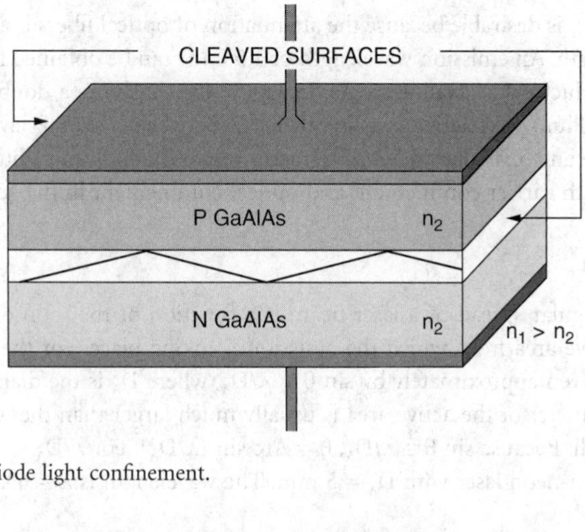

FIGURE 119.60 Laser diode light confinement.

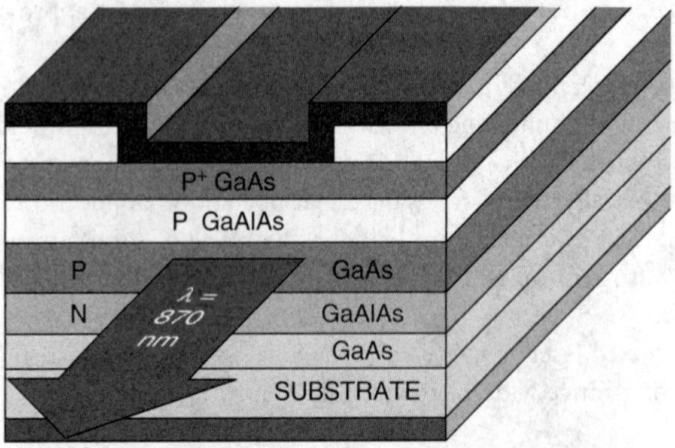

FIGURE 119.61 Light emission from the active layer of a laser diode.

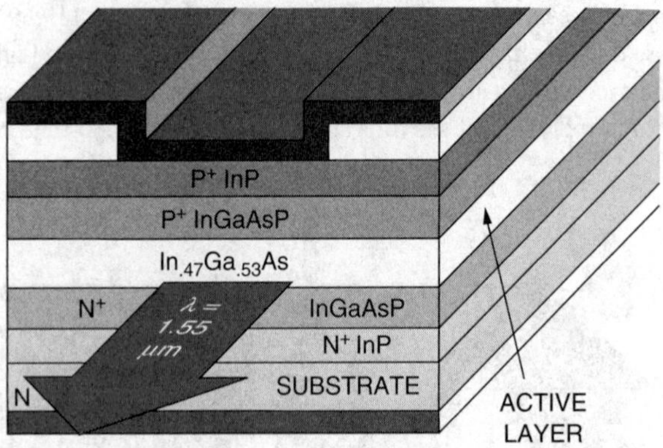

FIGURE 119.62 Double-heterostructure laser diode with an InGaAs active layer.

laser diode chip are cleaved mirror smooth surfaces, so there is a substantial amount of reflection at these faces. This forms a Fabry-Perot resonator. The other faces are roughened surfaces so that there will be no buildup of lasing modes in the perpendicular direction. The reflection coefficient at the GaAs/air interface for normal incidence is 0.33, so that a substantial amount of light will exit the cleaved faces.

A more detailed look at the structure of laser diodes is shown in Figure 119.61. The laser radiation is emitted from the P-type GaAs active layer that has an energy gap $E_G \cong 1.43$ eV. The corresponding wavelength is $\lambda = 1.24$ eV-μm/1.43 eV = 0.87 μm = 870 nm. For long-distance fiber optic applications, a wavelength of 1.55 μm is desirable because the attenuation of optical fibers is a minimum of about 0.2 dB/km at that wavelength. An emission wavelength of 1.55 μm can be obtained by using the ternary III-V compound semiconductor $In_{0.47}Gallium_{0.53}As$. In Figure 119.62 shows a double-heterostructure laser diode with an $In_{0.47}Gallium_{0.53}As$ active layer sandwiched between InGaAsP layers. The InGaAsP layer has a larger energy gap and a smaller index of refraction than the indium gallium arsenide active layer. Therefore, there are both carrier confinement and optical confinement in the active layer.

Laser Beamwidth

The beam width or angular spread of a laser beam is a function of the ratio of the wavelength to the dimensions of the active area from which the emission is taking place. For the case of $D >> \lambda$, as in Figure 119.63, this is given approximately by $\sin \theta = \lambda/D_A$, where D_A is the diameter of the active area. With gas lasers the diameter of the active area is usually much larger than the wavelength, so the beam width will be very small. Because $\sin \theta \cong \lambda/D_A$, $\theta = \text{Arc sin } (\lambda/D_A)$. For $\lambda/D_A << 1$, $\theta \cong \lambda/D_A$. Consider the example of a helium-neon laser with $D_A = 3$ mm. The wavelength is $\lambda = 633$ nm = 0.63 μm, which

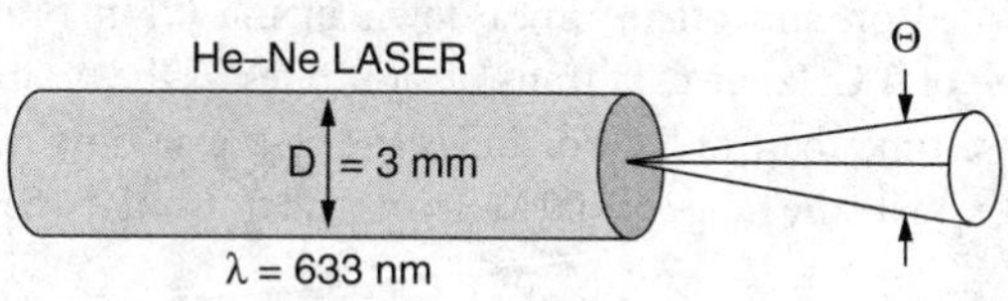

FIGURE 119.63 He-Ne laser beam width.

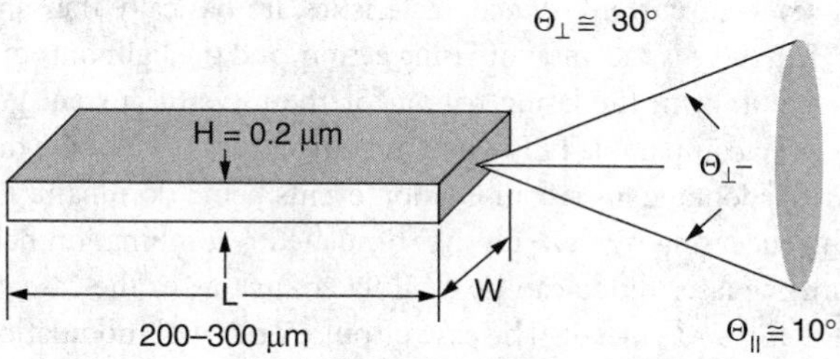

FIGURE 119.64 Laser diode beam width.

is in the red part of the visible spectrum. Assume that λ/D_A = 633 nm/3 mm, which gives $\lambda/D_A = 211 \cdot 10^{-6} = 2.11 \cdot 10^{-4}$. We thus obtain $\theta = 2.11 \cdot 10^{-4}$ radians. This corresponds to only 0.012°, so the beam divergence is very small indeed.

We now consider a laser diode, of which only the active layer is shown in Figure 119.64. The length of the active layer is typically in the 200- to 300-μm range. We typically have H ≅ 0.2 μm for the height or thickness of the active layer, and W in the range of 10 to 50 μm for the width. Since the thickness H of the active layer is actually smaller than the wavelength λ, the equation presented earlier for the beam divergence cannot be used. As a result of the small dimensions of the active layer in the laser diode, the angular spread of the beam will be much wider, especially in the direction perpendicular to the narrow dimension. The beam width in the direction perpendicular to the active layer is typically around 30°. The beam width in the direction parallel to the active layer is typically around 10°.

Although the laser diode does have a rather large beam width, the small dimensions of the active area and the coherent and nearly monochromatic nature of the light makes it possible to use lenses to produce well-collimated beams. It is also possible to focus the light in a very small spot size, as required for example in compact disc and laser disc applications. As a result of the small dimensions of the light emitting active area, the light output can be efficiently coupled to optical fibers as shown in Figure 119.65. The active area of a laser diode is usually comparable to or smaller than the area of an optical fiber, so a large fraction of the light output can be coupled into the fiber. Indeed, the coupling efficiency of light from a laser diode into a small diameter optical fiber can be of the order of 100 times greater than that of an LED.

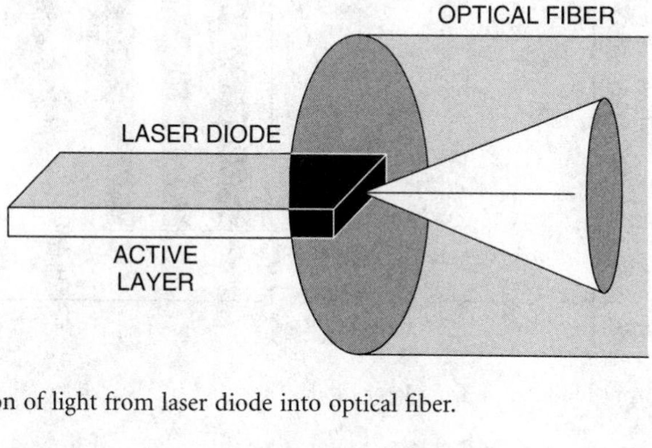

FIGURE 119.65 Emission of light from laser diode into optical fiber.

Laser diodes can be turned on and off in times down in the 0.1 to 1 ns range, corresponding to bandwidths up to the range of 2 GHz, or data transmission rates in the 2-Gbps range. Laser diodes also have a much faster response time than LEDs. For LEDs the response time is in the 3- to 300-ns range, corresponding to bandwidths up to the range of 100 MHz, or data transmission rates in the 100-Mbps range.

Light Output versus Forward Current

A graph of the light output (photon flux) ϕ versus the forward current for a laser diode is shown in Figure 119.66. At lower levels of current, the characteristics are basically the same as an LED. At some threshold current value, I_{TH}, there is the onset of lasing action, and the light output starts to increase much more rapidly. The power output in the lasing region for the forward current greater than I_{TH} is much greater than it would be for a comparable LED. For drive currents below I_{TH}, the diode operates essentially like an ordinary LED, with spontaneous recombination events being dominant. For drive currents above I_{TH}, the diode operates in the lasing mode, with the stimulated recombination being dominant.

The light output from the laser diode can be modulated, just as in the case of an LED, by variation of the driving current the forward current. The case of pulse or digital modulation, in which the driving current is a rectangular pulse waveform is shown in Figure 119.67. To take advantage of the much faster response time and high power output of the diode when operating in the lasing mode, the device is usually biased at, or slightly above the threshold current I_{TH}. This keeps the device operating in the lasing mode all of the time. The output is in the form of a series of pulses of light. The most commonly used

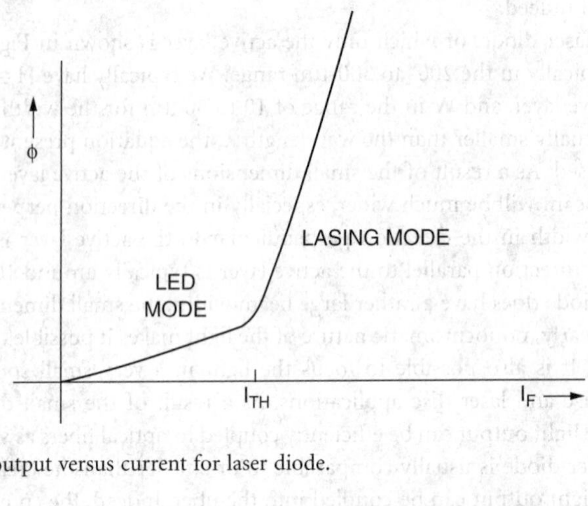

FIGURE 119.66 Light output versus current for laser diode.

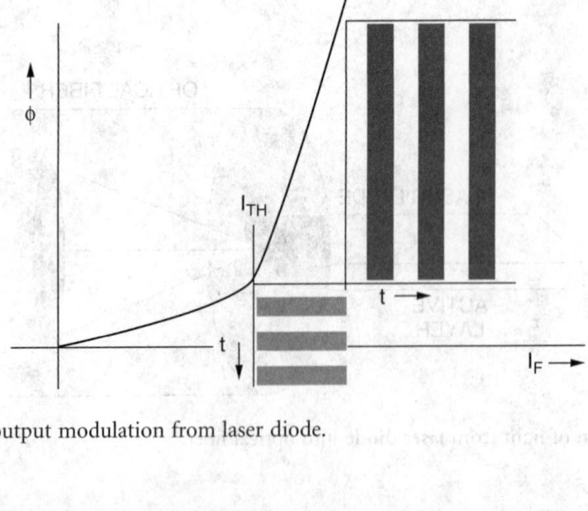

FIGURE 119.67 Light output modulation from laser diode.

type of pulse modulation is pulse code modulation or PCM in which the output is a series of off and on pulses representing a binary digital code of 1's and 0's. Another type of pulse modulation that can be used is pulse width modulation, or PWM. In this case the width of the pulses are proportional to the amplitude of an analog input signal.

Laser diodes offer several advantages over LEDs. The efficiency of a laser diode can be much greater than for an LED. The maximum frequency of modulation is much higher, the spectral width is much smaller, the emitting area is much smaller, and the output has phase coherence. The output of a laser diode can thus be easily turned into a well-collimated beam by a lens, and can also be focused down to a very small spot size. These factors make laser diodes especially useful for such applications as long-distance fiber optic systems and compact disc and laser disc systems. Laser diodes are, however, much more expensive than LEDs, and they require a more complex biasing circuit.

119.11 Conclusion

Some of the most important optoelectronic devices have been briefly described here. More detailed descriptions of their physics and properties are available in the works listed in the references.

Defining Terms

Avalanche photodiode — A photodiode in which the avalanche multiplication process increases the response of the device.

Critical wavelength — The long wavelength limit of photosensor devices, corresponding to photon energies that are just equal to the energy gap.

Current responsivity — The ratio of the current produced by illumination (*photocurrent*) to the incident optical power.

Heterojunction — Junction between two different semiconductor materials. Also known as a heterostructure.

Laser diode — A light-emitting diode that is especially constructed such that when operated above a threshold current level, will emit a coherent radiation of photons.

Light-emitting diode (LED) — A forward-biased diode that produces light as a result of the electron–hole pair recombination process. LEDs usually are made of III-V compound semiconductors such as gallium arsenide.

Photocell — Light-sensing device based on principle of photoconductivity. Also known as a *photoconductive cell* or *photoresistor.*

Photoconductive mode — Region of operation of a photodiode in which the PN junction is reverse biased. Photodiodes used for optical communications are usually operated in the photoconductive mode.

Photoconductivity — Change in conductivity of a semiconductor as a result of photons absorbed in the material.

Photodiode — A diode in which the reverse bias current is increased by the absorption of photons. A light-sensitive diode.

Phototransistor — An integrated combination of a photodiode and a transistor on one chip of silicon. The photocurrent of the photodiode is multiplied by the current gain of the transistor.

Photovoltaic mode — Region of operation of a photodiode in which the PN junction is forward biased. Solar cells operate in the *photovoltaic mode.*

PIN photodiode — A high-speed photodiode with a wide lightly-doped region that is fully depleted under operating conditions.

Radiative recombination — Recombination of electrons and holes in a semiconductor resulting in the emission of photons.

Schottky photodiode — A diode with enhanced short-wavelength response that uses a metal–semiconductor Schottky barrier instead of a PN junction.

Solar cell — A large-area photodiode especially constructed for the purpose of converting sunlight to electrical power.

Stimulated emission — Emission of light in which photons initiate the emission of other photons. The result is that the photons all have the same phase, frequency, and direction of propagation. This is called *coherent* emission of light.

References

Agrawal, G. P. and Dutta, N. K. 1993. *Semiconductor Lasers*, 2nd ed. Kluwer, Dordrecht.

Bhattacharya, P. 1997. *Semiconductor Optoelectronic Devices.* Prentice Hall, Englewood Cliffs, NJ.

Booth, K. M. and Hill, S. L. 1998. *The Essence of Optoelectronics (Essence of Engineering).* Prentice Hall, Englewood Cliffs, NJ.

Donati, S. 1999. *Photodetectors: Devices, Circuits and Applications.* Prentice Hall, Englewood Cliffs, NJ.

Jha, A. R. 2000. *Infrared Technology: Applications to Electro-Optics, Photonic Devices and Sensors.* John Wiley & Sons, New York.

Kasap, S. O. 2001. *Optoelectronics and Photonics: Principles and Practices.* Prentice Hall, Englewood Cliffs, NJ.

Martin, V. D. and Desmarais, L. 1997. *Optoelectronics*, Vols. 1, 2, and 3. Delmar Learning, Clifton Park, NY.

Piprek, J. 2003. *Semiconductor Optoelectronic Devices: Introduction to Physics and Simulation.* Academic Press, New York.

Rosencher, E. and Vinter, B. 2002. *Optoelectronics.* Cambridge University Press, London.

Saleh, B. E. A. and Teich, M. C. 1991. *Fundamentals of Photonics.* John Wiley & Sons, New York.

Smith, S. D. 1995. *Optoelectronic Devices.* Prentice Hall, Englewood Cliffs, NJ.

Uiga, E. 1995. *Optoelectronics.* Prentice Hall, Englewood Cliffs, NJ.

Wilson, J. and Hawkes, J. F. B. 1998. *Optoelectronics: An Introduction*, 3rd ed. Prentice Hall, Englewood Cliffs, NJ.

Zimmermann, H. 2000. *Integrated Silicon Optoelectronics.* Springer-Verlag, Heidelberg.

Further Information

Forrest, S. R. 1986. Optical detectors: three contenders. *IEEE Spectrum*, 23, pp. 76–84.

120

Power Electronics

Kaushik S. Rajashekara
Delphi Energy & Engine Management Systems

Timothy L. Skvarenina
Purdue University

Power electronics implies the use of electronics to convert electrical current and voltage into a more usable form for a given load. This typically involves the use of solid-state electronic switches of some sort, a converter circuit, and a control circuit. A variety of power electronic switches are available with a wide range of voltage and current ratings. Thus, power electronic circuits are used for low-power applications of only a few watts to very high-power applications using tens of megawatts. Power switches are discussed in the first section of this chapter.

Since the input and the output of the power electronic circuit can be AC or DC, there are four possible types of conversion:

AC–DC (rectification)
DC–AC (inversion)
AC–AC (change of voltage magnitude and/or frequency)
DC–DC (change of voltage magnitude)

Circuits that perform these functions are often referred to as converters, and within each class of converter there are a number of possible circuit topologies. Section two of this chapter discusses a number of converter circuits.

Applications for power electronics include power supplies for computers and other electronic devices, motor drives, uninterruptible power supplies, induction heaters, automotive ignition systems, battery chargers, flexible AC transmission system devices, electrically driven ships and vehicles, electrically controlled aircraft, and a host of other devices. Power electronics allows us to use electricity in the form (i.e., the voltage and frequency) that is most useful to the task at hand.

120.1 Power Semiconductor Devices

Power electronic devices require power switches to do their job. When we think of a switch, we normally think of a device that opens and closes. We also usually assume that the switch has infinite resistance when it is open and zero resistance when it is closed. Figure 120.1(a) shows an idealized characteristic (the shaded bars on the V and I axes) for a switch that conducts current in either direction with no

0-8493-1586-7/05/$0.00+$1.50

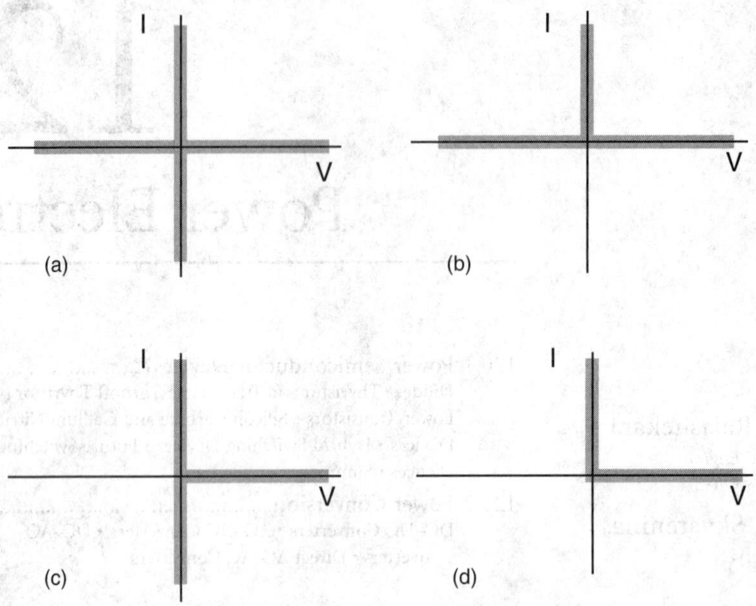

FIGURE 120.1 Ideal switch characteristics: (a) four-quadrant switch; (b) two-quadrant, voltage-blocking switch; (c) two-quadrant, reverse-conducting switch; and (d) one-quadrant switch.

voltage drop across the switch or holds off voltage of either polarity, without conducting current. A switch that can operate in this manner is a four-quadrant switch, since it can operate in any quadrant on the I-V axes. Most switches, however, are not capable of operating in all four quadrants; so one way of classifying switches is by the quadrants in which they operate. As shown in Figures 120.1(b), (c), and (d), switches may only operate in one or two quadrants. Figure 120.1(b) shows a two-quadrant switch that is capable of withstanding voltage in either direction, but can only conduct current in one direction, while Figure 120.1(c) shows a two-quadrant switch that conducts current in either direction but can only block voltage in the forward direction. Finally, Figure 120.1(d) shows a single-quadrant switch that can only block forward voltage or conduct positive current.

Of course, when one thinks about the operation of a switch, one assumes that it can be turned on or off, essentially at will. However, some power electronic switches are capable of closing but not opening when current is flowing through them. Thus, switches can also be categorized by how they operate. In particular, for an uncontrolled switch, such as a diode, the change between the on and off states is determined by the power circuit. Semi-controlled switches, such as the thyristor, are turned on by a control signal, but must be turned off by the operation of the power circuit. Finally, controlled switches can be turned on or off by application of an appropriate control signal, which may be a voltage or a current, depending on the switch. A number of popular semiconductor switches are described in this section.

Diodes

The diode consists of a P-N junction. Figure 120.2(a) shows a simple half-wave rectifier circuit containing a diode. The ideal diode blocks voltage when reversed biased and conducts when forward biased, with zero voltage drop across the diode. An ideal diode would not conduct any current when reverse biased and would provide any level of desired current with no voltage drop when forward biased, as shown in Figure 120.2(c). A real diode approximates that, as shown by the diode characteristic in Figure 120.2(d). When forward biased, there is a small voltage drop across the diode, typically about 0.7 to 1.0 volt for a silicon diode. This means the diode will dissipate power when current flows through it, and some type of cooling is necessary for high-power operation. When the diode is reverse biased, it allows a small leakage

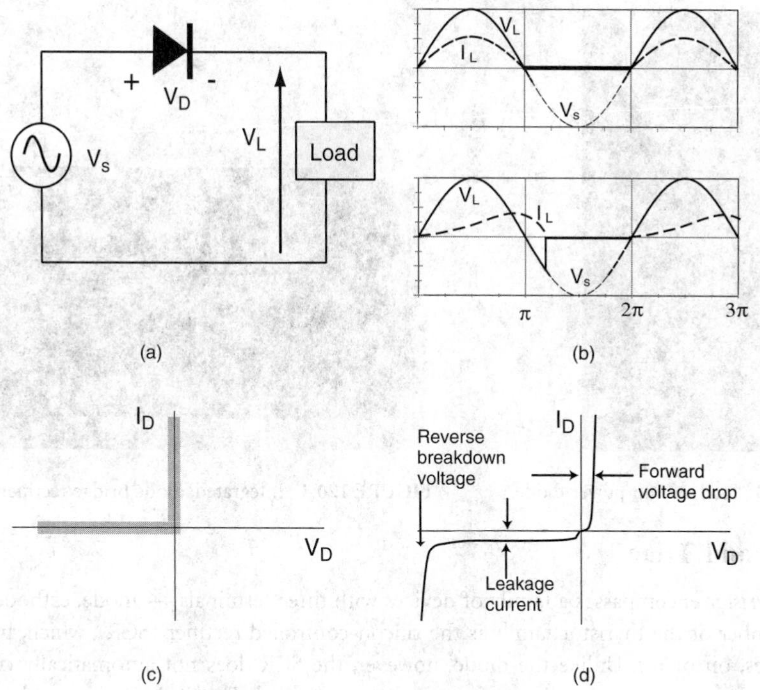

FIGURE 120.2 (a) Diode half-wave rectifier circuit. (b) Voltage and current waveforms for circuit of part (a) resistive load and R-L load. (c) Ideal diode volt-amp characteristic. (d) Actual diode volt-amp characteristic.

current until the reverse breakdown voltage is reached. When too much negative voltage is applied to the diode, it reaches a point where it begins conducting large amounts of current with little additional change in voltage. If the current is not limited by a series resistance, the diode would be rapidly destroyed due to the high-power dissipation. Clearly, a large reverse recovery voltage is desirable for high-power applications.

The principles of operation of the diode are demonstrated by the curves in Figure 120.2(b). The load is resistive for the top curves and resistive-inductive for the bottom curves. With a resistive load, the diode conducts during the positive half-cycle of the source voltage and blocks the negative half-cycle. Since the load is resistive, the current has the same wave shape as the voltage, except for the magnitude. For the lagging load, the diode cannot shut down while current is still flowing. Due to the inductance in the load, the current cannot change instantaneously. The current begins building up at the start of a positive half-cycle of voltage and is not back to zero when the source voltage goes negative. Thus, the diode continues conducting and the load voltage becomes negative as shown in Figure 120.2(b) until the current returns to zero. At that point the diode shuts off and the load voltage returns to zero until the next positive half-cycle.

It is important to realize, however, that a diode does not turn off instantaneously. Due to the charges within the diode, there is a small reverse recovery current that limits the speed of operation. For 60-Hz power applications the reverse recovery doesn't have a serious effect, but high-frequency applications may require fast-recovery diodes. Line frequency diodes, suitable for operation up to several hundred hertz, are available with ratings up to 6500 V and 8000 A. Fast recovery diodes have very low recovery times and are capable of operating up to several hundred kilohertz at 1 kV and 100 A. At 3 kHz, fast recovery diodes can operate up to 3300 V and 1400 A. Finally, Schottky diodes are used for low voltages (below 150 V) and very high-speed operation (up to 1 MHz). Figure 120.3 shows a picture of a stud-mounted rectifier rated for 150 A and 600 V. Manufacturers also provide integrated packages for standard configurations. Figure 120.4 shows a bridge rectifier module containing six diodes; the module is about 4" wide, with a peak reverse voltage of 1600 volts and a DC output current of 150 amps.

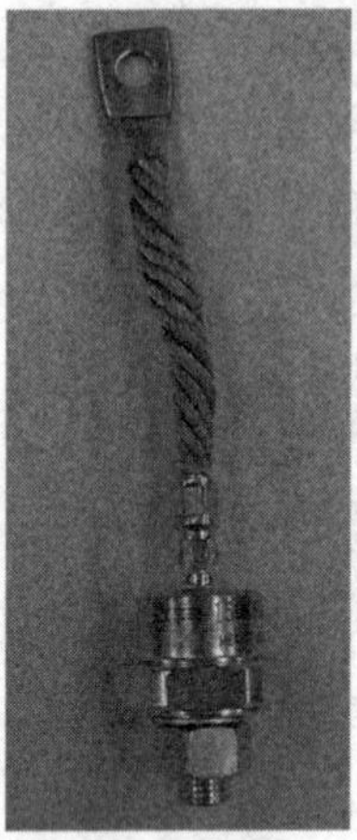

FIGURE 120.3 Stud-mount power diode.

FIGURE 120.4 Integrated, diode bridge rectifier.

Thyristor and Triac

The term *thyristor* encompasses a family of devices with three terminals — anode, cathode, and gate. A popular member of the thyristor family is the silicon-controlled rectifier (SCR), which, like the diode, has two states, on or off. Unlike the diode, however, the SCR does not automatically conduct when forward biased. As the name implies, turn-on can be controlled. The SCR was invented and named by the General Electric Company in the late 1950s. The thyristor was invented as a solid-state replacement for the thyratron, which is a tube device that is filled with mercury vapor.

Figure 120.5(a) shows a half-wave rectifier circuit using an SCR. Note that the symbol for the SCR is similar to the diode, with the addition of a gate. Unlike the diode, the SCR has the ability to block voltage in either the forward or reverse direction. The ideal volt-ampere characteristic for an ideal SCR is shown in Figure 120.5(b). Without a gate signal, the SCR will not conduct when forward biased, unless the forward breakdown voltage is exceeded. If a voltage of proper magnitude and duration is applied to the gate of the SCR, then forward conduction can begin at voltages below the forward breakdown voltage. The forward voltage at which conduction begins is determined by the magnitude of the gate current. Higher gate current causes lower forward breakdown voltage. Once the SCR is in the conducting mode, the gate signal is no longer needed. The thyristor has a low forward voltage drop while in the conducting mode, typically about 2 volts. Exceeding the forward breakdown voltage, applying a voltage with an excessively high DV/DT, or too much thermal stress, can also trigger a thyristor; however, these methods are not recommended. To preclude excessive DV/DT or DI/DT, snubber circuits are required for the thyristor.

Like the diode, the SCR conducts until the current drops nearly to zero. The value at which the SCR turns off is called the holding current. For purposes of describing the operation of various circuits, however, the holding current is often assumed to be zero. Figure 120.5(c) shows the voltage and current waveforms for the half-wave rectifier, again for a resistive load and for an R-L load. In the upper curves of Figure 120.5(c), the turn-on of the SCR has been delayed by an angle α, which is determined by the control circuit that generates the gate signal. Since the load is resistive, the current looks exactly like the voltage, except for the magnitude. When the voltage and current reach zero, at π, the SCR shuts off, until the next positive half-cycle. Essentially, the SCR chops out a portion of the positive half-cycle and the entire negative half-cycle of voltage. When the load has an inductive component, as in the lower curves of Figure 120.5(c), the current can no longer change instantaneously, so when the SCR is triggered to apply voltage to the load, the current starts building up, lagging behind the voltage. Thus, when the voltage reaches zero, there is still current flowing in the circuit and the SCR does not shut down, resulting in a negative load voltage. The result is a lower (DC) load voltage than was obtained for the resistive case. The RMS load voltage is actually higher, however, since the SCR conducts for a longer period.

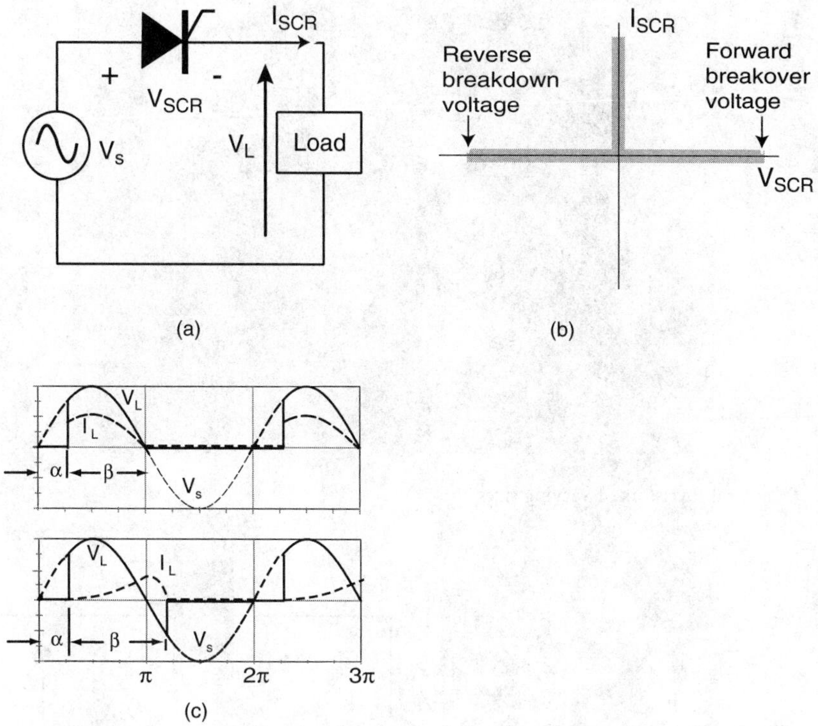

FIGURE 120.5 (a) Thyristor half-wave rectifier circuit. (b) Ideal thyristor volt-amp characteristic. (c) Voltage and current waveforms for circuit of part (a) resistive load and R-L load.

There are two types of thyristors. Phase-controlled, or converter, grade thyristors are used for rectifying line frequency voltage and current and offer very high voltage and current ratings with a low on-state voltage drop. Inverter grade thyristors are capable of faster operation and, as the name implies, are used in inverter circuits. However, since they cannot turn off, external commutation circuitry is required to artificially force the thyristor current to zero so it can shut off. Because of the need for commutating circuitry, other devices have largely replaced thyristors for inverter operation.

Some examples of high-end ratings for thyristors are 1500 A at 12 kV, 4000 A at 7 kV, and 7000 A at 6.5 kV. Figure 120.6 shows a photo of several thyristors of various sizes and styles. Thyristors can be connected in series for higher voltage ratings. Figure 120.7 shows a thyristor switch built from a press-pack stack of thyristors. The switch is rated at 50 kV and 1200 A and is about 30 in. high.

Diodes and SCRs allow rectification of AC voltage to DC; however, for some applications we want to control both half-cycles of the AC voltage. One example would be a dimmer switch for an incandescent light. An SCR-controlled, half-wave rectifier SCR would not allow the lamp to reach full brightness, because the negative half-cycle of voltage would always be blocked. Similarly, for some AC motors, we would want to control both the negative and positive half-cycles of the voltage, rather than rectify the voltage. Figure 120.8(a) shows a triac switch, which is essentially two SCRs placed back to back. The triac allows the control of AC voltage in both directions, and by delaying the firing of the triac in both directions we can reduce the RMS value of the voltage applied to the load. The triac tends to operate a bit slower than the SCR, so it is primarily useful for low-frequency applications.

Figure 120.8(b) shows the voltage and current waveforms for a simple triac AC voltage controller. In the upper curves, the load is purely resistive, and the voltage and current have the same wave shape. The triac effectively chops out a portion of each half-cycle of voltage. Intuitively, it makes sense that the larger the delay angle, α, the less will be the RMS voltage. The lower curves show an inductive load. In this

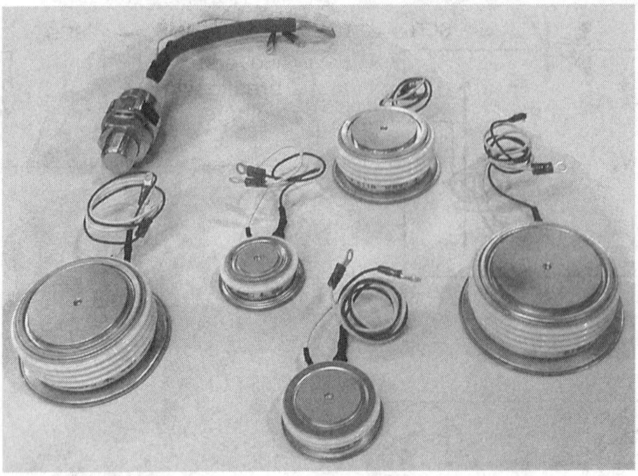

FIGURE 120.6 Several thyristors of varying sizes.

FIGURE 120.7 High-voltage switch built of stacked thyristors.

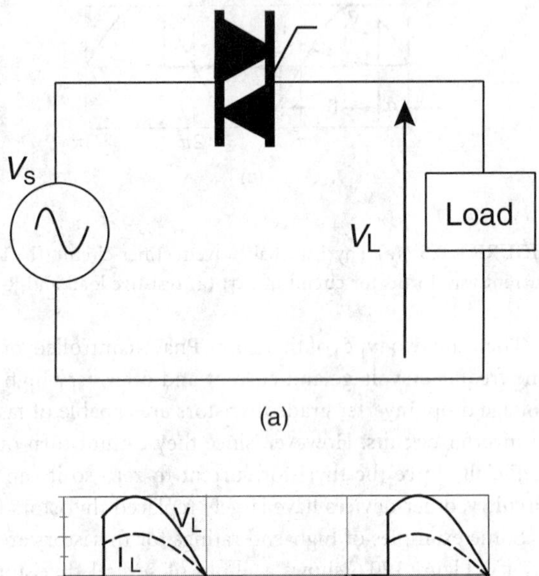

FIGURE 120.8 (a) Triac voltage-control circuit. (b) Voltage and current waveforms for circuit of part (a) resistive load and R-L load.

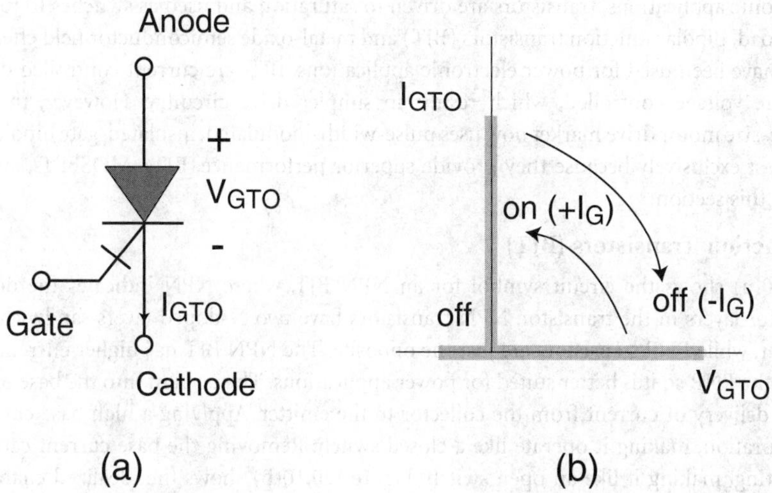

FIGURE 120.9 (a) GTO schematic element. (b) Ideal GTO volt-amp characteristic.

case, the current has not reached zero when the voltage does, and the voltage reverses polarity until the current reaches zero. If the inductance is large enough or the delay angle is small enough, the entire voltage waveform will be applied to the load.

The components discussed so far cannot interrupt a current; therefore, they continue delivering voltage to the load until the current reaches zero. In many applications, it is necessary to interrupt the current. Fortunately, several power electronic devices can turn off, as well as turn on.

Gate-Turnoff Thyristor (GTO)

As discussed above, the thyristor can be turned on with a positive gate current, but does not turn off until the current naturally goes to zero or is forced to zero by a commutating circuit. The gate-turnoff thyristor (GTO) was developed in the early 1980s and overcomes that limitation. A GTO operates much the same as a thyristor with respect to turn-on; however, a negative gate current can turn it off. The GTO has been used in motor drives and other applications, especially at high-power levels. With ratings of up to 6500 volts blocking and 4000 amps conducting, the GTO was largely responsible for the development of medium-voltage drives and megawatt inverters. Figure 120.9(a) shows the schematic symbol for the GTO and figure 120.9(b) shows its ideal volt-amp characteristic. In particular, GTOs generally have a low reverse breakdown voltage. For example, one manufacturer's GTO rated with a 4500-volt forward operating voltage has only a 19-volt reverse breakdown voltage. Some symmetrical GTOs are able to block voltages in both directions, however.

Compared to the SCR, the GTO has a similar turn-on time, but a faster turn-off time, making it suitable for higher-speed applications (typically a few kilohertz). The GTO has a higher forward voltage drop than the SCR when conducting (typically 3 volts), which means higher losses. The turn-on gate current for the GTO is somewhat higher than for the SCR, but the turn-off gate current is dependent on the current to be turned off and can be quite high, especially for GTOs with high blocking voltages. Thus, at GTO may require a large, expensive control circuit. Like standard thyristors, GTOs require snubber circuits, which limit the rate of rise of voltage across the switch or current through the switch. At high power levels, the snubbers can be very expensive.

Power Transistors

The transistor was invented in the 1940s, and began replacing the vacuum tube in amplifiers and other electronic applications. In those applications, the transistor operates in the active or linear region. For power electronic applications, transistors are driven to saturation and used as switches to turn power on and off to a load. Bipolar junction transistors (BJT) and metal-oxide semiconductor field effect transistors (MOSFET) have been used for power electronic applications. BJTs are current controlled devices, while MOSFETs are voltage controlled, which results in simpler drive circuitry. However, the small- and intermediate-size motor drive market now uses pulse-width-modulated, insulated-gate bipolar transistors (IGBT) almost exclusively because they provide superior performance. BJTs, MOSFETs, and IGBTs are described in this section.

Bipolar Junction Transistors (BJT)

Figure 120.10(a) shows the circuit symbol for an NPN BJT, where NPN indicates the doping of the semiconductor layers in the transistor. NPN transistors have two N-doped layers sandwiched around a P-doped layer, while PNP transistors are just the opposite. The NPN BJT has higher current and voltage ratings than the PNP, so it is better suited for power applications. The current into the base of a transistor controls the delivery of current from the collector to the emitter. Applying a high base current drives a BJT into saturation, making it operate like a closed switch. Removing the base current causes a BJT to stop conducting, making it like an open switch. Figure 120.10(b) shows the idealized characteristic for the power BJT. In reality, there is a drop of about 2 volts from the collector to the emitter when the transistor is conducting. Unlike the diode or SCR, the BJT does not block reverse voltage; it's reverse voltage rating may only be 5 to 10 volts. Thus, transistors are not used for AC applications unless they are protected by additional circuitry. They are available with ratings of 500 volts and up to 300 amps and have been used in devices up to several hundred kilowatts.

The main disadvantage of the power BJT is very low gain (I_C/I_B) — less than 10 for high voltage/power devices. A gain of 10 means that the base current must be 10% of the current that is to be switched, which, for high-power applications, would require a large power circuit to create the base current. A way to reduce the required base current is by using essentially a two-stage transistor, known as the Darlington connection, shown in Figure 120.10(c). The Darlington connection can be built from discrete components or as a single integrated circuit. In either case, the first transistor amplifies the base current into the second, and the overall gain is the product of the two individual gains. However, the Darlington connection has a higher forward voltage drop and longer turn-on time than the single BJT, which means higher losses and a lower frequency of operation. Darlington transistors are available with V_{CE} ratings

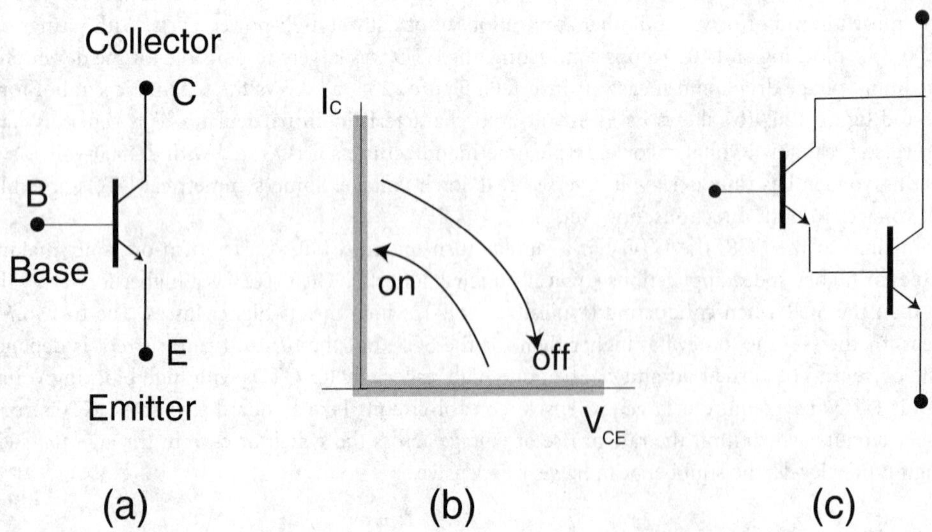

FIGURE 120.10 (a) BJT schematic element. (b) Ideal BJT volt-amp characteristic. (c) Darlington connection.

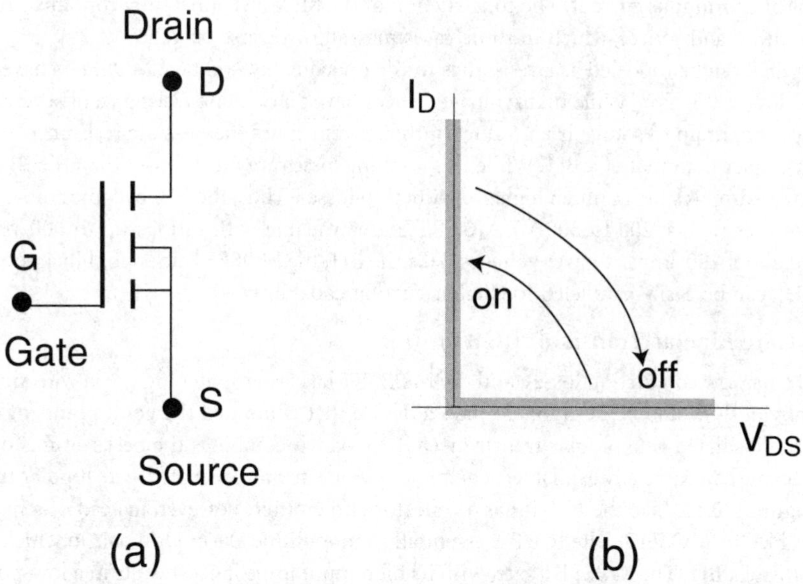

FIGURE 120.11 (a) MOSFET schematic element. (b) Ideal MOSFET volt-amp characteristic.

up to 1200 volts and current ratings up to 600 amps, although both ratings would not be available in a single device.

BJTs are difficult to operate in parallel and are limited to less than 10 kHz. As a result, BJTs have been largely replaced in new products by the MOSFET at low voltage and the IGBT at high voltage.

Metal-Oxide Semiconductor Field Effect Transistor (MOSFET)

Figure 120.11(a) shows the schematic symbol for a MOSFET, which is controlled by voltage rather than current. Figure 120.11(b) shows the ideal characteristic for the power MOSFET. Applying relatively small voltage (about 15 volts) between the gate and source turns on the MOSFET; applying zero volts turns the MOSFET off. Because of the low voltage and high gate input impedance, the gate current is very low. Thus, drive circuits for MOSFETs are relatively simple and can even be integrated with the MOSFET on a single power chip, greatly simplifying the system. MOSFETs can operate at much higher frequencies than BJTs, greater than 100 kHz, but are limited to low-power applications (a few kilowatts). Figure 120.12 shows a picture of some small MOSFETs. The largest one in the photo (at the left) is rated for 50 volts

FIGURE 120.12 Power MOSFETs.

and 20 amps of continuous current. The construction of the MOSFET inherently contains a reverse diode between the drain and source, which in some cases may allow reverse current.

The MOSFET can be modeled as a resistance in the on state. Low-voltage MOSFETs have an on-state resistance as low as 0.1 ohm, while high-voltage devices have an on-state resistance of several ohms. The voltage drop from drain to source is a product of the current times the on-state resistance, which can be significantly higher than that of a BJT. While its switching losses are much lower than the BJT, the higher forward voltage drop results in much higher conduction losses. Thus, the BJT becomes more economical at higher voltages (above 200 to 300 V). MOSFETs are available with ratings up to 600 volts at lower currents and about 200 amps at lower voltages. Like the BJT, the MOSFET does not block reverse voltage. The MOSFET can be easily paralleled for higher current capability.

Insulated Gate Bipolar Transistor (IGBT)

Since the BJT has low conduction losses, and the MOSFET has faster switching speed with simple control circuits, it obviously would be desirable to have a device that combined the good points of each. In the late 1980s, the insulated gate bipolar transistor (IGBT), which combines the best features of MOSFETs and BJTs, became commercially available. Schematic symbols for the insulated-gate bipolar transistor are shown in Figure 120.13. Like the BJT, it has a collector and emitter; however, instead of a base terminal, it has a gate, like the MOSFET. The IGBT is essentially a monolithic Darlington pair in which a MOSFET is used to control a BJT. The MOSFET gate, with its high input impedance, results in a low-power voltage control for the device with relatively fast switching times, although slower than a MOSFET. The BJT portion, which conducts the main current, provides lower conduction losses than a MOSFET. During the 1990s, the IGBT became the device of choice for variable-speed AC motor drives up to several hundred horsepower. By the end of the 1990s, IGBTs were pushing the lower limits of applications for GTOs, as higher blocking voltage ratings became available. IGBTs are now available with ratings up to 1200 amps at 3300 volts or 900 amps at 4500 volts.

Standard IGBTs are not capable of operating at as high a frequency as MOSFETs, although some manufacturers offer IGBTs that are optimized for speed and can approach the switching speed of a MOSFET. There is a tradeoff because the on-state power loss of an IGBT increases as the frequency of operation increases. For example, one manufacturer quotes a forward voltage drop of 1.3 volts for a standard IGBT, 1.5 volts for a fast IGBT, and 1.9 volts for an ultra-fast IGBT. Thus, for the same current level, the ultra-fast device would dissipate about 50% more power than the standard device.

IGBTs can be purchased as discrete modules, but are often sold in packaged assemblies with built-in protective circuitry. Figure 120.14 shows a discrete IGBT on the right and a cut-away of the same

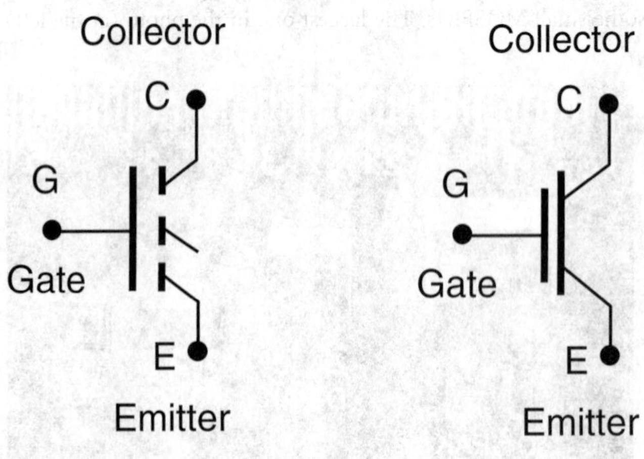

FIGURE 120.13 Two alternative IGBT schematic elements.

FIGURE 120.14 IGBT cut-away and packaged.

FIGURE 120.15 IGBT power module.

FIGURE 120.16 Integrated IGBT inverter module.

type on the left. Figure 120.15 shows an IGBT module rated for 200 amps and Figure 120.16 shows a complete inverter package. The three devices at the top of the package are the IGBTs and the three devices at the bottom are freewheeling diodes. A significant advantage of the IGBT is that it does not require snubber circuits.

Silicon Carbide and Gallium Nitride Devices

Nonsilicon-based, large band-gap semiconductors, silicon carbide (SiC) and gallium nitride (GaN), are two promising materials for high-temperature power electronics applications. Silicon carbide technology is more mature than gallium nitride. SiC is currently the most practical high-temperature wide band gap material for advanced power conversion. The large band gap enables operation with junction temperatures higher than 600°C and a high breakdown electric field. Also, high thermal conductivity of SiC enables higher power density, thus leading to compact packaging. In addition, the high temperature capability of SiC greatly reduces cooling requirements, thus reduces the total size of the power electronics unit. Some important material properties of silicon, SiC, and gallium nitride materials are shown in Table 120.1. Significant cost challenges have prevented these devices from being commercially successful. Crystal growth and material defect control are two of the primary issues that need to be resolved before cost can be competitive with silicon devices.

Presently, SiC Schottky diodes are commercially available from several suppliers. These devices can operate at high voltages and high temperatures. SiC and GaN diodes are best suited in high-temperature applications above 600 V.

TABLE 120.1 Material Properties of Si, SiC, and GaN

Property	Si	4H-SiC	GaN
Band gap (Ev)	1.1	3.26	3.49
Electron mobility at 300 K (cm^2/Vs)	1500	700	900
Critical breakdown field (MV/cm)	0.3	2.0	3.3
Thermal conductivity (W/cm-K)	1.5	4.5	>1.7
Relative dielectric constant (εr)	11.8	10	9.0

Hybrid Switching Devices

MOS-Controlled Thyristor (MCT)

Thyristors have lower on-state losses than BJTs and IGBTs, but cannot easily be turned off. Thus, a thyristor device that is controlled by a MOSFET would be very desirable. The MOS-controlled thyristor (MCT) is a promising device that became available in lower ratings during the late 1990s. The MCT has lower on-state losses than the IGBT for a given turn-off switching time, but the maximum switching frequency for the MCT is lower than that of the IGBT. Research has demonstrated MCTs with blocking voltages of 3000 volts, with low on-state voltage drop; however, the manufacturing of the MCT is more complex than other power electronic devices. Thus, the initial devices that came to market were limited to 75 amps and 600 volts.

Gate Commutated Thyristor (GCT)/Integrated Gate Commutated Thyristor (IGCT)

The thyristor offers a low voltage drop while conducting resulting in low losses while in the on state. The GTO, as described above, made possible the development of multi-megawatt inverters; however, the turn-off of the GTO is relatively slow, which causes it to have relatively high turn-off losses. BJTs, on the other hand, offer faster, less loss in turn-off, but consume more power while conducting. Clearly, a device that could conduct as a thyristor and turn off as a transistor would be superior to a GTO. The gate commutated thyristor (GCT), which is such a device, was developed in the 1990s and introduced commercially in 1997. Since then it has begun to replace the GTO in high-power applications.

The GCT is based on the GTO, but is designed so that the cathode emitter shuts off instantly. As a result, the cathode current is transferred to the gate, causing the thyristor to become a transistor during shut-off, resulting in up to 40% less loss than in a GTO. Another advantage of the GCT is that it does not need DV/DT snubbers, which creates a significant decrease in parts count and thus cost for an inverter. Three types of GCTs are available: asymmetric, reverse blocking, and reverse conducting. Figure 120.17 shows circuit symbols for the asymmetric and reverse blocking in 120.17(a) and for reverse conducting in 120.17(b). The reverse-conducting GCT requires a freewheeling diode, which may be built on the same silicon wafer as the GCT in some cases. GCTs have been built using wafers up to 6 in. in diameter, with ratings up to 6000 amps at 6 kV for asymmetrical devices. Figure 120.18 shows a picture of a 4 in. GCT wafer. The radial lines that are connected by rings comprise the gate circuit, while the ring midway from the center to the edge of the wafer is the freewheeling diode.

Like a thyristor, the GCT has a low on-state voltage of about 2 volts at 4000 amps. To reduce the turn-off losses, the device current must be turned off in 1 microsecond. Since the voltage applied to the gate is only 20 volts or so, the circuit inductance must be extremely low. For example, to turn off a current of 3000 amps in 1 microsecond, with 20 volts, requires that the inductance in the circuit must be less than about 6 nH. Achieving such a low gate circuit inductance requires special design considerations. For this reason, many GCTs are sold integrated with a gate circuit. The integrated unit is thus called an integrated gate commutated thyristor (IGCT). Figure 120.19 shows a picture of an IGCT package, which includes the GCT, freewheeling diode, and gate driver. The input circuit merely requires a 20-volt power supply and an infrared fiber optic link to deliver the turn-on and turn-off commands.

GCTs seem poised to take over virtually all applications that were previously done by GTOs. They currently offer a range of 0.3 MW to 300 MW at a cost per switched KW that is comparable to the IGBT. Due to its faster turn-off, the GCT can operate at higher frequencies than the GTO.

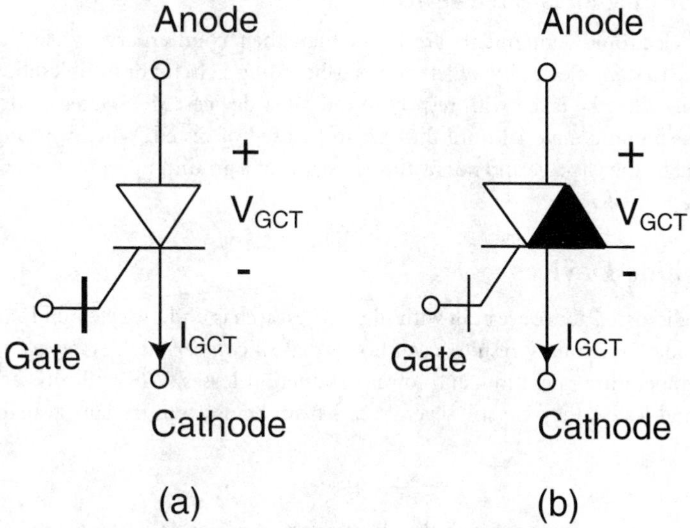

FIGURE 120.17 Circuit symbols for the GCT: (a) asymmetric and reverse blocking, and (b) reverse conducting.

FIGURE 120.18 GCT device wafer.

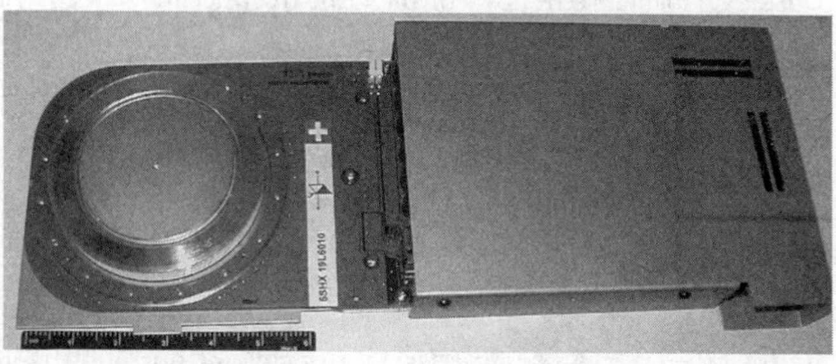

FIGURE 120.19 IGCT module.

Handling and Testing MOS-gated devices

Although power electronic components are larger than their counterparts in electronic circuits, they remain susceptible to static electricity. Whether bread-boarding a circuit or troubleshooting a commercial system, proper care must be taken with respect to the MOS devices. MOS-gated devices should be kept in their antistatic shipping material until they are to be used or tested. When working with them, they should be handled by the package and not by the leads. Use of a grounding wrist strap, while not essential, is a good practice as well.

Future Switching Devices

All of the devices discussed above are built with silicon. Research is underway to build various components from silicon carbide. Preliminary results have shown that silicon carbide devices offer higher switching speeds, higher temperature operation, and lower conduction losses. While silicon carbide devices have been fabricated and tested, they are still several years from being commercially available.

Summary

Table 120.2 shows a summary of some of the characteristics of power electronic switches.

120.2 Power Conversion

Power converters are generally classified as follows:

1. DC–DC converters (choppers, buck and boost converters)
2. AC–DC converters (phase-controlled converters)
3. DC–AC converters (inverters)
4. Direct AC–AC converters (cycloconverters)

DC–DC Converters

DC–DC converters are used to convert unregulated DC voltage to regulated or variable DC voltage at the output. They are widely used in switch-mode DC power supplies and in DC motor drive applications. In DC motor control applications they are called *chopper-controlled drives.* The input voltage source is usually a battery or is derived from AC supply using a diode bridge rectifier. These converters are generally either hard-switched PWM types or soft-switched resonant-link types. There are several DC–DC converter topologies, the most common ones being buck converter, boost converter, and buck-boost converter.

Buck Converter

A buck converter is also called a *step-down converter.* Its principle of operation is illustrated in Figure 120.20(a). The IGBT acts as a high-frequency switch. The IGBT is repetitively closed for a time t_{on} and opened for a time t_{off}. During t_{on} the supply terminals are connected to the load and power flows from supply to the load. During t_{off} load current flows through the freewheeling diode D_1, and the load voltage is ideally zero. The average output voltage is given by

$$V_{out} = D \cdot V_{in}$$

where D is the duty cycle of the switch and is given by $D = t_{on}/T$, where T is the time for one period. The term $1/T$ is the switching frequency of the power device IGBT.

Boost Converter

A boost converter is also called a *step-up converter.* Its principle of operation is illustrated by Figure 120.20(b). This converter is used to produce higher voltage at the load than the supply voltage. When

TABLE 120.2 Characteristics of Thyristor and Transistor Family Members for Power Electronics Applications

	Thyristors			Transistors		
	SCR	GTO	IGCT	BJT	MOSFET	IGBT
Approximate year introduced	Late 1950s	1980s	1997	1970s	Early 1980s	Late 1980s
Max forward voltage/volts	12,000	6500	6000	500	600	4500
Max reverse voltage	12,000	~20	~20 asymmetric 6000 reverse blocking	0	0	0
Max current amps	7000	4000	6000	300	200	1200
Max switching frequency	NA	<1 KHz	1 KHz	5–10 KHz	up to 1 MHz	50–100 KHz
Gate drive complexity	Simple turn on only	Expensive, high current for turn off	Simple, voltage control	Expensive, requires large base current when on	Simple voltage control	Simple voltage control
Forward voltage drop (volts) while conducting	2	2–3	2–3	1–2	Variable I * R*on*	2–6
Snubbers required	Sometimes	Yes	Turn on only	Yes	Less than BJT	Not usually
Status/prognosis	Useful for uncontrolled rectifiers and line commutated inverters	Mature technology, being replaced by IGCT and other variants	Good for high-power applications; technology still developing	Mature technology largely replaced by IGBTs in new products	Good for high-frequency low power (e.g., switched power supplies)	Technology still developing; widely used in low- and medium-power applications

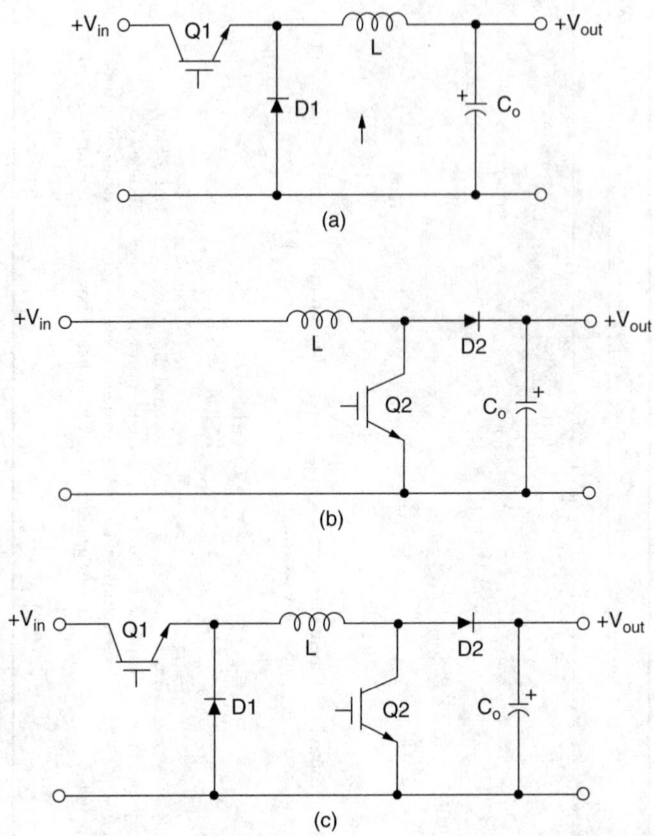

FIGURE 120.20 DC–DC converter configurations: (a) buck converter; (b) boost converter; (c) buck-boost converter. (*Source:* Rajashekara, K. S., Bhat, A. K. S., and Bose, B. K. 1997. Power electronics. In *The Electrical Engineering Handbook*, 2nd ed. R. C. Dorf, Ed., CRC Press, Boca Raton, FL, p. 709. With permission.)

the power switch is on, the inductor is connected to the DC source and the energy from the supply is stored in it. When the device is off, the inductor current is forced to flow through the diode and the load. The induced voltage across the inductor is negative. The inductor voltage adds to the source voltage of force the inductor current into the load. The output voltage is given by

$$V_{\text{out}} = V_{\text{in}} / (1 - D)$$

Thus, for variation of D in the range $0 < D < 1$, the load voltage V_o will vary in the range

$$V_{\text{in}} < V_{\text{out}} < \infty$$

Buck-Boost Converter

A buck-boost converter can be obtained by the cascade connection of the buck and the boost converter. The output voltage V_o, is given by

$$V_o = V_{\text{in}} \cdot D / (1 - D)$$

The output voltage is higher or lower than the input voltage based on the duty cycle D. A typical buck-boost converter topology is shown in Figure 120.20(c). When the power devices are turned on, the input provides energy to the inductor and the diode is reverse biased. When the devices are turned off, the energy stored in the inductor is transferred to the output. No energy is supplied by the input during this interval. In DC power supplies, the output capacitor is assumed to be very large, which results in a constant output voltage. The buck and boost converter topologies enable the four-quadrant operation of a DC motor. In DC drive systems, the chopper is operated in step-down mode during motoring and in step-up mode during regeneration operation.

Resonant-Link DC–DC Converters

The use of resonant converter topologies in power supplies would help to reduce the switching losses in DC–DC converters and enable the operation at switching frequencies in the megahertz range, which results in reduced size, weight, and cost of the power supplies. Other advantages of resonant converters are that the leakage inductances of the high-frequency transformers and the junction capacitances of semiconductors can be used to further increase the power density of the power supplies. An added advantage is the significant reduction of RFI/EMI. The major disadvantage of resonant converters is increased peak current or voltage stress.

Single-Ended Resonant Converters

These types of converters are referred to as *quasi-resonant converters*. Quasi-resonant converters (QRC) do exhibit a resonance in their power section, but, instead of the resonant elements being operated in a continuous fashion, they are operated for only one-half of a resonant sine wave at a time. The topologies within the quasi-resonant converters are simply resonant elements added to many of the basic PWM topologies as shown in Figures 120.21 and 120.22. The QRCs can operate with zero-current switching or zero-voltage switching or both. The power switch in quasi-resonant converters connects the input voltage source to the tank circuit and is turned on and off in the same step fashion as in PWM switching power supplies. The conduction period of the devices is determined by the resonant frequency of the tank circuit. The power switch turns off after the completion of one-half of a resonant period. So, the current at turn-on and turn-off transitions is zero, thus eliminating the switching loss within the switch. In zero-voltage switching ORCs, at the turn-on and turnoff of the power devices, the voltage across the device is zero, thus again reducing the switching loss.

The major problems with the zero-current switching QRC are the high-peak currents through the switch and capacitive turn-on losses. The zero-voltage switching QRCs suffer from increased voltage stress on the power device. The full-wave mode-zero-voltage switching circuit suffers from capacitive turn-on losses.

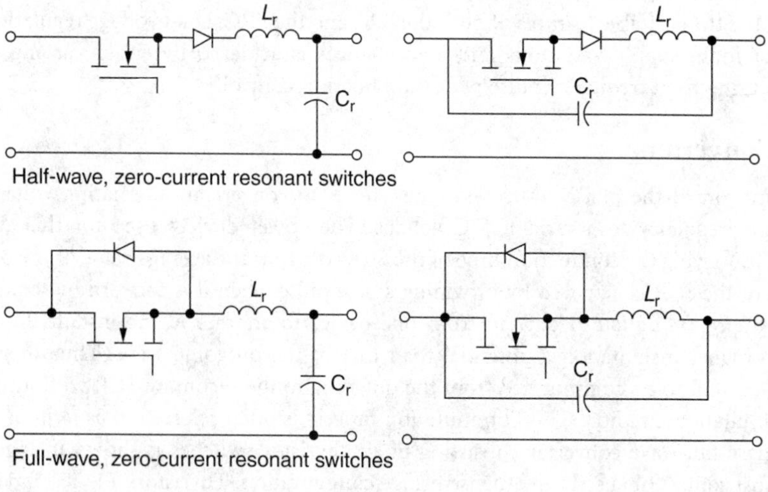

FIGURE 120.21 Zero-current resonant switches.

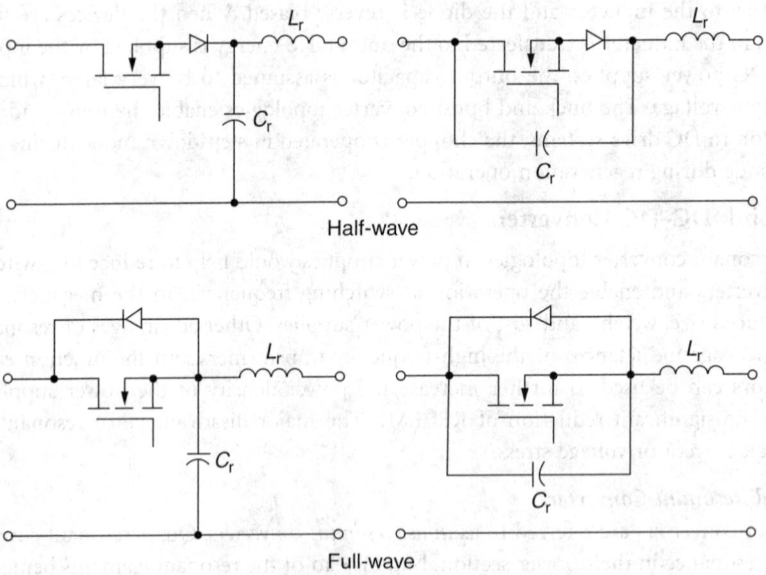

FIGURE 120.22 Zero-voltage quasiresonant switches.

Double-Ended Resonant Converters

These converters use full-wave rectifiers at the output, and they are generally referred to as resonant converters. A number of resonant converter configurations are realizable by using various resonant tank circuits; the three most popular configurations are the series resonant converter (SRC), the parallel resonant converter (PRC), and the series-parallel resonant converter (SPRC), as shown in Figure 120.23.

Series resonant converters have a high efficiency from full load to part load. Transformer saturation is avoided due to the series-blocking resonating capacitor. The major problems with the SRC are that it requires a very wide change in switching frequency to regulate the load voltage and that the output filter capacitor must carry high-ripple current.

Parallel resonant converters are suitable for low–output voltage, high–output current applications due to the use of filter inductance at the output with low-ripple current requirements for the filter capacitor. The major disadvantage of the PRC is that the device currents do not decrease with the load current, resulting in reduced efficiency at reduced load currents.

The SPRC has the desirable features of both the SRC and the PRC. Load voltage regulation in resonant converters for input supply variations and load changes is achieved by either varying the switching frequency or using fixed frequency pulse width modulation control.

AC–DC Converters

The basic function of the phase-controlled converter is to convert an alternating voltage of variable amplitude and frequency to a variable DC voltage. The power devices used for this application are generally SCRs. Varying the conduction time of the SCRs controls the average value of the output voltage. The turn-on of the SCR is achieved by providing a gate pulse when it is forward biased. The turnoff is achieved by the *commutation* of current from one device to another at the instant the incoming AC voltage has a higher instantaneous potential than that of the outgoing wave. Thus there is a natural tendency for current to be commutated from the outgoing to the incoming SCR without the aid of any external commutation circuitry. This commutation process is often referred to as *natural commutation.*

A three-phase full-wave converter consisting of six thyristor switches is shown in Figure 120.24(a). This is the most commonly used three-phase bridge configuration. Thyristors T1, T3, and T5 are turned on during the positive half-cycle of the voltages of the phases to which they are connected, and thyristors, T2, T4, and T6 are turned on during the negative cycle of the phase voltages. The reference for the angle

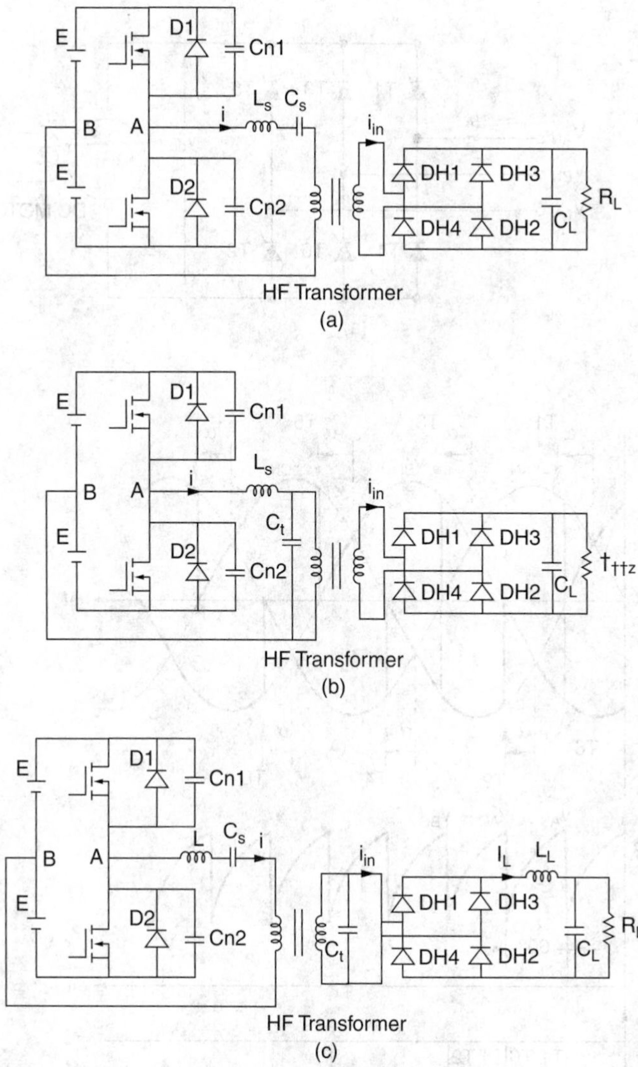

FIGURE 120.23 High-frequency resonant converter (half-bridge version) configurations suitable for operation above resonance. (a) Series resonant converter. Leakage inductances of the high-frequency (HF) transformer can be part of resonant inductance. (b) Parallel resonant converter. (c) Series-parallel resonant converter with capacitor *Ct* placed on the secondary side of the HF transformer. (Adapted from Rajashekara, K. S., Bhat, A. K. S., and Bose, B. K. 1997. Power electronics. In *The Electrical Engineering Handbook*, 2nd ed. R. C. Dorf, CRC Press, Boca Raton, FL, p. 788. With permission.)

in each cycle is at the crossing points of the phase voltages. The ideal output voltage, output current, and input current waveforms are shown in Figure 120.24(b). Controlling the firing angle α controls the output DC voltage. If the load is a DC motor, varying the firing angle α varies the speed of the motor.

The average output voltage of the converter at a firing angle α is given by

$$v_{d\alpha} = (3\sqrt{3}/\pi)E_m \cos\alpha$$

where E_m is the peak value of the phase voltage.

At $\alpha = 90°$ the output voltage is zero. For $0° < \alpha < 90°$, $v_{d\alpha}$ is positive and power flows from the AC supply to the DC load. For $90° < \alpha < 180°$, $v_{d\alpha}$ is negative and the converter operates in the inversion

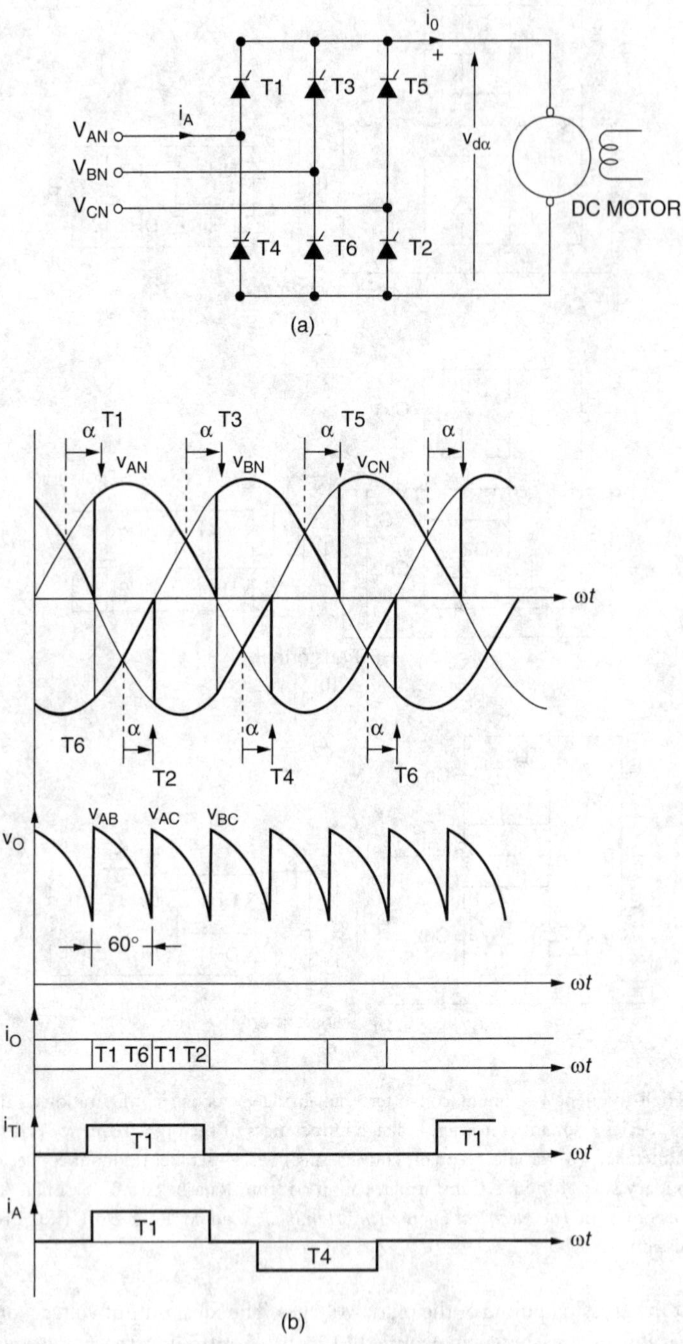

FIGURE 120.24 (a) Three-phase thyristor full-bridge configuration. (b) Output voltage and current waveforms. (Adapted from Rajashekara, K. S., Bhat, A. K. S., and Bose, B. K. 1997. Power electronics. In *The Electrical Engineering Handbook, 2nd ed.*, R. C. Dorf, Ed., CRC press, Boca Raton, FL, p. 773. With permission.)

mode. Thus the power can be transferred from the motor to the AC supply, a process known as *regeneration.*

In Figure 120.24(a), the top or the bottom thyristors could be replaced by diodes. The resulting topology is called *thyristor semiconverter.* With this configuration, the input power factor is improved but the regeneration is not possible.

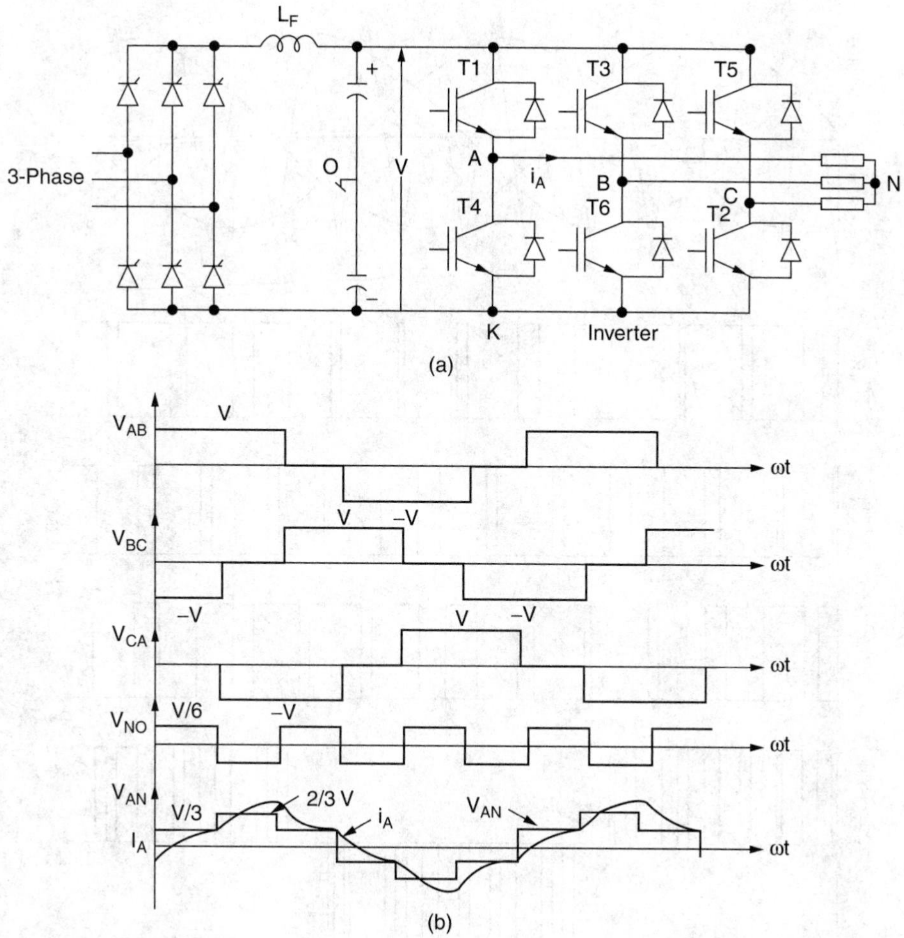

FIGURE 120.25 (a) Three-phase converter and voltage-source inverter configuration. (b) Square-wave inverter waveforms. (*Source:* Rajashekara, K. S., Bhat, A. K. S., and Bose, B. K. 1997. Power electronics. In *The Electrical Engineering Handbook, 2nd ed.*, R. C. Dorf, Ed., CRC Press, Boca Raton, FL, p. 774. With permission.

DC–AC Converters

The DC–AC converters are generally called *inverters*. The AC supply is first converted to DC, which is then converted to a variable-voltage and variable-frequency power. The power conversion stage generally consists of a three-phase bridge converter connected to the AC power source, a DC link with a filter, and the three-phase inverter bridge connected to the load, as shown in Figure 120.25(a). In the case of battery-operated systems there is no intermediate DC link. An inverter can be classified as a voltage source inverter (VSI) or a current source inverter (CSI). A voltage source inverter is fed by a stiff DC voltage, whereas a current source inverter is fed by a stiff current source. A voltage source can be converted to a current source by connecting a series inductance and then varying the voltage to obtain the desired current. A VSI can also be operated in current-controlled mode and, similarly, a CSI can also be operated in the voltage control mode. These inverters are used in variable-frequency AC motor drives, uninterrupted power supplies, induction heating, static VAR compensators, and so on.

Voltage Source Inverter

The inverter configuration shown in Figure 120.25(a) is the voltage source inverter. The voltage source inverters are controlled either in square wave mode or in pulse width modulation (PWM) mode. In a square wave mode the frequency of the output is controlled within the inverter; the devices are being

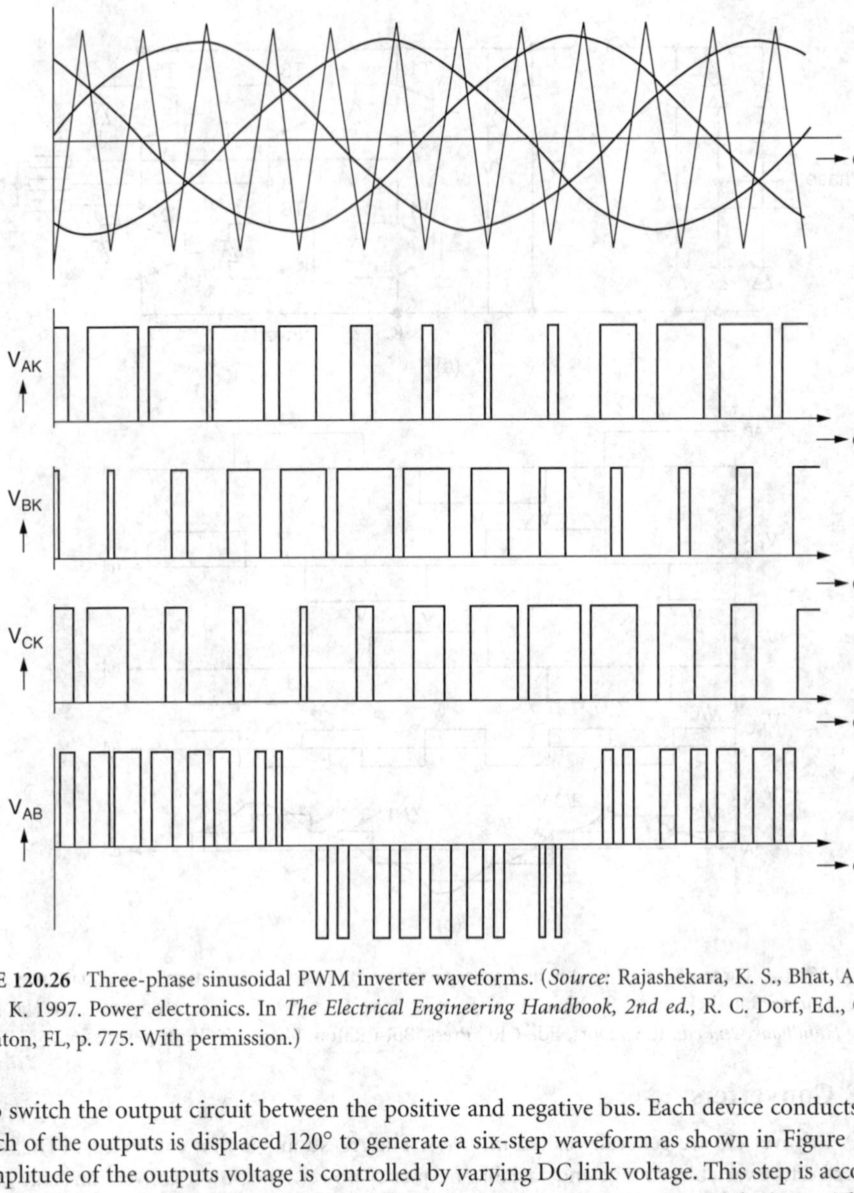

FIGURE 120.26 Three-phase sinusoidal PWM inverter waveforms. (*Source:* Rajashekara, K. S., Bhat, A. K. S., and Bose, B. K. 1997. Power electronics. In *The Electrical Engineering Handbook, 2nd ed.*, R. C. Dorf, Ed., CRC Press, Boca Raton, FL, p. 775. With permission.)

used to switch the output circuit between the positive and negative bus. Each device conducts for 180°, and each of the outputs is displaced 120° to generate a six-step waveform as shown in Figure 120.25(b). The amplitude of the outputs voltage is controlled by varying DC link voltage. This step is accomplished by varying the firing angle of the thyristors of the three-phase bridge converter at the input. The six-step output is rich in harmonics and hence needs heavy filtering.

In PWM inverters the output voltage and frequency are controlled within the inverter by varying the width of the output pulses. Hence, at the front end, instead of a phase-controlled thyristor converter, a diode bridge rectifier can be used. A very popular method of controlling the output voltage and frequency is by sinusoidal pulse width modulation. In this method a high-frequency triangle carrier wave is compared with a three-phase sinusoidal waveform as shown in Figure 120.26. The power devices in each phase are switched on at the intersection of the sine and triangle waves. The amplitude and frequency of the output voltage are varied by varying the amplitude and frequency, respectively, of the reference sine waves. The ratio of the amplitude of the sine wave to the amplitude of the carrier wave is called the *modulation index*. The harmonic components in a PWM wave are easily filtered because they are shifted to a higher-frequency region. It is desirable to have a high ratio of carrier frequency to fundamental frequency to reduce the harmonics of lower-frequency components. There are several other PWM

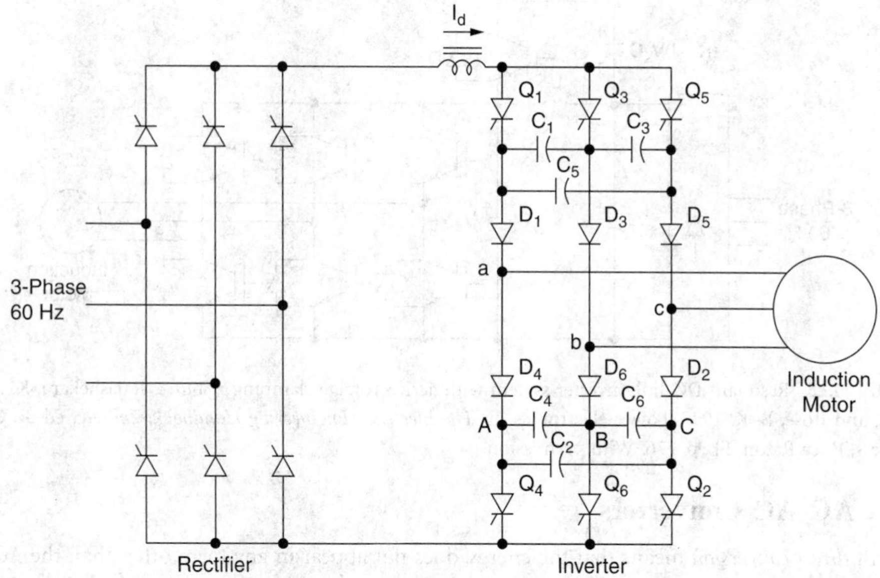

FIGURE 120.27 Force-commutated current-fed inverter control of an induction motor. (*Source:* Rajashekara, K. S., Bhat, A. K. S., and Bose, B. K. 1997. Power electronics. In *The Electrical Engineering Handbook, 2nd ed.*, ed. R. C. Dorf, CRC Press, Boca Raton, FL, p. 801. With permission.)

techniques mentioned in the literature. The most notable ones are selected harmonic elimination, hysteresis control, and the space vector PWM technique.

Current Source Inverter

Unlike the voltage source inverter — in which the voltage of the DC link is imposed on the motor windings — the current source inverter shown in Figure 120.27 imposes current into the motor. Here, the amplitude and phase angle of the motor voltage depend on the load conditions of the motor. The capacitors and series diodes help commutation of the thyristors. One advantage of this drive system is that regenerative braking is easily achieved because the rectifier and inverter can reverse their operation modes. Six-step machine current causes large harmonic heating and torque pulsation, which may be quite harmful at low-speed operation. Another disadvantage is that the converter system cannot be controlled in open loop like in a voltage source inverter.

Resonant-Link Inverter

The use of resonant switching techniques can be applied to inverter topologies to reduce the switching losses in the power devices. They also permit high switching frequency operation to reduce the size of the magnetic components in the inverter unit. In the resonant DC link inverter shown in Figure 120.28, a resonant circuit is added at the inverter input to convert a fixed DC to a pulsating DC voltage. This resonant circuit enables the devices to be turned on and off during the zero-voltage interval. Zero-voltage switching (ZVS) or zero-current switching (ZCS) is often called *soft switching*. Under soft switching, the switching loss in the power devices is almost eliminated. The EMI problem is less severe because resonant voltage pulses have lower DV/DT compared to those of hard-switched PWM inverters. Also, the machine insulation is less stretched because of lower DV/DT resonant voltage pulses. In Figure 120.28 all the inverter devices are turned on simultaneously to initiate a resonant cycle. The commutation from one device to another is initiated at the zero–DC link voltage. The inverter output voltage is formed by the integral numbers of quasi-sinusoidal pulses. The circuit consisting of devices Q, D, and the capacitor C acts as an active clamp to limit the DC link voltage to about 1.4 times the diode rectifier voltage V_s. It is possible to use passive clamp circuits instead of the active clamp.

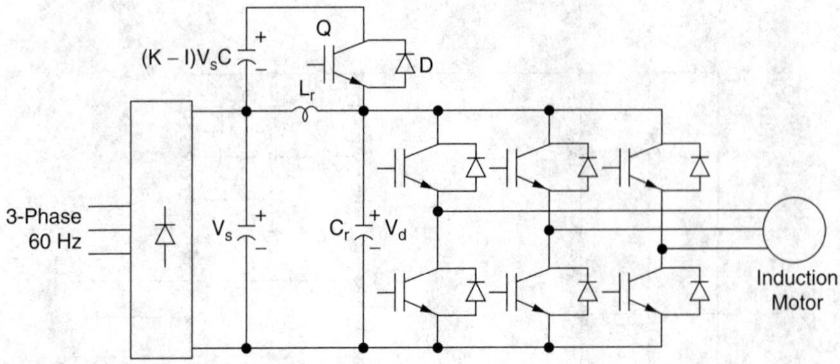

FIGURE 120.28 Resonant DC-link inverter system with active voltage clamping. (*Source:* Rajashekara, K. S., Bhat, A. K. S., and Bose, B. K. 1997. Power electronics. In *The Electrical Engineering Handbook, 2nd ed.*, ed. R. C. Dorf, CRC Press, Boca Raton, FL, p. 776. With permission.)

Direct AC–AC Converters

The term *direct conversion* means that the energy does not appear in any form other than the AC input or AC output. The cycloconverters are direct AC-to-AC frequency changers. The three-phase full-wave cycloconverters configuration is shown in Figure 120.29. Two antiparallel, phase-controlled, six-pulse thyristor bridge converters supply each phase of the three-phase motor. The output frequency is lower than the input frequency and is generally an integral multiple of the input frequency. The cycloconverter permits energy to be fed back into the utility network without any additional measures. Also, the phase sequence of the output voltage can be easily reversed by the control system. Cycloconverters have found applications in aircraft systems and industrial drives for controlling synchronous and induction motors.

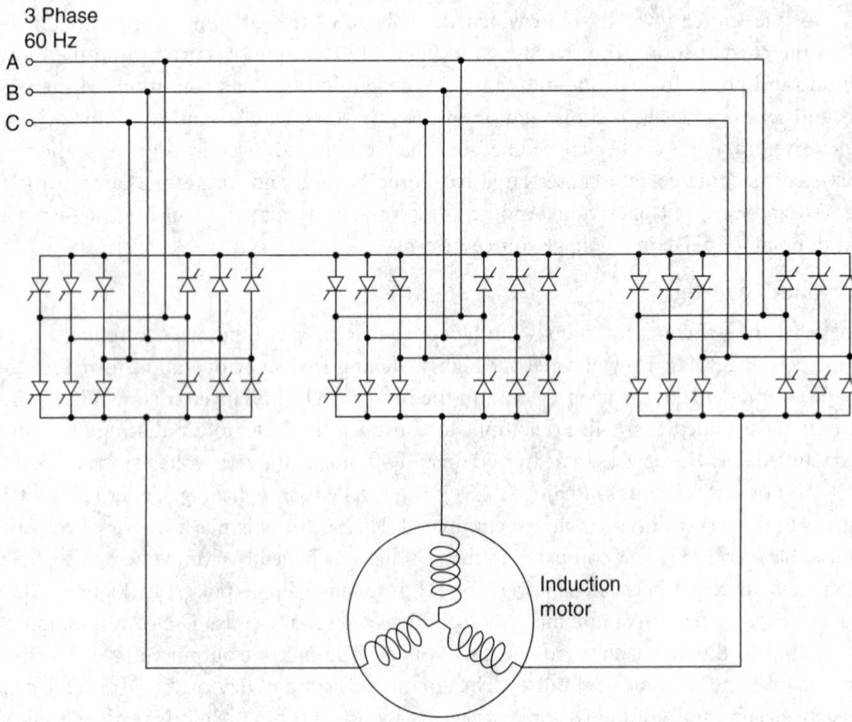

FIGURE 120.29 Cycloconverter control of an induction motor.

Defining Terms

Commutation — Process of transferring the current from one power device to another.
Duty cycle — Ratio between on time of a switch and the switching period.
Forward voltage — The voltage across the device when the anode is positive with respect to the cathode.
Full-wave control — Both the positive and negative half-cycle of the waveform is controlled.
Hysteresis control — A method of controlling current in which the instantaneous current can vary within a band.
Inverter — Circuit that converts DC to AC.
Isolated — A power electronic circuit or device having ohmic isolation between the input source and load circuit.
Rectifier — Circuit that converts AC to DC.
Snubber — An auxiliary circuit used to protect a power electronic switch from current or voltage transients. An R-L circuit in series can limit the rate of rise of current through the device; an R-C circuit in parallel with a switch limits the rate of rise of voltage across the device.
Switching frequency — The frequency at which the devices are turned on and turned off. Switching frequency = $1/(t_{on} + t_{off})$.

References

Ahmed, A. 1999. *Power Electronics for Technology.* Prentice Hall, Englewood Cliffs, NJ.
Bhat, A. K. S. 1991. A unified approach for the steady-state analysis of resonant converters. *IEEE Trans. Ind. Electron.*, 38:251–259.
Bose, B. K. 1992. *Modern Power Electronics.* IEEE Press, Piscataway, NJ.
Bose, B. K. 2002. *Modern Power Electronics and AC Drives.* Prentice Hall, New York.
Brown, M. 1990. *Practical Switching Power Supply Design.* Academic Press, New York.
Carroll, E. and Siefken, J. 2002. *IGCTs: moving on the right track. Power Electron. Technol.*, 28:16–18.
Kemerly, R. T., Wallace, H. B., and Yoder, M. N. 2002. *Impact of Wide Band Gap Microwave Devices on DoD Systems. Proc. IEEE*, 90:1059–1064.
Krein, P. 1998. *Elements of Power Electronics.* Oxford University Press, Oxford.
Mohan, N., Undeland, T., and Robbins, W. P. 1995. *Power Electronics: Converters, Applications, and Design.* John Wiley & Sons, New York.
Rajashekara, K. S., Bhat, A. K. S., and Bose, B. K. 1993. Power electronics. In: Dorf, R. C., *The Electrical Engineering Handbook,* CRC Press, Boca Raton, FL, pp. 694–737.
Rashid, M. H. 1993. *Power Electronics, Circuits, Devices and Applications,* 2nd ed. Prentice Hall, Englewood Cliffs, NJ.
Stephani, D. Prospects of SiC Power Devices from the State of the Art to Future Trends. In: PCIM Proceedings, Nuremberg, Germany, May 12–14.
Venkataramanan, G. and Divan, D. 1990. Pulse Width Modulation with Resonant DC Link Converter. Paper presented at IEEE Industry Applications Society annual meeting, pp. 984–990.

Further Information

Bose, B. K. 1986. *Power Electronics and AC Drives.* Prentice Hall, Englewood Cliffs, NJ.
Mohan, N. and Undeland, T. 1995. *Power Electronics: Converters, Applications, and Design*, 2nd ed. John Wiley & Sons, New York.
Murphy, J. M. D. and Turnbull, F. G. 1988. *Power Electronic Control of AC Motors.* Pergamon Press, New York.
Sen, P. C. 1981. *Thyristor DC Drives.* John Wiley & Sons, New York.
Sum, K. K. 1988. *Recent Developments in Resonant Power Conversion.* Intertech Communications, Ventura, CA.
Skvarenina, T. L., Ed. 2002. *The Power Electronics Handbook.* CRC Press, Boca Raton, FL.

121

A/D and D/A Converters

Rex T. Baird
Silicon Laboratories, Inc.

Jerry C. Hamann
University of Wyoming

Analog-to-digital (A/D) and digital-to-analog (D/A) converters provide the fundamental interface between continuous-time signals and digital-processing circuitry. These devices find wide application in consumer electronics, digital audio, instrumentation, data acquisition, signal processing, communication, automatic control, and related areas. The A/D converter provides a discrete-time, discrete-valued **digital encoding** of a real-world signal, typically a voltage. The D/A converter reverses this process, producing a continuous-time, or analog, signal corresponding to a given digital encoding.

The block diagram for an example application including both A/D and D/A functions is given in Figure 121.1. Here, the digital-processing block might be a dedicated digital signal processor (DSP) or a general purpose microprocessor or microcontroller that implements an application-specific signal-processing task. The input and output **signal conditioning** blocks provide functionality that might include input bandwidth limiting to prevent aliasing, output frequency shaping and bandwidth limiting to smooth the analog output, multiplexing of multiple input or output signals, or a **sample-and-hold** function. Component vendors provide some of the input and output signal conditioning functions within the A/D or D/A to simplify system integration.

Several conversion methods exist for A/D and D/A conversion. Competing methods co-exist primarily due to trade-offs in four major specifications; resolution (number of binary digits *n* in the conversion representation), **conversion rate** (the time required to complete a conversion), power consumption, and cost. Table121.1 and Table 121.2 provide samples of representative converter integrated circuits (ICs) and their associated parameters. Refer to the manufacturer datasheets listed in the chapter references for further details and additional examples. More complete coverage of A/D and D/A principles and circuit design can be found in Razavi [1995] or Van De Plassche [1994].

The Fundamentals of A/D Converters

Many methods exist for implementing the A/D conversion function; **integrating**, **sigma-delta**, **successive approximation register** (SAR), **pipelined**, and **flash** compose the bulk of competing architectures. These methods offer different trade-offs between resolution, conversion rate, power consumption, and cost. Although these techniques differ in the manner in which they arrive at a digital encoding of the continuous-time signal, many aspects of A/D conversion process are architecture independent.

Figure 121.2 shows a diagram of three important steps in the A/D quantization process, with the first two steps being signal conditioning. The first step is input bandwidth limiting, provided by the anti-alias

0-8493-1586-7/05/$0.00+$1.50

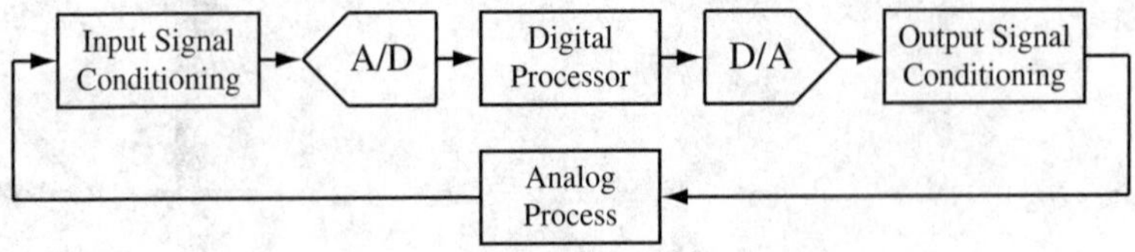

FIGURE 121.1 Example A/D and D/A application block diagram.

TABLE 121.1 Example A/D Integrated Circuits

A/D Integrated Circuit	Resolution	Conversion Rate	Conversion Method
Analog Devices AD7791A	24 bit	60,200 μs	Sigma-delta
Maxim MAX1186	10 bit	25 ns	Pipelined
Maxim MAX105	6 bit	1.25ns	Flash
National ADC12030	13 bit	8.8 μs	Successive approximation
Texas Instruments TLC7135	4.5 bit	333,000 μs	Dual-slope-integrating

TABLE 121.2 Example D/A Integrated Circuits

D/A Integrated Circuit	Resolution	Settling Time	Conversion Method
Analog Devices DAC08	8 bit	85 ns	Multiplying
Analog Devices AD1871	24 bit	N/A, 96 Hz sample rate	Sigma-delta
Maxim MAX5207	16 bit	25 μs	Multiplying
Texas Instruments DAC2902	12 bit	30 ns	Fixed precision reference

filter. See graphs (a) and (b) of Figure 121.2. Basically, this attenuates high frequency signal and noise spectra so that the A/D output is not corrupted by high frequency spectra which modulate (i.e., alias) to low frequency by the sampling process. Filter requirements are set by the input signal spectra, the sample rate and the conversion method. For example, an integrating A/D usually has strict filter requirements, but a sigma-delta converter used in the same application may relax filter requirements to the point that inherent bandwidth limiting of the input signal is sufficient, and no explicit anti-alias filter is required.

The second step in the conversion process is the sample-and-hold operation. Graph (c) Figure 121.2 shows the detail of this operation. During the acquisition phase (which may be very short as shown here) the sample-and-hold output transitions to match the input signal. During the hold phase the output value is frozen at that value until the beginning of the next acquisition phase. The sample-and-hold acts to sample the input at exact increments of time, namely, the end of the acquisition phase. The sample-and-hold reduces errors in the A/D output by allowing the encoder to act upon a static input during each conversion process. Component vendors generally integrate the sample-and-hold along with

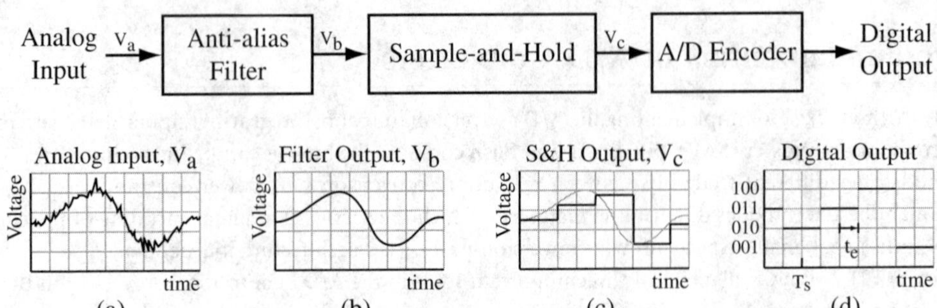

FIGURE 121.2 Steps in the A/D process.

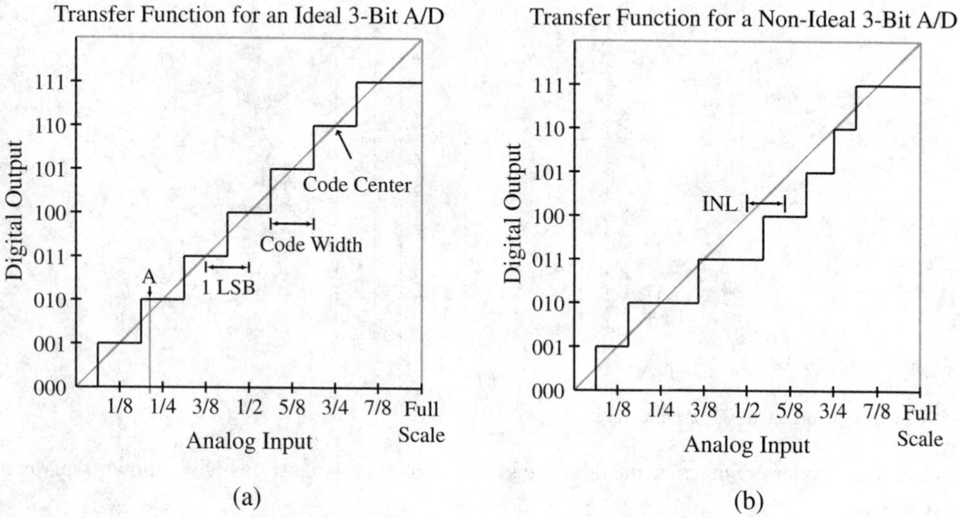

FIGURE 121.3

the A/D converter. In topologies such as sigma-delta and SAR A/Ds the sample-and-hold function is inherent in the circuit architecture.

The third step in the conversion process is quantization, detailed in graph (d) of Figure 121.2. During each hold phase of the sample-and-hold the A/D encoder input will be encoded to one of 2^n (n is the number of bits of the A/D converter) discrete digital encoding levels and represented by this level for T_s sec. There is an aperture delay time, t_e, from the beginning of the sample-and-hold acquisition phase to when it is available at the encoder output. The aperture delay time depends largely upon the conversion method.

The signal conditioning blocks of Figure 121.2 suppress two main sources of noise and distortion, aliasing and signal variation during the encoding process. The encoder introduces additional noise that should be considered. Figure 121.3a shows the transfer function for an ideal 3-bit A/D. The finite quantization introduces an unavoidable quantization error. For example, if the analog input is at point A, there is no digital code exactly on the diagonal line, so it will be represented by 010 and there is a quantization error of –(1 LSB)/4 where an LSB is change in input voltage for a least significant bit. The quantization noise causes a fundamental limit to the signal-to-noise ratio of a converter. Increasing the converter resolution increases the signal-to-noise ratio. See Razavi [1995] or Oppenheim and Schafer [1998] for more coverage of quantization principles.

The non-ideal transfer function of Figure 121.3b shows the final source of error to be discussed, inaccurate quantization levels. Ideally, the digital output will change each time the analog input changes by 1 LSB, called the code-width. The deviation in each code-width from the ideal 1 LSB is called the differential non-linearity (DNL). The largest deviation from any code center to the ideal diagonal line is referred to as the integral non-linearity (INL). Code-width errors will cause both noise and distortion in the digitally encoded waveform. There are many more error sources in actual circuit implementations. The Analog Devices application note is a good source for further exploration [AN-282].

To investigate how A/D quantization might be undertaken, consider the block diagram of a flash architecture given in Figure 121.4 for an n-bit converter. A resistive divider with 2^n resistors provides a set of reference voltages to an array of 2^n-1 comparators. The reference voltage of each comparator is 1 LSB greater than the reference voltage of the comparator immediately below it in the array. The bottom comparator has a reference voltage of 1/2 LSB. Each comparator produces a "1" when the analog input is greater than its reference voltage and a "0" otherwise. Generally, clocked comparators are used so the output only changes at the rising edge of the clock and then remains static until the next rising edge. In

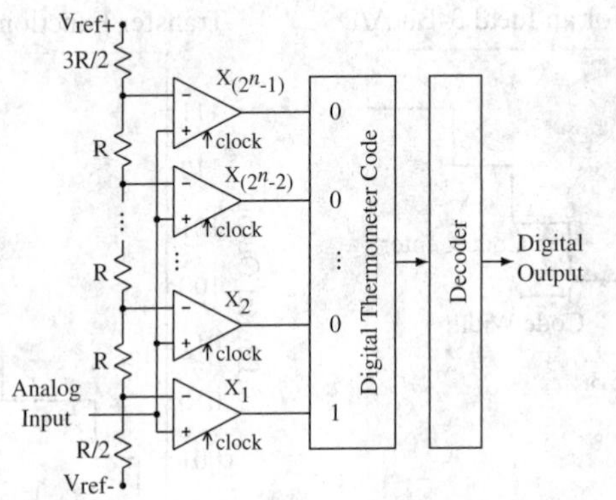

FIGURE 121.4 *n*-bit flash A/D architecture.

this case, the encoder architecture includes the sample-and-hold circuitry which is shown separately in Figure 121.2.

With this arrangement all comparators having a reference voltage less then the analog input will produce a "1" and all the remaining comparators will produce a "0." This type of encoding is referred to as a thermometer code. The thermometer code is usually decoded to a binary output code.

The Fundamentals of D/A Converters

The purpose of a D/A converter is to produce a continuous-time analog voltage or current representing a digital input. Just as in the case of A/D conversion, many methods exist for implementing the D/A function; **multiplying**, **fixed precision reference**, hybrid, and **sigma-delta** compose the bulk of competing architectures. This discussion focuses on fundamental concepts that are largely independent of the conversion method rather than discuss details of each method.

Figure 121.5 shows a block diagram of the steps important to D/A conversion. Note that the last two blocks, the signal conditioning blocks, are the same as for A/D conversion (Figure 121.2) but they serve a much different purpose. The first step is the actual D/A converter, which converts a digital word into an analog signal, usually a voltage as shown in graphs (a) and (b) of Figure 121.5.

Figure 121.6 shows example circuitry to accomplish the D/A conversion task. The input register stores a new digital input at each rising edge of the sample clock. The output of this register controls an array of switches connecting an array of weighted current sources to a load resistor. Each current source is twice the size of the previous current source, just as each bit of the digital word represents a value that is twice the previous bit. If a bit is "1" it closes the switch. The current in that branch will then flow through the load resistor and create a voltage weighted by that bit of the data word. If a bit is "0" the

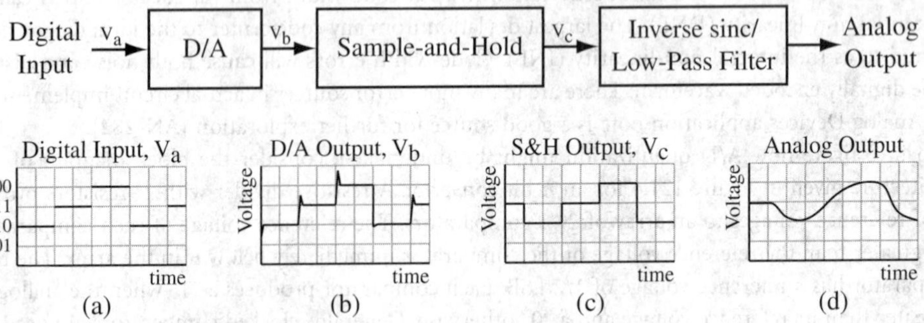

FIGURE 121.5 Steps in the D/A process.

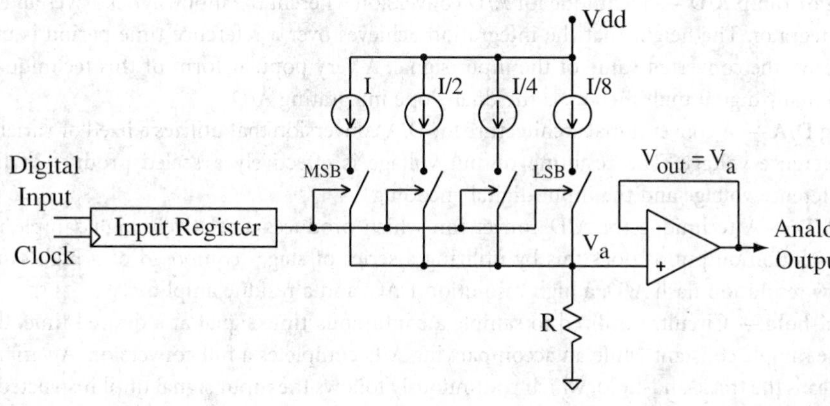

FIGURE 121.6 D/A architecture.

switch remains open. The current will then be prevented from flowing to the load resistance and it will not add to the output voltage. The currents from all bits which are "1" will sum in the load resistance and produce a voltage V_a corresponding to the value of the digital input word. The buffer amplifier drives the converter output to $V_{out} = V_a$ so that the output load will not affect the generated voltage.

As in the example just discussed, there is usually a sample clock which generates a new output upon each rising clock edge. This is called synchronization and allows precise control of when the output is generated. This is necessary because there may be significant skew between the bits of the input word arriving at the input of the D/A converter. But even with synchronization, there is often enough skew within the circuitry of the D/A converter that the output may momentarily go through a variety of output states before the final state is reached. This produces a "**glitch**" in the analog output each time a new output is generated, as shown in graph (b) of Figure 121.5. These glitches can add significant noise and distortion to the output waveform. The sample-and-hold block of Figure 121.5 eliminates glitches from the D/A converter by sampling the analog output after each glitch has settled. This result is shown in graph (c). The final block, the inverse sinc/analog low-pass filter, provides frequency shaping and bandwidth limiting to smooth the analog output as evident in graph (d). The inverse sinc filter is necessary to compensate for the sinc envelope present in the output spectra of the D/A.

Defining Terms

Conversion rate — The time elapsed between the start and completion of the conversion task. Typically applied in the context of A/D conversion. For D/A conversion, refer to settling time.

Digital-encoding — A variety of binary codes are utilized in conversion hardware. Examples include natural binary, binary-coded decimal (BCD), Gray code, sign-magnitude, offset binary, twos complement, and ones complement. Some converters allow for use of one or more of these coding schemes.

Fixed precision reference D/A — A converter architecture that utilizes either a fixed precision internal or external reference so the user cannot scale the output voltage by changing the reference.

Flash A/D — Technique for A/D conversion that utilizes 2^n-1 comparators, operating in parallel, to provide an n-bit digitization. This technique provides extremely fast conversion rates, while requiring a large component count in the IC realization.

Glitch — When a D/A converter receives a new output request, the asynchronous nature of the switching circuitry can result in momentary spikes or anomalous output values as the converter switches to the new state described by the digital encoding. These spurious changes in the output value are referred to as glitches, and the circuitry employed to remove or reduce their effects is said to deglitch the output.

Integrating or ramp A/D — Technique for A/D conversion wherein the input signal drives an electronic integrator. The height that the integration achieves over a reference time period is utilized to derive the converted value of the input signal. A very popular form of this technique utilized in many digital multimeters, is the dual-slope integrating A/D.

Multiplying D/A — A conventional architecture for D/A conversion that utilizes a fixed or variable-input reference voltage. The resulting output voltage is effectively a scaled product of the input reference voltage and the input digital encoding.

Pipelined A/D — A technique for A/D conversion which produces a relatively high sample rate high resolution output. It does this by utilizing a series of stages composed of a high sample rate low resolution flash A/D, a high resolution DAC, and a residue amplifier.

Sample-and-hold — Circuitry utilized to sample a continuous-time signal at a desired time, then hold the sample constant while an accompanying A/D completes a full conversion. A variant of this idea is the track-and-hold, which continuously follows the input signal until instructed to hold.

Settling time — With reference to the D/A task, the time elapsed between the change of the input digital encoding and the settling of the output analog signal to within a stated tolerance.

Sigma-delta A/D (also called delta-sigma A/D) — Technique for A/D conversion wherein the analog input signal level is compared to an approximate reconstruction of the input at each sampling instance. The repeated comparison results in a serial bit stream where the value of each bit denotes whether the input was greater than or less than the approximation at the given sample time. This technique typically involves a very high sampling rate that, when combined with a filtering of the low-resolution comparison results, yields a high precision moderate-rate conversion. This technique finds wide application in digital audio.

Sigma-delta D/A — Analogous to a sigma-delta A/D except that in this case the input is a multi-bit digital data stream and the output is an analog signal.

Signal conditioning — The general term for circuitry that preconditions the analog signal entering the A/D converter and similarly filters or smooths the output of the D/A.

Successive approximation register A/D — Technique for A/D conversion wherein a stepwise approximation of the input signal is constructed via a serial hardware comparator algorithm. The algorithm converges to an n-bit digitization of the input in n steps by successively assigning the most significant bits based upon a comparison of the input with the analog equivalent of the bits already assigned.

References

Analog Devices Inc. *Fundamentals of Sample Data Systems*, Application note AN-282. Analog Devices Inc., Norwood, MA.

Analog Devices Inc. Data converter application notes and datasheets at the website http://www.analog.com. Analog Devices Inc., Norwood, MA.

Demler, M. J. 1991. *High-Speed Analog-to-Digital Conversion*. Academic Press, San Diego, CA.

Maxim Integrated Products, Inc. Data converter application notes, datasheets and design guides at the website http://www.maxim-ic.com. Maxim Integrated Products Inc., Sunnyvale, CA.

National Semiconductor Corp. Data converter application notes and datasheets at the website http://www.national.com. National Semiconductor Corp., Santa Clara, CA.

Oppenheim, A. V., Schafer, R. W., and Buck, J. R. 1998. *Discrete-Time Signal Processing*, 2nd ed. Prentice Hall, Englewood Cliffs, NJ.

Razavi, B. 1995. *Principles of Data Conversion System Design*. IEEE Press, Piscataway, NJ.

Texas Instruments Inc. Data converter application notes and datasheets at the website http://www.ti.com. Texas Instruments Inc., Dallas, TX.

Van De Plassche, R. 1994. *Integrated Analog-to-Digital and Digital-to-Analog Converters*. Kluwer Academic Publishers, New York.

Further Information

Innovative uses of A/D and D/A hardware are broadly disseminated in trade journals, including *Electrical Design News (EDN)*, *Electronic Design*, and *Computer Design*. For application of conversion hardware in digital audio arenas, refer to *AES: Journal of the Audio Engineering Society*.

Manufacturers of conversion hardware often provide detailed application notes and supporting documentation for designing with and utilizing converters. Notable among these are the extensive publications of Analog Devices Inc. and Maxim Integrated Products Inc.

The frontier of conversion hardware technology is growing rapidly. Contemporary techniques for device fabrication and interfacing can often be found in the *IEEE Journal of Solid-State Circuits*, *IEEE Transactions on Instrumentation and Measurement*, and *IEEE Transactions on Consumer Electronics*. For a systems and signal-processing emphasis, refer to the *IEEE Transactions on Acoustics, Sound and Signal Processing* and the *IEEE Transactions on Circuits and Systems*.

122 Superconductivity[1]

Kevin A. Delin
Jet Propulsion Laboratory

Terry P. Orlando
Massachusetts Institute of Technology

122.1 Introduction

The fundamental ideal behind all of a superconductor's unique properties is that **superconductivity** is a quantum mechanical phenomenon on a macroscopic scale created when the motions of individual electrons are correlated. According to the theory developed by John Bardeen, Leon Cooper, and Robert Schrieffer (BCS theory), this correlation takes place when two electrons couple to form a Cooper pair. For our purposes, we may therefore consider the electrical charge carriers in a superconductor to be Cooper pairs (or more colloquially, superelectrons) with a mass m^* and charge q^* twice those of normal electrons. The average distance between the two electrons in a Cooper pair is known as the coherence length, ξ. Both the coherence length and the binding energy of two electrons in a Cooper pair, 2Δ, depend upon the particular superconducting material. Typically, the coherence length is many times larger than the interatomic spacing of a solid, and so we should not think of Cooper pairs as tightly bound electron molecules. Instead, there are many other electrons between those of a specific Cooper pair allowing for the paired electrons to change partners on a time scale of $h/(2\Delta)$, where h is Planck's constant.

If we prevent the Cooper pairs from forming by ensuring that all the electrons are at an energy greater than the binding energy, we can destroy the superconducting phenomenon. This can be accomplished, for example, with thermal energy. In fact, according to the BCS theory, the critical temperature, T_c, associated with this energy is

$$\frac{2\Delta}{k_B T_c} \approx 3.5 \tag{122.1}$$

where k_B is Boltzmann's constant. For low critical temperature (conventional) superconductors, 2Δ is typically on the order of 1 meV, and we see that these materials must be kept below temperatures of about 10 K to exhibit their unique behavior. Superconductors with high critical temperature, in contrast, will superconduct up to temperatures of about 100 K, which is attractive from a practical view because the materials can be cooled cheaply using liquid nitrogen. A second way of increasing the energy of the electrons is electrically driving them. In other words, if the critical current density, J_c, of a superconductor

[1] This chapter is modified from Delin, K. A. and Orlando, T. P. 1993. Superconductivity. In *The Electrical Engineering Handbook*, ed. R. C. Dorf, pp. 1114–1123. CRC Press, Boca Raton, FL.

0-8493-1586-7/05/$0.00+$1.50

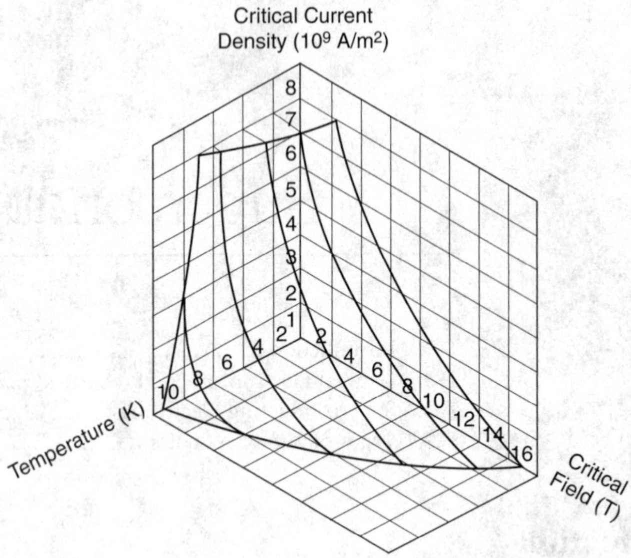

FIGURE 122.1 The phase space for the superconducting alloy niobium-titanium. The material is superconducting inside the volume of phase space indicated. [*Source:* Orlando, T. P. and Delin, K. A. 1991. *Foundations of Applied Superconductivity.* p. 10. Addison-Wesley, Reading, MA. With permission. (As adapted from Wilson, 1983.)]

is exceeded, the electrons have sufficient kinetic energy to prevent the formation of Cooper pairs. The necessary kinetic energy can also be generated through the induced currents created by an external magnetic field. As a result, if a superconductor is placed in a magnetic field larger than its critical field, H_c, it will return to its normal metallic state. To summarize, superconductors must be maintained under the appropriate temperature, electrical current density, and magnetic field conditions to exhibit its special properties. An example of this phase space is shown in Figure 122.1.

122.2 General Electromagnetic Properties

The hallmark electromagnetic properties of a superconductor are its ability to carry a static current without any resistance and its ability to exclude a static magnetic flux from its interior. It is this second property, known as the Meissner effect that distinguishes a superconductor from merely being a perfect conductor (which conserves the magnetic flux in its interior). Although superconductivity is a manifestly quantum mechanical phenomenon, a useful classical model can be constructed around these two properties. In this section we will outline the rationale for this classical model, which is useful in engineering applications such as waveguides and high-field magnets.

The zeros DC resistance criterion implies that the superelectrons move unimpeded. The electromagnetic energy density, w, stored in a superconductor is therefore

$$w = \frac{1}{2}\varepsilon\,\mathbf{E}^2 + \frac{1}{2}\mu_o\mathbf{H}^2 + \frac{n^*}{2}m^*\mathbf{v}_s^2 \tag{122.2}$$

where the first two terms are the familiar electric and magnetic energy densities, respectively. (Our electromagnetic notation is standard: ε is the permittivity, μ_o is the permeability, $\mathbf{E}$ is the electric field, and the magnetic flux density, $\mathbf{B}$, is related to the magnetic field, $\mathbf{H}$, via the constitutive law $\mathbf{B} = \mu_o\,\mathbf{H}$.) The last term represents the kinetic energy associated with the undamped superelectrons' motion (n^* and $\mathbf{v}_s$ are the superelectrons' density and velocity, respectively). Because the supercurrent density, $\mathbf{J}_s$, is related to the superelectron velocity by $\mathbf{J}_s = n^*q^*\,\mathbf{v}_s$, the kinetic energy term can be rewritten

$$n^*\left(\frac{1}{2}m^*\mathbf{v}_s^2\right)=\frac{1}{2}\Lambda\mathbf{J}_s^2 \qquad (122.3)$$

where Λ is defined as

$$\Lambda=\frac{m^*}{n^*(q^*)^2} \qquad (122.4)$$

Assuming that all the charge carriers are superelectrons, there is no power dissipation inside the superconductor, and so Poynting's theorem over a volume V may be written

$$-\int_V \nabla\cdot(\mathbf{E}\times\mathbf{H})\,dv=\int_V \frac{\partial w}{\partial t}\,dv \qquad (122.5)$$

where the left side of the expression is the power flowing into the region. By taking the time derivative of the energy density and appealing to Faraday's and Ampère's laws to find the time derivatives of the field quantities, we find that the only way for Poynting's theorem to be satisfied is if

$$\mathbf{E}=\frac{\partial}{\partial t}(\Lambda\mathbf{J}_s) \qquad (122.6)$$

This relation, known as the *first London equation* (after the London brothers, Heinz and Fritz), is thus necessary if the superelectrons have no resistance to their motion.

Equation (122.6) also reveals that the superelectrons' inertia creates a lag between their motion and that of the electric field. As a result, a superconductor can support a time-varying voltage drop across itself. The impedance associated with the supercurrent, therefore, is an inductor, and it will be useful to think of Λ as an inductance created by the correlated motion of the superelectrons.

If the first London equation is substituted into Faraday's law, $\nabla\times\mathbf{E}=-(\partial\mathbf{B}/\partial t)$, and integrated with respect to time, the *second London equation* results:

$$\nabla\times(\Lambda\mathbf{J}_S)=-\mathbf{B} \qquad (122.7)$$

where the constant of integration has been defined to be zero. This choice is made so that the second London equation is consistent with the Meissner effect, as we now demonstrate. Taking the curl of the quasi-static form of Ampère's law, $\nabla\times\mathbf{H}=\mathbf{J}_s$, results in the expression $\nabla^2\mathbf{B}=-\mu_o\nabla\times\mathbf{J}_s$, where a vector identity, $\nabla\times\nabla\times\mathbf{C}=\nabla(\nabla\cdot C)-\nabla^2\mathrm{C}$; the constitutive relation, $\mathbf{B}=\mu_o\mathbf{H}$; and Gauss's law, $\nabla\cdot\mathbf{B}=0$, have been used. By now appealing to the second London equation, we obtain the vector Helmholtz equation

$$\nabla^2\mathbf{B}-\frac{1}{\lambda^2}\mathbf{B}=0 \qquad (122.8)$$

where the penetration depth is defined as

$$\lambda\equiv\sqrt{\frac{\Lambda}{\mu_o}}=\sqrt{\frac{m^*}{n^*(q^*)^2\mu_o}} \qquad (122.9)$$

From Equation (122.8) we find that a flux density applied parallel to the surface of a semi-infinite superconductor will decay away exponentially from the surface on a spatial length scale of order λ. In other words, a bulk superconductor will exclude an applied flux as predicted by the Meissner effect.

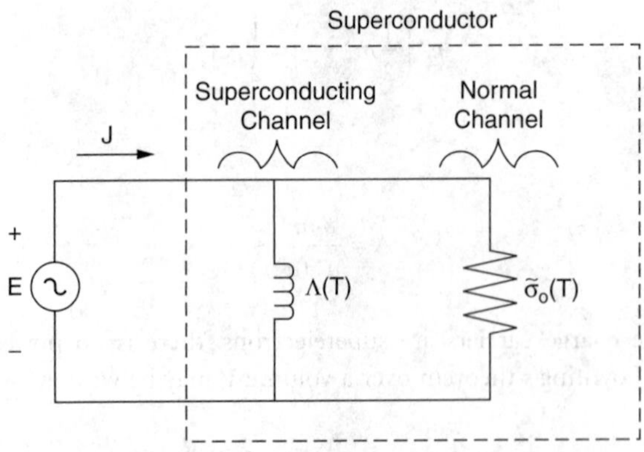

FIGURE 122.2 A lumped element model of a superconductor.

The London equations reveal that there is a characteristic length λ over which electromagnetic fields can change inside a superconductor. This penetration depth is different from the more familiar skin depth of electromagnetic theory, the latter being a frequency-dependent quantity. Indeed, the penetration depth at zero temperature is a distinct material property of a particular superconductor.

Notice that λ is sensitive to the number of correlated electrons (the superelectrons) in the material. As previously discussed, this number is a function of temperature, and so only at $T = 0$ do *all* the electrons that usually conduct ohmically participate in the Cooper pairing. For intermediate temperatures, $0 < T < T_c$, there are actually two sets of interpenetrating electron fluids: the uncorrelated electrons providing ohmic conduction and the correlated ones creating supercurrents. This two-fluid model is a useful way to build temperature effects into the London relations.

Under the two-fluid model, the electrical current density, **J**, is carried by both the uncorrelated (normal) electrons and the superelectrons: $\mathbf{J} = \mathbf{J}_n + \mathbf{J}_s$, where $\mathbf{J}_n$ is the normal current density. The two channels are modeled in a circuit, as shown in Figure 122.2, by a parallel combination of a resistor (representing the ohmic channel) and an inductor (representing the superconducting channel). To a good approximation, the respective temperature dependences of the conductor and inductor are

$$\tilde{\sigma}_o(T) = \sigma_o(T_c)\left(\frac{T}{T_c}\right)^4 \quad \text{for } T \le T_c \tag{122.10}$$

$$\Lambda(T) = \Lambda(0)\left(\frac{1}{1-(T/T_c)^4}\right) \quad \text{for } T \le T_c \tag{122.11}$$

where σ_o is the DC conductance of the normal channel. (Strictly speaking, the normal channel should also contain an inductance representing the inertia of the normal electrons, but typically such an inductor contributes negligibly to the overall electrical response.) Since the temperature-dependent penetration depth is defined as $\lambda(T) = \sqrt{\Lambda(T)/\mu_o}$, the effective conductance of a superconductor in the sinusoidal steady state is

$$\sigma = \tilde{\sigma}_o + \frac{1}{j\omega\mu_o\lambda^2} \tag{122.12}$$

where the explicit temperature dependence notation has been suppressed.

Most of the important physics associated with the classical model is embedded in Equation (122.12). As is clear from the lumped element model, the relative importance of the normal and superconducting channels is a function not only of temperature but also of frequency. The familiar L/R time constant, here equal to $\Lambda\tilde{\sigma}_o$, delineates the frequency regimes where most of the total current is carried by $\mathbf{J}_n$ (if

TABLE 122.1 Lumped Circuit Element Parameters Per Unit Length for Typical Transverse Electromagnetic Parallel Plate Waveguides*

Transmission Line Geometry	L_0	C_0	R_0
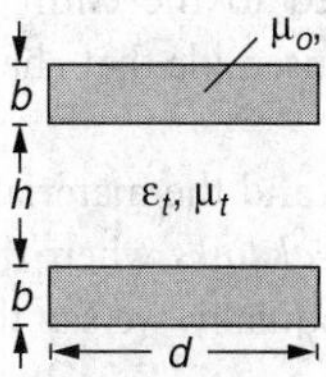 Two identical, thin $\lambda \gg b$ superconducting plates	$\frac{\mu_t h}{d} + \frac{2\mu_0 \lambda^2}{db}$	$\frac{\varepsilon_t d}{h}$	$\frac{8}{db\tilde{\sigma}_o}\left(\frac{\lambda}{\delta}\right)^4$
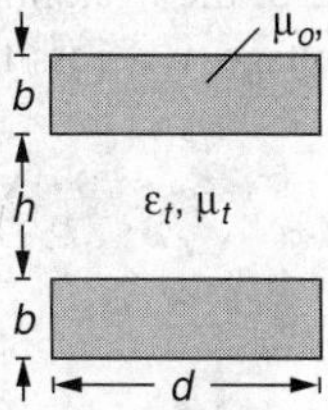 Two identical, thick $\lambda \ll b$ superconducting plates	$\frac{\mu_t h}{d} + \frac{2\mu_0 \lambda}{d}$	$\frac{\varepsilon_t d}{h}$	$\frac{4}{d\delta\tilde{\sigma}_o}\left(\frac{\lambda}{\delta}\right)^3$
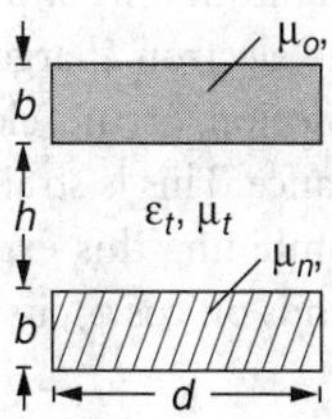 One thick $\lambda \ll b$ superconducting plate and one thick $(\delta_n \ll b)$ ohmic plate	$\frac{\mu_t h}{d} + \frac{\mu_0 \lambda}{d} + \frac{\mu_n \delta_n}{2d}$	$\frac{\varepsilon_t d}{h}$	$\frac{1}{d\delta_n \sigma_{o,n}}$

* The subscript *n* refers to parameters associated with a normal (ohmic) plate. Using these expressions, line input impedance, attenuation, and wave velocity can be calculated.

Source: Orlando, T. P. and Delin, K. A. 1991. *Foundations of Applied Superconductivity,* p. 171. Addison-Wesley, Reading, MA. With permission.

$\omega\Lambda\tilde{\sigma}_o >> 1$) or $\mathbf{J}_s$ (if $\omega\Lambda\tilde{\sigma}_o << 1$). This same result can also be obtained by comparing the skin depth associated with the normal channel, $\delta = \sqrt{2/(\omega\mu_o\tilde{\sigma}_o)}$, to the penetration depth to see which channel provides more field screening. In addition, it is straightforward to use Equation (122.12) to rederive Poynting's theorem for systems that involve superconducting materials:

$$-\int_V \nabla\cdot(\mathbf{E}\times\mathbf{H})\,dv = \frac{d}{dt}\int_V\left(\frac{1}{2}\varepsilon\mathbf{E}^2 + \frac{1}{2}\mu_o\mathbf{H}^2 + \frac{1}{2}\Lambda(T)\mathbf{J}_s^2\right)dv + \int_V \frac{1}{\tilde{\sigma}_o(T)}\mathbf{J}_n^2\,dv \qquad (122.13)$$

Using this expression, it is possible to apply the usual electromagnetic analysis to find the inductance (L_o), capacitance (C_o), and resistance (R_o) per unit length along a parallel plate transmission line. The results of such analysis for typical cases are summarized in Table 122.1.

122.3 Superconducting Electronics

The macroscopic quantum nature of superconductivity can be usefully exploited to create a new type of electronic device. Because all the superelectrons exhibit correlated motion, the usual wave-particle duality normally associated with a single quantum particle can now be applied to the entire ensemble of superelectrons. Thus, there is a spatiotemporal phase associated with the ensemble that characterizes the supercurrent flowing in the material.

Naturally, if the overall electron correlation is broken, this phase is lost and the material is no longer a superconductor. There is a broad class of structures, however, known as *weak links*, where the correlation is merely perturbed locally in space rather than outright destroyed. Colloquially, we say that the phase "slips" across the weak link to acknowledge the perturbation.

The unusual properties of this phase slippage were first investigated by Brian Josephson and constitute the central principles behind superconducting electronics. Josephson found that the phase slippage could be defined as the difference between the macroscopic phases on either side of the weak link. This phase difference, denoted as ϕ, determined the supercurrent, i_s, through and voltage, v, across the weak link according to the Josephson equations,

$$i_s = I_c \sin\phi \tag{122.14}$$

$$v = \frac{\Phi_o}{2\pi}\frac{\partial\phi}{\partial t} \tag{122.15}$$

where I_c is the critical (maximum) current of the junction and Φ_o is the quantum unit of flux. (The flux quantum has a precise definition in terms of Planck's constant, h, and the electron charge, e: $\Phi_o \equiv h/(2e) \approx 2.068 \times 10^{-15}$ Wb). As in the previous section, the correlated motion of the electrons, here represented by the superelectron phase, manifests itself through an inductance. This is straightforwardly demonstrated by taking the time derivative of Equation (122.14) and combining this expression with Equation (122.15). Although the resulting inductance is nonlinear (it depends on cos ϕ), its relative scale is determined by

$$L_j = \frac{\Phi_o}{2\pi I_c} \tag{122.16}$$

a useful quantity for making engineering estimates.

A common weak link, known as the Josephson tunnel junction, is made by separating two superconducting films with a very thin (typically 20 Å) insulating layer. Such a structure is conveniently analyzed using the resistively and capacitively shunted junction (RCSJ) model shown in Figure 122.3. Under the RCSJ model an ideal lumped junction [described by Equation (122.14) and Equation (122.15)] and a resistor R_j represent how the weak link structure influences the respective phases of the super and normal electrons, and a capacitor C_j represents the physical capacitance of the sandwich structure. If the ideal lumped junction portion of the circuit is treated as an inductor-like element, many Josephson tunnel junction properties can be calculated with the familiar circuit time constants associated with the model. For example, the quality factor Q of the RCSJ circuit can be expressed as

$$Q^2 = \frac{R_j C_j}{L_j / R_j} = \frac{2\pi I_c R_j^2 C_j}{\Phi_o} \equiv \beta \tag{122.17}$$

where β is known as the Stewart-McCumber parameter. Clearly, if $\beta \gg 1$, the ideal lumped junction element is underdamped in that the capacitor readily charges up, dominates the overall response of the

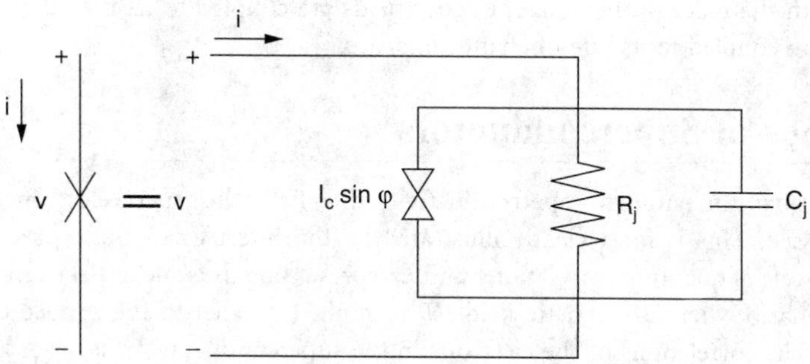

FIGURE 122.3 A real Josephson tunnel junction can be modeled using ideal lumped circuit elements.

circuit, and therefore creates a hysteretic *i-v* curve as shown in Figure 122.4(a). In the case when the bias current is raised from zero, no time-averaged voltage is created until the critical current is exceeded. At this point the junction switches to the voltage $2\Delta/e$ with a time constant $\sqrt{L_j C_j}$. Once the junction has latched into the voltage state, however, the bias current must be lowered to zero before it can again be steered through the superconducting path. Conversely, $\beta \ll 1$ implies that the L_j/R_j time constant dominates the circuit response, so that the capacitor does not charge up and the *i-v* curve is not hysteretic [Figure 122.4(b)].

Just as the correlated motion of the superelectrons creates the frequency-independent Meissner effect in a bulk superconductor through Faraday's law, so too the macroscopic quantum nature of superconductivity allows the possibility of a device whose output voltage is a function of a static magnetic field. If two weak links are connected in parallel, the lumped version of Faraday's law gives the voltage across the second weak link as $v_2 = v_1 + (d\Phi/dt)$, where Φ is the total flux threading the loop between the links. Substituting Equation (122.15) and integrating with respect to time yields

$$\phi_2 - \phi_1 = (2\pi\,\Phi)/\Phi_o \tag{122.18}$$

showing that the spatial change in the phase of the macroscopic wavefunction is proportional to the local magnetic flux. The structure described is known as a *superconducting quantum interference device* (*SQUID*) and can be used as a highly sensitive magnetometer by biasing it with current and measuring the resulting voltage as a function of magnetic flux. Such SQUID structures have also been proposed for quantum bits in quantum computing. From this discussion, it is apparent that a duality exists in how

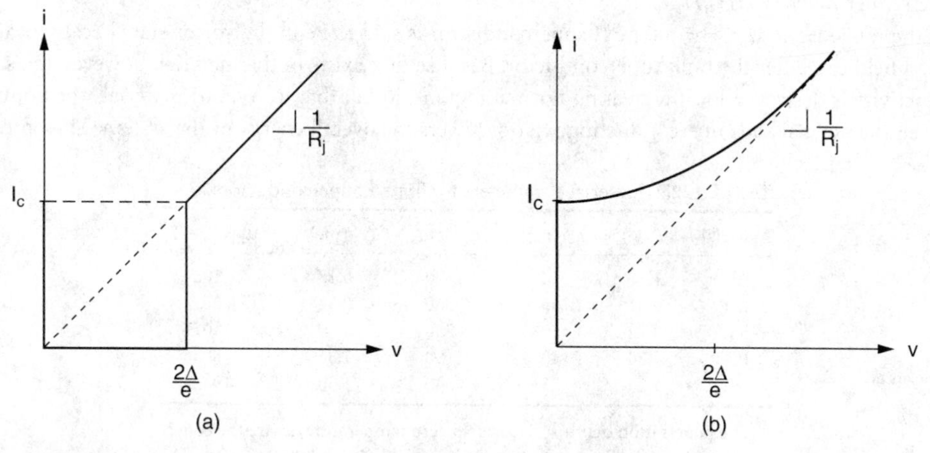

FIGURE 122.4 The *i-v* curves for a Josephson junction: (a) $\beta \gg 1$, (b) $\beta \ll 1$.

fields interact with the macroscopic phase: electric fields are coupled to its rate of change in time and magnetic fields are coupled to its rate of change in space.

122.4 Types of Superconductors

The macroscopic quantum nature of superconductivity also affects the general electromagnetic properties previously discussed. This is most clearly illustrated by the interplay of the characteristic lengths ξ, representing the scale of quantum correlations, and λ, representing the scale of electromagnetic screening. Consider the scenario where a magnetic field, H, is applied parallel to the surface of a semi-infinite superconductor. The correlations of the electrons in the superconductor must lower the overall energy of the system or else the material would not be superconducting in the first place. Because the critical magnetic field H_c destroys all the correlations, it is convenient to define the energy density gained by the system in the superconducting state as $(\frac{1}{2})\mu_o H_c^2$. The electrons in a Cooper pair are separated on a length scale of ξ, however, and so the correlations cannot be fully achieved until a distance roughly ξ from the boundary of the superconductor. There is thus an energy per unit area, $(\frac{1}{2})\mu_o H_c^2 \xi$, that is lost because of the presence of the boundary. Now consider the effects of the applied magnetic field on this system. It costs the superconductor energy to maintain the Meissner effect, $\mathbf{B} = 0$, in its bulk; in fact, the energy density required is $(\frac{1}{2})\mu_o H^2$. However, since the field can penetrate the superconductor a distance roughly λ, the system need not expend an energy per unit area of $(\frac{1}{2})\mu_o H^2 \lambda$ to screen over this volume. To summarize, more than a distance ξ from the boundary, the energy of the material is lowered (because it is superconducting), and more than a distance λ from the boundary the energy of the material is raised (to shield the applied field).

Now, if $\lambda < \xi$, the region of superconducting material greater than λ from the boundary, but less than ξ, will be higher in energy than that in the bulk of the material. Thus, the surface energy of the boundary is positive and so costs the total system some energy. This class of superconductors is known as type I. Most elemental superconductors, such as aluminum, tin, and lead, are type I. In addition to having $\lambda < \xi$, type I superconductors are generally characterized by low critical temperatures (~5 K) and critical fields (~0.05 T). Typical type I superconductors and their properties are listed in Table 122.2.

Conversely, if $\lambda < \xi$, the surface energy associated with the boundary is negative and lowers the total system energy. It is therefore thermodynamically favorable for a normal–superconducting interface to form inside these type II materials. Consequently, this class of superconductors does not exhibit the simple Meissner effect as do type I materials. Instead, there are now two critical fields: for applied fields below the lower critical field, H_{c1}, a type II superconductor is in the Meissner state, and for applied fields greater than the upper critical field, H_{c2}, superconductivity is destroyed. The three critical fields are related to each other by $H_c \approx \sqrt{H_{c1} H_{c2}}$.

In the range $H_{c1} < H < H_{c2}$, a type II superconductor is said to be in the vortex state because now the applied field can enter the bulk superconductor. Because flux exists in the material, however, the superconductivity is destroyed locally, creating normal regions. Recall that for type II materials the boundary between the normal and superconducting regions lowers the overall energy of the system. Therefore, the

TABLE 122.2 Material Parameters for Type I Superconductors*

Material	T_c(K)	λ_o (nm)	ξ_o (nm)	Δ_o (meV)	$\mu_o H_{co}$ (mT)
Al	1.18	50	1600	0.18	10.5
In	3.41	65	360	0.54	23.0
Sn	3.72	50	230	0.59	30.5
Pb	7.20	40	90	1.35	80.0
Nb	9.25	85	40	1.50	198.0

* The penetration depth λ_o is given at zero temperature, as are the coherence length ξ_o, the thermodynamic critical field H_{co}, and the energy gap Δ_o.

Source: Donnelly, R. J. 1981. Cryogenics. In *Physics Vade Mecum*, ed. H. L. Anderson. American Institute of Physics, New York. With permission.

TABLE 122.3 Material Parameters for Conventional Type II Superconductors*

Material	T_c(K)	λ_{GL} (0)(nm)	ξ_{GL} (0)(nm)	Δ_o (meV)	$\mu_o H_{c2,o}$ (T)
Pb-In	7.0	150	30	1.2	0.2
Pb-Bi	8.3	200	20	1.7	0.5
Nb-Ti	9.5	300	4	1.5	13
Nb-N	16	200	5	2.4	15
$PbMo_6S_8$	15	200	2	2.4	60
V_3Ga	15	90	2–3	2.3	23
V_3Si	16	60	3	2.3	20
Nb_3Sn	18	65	3	3.4	23
Nb_3Ge	23	90	3	3.7	38

* The values are only representative because the parameters for alloys and compounds depend on how the material is fabricated. The penetration depth $\lambda_{GL}(0)$ is given as the coefficient of the Ginzburg-Landau temperature dependence as $\lambda_{GL}(T) = \lambda_{GL}(0)(1 - T/T_c)^{-1/2}$; likewise for the coherence length where $\xi_{GL}(T) = \xi_{GL}(0)(1 - T/T_c)^{-1/2}$. The upper critical field $H_{c2,o}$ is given at zero temperature as well as the energy gap Δ_o.

Source: Donnelly, R. J. 1981. Cryogenics. In *Physics Vade Mecum*, ed. H. L. Anderson. American Institute of Physics, New York. With permission.

flux in the superconductor creates as many normal-superconducting interfaces as possible without violating quantum criteria. The net result is that flux enters a type II superconductor in quantized bundles of magnitude Φ_o known as *vortices* or *fluxons* (the former name derives from the fact that current flows around each quantized bundle in the same manner as a fluid vortex circulates around a drain). The central portion of a vortex, known as the core, is a normal region with an approximate radius of ξ. If a defect-free superconductor is placed in a magnetic field, the individual vortices, whose cores essentially follow the local average field lines, form an ordered triangular array, or flux lattice. As the applied field is raised beyond H_{c1} (where the first vortex enters the superconductor), the distance between adjacent vortex cores decreases to maintain the appropriate flux density in the material. Finally, the upper critical field is reached when the normal cores overlap and the material is no longer superconducting. Indeed, a precise calculation of H_{c2} using the phenomenological theory developed by Vitaly Ginzburg and Lev Landau yields

$$H_{c2} = \frac{\Phi_o}{2\pi\mu_o\xi^2} \tag{122.19}$$

which verifies out simple picture. The values of typical type II material parameters are listed in Table 122.3 and Table 122.4.

TABLE 122.4 Type II (Non-Conventional and High-Temperature Superconductors)

Material	T_c(K)
$BA_{1-x}K_xBi\ O_3$	30
Rb_3C_{60}	33
M_gB_2	39
$YBa_2Cu_3O_7$	95
$Bi_2Sr_2CaCu_2O_8$	85
$Bi_2Sr_2Ca_2Cu_3O_{10}$	110
$TlBa_2Ca_2Cu_3O_{10}$	125
$HgBa_2Ca_2\ Cu_3O_8$	131

Type II superconductors are of great technical importance because typical H_{c2} values are at least an order of magnitude greater than the typical H_c values of type I materials. It is, therefore, possible to use type II materials to make high-field magnet wire. Unfortunately, when current is applied to the wire, there is a Lorentz-like force on the vortices, causing them to move. Because the moving vortices carry flux, their motion creates a static voltage drop along the superconducting wire by Faraday's law. As a result, the wire no longer has a zero DC resistance, even though the material is still superconducting. To fix this problem, type II superconductors are usually fabricated with intentional defects, such as impurities or grain boundaries, in their crystalline structure to pin the vortices and prevent vortex motion. The pinning as created because the defect locally weakens the superconductivity in the material, and it is thus energetically favorable for the normal core of the vortex to overlap the nonsuperconducting region in the material. Critical current densities usually quoted for practical type II materials, therefore, really represent the depinning critical current density where the Lorentz-like force can overcome the pinning force. (The depinning critical current density should not be confused with the depairing critical current density, which represents the current when the Cooper pairs have enough kinetic energy to overcome their correlation. The depinning critical current density is typically an order of magnitude less than the depairing critical current density, the latter of which represents the theoretical maximum for J_c.)

By careful manufacturing, it is possible to make superconducting wire with tremendous amounts of current-carrying capacity. For example, standard copper wire used in homes will carry about 10^7 A/m^2, whereas a practical type II superconductor like niobium-titanium can carry current densities of 10^{10} A/m^2 or higher even in fields of several teslas. This property, more than a zero DC resistance, is what makes superconducting wire so desirable.

Defining Terms

Superconductivity — A state of matter whereby the correlation of conduction electrons allows a static current to pass without resistance and a static magnetic flux to be excluded from the bulk of the material.

References

Delin, K. A. and Kleinsasser, A. W. 1996. "Stationary Properties of High-Critical-Temperature Proximity Effect Josephson Junctions," *Superconductor Science and Technology* **9**, 227.

Donnelly, R. J. 1981. Cryogenics. In *Physics Vade Mecum*, ed. H. L. Anderson. American Institute of Physics, New York.

Foner, S and Schwartz, B. B. 1974. *Superconducting Machines and Devices.* Plenum Press, New York.

Foner, S. and Schwartz, B. B. 1981. *Superconducting Materials Science.* Plenum Press, New York.

Orlando, T. P. and Delin, K. A. 1991. *Foundations of Applied Superconductivity.* Addison-Wesley, Reading, MA.

Ruggiero, S. T. and Rudman, D. A. 1990. *Superconducting Devices.* Academic Press, Boston, MA.

Schwartz, B. B. and Foner, S. 1977. *Superconducting Applications: SQUIDs and Machines.* Plenum Press, New York.

Van Duzer, T. and Turner, C. W. 1999. *Principles of Superconductive Devices and Circuits*, 2nd ed. Prentice Hall PTR, New Jersey.

Wilson, M. N. 1983. *Superconducting Magnets.* Oxford University Press, Oxford, UK.

Further Information

Every two years an Applied Superconductivity Conference is held devoted to practical technological issues. The proceedings of these conferences have been published every other year from 1977 to 1991 in the *IEEE Transactions on Magnetics.*

In 1991 the *IEEE Transactions on Applied Superconductivity* began publication. This quarterly journal focuses on both the science and the technology of superconductors and their applications, including materials issues, analog and digital circuits, and power systems. The proceedings of the Applied Superconductivity Conference now appear in this journal.

123

Embedded Systems-on-Chips

Wayne Wolf
Princeton University

123.1 Introduction

Advances in VLSI technology now allow us to build systems-on-chips (SoCs), also known as systems-on-silicon (SoS). SoCs are complex at all levels of abstraction; they contain hundreds of millions of transistors; they also provide sophisticated functionality, unlike earlier generations of commodity memory parts. As a result, SoCs present a major productivity challenge.

One solution to the SoC productivity problem is to use embedded computers.[1] An embedded computer is a programmable processor that is a component in a larger system that is not a general-purpose computer. Embedded computers help tame design complexity by separating (at least to some degree) hardware and software design concerns. A processor can be used as a predesigned component — known as intellectual property (IP) — that operates at a known speed and power consumption. The software required to implement the desired functionality can be designed somewhat separately.

In exchange for separating hardware and software design, some elements traditionally found in hardware design must be transferred to software design. Software designers have traditionally concentrated on functionality while hardware designers have worried about critical delay paths, power consumption, and area. Embedded software designers must worry about real-time deadlines, power consumption, and program and data size. As a result, embedded SoC design disciplines require a blending of hardware and software skills.

This chapter considers the characteristics of SoCs built from embedded processors. The next section surveys the types of requirements that are generally demanded from embedded SoCs. Section 123.3 surveys the characteristics of components used to build embedded systems. Section 123.4 introduces the types of architectures used in embedded systems. Section 123.5 reviews design methodologies for embedded SoCs.

0-8493-1586-7/05/$0.00+$1.50

123.2 Requirements on Embedded SoCs

A digital system typically uses embedded processors to meet a combination of performance, complexity, and possibly design time goals. If the system's behavior is very regular and easy to specify as hardware, it may not be necessary to use embedded software. An embedded processor becomes more attractive when the behavior is too complex to be easily captured in hardwired logic.

Using embedded processors may reduce design time by allowing the design to be separated into distinct software and hardware units. In many cases, the CPU will be predesigned; even if the CPU and associated hardware is being designed for the project, many aspects of the hardware design can be performed separately from the software design. (Experience with embedded system designs does show, however, that the hardware and software designs are intertwined and that embedded software is prone to some of the same scheduling problems as mainframe software projects.)

But even if embedded processors seem attractive by reducing much of the design to "just programming," it must be remembered that embedded software design is much more challenging than typical applications programming for workstations or PCs. Embedded software must be designed to meet not just functional requirements — the software's input and output behavior — but also stringent nonfunctional requirements. Those nonfunctional requirements include:

- *Performance* — Although all programmers are interested in speed of execution, performance is measured much more precisely in the typical embedded system. Many embedded systems must meet real-time deadlines. The deadline is measured between two points in the software: if the program completely executes from the starting point to the end point by the deadline, the system malfunctions.
- *Energy/power* — Traditional programmers don't worry about power or energy consumption. However, energy and power are important to most embedded systems. Energy consumption is of course important in battery-operated systems, but the heat generated as a result of power consumption is increasingly important to wall-powered systems.
- *Size* — The amount of memory required by the embedded software determines the amount of memory required by the embedded system. Memory is often one of the major cost components of an embedded system.

Embedded software design resembles hardware design in its emphasis on nonfunctional requirements such as performance and power. The challenge in embedded SoC design is to take advantage of the best aspects of both hardware and software components to quickly build a cost-effective system.

123.3 Embedded SoC Components

CPUs

As shown in Figure 123.1, a CPU is a programmable instruction set processor. Instructions are kept in a separate memory — a program counter (PC) that points to the current instruction. This definition does not consider reconfigurable logic to be a programmable computer, because it does not have a separate instruction memory and a PC. Reconfigurable logic can be used to implement sequential machines, and so a CPU could be built in reconfigurable logic. But the separation of CPU logic and memory is an important abstraction for program design.

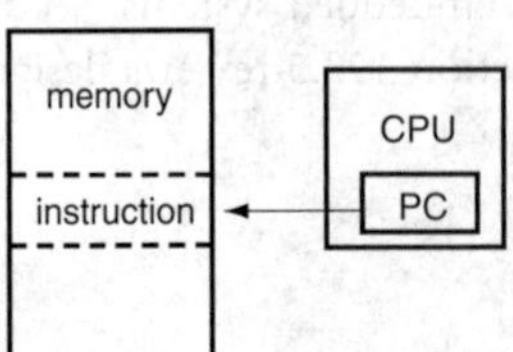

FIGURE 123.1 A CPU and memory.

An embedded processor is judged by several characteristics:

- *Performance* — The overall speed of execution may be important in some systems, but in many cases we particularly care about the CPU's performance on critical sections of code.
- *Energy and power* — Processors provide different mechanisms to manage power consumption.
- *Area* — The area of the processor contributes to the total implementation cost of the SoC. The area of the memory required to store the program also contributes to implementation cost.

These characteristics are judged relative to the embedded software they are expected to run. A processor may exhibit very different performance or energy consumption on different applications.

RISC processors are commonly used in embedded computing. ARM[2] and MIPS[3] processors are examples of RISC processors that are widely used in embedded systems. A RISC CPU uses a pipeline to increase CPU performance. Many RISC instructions take the same amount of time to execute, simplifying performance analysis. However, many RISC architectures do have exceptions to this rule. An example is the multiple-register feature of the ARM processor: an instruction can load or store a set of registers, for which the instruction takes one cycle per instruction.

Most CPUs used in PCs and workstations today are superscalar processors. A superscalar processor builds on RISC techniques by adding logic that examines the instruction stream and determines, based on what CPU resources are needed, when several instructions can be executed in parallel. Superscalar scheduling logic adds quite a bit of area to the CPU in order to check all the possible conflicts between combinations of instructions; the size of a superscalar scheduler grows as n^2, where n is the number of instructions that are under consideration for scheduling. Many embedded systems, and in particular SoCs, do not use superscalar processors and instead stick with RISC processors. Embedded system designers tend to use other techniques, such as instruction-set optimization caches, to improve performance. Because SoC designers are concerned with overall system performance, not just CPU performance, and because they have a better idea of the types of software run on their hardware, they can tackle performance problems in a variety of ways that may use the available silicon area more cost-effectively.

Some embedded processors are known as digital signal processors (DSPs). The term DSP was originally used to mean one of two things: either a CPU with a Harvard architecture that provided separate memories for programs and data; or a CPU with a multiply-accumulate unit to efficiently implement digital filtering operations. Today, the meaning of the term has blurred somewhat. For instance, version 9 of the ARM architecture is a Harvard architecture to better support digital signal processing. Modern usage applies the term DSP to almost any processor that can be used to efficiently implement signal processing algorithms.

The application-specific integrated processor (ASIP)[4] is one approach to improving the performance of RISC processors for embedded application. An ASIP's instruction set is designed to match the requirements of the application software it will run. On the one hand, special-purpose function units and instructions to control them may be added to speed up certain operations. On the other hand, function units, registers, and busses may be eliminated to reduce the CPU's cost if they do not provide enough benefit for the application at hand. The ASIP may be designed manually or automatically based on profiling information. One advantage of generating the ASIP automatically is that the same information can be used to generate the processor's programming environment: a compiler, assembler, and debugger are necessary to make the ASIP useful building blocks.

Another increasingly popular architecture for embedded computing is very long instruction word (VLIW). A VLIW machine can execute several instructions simultaneously but, unlike a superscalar processor, relies on the compiler to schedule parallel instructions at compilation time. A pure VLIW machine uses slots in the long, fixed-length instruction word to control the CPU's function units, with NOPs used to indicate slots that cannot be used for useful work by the compiler. Modern VLIW machines, such as the TI C6000[5] and the Motorola/Agere StarCore,[6] group single-operation instructions into execution packets; the packet's length can vary depending on the number of instructions that the compiler was able to schedule for simultaneous operation. VLIW machines provide instruction-level parallelism with a much smaller CPU than is possible in a superscalar system; however, the compiler must be able to extract parallelism at compilation time to be able to use the CPU's resources. Signal processing applications often have parallel operations that can be exploited at compilation time. For example, a parallel set of filter banks runs the same code on different data; the operations for each channel can be scheduled together in the VLIW instruction group.

Interconnect

Embedded SoCs may connect several CPUs, on-chip memories, and devices on a single chip. High-performance interconnect systems are required to meet the system's performance demands. The interconnection systems must also comply with standards so that existing components may be connected to them.

Busses are still the dominant interconnection scheme for embedded SoCs. Although richer interconnection schemes could be used on-chip, where they are not limited by pinout as in board-level systems, many existing architectures are still memory-limited and not interconnect-limited. However, future generations of embedded SoCs may need more sophisticated interconnection schemes.

A bus provides a protocol for communication between components. It also defines a memory space and the uses of various addresses in that memory space, for example, the address range assigned to a device connected to the bus. Busses for SoCs may be designed for high-performance or low-cost operation. A high-performance bus uses a combination of techniques — advanced circuits, additional bus lines, efficient protocols — to maximize transaction performance. One common protocol used for efficient transfers is the block transfer, in which a range of locations is transferred based on a single address, eliminating the need to transfer all the addresses on the bus. Some recent busses allow split transactions — the data request and data transfer are performed on separate bus cycles, allowing other bus operations to be performed while the original request is serviced. A low-cost bus design provides modest performance that may not be acceptable for instruction fetching or other time-critical operations. A low-cost bus is designed to require little hardware in the bus itself and to impose a small hardware and software overhead on the devices connecting to the bus. A system may contain more than one bus; a bridge can be used to connect one bus to another.

The ARM AMBA bus specification[7] is an example of a bus specification for SoCs. The AMBA spec actually includes two busses: the high-performance AMBA high-performance bus (AHB) and the low-cost AMBA peripherals bus (APB). The Virtual Sockets Interface committee has defined another standard for interconnecting components on SoCs.

Memory

One of the great advantages of SoC technology is that memory can be placed on the same chip as the system components that use the memory. On-chip memory both increases performance and reduces power consumption because on-chip connections present less reactive load than do pins and traces between chips; however, an SoC may still need to use separate chips for off-chip memory.

Although on-chip embedded memory has many advantages, it still is not as good as commodity memory. A commodity SRAM or DRAM's manufacturing process has been carefully tuned to the requirements of that component. In contrast, an on-chip memory's manufacturing needs must be balanced against the requirements of the logic circuits on the chip. The transistors, interconnections, and storage nodes of on-chip memories all have somewhat different needs than logic transistors.

Embedded DRAMs suffer the most because they need quite different manufacturing processes than do logic circuits. The processing steps required to build the storage capacitors for the DRAM cell are not good for small-geometry transistors. As a result, embedded DRAM technologies often compromise both the memory cells and the logic transistors, with neither being as good as they would be in separate, optimized processes. Although embedded DRAM has been the subject of research for many years, its limitations have kept it from becoming a widely used technology at the time of this writing.

SRAM circuits' characteristics are closer to those of logic circuits and so can be built on SoCs with less penalty. SRAM consumes more power and requires more chip area than does DRAM, but SRAM does not need refreshing, which noticeably simplifies the system architecture.

Software Components

Software elements are also components of embedded systems. Just as pre-designed hardware components are used to both reduce design time and to provide predictions of the characteristics of parts of the system, software components can also be used to speed up software implementation time and to provide useful data on the characteristics of systems.

CPU vendors often supply software libraries for their processors. These libraries generally supply code for two types of operations. First, they provide drivers for input and output operations. Second, they provide efficient implementations of commonly-used algorithms. For example, libraries for DSPs generally include code for digital filtering, fast Fourier transforms, and other common signal processing algorithms. Code libraries are important because compilers are still not as adept as expert human programmers at creating code that is both fast and small.

The real-time operating system (RTOS) is the second major category of software component. Many applications perform several different types of operations, often with their own performance requirements. As a result, the software is split into processes that run independently under the control of an RTOS. The RTOS schedules the processes to meet performance goals and efficiently utilize the CPU and other hardware resources. The RTOS may also provide utilities, such as interprocess communication, networking, or debugging. An RTOS's scheduling policy is necessarily very different from that used in workstations and mainframes, because the RTOS must meet real-time deadlines. A priority-driven scheduling algorithm such as rate-monotonic scheduling (RMS)[9] is often used by the RTOS to schedule activity in the system.

123.4 Embedded System Architectures

The hardware architecture of an embedded SoC is generally tuned to the requirements of the application. Different domains, such as automotive, image processing, and networking all have very different characteristics. In order to make best use of the available silicon area, the system architecture is chosen to match the computational and communication requirements of the application. As a result, a much wider range of hardware architectures is found for embedded systems as compared with traditional computer systems.

Figure 123.2 shows one common configuration, a bus-based uniprocessor architecture for an embedded system. This architecture has one CPU, which greatly simplifies the software architecture. In addition to I/O devices, the architecture may include several devices known as accelerators designed to speed up computations. (Though some authors refer to these units as co-processors, we prefer to reserve that term for units that are dispatched by the CPU's execution unit.) For example, a video operation's inner loops may be implemented in an application-specific IC (ASIC) so that the operation can be performed more

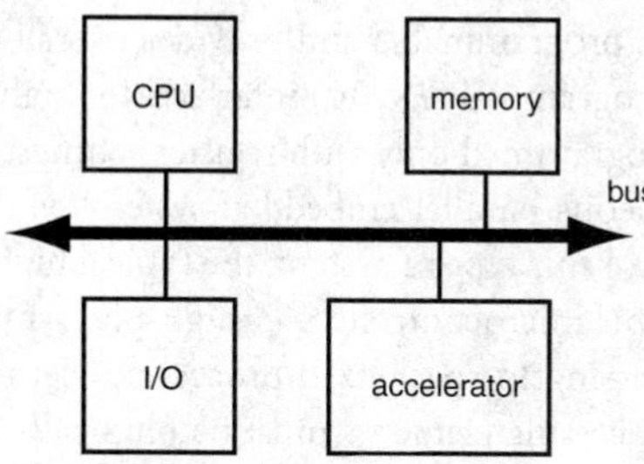

FIGURE 123.2 A bus-based, single-CPU embedded system.

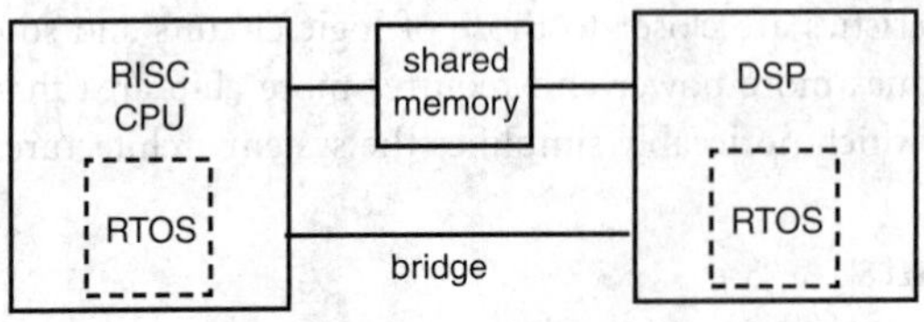

FIGURE 123.3 The TI OMAP architecture.[10]

quickly than would be possible on the CPU. An accelerator can achieve performance gains through several mechanisms: by implementing some functions in special hardware that takes fewer cycles than is required on the CPU, by reducing the time required for control operations that would require instructions on the CPU, and by using additional registers and custom data flow within the accelerator to more efficiently implement the available communication. The single-CPU/bus architecture is commonly used in applications that do not have extensive real-time characteristics and ones that need to run a wider variety of software. For example, many PDAs use this type of architecture. A single-CPU system simplifies software design and debugging since all the work is assumed to happen on one processing element. The single CPU system is also relatively inexpensive.

In general, however, a high-performance embedded system requires a heterogeneous multiprocessor — a multiprocessor that uses more than one type of processing element and/or a specialized communication topology. Scientific parallel processors generally use a regular architecture to simplify programming. Embedded systems use heterogeneous architectures for several reasons:

- *Cost* — A regular architecture may be much larger and more expensive than a heterogeneous architecture, which freed from the constraint of regularity, can remove resources from parts of the architecture where they are not needed and add them to parts where they are needed.
- *Real-time performance* — Scientific processors are designed for overall performance but not to meet deadlines. Embedded systems must often put processing power near the I/O that requires real-time responsiveness; this is particularly true if the processing must be performed at a high rate. Even if a high-rate, real-time operation requires relatively little computation on each iteration, the high interrupt rate may make it difficult to perform other processing tasks on the same processing element.

Many embedded systems use heterogeneous multiprocessors. One example comes from telephony. A telephone must perform both control- and data-intensive operations: both the network protocol and the user interface require control-oriented code; the signal-processing operations require data-oriented code. The Texas Instruments OMAP architecture, shown in Figure 123.3, is designed for telephony: the RISC processor handles general-purpose and control-oriented code while the DSP handles signal processing. Shared memory allows processes on the two CPUs to communicate, as does a bridge. Each CPU has its own RTOS that coordinates processes on the CPU and also mediates communication with the other CPU.

The C-Port network processor,[11] whose hardware architecture is shown in Figure 123.4, provides an example of a heterogeneous multiprocessor in a different domain. The multiprocessor is a high-speed bus. The RISC executive processor is C programmable and provides overall control, initialization, etc. Each of the 16 HDLC processors is also C programmable. Other interfaces for higher-speed networks are not general-purpose computers and can be programmed only with register settings.

Another category of heterogeneous parallel embedded systems is the networked embedded system. Automobiles are a prime example of this type of system: the typical high-end car includes over a hundred microprocessors ranging from 4-bit microcontrollers to high-performance 32-bit processors. Networks help to distribute high-rate processing to specialized processing elements, as in the HP DesignJet, but they are most useful when the processing elements must be physically distributed. When the processing elements are sufficiently far apart, busses designed for lumped microprocessor systems do not work well. The network is generally used for data transfer between the processing elements, with each processing element maintaining its own program memory as well as a local data memory. The processing elements

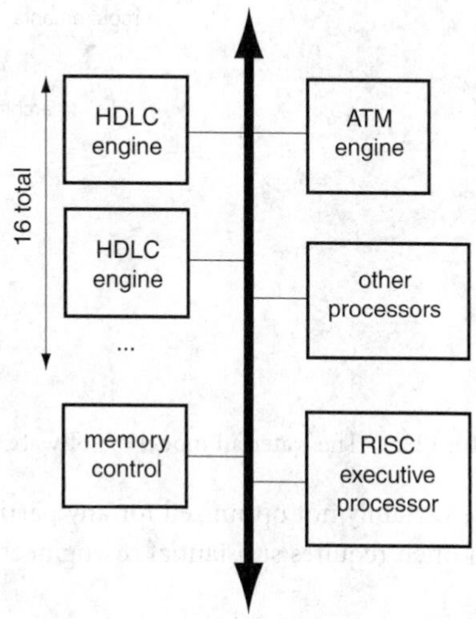

FIGURE 123.4 Block diagram of the C-Port network processor.[11]

communicate data and control information as required by the application. I^2C and CAN are two widely used networks for distributed systems.

123.5 Embedded SoC Design Methodologies

Specifications

As described in Section 123.2, embedded computers are typically used to build systems with complex functionality. Therefore, capturing a functional description of the system is an important part of the design process. A variety of specification languages have been developed. Many of these languages were developed for software systems, but several languages have been developed over the past decade with embedded systems in mind.

Specification languages are generally designed to capture particular styles of design. Many languages have been created to describe control-oriented systems. An early example was Statecharts,[12] which introduced hierarchical states that provided a structured description of state machines. The SDL language[13] is widely used to specify protocols in telecommunications systems. The Esterel language[14] describes a reactive system as a network of communicating state machines.

Data-oriented languages find their greatest use in signal processing systems. Dataflow process networks[15] are one example of a specification language for signal processing. Object-oriented specification and design have become very popular in software design. Object-oriented techniques mix control and data orientation. Objects tend to reflect natural assemblages of data; the data values of an object define its state and the states of the objects define the state of the system. Messages providing communication and control. The real-time object-oriented Methodology (ROOM)[16] is an example of an object-oriented methodology created for embedded system design.

In practice, many systems are specified in the C programming language. Many practical systems combine control and data operations, making it difficult to use one language that is specialized for any type of description. Algorithm designers generally want to prototype their algorithms and verify them through experimentation; as a result, an executable program generally exists as the golden standard with which the implementation must conform. This is especially true when the product's capabilities are defined by standards committees, which typically generate one or more reference implementations, usually in C. Once a working piece of code exists in C, there is little incentive to rewrite it in a different specification language; however, the C specification is generally a long way from an implementation. Algorithmic designs are usually written for uniprocessors and ignore many aspects of I/O, whereas embedded systems must perform real-time I/O and often distribute tasks among several processing elements. Algorithm designers often do not optimize their code for any particular platform, and their

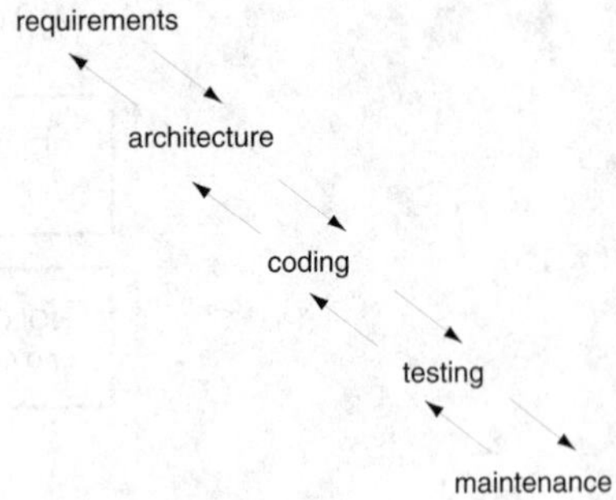

FIGURE 123.5 The waterfall model of software development.

code is certainly not optimized for any particular embedded platform. As a result, a C language specification often requires substantial re-engineering before it can be used in an embedded system.

Design Flows

In order to better understand modern design methodologies for embedded SoCs, we can start with traditional software engineering methodologies. The waterfall model, one of the first models of software design, is shown in Figure 123.5. The waterfall model is a top-down model with only local feedback. Other software design models, such as the spiral model, try to capture more bottom-up feedback from implementation to system design; however, software design methodologies are designed primarily to implement functionality and to create a maintainable design. Embedded SoCs must, as mentioned in Section 123.2, satisfy performance and power goals as well. As a result, embedded system design methodologies must be more complex.

The design of the architecture of an embedded SoC is particularly important because the architecture defines the capabilities that will limit both the hardware and software implementations. The architecture must, of course, be cost effective, but it must also provide the features necessary to do the job. Because the architecture is custom designed for the application, it is quite possible to miss architectural features that are necessary to efficiently implement the system. Retrofitting those features back into the architecture may be difficult or even impossible if the hardware and software design efforts do not keep in sync.

Important decisions about the hardware architecture include:

- How many processing elements are needed?
- What processing elements should be programmable and which ones should be hardwired?
- How much communication bandwidth is needed in the system and where is it needed?
- How much memory is needed and where should it go in the system?
- What types of components will be used for processors, communication, and memory?

The design of the software architecture is just as important and goes hand-in-hand with the hardware architecture design. Important decisions about the software architecture include:

- How should the functionality be split into processes?
- How are input and output performed?
- How should processes be allocated to the various processing elements in the hardware architecture?
- When should processes be scheduled?

In practice, information required to make these decisions comes from several sources. One important source is previous designs. Though technology and requirements both change over time, similar designs can provide valuable lessons on how to (and not to) design the next system. Another important source is implementation. Some implementation information can come from pre-designed hardware or software components, which is one reason why intellectual-property-based design is so important. Implementation can also come from early design efforts.

A variety of CAD algorithms have been developed to explore the embedded system design space and to help automate system architectures. Vulcan[21] and Cosyma[22] were early hardware/software partitioning systems that implemented a design using a CPU and one or more custom ASICs. Other algorithms target more general architectures.[23,24]

Once the system architecture has been defined, the hardware and software must be implemented. Hardware implementation challenges include:

- finding efficient protocols to connect together existing hardware blocks,
- memory design,
- clock rate optimization,
- power optimization.

Software implementation challenges include:

- meeting performance deadlines,
- minimizing power consumption,
- minimizing memory requirements.

The design must be verified throughout the design process. Once the design progresses to hardware and software implementation, simulation becomes challenging because the various components operate at very different levels of abstraction. Hardware units are modeled at the clock-cycle level. Software components must often be run at the instruction level or in some cases at even higher levels of abstraction. A hardware/software co-simulator[19] is designed to coordinate simulations that run at different time scales. The co-simulator coordinates multiple simulators — hardware simulators, instruction-level simulators, behavioral software processes — and keeps track of the time in each simulation. The co-simulator ensures that communications between the simulators happen at the right time for each simulator.

Design verification must include performance, power, and size as well as functionality. Although these sorts of checks are common in hardware design, they are relatively new to software design. Performance and power verification of software may require cache simulation. Some recent work has developed higher-level power models for CPUs.

Platform-Based Design

One response to the conflicting demands of SoC design has been the development of platform-based design methodologies. On the one hand, SoCs are becoming very complex. On the other hand, they must be designed very quickly to meet the electronics industry's short product lifecycles.

Platform-based design tries to tackle this problem by dividing the design process into two phases. In the first phase, a platform is designed. The platform defines the hardware and software architectures for the system. The degree to which the architecture can be changed depends on the needs of the marketplace. In some cases, the system may be customizable only by reprogramming. In other cases, it may be possible to add or delete hardware components to provide specialized I/O, additional processing capabilities, etc. In the second phase, the platform is specialized into a product. Because much of the initial design work was done in the first phase, the product can be developed relatively quickly based on the platform.

Platform-based design is particularly well suited to products derived from standards. On the one hand, all products must meet the minimum requirements of the standard. On the other hand, standards committees generally leave room for different implementations to distinguish themselves: added features, lower power, etc. Designers will generally want to modify their design to add features that differentiate their product in the marketplace.

Platform-based design also allows designers to incorporate design experience into products. Each product derived from the platform will teach something: how to better design part of the system, unexpected needs of customers, etc. The platform can be updated as new products are developed from it so that successive designs will be easier.

Platforms are usually designed within a technology generation. A new VLSI technology generally changes enough design decisions that platforms must be rethought for each new generation of technology. Therefore, the platform itself must be designed quickly and each product based on the platform must be completed quickly in order to gain effective use of the platform design effort in the 18-month lifecycle of a manufacturing technology.

Software Performance Analysis and Optimization

Although methods for hardware performance analysis and optimization are well known, software techniques for optimizing performance have been developed only recently to meet the demands of embedded design methodologies.

The performance of an embedded system is influenced by several factors at different levels of abstraction. The first is the performance of the CPU pipeline itself. RISC design techniques tend to provide uniform execution times for instructions, but software performance is not always simple to predict. Register forwarding, which is used to enhance pipeline performance, also makes execution time less predictable. Branch prediction causes similar problems.

Superscalar processors, because they schedule instructions at execution time based upon execution data, provide much less predictable performance than do either RISC or VLIW processors. This is one reason why superscalar processors are not frequently used in real-time embedded systems.

The memory system is often an even greater source of uncertainty in embedded systems. CPUs use caches to improve average memory response time, but the effect of the cache on a particular piece of software requires complex analysis. In pathological cases, the cache can add uncertainty to execution times without actually improving the performance of critical software components. Cache simulation is often used to analyze the behavior of a program in a cache. Analysis must take into account both instructions and data. Unlike in workstation CPUs, in which the cache configuration is chosen by the CPU architect based on benchmarks, the designer of an embedded SoC can choose the configurations of caches to match the characteristics of the embedded software. Embedded system designers can choose between hardware and software optimizations to meet performance goals.

Analyzing the performance of a program requires determining both the execution path and the execution time of instructions along that path.[18] Both are challenging problems.[20] The execution path of a program clearly depends on input data values. To ensure that the program meets a deadline, the worst-case execution path must be determined. The execution time of instructions along the path depend on several factors: data values, interactions between instructions, and cache state.

Energy/Power Analysis and Optimization

Many embedded systems must also meet energy and power goals as well as performance goals. The specification may impose several types of power requirements: peak power consumption, average power consumption, energy consumption for a given operation.

To a first-order, high-performance design is low-power design. Efficient implementations that run faster also tend to reduce power consumption, but trade-offs between performance and power in embedded system design. For example, the power consumption of a cache depends on both its size and the memory system activity.[25] If the cache is too small, too many references require expensive main memory accesses. If the cache is too large, it burns too much static power. Many applications exhibit a sweet spot at which the cache is large enough to provide most of the available performance benefit while not burning too much static power. Techniques have been developed to estimate hardware/software power consumption.[26]

System-level approaches can also help reduce power consumption.[27] Components can be selectively turned off to save energy; however, because turning a component on again may consume both time and energy, the decision to turn it off must be made carefully. Statistical methods based on Markov models can be used to create effective system-level power management methodologies.

123.6 Summary

Embedded computers promise to solve a critical design bottleneck for SoCs. Because we can design CPUs relatively independently of the programs they run and reuse those CPUs design across many chips, embedded computers help to close the designer productivity gap. Embedded processors, on the other hand, require that many design techniques traditionally reserved for hardware — deadline-driven performance, power minimization, size — must now be applied to software as well. Design methodologies for embedded SoCs must carefully design system architectures that will allow hardware and software components to work together to meet performance, power, and cost goals while implementing complex functionality.

References

1. Wayne Wolf, *Computers as Components: Principles of Embedded Computer System Design,* San Francisco: Morgan Kaufman, 2000.
2. http://www.arm.com.
3. http://www.mips.com.
4. G. Goossens, J. van Praet, D. Lanneer, W. Geurts, A. Kifli, C. Liem, and P. G. Paulin, "Embedded software in real-time signal processing systems: design technologies," *Proceedings of the IEEE,* 85(3), March 1997, pp. 436–453.
5. http://www.ti.com.
6. http://www.lucent.com/micro/starcore/motover.htm.
7. ARM Limited, *AMBA(TM) Specification (Rev 2.0),* ARM Limited, 1999.
8. http://www.vsi.com.
9. C. L. Liu and J. W. Layland, "Scheduling algorithms for multiprogramming in a hard real-time environment," *Journal of the ACM,* 20(1), 1973, pp. 46–61.
10. http://www.ti.com/sc/docs/apps/wireless/omap/overview.htm.
11. http://www.cportcorp.com/products/digital.htm.
12. D. Harel, "Statecharts: a visual formalism for complex systems," *Science of Computer Programming,* 8, 1987, pp. 231–274.
13. Anders Rockstrom and Roberto Saracco, "SDL — CCITT specification and description language," *IEEE Transactions on Communication,* 30(6), June 1982, pp. 1310–1318.
14. Albert Benveniste and Gerard Berry, "The synchronous approach to reactive real-time systems," *Proceedings of the IEEE,* 79(9), September 1991, pp. 1270–1282.
15. E. A. Lee and T. M. Parks, "Dataflow process networks," *Proceedings of the IEEE,* 83(5), May 1995, pp. 773–801.
16. Bran Selic, Garth Gullekson, and Paul T. Ward, *Real-Time Object-Oriented Modeling,* New York: John Wiley & Sons, 1994.
17. Henry Chang, Larry Cooke, Merrill Hunt, Grant Martin, Andrew McNelly, and Lee Todd, *Surviving the SOC Revolution: A Guide to Platform-Based Design,* Kluwer Academic Publishers, 1999.
18. Chang Yun Park and Alan C. Shaw, "Experiments with a program timing tool based on source-level timing scheme," *IEEE Computer,* 24(5), May 1991, pp. 48–57.
19. David Becker, Raj K. Singh, and Stephen G. Tell, "An engineering environment for hardware/software co-simulation," in *Proceedings, 29th Design Automation Conference,* IEEE Computer Society Press, 1992, pp. 129–134.
20. Yau-Tsun Steven Li, Sharad Malik, and Andrew Wolfe, "Performance estimation of embedded software with instruction cache modeling," in *Proceedings, ICCAD-95,* IEEE Computer Society Press, 1995, pp. 380–387.
21. Rajesh K. Gupta and Giovanni De Micheli, "Hardware-software cosynthesis for digital systems," *IEEE Design & Test,* 10(3), September 1993, pp. 29–41.
22. R. Ernst, J. Henkel, and T. Benner, "Hardware-software cosynthesis for microcontrollers," *IEEE Design & Test,* 10(4), December 1993, pp. 64–75.

23. Wayne Wolf, "An architectural co-synthesis algorithm for distributed, embedded computing systems," *IEEE Transactions on VLSI Systems,* 5(2), June 1997, pp. 218–229.
24. Asawaree Kalavade and Edward A. Lee, "The extended partitioning problem: Hardware/software mapping, scheduling, and implementation-bin selection," *Design Automation for Embedded Systems,* 2(2), March 1997, pp. 125–163.
25. Yanbing Li and Joerg Henkel, "A framework for estimating and minimizing energy dissipation of embedded HW/SW systems," in *Proceedings, 35th Design Automation Conference,* ACM Press, 1998, pp. 188–194.
26. W. Fornaciari, P. Gubian, D. Sciuto, and C. Silvano, "Power estimation of embedded systems: a hardware/software codesign approach," *IEEE Transactions on VLSI Systems,* 6(2), June 1998, pp. 266–275.
27. L. Benini, A. Bogliolo, and G. De Micheli, "A survey of design techniques for system-level dynamic power management," *IEEE Transactions on VLSI Systems,* 8(3), June 2000, pp. 299–316.

124

Electronic Data Analysis Using PSPICE and MATLAB

John Okyere Attia

Prairie View A&M University

SPICE is an of the industry standard software for circuit simulation [1]. It can perform DC, AC, transient, Fourier, and Monte Carlo analyses. There are several SPICE-derived simulations packages, including Orcad PSPICE, Meta-software HSPICE, and Intusoft IS-SPICE.

PSPICE contains more features than classical SPICE. Some of the most useful additional features are a post-processor program, PROBE, that can be used for interactive graphical display of simulation results; and an analog behavioral model facility that allows modeling of analog circuit functions by using mathematical equations, tables, and transfer functions [2,3].

MATLAB is primarily a tool for matrix computations [4,5]. MATLAB has a rich set of plotting capabilities. The graphics are integrated in MATLAB. Since MATLAB is also a programming environment, a user can extend the functional capabilities by writing new modules (*m-files*). This chapter shows how the strong features of PSPICE and the powerful functions of MATLAB can be used to perform extensive and complex data analysis of electronic circuits [3,6].

0-8493-1586-7/05/$0.00+$1.50

TABLE 124.1 Element Name and Corresponding Element

First Letter of Element Name	Circuit Element, Sources, and Subcircuit
B	GaAs MES field-effect transistor
C	Capacitor
D	Diode
E	Voltage-controlled voltage source
F	Current-controlled current source
G	Voltage-controlled current source
H	Current-controlled voltage source
I	Independent current source
J	Junction field-effect transistor
K	Mutual inductors (transformers)
L	Inductor
M	MOS field-effect transistor
Q	Bipolar junction transfer
R	Resistor
S	Voltage-controlled switch
T	Transmission line
V	Independent voltage source
X	Subcircuit

124.1 PSPICE Fundamentals

A general SPICE program consists of the following components: (a) title; (b) element statements; (c) control statements, and (d) end statement. The following two sections will discuss the element and control statements.

Element Statements

Element statements specify the elements in the circuit. An element statement contains the (a) element name; (b) the circuit nodes to which each element is connected; and (c) the values of the parameters that electrically characterize the element. The element name must begin with a letter of the alphabet that is unique to a circuit element, source, or subcircuit. Table 124.1 shows the beginning alphabet of an element name and the corresponding element.

Circuit nodes are positive integers. The nodes in the circuit need not be numbered sequentially. Node 0 is predefined for ground of a circuit. Element values can be integer, floating point number, and floating point followed by an exponent.

Resistors

The general format for describing resistors is

```
Rname N+ N- value*
```

where the name must start with the letter *R*. *N*+ and *N*– are the positive and negative nodes of the resistor, respectively.

Value specifies the value of the resistor. The latter may be positive or negative, but not zero.

Inductors

The general format for describing linear inductors is

```
Lname N+ N- value [IC = initial_current]
```

where the inductor name must start with the letter *L*, *N*+ and *N*– are positive and negative nodes of the inductor, respectively, *Value* specifies the value of the inductance, and the initial condition for transient analysis is assigned using *IC* = *initial_current*, to specify the initial current.

Capacitors

The general format for describing linear capacitors is

```
Cname N+ N- value [IC = initial_voltage]
```

where the capacitor name must start with the letter *C*, *N*+ and *N*– are the positive and negative nodes of the capacitor, respectively, *Value* indicates the value of the capacitance, and the initial condition for transient analysis is assigned using *IC* = *initial_voltage* on the capacitor.

Independent Voltage Source

The general format for describing independent voltage source is

```
Vname N+ N- [DC value] [AC magnitude phase]
[PULSE V1 V2 td tr tf pw per]
```

or

```
[SIN V0 Va freq td df phase]
```

or

```
[EXP V1 V2 td1 t1 td2 t2]
```

or

```
[PWL t1 V1 t2 V2 . . . tn, Vn]
```

or

```
[SFFM V0 Va freq md fs]
```

where the voltage source must start with letter *V*, *N*+ and *N*– are the positive and negative nodes of the source, respectively, and sources can be assigned values for DC analysis [*DC value*], AC analysis [*AC magnitude phase*], and transient analysis. Only one of the transient response source options (*PULSE*, *SIN*, *EXP*, *PWL*, *SFFM*) can be selected for each source. The AC phase angle is in degrees. Discussion on the transient signal generators (*PULSE*, *SIN*, *EXP*, *PWL*, *SFFM*) can be found in Al-Hashimi [2], Attia [3], or Roberts and Sedra [7].

Independent Current Source

The general format for describing independent current source is

```
Iname N+ N- [DC value] [AC magnitude phase]
[PULSE V1 V2 td tr tf pw per]
```

or

```
[SIN V0 Va freq td df phase]
```

or

```
[EXP V1 V2 td1 t1 td2 t2]
```

or

```
[PWL t1 V1 t2 V2 . . . tn, Vn]
```

or

[SFFM V_0 V_a freq md fs]

where the current source must start with letter *I*, *N*+ and *N*– are the positive and negative nodes of the source, respectively. Current flows from a positive node to the negative node, and independent current sources can be assigned values for DC analysis [*DC value*], AC analysis [*AC magnitude phase*], and transient analysis. Only one of the transient response source options (*PULSE, SIN, EXP, PWL, SFFM*) can be selected for each source. The AC phase angle is in degrees.

124.2 Control Statements

Circuit Title

The circuit title must be the first statement in the SPICE program or circuit netlist. If this is not done, the program will assume that the first statement is the circuit title.

DC Analysis (.DC)

The .DC control statement specifies the values that will be used for DC sweep or DC analysis. The general format for the .DC statement is

```
.DC SOURCE_NAME START-VALUE STOP_VALUE INCREMENT_VALUE
```

where SOURCE_NAME is the name of an independent voltage or current source and START_VALUE, STOP_VALUE and INCREMENT_VALUE represent the starting, ending, and increment values of the source, respectively.

Transient Analysis (.TRAN)

The .TRAN control statement is used to perform transient analysis on a circuit. The general format of the .TRAN statement is:

```
.TRAN TSTEP TSTOP <TSTART> <TMAX> <UIC>
```

where the terms inside the angle brackets are optional, *TSTEP* is the printing or plotting increment, *TSTOP* is the final time of the transient analysis, *TSTART* is the starting time for printing out the results of the analysis. If it is omitted, it is assumed to be zero. The transient analyses always start at time zero, *TMAX* is the maximum step size that PSPICE uses for the purposes of computation. If TMAX is omitted, the default is the smallest value of either TSTEP or (TSTOP – TSTART)/50, and *UIC* (Use Initial Conditions) is used to specify the initial conditions of capacitors and inductors.

There are five SPICE-supplied sources that can be used for transient analysis: (1) PULSE (for periodic pulse waveform); (2) EXP (for exponential waveform); (3) PWL (for piece-wise linear waveform); (4) SIN (for a sine wave); and (5) SFFM (for frequency-modulated waveform). The format for specifying the above sources for transient analysis can be obtained from Al-Hashimi [2], Attia [3], or Roberts and Sedra [7].

AC Analysis (.AC)

The .AC control statement is used to perform AC analysis on a circuit. The general format of the .AC statement is

```
.AC FREQ_VAR NP FSTART FSTOP
```

where *FREQ_VAR* is one of three keywords that indicates the frequency variation by decade (DEC), octave (OCT), or linearly (LIN), *NP* is the number of points; its interpretation depends on the keyword

TABLE 124.2 Name Types for AC Output Variable

Output Variable	Meaning
V or I	Magnitude of V or I
VR or IR	Real part of complex value V or I
VI or II	Imaginary part of complex number V or I
VM or IM	Magnitude of complex number V or I
VDB or IDB	Decibel value of magnitude, i.e., 20 log $10\|V\|$ or 20 log$10\|I\|$

(DEC, OCT, or LIN) in the FREQ. DEC - NP is the number of points per decade. OCT - NP is the number of points per octave. LIN - NP is the total number of points spaced evenly from frequency FSTART and ending at FSTOP, *FSTART* is the starting frequency; FSTART cannot be zero, and *FSTOP* is the final or ending frequency.

Printing Command (.PRINT)

The .PRINT control statement is used to print tabular outputs. The general format of the .PRINT statement is

```
.PRINT ANALYSIS_TYPE OUTPUT_VARIABLE
```

where *ANALYSIS_TYPE* can be *DC, AC, TRAN.* Only one analysis type must be specified for .PRINT statement and *OUTPUT-VARIABLE* can be voltages or currents. Up to eight output variables can accompany one .PRINT statement. If more than eight output variables are to be printed, additional .PRINT statements can be used.

The output variable may be node voltages and current through voltage sources. PSPICE allows one to obtain current flowing through passive elements. The voltage output variable has the general form

```
V(node 1, node 2) or V(node 1) if node 2 is node "0."
```

The current output variable has the general form

```
I (Vname)
```

where *Vname* is an independent-voltage source specified in the circuit netlist.

For PSPICE, the current output variable can also be specified as *I(Rname)*, where Rname is resistance defined in the input circuit. For AC analysis, output voltage and current variables may be specified as real or imaginary magnitude and phase. Table 124.2 shows the name types for AC output variables.

Initial Conditions (.IC, UIC)

The .IC statement is only used when the transient analysis statement, .TRAN, includes the "*UIC*" option. The initial voltage across a capacitor or the initial current flowing through an inductor can be specified as part of capacitor or inductor component statement. For example, for a capacitor we have

```
Cname N+ N- value IC = initial voltage
```

and for an inductor, we use the statement

```
Lname N+ N- value IC = initial current
```

It should be noted that the initial conditions on an inductor or capacitor are used provided that .TRAN statement includes the "UIC" option.

Several SPICE control commands are shown in Table 124.3. Details of using the control statements can be found in Vladimirescu [1], Al-Hashimi [2], and Attia [3].

TABLE 124.3 Other PSPICE Control Statements

Control Statement	Description
*	* in the first column indicates a comment line.
.FOUR	Allows the user to perform Fourier analysis. Fourier components from DC to the *n*th harmonic are calculated.
.MC	Used to vary device parameter and to observe the overall system for variation in circuit parameters.
.NODESET	Used to set the operating point at specified nodes of a circuit during the initial run of a transient analysis.
.OP	Used to obtain the nodal voltages and the current flowing through independent voltage sources.
.SENS	Performs DC sensitivity of circuit element values and variation of model parameters.
.TEMP	Used to change the temperature at which a simulation is performed.
.TF	Used to obtain the small-signal gain, DC input resistance, and DC output resistance.
.WCASE	Causes sensitivity and worst-case analysis to be performed.

PSPICE Probe Statement (.PROBE)

Probe is a PSPICE interactive graphics processor that allows the user to display SPICE simulation results in graphical format on a computer monitor. Probe has facilities that allow the user to access any point on a displayed graph and obtain its numerical values. In addition, Probe has many built-in functions that enable a user to compute and display mathematical expression that models aspects of circuit behavior. The general format for specifying the probe statement is

```
.PROBE OUTPUT_VARIABLES
```

where *OUTPUT_VARIABLES* can be node voltages and/or devices currents. If no OUTPUT_VARIABLE is specified, Probe will save all node voltages and device currents.

124.3 MATLAB Fundamentals

The Colon Symbol

The colon symbol (:) is one of the most important operators in MATLAB. It can be used (a) to create vectors and matrices; (b) to specify sub-matrices and vectors; and (c) to perform iterations.

Creation of Vectors and Matrices

The statement

```
j1 = 1:9
```

will generate a row vector containing the numbers from 1 to 9 with unit increment. Nonunity, positive, or negative increments may be specified.

Specifying Submatrices and Vectors

Individual elements in a matrix can be referenced with subscripts inside parentheses. For example, j2(4) is the fourth element of vector j2. Also, for matrix j3, j3(2, 3) denotes the entry in the second row and third column. Using the colon as one of the subscripts denotes all of the corresponding row or column. For example, j3(:,4) is the fourth column of matrix j3. Also, the statement j3(2,:) is the second row of matrix j3. If the colon exists as the only subscript, such as j3(:), the latter denotes the elements of matrix j3 strung out in a long column vector.

Iterative Uses of Colon Command

The iterative uses of the colon command are discussed in the next section.

TABLE 124.4 Relational Operators

Relational Operator	Meaning
<	Less than
<=	Less than or equal
>	Greater than
>=	Greater than or equal
= =	Equal
~=	Not equal

FOR Loops

FOR loops allow a statement or group of statements to be repeated a fixed number of times. The general form of a FOR loop is

```
for index = expression
statement group C
end
```

The expression is a matrix and the statement group *C* is repeated as many times as the number of elements in the columns of the expression matrix. The index takes on the elemental values in the matrix expression. Usually, the expression is something like

```
m:n or m:i:n
```

where *m* is the beginning value, *n* the ending value, and *i* the increment.

IF Statements

The general form of the *simple IF statement* is

```
if logical expression 1
statement group G1
end
```

In the case of a simple IF statement, if the logical expression *1* is true, the statement group *G1* is executed. However, if the logical expression is false, the statement group G1 is bypassed and the program control jumps to the statement that follows the end statement. Several variations of the IF statement are described in Etter [5] and Attia [6].

IF statements use relational or logical operations to determine what steps to perform in the solution of a problem. The relational operators in MATLAB for comparing two matrices of equal size are shown in Table 124.4.

When any one of the above relational operators is used, a comparison is done between the pairs of corresponding elements.

Graph Functions

MATLAB has built-in functions that allow you to generate x-y and 3-D plots. MATLAB also allows you to give titles to graphs, label the x- and y- axes, and add grids to graphs.

X-Y Plots and Annotations

If x and y are vectors of the same length, then the command

```
plot(x, y)
```

plots the element of x (x-axis) versus the elements of y (y-axis).

To plot multiple curves on a single graph, one can use the *plot* command with multiple arguments, such as

```
plot(x1, y1, x2, y2, x3, y3, …, xn, yn)
```

The variables x1, y1, x2, y2, and so on are pairs of vector. Each x-y pair is graphed, generating multiple lines on the plot. The above plot command allows vectors of different lengths to be displayed on the same graph. MATLAB automatically scales the plots.

When a graph is drawn, one can add a grid, title, label, and x- and y- axes to the graph. The commands for grid, title, x-axis label, and y-axis label are *grid* (grid lines), *title* (graph title), *xlabel* (x-axis label), and *ylabel* (y-axis label), respectively.

Logarithmic and Plot3 Functions

Logarithmic and semilogarithmic plots can be generated using the commands *loglog*, *semilogx*, and *semilogy*. Descriptions of these commands follow:

loglog(x,y) – Generates a plot of $\log_{10}$(x) versus $\log_{10}$(y)
semilogx(x, y) – Generates a plot of $\log_{10}$(x) versus linear axis of y
semilogy(x, y) – Generates a plot of linear axis of x versus $\log_{10}$(y)

The *plot3* function can be used to do three-dimensional line plots. The function is similar to the two-dimensional plot function. The plot3 function supports the same line size, line style, and color options that are supported by the plot function. The simplest form of the plot3 function is

```
plot(x, y, z)
```

where *x*, *y*, and *z* are equal-sized arrays containing the locations of the data points to be plotted.

Subplot Function

The graph window can be partitioned into multiple windows. The *subplot* command allows the user to split the graph into two subdivisions or four subdivisions. The general form of the subplot command is

```
subplot(i j k)
```

The digits *i* and *j* specify that the graph window is to be split into an i-by-j grid of smaller windows, arranged in *i* rows and *j* columns. The digit *k* specifies the *kth* window for the current plot. The subwindows are numbered from left to right, top to bottom.

Fprintf

The *fprintf* command can be used to print both text and matrix values. The format for printing the matrix can be specified, and line feed can also be specified. The general form of this command is

```
fprintf('text with format specification', matrices)
```

The format specifier, such as *%7.3e*, is used to show where the matrix value should be printed in the text. 7.3e indicates that the value should be printed with an exponential notation of 7 digits, 3 of which should be decimal digits.

Other format specifiers are %c (single character); %d (signed decimal notation); %e (exponential notation); %f (fixed-point notation); and %g (signed decimal number in either %e or %f format, whichever is shorter). The text with format specification should end with *\n* to indicate the end of line.

M-files

MATLAB is capable of processing a sequence of commands that are stored in files with extension *m*. MATLAB files with extension *m* are called *m-files*. The latter are ASCII text files and are created with a text editor or word processor. M-files can either be scripts or functions. Script and function files contain a sequence of commands. However, function files take arguments and return values.

Script Files

Script files are especially useful for analysis and design problems that require long sequences of MATLAB commands. With a script file written using a text editor or word processor, the file can be invoked by entering the name of the m-file without the extension. Statements in a script file operate globally on the workspace data.

Function Files

Function files are m-files that are used to create new MATLAB functions. Variables defined and manipulated inside a function file are local to the function, and they do not operate globally on the workspace. However, arguments may be passed into and out of a function file.

The general form of a function file is

```
function variable(s) = function_name (arguments)
% help text in the usage of the function
.
.
end
```

Data Analysis Functions

In MATLAB, data analyses are performed on column-oriented matrices. Variables are stored in the individual column cells, and each row represents different observation of each variable. Functions act on the elements in the column. Table 124.5 gives a brief description of various MATLAB functions for performing data analysis.

TABLE 124.5 Data Analysis Functions

Function	Description
corrcoef(x)	Determines correlation coefficients.
diff(x)	Computes the differences between elements of an array x. It approximates derivatives.
max(x)	Obtains the largest value of x. If x is a matrix, max(x) returns a row vector containing the maximum elements of each column.
[y, k]=max(x)	Obtains the maximum value of x and corresponding locations (indices) of the first maximum value for each column of x.
mean(x)	Determines the mean or the average value of the elements in the vector. If x is a matrix, mean(x) returns a row vector that contains the mean value of each column.
median(x)	Finds the median value of elements in vector x. If x is a matrix, this function returns a row vector containing the median value of each column.
min(x)	Finds the smallest value of x. If x is a matrix, min(x) returns a row vector containing the minimum values from each column.
[y, k]=min(x)	Obtains the smallest value of x and the corresponding locations(indices) the first minimum value from each column of x.
sort(x)	Sort the rows of matrix x in ascending order.
std(x)	Calculates and returns the standard deviation of x if it is a one-dimensional array. If x is a matrix, a row vector containing the standard deviation of each column is computed and returned.
sum(x)	Calculates and returns the sum of the elements in x. If x is a matrix, this function calculates and returns a row vector that contains the sum of each column.
trapz(x,y)	Trapezoidal integration of the function y = f(x).

TABLE 124.6 Save Command Options

Option	Description
-mat	Save data in MAT file format (default)
-ascii	Save data using 8-digit ASCII format
-ascii -double	Save data using 16-digit ASCII format
-ascii -double -tab	Saves data using 16-digit ASCII format with tabs

Curve Fitting (Polyfit, Polyval)

The MATLAB *polyfit* function is used to compute the best fit of a set of data points to a polynomial with a specified degree. The general form of the function is

```
poly_xy = polyfit(x, y, n)
```

where *x* and *y* are the data points, *n* is the n-th degree polynomial that will fit the vectors x and y, and *poly_xy* is a polynomial that fits the data in vector y to x in the least-squares sense. poly_xy returns (n + 1) coefficients in descending powers of x.

Thus, the polynomial fit to vectors x and y is given as

$$poly\_xy(x) = a_1 x^n + a_2 x^{n-1} + \ldots a_m$$

The degree of the polynomial is *n* and the number of coefficients $m = n + 1$. The coefficients (a_1, a_2, ..., a_m) are returned by the MATLAB *polyfit* function

Save, Load, and Textread Functions

Save and Load Commands

The *save* command saves data in the MATLAB workspace to disk. The save command can store data either in memory-efficient binary format, called a *MAT-file*, or *ASCII file*. The general form of the save command is

```
save filename [List of variables] [options]
```

where *save* (without filename, list of variables, and options) saves all the data in the current workspace to a file named *matlab.mat* in the current directory. If a filename is included in the command line, the data will be saved in file "*filename.mat*." If a list of variables in included, only those variables will be saved.

Options for the save command are shown in Table 124.6.

MAT-files are preferable for data that are generated to be used by MATLAB. MAT-files are platform independent. The files can be written and read by any computer that supports MATLAB. The *ASCII files* are preferable if the data are to be exported or imported to programs other than MATLAB.

The *load* command will load data from MAT-file or ASCII file into the current workspace. The general format of the *load* command is

```
load filename [options]
```

where *load* (by itself without filename and options) will load all the data in file *matlab.mat* into the current workspace and *load filename* will load data from the specified filename.

Options for the load command are shown in Table 124.7.

TABLE 124.7 Load Command Option

Option	Description
-mat	Load data from MAT file (default in file extension is mat)
-ascii	Load data from space-separated file

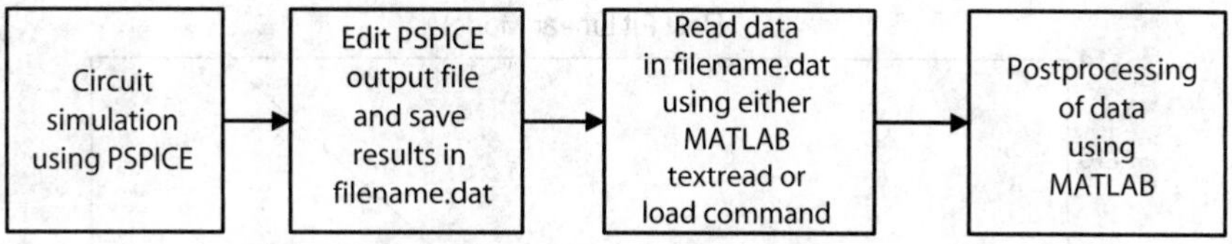

FIGURE 124.1 Flowchart of circuit simulation using PSPICE and post-processing of PSPICE.

TABLE 124.8 Voltage versus Current of a Diode

Forward-Biased Voltage, V	Forward Current, A
0.1	1.33e-13
0.2	1.79e-12
0.3	24.02e-12
0.4	0.321e-9
0.5	4.31e-9
0.6	57.69e-9
0.7	7.72e-7

The Textread Function

The *textread* command can be used to read ASCII files that are formatted into columns of data, where values in each column might be a different type. The general form of the textread command is

```
[a, b, c,…] = textread(filename, format, n)
```

where *filename* is the name of file to open. The filename should be in quotes, as in 'filename', *format* is a string containing a description of the type of data in each column. The format list should be in quotes. Supported functions include *%d* (read a signed integer value); *%u* (read an integer value); *%f* (read a floating point value); and *%s* (read a whitespace separated string), *n* is the number of lines to read. If n is missing, the command reads to the end of the file, and *a,b,c…* are the output arguments. The number of output arguments must match the number of columns that are being read.

124.4 Interfacing SPICE to MATLAB

To exploit the best features of PSPICE and MATLAB, circuit simulation is performed by using PSPICE. The PSPICE results, which are written into a file, *filename.out*, are edited using a text editor or a word processor and the data will be saved as *filename.dat*. The data are read using either MATLAB *textread* or *load* commands. Further processing of the data is performed using MATLAB. The methodology is shown in Figure 124.1.

In the following examples, the methodology described in this section will be used to analyze data obtained from electronic circuits.

Best-Fit Linear Model of a Diode

A forward biased diode has the corresponding voltage and current shown in Table 124.8. (a) Draw the equation of best fit for the diode data. (b) For the voltage of 0.64 V, what is the diode current?

The MATLAB script for analyzing the diode data follows:

```
% Example of Best-fit linear model
vt = 25.67e-3;
vd = [0.1 0.2 0.3 0.4 0.5 0.6 0.7];
id = [1.33e-13 1.79e-12 24.02e-12…
0.321e-9 4.31e-9 57.69e-9 7.72e-7];%
lnid = log(id);% Natural log of current
```

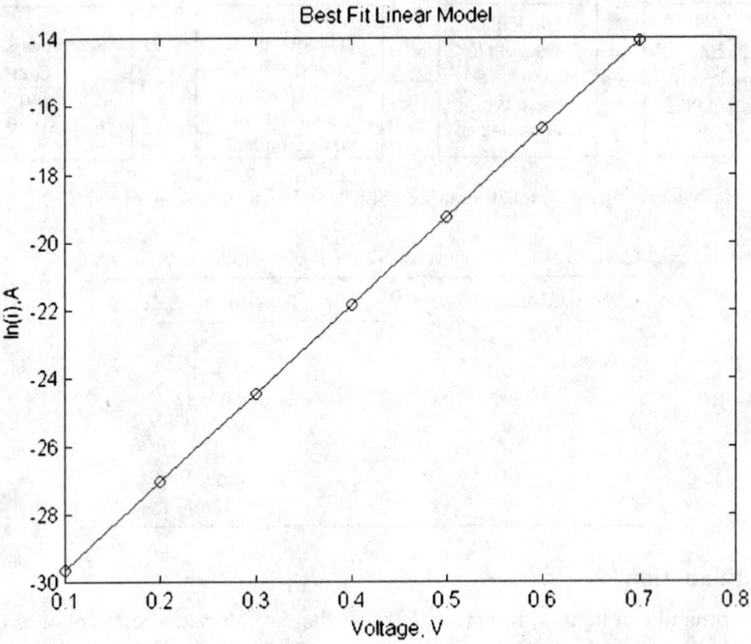

FIGURE 124.2 Best fit linear model of a diode.

```
% Determine coefficients
pfit = polyfit (vd, lnid, 1);% Curve fitting
% Linear equation is y = mx + b
b = pfit(2);
m = pfit(1);
ifit = m*vd + b;
% Calculate current when diode voltage is 0.64 V
Ix = m*0.64 + b;
I_64v = exp(Ix);
% Plot v versus ln(i) and best-fit linear model
plot(vd, ifit, 'b', vd, lnid, 'ob')
xlabel ('Voltage, V')
ylabel ('ln(i),A')
title ('Best-Fit Linear Model')
fprintf('Diode current for voltage of 0.64V is %9.3e\n', I_64v)
```

The plot is shown in Figure 124.2. Based on MATLAB, the diode current for voltage of 0.64V is 1.629e-007 A.

Op Amp Circuit with Series-Shunt Feedback Network

Figure 124.3 shows an op amp circuit with series-shunt feedback network. RS = 1 KΩ, RL = 10 KΩ, and R1 = 5 KΩ. Find the gain, V_0/V_S if RF varies from 10 KΩ to 100 KΩ. Plot voltage gain with respect to RF. Assume that the op amp is UA741 and the input voltage VS is sinusoidal waveform with a frequency of 5 KHz and a peak voltage of 1 mV.

PSPICE program for determining the gain follows:

```
* Example - OP AMP CIRCUIT WITH SERIES-SHUNT FEEDBACK
VS    1    0    AC    1E-3  0
RS    1    2    1K
```

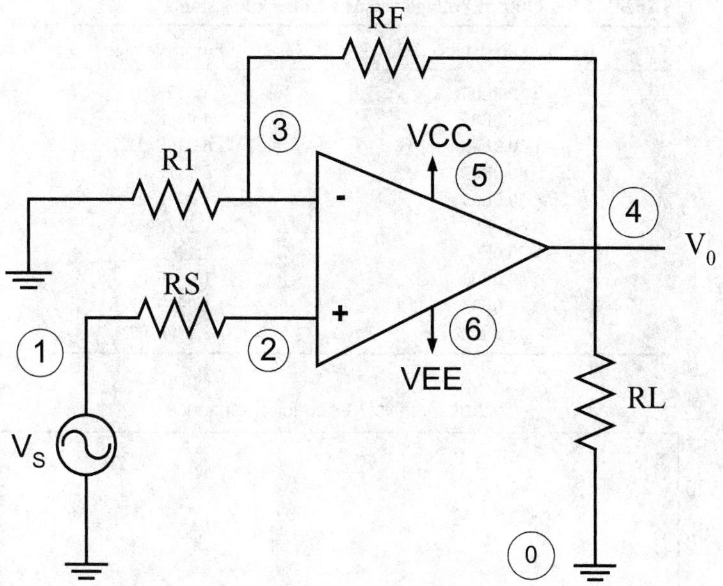

FIGURE 124.3 Op amp circuit with series-shunt feedback network.

```
R1    3    0    5K
RL    4    0    10K
VCC   5    0    15V
VEE   6    0    -15V
RF    4    3    RMOD  1
.MODEL  RMOD  RES(R=1)
.STEP RES   RMOD(R)  10.0E3  100.0E3  10.0E3
X1    2    3    5    6     4 UA741;UA741  OP  AMP
* +INPUT;  -INPUT;  +VCC;  -VEE;  OUTPUT;  CONNECTIONS  FOR  UA741
** ANALYSIS  TO  BE  DONE
.LIB  NOM.LIB
* UA741  OP  AMP  MODEL  IN  PSPICE  LIBRARY  FILE  NOM.LIB
** OUTPUT
.AC   LIN   1    5K    5K
.PRINT  ACV(4)
.END
```

PSPICE results are shown in Table 124.9. The data shown in Table 124.10 were stored in file *sol_52ps.dat*. The data from the latter file are used by MATLAB. The MATLAB script for solving the problem follows:

```
% Example
% Load data
load 'sol_52ps.dat' -ascii;
rf = sol_52ps(:,1);
gain = 1000*sol_52ps(:,2);
% Plot data
plot(rf, gain, rf, gain,'ob')
xlabel('Feedback Resistance, Ohms')
ylabel('Voltage Gain')
title('Voltage Gain vs. Feedback Resistance')
```

TABLE 124.9 Output Voltage versus Feedback Resistance

Feedback Resistance, RF	Output Voltage
10.0E03	3.000E-03
20.0E03	4.998E-03
30.0E03	6.996E-03
40.0E03	8.991E-03
50.0E03	1.098E-02
60.0E03	1.297E-02
70.0E03	1.496E-02
80.0E03	1.694E-02
90.0E03	1.892E-02
100.0E03	2.089E-02

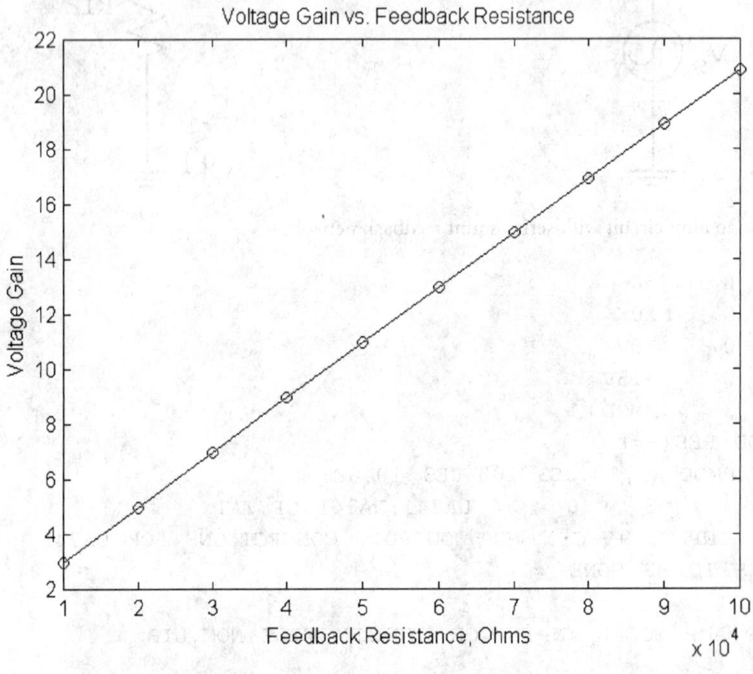

FIGURE 124.4 Voltage gain versus feedback resistance of an op amp circuit.

The gain versus feedback resistance is shown in Figure 124.4.

124.5 Conclusions

SPICE is an industry-standard software for circuit simulation. MATLAB is primarily a tool for matrix computations. It has numerous functions for data processing and analysis. PSPICE can be used to perform DC, AC, transient, Fourier, temperature and Monte Carlo analysis of electronic circuits with device models and subsystem subcircuits. MATLAB can then be used to perform calculation of device parameters, curve fitting, statistical analysis, and two-dimensional and three-dimensional plots. The strong features of PSPICE and powerful functions of MATLAB, when used in conjunction, allow extensive and complex data analysis of electronic circuits.

References

1. Vladimirescu, A., *The SPICE Book*, John Wiley & Sons, New York, 1994.
2. Al-Hashimi, B., *The Art of Simulation Using PSPICE, Analog, and Digital*, CRC Press, Boca Raton, FL, 1994.
3. Attia, J.O., *PSPICE and MATLAB for Electronics: An Integrated Approach*, CRC Press, Boca Raton, FL, 2002.
4. Chapman, S.J., *MATLAB Programming for Engineers*, 2nd ed., Brook, Cole Thompson, Learning, Pacific Grove, CA, 2002.
5. Etter, D.M., *Engineering Problem Solving with MATLAB*, 2nd ed., Prentice Hall, Upper Saddle River, NJ, 1997.
6. Attia, J.O., *Electronics and Circuit Analysis Using MATLAB*, CRC Press, Boca Raton, FL, 1999.
7. Roberts, G.W. and Sedra, A.S., *SPICE for Microelectronic Circuits*, Saunders College Publishing, Fort Worth, TX, 1992.

125

Electronic Packaging

Glenn Blackwell

Purdue University

Much like Moore's Law (3) is still true for semiconductor fabrications, the drive for more complexity in less space also continues in packaged assemblies. Smaller passive and integrated circuit (IC) packages and higher I/O counts, along with thinner printed circuit boards (PCBs) and the use of high-density interconnect (HDI) technologies continue to be driven by the desire for smaller, lighter, and higher-performing electronic products. This chapter is intended to provide an overview of currently used electronic packaging technologies, both of IC packages and of circuit board/circuit "packaging" or assemblies. Due to its brevity, the chapter is not intended to provide a comprehensive solution to any particular packaging problem, but rather to only introduce the reader to the breadth of solutions that are available, and to references that will allow the reader to delve further into the issues presented here. It also assumes that the reader is somewhat familiar with basic IC and PCB/PCA terminology, as well as with the three basic surface mount technology (SMT) assembly steps of solder paste deposition, parts placement, and reflow soldering.

IC packages may be developed by a packaging company, or by an IC manufacturer to use for their product. As an example of the proliferation of packages, consider the variety of packages that one manufacturer, Amkor Technology, makes available in several areas of current packaging (1):

- Ball grid array (BGA)
  - Plastic ball grid array (PBGA), the standard BGA package, 119 to 1156 ball count available
  - High-performance BGA (HPBGA), with enhanced thermal performance 276 to 1036 ball count
  - Multichip module BGA (MCM-PBGA), designed to use multiple die in one package, 119 to 1156 ball count
  - SuperBGA (SBGA), with high thermal performance
- Exposed pad technology ICs, with bottom solderable pads (die paddle) that are soldered directly to a corresponding pad on the PCB, for enhanced thermal performance
  - ExposedPad small-outline IC (SOIC), an SOIC with pad, 8 to 28 lead counts available

0-8493-1586-7/05/$0.00+$1.50

    - ExposedPad TQFP, a thin quad flat pack with pad, 32 to 256 lead counts available
    - ExposedPad TSSOP, thin small shrink outline package with pad, 8 to 80 pads available
- Tape packages from Amkor show four variations
- Chip scale packages from Amkor show nine variations

Many other package styles are available from Amkor and from other manufacturers. Examples are shown at websites and data books from Intel (2), Maxim (4), National Semiconductor (5), and others. Because of the proliferation of packages, it is imperative that the user always consult the latest manufacturer data.

Hundreds of published documents and standards exist that are related to many aspects of electronic packaging. Some of the publishers of these documents and standards follow:

- The **JEDEC** (Joint Electron Device Engineering Council) Solid State Technology Association (*www.jedec.org*) is the semiconductor engineering standardization body of the Electronic Industries Alliance (EIA). JEDEC/EIA publishes over 220 documents related to semiconductor packages and packaging. JEDEC also publishes a number of documents related to the testing of semiconductors.
- The Association Connecting Electronics Industries/IPC (*www.ipc.org*) publishes over 120 documents related to bare PCB design and manufacturing, and PCA assembly, test, and cleaning. Example IPC standards include:
    - IPC-2220, Design Standard Series
    - IPC7095, Design and Assembly Process Implementation for BGAs
    - J-STD-012,- Implementation of Flip Chip and Chip Scale Technology

Other groups that work in the area of packaging include:

- International Microelectronics and Packaging Society (formerly International Society for Hybrid Microelectronics, ISHM) (*www.imaps.org*)
- Surface Mount Technology Assn. (*www.smta.org*)

IC packaging technologies can be divided into the following types:

- Through-hole technology (THT) packaging
- Surface mount technology (SMT) packaging, which includes the following subtypes:
    - Leaded SMT devices
    - Leadless SMT devices
    - Chip scale packages: packaged ICs for which the package size does not exceed the die dimensions by more than 20%
    - Flip chip packages: die that have been passivated and bumped (described later), and are suitable for direct attach without a required protective covering
    - Tape packages, using a flexible tape to connect the die to the package
- Bare die that are suitable for chip-on-board (COB) assembly using ball and/or wedge wire bonding techniques. Protective covering/coating of some type is typically required when bare die are placed on PCAs.

Discussions of the various types of IC packages include the pitch of package terminations. Pitch is defined as the center-to-center dimension of adjacent leads or, for leadless devices, the center-to-center dimension of adjacent terminations.

125.1 IC Packaging

THT packages can be used with the standard SMT solder paste and reflow soldering processes. This "pin in paste" or "intrusive reflow" process is described in several articles (11,12,24,26,31). As THT ICs made up less than 15% of worldwide IC production in 2002, no further specific discussion of these packages is provided in this chapter. The reader is referred to the Intel and National websites, as well as to *The Electronic Packaging Handbook* (14).

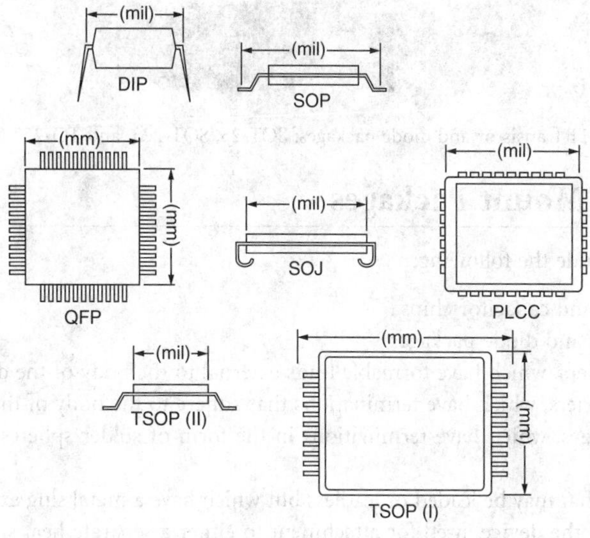

FIGURE 125.1 SMT package examples (lead pitch varies).

SMT IC packages undergo the standard manufacturing steps of die mount, wire bonding, and encapsulation. The final lead/termination configuration may be of several styles. These include the j-lead, gull-wing lead, unformed flat pack lead, and leadless terminations. The SMT IC may be in a SOIC, a plastic leaded chip carrier (PLCC), a BGA, a leadless ceramic chip carrier (LLCC), or one of several other SMT package styles. Unfortunately for the engineer and PCB layout designer, there is a lack of standardization in SMT packaging. Packages that are virtually identical in appearance may have a "controlling dimension" in inches or in millimeters. This is also true of passive SMT parts, where a "1206" package capacitor measures 0.120" × 0.060", but a "3216" capacitor measures 3.2 mm × 1.6 mm, making it almost, but not quite, equivalent to a 1206. The 3216 metric capacitor would actually be 0.126" × 0.063". An excellent discussion of passive part types and dimensions can be found in Prasad (27). As of this writing, passive parts are available as small as 0201s (0.020" × 0.010") (15). Examples of SMT IC packages are shown in Figure 125.1.

In the 1980s and 1990s, a number of IC assembly technologies developed to utilize either bare die or very small packages that approximated the size of bare die. Flip chips use the same bare die as COB, but the bond pads are placed toward the board or "face" down. Also placed face down are chip scale packages (CSPs), which are die enclosed in a package that is no more than 120% of the bare die X-Y dimensions. All of these die-size components are responses to the increasing demands for higher-density, higher-speed performance, and increased functionality.

125.2 PCB Design

The process steps from the design of a schematic to the successful design of a printed circuit board suitable for assembly are many, such as the following:

- Power and ground planes
- Issues relating to RF or fast risetime circuits
- Issues relating to analog and digital circuits on the same board
- Issues relating to testability, that is, design for test (DFT)
- Issues relating to manufacturability, that is, design for manufacturability (DFM)
- Issues relating to overall product packaging

While all of these are very important issues, they are not covered in this chapter and are mentioned only as issues that relate to the choice of both IC and PCB packaging techniques.

FIGURE 125.2 Leaded SMT transistor and diode packages: SOT-23, SOT-223, and TO-252 (DPAK).

125.3 Surface Mount Packages

SMT package types include the following:

- Leadless resistor and capacitor chips
- Leaded transistor and diode packages
- Leaded chip carriers, which have formable leads external to the body of the device
- Leadless chip carriers, which have terminations that adhere to the body of the device
- Area array packages, which have terminations in the form of solder spheres under the body of the device
- Power packages that may be leaded or leadless but which have a metal slug exposed on either the top or bottom of the device, used for attachment to either a separate heat sink or a copper pad on the surface of the PCB

Passive packages are not be discussed further in this chapter.

125.4 Transistor Packages

Leaded SMT packages are used for transistors, diodes, and integrated circuits. Transistors are commonly in a variety of SOT packages, including SOT-23 (also used for diodes), SOT-89, SOT-223, and TO-252 (Motorola D-PAK) packages. Packages such as the SOT-89, SOT-223, and TO-252 are used for high-power devices since they include relatively large solder tabs to improve heat transfer to the board. The SOT-23 package has been modified by several manufacturers to increase the lead count to a total of five or six, and to use them for IC packaging, such as op amps. Overall body size remains the same (Figure 125.2).

125.5 Leaded IC Packages

Selected leaded SMT IC packages are described below. This list should not be considered all inclusive, as new packages and package modifications are released by IC manufacturers on a regular basis.

- The modified SOT-23 package noted above with 25- and 50-mil pitch gull-wing leads.
- PLCCs with 50-mil pitch j-leads. This is the only SMT IC that is easily socketable (Figure 125.3).
- Small outline ICs, with 50-mil (1.27-mm) pitch gull-wing leads. Variations on the SOIC include the thin small outline package (TSOP), and the thin shrink small outline packages (TSSOP) with a 25-mil or 0.65-mm lead pitch (Figure 125.4).
- Quad flat packs (QFPs), with 25-mil (0.65-mm pitch) gull-wing leads (Figure 125.5).
- Flat packs with leads straight out from the four sides of the body. Typically these leads will be formed at the time of part placement.

125.6 Leadless IC Packages

Many packages are available with terminations that wrap around the bottom edge of the IC body, and do not have a formable lead external to the body. These include the leadless ceramic chip carrier (LLCCC) and the "no-lead" quad pack, with the common designation of a QFN package. The QFN, used by Maxim and TI, among others (10), is also available as an exposed pad power package (PQFN), with a solderable

FIGURE 125.3 PLCC-20 with J-leads.

FIGURE 125.4 Small outline IC: SOW-20 with gull-wing leads.

FIGURE 125.5 QFP-132 with gull-wing leads.

die pad under the device to be soldered to a corresponding pad on the PCB to allow for heat transfer to the board.

125.7 Area Arrays

The generic surface mount area array is known as a pad grid array or land grid array. The two most common types of area arrays are the BGA and the column grid array (CGA). Area arrays have a series of pads, with or without attachment interconnections, distributed across the bottom of the package. The pad array may be continuous across the entire surface, it may not be present directly under the die, or there may be a series of thermal pads directly under the die that are not used for I/O but rather for creating thermal connections to the PCB.

The BGA package is a style for which there is an overwhelming amount of information. A search of the Compendex Engineering Index for "ball grid array" results in 1145 articles being returned. Therefore, information about all aspects of BGAs can be found in virtually any technical journal related to the field.

BGAs are SMPs using an array of solder balls on the pads as the interconnect medium (Figure 125.6 and Figure 125.7). A corresponding array of solder lands is present on the board where the BGA will be mounted. The balls will be fused to the package pads by the IC supplier. The balls may be eutectic or near-eutectic solder and are commonly 0.75 mm (0.030") in diameter. Coplanarity of the balls is very important for acceptable soldering, with a requirement of at least 4 mils for standard 1.27-mm/50-mil pitch parts. BGAs and the smaller micro-BGA packages may have ball pitches of 1.27mm/50 mils, 1 mm/40 mils, down to 0.5 mm/20 mils for some micro BGA (μBGA) packages.

The BGA may be placed on a substrate with only flux present on the substrate pads or with additional solder paste on the pads. During reflow the balls will collapse to about 80% of their pre-reflow diameter, and during liquidus they will provide some self-centering on the pads. Since the balls collapse during the process, BGA assembly with additional solder paste is robust with regard to reasonable coplanarity issues on the board and of the balls themselves.

Most BGAs are in overmolded plastic packages. As such they are considered a class 3 component, susceptible to moisture absorption, and as such must be placed and reflowed within 24 hours of their

FIGURE 125.6 Ball grid array. BGA-256 with center ball depletion.

removal from dry pack, unless stored in a low-humidity environment. This cannot be emphasized enough, since the molding compound used for most BGAs absorbs moisture at a higher rate than the compound used for other IC packages.

BGAs are considered a viable alternative to QFPs when lead counts exceed 200, or when placement issues of 0.65-mm (0.025") and smaller-pitch QFPs override inspection issues of BGAs. With their higher pitch, BGAs can typically be placed with standard SMT placement systems, requiring less X-Y accuracy than QFPs with their 0.65-mm (0.025") and 0.5-mm (0.020") pitches. Their biggest disadvantage is that only the outer ring of balls/pads can be inspected without x-ray, and there is no reliable way to "touch up" a solder joint. Most manufacturers recommend that any rework be accomplished by removal and replacement of the BGA package.

CGAs have high-temperature columns of solder attached to pads on the bottom of the package. Eutectic solder paste is placed on the pad array of the substrate. When the system is reflowed, only the eutectic solder reflows, leaving the package with a controlled standoff from the substrate. Due to the height of the columns, the package accommodates well to thermal cycling, but uses a less robust placement/assembly procedure than BGAs.

125.8 Chip on Board

COB refers to the placement of bare IC die onto the substrate, with bond pads up. Since most IC die are available in bare die configuration, this assembly technique is usable by any manufacturer willing to make the investment in wire bonding and test equipment (Figure 125.8). In both COB and flip chips, discussed in the next section, multiple die/chips may be combined in a single package resulting in a system in a package (SIP).

In COBs, the die is placed into epoxy on the substrate, and the bond pads are then connected to the substrate using standard IC wire-bonding techniques. Once the bare die are wire bonded, the die and its associated wires are covered with epoxy for both mechanical and environmental protection. Die bond-pad pitch is commonly 0.25 mm (0.010"), with a corresponding requirement for very accurate placement. COB is attractive when space is extremely limited, such as in PCMCIA cards, and when maximum interconnect density is required. This technique is also attractive for low-cost high-volume consumer products, and is used extensively in digital watches and calculators. The die used for COBs are commonly known as "back-bonded" semiconductor chips, since the back of the chip is bonded to the substrate. This results in very good heat transfer to the substrate.

Any operation that uses bare IC die is faced with the issue of known good die (KGD) (9,25). IC manufactures test bare die for certain performance parameters, but do not do full tests until the die is

FIGURE 125.7 BGA-225 with full ball coverage.

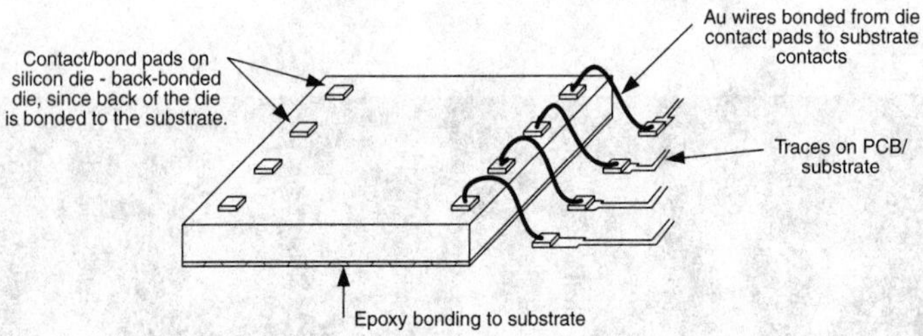

FIGURE 125.8 Example of chip on board.

packaged in its final form. Assemblers considering any of the assembly options discussed in this chapter will need to determine the "goodness" of the bare die they are using. While some IC manufacturers have put KGD testing into place, not all have done so.

One approach to the use of bare die in an assembly is for the user to put into place all the necessary new equipment for not only die mounting and wire bonding, but also for full test and burn in of the bare die. For these purposes, die probe and test carriers are available, and burn-in carrier and socket assemblies have been developed. These allow temporary connection with the die without damaging the die pad area for any subsequent assembly operations.

The EIA through JEDEC has put forth a procurement standard, EIA/JEDEC Standard #49, "Procurement Standard for Known Good Die (KGD)." The intent of this standard is to make suppliers of die aware of "the high levels of as-delivered performance, quality and long term reliability expected of this product type." The die procured may be used in multichip modules (MCMs) or hybrid assemblies (14,22), or in board-level assemblies such as COB or flip chip components. The standard applies to both military and commercial uses of KGD. It notes that KGD users must recognize that the levels of quality and reliability users have come to expect from packaged devices cannot be assured with bare die, although it also notes that KGD are "intended" to have equivalent or better performance than equivalent packaged parts. It also notes that close cooperation between supplier and user of KGD is a necessity. The standard notes that the user accepts the responsibility for final hermetic sealing and/or plastic encapsulation.

125.9 Flip Chip

Flip chips are passivated IC die placed on the substrate with bond pads down. The lands on the substrate must match the bond pad geometry on the chip. As in COBs, die bond-pad pitch is commonly 0.25 mm (0.010"), with a corresponding requirement for very accurate placement. The die used for flip chip assembly are face-bonded chips. Rather than the "bare" bond pads used on back-bonded chips, face-bonded chips must have prepared bond pads suitable for direct placement and soldering to the substrate lands. Typical pad preparations include

- Solder-coated copper ball
- Bond pad with bump ("bumped" chip)
- Substrate land with bump (Figure 125.9)

Pad bumping is done with solder at each bond pad while the die are still in wafer form, requiring special processing by the die manufacturer. A variety of bumping techniques can be used, and should be

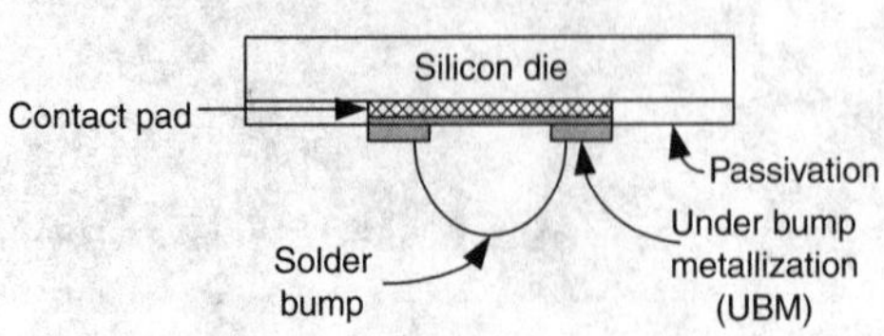

FIGURE 125.9 Example of flip-chip solder bump.

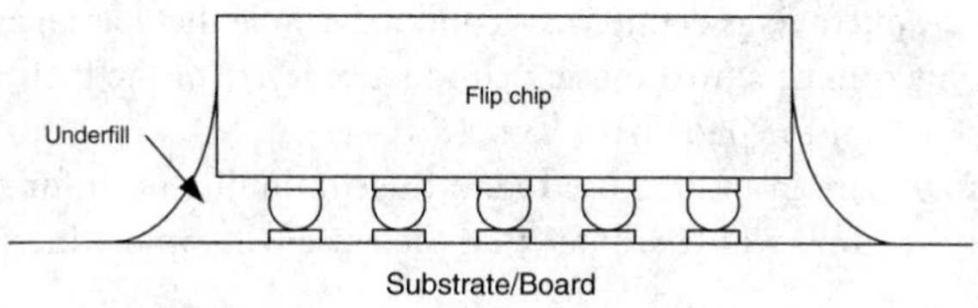

FIGURE 125.10 Flip chip with underfill.

understood by the user (27). After individual die are sliced from the wafer, the chip will be placed on the substrate lands and reflowed in a conventional reflow oven.

Unprotected flip chips are prone to thermally induced cracking, so standard assembly practice is to fill the area under the chip with adhesive in order to equalize the stresses across the chip (Figure 125.10) (13,14).

Bare die, whether wire bonded or flipped, can be placed in a common IC package to create a multichip module (MCM) (14). If the functions of the die combine to form a complete operating unit, this may also be known as a SIP (23,30).

125.10 Chip Scale Packages

CSPs have been defined by the IPC in PAS-62084, "Implementation of Flip Chip and Chip Scale Technology" as IC die that have been enclosed in a package that is no larger than 1.2 times the dimensions of the bare die itself. The packages have pitch dimensions as small as 0.25 mm/10 mil, making them valuable in space-limited applications such as memory products, PC cards, and MCMs.

Substrate, leadframe, and wafer-scale CSPs (some manufacturers also use the term μBGA) pack more functions into less space than many previous packages (21,26,29). CSPs are available in pin counts from 4 to 48 leads and more, are thin, and have very low parasitics. Most CSPs can be handled by standard placement machines. Current solder deposition and reflow systems are capable of soldering CSPs, most of which have a 0.5-mm lead pitch. Reliability can be an issue, although manufacturers subject these packages to the same reliability criteria applied to other semiconductor packaging styles.

125.11 High-Density Interconnects

In PCB terminology, a through hole is a conductive hole in the board that is used both for interlayer connectivity as well as accommodating a part (THT) lead (Figure 125.11). A via is a conductive hole for interlayer connectivity that may extend through the board or may connect only certain layers. For both through holes and vias, a defining measurement is the "aspect ratio" that relates the diameter of the hole

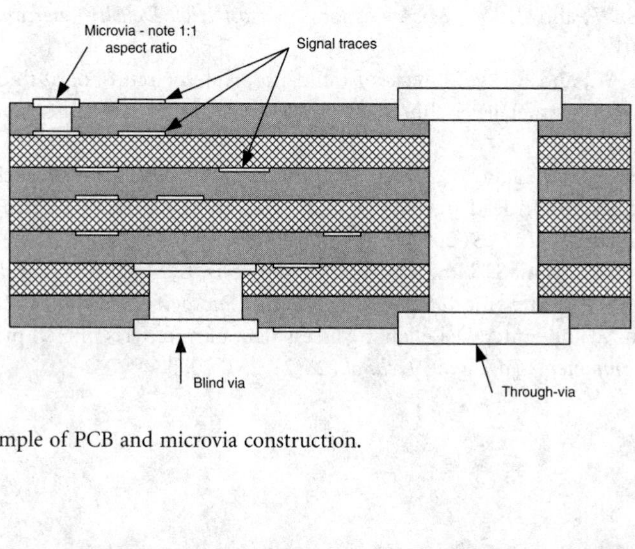

FIGURE 125.11 Example of PCB and microvia construction.

to the length of the hole. A microvia is defined as a conductive hole that has an aspect ratio of 1:1, which means that a microvia only connects two adjacent conductive layers of the PCB, and the diameter of the hole is equal to the thickness of the one board layer (6,16–19).

While HDI and microvias are in limited use today, the continuing push for smaller circuitry, higher clock speeds, and faster rise times will guarantee that their use will rapidly increase (7,8,18,23).

References

1. Amkor Technology, Inc., available at: *www.amkor.com*.
2. Intel Corp., Intel Packaging Handbook, available at: *www.intel.com/design/packtech/packbook.htm*.
3. Intel Corp., available at: *www.intel.com/research/silicon/mooreslaw.htm*.
4. Maxim Integrated Products, available at: *www.maxim-ic.com/cgi-bin/packages*.
5. National Semiconductor Packaging Information, available at: *www.national.com/packaging/*.
6. Association Connecting Electronics Industries/IPC, IPC 2226 — Sectional Design Standard for High Density Interconnect (HDI) Boards, IPC, Northbrook, IL.
7. Association Connecting Electronics Industries/IPC, IPC 6016 — Qualification & Performance Specification for High Density Interconnect (HDI) Layers or Boards, IPC, Northbook, IL.
8. Association Connecting Electronics Industries/IPC, IPC 6801 — IPC/JPCA Terms & Definitions, Test Methods, and Design Examples for Build-Up/High Density Interconnect (HDI) Printed Wiring Boards, IPC, Northbrook, IL.
9. Joint Electronic Device Engineering Council/JEDEC Solid State Technology Association, Standard 49, Procurement Standard for Known Good Die, JEDEC, Arlington, VA.
10. Maxim Integrated Products, Thermal Considerations of QFN and Other Exposed-Paddle Packages. Application Note HFAN-08.1, Maxim Integrated Products, Sunnyvale, CA, 2001.
11. Tyco Electronics/AMP, Solder Volumes for Through-Hole Reflow-Compatible Connectors, available at: *www.amp.com/products/technology/articles/dd66c.stm*.
12. Balchunas, J. and Hagio, A. X-ray Verifies SMT Process for Through-Hole Connectors. *Evaluation Eng.*, 33 (9), 1994.
13. Beddingfield, C. Flip Chip Process Challenges for SMT Assembly. *Electron. Prod. Packaging*, March 2003, p. 24.
14. Blackwell, G. R., Ed. *The Electronic Packaging Handbook*. CRC Press, Boca Raton, FL, 2000.
15. Butterfield, A., Guilford, M., and Pieper, K. Manufacturing with 0201s: The Latest Developments. *Circuits Assembly*, 14 (5), 24, 2003.
16. Dunsky, C. High-Speed Microvia Formation with UV Solid-State Lasers. Proceedings of the IEEE, 90(10), 1670, 2002.
17. Holden, H. Microviasí effect on high-frequency signal integrity. *Circuit World*, 28(3), 10, 2002.
18. Larson, S. HDI-buildup technology and microvias: no longer a mystery. *EDN*, 48(5), 111, 2003.
19. Lau, J. H., Lee, S. W., and Ricky, . *Microvias for Low Cost High Density Interconnects*. McGraw-Hill, New York, 2001.
20. Lau, J. H., Lee, S. W., and Ricky, . Effects of build-up printed circuit board thickness on the solder joint reliability of a wafer level chip scale package (WLCP). *IEEE Trans. Components Packaging Technol.*, 25(1), 3, 2002.
21. Lau, J. H., Pan, S. H., and Chang, C. Creep analysis of wafer level chip scale package (WLCSP) with 96.5Sn-3.5 Ag and 100In lead-free solder joints and microvia build-up printed circuit board. *J. Electron. Packaging*, 124(2), 69, 2002.
22. Lee, C.-H., Sutono, A., Han, S., Lim, K., Pinel, S., Tentzeris, E. M., and Laskar, J. A compact LTCC-based Ku-band transmitter module. *IEEE Trans. Adv. Packaging*, 25(3), 374, 2000.
23. Liu, F., Sundaram, V., and Sutter, D. Reliability assessment of microvias in HDI printed circuit boards. *IEEE Trans. Components Packaging Technol.*, 25(2), 254, 2000.

24. Manessis, D., Whitmore, M., Adriance, J. H., and Westby, G. R. Evaluation study of ProFlow system for stencil printing of thick boards (0.125 inches) in the alternative assembly and reflow technology (AART) or pin-in-paste process. *Proceedings of the Technical Program, National Electronic Packaging & Production Conference*, 1, 416, 1999.
25. Novellino, J. What is known — good die and how do you get there? *Electronic Design*, 445(17), 1997.
26. Monticelli, D. Chip-scale packaging is ready for prime time. *Electronic Design*, 49(3), 46, 2001.
27. Patterson, D. S., Elenius, P., and Leal, J. A. Wafer Bumping Technologies — A Comparative Analysis of the Solder Deposition Processes and Assembly Considerations. Paper presented at INTERPack conference, 1997.
28. Prasad, R. P. *Surface Mount Technology Principles and Practices*, 2nd ed. Chapman & Hall, New York, 1997.
29. Ramakrishnan, S., Srihari, K., and Westby, G. R. Decision support system for the alternative assembly and reflow technology process. *Comput. Ind. Eng.*, 35(1–2), 61, 1998.
30. Shetty, S., Lehtinen, V., and Dasgupta, A. Fatigue of chip scale package interconnects due to cyclic bending. *J. Electron. Packaging*, 123(3), 302, 2001.
31. Sturcken, K., Konecke, S., and Mason, K. Bare chip stacking. *Adv. Packaging*, Apr. 2003, p. 17.
32. Tonapi, Sandeep, S., and Srihari, K. Implementation of the pin-in-paste process in a contract manufacturing environment. *Int. J. Prod. Res.*, 38(17), 4535, 2000.

Further Information

Journals, available in hard copy and online
Advanced Packaging, www.apmag.com.
Circuits Assembly, www.circuitsassembly.com.
Electronics Cooling, www.electronics-cooling.com.
Electronics Packaging and Production, www.epp.com.
IEEE Transactions on Components and Packaging Technologies, www.cpmt.org/trans/trans-cpt.html.
IEEE Transactions on Electronics Packaging Manufacturing, www.cpmt.org/trans/trans-epm.html.
Printed Circuit Design and Manufacture, www.padandm.com.
SMT Magazine, www.smtmag.com.
Organizations related to packaging
Association Connecting Electronics Industries/IPC, *www.ipc.org.*
Joint Electron Device Engineering Council/JEDEC, *www.jedec.org.*
International Microelectronics and Packaging Society (IMAPS), *www.imaps.org.*
Surface Mount Technology Assn. (SMTA), *www.smta.org.*

126

Microwave and RF Engineering

Mike Golio
Golio Consulting

The fields of RF and microwave engineering involve the exploitation of the electromagnetic spectrum in the hundreds of megahertz to hundreds of gigahertz region. A precise and standard definition of RF signals and microwave signals is not agreed upon, but microwave engineering is identified by certain physical characteristics of circuits, systems, and measurement procedures that are uniquely important at frequencies above several hundred megahertz but below light waves.

Figure 126.1 illustrates schematically the electromagnetic spectrum from audio frequencies through cosmic rays. As seen in the illustration, the RF spectrum is defined loosely as the electromagnetic waves from approximately 300 MHz to 1 GHz, while the microwave spectrum includes frequencies in the 1-GHz to 100-GHz range. Alternatively, the RF and microwave spectrum can be defined in terms of the wavelength, λ_0, according the expression

$$\lambda_0 = c/f \tag{126.1}$$

where c = velocity of light and f = frequency.

The RF and microwave spectra are further divided into frequency bands as illustrated in Figure 126.2. Although frequency band designation standards using letters to define frequency ranges are presented in the technical literature, competing standard definitions exist and the use of lettered band designations can be misleading. Diagonal hashing at the ends of the frequency bands in Figure 126.2 indicates variations in the definitions by different sources in the literature. The double-ended arrows appearing above some of the bands indicate the definitions by the Institute of Electrical and Electronics Engineers (IEEE) for these bands. Both the IEEE and U.S. military have established letter designations for RF and microwave bands, but the designations are highly dissimilar. Bands identified as D, E, G, and L, for example, exist in both designations but do not identify the same spectra.

0-8493-1586-7/05/$0.00+$1.50

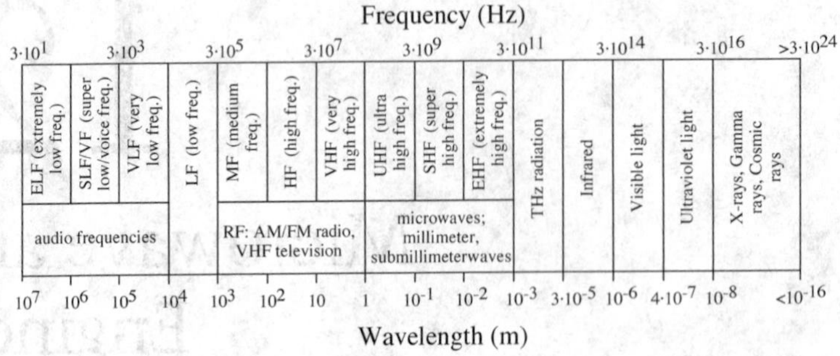

FIGURE 126.1 Electromagnetic frequency spectrum and associated wavelengths. (*Source:* Golio, M., Ed., *RF and Microwave Handbook*, CRC Press, Boca Raton, FL, 2000.)

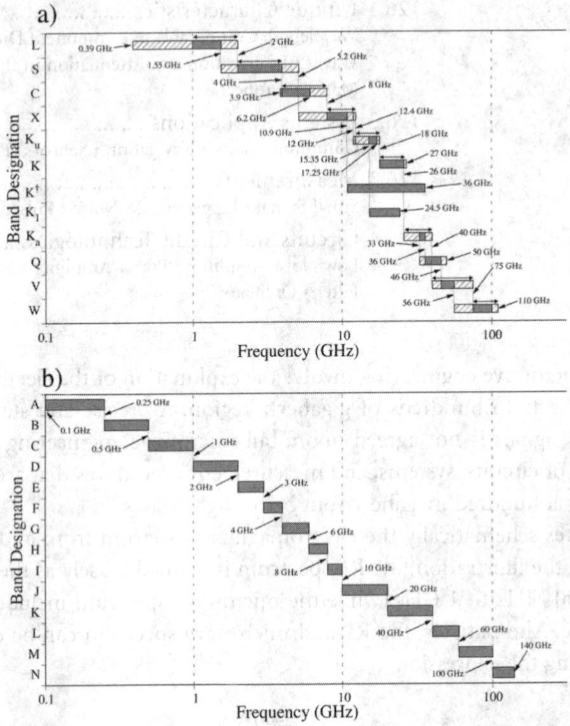

FIGURE 126.2 Microwave and RF frequency band designations. (a) Industrial and IEEE designations. Diagonal hashing indicates variation in the definitions found in the literature; dark regions in the bars indicate frequencies for which there is widespread agreement. Double-ended arrows appearing on bands indicate the current IEEE definitions for these bands where they exist, and K† denotes an alternative definition for K-band. (b) U.S. military frequency band designations. (*Source:* Golio, M., Ed., *RF and Microwave Handbook*, CRC Press, Boca Raton, FL, 2000.)

126.1 Unique Characteristics

Wavelength Comparable to Component Dimensions

At lower frequencies where the wavelength of the electromagnetic energy is several orders of magnitude larger than the dimensions of the circuit or system being examined, performance analysis using node voltages or loop currents is accurate. Consideration of propagation effects is not required. There is effectively no delay between the cause and effect of current and voltage changes in the system. As the frequency increases, however, the wavelength shrinks until the components (e.g., lumped inductors,

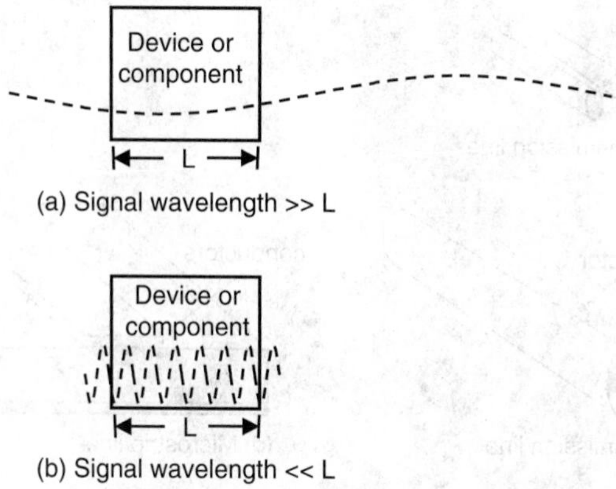

FIGURE 126.3 Illustration of the relationship between device or component dimensions, L, and signal wavelength, λ.

capacitors, or transistors) are of comparable size to the wavelength. Figure 126.3 illustrates this phenomenon. As these frequencies are approached, signal propagation effects become critical to the performance analysis of the circuit or system. Guided waves, signal transmission, signal reflection, and standing waves are key concepts involved in RF and microwave engineering.

Guided Waves

At higher frequencies it is often necessary to consider an electronic signal as an electromagnetic wave and the structure that guides this wave as a transmission line. Thus, wires become transmission lines for the RF and microwave engineer. The properties of transmission lines can be exploited to the advantage of the RF/microwave designer to realize circuit functions with very low loss. The physical realization of high frequency transmission lines come in several forms. Figure 126.4 illustrates some of the most important realizations of transmission lines in RF and microwave systems.

Transmission Lines

Figure 126.4(a) illustrates a two-wire transmission line. This sort of line is common at lower frequencies (television antenna cables often utilize this type of transmission line), but tends to become very lossy and ill behaved electrically at RF and microwave frequencies. The diameter of the wire and separation between the wires are critical parameters in determining the propagation properties of this type of line.

A coaxial cable is illustrated in Figure 126.4(b). This type of cable is very common in RF and microwave systems of all types. In the coaxial cable, a center conductor runs through a dielectric cylinder surrounded by a metallic outer shield. Both the inner and outer conductor diameters as well as the dielectric material properties determine the propagation characteristics of this type of cable. The outer conductor shields the electronic signal from interference and prevents losses due to unwanted signal radiation.

A section of rectangular waveguide is illustrated in Figure 126.4(c). Waveguide structures are used in many legacy microwave systems, but are also important for many high power and millimeter wave systems today. Waveguides consist of a closed conducting structure with a fixed cross section. Waveguides are often air filled but can also be filled with dielectric material. The dimensions of the guide and the dielectric properties determine the propagation characteristics of waveguides. One characteristic that distinguishes the metal waveguide from the other structures of Figure 126.4 is that it consists of a single conductor. This characteristic results in a number of different possible propagation modes for the waveguide. Each propagation mode has distinct velocity and attenuation characteristics. For most applications, single-mode propagation is desired to avoid signal distortion and attenuation. By appropriate selection of the cross-sectional dimensions of the waveguide, a dominant mode can be established over a specific frequency band. At lower frequencies, the wave will not propagate through the guide and at higher frequencies the wave will propagate in undesirable and multiple modes.

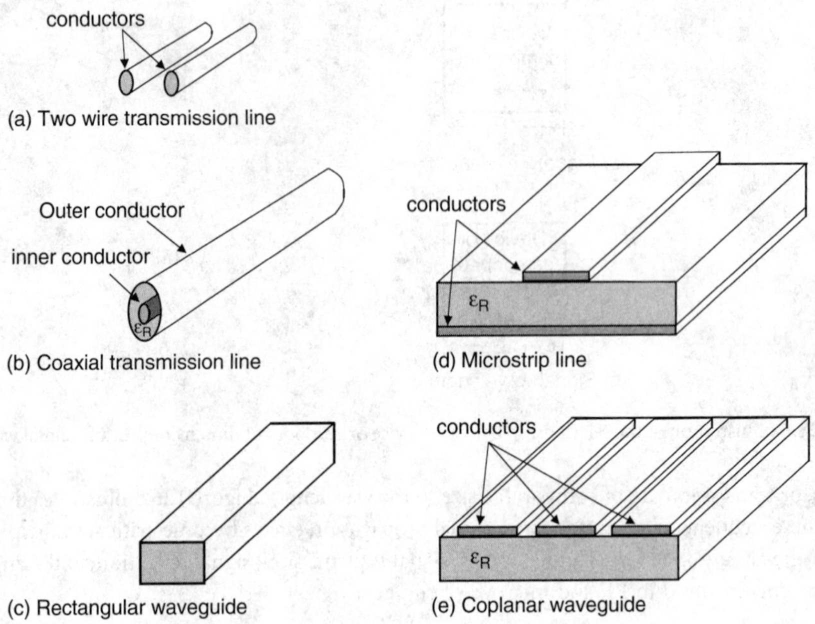

FIGURE 126.4 Physical representations of several important guided wave structures.

Figures 126.4(d) and (e) illustrate two realizations of planar-guided wave structures. These types of structures are extremely important in the development of RF and microwave integrated circuits. Since integrated circuits are fabricated as part of the semiconductor wafer production process, the transmission lines that are part of those circuits must be compatible with those planar processes. The microstrip and coplanar waveguide lines are easily fabricated on a semiconductor wafer using normal metal deposition and patterning processes. Both of the planar transmission lines of Figures 126.4(d) and (e) have properties similar to coaxial cable. In the case of microstrip lines, the upper conductor is analogous to the center conductor in the coax while the lower conductor is analogous to the outer conductor. The dielectric thickness and upper conductor line width along with the dielectric properties are the critical parameters that determine the propagation characteristics of microstrip line. In the case of coplanar waveguide, the center conductor is analogous to the center conductor of the coax while the two outer conductors are equivalent to the outer conductor of the coax. Critical parameters that determine propagation characteristics of coplanar waveguide include the width of the center conductor, the spacing between the center and outer conductors and the dielectric properties of the substrate.

All of the transmission line types illustrated in Figure 126.4 can be analyzed and modeled using the equivalent circuit illustrated in Figure 126.5. A differential length of transmission line Δz is associated with the following four characteristic parameters:

R, resistance per unit length of the conductors (Ω/m),
L, inductance per unit length of the conductors (H/m),
G, conductance per unit length (S/m),
C, capacitance per unit length (F/m).

Analysis of the equivalent circuit results in the definition of important transmission line characteristics. The *characteristic impedance* of a transmission line given in terms of the equivalent circuit is

$$Z_0 = \sqrt{\frac{(R + j\omega L)}{(G + j\omega C)}} \tag{126.2}$$

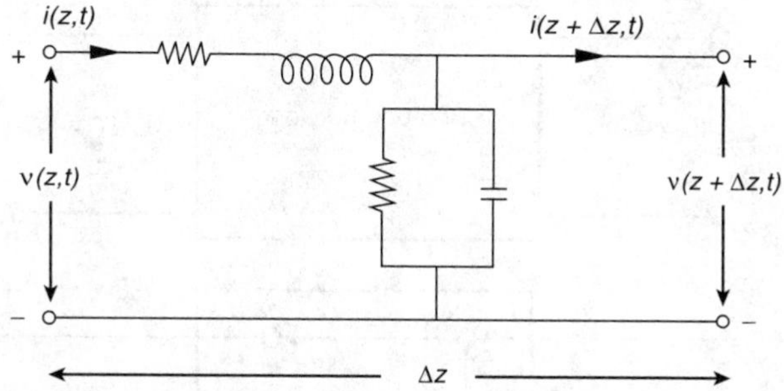

FIGURE 126.5 Distributed equivalent circuit model for a transmission line. (*Source:* Golio, M., Ed., *RF and Microwave Handbook*, CRC Press, Boca Raton, FL, 2000.)

The characteristic impedance of a transmission line is the impedance that when placed at the termination of the line results in no reflected energy. Such a transmission line is called "matched."

The *propagation constant* for the transmission line can be written as

$$\gamma = \alpha + j\beta = \sqrt{(R + j\omega L)(G + j\omega C)} \tag{126.3}$$

The *phase velocity* or *propagation velocity* of the electromagnetic energy is expressed as

$$v = \frac{\omega}{\beta} \tag{126.4}$$

Phase velocity is the speed at which energy or information moves along the transmission line.

S-Parameters

To analyze microwave transmission and circuit design problems, generalized parameters are desirable. For low frequency circuits these parameters are often defined in terms of node voltages and currents. Y-parameters, z-parameters and h-parameters are all examples of generalized network parameters that are defined in terms of voltages and currents. At RF and microwave frequencies, however, these parameters are very difficult to measure. The y-, z-, and h-parameters all involve parameter definitions that require that short circuits or open circuits be applied to the network. But at high frequencies, short circuits and open circuits are extremely difficult to realize. Conventional short and open circuits will exhibit stray capacitance and radiation characteristics at high frequencies that will dominate the performance of the structure. In addition, an attempt to provide a short or open circuit to an active device can result in oscillations at unwanted frequencies.

For RF and microwave circuits and systems, it is more convenient to develop parameters based on energy flow through multiport networks. Figure 126.6(a) illustrates a two-port network. A flow graph can be drawn to analyze the energy flow of a two-port network as shown in Figure 126.6(b). The flow graph has two nodes for each port — an incident wave node and a reflected wave node. The incident wave is the *a* node while the reflected wave is the *b* node. The numeral suffix on the *a* and *b* values denotes the port number.

Signal flow through the two-port can be described as follows. A wave incident on port 1 is represented as *a*1. Part of that wave will be reflected through the S_{11} path and appear as b_1. The remainder of that wave goes through the S_{21} path and appears at the output as b_2. Similarly, a wave incident at port 2 appears as *a*2 and is partially reflected through the S_{22} path and partially transmitted through the S_{12} path.

For a two-port device, the four s-parameters can therefore be written as

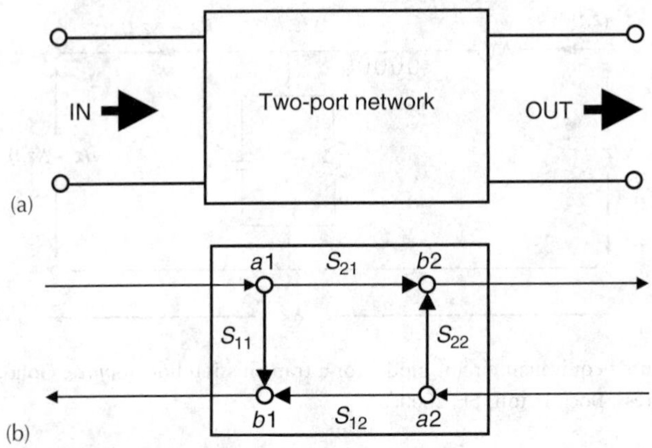

FIGURE 126.6 (a) Simple representation of a two-port network. (b) Flow graph of a two-port network.

$$S_{11} = \left.\frac{b1}{a1}\right|_{Z_0 load\_on\_port\_2} \tag{126.5}$$

$$S_{12} = \left.\frac{b1}{a2}\right|_{Z_0 load\_on\_port\_1} \tag{126.6}$$

$$S_{21} = \left.\frac{b2}{a1}\right|_{Z_0 load\_on\_port\_2} \tag{126.7}$$

$$S_{22} = \left.\frac{b2}{a2}\right|_{Z_0 load\_on\_port\_1} \tag{126.8}$$

These expressions also suggest the means of measuring the s-parameters. S_{11} can be measured when port 2 is terminated with its characteristic impedance by measuring the ratio of the reflected and incident wave. S_{21} is measured with the characteristic impedance termination on port 2 by measuring the ratio of the transmitted to incident wave. S_{22} and S_{12} are similarly measured with port 1 terminated in its characteristic impedance.

The s-parameters defined by Equation (126.5) through Equation (126.8) are complex quantities normally expressed as magnitude and phase. Several scalar quantities that can be defined in terms of s-parameters are also commonly utilized by microwave engineers. Note that S_{11} and S_{22} can be thought of as complex reflection ratios since they represent the magnitude and phase of waves reflected from ports 1 and 2, respectively. It is common to evaluate the quality of the match between components using the *reflection coefficient* defined as

$$\Gamma = |S_{11}| \tag{126.9}$$

for the input reflection coefficient of a two-port network, or

$$\Gamma = |S_{22}| \tag{126.10}$$

for the output reflection coefficient.

Reflection coefficient is often expressed in dB and referred to as *return loss* expressed as

$$L_{return} = -20\log(\Gamma) \tag{126.11}$$

Analogous to the reflection coefficient, both a forward and reverse transmission coefficient can be defined. The forward transmission coefficient is given as

$$T = |S_{21}| \tag{126.12}$$

while the reverse transmission coefficient is expressed as

$$T = |S_{12}| \tag{126.13}$$

As in the case of reflection coefficient, transmission coefficients are often expressed in dB and referred to as *gain* given by

$$G = 20\log(T) \tag{126.14}$$

Another commonly measured and calculated parameter is the *standing wave ratio* or the *voltage standing wave ratio* (VSWR). This quantity is the ratio of maximum to minimum voltage at a given port. It is commonly expressed in terms of a reflection coefficient as

$$(VSWR) = \frac{1+\Gamma}{1-\Gamma} \tag{126.15}$$

Propagation and Attenuation in the Atmosphere

Electromagnetic signals are attenuated by the atmosphere as they propagate from source to target. In general, atmospheric attenuation increases with increasing frequency. As shown in Figure 126.7, however, there is significant structure in the atmospheric attenuation versus frequency plot. If only attenuation is considered, it is clear that low frequencies would be preferred for long-range communications systems in order to take advantage of the low attenuation of the atmosphere. If high data rates or large information content is required, however, higher frequencies are needed. In addition to the atmospheric attenuation, the wavelengths of microwave systems are small enough to become affected by water vapor and rain. Above 10 GHz these effects become important. Above 25 GHz, the effect of individual gas molecules

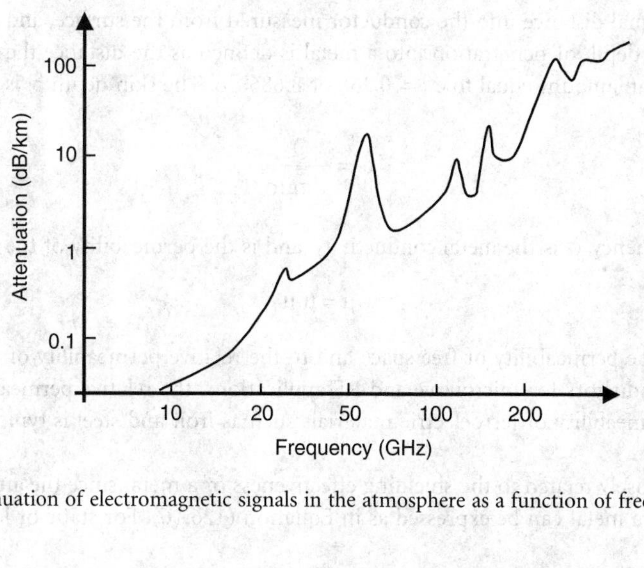

FIGURE 126.7 Attenuation of electromagnetic signals in the atmosphere as a function of frequency.

becomes important. Water and oxygen are the most important gases. These have resonant absorption lines at ~23, ~69, and ~120 GHz. In addition to absorption lines, the atmosphere also exhibits "windows" that may be used for communication, notably at ~38 GHz and ~98 GHz.

RF and microwave signal propagation is also affected by objects such as trees, buildings, towers, and vehicles in the path of the wave. Indoor systems are affected by walls, doors, furniture, and people. The interaction events of electromagnetic signals with objects can be categorized as reflection, refraction, diffraction, and scattering. Reflection occurs when a wave meets a planar object. The wave is reflected back with minimal distortion. Refraction occurs when a wave encounters a medium with a different wave speed (change in dielectric constant). The direction and speed of the wave is altered by refraction events. Diffraction occurs when a wave encounters a corner and is bent around the corner. Diffraction affects are diminished at higher frequencies but can be very important at lower frequencies. Finally, the term *scattering* refers to events that redirect the wave but are too complex to consider only as reflection, refraction, or diffraction.

As a result of the interaction of electromagnetic signals with objects, the propagation channel for wireless communication systems consists of multiple paths between the transmitter and receiver. Each path will experience different attenuation and delay. Some transmitted signals may experience a deep fade (large attenuation) due to destructive multipath cancellation. Similarly, constructive multipath addition can produce signals of large amplitude. Shadowing can occur when buildings or other objects obstruct the line-of-site path between transmitter and receiver.

The design of wireless systems must consider the interaction of specific frequencies of RF and microwave signals with the atmosphere and with objects in the signal channel that can cause multipath effects.

Material Properties

Electrical properties of metals, ceramics, and ferromagnetic materials are different at higher frequencies than they are at DC or low frequencies. The unique high-frequency properties of materials can result in both limitations and opportunities for exploitation to the microwave engineer.

Metals

The conductivity of metal objects is affected dramatically by the penetration of high-frequency electromagnetic fields into a conductor. The field amplitude decays exponentially from its surface value according to

$$A = e^{-x/\delta_s} \tag{126.16}$$

where x is the normal distance into the conductor measured from the surface, and δ_s is the skin depth. The skin depth or depth of penetration into a metal is defined as the distance the wave must travel in order to decay by an amount equal to e^{-1} = 0.368 or 8.686 dB. The skin depth δ_s is given by

$$\delta_s = \frac{1}{\sqrt{\pi f \mu \sigma}} \tag{126.17}$$

where f is the frequency, σ is the metal conductivity, and is the permeability of the metal is given as

$$\mu = \mu_0 \mu_r \tag{126.18}$$

with μ_0 equal to the permeability of free space and μ_r the relative permeability of the metal. For most metals used as conductors for microwave and RF applications, the relative permeability, μ_r, is equal to 1. The relative permeability of ferroelectric materials such as iron and steel is typically on the order of several hundred.

Skin depth is closely related to the shielding effectiveness of a metal since the attenuation of electric field strength into a metal can be expressed as in Equation (126.16). For static or low-frequency fields, the only method of shielding a space is by surrounding it with a high-permeability material. For RF and microwave frequencies, however, a thin sheet or screen of metal serves as an effective shield from electric fields.

Skin depth can be an important consideration in the development of guided wave and reflecting structures for high-frequency work. For the best conductors, skin depth is on the order of microns for 1-GHz fields. Since electric fields cannot penetrate very deeply into a conductor, all current is concentrated near the surface. As conductivity or frequency approach infinity, skin depth approaches zero and the current is contained in a narrower and narrower region near the surface of the conductor. For this reason, only the properties of the surface metal affect RF or microwave resistance. A poor conductor with a thin layer of high-conductivity metal will exhibit the same RF conduction properties as a solid high-conductivity structure.

Dielectrics

The dielectric permittivity of ceramics and other commonly used materials varies from DC into the microwave and optical range. Over this range a number of different loss mechanisms must be considered. Dielectric permittivity of a material can be expressed as

$$\varepsilon = \varepsilon' - j^*\varepsilon'' \tag{126.19}$$

where ε' is the real part of the dielectric permittivity. The existence of losses is explicitly included by the imaginary quantity, ε''. Both the real and imaginary parts of Equation (126.19) are a function of frequency.

The general trend of the dielectric permittivity as a function of frequency is shown in Figure 126.8. There are several regimes in which the dielectric permittivity is changing. Each transition in the permittivity represents a distinct interaction mechanism between the field and the material.

In Figure 126.8, the high value of permittivity at low frequency is attributable to interface effects. As the frequency increases, the field polarity begins to move more quickly than the charge in the interface states can respond, and the dielectric permittivity settles to a bulk material value — the value that the material would exhibit if absent of interface charges.

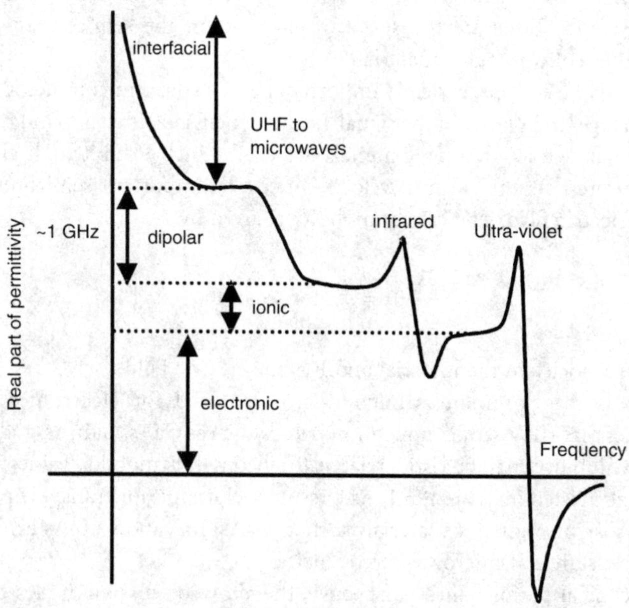

FIGURE 126.8 The real part of permittivity, ε', as a function of frequency. (*Source:* Golio, M., Ed., *RF and Microwave Handbook*, CRC Press, Boca Raton, FL, 2000.)

As the frequency is further increased, the next polarizability is associated with the material. Perovskites have oxygen vacancies that will exhibit a dipolar polarization. This effect can exhibit resonant frequencies as low as 100 MHz. It can therefore play a role in chip capacitors, which are typically made from mixed or doped perovskites. If appropriate attention is not paid to control the number of vacancies during the manufacturing process, frequency-dependent effects can appear in these devices at relatively low frequencies.

In the frequency range above 1 GHz, there will be contributions to the permittivity due to the ionic separation of the components of a material. Changes in the electromagnetic field at frequencies in this range will result in a decrease in the dielectric permittivity to a new plateau.

A final polarization at extremely high frequencies is the electronic contribution, that is, the distortion induced in the electronic cloud of the material by the external field.

The dielectric behavior presented in Figure 126.8 is somewhat idealized. Typically there are multiple mechanisms and resonances resulting in a distribution of relaxation times. This in turn causes the transitions of Figure 126.8 to be less distinct.

Ferrites

A class of magnetic materials known as ferromagnetic or ferrites has important properties at microwave frequencies, which allows them to be used to advantage for certain circuit functions. Internal magnetic moments in these materials spontaneously align parallel to each other to form domains, resulting in permeabilities considerably higher than unity. RF energy propagating through a ferromagnetic material will interact with the electrons of the material. The interaction results in a resonance behavior in the permeability of the material. The resonance (and therefore the permeability) can be affected by the application of a DC magnetic field. The use of this effect is exploited in the realization of a passive isolator device. Such a device allows a signal to flow nearly unattenuated and undistorted from port 1 to port 2, while a signal flowing from port 2 to port 1 is attenuated significantly. Microwave circulators are also realized using ferromagnetic materials.

Semiconductors

Although semiconductor materials are exploited in virtually all electronics applications today, the unique characteristics of RF and microwave signals require that special attention be paid to specific properties of semiconductors that are often neglected or of second-order importance for other applications. Two critical issues to RF applications are the speed of electrons in the semiconductor material, and the breakdown field of the semiconductor material.

The first issue, speed of electrons, is clearly important because the semiconductor device must respond to high-frequency changes in polarity of the signal. Improvements in efficiency and reductions in parasitic losses are realized when semiconductor materials are used which exhibit high electron mobility and velocity. Figure 126.9 presents the electron velocity of several important semiconductor materials as a function of applied electric field. The carrier mobility is given by

$$\mu_c = \frac{v}{E} \quad \text{for small E} \tag{126.20}$$

where v is the carrier velocity in the material and E is the electric field.

Although silicon is the dominant semiconductor material for electronics applications today, Figure 126.9 illustrates that III-V semiconductor materials such as GaAs, GaInAs, and InP exhibit superior electron velocity and mobility characteristics relative to silicon. Bulk mobility values for several important semiconductors are also listed in Table 126.1. As a result of the superior transport properties, transistors fabricated using III-V semiconductor materials such as GaAs, InP, and GaInAs exhibit higher efficiency and lower parasitic resistance at microwave frequencies.

From a purely technical performance perspective, the above discussion argues primarily for the use of III-V semiconductor devices in RF and microwave applications. These arguments are not complete, however. Most commercial wireless products also have requirements for high-yield, high-volume, low-cost, and rapid product development cycles. These requirements can overwhelm the material selection

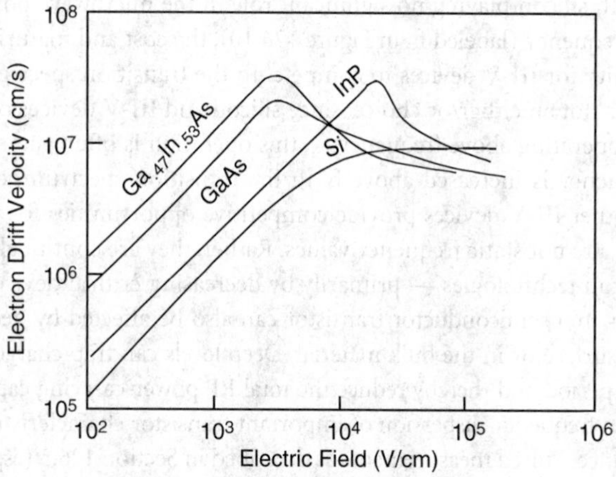

FIGURE 126.9 The electron velocity as a function of applied electric field for several semiconductor materials that are important for RF and microwave applications.

TABLE 126.1 Mobility and Breakdown Electric Field Values for Several Semiconductors Important for RF and Microwave Transmitter Applications

Property	Si	SiC	InP	GaAs	GaN
Electron mobility (cm^2/Vs)	1900	40 to 1000	4600	8800	1000
Breakdown field (V/cm)	3×10^5	20×10^4 to 30×10^5	5×10^5	6×10^5	$>10 \times 10^5$

Source: Golio, M., Ed., *RF and Microwave Handbook*, CRC Press, Boca Raton, FL, 2000.

process and they favor mature processes and high-volume experience. The silicon high-volume manufacturing experience base is far greater than that of any III-V semiconductor facility.

The frequency of the application becomes a critical performance characteristic in the selection of device technology. Figure 126.10 illustrates the relationship between frequency of the application and device technology choice. Because of the fundamental material characteristics illustrated in Figure 126.9, silicon technologies will always have lower theoretical maximum operation frequencies than III-V technologies. The higher the frequency of the application, the more likely the optimum device choice will be a III-V transistor over a silicon transistor. Above some frequency (labeled $f_{III\text{-}V}$ in Figure 126.10), III-

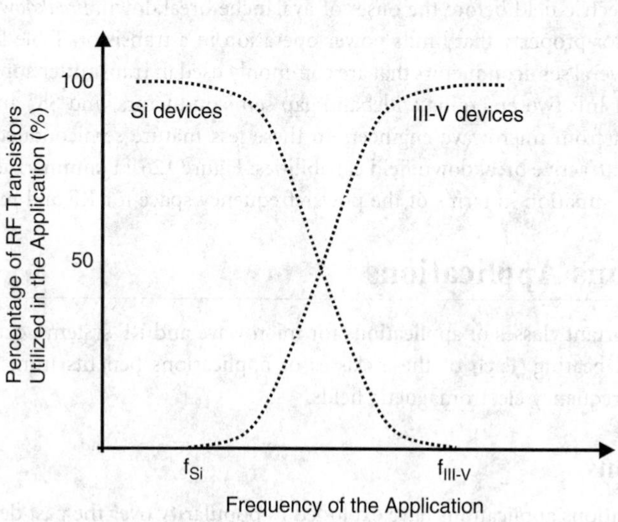

FIGURE 126.10 An illustration of the relationship between frequency of application and device technology of choice.

V devices dominate with silicon playing no significant role in the microwave portion of the product. In contrast, below some frequency (labeled f_{Si} in Figure 126.10), the cost and maturity advantages of silicon provide little opportunity for III-V devices to compete. In the transition spectrum between f_{Si} and f_{III-V}, the device technology is not an either/or choice, since silicon and III-V devices coexist. Although silicon devices are capable of operating above frequency f_{Si}, this operation is often gained at the expense of DC current drain. As frequency is increased above f_{Si} in the transition spectrum, efficiency advantages of gallium arsenide and other III-V devices provide competitive opportunities for these parts. The critical frequencies, f_{Si} and f_{III-V}, are not static frequency values. Rather, they are continually being moved upward by the advances of silicon technologies — primarily by decreasing critical device dimensions.

The speed of carriers in a semiconductor transistor can also be affected by deep levels (traps) located physically either at the surface or in the bulk material. Deep levels can trap charge for times that are long compared to the signal period and thereby reduce the total RF power-carrying capability of the transistor. Trapping effects result in frequency dispersion of important transistor characteristics such as transconductance and output resistance. Pulsed measurements as described in Section 126.3 (especially when taken over temperature extremes) can be a valuable tool to characterize deep-level effects in semiconductor devices. Trapping effects are more important in compound semiconductor devices than in silicon technologies.

The second critical semiconductor issue mentioned above is breakdown voltage. The constraints placed on the RF portion of radio electronics are fundamentally different from the constraints placed on digital circuits in the same radio. For digital applications, the presence or absence of a single electron can theoretically define a bit. Although noise floor and leakage issues make the practical limit for bit signals larger than this, the minimum amount of charge required to define a bit is very small. The bit charge minimum is also independent of the radio system architecture, the radio transmission path or the external environment. If the amount of charge utilized to define a bit within the digital chip can be reduced, then operating voltage, operating current or both can also be reduced with no adverse consequences for the radio.

In contrast, the required propagation distance and signal environment are the primary determinants for RF signal strength. If 1 watt of transmission power is required for the remote receiver to receive the signal, then reductions in RF transmitter power below this level will cause the radio to fail. Modern radio requirements often require tens, hundreds, or even thousands of watts of transmitted power in order for the wireless system to function properly. Unlike the digital situation where any discernable bit is as good as any other bit, the minimum RF transmission power must be maintained. A watt of RF power is the product of signal current, signal voltage, and efficiency, so requirements for high power result in requirements for high voltage, high current, and high efficiency.

The maximum electric field before the onset of avalanche breakdown, *breakdown field*, is the fundamental semiconductor property that limits power operation in a transistor. Table 126.1 presents breakdown voltages for several semiconductors that are commonly used in transmitter applications. In addition to silicon, GaAs and InP, two emerging wideband gap semiconductors, and SiC and GaN are included in the table. Interest from microwave engineers in these less mature semiconductors is driven almost exclusively by their attractive breakdown field capabilities. Figure 126.11 summarizes the semiconductor material application situation in terms of the power-frequency space for RF and microwave systems.

126.2 Systems Applications

There are four important classes of applications for microwave and RF systems: communications, navigation, sensors, and heating. Each of these classes of applications benefits from some of the unique properties of high-frequency electromagnetic fields.

Communications

Wireless communications applications have exploded in popularity over the past decade. Pagers, cellular phones, radio navigation, and wireless data networks are among the RF products that consumers are

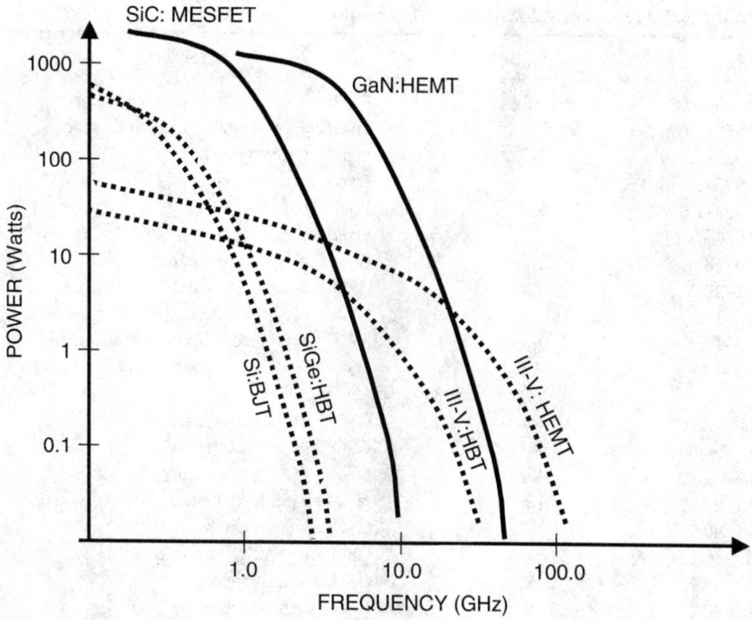

FIGURE 126.11 Semiconductor choices for RF applications are a strong function of the power and frequency required for the wireless application.

likely to be familiar with. Prior to the growth of commercial wireless communications, RF and microwave radios were in common usage for communications satellites, commercial avionics communications, and many government and military radios. All of these systems benefit from the high frequencies that offer greater bandwidth than low-frequency systems, while still propagating with relatively low atmospheric losses compared to higher-frequency systems.

Cellular phones are among the most common consumer radios in use today. Analog cellular (first generation or 1G cellular) operates at 900-MHz bands and was first introduced in 1983. Second-generation (2G) cellular technology using TDMA, GSM TDMA, and CDMA digital modulation schemes came into use more than 10 years later. The 2G systems were designed to get greater use of the 1.9-GHz frequency bands than their analog predecessors. Emergence of 2.5-G and 3-G systems operating in broader bands as high as 2.1 GHz is happening slowly. These systems make use of digital modulation schemes adapted from 2G GSM and CDMA systems. With each advance in cellular phones, requirements on the microwave circuitry have increased. Requirements for broader bandwidths, higher efficiency, and greater linearity have been coupled with demands for lower cost, lighter, smaller products, and increasing functionality. The microwave receivers and transmitters designed for portable cellular phones represent one of the highest volume manufacturing requirements of any microwave radio. Fabrication of popular cell phones has placed an emphasis on manufacturability and yield for microwave radios that was unheard of prior to the growth in popularity of these products.

Another microwave-based consumer product that is growing dramatically in popularity is the wireless local area network (WLAN) or Wi-Fi. These short-range systems offer data rates more than five times higher than cellular-based products using bandwidth at 2.4 GHz and 5 GHz. Although the volume demands for Wi-Fi components are not as high as for cellular phones, the emphasis on cost and manufacturability is still critical to these products.

Commercial communications satellite systems represent a microwave communications product that is less conspicuous to the consumer, but continues to experience increasing demand. Although the percentage of voice traffic carried via satellite systems is rapidly declining with the advent of undersea fiber-optic cables, new video and data services are being added over existing voice services. Today satellites provide worldwide TV channels, global messaging services, positioning information, communications

TABLE 126.2 Frequency Allocations for Communications Satellites

Frequency Band (GHz)	Band Designation	Service
2.500–2.655	S	Fixed regional systems, space to earth
2.655–2.690		Fixed regional systems, earth to space
3.625–4.200	C	Fixed, space to earth
5.925–6.245		Fixed, earth to space
11.700–12.500	Ku	Fixed, space to earth
14.000–14.500		Fixed, earth to space
17.8–18.6	Ka	GSO FSS, space to earth
19.7–20.2	Ka	GSO FSS, space to earth
18.8–19.3	Ka	NGSO FSS, space to earth
19.3–19.7	Ka	NGSO MSS feeder links, space to earth
27.5–28.35	Ka	GSO FSS, earth to space (coordination required with LMDS)
28.35–28.6	Ka	GSO FSS, earth to space
29.5–30.0	Ka	GSO FSS, earth to space
28.6–29.1	Ka	NGSO FSS, earth to space
29.1–29.5	Ka	NGSO MSS feeder links, earth to space
38.5–40.5	V	GSO, space to earth
37.5–38.5	V	GSO and NGSO, space to earth
47.2–48.2	V	GSO, earth to space
49.2–50.2	V	GSO, earth to space
48.2–49.2	V	GSO and NGSO, earth to space

Source: Golio, M., Ed., *RF and Microwave Handbook*, CRC Press, Boca Raton, FL, 2000.

from ships and aircraft, communications to remote areas, and high-speed data services including Internet access. Table 126.2 presents the frequency allocations for satellite communications systems in operation today. These allocations cover extremely broad bandwidths compared to many other communications systems and have spectrums from as low as 2.5 GHz to almost 50 GHz. Future allocations will include even higher frequency bands. In addition to the bandwidth and frequency challenges, microwave components for satellite communications are faced with reliability requirements that are far more severe than any earth-based systems.

Avionics applications include subsystems that perform communications, navigation, and sensor applications. Avionics products typically require functional integrity and reliability that are orders of magnitude more stringent than most commercial wireless applications. The rigor of these requirements is matched or exceeded only by the requirements for space and/or certain military applications. Avionics must function in environments that are more severe than most other wireless applications as well. Quantities for this market are typically very low when compared to commercial wireless applications; for example, the number of cell phones manufactured every single working day far exceeds the number of aircraft that are manufactured in the world in a year. Wireless systems for avionics applications cover an extremely wide range of frequencies, function, modulation type, bandwidth, and power. Due to the number of systems aboard a typical aircraft, *electromagnetic interference* (*EMI*) and *electromagnetic compatibility* (*EMC*) between systems are major concerns, and EMI/EMC design and testing are a major factor in the flight certification testing of these systems. RF and microwave communications systems for avionics applications include several distinct bands between 2 and 400 MHz and output power requirements as high as 100 watts.

In addition to commercial communications systems, military communications comprise an extremely important application of microwave technology. Technical specifications for military radios are often extremely demanding. Much of the technology developed and exploited by existing commercial communications systems today was first demonstrated for military applications. The requirements for military radio applications are varied, but will cover broader bandwidths, higher power, more linearity, and greater levels of integration than most of their commercial counterparts. In addition, reliability requirements for these systems are stringent. Volume manufacturing levels, of course, tend to be much lower than commercial systems.

Navigation

Electronic navigation systems represent a unique application of microwave systems. In this application, data transfer takes place between a satellite and a portable radio on earth. The consumer portable product consists of only a receiver portion of a radio. No data or voice signals are transmitted by the portable navigation unit. In this respect, electronic navigation systems resemble a portable paging system more closely than they resemble a cellular phone system. The most widespread electronic navigation system is geographic positioning system (GPS). The nominal GPS constellation is composed of 24 satellites in six orbital planes (four satellites in each plane). The satellites operate in circular 20,200-km-altitude (26,570 km radius) orbits at an inclination angle of 55°. Each satellite transmits a navigation message containing its orbital elements, clock behavior, system time, and status messages. The data transmitted by the satellite is sent in two frequency bands at 1.2 GHz and 1.6 GHz. The portable terrestrial units receive these messages from multiple satellites and calculate the exact location of the unit on the earth. In addition to GPS, other navigation systems in common usage include NAVSTAR, GLONASS, and LORAN.

Sensors (Radar)

Microwave sensor applications are addressed primarily with various forms of radar. Radar is used by police forces to establish the speed of passing automobiles, by automobiles to establish vehicle speed and danger of collision, by air traffic control systems to establish the locations of approaching aircraft, by aircraft to establish ground speed, altitude, other aircraft, and turbulent weather, and by the military to establish a multitude of different types of targets.

The receiving portion of a radar unit is similar to other radios. It is designed to receive a specific signal and analyze it to obtain desired information. The radar unit differs from other radios, however, in that the received signal is typically transmitted by the same unit. By understanding the form of the transmitted signal, the propagation characteristics of the propagation medium and the form of the received (reflected) signal, various characteristics of the radar target can be determined, including size, speed, and distance from the radar unit. As in the case of communications systems, radar applications benefit from the propagation characteristics of RF and microwave frequencies in the atmosphere. The best frequency to use for a radar unit depends upon its application. Like most other radio design decisions, the choice of frequency usually involves tradeoffs among several factors including physical size, transmitted power, and atmospheric attenuation.

The dimensions of radio components used to generate RF power and the size of the antenna required to direct the transmitted signal are, in general, proportional to wavelength. At lower frequencies where wavelengths are longer, the antennae and radio components tend to be large and heavy. At the higher frequencies where the wavelengths are shorter, radar units can be smaller and lighter.

Frequency selection can indirectly influence the radar power level because of its impact on radio size. Design of high-power transmitters requires that significant attention be paid to the management of electric field levels and thermal dissipation. Such management tasks are made more complex when space is limited. Since radio component size tends to be inversely proportional to frequency, manageable power levels are reduced as frequency is increased.

As in the case of all wireless systems, atmospheric attenuation can reduce the total range of the system. Radar systems designed to work above about 10 GHz must consider atmospheric loss at the specific frequency being used in the design.

Automotive radar represents a large class of radars that are used within an automobile. Applications include speed measurement, adaptive cruise control, obstacle detection, and collision avoidance. Various radar systems have been developed for forward-looking, rear-looking and side-looking applications.

V-band frequencies are exploited for forward-looking radars. Within V-band, different frequencies have been used in the past decade, including 77 GHz for U.S. and European systems, and 60 GHz in some Japanese systems. The choice of V-band for this application is dictated by the resolution requirement, antenna size requirement, and the desire for atmospheric attenuation to ensure that the radar is short range. The frequency requirement of this application has contributed to a slow emergence of this product into mainstream use, but the potential of this product to have a significant impact on highway safety continues to keep automotive radar efforts active.

As in the case of communications systems, avionics and military users also have significant radar applications. Radar is used to detect aircraft both from the earth and from other aircraft. It is also used to determine ground speed, establish altitude, and detect weather turbulence.

Heating

The most common heating application for microwave signals is the microwave oven. These consumer products operate at a frequency that corresponds to a resonant frequency of water. When exposed to electromagnetic energy at this frequency, all water molecules begin to spin or oscillate at that frequency. Since all foods contain high percentages of water, electromagnetic energy at this resonant frequency interacts with all foods. The energy absorbed by these rotating molecules is transferred to the food in the form of heat.

RF heating can also be important for medical applications. Certain kinds of tumors can be detected by the lack of electromagnetic activity associated with them, and some kinds of tumors can be treated by heating them using electromagnetic stimulation.

The use of RF/microwaves in medicine has increased dramatically in recent years. RF and microwave therapies for cancer in humans are presently used in many cancer centers. RF treatments for heartbeat irregularities are currently employed by major hospitals. RF/microwaves are also used in human subjects for the treatment of certain types of benign prostrate conditions. Several centers in the United States have been utilizing RF to treat upper airway obstruction and alleviate sleep apnea. New treatments such as microwave-aided liposuction, tissue joining in conjunction with microwave irradiation in future endoscopic surgery, enhancement of drug absorption, and microwave septic wound treatment are continually being researched.

126.3 Measurements

The RF/microwave engineer faces unique measurement challenges. At high frequencies, voltages and currents vary too rapidly for conventional electronic measurement equipment to gauge. Conventional curve tracers and oscilloscopes are of limited value when microwave component measurements are needed. In addition, calibration of conventional characterization equipment typically requires the use of open- and short-circuit standards that are not useful to the microwave engineer. For these reasons, most commonly exploited microwave measurements focus on the measurement of power and phase in the frequency domain as opposed to voltages and currents in the time domain.

Small Signal

Characterization of the linear performance of microwave devices, components, and boards is critical to the development of models used in the design of the next higher level of microwave subsystem. At lower frequencies, direct measurement of y-parameters, z-parameters, or h-parameters is useful to accomplish linear characterization. As discussed previously, however, RF and microwave design utilizes s-parameters for this application. Other small signal characteristics of interest in microwave design include impedance, VSWR, gain, and attenuation. Each of these quantities can be computed from s-parameter data.

The vector network analyzer (VNA) is the instrument of choice for small signal characterization of high-frequency components. Figure 126.12 illustrates a one-port vector network analyzer measurement. These measurements use a source with a well-defined impedance equal to the system impedance and all ports of the device under test (DUT) are terminated with the same impedance. This termination eliminates unwanted signal reflections during the measurement. The port being measured is terminated in the test channel of the network analyzer that has an input impedance equal to the system characteristic impedance. Measurement of system parameters with all ports terminated minimizes the problems caused

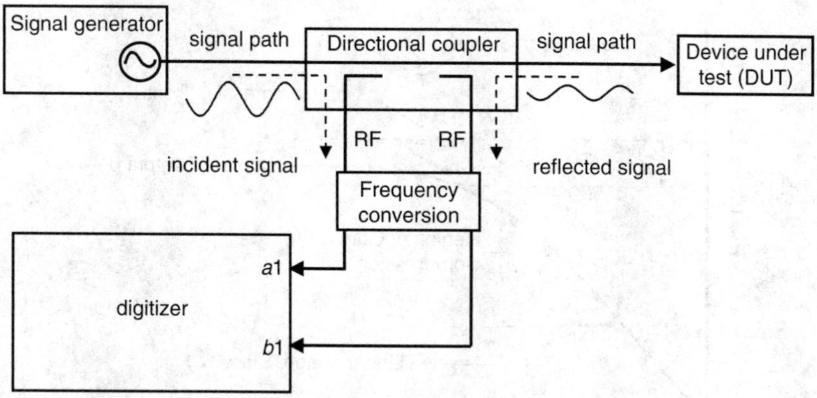

FIGURE 126.12 Vector network analyzer measurement configuration to determine s-parameters of a high-frequency device, component, or subsystem.

by short circuit, open circuit, and test circuit parasitics that cause considerable difficulty in the measurement of y- and h-parameters at very high frequencies. If desired, s-parameters can be converted to y- and h-parameters using analytical mathematical expressions.

The directional coupler shown in Figure 126.12 is a device for measuring the forward and reflected waves on a transmission line. During the network analyzer measurement, a signal is driven through the directional coupler to one port of the DUT. Part of the incident signal is sampled by the directional coupler. On arrival at the DUT port being measured, some of the incident signal will be reflected. This reflection is again sampled by the directional coupler. The sampled incident and reflected signals are then down-converted in frequency and digitized. The measurement configuration of Figure 126.12 shows only one-half of the equipment required to make full two-port s-parameter measurements. Figure 126.13 adds to this configuration to illustrate how two-port s-parameter measurements are accomplished. The s-parameters as defined in Equation (126.5) through Equation (126.18) are determined by analyzing the ratios of the digitized signal data.

For many applications, knowledge of the magnitude of the incident and reflected signals is sufficient. In these cases, the scalar network analyzer can be utilized in place of the vector network analyzer. The cost of the scalar network analyzer equipment is much less than VNA equipment and the calibration required for making accurate measurements is easier when phase information is not required. The scalar network analyzer measures reflection coefficient as defined in Equation (126.9) and Equation (126.10).

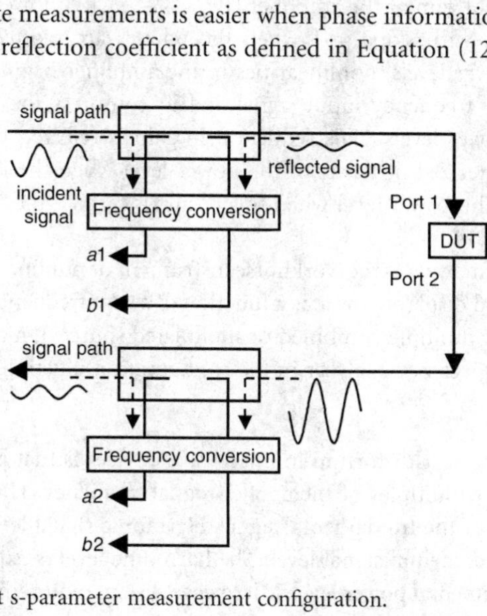

FIGURE 126.13 Full two-port s-parameter measurement configuration.

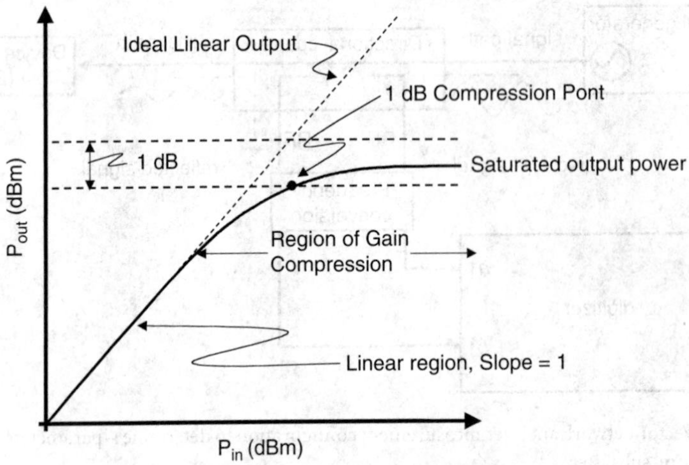

FIGURE 126.14 Output power versus input power at the fundamental frequency for a nonlinear circuit.

Large Signal

Virtually all physical systems exhibit some form of nonlinear behavior and microwave systems are no exception. Although powerful techniques and elaborate tools have been developed to characterize and analyze linear RF and microwave circuits, it is often the nonlinear characteristics that dominate microwave engineering efforts. Nonlinear effects are not all undesirable. Frequency conversion circuitry, for example, exploits nonlinearities in order to translate signals from one frequency to another. Nonlinear performance characteristics of interest in microwave design include harmonic distortion, gain compression, intermodulation distortion, phase distortion, and adjacent channel power. Numerous other nonlinear phenomena and nonlinear figures of merit are less commonly addressed, but can be important for some microwave systems.

Gain Compression

Figure 126.14 illustrates gain compression characteristics of a typical microwave amplifier with a plot of output power as a function of input power. At low power levels, a single frequency signal is increased in power level by the small signal gain of the amplifier ($P_{out} = G * P_{in}$). At lower power levels, this produces a linear P_{out} versus P_{in} plot with slope = 1 when the powers are plotted in dB units as shown in Figure 126.14. At higher power levels, nonlinearities in the amplifier begin to generate some power in the harmonics of the single frequency input signal and to compress the output signal. The result is decreased gain at higher power levels. This reduction in gain is referred to as *gain compression*. Gain compression is often characterized in terms of the power level when the large signal gain is 1 dB less than the small signal gain. The power level when this occurs is termed the *1dB compression point* and is also illustrated in Figure 126.14.

The microwave spectrum analyzer is the workhorse instrument of nonlinear microwave measurements. The instrument measures and displays power as a function of swept frequency. Combined with a variable power level signal source (or multiple combined or modulated sources), many nonlinear characteristics can be measured using the spectrum analyzer in the configuration illustrated in Figure 126.15.

Harmonic Distortion

A fundamental result of nonlinear distortion in microwave devices is that power levels are produced at frequencies which are integer multiples of the applied signal frequency. These other frequency components are termed *harmonics* of the fundamental signal. Harmonic signal levels are usually specified and measured relative to the fundamental signal level. The harmonic level is expressed in dBc, which designates dB relative to the fundamental power level. Microwave system requirements often place a maximum

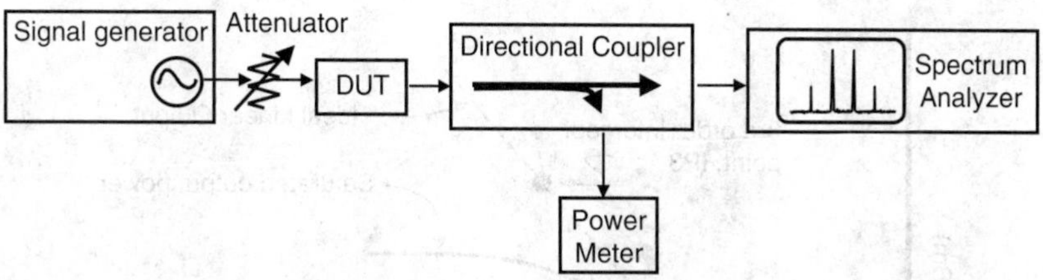

FIGURE 126.15 Measurement configuration to characterize gain compression and harmonic distortion. By replacing the signal generator with two combined signals at slightly offset frequencies, the configuration can also be used to measure intermodulation distortion.

acceptable level for individual harmonics. Typically, third and second harmonic levels are critical, but higher-order harmonics can also be important for many applications. The measurement configuration illustrated in Figure 126.15 can be used to directly measure harmonic distortion of a microwave device.

Intermodulation Distortion

When a microwave signal is comprised of power at multiple frequencies, a nonlinear circuit will produce *intermodulation distortion* (*IMD*). The IMD characteristics of a microwave device are important because they can create unwanted interference in adjacent channels of a radio or radar system. The intermodulation products of two signals produce distortion signals not only at the harmonic frequencies of the two signals, but also at the sum and difference frequencies of all of the signal's harmonics. If the two signal frequencies are closely spaced at frequencies f_c and f_m, then the IMD products located at frequencies $2f_c$-f_m and $2f_m$-f_c will be located very close to the desired signals. This situation is illustrated in the signal spectrum of Figure 126.16. The IMD products at $2f_c$-f_m and $2f_m$-f_c are third order products of the desired signals, but are located so closely to f_c and f_m that filtering them out of the overall signal is difficult.

The spectrum of Figure 126.16 represents the nonlinear characteristics at a single power level. As power is increased and the device enters gain compression, however, harmonic power levels will grow more quickly than fundamental power levels. In general, the nth order harmonic power level will increase at n times the fundamental. This is illustrated in the P_{out} versus P_{in} plot of Figure 126.17 where both the fundamental and the third order product are plotted. As in the case of the fundamental power, third-order IMD levels will compress at higher power levels. Intermodulation distortion is often characterized and specified in terms of the *third-order intercept point, IP3*. This point is the power level where the slope of the small signal gain and the slope of the low-power-level, third-order product characteristics cross as shown in Figure 126.17.

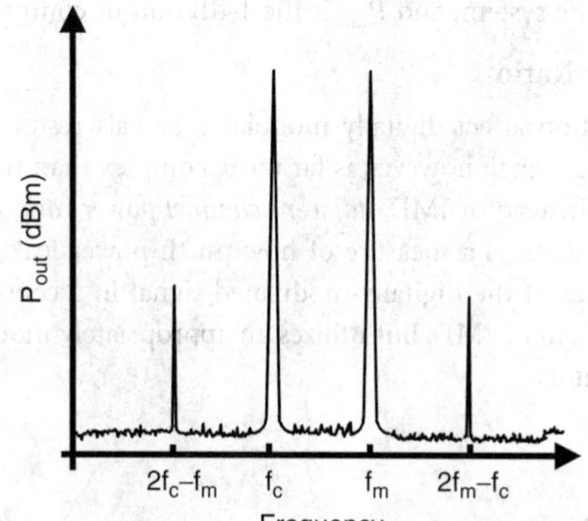

FIGURE 126.16 An illustration of signal spectrum due to intermodulation distortion from two signals at frequencies f_c and f_m.

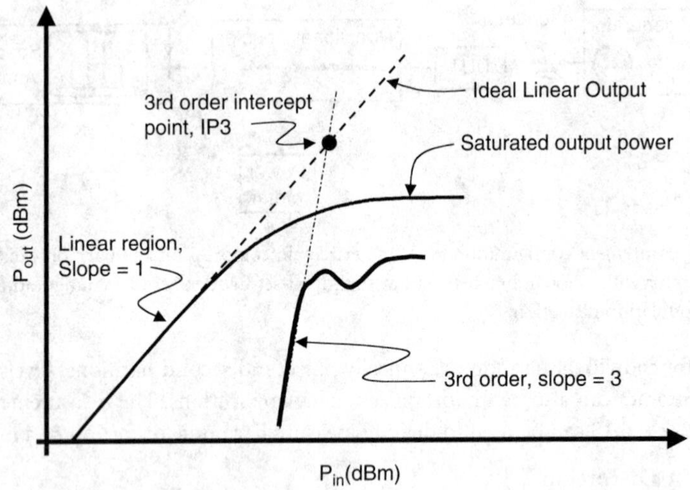

FIGURE 126.17 Relationship between signal output power and intermodulation distortion product levels.

Phase Distortion

Reactive elements in a microwave system give rise to time delays that are nonlinear. Such delays are referred to as *memory effects* and result in *AM-PM distortion* in a modulated signal. AM-PM distortion creates sidebands at harmonics of a modulating signal. These sidebands are similar to the IMD sidebands, but are repeated for multiple harmonics. AM-PM distortion can dominate the out-of-band interference in a radio. At lower power levels, the phase deviation of the signal is approximately linear and the slope of the deviation, referred to as the modulation index, is often used as a figure of merit for the characterization of this nonlinearity. The *modulation index* is measured in degrees per volt using a vector network analyzer. The phase deviation is typically measured at the 1-dB compression point in order to determine the modulation index. Because the vector network analyzer measures power, the computation of modulation index, k_ϕ, uses the formula

$$k_\phi = \frac{\Delta\Phi(P_{1dB})}{2Z_0\sqrt{P_{1dB}}} \tag{126.21}$$

where $\Delta\Phi(P_{1dB})$ is the phase deviation from small signals at the 1-dB compression point, Z_0 is the characteristic impedance of the system, and P_{1dB} is the 1-dB output compression point.

Adjacent Channel Power Ratio

Amplitude and phase distortion affect digitally modulated signals resulting in gain compression and phase deviation. The resulting signal, however, is far more complex than the simple one- or two-carrier results presented previously. Instead of IMD, *adjacent channel power ratio* (*ACPR*) is often specified for digitally modulated signals. ACPR is a measure of how much power leaks into adjacent channels of a radio due to the nonlinearities of the digitally modulated signal in a central channel. Measurement of ACPR is similar to measurement of IMD, but utilizes an appropriately modulated digital signal in place of a single tone signal generator.

Noise

Noise is a random process that can have many different sources such as thermally generated resistive noise, charge crossing a potential barrier, and generation-recombination (G-R) noise. Understanding noise is important in microwave systems because background noise levels limit the sensitivity, dynamic range, and accuracy of a radio or radar receiver.

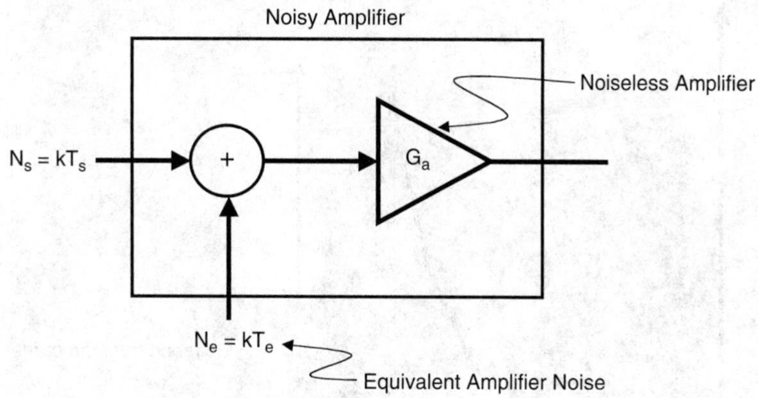

FIGURE 126.18 System view of amplifier noise.

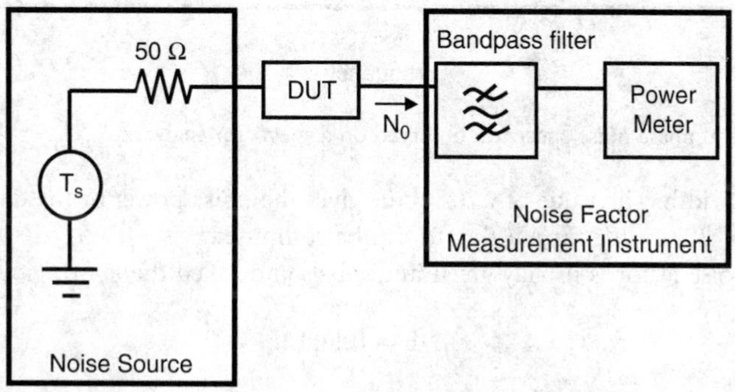

FIGURE 126.19 Measurement configuration for noise factor measurement.

Noise Figure

At microwave frequencies, noise characterization involves the measurement of noise power. The noise power of a linear device can be considered as concentrated at its input as shown in Figure 126.18. The figure considers an amplifier, but the analysis is easily generalized to other linear devices.

All of the amplifier noise generators can be lumped into an equivalent noise temperature with an equivalent input noise power per hertz of $N_e = kT_e$, where k is Boltzmann's constant $\cong 1.38 \times 10^{-23}$ JK^{-1} and T_e is the equivalent noise temperature. The noise power per hertz available from the noise source is $N_S = kT_S$ as shown in Figure 126.18. Since noise limits the system sensitivity and dynamic range, it is useful to examine noise as it is related to signal strength using a *signal-to-noise ratio* (*SNR*). A figure of merit for an amplifier, *noise factor* (*F*), describes the reduction in SNR of a signal as it passes through the linear device illustrated in Figure 126.18. The noise factor for an amplifier is derived from the figure to be

$$F = \frac{SNR_{IN}}{SNR_{OUT}} = 1 + \frac{T_e}{T_s} \tag{126.22}$$

Device noise factor can be measured as shown in Figure 126.19. To make the measurement, the source temperature is varied resulting in variation in the device noise output, N_0. The device noise contribution, however, remains constant. As T_S changes, the noise power measured at the power meter changes according to

$$N_0(T_S) = (kT_S G + kT_e G)B \tag{126.23}$$

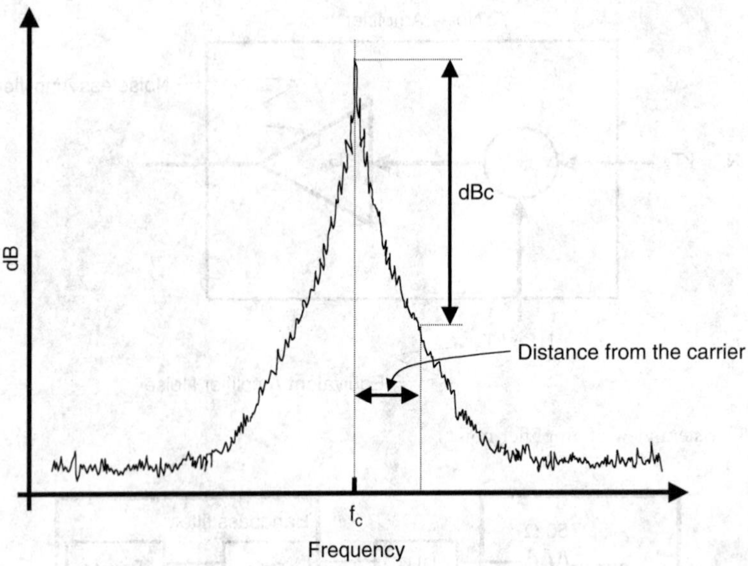

FIGURE 126.20 Typical phase noise spectrum observed on a spectrum analyzer.

where B is the bandwidth. The value of $N_0(T_S = 0)$ gives the noise power of the device alone. By using two known values of T_S, a value for $N_0(T_S = 0)$ can be computed.

In practice, the noise factor is usually given in decibels and called the *noise figure*, as follows:

$$NF = 10\log F \quad (126.24)$$

Phase Noise

When noise is referenced to a carrier frequency it modulates that carrier and causes amplitude and phase variations known as phase noise. Oscillator phase modulation noise (PM) is much larger than amplitude modulation noise (AM). The phase variations caused by this noise result in *jitter*, which is critical in the design and analysis of digital communication systems.

Phase noise is most easily measured using a spectrum analyzer. Figure 126.20 shows a typical oscillator source spectrum as measured directly on a spectrum analyzer. Characterization and analysis of phase noise are often described in terms of the power ratio of the noise at specific distances from the carrier frequency. This is illustrated in Figure 126.20.

Pulsed I-V

Although most of the measurements commonly utilized in RF and microwave engineering are frequency domain measurements, pulsed measurements are an important exception used to characterize high-frequency transistors. At RF and microwave frequencies, mechanisms known as *dispersion effects* become important to transistor operation. These effects reveal themselves as a difference in I-V characteristics obtained using a slow sweep as opposed to I-V characteristics obtained using a rapid pulse. The primary physical causes of I-V dispersion are thermal effects and carrier traps in the semiconductor. Figure 126.21 illustrates the characteristics of a microwave transistor under DC (solid lines) and pulsed (dashed lines) stimulation. In order to characterize dispersion effects, pulse rates must be shorter than the thermal and trapping time constants that are being monitored. Typically, for microwave transistors, that requires a pulse on the order of 100 ns or less. Similarly, the quiescent period between pulses should be long compared to the measured effects. Typical quiescent periods are on the order of 100 ms or more. The discrepancy between DC and pulsed characteristics is an indication of how severely the semiconductor traps and thermal effects will impact device performance.

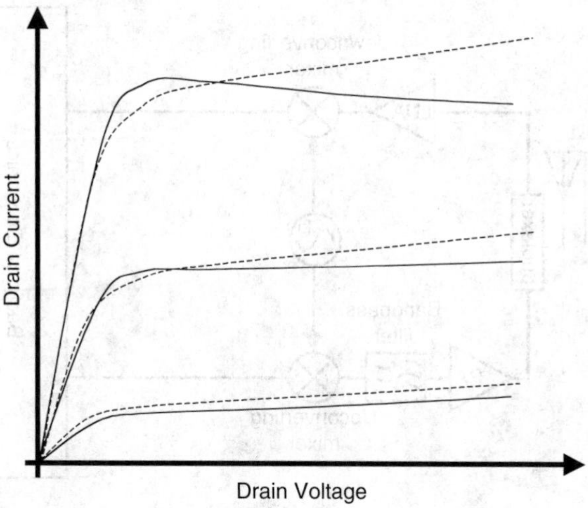

FIGURE 126.21 Pulsed I-V characteristics of a microwave FET. Solid lines are DC characteristics while dashed lines are pulsed.

Another use for pulsed I-V measurement is the characterization of high-power transistors. Many high-power transistors (greater than approximately 30 watts) are only operated in a pulsed mode or at a bias level far below their maximum currents. If these devices are biased at higher current levels for a few milliseconds, the thermal dissipation through the transistor will cause catastrophic failure. This is a problem for transistor model development, since a large range of I-V curves — including high current settings — is needed to extract an accurate model. Pulsed I-V data can provide input for model development while avoiding unnecessary stress on the part being characterized.

126.4 Circuits and Circuit Technologies

Figure 126.22 illustrates a generalized radio architecture that is typical of the systems used in many wireless applications today. A signal is received by the antenna and routed via the duplexer to the receive path of the radio. A low noise amplifier (LNA) amplifies the signal before a mixer down-converts it to a lower frequency. The down-conversion is accomplished by mixing the received signal with an internally generated local oscillator (LO) signal. The ideal receiver rejects all unwanted noise and signals. It adds no noise or interference and converts the signal to a lower frequency that can be efficiently processed without adding distortion.

On the transmitter side, a modulated signal is first up-converted, and then amplified by the power amplifier (PA) before being routed to the antenna. The ideal transmitter boosts the power and frequency of a modulated signal to that required for the radio to achieve communication with the desired receiver. Ideally, this process is accomplished efficiently (minimum DC power requirements) and without distortion. It is especially important that the signal broadcast from the antenna include no undesirable frequency components.

To accomplish the required transmitter and receiver functions described above, RF and microwave components must be developed either individually or as part of an integrated circuit.

Low Noise Amplifier

The low noise amplifier is often most critical in determining the overall performance of the receiver chain of a wireless radio. The noise figure of the LNA has the greatest impact of any component on the overall receiver noise figure and receiver sensitivity. The LNA should minimize the system noise figure, provide sufficient gain, minimize nonlinearities, and ensure stable 50-ohm impedance with low power consump-

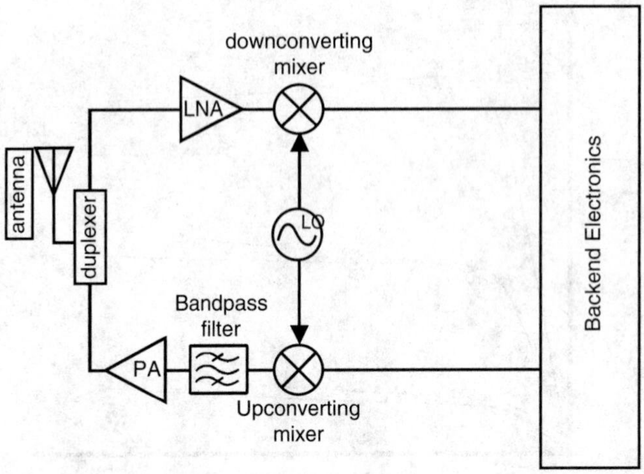

FIGURE 126.22 Generalized microwave radio architecture illustrating microwave components in both receiver and transmitter path. At present, most of the digital systems described in this section of the handbook are implemented in CMOS (complementary metal oxide semiconductor) technology. This technology allows for the fabrication of chips with wiring that is several times thinner than a human hair. Each system is packaged into an integrated circuit. The various layers of silicon and metal defining the circuit are then sent for fabrication. Copies of the same circuit, and sometimes different circuits, are fabricated on the same wafer, as shown here. (Photo courtesy of IBM.)

tion. The two performance specifications of primary importance to determine LNA quality are gain and noise figure.

In many radios, the LNA is part of a single chip design that includes a mixer and other receiver functions as well as the LNA. In these applications, the LNA may be realized using silicon, SiGe, GaAs, or another semiconductor technology. Si bipolar junction transistors and SiGe heterojunction bipolar transistors (HBTs) dominate the LNA business at frequencies below a couple of gigahertz because of their tremendous cost and integration advantages over compound semiconductor devices. Compound semiconductors are favored as frequency increases and noise figure requirements decrease. For applications that require extremely low noise figures, cooled compound semiconductor HEMTs are the favored device.

Power Amplifier

A power amplifier (PA) is required at the output of a transmitter to boost the signal to the power levels necessary for the radio to achieve a successful link with the desired receiver. PA components are often the most difficult and expensive part of microwave radio design. At high power levels, semiconductor nonlinearities such as breakdown voltage become critical design concerns. Thermal management issues related to dissipating heat from the RF transistor can dominate the design effort. Especially in the case of portable radio products, efficiency of the amplifier is critical. PA efficiency is essential to obtain long battery lifetime in a portable product. Critical primary design specifications for PAs include output power, gain, linearity, and efficiency.

Except in the case of very short-range, low-power radios, PA components tend to be discrete devices with minimal levels of integration. The unique semiconductor and thermal requirements of power amplifiers dictate the use of unique fabrication and manufacturing techniques in order to obtain required performance. The power and frequency requirements of the application typically dictate what device technology is required for the power amplifier. At frequencies as low as 1 GHz and power levels of 1 watt, compound semiconductor devices often compete with silicon and SiGe for power amplifier devices. As power and frequency increase from these levels, compound semiconductor HEMTs dominate in this application. Vacuum tube technology is still required to achieve performance for some extremely high-power or high-frequency applications.

Mixer

A mixer is essentially a multiplier and can be realized with any nonlinear device. If at least two signals are present in a nonlinear device, their products will be produced at the device output. The mixer is a frequency-translating device. Its purpose is to translate the incoming signal at frequency, f_{RF}, to a different outgoing frequency, f_{IF}. The local oscillator (LO) port of the mixer is an input port and is used to *pump* the RF signal and create the IF signal.

Mixer characterization normally includes the following parameters:

- Input match (at the RF port)
- Output match (at the IF port)
- LO to RF leakage (from the LO to RF port)
- LO to IF leakage (from LO to IF port)
- Conversion Loss (from the RF port to the IF port)

The first four parameters are single frequency measurements similar to s-parameters S_{11}, S_{22}, S_{13}, and S_{23}. Conversion loss is similar to s-parameter S_{21}, but is made at the RF frequency at the input port and at the IF frequency at the output port.

Although a mixer can be made from any nonlinear device, many RF/microwave mixers utilize one or more diodes as the nonlinear element. FET mixers are also used for some applications. As in the case of amplifiers, the frequency of the application has a strong influence on whether silicon or compound semiconductor technologies are used.

RF Switch

RF switches are control elements required in many wireless applications. They are used to control and direct signals under stimulus from externally applied voltages or currents. Phones and other wireless communications devices utilize switches for duplexing and switching between frequency bands and modes.

Switches are ideally a linear device so can be characterized with standard s-parameters. Since they are typically bidirectional, $S_{21} = S_{12}$. Insertion loss (S_{21}) and reflection coefficient (S_{11} and S_{22}) are the primary characteristics of concern in an RF switch. Switches can be reflective (high impedance in the off state) or absorptive (matched in both on and off state).

The two major classes of technologies used to implement switches are PIN diodes and FETs. PIN diode switches are often capable of providing superior RF performance to FET switches but the performance can come at a cost of power efficiency. PIN diodes require a constant DC bias current in the on state while FET switches draw current only during the switching operation. Another important emerging technology for microwave switching is the micro-electro-mechanical systems (MEMS) switch. These integrated circuit devices use mechanical movement of integrated features to open and close signal paths.

Filter

Filters are frequency selective components that are central to the operation of a radio. The airwaves include signals from virtually every part of the electromagnetic spectrum. These signals are broadcast using various modulation strategies from TV and radio stations, cell phones, and base stations, wireless LANs, police radar, and so on. An antenna at the front end of a radio receives all these signals. In addition, many of the RF components in the radio are nonlinear, creating additional unwanted signals within the radio. In order to function properly, the radio hardware must be capable of selecting the specific signal of interest while suppressing all other unwanted signals. Filters are a critical part of this selectivity. An ideal filter would pass desired signals without attenuation while suppressing signals at all other frequencies to elimination.

Although Figure 126.22 shows only one filter in the microwave portion of the radio, filters are typically required at multiple points along both the transmit and receive signal paths. Further selectivity is often accomplished by the input- or output-matching circuitry of amplifier or mixer components.

Filter characteristics of interest include the bandwidth or passband frequencies and the insertion loss within the passband of the filter, as well as the signal suppression outside of the desired band. The quality factor, Q, of a filter is a measure of how sharply the performance characteristics transition between passband and out of band behavior.

At lower frequencies, filters are realized using lumped inductors and capacitors. Typical lumped components perform poorly at higher frequencies due to parasitic losses and stray capacitances. Special manufacturing techniques must be used to fabricate lumped inductors and capacitors for microwave applications. At frequencies above about 5 or 6 GHz, even specially manufactured lumped element components are often incapable of producing adequate performance. Instead, a variety of technologies are exploited to accomplish frequency selectivity. Open- and short-circuited transmission line segments are often realized in stripline, microstrip, or coplanar waveguide forms to achieve filtering. Dielectric resonators, small pucks of high dielectric material, can be placed in proximity to transmission lines to achieve frequency selectivity. Surface acoustic wave (SAW) filters are realized by coupling the electromagnetic signal into piezoelectric materials and tapping the resulting surface waves with appropriately spaced contacts. Bulk acoustic wave (BAW) filters make use of acoustic waves flowing vertically through bulk piezoelectric material. MEMS are integrated circuit devices that combine both electrical and mechanical components to achieve both frequency selectivity and switching.

Oscillator

Oscillators deliver power either within a narrow bandwidth, or over a frequency range (i.e., they are tunable). Fixed oscillators are used for everything from narrowband power sources to precision clocks. Tunable oscillators are used as swept sources for testing, FM sources in communication systems, and the controlled oscillator in a phase locked loop (PLL). Fixed tuned oscillators will have a power supply input and oscillator output, while tunable sources will have one or more additional inputs to change the oscillator frequency. The output power level, frequency of output signal and power consumption are primary characteristics that define oscillator performance. The quality factor, Q, is an extremely important figure of merit for oscillator resonators. Frequency stability (jitter) and tunability can also be critical for many applications.

The performance characteristics of an oscillator depend on the active device and resonator technologies used to fabricate and manufacture the component. Resonator technology primarily affects the oscillators cost, phase noise (jitter), vibration sensitivity, temperature sensitivity, and tuning speed. Device technology mainly affects the oscillator maximum operating frequency, output power, and phase noise (jitter).

Resonator choice is a compromise of stability, cost, and size. Generally the quality factor, Q, is proportional to volume, so cost and size tend to increase with Q. Technologies such as quartz, SAW, yttrium-iron-garnet (YIG), and dielectric resonators allow great reductions in size while achieving high Q by using acoustic, magnetic, and dielectric materials, respectively. Most materials change size with temperature, so temperature stable cavities have to be made of special materials. Quartz resonators are an extremely mature technology with excellent Q, temperature stability, and low cost. Most precision microwave sources use a quartz crystal to control a high-frequency tunable oscillator via a phase locked loop (PLL). Oscillator noise power, and jitter, is inversely proportional to Q^2, making high resonator Q the most direct way to achieve a low noise oscillator.

Silicon bipolar transistors are used in most low noise oscillators below about 5 GHz. HBTs are common today and extend the bipolar range to as high as 100 GHz. These devices exhibit high gain and superior phase noise characteristics over most other semiconductor devices. For oscillator applications, complementary metal oxide semiconductor (CMOS) transistors are poor performers relative to bipolar transistors, but offer levels of integration that are superior to any other device technology. Above several GHz, compound semiconductor MESFETs and HEMTs become attractive for integrated circuit applications. Unfortunately, these devices tend to exhibit high phase-noise characteristics when used to fabricate oscillators. Transit time diodes are used at the highest frequencies where a solid-state device can be used. IMPATT and Gunn diodes are the most common types of transit time diodes available. The IMPATT diode produces power at frequencies approaching 400 GHz, but the avalanche breakdown mechanism inherent to its operation causes the device to be very noisy. In contrast, Gunn diodes tend to exhibit very clean signals at frequencies as high as 100 GHz.

References

1. Golio, M., Ed., *The RF and Microwave Handbook*, CRC Press, Boca Raton, FL, 2000.
2. Collin, R. E., *Foundations for Microwave Engineering*, McGraw-Hill, New York, 1992, 2.
3. Adam, Stephen F., *Microwave Theory and Applications*, Prentice Hall, Englewood Cliffs, NJ, 1969.
4. Halliday, D. and Resnick, R., *Fundamentals of Physics*, John Wiley & Sons, New York, 1970.
5. Schroder, D. K., *Semiconductor Material and Device Characterization*, John Wiley & Sons, New York, 1990.
6. Plonus, M. A., *Applied Electromagnetics*, McGraw-Hill, New York, 1978.
7. Jonscher, A. K., *Dielectric Relaxation in Solids*, Chelsea Dielectric Press, London, 1983.
8. Kittel, Charles, *Introduction to Solid State Physics*, 3rd ed., John Wiley & Sons, New York, 1967.
9. Tsui, J. B., *Microwave Receivers and Related Components*, Peninsula Publishing, Los Altos, CA, 1985.
10. Sze, S. M., *Physics of Semiconductor Devices*, 2nd ed., John-Wiley & Sons, New York, 1981.
11. Bahl, I. and Bhartia P., *Microwave Solid State Circuit Design*, John Wiley & Sons, New York, 1988.
12. Lee, W. C. Y., *Mobile Communications Engineering*, McGraw-Hill, New York, 1982.
13. Jakes, W. C. Jr., Ed., *Microwave Mobile Communications*, John Wiley & Sons, New York, 1974.
14. Evans, J., Network interoperability meets multimedia, *Satellite Communications*, February 2000, p. 30.
15. Laverghetta, T. S., *Modern Microwave Measurements and Techniques*, Artech House, Dedham, MA, 1989.
16. Ramo, S., Winnery, J. R., and Van Duzer, T., *Fields and Waves in Communications Electronics*, 2nd ed., John Wiley & Sons, New York, 1988.
17. Matthaei, G. L., Young L., and Jones, E. M. T., *Microwave Filters, Impedance-Matching Networks, and Coupling Structures*, Artech House, Dedham, MA, 1980.
18. Ambrozy, A., *Electronic Noise*, McGraw-Hill, New York, 1982.
19. Watson, H. A., *Microwave Semiconductor Devices and Their Circuit Applications*, McGraw-Hill, New York, 1969.
20. Vendelin, G., Pavio, A. M., and Rohde, U. L., *Microwave Circuit Design*, John Wiley & Sons, New York, 1990.

XIX

Digital Systems

127

Logic Devices

Richard S. Sandige
University of Wyoming

Logic devices as discussed in this chapter are of the electronic type and are also called **digital circuits and switching circuits**. Electronic logic devices that are manufactured today are parimarily transistor-transistor logic (TTL) circuits and complementary metal oxide semiconductor (CMOS) circuits. Specialized electronic design engineers called *digital circuit designers* design TTL and CMOS switching circuits. *Logic* and *system designers* gain familiarity with available logic devices via manufacturer's catalogs.

The AND gate, the OR gate, the INVERTER circuit, the NAND gate, and the NOR gate are the fundamental building blocks of practically all electronic manufactured switching circuits. From these fundamental logic devices, larger blocks such as counters, state machines, microprocessors, and computers are constructed. Construction of larger logic devices is accomplished using the mathematics introduced in 1854 by an English mathematician named George Boole [Boole, 1954]. The algebra he invented is appropriately called *Boolean algebra*. It was not until 1938 that Claude Shannon [Shannon, 1938], a research assistant at Massachusetts Institute of Technology (MIT), showed the world how to use Boolean algebra to design switching circuits.

In the following sections the building blocks for the fundamental logic device are discussed, and their truth table functions and Boolean equations are presented.

127.1 AND Gates

The AND logic device is referred to as an *AND gate*. The function it performs is the AND function. The function is most easily represented by a table of truth called a *truth table*, as illustrated in Table 127.1.

The values in the truth table are ones and zeroes, but they could just as easily be trues and falses, highs and lows, ups and downs, or any other symbolic entries that are easily distinguished. The output F is 1 only when both inputs A and B are 1, as illustrated in Table 127.1. For all other input values of A and B the output values of F are 0. Figure 127.1(a) shows the logic symbol for an AND gate with two inputs [ANSI/IEEE Std 91-1984]. The Boolean equation that represents the function performed by the AND gate is written as

$$F = A \text{ AND } B \quad \text{or} \quad F = A \cdot B \tag{127.1}$$

0-8493-1586-7/05/$0.00+$1.50

TABLE 127.1 AND Gate Truth Table Function

A	*B*	*F*
0	0	0
0	1	0
1	0	0
1	1	1

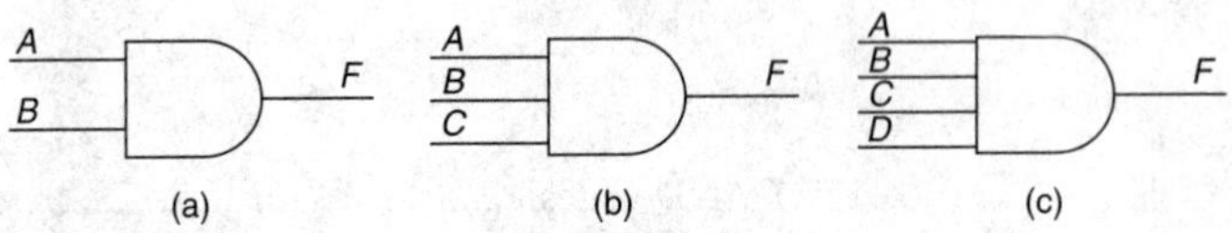

FIGURE 127.1 (a) AND gate logic symbol with two inputs; (b) logic symbol for a 3-input AND gate; (c) logic symbol for a 4-input AND gate.

The raised dot represents the AND operator and signifies the operation to be performed on the input variables *A* and *B*, with the result assigned to *F*. For *n* input **switching variables** there are 2^n rows in the truth table and output values for the switching variable *F*. An AND gate can consist of 2 or more inputs. Standard off-the-shelf TTL AND gate devices [Texas Instruments, 1984] have as many as 4 inputs. **Programmable logic devices** (PLDs) that are used to configure larger circuits [Advanced Micro Devices, 1988] either by fuses or by stored charge movement electrically have AND gates with 32 or more inputs. Figure 127.1(b) and Figure 127.1(c) show the logic symbols for a 3-input AND gate and a 4-input AND gate, respectively.

127.2 OR Gates

The OR logic device is referred to as an *OR gate*. The function it performs is the OR function. The truth table function for a 2-input OR gate is shown in Table 127.2.

Like the AND gate, the OR gate can have many inputs. The output *F* for the OR gate is 1 any time one of the inputs *A* or *B* is 1, as illustrated in Table 127.2. When the input values of *A* and *B* are both 0, the output value of *F* is 0. Figure 127.2(a) shows the logic symbol for an OR gate with two inputs. The Boolean equation that represents the function performed by the OR gate is written as

$$F = A \text{ OR } B \quad \text{or} \quad F = A + B \tag{127.2}$$

The plus sign represents the OR operator and signifies the operation to be performed on the input variables *A* and *B*, with the result assigned to *F*. Standard off-the-shelf TTL AND gate devices are generally limited to 2 inputs. Programmable logic devices (PLDs) that are used to configure larger circuits either by fuses or by stored charge movement electrically have OR gates with as many as 9 or more inputs. Figure 127.2(b) shows a logic symbol for a 9-input OR gate. Figure 127.2(c) shows a simplified logic symbol for a 9-input OR gate.

TABLE 127.2 OR Gate Truth Table Function

A	*B*	*F*
0	0	0
0	1	1
1	0	1
1	1	1

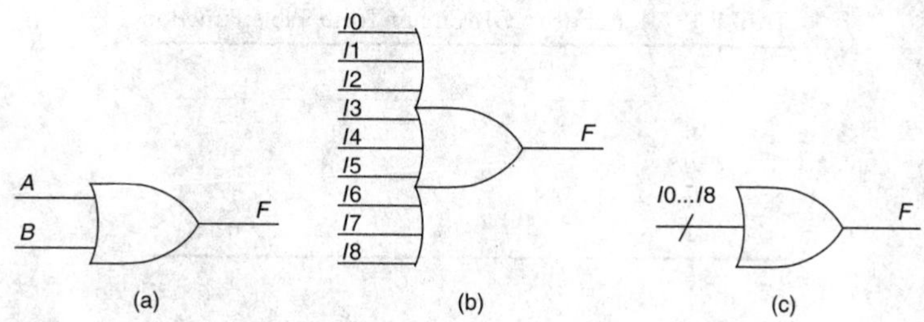

FIGURE 127.2 (a) OR gate logic symbol with two inputs; (b) logic symbol for a 9-input OR gate; (c) simplified logic symbol for a 9-input OR gate.

127.3 INVERTER Circuits

The INVERTER circuit, sometimes called the NOT circuit, is the simplest circuit of the fundamental logic devices because it only has one input. Its truth table function is shown in Table 127.3.

Notice in Table 127.3 that each output value of F is just the opposite of each input value for A. This inverting property provides a simple means of establishing a 1 output when a 0 input is supplied and a 0 output when a 1 input is supplied.

Larger circuits are often formed by connecting the outputs of two or more AND gates to the inputs of an OR gate. This form of configuration is called a **sum-of-products (SOP) form** since the AND gates perform a logic product and the OR gate performs a logic sum — hence the name *sum of products*. Figure 127.3(a) shows the sum-of-products form for a very common logic circuit that performs **modulo 2 addition** called an *exclusive OR*. Observe that the logic symbol for the INVERTER circuit — that is, the triangular symbol with the small circle at the output — is used in two places in the circuit for the exclusive OR. The switching variable A is inverted to obtain $\overline{A}$ by one INVERTER circuit, and B is inverted to obtain $\overline{B}$ by the second INVERTER circuit. The logic symbol for the exclusive OR circuit is shown in Fig. 127.3(b). The truth table function for the Exclusive OR circuit is illustrated in Table 127.4.

The Boolean equation for the exclusive OR circuit is

$$F = A\cdot\overline{B} + \overline{A}\cdot B \quad \text{or} \quad F = A \oplus B \tag{127.3}$$

The circled + symbol is the exclusive OR operator, which allows the functions to be expressed in a simpler form than the first form, which uses the AND operator (the raised dot), the OR operator (the plus symbol), and the INVERTER operator (the overbar).

TABLE 127.3 INVERTER Circuit Truth Table Function

A	F
0	1
1	0

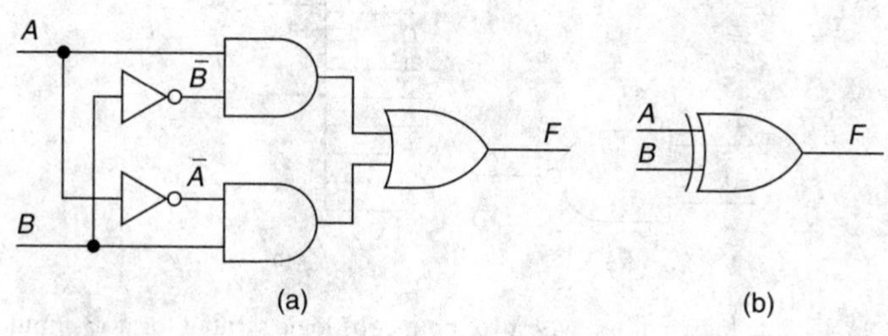

FIGURE 127.3 (a) Exclusive OR circuit in SOP form; (b) logic symbol for the exclusive OR circuit.

TABLE 127.4 Exclusive OR Circuit Truth Table Function

A	*B*	*F*
0	0	0
0	1	1
1	0	1
1	1	0

127.4 NAND Gates

The NAND gate — where NAND stands for NOT AND — is the basic gate perhaps most frequently used. It consists of an AND gate followed by an INVERTER circuit. An off-the-shelf NAND gate is simpler to design, costs less to manufacture, and provides less **propagation delay time** than the equivalent connection of an AND gate followed by an INVERTER circuit. The NAND gate performs the NAND function. The truth table function for a NAND gate with two inputs is shown in Table 127.5.

The output F is 0 only when both inputs A and B are 1, and for all other input values of A and B the output values of F are 1. As might be expected, the output values of F in the truth table function for the NAND gate, Table 127.5, are each inverted from those in the truth table function for the AND gate, Table 127.1. The Boolean equation for the NAND gate, Equation (127.4), is the same as the Boolean equation for the AND gate, Equation (127.1), with an added overbar to indicate that the output is inverted:

$$F = A \text{ NAND } B \quad \text{or} \quad F = \overline{A \cdot B} \tag{127.4}$$

Figure 127.4(a) illustrates the logic symbol for a NAND gate with two inputs. Notice that the NAND gate logic symbol simply consists of the AND gate logic symbol with a small circle attached to its output. The small circle indicates inversion like the small circle used in the logic symbol for the INVERTER circuit.

Off-the-shelf TTL NAND gates are available with up to 13 inputs, as illustrated by the logic symbol shown in Figure 127.4(b) for a 13-input NAND gate.

TABLE 127.5 NAND Gate Truth Table Function

A	*B*	*F*
0	0	1
0	1	1
1	0	1
1	1	0

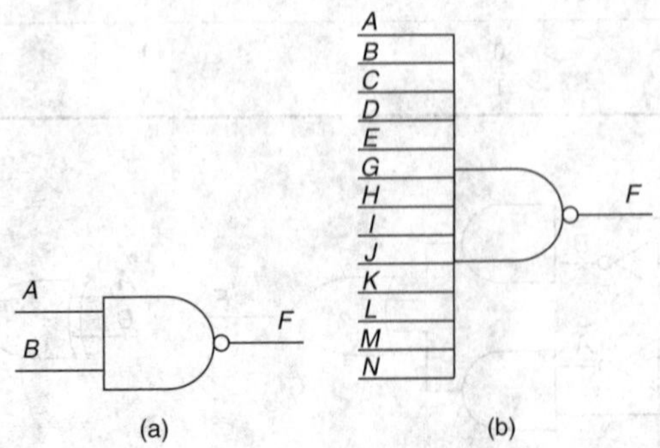

FIGURE 127.4 (a) NAND gate logic symbol with two inputs; (b) logic symbol for a 13-input NAND gate.

TABLE 127.6 NOR Gate Truth Table Function

A	*B*	*F*
0	0	1
0	1	0
1	0	0
1	1	0

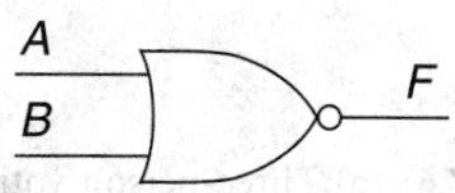

FIGURE 127.5 (a) NOR gate logic symbol with two inputs.

127.5 NOR Gates

The NOR gate consists of an OR gate followed by an INVERTER circuit. NOR is a contraction of NOT OR. Like the NAND gate, an off-the-shelf NOR gate is simpler to design, costs less to manufacture, and provides less propagation delay time than the equivalent connection of an OR gate followed by an INVERTER circuit. The NOR gate truth table function for two inputs is represented in Table 127.6.

The output *F* is 1 only when both inputs *A* and *B* are 0, and for all other input values of *A* and *B* the output values of *F* are 0. The output values of *F* in the truth table function for the NOR gate, Table 127.6, are each inverted from those in the truth table function for the OR gate, Table 127.2. The Boolean equation for the NOR gate, Equation (127.5), is the same as the Boolean equation for the OR gate, Equation (127.2), with an added overbar to indicate that the output is inverted.

$$F = A \text{ NOR } B \quad \text{or} \quad F = \overline{A + B} \tag{127.5}$$

The logic symbol for a NOR gate with two inputs is shown in Figure 127.5. The NOR gate logic symbol is represented by the OR gate logic symbol with a small circle attached to its output. As mentioned earlier, the small circle indicates inversion. Off-the-shelf TTL NOR gates are available with up to 5 inputs.

Example. Suppose a three-person digital voting circuit is needed to allow three judges to make a decision on a proposition, and their decision must be indicated as soon as all votes are made. For three judges there must be $2^3 = 8$ rows in the truth table such that each row represents a different combination of judges' votes. The entry 101 would represent judge 1 voting yes, judge 2 voting no, and judge 3 voting yes; the decision for this row would be 1, showing a majority ruling, and the proposition would pass. For the case of 001 the decision for the row would be a 0, showing that a majority was not achieved, and the proposition would fail. The complete truth table function for a three-person voting circuit is shown in Table 127.7.

TABLE 127.7 Truth Table Function for a Three-Person Voting Circuit

Judge 1	Judge 2	Judge 3	Decision
0	0	0	0
0	0	1	0
0	1	0	0
0	1	1	1
1	0	0	0
1	0	1	1
1	1	0	1
1	1	1	1

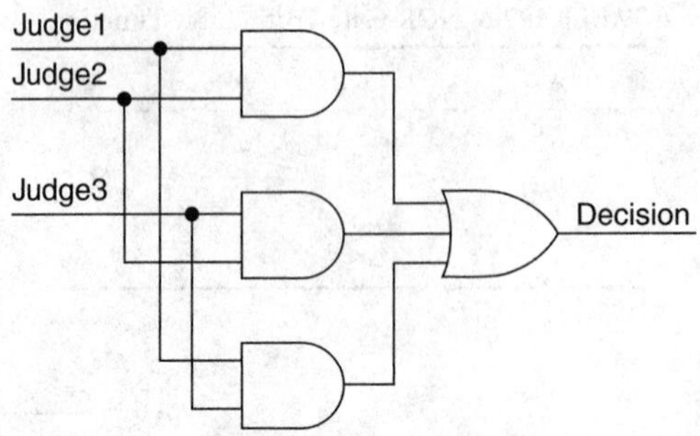

FIGURE 127.6 (a) Three-person voting circuit in SOP form.

The truth table function for the three-person voting circuit cannot be represented by one of the fundamental logic devices. The circuit requires a combination of the fundamental logic devices. By closely observing the truth table entries, we can write the following Boolean equation for the decision in terms of the three judges' inputs:

$$\begin{aligned}\text{Decision} &= (\text{judge 2 AND judge 3}) \text{ OR } (\text{judge 1 AND judge 3}) \\ &\quad \text{OR } (\text{judge 1 AND judge 2})\end{aligned} \tag{127.6}$$

which can be written more simply as

$$\text{Decision} = (\text{judge } 2 \cdot \text{judge } 3) + (\text{judge } 1 \cdot \text{judge } 3) + (\text{judge } 1 \cdot \text{judge } 2) \tag{127.7}$$

The circuit for this Boolean equation in SOP form is shown in Figure 127.6. Each input to the circuit can be supplied by an on/off switch that represent a yes/no vote, respectively, and the output can be set up to drive a light *on* (lighted) or *off* (not lighted) to indicate the decision of the voting — where *on* represents a proposition has passed and *off* represents a proposition has failed. Other circuit designs are carried out in a similar manner by logic designers and system designers to build larger circuits and systems.

Logic devices are operated with voltages, so voltage values are sometimes used for the two distinct values, that is, the 1 and 0 discussed earlier. Each symbolic entry in a truth table thus represents a range of voltages. Logic circuits of the same family, such as TTL or CMOS, are designed to be used in combination with other devices of the same family without the user worrying about actual voltages required other than **V_{cc} and GND**. Most switching circuits today are operated from a 5-volt supply source; however, newer switching circuits are rapidly appearing on the market requiring a 3-volt supply source. Logic devices utilizing a smaller voltage require less power to operate, and portable equipment manufacturers are the first to benefit from this trend in device technology.

Defining Terms

Digital circuits and switching circuits — A class of electronic circuits that operates with two distinct levels.

Modulo 2 addition — Binary addition of two single binary digits, as expressed in Table 118.4.

Programmable logic devices — Logic devices that are either one time programmable by blowing or leaving fuses intact or many times programmable by moving stored charges electrically.

Propagation delay time — Delay time through a logic device from input to output.

Sum-of-products (SOP form) — Very popular form of Boolean equation representation for a switching circuit, also called AND/OR form, since AND gates feed into an OR gate.

Switching variable — Another name for an input variable, output variable, or signal name that can take on only two distinct values.

V_{cc} and GND — Power supply connections required of all electronic devices. For a 5-volt logic device, the 5-volt power supply terminal is connected to the V_{cc} terminal of the device and the power supply ground terminal is connected to the GND terminal of the device.

References

Advanced Micro Devices. 1988. *PAL Device Data Book*. Advanced Micro Devices, Sunnyvale, CA.

ANSI/IEEE Std 91-1984. IEEE *Standard Graphic Symbols for Logic Functions*. Institute of Electrical and Electronic Engineers, New York.

ANSI/IEEE Std 991-1986. *IEEE Standard for Logic Circuit Diagrams*. Institute of Electrical and Electronic Engineers. New York.

Boole, G. 1954. *An Investigation of the Laws of Thought*. Dover, New York.

Roth, C. H., Jr. 1985. *Fundamentals of Logic Design*, 3rd ed. West, St. Paul, MN.

Sandige, R. S. 1990. *Modern Digital Design*. McGraw-Hill, New York.

Shannon, C. E. 1938. A symbolic analysis of relay and switching circuits. *Trans. AIEE*. 57:713–23.

Texas Instruments. 1984. *The TTL Data Book Volume 3 (Advanced Low-Power Schottky, Advanced Schottky)*. Texas Instruments, Dallas, TX.

Wakerly, J. F. 1994. *Digital Design Principles and Practices*, 2nd ed. Prentice Hall, Englewood Cliffs, NJ.

Further Information

IEEE Transactions on Education published quarterly by the Institute of Electrical and Electronic Engineers.

Dorf, R. C. (Ed.) 1993. *The Electrical Engineering Handbook*. CRC Press, Boca Raton, FL.

128

Counters and State Machines (Sequencers)

Barry Wilkinson
University of North Carolina, Charlotte

A *counter* is a logic circuit whose outputs follow a defined repeating sequence. After the final number in the sequence is reached, the counter outputs return to the first number. To cause the outputs to change from one number in the sequence to the next, a clock signal is applied to the circuit. A *clock signal* is a logic signal that changes from a logic 0 to a logic 1 and from a logic 1 to a logic 0 at regular intervals. The counter outputs change at one of the transitions of the clock signal, either a 0-to-1 transition or a 1-to-0 transition, depending on the design of the counter and logic components used. Counters are widely used in logic systems to generate control signal sequences and to count events.

128.1 Binary Counters

A common counter is the *binary counter*, whose outputs follow a linearly increasing or decreasing binary number sequence. A *binary-up counter* has outputs that follow a linearly increasing sequence. For example, a 4-bit binary-up counter (a binary-up counter with four outputs) has outputs that follow the sequence: 0000, 0001, 0010, 0011, 0100, 0101, 0110, 0111, 1000, 1001, 1010, 1011, 1100, 1101, 1110, 1111, 0000, … (i.e., 0, 1, 2, 3, 4, 5, 6, 7, 8, 9, 10, 11, 12, 13, 14, 15, 0, … in decimal). A *binary-down counter* is a counter whose outputs follow a counting sequence in reverse order, such as 1111, 1110, 1101, 1100, 1011, 1010, 1001, 1000, 0111, 0110, 0101, 0100, 0011, 0010, 0001, 0000, 1111, … (i.e., 15, 14, 13, 12, 11, 10, 9, 8, 7, 6, 5, 4, 3, 2, 1, 0, 15, … in decimal). A *bidirectional counter* is a counter that can count upwards or downwards depending on control signals applied to the circuit. The control signals specify whether the counter is to count up or down.

128.2 Arbitrary Code Counters

Counters can also be designed to follow other sequences. Examples include binary counters that count up to a specific number. For example, a 4-bit binary counter could be designed to count from 0 to 4 repeatedly, rather than from 0 to 15 repeatedly. Such counters can be binary counter designs with additional circuitry to reset the counter after the maximum value required has been reached. A *ring counter*

0-8493-1586-7/05/$0.00+$1.50

generates a sequence in which a 1 moves from one position to the next position in the output pattern. For example, the outputs of a 5-bit ring counter follow the sequence: 10000, 01000, 00100, 00010, 00001, 10000, …. A *Johnson counter* or *twisted ring counter* is a ring counter whose final output is "twisted" before being fed back to the first stage, so that when the final output is a 1 the next value of the first output is a 0, and when the final output is a 0 the next value of the first output is a 1. The sequence for a 5-bit Johnson counter is 10000, 11000, 11100, 11110, 11111, 01111, 00111, 00011, 00001, 10000, …. Johnson counters have the characteristic that only one digit changes from one number in the sequence to the next number. This characteristic is particularly convenient in eliminating *logic glitches* that can occur in logic circuits using counter outputs. Logic glitches are unwanted logic pulses of very short duration.

128.3 Counter Design

Most counters are based upon *D*-type or *J-K flip-flops*. *J-K* flops are particularly convenient as they can be made to "toggle" (i.e., change from 0 to 1, or from 1 to 0) if a logic 1 is applied permanently to the *J* output and to the *K* input of the flip-flop. Hence a single *J-K* flip-flop can behave as a one-bit binary counter whose output follows the sequence 0, 1, 0, … as shown in Figure 128.1. We will describe counter designs using the toggle action of *J-K* flip-flops. In Figure 128.1 the flip-flop output changes on a 1-to-0 transition of the clock signal — that is, a negative *edge-triggered* flip-flop is being used. We shall assume such flip-flops in our designs.

Figure 128.2 shows idealized waveforms of a binary-up counter whose outputs are *A*, *B*, *C*, and *D*. Such counter outputs can be created in one of several ways. There will be one flip-flop for each output and four flip-flops for a 4-bit counter.

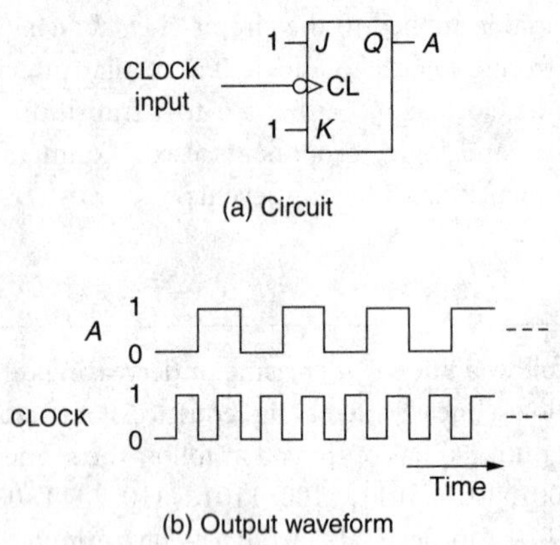

FIGURE 128.1 One-bit counter using a *J-K* flip-flop: (a) circuit; (b) output waveform.

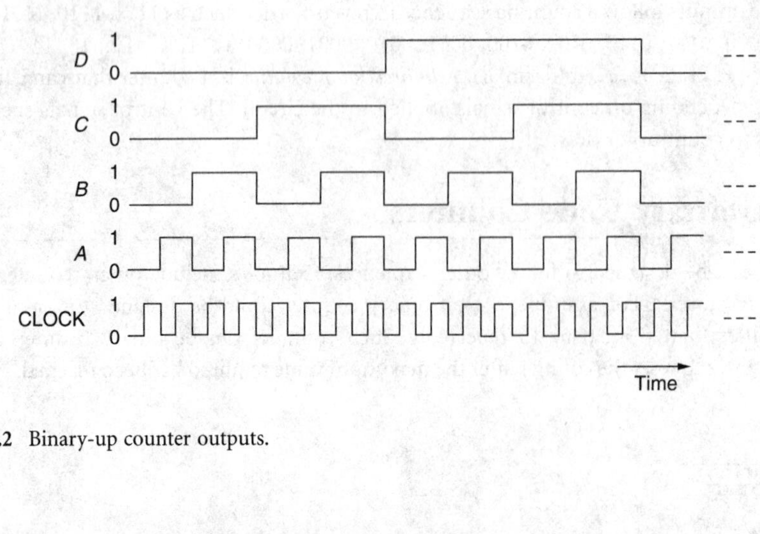

FIGURE 128.2 Binary-up counter outputs.

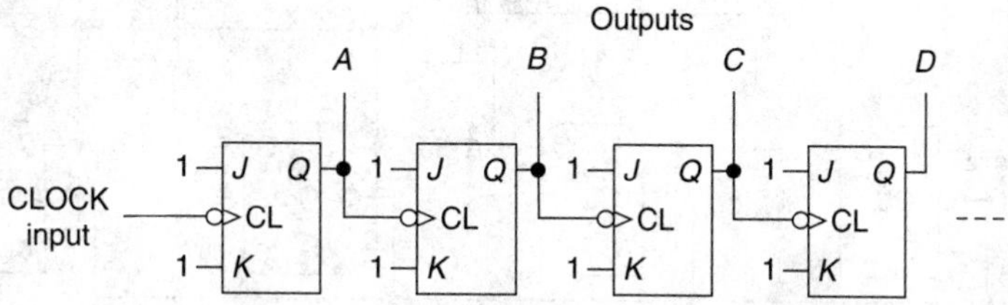

FIGURE 128.3 Asynchronous binary counter.

Counter designs can be classified as synchronous or asynchronous. The key characteristic of any *asynchronous counter* is that the output of one stage is used to activate the next stage. A 4-bit asynchronous binary (up) counter is shown in Figure 128.3. Output *A* is the least significant bit of the sequence, and output *D* is the most significant bit. We can understand the operation of this counter by referring to the required counter waveforms shown in Figure 128.2. Output *A* is to change when there is a 1-to-0 transition on the clock input signal. Output *B* is to change when there is a 1-to-0 transition on output *A*. Output *C* is to change when there is a 1-to-0 transition on output *B*. Output *D* is to change when there is a 1-to-0 transition on output *C*. These transitions are achieved with the connections shown.

The circuit arrangement can be extended to any number of stages. The counter is asynchronous in nature because each output depends on a change in the previous output. Consequently, there will be a small delay between changes in successive outputs. The delays are cumulative, and the overall delay between the first output change and the last output change could be significant, for example, from 1111 to 0000. This delay will limit the rate at which clock pulses can be applied (the maximum frequency of operation). Asynchronous binary counters are sometimes called *ripple counters* because the clock "ripples" through the circuit.

In the *synchronous counter*, we try to avoid delays between successive outputs by synchronizing flip-flop actions. This step is done by connecting the clock signal to all the flip-flop clock inputs. Consequently, when outputs of flip-flops change, the changes occur simultaneously (ignoring variations between flip-flops). Logic is attached to the flip-flop inputs to produce the required sequence. Figure 128.4 shows one design of a synchronous binary (up) counter. We can understand this design also by examining the counter waveforms of Figure 128.2. The first stage is the same as the asynchronous counter, and output changes occur on every activating clock transition. The second stage should toggle if, at the time of the activating clock transition, the *A* output is a 1. The third stage should toggle if, at the time of the activating clock transition, the *A* output and the *B* output are both 1. The fourth stage should toggle if, at the time of the activating clock transition, the *A* output, the *B* output, and the *C* output are all 1. These transitions are achieved by applying *A* to the *J*-*K* inputs of the second stage, the logic functions $A \cdot B$ (i.e., *A* AND *B*) to the third stage, and the logic function $A \cdot B \cdot C$ (i.e., *A* AND *B* AND *C*) to the fourth stage. In each case, an AND gate is used in Figure 128.4. The toggle action for a fifth stage occurs when all of *A*, *B*, *C*, and *D* are a 1, and hence requires a four-input AND gate. Similarly, the sixth stage requires a five-input AND gate, and so on. An alternative implementation that reduces the number of inputs of the gates to two inputs is shown in Figure 128.5. However, this particular implementation has the disadvantage that

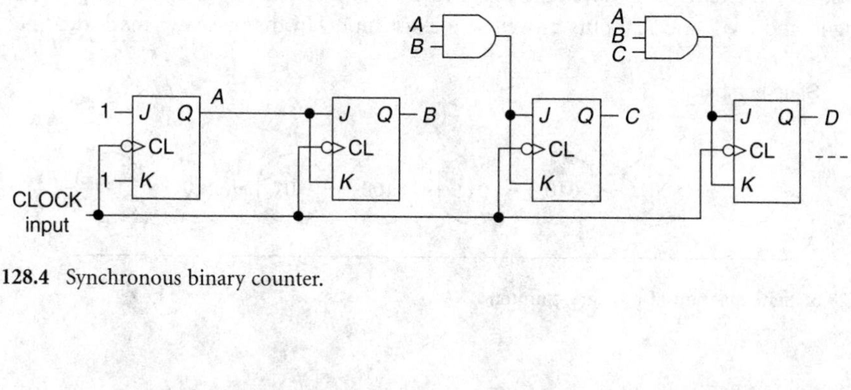

FIGURE 128.4 Synchronous binary counter.

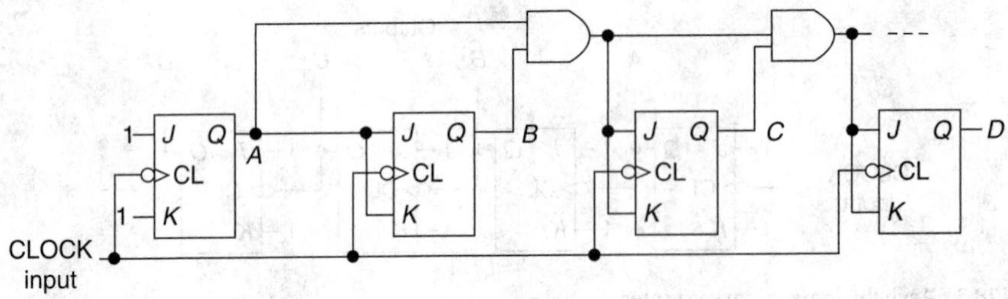

FIGURE 128.5 Synchronous binary counter using two-input AND gates.

signals may have to pass through several gates to reach the inputs of flip-flops, which will limit the speed of operation of the counter.

128.4 State Machines

Counters are in the logic classification called *sequential circuits*. The outputs of sequential circuits depend on present inputs and past output values, as opposed to *combinational circuits*, in which the outputs depend only on a particular combination of the present input values. Sequential circuits exist in defined *states*. The state of the circuit changes to another state on application of new input values. Such circuits are called *state machines*. Since in all practical circuits there will be a finite number of states, practical state machines are more accurately called *finite state machines*.

A counter is clearly a circuit that exists in defined states; each state corresponds to one output pattern. Other sequential circuits include circuits that detect particular sequences of input patterns (*sequence detectors*). As parts of the required input pattern are received, the circuit state changes from one state to the next, finally reaching the state corresponding to the complete input pattern being received.

Using *state variables*, each state is assigned a unique binary number, which is stored internally. An *n*-bit state variable can be used to represent 2*n* states. In *one-hot assignment* each state is assigned one unique state variable, which is a 1 when the system is in that state. All other state variables are zero for that state. Hence, with *n* states, *n* state variables would be needed. Although the one-hot assignment leads to more state variables, it usually requires less complicated logic.

128.5 State Diagrams

A *state diagram* is a graphical representation of the logical operation of circuits, such as counters, and sequential circuits in general. A state diagram indicates the various states that the circuit can enter and the required input/output conditions necessary to enter the next state. A state diagram of a 3-bit binary (up) counter is shown in Figure 128.6. States are shown by circles. The state number for the state, as given by the state variables, is shown inside each circle. In a counter, the state variables and the counter outputs are the same. Lines between the state circles indicate transitions from one state to the next state, which occurs for a counter circuit on application of the activating clock transition.

Other sequential circuits may have data inputs. For example, a *J-K* flip-flop is a sequential circuit in its own right, and has *J* and *K* inputs as well as a clock input. In that case, we mark the lines between

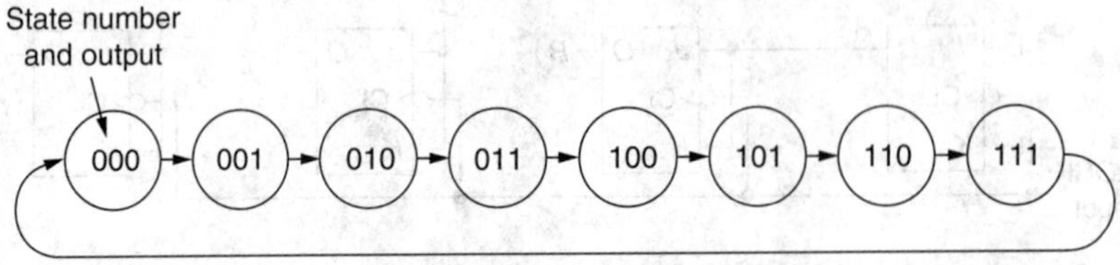

FIGURE 128.6 State diagram of a binary counter.

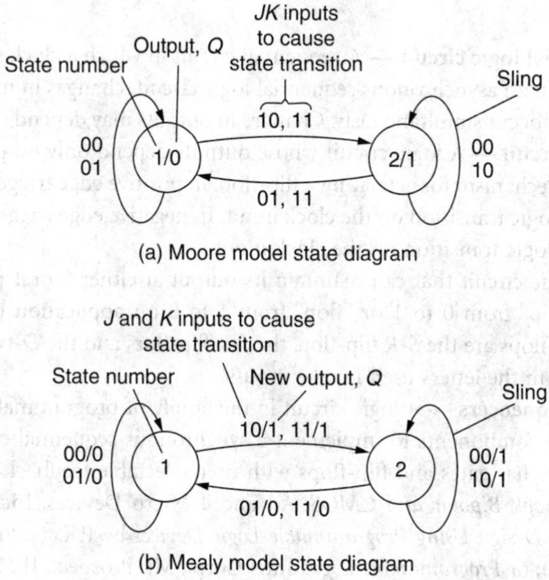

FIGURE 128.7 State diagrams of a *J-K* flip-flop: (a) Moore model state diagram; (b) Mealy model state diagram.

the states in the state diagram with the required input values to cause the state transition. There are two common forms of state diagram for circuits incorporating data inputs — the *Moore model state diagram* and the *Mealy model state diagram*. In both models, the inputs that cause a new state are shown next to the lines. The effects of all combinations of input values must be considered in each state. A line that returns to the same state forms a "sling" and indicates that the state does not change with the particular input values. In the Moore model, the outputs are associated with the states and are shown inside the circles. In the Mealy model, the outputs are associated with input values and resultant states and are shown next to the transition lines between states. Figure 128.7 shows the Moore model state diagram and the Mealy model state diagram for a *J-K* flip-flop. The fundamental difference between the Moore model state diagram and the Mealy model state diagram is that, in the Moore model state diagram, each state always generates a defined output, whereas in the Mealy model state diagram, states can generate various outputs depending upon present input values. Each model can represent any sequential circuit, although the number of states may be greater in the Moore model state diagram than in the Mealy model state diagram. The information given in the state diagram can be produced in a table called a *state table.*

128.6 State Diagrams Using Transition Expressions

An alternative form of state diagram that is particularly suitable for a state machine with a large number of states marks each line with a Boolean expression that must be true for the transition to occur. All expressions associated with each node must be exclusive of each other (i.e., no combinations of values must make more than one expression true). Also, the set of expressions must cover all combinations of variables. This form of state diagram for a *J-K* flip-flop is shown in Figure 128.8. In this diagram we can see clearly that the transition from state 1 to state 2 depends only on *J* being a 1, and the transition from state 2 to state 1 depends only on *K* being a 1.

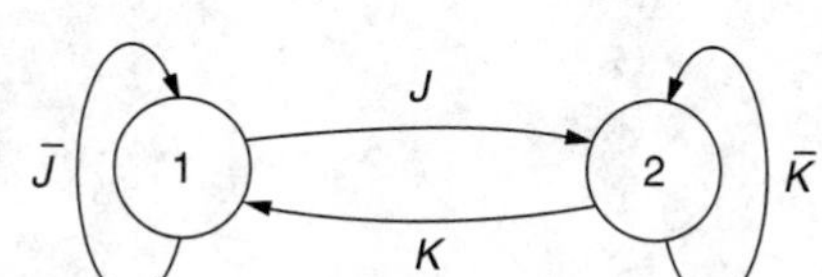

FIGURE 128.8 State diagram of a *J-K* flip-flop using transition expressions.

Defining Terms

Asynchronous sequential logic circuit — A sequential circuit in which a clock signal does not synchronize changes. In an asynchronous sequential logic circuit, changes in more than one output do not necessarily occur simultaneously. Changes in outputs may depend on other output changes.

Combinational logic circuit — A logic circuit whose outputs depend only on present input values.

Edge triggering — A mechanism for activating a flip-flop. In positive edge triggering the output changes after a 0-to-1 logic transition on the clock input. In negative edge triggering the output changes after a 1-to-0 logic transition on the clock input.

Flip-flop — A basic logic circuit that can maintain its output at either 0 or 1 permanently, but whose output can "flip" from 0 to 1 or "flop" from 1 to 0 on application of specific input values. Common flip-flops are the *S-R* flip-flop, the *J-K* flip-flop, and the *D*-type flip-flop. The names are derived from the letters used for the inputs.

Programmable logic sequencers — A logic circuit in the family of programmable logic devices (PLDs) containing the components to implement a synchronous sequential circuit. A programmable logic sequencer has gates and flip-flops with user-selectable feedback connections. See *PAL@ Device Data Book Bipolar and CMOS*, Advanced Micro Devices, Inc., Sunnyvale, CA. 1990; *Digital System Design Using Programmable Logic Devices* by P. Lala, Prentice Hall, Englewood Cliffs, NJ, 1990; or *Programmable Logic*, Intel Corp., Mt. Prospect, IL, 1994, for further details.

Propagation delay — The very short time period in a basic logic circuit between the application of a new input value and the generation of the resultant output.

Sequential logic circuit — A logic circuit whose outputs depend on previous output values and present input values. Sequential circuits have memory to maintain the output values.

Synchronous sequential logic circuit — A sequential circuit in which a clock signal initiates all changes in state.

References

Kats, R. H. 1994. *Contemporary Logic Design.* Benjamin/Cummings, Redwood City, CA.

Wakerly, J. F. 1990. *Digital Design Principles and Practices.* Prentice Hall, Englewood Cliffs, NJ.

Wilkinson, B. 1992. *Digital System Design,* 2nd ed. Prentice Hall, Hemel Hempstead, England.

Further Information

The topic of counters and state machines can be found in most logic design books, a sample of which is listed in the reference section. *Contemporary Logic Design* by Katz is very readable and thorough for the electrical engineer. Other logic design books include:

Prosser, F. P. and Winkel, D. E. 1987. *The Art of Digital Design,* 2nd ed. Prentice Hall, Englewood Cliffs, NJ.

McCalla, T. R. 1992. *Digital Logic and Computer Design.* Macmillan, New York.

Some books, such as the book by McCalla, integrate logic design with computer design because logic design is used in the design of computers.

129

Microprocessors and Microcontrollers

Michael A. Soderstrand
University of California, Davis, and Oklahoma State University

Every year since 1928, *Time Magazine* has announced in its first issue of the new year the "Man of the Year" for the previous year. Occasionally the "Man of the Year" was a group: *U.S. Fighting Men* in 1951, *U.S. Scientists* in 1961, *Those Under 25* in 1967, *Middle Americans* in 1970, and *American Women* in 1976 [1]. But in the June 3, 1983 issue, *Time Magazine* broke with all tradition with the announcement that the "Man of the Year" for 1982 was not a person at all, but rather *The Machine of the Year: The Computer Moves In* [2]. This was an extraordinary testimonial to a revolution in the making that had its roots in the astonishing development of the microprocessor chip that has revolutionized life throughout the world and invaded, for good or for bad, every aspect of human life. The article quotes an Apple Computer advertisement of the time:

"The ad provides not merely an answer, but 100 of them. A personal computer, it says, can send letters at the speed of light, diagnose a sick poodle, custom-tailor an insurance program in minutes, test recipes for beer. Testimonials abound. Michael Lamb of Tucson figured out how a personal computer could monitor anesthesia during surgery; the rock group Earth, Wind and Fire uses one to explode smoke bombs onstage during concerts; the Rev. Ron Jaenisch of Sunnyvale, Calif., programmed his machine so it can recite an entire wedding ceremony" [2].

The article also reports on a telephone survey of 1019 registered voters conducted December 8 and 9, 1982, with a sampling error of plus or minus 3%:

"A new poll for TIME by Yankelovich, Skelly and White indicates that nearly 80% of Americans expect that in the fairly near future, home computers will be as commonplace as television sets or dishwashers.

0-8493-1586-7/05/$0.00+$1.50

Although they see dangers of unemployment and dehumanization, solid majorities feel that the computer revolution will ultimately raise production and therefore living standards (67%), and that it will improve the quality of their children's education (68%)" [2].

Events since 1983 have not only confirmed, but exceeded the expectations outlined in the *Time Magazine* poll. The microprocessor that was and is the heart of the personal computers of 1982 has now expanded into every aspect of life, not just in the United States, but all over the world. Microprocessors, microcomputers, and microcontrollers are now everywhere, and it is virtually impossible to escape their influence. In this brief discussion, we cannot hope to cover but a very small part of the microprocessor story. However, we will provide an introduction to the topic through some critical definitions, a brief history, a review of the key features of microprocessors, a review of microprocessor applications, and finally a few words on where to go to get further information on this important topic.

129.1 Definitions

Figure 129.1 shows the basic architecture (i.e., basic components and their layout) for any computer system. There are four basic components in Figure 129.1: central processing unit (CPU), arithmetic logic unit (ALU), random-access memory (RAM), and input/output devices (I/O).

The CPU is essentially a controller for the operation of the computer. The CPU receives a clock signal (square wave in voltage, digitally this is alternating ones and zeros at the CPU clock frequency). On each cycle of the clock signal, the CPU steps through a standard sequence of operations as follows:

1. Fetch the next instruction from RAM memory.
2. Decode the instruction (i.e., determine what the instruction is asking the computer to do.)

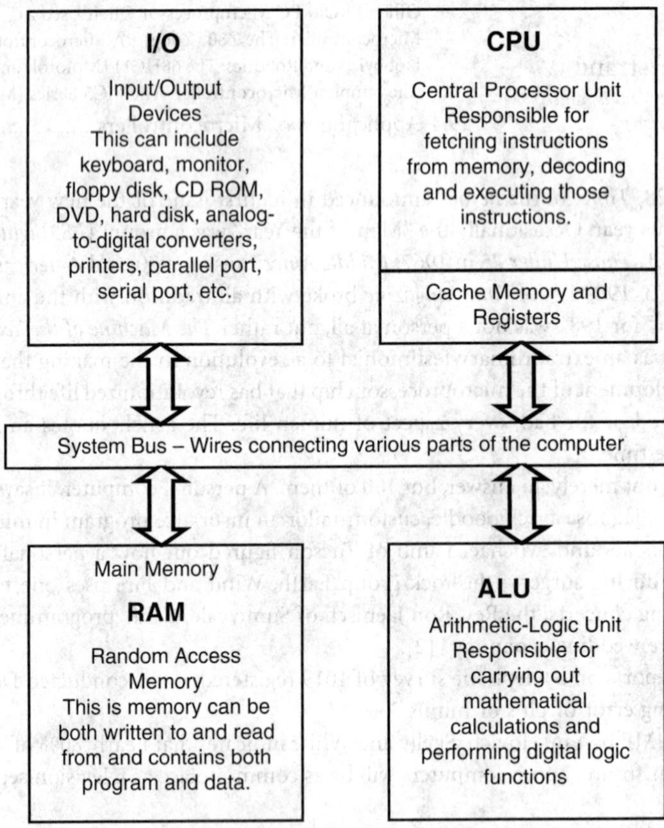

FIGURE 129.1 Architecture of a typical computer.

3. Fetch any data needed for the execution of the instruction from RAM memory and/or from the I/O devices. These data are stored in the CPU registers and/or the ALU.
4. Execute the instruction using the ALU if necessary.
5. Store the results of executing the instruction in RAM memory or send it to the I/O devices. (Note: Some instructions may only affect the CPU, registers or ALU.)

These basic five steps are repeated over and over again. The power of the computer is in the fact that we can store these instructions in RAM memory; hence, we can create very complicated programs that can do significant computing and this computing can be done very fast. For example, modern CPU clocks run in the low gigahertz range (1 billion cycles per second). With a 5-GHz CPU clock, the complete cycle of five steps can be completed 1 billion times every second!

The simple computer architecture of Figure 129.1 represents virtually all computers over the last 5 decades [3]. While modern computers do optimize the various parts of the computer, the basic structure remains intact. There are two significant improvements to this architecture that are worth mentioning, however. The first is the called a reduced instruction set computer (RISC). Initially these computers tried to reduce the number of instructions that the computer had to perform in order to optimize the hardware for fast execution of these instructions. However, the real advantage of RISC turned out to be in regularizing the instructions so that hardware could be designed more efficiently. The opposite of a RISC computer is a complex instruction set computer (CISC). The primary difference between modern RISC and CISC computers is not in the number of instructions, but rather that RISC computers generally take exactly the same amount of time to execute each instruction while CISC computer instructions generally take different amounts of time to execute. A more significant improvement in computer performance comes from the idea of pipelining the steps that the CPU goes through. Pipelining is like an assembly line where parallel hardware executes each of the five steps in assembly line fashion such that it would take five CPU clock cycles for the first instruction to be completed, but after that one instruction per CPU clock cycle will be completed as the other instructions were entered into the pipeline on each subsequent CPU clock. This concept is at the heart of the superscalar computers, which tend to be the fastest-executing modern computers.

Definition of a Microcomputer

In the old days, there were three major categories of computer: mainframe computer, minicomputer, and microcomputer. Generally, mainframe computers were made from discrete components, minicomputers made significant use of very large-scale integrated (VLSI) circuits, and microcomputers used single-chip VLSI circuits for the CPU. No one speaks much of mainframe or minicomputers anymore. Modern computers are more likely to be categorized as supercomputers, workstations, or microcomputers. Supercomputers are special-purpose computers designed for scientific applications (e.g., number crunching) and are typically the fastest computers available today. Workstations are used as servers and high-end personal computers. Microcomputers generally fall into two classes: personal computers designed for use primarily by a single user and special-purpose computers designed to do a particular function, such as control a home appliance or a scientific instrument [4].

Definition of a Microprocessor

A microprocessor is a single-chip VLSI circuit that contains the CPU, registers, cache memory, and part of all of the ALU of Figure 129.1 on a single chip. By definition, a microprocessor does not contain memory or I/O [4].

Definition of a Microcontroller

A single-chip VLSI circuit that contains all of the elements of Figure 129.1 is called a microcontroller. A microcontroller is sometimes defined as a microprocessor with significant amounts of RAM memory and I/O on chip. Microcomputers consist of personal computers (PCs) and microcontrollers. PCs usually are built from microprocessors with the RAM memory and I/O off chip. Microcontrollers, in contrast, are complete microcomputers of Figure 129.1 on a single VLSI chip [4].

Microcontrollers are single-chip computers that are designed for use in controlling some process. The modern world is full of microcontrollers. They are used in many of our home appliances, home electronics, automobiles, scientific instruments, airplanes, and many other places. On a single chip, modern microcontrollers contain the following:

1. The CPU and ALU or microprocessor
2. Read-only memory (ROM, PROM, or EPROM)
3. Read/write memory (RAM)
4. Input/output devices (serial port, parallel port, analog-to-digital converters [ADCs], digital-to-analog converters [DACs], and so on)
5. On-chip timers
6. Interrupt controller

129.2 Historical Development

In 1971, Intel introduced the Intel 4004 4-bit microprocessor on a single chip. This is generally recognized as the first microprocessor [3]. It was not very powerful; all it could do was add and subtract and it did that with only 4-bit numbers. But it did become the basis for a hand-held portable four-function electronic calculator. In 1972, Intel introduced the 8008, an 8-bit version of the Intel 4004. It too could only add and subtract, but now with 8-bit numbers. However, the really significant breakthrough was the introduction in 1974 of the Intel 8080 microprocessor. The 8080 did not offer much advancement in arithmetic functions over the 8008. It could still only add and subtract 8-bit numbers. But what it did add was a complete computer instruction set including memory reference, input/output control, and bit manipulation and logic functions. The Intel 8080 was to revolutionize the computer industry.

The First Personal Computer — The Mark-8

Researchers at the University of California, Davis, Computer Museum believe that the first personal computer was constructed with the Intel 8008 microprocessor. This computer was called the Mark 8 Computer with 256 bytes of RAM. About 1000 to 2000 of these were marketed starting in July 1974 [5]. The July 1974 issue of *Radio Electronics* featured the Mark 8 Computer on its cover [6,7].

The first portable computer was probably the IBM 5100 which was marketed primarily for business use and introduced by IBM in 1975 [8]. The IBM 5100 had 64 Kbytes of memory, and an integrated 5-inch CRT that could display 16 lines of 64 characters. For about $20,000, the IBM 5100 included an IBM 5103 dot-matrix printer and an IBM 5106 external tape drive for program storage. It could be programmed in the IBM APL language or in Basic. It had a serial I/O port and a communications adaptor. The CPU was based on an IBM controller called the PALM (put all logic in microcode). The PALM was technically not a microprocessor, but rather a computer controller that executed microcode in order to interpret a higher-level language like APL or Basic [9].

The First Commercially Successful Personal Computer

The first commercially successful personal computer was the Altair 8800 invented and marketed by Ed Roberts, the father of personal computing [8,10,11]. Roberts was a graduate of Oklahoma State University and a personal acquaintance of the author of this article. When the July 1974 issue of *Radio Electronics* featured the Mark-8 computer, Les Solomon, editor of *Popular Electronics,* was desperate to get something on homebrew computers into the magazine. He received a lot of potential articles, but he wanted to feature a personal computer built around the more powerful Intel 8080 or Motorola 6800 microprocessor. Roberts, through his company Micro Instrumentation Telemetry Systems (MITS), was making calculator kits, but exploring the potential of marketing a computer kit based on the Intel 8080. He had a great deal with Intel getting the $360 Intel 8080 microprocessor for only $75 in bulk. MITS initially marketed the basic computer kit with an Intel 8080 microprocessor, 256 bytes of RAM, front panel switches and LEDs, and power supply for $400, only $40 more than Intel's price for the microprocessor alone. Solomon's daughter Lauren actually provided the name Altair after the name of a galaxy that the "Star Trek" crew was headed for in one episode. The January 1975 issue of *Popular Electronics* featured the MITS Altair 8800 on the front cover. Roberts hoped to sell about 200 units, but was overwhelmed with orders, ultimately selling over 10,000 Altair 8800 computers in both kit and assembled form. It was this article in *Popular Electronics* that caught the attention of Bill Gates and Steve Jobs, who decided to leave Harvard and come to Albuquerque, New Mexico, to produce software for the Altair 8800. At the time, Roberts had Paul Allen working on software for the Altair, so when Bill Gates showed up in Albuquerque (Jobs stayed in Massachusetts), Ed put all three on the task of building an operating system and a Basic language for the Altair 8800. This eventually led to Jobs forming Apple Computers and Gates and Allen forming Microsoft [11].

Apple and Microsoft

Apple Computers was founded by Steve Jobs, Steve Wozniak, and Ron Wayne in April 1976. The Apple I computer, based on the MOStek 6502 microprocessor, was first introduced in kit form at the homebrew computer club in Palo Alto, California, in May 1976. Paul Terrell of The Byte Shop was so impressed he ordered 50 units at $500 apiece [12].

In early 1977, plans were made for the first West Coast Computer Faire, originally scheduled for a small venue in San Francisco. As preregistration increased, the show venue was moved twice, first to a major San Francisco hotel, and then when it outgrew those facilities, to the San Francisco convention center. These moves were all necessary to keep up with the companies that wanted to show their computers at the show and the number of people who wanted to attend. Apple and Commodore were the stars of the show. Apple introduced the Apple II, an extraordinary computer for 1977. It featured a built-in keyboard, a color graphics display (yes, color!), eight expansion slots, and Basic built into a ROM with built-in high-level language graphics commands [8]. The Commodore PET (Personal Electronic Transactor) was also introduced at the 1977 first West Coast Computer Faire. This was the first of a long line of inexpensive computers from Commodore that brought computers to the masses. The Commodore VIC-20 was the first personal computer to sell 1 million units [8].

The IBM-PC and the Information Age

The landmark decision of IBM, the leading business mainframe computer manufacturer, to enter the personal computer market with the announcement of the IBM-PC in 1981 is generally considered to be the dawn of the information age [8]. The IBM-PC made desktop computing a legitimate and eventually essential part of business computing. The original IBM-PC was based on the Intel 8088 chip, a hybrid chip with a 16-bit internal bus and an 8-bit external bus. IBM obtained their operating system for the IBM-PC from Microsoft and bundled in Lotus 1-2-3 business software as part of the package. As was noted in the 1983 *Time Magazine* article [2], this was a significant turning point in the personal computer market and in the dawning of the information age.

Microprocessor Advancement from 1970 to Present

Table 129.1 through Table 129.4 provide data on the advance of microprocessors used by the two major personal computer manufacturers, Apple and Intel. Intel, of course, produced its own chips that were used in IBM-PCs and PC clones (personal computers made by other manufacturers that were compatible with the IBM-PC). Apple, on the other hand, produced some chips of its own, but also purchased chips from various manufacturers. The microprocessors of the 1970s were primarily 8-bit machines implemented in three-micron integrated circuit (IC) technology with clock speeds in the low megahertz range and capable of only about 300,000 to 600,000 instructions per second. The microprocessors of the 1980s

TABLE 129.1 Microprocessors of the 1970s

Date	PC/Apple	Processor	Transistors	Microns	Clock Speed	Data Width	MIPS
1971	Intel	Intel 4004	2.3 k	10 μ	0.1 MHz	4-bits	
1972	Intel	Intel 8008	3.5 k	10 μ	0.2 MHz	8-bits	
1974	Intel	Intel 8080	4.5 k	6 μ	2 MHz	8-bits	0.64
1976	Intel	Intel 8085	6.5 k	3 μ	5 MHz	8-bits	
1976	Apple	MOStek 6502			1 MHz	8-bits	
1978	Intel	Intel 8086	29 k	3 μ	5–10 MHz	16-bits 16-bit bus	0.33
1979	Apple	Syner Tek 6502			1 MHz	8-bits	
1979	Intel	Intel 8088	29 k	3 μ	5–8 MHz	16-bits 8-bit bus	0.33

Sources: Information on Intel microprocessors from Intel Corporation, *Intel Microprocessor Quick Reference Guide*, Intel Corporation Online Series, available at: *www.intel.com/pressroom/kits/quickreffam.htm*; Apple microprocessors, Glen Sanford, Apple-History.Com, available at: *www.apple-history.com*; and MIPS, TSCP, TSCP Benchmark Scores, available at: *http://home.attbi.com/~tckerrigan/bench.html.*

TABLE 129.2 Microprocessors of the 1980s

Date	PC/Apple	Processor	Transistors	Microns	Clock Speed	Data Width	MIPS
1982	Intel	Intel 80186			10–12 MHz	16-bits	
1982	Intel	Intel 80286	134 k	1.5 μ	6–12 MHz	16-bits	1
1983	Apple	Motorola MC68000			5 MHz	16-bits	
1984	Apple	Syner Tek 65C02			4 MHz	8-bits	
1985	Intel	Intel 386	275 k	1 μ	16–33 MHz	32-bits	5
1986	Apple	WDS 65SC816			2.8 MHz	16-bits	
1987	Apple	Motorola MC68020			16 MHz	16-bits	
1988	Apple	Motorola MC68030			16 MHz	32-bits	
1989	Intel	Intel 486	1200 k	0.8 μ	25–50 MHz	32-bits	20

Sources: Information on Intel microprocessors from Intel Corporation, *Intel Microprocessor Quick Reference Guide*, Intel Corporation Online Series, available at: *www.intel.com/pressroom/kits/quickreffam.htm*; Apple microprocessors, Glen Sanford, Apple-History.Com, available at: *www.apple-history.com*; and MIPS, TSCP, TSCP Benchmark Scores, available at: *http://home.attbi.com/~tckerrigan/bench.html.*

were primarily 16-bit machines implemented in one-micron technology with clock speeds in the 5- to 50-MHz range and capable of 1 to 20 million instructions per second. The microprocessors of the 1990s were primarily 32-bit microprocessors, often with a 64-bit bus, implemented in submicron technology with clock speeds of 50 MHz to 1 GHz and capable of 100 million to a billion instructions per second. The microprocessors of the early 2000s are 64-bit processors in 0.13-micron technology running at clock speeds in the 2- to 5-GHz range and capable of nearly 3 billion instructions per second. The number of transistors on the microprocessor chip increased dramatically from only a few thousand in the early 1970s to 77 million in the early 2000s [13–15].

Microprocessors and Microcontrollers

From Table 129.1 through Table 129.4, we see that the PC market is continually striving for improved performance in microprocessors. These microprocessors for the PC market tend to be relatively expensive ($500 price range) when they first come out, but then as they are replaced by more powerful microprocessors the prices plummet (usually to less than $5). Initially, the microcontroller market was comprised primarily of older microprocessors whose values had plummeted so that the price of the microprocessor was so low that they became much more economical in control applications than discrete logic. Applications for the outdated Intel 8080, 8086, and the Motorola 6800, among others, soared as the cost of the processors plummeted. Applications were found in home appliances, intercoms, telephone systems, security systems, garage door openers, answering machines, FAX machines, TVs, cable TV tuners, VCRs,

TABLE 129.3 Microprocessors of the 1990s

Date	PC/Apple	Processor	Transistors	Microns	Clock Speed	Data Width	MIPS
1991	Apple	Motorola MC68040			25 Mhz	32-bits	
1991	Apple	Motorola MC68HC000			16 MHz	16-bits	
1993	Intel	Pentium	3100 k	0.8 μ	66 MHz	32-bits 64-bit bus	100
1993	Apple	Motorola MC68LCO40			20 MHz	32-bits	
1994	Apple	Power PC 601			120 MHz	64-bits	
1995	Apple	Power PC 601+			100 MHz	64-bits	
1995	Apple	Power PC 603/603e			120 MHz	64-bits	
1995	Apple	Power PC 604			180 MHz	64-bits	
1995	Intel	Intel Pentium Pro	5500 k	0.35 μ	200 MHz	32-bits 64-bit bus	250
1997	Apple	Power PC 750			250 MHz	64-bits	
1997	Intel	Pentium II	7500 k	0.35 μ	233 MHz	32-bits 64-bit bus	300
1997	Intel	Pentium MMX	4500 k	0.25 μ	233 MHz	32-bits 64-bit bus	300
1998	Intel	Celeron	19000 k	0.25 μ	333 MHz	64-bits	
1998	Intel	Pentium II Xeon	7500 k	0.25 μ	450 MHz	32-bits 64-bit bus	465
1999	Apple	Power PC 7400			500 MHz	64-bits	
1999	Intel	Pentium III Xeon	28000 k	0.18 μ	733 MHz	32-bits 64-bit bus	500
1999	Intel	Mobile Celeron	18900 k	0.25 μ	466 MHz	64-bits	541
1999	Intel	Mobile Pentium III	28000 k	0.18 μ	500 MHz	32-bits 64-bit bus	800
1999	Intel	Pentium III	28000 k	0.18 μ	733 MHz	32-bits 64-bit bus	825

Sources: Information on Intel microprocessors from Intel Corporation, *Intel Microprocessor Quick Reference Guide*, Intel Corporation Online Series, available at: *www.intel.com/pressroom/kits/quickreffam.htm*; Apple microprocessors, Glen Sanford, Apple-History.Com, available at: *www.apple-history.com*; and MIPS, TSCP, TSCP Benchmark Scores, available at: *http://home.attbi.com/~tckerrigan/bench.html.*

TABLE 129.4 Microprocessors of the 2000s

Date	PC/Apple	Processor	Transistors	Microns	Clock Speed	Data Width	MIPS
2000	Intel	Pentium 4	42000 k	0.18 μ	1.5 GHz	32-bits 64-bit bus	1700
2001	Apple	Power PC 7410			500 MHz	64-bits	
2001	Apple	Power PC 7450			733 MHz	64-bits	
2001	Intel	Xeon	42000 k	0.18 μ	2 GHz	64-bits	2350
2001	Intel	Itanium	25000 k	0.18 μ	800 MHz	64-bits	
2002	Apple	Power PC 7455			1 GHz	64-bits	
2002	Intel	Mobile Pentium 4	55000 k	0.13 μ	2.2 GHz	32-bits 64-bit bus	2750
2003	Intel	Pentium M (Merced)	77000 k	0.13 μ	1.6 GHz	64-bits	

Sources: Information on Intel microprocessors from Intel Corporation, *Intel Microprocessor Quick Reference Guide*, Intel Corporation Online Series, available at: *www.intel.com/pressroom/kits/quickreffam.htm*; Apple microprocessors, Glen Sanford, Apple-History.Com, available at: *www.apple-history.com*; and MIPS, TSCP, TSCP Benchmark Scores, available at: *http://home.attbi.com/~tckerrigan/bench.html.*

camcorders, remote controls, video games, cellular phones, musical instruments, sewing machines, lighting controls, pagers, cameras, pinball machines, exercise equipment, microwaves, copiers, printers, automobile trip counters, automobile engine controls, air bags, instrumentation, electronic entertainment equipment, climate control, keyless entry, and a host of others. Microprocessor manufacturers quickly realized that there was a huge market for microcontrollers and quickly modified some of their older

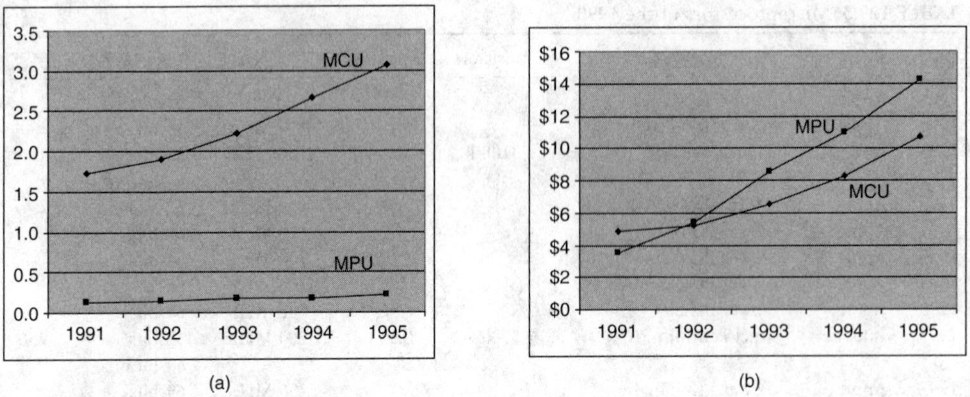

FIGURE 129.2 Microprocessor (MPU) and cicrocontroller (MCU) markets. (a) MPU vs. MCU unit sales in billions of units. (b) MPU vs. MCU dollar sales in billions of dollars.

microprocessors to include on-chip RAM and ROM, serial and parallel I/O, timers, A/D and D/A converters and other support for control applications revamping the old microprocessors into microcomputers marketed as microcontrollers for embedded systems. Figure 129.2 shows a comparison of microprocessor (MPU) and microcontroller (MCU) sales during the 1990s. Figure 129.2(a) depicts worldwide dollar sales, and Figure 129.2(b) worldwide unit sales. Several things become very clear from Figure 129.2. First, Figure 129.2(a) shows that microprocessor (MPU) sales have grown only slightly, hovering around 200 million units per year while microcontroller (MCU) sales have grown dramatically from 1.7 billion units in 1991 to over 3 billion units in 1995. In 1995, about 15 microcontrollers were sold for every microprocessor sold. Of course, the microprocessors are much more expensive and hence constitute significant dollar sales. Figure 129.2(b) compares the dollar sales of microprocessors (MPU) vs. microcontrollers (MCU). Despite the soaring unit price of microprocessors and the relatively low price of microcontrollers, the dollar sales track very well with over $10 billion in sales of microcontrollers in 1995 and around $14 billion sales in microprocessors [17].

Figure 129.3 compares the types of microcontrollers sold during the 1990s. Figure 129.3(a) plots the total sales of microprocessors plus microcontrollers, increasing from $6 billion in 1990 to $18 billion in 2000. Figure 129.3(b) looks at microcontroller dollar sales for 4-bit, 8-bit, and 16/32-bit microcontrollers. One sees a relatively constant market for the very inexpensive 4-bit processors (just under $2 billion per year), strong growth in the 8-bit microcontrollers (from $2 billion in 1990 to $10 billion in 2000) and

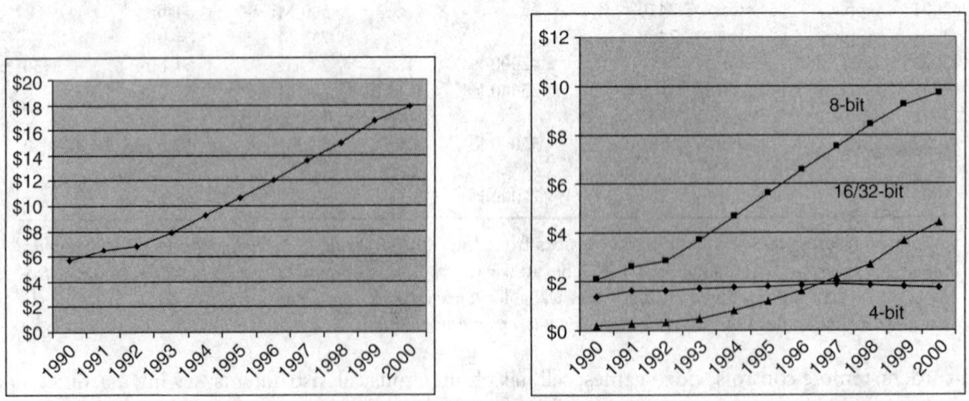

FIGURE 129.3 MPU and MCU sales: (a) total sales of MPU and MCU in billions of dollars; (b) sales of 4-bit, 8-bit and 16/32-bit MCUs's in billions of dollars.

an exponential growth in 16/32-bit processors starting from nearly nothing in 1990 and growing to nearly $5 billion in 2000 [18].

129.3 Key Features of Modern Microprocessors and Microcontrollers

Both microprocessors and microcontrollers implement part or all of Figure 129.1. However, the microprocessors emphasize the CPU and ALU functions, while microcontrollers balance CPU and ALU functions against RAM, ROM, and particularly I/O functions. Table 129.1 through Table 129.4 chart the progress of microprocessors over 4 decades. The clear trend is increasing power of the processor with faster CPU clocks, larger data widths, powerful floating-point ALUs, and more instructions per second. On the other hand, the most popular microcontrollers are based on older microprocessors such as the 8080, 8086, or 80186. By adding such things as a ROM containing a basic interpreter, RAM for program and data storage, A/D, D/A, serial and parallel I/O, and a set of timers, these older microprocessors are transformed into microcontrollers that are very low priced (average $3 per unit) and very well suited for embedded processor applications. Table 129.5 lists some of the major producers of microcontrollers and their 4-bit, 8-bit, 16-bit, and 32-bit lines.

The Most Popular Microcontroller: The 8051 (Intel and Others)

The 8051 was originally released by Intel in 1980 and quickly became the most popular microcontroller for embedded applications. Today many variants of the 8051 are available from multiple manufacturers. The 8051 is an 8-bit microprocessor with reasonable power at very low price. The 8051 is a modified Harvard architecture that has a separate memory for data and instructions allowing a single-cycle fetch of both instruction and data. The 8051 has a CPU with a Boolean processor, features five or six interrupts and two or three 16-bit timer counters. The I/O contains a full-duplex serial port and four 8-bit parallel ports. It comes with both on-chip RAM and EPROM. Strong points of the 8051 are its simple interrupt structure and the Boolean processor that allows for very simple bit testing and manipulation. The addressing structure is somewhat confusing, but very powerful.

The 8051 has the widest range of variants of any microcontroller. Atmel makes the smallest device, Dallas makes the fastest device and Siemens makes the most powerful. Unit costs for 8051 variants range from as low as $3.30 to as high as $40 depending on the features. The 8051 supports four separate memories: on-board RAM (only 128 or 256 bytes); external RAM (up to 64 Kbytes); on board EPROM (4 Kbytes or 8 Kbytes); and external EPROM (64 Kbytes).

Old PC Microprocessor Often Used as Microcontroller: The Z80 (Zilog)

In July 1976, Zilog introduced the Z80 microprocessor chip to compete with Intel's 8080 and 8085. The Z80 can execute 8080 code but includes powerful extension to the code. Many personal computers of the late 1970s and early 1980s were based on the Z80. Today the Z80 and its family (Z8 and Z180) are very popular as microcontrollers because of their very low cost (less than $1) and powerful instruction set. The Z80 is an 8-bit CPU with 16-bit addresses that can address 64 Kbytes of memory. Zilog's Z180, 84013, and 14015, Hitachi's 64180, and Toshiba's 84013 and 84015 are enhanced versions of the Z80 that integrate many of the traditional peripheral functions onto the chip while maintaining upgrade compatibility with the Z80 but with much lower power requirements. They also add additional features, such as watchdog timers, clock generators, and additional 16-bit parallel I/O ports.

A Microcontroller for Hobbyists and Robotics: The 68HC11 (Motorola and Others)

Motorola introduced the 6800 series of 8-bit microprocessors back in 1975 to compete with the Intel 8080. The 68HC11 is an upgraded version of the 6800 series that integrates many peripheral functions

TABLE 129.5 Microcontrollers

Manufacturer	4-Bit	8-Bit	16-Bit	32-Bit
Atmel	MARC4	AVR RISC		Rad Hard DSP
		AT91 ARM		
		8051		
		CAN		
		USB		
Cypress		Neuron		
Dallas/Maxim		Secure		
		CAN		
		8051		
		Mixed Signal		
Intel		8058	80186	80386
		8051	MCS 96/296	
		MCS 51/251		
Intersil		1802	8086	
		8088	80286	
Infineon		C500	C166	Tricore DSP
		8051		
Microchip		PIC12		
		PIC14		
		PIC16		
		PIC17		
		PIC18		
Motorola		68HC05	HCS12	68000
		68HC08	M68HC12	M683XX
		68HC11	M68HC16	MPC500
			DSP56800	
National	CR4	CR8	CR16	CR32/64
		COP8	8086 (GX)	
NEC	75XL	KOS	K4	V850
		KO		
Philips		8051	XA	ARM7TDMI
Renesas	HMCS400	7600	M16	Super H
(Hitachi and Mitsubishi)		38000	H8S	M32R
		740/7450/7470	H8/300H	SH-Mobile
		H8/300	H8/500	H8SX
		H8/300L	7900	
			7700/7751	
Sharp			LH75400	LH79520
STMicroelectronics		ST7	ST9	
Texas Instruments			TMS3201X	TMS3203X
			TMS3202X	TMS3204X
			TMS3205X	TMS3208X
Toshiba	47 series	870 series	900 series	TX system
Zilog		eZ80	Z89321	
		Z180		
		Z8		
		Z80		

onto the chip. Like the Intel 8051, the 68HC11 has an 8-bit CPU with 16-bit address lines that can address 64 Kbytes of memory. However, the 68HC11 has a common memory structure where data, instructions, I/O and timers all share the same memory space. Part of the reason for the popularity of the 68HC11 with hobbyists and robotics is the fact that various models of the microprocessor come with EPROM, RAM, I/O timers, A/D converters, and PWM generators are all on chip. Typical current draw for the 68HC11 is a low 20 ma.

The Simplest Microcontroller: The PIC5 Series (Microchip)

The roots of the PIC microprocessor go back to 1975 and a Department of Defense (DOD) contract with Harvard University that developed the Harvard architecture. However, the DOD went with a simpler and at the time more reliable single memory architecture proposed by Princeton University. In 1978, Signetics picked up the basic architecture and used it in the Signetics 8x300 Peripheral Interface Controller (PIC). Around 1985 Signetics spun off its microelectronics division to Arizona-based Microchip Technologies with their main product the PIC microcontroller. The PIC was the first RISC microcontroller featuring only 33 instructions compared to over 90 in the Intel 8048. Microchip has developed a line of PIC microprocessors with a large variety of features integrated onto the chip while maintaining extremely low cost ($1 to $10 per unit).

129.4 Applications of Microcontrollers

Microcontrollers are primarily used in embedded systems. This means that the microcontroller is connected directly to sensory inputs and control outputs that allow it to sense information about its environment and exert control over a device through its control outputs. An example in home electronics might be the microcontroller in your microwave that responds to the buttons pressed on the console and then performs the function desired including such tasks as thawing frozen items by adjusting the microwave energy and sensing the temperature of the food to adjust the microwave power levels. An example in automobile electronics might be the cruise control that senses the speed and then adjusts the throttle to maintain the speed or to increase and decrease the speed in response to user input to the cruise control buttons. An example of a hobbyist project might be the design of a robot to search through a model house for a flame and then extinguish that flame. In each of these examples, the microcontroller is responsible for the control of the system and is embedded within the system as an integral part of the product.

The popularity of the microcontroller families discussed in the previous section is primarily due to the availability of excellent development tools for these microcontrollers. Figure 129.4 shows the setup of the 8052-BASIC development system that is very popular for use with the 8081 line of microcontrollers. The development system consists of a board with the 8052 microprocessor with a BASIC interpreter in its on-chip ROM, with either on-chip or external RAM, peripheral chips to support the 8052, and an RS-232 serial port to connect to a host computer. The host computer is used to download instructions in BASIC to the RAM where they are then interpreted by the BASIC interpreter and cause the 8052 to perform the desired function. The RS-232 serial port provides the host computer full access to the 8052 to examine registers, debug software code, set and test various features of the chip, and view what is

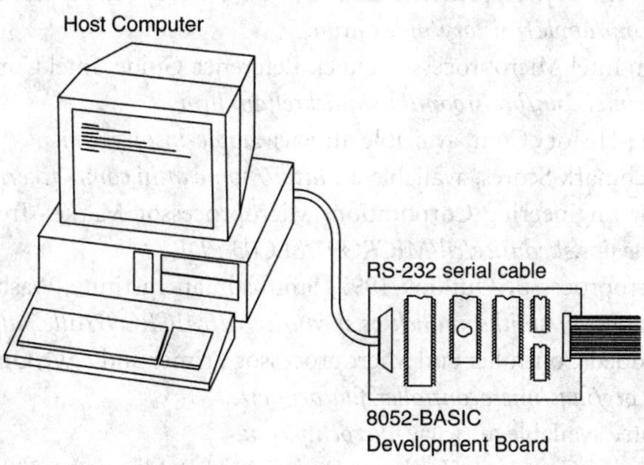

FIGURE 129.4 Typical microcontroller development system.

stored in all the registers and the memory of the 8052 chip. Once the program has been written and debugged to carry out the desired functions, the host computer can download the final program to EPROM (usually not to RAM as it would be lost should the power fail) and then the RS-232 cable can be removed and the 8052 can function independently of the host computer. Often the 8052-BASIC development board can be directly embedded into the system and become part of the integrated system. In other instances, the 8052-BASIC development system is used only to program and debug the 8052 microcontroller and then the microcontroller is unplugged from the development system and plugged into the embedded system where it will perform the desired function. There are lots of excellent development tools available both from commercial sources and in the form of freeware that make the use of microcontrollers in embedded systems very easy. Most designers find that they can learn very quickly the process to program the microcontroller and use it in their system.

References

1. *Time Magazine*, The Past Men of the Year, Time Magazine Online Webpage, available at: *www.time.com/time/special/moy/moypast.html.*
2. Friedrich, O., Mortiz, M., Nash, J. M., and Stoler, P., Machine of the year: the computer moves in, *Time Magazine*, January 3, 1983.
3. Brain, M., How microprocessors work, How Stuff Works Online Series, available at: *http://electronics.howstuffworks.com/microprocessor.htm/printable.*
4. Rongen, H., Introduction to Microprocessors and Microcontrollers, Zentrallabor fur Elektronik, Forschungszentrum Jülich, available at: *http://zelweb.zel.kfa-juelich.de/projects/dsp/training/MicroContr.PDF.*
5. Walters, R. Mark 8 and IBM 5100 Computers, UC Davis Computer Museum, available at: *wwwcsif.cs.ucdavis.edu/~csclub/museum/items/merk_8_ibm_5100.html.*
6. Apple 2 History Museum, Build your own Mark 8 personal computer, Apple 2 history, available at: *http://apple2history.org/museum/magazines/radioelect7407.html.*
7. Ritter, T., The Mark 8 experience, available at: *www.ciphersbyritter.com/MARK8/MARK8.HTM.*
8. Long, D. J., A Brief History of Personal Computing, Prepared for the 20th Anniversary of CHIPS, available at: *www.ciphersbyritter.com/MARK8/MARK8.HTM.*
9. Smith, E., IBM 5100 Personal Computer, Eric Smith Online Resources, Milpitas, CA, available at: *www.brouhaha.com/~eric/retrocomputing/ibm/5100/.*
10. Roberts, E., An interview with Ed Roberts, *Historically Brewed Magazine*, available at: *http://virtualaltair.com/.*
11. Burke, M., *The Pirates of Silicon Valley*, Warner Brothers Video, TNT Original, T6593.
12. Owad, T., Apple History, Applefritter.com Online History of Apple Computer, available at: *www.applefritter.com/apple1/history/index.html.*
13. Intel Corporation, Intel Microprocessor Quick Reference Guide, Intel Corporation Online Series, available at: *www.intel.com/pressroom/kits/quickreffam.htm.*
14. Sanford, G., Apple-History.Com, available at: *www.apple-history.com.*
15. TSCP, TSCP Benchmark Scores, available at: *http://home.attbi.com/~tckerrigan/bench.html.*
16. Integrated Circuit Engineering Corporation, Microprocessor Market Trends, 1997, available at: *http://smithsonianchips.si.edu/ice/cd/MICRO97/SEC03.PDF.*
17. McClean, B., Microprocessor Outlook 1997, Smithsonian Institute, Washington, DC, Sections 2 and 3, available at: *http://smithsonianchips.si.edu/ice/cd/MICRO97/title.pdf.*
18. Hersch, R., Embedded Controller and Microprocessor Primer and FAQ, Online Newsgroup, available at: *www.faqs.org/faqs/microcontroller-faq/primer/.*
19. Microchip web site, available at: *www.microchip.com.*
20. Arnold, K., *Embedded Controller Hardware Design*, LLH Publications, 2001 (includes CD-ROM).
21. Berger, A. S., *Embedded System Design: An Introduction to Process, Tools and Techniques*, CMP Books, 2001.

22. Barr, M., *Programming Embedded Systems in C and C++*, OíReilly and Associates, 1999.
23. Van Sickle, T., *Programming Microcontrollers in C*, LLH Publications, 2000.
24. Fox, T., *Programming and Customizing the HC11 Microcontroller*, McGraw-Hill, New York, 1999.
25. Huang, H.-W., *MC68HC11: An Introduction*, Delmar Learning, 2000.

Further Information

This chapter is intended to be a brief overview of microprocessors and microcontrollers. While it is not feasible to provide in this brief chapter the details needed to use these microprocessors and microcontrollers, this introduction should prepare the reader well for taking the next step of obtaining a microcontroller development system such as the one of Figure 129.4 and taking the first step in programming a microcontroller for an embedded application. The Internet is an excellent source of information on how to do this and is also a source for both commercial and freeware that can be used to program or to simulate various microprocessors. A good overview and introduction to microprocessors and microcontrollers can be found in Brain [3] and Rongen [4]. A particularly good starting point for the beginner wishing to build an embedded system is the Microchip PIC website, which does a pretty good job of leading one by the hand through the steps necessary to choose the correct PIC microprocessor and associated development tools and then how to use those tools to develop an embedded system [19]. There are also a very large number of books available aimed at both the beginner and the more advanced designer [20–25].

130

Memory Systems

Richard S. Sandige
University of Wyoming

This chapter deals primarily with relatively fast electronic memory systems. Slower magnetic media, other storage media, and input/output (I/O) interfaces are only briefly discussed. Read/write memory, read-only memory, static memory, dynamic memory, cache memory, volatility of memory, relative speed of memory, and applications of memory are the major topics discussed. Tradeoffs, advantages, and disadvantages are considered for various memory systems in the following sections. This approach provides insight as to the applications for which each type of memory system is used and why.

130.1 CPU, Memory, and I/O Interface Connections

Most applications that employ memory systems contain a processor or a central processing unit (CPU). The CPU carries out basic instructions that require moving data between internal registers and memory or between internal registers and an I/O interface. Accomplishing this task usually requires a minimum of three types of buses, as illustrated in Figure 130.1. The CPU provides the locations or addresses where the data must come from during a memory-read cycle or go to during a memory-write cycle. The bus that conveys these locations to the memory and I/O interface is called the *address bus.* For an address bus of n bits there are $2n$ storage locations available. When the memory and I/O interface are both accessed via the same address lines and the same control lines, as illustrated in Figure 130.1, the interface is called *memory-mapped I/O.*

When the I/O interface is accessed as an address space separate from the memory address space, the address bus lines are interpreted as port addresses of the CPU, and separate control bus lines exist between the CPU and the I/O interface to direct data into or out of requested ports on the CPU via special I/O instructions. In the case of processors that have separate I/O and memory control signals, memory-mapped I/O can always be used by ignoring both the separate I/O control signals and the special I/O instructions and using only those processor instructions that relate to memory. Using only memory instructions to access both memory and I/O in the memory space of the CPU generally increases the flexibility of the overall system and also makes the system easier to program. A necessary requirement for the programmer of a memory-mapped I/O memory system is a **memory map** that shows where all device locations exist.

Data, the contents of an address in a memory-mapped I/O system as illustrated in Figure 130.1, are read from the memory space on a CPU memory-read cycle and written to the memory space on a CPU

0-8493-1586-7/05/$0.00+$1.50

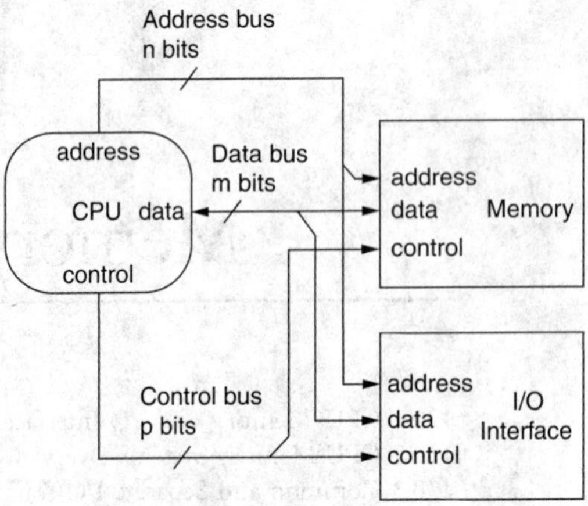

FIGURE 130.1 CPU, memory, and I/O interface for a memory-mapped I/O architecture.

memory-write cycle. An example of a memory-read cycle and a memory-write cycle timing diagram for a static RAM device will be presented later. Data travel on the **bidirectional data bus** in one direction at a time. The number of bits (data bus lines) associated with the data bus is dependent on the design of the CPU and ranges from a minimum of 4 bits for early microprocessors to as many as 32 bits for current microprocessors. Since heavy competition exists between major companies that manufacture microprocessors such as Motorola, Intel, IBM, Hewlett-Packard, and others, the push for 64 data bits and beyond for newer designs is underway. A larger number of data bits usually allows programs to be written more efficiently and also provides speed improvements in program execution since more bytes of data can be moved with fewer memory-read or -write cycles.

The control bus lines in a memory-mapped I/O system are routed to the memory and the I/O interface as illustrated in Figure 130.1. The I/O interface provides a data path for reading and writing data to the outside world that is beyond the system's existing memory system. The number of bits associated with the control bus is dependent on the architectural design of the complete system.

130.2 CPU Memory Systems Overview

A few of the various types of memory devices that can be connected to a CPU are illustrated in the memory systems connected to the CPU in Figure 130.2. The interconnections (all possible interconnections are not shown) must each have an address bus, a data bus, and a control bus, as discussed earlier. The **cache memory** block closest to the CPU usually contains static random access memory (RAM). Static RAM can also be used for primary memory; however, primary memory is usually designed with

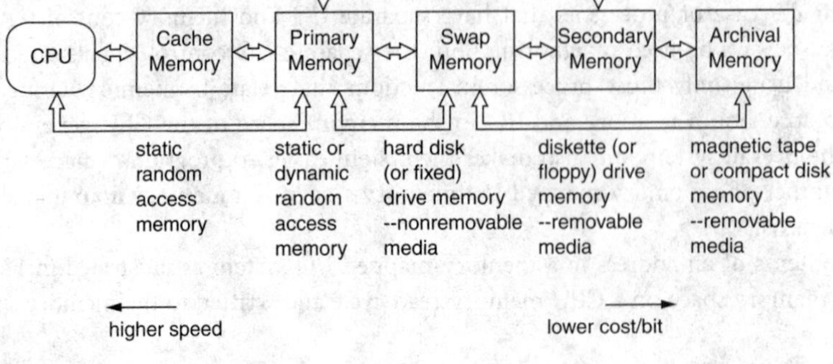

FIGURE 130.2 Overview of a CPU memory system.

dynamic RAM, which operates at a slower speed and is less expensive than static RAM. Dynamic RAM must be constantly refreshed (about every 2 to 8 milliseconds) or stored data are lost. Only one transistor per storage location is required in a dynamic RAM, compared to several transistors per storage location in a static RAM, where refresh is not required. Figure 130.2 also provides a general indication of where other types of memory such as hard (or fixed) drive memory, diskette (or floppy) drive memory, and magnetic tape or compact disk memory fit into a memory system relative to speed and cost/bit. Both static and dynamic RAMs are volatile types of memory. When power is removed, the content of the data in all storage locations is lost. The memory devices used for swap memory, secondary memory, and archival memory shown in Figure 130.2 are all nonvolatile memory. When power is removed from nonvolatile memory, the content of the data in all storage locations is preserved. Swap, secondary, and archival memories are all slower and cost less per bit than static and dynamic memories. This chapter will concentrate only on the faster or higher-speed types of static and dynamic memories that are contained in blocks closer to the CPU.

130.3 Common and Separate I/O Data Buses

Memory system blocks that consist of RAMs either utilize RAM devices that provide the input and output data lines via a common I/O data bus, as shown in Figure 130.3(a), or utilize RAM devices that provide the input and output data lines on separate I/O data buses, as shown in Figure 130.3(b). Memory systems that use memory blocks with a common I/O data bus have the advantage of costing less and taking up less printed circuit (PC) board space than those that use memory blocks with separate I/O data buses. Systems that use memory blocks with separate I/O data buses, however, have the advantage that **three-state buffers** can be added to the separate output data bus to eliminate **bus contention**, a problem in high-speed designs. Memory blocks that use either a common I/O bus or separate I/O buses are referred to as *single-port memory blocks* since data can only be accessed sequentially via a memory-read cycle or a memory-write cycle.

Some memory system blocks utilize dual-port memory devices, which allow data to be written to an input port while other data are read at a separate output port at the same time. The video memory of fast computer workstations is often designed using dual-port static memory devices called *VRAMs* (video random access memory). VRAMs have two ways to access data simultaneously. Data can be accessed

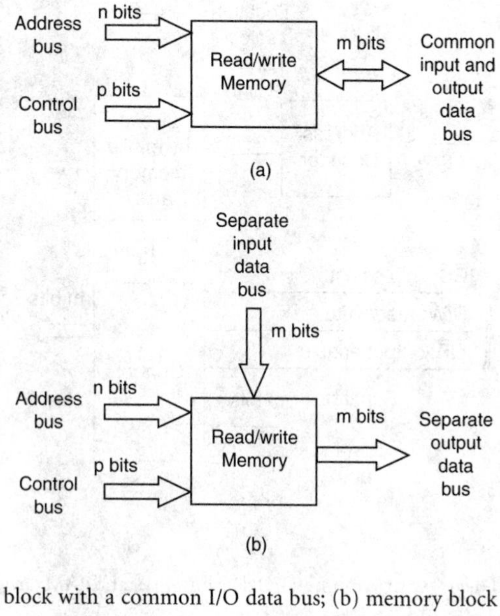

FIGURE 130.3 (a) Memory block with a common I/O data bus; (b) memory block with separate I/O data buses.

through a parallel RAM port and through a serial access memory (SAM) port. VRAMs are specialized devices that are specifically designed for use as display memories in **bit-mapped** graphic systems. Because of their multiple-access capability and functional flexibility, VRAM devices are usually the most expensive memory devices. It is also common in high-speed designs to have a dedicated video processor that issues video commands to speed up the video memory system.

130.4 Single-Port RAM Devices

Figure 130.4(a) shows a block diagram of a static RAM device with a common I/O data bus. A block diagram of a static RAM device with separate I/O data buses is shown in Figure 130.4(b). In both types of RAM devices, an address **decoder** is required to decode the *n* bits of address to determine one among the 2*n* location in the memory array where data are stored (written) or retrieved (read). Both types of RAM devices also can contain a chip select line, a read/write line, and an output enable line. Output

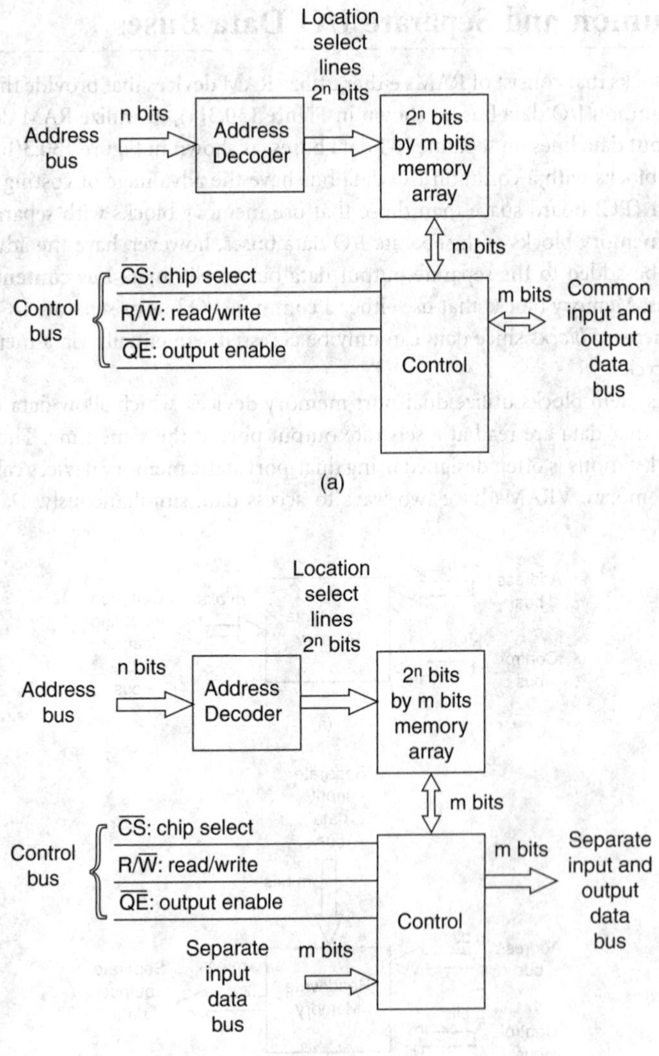

FIGURE 130.4 (a) Static RAM device with a common I/O data bus; (b) static RAM device with separate I/O data buses.

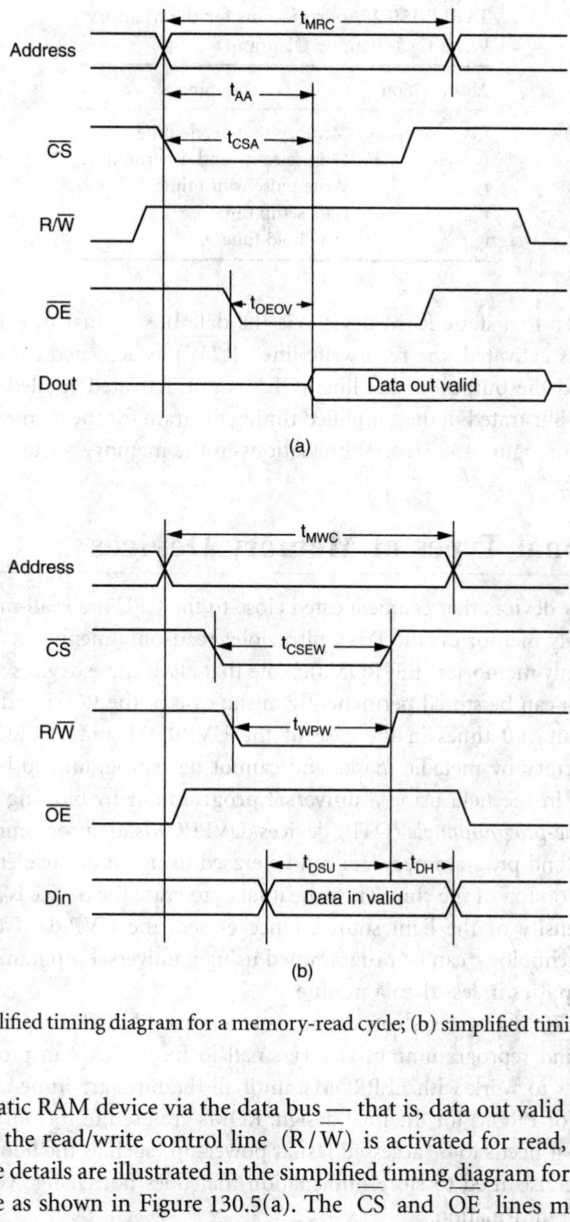

FIGURE 130.5 (a) Simplified timing diagram for a memory-read cycle; (b) simplified timing diagram for a memory-write cycle.

data are read from a static RAM device via the data bus — that is, data out valid — when the chip select line ($\overline{CS}$) is activated, the read/write control line ($R/\overline{W}$) is activated for read, and output enable line ($\overline{OE}$) is activated. These details are illustrated in the simplified timing diagram for the memory-read cycle of a static RAM device as shown in Figure 130.5(a). The $\overline{CS}$ and $\overline{OE}$ lines must both be pulled to a low-voltage logic state to be activated, since they both contain an overbar. The $R/\overline{W}$ line must be pulled to a high-voltage logic state to be activated for read (no overbar over R). Abbreviations in the memory-read cycle timing diagram are provided in Table 130.1.

TABLE 130.1 Abbreviations for the Memory-Read Cycle Timing Diagram

Abbreviation	Meaning
t_{MRC}	Memory-read cycle time
t_{AA}	Address access time
t_{CSA}	Chip select access time
t_{OEOV}	Output enable to output valid time

TABLE 130.2 Abbreviations for the Memory-Write Cycle Timing Diagram

Abbreviation	Meaning
t_{MWC}	Memory-write cycle time
t_{CSEW}	Chip select to end of write time
t_{WPW}	Write pulse width time
t_{DSU}	Data setup time
t_{DH}	Data hold time

Input data are written to a static RAM device via the data bus — that is, data in valid — when the chip select line ($\overline{CS}$) is activated, the read/write line ($R/\overline{W}$) is activated for write (pulled to a low-voltage logic state), and the output enable line ($\overline{OE}$) is not activated (pulled to a high-voltage logic state). These details are illustrated in the simplified timing diagram for the memory-write cycle of a static RAM device as shown in Figure 130.5(b). Abbreviations in the memory-write cycle timing diagram are provided in Table 130.2.

130.5 Additional Types of Memory Devices

Other types of memory devices that can be located close to the CPU are **read-only memories** (ROMs), programmable read-only memories (PROMs), ultraviolet read-only memories (UVPROMs), and electrically erasable read-only memories (EEPROMs). Note that all of these devices are read-only devices in normal operation. Data can be stored permanently in the case of the ROM and PROM devices, or can be restored up to about 100 times in the case of the UVPROM and EEPROM devices. ROMs are programmed at the factory by metallic masks and cannot be reprogrammed by the user. PROMs are generally programmed in the field using a **universal programmer** by blowing or leaving fuses intact; these are called *one-time-programmable* (OTP) devices. UVPROMs are programmed in the field using a universal programmer, and programmed data can be erased using an ultraviolet light source through a quartz window located on top of the chip. The time it takes to erase the device is about 15 to 45 minutes, depending on the intensity of the light source. Once erased, the UVPROM can be reprogrammed. EEPROMs, the latest technology, can be programmed using a universal programmer, and then they can be erased by electrical pulses in less than a minute.

When developing a ROM-based system, engineers often use EEPROMs, since the turnaround time (the time for erasing and reprogramming) is very small to fix mistakes in programmed bit patterns. Engineers usually prefer to work with EEPROMs, until all the bugs are ironed out of a design, before committing to a ROM or PROM for the final design. ROMs are used in memory systems in computers to store information that needs to be accessed fast at power-up (such as the boot ROM that starts a PC system). EEPROMs are also used to store information that does not change very often, such as date, time, and configuration information.

ROMs and PROMs are permanently nonvolatile, whereas UVPROMs and EEPROMs are nonvolatile for approximately 10 years after being programmed. ROMs are very inexpensive per part after an up-front high cost is paid to the manufacturer for making the mask. To program PROMs, UVPROMs, and EEPROMs requires the user to purchase a universal programmer. A high-end universal programmer can run a few thousand dollars, whereas a programmer that programs only certain types of PROMs can cost as little as \$400 or \$500. PROMs are a little more expensive per part compared to ROMs and require that the user purchase a programmer. The next most expensive devices are UVPROMs and EEPROMs. Erasing UVPROMs requires that the user purchase an ultraviolet light source in addition to a universal programmer. EEPROMs can be programmed and erased with a universal programmer.

Speeds of the various devices are generally given with the ROM as the fastest, the PROM the next fastest, and the UVPROM and EEPROM much slower. Table 130.3 provides a quick reference concerning cost/part, speed, and so forth for ROM and PROM-type devices.

TABLE 130.3 Summary of ROM and PROM-Type Devices

Device	Cost/Part	Speed	Nonvolatile	Programmer
ROM	Lowest[a]	Fastest	Permanently	N/A
PROM	Low	Fast	Permanently	Required
UVPROM	Higher	Slow	Semipermanently[b]	Required[c]
EEPROM	Higher	Slow	Semipermanently[b]	Required

[a] Up-front high cost must be paid to manufacturer for making the mask.
[b] Stores data for approximately 10 years.
[c] An ultraviolet light source is also required to erase these devices.

The timing diagram of interest for ROM users and the various PROM devices is the memory-read cycle. The universal programmer provides the required memory-write cycle for data storage for PROM devices. The memory-read cycle is basically the same as the memory-read cycle shown in Figure 130.5(a). The $R/\overline{W}$ signal for PROM devices is replaced by a program signal $\overline{PGM}$ used to perform a memory-write cycle. After the device is programmed the program signal $\overline{PGM}$ is disabled or not activated (pulled to a high-voltage logic state) so that the device can only be read.

130.6 Design Examples

In general, a cache memory system and a primary memory system often require a larger number of address bits and data bits than single-chip RAM devices can supply. This condition requires that several RAM devices be connected together to form the design for a cache or primary memory system.

Figures 130.6 and 130.7 illustrate the designs for two separate memory system blocks using static RAM devices. The first memory system design shown in Figure 130.6 addresses 1024 locations each 8 bits wide. The RAM devices used in this design have only 256 addresses with a common I/O 4-bit-wide data bus. This situation requires using a decoder to decode four ranges of addresses and two RAM devices simultaneously. The decoder requires 2 bits from the address bus to select the four different address ranges. Output enable for the two RAMs that are selected for a memory-read cycle is obtained by ANDing the read/write signal with the decoder output that selects the required address range. Each RAM device requires 8 bits from the address bus to select among the 256 locations in its memory. Each RAM device only provides 4 bits of content at each memory location from 0 through 1023, thus requiring two RAM devices for each address.

Figure 130.7 illustrates a simpler memory system design that also addresses 1024 locations each 8 bits wide. In this design, 1024-bit RAM devices organized as 1024 words of 1 bit are used. Eight RAM devices must be used to obtain 8 bits of data. The address bus supplies all the RAM devices simultaneously without the need for an external address decoder, thus simplifying the design. Expanding the memory to 8192 addresses using instances of this design requires an external address decoder for chip select logic.

The simplest design for a memory system that addresses 1024 locations each 8 bits wide uses a single 8192-bit RAM device organized as 1024 words of 8 bits. In this design, the address bus and the data bus are connected directly to the RAM device. Cost, speed, printed circuit board space, and so forth are factors that need to be considered in choosing the best memory system design for a particular application.

Defining Terms

Bus contention — A condition that occurs on a bus when two or more devices try to output opposite logic levels on the same bus line.

Bidirectional data bus — A bus that allows data to flow in either direction.

Bit mapped — A graphic display that has 1 bit (or more) of display memory for each possible pixel or dot on the display.

Cache memory — A high-speed memory that can contain small segments of a program that can be accessed faster than primary or bulk memory.

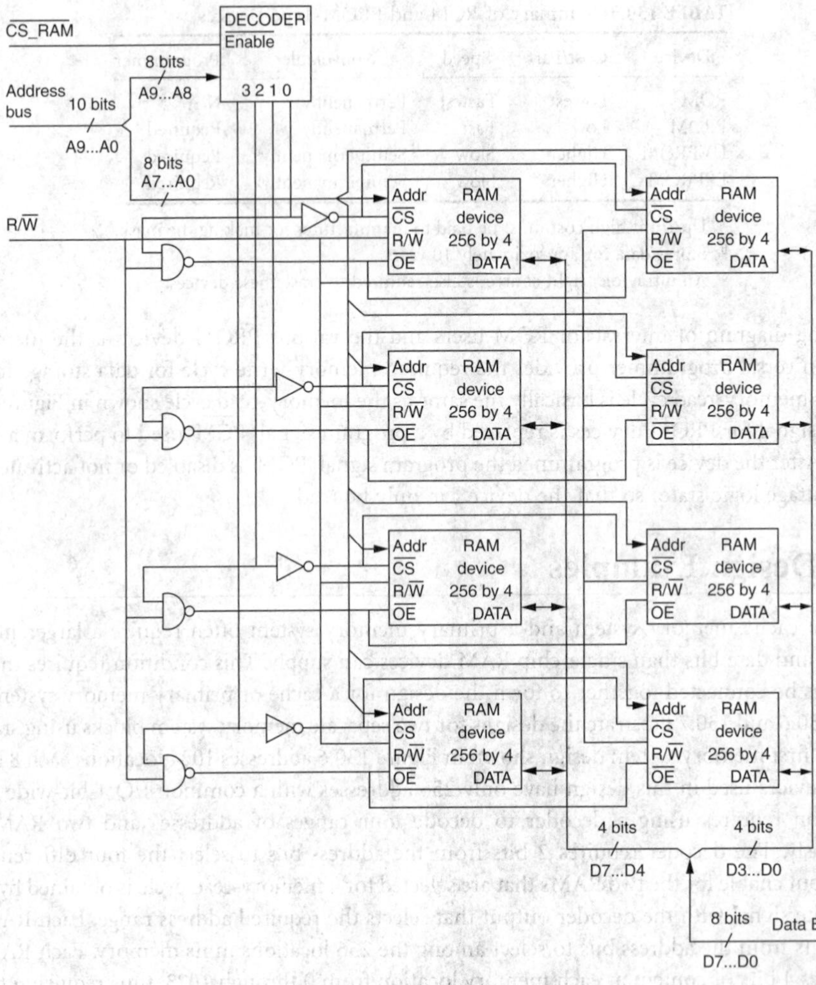

FIGURE 130.6 A 1024 × 8-bit static RAM memory system using 256 × 4-bit RAM devices with a common I/O data bus.

Decoder — A logic device that converts a binary code applied to *i*-input lines to 2^i different output lines.

Memory map — A table that specifies the location or address of every memory or I/O device that a CPU can access.

Read-only memories — Memories that are preprogrammed with data that can only be read under normal operation.

Three-state buffers — Buffers that have a third output state — that is, a state that is off, or high impedance, to effectively disconnect the outputs — in addition to the normal high- and low-voltage output states.

Universal programmer — A programming unit that can program various types of PROMs, erase EEPROMs, and program programmable logic array and programmable array logic devices.

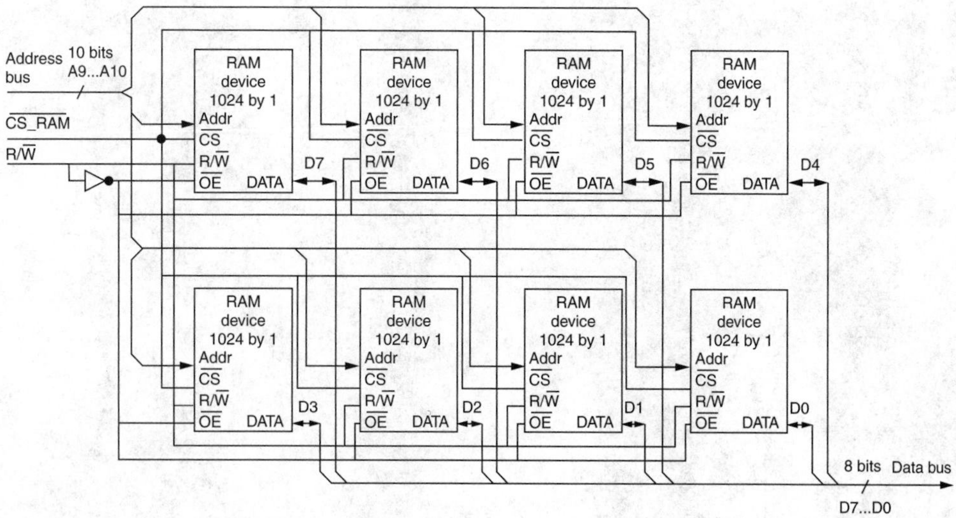

FIGURE 130.7 A 1024 × 8-bit static RAM memory system using 1024 × 1-bit RAM devices with a common I/O data bus.

References

Gaonkar, R. S. 1984. *Microprocessor Architecture, Programming and Applications with the 8085/8080A.* Merrill, Columbus, OH.

Gibson, G. A. 1991. *Computer System Concepts and Design.* Prentice Hall, Englewood Cliffs, NJ.

Greenfield, J. D. and William, C. W. 1988. *Using Microprocessors and Microcomputers: The Motorola Family,* 2nd ed. John Wiley & Sons, New York.

Hayes, J. P. 1988. *Computer Architecture and Organization,* McGraw-Hill, New York.

Hitachi. *Hitachi IC Memories Data Book.* #M11. Hitachi America, Ltd., San Jose, CA.

Motorola. 1990. *Motorola Memories.* DL113 REV 6. Motorola, Phoenix, AZ.

Pollard, H. L. 1990. *Computer Design and Architecture.* Prentice Hall, Englewood Cliffs, NJ.

Slater, M. 1989. *Microprocessor-Based Design: A Comprehensive Guide to Effective Hardware Design.* Prentice Hall, Englewood Cliffs, NJ.

Wear, L. L., Pinkert, J. R., Wear, C. W., and Land, W. G. 1991. *Computers: An Introduction to Hardware and Software Design.* McGraw-Hill, New York.

Further Information

Dorf, R. C., Ed. 1993. *The Electrical Engineering Handbook.* CRC Press, Boca Raton, FL.

IEEE Transactions on Education, published quarterly by the Institute of Electrical and Electronic Engineers.

Rigby, W. H. and Dalby, T. 1995. *Computer Interfacing: A Practical Approach to Data Acquisition and Control.* Prentice Hall, Englewood Cliffs, NJ.

131
Computer-Aided Design and Simulation

Michael D. Ciletti
University of Colorado, Colorado Springs

The size and complexity of very large-scale integrated (VLSI) circuits preclude manual design. Designers of VLSI circuits typically use specialized software tools in a workstation-based interactive environment. This chapter reviews important aspects of computer-aided circuit design and simulation and presents an introduction of the use of hardware description languages for design description and design simulation/verification of digital circuits. Digital simulation of analog circuits is also introduced.

131.1 Design Flow

The design flow of a methodology for designing VLSI circuits consists of a structured sequence of steps, beginning with design entry and culminating in the generation of a database containing geometric detail of the masks that will be used to fabricate the design. These steps can be summarized as follows:

1. Create a description of the design (design entry).
2. Establish testability of the design.
3. Verify the functionality of the design (simulation).
4. Develop a gate-level realization of the design.
5. Verify the timing specifications.
6. Place and route the design.
7. Evaluate the timing performance of the routed design.
8. Produce database for mask generation.

Multiple passes may be necessary through all or part of this flow. Other design flows are possible, and this design flow can be modified depending on the particular technology that is being used. For example, formal verification tools may be used in place of simulation, and a synthesis tool may be used to produce the gate-level realization.

Design specifications summarize the functional behavior and timing requirements of the design. They may include power and area constraints and additional information that is relevant to the task.

0-8493-1586-7/05/$0.00+$1.50

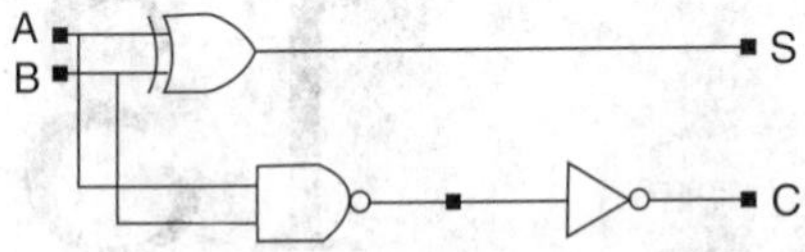

FIGURE 131.1 Schematic of a half-adder.

Design entry is the step of encapsulating a representation of the design. This representation may be in a variety of forms, such as a schematic. The testability of the design is generally addressed early in the design process to allow detection of untestable circuitry. The addition of hardware might be required, such as the insertion of a scan path into a sequential circuit. If the design is already in a form that is bound to a particular hardware realization, the timing specifications of the design can be verified by a static analysis of the paths in the circuit or by simulation. The place and route step may involve a full-custom layout of high-performance circuitry or semicustom layout of standard cells or gate arrays (field- or mask-programmable). In either case, the physical layout must be re-verified to confirm that the implementation not only realizes the desired functional behavior but also meets the externally imposed timing constraints of the design and the timing constraints imposed by storage elements (flip-flops) used in the design itself.

This chapter will focus on two steps in the overall design flow: design entry and simulation. *Design entry* is the process by which a description of the design is encapsulated in a database that serves all subsequent steps in the design flow. A designer may perform this step by drawing paper-and-pencil schematics, using a schematic entry tool, using a *hardware description language* (HDL), or selecting and interconnecting high-level macros representing hardware functional components. The two modes of entry that are of interest here are schematics and hardware description languages.

131.2 Schematic Entry

A schematic entry tool allows the designer to select and interconnect schematic symbols (icons) representing hardware components. Connections are made by graphical wires and buses representing physical signals in a circuit. The schematic entry software creates and manages a database containing the topological and incidental information created at the schematic, and this software also creates interfaces that will allow other design flow steps to access the database. For example, the information in the topological description can be used by a simulation tool to establish a database for simulating the design represented by the schematic. Figure 131.1 shows a simple schematic of a half-adder logic circuit generated by a schematic entry tool. Schematic entry tools have become popular during the past 10 years as the computational power of desktop interactive computer graphics has become capable of supporting complex design tasks at reasonable speeds.

131.3 Hardware Description Languages

Schematic entry focuses attention on structural detail of the design and is an appealing mode of entry because engineers are familiar with this traditional visual format. The objects in a schematic representation are high-level blocks and/or gate-level circuits. Structural modeling consists of interconnecting these objects to create a structure that has a desired behavior. An HDL is a computer-based programming language having special constructs and semantics to model, represent, and simulate the functional behavior and timing of digital hardware. An HDL supports structural, behavioral, and mixed descriptions of a design. Unlike schematic-based design, behavioral modeling uses HDL-based constructs and/or procedural code to describe the desired behavior without explicit binding to hardware.

Recently developed HDLs, such as Verilog [Thomas and Moorby, 1991; Sternheim, 1993] and VHDL [Navabi, 1993], provide an alternative mode of design entry by allowing the designer to create a text description of the circuit without relying on a schematic. The text itself can be generated on an ordinary terminal and is very portable. Some examples of Verilog descriptions are given as follows:

```
module Flip_flop(q,data_in,clk,rst);
input data_in,clk,rst;
output q;
always @ (posedge clk)
 begin
 if (rst == 1) q = 0;
 else q = data_in;
 end
endmodule
```

The basic element of design encapsulation in Verilog is a "module." The code in the flip-flop module declared here updates the value of the output *q* whenever the clock has a positive (rising) edge, provided that the reset line is not asserted.

Verilog also supports a *register transfer logic* (RTL) description with several built-in-language operators. The use of RTL is illustrated as follows:

```
module bitwise_or (y,A,B); // RTL model
 input [7:0] A,B;
 output [7:0] y;
 assign y = A | B;
endmodule
```

This bitwise_or module uses the built-in operator, |, to implement the "bitwise or" of the data words. The "assign" keyword effects an event-scheduling rule that updates the value of *y* whenever *A* or *B* changes. An event is said to occur whenever a signal changes value. Note that the Verilog language operators used in RTL design (such as the | operator) have implied logic. Their use simplifies the task of writing Verilog descriptions of behavior.

The text that follows declares a model of a half-adder circuit (see Figure 131.1) as a structural connection of CMOS standard cells xorf201, nanf201, and invf101.

```
module Add_half_structural (S,C,A,B,);
 output S,C;
 input A, B;
 wire C_bar;
 xorf201 G1 (S,A,B);
 nanf201 G2 (C_bar,A,B);
 invf101 G3 (C,C_bar);
endmodule
```

The description declares the name of the module, the input and output ports of the module, a wire that is used internally for a connection, and a list of instantiations of the library cells/modules. The arguments of the instantiated modules correspond directly to the physical wires that would interconnect their hardware counterparts.

HDLs allow a design to be represented abstractly or behaviorally without any binding to particular hardware elements. The fragments of text that follow show two alternative descriptions of a half-adder. The first fragment uses built-in language operators in an RTL style to implement a 4-bit-slice adder. (Here {} denotes a concatenation operator, which in this example creates a 5-bit wide data path from operations on the 4-bit buses and the *c_in* bit.) Verilog features built-in data types and operators that make this style of modeling very easy to implement.

```
module adder_4_RTL (sum, c_out, a,b,c_in);
 output [3:0] sum;
 output c_out;
 input [3:0] a,b;
```

```
  assign {c_out,sum} = a + b + c_in
endmodule
```

The next description is a fragment of procedural code that implements a behavioral model of a carry look-ahead adder [Sternheim, 1993].

```
module add_4_CLA (sum, c_out, a,b,c_in);
  output [3:0] sum;
  output c_out;
  input [3:0] a,b;
  reg [3:0] carrychain;
  wire [3:0] gen = a & b; / /bitwise and (carry generate);
  wire [3:0] prop = a ^ b;// bitwise xor (carry propagate);
  always @ a or b or c_in) // event "or"
    begin: carry_generation_block
      integer i;
      carrychain[0] = gen[0] + (prop[0] & c_in);
      for(i = 1; i <= 3; i = i + 1)
      begin
      #0 carrychain[i] = gen[i] + (prop[i] & carrychain[i-1]);
      end
    end
  wire [4:0] shiftedcarry = {carrychain, c_in};
  wire [3:0] sum = prop ^ shiftedcarry;
  wire c_out = shiftedcarry[4];
endmodule
```

A module that is modeled by procedural code has no predetermined binding to hardware. Admittedly, built-in language operators, such as +, have an implicit binding to hardware, but this binding can be deferred to later in the design flow. This allows the designer to focus on functionality rather than implementation. Both Verilog and VHDL support algorithmic description of behavior through the mechanism of procedural code, which executes serially.

CAD/CAE software vendors provide development and debug environments supporting the designer of HDL-based descriptions of a design. These may include text editors and interactive debuggers.

VHDL, the VHSIC (very high-speed integrated circuit) HDL, was created under the auspices of the Department of Defense and is now an IEEE standard (IEEE 1076-1987). Verilog was created as a proprietary language, became very popular as a widely used industry tool, and then was placed in the public domain in 1990. It is presently in the final stages of becoming an IEEE standard [IEEE 1364]. Both languages support high-level abstract descriptions of digital systems; Verilog also has built-in gate-level and switch-level functional primitives.

Simulation of HDL-Modeled Circuits

Simulators are available for simulating the behavior of digital circuit described by either Verilog or VHDL. These simulators provide a fast, efficient, visual representation of the behavior of a digital circuit. Logic simulation is usually done on event-driven simulators, which exploit the topological latency that is characteristic of digital circuits.

Simulation of a circuit that has been modeled by an HDL requires that a *design unit test bench* (DUTB) be developed to apply stimulus to the unit under test (UUT). A test bench for a nand latch (cross-coupled nand primitives) module is given in the following:

```
module DUTB_Nand_latch;
  reg preset, clear;
```

```
  wire q, qbar;
  Nand_latch(q, qbar, preset, clear); // Instantiate the UUT
  initial
  begin // Create stimulus to UUT
  $monitor ($time,,"preset =%b clear =%b q =%b qbar =%b,"
  preset, clear,q,qbar);
    #10 preset = 0; clear = 1;
    #10 preset = 1;
    #10 clear = 0;
    #10 clear = 1;
    #10 preset = 0;
    #10 $finish;
  end
endmodule
```

This DUTB applies stimuli to the inputs of the nand latch at intervals of ten simulation units. The $monitor system task effects a listing of the output shown below. (Note: In this example the nand latch has a unit delay between an event on one of its inputs and the resulting event on its output. The value *x* denotes an unknown logic value.)

```
0    preset = x   clear = x   q = x   qbar = x
10   preset = 0   clear = 1   q = x   qbar = x
11   preset = 0   clear = 1   q = 1   qbar = x
12   preset = 0   clear = 1   q = 1   qbar = 0
20   preset = 1   clear = 1   q = 1   qbar = 0
30   preset = 1   clear = 0   q = 1   qbar = 0
31   preset = 1   clear = 0   q = 1   qbar = 1
32   preset = 1   clear = 0   q = 0   qbar = 1
40   preset = 1   clear = 1   q = 0   qbar = 1
```

In addition to providing the standard output listing, a variety of tools offered by vendors of CAD/CAE software provide graphical output of simulation results.

131.4 Tradeoffs between HDLs and Schematic Entry

There are some tradeoffs that can be noted between schematic entry and HDL-based design entry. From the standpoint of support, a schematic-driven paradigm requires a color-graphic workstation (or suitably enhanced PC); language-based entry is easily done at a terminal, and an engineer can work at a remote site without requiring local support.

Editing a design that is described by an HDL can be shorter and simpler than the task of editing a design described by schematics. The task of removing, relocating, and rebinding schematic objects can be time consuming if the schematic itself is very dense and/or complex.

HDLs support a higher level of abstraction than can be described by schematics. A schematic can certainly have a symbol of any functional unit, but this must eventually be represented in terms of lower-level detail that is ultimately expressed as a structural description. An HDL can embody a behavior with no reference whatsoever to structural detail.

Schematic entry focuses the designer's attention on structural detail; a designer using an HDL can focus on structural detail, functional behavior, or a mixture of the two. HDLs support a top-down design methodology (TDM), in which a design is hierarchically decomposed in simpler, hierarchically organized functional units. In Verilog, the nested instantiation of modules is the mechanism by which hierarchical decomposition of a design is accomplished. The actual partitioning may be done according to functionality, but no restriction is implied by the language itself. In a TDM, the design is created in a top-down fashion; it is verified in a bottom-up sequence, beginning with verification of the lowest levels of the hierarchy and proceeding to verification of the integrated design. The hierarchical decomposition of a 4-bit adder is given below. The top-level module contains four instantiations of full adders, which themselves contain instantiations of half-adders and glue logic; the half-adders are defined in terms of modules declared in the cell library.

```
module Add_rca_4 (Sum, C_out, A, B, C_in);
 output Sum,C_out;
 input A,B,C_in;
 wire [3:0] Sum,A,B;
 wire C_out,C_in4,C_in3,C_in2,C_in;
 Add_full G1 (Sum[3],C_out,A[3],B[3],C_in4);
 Add_full G2 (Sum[2],C_in4,A[2],B[2],C_in3);
 Add_full G2 (Sum[1],C_in3,A[1],B[1],C_in2);
 Add_full G4 (Sum[0],C_in2,A[0],B[0],C_in);
endmodule
module Add_full (S,C_out,A,B,C_in);
 output S,C_out;
 input A,B,C_in;
 wire S1,C1,C2,C_out_bar;
 Add_half G1 (S1,C1,A,B);
 Add_half G2 (S,C2,S1,C_in);
 norf201 G3 (C_out_bar,C1,C2);
 invf101 G4 (C_out,C_out_bar);
endmodule
module Add_half(S,C,A,B);
 output S,C;
 input A,B;
 wire C_bar;
 xorf201 G1 (S,A,B);
 nanf201 G2 (C_bar,A,B);
 invf101 G3 (C,C_bar);
endmodule
```

HDLs support rapid prototyping of a design by allowing a designer to focus attention on the functionality of a design rather than its physical/structural implementation. This factor dramatically shortens the time required to create and verify a design. Shortening the design cycle allows more changes to be made in less time, thereby increasing the likelihood that a design error will be found before its effects become widespread. The ease of considering design alternatives in an HDL context can stimulate and encourage consideration of design alternatives.

Tools now exist that automatically create a schematic from the HDL description, resulting in another attractive feature of HDLs. Thus, a schematic is actually a by-product of the HDL design flow. This approach to design greatly shortens the design cycle. The designer must, however, conform to a style that ensures synthesizable results.

131.5 HDLs and Synthesis

The Verilog HDL is a key element in modern design flows that incorporate logic synthesis tools. These tools begin with a behavioral description of the functionality of the design and then create an optimal logic-level description. This description can then be mapped onto a particular technology to meet timing and area constraints.

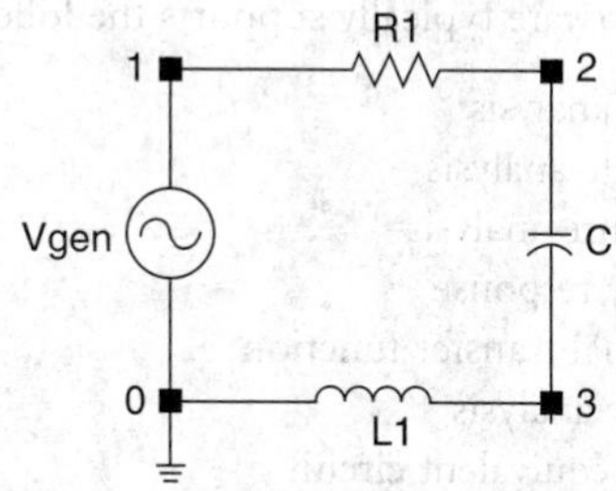

FIGURE 131.2 Low-pass filter section.

131.6 Transistor-Level Design and Simulation

Digital simulation of a circuit's analog waveforms has been popular for decades, especially since the creation of SPICE [Tuinenga, 1988]. This tool uses mathematical descriptions of active and passive circuit elements such as transistors and other components. These models provide a closer approximation of the actual analog waveforms that will be generated in the physical circuit. This additional resolution/detail comes at the price of greatly increased simulation times and memory requirements. The underlying models of the devices themselves are typically more complex than those used to support digital simulation.

Transistor-level simulators rely on numerical integration techniques, such as the Newton–Raphson method, and exploit the sparsity of the underlying matrices that result from applying Kirchhoff's laws to the possibly nonlinear circuit. In practice these simulators are used effectively to verify the timing performance of critical subcircuits, such as a detailed transient analysis of a memory design, rather than that of an entire system. The text that follows shows a fragment of SPICE code that describes the simple double-pole low-pass filter section shown in Figure 131.2.

```
Vgen 1 0 AC 1
R1 1 2 500
L1 2 3 10 mH
C1 3 0 1 uF
.AC DEC 100 100 HZ 10 Khz
.PROBE
.END
```

The first line of the code specifies an independent AC voltage source connected between node 1 and the reference node 0. This source has a level of 1. Each of the next three lines begins with a label specifying the type and identity index of a component, the indices of a pair of nodes to which the component is connected, and the value of the component in default or specified physical units. Then the AC source is specified to have a logarithmic sweep in decades from 100 Hz to 10 kHz in steps of 100 Hz. The .PROBE command is a PSpice [Tuinenga, 1988] statement that creates a database from simulation of the circuit. This database can then be examined to view various features of the response of the circuit. In this example, a Bode plot of the filter response will be created.

The following code describes a transient analysis of the same circuit that was considered earlier:

```
Vgen 1 0 pwl(0,0.1m,1 5.1m, 0)
R1 1 2 500
L1 2 3 10 mH
C1 3 0 1 uF
.TRAN 1m 10m
.PROBE
.END
```

Here, the source is programmed as a piecewise linear waveform describing a pulse that has a 0.1-ms transition from 0 to 1 and lasts for 5 ms. The .TRAN statement specifies that the transient waveform should be plotted in steps of 1 ms beginning at time = 0 (default) and ending at time 10 ms.

SPICE-like software typically supports the following analysis tasks:

- Transient analysis
- Steady state analysis
- Temperature analysis
- Frequency response
- Small-signal transfer function
- Sensitivity analysis
- Thevenin equivalent circuit
- Monte Carlo analysis
- Group delay analysis

Many variations of SPICE are available from CAD/CAE software vendors.

131.7 Conclusions

Powerful software tools exist to support circuit designers. These tools offer substantial gains in productivity in the overall engineering effort.

Defining Terms

Hardware description language (HDL) — A computer-based programming language having special constructs and semantics to model, represent, and simulate the functional behavior and timing of digital hardware.

SPICE — A software language used to create digital simulations of analog circuits. The acronym stands for *simulation program with integrated circuit emphasis.* SPICE was developed at the University of California, Berkeley; it is a public domain tool.

Verilog — An HDL that is widely used in industry to describe and simulate digital systems. It is in the final stages of becoming an IEEE standard. It supports hierarchical decomposition, switch- and gate-level structural modeling, register transfer logic/data flow modeling, and procedural modeling, including concurrent activity flows. It has built-in data types.

VHDL — An HDL that was developed under the support of the Department of Defense and is an IEEE standard. It supports hierarchical decomposition, data flow, and procedural modeling. It has user-defined data types.

References

Navabi, Z. 1993. *VHDL Analysis and Modeling of Digital Systems.* McGraw-Hill, New York.

Sternheim, E. 1993. *Digital Design and Synthesis with Verilog HDL.* Automata, San Jose, CA.

Thomas, D. E. and Moorby, P. 1991. *The Verilog Hardware Description Language.* Kluwer Academic, Boston, MA.

Tuinenga, P. W. 1988. *SPICE: A Guide to Circuit Simulation & Analysis Using PSpice.* Prentice Hall, Englewood Cliffs, NJ.

Further Information

For additional information and examples of design using the Verilog HDL, see Sternheim, *Digital Design and Synthesis with Verilog HDL.* For comprehensive information about VHDL, see the *IEEE Standard VHDL Language Reference Manual* (IEEE, New York, 1988). For additional information about recent research in computer-aided design, including simulation tools, see the *IEEE Transactions on Computer-Aided Design of Circuits and Systems.* For information on the mathematical models in SPICE-like software, see L. O. Chua and P.-M. Lin, *Computer-Aided Analysis of Electronic Circuits: Algorithms and Computational Techniques* (Prentice Hall, Englewood Cliffs, NJ, 1975).

132

Logic Analyzers

Samiha Mourad
Santa Clara University

Mary Sue Haydt
Santa Clara University

Several types of equipment are used to verify and test digital circuits. The type depends on the design itself and the stage at which it is tested. Logic analyzers are "testers" that combine general-purpose characteristics with ease of use. They allow engineers to measure digital signals in a fashion similar to oscilloscopes; the x axis represents the time, and the y axis the voltage sampled. Logic analyzers evolved from oscilloscopes in the early 1970s; in a later section, we will distinguish between the use of analyzers and the use of oscilloscopes. Logic analyzers come in various sizes and with a variety of features. They may be used to measure one signal or several signals simultaneously. To explain how they work, it is important to present some basic concepts. The following sections will first describe the nature of digital signals, and then explain what an analyzer is and how it is used in testing digital circuits.

132.1 Nature of Digital Signals

Digital signals are waveforms that assume either a low-limit V_L (logic state 0) or a high-limit V_H (logic state 1). The exact values of these two limits are technology dependent. An example of a digital signal is shown in Figure 132.1. In its ideal form it is a square wave but not necessarily periodic. Due to delays within the devices and interconnect, the rise and fall of the signal are not instantaneous. For a typical pulse, the rate of rise (or fall) is usually called the *slew rate*. The higher the slew rate is, the faster the operation of the circuit. Proper operation of digital circuits is dependent on the arrival of the waveform at various nodes of the circuit at the appropriate time.

Verifying the correctness of a digital signal involves state and timing measurement. The state is the value of the signal at any time, either state 0 or state 1. If the signal is above a certain reference V, it is logic 1; if it is below this reference, it is logic 0. This is illustrated in Figure 132.2. For the logic analyzer to reconstruct the waveform examined, measurements need to be taken at various time instants. The collection of such measurements forms a sample that is interpreted by the analyzer, displayed, and possibly stored for further analysis. This process is called **sampling** and is done at various sampling points. In a sense, sampling is like taking snapshots of the waveform. Sampling also serves in time analysis. In the remainder of this chapter, we describe how the analyzer works in both state and timing capacities, but first let us discuss sampling and its importance in logic analyzer operations.

0-8493-1586-7/05/$0.00+$1.50

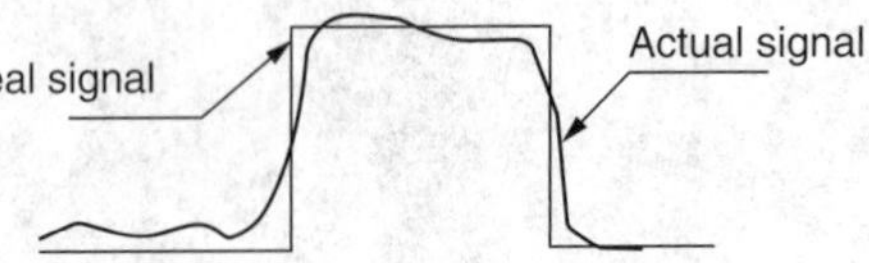

FIGURE 132.1 Digital signals: ideal vs. actual.

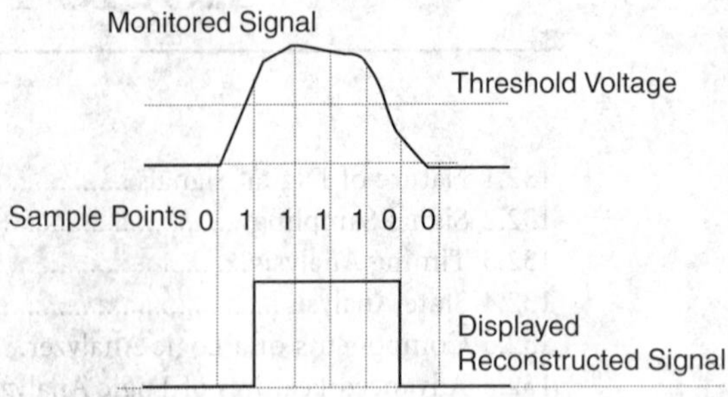

FIGURE 132.2 Sampling a waveform and reconstructing it.

132.2 Signal Sampling

Figure 132.2 shows a digital waveform, the sampling points, and the possible displays of the analyzer. The collected sample is stored in a buffer and used to reconstruct the waveform as illustrated in the figure. Sampling is a very important process in the operation of logic analyzers. If the signal is changing at 1 MHz, then it alternates between high and low every $T = 0.5$ μs. To accurately reconstruct this signal, we must sample it at least every 0.5 μs. That is, the sampling rate has to be at least double that of the rate at which the signal is changing. The size of the sample is limited to the size of the buffer. A normal buffer size contains 1 K (1024) sample points. For a sample rate of 2 megasamples per second, the buffer will hold data for 512 μs.

The logic analyzer is always sampling. When the user triggers it for display, it will show the waveform before and after the **trigger**. The triggering is done in one of two modes: level and edge. Both modes will be described later in conjunction with timing and state analysis.

132.3 Timing Analysis

Time analysis is very crucial to the performance of digital circuits. For proper functioning of a circuit, the signal must arrive at various nodes at precise times within a tolerance. Whenever the state changes from 0 to 1 or vice versa for two consecutive sampling points, the analyzer recognizes that the signal went through a transition during the period between these two points. This transition is then interpreted as having occurred at the second sampling instant, as illustrated in Figure 132.3. However, there is uncertainty about where the transition has really occurred. The first transition, 0 to 1, occurred just after sample point 1, but it is only recorded at sample point 2. Similarly, the transition from 1 to 0 occurred right before sample point 4, where it was recorded. Since $\delta_1 > \delta_2$, the recording at sample point 4 is more accurate than at point 2.

The interval of time between two sample points, the *sampling period*, denoted in the figure by Δ, is the maximum uncertainty in timing measurement. Thus, the shorter the sampling period is, the higher is the resolution. This period has to be smaller than any pulse width measured by the logic analyzer. Otherwise, the pulse may not be recorded at all, or it might be recorded as having the sampling period as its width. Both cases are shown in Figure 132.3 for the pulses b and c.

Increasing the sampling rate for a certain **buffer depth** results in a narrower window. Alternately, keeping the same window requires a *deeper buffer*. It is possible to reduce the buffer depth by storing

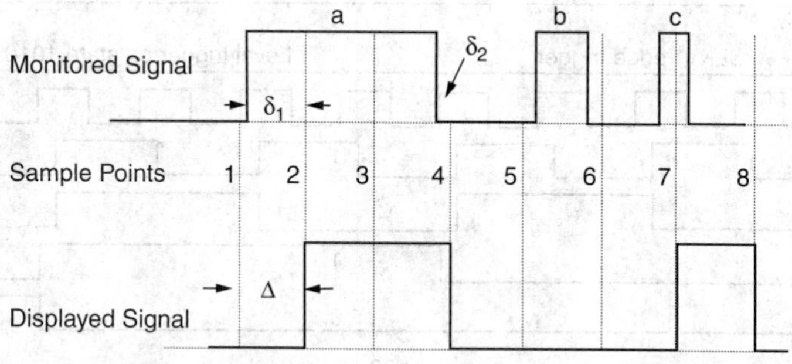

FIGURE 132.3 Sampling accuracy.

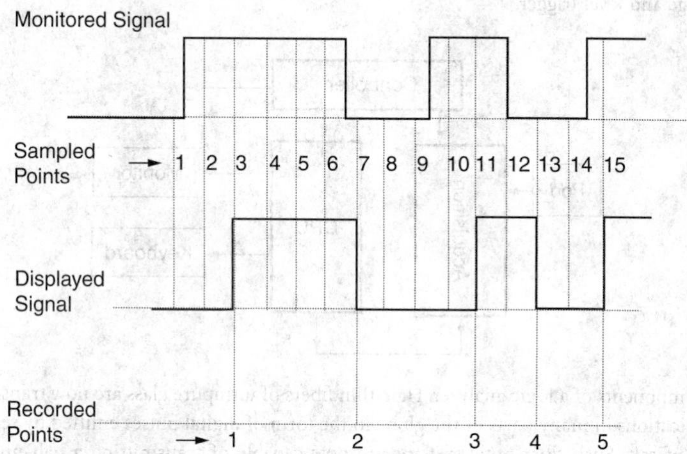

FIGURE 132.4 Transitional sampling.

only the sample points following transitions, as illustrated in Figure 132.4. Of all 15 sample points, only five are stored. Since the analyzer is sampling digital data, only the transition is relevant. This scheme is called **transitional sampling.**

Logic analyzers are not intended as parametric measuring instruments. They are used to determine timing relationship. We will consider such use of logic analyzers later in the chapter. Next, we examine the second type of analysis — state analysis.

132.4 State Analysis

In an earlier section, we used the term *state* to refer to the value of the signal on a certain node of the circuit. However, the advantage of the logic analyzer is in providing a means to determine the signal value on several nodes of the circuit. The **state** of a certain circuit is the collection of signals (0 or 1) on some lines of the circuit. The set of nodes may be the lines of a bus or the output of a counter.

The operation of the majority of digital circuits is synchronized by one signal called the *clock*. As the clock transitions from 0 to 1 (positive edge) or from 1 to 0 (negative edge), the signals propagate through the circuit and stabilize before the next clock edge. The clock is usually periodic, with duty cycle less than 50%. Thus, to observe the state of a circuit, it is convenient to use the system clock as the reference. The logic analyzer will capture the state of the circuit as the edge of the clock occurs. This mode of operation is called *edge triggering*.

Consider the example shown in Figure 132.5 for a 4-bit binary counter. The count is read on $Q_1Q_2Q_3Q_4$. These nodes form the state of the counter. Displayed also are the system clock and an asynchronous clear that returns the counter to 0 whenever asserted.

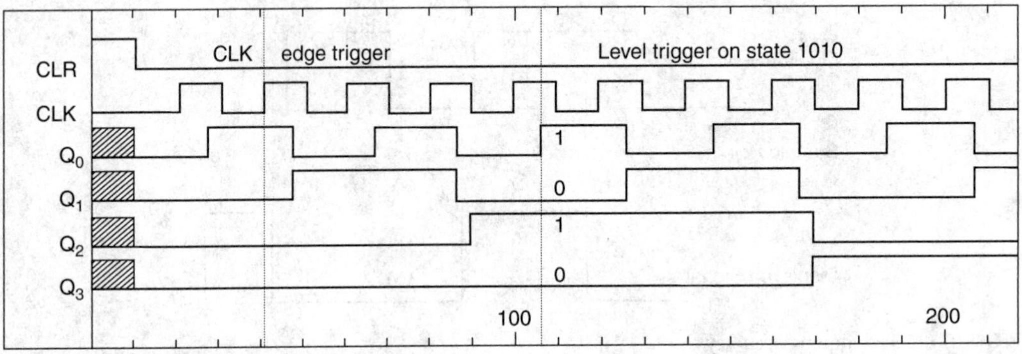

FIGURE 132.5 Edge and level trigger.

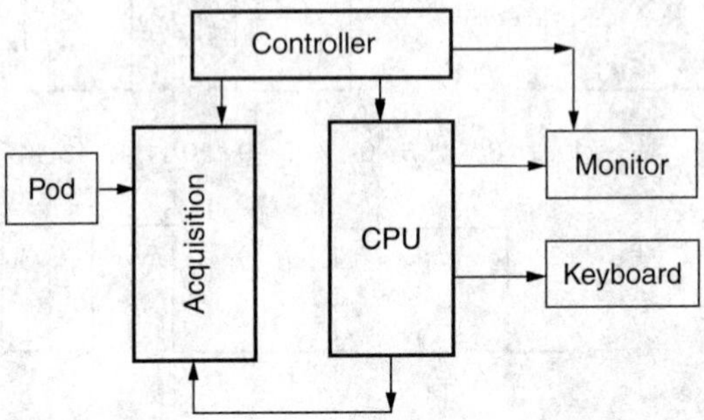

FIGURE 132.6 Components of a logic analyzer. Hair-thin fibers of ultrapure glass are now transmitting voice, data, and video communications in many parts of the globe in the form of digital pulses emitted by semiconductor lasers the size of a grain of salt. Such fiber-optic systems are now capable of transmitting a half-million simultaneous conversations. (Photo courtesy of AT&T Bell Laboratories.)

It is also possible to request the data when the circuit is in a certain state, say 1010. In such a case, a *level trigger* — specification of a level to start the capture of the data — is used. Every signal of the state — here, 4 — is measured by a channel. Besides observing the display, users may need to examine the values. For this 4-bit counter, using binary notations is manageable. It would be difficult, though, to use this representation for a 32-bit bus; instead, hexadecimal numbers would be used.

Now that we are familiar with the operation of logic analyzers, we briefly describe its components.

132.5 Components of a Logic Analyzer

A logic analyzer consists of the functional blocks shown in Figure 132.6: data acquisition block, data analyzer block that is part of the CPU, memory, display, and several ports that allow connection to a PC, printer, and other instruments. The interface between the data acquisition and the unit under test (UUT) is called the *pod*. The CPU is microprocessor based, but the data acquisition is usually an application-specific integrated circuit (ASIC). As IC technology advances the size and the cost of analyzers decreases. Portable analyzers may be used for on-site testing.

132.6 Advanced Features of Logic Analyzers

- Modern logic analyzers interface with PCs for postprocessing of the data. These data can be edited by the PC editor and also exported to a spreadsheet for further analysis.

- A logic analyzer can be bundled with a digitizing oscilloscope and thus work as three instruments in one, for timing analysis, state analysis, and parametric measurements. For intermodule analysis the data are viewed simultaneously by the three modules.
- Utilities are also available to translate state analysis files into test-pattern generation files to minimize time-consuming data entries.

132.7 Applications of Logic Analyzers

Although logic analyzers bear resemblance to digitizing oscilloscopes, their applications are not identical. Scopes are usually used when high fidelity is necessary in replicating every variation, no matter how small, in the waveform, and when timing between two or more events needs to be measured with high accuracy. The scope is suitable for parametric measurements.

Logic analyzers, on the other hand, are not as accurate in measuring time and voltages. They do have, however, many other equally important assets that have helped advance the state of the art of digital design. In the digital domain the signal is either high or low; the ripples in the waveform profile are of no consequence to proper operation of the product. It is more important to show that the actual waveforms of the circuit are as the designer expected. Logic analyzers are also useful in determining timing relationships among data lines of a bus or any group of various nodes in the UUT. This fact was illustrated earlier with the example of the 4-bit counter.

Logic analyzers are used for design verification during the development and integration cycle of digital systems. They are also useful in production testing. The product itself may be a microprocessor-based circuit or an ASIC.

The effectiveness of the logic analyzer depends on its characteristics. For example, analyzers with only a few channels are not efficient for testing a large circuit. If the maximum speed of an analyzer is under that of the UUT, it is useless for such a UUT. Also, logic analyzers must have sufficient capacity for data acquisition.

A Paradox

Logic analyzers are built of digital ICs that need to function at a speed higher than that of the UUT. They are used to develop tomorrow's ICs that are supposed to have higher performance than the products of today. To build a design tool for a customer who uses state-of-the-art devices, one must have the next-generation devices to make the tool effective for the present generation. And that is physically impossible.

Defining Terms

Buffer depth — The size of the acquisition memory.

Sampling — The process of recording signals at specific moments, called *sample points*, over a period of time.

State — The logic values of a collection of nodes in the circuit that depend on the transition in the clock of the circuit.

Transitional sampling — Storing samples indicating change of level.

Trigger — Signal entered by the user to flag the acquisition.

References

Hewlett-Packard. 1988. *Feeling Comfortable with Logic Analyzers,* Part 5954-2686. Hewlett-Packard, Palo Alto, CA.

Miner, G. F. and Comer, D. J. 1992. *Physical Data Acquisition for Digital Processing: Components, Parameters, and Specifications.* Prentice Hall, Englewood Cliffs, NJ.

Further Information

Bauer, N. 1989. Logic analyzers vs. IC design verification systems, *Test Meas. World*, p. 21.

DeSena, A. 1980. Logic analyzers, new capabilities, and challenges. *Electron. Test.*, pp. 24–27.

Editorial Staff. 1991. Designing a logic analyzer to cost and market needs. *Electron. Eng.*, pp. 41–44.

Hewlett-Packard. 1986. *Bandwidth and Sampling Rate in Digitizing Oscilloscopes*, Application note 344. Hewlett-Packard, Palo Alto, CA.

Jacob, G. 1990. Versatile trigger and data interpretation. *Eval. Eng.*, October.

Jacob, G. 1991. Faster processors place demands on logic analyzers. *Eval. Eng.*, October.

Jacob, G. 1992. Analyzers meet continuing "wider-deeper-faster" demands. *Eval. Eng.*, October.

XX

Communications and Signal Processing

133

Transforms and Fast Algorithms

Alexander D. Poularikas
University of Alabama, Huntsville

133.1 Fourier Transforms

One method used extensively calls for replacing the continuous Fourier transform by an equivalent *discrete Fourier transform* (DFT) and then evaluating the DFT using the discrete data. But evaluating a DFT with 512 samples (a small number in most cases) requires more than $1.5 \cdot 10^6$ mathematical operations. It was the development of the **fast Fourier transform (FFT)**, a computational technique that reduces the number of mathematical operations in the evaluation of the DFT to $N\log_2(N)$ (approximately $2.5 \cdot 10^4$ operations for the 512-point case mentioned above), that made DFT an extremely useful tool in almost all fields of science and engineering.

A data sequence is available only within a finite time window from $n = 0$ to $n = N - 1$. The transform is discretized for N values by taking samples at the frequencies $2\pi/NT$, where T is the time interval between sample points. Hence, we define the DFT of a sequence of N samples for $0 \le k \le N - 1$ by the relation

$$\begin{aligned} F(k\Omega) \doteq \mathrm{F}_d\{f(nT)\} &= T\sum_{n=0}^{N-1} f(nT)e^{-j2\pi nkT/NT} \\ &= T\sum_{n=0}^{N-1} f(nT)e^{-j\Omega Tnnk}, \qquad n = 0,1,\ldots,N-1 \end{aligned} \tag{133.1}$$

where

N = number of sample values
T = sampling time interval
$(N-1)T$ = signal length
$f(nT)$ = sampled form of $f(t)$ at points nT

0-8493-1586-7/05/$0.00+$1.50

$\Omega = \frac{2\pi}{T}\frac{1}{N} = \frac{\omega_s}{N}$ = the frequency sampling interval

$e^{-j\Omega T}$ = Nth principal root of unity

$j = \sqrt{-1}$

The inverse IDFT is given by

$$f(nT) \doteq \mathrm{F}_d^{-1}\{F(k\Omega)\} = \frac{1}{NT}\sum_{k=0}^{N-1} F(k\Omega)e^{j2\pi nkT/NT}$$
$$= \frac{1}{NT}\sum_{k=0}^{N-1} F(k\Omega)e^{j\Omega Tnk} \tag{133.2}$$

The sequence $f(nT)$ can be viewed as representing N consecutive samples $f(n)$ of the continuous signal, whereas the sequence $F(k\Omega)$ can be considered as representing N consecutive samples $F(k)$ in the frequency domain. Therefore, Equation (133.1) and Equation (133.2) take the compact form

$$F(k) \doteq \mathrm{F}_d\{f(n)\}$$
$$= \sum_{n=0}^{N-1} f(n)e^{-j2\pi nk/N} = \sum_{n=0}^{N-1} f(n)W_N^{nk}, \qquad k = 0,\ldots,N-1 \tag{133.3}$$

$$f(n) \doteq \mathrm{F}_d^{-1}\{F(k)\}$$
$$= \frac{1}{N}\sum_{k=0}^{N-1} F(k)e^{j2\pi nk/N} = \sum_{k=0}^{N-1} F(k)W_N^{-nk}, \qquad n = 0,\ldots,N-1 \tag{133.4}$$

where $W_N = e^{-j2\pi/N}$, and $j = \sqrt{-1}$. An important property of the DFT is that $f(n)$ and $F(k)$ are uniquely related by the transform pair in Equation (133.3) and Equation (133.4).

We observe that the functions W^{kn} are N-periodic, that is,

$$W_N^{kn} = W_N^{(k+N)n} = W_N^{k(n+N)}, \qquad k,n = 0,\pm 1,\pm 2,\ldots \tag{133.5}$$

As a consequence, the sequence $f(n)$ and $F(k)$ as defined by Equation (133.3) and Equation (133.4) are also N-periodic.

It is generally convenient to adopt the convention

$$\{f(n)\} \longleftrightarrow \{F(k)\} \tag{133.6}$$

to represent the transform pair in Equations (133.3) and (133.4).

Properties of the DFT

A detailed discussion of the properties of DFT can be found in the references cited at the end of this chapter. The following list gives a few of these properties that are of value for the development of the fast Fourier transform.

1. *Linearity*

$$\{af(n) + by(n)\} \longleftrightarrow \{aF(k) + bY(k)\} \tag{133.7a}$$

2. *Complex conjugate.* If $N/2$ is an integer and $\{f(n)\} \longleftrightarrow \{F(k)\}$, then

$$F\left(\frac{N}{2}+\ell\right) = F^*\left(\frac{N}{2}-\ell\right), \qquad \ell = 0,1,\ldots,\frac{N}{2} \tag{133.7b}$$

where $F^*(k)$ denotes the complex conjugate of $F(k)$. This identity shows the folding property of the DFT.

3. *Reversal*

$$\{f(-n)\} \longleftrightarrow \{F(-k)\} \tag{133.8}$$

4. *Time shifting*

$$\{f(n+\ell)\} \longleftrightarrow \{W^{-\ell k}F(k)\} \tag{133.9}$$

5. *Convolution of real sequences.* If

$$y(n) = \frac{1}{N}\sum_{\ell=0}^{N-1} f(\ell)h(n-\ell), \qquad n = 0,1,\ldots,N-1 \tag{133.10}$$

then

$$\{y(n)\} \longleftrightarrow \{F(k)H(k)\} \tag{133.11}$$

6. *Correlation of real sequences.* If

$$y(n) = \frac{1}{N}\sum_{\ell=0}^{N-1} f(\ell)h(n+\ell), \qquad n = 0,1,\ldots,N-1 \tag{133.12}$$

then

$$\{y(n)\} \longleftrightarrow \{F(k)H^*(k)\} \tag{133.13}$$

7. *Symmetry*

$$\left\{\frac{1}{N}F(n)\right\} \longleftrightarrow \{f(-k)\} \tag{133.14}$$

8. *Parseval's theorem*

$$\sum_{n=0}^{N-1} f^2(n) = \frac{1}{N}\sum_{k=0}^{N-1} |F(k)|^2 \tag{133.15}$$

where $|F(k)| = F(k)F^*(k)$.

Example 133.1 Verify Parseval's theorem for the sequence $\{f(n)\} = \{1, 2, -1, 3\}$.
Solution. With the help of Equation (133.3) we obtain

$$F(k)\big|_{k=0} = F(0) = \sum_{n=0}^{3} f(n)e^{-j\frac{2\pi}{4}kn}\Bigg|_{k=0}$$

$$= (1e^{-j\frac{\pi}{2}\cdot 0\cdot 0} + 2e^{-j\frac{\pi}{2}\cdot 0\cdot 1} - e^{-j\frac{\pi}{2}\cdot 0\cdot 2} + 3e^{-j\frac{\pi}{2}\cdot 0\cdot 3}) = 5$$

Similarly, we find

$$F(1) = 2 + j;\ F(2) = -5;\ F(3) = 2 - j$$

Introducing these values in Equation (133.15) we obtain

$$1^2 + 2^2 + (-1)^2 + 3^2 = \frac{1}{4}[5^2 + (2+j)(2-j) + 5^2 + (2-j)(2+j)] \text{ or } 15 = \frac{60}{4}$$

which is an identity as it should have been.

Relation between DFT and Fourier Transform

The sampled form of a continuous function $f(t)$ can be represented by the N equally spaced sampled values $f(n)$ such that

$$f(n) = f(nT),\ n = 0, 1, \ldots, N-1 \tag{133.16}$$

where T is the sampling interval. The length of the continuous function is $L = NT$, where $f(N) = f(0)$.

We denote the sampled version of $f(t)$ by $f_s(t)$, which can be represented by the expression

$$f_s(t) = \sum_{n=0}^{N-1} [Tf(n)]\delta(t - nT) \tag{133.17}$$

where $\delta(t)$ is the Dirac or impulse function.

Taking the Fourier transform of $f_s(t)$ in Equation (133.17) we obtain

$$\begin{aligned} F_s(\omega) &= T\int_{-\infty}^{\infty} \sum_{n=0}^{N-1} f(n)\delta(t - nT)e^{-j\omega t}dt \\ &= T\sum_{n=0}^{N-1} f(n)\int_{-\infty}^{\infty} \delta(t - nT)e^{-j\omega t}dt = T\sum_{n=0}^{N-1} f(n)e^{-j\omega nT} \end{aligned} \tag{133.18}$$

Equation (133.18) yields $F_s(\omega)$ for all values of ω. However, if we are interested only in values of $F_s(\omega)$ at a set of discrete equidistant points, then Equation (133.18) is expressed in the form [see also Equation (133.1)]

$$F_s(k\Omega) = T\sum_{n=0}^{N-1} f(n)e^{-jkn\Omega T}, \quad k = 0, \pm 1, \pm 2, \ldots, \pm\frac{N}{2} \tag{133.19}$$

where $\Omega = 2\pi/L = 2\pi/NT$. Therefore, comparing Equation (133.3) and Equation (133.19), we observe that we can find $F(\omega)$ from $F_s(\omega)$ using the relation

$$F(k) = F_s(\omega)\big|_{\omega=k\Omega} \tag{133.20}$$

Power, Amplitude, and Phase Spectra

If $f(t)$ represents voltage or current waveform supplying a load of 1 ohm, the left-hand side of Parseval's theorem [Equation (133.15)] represents the power dissipated in the 1-ohm resistor. Therefore, the right-hand side represents the power contributed by each harmonic of the spectrum. Thus, the DFT **power spectrum** is defined as

$$P(k) = F(k)F^*(k) = |F(k)|^2, \quad k = 0,1,\ldots,N-1 \tag{133.21}$$

For real $f(n)$, there are only $(N/2 + 1)$ independent DFT spectral points as the complex conjugate property shows [Equation (133.7)]. Hence, we write

$$P(k) = |F(k)|^2, \quad k = 0,1,\ldots,\frac{N}{2} \tag{133.22}$$

The *amplitude spectrum* is readily found from that of a power spectrum, and it is defined as

$$A(k) = |F(k)|, \quad k = 0,1,\ldots,N-1 \tag{133.23}$$

The power and amplitude spectra are invariant with respect to shifts of the data sequence $\{f(n)\}$.

The **phase spectrum** of a sequence $\{f(n)\}$ is defined as

$$\varphi_f(k) = \tan^{-1}\left[\frac{\operatorname{Im}\{F(k)\}}{\operatorname{Re}\{F(k)\}}\right], \quad k = 0,1,\ldots,N-1 \tag{133.24}$$

As in the case of the power spectrum, only $(N/2 + 1)$ of the DFT phase spectral points are independent for real $\{f(n)\}$. For a real sequence $\{f(n)\}$, the power spectrum is an *even function* about the point $k = N/2$ and the phase spectrum is an *odd function* about the point $k = N/2$.

Observations

1. The frequency spacing $\Delta\omega$ between coefficients is

$$\Delta\omega = \Omega = \frac{2\pi}{NT} = \frac{\omega_s}{N} \quad \text{or} \quad \Delta f = \frac{1}{NT} = \frac{f_s}{N} = \frac{1}{T_o} \tag{133.25}$$

2. The reciprocal of the record length defines the frequency resolution.
3. If the number of samples N are fixed and the sampling time is increased, the record length and the precision of frequency resolution is increased. When the sampling time is decreased, the opposite is true.
4. If the record length is fixed and the sampling time is decreased (N increases), the resolution stays the same and the computed accuracy of $F(n\Omega)$ increases.
5. If the record length is fixed and the sampling time is increased (N decreases), the resolution stays the same and the computed accuracy of $F(n\Omega)$ decreases.

Data Windowing

To produce more accurate frequency spectra, it is recommended that the data are weighted by a **window** function $\{w(n)\}$. Hence, the new data set will be of the form $\{f(n)w(n)\}$. The following are the most commonly used windows:

1. *Triangle (Fejer, Bartlet) window*

$$w(n)=\begin{cases}\dfrac{n}{N/2}, & n=0,1,\ldots,\dfrac{N}{2}\\ w(N-n), & n=\dfrac{N}{2},\ldots,N-1\end{cases} \tag{133.26}$$

2. $\cos^{\alpha}(x)$ *window*

$$\begin{aligned} w(n) &= \sin^2\left(\frac{n}{N}\pi\right), \quad n=0,1,\ldots,N-1, \quad \alpha=2 \\ &= 0.5\left[1-\cos\left(\frac{2n}{N}\pi\right)\right] \end{aligned} \tag{133.27}$$

   This window is also called the *raised cosine* or *Hann window.*

3. *Hamming window*

$$w(n)=0.54-0.46\cos\left(\frac{2\pi}{N}n\right), \quad n=0,1,\ldots,N-1 \tag{133.28}$$

4. *Blackman window*

$$w(n)=\sum_{m=0}^{K}(-1)^m a_m \cos\left(2\pi m\frac{n}{N}\right), \quad n=0,1,\ldots,N-1, \quad K\le\frac{N}{2} \tag{133.29}$$

   For $K = 2$, $a_0 = 0.42$, $a_1 = 0.50$, and $a_2 = 0.08$.

5. *Blackman-Harris window.* Harris used a gradient search technique to find three- and four-term expansions of Equation (133.29) that either minimized the maximum side-lobe level for fixed main-lobe width or traded main-lobe width for minimum side-lobe level (see Table 133.1).

6. *Centered Gaussian window*

$$w(n)=\exp\left[-\frac{1}{2}\alpha\left(\frac{n}{N/2}\right)^2\right], \quad 0\le|n|\le\frac{N}{2}, \quad \alpha=2,3,\ldots \tag{133.30}$$

   As α increases, the main lobe of the frequency spectrum becomes broader and the side-lobe peaks become lower.

TABLE 133.1 Blackman–Harris Window Parameters

Number of Terms in Equation (133.29)	Parameter Values			
	3	3	4	4
Minimum side lobe (dB)	−70.83	−62.05	−92	−74.39
Parameter				
a_0	0.42323	0.44959	0.35875	0.40217
a_1	0.49755	0.49364	0.48829	0.49703
a_2	0.07922	0.05677	0.14128	0.09892
a_3	—	—	0.01168	0.00188

7. *Centered Kaiser-Bessel window*

$$w(n) = I_o\left[\pi\alpha\sqrt{1.0-\left(\frac{n}{N/2}\right)^2}\right] / I_o[\pi\alpha], \quad 0 \le |n| \le \frac{N}{2} \tag{133.31}$$

where

$$\begin{aligned} I_o(x) &= \text{zero-order modified Bessel function} \\ &= \sum_{k=0}^{\infty}\left(\frac{(x/2)^k}{k!}\right)^2 \\ k! &= 1\times 2\times 3\times\cdots\times k \\ \alpha &= 2, 2.5, 3 \quad \text{(typical values)} \end{aligned} \tag{133.32}$$

Fast Fourier Transform

One of the approaches to speed the computation of the DFT of a sequence is the *decimation-in-time method.* This approach involves breaking the *N*-point transform into two $(N/2)$-point transforms, then breaking each $(N/2)$-point transform into two $(N/4)$-point transforms, and continuing the above process until the two-point transform is obtained. We start with the DFT expression and factor it into two DFTs of length $N/2$:

$$\begin{aligned} F(k) &= \sum_{n=0}^{N-2} f(n)W_N^{kn} \quad n \text{ even} \\ &+ \sum_{n=1}^{N-1} f(n)W_N^{kn} \quad n \text{ odd} \end{aligned} \tag{133.33}$$

Letting $n = 2m$ in the first sum and $n = 2m + 1$ in the second, Equation (133.33) becomes

$$F(k) = \sum_{m=0}^{(N/2)-1} f(2m)W_N^{2mk} + \sum_{m=0}^{(N/2)-1} f(2m+1)W_N^{(2m+1)k} \tag{133.34}$$

However, because of the following identities,

$$W_N^{2mk} = (W_N^2)^{mk} = e^{-\frac{2\pi}{N}2mk} = e^{-j\frac{2\pi}{N/2}mk} = W_{N/2}^{mk} \tag{133.35}$$

and the substitution $f(2m) = f_1(m)$ and $f(2m+1) = f_2(m)$, $m = 0, 1, \ldots, N/2 - 1$, Equation (133.33) takes the form

$$\begin{aligned} F(k) &= \sum_{m=0}^{(N/2)-1} f_1(m)W_{N/2}^{mk} \qquad (N/2) = \text{point DFT of even indexed sequence} \\ &+ W_N^k \sum_{m=0}^{(N/2)-1} f_2(m)W_{N/2}^{mk} \quad (N/2) = \text{point DFT of odd indexed sequence} \\ &k = 0, \ldots, N/2-1 \end{aligned} \tag{133.36}$$

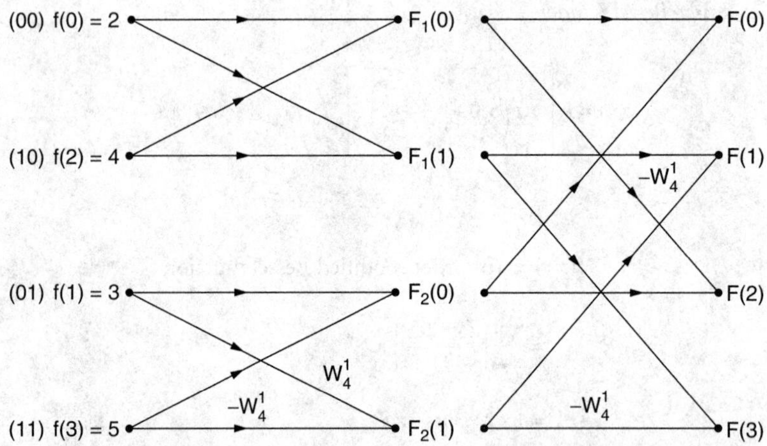

FIGURE 133.1 Illustration of Example 133.1. (*Source:* Dorf, R. C., Ed. 1993. *The Electrical Engineering Handbook*. CRC Press, Boca Raton, FL.)

We can also write Equation (133.36) in the form

$$F(k)=F_1(k)+W_N^kF_2(k), \qquad k=0,1,\ldots,N/2-1$$
$$F\left(k+\frac{N}{2}\right)=F_1(k)+W_N^{k+N/2}F_2(k) \tag{133.37}$$
$$=F_1(k)-W_N^kF_2(k), \qquad k=0,1,\ldots,N/2-1$$

where $W_N^{k+N/2}=-W_N^k$ and $W_{N/2}^{m(k+N/2)}=W_{N/2}^{mk}$. Since the DFT is periodic, $F_1(k)=F_1(k+N/2)$ and $F_2(k)=F_2(k+N/2)$.

Next, we apply the same procedure to each $N/2$ sample, where $f_{11}(m)=f_1(2m)$ and $f_{21}(m)=f_2(2m+1)$, $m=0,1,\ldots,(N/4)-1$. Hence,

$$F_1(k)=\sum_{m=0}^{(N/4)-1}f_{11}(m)W_{N/4}^{mk}+W_N^{2k}\sum_{m=0}^{(N/4)-1}f_{21}(m)W_{N/4}^{mk}, \qquad k=0,1,\ldots,\frac{N}{4}-1 \tag{133.38}$$

or

$$F_1(k)=F_{11}(k)+W_N^{2k}F_{21}(k)$$
$$F_1\left(k+\frac{N}{4}\right)=F_{11}(k)-W_N^{2k}F_{21}(k), \qquad k=0,1,\ldots,\frac{N}{4}-1 \tag{133.39}$$

Therefore, each of the sequences f_1 and f_2 have been split into two DFTs of length $N/4$.

Example 133.2 To find the EFT of the sequence {2, 3, 4, 5} we first bit-reverse the elements from position {00, 01, 10, 11} to position {00, 10, 01, 11}. The new sequence is {2, 4, 3, 5} (see also Figure 133.1). Using Equation (133.36) and Equation (133.37) we obtain

$$F_1(0) = \sum_{m=0}^{1} f_1(m)W_2^{m\cdot 0} = f_1(0)W_2^0 + f_1(1)W_2^0 = f(0)\cdot 1 + f(2)\cdot 1$$

$$F_1(1) = \sum_{m=0}^{1} f_1(m)W_2^{m\cdot 1} = f_1(0)W_2^{0\cdot 1} + f_1(1)W_2^1 = f(0) + f(2)(-j)$$

$$F_2(0) = W_4^0 \sum_{m=0}^{1} f_2(m)W_2^{m\cdot 0} = f_2(0)W_2^0 + f_2(1)W_2^0 = f(1) + f(3)$$

$$F_2(1) = W_4^1 \sum_{m=0}^{1} f_2(m)W_2^{m\cdot 1} = W_4^1[f(1)W_2^0 + f(3)W_2^1]$$

$$= W_4^1 f(1) - W_4^1 f(3)$$

From Equation (133.37) the output is $F(0) = F_1(0) + W_4^0 F_2(0)$, $F(1) = F_1(1) + W_4^1 F_2(1)$, $F(2) = F_1(0) - W_4^0 F_2(0)$, and $F(3) = F_1(1) - W_4^1 F_2(1)$ Figure 133.2 shows an eight-point decimation-in-time FFT. (See also Table 133.2.)

Computation of the Inverse DFT

To find the inverse of FFT using an FFT algorithm, we use the relation

$$f(n) = [\mathrm{FFT}(F^*(k))]^*/N \tag{133.40}$$

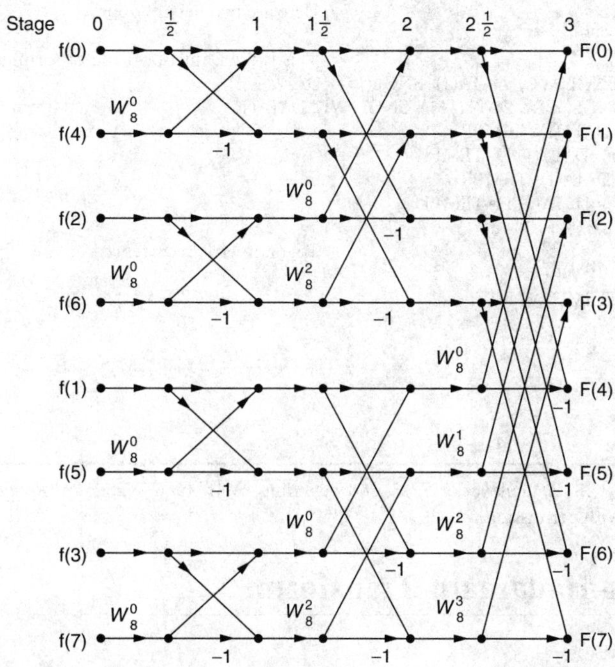

FIGURE 133.2 An eight-point decimation-in-time fast Fourier transform.

TABLE 133.2 Fast Fourier Transform Subroutine

```
SUBROUTINE FOUR1 (DATA, NN, ISIGN)
   Replaces DATA by its discrete Fourier transform, if ISIGN is input as 1; or replaces DATA by NN times its inverse discrete
   Fourier transform, if ISIGN is input as −1. DATA is a complex array of length NN or equivalently, a real array of length
   2*NN. NN must be an integer power of 2.
REAL*8 WR,WI,WPR, WPI,WTEMP,THETA              Double precision for the trigonometric recurrences.
DIMENSION DATA(2*NN)
N=2*NN
J=1
DO 11 I – 1, N, 2                              This is the bit-reversal section of the routine.
   IF (J.GT.I) THEN
      TEMPR=DATA(J)                            Exchange the two complex numbers.
      TEMPI=DATA(J+1)
      DATA(J)=DATA(I)
      DATA(J+1)=DATA(I+1)
      DATA(I)=TEMPR
      DATA(I+1)=TEMPI
   ENDIF
   M=N/2
1  IF ((M.GE.2).AND. (J.GT.M)) THEN
      J=J–M
      M=M/2
   GO TO 1
   ENDIF
   J=J+M
   11 CONTINUE
MMAX=2                                         Here begins the Danielson-Lanczos section of the routine.
2  IF (N.GT.MMAX) THEN                         Outer loop executed log2 NN times.
   ISTEP=2*MMAX
   THETA=6.28318530717959D0/(ISIGN*MMAX)       Initialize for trigonometric recurrence.
   WPR=–2.D0*DSIN(0.5D0*THETA)**2
   WPI=DSIN(THETA)
   WR=1.D0
   WI=0.D0
   D0 13 M=1,MMAX,2                            Here are two nested loops.
      D0 12 I=M, N, ISTEP
            J=I+MMAX                           This is the Danielson-Lanczos formula:
            TEMPR=SNGL(WR)*DATA(J)–SNGL(WI)*DATA(J+1)
            TEMPI=SNGL(WR)*DATA(J+1)+SNGL(WI)*DATA(J)
            DATA(J)=DATA(I)–TEMPR
            DATA(J+1)=DATA(I+1)–TEMPI
            DATA(I)=DATA(I)+TEMPR
            DATA(I+1)=DATA(I+1)+TEMPI
         12 CONTINUE
      WTEMP=WR                                 Trigonometric recurrence.
      WR=WR*WPR–WI*WPI+WR
      WI=WI*WPR+WTEMP*WPI+WI
      13 CONTINUE
   MMAX=ISTEP
GO TO 2
ENDIF
RETURN
END
```

Source: W. H., Flannery, B. P., Teukolosky, S. A., and Vetterling, W. T. 1986. *Numerical Recipes*. Cambridge University Press, Cambridge, UK. With permission.

133.2 Walsh–Hadamard Transform

Walsh Functions

In 1923, Walsh developed a complete orthogonal set of rectangular functions known as Walsh functions. The first eight functions are shown in Figure 133.3. These signals have been widely used to perform

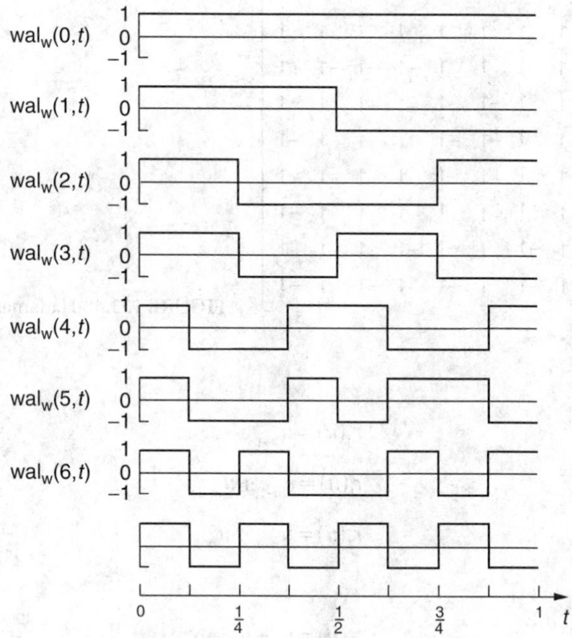

FIGURE 133.3 Walsh functions.

nonsinusoidal orthogonal transforms in various digital-signal processing applications since they can essentially be computed using addition and subtraction only.

Every signal $f(t)$ absolutely integrable in $0 \le t \le 1$ can be equated in a series of the form

$$f(t) = \sum_{k=0}^{\infty} d_k \mathrm{Wal}_w(k,t) \tag{133.41}$$

where

$$d_k = \int_0^1 f(t)\mathrm{Wal}_w(k,t)dt, \qquad k = 0,1,2,\ldots \tag{133.42}$$

The above series converges uniformly to $f(t)$ if it is continuous in $0 \le t \le 1$ and converges in the mean where $f(t)$ is discontinuous.

The Walsh–Hadamard transforms are analogous to the discrete Fourier transform. The basis functions are sampled Walsh functions, which can be expressed in terms of the Hadamard matrices $\underline{H}_w(n)$. An $\underline{H}_w(3)$ is shown in Figure 133.4, where $n = \log_2 N$.

Let u_i and v_i be the ith bits in the binary representation of the integers u and v, respectively; then

$$(u)decimal = (u_{n-1}, u_{n-2}, \ldots, u_1 u_0)binary$$

$$(v)decimal = (v_{n-1}, v_{n-2}, \ldots, v_1 v_0)binary$$

The Walsh-ordering Hadamard matrix elements are given by

$$h_{uv}^{(w)} = (-1)^{\sum_{i=0}^{n-2} r_i(u) v_i}, \qquad u,v = 0,1,\ldots,N-1 \tag{133.43}$$

$$\underline{H}_w(3) = \begin{bmatrix} 1 & 1 & 1 & 1 & 1 & 1 & 1 & 1 \\ 1 & 1 & 1 & 1 & -1 & -1 & -1 & -1 \\ 1 & 1 & -1 & -1 & -1 & -1 & 1 & 1 \\ 1 & 1 & -1 & -1 & 1 & 1 & -1 & -1 \\ 1 & -1 & -1 & 1 & 1 & -1 & -1 & 1 \\ 1 & -1 & -1 & 1 & -1 & 1 & 1 & -1 \\ 1 & -1 & 1 & -1 & -1 & 1 & -1 & 1 \\ 1 & -1 & 1 & -1 & 1 & -1 & 1 & -1 \end{bmatrix}$$

FIGURE 133.4 Hadamard matrix.

where

$$\begin{aligned} r_0(u) &= u_{n-1} \\ r_1(u) &= u_{n-1} + u_{n-2} \\ r_2(u) &= u_{n-2} + u_{n-3} \\ &\vdots \\ r_{n-1}(u) &= u_1 + u_0 \end{aligned}$$

The Hadamard matrices have the following properties:

1. $\underline{H}_w(k)$ is a symmetric matrix:

$$\underline{H}_w(k) = \underline{H}_w(k)^T \tag{133.44}$$

T stands for transpose.

2. $\underline{H}_w(k)$ are orthogonal:

$$\underline{H}_w(k)\underline{H}_w(k)^T = 2^k \underline{I}(k) \tag{133.45}$$

where $\underline{I}(k)$ is a $(2^k \times 2^k)$ identity matrix.

3. The inverse of $\underline{H}_w(k)$ is proportional to itself:

$$[\underline{H}_w(k)]^{-1} = (1/2^k)\underline{H}_w(k) \tag{133.46}$$

where $[\underline{H}_w(k)]^{-1}$ defines the inverse of $\underline{H}w(k)$.

Walsh-Ordered Walsh–Hadamard Transform (WHT_w)

The WHT_w of the data sequence $\{x(n)\} = \{x(0), x(1), \ldots, x(N-1)\}$ is defined by

$$\underline{X}(n) = \tfrac{1}{N}\underline{H}_w(n)\underline{x}(n) \tag{133.47}$$

where $\underline{X}(n)$ is the kth WHT_w coefficient and

$$\underline{X}(n)^T = [X(0), X(1), \ldots, X(N-1)] \tag{133.48}$$

Since $\underline{H}_w(n)$ is orthogonal and symmetric, we find that the inverse Walsh-ordered Hadamard transform $IWHT_w$ is given by

$$x(n) = H_w(n)X(n) \tag{133.49}$$

Example 133.3 Let $\{x(n)\} = \{1, 2, 2, 1\}$. To evaluate $\underline{X}(n)$ for $k = 0, 1, 2, 3$ we use Equation (133.47). Hence,

$$\begin{bmatrix} X(0) \\ X(1) \\ X(2) \\ X(3) \end{bmatrix} = \frac{1}{4}\begin{bmatrix} 1 & 1 & 1 & 1 \\ 1 & 1 & -1 & -1 \\ 1 & -1 & -1 & 1 \\ 1 & -1 & 1 & -1 \end{bmatrix}\begin{bmatrix} 1 \\ 2 \\ 2 \\ 1 \end{bmatrix} = \begin{bmatrix} 6/4 \\ 0 \\ -2/4 \\ 0 \end{bmatrix}$$

From the above example, we observe that N^2 additions and/or subtractions are required to compute the WHT_w coefficients $\underline{X}(n)$, $n = 0, 1, \ldots, N-1$.

Fast Walsh-Ordered Walsh–Hadamard Transform (FWHT_w)

Manz [1972] introduced an FWHT_w that has the following steps:

1. Bit-reverse the input sequence and order it in ascending index. For example, if $\{x(n)\} = \{x(1), x(2), x(3)\} = \{1, 2, 3\}$, the sequence becomes $\{\hat{x}(n)\} = \{1, 3, 2\} = \{\hat{x}(0), \hat{x}(1), \hat{x}(2)\}$.
2. Define a reversal, which is best illustrated by the simple case shown in Figure 133.5.

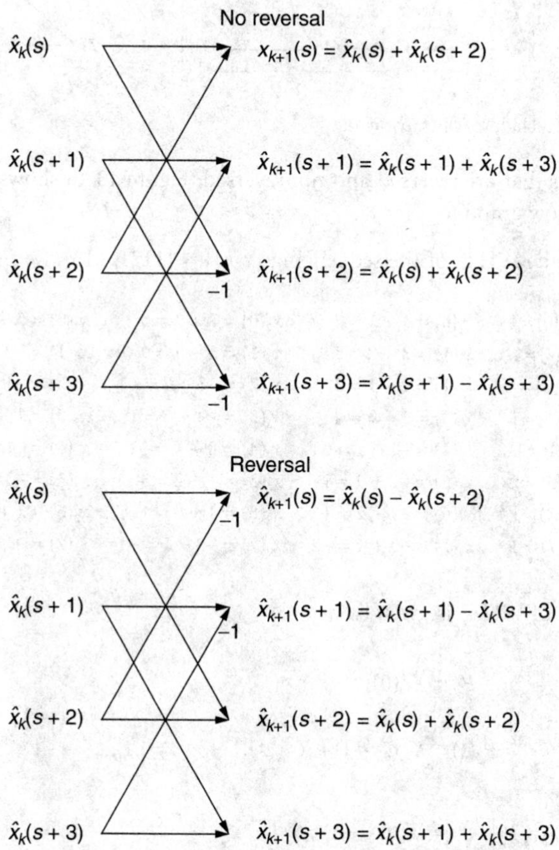

FIGURE 133.5 Reversal and nonreversal steps.

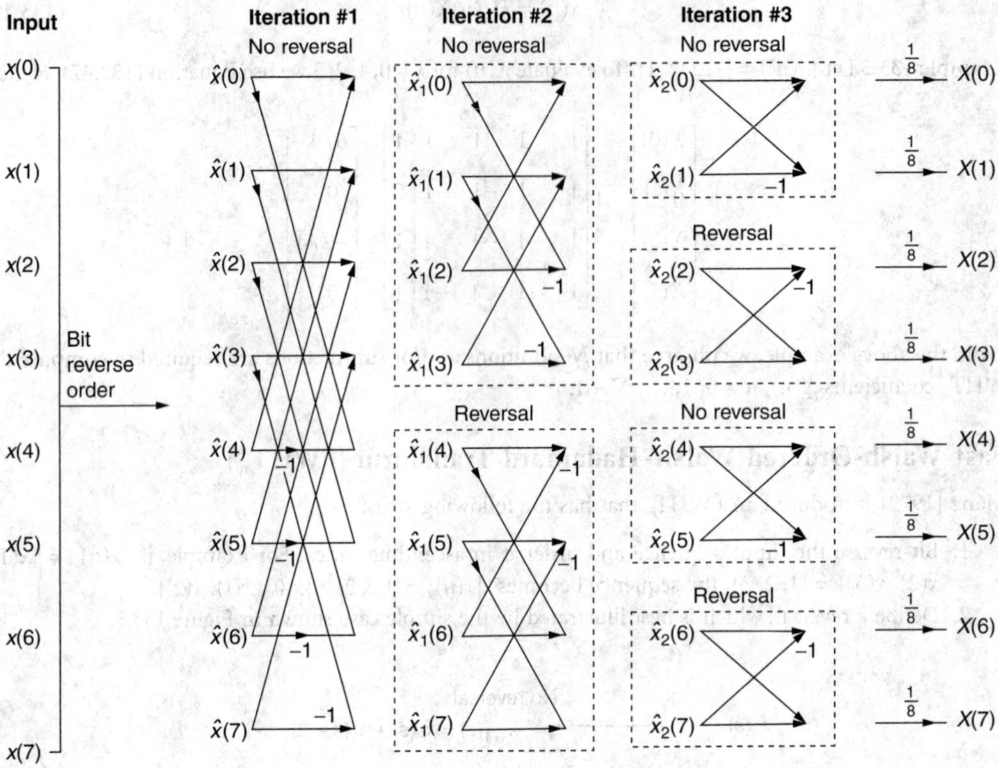

FIGURE 133.6 Fast Walsh–Hadamard transform.

3. Define the blocks that are reversed and nonreversed. Figure 133.6 shows the positions of blocks for an $FWHT_w$ flow graph for $N = 8$.

Example 133.4 To find the WHT_w of the sequencing $\{x(n)\} = \{11213213\}$, we proceed as follows:

$x(0) = 1$	$\hat{x}(0)[000] = x(0) = 1$	$x_1(0) = 1 + 1 = 2$	$x_2(0) = 2 + 3 = 5$	$x_3(0) = 5 + 9 = 14$	$\frac{1}{8}X(0) = \frac{14}{8}$
$x(1) = 1$	$\hat{x}(1)[100] = x(4) = 1$	$x_1(1) = 3 + 2 = 5$	$x_2(1) = 5 + 4 = 9$	$x_3(1) = -9 + 5 = -4$	$\frac{1}{8}X(1) = -\frac{4}{8}$
$x(2) = 2$	$\hat{x}(2)[010] = x(2) = 2$	$x_1(2) = 2 + 1 = 3$	$x_2(2) = -3 + 2 = -1$	$x_3(2) = -1 - 1 = -2$	$\frac{1}{8}X(2) = -\frac{2}{8}$
$x(3) = 1$	$\hat{x}(3)[110] = x(6) = 1$	$x_1(3) = 1 + 3 = 4$	$x_2(3) = -4 + 5 = 1$	$x_3(3) = 1 - 1 = 0$	$\frac{1}{8}X(3) = \frac{0}{8}$
$x(4) = 3$	$\hat{x}(4)[001] = x(1) = 1$	$x_1(4) = -1 + 1 = 0$	$x_2(4) = 0 - 1 = -1$	$x_3(4) = -1 + 3 = 2$	$\frac{1}{8}X(4) = \frac{2}{8}$
$x(5) = 2$	$\hat{x}(5)[101] = x(5) = 2$	$x_1(5) = -2 + 3 = 1$	$x_2(5) = 1 + 2 = 3$	$x_3(5) = -3 - 1 = -4$	$\frac{1}{8}X(5) = -\frac{4}{8}$
$x(6) = 1$	$\hat{x}(6)[011] = x(3) = 1$	$x_1(6) = -1 + 2 = 1$	$x_2(6) = 1 + 0 = 1$	$x_3(6) = 1 + 1 = 2$	$\frac{1}{8}X(6) = \frac{2}{8}$
$x(7) = 3$	$\hat{x}(7)[111] = x(7) = 3$	$x_1(7) = -3 + 1 = -2$	$x_2(7) = -2 + 1 = -1$	$x_3(7) = -1 + 1 = 0$	$\frac{1}{8}X(7) = \frac{0}{8}$

The power spectrum of WHT_w is given by

$$P_w = X^2(0)$$

$$P_w(s) = X^2(2s-1) + X^2(2s), \qquad s = 1, 2, \ldots, \frac{N}{2} - 1$$

$$P_w\left(\frac{N}{2}\right) = X^2(N-1)$$

Defining Terms

Fast Fourier transform (FFT) — A computational technique that reduced the number of mathematical operations in the evaluation of the discrete Fourier transform (DFT) to $N \log_2(N)$.

Phase spectrum — All phases associated with the spectrum harmonics constitute the phase spectrum.

Power spectrum — A power contributed by each harmonic of the spectrum.

Window — Any appropriate function that multiplies that data with the intent of minimizing the distortion of the Fourier spectra.

References

Ahmed, A. and Rao, K. R. 1975. *Orthogonal Transforms for Digital Signal Processing*. Springer-Verlag, New York.

Blahut, E. R. 1987. *Fast Algorithms for Digital Signal Processing*. Addison Wesley, Reading, MA.

Bringham, E. O. 1974. *The Fast Fourier Transform*. Prentice Hall, Englewood Cliffs, NJ.

Dorf, R. C., Ed. 1993. *The Electrical Engineering Handbook*. CRC Press, Boca Raton, FL.

Elliot, F. D. 1982. *Fast Transforms, Algorithms, Analysis, Application*. Academic Press, New York.

Harmuth, H. F. 1969. *Transmission of Information by Orthogonal Functions*. Springer-Verlag, New York.

Manz, J. W. 1972. A sequence-ordered fast Walsh transform. *IEEE Trans. Audio Electroacoustics*, AU-20:204–205.

Nussbaumer, H. J. 1982. *Fast Fourier Transform and Convolution Algorithms*. Springer-Verlag, New York.

Poularikas, A. D. and Seely, S. 1993. *Signals and Systems,* 2nd ed. Krieger, Melbourne, FL.

Press, W. H., Flannery, B. P., Teukolosky, S. A., and Vetterling, W. T. 1986. *Numerical Recipes*. Cambridge University Press, Cambridge, UK.

Further Information

A historical overview of the fast Fourier transform can be found in the following article: Cooley, J. W., Lewis, P. A. W., and Welch, P. D. 1967. Historical notes on the fast Fourier transform. *IEEE Trans. Audio Electroacoustics,* AU-15:76–79.

Fast algorithms appear frequently in the monthly magazine *Signal Processing*, which is published by the Institute of Electrical and Electronics Engineering. For subscriptions or ordering, contact: IEEE Service Center, 445 Hoes Lane, P.O. Box 1331, Piscataway, NJ 08855-1931.

134
Digital Filters

Bruce W. Bomar
University of Tennessee Space Institute

L. Montgomery Smith
University of Tennessee Space Institute

Digital filtering is concerned with the manipulation of **discrete data sequences** to remove noise or extract certain desired information. Discrete or digital signals are somewhat intuitive to human beings despite their rarity in nature. That is, people tend to make and tabulate measurements of physical entities in sets of data values with limited precision, while the entities being measured vary continuously both in their functional dependence and their value.

Although an infinite number of numerical manipulations can be applied to discrete data (e.g., finding the mean value, forming a histogram), the primary objective of digital filtering is to form a discrete output sequence $y(n)$ from a discrete input sequence $x(n)$. In some manner, each output data sample is computed from the input data sequence — not just from any one sample, but from many, in fact, possibly from all the input samples.

The reader may already be familiar with a few processing schemes. For example, computing a moving average is sometimes used as a means of reducing measurement uncertainty or noise in data. In a three-point symmetric scheme the output sequence is found by

$$y(n) = \frac{x(n-1) + x(n) + x(n+1)}{3}$$

Note that the output sequence in this case is a sum of products of the input sample values with certain coefficients. (In this case the coefficients all have the same value of 1/3.)

To make some headway in analyzing digital filters, it is necessary to restrict in two ways the types of filters being considered. First, a given filter is restricted to being linear. This means that the response of the filter to a sum of inputs is the sum of the outputs corresponding to the inputs individually and that, if the input to the filter is scaled by a multiplicative factor, the output is scaled by the same factor.

The second restriction placed on the types of filters being analyzed is that they be shift-invariant. This property states that for any given input $x(n)$ producing output $y(n)$, the response of the filter to the shifted input $x(n - n_0)$ produces the same output sequence of values shifted by the same amount, namely, $y(n - n_0)$. Note that the moving average processing scheme satisfies these restrictions. However, a much larger class of processing algorithms also falls into the category of linear shift-invariant filters.

The output of a digital filter to any input can be determined from a formula known as the *convolution sum*. To derive this relation, consider first the unit impulse sequence defined as

0-8493-1586-7/05/$0.00+$1.50

$$\delta(n) = \begin{cases} 1 & \text{for } n = 0 \\ 0 & \text{otherwise} \end{cases}$$

By use of this special sequence any discrete sequence can be written

$$x(n) = \sum_{m=-\infty}^{\infty} x(m)\delta(n-m)$$

Suppose this sequence were input to a linear shift-invariant digital filter. Since the filter is linear, the output will be a linear combination of the outputs resulting from shifted unit impulse inputs. Since the filter is shift-invariant, the outputs will also be shifted by the same amounts. Thus, if the output of the filter in response to a single unit impulse at $n = 0$ is the sequence $h(n)$, the output of the filter to the input $x(n)$ will be the linear combination of shifted $h(n)$ sequences weighted with the coefficients of the $x(n)$ sequence:

$$y(n) = \sum_{m=-\infty}^{\infty} x(m)h(n-m)$$

By a change of dummy index of summation, it is easily shown that this can also be written as

$$y(n) = \sum_{m=-\infty}^{\infty} h(m)x(n-m)$$

These equations are known as the **convolution summation** and describe the output of the filter to any arbitrary input. Thus, in theory, the response of a given filter to any input can be computed from knowledge of the sequence $h(n)$, which is referred to as the *impulse response* of the filter. It is a sequence that is of extreme importance in digital filtering, since it completely describes the filter.

A digital filter is often conveniently described in terms of its frequency characteristics, which are given by the Fourier transform of its impulse response. The impulse response and frequency response make up the Fourier transform pair,

$$H(e^{j\omega}) = \sum_{n=-\infty}^{\infty} h(n)e^{-j\omega n}, \qquad -\pi \le \omega \le \pi$$

$$h(n) = \frac{1}{2\pi}\int_{-\pi}^{\pi} H(e^{j\omega})e^{j\omega n}d\omega, \qquad -\infty \le n \le \infty$$

where ω is the digital radian frequency. The discrete-time Fourier transform is expressed as a function of $e^{j\omega}$ rather than simply as a function of ω to distinguish the discrete-time transform from its continuous-time counterpart. The radian frequency ranging from $\omega = 0$ to $\omega = \pi$ corresponds to the "real-world" frequency range from 0 to $F_s/2$, where F_s is the sample rate of the discrete-time data. For example, if the sample rate of data into a digital filter is 2 kHz, then the frequency 100 Hz corresponds to $\omega = 0.1\ \pi$.

Just as in the analysis of continuous-time systems, the operation of convolution is equivalent to the product of Fourier transforms in the frequency domain, $Y(e^{j\omega}) = H(e^{j\omega})X(e^{j\omega})$. Therefore, $H(e^{j\omega})$ may be treated as the *transfer function* of the digital filter since it relates the input Fourier transform to the Fourier transform of the output, specifying how each frequency component in the filter input is altered by the filter.

Closely related to the Fourier transform of $h(n)$ is the z transform defined by

$$H(z) = \sum_{n=-\infty}^{\infty} h(n) z^{-n}$$

$H(z)$ is referred to as the z-domain transfer function of the filter. The Fourier transform is then the z transform evaluated on the unit circle in the z plane ($z = e^{j\omega}$). An important property of the z transform is that $z^{-1}H(z)$ corresponds to $h(n-1)$, so z^{-1} represents a one-sample delay, termed a *unit delay*.

In this chapter, the focus will be restricted to **filter design** and **filter implementation** of frequency-selective filters. These filters are intended to pass frequency components of the input sequence in a given band of the spectrum while blocking the rest. Typical frequency-selective filter types are low-pass, high-pass, band-pass, and band-reject filters. Other special-purpose filters exist; however, their design is an advanced topic that will not be addressed here. In addition, special attention is given to causal filters for which the impulse response is identically zero for negative n and which can thus be implemented in real time.

134.1 Finite Impulse Response Filter Design

The objective of **finite impulse response (FIR) digital filter** design is to determine $N + 1$ coefficients

$$h(0), h(1), \ldots, h(N)$$

so that the transfer function $H(e^{j\omega})$ approximates a desired frequency characteristic $H_d(e^{j\omega})$. All other impulse response coefficients are zero. An important property of FIR filters for practical applications is that they can be designed to be *linear phase*; that is, the transfer function has the form

$$H(e^{j\omega}) = A(e^{j\omega})e^{-j\omega N/2}$$

where the amplitude $A(e^{j\omega})$ is a real function of frequency. The desired transfer function can be similarly written

$$H_d(e^{j\omega}) = A_d(e^{j\omega})e^{-j\omega N/2}$$

where $A_d(e^{j\omega})$ describes the amplitude of the desired frequency-selective characteristics. For example, the amplitude frequency characteristics of an ideal low-pass filter are given by

$$A_d(e^{j\omega}) = \begin{cases} 1 & \text{for } |\omega| \le \omega_c \\ 0 & \text{otherwise} \end{cases}$$

where ω_c is the *cutoff frequency* of the filter.

A linear phase characteristic ensures that a filter has a constant delay independent of frequency. Thus, all frequency components in the signal are delayed by the same amount, and the only signal distortion introduced is that imposed by the filter's frequency-selective characteristics.

Since an FIR filter can only approximate a desired frequency-selective characteristic, some measures of the accuracy of approximation are needed to describe the quality of the design. These are the *passband ripple* δ_p, the *stopband attenuation* δ_s, and the *transition bandwidth* $\Delta\omega$. These quantities are illustrated in Figure 134.1 for a prototype low-pass filter. The passband ripple gives the maximum deviation from the desired amplitude (typically unity) in the region where the input signal spectral components are desired to be passed unattenuated. The stopband attenuation gives the maximum deviation from zero in the region where the input signal spectral components are desired to be blocked. The transition bandwidth

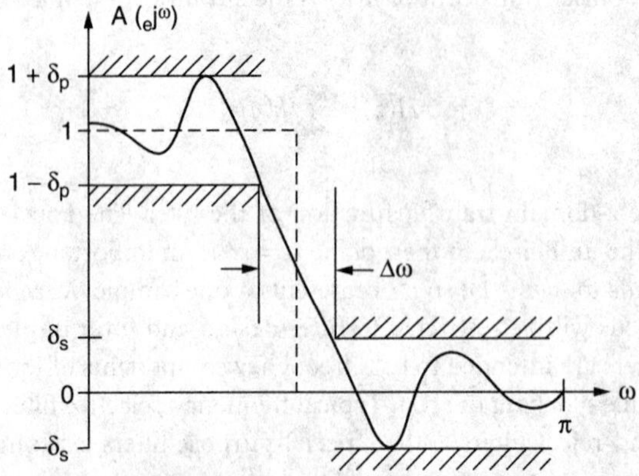

FIGURE 134.1 Amplitude frequency characteristics of an FIR low-pass filter showing definitions of passband ripple δ_p, stopband attenuation δ_s, and transition bandwidth $\Delta\omega$. (*Source:* Dorf, R. C., Ed. 1993. *The Electrical Engineering Handbook*. CRC Press, Boca Raton, FL, p. 240.)

gives the width of the spectral region in which the frequency characteristics of the transfer function change from the passband to the stopband values. Passband ripple and stopband attenuation are often specified in decibels, in which case their values are related to the quantities, δ_p and δ_s by the following:

$$\text{Passband ripple in dB} = P = -20\log_{10}(1-\delta_p)$$

$$\text{Stopband attenuation in dB} = S = -20\log_{10}\delta_s$$

FIR Filter Design by Windowing

The windowing design method is a computationally efficient technique for producing nonoptimal filters. Filters designed in this manner have equal passband ripple and stopband attenuation:

$$\delta_p = \delta_s = \delta$$

The method begins by finding the impulse response of the desired filter from

$$h_d(n) = \frac{1}{2\pi}\int_{-\pi}^{\pi} A_d(e^{j\omega})e^{j\omega(n-N/2)}\,d\omega$$

For ideal low-pass, high-pass, band-pass, and band-reject frequency-selective filters, the integral can be solved in closed form. The impulse response of the filter is then found by multiplying this ideal impulse response with a window $w(n)$ that is identically zero for $n < 0$ and for $n > N$:

$$h(n) = h_d(n)w(n),\ n = 0, 1, \ldots, N$$

Some commonly used windows are defined as follows:

Rectangular (truncation)

$$w(n) = \begin{cases} 1 & \text{for } 0 \le n \le N \\ 0 & \text{otherwise} \end{cases}$$

Hamming

$$w(n)=\begin{cases}0.54-0.46\cos(2\pi n/N) & \text{for } 0\le n\le N\\ 0 & \text{otherwise}\end{cases}$$

Kaiser

$$w(n)=\begin{cases}I_0\left(\beta\sqrt{1-[(2n-N)/N]^2}\right)/I_0(\beta) & \text{for } 0\le n\le N\\ 0 & \text{otherwise}\end{cases}$$

In general, windows that slowly taper the impulse response to zero result in lower passband ripple and a wider transition bandwidth. Other windows (e.g., Hanning, Blackman) are also sometimes used, but not as often as those shown above.

Of particular note is the Kaiser window where $I_0(.)$ is the 0th-order modified Bessel function of the first kind, and β is a shape parameter. The proper choice of N and β allows the designer to meet given passband ripple, stopband attenuation, and transition bandwidth specifications. Specifically, using S, the stopband attenuation in dB, the filter order must satisfy

$$N=\frac{S-8}{2.285\Delta\omega}$$

Then the required value of the shape parameter is given by

$$\beta=\begin{cases}0 & \text{for } S<21\\ 0.5842(S-21)^{0.4}+0.078\,86(S-21) & \text{for } 21\le S\le 50\\ 0.1102(S-8.7) & \text{for } S>50\end{cases}$$

Consider as an example of this design technique a low-pass filter with a cutoff frequency of $\omega_c = 0.4\pi$. The ideal impulse response for this filter is given by

$$h_d(n)=\frac{\sin[0.4\pi(n-N/2)]}{\pi(n-N/2)}$$

Choosing $N = 8$ and a Kaiser window with shape parameter of $\beta = 0.5$ yields the following impulse response coefficients:

$$h(0)=h(8)=-0.075\,682\,67$$

$$h(1)=h(7)=-0.062\,365\,96$$

$$h(2)=h(6)=0.093\,548\,92$$

$$h(3)=h(5)=0.302\,730\,70$$

$$h(4)=0.400\,000\,00$$

Design of Optimal FIR Filters

The accepted standard criterion for the design of optimal FIR filters is to minimize the maximum value of the error function

$$E(e^{j\omega}) = W_d(e^{j\omega})\left|A_d(e^{j\omega}) - A(e^{j\omega})\right|$$

over the full range of $-\pi \leq \omega \leq \pi$. $W_d(e^{j\omega})$ is a desired weighting function used to emphasize specifications in a given frequency band. The ratio of the deviation in any two bands is inversely proportional to the ratio of their respective weighting.

A consequence of this optimization criterion is that the frequency characteristics of optimal filters are *equiripple*: although the maximum deviation from the desired characteristic is minimized, it is reached several times in each band. Thus, the passband and stopband deviations oscillate about the desired values with equal amplitude in each band. Such approximations are frequently referred to as *mini-max* or *Chebyshev* approximations. In contrast, the maximum deviations occur near the band edges for filters designed by windowing.

Equiripple FIR filters are usually designed using the *Parks–McClellan* computer program [Antoniou, 1979], which uses the *Remez exchange algorithm* to determine iteratively the *extremal frequencies* at which the maximum deviations in the error function occur. A listing of this program along with a detailed description of its use is available in several references, including Parks and Burrus [1987] and DSP Committee [1979]. The program is executed by specifying as inputs the desired band edges, gain for each band (usually 0 or 1), band weighting, and filter impulse response length. If the resulting filter has too much ripple in some bands, those bands can be weighted more heavily and the filter redesigned. Details on this design procedure are discussed in Rabiner [1973] along with approximate design relationships that aid in selecting the filter length needed to meet a given set of specifications.

Although we have focused on the design of frequency-selective filters, other types of FIR filters exist. For example, the Parks–McClellan program will also design linear-phase FIR filters for differentiating broadband signals and for approximating the Hilbert transform of such signals.

For practice in the principles of equiripple filter design, consider an eighth-order low-pass filter with a passband $0 \leq \omega \leq 0.3\pi$, a stopband $0.5\pi \leq \omega \leq \pi$, and equal weighting for each band. The impulse response coefficients generated by the Parks–McClellan program are as follows:

$$h(0) = h(8) = -0.063\ 678\ 59$$

$$h(1) = h(7) = -0.069\ 122\ 76$$

$$h(2) = h(6) = 0.101\ 043\ 60$$

$$h(3) = h(5) = 0.285\ 749\ 90$$

$$h(4) = 0.410\ 730\ 00$$

These values can be compared to those for the similarly specified filter designed in the previous section using the windowing method.

134.2 Infinite Impulse Response Filter Design

An **infinite impulse response (IIR) digital filter** requires less computation to implement than an FIR digital filter with a corresponding frequency response. However, IIR filters cannot generally achieve a perfect linear phase response.

Techniques for the design of infinite impulse response analog filters are well established. For this reason, the most important class of IIR digital filter design techniques is based on forcing a digital filter to behave like a reference analog filter. For frequency-selective filters, this is generally done by attempting to match frequency responses. This task is complicated by the fact that the analog filter response is defined for an infinite range of frequencies ($\Omega = 0$ to ∞), whereas the digital filter response is defined for a finite range of frequencies ($\omega = 0$ to π). Therefore, a method for mapping the infinite range of analog frequencies, Ω, into the finite range from $\omega = 0$ to π known as the *bilinear transform* is employed.

Let $H_a(s)$ be the Laplace transform transfer function of an analog filter with frequency response $H_a(j\Omega)$ (Ω in radians/s). The bilinear transform method obtains the digital filter transfer function $H(z)$ from $H_a(s)$ using the substitution

$$s = \frac{2}{T}\frac{1-z^{-1}}{1+z^{-1}}$$

that is,

$$H(z) = H_a(s)\Big|_{s=(2/T)(1-z^{-1})/(1+z^{-1})}$$

This transformation maps analog frequency Ω to digital frequency ω according to

$$\omega = 2\tan^{-1}\left(\frac{\Omega T}{2}\right)$$

thereby warping the frequency response $H_a(j\Omega)$ and forcing it to lie between 0 and π for $H(e^{j\omega})$. Therefore, to obtain a digital filter with a cutoff frequency of ω_c, it is necessary to design an analog filter with cutoff frequency

$$\Omega_c = \frac{2}{T}\tan\left(\frac{\omega_c}{2}\right)$$

This process is referred to as *prewarping* the analog filter frequency response to compensate for the warping of the bilinear transform. Applying the bilinear transform substitution to this analog filter will then give a digital filter that has the desired cutoff frequency.

Analog filters and hence IIR digital filters are typically specified in a slightly different fashion than are FIR filters. Figure 134.2 illustrates how analog and IIR digital filters are usually specified. Note that by

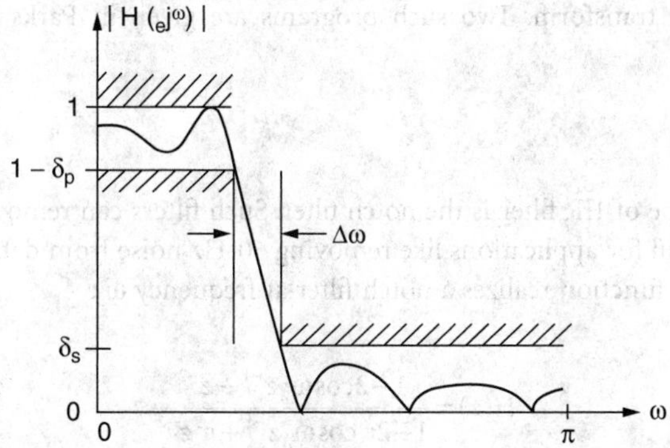

FIGURE 134.2 Frequency characteristics of an IIR digital low-pass filter showing definitions of passband ripple δ_p, stopband attenuation δ_s, and transition bandwidth $\Delta\omega$. (*Source:* Dorf, R. C., Ed. 1993. *The Electrical Engineering Handbook*. CRC Press, Boca Raton, FL, p. 244.)

comparing to Figure 134.1 that the passband ripple in this case never goes above unity, where in the FIR case the passband ripple is specified about unity.

Four basic types of analog filters are generally used to design digital filters: (1) Butterworth filters, which are maximally flat in the passband and decrease monotonically outside the passband; (2) Chebyshev filters, which are equiripple in the passband and decrease monotonically outside the passband; (3) inverse Chebyshev filters, which are flat in the passband and equiripple in the stopband; and (4) elliptic filters, which are equiripple in both the passband and stopband. Techniques for designing these analog filters are covered elsewhere [e.g., Van Valkenberg (1982)] and will not be considered here.

To illustrate the design of an IIR digital filter using the bilinear transform, consider the design of a second-order Chebyshev low-pass filter with 0.5 dB of passband ripple and a cutoff frequency of $\omega_c = 0.4\pi$. The sample rate of the digital filter is to be 5 Hz, giving $T = 0.2$ seconds. To design this filter, we first design an analog Chebyshev low-pass filter with a cutoff frequency of

$$\Omega_c = \frac{2}{0.2}\tan(0.2\pi) = 7.2654 \text{ rad/s}$$

This filter has a transfer function

$$H(s) = \frac{0.9441}{1 + 0.1249s + 0.01249s^2}$$

Substituting

$$s = \frac{2}{0.2}\frac{z-1}{z+1}$$

gives

$$H(z) = \frac{0.2665(z+1)^2}{z^2 - 0.1406z + 0.2695}$$

Computer programs are available that accept specifications on a digital filter and carry out all steps required to design the filter, including prewarping the frequencies, designing the analog filter, and performing the bilinear transform. Two such programs are given in Parks and Burrus [1987] and Antoniou [1979].

Notch Filters

An important special type of IIR filter is the notch filter. Such filters can remove a very narrow band of frequencies and are useful for applications like removing 60-Hz noise from data. The following second-order z-domain transfer function realizes a notch filter at frequency ω_0:

$$H(z) = \frac{1 - 2\cos\omega_o z^{-1} + z^{-2}}{1 - 2r\cos\omega_o z^{-1} + r^2 z^{-2}}$$

The parameter r is restricted to the range $0 < r < 1$. Substituting $z = e^{j\omega}$, multiplying the numerator and denominator by $e^{j\omega}$, and applying some trigonometric identities yields the following frequency response for the filter:

$$H(e^{j\omega}) = \frac{2\cos\omega - 2\cos\omega_o}{(1+r^2)\cos\omega - 2r\cos\omega_o + j(1-r^2)\sin\omega}$$

From this expression it can easily be seen that the response of the filter at $\omega = \omega_0$ is exactly zero. Another property is that, for $r \cong 1$ and $\omega \neq \omega_0$, the numerator and denominator are very nearly equal, and so the response is approximately unity. Overall, this is a very good approximation of an ideal notch filter as $r \rightarrow 1$.

134.3 Digital Filter Implementation

For FIR filters the convolution sum represents a computable process, and so filters can be implemented by directly programming the arithmetic operations

$$y(n) = h(0)x(n) + h(1)x(n-1) + \cdots + h(N)x(n-N)$$

However, for an IIR filter the convolution sum does not represent a computable process. Therefore, it is necessary to examine the general transfer function, which is given by

$$H(z) = \frac{Y(z)}{X(z)} = \frac{\gamma_0 + \gamma_1 z^{-1} + \gamma_2 z^{-2} + \cdots + \gamma_M z^{-M}}{1 + \beta_1 z^{-1} + \beta_2 z^{-2} + \cdots + \beta_N z^{-N}}$$

where $Y(z)$ is the z transform of the filter output, $y(n)$, and where $X(z)$ is the z transform of the filter input, $x(n)$. The unit-delay characteristic of z^{-1} then gives the following **difference equation** for implementing the filter

$$y(n) = \gamma_0 x(n) + \gamma_1 x(n-1) + \cdots + \gamma_M x(n-M) - \beta_1 y(n-1) - \cdots - \beta_N y(n-N)$$

In calculations of $y(0)$, the values of $y(-1)$, $y(-2)$, …, $y(-N)$ represent initial conditions on the filter. If the filter is started in an initially relaxed state, then these initial conditions are zero.

Defining Terms

Convolution summation — A possibly infinite summation expressing the output of a digital filter in terms of the impulse response coefficients of the filter and present and past values of the input to the filter. For finite-impulse-response filters, this summation represents a way of implementing the filter.

Difference equation — An equation expressing the output of a digital filter in terms of present and past values of the filter input and past values of the filter output. For infinite-impulse-response filters, a difference equation must be used to implement the filter since the convolution summation is infinite.

Discrete data sequence — A set of values constituting a signal whose values are known only at distinct sampled points.

Filter design — The process of determining the impulse response coefficients or coefficients of a difference equation to meet a given frequency or time response characteristic.

Filter implementation — The numerical method or algorithm by which the output sequence of a digital filter is computed from the input sequence.

Finite impulse response (FIR) digital filter — A filter whose output in response to a unit impulse function is identically zero after a given bounded number of samples.

Infinite impulse response (IIR) digital filter — A filter whose output in response to a unit impulse function remains nonzero for an indefinite number of samples.

References

Antoniou, A. 1979. *Digital Filters: Analysis and Design*. McGraw-Hill, New York.

Digital Signal Processing Committee, Eds. 1979. *Programs for Digital Signal Processing*. IEEE Press, New York.

Parks, T. W. and Burrus, C. S. 1987. *Digital Filter Design*. John Wiley & Sons, New York.

Rabiner, L. R. 1973. Approximate design relationships for low-pass FIR digital filters. *IEEE Trans. Audio Electroacoust.*, AU-21:456–460.

Van Valkenberg, M. E. 1982. *Analog Filter Design*. Holt, Rinehart & Winston, New York.

Further Information

The monthly journals *IEEE Transactions on Circuits and Systems II* and *IEEE Transactions on Signal Processing* routinely publish articles on the design and implementation of digital filters. The bimonthly journal *IEEE Transactions on Instrumentation and Measurement* also contains related information. The use of digital filters for integration and differentiation is discussed in the December 1990 issue (pp. 923–927).

135

Analog and Digital Communications

Tolga M. Duman
Arizona State University

135.1 Introduction

The objective in communications is to reproduce a source output (or, message signal) at a destination. Depending on the type of the message signal and the modulation scheme employed, communication systems can be classified as analog or digital. Examples of analog communication systems include commercial radio and television broadcast, first generation cellular systems such as advanced mobile phone service (AMPS) in the U.S., among others. On the other hand, most current telephone systems, second and third generation cellular systems such as GSM, and magnetic recording systems represent examples of digital communications systems.

The basic block diagram of an analog communication system is shown in Figure 135.1. The message signal is modulated, and sent over the channel by the transmitter. There are several objectives in modulation. For example, the channel may not be suitable for the transmission of the original message signal directly either due to the physical channel characteristics, or due to the frequency allocation mandated by a regulatory body. Furthermore, through modulation, multiple messages can be transmitted over the same medium, and also, depending on the specific modulation scheme, the signal quality can be improved at the receiver.

The channel block represents the physical medium over which transmission takes place. Examples include the atmosphere, telephone lines, optical fiber, and coaxial cables. The channel corrupts the transmitted signal by introducing attenuation, interference, multipath fading, additive noise, etc. The simplest, but useful, model to capture some of these effects is that of an additive white Gaussian noise (AWGN) channel where the transmitted signal is attenuated in a deterministic manner, and white Gaussian noise is added on it. For simplicity, we limit our attention to this channel model throughout the chapter. At the receiver side, the channel output is demodulated to obtain an estimate of the original message signal.

0-8493-1586-7/05/$0.00+$1.50

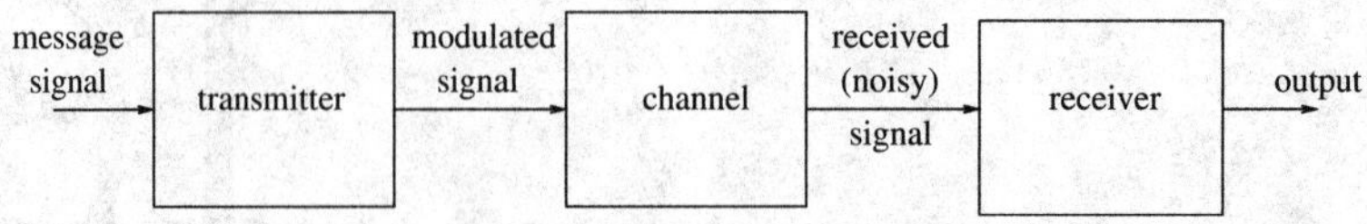

FIGURE 135.1 Block diagram of an analog communication system.

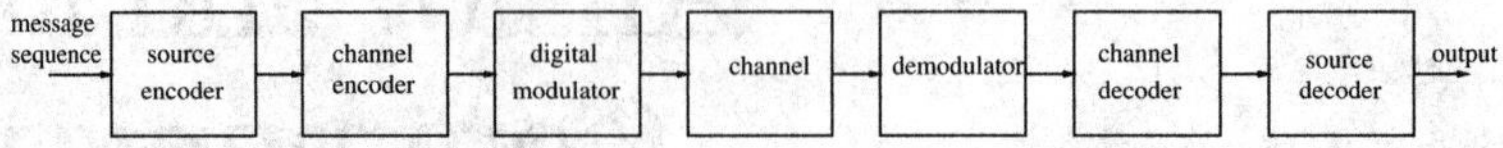

FIGURE 135.2 Block diagram of a digital communication system.

Some messages, such as computer files and text messages, are by nature digital, meaning that they are represented by a sequence of M-ary symbols. Furthermore, if an analog message signal is bandlimited, which is the case for almost all message signals, it can be sampled and quantized, and thus converted to a digital form. Therefore, it is also of importance to address the transmission of digital messages. In fact, digital communications offer various advantages over many analog communication techniques, for example, in terms of the performance in noise, cost, power, and bandwidth requirements as will become clear later in the chapter.

The block diagram of a general digital communication system is illustrated in Figure 135.2. The message sequence is first passed to a source encoder which removes the redundancy in the sequence, i.e., compresses it. This is very similar to the operation of the "zip" programs used to compress computer files. After the redundancy is removed, the information sequence is fed to a channel encoder block, which effectively adds "controlled" redundancy to protect the transmitted data against the channel impairments. The channel coded data is then modulated using a digital modulation technique which converts the sequence of symbols to a form that can be transmitted over the channel, for example, to an electrical signal for the case of telephony.

At the receiver side, the channel output is first demodulated and estimates of the channel coded symbols are obtained. Then, these symbols are passed through the channel decoder and the source decoder to obtain the message sequence at the destination.

This chapter reviews the basic blocks of both analog and digital communication systems. Emphasis is given to the description of the various modulation schemes. The chapter is organized as follows. Section 135.2 deals with analog modulation. Both amplitude modulation and angle modulation schemes are discussed, and the effects of noise on the performance are explained. Section 135.3 presents the basics of digital modulation techniques. Section 135.4 deals with the transmission of different messages over the same medium, and finally, Section 135.5 provides a brief summary.

135.2 Analog Communications

Before we present the details of various analog modulation techniques, we start with a review of the concept of frequency and present basic mathematical tools that will be necessary. We then review the amplitude modulation schemes where the signal is embedded in the amplitude of a carrier signal. We next describe the angle modulation schemes where the information is embedded in the phase or the frequency of the carrier. We conclude the section with a review of the effects of noise in analog modulation schemes.

Concept of Frequency and Fourier Transform

Frequency of a signal is a measure of how fast the signal is changing. In Figure 135.3, three sample signals are given. The first one is clearly changing more slowly than the second, and the second one is changing more slowly than the third. This is illustrated by the statement that the first signal is a "low" frequency signal, the second one is "medium" frequency signal, and the last one is a "high" frequency signal. We

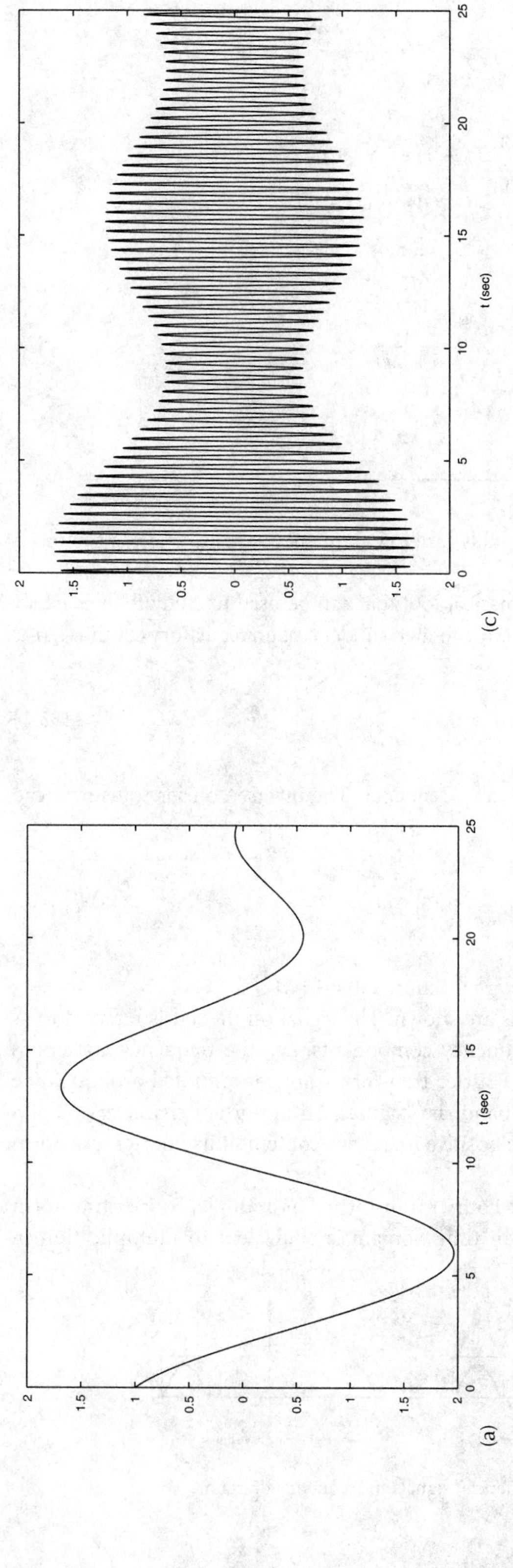

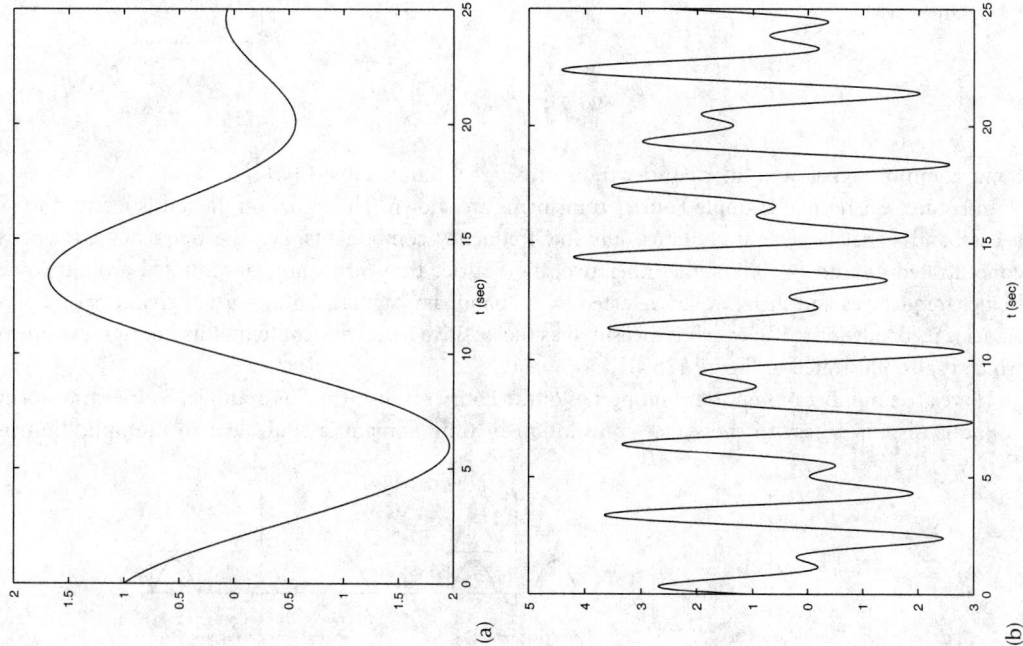

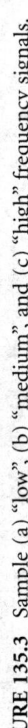

FIGURE 135.3 Sample (a) "low", (b) "medium", and (c) "high" frequency signals.

TABLE 135.1 Common Fourier Transform Pairs

Time Domain ($x(t)$)	Frequency Domain ($X(f)$)
$\delta(t)$	1
1	$\delta(f)$
$\delta(t-t_0)$	$e^{-j2\pi f t_0}$
$e^{j2\pi f_0 t}$	$\delta(f-f_0)$
$\cos(2\pi f_0 t)$	$\frac{1}{2}\delta(f-f_0)+\frac{1}{2}\delta(f+f_0)$
$\sin(2\pi f_0 t)$	$\frac{1}{2j}\delta(f-f_0)-\frac{1}{2j}\delta(f+f_0)$
$\Pi(t)=\begin{cases}1 & \lvert t\rvert \le 1/2 \\ 0 & \text{otherwise}\end{cases}$	$\frac{\sin(\pi f)}{\pi f}$
$\sin(\pi t)/\pi t$	$\Pi(f)$
$e^{-\alpha t}u_{-1}(t), \alpha>0$	$\frac{1}{\alpha+j2\pi f}$
$\text{sgn}(t)=\begin{cases}1 & t>0 \\ -1 & t<0 \\ 0 & t=0\end{cases}$	$\frac{1}{j\pi f}$

note that usually the message signals (e.g., speech signals) are low frequency signals, whereas the modulated signals are "high" frequency ones.

Fourier transform represents a convenient mathematical tool that can be used to study the frequency content of a signal. The Fourier transform of a signal $x(t)$ is defined (under certain regulatory conditions) as

$$X(f)=\int_{-\infty}^{\infty} x(t)e^{-j2\pi ft}\,dt \tag{135.1}$$

where $j=\sqrt{-1}$ and the dummy variable "f" denotes the frequency. The inverse Fourier transform can be obtained as

$$x(t)=\int_{-\infty}^{\infty} X(f)e^{j2\pi ft}\,df \tag{135.2}$$

Some common signals and their Fourier transforms are given in Table 135.1

In Figure 135.4, three example Fourier transforms are shown. The signal on the left is referred to as a baseband signal because it contains only low-frequency components, i.e., the frequency content is concentrated around $f = 0$. For the other two, the Fourier transforms are concentrated around some higher frequencies and therefore are referred to as "bandpass" signals. An important characteristic of a signal is its bandwidth (W), which is measured as the positive frequency content of its Fourier transform which is also illustrated in Figure 135.4.

There are a number of important properties of the Fourier transform. For example, Fourier transform is linear. Also, it is easy to show that convolution in time domain is equivalent to multiplication in

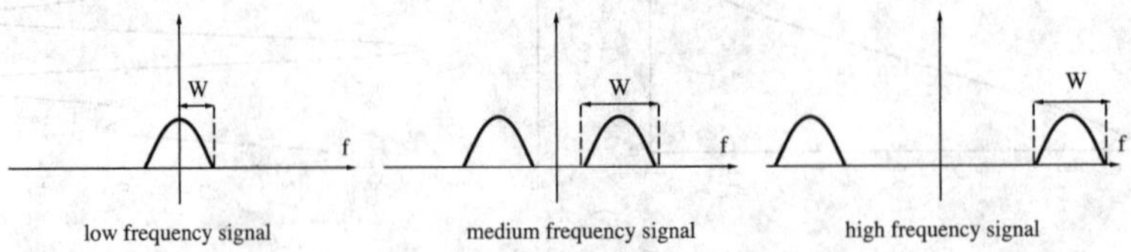

FIGURE 135.4 Sample "low", "medium", and "high" frequencey signals in frequency domain.

frequency domain, and vice versa. Furthermore, shift in time results in a multiplication by a complex exponential in the frequency domain. For a detailed discussion of these and other properties of the Fourier transform, the reader is referred to the references at the end of the chapter.

The response of a system to a given input is important to study various aspects of communications. A system is called linear if

$$\mathcal{L}\{ax_1(t)+bx_2(t)\}=a\mathcal{L}\{x_1(t)\}+b\mathcal{L}\{x_2(t)\} \tag{135.3}$$

for all scalar *a* and *b*. And, it is called time invariant if

$$y(t-t_0)=\mathcal{L}\{x(t-t_0)\} \tag{135.4}$$

for any t_0, where $y(t)$ is the response of the system to the input $x(t)$.

Linear time invariant (LTI) systems are useful to describe the effects of many communications channels, as well as the filters that are needed in the implementation of communication systems. An LTI system is completely characterized by its impulse response, denoted by $h(t)$, which is nothing but the response of the system to an impulse applied at $t = 0$. The output of the LTI system to an input $x(t)$ is given by the convolution integral

$$y(t)=x(t)*h(t) \tag{135.5}$$

$$=\int_{-\infty}^{\infty}x(\tau)h(t-\tau)d\tau \tag{135.6}$$

The Fourier transform of the impulse response, $H(f)$, is referred to as the transfer function of the system. With this definition, the input/output relationship of the LTI system (in the frequency domain) can be conveniently written as

$$Y(f)=X(f)H(f)\cdot \tag{135.7}$$

Amplitude Modulation

In amplitude modulation (AM), the message signal $m(t)$ is embedded in the amplitude of the carrier $c(t)=A_c\cos(2\pi f_c t+\phi_c)$ where A_c is the amplitude, ϕ_c is the phase and f_c is the frequency of the carrier signal. There are several amplitude modulation techniques, namely, double sideband suppressed carrier (DSB-SC) AM, conventional AM, single sideband (SSB) AM, and vestigial sideband (VSB) AM.

DSB-SC AM

For DSB-SC AM, the modulated signal is given by

$$u(t)=m(t)c(t) \tag{135.8}$$

$$=A_c m(t)\cos(2\pi f_c t+\phi_c)\cdot \tag{135.9}$$

It is easy to show that in the frequency domain we have

$$U(f)=M(f)*C(f) \tag{135.10}$$

$$=\frac{A_c}{2}e^{j\phi_c}M(f-f_c)+\frac{A_c}{2}e^{-j\phi_c}M(f+f_c)\cdot \tag{135.11}$$

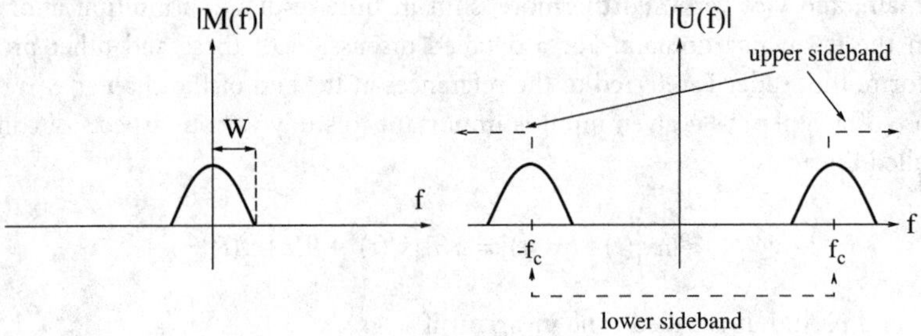

FIGURE 135.5 A typical message signal and its DSM AM modulated version.

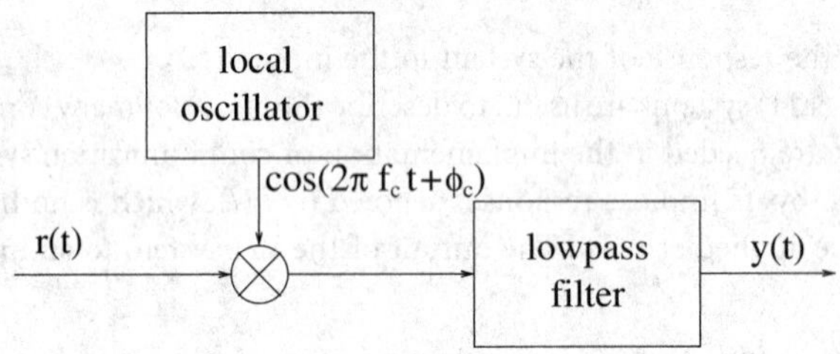

FIGURE 135.6 Demodulation of DSB-SC AM modulated signals.

Figure 135.5 illustrates the magnitude spectrum of a typical message signal, and the DSB-SC AM modulated signal. It is clear that the modulator simply "translates" the frequency content of the (baseband) message signal to the carrier frequency, which is selected to make sure that the transmitted signal matches the channel characteristics. In this case, both upper and lower sidebands of the modulated signal (shown in the figure) are transmitted. Therefore, the bandwidth of the modulated signal (i.e., the transmission bandwidth necessary) is twice the bandwidth of the message signal (W).

The demodulation of a DSB-SC AM modulated signal is performed as follows. A reference carrier signal is obtained using a phase locked loop circuit, then the received signal is mixed with the reference carrier signal, the resulting signal is passed through a low-pass filter whose bandwidth is the same as the original message bandwidth. The reference signal can also be obtained through the transmission of a small carrier component (i.e., pilot tone) along with the message signal. Figure 135.6 illustrates the demodulation process. In the absence of channel noise, i.e., when the received signal is identical to the transmitted signal, assuming that there is no estimation error in the phase of the carrier, it is easy to see that the receiver output is proportional to $m(t)$, thus the message signal is recovered.

Conventional AM

For DSB-SC AM, a phase-locked loop is needed to reproduce the carrier signal to be used in the demodulation process. However, this circuit is usually expensive, and thus not appropriate for certain applications such as broadcasting where a very cheap receiving circuit is required. With this motivation, in conventional AM, a large carrier component along with the amplitude modulated signal is transmitted. Mathematically, the modulated signal is given by

$$u(t) = A_c(1 + am_n(t))\cos(2\pi f_c t + \phi_c) \tag{135.12}$$

where the $m_n(t)$ is the normalized message signal, i.e.,

$$m_n(t) = \frac{m(t)}{\max|m(t)|}$$

and a is the modulation index. We usually select $a \leq 1$, otherwise, the signal is over-modulated.

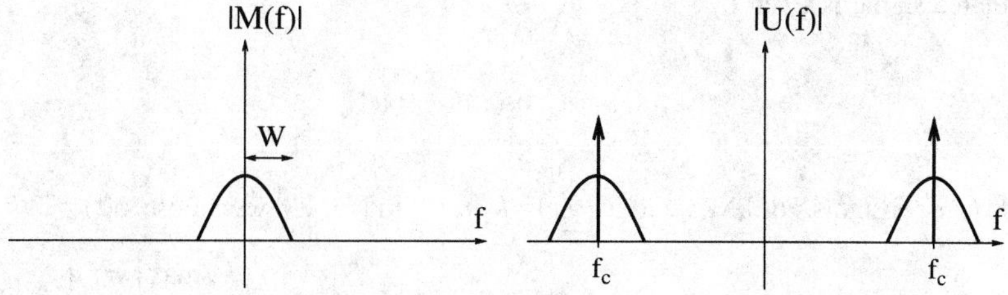

FIGURE 135.7 A typical message signal and its conventional AM modulated version.

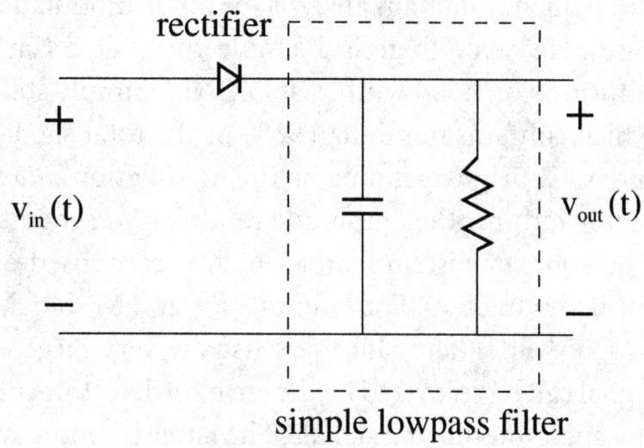

FIGURE 135.8 Envelop detector used to demodulate conventional AM modulated signals.

In the frequency domain, we have

$$U(f) = \frac{A_c}{2} e^{j\phi_c}(M(f - f_c) + \delta(f - f_c)) + \frac{A_c}{2} e^{-j\phi_c}(M(f + f_c) + \delta(f + f_c)). \tag{135.13}$$

Figure 135.7 illustrates a typical conventional AM modulated signal in the frequency domain. The bandwidth required for transmission is the same as the one for DSB-SC AM, i.e., $2W$.

As long as the signal is not over-modulated, a simple envelop detector circuit can be used for demodulation, which is nothing but a rectifier and a low-pass filter as shown in Figure 135.8. Clearly, a synchronous demodulator can also be used; however, this will unnecessarily complicate the implementation.

SSB and VSB AM

For DSB-SC AM, both the lower sideband and the upper sideband of the modulated signal are transmitted as shown in Figure 135.5. However, only one of the sidebands is sufficient for the proper demodulation of the message signal at the receiver. SSB AM transmits only one of the sidebands to reduce the bandwidth requirement to W (as compared to $2W$). The SSB AM signal can simply be obtained by bandpass filtering the DSB AM signal, and can be demodulated in exactly the same way.

In practice, the strict requirements for the bandpass filter make the implementation of the SSB AM expensive. As a remedy, another filter that keeps a "vestige" of the other sideband along with the sideband being transmitted can also be used. The resulting modulation scheme is the VSB AM. In this case, the bandwidth required to transmit the modulated signal is only slightly larger than W, and the scheme is much easier to implement. As long as the filter is selected such that $H(f - f_c) + H(f + f_c)$ is constant for $|f| \leq W$, the message signal can be demodulated in exactly the same way as in DSB-SC AM and SSB AM. VSB AM finds applications in, for example, TV broadcasting.

Angle Modulation

In angle modulation, the message signal is transmitted in the "angle" (i.e., frequency or phase) of the carrier signal as opposed to its amplitude. There are two main angle modulation schemes, namely, frequency modulation (FM) and phase modulation (PM).

The modulated signal is given by

$$u(t) = A_c \cos(2\pi f_c t + \phi(t)) \tag{135.14}$$

where *kp* and *kf* are constants. For PM, we select $\phi(t) = k_p m(t)$, and for FM we choose $\phi(t) = 2\pi k_f \int_{-\infty}^{t} m(\tau)d\tau$.

The modulation index is defined as $\beta_p = k_p \max|m(t)|$ for PM, and $\beta_f = \dfrac{k_f \max|m(t)|}{W}$ for FM, where k_p and k_f are constants.

Frequency domain analysis for angle modulation schemes is very difficult except for some very simple cases. However, there is a simple rule called Carson's rule that can be used to determine the required transmission bandwidth. Carson's rule simply states that the effective bandwidth of the modulated signal which contains more than 98% of the total signal power is $B_c = 2(\beta + 1)W$ where β is the modulation index. Clearly, depending on the modulation index used, this may be significantly more than the required bandwidth for the amplitude modulation schemes.

Frequency discriminator which is composed of a differentiator and an envelop detector can be used for demodulation. For example, for an FM signal, the output of the differentiator is simply proportional to $f_c + k_f m(t)$, and since f_c is usually very large, this quantity is always non-negative, and the message signal can be recovered by the envelop detector. One can also use a phase locked loop circuit to demodulate the angle-modulated signals. The latter is more expensive to implement, but it offers a better performance (when there is noise present).

Noise in Analog Communications

As mentioned in the introductory section, the received signal is corrupted by channel noise which degrades the demodulated signal quality at the output of the receiver. Signal to noise ratio (SNR) is usually used as such a measure. A detailed mathematical treatment of noise and the resulting SNR is beyond the scope of this chapter; however, some observations are in order.

Amplitude modulation schemes cannot improve the signal quality (compared to the baseband system, (i.e., no modulation). Specifically, it can be shown that DSB-SC AM and SSB AM result in the same output SNR as the reference baseband system. SNR for the VSB AM depends on the characteristics of the filter used, however, it is very similar to the SSB AM case. On the other hand, for conventional AM together with the message signal, a large carrier component – which contains no information – is transmitted. The modulation index should be selected as close to 1 as possible to maximize the SNR; however, even for $a = 1$, at least 50% of the power is spent on the carrier component. Therefore, in order to achieve the same output SNR, the transmission power must be increased considerably, resulting in an inferior output SNR compared to that of the baseband system, and the other amplitude-modulation schemes.

On the other hand, the significant bandwidth expansion introduced by the angle modulation schemes helps improve the signal quality at the receiver output considerably. In general, the larger the modulation index (for the angle-modulation scheme), the higher the resulting output SNR. However, the modulation index cannot be made arbitrarily large to increase the output SNR, as a phenomenon called "threshold effect" occurs. If the baseband SNR reduces below a threshold (which increases with the modulation index), the receiver output cannot be distinguished from noise. Therefore, for a given background noise level, there is an upper limit on the modulation index that can be used. The output signal fidelity can further be improved by the use of pre-emphasis/de-emphasis filtering, which effectively uses FM for the low-frequency components of the message signal, and PM for the high-frequency components. Finally, we note that the demodulation scheme being used affects the output signal-to-noise ratio that can be obtained. The receivers which have feedback (e.g., a phase-locked loop) result in a significantly larger output SNR compared to the other alternatives (e.g., frequency discriminator).

Regardless of the specific modulation scheme employed, transmitted signal gets attenuated over the communication link. For example, for free space, the received signal power is inversely proportional to the square of the transmitter/receiver separation. Therefore, if the transmitter/receiver separation is large, the signal power at the receiver input is small, resulting in a significant loss in the SNR. One way to combat attenuation is to boost the transmit signal power; however, this is not an efficient way. For example, for free space, if the separation is increased tenfold, the transmit power should be increased 100 times to ensure the same signal-to-noise ratio is achieved. A more efficient way to counteract the harmful effects of signal attenuation is to use "analog repeaters" between the source and the destination. An analog repeater simply amplifies the signal it receives, and relays it to the destination. It can be shown that even though amplifiers used may have an internal noise that "corrupts" the signal before retransmission, they can be very effective in reducing the overall power requirements of the transmitter.

135.3 Digital Communications

This section deals with the transmission of digital information as opposed to the analog sources considered in the previous section. Before we proceed, we should note that there are many advantages of digital communications over analog communication techniques. For example, with digital communications, the received signal fidelity can be improved significantly. Also, digital transmitters require less power consumption, less bandwidth (by exploiting the redundancy in the data), and are inexpensive to build. Furthermore, data transmission is possible, and also secure communications can be accomplished with relative ease.

The section is organized as follows. We first review digital information sources in brief, and consider analog to digital conversion. We then give an overview of various baseband and carrier modulation techniques. We consider the optimal receiver in the presence of additive white Gaussian noise, and comment on the effects of noise on the performance of the system.

Digital Sources and Analog to Digital Conversion

As opposed to analog sources such as a speech signal, many information sources, such as text messages or data files, are digital. That is, they are represented by sequences of M-ary symbols. For the case of $M = 2$, the resulting source is said to be a binary information source, and the alphabet used is simply $\{0,1\}$. This is probably the most common way of representing digital information sources.

Even though a source output may be analog to start with, it can be converted to digital with negligible loss of fidelity provided that the analog to digital (A/D) conversion is carried out properly. This conversion is done by sampling and quantization. Sampling theorem states that, if the message signal is bandlimited to W, we can reconstruct it from its samples provided that the samples are taken at a rate greater than or equal to the Nyquist rate, i.e., $2W$. For example, a telephone-quality speech signal is bandlimited to $W = 3.4$ KHz. Therefore, if we sample the signal using a sampling frequency of 6.8 KHz (i.e., 6800 samples per second) or more, the samples are theoretically sufficient to represent the original message signal. We note that there is no loss in signal fidelity in this step.

The second step in A/D conversion is quantization. The samples of the analog message signal usually take on a continuum of values, thus it is impossible to represent them with infinite accuracy digitally. Instead, each sample is represented by a finite number of bits by employing a quantizer which maps the signal value to the closest allowable quantization level. For example, for a speech source, if 8 bit (256 level) quantization is used, and 8 KHz sampling frequency is employed, the original source is represented by 64 K bits/sec. We note that the quantization operation results in a "loss" in signal quality. However, this loss may be reduced by selecting the number of quantization levels sufficiently large.

Pulse Code Modulation

Pulse Code Modulaton (PCM) whose block diagram is given in Figure 135.9 represents a popular method of analog-to-digital conversion. The analog signal is first sampled at a rate greater than the Nyquist rate,

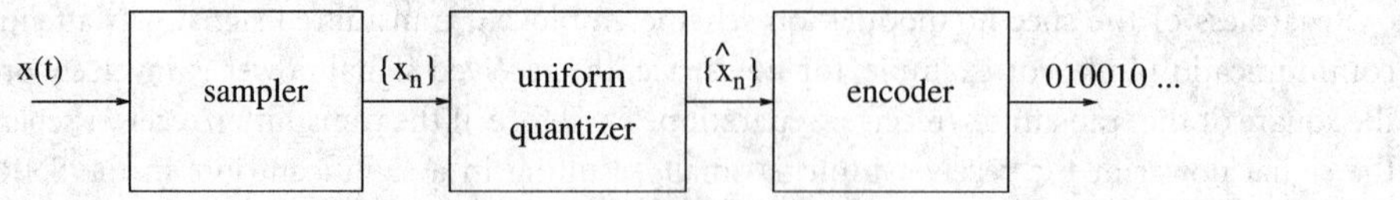

FIGURE 135.9 Block diagram of a uniform pulse code modulation scheme.

the samples are quantized using a uniform quantizer, and then represented by using binary digits. Uniform quantization means that the quantization regions are selected to be of equal length. The number of quantization levels is usually selected to be a power of 2 so that a ν-bit quantizer can be used effectively. A popular application of PCM is digital telephony.

In order to transmit one bit per second, the minimum bandwidth needed (with binary modulation) is $^1/_2$ Hz (although, in practice a higher bandwidth may be necessary due to practical limitations). If the source is sampled at a rate f_s samples/sec, the data rate is $f_s\,\nu$ bits/sec. Therefore, the minimum bandwidth needed to transmit the PCM signal is $f_s\,\nu/2$ Hz. Furthermore, if the sampling frequency is selected to be equal to the Nyquist rate (i.e., $f_s = 2\,W$), the minimum bandwidth needed is found to be $W\nu$. Therefore, if the number of quantization levels is increased in an attempt to make the quantization loss small, a larger transmission bandwidth will be required. For example, increasing the number of bits per sample by 1 increases the signal to quantization noise ratio by approximately 6 dB while requiring an additional bandwidth of W Hz for transmission.

For many sources, the source outputs are not uniformly distributed on their full range. For example, for speech sources, the low amplitude levels are more likely than the higher amplitude levels. Therefore, it will be beneficial if, effectively, a larger number of quantization levels (resulting in finer quantization) are used for the more likely amplitude levels. One way to accomplish this is to use a nonuniform quantizer instead of the uniform one in the PCM scheme. However, this is unnecessarily expensive. This motivates the use of a nonuniform PCM which is obtained by using a compressor/expander pair (usually referred as a compander) together with a uniform PCM scheme. A compressor is an invertible nonlinear mapping that maps the range of the more likely outputs to a wider region, and compresses the less likely ones. A compressor is used before sampling the analog signal, and its inverse (corresponding expander) is used at the receiver end after the signal is converted to analog. A μ-law compander and an A-law compander are used for digital telephony in North America and in Europe, respectively. The use of nonuniform PCM results in significant increase in the signal quality.

By exploiting the properties of typical analog sources (e.g., speech sources), other variations of "waveform coding" techniques, such as differential PCM, Delta Modulation, and Adaptive Delta Modulation are also used in practice.

Digital Modulation Techniques

Many different modulation methods can be used to convert the digital information (i.e., the sequence of M-ary symbols) to a form suitable for transmission over the channel. In this section, we review several important digital baseband and carrier modulation schemes.

Digital Baseband Transmission

An example of a digital baseband modulation scheme is the baseband pulse amplitude modulation (PAM) where the M–ary symbols are transmitted in the amplitude of a basic pulse. More specifically, the transmitted signal corresponding to the symbol m is given by

$$s_m(t) = A_m g(t) \leq t \leq T \tag{135.15}$$

where $A_m = 2m + 1 - M,\ m \in \{0,1,2,\cdots,M-1\}$, and $g(t)$ is the baseband pulse. For the case of binary transmission ($M = 2$), "0" is transmitted using $s_0(t) = -g(t)$, and "1" is transmitted with $s_1(t) = g(t)$.

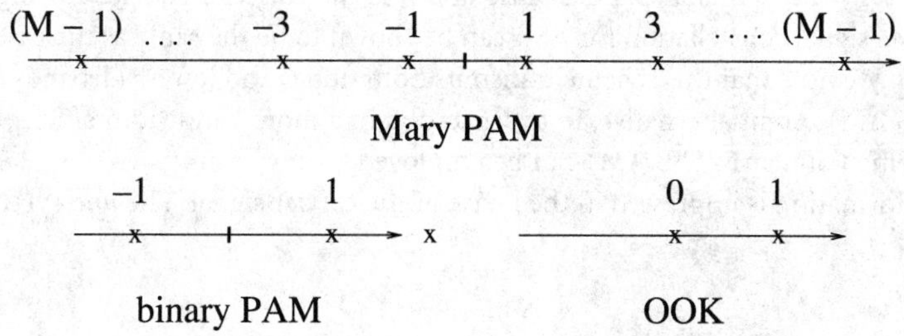

FIGURE 135.10 Signal constellations for PAM and OOK.

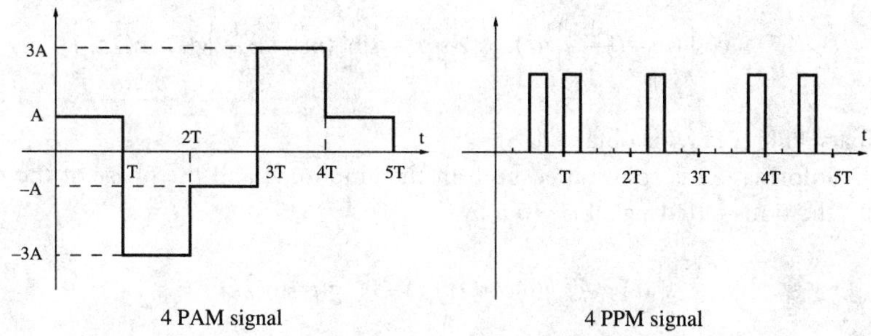

FIGURE 135.11 Transmitted 4-ary PAM and 4-ary PPM signals corresponding to the sequence {2,0,1,3,2}.

Another baseband modulation scheme is on-off keying (OOK) where "1" is transmitted using a basic pulse $s_1(t) = g(t)$, "0" is transmitted using $s_0(t) = 0$.

If $g(t)$ is used as a basis function, the PAM- and OOK-modulated signals can conveniently be represented in vector form. For PAM the vector representation is then $\underline{s}_m = A_m$, and for OOK, $\underline{s}_0 = 0$ and $\underline{s}_1 = 1$. This representation leads to a convenient representation for the signal constellation for PAM and OOK as shown in Figure 135.10.

Pulse position modulation (PPM) is another method of digital baseband modulation. As opposed to the PAM and OOK, in this case, the information is transmitted in the relative position of a pulse within the bit duration. For example, if a rectangular pulse is used and the bit duration is T, the symbol m is transmitted using

$$s_m(t) = \begin{cases} 1 & \text{if } \dfrac{mT}{M} \le t \le \dfrac{(m+1)T}{M} \\ 0 & \text{otherwise} \end{cases} \quad (135.16)$$

As an example, consider a 4-ary communication scheme. The transmitted signal with 4-ary PAM (using a rectangular pulse), and 4-ary PPM are shown in Figure 135.11 for the information sequence of $\{2,0,1,3,2\}$.

Digital Carrier Modulation

For many communication systems, digital baseband transmission is not possible due to the physical characteristics of the channel or due to regulatory reasons. Therefore, suitable digital carrier modulation techniques are also developed. Different carrier modulation techniques include amplitude shift keying (ASK), phase shift keying (PSK), frequency shift keying (FSK), and quadrature amplitude modulation (QAM).

ASK (also called carrier modulated PAM) is very similar to baseband PAM, that is, information is embedded in the amplitude of a basic signal. The mth symbol is transmitted using

$$s_m(t) = A_m g(t)\cos(2\pi f_c t) \quad 0 \le t \le T \quad (135.17)$$

where $A_m = 2m + 1 - M$, $g(t)$ is a shaping pulse and f_c is the carrier frequency. Using a similar vector representation, the signal constellation for ASK can be shown to be the same as the case of PAM (shown in Figure 135.10). We note that this scheme transmits both upper and lower sidebands of the modulated signal (as in DSB AM). As an alternative, in order to obtain a more bandwidth efficient version of ASK, an SSB ASK scheme (similar to SSB AM) can be employed.

In PSK, the information is impressed in the phase of the carrier signal. The *mth* symbol is transmitted using

$$s_m(t) = g(t)\cos\left(2\pi f_c t + (m-1)\frac{2\pi}{M}\right) \quad (135.18)$$

$$= \cos\left((m-1)\frac{2\pi}{M}\right)g(t)\cos(2\pi f_c t) - \sin\left((m-1)\frac{2\pi}{M}\right)g(t)\sin(2\pi f_c t) \quad (135.19)$$

For $M = 2$, binary PSK (BPSK) is obtained.

In QAM, the information is transmitted both in the amplitude and the phase of the carrier signal. Mathematically, the transmitted signal is given by

$$s_m(t) = A_{mc}g(t)\cos(2\pi f_c t) - A_{ms}g(t)\sin(2\pi f_c t) \quad (135.20)$$

where A_{mc} and A_{ms} are selected as the amplitudes of the in-phase and quadrature components of the transmitted signal.

For PSK and QAM, two orthogonal functions $g(t)\cos(2\pi f_c t)$ and $-g(t)\sin(2\pi f_c t)$ can be used as the basis functions to represent the transmitted signals in vector form. For PSK, we can write $\underline{s}_m = \left[\cos\left((m-1)\frac{2\pi}{M}\right) \quad \sin\left((m-1)\frac{2\pi}{M}\right)\right]$. Similarly, for QAM, we obtain $\underline{s}_m = [A_{mc} \quad A_{ms}]$. In Figure 135.12, the signal constellations for 4 PSK, 8 PSK and (rectangular) 16 QAM are shown.

Finally, in FSK, different signals are transmitted using carriers with different frequencies as opposed to the different amplitudes and phases.

Bandwidth Requirements

Roughly speaking, a pulse of duration T occupies an approximate bandwidth of $1/T$. Since with PAM each symbol is transmitted using exactly one of these pulses, in order to transmit at a rate of 1 symbol/sec, a bandwidth of 1 Hz is needed.[1] This is the same for the two-dimensional modulation schemes (e.g., PSK

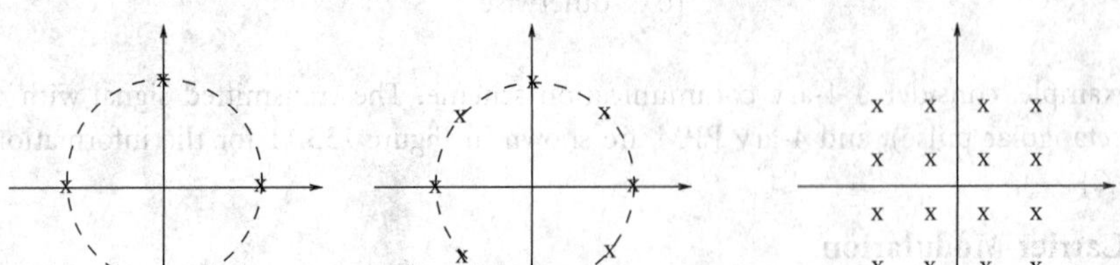

FIGURE 135.12 4 PSK, 8 PSK, and 16 QAM signal constellations.

[1]In fact, if single side band modulation is used, each bit would require $1/2$ Hz.

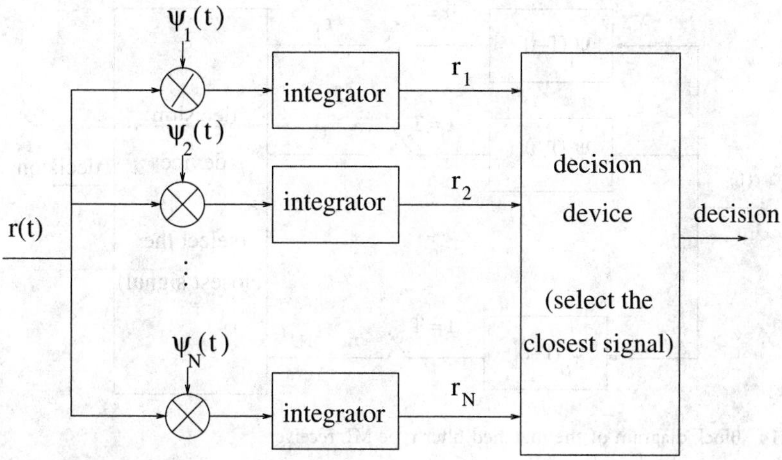

FIGURE 135.13 Block diagram of the ML receiver.

and QAM) described earlier since the in-phase and quadrature components of the signal occupy the same frequency band. Therefore, with PAM, PSK, and QAM, the spectral efficiency achieved is $log_2 M$ bits/sec/Hz. That is, when the constellation size is increased, a more efficient use of the frequency band is achieved. This improvement is obtained at the expense of an increase in the error rate, as we will see shortly.

Receiver Design in AWGN

For an AWGN channel (which is one of the simplest channel models for practical communication systems), ignoring the channel attenuation, the received signal is nothing but the transmitted signal corrupted by the additive white Gaussian noise. That is,

$$r(t) = s(t) + \eta(t) \tag{135.21}$$

where $s(t)$ is the transmitted signal, and $\eta(t)$ is a white Gaussian noise process. The receiver has to make a decision on the symbol transmitted based on this noisy observation. We now describe this process in some detail.

Assume that the transmitted signal is represented in vector form by $\underline{s} = [s_1 s_2 \cdots s_N]$ using orthonormal basis functions[1] denoted by $\{\psi_1(t), \psi_2(t), \cdots, \psi_N(t)\}$. Further assume that all the symbols are equally likely to be transmitted, and are independent from one symbol to the next. With this assumption (which characterizes "uncoded" communication systems appropriately), symbol by symbol maximum likelihood (ML) decoding becomes optimal, i.e., minimizes the probability of symbol error.

For ML decoding, the receiver first demodulates the received signal $r(t)$ using a bank of correlators, that is, it obtains the component of the received signal in the signal space denoted by $\underline{r}$. $\underline{r}$ is nothing but the vector representation of $r(t)$ with respect to the orthonormal basis functions defined in the previous paragraph. Although $r(t)$ may not lie in that space completely, its component outside of this space is not important for making a decision on the transmitted symbol. Then, this vector is fed to a decision device which declares the closest signal vector (i.e., the corresponding symbol) to the received vector in the Euclidean distance sense as its decision. This process is illustrated in Figure 135.13.

Each correlator can be replaced by a filter matched to the corresponding basis function $\psi(t)$ and a sampler sampling at $t = T$. The impulse response of the filter matched to $\psi(t)$ is given by $\psi(T-t)$. With this modification, the matched filter type receiver is obtained as shown in Figure 135.14.

---

[1]That is, the basis functions are normalized and orthogonal to each other.

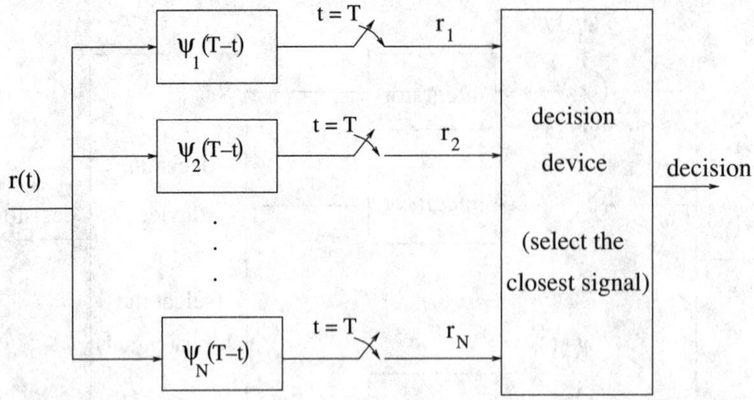

FIGURE 135.14 Block diagram of the matched filter type ML receiver.

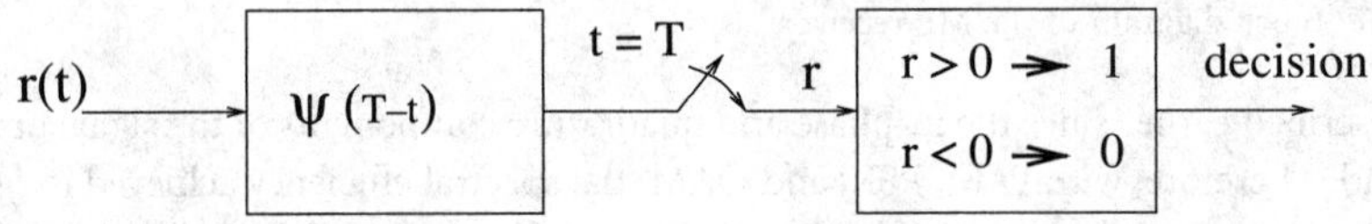

FIGURE 135.15 ML receiver for BPSK modulation.

Furthermore, the ML receiver can be simplified considerably for the specific modulation schemes described in the previous section. As an example, let us consider the case of BPSK modulation. In this case, the (matched filter type) ML receiver can be obtained by a single matched filter, and a simple threshold detector as illustrated in Figure 135.15.

Effects of Noise in Digital Transmission

Due to the additive noise present in the system, the receiver will make some errors in its decisions. A thorough treatment of the noise in digital communications is beyond the scope of this chapter; however, we will make some observations.

Assume that a certain modulation scheme with a given constellation is employed, and a specific constellation point (i.e., the corresponding symbol or signal) is transmitted. The receiver will fail to make the correct decision if the received vector is closer to another constellation point due to the channel noise. The most common types of errors will be due to the signal "closest" to the transmitted symbol (in the Euclidean distance sense). This is particularly true if the noise variance is small, or the signal-to-noise ratio is high, meaning that the performance of the modulation scheme is determined by the distance between the closest constellation points for large SNRs. Therefore, in order to compare the performance of two different modulation schemes, assuming that the energy of the constellation is normalized, we can use this minimum distance as a metric. The smaller the minimum distance, the more the modulation scheme is error prone.

As an example, let us compare the performance of BPSK with OOK. To maintain the same distance between the two constellation points in both schemes, it is easy to see that the energy of the OOK scheme should be doubled. That is, the signal-to-noise ratio should be increased by a factor of 2, i.e., approximately 3 dB. Therefore, we say that BPSK is 3 dB better than OOK in terms of power efficiency. Similar comparisons can be made for other modulation schemes as well. In general, when the constellation size is increased in an attempt to improve the spectral efficiency of the system, the power efficiency of the system reduces. In other words, there is a trade-off between the power consumption and the bandwidth efficiency.

In order to combat the effects of significant signal attenuation with the distance (both in wireless and wireline signal transmission), just as in the case of analog communication, repeaters can be employed.

For the case of digital transmission, the repeaters can be used to "amplify and forward" the intermediate signal, or to demodulate the transmitted digital information, and retransmit it. It can be shown that the latter scheme (also called regenerative repeaters) is a better choice in terms of the bit error rates achievable by the end-to-end system.

135.4 Multiplexing and Multiple Access

In many analog or digital communication systems, the communication medium is used to transmit different messages. This is accomplished through "multiplexing techniques" if the messages belong to the same user, and through "multiple-access techniques" if the messages belong to different users.

A straightforward multiplexing technique is what is called "quadrature carrier multiplexing." In this case, two messages (of bandwidth W) are transmitted using a double sideband AM modulation scheme employing two orthogonal carriers of the same frequency, i.e., $\cos(2\pi f_c t)$ and $\sin(2\pi f_c t)$. Since both messages occupy the same frequency band, two messages are transmitted using a bandwidth of $2W$, resulting in the same spectral efficiency as the single sideband AM.

Frequency division multiplexing (FDM) and frequency division multiple access (FDMA) use different nonoverlapping frequency bands to transmit the different signals over the same medium as illustrated in Figure 135.16. The underlying transmission may be analog or digital. Since the transmitted signals do not overlap in frequency domain, they can be easily recovered at the receiver by the help of bandpass filters. Commercial AM, FM radio transmission, and TV broadcast are all communication systems employing FDMA; cable TV is a good example of a system employing FDM.

Time division multiplexing (TDM), or time division multiple access (TDMA) use different time "intervals" to accommodate the transmission of different messages. A typical time frame (suitable for the transmission of K different messages) is shown in Figure 135.17. In this case, the messages do not overlap in time allowing proper communication. Clearly, unlike frequency division multiplexing or multiple access, TDM or TDMA are only suitable for digital communications. A good example of a TDM system is digital telephony where the "bits" of different users are multiplexed by the "central office" before transmission over a common medium. A good example of a TDMA scheme is the GSM system commonly used for second and third generation cellular telephony in the U.S. and Europe.

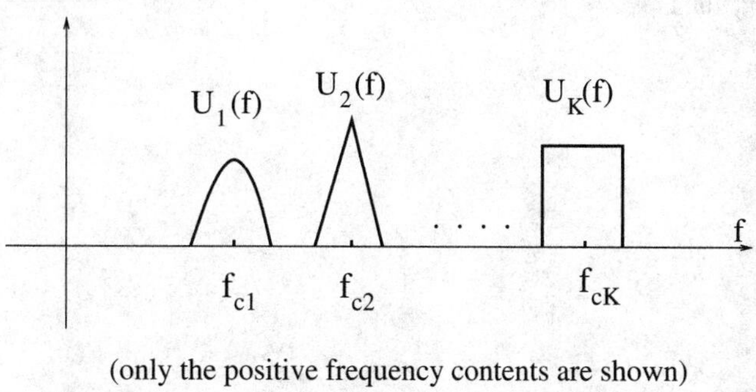

FIGURE 135.16 FDM or FDMA illustrated in the frequency domain.

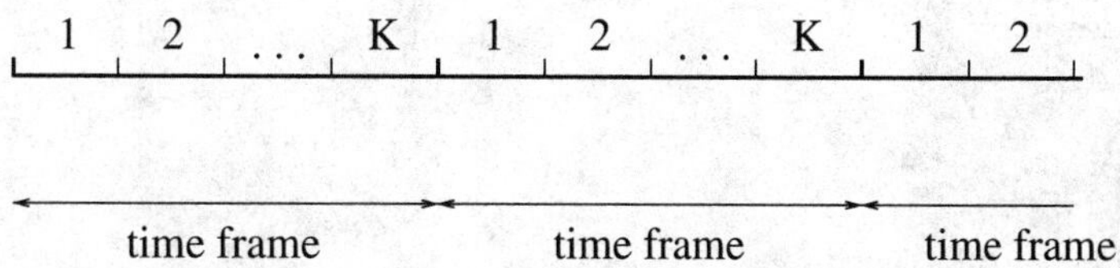

FIGURE 135.17 TDM or TDMA time frame.

There are other multiple access schemes such as code division multiple access (CDMA) or space division multiple access (SDMA) that are beyond the scope of this chapter.

135.5 Summary and Further Reading

We have presented a brief overview of analog and digital communication techniques. We have illustrated some of the main communications principles without going into the mathematical details involved.

In addition to the details of analog and digital modulation techniques not presented here, there are many issues that remain in particular for digital communications. For example, we have not talked about the timing and synchronization issues, detailed descriptions of source and channel coding techniques, signaling for other practical communication channels such as channels with intersymbol interference (ISI), and fading channels, other modulation schemes including spread spectrum communications, and multiple access techniques such as code division multiple access (CDMA). For these important more advanced topics and the details of analog and digital communication systems the reader is referred to the textbooks on communications such as References [1] through [8].

References

[1] Proakis J. G., *Digital Communications, Fourth Edition*, McGraw Hill, New York, 2001.

[2] Proakis J. G. and Salehi M., *Communication Systems Engineering, Second Edition*, Prentice Hall, Englewood Cliffs, NJ, 2002.

[3] Haykin S., *Communication Systems, Fourth Edition*, John Wiley & Sons, New York, 2001.

[4] Couch L. W., *Digital and Analog Communication Systems, Sixth Edition*, Prentice Hall, Englewood Cliffs, NJ, 2001.

[5] Sklar B., *Digital Communications: Fundamentals and Applications, Second Edition*, Prentice Hall, Englewood Cliffs, NJ, 2001.

[6[Gibson J. D., *Principles of Digital and Analog Communications, Second Edition*, Macmillan, 1993.

[7] Lee. E. A and Messerschmitt D. G., *Digital Communication, Second Edition*, Kluwer Academic Publishers, Dordrecht, the Netherlands, 1994.

[8] Carlson A. B., Crilly P. B. and Rutledge J. C., *Communication Systems, Fourth Edition*, McGraw-Hill, New York, 2002.

136

Coding

Scott L. Miller
University of Florida

Leon W. Couch II
University of Florida

Error correction coding involves adding redundant symbols to a data stream in order to allow for reliable transmission in the presence of channel errors. Error correction codes fall into two main classes: block codes and convolutional codes. In block codes, the code word is only a function of the current data input, whereas in convolutional codes the current output is a function of not only the current data input but also of previous data inputs. Thus, a convolutional code has memory, whereas a block code is memoryless. Codes can also be classified according to their code rate, symbol alphabet size, error detection/correction capability, and complexity. The code rate determines the extra bandwidth needed and also the reduction in energy per transmitted symbol relative to an uncoded system. The alphabet size is usually either two (binary) or a power of two. When the alphabet size is binary, the symbol is called a bit. Codes can be made either to detect or to correct errors or some combination of both and also can be made to detect/correct either random errors or burst errors. The complexity of a code is usually a function of the decoding procedure that is used, so the most popular codes are those that can be decoded in a relatively simple manner and still provide good error correction capability.

136.1 Block Codes

An (n, k) block code is a mapping of a k symbol information word into an n symbol code word. The code is said to be systematic if the first k symbols in the code word are the information word. In an (n, k) code, $n - k$ symbols are redundant and are often referred to as *parity symbols.* The code rate, R, is defined as $R = k/n$. An important parameter of a block code is its minimum distance, $d_{\min}$, which is defined as the minimum Hamming distance between any two code words. (Hamming distance is the number of places in which two code words disagree.) A code with a minimum distance of $d_{\min}$ can guarantee to correct any pattern of $t = \lfloor (d_{\min} - 1)/2 \rfloor$ or fewer errors. Alternatively, the code can guarantee to detect any error pattern consisting of $s = d_{\min} - 1$ or fewer errors. If simultaneous error detection and correction is desired, the minimum distance of the code must be such that $d_{\min} \geq s + t + 1$.

Most block codes used in practice are linear (or group) codes. In a linear code the set of code words must form a mathematical group. Most importantly, the sum of any two code words must itself be a code word. Encoding of linear block codes is accomplished by multiplying the k symbol information word by an $n \times k$ matrix (known as a generator matrix) forming an n symbol code word. To decode, a received word is multiplied by an $n \times (n - k)$ matrix (known as a *parity check matrix*) to form an $n - k$ symbol vector known as the *syndrome.* A look-up table is then used to find the error pattern that corresponds to the given syndrome and that error pattern is subtracted from the received word to form the decoded word. The parity (redundant) symbols are then removed to form the decoded information

0-8493-1586-7/05/$0.00+$1.50

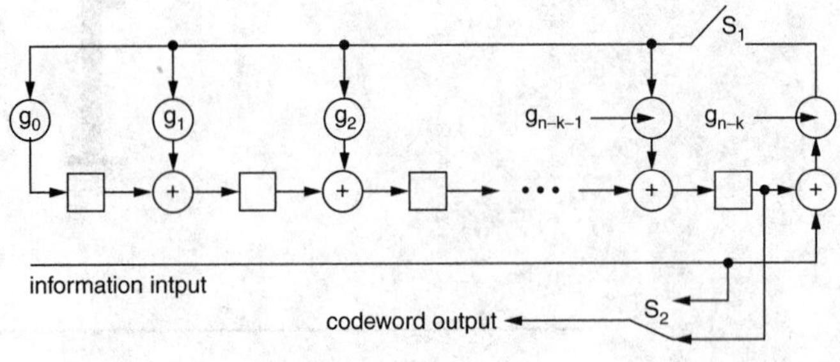

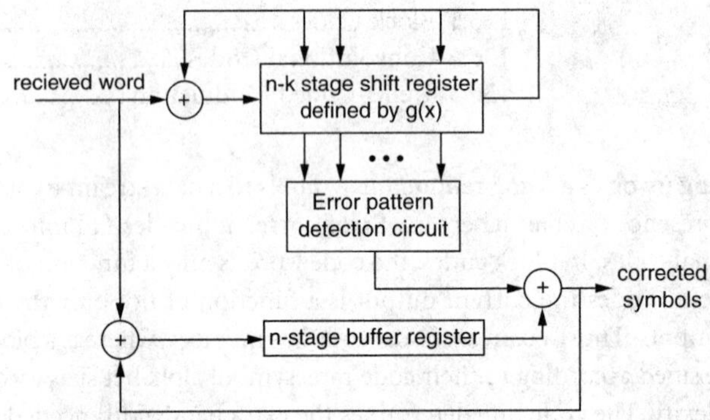

FIGURE 136.1 (a) Encoder for a cyclic group code; (b) Meggitt decoder for cyclic group codes.

word. In general, unless the number of parity symbols $(n - k)$ of a linear code is very small, syndrome decoding is too complicated and so codes with a simpler decoding procedure are used.

A cyclic code is a group code that has the additional property that any cyclic shift of a code word produces another code word. A cyclic code is specified by a generator polynomial, which is a polynomial of order $n - k$ of the form $g(x) = g_0 + g_1x + \cdots + g_{n-k}x^{n-k}$, or by its parity check polynomial, $h(x)$, which is related to $g(x)$ by $g(x)h(x) = x^n - 1$. The code is encoded with a division circuit (a linear feedback shift register) as shown in Figure 136.1(a). In that circuit, switches S_1 and S_2 are in the closed and up positions, respectively, until the k data symbols are clocked in. After that, the two switches are flipped and the parity symbols are read out. The received sequence can be decoded using a similar circuit known as the *Meggitt decoder*, shown in Figure 136.1(b). Most of the complexity of the decoder is in the error pattern detector, which must recognize all correctable error patterns with errors in the last position. For a single-error correcting code this can be implemented with a multiple-input AND gate, but for codes with large error correction capability the number of error patterns that must be recognized becomes unfeasible.

A technique known as *error-trapping decoding* can be used to simplify the error pattern detector of the Meggitt decoder, but again, this technique only works for short codes with low error-correcting capability. Error trapping is quite effective for decoding burst error correcting codes and also works nicely on the well-known Golay code. When longer codes with high error correction capability are needed, it is common to consider various subclasses of cyclic group codes that lead to implementable decoding algorithms. One such subclass that is extremely popular comprises the BCH codes. Since BCH codes are cyclic codes, they can be encoded with the same general encoder shown in Figure 136.1(a). In addition, a well-known technique called the *Berlekamp algorithm* can be used to decode BCH codes. The Berlekamp algorithm is fast and easy to implement, but it is not a maximum likelihood decoding technique, and so it does not always take advantage of the full error-correcting capability of the code. In addition, it is often desirable for a demodulator to feed soft decisions on to the decoder. This will usually enable the decoder to make a more reliable decision about which symbols are in error. A demodulator is said to make soft decisions when it quantizes its output into an alphabet size larger than the size of the transmitted symbol set. All of the standard techniques for decoding block codes suffer from the fact that they cannot be used on soft decisions; this is the main reason why convolutional codes are often preferred.

For a memoryless channel with an error rate of p, the probability of decoding error, P_d, for an (n, k) t-error correcting code is given by

$$P_d = \sum_{i=t+1}^{n} \binom{n}{i} p^i (1-p)^{n-i} = 1 - \sum_{i=0}^{t} \binom{n}{i} p^i (1-p)^{n-i}$$

Finding exact expressions for the probability of a symbol error, P_s (i.e., probability of bit error or bit-error rate for the case of binary symbols) is generally not tractable, but the following bounds work quite well:

$$\frac{d_{\min}}{n} P_d \le P_s \le \sum_{i=t+1}^{n} \binom{n}{i} \frac{i+t}{n} p^i (1-p)^{n-i}$$

Another method of specifying the performance of an error correction coding scheme focuses on coding gain. Coding gain is the reduction in dB of required signal-to-noise ratio needed to achieve a given error rate relative to an uncoded system. Specifically, if an uncoded system needs an energy per bit of $E_u(P_s)$ to achieve a symbol error rate of P_s, and a coded system requires $E_c(P_s)$, the coding gain is

$$\gamma(P_s) = 10\log\{[E_u(P_s)]/[E_c(P_s)]\} \text{ dB}$$

The asymptotic coding gain is the coding gain measured in the limit as P_s goes to zero and is given for a Gaussian noise channel by

$$\gamma = \lim_{P_s \to 0} \gamma(P_s) = 10\log\left(\frac{k}{n}(t+1)\right) \text{ dB}$$

136.2 Convolutional Codes

In a convolutional code, as in a block code, for each k information symbols, $n - k$ redundant symbols are added to produce n code symbols and the ratio of k/n is referred to as the code rate, R. The mapping, however, is not memoryless, and each group of n code symbols is a function of not only the current k input symbols but also the previous $K - 1$ blocks of k input symbols. The parameter K is known as the *constraint length*. The parameters n and k are usually small and the code word length is arbitrary (and potentially infinite). As with block codes, the ability to correct errors is determined by $d_{\min}$, but in the case of convolutional codes this quantity is generally referred to as the *free distance*, d_{free}. Although the error correction capability, t, has the same relationship to d_{free} as in block codes, this quantity has little significance in the context of convolutional codes.

A convolutional code is generated using a feed-forward shift register. An example of the structure for the convolutional encoder is shown in Figure 136.2(a) for a code with $R = 1/2$ and $K = 3$. In general, for each group of k symbols input to the encoder, n symbols are taken from the output. A convolutional code can be described graphically in terms of a state diagram or a trellis diagram, as shown in Figure 136.2(b) and Figure 136.2(c) for the encoder of Figure 136.2(a). The state diagram represents the operation of the encoder, whereas the trellis diagram illustrates all the possible code sequences. The states in both the state and trellis diagrams represent the contents of the shift register in the encoder, and the branches are labeled with the code output corresponding to that state transition.

The optimum (maximum likelihood) method for decoding convolutional codes is the Viterbi algorithm, which is nothing more than an efficient way to search through the trellis diagram for the most likely transmitted code word. This technique can be used equally well with either hard or soft

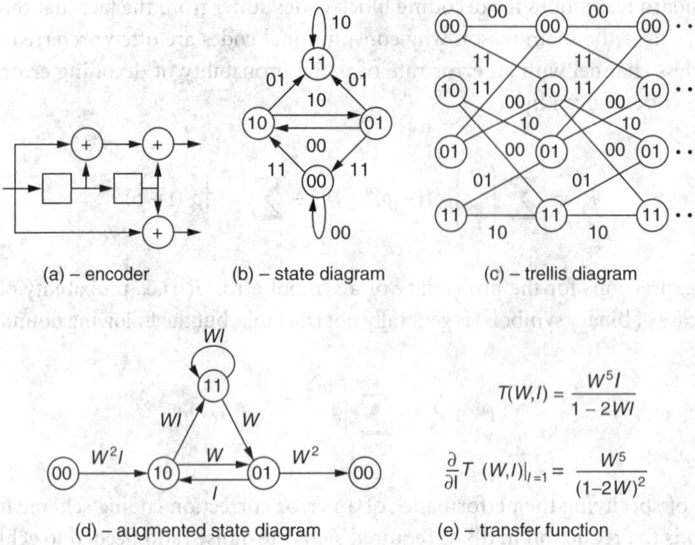

FIGURE 136.2 A rate 1/2, $K = 3$ convolutional encoder.

channel decisions and has a complexity that is exponential in the constraint length. As a result, the Viterbi algorithm can be used only on codes with a moderate constraint length (currently about $K \leq 10$). For most applications, this limit on K is not too severe and so Viterbi decoding is used almost exclusively. If a larger constraint length is needed, several suboptimal decoding algorithms exist whose complexities do not depend on the constraint length. Sequential decoding techniques such as the Stack and Fano algorithms are generally nearly optimum in terms of performance and faster than Viterbi decoding, whereas majority logic decoding is very fast, but generally gives substantially worse performance.

The performance of a convolutionally encoded system is specified in terms of the decoded symbol error probability or in terms of a coding gain, as defined in Section 136.1. An exact result for the error probability cannot be easily found but a good upper bound is provided through the use of transfer functions. To find the transfer function for a given code, an augmented state diagram is produced from the original state diagram of the code by splitting the all-zero state into a starting and ending state and labeling each branch with $W^w I^i$, where w is the Hamming weight of the output sequence on that branch, i is the Hamming weight of the input that caused that transition, and W and I are dummy variables. The transfer function of the code is then given by applying Mason's gain formula to the augmented state diagram. The augmented state diagram and the transfer function for the encoder in Figure 136.2(a) are shown in Figure 136.2(d) and Figure 136.2(e). The probability of decoded symbol error can be found from the transfer function by

$$P_e \leq \frac{1}{k}\frac{\partial}{\partial I}T(W,I)\bigg|_{I=1,W=Z}$$

The parameter Z that appears in this equation is a function of the channel being used. For a binary symmetric channel (BSC), $Z = \sqrt{4p(1-p)}$, where p is the crossover probability of the channel; for an additive white Gaussian noise (AWGN) channel with antipodal signaling, $Z = \exp(-E_s/N_0)$, where E_s is the average energy per symbol, and $N_0/2$ is the two-sided power spectral density of the noise.

Although the given bound is convenient because it is general, it is often not very tight. A tighter bound can be obtained for various channel models. For example, for the AWGN channel, a tighter bound is

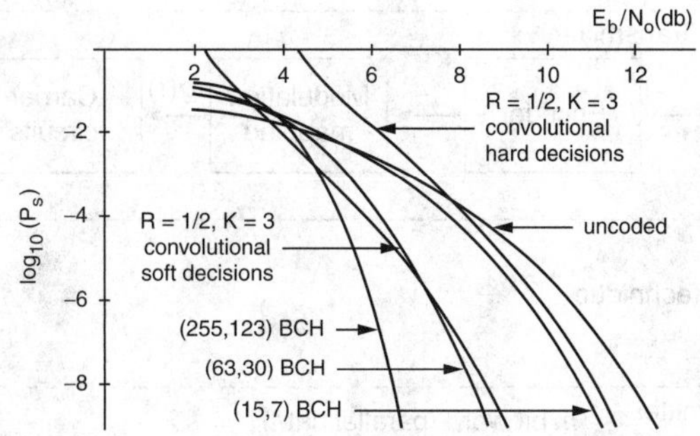

FIGURE 136.3 Bit-error-rate comparison of block and convolutional codes.

$$P_e \le \frac{1}{k}\exp\left(\frac{d_{\text{free}}E_s}{N_0}\right)Q\left(\sqrt{\frac{2d_{\text{free}}E_s}{N_0}}\right)\frac{\partial}{\partial I}T(W,I)\Bigg|_{I=1,W=\exp(-E_s/N_0)}$$

This bound generally gives a good estimate of the true error probability, at least for high values of E_s/N_0. For complex codes (long constraint length), calculating the transfer function can be tedious. In that case, it is common to use a lower bound that involves just the first term in the Taylor series expansion of the transfer function, resulting in

$$P_e \ge \frac{B_{d_{\text{free}}}}{k}Q\left(\sqrt{\frac{2d_{\text{free}}E_s}{N_0}}\right)$$

where $B_{d_{\text{free}}}$ is the sum of the input Hamming weights of all paths with an output Hamming weight of d_{free}. In many cases $B_{d_{\text{free}}} = 1$, and this can always be used as a lower bound. This expression can also be used to obtain the expression for asymptotic coding gain:

$$\gamma = 10\log\left(\frac{kd_{\text{free}}}{2n}\right)\text{ dB}$$

This result is very similar to the expression for asymptotic coding gain for block codes.

Finally, Figure 136.3 shows a performance comparison of the rate 1/2, $K = 3$ convolutional code depicted in Figure 136.2 with several block codes whose code rates are nearly equal to 1/2. The bit-error rate is plotted as a function of E_b/N_0 (energy per bit divided by noise spectral density). For all cases the assumed modulation is BPSK. Note that even this simple convolutional code with soft decisions can perform as well as a fairly long (and hence complex) block code.

136.3 Trellis-Coded Modulation

In both block and convolutional coding a savings in power (coding gain) is obtained at the cost of extra bandwidth. In order to provide the redundancy, extra symbols are added to the information stream. Accommodating these symbols without sacrificing information rate requires a shorter symbol duration or, equivalently, a larger bandwidth. The bandwidth expansion relative to an uncoded system is equal to $1/R = n/k$. Thus, coding is used when power is at a premium but bandwidth is available — that is, a power-limited environment. When power savings is not so crucial and bandwidth is at a premium (a bandwidth-limited environment), spectrum can be conserved by using a multilevel signaling scheme (some form of M-ary PSK or M-ary quadrature amplitude modulation is popular in this case). When

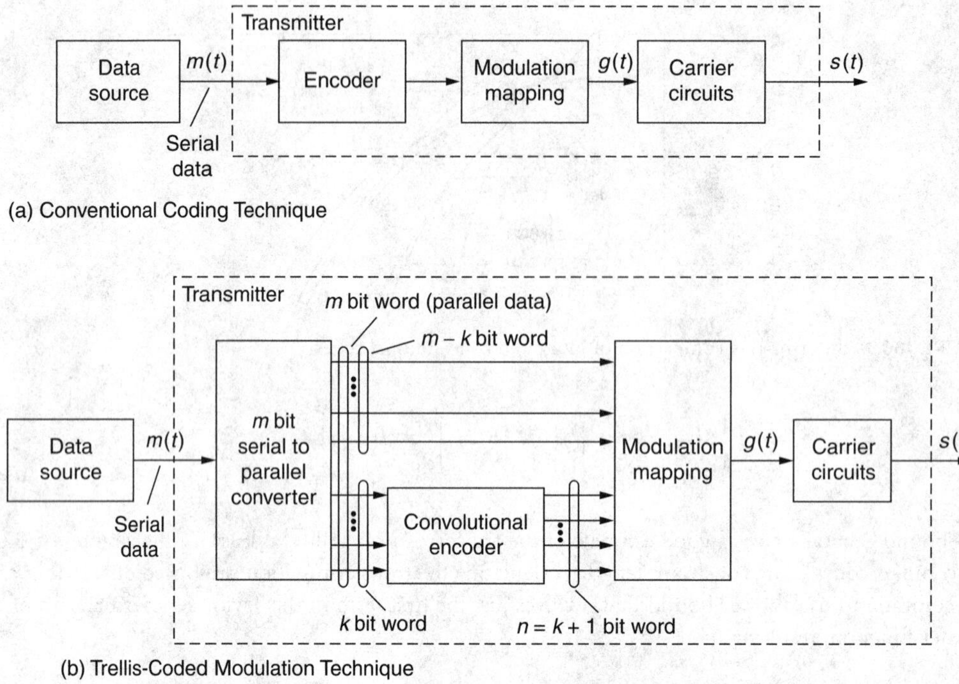

FIGURE 136.4 Block diagram of transmitter for TCM. (*Source:* Couch, L. W. 1993. *Digital and Analog Communication Systems*, 4th ed. Macmillan, New York. With permission.)

both bandwidth and power are tightly constrained, trellis-coded modulation (TCM) provides power savings with no bandwidth expansion. Redundancy is added to the data stream through the expansion of the size of the signal set. For example, binary data could be sent using a QPSK signal set. Since a QPSK constellation has four points and thus could represent two bits, each QPSK signal would represent one data bit and one parity bit (a rate 1/2 code). Alternatively, one could use 8-PSK with a rate 2/3 code and send two bits of information per symbol. As a general rule of thumb, if it is desired to send m bits of information per channel symbol, the size of the signal set is doubled and a rate $m/(m+1)$ code is used.

A general block diagram of a TCM encoder is shown in Figure 136.4. Data bits are grouped into m bit blocks. These blocks are then input to a rate $m/(m+1)$ binary convolutional encoder. The $m+1$ output bits are then used to select a signal from an $M = 2^{m+1}$-ary constellation. The constellation is usually restricted to being one- or two-dimensional in order to save bandwidth. The encoding is often done in a systematic manner as indicated in Figure 136.4, where $m - k$ of the input bits are left uncoded and the remaining k bits are encoded using a rate $k/(k+1)$ encoder. As with convolutional codes, the TCM scheme is designed to maximize the free distance of the code, but in this case distance is measured as Euclidean distance in the signal constellation. Each path through the trellis diagram now represents a sequence of signals from the constellation, and performance is mainly determined by the minimum Euclidean distance between any two distinct sequences of signals.

At the receiver, the received signal is demodulated using soft decisions and then decoded using the Viterbi algorithm. Performance is determined in a manner very similar to that for a convolutional code. Bounds on the bit-error rate can be found using transfer function techniques, although the method to obtain these transfer functions can be more complicated if the signal constellation does not exhibit a great deal of symmetry. Also, a coding gain relative to an uncoded system can be calculated as

$$\gamma = 20\log\left(\frac{d_{\text{free}}}{d_{\text{uncoded}}}\right)\ \text{dB}$$

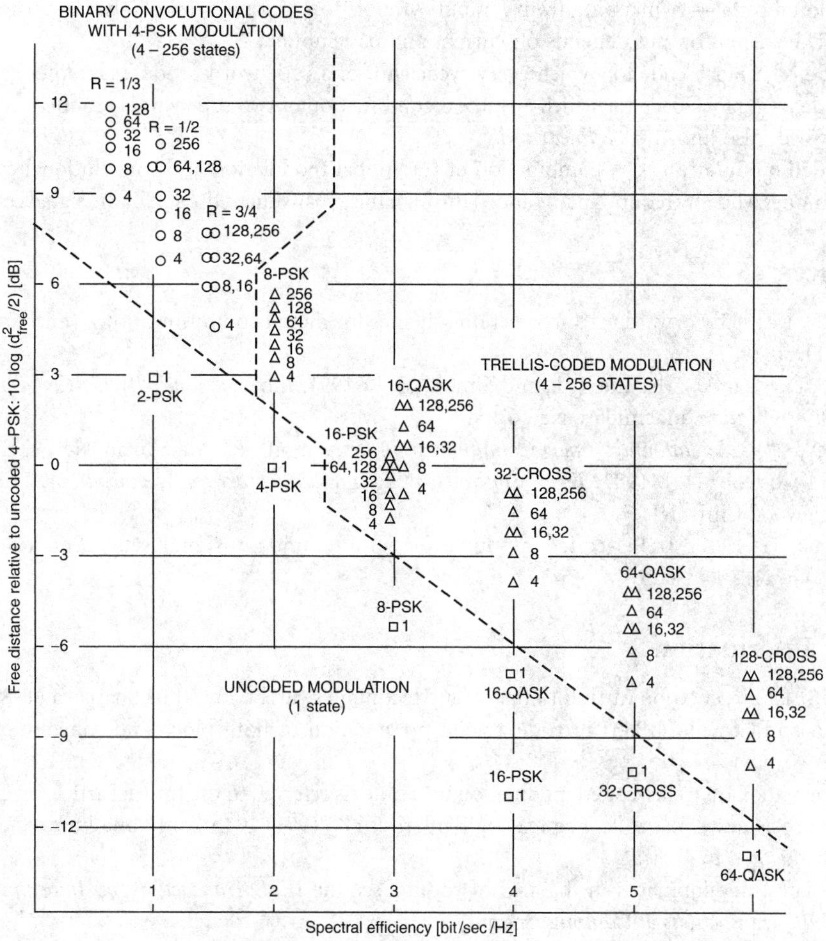

FIGURE 136.5 Free distance of binary convolution codes with 4-PSK modulation, and TCM with a variety of two-dimensional modulation schemes. (*Source:* Ungerboeck, G. 1987. Trellis-coded modulation with redundant signal sets, Part I: Introduction. *IEEE Comm. Mag.*, 25(2):5–11. ©1987 IEEE. With permission.)

where d_{uncoded} is the Euclidean distance between nearest neighbors in the uncoded systems signal constellation. Figure 136.5 shows some coding gains that can be achieved using TCM. To calculate coding gain relative to an uncoded system using this figure, pick the desired spectral efficiency, r (measured in bits/second/Hz), on the horizontal axis and then measure the vertical distance between the uncoded scheme and the TCM scheme. For example, if it is desired to send three bits per channel symbol, an uncoded 8-PSK signal constellation could be used. Using TCM with 16-QASK (i.e., 16-QAM) would give about 4.4 to 7.4 dB coding gain depending on the number of states in the TCM scheme, and the spectral efficiency would be 3 bits/second/Hz in both cases.

Defining Terms

Block code — A memoryless mapping from k input symbols to n output symbols.

Code rate — The ratio of the number of input symbols to the number of output symbols for a block or convolutional encoder.

Constraint length — The number of input blocks that affect a current output block in a convolutional encoder. It is also a measure of the complexity of the code.

Convolutional code — A mapping from k input symbols to n output symbols that is not memoryless. The current output depends on current and past inputs.

Cyclic code — A block code for which every cyclic shift of a code word produces another code word.

Group code — A block code for which any linear combination of two code words produces another code word. Also known as a *linear code.*

Trellis-coded modulation — A combination of traditional modulation and convolutional coding techniques whereby redundancy is added through the expansion of the size of the signal constellation.

References

Bhargava, V. K. 1983. Forward error correction schemes for digital communications. *IEEE Comm. Mag.*, 21(1):11–19.

Biglieri, E., Divsalar, D., McLane, P. J., and Simon, M. K. 1991. *Introduction to Trellis-Coded Modulation with Application.* Macmillan, New York.

Couch, L. W. 1993. *Digital and Analog Communication Systems,* 4th ed. Macmillan, New York.

Lin, S. and Costello, D. J. 1983. *Error Control Coding: Fundamentals and Applications.* Prentice Hall, Englewood Cliffs, NJ.

Ungerboeck, G. 1987. Trellis-coded modulation with redundant signal sets. *IEEE Comm. Mag.*, 25(2):5–21.

Further Information

A good tutorial presentation of traditional coding techniques can be found in Bhargava (1983). For full details, Lin and Costello (1983) provide a good presentation of both block and convolutional coding techniques.

For information on trellis-coded modulation the reader is referred to the tutorial article by Ungerboeck (1987), which is quite readable. The text by Biglieri et al. (1991) is the only one known to us that is completely devoted to TCM.

For the latest developments in the area of coding, see the *IEEE Transactions on Information Theory* and the *IEEE Transactions of Communications.*

137

Computer Communication Networks

J. N. Daigle
University of Mississippi

Over the last decade, computer communication networks have evolved from the domain of research and business tools for the few into the mainstream of public life. We have seen continued explosive growth of the Internet, a worldwide interconnection of computer communication networks that allows low-latency person-to-person communications on a global basis. Advancements in *firewall* technology have made it possible to conduct business using the Internet with significantly reduced fear of compromise of important private information, and broad deployment of *World Wide Web*, or simply *Web*, technology has facilitated the creation of new network-based businesses that span the globe. Significant steps have been taken to extend networking services to the mobile consumer. Perhaps more significantly, the introduction of high-speed Internet access to homes and the introduction of the point-to-point protocol (PPP) have made it possible to extend the Internet to every home that has at least a basic twisted-pair telephone line.

The potential for using networking technology as a vehicle for delivering multimedia — voice, data, image, and video — presentations, and indeed multiparty, multimedia conferencing service, has been demonstrated, and the important problems that must be solved in order to realize this potential are rapidly being defined and focused upon. User-friendly applications that facilitate navigation within the Web have been developed and made available to networking users on a non-fee basis via network servers, thus facilitating virtually instantaneous search and retrieval of information on a global basis. For example, it is now common for users to access broadcasts of exciting events, such as Le Tour de France, in real time over the Internet using only a dial-up telephone connection. Access to details concerning events of all kinds to all people is unprecedented in history.

By definition, a *computer communication network* is a collection of applications hosted on different machines and interconnected by an infrastructure that provides communications among the communicating entities. While the applications are generally understood to be computer programs, the generic model includes the human being as an application, an example being the people involved in a telephone call.

This article summarizes the major characteristics of computer communication networks. Our objective is to provide a concise introduction that will allow the reader to gain an understanding of the key distinguishing characteristics of the major classes of networks that exist today and some of the issues involved in the introduction of emerging technologies.

0-8493-1586-7/05/$0.00+$1.50

There are a significant number of well-recognized books in this area. Among these are the excellent texts by Schwartz [1987] and Spragins [1991], and more recently Kurose and Ross [2001], which have enjoyed wide acceptance both by students and practicing engineers and cover most of the general aspects of computer communication networks. Other books found to be especially useful by many practitioners are Rose [1990], Black [1991], and Bertsekas and Gallagher [1987].

The latest developments are, of course, covered in the current literature, conference proceedings, formal standards documents, and the notes of standards meetings. A pedagogically oriented magazine that specializes in computer communications networks is *IEEE Network*, but *IEEE Communications* and *IEEE Computer* often contain interesting articles in this area. *ACM Communications Review*, in addition to presenting pedagogically oriented articles, often presents very useful summaries of the latest standards activities. Major conferences that specialize in computer communications include the IEEE INFOCOM and ACM SIGCOMM series, which are held annually. It is becoming common at this time to have more and more discussion about personal communication systems, and the mobility issues involved in communication networks are often discussed in *IEEE Network* and *IEEE Personal Communication Systems*. There are also numerous journals, magazines, and conferences specializing in subsets of computer communications technologies.

We begin our discussion with a brief statement of how computer networking came about and a capsule description of the networks that resulted from early efforts. Networks of this generic class, called *wide area networks* (WANs) are broadly deployed today and there are still a large number of unanswered questions with respect to their design. The issues involved in the design of those networks are basic to the design of most networks, whether wide area or otherwise. In the process of introducing these early systems, we describe and contrast three basic types of communication switching: circuit, message, and packet.

We next turn to a discussion of computer communication *architecture*, which describes the structure of communication-oriented processing software within a communication-processing system. We introduce the International Standards Organization/Open Systems Interconnection (ISO/OSI) reference model (ISORM). Initially, the ISORM model was intended to guide the development of protocols, but its most important contribution seems to have been the provision of a framework for discussion of issues and developments across the communications field in general and communication networking in particular. We then present the Internet reference model, which parallels the Internet protocol stack that has evolved, and we base our discussion of protocols on selections from the Internet protocol suite. This discussion is necessarily simplified in the extreme, as thorough coverage requires hundreds of pages, but we hope that our brief description will enable the reader to appreciate some of the issues.

Having introduced the basic architectural structure of communication networks, we next turn to a discussion of an important variation on this architectural scheme: the *local area network* (LAN). Discussion of this topic is important because it helps to illustrate what a reference model is and what it is not. Specifically, early network architectures anticipate networks in which individual node pairs are interconnected via a single link, and connections through the network are formed by concatenating node-to-node connections.

LAN architectures, on the other hand, anticipate all nodes being interconnected in some fashion over the same communication link (or medium). This, then, introduces the concept of adaption layers in a natural way. It also illustrates that if the services provided by an architectural layer are carefully defined, then the services can be used to implement virtually any service desired by the user, possibly at the price of some inefficiency.

Next, we discuss asynchronous transfer mode and frame relay, two important link-layer technologies. We conclude with a brief discussion of recent developments.

137.1 General Networking Concepts

Data communication networks have existed since about 1950. The early networks existed primarily for the purpose of connecting users of a large computer to the computer itself, with additional capability to provide communications between computers of the same variety and having the same operating software. The

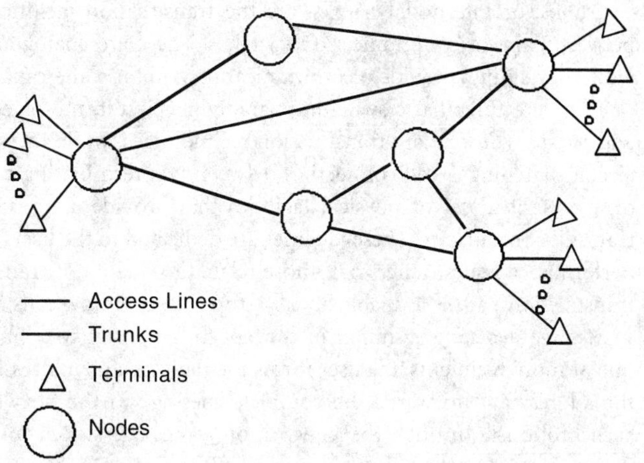

FIGURE 137.1 Generic computer communication network.

lessons learned during the first 20 or so years of operation of these types of networks have been valuable in preparing the way for modern networks. For the purposes of our current discussion, however, we think of communication networks as being networks whose purpose is to interconnect a set of applications that are implemented on hosts manufactured by possibly different vendors and managed by a variety of operating systems. Networking capability is provided by software systems that implement standardized interfaces specifically designed for the exchange of information among heterogeneous computers.

The earliest effort to develop large-scale, general-purpose networking capability based on packet switching was lead by the Advanced Research Projects Agency (ARPA) of the Department of the Army in the late 1960s; this effort resulted in the computer communication network called ARPANET. The end results of the ARPA networking effort, its derivatives, and the early initiatives of many companies such as AT&T, DATAPOINT, DEC, IBM, and NCR have been far-reaching in the extreme. We will concentrate on the most visible product of these efforts, which is a collection of programs that allows applications running on different computers to intercommunicate. Before turning to our discussion of the software, however, we provide a brief description of a generic computer communication network.

Figure 137.1 shows a diagram of a generic computer communication network. The most visible components of the network are the *terminals*, *access lines*, *trunks*, and *switching nodes.* Work is accomplished when the users of the network, the terminals, exchange messages over the network.

The terminals, which are usually referred to as end systems, represent the set of communication-terminating equipment communicating over the network. Equipment in this class includes, but is not limited to, user terminals, general-purpose computers, and database systems. This equipment, either through software or human interaction, provides the functions required for information exchange between pairs of application programs or between application programs and people. The functions include, but are not limited to, call setup, session management, and message transmission control. Examples of applications include electronic mail transfer, WWW browsing, and playback of audio streams. An extensive discussion of applications is provided in Kurose and Ross [2001].

Access lines provide for data transmission between the terminals and the network switching nodes. These connections may be set up on a permanent basis or they may be switched connections, and there are numerous transmission schemes and protocols available to manage these connections. The essence of these connections, however, from our point of view is a channel that provides data transmission at some number of bits per second (bps), called the channel capacity, C. The access line capacities may range from a few hundred bps to in excess of millions of bps, and they are usually not the same for all terminating equipments of a given network. The actual information-carrying capacity of the link depends on the protocols employed to effect the transfer; the interested reader is referred to Bertsekas and Gallagher [1987], especially chapter 2, for a general discussion of the issues involved in data transmission over communication links.

Trunks, or internodal trunks, are the transmission facilities that provide for transmission of data between pairs of communication switches. These are analogous to access lines, and from our point of view, they simply provide a communication path at some capacity specified in bps.

There are three basic switching paradigms: circuit, message, and packet switching. *Circuit switching* and *packet switching* are transmission technologies, while message switching is a service technology. In circuit switching, a call connection between two terminating equipments corresponds to the allocation of a prescribed set of physical facilities that provide a transmission path of a certain bandwidth or transmission capacity. These facilities are dedicated to the users for the duration of the call. The primary performance issues, other than those related to quality of transmission, are related to whether or not a transmission path is available at call setup time and how calls are handled if facilities are not available.

Message switching is similar in concept to the postal system. When a user wants to send a message to one or more recipients, the user forms the message and addresses it. The message switching system reads the address and forwards the complete message to the next switch in the path. The message moves asynchronously through the network on a message switch-to-message switch basis until it reaches its destination. Message switching systems offer services such as mailboxes, multiple destination delivery, automatic verification of message delivery, and bulletin boards. Communication links between the message switches may be established using circuit or packet switching networks as is the case with most other networking applications. An example of a message-switching protocol that has been used to build message-switching systems is the simple mail-transfer protocol (SMTP).

In the circuit-switching case, there is a one-to-one correspondence between the number of trunks between nodes and the number of simultaneous calls that can be carried. That is, a trunk is a facility between two switches that can service exactly one call, and it does not matter how this transmission facility is derived. Major design issues include the specification of the number of trunks between node pairs and the routing strategy used to determine the path through a network in order to achieve a given call-blocking probability. When blocked calls are queued, the number of calls that may be queued is also a design question.

A packet-switched communication system exchanges messages among users by transmitting sequences of packets comprising the messages. That is, the sending terminal equipment partitions a message into a sequence of packets, the packets are transmitted across the network, and the receiving terminal equipment reassembles the packets into messages. The transmission facility interconnecting a given node pair is viewed as a single trunk, and the transmission capacity of this trunk is shared among all users whose packets traverse both nodes. While the trunk capacity is specified in bps, the packet-handling capacity of a node pair depends both upon the trunk capacity and the nodal-processing power.

In some packet-switched networks, the path traversed by a packet through the network is established during a call-setup procedure, and the network is referred to as a virtual circuit-packet–switching network. Other networks provide datagram service, which allows users to transmit individually addressed packets without the need for call setup. Datagram networks have the advantage of not having to establish connections before communications take place, but have the disadvantage that every packet must contain complete addressing information. Virtual-circuit networks have the advantage that addressing information is not required in each packet, but have the disadvantage that a call setup must take place before communications can occur. The datagram is an example of *connectionless service*, while virtual circuit is an example of *connection-oriented service.*

Prior to the late 1970s, signaling for circuit establishment was in-band. That is, in order to set up a call through the network, the call-setup information was sent sequentially from switch to switch using the actual circuit that would eventually become the circuit used to connect the end users. In an extreme case, this amounted to trying to find a path through a maze, sometimes having to retrace one's steps before finally emerging at the destination or just simply giving up when no path could be found. This had two negative characteristics: first, the rate of signaling-information transfer was limited to the circuit speed, and second, the circuits that could have been used for accomplishing the end objective were being consumed simply to find a path between the endpoints. This resulted in tremendous bottlenecks on major holidays, which were solved by virtually disallowing alternate routes through the toll-switching network.

An alternate out-of-band signaling system, usually called *common channel interoffice signaling* (CCIS), was developed primarily to solve this problem. Signaling now takes place over a signaling network that is partitioned from the network that carries the user traffic. This principle is incorporated into the concept of integrated services digital networks (ISDNs), which is described thoroughly in Helgert [1991]. The basic idea of ISDN is to offer to the user some number of 64 Kbps access lines plus a 16-Kbps access line through which the user can describe to an ISDN how the user wishes to use each of the 64-Kbps circuits at any given time. The channels formed by concatenating the access lines with the network inter-switch trunks having the requested characteristics are established using an out-of-band signaling system, the most modern of which is known as *signaling system#7* (SS#7).

In either virtual circuit or *datagram networks*, packets from a large number of users may simultaneously need transmission services between nodes. Packets arrive to a given node at random times. The switching node determines the next node in the transmission path, and then places the packet in a queue for transmission over a trunk facility to the next node. Packet-arrival processes tend to be bursty, that is, the number of packet arrivals over fixed-length intervals of time has a large variance. Because of the burstiness of the arrival process, packets may experience significant delays at the trunks. Queues may also build due to the difference in transmission capacities of the various trunks and access lines and delays result. Processing is also a source of delay, and the essence of packet-switching technology is to trade delay for efficiency in resource utilization.

Protocol design efforts, which seek to improve network efficiencies and application performance, are frequent topics of discussion at both general conferences in communications and those specialized to networking. The reader is encouraged to consult the proceedings of the conferences mentioned earlier for a better appreciation of the range of issues and the diversity of the proposed solutions to the issues.

137.2 Computer Communication Network Architecture

In this section, we begin with a brief, high-level, definition of the ISORM, which is discussed in significant detail in Black [1991], and then we turn to a more detailed discussion of a practical Internet reference model (PIRM). The ISORM has seven layers, none of which can be bypassed conceptually. In general, a layer is defined by the types of services it provides to its users and the quality of those services. For each layer in the ISO/OSI architecture, the user of a layer is the next layer up in the hierarchy, except for the highest layer for which the user is an application. Clearly, when a layered architecture is implemented under this philosophy, then the quality of service obtained by the end user, the application, is a function of the quality of service provided by all of the layers.

Figure 137.2, adapted from Spragins [1991], shows the basic structure of the OSI architecture and how this architecture is envisaged to provide for exchange of information between applications. As shown in the figure, there are seven layers: application, presentation, session, transport, network, data link, and physical. Brief definitions of the layers are now given, but the reader should bear in mind that substantial further study will be required to develop an understanding of the practical implications of the definitions.

Physical layer: Provides electrical, functional, and procedural characteristics to activate, maintain, and deactivate physical data links that transparently pass the bit stream for communication between data link *entities.*

Data link layer: Provides functional and procedural means to transfer data between network entities; provides for activation, maintenance, and deactivation of data link connections, character and frame synchronization, grouping of bits into characters and frames, error control, media access control, and flow control.

Network layer: Provides switching and routing functions to establish, maintain, and terminate network layer connections and transfer data between transport layers.

Transport layer: Provides host-to-host, cost-effective, transparent transfer of data, end-to-end flow control, and end-to-end quality of service as required by applications.

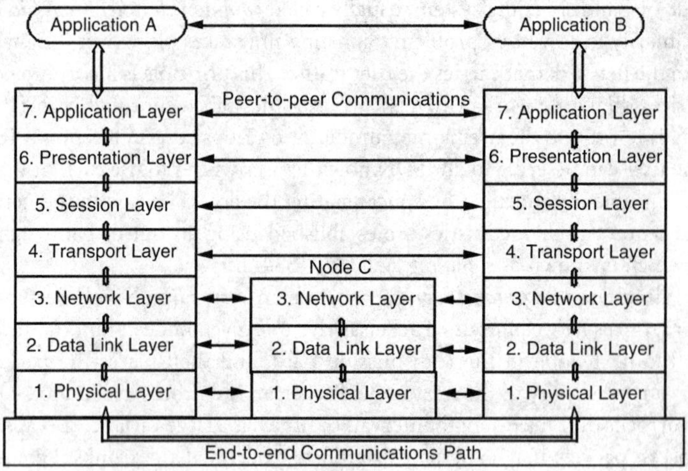

FIGURE 137.2 ISO reference model. (Adapted from Spragins, J. D. *Telecommunications: Protocols and Design.* Addison-Wesley, Reading, MA, 1991. With permission.)

Session layer: Provides mechanisms for organizing and structuring dialogues between application processes.

Presentation layer: Provides for independent data representation and syntax selection by each communicating application and conversion between selected contexts and the internal architecture standard.

Application layer: Provides applications with access to the ISO/OSI communication stack and certain distributed information services.

As we have mentioned previously, a layer is defined by the types of services it provides to its users. In the case of a request or a response, these services are provided via invocation of *service primitives* of the layer in question by the layer that wants the service performed. In the case of an indication or a confirm, these services are provided via invocation of service primitives of the layer in question by the same layer that wants the service performed.

The process of invoking a service primitive is not unlike a user of a programming system calling a subroutine from a scientific subroutine package in order to obtain a service, say, matrix inversion or memory allocation. For example, a request is analogous to a CALL statement in a FORTRAN program, and a response is analogous to the RETURN statement in the subroutine that has been called. The requests for services are generated asynchronously by all of the users of all of the services and these join (typically prioritized) queues along with other requests and responses while awaiting servicing by the processor or other resource such as a transmission line.

The service primitives fall into four basic types, which are as follows: request, indication, response, and confirm. These types are defined as follows:

Request: A primitive sent by layer $(N + 1)$ to layer N to request a service.

Indication: A primitive sent by layer N to layer $(N + 1)$ to indicate that a service has been requested of layer N by a different layer $(N + 1)$ entity.

Response: A primitive sent by $(N + 1)$ to layer N in response to an *indication primitive.*

Confirm: A primitive sent by layer N to layer $(N + 1)$ to indicate that a response to an earlier *request* primitive has been received.

To understand the basic ideas of a network, it is useful to first ask the question: "What functionality would be needed in a point-to-point connection between two computers that are hard-wired together?" We would find that we need everything except the functions provided by layer 3; there is no need for switching and routing functions. Thus, it is natural to think of the networking aspect of a computer

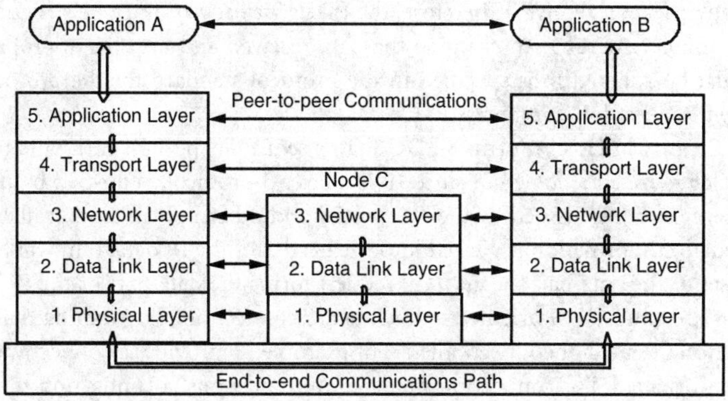

FIGURE 137.3 Practical Internet reference model.

communication network as the interconnection of layer three entities. In fact, we can think of the network as the collection of layer 3 entities. Everything above layer 3 facilitates activities that are taking place over the network, and everything below layer 3 is there simply for the purpose of connecting the layer 3 entities together.

As a practical matter, networking has evolved such that the more realistic reference model has five, rather than seven, layers. In essence, the session and presentation layers are absorbed into the application layer, with the resulting reference model as shown in Figure 137.3.

In order to be more specific about how communications takes place, we now turn to a brief discussion of Internet protocol (IP), which is the network layer protocol of the Internet. The IP protocol provides a datagram delivery service to the transport layer. In order to do this, IP provides exactly one primitive to its upper layer protocol (ULP): SEND.

As mentioned earlier, an upper layer protocol issuing a primitive is equivalent to a computer program calling a subroutine or procedure. A *primitive* has an associated set of *formal parameters*, which are analogous to the formal parameters of a procedure in a programming language. In this case, the parameters for the SEND.request primitive are the source address, the destination address, the type-of-service parameter set, the data length, options parameters, and the data. These parameters provide all the information needed by IP to deliver data to the required destination. The data itself is the entity that the ULP wants sent. The source address is the Internet address of the ULP requesting the SEND service and the destination address is the Internet address of the ULP to which the data is to be sent. The type of service and options parameters give IP information about how the delivery is to be handled, such as the priority of the send action. The data length parameter tells IP how much data it has to send.

In general, communications take place between peer layer protocols by the exchange of *protocol data units* (PDUs), which contain all of the information required for the receiving protocol *entity* to provide the required service. For the PIRM, we say that applications exchange *messages*, layer 4 entities exchange *segments*, layer 3 entities exchange *datagrams*, layer 2 entities exchange *frames*, and layer 1 protocols exchange *1-PDUs*. It should be mentioned that datagrams are packets that traverse networks using connectionless layer 3 protocols. Equivalently, we can say that the packet-switched Internet is a datagram network because IP is the only layer 3 protocol, and it provides connectionless service.

In order to exchange PDUs, entities of a given layer use the services of the next lower layer. To make things concrete, assume that the ULP sending the information is a transmission control protocol (TCP) entity in Host A and the ULP to which the information is being sent is the TCP entity in Host B. In the case of the SEND discussed previously, the TCP entity in Host A is using the SEND service to send the information in the data field to its peer destination TCP entity in Host B; the content of this data field is the PDU and it is called a segment since TCP is a layer 4 protocol. At the destination, Host B, IP uses its DELIVER primitive to deliver the packet to the TCP entity in Host B. Thus, the TCP's packet is the data contained in the data field of its SEND request. The destination TCP entity must be able to determine what to do with any packet it receives. Therefore, the PDUs, or packets, must be structured in a standard way so that a computer program can examine the data received and take the appropriate actions. The structure that must be adhered to is specified in the protocol standard for the protocol. For example, TCP is specified in Request for Comments RFC 793.

Similarly, the IP entity in Host A forms a PDU that is sent to its peer IP entity in another mode. The IP layer PDU, which is called a *datagram* since IP is a layer 3 protocol, is formed by using information provided in the service request issued by the TCP entity and other information available in the IP layer entity itself. Instructive information is packed into the header of the IP datagram while the data part, or sometimes a subset of the data part, of the TCP request forms the data part of the IP datagram.

Some protocols are connection oriented and some are connectionless, meaning that the service they provide is connection oriented or connectionless. For example, IP provides a connectionless service while TCP is connection oriented. For connection mode communications, a connection must be established between two peer entities before they can exchange PDUs. Therefore, connection-oriented protocols must provide primitives that facilitate establishment of a connection. For example, TCP has several forms of OPEN primitives that an application can use initiate connections, and TCP also has CLOSE and ABORT primitives that an application can use to terminate connections.

Once a connection is established, data exchange between the two application layer entities can take place; that is, the entities can exchange PDUs. For example, if an application layer entity in Host A wishes to send a PDU to an application layer entity in Host B, the application layer entity in Host A would issue a T_SEND.request to the appropriate transport link layer entity in Host A. This entity would package the PDU together with appropriate control information into a transport service data unit (TSDU) and send it to its peer transport layer entity in Host B. In order to do this, the transport layer would invoke the services of the network layer. In turn, the network layer would invoke the services of the data link layer and so on. Specifically, the network layer entity in Host A would issue a DL_DATA.request to the appropriate data link layer entity in Host A. This entity would package the PDU together with appropriate control information into a data link service data unit (DLSDU) and send it to its peer at C. The peer data link entity at C would extract the layer-3 PDU deliver it to the network entity at C, which would forward it to the data link entity in C providing the connection to Host B. This entity would then send the DLSDU to its peer in Host B, and this data link entity would extract the layer 3 PDU and pass it to the Host B network entity via a DL_DATA.indication. The receiving network layer entity would extract the SDU and forward it to the transport layer in Host C using the N_Data.indication.

Now, network layer PDUs are called packets and DL layer PDUs are called frames. But, the data link layer does not know that the information it is transmitting is a packet; to the DL layer entity, the packet is simply user information. From the perspective of a data link entity, it is not necessary to have a network layer. The network layer exists to add value for the user of the network layer to the services provided by the DL layer. In the example above, the network layer added value by providing a relaying capability since Hosts A and C were not directly connected. Similarly, the DL layer functions on a hop-by-hop basis, each hop being completely unaware that there are any other hops involved in the communication. We will see later that the data link need not be limited to a single physical connection.

We now turn to a discussion of LANs, which have inherent properties that make the use of sublayers particularly attractive.

137.3 Local Area Networks and Internets

In this section, we discuss the organization of communications software for LANs. In addition, we introduce the idea of internets, which were brought about to a large extent by the advent of LANs. We discuss the types of networks only briefly and refer the reader to the many excellent texts on the subject. Layers 4 and 5 for local area communications networks are identical to those of wide area networks. However, because the hosts communicating over a LAN share a single physical transmission facility, the routing functions provided by the network layer, layer 3, are not necessary. Thus, the functionality of a layer 3 within a single LAN can be substantially simplified without loss of utility. On the other hand, a

DL layer entity must now manage many simultaneous DL layer connections because all connections entering and leaving a host on a single LAN do so over a single physical link. Thus, in the case of connection-oriented communications, the software must manage several virtual connections over a single physical link.

There were several basic types of transmission schemes used in early local area networks. Three of these received serious consideration for standardization: the *token ring*, *token bus*, and *carrier sense multiple access* (CSMA). All three access methods became IEEE standards (the IEEE 802 series), and eventually became ISO standards (ISO 8802 series) because all merited standardization. On the other hand, all existed for the express purpose of exchanging information among peers, and it was recognized at the outset that the upper end of the data link layer could be shared by all three access techniques. The decision to use a common logical link control (LLC) sublayer for all of the LAN protocols and develop a separate sublayer for each of the different media apparently ushered in the idea of adaption sublayers, in this case, the media access control (MAC) sublayer. The most dominant today is the CDMA-based version, which is known as Ethernet or IEEE 802.3.

The idea of sublayering has proven to be valuable, as new types of technologies have become available. For example, the new fiber-distributed digital interface (FDDI) uses the LLC of all other LAN protocols, but its MAC is completely different from the token ring MAC even though FDDI is a token ring protocol. A thorough discussion of FDDI and related technologies is given in Jain [1994].

Metropolitan area networks (MANs) have been deployed for the interconnection of LANs within a metropolitan area. The primary media configuration for MANs is a dual bus configuration, and it is implemented via the distributed queue, dual bus (DQDB) protocol, also known as IEEE 802.6. The net effect of this protocol is to use the dual bus configuration to provide service approaching the FCFS service discipline to the traffic entering the FDDI network, which is remarkable considering that the LANs being interconnected are geographically dispersed. Interestingly, DQDB concepts have recently also been adapted to provide wide area communications. Specifically, structures have been defined for transmitting DQDB frames over standard DS-1 (1.544 megabits per second) and DS-3 (6.312 megabits per second) facilities, and these have been used as the basis for a service offering called switched multimegabit data services (SMDS).

One of the more interesting consequences of the advent of local area networking is that many traditional computer communication networks became internets overnight. LAN technology was used to connect stations to a host computer, and these host computers were already on a WAN. It was then a simple matter to provide a relaying, or bridging, service at the host in order to provide wide area interconnection of stations on LANs to each other. In short, the previously established WANs became networks for interconnection of LANs; that is, they were interconnecting networks rather than stations. Internet performance suddenly became a primary concern in the design of networks; a new business developed to provide specialized equipment, routers, to provide interconnection among networks.

Over the last decade, wireless LANs, which are local area networks in which radio or photonic links serve as cable replacements, have become common. Indeed, wireless LAN technology is now widely deployed in homes as well as in offices. Most of this technology features spread-spectrum wireless transmission with the media access technology being in the IEEE 802.11 family.

We now describe two broadly deployed technologies that provide link-layer connections in communication networks: *frame relay* (FR) technology, which is described in Braun [1994] and *asynchronous transfer mode* (ATM) technology, which is described in McDysan and Spohn [1994] and Pildush [2000].

As we have mentioned previously, there is really no requirement that the physical media between two adjacent data link layers be composed of a single link. In fact, if a path through the network is initially established between two data link entities, there is no reason that DLC protocols need to be executed at intermediate nodes. Through the introduction of adaption layers and an elementary routing layer at the top of the DL layer, DLC frames can be relayed across the physical links of the connection without performing the error checking, flow control, and retransmission functions of the DLC layer on a link-by-link basis. The motivation is that since link transmission is becoming more reliable, extensive error checking and flow control are not needed across individual links; an end-to-end check should be sufficient. Meanwhile, the savings in processing due to not processing at the network layer can be applied to frame processing, which allows interconnection of the switches at higher line speeds. Since bps costs decrease with increased line speed, service providers can offer savings to their customers through FRNs. Significant issues are frame loss probability and retransmission delay. Such factors will determine the retransmission strategy deployed in the network. The extensive deployment of FR technology at this time suggests that this technology provides improvements over standard packet technology.

Another recent innovation is the ATM. The idea of ATM is to partition a user's data into many small segments, called cells, for transmission over the network. Independent of the data's origin, the cell size is 53 octets, of which 5 octets are for use by the network itself for routing and error control. Users of the ATM are responsible for segmentation and reassembly of their data. Any control information required for this purpose must be included in the 48 octets of user information in each cell. In the usual case, these cells would be transmitted over networks that provide users with 135 Mbps or more data transmission capacity (with user overhead included in the capacity).

The segmentation of units of data into cells introduces tremendous flexibility for handling different types of information, such as voice, data, image, and video, over a single transmission facility. As a result, there has been a tremendous investment in developing implementation agreements that will enable a large number of vendors to independently develop interoperable equipment. This effort is focused primarily in the ATM Forum, a private, not-for-profit consortium of over 500 companies of which more than 150 are principal members and active contributors.

LANs, WANs, and MANs based on the ATM paradigm are being deployed as described in Pildush [2000], for example. ATM is an excellent technology for providing worldwide Ethernet LAN interconnection at a full rate of 10 or 100 megabits per second. Originally, numerous vendors were planning to have ATM capabilities at the back plane in much the same way that Ethernet has been provided, but the bulk of deployment of ATM today is at the link layer between IP routers.

137.4 ATM and Frame Relay

ATM technology was initially developed in hopes of extending ISDN technology toward the concept of a broadband integrated services data network that would provide bandwidth on demand to end users. The ATM architecture consists of three sublayers: the ATM adaption layer (AAL), the ATM layer, and the physical media dependent (PMD) layer. It is convenient for the purposes of this discussion to think of services as falling into two categories: circuit mode and packet mode, where a circuit mode service, such as voice, is a service that is naturally implemented over a circuit switched facility and a packet-mode service, such as e-mail, is a service that is more naturally implemented over a packet-switched connection. From many perspectives, it is natural to implement circuit-mode services directly over ATM while it is more natural to implement packet-mode services at the Internet (or packet) layer.

The implication of this partitioning of service types is that any service that has been developed for deployment over an IP network would naturally be deployed over an ATM network by simply using the ATM network as a packet delivery network. Each packet would traverse the network as a sequence of cells over an end-to-end virtual connection. If the flow control and resource management procedures can be worked out, the net effect of this deployment strategy would be, for example, that an application designed to be deployed over an Ethernet segment could be deployed on a nationwide (or even global) network without a noticeable performance degradation. The implications of this type of capability are obvious as well as mind-boggling.

137.5 Recent Developments

Broad deployment of networking technologies and wireless communications devices are the big story in recent times. A typical home deployment of networking technology today includes a router, which is at the interface of the home network and the outside world. Typically, the router has an IEEE 802.3

(Ethernet) port of the 100-Mbps variety as well as an IEEE 802.11 wireless LAN port facing the in-home side of the network. The connection from the home may be one of many technologies, ranging from a traditional relatively low-speed dialed connection over a phone line to relatively high-speed connection based on some form of digital subscriber loop (DSL) technology, such as asymmetric DSL (ADSL), which delivers service to the home in the range of hundreds of kilobytes per second to low megabytes per second. Connected together to form a network within a home may be a number of computers and a print server having an attached printer. A number of additional computers may be attached to the same network via wireless IEEE 802.11 connections. The router itself is usually connected to an Internet service provider (ISP) over the point-to-point protocol (PPP). All Internet applications running on all end systems (computers) in the home share the same PPP link-layer connection to the ISP port to the Internet.

Many of the services that network users enjoy today are the result of enhancements to the development of multimedia applications and protocols that are integrated into Web browsers. The Web browsers allow users to access search engines through which they can find interesting Web content. Once the content is found, the user can generally access the content quickly by simply clicking on a URL. When the object arrives at the user's system, the Web browser automatically invokes the services of a helper application, which renders the object to the user. For example, in the event that the object is a JPEG file, the Web browser will invoke an application that can display a picture that is stored in JPEG format.

At the time of this writing, networking technologies are being extended to the wireless domain as discussed in Lin and Chlamtac [2001] and Garg [2002]. The intention is to make all networking services available to mobile end systems using essentially the same infrastructure as the cellular telephone system. Alternatively stated, the intention is to evolve the wireless cellular phone system to a point where it is just part of the Internet. Meanwhile the intention is to evolve the Internet to the point where it handles all types of communications services. Thus, over the long term it is expected that communications types will be available over the Internet and that access to the Internet will be available to users at any time and place, whether the user is mobile or stationary. We are all looking forward to the end objective, the universal personal communications services (UPCS).

Defining Terms

Access line — A communication line that connects a user's terminal equipment to a switching node.

Adaption sublayer — Software that is added between two protocol layers to allow the upper layer to take advantage of the services offered by the lower layer in situations where the upper layer is not specifically designed to interface directly to the lower layer.

Architecture — The set of protocols defining a computer communication network.

Asynchronous transfer mode (ATM) — A mode of communication in which communication takes place through the exchange of tiny units of information called *cells*.

Broadband Integrated Services Digital Network (B-ISDN) — A generic term that generally refers to the future network infrastructure that will provide ubiquitous availability of integrated voice, data, imagery, and video services.

Carrier sense multiple access — A random access method of sharing a bus-type communication medium in which a potential user of the medium listens before beginning to transmit.

Circuit Switching — A method of communication in which a physical circuit is established between two terminating equipments before communication begins to take place. This is analogous to an ordinary phone call.

Common channel interoffice signaling — Use of a special network dedicated to signaling to establish a path through a communication network, which is dedicated to the transfer of user information.

Computer communication network — Collection of applications hosted on different machines and interconnected by an infrastructure that provides intercommunications.

Connectionless service — A mode of packet switching in which packets are exchanged without first establishing a connection. Conceptually, this is very close to message switching, except that if the destination node is not active, then the packet is lost.

Connection-oriented service — A mode of packet switching in which a call is established prior to any information exchange taking place. This is analogous to an ordinary phone call, except that no physical resources need be allocated.

Entity — A software process that implements a part of a protocol in a computer communication.

Formal parameters — The parameters passed during the invocation of a service primitive; similar to the arguments passed in a subroutine call in a computer program.

Fast packet networks — Networks in which packets are transferred by switching at the frame layer rather than the packet layer. Such networks are sometimes called frame relay networks. At this time, it is becoming vogue to think of frame relay as a service, rather than transmission, technology.

Firewall — Computer communication network hardware and software introduced into an internet at the boundary of the public network and a private network for the purpose of protecting the confidential information and network reliability of the private network.

International Standards Organization Reference Model — A model, established by the International Standards Organization (ISO), that organizes the functions required by a complete communication network into seven layers.

Internet — A network formed by the interconnection of networks.

Local area networks — A computer communication network spanning a limited geographic area, such as a building or college campus.

Media access control — A sublayer of the link layer protocol whose implementation is specific to the type of physical medium over which communication takes place and controls access that medium.

Message switching — A service-oriented class of communication in which messages are exchanged among terminating equipments by traversing a set of switching nodes in a store and forward manner. This is analogous to an ordinary postal system. The destination terminal need not be active at the same time as the originator in order that the message exchange take place.

Metropolitan area networks — A computer communication network spanning a limited geographic area, such as a city; sometimes features interconnection of LANs.

Packet switching — A method of communication in which messages are exchanged between terminating equipments via the exchange of a sequence of fragments of the message called packets.

Protocol data unit — The unit of exchange of protocol information between entities. Typically, a protocol data unit (PDU) is analogous to a structure in C or a record in Pascal; the protocol is executed by processing a sequence of PDUs.

Service primitive — The name of a procedure that provides a service; similar to the name of a subroutine or procedure in a scientific subroutine library.

Switching node — A computer or computing equipment that provides access to networking services.

Token bus — A method of sharing a bus-type communications medium that uses a token to schedule access to the medium. When a particular station has completed its use of the token, it broadcasts the token on the bus, and the station to which it is addressed takes control of the medium.

Token ring — A method of sharing a ring-type communications medium that uses a token to schedule access to the medium. When a particular station has completed its use of the token, it transmits the token on the bus, and the station that is physically next on the ring takes control.

Trunk — A communication line between two switching nodes.

Wide area networks — A computer communication network spanning a broad geographic area, such as a state or country.

World Wide Web — A collection of hypertext-style servers interconnected via Internet services.

References

Bertsekas, D. and R. Gallagher. *Data Networks*, 2nd ed. Prentice Hall, Englewood Cliffs, NJ, 1987.

Black, U. D. *OSI: A Model for Computer Communication Standards.* Prentice Hall, Englewood Cliffs, NJ, 1991.

Braun, E. *The Internet Directory.* Fawcett Columbine, New York, 1994.
Garg, V. K. *Wireless Network Evolution: 2 G to 3G.* Prentice Hall, Upper Saddle River, NJ, 2002.
Hammond, J. L. and P. J. P. O'Reilly. *Performance Analysis of Local Computer Networks.* Addison-Wesley, Reading, MA, 1986.
Helgert, H. J. *Integrated Services Digital Networks.* Addison-Wesley, Reading, MA, 1991.
Jain, R. *Handbook: High-Speed Networking Using Fiber and Other Media.* Addison-Wesley, Reading, MA, 1994.
Kurose, J. F. and K. W. Ross. *Computer Networking: A Top-Down Approach Featuring the Internet.* Addison-Wesley, Reading, MA, 2001.
Lin, Y.-B. and I. Chlamtac. *Wireless and Mobile Network Architectures*, John Wiley & Sons, New York, 2001.
McDysan, D. E. and D. E. Spohn. *ATM: Theory and Application.* McGraw-Hill, New York, 1994.
Pildush, G. D. *Cisco ATM Solutions: Master ATM Implementation of Cisco Networks.* Cisco Systems Press, Indianapolis, 2000.
Rose, M. *The Open Book.* Prentice Hall, Englewood Cliffs, NJ, 1990.
Schwartz, M. *Telecommunications Networks: Protocols, Modeling and Analysis.* Addison-Wesley, Reading, MA, 1987.
Spragins, J. D. *Telecommunications: Protocols and Design.* Addison-Wesley, Reading, MA, 1991.
Stallings, W. *Handbook of Computer-Communications Standards: The Open Systems Interconnection (OSI) Model and OSI-Related Standards.* Macmillan, New York, 1990.

Further Information

There are many conferences and workshops that provide up-to-date coverage in the computer communications area. Among these are the IEEE INFOCOM and ACM SIGCOMM conferences and the IEEE Computer Communications Workshop, which are specialized in computer communications and are held annually. In addition, IEEE GLOBCOM (annual), IEEE ICC (annual), IFIPS ICCC (biannual), and the International Telecommunications Congress (biannual) regularly feature a substantial number of paper and panel sessions in networking.

The *ACM Communications Review*, a quarterly, specializes in computer communications and often presents summaries of the latest standards activities. *IEEE Network*, a bimonthly, specializes in tutorially oriented articles across the entire breadth of computer communications and includes a regular column on books related to the discipline. Additionally, monthly magazines *IEEE Communications* and *IEEE Computer* frequently have articles on specific aspects of networking. Also, see *IEEE Personal Communication Systems*, a quarterly magazine, for information on wireless networking technology.

For those who wish to be involved in the most up-to-date activities, there are many interest groups on the Internet that specialize in some aspect of networking. Searching for information on the Internet has become greatly simplified with the advent of the World Wide Web and the deployment of a number of search engines. For example, you can go to *www.google.com* and then enter *mobile IP* as the search keyword, and Google will return ten pages of references on the topic, including the set of standards relating to the topic.

138

Satellites and Aerospace

Samuel W. Fordyce
Advanced Technology Mechanization Company, ATMco

William W. Wu
Advanced Technology Mechanization Company, ATMco

138.1 Communications Satellite Services and Frequency Allocation

An agency of the United Nations, the International Telecommunications Union (ITU), issues the radio regulations that have treaty status among the ITU members. The services of interest in satellite communications include the fixed satellite service, mobile satellite service (including maritime, aeronautical, and land mobile vehicles), broadcast satellite service, and intersatellite service. Most commercial communications satellites operating in the fixed satellite service use the C band (6 GHz up, 4 GHz down) or the Ku band (14 GHz up, 12 GHz down). At higher frequencies, the 30-GHz (Ka band) and 50-GHz satellites are under development. The mobile satellite services operate primarily in the L band (1.6 GHz up, 1.5 GHz down). Exact frequencies and permitted signal characteristics are contained in the radio regulations and tables of frequency allocations issued by the ITU in Geneva, Switzerland.

138.2 Information Transfer and Link Margins — Ground to Space (Up-Link)

A transmitter with an output power of P_t, transmitting through an antenna with a gain of G_t, will provide a power flux density, ϕ, at a range of r_u, according to the formula

$$\phi = \frac{P_t G_t}{4\pi r_u^2} \tag{138.1}$$

The satellite receiver antenna with an effective aperture of A_r located at a range from the ground transmitter will have a received power level, P_r, given by

0-8493-1586-7/05/$0.00+$1.50

$$P_r = \phi A_r = \frac{\phi G_u \lambda_u^2}{4\pi} \tag{138.2}$$

where λ_u is the wavelength of the up-link, and G_u is the gain of the receiving antenna.

The signal is received in the presence of thermal noise from the receiver plus external noise. The total noise power density is $N_0 = kT_s$, where $k = 1.38 \cdot 10^{-23}$ J/K (joules per kelvin) is Boltzmann's constant, and T_s is the system temperature. The noise power in the radio frequency (RF) transmission bandwidth, B, is $N = N_0 B$.

The up-link signal-to-noise power ratio is

$$P_u/N_u = (P_t G_t G_u / kT_s B)(\lambda / 4\pi r)^2 \tag{138.3}$$

In decibels,

$$\begin{aligned} P_u / N_u &= 10\log P_t G_t - 20\log(\lambda / 4\pi r_u) - 10\log B \\ &+ 10\log(G/T) - 10\log k \end{aligned} \tag{138.4}$$

The terms of this equation represent the earth station effective isotropic radiated power (EIRP) in dB W; free space loss in dB; RF channel bandwidth in dB Hz; satellite G/T ratio in dB/K, and Boltzmann's constant, which is –228.6 dBW/Hz K.

Similarly, the signal-to-noise power ratio of the space-to-earth link, or down-link (P_d/N_d), yields

$$P_d/N_d = (P_s G_s G_e / kT_e B)(\lambda_d / 4\pi r_d)^2 \tag{138.5}$$

where P_s is the power of the satellite transmitter, G_s is the gain of the satellite transmitting antenna, G_e is the gain of the earth station receiving antenna, k is Boltzmann's constant, T_e is the temperature of the earth station receiver, B is the bandwidth, λ_d is the wavelength of the down-link, and r_d is the range of the down-link.

The overall system power-to-noise ratio (P_s/N_s) is given by

$$\frac{1}{P_s / N_s} = \frac{1}{P_u / N_u} + \frac{1}{P_d / N_d} \tag{138.6}$$

Signals relayed via communications satellites include voice channels (singly, or in groups or supergroups), data channels, and video channels. Signals from multiple sources are multiplexed onto a composite baseband signal. The signals relayed via communications satellites include voice channels (singly, or in groups or supergroups), data techniques used to modulate these signals to that they can be kept separate use a physical domain such as frequency, time, space (separate antenna beams), or encoding. Access can be preassigned or assigned on demand. The three principal modes of multiple access include the following:

Frequency-demand multiple access (FDMA) isolates signals by filtering different frequencies.
Time-demand multiple access (TDMA) isolates signals by switching time slots.
Code-division multiple access (CDMA) isolates signals by correlation.

138.3 Communication Satellite Orbits

Earth-orbiting satellite trajectories are conic sections with the earth's center of mass located at one focus. The simplest orbit is circular, wherein the centrifugal force on the satellite is balanced by the gravitational force:

$$F_c = F_g \tag{138.7}$$

$$\frac{mv^2}{r} = \frac{GMm}{r^2} \tag{138.8}$$

where v is the satellite's velocity, r is the distance from the earth's center to the satellite, m is the satellite's mass, M is the earth's mass, and G is the constant of universal gravitation. Solving for v gives

$$v = \sqrt{\frac{GM}{r}} \tag{138.9}$$

$$(GM = 3.9858 \cdot 10^5 \text{km}^3/\text{s}^2) \tag{138.10}$$

If the satellite is in an elliptical orbit, with the semimajor axis of the ellipse given by a, the velocity at any point is given by the "vis-viva" equation:

$$v = \sqrt{GM\left(\frac{2}{r} - \frac{1}{a}\right)} \tag{138.11}$$

The maximum distance from the earth's center is known as the *apogee radius* (A), and the minimum distance is known as the *perigee radius* (P). The eccentricity (e) of the ellipse is given by

$$e = \frac{A - P}{A + P} \tag{138.12}$$

The velocity at apogee v_A is give by

$$v_A = \sqrt{\frac{GM}{A}(1 - e)} \tag{138.13}$$

and at perigee

$$v_P = \sqrt{\frac{GM}{P}(1 + e)} \tag{138.14}$$

The orbit period (T) is given by

$$T = 2\pi\sqrt{\frac{a^3}{GM}} \tag{138.15}$$

Most communications satellites are in geostationary orbits, which are in the equatorial plane with a period equal to one (sidereal) day, which is approximately 4 minutes shorter than a solar day. These satellites have an altitude of approximately 36,000 km.

Geostationary satellites appear to be stationary to observers (and antennas) on earth. The antennas are not required to track such a stationary target. Low-gain antennas with broad beams do not need to track satellites, and many of the satellites designed to operate with small mobile terminals use orbits with lower altitudes than the geostationary satellites. These orbits usually have high inclinations to the equatorial plane to provide coverage of the earth's surface to high latitudes.

TABLE 138.1 Launch Site Locations

Site	Latitude	Longitude
Kourou, French Guiana	5°N	53°W
Cape Canaveral, Florida	28°N	81°W
Vandenberg Air Force Base, California	35°N	121°W
Baikonur (Tyuratam), Kazakhstan	46°N	63°E
Plesetsk, Russia	63°N	35°E
Xiachang, People's Republic of China	28°N	102°E
Tanegashima, Japan	30°N	131°E

138.4 Launch Vehicles

Launch services that were once the province of government organizations have become commercial enterprises. Practically all commercial launches to date have been to geostationary orbit.

Recently, the most popular launch provider has been Arianespace, which launches from Kourou, French Guiana, on the Atlantic coast of South America. Three launch vehicles used for commercial launches in the U.S. are McDonnell Douglas's Delta and Martin Marietta's Atlas and Titan. In the U.S., all eastward launches are conducted from Cape Canaveral; the polar and nearpolar launches use Vandenberg Air Force Base in California. Russia provides launches using the Proton and Zenit launch vehicles from Baikonur (Tyuratam) in Kazakhstan and from Plesetsk in Russia. China's Great Wall Trading Co. provides launches on the Long March launch vehicle from Xichang, China. Japan launches from Tanegashima, Japan, using the H-1 and H-2 launch vehicles. Locations of these launch sites are given in Table 138.1.

The velocity increment (v) gained by a launch vehicle when a propellant is burned to depletion is given by the equation

$$v = c\log_{e} M \tag{138.16}$$

The term c is the characteristic velocity of the propellants and is often expressed as

$$c = I_{sp}\, g \tag{138.17}$$

where I_{sp} is the specific impulse of the propellants (typically, I_{sp} = 300 s for lox/kerosene propellants in space) and g is the acceleration due to gravity. The mass ratio (M) can be expressed as

$$M = \frac{\text{Propellents} + \text{structure} + \text{payload}}{\text{Structure} + \text{payload}} \tag{138.18}$$

This velocity increment assumes that the vehicle doesn't change altitude appreciably during the propellant burn or experience significant aerodynamic drag.

138.5 Spacecraft Design

The primary subsystem in a communications satellite is the payload, which is the communications subsystem. The supporting subsystems include structure; electrical power; thermal control; attitude control; propulsion; and telemetry, tracking, and control (TT&C).

The communications subsystem includes the receiving antennas, receivers, transponders, and transmitting antennas. This payload makes up 35 to 60% of the mass of the spacecraft. The communications capability of the satellite is measured by the antenna gain (G), receiver sensitivity (T_r), transmit power (P_t)$_s$, and signal bandwidth (B). Sensitivity is often expressed as G_r/T_r, and radiated power (EIRP) as G_tP_t.

Early models of communications satellites used antennas with broad beams and little directivity. These beams were spread over the whole earth and required large, highly directional antennas on the ground to pick up the weak satellite signals.

Improvements in satellite attitude control subsystems and in launch vehicle capabilities have enabled satellites to carry large antennas with multiple feeds to provide narrow beams that "spotlight" the desired coverage area ("footprint") on earth. These footprints can be focused and contoured to cover designated areas on earth. The narrow beams also permit multiple reuse of the same frequency allocations.

Antenna gain G is given by

$$G = 4\pi A\eta/\lambda^2 \tag{138.19}$$

where A is the effective cross-sectional area, η is the antenna efficiency (typically 55 to 80%), and λ is the wavelength. At the Ku band down-link (12 GHz), where $\lambda = 2.5$ cm, a 1-m diameter antenna can provide a gain of 40 dB. The half-power beam width in degrees θ is given approximately by

$$\theta \approx 21/fD \tag{138.20}$$

where f is the frequency in gigahertz and D is the antenna diameter in meters. In this example the beam width $\theta = 1.75°$. Using the same antenna reflector on the Ku band up-link (14 GHz), the beam width would be slightly narrower (1.5°) if the reflector is fully illuminated.

Transponders are satellite-borne microwave repeaters. A typical C-band transponder is composed of filters and low-noise amplifiers to select and amplify the received (6 GHz) signal. Local oscillators are fed into mixers along with the incoming signal to produce intermediate frequency (IF) signals, which are further amplified before mixing with another local oscillator to produce the down-link signal. This signal is fed to a high-power amplifier (HPA). Traveling wave tubes were used originally as HPAs but have been replaced by solid-state power amplifiers (SSPAs) at C band. The amplified signal is fed to the transmitting antenna for transmission to earth on the down-link (4 GHz).

Geostationary communications satellites operating at C band usually carry 24 transponders with 40 MHz of separation between them. Dual polarizations permit double use of the 500-MHz frequency allocation.

Without onboard processing the transponders are usually "transparent," in that they receive the incoming (up-link) signals, amplify them, and change the carrier frequency before transmitting the down-link signals. Transparent transponders provide no signal processing other than amplification and changing the carrier frequency (heterodyning). These transponders can relay any signals that are within the transponders' bandwidth. Other types of transponders can demodulate the incoming signals and remodulate them on the down-link carrier frequencies.

The spacecraft structure must support the payload and the subsystems through the propulsion phases as well as in orbit. The accelerations can be as high as 6 to 8 g during the launch phases. In orbit the structure must permit deployment and alignment of solar cell arrays and antenna reflectors. The spacecraft utilization factor, U, is defined as

$$U = \frac{M_u}{M} \tag{138.21}$$

where M_u is the mass of the communications payload and power subsystems, and M is the total spacecraft mass in orbit.

Representative values of U range from 0.35 to 0.60. Typical large geostationary communications satellites have a mass of 2000 kg in orbit.

The electric power subsystem uses solar cell arrays to provide electric power. The transmitters of the communications subsystem consume most of this power. Increasing demands for electric power led to large deployable solar arrays composed of silicon solar cells. Typically, power demand is for 2 kW of 28 V DC power. Power-conditioning units maintain the voltage levels within prescribed specifications. Rechargeable batteries (nickel-cadmium or nickel-hydrogen) provide power in emergencies and during solar eclipses.

The thermal control subsystem must maintain the specified temperature for all components of the spacecraft. Heat sources include incident sunlight and internal electrical heat sources (especially the HPAs). The only way to eliminate heat is to radiate it into space. The Stefan–Boltzmann law of radiation gives the radiated heat (q) as $q = \varepsilon A\sigma T^4$ where ε is the emissivity, which is between 0 and 1 (a black body has $\varepsilon = 1$); A is the surface area; σ is the Stefan–Boltzmann constant ($= 5.760 \pm 0.007 \cdot 10^{-8}\ W/m^2K^4$); and T is the absolute temperature.

The temperature of a passive black sphere in geostationary orbit around the earth in sunlight is between 275 and 280 K. The incident solar flux density (S) is approximately 1.37 kW/m². Preventing absorption of this heat energy requires the spacecraft to have a low absorptivity (α). A designer can control the temperature by controlling the absorptivity/emissivity ratio and by using spacecraft radiators.

The satellite's average temperature is given by

$$T = \frac{1}{\sigma}\left(\frac{\alpha a S}{\varepsilon A} + \frac{Q}{\varepsilon A}\right) \tag{138.22}$$

where a is the projected area (facing the sun), A is the total surface area, and Q is the internal heat dissipation.

Satellites need attitude control and station-keeping subsystems to maintain their orientation so that the antenna beams will illuminate the desired coverage areas on earth, the solar arrays will intercept the sun's rays, and velocity increments from the onboard propulsion subsystem will keep the satellite in its desired location. Onboard sensors can be used to detect the earth's horizon, the sun, and reference stars. Radio-frequency sensors are also used to detect beacons from earth stations.

Inertial measurements from onboard gyroscopes are used to detect attitude changes. These changes can be made using onboard reaction wheels, which transfer angular momentum between the spacecraft and the wheel, or by the reaction jets on the attitude control system.

Communications satellites are usually held to accuracies of approximately 0.1° or less in all three axes (yaw, pitch, and roll). The onboard propulsion system can provide velocity for the initial insertion in orbit and subsequent station-keeping velocity increments, as well as attitude control.

A typical communications satellite has propellant tanks containing monopropellant hydrazine (N_2H_4). On command, this fuel will flow through a valve to a thruster that contains a catalyst bed, combustion chamber, and nozzle. The resulting force from the thruster is used in velocity corrections needed for station keeping or for attitude control. This force is given by

$$F = \dot{W} I_{sp} \tag{138.23}$$

where $\dot{W}$ is the propellant weight flow rate, and I_{sp} is the specific impulse, which is about 230 s for the example described.

By using electrically heated thrusters and bipropellant systems, the specific impulse can be raised to 300 s or more. Some communications satellites have such large propellant tanks that velocity increments of several kilometers per second can be achieved.

TT&C subsystems are provided for satellites in order to determine the status, performance, position, and velocity of the satellite, as well as to control it. Data from onboard sensors are transmitted by telemetry transmitters to ground control for monitoring the status of the satellite. Beacons are tracked in range and angle to determine position and velocity. Ground commands are received, demodulated, and processed by command receivers onboard the satellite. TT&C subsystems include high-power transmitters and omnidirectional antennas that can provide communications even if attitude control is lost and the spacecraft is tumbling. Link margins are similar to those described for the communications payload, but with more robust signal-to-noise ratios.

138.6 Propagation

Free space is transparent to electromagnetic waves; however, waves undergo refraction and absorption when passing through the troposphere and the ionosphere. The most important effect on satellite communications is the clear sky attenuation caused by the molecular resonance absorption bands. The

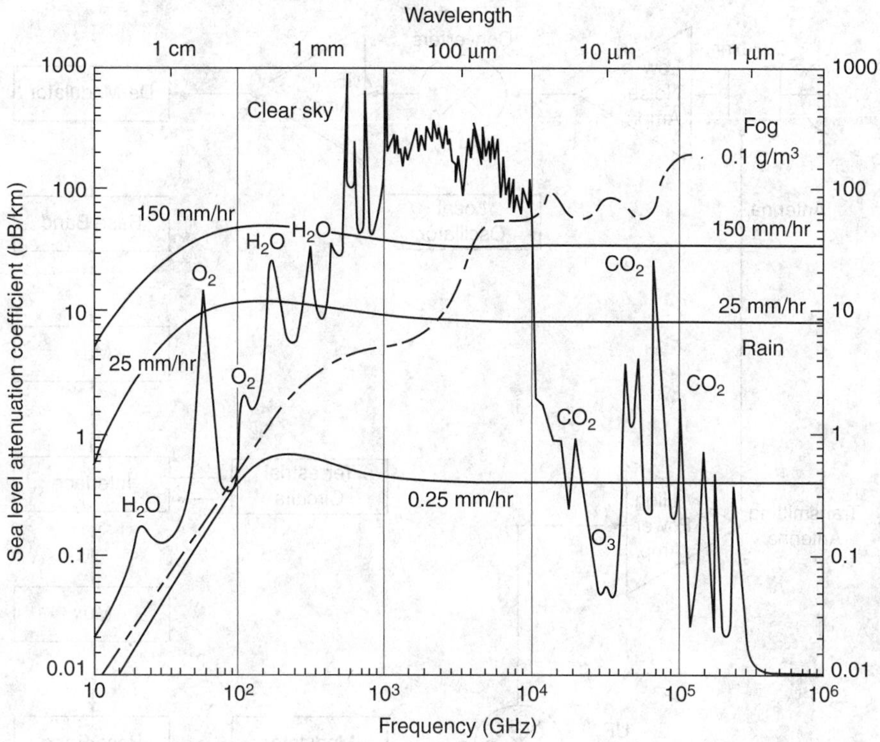

FIGURE 138.1 Atmospheric path losses at sea level.

clear sky attenuation per kilometer is plotted versus frequency in Figure 138.1. Resonance of water vapor molecules reaches 0.15 dB/km at 25 GHz. During heavy rainstorms, this attenuation can increase significantly, as shown in Figure 138.1.

Below 10 GHz, the atmosphere is transparent, but the resonant peaks shown for various molecules in the atmosphere can cause serious attenuation in the K band. The use of this band may be hampered by heavy rainstorms, which can cause depolarization as well as attenuation. One technique used to overcome this problem is to use diverse ground stations, spaced 5 to 10 km apart, to avoid transmission through localized thunderstorms.

138.7 Earth Stations

Earth stations are located on the ground segment of satellite communications systems. They provide communications with the satellites and interconnections to terrestrial communications systems. A simplified block diagram is shown in Figure 138.2.

The antennas, the most prominent subsystems, focus beams to enhance the sensitivity of the receivers, enhance the radiated power of the transmitters, and discriminate against radio interference from transmitters located outside the narrow antenna beam.

Both fixed and mobile earth stations have been developed for lower- and higher-frequency satellites.

138.8 Ka Band Satellites

With wider bandwidth availability, higher-frequency satellites have been conceived. In 2003 there were about 95 orbital slots assigned and planned Ka band satellites operated in the 20- to 30-GHz region. However, some satellite systems came and went before become operational. As mentioned in Section 138.6 and Figure 138.1, attenuation increases as frequency increase. Thus, transmission characteristics

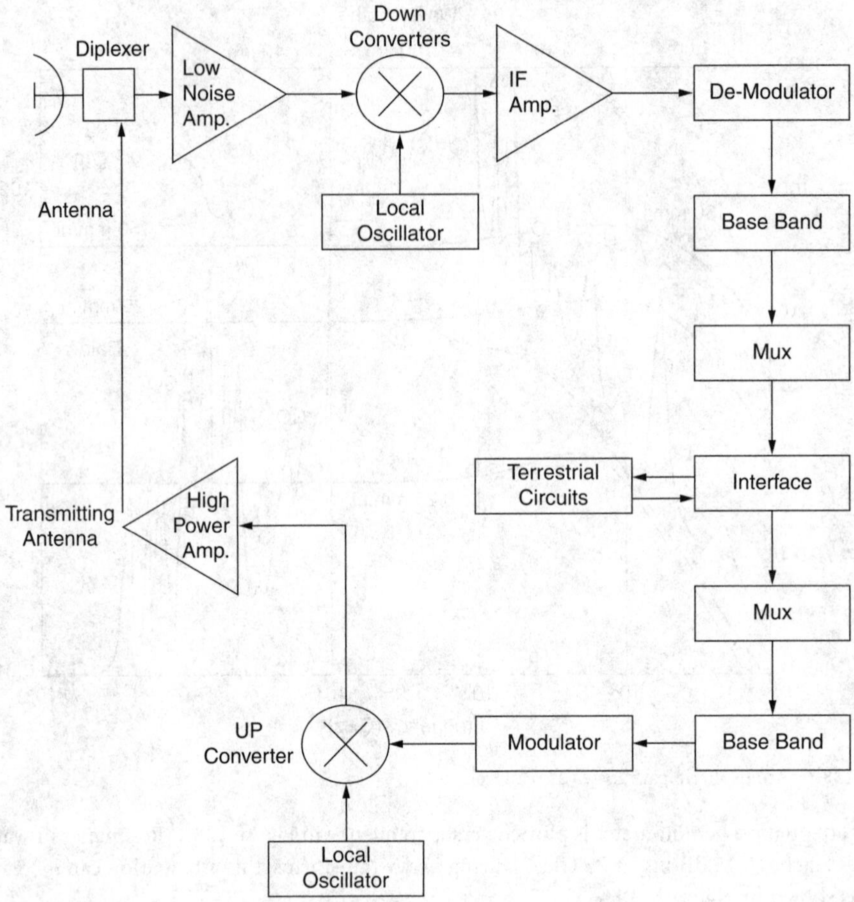

FIGURE 138.2 Earth station block diagram.

of such satellite operations become significant particular for small mobile terminals (Christopher, 1999; Pinck and Rice, 1995; Slack and Rice, 1996).

138.9 Satellites and the Internet

"Internet in the sky" has been announced by a number of satellite operators or would-be operators. In general, Internet and satellites are not compatible. Even with onboard processing, satellites are bottlenecks and cannot provide the optimal freedom of multiple-path flexibility as in terrestrial Internet operations. For recent developments in this area, see Hu and Li (2001), Allman et al. (1997), and publications by the Internet Protocols over Satellite Working Group.

Internet Protocols over Satellite (IpoS) Working Group is the concise name for the TIA-SCD-CIS-Internet Protocol Working Group (TR34.1.1), which is within the Telecommunications Industry Association-Satellite Communications Division (TR43), Communications and Interoperability Section (TR43.1). The IpoS Group is concerned with developing enhancements of Internet protocols to facilitate integration between Internet and satellite technologies.

138.10 Consultative Committee for Space Data Systems

Originated by NASA, the Consultative Committee for Space Data Systems (CCSDS) is now a multinational satellite and space communications standards organization that addresses both space and ground segments. The space segments consist of spacecraft platforms, onboard systems, and space-qualified devices. The ground segments consist of ground networks and command, and telemetry data processing. Active members include the British National Space Centre (BNSC)/United Kingdom; Canadian Space Agency; Centre National d'Etudes Spatiales (CNES)/France; Deutsches Zentrum fur Luft- und Raumfahrt e.V (DLR)/Germany; European Space Agency (ESA); Instituto Nacional de Pesquisas Espaciais (INPE)/ Brazil; NASA; National Space Development Agency of Japan (NASDA); and Russian Space Agency (RSA)/ Russian Federation. In addition, there are 25 satellite- and space-related organizations worldwide that belong to CCSDS as observer members.

References

Allman, M., Hayes, C., Kruse, H., et al. TCP Performance over Satellite Link. Paper presented at 5th International Conference on Telecommunication Systems, 1997.

Christopher, P. World Wide Millimeter Wave Attenuation Functions from Barbaliscia's 49/22 GHz Observations. Proceedings of the Ka Band Utilization Conference, 1999.

Hu, Y. and Li, O. K. Satellite-based Internet: a tutorial. *IEEE Commun. Mag.*, March 2001.

Pinck, D. and Rice, M. K/Ka–Band Channel Characterization for Mobile Satellite Systems. International Mobile Satellite Conference Proceedings, 1995.

Slack, J. and Rice, M. Finite State Markov Model for Error Bursts on the ACTS Land Mobile Satellite Channel. Department of Electrical Engineering, Brigham Young University, May 1996.

Further Information

More complete explanations of this material are contained in the following:

Literature

Battin, R. H. *An Introduction to the Mathematics and Methods of Astrodynamics*. American Institute of Aeronautics and Astronautics, New York.

Griffin, M. D. and French, J. R. *Space Vehicle Design*. American Institute of Aeronautics and Astronautics, Washington, DC.

Ippolito, L. Jr. *Radiowave Propagation in Satellite Communications*. Van Nostrand-Reinhold, New York, 1986.

Isakowitz, S. J. *Reference Guide to Space Launch Systems*. American Institute of Aeronautics and Astronautics, Washington, DC, 1991.

Morgan, W. and Gordon, G. *Communications Satellite Handbook*, John Wiley & Sons, New York, 1989.

Pratt, T. and Bostian, C. *Satellite Communications*. John Wiley & Sons, New York, 1986.

Pritchard, W., Suyderhoud, H., and Nelson, R. *Satellite Communication Systems Engineering*. Prentice Hall, Englewood Cliffs, NJ, 1993.

System Alternatives, Analyses, and Optimization, vol. 1, 1984, and *Channel Coding and Integrated Services Digital Satellite Networks*, vol. 2, 1985, of *Elements of Digital Satellite Communications*. Computer Science Press, Rockville, MD.

Symposia

Developments in satellite communications are discussed in many symposia, including those sponsored by the Institute of Electrical and Electronics Engineers and the American Institute of Aeronautics and Astronautics.

Websites

Consultative Committee for Space Data Systems, *www.ccsds.org*.

Internet Protocols over Satellite Working Group (IpoS), *www.isr.umd.edu*.

139

Mobile and Portable Radio Communications

Rias Muhamed
SBC Laboratories Inc.

Michael Buehrer
Virginia Polytechnic Institute & State University

Anil Doradla
SBC Laboratories Inc.

Theodore S. Rappaport
University of Texas at Austin

All over the world, wireless communications services have enjoyed phenomenal growth over the past two decades. It was only in late 1983 that the first commercial cellular telephone system in the U.S. was deployed by Ameritech in the Chicago area. This was the analog service called Advanced Mobile Telephone System (AMPS). Today, cellular telephone services, most of them digital, are available throughout the world, and are beginning to rival fixed-line telephone services both in terms of availability and number of users. In a span of less than 20 years, the number of mobile wireless subscribers worldwide has grown from zero to over a billion users. This remarkable growth demonstrates not only the strong desire of people around the world to communicate to one another while on the move, but also the tremendous strides that technology has made in both fulfilling and further fueling this need. While the ability to communicate using radio signals was demonstrated by Marconi more than a century ago, it was the remarkable developments during the past few decades in RF circuit fabrication, digital signal processing, and several miniaturization technologies that made it possible to deploy and deliver wireless communication services at the scale and scope that we see today.

While a vast majority of subscribers today use wireless systems for voice telephony, we have begun to see the adoption of wireless data services as well. Short Messaging Service (SMS), which allows users to send text messages of size 160 characters or less using their cellular handsets, was first deployed in Europe in 1991, and has risen to a worldwide volume of several billions of messages per month in less than a decade. Operators are now beginning to offer Multi-media Messaging Service (MMS), which allows users to add pictures, sound, and other contents to their messages.

In addition to messaging, wireless data services include information services such as the delivery of news, stock quotes, weather, directions, etc. Due to the limitations in data rate and available space for display in handheld devices, specialized technologies such as the Wireless Access Protocol (WAP) and i-mode were developed to tailor and deliver Internet content to handheld devices. WAP is an open global specification that has so far had limited success, and i-mode is a proprietary technology developed by NTT DoCoMo that has enjoyed greater success in Japan since its introduction in February 1999.

0-8493-1586-7/05/$0.00+$1.50

Wireless data applications also include wireless access to the Internet and corporate Intranets. These applications are typically experienced in a laptop computer environment, where wireless connectivity is provided through a wireless adapter card. Currently available wireless Internet access services are limited to data rates in the order of tens of kilobits per second, with user experience comparable to the traditional dial-up services available over telephone lines. Future wireless services such as the third generation (3G) services promises greater throughputs on the order of several hundred kilobits per second. The evolution of newer and richer wireless data applications is accompanied by a migration of cellular wireless networks from a traditional circuit switched platform toward an all-IP packet switched network.

In this chapter we cover the fundamental aspects of wireless communications technology and provide a brief introduction to several existing and emerging wireless communication systems and standards. We begin by outlining the major technical challenges to providing wireless communications services, and provide an overview of the various techniques that have been developed to meet these challenges. Since the wireless industry has evolved to support a multiplicity of standards and systems, we attempt to cover the prominent ones such as GSM and cdma2000 in brief detail in this chapter. While most of the focus is on wide area mobile communications systems, Wireless Local Area Networks (WLAN) is also briefly covered. While WLAN systems do not provide the ubiquitous wide area coverage or high-velocity mobility support that traditional cellular and personal communications networks do, they are an important component of the wireless connectivity solution for the end-users since they are capable of much greater data rates and provide untethered access to communication networks. We also envision the possibility of seamless roaming across multiple heterogeneous networks such as the cellular packet data networks and wireless LAN networks. A number of wireless LAN systems and technologies that are standardized in the IEEE 802.11 group are covered in this chapter.

There is a lot of research activity currently underway in the area of Personal Area Networking (PAN) that provides interconnectivity and networking among the devices carried by a user within a Personal Operating Space (POS) using short-range low-power radio technology. Bluetooth® is an example of a Personal Area Networking technology. Personal Area Networks are being standardized in the IEEE 802.15 standards group. Task Group 3a within the 802.15 body is exploring the use of ultra-wideband wireless technology to develop extremely high-data rate short-range connectivity solutions.

While the focus of this chapter is on mobile and portable radio communications, fixed wireless communications is also gaining a lot of traction. The IEEE 802.16 standards group has ratified a new standard for wireless broadband metropolitan area networking that provides a wireless alternative to wired solutions such as cable modems and Digital Subscriber Lines (DSL) for broadband access. These systems provide multi-megabits per second connectivity over distances of a few miles for stationary applications.

139.1 Technical Challenges to Wireless Communications

In this section we identify the major technical challenges that make the design of a wireless communication system uniquely different from that of traditional wired communication systems, and provide an overview of the various approaches that have been explored to overcome these challenges.

In wireless communications systems the transmission channel is simply the intervening space between the transmitter and the receiver. Unlike in wired communications channels, there is no physical connection (a copper wire or fiber optic cable) to guide the signal from one point to another. Instead wireless systems rely on the ability of radiowaves to propagate through space. The first challenge comes from the fact that the radio channel is a difficult and unpredictable media for communication. This is especially true of mobile radio channels where the end-user is moving at high-speeds. Developing a communication system that can reliably transport signals between a mobile user terminal and a fixed base station through this harsh media is a challenge that continues to excite wireless researchers.

The second challenge comes from the fact that the radio spectrum is a scarce resource. Government regulatory bodies around the world have allocated only a limited amount of spectrum for commercial use. While there is a large amount of spectrum for use in the low-power license-exempt bands, this is not the case with licensed frequency bands. Being a scarce resource, operators around the world have paid billions of dollars in licensing fees to gain rights to use these frequency bands. The need to accommodate an ever-increasing number of users and providing them with applications that demand larger amounts of bandwidth within the allocated spectrum challenges the system designer to continuously search for solutions that use the spectrum more efficiently. Developing methods to improve the spectral efficiency of wireless communication systems is an area of keen interest for wireless researchers.

One particular challenge that can be thought of as resulting from the scarcity of spectral resources, but requiring special mention is the issue of multi-user channel access. In addition to the need to ensure that each user utilizes the spectrum as efficiently as possible, there is also the need to devise methods to share the resources among the multiple users in the most effective manner. Since spectrum constraints do not always permit the luxury of providing dedicated resources to each user, methods need to be developed to divide and allocate the spectrum resources to different users on demand and in an optimal manner.

The next two challenges result directly from the two fundamental differentiating value propositions that wireless communications systems attempt to bring to the end-user: mobility and portability. Mobility implies that (1) a user can send and receive data packets or voice calls wherever they are, and (2) a user can maintain a communication session without interruption while on the move. The first implication requires that the system provide the ability to identify the location of each user at any given time and develop routing and roaming methods. The second implication requires that the system provide a method of seamless handover between base stations as the user moves around the coverage area.

Portability implies that the communications device used by the end-user is battery powered and lightweight. Since terminal devices are battery powered, it is important that they consume as little power as possible. Given that in many industrialized countries the total power consumed by wireless networks and terminals is turning out to be a non-significant portion of the total national power consumption, power saving has also become a national macro-economic need. The requirement of low-power consumption challenges designers to look for power efficient transmission schemes, power saving protocols, and battery technologies with longer life.

Superimposed on all these challenges is the need to provide these services at a low cost in order to ensure mass market adoption. It is therefore imperative to understand that the wireless system designer often does not have the luxury of choosing the best possible technical solution, but is instead constrained by several practical and economic realities.

In the following subsection we discuss each of these challenges in more detail. We begin with the challenges imposed by the radio channel.

Radio Channel Impairments

Radio signals travel most effectively when there is a clear line-of-sight between the transmitter and receiver, and if the intervening area is free of clutter. In clear line-of-sight conditions, a radio signal simply experiences an exponential decay of power with distance. The requirements of modern communications systems such as cellular systems and wireless LAN systems, however, do not allow the luxury of line-of-sight communications, and have to work under much more unforgiving conditions. The several large and small obstructions in the channel, undulations in the terrain, relative motion between the transmitter and receiver, interference from other signals, and several other complicating factors together not only weaken but also delay and distort the signal in an unpredictable and time-varying fashion. Instead of a clean copy of the original signal, what is received at the far end is an unpredictable mess that often bears only a poor resemblance to the signal transmitted. Under these conditions it is a challenge not only to design signals that can withstand the vagaries of the wireless channel but also to build receivers capable of extracting the transmitted information buried deep within the distorted received signal. Any attempt to design a wireless system and build resilient signal sets and good receivers must begin with a thorough understanding of the wireless channel.

It is instructive to characterize the radio channel at three different levels of spatial scale. The first level of characterization is to develop a model that describes the median distance-dependent decay in power that the signal undergoes as it traverses the channel. These large-scale models provide the systems engineer a rough estimate of the area that can be covered by a given radio transmitter. Since radio signal power tends to decay exponentially with distance, these models are typically linear on a logarithmic decibel scale with a slope and intercept that depend on the overall terrain and clutter environment. Widely-used large scale propagation models derived from empirical measurements include the Okumura-Hata Model [Hata, 1980], and the COST-231 Model [EURO-COST, 1991].

The next level of characterization is the variation in received signal power from the median distance-dependant value. These variations called large-scale fading are often caused by undulations in the terrain and large obstructions such as buildings, and depend largely on the diffraction losses that radio signals suffer as they get around these obstacles. Measurements have shown that these large-scale variations tend to have a log-normal distribution around the median value. The variance of this log-normal distribution dictates the amount of power margin that the system engineer needs to accommodate in the link budget to ensure adequate signal coverage at over 90–99% of the areas that need to be covered. This margin is often referred to as the shadow margin.

The final level of spatial scale at which the radio channel can be characterized is the variation in signal strength observed over small areas. Due to the phenomenon of multi-path propagation the amplitude of the received radio signal can vary significantly (several tens of dBs) over very small distances (on the order of wavelengths). These variations are called small-scale fading. A good understanding of the nature of small-scale fading is critical to the design of wireless modems.

Figure 139.1 illustrates the phenomenon of multi-path propagation experienced by signals traveling from a transmitter to a receiver via several reflecting and scattering objects. Each of the multiple paths experiences a different attenuation and time delay (and resulting phase offset). What is seen at the receiver is the superposition of these multiple waves. This superposition can cause constructive or destructive interference depending on the phase relationship among the different waves. Therefore the amplitude of the resultant received signal will depend on the particular multipath structure seen at the receiver. As the mobile receiver moves in space, the radio paths between the transmitter and receiver will change giving rise to a different multipath structure. This change in multipath structure changes the relative phase and amplitude of the received waves, and thereby changes the amplitude of the resultant signal. Therefore, as the receiver moves in space, the amplitude of the received signal varies depending on variation in the multipath structure. Thus, spatial variation in multipath structure is manifested as temporal variation in the received signal strength of a mobile receiver. This variation in received signal strength caused by multipath propagation is called **multipath fading**. Depending on the absence or presence of a dominant path (e.g., direct line of sight), the amplitude variation in the received signal caused by multi-path fading is often found to follow either a Rayleigh or Rician probability distribution, and therefore referred to as *Rayleigh fading* or *Rician fading*, respectively.

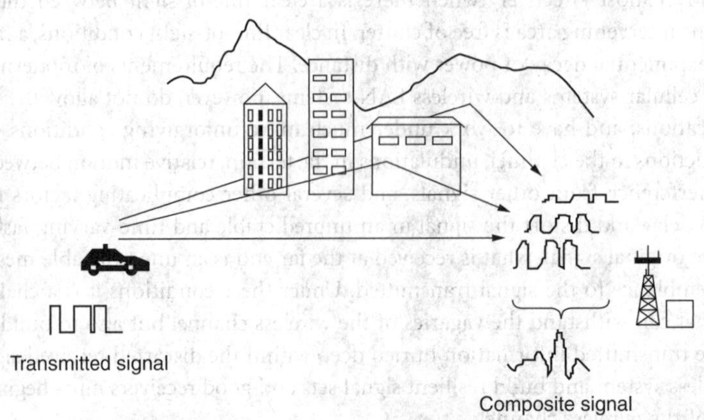

FIGURE 139.1 Illustration of multipath propagation in a wireless channel.

TABLE 139.1 Summary of Techniques to Mitigate Radio Channel Impairments

Type of Fading	Mitigation Technique
Flat and Slow Fading	Obtain multiple uncorrelated copies of the same signal using some type of diversity. Examples include: Antenna or Space Diversity Error Correction Coding Time Diversity using Interleaving Frequency Diversity Polarization Diversity Other techniques include: Adaptive Modulation Automatic Repeat Request Increase Fade Margin in Link Budget
Frequency Selective Fading	Adaptive Equalization Multi-carrier modulation schemes such as Orthogonal Frequency Division Multiplexing (OFDM) Direct Sequence Spread Spectrum with RAKE Frequency Hopped Spread Spectrum Pilot Signal Antenna solutions such as distributed antenna systems, small cells, and directional antennas to reduce delay spread
Fast Fading	Non-coherent or Differentially Coherent Modulation Error Correction Coding Interleaving Increase Symbol Rate by adding signal redundancy

While multi-path propagation is inherent to a mobile radio channel, the effect of it on a radio signal depends on the signal characteristics. The two main effects of multi-path propagation are *flat fading* and *frequency selective fading*. If the signal bandwidth is less than the channel bandwidth (band of frequencies over which the frequency response of the channel is nearly constant), flat fading results. Flat fading is manifested as large variations in signal amplitude, but does not cause signal distortion. If the signal bandwidth is larger than the channel bandwidth, frequency-selective fading results. Frequency-selective fading is manifested as time dispersion, which causes inter-symbol interference (ISI). Time dispersion caused by multipath delays is often quantified by a parameter called *rms delay spread*. Measurement results show that in outdoor cellular environments delay spread values are in the range of hundreds of nanoseconds to a few microseconds, and in indoor environments they range from a few nanoseconds to a few hundred nanoseconds.

The time-variation in the channel characteristics caused by the relative motion between the transmitter and receiver leads to carrier frequency dispersion called *Doppler Spread*. Doppler Spread is directly related to vehicle speed and carrier frequency. When the time-variations in channel conditions caused by motion is faster than the time-variations in the transmitted information signal, the effect is called *fast fading*. And conversely, when the channel characteristics vary slower than the information signal, the effect is called *slow fading*.

There are a variety of processing techniques that can be used to mitigate the effects of small-scale fading. Some of these techniques require processing at the transmitter, some at the receiver, some both. Channel coding, equalization, antenna diversity combing, etc. are examples of techniques to mitigate multipath effects. Some of the more advanced techniques do not seek to mitigate the challenges posed by multipath fading, instead see multipath as offering a richly diverse environment that can be cleverly exploited to gain capacity enhancements. Examples of such techniques include space division multiplexing using intelligent antennas and multi-user diversity that exploits the uncorrelated fades experienced by different users to make capacity gains. Table 139.1 summarizes several of these signal processing techniques to mitigate multipath fading.

Scarcity of Licensed Frequency Spectrum

Radio frequency spectrum is a scarce and expensive resource that needs to be managed carefully. This is especially true of the licensed spectrum allocated to provide cellular mobile voice and data services. As demand for wireless services have grown, a number of techniques have been developed and adopted to utilize the scarce spectrum resources more and more efficiently. One of the most basic techniques for efficiently managing the limited spectrum resource is called **frequency reuse**, which is the key to providing capacity and accommodating a growing subscriber base in cellular systems [MacDonald, 1979]. Frequency reuse is fundamental to the cellular radio concept.

Cellular and Frequency Reuse

Cellular radio systems rely on an intelligent allocation and reuse of the radio frequency spectrum throughout a coverage region. A subset of the spectrum is assigned to a small geographic area called a cell. Each cell is served by a base station that uses channels in the spectrum allocated to that cell. The power radiated by a base station is deliberately kept low, and antennas are located so as to achieve coverage within the particular cell. By limiting the coverage area within a cell, the same group of channels can be used to cover various cells that are separated from one another by distances large enough to keep the interference between them (termed co-channel interference) within tolerable limits. This replication of channels in separated geographic areas is termed frequency reuse.

Figure 139.2 shows a cellular layout where cells labeled with the same letter use the same group of channels (i.e., spectrum). Due to random propagation effects, actual cell coverage areas are amorphous in nature. It is, however, useful to visualize cells as hexagons for system design purposes.

To better understand the frequency reuse concept, consider a cellular system that has a total of S duplex channels available for use. If each cell is allocated a group of k channels ($k < S$), and if the S channels are divided among N cells into unique and disjoint channel groups with the same number of channels, the total number of available radio channels can be expressed as

$$S = kN \tag{139.1}$$

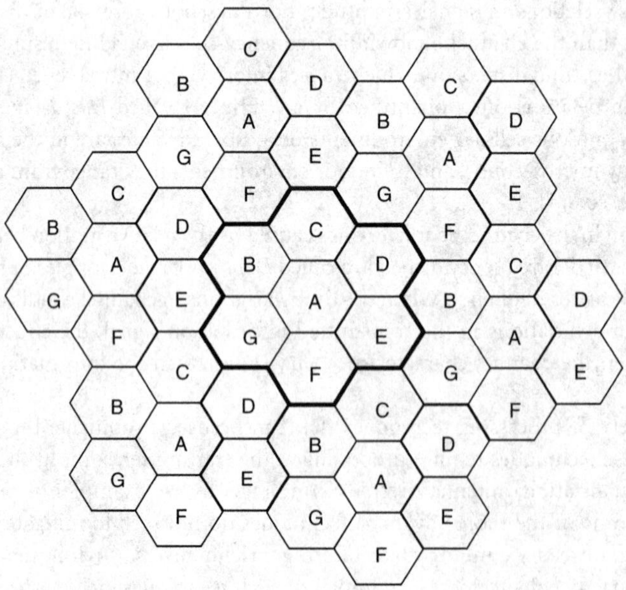

FIGURE 139.2 Illustration of the cellular concept. Cells labeled with the same letter use the same set of frequencies. A cell cluster is outlined in bold and is replicated over the coverage area. In this example the cluster size, N, is equal to 7, and each cell contains $1/7$ of the total number of available channels.

The factor N is called the cluster size and is typically equal to or less than 7. The N cells that use the complete set of frequencies are collectively called a cluster. If a cluster is replicated M times within the system, the total number of duplex channels, C, available to the system is given by

$$C = MkN = MS \tag{139.2}$$

As seen from Equation (139.2), the capacity of a cellular system is directly proportional to the number of times a cluster is replicated in a given service area. If the cluster size N is reduced, more clusters are used to cover a given area and hence more capacity (larger value of C) is achieved. The choice of N depends on the co-channel interference level that can be tolerated. For example, GSM systems typically can work well at $N = 4$, while CDMA systems can operate at $N = 1$.

To this point we have considered the use of omni-directional antennas at each base station. The co-channel interference of a cellular system can be further reduced (and thus capacity can be increased) if directional antennas are used at the base station. Each directional antenna covers a disjoint section of the cell and is assigned a fraction of the cellís channels. This technique is termed sectorization. Due to the use of directional antennas, a given sector will now interfere with (and be interfered by) a fraction of the co-channel cells. This fraction depends on the amount of sectoring used. Normally, sectoring is done either in thirds or sixths. As a result of the reduced co-channel interference, a smaller cluster size becomes allowable. This reduced cluster size results in higher capacity.

When sectorization is employed, the frequencies used in a particular cell are broken down into three or six groups and used only within a particular sector. The implementation is accomplished using fixed directional antennas. In the case of 120° sectors, the number of interferers in the first level is reduced from six to two. This is because only two of the six co-channel cells will be directing a particular frequency group towards the cell. The penalty paid for this improvement is an increased number of antennas as well as an increased number of necessary hand-offs. Additionally, the trunking efficiency is decreased.

Dynamic Channel Allocation

For efficient utilization of the radio spectrum, a frequency reuse scheme that is consistent with the objectives of increasing capacity and minimizing interference is required. In other words, a system designer needs to devise efficient channel assignment strategies. Channel assignment strategies can be classified as either fixed or dynamic. The particular type of channel assignment employed affects the performance of the system, particularly in how calls are managed when a mobile user travels from one cell to another [Tekinay and Jabbari, 1991].

The preceding discussion on the number of channels per cell assumes that traffic is uniformly distributed in the service area. In real systems, however, traffic is not uniformly distributed. Certain areas such as shopping malls or stadiums may experience higher traffic volume as compared to other areas. Such hot spots can be alleviated through cell-splitting or the use of micro-cells. However, if the traffic volume increases are temporary (e.g., during special events or during rush hour) a more effective means of servicing the increased demand is the use of dynamic channel allocation. A fixed channel allocation scheme permanently assigns channels (channel definitions depend on the multiple access scheme as will be discussed later) to a given sector or cell whereas a dynamic channel allocation scheme will associate channels and cells based on demand. Such a scheme requires careful assignment strategies to ensure reuse distances are not violated. However, the advantage is that it allows a system to adjust to non-uniform traffic patterns. Typically such schemes allow heavily loaded sectors (or cells) to borrow channels from neighboring cells which are lightly loaded.

In a fixed channel assignment strategy, each cell is allocated a predetermined set of channels. Any call attempt within the cell can be served only by the unoccupied channels in that particular cell. If all the channels in that cell are occupied, the call is blocked and the subscriber does not receive service. In dynamic channel assignment, channels are not allocated to various cells permanently. Instead, each time a call is attempted, the cell base station requests a channel from the MTSO, which allocates a channel based on an algorithm that minimizes the cost of channel allocation.

Intelligent Antennas

As mentioned in a previous section, directional antennas can increase system capacity by reducing co-channel interference thus allowing smaller reuse patterns. Theoretically, the capacity could be continually increased by further sectorization. However, practical constraints limit the amount of sectorization possible. For example, increased sectorization results in an increased handoff overhead and a higher risk of dropped calls. One potential solution to this situation is the concept of intelligent antennas [Buehrer et al., 1999]. Intelligent antennas can allow for reduced reuse patterns without increasing hand-offs. This is accomplished by using phased arrays or switched beams to direct energy over a more narrow part of the cell or sector without dividing the spectrum. By monitoring the received signal the antenna adapts to mobile movement within a sector or cell to maintain a narrow beam without requiring handoff. The downside to such a system is that it requires additional antenna hardware to form the narrow beams (either through antenna arrays or butler matrices). Unfortunately, RF hardware is among the more expensive components of the base station. Additionally, intelligent antenna systems require additional ASICs to run the appropriate algorithm which either adapts the phased array weights or switch the beams.

Multiple Access Strategies

A basic component of any wireless (or wired) communication system that involves multiple connections is the method by which different users access the medium. Since users must share a common medium in order to communicate, there must be some known method for accomplishing this. This is termed **multiple access.** We have already discussed the way in which geography is used in cellular systems to create isolated channels through frequency reuse patterns. Now we discuss how within a given cell (and thus specific frequency band) users must have some method of accessing or sharing the medium.

Multiple access methods are generally divided into two basic categories: random access and channelized access. In the former method there are no pre-assigned channels for each user. Instead, users must compete for the medium each time they have information to send. In the latter method users are assigned a specific channel which is dedicated for their use as long as they have information to transmit. Such a scheme requires significant overhead to set up the channels and thus is most applicable to constant rate applications such as voice. However, setting up dedicated channels is not particularly efficient when the information source is bursty (such as in computer data networks). In such cases it is much more efficient to allow users to randomly access the channel.

There are several general random access methods that are employed for data networks including ALOHA, slotted ALOHA, and carrier sense techniques. In each of these techniques users access a single channel whenever they have data to send. In the ALOHA system [Abramson, 1970], users simply transmit data as soon as it arrives. They can then listen (either to the channel or via feedback) and determine whether or not their data was received successfully. When two or more users access simultaneously, there will be a collision and the data will not be received correctly. However, if traffic is bursty and the number of users is not large, the frequency of collisions should be small. Further, if transmitters know when a collision occurs they can retransmit using a channel access protocol. When users are restricted to begin transmitting at fixed time intervals the system is called *Slotted ALOHA*. Slotted ALOHA, while requiring synchronization between users, provides an improvement in throughput, particularly at moderate to high offered traffic loads. This is because the probability of a collision is reduced by synchronizing transmissions.

The throughput of ALOHA and Slotted ALOHA is not particularly good since transmitters listen to the channel only after they have transmitted. Collisions can be reduced by listening before transmitting (a "look before you leap" approach). This is termed *Carrier Sense Multiple Access* or CSMA. In CSMA users simply monitor the channel and refrain from transmitting if there is current traffic. Collisions will still occur when two or more users sense the idle channel and transmit simultaneously. The throughput of CSMA can be much higher than ALOHA especially if it is used in conjunction with *Collision Detection* (CSMA/CD). In CSMA/CD users also listen for collisions and when a collision is detected a special jam signal is transmitted which informs all users that they should cease transmitting. Such a procedure reduces

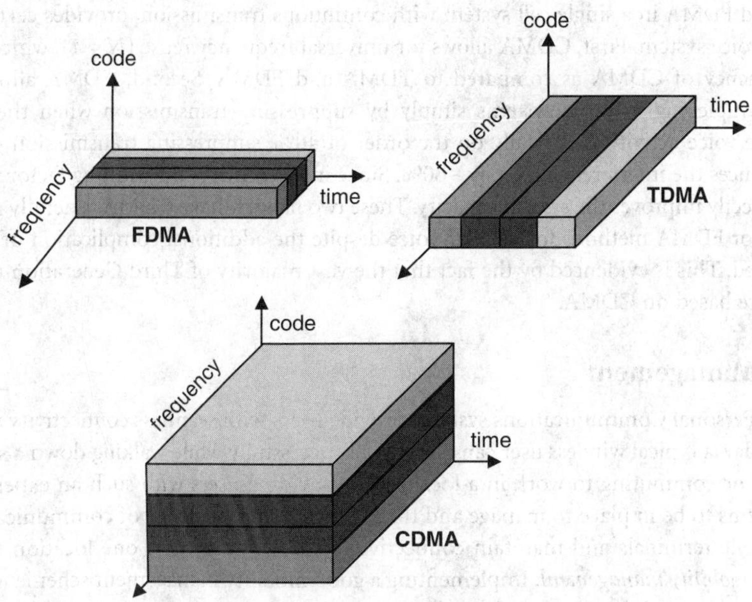

FIGURE 139.3 Illustration of the three multiplexing techniques, and how multiple channels are provided without interfering with each other: (a) FDMA, (b) TDMA, and (c) CDMA.

the downtime of the channel and thus improves throughput. As discussed in the WLAN section CSMA/CD has limitations in a wireless channel, and therefore a Collision Avoidance (CA) scheme is used instead.

In voice systems, random access methods are not particularly useful since they cannot guarantee that a user will be able to transmit when necessary. Voice systems thus use channelized multiple access strategies where each user requests a channel and is given one for the duration of the call. There are three basic methods for creating channels in a communication system: frequency channels, time channels, and code channels. These correspond directly to *Frequency Division Multiple Access* (FDMA), *Time Division Multiple Access* (TDMA), and *Code Division Multiple Access* (CDMA). In FDMA systems the medium is divided into frequency slots which are then used for specific channels. Each transmitter-receiver pair must know which frequency band to use and must then tune their receiver to detect that band. In a TDMA system all users share a common frequency band, but are restricted to transmitting in a particular time slot. Thus, only a single frequency band is needed, but the receiver must have accurate knowledge of the timing of the different transmissions in order to correctly receive the signal. Such a system has some advantages compared to FDMA when multiple data rates are used in the system. In many wireless systems, the medium is first divided into several frequency channels using FDMA, and then each frequency channel is further divided among multiple users using TDMA. Figure 139.3 provides an illustration of the three multiple access schemes.

CDMA is significantly different than FDMA and TDMA. CDMA systems rely on spread spectrum techniques, where all users can transmit in the same frequency band and at the same time. However, each transmitter-receiver pair uses a specific spreading waveform which is modulated by the information. The spreading waveforms of different users have very low cross correlation. Thus, while TDMA and FDMA provided ideally orthogonal multiple access schemes, CDMA is inherently interference limited. While it is theoretically possible to create an orthogonal CDMA system, it would require synchronous reception of all signals on both the uplink and downlink. This is clearly possible on the downlink, but extremely difficult on the uplink. Further, when there are multipath delays which are greater than a chip[1] duration, maintaining orthogonal reception upon synchronization is not possible. CDMA, while inferior

[1] A 'chip' is one symbol in a direct sequence spread spectrum signal.

to TDMA and FDMA in a single cell system with continuous transmission, provides certain advantages in a cellular voice system. First, CDMA allows for universal frequency reuse ($N = 1$), which improves the spectral efficiency of CDMA as compared to TDMA and FDMA. Second, CDMA allows for natural statistical multiplexing in voice systems simply by suppressing transmission when there is no voice present. Since voice activity is typically on the order of 40%, suppressing transmission during periods of silence reduces the interference level by ~60%. Since interference is the limiting factor in CDMA, this technique directly improves the system capacity. These two factors make CDMA generally more attractive than TDMA or FDMA methods for cellular voice despite the additional complication of power control that is required. This is evidenced by the fact that the vast majority of Third Generation (3G) standards for cellular are based on CDMA.

Mobility Management

Cellular and Personal Communications systems provide users with seamless connectivity while being on the move. Today, a typical wireless user can place a call successfully while walking down a street, traveling in his vehicle or commuting to work in a local train. Providing users with such an experience requires complex systems to be in place to manage and track the call. The challenge of communication networks to locate mobile terminals and maintain connectivity as they move from one location to the other is referred to as *mobility management*. Implementing a good mobility management scheme is critical to the overall performance of the network and providing a good user experience. As a result, the area of mobility management has gained considerable attention since the early days of the first wireless systems.

Functionally, mobility management may be divided into two categories — location management and handoff management. Location management protocols focus on issues like mobile registration, location updating, and paging while handoff management protocols allow users to maintain a seamless connection with the network as they move and change their location.

Location management involves both the identification of the mobile location and successful call delivery. In the identification stage, the mobile periodically registers with the network to identify its presence and location within the network. In the call delivery stage, the network is queried on the current location of the terminal device. In a nutshell, location management within the wireless network is a combination of several database queries and responses. These databases include Home Location Registers (HLR) and Visitor Location Registers (VLR), which store information about the local home users and the visiting users, respectively. For larger networks, the location of databases and network elements participating in the location management process is critical to the overall performance.

Today, there are two main standards that provide location management in cellular and personal communications systems. They are GSM's Mobile Application Part (MAP) and the EIA/TIA's Interim Standard IS-41. GSM MAP was developed as part of the GSM standard and the IS-41 protocol was primarily developed for wireless systems in North America.

Handoff management is the second aspect of mobility management. A proper handoff management mechanism is crucial to the overall performance of a wireless system that supports roaming and mobility. **Handoff** refers to the process where a data session or voice call is transferred from one base station to another. A successful handoff is executed by the close co-ordination and involvement of several network elements and multiple protocols within the wireless system. For example, in a GSM system, handoffs between two base stations involve the base stations, Base Station Controllers (BSC) and Mobile Switching Centers (MSC) [3GPP TS 05.08, 1997].

Processing handoffs is an important task in any cellular mobile system. Many system designs prioritize handoff requests over call initiation requests. It is required that every handoff be performed successfully and that they happen as infrequently and imperceptibly as possible. In cellular systems the signal strength on either the forward or reverse channel link is continuously monitored and, when the mobile signal begins to decrease (e.g., when the reverse channel signal strength drops to below between −90 dBm and −100 dBm at the base station), a handoff occurs. In first-generation analog cellular systems, the Mobile Telephone Switching Office (MTSO) monitored the signals of all active mobiles at frequent intervals to obtain a rough estimate of their location and decide if a handoff was necessary. In second-generation systems handoff decisions are mobile assisted. In a mobile-assisted handoff (MAHO) the mobile stations take measurements of the received power from several surrounding base stations and continually report the results of these measurements to the base station in use, which initiates the handoff. The MAHO method provides faster handoff than first-generation analog cellular systems since the MTSO is not burdened with additional computation.

Handoffs may be classified based on the link layer establishment and termination process between the mobile terminal and the base station. Accordingly, there are two types that are referred to as "break-before-make," or "make-before-break." In a break-before-make handoff, the mobile breaks the link layer from one base station before establishing it with another. During the time of switchover between the base stations, the mobile terminal, for a fraction of a second, is not connected to either base station. Examples of wireless protocols supporting the break-before-make handoffs include AMPS, GSM, and IS-136. In the make-before-break handoff, the mobile terminal establishes concurrent link layer connections with all the base stations taking part in the handoff process. IS-95 and 1XRTT are examples of systems that support make-before-break handoffs.

As mobility management protocols provide end to end solutions, they operate across multiple physical and link layer connections (Layer 1 and 2). Hence, these protocols are generally designed to operate around the network layer (Layer 3). For example, in GSM mobility management is distributed across different network elements. Depending on the situation, different network elements participate in providing mobility management to the terminal. Examples of mobility management in GSM include GSM-MAP between the HLR and Mobile Switching Center (MSC) VLR, ISUP between MSCs, and BSSMAP between BSCs and MSCs.

The convergence of wireless technologies with the Internet has created a special need for providing mobility support in an Internet Protocol (IP) environment. Since the original Internet protocols were designed for stationary devices and most of the existing mobility protocols were meant for non-IP networks, newer protocols, such as Mobile IP (MIP), had to be developed to support mobility for an IP enabled device in a wireless and fixed-line environment.

Mobile IP (MIP) is a protocol that has gained great attention over the past few years. The protocol is being specified under the auspices of the Mobile IP Working Group (WG) at the Internet Engineering Task Force (IETF). The protocol attempts to offer a simple solution to the mobility problem for IP devices [Perkins, 2002]. IETF MIP WG has specified IP mobility solutions for networks based on IPv4 and IPv6. Since its inception, the WG has completed the development of the IPv4 based mobility protocol and is currently working on the IP-v6 based mobility protocol. At the time of this writing, the IETF WG was working on the 21st draft for MIPv6. Some of the issues that have been addressed by the protocols include mobility management across heterogeneous IP based networks, triangular routing, and route optimization. As a consequence of MIP implementation in networks, we can expect to see devices successfully roaming across different wireless technologies. For example, a user can expect to surf the Internet with an Internet enable PDA device on an 802.11b network and handover the session to an UMTS network. To the user, the transition between the networks will be seamless and the application will not be interrupted.

Another important benefit that mobility management techniques offer wireless operators is the ability to off-load traffic from one network to the other. An operator may avoid congestion to its GSM/GPRS network by off-loading some of its traffic to the EGPRS or UMTS network. Of course, we are assuming that the mobile terminal has the ability to maintain link layer connectivity across all the technologies participating in the handover.

Low Power Requirements

Since wireless communications systems typically involve mobile and portable devices, it is of utmost importance that the systems be designed with low power consumption requirements of the user terminal device in mind. Greater efficient use of power in portable wireless devices not only increases the usage time of the devices, but also enhances the user experience. Charging one's wireless device often can be the most frustrating experience to any user. Since the early days of wireless communication, low power consumption for terminal devices has been an area of great focus and attention. It used to be that designers could allow for greater power consumption in order to get greater performance, but researchers are now looking to save power without compromising performance. Greater power savings may be achieved by using a combination of efficient signal processing techniques, better physical layer design, improvements in protocols, and superior materials in the device.

In wireless handsets, battery quality and performance plays an important role in the overall performance and usability of handsets. This issue becomes more critical with wireless terminals transmitting and receiving data. Keeping other parameters such as modulation, coding and channel conditions constant, the power consumed by the terminal device is directly proportional to the bit rate with which it is transmitting.

Development of high performance, low power wireless systems require designers to focus on coupling different areas of communications. Some of the most important areas include communications theory, integrated circuit fabrication, low power circuit design, digital signal processing techniques, and wireless protocol design.

Low-power consumption requirement drives physical layer design toward the direction of using power-efficient modulation schemes — signal sets that can be detected and decoded at lower signal levels. Unfortunately, power-efficient modulation and coding schemes tend to be less spectrally efficient. Since spectral efficiency is also a very important requirement for many wireless systems, it becomes challenging to pick the appropriate trade-off between power efficient and spectral efficient signal sets.

Protocol design is another important area which can be utilized to gain power efficiencies. Ensuring that the transmitter circuitry is turned on only when required, and on a demand basis, incorporating low-power sleep modes with methods to wake up when required, etc. are tricks that can be played to reduce the overall power consumption. Bluetooth is an example of a protocol that is designed with low-power consumption requirement being among the most important consideration.

Along with hardware designs that employ power efficient architectures, battery technologies play a vital role in the performance of terminals. Battery technologies that offer higher power densities over greater periods of time can compensate for some of the limitations in the hardware and software systems.

Most of the earlier handsets used Nickel Cadmium (NiCd) batteries. Although these batteries were widely used, and to a large extent are being currently used, they had a problem related to their charging levels. When a NiCd battery is charged from a partially charged level, it assumes its new uncharged level as the partially charged level at which it begins charging. This memory effect in the battery results in a reduction of the total power delivered with each charge. With the introduction of Nickel Metal Hydride (NiMH) batteries, the problem related to the memory-effect in batteries was solved. These batteries also had higher power levels in comparison with NiCd batteries. The latest battery technology involves Lithium Ion and contains a built in electronic charge and discharge circuitry. These batteries have superior performance characteristics and are more expensive than their NiCd and NiMH counterparts.

139.2 Evolution of Wireless Services and Standards

Wireless communication systems have come a long way since Marconi first showed how information could be transmitted over the wireless link. Since the early 1970s, commercial wireless systems have evolved considerably. Advances in digital signal processing, semiconductor technologies, and software systems have enabled the development of advanced wireless technologies that support services ranging from voice applications to real-time video applications. Presently, there are several wireless standards available; each supporting multiple services. Most of the countries around the world are at a stage of wireless maturity where wireless subscribers expect their handsets to work wherever they go. To a casual subscriber, this may not be a great achievement, but on closer look, one can easily appreciate the amount of effort and time that was put in to develop these global standards. Although today most of the standardization is achieved with the collective effort of companies and government bodies on a global scale, this was not so in the early stages of the development of wireless systems.

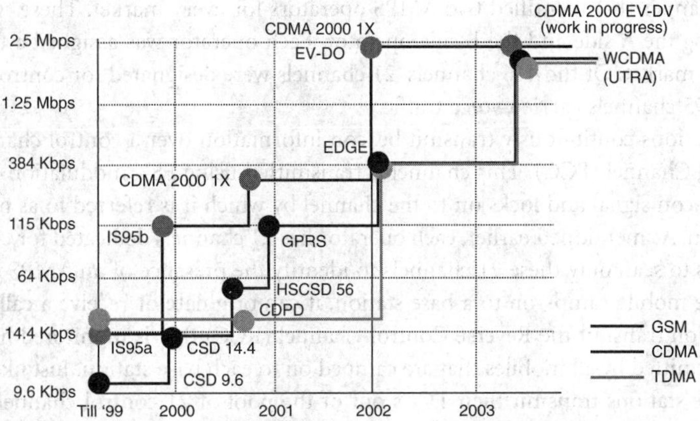

FIGURE 139.4 Evolution of the various mobile wireless technology standards and their data rate capabilities.

United States, Japan, and parts of Europe led the development of the first generation of wireless systems. The first generation systems were characterized by their analog modulation schemes and were geared primarily towards delivering voice. Examples of these technologies include the Advanced Mobile Phone System (AMPS) from the U.S. and the Nordic Mobile Telephone (NMT) from Norway, Sweden and Finland.

Improvements in processing abilities of hardware platforms over time enabled the development of the second generation (2G) wireless systems. These systems were also aimed primarily towards the voice market, but unlike the first generation systems, they used digital modulation schemes. Examples of 2G systems include the Global System of Communications (GSM), IS-95, and IS-136. GSM was adopted all over Europe and most of Asia, while North American market was fragmented between IS-95, IS-136, and GSM.

Although 2G systems offered consumers with superior voice quality over their first generation counterparts, they were aimed primarily towards delivering voice service. There was still the need to deliver data services over wireless. Keeping in mind that large quantities of money were spent in deploying the 2G networks, it was essential to upgrade 2G systems, at a minimal cost, to provide data. Such upgrades to 2G systems to support wireless data came to be commonly referred to as being the 2.5G systems. Examples of 2.5G systems include GPRS and IS-95b.

The International Telecommunications Union (ITU) in the 1990s invited contributions for third generation wireless systems, or 3G systems. Some of the main criteria to qualify as a 3G system included the ability to provide 144 kbps in a vehicular environment, 384 kbps for pedestrian users and 2 Mbps for stationary terminals. As part of its standardization process, ITU also identified spectrum segments for the use by 3G systems. Examples of 3G systems include WCDMA and cdma2000. One of the most interesting aspects of the 3G contributions was the choice of CDMA as the preferred access technique for the majority of 3G systems. Not only did the IS-95 camp propose evolution towards a CDMA based 3G technology, but the GSM camp offered its own version of CDMA, called W-CDMA. Figure 139.4 shows the evolutionary path for wireless data systems.

Advanced Mobile Phone Service (AMPS)

The Advance Mobile Phone Service, or AMPS, was developed by AT&T Bell Labs in the late 1970s and was first deployed commercially in 1983 in Chicago and its nearby suburbs. Unlike the cellular systems of today, the first system used large cell areas and omni-directional base station antennas. The system covered 2,100 square miles with only 10 base stations. Each base station had a height between 150 ft and 550 ft. A total of 136 voice channels were deployed unevenly among the 10 base stations on the basis of expected levels of traffic. The earliest systems were designed for a Carrier to Interference Ratio (CIR) of 18 dB for satisfactory voice performance, and were deployed in a 7 cell frequency reuse pattern with 3 sectors per cell.

Since its first deployment in Chicago, AMPS has been deployed in several countries in South America, Asia and North America. In the United States, the Federal Communications Commission assigned spectrum in a manner that identified two AMPS operators for every market. These two operators were identified as being the A side and the B side operator. Each operator was assigned a total of 416 AMPS channels in each market. Of the 416 channels, 21 channels were designated for control information and the remaining 395 channels carried voice traffic.

AMPS base stations continuously transmit beacon information over a control channel known as the Forward Control Channel (FCC). This channel is transmitted using FSK modulation. An AMPS mobile listens to this beacon signal and locks on to the channel by which it is referred to as being "camped on" to the base station. As mentioned earlier, each operator has 21 channels dedicated for control. As a result, the mobile needs to scan only these 21 channels to identify the presence of an AMPS signal from a base station. Once the mobile camps-on to a base station, it can originate or receive a call. Mobiles camped to each base station transmit the Reverse Control Channel (RCC) that is monitored by the base station. The RCC is transmitted by all mobiles that are camped on to each base station. Just like the reverse link's RCC, all the base stations transmit their FCCs out of the pool of 21 control channels. Hence, it is the responsibility of network providers to see that FCCs allocated to adjacent base stations are not a source of interference.

Once the mobile associates itself with a base station, it may originate a call by transmitting an origination message over the RCC. Details of the subscriber such as the Electronic Serial Number (ESN), Station Class Mark (SCM) and Mobile Identification Number (MIN) are also transmitted to the base station. On receiving these details of the mobile subscriber, the base station verifies the authenticity of the user by checking with the Mobile Telephone Switching Office (MTSO). On getting an approval from the MTSO, the call is routed to the Public Switched Telephone Network (PSTN) and on to the final destination.

During the course of an AMPS call, supervisory signals are used to make sure that the call between the mobile subscriber and base station is properly connected. The supervisory signals in the AMPS systems are the Supervisory Audio Tone (SAT) and the Signaling Tone (ST). The SAT is a range of three tones that are transmitted at 5970, 6000, and 6030 Hz. Each base station uses one of these three tones during a voice call. This signal is superimposed over the voice signal in the forward link and reverse link and is transmitted at levels that are barely audible to a user. In geographic areas where mobiles receive multiple co-channel signals from multiple base stations, the SAT provides a way of identifying the base station that is handling the call.

The ST is a burst of 10 kbps data stream that consists of a pattern of alternating 1's and 0's. The main function of the ST is to indicate the end of a call and is transmitted along with the SAT. When the mobile terminates a call, it sends the ST to the base station. On its reception, the base station is alerted of the call being terminated by the mobile station.

Though all operators have already deployed 2G digital systems, AMPS continues to be used in North America as a common fallback service available throughout the geography, and is often used in the context of providing roaming between different operator networks.

GSM and its Evolution

In the early 1990s, many European countries came together under the auspices of the Conference of European Posts and Telegraphs (CEPT) to develop and standardize a pan-European system for mobile services. The group was called the Groupe Spécial Mobile (GSM) and their main charter was to develop a system that can deliver inexpensive wireless voice services, and work all across Europe.

By 1989, the European Telecommunications Standards Institute (ETSI) took over the development of the GSM standard. The first version, called GSM Phase I, was released by the early 1990s and in a short period of time was deployed by several European operators. As GSM began gaining acceptance beyond Europe, the standard was renamed as the Global System for Mobile Communications.

Since the first commercial GSM deployment in Europe in 1991, GSM has gained widespread acceptance across the globe. Today GSM stands out from the rest of the technologies as the most popular standard in terms of volumes of terminals and operators deploying it.

TABLE 139.2 Summary of GPRS Radio Specifications

Parameter	Specification
Downlink Frequency Band	824–849 MHz, 1850–1910 MHz (U.S.) 880–915 MHz, 1710–1785 MHz (Europe)
Uplink Frequency Band	869–894 MHz, 1930–1990 MHz (U.S.) 925–960 MHz, 1805–1880 MHz (Europe)
Transmitter/Receiver Frequency Separation	45 MHz, 80 MHz (U.S.) 45 MHz, 95 MHz (Europe)
Multiple Access Technique	TDMA/FDMA
Duplexing Technique	FDD
Modulation Data Rate	270.83333 Kbps
Channel Spacing	200 kHz
Time Slot Period	.576 ms
Frame Period	4.615 ms
Number of Timeslots per Frame	8
Bit Period Duration	3.692 μs
Modulation	GMSK

In Europe and other parts of the world GSM has been deployed in the 900 MHz and 1800 MHz bands while in the United States it has been deployed in the 850 MHz and 1900 MHz frequency bands. Table 139.2 shows details of the radio interface of a GSM system.

By the mid-1990s, ETSI introduced the GSM Packet Radio Systems (GPRS) as an evolutionary step for GSM systems toward higher data rates. Under favorable conditions, these systems can provide a maximum data rate of 160 kbps. GPRS and GSM systems share the same frequency bands, time slots, and signaling links. Figure 139.5 shows the architecture of a GSM/GPRS system. As shown in the diagram, a GSM system may be upgraded to a GPRS system by introducing new elements, such as the Serving GPRS Support Node (SGSN) and Gateway GPRS Support Node (GGSN), and upgrading existing network elements such as the BTS and HLR. SGSN provides mobility management and may be thought of as equivalent to the MSC for voice systems. GGSN provides the IP router functionality and connects the GPRS network to the Internet and other IP networks.

The GSM standard got a further boost in its data handling capabilities with the introduction of Enhanced Data Rate for GSM systems, or EDGE, in the early part of 1997. Although EDGE was developed for existing GSM systems, it can also be implemented as a stand-alone system that can be deployed by a non-GSM operator. Under favorable conditions, EDGE systems can provide users with a maximum data rate in the range of 400 kbps.

In the late 1990s ETSI submitted a proposal for a 3G system called Universal Mobile Telephone System (UMTS). UMTS included a CDMA-based air interface, called Wide-band CDMA (W-CDMA). Although the air interface was different from the air interface of GSM and EDGE, its back end network elements such as MSC, SGSN, and GGSN was backward compatible with the 2G and 2.5G system.

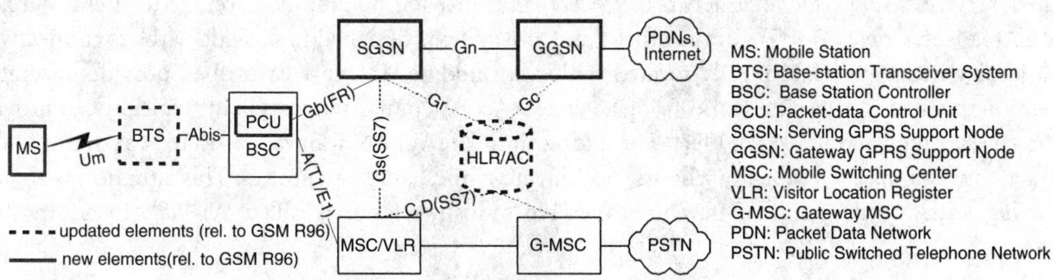

FIGURE 139.5 Block diagram representation of the GSM/GPRS network architecture.

IS-95 and its Evolution

The IS-95 standard based on CDMA for cellular telephony was introduced in 1993 as a voice-only system. The standard uses universal frequency reuse ($N = 1$) with direct sequence spread spectrum modulation using 64-ary Orthogonal Walsh codes on the uplink and QPSK on the downlink. Unlike TDMA/FDMA-based systems, CDMA requires strict power control. Thus, IS-95 includes fast (800Hz) power control on the reverse link and 50 Hz power control on the forward link. The original (IS-95A) system supported 9.6 kbps voice channels[1] and a single dedicated data channel at 9.6 kbps. The introduction of IS-95B allowed for multiple 9.6 kbps channels to be used simultaneously for a peak data rate of 76.8 kbps albeit at the cost of 8 voice channels. Further, in order to improve the efficiency of data transmission, burst (or packet) mode was introduced. The key benefits of CDMA-based cellular systems are universal frequency use and statistical multiplexing of voice traffic. The former benefit is directly enabled by the use of spread spectrum modulation. The latter benefit is the result of voice activity as discussed earlier. A third advantage of CDMA is the direct benefit of sectorized antennas.

In recent years the IS-95 family of standards (branded cdmaOne by the CDMA Development Group) has been upgraded via the Third Generation Partnership Project 2 to what is termed cdma2000. cdma2000 (also called 3G1X for third generation with one times the chip rate of IS-95) improves the capabilities of IS-95 to meet the standards set forth by the International Telecommunications Union (ITU) for International Mobile Telecommunications 2000 (IMT-2000). Specifically, the data capabilities were enhanced by adding separate logical channels termed *supplemental channels*. Each link can support a single fundamental channel (at 9.6 kbps) and multiple supplemental channels (up to 307 kbps). Further, the data rate can be increased up to 2 Mbps through the use of multiple carriers (3G3X). The link quality was improved by upgrading the uplink through the use of coherent modulation and the downlink through the addition of fast (800 Hz) power control to match the uplink. Advanced antenna capabilities were also integrated into the new standard through options for transmit diversity as well as supplemental pilot options for beam-steering. A key to these upgrades is that they are backward compatible. cdma2000 and IS-95A/B may be deployed on the same carrier allowing gradual deployment.

In order to both achieve higher data rates (up to 2 Mbps) as well as improve overall system throughput for packet data scenarios the 3G1X standard was also evolved to 3G1X-EVDO (*EV*olution, *D*ata *O*nly). As the name implies, the standard is applicable to data traffic only since there is no guarantee of real time service. The downside to this standard is that although it is backward compatible with the IS-95 radio characteristics, it cannot be deployed on the same carrier as 3G1X RTT or IS-95. This requires service providers to dedicate a single carrier to data services in order to deploy data. In order to allow providers to use the same carrier for both data and voice the IS-95 family of standards also includes 3G1x-EVDV (*EV*olution, *D*ata and *V*oice). This standard combines EVDO packet data service with voice service on a single carrier.

Wireless Local Area Networks

The wireless standards and systems covered thus far are for providing cellular and personal communications services using wide area networks (WAN) that provide ubiquitous coverage to users. Wireless Local Area Networks (WLAN), on the other hand, are systems designed to provide wireless connectivity to a local wired network, typically located within a building. In most cases they provide a wireless extension to wired Ethernet networks using a wireless Access Point. An Access Point typically can provide radio coverage for about 100–200 feet. For a long time, the WLAN use was restricted to vertical niche market applications such as factory floors, health care clinics, and warehouses. This situation is rapidly changing with the adoption of new interoperable industry standard called Wi-Fi™. In addition to

[1] This was termed the fundamental rate and was dependent on the vocoder used. Some systems used a 14.4 kbps vocoder which meant that the data channel was also available at 14.4 kbps.

enterprise wireless networking, WLAN systems are now being used for applications such as home networking and wireless data access at targeted public locations.

The IEEE 802.11 is the predominant wireless LAN standard in the US. It took several years of work at the IEEE to complete its first version of the standard, thereby delaying the adoption of WLAN systems. After completing their first standard in June 1997, IEEE 802.11 committee worked very hard to quickly complete several revisions to the standard. The first one is the IEEE 802.11b, which is an 11 Mbps extension to the original IEEE 802.11, and operates in the 2.4 GHz ISM band. This is the standard that is currently trademarked as Wi-Fi™. The second one, the IEEE 802.11a is a higher speed alternative that operates in a 5 GHz U-NIII band. Both of these standards were completed in a very short time. We now also have a new standard called IEEE 802.11g, which can be thought of as essentially IEEE 802.11a operating in the 2.4 GHz band, while at the same time also including backward compatibility with IEEE 802.11b. While the physical layer standard for each of these variations is different, they all have the same medium access control (MAC) layer specification.

The MAC layer is based on a variation of the Carrier Sense Multiple Access (CSMA) system used in Ethernet. While the traditional wired Ethernet uses Collision Detection (CD) along with CSMA, the 802.11 standard uses a collision avoidance (CA) mechanism. Collision Avoidance is used since Collision Detection cannot work properly in the wireless environment. Collision Detection requires that every node be able to hear every other node in the network. In a wireless environment, it is quite possible that a node A can hear a node B, and node B can hear a node C, while A and C cannot hear each other. This problem is often referred to as the "hidden node" problem. Collision Avoidance overcomes the hidden node problem, by using a simple Request to Send (RTS), Clear to Send (CTS) protocol that ensures that collisions do not occur very often. The MAC layer design philosophy is premised on supporting only best effort services. There are no good provisions made for QoS. There is, however, a new MAC layer standard called IEEE 802.11e being specified that will incorporate methods to accommodate service prioritization and QoS support.

The MAC layer also supports a rudimentary authentication and encryption scheme called Wired Equivalent Privacy (WEP). Efforts to improve the security offered in wireless LAN is underway in a task group within the IEEE (called IEEE 802.11i).

The IEEE 802.11b uses a Direct Sequence Spread Spectrum (DSSS) physical layer and operates in the 2.4 GHz ISM band. The DSSS signal occupies a relatively wide 22 MHz channel. There are a total of 11 channels defined in the IEEE 802.11b standard. Given that there is only a total of 80 MHz in the 2.4 GHz ISM band, out of the eleven channels one can only pick sets of three non-overlapping channels at a time. When two IEEE 802.11b systems operate in close proximity, it is important to ensure that they are operating in non-overlapping channels. Operating in the same channel or overlapping channel can degrade the throughput performance.

The IEEE 802.11b standard supports multiple data rates: 1, 2, 5.5, and 11 Mbps. Signaling at 1 Mbps and 2 Mbps is achieved using BPSK and QPSK modulation, respectively, and are fully backward compatible with the original 802.11 DSSS standard. The 5.5 Mbps and 11 Mbps signaling are achieved by adding a new coding scheme called complementary code keying (CCK) to the standard. CCK is able to deliver 5.5 Mbps and 11 Mbps without changing the bandwidth of the signal. CCK uses an "eight chip" code-spreading sequence and can operate at 11 Mbps in environments with delay spreads of 100 ns, and at 5.5 Mbps in environments with 250 ns delay spread. CCK does not actually change the modulation scheme used; it merely replaces the Barker code sequences used in the original IEEE 802.11 standard with a new set of CCK codes. Because there are 64 unique CCK code words that can be used to encode the signal, up to 6 bits can be represented by any one particular code word (instead of the 1 bit represented by a Barker symbol). The CCK code word is then modulated with the QPSK technology used in 2 Mbps wireless DSSS radios.

The need for higher speeds and the desire to move away from the crowded 2.4 GHz band led to the development of IEEE 802.11a standard. This standard offers up to 54 Mbps in the newly allocated 5 GHz U-NII band. The physical layer is not only at a different frequency, but also uses a completely different modulation scheme. Orthogonal Frequency Division Multiplexing (OFDM) is the radio technology adopted by the IEEE 802.11a standard. OFDM offers superior performance both in terms of spectral efficiency (how many bits can be squeezed in a given bandwidth) and immunity to multi-path interference. IEEE 802.11a breaks the frequency band of 5.15–5.35 GHz into eight 20 MHz channels. Each of these 20 MHz channels is composed of 52 narrow-band carriers of 300 kHz bandwidth. OFDM sends data in parallel across all of these carriers and aggregates the throughput. Each carrier can transport 125 kbps of data. Of the 52 carriers, four are designated as pilot (for synchronization and control), and 48 are used for data transport. Thus, each 20 MHz channel yields an aggregate data rate of 6 Mbps. Now, using higher order modulation (64QAM) the channel capacity can be increased up to 54 Mbps.

Wireless LAN/WAN Integration

Telecommunications service providers and mobile operators around the world are recognizing the potential for a new high-speed wireless data service at targeted public locations called "hotspots" using wireless LAN technology. The rapid adoption and proliferation of Wi-Fi™ offers a compelling case to deploy broadband wireless services using this technology. Given that the range of these wireless systems is only 100–200 feet, it is not possible to provide ubiquitous wide area cellular-like coverage using this technology. It is, however, eminently possible (with significantly lower capital outlay) to provide service at targeted islands of coverage, called "hotspots," which may be chosen carefully depending on the target market and type of service offering. Plausible hotspots include airports, hotels, convention centers, cafes, restaurants, bookstores, truck stops, MDUs, and other locations where people tend to congregate and/or are captive for a certain period of time. Some operators see the initial target market as the so-called "road-warrior" segment — business travelers with laptops — and offer services in airports and hotels. Other operators target local travelers like the real-estate agents and the general tech-savvy consumers and are offering this service at places such as cafes and restaurants.

As wireless LAN systems are used to deploy public communications service, there is an interest in integrating them with the existing and emerging wide area wireless data systems such as GPRS, and cdma2000 to provide users a seamless connectivity experience. Having a WLAN-based network in targeted locations can enable an operator to provide higher-speed services to customers at those locations, and could help off-load the traffic on the wide area network. Using WLAN-based networks for providing high-speed services in areas of high-traffic concentration, and within buildings appears very attractive. Providing a wireless LAN-based service can complement the mobility and ubiquity of the cellular networks with the offer of high bandwidth in a relatively limited and stationary or portable environment.

The LAN/WAN integration could happen at various levels. At one level, it would simply be a common authentication and billing between the two services. At the deeper level, the system would support seamless mobility between the two using Mobile IP–based technologies. Even tighter coupling between the two services will require developing interworking functions between the two. It is conceivable that future third-generation wireless data networks and wireless LAN networks could get fully integrated into a single IP-based network providing seamless handover functionality between the two.

139.3 Summary and Conclusions

Mobile and portable radio communication systems and services have enjoyed tremendous growth in the past two decades. Remarkable strides have been made in understanding the challenges to wireless mobile and portable communications and developing solutions to meet those challenges. In this chapter, we provided a brief introduction to wireless communications, and outlined some of the major challenges. While good quality mobile voice communications and low data rate wireless data communications are a reality today, there is still work to be done in wireless communications. Mobile radio communications continue to excite researchers around the world, who are driven by the powerful vision of providing affordable broadband multimedia communications to mobile users anywhere, anytime, and in a manner that is secure and easy to use.

Defining Terms

Multipath Fading — The variations in received signal level caused by the superimposition of multiple propagating radio waves impinging on the receiver.

Frequency reuse — The use of radio channels on the same carrier frequency to cover various areas that are separated from one another so that cochannel interference is not objectionable.

Multiple Access — The process and techniques by which a multiplicity of users gain access to the available channel resources from transmission and reception. The channel resources are typically shared among all the users on a demand basis.

Handoff — The process of transferring a mobile station from one channel to another. This transfer could be across adjacent sectors or base stations.

References

Abramson, N. 1970. The ALOHA System — Another Alternative for Computer Communications. *Fall Joint Computing. Conf., AFIDS Conf. Proc.*, vol. 37, pp. 281–285, AFIDS Press, Montvale, NJ.

R. M. Buehrer, A. G. Kogiantis, S.-C. Liu, J.-A. Tsai, and D. Uptegrove. 1999. Intelligent Antennas for Wireless Communications — Uplink. *Bell Labs Technical Journal*, vol. 4, no. 3, pp. 73–103, July–September 1999.

Hata, M. 1980. Empirical Formula for Propagation Loss in Land Mobile Radio Services. *IEEE Transactions on Vehicular Technology* 29(3):317–325.

EURO-COST 231. 1991. Urban Transmission Loss Models for Mobile Radio in the 900 and 1800 MHz Bands. European Cooperation in the Field of Scientific and Technical Research, The Hague.

MacDonald, V. H. 1979. The Cellular Concept. *The Bell Systems Tech. J.* 58(1):15–43.

Tekinay, S. and Jabbari, B. 1991. Handover and Channel Assignment in Mobile Cellular Networks. *IEEE Comm. Mag.* November, pp. 42–46.

3GPP TS 05.08 V6.9.0. 1997. Digital Cellular Telecommunications System (Phase 2+); Radio Subsystem Link Control (Release 1997)

Perkins, C. 2002. IP Mobility Support for IPv4. Internet Engineering Task Force (IETF) RFC3344.

Further Information

A detailed treatment of wireless communications technology, systems, and standards is presented in *Wireless Communications — Principles and Practice*, by T. S. Rappaport, Prentice-Hall, Second Edition 2002.

The *IEEE Communications Magazine* and the *IEEE Wireless Communications Magazine* are good sources of information for the latest developments in the field.

140

Optical Communications

Joseph C. Palais
Arizona State University

Electronic communications over conducting wires or by atmospheric radio transmission began in the latter part of the 19th century and was highly developed by the mid-20th century. Widespread communication via beams of light traveling over thin glass fibers is a relative newcomer, beginning in the 1970s, reaching acceptance as a viable technology in the early 1980s, and continuing to evolve since then [Chaffee, 1988]. Fibers now form a major part of the infrastructure for a national telecommunications information highway in the U.S. and elsewhere.

The fundamentals of optical communications are covered in many textbooks [e.g., Palais, 1998; Keiser, 2000; Agrawal, 2002], which elaborate on the information presented in this chapter.

Optical communications refers to the transmission of messages over carrier waves that oscillate at optical frequencies. The frequency spectrum of electromagnetic waves can range from DC to beyond 10^{21} Hz. (A tabulation of units and prefixes used in this chapter appears in Tables 140.1 and 140.2.) Optical waves oscillate much faster than radio waves or even microwaves. Their characteristically high frequencies (on the order of $3 \cdot 10^{14}$ Hz) allow vast amounts of information to be carried. An optical channel utilizing a bandwidth of just 1% of this center frequency would have an enormous bandwidth of $3 \cdot 10^{12}$ Hz. Numerous schemes exist for taking advantage of the vast bandwidths available. These include wavelength-division multiplexing (WDM) and optical frequency-division multiplexing (OFDM), which allocate various information channels to bands within the optical spectrum.

For historical reasons optical waves are usually described by their wavelengths rather than their frequencies. The two characteristics are related by

$$\lambda = c/f \tag{140.1}$$

where f is the frequency in hertz, λ is the wavelength, and c is the velocity of light in empty space ($3 \cdot 10^8$ m/s). A frequency of $3 \cdot 10^{14}$ Hz corresponds to a wavelength of 10^{-6} meters (or 1 μm, often called a *micron*). Wavelengths of most interest for optical communications cover the range from 0.8 to almost 1.7 microns.

The Gaussian beam profile is common in optical systems. Its intensity pattern in a plane transverse to the direction of wave travel is described by

0-8493-1586-7/05/$0.00+$1.50

TABLE 140.1 Units Used in Optical Communications

Unit	Symbol	Measure
Meter	m	Length
Second	s	Time
Hertz	Hz	Frequency
Bits per second	b/s	Data rate
Watt	W	Power

TABLE 140.2 Commonly Used Prefixes

Prefix	Symbol	Multiplication Factor
Tera	T	10^{12}
Giga	G	10^{9}
Mega	M	10^{6}
Kilo	k	10^{3}
Milli	m	10^{-3}
Micro	μ	10^{-6}
Nano	n	10^{-9}

$$I = I_o e^{-2(r/w)^2} \tag{140.2}$$

where r is the polar radial coordinate and the factor w is called the *spot size*. This is the light pattern emitted by many lasers and is (approximately) the intensity distribution in a single-mode fiber.

Optical transmission through the atmosphere is possible, but it suffers serious liabilities. A fundamental limit is imposed by diffraction of the light beam as it propagates away from the transmitting station. The full divergence angle of a Gaussian beam is given by

$$\theta = 2\lambda/\pi w \tag{140.3}$$

Because of the continual beam enlargement with propagation distance, the amount of light captured by a finite receiving aperture diminishes with increasing path length. The resultant low received power limits the distances over which atmospheric optical systems are feasible.

The need for an unobstructed line-of-sight connection between transmitter and receiver and a clear atmosphere also limits the practicality of atmospheric optical links. Although atmospheric applications exist, the vast majority of optical communications is conducted over glass fiber.

A key development leading to fiber communications was the demonstration of the first laser in 1960. This discovery was quickly followed by plans for numerous laser applications. Progress on empty-space optical systems in the 1960s laid the groundwork for fiber communications in the 1970s. The first low-loss optical waveguide, the glass fiber, was produced in 1970. Soon after, multimode fiber transmission systems were being designed, tested, and installed. The mass production commercialization of the single-mode fiber in 1983 brought about the fiber revolution in the communications industry. Fibers are practical for a range of path lengths, from under a meter to as long as required on the earth's surface and beneath its oceans (e.g., almost 10,000 kilometers for transpacific links).

Fiber systems are limited in length by the bandwidth of their components (a fiber's bandwidth decreases with length) and by component losses (a fiber's loss increases with length). Loss is usually expressed in the decibel scale, which compares two power levels and is defined by

$$\text{dB} = 10\log(P_2/P_1) \tag{140.4}$$

140.1 Optical Communications Systems Topologies

Fibers find their greatest use in telephone networks, local area networks (LANs), and cable television networks. They are also useful for short data links, closed-circuit video links, and elsewhere.

A block diagram of a point-of-point fiber optical communications system appears in Figure 140.1. This diagram represents the architecture typical of the telephone network. The fiber telephone network is digital, operating at levels indicated in Table 140.3. Single fibers operating as high as 10 Gb/s are commonplace. At this rate, almost 130,000 digitized voice channels (each operating at 64 kb/s) can be transmitted along a single fiber using time-division multiplexing (TDM). Fiber systems operating at 40 Gb/s have been developed, further increasing the capacity of a single fiber operating at a single carrier

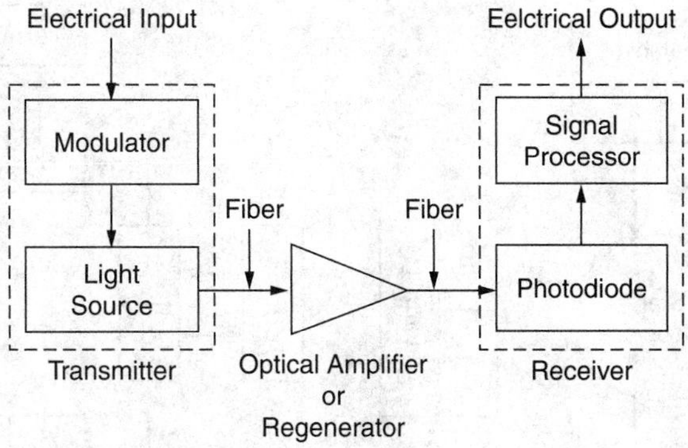

FIGURE 140.1 Point-to-point fiber transmission system.

wavelength. When the number of carrier wavelengths is increased using WDM (tens or even hundreds of channels), capacities of a single fiber enter the terabit/s range.

Because fiber cables may contain more than one fiber (in fact, some cables contain hundreds of fibers), the capacity for voice and data transmission within a single cable is enormous.

Telephone applications may be broken down into several distinct areas: transmission between telephone exchanges, long-distance links, undersea links, and distribution in the local loop (i.e., to subscribers). Although similarities exist among these systems, requirements are somewhat different. Between telephone exchanges, large numbers of calls must be transferred over moderate distances. Because of the moderate path lengths, optical amplifiers or regenerators are not required. Long-distance links, such as between major cities, require signal boosting of some sort (either regenerators or optical amplifiers). Undersea links (such as transatlantic or transpacific) require multiple boosts in the signal because of the long path lengths involved [Thiennot et al., 1993].

TABLE 140.3 Digital Transmission Rates

Designation	Data Rate (Mb/s)
DS-1	1.5444
DS-3	44.736
DS-4	274.175
OC-1	51.84
OC-3	155.52
OC-12	622.08
OC-24	1244.16
OC-48	2488.32
OC-192	9953.28
OC-768	39813.12

Note: The DS levels refer to U.S. telephone signaling rates, whereas the OC levels are those of the SONET standard.

Fiber to the home (for broadband services, such as cable television distribution) does not involve long path lengths but does include division of the optical power in order to share fiber transmission paths over all but the last few tens of meters into the subscriber's premises. In many subscriber networks, the fibers terminate at optical-to-electrical conversion units located close to the subscriber. From that point, copper wires transmit the signals over the remaining short distance to the subscriber. Because of the power division, optical amplifiers are needed to keep the signal levels high enough for proper reception.

Cable television distribution remained mostly coaxial cable based for many years, which was partly due to the distortion produced by optical analog transmitters. Production of highly linear laser diodes (such as the distributed feedback [DFB] laser diode) permitted the design of practical television fiber distribution links.

Some applications, such as LANs, require distribution of the signals over shared transmission fiber. Topologies include the passive star, the active star, and the ring network [Hoss, 1990]. The passive star and the active ring are illustrated in Figure 140.2 and Figure 140.3. LANs are *broadcast systems* that allow all terminals to receive messages sent from any one transmitting terminal.

The star coupler distributes optical signals from any input port to all output ports. Its loss is simply

$$L = 10\log(1/N) \tag{140.5}$$

where N is the number of output ports on the star coupler.

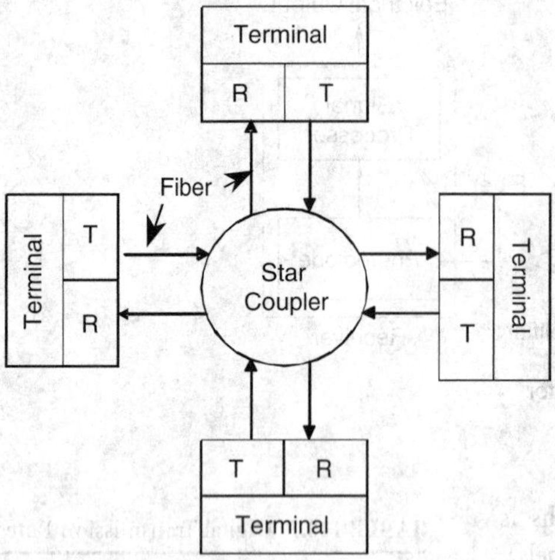

FIGURE 140.2 Star topology. Blocks T and R are, respectively, the optical transmitter and optical receiver.

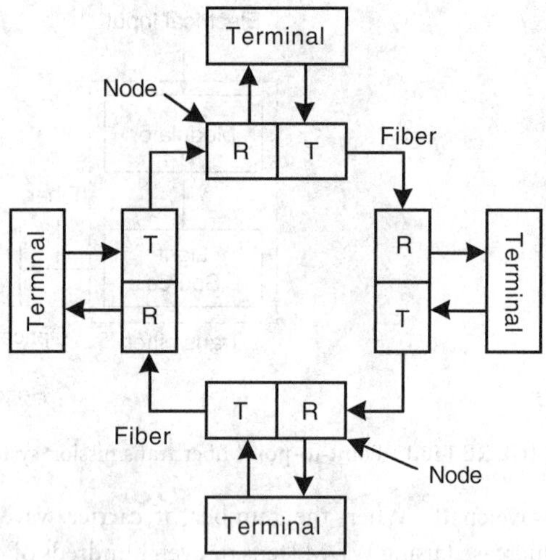

FIGURE 140.3 Ring topology. Blocks T and R are, respectively, the optical transmitter and optical receiver.

In the ring each node acts as a regenerator. The ring is simply a series of connected point-to-point networks.

The major components found in optical communications systems are modulators, light sources, fibers, photodetectors, connectors, splices, star couplers, regenerators, and optical amplifiers.

140.2 Fibers

Silica glass fibers make up the majority of optical transmission lines. The wavelength regions around 850 nm and in the range from 1260 to 1675 nm have been heavily utilized because of their low losses. Losses are lower in the longer wavelengths, making them more suitable for long transmission paths. The nomenclature for the longer wavelength bands appears in Table 140.4. Selected properties of fibers are tabulated in Table 140.5.

Dispersion causes pulse spreading, leading to intersymbol interference. This interference limits the fiber's allowable data rate. The amount of pulse spreading is given by

$$\Delta\tau = ML\Delta\lambda \tag{140.6}$$

where M is the dispersion (a few values are given in Table 140.5), L is the fiber length, and $\Delta\lambda$ is the spectral width of the light source emission. Dispersion is created by the material and the waveguide structure, both of which have a wavelength-dependent pulse velocity.

Multimode fibers allow many modes to simultaneously traverse the fiber. This produces *multimode distortion* (resulting in additional pulse spreading) because the various modes travel at different speeds. For this reason, multimode fibers can only be used for applications in which the product of data rate (or modulation frequency) and path length is not high.

Because single-mode fibers eliminate multimode spreading, they have larger bandwidths and are used for all long-distance, high-rate links. Multimode fibers have relatively high loss and large dispersion in the 850-nm first window region. Applications are, therefore,

TABLE 140.4 Spectral Band Nomenclature

Band	Descriptor	Range (nm)
0-band	Original	1260–1360
E-band	Extended	1360–1460
S-band	Short wavelength	1460–1530
C-band	Conventional	1530–1565
L-band	Long wavelength	1565–1625
U-band	Ultra-long wavelength	1625–1675

TABLE 140.5 Typical Fiber Properties

Wavelength (nm)	Loss (dB/km)	Dispersion (ps/nm · km)
800–900	3.0	120
1290–1330	0.35	Nearly zero
1520–1570	0.22	15

restricted to moderately short part lengths (typically less than a kilometer). Components in this window tend to be cheaper than those operating at the longer wavelengths.

At wavelengths close to 1300 nm, fibers exhibit low losses and nearly zero dispersion. Single-mode, nonrepeatered paths up to 70 km or so are attainable at this wavelength. Here, multimode fiber is feasible for modest lengths required by LANs and campus-based networks, while single-mode fiber is better for longer point-to-point links.

Fiber systems in the 1550-nm region operate at the highest rates and longest unamplified, nonrepeatered distances because of the low fiber attenuation. Unamplified lengths over 100 km are possible. Typically, only single-mode fibers are used in this region to eliminate capacity-limiting multimode distortion. Dispersion is managed using specially designed dispersion-shifted fibers that reduce it below the level given in Table 140.5.

140.3 Other Components

Most systems utilize semiconductor laser diodes (LD) or light-emitting diodes (LEDs) for the light source. These sources are typically modulated by controlling their driving currents. The conversion from current, *i*, to optical power, *P*, is given by

$$\begin{aligned} P &= a_1(i - I_{th}), \qquad & i > I_{th} \\ P &= 0, \qquad & i < I_{th} \end{aligned} \tag{140.7}$$

where a_1 is a constant and I_{th} is the turn-on threshold current for the diode. The threshold current for LEDs is zero and in the range of a few milliamperes to a few tens of milliamperes for LDs. For example, a_1 may be 0.1 mW/mA. Thus, the optical power waveform is a replica of the modulation current if the light source is operated above its threshold current.

Laser diodes are more coherent (i.e., they have smaller spectral widths) than LEDs, and thus produce less dispersion. In addition, LDs can be modulated at higher rates (above 10 GHz), whereas LEDs are limited to rates of just a few hundred MHz. LEDs have the advantage of lower cost and simpler circuitry requirements. For high modulation rates, external modulators replace direct modulation of the light source. These modulators follow the light source, rather than precede it as in Figure 140.1. These modulators use the electrooptic effect in crystals or electroabsorption effects in semiconductors and take the form of integrated optic devices.

The photodetector converts the light beam back into an electrical signal. Semiconductor PIN photodiodes and avalanche photodiodes are normally used. The conversion for the PIN diode is given by the linear equation

$$i = \rho P \tag{140.8}$$

where i is the detected current, P is the incident optical power, and ρ is the photodetector's *responsivity*. Typical values of the responsivity are on the order of 0.5 A/W. The avalanche photodiode response follows this same equation but includes an amplification factor that can be as high as several hundred.

140.4 Signal Quality

Signal quality is measured by the signal-to-noise (S/N) ratio in analog systems and by the bit error rate (BER) in digital links. High-quality analog video links may require S/N ratios on the order of a hundred thousand (50 dB) or more. Good digital systems operate at error rates of 10^{-9} or better.

In a thermal noise-limited system the probability of error, P_e (which is the same as the bit error rate) is

$$P_e = 0.5 - 0.5\,\mathrm{erf}\,(0.354\sqrt{S/N}) \tag{140.9}$$

where erf is the *error function*, tabulated in many references and available electronically as part of several math packages. An error rate of 10^{-9} requires a signal-to-noise ratio of nearly 22 dB (S/N = 158.5).

Defining Terms

Avalanche photodiode — Semiconductor photodetector having internal gain.

Bit error rate — Probability of error.

Dispersion — Wavelength-dependent pulse group velocity caused by the material and the fiber structure. Results in pulse spreading due to the nonzero spectral emission widths of the light source.

Laser — Source producing highly coherent light.

Laser diode — Semiconductor laser. Spectral emission widths typically in the range of 1 to 5 nm. Special devices are available with spectral widths of 0.1 nm or less.

Light emitting diode — Semiconductor emitter having spectral emission widths much larger than those of the laser diode. Typical emission widths are in the range of 20 to 100 nm.

Optical frequency-division multiplexing (OFDM) — Transmission of closely spaced carrier wavelengths (a few tenths of a nanometer, or less) along a single fiber. Demultiplexing usually requires an optical coherent receiver.

PIN photodiodes — Semiconductor photodetector.

Signal-to-noise ratio — Ratio of signal power to noise power.

Single-mode fiber — Fiber that allows only one mode to propagate; has larger bandwidth which allows greater data rate than multimode fibers.

Wavelength-division multiplexing (WDM) — Simultaneous transmission of numerous carrier wavelengths to increase the capacity of a single fiber [Agrawal, 2002]. The individual carriers are spaced from a fraction of a nanometer to several hundred nm, permitting the number of independent channels to range from just a few to over 100.

References

Agrawal, G. P. 2002. *Fiber-Optic Communication Systems.* John Wiley & Sons, New York.

Chaffee, C. D. 1988. *The Rewiring of America.* Academic Press, Orlando, FL.

Hoss, R. J. 1990. *Fiber Optic Communications.* Prentice Hall, Englewood Cliffs, NJ.

Keiser, G. 2000. *Optical Fiber Communications.* McGraw-Hill, New York.

Palais, J. C. 1998. *Fiber Optic Communications.* Prentice Hall, Upper Saddle River, NJ.

Thiennot, J., Pirio, F., and Thomine, J.-B. 1993. Optical undersea cable systems trends. *Proc. of the IEEE.*, 81:1610–1611.

Further Information

Information on optical communications is included in several professional society journals, including the *IEEE Journal of Lightwave Technology* and *IEEE Photonics Technology Letters*. Valuable information is also contained in several trade magazines such as *Lightwave* and *Laser Focus World.*

141

Digital Image Processing

Jonathon Randall
University of Sydney

Ling Guan
Ryerson University, Canada

Digital image processing is one of the most important areas of information processing and communications. It forms the basis of image and video processing, computer vision, and multimedia with a long list of applications in media arts and communications, preservation of cultural heritage, virtual digital museums, film-making, medical decision support, e-health care, distance education, e-commerce, telepresence, security/surveillance, training of athletes, and many more.

A digital image is an array of numbers, each representing a color or gray level. Each entry in the array is termed a *picture element* (pel) or *pixel*. The array consists of an organization or structure that when viewed by a human observer may exhibit some meaning greater than just an array of numbers.

The information revolution of the past two decades has relied mainly on digital processing techniques. Computer calculations rely on digital representations (numbers, concepts, and memory are represented by combinations of 1's and 0's, and communication channels rely largely on digital transmission), and they relay streams of 0's and 1's or bits. In addition, with the reduced cost and high availability of digital scanners, most image and document archives are currently being converted to digital format.

Most digital images are captured on charge-coupled devices (CCDs), which are used in digital cameras, scanners, photocopiers, electron microscopes, bar code readers, and so on. The CCD can either capture the image directly as with a digital camera, or can be used in converting images from analog to digital form, as with image scanners. A CCD consists of two fundamental elements: (1) a solid-state array, which captures the energy (energy is directly related to wavelength) of the incident light and translates it into a voltage; and (2) an analog-to-digital converter that converts the voltage into a digital signal.

141.1 Color and Gray Level

Digital images can be color or black and white. Black-and-white images are referred to as gray-level images, as they are not really black and white, but rather are made up of different shades of gray. In digital image processing, the gray levels are commonly represented by 8-bit values. Thus, there are 256

0-8493-1586-7/05/$0.00+$1.50

FIGURE 141.1 The gray square on the *left* appears "darker" to the human eye than the gray square on the *right*, even though both squares possess equivalent gray-level values.

different gray levels. The human eye can adapt to a large range of gray levels, but can only distinguish a very limited range of intensity levels simultaneously. In Figure 141.1, we note that the right square appears brighter than the left square, even though both squares are of exactly the same intensity.

Color is represented by a three-dimensional space, usually the primary colors red, green, and blue termed the *RGB* color space, but other combinations such as cyan, yellow, and magenta can be used. Each value is given by an 8-bit binary number, resulting in 24 bits for most color representations and 16,777,216 different colors! Another common representation is the *YIQ* color space. The *Y* value represents the luminance and the *I* (in phase) and *Q* (quadrature) values give the color information or chrominance. This representation was introduced for color television, since it sets up compatibility between color television and so-called "black and white" television. While the color television receives all three values, the black and white television will only receive the luminance value. This color space also has the advantage that the eye is more sensitive to luminance; thus the *I* and *Q* values are often down-sampled at a rate of 2:1 in each direction with little effect on the overall quality of the video. The *RGB* to *YIQ* conversion is given by the following equation:

$$\begin{bmatrix} Y \\ I \\ Q \end{bmatrix} = \begin{bmatrix} 0.299 & 0.587 & 0.114 \\ 0.596 & -0.275 & -0.321 \\ 0.212 & -0.523 & 0.311 \end{bmatrix} \begin{bmatrix} R \\ G \\ B \end{bmatrix} \tag{141.1}$$

The last color models worth mentioning are the *HSV* and *HSI*. The HSV and HSI (hue, saturation, value/intensity) are closely related to the way in which colors are perceived by human beings, and are thus widely used in computer vision applications. The hue gives the actual color, such as orange, red, blue, and so on. The saturation gives the purity of the color, such as how orange something is. The intensity or value gives the brightness of the color, such as dark orange as opposed to light orange.

141.2 Point Operations

Point operations are those procedures that modify a point or entry in the digital array, including *contrast stretching, brightening,* and *thresholding*. Contrast stretching and brightening are used to enhance the appearance of an image. Thresholding is used to obtain important aspects of image structure.

Contrast stretching is achieved with histogram equalization. The grayscale histogram measures the frequency of gray levels in an image. The histogram for an image with N gray levels is a discrete function $f(j) = n_j$, such that n_j is the number of pixels in the image with gray level j, $j = 0, 1, \ldots, N$. This involves transforming the histogram of the image, which defines the probability distribution of the gray levels in the image to a uniform probability distribution, which can be achieved by considering the cumulative distribution of the image gray levels. The procedure is shown below, and transforms all pixels having gray level k to pixels with gray level b_k.

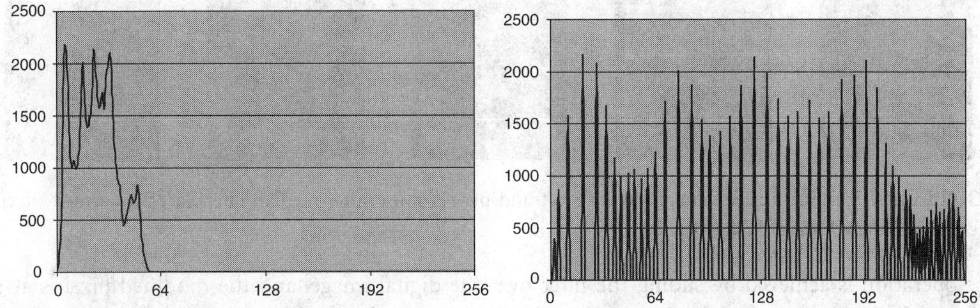

FIGURE 141.2 The histogram of Lena (darkened), and the histogram after equalization.

FIGURE 141.3 The darkened image Lena, and the image Lena after histogram equalization.

1. Calculate the image histogram. For an 8-bit image we will have values defined in the range [0, 255]. Define the "new" histogram value as

$$b_k = 255 * \sum_{j=0}^{k} \frac{n_j}{n}$$

where n is the total number of pixels in the image and n_j is the number of pixels with value j, j = 0, 1, 2, …., 255 and k = 0, 1, 2, …, 255.

Figure 141.2 shows the histogram of the original image (left) and after equalization (right). The corresponding images are illustrated in Figure 141.3.

141.3 Image Filtering

Image filtering is most commonly performed by using a 3 × 3 window and modifying the central pixel in the 3 × 3 window according to the filter coefficients, $a_1 - a_9$. A 3 × 3 filter is shown in the Equation (141.2).

$$\begin{bmatrix} \mathrm{a}_1 & a_3 & a_3 \\ a_4 & a_5 & a_6 \\ a_7 & a_8 & a_9 \end{bmatrix} \tag{141.2}$$

FIGURE 141.4 Original image Lena, Lena with salt-and-pepper noise, after median filtering, after average filtering. The median filtered image is sharpened.

The operation is achieved by sliding the filter over the digital image, and the modified pixel is at the point in the image corresponding to a_5 when the filter is passed over this point. Thus,

$$x_5' = \sum_{i=1}^{9} a_i x_i \tag{141.3}$$

where x_5' represents the modified image pixel.

If all the coefficients are positive and sum to 1, the filter is a low-pass filter. If the center coefficient is positive and all other coefficients are less than or equal to zero, then the filter is a high-pass filter, and if all coefficients sum to zero, the filter is an edge detector. Low-pass filtering is used for smoothing and noise reduction, while high-pass filtering is used for sharpening image features. Low-pass operations also include *median filtering*. The median is defined as the value for which half the pixels are less than and half the pixels are greater than. A median filter sets the new center pixel equal to the median of the 9 pixels in the window. This works better than averaging when the goal is to eliminate impulse noise. Impulse noise usually consists of pixels with values at one of the two extremes; for an 8-bit image the "noise" pixels will be close to 0 and 255. The effect of one of the "noise" pixels will be to raise or lower the average.

A common method for image sharpening is employment of a high-boost filter. A high-boost filter adds a high-pass filtered image to the original image to produce much sharper image details. Image sharpening can be used to reduce the blurring effects of a median filter. Figure 141.4 shows an image after undergoing different low-pass and high-pass filtering operations. Note that although the sharpened image is numerically different from the original one, it is visually much closer than the others.

Filters of other sizes, such as 7×7 or 25×25, may be used; the only requirement is that filter dimensions are odd numbers that will result in a well-defined central point.

141.4 Edge Detection

An edge is given as the boundary between two regions of differing gray levels. At an edge the gradient of the gray level will be maximal; thus, an edge can be detected with a first-derivative filter. Second-derivative filters may also be used. The second-derivative will experience a zero crossing at a maximum point in the first derivative.

One of the simplest and most popular edge detectors is the Sobel filter, which detects the gradient of the image in the x and y directions. The structure of the Sobel filter is shown in Equation (141.4). The image is filtered with each of the filters. The absolute values of the two images obtained from the two filters are then added. An example of edge detection by the Sobel filter is shown in Figure 141.5.

FIGURE 141.5 Sobel image.

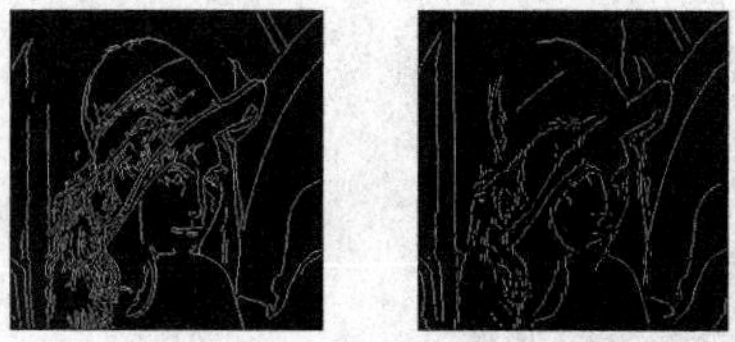

FIGURE 141.6 Canny image with small Gaussian blur and large Gaussian blur.

$$\begin{bmatrix} -1 & 0 & 1 \\ -2 & 0 & 2 \\ -1 & 0 & 1 \end{bmatrix} \tag{141.4}$$

A more sophisticated edge detector is the Canny edge detector. This detector starts with a Gaussian smoothing filter, which helps in the detection of global edges, and then incorporates hysteresis to locate secondary edge points that are connected to primary edge points. Figure 141.6 illustrates the Canny detector with different Gaussian blurs. Other edge detectors include the Prewitt operator, Roberts operator, and the Gaussian Laplacian.

141.5 Digital Image Restoration

Image restoration techniques rely on modeling the degradation function and then trying to reconstruct the image through an inverse process. Some common degradation types that are easily restored include constant motion relative to the camera lens, lens out of focus, and atmospheric turbulence. The degradation model is described by the following equation:

$$g(x, y) = (h \bullet f)(x, y) + \eta(x, y) \tag{141.5}$$

where $g(x, y)$ is the degraded image, $f(x, y)$ is the original image, $h(x, y)$ is the degradation system, $\eta(x, y)$ represents the additive noise, and $\bullet$ represents the convolution operator.

In the following, we discuss two common types of restoration: inverse filtering and the Wiener filter.

Inverse filtering

Taking the Fourier transform of Equation (141.5), we obtain

$$G(u, v) = H(u, v)F(u, v) + N(u, v) \tag{141.6}$$

Rearranging this we obtain

$$F(u, v) = \frac{G(u, v)}{H(u, v)} - \frac{N(u, v)}{H(u, v)} \tag{141.7}$$

Assuming that the noise is small — the degradation being caused by the system $h(x, y)$ — we can neglect the noise term. The zeros of $H(u, v)$ will be few enough to ignore, but we must be careful because $H(u, v)$ will tend to decay faster than $N(u, v)$ as we move away from the origin in the *uv* plane. Reasonable results can be obtained by limiting the values of u and v close to the origin.

Obtaining $H(u, v)$ can be done by testing the system response to an impulse or point light source, that is, setting $f(x, y)$ to $\delta(x, y)$, where $\delta(x, y)$ is the Dirac delta function, as defined in Equation (141.8) and shown in Figure 141.7. An example of a test image is given in the figure. For motion relative to camera lens, lens out of focus, and atmospheric turbulence, the form of $H(u, v)$ is well known:

FIGURE 141.7 The impulse.

$$\delta(x,y)=\begin{cases}1, \text{ if } x=0 \text{ and } y=0\\ 0, \text{ otherwise}\end{cases} \tag{141.8}$$

Suppose an object traveling at a constant speed *V* in the *x*-direction is photographed with the shutter speed set so that the shutter is open for time *T*. Then the Fourier transform of the degradation is given by [2]

$$H(u,v)=\frac{\sin(\pi VTu)}{\pi Vu} \tag{141.9}$$

For a thin lens out of focus, the degradation is given by [2]

$$H(u,v)=\frac{J_1(a(u^2+v^2))}{a(u^2+v^2)} \tag{141.10}$$

where J_1 is a Bessel function of the first order, and *a* is the displacement.

Atmospheric turbulence mostly arises in remote sensing applications, such as taking terrestrial photographs from a satellite positioned thousands of miles above the earth. This is due to heat effects in the atmosphere. The refractive index of the medium will vary with temperature. When driving along a road on a really hot dry day, one sometimes notices the illusion of water on the road ahead. This is actually the sky. Because the sun heats the road, the air closer to the surface of the road is much warmer; thus the path of light from the sky to your eye bends toward the road. These refraction effects will alter the path of the light from the terrestrial surface to the satellite. This phenomenon is known as atmospheric turbulence and is modeled by the following equation [2]:

$$H(u,v)=e^{-c(u^2+v^2)^{5/6}} \tag{141.11}$$

where *c* is a constant that can be determined experimentally.

Wiener filter

The Wiener filter minimizes the mean square error, e^2, between the estimated image, $\hat{f}$, and the original image *f*. The error takes the following form:

$$e^2=\left\{\left[f-\hat{f}\right]^2\right\} \tag{141.12}$$

The Fourier transform of the Wiener filter is given as

$$W(u,v)=\frac{1}{H(u,v)}\frac{|H(u,v)|^2}{|H(u,v)|^2+\left[S_\eta(u,v)/S_g(u,v)\right]} \tag{141.13}$$

FIGURE 141.8 The image on the left is Lena corrupted by Gaussian noise, and the image on the right is Lena restored with the Wiener filter.

FIGURE 141.9 Lena after histogram thresholding.

where $S_g(u, v)$ and $S_\eta(u, v)$ are the power spectrums of the degraded image and noise components, respectively. These are often unknown, and it is often useful to approximate the Wiener filter as

$$W(u,v) = \frac{1}{H(u,v)} \frac{|H(u,v)|^2}{|H(u,v)|^2 + K} \tag{141.14}$$

where K is a real constant. Figure 141.8 demonstrates restoration of a noise-corrupted image by the Wiener filter.

141.6 Digital Image Segmentation

Image segmentation refers to grouping image regions. This can be done based on color, texture, or image structure. The simplest form of image segmentation is image thresholding as described in Section 141.2. Thresholding converts a gray-level image into a binary image, thus dividing the image into regions of black and regions of white. Thresholding is performed on a gray-level histogram, where object points are described as those above or those below a certain gray level or threshold, T. The output image $g(x, y)$ is defined as follows:

$$g(x,y) = \begin{cases} 255 \text{ if } f(x,y) > T \\ 0 \quad \text{otherwise} \end{cases} \tag{141.15}$$

An optimal threshold can be found by modeling the histogram as a mixture of two Gaussians and taking the threshold as the point of intersection of the Gaussians. This method may fail in the case of overlapping histograms between the object and the background. Figure 141.9 shows segmentation of an image by histogram thresholding.

Other methods of segmentation include *region growing by pixel aggregation* and *quad-tree split and merge*. Region growing by pixel aggregation starts with a scattered subset of pixels or "seeds," and combines them with neighboring pixels based on some property such as color or texture. One of the problems that this method presents is seed selection. We would like to select seeds that lie within regions of interest, which requires some kind of knowledge of the application. For example, if we wish to segment only bananas we may select pixels corresponding to the color yellow.

Quad-tree split and merge works as follows:

1. Split the image into four equal parts.
2. Test each region to see if it satisfies some criterion, such as having a small gray-level variance, that is,

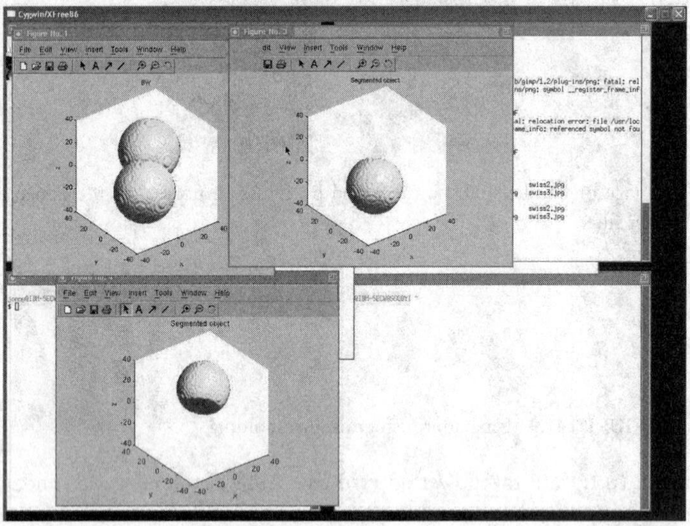

FIGURE 141.10 Watershed is used to split the two three-dimensional binary globes. The color in this example is used to capture the three-dimensional nature of the image, but the image is actually binary.

$$\frac{1}{N}\sum_{i=1}^{N}\left|x_i-\mu\right|^2 < T$$

If the criterion is satisfied, then no more splitting is required. If the region combined with one of its neighbors satisfies the criterion, then merge the two regions.

3. Start from Step 1 for regions that do not satisfy the criterion of Step 2.

Watershed

Watershed segmentation operates on two- and three-dimensional binary images, and is used for separating convex objects that slightly overlap. The algorithm starts with a distance map to find the fattest parts of the object (the peaks), and dilates (mathematical morphology) them until it reaches the edge of an object, or until it reaches the edge of another object. Significantly overlapping features may not be detected. An example of watershed segmentation is given in Figure 141.10.

141.7 Digital Image Compression

With most digital cameras today offering resolution in about the 2048 × 2048 range or 4 megapixels, for a color image at 24 bits per pixel (or 3 bytes per pixel) a single image will occupy 12 Mb of data space. The average computer screen has a resolution of about 1 Mpixel, so why do we need digital cameras with such high resolution? If we want to make 7" × 7" photo-quality prints at 300 dpi (dots per inch), then a 1 Mpixel resolution will require four dots per pixel, which will give us a blocky effect. In this raw format, 40 photographs of our last holiday will take up half a gigabyte of memory, and 5 minutes of video at 30 frames per second will occupy over 100 Gb of memory, which presents problems for both transmission and storage.

Image compression may be *lossless*, meaning that the decompressed image is exactly the same as the compressed image, or it may be *lossy*, meaning that some information is lost in the decompressed image. Lossless methods deliver, at best, compression ratios of about 3:1, while lossy methods may typically result in compression ratios of more than 10:1. Images usually contain large amounts of redundancy. Types of redundancy are coding redundancy, which occurs when more bits are used to code the pixel

FIGURE 141.11 The original image Lena (233 gray levels), and Lena with 64 gray levels.

values than absolutely necessary, psychovisual redundancy, which applies to information that may be removed without significantly degrading the performance of the image as perceived by the human observer, and interpixel redundancy, which occurs when pixels can be easily predicted from the values of their neighbors. For video images, interpixel redundancy can be extended to interframe redundancy. Coding redundancy occurs when more bits are used to code the pixel values than absolutely necessary.

Lossless compression takes advantage of both coding and interpixel redundancies. Coding redundancy is usually based on variable length coding (VLC). The most well-known example of VLC is the Morse code. The length of the codeword for each letter is based on the frequency with which that letter appears in the English language. For example the most common letter, the letter "e," is given by a dot, while a letter less likely to appear, the letter "q," is given by two dashes, a dot, and a dash. A pause is implemented between two letters so that one may recognize the finish of one letter and the start of the next one. This is not feasible for coding of digital images; thus, the VLC must be uniquely decidable (UD). A code is UD if and only if no codeword is a concatenation of two other codewords. Assume that we decide to assign the codeword 0 to the letter "e," and the codeword 1 to the letter "t." Then if the letter "q" is given by 1101, we will not be able to tell the difference between ttet and q. The best-known VLC for a zero memory source is the Huffman code.

Interpixel redundancies can be implemented in lossless coding in the following manner. Assume we have some kind of reasonable predictor to calculate the value of a pixel based on the value of the previous pixel. If the predictor is designed well, for a large proportion of the image pixels the error will be close to zero or zero. For an image with 256 gray levels, we only need to code the error, which will have a high probability of being close to zero. Thus, applying a Huffman code to the error will result in much higher compression than it would if applied to the original image. Lossless compression results in, at best, compression ratios of 3:1. We can achieve much higher compression ratios if we use lossy compression.

Lossy compression takes advantage of psychovisual redundancy, as well as coding and interpixel redundancies; thus, information "lost" due to compression should not infringe too heavily on image quality. The common methods of achieving this are scalar or vector quantizations. As discussed above, the amount of gray levels the human eye can distinguish at once is limited. Thus, we may reduce the number of bits used to encode the image without significantly degrading the quality as seen by the human observer. A look-up table is used to decode the image. In Figure 141.11, the image on the left contains 233 gray levels and the image on the right contains 64 gray levels.

Another common type of lossy compression is based on transform coding. This involves using a reversible transformation, such as the Fourier transform, and then coding the coefficients. For most natural images, most of the important information will be contained in only a few coefficients; thus, coefficients that make a smaller contribution to the image may be coded with fewer bits than those that make a significant contribution to the image. The most common lossy code is the JPEG (Joint Photographic Expert Group) standard, which relies on a discrete cosine transform (DCT). However, the most recent version of JPEG Standard, JPEG-2000, uses the discrete wavelet transform (DWT) to achieve higher compression ratios.

141.8 Image Description and Analysis

Used to describe and analyze images for applications, such as image object recognition, three-dimensional reconstruction, and automatic navigation, this process often relies on higher-level features, such as edge

FIGURE 141.12 Image Lena after morphological operations.

alignment, edge continuity, the shape of a given region, and region convexity. Methods include mathematical morphology, graph methods, and gestalt psychology.

Mathematical morphology can be used to *dilate* and *erode* image regions. Morphology consists of transforming the image based on its set relation to a structuring element. Basic morphological operations include *dilation, erosion, opening*, and *closing*. Figure 141.12 shows a thresholded image after the closing operation.

A graph is a data structure in which image elements are described by nodes, and relationships between the nodes are described by arcs, or connections. One of the simplest forms of such a data structure is the *region adjacency graph*. In such a graph, each region is represented by the node, and connections are formed with adjacent regions. The value of the connection may be given by some property such as the length of the common boundary, or a description of the smoothness of the boundary. Gestalt psychology derives such features as smoothness, continuity, and parallelism, all of which may be used to describe graphs.

141.9 Conclusion

This chapter presented a brief introduction to digital image processing. For readers interested in learning more about digital image processing, numerous references are available. Gonzalez and Woods [1] and Sonka et al. [2] are two of the best. Gonzalez and Woods [1] provide a thorough mathematical and theoretical outline of the subject, while Sonka et al. [2] are geared toward a more practical treatment of digital image processing.

References

[1] R. C. Gonzalez and R. E. Woods. *Digital Image Processing*. Addison-Wesley, Reading, MA, 1993.

[2] M. Sonka, V. Hlavac, and R. Boyle. *Image Processing, Analysis and Machine Vision*. Chapman & Hall, New York, 1993.

142

Complex Envelope Representations for Modulated Signals*

Leon W. Couch, II
University of Florida

142.1 Introduction

What is a general representation for bandpass digital and analog communication signals? How do we represent a **modulated signal**? How do we evaluate the spectrum and the power of these signals? These are some of the questions that are answered in this chapter.

A *baseband* waveform has a spectral magnitude that is nonzero for frequencies in the vicinity of the origin (i.e., $f = 0$) and negligible elsewhere. A *bandpass* waveform has a spectral magnitude that is nonzero for frequencies in some band concentrated about a frequency $f = \pm f_c$ (where $f_c >> 0$), and the spectral magnitude is negligible elsewhere. f_c is called the *carrier frequency.* The value of f_c may be arbitrarily assigned for mathematical convenience in some problems. In others, namely, **modulation** problems, f_c is the frequency of an oscillatory signal in the transmitter circuit and is the assigned frequency of the transmitter, such as 850 kHz for an AM broadcasting station.

In communication problems, the information source signal is usually a baseband signal, for example, a transistor–transistor logic (TTL) waveform from a digital circuit or an audio (analog) signal from a microphone. The communication engineer has the job of building a system that will transfer the information from this source signal to the desired destination. As shown in Figure 142.1, this usually requires the use of a bandpass signal, $s(t)$, which has a bandpass spectrum that is concentrated at $\pm f_c$, where f_c is selected so that $s(t)$ will propagate across the communication channel (either a wire or a wireless channel).

Modulation is the process of imparting the source information onto a bandpass signal with a carrier frequency f_c by the introduction of amplitude and/or phase perturbations. This bandpass signal is called

Source: Couch, L. W., II. *Digital and Analog Communication Systems,* 6th ed., ©2001, Prentice Hall, reprinted by permission of Pearson Education, Inc., Upper Saddle River, NJ.

0-8493-1586-7/05/$0.00+$1.50

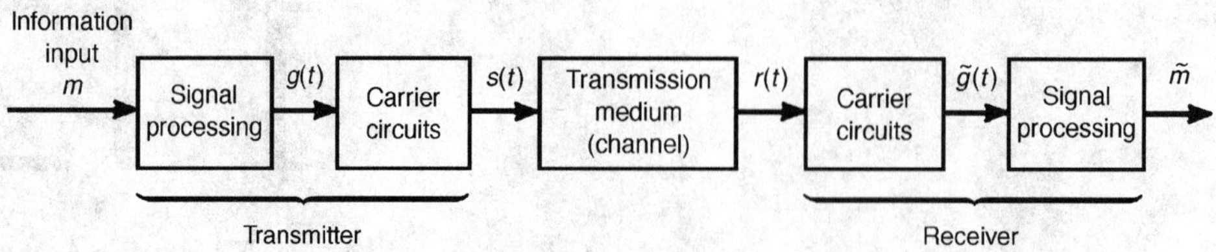

FIGURE 142.1 Bandpass communication system. *Source:* Couch, L. W., II. *Digital and Analog Communication Systems,* 6th ed., ©2001, Prentice Hall, reprinted by permission of Pearson Education, Inc., Upper Saddle River, NJ.

the *modulated* signal $s(t)$, and the baseband source signal is called the *modulating* signal $m(t)$. Examples of exactly how modulation is accomplished are given later in this chapter. This definition indicates that modulation may be visualized as a mapping operation that maps the source information onto the bandpass signal $s(t)$ that will be transmitted over the channel.

As the modulated signal passes through the channel, noise corrupts it. The result is a bandpass signal-plus-noise waveform that is available at the receiver input, $r(t)$, as illustrated in Fig. 1.1. The receiver has the job of trying to recover the information that was sent from the source; $\tilde{m}$ denotes the corrupted version of m.

142.2 Complex Envelope Representation

All bandpass waveforms, whether they arise from a modulated signal, interfering signals, or noise, may be represented in a convenient form given by the following theorem. $v(t)$ will be used to denote the **bandpass waveform** canonically. That is, $v(t)$ can represent the signal when $s(t) \equiv v(t)$, the noise when $n(t) \equiv v(t)$, the filtered signal plus noise at the channel output when $r(t) \equiv v(t)$, or any other type of bandpass waveform.*

Theorem 142.1 *Any physical bandpass waveform can be represented by*

$$v(t) = \text{Re}\{g(t)e^{j\omega_c t}\} \tag{142.1a}$$

Re{·} denotes the real part of {·}. g(t) is called the complex envelope of v(t), and f_c is the associated carrier frequency (hertz) where $\omega_c = 2\pi f_c$. Furthermore, two other equivalent representations are

$$v(t) = R(t)\cos[\omega_c t + \theta(t)] \tag{142.1b}$$

and

$$v(t) = x(t)\cos\omega_c t - y(t)\sin\omega_c t \tag{142.1c}$$

where

$$g(t) = x(t) + jy(t) = |g(t)|e^{j\angle g(t)} \equiv R(t)e^{j\theta(t)} \tag{142.2}$$

$$x(t) = \text{Re}\{g(t)\} \equiv R(t)\cos\theta(t) \tag{142.3a}$$

$$y(x) = \text{Im}\{g(t)\} \equiv R(t)\sin\theta(t) \tag{142.3b}$$

$$R(t) \triangleq |g(t)| \equiv \sqrt{x^2(t) + y^2(t)} \tag{142.4a}$$

*The symbol ≡ denotes an equivalence and the symbol $\triangleq$ denotes a definition.

$$\theta(t) \triangleq \angle g(t) = \tan^{-1}\left(\frac{y(t)}{x(t)}\right) \tag{142.4b}$$

The waveforms $g(t)$, $x(t)$, $y(t)$, $R(t)$, and $\theta(t)$ are all **baseband waveforms**, and, except for $g(t)$, they are all real waveforms. $R(t)$ is a nonnegative real waveform. Equation (142.1) is a low-pass-to-bandpass transformation. The $e^{j\omega_c t}$ factor in Equation (142.1a) shifts (i.e., translates) the spectrum of the baseband signal $g(t)$ from baseband up to the carrier frequency f_c. In communications terminology, the frequencies in the baseband signal $g(t)$ are said to be *heterodyned* up to f_c. The **complex envelope**, $g(t)$, is usually a complex function of time and it is the generalization of the phasor concept. That is, if $g(t)$ happens to be a complex constant, then $v(t)$ is a pure sine wave of frequency f_c and this complex constant is the phasor representing the sine wave. If $g(t)$ is not a constant, then $v(t)$ is not a pure sine wave because the amplitude and phase of $v(t)$ varies with time, caused by the variations of $g(t)$.

Representing the complex envelope in terms of two real functions in Cartesian coordinates, we have

$$g(x) \equiv x(t) + jy(t) \tag{142.5}$$

where $x(t) = \text{Re}\{g(t)\}$ and $y(t) = \text{Im}\{g(t)\}$. $x(t)$ is said to be the *in-phase modulation* associated with $v(t)$, and $y(t)$ is said to be the *quadrature modulation* associated with $v(t)$. Alternatively, the polar form of $g(t)$, represented by $R(t)$ and $\theta(t)$, is given by Equation (142.2), where the identities between Cartesian and polar coordinates are given by Equation (142.3) and Equation (142.4). $R(t)$ and $\theta(t)$ are real waveforms, and, in addition, $R(t)$ is always nonnegative. $R(t)$ is said to be the *amplitude modulation* (AM) on $v(t)$, and $\theta(t)$ is said to be the *phase modulation* (PM) on $v(t)$.

The usefulness of the complex envelope representation for bandpass waveforms cannot be overemphasized. In modern communication systems, the bandpass signal is often partitioned into two channels, one for $x(t)$ called the I (in-phase) channel and one for $y(t)$ called the Q (quadrature-phase) channel. In digital computer simulations of bandpass signals, the sampling rate used in the simulation can be minimized by working with the complex envelope, $g(t)$, instead of with the bandpass signal, $v(t)$, because $g(t)$ is the baseband equivalent of the bandpass signal [1].

142.3 Representation of Modulated Signals

Modulation is the process of encoding the source information $m(t)$ (modulating signal) into a bandpass signal $s(t)$ (modulated signal). Consequently, the modulated signal is just a special application of the bandpass representation. The *modulated signal* is given by

$$s(t) = \text{Re}\{g(t)e^{j\omega_c t}\} \tag{142.6}$$

where $\omega_c = 2\pi f_c$. f_c is the carrier frequency. The complex envelope $g(t)$ is a function of the modulating signal $m(t)$. That is,

$$g(t) = g[m(t)] \tag{142.7}$$

Thus $g[\cdot]$ performs a mapping operation on $m(t)$. This was shown in Figure 142.1.

Table 142.1 gives an overview of the *big picture* for the modulation problem. Examples of the mapping function $g[m]$ are given for amplitude modulation (AM), double-sideband suppressed carrier (DSB-SC), phase modulation (PM), frequency modulation (FM), single-sideband AM suppressed carrier (SSB-AM-SC), single-sideband PM (SSB-PM), single-sideband FM (SSB-FM), single-sideband envelope detectable (SSB-EV), single-sideband square-law detectable (SSB-SQ), and quadrature modulation (QM). For each $g[m]$, Table 142.1 also shows the corresponding $x(t)$ and $y(t)$ quadrature modulation components and

TABLE 142.1 Complex Envelope Functions for Various Types of Modulation[a]

Type of Modulation	Mapping Function $g(m)$	Corresponding Quadrature Modulation $x(t)$	$y(t)$
AM	$A_c[1+m(t)]$	$A_c[1+m(t)]$	0
DSB-SC	$A_c m(t)$	$A_c m(t)$	0
PM	$A_c e^{jD_p m(t)}$	$A_c \cos[D_p m(t)]$	$A_c \sin[D_p m(t)]$
FM	$A_c e^{jD_f \int_{-\infty}^{\infty} m(\sigma)d\sigma}$	$A_c \cos\left[D_f \int_{-\infty}^{t} m(\sigma)d\sigma\right]$	$A_c \sin\left[D_f \int_{-\infty}^{t} m(\sigma)d\sigma\right]$
SSB-AM-SC[b]	$A_c[m(t) \pm j\hat{m}(t)]$	$A_c m(t)$	$\pm A_c \hat{m}(t)$
SSB-PM[b]	$A_c e^{jD_p[m(t) \pm j\hat{m}(t)]}$	$A_c e^{\mp D_p \hat{m}(t)} \cos[D_p m(t)]$	$A_c e^{\mp D_p \hat{m}(t)} \sin[D_p m(t)]$
SSB-FM[b]	$A_c e^{jD_f \int_{-\infty}^{\infty} [m(\sigma) \pm j\hat{m}(\sigma)]d\sigma}$	$A_c e^{\mp D_f \int_{-\infty}^{t} \hat{m}(\sigma)d\sigma} \cos\left[D_f \int_{-\infty}^{t} m(\sigma)d\sigma\right]$	$A_c e^{\mp D_f \int_{-\infty}^{t} \hat{m}(\sigma)d\sigma} \sin\left[D_f \int_{-\infty}^{t} m(\sigma)d\sigma\right]$
SSB-EV[b]	$A_c e^{\{\ln[1+m(t)] \pm j\hat{\ln}\lvert 1+m(t)\rvert\}}$	$A_c[1+m(t)]\cos\{\hat{\ln}[1+m(t)]\}$	$\pm A_c[1+m(t)]\sin\{\hat{\ln}[1+m(t)]\}$
SSB-SQ[b]	$A_c e^{(1/2)\{\ln[1+m(t)] \pm j\hat{\ln}\lvert 1+m(t)\rvert]}$	$A_c\sqrt{1+m(t)}\cos\left\{\frac{1}{2}\hat{\ln}[1+m(t)]\right\}$	$\pm A_c\sqrt{1+m(t)}\sin\left\{\frac{1}{2}\hat{\ln}[1+m(t)]\right\}$
QM	$A_c[m_1(t)+jm_2(t)]$	$A_c m_1(t)$	$A_c m_2(t)$

Type of Modulation	Corresponding Amplitude and Phase Modulation $R(t)$	$\theta(t)$	Linearity	Remarks
AM	$A_c\lvert 1+m(t)\rvert$	$\begin{cases} 0, & m(t) > -1 \\ 180°, & m(t) < -1 \end{cases}$	L[c]	$m(t) > -1$ required for envelope detection
DSB-SC	$A_c\lvert m(t)\rvert$	$\begin{cases} 0, & m(t) > 0 \\ 180°, & m(t) < 0 \end{cases}$	L	Coherent detection required
PM	A_c	$D_p m(t)$	NL	D_p is the phase deviation constant (rad/volt)
FM	A_c	$D_f \int_{-\infty}^{t} m(\sigma)d\sigma$	NL	D_f is the frequency deviation constant (rad/volt-sec)
SSB-AM-SC[b]	$A_c\sqrt{[m(t)]^2+[\hat{m}(t)]^2}$	$\tan^{-1}[\pm\hat{m}(t)/m(t)]$	L	Coherent detection required
SSB-PM[b]	$A_c e^{\mp D_p \hat{m}(t)}$	$D_p m(t)$	NL	
SSB-FM[b]	$A_c e^{\mp D_f \int_{-\infty}^{t} \hat{m}(\sigma)d\sigma}$	$D_f \int_{-\infty}^{t} m(\sigma)d\sigma$	NL	
SSB-EV[b]	$A_c\lvert 1+m(t)\rvert$	$\pm\hat{\ln}[1+m(t)]$	NL	$m(t) > -1$ is required so that the $\ln(\cdot)$ will have a real value
SSB-SQ[b]	$A_c\sqrt{1+m(t)}$	$\pm\frac{1}{2}\hat{\ln}[1+m(t)]$	NL	$m(t) > -1$ is required so that the $\ln(\cdot)$ will have a real value
QM	$A_c\sqrt{m_1^2(t)+m_2^2(t)}$	$\tan^{-1}[m_2(t)/m_1(t)]$	L	Used in NTSC color television; requires coherent detection

Source: Couch, L. W., II. *Digital and Analog Communication Systems*, 6th ed., ©2001, Prentice Hall, reprinted by permission of Pearson Education, Inc., Upper Saddle River, NJ, pp. 235–236.

[a] $A_c > 0$ is a constant that sets the power level of the signal as evaluated by use of Eq. (1.11); L, linear; NL, nonlinear; and $[\hat{\cdot}]$ is the Hilbert transform (a –90° phase-shifted version of $[\cdot]$). For example, $\hat{m}(t) = m(t) * \frac{1}{\pi t} = \frac{1}{\pi}\int_{-\infty}^{\infty} \frac{m(\lambda)}{t-\lambda} d\lambda$.

[b] Use upper signs for upper sideband signals and lower signals for lower sideband signals.

[c] In the strict sense, AM signals are not linear because the carrier term does not satisfy the linearity (superposition) condition.

the corresponding $R(t)$ and $\theta(t)$ amplitude and phase modulation components. Digitally modulated bandpass signals are obtained when $m(t)$ is a digital baseband signal, for example, the output of a transistor transistor logic (TTL) circuit.

Obviously, it is possible to use other $g[m]$ functions that are not listed in Table 142.1. The question is: are they useful? $g[m]$ functions are desired that are easy to implement and that will give desirable spectral properties. Furthermore, in the receiver, the inverse function $m[g]$ is required. The inverse should be single valued over the range used and should be easily implemented. The inverse mapping should suppress as much noise as possible so that $m(t)$ can be recovered with little corruption.

142.4 Generalized Transmitters and Receivers

A more detailed description of transmitters and receivers, as first shown in Figure 142.1, will now be illustrated.

There are two canonical forms for the generalized transmitter, as indicated by Equation (142.1b) and Equation (142.1c). Equation (142.1b) describes an AM-PM type circuit, as shown in Figure 142.2. The baseband signal processing circuit generates $R(t)$ and $\theta(t)$ from $m(t)$. The R and θ are functions of the modulating signal $m(t)$, as given in Table 142.1, for the particular modulation type desired. The signal processing may be implemented either by using nonlinear analog circuits or a digital computer that incorporates the R and θ algorithms under software program control. In the implementation using a digital computer, one analog-to-digital converter (ADC) will be needed at the input of the baseband signal processor, and two digital-to-analog converters (DACs) will be needed at the output. The remainder of the AM-PM canonical form requires radio frequency (RF) circuits, as indicated in the figure.

Figure 142.3 illustrates the second canonical form for the generalized transmitter. This uses in-phase and quadrature-phase (IQ) processing. Similarly, the formulas relating $x(t)$ and $y(t)$ to $m(t)$ are shown in Table 142.1, and the baseband signal processing may be implemented by using either analog hardware or digital hardware with software. The remainder of the canonical form uses RF circuits as indicated. Modern transmitters and receivers, such as those used in cellular telephones, often use IQ processing.

Analogous to the transmitter realizations, there are two canonical forms of receiver. Each one consists of RF carrier circuits followed by baseband signal processing, as illustrated in Figure 142.1. Typically, the carrier circuits are of the superheterodyne-receiver type which consist of an RF amplifier, a down converter (mixer plus local oscillator) to some intermediate frequency (IF), an IF amplifier, and then detector circuits [1]. In the first canonical form of the receiver, the carrier circuits have amplitude and phase detectors that output $\tilde{R}(t)$ and $\tilde{\theta}(t)$, respectively. This pair, $\tilde{R}(t)$ and $\tilde{\theta}(t)$, describe the polar form of the received complex envelope, $\tilde{g}(t)$. $\tilde{R}(t)$ and $\tilde{\theta}(t)$ are then fed into the signal processor, which uses the inverse functions of Table 142.1 to generate the recovered modulation, $\tilde{m}(t)$. The second canonical

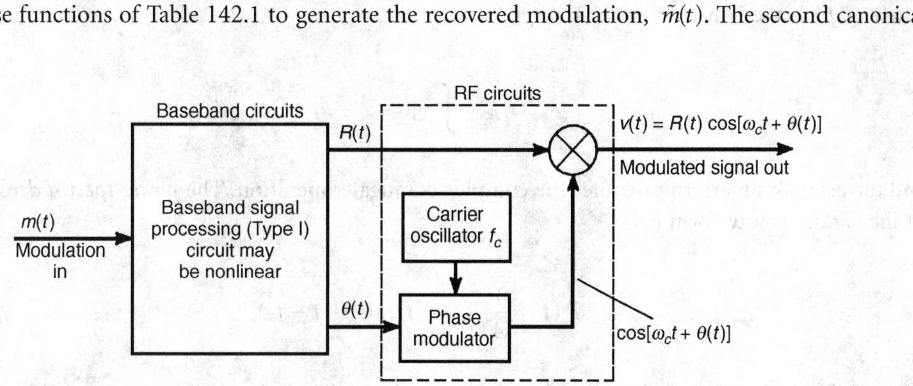

FIGURE 142.2 Generalized transmitter using the AM-PM generation technique. *Source:* Couch, L. W., II. *Digital and Analog Communication Systems,* 6th ed., ©2001, Prentice Hall, reprinted by permission of Pearson Education, Inc., Upper Saddle River, NJ, p. 282.

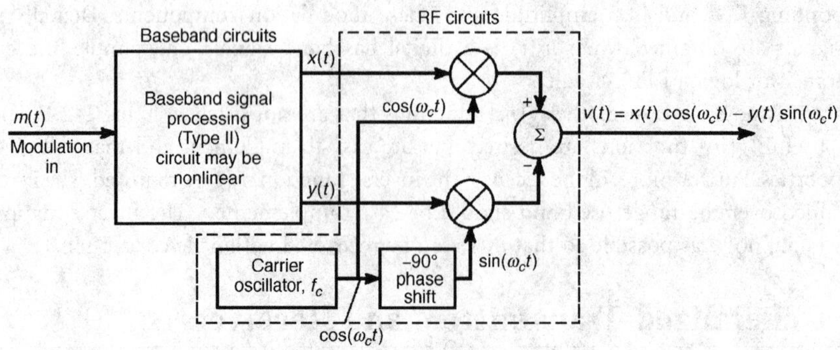

FIGURE 142.3 Generalized transmitter using the quadrature generation technique. *Source:* Couch, L. W., II. *Digital and Analog Communication Systems,* 6th ed., ©2001, Prentice Hall, reprinted by permission of Pearson Education, Inc., Upper Saddle River, NJ, p. 283.

form of the receiver uses quadrature product detectors in the carrier circuits to produce the Cartesian form (IQ processing) of the received complex envelope, $\tilde{x}(t)$ and $\tilde{y}(t)$. $\tilde{x}(t)$ and $\tilde{y}(t)$ are then inputted to the signal processor, which generates $\tilde{m}(t)$ at its output.

Once again, it is stressed that any type of signal modulation (see Table 142.1) may be generated (transmitted) or detected (received) by using either of these two canonical forms. Both of these forms conveniently separate baseband processing from RF processing. Digital signal processing (DSP) techniques are especially useful to realize the baseband processing portion. Furthermore, if DSP circuits are used, any desired modulation type can be realized by selecting the appropriate software algorithm. This is the basis for software radios [1].

142.5 Spectrum and Power of Bandpass Signals

The spectrum of the bandpass signal is the translation of the spectrum of its complex envelope. Taking the **Fourier transform** of Equation (142.1a), the spectrum of the bandpass waveform is [1]

$$V(f)=\frac{1}{2}[G(f-f_c)+G^*(-f-f_c)] \tag{142.8}$$

where $G(f)$ is the Fourier transform of $g(t)$,

$$G(f)=\int_{-\infty}^{\infty} g(t)e^{-j2\pi ft}\,dt$$

and the asterisk superscript denotes the complex conjugate operation. The *power spectra density* (PSD) of the bandpass waveform is [1]

$$P_v(f)=\frac{1}{4}[P_g(f-f_c)+P_g(-f-f_c)] \tag{142.9}$$

where $P_g(f)$ is the PSD of $g(t)$.

The average power dissipated in a resistive load is V_{rms}^2/R_L or $I_{\text{rms}}^2 R_L$, where V_{rms} is the rms value of the voltage waveform across the load and I_{rms} is the rms value of the current through the load. For bandpass waveforms, Equation (142.1) may represent either the voltage or the current. Furthermore, the rms values of $v(t)$ and $g(t)$ are related by [1]

$$v_{\text{rms}}^2 = \langle v^2(t) \rangle = \frac{1}{2}\langle |g(t)|^2 \rangle = \frac{1}{2} g_{\text{rms}}^2 \tag{142.10}$$

where $\langle \cdot \rangle$ denotes the time average and is given by

$$\left\langle \left[\quad \right] \right\rangle = \lim_{t\to\infty} \frac{1}{T} \int_{-T/2}^{T/2} \left[\quad \right] dt$$

Thus, if $v(t)$ of Equation (142.1) represents the bandpass voltage waveform across a resistive load, the average power dissipated in the load is

$$P_L = \frac{v_{\text{rms}}^2}{R_L} = \frac{\langle v^2(t) \rangle}{R_L} = \frac{\langle |g(t)|^2 \rangle}{2R_L} = \frac{g_{\text{rms}}^2}{2R_L} \tag{142.11}$$

where g_{rms} is the rms value of the complex envelope, and R_L is the resistance of the load.

142.6 Amplitude Modulation

Amplitude modulation (AM) will now be examined in more detail. From Table 142.1, the complex envelope of an AM signal is

$$g(t) = A_c[1 + m(t)] \tag{142.12}$$

so that the spectrum of the complex envelope is

$$G(f) = A_c\delta(f) + A_c M(f) \tag{142.13}$$

Using Equation (142.6), we obtain the AM signal waveform

$$s(t) = A_c[1 + m(t)]\cos\omega_c t \tag{142.14}$$

and, using Equation (142.8), the AM spectrum

$$S(f) = \frac{1}{2} A_c[\delta(f - f_c) + M(f - f_c) + \delta(f + f_c) + M(f + f_c)] \tag{142.15}$$

where $\delta(f) = \delta(-f)$ and, because $m(t)$ is real, $M^*(f) = M(-f)$. Suppose that the magnitude spectrum of the modulation happens to be a triangular function, as shown in Figure 142.4(a). This spectrum might arise from an analog audio source where the bass frequencies are emphasized. The resulting AM spectrum, using Equation (142.15), is shown in Figure 142.4(b). Note that because $G(f - f_c)$ and $G^*(-f - f_c)$ do not overlap, the magnitude spectrum is

$$|S(f)| = \begin{cases} \frac{1}{2}A_c\delta(f - f_c) + \frac{1}{2}A_c|M(f - f_c)|, & f > 0 \\ \frac{1}{2}A_c\delta(f + f)_c + \frac{1}{2}A_c|M(-f - f)_c|, & f < 0 \end{cases} \tag{142.16}$$

The 1 in

$$g(t) = A_c[1 + m(t)]$$

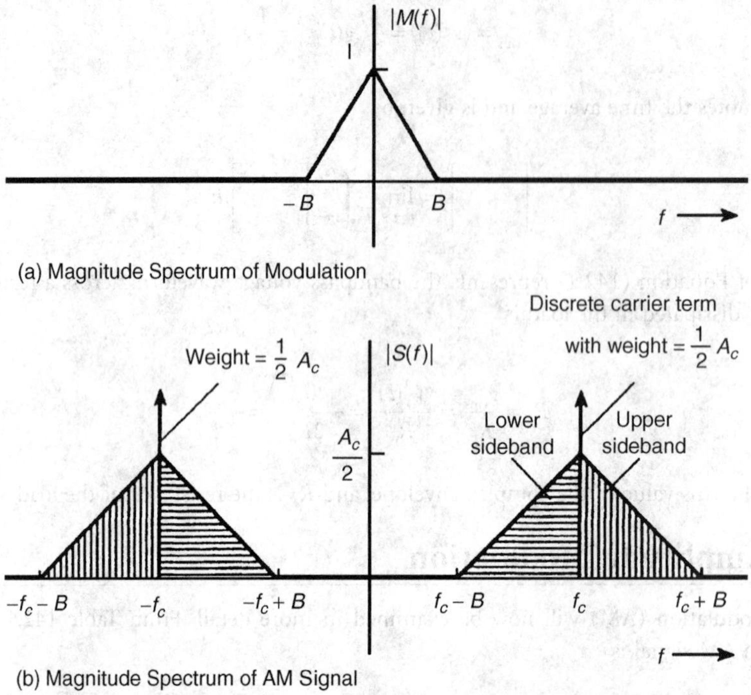

FIGURE 142.4 Spectrum of an AM signal. *Source:* Couch, L. W., II. *Digital and Analog Communication Systems,* 6th ed., ©2001, Prentice Hall, reprinted by permission of Pearson Education, Inc., Upper Saddle River, NJ, p. 239.

causes delta functions to occur in the spectrum at $f = \pm f_c$, where f_c is the assigned carrier frequency. Also, from Figure 142.4 and Equation (142.16), it is realized that the bandwidth of the AM signal is 2B. That is, the bandwidth of the AM signal is twice the bandwidth of the baseband modulating signal.

The average power dissipated into a resistive load is found by using Equation (142.11).

$$P_L = \frac{A_c^2}{2R_L}\left\langle |1+m(t)|^2 \right\rangle = \frac{A_c^2}{2R_L}[1 + 2\langle m(t)\rangle + \langle m^2(t)\rangle]$$

If we assume that the *dc* value of the modulation is zero, $\langle m(t)\rangle = 0$, then the average power dissipated into the load is

$$P_L = \frac{A_c^2}{2R_L}\left\langle 1 + m_{\text{rms}}^2 \right\rangle \tag{142.17}$$

where m_{rms} is the rms value of the modulation, $m(t)$. Thus, the average power of an AM signal changes if the rms value of the modulating signal changes. For example, if $m(t)$ is a sine wave test tone with a peak value of 1.0 for 100% modulation,

$$m_{\text{rms}} = 1/\sqrt{2}$$

Assume that $A_c = 1000$ volts and $R_L = 50$ ohms, which are typical values used in AM broadcasting. Then, the average power dissipated into the 50 Ω load for this AM signal is

$$P_L = \frac{(1000)^2}{2(50)}\left[1 + \frac{1}{2}\right] = 15{,}000 \text{ watts} \tag{142.18}$$

The Federal Communications Commission (FCC) rated carrier power is obtained when $m(t) = 0$. In this case, Equation (142.17) becomes $P_L = (1000)^2/100 = 10{,}000$ watts, and the FCC would rate this as a 10,000 watt AM station. The sideband power for 100% sine wave modulation is 5000 watts.

Now let the modulation on the AM signal be a binary digital signal such that $m(t) = \pm 1$ where +1 is used for a binary one and −1 is used for a binary 0. Referring to Equation (142.14), this AM signal becomes an *on–off keyed* (OOK) digital signal where the signal is on when a binary one is transmitted and off when a binary zero is transmitted. For $A_c = 1000$ and $R_L = 50\ \Omega$, the average power dissipated would be 20,000 watts since $m_{\text{rms}} = 1$ for $m(t) = \pm 1$.

142.7 Phase and Frequency Modulation

Phase modulation (PM) and *frequency modulation* (FM) are special cases of angle-modulated signalling. In angle-modulated signalling, the complex envelope is

$$g(t) = A_c e^{j\theta(t)} \tag{142.19}$$

Using Equation (142.6), the resulting *angle-modulated* signal is

$$s(t) = A_c \cos[\omega_c + \theta(t)] \tag{142.20}$$

For PM, the phase is directly proportional to the modulating signal:

$$\theta(t) = D_p m(t) \tag{142.21}$$

where the proportionality constant D_p is the phase sensitivity of the phase modulator, having units of radians per volt [assuming that $m(t)$ is a voltage waveform]. For FM, the phase is proportional to the integral of $m(t)$:

$$\theta(t) = D_f \int_{-\infty}^{t} m(\sigma)d\sigma \tag{142.22}$$

where the frequency deviation constant D_f has units of radians/volt-second. These concepts are summarized by the PM and FM entries in Table 142.1.

By comparing the last two equations, it is seen that if we have a PM signal modulated by $m_p(t)$, there is *also* FM on the signal corresponding to a *different* modulating waveshape that is given by

$$m_f(t) = \frac{D_p}{D_f}\left[\frac{dm_p(t)}{dt}\right] \tag{142.23}$$

where the subscripts f and p denote frequency and phase, respectively. Similarly, if we have an FM signal modulated by $m_f(t)$, the corresponding phase modulation on this signal is

$$m_p(t) = \frac{D_f}{D_p}\int_{-\infty}^{t} m_f(\sigma)d\sigma \tag{142.24}$$

By using Equation (142.24), a PM circuit may be used to synthesize an FM circuit by inserting an integrator in cascade with the phase modulator input.

Other properties of PM and FM are that the **real envelope**, $R(t) = |g(t)| = A_c$, is a constant, as seen from Equation (142.19). Also, $g(t)$ is a *nonlinear* function of the modulation. However, from Equation (142.21) and Equation (142.22), $\theta(t)$ is a linear function of the modulation $m(t)$. Using Equation (142.11), the average power dissipated by a PM or FM signal is the constant

$$P_L = \frac{A_c^2}{2R_L} \tag{142.25}$$

That is, the average power of a PM or FM signal does not depend on the modulating waveform $m(t)$.

The *instantaneous frequency deviation* for an FM signal from its carrier frequency is given by the derivative of its phase $\theta(t)$. Taking the derivative of Equation (142.22), the *peak frequency deviation* is

$$\Delta F = \frac{1}{2\pi} D_f M_p \text{ Hz} \tag{142.26}$$

where $M_p = \max[m(t)]$ is the peak value of the modulation waveform and the derivative has been divided by 2π to convert from radians/sec to Hz units.

For FM and PM signals, Carson's rule estimates the transmission bandwidth containing approximately 98% of the total power. This FM or PM signal bandwidth is

$$B_T = 2(\beta + 1)B \tag{142.27}$$

where B is bandwidth (highest frequency) of the modulation. The modulation index β, is $\beta = \Delta F/B$ for FM and $\beta = \max[D_p m(t)] = D_p M_p$ for PM.

The AMPS (Advanced Mobile Phone System) analog cellular phones use FM signalling. A peak deviation of 12 kHz is specified with a modulation bandwidth of 3 kHz. From Equation (142.27), this gives a bandwidth of 30 kHz for the AMPS signal and allows a channel spacing of 30 kHz to be used. To accommodate more users, narrow-band AMPS (NAMPS) with a 5 kHz peak deviation is used in some areas. This allows 10 kHz channel spacing if the carrier frequencies are carefully selected to minimize interference to used adjacent channels. A maximum FM signal power of 3 watts is allowed for the AMPS phones. However, hand-held AMPS phones usually produce no more than 600 mW, which is equivalent to 5.5 volts rms across the 50 Ω antenna terminals.

The GSM (Group Special Mobile) digital cellular phones use FM with *minimum frequency-shift-keying* (MSK) where the peak frequency deviation is selected to produce orthogonal waveforms for binary one and binary zero data. (Digital phones use a speech codec to convert the analog voice source to a digital data source for transmission over the system.) Orthogonality occurs when $\Delta F = 1/4R$ where R is the bit rate (bits/sec)[1]. Actually, GSM uses Gaussian shaped MSK (GMSK). That is, the digital data waveform (with rectangular binary one and binary zero pulses) is first filtered by a low-pass filter having a Gaussian shaped frequency response (to attenuate the higher frequencies). This Gaussian filtered data waveform is then fed into the frequency modulator to generate the GMSK signal. This produces a digitally modulated FM signal with a relatively small bandwidth.

Other digital cellular standards use QPSK signalling, as discussed in the next section.

142.8 QPSK, π/4 QPSK, QAM, and OOK Signalling

Quadrature phase-shift-keying (QPSK) is a special case of quadrature modulation, as shown in Table 142.1, where $m_1(t) = \pm 1$ and $m_2(t) = \pm 1$ are two binary bit streams. The complex envelope for QPSK is

$$g(t) = x(t) + jy(t) = A_c[m_1(t) + jm_2(t)]$$

where $x(t) = \pm A_c$ and $y(t) = \pm A_c$. The permitted values for the complex envelope are illustrated by the QPSK **signal constellation** shown in Figure 142.5a. The *signal constellation* is a plot of the permitted values for the complex envelope $g(t)$. QPSK may be generated by using the quadrature generation technique of Figure 142.3, where the baseband signal processor is a serial-to-parallel converter that reads in two bits of data at a time from the serial binary input stream, $m(t)$, and outputs the first of the two

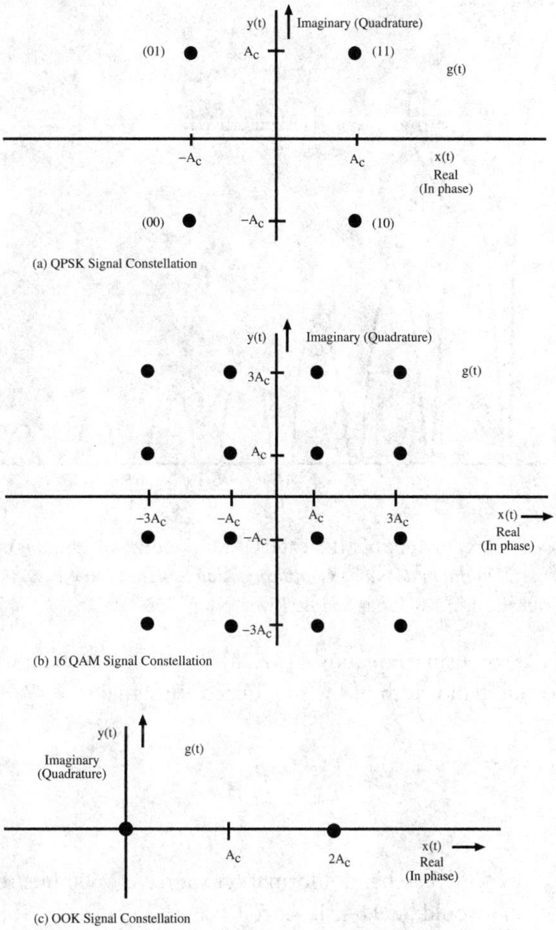

FIGURE 142.5 Signal constellations (permitted values of the complex envelope).

bits to $x(t)$ and the second bit to $y(t)$. If the two input bits are both binary ones, (11), then $m_1(t) = +A_c$ and $m_2(t) = +A_c$. This is represented by the top right-hand dot for $g(t)$ in the signal constellation for QPSK signalling in Figure 142.5a. Likewise, the three other possible two-bit words, (10), (01), and (00), are also shown. The QPSK signal is also equivalent to a four-phase phase-shift-keyed signal (4PSK) since all the points in the signal constellation fall on a circle where the permitted phases are $\theta(t) = 45°, 135°, 225°$, and 315°. There is no amplitude modulation on the QPSK signal since the distances from the origin to all the signal points on the signal constellation are equal.

For QPSK, the spectrum of $g(t)$ is of the sin x/x type since $x(t)$ and $y(t)$ consist of rectangular data pulses of value $\pm A_c$. Moreover, it can be shown that for equally likely independent binary one and binary zero data, the power spectral density of $g(t)$ for digitally modulated signals with M point signal constellations is [1]

$$P_g(f) = K\left(\frac{\sin \pi f \ell T_b}{\pi f \ell T_b}\right)^2 \tag{142.28}$$

where K is a constant, $R = 1/T_b$ is the data rate (bits/sec) of $m(t)$, and $M = 2^\ell$. M is the number of points in the signal constellation. For QPSK, $M = 4$ and $\ell = 2$. This PSD for the complex envelope, $P_g(f)$, is plotted in Figure 142.6. The PSD for the QPSK signal ($\ell = 2$) is given by translating $P_g(f)$ up to the carrier frequency as indicated by Equation (142.9).

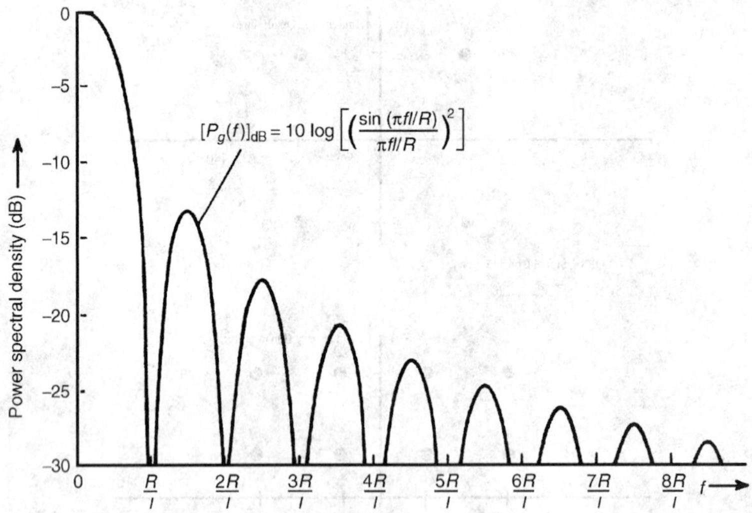

FIGURE 142.6 PSD for the complex envelope of MPSK and QAM where $M = 2^\ell$ and R is bit rate (positive frequencies shown). *Source:* Couch, L. W., II. *Digital and Analog Communication Systems,* 6th ed., ©2001, Prentice Hall, reprinted by permission of Pearson Education, Inc., Upper Saddle River, NJ, p. 358.

Referring to Figure 142.6 or using Equation (142.28), the first-null bandwidth of $g(t)$ is R/ℓ Hz. Consequently, the null-to-null bandwidth of the modulated RF signal is

$$B_{null} = \frac{2R}{\ell} \text{ Hz} \qquad (142.29)$$

For example, if the data rate of the baseband information source is 9600 bits/sec, then the null-to-null bandwidth of the QPSK signal would be 9.6 kHz since $\ell = 2$.

Referring to Figure 142.6, it is seen that the sidelobes of the spectrum are relatively large, so, in practice, the sidelobes of the spectrum are filtered off to prevent interference to the adjacent channels. This filtering rounds off the edges of the rectangular data pulses and this causes some amplitude modulation on the QPSK signal. That is, the points in the signal constellation for the filtered QPSK signal would be fuzzy since the transition from one constellation point to another point is not instantaneous because the filtered data pulses are not rectangular. QPSK is the modulation used for digital cellular phones with the IS-95 Code Division Multiple Access (CDMA) standard.

$\pi/4$ QPSK is generated by alternating symbols between a signal constellation shown by Figure 142.5a and a signal constellation (of Figure 142.5a) rotated by 45°. $\pi/4$ QPSK has the same spectrum as QPSK. However, the peak-to-average power ratio of filtered $\pi/4$ QPSK is much lower than that for filtered QPSK because there are no 180° phase transitions (from symbol to symbol) in the $\pi/4$ QPSK signal. $\pi/4$ QPSK is the modulation technique used for IS-136 TDMA digital cellular phones.

Equation (142.28) and Figure 142.6 also represent the spectrum for *quadrature modulation amplitude modulation* (QAM) signalling. QAM signalling allows more than two values for $x(t)$ and $y(t)$. For example, QAM where $M = 16$ has 16 points in the signal constellation with 4 values for $x(t)$ and 4 values for $y(t)$ such as, for example, $x(t) = +A_c, -A_c, +3A_c, -3A_c$ and $y(t) = +A_c, -A_c, +3A_c, -3A_c$. This is shown in Figure 142.5b. Each point in the $M = 16$ QAM signal constellation would represent a unique four-bit data word, as compared with the $M = 4$ QPSK signal constellation shown in Figure 142.5a, where each point represents a unique two-bit data word. For an $R = 9600$ bits/sec information source data rate, an $M = 16$ QAM signal would have a null-to-null bandwidth of 4.8 kHz since $\ell = 4$.

For OOK signalling, as described at the end of Section 142.6, the signal constellation would consist of $M = 2$ points along the x axis where $x = 0, 2A_c$ and $y = 0$. This is illustrated in Figure 142.5c. For an $R = 9600$ bit/sec information source data rate, an OOK signal would have a null-to-null bandwidth of 19.2 kHz since $\ell = 1$.

Defining Terms

Bandpass waveform: The spectrum of the waveform is nonzero for frequencies in some band concentrated about a frequency $f_c >> 0$; f_c is called the carrier frequency.

Baseband waveform: The spectrum of the waveform is nonzero for frequencies near $f = 0$.

Complex envelope: The function $g(t)$ of a bandpass waveform $v(t)$ where the bandpass waveform is described by

$$v(t) = \text{Re}\{g(t)e^{j\omega_c t}\}$$

Fourier transform: If $w(t)$ is a waveform, then the Fourier transform of $w(t)$ is

$$W(f) = \Im[w(t)] = \int_{-\infty}^{\infty} w(t)e^{-j2\pi ft}\,dt$$

where f has units of hertz.

Modulated signal: The bandpass signal

$$s(t) = \text{Re}\{g(t)e^{j\omega_c t}\}$$

where fluctuations of $g(t)$ are caused by the information source such as audio, video, or data.

Modulation: The information source, $m(t)$, that causes fluctuations in a bandpass signal.

Real envelope: The function $R(t) = |g(t)|$ of a bandpass waveform $v(t)$ where the bandpass waveform is described by

$$v(t) = \text{Re}\{g(t)e^{j\omega_c t}\}$$

Signal constellation: The permitted values of the complex envelope for a digital modulating source.

Reference

1. Couch, L. W., II, *Digital and Analog Communication Systems,* 6th ed., Prentice Hall, Upper Saddle River, NJ, 2001.

Further Information

1. Bedrosian, E., The analytic signal representation of modulated waveforms. *Proc. IRE,* vol. 50, October, 2071–2076, 1962.
2. Couch, L. W., II, *Modern Communication Systems: Principles and Applications,* Macmillan Publishing, New York (now Prentice Hall, Upper Saddle River, NJ), 1995.
3. Dugundji, J., Envelopes and pre-envelopes of real waveforms. *IRE Trans. Information Theory,* vol. IT-4, March, 53–57, 1958.
4. Voelcker, H. B., Toward the unified theory of modulation—Part I: Phase-envelope relationships. *Proc. IRE,* vol. 54, March, 340–353, 1966.

5. Voelcker, H. B., Toward the unified theory of modulation—Part II: Zero manipulation. *Proc. IRE,* vol. 54, May, 735–755, 1966.
6. Ziemer, R. E. and Tranter, W. H., *Principles of Communications,* 4th ed., John Wiley & Sons, New York, 1995.

XXI

Computers

143 Computer Organization: Architecture

Vojin G. Oklobdzija
University of California, Davis

On 7 April 1964, the term **computer architecture** was first defined by Amdahl, Blaauw, and Brooks [1964] of IBM Corporation in the paper announcing the IBM System/360 computer family. On that day IBM Corporation introduced, in the words of an IBM spokesman, "the most important product announcement that this corporation has made in its history." There were six models introduced originally, ranging in performance from 25 to 1. Six years later this performance range was increased to about 200 to 1. This was the key feature that prompted IBM's effort to design an architecture for a new line of computers that were to be code compatible with each other. The recognition that architecture and **computer implementation** could be separated and that one need not imply the other led to establishment of a common System/360 machine architecture implemented in the range of models.

In their milestone paper Amdahl, Blaauw, and Brooks identified three interfaces: architecture, implementation, and realization. They defined computer *architecture* as the attributes of a computer seen by the machine language programmer, as described in the **principles of operation**. IBM referred to the principles of operation as a definition of the machine that enables the machine language programmer to write functionally correct, time-independent programs that run across a number of implementations of that particular architecture. Therefore, the architecture specification covers all functions of the machine that are observable by the program [Siewiorek et al. 1982]. On the other hand, the principles of operation are used to define the functions that the implementation should provide. In order to be functionally correct, the implementation must conform to the principles of operation. Accordingly, for the first time in the history of computer development, IBM has separated *machine definition* from *machine implementation*, thus enabling the company to develop several machine implementations in a wide price and performance range that has reached — 22 years after the introduction of System/360 — 2000 to 1.

The principles of operation define computer architecture, which includes the following components:

- Instruction set
- Instruction format
- Operation codes
- Addressing modes

0-8493-1586-7/05/$0.00+$1.50

- All registers and memory locations that may be directly manipulated or tested by a machine language program
- Formats for data representation

Machine implementation was defined as the actual system organization and hardware structure encompassing major functional units, data paths, and control. **Machine realization** includes issues such as logic technology, packaging, and interconnections.

An example of simple architecture of an 8-bit processor that uses a 2s complement representation to represent integers and contains 11 instructions is shown in Figure 143.1. The figure contains all necessary information for the architecture to be defined.

Separation of the machine architecture from implementation enables several embodiments of the same architecture to be built. Operational evidence proved that architecture and implementation could be separated and that one need not imply the other. This separation made it possible to transfer programs routinely from one model to another and expect them to produce the same result, which defined the notion of **architectural compatibility**. Implementation of the whole line of computers according to a

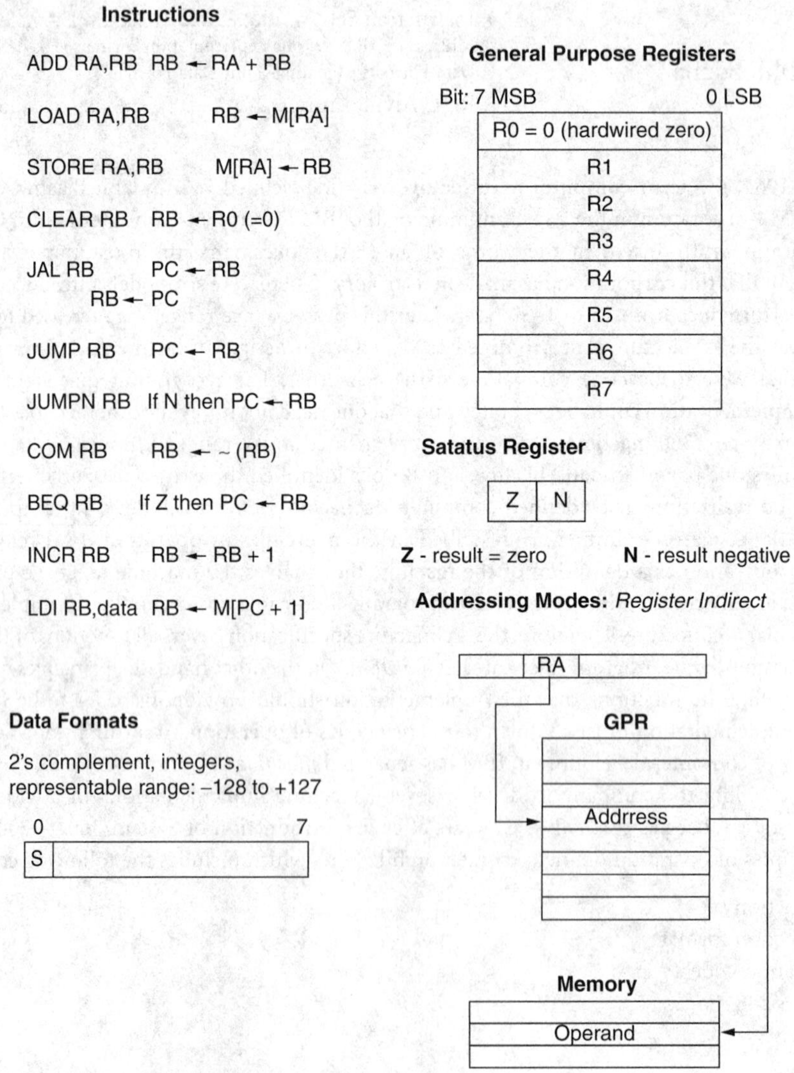

FIGURE 143.1 Example of minimal architecture: PRISC.

common architecture requires unusual attention to details and some new procedures, which are described in the architecture control procedure. The design and control of system architecture is an ongoing process whose objective is to remove ambiguities in the definition of the architecture and, in some cases, to adjust the functions provided.

Definition of an architecture facilitated future development and introduction of not only new models but new **upwardly compatible architectures**. The architecture is upwardly compatible if the user's programs written for the old architecture run efficiently on the new models without modifications to the program. The limitations to upward compatibility are (1) that new systems have the same or equivalent facilities, and (2) that the programs have no time dependence, use only model-independent functions defined in the principles of operation, and do not use unassigned formats and operation codes [Case and Pedegs, 1978]. An example of upward compatibility is IBM System/360, introduced in June 1970.

143.1 Instruction Set

An instruction set defines a basic set of operations, as specified by the architecture, that a particular implementation of that architecture is required to perform. An *instruction* of the instruction set defines an atomic operation that may alter data or the machine state or may perform an I/O operation. In terms of the operation performed, instructions of the instruction set are broadly classified in one of the four general categories:

1. Instructions performing transformation of data
2. Instructions altering the program flow
3. Instructions performing data movement
4. System instructions

The first category includes instructions performing arithmetic and logical operations. The operations can be arithmetic, string, logical, or floating point. They are performed in the appropriate functional units of the particular implementation of the architecture.

Instructions affecting the flow of the program and/or machine state are branches, calls, and returns as well as loop control instructions.

The third category of instructions performs data movement across various functional units of the machine. Examples of such instructions are the load instruction, which loads a content of a memory location to a particular register in the general purpose register file (GPR), and the store instruction, which does the opposite. The move instruction moves a block of data from one memory location to another, or to and from the stack or GPR.

The system instructions change the system's mode and are not generally visible by the programmer who programs in the *problem state*. The problem state is the domain of the machine visible to a programmer executing a general-purpose program, as opposed to the *system state*, which is visible to the operating system. An example of the instruction set specified in the IBM System/360 architecture is given in Figure 143.2.

We can further classify instructions in terms of the number of *explicit operands, operand locations,* and *type* and *size* or the operands. Instructions architecture that specifies no explicit operands is better known as *stack architecture*. In stack architecture all operations are performed on the data that are on the top of the stack. Examples of stack architecture are HP 3000/70 by Hewlett-Packard and B5500 by Burroughs. In the **accumulator architecture**, all operations are performed between the operand specified in the instruction and the *accumulator,* which is a special register. An example of accumulator architecture is the PRISC processor shown in Figure 143.1. One of the well-known accumulator-based architectures is PDP-8, by Digital Equipment Corporation. Almost all modern machines have a repertoire of available general-purpose registers whose numbers range from 16 to 32 and in some cases even more than 32 (SPARC). The number of operands explicitly specified in the instructions of a modern architecture today can be two or three. In the case of three operands, an instruction explicitly specifies the location of both operands and the location where the result is to be stored. In some architectures (e.g., IBM System/360),

RR Format

xxxx	*Branching and status switching 0000xxxx*		*Fixed-point fullword and logical 0001xxxx*		*Floating-point long 0010xxxx*		*Floating-point short 0011xxxx*	
0000			LPR	LOAD POSITVE	LPDR	LOAD POSITIVE	LPER	LOAD POSITIVE
0001			LNR	LOAD NEGATIVE	LNDR	LOAD NEGATIVE	LNER	LOAD NEGATIVE
0010			LTR	LOAD AND TEST	LTDR	LOAD AND TEST	LTER	LOAD AND TEST
0011			LCR	LOAD COMPLEMENT	LCDR	LOAD COMPLEMENT	LCER	LOAD COMPLEMENT
0100	SPM	SET PROGRAM MASK	NR	AND	HDR	HALVE	HER	HALVE
0101	BALR	BRANCH AND LINK	CLR	COMPARE LOGICAL				
0110	BCTR	BRANCH ON COUNT	OR	OR				
0111	BCR	BRANCH/CONDITION	XR	EXCLUSIVE OR				
1000	SSK	SET KEY	LR	LOAD	LDR	LOAD	LER	LOAD
1001	ISK	INSERT KEY	CR	COMPARE	CDR	COMPARE	CER	COMPARE
1010	SVC	SUPERVISOR CALL	AR	ADD	ADR	ADD N	ALR	ADD N
1011			SR	SUBRRACT	SDR	SUBRACT N	SER	SUBRACT N
1100			MR	MULTIPLY	MDR	MULTIPLY	MER	MULTIPLY
1101			DR	DIVIDE	DDR	DIVIDE	DER	DIVIDE
1110			ALR	ADD LOGICAL	AWR	ADD U	AUR	ADD U
1111			SLR	SUBRACT LOGICAL	SWR	SUBRACT U	SUR	SUBRACT U

RX Format

xxxx	*Fixed-point halfword and branching 0100xxxx*		*Fixed-point fullword and logical 0101xxxx*		*Floating-point long 0110xxxx*		*Floating-point short 0111xxxx*	
0000	STH	STORE	ST	STORE	STD	STORE	STE	STORE
0001	LA	LOAD ADDRESS						
0010	STC	STORE CHARACTER						
0011	IC	INSERT CHARACTER						
0100	EX	EXECUTE	N	AND				
0101	BAL	BRANCH AND LINK	CL	COMPARE LOGICAL				
0110	BCT	BRANCH AND LINK	O	OR				
0111	BC	BRANCH/CONDITION	X	EXCLUSIVE OR				
1000	LH	LOAD	L	LOAD	LD	LOAD	LE	LOAD
1001	CH	COMPARE	C	COMPARE	CD	COMPARE	CE	COMPARE
1010	AH	ADD	A	ADD	AD	ADD N	AE	ADD N
1011	SH	SUBTRACT	S	SUBTRACT	SD	SUBTRACT N	SE	SUBTRACT N
1100	MH	MULTIPLY	M	MULTIPLY	MD	MULTIPLY	ME	MULTIPLY
1101			D	DIVIDE	DD	DIVIDE	DE	DIVIDE
1110	CVD	CONVET-DECIMAL	AL	ADD LOGICAL	AW	ADD U	AU	ADD U
1111	CVB	CONVERT-BINARY	SL	SUBRACT LOGICAL	SW	SUBTRACT U	SU	SUBTRACT U

FIGURE 143.2 IBM System/360 instruction set. (*Source*: Blaauw, G. A. and Brooks, F. P. 1964. The structure of system/360. *IBM Syst. J.*, 3:119–135. With permission. Copyright 1964 by International Business Machines Corporation.)

RS, S1 Format

xxxx	*Branching status switching and shifting* 1000xxxx		*Fixed-point logical and input/output* 1001xxxx		1010xxxx	1011xxxx
0000	SSM	SET SYSTEM MASK	STM	STORE MULTIPLE		
0001			TM	TEST UNDER MASK		
0010	LPSW	LOAD PSW	MVI	MOVE		
0011		DIAGNOSE	TS	TEST AND SET		
0100	WRD	WRITE DIRECT	NI	AND		
0101	RDD	READ DIRECT	CLI	COMPARE LOGICAL		
0110	BXH	BRANCH/HIGH	OI	OR		
0111	BXLE	BRANCH/LOW-EQUAL	XI	EXCLUSIVE OR		
1000	SRL	SHIFT RIGHT SL	LM	LOAD MULTIPLE		
1001	SLL	SHIFT LEFT SL				
1010	SRA	SHIFT RIGHT S				
1011	SLA	SHIFT LEFT S				
1100	SRDL	SHIFT RIGHT DL	SIO	START I/O		
1101	SLDL	SHIFT LEFT DL	TIO	TEST I/O		
1110	SRDA	SHIFT RIGHT D	HIO	HALT I/O		
1111	SLDA	SHIFT LEFT D	TCH	TEST CHANNEL		

SS Format

xxxx	1100xxxx	*Logical* 1101xxxx		1110xxxx	*Decimal* 1111xxxx	
0000						
0001		MVN	MOVE NUMERIC		MVO	MOVE WITH OFFSET
0010		MVC	MOVE		PACK	PACK
0011		MVZ	MOVE ZONE		UNPK	UNPACK
0100		NC	AND			
0101		CLC	COMPARE LOGICAL			
0110		OC	OR			
0111		XC	EXCLUSIVE OR			
1000					ZAP	ZERO AND ADD
1001					CP	COMPARE
1010					AP	ADD
1011					SP	SUBTRACT
1100		TR	TRANSLATE		MP	MULTIPLY
1101		TRT	TRANSLATE AND TEST		DP	DIVIDE
1110		ED	EDIT			
1111		EDMK	EDIT AND MARK			

NOTE: N = NORMALIZED, SL = SINGLE LOGICAL, DL = DOUBLE LOGICAL, U = UNNORMALIZED, S = SINGLE, D = DOUBLE

FIGURE 143.2 *Continued.*

only two operands are *explicitly specified* in order to save the bits in the instruction. As a consequence, one of the operands is always replaced by the result and its content is destroyed. This type of instruction is sometimes referred to as *diadic instruction.*

In terms of the operand locations, instructions can be classified as (1) register-to-register (R-R) instructions; memory-to-register (R-M) instructions; and memory-to-memory (M-M) instructions. The addresses of the operands are specified within the instruction. From the information contained in the particular operand field of the instruction, the address of the particular operand can be formed in various ways, described in the next section.

Addressing Modes

The way in which the address of the operand is formed depends on the location of the operand as well as choices given in the instruction architecture. It is obvious that in the case of stack or accumulator architecture, the address of the operand is implied and there is no need to specify the address of the operand. If the operand is in one of the GPRs, the operand field in the instruction contains the number (address) of that particular register. This addressing mode is known as *register direct addressing*, and is one of the simplest ways of pointing to the location of the operand. The addressing of an operand can be even simpler, as in the case where the operand is contained within the instruction. This mode is called the *immediate addressing mode*. The location pointed to by the address formed from the information contained in the operand field of an instruction can contain the operand itself or an address of the operand. The latter case is referred to as *indirect addressing*. Examples of several ways of forming an address of the operand are presented in Figure 143.3.

Data Representation Formats

Another important issue in computer architecture is the determination of data formats. Data formats, along with instruction formats, were formerly of much influence in determining **word size.** Today it is commonly assumed that most of the machines use a 32-bit word size (which is gradually shifting toward 64 bits). This standard was not common in the past, and there was not a common word size used by the majority of the machines. A 36-bit word size was quite common (e.g., IBM's 701 and 704 series), and word sizes of 12, 18, and 60 bits were represented as well (PDP-8, CDC 6600). In the early days of computer development, interaction with the operator was done mainly via the teletype machine, which used 6 bits to represent each character. Therefore, the word sizes of the machines of that period were determined with the objective of being able to pack several characters in the machine word. The size of I/O interfaces was commonly 12 bits (two characters). Anticipation of the new standard for the representation of digits (USASCII-8) prompted IBM to introduce an 8-bit character (EBCDIC) in their introduction of System/360 architecture, which was also its reason for switching from the 36-bit to a new 32-bit word size. Since then (and until today), 32-bit word size and the multiples of the 8-bit quantity (**byte**) have been the most common data formats among various computer architectures. The new standard for representation of digits, USASCII-8, however, did not materialize. Instead, a 7-bit standard for data representation, ASCII, has been commonly used almost everywhere, except by IBM, which could not diverge from the 8-bit character representation defined in its architecture.

Every architecture must specify its representation of characters, integers, floating-point numbers, and logical operands. This representation must specify the number of bits used for every particular field, their order in the computer word, meaning of special bits, interpretation, and total length of data. Data types and data formats as defined in the Digital Equipment Corporation VAC 11/780 architecture are shown in Figure 143.4.

Fixed-Point Data Formats

Fixed-point data forms are used to represent integers. Full-word (32-bit), half-word (16-bit), or double-word (64-bit) quantities are used for representation of integers. They can be signed or unsigned positive

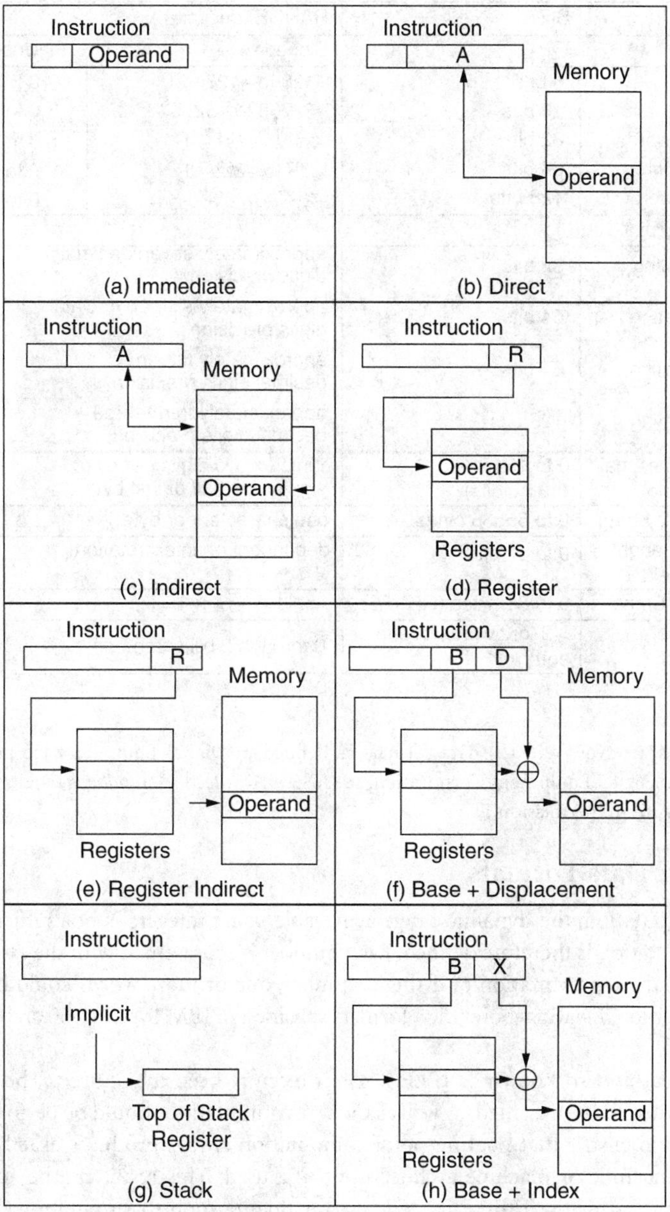

FIGURE 143.3 Example of addressing modes. (Adapted from Stallings, W. 1993. *Computer Organization and Architecture.* Macmillan, New York.)

integers. In the case of signed integers, one bit is used for representation of the sign to represent a range of positive and negative integers. The most common representation of integers is 2s complement format. Another not-so-common representation of integers is binary-coded decimal representation (BCD), used to represent integers as decimal numbers. Each digit position is represented by 4 bits. The coding is straightforward for the numbers from 0 to 9, and the unused bit combinations are used to represent the sign.

For the *logical operand*, a word is treated as a collection of individual bits, where each bit is assigned a Boolean value. A *variable bit field* can also be defined in cases where the field can be treated as a signed or unsigned field of bits.

DATA TYPE	SIZE	RANGE (decimal)	
Integer		Signed	Unsigned
Byte	8 bits	−128 to +127	0 to 255
Word	16 bits	−32768 to +32767	0 to 65535
Longword	32 bits	-2^{31} to $+2^{31} - 1$	0 to $2^{32} - 1$
Quadword	64 bits	-2^{63} to $+2^{63} - 1$	0 to $2^{64} - 1$
Octaword	128 bits	-2^{127} to $+2^{127} - 1$	0 to $+2^{128} - 1$
Floating Point			
F floating	32 bits	approximately seven decimal digits precision	
D floating	64 bits	approximately sixteen decimal digits precision	
G floating	64 bits	approximately fifteen decimal digits precision	
H floating	128 bits	approximately thirty-three decimal digits precision	
Packed Decimal String	0 to 16 bytes (31 digits)	numeric, two digits per byte sign in low half of last byte	
Character String	0 to 65535 bytes	one character per byte	
Variable-lenght Bit Field	0 to 32 bits	dependent on interpretation	
Numeric String	0 to 31 bytes (DIGITS)	$-10^{31}-1$ to $+10^{31}-1$	
Queue	> 2 longwords/queue entry	0 through 2 billion entries	

(a)

FIGURE 143.4 (a) Data types and (b) data formats as defined in Digital Equipment Corporation VAX 11/780 architecture. (*Source:* Digital Equipment Corporation. 1981. *VAX Architecture Handbook*. Digital Equipment Corporation, Maynard, MA. With permission.) *Continued.*

Floating-Point Data Formats

For scientific computation, the dynamic range achievable using integers is not sufficient, and **floating-point** data representation is therefore defined. Each number is represented with the *exponent* and *fraction* (or mantissa). For the representation of a single number, one or more words could be used if required by the desired precision. Floating-point data formats specified in IBM System/360 architecture are shown in Figure 143.5(a).

A floating-point standard known as IEEE 754 has recently been introduced. The standard specifies the way that data are to be represented, as well as the way computation should be performed. The purpose of this standard is to ensure that floating-point computation always produces exactly the same results, regardless of the machine or machine architecture being used. This result can be achieved only if the architecture complies with the IEEE 754 standard for floating-point computation. Data formats prescribed by IEEE 754 are shown in Figure 143.5(b).

143.2 RISC Architecture

A special place in computer architecture has been given to RISC. RISC architecture was developed as a result of the 801 project, which started in 1975 at the IBM T. J. Watson Research Center and was completed by the early 1980s [Rading, 1982]. This project was not widely known to the world outside of IBM and people involved in two other projects with similar objectives started in the early 1980s at the University of California, Berkeley, and Stanford University [Patterson and Sequin, 1982; Hennessy, 1984]. The term **RISC** (reduced instruction set architecture), used for the Berkeley research project, is the term under which this architecture became widely known and recognized today.

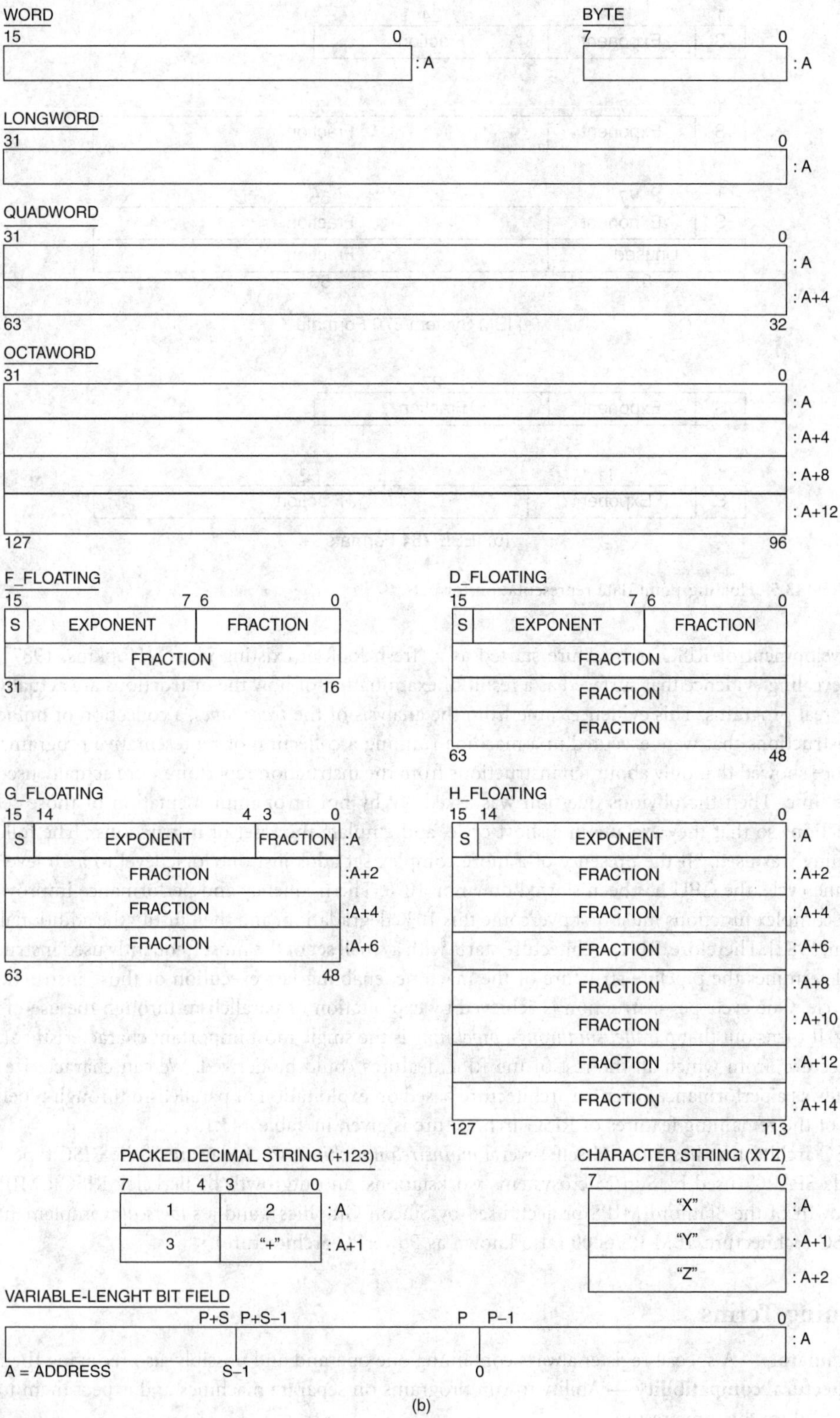

FIGURE 143.4 *Continued.*

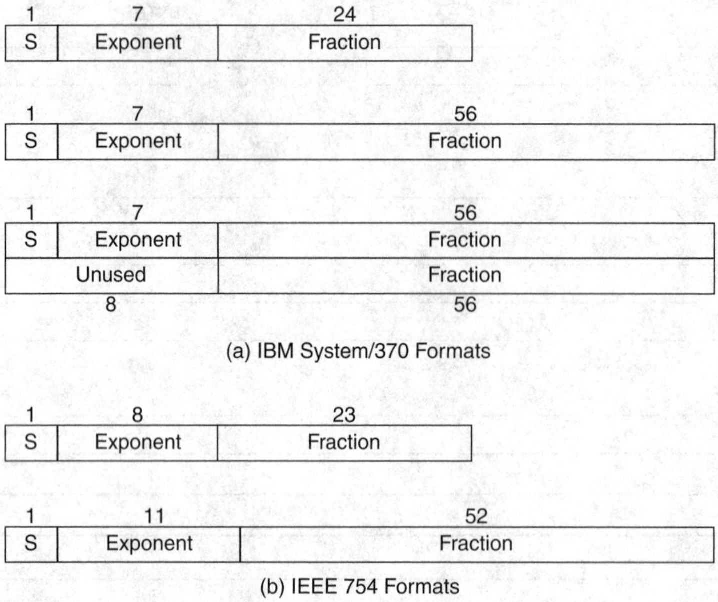

FIGURE 143.5 Floating-point data representation formats.

Development of RISC architecture started as a "fresh look at existing ideas" [Hopkins, 1987], after first revealing evidence that surfaced as a result of examination of how the instructions are actually used in the real programs. This evidence came from the analysis of the *trace tapes*, a collection of millions of the instructions that were executed in a machine running a collection of representative programs. This evidence showed that only about ten instructions from the instruction repertoire were actually used 90% of the time. Then the obvious question was asked: "Why not favor implementation of those selected instructions so that they execute in a short cycle, and emulate the reset of instructions." The following reasoning was used: "If the presence of a more complex set adds just one logic level to a 10 level basic machine cycle, the CPU has been slowed down by 10%. The frequency and performance improvement of the complex functions must first overcome this 10% degradation, and then justify the additional cost" [Radin, 1982]. Therefore, RISC architecture starts with a small set of the most frequently used instructions that determines the pipeline structure of the machine, enabling fast execution of those instructions in one cycle. One cycle per instruction is achieved by exploitation of parallelism through the use of **pipelining**. It turns out that *parallelism through pipelining* is the single most important characteristic of RISC architecture, from which all the rest of the RISC features could be derived. We can characterize RISC basically as a performance-oriented architecture based on exploitation of parallelism through pipelining. A list of the remaining features of RISC architecture is given in Table 143.1.

RISC architecture has proven itself; several *mainstream architectures* today are of the RISC type. These include SPARC (used by Sun Microsystems workstations, an outgrowth of Berkeley RISC); MIPS (an outgrowth of the Stanford MIPS project, used by Silicon Graphics); and a *superscalar* implementation of RISC architecture, IBM RS/6000 (also known as PowerPC architecture).

Defining Terms

Accumulator — A special register always containing one operand and possibly also receiving the result.
Architectural compatibility — Ability to run programs on separate machines and expect them to produce the same results.
Byte — An 8-bit quantity being treated as a unit.

TABLE 143.1 Features of RISC Architecture

Feature	Characteristic
Load/store architecture	All operations are register to register. In this way, *operation* is decoupled from the *access to memory.*
Carefully selected subset of instructions	Control is implemented in hardware. There is no microcoding in RISC. In addition, this set of instructions is not necessarily small.[a]
Simple addressing modes	Only the most frequently used addressing modes are used. It is also important that they fit into the existing pipeline.
Fixed-size and fixed-field instructions	This is necessary to be able to decode instruction and access operands in one cycle (although there are architectures using two sizes for the instruction format, IBM PC-RT).
Delayed branch instruction (known also as *branch and execute*)	The most important performance improvement through instruction architecture.
One instruction per cycle execution rate, CPI = 1.0	Possible only through the use of pipelining.
Optimizing compiler	Close coupling between the architecture and the compiler. Compiler "knows" about the pipeline.
Harvard architecture	Separation of instruction and data cache, resulting in increased memory bandwidth.

[a] IBM PC-RT instruction architecture contains 118 instructions, whereas IBM RS/6000 (PowerPC) contains 184 instructions. This should be contrasted to the IBM System/360, containing 143 instructions, and IBM System/370, containing 208. The first two are representative of RISC architecture, whereas the latter two are not.

Computer architecture — The attributes of a computer, as seen by the machine language programmer that enable this programmer to write functionally correct, time-independent programs.

Computer implementation — System organization and hardware structure.

Computer organization — Hardware structure encompassing the major functional units, data paths, and control.

Fixed point — Positive or negative integer.

Floating point — A number format, containing a fraction and an exponent, used for representation of numbers covering a wide range of values. Used for scientific computation where the range is important.

Pipelining — The technique used to initiate one operation in every cycle without waiting for the final result to be produced or completion of previously initiated operations.

Principles of operation — A definition of the machine. Term used for computer architecture by IBM.

RISC — Reduced instruction set computer.

Superscalar — Implementation of an architecture capable of executing more than one instruction in the same cycle.

Upwardly compatible architectures — Ability to efficiently run user programs written for the old architecture on the new models without modifications to the program; however, the capacity to do the reverse does not exist.

Word size — A quantity defined as the number of bits being operated on as a unit.

References

Amdahl, G. M., Blaauw, G. A., and Brooks, F. P. 1964. Architecture of the IBM System/360. *IBM J. Res. Dev.*, 8:87–101.

Blaauw, G. A., and Brooks, F. P. 1964. The structure of System/360. *IBM Syst. J.*, 3:119–135.

Case, R. P. and Padegs, A. 1978. Architecture of the IBM System/370. *Commun. ACM*, 21:73–96.

Digital Equipment Corporation. 1981. *VAX Architecture Handbook.* Digital Equipment Corporation, Maynard, MA.

Hennessy, J. L. 1984. VLSI processor architecture. *IEEE Trans. Comput.*, C-33(12).

Hopkins, M. E. 1987. A perspective on the 801/reduced instruction set computer. *IBM Syst. J.*, 26(1).

Patterson, D. A. and Sequin, C.H. 1982. A VLSI RISC. *IEEE Comput. Mag.*
Radin, G. 1982. The 801 minicomputer. *SIGARCH Comput. Architecture News*, 10(2):39–47.
Siewiorek, D. P., Bell, C. G., and Newell, A. 1982. *Computer Structures: Principles and Examples.* McGraw-Hill, New York.
Stallings, W. 1993. *Computer Organization and Architecture.* Macmillan, New York.

Further Information

A good introductory text for computer architecture is William Stalings, *Computer Organization and Architecture* (Macmillan, New York, 1993).

For the advanced reader, more information on computer hardware, design, and performance analysis can be found in David A. Patterson and John L. Hennessy, *Computer Organization and Design: The Hardware/Software Interface* (Morgan Kaufmann, San Francisco, 1994). For quantitative analysis of instruction usage and various factors affecting performance, as well as insight into RISC architecture, *Computer Architecture: A Quantitative Approach*, by the same authors and publisher, is highly recommended.

An important historical insight into the development of computer architecture is provided in an interview with Richard Case and Andris Padegs, "Case Study: IBM's System/360-370 Architecture," conducted by editors David Gifford and Alfred Spector in *Communications of ACM* (30, April 1987), as well as the paper "The Architecture of IBM's Early Computers," published in the *IBM Journal of Research and Development* (25, September 1981). The first chapter of David J. Kuck, *The Structure of Computers and Computation* (John Wiley & Sons, New York, 1978) contains an excellent overview of the history of computer development.

Various useful articles on computer architecture, performance, and systems can be found in *Computer Magazine*, published by the Computer Society of the Institute of Electrical and Electronics Engineers (IEEE). More advanced articles on the subject of computer architecture, performance, and design can be found in *IEEE Transactions on Computers*, published by the IEEE. For subscription information on IEEE publications, contact IEEE Service Center, 445 Hoes Lane, P. O. Box 1331, Piscataway, NJ 08855-1331, or phone (800)678-IEEE.

144

Operating Systems

Yingxu Wang
University of Calgary

An operating system is a type of system software that manages and controls the resources and computing capability of a computer or a computer network, and provides users a logical interface for accessing the physical computer to execute applications. This chapter describes typical architectures and generic functions of operating systems. First, the basic concepts and brief history of operating system technologies are introduced. Next is a description of the conceptual and typical commercial architectures of operating systems. Common functions of operating systems, such as process and thread management, memory management, file system management, I/O system management, and network/communication management are then presented. Finally, real-time operating systems are described. with illustrations of the RTOS+ operating system.

144.1 Introduction

Operating system technologies have evolved from rather simple notions of managing the hardware on behalf of a single user or sequentially scheduled users to multiuser time-sharing systems and then to networked and distributed systems. Most modern operating systems are based on multiprogrammed timesharing technologies [Brinch-Hansen, 1971; Peterson and Silberschatz, 1985; Silberschatz et al., 2003; Tanenbaum, 2001].

An *operating system* is a type of system software that manages and controls the resources and computing capability of a computer or a computer network, and provides users a logical interface for accessing the physical computer to execute applications. Almost all general-purpose computers need an operating system before any specific application may be installed and executed by users. The role of an operating system as a conceptual model of a physical computer is shown in Figure 144.1.

Operating systems have evolved from being one-of-a-kind programs useful for only one type of hardware configuration to being portable programs that can be made to operate in a homogeneous family

0-8493-1586-7/05/$0.00+$1.50

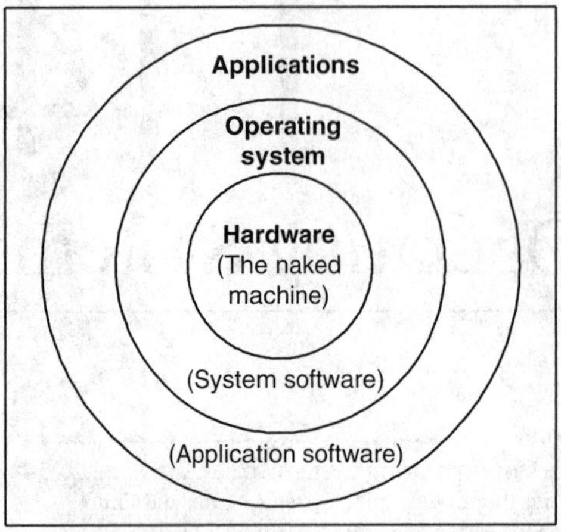

FIGURE 144.1 Role of an operating system in a general-purpose computer.

of hardware configurations and even in heterogeneous hardware platforms. The first operating system that operated on several different kinds of hardware configurations supporting many different kinds of devices was called OS/360. OS/360 was developed by IBM in the early 1960s as a computing platform for commercial applications. OS/360 enabled, for the first time, the support of different generations of hardware devices in a continuous manner, without having to modify the software that ran in the systems and avoiding the creation of completely new operating systems to accommodate new devices. The layered structure of operating systems was first proposed by Dijkstra in his THE multiprogramming system [Dijkstra, 1968].

The first commercial operating system, IBM VM/370, was developed in 1968. The first operating system to operate in a wide variety of hardware platforms was Unix, introduced by AT&T Bell Laboratories in the mid-1970s as a computing platform for research and development of engineering applications [Earhart, 1986]. Unix adopted the multitasking operating system technology, by which users may have several different kinds of concurrent activities to make better use of time and resources while getting work done. The size of Unix was small, and thus it could operate in inexpensive computers used for small groups of people. The first operating system to be used massively was MS-DOS, introduced by Microsoft in the early 1980s as a computing platform for personal or individual use [Microsoft, 1986].

The general-purpose operating systems can be classified into four types: the batch systems, time-sharing systems, real-time systems, and distributed systems. A *batch system* is an early type of operating systems that runs similar jobs sorted by an operator as a batch through an operation console. A *time-sharing system* is a type of multitasking operating systems that executes multiple jobs by automatically switching among them with predefined time slice. A *real-time operating system* is a type of special-purpose operating systems that is designed for dealing with highly time-constraint events of I/Os and processes for control systems. A *distributed operating system* is a type of operating systems that supports networking, communication, and file sharing among multiple computers via a network protocol [Sloane, 1993; Tanenbaum, 1994].

144.2 Architectures of Operating Systems

The architectures of operating systems have evolved over the years from being a monolithic set of system services whose boundaries were difficult to establish to being a structured set of system services. Today, for example, some of these services, like the file system in charge of administering the file data stored by users, have boundaries, or software interfaces, that are standards regulated by organizations that are not related to any computer vendor. This modularization of operating systems into well-understood components provides the possibility to "mix and match" components from different vendors in one system. This trend toward pluggable system parts is accelerating with the formation of various consortia whose charters are to specify common, appropriate, and widely accepted interfaces for various system services.

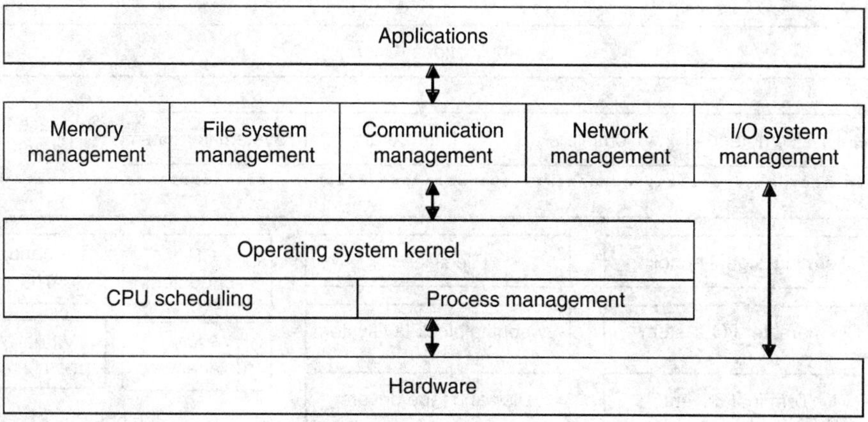

FIGURE 144.2 Generic architecture of operating systems.

Current operating systems are all based on the idea of building higher-level hardware abstraction from lower-level hardware-oriented functions. In other words, current operating systems build higher-level abstractions from lower-level functions that are closer to the raw hardware. Thus, all kinds of hard disks, for example, are made to look and operate in the same manner by their low-level device drivers. Then, in turn, the operating system presents, with all other services in the system (such as the file system), a uniform, common view of a hard disk. This process of successive layers of abstraction is also followed within other services [Dijkstra, 1968]. In file systems, for example, a layer providing the abstraction of continuous storage that does not have device boundaries is built from the abstraction provided by the individual storage in several disks.

Generic Architecture of Operating Systems

An operating system may be perceived as an agent between the hardware and computing resources of a computer or a computer network and the applications and users as shown in Figure 144.2. A generic operating system may be divided into two parts: the kernel and the resource management subsystems [Peterson and Silberschatz, 1985; Silberschatz et al., 2003; Tanenbaum and Tanenbaum, 2001]. The former is a set of central components for computing, including CPU scheduling and process management. The latter is a set of individual supporting software for various system resources and user interfaces.

The kernel is the most basic set of computing functions needed for an operating system. The kernel contains the interrupt handler, the task manager, and the interprocess communication manager. It may also contain the virtual memory manager and the network subsystem manager. With these services, the system can operate all the hardware present in the system and also coordinate the activities between tasks. The services provided by an operating system can be organized in categories. Four possible categories are task control, file manipulation, device control, and information maintenance.

Unix and Linux Operating Systems

The history of Unix goes back to 1969 based on Ken Thompson, Dennis Ritchie, and others' work [Thomas et al., 1986]. The name "Unix" was intended as a pun on Multics (UNiplexed Information and Computing System). The development of Unix was essentially confined to Bell Labs for DEC's PDP-11 (16 bits) and later VAXen (32 bits) [Earhart, 1986]. But much happened to Unix outside AT&T, especially at Berkeley [CSRG, 1986]. Major vendors of workstations, such as SUN's NFS, also contributed to its development.

The architecture of Unix is shown in Figure 144.3, which can be divided into the kernel and the system programs. The Unix kernel consists of system resource management, interfaces, and device drivers, such as the CPU scheduling, file system, memory management, and I/O management.

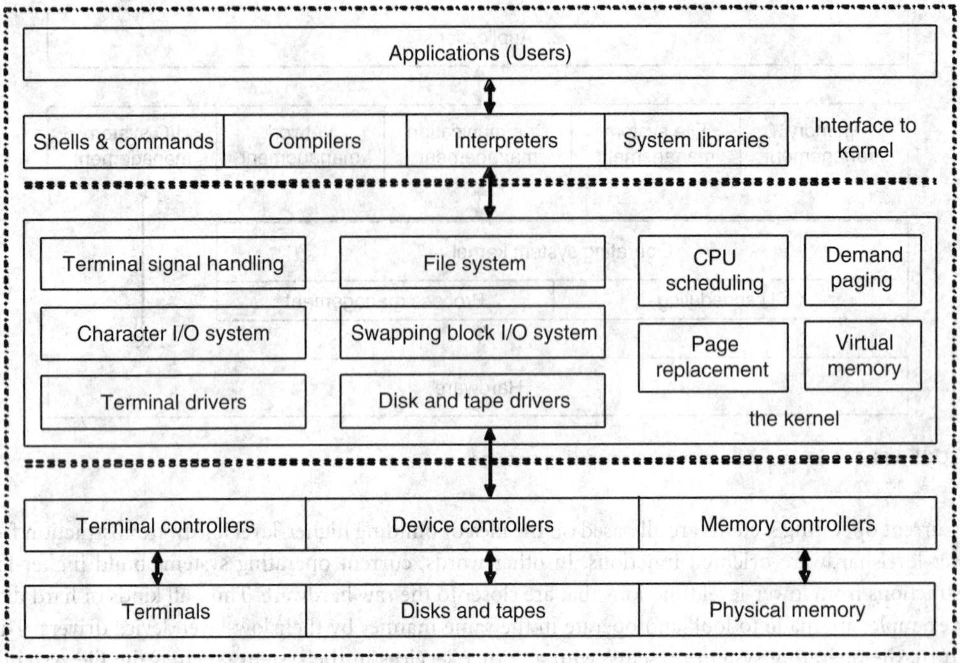

FIGURE 144.3 Architecture of Unix.

Linux is a complete Unix clone for Intel 386/486/Pentium machines [Siever et al., 2003]. Linux is an operating system, which acts as a communication service between the hardware and the software of a computer system. The Linux kernel contains all of the features that one would expect in any operating system. Some of these features are:

- Multitasking (a technique for sharing a single processor between several independent jobs)
- Virtual memory (allows repetitive, extended use of the computer's RAM for performance enhancement)
- Fast transmission control protocol (TCP)/Internet protocol (IP) drivers (for speedy communication)
- Shared libraries (enable applications to share common code)
- Multiuser capability (this means hundreds of people can use the computer at the same time, either over a network, the Internet, or on laptops/computers or terminals connected to the serial ports of those computers)
- Protected mode (allows programs to access physical memory, and protects stability of the system)

Windows XP Operating System

Windows XP is a multitasking operating system built on enhanced technologies that integrate the strengths of Windows 2000, such as standard-based security, manageability, and reliability, with the best features of Windows 98 and Windows Me, such as plug and play, and easy-to-use user interfaces.

The architecture of Windows XP is shown in Figure 144.4. Windows XP adopts a layered structure that consists of the hardware abstraction layer, the kernel layer, the executive layer, the user mode layer, and applications.

Each kernel entity of Windows XP is treated as an object that is managed by an object manager in the executive. The kernel objects can be called by the user-mode applications via an object handle in a process. The use of kernel objects to provide basic services, and the support of client-server computing, enable Windows XP to support a wide variety of applications. Windows XP also provides virtual memory, integrated caching, preemptive scheduling, stronger security mode, and internationalization features.

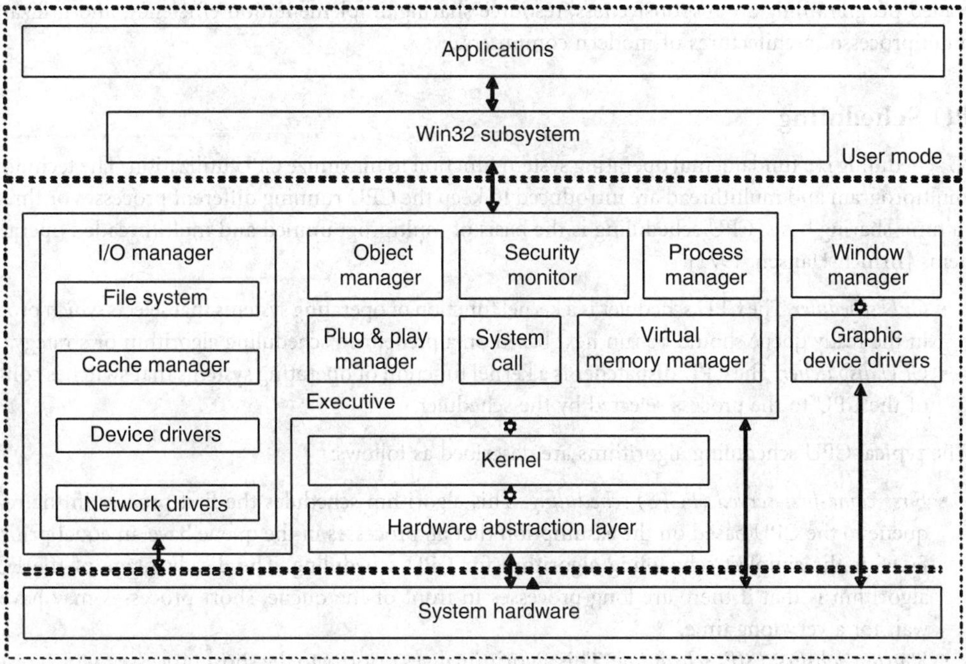

FIGURE 144.4 Architecture of Windows XP.

144.3 Functions of Operating Systems

The basic functions of operating systems can be classified as process and thread management, memory management, file system management, I/O system management, and network/communication management. This section describes the fundamental technologies of main operating system functions in modern operating systems.

Process Management

A *process* is an execution of a program on a computer under the support of an operating system. A process can be a system process or user process. The former executes system code, and the latter runs a user's application. Processes may be executed sequentially or concurrently depending on the type of operating systems.

The operating system carries out process management by the following activities:

- Detection of process requests
- Creation of processes by individual process control blocks (PCBs)
- Allocation of system resources to processes, such as CPU time, memory, files, and I/O devices
- Scheduling of processes based on a predefined process state transition diagram
- Termination of processes

A typical process state transition diagram of a real-time operating system, RTOS+ [Wang and Ngolah, 2003], is shown in Figure 144.7.

Threads are an important concept of process management in operating systems [Anderson et al., 1989; Lewis and Berg, 1998]. A *thread* is a basic unit of CPU utilization, or a flow of control within a process, supported by a thread control block (TCB) with a thread ID, a program counter, a set of registers, and a stack. Conventional operating systems are single-thread systems. Multithreaded systems enable a process to control a number of execution threads. The benefits of multithreaded operating system and multithreaded programming are responsiveness, resource sharing, implementation efficiency, and utilization of multiprocessor architectures of modern computers.

CPU Scheduling

CPU scheduling is a fundamental operating system function to maximize CPU utilization. The techniques of multiprogram and multithread are introduced to keep the CPU running different processes or threads on a time-sharing basis. CPU scheduling is the basis of multiprogrammed and multithreaded operating systems [Brinch-Hansen, 1971].

- *CPU scheduler.* The CPU scheduler is a kernel function of operating systems that selects which process in the ready queue should be run next based on a predefined scheduling algorithm or strategy.
- *CPU dispatcher.* The CPU dispatcher is a kernel function of operating systems that switches control of the CPU to the process selected by the scheduler.

The typical CPU scheduling algorithms are described as follows:

- *First-come-first-served (FCFS) scheduling.* This algorithm schedules the first process in the ready queue to the CPU based on the assumption that all processes in the queue have an equal priority. FCFS is the simplest scheduling algorithm for CPU scheduling. The disadvantage of the FCFS algorithm is that if there are long processes in front of the queue, short processes may have to wait for a very long time.
- *Shortest-job-first (SJF) scheduling.* This algorithm gives priority to the short processes, which results in the optimal average waiting time. But the predication of process length seems to be difficult when using the SJF strategy .
- *Priority (PR) scheduling.* This algorithm assigns different priorities to individual processes. Based on this, CPU scheduling will be carried out by selecting the process with the highest priority. The drawback of the priority algorithm is *starvation*, a term that denotes the indefinite blocking of low-priority processes under high CPU load. To deal with starvation, the *ageing* technique may be adopted that increases the priority levels of low-priority processes periodically, so that the executing priorities of those processes will be increased automatically while waiting in the ready queue.
- *Round-robin (RR) scheduling.* This algorithm allocates the CPU to the first process in the FIFO ready queue for only a predefined time slice, and then it is put back at the tail of the ready queue if it has not yet completed.
- *Multiprocessor scheduling.* This algorithm schedules each processor individually in a multiprocessor operating system on the basis of a common queue of processes. In a multiprocessor operating system, processes that need to use a specific device have to be switched to the right processor that is physically connected to the device.

Memory Management

Memory management is one of the key functions of operating systems because memory is both the working space and storage of data or files. Common memory management technologies of operating systems are contiguous allocation, paging, segmentation, and combinations of these methods [Silberschatz et al., 2003].

- *Contiguous memory allocation.* This method is used primarily in a batch system where memory is divided into a number of fixed-sized partitions. The contiguous allocation of memory may be carried out by searching a set of holes (free partitions) that best fit the memory requirement of a process. A number of algorithms and strategies were developed for contiguous memory allocation such as the first-fit, best-fit, and worst-fit algorithms [Tanenbaum and Tanenbaum, 2001].
- *Paging.* Paging is a dynamic memory allocation method that divides the logical memory into equal blocks known as pages corresponding to physical memory frames. In a paging system, each logical

address contains a page number and a page offset. The physical address is generated via a page table where the base address of an available page is provided. Paging technology is used widely in modern operating systems to avoid the fragmentation problem as found in the early contiguous memory allocation techniques.

- *Segmentation*. This is a memory management technique that uses a set of segments (logical address spaces) to represent a user's logical view of memory independent of the physical allocation of system memory. Segments can be accessed by providing their names (numbers) and offsets.
- *Virtual memory*. When the memory requirement of a process is larger than physical memory, an advanced technique needs to be adopted known as the virtual memory, which enables the execution of processes that may not be completely in memory. The main approach to implement virtual memory is to separate the logical view of system memory from its physical allocation and limitations. Various technologies have been developed to support virtual memory, such as the demand paging and demand segmentation algorithms [Silberschatz et al., 2003].

In memory-sharing systems, the sender and receiver use a common area of memory to place the data that is to be exchanged. To guarantee appropriate concurrent manipulation of these shared areas, the operating system has to provide synchronization services for mutual exclusion. A common synchronization primitive is the *semaphore*, which provides mutual exclusion for two tasks using a common area of memory. In a shared memory system the virtual memory subsystem must also collaborate to provide the shared areas of work.

File System Management

The file system is the most used function of operating systems for nonprogramming users. A file is a logical storage unit of data or code separated from its physical implementation and location. Types of files can be text, source code, executable code, object code, word processor formatted, or system library code. The attributes of files can be identified by name, type, location (path of directory), size, date/time, user ID, and access control information. Logical file structures can be classified as sequential and random files. The former are files that organize information as a list of ordered records; while the latter are files with fixed-length logical records accessible by block number.

Typical file operations are reading, writing, and appending. Common file management operations are creating, deleting, opening, closing, copying, and renaming.

The file system of an operating system consists of a set of files and a directory structure that organizes all files and provides detailed information about them. The major function of a file management system is to map logical files onto physical storage devices such as disks or tapes. Most file systems organize files by a tree-structured directory. A file in the file system can be identified by its name and detailed attributes provided by the file directory. The most frequently used method for directory management is the *hash table*. Although it is fast and efficient, backup is always required to recover a hash table from unpredicted damage.

A physical file system can be implemented by contiguous, linked, and indexed allocation. Contiguous allocation can suffer from external fragmentation. Direct access is inefficient with linked allocation. Indexed allocation may require substantial overheads for its index block.

I/O System Management

I/O devices of a computer system encompass a variety of generic and special hardware and interfaces. Typical I/O devices that an operating system deals with are shown in Table 144.1.

I/O devices are connected to the computer through buses with specific ports or I/O addresses. Usually, between an I/O device and the bus, there is a device controller and an associated device driver program. The I/O management system of an operating system is designed to enable users to use and access system I/O devices seamlessly, harmoniously, and efficiently.

I/O management techniques of operating systems can be described as follows:

TABLE 144.1 Types of I/O Devices

Types of I/O devices	Examples
System devices	System clock, timer, interrupt controller
Storage devices	Disks, CD drivers, tapes
Human interface devices	Keyboard, monitor, mouse
Communication devices	Serial/parallel buses, network cards, DMA controllers, modems
Special devices	Application-specific control devices

- *Polling*. Polling is a simple I/O control technique by which the operating system periodically checks the status of the device until it is ready before any I/O operation is carried out.
- *Interrupt*. Interrupt is an advanced I/O control technique that lets the I/O device or control equipment notifies the CPU or system interrupt controller whenever an I/O request occurs or a waiting event is ready. When an interrupt is detected, the operating system saves the current execution environment, dispatches a corresponding interrupt handler to process the required interrupt, and then returns to the interrupted program. Interrupts can be divided into different priorities on the basis of processor structures in order to handle complicated and concurrent interrupt requests.
- *DMA*. Direct memory access (DMA) is used to transfer a batch of large amount of data between the CPU and I/O devices, such as disks or communication ports. A DMA controller is handled by the operating system to carry out a DMA data transfer between an I/O device and the CPU.
- *Network sockets*. Most operating systems use a socket interface to control network communications. When requested in networking, the operating system creates a local socket and asks the target machine to be connected to establish a remote socket. Then, the pair of computers may communicate by a given communication protocol.

Communication Management

A fundamental characteristic that may vary from system to system is the manner of communication between tasks. The two manners in which this is done are via messages sent between tasks or via the sharing of memory where the communicating tasks can both access the data. Operating systems can support either. In fact, both manners can coexist in a system. In message-passing systems, the sender task builds a message in an area that it owns and then contacts the operating system to send the message to the recipient. There must be a location mechanism in the system so that the sender can identify the receiver. The operating system is then put in charge of delivering the message to the recipient. To minimize the overhead of the message delivery process, some systems try to avoid copying the message from the sender to the kernel and then to the receiver, but to provide means by which the receiver can read the message directly from where the sender wrote it. This mechanism requires the operating system intervene if the sender wants to modify the contents of a message before the recipient has gone through its content.

In memory-sharing systems, the sender and receiver use a common area of memory to place the data that is to be exchanged. To guarantee appropriate concurrent manipulation of these shared areas, the operating system has to provide synchronization services for mutual exclusion. A common synchronization primitive is the *semaphore*, which provides mutual exclusion for two tasks using a common area of memory. In a shared memory system, the virtual memory subsystem must also collaborate to provide the shared areas of work.

There are systems in which the amount of data that can be shared or sent between tasks is minimal. In the original Unix, for example, different tasks would not have shared memory, and only a very limited form of message passing was provided. The messages were constrained to be event notification. In this environment, data shared by processes have to be written into a common file. To support file sharing, some systems provide a locking service that is used to synchronize accesses to files by different tasks.

The ISO open systems interconnection (OSI) reference model was developed in 1983 [Day and Zimmermann, 1983] for standardizing data communication protocols between different computer sys-

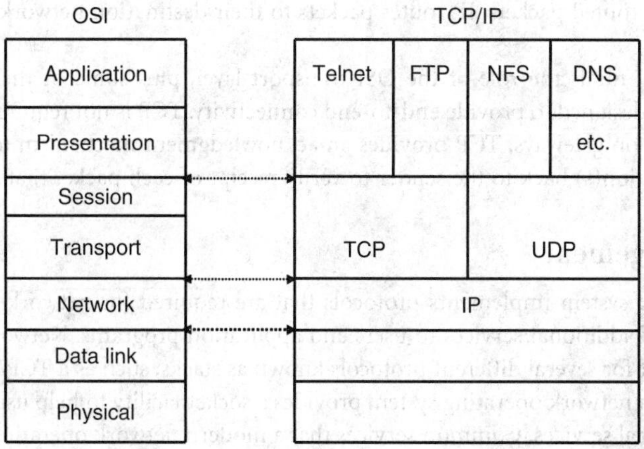

FIGURE 144.5 OSI reference model and TCP/IP.

tems and networks. The OSI reference model is an important protocol framework for regulating multi-vendor multi-OS computers interconnection in local area network (LAN) and wide area network (WAN) environments [Stallings, 2000]. From the bottom up, the seven layers are physical, data link, network, transport, session, presentation, and application, as shown in Figure 144.5.

- *Physical layer.* This layer provides a real communication channel for transmitting bit stream data. Mechanical, electrical, and procedural interfaces for different physical transmission media are specified on this layer.
- *Data link layer.* This layer implements data frames in the sender end and acknowledges frames in the receiver end, and provides retransmission mechanisms when a frame is not successfully transmitted. The data link layer converts a raw transmission line on the physical layer into a line that is free of transmission errors to the network layer.
- *Network layer.* This layer controls the operation of a subnet by determining how message packets are routed from source to destination. It also provides congestion control, information flow counting, and packet size conversion between different networks.
- *Transport layer.* This layer accepts data from the session layer, splits it up into smaller units for adapting to the network layer, and provides an interface between the session layer and the different implementations of the lower-layer protocol and hardware.
- *Session layer.* This layer provides transport session establishment, synchronization, and data flow control between different machines modeled by the transport layer.
- *Presentation layer.* This layer converts the original data represented in vendor-dependent format into an abstract data structure at the sender end and vice versa at the receiver end. The presentation layer enables data to be compressed or encoded for transmitting on the lower layers.
- *Application layer.* This layer adapts a variety of terminals into a unified network virtual terminal interface for transferring different files between various file systems.

Figure 144.5 also contrasts the functional equivalency between the OSI model and TCP/IP [Day and Zimmermann, 1983]. The TCP/IP design philosophy [Comer and Stevens, 1996; Comer, 2000] was intended to provide universal connectivity with connection-independent protocols at the network layer. Thus, TCP/IP does not address the data link and physical layers that determine the communication channels. There are no separate application, presentation, and session layers in TCP/IP; instead, a combined application layer is provided in TCP/IP, which has the functions of those layers.

The IP protocol is approximately equivalent to the OSI network layer. In a WAN, IP is presented on every node in the network. The role of IP is to segment messages into packets (up to 64 Kbytes) and then route and pass the packets from one node to another until they reach their destinations. IP uses packet switching as its fundamental transmission algorithm. A message is transmitted from gateway to gateway by dynamic routed packets. IP routes packets to their destination network, but final delivery is left to TCP.

The TCP protocol fulfils the role of the OSI transport layer, plus some of the functionality of the session layer. TCP is designed to provide end-to-end connectivity. TCP is not required for packet routing, so it is not included on gateways. TCP provides an acknowledgment mechanism to enable messages to be sent from destination(s) back to the sender to verify receipt of each packet that makes up a message.

Network Management

A network operating system implements protocols that are required for network communication and provides a variety of additional services to users and application programs. Network operating systems may provide support for several different protocols known as stacks, such as a TCP/IP stack and an IPX/SPX stack. A modern network operating system provides a socket facility to help users to plug-in utilities that provide additional services. Common services that a modern network operating system can provide include the following:

- *File services.* File services transfer programs and data files from one computer on the network to another.
- *Message services.* Message services allow users and applications to pass messages from one computer to another on the network. The most familiar application of message services consists of email and intercomputer talk facilities.
- *Security and network management services.* These services provide security across the network and allow users to manage and control the network.
- *Printer services.* Printer services enable sharing of expensive printer resources in a network. Print requests from applications are redirected by the operating system to a network workstation that manages the requested printer.
- *Remote procedure calls (RPCs).* RPCs provide application program interface services to allow a program to access local or remote network operating system functions.
- *Remote processing services.* These services allow users or applications to remotely log in to another system on the network and use its facilities for program execution. The most familiar service of this type is Telnet, which is included in the TCP/IP protocol suite [Comer and Stevens, 1996; Comer, 2000] and many other modern network operating systems.

144.4 Real-Time Operating Systems

A real-time operating system (RTOS) is essential to implement embedded and/or real-time control systems. An RTOS is an operating system that guarantees timely processing of external and internal events of real-time systems. Problems often faced by RTOSs are CPU and tasks scheduling, timing/event management, and resource management. An RTOS requires multitasking, process threads, and explicit interrupt levels to deal with real-time events and interrupts.

Various RTOS models were developed [Labrosse, 1999; Laplante, 1977; Liu, 2000]. Existing RTOSs are target machine specific, implementation dependent, and not formally described. Therefore, they are usually not portable as a generic RTOS to be seamlessly incorporated into real-time or embedded system solutions.

An extended RTOS, RTOS+, is described in this section to address the above issues [Wang and Ngolah, 2003]. RTOS+ is a portable and multitask/multithread operating system capable of handling event-, time-, and interrupt-driven processes in a real-time environment. RTOS+ is also specified by real-time process algebra (RTPA), a formal specification language for real-time architecture and behavior description, particularly for the explicit description of the dynamic behaviors of real-time software systems [Wang, 2002/2003].

RTOS+ Architecture

The architecture of RTOS+ is described as shown in Figure 144.6, where interactions between system resources, components, and internal control models are illustrated. There are four subsystems in RTOS+:

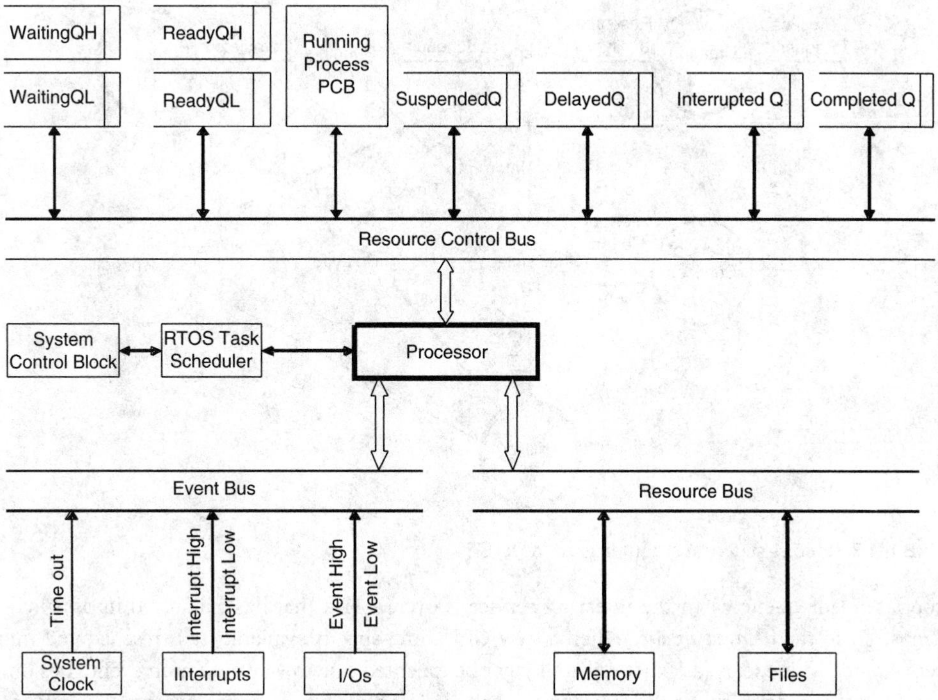

FIGURE 144.6 Architecture of RTOS+.

processor and task scheduler, resource controller, event handler, and system resources. The task scheduler is the innermost operating system kernel that directly controls the CPU and all other system resources by system control blocks. The resources of RTOS+, supplement to the CPU, are mainly the memory, system clock, I/O ports, interrupt facilities, files, and internal control models such as queues and task/resource control blocks as shown in Figure 144.6.

The task scheduling of RTOS+ is priority based. We use a fixed-priority scheduling algorithm where the priority of a task is assigned based on its importance when it is created. Tasks are categorized into four priority types: high- and low-priority interrupts, as well as base and high-priority processes. A process, when created, is placed in a proper queue corresponding to its predefined priority levels.

RTOS+ Task Scheduler

The *Task Scheduler* is the kernel of RTOS+. Its behaviors can be modeled by a state transition diagram as shown in Figure 144.7. The task scheduler of RTOS+ is designed for handling event-, time-, and interrupt-driven processes, in which the CPU is allocated by a fixed time slice for executing a given process.

Process requests are handled by the task scheduler for creating a process. When a new process is generated, it is first put into the *waiting state* with a process control block (PCB) and a unique task ID (Figure 144.7). The system uses a resource control block (RCB) (Figure 144.6) to manage system resources. Each task in the waiting state must be checked to see if there are enough resources for its execution. If resources are available, it is transferred into the *ready state*; otherwise, it has to be re-queued at the tail of the waiting queue until resources are available.

The task scheduler continuously checks the ready queue for any ready tasks. If there are ready tasks, it executes the first task in the queue until it is completed (State 4) or is suspended. There are three conditions that may cause a running task to be suspended during execution: interrupted by a task or event with higher priority, time-out for a scheduled time-slice of CPU, and waiting for a specific event. The scheduler may remove the CPU from a running task if a higher priority interrupt request occurs. Such interruption will cause the running task to go into the *interrupted queue* and later return to the

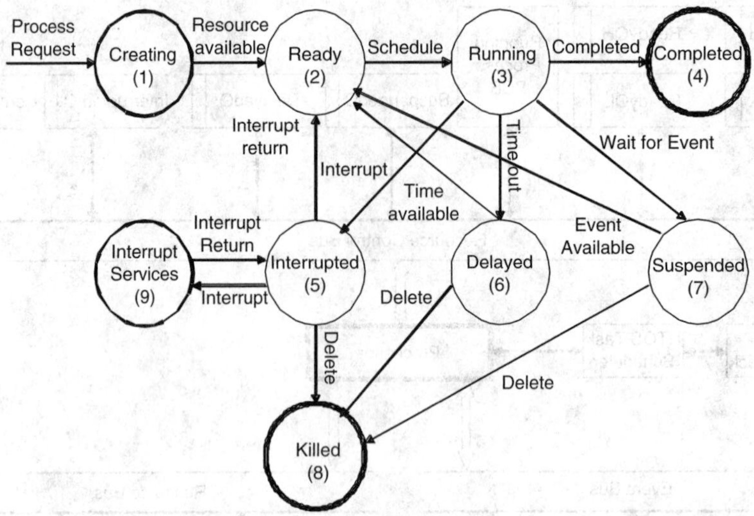

FIGURE 144.7 Process state-transition diagram of RTOS+.

appropriate ready queue when the interrupt service is over. A task that has exhausted its assigned time slice must go to the *delayed queue*. When a new CPU time-slice is available, it is rescheduled into the appropriate ready queue. A task that can no longer be executed due to waiting for an event goes into the *suspended queue*, and returns to the appropriate ready queue once the event has occurred. In any of these suspended cases, the task is put into a corresponding queue in States 5, 6, or 7, respectively. The task will be rescheduled into ready state when the cause of the suspension is no longer true.

A task suspended may be cancelled (State 8) by the scheduler from the queues of States 5, 6, or 7 in case there is a lack of resources or under the request of users.

Formal Description of RTOS+

In the previous subsection, the conceptual model of RTOS+ was established. To further refine the design of RTOS+, real-time process algebra (RTPA) is adopted as a formal specification means [Wang, 2002/2003]. RTPA is a mathematical notation capable of specifying and describing both architectures and behaviors of real-time systems.

The dynamic behaviors of RTOS+ can be described by the interactions of parallel processes between *TaskScheduling*, *TimeManagement*, *ResourceManagement*, and *SystemControlUpdate* as shown in Figure 144.8. RTOS+ runs the *TaskManagment* routine continuously by updating and dispatching various processes in different queues. If an interrupt occurs during run-time, the interrupt handling process (*SysClock100msInst*) saves the current executing environment, switches control to the interrupting process, and then automatically returns to the interrupted process following completion of a higher-priority operation. *SysClock100msInt* handles low-level interrupt events, such as system ResourceManagement at 100 ms intervals.

Corresponding to the state transition diagram as shown in Figure 144.7, the task scheduler of RTOS+ is specified by RTPA in Figure 144.9. RTPA describes a finite state machine as an event-driven process dispatching system, where a specific event is used to trigger a predefined process. *Process dispatch* in RTPA models dynamic process relations behaving at run-time. The process dispatching mechanism of RTPA is used to formally describe the RTOS+ dynamic behaviors by a set of event-driven relationships between system events and functional processes of the operating system kernel.

The core operations of RTOS+, such as task scheduling, and time, event, and resource management are rigorously described for better real-time performance and improved resource utilization. On the basis of the formal specification of RTOS+ by RTPA, architectural and behavioral consistency and correctness

```
RTOS. Process Deployment ≙
{ // Basic level processes
    @System Initial
      ↳ ( SysInitial
            @SysShut Down = T
          ↳    R     Task Management
              ≥1
            → ⊠
        )
|| // High-interrupt level processes
   ⊙ @SysClock1msInt
      ↗ ( SysClock
            ↳ Task Scheduling
            ↳ System Control Block Update
            ↳ Interrupt Management
            ↳ Time Management
            ↳ High Priority Event Scan
          )
      ↘⊙
|| // Low-interrupt level processes
   ⊙ @SysClock 100ms Int
      ↗ (
          Base Event Scane
          ↳ Resource Management)
      ↘⊙
}
```

FIGURE 144.8 Real-time process deployment in RTOS+.

```
TaskScheduler ≙
{   ? ⓢNewProcRequestBL = T
        → CreatePCB (<I::ProcIDN>; <O:: ()>
        → ProcStateN = 1
    | ? ⓢ PCB.ProcStateN ≥ 1
        ? ProcStateN
        | 1 → Waiting (<I::(ProcIDN); O::()>)
        | 2 → Ready (<I::(ProcIDN); O::()>)
        | 3 → Running (<I::(ProcIDN); O::()>)
        | 4 → Completed (<I::(ProcIDN); O::()>)
        | 5 → Interrupted (<I::(ProcIDN); O::()>)
        | 6 → Delayed (<I::(ProcIDN); O::()>)
        | 7 → Suspended (<I::(ProcIDN); O::()>)
        | 8 → Killed (<I::(ProcIDN); O::()>)
        | 9 → InterruptServices (<I::(ProcIDN); O::(IntReturnBL)>)
        | ? ~ → ∅
}
```

FIGURE 144.9 Dynamic behaviors of the RTOS+ task scheduler.

of RTOS+ can be improved. RTOS+ has been found useful in a wide range of applications in real-time and embedded system design.

144.5 Conclusions

An operating system has been perceived as the agent between the hardware and computing resources of a computer or a computer network and the applications and users that manages and controls the resources and computing capability, and provides users a logical interface for accessing the physical computer in executing computing applications. This chapter has described typical architectures and generic functions of operating systems. The basic concepts and brief history of operating system technologies have been reviewed. The conceptual and typical commercial architectures of operating systems have been illustrated. The common functions of operating systems, such as process and thread management, memory management, file system management, I/O system management, and network/communication management have been described. A real-time operating system, RTOS+, has been presented with a formal description by using real-time process algebra (RTPA).

Acknowledgment

The author would like to thank L.F. Cabrera and C.F. Ngolah for their work and comments on this chapter.

References

Anderson, T.E., E.D. Lazowska, and H.M. Levy. 1989. The performance implications of thread management alternatives for shared-memory multiprocessors, *IEEE Trans. Comput.*, 38:1631–1644.

Brinch-Hansen, P. 1971. Short-term scheduling in multiprogramming systems, *Proceedings of the Third ACM Symposium on Operating Systems Principles*, October, pp. 103–105.

Comer, D.E. 2000. *Internetworking with TCP/IP, Vol. I: Principles, Protocols, and Architecture*, 4th ed., Prentice Hall, Englewood Cliffs, NJ.

Comer, D.E. and Stevens, D. L. 1996. *Internetworking with TCP/IP, Vol. III: Client-Server Programming and Application*, Prentice Hall, Englewood Cliffs, NJ.

Computer Systems Research Group, University of California at Berkeley. 1986. *BSD UNIX Reference Manuals*, USENIX Association, California.

Day, J.D. and H. Zimmermann 1983. The OSI reference model, *Proc. IEEE*, 71:1334–1340.

Deitel, H.M. and Kogan, M.S. 1992. *The Design of OS/2*, Addison-Wesley, Reading, MA.

Dijkstra, E.W. 1968. The structure of the THE multiprogramming system, *Commun. ACM*, 11:341–346.

Earhart, S.V., Ed. 1986. *AT&T UNIX Programmer's Manual*, Holt, Rinehart, and Winston, New York.

Labrosse, J.J. 1999. *MicroC/OS-II, The Real-Time Kernel*, 2nd ed., R&D Books, Gilroy, CA.

Laplante, P.A. 1977. *Real-Time Systems Design and Analysis*, 2nd ed., IEEE Press, New York.

Lewis, B. and D. Berg 1998. *Multithreaded Programming with Pthreads*, Sun Microsystems Press, Upper Saddle River, NJ.

Liu, J. 2000. *Real-Time Systems*, Prentice Hall, Upper Saddle River, NJ.

Microsoft Corp. 1986. *Microsoft MS-DOS User's Reference and Microsoft MS-DOS Programmer's Reference*, Microsoft Press, Redmond, WA.

Peterson, J. L. and Silberschatz, A. 1985. *Operating System Concepts*. Addison-Wesley, Reading, MA.

Silberschatz, A., P. Galvin, and G. Gagne 2003. *Applied Operating System Concepts*, John Wiley & Sons, New York.

Siever, E., A. Weber, S. Figgins, and A. Oram 2003. *Linux in a Nutshell*, 4th ed., O'Reilly & Associates, Sebastapol, CA.

Sloane, A. 1993. *Computer Communications: Principles and Business Applications*, McGraw-Hill, New York.

Stallings, W. 2000. *Local and Metropolitan Area Networks*, 6th ed., Prentice Hall, Englewood Cliffs, NJ.

Tanenbaum, A.S. 1994. *Distributed Operating Systems*, Prentice Hall, Englewood Cliffs, NJ .

Tanenbaum, A.S. and A. Tanenbaum. 2001. *Modern Operating Systems*, Prentice Hall, Englewood Cliffs, NJ.

Thomas, R., L.R. Rogers, and J.L. Yates. 1986. *Advanced Programmer's Guide to Unix System V*, Osborne McGraw-Hill, Berkeley, CA.

Wang, Y. 2002. The real-time process algebra (RTPA), *Ann. Software Eng. Int. J.*, 14:235–274.

Wang, Y. 2002. A new approach to real-time systems specification, *Proceedings of the 2002 IEEE Canadian Conference on Electrical and Computer Engineering (CCECE'02)*, Winnipeg, Manitoba, Canada, May, vol. 2, pp. 663–668.

Wang, Y. 2003. Using process algebra to describe human and software behaviors, *Transdisciplinary J. Neurosci. Neurophilosophy*, 4(2):199–213.

Wang, Y. and C.F. Ngolah. 2003. Formal description of real-time operating systems using RTPA, *Proceedings of the 2003 Canadian Conference on Electrical and Computer Engineering (CCECE'03)*, Montreal, Canada, May, pp. 35.3.1–4.

145
Programming Languages

Jens Palsberg
University of California, Los Angeles

145.1 Introduction

The goal of a programming language is to make it easier to build software. A programming language can help make software flexible, correct, and efficient. How good are today's languages and what hope is there for better languages in the future?

Many programming languages are in use today, and new languages are designed every year; however, the quest for a language that fits all software projects has so far come up short. Each language has its strengths and weaknesses, and offers ways of dealing with some of the issues that confront software development today. The purpose of this chapter is to highlight and discuss some of those issues, to give examples from languages in which the issues are addressed well, and to point to ongoing research that may improve the state of the art. It is *not* our intention to give a history of programming languages [42, 10].

We will focus on flexibility, correctness, and efficiency. While these issues are not completely orthogonal or comprehensive, they will serve as a way of structuring the discussion. We will view a programming language as more than just the programs one can write; it will also include the implementation, the programming environment, various programming tools, and any software libraries. It is the combination of all these things that makes a programming language a powerful means for building software.

Two particular ideas are used for many purposes in programming languages: compilers and type systems. Traditionally, the purpose of a compiler is to translate a program to executable machine code, and the purpose of a type system is to check that all values in a program are used correctly. Today, compilers and type systems serve more needs than ever, as we will discuss along the way.

145.2 Flexibility

We say that software is flexible when it can run on a variety of platforms and is easy to understand, modify, and extend. Software flexibility is to a large extent achieved by good program structure. A

0-8493-1586-7/05/$0.00+$1.50

programming language can help with the structuring of programs by offering language constructs for making the program structure explicit. Many such language constructs have been designed, and some of them will be highlighted in the following discussion of several aspects of software flexibility.

Platform Independence

Goal: Write once, run everywhere.

The C programming language [27] was invented in the 1970s. Over the years, it became the "gold standard" for platform independence: most general-purpose processors, supercomputers, etc. have their own C compiler, which, in principle, enables any C program to run on any such computer. This is much more flexible than rewriting every program from scratch, perhaps in machine code, for each new computer architecture.

In practice, most C programs will have some aspects that tie them closely to a particular architecture. Programmers are well aware of this and tend to design their programs such that porting software from one computer to the next entails minimal changes to the C source code. The main challenges to porting software lie partly in finding out the places in the C program where changes have to be made, and partly in understanding the differences between the two architectures in question.

In the 1990s, the Java programming language [23] popularized the use of virtual machines. Virtual machines increase platform independence by being an intermediate layer between high-level languages such as Java, and low-level languages such as machine code. A Java compiler will translate a Java program to virtual machine code, which the virtual machine executes. Each kind of computer will have its own implementation of the virtual machine, much like it will have its own C compiler. The virtual machine offers capabilities such as event handling in a platform-independent way, which in C would be platform dependent. This is a step towards the ultimate goal of "write once, run everywhere." Intuitively, we have a spectrum

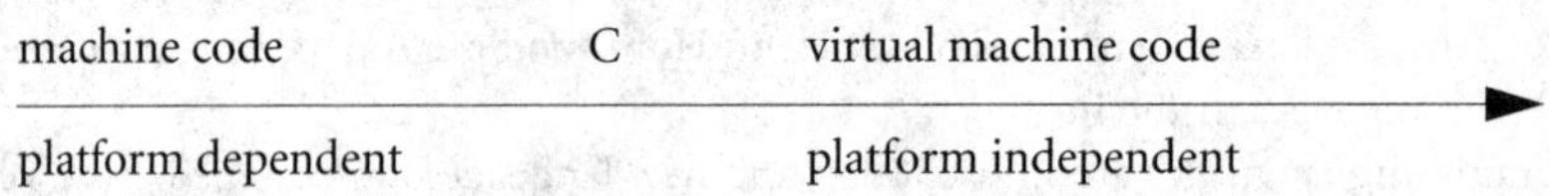

Virtual machines such as those for Java and the .NET Common Language Runtime [30] are slowly replacing C as "portable assembly code," that is, the preferred target for compiling languages such as Java and C# [29]. Some virtual machines work in part by translating virtual machine code to C.

Platform independence is particularly important for mobile code that can move from one computer to another while it is executing. If the computers all run the same kind of virtual machine, then such movement is greatly simplified.

Platform independence via virtual machines comes with a price: it is more difficult to execute virtual machine code efficiently than it is to execute C code efficiently. Efficient implementation of virtual machines remains an active research area [2].

Abstraction

Goal: Never duplicate code.

When a programmer faces a programming task, it is tempting to program the needed functionality from scratch. It gives complete control over the code and minimizes the amount of attention that needs to be paid to other code. But it also runs the risk of duplicating work already done by another programmer.

A better scenario for the programmer would be to come across some portion of an already written program that does almost exactly the same as what needs to be programmed. The programmer can now copy and paste the code and perhaps customize it to fit the particular situation. This increases the amount of reading of code by the programmer, but greatly decreases the amount of writing of code. Thus, the programmer is likely to finish the job in less time. However, the copied code may have a problem or defect, and if the original code gets fixed some day, the copy will stay the same. In other words, improvements cannot be done once and for all. The problem is that the code "lives" in several duplicates.

A vastly superior technique is to use abstraction to isolate and harness a piece of code that is useful in many scenarios. A popular abstraction mechanism is that of procedures which allow a code sequence to be called from many places. For example, suppose we have written code for sorting a list of data. In a language like C, we can abstract it into a procedure

```
void sort(int a[ ], int n) { ... }
```

which sorts the argument array a from positions 1 to n, whenever it is called. The devices of passing arguments and possibly returning results increase the usefulness of procedures as an abstraction mechanism.

An additional advantage of a procedure is that it creates a natural interface between the user and the implementer of the procedure. The implementer is free to change the implementation of the procedure as long as the new implementation has the same behavior. Thus, if the current implementation has a problem, then it can be fixed once and for all, to the benefit of all call sites.

Another abstraction mechanism is that of data abstraction, for example in the form of classes and objects. The idea is to let the programmer define a new data type with a well-defined set of operations and a hidden representation. Such a data type can be used in many places without duplication of code. Similar to the case of procedures, one can change the representation of a data type without affecting the clients.

Constructs for procedural and data abstraction are an integral part of object-oriented languages such as Java and C++. The highly successful and widely-used libraries that come with these languages were to a large extent enabled by these abstraction mechanisms.

The use of abstractions, particularly in libraries, switches some of the work of a programmer from writing to reading code. If a good abstraction can be located, perhaps the programmer needs to do little more than use it. The better the documentation, the more likely the programmer is to find something useful in a library.

Programmers continue to develop useful abstractions that cannot be easily expressed with the abstraction mechanisms of today's programming languages. Instead they are expressed and documented more informally, for example, in the form of design patterns [21]. As a result, the code for a design pattern is often duplicated to each place it is used. Developing new abstraction mechanisms that can capture some of the growing list of design patterns remains an active research area [8, 13].

Code Management

Goal: Separation of concerns.

A large program tends to have other big features than many lines of code. It may have a large state space, represented by many variables, and it may have several rather independent subtasks that are handled by quite separate parts of the code. A programmer will manage the code complexity by trying to keep separate things separate. For example, a classical system design is to have three layers: a user interface, the business logic, and a database. If the code for each of the layers is kept separate and has minimal and well-defined interaction with the other layers, then the overall system will be easier to build, understand, and maintain.

A programming language can help enforce separation of concerns by offering ways of creating separate name spaces. For that purpose, Java has packages, ML [32] has modules, and C++ has so-called namespaces. In each case, the idea is, for example, to enable the name space of the user interface to be separate from that of the database. The advantage is that the programmer of the user interface will not accidentally or purposefully manipulate the internals of the database. Furthermore, the programmer of the database can choose names of variables, procedures, etc. without worrying about clashes with names used in the user interface code.

Good constructs for code management greatly enhance software engineering by teams of programmers where each programmer is responsible for a module. As long as the interfaces between the modules are well defined, each programmer can work independently.

Java packages are a simple mechanism for separation of concerns. Each package has a name and a separate name space, and it can access names from other packages only by explicitly importing them. For example, in Java, if we write

```
import java.util.*;
```

then we get access to all classes and interfaces defined in the package java.util.

The import statements make explicit the relationships with other modules and therefore avoid name clashes. The collection of Java classes is flat and unstructured; a Java package cannot be nested in another package. Moreover, a Java package is not a value; it cannot be stored in a variable, passed as an argument, or returned as a result.

In ML, a module is a first-class entity which can be stored and passed around in much the same way as a basic value such as an integer. There is a layering, though: functions that take modules as arguments are functors, not functions. So, functors are functions from modules to modules, while normal functions are functions from normal values to values.

ML modules and functors blend mechanisms for abstraction and for separation of concerns. Further blending of modules with mechanisms of object-oriented programming remains an active research area [20].

Concurrency Control

Goal: Control access to shared resources.

At the hardware level, synchronization of concurrent processes can be supported in a variety of ways. For example, there can be an atomic test-and-set operation that allows a shared register to be simultaneously read and written. Another example is an atomic operation for swapping the contents of two registers. Such operations enable a register to be a *lock* for a shared resource. With either of the two mentioned operations, one can ask for the lock, and possibly get it, in one computation step. If it had happened in two computation steps, then it is possible that two different concurrent processes simultaneously would ask for the lock, both find that the lock is not currently held, and then both processes take the lock. This would be a programming error.

Representing and operating on locks with atomic operations is fairly cumbersome and highly machine dependent. Programming languages tend to provide more convenient constructs for concurrency control. Those constructs can often be easily implemented using test-and-set, etc. A classical example is that of a semaphore [19], which is a data structure with just two operations: wait and signal. We can think of a semaphore as an abstraction of a lock. When a process wants access to a shared resource, it issues a wait to the semaphore that guards the resource. If no other process is currently accessing the resource, then the process is granted access right away and is now holding the lock. When the process is done with the access, it issues a signal to the semaphore, thereby releasing the lock. If the wait is issued while another process holds the lock, then the process will wait until the lock is released. In general, the semaphore may have a queue of processes waiting to access the shared resource.

A different construct for concurrency control is that of monitors [24]. A monitor is a data structure where we can think of all the operators as being guarded by a single semaphore. Whenever an operator is called, a wait to the semaphore is automatically issued, and when an operation is about to return, a signal is issued. The effect is that the data structure is accessed by at most one concurrent process at a time.

In Java, any variable can be used as a lock, and any block of code can be guarded by a lock. For example, we can write

```
synchronized(lock){ balance = balance + x; }.
```

Here we can think of lock as a semaphore on which a wait is automatically issued on entry to the guarded block of code, and on which a signal is automatically issued on exit. If every method of a class is guarded by *synchronized* on the same lock, then each object of that class is effectively a monitor.

Much research has gone into finding efficient implementations of the various constructs for concurrency control [5]. A radical idea is to let a compiler prove that a certain lock can be held by at most one process at a time. In such a case, the lock is unnecessary and the calls to be operations on it can be eliminated [1].

145.3 Correctness

We say that software is correct when it does what it is supposed to do. Most software has complex behavior and one can rarely get a waterproof guarantee that software is fully correct. However, a programming language can help establish partial correctness by offering ways of specifying and checking key properties. A variety of approaches to software verification and validation have been devised, and some of them will be highlighted in the following discussion of several aspects of software correctness.

Memory Safety

Goal: Avoid jumps to data addresses and data manipulation of code addresses.

One of the most dreaded software errors is reported at run time as something like "illegal address—core dump." Such errors preclude graceful recovery and continuation of the computation, and they can be difficult to troubleshoot. They typically arise from careless manipulation of memory addresses which can look innocent in a machine code program. For example, a program may by mistake jump to a data address, or it may add a number to a code address and thereby get an illegal jump target. The idea of memory safety is to avoid all such errors.

For a given machine code program, it can be difficult to determine whether it is memory safe. Extensive software testing can increase confidence in memory safety, but, as always, testing can only show the presence of bugs, not prove the absence of bugs.

In high-level languages, such as Java, ML, and Scheme [15], memory safety is guaranteed via a type system. In these languages, data and code addresses are not manipulated directly; rather, the programmer works with convenient abstractions such as objects and functions. The type system enforces that those abstractions are used correctly. From the programmer's point of view, data and code manipulation errors no longer manifest themselves as core dumps at the hardware level; instead they show up as type errors that are reported in a more intelligible way by the run-time system of the language. For example, in Scheme, each data value is tagged with information about its type. The tag can be "integer," "procedure," "heap address," etc. Whenever a value is used, the run-time system checks whether the value is of the appropriate type. The run-time system will allow a number to be added to another number, but will disallow a number to be added to a procedure. A type error in Scheme will terminate the execution, but in a graceful way. An alternative is that the run-time system throws a type exception which can then, perhaps, be caught by an exception handler and thereby allow the computation to continue.

The Scheme style of type checking is known as dynamic type checking because it takes place at run time. It comes with an overhead in time and space because each value must be tagged and because the tags must be checked each time values are used. In contrast, ML has entirely static type checking. The idea is that before the program is run, a type checker will determine whether all data manipulation and procedure calls will operate on values of the correct types. If not, then the program is not executed at all. While this may seem restrictive, it has the benefit of saving time and space at run time, and of enforcing the program abstractions in a static way, that is, reporting problems early — even before running the program. Static type checking takes time itself, but is typically done along with compilation of the program and tends to take less time that the rest of the compilation process. Static type checking is a win-win technology: (1) problems are caught early, making it possible to complete software testing faster, and (2) it comes with no overhead for the running program.

In an object-oriented language such as Java, a data type can be a subtype of another type. For example, Plane and Helicopter can be two subtypes of Aircraft. The idea is that every Plane is an Aircraft, but not vice versa. If we have a variable of type Aircraft, the actual object residing in the variable may be a Plane or a Helicopter. Since the type system does not know which one, it would be a type error to do a Plane-specific operation on the variable. This is where type casts come in: we can cast the variable to be a Plane before operating on it. If the variable actually contains a Plane object, then all is well; otherwise a type cast exception is thrown. If the exception is not caught, then the effect is the same as a dynamic type error in Scheme.

The quest of type systems research is to strike a balance between expressiveness, that is, the number of programs that can be type checked, and efficiency, that is, the time it takes to do the type checking. One of the current active research areas is to combine static type systems for object-oriented languages with a notion of generic types [11].

Proof

Goal: A guarantee that the program meets the specification.

Mathematical proofs of program properties are based on a formal description of program behavior. A description of program behavior is also known as the semantics of a programming language. There are several approaches to formal semantics. Operational semantics [40] models a computation as a step-by-step process in which commands are executed one after the other. The execution process is modeled on a mathematical, high-level representation of the bits and bytes in an actual computer. Denotational semantics [34] formalizes a program as simply its input-output behavior, that is, in the simplest case, how it maps an initial state to a final state. Axiomatic semantics [25] specifies relationships between program states, such as if a predicate is true of a state, then after executing a command, a somewhat different predicate is true of the next state. Once a formal semantics is in place, we have a foundation for mathematical reasoning about programs. The properties we want to prove must also be stated formally.

In general, proving that a program has a property can be difficult. The ultimate goal of proving that a program meets its entire specification remains elusive and is getting increasingly problematic as software grows in size and complexity. One kind of program has received particular attention when it comes to proving correctness: compilers. The motivation is that compilers are an important part of today's computational infrastructure; if the compiler is not working correctly, then all bets are off. Moreover, many aspects of compilers are well understood which increases the chance that good proof techniques can be found [17]. One aspect of compiler correctness can be stated as follows:

If a program p is compiled to code c, and evaluating p gives result r, then evaluating c gives a result which is a machine representation of r.

Notice that the semantics of both the source language and the target language play a crucial role in the statement of compiler correctness.

One of the goals of modern programming language design is to ensure that *all* programs in the language have a certain correctness property. One example is type soundness [31], which states:

Well-typed programs cannot go wrong.

Here, "well-typed" simply means that the program type checks, and "go wrong" is, intuitively, an abstraction of the error "illegal address — core dump" that one can encounter at the hardware level. The standard way of proving type soundness is to first prove two lemmas [37, 43]:

Preservation: If a well-typed program state takes a step, then the new program state is also well typed.

Progress: A well-typed program state is either done with its computation, or it can take a step.

The Preservation lemma says that if we have type checked a program, then during the computation, all reachable program states will also type check. The Progress lemma says that if we have reached a program state that type checks, then we cannot go wrong in the next step of computation. Together, the two lemmas are sufficient to prove type soundness: when we have type checked a program, Preservation ensures that we will only reach typable program states, and Progress ensures that in those states, the computation cannot go wrong.

The field of program synthesis [7] is concerned with generating programs from specifications. This obviates the need for proofs altogether; if we can prove the correctness of the program generator, of course!

An active research area is that of proving the correctness of optimizing compilers [28]. Some optimizations radically change the code and remain a challenge to prove correct.

Certification

Goal: Verify a certificate before executing the program.

Clicking on a link may lead to the downloading and execution of Java bytecode. In addition, the bytecodes will be verified before execution, giving the same kind of guarantee as we get from type checking source code. This can be done easily because Java bytecode contains type information that guides the verifier. We can view this extra type information as a certificate which is checked by the bytecode verifier. If the raw bytecode does not match the certificate, then the bytecode will not be executed. The combination of code and certificate achieves a level of tamperproofing: one can try to tamper with either the code, the certificate, or both, but if the end result does not verify, then it will not be executed. In Java, bytecode verification first and foremost guarantees memory safety. Even if an attacker manages to change the code and the certificate in such a way that the result verifies, the code will be memory safe. It may no longer compute factorial, but will not crash your computer!

A weakness of the Java bytecode verification model is that after the verification is done, it may be necessary to compile the bytecode in order to achieve efficient execution. This entails that the compiler is part of the trusted computing base: if the compiler is faulty, then the guarantees obtained at the bytecode level are worthless. This raises the challenge of whether we can avoid trusting the compiler and have a certificate for machine code. The idea is that, instead of downloading Java bytecodes, we would like to download machine code plus a certificate which can be verified before execution. The certificate will be issued by the code producer, and there will be nothing to compile on the receiving end, thus obviating the need for a compiler in the trusted computing base.

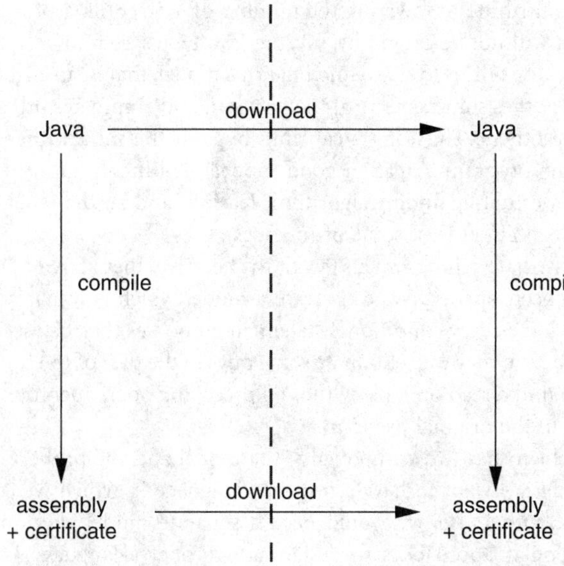

The obvious place to put the burden of producing the certificate is with the compiler. This leads to the concept of a certifying compiler, which not only produces memory-safe code, but also a certificate that the code is memory safe.

One approach to low-level certified code is typed assembly language [33]. The idea is the same as for Java bytecode: extra type annotations in the code guide the type checker, and type checking guarantees memory safety. One of the main challenges in designing a typed assembly language is that high-level constructs and values, such as procedures and objects, at the assembly level are represented by a multitude of individual entities and instructions. For example, there may be one type rule for a Java method call, but at the machine level, the method call is represented by several instructions, each of which must type check. Another challenge is to design a type system that will be a convenient target for the compilation of a wide variety of source languages. The ultimate typed assembly language would enable easy compilation of all source languages. This goal seems elusive, and it may be that in the future there will be several competing type systems for the same assembly language.

While certified assembly code does eliminate the compiler from the trusted computing base, it does introduce the need for having a verifier in the trusted computing base. The good news is that a verifier tends to be much simpler than a compiler.

If we allow certificates to be written as proofs in a full-blown logic, then the certified code is known as proof-carrying code [36]. This is vastly more powerful and flexible than typed assembly language, but it is also more difficult to produce the proofs. It remains an active research area to develop industrial-strength certifying compilers that produce proof-carrying code. Another active research area concerns producing better certified bytecode representations of high-level programs [3].

Bug Finding

Goal: Find bugs faster than via software testing.

In most software engineering efforts, more time is spent on testing software than on writing software. The goal of software testing is to find as many bugs as possible. It is widely believed that no large piece of software can be bug free, even after extensive testing. Part of the testing process can be automated, but testing remains labor intensive and therefore costly. It is particularly time consuming to invent test cases and to figure out what the correct output should be.

Modern programming languages help with bug finding, even without running the program. Structured programming, variables instead of registers, static type checking, automatic memory management, and certifying compilers all contribute to lowering the number of bugs. Still, a program can contain errors in the program logic that will not be caught by, say, the Java type system.

The idea of model checking [14] is to try to find bugs in a model, that is, an abstraction of the program. If the model faithfully preserves some aspects of the program and eliminates others, then it will preserve some bugs and eliminate others. Thus, if we can find bugs in the model, then those bugs will likely correspond to bugs in the program. Creating good models is difficult. On one hand, we want small models that will make powerful bug finding algorithms feasible, and on the other hand, we want models that are large enough to contain at least some of the bugs.

For example, suppose we have a program and want to check whether all read operations on a file only happen after the file has been opened. We can create a model which eliminates all operations on data other than files, and which for every conditional statement embodies the abstraction that both branches are executable. This abstraction may well eliminate some bugs in the part of the code that is not concerned with files. However, if we find a read on a file without a preceding open operation, then chances are that we have identified a bug in the original program.

A model checking problem has two components: the model and the property we want to check. To enable flexible bug finding, we want to have a property language in which we can express a variety of properties to be checked. Otherwise, we would need a separate model checking algorithm for every property. One of the popular approaches to designing a property language is to base it on regular expressions. For example, we might choose the alphabet {open, read, write, close}, and check the property:

open · read*

We can understand the property as stating that a file must be opened exactly once, and only after that can it be read any number of times. A somewhat different property could be:

open · (read | write)* · close

The same general model checking algorithm would handle both properties and possibly report bugs.

While it is easiest to build the model independently of the desired property, it can lead to faster model checking if the model is property driven. For example, the first property above has no need for a model with write and close operations.

Model checking has been immensely successful at bug finding for hardware. Software model checking has turned out to be more difficult and remains an active research area [16, 6].

145.4 Efficiency

We say that software is efficient when it can do its job using the available resources. A variety of resources can be available to software, including time, space, power, databases, and networks. When resources are constrained, a programming language can help with using them judiciously by offering resource-aware language constructs and compilers. A variety of approaches to resource awareness have been devised, and some of them will be highlighted in the following discussion of several aspects of efficiency.

Execution Time

Goal: Complete the task faster.

For desktop computing, users want spreadsheets that update the fields quickly after changes, they want internet searches to complete quickly, they want games with fancy graphics and computer players that move quickly, and so on. High speed is in practice achieved by a combination of good algorithms, careful coding, good compilers, etc.

A good programming language can contribute to fast execution time in at least two ways. First, the language can make it easier to express efficient algorithms. For example, a recursive algorithm such as quicksort is easier to express in a language such as C than in assembly language because recursion is supported in C. Second, the compiler can have a major impact by doing good instruction selection, register allocation, etc. The difference between a simple compiler and a highly optimizing compiler can be several factors when measuring execution speed.

The traditional role of a compiler is to do its job "way ahead of time" (WAOT), that is, before the compiled code is run. A WAOT compiler can take as long time as it likes, as long as it generates efficient code. Usually, a highly-optimizing compiler has the following structure:

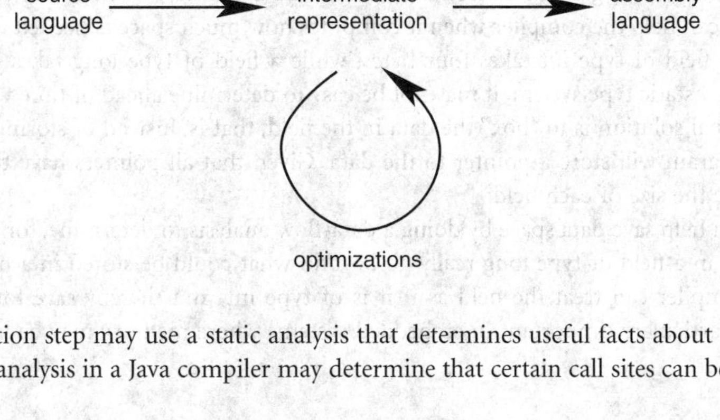

Each optimization step may use a static analysis that determines useful facts about the program. For example, a static analysis in a Java compiler may determine that certain call sites can be inlined, and the optimizer may then go ahead and inline the calls. In more detail, consider a Java call site x.m() where x is variable and m is a method name. From the syntax alone, we cannot determine which method will be called. However, a static analysis might determine that only objects of class C will be assigned to x, so only the m method in C can be called, and therefore we can inline it [22].

In Java, classes can be loaded and used while a program is running. If we want newly loaded code to execute fast, then we need to compile "just in time" (JIT), that is, after the code is loaded but before it is executed. A JIT compiler is executed at run time, so the compilation time directly impacts the run time. This creates a trade off between compilation time and run time. For example, an inlining optimization for Java may take 4 seconds to execute on a large program and make the program execute 3 seconds faster. This is fine for a WAOT compiler. In a JIT compiler, we would prefer a simpler inlining optimization that takes just 1 second to execute and makes the program execute 2 seconds faster. Each optimization needs to "pay for itself" by creating a larger speed up than the time it takes to do the optimization.

Some program analyses and optimizations work on a per-procedure basis. That property make them good candidates for JIT compilers. Other program analyses are inherently whole-program analyses. Those analyses do not work so well for languages such as Java because the program can change after the analysis is done, due to dynamic class loading.

Some program analyses take advantage of being applied to programs that have already been type checked. Such analyses are called type-base analyses [38] and are the topic of much current research. Another active research area is to find fast and effective program analyses for use in JIT compilers [18, 41].

Memory Management

Goal: Use less space.

Many efficient algorithms require the use of pointers and dynamically allocated data structures. In languages such as C, programs manage a memory space known as the "heap" for dynamically allocated data. A program can allocate and deallocate heap space, and it can access and update heap objects. The "do-it-yourself" style of heap management can lead to efficient programs but also to nasty bugs. For example, a program may reference a memory area after it has been deallocated, a bug known as the "dangling pointer" error. A program might also keep allocating heap space indefinitely, a bug known as a "memory leak." Some of these problems can be avoided by dealing with the heap at a higher level of abstraction and leaving the details to a memory management system. The memory manager is a part of the run-time system which is present during the execution of a program.

For example, in Java all dynamic memory allocation is done with expressions of the form

```
new C()
```

or something similar, where C is a class name. The memory manager will allocate space for a C object, that is, sufficient heap space for representing the fields and methods of a C object. When the program can no longer reach the object, then the memory manager will automatically deallocate it. The automatic deallocation, also known as garbage collection, ensures that dangling-pointer errors cannot occur.

A type system can help the compiler when it computes how much space is needed for the fields of an object. In Java, a field of type int takes four bytes, while a field of type long takes eight bytes. For a language without a static type system, it may not be easy to determine ahead of time what will be stored in a field. The usual solution is to "box" the data in the field, that is, instead of storing the data itself in the field, the program will store a pointer to the data. Given that all pointers have the same size, it is easy to determine the size of each field.

A compiler can help save data space by doing a data-flow analysis to determine, for example, whether the values stored in a field of type long really go beyond what could be stored in a field of type int. If not, then the compiler can treat the field as if it is of type int, and thereby save four bytes. Such an optimization is particularly important in embedded systems where memory may be limited [4].

The use of dynamic memory allocation can make it difficult to determine an upper bound on the need for heap space. As a consequence, some embedded software refrains entirely from dynamic memory allocation and instead allocates all data in a global area or on a stack. Even without dynamic memory allocation, it can be difficult to determine an upper bound on the need for stack space [12]. In particular, recursion makes it difficult to find such upper bounds. So, some embedded software also refrains from using recursion.

A compiler can help save stack space by inlining procedure calls. If a procedure is called more than once, such inlining may in turn increase the code size, creating a trade off between stack size and code size.

Current research includes developing memory managers that can allocate and deallocate data more efficiently and with smaller overhead [9]. Another active area is the development of methods for statically determining the need for heap space in recursive programs with dynamic memory allocation [26]. Yet another active area is resource-aware compilation [35, 39], where the compiler is told up front about resource constraints, such as memory limits, and the compiler then generates code that meets the requirements, or tells the programmer that it cannot be done.

145.5 Concluding Remarks

As the need for software grows, so does the need for better programming languages. New directions include power-aware compilers that can help save energy in embedded systems, and XML processing languages for programming of web services.

Acknowledgments

Thanks to Mayur Naik, Vidyut Samanta, and Thomas VanDrunen for helpful comments of a draft of the chapter.

Defining Terms

Flexibility — Software is flexible when it can run on a variety of platforms and is easy to understand, modify, and extend.

Correctness — Software is correct when it does what it is supposed to do.

Efficiency — Software is efficient when it can do its job using the available resources.

Compiler — The purpose of a compiler is to translate a program to executable machine code.

Type system — The purpose of a type system is to check that all values in a program are used correctly.

References

[1] J. Aldrich, C. Chambers, E. Gün Sirer, and S. J. Eggers. Static analyses for eliminating unnecessary synchronization from Java programs. In *Proceedings of SAS'99, 6th International Static Analysis Symposium*, pp. 19–38. Springer-Verlag (*LNCS* 1694), 1999.

[2] B. Alpern, C. R. Attanasio, J. J. Barton, M. G. Burke, P. Cheng, J.-D. Choi, A. Cocchi, S. J. Fink, D. Grove, M. Hind, S. F. Hummel, D. Lieber, V. Litvinov, M. F. Mergen, T. Ngo, J. R. Russell, V. Sarkar, M. J. Serrano, J. C. Shepherd, S. E. Smith, V. C. Sreedhar, H. Srinivasan, and J. Whaley. The Jalapeno virtual machine. *IBM Sys. J.*, 39(1), 2000.

[3] W. Amme, N. Dalton, M. Franz, and J. Von Ronne. SafeTSA: A type safe and referentially secure mobile-code representation based on static single assignment form. In *Proceedings of PLDI'01, ACM SIGPLAN Conference on Programming Language Design and Implementation*, pp. 137–147, 2001.

[4] C. S. Ananian and M. Rinard. Data size optimizations for Java programs. In *LCTES'03, Languages, Compilers, and Tools for Embedded Systems*, 2003.

[5] D. F. Bacon, R. B. Konuru, C. Murthy, and M. J. Serrano. Thin locks: Featherweight synchronization for Java. In *Proceedings of PLDI'98, ACM SIGPLAN Conference on Programming Language Design and Implementation*, pp. 258–268, 1998.

[6] T. Ball, R. Majumdar, T. Millstein, and S. Rajamani. Automatic predicate abstraction of C programs. In *Proceedings of PLDI'01, ACM SIGPLAN Conference on Programming Language Design and Implementation*, pp. 203–213, 2001.

[7] D. A. Basin. The next 700 synthesis calculi. In *Proceedings of FME'02, International Symposium of Formal Methods Europe*, p. 430. Springer-Verlag (*LNCS* 2391), 2002.

[8] G. Baumgartner, K. Läufer, and V. F. Russo. On the interaction of object-oriented design patterns and programming languages. Technical Report CSD-TR-96-020, 1998. citeseer.nj.nec.com/baumgartner96interaction.html.

[9] E. D. Berger, B. G. Zorn, and K. S. McKinley. Reconsidering custom memory allocation. In *Proceedings of OOPSLA'02, ACM SIGPLAN Conference on Object-Oriented Programming Systems, Languages and Applications*, pp. 1–12, 2002.

[10] T. J. Bergin and R. G. Gibson. *History of Programming Languages (Proceedings)*. Addison-Wesley, New York, 1993.

[11] G. Bracha, M. Odersky, D. Stoutamire, and P. Wadler. Making the future safe for the past: Adding genericity to the Java programming language. In *Proceedings of OOPSLA'98, ACM SIGPLAN Conference on Object-Oriented Programming Systems, Languages and Applications*, pp. 183–200, 1998.

[12] D. Brylow, N. Damgaard, and J. Palsberg. Static checking of interrupt-driven software. In *Proceedings of ICSE'01, 23rd International Conference on Software Engineering*, pp. 47–56, Toronto, May 2001.

[13] C. Chambers, B. Harrison, and J. M. Vlissides. A debate on language and tool support for design patterns. In *Proceedings of POPL'00, 27nd Annual SIGPLAN–SIGACT Symposium on Principles of Programming Languages*, pp. 277–289, 2000.

[14] E. Clarke, O. Grumberg, and D. Peled. *Model Checking*. MIT Press, Cambridge, MA, 2000.

[15] W. Clinger and J. Rees, Eds. *Revised Report on the Algortihmic Language Scheme*, November 1991.

[16] J. C. Corbett, M. B. Dwyer, J. Hatcliff, S. Laubach, C. S. Pasareanu, Robby, and H. Zheng. Bandera: Extracting finite-state models from Java source code. In *Proceedings of ICSE'00, 22nd International Conference on Software Engineering*, pp. 439–448, 2000.

[17] J. Despeyroux. Proof of translation in natural semantics. In *LICS'86, First Symposium on Logic in Computer Science*, pp. 193–205, June 1986.

[18] D. Detlefs and O. Agesen. Inlining of virtual methods. In *Proceedings of ECOOP'99, European Conference on Object-Oriented Programming*, pp. 258–278. Springer-Verlag (*LNCS* 1628), 1999.

[19] E. W. Dijkstra. The structure of the t.h.e. multiprogramming system. *Communications of the ACM*, 11(5):341–346, May 1968.

[20] K. Fisher and J. Reppy. Extending Moby with inheritance-based subtyping. In *Proceedings of ECOOP'00, European Conference on Object-Oriented Programming*, pp. 83–107. Springer-Verlag (*LNCS* 1850), 2000.

[21] E. Gamma, R. Helm, R. Johnson, and J. Vlissides. *Design Patterns: Elements of Reusable Object-Oriented Software*. Addison-Wesley, New York, 1995.

[22] N. Glew and J. Palsberg. Type-safe method inlining. In *Proceedings of ECOOP'02, European Conference on Object-Oriented Programming*, pp. 525–544. Springer-Verlag (*LNCS* 2374), Malaga, Spain, June 2002.

[23] J. Gosling, B. Joy, and G. Steele. *The Java Language Specification*. Addison-Wesley, New York, 1996.

[24] C. Hoare. Monitors: An operating system structuring concept. *Communications of the ACM*, 17(10):549–557, October 1974.

[25] C. A. R. Hoare. An axiomatic basis for computer programming. *Communications of the ACM*, 12(10):576–580, 1969.

[26] M. Hofmann and S. Jost. Static prediction of heap space usage for first-order functional programs. In *Proceedings of POPL'03, SIGPLAN–SIGACT Symposium on Principles of Programming Languages*, pp. 185–197, 2003.

[27] B. W. Kernighan and D. M. Ritchie. *The C Programming Language*. Prentice-Hall, Englewood Cliffs, NJ, 1978.

[28] D. Lacey, N. D. Jones, E. Van Wyk, and C. C. Frederiksen. Proving correctness of compiler optimizations by temporal logic. In *Proceedings of POPL'02, SIGPLAN–SIGACT Symposium on Principles of Programming Languages*, pp. 283–294, 2002.

[29] Microsoft. Microsoft Visual C#. http://msdn.microsoft.com/vcsharp.

[30] Microsoft. The .NET common language runtime. http://msdn.microsoft.com/net.

[31] R. Milner. A theory of type polymorphism in programming. *J. Comp. Sys. Sci.*, 17:348–375, 1978.

[32] R. Milner, M. Tofte, and R. Harper. *The Definition of Standard ML*. MIT Press, Cambridge, MA, 1990.

[33] G. Morrisett, D. Walker, K. Crary, and N. Glew. From System F to typed assembly language. *ACM Transactions on Progamming Languages and Systems*, 21(3):528–569, May 1999.

[34] P. D. Mosses. Denotational semantics. In J. van Leeuwen, A. Meyer, M. Nivat, M. Paterson, and D. Perrin, Eds., *Handbook of Theoretical Computer Science*, volume B, chapter 11, pp. 575–631. Elsevier Science Publishers, Amsterdam and MIT Press, Cambridge, MA, 1990.

[35] M. Naik and J. Palsberg. Compiling with code-size constraints. In *LCTES'02, Languages, Compilers, and Tools for Embedded Systems Joint with SCOPES'02, Software and Compilers for Embedded Systems*, pp. 120–129, Berlin, Germany, June 2002.

[36] G. Necula. Proof-carrying code. In *Proceedings of POPL'97, 24th Annual SIGPLAN–SIGACT Symposium on Principles of Programming Languages*, pp. 106–119, 1997.

[37] F. Nielson. The typed lambda-calculus with first-class processes. In *Proceedings of PARLE*, pp. 357–373, April 1989.

[38] J. Palsberg. Type-based analysis and applications. In *Proceedings of PASTE'01, ACM SIGPLAN/SIGSOFT Workshop on Program Analysis for Software Tools*, pp. 20–27, Snowbird, UT, June 2001. Invited paper.

[39] J. Palsberg and D. Ma. A typed interrupt calculus. In *FTRTFT'02, 7th International Symposium on Formal Techniques in Real-Time and Fault Tolerant Systems*, pp. 291–310. Springer-Verlag (*LNCS* 2469), Oldenburg, Germany, September 2002.

[40] G. D. Plotkin. A structural approach to operational semantics. Technical Report DAIMI FN-19, Computer Science Department, Aarhus University, September 1981.

[41] M. Poletto and V. Sarkar. Linear scan register allocation. *ACM Transactions on Progamming Languages and Systems*, 21(5):895–913, 1999.

[42] R. L. Wexelblat. *History of Programming Languages (Proceedings)*. Academic Press, New York, 1981.

[43] A. Wright and M. Felleisen. A syntactic approach to type soundness. *Information and Computation*, 115(1):38–94, 1994.

146

Input /Output Devices

Chih-Kong Ken Yang
University of California at Los Angeles

Input/output (I/O) devices are the elements that allow digital processing units to interact with the environment. The connection that transfers data between the central processor and the I/O devices is the I/O port. Through the I/O ports, a user can program a processor, access information, and store information. The I/O port is a connection that may comprise segments of physical connections with different signal environment and protocols. Dedicated processors that translate between the protocols bridge the segments or channels.

Performance measurement for an I/O device or I/O port is generically determined by the impact on the overall throughput of a processor. System throughput is often not a viable metric since it depends on many additional factors. More isolated metrics are commonly used for I/O channels such as (1) bandwidth, the amount of information that can be communicated per second (bits-per-second or baud-per-second); (2) latency, the delay of a response to a query; and (3) reliability, the number of erroneous data transfers (bit error rate). In this increasingly portable and environmentally aware world, some secondary metrics are the size/area and power associated with the I/O channel and I/O device.

As an example, Figure 146.1 illustrates the common I/O connections of a central processor (CPU). As data rates of digital processors increase, the system throughput for many applications is increasingly dependent on the performance of the I/O. Particularly in storage, memory access, and network interface, I/O performance is designed to scale with the processor clock rate. This chapter first describes various I/O devices, followed by how a CPU interfaces with an I/O subsystem. The chapter ends with a discussion of selected I/O channel specifications. As the system demands higher performance, specifications has become more numerous and more difficult.

146.1 I/O Devices

I/O devices can be classified based on the type of agent, human or machine. Further differentiation can be based on whether the device is input only, output only, or both. The grouping can also be based on the application: memory access, storage, networking, video, audio, pointer, and so on. The following section briefly discusses a sampling of the common I/O devices.

0-8493-1586-7/05/$0.00+$1.50

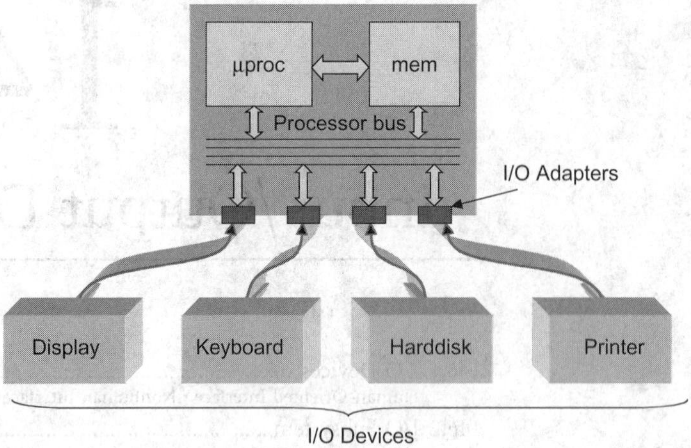

FIGURE 146.1 I/O connections of a central processor.

Human-Oriented Interface

Our increasing reliance on computers spawns a diversity of human–machine interfaces. A short list of categories would include video displays, printers, keyboards, pointing devices, scanners, and digital cameras. Each category includes many implementations. For instance, displays include cathode ray tubes (CRTs), light-emitting diodes (LED), liquid crystal displays (LCDs, passive or active), and plasma displays. Output devices have predominantly involved only two of our senses, vision and hearing, in the form of displays and speakers. The input device that can be most easily interpreted by the computer is still the alphanumeric keyboard aided by a pointing device. Recent advances in pen-based entry on touch screens have enabled a growing set of small portable computers. Enabling accurate electronic interpretation of pictorial and audio inputs is an area of very active research that may eventually make scanners and voice recognition the norms of human–machine interface. Until then, this discussion is limited to the more common set of devices: displays, printer, keyboard, and mouse.

Displays

Video displays based on CRTs have existed for over a hundred years. The displays require large voltages in order to bend an electron beam so that it scans horizontally across the screen one line at a time. Physical depth is needed for a good control of the scan. A memory or frame buffer, contains a byte or word for the intensity of each color of each point in the picture (pixel). Computer displays have standardized resolution of 640 × 480, 800 × 600, 1280 × 1024, or 1600 × 1200 (pixels). As one would expect, the higher resolution results in a larger frame buffer. The digital values in the frame buffer control the intensity of the electron beam as it scans. With a sufficiently fast scan rate, a picture is painted on the screen. The entire picture is refreshed 50 to 80 times per second. A display controller interfaces with the graphics processor to load the contents of the frame buffer.

The LCD is rapidly replacing the CRT; its primary advantages are smaller space requirements and lower power. The LCD consists of two thin sheets of glass that are transparent to opposite polarization and contain transparent electrodes and (for *active* displays) switches. Light (either backlit or reflected) illuminates the glass. The sheets sandwich a semisolid gel that is transparent and rotates the light polarization. The rotation changes with an applied voltage, which would cause the gel to be opaque. The opacity depends on the applied voltage. Each pixel has two electrodes and appears as a capacitance. For high density, one glass sheet contains horizontal conducting traces connecting rows of electrodes and the second sheet contains vertical traces connecting columns. Each of the three primary colors are connected separately. The minimum pitch (0.25 mm) of each dot (all three primary colors) limits the screen resolution and varies between manufacturers. The pixels are addressed via the rows and columns. Similar to a CRT, each pixel is activated serially by first converting the byte/word in the frame buffer for

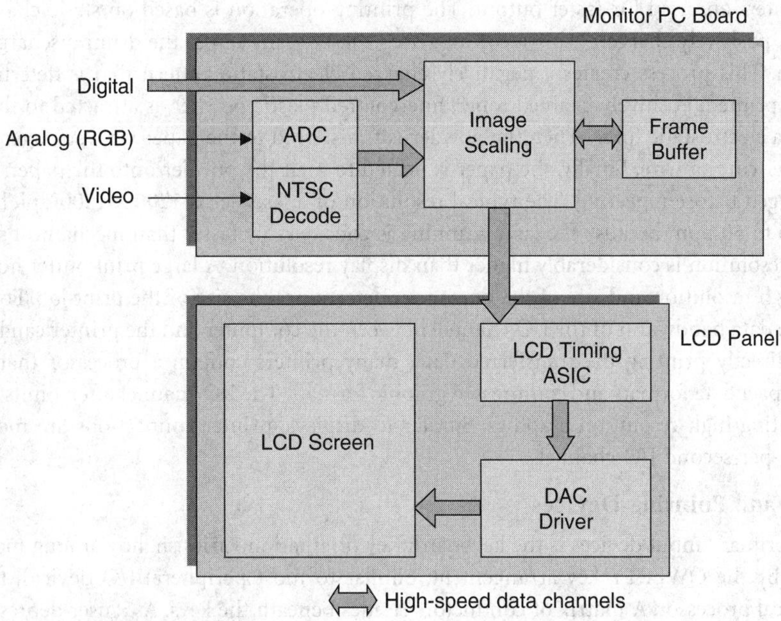

FIGURE 146.2 Liquid crystal display subsystem.

each color into an analog voltage and then driven onto the desired column. An LCD system is shown in Figure 146.2. Because of the capacitance of each pixel, driving a conductor loaded with pixels can be slow. This manifests as a shadow trailing behind a moving object on the screen. To improve the speed, manufacturers include a small transistor (thin-film transistor) for each color of a pixel. The transistor has little capacitance and can be switched quickly. The displays with the active switching are known as active-pixel LCD. Particularly for larger screens, LCDs are more costly than CRTs. Large LCDs have a low manufacturing yield because any processing defects result in dead pixels.

The connection between a display and a computer is a high-speed connection. A raw calculation of the data rate needed to load the frame buffer every refresh cycle would require several gigabits per second of transfer rate. Fortunately, not every pixel needs updating. The display controller substantially reduces the data bandwidth. However, the data rate still scales with screen size for a given resolution. I/O channels of multiple gigabits per second are used in display connections to support sufficient data bandwidth.

Printers

Another common output device is the printer. Printer technology can be classified into two main categories: impact and nonimpact. Impact printers are similar to a typewriter where small pins strike the paper through an ink ribbon. The resolution is limited by the size of the pins. Instead of using a pin, character printers limit the symbol set to predefined characters. Very fast and high-resolution character images can be produced but the application is limited. Impact technology has been, for the most part, replaced by nonimpact technology where the printer does not touch the paper. The dominant nonimpact printers are the inkjet printer and the laser printer.

Because of its low price, the inkjet printer is the most popular. Cartridges contain liquid black or multicolored ink and are attached to a print head. Very small droplets (50 to 60 mm) are dropped onto the paper, achieving resolution of up to 4800 × 1200 dots per inch (dpi). A droplet that has a well-controlled size is formed on a nozzle by various patented techniques such as vibration or heating. Each print head contains hundreds of nozzles. The printer head is scanned across the page and the paper is advanced. Both the scanning and paper feeding use accurate stepper motors. For color, several print heads are aligned and applied simultaneously. Because of mechanical scanning, the print speed is less than 20 pages per minute (ppm).

Laser printers are used for faster output. The printing operation is based on static charge. A rolling drum is first positively charged using a corona wire. A laser scans across the drum discharging portions of the drum. This process creates a negatively charged electrostatic pattern that is determined by the image to be printed. Positively charged toner, fine-colored plastic powder, is attracted to the pattern on the drum via electrostatic force. Then the powder is transferred to the paper via a stronger electric field using another corona wire. Finally, the paper is heated to melt the powder onto the paper. The drum is fully discharged before repeating. The typical resolution of a laser jet is 1200 × 1200 dpi, but can print at speeds up to 60 ppm because the laser scanning is considerably faster than mechanical scanning.

Printing resolution is considerably higher than display resolution. A large print buffer holds the print data. The high resolution and size of the document determines the size of the print job. For high-speed printing, the data bandwidth of the I/O channel between the computer and the printer can be very large. Instead of directly printing the transferred data, many printers contain a processor that can decode various compact data formats and compressed graphic formats. The I/O channel often limits the throughput for printing high-resolution graphics. Similar to displays, printer connections are moving toward multigigabit-per-second I/O channels.

Keyboards and Pointing Devices

One of the primary input devices is the keyboard. Key orientation varies slightly among manufacturers. Most abide by the QWERTY key arrangement. Similar to most peripheral I/O devices, the keyboard embeds a local processor. A matrix of conductors resides beneath the keys. As a user depresses each key, either the capacitance increases for that conductor, or a physical connection is made between two conductors beneath the key. The processor senses the change and interprets the meaning. The processor must handle multiple simultaneous depressions. Since key press may bounce, the processor must also filter the input and distinguish a bounce from a rapid series of presses. The keyboard processor transfers an encoded, 8-bit value of the keystroke to the computer.

Accompanying the keyboard is typically a mechanism for pointing to a location on a screen. A touch screen directly translates the pen location to the screen. The mechanism is very similar to that of a keyboard. By using additional layers of transparent conductors on the screen, the tap on a screen can be sensed as a change in the electrical resistance of the conductors. The resistance drops at the x and y location of the touch. Capacitive sensing is also possible through either using a mesh or charging a glass plate. When a user touches the screen, sensors measure a change in charge at the edges of the screen and estimate the location by comparing the differences between the sensors.

The most common pointing tool for computers is still the mouse. The introduction of the mouse dates back to 1960s and has become ubiquitous in the 1980s. The pointer on the screen, the cursor, has a reference origin chosen by the software. The mouse movements introduce Δx and Δy that adjust the current cursor position. The mouse contains a small processor that transfers the position adjustment periodically to the computer. A typical transfer rate is 40 reports per second or 1200 bits per second. The mouse consists of a rubber ball that drives an x shaft and a y shaft. Each shaft connects to a wheel that has slots on the surface. For each direction, an LED and optical sensor detects the amount of rotation with the number of pulses and the speed of rotation with the rate of the pulses. The mouse processor interprets the pulses into a binary adjustment to the registers that store the position.

In 1999, Agilent introduced an improved mouse based on optical sensing. Older models of optical mice require a special pad with a reflecting surface and a grid for the optics to determine the cursor position. The new optical mice rely on sophisticated digital signal processing (DSP) with an LED and a sensor array. The mouse can detect an image of the area beneath the mouse at 1500 images per second. Since processing chips are inexpensive, a DSP detects changes in the image pattern to determine Δx, Δy, and rate of motion. With no moveable parts that may fail or collect dust, the optical mouse is rapidly replacing the wheeled mice.

For portability, laptops use touch pads and strain gauges. Touch pads are very similar to touch screens without needing the absolute position. Strain gauges are small knobs situated within the keyboard. The cursor direction and speed are based on the amplitude of strain in both the x- and y-axes.

TABLE 146.1 Human-to-Machine Data Rates

Human	Data Rate	Machine
Reading text	30 to 375 B/s	Alphanumeric display
Typing at 100 wpm	10 to 20 B/s	Keyboard
Pointing/selecting	1 kB/s	Mouse; TrackPoint II™; tablet
Hearing	8 to 60 kB/s	Speakers
Voice	3 to 15 B/s	Microphone

Human-oriented input devices typically do not limit a CPU's throughput because they require little data bandwidth. Table 146.1 summarizes the data rate for various human–machine interfaces.

Nonhuman Interface

Many I/O devices do not have a human interface. Some are machines that are slave devices to a CPU and responds to queries from the CPU such as memories and storage. Other I/O devices serve as communication between peers such as modems and multiprocessor interconnections. In fact, many human-oriented I/O devices have internal processors that run built-in operating systems and communicate with the CPU. This section briefly discusses these devices.

Main Memory

The memory storage for CPU is organized hierarchically with the low-latency but small-capacity level residing on chip as the first and second levels. The bulk of the memory for CPUs is off-chip dynamic RAMs (DRAMs) configured as a multidrop bus module. Although memory modules are often not considered to be a computer I/O, it is an important I/O port for the CPU. Each module contains up to eight ICs and can currently provide 1 Gb of storage capacity. The DRAM IC utilizes a dense capacitor arrays. Each capacitor dynamically stores charge and the amount of charge corresponds to a binary bits. Specialized processing of the capacitor is needed to achieve extremely high density while maintaining roughly 40 fF of capacitance. The minimum capacitance guarantees reliable charge storage that can withstand leakage, noise, and radiation. The module is connected to the CPU through a memory controller. The interface has a significant impact on system throughput and has very high data bandwidths and low latencies. A high-speed memory channel has bandwidth of 1.6 Gb/s with a latency of 40 ns.

Storage

Disks provide even larger storage capacity at low cost. Disks vary depending on their storage mechanism. The most common are magnetic disks (hard drives, tapes) and optical disks (CDs and DVDs). Data are stored on a platter or tape. Access times are significantly slower than DRAMs and depend on the physical mechanism of moving a sensing head to the appropriate location on the platter. Many architectures use multiple heads to speed along the search. Sequential access rate is considerably faster but are still slower than a DRAM due to the signal processing needed to recover the bit.

Magnetic disks are based on a magnetic recording medium in which the direction of the magnetic field over a small domain can be programmed. Writing data onto the platter involves moving a magnetic coil over the surface. With a sufficient magnetic field, the magnetic domain retains the direction. Reading involves using a sensitive magnetoresistive material on a moving head that senses the magnetic field from the domain. Hard disks comprised of multiple platters accessed in parallel with multiple heads. The platter spins under the head at speeds of 7200 to 10,000 rotations per minute. Data can be continuously read while the platter spins. The head moves along the radius to read different areas of the platter.

Optical disks operate in a similar manner except the storage mechanism is different. Instead of a magnetic domain, the platter is marked with very small bumps (1.6 μm in pitch between bumps for CDs and 0.74 μm for DVDs). A laser scans over the bumps to detect a difference in the reflection. A light sensor outputs pulses to indicate the data bits. To maintain constant bump density and constant sampling rate across the disk, the rotation slows as the head moves radially outward. Because of the extraordinary density of CDs and DVDs, magnetic tapes have become much less common mediums.

The data rate for hard drives or CDs depends not only on platter size and rotation speed but also on the number of storage drives. Due to parallel access, the data rate for disks is not limited by the access latency of a single platter or disk.

Flash Memory

Recent advancements in electrically erasable and programmable ROMs have enabled dense, solid state nonvolatile memories known as flash ROMs or flash memories (with brand names such as Memory Stick, SmartMedia, etc.). Each IC can store several gigabits and several can be combined to form a very small memory card ranging from 32 to 512 Mb of capacity. Flash memories use specially processed transistors that behave as programmable switches. Each switch can be programmed to conduct or not conduct by using a large voltage (10 V). The initial write can take up to 7 μs but subsequent serial writes can achieve rates near that of DRAMs of 50 ns. With considerably faster access times, the form factor of a quarter and no moving parts, flash memory cards serve as the ultimate portable electronic file. The technology is more costly than disks and is limited in the number of write/erase cycles.

Networking

Another type of I/O device enables peer-to-peer communication between computers, also known as computer networking. The physical distance between computers may range from a few meters to hundreds of kilometers. The data are not directly transmitted from one computer to the next. Depending on distance, the data may travel over many segments with differing mediums and protocols. The mediums may be optical fiber, shielded cables, unshielded cables, or wireless. The protocols may include transmission by packets (inserting headers and addresses), flow control, error correction, data encoding, data modulation, and so on. For a better understanding of networking layers and protocols, see the excellent texts appearing in the Bibliography section.

For a computer, the I/O device is a modem or network interface card (NIC). The NIC is an intelligent adapter that converts the data from the CPU into a format that can be transmitted over long distances. The transmitted signals are often analog in nature and modulated/encoded to ensure reliable transmission. Data flows back and forth so that erroneous data can be detected and resent. The transmitted signals have long latencies due to distance, so latency within the modem is typically not a concern.

Common interface standards are Ethernet (data rates of 10, 100, and 1000 Mbps), dial-up modems (300 to 56 Kbps), and xDSL (256 Kb to 20 Mbps). Although Ethernet has the highest transmission rates, it requires a high-quality connection with a shielded twisted-pair cable (CAT-5) for each segment. Cables from each computer are concentrated at a switch that redirects the data between the inputs. Each computer is assigned an address (unique name) and the information is transmitted in packets. The switching depends on the target computer's address associated with the data packet. Because homes are typically wired with high-performance cables, as an alternative, computers communicate from homes using the existing telephone cables. Computers can *dial up* a network provider and transmit data over the frequency bandwidth allotted for voice communication. Advances in the technology have reach data rates of 56 Kbps which very near the theoretical limit of data transmission in the medium. For further increase in bandwidth, telephone companies must relax the frequency bandwidth requirement with an explicit modem at the local switchbox. xDSL (digital subscriber line) are standards that can extend the home connection upward to 20 Mbps.

Table 146.2 and Table 146.3 list some of the performance and characteristics of the nonhuman interfaces.

146.2 I/O Subsystem

Most I/O events are not synchronous to the operation of the CPU. The data format to/from the I/O is not compatible with the CPU and hence an I/O cannot directly connect to the processor. In order to

TABLE 146.2 Storage I/O

Device	Storage	Transfer Rate	Latency	Seek Time
Floppy disk	1.44 Mb	0.125–1 Mb/s	50–150 ms	100–200 ms[a]
Disk drive	—[b]	—	2–16 ms	8–10 ms
CD ROM	0.6–1.8 Gb	0.15–6 Mb/s	80 ms	70–160 ms[c]
DVD ROM	7 Gb	1.35–21.6 Mb/s	95 ms	80–240 ms
Flash memory	32–512 Mb	—	40 ns[d]	—

[a] 500 ms spin-up time.
[b] Disk arrays can have hundreds of gigabytes and a transfer rate that depends on ATAPI.
[c] 6 s spin up time.
[d] 7 μs initial access.

TABLE 146.3 Networking I/O

Medium	Bandwidth	Mode
Telephone modem	56 kb/s	C
xDSL	256 kb/s–10 Mb/s	C
Wireless LAN	200 kb/s	P
Ethernet	10–1000 Mb/s	P

Note: C, connection; P, packet.

TABLE 146.4 Processor Buses

Bus	Data Width	Transfer Data Rate
ISA	8, 16	8.33 Mb/s
EISA	32	33
PCI	32, 64	133
PCI express	1, 4, 12	625–7500

Note: ISA, industry standard architecture; PCI, peripheral component interconnect; EISA, extended ISA; PCI express, a point-to-point PCI.

query a CPU or respond to a query, the data must pass through an I/O adapter that accompanies the device to "translate." Once connected, there are several modes by which an I/O device can request the attention of a CPU with differing levels of intrusion and inefficiency. A device driver, or software that is part of the operating system specific to the I/O device, regulates the interface. In its entirety, the I/O subsystem appears to a computer user as a combination of both software and hardware. This section provides an overview of this subsystem.

I/O Adapters

As shown in the previous section, each I/O device connects to the computer with differing connectors and protocols. In order for each device to interface uniformly with the CPU, an *I/O adapter* must be used. Unlike cable adapters which only change the form factor of the connector, an I/O adapter converts the electrical signals and the signaling protocol into one that a CPU and memory can understand. Interestingly, this is similar to a network. Because of the short distances and latency concerns, the I/O adapter performs much less signal processing and error correction than typical communication networks.

CPU interfaces have evolved over time. Table 146.4 shows the transfer rates for various processor buses starting from the early ISA bus to the newly specified PCI Express. The most common interface for current personal computers is the peripheral component interconnect (PCI) bus. As illustrated in Figure 146.3, PCI is a multidrop bus that allows connections to I/O devices, a memory controller, a storage controller (ATA), a graphics port (AGP), and bridges to older interfaces such as ISA. Because PCI is a shared bus, the I/O adapter is responsible for placing the data onto the bus with the appropriate timing. In addition to behaving as a bridge between the I/O device and the CPU, the I/O adapter can be designed to have significant processing and storage capability. It can often serve to detect errors in the data transmission, collect status information, and buffer data. A graphics card is one such example. Many personal computer motherboards use a single "super I/O processor" to handle most of the standard I/O interfaces such as keyboard, mouse, audio, and even simple networking.

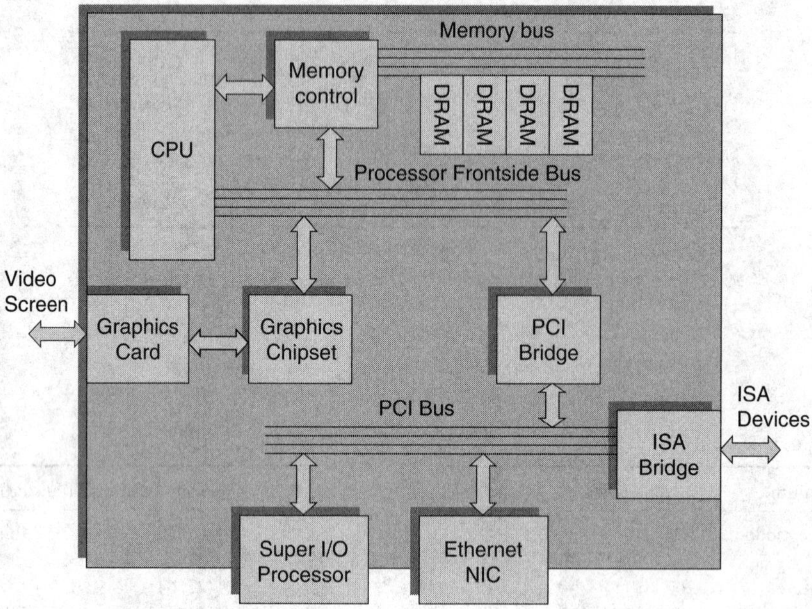

FIGURE 146.3 CPU I/O bus.

CPU Interface Modes

Once the physical connection has been established, depending on the design, different I/O devices require varying degrees of attention from the CPU. These interface modes can be classified into three main categories: programmed I/O, interrupt-driven I/O, and direct memory access. Typically, the CPU supports all three modes and each is needed by a particular I/O device and adapter.

In programmed I/O, the software code directly controls the CPU with specific I/O access commands. The CPU is directly programmed to read and write data, control, and status information to and from the I/O device. There is very little processing done in the I/O adapter. Since an I/O's bandwidth requirement varies, in a simple implementation the CPU periodically checks for an I/O event. This is known as *polling*. If the I/O device responds positively, the CPU acts to grab the new information. Because of the continuous attention from the CPU, the mode sacrifices throughput.

With a more intelligent I/O adapter, the CPU can offload the task of the continuously polling. However, the I/O adapter would interrupt the CPU upon new information or a hardware event. An interrupt handler determines the appropriate sequence of operations. The CPU must halt its execution, save the state of the execution, and service the interrupt. The original operation resumes after the interrupt is serviced. Interrupt handling must deal with the importance of an interrupt and simultaneous interrupt requests. When the CPU is running an important operation, the handler may need to disable the interrupt system. Otherwise, interrupts can be prioritized based on a priority mask that accompanies the interrupt. A particular interrupt can be disarmed or masked to ignore or postpone its service depending on the priority. Important interrupts, such as impending system failure, would be flagged as unmaskable. Interrupt handlers are increasingly complex because current CPUs are deeply pipelined (requiring more than ten clock cycles for a single operation), and can handle multiple streams of execution. Both software and hardware scheduling can help determine if the interrupt requires the CPU to fully halt its execution or to simply treat it as another stream.

An interrupt-driven I/O typically informs the CPU of an I/O event, and then the CPU proceeds to load or store the I/O data. If the data block is large, the transfer occupies the CPU and the system throughput decreases. To minimize the impact of large data transfers, some I/O adapters are given *direct memory access* (DMA). A specific location and size in either the physical memory or the virtual memory is allotted to the I/O. A DMA controller writes the block of data from the I/O sequentially into the memory. Traditionally, a count is decremented starting from the desired block size. The transfer stops when the count decreases to zero. For some data transfers, the block size can be varied and even dynamically adjusting during data transfer. Once the desired block size is reached, the controller flags an interrupt to the CPU. Depending on the I/O device, the DMA controller can reside either in the I/O adapter (first-party DMA transfer) or externally in the memory controller (third-party DMA transfer). Whereas an interrupt-driven I/O mode would work well for the low data rate 56 K modem, DMA transfers are commonly used for storage devices or high-bandwidth network connections.

Device Drivers

The final part of the I/O subsystem links the software application with the I/O adapter and interrupt handler. The computer's operating system must recognize the I/O device; thus, an additional piece of software known as the *device driver* is installed into the operating system along with the installation of a new device. The driver provides a device name and memory location that the software can address. Instead of burdening a software programmer with knowing the specifics of an I/O device, the driver manages the I/O events and provides some degree of user programmability. The driver deals with resetting and arming interrupts, initializing the DMA controller, starting the I/O adapter, accessing status bits, handling errors, and so on.

146.3 I/O Channel

As seen in the previous section, the I/O connection between the I/O device and the CPU is not direct. It consists of the communication between three different processors: the processor resident on the device, the I/O adapter or I/O processor, and the CPU. The arrangement is not unlike a small network. As is typical of a network, different segments have different interface standards and bridges (I/O processors) convert between the standards. This section focuses on the segments and looks at some of the common physical mediums and specifications of the I/O channel.

Cables

The quality of a transmission medium is determined by signal loss. A signal propagating along a conductor establishes electric and magnetic fields with a nearby conductor (or conductors). Although the nearby conductor(s) may be grounded and do not carry a signal, they provide a path for the signal's return current that results from the terminating fields. The signal propagation depends on both conductors. Series resistance of the conductors appears as signal loss. Due to a phenomenon called *skin effect*, the series resistance increases with signal frequency. Nonuniform physical arrangements of the conductors from manufacturing imperfections result in varying fields, which also cause additional signal loss at higher frequencies. Therefore, transmission mediums such as cables or traces on a printed circuit board (PCB) appear as low-pass filters. The attenuation is proportional to the length of the transmission medium. High-quality cables use a metal layer around the signaling conductors to shield the signals and provide a uniform return path. Unshielded cables may use several conductors in parallel to provide the signal's return path. Pairs of wires are often twisted together (twisted pairs) to and driven differentially to guarantee their physical and electrical relationship.

A channel that connects between two IC can be described in many ways. Signals may be single ended or differential (where two signals with opposing polarity carry the data). The connection may be a point-to-point link or multidrop bus (where multiple ICs hang from each PCB trace). The channel may be parallel (where many single-ended or differential signals propagate the data) or serial (where only one or few signals are used). The connection may be simplex (only one direction of signal propagation); half-duplex (both directions of signal propagation are possible at different times); full duplex (both directions

TABLE 146.5 RS-232 Pin Assignments

D-type-25 Pin no.	D-type-9 Pin no.	Abbreviation	Full Name
Pin 2	Pin 3	TD	Transmit data
Pin 3	Pin 2	RD	Receive data
Pin 4	Pin 7	RTS	Request to send
Pin 5	Pin 8	CTS	Clear to send
Pin 6	Pin 6	DSR	Data set ready
Pin 7	Pin 5	SG	Signal ground
Pin 8	Pin 1	CD	Carrier detect
Pin 20	Pin 4	DTR	Data terminal ready
Pin 22	Pin 9	RI	Ring indicator

of signal propagation on different wires), or simultaneous bidirectional (both directions on the same physical wire).

There is a myriad of different cables and connectors in a computer system. Over short distances, the connection is through the PCB. For instance, the motherboard of a computer provides the connection between the CPU and I/O adapters through the PCI bus. The board is made with multiple insulating layers sandwiching patterned metal traces. Internal layers are often used as the controlled return path for the traces above it. To connect between PCBs within a machine's chassis, many use ribbon cables, a broad linear array of wires in parallel. The ribbon cable's poor frequency characteristic is not an issue over such short distances. Also the weak mechanical integrity of the cable and connector are protected within a computer's casing.

Many connections to the computer chassis use shielded cables and D-type connectors. A standard connection is the 25-pin or 9-pin connector for the RS-232 serial port. Table 146.5 shows the pin specifications for an RS-232 port. The protocol transmits data with two wires: one to transmit, and one to receive. The connection is named serial in contrast to using many wires in parallel to carry the data. The simple connection at 20 Kbps can be used over several tens of meters. The RS-232 protocol has been widely used for printers, scanners, and even single peer-to-peer computer connection. The 50-pin D connector is often used for the small computer systems interface (SCSI), a parallel connection for higher bandwidth. Similar to the D-type connector is the VGA connector with three rows of pins for a high-speed display connection. Small and round Deutsche Industrie Norm (DIN) connectors are the standard keyboard connectors. Telephone cables are unshielded twisted pair with the small rectangular, RJ-11, connector jack. Similar but slightly larger in shape is the commonly used thin Ethernet connector, RJ-45. To support a high data rate, the Ethernet cable, CAT-5, consists of four shielded, twisted pairs.

New standards are slowly replacing many of the I/O channels. As systems require higher data bandwidths, faster connections are needed. For example, the PCI bus has evolved from 33-MHz data frequency using 84 pins/traces (32 bits in parallel) to 133-MHz data frequency across 150 pins/traces (64 bits in parallel). Even at its fastest, the aggregate data bandwidth is approximately 1 Gb/s. Instead of continuing this trend of wider multidrop data buses, computer system manufacturers are moving toward an even faster connection with much fewer pins in parallel and point-to-point connections called PCI express. Using as few as 90 pins, the CPU connects to each I/O adapter with up to 12 parallel full-duplex connections. Each point-to-point connects at 2.5 Gbps, hence providing for an aggregate data rate of 60 Gbps for future system generations. In another example, the hard-drive interface known as ATAPI (AT attachment packet interface) had evolved to a 40-pin ribbon cable that has a total 133-Mbps transfer rate. Serial ATA is a new serial standard that will allow a 600 Mbps transfer rate over a single differential line. Finally, universal serial bus (USB) is a generic point-to-point serial connection with a pair of differential full-duplex signals that can operate up to 480 Mbps. It is rapidly replacing the connection to most peripheral devices such as keyboard, mouse, printers, scanners, flash cards, cameras, and so on. For each of these serial interfaces, an I/O adapter would serve as a bridge or hub that translates between data rates and signaling protocols.

Channel Specification

The specifications of a high-performance I/O channel contain the information needed to implement an I/O device that can directly plug in to the computer. The example that follows is based on Infiniband (part of the PCI express) specification.

Each pin of the interface is defined with the appropriate signal. The channel consists of 4 to 48 data signaling pins (1, 4, or 12 differential duplex channels). For the motherboard, an additional 24 pins are allotted for the return path for the shield of the cable. The 13 pins provide the power and power control signals. Five more pins manage the link. Only the 48 data pins operate at 2.5 Gbps. Management signals operate at kilohertz data rates. Each pair of data pins (a lane) transmits differentially in a single direction.

Each lane serially transmits a byte. Bytes are sequentially transmitted across 1, 4, or 12 lanes. Each byte is encoded using 8 bits/10 bits (encoding 8 bits using 10 bits). The encoding removes the DC component in the signal and also allows special control bytes that do not represent valid data. The data or control blocks are broken into packets that use special bytes as headers and trailers. Errors are checked per packet using a cyclic redundancy checker (CRC).

The binary 2.5-Gbps signals are carefully defined, including not only the voltage levels but also the signal transition times. There is a maximum tolerable skew between lanes. Also defined are the noise requirements for both voltage and timing. For instance, the receiver must be able to receive minimum signal amplitude of 175 mV, hence limiting the maximum internal noise of the receiver. Figure 146.4 illustrates the *mask* for the received data eye. The eye is the result of folding and overlapping a long data waveform over the time of one bit. The mask indicates where the signal voltage may transition. The opening indicates the location where the data can be reliably sampled with little error.

As data processing speeds increase, higher performance I/O channels will be developed to improve system throughput. As optical cables and attachments become more affordable, more optical I/O channels can be expected. At the same time, as we increasingly depend on computing, new I/O devices will allow more seamless interaction with machines. For instance, instead of being bound to the computer with cables, wireless protocols for input devices will be forthcoming. Ultra-wideband (UWB) is a developing

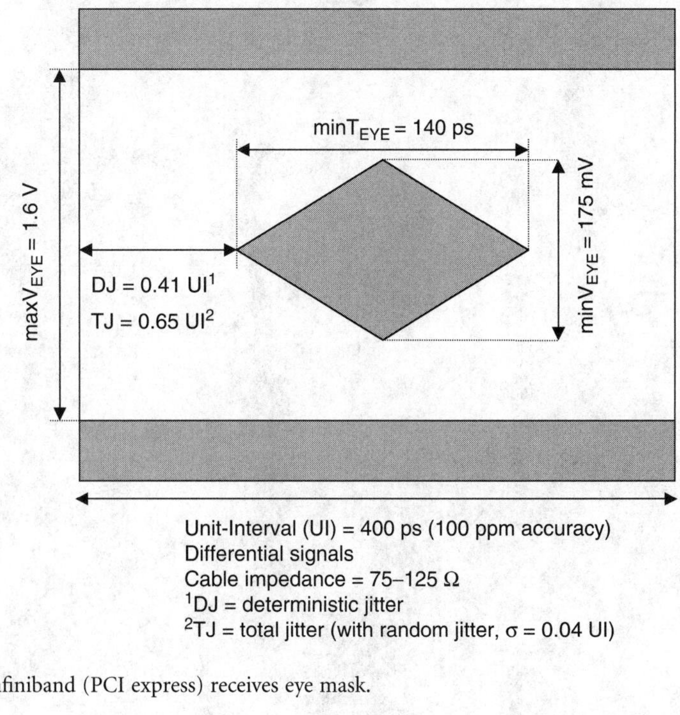

FIGURE 146.4 Infiniband (PCI express) receives eye mask.

standard that can provide low-data-rate, short-distance connectivity. The recently available 60-GHz wireless frequencies may even provide high data rate wireless.

References

Holst, G. C. 1998. *CCD Arrays, Cameras, Displays*, 2nd ed. International Society for Optical Engineering, Bellingham, WA.

Goupille, P.-A., trans. Loadwick, G. M. 1993. *Introduction to Computer Hardware and Data Communications*. Prentice Hall, Englewood Cliffs, NJ.

Ashar, K. G. 1997. *Magnetic Disk Drive Technology: Heads, Media, Channel, Interfaces, and Integration.* IEEE Press, Piscataway, NJ.

Dally, W. J., and Poulton J. W. 1998. *Digital Systems Engineering*. Cambridge University Press, Cambridge, UK.

Hennessy, J. L. and Patterson, D. A. 2002. *Computer Architecture: A Quantitative Approach*, 3rd ed. Morgan Kaufmann, San Mateo, CA.

Tanenbaum, A. S. 1996. *Computer Networks*, 3rd ed. Prentice Hall, Englewood Cliffs, NJ.

Further Information

In addition to the reference material above, see manufacturers' detailed information on I/O devices, often on the company websites. In particular, recent device drivers are often updated. Driver settings for a new display in an O/S such as Linux can be a headache. Such helpful information is available in newsgroups. Technical information on various I/O channels and their specifications can best be found through contacting the working group websites (i.e., *www.pcisig.com* for PCI standards). General magazines that are useful are *IEEE Spectrum*, *Computer*, *IEEE Micro*, and *PC Magazine*.

147

Memory and Storage Systems

Peter J. Varman
Rice University

Memory and storage devices are ubiquitous components of digital electronic systems ranging from PDAs and MP3 players, to high-end supercomputers driving businesses and scientific research. At a fundamental level such devices simply remember encoded information stored in them for recall at a later time. For instance, digitally encoded video written onto the surface of a video disk by the manufacturer can be repeatedly read using a DVD player that retrieves the stored information. Cassette tapes, audio CDs, smart cards, memory sticks, memory cards, hard disks, and random access memory (RAM) are some of the many storage devices found in everyday applications.

147.1 Storage Basics

Digital storage devices store information encoded as strings of binary digits or bits. A bit is stored by changing some distinguishing physical property of the storage medium to represent its two values. Solid-state DRAM chips for instance, alter the electrical charge on a capacitor to represent a bit, hard disks flip the polarization state of tiny magnetic particles on its surface, optical WORM devices create tiny pits on the medium to store a bit, and future biologically inspired memory devices may use different molecular sequences to encode information.

In a write-once memory (also referred to as a read-only memory) like a WORM device or a ROM, the change of state is irreversible. Once the information has been written it cannot be changed. In contrast, read/write memory devices, such as DRAM and SRAM chips or magnetic disks, can be rewritten at will, with the new information replacing the previous contents. A read returns the last written information. In between these extremes are erasable memories such as flash memory, which can be rewritten several times up to some limit.

A nonvolatile memory maintains the stored state even when power is removed. Write-once devices are typically nonvolatile, as are some read/write memory like magnetic disks and tapes, and erasable devices such as flash. Nonvolatile storage is also referred to as persistent storage. In contrast, volatile memory devices revert to a neutral state when power is removed, effectively erasing any stored informa-

0-8493-1586-7/05/$0.00+$1.50

tion. DRAM and SRAM chips are volatile. Persistence in systems built from volatile devices can be obtained by using redundant uninterruptible power supplies.

Two fundamental characteristics of memory systems are speed and cost. Speed reflects how long it takes the device to read or to write information. The access time or read latency of a device is the time required for it to respond to a read request. The throughput or bandwidth of the device is the rate at which it transfers information. The relation between these terms can be illustrated by analogy with a photocopy machine making multiple copies of a single page. The time required for the first copy to be output is the latency, which includes the time required to scan the page and for internal processing. Subsequent copies of the page are output faster by the machine, because it can avoid the initial scan and can overlap processing steps required for successive copies. The throughput reflects the rate at which successive copies are produced. Device speeds vary greatly. Solid-state SRAMs and DRAMs have latencies of a few to tens of nanoseconds, respectively, while mechanically driven devices like magnetic tapes can have latencies on the order of seconds. A magnetic disk has a latency of tens of milliseconds, but can transfer data at the rate of several million bits per second. In a computer system, the term memory usually refers to fast devices that are directly accessed by the processor, while the term storage refers to slower devices further from the processor that are accessed using specialized hardware controllers under the supervision of operating system software.

Sequential memory devices can be read only in a fixed order, and consequently incur a high latency when accessing arbitrary sections of storage. Random access devices, in contrast, permit efficient access to any portion of the stored information. Audio CDs and audiocassette tapes are easily recognizable examples of random access and sequential memory devices respectively.

The cost of a memory or storage device is often expressed in cents per bit. The capacity of a device is the number of bits it can store. Speed, capacity, and cost are related; not surprisingly, slow high-capacity devices have a lower cost than low-capacity, high-speed devices. Computer systems are built with a combination of devices organized in a hierarchy to optimize cost and performance characteristics: the goal is to achieve performance close to the fastest devices in the hierarchy at a cost approaching that of the cheapest. A typical memory hierarchy ordered from fastest to slowest consists of processor registers, primary or L1 cache, L2 and L3 caches, main memory, magnetic disks, and magnetic tape. Computer system designers have engineered ingenious schemes using caching and virtual memory techniques, to give programs the illusion of having access to unlimited amounts of fast memory. The program image is stored in its entirety on slow disks; active portions of a program are automatically moved to fast memory on demand. By multiplexing the fast memory among the currently active segments of a program, these techniques achieve favorable cost/performance characteristics.

The reliability of a storage device is usually expressed in terms of the mean time to failure (MTTF). This figure of merit informally refers to the length of the continuous up-time of the device. Transient disturbances can result in the corruption of data stored on a device. To protect against random bit flips that destroy data integrity, error detection or error correcting codes (ECCs) are often employed. By adding a few extra bits to each fixed-length sequence of bits, it is possible to detect whether one or more bits have been corrupted and to reconstruct the original data from the corrupted bit sequence.

147.2 Memory Systems

This section introduces some of the major types of memory and discusses their operation in more detail. We begin with a description of RAM-based memory system organization, followed by discussions of RAM and DRAM architectures, associative memory, and nonvolatile memory devices.

Random Access Memory

Random access memory (RAM) refers to read/write memory whose contents can be accessed in arbitrary order. Logically, a random access memory system can be viewed as a table of uniquely addressed words. Each word is made up of a fixed number of cells; a cell stores a single bit of information. The number of bits (cells) in a word is the memory width. The memory is accessed using an n-bit address, which can uniquely address any of 2^n words. To perform a read the processor provides the n-bit address to the memory system on a set of signal lines known as the address bus; following the read latency, the bits in the word are available to the processor on the data bus. To perform a write, the processor provides the memory system with the address and the bits to be stored on the address and data buses, respectively; the memory system stores the bits present on the data bus in the cells at the addressed word.

A RAM system is implemented using memory devices such as SRAM (static RAM) and DRAM (dynamic RAM). Early systems used individual SRAM- or DRAM-integrated circuit chips in dual inline packages (DIP) directly in implementing the memory system. Nowadays, memory devices are packaged as industry-standard modules such as SIMMs (single inline memory modules) and DIMMs (dual inline memory modules). Memory modules consist of several memory chips assembled on a single standardized small circuit board, thereby easing implementation and increasing reliability.

From the viewpoint of the processor, a DRAM chip appears as a scaled-down version of the memory system. Logically a DRAM chip may be viewed as having 2^m, B-bit-wide memory locations. The memory system is built by arranging the appropriate number of memory chips in a two-dimensional matrix; each row will implement a word of the memory. Early DRAM chips had a width of 1 bit (B = 1) to minimize the number of data pins on the chip. For instance, using individual 4K × 1 DRAM chips (m = 12, B = 1), a 1M-word, 32-bit-wide memory system would be constructed as a 256 × 32 matrix. The 20-bit word address is decoded as follows: 8 bits of the address are used to select one of the 256 rows of the matrix. Each chip in the selected row is activated, and the 12 remaining bits of the address are used to access a location in each chip in this row. The 32 individual bits, 1 bit per chip in the selected row, make up the desired word. Writing a word of memory is similar; one bit of the 32-bit word is stored in each chip of the activated row, at the same location in each chip.

Memory packaging like SIMM and DIMM incorporate several chips into the same module. It would take just 256 modules, each made up of 32 4 K × 1 DRAM chips, to implement the memory system described above, with each module providing an entire 32-bit word. Later generations of DRAM and SRAM chips were able to devote more pins for data, and memory widths of 4, 8, or 16 bits are common. With wider memory chips, fewer of them are needed to build a module of a given width, thereby improving power consumption and reliability, while reducing system complexity. Some SIMMs and DIMMs may also provide parity and error-correction bits for fault tolerance. Single-bit error detection is accomplished by having one extra bit beyond the required width of the module; error-correcting modules such as a 32/36-DIMM module have a 32-bit data width and 4 additional bits for error correction.

SRAM and DRAM Organization

Internally, a RAM chip is organized as a two-dimensional grid with a memory cell at each of the grid points. A memory cell stores 1 bit. The geometry of the grid is governed by physical constraints, and is usually different from the logical view of the system designer. A 1-K × 1 RAM chip for instance may be physically organized as a 32 × 32 grid, rather than a single column of 1024 cells. The 10-bit address in this case is split into two halves, referred to as the row address and column address, respectively: the five high-order bits are used to address the rows of the grid, and the five low-order bits selects one of the columns of the grid. The cell at the intersection of the activated row and selected column is the one that is accessed.

The differences between a DRAM and SRAM stem fundamentally from different implementations of the memory cell. In an SRAM, the cell consists of a cross-coupled logic circuit built from four to six transistors; a SRAM memory cell will indefinitely remember the bit written to it as long as it is powered. In contrast, a DRAM cell is made from a single transistor and capacitor, and its memory is transient. The DRAM cell is written by charging its capacitor. This charge slowly leaks out of the device in the form of an electrical current so that without special intervention the information stored in the cell is irrevocably lost. As a consequence, DRAM cells need to be periodically refreshed every few milliseconds or so. To refresh a cell, its contents are read and then written back again, thereby replenishing the charge on the capacitor as needed. The need for refresh circuitry complicates the design and operation of the DRAM interface, and impacts its operating speed since some portion of the time is used by the chip for refresh.

However, because of the small size of each cell, DRAM chips are much denser than SRAM and have four to six times the capacity. SRAM, by contrast, is much faster and consumes less power than the DRAM, but holds fewer bits per chip and is considerably more expensive than DRAM.

In order to conserve pins on the chip, the row and column addresses on a DRAM chip are multiplexed on the same set of pins on consecutive cycles. The DRAM controller first presents the row address to the chip, which stores it in an internal register abbreviated as RAS (row address select), and uses it to activate the addressed row. Next the column address is sent on the same pins; the chip stores the address in the internal CAS (column access select) register, and uses it to select the cell in the addressed column of the previously activated row. SRAM chips being less dense typically have separate pins for both row and column addresses and do not require the two-cycle addressing strategy of a DRAM.

During a read, the entire row of the DRAM addressed by the RAS register is read internally, and then written back again, thereby refreshing all the cells in that row. The DRAM controller performs refresh cycles in a periodic pattern; each refresh cycle refreshes one row. The controller cycles through the row addresses in order using either a distributed or block refresh mode. In the former, the refresh cycles are evenly spaced, with normal memory accesses occurring between consecutive row refresh operations. In block refresh, all refresh cycles are performed together without any intervening memory accesses. The smaller the number of rows the fewer the number of refresh cycles needed, favoring organizations with more columns than rows.

High-Performance DRAM

With modern processors operating at gigahertz clock rates, DRAM access latencies of tens of nanoseconds is a significant performance bottleneck. Consequently, a number of techniques spanning the device, architecture, compiler, and application levels have been developed to mitigate this problem. At the device level, evolution of the basic DRAM technology and architecture have resulted in current industry-standard DRAMs such as fast page mode (FPM) DRAM; extended data out (EDO) DRAM; synchronous DRAM (SDRAM); double data rate (DDR) SDRAM; and direct RAMBUS DRAM (DRDRAM).

Current DRAM designs have been optimized for burst modes of access, in which a number of adjacent memory words are read in succession. A FPM DRAM is organized so that whenever a cell is read, adjacent cells in the same row (or page) may be read quickly. Since these locations all have the same row address, accessing them only requires the column address to be incremented, decreasing their access latency. An EDO DRAM also allows FPM capability; in addition, it pipelines internal operations so that the period of time that data remain available on the data bus is increased, providing greater flexibility to the memory controller.

SDRAM, the current-volume industry-standard DRAM has several advantages over FPM/EDO. First, SDRAM operates synchronously with the system clock increasing performance, and provides flexible programmable block access modes to control burst length and column access order. In addition, the internal architecture of SDRAM supports a multiple bank organization. The separate banks can be controlled so that while one bank is being read, another can be primed by activating its RAS, an operation known as precharging. Consequently, different rows can also be accessed in burst mode as long as they reside in different banks, resulting in high memory bandwidth. DDR SDRAM is similar to SDRAM, but doubles the memory bandwidth by transferring data twice per clock cycle, at both the rising and falling edges of the clock.

Rambus memory is a proprietary memory technology invented by Rambus Inc. to provide high memory bandwidth. Rambus memory chips are called RDRAMs (Rambus DRAMs), and the memory modules are called RIMM. RDRAM is considered to be a revolutionary rather than evolutionary memory technology and necessitates a totally new memory architecture.

Content Addressable Memory

Unlike memory devices in which information is accessed by specifying the address or location of the desired information, content-addressable memories (CAMs) or associative memories access information by value. Each block of memory contains additional logic in the form of a comparator that compares an input search value against a stored tag in the block. If the tag matches the input, the information associated with the selected block is retrieved.

Associative memories are used as main memory caches within a computer system. A cache is a small high-speed memory (usually SRAM) that stores a copy of frequently accessed main memory (DRAM) locations. Since many possible main memory locations may map to the same cache block, there must to be some way of distinguishing the actual memory words that are currently stored in a cache block. The cache is organized as an associative memory whose search key is the main memory address of a word, and the data are the actual contents of the word. When the processor presents the address of a main memory location, the cache is associatively searched. If the desired address is found in the cache, the associated data can be immediately returned without the need to access the much slower DRAM memory.

Processor-in-memory architectures, which attempt to migrate processing logic closer to the memory, generalize the concept underlying associative memories and are an area of current research.

Nonvolatile Memory

Nonvolatile memory devices retain stored information even when power is removed. The basic solid-state nonvolatile memory is the read only memory (ROM) chip. ROM has evolved to include several variants, such as programmable ROM (PROM), erasable PROM (EPROM), electrically erasable PROM (EEPROM), and the ubiquitous flash memory. Unlike RAM, which can be read and written with equal ease, ROM devices are either unchangeable or require special operations to change the stored information.

Like RAM chips, the family of ROM devices also consists of an orthogonal grid of wires, with a bit stored at each grid point. Members of the family differ according to the type of circuit element at each grid point. In a ROM, the presence or absence of a diode between the row and column wires fixes the bit value stored at a grid point. A diode permits current flow between the column and row when energized, while the two wires are isolated if there is no diode at the grid point. A PROM allows the user to program the pattern of 0's and 1's at the grid points, rather than building it in at the time of manufacture. Equipment known as a PROM programmer is used to selectively blow out fuses between the column and rows at each grid point. Once programmed, the PROM cannot be altered. An EPROM permits the stored information to be altered, by first erasing the information using ultraviolet light, and then reprogramming the grid. Each grid point contains a two-transistor device that exploits quantum tunneling to coerce electrons to cross an insulating barrier in the presence of an applied voltage. When the voltage is removed, the electrons are trapped. Erasure using ultraviolet light causes the entire memory to be reset. An EEPROM permits selective erasure of the device by the application of a suitable electrical voltage. This allows the EEPROM to be altered in place without removing the chip, and without the need for specialized equipment. However, EEPROMs are too slow for applications that must make rapid changes to the stored data.

Flash memory is the latest generation of nonvolatile solid-state memory. Flash is used in the BIOS chip of PCs, digital cameras, memory sticks, PCMCIA Type I and Type II memory cards used as solid-state disks in laptop computers, and memory cards for video game consoles. Flash memory is similar to the EEPROM described above, but can be erased in units known as blocks, by the application of a high voltage using in-circuit wiring. Flash memory has a maximum number of erase cycles depending on the process, which can be a limitation in some applications. As the price of flash memory drops, it offers an alternative to hard disk technology in environments where reliability, speed, and power consumption are the dominant considerations.

A number of nonvolatile memory technologies are currently under research and development. Polymer memory uses the polarization state of a polymer chain instead of a transistor to record a bit of information. The advantages of the technology are low power, high capacity, and low cost. Its disadvantages are its slower speed compared to electronic memory, and destructive reads that necessitate rewriting circuitry for refresh. Polymer memory is thought to be a good candidate for portable devices such as MP3 players, digital cameras, PDAs, and other hand-held devices.

Magnetoresistive RAM (MRAM) technology is based on altering the magnetic polarity and consequent change of resistance in response to an electrical current. Its promise is density, speed, and nonvolatility, and is considered an alternative to flash memory. Another nascent technology, OUM, uses a thin film of chalcogenide alloy used in rewritable CDs and DVDs between electrodes, and electronically switches the material from crystalline to amorphous. Potential advantages include high density, low power and cost, ease of integration with existing technology, a nondestructive read cycle, and a lifetime of an estimated trillion writes.

147.3 Storage Systems

In this section, we focus on mass storage systems built from hard disks and magnetic tapes. The most common mass storage devices are hard disks, while magnetic tapes are used primarily for backup and archival storage. The current trend in storage systems is toward decentralized, networked storage decoupled from the server.

Magnetic Disks

Magnetic disks are the most common devices used in online mass-storage systems. Magnetic disks record information in the form of magnetic polarization of particles on its surface. A disk system consists of a number of platters mounted on a common spindle, with coupled read/write heads for each platter. A platter is made up of a number of concentric tracks, divided into smaller units known as sectors. A common sector size is 512 bytes, and is the smallest unit of access. The tracks in different platters line up to form a cylinder. Disk systems have shown remarkable improvements in capacity and performance over the past decade, resulting in dramatically decreasing costs. Innovations in technology have resulted in higher recording densities, smaller form factors, increased spindle speeds, and optimized transfer rates by the use of multizoned disks.

Reading and writing a disk is slow since it requires moving the head till it is positioned on the correct track, and then waiting for the correct sector to rotate under the head. The mechanical and rotational delay result in long access latencies. Reading or writing several consecutive sectors together in a block can amortize the positioning overheads. Software and hardware optimizers therefore chain together accesses to consecutive sectors to improve performance. For the same reason, I/O intensive applications like database systems place data so that it can be accessed in units of a track or cylinder. Low level hardware enhancements like track buffering store the contents of the entire track in an internal buffer as the disk rotates, avoiding rotational latency for accesses to different sectors on the same cylinder.

RAID Systems

RAID (redundant arrays of independent disks) systems provide increased storage capacity and higher bandwidth by incorporating several multiple disk drives within a single storage unit. Fault tolerance techniques are employed to cope with the increased probability of system failure arising from the use of multiple devices. RAID organizations, referred to as RAID level 0 through RAID 5, use different redundancy techniques for fault tolerance, similar to the use of parity and ECC codes in DRAM memory. RAID 1 uses data mirroring whereby the entire disk contents are replicated on a second disk. RAID 4 and RAID 5 employ the concept of a parity block for fault tolerance. The multiple disk system is viewed as a collection of stripes, composed of a block from each disk. One block of each stripe is designated as the parity block, and stores the bit wise exclusive or of the corresponding bits of the remaining blocks in that stripe. In the event of a single-disk failure, the blocks on the failed disk can be reconstructed from the blocks in the same stripe on the working disk. The storage overhead is much less than the 100% redundancy of RAID 1 systems. The penalty is an increase in write time since a write to a block requires a read-modify-write operation on the parity block as well. A RAID 4 system uses a single designated disk as the parity disk for all stripes. In RAID 5, the use of roving parity that associates different parity disks for different stripes, alleviates the parity-disk bottlenecking of a RAID 4 design. Other RAID organizations have since been proposed. RAID 6 permits the failure of up to two disks without data loss, by employing either two parity blocks with differently computed parities, or employing a two-dimensional arrangement of disks with associated row and column parities. RAID 0 does not provide fault tolerance, but allows data to be striped across the disks, thereby facilitating high-bandwidth data transfers. Hybrid combinations such as RAID 10 and RAID53 attempt to combine the advantages of different RAID levels.

Networked Storage

Storage architecture and the interconnection of storage devices is one of the rapidly changing characteristics of modern computing systems. Traditional server-centric storage architectures use bus-based interconnects such as IDE (interface data exchange) or SCSI. These direct-attached architectures are finding it difficult to meet the growing demand for flexible, high-performance storage. To achieve desired scalability, decentralized storage architectures networked together to create a virtual storage pool have gained popularity.

A SAN or storage area network is a dedicated network connecting servers and storage devices. Storage subsystems such as disk arrays or magnetic tape backup units are accessed over this backend network without loading of the LAN. SANs facilitate sharing stored data among multiple servers by avoiding the three-step process (read I/O, network transfer, write I/O) required in transferring data on traditional server-hosted I/O architectures. Furthermore, they permit autonomous data transfer between devices simplifying backup and data replication for performance or reliability, and encourage the spatial distribution of devices on the network, while maintaining the capability for centralized management.

In a typical SAN, dedicated protocols such as fiber channel protocol (FCP) are used to transport SCSI commands from a host bus adapter (HBA) in the server, over the fiber channel optical (or copper) interconnection fabric, to storage units connected to the SAN. The high costs of fiber channel switch hardware, training, and support have spurred interest in an alternative SAN technology known as iSCSI. The iSCSI protocol uses TCP/IP (running over Gigabit Ethernet for instance) as the transport for SCSI commands, reducing costs by leveraging the networking infrastructure already developed for Ethernet-based systems. Furthermore, the solution naturally scales to wide area networks due to the use of Internet protocols like TCP/IP as the transport mechanism.

A network-attached storage (NAS) device allows many of the server functions to be offloaded directly to the device. Once a request is authenticated by the server and forwarded to the device, data transfer to the network proceeds independently without further involvement of the server. In principle, a NASD can be directly connected to the local area network or may serve as an independent module in a backend SAN.

As storage and memory systems get more complex with increased numbers of concurrent activities, automated management techniques for data allocation, replication, backup, compression, scheduling, and load balancing are becoming necessary for cost-effective deployment.

147.4 Summary

Memory and storage are fundamental components of all digital electronic devices. The range of technologies, devices, and system architectures employed is astounding. There will continue to be a push for innovative solutions to keep pace with ever-faster processor speeds, exponentially increasing data volumes, need for reliability for critical data, protection techniques for security, complexity, manageability, and performance. Research into new MEMS-based (micro electromechanical) storage, miniature devices based on nanotechnology, and biologically inspired memory hold the promise for revolutionary future developments.

List of Acronyms

BIOS	Basic input/output system
CAM	Content addressable memory
CAS	Column address select

CD	Compact disk
DDR SDRAM	Double data rate SDRAM
DIP	Dual inline package
DIMM	Dual inline memory module
DRAM	Dynamic random access memory
DRDRAM	Direct RAM bus DRAM
DVD	Digital video disk
ECC	Error-correcting codes
EDO DRAM	Extended data out DRAM
EEPROM	Electrically erasable PROM
EPROM	Erasable PROM
FCP	Fiber channel protocol
FPM DRAM	Fast-page mode DRAM
HBA	Host bus adapter
IDE	Interface data exchange
ISCSI	SCSI over TCP/IP
L1 cache	Level 1 (or primary) cache
L2, L3 cache	Level 2 (level 3) or secondary cache
MEMS	Micro electromechanical systems
MP3	A popular encoding and compression scheme for music
MTTF	Mean time to failure
NAS	Network attached storage
PDA	Personal digital assistant
PROM	Programmable read-only memory
RAID	Redundant array of independent disks
RAM	Random access memory
RAS	Row address select
ROM	Read-only memory
SAN	Storage area network
SCSI	Small computer system interface
SDRAM	Synchronous DRAM
SIMM	Single inline memory module
SRAM	Static random access memory
WORM	Write once/read many

References

J. Hennessy and D. A. Patterson, *Computer Architecture: A Quantitative Approach*, Morgan Kaufmann, San Francisco, 2002.

J. F. Wakerly, *Digital Design Principles and Practices*, Prentice Hall, New York, 2000.

Further Information

In the United States, the most relevant professional organizations that sponsor publications and conferences related to memory and mass storage systems are the Institute of Electrical and Electronics Engineers (IEEE) and the Association for Computing Machinery (ACM). In particular, *IEEE Transactions on Computers, ACM Transactions on Computer Systems, IEEE Journal of Solid-State Circuits,* IEEE/NASA Symposia on Mass Storage Systems and Technologies, USENIX Conferences on File and Storage Technologies, IEEE/ACM International Symposia on Computer Architecture, ACM Symposium on Parallel Algorithms and Architectures, IEEE/ACM International Conferences on Architectural Support for Programming Languages and Operating Systems, and ACM Symposia on Operating Systems Principles are excellent sources for up-to-date technical information on designing, building, and using memory and mass storage systems.

Selected websites with information on memory and storage products are listed below.

www.micron.com
www.intel.com/technology/memory
www.intel.com/network/connectivity
www.hpl.hp.com/research
www.insic.org
www.rambus.com
www.sun.com/storage
www.netapp.com
www.storage.ibm.com/
http://h18006.www1.hp.com/storage/

148

Nanocomputers, Nanoarchitectronics, and NanoICs*

Sergey Edward Lyshevski
Rochester Institute of Technology

This chapter discusses far-reaching developments in design, analysis, modeling, and synthesis of nanocomputers using nanotechnology. In particular, nanoICs can be applied to synthesize and design nanocomputers utilizing two- and three-dimensional architectures. The nanoarchitectronics paradigm allows one to develop high-performance nanoICs utilizing recently developed nanoelectronics components and devices, including nanoscale transistors, switches, capacitors, etc. Nanotechnology offers benchmarking opportunities to fabricate the components needed, and to date, different nanotechnologies are under developments. Though molecular-, single-electron, carbon-, silicon-, GaAs-, and organic-based electronics has been developed to fabricate nanodevices, and promising pioneering results have been reported, it is difficult to accurately predict and envision the most promising high-yield affordable technologies and processes. Correspondingly, while covering some nanotechnologies to fabricate nanodevices and nanoICs, this chapter focuses on the advanced concepts in fundamental and applied research in nanocomputers synthesis, design, and analysis. Mathematical models are developed using nanodevices and nanoICs behavior description. The models proposed are straightforwardly applied to perform analysis, design, simulation, and optimization of nanocomputers. Novel methods in design of nanocomputers and their components are documented. Fundamental and applied results documented expand the horizon of nanocomputer theory and practice. It is illustrated that high-performance nanocomputers can be devised and designed using nanoICs integrated within the *nanoarchitectronics* concept to synthesize two- and three-dimensional computer topologies with novel robust fault-tolerant organizations.

*Parts of this chapter were published in: S. E. Lyshevski, Nanocomputer Architectronics and Nanotechnology, in *Handbook of Nanoscience, Engineering, and Technology,* W. A. Goddard, D. W. Brenner, S. E. Lyshevski, and G. J. Iafrate, Eds., CRC Press, Boca Raton, FL, 2003.

0-8493-1586-7/05/$0.00+$1.50

148.1 Introduction

Recently meaningful applied and experimental results in nanoelectronic have been reported [1–15]. Though fundamental research has been performed in analysis of nanodevices, limited number of feasible technologies has been developed to date to design applicable nanodevices operating at room temperature. However, the progress made in integrated carbon/silicon-based technology provides the confidence that nanoscale devices will be mass-produced in the nearest future. Unfortunately, no perfect nanodevices and large-scale nanodevice assemblies (matrices) can be fabricated. Correspondingly, novel adaptive and robust computing architectures, topologies, and organizations must be devised and developed. Our goal is to further expand and apply fundamental theory of nanocomputers, develop the basic research towards the sound nanocomputer theory and practice, as well as report the application of nanotechnology to fabricate nanocomputers and their components.

Several types of nanocomputers have been proposed. For last thousand years the synthesis and design of computers have been performed as a sequence of different paradigms and changes, e.g., from gears to relays, then to valves and to transistors, and recently (in 1980s) to integrated circuits [16]. Currently, the ICs technology progressed to the point that hundreds of millions of transistors are fabricated on 1 cm^2 silicon chip. The transistors are downsized to hundreds of nanometers using 50- and 70-nm fabrication technologies. However, the silicon-based technology reaching the limit, and scaling laws cannot be applied due to the fundamental limitations and basic physics. Nanotechnology allows to fabricate nanoscale transistors that operate based on the quantum mechanics paradigms. Furthermore, computing can be performed utilizing quantum principles. Digital computers use bits (charged is 1 and not charged is 0). A register of 3 bits is represented by eight numbers, e.g., 000, 001, 010, 100, 011, 101, 110, and 111. In the quantum state (quantum computing), an atom (one bit) can be in two places at once according to the laws of quantum physics. Therefore, two atoms (quantum bits or qubits) can represent eight numbers at any given time. For x number of qubits, 2^x numbers are stored. Parallel processing can take place on the 2^x input numbers, performing the same task that a classical computer would have to repeat 2^x times or use 2^x processors working in parallel. Therefore, quantum computer offers an enormous gain in the use of computational resources in terms of time and memory. Unfortunately, limited progress has been made to date in practical application of this concept. Though conventional computers require exponentially more time and memory to match the power of a quantum computer, it is unlikely that quatum computers become the reality in the near observable future. Correspondingly, experimental, applied, and fundamental research in nanocomputers has been performed to significantly increase the computational and processing speed, memory, and other performance characteristics utilizing novel nanodevices trough application of nanotechnology (rather than solely rely on the quantum computing and possible quantum computers due to existing formidable challenges). Nanocomputer will allow one to compute and process information thousands times faster than advanced computers.

Bioelectrochemical nanocomputers store and process information utilizing complex electrochemical interactions and changes. Bioelectrochemical nanobiocircuits that store, process, and compute exist in nature in enormous variety and complexity evolving through thousands years of evolution. The development and prototyping of bioelectrochemical nanocomputer and three-dimensional circuits have been progressed through engineering bioinformatic and biomimetic paradigms. Mechanical nanocomputers can be designed using tiny moving components (nanogears) to encode information. These nanocomputers can be considered as evolved Babbage's computer (Charles Babbage in 1822 built 6-digit calculator that performed mathematical operations using gears). Complexity of mechanical nanocomputers is a major drawback, and this deficiency unlikely can be overcome.

First-, second-, third-, and fourth-generations of computers emerged, and tremendous progress has been achieved. The Intel® Pentium® 4 (3 GHz) processor was built using advanced Intel® NetBurst™ microarchitecture. This processor ensures high-performance processing, and is fabricated using 0.13 micron technology. The processor is integrated with high-performance memory systems, e.g., 8KB L1 data cache, 12 K μops L1 Execution Trace Cache, 256 KB L2 Advanced Transfer Cache and 512 KB Advance Transfer Cache. Currently, 70 and 50 nm technologies are emerged and applied to fabricate

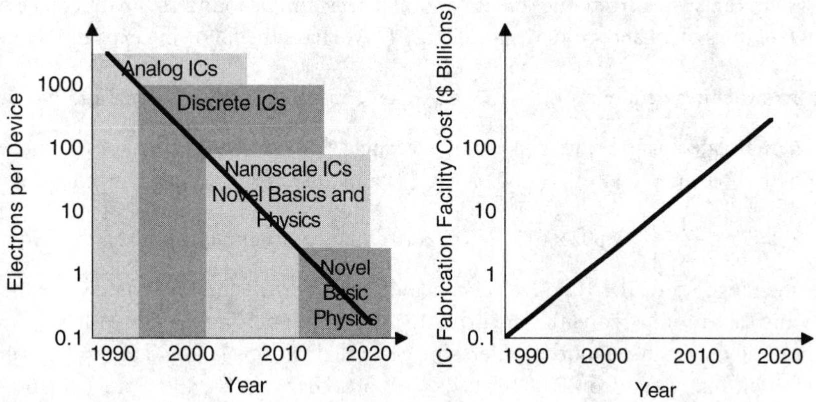

FIGURE 148.1 Moore's laws and nanotechnologies.

high-yield high-performance ICs with billions of transistors on a single 1 cm^2 die. Further progress is needed, and novel developments are emerging. The fifth generation of computers will be built using emerging nanoICs. This chapter studies the application of nanotechnology to design nanocomputers with nanoICs. Synthesis, integration, and implementation of new affordable high-yield nanoICs are critical to meet Moore's first law. Figure 148.1 illustrates the first and second Moore laws (reported data and foreseen trends can be viewed as controversial and subject to adjustments, however, the major trends and tendencies are obvious, and most likely cannot be seriously argued and disputed) [10].

Nanotechnology and far-reaching fundamental progress eventually will lead to three-dimensional nanoICs. This will result in synthesis and design of nanocomputers with novel computer architectures, topologies, and organizations. These nanoICs and nanocomputers will guarantee superior level of overall performance. In particular, compared with the existing most advanced computers, in nanocomputers the execution time, switching frequency, bandwidth, and size will be decreased by the order of millions, while the memory capacity will be increased by the order of millions. However, significant challenges must be overcome particularly in synthesis and design of three-dimensional nanodevices, nanoelectronics, and nanoICs. Many problems (novel nanocomputer architectures, advanced organizations, robust adaptive topologies, high-fidelity modeling, data-intensive analysis, heterogeneous simulations, optimization, reconfiguration, self-organization, robustness, utilization, and other) must be addressed, researched, and solved. Many of the above-mentioned problems have not been even addressed yet. Due to tremendous challenges, much effort must be focused to solve these problems. This chapter formulates and solves some long-standing fundamental and applied problems in synthesis, design, analysis, and optimization of nanocomputers utilizing nanoICs. The fundamentals of nanocomputer *architectronics* are reported, and the basic organizations and topologies are examined progressing from the general system-level consideration to the nanocomputer subsystem/unit/device-level study.

148.2 Nanoelectronics and Nanocomputer Fundamentals

In general, reversible and irreversible nanocomputers can be designed. Now days, all existing computers are irreversible. The system is reversible if it is deterministic in the reverse and forward time directions. The reversibility implies that no physical state can be reached by more than one path (otherwise, reverse evolution would be nondeterministic). Current computers constantly irreversibly erase temporary results, and thus, the entropy changes. The average instruction execution speed (in millions of instructions executed per second I_{PS}) and cycles per instruction are related to the time required to execute instructions as given by $T_{inst} = 1/f_{clock}$. The clock frequency f_{clock} depends mainly on the ICs or nanoICs used as well as upon the fabrication technologies (for example, the switching frequency 3 GHz was achieved in the existing Intel Pentium processors). The quantum mechanics implies an upper limit on the frequency at which the system can switch from one state to another. This limit is found as the difference between the total energy E of the system and ground state energy E_0. We have the following explicit inequality to find the maximum switching frequency $f_l \le \frac{4}{h}(E - E_0)$. Here, h is the Planck constant, and $h = 6.626 \times 10^{-34}$ J-sec or J/Hz. An isolated nanodevice, consisting of a single electron at a potential of 1 V above its ground state, contains 1 eV of energy (1 eV = 1.602×10^{-19} J) and, therefore, cannot change its state faster than $f_l \le \frac{4}{h}(E - E_0) = \frac{4}{6.626 \times 10^{-34}} 1.602 \times 10^{-19} \approx 0.97 \times 10^{15}$ Hz. For example, the switching frequency can be achieved in the range up to 1×10^{15} Hz. We conclude that the switching frequency of nanoICs can be significantly increased compared with the currently used CMOS-technology-based ICs.

In asymptotically reversible nanocomputers, the generated entropy is found as $S = b/t_t$, where b is the entropy coefficient that varies from 1×10^7 to 1×10^6 bits/GHz for ICs, and from 1 to 10 bits/GHz for quantum FETs; t_t is the length of time over which the operation is performed. Hence, the minimum entropy and processing (operation) rate for quantum nanodevices are $S = 1$ bit/operation and $r_e = 1 \times 10^{26}$ operation/sec-cm^2, while CMOS technology allows one to fabricate devices with $S = 1 \times 10^6$ bits/operation and $r_e = 3.5 \times 10^{16}$ operation/sec-cm^2.

The nanocomputer architecture integrates the functional, interconnected, and controlled hardware units and systems that perform propagation (flow), storage, execution, and processing of information (data). Nanocomputer accepts digital or analog input information, processes and manipulates it according to a list of internally stored machine instructions, stores the information, and produces the resulting output. The list of instructions is called a program, and internal storage is called memory.

In general, nanocomputer architecture can be synthesized utilizing the following major systems: (1) input – output, (2) memory, (3) arithmetic and logic, and (4) control units.

The input unit accepts information from electronic devices or other computers through the cards (electromechanical devices, such as keyboards, can be also interfaced). The information received can be stored in the memory, and then, manipulated and processed by the arithmetic and logic unit (ALU). The results are output using the output unit. Information flow, propagation, manipulation, processing and storage (memory) are coordinated by the control unit. The arithmetic and logic unit, integrated with control unit, is called the processor or central processing unit (CPU). Input and output systems are called the input-output unit (I/O unit). The memory unit integrates memory systems that stores programs and data. There are two main classes of memory called *primary* (main) and *secondary* memory. The primary memory is implemented using nanoICs that can consist of billions of nanoscale storage cells (each cell can store one bit of information). These cells are accessed in groups of fixed size called words. The main memory is organized such that the contents of one word can be stored or retrieved in one basic operation called a memory cycle. To provide a consistent direct access to any word in the main memory in the shortest time, a distinct address number is associated with each word location. A word is accessed by specifying its address and issuing a control command that starts the storage or retrieval process. The number of bits in a word is called the word length. Word lengths vary, for example from 16 to 64 bits. Personal computers and workstations usually have a few million words in the main memory, while nanocomputers can have hundreds of millions of words with the time required to access a word for reading or writing within psec range. Although the main memory is essential, it tends to be expensive and volatile. Therefore, nanoICs can be effectively used to implement the additional memory systems to store programs and data forming secondary memory.

Using the number of instructions executed (N), the number of cycles per instruction (C_{PI}) and the clock frequency (f_{clock}), the program execution time is found to be $T_{ex} = \frac{N \times C_{PI}}{f_{clock}}$. In general, the hardware defines the clock frequency f_{clock}, the software influences the number of instructions executed N, while the nanocomputer architecture defines the number of cycles per instruction C_{PI}.

One of the major performance characteristics is the time that takes to execute a program. Suppose N_{inst} is the number of the machine instructions needed to be executed. A program is written in high-

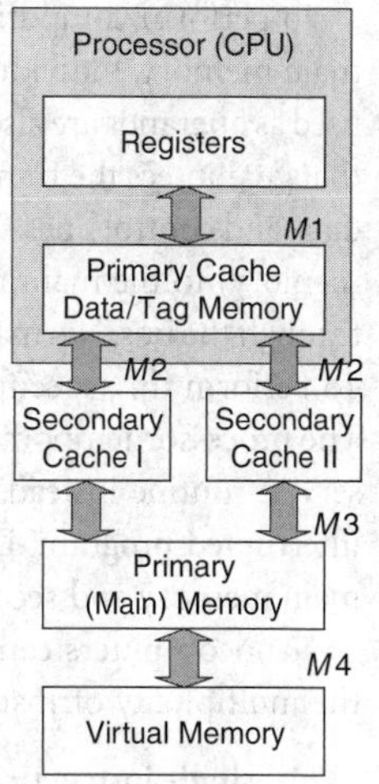

FIGURE 148.2 Processor and memory hierarchy in nanocomputer with cache (primary and secondary), primary (main) memory, and virtual memories.

level language, translated by compiler into machine language, and stored. An operating system software routine loads the machine language program into the main memory for execution. Assume that each machine language instruction requires N_{step} basic steps for execution. If basic steps are executed at the constant rate of R_T [steps/sec], then, the time to execute the program is $T_{ex} = \frac{N_{inst} \times N_{step}}{R_T}$. The main goal is to minimize T_{ex}. Optimal memory and processor design allows one to achieve this goal. The access to operands in processor registers is significantly faster than access to the main memory. Suppose that instructions and data are loaded into the processor. Then, they are stored in a small and fast cache memory (high-speed memory for temporary storage of copies of the sections of program and data from the main memory that are active during program execution) on the processor. Hence, instructions and data in cache are accessed repeatedly and correspondingly. The program execution will be much faster. The cache can hold small parts of the executing program. When the cache is full, its contents are replaced by new instructions and data as they are fetched from the main memory. A variety of cache replacement algorithms are used. The objective of these algorithms is to maximize the probability that the instructions and data needed for program execution can be found in the cache. This probability is known as the cache hit ratio. High hit ratio means that a large percentage of the instructions and data are found in the cache, and the requirement to access the slower main memory is reduced. This leads to decreasing in the memory access basic step time components of N_{step}, and this results in a smaller T_{ex}. The application of different memory systems results in a memory hierarchy concept as will be studied later. The nanocomputer memory hierarchy is shown in Figure 148.2. As was emphasized, to attain efficiency and high performance, the main memory should not store all programs and data. Specifically, caches are used. Furthermore, virtual memory, which has the largest capacity but the slowest access time, is used. Segments of a program are transferred from the virtual memory to the main memory for execution. As other segments are needed, they may replace the segments existing in the main memory when the main memory is full. The sequential controlled movement of large program and data between the cache, main, and virtual memories, as programs execute, is managed by a combination of operating system software and control hardware. This is called the memory management.

Using the memory hierarchy illustrated in Figure 148.2, the CPU communicates directly only with $M1$, and $M1$ communicates with $M2$, and so on. Therefore, for the CPU to assess the information, stored in the memory M_j, requires the sequence of j data transfer as given by

$$M_{j-1}:=M_j,\ M_{j-2}:=M_{j-1},\ \ldots\ M_1:=M_2,\ \text{CPU}:=M_1.$$

However, the memory bypass can be implemented and effectively used.

To perform computing, specific programs consisting of a set of machine instructions are stored in the main memory. Individual instructions are fetched from the memory into the processor for execution. Data used as operands are also stored in the memory. The connection between the main memory and the processor that guarantees the transfer of instructions and operands is called the bus. A bus consists of a set of address, data, and control lines. The bus permit transfer program and data files from their long-term location (virtual memory) to the main memory. Communication with other computers is ensured by transferring the data through the bus. Normal execution of programs may be preempted if some IO device requires urgent control. To perform this, specific programs are executed, and the device sends an interrupt signal to the processor. The processor temporarily suspends the program that is being executed, and executes the special interrupt service routine instead. After providing the required interrupt service, the processor switches back to the interrupted program. During program loading and execution, the data should be transferred between the main memory and secondary memory. This is performed using the direct memory access.

Nanocomputers can be classified using different classification principles. For example, making use of the multiplicity of instruction and data streams, the following classification can be applied:

1. single instruction stream/single data stream – conventional word-sequential architecture including pipelined nanocomputers with parallel arithmetic logic unit (ALU)
2. single instruction stream/multiple data stream – multiple ALU architectures, e.g., parallel-array processor (ALU can be either bit-serial or bit-parallel)
3. multiple instruction stream/single data stream
4. multiple instruction stream/multiple data stream – the multiprocessor system with multiple control unit

The execution of most operations is performed by the ALU. In the ALU, the logic nanogates and nanoregisters used to perform the basic operations (addition, subtraction, multiplication, and division) of numeric operands, and the comparison, shifting, and alignment operations of general forms of numeric and nonnumeric data. The processors contain a number of high-speed registers, which are used for temporary storage of operands. Register, as a storage device for words, is a key sequential component, and registers are connected. Each register contains one word of data and its access time at least 10 times faster than the main memory access time. A register-level system consists of a set of registers connected by combinational data-processing and data-processing nanoICs. Each operation is implemented as given by the following statement

$$\texttt{cond}: X:=f(x_1, x_2, \ldots, x_{i-1}, x_i).$$

Thus, when the condition cond holds, compute the combinational function of f on $x_1, x_2, \ldots, x_{i-1}, x_i$ and assign the resulting value to X. Here, cond is the control condition prefix which denotes the condition that must be satisfied; $X, x_1, x_2, \ldots, x_{i-1}, x_i$ are the data words or the registers that store them; f is the function to be performed within a single clock cycle.

Suppose that two numbers located in the main memory should be multiplied, and the sum must be stored back into the memory. Using instructions, determined by the control unit, the operands are first fetched from the memory into the processor. They are then multiplied in the ALU, and the result is stored back in memory. Various nanoICs can be used to execute data-processing instructions. The complexity of ALU is determined by the arithmetic instruction implementation. For example, ALUs that perform fixed-point addition, subtraction, and word-based logics can be implemented as combinational nanoICs. The floating-point arithmetic requires complex implementation, and arithmetic coprocessors to perform complex numerical functions needed. The floating-point numerical value of a number X is (X_m, X_e), where X_m is the mantissa and X_e is the fixed-point number. Using the base b (usually, $b = 2$), we have $X = X_m \times b^{X_e}$. Therefore, the general basic operations are quite complex and some problems (biasing, overflow, underflow, etc.) must be resolved.

Memory–processor interface — A memory unit that integrates different memory systems stores the information (data). The processor accesses (reads or loads) the data from the memory systems, performs

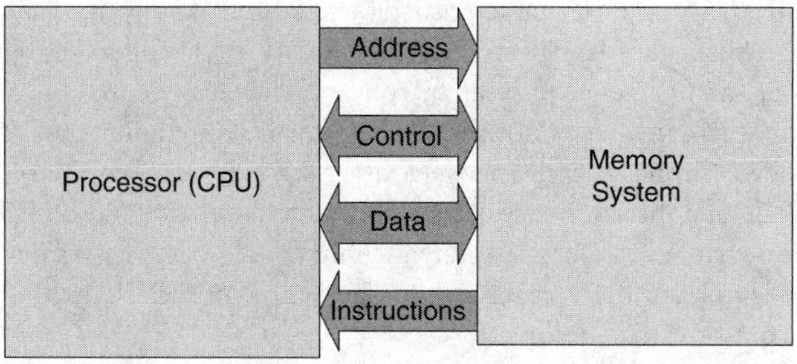

FIGURE 148.3 Memory–processor interface.

computations, and stores (writes) the data back to memory. The memory system is a collection of storage locations. Each storage location (memory word) has a numerical address. A collection of storage locations forms an address space. Figure 148.3 documents the data flow and its control, representing how a processor is connected to a memory system via address, control and data interfaces. High-performance memory systems should be capable to serve multiple requests simultaneously, particularly for vector nanoprocessors.

When a processor attempts to load or read the data from the memory location, the request is issued, and the processor stalls while the request returns. While nanocomputers can operate with overlapping memory requests, the data cannot be optimally manipulated if there are long memory delays. Therefore, a key performance parameter in the design of nanocomputer is the effective speed of its memory. The following limitations are imposed on any memory systems: the memory cannot be infinitely large, cannot contain an arbitrarily large amount of information, and cannot operate infinitely fast. Hence, the major characteristics are speed and capacity.

The memory system performance is characterized by the latency (τ_l) and bandwidth (B_w). The memory latency is the delay as the processor first requests a word from memory until that word arrives and is available for use by the processor. The bandwidth is the rate at which information can be transferred from the memory system. Taking note of the number of requests that the memory can service concurrently $N_{request}$, we have $B_w = \dfrac{N_{request}}{\tau_l}$. Using nanoICs, it become feasible to design and build superior memory systems with desired capacity, low latency, and high bandwidth approaching the physical limits. Using nanoICs, it will be possible to match the memory and processor performance characteristics.

Memory hierarchies provide decreased average latency and reduced bandwidth requirements, whereas parallel memories provide higher bandwidth. As was emphasized, nanocomputer architectures that utilize a small and fast memory located in front of a large but relatively slow memory, can tremendously enhance speed and memory capacity by making use of nanoICs. This results in application of registers in the CPU, and most commonly accessed variables should be allocated at registers. A variety of techniques, employing either hardware, software, or a combination of hardware and software, are employed to ensure that most references to memory are fed by the faster memory. The locality principle is based on the fact that some memory locations are referenced more often than others. The implementation of spatial locality, due to the sequential access, provides one with the property that an access to a given memory location increases the probability that neighboring locations will soon be accessed. Making use of the frequency of program looping behavior, temporal locality ensures the access to a given memory location increasing the probability that the same location will be accessed again soon. It is evident that if a variable was not referenced for a while, it is unlikely that this variable will be needed soon.

As illustrated in Figure 148.2, the top of the memory hierarchy level are the superior speed CPU registers. The next level in the hierarchy is a high-speed cache memory. The cache can be divided into multiple levels, and nanocomputers will likely have multiple caches that can be fabricated on the CPU nanochip. Below the cache memory is the slower but lager main memory, and then, the large virtual memory which is slower than the main memory. These level can be designed and fabricated utilizing different naoICs. Three performance characteristics (access time, bandwidth, and capacity) and many factors (affordability, robustness, adaptability, reconfigurability, etc.) support the application of multiple levels of cache memory and the memory hierarchy concept. The time needed to access the primary cache should match with the clock frequency of the CPU, and the corresponding nanoICs must be used. We place a smaller first-level (primary) cache above a larger second-level (secondary) cache. The primary cache is accessed quickly, and the secondary cache holds more data close to the CPU. The nanocomputer architecture, build using nanoelectronics integrated within nanoICs assemblies, depends on the technologies available. For example, primary cache can be fabricated on the CPU chip, while the secondary caches can be on- or out-of-chip solution.

Size, speed, latency, bandwidth, power consumption, robustness, affordability, and other performance characteristics are examined to guarantee the desired overall nanocomputer performance based upon the specifications imposed. The performance parameter, which can be used to quantitatively examine different memory systems, is the effective latency τ_{ef}. We have

$$\tau_{ef} = \tau_{hit} R_{hit} + \tau_{miss}(1 - R_{hit}),$$

where τ_{hit} and τ_{miss} are the hit and miss latencies; R_{hit} is the hit ratio, $R_{hit} < 1$.

If the needed word is found in a level of the hierarchy, it is called a hit. Correspondingly, if a request must be sent to the next lower level, the request is said to be a miss. The miss ratio is given as $R_{miss} = (1 - R_{hit})$. These R_{hit} and R_{miss} are strongly influenced by the program being executed and influenced by the high-/low-level memory capacity ratio. The access efficiency E_{ef} of multiple-level memory ($i - 1$ and i) is found using the access time, hit and miss ratios. In particular,

$$E_{ef} = \frac{1}{\dfrac{t_{\text{access time}\,i-1}}{t_{\text{access time}\,i}} R_{miss} + R_{hit}}$$

The hardware can dynamically allocate parts of the cache memory for addresses likely to be accessed soon. The cache contains only redundant copies of the address space. The cache memory can be associative or content-addressable. In an associative memory, the address of a memory location is stored along with its content. Rather than reading data directly from a memory location, the cache is given an address and responds by providing data which might or might not be the data requested. When a cache miss occurs, the memory access is then performed from main memory, and the cache is updated to include the new data. The cache should hold the most active portions of the memory, and the hardware dynamically selects portions of main memory to store in the cache. When the cache is full, some data must be transferred to the main memory or deleted. Therefore, a strategy for cache memory management is needed. Cache management strategies are based on the locality principle. In particular, spatial (selection of what is brought into the cache) and temporal (selection of what must be removed) localities are embedded. When a cache miss occurs, hardware copies a contiguous block of memory into the cache, which includes the word requested. This fixed-size memory block can be small (bit or word) or hundreds of bytes. Caches can require all fixed-size memory blocks to be aligned. When a fixed-size memory block is brought into the cache, it is likely that another fixed-size memory block must be removed. The selection of the removed fixed-size memory block is based on effort to capture temporal locality. In general, this is difficult to achieve. Correspondingly, viable methods are used to predict future memory accesses. A least-recently-used concept can be the preferred choice for nanocomputers.

The cache can integrate the data memory and the tag memory. The address of each cache line contained in the data memory is stored in the tag memory (the state can also track which cache lines is modified). Each line contained in the data memory is allocated by a corresponding entry in the tag memory to indicate the full address of the cache line. The requirement that the cache memory be associative (content-addressable) complicates the design because addressing data by content is more complex than by its address (all tags must be compared concurrently). The cache can be simplified by embedding a mapping of memory locations to cache cells. This mapping limits the number of possible cells in which a particular line may reside. Each memory location can be mapped to a single location in the cache through direct mapping. Although there is no choice of where the line resides and which line must be replaced, however, poor utilization results. In contrast, a two-way set-associative cache maps each memory location into either of two locations in the cache. Hence, this mapping can be viewed as two identical directly-mapped caches. In fact, both caches must be searched at each memory access, and the appropriate data selected and multiplexed on a tag match — hit and on a miss. Then, a choice must be made between two possible cache lines as to which is to be replaced. A single least-recently-used bit can be saved for each such pair of lines to remember which line has been accessed more recently. This bit must be toggled to the current state each time. To this end, an M-way associative cache maps each memory location into M memory locations in the cache. Therefore, this cache map can be constructed from M identical direct-mapped caches. The problem of maintaining the least-recently-used ordering of M cache lines is primarily due to the fact that there are $M!$ possible orderings. In fact, it takes at least $\log_2 M!$ bits to store the ordering. It can be envisioned that two-, three- or four-way associative cache will be implemented in nanocomputers.

Consider the write operation. If the main memory copy is updated with each write operation, then, write-through or store-through technique is used. If the main memory copy is not updated with each write operation, write-back or copy-back or deferred writes algorithm is enforced. In general, the cache coherence or consistency problem must be examined due to implementation of different bypass techniques.

Parallel Memories — Main memories can comprise a series of memory nanochips or nanoICs on a single nanochip. These nanoICs form a *nanobank.* Multiple memory *nanobanks* can be integrated together to form a parallel main memory system. Since each *nanobank* can service a request, a parallel main memory system with N_{mb} *nanobanks* can service N_{mb} requests simultaneously, increasing the bandwidth of the memory system by N_{mb} times the bandwidth of a single *nanobank*. The number of *nanobank* is a power of two, that is, $N_{mb} = 2^p$. An n-bit memory word address is broken into two parts: a p-bit *nanobank* number and an m-bit address of a word within a *nanobank*. The p bits used to select a *nanobank* number could be any p bits of the n-bit word address. Let us use the low-order p address bits to select the *nanobank* number, and the higher order $m = n - p$ bits of the word address is used to access a word in the selected *nanobank*.

Multiple memory *nanobank* can be connected using *simple paralleling* and *complex paralleling.* Figure 148.4 shows the structure of a simple parallel memory system where m address bits are simultaneously supplied to all memory *nanobanks*. All *nanobanks* are connected to the same read/write control line. For a read operation, the *nanobanks* perform the read operation and deposit the data in the latches. Data can then be read from the latches one by one by setting the switch appropriately. The *nanobanks* can be accessed again to carry out another read or write operation. For a write operation, the latches are loaded one by one. When all latches have been written, their contents can be written into the memory *nanobanks*

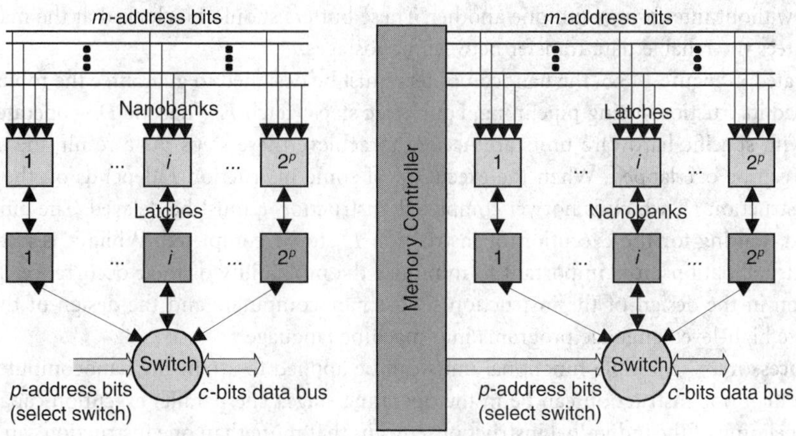

FIGURE 148.4 Simple and complex parallel main memory systems.

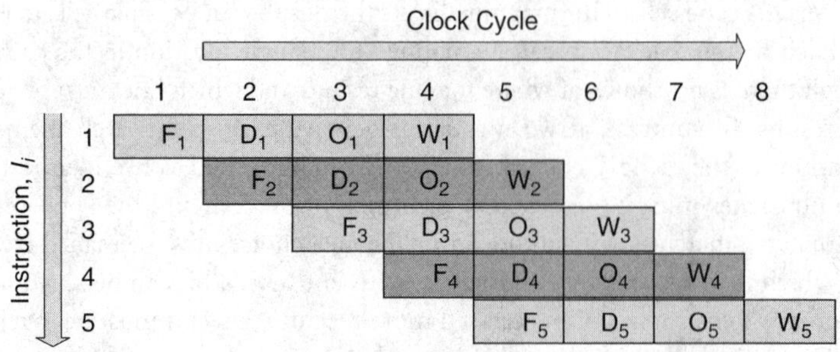

FIGURE 148.5 Pipelining of instruction execution.

by supplying *m* bits of address. In a simple parallel memory, all *nanobanks* are cycled at the same time. Each *nanobank* starts and completes its individual operations at the same time as every other *nanobank*, and a new memory cycle starts for all *nanobanks* once the previous cycle is complete.

A complex parallel memory system is documented in Figure 148.4. Each *nanobank* is set to operate on its own independent of the operation of the other *nanobanks*. For example, *i*th *nanobank* performs a read operation on a particular memory address, while $(i + 1)$th *nanobank* performs a write operation on a different and unrelated memory address. Complex paralleling is achieved using the address latch and a read/write command line for each *nanobank*. The *memory controller* handles the operation of the complex parallel memory. The processing unit submits the memory request to the memory controller, which determines which *nanobank* needs to be accessed. The controller then determines if the *nanobank* is busy by monitoring a busy line for each *nanobank*. The controller holds the request if the *nanobank* is busy, submitting it when the *nanobank* becomes available to accept the request. When the *nanobank* responds to a read request, the switch is set by the controller to accept the request from the *nanobank* and forward it to the processing unit. It can be foreseen that complex parallel main memory systems will be implemented is in vector nanoprocessors. If consecutive elements of a vector are present in different memory *nanobank*, then the memory system can sustain a bandwidth of one element per clock cycle. Memory systems in nanocomputers can have hundreds of *nanobanks* with multiple memory controllers that allow multiple independent memory requests at every clock cycle.

Pipelining — Pipelining is a technique to increase the processor throughput with limited hardware in order to implement complex *datapath* (data processing) units (multipliers, floating-point adders, etc.). In general, a pipeline processor integrates a sequence of *i* data-processing nanoICs (nanostages) which cooperatively perform a single operation on a stream of data operands passing through them. Design of pipelining nanoICs involves deriving multistage balanced sequential algorithms to perform the given function. Fast buffer registers are placed between the nanostages to ensure the transfer of data between nanostages without interfering with one another. These buffers should be clocked at the maximum rate that guarantees the reliable data transfer between nanostages.

As illustrated in Figure 148.5, the nanocomputers must be designed to guarantee the robust execution of overlapped instructions using pipelining. Four basic steps (fetch F_i – decode D_i – operate O_i – and – write W_i) with specific hardware units are needed to achieve these steps. As a result, the execution of instruction can be overlapped. When the execution of some instruction I_i depends on the results of a previous instruction I_{i-1} which is not yet completed, instruction I_i must be delayed. The pipeline is said to be stalled, waiting for the execution of instruction I_{i-1} to be completed. While it is not possible to eliminate such situations, it is important to minimize the probability of their occurrence. This is a key consideration in the design of the instruction set for nanocomputers and the design of the compilers that translate high-level language programs into machine language.

Multiprocessors — Multiple functional units can be applied to attain the nanocomputer operation when more than one instruction can be in the operating stage. The parallel execution capability, when added to pipelining of the individual instructions, means that more than one instruction can be executed per basic step. Thus, the execution rate can be increased. This enhanced processor performance is called

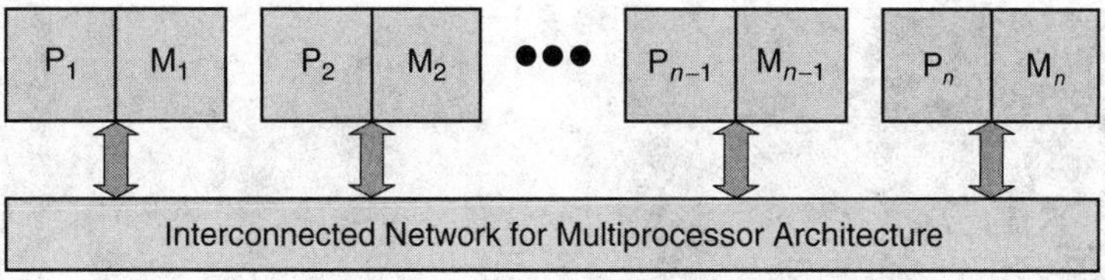

FIGURE 148.6 Multiprocessor architecture. The basic parallel organization for nanocomputers is represented in Figure 148.7.

superscalar processing. The rate R_T of performing basic steps in the processor depends on the processor clock rate. This rate is in the order of billions steps per second in current high-performance nanoICs. It was illustrated that physical limits prevent single processors from being speeded up indefinitely. Nanocomputers with multiprocessors will speed up the execution of large programs by executing subtasks in parallel. The main difficulty in achieving this is decomposition of given task into its parallel subtasks and ordering these subtasks to the individual processors in such a way that communication among the subtasks will performed efficiently and robustly. Figure 148.6 documents a block diagram of a multiprocessor system with the interconnection network needed for data sharing among the processors P_i. Parallel paths are needed in this network in order to parallel activity to proceed in the processors as they access the global memory space represented by the multiple memory units M_i.

Nanocomputer Architectronics — The theory of computing, computer architecture, and networking are the study of efficient robust processing and communication, modeling, analysis, optimization, adaptive networks, architecture and organization synthesis, as well as other problems of hardware and software design. These studies have emerged as a synergetic fundamental discipline (computer engineering and science), and many problems have not been solved yet. Correspondingly, a large array of questions is still unanswered. Nanocomputer *architectonics* is the theory of nanocomputers devised and designed using fundamental theory and applying nanoICs. Our goal is to develop the nanocomputer *achitectronics* basics, e.g., fundamental methods in synthesis, design, analysis, modeling, computer-aided-design, etc. Nanocomputer *achitectronics*, which is a computer science and engineering frontier, will allow one to solve a wide spectrum of fundamental and applied problems. Applying the nanocomputer *achitectronics* paradigm, this chapter spans the theory of nanocomputers and related areas, e.g., information processing, communication paradigms, computational complexity theory, combinatorial optimization, architecture synthesis, optimal organization, and their applications. Making of the nanocomputer *achitectronics* paradigms, one can address and study fundamental problems in nanocomputer architecture synthesis, design and optimization applying three-dimensional organization, application of nanoICs, multithreading, error recovery, massively parallel computing organization, shared memory parallelism, message passing parallelism, etc. The nanocomputer fundamentals, operation, and functionality can be devised through neuroscience. The key to understand learning, adaptation, control, architecture, hierarchy, organization,

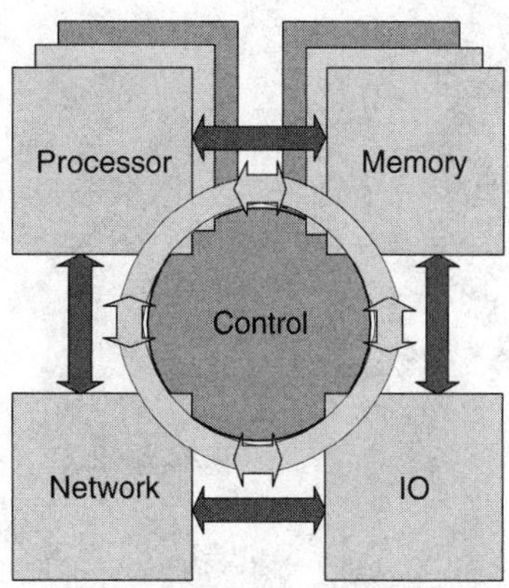

FIGURE 148.7 Parallel nanocomputer organization to maximize the computer concurrency.

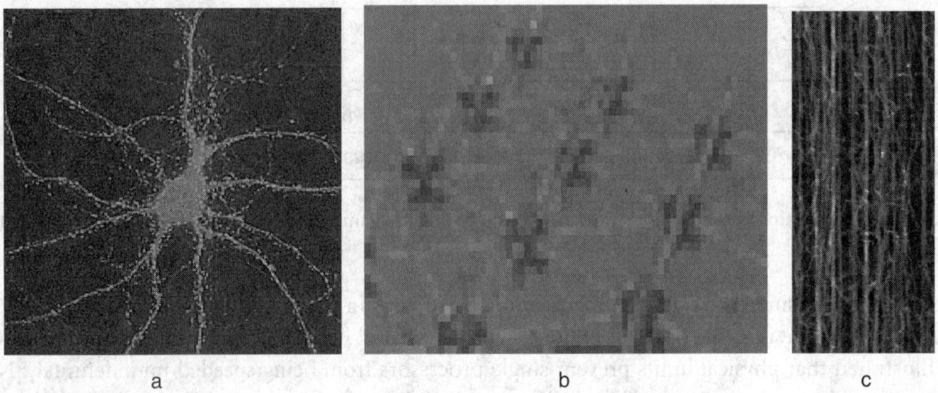

FIGURE 148.8 *Vertebrate neuron* (soma, axon with synaptic terminals, dendrites, and synapses), three-dimensional nanoICs, aligned carbon nanotubes, and single C_{60} buckyballs on Si.

memory, intelligence, diagnostics, self-organization, computing and other system-level basics lies in the study of phenomena and effects in the central nervous system, its components (brain and spinal cord), and the fundamental building blocks, e.g., neurons. Neuronal cells have a large number of synapses. A typical nerve cell in the human brain has thousands of synapses which establish communication and information processing. The communication and processing are not fixed, and constantly change and adapt. Neurons function in the hierarchically distributed robust adaptive network manner. During information transfer, some synapses on a cell are selectively triggered to release specific neurotransmitters, while other synapses remain passive. Neurons consist of a cell body with a nucleus (soma), axon (which transmits information from the soma), and dendrites (which transmit information to the soma), see Figure 148.8. It becomes possible to implement three-dimensional nanoICs structures using biomematic analogies as illustrated in Figure 148.8. For example, the complex inorganic dendrite-like trees can be implemented using the carbon nanotube-base technology (the Y-junction branching carbon nanotube networks ensure robust ballistic switching and amplification behavior desired in nanoICs. Other nanodevices are emerged. For example, single isolated C_{60} molecules and multi-layered C_{60} can form nanodevices (superconducting transitions), see Figure 148.8. Unfortunately, due to low temperature, the buckyball C_{60} solution in nanodevices is not feasible. However, the carbon-based technology is progressed to possible application in nanoICs within three-dimensional architectures.

Two- and three-dimensional topology aggregation — There is a critical need to study the topology aggregation in the hierarchical distributed nanocomputer that will utilize two- and three-dimensional nanoICs. One can examine how a hierarchical network with multiple paths and routing domain functions. Figure 148.9 shows the principle of topology aggregation of nanodevices in nanoICs for nanocomputers. For example, the documented network consists of four routers (buffers) numbered 1, 2, 3, and 4, and

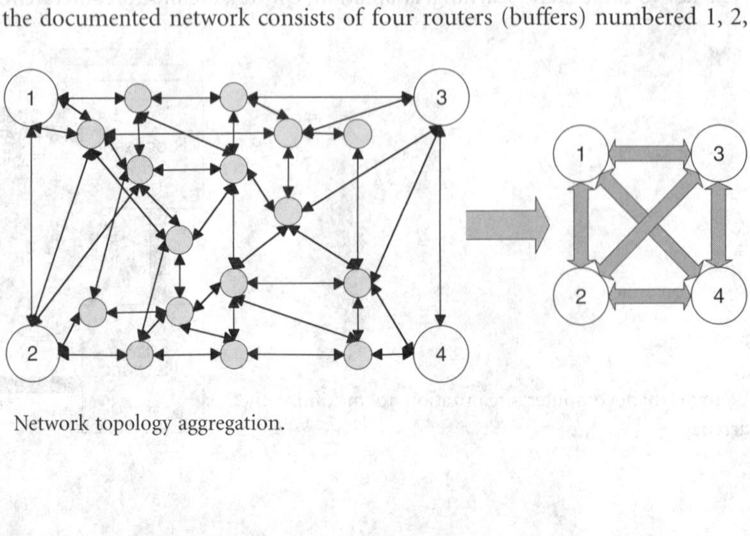

FIGURE 148.9 Network topology aggregation.

interior (core) routers are illustrated. After the aggregation, a meshed network with four nodes represents the same topology. The difficulty in topology aggregation is to calculate the link metrics to attain the accuracy and to synthesize optimized network topologies. The aggregated link metrics can be computed using the methods used for multiplayer artificial networks. The bandwidth and delay are integrated to optimize the hierarchically distributed networks. We believe that the *distributed-memory* nanocomputers can be adaptively configured and reconfigured based upon *processing*, *communication*, and *control* (instruction) *parallelisms*.

148.3 Nanocomputer Architecture

System-level performance analysis is based on mathematical models used to examine nanocomputers and optimize their architectures, organization, topologies, and parameters, as well as to perform optimal hardware-software codesign. Computers process inputs, producing outputs accordingly. To examine the performance, one should develop and apply the mathematical model of a nanocomputer which comprises central processing, memory, input – output, control, and other units. However, one cannot develop the mathematical model without explicitly specifying the nanocomputer architecture, identifying the optimal organization, and synthesizing the topologies. There are different levels of abstraction for nanocomputer architectures, organizations, models, etc.

Advanced nanocomputer architectures (beyond Von Neumann computer architecture) can be devised to guarantee superior processing and communication speed. Novel nanodevices that utilize new effects and phenomena to perform computing and communication are sought. For example, through the quantum computing, the information can be stored in the phase of wavefunctions, and this concept leads to utilization of massive parallelism of these wavefunctions to achieve enormous processing speed.

The central processing unit (CPU) executes sequences of instructions and operands, which are fetched by the program control unit (PCU), executed by the data processing unit (DPU), and, then, placed in memory. In particular, caches (high speed memory where data is copied when it is retrieved from the random access memory improving the overall performance by reducing the average memory access time) are used. The instructions and data form instruction and data streams which flow to and from the processor. The CPU may have two or more processors and coprocessors with various execution units and multi-level instruction and data caches. These processors can share or have their own caches. The CPU *datapath* contains nanoICs to perform arithmetic and logical operations on words such as fixed- or floating-point numbers. The CPU design involves the trade-offs analysis between the hardware, speed, and affordability. The CPU is usually partitioned on the control and *datapath* units, and the control unit selects and sequence the data-processing operations. The core interface unit is a switch that can be implemented as autonomous cache controllers operating concurrently and feeding the specified number (32, 64, or 128) of bytes of data per cycle. This core interface unit connects all controllers to the data or instruction caches of processors. Additionally, the core interface unit accepts and sequences information from the processors. A control unit is responsible for controlling data flow between controllers that regulate the *in* and *out* information flows. There is the interface to input/output devices. On-chip debug, error detection, sequencing logic, self-test, monitoring, and other units must be integrated to control a pipelined nanocomputer. The computer performance depends on the architecture and hardware components (which are discussed in this Chapter). Figure 148.10 illustrates the possible nanocomputer organization.

In general, nanodevices ensure high density, superior bandwidth, high switching frequency, low power, etc. It is envisioned that in the near future nanocomputers will allow one to increase the computing speed by a factor of millions compared with the existing CMOS ICs computers. Three-dimensional multiple-layered high-density nanoIC assemblies, shown in Figure 148.11, are envisioned to be used. Unfortunately, the number of formidable fundamental, applied, experimental, and technological challenges arise, e.g., robust operation and characteristics of nanoICs are significantly affected by the "second-order" effects (gate oxide and bandgap tunneling, energy quantization, electron transport, etc., and, furthermore, the operating principles for nanodevices can be based on the quantum effects), noise vulnerability, complexity, etc. It is well-known that high-fidelity modeling, data-intensive analysis, het-

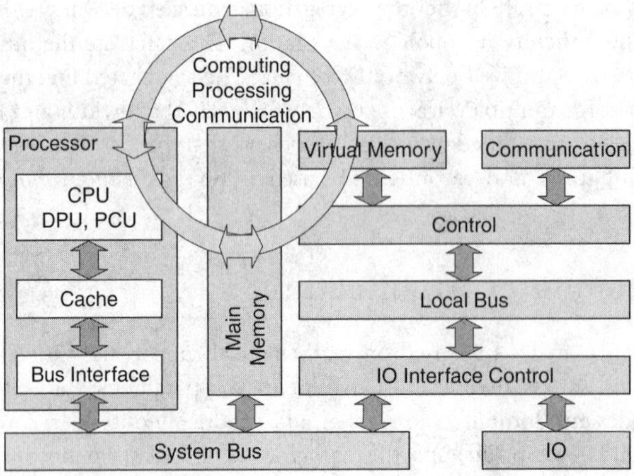

FIGURE 148.10 Nanocomputer organization.

erogeneous simulation, and other fundamental issues even for a single nanodevice are not completely solved yet. In addition, the currently existing fabrication processes and technologies do not allow one to fabricate ideal nanodevices and nanostructures. In fact, even molecular wires are not perfect. Different fabrication technologies, processes, and materials have been developed to attain self-assembly and self-ordered features in fabrication of nanoscale devices [7–9], see Figure 148.11. As example, the self-assembled and aligned carbon nanotubes array is illustrated in Figure 148.11.

One of the basic component of the current computers is CMOS transistors fabricated using different technologies and distinct device topologies. However, the CMOS technology and transistors fabricated by using even most advanced CMOS processes reach the physical limits. Therefore, the current research developments have been concentrated on the alternative solutions, and leading companies (IBM, Intel, Hewlett-Packard, etc.), academia, and government laboratories develop novel nanotechnologies. We study computer hardware. Nanodevices (switches, logics, memories, etc.) can be implemented using the illustrated in Figure 148.11 three-dimensional nanoelectronics arrays. It must be emphasized that the

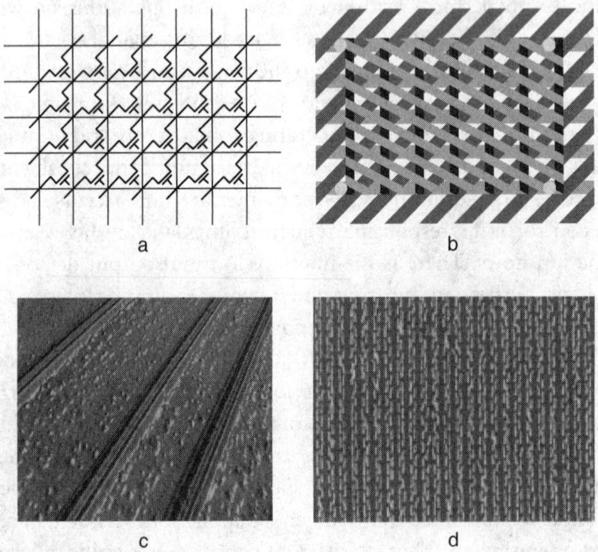

FIGURE 148.11 Three-dimensional multiple-layered high-density nanoIC assemblies (crossbar switching, logic, or memory arrays), 3 nm wide parallel (six-atom-wide) erbium disilicide ($ErSi_2$) nanowires (Hewlett-Packard), and carbon nanotube array.

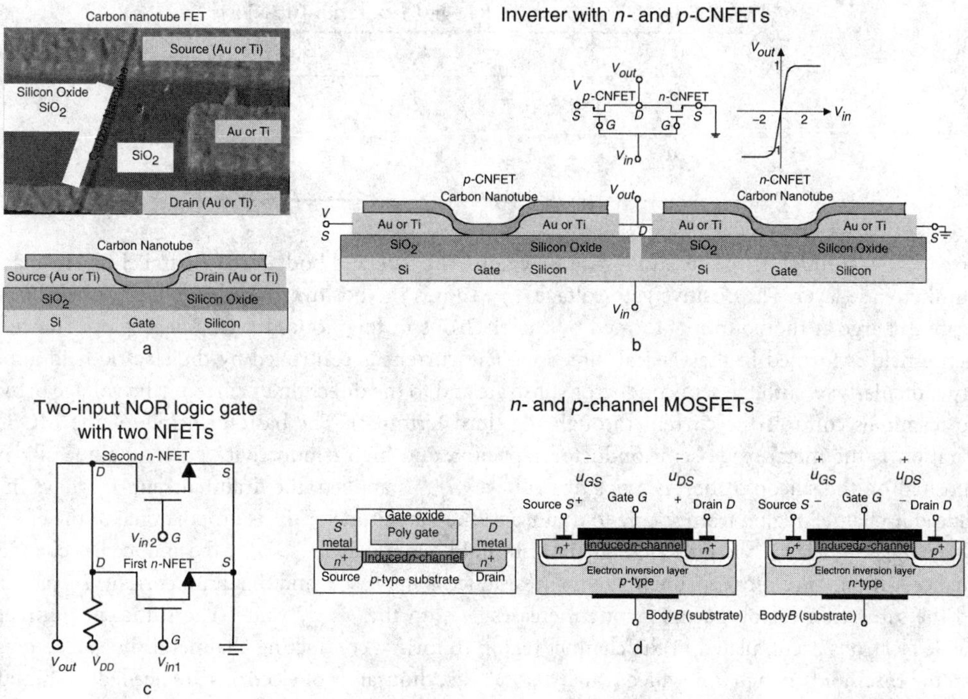

FIGURE 148.12 (a) Carbon nanotube FETs; (b) inverter with CNFETs; (c) NOR gate with NFETs; (d) *n*- and *p*-channel MOSFETs.

extremely high-frequency logic gates can be fabricated using carbon nanotubes, which are from 1 to 10 nm in diameter (100,000 times less then the diameter of human hair). *P*- and *n*-type carbon nanotube field-effect transistors (CNFETs) with single- and multi-wall carbon nanotubes as the channel were fabricated and tested [11]. The atomic force microscope image of a single-wall CNFET (50 nm total length) and CNFET are documented in Figure 148.12.a.

It should be emphasized that the two-dimensional carbon nanotube structure can be utilized to devise and built different transistors with distinct characteristics utilizing different phenomena [10, 11]. For example, twisted carbon nanotubes can be used. Carbon nanotubes can be grown on the surface using chemical vapor deposition, deposited on the surface from solvent, etc. Photolithography can be used to attain the device-level structural and functional integration connecting source, drain, gate, etc. One concludes that different transistors topologies and configurations are available, and these results are reported in [11–13]. Taking note of this fact, we use nanoscale field-effect transistors (NFET) to synthesize and analyze the nanoICs. The carbon nanotube inverter, formed using the series combination of *p*- and *n*-CNFETs, is illustrated in Figure 148.12.b. The gates and drains of two CNFETs are connected together to form the input and output. The voltage characteristics can be examined studying the various transistor bias regions. When the inverter input voltage V_{in} is either a logic 0 or a logic 1, the current in the circuit is zero because one of the CNFETs is cut off. When the input voltage varies in the region $V_{threshold} < V_{in} < V - |V_{threshold}|$, both CNFETs are conducting and a current exists in the inverter.

The current-control mechanism of the field-effect transistors is based on an electrostatic field established by the voltage applied to the control terminal. Figure 148.12.d shows an *n*- and *p*-channel enhancement-type MOSFETs with four terminals (the gate, source, drain, and base (substrate) terminals are denoted as G, S, D, and B). Consider the *n*-channel enhancement-type MOSFETs. A positive voltage u_{GS}, applied to the gate, causes the free positively charged holes in the *n*-channel region. These holes are pushed downward into the *p*-base (substrate). The applied voltage u_{GS} attracts electrons from the source and drain regions into the channel region. The voltage is applied between the drain and source, and the

TABLE 148.1 Two-input NOR Logic Circuit with Two NFETs

V_{in1}	V_{in2}	V_{out}
0	0	1
1	0	0
0	1	0
1	1	0

current flows through the induced *n*-channel region. The gate and body form a parallel-plate capacitor with the oxide layer. The positive gate voltage u_{GS} causes the positive charge at the top plate, and the negative charge at the bottom is formed by the electrons in the induced *n*-channel region. Hence, the electric field is formed in the vertical direction. The current is controlled by the electric field applied perpendicularly to both the semiconductor substrate and to the direction of current (the voltage between two terminals controls the current through the third terminal). The basic principle of the MOSFET operation is the metal-oxide-semiconductor capacitor, and high-conductivity polycrystalline silicon is deposited on the silicon oxide. As a positive voltage u_{DS} is applied, the drain current i_D flows in the induced *n*-channel region from source to drain, and the magnitude of i_D is proportional to the effective voltage $u_{GS} - u_{GSt}$. If u_{GS} is greater than the threshold value u_{GSt}, $u_{GS} > u_{GSt}$, the induced *n*-channel is enhanced. If one increases u_{DS}, the *n*-channel resistance is increased, and the drain current i_D saturates, and the saturation region occurs as one increases u_{DS} to the u_{DSsat} value. The sufficient number of mobile electrons accumulated in the channel region to form a conducting channel if the u_{GS} is greater than the threshold voltage u_{GS}, which usually is 1 V, e.g., thousands of electrons are needed. It should be emphasized that in the saturation region, the MOSFET is operated as an amplifier, while in the triode region and in the cut off region, the MOSFET can be used as a switch. A *p*-channel enhancement-type MOSFETs are fabricated on the *n*-type substrate (body) with p^+ regions for the drain and source. Here, the voltages applied u_{GS} and u_{DS}, and the threshold voltage u_{GSt} are negative.

We use and demonstrate the application of nanoscale field-effect transistors (NFETs) to design the logic nanoICs. The NOR logic can be straightforwardly implemented when the first and second *n*-NFETs are connected in parallel, see Figure 148.12c. Different flip-flops usually are formed by cross-coupling NOR logic gates. If two input voltages are zero, both *n*-NFET are cut off, and the output voltage V_{out} is high. Specifically, $V_{out} = V_{DD}$, and if $V_{DD} = 1$ V, then $V_{out} = 1$ V. If $V_{in1} \neq 0$ (for example, $V_{in1} = 1$ V) and $V_{in2} = 0$, the first *n*-NFET turns on and the second *n*-NFET is still cut off. The first *n*-NFET is biased in the nonsaturation region, and V_{out} reaches the low value ($V_{out} = 0$). By changing the input voltages such as $V_{in1} = 0$ and $V_{in2} \neq 0$ ($V_{in2} = 1$ V), the first *n*-NFET becomes cut off, while the second *n*-NFET is biased in the nonsaturation region. Hence, V_{out} has the low value, $V_{out} = 0$. If $V_{in1} \neq 0$ and $V_{in2} \neq 0$ ($V_{in1} = 1$ V and $V_{in2} = 1$ V), both *n*-NFET become biased in the nonsaturation region, and V_{out} is low ($V_{out} = 0$). Table 148.1 summarizes the result and explains how the NFET can be straightforwardly used in nanocomputer units.

The series-parallel combination of the NFETs is used to synthesize complex logic gates. As an example, the resulting carbon nanotube circuitry using *n*-NFETs to implement the Boolean output function $V_{out} = \overline{f(A \cdot B + C)}$ is illustrated in Figure 148.13. The *n*-NFETs executive OR logic gate $V_{out} = A \otimes B$ can be made, see Figure 148.13. If $A = B =$ logic 1, the path exists from the output to ground trough NFET

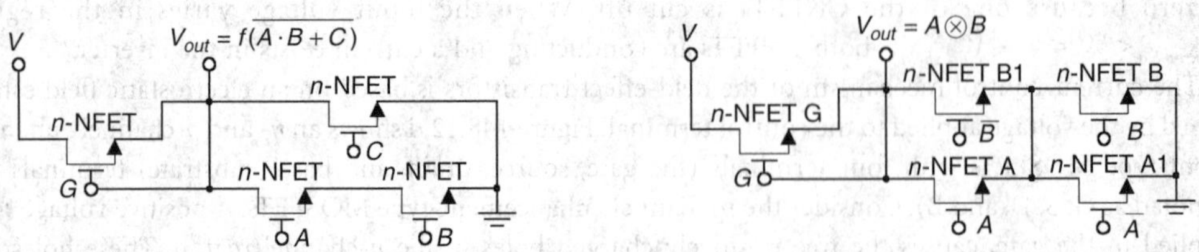

FIGURE 148.13 Static logic gates $V_{out} = \overline{f(A \cdot B + C)}$ and $V_{out} = A \otimes B$ synthesized using *n*- and *p*-NFETs.

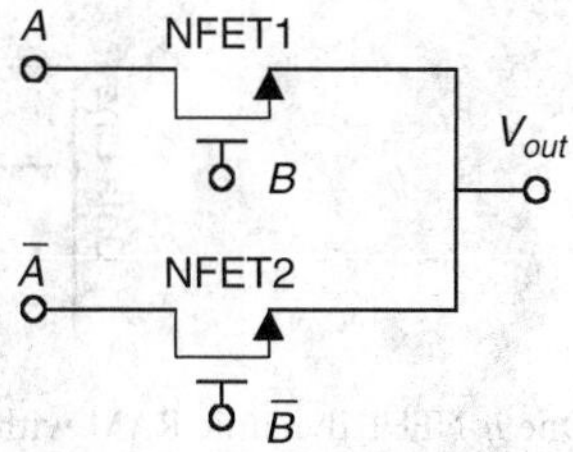

FIGURE 148.14 NanoICs pass network.

TABLE 148.2 Two-input NOR Logic Circuit with Two NFETs

	Input Gate Control						
State	A	$\overline{A}$	B	$\overline{B}$	NFET1	NFET2	V_{out}
1	0	1	0	1	off	on	1
2	1	0	0	1	off	on	0
3	0	1	1	0	on	off	0
4	1	0	1	0	on	off	1
5	0	1	0	1	off	on	1
6	1	0	0	1	off	on	0
7	0	1	1	0	on	off	0
8	1	0	1	0	on	off	1

A and NFET B transistors, and the output goes low. If $A = B =$ logic 0 ($\overline{A} = \overline{B} =$ logic 1), the path exists from the output to ground trough NFET A1 and NFET B1 transistors, and the output goes low. For all other input logic signal combinations, the output is isolated from ground, and, hence, the output goes high. Two logic gates $V_{out} = f(A \cdot B + C)$ and $V_{out} = A \otimes B$ synthesized are the static nanoICs (the output voltage is well-defined and is never left floating). The static nanoICs can be redesigned adding the clock.

The nanoICs pass networks can be easily implemented. Consider the nanoICs with two n-NFETs documented in Figure 148.14. The nonoICs output V_{out} is determined by the conditions listed in Table 2. In states 1 and 2, the transmission gate of NFET2 is biased in its conduction state (NFET2 is on), while NFET1 is off. For state 1, $\overline{A}$ =logic 1 is transmitted to the output, and $V_{out} = 1$. For state 2, $\overline{A} =$ logic 0, and though $A =$ logic 1, due to the fact that NFET1 is off, $V_{out} = 0$. In states 3 and 4, the transmission gate of NFET1 is biased, and NFET2 is off. For state 3, $A =$ logic 0, we have $V_{out} = 0$ because NFET2 is off ($\overline{A} =$ logic 1). In contrast, in state 4, $A =$ logic 1 is transmitted to the output, and $V_{out} = 1$. For other states the results are reported in Table 148.2. We conclude that the output V_{out} is a function of two variables, e.g., gate control and input (logic) signals.

The clocked nanoICs are dynamic circuits that in general precharge the output node to a particular level when the clock is at a logic 0. The generalized clocked nanoICs is illustrated in Figure 148.15,

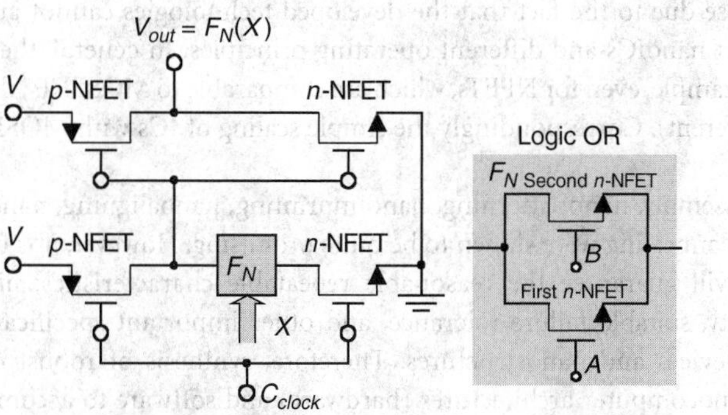

FIGURE 148.15 Dynamic generalized clocked nanoIC with logic function $F_N(X)$.

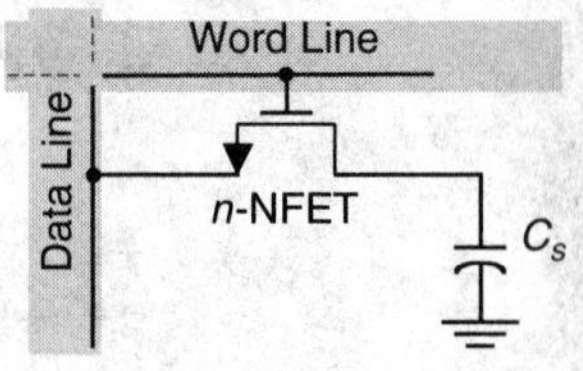

FIGURE 148.16 One n-NFET dynamic RAM with the storage capacitor.

where F_N is the NFETs network that performs the particular logic function $F_N(X)$ of i variables, here $X = (x_1, x_2, \ldots, x_{i-1}, x_i)$. The set of X inputs to the logic nanoICs $F_N(X)$ is derived from the outputs of other static and dynamics nanoICs. When C_{clock} = logic 0, the outputs of all inverters are a logic 0 during the precharged cycle, and during the precharged cycle all x variables of $X = (x_1, x_2, \ldots, x_{z-1}, x_z)$ are logic 0. During the precharge phase, all NFETs are cut off, and the transistor outputs are precharged to V. The transitions are possible only during the evaluation phase. The output of the nonoICs buffer change from 0 to 1. Specifically, the logic OR function is illustrated to demonstrate the generalized clocked nanoICs in Figure 148.15.

Combination of logic gates is used to perform logic functions, addition, subtraction, multiplication, division, multiplexing, etc. To store the information, the memory cell nanoICs are used. A systematic arrangement of memory cells and peripheral nanoICs (to address and write the data into the cells as well as to delete data stored in the cells) constitute the memory. The NFETs can be used to build superior static and dynamic random access memory (RAM is the read-write memory in which each individual cell can be addressed at any time), programmable and alterable read-only memory (ROM is commonly used to store instructions of a system operating systems). The static RAM (implemented using six NFETs) consists a basic flip-flop nanoICs with two stable states (0 and 1), while dynamic RAM (implemented using one NFET and storage capacitor) stores one bit of information charging the capacitor. As an example, the dynamic RAM (DRAM) cell is documented in Figure 148.16. In particular, the binary information is stored as the charge on the storage capacitor C_s (logic 0 or 1). The DRAM cell is addressed by turning on the pass n-NFET via the world line signal S_{wl} and charge is transferred into and out of C_s on the data line (C_s is isolated from the rest of the nanoICs when the n-NFET is off, but the leakage current through the n-NFET requires the cell refreshment to restore the original signal).

Dynamic shift registers are implemented using transmission gates and inverters, flip-flops are synthesized by cross-coupling NOR gates, while delay flip-flops are built using transmission gates and feedback inverters.

Though for many nanoICs (we do not consider the optical, single-electron, and quantum nonoICs) synthesis and design can be performed using the basic approaches, methods and computer-aided design developed for conventional ICs (viable results exist for MOSFETs), the differences must be outlined. The major problems arise due to the fact that the developed technologies cannot guarantee the fabrication of high-yield perfect nanoICs and different operating principles. In general, the secondary effects must be integrated (for example, even for NFETs, which are comparable to MOSFETs, the switching and leakage phenomena are different). Correspondingly the simple scaling of ICs with MOSFETs to nanoICs cannot be performed.

The direct self-assembly, nanopatterning, nanoimprinting, nanoaligning, nanopositioning, overlayering, margining, and annealing were shown to be quite promising. However, it is unlikely that near-future nanotechnologies will guarantee the reasonable repeatable characteristics, high-quality, satisfactory geometry uniformity, suitable failure tolerance, and other important specifications and requirements imposed on nanodevices and nanostructures. Therefore, synthesis of robust defect-tolerant adaptive (reconfigurable) nanocomputer architectures (hardware) and software to accommodate failures, inconsistence, variations, nonuniformity, and defects is critical [10].

148.4 Hierarchical Finite-State Machines and Their Use in Hardware and Software Design

Simple register-level systems perform a single data-processing operations, e.g., summation $X{:=}x_1{+}x_2$, subtraction $X{:}x_1{-}x_2$, etc. To do different complex data-processing operations, multifunctional register-level systems should be synthesized. These multifunctional register-level systems are partitioned as a data-processing unit (*datapath*) and a controlling unit (control unit). The control unit is responsible for collecting and controlling the data-processing operations (actions) of the *datapath*. To design the register-level systems, one studies a set of operations to be executed, and then, designs nanoICs using a set of register-level components that implement the desired functions satisfying the affordability and performance requirements. It is very difficult to impose meaningful mathematical structures on register-level behavior or structure using Boolean algebra and conventional gate-level design. Due to these difficulties, the heuristic synthesis is commonly accomplished as sequential steps listed below:

1. define the desired behavior as a set of sequences of register-transfer operations (each operation can be implemented using the available components) comprising the algorithm to be executed
2. examine the algorithm to determine the types of components and their number to attain the required *datapath*. Design a complete block diagram for the *datapath* using the components chosen
3. examine the algorithm and *datapath* in order to derive the control signals with ultimate goal to synthesize the control unit for the found *datapath* that meet the algorithm's requirements
4. analyze and verify performing modeling and detail simulation (VHDL, Verilog, ModelSim, and other environments are commonly used)

Let us perform the synthesis of virtual control units that ensures extensibility, flexibility, adaptability, robustness, and reusability. The synthesis will be performed using the hierarchic graphs (HGs). A most important problem is to develop straightforward algorithms which ensure implementation (nonrecursive and recursive calls) and utilize hierarchical specifications. We will examine the behavior, perform logic synthesis, and implement reusable control units modeled as hierarchical finite-state machines with virtual states. The goal is to attain the top-down sequential well-defined decomposition in order to develop complex robust control algorithm step-by-step.

We consider *datapath* and control units. The *datapath* unit consists memory and combinational units. A control unit performs a set of instructions by generating the appropriate sequence of micro instructions that depend on intermediate logic conditions or on intermediate states of the *datapath* unit. To describe the evolution of a control unit, behavioral models were developed [27, 28]. We use the direct-connected HGs containing nodes. Each HG has an entry (*Begin*) and an output (*End*). Rectangular nodes contain micro instructions, macro instructions, or both.

A micro instruction U_i includes a subset of micro operations from the set $U = \{u_1, u_2, \ldots, u_{u-1}, u_u\}$. Micro operations $\{u_1, u_2, \ldots, u_{u-1}, u_u\}$ force the specific actions in the *datapath* (see Figure 148.17 and Figure 148.19). For example, one can specify that u_1 pushes the data in the local stack, u_2 pushes the data in the output stack, u_3 forms the address, u_4 calculates the address, u_5 pops the data from the local stack, u_6 stores the data from the local stack in the register, u_7 pops the data from the output stack to external output, etc.

A micro operation is the output causing an action in the *datapath*. Any macro instruction incorporates macro operations from the set $M = \{m_1, m_2, \ldots, m_{m-1}, m_m\}$. Each macro operation is described by another lower level HG. Assume that each macro instruction includes one macro operation. Each rhomboidal node contains one element from the set $L \cup G$. Here, $L = \{l_1, l_2, \ldots, l_{l-1}, l_l\}$ is the set of logic conditions, while $G = \{g_1, g_2, \ldots, g_{g-1}, g_g\}$ is the set of logic functions. Using logic conditions as inputs, logic functions are calculated examining predefined set of sequential steps that are described by a lower level HG. Directed lines connect the inputs and outputs of the nodes.

Consider a set $E = M \cup G$, $E = \{e_1, e_2, \ldots, e_{e-1}, e_e\}$. All elements $e_i \in E$ have HGs, and e_i has the corresponding HG Q_i which specifies either an algorithm for performing e_i (if $e_i \in M$) or an algorithm

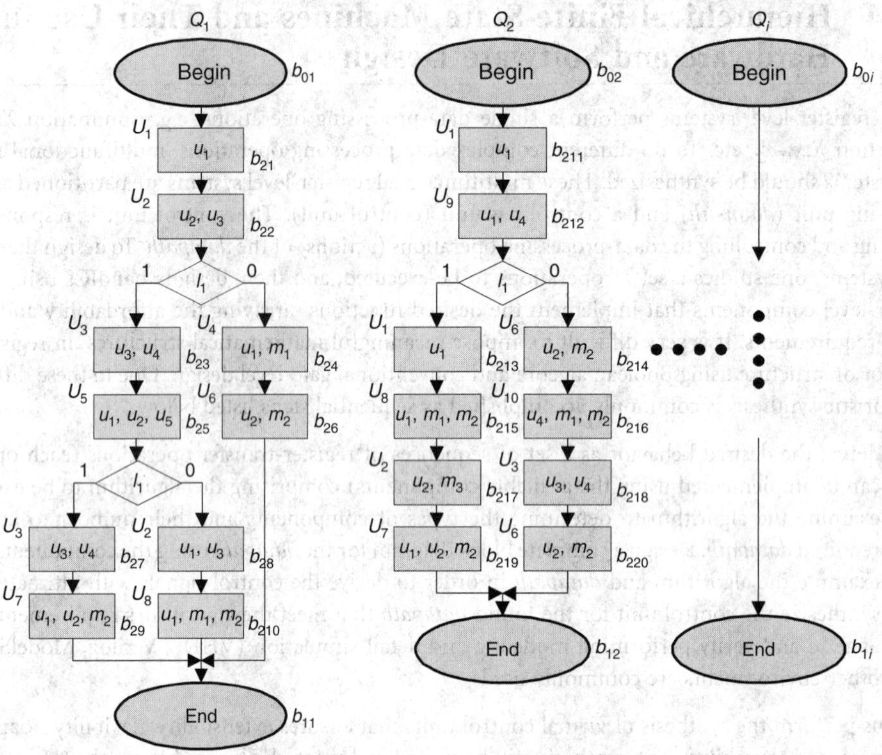

FIGURE 148.17 Control algorithm represented by HGs $Q_1, Q_2, \ldots, Q_i$.

for calculating e_i (if $e_i \in G$). Assume that $M(Q_i)$ is the subset of macro operations and $G(Q_i)$ is the subset of logic functions that belong to the HG Q_i. If $M(Q_i) \cup G(Q_i) = \varnothing$, the well-known scheme results [10, 27, 28]. The application of HGs enables one to gradually and sequentially synthesize complex control algorithm concentrating the efforts at each stage on a specified level of abstraction because specific elements of the set E are used. Each component of the set E is simple and can be checked and debugged independently. Figure 148.17 reports HGs $Q_1, Q_2, \ldots, Q_i$ which describe the control algorithm.

The execution of HGs is examined studying complex operations $e_i = m_j \in M$ and $e_i = g_j \in G$. Each complex operation e_i that is described by a HG Q_i must be replaced with a new subsequence of operators that produces the result executing Q_i. In the illustrative example, shown in Figure 148.18, Q_1 is the first HG at the first level Q^1, the second level Q^2 is formed by Q_2, Q_3, and Q_4, etc. We consider the following hierarchical sequence of HGs $Q_{1\ (\text{level } 1)} \Rightarrow Q^2_{(\text{level } 2)} \Rightarrow \ldots \Rightarrow Q^{q-1}_{(\text{level } q-1)} \Rightarrow Q^q_{(\text{level } q)}$. All $Q_{i\ (\text{level } i)}$ have the corresponding HGs. For example, Q^2 is a subset of the HGs that are used to describe elements from the set $M(Q_1) \cup G(Q_1) = \varnothing$, while Q^3 is a subset of the HGs that are used to map elements from the sets $\cup_{q \in Q^2} M(q)$ and $\cup_{q \in Q^2} G(q)$. In Figure 148.18, $Q^1 = \{Q_1\}$, $Q^2 = \{Q_2, Q_3, Q_4\}$, $Q^3 = \{Q_2, Q_4, Q_5\}$, etc.

Micro operations u^+ and u^- are used to increment and to decrement the stack pointer (SP). The problem of switching to various levels can be solved using a stack memory, see Figure 148.18. Consider an algorithm for $e_i \in M(Q_1) \cup G(Q_1) = \varnothing$. The SP is incremented by the micro operation u^+, and a new register of the stack memory is set as the current register. The previous register stores the state when it was interrupted. New Q_i becomes responsible for the control until terminated. After termination of Q_i, the micro operation u^- is generated to return to the interrupted state. As a result, control is passed to the state in which Q_f is called.

The synthesis problem can be formulated as: for a given control algorithm $\boldsymbol{A}$, described by the set of HGs, construct the FSM that implements $\boldsymbol{A}$.

In general, the synthesis includes the following steps:

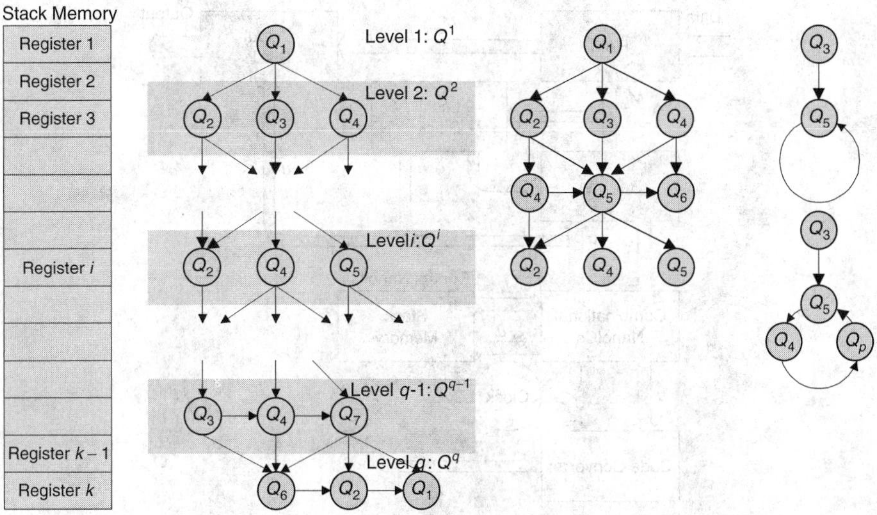

FIGURE 148.18 Stack memory with multilevel sequential HGs with illustration of recursive call.

- transforming the HGs to the state transition table
- state encoding
- combinational logic optimization
- final structure design

The first step is divided into substeps as: (1) marking the HGs with labels b, see Figure 148.18; (2) recording all transitions between the labels in the extended state transition table; and (3) convert the extended table to ordinary form.

The labels b_{01} and b_{11} are assigned to the nodes *Begin* and *End* of the Q_1. The label b_{02}, …, b_{0i} and b_{12}, …, b_{1i} are assigned to nodes *Begin* and *End* of Q_2, …, Q_i, respectively. The labels b_{21}, b_{22}, …, b_{2j} are assigned to other nodes of HGs, inputs and outputs of nodes with logic conditions, etc. Repeating labels are not allowed. The labels are considered as the states. The extended state transition table is designed using the state evolutions due to inputs (logic conditions) and logic functions which cause the transitions from $\boldsymbol{x}(t)$ to $\boldsymbol{x}(t+1)$. All evolutions of the state vector $\boldsymbol{x}(t)$ are recorded, and the state $x_k(t)$ has the label k. It should be emphasized that the table can be converted from the extended to the ordinary form. To program the Code Converter, one records the transition from the state x_1 assigned to the *Begin* node of the HG Q_1, e.g., $x_{01} \Rightarrow x_{21}(Q_1)$. The transitions between different HGs are recorded as $x_{ij} \Rightarrow x_{nm}(Q_j)$. For all transitions, the data-transfer instructions are synthesized. The hardware implementation is illustrated in Figure 148.19.

Robust algorithms are synthesized and implemented using the HGs utilizing the hierarchical behavior specifications and top-down decomposition. The reported method guarantees exceptional adaptation and reusability features through reconfigurable hardware and reprogrammable software.

148.5 Reconfigurable Nanocomputers

To design nanocomputers, specific hardware and software solutions must be developed and implemented. For example, ICs are designed by making use of hardware description languages, e.g., Very High Speed Integrated Circuit Hardware Description Language (VHDL) and Verilog. Making the parallel to the conventional ICs, the programmable gate arrays (PGAs) developed by Xilinx, Altera, Actel, and other companies, can serve as the meaningful inroad in design of reconfigurable nanocomputers. These PGAs lead one to the on-chip reconfigurable logic concept. The reconfigurable logic can be utilized as a functional unit in the *datapath* of the processor, having access to the processor register file and to on-

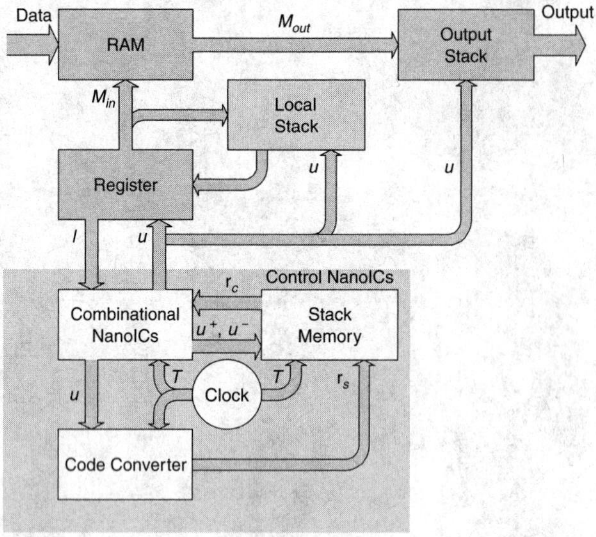

FIGURE 148.19 Hardware implementation.

chip memory ports. Another approach is to integrate the reconfigurable part of the processor as a coprocessor. For this solution, the reconfigurable logic operates concurrently with the processor. Optimal design and memory port assignments can guarantee the coprocessor reconfigurability and concurrency.

In general, the reconfigurable nanocomputer architecture synthesis emphasizes a high-level design, rapid prototyping and reconfigurability in order to reduce time and cost improving performance and reliability. The goal is to device, design, and fabricate affordable high-performance high-yield nanoICs arrays and application-specific nanoICs (ASNICs). These ASNICs should be testable to detect the defects and faults. The design of ASNICs involves mapping application requirements into specifications implemented by nanoICs. The specifications are represented at every level of abstraction including the system, behavior, structure, physical, and process domains. The designer should be able to differently utilize the existing nanoICs to meet the application requirements. User-specified nanoICs and ASNICs must be developed to attain affordability and superior performance.

The PGAs can be used to implement logic functions utilizing millions of gates. The design starts by interpreting the application requirements into architectural specifications. As the application requirements are examined, the designer translates the architectural specifications into behavior and structure domains. Behavior representation means the functionality required as well as the ordering of operations and completion of tasks in specified times. A structural description consists of a set of nanodevices and their interconnection. Behavior and structure can be specified and studied using hardware description languages. This nanoICs Hardware Description Language (NHDL) should manage efficiently very complex hierarchies which can include millions of logic gates. Furthermore, NHDLs should be translated into net-lists of library components using synthesis software. The NHDLs software, which is needed to describe hardware and must permit concurrent operations, should perform the following major functions:

- translate text to a Boolean mathematical representation,
- optimize the representation based specified criteria (size, delays, optimality, reliability, testability, etc.),
- map the optimized mathematical representation to a technology-specific library of nanodevices.

Reconfigurable nanocomputers should use reprogrammable logic units (e.g., PGAs) to implement a specialized instruction set and arithmetic units to optimize the performance. Ideally, reconfigurable nanocomputers can be reconfigured at real-time (runtime), enabling the existing hardware to be reused

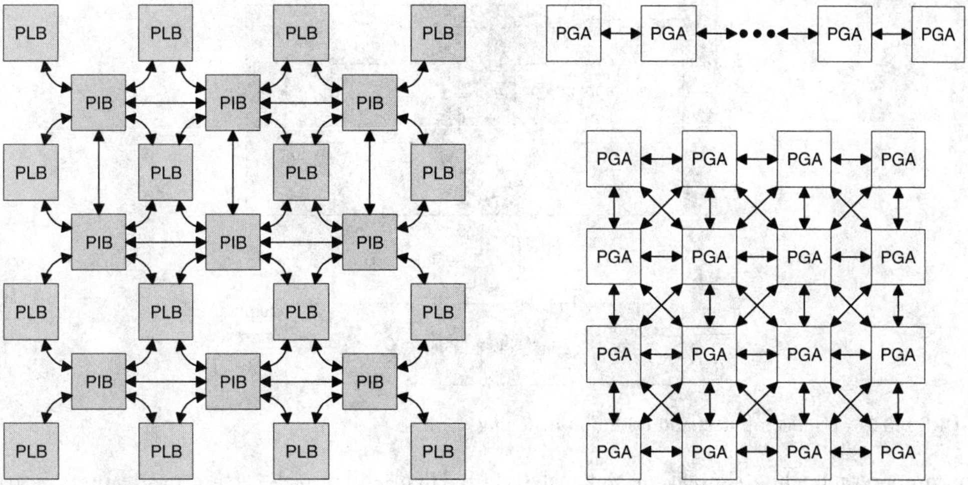

FIGURE 148.20 Programmable gate arrays and multiple PGAs architectures.

depending on its interaction with external units, data dependencies, algorithm requirements, faults, etc. The basic PGAs architecture is built using the programmable logic blocks (PLBs) and programmable interconnect blocks (PIBs), see Figure 148.20. The PLBs and PIBs will hold the current configuration setting until adaptation will be accomplished. The PGA is programmed by downloading the information in the file through a serial or parallel logic connection. The time required to configure a PGA is called the configuration time (PGAs could be configured in series or in parallel). Figure 148.20 illustrates the basic architectures from which most of multiple PGAs architectures can be derived (pipelined architecture with the PGAs interfaced one to other is well fit for functions that have streaming data at specific intervals, while arrayed PGAs architecture is appropriate for functions that require a systolic array). A hierarchy of configurability is different for the different PGAs architectures.

The goal is to design reconfigurable nanocomputer architectures with corresponding software to cope with less-than-perfect, entirely or partially defective and faulty nanoscale devices, structures, and connects (e.g., nanoICs) encountered in arithmetic and logic, control, input-output, memory, and other units. To achieve our objectives, the redundant nanoICs units can be used (the redundancy level is determined by the nanoscale ICs quality and software capabilities). Hardware and software evolutionary learning, adaptability and reconfigurability can be achieved through decision-making, diagnostics, health-monitoring, analysis, and optimization of software, as well as pipelining, rerouting, switching, matching, and controlling of hardware. We concentrate our research on how to devise – design – optimize – build – test — configurate nanocomputers. The overall objective can be achieved guaranteeing the evolution (behavior) matching between the ideal (C_I) and fabricated (C_F) nanocomputer, their units, systems, or components. The nanocompensator (C_{F1}) can be designed for the fabricated C_{F2} such that the response of the cf. will match the evolution of the C_I, see Figure 148.21. The C_I gives the reference ideal evolving model which analytically and/or experimentally maps the ideal (desired) input-output behavior, and the nanocompensator C_{F1} should modify the evolution of C_{F2} such that cf. described by $C_F = C_{F1} \circ C_{F2}$ (series architecture), matches the C_I behavior. Figure 148.21 illustrates the concept. The necessary and sufficient conditions for strong and weak evolution matching based on C_I and C_{F2} must be derived.

148.6 Mathematical Models for Nanocomputers

Six-Tuple Nanocomputer Model

To address analysis, control, diagnostics, optimization, and design problems, the explicit mathematical models of nanocomputers must be developed and applied. There are different levels of abstraction in

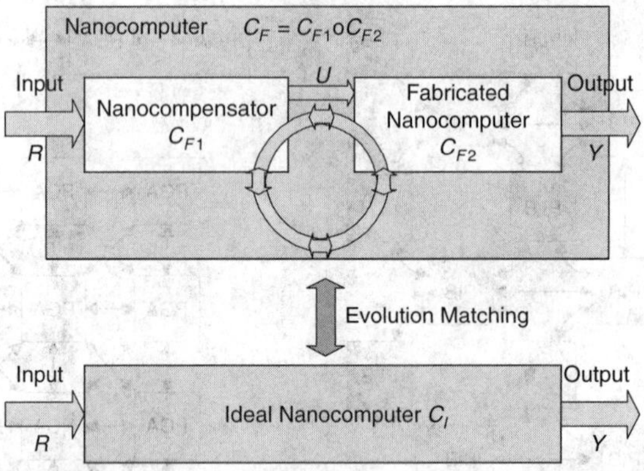

FIGURE 148.21 Nanocomputers and evolution matching.

nanocomputer modeling, simulation, and analysis [10]. High-level models can accept streams of instruction descriptions and memory references, while the low-level (device-level) logic gates or memory modeling can be performed by making use of streams of input and output signals and nonlinear transient behavior of devices. The system- and subsystem-level modeling (medium-level) also can be formulated and performed. It is evident that one subsystem can contain millions of nanodevices. For example, computer systems can be modeled as queuing networks. Different mathematical modeling frameworks exist and have been developed for each level. In this section we concentrate on the high-, medium-, and low-level systems modeling of nanocomputer systems using the finite state machine concept.

Computer accepts the input information, processes it according to the stored instructions, and produces the output. There exist numerous computer models, e.g., Boolean models, polynomial models, information-based, etc. However, all mathematical models are the mathematical idealization based upon the abstractions, simplifications and hypotheses made. In fact, it is virtually impossible to develop and apply the complete mathematical model due to complexity and numerous obstacles.

It is possible to concurrently model nanocomputers by the six-tuple

$$C = \{X, E, R, Y, F, X_0\},$$

where $\boldsymbol{X}$ is the finite set of states with initial and final states $\boldsymbol{x}_0 \in \boldsymbol{X}$ and $\boldsymbol{x}_f \subseteq \boldsymbol{X}$; $\boldsymbol{E}$ is the finite set of events (concatenation of events forms a string of events); $\boldsymbol{R}$ and $\boldsymbol{Y}$ are the finite sets of the input and output symbols (alphabets); $\boldsymbol{F}$ are the transition functions mapping from $X \times E \times R \times Y$ to $\boldsymbol{X}$ (denoted as $\boldsymbol{F_X}$), to $\boldsymbol{E}$ (denoted as $\boldsymbol{F_E}$) or to $\boldsymbol{Y}$ (denoted as $\boldsymbol{F_Y}$), $F \subseteq X \times E \times R \times Y$ (we assume that $\boldsymbol{F} = \boldsymbol{F_X}$, e.g., the transition function defines a new state to each quadruple of states, events, references, and outputs, and $\boldsymbol{F}$ can be represented by a table listing the transitions or by a state diagram).

The nanocomputer evolution is due to inputs, events, state evolutions, and parameter variations (as explained in the end of this subsection), etc.

We formulate two useful definitions.

Definition 1. A vocabulary (or an alphabet) $\boldsymbol{A}$ is a finite nonempty set of symbols (elements). A world (or sentence) over $\boldsymbol{A}$ is a string of finite length of elements of $\boldsymbol{A}$. The empty (null) string is the string which does not contain symbols. The set of all words over $\boldsymbol{A}$ is denoted as $\boldsymbol{A_w}$. A language over $\boldsymbol{A}$ is a subset of $\boldsymbol{A_w}$.

Definition 2. A finite-state machine with output $C_{FS} = \{X, A_R, A_Y, F_R, F_Y, X_0\}$ consists of a finite set of states $\boldsymbol{S}$, a finite input alphabet $\boldsymbol{A_R}$, a finite output alphabet $\boldsymbol{A_Y}$, a transition function $\boldsymbol{F_Y}$ that assigns a new state to each state and input pair, an output function $\boldsymbol{F_Y}$ that assigns an output to each state and input pair, and initial state $\boldsymbol{X_0}$.

Using the input-output map, the evolution of $\boldsymbol{C}$ can be expressed as

$$E_C \subseteq R \times Y .$$

That is, if the computer in state $x \in X$ receives an input $r \in R$, it moves to the next state $f(x, r)$, and produces the output $y(x, r)$. Nanocomputers can be represented as the state tables which describe the state and output functions. In addition, the state transition diagram (direct graph whose vertices correspond to the states and edges correspond to the state transitions, and each edge is labeled with the input and output associated with the transition) is frequently used.

Nanocomputer Modeling With Parameters Set

Nanocomputers can be modeled using the parameters set P. Designing reconfigurable fault-tolerant nanocomputer architectures, sets P and P_0 should be integrated, and we have

$$C = \{X, E, R, Y, P, F, X_0, P_0\}.$$

It is evident that the nanocomputer evolution depends upon P and P_0. The optimal performance can be achieved through adaptive synthesis, reconfiguration and diagnostics. For example, one can vary F and variable parameters P_v to attain the best possible performance.

The nanocomputer evolution, considering states, events, outputs, and parameters, can be expressed as

$$(x_0, e_0, y_0, p_0) \overset{\text{evolution 1}}{\Rightarrow} (x_1, e_1, y_1, p_1) \overset{\text{evolution 2}}{\Rightarrow} \cdots \overset{\text{evolution } j-1}{\Rightarrow} (x_{j-1}, e_{j-1}, y_{j-1}, p_{j-1}) \overset{\text{evolution } j}{\Rightarrow} (x_j, e_j, y_j, p_j).$$

The input, states, outputs, events, and parameter sequences are aggregated within the model $C = \{X, E, R, Y, P, F, X_0, P_0\}$. The concept reported allows us to find and apply the minimal — but – complete functional description of nanocomputers. The minimal subset of state, event, output, and parameter evolutions (transitions) can be used. That is, the partial description $C_{partial} \subset C$ results, and every essential nanocomputer quadruple (x_i, e_i, y_i, p_i) can be mapped by $(x_i, e_i, y_i, p_i)_{partial}$. This significantly reduces the complexity modeling, simulation, analysis, and design problems.

Let the transition function F maps from $X \times E \times R \times Y \times P$ to X, e.g., $F: X \times E \times R \times Y \times P \to X$, $F \subseteq X \times E \times R \times Y \times P$. Thus, the transfer function F defines a next state $x(t + 1) \in X$ based upon the current state $x(t) \in X$, event $e(t) \in E$, reference $r(t) \in R$, output $y(t) \in Y$ and parameter $p(t) \in P$. Hence,

$$x(t+1) = F(x(t), e(t), r(t), y(t), p(t)) \text{ for } x_0(t) \in X_0 \text{ and } p_0(t) \in P_0.$$

The robust adaptive control algorithms must be developed. The control vector $u(t) \in U$ is integrated into the nanocomputer model. We have $C = \{X, E, R, Y, P, U, F, X_0, P_0\}$.

Nanocompensator synthesis procedure is reported in the following sections. This synthesis approach is directly applicable to design nanocomputers with two- and three-dimensional nanoICs.

148.7 Nanocompensator Synthesis and Design Aspects

In this section we will design the nanocompensator. Two useful Definitions which allow one to precisely formulate and solve the problem are formulated.

Definition 3. The strong evolutionary matching $C_F = C_{F1} \circ C_{F2} =_B C_I$ for given C_I and cf. is guaranteed if $E_{C_F} = E_{C_I}$. Here, $C_F =_B C_I$ means that the behaviors (evolution) of C_I and cf. are equivalent.~

Definition 4. The weak evolutionary matching $C_F = C_{F1} \circ C_{F2} \subseteq_B C_I$ for given C_I and cf. is guaranteed if $E_{C_F} \subseteq E_{C_I}$. Here, $C_F \subseteq_B C_I$ means that the evolution of cf. is contained in the behavior C_I. ~

The problem is to derive a nanocompensator $C_{F1} = \{X_{F1}, E_{F1}, R_{F1}, Y_{F1}, F_{F1}, X_{F10}\}$ such that for given $C_I = \{X_I, E_I, R_I, Y_I, F_I, X_{I0}\}$ and $C_{F2} = \{X_{F2}, E_{F2}, R_{F2}, Y_{F2}, F_{F2}, X_{F20}\}$ the following conditions

$$C_F = C_{F1} \circ C_{F2} =_B C_I \text{ (strong behavior matching)}$$

or

$$C_F = C_{F1} \circ C_{F2} \subseteq_B C_I \text{ (weak behavior matching)}$$

are satisfied.

Here we assume that the following conditions are satisfied:

- output sequences generated by C_I can be generated by C_{F2};
- the C_I inputs match the C_{F1} inputs.

It must be emphasized that the output sequences means the state, event, output, and/or parameters vectors, e.g., the triple (x, e, y, p).

Lemma 1. If there exists the state-modeling representation $\gamma \subseteq X_I \times X_F$ such that $C_I^{-1} \times {}_B^{\gamma}C_{F2}^{-1}$ (if $C_I^{-1} \times {}_B^{\gamma}C_{F2}^{-1}$, then $C_I \times {}_B^{\gamma}C_{F2}$), then the evolution matching problem is solvable. The nanocompensator C_{F1} solves the strong matching problem $C_F = C_{F1} \circ C_{F2} =_B C_I$ if there exist the state-modeling representations $\beta \subseteq X_I \times X_{F2}$, $(X_{I0}, X_{F2\,0}) \in \beta$ and $\alpha \subseteq X_{F1} \times \beta$, $(X_{F1\,0}, (X_{I0}, X_{F2\,0})) \in \alpha$ such that $C_{F1} = {}_B^{\alpha}C_I^{\beta}$ for $\beta \in \Gamma = \{\gamma | C_I^{-1} \times {}_B^{\gamma}C_{F2}^{-1}\}$. Furthermore, the strong matching problem is tractable if there exist C_I^{-1} and C_{F2}^{-1}.

The nanocomputer can be decomposed using algebraic decomposition theory which is based on the closed partition lattice. For example, consider the fabricated nanocomputer C_{F2} represented as $C_{F2} = \{X_{F2}, E_{F2}, R_{F2}, Y_{F2}, F_{F2}, X_{F20}\}$.

A partition on the state set for C_{F2} is a set $\{C_{F2\,1}, C_{F2\,2}, \ldots, C_{F2\,i}, \ldots, C_{F2\,k-1}, C_{F2\,k}\}$ of disjoint subsets of the state set X_{F2} whose union is X_{F2}, e.g., $\bigcup_{i=1}^{k} C_{F2i} = X_{F2}$ and $C_{F2i} \bigcap C_{F2j} = \varnothing$ for $i \neq j$. Hence, we can design and implement the nanocompensators (hardware) $C_{F1\,i}$ for given $C_{F2\,i}$.

References

1. J. Appenzeller, R. Martel, P. Solomon, K. Chan, P. Avouris, J. Knoch, J. Benedict, M. Tanner, S. Thomas, L. L. Wang, and J. A. Del Alamo, "A 10 nm MOSFET concept," *Microelectronic Engineering*, vol. 56, no. 1–2, pp. 213–219, 2001.
2. Y. Chen, D. A. A. Ohlberg, G. Medeiros-Ribeiro, Y. A. Chang, and R. S. Williams, "Self-assembled growth of epitaxial erbium disilicide nanowires on silicon (001)," *Appl. Phys. Lett.*, vol. 76, pp. 4004–4006, 2000.
3. V. Derycke, R. Martel. J. Appenzeller, and P. Avouris, "Carbon nanotube inter- and inramolecular logic gates," *Nano Letters*, 2001.
4. J. C. Ellenbogen and J. C. Love, "Architectures for molecular electronic computers: Logic structures and an adder designed from molecular electronic diodes," *Proc. IEEE*, vol. 88, no. 3, pp. 386–426, 2000.
5. W. L. Henstrom, C. P. Liu, J. M. Gibson, T. I. Kamins, and R. S. Williams, "Dome-to-pyramid shape transition in Ge/Si islands due to strain relaxation by interdiffusion," *Appl. Phys. Lett.*, vol. 77, pp. 1623–1625, 2000.
6. S. C. Goldstein, "Electronic nanotechnology and reconfigurable computing," *Proc. Computer Society Workshop on VLSI*, pp. 10–15, 2001.

7. T. I. Kamins and D. P. Basile, "Interaction of self-assembled Ge islands and adjacent Si layers grown on unpatterned and patterned Si (001) substrates," *J. Electronic Materials*, vol. 29, pp. 570–575, 2000.
8. T. I. Kamins, R. S. Williams, Y. Chen, Y. L. Chang, and Y. A. Chang, "Chemical vapor deposition of Si nanowires nucleated by $TiSi_2$ islands on Si," *Appl. Phys. Lett.*, vol. 76, pp. 562–564, 2000.
9. T. I. Kamins, R. S. Williams, D. P. Basile, T. Hesjedal, and J. S. Harris, "Ti-catalyzed Si nanowires by chemical vapor deposition: Microscopy and growth mechanism," *J. Appl. Phys.*, vol. 89, pp. 1008–1016, 2001.
10. S. E. Lyshevski, *Nanocomputer Architectronics and Nanotechnology*, in Handbook of Nanoscience, Engineering, and Technology, Edited by W. A. Goddard, D. W. Brenner, S. E. Lyshevski, and G. J. Iafrate, CRC Press, Boca Raton, FL, pp. 6-1–6-38, 2003.
11. R. Martel, H. S. P. Wong, K. Chan, and P. Avouris, "Carbon nanotube field effect transistors for logic applications," *Proc. Electron Devices Meeting, IEDM Technical Digest*, pp. 7.5.1–7.5.4, 2001.
12. P. L. McEuen, J. Park, A. Bachtold, M. Woodside, M. S. Fuhrer, M. Bockrath, L. Shi, A. Majumdar, and P. Kim, "Nanotube nanoelectronics," *Proc. Device Research Conf.*, pp. 107–110, 2001.
13. R. Saito, G. Dresselhaus and M. S. Dresselhaus, *Physical Properties of Carbon Nanotubes*, Imperial College Press, London, 1999.
14. W. T. Tian, S. Datta, S. Hong, R. Reifenberger, J. I. Henderson, and C. P. Kubiak, "Conductance spectra of molecular wires," *Int. J. Chemical Physics*, vol. 109, no. 7, pp. 2874–2882, 1998.
15. K. Tsukagoshia, A. Kanda, N. Yoneya, E. Watanabe, Y. Ootukab, and Y. Aoyagi, "Nano-electronics in a multiwall carbon nanotube," *Proc. Microprocesses and Nanotechnology Conf.*, pp. 280–281, 2001.
16. E. K. Drexler, *Nanosystems: Molecular Machinery, Manufacturing, and Computations*, Wiley-Interscience, New York, 1992.
17. J. Carter, *Microprocessor Architecture and Microprogramming, a State Machine Approach*, Prentice-Hall, Englewood Cliffs, NJ, 1996.
18. V. C. Hamacher, Z. G. Vranesic, and S. G. Zaky, *Computer Organization*, McGraw-Hill, New York, 1996.
19. J. P. Hayes, *Computer Architecture and Organizations*, McGraw-Hill, Boston, MA, 1998.
20. J. L. Hennessy and D. A. Patterson, *Computer Architecture: A Quantitative Approach*, Morgan Kaufman, San Mateo, CA, 1990.
21. K. Hwang, *Computer Arithmetic*, Wiley, New York, 1978.
22. K. Likharev, "Riding the crest of a new wave in memory," *IEEE Circuits and Devices Magazine*, vol. 16, no. 4, pp. 16–21, 2000.
23. D. A. Patterson and J. L. Hennessey, *Computer Organization and Design — The Hardware/Software Interface*, Morgan Kaufman, San Mateo, CA, 1994.
24. L. H. Pollard, *Computer Design and Architecture*, Prentice-Hall, Englewood Cliffs, NJ, 1990.
25. A. S. Tanenbaum, *Structured Computer Organization*, Prentice-Hall, Englewood Cliffs, NJ, 1990.
26. R. F. Tinder, *Digital Engineering Design: A Modern Approach*, Prentice-Hall, Englewood Cliffs, NJ, 1991.
27. S. Baranov, *Logic Synthesis for Control Automata*. Kluwer, Norwell, MA, 1994.
28. V. Sklyarov, *Synthesis of Finite State Machines Based on Matrix LSI*. Science, Minsk, Belarus, 1984.

149

Software Engineering

Phillip A. Laplante
Penn State University

149.1 Classification of Software Qualities

Software can be characterized by any of a number of qualities. External qualities are visible to the user, such as usability and reliability, and are of concern to the end user. Internal qualities may not be necessarily visible to the user, but help developers to achieve improvement in external qualities. For example, good requirements and design documentation might not be seen by the typical user, but these are necessary to achieve improvement in most of the external qualities. A specific distinction between whether a particular quality is external or internal is not often made because they are so closely tied. Moreover, the distinction is largely a function of the software itself and the kind of user involved.

While it is helpful to describe these qualities, it is equally desirable to quantify them. Quantification of these characteristics of software is essential in enabling users and designers to talk succinctly about the product and for software process control and project management.

Reliability

Reliability is a measure of whether a user can depend on the software. This notion can be informally defined in a number of ways. For example, one definition might be "a system that a user can depend on." Other loose characterizations of a reliable software system include:

- The system "stands the test of time."
- There is an absence of known catastrophic errors, that is, errors that render the system useless.
- The system recovers "gracefully" from errors.
- The software is robust.

For mission critical systems, other informal characterizations of reliability might include:

- Downtime is below a certain threshold.
- The accuracy of the system is within a certain tolerance.
- Real-time performance requirements are met consistently.

0-8493-1586-7/05/$0.00+$1.50

While all of the above are desirable in any system, these informal characteristics are difficult to measure. Moreover, they are not truly measures of reliability, but of other attributes of the software.

Specialized literature on software reliability exists that defines this quality in terms of statistical behavior, that is, the probability that the software will operate as expected over a specified time interval. These characterizations generally take the following approach. Let S be a software system, and let T be the time of system failure. Then the reliability of S at time t, denoted $r(t)$, is the probability that T is greater than t; that is,

$$r(t) = P(T > t) \tag{149.1}$$

this is the probability that a software system will operate without failure for a specified period of time.

Thus, a system with reliability function $r(t) = 1$ will never fail. However, it is unrealistic to have such expectations. Instead, some reasonable goal should be set, for example, in mission critical systems that the failure probability be no more than 10^{-9} per hour. This represents a reliability function of $r(t) = (0.99999999)^t$ with t in hours. Note that as $t \rightarrow \infty$, $r(t) \rightarrow 0$.

Another way to characterize software reliability is in terms of a real-valued failure function. One failure function uses an exponential distribution where the abscissa is time and the ordinate represents the expected failure intensity at that time (Equation 149.2).

$$f(t) = \lambda e^{-\lambda t} \qquad t \geq 0 \tag{149.2}$$

Here the failure intensity is initially high, as would be expected in new software as faults are detected during testing. However, the number of failures would be expected to decrease with time, presumably as failures are uncovered and repaired (Figure 149.1). The factor λ is a system-dependent parameter.

A second failure model is given by the "bathtub curve" shown in Figure 149.2. Brooks notes that while this curve is often used to describe the failure function of hardware components, it might also be useful in describing the number of errors found in a certain release of a software product.[1]

The interpretation of this failure function is clear for hardware — a certain number of product units will fail early due to manufacturing defects. Later, the failure intensity will increase as the hardware ages

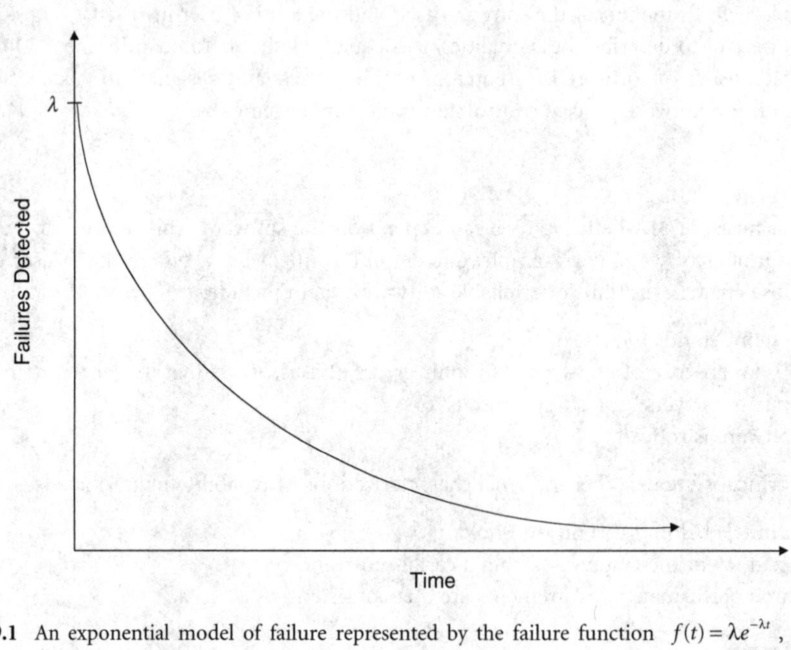

FIGURE 149.1 An exponential model of failure represented by the failure function $f(t) = \lambda e^{-\lambda t}$, $t \geq 0$. λ is a system-dependent parameter.

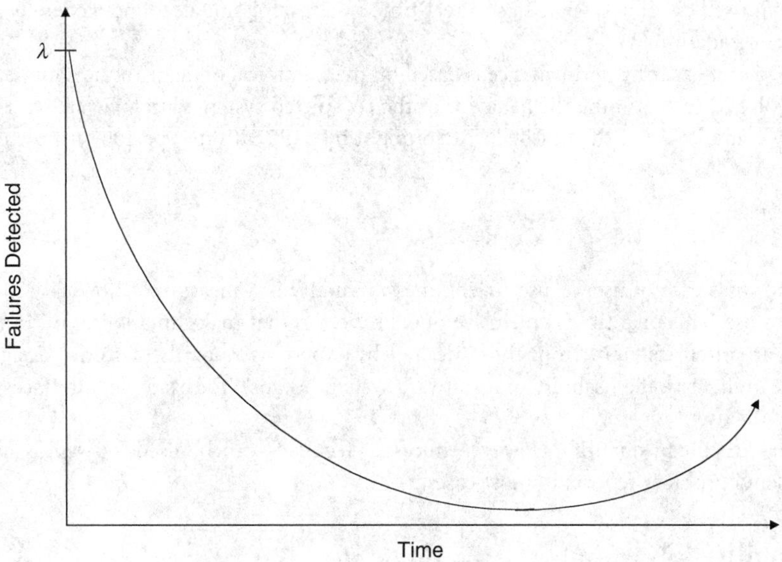

FIGURE 149.2 A software failure function represented by the bathtub curve.

and wears out. But software does not wear out. If systems seem to fail according to the bathtub curve, then there has to be some plausible explanation.

It is clear that a large number of errors will be found in a particular software product early, just as in the exponential model of failure. But why would the failure intensity increase much later? There are at least three possible explanations. The first is that the errors are due to effects of patching the software for various reasons. The second reason is that late software failures are really due to failure of the underlying hardware. Finally, additional failures could appear because of the increased stress on the software by expert users. That is, as users master the software and begin to expose and strain advanced features, it is possible that certain poorly tested functionality of the software is beginning to be used.

Often the traditional quality measures of mean time to first failure (MTTF) or mean time between failures (MTBF) are used to stipulate reliability in the software requirements specification. This approach to failure definition places great importance on the effective elicitation (gathering) and specification of functional requirements, because requirements define software failure.

Correctness

Software correctness is closely related to reliability and the terms are often used interchangeably. The main difference is that minor deviation from the requirements is strictly considered a failure, and hence means that the software is incorrect. However, a system may still be deemed reliable if only minor deviations from the requirements are experienced. Correctness is measured in terms of number of failures detected over time.

Performance

Performance is a measure of some required behavior. For example, an imaging system might be required to display a filtered image at a rate of 30 frames per second. A photo reproduction system might be required to digitize, clean, and output color copies at a rate of 1 every 2 seconds.

Many software systems are real-time systems, that is, performance satisfaction is based on both the correctness of the outputs, as well as the timeliness of those outputs. Hard real-time systems are those in which missing even a single deadline will lead to total system failure. Firm real-time systems can tolerate a few missed deadlines, while in soft real-time systems, missing deadlines generally leads only to performance degradation.

One method of measuring performance is based on mathematical or algorithmic complexity. Another approach involves directly timing the behavior of the completed system with logic analyzers and similar tools. Finally, a simulation of the finished system might be built with the specific purpose of estimating performance.

Usability

Often referred to as ease of use, or user friendliness, usability is a measure of how easy the software is for humans to use. This quantity is an elusive one. Properties that make an application user-friendly to novice users are often different from those desired by expert users or the software designers. Use of prototyping can increase the usability of a software system because, for example, interfaces can be built and tested by the user.

Usability is difficult to quantify. However, informal feedback can be used, as well as user feedback from surveys and problem reports in most cases.

Interoperability

This quality refers to the ability of the software system to coexist and cooperate with other systems. For example, in certain systems the software must be able to communicate with various devices using standard bus structures and protocols.

A concept related to interoperability is that of an open system. An open system is an extensible collection of independently written applications that cooperate to function as an integrated system. Open systems differ from open source code, which is source code that is made available to the user community for moderate improvement and correction.

An open system allows the addition of new functionality by independent organizations through the use of interfaces whose characteristics are published. Any applications developer can then take advantage of these interfaces, and thereby create software that can communicate using the interface. Open systems allow different applications written by different organizations to interoperate.

Interoperability can be measured in terms of compliance with open system standards.

Maintainability

Anticipation of change is a general principle that should guide the software engineer. A software system in which changes are relatively easy to make has a high level of maintainability. In the long run, design for change will significantly lower software life-cycle costs and lead to an enhanced reputation for the software engineer, the software product, and the company.

Maintainability can be decomposed into two contributing properties: evolvability and repairability. Evolvability is a measure of how easily the system can be changed to accommodate new features or modification of existing features. Software is repairable if it allows for fixing defects.

Measuring these qualities of software is not always easy, and is often based on anecdotal observation only. This means that changes and the cost of making them are tracked over time. Collecting these data has a twofold purpose. First, the costs of maintenance can be compared to other similar systems for benchmarking and project management purposes. Second, the information can provide experiential learning that will help to improve the overall software production process and skills of the software engineers.

Portability

Software is portable if it can easily run in different environments. The term *environment* refers to the hardware on which the system runs, operating system, or other software with which the system is expected

TABLE 149.1 Selected Software Properties and Measurement Approaches

Software Quality	Possible Measurement Approach
Correctness	Probabilistic measures, MTBF, MTFF
Interoperability	Compliance with open standards
Maintainability	Anecdotal observation of resources spent
Performance	Algorithmic complexity analysis, direct measurement, simulation
Portability	Anecdotal observation
Reliability	Probabilistic measures, MTBF, MTFF, heuristic measures
Usability	User feedback from surveys and problem reports
Verifiability	Software monitors

Note: MTBF, mean time between failures; MTTF, mean time to first failure.

to interact. Because of the specialized hardware with which they interact, special care must be taken in making imaging systems portable.

Portability is achieved through a deliberate design strategy in which hardware-dependent code is confined to the fewest code units as possible. This strategy can be achieved using either object-oriented or procedural programming languages and through object-oriented or structured approaches. Both of these will be discussed below.

Portability is difficult to measure, other than through anecdotal observation. Person months required to perform the port are the standard measure of this property.

Verifiability

A software system is verifiable if its properties, including all of those previously introduced, can be verified easily. One common technique for increasing verifiability is through the insertion of software code that is intended to monitor various qualities such as performance or correctness. Modular design, rigorous software engineering practices, and the effective use of an appropriate programming language can also contribute to verifiability.

Summary of Software Properties and Associated Metrics

Thus far, it has been emphasized that measurement of software properties is essential throughout the software life cycle. A summary of the software qualities just discussed and possible ways to measure them is shown in Table 149.1.

149.2 Basic Software Engineering Principles

Software engineering has been criticized for not having the same kind of underlying rigor as other engineering disciplines. And while it may be true that there are few formulaic principles, there are many fundamental rules that form the basis of sound software engineering practice. The following sections describe the most general and prevalent of these.

Rigor and Formality

Because software development is a creative activity, there is an inherent tendency toward informal ad hoc techniques in software specification, design, and coding. But the informal approach is contrary to good software engineering practice.

Rigor in software engineering requires the use of mathematical techniques. Formality is a higher form of rigor in which precise engineering approaches are used.

For example, imaging systems require the use of rigorous mathematical specification in the description of image acquisition, filtering, enhancement, and so on. But the existence of mathematical equations in the requirements or design does not imply an overall formal software engineering approach. In the case of imaging systems, formality further requires that there be an underlying algorithmic approach to the specification, design, coding, and documentation of the software.

Separation of Concerns

Separation of concerns is a kind of divide-and-conquer strategy that software engineers use. There are various ways in which separation of concerns can be achieved. In terms of software design and coding, it is found in modularization of code and in object-oriented design. There may be separation in time, such as developing a schedule for a collection of periodic computing tasks with different periods.

Yet another way of separating concerns is in dealing with qualities. For example, it may be helpful to address the fault tolerance of a system while ignoring other qualities. However, it must be remembered that many of the qualities of software are interrelated, and it is generally impossible to affect one without affecting the other, possible adversely.

Modularity

Some separation of concerns can be achieved in software through modular design. Modular design involves the decomposition of software behavior in encapsulated software units, and can be achieved in either object-oriented or procedurally oriented programming languages.

Modularity is achieved by grouping together logically related elements, such as statements, procedures, variable declarations, object attributes, and so on in increasingly fine-grained levels of detail (Figure 149.3).

The main objectives in seeking modularity is to foster high cohesion and low coupling. With respect to the code units, cohesion represents intramodule connectivity and coupling represents intermodule connectivity. Coupling and cohesion can be illustrated informally as in Figure 149.4, which shows software

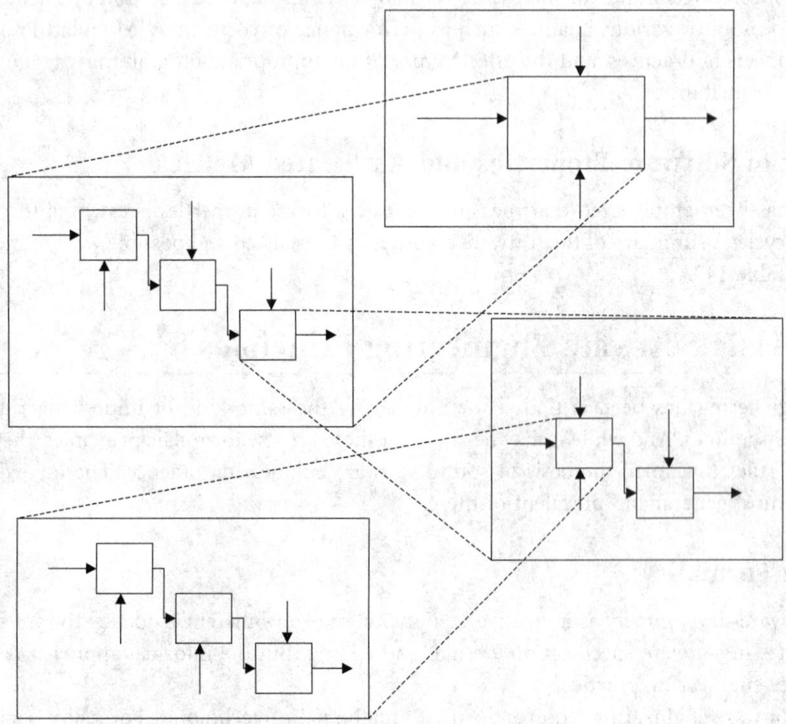

FIGURE 149.3 Modular decomposition of code units. *Arrows* represent inputs and outputs in the procedural paradigm. In the object-oriented paradigm, *arrows* represent associations or messages. The *boxes* represent encapsulated data and procedures in the procedural paradigm. In the object-oriented paradigm they represent classes.

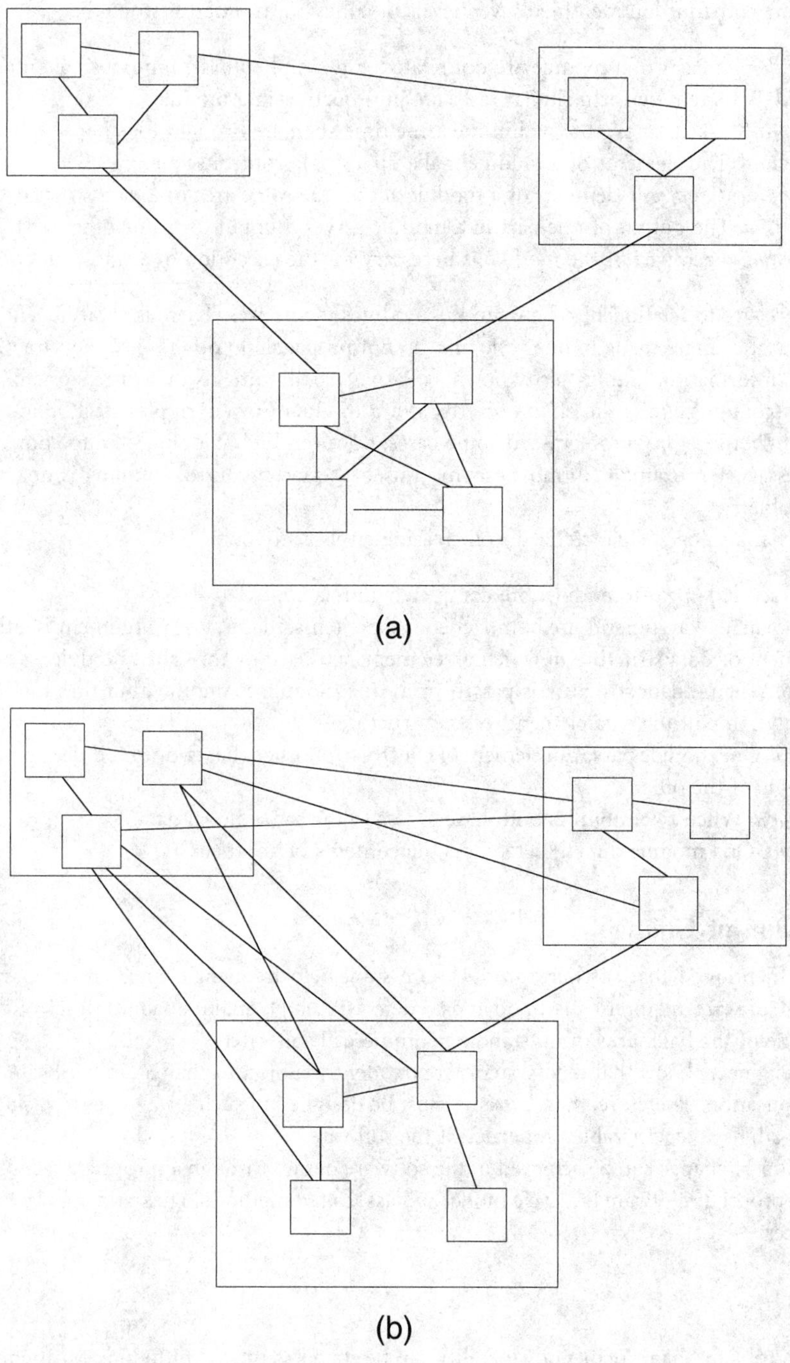

FIGURE 149.4 Software structures with high cohesion and low coupling (a) and low cohesion and high coupling (b). The inside *squares* represent statements or data, and *arcs* indicate functional dependency.

structures with high cohesion and low coupling, and low cohesion and high coupling. The inside squares represent statements or data, and arcs indicate functional dependency.

Cohesion relates to the relationship of the elements of a module. High cohesion implies that each module represents a single part of the problem solution. Therefore, if the system ever needs modification, then that part that needs to be modified exists in a single place, making it easier to change.

Constantine and Yourdon identified seven levels of cohesion in order of strength:[4]

- *Coincidental.* Parts of a module are not related, but simply bundled into a single module.
- *Logical.* Parts that perform similar tasks are put together in a module.
- *Temporal.* Tasks that execute within the same time span are brought together.
- *Procedural.* The elements of a module make up a single control sequence.
- *Communicational.* All elements of a module act on the same area of a data structure.
- *Sequential.* The output of one part in a module serves as input for some other part.
- *Functional.* Each part of the module is necessary for the execution of a single function.

Coupling relates to relationships between the modules themselves. There is great benefit in reducing coupling so that changes made to one code unit do not propagate to others (i.e., they are hidden). This principle of "information hiding," also known as Parnas partitioning, is the cornerstone of all software design.[3] Low coupling limits the effects of errors in a module (lower "ripple effect"), and reduces the likelihood of data integrity problems. In some cases, however, high coupling due to control structures may be necessary. For example, in most graphical user interfaces, control coupling is unavoidable, and indeed desirable.

Coupling has also been characterized in increasing levels as follows:

1. *No direct coupling.* All modules are completely unrelated.
2. *Data.* When all arguments are homogeneous data items, that is, every argument is either a simple argument or data structure in which all elements are used by the called module.
3. *Stamp.* When a data structure is passed from one module to another, but that module operates on only some of the data elements of the structure.
4. *Control.* One module passes an element of control to another, that is, one module explicitly controls the logic of the other.
5. *Common.* When two modules both have access to the same global data.
6. *Content.* One module directly references the contents of another.

Anticipation of Change

It has been mentioned that software products are subject to frequent change either to support new hardware or software requirements or to repair defects. A high maintainability level of the software product is one of the hallmarks of outstanding commercial software.

Software engineers know that their systems are frequently subject to changes in hardware, algorithms, and even application. Therefore, these systems must be designed in such a way so as to facilitate changes without degrading other desirable properties of the software.

Anticipation of change can be achieved in the software design through appropriate techniques, adoption of an appropriate software life-cycle model and associated methodologies, and appropriate management practices.

Generality

In solving a problem, the principle of generality can be stated as the intent to look for the more general problem that may be hidden behind it. For instance, designing the visual inspection system for a specific application is less general than designing it to be adaptable to a wide range of applications.

Generality can be achieved through a number of approaches associated with procedural and object-oriented paradigms. For example, in procedural languages, Parnas's information hiding can be used. In object orientation, the Liskov substitution principle can be used.

Although generalized solutions may be more costly in terms of the problem at hand, in the long run, the costs of a generalized solution may be worthwhile.

Incrementality

Incrementality involves a software approach in which progressively larger increments of the desired product are developed. Each increment provides additional functionality, which brings the product closer to the final one. Each increment also offers an opportunity for demonstration of the product to the customer for the purposes of gathering requirements and refining the look and feel of the product.

Traceability

Traceability is concerned with the relationships among requirements, their sources, and system design. Regardless of the process model, documentation and code traceability are paramount. A high level of traceability ensures that the software requirements flow down through the design and code and then can be traced back up at every stage of the process. This would ensure, for example, that a coding decision could be traced back to a design decision to satisfy a corresponding requirement.

Traceability is particularly important in embedded systems because often design and coding decisions are made to satisfy hardware constraints that may not be easily associated with a requirement. Failure to provide a traceable path from such decisions through the requirements can lead to difficulties in extending and maintaining the system.

Generally, traceability can be obtained by providing links between all documentation and the software code. In particular, the following links should be in place:

- From requirements to stakeholders who proposed these requirements
- Between dependent requirements
- From the requirements to the design
- From the design to the relevant code segments
- From requirements to the test plan
- From the test plan to test cases

One way to achieve these links is through the use of an appropriate numbering system throughout the documentation. For example, a requirement numbered 3.2.2.1 would be linked to a design element with a similar number (the numbers do not have to be the same so long as the annotation in the document provides traceability). These linkages are depicted in Figure 149.5. Although the documents shown in Figure 149.5 have not been introduced yet, the point to be made is that the documents are all connected through appropriate referencing and notation.

Figure 149.5 is simply a graphical representation of traceable links. In practice, a traceability matrix is constructed to help cross-reference documentation and code elements (Table 149.2). The matrix is constructed by listing the relevant software documents and the code unit as columns, and then each software requirement in the rows. Constructing the matrix in a spreadsheet software package allows for providing multiple matrices sorted and cross referenced by each column as needed. For example, a traceability matrix sorted by test case number would be an appropriate appendix to the text plan.

The traceability matrices are updated at each step in the software life cycle. For example, the column for the code unit names (e.g., procedure names, object class) would not be added until after the code is developed.

Finally, a way to foster traceability between code units is through the use of data dictionaries.

149.3 Role of Software Engineer

Software production is a problem-solving activity that is accomplished by modeling. As a problem-solving, modeling discipline, software engineering is a human activity that is biased by previous experience and subject to human error.

Modeling is a translation activity. The software product concept is translated into requirements specification. The requirements are converted into a design. The design is then converted into code, which

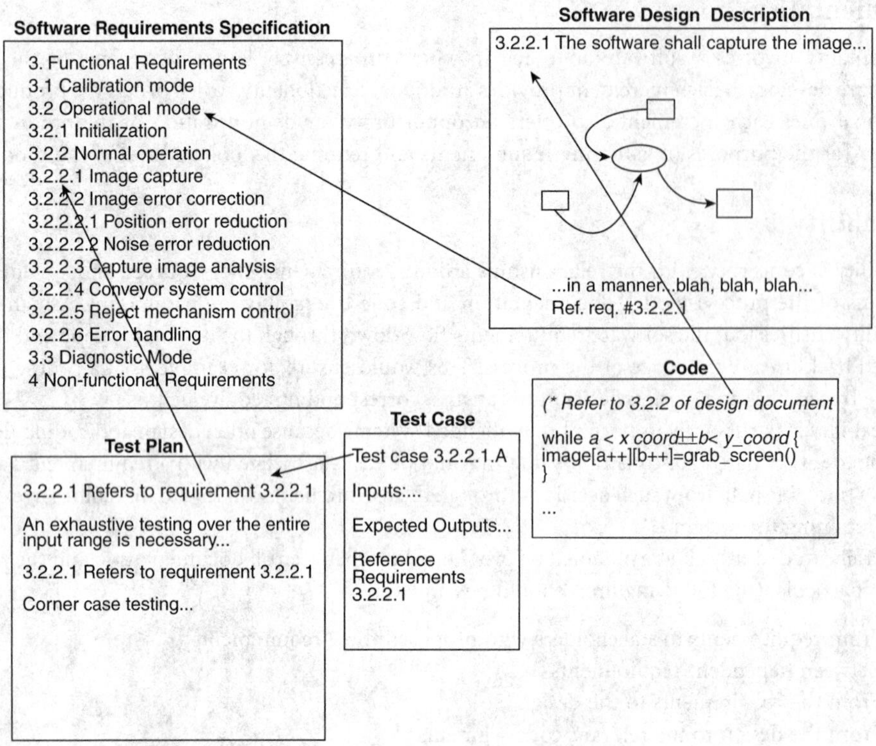

FIGURE 149.5 Linkages between software documentation and code. In this case, the links are achieved through both similarity in numbering and through specific reference to the related item in the appropriate document.

TABLE 149.2 Traceability Matrix Corresponding to Figure 149.5 Sorted by Requirement Number

Requirement Number	Software Design Document Reference Number(s)	Test Plan Reference Number(s)	Code Unit Name(s)	Test Case Number(s)
3.1.1.1	3.1.1.1, 3.2.4	3.1.1.1, 3.2.4.1, 3.2.4.3	Simple_fun	3.1.1.A, 3.1.1.B
3.1.1.2	3.1.1.2,	3.1.1.2	Kalman_filter	3.1.1.A, 3.1.1.B
3.1.1.3	3.1.1.3	3.1.1.3	Under_bar	3.1.1.A, 3.1.1.B, 3.1.1.C

is automatically translated by compilers and assemblers, which produce machine executable code. In each of these translation steps, however, errors are likely to be introduced either by the humans involved or the tools they use. Thus, software engineer must strive to identify these likely errors and avoid or fix them.

Software engineers should also strive to develop code that is built to be tested, designed for reuse, and ready for inevitable change. Anticipation of problems can only come from experience and from drawing on a body of software practice experience from the past 50-plus year.

Software engineers probably spend less than 10% of their time writing code. The other 90% of the time is involved other activities that are generally more important than writing the code. These activities include:

1. Eliciting requirements.
2. Analyzing requirements.
3. Writing software requirements documents.
4. Building and analyzing prototypes.
5. Developing software designs.
6. Writing software design documents

7. Researching software engineering techniques or obtaining information about the application domain.
8. Developing test strategies and test cases.
9. Testing the software and recording the results.
10. Isolating problems and solving them.
11. Learning to use or installing and configuring new software and hardware tools.
12. Writing documentation such as user manuals.
13. Attending meetings with colleagues, customers, and supervisors.
14. Archiving software or preparing it for distribution.

This is only provides a partial list of software engineering activities. These activities are not necessarily sequential and not all encompassing. Finally, most of these activities can recur throughout the software life cycle and in each new minor or major software version. Many software engineers specialize in a small subset of these activities, for example, software testing.

Misconceptions about Software Engineering

There are many misconceptions about what software engineering is and what it is intended to do. Often practitioners of "hard" engineering disciplines such as mechanical or electrical engineering view software engineering dimly or not as an engineering discipline at all. This perception could exist partly because there are no fundamental physical laws governing the practice of software engineering, as there are in mechanical, electrical, civil, and other types of engineering. More likely, however, the reason that software engineering does not always get the respect it deserves is that it is often not practiced as an engineering discipline and the barriers to "practicing" it are considered to be too low. Anyone can call herself or himself a software engineer if she or he writes code, but these individuals are not usually practicing software engineering. (As of this writing, one notable exception involves the requirement of professional licensure for individuals practicing software engineering in New York State.)

Some misconceptions about software engineering and brief rebuttals follow:

1. Software system development is primarily concerned with programming.

As mentioned before, 10% or less of the software engineer's time is spent writing code. Someone who spends the majority of his or her time generating code is more aptly called a "programmer." Just as wiring a circuit designed by an electrical engineer is not engineering, writing code designed by a software engineer is not an engineering activity.

2. Software tools and development methods can solve most or all of the problems pertaining to software engineering.

This is a dangerous misconception. Tools, software or otherwise, are only as good as the user. Bad habits and flawed reasoning can just as easily be amplified by tools as corrected. While software engineering tools are essential and provide significant advantages, to rely on them to remedy process or engineering deficiencies is naive.

3. Software productivity is a function of system complexity.

While it is certainly the case that system complexity can degrade productivity, there are many other factors that affect productivity. Requirements stability, engineering skill, quality of management, and availability of resources are just a few of the factors that affect productivity.
Once software is delivered, the job is finished.

4. Of course, this is not true. At the very least, some form of documentation of the end product as well as the process used needs to be written. More likely, the software product will now enter a maintenance mode after delivery in which it will experience many recurring life cycles as errors are detected and corrected and features are added.

5. Errors are an unavoidable side effect of software development.

While it is unreasonable to expect that all errors can be avoided (as in every discipline involving humans), good software engineering techniques can minimize the number of errors that are delivered to a customer. The attitude that errors are inevitable can be used to excuse sloppiness or complacency, whereas an approach to software engineering that is intended to detect every possible error, no matter how unrealistic this goal may be, will lead to a culture that will encourage engineering rigor and high software quality.

Acknowledgments

This chapter is adapted from *Software Engineering for Image Processing*, by Phillip A. Laplante, CRC Press, Boca Raton, FL, 2003, and from *The Computer Science and Engineering Handbook*, edited by Alan Tucker, CRC Press, Boca Raton, FL, 1996.

Defining Terms

Software design — A phase of the software development life cycle that maps what the system is supposed to do into how the system will do it in a particular hardware/software configuration.

Software engineering — Systematic development, operation, maintenance, and retirement of software.

Software evolution — The process that adapts the software to changes of the environment in which it is used.

Software reengineering — The reverse analysis of an old application to conform to a new methodology.

References

1. Brooks, F. *The Mythical Man-Month*, 2nd ed. Addison-Wesley, Reading, MA, 1995.
2. Laplante, P, *Software Engineering for Image Processing*, CRC Press, Boca Raton, FL, 2003.
3. Parnas, D. L., Designing Software for Ease of Extension and Contraction, *IEEE Trans. Software Eng.*, 1979;SE-5:128–138.
4. Pressman, R. S. *Software Engineering: A Practitioner's Approach*, 5th ed., McGraw-Hill, New York, 2000.
5. Tucker, Allen B. Jr., Ed., *The Computer Science and Engineering Handbook*, CRC Press, Boca Raton, FL, 1996.

150 Human–Computer Interface Design

Mansour Rahimi
University of Southern California

Jennifer Russell
University of Southern California

Greg Placencia
University of Southern California

Human–computer Interface (HCI) design is a new field of scientific inquiry. We now know that the idea of "slapping on an interface" after the software has been programmed could be costly and dangerous if the software is used in safety-critical systems. This chapter focuses on interface design and evaluation, mostly from the user perspective. We will describe the elements of the HCI first. Then, we will present the tools and techniques used to design and evaluate interfaces. Overall, we hope to show that HCI design is a necessary and critical component of any software development in the early design stages.

150.1 A Definition for HCI

HCI is the study of the human and the computer interacting through a medium called an interface. Computers and humans speak different languages, and in order to bridge the gap between them, a common language (or translator) is needed. Given the limitations of humans and the ever-expanding capabilities of computers, it is important to design for this cross-communication through an interface.

150.2 Why Is HCI Design Important?

Most users are familiar with common commercial interfaces using the windows, icons, menus, and pointers (WIMP) design from OpenWindows, KDE, Microsoft Windows, or the MacOS. They most likely can recall receiving a cryptic error message after performing an illegal action, such as a 404 error in Internet Explorer. While these messages convey to the user that there is something wrong, they do not

0-8493-1586-7/05/$0.00+$1.50

inform the user about the exact problem nor how to begin fixing the problem at hand. This results in a frustrated user with limited options to perform the next set of tasks. At some time during both the design of the HTTP status code and the design of a compiler, decisions were made on how user feedback was generated. In the HTTP case, error messages have at least a code number, and are usually accompanied by a text description. But it is frequently the case that the messages do not describe the exact problem, especially for those who are unfamiliar with HTTP status codes.

While the idea of providing intuitive feedback to the user seems simple, unfortunately a number of complex issues need to be addressed before the entire system can be rendered user-friendly. In many cases, an interface design that might support a novice user might not allow enough flexibility for more experienced users. For example, a person who programs HTTP code would certainly look at HTTP status codes differently from someone who only uses the web to read email. And anyone who has programmed even a few hundred lines of code knows that a long list of syntax errors could be the result of a single instruction error several lines away from the indicated location of the error. It is now believed that incorporating HCI principles and interface design guidelines have important ramifications on the product's success in today's information-intensive society. Moreover, in safety critical systems, well-designed interfaces could reduce the potential for death and destruction. For example, a fratricide incident in December 2001 in Afghanistan was caused when a soldier mistook his own position for the enemy's because of a poorly designed interface.

150.3 The Human

Computational power has increased by leaps and bounds since the first ENIAC computer was created at Penn State in 1946 [Weik, 1961]. Yet, our human capabilities, as the basis for understanding the computational universe, have remained unchanged over time. As a result, the need to "adapt" the computer to suit human end users has elevated the field of human factors to become an important component in software design. So, the premise here is to design software to fit and adapt to human capabilities by studying the human component of the system first.

Most computer systems require some form of user interaction. By considering the human as a component of system design, system developers are required to include users from the early stages of design. The interface can be viewed as the communication channel through which the human and computer provide each other inputs and outputs. The inputs and outputs need to be specified and protocols established that govern the interaction through the interface. The illustration in Figure 150.1 shows the

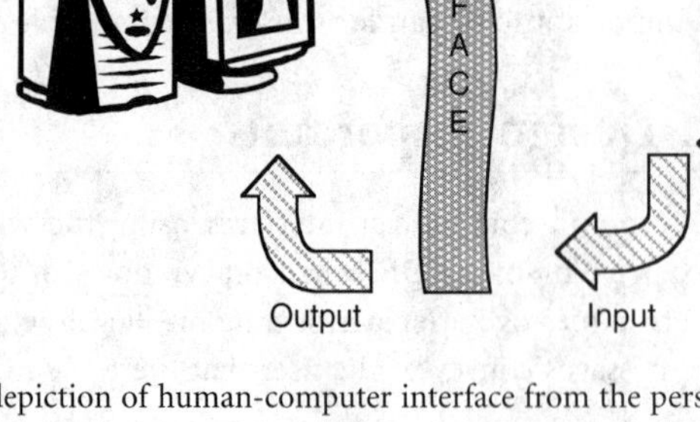

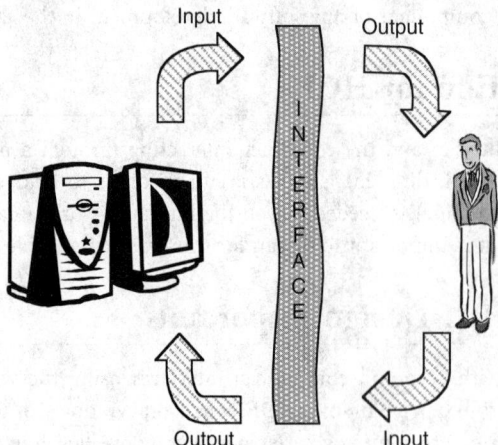

FIGURE 150.1 A graphical depiction of human-computer interface from the perspective of input/output communication channels.

human output information channels and the computer's input information channels in a closed system loop. The field of HCI attempts to optimize the compatibility among these input/output channels.

Human Physical Abilities

Human capabilities can be broken down into two major categories: physical and cognitive. There are five basic human senses that are used as input channels for humans: visual, auditory, haptic (or touch), olfactory (or smell), and taste. Human cognitive abilities are used to process and interpret these sensory inputs. Regardless of how technologically advanced our systems may be (e.g., how fast the processor processes, the software compiles, or the display shows rich colors), the range of physical and cognitive capabilities of human beings remains fairly constant across the general population. It seems that in terms of computer technologies, it is the human that is the limiting factor. It is also important to mention that human capabilities vary significantly among user populations, which makes HCI design a challenging multidisciplinary task.

Visual

Among the human senses, vision has been extensively studied and documented in the HCI literature. Vision is the ability to perceive size, colors, and movement of objects. For a visual display to consider these aspects of human perception, there are a number of guidelines, some of which are mentioned below.

Contrast and Color

Contrast is the distinction of light and dark between elements of the display. With proper contrast between the display elements, the user will be able to perceive the information presented on the screen with higher accuracy and lower eyestrain. "When contrast is low, objects must be larger to be equally discriminable" [Sanders and McCormick, 1993]. In addition to perceiving contrast, the human eye distinguishes three primary colors: red, green, and blue. All other colors are a resultant combination of these colors. Contrast is typically more important than color. In general, a dark background with light text is preferred for human perception. The dark background has an added benefit of decreasing the perception of screen flicker as the display is refreshed. However, color can attract attention to a particular area of an interface. When used correctly, color can add depth and aid in conveying importance for key elements of the display. An interface with highlighted text will attract the user's attention to that portion before any other aspect of the text. The application of color to an interface design can allow the designer to focus the user's attention to the areas of the designer's intent.

However, among the general population, 8% of males and 1% of females are colorblind, rendering color displays useless for these users. Color blindness is defined as the inability to perceive one of the three primary colors. Only rarely is a person totally color blind (monochromatic). So, if the interface is designed with contrast rather than color in mind, colorblind users will also be able to use the interface with high degree of accuracy. A quick test of the visibility of the design is to turn the monitor to black and white mode and to evaluate each of the elements of the display in grayscale. Any elements that are not easily detectable should be considered for presentation with higher contrast or a different color.

Movement, Size, and Depth

Humans view size, depth, dimensions, movement, and other distinguishing features of objects in a combinatorial fashion. Size, depth, and dimension are largely determined through the interpretation of the light or brightness of an object. Generally, objects that flash quickly indicate an emergency and attract the most immediate attention. In detecting and perceiving objects, our stored memory about an object also plays a large role. For example, if a person is looking at a pine tree that appears to be 6" tall, the human brain automatically interprets that the tree is far away from the viewer. This is because not only is the reflection of the light from the tree less intense than that of a nearby object, but also we have previous knowledge that pine trees are tens of feet tall. The combination of these two factors allows the perception that the tree might be hundreds of feet away. This principle holds true when viewing a display as well. An object that is known to the user will be interpreted relative to the other objects in the display. Therefore, a tree in the foreground should be larger than a tree in the background.

Auditory

Human auditory receptors can perceive sounds within a wide range of frequency and intensity. Based on these abilities, there are some key concepts that help in designing auditory interfaces.

Pitch and Intensity

Generally, auditory perception is a result of interpreting pitch, intensity, and duration of sound. Like visual perception, sounds should convey information consistent with existing knowledge (e.g., smoke detector sounds indicate an emergency). They should also be detectable and distinguishable from surrounding noise and other auditory signals. Humans can perceive even minute pitch or tone differences. On the other hand, auditory displays should avoid excessive auditory signals because they can become distractions rather than providing useful information to the user. When designing for an interface in an area with excessive ambient noise, a signal's auditory pitch and intensity must be determined such that it is not masked by environmental noise. Another consideration is the possibility of perceptual adaptation when the user "filter's out" a signal because it fails to provide further useful information after first encountered. Interrupted or variable signals are less likely to encounter this problem and will have a longer-lasting capability of transferring the required information.

In order to best differentiate warning sounds and signals used in an interface, the following guidelines are recommended [Sanders and McCormick, 1993]:

- Use frequencies between 200 and 5000 Hz, preferably between 500 and 3000 Hz.
- Use a modulated signal (1 to 8 beeps per second, or warbling sounds varying 1 to 3 times per second), since it is different enough from normal sounds.
- Use frequencies different from those that dominate any background noise to minimize masking.

Loudness

Sound intensity is measured in decibels, a measure of the physical compression of air in a sinusoidal waveform. It is important to recognize that both pitch and intensity affect the overall sound effect. Loudness is the human perception of this effect. Obviously, the perceived relative loudness of a sound varies by individual. Another consideration is the relative loudness of an auditory signal as it is perceived in the environment where the auditory interface is being used. For example, if an interface is to be used in the cargo area of a container ship, an auditory signal will be perceived as less loud compared to the same signal in an executive office area.

Haptic

The fingers and thumbs contain the most sensitive receptors to external pressure. Haptic interfaces take advantage of these sensitive receptors. However, the methods for transmitting complicated information through haptic senses require significant training (e.g., Braille). In its simplest form, computer users have been using their haptic sense through a keyboard, mouse, touch pad, and so on.

Haptic receptors on human hands can detect such things as mechanical vibration, electrical impulses, and physical pressure. The fingers can generally sense pressures with a two-point threshold of 0.25 cm, the thumb a bit less at 0.3 cm. Therefore, if a haptic element is planned for an interface, it is important to recognize the need to provide the user with at least that level of resistance so that the user perceives the interaction. The user will not perceive any response that is lower than the measured resistance. Touch screens are a prime example of a situation when tactile feedback contributes to the user's overall ability to control the system. In these systems, a user will manipulate the interface faster and with greater accuracy if the system provided tactile feedback that the input was received by the interface. A tactile feedback coupled with auditory cues is the best solution. An example of this is a standard keyboard. The keys provide resistance when depressed and a light tapping sound is registered, signaling that the key has been successfully engaged to convey information to the computer.

Olfactory/Taste

While these two senses are very important human receptors, there is no evidence that olfactory or taste sensations have been used successfully in human computer interface design. There is potential use for these senses as virtual reality becomes more developed. For example, a new research endeavor is being organized in "affective" computing. Affective computing attempts to provide human–computer interactions that contain feelings and emotions. In this respect, the above senses might play an important role in the future.

Human Cognitive Capabilities

In HCI, the designer must be aware of how humans acquire and process information from the interface. Therefore, human memory, reasoning (problem solving), and learning play key roles in how humans interact with computing systems. Limitations in these human abilities and large variations among them must also be understood in order to design interfaces properly.

Memory

In general, three types of human memory systems interact to give us the ability to store information. The *sensory* memory acts as a buffer for stimuli received through the senses. These memories are constantly overwritten by new information coming in on a sensory channel. Information is passed from sensory memory into *short-term* memory by attention. Attention filters the stimuli, leaving those that are of interest at a given time. This is the memory that helps us listen to a conversation while attending to other information in the background. From an HCI design perspective, an example for the use of this memory is to provide "closure" when a specific task is complete. A dialog box indicating: "Are you sure you want to exit this module …" brings closure to a subtask and allows the user to delete short-term memory storage and continue with other tasks required by the system. Lastly, a *long-term* memory is a permanent storage of factual information, experiential knowledge, procedural rules, and so on. Its characteristics are large amounts of storage, slow access and retrieval time, and a slow forgetting process. We store information in long-term memory through an elaborate network of identity connections (semantic, visual, procedural, etc.) and retrieve it by familiarity, structure, and concreteness. Two types of information retrieval have been extensively used in HCI design: recognition and recall. Information displays using memory recognition require knowledge that the information has been seen before. For example, a web page giving a choice of menus uses recognition memory. On the other hand, information displays using memory recall are based on the fact that the information is reproduced from memory. An example of this is a page that requires the user to enter a choice of airline in a data field for ticketing purposes.

Reasoning

Reasoning is the process by which we use the knowledge we have to draw conclusions or generate new inferences about the domain of interest. *Deductive* reasoning derives the logically necessary conclusions from the given set of premises. For example, the use of the word "Lassie" in a web page on dogs logically conjures up the image of the Lassie dog from a famous television serial. However, the same name in a different context might conclude a different entity. *Inductive* reasoning, on the other hand, is a generalization from the cases we have seen or experienced to infer information about cases that we have not seen or experienced. For example, the use of color red on a dial generally indicates the presence of a hazard. If we use the color red on an interface designed to show different levels of pressure, then, we should expect the users to make a similar inference. Of course, this inference may be unreliable in some other situations (e.g., the use of color red in a nuclear power control panel may denote safety rather than danger). The third type of reasoning is *abduction*, which derives a fact from the action or state that caused it. For example, in a word processor, closing a document assumes that the user has finished using or modifying the document. However, a proper design would include a dialog box asking the user if the system should save any unsaved portion.

Learning/Skill Acquisition

One of the distinguishing characteristics of human beings is the ability to learn. As such, interface designers can use this human potential to design adaptive systems. At the lowest common denominator, a designer must consider the "norm" behavior and abilities of the user population in allocating functions to any interface feature. In general, there are three basic levels of skills as users go through their learning process [Dix et al., 1998]: (1) the learner uses general-purpose rules, which interpret facts about a problem; (2) the learner develops rules specific to the task; and (3) the rules are tuned to speed up performance. From an interface design standpoint, a thorough investigation and understanding of the language and symbols of the user population can facilitate interface design because many of the rules to perform the tasks are already known as part of the user skill set. A prime example of learning gone wrong is the case of the QWERTY keyboard layout. The QWERTY keyboard layout is certainly not the optimal interface design for typing; however, through many years of learned behavior, this less-than-optimal design has become the standard for keyboard interface. Other keyboard layout designs that have proven to be faster and more accurate have not achieved market success because the relearning process would be cost prohibitive.

150.4 The Interaction

In an ideal human–computer interactive world, humans and computers would perfectly understand each other's intentions. Current popular software systems in use are quite distant from such a perfect communicative state. HCI designers must anticipate the needs of the end user to create programs that are natural to use and easy to learn. While human limitations must be considered when designing any HCI, there are a number of usability issues that help to account for user abilities and intentions in a more natural interactive fashion.

Paradigm Shifts

Perceptions about the use of computers have shifted drastically over the years. When the first electronic computer was introduced during the mid-1940s, interaction consisted of flipping a series of switches. It required weeks to correctly reprogram by changing the status of these manual switches. During the early 1970s, Xerox PARC developed the direct manipulation concepts of the WIMP and graphical user interface (GUI) paradigms that have been used by most HCI designers since the late 1980s. Since the mid-1990s, much research has been conducted in three-dimensional visual interfaces, natural language interfaces, and even direct neural interfaces — a clear paradigm shift to a more natural human–computer interaction.

Despite these shifts, it is still unclear as to what exactly constitutes good interaction. With each new paradigm there is an associated set of advantages and disadvantages. For example, while WIMP and GUI have proven extremely successful over the past 20 years, they still have many problems. The user must still be able to successfully navigate through a program to perform different tasks. For this reason most Windows programs have a basic set of tasks such as Open File, Close File, Save, and so on that are always located in the same relative place in the interface. However, beyond these basic tasks, each program has its own tasks with which a user must slowly and fitfully learn and use. For example, a paint program might require users to manipulate filters in order to modify a graphics file, whereas a word processor might require format changes in how a text file is printed. These may be considered as basic tasks within their respective program environments. While there are no obvious methods to instruct users how to perform tasks within each program, there are principles that can be used to aid program designer to convey their intent to the user. These are discussed in the subsequent sections.

150.5 Usability Principles

Over the years there have been many attempts to produce guidelines to help designers develop better interfaces. From these attempts, usability heuristics seem to be the most widely used approach. Jakob

Nielsen's *Ten Usability Heuristics* is a simple set of guidelines for HCI design and redesign improvements [Nielson, 1994]. These ten heuristics are listed below:

1. *Visibility of system status*
   The system should always keep users informed about what is going on, through appropriate feedback within reasonable time.
2. *Match between system and the real world*
   The system should speak the user's language with words, phrases, and concepts familiar to the user, rather than system-oriented terms. Follow real-world conventions, making information appear in a natural and logical order.
3. *User control and freedom*
   Users often choose system functions by mistake and will need a clearly marked "emergency exit" to leave the unwanted state without having to go through an extended dialogue (e.g., support "undo" and "redo").
4. *Consistency and standards*
   Users should not have to wonder whether different words, situations, or actions mean the same thing. Follow platform conventions.
5. *Error prevention*
   Even better than good error messages is a careful design that prevents a problem from occurring in the first place.
6. *Recognition rather than recall*
   Make objects, actions, and options visible. The user should not have to remember information from one part of the dialogue to another. Instructions for use of the system should be visible or easily retrievable whenever appropriate.
7. *Flexibility and efficiency of use*
   Accelerators — unseen by the novice user — may often speed up the interaction for the expert user such that the system can cater to both inexperienced and experienced users. Allow users to tailor frequent actions.
8. *Aesthetic and minimalist design*
   Dialogues should not contain information that is irrelevant or rarely needed. Every extra unit of information in a dialogue competes with relevant units of information and diminishes their relative visibility.
9. *Help users recognize, diagnose, and recover from errors*
   Error messages should be expressed in plain language (not code numbers), precisely indicate the problem, and constructively suggest a solution.
10. *Help and documentation*
    It is ideal if the system can be used without documentation. Nevertheless, in most systems it may be necessary to provide help and documentation. Any such information should be easy to search, focused on the user's task, list concrete steps to be carried out, and not be too lengthy.

One interesting feature of the above set of heuristics is that the ten points are mostly independent of each other and therefore can be applied separately to any interface design. And, these heuristics are general enough to be applied in various user interface domains and applications.

As an example, to increase the precision of using these heuristics, Rahimi [2002] has applied weighted scores to the heuristics to compare usability between a DOS and Windows version of a vehicle routing system. Each heuristic was used as an independent attribute and was assigned a score for each system. When the scores were summed, the DOS version received a higher overall usability score than the Windows version. This may have explained why users were consistently in favor of using the DOS routing system. This finding reinforces the notion of user preference for a simpler and more consistent interface over a sophisticated GUI innovation.

150.6 Models of Users in Design

From an HCI design perspective, there are two basic types of users: novice and expert. A basic premise is that the system must facilitate the transition from the novice level of use to expert level of use with minimum transition time and the least number of errors possible. One way of doing this is to allow the user to modify a program while minimizing the degree of task interference. Another approach is to let the system determine where the user is in the expertise spectrum and adapt itself to the user's dynamic performance, using intelligent support tools and interactive help systems.

An example for this transition is the use of "agents." The idea of agents is to create a simple, guided, and intelligent capability within the interface. Most current Microsoft products include subsidiary programs that help (novice) users perform any of a number of tasks. However, for many expert users, these agents tend to interfere with the more specialized tasks expert users typically perform. To counter this problem, expert users frequently decide to relax or remove agent-imposed constraints altogether. Microsoft Office's Paperclip is an example of a type of an agent that was designed to help users perform various tasks with Microsoft Office. However, it is considered unhelpful by many intermediate to expert users.

Modeling the User

There are several tools to analyze user interactions across an interface. One useful technique is the goal, operators, methods, selections (GOMS) model by Card et al. [1983] and its later derivatives. GOMS attempts to model and prescribe user interaction in a task domain. Typically, the focus is more on the physical interactions with the system than on a purely cognitive process. As a result, this technique has been used for analyzing how long a task should take for a group of users. For instance, a simple GOMS model of writing a letter on paper might be:

```
GOAL: WRITE-LETTER
    GOAL: LOCATE-PAPER
    GOAL: LOCATE-PEN
    GOAL: WRITE-WORDS-ON-PAPER
        WRITE-GREETING
        WRITE-THINGS-YOU-WANT-TO-SAY
        WRITE-ENDING
...
```

By assigning task times to each GOAL, one could estimate total performance time. There is an extensive literature on GOMS and other similar user modeling techniques [see Dix et al., 1998].

150.7 Task Analysis

One important tool for HCI design is *task analysis*. Dix et al. [1998] define task analysis as "the process of analyzing the way people perform their jobs: the things they do, the things they act on and the things they need to know." We use a brief example of checking for email using a computer web browser to illustrate this process:

In order to check email on a computer web browser:

find a usable computer and turn it on
start web browser
enter URL of web-based email program
enter username and password into program
check email

For each one of these tasks, one can create a more complex set of subtasks to explain the user performance in more detail. For example, we can expand "start web browser" into a series of steps such as:

find web browser icon on screen or menu
click web browser icon to start web browser program

There are various ways in which a task can be decomposed into its smaller components. We explain three classes of task analyses in this chapter:

1. Task decomposition examines the way in which a task can be broken down into subtasks and the order in which these can be performed.
2. Knowledge-based techniques examine what users need to know about the object and actions involved in performing a task and how that knowledge is organized.
3. Entity-relation-based analysis is an object-based approach that emphasizes identifying the actor and objects, the relationships between them, and the actions they perform.

Task Decomposition

As stated above, task decomposition requires breaking down a task into several subtasks and ordering them in such a way as to show how the subtasks are performed. One approach to performing task decomposition is *hierarchical task analysis* (HTA) [Shepherd, 1989]. HTA creates a listing of tasks and the order in which these tasks must be performed. We again use the email example above to illustrate how to make a simple HTA. We assume that there are no server-related problems while retrieving the e-mail. We also use a "plan" to denote the ordering of subtasks or any other special arrangements in regards to the tasks to be performed.

```
0.     In order to check email on a computer web browser
       1. If computer does not function find functional computer
       2. If computer is off turn it on
       3. If monitor is off turn it on
       4. Start web browser
       5. Check email
5.1 Enter email program URL into web browser address bar
5.2 Login to retrieve email
5.2.1 When email portal appears enter username into username datafield
5.2.2 When email portal appears enter password into password datafield
5.2.3 Send username and password
5.3 Read email
Plan 0:     do 1 before 2 - 3 - 4 - 5
do 2 - 3 in any order before 4 - 5
do 4 - 5 in that order
Plan 5:     do 5.1 - 5.2 - 5.3 in that order
Plan 5.2:   do 5.2.1 - 5.2.2 in any order before 5.2.3
            continue to do 5.2 until successfully logged in
```

Figure 150.2 converts the textual format outlined above into a graphical depiction.

A number of assumptions were made while performing this HTA. We assumed that the user knows how to operate a computer with enough proficiency to turn one on and to start a web browser program. Furthermore, we assumed that the user knows the URL of the email portal, and his or her username and password. Lastly, we assumed that the user knows how the web-based email program works. The key concept in designing any HTA is that tasks can be studied by breaking them down into smaller and often more manageable subcomponents. One of the key skills in creating an HTA, therefore, is developing an understanding about the degree of depth that meaningfully captures an actual user-task interaction. Keep in mind that any assumptions made by the modeler have significant ramifications on how a designer reflects these task interactions into the interface design requirements.

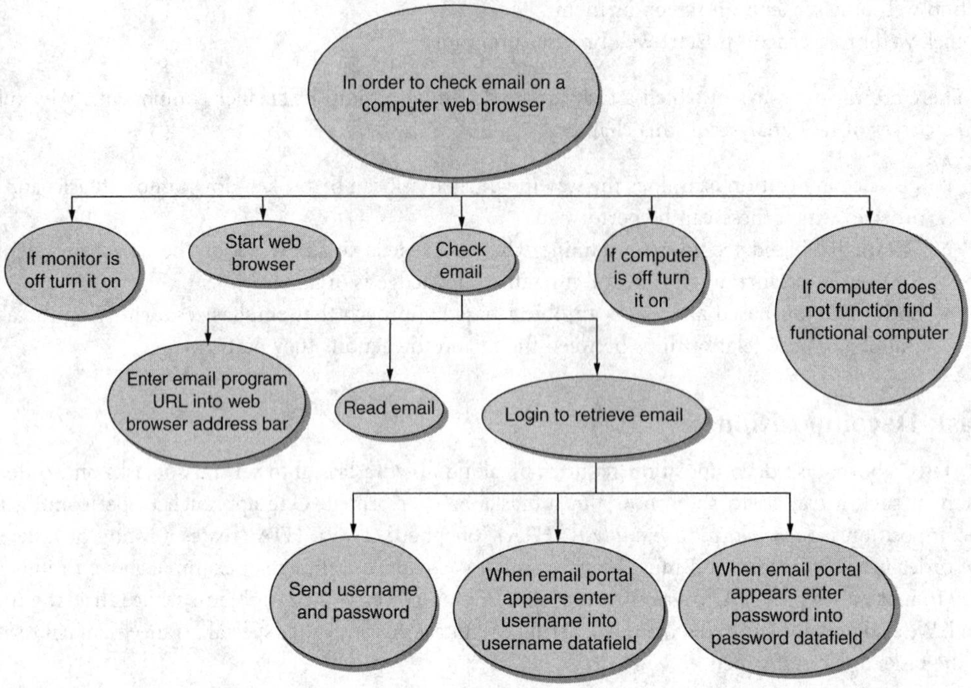

FIGURE 150.2 A graphical depiction of a Hierarchical Task Analysis for checking and reading an e-mail.

Knowledge-Based Analysis

Knowledge-based analysis is similar to task decomposition except that it creates a taxonomy or semantic-based relationship among objects and actions associated with a task. Instead of looking exclusively at how a task is performed, a knowledge-based analysis looks at how the multiple components of an interactive task fit together. One technique for performing knowledge-based analysis is task analysis for knowledge description (TAKD) [Diaper, 1989], which uses a special taxonomy called task description hierarchy (TDH).

TAKD uses a list of attributes needed to perform a task and logically relates them using TDH's three basic logical operators: AND, OR, and XOR. For instance, creating a knowledge-based analysis of the email example above, we have the following:

```
check email on computer AND
/____computer status AND
      /_______functional
      /_______turned on
/____monitor status AND
      /_______functional
      /_______turned on
/____web browser program AND
      /_______started
      /_______email URL entered
      /_______email username entered
      /_______email password entered
```

Note that unlike the task decomposition, there is no indication as to how the task is performed. Rather, the focus is on what actions and objects must be used to correctly check email on a computer. This method provides a more abstract approach to analyzing a task, thereby allowing the modeler to understand the general components required to perform a task with logical interconnections.

Entity-Relationship-Based Techniques

Entity-relationship-based analysis is an object-oriented approach. It stresses understanding the relationship of the entities to a task, rather than on the components of the task alone. It is similar in design and function to object-oriented programming. This approach shares the need to catalog actions and objects with the knowledge-based approach. For example, using the email illustration again we have:

```
Object email user human actor
     Actions:
          U1: Check if X works
          U2: Start X
          U3: Use X
          U4: Type X
          U5: Read X
Object computer simple
     Attribute:
          functional
          on
Object monitor simple
     Attributes:
          functional
          on
Object web browser simple
     Attributes:
          functional
          on
     Event:
          URL found
Relations: action-event
     before (U1, U2)
-device must work before it can be turned on
     before (U2, U3)
-device or program must be started before they can be used
     before (U2, U3)
-user must type in correct username and password before they can read
their email.
```

Unlike the previous example, the entity-relationship technique focuses on object interactions rather than on the task itself. Therefore, it is useful for creating models where less is known about a task and more is known about the objects performing a task.

150.8 Implementation Evaluation

Interface valuations are performed after a design is implemented. Therefore, these techniques center on examining how well users perform in a given environment. The evaluations could take one of several forms mentioned below.

Empirical methods frequently use statistical analysis to examine user performance. This requires gathering data indicating user performance attributes such as the time required for a user to finish a task or how many errors occur in performing a task. In its simplest form, a Student's *t*-test is performed to show differences in performance between two interface designs.

Observational methods involve watching and recording how users perform the required tasks in an interactive manner. These methods are generally employed when the users need to perform their tasks in a natural setting. Therefore, the investigators must pay special attention to the observation tools used so as not to interfere with the actual user tasks. An example for this type of technique is called verbal protocol analysis.

Query techniques "debrief" users after they have performed a task in order to understand the user's point of view. While this technique is similar to the observation method above, query techniques tend to be more informal and limit external interference. A formal survey or question/answer tool would be an excellent complement to some of the more specific task-modeling approaches mentioned previously.

150.9 Concluding Remarks

HCI design is a broad and interdisciplinary field. While we have attempted to briefly summarize key areas in this field, a vast amount of literature exists on different aspects of HCI design not covered here. We intended to explain HCI design in the context of software design process. We believe that interface design is an effort that should be addressed at all stages of software design. A proper interface design should support the tasks that users actually want to do, and forgive those that they inadvertently initiate as mistakes. Therefore, it is a design process that should consider the users as intelligent entities with specific tasks in mind, who want to use the interface in a way that is seamless with respect to their everyday jobs. In this respect, the interface as a critical component of the software should be designed to:

- Match the tasks that the users want to perform
- Be easy to use and adaptable to user-level skills and knowledge
- Provide feedback on system status and users' upcoming task options
- Provide information on task completion

Finally, the basis of this chapter seems to be users and their tasks, rather than computers and their ability to generate objects on the interface. This emphasis forces us to consider HCI as an iterative design process, since we do not have a sufficiently strong theory about human task performance. In this respect, we are dealing with a field that is both science and art.

Defining Terms

GUI — Graphical user interface.
GOMS — Goal, operators, methods, selections.
Human senses — Five essential human senses used for interacting with computers.
Paradigm shift — Perceptions about computer use in society.
Portal — A website or service that offers a broad array of resources and services, such as e-mail, forums, search engines, and online shopping malls.
Task analysis — The process of decomposing a user task into its meaningful components.
TAKD — Task analysis for knowledge description.
TDH — Task description hierarchy.
Usability — The process of evaluating software useage.
WIMP — Windows, icons, menus, and pointers.

References

Card, S., Moran, T., and Newell, A. 1983. *The Psychology of Human Computer Interaction*. Erlbaum, Hillsdale, NJ.

Diaper, D. 1989. Task analysis for knowledge descriptions (TAKD); the method and an example. In: D. Diaper, Ed., *Task Analysis for Human–Computer Interaction*. Ellis Horwood, Chichester, UK, chapter 1.

Dix, A., Finlay, J., Abowd, G., et al. 1998. *Human–Computer Interaction*, 2nd ed. Prentice Hall, London.

Nielsen, J. 1994. Heuristic evaluation. In: Nielsen, J., and Mack, R.L. Eds., *Usability Inspection Methods*. John Wiley & Sons, New York.

Rahimi, M. 2002. Computer-aided demand-response transit dispatching: a human–computer interaction perspective. Paper presented at 6th International Conference in Engineering Design and Automation, Maui, August 4–7.

Sanders, M., and McCormick, E. 1993. *Human Factors in Engineering and Design*, 7th ed. McGraw-Hill, New York.

Schneiderman, B. 1997. *Designing the User Interface: Strategies for Effective Human–Computer Interaction*, 3rd ed. Addison-Wesley, Reading, MA.

Shepherd, A. 1989. Analysis and training in information technology tasks. In: D. Diaper, Ed., *Task Analysis for Human–Computer Interaction*. Ellis Horwood, Chichester, UK, chapter 1.

Weik, M. 1961. *The ENIAC Story*. Available at: *http://ftp.arl.mil/~mike/comphist/eniac-story.html*, accessed on 3 January 2003.

Further Information

Galitz, W.O. 2002. *The Essential Guide to User Interface Design*, 2nd ed. John Wiley & Sons, New York.

Jack, J.A. and Sears, A. 2003. *The Human–Computer Interaction Handbook*. Lawrence Erlbaum Associates, Mahwah, NJ.

Molich, R. and Nielsen, J. 1990. Improving a human–computer dialogue, *Commun. ACM*, 33338–348.

Nielsen, J. 2000. *Designing Web Usability: The Practice of Simplicity*. New Rider Publishers, Indianapolis.

Schneiderman, B. 1998. *Designing the User Interface: Strategies for Effective Human–Computer Interaction*. Addison Wesley Longman, Reading, MA.

Stillings, N.A., Weisler, S.E., Chase, W.C., et al. 1995. *Cognitive Science: An Introduction*. MIT Press, Cambridge, MA.

XXII

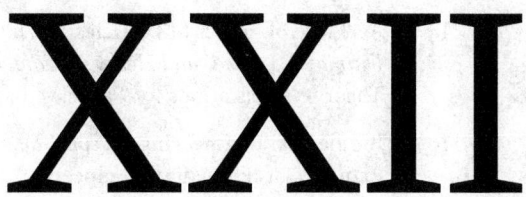

Measurement and Instrumentation

151

Sensors and Transducers

Rosemary L. Smith
University of California, Davis

Sensors are critical components in all measurement and control systems. Along with the always-present need for sensors in science and medicine, the demand for sensors in automated manufacturing and processing is rapidly growing. In addition, small, inexpensive sensors are finding their way into all sorts of consumer products, from children's toys to dishwashers to automobiles. Because of the vast variety of useful things to be sensed and sensor applications, sensor engineering is a multidisciplinary and interdisciplinary field or endeavor. This chapter introduces some basic definitions, concepts, and features of sensors and illustrates them with several examples. The reader is directed to the references and further information for more details and examples.

The terms *sensor, transducer, meter, detector,* and *gage* are often used synonymously. Generally, a transducer is defined as a device that converts energy from one form to another. However, the most widely used definition for sensor is that which has been applied to electrical transducers by the Instrument Society of America [ANSI, 1975]: "Transducer — A device which provides a usable output in response to a specified measurand." The measurand can be any physical, chemical, or biological property or condition to be measured. A usable output refers to an optical, electronic, or mechanical signal. Following the advent of the microprocessor, a usable output has come to mean to many an electronic output signal. For example, consider the common thermometer. In response to a temperature increase, the volume of mercury inside increases via thermal expansion. The change in height of the mercury column inside the glass tube is a usable output, although it is not an electrical signal. The change in mercury height is detected by the human eye, which acts as a secondary transducing element. In order to produce an electrical signal as output, the height of the mercury has to be converted, for example, by using optical or capacitive effects. However, there are more direct temperature-sensing methods in which an electrical output is produced in response to a change in temperature. An example is given in the next section on physical sensors.

Many sensors employ more than one transduction mechanism to produce a usable output signal. Sometimes sensors are classified as direct or indirect according to how many transduction mechanisms are required to produce the desired output signal. A thermocouple, used to measure temperature, is a direct sensor because it produces an electrical voltage directly as output in response to a temperature gradient. However, most flow rate sensors are indirect sensors. Flow rate is usually inferred from a secondary measurement, such as the displacement of an object placed in the flow stream (e.g., rotome-

0-8493-1586-7/05/$0.00+$1.50

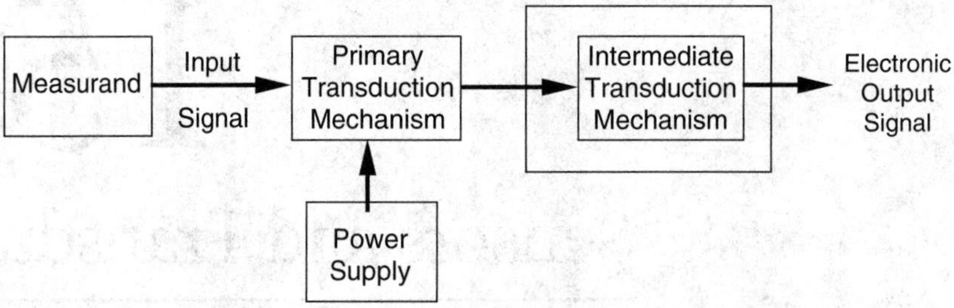

FIGURE 151.1 Sensor block diagram. Active sensors require input power to accomplish transduction. Many sensors employ multiple transduction mechanisms in order to produce an electronic output in response to the measurand.

ters), the temperature measured downstream from a hot element (anemometer), or the difference in pressure measured at two points along the flow path. Figure 151.1 depicts a typical sensor block diagram identifying the measurand and associated input signal, the primary and intermediate transduction mechanisms, and the electronic output signal. Indirect sensors typically require an electrical power source be applied in order to produce an electronic output signal. A piezoresistor, for example, is a resistor that changes value when strained. But in order to generate a corresponding electrical signal, a current is passed through the resistor to generate an output voltage signal, whose value is related to the resistance by Ohm's law: $V = I \cdot R$. Table 151.1 is a 6 × 6 matrix of some of the more commonly employed physical and chemical transduction mechanisms. Detailed descriptions of these mechanisms can be found in college-level physics textbooks.

In choosing a particular sensor for a given application, many factors must be considered. These deciding factors, or specifications, can be divided into three major categories: environmental factors, economic factors, and sensor performance. The most commonly encountered factors are listed in Table 151.2, although not all of these may be pertinent to a particular application. Environmental factors determine the *packaging* of the sensor — which refers to its encapsulation or insulation that provides protection or isolation — and the input/output, leads, connections, and cabling. Economic factors determine the type of manufacturing and materials used in the sensor, and to some extent, the quality of the materials. For example, a very expensive sensor may be cost-effective if it is used repeatedly, operates reliably for very long periods of time, or has exceptionally high performance. On the other hand, a disposable sensor, such as is desired in many medical applications, should be inexpensive. The performance requirements of the sensor are usually the specifications of primary concern. The most important of these are *sensitivity, stability,* and *repeatability.* Normally, a sensor is only useful if all three of these parameters are tightly specified for a given range of measurand and time of operation. For example, a highly sensitive device is not very useful if its output signal drifts greatly during the measurement time, and the data obtained are not reliable if the measurement is not repeatable. Other output signal characteristics, such as selectivity and linearity, can often be compensated for by using additional independent sensors or with signal-conditioning circuits. In fact, most sensors have a response to temperature, since most transduction mechanisms are temperature dependent.

Sensors are most often classified by the type of measurand they are to measure — that is, physical, chemical, or biological. This is a much simpler means of classification than by transduction mechanism or type of output signal (e.g., digital or analog), since many sensors use multiple transduction mechanisms and the output signal can always be processed, conditioned, or converted by a circuit so as to cloud the definition of output. A description of each classification and examples are given in the following sections. In Section 151.4, *microsensors* are introduced with a few examples.

151.1 Physical Sensors

Physical measurands include temperature, strain, force, pressure, displacement, position, velocity, acceleration, optical radiation, sound, flow rate, viscosity, and electromagnetic fields. In Table 151.1, note that

TABLE 151.1 Physical and Chemical Transduction Principles

	Secondary Signal					
Primary Signal	Mechanical	Thermal	Electrical	Magnetic	Radiant	Chemical
Mechanical	(Fluid) mechanical andacoustic effects (e.g., diaphragm, gravity balance, echo sounder)	Friction effects (e.g., frictioncalorimeter), cooling effects (e.g., thermal flowmeters)	Piezoelectricity; piezoresistivity; resistive, capacitive, andinductive effects	Magnetomechanical effects (e.g., piezomagnetic effect)	Photoelastic systems (stress-induced birefringence), interferometers, Sagnac effect, Doppler effect	
Thermal	Thermal expansion (bimetal strip, liquid in glass, and gas thermometers: resonantfrequency), radiometer effect (light mill)		Seebeck effect, thermoresistance, pyroelectricity, thermal (Johnson) noise		Thermo-optical effects (e.g., in liquid crystals), radiant emmission	Reaction activation (e.g., thermal dissociation)
Electrical	Electrokinetic and electromechanicaleffects (e.g., piezoelectricity, electrometer, Ampere's law)	Joule (resistive) heating, Peltier effect	Charge collectors, Langmuir probe	Biot–Savart's law	Electro-optical effects (e.g., Kerr effect, Pockels effect, electroluminescence)	Electrolysis, electromigration
Magnetic	Magnetomechanicaleffect (e.g., magnetorestriction, magnetometer)	Thermomagnetic effects (e.g., Righi-Leduc effect), galvanomagnetic effects (e.g., Ettingshausen effect)	Thermomagnetic effects (e.g., Ettingshausen–Nernst effect), alvano-magnetic effects (e.g., Hall effect, magnetoresistance)		Magneto-optical effects (e.g., Faraday effect, Cotton–Mouton effect)	
Radiant	Radiation pressure	Bolometer, thermopile	Photoelectric effects (e.g., photovoltaic effect, photoconductive effect)		Photorefractive effects, optical bistability	Photosynthesis, dissociation
Chemical	Hygrometer, electrodeposition cell, photoacousticeffect	Calorimeter, thermal conductivity cell	Potentiometry, conductimetry, amperometry, flameionization, Volta effect, gas sensitive field effect	Nuclear magnetic resonance	(Emission and absorption) spectroscopy, chemiluminiscence	

Source: Grandke, T. and Hesse, J. Fundamentals and general aspects. Vol. 1 of Gopel, W., Hesse, J., and Zemel, J. N., Eds., *Sensors: A Comprehensive Survey,* VCH, Weinheim, Germany. With permission.

TABLE 151.2 Factors for Considerations in Sensor Selection

Environmental Factors	Economic Factors	Sensor Performance
Temperature range	Cost	Sensitivity
Humidity effects	Availability	Range
Corrosion	Lifetime	Stability
Size	Performance	Repeatability
Over-range protection		Linearity
Susceptibility to EM interferences		Accuracy
Self-test capability		Response time
Ruggedness		Frequency response
Power consumption		

all but those transduction mechanisms listed in the chemical row are used in the design of physical sensors. Clearly, physical sensors compose a very large proportion of all sensors. It is impossible to illustrate all of them, but three measurands stand out in terms of their widespread application: temperature, displacement (or associated force), and optical radiation.

Temperature Sensors

Temperature is an important parameter in many control systems, most familiarly in environmental control systems. Several distinct transduction mechanisms have been employed. The mercury thermometer was mentioned earlier as a nonelectronic sensor. The most commonly used electronic temperature sensors are thermocouples, thermistors, and resistance thermometers. Thermocouples operate according to the Seebeck effect, which occurs at the junction of two dissimilar metal wires, such as copper and constantan. A voltage difference is generated at the hot junction due to the difference in the energy distribution of thermally energized electrons in each metal. A temperature gradient between the hot junction and the cooler ends produces the requisite, albeit minute, current that enables the measurement of the junction voltage between the ends of the two wires. The output voltage changes linearly with temperature over a given range, depending on the choice of metals. To minimize measurement error, the cool end of the couple must be kept at a constant temperature and the voltmeter must have a high input impedance. Connecting thermocouples in series creates a thermopile that produces an output voltage approximately equal to that produced from a single thermocouple voltage multiplied by the number of couples.

The resistance thermometer relies on the increase in resistance of a metal wire with increasing temperature. As the electrons in the metal gain thermal energy, they move about more rapidly and undergo more frequent collisions with each other and the atomic nuclei. These scattering events reduce the mobility of the electrons, and since resistance is inversely proportional to mobility, the resistance increases. Resistance thermometers usually consist of a coil of fine metal wire. Platinum wire gives the largest linear temperature range of operation. To determine the resistance, a constant current is supplied and the voltage is measured. A resistance measurement can be made by placing the resistor in the sensing arm of a Wheatstone bridge and adjusting the opposing resistor to "balance" the bridge, which produces a null output. A measure of the sensitivity of a resistance thermometer is its temperature coefficient of resistance: $\mathrm{TCR} = (\Delta R/R)(1/\Delta T)$, in units of percent resistance per degree of temperature.

Thermistors are resistive elements made of semiconductive materials. Semiconductors have a negative coefficient of resistance that can be up to 100 times higher in value than the TCR of a metal, which is positive. The mechanism governing the resistance change of a thermistor is the increase in the number of charge carriers with increase in temperature due to thermal generation. Electrons that are tightly bound to the nucleus by coulombic attraction, called the valence electrons, gain sufficient thermal energy to break away from the nucleus and become influenced by external electric fields. Thermistors are measured in the same manner as resistance thermometers, but are much more sensitive.

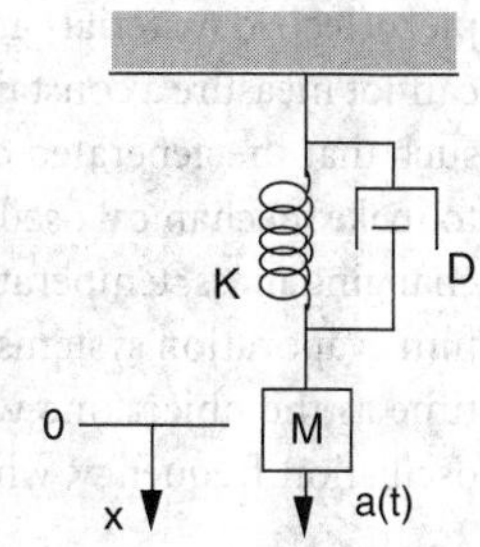

FIGURE 151.2 Schematic diagram of a mechanical oscillator made up of a mass at the end of a spring. The displacement of the mass, $x(t)$, in response to an acceleration, $a(t)$, is given by $F = M \cdot a(t) = M\ddot{x} + D\dot{x} + Kx$, where K is the spring constant, D is viscous drag, M is the mass, and F is the net force exerted on the mass. For a constant acceleration due to gravity, g, the steady-state solution to this differential equation yields $x = Mg/K$. The displacement of a known mass, therefore, can be used to determine its acceleration.

Displacement and Force Sensors

Many types of forces are sensed by the displacements they create. For example, the force due to acceleration of a mass at the end of a spring will cause the spring to stretch and the mass to move. This phenomenon is illustrated in Figure 151.2. The displacement, x, of the mass, M, from its zero acceleration position, is governed by the force generated by acceleration ($F = M \cdot a$) and the restoring force of the spring ($F = -K \cdot x$). For a constant acceleration due to gravity, g, the steady state displacement, x, is equal to $M \cdot g/K$. Therefore, the force due to acceleration can be determined by measuring the displacement of a known mass on which the force is acting.

Another example of force sensed by displacement is the displacement of the center of a deformable membrane due to a difference in pressure across it. Both the pressure sensor and the accelerometer examples use multiple transduction mechanisms to produce an electronic output. The primary mechanism converts force to displacement (mechanical to mechanical). An intermediate mechanism is used to convert displacement to an electrical signal (mechanical to electrical). Displacement can be determined by measuring the capacitance between two plates that move with respect to each other. The gap between the two plates is the displacement, and the associated capacitance is given by C = area × dielectric constant/gap length. The ratio of plate area to gap length must be greater than 100, since most dielectric constants are on the order of $1 \cdot 10^{-13}$ farads/cm and capacitance is readily resolvable to only about 10^{-11} farads. This is because measurement leads and contacts create parasitic capacitances of about 10^{-12} farads. If the capacitance is measured at the generated site by an integrated circuit (see discussion of microsensors), capacitances as small as 10^{-15} farads can be measured. Displacement is also commonly measured by the movement of a ferromagnetic core inside an inductor coil. The displacement produces a change in inductance that can be measured by placing the inductor in an oscillator circuit and measuring the change in frequency of oscillation.

The most commonly used force sensor is the strain gage. It consists of metal wires that are fixed to an immobile structure at one end and to a deformable element at the other. The resistance of the wire changes as it undergoes strain; that is, resistance changes in length, since the resistance of a wire is R = resistivity × length/cross-sectional area. The wire's resistivity is a bulk property of the metal, which is a constant for constant temperature. For example, a strain gage can be used to measure force by attaching both ends of the wire to a cantilever beam, with one end of the wire at the attached beam end and the other at the free end. The cantilever-beam free end moves in response to an applied force that produces strain in the wire and a subsequent change in resistance. The sensitivity of a strain gage is described by the unitless gage factor, $G = (\Delta R/R)/(\Delta L/L)$. For metal wires, gage factors typically range from 2 to 3. Semiconductor materials are known to exhibit *piezoresistivity*, a change in resistance in response to strain that involves a large change in resistivity in addition to the change in linear dimension. Piezoresistors have gage factors as high as 130. Piezoresistive strain gages are frequently used in silicon microsensors because they can be fabricated inside the same semiconductor material as the micromechanical device.

Piezoelectric sensors use materials that produce electric charge, measured as a voltage, at their surface in response to mechanical deformation, for example by application of pressure. The charge is generated by the deformation of dipoles that are a consequence of the molecular structure of the material itself. The inverse effect is also true, that is, application of a voltage will produce deformation. Examples of piezoelectric materials are polyvinyldifluoride (PVDF), quartz, and ZnO. Piezoelectric sensors generally cannot measure a constant pressure or strain, because the materials employed are not perfectly insulating, such that the generated charge eventually leaks away. Hence, most often, piezoelectric materials are used to make mechanical oscillators, which in turn can be used to sense several different measurands, including changing mass, temperature, and viscosity. For example, a very common film thickness monitor for thin film evaporation systems is one that uses a quartz oscillator, onto which the film is deposited at the same time as the object one wishes to coat. The added film increases the mass of the oscillator, changing its oscillation frequency, which is measured by an accompanying electronic circuit.

Optical Radiation Sensors

The intensity and frequency of optical radiation are parameters of ever-growing interest and utility in consumer products, such as the video camera and home security systems, and in optical communications systems. The conversion of optical energy to electronic signals can be accomplished by several mechanisms; however, the most commonly used is the photogeneration of electrons in semiconductors. Band-to-band generation requires that the energy of the incident photons, $E = h\nu$, is greater than the semiconductor bandgap. The most often used device for this is the semiconductor PN junction photodiode. The construction of this device is very similar to that of diodes used in electronic circuits as rectifiers. The photodiode is operated in reverse bias, where very little current normally flows. When light is incident on the structure and is absorbed in the semiconductor, energetic electrons are produced. These electrons flow in response to the electric field sustained internally across the junction, producing an externally measurable current. The current magnitude is proportional to the light intensity and also depends on the frequency of the light according to the semiconductor material used.

Sensing infrared radiation (IR) by means of photogeneration requires very small bandgap semiconductors, such as HgCdTe. Since electrons in semiconductors are also generated by thermal energy (e.g., thermistors), these IR sensors require cryogenic cooling in order to achieve reasonable sensitivity. Another means of sensing IR is to first convert the optical energy to heat inside an absorbing material, and then measure the accompanying temperature change. Accurate and highly sensitive measurements require passive temperature sensors, such as a thermopile or a pyroelectric sensor, since active devices can themselves generate enough joule heating to effect a temperature change.

151.2 Chemical Sensors

Chemical measurands include ion concentration, chemical composition, rate of reactions, reduction-oxidation potentials, and gas concentration. The last row of Table 151.1 lists some of the transduction mechanisms that have been, or could be, employed in chemical sensing. Two examples of chemical sensors are described here: the ion-selective electrode (ISE) and the gas chromatograph. These sensors were chosen because of their general use and availability and because they illustrate the use of a primary (ISE) versus a primary plus intermediate (gas chromatograph) transduction mechanism.

Ion-Selective Electrode

As the name implies, ISEs are used to measure the concentration of a specific ion concentration in a solution of many ions. To accomplish this, a membrane that selectively generates a potential dependent on the concentration of the ion of interest is used. The generated potential is an equilibrium potential, called the *Nernst potential*, which develops across the interface of the membrane and the solution under test. This potential is generated by the initial net flow of ions (charge) across the membrane in response to a concentration gradient. From that moment on, the diffusional force becomes balanced by an opposing electric force, and equilibrium is established. This potential is very similar to the built-in potential of a PN junction diode, except that ions are the charge carriers instead of electrons. The ion-selective membrane acts in such a way as to ensure that the generated potential is dependent mostly on a particular

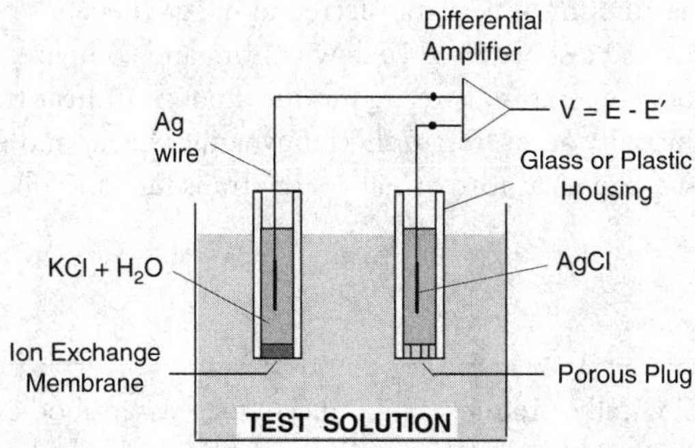

FIGURE 151.3 An electrochemical cell comprising an ion-selective electrode (ISE) and reference electrode in an ionic solution. The ion exchange membrane of the ISE generates a potential, *E*, that is proportional to the log of the concentration of the ion of interest.

ion of interest and negligibly on any other ions in solution. This condition is achieved by enhancing the rate at which the ion of interest is transported across the membrane relative to all others, so that it is the ion that generates and maintains the potential.

The most familiar ISE is the pH electrode. In this device the membrane is a thin, sodium-rich glass that exhibits a high exchange rate for H^+. The generated Nernst potential, *E*, is described by the expression: $E = E_0 + (RT/F)\text{In}[H^+]$, where E_0 is a constant for constant temperature, *R* is the gas constant, and *F* is the Faraday constant. The pH is defined as the negative of the $\log[H^+]$; therefore $\text{pH} = (E_0 - E)(2.3)F/RT$. One pH unit change corresponds to a ten-fold change in the molar concentration of H^+ and a 59 mV change in the Nernst potential at room temperature. Other ISEs exhibit the same form of voltage dependence with concentration, but the response is specific to a different ion, depending on the choice of membrane. Some ISEs employ ionophores trapped inside a polymeric membrane to generate selectivity. An ionophore is a molecule that selectively and reversibly binds with an ion. When placed inside a membrane, it provides a source (or sink), thereby creating a high exchange rate for that particular ion.

The typical ISE consists of a glass or plastic tube with the ion-selective membrane closing the end of the tube that is immersed into the test solution (see Figure 151.3). The Nernst potential is measured by making electrical contact to either side of the membrane. The solution side of the membrane is contacted by a reference electrode placed inside the same solution under test. The reference electrode is constructed in the same manner as the ISE, but it has a porous membrane or leaky plug that creates a liquid junction between its inner filling solution and the test solution. That junction is designed so that the potential across it is ideally invariant with changes in concentration of any ion in the test solution. The opposite side of the membrane is contacted by a fixed-concentration, conductive, filling solution and specially coated metal wire inside the tube. The reference electrode, the solution under test, and the ISE form an electrochemical cell. The reference electrode potential acts like the ground reference in electric circuits, and the ISE potential is measured between the two wires emerging from the respective electrodes. The details of the mechanisms of transduction in ISEs and conditions necessary to create a stable electrochemical cell are beyond the scope of this article. See Bard and Faulkner [1980] or Koryta [1975] for more details.

Gas Chromatograph

Molecules in gases have thermal conductivities that are dependent on their masses; therefore, a pure gas can be identified by its thermal conductivity. One way to determine the composition of a gas is to first separate it into its components and then measure the thermal conductivity of each. A gas chromatograph does exactly that. The gas mixture flows through a long narrow column, which is packed with an adsorbent solid (for gas-solid chromatography), wherein the gases are separated according to the retentive properties of the packing material for each gas. As the individual gases exit the end of the tube one at a time, they flow over a heated wire. The amount of heat transferred to the gas depends on its thermal conductivity. The gas temperature is measured a short distance downstream and compared to a known gas flowing in a separate sensing tube. The temperature is related to the amount of heat transferred and can be used to derive the thermal conductivity according to thermodynamic theory and empirical data. This sensor requires two transductions: a chemical-to-thermal energy transduction, followed by a thermal-to-electrical transduction.

151.3 Biosensors

Biosensors respond to biologically produced substances, such as antibodies, glucose, hormones, and enzymes. A biosensor is not the same as a biomedical sensor, which can be any type of sensor that is used in biomedical applications, such as a blood pressure sensor or hear rate monitor. Many biosensors are biomedical sensors; however, biosensors are also used in industrial applications, such as the monitoring and control of fermentation reactions or in drug discovery. Table 151.1 does not include biological signals as primary signals because they can usually be classified as either chemical or physical in nature. Biosensors are of special interest because of the very high selectivity associated with biological reactions and binding. However, detection of that reaction or binding is often elusive. A very familiar commercial biosensor is the in-home pregnancy test sensor, which detects the presence of human growth factor in urine. That device is a nonelectronic sensor since the output is a color change that your eye senses. In fact, most biosensors require multiple transduction mechanisms to arrive at an electrical output signal. Two examples are given below: immunosensor and enzyme sensor.

Immunosensor

Most commercial techniques for detecting antibody-antigen binding use optical fluorescence detection. For example, an optically fluorescent molecule can be attached to the species of interest by selective chemical attachment via a linking molecule. The fluorescent tag is introduced into the test solution where it teams up with the target species, if it is present. The complementary binding species is chemically bound to a glass substrate, or onto glass beads that are packed into a column. The solution containing the tagged species of interest, say the antibody, is passed over the antigen-coated surface, where the two selectively bind. The nonbound fluorescent molecules are washed away, and the antibody concentration is determined by the intensity of fluorescence emitted when the antigen-coated surface is excited by a UV (typical) light source. These sensing techniques are quite costly and bulky, and therefore other biosensing mechanisms are rapidly being developed and commercialized. One technique uses the change in the permittivity or index of refraction of the bound antibody-antigen complex in comparison to an unbound surface layer of antigen. The technique utilizes surface plasmon resonance as the mechanism for detecting the change in permittivity. Surface plasmons are charge density oscillations that propagate along the interface between a dielectric and a thin metal film. They can be generated by the transfer of energy from an incident, transverse magnetic (TM) polarized light beam. Plasmon resonance, or maximum energy transfer, occurs when the light beam is incident at a specific angle that depends on, and is extremely sensitive to, the dielectric permittivity at the interface on either side of a given metal film. At resonance, the optical beam intensity reflected from the metal film drops to a minimum; therefore, the resonance angle can be determined by recording reflected intensity (e.g., with a photodiode), versus incident light beam angle. The binding of antibody to an antigen layer attached to the metal film will produce a change in this resonance angle because of the minute change in permittivity produced at the interface.

Enzyme Sensor

Enzymes selectively react with a chemical substance to modify it, usually as the first step in a chain of reactions to release energy (metabolism). A well-known example is the selective reaction of glucose oxidase (enzyme) with glucose to produce gluconic acid and peroxide, according to the following formula:

$$C_6H_{12}O_6 + O_2 \xrightarrow{\text{glucose oxidase}} \text{gluconic acid} + H_2O_2 + 80 \text{ kilojoules heat}$$

An enzymatic reaction can be sensed by measuring the rise in temperature associated with the heat of reaction or by the detection and measurement of the reaction by-products. In the glucose example, the reaction can be sensed by measuring the local dissolved peroxide concentration. This is done via an electrochemical analysis technique called *amperometry* [Bard and Faulkner, 1980]. In this method, a potential is placed across two inert metal wire electrodes immersed in the test solution, and the current that is generated by the reduction/oxidation (redox) reaction of the species of interest is measured. The current is proportional to the concentration of the reducing/oxidizing species. A selective response is obtained if no other species in the vicinity of the electrodes has a lower redox potential. Because the selectivity of peroxide over oxygen is poor, some glucose-sensing schemes employ a second enzyme called *catalase*, which converts the peroxide to oxygen and hydroxyl ions. This produces a change in the local pH. As described earlier, an ISE can then be used to convert the pH to a measurable voltage. In this example, glucose sensing involves two chemical-to-chemical transductions followed by a chemical-to-electrical transduction mechanism.

151.4 Microsensors

Microsensors are manufactured using integrated circuit fabrication techniques and/or micromachining technology. Integrated circuits are fabricated using a series of process steps that are done in batch fashion, meaning that thousands of circuits are processed together at the same time in the same way. The patterns that define the components of the circuit are photolithographically transferred from a master template to a semiconducting substrate using a photosensitive organic coating, called photoresist. The coating pattern is then transferred into the substrate, or into a thin film sitting on top of the substrate through an etching process. The photolithographic transfer process is illustrated in Figure 151.4. This process is repeated for each additional thin film that is deposited, building up the respective layers comprising the circuit elements. Each template, called a *photomask,* can contain thousands of identical sets of patterns, with each set representing a circuit. This "batch" method of manufacturing is what makes integrated circuits so reproducible and inexpensive. In addition, photoreduction enables one to make extremely small features, on the order of microns, which is why this collection of process steps is referred to as *microfabrication.* The resulting integrated circuit is contained in only the top few microns of the semiconductor substrate and the submicron thin films on its surface. Hence, integrated circuit technology is said to consist of a set of planar microfabrication processes.

Micromachining refers to the set of processes that produce three-dimensional microstructures using technology that is similar to integrated circuit manufacturing methods. The third dimension refers to the height above or the depth into the substrate. Micromachining produces objects with third dimension in the range of 1 to 500 microns. The use of microfabrication to manufacture sensors promises the same

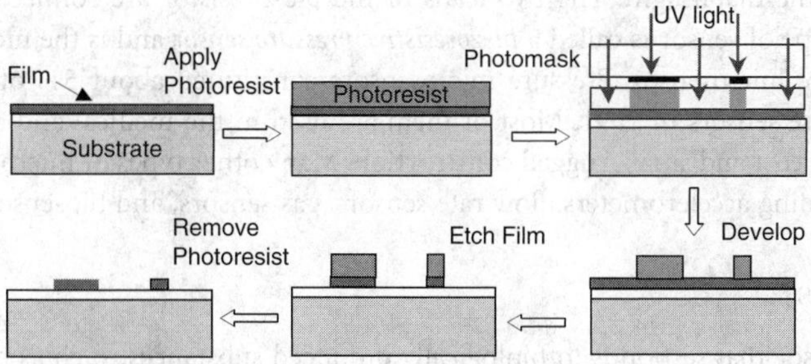

FIGURE 151.4 An illustration of the photolithographic process used to transfer a pattern from a master templete (photomask) to a thin solid film in integrated circuit manufacturing.

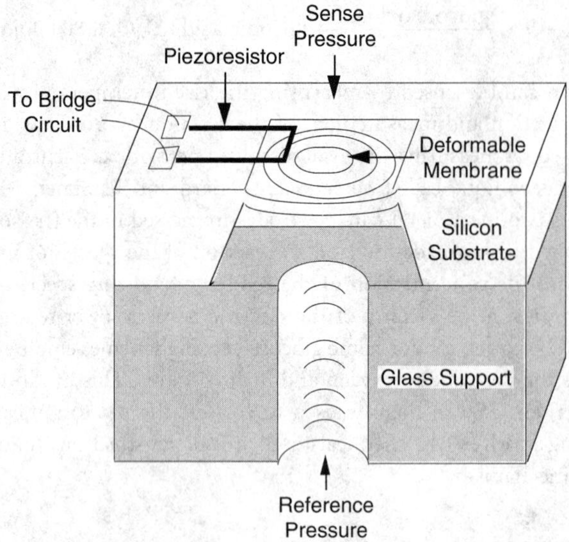

FIGURE 151.5 Schematic cross-section of a silicon piezoresistive pressure sensor. A differential pressure deforms the silicon diaphragm, producing strain in the integrated piezoresistor. The change in resistance is measured via a Wheatstone bridge.

benefits as it does for circuits: low cost per sensor, small size, and highly reproducible geometries. It also enables the integration of circuits, sensors, and actuators — that is, entire sensing and control systems — which provides a dramatic improvement in capability for very little increase in cost. For these reasons there is a great deal of research and development activity in microsensors and microsystems. The first microsensors were standard integrated circuit components, such as semiconductor resistors and PN junction diodes. The piezoresistivity of semiconductors and optical sensing by the photodiode were discussed earlier. Diodes can also be used as temperature-sensing devices. When forward biased with a constant diode current, the resulting diode voltage will increase approximately linearly with increasing temperature. The first micromachined microsensor to be commercially produced was the silicon pressure sensor. It was invented in the mid- to late 1950s at Bell Labs and commercialized in the 1960s by General Electric, Endevco, and Fairchild Control Division (now Foxboro/ICT, Inc.). This device contains a thin silicon diaphragm ($\approx$10 microns) that deforms in response to a pressure difference across it (Figure 151.5). The deformation produces two effects: a position-dependent displacement, which is maximum at the diaphragm center, and position-dependent strain, which is maximum near the diaphragm edge. Both of these effects have been used to produce an electrical output that is proportional to differential pressure. Membrane center displacement is sensed capacitively, as previously described, in one type of pressure sensor. In another, the strain is sensed by placing a piezoresistor, fabricated in the same silicon substrate, along one edge of the diaphragm. The two leads of the piezoresistor are connected to a Wheatstone bridge. The latter type of sensor is called a *piezoresistive pressure sensor* and is the more common type of commercial pressure microsensor. Pressure microsensors constituted about 5% of the total U.S. consumption of pressure sensors in 1991. Most of them are used in the medical and automotive industry because of their low cost and small, rugged construction. Many other types of microsensors are available commercially, including accelerometers, flow rate sensors, gas sensors, and biosensors.

Defining Terms

Biosensor — A sensor that responds to biologically produced substances, such as enzymes, antibodies, viruses, RNA, and hormones.

Micromachining — The set of processes that produce three-dimensional microstructures using sequential photolithographic pattern transfer and etching or deposition in a batch-processing method.
Microsensor — A sensor that is fabricated using integrated circuit and micromachining technologies.
Repeatability — The ability of a sensor to reproduce output readings for the same value of measurand, when applied consecutively and under the same conditions.
Sensitivity — The radio of the change in sensor output to a change in the value of the measurand.
Sensor — A device that produces a usable output in response to a specified measurand.
Stability — The ability of a sensor to retain its characteristics over a relatively long period of time.
Transducer — A device that converts one form of energy to another.

References

ANSI. 1975. *Electrical Transducer Nomenclature and Terminology.* ANSI Standard MC6.1-1975. (ISA S37.1). Instrument Society of America, Research Triangle Park, NC.

Bard, A. J. and Faulkner, L. R. 1980. *Electrochemical Methods: Fundamentals and Applications.* John Wiley & Sons, New York.

Carstens, J. R. 1993. *Electrical Sensors and Transducers.* Regents/Prentice Hall, Englewood Cliffs, NJ.

Cobbold, R. S. C. 1974. *Transducers for Biomedical Measurements: Principles and Applications.* John Wiley & Sons, New York.

Fraden, J., "Handbook of Modern Sensors," 2nd ed. American Institute of Physics Press, Woodbury, NY, 1996.

Grandke, T. and Ko, W. H. 1989. Fundamentals and general aspects. Vol. 1 of Gopel, W., Hesse, J., and Zemel, J. N., Eds., *Sensors: A Comprehensive Survey,* VCH, Weinheim, Germany.

Janata, J. 1989. *Principles of Chemical Sensors.* Plenum Press, New York.

Koryta, J. 1975. *Ion-Selective Electrodes.* Cambridge University Press, London.

Norton, H. N. 1989. *Handbook of Transducers.* Prentice Hall, Englewood Cliffs, NJ.

Further Information

Sze, S. M., Ed., *Semiconductor Sensors.* Wiley, New York, 1994.

Gopel, W., Hesse, J., and Zemel, J. N., Eds., *Sensors: A Comprehensive Survey.* VCH, Weinheim, Germany. Vols. 2, 3: Mechanical Sensors, 1991; Vol. 4: Thermal Sensors, 1990; Vol. 5: Magnetic Sensors, 1989; Vol. 6: Optical Sensors, 1991; Vol. 7, 8: Chemical and Biochemical Sensors, 1990.

Sensors and Actuators is a technical journal devoted to solid-state sensors and actuators. It is published bimonthly by Elsevier Press in two volumes: Volume A: Physical Sensors and Volume B: Chemical Sensors.

The International Conference on Solid-State Sensors and Actuators is held every 2 years, hosted in rotation by the United States, Japan, and Europe. It is sponsored in part by the Institute of Electrical and Electronic Engineers (IEEE). The *Digest of Technical Papers* is published by and available through IEEE, Piscataway, NJ.

152

Measurement Errors and Uncertainty

Steve J. Harrison
Queen's University

Ronald H. Dieck
Ron Dieck Associates Inc.

Engineers are increasingly being asked to monitor or evaluate the efficiency of a process or the performance of a device. When a measurement result is as expected, there is the temptation to claim that the data are "accurate." When unexpected results occur, often the claim is made that there must be major errors and engineers begin searching for them. What is needed is an objective method for assessing data quality or measurement uncertainty.

Results are often derived from the combination of values determined from a number of individual measurements. In the planning of an experiment, an attempt is usually made to minimize the error associated with the experimental measurements. Unfortunately, every measurement is subject to error, and the degree to which this error is minimized is a compromise between the (overall) accuracy desired and the expense required to reduce the error in the component measurements to an acceptable value.

Good engineering practice dictates that an indication of the error or uncertainty should be reported along with the derived results. Occasionally, the error in a derived result would be treated as the sum of the errors of the component measurements. Implicit in this assumption is that the worst-case errors will occur simultaneously in all the measurements and in the most detrimental fashion. This condition is unlikely, and therefore such an analysis usually results in an over prediction of the error in a derived result.

A realistic estimate of error cannot be achieved. Error cannot be known. It is possible only to estimate (with some confidence) the potential size of a particular error or combination of errors. This estimation process is called measurement uncertainty analysis. The estimation of these uncertainties and their combination process in a measurement process are detailed in two excellent sources. The first is *Guide to the Expression of Uncertainty in Measurement* [ISO, 1993]. It is especially suited to meteorologists. The second is the American Society of Mechanical Engineers' PTC 19.1 publication on test uncertainty [ASME, 1998]. Its approach is better suited for engineers and scientists. The following discussion is meant to provide insight into measurement uncertainty rather than a rigorous treatment of its theoretical basis. Readers are encouraged to consult the references noted dealing with these issues if they consider it necessary.

0-8493-1586-7/05/$0.00+$1.50

152.1 Measurement Errors and Uncertainty

Measurement error may be defined as the difference between the true value and the measured value of the quantity:

$$E = M - T$$

where E is the error in x, M is a measured or observed value of some physical quantity, and T is the actual value.

The errors that occur in an experiment are usually categorized as mistakes or recording errors (blunders), systematic or fixed errors, and accidental or random errors.

- Mistakes or recording errors are usually the result of blunders (e.g., the observer reads 10.1 instead of 11.1 units on the scale of a meter). It is assumed that careful experimental practices will minimize the occurrence of these blunders. They are not a part of an objective uncertainty analysis.
- Systematic errors (formerly called bias errors) are errors that persist and cannot be considered due entirely to chance. Systematic errors may result from the residual errors in instrument calibration and relate to instrument performance (the ability of the instrument to indicate the true value). These errors do not change for the duration of a test or experiment.
- Random errors cause readings to take random values on either side of some mean value. They may be due to the observer or the instrument and are revealed by repeated observations. They are disordered in incidence and variable in magnitude.
- If an error source causes scatter in test data, it is a random source. If not, it is systematic.

In measurement systems, accuracy generally refers to the closeness of agreement of a measured value and the true value. All measurements are subject to both systematic and random errors to differing degrees, and consequently the true value can only be estimated. To illustrate the above concepts, consider the case shown in Figure 152.1, where measurements of a fixed value are taken over a period of time.

As illustrated, the measured values are scattered around some mean value, x_m. In effect, if many measurements of the quantity were taken, we could calculate a mean value of the fixed quantity. If we further grouped the data into ranges of values, it would be possible to plot the frequency of occurrence in each range as a histogram (Figure 152.2) If a large number, N, of measurements were recorded (i.e., as N goes to infinity), a plot of the frequency distribution of the measurements, $f(X)$, could be constructed (Figure 152.3). Figure 152.3 is often referred to as a plot of the probability density function, and the area under the curve represents the probability that a particular value of X(the measured quantity) will occur. The total area under the curve has a value of 1, and the probability that a particular measurement will fall within a specified range (e.g., between X_1 and X_2) is determined by the area under the curve bounded by these values. Figure 152.3 indicates that there is a likelihood of individual measurements being close to X_m, and that the likelihood of obtaining a particular value decreases for values farther away from the mean value, X_m.

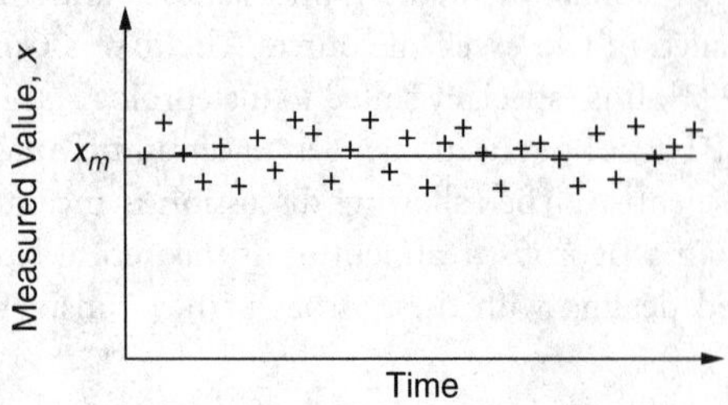

FIGURE 152.1 Repeated measurements of a fixed value.

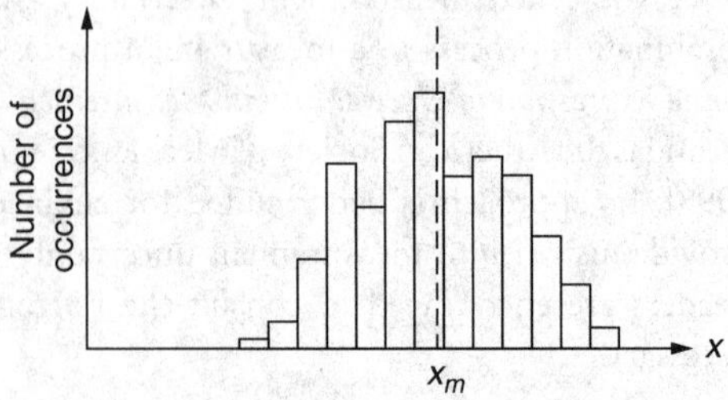

FIGURE 152.2 Histogram of measured values.

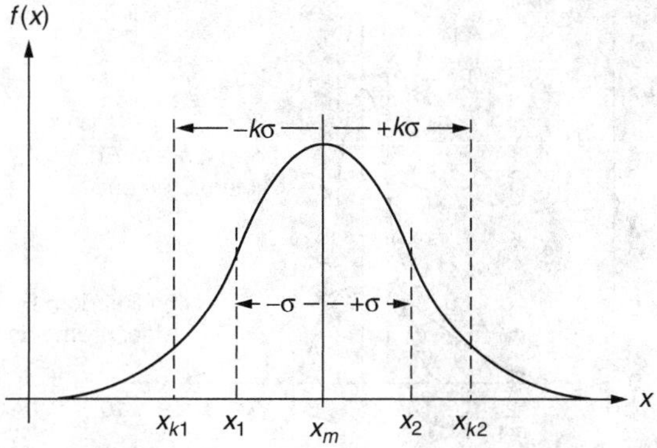

FIGURE 152.3 Frequency distribution of repeated measurements.

The frequency distribution shown in Figure 152.3 corresponds to a Gaussian or normal distribution curve [Bolz and Tuve, 1987], the form generally assumed to represent random measurement uncertainty. There is no guarantee that this symmetrical distribution, indicating an equal probability of measurements falling above or below the mean value, X_m, will occur, but experience has shown that the normal distribution is generally suitable for most measurement applications. Other distributions could be determined based on taking large numbers of measurements and plotting the results [Bevington, 1969], but generally this information (large numbers of measurements) is unavailable.

In analyzing these results, we may apply standard statistical tools to express our confidence in the determined value based on the probability of obtaining a particular result. If experimental errors follow a normal distribution, then a widely reported value is the standard deviation, σ [Bevington, 1969]. There is a 68% (68.27%) probability than an observed value x will fall within $\pm\sigma$ of X_m (Figure 152.3). Other values are often reported for $\pm 2\sigma$ (probability 95.45%) or $\pm 3\sigma$ (probability 99.73%). Values of probabilities are tabulated [Bolz and Tuve, 1987] for ranges from X_{k1} to X_{k2}, or $X_m \pm k\sigma$, as illustrated in Figure 152.3. It should also be noted that these confidences, based on one, two, or three sigma, are also based on an infinite population or set of data. In experimental work, we usually only have a data sample. We then cannot use sigma, but rather the sample standard deviation. We will discuss more details on this later.

In reporting measurements, an indication of the probable error in the result is often stated based on an absolute error prediction (e.g., a temperature of 48.3 ±0.1, based on a 95% probability) or on a relative error basis (e.g., voltage of 9.0 V ±2% (based on a 95% probability)]. The choice of probability value corresponding to the error limits is arbitrary, but a value of 95%, corresponding to $\pm 2\sigma$, appears to be widely used.

When considering the results in Figure 152.3, lacking other information, our best estimate of the true value of the measured quantity is the mean (or average) of the measured values. Based on the previous discussions we may characterize measurements and instrumentation as being of high or low random uncertainty. This is illustrated in Figure 152.4, where two probability distributions are plotted. As may be seen, the probability of obtaining values near the mean is greater for the low random uncertainty measurement (i.e., the measurements are highly repeatable or precise). The high random uncertainty measurements have a wider distribution and are characterized by a greater standard deviation, $\pm\sigma_{lH}$ compared with the low random uncertainty measurements, $\pm\sigma_L$.

It is worth noting that while the high random uncertainty measurements are not highly repeatable, if the mean value of a large number of measurements is calculated, then the value x_m should be the same as that determined from the low random uncertainty measurement. Therefore, the more measurements that we average, the closer to the true mean our sample mean will be. What is needed usually is not an estimate of the standard deviation of our measurements, but the standard deviation of our average. This

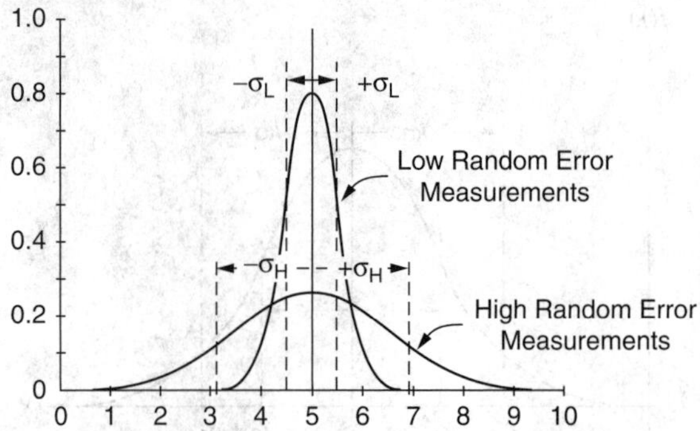

FIGURE 152.4 Frequency distributions for instruments of varying precision.

is obtained by dividing the standard deviation of our measurements by the square root of the number of measurements in the average as follows:

$$S_{\overline{X}} = \frac{S_X}{\sqrt{N}}$$

where S_X = the standard deviation of our measurements. Note that this is our estimate of σ, $S_{\overline{X}}$ = the standard deviation of our average result. This is also called the "random standard uncertainty, and N = the number of measurements in our average result. Note that we have now made the transition from the population estimates of random uncertainty, σ, to that of our sample of measurements, S_X.

The previous discussion, which illustrated the concept of random measurement error, did not address the effects of systematic or fixed errors. In reality, even though a high probability exists that an individual measurement will be close to the mean value, there is the guarantee that the value of the mean of the large sample of measurements will not be the true value (Figure 152.5). In effect, a particular measurement and measurement instrument may be highly repeatable (low random errors and uncertainty), but possess significant error (i.e., not accurate). This additional error source beyond that of random is called systematic. Systematic errors may arise from either the observer or the instrument. They are constant for the duration of a measurement process. When an error source introduces variable errors into a measurement, this results in observable data scatter and that error source is classified as random. Systematic error sources displace all measurements of a process from the true value by a fixed amount.

The estimate of the limits of a systematic error begins with assuming a distribution for those errors (usually Gaussian) and assigning a confidence (usually 95%). The term often applied here is the "systematic standard uncertainty," which we will define as: $B_{\overline{X}}$. $B_{\overline{X}}$ represents one standard deviation for the systematic errors affecting our average, $\overline{X}$.

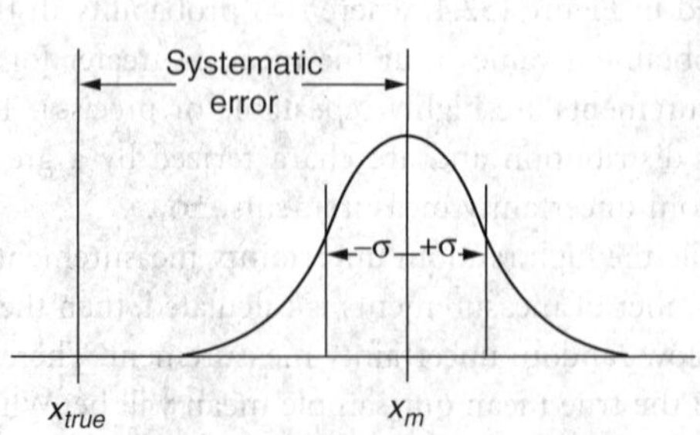

FIGURE 152.5 Measurements with bias errors in addition to random errors.

Note that trends in data or errors may be predictable and may alter test results. These trends need to be quantified. Then the error in that quantification can be estimated with either systematic or random uncertainty. Systematic errors may be significant but, by their nature, should be identifiable and their effects accounted for through careful calibration and comparison with other instruments. This will reduce their effects but not eliminate them. Recently, methods for combining systematic and random uncertainty estimates into an overall uncertainty estimate have been presented [ANSI/ASME, 1998; Dieck, 2002].

152.2 Estimating Measurement Uncertainty

In the previous discussion, we considered multisample experiments of a single parameter. Multisample experiments are those in which, for a given set of the independent experimental variables, the readings are taken many times. If we could repeat our tests many times, with many observers and a variety of instruments, we could apply statistics to determine the reliability of the results as in the previously discussed methods.

In engineering we are often forced to conduct "single-sample" experiments, and therefore we cannot estimate uncertainty by observing repeated tries. Single-sample experiments are those in which, for a given set of experimental conditions, the readings are taken only once. These are typical in engineering, where financial or time constraints limit the number of repetitions of a particular test. In all cases the uncertainty in the measurements can only be estimated.

In estimating the random uncertainty in a measurement, there is usually data from which to calculate the sample standard deviation, S_X, and the standard deviation of the average, $S_{\overline{X}}$. When there are no data, an estimate of S_X must be made from judgment or previous experience. For the systematic uncertainty estimates, the experimenter often must rely on judgment based on experience.

152.3 Propagation of Measurement Uncertainty

As previously stated, results are often derived from the combination of values determined from a number of individual measurements. The propagation of uncertainty is defined as the way in which uncertainties in the variables affect the uncertainty in the calculated results. Both the abovementioned ISO and ASME documents have presented a technique for determining the propagation of uncertainties into a derived result. In this technique, a best estimate of the uncertainty in each variable is made by the researcher assuming an equivalent 1S value, or 68% confidence, that a measured value will fall with the uncertainty interval. Therefore, assuming that a desired result is derived from n independent variables or simultaneous measurements (e.g., $R = R[X_1, X_2, X_3, \ldots, X_n]$) and letting $U_1, U_2, \ldots, U_n$ be the uncertainties in the independent variables, we can derive an expression for the uncertainty in the result, U_R. Note especially here that the "*U*" values are typical of the random and/or systematic components of an uncertainty analysis. They are either all random, that is, $S_{\overline{X}}$, or all systematic, that is, $B_{\overline{X}}$, in any one propagation equation. We cannot mix the random and systematic terms in one propagation equation, but must handle each separately. Later, the random and systematic uncertainties for a result will be combined for a result uncertainty.

If the same uncertainty estimate is used for all the component variables in the analysis (i.e., all the uncertainties must be standard uncertainties, systematic or random), the uncertainty in the result will be given by

$$u_R = \pm\left[\left(\frac{\partial R}{\partial X_1}u_1\right)^2 + \left(\frac{\partial R}{\partial X_2}u_2\right)^2 + \left(\frac{\partial R}{\partial X_3}u_3\right)^2 + \cdots + \left(\frac{\partial R}{\partial X_n}u_n\right)^2\right]^{1/2}$$

This equation is in engineering units or absolute uncertainty.

For systematic terms, the previous equation becomes:

$$B_{\overline{R}} = \pm\left[\left(\frac{\partial R}{\partial X_1}B_{\overline{X}_1}\right)^2 + \left(\frac{\partial R}{\partial X_2}B_{\overline{X}_2}\right)^2 + \left(\frac{\partial R}{\partial X_3}B_{\overline{X}_3}\right)^2 + \cdots + \left(\frac{\partial R}{\partial X_n}B_{\overline{X}_n}\right)^2\right]^{1/2}$$

For random terms, this equation becomes:

$$S_{\overline{X},R} = \pm\left[\left(\frac{\partial R}{\partial X_1}S_{\overline{X}_1}\right)^2 + \left(\frac{\partial R}{\partial X_2}S_{\overline{X}_2}\right)^2 + \left(\frac{\partial R}{\partial X_3}S_{\overline{X}_3}\right)^2 + \cdots + \left(\frac{\partial R}{\partial X_n}S_{\overline{X}_n}\right)^2\right]^{1/2}$$

In certain instances (such as a product form of the equation, i.e., $R = X_1^a \cdot X_2^b \cdot X_3^c \cdots X_n^m$), the former equation can be reduced to

$$\frac{u_R}{R} = \pm\left[\left(a \cdot \frac{u_1}{X_1}\right)^2 + \left(b \cdot \frac{u_2}{X_2}\right)^2 + \left(c \cdot \frac{u_3}{X_3}\right)^2 + \cdots + \left(m \cdot \frac{u_n}{X_n}\right)^2\right]^{1/2}$$

where U_R/R is the relative uncertainty in the result, R [Moffat, 1988]. For this equation as well, we can substitute the $B_{\overline{R}}$ and $S_{\overline{X}}$ terms for the systematic and random components of uncertainty.

152.4 Uncertainty of a Result

Once the systematic and random uncertainties have been obtained, it is necessary to combine them into the uncertainty of a result. This is done with the following equation:

$$U_R = \pm t_{95}\left[\left(B_{\overline{R}}\right)^2 + \left(S_{\overline{X},R}\right)^2\right]^{\frac{1}{2}}$$

where U_R = the result uncertainty at 95% confidence, $B_{\overline{R}}$ = the systematic uncertainty of the average result. It is a standard uncertainty, $S_{\overline{X},R}$ = the random uncertainty of the average result. It is a standard uncertainty, and t_{95} = Student's t for 95% confidence level. This is the most common confidence level used in uncertainty analysis.

Note that the Student's t_{95} depends on the degrees of freedom for U_R. For this presentation, we will assume all degrees of freedom are 30 or higher and that Student's t_{95}, therefore, is 2.

152.5 Uncertainty Analysis in Experimental Design

Possible experimental errors should be examined before conducting an experiment to ensure the best results. The experimenter should estimate the uncertainties in each measurement. Uncertainty propagation in the result depends on the squares of the uncertainties in the independent variables; thus the squares of the large uncertainty values dominate the result. Consequently, in designing an experiment, very little is gained by reducing the small errors (uncertainties). A series of measurements with relatively large uncertainties could produce a result with an uncertainty not much larger than that of the most uncertain measurement. Thus, to improve the overall experimental result, the large uncertainties must be reduced [Moffat, 1985, 1988].

Example

Consider an experiment to measure the thermal efficiency, η, of a solar collector for heating water. During testing, the solar collector is connected in series with an electric reference heater, and cooling water is circulated through both. Values of η are obtained by individual measurements of: the temperature rise across the solar collector, ΔT_c, and reference heater, ΔT_h; the power input to the reference heater, P_h; and the total solar energy incident on the collector surface, G_i. The thermal efficiency, η, is determined according to

$$\eta = \frac{P_h}{G_i} \cdot \left(\frac{\Delta T_c}{\Delta T_h}\right)$$

To estimate the relative uncertainty in the measured value of η, the uncertainties in the component measurements must be estimated. The electrical power input to the reference heater was measured with a power transducer with an estimated uncertainty of ±1% of reading random uncertainty and a ±1% of reading systematic uncertainty. The measurement of solar radiation introduced the largest uncertainty and was estimated at ±5% of reading random uncertainty and a ±5% of reading systematic uncertainty. Uncertainties in the measurement of temperature rise across the heater and collector were estimated at ±0.1°C for the random component and ±0.1% for the systematic component. These uncertainties are summarized below.

Measurement	Systematic Uncertainty	Random Uncertainty
Power, P_h	0.1%R	0.1%R
Radiation, G_i	5%	5%R
Temperature rise, ΔT	0.1°C	0.1°C

Therefore, the uncertainty values are

$$B_{\bar{P}_h} = S_{\bar{P}_h} = \pm 0.01 \times \mathrm{P_h}$$

$$B_{\bar{G}_i} = S_{\bar{G}_i} = \pm 0.05 \times \mathrm{G_i}$$

$$B_{\Delta\bar{T}} = S_{\Delta\bar{T}} = \pm 0.1°\mathrm{C}$$

These values are based on best estimates and represent the odds of 20 to 1 that measured values will fall within these limits.

Applying the analysis described above, the relative systematic uncertainty in determining η is

$$\frac{B_{\bar{\eta}}}{\eta} = \pm\left[\left(\frac{B_{\bar{P}_h}}{P_h}\right)^2 + \left(\frac{B_{\bar{G}_i}}{G_i}\right)^2 + \left(\frac{B_{\Delta\bar{T}_h}}{\Delta T_h}\right)^2 + \left(\frac{B_{\Delta\bar{T}_c}}{\Delta T_c}\right)^2\right]^{1/2}$$

Substituting typical test conditions, that is, $\Delta T_c \approx 3.5°C$ and $\Delta T_h \approx 6°C$, then

$$\frac{B_{\bar{\eta}}}{\eta} = \pm[(0.01)^2 + (0.05)^2 + (0.1/6)^2 + (0.1/3.5)^2]^{0.5} = \pm 0.06$$

Thus, the relative uncertainty in the measured value of solar collector efficiency would be ±0.06, or ±6% (based on 20-to-1 odds). The same uncertainty would result from the random terms.

We now must calculate the uncertainty for the result. Here we will assume over 30 degrees of freedom so that $t_{95} = 2$. We then have

$$U_R = \pm t_{95}\left[\left(B_{\overline{R}}\right)^2 + \left(S_{\overline{X},R}\right)^2\right]^{\frac{1}{2}} = \pm 2\left[(0.06)^2 + (0.06)^2\right]^{\frac{1}{2}} = \pm 0.17 or \pm 17\%$$

It is apparent from this analysis that the uncertainty in the solar radiation measurement dominates the uncertainty in the result; that is, decreasing the uncertainty in the other measurements has little effect on the result. It is also worth noting that as ΔT_c becomes smaller, the overall uncertainty in the result increases. This effect could be minimized by reducing the uncertainty in the ΔT_c measurement.

152.6 Additional Considerations

All the above analysis has been based on several simplifying assumptions, summarized below.

1. Degrees of freedom are all 30 or higher (noted in the text above).
2. All uncertainties are symmetrical about the measured average.
3. There are no correlated error sources.

Assumption 3 is particularly troublesome when there is error correlation. Nonsymmetrical uncertainties also must be considered in many experimental results. For a treatment of all these effects, consider consulting Dieck [2002].

Defining Terms

Accuracy — The closeness of agreement of a measured value and the true value. As there are many confusing definitions for "accuracy," its use is not recommended.

Bias error — The tendency of a deviation estimate in one direction from a true value. As there are many, conflicting definitions for the term "bias," its use is not recommended.

Frequency distribution — A description of the frequency of occurrence of the values of a variable.

Gaussian or normal distribution curve — The theoretical distribution function for the conceptual infinite population of measured values. This distribution is characterized by a limiting mean and standard deviation. The interval of the limiting mean plus or minus two standard deviations will include approximately 95% of the total scatter of measurements.

Measurement error — The difference between true and observed values.

Multisample experiments — Experiments in which uncertainties are evaluated by many repetitions and many diverse instruments. Such experiments can be analyzed by classical statistical means.

Precision — The repeatability of measurements of the same quantity under the same conditions. As there are many definitions for this term, its use is not recommended.

Random errors — Errors that cause readings to take random values on either side of some mean value. They may be due to the observer or the instrument and are revealed by repeated observations.

Random uncertainty — The uncertainty estimate that applies to random error sources. It estimates the limits to which a random error may go with some confidence.

Standard deviation — A measure of dispersion of a population. It is calculated as the square root of the average of the squares of the deviations from the mean (root mean square) deviation.

Standard uncertainty — An uncertainty, systematic or random, that is the equivalent of one standard deviation of the average for a group of measurements.

Systematic errors — Errors that persist and cannot be considered due to chance. Systematic errors may be due to instrument manufacture or incorrect calibration and relate to instrument accuracy (the ability of the instrument to indicate the true value).

Systematic uncertainty — The uncertainty estimate that applies to systematic error sources. It estimates the limits to which a systematic error may go with some confidence.

Uncertainty — The estimated error limit for a measurement of a result for given odds.

References

ANSI/American Society of Mechanical Engineers. 1998. *Instruments and Apparatus. Part I: Measurement Uncertainty.* PTC 19.1. American Society of Mechanical Engineers, New York.

Beckwith, T. G., Marangoni, R. D., and Lienhard, V. J. H. 1993. *Mechanical Measurements,* 5th ed. Addison-Wesley, Reading, MA.

Bevington, R. P. 1969. *Data Reduction and Error Analysis for the Physical Sciences.* McGraw-Hill, New York.

Bolz, R. E. and Tuve, G. L. 1987. *CRC Handbook of Tables for Applied Engineering Science,* 2nd ed. CRC Press, Boca Raton, FL.

Dieck, Ronald H. 2002. *Measurement Uncertainty, Methods and Appliations,* 3rd ed. Instrument Society of America, 2002.

International Standards Organization. 1993. *Guide to the Expression of Uncertainty in Measurement.* ISO.

Moffat, R. J. 1985. Using uncertainty analysis in the planning of an experiment. *J. Fluids Eng. Trans. ASME,* 107:173–178.

Moffat, R. J. 1988. Describing the uncertainties in experimental results. *Exp. Therm. Fluid Sci.*, 1:3–17.

Further Information

Extensive discussion of measurement instrumentation and an introduction to measurement errors and uncertainty analysis are presented in Beckwith et al. [1993] and J. P. Holman, 1994, *Experimental Methods for Engineers,* 6th ed., McGraw-Hill, New York.

Detailed discussions of the treatment of systematic and random errors for measurement systems are given in ASME Standard 19.1-1985 and the *ASHRAE Handbook of Fundamentals,* chapter 13, available from American Society of Heating, Refrigerating and Air-Conditioning Engineers, New York.

The use of uncertainty analysis in the planning of experiments is outlined with examples by Moffat [1985].

153

Signal Conditioning

Stephen A. Dyer
Kansas State University

Kelvin's first rule of instrumentation states, in essence, that the measuring instrument must not alter the event being measured. For the present purposes, we can consider the instrument (Figure 153.1) to consist of an input transducer followed by a signal-conditioning section, which in turn drives the signal and data processing and communication sections (the remainder of the instrument). We are using the term *instrument* in the broad sense, with the understanding that it may actually be a measurement subsystem within virtually any type of system.

Certain requirements are imposed upon the transducer if it is to reproduce an event faithfully: it must exhibit amplitude linearity, phase linearity, and adequate frequency response. But it is the task of the signal conditioner to accept the output signal from the transducer and from it produce a signal in the form appropriate for introduction to the remainder of the instrument.

Analog signal conditioning can involve strictly *linear* operations, strictly *nonlinear* operations, or some combination of the two. In addition, the signal conditioner may be called on to provide auxiliary services, such as introducing electrical isolation, providing a reference of some sort for the transducer, or producing an excitation signal for the transducer.

Important examples of linear operations include *amplitude scaling, impedance transformation, linear filtering,* and *modulation.*

A few examples of nonlinear operations include obtaining the *root-mean-square (rms) value, square root, absolute value,* or *logarithm* of the input signal.

A wide variety of building blocks are available in either modular or integrated circuit (IC) form for accomplishing analog signal conditioning. Such building blocks include operational amplifiers, instrumentation amplifiers, isolation amplifiers, and a plethora of nonlinear processing circuits such as comparators, analog multiplier/dividers, log/antilog amplifiers, rms-to-DC converters, and trigonometric function generators.

Also available are complete signal-conditioning subsystems consisting of various plug-in input and output modules that can be interconnected via universal backplanes that can be either chassis or rack mounted.

153.1 Linear Operations

Three categories of linear operations important to signal conditioning are amplitude scaling, impedance transformation, and linear filtering.

0-8493-1586-7/05/$0.00+$1.50

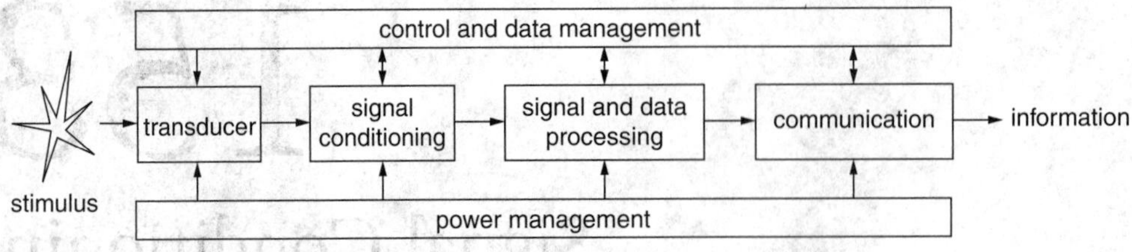

FIGURE 153.1 Basic block diagram of an electronic instrument.

Amplitude Scaling

The amplitude of the signal output from a transducer must typically be scaled — either amplified or attenuated — before the signal can be processed.

Amplification

Amplification is generally accomplished by an *operational amplifier*, an *instrumentation amplifier*, or an *isolation amplifier*.

Operational Amplifiers

A conventional operational amplifier (op amp) has a differential input and a single-ended output. An *ideal* op amp, used often as a first approximation to model a real op amp, has infinite gain, infinite bandwidth, infinite differential input impedance, infinite slew rate, and infinite common-mode rejection ratio (CMRR). It also has zero output impedance, zero noise, zero bias currents, and zero input offset voltage. Real op amps, of course, fall short of the ideal in all regards.

Important parameters to consider when selecting an op amp include:

1. DC voltage gain K_0.
2. Small-signal gain-bandwidth product (GBWP) f_T, which for most op amps is $f_T \approx K_0 f_l$, where f_l is the lower break frequency in the op amp's transfer function. The GBWP characterizes the closed-loop, high-frequency response of an op-amp circuit.
3. Slew rate, which governs the large-signal behavior of an op amp. Slew rates range from less than 1 v/μs to several thousand volts per microsecond.

Other parameters, such as input and output impedances, DC offset voltage, DC bias current, drift voltages and currents, noise characteristics, and so forth, must be considered when selecting an op amp for a particular application.

There are several categories of operational amplifiers. In addition to "garden-variety" op amps, there are many op amps whose characteristics are optimized for one or more classes of use. Some categories of op amps include:

- *Low-noise* op amps, which are useful in the portions of signal conditioners required to amplify very-low-level signals.
- *Chopper-stabilized* op amps, which are useful in applications requiring extreme DC stability.
- *Fast* op amps, which are useful when large slew rates and large GBWPs are required.
- *Power* op amps, which are useful when currents of greater than a few mA must be provided to the op amp's load.
- *Electrometer* op amps, which are used when very high ($>10^{13}$ Ω) input resistances and very low (<1 pA) input bias currents are required.

An introduction to op amps and basic circuit configurations can be found in essentially any modern text on circuit theory or electronics, and the reader can find detailed theoretical developments and many useful configurations and applications in Roberge (1975), Graeme et al. (1971), Graeme (1973, 1977), Horowitz and Hill (1989), and Stout and Kaufman (1976).

Instrumentation Amplifiers

Instrumentation amplifiers (IAs) are gain blocks optimized to provide high input impedance, low output impedance, stable gain, relatively high common-mode rejection (CMR), and relatively low offset and drift. They are well suited for amplification of outputs from various types of transducers such as strain gages, for amplification of low-level signals occurring in the presence of high-level common-mode voltages, and for situations in which some degree of isolation is needed between the transducer and the remainder of the instrument.

Although instrumentation amplifiers can be constructed from conventional op amps (a three-op-amp configuration is typically discussed; see, for example, Stout and Kaufman [1976]), they are readily available and relatively inexpensive in IC form. Some IAs have digitally programmable gains, whereas others are programmable by interconnecting resistors internal to the IA via external pins. More-basic IAs have their gains set by connecting external resistors.

Isolation Amplifiers

Isolation amplifiers are useful in applications in which a voltage or current occurring in the presence of a high common-mode voltage must be measured safely, accurately, and with a high CMR. They are also useful when safety from DC and line-frequency leakage currents must be ensured, such as in biomedical instrumentation.

The isolation amplifier can be thought of as consisting of three sections: an input stage, an output stage, and a power circuit. All isolation amplifiers have their input stages galvanically isolated from their output stages. Communication between the input and output stages is accomplished by modulation/demodulation.

An isolation amplifier is said to provide two-port isolation if there is a DC connection between its power circuit and its output stage. If its power circuit is isolated from its output stage as well as its input stage, then the amplifier is said to provide three-port isolation. Isolation impedances on the order of 10^{10} Ω are not atypical.

Isolation amplifiers are available in modular form with either two-port or three-port isolation. Both single-channel and multichannel modules are offered.

Attenuation

Although the majority of transducers are low-level devices such as thermocouples, thermistors, resistance temperature detectors (RTDs), strain gages, and so forth, whose outputs require amplification, there are many measurement situations in which the input signal must be attenuated before introducing it to the remainder of the system.

Voltage Scaling

Most typically, the signals to be attenuated take the form of voltages. Broadly, the attenuation is accomplished by either a *voltage divider* or a *voltage transformer.*

Voltage Dividers — In many cases, a simple chain divider proves adequate. The transfer function of a two-element chain of impedances $Z_1(s)$ and $Z_2(s)$ is

$$\frac{V_o(s)}{V_{\text{in}}(s)} = \frac{Z_1(s)}{Z_1(s) + Z_2(s)}$$

where the output voltage $V_o(s)$ is the voltage across $Z_1(s)$, and the input voltage $V_{\text{in}}(s)$ is the voltage across the two-element combination.

Of course, the impedances of the source (transducer) and the load (the remainder of the system) must be taken into account when designing the divider network.

Resistive Dividers — If the elements in the chain are resistors, then the divider is useful from DC up through the frequencies for which the impedances of the resistors have no significant reactive components. For $Z_1(s) = R_1$ and $Z_2(s) = R_2$,

$$\frac{V_o(s)}{V_{in}(s)} = \frac{R_1}{R_1 + R_2}$$

Other configurations are available for resistive dividers. One example is the Kelvin–Varley divider, which has several advantages that make it useful in situations requiring high accuracy. For a detailed description, see Gregory (1973).

Capacitive Dividers — If the elements in the chain divider are capacitors, then the divider has as its transfer function

$$\frac{V_o(s)}{V_{in}(s)} = \frac{C_2}{C_1 + C_2}$$

This form of divider is useful from low frequencies up through frequencies of several megahertz. A common application is in the scaling of large voltages.

Inductive Dividers — If the elements in the chain divider are inductors, then an autotransformer results. Inductive dividers are useful over frequencies from a few hertz to several hundred kilohertz. Errors in the parts-per-billion range are achievable.

Voltage Transformers — Voltage transformers constitute one of the most common means of accomplishing voltage scaling at line frequencies. Standard double-wound configurations are useful unless voltages above about 200 kV are to be monitored. For very high voltages, alternative configurations such as the *capacitor voltage transformer* and the *cascade voltage transformer* are employed (Gregory, 1973).

Current Scaling

Current scaling is typically accomplished via either a current shunt or a current transformer.

A *current shunt* is essentially an accurately known resistance through which the current to be measured is passed. The voltage developed across the shunt as a result of the current is the quantity measured. Shunts are useful at DC and frequencies through the audio range. Two disadvantages are: (1) the shunt consumes power, and (2) the measurement circuitry must be operated at the same potential as the shunt.

The *current transformer* overcomes the mentioned disadvantages of the current shunt. Typically, the current transformer consists of a specially constructed toroidal core upon which the secondary (sense) winding is wrapped and through which the primary winding is passed. A single-turn primary is commonly used, although multiturn primaries are available.

Other Attenuators

In addition to the aforementioned means of voltage and current scaling, there are attenuator pads that provide, in addition to voltage or power reduction, the ability to be matched in impedance to the source and load circuits between which it is connected. The common pads include the T, L, and Π types, either balanced or unbalanced. Resistive attenuator pads are discussed in most textbooks on circuit design (e.g., Cuthbert, 1983). They are useful from DC through several hundred megahertz.

Impedance Transformation

Oftentimes the impedance of the transducer must be transformed to a value more acceptable to the remainder of the measurement system. In many cases, maximum power must be transferred from the transducer's output signal to the remaining circuitry. In other cases, it is sufficient to provide buffering that presents very high impedance to the transducer, a very low impedance to the rest of the system, and a voltage gain of unity.

Matching transformers, passive matching networks such as attenuator pads, and unity-gain buffers are standard means of accomplishing impedance transformation. Unity-gain buffers are available in IC form.

Linear Filtering

Although, in general, digital signal processing offers many advantages over analog techniques for filtering signals, there are many relatively simple applications for which *frequency-selective analog filtering* is well suited.

Filters are used within signal conditioners (1) to reduce the effects of noise that corrupts the input signal; (2) as part of a demodulator; (3) to limit signal bandwidth; or (4) if the signal is to be sampled, to limit its bandwidth in order to prevent aliasing. These filters can be built either entirely of passive components or based on active devices such as op amps.

There are many good references that discuss methods of characterizing, specifying, and implementing frequency-selective analog filters. See Van Valkenburg (1960) for design of passive filters; for the design of active-RC filters, see Sedra and Brackett (1978) and Stephenson (1985).

153.2 Nonlinear Operations

A wide variety of nonlinear operations are useful to signal-conditioning tasks. Listed below are some typical nonlinear blocks along with brief descriptions. Most of the blocks are available as ICs.

1. *Comparator.* A comparator is a two-input device whose output voltage, V_o, takes on one of two stable values, V_{o0} and V_{o1}, as follows:

$$V_o = \begin{cases} V_{o0}, & \text{if } V_2 < V_1 \\ V_{o1}, & \text{otherwise} \end{cases}$$

   where V_1 and V_2 are the voltages at the two inputs.
2. *Schmitt trigger.* A Schmitt trigger is a comparator with hysteresis. It can be constructed from a comparator by applying positive feedback.
3. *Multiplier.* A two-input multiplier supplies an output voltage that is proportional to the product of its input voltages.
4. *Divider.* A two-input divider has as its output a voltage proportional to the ratio of its input voltages. The functions of multiplication and division are usually combined within a single device.
5. *Squarer.* A squarer has as its output a voltage proportional to the square of its input. Squarers can be constructed by a number of means: from multipliers, based on diode–resistor networks, based on FETs, and so forth.
6. *Square-rooter.* A square-rooter has as its output a voltage proportional to the square root of its input. A square-rooter can be built most easily from either a divider or a log/antilog amplifier.
7. *Logarithmic/antilogarithmic amplifier.* A log/antilog amplifier produces an output voltage proportional to the logarithm or the antilogarithm of its input voltage.
8. *True RMS-to-DC converter.* A true RMS-to-DC converter computes the square root of the average, over some interval of time, of the instantaneous square of the input signal. The averaging operation is generally accomplished via a simple low-pass filter whose capacitor is selected to give the desired interval.
9. *Trigonometric function generator.* Generators are available in IC form that produce as their outputs any of the standard trigonometric functions or their inverses, taken as functions of the differential voltage at the generator's inputs.
10. *Sample-and-hold and track-and-hold amplifiers.* A sample-and-hold amplifier (SHA) is a device that samples the signal at its input and holds the instantaneous value whenever commanded by a logic control signal. A track-and-hold amplifier is identical to an SHA but is used in applications

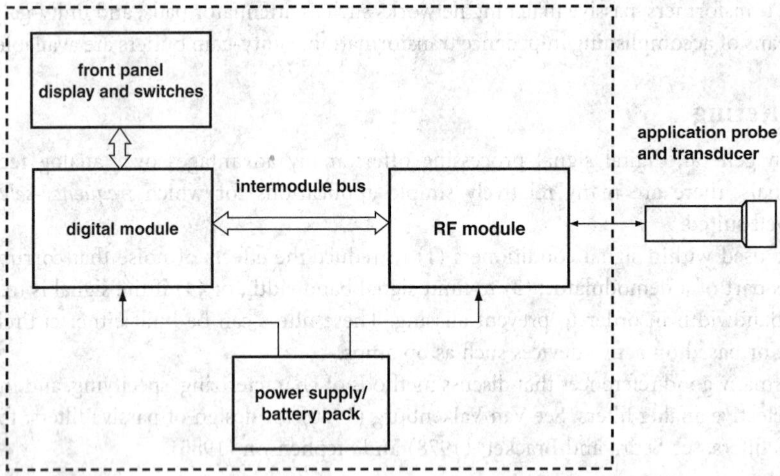

FIGURE 153.2 Basic block diagram of the therapeutic ultrasound unit discussed as an example.

where it spends most of its time tracking the input signal (i.e., in "sample" or "track" mode), in contrast to the SHA, which spends most of its time in "hold" mode.

11. *Precision diode-based circuits.* Circuits such as precision half-wave rectifiers, absolute-value circuits, precision peak detectors, and precision limiters are relatively easy to design and implement based on diodes and op amps. See Horowitz and Hill (1989), Stout and Kaufman (1976), and Graeme (1977).

A detailed description of these and other nonlinear circuit blocks can be found in Sheingold (1976).

153.3 Example

We provide briefly an example of a device that has embedded within it several signal-conditioning circuits. Figure 153.2 shows the basic block diagram of a therapeutic ultrasound unit, which finds widespread use in medicine.

The particular unit being discussed consists of five principal subsystems:

1. An application probe and ultrasound transducer, which imparts ultrasonic energy to the tissue being treated. *Note that this transducer is* not *an input transducer such as has been discussed in relation to signal conditioners.*
2. A radio-frequency (RF) module, which provides electrical excitation to the ultrasound transducer.
3. Front-panel display and switches, which allow communication between the unit and its operator.
4. A microprocessor-based digital module, which orchestrates the overall control of the ultrasound unit.
5. A power supply/battery pack, which provides operating power to the unit.

We focus now on the RF module, whose basic block diagram is shown in Figure 153.3. The module consists of a sine-wave oscillator that produces a signal at the resonant frequency of the transducer, a modulator that allows that signal to be pulse-modulated, and an amplifier with RF-voltage feedback. Incorporated in the amplifier are a power amplifier capable of driving the transducer and automatic-gain-control (AGC) circuitry required to adjust the output power to coincide with that selected by the operator. The AGC uses a standard feedback-control loop to maintain a constant-voltage envelope on the RF signal output from the power amplifier.

Some of the signal conditioners employed within the RF module include the following:

1. The RF-voltage pickoff at the output of the power amplifier. The pickoff employs a half-wave rectifier, followed by a simple capacitive chain divider for voltage scaling.

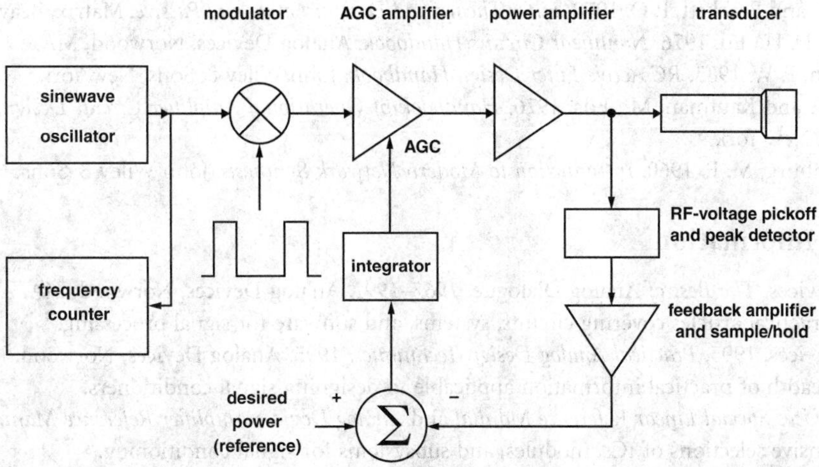

FIGURE 153.3 Simplified block diagram of the RF module used in the ultrasound unit of Figure 153.2.

2. A precision peak detector, which obtains the peak value of the output from the voltage divider during a modulation cycle and presents that value to the feedback loop.
3. An amplifier, having digitally selectable gain, which amplifies the output of the peak detector.
4. A sample-and-hold amplifier, used to hold the amplified output from the peak detector during the "off time" of the modulator. The SHA is needed since the time constant of the peak detector is not sufficient to prevent significant "droop" during the off-time of the modulator.
5. An integrator (an example of frequency-selective filtering), which develops the control voltage for the AGC loop from the output of the differencer.
6. A current shunt, not shown in Figure 153.3, which is used to monitor the DC current supplied to the power amplifier.

As can be seen from this simple example, several signal-conditioning functions may be employed within a single system, and the system itself might not even be an instrument!

Defining Terms

Common-mode rejection (CMR) — CMRR given in decibels. CMR = 20 log|CMRR|. CMR is a non-linear function of common-mode voltage and depends on other factors such as temperature.

Common-mode rejection ratio (CMRR) — The ratio of the differential gain to the common-mode gain of an amplifier.

Gain bandwidth product (GBWP) — The product of an amplifier's highest gain and its corresponding bandwidth. Used as a rough figure of merit for bandwidth.

Slew rate — The maximum attainable time rate of change of an amplifier's output voltage in response to a large step change in input voltage.

References

Cuthbert, T. R. 1983. *Circuit Design Using Personal Computers.* John Wiley & Sons, New York.

Graeme, J. G. 1973. *Applications of Operational Amplifiers.* McGraw-Hill, New York.

Graeme, J. G. 1977. *Designing with Operational Amplifiers.* McGraw-Hill, New York.

Graeme, J. G., Tobey, G. E. and Huelsman, L. P., Eds. 1971. *Operational Amplifiers.* McGraw-Hill, New York.

Gregory, B. A. 1973. *An Introduction to Electrical Instrumentation.* Macmillan, London.

Horowitz, P. and Hill, W. 1989. *The Art of Electronics*, 2nd ed. Cambridge University Press, New York.

Roberge, J. K. 1975. *Operational Amplifiers.* John Wiley & Sons, New York.

Sedra, A. S. and Brackett, P. O. 1978. *Filter Theory and Design: Active and Passive.* Matrix, Beaverton, OR.

Sheingold, D. H., Ed. 1976. *Nonlinear Circuits Handbook.* Analog Devices, Norwood, MA.

Stephenson, F. W. 1985. *RC Active Filter Design Handbook.* John Wiley & Sons, New York.

Stout, D. F. and Kaufman, M., Eds. 1976. *Handbook of Operational Amplifier Circuit Design.* McGraw-Hill, New York.

Van Valkenburg, M. E. 1960. *Introduction to Modern Network Synthesis.* John Wiley & Sons, New York.

Further Information

Analog Devices. *The Best of* Analog Dialogue, *1967–1991.* Analog Devices, Norwood, MA. A collection of practical articles covering circuits, systems, and software for signal processing.

Analog Devices. 1995. *Practical Analog Design Techniques.* 1995. Analog Devices, Norwood, MA. Offers a breadth of practical information applicable to designing signal conditioners.

Analog Devices Special Linear Reference Manual and *Analog Devices Amplifier Reference Manual.* Present extensive selections of ICs, modules, and subsystems for signal conditioning.

Barnes, J. R. 1987. *Electronic System Design: Interference and Noise Control Techniques.* Prentice-Hall, Englewood Cliffs, NJ. Concentrates on the task of controlling electronic noise, both as one aspect of signal conditioning and as an important part of the overall design process.

Dyer, S. A., Ed. 2001. *Survey of Instrumentation and Measurement.* John Wiley & Sons, New York. Gives encyclopedic coverage to the field of instrumentation and measurement, including articles relating specifically to signal conditioning.

Fowler, K. R. 1996. *Electronic Instrument Design: Architecting for the Life Cycle.* Oxford University Press, New York. Provides an excellent heuristic framework for developing electronic instrumentation and approaching the various aspects, such as signal conditioning, of a design.

IEEE Instrumentation & Measurement Magazine. Published quarterly by the Institute of Electrical and Electronics Engineers.

IEEE Transactions on Instrumentation and Measurement. Published bimonthly by the Institute of Electrical and Electronics Engineers.

IEEE Transactions on Circuits and Systems — II: Analog and Digital Signal Processing. Published monthly by the Institute of Electrical and Electronics Engineers.

O'Dell, T. H. 1991. *Circuits for Electronic Instrumentation.* Cambridge University Press, London. Takes an experimental approach to the development of analog circuits useful to signal conditioning.

Ott, H. W. 1988. *Noise Reduction Techniques in Electronic Systems,* 2nd ed. John Wiley & Sons, New York. Covers the practical aspects of noise suppression and control in electronic circuits and systems.

Pallás-Areny, R. and Webster, J. G. 1991. *Sensors and Signal Conditioning.* John Wiley & Sons, New York. Provides an excellent introduction to sensors and signal-conditioning circuits required by them.

Pallás-Areny, R. and Webster, J. G. 1999. *Analog Signal Processing.* John Wiley & Sons, New York. Gives broad coverage to various aspects of signal conditioning, which the authors refer to as analog signal processing.

Sheingold, D. H., Ed. 1980. *Transducer Interfacing Handbook.* Analog Devices, Norwood, MA. Covers signal-conditioning techniques applicable to temperature, pressure, force, level, and flow transducers.

154

Telemetry

Stephen Horan
New Mexico State University

Telemetry systems are found in a variety of applications, from automobiles, to hospitals, to interplanetary spacecraft. Although these examples represent a broad range of applications, they all have many characteristics in common: a natural parameter is measured by a sensor system, the measurement is converted to numbers or data, the data are transported to an analysis point, and an end user makes use of the data gathered. After all, the implication of telemetry is to "measure at a distance."

154.1 Data Measurements

The general purpose of a telemetry system is to gather information about a subject of interest and present that measurement to a user. The physical quantity or property that is being measured is known as the *measurand*. Each measurand is sampled by a sensor at a rate appropriate for the bandwidth of the signal. The sampled analog signals are then frequently digitized by using an analog-to-digital converter. This produces a *pulse-coded modulation* (PCM) data stream. This process is illustrated in Figure 154.1 where an analog signal is sampled as a function of time. The sampled signal is represented by a voltage at the sensor output. This voltage is converted to the PCM form in the analog-to-digital conversion process to produce one digital sample for each sample. The data user will reverse this process to estimate the measurand value as a function of time. To produce an accurate estimate of the measurand, the sensor and conversion process will need to be calibrated and the calibration coefficients incorporated in the inverse transformation.

The sampling rate for each sensor is determined by the signal's *Nyquist sampling rate*. If W is the signal bandwidth in hertz, then the Nyquist sampling rate, f_N, in samples per second, is given by

$$f_N = 2W \tag{154.1}$$

In practice, a minimal sampling rate of five times the signal bandwidth is typically used to accurately reconstruct the measurand.

154.2 Telemetry Systems

Measurement systems requiring the sampling of more than a single sensor require more complicated support and coordination to allow sampling of several measurands. These measurement systems are typically located at a distance from the eventual data user, which can be a person or a machine. The measurements are transmitted over some type of data channel, such as a radio link, fiber optic cable, or

0-8493-1586-7/05/$0.00+$1.50

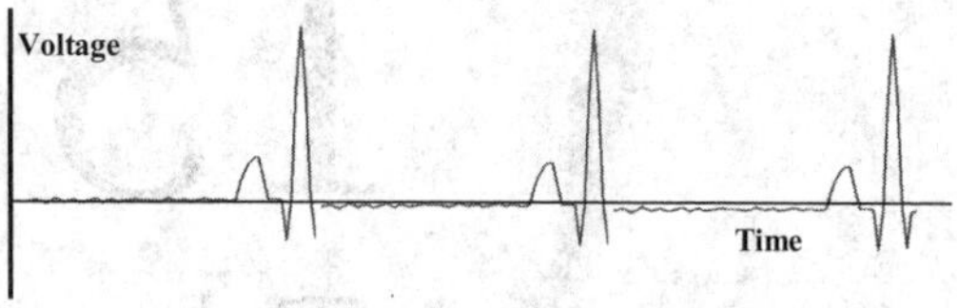

Analog Voltage Signal

Time	Volt	PCM
0.00	0.0509	257
0.05	-0.0712	254
0.10	0.0000	256
0.15	0.0734	257
0.20	-0.0554	254
0.25	-0.1558	252
0.30	0.0000	256
0.35	0.1620	259
0.40	0.0606	257

Digital Representation

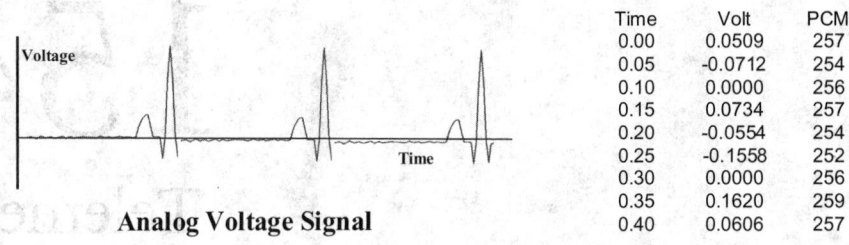

FIGURE 154.1 Sampling an analog voltage signal and converting it to a PCM digital signal.

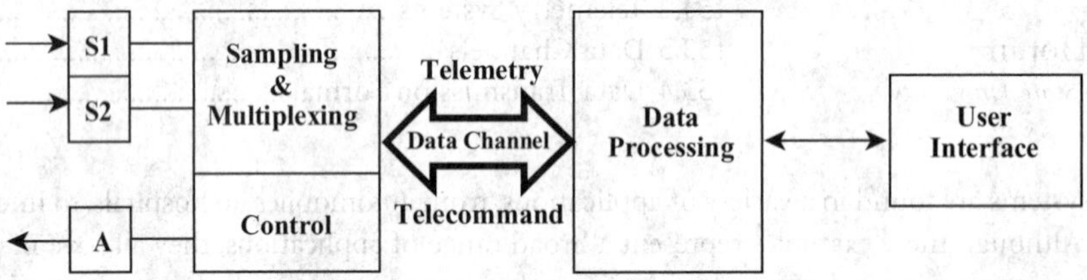

FIGURE 154.2 Overall data flow in a telemetry and telecommand system composed of sensors (S) and actuators (A).

telecommunications network. This architecture produces a telemetry system having the following characteristics of the standard telecommunications definition of telemetry [1]:

1. The use of telecommunication for automatically indicating or recording measurements at a distance from the measuring instrument.
2. The transmission of nonvoice signals for the purpose of automatically indicating or recording measurements at a distance from the measuring instrument.

The control functions in a telemetry system are accomplished through *actuators*, which are devices that respond to transmitted control signals to affect the measurement system characteristics or interact with the measurement environment. The control is achieved through a *telecommand* link which is the "use of telecommunication for the transmission of signals to initiate, modify, or terminate functions of equipment at a distance" [1]. The data flow in a telemetry/telecommand system is illustrated in Figure 154.2. The telemetry measurements from the sensors flow back to the user over the data channel. The sampling and multiplexing components format and sequence order the data for transmission over the channel. The data processing returns the measurement values to estimates of the actual measurand values for presentation to the user. In the user interface, the PCM signal will either be converted back to an analog waveform or left as a series of discrete measurements for use and analysis. The user interface may be a display screen, a set of gages, or a chart recorder. The user interface often contains data-logging capabilities to provide a permanent record of the measurements.

The user also enters commands at the user interface to control the actuators. The data processing prepares the user input into telecommands for transmission over the data channel. The control functions then interpret these commands and cause the actuators to respond. The design of the telemetry and telecommand data structures is generally individualized for each system by the system designers. The overall structure of the data structures can be classified as either frames or packets as discussed later.

153.3 Data Channels

The configuration and quality of the data channel will influence the design of the data transmission structure. Telemetry systems have two general channel configurations: common channel mode or distributed channel mode, as illustrated in Figure 154.3. In common channel mode, all telemetry and

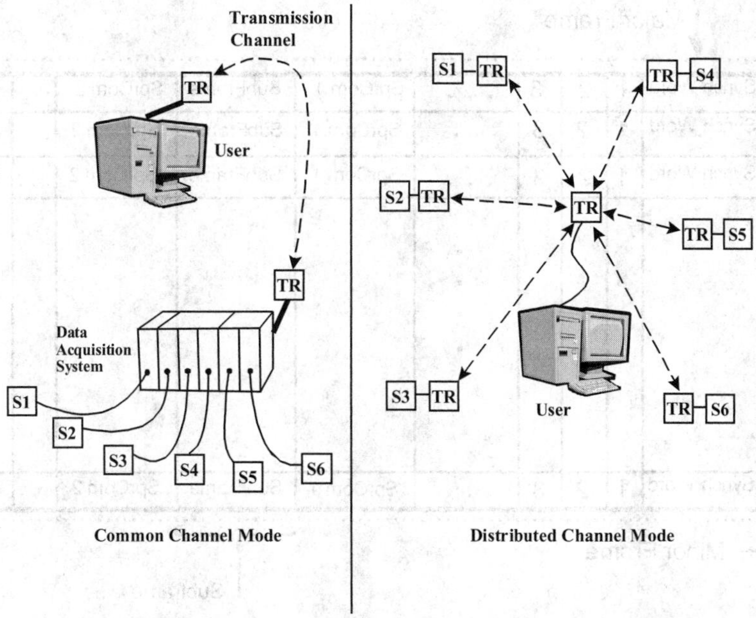

FIGURE 154.3 Common and distributed data channels for telemetry and telecommand systems. The data sources are the sensors (S) and the channel interface occurs at a transceiver (TR).

telecommand data traverse a single, logical link. This link may be either full duplex or parallel simplex links, depending on the nature of the physical channel and the access protocol. In this mode, all sensor data are combined at a common data acquisition system prior to transmission over the channel. Typical environments for this mode include rocket payloads, laboratory data acquisition systems, and remote vehicle telemetry systems. The common channel is frequently a radio link, a wire, or a local area network.

In the distributed mode channel, each sensor connects to the channel via a transceiver. For example, this transceiver can be a simple 802.11 interface that permits the sensors to be connected as part of a wireless network with a central computer receiving the data via a wireless access point. The distributed channel can use either polling techniques or code division multiple access techniques to give each sensor shared access to the channel. Typical environments for this mode include building temperature and fire sensing or factory process measurement and control.

153.4 Data Transmission Formats

The channel bandwidth and quality are limiting factors in the system because they restrict the volume of data that can be reliably sent and they determine the data synchronization methodology. The channel bandwidth will affect the speed at which each sensor can be sampled to ensure that the Nyquist sampling rate is satisfied. Because different sensors will have different sampling rates, a time-division multiplexing technique can be used to share the data channel. For channels where data loss due to data dropouts is a significant problem or where tight synchronization is required between the sensor system and the user interface, a continuous frame transmission data structure is used. For systems where the channel provides high reliability or there are many different sampling rates possible, a packet data structure may be preferred.

Frame telemetry is the traditional method for time multiplexing telemetry data from the source to the destination. The multiplexing structure repeats in fixed time intervals to allow the receiving side to synchronize to the data stream based on the contents of the data stream alone. Most frame systems use a format standard called IRIG 106 [2], developed by the military test ranges' Inter-Range Instrumentation Group (IRIG). This frame structure is used by all industries and not just for military systems. The frame structure is arranged like a matrix, as shown in Figure 154.4, with the *major frame* being one complete

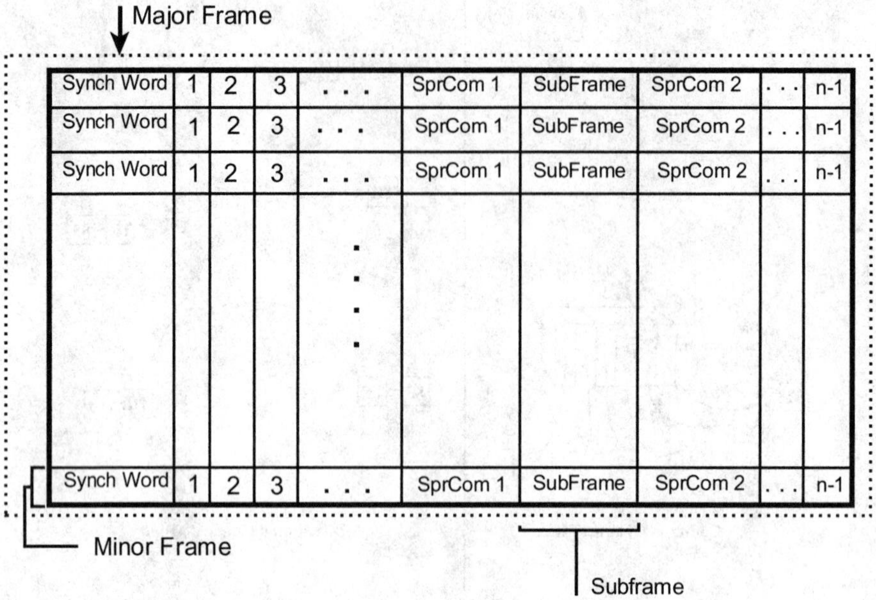

FIGURE 154.4 The standard IRIG telemetry frame structure.

cycle through the time-multiplexing sequence, during which time each sensor is sampled at least once. Each major frame is divided along rows of the matrix into *minor frames*, beginning with the synchronization word. The columns of the matrix in the major frame can represent either the output from sensors or management data repeating in each minor frame. A sensor's output reading stays in its assigned column from frame to frame. If the same sensor is sampled once per minor frame — for example, columns 1, 2, and 3 in Figure 154.4 — then it is sampled at the commutation rate for the system and is called *commutated data*. If a small number of sensors need to be sampled at a rate higher than the commutation rate, they are *supercommutated* sensors ("SprCom" locations in Figure 154.4) and take up more than one column in the matrix. Some columns in the matrix may represent a group of sensors rather than a single sensor. As shown in the column labeled "subframe" in Figure 154.4, the sensors in that group form a *subframe* within the major frame. The sensors are sampled at a rate below the commutation rate, so the data are *subcommutated*.

Standard synchronization codes are usually 16 or 32 bits long. For example, the standard 16-bit code is EB90 in hexadecimal, whereas the standard 32-bit code is FE6BA840 in hexadecimal. These codes are optimal in the sense of having the lowest autocorrelation values when shifted by 1 bit. This minimizes the probability of a false lock in the synchronization detection circuitry. All synchronization codes suffer the possibility of being mimicked by random data. Therefore, several occurrences of the code are required upon initial synchronization. The condition in which the synchronizing circuitry locks onto random data and not the synchronization word is called a false lock. The probability of this false lock condition occurring, P_{FL}, is given by

$$P_{\mathrm{FL}} = \frac{\sum_{i=o}^{k} \binom{N}{i}}{2^N} \tag{154.2}$$

Here N is the length of the synchronization code in bits and k is the number of differences allowed between the received code and the exact code value.

The synchronization code can also be missed if the data become corrupted in the channel. The probability of a missed synchronization code, P_M, due to channel errors is given by

$$P_M = \sum_{i=k+1}^{N} \binom{N}{i} p^i (1-p)^{N-i} \tag{154.3}$$

Here p is the channel bit-error rate while N and k are as before.

If the minor frame has a length of L bits, then with a channel BER of p, the frame data have a probability of correct reception, P_0, given by

$$P_0 = (1-p)^L \tag{154.4}$$

This can be used by the designer to estimate necessity for any error detection and correction coding to be applied to the transmission.

The synchronization process for a telemetry frame follows these steps:

1. Achieve lock onto the data signal from the data channel.
2. Find the individual bits in the data stream using a bit synchronizer circuit.
3. Find the occurrences of the frame synchronization word by using a correlator comparing the data with the desired stored pattern.
4. Once the synchronization marker is reliably found, ensure that it repeats at the minor frame rate and that any management information in the frame is intact by using a frame synchronizer.
5. Once the frame structure is fully identified, begin processing the data.

Packet telemetry systems are becoming more common, especially in systems where the data acquisition and data reception subsystems have high degrees of computational capability and the link between them can be viewed as reliable. Packet systems have several advantages over frame systems, with the main advantage being flexibility. With a packet system, instead of having a master commutation rate, the sampling rate for each sensor or sensor system can be individualized to the natural signal bandwidth. For example, battery voltages being measured may not change significantly over five minutes. With frame telemetry they may be sampled more frequently than once per minute due to the commutation rate specification. With packets the voltages may be sampled only as needed and then transmitted in a data packet. Packets also have the advantage of allowing the data to be more easily routed over a computer network for analysis and distribution to end users. Examples of packet telemetry formats are given in [3].

The general packet format is composed of a header, containing accounting and addressing information, followed by the actual data similar to the format in Figure 154.5. The packet may end with a trailer composed of error-checking codes or other administrative information. The addressing information in the packet identifies the sensor system originating the packet and the destination process for analysis. Other information included in the header might be counters to identify the sequence number or a time stamp to show when the packet was created. The header will often contain a size parameter to specify the length of the data field.

For actual transmission across the data channel, a link-layer packet may used. This link packet may multiplex the data packets from several subsystems together for efficient transport. If the data channel is synchronous, as with a radio channel, then it is common for the channel packets to be sent at regular

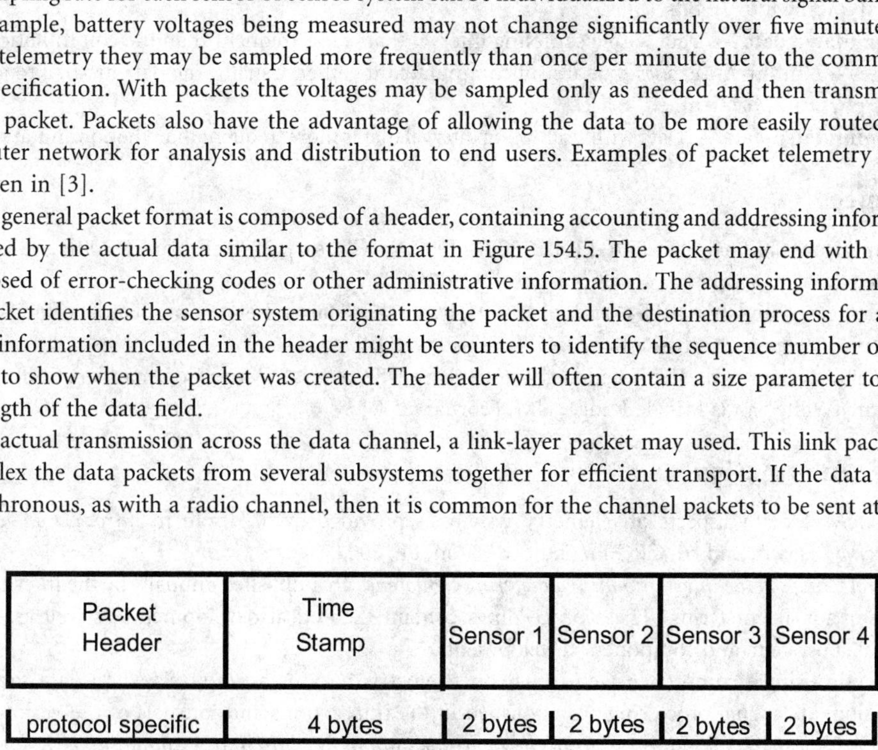

FIGURE 154.5 An example of a telemetry packet.

intervals to maintain transmission synchronization. When this is done, fill packets are used to keep the channel active if there are no actual data to be sent. The packet header will then have a special code to indicate that the packet is a fill packet and should not be processed.

The packet usually begins with a synchronization marker, just as frame telemetry does. The same synchronization codes can be used in packet systems as in frame systems. After synchronization the header is analyzed to identify the type of processing to be performed based on the source of the data. The probability of missing the synchronization marker due to channel errors is computed as in Equation (154.3) and the probability of receiving the packet correctly is computed as in Equation (154.4). Many commercial protocols have error detection as part of the protocol specification so this can be used to assess data quality in the packet.

Defining Terms

Commutated data — Data that are sampled once per main sampling interval. This main sampling interval is called the *commutation rate*. Minor frames, major frames, and subframes are tied to this commutation rate.

Major frame — The set of an integer number of minor frames where each sensor value is sampled at least once.

Measurand — The physical quantity or parameter being measured.

Minor frame — The set of sensor values, synchronization markers, and other management data between successive synchronization words.

Nyquist sampling rate — The minimum sampling rate for signal recovery. If a signal of limited bandwidth is sampled at twice this rate, then the signal can be reconstructed in principle. Most signals are sampled at a higher rate than the Nyquist rate to give better reconstruction.

Pulse-coded modulation (PCM) — Modulation in which each analog sensor value is converted to a digital number once per sampling interval. The number of bits used in the representation is typically 8 to 16 bits.

Subcommutated data — Data with a sampling interval that is less frequent than the commutation rate.

Subframe — A group of sensors that are subcommutated together. Usually, the data are tied to a single physical subsystem.

Supercommutated data — Data with a sampling interval that is more frequent than the commutation rate.

References

[1] American National Standard T1.523-2001, Telecom Glossary 2000, February 2001, available at: *www.its.bldrdoc.gov/projects/telecomglossary2000/*.

[2] Telemetry Standards, IRIG Standard 106-01, Part 1, Secretariat, Range Commanders Council, U.S. Army White Sands Missile Range, NM, February 2001.

[3] Telemetry Standards, IRIG Standard 106-01, Part 2, Secretariat, Range Commanders Council, U.S. Army White Sands Missile Range, NM, February 2001.

Further Information

An overview of many aspects of telemetry systems is provided by S. Horan in *Introduction to PCM Telemetering Systems*, 2nd ed.,CRC Press, Boca Raton, FL, 2002.

The *Proceedings of the International Telemetering Conference* are published annually by the International Foundation for Telemetering. These proceedings contain theoretical developments as well as system, individual subsystem, and component developments.

There is no single comprehensive journal on telemetry. Various aspects related to data sampling, transmission, and signal processing are published in the transactions and journals of the various IEEE societies. Additionally, the *Transactions of the Instrumentation, Systems, and Automation Society*, and the *Journal of the International Test and Evaluation Association* should be consulted.

155

Recording Instruments

Timothy M. Chinowsky
University of Washington

Recording instruments are needed wherever repetitive measurements must be made and stored for later review or analysis. Some general types of applications that require recording instruments include the following:

Measurements for longitudinal characterization and recordkeeping. Measurement and recording of industrial process parameters allow for process monitoring and characterization. Chart records are required by some industries to comply with regulations such as the Environmental Protection Agency (environmental monitoring), Food and Drug Administration (food and pharmaceutical), and North American Electric Reliability Council (power grid interchange). Flight data recorders record aircraft performance data to allow for diagnosis in the event of an adverse flight event. As in many applications of this type, the recorded data are generally only used when the rare problem occurs, but then the recorded data become crucial.

Measurements requiring rapid scanning and review. Medical applications such as monitoring of patient vital signs in critical care settings require that medical personnel have the ability to quickly scan and summarize a variety of patient data, such as electrocardiogram traces, blood pressure measurements, and oxygen saturation. Recorders that can measure various patient data and assemble them into a graphical presentation which can be quickly comprehended are essential for this application.

Fast measurements requiring further analysis. Measurements that possess sufficient speed and complexity to require human interpretation at a slower speed must be recorded to allow for this analysis. Whether the data are relatively simple, such as an electrocardiogram, or more complicated, such as high-speed telemetry, a data recording is necessary to make full use of the data.

Measurements of slow trends. On the other extreme, human operators cannot efficiently monitor quantities that change very slowly, due to the limited human attention span and prohibitive labor costs. Data recorders can provide cost-effective measurements of quantities that change over time periods ranging from nanoseconds to years.

Measurements in remote or unattended locations. When a measurement must be made at a location where an operator is not present, measurements must be recorded. Automated data recorders allow measurements to be made in the most appropriate location, whether or not that location is conveniently accessible to a human operator.

0-8493-1586-7/05/$0.00+$1.50

155.1 Characteristics of Data Recorders

A great variety of devices for data recording are available in the market today. To ensure an optimal choice of recording device, it must be matched to the application. General attributes that must be considered when choosing a recording device include the following:

Type of measurement. What type or types of signal is being recorded? Does the recorder use internal sensors or does it connect to external sensor inputs?
Number of channels. How many measurements must be recorded simultaneously?
Speed. How fast (measurements per second) is the data recorded?
Recording medium. Onto what medium are the data recorded? Are data recorded in an analog or a digital format?
Data storage. How much data can the recorder store internally? When the recorder is "full," how are data removed and stored? Is the storage archival?
Machine readability. Is the recorded data machine readable, or is intended solely for human interpretation?
Data display. Does the recorder display the data as they are recorded, or are they stored for later viewing? Does the recorder display only the current measurement, or a complete measurement history?
Location. Is the recorder located where the measurements are made, or are the measurements made remotely and transmitted to the recorder over a wired or wireless connection? Are the recorded data viewed and analyzed where they are recorded, or are the recorded data removed from the recorder and analyzed at some other location?
Other features. Modern computerized data recorders often are capable of performing many more functions than simple data recording and display, including interfacing to personal computers (PCs) and computer networks, built-in data analysis algorithms, and the ability to control as well as monitor a process. The user must determine what combination of features will best serve to meet the short- and long-term goals of the application.

155.2 Paper Recorders: Strip Charts and Circular Charts

The oldest form of recording instrument is the paper recorder. In an all-mechanical temperature recorder, for instance, a mechanical displacement generated by a bimetallic strip moves a pen across a slowly moving piece of paper. As the paper moves, a plot of temperature against time is created. More sophisticated electromechanical pen recorders measure a wide variety of inputs such as current, voltage, RTD, or thermocouple, and convert the sensor data into mechanical motion of a pen using a servomechanism.

Paper recorders typically use either a strip chart or circular chart format. A strip chart records on a long strip of paper, typically 100 mm to 250 mm wide and many meters long. The paper is stored in either a fanfold or roll format, and is fed across the writing mechanism at a rate appropriate to the measurement rate. The circular chart records data along the radius of a slowly rotating circular paper chart, typically between 150 mm and 300 mm in diameter. The distinctive feature of the circular chart is that the chart data are inherently organized according to the chart rotation period, which is typically set to be a convenient period such as 24 hours, 7 days, or 30 days. A circular chart with a period of 24 hours would therefore be convenient for measuring and presenting parameters (temperature, for instance) expected to have a periodic daily variation.

Pen-based recorders are much improved since their invention. Modern pen recorders such as the Love Controls 16R (Figure 155.1) incorporate several ink colors, allowing for easy trace identification, and pens can provide automatic alphanumeric chart annotation as well as measurement traces. In "multipoint" recorders, each pen is not associated with a fixed trace, and many simultaneous traces in various line styles may be plotted. The ABB SR250A (Figure 155.2) allows recording of up to 24 channels, each with a unique trace style, along with a wide variety of alphanumeric annotations.

Mechanical limitations of pen recorders such as ink depletion and pen slew rate are addressed by strip chart recorders based on thermal line printer technology. These recorders digitize measurement data,

FIGURE 155.1 Love Controls 1600R 4-trace pen recorder.

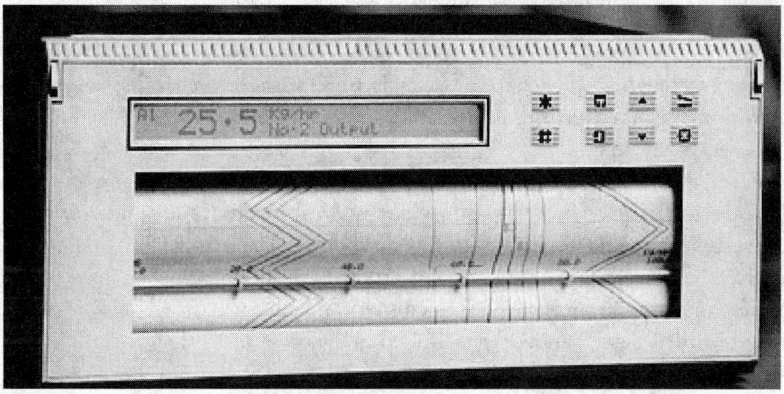

FIGURE 155.2 ABB SR250A multipoint strip chart recorder.

format the data for raster display, and plot the data on a dot-matrix thermal line printer. No physical pen motion is required, so measurement speed is limited only by measurement electronics, paper speed, and printer resolution rather than pen motion. The more sophisticated display capabilities of the thermal printer allow incorporation of information other than raw display traces onto the strip chart. Grid lines may be printed at the same time as the traces, eliminating the alignment difficulty found with pen-based recorders, in which pen traces are aligned with grid lines preprinted on the chart paper. Other graphic and alphanumeric information may also be printed on the chart as it is created. For instance, the RMS Instruments GP300 (Figure 155.3) prints on a 300-mm wide strip at a resolution of 300 dots per inch

FIGURE 155.3 RMS Instruments GP300 graphic printer/recorder.

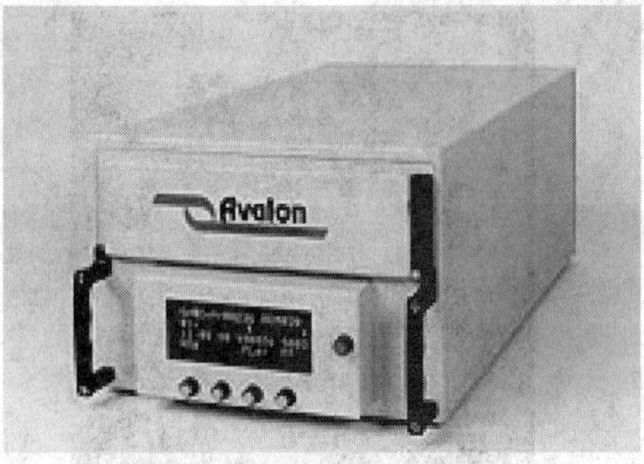

FIGURE 155.4 Avalon AE3000FL high-speed data recorder. Analog data of up to 12-MHz bandwidth is digitized and stored on conventional videotape.

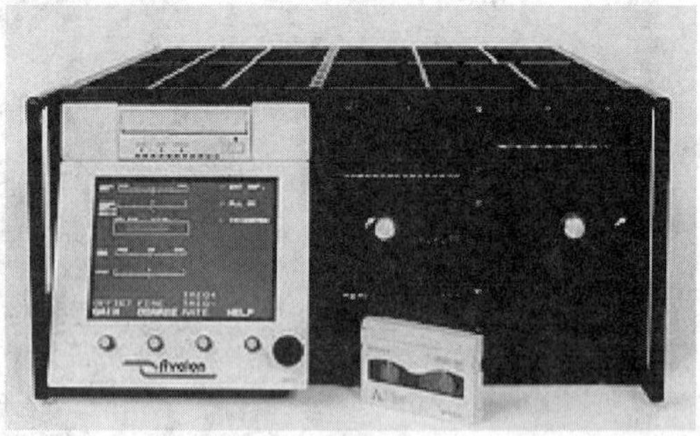

FIGURE 155.5 Avalon AE7000 hard-disk data recorder.

and paper feed speeds ranging up to 260 mm/sec. Up to 32 waveform channels can be plotted simultaneously, along with many other types of information, including 16 event (logic) channels, alphanumeric information, and grayscale images.

The use of paper as the primary storage medium is both the primary feature and main liability of paper recorders. On the positive side, paper provides an ideal medium for rapid viewing and review of data. Fanfold paper, in particular, allows the user to review a large quantity of data with little effort. Storage of the chart requires little special handling, and with proper treatment can be of archival quality. On the downside, paper is bulky and must be replenished, and paper-recorded data are not easily machine readable. Finally, recorded signal bandwidth is limited to the order of 10 Hz by printing resolution, paper speed, and paper consumption.

155.3 Magnetic Media Recorders

The bandwidth limitations, lack of machine readability, and bulky consumables of paper recorders are reduced by data recorders that use magnetic media. Removable media such as floppy disks, ZIP disks, videotape, and DAT provide for unlimited storage capability, while fixed hard disks offer high-performance random-access reading and writing of data. As the cost per gigabyte for hard disks continues to fall, removable/replaceable hard drives are an increasingly attractive option for high-density data storage, and are offered on some data recorders. Magnetic media may be used as an enhancement to a paper recorder, providing a machine-readable recording simultaneous with the paper record. The capabilities of magnetic media may be used to provide performance unobtainable with a paper record. The Avalon AE3000FL (Figure 155.4) is capable of recording analog signals with 18-MHz bandwidth on standard videotape; the hot-swappable-hard-disk based AE7000 (Figure 155.5) can record signals with 50-MHz bandwidth.

155.4 Semiconductor Memory Recorders

Increasing density and decreasing cost of nonvolatile semiconductor memory (flash memory) has made semiconductor data storage an attractive alternative to magnetic media. Packaged in various formats such as CompactFlash and PC-Card ATA, removable flash memory cartridges in capacities up to several gigabytes are increasingly used in data recorders as the storage medium. Battery-backed low-power static RAM is also widely used for semiconductor-based data storage, but battery life must be taken into account. Semiconductor memory is particularly attractive for data-logger applications in which a recorder is intended to operate unattended for a long period of time. The Onset StowAway Tidbit (Figure 155.6) is a 1.2" × 1.6" × 0.65", 0.8-oz. device that can make and store over 32,000 temperature measurements at a user-programmable rate ranging from seconds to hours; its encapsulated construction makes it usable up to 1000 feet underwater. Stored data can then be read out by a noncontact optical link. More general-

FIGURE 155.6 Onset Stowaway TidBit miniature encapsulated temperature data logger. Data are read from the device by a noncontact optical interface.

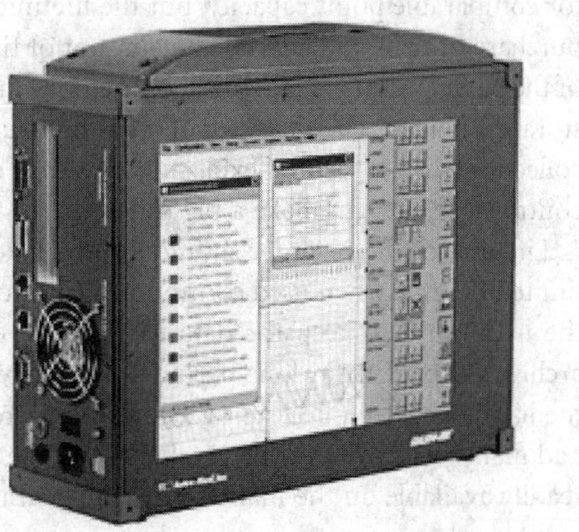

FIGURE 155.7 Astro-Med Dash 8x modular data recorder.

purpose data loggers provide self-contained versatile measurement, and data storage and retrieval capability for a wide variety of types of input signals and sensors.

155.5 Hybrid Recorders

Modern recording solutions typically offer a combination of the above features. Recorders that produce both a paper and machine-readable record are termed "hybrid" recorders and are increasingly ubiquitous. The trend is for the paper output, rather than the machine readable recording, to be an optional feature. The Astro-Med Dash 8x (Figure 155.7), for instance, incorporates a modular architecture that allows up to eight inputs of various types to be sampled at up to 200 kHz, with measurements saved to the recorder's internal hard disk. Data is displayed on an integral flat-panel touch-screen monitor. An optional printer may be attached to obtain conventional strip-chart output.

155.6 Personal Computer-Based Recorders

The ubiquity of personal computers has made personal computer (PC)-based solutions attractive for data recording applications. A typical office PC incorporates the display, storage, and printing capabilities required for most data recording applications, and requires only the addition of relatively low-cost measurement hardware and software to complete a data recording solution. Many companies have developed software and hardware for this application. National Instruments, in particular, has developed a large variety of sophisticated hardware and software programming environments for data acquisition. For instance, its VI Logger software combined with any of a variety of the company's measurement hardware enables rapid assembly of a data recording system. This approach is economical, as it makes use of general-purpose PCs, and also versatile, as software and hardware can be modified to add more sophisticated data acquisition and control capabilities. However, systems built using commercial, multipurpose hardware may be less rugged, less reliable, and more difficult to maintain than a self-contained piece of hardware dedicated to data recording.

155.7 Tradeoffs in Recorder Selection

The great variety of recording devices available makes it necessary to consider a large number of factors when choosing a recording device. Choosing between a paper-based recorder or a paperless recorder is one major decision. The initial cost of the paperless unit is higher than that of its paper-based counterpart for comparable point capacity, but the lifetime cost of ownership for a paper recorder must include the purchase of pens and charts. The amount of time needed to store data, the sample rate, and the capacity of the paper, magnetic tape, disk, or memory card tape are issues that must be resolved to ensure sufficient storage capacity. The question of what is required to enable the user to record data and view data after collection should also be explored. The ease of configuring the unit using its front-panel interface or control software can make a big difference in the usability of the recorder.

Lifetime of the recorder should also be taken into consideration. For instance, many displays have a limited lifetime when used continuously. If archival storage of data is necessary, one must consider what the user must do keep the data for several years. Many units require additional steps to maintain or archive stored data for long periods after the recorder has collected the electronic information. Magnetic media may have limited storage times compared with paper. Fortunately, the data recorder industry has had many years to refine its products, and the vast selection of reasonably priced recording instruments readily available on the market virtually guarantees that an acceptable solution can be found.

Further Information

A distributor of a wide variety of data recording devices is OMEGA Engineering, INC., P.O. Box 4047, One Omega Drive, Stamford, CT 06907-0047, *www.omega.com*.

Manufacturers mentioned in this chapter are listed below.

- ABB USA, P.O. Box 5308, 501 Merritt 7, Norwalk, CT 06856-5308, *www.abb.com*
- Astro-Med, Astro-Med Industrial Park, 600 East Greenwich Avenue, West Warwick, RI 02893, *www.astro-med.com*
- Avalon Electronics, Inc., 100 Bartow Municipal Airport, Bartow, FL 33830, *www.avalon-electronics.com*
- Dwyer Instruments Inc., P.O. Box 373, 102 Indiana Hwy. 212, Michigan City, IN 46361, *www.dwyer-inst.com*
- National Instruments, 11500 N. Mopac Expwy, Austin, TX 78759-3504, *www.ni.com*
- Onset Computer, 470 MacArthur Blvd., Bourne, MA 02532, *www.onsetcomp.com*
- RMS Instruments, 6877-1 Goreway Drive, Mississauga, Ontario, Canada L4V 1L9, *www.rmsinst.com*

156

Bioinstrumentation

Wolf W. von Maltzahn
University of Karlsruhe
Karsten Meyer-Waarden
University of Karlsruhe

Biomedical instruments measure, amplify, process, record, and store physiological quantities to help physicians diagnose illnesses, set up treatment plans, and restore lost functions. The scope of such instruments is enormous, both in complexity and the range of applications. Only a few can be covered in this chapter. Since a choice among topics must be made, this chapter focuses primarily on instruments that measure physiological signals; it does not cover devices that treat diseases or restore compromised physiologic functions. For further information, see the Bibliography section.

In contrast to technical systems, physiologic systems comprise living cells and organs. This requires that measurements do not adversely affect the living system by introducing toxic substances, destroying delicate tissues, or otherwise interfering with the chemical, electrical, or mechanical balance of the living cell or organ. Furthermore, many physiologic measurement sites are either inaccessible or direct measurements are only accessible invasively, that is, by puncturing the skin. Noninvasive biomedical instruments often use indirect measurement methods, even though these methods are usually less accurate, slower, and less informative than their direct counterparts.

Physiologic measurements seldom depend on one variable alone. The electrical impedance of tissue, for instance, depends not only on resistivity, cross-sectional area, and length, but also on electrolytes, enzymes, temperature, and other factors related to life processes. To extract meaningful information from impedance measurements, the system needs to focus on one of these factors and suppress the others. Physiologic events in living systems are unstationary, stochastic, time dependent, and noisy. It takes a lot of skill and ingenuity to measure them.

Biological cells pump primarily sodium and potassium ions across the cell membrane to maintain specific concentration differences. The resulting membrane potential of about –90 mV can be calculated from a modified Nernst equation. Nerve and muscle cells are excitable cells and generate action potentials by first changing the membrane permeability of sodium and then that of potassium. Action potentials influence the electric fields in the tissues surrounding the excitable cells all the way to the skin, from where they can be detected and recorded as surface biopotentials. These biopotentials are characterized by high source impedances, small signal voltages, significant interference voltages, and a modest frequency range. Special biopotential amplifiers convert these biopotentials into high-quality signals by amplifying them, suppressing interferences, and preparing them for further signal processing, display, or recording.

Biomedical instruments operate in hospital rooms, where they interact with other devices, work in the vicinity of other devices, or connect to patients. Such interconnections not only cause unpredictable interferences, but also provide current pathways that may endanger the lives of patients and/or operators. To ensure high-quality measurements and electrical safety, requirements must be imposed on biomedical

0-8493-1586-7/05/$0.00+$1.50

instruments, uncommon in typical industrial measurement systems. Biomedical measurements must do the following:

- Amplify the biosignal and suppress interferences
- Protect input stages against high voltages generated by other devices
- Be electrically safe for operator and subject
- Avoid adversely affecting the living system

156.1 Basic Bioinstrumentation Systems

Figure 156.1 shows a block diagram of a basic bioinstrumentation system. The physiologic variable to be measured, the measurand, may be the result of a molecular, cellular, or systemic event, and be mechanical, electrical, or chemical in nature. Blood pressure and blood flow, for instance, are important mechanical variables of the circulatory system. The electrocardiogram, electromyogram, and electroencephalogram provide information on the electrical activity of the heart, skeletal muscle, and brain, respectively. Partial pressures of oxygen or carbon dioxide reflect the status of the chemical balance in the blood. The sensing element in Figure 156.1 interacts with the measurand, directly or indirectly, and is designed with three main goals in mind: minimize the unavoidable disturbance of the measurand and its environment, avoid interference with life processes, and provide an output signal sensitive primarily to the measurand and insensitive to other parameters.

The transducer takes the output of the sensing element and transforms it into an electrical signal. The most common analog signal-conditioning elements are amplifiers, filters, rectifiers, triggers, comparators, and wave shapers, just like in other instrumentation systems. Appropriate amplification and filtering prepare the analog signal for analog-to-digital conversion (ADC). Once in digital form, a microcomputer further processes the signal and displays it on a CRT screen or on an LCD panel, generates a hard copy on a recorder, sends it to another device, or stores it in a mass storage device. Some measurements require external energy or stimulation applied to the subject across an isolation barrier.

Medical instruments contain special circuits to safeguard patients and operators against electrical shock, suppress interferences from the noisy hospital environment, and protect sensitive input stages. The most important of these is the isolation barrier, made of transformers, capacitors, optocouplers, or a combination of these. The isolation barrier keeps current densities in the body below the safe level of 100 μA/cm^2. Several national and international standards give limits for safe voltages and currents, describe methods to avoid potentially dangerous connections, and give detailed test procedures to ensure electrical safety [NFPA, 1990; AAMI, 1990; IEC, 1982].

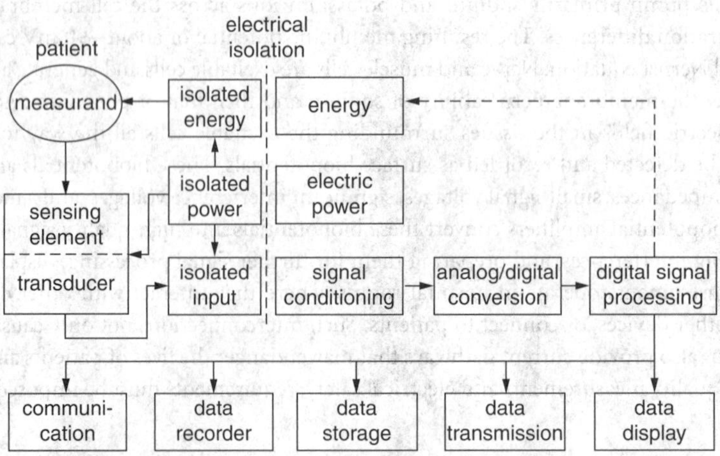

FIGURE 156.1 Schematic block diagram of a biomedical instrument.

A single hospital room may contain electrocardiographs, blood pressure monitors, x-ray machines, cardiac output monitors, respirators, electrosurgery devices, defibrillators, and other devices. The useful and beneficial output of one device may severely interfere with another. To function in the same environment, these devices must contain special hardware circuits or software algorithms that suppress undesired outputs from other devices.

156.2 Applications and Examples

Action potentials of nerve and muscle cells inside the body manifest themselves on the surface of the skin as small voltages between 10 μV and 100 mV with high source impedances and high levels of interfering noise signals. These surface biopotentials are detected with surface electrodes that function as transducers between the ionic currents inside the body and electronic currents in wires and amplifiers. Electrodes represent complicated electrochemical systems, as described in many bioinstrumentation books [Aston, 1990; Bronzino, 1986; Geddes and Baker, 1989; Norman, 1988; Profio, 1993; Webster, 1992]. Most electrodes for recording surface biopotentials are made of silver or silver chloride because they provide stable and relatively noise-free recordings.

The diagram in Figure 156.2 for obtaining a lead I electrocardiogram serves as an example of how to obtain a high-quality biopotential recording in general. Three electrodes attach to the subject — two for recording the biopotential and one for providing a reference potential. The electrodes on the left and right arms connect directly to the differential input of an isolated instrumentation amplifier; the reference electrode on the right leg connects to the floating ground terminal. This configuration separates the desired differential-mode biosignal e_S from the undesired common-mode noise e_N, as shown in the electric equivalent circuit of Figure 156.2. The instrumentation amplifier provides a high differential-mode gain and a low common-mode gain to suppress power line interferences and electrode potentials. The ratio between these two gains is called common-mode rejection ratio (CMRR). Although the CMRR of a good instrumentation amplifier may be as high as 130 dB, the CMRR of the overall circuit is seldom greater than 80 dB or 10,000:1. The primary cause for this large reduction in CMRR lies in the differences in electrode and skin impedances, represented by Z_1 and Z_2 in Figure 156.2. Some bioinstrumentation systems use additional amplifiers that set the voltage on the right leg equal to the mean voltage between the left and right arm ("driven right leg"), thereby decreasing common-mode voltages.

The most important and most investigated biopotential waveform is the *electrocardiogram* (ECG) [Webster, 1992]. Since the electrocardiogram represents the heart's electrical activity associated with cardiac contraction, it provides diagnostic insight into many heart functions. The curve in the left panel of Figure 156.3 represents a typical ECG waveform with the standard P, QRS, and T labels. The P wave and the QRS complex represent depolarization of cardiac muscle cells — the first one of atrial cells and the second one of ventricular cells. Although atrial repolarization cannot be seen in the ECG, the T wave represents ventricular repolarization. These P waves and QRS complexes precede cardiac contraction and the pumping of blood, first of the atria and then of the ventricles.

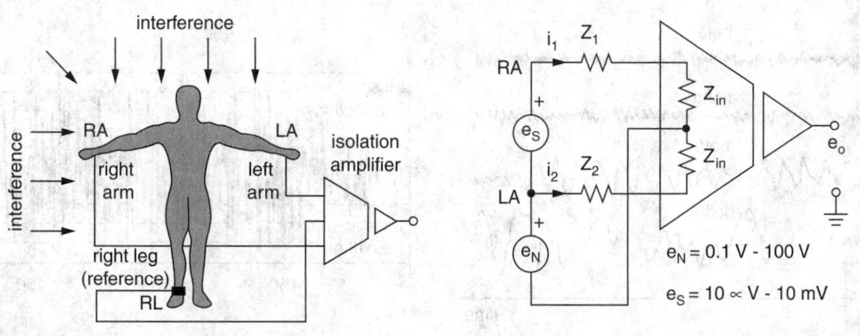

FIGURE 156.2 Schematic diagram for measuring biopotentials (left); electrical equivalent circuit (right).

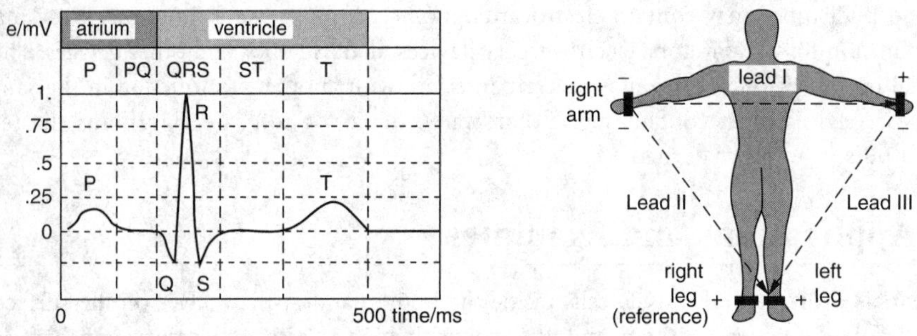

FIGURE 156.3 Typical lead II ECG recording (left). Einthoven triangle (right).

To enable comparisons of waveforms between subjects, certain standardized recording techniques have evolved. Einthoven, the pioneer of electrocardiography, developed an extremity lead system still in use today. As shown in the right panel of Figure 156.3, the three recording electrodes of the Einthoven triangle as placed on the left arm, the right arm, and the left leg, whereas the reference electrode is placed on the right leg. The voltage differences between the three recording electrodes are known as lead I, lead II, and lead III recordings. The lead II recording from the right arm to the left leg runs almost in parallel to the main axis of the heart and is therefore the preferred recording of a single-channel ECG. In addition to Einthoven's recordings, physicians use augmented limb leads and unipolar chest leads.

Recordings of the electrical activities of the brain, called *electroencephalograms* (EEGs), are more difficult to obtain than recordings of the ECG. The skull is a poor electrical conductor and EEG voltages on the scalp are in the μV range, as opposed to ECG voltages in the mV range. Furthermore, it is difficult to relate EEG voltages to specific neuronal activities; they are the net result of many different neurons firing seemingly independently of each other. With a subject in a relaxed state with closed eyes, alpha waves are easily recorded between two electrodes on the scalp. As shown in Figure 156.4, alpha waves are simple periodic waveforms in the frequency band of 8 to 13 Hz with slightly varying amplitudes. Alpha waves disappear when the subject's eyes are opened. The other waveforms shown in Figure 156.4 have clinical significance, particularly in diagnosing epilepsy and sleep disorders or in providing feedback during general anesthesia. The low-frequency theta waves between 4 and 8 Hz indicate sleep, whereas the high-frequency beta waves between 14 and 22 Hz appear during high states of alertness.

Recordings of the electrical activities of skeletal muscles are called *electromyograms* (EMGs). They can be obtained with surface electrodes similar to ECG electrodes or needle electrodes. The latter puncture the skin and are positioned close to the muscle group, from which they record. The mean amplitude and mean frequency of the EMG power spectrum reveal both muscle strength and muscle fatigue, which can provide feedback information in rehabilitation engineering and functional electrical stimulation [Phillips, 1991]. Table 156.1 summarizes amplitudes and frequency ranges of ECG, EEG, and EMG. Similarly, other biopotentials originating from other nerve or muscle cells can be recorded from the surface of the body.

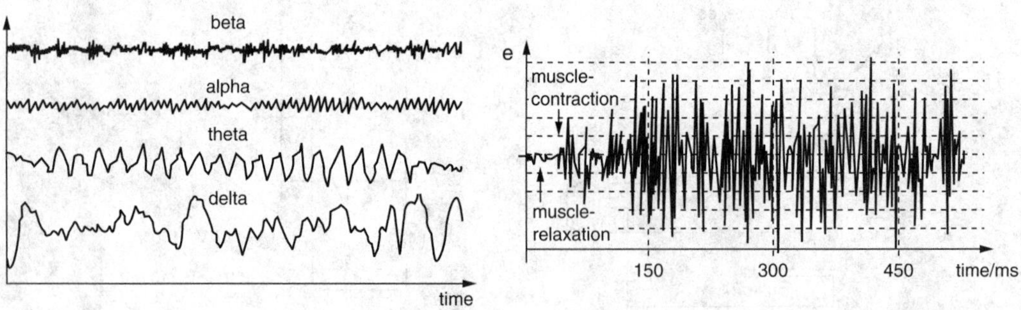

FIGURE 156.4 Typical EEG waveforms (left); typical EMG waveform (right).

TABLE 156.1 Amplitude and Frequency Ranges of ECG, EEG, and EMG

Origin of Electrical Activity	Name	Amplitude (mV)	Frequency Band (Hz)
Cardiac muscle	ECG	1–10	0.01–150
Skeletal muscle	EMG	5–100	10–5000
Brain	EEG	<0.2	1–30
	β	0.01	14–30
	α	0.03	8–13
	θ	0.05–0.1	4–7
	δ	0.1–0.15	1–4

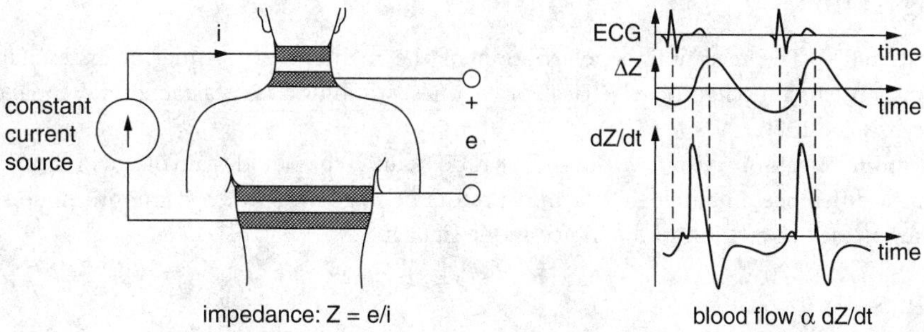

FIGURE 156.5 Electrode positions and typical waveforms of electrical impedance of thorax measurements.

Bioelectric impedance measurements provide qualitative and quantitative information about volume changes in the heart and in peripheral arteries and determine body characteristics such as percent body fat, total body fluid volume, and cell volume. They are also used in sleep apnea monitoring, especially in infants, and in the detection of venous thrombus. Multiple bioelectric impedance measurements lead to computed cross-sectional images of the body, so-called *computed impedance tomograms.*

Figure 156.5 shows four band electrodes for the measurement of respiration, stroke volume, and cardiac output. A constant sinusoidal current with an amplitude in the range of 0.5 to 4 mA and a frequency between 50 and 100 kHz flows from the top electrode through the thorax to the bottom electrode. This current is maintained independent of skin or tissue impedances. The voltage drop between the two middle electrodes is amplified by a voltage amplifier with a high-input impedance. The output voltage e_0 is proportional to impedance changes ΔZ that arise from thoracic volume changes due to respiration and from blood volume changes due to the pumping heart. Figure 156.4 shows the curves for ΔZ and dZ/dt, the former being proportional to respiratory volume changes and the latter to blood flow.

The noninvasive measurement of blood pressure is important in the diagnosis of many cardiovascular problems. Blood pressure varies — often within seconds — to meet the physiologic demands of organs and muscles. Although single blood pressure readings are valuable for entry-level screening, they do not provide as much information as monitoring over a specified length of time, usually 24 hours. The most common body-worn instrument for monitoring blood pressure noninvasively is a battery-powered oscillometric blood pressure monitor. In this monitor, an inflatable cuff fits snugly around the upper or lower arm and connects to a pump and pressure transducer. The microcontroller (CPU) triggers the pump periodically to inflate the cuff up to a pressure slightly greater than systolic arterial pressure. It then controls the valve to release the cuff pressure at a rate of about 4 mmHg/s and measures the cuff pressure. The resulting cuff pressure curve consists of a descending curve superimposed by small periodic oscillations. These oscillations start just before the cuff pressure equals systolic pressure, increase rapidly, reach the maximum amplitude at mean arterial pressure, and then gradually decrease. A special algorithm suppresses motion artifacts, extracts systolic and mean pressure (P_s and P_m) from the pressure curve, and calculates the diastolic arterial pressure. The oscillations also permit the calculation of heart rate. Blood pressure data, heart rate, date, and time are displayed on an LCD panel and stored in a data logger. The serial output permits uploading data to another computer for further evaluation. Figure 156.6 shows the block diagram of this device and an annotated typical oscillometric waveform.

156.3 Summary

This chapter has focused primarily on electrical measurement techniques and examples thereof. Many interesting subjects and instruments could not be covered, such as blood flow and blood volume, partial pressures of oxygen and carbon dioxide, oxygen saturation, glucose, and enzyme concentrations, to mention only a few. The emerging technologies related to home care could not even be mentioned. For further reading, see the texts listed in the References section.

Defining Terms

Action potential — The reversible depolarization of the membrane potential of an excitable cell in response to a mechanical, electrical, or chemical stimulus. The peak action potential of a single cell is about 70 mV.

Common-mode rejection ratio (CMRR) — The ratio of difference-mode gain over common-mode gain in a difference amplifier. It is a measure of the degree to which common-mode signals are suppressed in relation to difference-mode signals.

References

AAMI. 1990. *Design of Clinical Engineering Quality Assurance and Risk Management Programs.* Association for the Advancement of Medical Instrumentation, Arlington, VA.

Aston, R. 1990. *Principles of Biomedical Instrumentation and Measurement.* Merrill, Columbus, OH.

Bronzino, J. D. 1986. *Biomedical Engineering and Instrumentation: Basic Concepts and Applications.* PWS Engineering, Boston, MA.

Carr, J. J. and Brown, J. M. 1993. *Introduction to Biomedical Equipment Technology,* 2nd ed. Prentice Hall, Englewood Cliffs, NJ.

Geddes, L. A. and Baker, L. A. 1989. *Principles of Applied Biomedical Instrumentation,* 3rd ed. John Wiley & Sons, New York.

International Electrical Commission. 1982. *Regulations for Electro-Medical Devices.* IEC-601. International Electrical Commission, Beuth Verlag GmbH, Berlin and Cologne, Germany.

National Fire Protection Association. 1990. *Standard for Health Care Facilities.* NFPA-99. National Fire Protection Association, Publication Sales Department, Batterymarch Park, Quincy, MA 02269.

Norman, R. A. 1988. *Principles of Bioinstrumentation.* John Wiley & Sons, New York.

Phillips, C. A. 1991. *Functional Electrical Rehabilitation.* Springer Verlag, New York.

Profio, E. A. 1993. *Biomedical Engineering.* John Wiley & Sons, New York.

Webster, J. G., Ed. 1992. *Medical Instrumentation: Application and Design,* 2nd ed. Houghton Mifflin, Boston, MA.

Further Information

The monthly journal *IEEE Transactions on Biomedical Engineering* publishes research articles on recent advances in biomedical instrumentation. For subscription, contact IEEE Service Center, 445 Hoes Lane, P.O. 1331, Piscataway, NJ 08855-1331; phone (800)678-IEEE.

The monthly journal *Annals of Biomedical Engineering* also publishes research articles on recent advances in biomedical instrumentation. For subscription, contact Biomedical Engineering Society, P.O. Box 2399, Culver City, CA 90230; phone (310)618-9322.

The Emergency Care Research Institute evaluates medical devices, collects information about medical devices, and publishes periodic reports. Contact Emergency Care Research Institute, 5200 Butler Pike, Plymouth Meeting, PA 19462-1298; phone (215)825-6000; fax (215)834-1276.

157

G (LabVIEW™) Software Engineering

Christopher G. Relf
National Instruments Certified LabVIEW Developer

The LabVIEW (Laboratory Virtual Instrument Engineering Workbench) development system has certainly matured beyond its original role of simple laboratory bench experiment automation and data collection — it has become a full-featured programming language (known as *G*) in its own right. Although National Instruments (the creators of LabVIEW) continue to push its rapid prototyping and code development virtues, an increasing number of G software developers are relying on advanced programming techniques, far beyond those initially apparent.

The documentation that ships with a standard LabVIEW installation is sufficient to assist the development of simple applications by users not initially familiar with LabVIEW, and the various online documents provided by National Instruments and third-party vendors extend that learning experience. This chapter attempts to describe some of the features, methods, and more obscure tidbits of programming expertise that are of use to the advanced G software engineer.

157.1 Data Types

Like most modern programming languages, G code uses various data types. These data types allow for efficient data storage, transmission, manipulation, and processing of information. Simple data types include the integer, floating point, strings, paths, arrays, clusters, and reference numbers (*refnums*). The

0-8493-1586-7/05/$0.00+$1.50

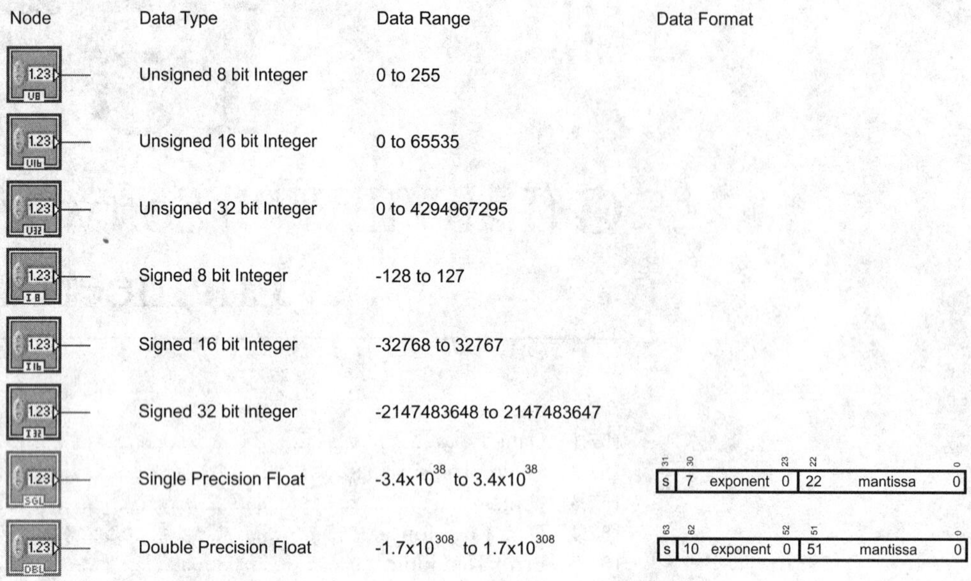

Node
Data Type
Data Range
Data Format
Unsigned 8 bit Integer
0 to 255
Unsigned 16 bit Integer
0 to 65535
Unsigned 32 bit Integer
0 to 4294967295
Signed 8 bit Integer
-128 to 127
Signed 16 bit Integer
-32768 to 32767
Signed 32 bit Integer
-2147483648 to 2147483647
Single Precision Float
-3.4x10^38 to 3.4x10^38
s 7 exponent 0 22 mantissa 0
Double Precision Float
-1.7x10^308 to 1.7x10^308
s 10 exponent 0 51 mantissa 0

FIGURE 157.1 Numerical data types.

method G uses to store data depends on its type, and it is important to understand the way data will reside in memory. Savvy use of data types decreases the overall memory footprint of an application, minimizes data coercion errors, and improves system performance.

Figure 157.1 lists representations of the standard numerical data types available to G. Others exist, including the extended-precision float, although its representation is platform dependent, as shown in Figure 157.2. (The representation of an extended-precision float is identical to a double-precision float when using HP-UX.)

Unlike numerical values, strings are represented by dimension-defined arrays (Figure 157.3). Each array consists of a 4-byte integer that contains the number of characters required to represent the string, and subsequent byte integers that represent each character. Therefore, as the highest number that will fit into a 4-byte word is 4,294,967,296, the longest string that can be continuously represented has 4,294,967,296 characters.

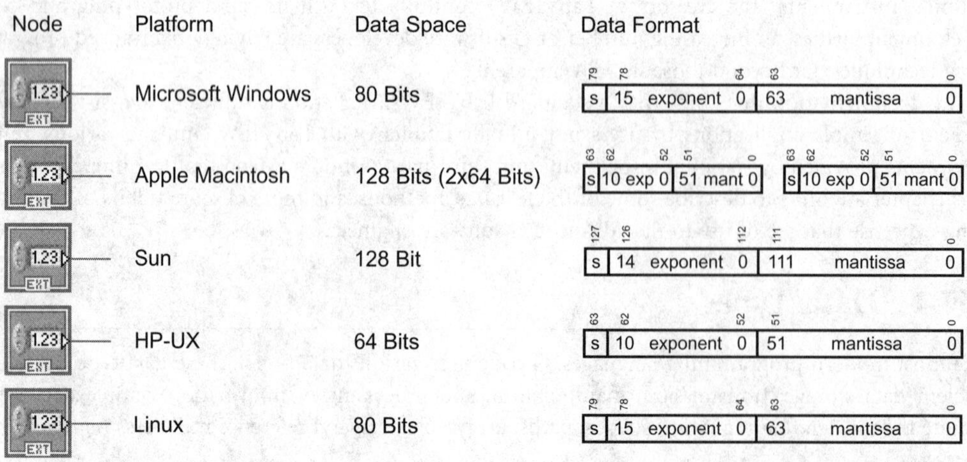

Node
Platform
Data Space
Data Format
Microsoft Windows
80 Bits
s 15 exponent 0 63 mantissa 0
Apple Macintosh
128 Bits (2x64 Bits)
s 10 exp 0 51 mant 0
s 10 exp 0 51 mant 0
Sun
128 Bit
s 14 exponent 0 111 mantissa 0
HP-UX
64 Bits
s 10 exponent 0 51 mantissa 0
Linux
80 Bits
s 15 exponent 0 63 mantissa 0

FIGURE 157.2 Extended precision data type.

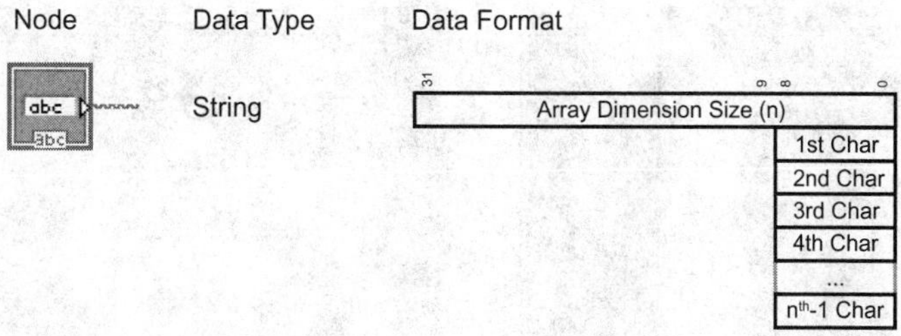

FIGURE 157.3 String data type.

TABLE 157.1 Path Type Descriptions

Path Type	Description	Microsoft Windows Examples
0	Absolute	C:\DATA\TEMP.DAT
1	Relative	\DATA\TEMP.DAT
2	Undefined	
3	Universal naming convention	\\PC_NAME\DATA\TEMP.DAT
4→∞	Undefined	

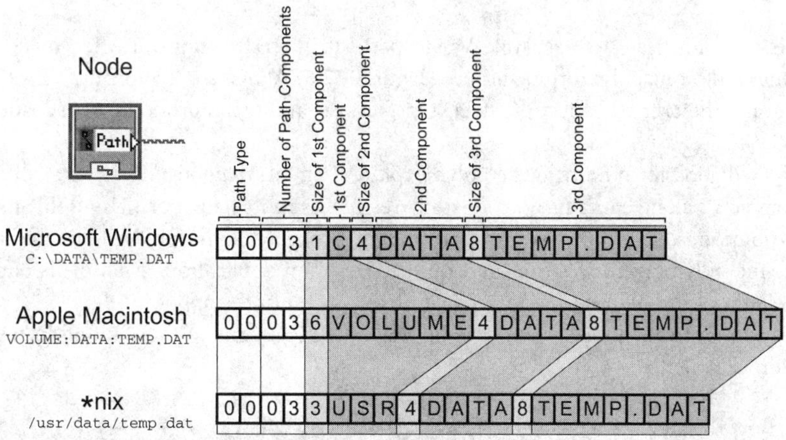

FIGURE 157.4 Path data type representations.

Paths are also stored as arrays, although the format is different from strings. The first element of a path array data format describes the *path type*, which is determined as shown in Table 157.1.

The second element of the path array data format defines the *number of path components*. For example, the following path,

```
\\PC_NAME\DATA\TEMP.DAT
```

contains three components: the PC name, then a folder, and a file. Each of the components is then described in the array as a double-element pair — the first element is a byte that represents the character length of the component, and the second element is the component. As might be expected, path representations are platform dependent, as shown in Figure 157.4.

157.2 Polymorphism

Considered a generally underrated technique, polymorphic virtual instruments (VIs) are simply a collection of related VIs that can perform functions on differing data types. For example, a VI that loads a waveform from a file could return the result as an array of values or the waveform data type (Figure 157.5).

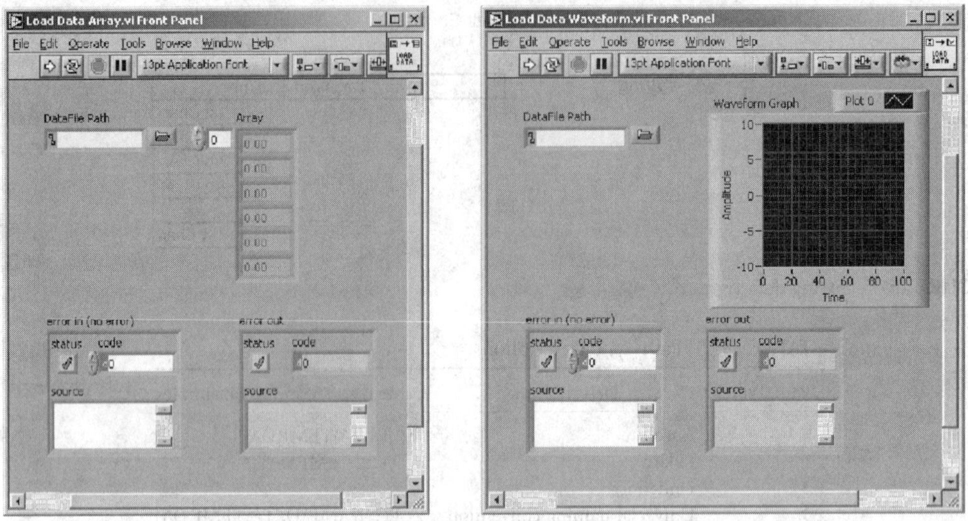

(a) (b)

FIGURE 157.5 (a) Load data array, 5; (b) load data waveform.

While it is valid to create two separate VIs to perform these functions, it often makes more sense combining them into one polymorphic library (Figure 157.6). When a polymorphic VI is dropped onto a wiring diagram, the user can select the data "type" to use, and the appropriate VI is loaded (as shown in Figure 157.7).

It is considered good form to include a polymorphic VI and its daughter VIs in one LabVIEW library, as a polymorphic VI alone contains no code, only references to its daughters. Using this method assists in code distribution, and minimizes the incidence of broken polymorphic VIs unable to find their daughters. Using polymorphic VIs can increase the size of an application, as all of its daughters, irrespective of whether they are used in the application, are saved in the application.

Polymorphism is already used in several vi.lib VIs, including many of those found in the data acquisition function subpalette.

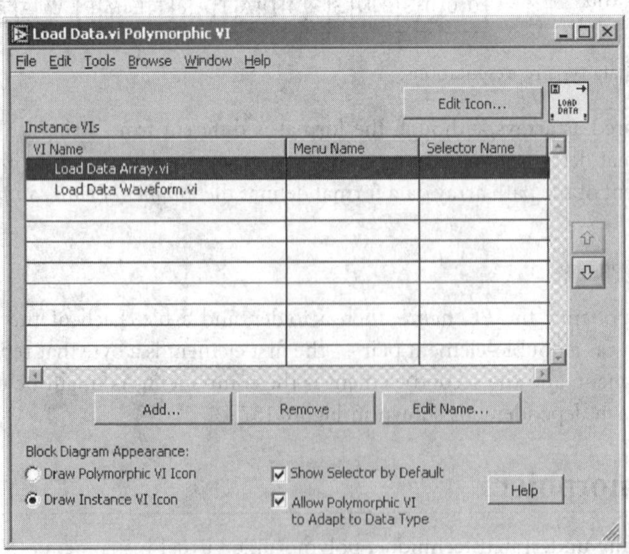

FIGURE 157.6 Creating a polymorphic VI.

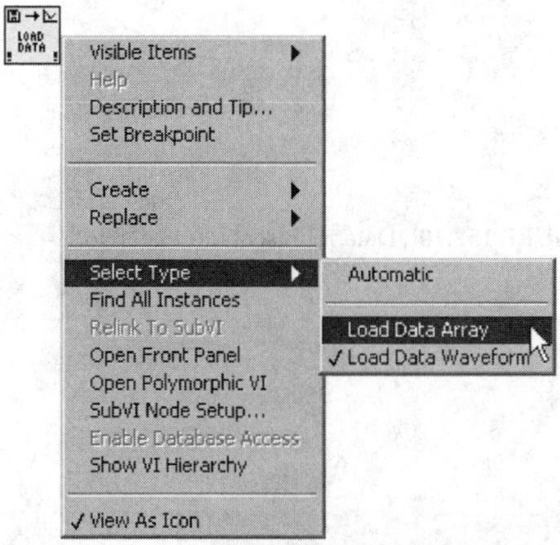

FIGURE 157.7 Select polymorphic type.

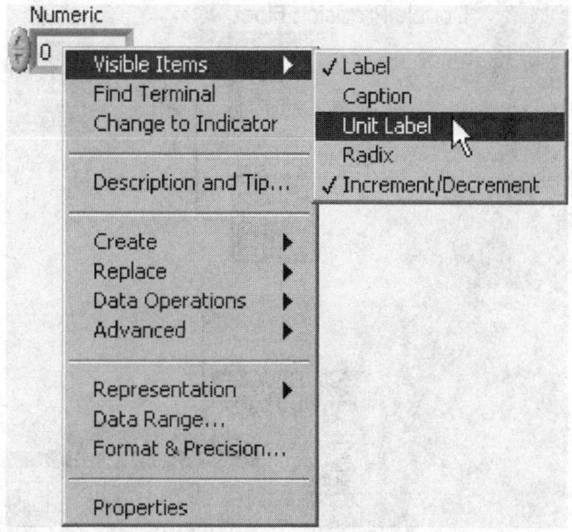

FIGURE 157.8 Selecting unit label.

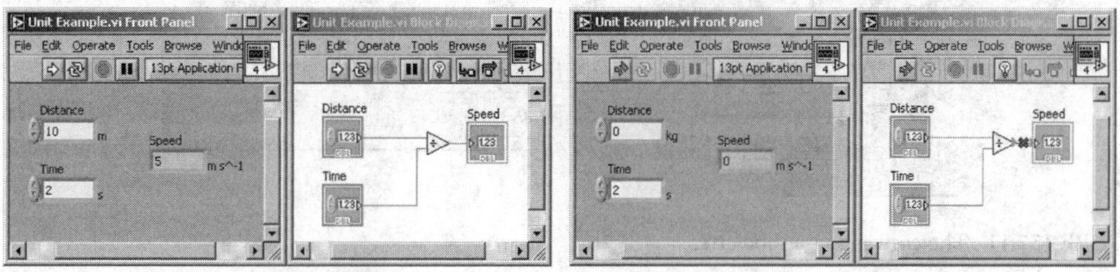

(a) (b)

FIGURE 157.9 (a) Unit example OK; (b) unit example not OK.

157.3 Units

Associating units to raw data adds an additional level of numerical checking when evaluating expressions and formulas. Incorporating units also adds the ability to display data to the user in dynamically controlled units, without the need for code to perform conversions.

A data node's unit association can be found by right-clicking on the node, and selecting Visible Items→Units (as shown in Figure 157.8). Once the unit label is displayed, the user can enter a unit type (a comprehensive list can be found in the LabVIEW online help), such as g for "grams." Now that a unit is associated with the node, all data to that node are converted to the unit specified, and all data from it bears its unit type. VIs that do not respect mathematical unit relationships are broken, and will not run. Figure 157.9(a) shows a VI where a distance (m) value is divided by a time (s) value, resulting in a speed (ms^{-1}) value. If the distance control's unit is changed to an incompatible type (e.g., kilogram in Figure 157.9(b)), the G development system reports a bad wire between the division primitive and the ms^{-1} indicator, suggesting an incompatible unit.

The LabVIEW built-in functions, such as Add and Multiply, are polymorphic with respect to units, so they automatically handle different units. In order to build a subVI with the same polymorphic unit capability, you must create a separate daughter VI to deal with each unit configuration.

157.4 Data Coercion

When performing functions on data, it is possible to combine differing representations in the same structure. When this occurs, LabVIEW must decide on how to convert the incident types to a common

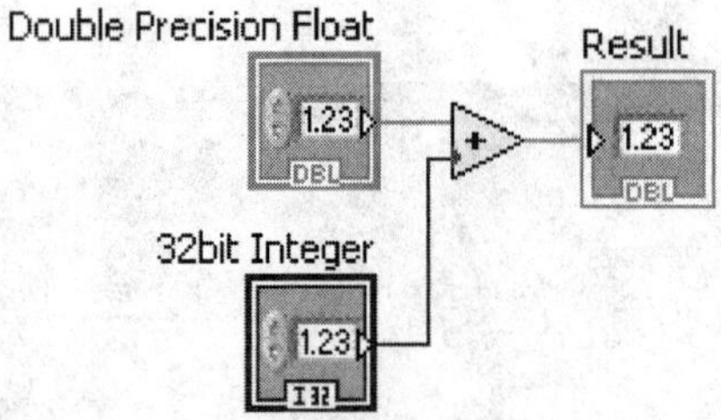

FIGURE 157.10 Data representation coercion.

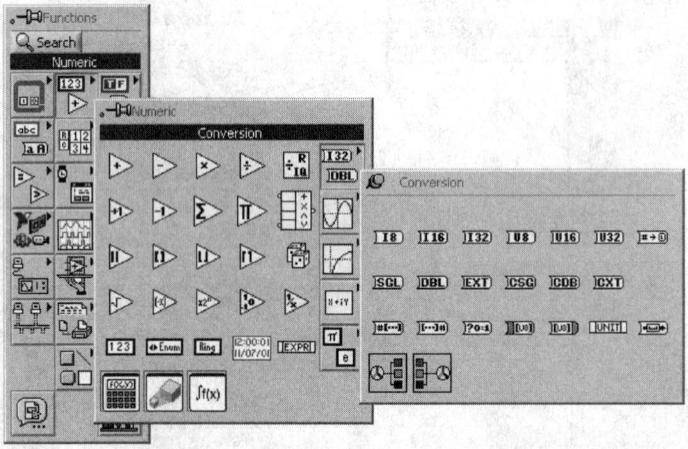

FIGURE 157.11 The conversion subpalette.

format before executing the function. As shown in Figure 157.10, the widest data representation is chosen for the operation and subsequent display and/or manipulation.

The coercion of the 32-bit integer is demonstrated by the coercion dot on the input terminal of the function, indicating that the data type of the data input at this terminal is changed before the function is executed. Although in the example shown in Figure 157.10 no data are lost (the narrower data type is expanded), data coercion can lead to inefficient memory usage (storing numbers in a data format beyond what is required to accurately represent the physical data) and occasionally the loss of data (coercing data to a type narrower than the information encoded in it, thus clipping the data to its highest permitted value, or value wrapping when using signed/unsigned conversions). It is considered good form to hunt down and remove coercion dots from one's code, as they can represent coercion of data that are either beyond the software engineer's control and/or knowledge. Explicitly performing data coercion using the Conversion subpalette (Figure 157.11) of the Numerical function palette minimizes the incidence of poor block diagram comprehension.

Data coercion can be very important when considering application memory management. For example, if a floating point is inputted to the *N* node of a *for loop*, its data is coerced to a 32-bit integer format. If an upstream calculation causes the floating point to become *NaN* (not a number), it coerces to a very large number (in the order of 2×10^9), resulting in a large number of for loop iterations. LabVIEW will attempt to allocate an appropriate amount of memory to execute the loop (and handle any data inputs and outputs), and often crash trying. This issue may seem intermittent, as it will only occur when an upstream calculation fails. (This example was provided by Michael Porter, Porter Consulting, LLC.)

157.5 Error Handling

Writing VIs and SubVIs that incorporate error handling is considered good form, not only to allow the user indication of abnormal software execution, but also to allow the software to make decisions based

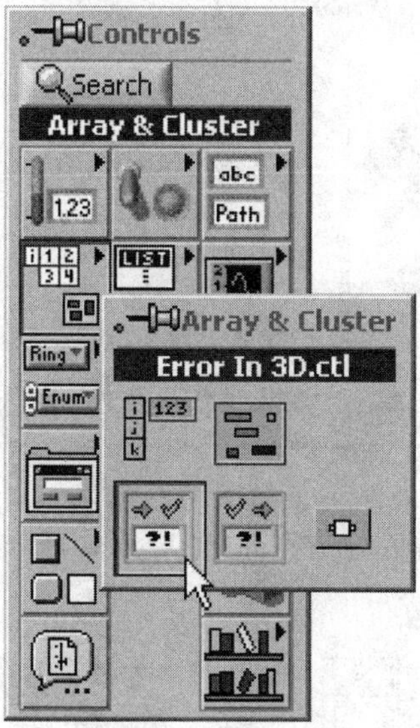

FIGURE 157.12 Error cluster control palette.

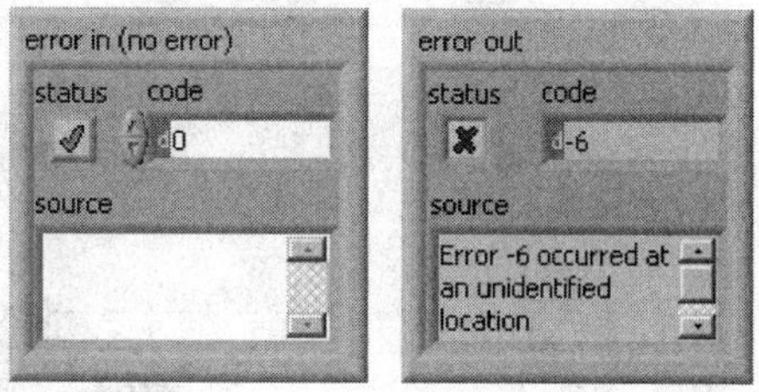

FIGURE 157.13 Front panel error clusters.

on the status of previous operations. Including error clusters in your code assists in troubleshooting, modularity, and user friendliness.

The Error Cluster

The error cluster is a special predefined LabVIEW cluster that is used to contain error status information (Figures 157.12 and 157.13). The cluster contains the following three components:

Name	Data Type	Description
Status	Boolean	Indicates if an error has occurred (TRUE = error, FALSE = no error).
Code	32-Bit integer	A standardized error code specific to the particular error. LabVIEW has a table of default error codes, although the user is able to define custom error codes. See below for more information.
Source	String	Textual information often describing the error, the VI it occurred within, and the call chain.

Usage

Although one of the most obvious methods of conditional code execution might be to unbundle the *Status* Boolean of the error cluster, and feed it to the conditional terminal of a case structure (Figure 157.14(a)), the complete cluster can be wired to it instead (Figure 157.14(b)).

The functionality of the code in Figure 157.14(a) and (b) is identical, although its readability is vastly improved as the second example colors the case structure green for the *No Error* case, and red for the *Error* case. Wrapping the entire code of a SubVI within a conditional case structure based on the error input allows VIs at higher levels to continue functioning, without executing code that could be useless or even dangerous when an error has occurred. Consider the simple report generation example shown in Figure 157.15.

A report is initially created, a file is then attached, the report is printed, and finally destroyed. The dataflow link between these SubVIs is the report's reference number (*refnum*), which ensures each execute in a predefined order. If each of the SubVIs execute without error, then the process completes successfully. Conversely, if one of the SubVIs encounters an error, subsequent SubVIs are unaware of the problem and attempt to execute regardless. In the example above, an error in the *Attach a File to the Report* SubVI will not cause the *Print the Report* SubVI to fail, resulting in a blank report print. If effective error handling is introduced (Figure 157.16), the Print the Report SubVI will know if an error has occurred before its

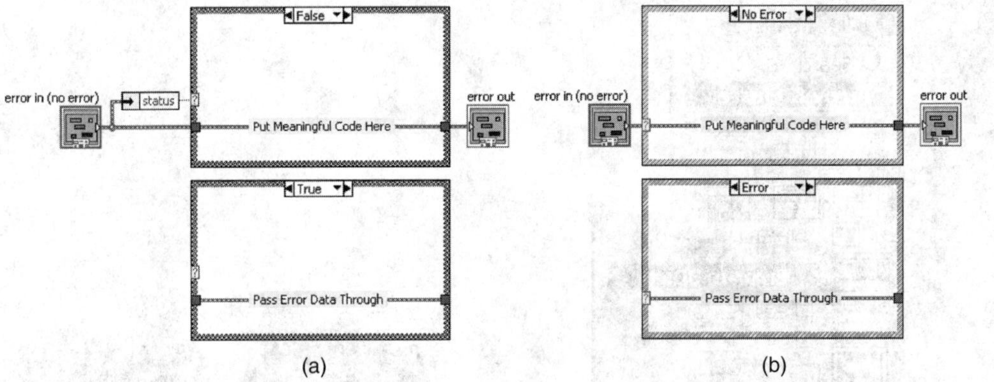

FIGURE 157.14 (a) Wire error cluster status into the conditional node 14; (b) wire error cluster directly into the conditional node.

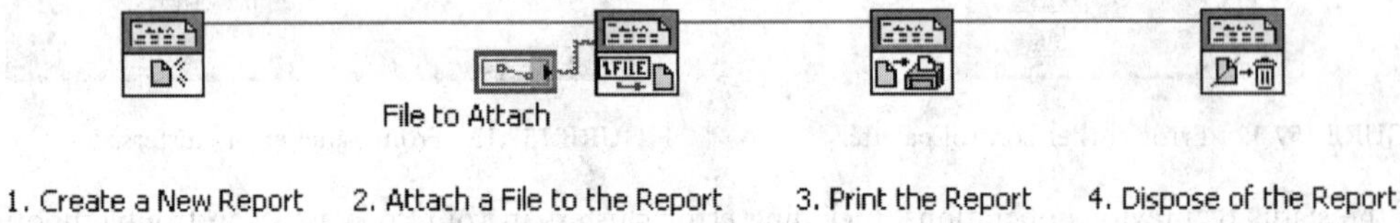

FIGURE 157.15 Create and print report without error handling.

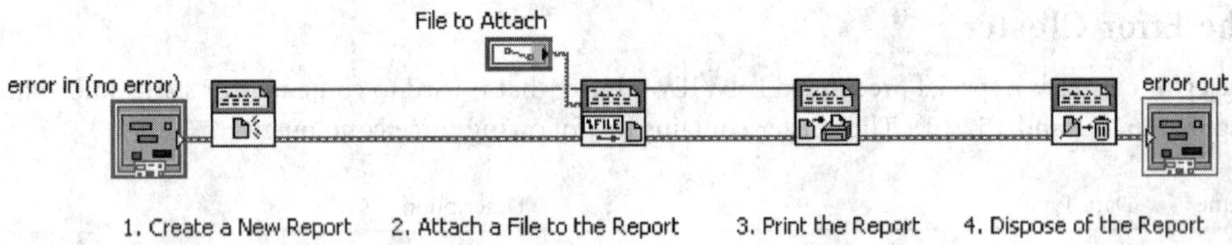

FIGURE 157.16 Create and print report with error handling.

execution is requested. If the functional code inside the *Print the Report* SubVI is enclosed within a conditional case structure based on the error input, the printing code is bypassed, and the data in the error cluster is passed on to the next SubVI.

To attain a higher level of user interaction, standard SubVIs exist to alert the user to an error on the error cluster, and prompt for conditional actions. The Simple Error Handler.vi (Figure 157.17) allows for a basic level of error cluster status reporting, displaying detected errors in a dialog box, and prompting the user for an action based on the *type of dialog* input (for example, OK, Cancel, Stop, etc.).

The Simple Error Handler.vi is a wrapper for the lower level General Error Handler.vi. The latter (Figure 157.18) is more configurable, and permits the dynamic definition of custom error codes, and error exception handling.

LabVIEW 7.0 brings with it an addition to the error handling function palette, Clear Errors.vi (Figure 157.19). This VI is simply an *error in* control and an *error out* indicator, which are not linked on the wiring diagram, causing any errors on the wired error link to be cancelled. This VI can be useful

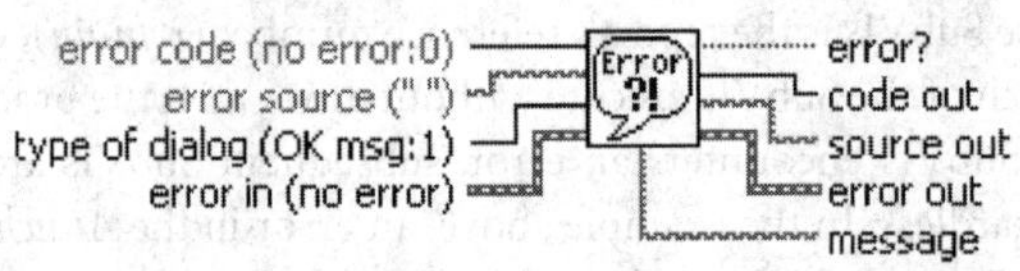

FIGURE 157.17 Simple error handler.

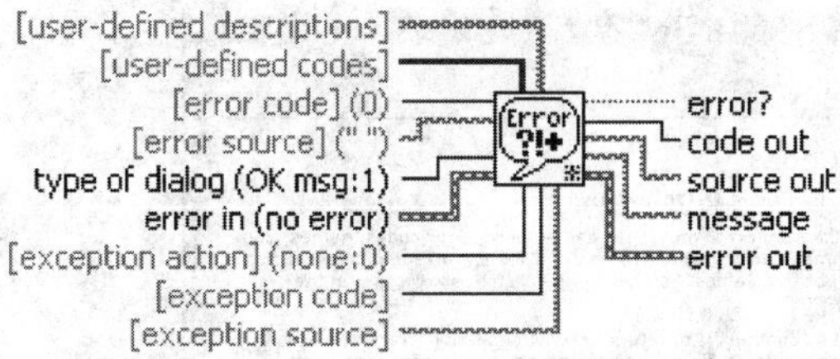

FIGURE 157.18 General error handler.

FIGURE 157.19 Clear errors.

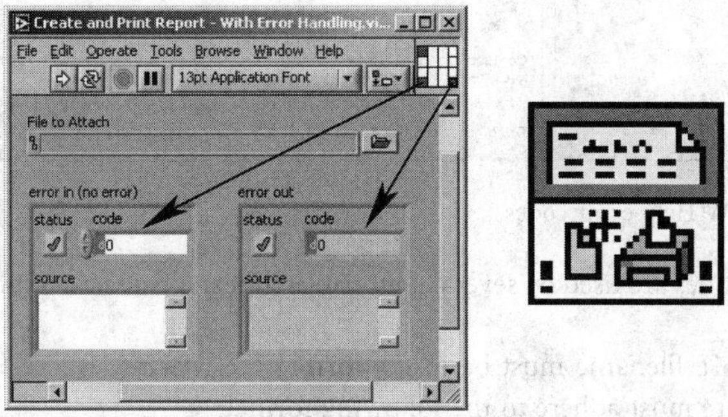

FIGURE 157.20 SubVI error cluster connector pane placement.

when constructing custom error-handling VIs, including dialog boxes allowing user interaction that is not covered by the simple and general error-handling VIs, but should not be used alone. Although dumping the errors from the error cluster may be tempting, one *must* incorporate appropriate code to handle them.

Wiring Errors into a SubVI Connector Pane

As described above, SubVIs have an associated connector pane, and the placement of error cluster inputs and output is generally in the lower quadrants (*error in* on the left, and *error out* on the right), with corresponding exclamation marks over the connectors on the SubVI's icon (as shown in Figure 157.20).

Default Error Codes

The LabVIEW execution system contains a large list of standard and specific errors, encouraging the reuse of generic codes across applications. A list of generic errors codes can be found under the LabVIEW *Help* menu (Figure 157.21).

Another method of parsing a small number of errors is to select the *Help→Explain Error* menu item. This will launch an interface that allows the user to enter an error code, and a brief explanation is displayed (Figure 157.22). This interface is also accessible by right-clicking on a front panel error cluster, and selecting *Explain Error*.

Custom Error Codes

National Instruments has set several error codes aside for custom use. If an existing error code does not adequately describe the error condition, the user can define custom codes that are specific to the application. Codes between 5000 through to 9999 are available for use, and do not need to conform to any other application. Although General Error Handler.vi can be used to define custom error codes, one can also create an XML file in the labview\user.lib\errors directory that contains custom error codes and their descriptions. This method is particularly useful if the user requires custom error codes to apply to several

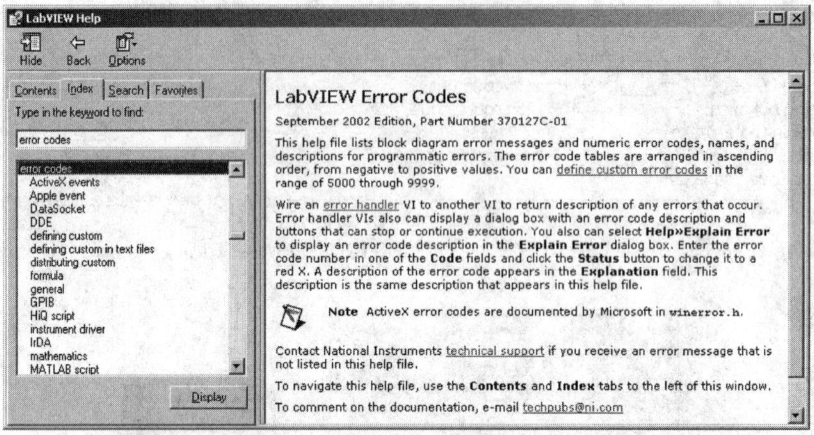

FIGURE 157.21 LabVIEW Help error codes.

applications, or if the codes are used by several software engineers in a team or are to be distributed with an application.

For example, the XML filename must be in the format *-errors.txt (where * is user definable), and the internal file structure must adhere to the following format:

```
<?XML Version="1.0">
<nidocument>
<nicomment>
This is a custom error code definition file for my application.
</nicomment>
<nierror code="5000">
User Access Denied!
Contact the Security Department to gain clearance to perform this function.
</nierror>
<nierror code="5001">
User Unknown.
Contact the People Development Department.
</nierror>
<nierror code="5100">
Driver Unable to Contact Instrument Database.
Contact the Software Engineering Department Helpdesk.
</nierror>
<nierror code="5200">
Plug-In Module in R&D mode - not to be used in Production Environment.
Contact the New Product Development Group.
</nierror>
</nidocument>
```

As can be seen, a file comment can be created within the <nicomment> tag space. Each custom error is defined as an <nierror> with it's associated error code, and its error message is then placed inside the <nierror> tag space. Although hand coding a custom error code XML file is possible, the Error Code File Editor (**Tools→Advanced→Edit Error Codes**) provides a simple GUI for file creation and editing (see Figure 157.23). Once custom error code files have been created and/or altered, LabVIEW must be restarted for the changes to take effect.

It is often useful to define error code bands during the project planning stage, setting aside bands for specific related error groups.

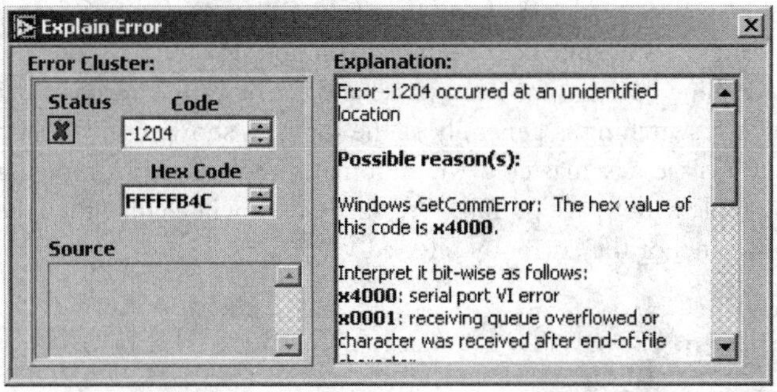

FIGURE 157.22 Explain error.

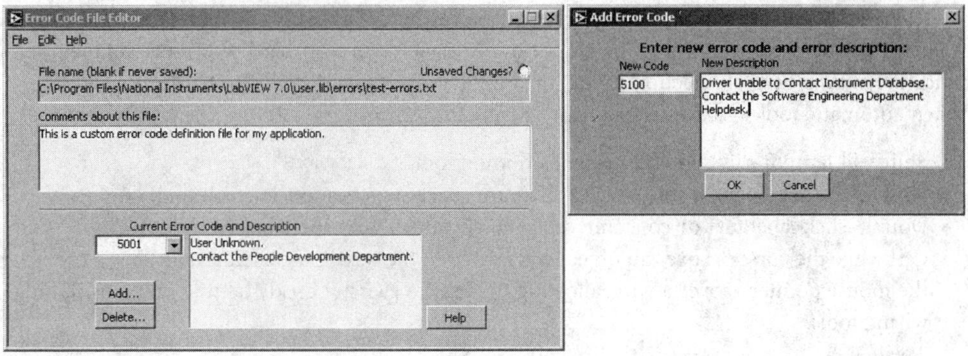

FIGURE 157.23 Error code file editor.

157.6 Shortcuts

Keyboard Shortcuts*

Most developers soon realize that their mouse or pointing device is not the only way to perform tasks; in fact some actions can only be achieved through keyboard sequences (or *shortcuts*). Once committed to memory, keyboard shortcuts are often a faster way to achieve tasks that might otherwise take several mouse maneuvers. Appearing below are most of the common and some of the obscure keyboard shortcuts that LabVIEW offers. Occasionally the only way to achieve a task is to use the keyboard in combination with the mouse, and these cases have also been included for completeness.

LabVIEW offers the developer a wealth of useful keyboard shortcuts. The shortcuts presented here refer to LabVIEW installed on the Windows operating system. Substitutions for the Ctrl key apply for other operating systems, as detailed in Table 157.2.

TABLE 157.2 Control Key Substitutions across LabVIEW-Supported Platforms

Operating System	Control Key
Windows™	Ctrl
Macintosh™	Option or Command (context dependent)
Sun™/UNIX®	Meta
Linux™/HP-UX™	Alt

*This information on shortcuts was provided by Peter Badcock, engineering consultant, Australia.

The most commonly encountered tasks faced by a LabVIEW programmer focus on constructing and wiring the block diagram of a VI, while the next most common set of tasks is the manipulation of objects on the front panel. Switching between the front panel and the block diagram is a regular task. A shortcut to achieve panel/diagram switching is generally at the top of a shortcut enthusiast's "commonly used" list (assuming the enthusiast has mastered the automatic tool selection mode, and is therefore not continually switching tools with Space or Tab). Pressing Ctrl-E will change the focus between the block diagram and the front panel of the currently selected VI.

Tool Palette Shortcuts

An automatic tool selection tool palette mode was introduced with LabVIEW 6.1, where the active tool automatically changes to a selection that the development system considers the most useful as the cursor is moved over different parts of the front panel or block diagram. For the LabVIEW programmer who is accustomed to manually selecting the appropriate tool (whether by using the Tools Palette, or a keyboard shortcut), this takes some familiarization. Committing certain keyboard shortcuts to memory will enhance the automatic tool selection mode, making it a more useful (and less frustrating!) aid.

When automatic tool selection is enabled:

- Shift will temporarily switch to the positioning tool.
- Shift-Ctrl while cursor over panel or diagram space switches to the scroll tool.
- Double-click label text or constant values to edit them with the labeling tool.
- Ctrl while the cursor is over an object will switch to the next most useful tool. This is useful when the mouse pointer is over a wire, allowing the user to switch quickly between the positioning and wiring tools.

Note that Shift-Tab enables and disables the automatic tool selection mode.

When automatic tool selection is disabled:

- Tab selects the next tool on the tools palette.
- Space toggles between the positioning tool and the scrolling tool on the front panel, or the positioning tool and the wiring tool on the block diagram.

Miscellaneous Front Panel and Block Diagram Shortcuts

To create additional space in the selection, Ctrl-click (for operating systems other than Windows, use Shift+Ctrl+Click) from a clear point on the panel or diagram and select an area. The cursor keys (←, ↑, →, ↓) nudge selections one pixel in the direction chosen, whereas simultaneously holding down Shift while moving a selection with a cursor key increases the distance moved to eight pixels. Other miscellaneous shortcuts include those listed in Table 157.3.

TABLE 157.3 Other Miscellaneous Shortcuts

Shortcut	Description
Ctrl-double click	On a clear point, places text (otherwise known as a free label).
Shift-right click	Displays the tool palette.
Ctrl-double click	On a sub VI, opens its block diagram.
Ctrl-click-drag	An object with the positioning tool to duplicate it.
Shift-click	The block diagram to remove the last wiring point.
Esc	Cancels a wire route (dashed).
Ctrl-B	Removes broken wires. (Beware: nonvisible broken wires may exist.)
Drag-space	Activate autowiring when moving an element on the block diagram. (Autowiring only occurs when first placing an element, or a copy of an element, on the block diagram.)

TABLE 157.4 Common Shortcuts

Shortcut	Description	Comments
Ctrl-S	Saves the currently selected VI.	
Ctrl-Q	Quits LabVIEW.	The user is prompted to save any open unsaved VIs.
Ctrl-C	Copy a selection to the clipboard.	
Ctrl-X	Cut a selection and place it on the clipboard.	
Ctrl-V	Paste the data from the clipboard.	Pasting data from a previous LabVIEW session will insert an image representing the data, and not the data itself.
Ctrl-F	Find object or text.	
Ctrl-R	Run the currently selected VI.	
Ctrl-.	Aborts a running VI.	Only available when Show Abort Button option in VI Properties»Windows Appearance»Show Abort Button is selected.
Ctrl-E	Toggle focus between the front panel and block diagram.	
Ctrl-H	Display context help.	
Ctrl-W	Close the current window.	The block diagram is automatically closed when a front panel is closed using this shortcut.
Ctrl-Z	Undo the last action.	

General Shortcuts

A complete list of keyboard shortcuts for the main pull-down menus can be found on the LabVIEW Quick Reference Card (shipped with LabVIEW and accessible in the online help). Some commonly used keyboard shortcuts include those shown in Table 157.4.

157.7 GOOP

G is well known for its speed of code development that provides engineers with a rapid prototyping advantage. This advantage is particularly evident when working with small projects with flat structures. However, as project complexity grows, initial stages of software architecture planning become vitally important.

GOOP (graphical object-oriented programming) is a method for developing object-oriented code, and is created using an external object model (often referred to as "LabVIEW external object" oriented programming, or LEO). Using GOOP has three main advantages to application development: code maintainability, scalability, and reusability. GOOP is the most common form of OOP with LabVIEW, and has been used successfully in many commercially available applications developed by G software engineers.

By implementing GOOP classes, you can take advantage of standard OOP analysis and design tools, including Microsoft Visual Modeler, and apply principles that describe object-oriented design and development. The OOP component-based development approach to developing applications enhances the traditional data-flow programming paradigm used in G applications, and using tools including the GOOP Wizard (access *www.ni.com*, and enter "GOOP" in the Search box for more information) allows you to create and manage OOP classes easily (Figure 157.24).

Once the fundamental concepts are understood, GOOP is easy to use, allows encapsulation, and permits you to create, control, and destroy as many objects as required. Both functions and data can be accessed using LabVIEW refnums, which simplifies wiring diagrams and encourages dynamic and modularized code.

157.8 Code Distribution

Distributing code to encourage multiple programmer input and commercial reward can be an important part of software development, and hence should not be considered only in the final moments of a project. Depending on the purpose of distribution, the package format created can vary significantly.

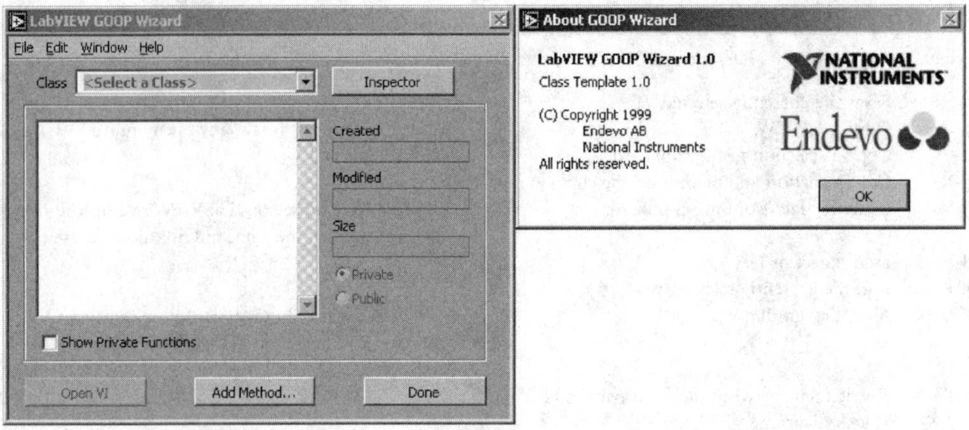

FIGURE 157.24 GOOP wizard.

llbs

A VI library (*.llb) is a National Instruments proprietary file that can contain several VIs and some of their settings. Files that are saved into an llb are lightly compressed, and can be set as *top level* (denotes which VIs should be launched when their respective llb is opened) or *companion*. Other than that, an llb is formless; all of the files reside at one level (i.e., you cannot create subdirectories or sub-llbs in a VI library). Although a library has a flat structure, it is possible to assign one of two levels to a VI in a library. A top-level VI will launch automatically when opened from your operating system's file explorer, whereas a normal-level VI will only open if explicitly opened from within the library. (A new feature to LabVIEW 7.0 allows Windows Explorer to navigate inside G libraries, effectively bypassing top-level assignments.) It is also acceptable to assign more than one VI in a library as top level, although this can make window negotiation difficult, as they all will be launched when the library is opened. Figure 157.25 demonstrates a library with one top-level VI.

Development Distribution Library

A development distribution contains all custom VIs and controls that have been created in the project, so that it can be successfully opened and executed on a PC with a similar LabVIEW configuration installed. VIs from vi.lib and add-on toolkits are not included in a development library, as it is assumed that environments used for subsequent openings of the library will have these installed.

Application Distribution Library

An application library distribution is an llb file that contains all of the VIs listed in a development library, and all others that are used in the application. VIs from vi.lib and toolkits are included, as are all custom VIs and controls. Distributing an application library is useful when archiving an application, including all of the components used to execute it.

Diagram Access

Altering a user's access to a VI's diagram can minimize code theft and ensure that the VIs are opened with an appropriate version of LabVIEW. Setting a password to a VI's block diagram prevents users from accessing the code diagram and understanding hierarchical dependences between modules. To change a VI's status, access the Security VI property (Figure 157.26), and select one of the settings shown in Table 157.5.

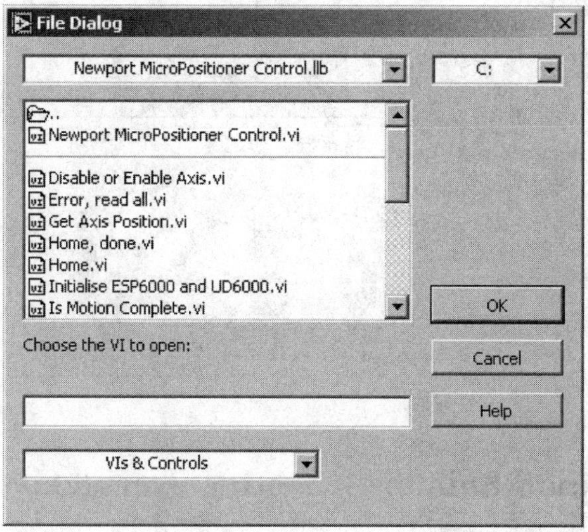

FIGURE 157.25 A library with a top-level VI.

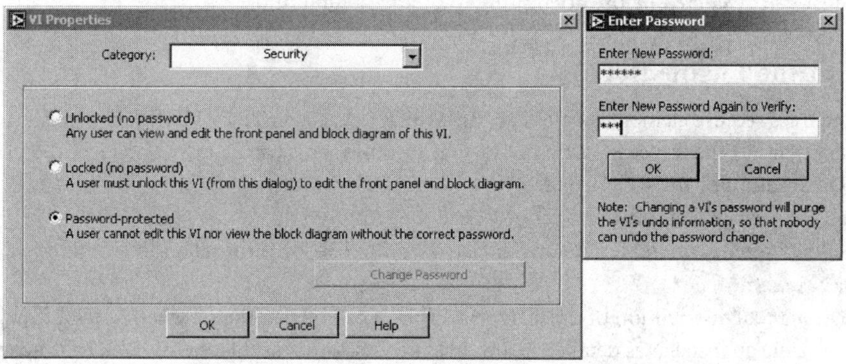

FIGURE 157.26 Set block diagram password.

TABLE 157.5 Block Diagram Security Access Levels

Security Level	Description
Unlocked (no password)	The default for a new VI. Permits read/write access to the block diagram.
Locked (no password)	Access to the block diagram is prohibited until the user changes the security to *Unlocked*. A password is not required to change security settings.
Password protected[a]	The highest level of block diagram security; access is prohibited unless the user enters the correct password.

[a] Due to the nature of LabVIEW VIs, the password is not recoverable from the file. If you forget a password, you are unable to parse the file to "crack" it.

An extreme method of code protection is to remove the block diagram altogether, and can be achieved by performing a custom save (Figure 157.27), using File → Save with Options.

When upgrading a VI from a previous version of LabVIEW, the diagram is recompiled; therefore, removing the diagrams from your VIs will prevent them from being opened with a version of LabVIEW other than the version used to create the VI. This minimizes conversion errors that may occur between versions (e.g., if your VI uses functions from vi.lib), and enhances commercial code control (limiting the lifetime of a VI to a particular version of LabVIEW, requiring users to obtain new versions when migrating between major versions of LabVIEW).

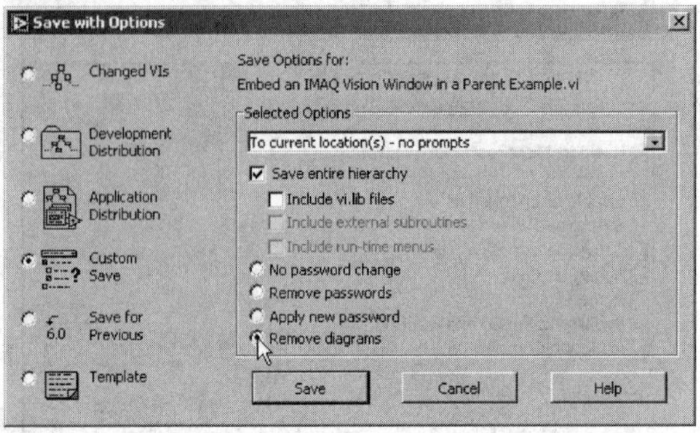

FIGURE 157.27 Remove diagrams.

157.9 Application Building (Creating Executables)

Often left to the last minute in application development, building stand-alone executable code may not be as straightforward as one might expect. External code libraries, target operating systems, engines, and licenses must be considered before attempting to build an application.

The Runtime Engine (RTE)

Built executables require elements of the LabVIEW RTE to execute. The RTE should be distributed with your applications, as the target PC may not have it installed, and your applications will therefore fail to execute. The location of the RTE installed on a LabVIEW development PC is

```
..\National Instruments\Shared\LabVIEW Run-Time\Version\
```

and the following redistributable version is located on the LabVIEW installation CD-ROM:

```
..\LVRunTimeEng\
```

Although manual distribution of the RTE is possible, it can be automated when distributing your built software, by using an installer as detailed below. In some cases, the installation of the RTE is not necessary. (This method is not authorized or supported by National Instruments and does not support DataSocket, NI-Reports, or 3D Graphs.) If an executable is a simple one, it may not need the RTE to be installed on the target machine; all one needs to do is include some of the engine's files with the exe for it to work. To determine whether this method will work with a particular built executable, copy the following files and folders into the folder containing the exe:

```
..\National Instruments\shared\nicontdt.dll
..\National Instruments\shared\nicont.dll
..\National Instruments\shared\LabVIEW Run-Time\ver\* (including all
subdirectories)
```

Using the LabVIEW 6.1 development system as an example, the exe's directory would look something like the following:

```
..\AppDirectory\MyApp.exe
..\AppDirectory\MyApp.ini
..\AppDirectory\My_Apps_DLLs (if_any)
..\AppDirectory\nicontdt.dll
..\AppDirectory\nicont.dll
..\AppDirectory\lvapp.rsc
..\AppDirectory\lvjpeg.dll
..\AppDirectory\lvpng.dll
```

```
..\AppDirectory\lvrt.dll
..\AppDirectory\mesa.dll
..\AppDirectory\serpdrv
..\AppDirectory\models\*
..\AppDirectory\errors\*
..\AppDirectory\script\*
```

When the executable is launched on a PC without the RTE installed, it should find all of the RTE files it requires in the application's root directory, and execute normally. This method can be particularly useful when distributing autorun presentations on CD-ROMs.

Reverse Engineering Built Executables

As built executables have a similar structure to VI libraries, it is possible to extract and use the diagramless VIs within. Simply change the extension of the executable file to llb, and open the library within LabVIEW. One should be careful when designing the architecture of a commercial application, as this G executable characteristic can allow users to reverse engineer an executable, effectively accessing the functionality of the SubVIs within the executable.

Run Time Licenses (RTLs)

Distributing a simple executable to the world is free; the software engineer developed the code, so he or she may sell and distribute the resulting code desired. When the software contains specialist code that the software engineer did not write, then the users may need to pay the author a small amount for each instance of their code that is on-sold as a portion of the application. Products that currently require the inclusion of a runtime license include the National Instruments Vision toolkit, meaning that any applications released that use Vision VIs must be accompanied by an official NI Vision toolkit RTL, available from NI. Always check with the vendor of any software tools that you are planning to incorporate in an application, indicating the RTL costs at the project proposal stage.

Installers

Rather than providing a single executable file to customers, it is almost always preferable to pack your application into an installer. An installer is a mechanism that encourages the simple installation of the developer's application, support files (including RTEs is required), registration of external components (ActiveX controls, DLLs, documentation, links to web pages, etc.), and manipulation of the operating system (registry alteration). Installers provide intuitive graphical user interfaces that assist end users to customize your application installation to their needs (including the selection of components, installation destination locations, readme file display, etc.) and provide a professional finish to your product. Modern installers also have automatic support for damaged installations, and a facility to uninstall your product cleanly, returning the end user's system to a state comparable to that before installation. The NI LabVIEW Application Builder contains a simple installer that is often sufficient for small applications that do not require advanced operating system alterations, whereas there are several commercially available installers available (Figure 157.28). One such installer, Setup2Go, fulfills many of the advanced installation requirements, and has a very simple interface to package your software product. More information on Setup2Go can be found at *http://dev4pc.com/setup2go.html*.

157.10 Open Source G: Distributed Development

One of the most powerful tools available to software engineers today is open source project development. Unlike other engineering branches (including structural, electrical, etc.), software engineering projects can be easily transported to all corners of the earth in seconds using the Internet. Using this unique ability, projects can be developed at many different sites by several professionals simultaneously, ultimately

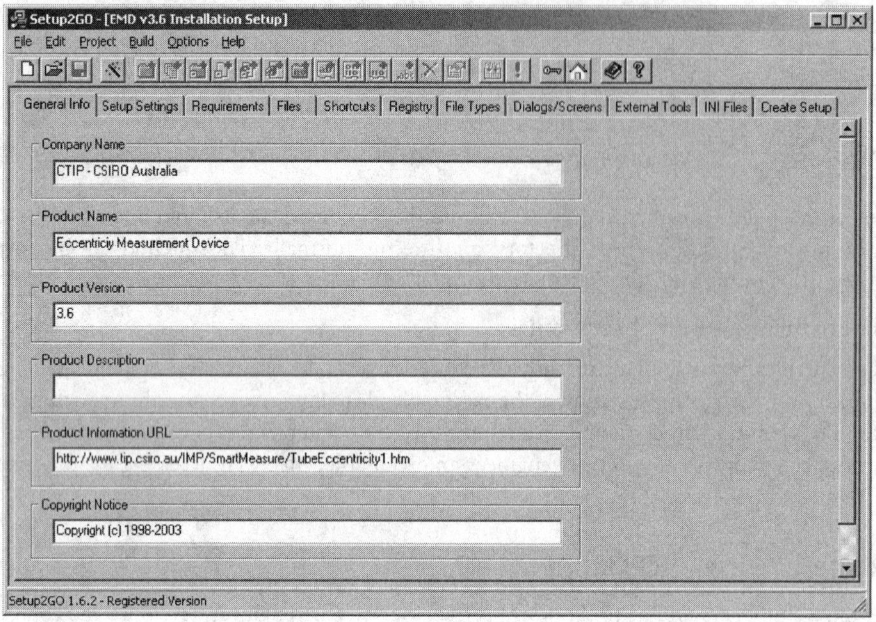

FIGURE 157.28 Custom InstallerSetup2Go.

leading to better tested software with extended features and more robust code. Sharing the burden of developing a software project allows engineers a much wider range of experience and expertise, and thus a more stable and useful application.

As you might imagine, coordinating such projects can be time intensive and legally difficult, which is why the OpenG community has been created. Spearheaded by the longtime LabVIEW expert Jim Kring, *OpenG.org* (Figure 157.29) is the home of our Open Source LabVIEW development community, hosting the largest registry of Open Source LabVIEW code and applications available on the Internet. OpenG.org is a not-for-profit venture, and provides each project a web page, where you can embed HTML, allowing you to link to software and websites, as well as embed images and other content hosted remotely.

All OpenG.org projects are covered by the *LGPL software license*, which protects source code, so that it remains Open Source, but it allows using protected code in commercial, proprietary, closed applications. Most people using LabVIEW are using it directly in commercial or industrial applications and systems that are sensitive because they contain intellectual property in the form of valuable trade secrets. The ideals of the OpenG community promote the adoption of open source everywhere that LabVIEW is used.

More information on the OpenG project, and how to develop software using open source principles is available at the OpenG website (*www.openg.org*) (Figure 157.29).

Defining Terms

Block (wiring) diagram — The coding portion of a VI, where the software engineer writes the G code.

Data type — The format the G development system uses to store a piece of information in memory. Using appropriate data types decreases application execution time, decreases memory footprint, and eradicates coercion errors.

Error cluster — A special cluster containing an error code, a Boolean indicating if an error has occurred, and the source of that error. The error cluster format is consistent throughout the G development system.

Front panel — The graphical user interface of a VI, including a method in which data are passed to and from the VI.

G — The programming language used to develop applications using the LabVIEW Development System.

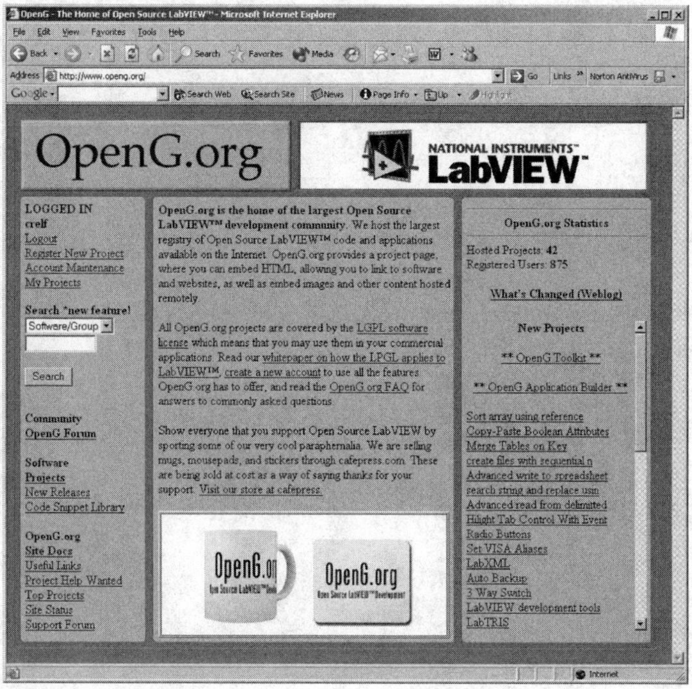

FIGURE 157.29 OpenG website.

Installer — Software that facilitates the automatic installation, upgrade, repair, or uninstallation of third-party software.

LabVIEW — The development system used to develop code in the G programming language.

Polymorphic VI — A collection of VIs with identical missions, operating on differing data types. A polymorphic VI exists on disk as a special form of a VI library.

Runtime engine (RTE) — The system under which compiled G code runs, without the presence of the G development system.

Runtime license (RTL) — A "user pays" system of charging for code portion development.

Unit — An attribute associated with a numerical value, giving it physical significance.

Further Information

National Instruments, *www.ni.com/labview* (G development system vendor)
National Instruments Alliance Members (system integrators), *www.ni.com/alliance*
The LabVIEW Zone, *www.ni.com/labviewzone* (developer forum)
OpenG — Open Source G Development, *www.openg.org* (developer forum)
InfoLabVIEW Mailing List, *www.info-labview.org* (developer forum)
LabVIEW Technical Resource, *www.ltrpub.com* (journal)

158

Sensors

Halit Eren
Polytechnic University, Hong Kong

Our capacity to interact with the environment depends on the knowledge about it. Although our physical senses help us to understand and comprehend much of the world around us, there are many situations where better understanding can be gained by using manufactured sensors and instruments. The industrial plant variables, ocean streams, and electric and magnetic fields are some examples of quantities that are conveniently sensed and measured by using sensors.

Sensors can be thought of as extensions of human senses. Sensors usually generate electrical signals that can easily be processed and transmitted by manufactured devices or systems of devices. In recent years, the use of sensors has undergone explosive growth due to the availability of processing devices to manipulate output signals and progress in communication technology. A schematic representation of a sensor is illustrated in Figure 158.1.

The output signal, $y_o(t)$, is produced in response to the input signal, $x_i(t)$, thus characterizing the state of the measured system (the *measurand*). An ideal sensor should not respond to any parameters other

0-8493-1586-7/05/$0.00+$1.50

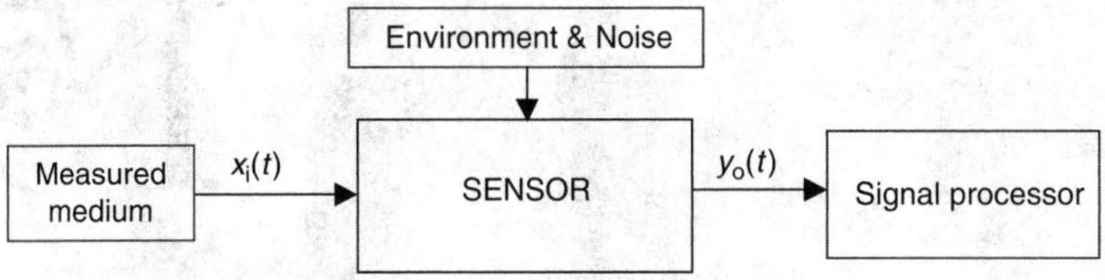

FIGURE 158.1 Schematic representation of a sensor.

than what it is intended for. Unfortunately, this is a simplified assumption and many sensors are unintentionally sensitive to a number of other parameters as well.

For an ideal sensor, the relation between the system input and output can be expressed by

$$y_o(t) = f(x_i(t)) \tag{158.1}$$

In practice, the output depends on various parameters affecting the system and the corresponding equations are much more complex.

158.1 Recent Trends in Sensors and Sensor Technology

Sensors and sensing technologies are well established and are widely applied in everyday life and industrial applications. In recent years, with the use of micro and nano technologies new families of semiconductor sensors are emerging such as *smart* and *soft sensors* and *microelectromechanical* sensors. In the near future, the new generation of sensors will change measurement technologies considerably, leading to new approaches in instruments and instrumentation.

Intelligent (Smart) and Soft Sensors

At present, many instruments use intelligent sensors that contain the sensor, signal processors, and intelligence capabilities in a single chip. They appear in the marketplace as pressure sensors and accelerometers, biosensors, chemical sensors, optical sensors, and magnetic sensors, among other types. Some of these sensors are manufactured with neural networks and other sophisticated intelligence techniques on board the chip. Many types of intelligent sensors are currently in use, such as neural processors, intelligent vision systems, and intelligent parallel processors.

Today's technology allows the development of *soft sensors*. A soft sensor is a device that does not include any sensing hardware but involves a microprocessor that processes data acquired from a number of devices and combines them by using a mathematical model to produce an estimation of the parameter of interest. Either the traditional models, such as the ARMAX, and NARMAX, or the more exotic neural, fuzzy, and neuro-fuzzy models, are used to obtain the required estimation capability.

Standards for Intelligent Sensors

Parallel to rapid developments in hardware aspects of modern sensors, some new standardized techniques are emerging for software, communications, and networking. This is making a revolutionary contribution to the development of electronic instruments. For example, the IEEE-1451 is a set of standards that define interface network-based data acquisition and control of sensors.

The IEEE-1451 standard aims to make it easy to create solutions using existing networking technologies, standardized connections, and common software architectures. The standards allow application software, field network, and transducer decisions to be made independently. It offers flexibility to choose the products and vendors that are most appropriate for a particular application.

Micro Electromechanical Systems

Developments of silicon-based sensors are enhanced considerably due to advances in embedded controllers, microelectronics, and micromachining. Micromachining technology combined with semiconductor processing technology provides variety of sensors. Micro Electromechanical Systems (MEMS) comprise one of the most active sensor research fields. MEMS largely use integrated circuit (IC) technology to obtain micro electromechanical systems.

Microscale physical principles can be used to build sensors with characteristics that cannot normally be obtained with traditional techniques. Therefore, IC technology provides the opportunity to construct monolithic hybrid systems that contain sensors, signal processing circuits, analog-to-digital converters, compensation circuits, and microcontrollers in the same chip. Examples of such sensors some are discussed in subsequent sections.

Sensor Communications

Communications capabilities of electronic instruments is enhancing their usability in many types of applications. There are two levels of communications: communication at the sensor level, which is mainly realized by intelligent sensors; and communication at the device level, which takes place either in wired form or infrared with RF or microwave techniques.

In the implementation of wireless communication features, the Bluetooth technology and near-field communication protocols are gaining wide acceptance. Bluetooth is considered to be a low-cost and short-range wireless technology that provides communication functionalities, ranging from wire replacements to simple personal area networks. Initially, it was aimed to bring short-distance wireless interfaces mainly to consumer products on a large scale. As the number of products incorporating Bluetooth wireless technology increases, the development of various types of instruments for a diverse range of applications becomes more important.

If the data need to be transmitted for long distances in which transmitter and receivers are out of range, wireless bridges are employed, which provide long-range point-to-point or point-to-multipoint links. Some of these devices use direct-sequence-spread spectrum (DSSS) radio technology operating at typical frequencies (i.e., 2.4 GHz). Data may be transmitted at speeds up to 11 Mbps. These bridges comply with standards, such as IEEE-802.11b, while connecting one or more remote sites to a central server or Internet connection.

Built-in Test Sensors

RF data communication is extensively applied in built-in tests (BITs) in which the sensors are embedded in operational systems, such as rotating machinery, industrial systems, or even implanted in living organisms. The information gathered from the sensors is transmitted by a built-in RF transmitter to a nearby receiver. BIT sensor quality is improving steadily through the availability of intelligent and other sophisticated sensors, instrument software, interoperability, and self-test effectiveness.

Sensor Networks

Geographically distributed sensors can be networked as a distributed sensor network (DSN). The networking requires intelligent sensors that may obey some form of hierarchical structure.

In recent years, the IEEE-1451 compatible interface between Internet/Ethernet serial port web sensors has been developed. These sensors have direct Internet addresses. The interface is realized in IEEE-451 network-capable application processors (NCAP). NCAPs connect the Internet through Ethernet. NCAP is a communication board capable of receiving and sending information using the standard TCP/IP format. The sensor data are formatted to and from the serial port by one of the following: RS232, RS485, TII, Microlan/1-wire, Esbus, or I^2C.

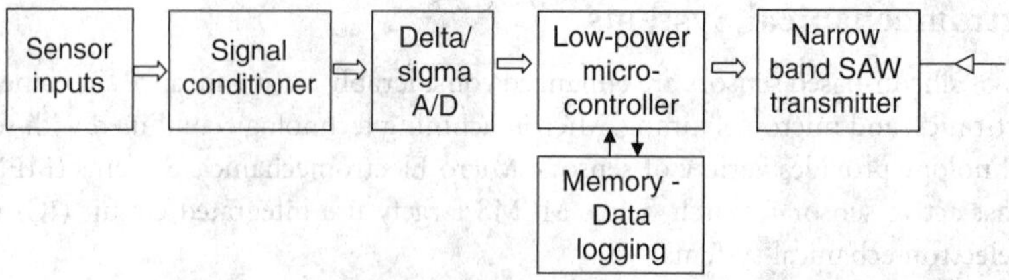

FIGURE 158.2 Wireless websensor.

The useful lifetime of a particular sensor will probably be shaped by how it interfaces to the network to share information. The importance of sensor networks has been realized by many vendors; hence, they promote proprietary solutions to connect smart sensors into TCP/IP-based networks. A typical wireless web sensor node arrangement is illustrated in Figure 158.2. Recent developments are focused on low- and high-level interfaces.

158.2 Classification of Sensors

Sensors and transducers produce outputs in response to physical variations. These outputs are processed suitably by analogue and digital methods to obtain desired information from the physical process. Therefore, there are diverse approaches to classifying sensors based on physical process properties, sensor operational principles, or both.

One approach is by referring to the nature of the sensed parameter (temperature, pressure, etc.). This approach is useful for people who want to observe particular parameters of physical variables and select suitable devices for monitoring them. A second approach may be the classification of sensors by referring to the nature of their output signals (resistive, voltage generating, etc.). This approach is useful for people who want to design a measuring system and process sensor outputs by using appropriate hardware and software. A third approach begins with the energy input and output. From this perspective, there are three fundamental types of sensors: modifiers, self-generators, and modulators.

In this chapter, sensors are classified from the applications point of view, but basic operational principles are explained whenever possible. Sensors in the following areas are discussed:

- Voltage and current
- Magnetic and electromagnetic
- Capacitive
- Acoustic
- Temperature and heat
- Light and radiation
- Chemical and gas
- Environmental and biological
- Mechanical variables

158.3 Voltage and Current Sensors

Voltage sensing is very important in many types of instruments, since the signals generated by most sensors are in voltage form. The voltage is related to electric potential difference. Four basic types of voltage sensors are commonly used in voltage measurements: inductive, thermal, capacitive, and semiconductor.

Inductive voltage sensors are based on the characteristics of magnetic fields. They make use of voltage transformers, AC voltage inductive coils, eddy currents, and so on. Classical *electromechanical devices* are typical examples of such sensors. They are based on the mechanical interaction between currents and magnetic fields. This interaction generates a mechanical torque proportional to the voltage or the squared voltage to be measured (voltmeters), or proportional to the current or the squared current to be measured (ammeters). A restraining torque, usually generated by a spring, balances this torque. The spring causes the instrument pointer to be displaced by an angle proportional to the driving torque, and hence to the quantity, or squared quantity, to be measured. The value of the input voltage or current is therefore given by the reading of a pointer displacement on a graduated scale.

Thermal voltage sensors are based on the thermal effects of a current flowing into a conductor. The sensor output is proportional to the squared input voltage or current.

Capacitive voltage sensors are based on the characteristics of electric fields. These sensors detect voltages by different methods, such as electrostatic force, Kerr or Pockels effects, Josephson effect, and change of reflective index of the optic fibers. Capacitive voltage sensors are generally used in low-frequency, high-voltage measurements. In high-voltage applications, the capacitive dividers are used to reduce the voltages to low levels.

Semiconductor voltage sensors constitute a wide range of voltmeters and ammeters. They are based on purely electronic circuits, and attain the required measurement by processing the input signal via semiconductor devices. The method employed to process the input signal can be either analogue or digital.

Types of Current Sensors

Current flowing through a circuit element can also be related to the properties of the circuit element and the voltage across its terminals by Ohm's law:

$$V = RI \tag{158.2}$$

where R is a quantity representing the electric behavior of the circuit element. Under DC conditions, this quantity is called "resistance" and its measurement unit is, in the SI, the "ohm" (Ω).

Fundamentally, the properties of magnetic fields, electric fields, and heat can be used for currents sensing. In addition, any magnetic field sensor can be configured for current measurements. An example is the *Rogowski coil*, illustrated in Figure 158.3. The Rogowski coil is a solenoid air core winding of a small cross-section looped around a conductor carrying the current. The voltage induced across the terminals of the coil is proportional to the derivative of the current.

One of the simplest methods of current sensing is the current-to-voltage conversion by means of resistors. In current measurements, the resistor is called the shunt resistor despite the fact that it is connected in series with the load. The shunt produces a voltage output that can be presented by a variety of secondary meters, such as analogue meters, digital meters, and oscilloscopes.

Hall-effect current sensors are typical semiconductor devices. They operate on the principle that the voltage difference across a thin conductor carrying a current depends on the intensity of magnetic field applied perpendicular to the direction of the current flow (Figure 158.4). Electrons moving through a magnetic field experience Lorentz force perpendicular to both the direction of motion and to the direction

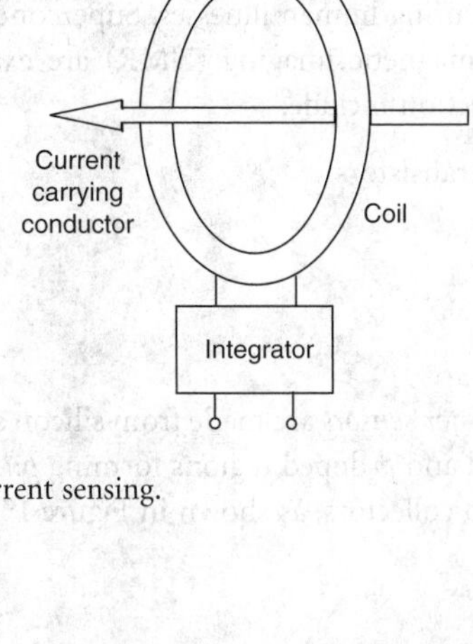

FIGURE 158.3 Rogowski coil for current sensing.

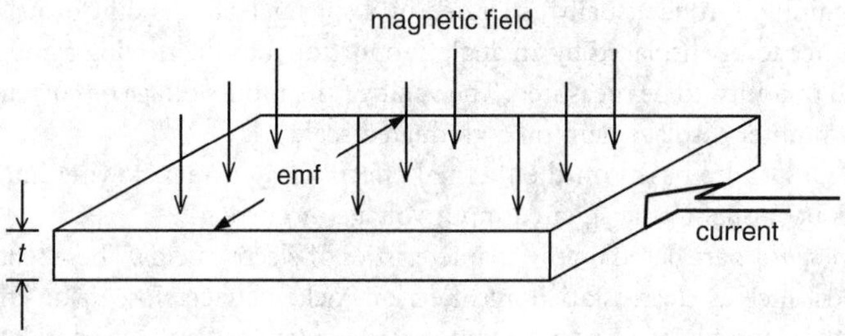

FIGURE 158.4 Hall-effect sensor.

of the field. The response of electrons to Lorentz force creates *Hall voltage.* If a current I flows through the sensor, the Hall voltage can mathematically be expressed by

$$V = R_H IB/t \tag{158.3}$$

where R_H is the Hall coefficient ($m^3/°C$), B is the flux density (T), and t is the thickness of the sensor (m).

Hall-effect sensors are made from metals or silicon, but mainly semiconductors with high electron mobility such as indium antimonite. They are usually manufactured in the form of probes with sensitivity down to 100 μT. These sensors have good temperature characteristics, varying from 200°C to near absolute zero.

158.4 Magnetic Sensors

Magnetic sensors find many applications in everyday life and in industry. They provide convenient, noncontact, simple, rugged, and reliable operational devices compared to many other sensors.

Generally, magnetic sensors are based on sensing the properties of magnetic materials, which can be done in many ways. For example, magnetization, which is the magnetic moment per volume of materials, is used in some systems by sensing force, induction, field methods, and superconductivity. However, most sensors make use of the relationship between magnetic and electric phenomena.

Many types of magnetic sensors are available, but they can broadly be classified as *primary* or *secondary sensors.* In the primary sensors, also known as magnetometers, the physical parameters to be measured are external magnetic fields. In the secondary sensors, the external parameters are derived from other physical variables such as force and displacement.

Primary Magnetic Sensors

Primary sensors are based on measurement of external magnetic fields. They find applications in many areas from biological and geophysical measurements to the determination of characteristics of extraterrestrial objects and stars. They also find applications where sensitivity is of utmost importance, as in the case of devices for diagnosing and curing human illnesses. Superconducting quantum interface devices (SQUIDs) and nuclear resonance magnetic imaging (NMR) are examples of such devices. Types of primary sensors discussed in this section include:

- Magnetodiode and magnetotransistors
- Magnetoresistive sensors
- Magneto-optical sensors
- Magnetic thin films
- Hall-effect sensors

Magnetodiodes and magnetotransistor sensors are made from silicon substrates with undoped areas that contain the sensor between *n*-doped and *p*-doped regions forming *pn*, *npn*, or *pnp* junctions. In the case of magnetotransistors, there are two collectors, as shown in Figure 158.5.

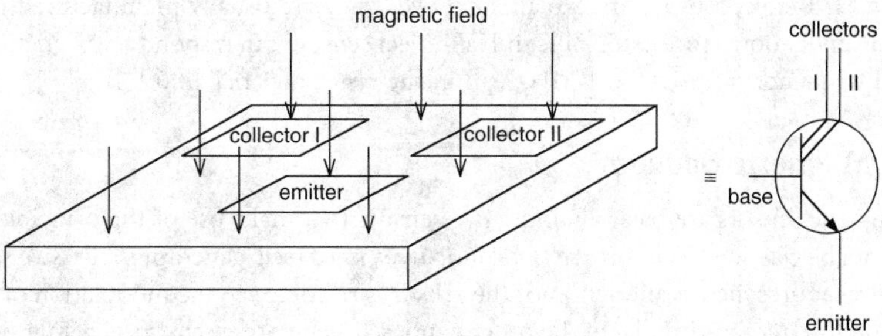

FIGURE 158.5 Semiconductor magnetotransistor.

Depending on the direction, an external magnetic field deflects electron flow between emitter and collector in favor of one of the collectors. The two collector voltages are sensed and related to the current or the applied magnetic field.

In *magnetoresistive sensors*, the resistivity of some current carrying material changes in the presence of a magnetic field mainly due to the inhomogeneous structure of some materials. For example, the resistance of bismuth can change by a factor 10^6. Most conductors have a positive magnetoresistivity. Magnetoresistive sensors are largely fabricated from permalloy stripes positioned on a silicon substrate. Each strip is arranged to form one arm of a Wheatstone bridge so that the output of the bridge can directly be related to the magnetic field strength.

As in the case of Hall-effect sensors, the basic cause of magnetoresistivity is the Lorentz force, which causes electrons to move in curved path between collisions. One advantage is that for small values of magnetic field, the change in resistance is proportional to the square of the magnetic field strength, thereby giving better sensitivity.

Magneto-optical sensors constitute an important component of magnetometers. In recent years, highly sensitive magneto-optical sensors have been developed. These sensors are based on various technologies, such as fiber optics, polarization of light, Moire effect, and Zeeman effect. These type of sensors lead to highly sensitive devices and are used in applications requiring high resolution, such as human brain function mapping and magnetic anomaly detection.

Magnetic thin-film sensors are an important part of superconducting instrumentation, sensors, and electronics in which active devices are made from deposited films. The thin films are usually made from amorphous alloys, amorphous gallium, and the like. As an example, a thin-film Josephson junction is given in Figure 158.6. The deposition of thin films can be done by several methods, such as thermal evaporation, electroplating, sputter deposition, or by some chemical means. The choice of technology depends on the characteristics of the sensors. For example, thin-film superconductors require low-temperature operations, whereas common semiconductors normally operate well at room temperatures.

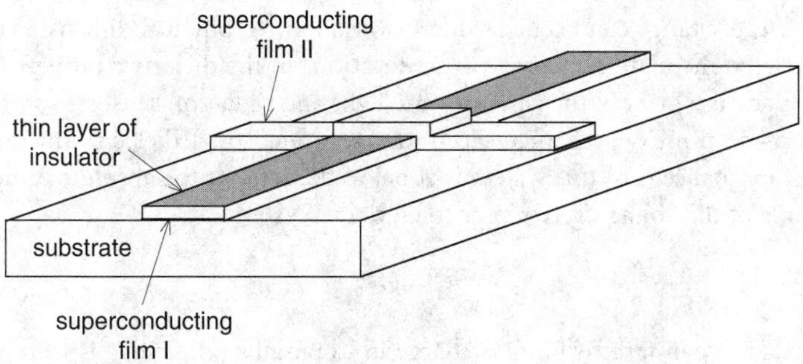

FIGURE 158.6 Magnetic thin-film Josephson junction.

Principles of Hall-effect sensors are explained above. They are usually manufactured in the form of probes with sensitivity down to 100 μT. Silicon Hall-effect sensors can respond to magnetic fluxes having an operational frequency from DC to 1 MHz within the range of 1 mT to 0.1 T.

Secondary Magnetic Sensors

Secondary magnetic sensors are basically inductive sensors that make use of the principles of magnetic circuits. They can be classified as *passive sensors* or *active sensors* (self-generating). Passive sensors require an external power source; hence, the action of the sensor is restricted to the modulation of the excitation signal in relation to external stimuli. On the other hand, self-generating sensors generate signals by using the electrical generation principle based on Faraday's law of induction. That is, "when there is a relative motion between a conductor and a magnetic field, a voltage is induced in the conductor," or a varying magnetic field linking a stationary conductor produces voltage in the conductor. This relationship can be expressed as

$$e = -d\Phi/dt \tag{158.4}$$

where Φ is the magnetic flux, and t is time.

There are many different types of inductive sensors, such as linear and rotary variable-reluctance sensors; synchros; microsyn, linear variable inductors; induction potentiometers; linear variable differential transformers; and rotary variable-differential transformers. Two of these methods are briefly discussed below.

Single-coil linear variable-reluctance sensors consist of three elements: a ferromagnetic core in the shape of a semicircular ring, a variable air gap, and a ferromagnetic plate. Their performance depends on the core geometry and permeability of the materials; therefore, they can be highly nonlinear. Despite this nonlinearity, they find applications in many areas, such as force measurements. The coil usually forms one of the components of an LC oscillator circuit whose output frequency varies with applied force. Hence, the coil modulates the frequency of the local oscillator.

Linear variable-differential transformers (*LVDT*) make use of the principle of transformer action, that is, magnetic flux created by one coil links with the other coil to induce voltages. The LVDT is a passive inductive transducer with many applications. It consists of a single primary winding positioned between two identical secondary windings wound on a tubular ferromagnetic former. The primary winding is energized by a high-frequency 50-Hz to 20-kHz AC voltages. The two secondaries are made identical by having an equal number of turns. They are often connected in series opposition so that the induced output voltages oppose each other.

158.5 Capacitive and Charge Sensors

Many sensors are based on the characteristics of electrical charges and associated electric field properties. The electrical charge and field are related to changes in capacitances in response to physical variations. The changes in the capacitance can occur as due to variations in physical dimensions such as the area or the distance between the plates. In some cases, variations in the dielectric properties of the material between plates is made use of as in the case of some light and radiation sensors.

Capacitors are made from two charged electrodes separated by a dielectric material, as shown in Figure 158.7. The capacitance C of this system is equal to the ratio of the absolute value of the charge Q to the absolute value of the voltage between charged bodies, as in:

$$C = Q/V \tag{158.5}$$

where C is the capacitance in farads (F), Q is charge in Coulombs (C), and V is voltage (V).

The capacitance C depends on the size and shape of charged bodies and their positioning relative to each other. In many sensors, it is necessary to deal with the useful capacitances as well as the stray

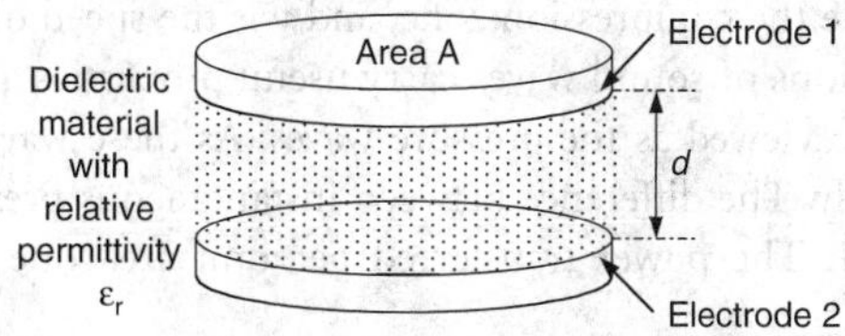

FIGURE 158.7 Typical capacitor.

capacitances. Stray capacitances may be introduced externally or internally whenever two charged bodies exist side by side.

Integrated circuit capacitive sensors find wide applications, particularly in MEMs, and they are made as MOS integrated circuits; monolayer versions contain tantalum or other suitable deposits. In a typical sensor, two heavily doped polysilicon layers formed on a thick layer of oxide generally comprise the plates of IC capacitors. The dielectric is usually made from a thin layer of silicon oxide. Important parameters for IC capacitors are the tolerances, voltage coefficients, temperature coefficients, and capacitance values. These capacitors are temperature stable with about 20 ppm/°C temperature coefficients. The voltage coefficients are usually less than 50 ppm/V. IC capacitive sensors are achieved by incorporating a dielectric material sensitive to physical variables.

Many capacitive IC sensors have typically one plate fixed and one that moves as a result of the applied measurand. Silicon technology combined with the surface micromachining allows the capacitance between interdigitized fingers to measure acceleration as well as other inputs. The value of nominal capacitance is 100fF to 1pF and the variations in capacitance are on the order of femtofarads.

Piezoelectric sensors can be classified as capacitive or charge sensors. They produce an electric charge when a force is applied on to it. This effect exists in natural crystals such as quartz (SiO_2), or manmade artificially from polarized ceramics and polymers. A piezoelectric sensor can be viewed to have a crystal structure with a dielectric material. When this dielectric material is arranged like a capacitor with electrodes, it behaves like an electric charge generator when stress is applied on it. Therefore, a resulting voltage, V, across the capacitor appears as a result of stress. The piezoelectric effect is a reversible physical phenomenon; that is, applying a voltage across the terminals the crystal produces a mechanical strain.

Most modern piezoelectric sensors contain lead zirconate titanate (PZT). This ceramic-based sensor is used to construct biomorph sensors for force, sound, motion, vibration, and acceleration sensing. The biomorph consists of two layers of different PZT formulations that are bonded together in rectangular strips. The rectangular shape structure can conveniently be mounted as a cantilever on structures and the motions perpendicular to the surface of the sensor generates signals of measurements.

A variation in piezoelectric technology is piezoresistive sensors, which are the most common micromachined sensors. Their output signals are highly uniform and predictable; therefore, these signals can easily be conditioned.

158.6 Acoustic Sensors

Sound is defined as the vibrations of a solid, liquid, or gaseous medium in the frequency range of 20 Hz to 20 kHz that can be detected by the human ear. Sound travels in a media by obeying the laws of shear and longitudinal forces. In contrast to solids, liquid, or gaseous media cannot transmit shear forces; therefore, in these media the sound waves are always longitudinal, that is the particles move in the direction of the propagation of the wave. As the sound waves travel in medium, the medium compresses or expands; hence, its volume changes from V to $V - \Delta V$. As a result, the *bulk modulus* of medium elasticity, which is the ratio of change in pressure Δp relative to change in volume, can be written as

$$B = \frac{\Delta p}{\Delta V / V} = \rho_0 v^2 \qquad (158.6)$$

where ρ_0 is the density outside the compression zone, and v is the speed of sound in the medium.

In practice, pressure variations of sound waves carry useful properties and find wide applications. In this case, sound waves can be viewed as the pressure waves. As these waves travel, the pressure at any given point changes constantly. The difference between instantaneous pressure and average pressure is called the *acoustic pressure*, *P*. The power transferred per unit area due to acoustic pressure can be expressed as

$$I = \frac{P^2}{Z} \tag{158.7}$$

where Z is the acoustic impedance that can be determined from acoustic pressure and instantaneous velocity.

Pressure levels are often expressed in decibels, as

$$P_L = 20\log_{10}\frac{p}{p_0} \tag{158.8}$$

where p_0 is the lowest pressure level that the human ear can detect (2×10^{-5} N/m^2).

The sound pressure is almost always the only parameter sensed directly. All other parameters, such as sound power, particle velocity, reverberation time, and directivity, are derived from the pressure measurements. The sound pressure measurements are performed by microphones in the gaseous media and hydrophones in the liquid media. Microphone types are listed below.

- Capacitive
- Piezoelectric and electret
- Fiber optic
- Carbon
- Moving iron (variable reluctance)
- Moving coil

The most commonly used microphones for sound measurements are capacitive, piezoelectric, electret, and fiber optic microphones. They tend to have good stability, are sensitive and precise, and possess excellent temperature properties.

Piezoelectric and electret microphones use piezoelectric materials such as lead zirconate titanate (PZT). They have very high impedance levels and very broad operating frequencies, and are suitable for ultrasonic applications.

The operational principles of electret microphones are similar to piezoelectric microphones. Electret principles are similar to piezoelectric and pyroelectric materials. Electret is permanently polarized crystalline dielectric material. An electret microphone consists of a metallized diaphragm and backplane separated from the diaphragm by an air gap as shown in Figure 158.8.

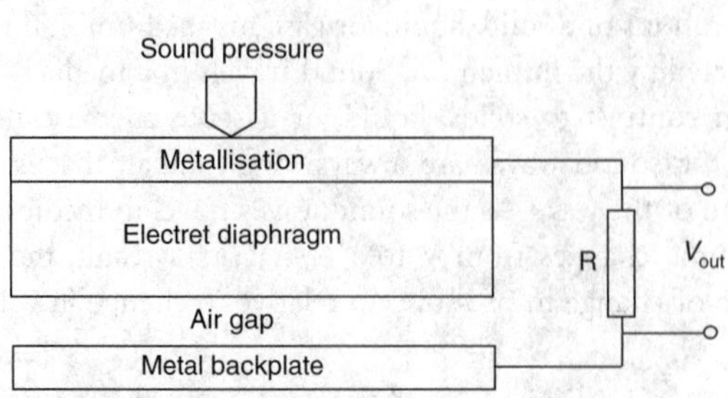

FIGURE 158.8 Electret microphone.

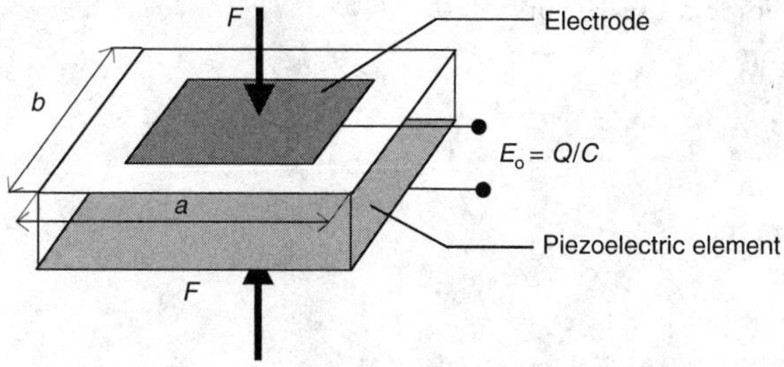

FIGURE 158.9 Piezoelectric ultrasonic detection system.

Fiber optic microphones make use of reflection of laser beams from a vibrating diaphragm. The diaphragm vibrates in response to sound pressure, and the interference between incident light and the reflected light beams is sensed by a suitable interferometer (e.g., Michelson interferometer).

Optical microphones are suitable for hostile environments, such as monitoring the noise in rocket and jet engines. In some applications, extreme measures may be required for good sensitivity, such as cooling the temperature-sensitive components with water or other suitable fluids.

Ultrasonic sensors are used for acoustic waves of frequencies higher than 20,000 Hz, and they find many medical and industrial applications. Ultrasonics are generated naturally or by transducers, which are available over a wide range of frequencies in many sizes and shapes. An ultrasonic sensor is illustrated in Figure 158.9. A common method of transduction is piezoelectric material, but in some applications magnetorestrictive materials are used as well. Both piezoelectric and magnetorestrictive transducers can simultaneously function as transmitters and receivers through air, solids, or liquids.

Ultrasonic sensors in acoustic applications are commonly used in the following areas:

- Detection of defects
- Fluid and gas flow measurements
- Ultrasonic cleaner
- Security devices
- Elastic properties of solids
- Porosity and size estimation
- Acoustic microscopy
- Medical applications
- Motion, displacement, and range measurements
- Thickness measurements in process control and other applications

158.7 Temperature and Heat Sensors

Temperature is a measure of heat intensity. The simplest way of measuring temperature is the thermometer, which makes use of thermal expansion of materials, such as liquid in glass. For electronic measurements, there are many different sensors available, such as resistive sensors, thermoelectric sensors, thermocouples, thermostats, IC temperature sensors, and piezoelectric sensors.

Heat is a form of energy known as the thermal energy. The quantity of heat contained in an object cannot easily be measured, but the changes in heat can be measured as the temperature. In temperature measurements, a small portion of thermal energy of the object is transmitted to the sensors for conversion to electrical signals. The operational principles of temperature sensors depend on the heat exchange that takes place in the materials. The heat exchange can be conductive, convective, and radiative. Sensors for thermal energy measurement are listed below.

- Thermocouples
- Resistive temperature sensors
- Silicon resistive temperature sensors

- Semiconductor temperature sensors
- Pyroelectric sensors
- Thermistors
- Infrared sensors
- Optical temperature sensors
- Fiber optic thermometers
- Fluoroptic sensors
- Interferometric sensors
- Acoustic temperature sensors
- Thermal conductivity sensors
- Bimetallic thermometers
- Liquid or gas expansion sensors

The operational principles and characteristics of some of these sensors are explained below.

Thermocouples are contact-type temperature sensors made from the bonding of two different metals in the form of two symmetrical junctions. One of the junctions is kept at a constant temperature (e.g., room temperature) to act as a reference point. The output voltage between the junctions is on the order of tens of millivolts; hence the output must be amplified, usually by FET amplifiers. The most popular thermocouples that are used in portable instruments are copper-constantan (–190°C to 400°C); iron-constantan (–190°C to 870°C); chromel-alumel (–200°C to 1000°C); and platinum-platinum-rhodium (–0°C to 1760°C).

An electric potential is always produced at the junctions of two dissimilar metals bonded together. The junction potential usually exhibits temperature dependence with magnitudes varying in accordance withcharacteristics of the selected metals. The potential difference between the junctions, known as the *Thompson effect*, is described by

$$E_j = \alpha_1 (T_1 - T_2) + \alpha_2 (T_1^2 - T_2^2) \tag{158.9}$$

where E_j is the voltage (in μV) across the junctions, α_1 is the *Seebeck coefficient*, α_2 is the *Thomson coefficient*, and T_1 and T_2 are the absolute junction temperatures in Kelvin. The coefficients α_1 and α_2 are dependent on thermoelectric and bulk properties of the materials. They are also independent of homogeneous lengths and shapes of materials that form the thermocouple; hence, they have constant values. Depending on the materials used, Equation (158.9) can be expanded as a third-order or higher-order equation as a function of temperature gradient.

The component of the voltage proportional to junction temperature $[\alpha_1(T_1 - T_2)]$ is known as the *Peltier effect*. The Peltier effect is considered to be localized at the junctions. The component of the voltage proportional to the square of the junction temperature $[\alpha_2(T^2_1 - T^2_2)]$ is the *Thomson effect* and it is considered to be distributed along each conductor between the junctions.

For a single junction, the contact potential is not measurable. A voltage can be detected when junctions are subjected to different temperatures. A two-junction arrangement is illustrated in Figure 158.10. With the use of certain suitable physical arrangements and proper electronic compensation techniques, the temperature of the object can be measured linearly.

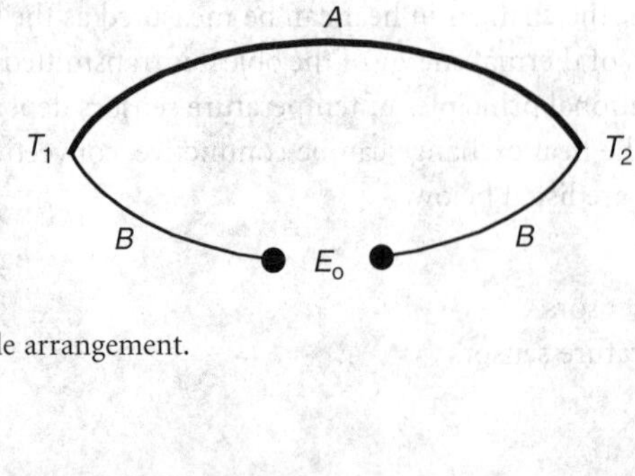

FIGURE 158.10 Thermocouple arrangement.

Thermoresistive sensors are based on the changes in electrical resistance with respect to temperature. Platinum is a commonly used element because of its good corrosive and resistive properties. Generally, thermoresistive sensors require simple signal-processing circuits. They have good sensitivity and stability. These sensors can be divided into three main groups: the resistance temperature detectors (RTDs), silicon-resistive sensors also known as *pn*-junction sensors, and thermistors.

RTDs are fabricated either in the form of wire or thin film. The resistivities of virtually all metals are dependent on temperature, but some metals and alloys are exclusive in their responses, stability, and durability. Platinum and its alloys are typical examples of such metals. Another popular metal is tungsten, which is used at high temperatures (over 600°C).

In general, all metallic conductors exhibit changes in resistivity when subjected to temperature variation. The resulting change in conductor resistance can be expressed by

$$R_\theta = R_0(1 + \alpha\theta) \tag{158.10}$$

where R_θ is resistance at temperature θ 0°C. R_0 is the resistance at 0°C, and α is the temperature coefficient of resistance at °C.

Silicon resistive sensors make use of the conductive properties of bulk silicon. Pure silicon has intrinsically a negative temperature coefficient (NTC); that is, resistance decreases as temperature increases. However, when doped with an *n*-type impurity, its temperature coefficient becomes positive at a certain temperature range. Therefore, resistance/temperature characteristics of these sensors are highly nonlinear. However, in some ranges, the characteristics transfer function can be approximated by a second-order polynomial as

$$R_T = R_0\left[1 + A(T - T_0) + B(T - T_0)^2\right] \tag{158.11}$$

where R_0 and T_0 are resistance and temperature (K), respectively, at a reference point. Silicon resistive sensors are used inside IC chips to detect internal temperatures.

Thermistors are based on semiconductors and often used as a part of transistor switches or as an operational amplifier inside the chips. Thermistors are a mixture of semimetal oxides (rare earths) and they can be classified into three major groups: *bead-type* thermistors, *chip* thermistors, and *semiconductive* thermistors. Each one of these types is manufactured in various sizes and shapes, as illustrated in Figure 158.11.

Thermistors can be made from NTC materials or positive temperature coefficient (PTC) materials, where resistance increases as temperature increases. Most metal-oxide thermistors have an NTC. The physical dimensions and resistivity of the material determine the value of resistance. The relation between temperature and resistance can be highly nonlinear, such as an exponential function. This relationship can be expressed as

$$R_T = R_0 e^{\beta\left(\frac{1}{T} - \frac{1}{T_0}\right)} \tag{158.12}$$

where R_T is the resistance at temperature T measured in K, and β is the characteristic temperature coefficient of the thermistor.

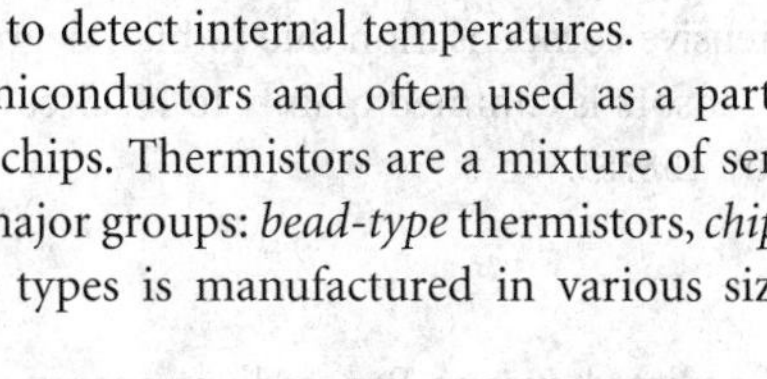

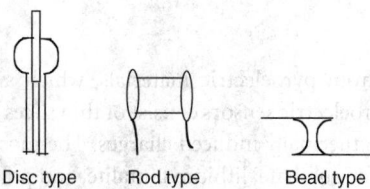

FIGURE 158.11 Various types of thermistors.

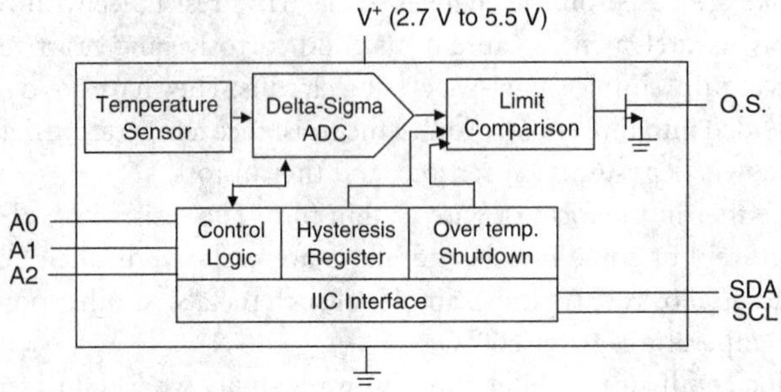

FIGURE 158.12 Typical IC temperature sensor.

PTC thermistors are generally made from polycrystalline ceramic materials such as barium titanate and strontium titanate. PTC thermistors are capable of maintaining their temperature coefficients constant.

Semiconductor temperature sensors consist of a *pn*-junction diode, and bipolar transistors exhibit strong thermal dependence. For example, the current-to-voltage equation of a diode can be expressed as

$$I = I_0 e^{\left(\frac{qV1}{2kT}\right)} \tag{158.13}$$

where I_0 is the saturation current, q is the charge of an electron, k is the Boltzman constant, and T is the temperature in K.

In principle, the transistors themselves can be used as sensing elements, since they exhibit exponential negative temperature characteristics. Indeed, an inherent amplification property of transistors makes them very attractive for temperature measurements. Particularly, the use of temperature sensitivity of base-emitter voltage becomes a good proposal since the output can be taken in amplified form from the collector. However, this needs extensive compensation due to high degree of nonlinearity. Therefore, the use of transistors as temperature sensors is confined mainly to IC circuits. A transistor-based IC temperature sensor is illustrated in Figure 158.12.

Heat Sensors

Heat can be viewed as a form of radiated energy. But, radiant energy can take many other forms, such as light, radio waves, or ionizing radiation. Therefore, the operational principles and manufacturing techniques of sensors for radiant energy differ according to the type of radiation. For example, black body radiation is sensed by bolometers, light is sensed by photometers, radio waves are sensed by antenna, and particles are measured by various types of radiation sensors.

Thermal sensors known as *pyrometers* are based on the conversion of thermal radiation into heat, and then converting the heat into electrical signals. All thermal radiation sensors can be divided into two groups: *passive* and *active.* Passive sensors absorb incoming radiation and convert it into heat. Active sensors, on the other hand, emit thermal radiation directed toward the object under investigation. There are many different types of thermal sensors, including the following:

- Pyroelectric
- Thermopile
- Infrared

Pyroelectric sensors are made from pyroelectric materials, which are crystals capable of generating charge in response to heat flow. Pyroelectric sensors consist of thin slices or films with electrodes deposited on the opposite side to collect the thermally induced charges. The most common pyroelectric materials are lead zirconate titanate, triglycine sulphate, lithium tantalite, and polyvinyl fluoride. Most commercial pyroelectric sensors are made from single crystals, such as triglycine sulphate, PZT ceramics, or $LiTaO_3$. The polyvinyl fluoride-based sensors exhibit high-speed responses and good lateral resolutions.

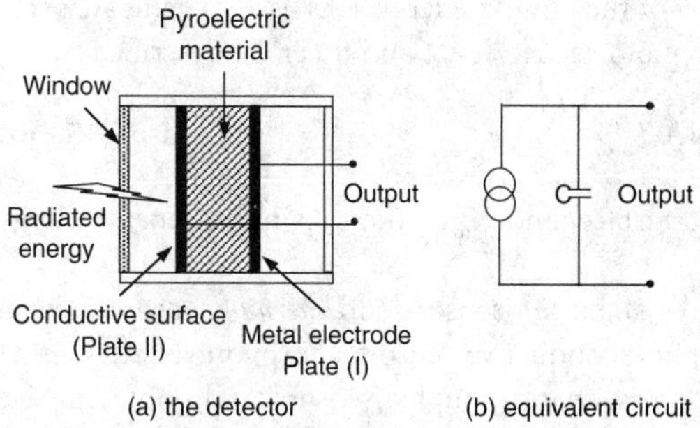

FIGURE 158.13 Pyroelectric detector and its equivalent circuit.

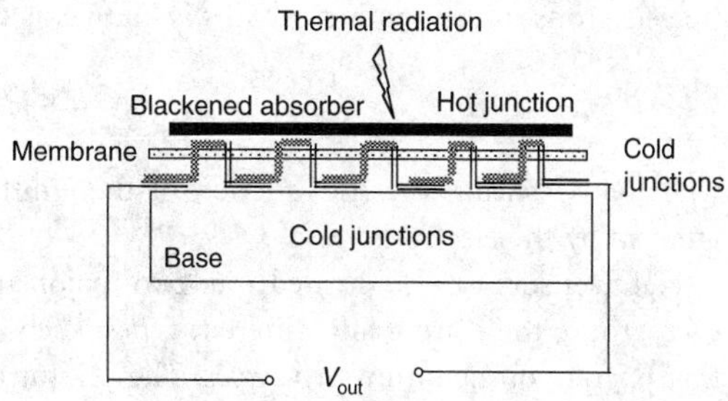

FIGURE 158.14 Thermopile.

Pyroelectric sensors are manufactured by having metallized surfaces and a protective window, as shown in Figure 158.13(a). In the case of infrared (IR) detectors, the thin-slice films are made from the dielectric materials whose surfaces are electrically charged by the incoming radiation. Therefore, these detectors are primarily capacitive types with very high impedances. The equivalent circuit of a pyroelectric sensor is illustrated in Figure 158.13(b). Although many plastic films are suitable, in most applications, a lithium tantalate film forms the basic material of passive IR detectors. These detectors are constructed just like capacitors. A metal plate forms one of the electrodes, and the other side of the pyroelectric material contains a conductive material that acts as the second plate. The voltages across the plates alter since the radiated energy alters the charge of the pyroelectric material. In practical applications, most pyroelectric detectors and signal processing circuits are integrated in a MOSFET structure.

Thermopiles are passive infrared sensors. They operate on the same principles as thermocouples; that is, they have a number of serially connected hot junctions (Figure 158.14) with improved absorptivity and a cold junction that gives the temperature difference between the two. The sensor is hermetically sealed in a metal can with a hard, infrared transparent window such as silicon, germanium, or zinc selenite. The operating frequency is determined by the thermal capacity, and thermal conductivity of the membrane is sandwiched between the hot junctions and the cold junction. These sensors have excellent noise properties.

158.8 Light and Radiation Sensors

Light is basically an electromagnetic radiation that consists of an electric field and a magnetic field component. Compared to radio waves, light has a short wavelength, and hence a very high frequency. Many different properties of light are used for instrumentation and measurement purposes. Applications vary from photographic imaging to high-speed data transmission via fibers. Once the light is generated and propagated from a source, it can be expanded, condensed, collimated, reflected, polarized, filtered, diffused, absorbed, refracted, and scattered to develop sensors and measurement systems. Some of these properties of light are manipulated on purpose to serve particular application needs, and sometimes manipulations are not necessary since they happen naturally due to the optical properties of the media and physical characteristics of light.

Like the other electromagnetic waves in the electromagnetic range, light has a velocity, c, in space 3×10^8 mps so that the relationship between frequency, f, and wavelength, λ, can be written as

$$c = \lambda \times f \tag{158.14}$$

Also, the energy is carried by photons. The energy of a single photon is given by

$$E = h \times f \tag{158.15}$$

where h is the Plank's constant ($h = 6.63 \times 10^{-34}$ J or 4.13×10^{-15} eV.s)

When a photon strikes a surface of a conductor, part of the photon energy, E, is used to generate some free electrons in the conductor. This is known as the photoelectric effect and it can be described by

$$h \times f = \varphi + K_{\mathrm{m}} \tag{158.16}$$

where φ is called the work function of the emitting surface, and K_{m} is the maximum kinetic energy attained by the electron.

Light sensors can be divided into two major groups: *quantum* sensors and *thermal* sensors. Within each group, there are many different types, such as photoconductive sensors and photovoltaic sensors that operate on quantum principles. The phototransistors, charge-coupled sensors, and photoemissive sensors are other types that operate on thermal principles. These sensors are available in many different configurations, some built with powerful lenses and high-speed responses to monochromatic light, and others having different color sensitivities to chromatic light.

Photosensors

Photosensors are essentially photon detectors that are sensitive to photons with energies greater than a given intrinsic gap energy of the sensor. Some of the most common photon detectors make use of properties of cadmium and lead, such as cadmium sulphides (CdS) and cadmium selenite (CdSe), lead sulphide (PbS) and lead selenite (PbSe) devices. Their response times can be less than 5 μs. The main classes of photosenors are photodiodes/phototransistors, photovoltaic cells, and photoresistors/photoconductors.

Photodiodes and *phototransistors* are semiconductor devices that are sensitive to light. In a photodiode, as the incident light falls on a reverse-biased *pn* junction, the photonic energy carried by the light creates an electron-hole pair on both sides of the junction causing current to flow in a closed circuit (Figure 158.15). The output voltage of photodiodes may be highly nonlinear, thus requiring suitable linearization and amplification circuits.

Photodiodes are constructed like any other diodes, but without the opaque coating, so that light can fall onto the junction. The current generated at the junction can be linearly related to the amount of illumination falling on the junction. The response time of photodiode sensors is very short, typically 250 ns. The short response time of photodiodes allows these devices to be used in modulated light-beam applications.

Wide ranges of photodiodes are available such as *pn* photodiodes, *PIN* photodiodes, Schottky photodiodes, and avalanche photodiodes.

Phototransistors are the most used light sensors. In addition to converting photons into charge carriers, they have current gain properties. Particularly, Darlington phototransistors possess high sensitivity and high current gain.

A phototransistor (Figure 158.16) is a form of transistor in which the base-emitter junction is sealed with a transparent package. As in the case of photodiodes, this junction is affected by incident light and acts as a diode. The light at the junction promotes electrons from valence to conduction band. The current in this junction is then amplified by the normal transistor action to provide a much larger

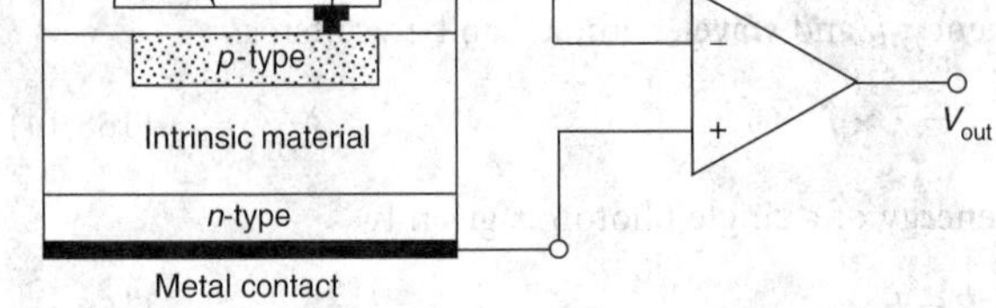

FIGURE 158.15 Typical structure of a photodiode.

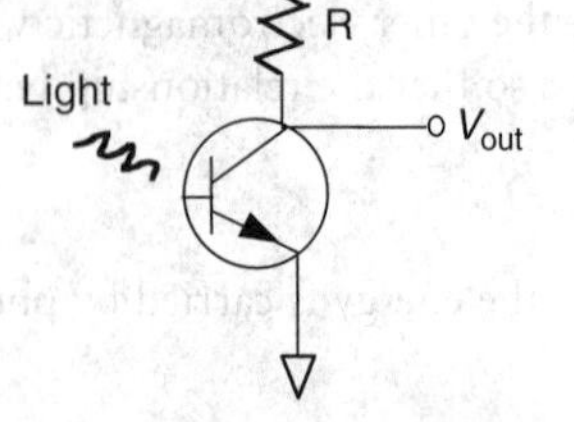

FIGURE 158.16 Phototransistor for sensing light.

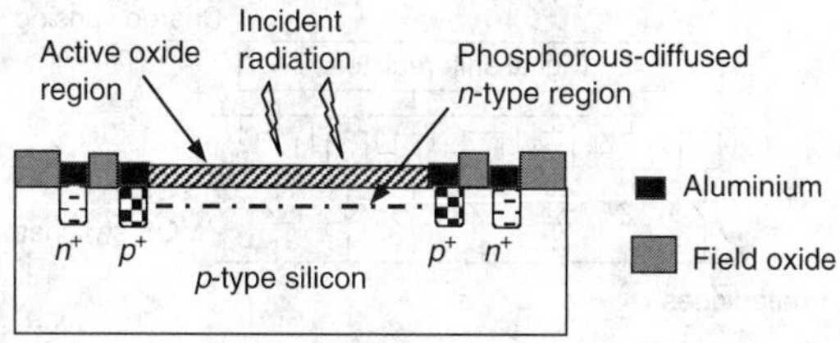

FIGURE 158.17 Si-photodiode.

collector current. The spectral responses of these devices are limited to red and near-infrared portions of the spectrum. The response of the phototransistor can be adjusted by applying an appropriate bias voltage to the base. The response times of phototransistors are much longer than photodiodes (microseconds). Phototransistors are manufactured in the IC forms, containing light sensors integrated with the amplifiers and other processing circuits.

Photovoltaic sensors are based on silicon cells that generate voltages in response to a beam of light. The response time of these sensors is slow and it takes a long time to stabilize the output (up to 20 s). The wavelength response of a photovoltaic silicon cell covers the entire visible spectrum, which makes them very useful for environmental light measurements.

The first form of photovoltaic device was the selenium cell, which produced voltage across the cell in proportion to the illumination. Modern photovoltaic devices are constructed from silicon, thus acting as a silicon photodiode with a large area junction and used without bias (Figure 158.17). These devices find applications in camera exposure controls and control of light levels in manufacturing processes, because with some suitable filtering the response curves can be made very similar to the response of the human eye to the same light.

Photoresistors/photoconductors make use of changes in resistance of many materials when exposed to light. The materials are known as the light-dependent resistors, such as CdS and CdSe. In photoresistors, the material (e.g., cadmium sulphide) is deposited as a thread pattern on an insulator in a zigzag form. The cell is then encapsulated in a transparent resin or encased in glass to protect the material from the contamination by the atmosphere. A typical sensor has a spectral response of 610 nm corresponding to a color in the yellow-orange region, the dark resistance is 10 MΩ and the cell resistance decreases typically from 2.4 kΩ at 50 lux to 130 Ω at 1000 lux. However, the changes in the resistances against illumination are not linear, thus requiring additional linearization circuitry.

The response time of photoconductors is fairly slow and the resistance changes in response to light might take a few seconds. Consequently, these sensors are unsuitable in applications where fast responses are required.

The *image sensor* is a device that transforms an optical image into electrical signals. A standard image sensor converts the light photons falling onto the image plane into the corresponding spatial distribution of electric charges. The accumulated charge at the point of generation is transferred suitably and converted to useable voltage signals. Each of these functions can be accomplished by a variety of approaches, such as vacuum tube and solid-state technologies. Solid-state sensors are primarily based on photodiodes and charge-coupled devices.

The simplest image sensors are the parallel types, in which a matrix of photodiodes is used in small-scale applications such as optical character recognition. A photodiode consists of a thin surface region of *p*-type silicon formed on an *n*-type silicon substrate. A negative voltage applied to a surface electrode reverses the biases of the *p-n* junction. This creates a depletion region in the *n* silicon, which contains only an immobile positive charge. Light penetrating into the depletion region creates electron-hole pairs, which discharges the capacitor linearly in time. Solid-state image sensors are quite complex and made in IC forms; there are two basic types — the serially switched photodiode arrays (Figure 158.18) and the charge-coupled photodiode arrays. In these arrays, the basic principle is to use light intensity to charge a capacitor and then to read the capacitor voltage by shifting it through the register. These arrays have long life and high reliability, small in size, and immune to damage from bright lights.

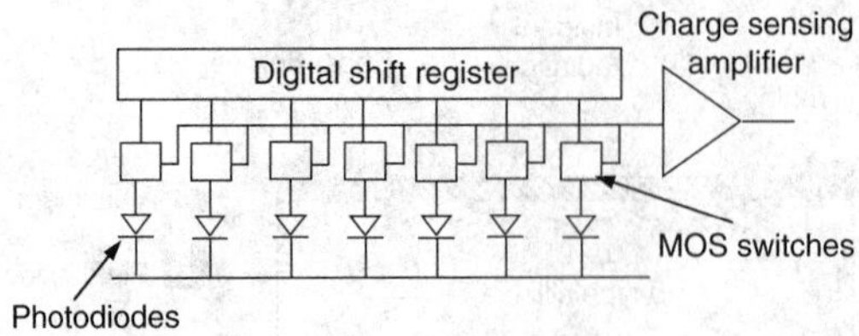

FIGURE 158.18 IC image transducer.

Optical fiber sensors are used in varying forms, such as temperature sensors and optical gyroscopes. They are based on the properties of light as well as those of optical materials that are constructed suitably for specific applications. Operational principles of the sensors vary considerably. Some make use of the scattering properties of light while others make use of attenuations and losses. There are three basic types of fibers: single-mode step-index, multimode step index, and multimode graded index.

Photoemissive sensors are vacuum tubes with treated metal surfaces. Light falling on the surface causes electrons to be ejected to be picked up by an anode. Photoemissive sensors are available with many different spectral sensitivities. These sensors have low power demands but require a high voltage. The most commonly used photoemissive device is the photomultiplier.

Radiation Sensors

Because of the health and safety implications to humans and other living organisms, the correct sensing of radiation has been taken very seriously. Therefore, a wide range of sensors are available. Many radiation sensors use ionization as the basic physical principle behind operations. There are two types of ionizing radiation: particle radiation such as alpha particles, beta particles, neutrons, and cosmic rays; and electromagnetic radiation such as X-rays and gamma rays.

There are four basic types of radiation sensors: gas filled, scintillation, semiconductor, and thin film. The film detectors primarily consist of chemical thin films, which are sensitive to incoming radiation. The intensity of the darkening of the film indicates the amount of radiation that a person who is wearing the detector has been exposed to.

A *gas-filled sensor* consists of a cylindrical container filled with an inert gas (e.g., argon). A voltage is applied in a wire placed in the cylinder and the cylinder wall. Radiation entering the container ionizes some of the gas atoms to give negatively charged and positively charged ions. Free electrons travel to the anode (the wire) and the cathode (the cylinder wall). The resulting electronic pulse is amplified and recorded as a count. In some gas-filled detectors, electrons traveling to the anode can cause further ionization, magnifying the incoming radiation. Instruments based on these principles can be very sensitive to small amounts of incoming radiation.

A typical and widely used example of gas-filled detectors is the Geiger–Müller detector, also known as the Geiger counter, shown in Figure 158.19. In this arrangement, the anode is a tungsten or platinum wire, while the cylindrical tube acts the cathode for the circuit. The tube is filled with argon with a slight amount of hydrocarbon gas. Internal pressure is kept at slightly less than atmospheric pressure. The radiation particle is transmitted through the cathode material. Interaction of the particle produces ionization of the gas molecules, thus causing a voltage pulse for each particle.

Semiconductor radiation sensors are made from semiconductor materials. When an electric field is placed across the semiconductor, a region is created such that no free electrons exist. An incoming radiation ionizes atoms in this region producing free electrons, which can be collected and counted. The semiconductor sensors are highly sensitive, accurate, and have high resolution compared to gas-filled counterparts. They can distinguish between radiation particles that are close together in energy content.

There are many different types of semiconductor radiation sensors: diffused junction, ion implanted, strip types, drift types, CCD sensors, and so on. As a typical semiconductor radiation sensor, a position-sensitive silicon-strip type of sensor is illustrated in Figure 158.20. This sensor is

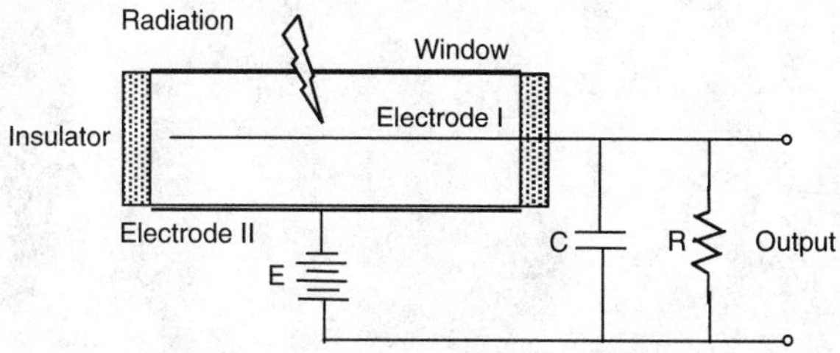

FIGURE 158.19 Geiger–Müller radiation detector.

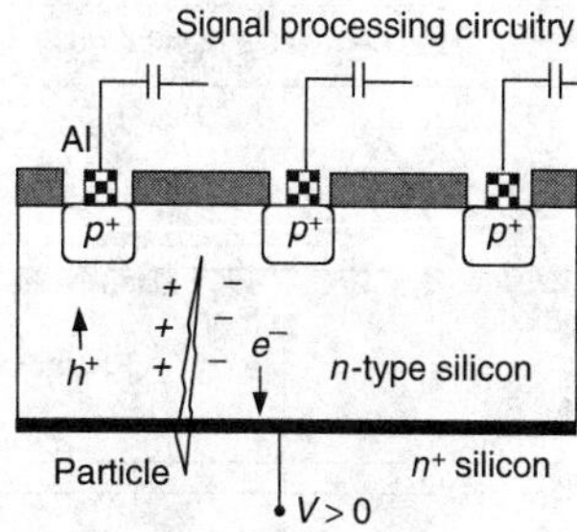

FIGURE 158.20 Semiconductor radiation detector.

suitable for detecting radioactive particles with greater than 5 μm localized hit rates. They are often used in particle physics experiments.

158.9 Chemical and Gas Sensors

Chemical sensors are extensively used for identifying chemical compounds and elements in industrial, environmental, food processing, and domestic applications. Some of the applications are air and water quality measurements, pollution level determination, detection of chemical and gas leaks, determination of toxicity levels, prospecting of minerals in mining and metallurgy industries, finding explosives, detecting drugs, fire warning, and many other types of domestic, medical, industrial, and health- and safety-related applications.

Chemical sensors can be classified as *contact types* and *noncontact types*. The contact types involve the physical interaction of the sensor with the chemicals under observation. The chemicals can be in gas, liquid, solid, or mixture forms. The noncontact types use remote sensing techniques that are based on the analysis of the gaseous part of the chemical substances under investigation.

Some of the many methods of sensing chemicals include catalytic sensors, enzyme sensors, chemosensors, chromatographic sensors, electrochemical sensors such as potentiometric and amperometric sensors, radioactive sensors, thermal sensors, optical sensors, mass sensors, concentration sensors, titration sensors, and semiconductor sensors. Out of this vast repertoire of sensors only a few are discussed here.

Enzyme sensors consist of a special type of catalyst, proteins found in living organisms. Enzymes exist in aqueous environment in the form of immobilization matrices as gels or hydrogels. As sensors, enzymes tend to be very effective in increasing the rate of certain chemical reactions and also are strongly selective to a given substrate.

Figure 158.21 illustrates a typical enzyme sensor. The sensing element can be a heated probe, electrochemical, or optical sensor. The enzyme, acting as a catalyst, is immobilized inside the layer into which the substrate diffuses; hence, it reacts with the substrate and the product is diffused out of the layer into the sample solution.

Radioactive chemical sensors are most commonly used in home smoke alarms. In this arrangement, two metal plates exist inside the chamber, and a small radioactive source emits alpha radiation that ionizes the air in between the metal plates and allows a current to flow. If chemicals/particles in the air enter

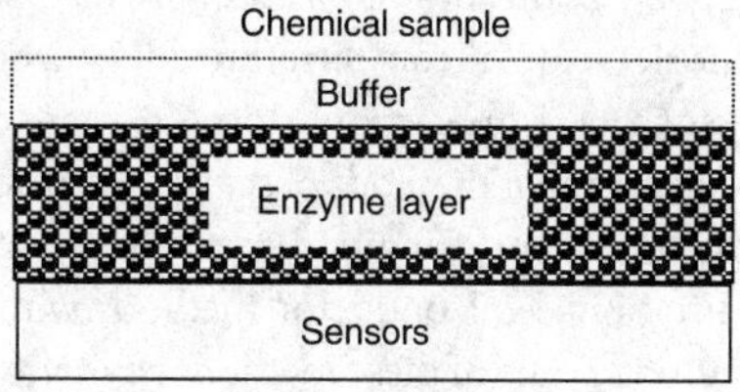

FIGURE 158.21 Arrangement of an enzyme sensor.

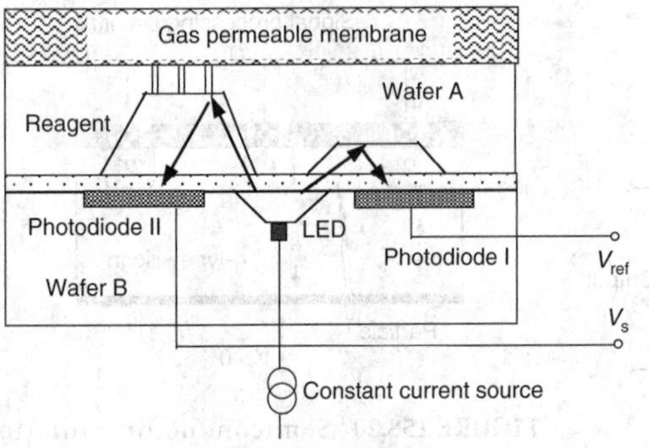

FIGURE 158.22 Optical semiconductor chemical sensors.

the space between the plates and bind with the ions, then the amount of normal current is altered, thus triggering an alarm. This method may be applied only to certain analytes.

Chemosensors are based on molecules that can bind selectively and reversibly to the analyte of interest. A chemosensor binds and/or unbinds easily with the dissolved gas molecules of interest. The constituents and physical properties of a chemosensor may be unique to each analyte, depending on chemical composition; that is, a different chemosensor will be required for each analyte. The analyte can be made from various compounds; for instance methane dissolves into the chemosensor solution and is randomly bound to the chemosensor.

Sensors with light and properties of optics are based on the interaction of electromagnetic radiation with matter. As a result of this interaction, some property of light is altered, such as intensity, polarization, and velocity.

Figure 158.22 illustrates a typical semiconductor-based optical chemical sensor. This sensor consists of two chambers, one acting as a reference containing a reference concentration chemical, and the other acting as the sensor. These chambers are illuminated by a common light-emitting diode (LED). The surfaces of chambers are metallized to improve internal reflectivity, and the bottom of chambers are covered by glass. One of the chambers has slots covered with a gas-permeable membrane. The slots allow the penetration of gas to be measured (e.g., CO_2) into the chamber. Wafers A and B form optical waveguides. The chamber filled with reagent is used to monitor the optical absorbency for comparison with that obtained from the reference chamber.

Absorption and emission sensors act as a binding agent of analytes to some chemosensor. The absorption and emission spectrum gives distinctive bands of colors that may be analyzed and compared against known spectra for determining the chemical composition. This method is also used in the analysis of gases, such as oxygen.

Gas Sensors

Gas sensors could be included in the section of chemical sensors since gas, in chemical terms, is any matter that is in the gaseous state characterized by high molecular kinetic energy and a tendency to expand out of an open container. In practice, the word *gas* takes distinct and separate meanings. For example, natural gas is a fossil fuel that is collected in the gas phase, and gasoline, which is also called gas, is a liquid mixture of hydrocarbon fuels that becomes gaseous in the combustion chamber of an engine. In this article, gas will be taken in the chemical sense. For the purpose of detecting gases, there are effectively two types of gases: *inorganic*, such as sulfur dioxide, hydrogen sulphide, nitrogen oxides, hydrochloric acid, silicon tetrafluoride, carbon monoxide and carbon dioxide, ammonia, ozone; and *organic*, such as hydrocarbons, terpins, mercaptans, formaldehyde, dioxin, and fluorocarbons.

The most popular methods of sensing gases are infrared and fluorescent spectrometry and light absorption methods, colorimetry or photometry, chromatographic methods, flame ionization, indicator tubes, test papers, titration, semiconductor methods with electric conductivity and amperimetric methods, and ring oven methods. Some of these methods have already been discussed, and others will briefly be discussed below.

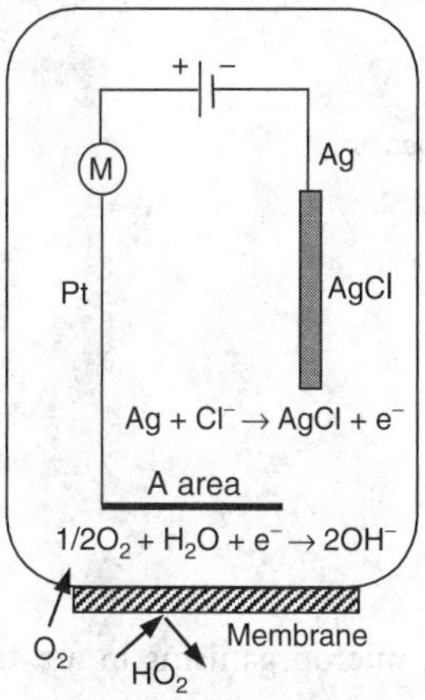

FIGURE 158.23 Amperimetric oxygen sensor.

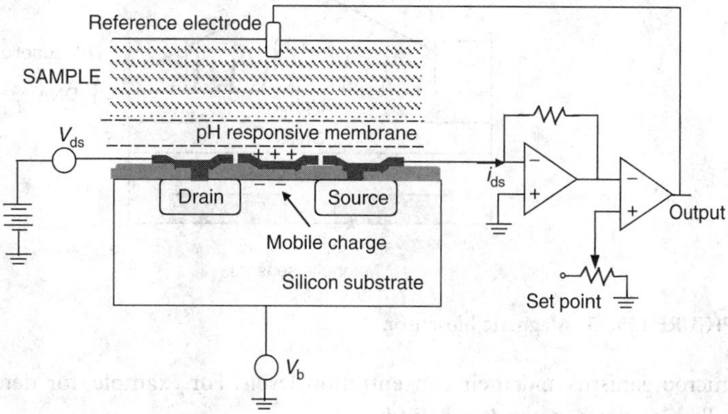

FIGURE 158.24 A pH sensor based on ion-selective FET.

Amperimetric chemical and gas sensors are based on the generation of electric current or potential by free elements or compounds. For example, sulfur dioxide and other reducing gases in an air sample react with electrically generated free iodine or bromine in a detection cell. The cell has an anode and a cathode electrode and contains a constant titrant concentration. Any sulfur compound introduced into the cell reacts with titrant and changes the concentration to produce a potential change in the solution to be picked up by electrodes. The range of the instrument is 0.0001 ppm to 0.2 ppm.

In amperimetric sensors, the chemical reaction occurs at the active electrodes by means of an exchange of electrodes leading to the flow of current. A typical amperimetric cell contains three electrodes inserted in the solution to be measured. The desired reaction takes place in the working electrodes (WE). The WE are usually made from materials, such as mercury, platinum, carbon, or gold, that show different responses to specific chemicals. The potential of the WE relative to the solution is measured by a reference electrode (RE). The WE carries current but the RE does not. In some cases, the chemical reaction is encouraged by setting a potential difference between the electrodes by counter electrodes, called the *potentiostat.*

Basic components of amperimetric sensors are illustrated in Figure 158.23. The cathode is made from platinum. The reference electrode, called the Clark oxygen sensor, is made from AgCl, which results in the generation of electrons. The cathode is maintained at a fixed reference condition by establishing constant chloride ion concentration. The cathode is separated from the sensing fluid by an oxygen-permeable membrane made from silicone polymer.

Potentiometric chemical and gas sensors act as an electric battery. A typical example of potentiometric sensors based on semiconductors is given in Figure 158.24. These sensors are based on the use of ion-selective field-effect transistors (ISFETs). They are derived from metal-oxide semiconductor FETs or MOSFETs with hydrogen ion-sensitive elements. These silicon chips contain a pH-responsive membrane much like glass electrodes. The signals are amplified and processed on the same chip.

In these sensors, the potential or current signals generated by the electrochemical process between the solution and the sensors are processed. Potentiometric sensors have two important characteristics: very small current needs to be drawn during measurements, thus implying alterations in the characteristics; and the response is not linear but usually logarithmic.

158.10 Biological and Environmental Sensors

Biological Sensors

Traditionally, biological detectors require human intervention in a laboratory environment. However, in recent years, automatic devices and robots are involved in biological applications, such as detection of

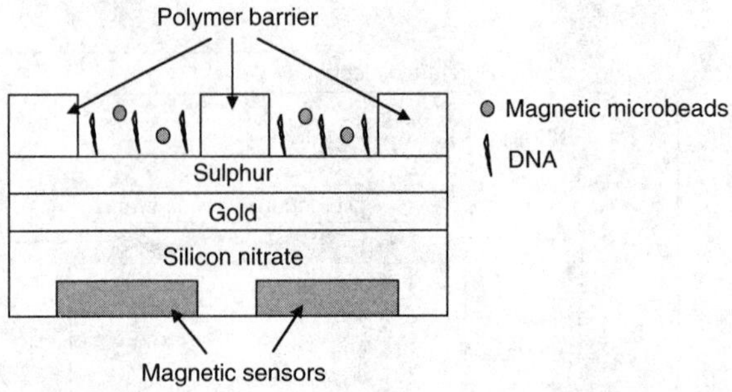

FIGURE 158.25 Magnetic biosensor.

microorganisms and their concentration levels. For example, for detecting microorganisms in air, the following three methods are used:

1. *Biochemical*, which detects a DNA sequence and protein that are unique to a bioagent through its interaction with test modules
2. *Chemical*, for example, mass spectrometry, which works by breaking down a sample into its components such as amino acids and then comparing their weights with those of known bioagents and other molecules
3. *Biological tissue-based systems*, in which a bioagent or biotoxin affects live mammalian cells, causing them to undergo some measurable response

Some classes of biological sensors rely on comparing the DNA taken from a microorganism with the DNA of a known agent. These sensors are built on prior knowledge of the bioagent. In automatic systems, a sensor collects samples and then multiple copies of the DNA are made by using a polymerize chain reaction (PCR). In this method, the samples are heated and cooled periodically to produce copies of the DNA. As the DNA is copied, the resulting strands are mixed with fluorescent DNA probes. The probe bound with the bioagent glows under ultraviolet light. By using different probe markers, several PCR reactions can run at the same time, thus enabling detection of more than one agent at a time.

There are microfluidic devices, which contain tiny channels, valves, and chambers in a single chip. Once the microorganisms are in the chip, their cells are cracked open ultrasonically. The PCR is applied by means of small thin-film heaters.

Apart from optical and fluorescent methods, magnetic methods, illustrated in Figure 158.25, are also used to detect DNA. These devices comprise an array of wire-like magnetic field microsensors. These sensors are coated with single-stranded DNA probes specific to a gene from a bioagent. Once a strand of bioagent DNA in a sample binds with a probe, the resulting double strand binds a single magnetic microbead. When a magnetic bead is present above a sensor, the resistance of the sensor decreases in proportion to the number of microbeads.

Some sensors use fluorescent antibodies to bind to bacterial cells. In these devices, the sample passes through a portable flow cytometer, which counts cells by measuring their fluorescent or other properties as they move in a liquid.

Another commonly used method in biological sensors uses live tissues. Many toxins trigger measurable or differentiable reactions in living cells. Mammalian cells, such as heart cells, are cultured in a lab, and then seeded into a cartridge containing a microelectrode array. When a biotoxin is introduced, the cells create voltages detectable in millivolts at the electrodes.

Environmental Sensors

Humidity and moisture measurements are very important in agricultural applications and international food commodity trading. Moisture is the amount of water, in some cases the amount of liquid, in materials. The presence of moisture in a gas is termed as humidity. The absolute humidity of a gas is the mass of water per unit mass of gas. The maximum humidity that can be attained is called the saturation humidity. The saturation humidity heavily depends on temperature. For many purposes, relative humidity is important. The relative humidity is the ratio of absolute humidity to saturation humidity at a particular temperature.

Another measure of relative humidity is the dew point. When a surface is cooled and in contact with gas, a temperature will be reached such that gas-containing water deposits the water onto the surface by condensation. This surface temperature is called the dew-point temperature.

Many methods are available for moisture and humidity sensing. These methods range from a change of electrical properties to changes in optical properties or even changes in mechanical characteristics of the sensors. Diverse techniques may be designed and developed for specific application in a particular process.

The oldest method of measuring humidity is the *hygrometer.* A human hair, free from oil or grease, is strongly affected by humidity, shrinking in dry conditions and expanding in moist conditions. Relative humidity sensors are constructed by slightly tensioned hair along with some method of detecting changes in the length of materials. Some of these methods include the use of linear potentiometers, LVDTs, capacitive techniques, or some forms of oscillatory circuits.

Use of *lithium chloride* for moisture measurement is very common. Lithium chloride has very high resistance in its dry state, but resistance drops considerably in the presence of water with a reasonable linearity in proportion to the amount of water. The lithium chloride cell is made part of a measuring bridge, and the output of the bridge is appropriately processed.

A common method for measuring humidity and moisture content is by using *microwave energy.* Microwave energy at various frequencies (e.g., 2.45 GHz) is absorbed strongly by water. The attenuations and phase shifts of the microwave across the moist media become a good way measuring water content of the media. Microwave techniques are extensively used in agricultural products such as grain moisture determinations.

Moisture of solids is usually sensed in terms material conductivity, changes in permittivity, or microwave absorption. For moisture content in materials of fixed composition, resistance measurements between the conductors set at fixed distances is sufficient, but need lengthy calibration procedures.

Conductive humidity and moisture sensors make use of changes in resistance of materials doped in semiconductor structures. The sensing elements are made from many different materials such as polystyrene films (Pope cells) treated with sulfuric acid. In some cases, solid polyelectrolides are used. Solid-state humidity sensors are fabricated from silicon substrate as illustrated in Figure 158.26.

Capacitive humidity and moisture sensors find extensive applications. In capacitive methods, changes in permittivity of atmospheric air as well as changes in permittivity of many solid materials are used. The permittivity of most dielectric materials is a function of moisture content and temperature. The capacitive humidity sensor is based on changes in permittivity of the dielectric material between plates of capacitors.

Thin-film capacitive sensors can be fabricated on silicon substrates as shown in Figure 158.27. In this construction, a SiO_2 layer is deposited on the *n*-type silicon substrate. The two metal electrodes, shaped in interdigitized patterns, are located on the SiO_2. These electrodes may be made from aluminium, tantalum, chromium, or phosphorous-doped polysilicon. Since there are many types of capacitive humidity sensors operating on similar principles, only the aluminium and tantalum types will be introduced as typical examples.

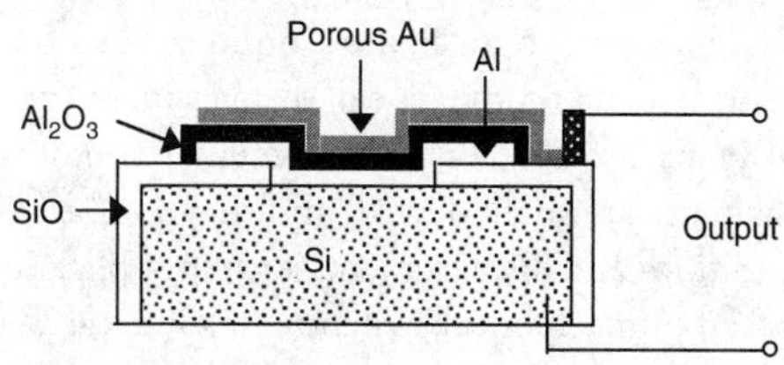

FIGURE 158.26 Aluminium oxide thin-film conductive moisture sensor.

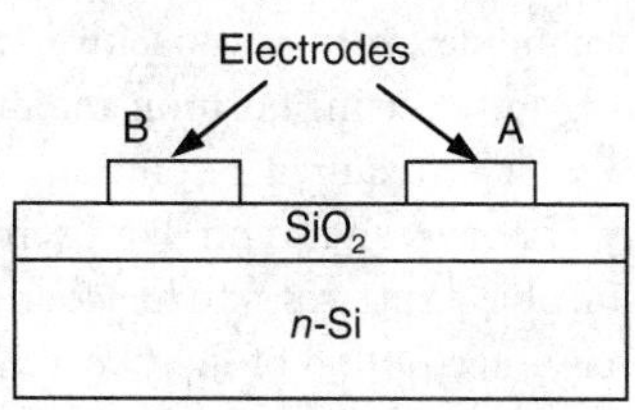

FIGURE 158.27 Capacitive thin-film humidity sensor.

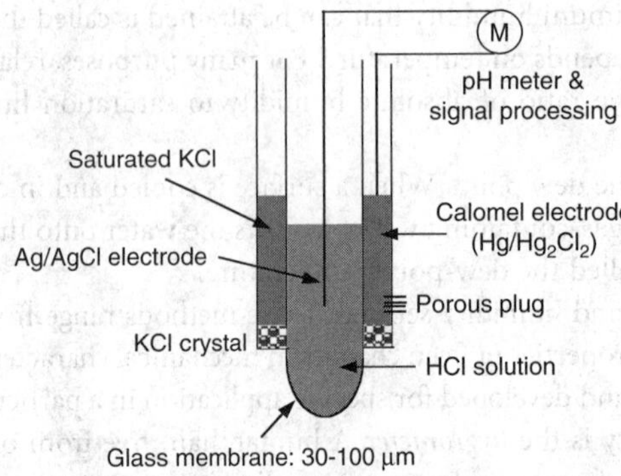

FIGURE 158.28 A pH sensor.

Aluminium-type capacitive humidity sensors constitute the majority of capacitive humidity sensors. In these types of sensors, high-purity aluminium is chemically oxidized to produce a prefilled insulating layer of partially hydrated aluminium oxide, which acts as the dielectric. A water-permeable but conductive gold film is deposited onto the oxide layer, usually by vacuum deposition, which forms the second electrode of the capacitor.

Acidity and Alkalinity Sensors

Acidity and alkalinity of water are important factors in monitoring water supplies for industrial and domestic water use, at hydroelectric power-generation stations, and in various agriculture and horticulture applications. The acidity and alkalinity of water are measured on the pH scale, which is based on free hydrogen ions in the water. Natural water has a pH value of 7, fairly strong acid solutions have a pH of 2, and fairly strong alkaline solutions have a pH of 12.

The sensing of pH makes use of changes in ionization. Natural water with 7.0 pH levels has a very high resistivity, but ionization causes the resistivity to drop very sharply. The standard electrical system of sensing acidity depends on the glass electrode system as shown in Figure 158.28. In this arrangement, a thin glass bulb contains a mildly acidic solution with good conducting properties. A suitable reference electrode, calomel, and a sensing element inside the bulb provide electrical signals for pH levels.

158.11 Mechanical Variables Sensors

Mechanical variable measurements can be divided into two broad categories: solids and fluids. Solid variable measurements consist of stress and strain, mass and weight, density, acceleration, velocity, force, torque, and power measurements. Variables for fluids can be categorized as pressure, velocity, flow, level, viscosity, and surface tension measurements. In this section, only a few sensors associated with mechanical variables will be discussed.

Position, Displacement, and Proximity Sensors

The measurement of distances or displacements is an important aspect in many industrial, scientific, and engineering applications. Sensors for position and displacement sensing may be based on many different techniques, such as capacitive, inductive, resistive, magnetic, optical, ultrasound, and so on.

Designing and selecting position and displacement sensors depend on the type of displacement (linear, nonlinear, rotary), required resolution, type of material being applied for, environmental conditions, availability of power, and so on. For example, the microswitches are small devices and they can be easily mounted on object surfaces, and hence are used in many position- and displacement-sensing applications. Alternatively, interruption of light beams can be used instead of mechanical switches. In some cases, the location of a reflected beam can be used to measure the distance of objects. Other methods may include potentiometers, ferrite core inductors with movable cores, or variable capacitors to track the movements of an attached object.

Capacitive position, displacement, and proximity sensors consist of two simple electrodes with capacitance *C*. There are three basic methods for realizing such sensors: varying the distance between the plates, varying the area between plates, and making use of dielectric properties of the material between the plates.

Magnetic position, displacement, and proximity sensors are based on magnetic properties of materials and circuits. The many types include the following:

- Inductive
- Hall effect
- Magnetoresistive
- Eddy current proximity

Inductive proximity sensors operate on the principle of change inductance of a coil in the presence of a metal in the proximity of the core. For signal processing, the coil can be configured as part of a bridge circuit or as a part of a tuned circuit. When used as part of a bridge, the presence of metal near the coil forces the bridge output to be off balance, thus registering a voltage output. In the case of tuned circuits, change in the tuned frequency due to variations in the magnetic properties can easily be processed.

Hall-effect sensors are usually fabricated as monolithic silicon chips and encapsulated into epoxy or ceramic packages. There are two types of Hall-effect sensors: linear and threshold sensors. A linear sensor incorporates an amplifier and other electronic circuits for interfacing with peripherals. Compared to basic Hall-effect sensors, they tend to be more stable and operate over wider voltage ranges. The threshold sensors have a Schmitt trigger with a built-in hysteresis to eliminate spurious oscillations.

Magnetoresistive sensors are based on the change in resistance of some materials such as permalloys in the presence of magnetic fields. In these materials, current passing through the material magnetizes the material in a particular magnetic orientation. The resistance is highest when the magnetization is parallel to the current and lowest when it is perpendicular to the current. Hence, depending on the intensity of the external magnetic field, the resistance of the permalloy changes in proportion. The magnetoresistive sensors are manufactured as thin films and are usually integrated as part of an appropriate bridge circuit. They have good linearity and low-temperature coefficients. These devices have sensitivity ranging from 10^{-6} T to 50 mT. By choosing good electronic components and suitable feedback circuits, sensitivity can be as low as 10^{-10} T. They can operate from DC to frequencies in several gigahertz.

Eddy current sensors are inductive-type sensors that contain two coils in the shape of probes. One of the coils, known as the active coil, is influenced by the presence of a target with conductive properties. The second coil, known as the balance coil, serves as an arm of a bridge that balances the circuit. This also provides temperature compensation. The magnetic flux from the active coil passes into the conductive target from the probe. When the probe is brought close to the target, the flux from the probe links with the target, producing eddy currents within the target. The eddy currents change the magnetic properties of the sensors, thus yielding unbalanced conditions of the bridge.

Many other sensors are used for position, displacement, and proximity sensing. Most of these sensors can also be used as motion detectors by the use of suitable signal processing and display techniques. Some of these sensors include:

- Position-sensitive detectors
- Proximity sensors with polarized light
- Photo beam sensors
- Fiber optic sensors
- Grating sensors such as encoded discs
- Ultrasonic sensors
- Microwave and Doppler radar sensors

Some of these are briefly discussed below.

Position-sensitive detector (PSD) optical systems operating at the near-infrared region can be a very effective way of sensing short- and long-range displacements and positions. As in the case of cameras, PSDs are used in conjunction with LEDs. A PSD operates on the principle of photoelectric effect. The

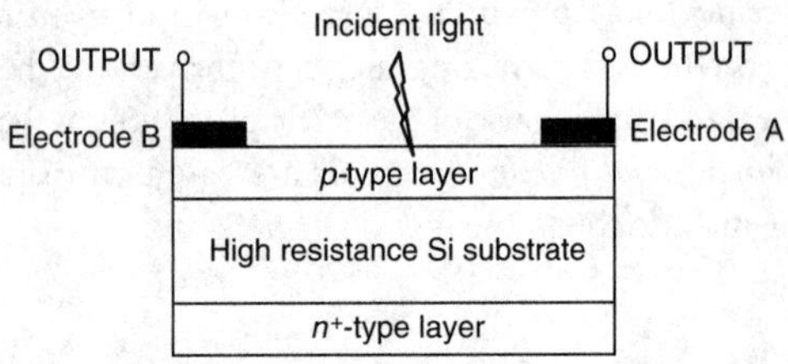

FIGURE 158.29 One-dimensional silicon position-sensitive detectors.

surface resistance of a silicon photodiode changes depending on the light falling on it. These sensors can be single dimensional or two dimensional. A single-dimensional sensor, illustrated in Figure 158.29, is fabricated with high resistance silicon with *p*- and n^+-type layers. It has two electrodes to form contact with the *p*-type layer. The photoelectric effect occurs in the upper *pn* junction.

Photo beam sensors, or optical position, displacement, and proximity sensors of this type mostly use a reflection technique — the transmitted beam reflects from the stationary or moving object. The light source is usually an LED working in the visible red or invisible infrared region. Visible red makes sensor configuration easier, but the infrared type is less affected by interference from other light sources. These sensors make use of pulse-modulated signals and polarizing filters to avoid stray light and multiple reflection effects.

Ultrasonic sensors mainly make use of reflected energy from the target or time of flight of the ultrasonic beam between the transmitter and receiver. They operate on diffuse-scan or through-scan modes. The diffuse-scan mode uses a single transducer as transmitter and receiver, whereas through-scan mode uses two transducers. In proximity detection applications, using an ultrasonic frequency of 215 kHz is an industrial norm. The transmitted signal is modulated, using frequencies in the range of 30 to 360 Hz to minimize interference.

Pressure Sensors

Pressure is defined as the normal component of the force exerted on the unit surface area of an object. Most pressure sensors operate on the principle of converting pressure to some form of physical displacement, such as compression of a spring or deformation of an object. If displacement is used, there are many options for converting displacement into electrical signals. However, because of heavy reliance on physical displacement, many pressure sensors are sensitive to vibration and shock, which introduce a drawback to their use in many applications.

Pressure can be sensed indirectly or directly. Indirect methods rely on the action of pressure to cause displacement of a diaphragm, piston, or other device, so that the electronic sensing of displacement can be related to the pressure. The most commonly employed method is indirect pressure sensing.

Indirect pressure sensors are commonly known as aneroid barometers. There are many different versions of aneroid barometers, defined by the sensing elements of the electrical components. Figure 158.30 illustrates a semiconductor, piezoresistive indirect-pressure sensor. In this sensor, *p*-type doping material is introduced into an *n*-type silicon diaphragm using ion implementation technology. Another method of sensing pressure is using piezoelectric crystals such as the PZT. These sensors are useful in measuring pressure that varies in time.

Force Sensors

Force sensors make use of physical behavior of the body under external force. Force can be measured by the following methods:

- Balancing the force against a standard mass through a system of levers
- Measuring the acceleration of a known mass

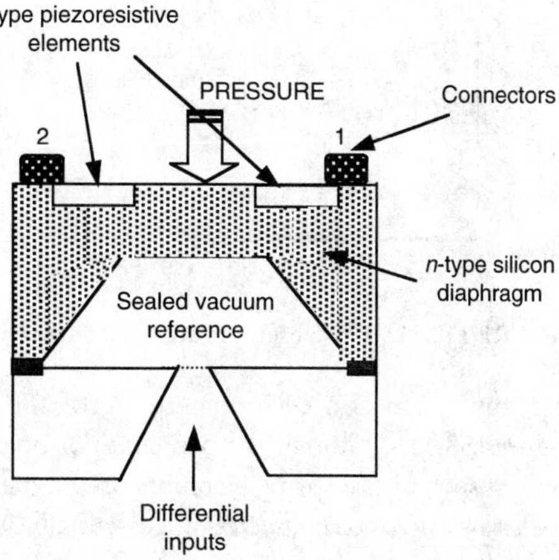

FIGURE 158.30 Semiconductor piezoresistive pressure sensor.

- Equalizing the force to magnetic force by suitable coil and magnet arrangements
- Measuring the pressure caused by the force on a surface
- Converting the force into the deformation of an elastic element

Consequently, a vast range of sensors is suitable for force measurements:

- Accelerometers
- Strain gauge load cells
- Piezoelectric sensors
- Resistive sensors
- Force sensing polymers
- Magnetoresistive sensors
- Inductive sensors
- Magnetoelastic sensors
- Piezotransistors
- Capacitive sensors

The basic operational principles of most of these sensors are explained above. Because of diverse range and large volume of applications, only selected examples of accelerometers are discussed below.

Accelerometers

Acceleration sensors, also called accelerometers, are extensively used in absolute motion measurements and vibration and shock sensing. Modern accelerometers are quite advanced; they are manufactured as intelligent sensors, providing good examples of advanced IC technologies. Accelerometers find a diverse range of applications and they are commercially available in a wide variety of ranges and types to meet specific requirements. They are manufactured to be small in size, light in weight, and rugged and robust to operate in harsh environments. They can be configured as active or passive sensors. An active accelerometer (e.g., piezoelectric) provides an output without the need for an external power supply, while a passive accelerometer only changes its electric properties (e.g., capacitance) and requires an external electrical power source.

Accelerometers can be classified in a number of ways, such as deflection or null-balance types, mechanical or electrical types, and dynamic or kinematic types. The majority of industrial accelerometers can be classified as deflection or null-balance types. Accelerometers used in vibration and shock measurements are usually the deflection types, whereas those used for measuring vehicle motion (automotive, aircraft, and others) for navigation purposes may be of either type. In general, null-balance types are used when extreme accuracy is needed.

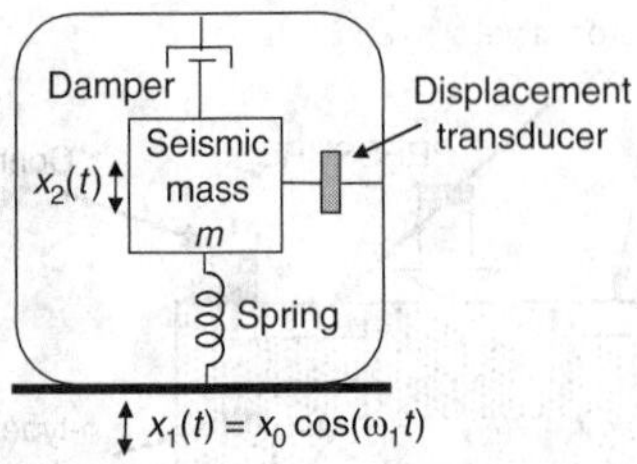

FIGURE 158.31 Typical seismic accelerometer.

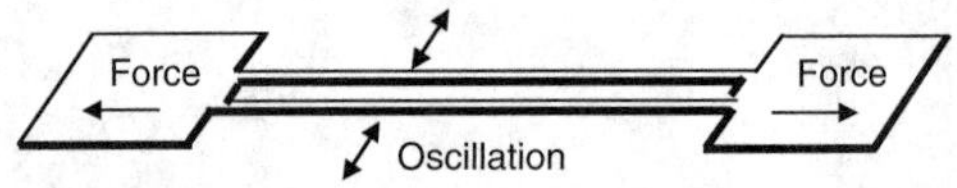

FIGURE 158.32 Double-ended tuning fork accelerometer.

Many practical accelerometers are of the deflection type; the general configuration is shown in Figure 158.31. Although the principles of operation are similar, deflection-type models differ in minor details, such as the spring elements used, types of damping provided, and types of relative motion.

Piezoresistive accelerometers are essentially semiconductor strain gauges with large gauge factors. High gauge factors are obtained since the material resistivity is dependent primarily on the stress, not only on the dimensions of the device. This effect can be greatly enhanced by appropriate doping of semiconductors such as silicon. The increased sensitivity is critical for vibration measurements since it allows miniaturization of the accelerometer. Most piezoresistive accelerometers use two or four active gauges arranged in a Wheatstone bridge form. Extra-precision resistors are used, as part of the circuit, in series with the input to control the sensitivity, balancing, and offsetting the temperature effects.

Piezoelectric accelerometers are used widely for general-purpose acceleration, shock, and vibration measurements. They are basically motion transducers with large output signals and comparatively small size. These accelerometers are available with very high natural frequencies, and are therefore suitable for high-frequency applications and shock measurements.

Microaccelerometers come in many different sizes and shapes. An example of microaccelerometers is the vibrating-beam accelerometer, also termed resonant-beam force transducer. This type is made in such a way that acceleration along a positive input axis places the vibrating beam in tension. Thus, the resonant frequency of the vibrating beam increases or decreases with the applied acceleration. A mechanically coupled beam structure, also known as a double-ended tuning fork (DETF), is shown in Figure 158.32.

In DETF, an electronic oscillator capacitively couples energy into two vibrating beams to keep them oscillating at their resonant frequency. The beams vibrate 180° out of phase to cancel reaction forces at the ends. The dynamic cancellation effect of the DETF design prevents energy from being lost through the ends of the beam. Hence, the dynamically balanced DETF resonator has a high Q factor, which leads to a stable oscillator circuit. The acceleration signal appears as an output from the oscillator in the form of a frequency-modulated square wave, which can be used in a digital interface.

Stress and Strain Sensors

Strain is a fractional change in the dimensions of an object. An example of strain is the change in the length of an object divided by its original length; therefore, strain does not have any dimensions. Stress, on the other hand, is the force applied on an object divided by the area. Consequently, strain takes place as a result of stress. Tensile stress is applied force divided by the area over which it is applied, whereas bulk stress is force per unit of area. The most common strain sensors are produced for tensile strain measurements. Strain measurement also allows calculation of the amount of stress by using knowledge of the object's elasticity, using elastic modules such as Young's modulus, shear (twisting) modulus, and bulk (pressure) modulus in the calculations.

Sensing tensile strain involves the measurement of small changes in the length of a sample. The commonest form of strain measurement uses resistive strain gauges, semiconductor strain gauges, piezoelectric strain gauges, and optical interferometry. The basic principles behind the resistive, semiconductor, and piezoelectric types are explained elsewhere in this chapter. An advanced method for measuring the strain is achieved by interferometry and fiber optic sensors. In these methods, generally high sensitivities are achieved when lasers are used. A laser consists of a coherent light beam resulting in a good interference effect. Other methods involve various types of fiber optic sensors.

Density Sensors

Density of a substance is defined as the mass per unit volume ($\rho = m/V$) under fixed conditions. The term is applicable to solids, liquids, and gases. Density depends on temperature and pressure, which is much greater in gases. Although there are many different units in use, usually density values are given in terms of grams per cubic centimeter.

There is no single universally applicable density measurement technique available. Different methods must be employed for solids, liquids, and gases. The measurement of fluid densities is much more complex than for solids; therefore, diverse techniques have been developed. Hydrometers, pycnometers, hydrostatic weighing, flotation methods, drop methods, radioactive methods, and optical methods are examples of liquid density measurement approaches. Flask methods, gas balance methods, optical methods, and X-ray methods are typical techniques employed for gas density measurements. Some of these sensors are briefly discussed below.

Magnetic methods are used for both liquids and gases. These methods allow the determination of effects of pressures and temperatures down to the cryoscopic range. Basically, these devices contain a small ferromagnetic cylinder encased in a glass jacket. The jacket and ferromagnetic material combination constitutes a buoy or float. The cylinder is held at a precise height within the medium by means of solenoid that is controlled by a servo system integrated with a height sensor. The total magnetic force on the buoy is the product of the induced magnetic moment by the solenoid and the field gradient in the vertical direction. The total magnetic force at a particular distance in the vertical direction in the solution compensates for the difference in the opposing forces of gravity (downward) and buoyancy (upward) exerted by the medium through the Archimedes principle. The magnetic force is directly proportional to the square of the current in the solenoid. If the buoyant force is sufficient to make the ferromagnetic assembly float on the liquids of interest, the force generated by the solenoid must be downward to add to the force of gravity for equilibrium.

Vibrational methods make use of the changes in natural frequency of vibration of a body containing or surrounded by fluid. The natural frequency of the vibrating body is directly proportional to the stiffness and inversely proportional to the combined masses of the body and the fluid. It also depends on the shape size and elasticity of materials, induced stresses, and the total mass and mass distribution of the body. Vibrational methods include vibrating tube densitometers, vibrating cylinder densitometers, and tuning fork densitometers, all of which make use of the natural frequency of a low-mass tuning fork. In some cases, the liquid or gas is taken into a small chamber in which the electromechanically driven forks are situated. In the other cases, the fork is inserted direct into the liquid. Calibration is necessary in each application.

Hydrometers are direct reading instruments, and are the most commonly used devices for measuring liquid density. They are so common that specifications and procedures of use are described by national and international standards such as ISO 387. The buoyancy principle is used as the main technique of operation. Almost all hydrometers are made from high-grade glass tubing. The volume of fixed mass is converted to a linear distance by a sealed-bulb, shaped glass tube containing a long-stem measurement scale. The bulb is ballasted with a lead shot and pitch, the mass of which is dependent on the density range of the liquid to be measured. The bulb is simply placed into the liquid and the density is read from the scale. The scale may be graduated in density units such as kilograms per cubed meter. Hydrometers can be calibrated at different ranges for surface tensions and temperatures.

Absorption techniques are also used for density measurements in specific applications. X-rays, visible light, ultraviolet (UV), and sonic absorptions are typical examples of this method. Essentially, attenuation and phase shift of a generated beam going through the sample are sensed and related to the sample density. Most absorption-type densitometers are custom designed for applications having particular characteristics. Two typical examples are UV absorption or X-ray absorption used for determining local densities of mercury deposits in arc discharge lamps; and ultrasonic density sensors used in connection with difficult density measurements such as density measurement of slurries. Lime slurry, for example, is a very difficult material to handle. It has a strong tendency to settle out and to coat all equipment it comes into contact with. An ultrasonic density control sensor can fully be emerged into agitated slurry, thus avoiding the problems of coating and clogging.

Rotation Sensors

Sensing and measurement of rotational movement are necessary in many industrial applications since many industrial machines have rotating shafts. The simplest form of sensing angular velocity is by using AC or DC generators, commonly known as *tacho-generators.*

In some applications, signals are generated for each revolution of the wheel or shaft by optical or piezoelectric or magnetic-pulsing arrangements. These pulses enable the determination of the angular velocities as well as positions of shafts or wheels.

Angular methods for displacement measurements can be grouped to be small or large angular measurements. Small displacements are measured by the use of capacitive, strain gauge, or piezo sensors. For larger displacements, potentiometric or inductive techniques are more appropriate. Rotary digital encoders also find wide applications.

References

Bentley, J. P. 1988. *Principles of Measurement Systems*, 2nd ed. Wiley, New York.

Brignell, J. and N. White. 1996. *Intelligent Sensor Systems*, rev. ed. Institute of Physics Pub., Bristol, UK.

Brooks, T. 2001. Wireless technology for industrial sensor and control networks, Proceedings of the First ISA/IEEE Sensors for Industry Conference, pp. 73–77.

Coombs, C. F. Jr. 2000. *Electronic Instrument Handbook*, 3rd ed. McGraw-Hill, New York.

Dunbar, M. 2001. Plug-and-play sensors in wireless networks. *IEEE Instrum. Meas. Mag.*, March, pp. 19–23.

Dyer, S. A., Ed. 2001. *Survey of Instrumentation and Measurement*, John Wiley & Sons, New York.

Eren, H., 1999. Inductive displacement sensors. In: Webster, J. G., Ed., *The Measurement, Instrumentation, and Sensors Handbook.* CRC Press, Boca Raton, FL.

Eren, H. 1999. Capacitive sensors. In: Webster, J. G., Ed., *The Measurement, Instrumentation, and Sensors Handbook.* CRC Press, Boca Raton, FL.

Eren, H. 1999. Acceleration, vibration and shock. In: Webster, J. G., Ed., *The Measurement, Instrumentation, and Sensors Handbook.* CRC Press, Boca Raton, FL.

Eren, H. 1999. Capacitance and capacitance measurement. In: Webster, J. G., Ed., *The Measurement, Instrumentation, and Sensors Handbook.* CRC Press, Boca Raton, FL.

Eren, H. 2000. Density measurement. In: Webster, J. G., Ed., *Mechanical Variables Measurement: Solid, Fluid, and Thermal.* CRC Press, Boca Raton, FL.

Eren, H. 2001. Instruments. In: Dyer, S. A., Ed., *Survey of Instrumentation and Measurement.* Wiley, New York, pp. 1–14.

Eren, H. 2001. Magnetic sensors. In: Dyer, S. A., Ed., *Survey of Instrumentation and Measurement.* Wiley, New York, pp. 46–60.

Eren, H. 2001. Displacement measurement. In: Dyer, S. A., Ed., *Survey of Instrumentation and Measurement.* Wiley, New York, pp. 509–521.

Eren, H. 2001. Acceleration measurements. In: Bishop, R. H., Ed., *The Mechatronics Handbook.* CRC Press, Boca Raton, FL.

Ewing, R. L. and A. A. Zohdy. 2001. Design Approach for Biomedical Smart Sensors. Proceedings of the 44th Symposium on Circuits and Systems, vol. 1, pp. 474–477.

Ferrari, P., A. Flaminni, D. Marioli, and A. Taroni. 2002. A Low-Cost Smart Sensor with Java Interface. IEEE Sensors for Industry Conference Proceedings, Houston, TX, pp. 161–167.

Ferri, G., M. Faccio, G. Stochino, A. D'Amico, D. Rossi, and G. Ricotti. 1999. A very low voltage bipolar op-amp for sensor applications. *Analogue Integrated Circuits Signal Process.*, 20:11–23.

Filipov A., N. Srour, and M. Falco. 2002. Networked Microsensor Research at ARL and the ASCTA. IEEE Sensors for Industry Conference Proceedings, Houston, TX, pp. 212–218.

Frenzel, L. E. 2002. After a slow start, Bluetooth shows its colors. *Electron. Design*, 50:68–74.

Gardner, J. L. 1994. *Microsensors: Principles and Applications*. Wiley, New York.

Gotz, A., I. Gracia, J. A. Plaza, C. Cane, P. Roetsch, H. Bottner, and K. Seibert. 2001. Novel methodology for the manufacturability of robust CMOS semiconductor gas sensor arrays. *Sensors Actuators*, B77:395–400.

Lee K. 2000. IEEE 1451: A Standard in Support of Smart Transducer Networking. Proceedings of the 17th IEEE Instrumentation and Measurement Technology Conference, vol. 2, pp. 525–528.

Lee, K. B. and R. D. Schneeman. 2000. Distributed measurement and control based on the IEEE 1451 smart transducer interface standards. *IEEE Trans. Instrum. Meas.*, vol. 49(3):621–627.

Lipták, B. 2002. *Instrument Engineers' Handbook*, 3rd ed. CRC Press, Boca Raton, FL.

Pallás-Areny, R. and J. G. Webster. 2001. *Sensors and Signal Conditioning*, 2nd ed., John Wiley & Sons, New York.

Potter, D. 2002. Smart plug and play sensors. *IEEE Instrum. Meas. Mag.*, 5:28–30.

Randy, F. 2000. *Understanding Smart Sensors*, 2nd ed. Artech House, Boston.

Soloman, S. 1999. *Sensors Handbook*. McGraw-Hill, New York.

Sze, S. M. 1994. *Semiconductor Sensors*. Wiley, New York.

Usher, M. J. and D. A. Keating. 1996. *Sensors and Transducers: Characteristics, Applications, Instrumentation, Interfacing*, 2nd ed. MacMillan, London.

Webster, J. G., Ed. 2000. *Mechanical Variables Measurement: Solid, Fluid, and Thermal*. CRC Press, Boca Raton, FL.

Webster, J. G., Ed. 1999. *Wiley Encyclopedia of Electrical and Electronics Engineering*, John Wiley & Sons, New York.

Webster, J. G., Ed. 1999. *The Measurement, Instrumentation, and Sensors Handbook*, CRC Press, Boca Raton, FL.

Wise, K. D., Ed. 1998. Special issue on integrated sensors, microactuators, and microsystems (MEMS). *Proc. IEEE*, 86:1531–1811.

159

AC Electrokinetics of Particles

Michael Pycraft Hughes
University of Surrey

Kai F. Hoettges
University of Surrey

Fatima H. Labeed
University of Surrey

Henry O. Fatoyinbo
University of Surrey

Certain types of particles, when subjected to an electric field, will polarize; charges associated with given particles will move to form a dipole or higher-order pole. These poles then interact with the electric field and generate electrostatic forces. If the field is nonuniform, and the particle carries no net charge, the greater electric field strength across one side of the particle means that the forces generated by Coulombic interactions between charge and field will be greatest on the charges where the field is highest, so that the particle will experience a net force toward or away from the field gradient, according to the sign of the charge where the field is greatest. Since the dipole is induced, reversing the orientation of the field will also reverse the orientation of the dipole, so that the direction of force is independent of the direction of the applied electric field. This force is known as dielectrophoresis (shown schematically in Figure 159.1), and was first reported by Herbert Pohl in 1951 [1], although previous observations of the effect do exist [2].

The orientation of the dipole in particles on the scale to which dielectrophoresis is most often applied — from tens of nanometers up to hundreds of micrometers — is usually governed by the relative contributions of charge induced in the dipole formed in the particle, and the relative amount of countercharge formed at the interface of the particle, supplied from the material that the particle is suspended in (typically an aqueous electrolyte). The relative amounts of charge supplied to this interface by particle and medium will affect the net charge of the dipole and hence the direction of the dipole, which in turn governs the direction of the force, as shown in Figure 159.2. If there is greater charge supplied from the particle (i.e., it is more polarizable than the medium), the dipole will oppose the field and the particle will be attracted up the field gradient. If the medium is more polarizable than the particle, then the dipole will align with the field and the particle will be repelled and move down the field gradient. These phenomena are termed "positive dielectrophoresis" and "negative dielectrophoresis," respectively. Since the polarizibility of the particle and medium are dependent on the electrical impedance of those materials (and hence conductance and capacitance), which are frequency dependent, the net dielectrophoretic behavior of the particle will depend on the frequency of the applied electric field; the magnitude and direction of the force on a given particle can change with the frequency of the applied electric field. Homogeneous particles typically exhibit one *dielectric dispersion*, a frequency region where the dielectric behavior of the particle changes from being medium dominated to being particle dominated. Complex dielectric particles containing more than one dielectric material can exhibit more than

0-8493-1586-7/05/$0.00+$1.50

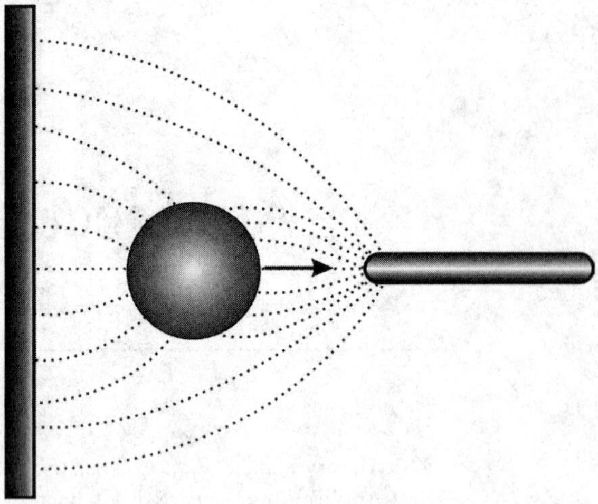

FIGURE 159.1 A schematic of a polarizable particle suspended within a point-plane electrode system. When the particle polarizes, the interaction between the dipolar charges and the local electric field produces a force. Due to the inhomogeneous nature of the electric field, the force is greater in the side facing the point than that on the side facing the plane, and there is net motion toward the point electrode. This effect is called *positive dielectrophoresis.* If the particle is less polarizable than the surrounding medium, the dipole will align counter to the field and the particle will be repelled from the high field regions, which is called *negative dielectrophoresis.*

one transition between behaviors, such as exhibiting negative dielectrophoresis at low and high frequencies with a window of positive dielectrophoresis in between. While this variation in behavior is the main advantage in using the AC rather than DC field, there are other advantages including a reduction in the effects of electrode screening and an elimination of DC (electrophoretic) forces due to net charge on the particle.

Pohl's work advanced the use of dielectrophoresis for investigating the properties of suspensoids, and for providing a means of separating particles from suspension. Similar investigations have been conducted using a frequency-based examination of dielectrophoretic response of cell populations [3,4], yeast [5,6], and bacteria [7], including work by Nobel laureate Albert Szent-Györgyi. Practical applications of dielectrophoresis have included the collection of cells for cellular fusion in biological experiments [8–10]. The use of positive and negative dielectrophoresis has been used to separate mixtures of viable and nonviable yeast cells [5,11] and mixtures of healthy and leukemic blood cells [12]. Work by Rousselet et al. [13] and others applied dielectrophoresis to the induction of continuous linear motion of particles, expanding on the basic concept of dielectrophoresis as a means of trapping particles in a specific region in space. An important class of electrokinetic particle manipulator is the levitator — a device used to propel a particle against gravity — resulting in it hovering in mid-solution (or mid-air) at a height governed by its dielectric properties, allowing those properties to be measured, and allowing those particles to be selected and trapped [14,15]. Early experiments used electric fields generated by (relatively) large electrodes and high voltages to trap particles; more recently, electrode structures have been fabricated using techniques borrowed from the computer industry [6,11,13] to manipulate much smaller particles at much lower voltages.

A related electrokinetic effect is observed when the electric field contains a *phase* gradient instead of a *magnitude* gradient. Although this phenomenon produces quite different particle behavior to dielectrophoresis, the two are closely related in origin [16,17], and produces phenomena where particles rotate *in situ* (electrorotation) or move along the direction of increasing phase gradient (traveling-wave dielectrophoresis). In these phenomena, the fundamental principle is the same; the dipole takes a finite time (relaxation time) to form, by which time the electric field has rotated slightly. There is thus a lag between the orientation of the electric field and that of the dipole moment, and a torque or force is induced as the dipole moves to reorient itself with the electric field. Due to the continuous rotation of the electric

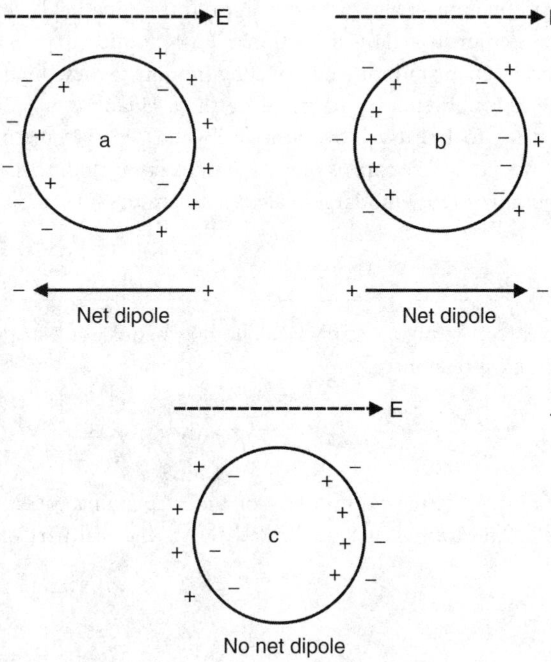

FIGURE 159.2 A lossy dielectric sphere in a lossy dielectric medium, and exposed to an electric field, will experience interfacial polarization. The amount of charge on each side of the interface will depend on the relative capacitances and conductances of the particle and medium, leading to a net dipole across the particle. (a) Particle more polarizable than medium. (b) Particle less polarizable than medium. (c) Particle and medium equally polarizable.

field, the torque is induced continually and the cell rotates, in the case of electrorotation, or translates, in the case of traveling-wave dielectrophoresis. The direction of movement is determined by the displacement between the dipole moment and the electric field. For example, in electrorotation, if the phase lag is less than 180°, the particle rotation will follow that of the applied field, referred to as co-field rotation. If the phase angle is greater than 180°, the shortest path for the dipole to align with the electric field is by rotating in a contrary fashion to that of the electric field, and hence particle rotation will act in this direction (anti-field rotation). As with dielectrophoresis, the rate and direction of cell rotation are related to the dielectric properties of both the particle and the suspending medium. Traveling-wave dielectrophoresis is effectively an extension of the principle of electrorotation to include a linear case. An AC electric field is generated which travels linearly along a series of electrodes. Particles suspended within the field establish dipoles which, due to the relaxation time, are displaced from the regions of high electric field. This induces a force in the particle as the dipole moves to align with the field. If the dipole lags within half a cycle of the applied field net motion acts in the direction of the applied field, while a lag greater than this results in motion counter to the applied field.

Cell rotation was observed and reported by experiments on AC dielectrophoresis [18], and was later suggested to be the result of the dipole-dipole interaction of neighboring cells [19]. This led Arnold and Zimmermann [20] to the principle of suspending single particles in a rotating field, and thus to a more amenable means of studying the phenomenon. Electrorotation has been used to study the dielectric properties of matter, such as the interior properties of biological cells and biofilms [21,22]. Traveling-wave dielectrophoresis was first reported by Batchelder [23] and subsequently by Masuda et al. [24,25], where the electric fields "travel" along a series of bar-shaped electrodes where low-frequency (0.1 Hz to 100 Hz) sinusoidal potentials, advanced 120° for each successive electrode, were applied. This was found to induce controlled translational motion in lycopodium particles [25] and red blood cells [17]. At low frequencies, the translational force was largely electrophoretic, and it was proposed that such traveling fields could eventually find application in the separation of particles according to their size or electrical charge. However, later work by Fuhr et al. [26], using applied traveling fields at much higher frequency ranges (10 kHz to 30 MHz), demonstrated the induction of linear motion in pollen and cellulose particles, and demonstrated that the mechanism inducing traveling motion at these higher frequencies is dielectrophoretic, rather than electrophoretic, in origin. Since then, Huang et al. [27] and others have, for example, used traveling fields to linearly move yeast cells and separate them from a heterogeneous population of yeast and bacteria. A large corpus of work now exists, including theoretical studies [27,28], devices for electrostatic pumping [29], and large-scale cell separators [30].

159.1 Theory

The dielectrophoretic force, F_{DEP}, acting on a spherical, homogeneous body suspended in a local electric field gradient is given by the expression:

$$F_{DEP} = 2\pi r^3 \varepsilon_m Re[K(\omega)]\nabla E^2 \tag{159.1}$$

where r is the particle radius, ε_m is the permittivity of the suspending medium, ∇ is the Del vector (gradient) operator, and E is the *rms* electric field. $Re[K(\omega)]$ is the real part of the Clausius–Mossotti factor, given by

$$K(\omega) = \frac{\varepsilon_p^* - \varepsilon_m^*}{\varepsilon_p^* + 2\varepsilon_m^*} \tag{159.2}$$

where ε_m^* and ε_p^* are the complex permittivities of the medium and particle, respectively; and $\varepsilon^* = \varepsilon - \frac{j\sigma}{\omega}$ with σ the conductivity, ε the permittivity, and ω the angular frequency of the applied electric field. The limiting (DC) case of Equation (159.2) is

$$K(\omega = 0) = \frac{\sigma_p - \sigma_m}{\sigma_p + 2\sigma_m} \tag{159.3}$$

The frequency dependence of $Re[K(\omega)]$ indicates that the force acting on the particle varies with the frequency. The magnitude of $Re[K(\omega)]$ varies depending on whether the particle is more or less polarizable than the medium. If $Re[K(\omega)]$ is positive, then particles move to regions of highest field strength (positive dielectrophoresis); the converse is negative dielectrophoresis where particles are repelled from these regions. The frequency dependence of $K(\omega)$, both real and imaginary parts, is shown in Figure 159.3 for a homogeneous, solid, spherical latex particle of diameter 216 nm in a dilute electrolyte.

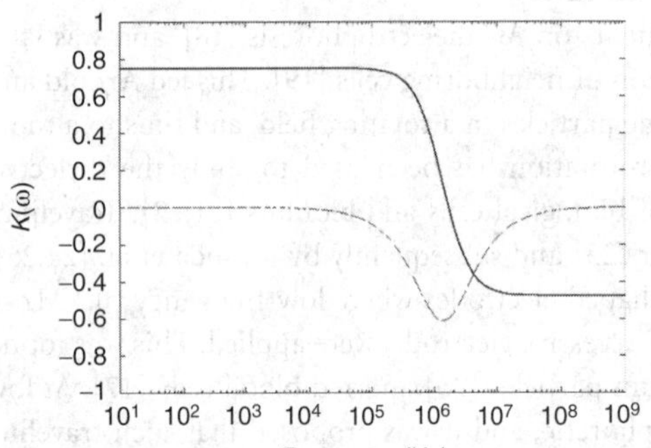

FIGURE 159.3 A plot of the real (solid line) and imaginary (dotted line) parts of the Clausius–Mossotti factor calculated for a 216-nm latex bead in a 1 mSm^{-1} solution, neglecting surface charge effects. The magnitude and signs of the real and imaginary parts govern the magnitude and direction of the dielectrophoretic force and electrorotational torque, respectively.

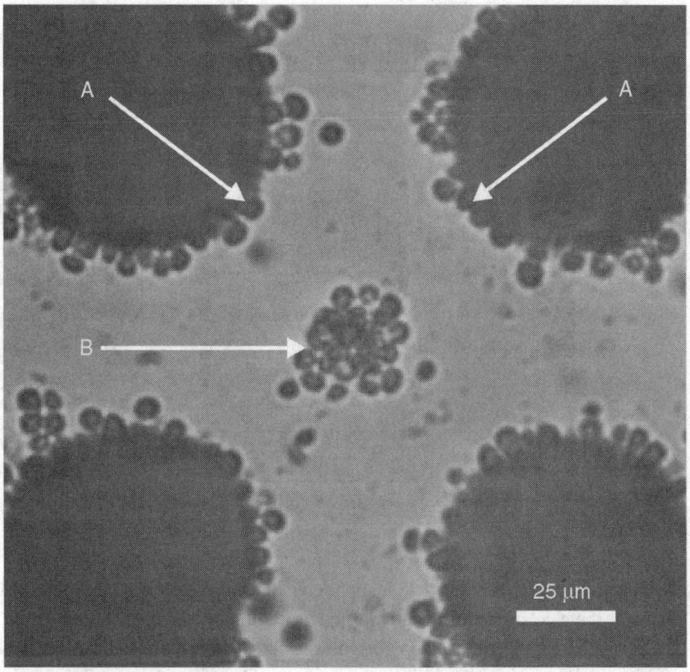

FIGURE 159.4 A mixture of live and dead (heat-treated) yeast cells undergoing dielectrophoresis in a quadrupolar electrode array; the live cells are collected at regions of high electric field (A) at the electrode edges, while the dead cells are repelled from these regions and collect in the interelectrode space (B).

By careful construction of the electrode geometry that creates the electric field, it is possible to create electric field morphologies so that potential energy minima are bounded by regions of increasing electric field strengths. An example of yeast cells in such an array (often called a *quadrupolar* or *polynomial* electrode array) is shown in Figure 159.4. In such electrodes, particles experiencing positive dielectrophoresis are attracted to the regions of highest electric field (typically the electrode edges, particularly where adjacent electrodes are close), while particles experiencing negative dielectrophoresis are trapped in an isolated field minimum at the center of the array.

The second type of AC electrokinetic phenomenon described above depends on the interactions between an *out-of-phase* dipole with a spatially moving electric field. Since an induced dipole may experience a force with both an in-phase and out-of-phase component simultaneously, the induced forces due to these components will be experienced at the same time, with the respective induced forces superimposed. If our particle is suspended in a rotating electric field, the induced dipole will form across the particle and should rotate in synchrony with the field. However, if the angular velocity of the field is sufficiently large, the time taken for the dipole to form (the *relaxation time* of the dipole) becomes significant and the dipole will lag behind the field. This results in a nonzero angle between field and dipole, which induces a torque in the body and causes it to rotate asynchronously with the field; the rotation can be with or against the direction of rotation of the field, depending on whether the lag is less or more than 180°. This phenomenon was called electrorotation by Arnold and Zimmermann [20], and is shown schematically in Figure 159.5. The general equation for time-averaged torque Γ experienced by a spherical polarizable particle of radius r suspended in a rotating electric field E is given by:

$$\Gamma = -4\pi\varepsilon_m r^3 Im[K(\omega)]\mathrm{E}^2 \tag{159.4}$$

where $Im[K(\omega)]$ represents the imaginary component of the Clausius–Mossotti factor shown in Figure 159.4. The minus sign indicates that the dipole moment lags the electric field. When viscous drag is accounted for, the rotation rate $\mathrm{R}(\omega)$ of the particle is given by

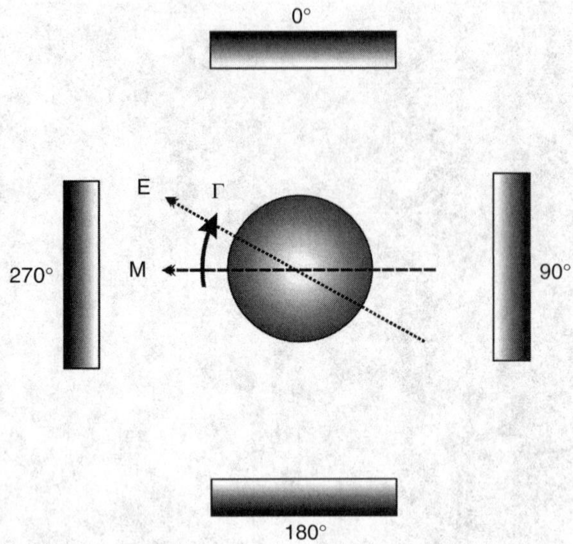

FIGURE 159.5 A schematic of a polarizable particle suspended in a rotating electric field generated by four electrodes with 90° advancing phase. If the electric field E rotates sufficiently quickly, the induced dipole M will lag behind the electric field by an angle related to the time taken for the dipole to form (the relaxation time). The interaction between the electric field and the lagging dipole induces a torque Γ in the particle, causing the particle to rotate. This effect is known as *electrorotation*.

$$R(\omega) = -\frac{\varepsilon_m Im[K(\omega)]E^2}{2\eta} \tag{159.5}$$

where η is the viscosity of the medium. Note that unlike the dielectrophoretic force Equation (159.1), the relationship with the electric field is as a function of the square of the electric field rather than of the *gradient* of the square of the electric field. Furthermore, the torque depends on the *imaginary* rather than the real part of the Clausius–Mossotti factor. A particle may experience both dielectrophoresis and electrorotation simultaneously, and the magnitudes and directions of both are related to the interaction between the dielectric properties of particle and medium; the relative magnitudes of force and torque are proportional to the real and imaginary parts of the Clausius–Mossotti factor.

Where the electric field has a phase gradient that moves in space rather than induces rotation, the particles within that field will experience a translational force rather than torque in particles, an effect known as *traveling wave* dielectrophoresis. This shares with dielectrophoresis the phenomenon of translational induced motion, but rather than acting toward a specific point (that of the highest electric field strength), the force acts to move particles *along* an electrode array in the manner of an electrostatic "conveyor belt" as shown in Figure 159.6.

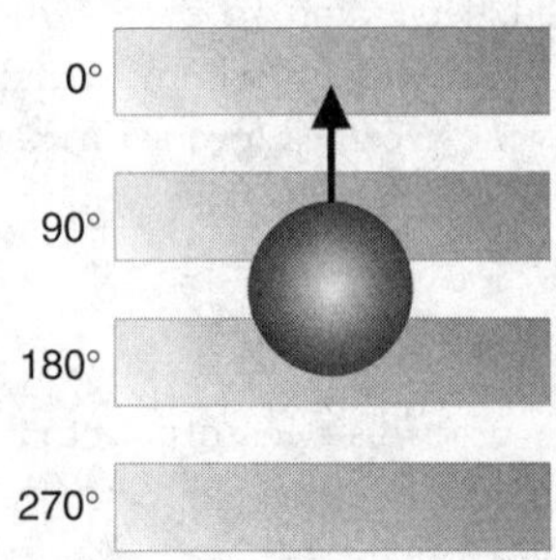

FIGURE 159.6 A schematic showing a polarisable particle suspended in a traveling electric field generated by electrodes on which the applied potential is 90° phase-advanced with respect to the electrode above. If the electric field moves sufficiently quickly, the induced dipole M will lag behind the electric field, inducing a force in the particle. This causes the particle to move along the electrodes, a phenomenon known as *traveling-wave dielectrophoresis.*

Consider a particle in a sinusoidal electric field that travels — that is, rather than merely changing magnitude, the field maxima and minima move through space, like waves on the surface of water. These waves move across a particle, and a dipole is induced by the field. If the speed at which the field crosses the particle is great enough, then there will be a time lag between the induced dipole and the electric field, in much the same way as there is an angular lag in a rotating field that causes electrorotation. This physical lag between dipole and field induces a force on the particle, resulting in induced motion; the degree of lag, related to the velocity (and hence the frequency) of the wave, will dictate the speed and direction of any motion induced in the particle. The underlying principle is closely related to electrorotation; it could be argued that the name "traveling wave dielectrophoresis" is misleading because the origin of the effect is not dielectrophoretic; that is, it does not involve the interaction of dipole and field gradient. Instead, the technique is a linear analogue of electrorotation, in a similar manner to the relationship between rotary electric motors and the linear electric motors used to power magnetically-levitated trains. As with the rotation of particles, the movement is asynchronous with the moving field, with rates of movement of 100 μmps being reported. The value of the force F_{TWD} is given by [27]:

$$F_{TWD} = \frac{-4\pi\varepsilon_m r^3 Im\left[K(\omega)\right]E^2}{\lambda} \tag{159.6}$$

where λ is the wavelength of the traveling wave, and is usually equal to the distance between electrodes that have signals of the same phase applied.

It is possible to extend the model of dielectric behavior to account for more complex particle structure. Two cases of this will be considered here; the first is that of nonhomogeneous (shelled) particles, and the second case is that of nonspherical ellipsoids. In the first case, we can extend the models described above by replacing the Clausius–Mossotti factor for a homogeneous sphere with a more complex term representing the many dielectric materials in a layered or "shelled" object (such as a cell, where the inner cytoplasm is enclosed by an enveloping membrane). The model most widely used now was developed by Irimajiri et al. [31], and works by considering each layer as a homogeneous particle suspended in a medium, where that medium is in fact the layer surrounding it. So, starting from the core we can determine the dispersion at the interface between the core and the layer surrounding it, which we will call shell 1. This combined dielectric response is then treated as a particle suspended in shell 2, and a second dispersion due to that interface is determined; then a third dispersion due to the interface between shells 2 and 3 is determined, and so on. In this way, the dielectric properties of all the shells combine to give the total dielectric response for the entire particle. This is illustrated schematically in Figure 159.7. In order to examine this mathematically, let us consider a spherical particle with N shells surrounding a central core. To each layer *i* we assign an outer radius a_i, with a_1 being the radius of the core and a_{N+1} being the radius of the outer shell (and therefore the radius of the entire particle). Similarly, each layer has its own complex permittivity given by

$$\varepsilon_i^* = \varepsilon_i - j\frac{\sigma_i}{\omega} \tag{159.7}$$

where i has values from 1 to N+1. In order to determine the effective properties of the whole particle, we first replace the core and the first shell surrounding it with a single, homogeneous "core." This new core has a radius a_2 and a complex permittivity given by

$$\varepsilon_{1eff}^* = \varepsilon_2^* \frac{\left(\frac{a_2}{a_1}\right)^3 + 2\frac{\varepsilon_1^* - \varepsilon_2^*}{\varepsilon_1^* + 2\varepsilon_2^*}}{\left(\frac{a_2}{a_1}\right)^3 - \frac{\varepsilon_1^* - \varepsilon_2^*}{\varepsilon_1^* + 2\varepsilon_2^*}} \tag{159.8}$$

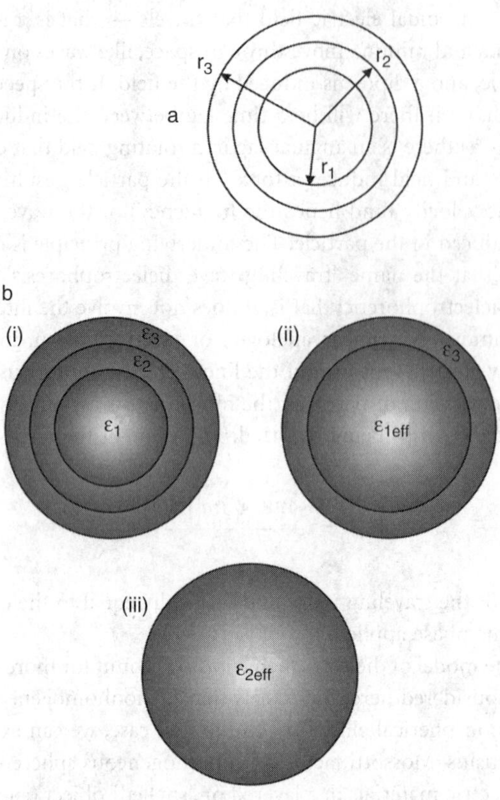

FIGURE 159.7 (a) A sphere comprises a core and inner and outer shells, with radii r1, r2, and r3. (b) These three layers have complex permittivity ε_1^*, ε_2^*, and ε_3^*. We can find the total polarizibility of the particle by successively combining the two innermost layers to find the effective combined complex permittivity (i to iii).

We now have a "core" surrounded by N – 1 shells. We then proceed by repeating the above calculation, but combining the "new" core with the second shell:

$$\varepsilon_{2eff}^* = \varepsilon_3^* \frac{\left(\dfrac{a_3}{a_2}\right)^3 + 2\dfrac{\varepsilon_{1eff}^* - \varepsilon_3^*}{\varepsilon_{1eff}^* + 2\varepsilon_3^*}}{\left(\dfrac{a_3}{a_2}\right)^3 - \dfrac{\varepsilon_{1eff}^* - \varepsilon_3^*}{\varepsilon_{1eff}^* + 2\varepsilon_3^*}} \tag{159.9}$$

If this procedure is repeated a further N – 2 times, then the final step will replace the final shell and the particle will be replaced by a single homogeneous particle with effective complex permittivity ε_{Peff}^* given by

$$\varepsilon_{Peff}^* = \varepsilon_{Neff}^* \frac{\left(\dfrac{a_{N+1}}{a_N}\right)^3 + 2\dfrac{\varepsilon_{N-1eff}^* - \varepsilon_{N+1}^*}{\varepsilon_{N-1eff}^* + 2\varepsilon_{N+1}^*}}{\left(\dfrac{a_{N+1}}{a_N}\right)^3 - \dfrac{\varepsilon_{N-1eff}^* - \varepsilon_{N+1}^*}{\varepsilon_{N-1eff}^* + 2\varepsilon_{N+1}^*}} \tag{159.10}$$

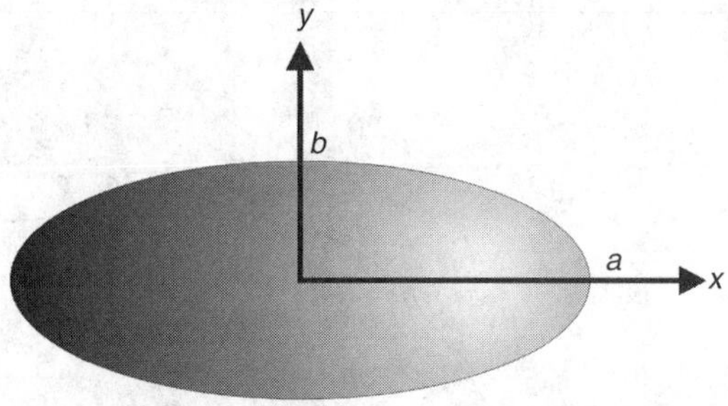

FIGURE 159.8 A schematic diagram of an elliptical particle, showing axes *x* and *y*, along which the particle extends by distances *a* and *b*. The particle extends along the *z* axis (beyond the page) by length *c*. If $c = b$, the particle is prolate; if $c = a$, the particle is oblate.

This value provides an expression for the combined complex permittivity of the particle at any given frequency ω. It can also be combined with the complex permittivity of the medium to calculate the Clausius–Mossotti factor, as demonstrated by Huang et al. [29] for yeast cells.

Moving to the behavior of ellipsoidal particles, we find that the dipole moment exerted along the different axes has different dispersion frequencies, so that for example, a prolate (football or rugby ball shaped) ellipsoid will have dispersion along its long axis of different relaxation frequency to the dispersion across its short (but equal) axes. The dispersion frequency of the dipole formed along the long axis will be of lower frequency than that formed across the shorter axes (because the charges have farther to travel from end to end in an alternating field), but the magnitude of the dipole formed will be greater due to the greater separation between the charges.

Consider an elliptical particle such as that shown in cross-section in Figure 159.8. It consists of two axes in projection, *x* and *y*, plus a third axis projecting from the page, *z*. The radii of the object along these axes are *a*, *b*, and *c*, respectively. It can be demonstrated [32,33] that the particle will undergo three dispersions at different frequencies according to the thickness of the ellipsoid along each axis. However, in addition to the dielectrophoretic force experienced by the particle, it will also experience a torque acting so as to align the longest nondispersed axis with the field. This phenomenon, often observed in practical dielectrophoresis, is *electro-orientation* [32,33]. When a nonspherical object is suspended in an electric field (for example, but not solely, when experiencing dielectrophoresis), it rotates such that the dipole along the longest nondispersed axis aligns with the field. Since each axis has a different dispersion, the particle orientation will vary according to the applied frequency. For example, at lower frequencies, a rod-shaped particle experiencing positive dielectrophoresis will align with its longest axis along the direction of electric field; the distribution of charges along this axis has the greatest moment and therefore exerts greatest torque on the particle to force it into alignment with the applied field. As the frequency is increased, the dipole along this axis reaches dispersion, but the dipole formed *across* the rod does not and the particle will rotate 90° and align perpendicular to the field. This smaller axis has a shorter distance for charges to travel between cycles and so the dispersion frequency will be higher; however, the shorter distance means the dipole moment is smaller. This will result in the force experienced by the particle being smaller in this mode of behavior.

When aligned with one axis parallel to the applied field, a prolate ellipsoid experiences a force given by the equation

$$F_{DEP} = \frac{2\pi abc}{3}\varepsilon_m Re\left[X(\omega)\right]\nabla E^2 \tag{159.11}$$

where

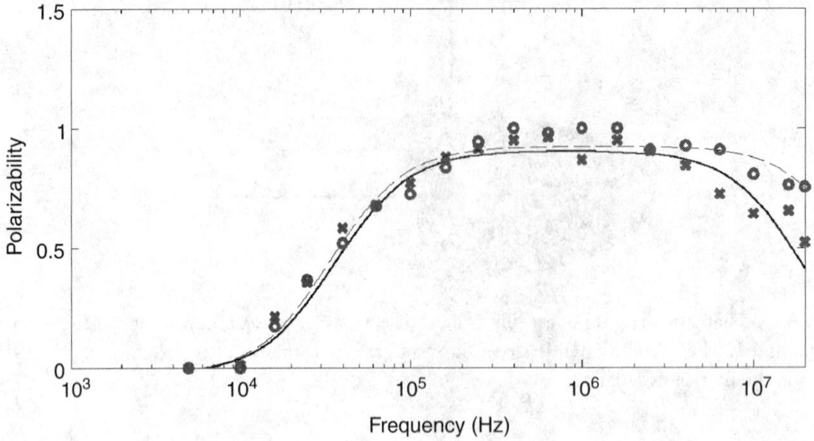

FIGURE 159.9 The DEP collection spectrum for cancer cell lines K562 (drug sensitive, open circles) and K562AR (drug resistant, closed circles) scaled to fit the polarization model. The simulation models are denoted by solid and broken lines, respectively. K562 collection starts from 10^4 Hz and starts to decrease at approximately 3 MHz; K562AR collection also starts around 10^4 Hz, but starts decreasing at about 8 MHz, indicating a higher internal conductivity.

$$X(\omega) = \frac{\varepsilon_p^* - \varepsilon_m^*}{(\varepsilon_p^* - \varepsilon_m^*)A_\alpha + \varepsilon_m^*} \tag{159.12}$$

where α represents the x, y, or z axis and A is the *depolarization factor*, which represents the different degrees of polarization along each axis and takes a value between 0 and 1.

159.2 Applications for Particle Characterization and Separation

As can be seen from Equations (159.1) and (159.2), the frequency-dependent dielectrophoretic response is governed by the Clausius–Mossotti factor. If we can determine the frequency-dependent variation of force or torque, we may then be able to directly determine the value of $K(\omega)$. In fact, this is only truly possible for solid homogeneous spheres or ellipsoids, for which a unique solution of Equation (159.2) exists. Where a particle consists of one or more shells surrounding a central core, the solution can often only be known when the dimensions of each layer are known accurately. Even then, a number of different possible outcome sets often exists, or the solution is insensitive to infinite variation of a particular parameter in the inner core, particularly where there is more than a single shell. In order to overcome this problem, it is often best to approximate the system to a single shell, such as modeling the response of a cell as a membrane surrounding the cytoplasm.

Analysis of the frequency-dependent value of $K(\omega)$ can be performed using dielectrophoresis, usually by measuring the rate at which a population of particles is collected at an electrode array at different frequencies. Examples of dielectrophoretic collection spectra are shown in Figure 159.9, where cancer cells (K562 leukemic cells) have been collected at an electrode edge. The plot shows two sets of data with accompanying best-fit lines according to Equation (159.8); one set was collected for drug-sensitive cells, the other for a drug-resistant counterpart. The model indicates that the resistant cell line has similar membrane properties to the parental line (as indicated by the similar behavior at low frequencies), but has a considerably higher cytoplasmic conductivity, as shown by the continued plateau of high collection extending to higher frequencies before beginning to reduce. The principal drawback with this method of dielectrophoretic data analysis is that it cannot be used to determine the response of a single cell; this would entail returning the cell to an identical position in the electrode array and applying fields of different frequency in order to monitor the force exerted on that cell. For the analysis of single cells the preferred electrokinetic method is electrorotation, since electrorotation does not exert a translational force on the particle and therefore the induced rotation as a function of frequency can directly give an indication of the imaginary part of the Clausius–Mossotti factor.

Dielectrophoretic methods have been used to analyze the properties of a wide range of particle types. In recent years, the majority of such research has been directed at the study of bioparticles such as cells and viruses, since AC electrokinetic methods offer the opportunity to probe the interior cytoplasm of cells noninvasively and to study the cell interior over time. Common cell types include yeast cells that have been examined in live and dead states [6], and that have been examined across the frequency spectrum to over 1 GHz [34]. Other common cell types include human cells such as blood cells [35] and cancers [36]. Since electrokinetic means are highly sensitive to interior changes, the method has allowed the study of changes occurring in oocytes following fertilization [37], in kidney cells following infection with Herpes viruses [38], and the discrimination among different forms of parasite spores [39]. Dielectrophoretic analysis also offers potential for the analysis of the electrical characteristics of nanoparticles [40,41].

It is possible that in a heterogeneous mixture of particles, groups of particles with different crossover frequencies will experience forces of different sign in the frequency window bounded by the crossover frequencies of the two populations, causing one population to be attracted to high field regions (by ascending the field gradient by positive dielectrophoresis), while the other is repelled (down the field gradient by negative dielectrophoresis) when exposed to an electric field of a particular frequency, allowing the particles to be separated. Similarly, particles with differences in dielectric makeup can be driven in opposite directions along a traveling electric field. Furthermore, a more heterogeneous mix of particles may be fractionated by exploiting small differences in polarizibility (and hence force), or by using dielectrophoretic force in conjunction with other factors such as imposed flow or particle diffusion. All these phenomena allow the physical separation of particles according to electrical or other characteristics using dielectrophoresis as the means by which separation is realized. Dielectrophoretic separation devices are presently receiving a great deal of attention, particularly for applications in cell analysis; this is because they have the ability to noninvasively discriminate between cells with identical membrane characteristics but different interior properties. Examples of cell separation by dielectrophoresis include the separation of cancerous and healthy erythrocytes [42], viable and nonviable cells [43], leukemic cells from blood cells [44], species of bacteria [45], and placental cells from maternal blood [46]. Technological advancements have allowed smaller electrode structures to be constructed, which have in turn shown that dielectrophoretic separation of nanoparticles is also possible. Examples of such separations have included those of 92-nm diameter latex beads with varying surface charge [47]; later studies showed that modification of the surface allowed detection of the presence of antibodies [48].

As with cells, so dielectrophoretic separation can be applied to nano-scale bioparticles; for example, different species of virus [49], or components of the same virus [50], have been separated on an electrode array. Furthermore, dielectrophoresis is sensitive enough to detect variations in the state of virus particles when treated with antibiotics [51], with potential applications for drug discovery. Finally, dielectrophoretic separation has also been applied to biological macromolecules such as proteins and DNA fragments [52] by fractioning a mixture of protein and DNA molecules by a combination of field flow and dielectrophoresis.

Many practitioners of dielectrophoresis assert that the most likely future role of these phenomena is in the development of devices called "laboratories on a chip" [53]. Laboratories on a chip are devices wherein the components of a modern laboratory — fluid handling, reactors, heaters, pumps, separators and sensors — are integrated in miniature onto a single device, typically of a size between that of a postage stamp and a credit card. The original "labs on chips" were miniaturized chemistry devices such as capillary electrophoresis columns and liquid chromatographs, but since then the term has expanded to include a wide range of devices from DNA microarrays to miniaturized chemical assays to dielectrophoretic particle sorters. Implementations vary, but a common presentation is to enclose the lab-on-a-chip in a cartridge format that is inserted into another unit containing the ancillary electronics (power supplies, signal generation, optics for analysis, and so forth) and liquid handling (reservoirs and pumps). The liquid sample to be analyzed is inserted either into the control unit, or into the cartridge itself, where it is processed and analyzed. From a commercial perspective, the reduction in size, with benefits for mass production, transportability, and ease of use, mean that such devices have great potential for point-of-care diagnostics; portable water-screening equipment; and rapid cell, protein, and DNA analysis for rapid drug discovery. Such devices could, for example, potentially allow the testing of blood samples at the patient's bedside, or at least provide the potential for cheap analysis of blood or urine samples on a hospital-by-hospital basis. Another application is for the identification of pathogens such as toxic viruses used for biological warfare. The integrated nature of these devices, combining many analysis methods, lends itself to another general term — micro-total analytical systems, abbreviated microTAS or μTAS.

Since cells are easily manipulated using dielectrophoresis, and it is possible to discriminate between cell types on the basis of both surface and interior properties, much of the work on dielectrophoretic laboratory-on-a-chip devices has been aimed at the development of particle detectors for medical applications. However, most lab-on-a-chip devices have features in common, the principal one being that they are fabricated using semiconductor methods and operate on small samples (of the order of microliters) using channels etched into glass, photoresist, or some other polymer, through which material is pumped from an external source.

There have been a number of approaches to the application of dielectrophoretic techniques to the laboratory-on-a-chip concept. Perhaps the first example of the use of microengineering techniques to construct a dielectrophoretic laboratory on a chip is the "dielectrophoretic fluid integrated circuit" described by Washizu et al. [54,55]. This device was able to move cells around microfabricated channels and sort them into different outlets. Some researchers have described attempts to perform a range of functions using dielectrophoresis, from separation to trapping and analysis [56]; others have used dielectrophoresis as part of a broader system including electroporation or biochemical methods for cell detection, and using dielectrophoresis primarily as a method for isolating specific cells at a preliminary stage [57].

A complete laboratory-on-a-chip device requires the integration of many subsystems in order to perform useful functions such as particle separation and analysis. A number of workers in the field have constructed such systems, either relying totally on dielectrophoretic methods of particle manipulation, or integrating other methods. Both approaches have advantages; using only dielectrophoresis allows the entire device to be fabricated in a single operation, and since dielectrophoresis requires no extra material beyond what is used (except some ancillary devices for detection fluid handling), operation of the device is easier. However, by restricting the device to all-dielectrophoretic methods, other possibilities of particle discrimination according to factors that do not affect electrical properties — such as the presence of a certain gene — cannot be used, whereas methods borrowed from molecular biology, such as flow cytometry and polymerase chain reaction (PCR) are sensitive to these types of factors.

Where the system relies largely on dielectrophoresis to provide many different functions, there are additional requirements that are placed on the dielectrophoretic system; for example, particles entering a separator should be guided into the device in an appropriate place (where electric field gradient is greatest) or time. An example of this was presented by Cui et al. [58] as part of a traveling-wave dielectrophoresis-based particle separator. Traveling wave methods offer the possibility of highly sensitive particle fractionation, but only if the particles start from one end of the array at the same time; if they drift onto the array at different times, fractionating different types of particles is impossible. In order to avoid this, Cui et al. employed an extra pair of individually addressable electrodes at the beginning of the array. When the traveling field is applied to the remainder of the electrodes, the phases applied to the trapping electrodes can be changed so that they become part of the traveling array, ensuring that all particles start at the same point. Once moving, particles can be detected using laser interrogation of the particles by splicing an optical fiber into the path of the particles, allowing both detection and identification by using fluorescent dyes.

Another proposal for a traveling-wave-based lab-on-a-chip was proposed by Pethig et al. in 1998 [56]. However, unlike the Cui design, traveling wave dielectrophoresis is employed principally as a means of transporting particles between inlets, analysis electrodes and dielectrophoretic traps. However, a novel idea is proposed: by using a forked junction in the traveling wave "conveyor belt," energized at two different frequencies, particles could be encouraged into one of the two forward paths according to their dielectrophoretic properties, thereby allowing separation. These junctions could theoretically shift particles toward different dielectrophoretic traps, analysis electrodes, or outlets.

Other workers have used conventional dielectrophoresis as a means of sorting and analyzing particles as they are pumped around the microsystem by using a conventional external pump. In order to organize particles into a single-file sequence for analysis, devices such as dielectrophoretic funnels and concentrators were used [59,60], constructed of triangular electrode arrays placed across the flow. Electrodes are placed at the top and bottom of the array and energized so as to repel the oncoming particles. This forces them toward the constriction at the end of the array, where the gap between the electrodes is small enough to let only a single particle pass. The device was shown to be effective for flow rates up to 3.5 mm/sec^{-1}, allowing high-speed processing of cells to be achieved. Subsequent arrays provide dielectrophoretic trapping, analysis, and switching, the latter process operating by energizing one or two electrodes by negative dielectrophoresis to selectively move the particles toward one of two outlets. Other workers in the field have sought to integrate dielectrophoretic manipulation with other methods. For example, the U.S. company Nanogen has integrated dielectrophoretic trapping of cells with hybridization techniques for bacterial analysis [61–63]. Dielectrophoresis is used to provide initial cell sorting to isolate the bacteria from blood cells, and then trap them on electrodes where a high voltage is used to break apart the bacteria. The DNA then being allowed to drift free, PCR is used to identify the type of bacterium. Other workers have used dielectrophoresis with laser trapping to sort cells, with unwanted cells being trapped by positive dielectrophoresis, and an automated tracking system used to identify bacteria and move them to an outlet port [63]. Finally, some researchers are attempting to move away from traditional planar (semi-two-dimensional) electrode configurations. Our group at Surrey has developed fully three-dimensional electrode arrays using a laminate construction to create electrodes on the scale of those used in conventional dielectrophoresis electrodes, but arranged into tubes; this dramatically increases the throughput of the device, making it possible to analyze many milliliters of analyte in a relatively short time. Furthermore, the device has the advantage of being inexpensive to make, and is not reliant on microengineering principles used for the majority of dielectrophoresis electrodes. An example of this type of array is shown in Figure 159.10.

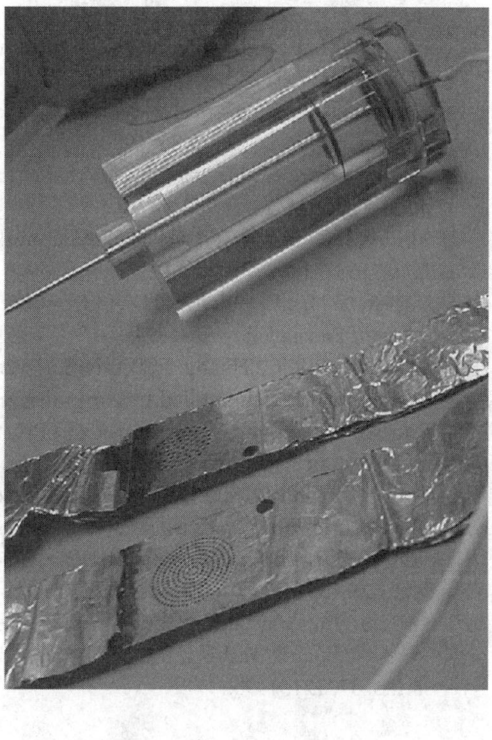

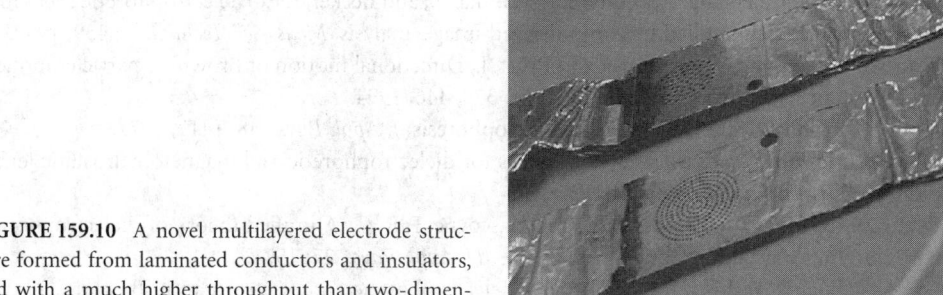

FIGURE 159.10 A novel multilayered electrode structure formed from laminated conductors and insulators, and with a much higher throughput than two-dimensional (planar) electrode arrays.

159.3 Conclusions

AC electrokinetic phenomena such as dielectrophoresis and electrorotation offer a number of benefits for manipulating particles on the micrometer and nanometer scales. In particular, the phenomena can be applied to the manipulation, separation, and analysis of bioparticles, including cells, viruses, and proteins, as well as allowing analysis of colloids and other particles.

Defining Terms

Dielectrophoresis — The manipulation of matter in nonuniform electric fields.
Electrorotation — The exertion of torque on dielectric particles in rotating electric fields.
Electrokinetics — Imparting forces using elecric fields.
Dielectric — Material in which an electrical dipole can be induced.

References

1. Pohl, H. A., The motion and precipitation of suspensoids in divergent electric fields, *J. Appl. Phys.*, 22, 869, 1951.
2. Hatschek, E. and Thorne, P. C. L., Metal sols in non-dissociating liquids. I. Nickel in toluene and benzene. *Proc. R. Soc.*, 103, 276, 1923.
3. Gascoyne, P. R. C., Pethig, R., Burt, J. P. H., and Becker, F. F., Membrane changes accompanying the induced differentiation of Friend murine erythroleukemia cells studied by dielectrophoresis, *Biochim. Biophys. Acta*, 1149, 119, 1993.
4. Kaler K. V. I. S. and Jones, T. B., Dielectrophoretic spectra of single cells determined by feedback-controlled levitation, *Biophys. J.*, 57, 173, 1990.
5. Pohl, H. A. and Hawk, I., Separation of living and dead cells by dielectrophoresis, *Science*, 152, 647, 1966.
6. Huang, Y., Hölzel, R., Pethig, R., and Wang X.-B., Differences in the AC electrodynamics of viable and non-viable yeast cells determined through combined dielectrophoresis and electrorotation studies, *Phys. Med. Biol.*, 37, 1499, 1992.
7. Hughes, M. P. and Morgan, H., Determination of bacterial motor force by dielectrophoresis, *Biotechnol. Prog.*, 245, 15, 1999.
8. Zimmermann, U., Vienken, J., and Scheurich, P., Electric field induced fusion of biological cells, *Biophys. Struct. Mech.*, 6, 86, 1980.
9. Zimmermann, U., Electric field-mediated fusion and related electrical phenomena, *Biochim. Biophys. Acta*, 694, 227, 1982.
10. Abidor, I. G. and Sowers, A. E., Kinetics and mechanism of cell-membrane electrofusion, *Biophys. J.*, 61, 1557, 1992.
11. Markx, G. H. and Pethig, R., Dielectrophoretic separation of cells: continuous separation, *Biotechnol. and Bioeng.*, 45, 337, 1994.
12. Gascoyne, P. R. C., Huang, Y., Pethig, R., Vykoukal, J., and Becker, F. F., Dielectrophoretic separation of mammalian cells studied by computerized image analysis, *Meas. Sci. Technol.*, 3, 439, 1992.
13. Rousselet, J., Salome, L., Ajdari, A., and Prost, J., Directional motion of Brownian particles induced by a periodic asymmetric potential, *Nature*, 370, 446, 1994.
14. Jones, T. B. and Bliss G. W., Bubble dielectrophoresis, *J. Appl. Phys.*, 48, 1412, 1977.
15. Lin, I. J. and Jones, T. B., General conditions for dielectrophoretic and magnetohydrostatic levitation, *J. Electrostatics*, 15, 53, 1984.
16. Wang, X.-B., Huang, Y., Becker, F. F., and Gascoyne, P. R. C., A unified theory of dielectrophoresis and traveling-wave dielectrophoresis, *J. Phys. D: Appl. Phys.*, 27, 1571, 1994.
17. Jones, T. B. and Washizu, M., Multipolar dielectrophoretic and electrorotation theory, *J. Electrostatics*, 37, 121, 1996.

18. Teixeira-Pinto, A. A., Nejelski, L. L., Cutler, J. L., and Heller, J. H., The behaviour of unicellular organisms in an electromagnetic field, *J. Exp. Cell Res.*, 20, 548, 1960.
19. Holzapfel, C., Vienken, J., and Zimmermann, U., Rotation of cells in an alternating electric-field — theory and experimental proof, *J. Membrane Biol.* 67, 13, 1982.
20. Arnold, W. M. and Zimmermann, U., Rotating-field-induced rotation and measurement of the membrane capacitance of single mesophyll cells of *Avena sativa, Z. Naturforsch.*, 37c, 908, 1982.
21. Arnold W. M. and Zimmermann, U., Electro-rotation: development of a technique for dielectric measurements on individual cells and particles, *J. Electrostatics*, 21, 151, 1988.
22. Zhou, X.-F., Markx, G. H., Pethig, R., and Eastwood, I. M., Differentiation of viable and non-viable bacterial biofilms using electrorotation, *Biochim. Biophys. Acta*, 1245, 85, 1995.
23. Batchelder, J. S., Dielectrophoretic manipulator, *Rev. Sci. Instrum.*, 54, 300, 1983.
24. Masuda, S., Washizu, M., and Iwadare, M., Separation of small particles suspended in liquid by nonuniform traveling field, *IEEE Trans. Ind. Appl.*, 23, 474, 1987.
25. Masuda, S., Washizu, M., and Kawabata, I., Movement of blood-cells in liquid by nonuniform traveling field, *IEEE Trans. Ind. Appl.*, 24, 217, 1988.
26. Fuhr, G., Hagedorn, R., Müller, T., Benecke, W., Wagner, B., and Gimsa, J., Asynchronous traveling-wave induced linear motion of living cells, *Stud. Biophys.*, 140, 79, 1991.
27. Huang, Y., Wang, X.-B., Tame, J., and Pethig, R., Electrokinetic behaviour of colloidal particles in travelling electric fields: studies using yeast cells, *J. Phys. D: Appl. Phys.*, 26, 312, 1993.
28. Hughes, M. P., Pethig, R., and Wang, X.-B., Forces on particles in traveling electric fields: computer-aided simulations, *J. Phys. D: Appl. Phys.*, 29, 474, 1996.
29. Fuhr, G., Schnelle, T., and Wagner, B., Travelling-wave driven microfabricated electrohydrodynamic pumps for liquids, *J. Micromech. Microeng.*, 4, 217, 1994.
30. Morgan, H., Green, N. G., Hughes, M. P., Monaghan, W., and Tan, T. C., Large-area travelling-wave dielectrophoresis particle separator, *J. Micromech. Microeng.*, 7, 65, 1997.
31. Irimajiri, A., Hanai, T., and Inouye, V., A dielectric theory of "multi-stratified shell" model with its application to lymphoma cell, *J. Theor. Biol.*, 78, 251, 1979.
32. Jones, T. B., *Electromechanics of Particles,* Cambridge University Press, Cambridge, 1995.
33. Hughes, M. P., *Nanoelectromechanics in Engineering and Biology,* CRC Press, Boca Raton, 2002.
34. Holzel, R., Electrorotation of single yeast cells at frequencies between 100 Hz and 1.6 GHz, *Biophys. J.*, 73, 1103, 1997.
35. Yang, J., Huang, Y., Wang X. J., Wang, X.-B., Becker, F. F., and Gascoyne P. R. C., Dielectric properties of human leukocyte subpopulations determined by electrorotation as a cell separation criterion, *Biophys. J.*, 76, 3307, 1999.
36. Huang, Y., Wang, X.-B., Holzel, R., Becker, F. F., and Gascoyne P. R. C., Electrorotational studies of the cytoplasmic dielectric-properties of friend murine erythroleukemia-cells, *Phys. Med. Biol.* 40, 1789, 1995.
37. Arnold, W. M., Schmutzler, R. K., Alhasani, S., Krebs, D., and Zimmermann, U., Differences in membrane-properties between unfertilized and fertilized single-rabbit oocytes demonstrated by electro-rotation: comparison with cells from early embryos, *Biochim. Biophys. Acta*, 979, 142, 1989.
38. Archer, S., Morgan, H., and Rixon, F. J., Electrorotation studies of baby hamster kidney fibroblasts infected with herpes simplex virus type 1, *Biophys. J.*, 76, 2833, 1999.
39. Gascoyne, P., Pethig, R., Satayavivad, J., Becker, F. F., and Ruchirawat, M., Dielectrophoretic detection of changes in erythrocyte membranes following malarial infection, *Biochim. Biophys. Acta*, 1323, 240, 1997.
40. Hughes, M. P., Morgan, H., and Flynn, M. F., The dielectrophoretic behaviour of submicron latex nanospheres: influence of surface conduction, *J. Coll. Int. Sci.*, 220, 454, 1999.
41. Hughes, M. P., Dielectrophoretic behaviour of latex nanospheres: low-frequency dispersion, *J. Coll. Int. Sci.*, 250, 291, 2002.
42. Gascoyne P. R. C., Huang, Y., Pethig, R., Vykoukal, J., and Becker, F. F., Dielectrophoretic separation of mammalian-cells studied by computerized image-analysis, *Meas. Sci. Technol.*, 3, 439, 1992.

43. Markx, G. H., Talary, M. S., and Pethig, R., Separation of viable and nonviable yeast using dielectrophoresis, *J. Biotechnol.*, 32, 29, 1994.
44. Becker, F. F., Wang, X.-B., Huang, Y., Pethig, R., Vykoukal, J., and Gascoyne, P. R. C., The removal of human leukemia-cells from blood using interdigitated microelectrodes, *J. Phys. D: Appl. Phys.*, 27, 2659, 1994.
45. Markx, G. H., Dyda, P. A., and Pethig, R., Dielectrophoretic separation of bacteria using a conductivity gradient, *J. Biotechnol.*, 51, 175, 1996.
46. Chan, K. L., Morgan, H., Morgan, E., Cameron, I. T., and Thomas, M. R., Measurements of the dielectric properties of peripheral blood mononuclear cells and trophoblast cells using AC electrokinetic techniques, *Biochim. Biophys. Acta,* 1500, 313, 2000.
47. Green, N. G. and Morgan, H., Dielectrophoretic separation on nano-particles, *J. Phys. D: Appl. Phys.*, 30, L41, 1997.
48. Hughes, M. P. and Morgan, H., Dielectrophoretic characterization and separation of antibody coated submicrometer latex spheres, *Anal. Chem.*, 71, 3441, 1999.
49. Morgan, H., Hughes, M. P., and Green, N. G., Separation of submicron bioparticles by dielectrophoresis, *Biophys. J.*, 77, 516, 1999.
50. Hughes M. P., Nanoparticle manipulation by electrostatic forces, in Goddard W., Breener D., Lyshevski S., Iafrate G. (Eds.) *The Nanoscience, Engineering and Technology Handbook*, CRC Press, Boca Raton, FL, 2002.
51. Hughes, M. P., Morgan, H., and Rixon, F. J., Measuring the dielectric properties of herpes simplex virus type 1 virions with dielectrophoresis, *Biochim. Biophys. Acta,* 1571, 1, 2002.
52. Washizu M., Suzuki, S., Kurosawa, O., Nishizaka, T., and Shinohara, T., Molecular dielectrophoresis of biopolymers, *IEEE Trans. Ind. Appl.,* 30, 835, 1994.
53. Effenhauser, C. S. and Manz, A., Miniaturizing a whole analytical laboratory down to chip size, *Am. Lab.,* 26, 15, 1994.
54. Washizu, M., Nanba, T., and Masuda, S., Novel method of cell fusion in field constriction area in fluid integrated circuit, *IEEE Trans. Ind. Appl.*, 25, 732, 1989.
55. Washizu, M., Nanba, T., and Masuda, S., Handling biological cells using a fluid integrated circuit, *IEEE Trans. Ind. Appl.*, 26, 352, 1990.
56. Pethig, R., Burt, J. P. H., Parton, A., Rizvi, N., Talary, M. S., and Tame, J. A., Development of biofactory-on-a-chip technology using excimer laser micromachining, *J. Micromech. Microeng.*, 8, 57, 1998.
57. Suehiro, J. and Pethig, R., The dielectrophoretic movement and positioning of a biological cell using a three-dimensional grid electrode system, *J. Phys. D: Appl. Phys.*, 31, 3298, 1998.
58. Cui, L., Holmes, D., and Morgan, H., The dielectrophoretic levitation and separation of latex beads in microchips, *Electrophoresis*, 22, 3893, 2001.
59. Fiedler, S., Shirley, S. G., Schnelle, T., and Fuhr, G., Dielectrophoretic sorting of particles and cells in a microsystem, *Anal. Chem.*, 70, 1909, 1998.
60. Müller, T., Gradl, G., Howitz, S., Shirley, S.,Schnelle, T., and Fuhr, G., A 3D microelectrode system for handling and caging single cells and particles, *Biosens. Bioelectron.*, 14, 247, 1999.
61. Cheng, J., Sheldon, E. L., Wu, L., Uribe, A., Gerrue, L. O., Carrino, J., Heller, M. J., and O'Connell, J. P., Preparation and hybridization analysis of DNA/RNA from E-coli on microfabricated bioelectronic chips, *Nat. Biotechnol.*, 16, 541, 1998.
62. Cheng, J., Sheldon, E. L., Wu, L., Heller, M. J., and O'Connell, J. P., Isolation of cultured cervical carcinoma cells mixed with peripheral blood cells on a bioelectronic chip, *Anal. Chem.*, 70, 2321, 1998.
63. Arai, F., Ichikawa, A., Ogawa, M., Fukuda, T., Horio, K., and Itoigawa, K., High-speed separation systems of randomly suspended single living cells by laser trap and dielectrophoresis, *Electrophoresis*, 22, 283, 2001.

160
Biomedical Engineering

Joseph Bronzino
Trinity College

Today, many of the problems confronting health professionals are of extreme interest to engineers because they involve the design and practical application of medical devices and systems, which are processes that are fundamental to engineering practice. These medically related design problems can range from very complex large-scale constructs, such as the design and implementation of automated clinical laboratories, multiphasic screening facilities (i.e., centers that permit many clinical tests to be conducted), and hospital information systems, to the creation of relatively small and "simple" devices, such as recording electrodes and biosensors, that may be used to monitor the activity of specific physiologic processes in either a research or clinical setting. They encompass the many complexities of remote monitoring and telemetry, including the requirements of emergency vehicles, operating rooms, and intensive care units. Today these problems also include bioprocessing systems, bioinformatics, and biorobotics critical to exploit the information provided by the human genome project. The U.S. healthcare system, therefore, encompasses many problems that represent challenges to *biomedical engineers.*

Although what is included in the field of biomedical engineering is considered by many to be quite clear, there are some disagreements about its definition. For example, consider the terms *biomedical engineering, bioengineering,* and *clinical* (or medical) *engineering*, which have been defined in Pacela (1992). While bioengineering has been defined as the broad umbrella term used to describe this entire field, bioengineering *is* usually defined as a basic research-oriented activity closely related to biotechnology and genetic engineering, that is, the modification of animal or plant cells, or parts of cells, to improve plants or animals or to develop new microorganisms for beneficial ends. In the food industry, for example, this has meant the improvement of strains of yeast for fermentation. In agriculture, bioengineers may be concerned with the improvement of crop yields by treatment of plants with organisms to reduce frost damage. It is clear that bioengineers of the future will have a tremendous impact on the quality of human life. The potential of this specialty is difficult to imagine. Consider the following activities of bioengineers:

- Development of improved species of plants and animals for food production
- Invention of new medical diagnostic tests for diseases
- Production of synthetic vaccines from clone cells
- Bioenvironmental engineering to protect human, animal, and plant life from toxicants and pollutants
- Study of protein–surface interactions
- Modeling the growth kinetics of yeast and hybridoma cells
- Research in immobilized enzyme technology
- Development of therapeutic proteins and monoclonal antibodies

0-8493-1586-7/05/$0.00+$1.50

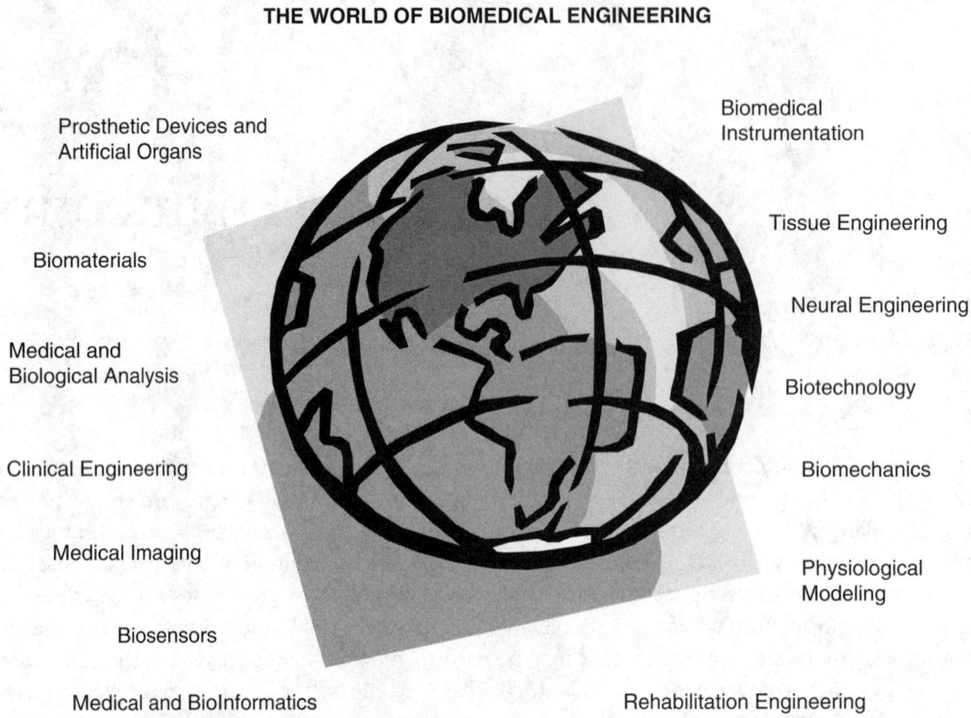

FIGURE 160.1 The world of biomedical engineering.

160.1 Activities of Biomedical Engineers

Biomedical engineers, on the other hand, apply electrical, mechanical, chemical, optical, and other engineering principles to understand, modify, or control biologic (i.e., human and animal) systems, as well as design and manufacture products that can monitor physiologic functions and assist in the diagnosis and treatment of patients. When biomedical engineers work within a hospital or clinic, they are more properly called clinical engineers.

The breadth of activity of biomedical engineers is significant (Figure 160.1). The field has moved significantly from being concerned primarily with the development of medical devices in the 1950s and 1960s to include a more wide-ranging set of activities. The field of biomedical engineering now includes many new career areas, including the following:

- Application of engineering system analysis (physiologic modeling, simulation, and control) to biologic problems
- Detection, measurement, and monitoring of physiologic signals (i.e., biosensors and biomedical instrumentation)
- Diagnostic interpretation via signal-processing techniques of bioelectric data
- Therapeutic and rehabilitation procedures and devices (rehabilitation engineering)
- Devices for replacement or augmentation of bodily functions (artificial organs)
- Computer analysis of patient-related data and clinical decision making (i.e., medical informatics and artificial intelligence)
- Medical imaging, that is, the graphic display of anatomic detail or physiologic function
- The creation of new biologic products (i.e., biotechnology and tissue engineering)

Typical pursuits of biomedical engineers, therefore, include:

- Research in new materials for implanted artificial organs

- Development of new diagnostic instruments for blood analysis
- Computer modeling of human heart function
- Writing software for analysis of medical research data
- Analysis of medical device hazards for safety and efficacy
- Development of new diagnostic imaging systems
- Design of telemetry systems for patient monitoring
- Design of biomedical sensors for measurement of human physiologic systems variables
- Development of expert systems for diagnosis of disease
- Design of closed-loop control systems for drug administration
- Modeling of the physiologic systems of the human body
- Design of instrumentation for sports medicine
- Development of new dental materials
- Design of communication aids for the handicapped
- Study of pulmonary fluid dynamics
- Study of the biomechanics of the human body
- Development of material to be used as replacement for human skin

Biomedical engineering, then, is an interdisciplinary branch of engineering that ranges from theoretical, nonexperimental undertakings to state-of-the-art applications. It can encompass research, development, implementation, and operation. Accordingly, like medical practice itself, it is unlikely that any single person can acquire expertise that encompasses the entire field. Yet, because of the interdisciplinary nature of this activity, there is considerable interplay and overlapping of interest and effort between them. For example, biomedical engineers engaged in the development of biosensors may interact with those interested in prosthetic devices to develop a means to detect and use the same bioelectric signal to power a prosthetic device. Engineers engaged in automating the clinical chemistry laboratory may collaborate with those developing expert systems to assist clinicians in making decisions based on specific laboratory data. The possibilities are endless.

Perhaps a greater potential benefit occurring from the use of biomedical engineering is identification of the problems and needs of our present healthcare system that can be solved using existing engineering technology and systems methodology. Consequently, the field of biomedical engineering offers hope in the continuing battle to provide high-quality health care at a reasonable cost; if properly directed toward solving problems related to preventive medical approaches, ambulatory care services, and the like, biomedical engineers can provide the tools and techniques to make our healthcare system more effective and efficient. For detailed examples of specific research and development activities in this field, see Bronzino (2000).

References

Pacela, A. F., Biomedical Careers Book: Careers in Biomedical or Clinical Engineering, Technology, Supervision and Management, Quest Pub. Co., Brea, CA, 1992.

Bronzino, J. D., *The Biomedical Engineering Handbook*, 2nd ed. CRC Press, Boca Raton, FL, 2000.

XXIII

Surveying

161

Quality Control

N. W. J. Hazelton
The Ohio State University

Boudewijn H. W. van Gelder
Purdue University

Surveyors collect measurements that are often critical in supporting the needs and decision-making processes of others. They serve the general public by providing information about property lines, and they serve the engineer by providing a proper stakeout of a subdivision, the distances between pillars of a bridge or the fine placement of machinery in a paper mill. They may provide a state department with an accurate location for a global positioning system (GPS) base station, to which the department ties all its geographic information system (GIS) data. A surveyor may provide geologists with regularly remeasured distances between two monuments on each side of a fault zone in order to monitor movement along that fault zone, or may provide the Federal Aviation Authority with proper coordinates of GPS beacons that aid automated landings of commercial aircraft. A surveyor may monitor the location of large-scale physical experimental apparatus, such as cyclotrons, to ensure that the system is meeting design specifications. In addition, a surveyor provides local departments of natural resources with proper heights for the mapping of flood plains.

Measurement science is one of the main foundations of surveying. Measurements rarely stand by themselves; they are usually part of a larger set, a measurement system. Design of the measurement system — from deciding what results are needed, through deciding on the particular measurements and how they are to be made, to knowing the quality of the measurements and the results and making sure that the required quality is achieved — is a fundamental part of a surveyor's work.

An area of measurement science that is currently considered very little in surveying literature is that of attribute measurement. With the growth of GIS, the collection of attribute data, nonspatial data that describe what is at a location, is becoming a larger part of a surveyor's work. How well can tree species, vegetation type, soil type, land value, land use, and innumerable other attributes be determined? The quality of these measurements has a significant impact on the quality of GIS and the decision making that GIS support. In many cases, working with these data is more difficult than the spatial measurement

0-8493-1586-7/05/$0.00+$1.50

data, as statistical tools for the latter are well developed and well understood. We cannot ignore the impact of the quality of nonspatial data in a GIS world.

The importance of the positions, heights, distances, or attributes resulting from various measurements makes the surveyor very careful about the quality of the measurement data, as well as the quality of the results computed from those measurements. Quality control is a central feature of measurement science.

161.1 Assumptions

When a surveyor measures a quantity, such as a distance or an angle, there are some important basic assumptions involved in the measurement processes. As these can have a significant impact on the quality of the results, both actual and perceived, we need to think about these assumptions. The most important assumptions are about independent measurements and the types and distribution of measurement errors.

Most raw measurements that surveyors collect are independent of all other measurements that are being made. A distance measured along one line has no relationship to a distance measured along a different line; an angle at one point is independent of an angle at another point. As such, the raw measurements are usually independent or uncorrelated with each other. But this is not always the case.

As technology becomes more complex with the aim of simplifying the measurement process, the independence assumption may no longer apply to the measurement values that a complex system produces. GPS measurements are very highly correlated to each other (both a point's coordinates and vectors between points). Total station results (e.g., coordinates, horizontal distances) often show correlations, owing to the internal processing of raw measurements. So we must be very careful about how we model our measurements.

For measurements that are independent, there is usually a distribution of random errors, which follows the normal or Gaussian distribution. This is the usual situation for most measurements, but derived quantities may have different distributions. For example, variances follow a chi-square (χ^2) distribution.

We must also define our terminology. Two terms often causing confusion are *accuracy* and *precision*. By precision we mean the repeatability of a measurement. If I measure a distance several times, what is the spread or range of measurements? We often use the standard deviation statistic to tell us something about precision. A large spread or standard deviation suggests poor precision, whereas a small spread or standard deviation suggests good precision. Related to this is how finely we can measure the quantity, but this is a very limited way of looking at precision, as it focuses at just one small part of the situation. By accuracy we mean how close a measurement is to the "true" or "real" value. But we can never know the true value; instead we can only approximate it by measuring the quantity, thereby incurring some error. So the term "accuracy" isn't used a great deal in its common sense, as it does not really tell you anything. However, it is used in a specialized and restricted way, as described in Section 161.6.

While precision can be interpreted as *repeatability*, we would like to narrow the definition of this term. We can consider repeatability to be how well I can repeat the measurement with the same equipment and in much the same circumstances, at about the same time. This will be the case if I measure a distance or angle several times at the same time I am at the site.

We can define the term *reproducibility* as a measure of how well someone else is able to measure the quantity, usually with different equipment, in different conditions and at a different time. This is important in work that must be repeated over time, such as deformation surveys, or that provides data for later work, such as control surveys. The up-to-date-ness (or currency) of the work is important here, as well as what changes might be happening to the objects being measured.

161.2 Errors

Even the most careful surveyor, like anybody else, makes errors — errors in judgment and, not necessarily more important, measurement errors. Some errors cannot be controlled or known at the moment of making the measurement. Some errors cannot be avoided, but should and can be foreseen and a measurement system designed accordingly. To distinguish among a variety of classes of errors (and this is not a game of semantics), one speaks of the precision of a measurement, the accuracy of a measurement, the reliability of a measurement, and, last but not least, the "up-to-dateness" of a measurement. In this respect, we have to view a measurement in a very loose manner: we may want to address the accuracy of the measurement itself or the accuracy of the variable directly or indirectly related to the actual measurement. In subsequent sections, each type will be addressed.

The classification of errors that will be used in this chapter is often found in the statistical literature (see Papoulis and Unnikrishna [2001], among many others) and the surveying literature:

- Stochastic or random errors
- Systematic errors, including periodic errors
- Blunders or gross errors

The usefulness of this classification is that it suggests causes, and thus ways to deal with and minimize or eliminate the different types of error.

Blunders or gross errors are usually caused by carelessness and can be kept to a minimum by relentless checking (particularly by building independent check measurements into the measurement system) and repeated measurements. For example, we may write down a digit incorrectly or sight to the wrong point. If we have the proper checks built into our entire measurement system, we would expect to detect these errors.

Systematic errors are the result of a mismatch between the model of the measurement process being used to understand and interpret the measurement and the reality of the measurement. For example, I may measure a distance between two points, but neglect to keep the tape horizontal. My model (that I am measuring a horizontal distance) and the reality (that I am measuring a slope distance) are not in agreement. An important characteristic of systematic errors is that a better model of the measurement system can be used to eliminate or minimize them.

Stochastic or random errors are intrinsic to the measurement process and cannot be eliminated by measuring more finely. As the name suggests, they have random individual behavior, but collectively they tend to follow definite statistical patterns. Because of this, we can model their general behavior and take steps to minimize their impact. As a simple example, I measure a distance several times, each measurement having some random error in it. If I compute the mean of the various measurements, this will give me a better estimate of the true value than any individual measurement could have done.

It is important to design the measurement system so that blunders and systematic errors are eliminated, or minimized to the point where these errors are indistinguishable from the stochastic or random errors. The statistical adjustment theory used to minimize the effect of the stochastic or random errors usually assumes that there are only random errors present, and can produce unreliable results if this assumption is not true.

The one constant in our world is change. We know that the very bedrock we always used for long-term stability of marks moves measurably over time, but many other things in the world change much more quickly. Nonspatial, descriptive, or attribute data can change suddenly, such as the owner of a property, or slowly over time, such as the salinity of groundwater. We need to include a time component in our work that acknowledges the fact of change, and if possible a rate of change. GPS Continuously Operating Reference Stations (CORS) include estimates of their movement over time in their general station descriptions, and it is important that this information is available for all types of survey data.

Stochastic errors and blunders first come to mind when talking about precision. Systematic errors are often viewed as an issue of accuracy. Reliability addresses all three types of errors. There are cases where a measurement is very precise but not accurate. In other cases, the measurement or related outcome may be very precise and accurate, but not reliable. In still other cases, a measurement or the resulting parameter may be very precise, very accurate, and very reliable, but not up-to-date, so that the overall quality is still low.

For nonspatial data that are measured, we may find that they do not follow a Gaussian error distribution, as they may be categorical or some other form of data. As Chrisman (2002) has pointed out, there are about ten different basic types of nonspatial data that may have to be dealt with, each requiring different statistical tools. Section 161.9 deals briefly with nonspatial data, but it is important to be aware that this is another major field of data collection and brings with it a new range of errors.

161.3 Precision

Precision in the classical sense deals with the repeatability of an experiment. A surveyor measures a distance over and over again, and the outcome differs each time. These are errors of a stochastic nature, over which the surveyor has no control. If the readings give the same outcome, the surveyor can make them a little more precise, for instance by reading the tape not to the nearest centimeter but to the nearest millimeter, to make the errors (deviations) appear more obviously random. The surveyor may compute the arithmetic mean of ten distance measurements, according to

$$\hat{\mu} = \frac{\sum_{i=1}^{n} l_i}{n} \tag{161.1}$$

where $\hat{\mu}$ is the estimate of the true mean μ based on the sample of n observations l_i.

Based on this sample of ten distance measurements l_i, with $i = 1, \ldots, 10$ ($n = 10$), the surveyor gets an estimate of the mean distance $\hat{l}$. Collecting an additional ten distance measurements would undoubtedly give a different mean. This is why we say that the calculated mean $\hat{l}$ is just an estimate of the mean based on that sample. So, for this example, we have

$$\hat{l} = \frac{\sum_{i=1}^{10} l_i}{10} = \frac{l_1 + l_2 + l_3 + \cdots + l_{10}}{10} \tag{161.2}$$

In statistics this average is known as the first moment.

A second interesting quantity to look at is the spread of the n observations. If the n observations are close together, we may call our measurement process precise. If the n values vary over a large range, we tend to think of them as the result of a less precise measurement process. To compute the spread, we compute the variations with respect to the estimated mean $\hat{l}$. We subtract the estimated mean from each individual observation to get a residual v_i:

$$v_i = l_i - \hat{l} \tag{161.3}$$

If we counted how many observations happen to fall within a specified range, we would obtain a histogram, as in Figure 161.1.

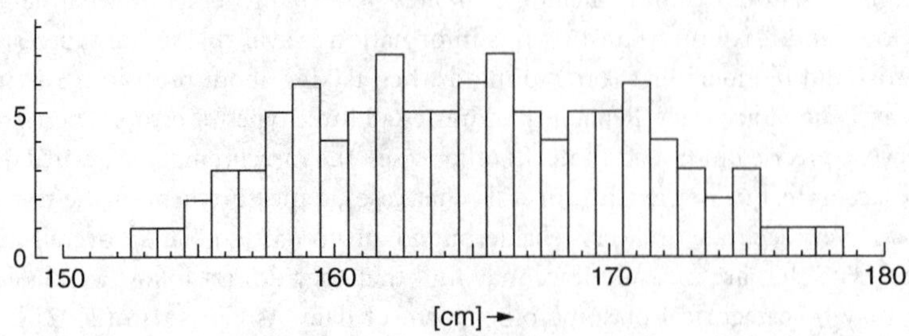

FIGURE 161.1 Histogram for n observations l_i.

If we add all the squared residuals, divide the sum by $n - 1$, and take the square root, we obtain an estimate of the standard deviation (we do not divide by n since we already extracted one piece of information from the sample of n observations: the estimate of the mean — and we used this estimate to compute the spread). The sample variance is the square of the sample standard deviation. We have for the sample standard deviation

$$\hat{\sigma} = \sqrt{\frac{\sum_{i=1}^{n} v_i^2}{n-1}} \tag{161.4}$$

and for the sample variance

$$\hat{\sigma}^2 = \frac{1}{(n-1)} \sum_{i=1}^{n} v_i^2 \tag{161.5}$$

In statistics the variance is also known as the second moment.

161.4 Law of Propagation of Errors

Often one is interested not so much in the standard deviation of the individual observations (say, those ten distance measurements), but in the standard deviation of the variable x computed from those n observations. Differentially small errors will propagate according to the rules of total and partial derivatives of multivariate calculus. An error dl_i in an observation l_i will propagate into the vector of variables x_j as an error dx_j according to

$$dx_j = \sum_{i=1}^{n} \frac{\partial x_j}{\partial l_i} dl_i \tag{161.6}$$

For all u variables x_j ($j = 1, \ldots, u$), we have

$$\begin{aligned} dx_1 &= \frac{\partial x_1}{\partial l_1} dl_1 + \cdots + \frac{\partial x_1}{\partial l_n} dl_n \\ dx_2 &= \frac{\partial x_2}{\partial l_1} dl_1 + \cdots + \frac{\partial x_2}{\partial l_n} dl_n \\ &\vdots \\ dx_u &= \frac{\partial x_u}{\partial l_1} dl_1 + \cdots + \frac{\partial x_u}{\partial l_n} dl_n \end{aligned} \tag{161.7}$$

or, in matrix form,

$$\vec{dx} = \mathbf{J}_{ji} \vec{dl} \tag{161.8}$$

where the matrix J_{ji} is the so-called Jacobian.

If we collect all the variances of the observations into a matrix V_l, the variances of the variables x_j are computed from

$$\begin{aligned} \mathbf{V}_x &= \vec{dx}\vec{dx}^{\mathrm{T}} \\ &= \mathbf{J}_{ji}\vec{dl}(\mathbf{J}_{ji}\vec{dl})^{\mathrm{T}} \\ &= \mathbf{J}_{ji}\vec{dl}\vec{dl}^{\mathrm{T}}\mathbf{J}_{ij} \\ &= \mathbf{J}\mathbf{V}_l\mathbf{J}^{\mathrm{T}} \end{aligned} \tag{161.9}$$

where J^T is the transpose of the Jacobian in Equation (161.8).

V_x represents the variance/covariance matrix of the variables. Consequently, for the observations l_i, the matrix V_l is

$$\mathbf{V}_l = \begin{bmatrix} \sigma_{l_1l_1} & \sigma_{l_1l_2} & \cdots & \sigma_{l_1l_n} \\ \sigma_{l_2l_1} & \sigma_{l_2l_2} & \cdots & \sigma_{l_2l_n} \\ \vdots & \vdots & \cdots & \vdots \\ \sigma_{l_nl_1} & \sigma_{l_nl_2} & \cdots & \sigma_{l_nl_n} \end{bmatrix} \tag{161.10}$$

In Equation (161.10), $\sigma_{l_il_i}$ is the variance, as in Equation (161.5). So we have

$$\sigma_{l_il_i} = \sigma_{l_i}^2 \tag{161.11}$$

In Equation (161.10) $\sigma_{l_il_j}$ is the covariance. From the covariance, the so-called correlation coefficient $\rho_{l_il_j}$ can be computed:

$$\rho_{l_il_j} = \frac{\sigma_{l_il_j}}{\sqrt{\sigma_{l_i}^2\sigma_{l_j}^2}} \tag{161.12}$$

The values of the correlation coefficient will range between −1 and 1. Values close to 1 indicate high positive correlation, or that an error in quantity l_i will cause an error of equal sign and size in quantity l_j. Likewise, values close to −1 indicate that an error in l_i will cause an error with opposite sign in quantity l_j. In general, we have

$$-1 \le \rho_{l_il_j} \le +1 \tag{161.13}$$

In surveying we strive for uncorrelated observations, meaning that $\sigma_{l_il_j}$ (and so $\rho_{l_il_j}$) is equal to zero. This results in a diagonal variance/covariance matrix V_l for the observations.

Example: Standard Deviation of the Sample Mean

To compute the standard deviation of the sample mean $\hat{\mu}$, we have to apply the law of propagation of errors. For the error in the sample mean, we have from Equation (161.1) and Equation (161.6),

$$d\hat{\mu} = \sum_{i=1}^{n} \frac{\partial\hat{\mu}}{\partial x_i} dx_i = \frac{1}{n}\sum_{i=1}^{n} dx_i \tag{161.14}$$

This leads to the variance of the sample mean

$$\hat{\sigma}_{\hat{\mu}}^2 = n\frac{1}{n^2}\hat{\sigma}_{x_i}^2 = \frac{1}{n}\hat{\sigma}_{x_i}^2 \tag{161.15}$$

or, for the standard deviation of the sample mean,

$$\hat{\sigma}_{\hat{\mu}} = \sqrt{\frac{\sum_{i-1}^{n} v_i^2}{n(n-1)}} \tag{161.16}$$

This tells us that the standard deviation of the mean of a set of observations is smaller than the standard deviation of the individual measurements from which it was derived, by a factor of the square root of the number of observations:

$$\hat{\sigma}_{\hat{\mu}} = \frac{1}{\sqrt{n}}\hat{\sigma}_{x_i} \tag{161.17}$$

The implication is that to improve the precision of our measurements, we can make more of them. Four times as many observations should reduce the standard deviation of the mean by half, but to improve the standard deviation by an order of magnitude (make it one-tenth of what it was before) would require 100 times as many observations. There is a law of diminishing returns at work here.

Monte Carlo Simulation

The above discussion of the theory of error propagation assumes that determining the Jacobian associated with the measurement process being modeled is fairly straightforward. In some cases involving complex processes, this may not be the case. An alternative is to consider Monte Carlo simulation for a simplified form of error propagation analysis.

With Monte Carlo simulation of error propagation, we take the measurement model and randomly assume a value for each observation, based on actual observation and expected precision. A random number generator is typically used for this purpose, with the results scaled to the appropriate precision and applied to the actual observation. We repeat the process some reasonable number of times, generating different values of each of the observations, based on the statistical behavior of the random errors associated with each observation, and using the random number generator. Modern spreadsheets often include random number generators that can simulate a number of different distributions, and it is very simple to generate a thousand normally distributed values based on the actual observation (as the mean) and the measurement precision (as the standard deviation) and apply these values to the measurement model.

The result is a large number of answers for the derived value, which have a statistical pattern of behavior. The standard deviation can be determined, and the surveyor can see if the specifications for precision of the derived quantities have been met, for example. It is also possible to adjust the various precisions of individual measurements, and thus use this technique as a planning tool to design the measurement system.

In complex measurement environments, it is possible to find points that are particularly sensitive, in that a small change in one variable can lead to a large change in the results. Sometimes this sensitivity is widespread; other times it is restricted to a small range of values. Monte Carlo simulation allows the surveyor to run a wide range of values through the proposed measurement system and undertake a sensitivity analysis of the measurement system as part of the design process.

Monte Carlo simulation is not a replacement for proper error propagation analysis, but it can be a very useful approximate tool for the measurement system design process.

161.5 Statistical Testing

We can test the individual observation against an arbitrary number or against the sample mean. Since the possible outcomes of an experiment, including the values of their sample means and sample standard deviations, are limitless, we work with normalized observations l_i^n:

$$l_i^n = \frac{l_i - \hat{\mu}}{\hat{\sigma}} \tag{161.18}$$

The advantage of the normalized observations is that the expected outcome of the mean is equal to zero, and the standard deviation of the normalized observations is equal to 1. Tables are available in the statistical literature to test any normalized observation against any value.

The probability that an observation lies in the interval $[a, b]$ is readily computable follows:

$$\begin{aligned} P(a \le l_i \le b) &= P\left(\frac{a-\hat{l}}{\hat{\sigma}} \le \frac{l_i - \hat{l}}{\hat{\sigma}} \le \frac{b-\hat{l}}{\hat{\sigma}}\right) \\ &= P\left(\frac{a-\hat{l}}{\hat{\sigma}} \le l_i^n \le \frac{b-\hat{l}}{\hat{\sigma}}\right) \\ &= P\left(l_i^n \le \frac{b-\hat{l}}{\hat{\sigma}}\right) - P\left(l_i^n \le \frac{a-\hat{l}}{\hat{\sigma}}\right) \end{aligned} \tag{161.19}$$

Assuming that the standard deviation of a distance measurement is 5 cm and the sample mean is 38.95 m, the probability that an individual observation will fall between 38.90 and 39.10 m is

$$\begin{aligned} P(38.90 \le l_i \le 39.05) &= P\left(\frac{38.90-38.95}{0.05} \le \frac{l_i - 38.95}{0.05} \le \frac{39.05-38.95}{0.05}\right) \\ &= P(-1 \le l_i^n \le +2) \\ &= P(l_i^n \le +2) - P(l_i^n \le -1) \\ &= 0.9772 - 0.1587 = 0.8185 \\ &= 81.85\% \end{aligned} \tag{161.20}$$

Similarly, the probability that the sample mean lies in an interval $[a, b]$ is computed according to Equations (161.16) and (161.19):

$$\begin{aligned} P(a \le \mu_i \le b) &= P\left(\frac{a-\hat{\mu}}{\sigma_{\hat{\mu}}} \le \frac{\mu_i - \hat{\mu}}{\sigma_{\hat{\mu}}} \le \frac{b-\hat{\mu}}{\sigma_{\hat{\mu}}}\right) \\ &= P\left(\frac{a-\hat{\mu}}{\sigma_{\hat{\mu}}} \le \mu_i^n \le \frac{b-\hat{\mu}}{\sigma_{\hat{\mu}}}\right) \\ &= P\left(\mu_i^n \le \frac{b-\hat{\mu}}{\sigma_{\hat{\mu}}}\right) - P\left(\mu_i^n \le \frac{a-\hat{\mu}}{\sigma_{\hat{\mu}}}\right) \end{aligned} \tag{161.21}$$

Given the example of ten distance measurements, using Equation (161.5), we first determine that the standard deviation of the mean is

$$\sigma_{\hat{\mu}} = \frac{\sigma_{l_i}}{\sqrt{n}} = \frac{5\text{ cm}}{\sqrt{10}} = 1.6\text{ cm} \tag{161.22}$$

Knowing that the standard deviation of the mean of the ten distance measurements is 1.6 cm and the sample mean is 38.95 m, the probability that the mean will fall between 38.94 and 38.97 m is

$$\begin{aligned} P(38.94 \le \mu_i \le 38.97) &= P\left(\frac{38.94-38.95}{0.016} \le \frac{\mu_i - 38.95}{0.016} \le \frac{38.97-38.95}{0.016}\right) \\ &= P(-0.625 \le \mu_i^n \le +1.250) \\ &= P(\mu_i^n \le +1.250) - P(\mu_i^n \le -0.625) \\ &= 0.8944 - 0.2659 = 0.6285 \\ &= 62.85\% \end{aligned} \tag{161.23}$$

Actually, various tests have to be distinguished, depending on the type of variable investigated and whether its standard deviation is considered to be known or unknown [Hamilton, 1964]. When the standard deviation is known, we invoke a normalized test, as in the example above. For tests of variables with an unknown sample standard deviation, we use the Student's *t*-test. Under these circumstances, the mean would fall within the specified limits with a probability of

$$\begin{aligned} P(38.94 \le \mu_i \le 38.97) &= P\left(\frac{38.94-38.95}{0.016} \le \frac{\mu_i - 38.95}{0.016} \le \frac{38.97-38.95}{0.016}\right) \\ &= P(-0.625 \le \mu_i^t \le +1.250) \\ &= P(\mu_i^t \le +1.250) - P(\mu_i^t \le -0.625) \\ &= 0.877 - 0.274 = 0.603 \\ &= 60.3\% \end{aligned} \tag{161.24}$$

To test individual observations and errors attached to them, two tests are available: the normalized residual test, as part of the B-method of statistical testing [Baarda, 1968], and the τ-test, as proposed by Pope [1976].

Normalized Residual Test

The first test computes the normalized residual according to

$$w_i = -\frac{v_i}{\sigma_{v_i}} \tag{161.25}$$

At a specified significance level α_0, the residuals, v_i of the observations l_i are tested in the following standardized way:

$$|w_i| < \sqrt{F_{1-\alpha_0;1,\infty}} \tag{161.26}$$

with F the related critical value of the F-statistic for one degree of freedom. It should be noted that Equation (161.25) only holds for uncorrelated observations. For the more general case of correlated observations, one is referred to Baarda [1968]. The multidimensional test on the sum of the squared residuals is

$$\frac{\hat{\sigma}_0}{\sigma_0} < \sqrt{F_{1-\alpha_0;r,\infty}} \tag{161.27}$$

where σ_0^2 and $\hat{\sigma}_0^2$ are the *a priori* and *a posteriori* variances of unit weight, respectively; r = the number of degrees of freedom (generally the difference between the number of observations and the number of parameters to be estimated from the observations); and α = the significance level of testing for r degrees of freedom.

For the estimation of the mean (one parameter) from ten distance measurements, we would have $r = n - 1 = 10 - 1 = 9$.

Sigma Zero

To take this type of testing to a practical situation, let us assume that we have adjusted a set of measurements using the method of least squares. Each observation, l_i, will have been adjusted by some amount, v_i, its residual or correction. It is instructive to compare the residuals with the standard deviations of each measurement and see what happened. A large residual compared to the standard deviation may point to a gross error in the observation.

It should be noted that the adjustment process will attempt to adjust out the large residual caused by a gross error over a region of the measurement network, so that adjacent observations may have significant residuals as well. A plot of the residuals mapped over the network can show a "peak" at the location of the gross error.

As a "first pass" at this type of error monitoring, it is a good move to compute a value commonly called *sigma zero* or *the standard deviation of a measurement of unit weight* or the *a posteriori standard deviation of unit weight*. We compute sigma zero as

$$\sigma_0 = \sqrt{\frac{\mathbf{v}^T \mathbf{V}_l^{-1} \mathbf{v}}{n-m}} \tag{161.28}$$

where $\mathbf{v}$ is the vector of residuals ($\mathbf{v}^T$ is that vector transposed), $\mathbf{V}_l^{-1}$ is the inverse of the variance-covariance matrix of the measurements, n is the number of observations and m is the number of unknowns.

In Equation (161.28), the numerator is a single value (actually a 1×1 matrix), which is a χ^2 variable with $n - m$ degrees of freedom. The expected value of such a variable is $n - m$, so Equation (161.28) simply normalizes it to unity.

With the expectation that σ_0 will be 1, we can examine it as a quick check on the quality of an adjustment. If σ_0 is significantly greater than 1, the possibilities are that there is at least one gross error in the observations, or that the assumed precisions of the measurements were too small (i.e., we thought we were better that we actually were). If all the residuals are larger than the standard deviations of the measurements by about the same factor, this may be the problem. Gross errors will usually show up as large corrections in a single region.

If σ_0 is significantly less than 1, it may mean that the standard deviations were estimated too large, or there has been preliminary adjustment of the measurements. The latter case destroys the assumption of uncorrelated observations, rendering the adjustment incorrect in any case.

In the statistical literature, σ_0 is sometimes used to re-scale the original variance-covariance matrix, $\mathbf{V}_l$, so that the precision of the derived values (usually coordinates of points) is corrected to allow for the *reestimated* precisions of measurements. However, while algebraically defensible, this is not always a reasonable or sensible thing to do. If the surveyor has made a reasoned assessment of the precision of the measurements, a value of σ_0 significantly different from 1 is a warning of possible problems in the measurement system or the adjustment and must be followed up. The author knows of only one case where this approach was reasonable and justifiable, and that was over 40 years ago.

What is "significantly different from 1" for σ_0? As a very rough guide, $0.8 \le \sigma_0 \le 1.3$ is a start, but better values for each particular case can be obtained using Equation (161.27).

Tau-Test

The τ-test avoids the cumbersome computations of the variance/covariance matrix of the residuals [Pope, 1976] needed in the denominator in Equation (161.25). The standard deviation of the residual σ_{v_i} is approximated by

$$\sigma_{v_i} \cong \sqrt{\frac{n-u}{n}}\sigma_{l_i} = \sqrt{\frac{r}{n}}\sigma_{l_i} \tag{161.29}$$

Testing of Variances

For tests of variables of a quadratic nature, such as the variance, we use the so-called χ^2-test. The probability that the standard deviation lies in the interval $[a, b]$ is readily computable from

$$\begin{aligned} P(a \le \sigma_{l_i}^2 \le b) &= P\left(\frac{1}{a} \ge \frac{1}{\sigma_{l_i}^2} \ge \frac{1}{b}\right) \\ &= P\left(\frac{r\hat{\sigma}^2}{a} \ge \frac{r\hat{\sigma}^2}{\sigma_{l_i}^2} \ge \frac{r\hat{\sigma}^2}{b}\right) \\ &= P(\chi^2_{r;\alpha_0/2} \ge \chi^2 \ge \chi^2_{r;1-\alpha_0/2}) \\ &= 1-\alpha_0 \end{aligned} \tag{161.30}$$

In our example, we compute that our standard deviation falls, with a probability of 95%, in the interval

$$\left[\sqrt{\frac{r\hat{\sigma}^2}{\chi^2_{r;\alpha_0/2}}} \le \sigma \le \sqrt{\frac{r\hat{\sigma}^2}{\chi^2_{r;1-\alpha_0/2}}}\right] \tag{161.31}$$

with the substituted values

$$\left[\sqrt{\frac{9\cdot 25}{19.02}} \le \sigma \le \sqrt{\frac{9\cdot 25}{2.70}}\right] \tag{161.32}$$

or

$$[3.44 \text{ cm} \le \sigma \le 9.13 \text{ cm}] \tag{161.33}$$

For testing ratios of two squared estimates of the same (squared) variable, we use the *F*-test. Consult Chapter 166 and the statistical literature for more details.

From statistical tables for normally distributed data (see the statistical literature for the definition of a normal distribution), it can be seen that, for instance, 95% of the observations fall in the region [–1.96σ, +1.96σ]. However, in 5% of the cases we may expect to have an observation that is more than 1.96σ away from its mean. This is a good observation (it belongs to the distribution just described) but will probably be removed from the data set. If we make such an "error," we speak of an *error of the first kind*, or the so-called producer's error (e.g., a TV manufacturer erroneously rejects a good although deviating specimen). However, if this data set contained a systematic error, such an outlier may be an observation belonging to a set of observations for which the systematic error had been removed. The surveyor is, in this sense, a producer as well: a producer of survey results who does not want to erroneously reject good data. On the other hand, the erroneous acceptance of false data is an *error of the second kind*, or the consumer's error (e.g., a person purchasing a TV accepts a faulty specimen that erroneously slipped through the manufacturer's testing procedures). The client of the surveyor should be most worried about this type of error. However, the surveyor generally serving the public or any other client has to design the measurement system in such a way as to minimize these types of errors, or at least the effect of errors of the second kind. The latter type of error brings us to the notions of accuracy and reliability.

161.6 Accuracy

Assume that a surveyor using a tape measures a distance very diligently (the proper tension is applied, temperature corrections are being applied, and so forth), but because of an earlier breakage the tape was repaired, resulting in an even 50 millimeters missing. All distances (assuming that all distances measured are shorter than one tape length) will have outcomes which are highly repeatable (the $\hat{\sigma}_{\hat{\mu}}$ is very small); however, all the distance measurements will be 50 millimeters too long. The surveyor got inaccurate results despite a very precise measurement procedure. Inaccuracy is the deviation of the measurement outcomes from the "real" or "true" value of the quantity the surveyor tried to measure. (The adjectives *real* and *true* are in quotes because it can be easily argued that the true value of, say, a distance will never be known, since any measurement process will only provide us with a sample mean, never the true value.)

A very dangerous but widespread practice is to increase the standard deviations of the measurements (remember, the standard deviations reflect precision or repeatability, not accuracy) by a factor of 2 to 10 in order to catch any inaccuracies caused by mismodeling, uncalibrated instrumental errors, setup errors, warming-up effects on a tripod that cause it to torque, personal errors, and so on. So we have

$$\sigma_{\text{accuracy}} = c \cdot \sigma_{\text{precision}} \tag{161.34}$$

with

$$\Sigma_{L_b} = \Sigma_{L_b(t)}(t) \tag{161.35}$$

This procedure is, in effect, an admission of the surveyor's inability to work to the standards expected with the equipment and situation, as well as a poor understanding of error propagation and analysis.

The preferred procedure is to apply any correction to the observations, such as the personal equation of the surveyor who experiences a consistent delay in the triggering of a chronometer between the moment a star crosses a cross wire and the moment of an audible time signal. Similarly, it is better to strive for the correct atmospheric setting on a total station, rather than increasing the standard deviation of the distance measurements.

It is also important that the surveyor has a good idea of the probable precision of survey measurements and their derived values, so that measurements of a lesser precision are noticed as early as possible. The surveyor must understand the error propagation process, both mathematically and intuitively, so that results can be checked for unexpected errors. In the example of the missing 50 mm in the tape, the measurements should have been checked by measuring enough additional lines to allow

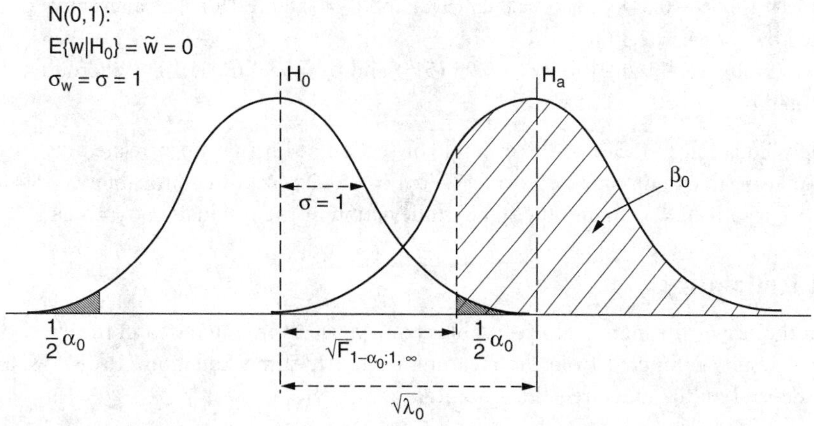

FIGURE 161.2 The one-dimensional test of the normal standardized variable *w*, shown for hypothesis H_0 and alternative hypothesis H_a.

a closing figure to be developed. Any misclosure in the figure would suggest something was wrong with the tape or the procedures.

Despite all precautions, the surveyor wants, on behalf of the client, to have insight into the size of any inaccuracies that may be avoided by following certain measurement procedures as part of an overall network of measurements. This leads us to the notion of reliability.

161.7 Reliability

Having calibrated the survey equipment, applied temperature corrections, and so on, the surveyor is able to assure himself and the client about the size of the error (inaccuracies) that can be detected with a certain probability under the given circumstances. This is often subject to a cost-versus-effect analysis.

Internal Reliability

To be able to detect distance measurement errors of 10 cm with a probability of 80%, one has to evaluate the distance measurements as an intrinsic part of the overall network design. Increasing the probability from 80% to 90% will increase the cost of the survey. Similarly, the ability to detect errors smaller than 10 cm with a probability of 80% will also considerably increase the cost of the survey. This is an issue of *internal reliability*: what size of measurement error is detectable with what probability? Such errors are referred to as *marginally detectable errors* (MDE). The marginally detectable error is denoted by a nabla (∇).

λ_0 denotes the distance between the normal distribution under the hypothesis (H_0) that the observation belongs to this distribution, and the distribution corresponding to the alternative hypothesis (H_a) (see Figure 161.2). In the statistical literature, one can find tables or graphs depicting λ_0 as a function of the level of significance α_0 and the power of the test β_0 [Baarda, 1968]. Common values for α_0 and β_0 are 5% and 80%, respectively. It can be shown that for uncorrelated observations the marginally detectable error is

$$\nabla_o l_i = \frac{\sigma_{l_i}^2}{\sigma_{v_i}} \sqrt{\lambda_0(\alpha_0, \beta_0)} \tag{161.36}$$

Following are three sets of examples for values of α_0 and β_0:

1. $\lambda_0 = 11.70$ [$(2.58 + 0.84)^2$] for $\alpha_0 = 0.01$ (1%) and $\beta_0 = 0.80$ (80%). The alternative distribution is shifted by $\sqrt{11.70} = 3.42$ units.

2. $\lambda_0 = 7.84$ [$(1.96 + 0.84)^2$] for $\alpha_0 = 0.05$ (5%) and $\beta_0 = 0.80$ (80%). The alternative distribution is shifted by $\sqrt{7.84} = 2.80$ units.
3. $\lambda_0 = 10.50$ [$(1.96 + 1.28)^2$] for $\alpha_0 = 0.05$ (5%) and $\beta_0 = 0.90$ (90%). The alternative distribution is shifted by $\sqrt{10.50} = 3.24$ units.

If we apply Equation (161.36) to the example of section 161.3 (ten uncorrelated distance measurements), we may be able to detect an error of 17.5 cm (= 25 · 2.8/4) with a probability of 80% (and level of significance 5%) if $\sigma_l = 5$ cm and the standard deviation of the residuals is $\sigma_v = 4$ cm.

External Reliability

Rather than the size of the marginally detectable error, we are more interested in the effect of the MDE on the actual results computed from the measurements. Repeating Equation (161.8), we assume that variables x_j depend on the measurements l_i according to

$$d\vec{x} = \mathbf{J}_{ji} d\vec{l} \tag{161.37}$$

The influence of a marginally detectable error of the size of $\nabla l_{(i)}$ on the variables x_j can be computed from

$$\nabla \vec{x} = \mathbf{J}_{ji} \nabla \vec{l}_{(i)} \tag{161.38}$$

It should be noted that the parentheses around the subscript i denote that the vector $\nabla l_{(i)}$ in Equations (161.36) and (161.38) contains all zeros except for the ith observation for which the influence on all variables x_j is computed.

Surveyors design networks to minimize the influence of MDEs in the observations. For classical terrestrial networks this leads to designs avoiding triangles with small angles. Survey resection problems with small top angles lead to large MDEs with large negative effects on the positional accuracy for the point to be resected. In modern surveying using the global positioning system (GPS), dilution of precision (DOP) (see Chapter 167 for an explanation) similarly lead to large MDEs in the GPS observables. In these cases precise and accurate measurements may lead to highly unreliable results because of the system's incapability to flag errors of small magnitude. Large errors in the GPS network positions may go undetected as a result (see [FGCC, 1984] and [FGCC, 1989]).

161.8 Up-to-date-ness or Currency

Precise, accurate, and reliable measurements and the resulting maps may be worthless if they tend to represent locations of a network in a highly dynamic environment. Leveled heights referring to some vertical datum in an area of large annual subsidence may be worthless. Maps showing property boundaries or a GIS land use database for an area where the land use rapidly changes soon become worthless also. We usually want current, up-to-date information, not historical information.

The vector $\vec{x}_j$ of results should be time tagged and reflect the epoch t_0 during which the situation is measured:

$$\vec{x}_j = \vec{x}_j(t_0) \tag{161.39}$$

Models may be available to represent the time history of the vector $\vec{x}_j$. Transformation parameters are capable of expressing the current result vector $\vec{x}_j(t)$ as a function of the result vector at reference epoch t_0. In this case, a Jacobian matrix $\mathbf{\Phi}$ will relate the state of $x_j(t)$ at t to the state of $x_j(t_0)$ at t_0:

$$\vec{dx}(t) = \frac{\partial \vec{x}(t)}{\partial \vec{x}(t_0)} \vec{dx}(t_0) \\ = \mathbf{\Phi}_{t,t_0} \vec{dx}(t_0) \tag{161.40}$$

with the state transition matrix being

$$\Phi(t,t_0) = \begin{pmatrix} \frac{\partial x_1(t)}{\partial x_1(t_0)} & \frac{\partial x_1(t)}{\partial x_2(t_0)} & \cdots & \frac{\partial x_1(t)}{\partial x_u(t_0)} \\ \frac{\partial x_2(t)}{\partial x_1(t_0)} & \frac{\partial x_2(t)}{\partial x_2(t_0)} & \cdots & \frac{\partial x_2(t)}{\partial x_u(t_0)} \\ \vdots & \vdots & \cdots & \vdots \\ \frac{\partial x_u(t)}{\partial x_1(t_0)} & \frac{\partial x_u(t)}{\partial x_2(t_0)} & \cdots & \frac{\partial x_u(t)}{\partial x_u(t_0)} \end{pmatrix} \tag{161.41}$$

Analytic expressions for the behavior of a set of coordinates as a function of time may be available. These functions may have the character of polynomials or may reflect more realistic (geo)physically or geometrically oriented time behavior.

The ability to represent changes in measurements over time is the essence of deformation surveying, which would also bring us back to the issues of repeatability and reproducibility. If the reproducibility of a set of measurements is significantly lower than its repeatability, it is important to look for causes. Movement of control points is a common cause, as are differences in measurement techniques.

For control and deformation work, the long-term viability of measurement quality includes the issue of stability of marks or monuments. If a surveyor determines the elevation of a mark to first-order standards, but the mark moves enough that the new elevation is now to third-order standards with respect to the measurement from the original mark, the work becomes third order, regardless of its internal quality. Any effort made to achieve higher results was a waste of time and money. So we must pay close attention to stability of control marks. For deformation surveys, the control marks must be assured of being stable, and the measured points must be able to be reoccupied exactly, if the surveyor wishes to ensure the reproducibility precision of the work meets required standards.

161.9 Attribute Data Errors

Nonspatial data collected for GIS are becoming an increasingly important part of a surveyor's work as GIS becomes ubiquitous. As surveyors have a responsibility to know something about the errors involved in the data they collect, it is important that they understand errors in the attribute data.

Attribute data of a location would be the information that a particular point represents a fire hydrant, that a particular line represents Main Street or a property boundary, or that a particular polygon represents Fred Smith's property, a school, or the State of Ohio.

Attribute data must be analyzed in ways that suit the nature of the data themselves. For continuous attribute data, we can often express the error level in ways very similar to spatial measurement data, such as with a standard deviation or confidence interval. However, some types of continuous attribute data encountered in GIS do not lend themselves to this approach, such as probabilities and fuzzy set membership values. It is critical to understand the particular data being handled, their meaning and statistical nature, before embarking on error analysis. Chrisman (2002) outlines some of the types of data encountered.

Categorical attribute data is usually discrete and refers to distinct types of features, such as soil types and land use classifications. While gross errors may be relatively few with careful data collection, a lot of the categorical data are used over a region, and it may be possible to classify a region to more than one category, such as river or vegetation, depending on the spatial resolution of the area being classified. This is a significant problem with remotely sensed data, especially satellite imagery.

TABLE 161.1 Misclassification Matrix

	Class on Ground			
Class in Database	A	B	C	D
A	—	—	—	—
B	—	—	—	—
C	—	—	—	—
D	—	—	—	—

One way to test the quality of categorical attribute data is with a misclassification matrix (also called a confusion or error matrix) (Table 161.1). The surveyor selects a number of points at random, using a suitable method of selection, and examines their attributes in the field and as recorded by previous measurements. The appropriate classification is determined in the field, and the matrix is completed for all of the points, with counts for each cell in the matrix being recorded.

Ideally, we want all the points to fall along the diagonal, indicating that all the points are correctly classified. The matrix should not be too large, or the classification process may induce additional errors which are not real errors.

Errors of omission occur when a point's classification on the ground is incorrectly recorded in the database. Errors of commission occur when the classification recorded in the database does not exist on the ground.

To summarize the matrix, we can simply state the number of points correctly classified as a percentage of all points checked. However, even in a totally random data set, we would expect some points to be correct purely by chance. We can adjust the percentage by using Cohen's kappa (sometimes also called the Kappa index of agreement or KIA), as described in Lillesand and Keifer (1999).

Kappa is determined using

$$\kappa = \frac{(d-q)}{(N-q)} \tag{161.42}$$

where d is the number of cases in diagonal cells, q is the number of cases expected in diagonal cells by chance, and N is the total number of cases. $\kappa = 1$ indicates a perfect result, while $\kappa = 0$ indicates agreement no better than chance.

Kappa is sensitive to the number of classes being tested, as well as the number of polygons whose classifications are to be tested. Fewer classes and polygons lead to a better result for kappa. Kappa may well vary over a large region, so it may be misleading to use a single kappa for a large area.

How many points should we test to determine attribute data quality? To some extent, this depends upon the expected level of error. For continuous attribute data, we can estimate the likely error, and call it an estimated standard deviation, s. We look up the standard normal statistical score, z, for the required confidence interval, e (e.g., $e = 95\%$ has $z = 1.96$), and compute the number of points required, n,

$$n = \frac{z^2 s^2}{(1-e)^2} \tag{161.43}$$

For categorical attribute data, standard deviations are meaningless, so we estimate p, the proportion of points that will be correct (as a number between 0 and 1), and hence $q = (1 - p)$. The number of points to be sampled, n, is computed by

$$n = \frac{z^2 pq}{(1-e)^2} \tag{161.44}$$

Examples: Number of Sampling Points

1. We want to test a continuously varying attribute, e.g., elevation, so that we can be 95% confident that the precision is 0.5. How many sample test points do we need to measure?
   For $e = 95\%$ (0.95), $z = 1.96$ (from statistical tables), and for $s = 0.5$ we have

$$n = \frac{1.96^2 \times 0.5^2}{(1-0.95)^2} = 384 \text{ points}$$

2. We want to test a categorical attribute, e.g., land use or soil type, so that we are 95% confident that no more than 10% of point are incorrect. How many sample test points do we need to measure?
   For $e = 95\%$ (0.95), $z = 1.96$, and for $p = 0.9$ and $q = 0.1$ we have

$$n = \frac{1.96 \times 0.9 \times 0.1}{(1-0.95)^2} = 138 \text{ points}$$

Methods of deciding which points to test on the ground come in three basic schemes: random, regular, or stratified. With random sampling, the surveyor selects the required number of points by some random method, such as a random number generator producing coordinates or sticking pins in a map. It is also possible to draw a regular grid of points over the area of interest and select sufficient points at random from that grid. For a regular selection of points, a grid of a suitable point density is laid over the map, so that all the points on the grid are measured to provide the required number of points. With a stratified scheme, the surveyor selects the areas where the most difficult cases are likely, and concentrates the points in those areas, although still being random in their final location within the area. For checking vegetation types, for instance, more points would be allocated to areas with small polygons of each classification, while proportionally fewer points would be allocated to areas with little variation in vegetation. For checking elevation data, more points would be allocated to areas with greater variation in elevation, and fewer points to those that are relatively flat. This allows better discrimination from a given number of points to be tested.

References

Baarda, W. 1968. A testing procedure for use in geodetic networks. *Publ. Geodesy.*, N.S. 2(5).

Chrisman, N. 2002. *Exploring Geographic Information Systems.* John Wiley & Sons, New York.

Federal Geodetic Control Committee. 1984. *Standards and Specifications for Geodetic Control Networks.* Federal Geodetic Control Committee, Rockville, MD.

Federal Geodetic Control Committee. 1989. *Geometric Geodetic Accuracy Standards and Specifications for Using GPS Relative Positioning Techniques.* Version 5.0 Federal Geodetic Control Committee. Rockville, MD.

Hamilton, W. C. 1964. *Statistics in Physical Science: Estimation, Hypothesis Testing, and Least Squares.* Ronald Press, New York.

Lillesand, T. M. and Kiefer, R. W. 1999. *Remote Sensing and Image Interpretation.* John Wiley & Sons, New York.

Papoulis, A. and Unnikrishna, S. 2001. *Probability, Random Variables and Stochastic Processes.* McGraw-Hill, New York.

Pope, A. J. 1976. *The Statistics of Residuals and the Detection of Outliers.* NOAA Technical Report NOS 65 NGS 1. National Oceanic and Atmospheric Administration, Rockville, MD.

Further Information

Textbooks and Reference Books

For additional reading and more background, from the very basic to the advanced level, you can consult specific chapters in a variety of textbooks on geodesy, satellite geodesy, physical geodesy, surveying, photogrammetry, or statistics itself.

Anderson, J. M. and Mikhail, E. M. 1997. *Surveying: Theory and Practice.* McGraw-Hill, New York.

Ch. 2: Survey Measurements and Adjustments
App. B: Introduction to Vector and Matrix Algebra
App. D: Introductory Probability and Statistics
App. E: Trigonometric Formulas and Statistical Tables

Bannister, A., Raymond, S., and Baker, R. 1998. *Surveying.* Addison-Wesley Longman, Harlow, UK.

Ch. 1: Introductory
Ch. 8: Analysis and adjustment of measurements

Bjerhammer, E. A. 1973. *Theory of Errors and Generalized Matrix Inverses.* Elsevier, New York.

Bomford, G. 1985. *Geodesy.* Clarendon Press, Oxford.

Ch. 1: Triangulation, Traverse, and Trilateration (Field Work)
Ch. 2: Computation of Triangulation, Traverse, and Trilateration
App. D: Theory of Errors

Buckner, R. B. 1983. *Surveying Measurements and Their Analysis.* Landmark Enterprises, Rancho Cordova, CA.

Burnside, C. D. 1991. *Electromagnetic Distance Measurement.* BSP Professional Books, Oxford.
Ch. 6: The Measurement Process and Calibration of Instruments

Escobal, P. R. 1976. *Methods of Orbit Determination.* John Wiley & Sons, New York.
App. IV: Minimum Variance Orbital Parameter Estimation

Heiskanen, W. A. and Moritz, H. 1978. *Physical Geodesy.* W.H. Freeman & Co., New York.
Ch. 7: Statistical Methods in Physical Geodesy

Hirvonen, R. A. 1979. *Adjustment by Least Squares in Geodesy and Photogrammetry.* Frederick Ungar Publishing, New York.

Hofmann-Wellenhof, B., Lichtenegger, H., and Collins, J. 2001. *GPS: Theory and Practice.* Springer-Verlag, New York.
Ch. 9: Data Processing

Kaula, W. M. 1966. *Theory of Satellite Geodesy: Applications of Satellites to Geodesy.* Blaisdell, New York. (reprinted 2001, by Dover Publications)

Ch. 4: Geometry of Satellite Observations
Ch. 5: Statistical Implications
Ch 6: Data Analysis

Koch, K. R. 1999. Parameter Estimation and Hypothesis Testing in Linear Models. Springer-Verlag, New York.

Kok, J. 1984. *On Data Snooping and Multiple Outlier Testing.* NOAA Technical Report NOS NGS 30. National Oceanic and Atmospheric Administration, Rockville, MD.

Kraus, K. 1993. *Photogrammetry.* Ferd. Dümmlers Verlag, Bonn.
App. 4.2-1: Adjustment by the Method of Least Squares

Leick, A. 1995. *GPS: Satellite Surveying*, 2nd ed. Wiley & Sons, New York.

Ch. 4: Adjustment Computations
Ch. 5: Least-Squares Adjustment Examples
Ch. 11: Network Adjustments
App. B: Linearization
App. C: One-Dimensional Distributions

McCormac, J. C., and Anderson, W. 1999. *Surveying.* John Wiley and Sons, New York.
Ch. 2: Introduction to Measurements

Mikhail, E. M. 1976. *Observations and Least Squares.* IEP-A Dun-Donnelley, New York.

Mikhail, E. M. and Gracie, G. 1997. *Analysis and Adjustment of Survey Measurements.* Van Nostrand Reinhold, New York.

Moffitt, F. H. and Bossler, J. D. 1999. *Surveying.* Addison-Wesley Longman, Menlo Park, CA.

Ch. 1-10: Errors and Mistakes
Ch. 1-11: Accuracy and Precision
Ch. 5: Random Errors
App. A: Adjustment of Elementary Surveying Measurements by the Method of Least Squares
App. B: The Adjustment of Instruments

Mueller, I. I. and Ramsayer, K. H. 1979. *Introduction to Surveying.* Frederick Ungar Publishing, New York.

Ch. 2: Nature of Errors and Measurements
Ch. 5: Adjustment Computation by Least Squares

Roberts, J. 1995. *Construction Surveying: Layout and Dimension Control.* Delmar, New York.
Ch. 6-2: Establishing Dimension Control Program Standards (ISO 6643)

Tienstra, J. M. 1966. *Theory of Adjustment of Normally Distributed Observations.* Argus, Amsterdam.

Uotila, U. A. 1985. Adjustment Computations Notes. Department of Geodetic Science and Surveying, Ohio State University, Columbus.

Vaníček, P. and Krakiwsky, E. J. 1982. *Geodesy: The Concepts.* North-Holland, Amsterdam.

Part I: Introduction
Ch. 3: Mathematics and Geodesy
Part III: Methodology
Ch. 10: Elements of Geodetic Methodology
Ch. 11: Classes of Mathematical Models
Ch. 12: Least-Squares Solution of Overdetermined Models
Ch. 13: Assessment of Results
Ch. 14: Formulation and Solving of Problems

Wolf, P. R. and Dewitt, B. A. 2000. *Elements of Photogrammetry with Applications in GIS.* McGraw-Hill, New York.
App. A: Random Errors and Least Squares Adjustment

Wolf, P. R. and Ghilani, C. D. 1997. *Adjustment Computations: Statistics and Least Squares in Surveying and GIS*, 3rd ed. Wiley & Sons, New York.

Wolf, P. R., and Ghilani, C. D. 1994. *Elementary Surveying: An Introduction to Geomatics.* Prentice Hall, Englewood Cliffs, NJ.

Ch. 2: Theory of Measurements and Errors
App. A: Instrument Testing and Adjusting
App. C: Propagation of Random Errors and Least-Squares Adjustment

Journals and Organizations

The latest results from research of statistical applications in geodesy, surveying, mapping, and photogrammetry are published in a variety of journals.

The following are two international magazines under the auspices of the International Association of Geodesy, both published by Springer-Verlag (Berlin/Heidelberg/New York): *Bulletin Géodésique* and *Manuscripta Geodetica.*

Geodesy- and geophysics-related articles can be found in the following publications:

American Geophysical Union, Washington, DC: *EOS* and *Journal of Geophysical Research*
Royal Astronomical Society, London: *Geophysical Journal International*

Statistical articles related to kinematic GPS can be found in:

Institute of Navigation: *Navigation*

Many national mapping organizations publish journals in which recent statistical applications in geodesy/surveying/mapping/photogrammetry are documented, including:

American Congress of Surveying and Mapping; Surveying and Land Information Science; Cartography and Geographic Information Systems
American Society of Photogrammetry and Remote Sensing: *Photogrammetric Engineering & Remote Sensing*
American Society of Civil Engineers: *Journal of Surveying Engineering*
Deutscher Verein für Vermessungswesen: *Zeitschrift für Vermessungswesen*, Konrad Wittwer Verlag, Stuttgart
The Canadian Institute of Geomatics: *Geomatica*
The Royal Society of Chartered Surveyors: *Survey Review*
Institution of Surveyors of Australia: *The Australian Surveyor*

Worth special mention are the following trade magazines:

GPS World, published by Advanstar Communications, Eugene, OR
P.O.B. (*Point of Beginning*), published by P.O.B. Publishing Company, Canton, MI
Professional Surveyor, published by American Surveyors Publishing Company, Arlington, VA
Geodetical Info Magazine (GIM), published by Geodetical Information & Trading Centre bv., Lemmer, the Netherlands

National mapping organizations such as the U.S. National Geodetic Survey (NGS) regularly make software available (free and at cost). Information can be obtained from: National Geodetic Survey, Geodetic Services Branch, National Ocean Service, NOAA, 1315 East-West Highway, Station 8620, Silver Spring, MD 20910-3282.

162

Elevation

Steven D. Johnson
Purdue University

Elevation is the distance above or below a specified reference surface. In engineering and surveying, the surface most commonly used to reference elevation is the geoid. The geoid is defined as an equipotential surface that closely approximates mean sea level. However, the geoid and mean sea level surfaces are not coincident. They may be separated by a meter or more at any specific location.

By definition, the potential due to gravity is equal at all points on the geoid, and the force of gravity is perpendicular to the geoid at all points. The geoid is an irregular surface that is affected by density variations within the Earth. The geoid is not readily defined by mathematical equations. The mathematical ellipsoid used as a datum surface for geodetic position can approximate the geoid, but the ellipsoid and geoid will be separated by up to 100 meters or more for a mean global fit. Figure 162.1 illustrates the general relationship between the geoid and some approximating geodetic ellipsoid.

162.1 Measures of Elevation and Height

The vertical distance referenced to the geoid is called an orthometric height (elevation), H. Orthometric height is measured along the plumb line. A height referenced to the ellipsoid is called an *ellipsoidal height, h.* Ellipsoidal height is measured along the normal to the ellipsoid. Geoid height, N, is the distance between the geoid and the ellipsoid measured along the normal to the ellipsoid. Neglecting the deviation between the plumb line and the ellipsoidal normal, the geoid height is related to the orthometric and ellipsoidal heights by the following equation:

$$h = H + N$$

Thus, Figure 162.2 shows a negative geoid height, which is typical in the United States. Values of N can be estimated using the geoid models published by the National Geodetic Survey, National Oceanic and Atmospheric Administration. The current model for the continental United States is GEOID99, and it is available from the National Geodetic Survey at the address given at the end of this chapter.

0-8493-1586-7/05/$0.00+$1.50

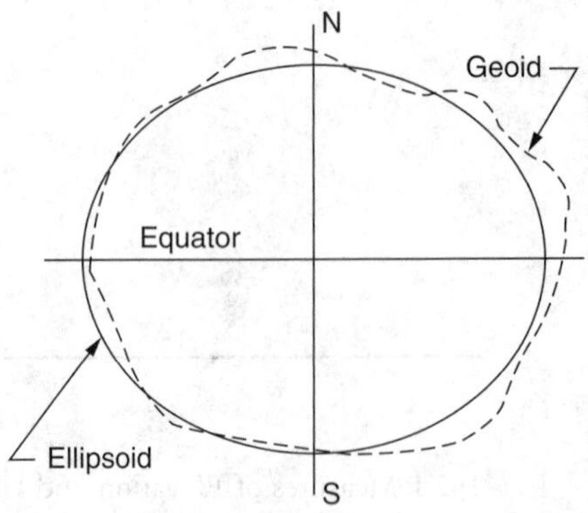

FIGURE 162.1 Relationship between geoid and approximating geodetic ellipsoid.

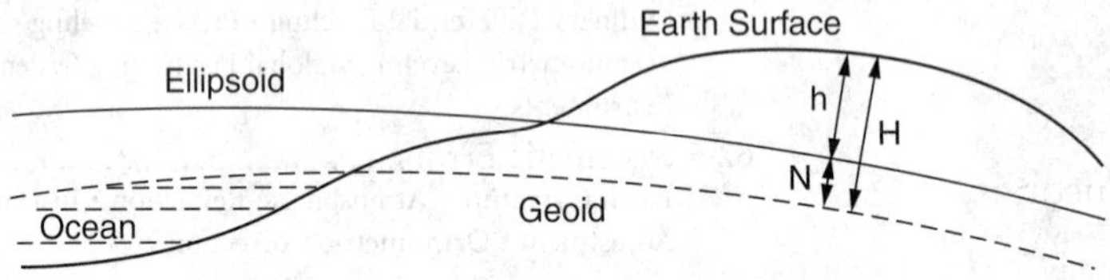

FIGURE 162.2 Illustration of negative geoid height.

Equipotential surfaces (level surfaces) are not parallel to one another. The distance between the surfaces decreases toward the Earth's north pole. Each level surface has a different gravity potential relative to the geoid that can be expressed as a geopotential number, C, measured in terms of geopotential units: 1 geopotential unit (gpu) = 1000 gal-meters. As an expression of height, the geopotential number (a potential) is constant for a given level surface, whereas the orthometric height decreases for a given level surface proceeding toward the pole. Geopotential numbers can be converted to distance units of height by dividing by a defined value of gravity. Helmert orthometric heights are computed using an estimated mean gravity value based on a constant crustal density, fixed free-air gravity gradient, and linear variation of gravity from geoid to the surface. Known as the Poincaré–Prey model for mean gravity, the calculation is

$$H_{Helmert} = \frac{C}{\bar{g}_H}$$

$H_{Helmert}$ is the Helmert orthometric height, C is the geopotential number on the level surface, and $\bar{g}_H$ is the mean value of gravity along the plumb line between the geoid and the surface. The change in orthometric height between two surface stations can be calculated using the formula found at the end of this chapter.

Another definition of height is dynamic height, H_D, given as

$$H_D = \frac{C}{g_{Constant}}$$

where H_D is the dynamic height, C is the geopotential number on the level surface, and $g_{Constant}$ is some constant value of gravity calculated on the ellipsoid, often at 45° latitude, using a standard gravity formula. The computation simply scales the geopotential number to a linear height value. In the dynamic height system, the elevation of a large body of water would be constant. In the Helmert orthometric height system, the height distance above the geoid at the north end of the water body would be smaller than height distance above the geoid at the south end of the water body.

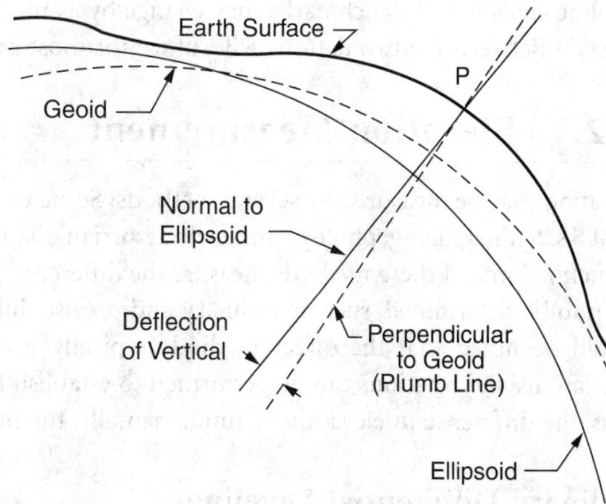

FIGURE 162.3 Definition of deflection of the vertical.

162.2 Deflection of the Vertical

A vertical or plumb line is perpendicular to the geoid and therefore parallel to the direction of gravity. A normal line is perpendicular to the ellipsoid. Because the geoid and ellipsoid are generally not parallel, a vertical line and a normal line are not coincident for most points on the Earth's surface. The deflection of the vertical is the angle between these two lines at a point on the Earth's surface. Figure 162.3 illustrates the definition of the deflection of the vertical. The deflection of the vertical is generally resolved into two components: ξ in the N-S direction and η in the E-W direction.

The deflection of the vertical has been modeled by the National Geodetic Survey. Program DEFLEC99 estimates the deflection components in the continental United States. The program is available from the National Geodetic Survey at the address given at the end of this chapter.

162.3 Vertical Datums

Vertical datums are defined for the purpose of specifying the elevation of points with respect to the geoid. There are two relevant continental vertical datums within the U.S.: the National Geodetic Vertical Datum of 1929 (NGVD 29) and the North American Vertical Datum of 1988 (NAVD 88).

The NGVD 29, a long-used reference for mean sea level in the U.S. was the product of a 1929 general adjustment of the U.S. and Canadian vertical control networks. The 1929 adjustment was based, in part, upon the assumption that the local mean sea level at the tide stations used in the adjustment was equal (same equipotential surface). This is not a valid assumption since the elevation of mean sea level varies from the Atlantic coast to the Pacific coast of the U.S. This distortion of the vertical datum caused the official name of the 1929 datum to change from "Sea Level Datum of 1929" to "National Geodetic Vertical Datum of 1929" in 1976. Other distortions, including those from upheaval and subsidence of the Earth's crust, are present in the NGVD 29; however, it remains a datum of reference for the U.S.

The most recent adjustment of the vertical datum for North America is known as the NAVD 88. It is a least-squares readjustment of over 600,000 benchmarks across the North American continent, resulting in a better approximation of the geoid. The project includes releveling of approximately 83,000 km of first-order vertical control within the U.S. Leveling data from Canada, Mexico, and Central America are included in the adjustment. The result of the NAVD 88 adjustment is a computer database of vertical control stations and elevations across the U.S. The published values are Helmert orthometric heights. This improved model of the geoid is beneficial for the determination of Helmert orthometric heights using global positioning system (GPS) derived heights above the ellipsoid.

The change in elevations from NGVD 29 to NAVD 88 varies, depending on the area of the country involved. The relative elevation between existing benchmarks changes only by a few millimeters. The absolute elevation of benchmarks may change by as much as a few decimeters. An elevation correction constant between the two datums will suffice for most project areas.

162.4 Elevation Measurement

Elevation may be measured by several methods. Some of these methods measure elevation directly, such as GPS satellite ranging, photogrammetric aerotriangulation, inertial surveying methods, and barometric altimetry. Some of these methods measure the difference in elevation from a reference benchmark to the point to be determined, such as ordinary and precise differential leveling and trigonometric leveling. It should be noted that the direct methods typically must also be referenced to benchmarks so that translations and rotations can be performed to establish the proper relationship to the elevation datum. Thus, the difference in elevation is, fundamentally, the important value to be measured.

Ordinary Differential Leveling

Differential leveling is a very simple process based on the measurement of vertical distances from a horizontal line. Elevations are transferred from one point to another through the process of using a leveling instrument to read a rod held vertically on, first, a point of known elevation and, then, on the point of unknown elevation.

A single-level setup is illustrated in Figure 162.4. The known elevation of the backsight point is transferred vertically to the line of sight by adding the known elevation and the backsight rod reading. The elevation of the line of sight is the height of instrument, HI. By definition, the line of sight generates a horizontal plane at the instrument location when the telescope is rotated on the vertical axis. The line-of-sight elevation is transferred down to the unknown elevation point by turning the telescope to the foresight, subtracting the rod reading from the height of instrument. Note that the difference in elevation from the backsight station to the foresight station is determined by subtracting the foresight rod reading from the backsight rod reading.

A level route consists of several level setups, each one carrying the elevation forward to the next foresight using the differential leveling method. A level route is typically checked by closing on a second known benchmark or by looping back to the starting benchmark. At the closing benchmark, closure = computed elevation − known elevation. Since differential leveling is usually performed with approximately equal setup distances between turning points, the level route is adjusted by distributing the closure equally to each setup:

$$\text{Adjustment} = \left(-\frac{\text{Closure}}{n}\right)\text{per setup}$$

where n is the number of setups in the route. When a network of interconnected routes is surveyed, a least-squares adjustment is warranted.

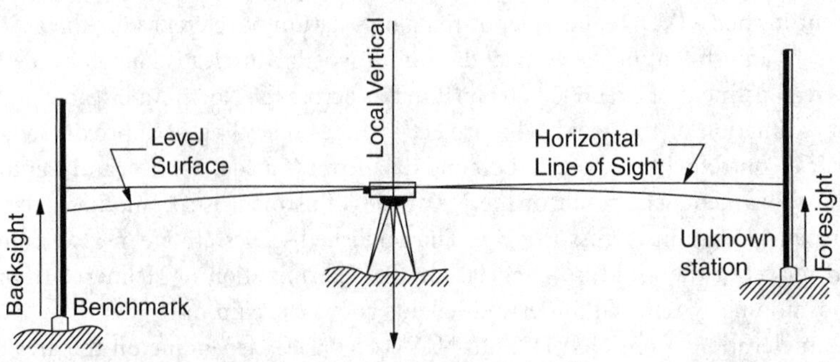

FIGURE 162.4 Using differential leveling to measure elevation.

Precise Leveling

Precise leveling methods are used when the highest accuracy is required for engineering and surveying work. Instruments used in precise leveling are specifically designed to obtain a high degree of accuracy in leveling. Improved optics in the telescope, improved level sensitivity, and carefully calibrated rod scales are all incorporated into the differential leveling process.

Typically when performing precise leveling, a method of leveling called *three-wire leveling* is used. This involves reading the center crosshair as well as the upper and lower "stadia" crosshairs. The basic process of leveling is the same as ordinary differential leveling, except that the three crosshair readings are averaged to improve the precision of each backsight and foresight value.

Another method that can be used to improve the precision of the level rod reading is to use an optical micrometer on the telescope. The optical micrometer is a rotating parallel-plate prism attached in front of the objective lens of the level. The prism enables the observer to displace the line of sight parallel with itself and set the horizontal crosshair exactly on the nearest rod gradation. The observer adds the middle crosshair rod reading and the displacement reading on the micrometer to obtain a precise rod reading to the nearest 0.1 millimeter.

Trigonometric Leveling

Trigonometric leveling is a method usually applied when a total station is used to measure the slope distance, S, and the vertical angle, α, to a point. Assuming the total station is set up on a station of known elevation and the height of instrument, HI, and reflector, HR, are measured, the elevation of the unknown station is

$$V = S \sin \alpha$$

$$P_{\text{Elev}} = A_{\text{Elev}} + \text{HI} + V - \text{HR}$$

The precision of trigonometric elevations is determined by the uncertainty in the vertical angle measurement and the uncertainty caused by atmospheric refraction effects. When the effects of Earth curvature and atmospheric refraction must be included, a useful reduction method for trigonometric leveling observations using short- and medium-range total station observations is

$$V = S\left[\sin \alpha + (c - r)_{radians} \cos \alpha\right]$$

where $(c - r)$ is the combined effect of curvature and refraction. Assuming an average Earth radius and standard atmospheric refraction of light, the effect may be approximated as 4.24 seconds per 1000 feet of distance or 2.055 radians per 1000 feet.

Global Positioning System Heights

The global positioning system (GPS) determines heights relative to the ellipsoid. The ellipsoid heights can be converted to orthometric heights as noted in Section 162.1. However, because of biases in horizontal and vertical datums, the absolute orthometric elevation may not be referenced to the desired datum. GPS survey networks should be adjusted to known benchmarks on the perimeter of the network. Detailed suggested survey and network adjustment procedures are published by the National Geodetic Survey. Results of GPS-determined heights using proper network geometry, adjustment methods, and the GEOID99 model to convert between NAD 83 GPS-derived ellipsoid heights and NAVD 88 Helmert orthometric heights can meet accuracy requirements for some applications.

Instruments

Altimeters

Surveying altimeters are precise aneroid barometers that are graduated in feet or meters. As the altimeter is raised in elevation, the barometer senses the atmospheric pressure drop. The elevation is read directly on the face of the instrument. Although the surveying altimeter may be considered to measure elevation directly, best results are obtained if a difference in elevation is observed by subtracting readings between a base altimeter kept at a point of known elevation and a roving altimeter read at unknown points in the area to be surveyed. The difference in altimeter readings is a better estimate of the difference in elevation, since the effects of local weather changes, temperature, and humidity that affect altimeter readings are canceled in the subtraction process. By limiting the distance between base and roving altimeters, accuracies of 3 to 5 feet are possible. Other survey configurations utilizing low- and high-base stations or leapfrogging roving altimeters can yield good results over large areas.

Level Bubble Instruments

Level bubble instruments, or spirit levels, contain a level vial with a bubble that must be centered to define a horizontal plane. Field instruments consist of three main components: a telescope to define a line of sight and magnify the object sighted, a level vial attached to the telescope to define the orientation of the instrument with respect to gravity, and a leveling head to tilt and orient the instrument.

All level bubble instruments are designed around the same fundamental relationships. These relationships are as follows:

- The axis of the level bubble (or compensator) should be perpendicular to the vertical axis.
- The line of sight should be parallel to the axis of the level bubble.

When these instrument adjustment relationships are true and the instrument is properly set up, the line of sight will sweep out a horizontal plane that is perpendicular to gravity at the instrument location.

Instruments that use a level bubble to orient the axes to the direction of gravity depend on the bubble's sensitivity for accuracy. Level bubble sensitivity is defined as the central angle subtended by an arc of one division on the bubble tube. The smaller the angle subtended is, the more sensitive the bubble is to dislevelment. A bubble division is typically 2 millimeters long, and bubble sensitivity typically ranges from 60 seconds to 1 second.

Builder's Level

The builder's level typically is less precise than other instruments in this category, but it is one of the most inexpensive and versatile instruments used by field engineers for construction layout. In addition to being able to perform leveling operations, it can be used to turn angles, and the scope can be tilted for inclined sights.

Transit

Although the primary functions of the transit are for angle measurement and layout, it can also be used for leveling because it has a bubble attached to the telescope. However, the field engineer should be aware that the transit may not be as sensitive and stable as a quality level.

Dumpy Level

The engineer's dumpy-type level has been the workhorse of leveling instruments for more than 150 years. Even with advancements in other leveling instruments, such as the automatic level and the laser, the dumpy may still be the instrument of choice in a construction environment because of its stability.

Automatic Compensator Levels

Compensators were developed about 50 years ago and incorporated into field levels. The compensator is a free-swinging pendulum arrangement in the optical path that maintains a fixed relationship between the line of sight and the direction of gravity. If the instrument is an adjustment, the line of sight will be maintained as a horizontal line. Compensator instruments are extremely fast to set up and level.

Laser Levels

A laser level uses a laser beam directed at a spinning optical reflector. The reflector is oriented so that the rotating laser beam sweeps out a horizontal reference plane. The level rod is equipped with a sensor to detect the rotating beam. By sliding the detector on the rod, a vertical reading can be obtained at the rod point. Laser levels are especially useful on construction sites. The spinning optics can also be oriented to produce a vertical reference plane.

Digital Levels

Digital levels are electronic levels that can be used to more quickly obtain a rod reading and make the reading process more reliable. The length scale on the level rod is replaced by a bar code. The digital level senses the bar code pattern and compares it to a copy of the code held in its internal memory. By matching the bar code pattern, a rod reading length can be obtained. Digital levels are available for ordinary and precise leveling applications.

Level Rods

In addition to the chosen leveling instrument, a level rod is required to be able to transfer elevations from one point to another. The level rod is a graduated length scale affixed to a rod and held vertically on a turning point or benchmark. The scale is read to obtain the vertical distance from the point to the line of sight.

Level rods are graduated in feet, inches, and fractions; feet, tenths, and hundredths, or meters and centimeters. Rods used in ordinary leveling may be multipiece extendable rods with graduations marked directly on the rod material or on a metal strip affixed to the rod for support. Rods used in precise leveling are one-piece rods with a stable invar metal graduated scale supported under constant tension by the rod. A precise rod can be calibrated for changes in length caused by temperature.

Accurate field leveling work is also aided by the use of rod targets, rod levels, and stable turning points when required.

162.5 Systematic Errors

Earth Curvature

The curved shape of the Earth results in the equipotential surface through the telescope departing from the horizontal plane through the telescope as the line of sight proceeds to the horizon. This effect makes actual level rod readings too large by the following approximate relation:

$$C = 0.0239D^2$$

where C is the error in the rod reading in feet and D is the sight distance in thousands of feet.

Atmospheric Refraction

The atmosphere refracts the horizontal line of sight downward, making the level rod reading smaller. The typical effect of refraction is equal to about 14% of the effect of Earth curvature. Thus, the combined effect of curvature and refraction is approximately

$$(C - r) = 0.0206D^2$$

Instrument Adjustment

If the geometric relationships defined in the preceding discussion are not correct in the leveling instrument, the line of sight will slope upward or downward with respect to the horizontal plane through the telescope. The test of the line of sight of the level to ensure that it is horizontal is called the "two-peg test." If the line of sight is inclined, the difference in elevation obtained from the two setups will not be equal. Either the instrument must be adjusted, or the slope of the line of sight must be calculated. The slope is expressed as a collimation factor, C, in terms of rod reading correction per unit of sight distance. It may be applied to each sight by the following:

$$\text{Corrected rod reading} = \text{Rod reading} + (C_{\text{Factor}} \cdot D_{\text{Sight}})$$

In ordinary differential leveling, these effects are canceled in the field procedure by always setting up so that the backsight distance and foresight distance are equal. The errors are canceled in the subtraction process. If long unequal sight distances are used, the rod readings should be corrected for curvature, refraction, and collimation error.

Orthometric Correction

When long, precise level routes are surveyed, it is necessary to account for the fact that the equipotential surfaces converge as the survey proceeds north. The correction to be applied for convergence of equipotential surfaces at different elevations can be calculated by

$$\text{Correction} = -0.0053\sin(2\phi)H\Delta\phi_{\text{rad}}$$

where ϕ is the latitude at the beginning point, H is the elevation at the beginning point, and $\Delta\phi$ is the change in latitude from the southerly station to the northerly station expressed in radians. A useful program has been developed by the National Geodetic Survey to assist with this solution. Program LVL_DH converts the published orthometric height difference between two NAVD 88 benchmarks into a leveled height difference by removing the orthometric correction from the published relative height. The program is available from the National Geodetic Survey at the address given at the end of this chapter.

Defining Terms

Benchmark (BM) — A benchmark is a permanent object having a mark of known elevation, such as, a cross-chiseled on a boulder or a concrete monument with an embedded brass disk.

Datum — Any quantity or set of such quantities that may serve as a reference or basis for calculation of other quantities.

Datum sea level — An equipotential surface passing through a specified point at mean sea level that is used as a reference for elevations; a surface passing through mean sea level at certain specified points to which elevations determined by leveling are referred. Note that, in general, the latter surface is not an equipotential surface.

Elevation — The distance, measured along the direction of gravity (plumb line), between a point and a reference equipotential surface, usually the geoid.

Height — The distance, measured along a perpendicular, between a point and a reference surface, such as the ellipsoidal height; the distance, measured along the direction of gravity, between a point and a reference surface of constant geopotential, such as the orthometric height. Note that the term *elevation* is preferred when the geoid is used as the reference surface.

Mean sea level — The arithmetic mean of elevations (heights) of the water's surface observed hourly over a specific 19-year cycle.

Vertical — The direction in which gravity acts.

References

Anderson, J. M. and Mikhail, E. M. 1998. *Surveying: Theory and Practice*, 7th ed. WCB/McGraw-Hill, New York.

Bomford, G. 1980. *Geodesy*, 4th ed. Clarendon, Oxford.

Leick, A. 1990. *GPS Satellite Surveying*. John Wiley & Sons, New York.

Moffitt, F. H. and Bossler, J. D. 1998. *Surveying*, 10th ed. Addison-Wesley, Reading, MA.

National Geodetic Survey. 1984. *Standards and Specifications for Geodetic Control Networks.* Federal Geodetic Control Committee, National Geodetic Survey, U.S. Department of Commerce, Silver Spring, MD.

National Geodetic Survey. 1986. *Geodetic Glossary.* National Geodetic Survey, U.S. Department of Commerce, Silver Spring, MD.

Schmidt, M. O. and Wong, K. W. 1985. *Fundamentals of Surveying,* 3rd ed. PWS Publishers, Boston.

Schwarz, C. R., Ed. 1989. *North American Datum of 1983.* NOAA Professional Paper NOS 2, January. National Oceanic and Atmospheric Administration, Washington, DC.

Torge, W. 1991. *Geodesy,* 2nd ed. De Gruyter, Berlin.

Vanicek, P. and Krakiwsky, E. J. 1986. *Geodesy: The Concepts,* 2nd ed. Elsevier, New York.

Wolf, P. R. and Ghilani, C. D. 2002. *Elementary Surveying: An Introduction to Geomatics,* 10th ed. Prentice Hall, Upper Saddle River, NJ.

Further Information

The material in this chapter is intended only as an overview of elevation reference systems and basic surveying methods. There are many textbooks dedicated exclusively to various aspects of surveying. For a more complete presentation of surveying theory, consult *Surveying: Theory and Practice,* 7th ed., by Anderson and Mikhail, WCB/McGraw-Hill, 1998, or *Elementary Surveying,* 10th ed., by Wolf and Ghilani, Prentice Hall, 2002. For more detailed information on the capabilities of various instruments and software, Business News Publishing prepares the trade magazine *P.O.B.* (Business News Publishing Company, 755 W. Big Beaver Rd., Suite 1000, Troy, MI 48084), and American Surveyors Publishing Company prepares the trade magazine *Professional Surveyor* (American Surveyors Publishing Company, Inc., Suite 501, 2300 Ninth Street South, Arlington, VA 22204). Each of these publications conducts annual reviews of surveying instruments and software. These listings allow the reader to keep up-to-date and compare "apples to apples" when analyzing equipment.

Survey control information, software, and many useful technical publications are available from the National Geodetic Survey (NGS). The address is NGS Information Services, NOAA, N/NGS12, National Geodetic Survey, SSMC-3, #9202, 1315 East-West Highway, Silver Spring, MD 20910-3282. Web address: *www.ngs.noaa.gov/INFO/NGSinfo.html.*

163

Distance Measurements

N. W. J. Hazelton
The Ohio State University

R. Ben Buckner
Surveying Education Consultant

When the word "distance" is used in surveying without qualification, it is usually construed to mean *horizontal distance* — that is, a distance either measured directly along a horizontal line or measured along a slope and then mathematically projected to the horizontal. If linear measurements along other alignments are being discussed, they are given specific designations, such as *slope distance* or *vertical distance* (elevation difference).

Since ancient times, direct distance measurements have been made using lines, cords, ropes, rods, chains, tapes, and other such devices. Distances have also been measured indirectly, employing stadia and other tacheometric methods, as well as trigonometrically, using a combination of distance and angle measurements and calculations. Technological advances during the last half of the 20th century have created changes in the methods used to measure many distances. During the 1960s and 1970s, most surveyors made a gradual shift from the chain and tape to electronic distance measurement systems that measure distances with light waves or microwaves. During the 1980s, this trend continued into using *total station* systems, which measure both distances and angles electronically and process the combined measurements into a Cartesian coordinate form for the survey points. Thus, many distances are determined by computations from coordinate positions rather than by direct measurements. Most recently, there has been a trend toward using the global positioning system (GPS) for precise determination of positions of points, and these positions are then used to determine indirectly the distances between survey points. Eventually, GPS positioning may replace both the tape and electronic distance instruments (including the total station) for most surveying measurements, including angles, elevations, and distances. GPS is covered in Chapter 167.

An overview of the concepts of direct and indirect distance measurements will be given here. Only the basic measuring procedures and the appropriate corrections needed to achieve optimum accuracy are discussed in this chapter; the geometric, electronic, and other physical principles of the measurement systems will not be discussed in detail. See Chapter 161 for discussion of quality control issues and the basic terminology used here for that aspect of distance measurement.

0-8493-1586-7/05/$0.00+$1.50

In a surveying world increasingly dominated by GPS and total stations, it may seem strange to spend as much time and space on taping as this chapter does. The reason for this is that taping manages to include most of the common forms of corrections and errors found in all distance measurements, while allowing them to be studied in a way that is easy to understand.

163.1 Fundamentals of Distance Measurement

The Need for Corrections

Measurements, whether of distance or any other quantity, are but estimates of the magnitude of the quantity. Few initial readings (observations) in surveying contain the accuracy desired without some correction for systematic errors. Much of the consideration today regarding measurements is not so much of how to operate the instruments and perform the calculations to reduce the data, but of how to identify and evaluate the magnitude of the corrections that must be applied to the readings. The "true value," theoretically, is always equal to the reading plus the corrections. That is, $T = R + \Sigma C$. If the true value is known (which is highly unusual, but a standard distance for calibration would be an example), the reading needed to achieve this is $R = T - \Sigma C$. We can also think of this as knowing a *required* value, such as for setting out points, and computing the actual reading needed to achieve this in the prevailing conditions. Application of this basic concept will be illustrated subsequently. The following sections discuss the typical methods of determining survey distances.

Taped Measurements

Most taped measurements are made using a steel tape, usually of 100-ft length. The tape is laid along a line between points and stretched tight. If the distance desired is longer than the length of the tape, more than one tape length is employed, the ends of the tape being marked with taping pins (sometimes called "arrows"). Alignment of the ends of the tape, to keep the tape on a straight line, is achieved either by hand signals from a rear tapeman or by a transit or theodolite centered over one end of the overall line.

The tape is either allowed to rest on the ground surface ("fully supported"), or suspended throughout its length ("end supported"), with the distance mark on the tape being transferred to the ground using plumb bobs at one or both ends. In the former case corrections for ground slope must be made, the slope having been measured using other instruments, such as hand levels, clinometers, or theodolites. In the latter case, the two ends of the tape are usually held near the same elevation as determined by hand levels or clinometers. End-supported taping requires corrections for the sag of the tape, but possibly not the slope. In either type of taping, corrections must usually be made (depending on the accuracy desired) for calibrated length, changes from calibration temperature to field temperature, and tension (if different from that used when calibrated).

Steel tapes are made in various lengths other than 100 ft, including 25 m, 50 m, 100 m, 200 ft, and so forth. The principles in the use and correction of readings is much the same for various tapes, regardless of length.

Tapes are also made of other materials, including woven fiber and invar (an iridium-nickel-vanadium-steel alloy). The fiber ("cloth") tapes are used only for rough measurements, and there is usually no attempt to apply corrections to the readings. In contrast, invar tapes are used for precise surveys, such as the establishment of calibration base lines for electronic distance measurements, and all applicable corrections must be made. The use of these more precise tapes and the less precise "cloth" tapes will not be covered here.

Tacheometric Measurements

The most common method in the tacheometry category uses the stadia. Such measurements are made through the telescope of a transit, theodolite, or level instrument. The measurement is achieved by observing where the two horizontal stadia hairs, viewed through the telescope, strike a graduated rod held vertically on a survey point. The difference between the two readings is a function of the separation of the stadia hairs, the focal length of the telescope, and the distance between the instrument and the rod. When the line of sight is other than horizontal, the vertical or zenith angle must also be considered. Stadia distances are generally accurate to only about 1 or 2 ft for normal sight distances, and thus there is little need to be concerned about corrections to the data, other than for slope. Stadia tacheometry has been largely superseded by total stations, but it is still used occasionally to get a quick, approximate distance, especially when leveling, and to check that level sight lengths are about equal.

The subtense bar is another tacheometric instrument. The principle is the precise measurement of the horizontal angle subtended between two distinct targets at the two ends of the bar. The bar is erected on a tripod so as to lie horizontally and perpendicular to the line of sight, centered over one end of the line. The theodolite is centered over the other end of the line. The separation between the two targets defining the bar is a known length (commonly 2 m). Using this known length and the measured angle, the horizontal distance is computed by trigonometry. For relatively short distances (up to perhaps 150 to 200 ft) this method can exceed the precision of both taping and electronic measurements, and thus it is very useful for measurements across busy streets or other small, inaccessible places. Because this method has no error caused by sloping lines, it is useful for measuring to high places, such as to the tops of buildings. Its application is overlooked nowadays because of the ease of using electronic instruments, even though it is often more precise for short measurements, has advantages similar to those of electronic distance measurement, and avoids slope corrections. We can consider the use of subtense bars for distance measurement to have been totally superseded by total stations.

Electronic Distance Measurements

The most common electronic distance measurement instruments (EDMs) employ an infrared beam reflected off a system of reflectors called *retroprisms*. The infrared beam is reflected onto the instrument for interpretation of the wavelengths and partial wavelengths comprising the double-slope distance between the instrument and the reflector. Older EDMs are individual units, measuring distance only. The most modern versions are part of the aforementioned total station systems, which also measure angles. Whether an individual unit or part of a more complete survey system, the principle of EDM operation is much the same.

A common fallacy, particularly when using total station systems, is that the measurements are free of errors. As has been mentioned, there is error in any measurement system. When measuring with an EDM, the surveyor must be concerned with calibration, just as with any distance-measuring system. For example, the electronic center of the instrument may not be located precisely along the same vertical line as the geometric center plumbed over the ground station. The instrument will usually have this small, constant instrument error, as well as a scale error that is proportional to the distance measured (often called the *parts per million* or *ppm correction*). Also, the reflector has a "constant," and the optical plummets in the tribrach mounting system can be out of adjustment. Some EDMs may have a "cyclic error," caused by errors in the determination of the phase differences between the incoming and outgoing signals, which can cause small differences in distance that vary in a cycle with distance. The magnitude and sign of these errors should be determined by field tests for best accuracy. EDMs are also affected by variations in atmospheric pressure and temperature. Microwave measurements are also affected by humidity.

All measurements made with EDMs are "slope" measurements between the center of the EDM and the reflector, and thus must be corrected to horizontal distances using measured vertical or zenith angles, together with instrument and reflector heights.

How Does EDM Work?

The basic idea is that a carrier beam is sent out by the main piece of equipment (the master instrument in microwave systems, or the instrument in visible light and infrared systems) and returned by the other piece of the equipment (the reflectors for visible light and infrared systems, or the remote instrument for microwave systems). Another frequency is superimposed on the carrier wave and it is this that is used to measure the distance. The superimposition can be done by amplitude modulation, AM (for visible and infrared systems), or frequency modulation, FM (for some of the later microwave equipment). Note that the frequency of the carrier wave remains practically constant, and the meteorological corrections are applied on the basis of the carrier frequency, not the modulating frequencies.

There have been several ways to measure the distance using the superimposed modulating frequencies. The first of these was used by the early Tellurometers, such as the MRA series. Here, the modulating frequencies were set up so that the difference between the different frequencies allowed the determination of parts of the overall distance, in such a manner that the transit time was not the critical factor in measuring the distance. See Bannister et al. (1998) for more detail.

A later method is discussed in Moffitt and Bossler (1999). Here the phase difference of the signal at different frequencies is used to build up the measurement, generally in powers of ten (meters or feet, or some other suitable unit); hence, the term *decade modulation*. Thus, the finest (highest frequency) modulation would yield a measurement of the parts of 10 m, the next would provide the parts of 100 m, the next parts of 1000 m, and so on. This was the approach adopted in the Tellurometer CA1000 and MA100, and the HP 3800 series. The user generally switched between the different modulation frequencies manually, and grouped the recorded measurements to produce the final measurement. The advantage of this method was that the oscillators in the EDM did not have to be good enough to measure the time for the beam to travel, they only had to give a good performance for phase difference measurement and calibration, and to be read out to two or three significant figures. This reduced the cost and weight of EDM.

The process of decade modulation suffered two problems. The first was that measurement was rather time consuming, and that the operator had to change the frequencies manually (and at both ends of the line for microwave systems). Rather than include a lot of servo or heavy electronic equipment, a new method was developed, mainly in infrared laser-based EDMs, of measuring using two modulations that were very close together in wavelength. One of these, set to a 10-m wavelength for example, would yield the parts of 10 m in the distance, generally to the nearest millimeter, that is, 1 in 10,000 precision. The other frequency would measure the distance with a 9.990 m wavelength (as an example only) as the modulation value. The difference between the resulting phase differences from the two frequencies would give the overall distance with no ambiguity up to 10 km. With high-speed electronics and high-quality oscillators in the EDMs, the required 1:10,000 measurement precision is generally well exceeded and two frequencies only are required. The EDM switches between the two to establish the overall distance and then measures a great many "fine" measurements to obtain the mean. The resulting measurement, obtained in a matter of a few seconds, will contain perhaps hundreds of actual measurements averaged to give the best results. Many modern machines continue to update their result as the machine stands, and you may see the last digit change from time to time. See Bannister et al. (1998).

The second problem with decade modulation was that you still needed fairly high-precision oscillators (for the time) in the EDM, and problems with these affected the overall measurements, as well as making the overall machine very expensive. For example, the Tellurometer MA100 (known as the Modlite) needed a small oven to be running to keep the oscillator at the correct temperature, which drained a lot of power. In addition, the decade system in the early days was susceptible to beam interruption, which caused problems with the whole measurement.

Despite all these difficulties, even the most inconvenient EDM was far quicker, a lot easier, and a great deal better than triangulation and taped baseline measurement. The equipment was expensive in the early days, and so out of the reach of most small businesses, but as it became cheaper, it revolutionized the way that surveying was done. It is a classical example of the effects of introducing a new and powerful technology.

The latest development in EDMs is pulsed lasers. These use a timing system for distance measurement, as this is possible with the improved oscillators now available. By making the laser more powerful and the detector more sensitive, you can dispense with the reflector over short distances. This allows you to measure to inaccessible objects easily. Using a visible spot, you can be sure you are measuring to the correct point. These EDM are rather more expensive than conventional systems, but they do allow you to do things you could not do before. See Bannister et al. (1998) for more details.

Today, the two-frequency system is almost universally used in EDMs, with the newer pulsed laser systems starting to appear in more specialized equipment. As the price of the latter falls, we can expect to see greater adoption of the newer technology. Of particular note is the absence of long-range EDM on the market. Distances over a few miles are more commonly measured using GPS today, so most available EDM equipment is capable of only relatively short ranges.

Indirect Computational Methods

Distances are commonly determined by indirect measurement using trigonometric principles. The most common of these methods are *intersection* and the coordinate *inverse*. Using intersection, a distance is measured (or computed from other measurements) and angles measured to a common point from the two ends of this *base line*, which forms an oblique triangle. The unknown sides of the triangle are solved using the sine law. Using the inverse, the computation starts with determination of the departure (difference in x coordinates) and the latitude (difference in y coordinates) of the line defined by the two coordinate points. The distance is the hypotenuse of a right triangle whose sides are the departure and latitude of the line; it is determined using the Pythagorean theorem. Many variations of these indirect methods occur in practice and usually involve some variation of an oblique triangle solution or the coordinate inverse. Some of these concepts are discussed in Section 163.2.

The usual method of coordinate inverse outlined above can be implemented as follows. If we know the coordinates of the two points in some orthogonal coordinate system, for example, (X_1, Y_1) and (X_2, Y_2), then the distance, d, of the line connecting them can be computed using

$$d = \sqrt{(X_2 - X_1)^2 + (Y_2 - Y_1)^2}$$

The above formula gives the horizontal distance between the two points. If the heights of the points, Z_i, are known, then the slope distance, l, between the points can be calculated using the three-dimensional (3D) extension of the Pythagorean theorem:

$$l = \sqrt{(X_2 - X_1)^2 + (Y_2 - Y_1)^2 + (Z_2 - Z_1)^2}$$

The azimuth of the line, θ, going from point 1 to point 2, based on whatever the zero direction of the coordinate system represents, can be computed using

$$\theta = \arctan\left(\frac{(X_2 - X_1)}{(Y_2 - Y_1)}\right)$$

Note that it is necessary to determine the correct quadrant in which the line lies. The ATAN2() function in the MS-Excel spreadsheet and Fortran can compute this automatically, based on the signs of the numerator and denominator in the formula above.

Photogrammetry is another method for determining distances. After proper orientation of a stereo-model, coordinates can be read from it, from which distances can be determined by the same inverse computation as discussed above. The science of photogrammetry is briefly explained in Chapter 165.

Approximate Methods

As mentioned, all measurements are estimates of an unknown quantity. The method used to determine a distance is chosen as dictated by the precision needed. The above methods are typical for measurements where precisions of a few millimeters to a fraction of a meter are desired. If less precision is needed, a range finder, a measuring wheel, pacing, the odometer on a bicycle or vehicle, map scaling, digitizing coordinates from maps (with subsequent "inverse" computations), or even visual estimation can be used.

Range finders are optical instruments that might be considered under the classification of tacheometric methods. These instruments and the calibrated measuring wheel or an odometer on a bicycle have precisions comparable to the stadia method. A car odometer can yield a precision of perhaps 50 to 100 ft in a mile, if calibrated. The accuracies of the other approximate methods depend on several factors, particularly the map scale and the map accuracy. It must be emphasized that digitizing from maps cannot yield high precisions of positions or distances. Even for a fairly large-scale map, the accuracy of such methods is seldom better than 10 ft.

The importance of approximate methods is that they provide a quick and easy check of the precise methods for gross errors. They can also be used to help find previously placed marks, to help with setting out points on construction and as an aid to data collection for topographic surveys. Skill in using approximate methods can make the overall process much quicker and easier, especially as modern equipment becomes more self-contained and less useful for quick, approximate measurements. For example, one cannot easily use a precise, survey-grade GPS receiver to help look for old monuments or to help draw up a recovery diagram for the surveyed point, but a tape is very useful.

163.2 Applications and Calculations

Taping

The variables affecting a horizontal distance using a tape, with their correction equations, are as follows:

1. *Calibration.* $C_l = l_t - l_r$ The correction per tape length is the actual ("true") length of the tape minus the nominal length (reading between the end marks). The actual length is determined by calibration at some observed temperature, tension, and support condition, comparing the tape's length with an accurate baseline.
2. *Temperature.* $C_t = K_t l(t_f - t_s)$. Temperature correction is given by the coefficient of thermal expansion multiplied by the nominal length of the tape times the difference between the field temperature and the calibration temperature. The correction is positive when $t_f > t_s$, since the tape expands when the temperature is warmer. The value for K_t is 0.000 006 45 units per unit per degree Fahrenheit (the constant is 0.000 011 6 if degrees Centigrade are used).
3. *Pull* (tension). $C_p = [(P_f - P_s)l]/AE$. This is the correction for tension if the field tension is different from the standardization tension. In this equation, A is the cross-sectional area of the tape and E is the modulus of elasticity, which is 29,000,000 lb/in^2. Care should be taken that the field tension appears first in the equation, subtracting the standardization tension from it.
4. *Sag.* $C_s = -(w^2 l)/(24P^2)$. This is the correction for sag, where w and l are the weight and length of the portion of the tape suspended between supports. The pull, P, is the actual tension on the tape. It has no relationship to the standardization pull.
5. *Slope.* $C_g = -(v^2)/(2s)$. This is the correction for grade or slope, which, when added to the slope length, yields the horizontal distance. This correction is applied when the tape ends are not at the same elevation. In this equation s is the slope distance being corrected and v is the elevation difference between the two tape ends at this length. This equation is approximate but is satisfactory for slopes less than about 10%. The Pythagorean theorem can be used in any case, solving the equation $h = \sqrt{s^2 - v^2}$ for the horizontal distance, h. In the event that vertical or zenith angles have been measured, h can be computed from $h = s\cos\alpha$ or $h = s\sin z$, where α is the measured vertical angle and z is the measured zenith angle. In this case, no correction is computed, but the horizontal distance is calculated directly.

The alignment error is generally not corrected in practice, but instead rendered negligible by the process of careful tape alignment. The correction for pull or tension can be avoided by always using the correct tension. It is common to combine the calibration and temperature correction, by computing the temperature for which the tape is the correct length, and using this as the standardized temperature. This reduces the number of computations, which in turn reduces the potential for gross errors in calculation.

For many errors the correction can usually be made for one tape length and then multiplied by the number of tape lengths, as long as the condition causing the error does not vary between tape lengths. This generally always applies to the calibration, temperature, and tension errors, and sometimes to the sag and slope errors.

Taping problems are of two types: (1) calculation of the horizontal distance between two established points; and (2) calculation of the reading to be observed to establish a given distance. The theory of systematic errors is applicable in making taping corrections. Solving for the true value $T = R + \Sigma C$ is the first type of problem. If "layout" is required, then the reading $R = T - \Sigma C$ is computed from the given value and the corrections.

The following problems will use a calibrated 100-ft tape, found to be 99.992 ft long at 70°F, 15 lb tension, fully supported. It has a cross-sectional area of 0.006 in^2 and weighs 2.2 lb. In the solutions, N is the number of tape lengths.

Example 163.1

A reading of 458.97 ft is observed between two points when the field temperature is 40°F, along a 4% slope. Find the correct horizontal distance.

Solution. There are three systematic errors to consider: calibration, temperature, and slope.

$$C_L = (l_t - l_r)\text{N} = (99.992 - 100.000) \times 4.59 = -0.037 \text{ ft}$$

$$C_t = K_t l (t_f - t_s)\text{N} = 0.000\,006\,45 \times (100) \times (40 - 70) \times 4.59 = -0.089 \text{ ft}$$

$$C_g = -\frac{v^2}{2l}\text{N} = [4^2 \div (2 \times 100)] \times (4.59) = -0.367 \text{ ft}$$

$$\sum C = -0.493 \text{ ft, from which } T = 458.97 - 0.49 = 458.48 \text{ ft.}$$

We could also set the standardized temperature for this tape to be the temperature when the tape is exactly 100 ft long:

$$C_t = K_t l (t_f - t_s)$$

$$t_f = \frac{C_t}{K_t l} + t_s$$

$$t_f = \frac{0.008}{(0.000006\,45 \times 100)} + 70$$

$$t_f = 82.4^\circ \text{F}$$

Consequently, in future we could apply the temperature correction and allow for the calibration of the tape at the same time, by adopting t_s = 82.4°F (or more reasonably 82°F). Note that this temperature will have to be updated with each calibration of the tape.

Example 163.2

A distance of 200 ft is to be laid out along a horizontal alignment. The tape must be suspended for 60 ft of one of its tape lengths. A tension of 30 lb is used for this portion of the layout; otherwise, 15 lb is used. Temperature is 70°F.

Solution. There are three systematic errors to consider: calibration, tension, and sag.

$$C_L = (l_t - l_r)\text{TL} = (99.992 - 100.000)2.00 = -0.016 \text{ ft}$$

$$C_p = (P_f - P_s)l \div AE = (30 - 15)60 \div (0.006 \times 29 \times 10^6) = 0.005 \text{ ft}$$

$$C_s = -\frac{w^2 l}{24P^2} = -(2.2 \times 0.6)^2 60 \div (24 \times 30^2) = -0.005 \text{ ft}$$

$$\sum C = -0.016 \text{ ft} \quad \text{and} \quad R = 200.00 - (-0.016) = 200.016 = 200.02 \text{ ft}$$

It is seen in this example that the added tension for the end-supported part of the taping compensated for the sag effect. This works in this case, but it is not recommended as a general practice, as it is necessary to compute the sag correction first, and then find the appropriate tension to correct that effect, when it would be a lot easier and less error-prone to use the correct tension and compute the corrections once after the fact.

What is the quality of taped distances? This depends very much on the care taken with the measurements, whether the tape is calibrated, and the effort put into determining the systematic errors, especially calibration of thermometers and spring balances. For conventional taping with calibrated tapes of about 100 m in length, applying all corrections and taking considerable care about the measurements, we would expect the precision of a single measured distance to be on the order of ±0.020 m + 20 ppm. This also assumes that the lines are not longer than about 1 km.

To achieve better results than this, we have to move to invar tapes and the processes used for baselines in the past. This allowed precisions of 10 to 12 parts per million over lines of up to 20 miles.

It can be seen that for high-precision distance measurement EDM is now usually quicker, cheaper, easier and better, but there is still a place for taping, especially if care is taken in its operation.

Stadia

Stadia hairs or lines are placed in most telescopes so that the *stadia interval factor, K*, equals 100. This makes it convenient to measure distance, merely computing the difference between the two stadia hair readings and multiplying by 100 to get the distance between the rod and the theodolite. Since the stadia hairs are each read with a precision of approximately ±0.01 ft at ordinary distances, the precision of a stadia distance is no better than ±1.0 ft, with variation according to how clearly the rod is seen and how carefully it is read. Elevations are good to no better than ±0.2 ft.

Old *external focusing* telescopes had a *stadia constant, C*, of about 1 ft. However, the instruments of recent generations (post-1920 for European instruments, post-1960 for many U.S. instruments), being *internally focusing*, eliminate this constant, that is, $C = 0$.

For a horizontal sighting $H = KI + C$, assume that K and C are as defined earlier and I is the rod intercept (difference between the upper and lower rod readings). When vertical angles are involved, $H = KI \cos^2 \gamma + C \cos \gamma$, where γ is the vertical angle.

For modern instruments, with a zenith angle of z, we can use

$$H = KI \cos^2 \gamma = KI \sin^2 z$$

$$EL = EL_{IP} + h_I + \frac{KI}{2} \cos 2z - r_C$$

where EL is the elevation of the point sighted, EL_{IP} is the elevation of the point over which the instrument is set, h_I is the height of the instrument's telescope above the point, and r_C is the reading on the rod beneath the center stadia wire (the horizontal crosshair).

Example 163.3

A modern theodolite has a stadia interval factor of 100. The reading on the upper stadia hair is 7.54 ft, and on the lower hair it is 3.66 ft. The zenith angle to the center crosshair is 96°36′36″. What is the horizontal distance, to the nearest foot?

Solution. Assuming the stadia constant to be zero, the appropriate values in the preceding equation are as follows:

$$H = 100 \times (7.54 - 3.66)\sin^2(96°36'30'') = 383 \text{ ft}$$

Subtense Bar

After the horizontal angle is measured between the two end targets on the bar, the horizontal distance is computed from $H = \frac{1}{2}b\cot\alpha/2$, where b is the bar length and α is the horizontal angle. Using a bar of the usual 2-meter length, the value of H is in meters, since half the bar length is 1 m.

Note that the distance is *always* horizontal, since the horizontal angle is the same regardless of the relative elevation of the two points. Thus, no slope corrections are ever required. In practice, a 1″ theodolite is generally used and several angles are measured, in order to achieve adequate precision.

Example 163.4

A 1″ theodolite is used to measure the angle between the targets on the ends of a 2-meter subtense bar. The mean of six independent readings of the angle is 0°45′46″. Compute the horizontal distance between the theodolite and the bar.

$$H = \frac{1}{2}b\cot\alpha/2 = \frac{1}{2}(2 \text{ meters})\cot(0°45'46''/2) = 75.110 \text{ m}$$

Electronic Distance Measurements

Although the instrument constant is, in practice, usually adjusted to zero whenever the instrument is serviced, calibration can discover a small instrument constant. Similarly, the reflector constant is usually keyed into the instrument by the surveyor and thus compensated, but a field test of reflector constants can discover slight discrepancies between what the manufacturer states the constant to be and what it actually is. Likewise, the atmospheric errors are generally keyed into the instrument after reading the temperature and pressure but are sometimes overlooked, and an old setting remains in the instrument. The surveyor should be aware of these possible error sources. The following example assumes that the atmospheric errors and reflector constant have been handled properly.

Example 163.5

An EDM has been calibrated using a four-station NGS base line, and errors are found as follows: C = +0.003 m and P = +0.000 004 56, where C is the constant correction and P is the "scale" correction, which may be expressed as 4.56 ppm. The zenith angle along the line is 88°34′42″. The observed slope distance is 1789.783 m. What is the corrected horizontal distance?

Solution. The ppm correction is +0.000 004 56 (1789.8 m) = +0.0082 m. The constant correction is +0.003 m. The corrected slope distance is

$$1789.783 + 0.008 + 0.003 = 1789.794 \text{ m}$$

$$H = 1789.794 \sin 88°34'42'' = 1789.243 \text{ m}$$

An additional correction might be necessary if the reflector is not set at the same height as the EDM because, in this case, the measured slope angle does not correspond to the slope of a line connecting the two ground points. The correction involves adjusting the measured zenith angle before calculating the horizontal distance as done previously.

What is the effect of the meteorological corrections? We find that for the common infrared EDMs, a 1 ppm error in distance is caused by a 1°C error in temperature, a 3.6-hPa error on atmospheric pressure, or a 25-hPa error in vapor pressure. As a consequence, we can generally ignore the effect of humidity for infrared instruments. (However, humidity is the biggest meteorological factor in microwave EDM and should also be allowed for when using infrared EDM in hot, humid conditions.) Thus, for good results from EDM, we must measure the temperature and pressure for each line, and apply the corrections. We should also ensure that the thermometer and barometer used for the EDM are properly calibrated, so that they produce accurate readings. In most cases, the temperature and pressure can be keyed into the total station, which makes the process very quick and simple.

Reflector constants are another source of concern. There are a number of different reflector systems on the market today. The "standard" reflector systems produced by Leica are designed to have their weight balanced, and so have a reflector constant of around –30 mm. Most other standard reflector systems are produced with a zero reflector constant. Leica's 360° prisms have a reflector constant of around –19 mm, while some of the other 360° systems have different reflector constants. This means that it is important to determine which reflectors are being used, and that the correct constant has been entered into the total station. Most organizations prefer to standardize on one particular type of reflector to avoid this problem, but sometimes it cannot be avoided and the reflector constants in use must be checked in detail.

With some of the reflector mounts currently available, it is possible to set the reflector in two positions: zero constant or –30 mm constant. It is a good idea to check that the reflectors are properly mounted for the settings that are to be used, and all the reflectors are mounted the same way.

With modern total stations that allow any reflector constant to be keyed in, as opposed to earlier EDM that was either pre-set at the factory or had a switch for just the two common settings, it is possible to include the EDM's offset or constant error in the reflector constant, thereby removing it from consideration. This underlines the importance of thinking about a total station and its reflectors as a single measurement system.

Most modern EDMs are able to achieve a high level of precision, which is commonly expressed using two numbers in the form $\pm x$ mm + y ppm. The first number, x, relates to the constant error of the instrument, plus the effect of centering by the user, plus any cyclic error. The second number, y, is a scale factor and includes the scale error and (mostly) uncertainty in the meteorological correction. EDMs commonly have manufacturers' claimed precisions of the order of ±10 mm + 10 ppm, but with the improvement in electronics and the shorter ranges more common today, some instrument makers claim ±5 mm + 5 ppm for some machines.

If the EDM is properly calibrated, the biggest component of the precision is the meteorological correction. A 10°C error in temperature can lead to a 10-ppm systematic error in the distance, which may exceed the manufacturer's figures substantially. This is why it is important to pay proper attention to collecting temperature and pressure data for each line.

Sea Level Correction

Distances in surveying are not only assumed to be horizontal distances, but for consistency are also reduced to the equivalent distance at sea level. This is because the distance between two locations on the Earth (expressed using some coordinate system) differs with the elevation of the points. For many applications, this correction is smaller than the other errors in the measurements and can safely be ignored. However, with EDM and GPS being able to span long distances, this correction is necessary to ensure consistency of distances, especially when the surveyor is working with state plane coordinates or trying to tie in to national control points.

The rigorous formula for sea level correction will not be given here, since it is mainly of interest for geodetic surveys, but two useful approximations follow:

$$s = d\left(1 - \frac{h_m}{R + h_m}\right) \quad \text{or}$$

$$s = d\left(1 - \frac{h_m}{6{,}365{,}400}\right)$$

where s is the sea level distance, d is the measured distance, reduced to a horizontal distance, h_m is the elevation of the midpoint of the line, and R is the radius of the Earth. The second, simplified formula can be used for most practical purposes when distances are in meters.

A note on the h_m term: if the h_m is the geoidal or orthometric height, s will be the geoidal distance. If h_m is the ellipsoidal height, then s will be an ellipsoidal distance.

Example 163.6

A surveyor measures a distance with EDM and reduces it to a horizontal distance. The horizontal distance is 5,607.234 m. The midpoint of the line is 348.23 m above the ellipsoid. What is the corrected ellipsoidal distance?

Solution. Substituting the values into the simplified formula above, we obtain

$$s = 5{,}607.234 \times \left(1 - \frac{348.23}{6{,}365{,}400}\right)$$

$$s = 5{,}606.928$$

We can see that the correction was –0.306 m, which is 54.5 parts per million. This is larger than most meteorological corrections and must be applied for all geodetic work, and for work that ties into a state, national, or international coordinate system.

Laser Scanning

A recent development in surveying equipment has been laser scanners. These work by using a steered pulsed laser beam to scan the field of view before the instrument at a fine resolution. The pulsed laser gives a range to any object that allows a return signal to be reflected, and knowing the direction that the beam was transmitted, the 3D location of the object can be calculated. The result is a large number of 3D points measured to various parts of objects in the field of view, out to a distance of perhaps 100 m.

These points are processed by software that searches for neighboring points and joins those that seem to go together by a series of surfaces. The resulting surfaces and the measurements points can then be downloaded to a suitable 3D CAD package, where a 3D map of the object could now be used as needed. For complex objects, it is possible to link together several scans, based on knowing the scanner's location and orientation at each observation location.

This equipment is capable of recording several thousand 3D points in a matter of minutes, allowing very rapid mapping of complex 3D objects with relatively little effort. It is a direct competitor to digital terrestrial photogrammetry, with the differences that the laser scanner can work well at night, while the photogrammetric equipment can measure longer distances and has a recording system that allows easier interpretation of objects at a later date (although you can always take photographs while running the laser scanner). Laser scanners are still quite expensive, but so is a terrestrial digital photogrammetric system.

Laser scanners have similar errors to pulsed laser total stations, in that distance and direction must be calibrated to ensure that the computed 3D coordinates are of sufficient quality. Measurement of a standardized test range is one method of calibration, with the advantage of being fairly realistic. Some of the autocalibration and in-flight calibration methods used in photogrammetry may be able to be used for laser scanner calibration.

163.3 Calibration of Distance Measuring Equipment

The equipment that the surveyor uses to measure distances must be regularly calibrated to ensure that it is providing results that are both accurate and precise. In some jurisdictions, tapes and EDM must be calibrated every 6 months, and most jurisdictions place a burden of proof of quality of equipment on the professional surveyor.

For most other distance measurement methods, calibration is rarely necessary. This is the case with stadia and estimated methods, although it is necessary to recalibrate one's pacing from time to time. However, tapes and EDM, which carry the work in most serious distance measurement jobs, need to be regularly calibrated.

As mentioned in Chapter 161, there are three main types of errors: gross, systematic, and random. Gross errors can be dealt with by independent checking and additional care. Random errors are intrinsic to the measurement process and can be handled statistically. Systematic errors are the result of a mismatch between the measurement process in reality and the abstract model in use to interpret the measurements. The purpose of calibration is to determine the systematic errors, so that they can be eliminated or minimized by making the measurement model better accord with the reality of the measurement process.

Tapes

Calibration of a tape is fairly simple, once one has established a base line of known length. It is often a good idea for surveyors to set up such a line close to their headquarters. The ideal situation is a flat, level surface for the full length of the tape, preferably in a completely shaded area. If all the tapes to be calibrated are nominally the same length, a zero point fixed in the surface is required at one end, while a second mark is fixed at around the correct distance. It is often handy to fix a small segment of a finely marked measuring tape at the second end, as it can make measuring the tapes a great deal easier.

The surveyor should have a tape that is used as a standard. In some jurisdictions, such tapes must be reserved for calibration purposes only and sent to a government office for regular recalibration. If there is no such facility, a tape should be set aside for calibration purposes only, and calibrated by EDM or some other means (e.g., subtense measurements).

The surveyor begins by allowing the tapes to gain the same temperature as the calibration area, and then measures the line with the standard tape. The advantage of having the piece of tape at the end of the line is that it is fairly simple to determine the point on the ground that corresponds with the correct nominal distance. To do this, the point at which the end of the standard tape falls is noted, and then the distance is corrected to allow for the various systematic errors, which should be only temperature and any calibration constant of the standard.

The tape to be calibrated is then compared to the line and its length noted against the piece of tape. It is then fairly simple to compute the actual length of the tape at that time, and using the temperature correction, compute the temperature at which it is correct.

With a simple calculator program or spreadsheet and some practice, this whole procedure can be done in little more than the time it takes to unroll and roll up the two tapes, and the tape's calibration values are then known.

It is important to ensure that the correct tension is being used during calibration. A little thought in the design of the base line will ensure that all the other corrections are zero, and that temperature is practically constant for the full length of the line. This will help achieve a high quality calibration of the tape.

Electronic Distance Meters

EDM equipment has a slightly different set of systematic errors that must be determined before quality results can be expected. The three systematic errors that are commonly determined by calibration are the zero, offset or constant error, the scale error, and the cyclic error.

The zero, offset or constant error, as mentioned above, is caused by the electronic center of the instrument not being vertically above the measurement mark. It can also be considered to include the reflector constant. It is characterized by appearing with the same sign and magnitude in all measurements that the instrument makes. In most cases, it remains fairly constant over time.

The zero, offset, or constant error can be simply determined by measuring a line in one piece, and then in two (or more) pieces. Each measurement will have the same constant error in it. It is a good idea to ensure that all the meteorological corrections are applied, and that the test line is level.

Example 163.7

A surveyor measures a line with an EDM from one end to the other and finds that it is 210.273 m long. The EDM is then moved to a point lined in on the same line between the end points, and the line measured in two parts from the new point. The two distances are found to be 142.916 m and 67.363 m. What is the zero, offset, or constant error of the EDM?

Solution. Each measurement that the EDM makes, d, can be considered to have two components: the "true" distance, t, and the constant error, e. So the measurement model is $d = t + e$.

For the whole line, the result is

$$d_w = t_w + e$$

For the partial line measurements, the results are

$$d_1 = t_1 + e$$

and

$$d_2 = t_2 + e$$

Now,

$$t_w = t_1 + t_2$$

So,

$$d_1 + d_2 = t_1 + t_2 + 2e = t_w + 2e$$

Therefore

$$(d_1 + d_2) - d_w = e$$

In this example,

$$e = (142.916 + 67.363) - 210.273 = 210.279 - 210.273 = 0.006 \text{ m}$$

The correction, being of the opposite sign to the error, is –0.006 m. Check:

$$t_1 = 142.916 - 0.006 = 142.910$$

$$t_2 = 67.363 - 0.006 = 67.357$$

$$t_w = 210.273 - 0.006 = 210.267 = 149.910 + 67.357$$

EDM scale error is caused by the measurement frequency that is superimposed on the carrier wave being incorrect. This leads to the wavelength of the measurement wave being incorrectly scaled by some factor, which in turn translates into an error that is directly proportional to the distance that the beam travels. The difficulty in determining this error is that it cannot be checked by the EDM alone, as a constant scale factor applies to all its measurements. The only way to determine this error outside a workshop is to measure lines whose lengths are precisely known. Naturally, the longer the line, the better, which means that the known lines need to be determined by a carefully calibrated high-precision EDM. The preferred machines for this are the Kern Mekometer 3000 and 5000, which use a different form of measurement compared to other EDM and can achieve consistent submillimeter precision.

To calibrate the EDM, it is run over several lines previously calibrated using a high-precision instrument, such as the Mekometer. Once the constant error has been eliminated, the data are used to determine the scale error. The alternative to this is calibration of the EDM's oscillators in an electronics workshop.

It is important to be able to monitor the behavior of the scale error over time. A known, constant scale error can be corrected, but one that varies indicates faulty oscillators, which must be replaced. This requires a workshop visit and takes the instrument out of operation.

Cyclic error in EDM is caused by variability in the reading of the phase differences used to determine the distance. The result is a small difference between the measured value and the correct value that varies in a sinusoidal pattern with a period equal to the wavelength of the measurement unit, commonly 10 m. This was more of a problem in earlier EDMs, and today could be expected to be less than 5 mm.

Cyclic error can be detected in two ways. The first is to place a tape, suitably tensioned, along a flat surface about as long as the wavelength of the measurement unit. The top of an even garden wall is good for this. A reflector is then moved along the tape and a measurement made using the EDM at regular intervals along the tape. The EDM distance and the tape distance are compared to see if there is any pattern to the differences between them. If a sinusoidal pattern of differences appears, this can indicate the presence of cyclic error.

The amplitude and phase offset of the cyclic error are noted, and the correction can then be applied to subsequent measurements. While it can be ignored for some measurements, it is necessary to allow for any cyclic error in the EDM for high-precision work. This means that the surveyor should know if any cyclic error exists in the machine.

The ideal form of calibration for EDM is a purpose-built test range. An example of good ranges can be found in the surveying literature. They commonly consist of up to seven concrete pillars spread over a distance of perhaps 1.2 km. The pillars are often offset, so that it is possible to place reflectors on all the pillars and measure them all at a single setup, without having to turn reflectors away to avoid interference.

The surveyor travels along the line placing reflectors, then returns along the line, setting up the EDM at each pillar in turn and measuring to all the pillars with reflectors. As a result, all combinations of lines are measured.

By careful design of the range, the pillars can be placed so that the various distances provide a complete and even coverage of the wavelength of the measurement unit, allowing the cyclic error to be assessed. Calibration of the range at regular intervals by a high-precision unit, such as a Mekometer, allows the distances to be known and so the scale error to be computed. The process of cumulative distances in Example 163.7 allows the constant error to be determined.

On some test ranges, a form is provided for the results, which are then computed as a service to the surveyor. A least squares adjustment provides the best estimate of the calibration values for the instrument, and a report and certificate of calibration may be issued. If the service also tracks each instrument's calibration history, it is possible to pick up stability problems in oscillators and other problems that may require repair or replacement of the unit.

Defining Terms

Distance — A linear value, either measured or computed. Unless qualified otherwise, it is understood to lie along the horizontal. The observed distance is considered inaccurate until corrected for systematic errors caused by instruments, nature, or other sources.

Electronic distance instrument — An instrument that measures distances using reflected light waves or microwaves.

Indirect measurement — A measurement that has been computed from other measurements, usually employing trigonometric principles.

Mekometer — A high-precision EDM developed at the UK's National Physical Laboratory and produced by the former Kern instrument company. Precision of ±0.2 mm + 0.2 ppm is claimed over distances up to 8 km.

Tacheometry — A distance-measuring procedure that involves measuring intervals between cross hairs on a rod or angles subtended between marks on a bar of known length, or a similar indirect method. The term derives from *fast measurement.*

Tape — A surveying instrument used to measure distance by stretching it, end for end, along a line between points. Also termed a *chain* or *band.*

References

Bannister, A., Raymond, S., and Baker, R. 1998. *Surveying.* Addison-Wesley Longman, Harlow, UK.

Buckner, R. B. 1983. *Surveying Measurements and Their Analysis.* Landmark Enterprises, Rancho Cordova, CA.

Moffitt, F. H. and Bossler, J. D. 1999. *Surveying.* Addison-Wesley Longman, Menlo Park, CA.

Further Information

Anderson, J. M. and Mikhail, E. M. 1997. *Surveying: Theory and Practice.* McGraw-Hill, New York.

McCormac, J. C. and Anderson, W. 1999. *Surveying.* John Wiley & Sons, New York.

Mueller, I. I. and Ramsayer, K. H. 1979. *Introduction to Surveying.* Frederick Ungar, New York.

Roberts, J. 1995. *Construction Surveying: Layout and Dimension Control.* Delmar, New York.

Wolf, P. R. and Ghilani, C. D. 1994. *Elementary Surveying: An Introduction to Geomatics.* Prentice Hall, Englewood Cliffs, NJ.

164

Directions

Bon A. Dewitt
University of Florida

In surveying, direction is the term used to denote the course or heading of a line. Here, a line is defined by its end points, giving it a magnitude (length) and direction, much like a vector. By convention, direction is specified in terms of angles and is separated into horizontal and vertical components. This chapter deals primarily with the horizontal component of direction within the context of plane surveys. In geodetic surveys, where the Earth's curvature is taken into account, the fundamental concepts of direction still apply, although their use in subsequent calculations is far more complex.

Traditionally, the basis for direction has been established through astronomic observations or compass readings. Although these traditional approaches may still be applicable in certain situations, the global positioning system is now preferred due to its accuracy and convenience.

164.1 Angles

Angles form the basis for quantification of direction. There are several conventions available for specification of angular units. In the U.S., the sexagesimal system is currently the most commonly used for surveying applications. In this system a full circle is divided into 360 degrees, with further subdivisions into minutes and seconds. Sixty minutes is equivalent to 1 degree and 60 seconds is equivalent to 1 minute. Thus, an angle can be expressed in degrees (°), minutes (′), and seconds (″), much like time is expressed in terms of hours, minutes, and seconds. As an alternative, angles can be expressed in terms of degrees and a decimal fraction thereof; however, this is not conventional surveying notation.

Other angular units are available, such as radians, grads, or mils (see equation below for unit equivalents). The radian is a dimensionless unit that utilizes the ratio of the length of a circular arc to its radius to express the magnitude of the subtended angle. The grad (or *gon*) is a unit of measure in the centesimal system (widely used in Europe) that corresponds to 1/400 of a full circle. The mil is an angular unit corresponding to 1/6400 of a full circle and is used by the U.S. military, primarily in artillery applications. These three alternate systems are decimal based, as opposed to the base-60 approach of the sexagesimal system.

$$\text{Right angle} = 90 \text{ degrees} = 100 \text{ grads} = 1600 \text{ mils} = \pi/2 \text{ radians} \tag{164.1}$$

Horizontal angles are measured in a plane perpendicular to the direction of gravity. There are many types of horizontal angles, such as angles to right (clockwise) or left (counterclockwise), interior angles,

0-8493-1586-7/05/$0.00+$1.50

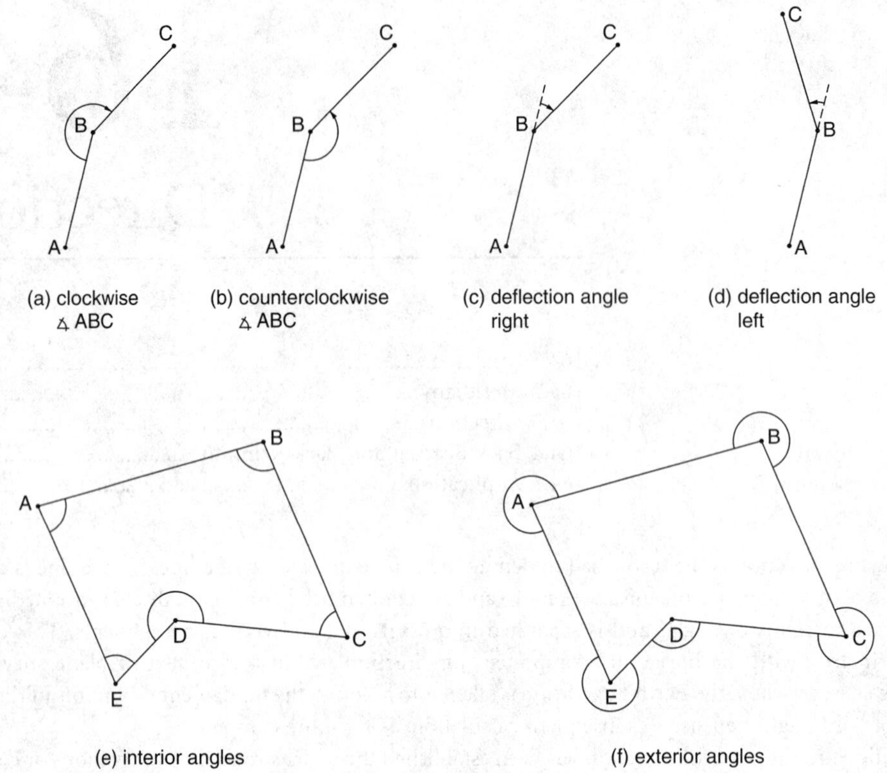

FIGURE 164.1 Illustration of various types of horizontal angles.

exterior angles, and deflection angles. Figure 164.1 shows the various types of angles. Angles to the right — with the backsight, occupied, and foresight points specified — are recommended for modern surveys due to their applicability to electronic data collectors and computer software. An illustration of this notation is given in Figure 164.1(a), where point A is the backsight, B is occupied, and C is the foresight point.

Vertical angles are measured in a plane perpendicular to the horizontal. There are two fundamental types of vertical angles: zenith angles and altitude angles. A zenith angle (or zenith distance), illustrated in Figure 164.2(a), is the angle from the observer's zenith direction to the target. Zenith angles range

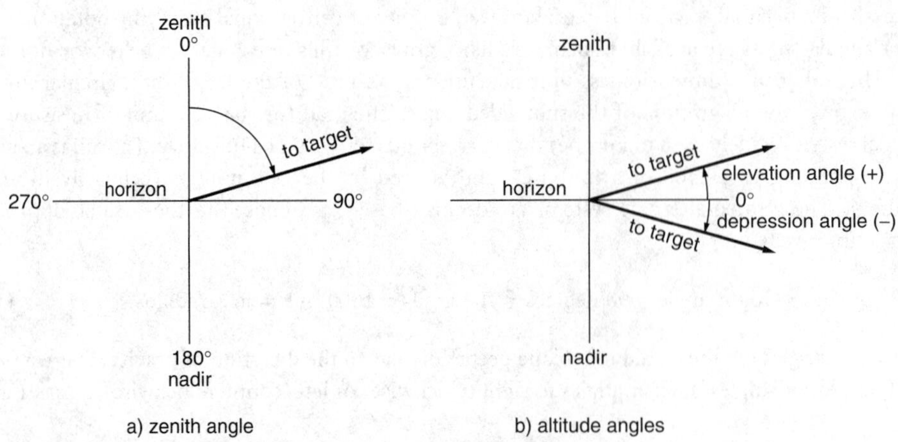

FIGURE 164.2 Profile view illustrating two types of vertical angles.

from 0° to 360°, with 90° and 270° corresponding to the direction of the horizon in the direct and reverse positions, respectively. An altitude angle, illustrated in Figure 164.2(b), is the angle from the observer's horizon to the target. Altitude angles above the horizon are called *elevation angles* and are considered positive by convention, whereas those below the horizon are called *depression angles* and are considered negative.

Surveyors determine angles with instruments known as *transits* or *theodolites*. Both instruments enable the user to perform the same basic functions, that is, to measure or establish horizontal and vertical angles. Generally, theodolites are more accurate and precise than transits, although this is not always the case. Many modern theodolites have electronic angle-reading systems and are incorporated into total stations, which also have the capability of measuring distances electronically. Total stations are particularly convenient due to their ability to transfer recorded data directly to a computer.

164.2 Meridians

In order to specify the horizontal component of direction, it is first necessary to specify the reference meridian. A meridian is an imaginary line that is selected as the nominal north–south indicator in the observer's horizon plane. It can be based on any one of several references: geodetic, astronomic, magnetic, grid, or assumed.

A geodetic (also called *geographic*) meridian is based on the north and south poles as defined by a particular latitude and longitude reference or graticule. It has been demonstrated that the rotational axis of the Earth changes slightly over time, so in essence the geographic graticule is a "snapshot" in time. This reference becomes standardized by virtue of published coordinates for a network of monumented points, based on its definition.

An astronomic meridian is based on the rotational axis of the Earth and the direction of gravity. It derives its name from the means by which it is typically established: astronomic observation. The angular difference between astronomic and geodetic meridians is expressed in terms of the Laplace equation. For most practical purposes in plane surveying, this difference is negligible and both meridians are often collectively referred to as the "true" meridian.

A magnetic meridian is based on the magnetic north and south poles of the Earth. These poles are distinct from the geographic poles and change appreciably over time. The angle from true north to magnetic north is called the *magnetic declination* and is a function of the observer's location with respect to the poles. The effect of magnetic declination can be quite large; for example, in parts of Alaska, magnetic declination is greater than 30° east.

A map projection is a distorted rendition of a portion of the curved Earth's surface projected onto a surface that can be laid out flat. The projection has an inherent x, y coordinate system, with y in the general direction of north. In the projection any line parallel to the y axis is a grid meridian. They are different from the three meridians previously mentioned, in that grid meridians are parallel to each other, whereas the others are not.

Assumed (arbitrary) meridians are chosen for convenience. Here, a direction is arbitrarily specified for a line connecting two survey monuments. This direction is often chosen so as to approximate some specific meridian; however, no actual observation of a meridian is performed. There is an inherent risk associated with assumed meridians. If one or both of the survey monuments is lost, the assumed meridian becomes unrecoverable.

164.3 Direction

In mathematics, polar coordinates are often used to specify the position of a point. Here, the direction of the line from the origin to the point is based on the angle from the positive x axis, with counterclockwise angles being positive. In surveying, the direction of a line can be expressed either in terms of bearing or azimuth. Both forms depend on the meridian definition mentioned earlier. No matter which approach is used, it is important to clearly specify the meridian upon which the direction is based.

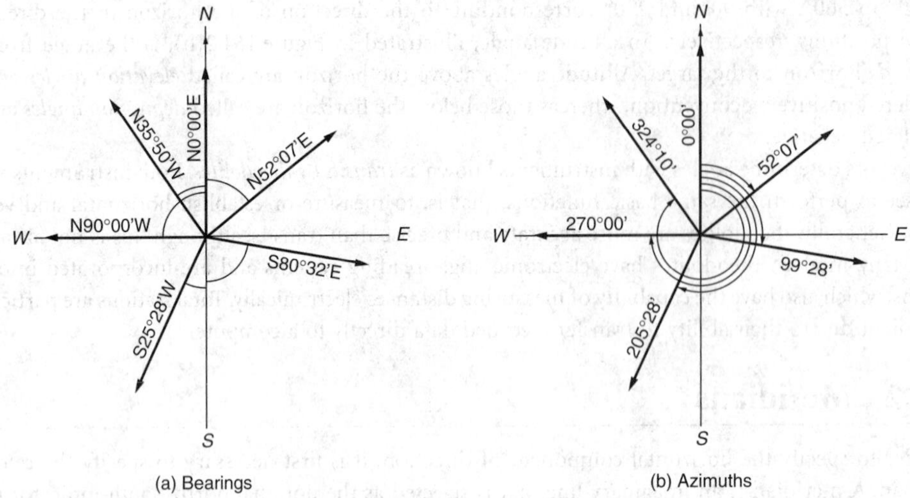

FIGURE 164.3 Examples of bearings of selected lines and their equivalent azimuths.

The bearing of a line is specified as an acute horizontal angle between the line and meridian, along with letters specifying the proper quadrant. It is expressed in the form of the letter N or S, followed by an angle (less than or equal to 90°), followed by the letter E or W.

Azimuth is specified as the clockwise horizontal angle from the meridian to the line. The angular value is positive and less than 360°. Azimuths are commonly specified from the north, although this is not a universally accepted standard. Some applications use azimuths that are referenced from the south. Due to this possible ambiguity, one should clearly indicate whether a north or south reference is implied. Figure 164.3 gives examples of bearings and azimuths (from north) for selected lines.

164.4 Back Bearing and Back Azimuth

Back bearings are expressions of the opposite direction of a line. The expression is formed by starting with the original (forward) bearing and then changing the "sense" of the letters. An N is changed to an S (or vice versa), and an E is changed to a W (or vice versa); however, the angular value remains the same. For example, if the bearing from point 1 to point 2 is N47°15′W, its back bearing (i.e., the bearing from point 2 to point 1) is S47°15′E.

Back azimuths are also expressions of the opposite direction of a line. The expression is formed by adding 180° to or subtracting 180° from the original (forward) azimuth, keeping in mind that the result must be in the range of 0 to 360°. For example, if the azimuth from point 1 to point 2 is 312°45′, its back azimuth (i.e., the azimuth from point 2 to point 1) is 132°45′.

The foregoing simple relations for back bearings and back azimuths are applicable only to plane surveys of a limited extent, where it can be assumed that all meridians are parallel. In surveys covering a large area, Earth curvature and meridian convergence are appreciable factors and therefore a more complicated relation must be used.

164.5 Applications

There are many applications in surveying that call for the use of directions. Property surveys, geodetic control surveys, transportation corridor (route) surveys, and topographic surveys are but a few. Most applications involving directions utilize plane trigonometry in the solution. Computations involving addition and subtraction of angles, sine and cosine laws, right triangle relationships, and sum of angles in a closed polygon are routinely performed.

Defining Terms

Azimuth — An expression for the direction of a line consisting of the clockwise horizontal angle (≥0° and <360°) from one end of the meridian. Azimuths from the north are generally used; however, some conventions employ azimuth from the south.

Bearing — An expression for the direction of a line consisting of the horizontal angle (≤90°) that the line makes with the meridian in conjunction with prefixed and postfixed letters that specify the quadrant.

Geodetic survey — A survey in which the Earth's true three-dimensional shape and gravity field are taken into account.

Global Positioning System — A system of satellites and ground receivers that enables users to determine geodetic coordinates of points to a high degree of accuracy. The system is under the control of the U.S. Department of Defense, but has been used for civilian applications since the 1980s.

Horizontal angle — An angle that is defined in a plane perpendicular to the direction of gravity.

Meridian — In a global context a meridian is the intersection of the plane containing the north pole, the south pole, and the observer's position with the spheroidal figure that approximates the Earth. In a local context, a meridian is a reference line that defines the north–south direction. This reference can be on a geodetic, astronomic, magnetic, grid, or assumed basis.

Plane survey — A survey of limited extent and accuracy in which the Earth's surface is assumed to be a plane. This assumption permits the use of plane trigonometry in computations involving coordinates and other parameters.

Total station — A device used in surveying that incorporates an electronic theodolite with an electronic distance-measuring instrument and a computer. The device can automatically read and record horizontal and vertical angles and slope distances.

Vertical angle — An angle that is defined in a plane parallel to the direction of gravity.

Reference

Wolf, P. R. and Ghilani, C. D. 2002. *Elementary Surveying: An Introduction to Geomatics*, 10th ed. Prentice Hall, Englewood Cliffs, NJ.

Further Information

Anderson, J. M. and Mikhail, E. M. 1998. *Surveying: Theory and Practice*, 7th ed. WCB/McGraw-Hill, New York. A good mathematical treatment of various surveying topics.

Bomford, G. 1980. *Geodesy*, 4th ed. Oxford University Press, New York. This book is considered to be a definitive text on the subject of geodesy, with a high degree of mathematical rigor.

Brinker, R. C. and Minnick, R., Eds. 1995. *The Surveying Handbook*, 2nd ed. Chapman and Hall, New York. A good general reference on virtually all topics of surveying.

165

Photogrammetry and Topographic Mapping

Sandra L. Arlinghaus (Second Edition)
University of Michigan

Robert F. Austin (Second Edition)
Austin Communication Education Services

Jim Bethel (First Edition)
Purdue University

The original introductory written material by Jim Bethel is included here, as well as selected original references [5,6,14,16,17,20,23]. In subsequent sections, Arlinghaus and Austin add their own approach to this material.

The term *photogrammetry* refers to the measurement of photographs and images for determining the size, shape, position, and other spatial attributes of features appearing in the images. The most common application of this technique is aerial photogrammetry, in which nominally vertical photographs are used to produce topographic maps, which are often used for engineering design and land development. Aerial cameras are made to precise tolerances; small systematic errors that may be present in such photographs can be modeled mathematically so that accurate ground positions and elevations can be inferred from measurements based on these photographs. The mathematical basis of the imaging equations that relate object points (three dimensional) to image points (two dimensional) is that of a perspective projection, with a point (actually two points) in the lens assembly serving as the perspective center(s). The geometric relationship between the image and object spaces may be modeled in *analog* fashion by optical rays or by mechanically gimbaled steel rods. Today it is more commonly modeled mathematically in an *analytical* instrument.

Historically, the production of topographic maps from aerial photographs has been a very labor-intensive operation, beginning with the establishment of *ground control points*, proceeding through *aerial triangulation* to densify the ground control, and culminating in the meticulous tracing of planimetric features from adjacent *stereo pairs* and the tracing of contour lines to depict the shape of land forms. In recent years, several trends have emerged which promise to make the mapping process more efficient. The use of the *global positioning system* (GPS), for ground control as well as for direct observation of the *exposure stations* in the aircraft, is greatly simplifying the preliminary steps in the photogrammetric

0-8493-1586-7/05/$0.00+$1.50

mapping process. High-level image processing of photographs converted to digital form is replacing the tedious operations of point selection, orientation and even feature extraction.

The following sections discuss the status of photogrammetric technique and suggest one approach to topographic mapping based on the use of GIS software.

165.1 Traditional Methodology

Planning

Typically, the first step in data collection is to obtain planning maps for the project area. These maps are used for the delineation of aircraft flight lines. The maps also are used for preliminary estimates of the numbers of aerial photographs and stereo models to be produced.

After defining the project area on the planning maps, analytical aerial triangulation ("analytics") technicians identify the location of the control points needed for proper registration and rectification of imagery. These points are noted on the map sheets, which then are delivered to the project surveyors.

During these preliminary stages of production, the surveyors conduct a review of the project's accuracy specifications, paying particular attention to benchmark and monument requirements. After receiving the annotated map sheets from the analytics technician, the surveyors identify and mark existing benchmarks and monuments within the project area. From these existing positions, the surveyor plans the traverse and bench lines required for horizontal and vertical control, respectively. To supplement the existing benchmarks and monuments, the surveyor establishes temporary benchmarks and temporary stations. These recoverable positions are supplemented, if necessary, to satisfy the requirements defined by analytics technicians.

Photography Acquisition

At this point in production, aerial photography is obtained. The aircraft pilot obtains the annotated planning maps from the surveyor and plans the required flight lines. In some instances, permission to fly over may be needed (particularly if military bases or civil airports are located nearby).

Aerial photography traditionally was collected using panchromatic, black-and-white film, primarily due to cost considerations. In recent years, color film has replaced black-and-white media for many applications due in part to reductions in the costs of the color film and film processing and in part to the benefits of color for image analysis. Nevertheless, certain special-purpose applications continue to use black-and-white film, particularly when budgets are restrictive. Whether color or black-and-white, the film emulsion used for a typical project should be fine grained, low distortion, and high resolution, and the entire project should be flown with one type of film.

Cameras of appropriate focal length (generally 3.5 or 6 in.), certified as calibrated to U.S. Geological Survey (USGS) specifications within the required period of time (most commonly, 3 years), are used. The camera should be equipped with a forward motion-compensating device and a high-resolution, distortion-free lens. A typical resolution for aerial photography would be between 6 in. per pixel and 3 ft per pixel: the smaller the pixel size, the greater the detail that can be seen.

The optimal times for taking aerial photography for topographic mapping are the autumn and spring seasons when there is a reduction of vegetation (i.e., "leaf-off") and the greatest frequency of clear weather conditions. Aerial photography is acquired after flight plan approval and ground control target placement (for image registration) as weather and ground conditions permit. Photography is done only when the sun angle is greater than 30° to minimize shadows and reduce contrast.

The aerial flyover should be performed with an airborne GPS/IMU (inertial measurement unit) system for direct georeferencing of the aerial photography during the flight. To maintain GPS data quality, aerial photography should be acquired only when at least four and preferably five satellites are observable with a positional dilution of precision (PDOP) of less than three and a cutoff angle/elevation mask greater than 20°.

The altitude at which photography is acquired depends in part on the intended purpose and in part on the project budget. For example, the National Aerial Photography Program (NAPP) that began in 1987 acquired photography on a 5- to 7-year cycle from an altitude of 20,000 feet. The predecessor National High Altitude Photography (NHAP) program acquired photographs at an altitude of 40,000 feet. Contemporarily, aerial photography intended for use in large-scale mapping typically would be captured at an average altitude of 7200', yielding a photonegative scale of 1" = 1200'.

The flight lines for data capture should be oriented in an east-west direction with approximately 7500' spacing, yielding a sidelap (side overlap) of approximately 30% between adjacent flight lines. The flight lines should be extended as required to provide stereo coverage of the designated mapping area and peripheral ground control points. The aerial photography typically provides 60% stereoscopic endlap (end overlap) between exposures in the flight line. The GPS navigation and flight management system are used to trigger the camera automatically with sufficient endlap. Camera tip, tilt, and crab should be limited to not more than 3°.

For quality control purposes, samples from each roll of film are checked for acceptable levels of base fog and density before the film is used. The aerial camera should directly title each frame of the aerial photography at the time of exposure. The title information includes the date, photo scale, camera serial number/focal length, site/project name, project number, and exposure number. Once processed, the aerial photography is checked for acceptable exposure, resolution, contrast, overlap, camera crab, tip and tilt level, and the absence of extraneous markings.

Survey Control

Ground-based survey points consist of targeted control points and checkpoints distributed throughout, and around the perimeter of, the target area. The combination of airborne and ground-based control points supports accurate aerotriangulation of the aerial photography and production of base mapping. Where possible, surveyors should use existing control monuments for the ground control network in place of newly surveyed points.

All new control points should have coordinates established on an appropriate horizontal coordinate system (e.g., NAD83) and elevations established on an appropriate vertical datum (e.g., NAVD88). The horizontal coordinates should be surveyed to at least second-order Class I (1:50,000) accuracy and elevations should be surveyed to at least third-order, 3-cm accuracy. Examples of appropriate new points include easily recognized points (e.g., the intersection of the edge of a bridge with the edge of a paved road) or points that can be made easily recognizable with limited preparation (e.g., painting selected manholes).

After photography is complete, the film is processed, typically by a firm or agency that specializes in aerial photography processing. Two sets of 9" × 9" prints and one set of diapositives, for internal use, typically would be produced in the photography lab. One set of prints generally is reserved for archiving, while the second set of prints is prepared for the field survey. The diapositives are reserved for processing in a later step.

The next production step is selecting photo control. The photographs are laid out and control points are selected and marked on the photographs. Four control points per stereo model (two per photograph) are required for vertical control. Three points are required for horizontal control. Enlargements of the high-altitude flight negatives centered on each section are used to identify horizontal control points.

After selecting and noting photo control, the photogrammetrist delivers the photographs to the surveyor. During the field survey, survey crews determine the elevations and horizontal coordinates for the control points. Description sheets of temporary benchmarks and temporary station recoverable positions are prepared. Depending on the project's specifications, complete field survey notebooks may be maintained for delivery to the client.

The surveyor prepares an overall map of points used (also known as a "control diagram"). The surveyor also performs the computations necessary to close traverses and bench lines. After the survey is completed, the surveyor posts the control point data on the back of the photographs. The photographs then are returned for analytical triangulation.

Fieldwork for a GPS/IMU survey typically involves the operation of one or more GPS base station receivers on the ground for the duration of the flight. The base station receiver(s) should be first-order-certified dual-frequency systems. Within the United States, the station locations should be coordinated with appropriate high-accuracy reference network (HARN) stations. The base stations should be positioned to within approximately 50 km of the aircraft receiver during the flight. Within the United States, National Geodetic Survey (NGS) Continuously Operating Reference Stations (CORS) in the area may be used to provide redundancy. Data from all base stations are processed so that the camera position at the time of exposure can be calculated independently for each station and compared to permit independent confirmation of position.

Analytical Triangulation

Analytical triangulation is performed in the following three phases: (1) photograph plan and layout; (2) point marking and measuring; and (3) data processing with bundle adjustments. With the advent of computers, many technicians use standard software packages for processing. Analytics technicians commit the model coordinates to a computer disk or drive, transfer these data to an appropriate computer, and reformat the data. Technicians then use appropriate computer hardware and software to manipulate the coordinate file to obtain X, Y, Z ground coordinates. The technicians reformat the coordinates and place them in a separate computer file used to achieve registration for digitizing.

For analytical photogrammetry, the positions of all control, orientation, and triangulation pass points (nine points per photo) and orthophoto orientation points (five points per photo) are marked on a set of contact prints. Technicians then "pug" (drill) into the set triangulation diapositives created previously using a calibrated, 1-micron accuracy point-marking-and-transfer device equipped with diamond tip drills. A calibrated analytical stereoplotter with appropriate pointing accuracy interfaced with a data collection station is used to record photo coordinates of the pugged points and fiducials, with the system performing a preliminary triangulation adjustment during the point reading to identify point reading/pugging errors.

The triangulation programs are block-and-bundle refinement-adjustment programs run on an appropriate workstation. The triangulation software accepts ground and/or airborne GPS control data and analytical triangulation pass point input. The software produces a simultaneous polynomial solution with RMS (root mean square) residual values as output.

For the completed triangulation, a triangulation report is prepared. The triangulation report contains summaries of the procedures used, results achieved, printouts of the triangulation input and output (including control points) and weighting factors, adjustment iterations, and triangulation point listings and residuals.

Traditional Photogrammetry

Following analytics processing, stereo compilation is performed. Using stereo compilation instruments, contours are compiled manually to represent contour (hypsographic) data. Pairs of overlapping images are examined using stereoscopic viewing instruments, which permit the perception of depth in the image, and are used to identify and draw lines of constant relative elevations. These lines are registered using the control points obtained from surveys and from analytics processing. These manuscripts then are submitted for board (vector) digitizing or scanning.

The process of board digitizing involves the use of highly accurate digitizing tables and digitizing cursors to trace the manuscript contours. In the case of scanning, the stereo compiled sheets are scanned at a relatively high resolution of 12 to 15 lines per millimeter. After scanning is complete, raster system operators edit the data to ensure that contour coalescence is eliminated. The pixel data array then is thinned and prepared for vectorization. The vectorizing process generates a file containing data whose number had been reduced using some standard point-reduction algorithm.

One fairly common adaptation of this traditional method of photogrammetry in later years involved retrofitting the stereo compilation instruments to permit direct interaction with a computer. In this method, contours were digitized directly during compilation, eliminating the manuscript step.

Softcopy Photogrammetry

Contemporary photogrammetry is performed increasingly, and in many cases exclusively, using softcopy techniques. Softcopy, or digital, photogrammetry is based on the use of digital (softcopy) images, as opposed to conventional photogrammetry, which is based on film negative or diapositive (hardcopy) images. "Digital systems are considered third generation photogrammetric instruments, after analytical (second generation) and analog (first generation) stereo plotters" [11: 1-1].

The digital photo imagery used for the project digital orthophotography should be produced by scanning aerial film directly. The scanning should be performed immediately after the film is processed and checked and while it is still in a relatively "pristine" state (i.e., free of surface scratches and dust). The scanning aperture used should be at least 1200 dots per inch (dpi). The scanned imagery should be resampled to the final pixel resolution needed.

For each selected exposure, the scanning covers the entire frame, including fiducials. The imagery initially is rasterized using a project-specific "lookup table" defined by the film type, photographic characteristics, and desired image characteristics. This process permits very fine adjustments of the color balance at the scanning stage. The scanner is recalibrated periodically to ensure proper color distribution.

The interior orientation parameters of each photographic exposure are computed automatically during the scanning phase. These measurements are saved for each image and used for production quality assurance testing as well as softcopy aerotriangulation and stereo compilation. If the root mean square error of the fiducial measurements exceeds predefined tolerances, the negative is scanned again.

Directly scanning the aerial film transparencies is preferable to scanning diapositives, especially if the imagery will subsequently be used to produce digital orthophoto imagery. Direct negative scanning provides better detail visibility in shadow and highlight areas. It also reduces film-handling time prior to scanning, reducing opportunities for dust particles, micro-scratches, and other anomalies that may affect the imagery. In general, direct scanning results in improved image quality, and less time and less cost for digital image production.

In this approach, softcopy analytical aerotriangulation of the aerial photography is used to densify and check the project control and generate orientation parameters as required for the image rectification and mapping work to be performed. The initial input to the triangulation process includes the accurate photo center coordinate data derived from the airborne GPS/IMU survey. This allows the process to converge to a final solution with minimal iteration.

A softcopy aerotriangulation workstation uses image correlation technology to join digital photo image pairs and automatically generate hundreds of pass points for each stereo pair. The generated pass points are run through a preliminary triangulation adjustment with a limited number of measured control points to determine the pass point residuals, with pass points having a greater than predefined residual filtered out. The final triangulation is performed with the accepted pass points and all control points.

Analog and analytical stereo plotters allowed technicians to view film images or photographic prints in stereo by merging adjacent overlapping images through a view piece similar in appearance to fixed-mount binoculars. In softcopy photogrammetry, adjacent digital images are merged in an overlapping fashion to simulate the three-dimensional display. Most commonly, stereo viewing is accomplished using special-purpose eyewear. "A flickerless view is achieved by splitting the high monitor vertical scan rate of 120 fields/sec into alternating left/right fields, each at 60 fields/sec. The viewing system presents to the left and right eye sequentially the corresponding perspective image (aerial camera view), in synchronization with the LCD shutter eyewear at 60 Hz/sec (twice the frequency of VHS video or television. The human brain does the rest" [11: 2–4].

Using this viewing system with a high-power computer workstation and a large monitor (typically 19" or larger), technicians can perform direct stereodigitizing of contours (and planimetric features) with extremely high accuracy. This approach has several advantages over traditional photogrammetry, including insensitivity to vibration, which permits placement on any floor of an office building, a smaller floor space footprint, and lower cost.

165.2 Recent Enhancements and Advances

Digital Orthophotography

Aerial photo rectification is a process in which the effects of camera crab, tip, and tilt on photographic images are corrected. As a camera points down on the Earth's surface, the image it collects is most accurate at the center of the frame. Rectification also corrects for the distortion of imagery away from the center of the frame. This process allows the assembly of multiple frames into a mosaic of images used as the basis for mapping.

Orthophoto rectification is a process that performs the same function as normal rectification, but also removes the effects of object displacements due to ground relief. This is done mathematically by moving the camera's perspective from the actual flight altitude to "infinity" much like an orthographic map projection. This is based on the use of a terrain model.

In simple terms, the orthophoto is modeled by draping the imagery over a terrain surface model (see Section 165.3), permitting localized adjustments to the rectification. Until recently, there would have been no further correction for objects on the terrain surface (e.g., buildings). However, new modes of data acquisition have made it possible to incorporate object elevations to produce a true continuous surface model.

During digital orthophoto production, compilers accomplish the stereocompilation using softcopy stereoplotting systems interfaced with photogrammetric data collection stations. Data are collected in three dimensions from parallax-cleared stereomodels of the triangulated aerial photography. Stereomodel orientation residuals should be limited to a predetermined error tolerance. The three-dimensional stereocompilation process ensures a correct representation of hydrographic feature elevations and "true" positioning of buildings and other vertical surface features.

The terrain model consists of a grid of spot elevations and "breaklines" compiled along significant breaks in grade. The spot elevations should be collected at an interval averaging 75 ft. The breakline data should generally include major road edges and/or centerlines, railroads, major driveways, major drainage features, large retaining walls, ridges, and other prominent linear discontinuities that would affect the image rectification. Where bridges and elevated overpasses are located, analysts collect breaklines representing the bridge decks, overpass structures, and roofs of taller buildings to support proper orthorectification of these features.

The orthophotography produced should be mosaicked, with consistent balance, tone, and contrast ranges within and between images, ensuring a smooth transition between images, although details in dark tone and highlight areas should be preserved. Mosaicking should be performed using a seaming process that eliminates image distortions caused by aboveground features mosaicked from adjacent photographs. With current technology, image processing and tiling procedures that previously took weeks to accomplish and review manually are now carried out in a matter of hours or days with minimal need for operator interaction.

Digital Image Acquisition

In 1999, Heier and Hinz observed that digital airborne cameras offered several advantages over traditional film-based aerial mapping cameras, including "higher radiometric resolution, reproducible color information, cost savings for film, cost savings for film processing, cost savings for scanning and immediate availability of the image data" [9: 2]. However, they also noted that the technology had not matured to the point that digital airborne cameras offered a complete replacement alternative to traditional camera systems.

As with ordinary digital cameras, the primary data collection mechanism in an airborne digital camera is an array of charge-coupled devices (CCD) that gather light reflected from the Earth's surface. The ideal sensor would be a single large CCD element with the same dimensions (or greater) of traditional camera film (240 mm × 240 mm). At present, technical and economic considerations preclude the use of such a sensor, hence an array of multiple smaller elements.

The greatest relative deficiency of airborne digital cameras is geometric resolution: "230 mm aerial film can be digitized . . . with a pixel size of 7 μm. This resolution would correspond to a sensor resolution of 32,800 pixels" [9: 2]. No airborne digital camera currently offers this resolution. Moreover, an uncompressed 12-bit color image would require approximately 4.5 gigabytes of storage and a complete roll of aerial film would require approximately 2.7 terabytes. While this volume of storage is not an insurmountable obstacle (at least on the ground), it does represent a technical issue to be overcome.

These deficiencies notwithstanding, digital airborne cameras have become an integral part of the cartographer's tool set. It is clear that the benefits of digital imagery will encourage continued technical evolution.

Light Detection and Ranging

Remote sensing is the science and art of collecting information about an object, area, or phenomenon through the analysis of data acquired by a device that is not in physical contact with the object, area, or phenomenon under investigation. On July 23, 1972, the U.S. National Aeronautics and Space Administration launched the first in a series of remote sensing satellites. Initially designated Earth Resources Technology Satellite-A (ERTS-A), this platform was renamed ERTS-1 after launch. Since that time, satellite remote sensing (originally relatively small-scale imagery) has evolved to the point that 0.5-m imagery is commercially available.

The small-scale imagery acquired by satellite platforms offers several advantages over aerial photography. The imagery provides a regional view (large areas) and provides repetitive views of the same areas. The imagery views a broader portion of the spectrum than the human eye and focuses on very specific bandwidths in an image simultaneously. Finally, the imagery provides georeferenced, digital data and the system operates in all seasons, in bad weather, and at night.

However, the comparatively small scale of the imagery can be a limitation for many mapping applications. In parallel with the development of increasingly sensitive satellite sensors, the remote sensing industry has adapted digital sensors that focus on the visible spectrum to aircraft platforms (i.e., digital airborne cameras). The industry also has adapted or developed sensors that focus on nonvisible portions of the electromagnetic spectrum and that offer benefits for a variety of mapping activities.

Light detection and ranging (LIDAR or lidar), perhaps the most significant innovation to date, is similar in many respects to radio detection and ranging (radar). Lidar can be used to measure characteristics of a remote target, including distance, speed, rotation, and chemical composition and concentration. Known initially as optical radar or laser radar, lidar can be thought of as a radar system that uses electromagnetic radiation at higher, optical frequencies — most commonly a laser.

In operation, a lidar instrument transmits light to a target. The transmitted light interacts with and is changed by the target. Some of this light is reflected and/or scattered back to the sensing instrument and analyzed. Change in the properties of the light enables some property of the target to be determined and the time that the light travels to the target and back is used to determine the distance to the target.

This last characteristic is particularly important for topographic data collection. Lidar systems perform an "echo time" distance conversion based on an assumed speed of light (c) equal to 3.0E8 mps as follows:

1 nanosecond	0.15 m	5.9 inches
1 microsecond	150 m	492 feet
10 microseconds	1.5 km	0.93 statute miles
100 microseconds	15 km	9.32 statute miles
1 millisecond	150 km	93.2 statute miles

The elevation of aircraft carrying lidar sensors can be fixed using GPS/IMU technology. As the aircraft flies over an area, laser pulses are emitted with high rapidity toward the ground. Pulses are reflected by the ground and/or objects on it (e.g., trees and buildings). The elapsed time between the emitted and returning signals is measured for each pulse, to compute a slant distance. The position and altitude of the aircraft are measured with airborne GPS and inertial measurement unit (IMU) systems.

After post-processing, the positional data are combined with the slant distance, information on atmospheric conditions, hardware characteristics, and other relevant parameters to generate an X, Y, Z coordinate triplet of a point on the ground. Millions of such points are captured as the mission progresses, providing a dense digital terrain model.

Frequently, the greatest problem posed during this post-processing is thinning of the data array. Lidar systems typically are transmitting 10,000 pulses per second and often achieve pulse rates in excess of 50,000. In this higher-resolution mode, lidar data have been used, for example, to support mapping of floodplains. The level of detail has been sufficient to support hydrographic and hydrologic modeling for the Federal Emergency Management Agency's (FEMA) digital flood insurance rate maps (DFIRM) program.

In effect, the measurement of terrain elevations using laser light produces an accurate data set for the creation of digital elevation models (DEMs) and contours. Lidar data sets generally are expected to have vertical accuracies of less than 1 ft, producing smaller contour intervals with minimal additional effort. By measuring the ground height and the height of buildings, lidar supports creation of low-cost, true continuous surface models for orthophoto production.

This extremely fine granularity in observations offers the possibility of intriguing new applications, including, for example, the ability to distinguish the heights of transmission cables over the ground below and the heights above the ground of tree canopies, the latter a function of the ability in many situations to penetrate the canopy and capture ground heights.

Another example of the use of this level of detail is offered by the wireless communications industry. The modeling requirements of the wireless market (especially the broadband wireless segment) include line-of-sight calculations for unobstructed transmission. These translate to technical specifications of 1-m (or better) vertical and horizontal accuracy, 1-m spatial DEM postings, separate terrain layers for structures, bare earth and vegetation, and good structure/building definition. Lidar data collection satisfies these criteria, and use of this technology is growing rapidly as the number of sensors in the commercial marketplace increases.

Synthetic Aperture Radar and Interferometric Synthetic Aperture Radar

Synthetic aperture radar (SAR) provides the means to obtain large-area imaging at high resolutions even in inclement weather or at night. SAR systems take advantage of the long-range propagation characteristics of radar signals to provide high-resolution imagery. SAR complements photographic and digital optical imaging capabilities by overcoming time-of-day and atmospheric constraints. Originally developed for military and large-scale economic applications (such as petroleum exploration), reductions in cost of SAR technology have combined to make it economically feasible for smaller commercial and public interest users.

Traditionally, SAR systems produced a two-dimensional image. One dimension, called the "range" (or cross track), is a measure of the "line-of-sight" distance from the radar to the target. As with other radar systems, range is determined by precisely measuring the time from the transmission of a pulse to the time of receiving the echo from a target.

The second dimension, called the "azimuth" (or along track), is perpendicular to the range. The resolution of the azimuth measurement is dependent on the concentration of the pulsed radar beam. For ground-based systems, large-aperture receivers can be used to improve image resolution, but even moderate SAR resolutions would require an antenna physically larger than reasonably could be carried aboard an aircraft. Airborne SAR systems overcome this limitation by simulating a large antenna with multiple observations along the flight light; the observations are combined (with appropriate consideration of Doppler effects) to simulate or "synthesize" a much larger antenna; hence the name for the technology.

FIGURE 165.1 Hypothetical dot scatter. Each dot has positional and elevation data associated with it.

A variation of SAR technology called Interferometric synthetic aperture radar (IFSAR) has been used to generate three-dimensional building models. High resolutions at pixel sizes down to 10 cm and single-pass interferometry support the geometric reconstruction of relatively small objects. Although the resolutions typically are less than those of lidar, ISSAR systems ultimately may represent an important commercial source of terrain information.

165.3 Topographic Mapping in a Geographic Information System

Data points collected either by field methods or by photogrammetric methods can be processed in a number of ways to make them useful in a variety of applications [14]. We focus on applications that employ GIS software to process them. Consider a hypothetical array of dot scatter representing collected data points and assume that elevation and positional data have been associated with each dot (Figure 165.1). The following methods of interpolating values between dots use ArcView 3.2 GIS (Environmental Systems Research Institute, Redlands, CA) with ArcView Spatial Analyst and ArcView 3D Analyst extensions loaded on the computer (both from ESRI). In addition, the Animal Movement and Home Range extension package is also loaded on the computer [10].

Use Animal Movement to calculate a triangulation of the data points (Figure 165.2). For any given set of points there is a unique network of triangles, with the points as vertices, that most closely satisfies the criterion that the triangles be as nearly equilateral as possible. This triangulation, called the Delaunay triangulation, has the property that it maximizes the minimum angle over all triangles in the triangulation. The triangulation may be calculated easily in ArcView [10]. That calculation is based on one of several algorithms, all of which lead to the same result [4,21].

Use the Home Range extension to calculate the minimum convex polygon containing the dot scatter (Figure 165.3) [10]. The convex hull (minimum convex polygon) is useful for constraining other calculations.

Use ArcView 3D Analyst (ESRI) to calculate a triangulated irregular network (TIN) from the dot scatter (Figure 165.4). As with the Delaunay triangulation, both may be easily done in ArcView in "blackbox" mode, without understanding the theory behind the software.

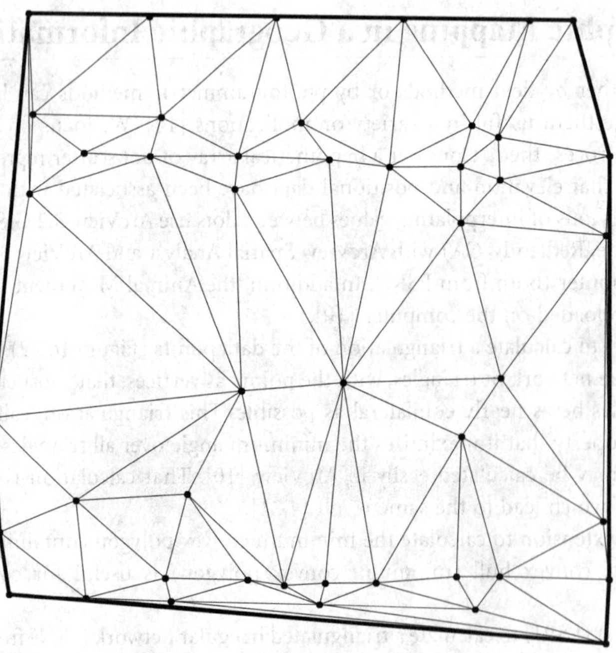

FIGURE 165.2 Delaunay triangulation of dot scatter in Figure 165.1.

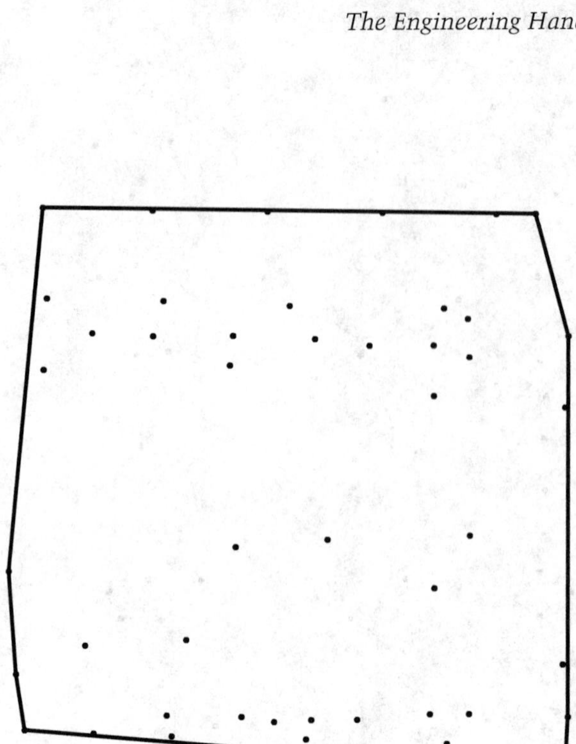

FIGURE 165.3 Minimum convex polygon containing the dot scatter.

FIGURE 165.4 Triangulated irregular network (TIN) calculated by the GIS using the dot scatter.

Superimpose the Delaunay triangulation on the TIN (Figure 165.5): Note that the Delaunay triangulation itself generates the TIN. Suitable partition of Delaunay triangles causes the surface of the TIN to appear to be three-dimensional. Hence, we reveal the need to understand that the Delaunay triangulation is the conceptual backing of the TIN and that the Watson geometric algorithm is one characterization of the triangulation [4,21].

Calculate contours from the TIN using ArcView 3D Analyst. In this case, the conceptual background is the calculus application of calculating level curves of the three-dimensional surface represented by the TIN: note that the contours (Figure 165.6) follow the TIN (Figures 165.4 and 165.5). Remove the TIN and the surface is contoured, as is a topographic map (Figure 165.6).

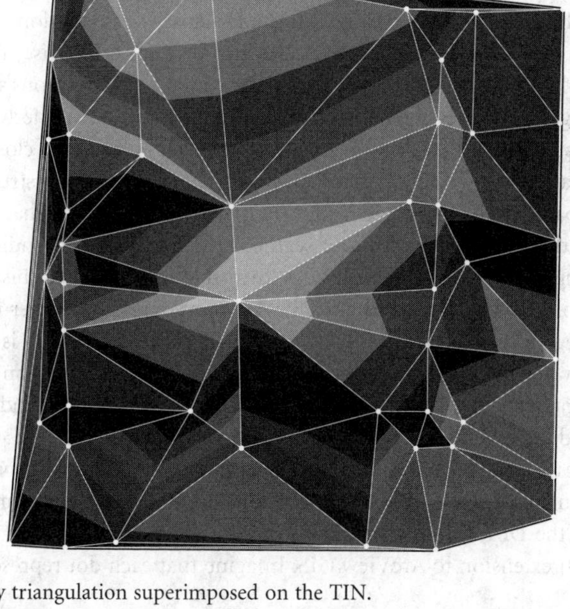

FIGURE 165.5 Delaunay triangulation superimposed on the TIN.

FIGURE 165.6 Remove the TIN to see the contours alone as a topographic map.

This sequence of examples is designed to illustrate the need to understand the theory behind practice. The strategy of using a geometric algorithm to generate a Delaunay triangulation, which is then used to create a TIN, which is then used to find level curves of the TIN as contours, which are then used on a topographic map, is a standard conceptual design that endures far beyond the life of any software package. It is a procedure that has been employed in mapping from the conventional to the modern, and it rests on a clear understanding of geometry and calculus. Software changes frequently, but theory does not. Thus, it is critical to have at least some theoretical grounding in methods if one wishes to keep current.

165.4 Delaunay Triangulation and Dirichlet Tessellation

Indeed, theory not only serves as an enduring underpinning for practice; it also serves to lead to other theoretical results. Such a situation is the case with the Delaunay triangulation and another partition of the dot scatter, known variously as a Dirichlet tessellation, Voronoi polygons, or Thiessen polygons.

In 1911, Thiessen and Alter [21] wrote on the analysis of rainfall using polygons surrounding rain gauges. Given a scatter of rain gauges, represented abstractly as dots, partition the underlying plane into polygons containing the dots in such a way that all points within any given polygon are closer to the rain gauge dot within that polygon than they are to any other gauge dot. The geometric construction usually associated with performing this partition of the plane into a mutually exclusive, yet exhaustive, set of polygons is performed by joining the gauge dots with line segments, finding the perpendicular bisectors of those segments, and extracting a set of polygons with sides formed by perpendicular bisectors. It is this latter set of polygons that has come to be referred to as "Thiessen polygons" (and earlier names such as Dirichlet region or Voronoi polygon; see Coxeter [7]). The construction using bisectors is tedious and difficult to execute with precision when performed by hand. Kopec (1963) [22] noted that an equivalent construction results when circles of radius of the distance between adjacent points are used. Indeed, that construction is but one case of a Euclidean general construction. Like Kopec, Rhynsburger (1973) [19] also sought easier ways to construct Thiessen polygons: Kopec through knowledgeable use of geometry and Rhynsburger through the development of computer algorithms. GIS software affords an opportunity to combine both [1].

Figure 165.7 shows the Dirichlet tessellation of the dot scatter in Figure 165.1 calculated using the Animal Movement [10] extension to ArcView GIS. Imagine that each dot represents a rain gauge. A cell

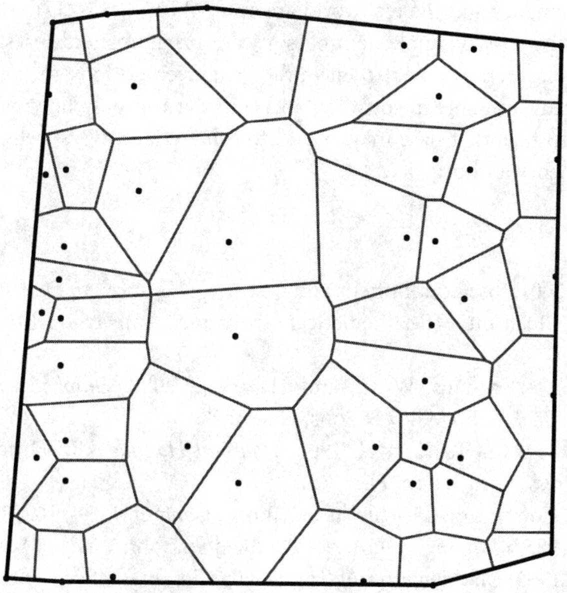

FIGURE 165.7 Dirichlet tessellation of the hypothetical dot scatter.

in the Dirichlet tessellation about a point (rain gauge) is the region that is closer to, or as close to, any other point (rain gauge).

The Dirichlet tessellation and the Delaunay triangulation are graph-theoretic duals; thus, one may be derived from the other [8]. To see this relationship, consider Figure 165.8. Replace each Dirichlet cell with a point interior to it (rain gauge). Join two points with an edge if there is a boundary of a cell of the Dirichlet tessellation separating these two points. Repeat the process for all pairs of points, creating the Delaunay triangulation.

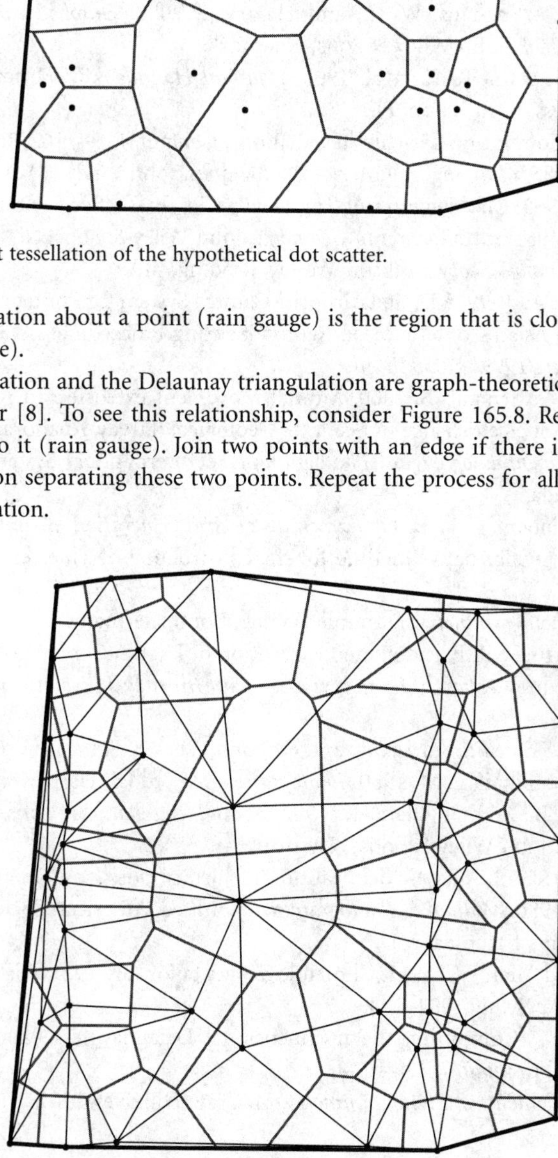

FIGURE 165.8 Graph-theoretic duality of Dirichlet tessellation and Delaunay triangulation.

Links between mathematics and the real world are everywhere [2,3,14,17]. All too often, though, one is caught up in the mechanics of implementing software programs and loses sight of the theoretical underpinnings of process. Here we seek to encourage engineers to look to the conceptual base of the wonderful tool kit already available in software, and consider not only how practice derives from that conceptual base but also how that base may extend to other concepts, as well. To do so is to be at the leading edge of both theory and practice.

References

1. Arlinghaus, S. L. 2001. Bisectors, buffers, and base maps. *Solstice: An Electronic Journal of Geography and Mathematics*. Institute of Mathematical Geography, Ann Arbor, MI. Available at: *www.imagenet.org*.
2. Arlinghaus, S. L., Arlinghaus, W. C., and Harary, F. 2002. *Graph Theory and Geography: An Interactive View eBook*, John Wiley & Sons, New York.
3. Arlinghaus, S. L., Austin, R. F., et al. 1995. *Practical Handbook of Digital Mapping: Terms and Concepts*. CRC Press, Boca Raton, FL.
4. Bowyer, A. 1981. Computing Dirichlet tessellations, *Comput. J.*, 24, 162–166.
5. Burnside, C. D. 1985. *Mapping from Aerial Photographs*. John Wiley & Sons, New York.
6. Chen, W.-F., Ed. 1995. *Civil Engineering Handbook*. CRC Press, Boca Raton, FL.
7. Coxeter, H. S. M. 1961. *Introduction to Geometry*. John Wiley & Sons, New York.
8. Harary, F. 1969. *Graph Theory*. Addison-Wesley, Reading, MA.
9. Heier, H. and Hinz, A., 1999. A Digital Airborne Camera System for Photogrammetry and Thematic Applications. 20th Asian Conference on Remote Sensing Proceedings. Available at: *www.gisdevelopment.net/aars/acrs/1999/ts7058.shtml*.
10. Hooge, P. N. and Eichenlaub B. 2000. Animal Movement extension to ArcView ver. 2.0. Alaska Science Center, Biological Science Office, U.S. Geological Survey, Anchorage, AK.
11. ISM, 1997. *Fundamentals of Digital Photogrammetry*. International Systemap Corporation, Vancouver, B.C., Canada.
12. Knupp, P. and Steinberg, S. 1993. *Fundamentals of Grid Generation*, Boca Raton, CRC Press.
13. Kopec, R. J. 1963. An alternative method for the construction of Thiessen polygons. *Prof. Geographer*, 15:24–26.
14. Kraus, K. 1993. *Photogrammetry*. Dummler Verlag, Bonn, Germany.
15. Ladak, A. and Martinez, R. B. Automated Derivation of High Accuracy Road Centrelines Thiessen Polygons Technique. Available at: *www.esri.com/library/userconf/proc96/TO400/PAP370/P370.HTM*.
16. Leick, A. 1995. *GPS Satellite Surveying*, 2nd ed. John Wiley & Sons, New York.
17. Moffitt, F. H. and Mikhail, E. M. 1980. *Photogrammetry*, 3rd ed. Harper & Row, New York.
18. Okabe, A.; Boots, B.; and, Sugihara, K. 1992. *Spatial Tessellations: Concepts and Applications of Voronoi Diagrams*. John Wiley & Sons, New York.
19. Rhynsburger, Dirk, 1973. Analytic delineation of Thiessen polygons. *Geogr. Anal.*, 5, 133–144.
20. Slama, C. C., Ed. 1980. *Manual of Photogrammetry*, 4th ed. American Society of Photogrammetry and Remote Sensing, Bethesda, MD.
21. Thiessen, A. H. and Alter, J. C. 1911. Climatological data for July 1911: District No. 10. *Great Basin Monthly Weather Rev.*, July:1082–1089.
22. Watson, D. F. 1981. Computing the n-dimensional Delaunay tessellation with application to Voronoi polytopes. *Comput. J.*, 24, 167–172.
23. Wolf, P. R. 1983. *Elements of Photogrammetry*. McGraw-Hill, New York.

Further Information

American Congress on Surveying and Mapping (ACSM), *www.acsm.net*
American Geographical Society, *www.amergeog.org*
American Society for Photogrammetry and Remote Sensing (ASPRS), *www.asprs.org*
American Society of Civil Engineers, *www.asce.org*
Association of American Geographers, *www.aag.org*
Environmental Systems Research Institute, ESRI, *www.esri.com*
UK Mapping Links, *www.geog.gla.ac.uk/~mshand/Ukmapping.htm*
United States Geological Survey (USGS), *www.usgs.gov*

Software Used

Adobe, Photoshop, 7.0
Environmental Systems Research Institute (ESRI), Redlands, CA, ArcView 3.2
ArcView Spatial Analyst Extension (ESRI)
ArcView 3D Analyst Extension (ESRI)
Animal Movement and Home Range Extensions to ArcView

166

Surveying Computations

Boudewijn H. W. van Gelder
Purdue University

Surveyors collect measurements (observations, data, etc.) for the purpose of determining a wide variety of variables (parameters, unknowns, etc.). These measurements may be distances, directions, angles, look angles, azimuths, elevation angles, height differences, time, and so on. From these observations, the surveyor deduces results in terms of parameters, such as coordinates of an intersection point of two boundary lines or the height of a marker on a bridge. Rather than identifying a variety of (classical) survey problems such as point positioning through traversing or height determinations through spirit leveling, this chapter follows a more general approach by discussing a general relationship between observational data and parameters. In Section 166.6 an example is presented.

166.1 Principles of Multivariate Calculus

A general relationship between two classes of variables may exist. One class of variables is denoted by l and the other by x. The two classes of variables refer to the two main classes that a surveyor deals with: observations or the input data (l) and the results or output data (x). The variable l represents a group of n variables l_i with $i = 1, \ldots, n$. The other variable x represents a group of u variables x_j with $j = 1, \ldots, u$. A functional relationship F exists between these two groups of variables. This functional relationship is nothing other than the mathematical model that expresses the assumed interdependency between the variables in question. Realize, though, that it is the surveyor who makes this enormously important judgment call on this mathematical relationship: like a car driver engages a gear, the surveyor engages at a certain moment a mathematical model, or at least follows the spiritual ancestor of the model provided in terms of a survey software package. The number of functional relationships does not need to be equal to n or u; we may have F_k ($k = 1, \ldots, r$) relationships. Summarizing, we have

0-8493-1586-7/05/$0.00+$1.50

$$F_k = F(l_i, x_j) \tag{166.1}$$

with

$$k = 1, \ldots, \gamma \tag{166.2}$$

$$i = 1, \ldots, n \tag{166.3}$$

$$j = 1, \ldots, u \tag{166.4}$$

A simple example will illustrate. Suppose that we observe four pairs of coordinates $\{X_l, Y_l\}$ that we assume lie on a circle. So the model F_k is the equation of a circle, centered at an arbitrary coordinate $\{X_c, Y_c\}$ with an arbitrary radius R:

$$F_k = (X_k - X_c)^2 + (Y_k - Y_c)^2 - R^2 \tag{166.5}$$

Geometry dictates that a circle is defined through three points (six coordinates), provided that those three points are not on a line. Thus, four sets of coordinates will not quite fit the "circle" model. Small adjustments to the eight coordinates will be necessary to end up with one unique circle. In addition, we may have to slightly adjust the unknown circle centered at $\{X_c, Y_c\}$ and its unknown radius R. We will probably have to slightly adjust any initial guess of these three variables $\{X_c, Y_c, R\}$ to find the optimum circle passing through those four pairs of coordinates. These small adjustments to the variables l and x, which we call dl and dx, will make the variables fit the circle model just right. Adding the small corrections dl and dx to l and x will make the equations F_k become equal to zero:

$$F_k = F(l_i + dl_i, x_j + dx_j) = 0 \tag{166.6}$$

The problem of finding these hopefully small corrections dl_i and dx_j will be made much easier if we linearize the model according to the rules of multivariate calculus:

$$F_k(l_i + dl_i, x_j + dx_j) = F_k(l_i, x_j) + \frac{\partial F_k}{\partial l_i} dl_i + \frac{\partial F_k}{\partial x_j} dx_j + \cdots = 0 \tag{166.7}$$

With the wealth of variables r equations F_k, n variables l_i, and u variables x_j in Equation (166.7) at hand, it is much more practical to use the tools (vectors and matrices) of linear algebra.

166.2 Principles of Linear Algebra

Equation (166.7) can be rewritten in terms of vectors and matrices. We define the vectors (shown below in transposed form) L, X, and W and the matrices A and B according to

$$\mathbf{L}^{\mathrm{T}} = [l_1, l_2, \ldots, l_i, \ldots, l_n]^{\mathrm{T}} \tag{166.8}$$

$$\mathbf{X}^{\mathrm{T}} = [x_1, x_2, \ldots, x_j, \ldots, x_u]^{\mathrm{T}} \tag{166.9}$$

$$\mathbf{W}^{\mathrm{T}} = [F_1, F_2, \ldots, F_k, \ldots, F_r]^{\mathrm{T}} \tag{166.10}$$

and

$$\mathbf{A} = \begin{bmatrix} \frac{\partial F_1}{\partial l_1} & \cdots & \frac{\partial F_1}{\partial l_n} \\ \vdots & \cdots & \vdots \\ \frac{\partial F_r}{\partial l_1} & \cdots & \frac{\partial F_r}{\partial l_n} \end{bmatrix} \tag{166.11}$$

$$\mathbf{B} = \begin{bmatrix} \frac{\partial F_1}{\partial x_1} & \cdots & \frac{\partial F_1}{\partial x_u} \\ \vdots & \cdots & \vdots \\ \frac{\partial F_r}{\partial x_1} & \cdots & \frac{\partial F_r}{\partial x_u} \end{bmatrix} \tag{166.12}$$

Matrix A is the Jacobian of the model equations F with respect to the variables l, and consists of r rows and n columns. Similarly, the matrix B is the Jacobian of the model equations F with respect to the variables x, and consists of r rows and u columns. The vector/matrix notation enables us to shorten Equation (166.7) to

$$\mathbf{W} + \mathbf{B}\,\mathbf{dL} + \mathbf{A}\,\mathbf{dX} + \cdots = 0 \tag{166.13}$$

Referring back to our circle fitting example, we have, for instance,

$$\mathbf{dL}^{\mathrm{T}} = [dX_1, dY_1, dX_2, dY_2, dX_3, dY_3, dX_4, dY_4]^{\mathrm{T}} \tag{166.14}$$

$$\mathbf{dX}^{\mathrm{T}} = [dX_c, dY_c, dR]^{\mathrm{T}} \tag{166.15}$$

The matrix element $A_{3,2}$ (third row, second column) is equal to

$$A_{(\text{row}=3,\ \text{col}=2)} = \frac{\partial F_3}{\partial Y_c} = -2\times(Y_3 - Y_c) \tag{166.16}$$

The matrix element $B_{3,5}$ (third row, fifth column) is equal to

$$B_{(\text{row}=3,\ \text{col}=5)} = \frac{\partial F_3}{\partial X_3} = 2\times(X_3 - X_c) \tag{166.17}$$

Matrix A is a full matrix; however, the B matrix is of the nature

$$\mathbf{B} = \begin{bmatrix} X & X & 0 & 0 & 0 & 0 & 0 & 0 \\ 0 & 0 & X & X & 0 & 0 & 0 & 0 \\ 0 & 0 & 0 & 0 & X & X & 0 & 0 \\ 0 & 0 & 0 & 0 & 0 & 0 & X & X \end{bmatrix} \tag{166.18}$$

with the X's denoting nonzero elements.

In this example, the running indices i, j, k (see Equations (166.3), (166.4), and (166.2), respectively) have ranges

$$i = 1,\ldots,8$$
$$j = 1,\ldots,3 \tag{166.19}$$
$$k = 1,\ldots,4$$

denoting eight observations and three parameters to be estimated, all being part of four equations.

166.3 Model of Two Sets of Variables, Observations, and Parameters: The Mixed Model

The linearized version of Equation (166.6), Equation (166.7), demands closer inspection of the definition of all variables involved. The surveyor has the benefit of observed values and approximate values for the parameters; however, they should not be mixed. With subscripts we indicate the nature of the variable. Subscript b denotes "observed value," subscript 0 denotes "approximate value," and subscript a denotes the value of the observable quantity or the (unknown) parameter that perfectly fits the model F. V denotes the residual, which is the value that must be added to the observed value L_b to obtain the value for L_a that perfectly fits the model F. Similarly, X denotes the correction that needs to be added to the approximate values of the parameter X_0 to obtain the value for the parameter X_a that perfectly fits the model F.

The following steps need to be taken to arrive at the linearized model (Equation (166.7) or (166.26)), starting from Equation (166.6) or (166.20), with the linearization around the Taylor point $\{L_0, X_0\}$:

$$F(\mathbf{L}_a, \mathbf{X}_a) = 0 \tag{166.20}$$

or

$$F(\mathbf{L}_b + \mathbf{V}, \mathbf{X}_0 + \mathbf{X}) = 0 \tag{166.21}$$

$$F(\mathbf{L}_0, \mathbf{X}_0) + \frac{\partial F}{\partial \mathbf{L}}(\mathbf{L}_a - \mathbf{L}_0) + \frac{\partial F}{\partial \mathbf{X}}(\mathbf{X}_a - \mathbf{X}_0) = 0 \tag{166.22}$$

$$\mathrm{W}_0 + \mathrm{B} \times (\mathbf{L}_a - \mathrm{L}_b + \mathrm{L}_b - \mathbf{L}_0) + \mathrm{A} \times (\mathrm{X}_a - \mathrm{X}_0) = 0 \tag{166.23}$$

$$\mathrm{W}_0 + \mathrm{B} \times (\mathrm{V} + \mathrm{L}) + \mathrm{A} \times \mathrm{X} = 0 \tag{166.24}$$

$$\mathrm{AX} + \mathrm{BV} + (\mathrm{W}_0 + \mathrm{BL}) = 0 \tag{166.25}$$

or, arriving at the linearized form of Equation (166.20),

$$\mathrm{AX} + \mathrm{BV} + \mathrm{W} = 0 \tag{166.26}$$

with

$$\begin{aligned} \mathrm{A} &= \frac{\partial F}{\partial \mathrm{X}} & \mathrm{X} &= \mathrm{X}_a - \mathrm{X}_0 \\ \mathrm{B} &= \frac{\partial F}{\partial \mathrm{L}} & \mathrm{V} &= \mathrm{L}_a - \mathrm{L}_b \\ & & \mathrm{L} &= \mathrm{L}_b - \mathrm{L}_0 \\ & & \mathrm{W} &= \mathrm{W}_0 + \mathrm{BL} \end{aligned} \tag{166.27}$$

In words, the variables denote

L_a: adjusted observations
L_b: observed values
V: residuals
L_0: approximate observations
L: residual observations
X_a: adjusted parameters
X_0: approximate parameters (initial guess)
X: unknown correction to parameters
W_0: misclosure vector
W: misclosure vector
A: design matrix (partial derivative matrix for the parameters)
B: partial derivative matrix for the observations

Equation (166.26) is a set of r equations with u unknowns. This is an inconsistent set of equations and cannot be solved since often $r > u$. The method of Lagrangian multipliers provides a set of u equations with u unknowns under the conditions that the sum of the squared residuals is minimum. This leads to a minimum variance estimate for the vector X (see, for example, Hamilton, 1964; Strang, 1986; and Strang, 1988). The unknown vector X can be computed from the u equations with u unknowns, also known as the "normal equations." Without derivation, the normal equations are

$$A^T M^{-1} A X + A^T M^{-1} W = 0 \tag{166.28}$$

or, in short,

$$NX + U = 0 \tag{166.29}$$

with

$$N = A^T M^{-1} A \tag{166.30}$$

$$U = A^T M^{-1} W \tag{166.31}$$

and

$$M = BB^T \tag{166.32}$$

The model that mixes the observations L_a and the parameters X_a can be further generalized, assuming a statistical model (variance/covariance matrix for the observations) (see also Chapter 161):

$$\Sigma_{L_b} = \begin{bmatrix} \sigma_{l_1 l_1} & \sigma_{l_1 l_2} & \cdots & \sigma_{l_1 l_n} \\ \sigma_{l_2 l_1} & \sigma_{l_2 l_2} & \cdots & \sigma_{l_2 l_n} \\ \vdots & \vdots & \cdots & \vdots \\ \sigma_{l_n l_1} & \sigma_{l_n l_2} & \cdots & \sigma_{l_n l_n} \end{bmatrix} \tag{166.33}$$

Factoring a common (standard unit weight) constant out, we get

$$\Sigma_{L_b} = \sigma_0^2 Q_{L_b} = \begin{bmatrix} q_{l_1 l_1} & q_{l_1 l_2} & \cdots & q_{l_1 l_n} \\ q_{l_2 l_1} & q_{l_2 l_2} & \cdots & q_{l_2 l_n} \\ \vdots & \vdots & \cdots & \vdots \\ q_{l_n l_1} & q_{l_n l_2} & \cdots & q_{l_n l_n} \end{bmatrix} \tag{166.34}$$

The Q matrix is called the weight coefficient matrix. In terms of the weight matrix P, we get

$$\Sigma_{L_b} = \sigma_0^2 P_{L_b}^{-1} = \begin{bmatrix} p_{l_1l_1} & p_{l_1l_2} & \cdots & p_{l_1l_n} \\ p_{l_2l_1} & p_{l_2l_2} & \cdots & p_{l_2l_n} \\ \vdots & \vdots & \cdots & \vdots \\ p_{l_nl_1} & p_{l_nl_2} & \cdots & p_{l_nl_n} \end{bmatrix}^{-1} \tag{166.35}$$

For weighted observations, the matrix M, Equation (166.32), is simply replaced by

$$M = BP^{-1}B^T = \frac{1}{\sigma_0^2} B\Sigma_{L_b} B^T \tag{166.36}$$

The least squares estimate for the solution vector can be obtained from

$$\begin{aligned} X_a &= X_0 + X \\ &= X_0 - [A^T M^{-1} A]^{-1} A^T M^{-1} W \\ &= X_0 - [A^T (BP^{-1}B^T)^{-1} A]^{-1} A^T (BP^{-1}B^T)^{-1} W \end{aligned} \tag{166.37}$$

The variance/covariance matrix of the parameter vector X or X_a, after applying the law of propagation of errors, can be shown to be equal to

$$\Sigma_{X_a} = \Sigma_X = \sigma_0^2 (A^T M^{-1} A)^{-1} \tag{166.38}$$

The relationship between the *a priori* variance of unit weight σ_0^2 and the a posteriori variance of unit weight $\hat{\sigma}_0^2$, the latter being computed from

$$\hat{\sigma}_0^2 = \frac{V^T PV}{r-u} \tag{166.39}$$

can be tested according to

$$\frac{\chi^2_{r-u;1-\alpha/2}}{r-u} < \frac{\hat{\sigma}_0^2}{\sigma_0^2} < \frac{\chi^2_{r-u;\alpha/2}}{r-u} \tag{166.40}$$

Equation (166.40) reflects the probability that the ratio of the variances will fall within the specified bounds, and is equal to $1 - \alpha$ (= 95% if $\alpha = 0.05$ or 5%). This test gives insight between the expected observational precision and the overall behavior of the residuals once a model has been "engaged." Rejection of the test may also lead to rejection of the particular model — for instance, in our example, if the observations were actually to lie on an ellipse rather than a circle.

From the "mixed model" $F(L_a, X_a) = 0$, two special models can be derived, which will be treated in the following two sections.

166.4 Observations as a Function of Parameters Only: The Model of Observation Equations

From the r (linearized) equations with u unknowns, a special case results if it so happens that each individual observation can be expressed as a function of the unknowns only. In this case we have $L_a = F(X_a)$, which can be derived directly rewriting Equation (166.26) as

$$-BL - BV = AX + W_0 \tag{166.41}$$

If B is assumed to be equal to a negative unit matrix (a square matrix with zeros and −1 as diagonal elements), Equation (166.41) becomes simply, with W_0 being absorbed in the L vector (see Equations (166.44) and (166.45)),

$$L + V = AX \tag{166.42}$$

with

$$\begin{aligned} A &= \frac{\partial F}{\partial X} \\ X &= X_a - X_0 \\ V &= L_a - L_b \\ L &= L_b - L_0 \\ L_0 &= F(X_0) \end{aligned} \tag{166.43}$$

When you start from the unlinearized model $L_a = F(X_a)$, we find similarly

$$\begin{aligned} L_a &= F(X_a) \\ L_b + V &= F(X_0 + X_a - X_0) \\ L_b + V &= F(X_0) + \frac{\partial F}{\partial X} \cdot (X_a - X_0) \\ L_b + V &= L_0 + A \cdot X \end{aligned} \tag{166.44}$$

Bringing the L_0 vector to the left side,

$$\begin{aligned} L_b - L_0 + V &= A \cdot X \\ L + V &= A \cdot X \end{aligned} \tag{166.45}$$

In words, the variables denote

- L_a: adjusted observations
- L_b: observed values
- V: residuals
- L_0: approximate observations
- L: residual observations
- X_a: adjusted parameters
- X_0: approximate parameters (initial guess)
- X: unknown correction to parameters
- A: design matrix (partial derivative matrix for the parameters)

The normal equations simplify to

$$NX + U = 0 \tag{166.46}$$

with

$$N = A^T PA \tag{166.47}$$

$$U = -A^T PL \tag{166.48}$$

The solution vector X_a is equal to

$$\begin{aligned} X_a &= X_0 + X \\ &= X_0 + [A^T PA]^{-1} A^T PL \end{aligned} \tag{166.49}$$

The variance/covariance matrix of the parameter vector X or X_a, applying the law of propagation of errors, can be shown to be equal to

$$\Sigma_{X_a} = \Sigma_X = \sigma_0^2 (A^T PA)^{-1} \tag{166.50}$$

The relationship between the *a priori* variance of unit weight σ_0^2 and the a posteriori variance of unit weight $\hat{\sigma}_0^2$, the latter being computed from

$$\hat{\sigma}_0^2 = \frac{V^T PV}{n-u} \tag{166.51}$$

Note that the denominator of Equation (166.51) represents the degrees of freedom, which are equal to $n - u$ since we deal with n (linearized) equations with u unknowns.

The method of observation equations is also known as "adjustment of indirect observations" (see, for instance, Mikhail, 1976).

Notation

The linearized observation equation, Equation (166.45),

$$L + V = AX \tag{166.52}$$

appears under a variety of notations in the literature. For instance, Mikhail [1976] and Mikhail and Gracie [1981] use

$$I + v = -B\Delta \tag{166.53}$$

In the statistics literature one often finds

$$y - \varepsilon = X\beta \tag{166.54}$$

or, as in Gelb [1974],

$$z - v = Hx \tag{166.55}$$

Also, a more tensor-oriented notation may be found, as in Baarda [1967]:

$$x^i + \varepsilon^i = a^i_\alpha Y^\alpha \tag{166.56}$$

166.5 All Parameters Eliminated: The Model of Condition Equations

From the r (linearized) equations with u unknowns, another special case may be derived by eliminating the u unknowns from the linearized model, Equation (166.26):

$$\mathrm{AX} + \mathrm{BV} + \mathrm{W} = 0 \tag{166.57}$$

After elimination, we obtain $r - u$ equations which reflect the mathematical relationship between the observations only. The equations are of the type

$$\mathrm{B'V} + \mathrm{W'} = 0 \tag{166.58}$$

They reflect the existing conditions between the observables — hence the name of the method. The classical example in surveying is that in a (not too large) triangle, the three measured angles have to sum to π. Another example concerns the loop closures between the leveled height differences, which have to sum to zero in each loop.

Starting from the nonlinearized model $F(\mathrm{L}_a) = 0$, we find

$$\begin{aligned}
F(\mathrm{L}_a) &= 0\\
F(\mathrm{L}_b + \mathrm{V}) &= 0\\
F(\mathrm{L}_0 + \mathrm{L}_b - \mathrm{L}_0 + \mathrm{V}) &= 0\\
F(\mathrm{L}_0 + \mathrm{L} + \mathrm{V}) &= 0\\
F(\mathrm{L}_0) + \frac{\partial F}{\partial \mathrm{L}} \cdot (\mathrm{L} + \mathrm{V}) &= 0\\
\mathrm{W}_0 + \mathrm{B} \cdot \mathrm{L} + \mathrm{B} \cdot \mathrm{V} &= 0\\
\mathrm{W} + \mathrm{BV} &= 0
\end{aligned} \tag{166.59}$$

with

$$\begin{aligned}
\mathrm{B} &= \frac{\partial F}{\partial \mathrm{L}}\\
\mathrm{V} &= \mathrm{L}_a - \mathrm{L}_b\\
\mathrm{L} &= \mathrm{L}_b - \mathrm{L}_0\\
\mathrm{W}_0 &= F(\mathrm{L}_0)\\
\mathrm{W} &= \mathrm{W}_0 + \mathrm{BL}
\end{aligned} \tag{166.60}$$

In words, the variables denote

L_a: adjusted observations
L_b: observed values
V: residuals
L_0: approximate observations
L: residual observations
W_0: misclosure vector

W: misclosure vector
B: partial derivative matrix for the observations

166.6 An Example: Traversing

Various methods are presented in the survey literature (see, for instance, Wolf and Brinker, 1995, chapters 12 and 13) to adjust data collected as part of a traverse. The (two-dimensional) least squares example discussed illustrates the formation of observation equations and can be adapted to reflect any traverse method. The example here involves the measurements of directions and distances along a traverse which stretches between two known points A and B, in coordinates $\{x_A, y_A\}$ and $\{x_B, y_B\}$. In these terminal points two closing directions are measured to two known azimuth markers, the points P and Q, respectively. The traverse is to solve for parameters such as the unknown coordinates of points 1 through n, the orientation unknowns for each point where directional measurements took place. Finally we assume the existence of an unknown scale factor λ between the distance measurement equipment and the distances implied by the known coordinates of points A, B, P, and Q.

Directional Measurements r_{ij}

In point A two directions are measured, a backsight direction to P and a foresight direction to point 1. In point 1 two directions are measured, to the previous point A and to point 2; in point 2 directions are measured to 1 and 3; and so on. In the next to last point n a backsight direction is measured to point $n - 1$ and a foresight direction to point B. In point B two directions are measured, to n and to the azimuth marker Q. The directional measurements can be written as a function of differences of azimuths Az_{ij} between points i and j and the unknown azimuths of the directional zero orientations o_i. The latter refer to the (unknown) azimuths of the "zero" reading on the horizontal circle of the theodolite. This so-called zero "reading" is not a reading on the circle at all, but the result of an analysis of a series of directional measurements taken in the point in question. The azimuths Az_{ij} are in turn a function of the coordinate unknowns $\{x_i, y_i\}$ and $\{x_j, y_j\}$.

The first two observation equations, Equations (166.61) and (166.62), generated by the two directional measurements in point A are

$$r_{AP} = \mathrm{Az}_{AP} - o_A = \arctan\left(\frac{x_P - x_A}{y_P - y_A}\right) - o_A \qquad (166.61)$$

Note that o_A in Equation (166.61) is the only unknown since we adopted the coordinates of A and P. However, the second (forward) direction in $A(r_{A1})$ is dependent for three unknowns, o_A, x_1, y_1:

$$r_{A1} = \mathrm{Az}_{A1} - o_A = \arctan\left(\frac{x_1 - x_A}{y_1 - y_A}\right) - o_A \qquad (166.62)$$

The next two directional measurements, Equations (166.63) and (166.64), in point 1 are dependent on five unknowns, o_1, x_1, y_1, x_2, and y_2:

$$r_{1A} = \mathrm{Az}_{1A} - o_1 = \arctan\left(\frac{x_A - x_1}{y_A - y_1}\right) - o_1 \qquad (166.63)$$

For the foresight direction,

$$r_{12} = \mathrm{Az}_{12} - o_1 = \arctan\left(\frac{x_2 - x_1}{y_2 - y_1}\right) - o_1 \tag{166.64}$$

In point *i*, in the middle of the traverse, we have the backsight direction

$$r_{i,\,i-1} = \mathrm{Az}_{i,i-1} - o_i = \arctan\left(\frac{x_{i-1} - x_i}{y_{i-1} - y_i}\right) - o_i \tag{166.65}$$

and the foresight direction

$$r_{i,\,i+1} = \mathrm{Az}_{i,i+1} - o_i = \arctan\left(\frac{x_{i+1} - x_i}{y_{i+1} - y_i}\right) - o_i \tag{166.66}$$

In the last point of the traverse, point *B*, we have

$$r_{Bn} = \mathrm{Az}_{Bn} - o_B = \arctan\left(\frac{x_n - x_B}{y_n - y_B}\right) - o_B \tag{166.67}$$

and the foresight direction to azimuth marker *Q*,

$$r_{BQ} = \mathrm{Az}_{BQ} - o_B = \arctan\left(\frac{x_Q - x_B}{y_Q - y_B}\right) - o_B \tag{166.68}$$

So far, these directional measurements have generated $2(n + 2)$ observational equations with $(n + 2)$ directional unknowns and $2n$ unknown coordinates, totaling $3n + 2$ unknown parameters. The problem would not be solvable: $2n + 4$ equations with $3n + 2$ unknowns, a fact known to a surveyor because of geometric considerations alone. The addition of distance measurements will make traversing an efficient survey tool.

Note that one may have to add or subtract multiples of 360° to keep the directional measurements r_{ij} between 0 and 360°.

Distance Measurements s_{ij}

The distance measurements with unknown (common) scale λ can be written in terms of the following functions. Since no distance measurements between point *A* and the azimuth marker *P* were assumed, we have in point *A* only one (forward) distance measurement

$$s_{A1} = \lambda[(x_1 - x_A)^2 + (y_1 - y_A)^2]^{1/2} \tag{166.69}$$

In point 1 we have a backsight distance s_{1A} and a foresight distance s_{12}, generating the following two observation equations:

$$s_{1A} = \lambda[(x_A - x_1)^2 + (y_A - y_1)^2]^{1/2} \tag{166.70}$$

and

$$s_{12} = \lambda[(x_2 - x_1)^2 + (y_2 - y_1)^2]^{1/2} \tag{166.71}$$

In point i, in the middle of the traverse, we have for the backsight distance

$$s_{i,i-1} = \lambda[(x_{i-1} - x_i)^2 + (y_{i-1} - y_i)^2]^{1/2} \tag{166.72}$$

and the foresight distance,

$$s_{i,i+1} = \lambda[(x_{i+1} - x_i)^2 + (y_{i+1} - y_i)^2]^{1/2} \tag{166.73}$$

In the last point of the traverse, in point B, we have for the backsight distance

$$s_{Bn} = \lambda[(x_n - x_B)^2 + (y_n - y_B)^2]^{1/2} \tag{166.74}$$

We assumed that no foresight distance measurement to azimuth marker Q took place.

The distance measurements added $2(n + 1)$ observation equations and only one additional unknown, the scale factor λ. The $2n$ coordinates $\{x_i, y_i\}$ were already included in the previous directional observation equations. Summing the equations for both the directions and distances, we have for this particular traverse $2(n + 2) + 2(n + 1) = 4n + 6$ observations with $(3n + 2) + 1 = 3n + 3$ unknowns. The degrees of freedom are in this case $(4n + 6) - (3n + 3) = n + 3$ for the traverse under mentioned measurement conditions.

166.7 Dynamical Systems

Modern survey techniques incorporate the element *time* in two different aspects: first of all, in classical survey systems the assumption is made that we deal with stationary systems. That is, the random behavior of, say, the residuals is invariant of time. Second, during a kinematic survey — for instance, having an aircraft make aerial photographs equipped with a GPS receiver — we have to deal with estimating a vector of unknowns, say the position of the GPS antenna fixed on top of the airplane's fuselage, which is not independent of time anymore. We get for the variance/covariance matrix of the observations, Equation (166.33), and the linearized version of Equation (166.52), respectively,

$$\Sigma_{L_b} = \Sigma_{L_b(t)}(t) \tag{166.75}$$

and

$$L(t) + V(t) = A(t)X(t) \tag{166.76}$$

The vector $X(t)$ reflects the state of the parameters at epoch t. At the same time, a different model may be at hand that describes the rate of change of this vector (vector velocity),

$$\dot{X}(t) = F(t)X(t) + \varepsilon' \tag{166.77}$$

One may similarly be able to write the state of the vector X at epoch $(t + dt)$ as a function of the state at t, according to

$$X(t + dt) = \mathbf{\Phi}(t + dt, t)X(t) + \varepsilon \tag{166.78}$$

The (Jacobian) matrix $\mathbf{\Phi}$ (see also Section 161.7 elsewhere in this book) is called the state transition matrix. A new estimate $X(t + dt)$ is computed from a new measurement through Equation (166.76) and through the use of the previous estimate $X(t)$ through Equation (166.77) or (166.78).

Equations (166.75) through (166.78) lead to dynamical estimation models. One of the better-known estimation (filtering) models has been developed by Kalman and others. These models are developed in the so-called time domain (as opposed to the frequency domain). The reader is referred to the vast literature in this area (see, for instance, Gelb, 1974).

References

Baarda, W. 1967. Statistical concepts in geodesy. *Publ. Geodesy.* N.S. 2(4).
Gelb, A. 1974. *Applied Optimal Estimation.* MIT Press, Cambridge, MA.
Hamilton, W. C. 1964. *Statistics in Physical Science: Estimation, Hypothesis Testing, and Least Squares.* Ronald Press, New York.
Mikhail, E. M. 1976. *Observations and Least Squares.* IEP-A Dun-Donnelley, New York.
Mikhail, E. M. and Gracie, G. 1981. *Analysis and Adjustment of Survey Measurements.* Van Nostrand Reinhold, New York.
Strang, G. 1986. *Introduction to Applied Mathematics.* Wellesley-Cambridge Press, Wellesley, MA.
Strang, G. 1988. *Linear Algebra and Its Applications*, 3rd ed. Saunders College Publishing, Fort Worth, TX.
Wolf, P. R. and Brinker, R. C. 1994. *Elementary Surveying.* HarperCollins College Publishers, New York.

Further Information

Textbooks and Reference Books

For additional reading and more background, from the very basic to the advanced level, consult specific chapters in a variety of textbooks on geodesy, satellite geodesy, physical geodesy, surveying, photogrammetry, or statistics itself.

Bjerhammer, E. A. 1973. *Theory of Errors and Generalized Matrix Inverses.* Elsevier, New York.
Bomford, G. 1980. *Geodesy.* Clarendon Press, Oxford.

Ch. 1: Triangulation, Traverse, and Trilateration (Field Work)
Ch. 2: Computation of Triangulation, Traverse, and Trilateration
App. D: Theory of Errors

Carr, J. R. 1995. *Numerical Analysis for the Geological Sciences.* Prentice Hall, Englewood Cliffs, NJ.
Davis, R. E., Foote, F. S., Anderson, J. M., and Mikhail, E. M. 1981. *Surveying: Theory and Practice.* McGraw-Hill, New York.

Ch. 2: Survey Measurements and Adjustments
App. B: Least-Squares Adjustment

Escobal, P. R. 1976. *Methods of Orbit Determination.* John Wiley & Sons, New York.
App. IV: Minimum Variance Orbital Parameter Estimation

Fraleigh, J. B. and Beauregard, R. A. 1987. *Linear Algebra.* Addison-Wesley, Reading, MA.

Ch. 5: Applications of Vector Geometry and of Determinants
Sec. 5.2: The Method of Least Squares

Heiskanen, W. A. and Moritz, H. 1967. *Physical Geodesy.* W. H. Freeman & Co., San Francisco.
Ch. 7: Statistical Methods in Physical Geodesy

Hirvonen, R. A. 1965. *Adjustment by Least Squares in Geodesy and Photogrammetry.* Frederick Ungar Publishing Co., New York.
Hofmann-Wellenhof, B., Lichtenegger, H., and Collins, J. 1995. *GPS: Theory and Practice.* Springer-Verlag, New York.
Ch. 9: Data Processing

Kaula, W. M. 1966. *Theory of Satellite Geodesy: Applications of Satellites to Geodesy.* Blaisdell, Waltham, MA.

Ch. 5: Statistical Implications
Ch. 6: Data Analysis

Koch, K. R. 1988. *Parameter Estimation and Hypothesis Testing in Linear Models.* Springer-Verlag, New York.

Kraus, K. 1993. *Photogrammetry.* Ferd. Dümmlers Verlag, Bonn, Germany.
App. 4.2-1: Adjustment by the Method of Least Squares

Leick, A. 1995. *GPS: Satellite Surveying*, 2nd ed. John Wiley & Sons, New York.

Ch. 4: Adjustment Computations
Ch. 5: Least-Squares Adjustment Examples
App. B: Linearization
App. C: One-Dimensional Distributions

McCormac, J. C. 1995. *Surveying.* Prentice Hall, Englewood Cliffs, NJ.

Ch. 2: Introduction to Measurements
Ch. 11: Traverse Adjustment and Area Computation

Menke, W. 1989. *Geophysical Data Analysis: Discrete Inverse Theory.* Academic Press, San Diego, CA.

Moffitt, F. H. and Bouchard, H. 1992. *Surveying.* HarperCollins, New York.

App. A: Adjustment of Elementary Surveying Measurements by the Method of Least Squares
App. B: The Adjustment of Instruments

Mueller, I. I. and Ramsayer, K. H. 1979. *Introduction to Surveying.* Frederick Ungar Publishing Co., New York.

Ch. 5: Adjustment Computation by Least Squares

Papoulis, A. 1985. *Probability, Random Variables and Stochastic Processes.* McGraw-Hill, New York.

Tienstra, J. M. 1966. *Theory of Adjustment of Normally Distributed Observations.* Argus, Amsterdam.

Uotila, U. A. 1985. Adjustment Computations Notes. Department of Geodetic Science and Surveying, Ohio State University, Columbus.

Vaníček, P. and Krakiwsky, E. J. 1982. *Geodesy: The Concepts.* North-Holland, Amsterdam.

Ch. 11: Classes of Mathematical Models
Ch. 12: Least-Squares Solution of Overdetermined Models
Ch. 13: Assessment of Results
Ch. 14: Formulation and Solving of Problems

Wolf, P. R. 1983. *Elements of Photogrammetry.* McGraw-Hill, New York.
App. A: Random Errors and Least Squares Adjustment

Wolf, P. R. 1987. *Adjustment Computations: Practical Least Squares for Surveyors.* Landmark Enterprises, Rancho Cordova, CA.

Wolf, P. R. and Brinker, R. C. 1994. *Elementary Surveying.* HarperCollins, New York.

Ch. 2: Theory of Measurement and Errors
App. C: Propagation of Random Errors and Least-Squares Adjustment

Journals and Organizations

The latest results from research of least squares applications in geodesy, surveying, mapping, and photogrammetry are published in a variety of journals.

Two international magazines under the auspices of the International Association of Geodesy, both published by Springer-Verlag (Berlin/Heidelberg/New York), are *Bulletin Geodésiqué Manuscripta* and *Geodetica*.

Geodesy- and geophysics-related articles can be found in:

American Geophysical Union, Washington, D.C.: *EOS* and *Journal of Geophysical Research*
Royal Astronomical Society, London: *Geophysical Journal International*

Statistical articles related to kinematic GPS can be found in Institute of Navigation: *Navigation*

Many national mapping organizations publish journals in which recent statistical applications in geodesy/surveying/mapping/photogrammetry are documented:

American Congress of Surveying and Mapping: Surveying and Land Information Systems and Cartography and Geographic Information Systems
American Society of Photogrammetry and Remote Sensing: *Photogrammetric Engineering & Remote Sensing*
American Society of Civil Engineers: *Journal of Surveying Engineering*
Deutscher Verein für Vermessungswesen: *Zeitschrift für Vermessungswesen*, Konrad Wittwer Verlag, Stuttgart
Canadian Institute of Geomatics: *Geomatica*
Royal Society of Chartered Surveyors: *Survey Review*
Institute of Surveyors of Australia: *Australian Surveyor*

Worth special mention are the following trade magazines:

GPS World, published by Advanstar Communications, Eugene, OR
P.O.B. (Point of Beginning), published by P.O.B. Publishing Co., Canton, MI
Professional Surveyor, published by American Surveyors Publishing Co., Arlington, VA
Geodetical Info Magazine, published by Geodetical Information & Trading Centre bv., Lemmer, the Netherlands

National mapping organizations such as the U.S. National Geodetic Survey (NGS) regularly make software available (free and at cost). Information can be obtained from: National Geodetic Survey, Geodetic Services Branch, National Ocean Service, NOAA, 1315 East-West Highway, Station 8620, Silver Spring, MD 20910-3282.

167

Satellite Surveying

Boudewijn H. W. van Gelder (First Edition)
Purdue University

Robert F. Austin (Second Edition)
Austin Communication Education Services

The distance to an object can be calculated by measuring the time that it takes for a transmitted signal to reach us. In the case of the global positioning system (GPS), we can use the time it takes for transmitted signals to reach us from four different satellites to calculate our position. GPS is a tool used widely both within and outside engineering.

Positioning has become possible with accuracies ranging from the subcentimeter level — for high-accuracy geodetic applications as used in state, national, and global geodetic networks, deformation analysis in engineering, and geophysics — to the hectometer level in navigation applications. Similar to the space domain, a variety of accuracy classes may be assigned to the time domain: GPS provides position and velocity determinations averaged over time spans from fractions of seconds (essentially instantaneous) to one or two days. Stationary applications of the observatory type are used in GPS tracking for orbit improvement.

167.1 A Satellite Orbiting the Earth

The path of an Earth-orbiting satellite is similar to that of a planet around the sun. In history the solution to the motion of planets around the sun was found before its explanation. Johannes Kepler discovered certain regularities in the motions of planets around the sun. Through the analysis of his own observations and those made by Tycho Brahe, he formulated the following three laws:

First law (1609):
The orbit of each planet around the sun is an ellipse, with the sun is at one focus.

Second law (1609):
A ray (vector) from the sun to a planet sweeps out equal areas in equal time periods.

0-8493-1586-7/05/$0.00+$1.50

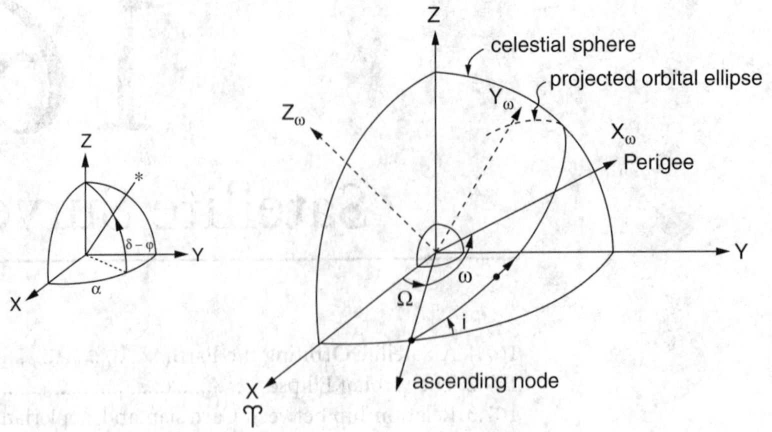

FIGURE 167.1 Celestial sphere with projected orbital ellipse and equator.

Third law (1611):

This ratio between the square of a planet's orbital period and the third power (cube) of its average distance from the sun is constant.

Kepler's third law leads to the famous equation,

$$n^2 a^3 = GM \tag{167.1}$$

where n is the average angular rate and a is the semimajor axis of the orbital ellipse.

In 1665–1666, Newton formulated his more fundamental laws of nature and showed that Kepler's laws follow from those fundamental laws (Newton, 1686).

167.2 The Orbital Ellipse

In a (quasi-) inertial frame, the ellipse of an Earth-orbiting satellite must be positioned: the focal point will coincide with the center of mass (CoM) of the Earth. Instead of picturing the ellipse itself, we project the ellipse on a celestial sphere centered at the CoM. On the celestial sphere we also project the Earth's equator (Figure 167.1).

The orientation of the orbital ellipse requires three orientation angles with respect to the inertial frame *XYZ*: two for the orientation of the plane of the orbit, Ω and *I*, and one for the orientation of the ellipse in the orbital plane, for which one refers to the point of closest approach, the perigee, ω (Satellite Observing).

Ω represents the right ascension (α) of the ascending node. The ascending node is the (projected) point where the satellite rises above the equator plane.

I represents the inclination of the orbital plane with respect to the equator plane.

ω represents the argument of perigee — the angle from the ascending node (in the plane of the orbit) to the perigee (for planets, the perihelion), which is that point where the satellite (planet) approaches the closest to the Earth (sun), or, more precisely, the center of mass of the Earth (sun).

Similar to the Earth's ellipsoid, we define the orbital ellipse by a semimajor axis a and the eccentricity e. In orbital mechanics it is unusual to describe the shape of the orbital ellipse by its flattening.

The position of the satellite in the orbital plane is depicted in Figure 167.2. The major variables in Figure 167.2 are defined as follows:

a = the semimajor axis of the orbital ellipse

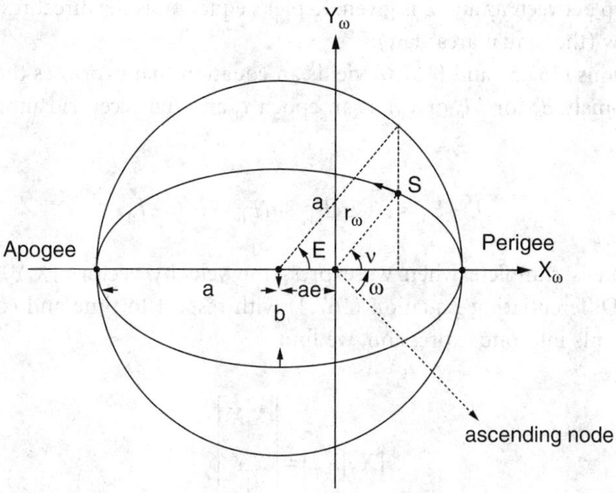

FIGURE 167.2 The position of the satellite (*S*) in the orbital plane.

b = the semiminor axis of the orbital ellipse
e = the eccentricity of the orbital ellipse, with

$$e^2 = \frac{a^2 - b^2}{a^2} \tag{167.2}$$

The relationship between the true anomaly and the eccentric anomaly can be derived as:

$$\tan\left(\frac{E}{2}\right) = \sqrt{\frac{1-e}{1+e}} \cdot \tan\left(\frac{\nu}{2}\right) \tag{167.3}$$

where ν = the true anomaly, sometimes denoted by *f* and E = the eccentric anomaly.

The Cartesian coordinates of the satellite position are

$$X_T = \begin{pmatrix} X \\ Y \\ Z \end{pmatrix} = \Re_3(-\Omega)\cdot\Re_1(-I)\cdot\Re_3(-\omega)\cdot\begin{pmatrix} a\cdot(\cos E - e) \\ a\cdot\sqrt{1-e^2}\cdot\sin E \\ 0 \end{pmatrix} \tag{167.4}$$

In Equation (167.4), the Cartesian coordinates are expressed in terms of the six so-called Keplerian elements: a, e, I, Ω, ω, and E. If we know the position of the satellite at an epoch t_0 through $\{a, e, I, \Omega, \omega, E_0\}$, we are capable of computing the position of the satellite at an arbitrary epoch t through Equation (167.4) if we know the relationship in time between E and E_0. In other words, how does the angle E increase with time?

We define an auxiliary variable (angle) M that increases linearly in time with the mean motion n [= $(GM/a^3)^{1/2}$] according to Kepler's third law. The angle M, the mean anomaly, may be expressed as a function of time by:

$$M = M_0 + n\cdot(t - t_0) \tag{167.5}$$

Through Kepler's equation,

$$M = E - e\cdot\sin E \tag{167.6}$$

the (time) relationship between M and E is given. Kepler's equation is the direct result of the enforcement of Kepler's second law (the "equal area" law).

Combining Equations (167.5) and (167.6) yields an equation that expresses the relationship between a given eccentric anomaly E_0 (or M_0 or ν_0) at an epoch t_0 and the eccentric anomaly E at an arbitrary epoch t:

$$E - E_0 = e\cdot(\sin E - \sin E_0) + n\cdot(t - t_0) \tag{167.7}$$

The transformation is complete when we express the velocity vector $\{\dot{X},\dot{Y},\dot{Z}\}$ in terms of those Keplerian elements. Differentiating Equation (167.4) with respect to time and combining the position and velocity components into one expression, we find:

$$[X_T|\dot{X}_T] = \begin{bmatrix} X|\dot{X} \\ Y|\dot{Y} \\ Z|\dot{Z} \end{bmatrix} \tag{167.8}$$

$$= \Re_3(-\Omega)\cdot\Re_1(-I)\cdot\Re_3(-\omega)\begin{bmatrix} a(\cos E - e) & | & -a\dot{E}\sin E \\ a\sqrt{1-e^2}\,\sin E & | & a\dot{E}\sqrt{1-e^2}\,\cos E \\ 0 & | & 0 \end{bmatrix} \tag{167.9}$$

The remaining variable $\dot{E}$ is obtained through differentiation of Equation (167.6):

$$\dot{E} = \frac{n}{1 - e\cdot\cos E} \tag{167.10}$$

Now all six Cartesian orbital elements (state vector elements) are expressed in terms of the six Keplerian elements.

167.3 Relationship between Cartesian and Keplerian Orbital Elements

To compute the inertial position of a satellite in a central force field, it is simpler to perform a time update in the Keplerian elements than in the Cartesian elements. The time update takes place through Equations (167.5), (167.6), and (167.7).

Schematically, the following procedure is to be followed:

$$\begin{array}{lcl} t_0: & \{X, Y, Z, \dot{X}, \dot{Y}, \dot{Z}\} & \\ & \downarrow & \text{Conversion to Keplerian elements} \\ t_0: & \{a, e, I, \Omega, \omega, E_0\} & \\ & \downarrow & \text{Equation of Kepler, Equation (167.7)} \\ t_1: & \{a, e, I, \Omega, \omega, E_1\} & \\ & \downarrow & \text{Conversion to Cartesian elements} \\ t_1: & \{X, Y, Z, \dot{X}, \dot{Y}, \dot{Z}\} & \end{array}$$

The conversion from Keplerian elements to state vector elements was treated in the previous section. For the somewhat more complicated conversion from position and velocity vector to Keplerian representation, the reader is referred to textbooks such as Escobal [1976]. Basically we "invert" Equations (167.8) and (167.9) by solving for the six elements $\{a, e, I, \Omega, \omega, E\}$ in terms of the six state vector elements.

167.4 Orbit of a Satellite in a Noncentral Force Field

The equations of motion for a real satellite are more difficult than implied by Equations (167.8) and (167.9). First of all, we do not deal with a central force field: the Earth is not a sphere, nor does it have a radial symmetric density. Second, we deal with other forces, including the gravity of the moon and the sun, atmospheric drag, and solar radiation pressure. Equations (167.8) and (167.9) get a more general meaning if we suppose that a potential function is being generated by the sum of the forces acting on the satellite:

$$V = V_c + V_{nc}^t + V_{\text{sun}}^t + V_{\text{moon}}^t + \cdots \tag{167.11}$$

where V_c is the central part of the Earth's gravitational potential,

$$V_c = \mu |X| \tag{167.12}$$

and V_{nc}^t is the noncentral and time-dependent part of the Earth's gravitational field. (The upper index t has been added to various potentials to reflect their time variance with respect to the inertial frame.)

The equations of motion to be solved are

$$\begin{aligned} \ddot{X} &= \nabla(V_c + V_{nc}^t + V_{\text{sun}}^t + V_{\text{moon}}^t + \cdots) \\ &= \nabla V_c + \nabla V_{nc}^t + \nabla V_{\text{sun}}^t + \nabla V_{\text{moon}}^t + \cdots \end{aligned} \tag{167.13}$$

For the Earth's gravitational field, we have (in an Earth-fixed frame):

$$V_c + V_{nc} = \frac{\mu}{r}\left[1 + \sum_{l=1}^{\infty}\sum_{m=0}^{l}\left(\frac{a_e}{r}\right)^l \cdot (C_{lm}\cos m\lambda + S_{lm}\sin m\lambda)\cdot P_{lm}(\sin\phi)\right] \tag{167.14}$$

With Equation (167.14), one is able to compute the potential at each point $\{\lambda, \phi, r\}$ necessary for the integration of the satellite's orbit. The coefficients C_{lm} and S_{lm} of the spherical harmonic expansion are in the order of 10^{-6} except for C_{20} ($l = 2$, $m = 0$), which is about 10^{-3}. This has to do with the fact that the Earth's equipotential surface at mean sea level can be approximated best by an ellipsoid of revolution. One must realize that the coefficients C_{lm} and S_{lm} describe the shape of the potential field and not the shape of the physical Earth, despite a high correlation between the two. P_{lm} (sin ϕ) are the associated Legendre functions of the first kind, of degree l and order m; a_e is some adopted value for the semimajor axis (equatorial radius) of the Earth. (For values of a_e, μ (= GM), and C_{20} (= $-J_2$), see IAG, 1971; IAG, 1980; IAG, 1984; IAG, 1988a; IAG, 1988b; DMA, 1988; McCarthy, 1992; and Cohen and Taylor, 1988.)

The equatorial radius a_e, the geocentric gravitational constant GM, and the dynamic form factor J_2 characterize the Earth as an ellipsoid of revolution with an equipotential surface.

If we restrict ourselves to the central part ($\mu = GM$) and the dynamic flattening ($C_{20} = -J_2$), then Equation (167.14) becomes:

$$V_c + V_{nc} = \frac{\mu}{r}\left[1 + \frac{J_2 a_e^2}{2r^2}\cdot(1 - 3\sin^2\phi)\right] \tag{167.15}$$

with

$$\sin\phi = \sin\delta = \frac{z}{r} \tag{167.16}$$

where ϕ is the latitude and δ the declination (Figure 167.2).

The solution expressed in Keplerian elements shows periodic perturbations and some dominant secular effects. An approximate solution using only the latter effects is (position only)

$$X_I = \Re_3[-(\Omega_0 + \dot{\Omega}\Delta t)] \cdot \Re_1(-I) \cdot \Re_3[-(\omega_0 + \dot{\omega}\Delta t)] \cdot X_\omega \tag{167.17}$$

with

$$\Delta t = t - t_0 \tag{167.18}$$

$$\dot{\Omega} = -\frac{3}{2}\frac{J_2 a_e^2}{a^2(1-e^2)^2} n \cos I \tag{167.19}$$

$$\dot{\omega} = \frac{3}{2}\frac{J_2 a_e^2}{a^2(1-e^2)^2} n(2 - 2\tfrac{1}{2}\sin^2 I) \tag{167.20}$$

$$n = n_0 \cdot \left[1 + \frac{3}{2}\frac{J_2 a_e^2 \sqrt{1-e^2}}{a^2(1-e^2)^2}(1 - 1\tfrac{1}{2}\sin^2 I)\right] \tag{167.21}$$

with

$$n_0 = \sqrt{\frac{GM}{a^3}} \tag{167.22}$$

Whenever

$I = 0°$	we have	an equatorial orbit
$0° < I < 90°$		a direct orbit
$I = 90°$		a polar orbit
$90° < I < 180°$		a retrograde orbit
$I = 180°$		a retrograde equatorial orbit

Equation (167.19) shows that the ascending node of a direct orbit slowly drifts to the west. For a satellite at about 150 km above the Earth's surface, the right ascension of the ascending node decreases about 9° per day. The satellites belonging to the global positioning system have an inclination of about 55°. Their nodal regression rate is about –0.04187° per day.

167.5 The Global Positioning System (GPS)

The first U.S. navigation satellite was the Transit, "launched on April 13, 1960 into a 51° inclination orbit with an apogee of 745 and a perigee of 373 km. The spacecraft operated for 3 months. By 1968 there were 23 Transit satellites operating in circular orbits of 850km" [Graham, 1995]. The Transit system was intended to update the inertial navigation systems in the Polaris submarine.

Since Transit, several GPS satellite clusters have been put into operation or are proposed for operation in the near future. The Navstar GPS was conceived in 1973 as a replacement for the Transit system. The first Navstar satellite was launched in 1978. The Navstar GPS comprises a constellation of 24 satellites that orbit the Earth at an altitude of 20,200 km, constantly broadcasting GPS information. These satellites complete their orbits every 12 hours in six orbital planes.

The system is operated and controlled by the 50th Space Wing, located at Schriever Air Force Base, Colorado. The master control station crew sends updated navigation information to GPS satellites through ground antennas using an S-band signal. The ground antennas are also used to transmit commands to satellites and to receive state-of-health data (telemetry).

GPS receivers on Earth calculate their positions by making distance measurements to four or more satellites. Individual distance measurements to each satellite are determined by analyzing the time it takes for a signal to travel from a satellite to a GPS receiver. Using some relatively simple geometry, the receiver determines its position. This process is termed "trilateration."

The Russian Global'naya Navigatsionnaya Sputnikovaya Sistema (Global Navigation Satellite System, or GLONASS) was proposed in 1976 to provide navigational and time reference data for U.S.S.R. military use. As with the Navstar system, the GLONASS constellation was intended to include 24 satellites (21 on-line plus 3 spares in orbit), although in three orbital planes rather than Navstar's six planes. The GLONASS constellation orbited at an altitude of 19,100 km at an inclination of 64.8° and an orbit time of approximately 11 hours and 15 minutes. The GLONASS network was made available to civilian users in the early 1990s. Since that time, Russia and the United States have cooperated on integrating the GLONASS and Navstar GPS systems, so that users can use both networks. The capabilities of the GLONASS network were diminished during the 1990s, because Russia was unable to replace the aging GLONASS satellites on a regular schedule. Only six GLONASS satellites were functioning in orbit when this chapter was revised in 2003.

In 1999, the European Commission announced plans for another satellite navigation system, named Galileo. The rationale for a third satellite survey and navigation system was in part political (i.e., Europe had no control over the U.S. and Russian systems) and part economic (it was designed to allow the European Union's members to benefit from the projected demand for value-added services and equipment for navigation systems). The Galileo system also was intended to allow the European Union to develop its own integrated transport and navigation systems. Europe's first effort with satellite navigation will be EGNOS, a system intended to improve the reliability of Navstar GPS and GLONASS for certain critical applications that is scheduled to come on line in 2004. The Galileo constellation will comprise 30 satellites when complete. The satellites are scheduled to be in orbit by 2006 and the complete system, including ground infrastructure, is scheduled to be operational in 2008. It is the intention of the developers to be interoperable with the Navstar GPS and GLONASS systems.

Back on the ground, prices for GPS have dropped steadily over time. Several personal location GPS receivers intended for hiking, fishing, and other recreational uses are priced below $100, while geodetic survey grade instruments can be purchased for less than $10,000. GPS consumer markets have been rapidly expanding. In the areas of land, marine, and aviation navigation; precise surveying; electronic charting, and time transfer, the deployment of GPS equipment seems to have become indispensable. As events of recent years have demonstrated, this holds true for military users as well as civilian users.

Positioning

Two classes of positioning are recognized: standard positioning service (SPS) and precise positioning service (PPS). In terms of positional accuracies, one must distinguish between SPS with and without selective availability (SA) on the one hand and PPS on the other. SA deliberately introduced clock errors and ephemeris errors in the data being broadcast by the Navstar satellites. The accuracy of the (civil) signal without SA was approximately 20 m to 40 m. With SA implemented, SPS accuracy was degraded to 100 m. On May 1, 2000, President Clinton announced that the United States would stop the intentional degradation of Navstar GPS signals beginning at midnight of that date. This decision was based on a recommendation by the Secretary of Defense in coordination with the Departments of State, Transportation and Commerce, the Central Intelligence Agency, and other executive branch departments and agencies, who realized that worldwide transportation safety, scientific, and commercial interests would best be served by discontinuation of SA.

In several applications, GPS receivers are interfaced with other positioning systems, such as inertial navigation systems (INS), hyperbolic systems, and even automatic braking systems in cars. GPS receivers operating in combination with other equipment can provide the answers to such general questions as [Wells and Kleusberg, 1990]:

Absolute positioning: Where am I? Where are you?
Relative positioning: Where am I with respect to you? Where are you with respect to me?
Orientation: Which way am I heading?
Timing: What time is it?

All these questions may refer to an observer either at rest (*static* positioning) or in motion (*kinematic* positioning). The questions may be answered immediately (*real-time* processing, often misnamed DGPS, which stands for differential GPS) or after the fact (*post*-processing).

Limiting Factors

The physics of the environment, instruments, broadcast ephemeris, and the relative geometry between orbits and networks are all limiting factors on the final accuracy of the results. Dilution of precision (DOP) is used as a scaling factor between the observational accuracy and positioning accuracy [Wells, 1986]. For reasons of safety and accuracy, one should avoid periods in which the DOP factor is larger than 5.

The atmosphere of the Earth changes the speed and the geometric path of the signals broadcast by GPS satellites. In the uppermost part of the atmosphere (the ionosphere), charged particles vary in number spatially as well as temporally. Ionospheric refraction errors may amount to several tens of meters. Because this effect is frequency dependent, the first-order effect can be eliminated for the most part by the use of dual-frequency receivers. The lower part of the atmosphere (the troposphere) causes refraction errors of several meters. Fortunately, the effect can be modeled rather well by measuring the atmospheric conditions at the measuring site.

GPS instruments are capable of measuring one or a combination of the following signals:

C/A code, with an accuracy of a few meters
P code, with an accuracy of a few decimeters
Carrier phase, with an accuracy of a few millimeters

In addition to this measurement noise, receiver clock errors must be modeled as to-be solved-for parameters. It is this synchronization parameter between satellite time and receiver time that makes it necessary to have at least four satellites in view to get a three-dimensional (3-D) fix.

Because of the high frequency of the GPS signals, multipath effects may hamper final accuracy; the signal arriving at the receiver through a reflected path may be stronger then the direct signal. By careful antenna design and positioning, multipath effects are reduced. The phase center of the antenna must be calibrated carefully with respect to a geometric reference point on the antenna assembly. However, because of the varying inclination angle of the incoming electromagnetic signals, effects of a moving phase center may be present at all times.

Information on the orbit of the satellite, as well as the orbital geometry relative to the network/receiver geometry, influences the overall positioning accuracy. The information that the satellite broadcasts on its position and velocity is necessarily the result of a process of prediction. Also, the on-board satellite clock is not free of errors. Orbital and satellite clock errors largely can be addressed by careful design of the functional model.

Differencing techniques are successfully used to eliminate a wide variety of errors, provided the receivers are to not far apart. In essence, two close-by receivers are influenced almost equally by (deliberate) orbital errors and by part of the atmosphere error. Differencing of the measurements of both receivers will cancel a large portion of the first-order effects of these errors.

Modeling and the GPS Observables

Developing well-chosen functional models *F*, relating the GPS measurements L to the modeled parameters *X*, enables users to fit GPS to their needs. A wide class of applications, from monitoring the subsidence of oilrigs in the open sea to real-time navigation of vehicles collecting geospatial information, belongs to the range of possibilities opened up by the introduction of GPS.

In satellite geodesy, one traditionally modeled the state of the satellite — a vector combining the positional (X) and the velocity ($\dot{X}$) information. GPS provides geodesists, and geoscientists in general, with a tool by which the state of the observer, also in terms of position (x) and velocity ($\dot{x}$), can be determined with high accuracy and often in real time. The GPS satellite geodetic model has evolved to:

$$\mathbf{L} = F(X, \dot{X}, x, \dot{x}, \mathbf{p}, t) \tag{167.23}$$

where L = C/A code, P code, or carrier phase observations, $i = 1, \ldots, n$, X = 3-D position of the satellite at epoch t, $\dot{X}$ = 3-D velocity of the satellite at epoch t, x = 1-D, 2-D, or 3-D position of the observer at epoch t, $\dot{x}$ = 1-D, 2-D, or 3-D velocity of the observer at epoch t, p = vector of modeled (known or unknown) parameters, $j = 1, \ldots, u$, and t = epoch of measurement taking.

Various differencing operators D^k, up to order three, are applied to the original observations to take full benefit of the GPS measurements. The difference operator D^k may be applied in the observation space spanned by the vector L [Equation (167.24)], or the D^k operator may be applied in the parameter space x (Equation (167.25)):

$$D^k[\mathbf{L}] = D^k[F(X, \dot{X}, x, \dot{x}, \mathbf{p}, t)] \tag{167.24}$$

$$\mathbf{L} = F(X, \dot{X}, D^1(x, \dot{x}), \mathbf{p}, t) \tag{167.25}$$

The latter method is sometimes referred to as "delta positioning." This is a difficult way of saying that one may either construct so-called derived observations from the original observations by differencing techniques, or model the original observations, compute parameters (e.g., coordinates) in this way, and subsequently start a differencing technique on the results obtained from the roving receiver and the base receiver.

Pseudo Range

We restrict the discussion to the C/A-code pseudo-range observables (pr). The ranges are called "pseudo" because they are calculated using time difference based on two independent clocks. The time offset δt_E^S between the satellite clock S and the receiver clock E yields one additional parameter to be solved for. Writing the observation equation in the Earth-fixed reference frame, we have:

$$pr = \sqrt{(x^S - x_E)^2 + (y^S - y_E)^2 + (z^S - z_E)^2} - c \cdot \delta t_E^S \tag{167.26}$$

Inspection of the partials,

$$\frac{\partial pr}{\partial x^S} = \frac{x^S - x_E}{pr} = -\frac{\partial pr}{\partial x_E} \tag{167.27}$$

$$\frac{\partial pr}{\partial y^S} = \frac{y^S - y_E}{pr} = -\frac{\partial pr}{\partial y_E} \qquad (167.28)$$

$$\frac{\partial pr}{\partial z^S} = \frac{z^S - z_E}{pr} = -\frac{\partial pr}{\partial z_E} \qquad (167.29)$$

$$\frac{\partial pr}{\partial (\delta t_E^S)} = -c \qquad (167.30)$$

reveals the following:

- The coordinates of the stations are primarily obtained in a frame determined by the satellites or, better, by their broadcast ephemeris.
- Partials evaluated for neighboring stations are practically identical, so the coordinates of one station need to be adopted.

Carrier Phase

For precise engineering applications, the phase of the carrier wave is measured. Two wavelengths are available in principle:

$$L_1{:}\ \lambda_1 = \frac{c}{f_1} \quad \text{with } f_1 = 1.57542 \text{ GHz} \qquad (167.31)$$
$$\cong 19.0 \text{ cm}$$

and

$$L_2{:}\ \lambda_2 = \frac{c}{f_2} \quad \text{with } f_2 = 1.22760 \text{ GHz} \qquad (167.32)$$
$$\cong 24.4 \text{ cm}$$

For phase measurements the following observation equation can be set up:

$$\text{Range} = \phi + N \cdot \lambda_l, \quad l = 1,\ldots,2 \qquad (167.33)$$

or

$$\Phi_E^S = \sqrt{(x^S - x_E)^2 + (y^S - y_E)^2 + (z^S - z_E)^2} - N_E^S \cdot \lambda_l \qquad (167.34)$$

where Φ_E^S is the phase observable in a particular *S*-*E* combination, and N_E^S is the integer multiple of wavelengths in the range: the ambiguity. Phase measurements can be done with probable 1% accuracy. This yields an observational accuracy — in case the ambiguity *N* can be properly determined — in the millimeter range!

Various differencing operators on the phase measurements are used in GPS positioning:

- Single differences
  - Between receiver differences, $\Delta\Phi$, eliminating or reducing satellite-related errors
  - Between satellite differences, $\nabla\Phi$, eliminating or reducing receiver-related errors
  - Between epoch differences, $\delta\Phi$, eliminating phase ambiguities per satellite-receiver combination

TABLE 167.1 Accuracy Grades of Civilian/Commercial GPS Receivers

Navigation grade	15–100 m	C/A code, in stand-alone mode
Mapping (GIS) grade	2–5 m	C/A code, in differenced mode
Surveying grade	1–2 cm within 10 km	C/A code + phase, differenced
Geodesy grade	5–15 mm over any distance	C/A + P code + phase, differenced

- Double differences
  - Between receiver/satellite differences, $\nabla\Delta\Phi$, eliminating or reducing satellite-and receiver-related errors, and so forth
- Triple differences
  - Between epoch/receiver/satellite differences, $\delta\nabla\Delta\Phi$, eliminating or reducing satellite/receiver-related errors, and ambiguities

Receivers that use carrier wave observations have, in addition to the electronic components that do the phase measurements, a counter that counts the complete cycles between selected epochs. GPS analysis software uses the triple differences to detect and possibly repair cycle slips occurring during loss of lock.

Design specifications and receiver selection are dependent on the specific project accuracy requirements. In the United States, the Federal Geodetic Control Committee has adopted various specifications [FGCC, 1989].

GPS Receivers

A variety of receivers are on the market. Basically, they can be grouped in the four classes listed in Table 167.1. The observations of the first three types of receivers are subjected to models that can be characterized as *geometric* models. The position of the satellite is considered known, based primarily on the information taken from the broadcast ephemeris. The "known" positions are not without error for several reasons. First, the positions are predicted and thus contain errors because of an extrapolation process in time. Second, the positions being broadcast may be corrupted by intentional errors (e.g., by SA if reinstated). Differencing techniques are capable of eliminating most of the errors if the separation between the base station and the roving receiver is not too large.

Millimeter-accurate observations from geodesy-grade receivers often are subjected to analysis through models of the *dynamic* type. Software packages containing dynamic models are very elaborate and allow for an orbit improvement estimation process.

GPS Base Stations

For achieving better accuracy (couple meter or better), additional GPS receiver(s) at known location(s) are required to take measurements in order to correct and reduce errors. This implies that for most applications of GPS in geodesy (e.g., surveying, mapping, photogrammetry, and GIS), one must have access to at least two GPS receivers. If one of the receivers occupies a known location during an acceptable minimum period, one may obtain accurate coordinates for the second receiver *in the same time frame*. In surveying/geodesy applications, it is preferable to include three stations with known horizontal coordinates and at least four with known vertical (orthometric) heights. In most GIS applications, a receiver is left at one particular site. This station serves as a *base station*. There are also GPS correction services that are available. Some are free and some are required to be subscribed. Instead of setting up a base station, a differential GPS survey can be conducted using other available base correction services.

The National Geodetic Survey (NGS), an agency of the National Oceanic and Atmospheric Administration, defines and manages the National Spatial Reference System (NSRS). The NSRS is a consistent coordinate system that defines latitude, longitude, height, scale, gravity, and orientation throughout the United States. The accuracy and accessibility of NSRS depends in large and growing measure on use of GPS.

The High Accuracy Reference Network (HARN) was designed to provide the surveying community with a network of highly reliable positional coordinates to serve as control for surveys. A framework of

GPS stations was established across the country and verified on several occasions by the NGS. The coordinates of the HARN stations were determined by very long baseline interferometry (VLBI), which served as the control for an A-Order survey called the "Eastern Strain Network Project" observed in 1987 and 1990. ESNP became the foundation of the HARN.

The HARN, which is homogeneous across the country, serves as a standard reference to define position accuracy. Relative position accuracy with respect to this reference network may be considered absolute positional accuracy for practical surveying purposes. The accuracy of classical horizontal control was on the order of one part in 100,000 (1 cm over 1 km). GPS is a survey tool with an accuracy of one part per 1,000,000 (1 cm over 10 km). Even for GIS applications where 0.5-m accuracies are claimed for the roving receivers, one may speak of 1-ppm surveys whenever those rovers operate at a distance of 500 km from their base station.

To simplify the process of relative GPS positioning, many government agencies established automated, permanently configured GPS base station facilities to continuously collect and record GPS data. The NGS refers to these automated reference stations as continuously operating reference stations (CORS). The NGS defined a nationwide CORS initiative to support of its mission of providing an accurate and consistent national coordinate system, that is, the NSRS. In effect, the NSRS consists of this CORS network, a network of permanently marked points, and a set of models describing dynamic, geophysical processes that affect spatial measurements.

GPS ties between the HARN and CORS make it possible to evaluate the HARN relative to the CORS, and the NGS has adjusted the HARN to reflect CORS accuracy. Improvements to the reference system began with the High-Precision GPS Network (HPGN) adjusted by NGS in 1990. Subsequent surveys resulted in the current network. The acronym HPGN evolved into the name High-Accuracy Reference Networks (HARN). The HARN is comprised of the NGS-maintained Federal Base Network (FBN), which features 100-km station spacing, and the volunteer-densified Cooperative Base Network (CBN), which typically features 25-km to 50-km spacing. Given its source, the HARN is occasionally referred to as the FBN/CBN.

It should not be forgotten that GPS is a geometric survey tool yielding results in terms of Earth-fixed coordinate differences. From these coordinate differences expressed in curvilinear coordinates, we obtain at best somewhat reproducible ellipsoidal height differences. These height differences are *not* easily converted to orthometric height differences of equal accuracy. The latter height differences are of interest in engineering and GIS applications, as discussed in the next section.

167.6 Gravity Field and Related Issues

One-Dimensional Positioning: Heights and Vertical Control

Some of the most accurate measurements surveyors are able to make are the determinations of height differences by spirit leveling. Because a leveling instrument's line of sight is tangent to the potential surface, one may say that leveling actually determines the height differences with respect to equipotential surfaces. If one singles out a particular equipotential surface at mean sea level (the so-called geoid), then the heights a surveyor determines are actually *orthometric heights* (Figure 167.3).

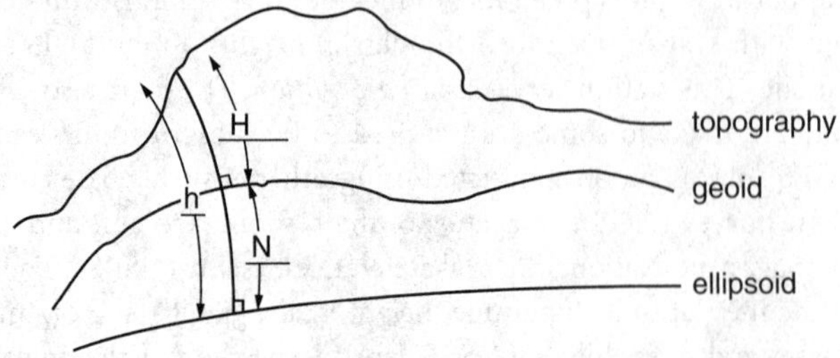

FIGURE 167.3 Orthometric heights.

Leveling in a closed loop is a check on the actual height differences, not in a metrical sense but in a potential sense: the distance between equipotential surfaces varies due to local gravity variations. In spherical approximation the potential at a point A is

$$V = -\frac{GM}{r} = -\frac{GM}{R+h} \tag{167.35}$$

The gravity is locally dependent on the change in the potential per height unit, or

$$\frac{dV}{dr} = g = \frac{GM}{r^2} \tag{167.36}$$

The potential difference dV between two equipotential surfaces is

$$dV = g \cdot dr \tag{167.37}$$

Consequently, if one levels in a loop, one has

$$\sum dV = \sum g \cdot dr = 0 \tag{167.38}$$

or

$$\oint dV = \oint g \cdot dr = 0 \tag{167.39}$$

This implies that for each metrically leveled height difference dr, one must multiply this difference by the local gravity. Depending on the behavior of the potential surfaces in a certain area and the diameter of one's project, one may require a gravimeter while leveling.

Variations in local gravity depend on the geology of the area. Variations on the order $10^{-7}g$ may yield errors as large as 10 mm for height differences in the order of several hundred meters. For precise leveling surveys (≤0.1 mm/km), gravity observations must be made with an interval of

2 to 3 km in relatively "flat" areas
1 to 2 km in hilly terrain
0.5 to 1.5 km in mountainous regions

For more design criteria on leveling and gravity surveys, see FGCC [1989] and Table 167.2.

TABLE 167.2 FGCC Vertical Control Accuracy Standards (Differential Leveling)

First Order	
Class I	$b^a < 0.5$
Class II	0.7
Second Order	
Class I	1.0
Class II	1.3
Third Order	
	2.0

[a] $b = S/\sqrt{d}$ (mm/$\sqrt{\text{km}}$), where S = standard deviation of elevation difference between control points (mm), and d = approximate horizontal distance along leveled route (km).

GPS surveys yield at best ellipsoidal height differences. These are rather meaningless from the engineering point of view. Therefore, extreme caution should be exercised when GPS height information, even after correction for geoidal undulations, is to be merged with height information from leveling. For two different points i and j,

$$h_i = H_i + N_i \tag{167.40}$$

$$h_j = H_j + N_j \tag{167.41}$$

Subtracting Equation (167.40) from (167.41), we find the ellipsoidal height differences h_{ij} (from GPS) in terms of the orthometric height differences H_{ij} (from leveling) and the geoidal height differences N_{ij} (from gravity surveys):

$$h_{ij} = H_{ij} + N_{ij} \tag{167.42}$$

where

$$h_{ij} = h_j - h_i \tag{167.43}$$

$$H_{ij} = H_j - H_i \tag{167.44}$$

$$N_{ij} = N_j - N_i \tag{167.45}$$

For instance, with the National Geodetic Survey's software program GEOID93, geoidal height differences are as accurate as 10 cm over 100 km for the conterminous U.S. For GPS leveling, this means that GPS may compete with third-order leveling as long as the stations are more than 5 km apart.

In principle, any equipotential surface can act as a vertical datum. The National Geodetic Vertical Datum of 1929 (NGVD29) is not a true mean sea-level datum. Problems may arise in merging GPS heights, gravity surveys and orthometric heights referring to NGVD29. Heights referring to the NGVD88 datum are more suitable for use with GPS surveys. More than 600,000 vertical control stations exist in the U.S.

Two-Dimensional Positioning: East/North and Horizontal Control

In classical geodesy, measurements of height were separated from the horizontal measurements (directions, angles, azimuths, and distances). To allow for the curvature of the Earth and the varying gravity field, the horizontal observations were reduced first to the geoid, taking into account the orthometric heights. Subsequently, it was desirable to take advantage of geometrical properties between the once-reduced horizontal observations, and the observations were reduced once more, from the geoid to the ellipsoid. An ellipsoid approximates the geoid up to 0.01%; variations of the geoid are not larger than 150 m. On the ellipsoid, which is a precise mathematical figure, one could check, for instance, whether the sum of the three angles equals a prescribed value.

In general, geodesists have relied on a biaxial ellipsoid of revolution. A semimajor axis a_e and a semiminor axis b_e define the dimensions of the ellipsoid. Rather than using this semiminor axis, one may specify the flattening of the ellipsoid:

$$f = \frac{a_e - b_e}{a_e} \approx \frac{1}{298.257\ldots} \tag{167.46}$$

For a semimajor axis of about 6378.137 km, this implies that the semiminor axis is 6378.137/298.257 ≈ 22 kilometers shorter than a_e.

Distance measurements must be reduced to the ellipsoid. Angular measurements made with theodolites, total stations, and other instruments must be corrected for the following effects:

- The direction of local gravity does not coincide with the normal to the ellipsoid.
- The direction of the first axis of the instrument coincides with the direction of the local gravity vector. Notwithstanding this effect, the Earth's curvature causes nonparallelism of first axes of one arcsecond for each 30 m.
- The targets aimed at for distance calculations generally do not reside on the ellipsoid.

TABLE 167.3 FGCC Horizontal Control Accuracy Standards (Classical Techniques)

First Order	
	1:100 000 (10 mm/km)
Second Order	
Class I	1:50 000 (20 mm/km)
Class II	1:20 000 (50 mm/km)
Third Order	
Class I	1:10 000 (100 mm/km)
Class II	1:5 000 (200 mm/km)

TABLE 167.4 Geographic (Spherical) Latitude as a Function of Geodetic Latitude

Geodetic Latitude			Geographic Latitude			Geodetic Minus Geographic Latitude		
Degrees	Minutes	Seconds	Degrees	Minutes	Seconds	Degrees	Minutes	Seconds
00	0	0.000	00	00	00.000	00	00	00.000
10	0	0.000	09	56	03.819	00	03	56.181
20	0	0.000	19	52	35.868	00	07	24.132
30	0	0.000	29	50	01.089	00	09	58.911
40	0	0.000	39	48	38.198	00	11	21.802
50	0	0.000	49	48	37.402	00	11	22.598
60	0	0.000	59	49	59.074	00	10	00.926
70	0	0.000	69	52	33.576	00	07	26.424
80	0	0.000	79	56	02.324	00	03	57.676
90	0	0.000	90	00	00.000	00	00	00.000

The noncoincidence of the gravity vector and the normal is called "deflection of the vertical." Proper knowledge of the behavior of the local geopotential surfaces is needed for proper distance and angle reductions. Consult Vanicek and Krakiwsky [1982], for example, for the mathematical background of these reductions. The FGCC adopted the accuracy standards given in Table 167.3 for horizontal control using classical geodetic measurement techniques [FGCC, 1984]. In the U.S., over 270,000 horizontal control stations exist.

3-D Positioning: Geocentric Positions and Full 3-D Control

Modern 3-D survey techniques, most noticeably GPS, allow for immediate 3-D relative positioning. 3-D coordinates are equally accurately expressed in ellipsoidal, spherical, or Cartesian coordinates. Care should be exercised to properly label curvilinear coordinates as spherical (geographic) or ellipsoidal (geodetic). Table 167.4 shows the large discrepancies between the two. At the mid-latitudes they may differ by more than 11′. This could result in a north-south error of 20 km. When merging GIS data sets, one should be aware of the meaning "LAT/LONG" in any instance. Consult Stem [1991] for the use of U.S. state plane coordinates. Curvilinear coordinates and their transformations and their use are discussed in a variety of textbooks; see the section "Further Information" for articles such as Soler [1976], Leick and van Gelder [1975], Soler and Hothem [1988], and Soler and van Gelder [1987].

Despite their 3-D characteristics, networks generated by GPS are the weakest in the vertical component, not only because of the lack of physical significance of the GPS determined heights, as described in the preceding subsection, but also because of the geometric distribution of satellites with respect to the vertical: no satellite signals are received from "below the network." This lopsidedness makes the vertical the worst determined component in 3-D.

Because of the inclination of the GPS satellites, there are places on Earth, most notoriously the mid-latitudes, where there is not an even distribution of satellites in the azimuth sense. For instance, in the

TABLE 167.5 Federal Geodetic Control Committee Three-Dimensional Accuracy Standards (Space System Techniques)

AA Order (global)	3 mm + 1:100 000 00 (1 mm/100 km)
A Order (primary)	5 mm + 1:10,000,000 (1 mm/10 km)
B Order (secondary)	8 mm + 1:1 000 00 (1 mm/km)
C Order (dependent)	10 mm + 1:100 000 (10 mm/km)

northern mid-latitudes we never have as many satellites to the north as we have to the south (see, for example, Santerre [1991]). This makes latitude the second-best determined curvilinear coordinate.

Acknowledgments

The assistance of Guangping He in the preparation of updates for the second edition is gratefully acknowledged.

References

Andrews Space & Technology. 2001. Space and Tech. GLONASS — Summary. Available at: *www.spaceandtech.com/spacedata/constellations/glonass_consum.shtml.*

Cohen, E. R. and Taylor, B. N. 1988. The fundamental physical constants. *Phys. Today*, 41:9–13.

Defense Mapping Agency (DMA). 1988. Department of Defense World Geodetic System: Its Definition and Relationships with Local Geodetic Systems. DMA Technical Report 8350.2 (revised 1 March). DMA, Washington, DC.

Escobal, P. R. 1976. *Methods of Orbit Determination.* John Wiley & Sons, New York.

Federal Geodetic Control Committee (FGCC). 1984. *Standards and Specifications for Geodetic Control Networks.* (Reprint version February 1991.) Federal Geodetic Control Committee. Rockville, MD.

Federal Geodetic Control Committee (FGCC). 1989. *Geometric Geodetic Accuracy Standards and Specifications for Using GPS Relative Positioning Techniques.* Version 5.0. Federal Geodetic Control Committee. Rockville, MD.

GPS World. 1994. Satellite almanac overview. *GPS World*, 5:60.

Graham, J. 1995. Navigation Satellites, Chapter 26 in *Space Exploration from Talisman of the Past to Gateway for the Future.* Available at: *www.space.edu/projects/book/chapter26.html.*

International Association of Geodesy (IAG). 1971. *Geodetic Reference System 1967.* Publication Spéciale No. 3. IAG, Paris.

International Association of Geodesy. 1980. Geodetic reference system 1980 (compiled by H. Moritz). *Bull. Géodésique*, 54395–405.

International Association of Geodesy. 1984. Geodetic reference system 1980 (compiled by H. Moritz). *Bull. Géodésique*, 58388–398.

International Association of Geodesy. 1988a. Geodetic reference system 1980 (compiled by H. Moritz). *Bull. Géodésique*, 62348–358.

International Association of Geodesy. 1988b. Parameters of common relevance of astronomy, geodesy, and geodynamics (compiled by B.H. Chovitz). *Bull. Géodésique*, 62359–367.

Leick, A. and van Gelder, B. H. W. 1975. On Similarity Transformations and Geodetic Network Distortions Based on Doppler Satellite Coordinates. Reports of the Department of Geodetic Science, No. 235. Ohio State University, Columbus.

McCarthy, D. D., Ed. 1992. *IERS Standards (1992).* IERS Technical Note 12. Central Bureau of International Earth Rotation Service, Observatoire de Paris.

Montgomery, H. 1993. City streets, airports, and a station roundup. *GPS World*, 416–19.

Newman, Yona. 2002. GPS Receiver Survey. Available at: *http://hona_n.tripod.com/gps/gps-survey.html.*

Newton, I. S. 1686. *Philosophiae Naturalis Principia Mathematica.*

North Atlantic Treaty Organization (NATO). 1988. Standardization Agreement on NAVSTAR Global Positioning System (GPS), System Characteristics — Preliminary Draft. STANAG 4294 (revision: 15 April).

Santerre, R. 1991. Impact of GPS satellite sky distribution. *Manuscripta Geodetica*, 6128–53.

Satellite Observing: Orbital Elements. Available at: *www.accesscom.com/~iburrell/sa/elements.html.*

Soler, T. 1976. On Differential Transformations between Cartesian and Curvilinear (Geodetic) Coordinates. Reports of the Department of Geodetic Science, No. 236. Ohio State University, Columbus.

Soler, T. and Hothem, L.D. 1988. Coordinate systems used in geodesy: basic definitions and concepts. *J. Surv. Eng.*, 11484–97.

Soler, T. and van Gelder, B. H. W. 1987. On differential scale changes and the satellite Doppler z-shift. *Geophys. J. R. Astron. Soc.*, 91:639–656.

Stem, J. E. 1991. State Plane Coordinate System of 1983. NOAA Manual NOS NGS 5. National Oceanic and Atmospheric Administration, Rockville, MD.

Wells, D., Ed. 1986. *Guide to GPS Positioning.* Canadian GPS Associates, Fredericton, New Brunswick.

Wells, D. and Kleusberg, A. 1990. GPS: a multipurpose system. *GPS World*, 160–63.

Further Information

Textbooks and Reference Books

For additional reading and more background, from the very basic to the advanced level, in geodesy, satellite geodesy, physical geodesy, mechanics, orbital mechanics and relativity, see the textbooks listed below.

Bomford, G. 1980. *Geodesy.* Clarendon Press, Oxford.

Goldstein, H. 1965. *Classical Mechanics.* Addison-Wesley, Reading, MA.

Heiskanen, W. A. and Moritiz, H. 1967. *Physical Geodesy.* W. H. Freeman, San Francisco.

Hofmann-Wellenhof, B., Lichtenegger, H. and Collins, J. 2001. *GPS: Theory and Practice*, 5th ed. Springer-Verlag, New York.

Jeffreys, H. 1970. *The Earth: Its Origin, History and Physical Constitution.* Cambridge University Press, Cambridge.

Kaula, W. M. 1966. *Theory of Satellite Geodesy: Applications of Satellites to Geodesy.* Blaisdell, Waltham, MA (reprinted 2000).

Lambeck, K. 1988. *Geophysical Geodesy: The Slow Deformations of the Earth.* Clarendon Press, Oxford.

Leick, A. 1995. *GPS: Satellite Surveying*, 2nd ed. John Wiley & Sons, New York.

Maling, D. H. 1993. *Coordinate Systems and Map Projections.* Pergamon Press, New York.

Melchior, P. 1978. *The Tides of the Planet Earth.* Pergamon Press, New York.

Moritz, H. 1990. *The Figure of the Earth: Theoretical Geodesy and the Earth's Interior.* Wichmann, Karlsruhe.

Moritz, H. and Mueller, I. I. 1988. *Earth Rotation: Theory and Observation.* Frederick Ungar Publishing Co., New York.

Mueller, I. I. 1969. *Spherical and Practical Astronomy, As Applied to Geodesy.* Frederick Ungar Publishing Co., New York.

Munk, W. H. and MacDonald, G. J. F. 1975. *The Rotation of the Earth: A Geophysical Discussion.* Cambridge University Press, Cambridge.

Seeber, G. 2000. *Satellite Geodesy: Foundations, Methods, and Applications*, 2nd ed. Walter de Gruyter, New York.

Soffel, M. H. 1989. *Relativity in Astrometry, Celestial Mechanics and Geodesy.* Springer-Verlag, New York.

Torge, W. 2001. *Geodesy*, 3rd ed. Walter de Gruyter, New York.

U.S. Air Force. USAF Fact Sheet: NAVSTAR Global Positioning System. Available at: *www.af.mil/news/factsheets/NAVSTAR_Global_Positioning_Sy.html.*

Vaníček, P. and Krakisky, E. J. 1982. *Geodesy: The Concepts.* North-Holland, Amsterdam.

Journals and Organizations

The latest results from research in geodesy are published in two international magazines under the auspices of the International Association of Geodesy, both published by Springer-Verlag (Berlin/Heidelberg/New York): *Bulletin Géodésique* and *Manuscripta Geodetica.*

Geodesy- and geophysics-related articles can be found in:

American Geophysical Union (Washington, DC): *EOS and Journal of Geophysical Research*
Royal Astronomical Society (London): *Geophysical Journal International*

Kinematic GPS-related articles can be found in:

Institute of Navigation: *Navigation*
American Society of Photogrammetry and Remote Sensing: *Photogrammetric Engineering & Remote Sensing*

Many national mapping organizations publish journals in which recent results in geodesy/surveying/mapping are documented:

American Congress of Surveying and Mapping: Surveying and Land Information Systems and Cartography and Geographic Information Systems
American Society of Civil Engineers: *Journal of Surveying Engineering*
Deutscher Verein für Vermessungswesen: *Zeitschrift für Vermessungswesen, Konrad Wittwer Verlag, Stuttgart*
The Canadian Institute of Geomatics: *Geomatica*
The Royal Society of Chartered Surveyors: *Survey Review*
Institute of Surveyors of Australia: *Australian Surveyor*

Worth special mention are the following trade magazines:

GPS World, published by Advanstar Communications, Eugene, OR
P.O.B. (*Point of Beginning*), published by P.O.B. Publishing Co., Canton, MI
Professional Surveyor, published by American Surveyors Publishing Co., Arlington, VA
Geodetical Info Magazine, published by Geodetical Information & Trading Centre bv., Lemmer, the Netherlands

National mapping organizations such as the U.S. National Geodetic Survey (NGS) regularly make geodetic software available (free and at cost). Information can be obtained from: National Geodetic Survey, Geodetic Services Branch, National Ocean Service, NOAA, 1315 East-West Highway, Station 8620, Silver Spring, MD 20910-3282.

168

Surveying Applications for Geographic Information Systems

Baxter E. Vieux
(Second Edition)
University of Oklahoma

James F. Thompson
(First Edition)
J.F. Thompson, Inc.

Surveying and mapping has been transformed by the explosive growth of geospatial data and the software systems designed to manage, analyze, and map this data. Distinctions have eroded between *computer aided design and drafting* (CADD) and *geographic information systems* (GIS) because both systems are able to use data formats designed for one another. However, distinct differences still exist between CADD and GIS in terms of the types of spatial analysis that can be performed. Automated mapping and facility management (AM/FM) that relies on a database of attributes and spatial location has become widespread and is referred to as an AM/FM system. Even though the integration of GIS data sources into engineering design and mapping is commonplace, understanding the differences in source, resolution, and accuracy between data collected by field surveys and GIS data is crucial to the effective use of such data.

This chapter provides guidance on engineering surveying and mapping standards for geospatial data. It is intended to guide the engineer in the selection and specification of data for use in various project documents, architectural and engineering drawings, planning maps, construction plans, hydraulic and hydrologic studies, and related GIS, CADD, and AM/FM products.

Data collected by traditional survey means, satellite remote sensing, or aerial platforms provide the base mapping that may be assembled in a CADD/GIS system on separate layers. Other sources include digitizing paper maps or converting CADD drawings into GIS data sets. Integration of these diverse data types, sources, and formats into a common database or CADD/GIS system should explicitly recognize the widely differing accuracy and precision. Of fundamental importance is the coordinate system used to plot position of features or themes. Assembly of data from disparate sources often requires transformation of the data into a common coordinate system.

A concurrent step is to document information about the data known as *metadata*, which are data *about* the data. Knowing the origin, lineage, and other aspects of the data you are using is essential to understanding their limitations and getting the most usefulness from the data. The metadata include projection system, datum, and units (metrication) of the original data and derivative products such as reprojection or conversion from meters to U.S. customary units. Documentation for GIS data in the United States has been set forth in the Federal Geographic Data Committee (FGDC) *Content Standards*

0-8493-1586-7/05/$0.00+$1.50

for Digital Geospatial Metadata (FGDC, 1998). FGDC-compliant metadata files contain detailed descriptions of the data sets, and include narrative sections describing the procedures used to produce the data sets in digital form. Even if detailed FGDC procedures are not followed for a CADD/GIS engineering project, keeping metadata about essential characteristics and sources of the data saves critical project resources by avoiding confusion or over-reliance on inaccurate data.

168.1 GIS Fundamentals

A GIS can be generally described as a computer-based system that stores and manages information and data that are geographically located and graphically displayed in map form. Various CADD/GIS themes are arranged by placing them on separate data layers for convenience in accessing the database or for turning on and off the display of each theme. Decisions regarding data collection in CADD or GIS format rest on the type of analysis required during the project or with future applications. Two major GIS data types are raster and vector. A raster data set has a regular grid of cells with a single attribute assigned to that cell. A vector data set is composed of points, lines, and polylines/polygons that may be assigned multiple attributes within a database. Both data types can be used and displayed in CADD and GIS alike.

Query, selection, and display of georeferenced data are functions supported by both CADD and GIS. Unlike CADD, GIS formats and GIS software permit spatial analysis where new information is derived from the overlay and intersection of two or more features or themes. However, a GIS allows new information to be obtained as the result of analyzing spatial relationships called *topology*. For example, knowing which counties have land areas within a series of watersheds can easily be obtained by overlaying the two themes, counties and watersheds, and getting the spatial intersection. While a CADD system can display the overlay, a new layer will not be produced that shows the counties which have land in a certain watershed. Figure 168.1 represents a simple GIS configuration where information of different types is stacked to form an organizational structure for the system itself. This graphic information is linked to one or more databases to form a relationship between the graphic information and supporting data. Within a GIS, every graphic element, together with its supporting data or attributes, has its own unique geographic coordinates. As such, a GIS can be used to query a database and find information on a geographic feature, element, or area. Many engineering, planning, and environmental projects can make use of both CADD and GIS formats.

Photogrammetric mapping for acquisition of features or topography is of particular importance to an engineering project. Photogrammetric mapping usage is generally divided into four categories: (1) topographic features, planimetric features, orthophotography, and land use. Topography acquisition requires stereo imagery with sufficient overlap, usually at least 60%, and stereo viewing equipment. Because increased precision is expensive, care should be taken to collect photogrammetric maps with appropriate precision and accuracy and with formats that can easily be assembled into CADD/GIS software packages. Orthophotography are aerial photographs that have been georeferenced and corrected for elevation differences, and generally referred to as *digital orthophoto quarter quadrangles* (DOQQs). One-meter resolution panchromatic photography is now available in most U.S. states. Both CADD and GIS software can readily use this data in common image formats such as the tagged image file format (TIFF). DOQQs provide an inexpensive background map for project mapping in CADD/GIS software packages.

Planimetric and topographic mapping most often forms the base data set in a CADD/GIS project. The accuracy of computations and queries made from these base mapping data sets is limited by the least accurate imagery or survey data collected. Collecting geospatial data from satellite or aerial sources involves several types of imagery for various purposes. Table 168.1 lists general types of imagery and associated usage. "Small scale" refers to a ratio scale, meaning that features appear small. An example of a small-scale map would be at 1:100,000. Conversely, large-scale maps are at 1:20,000 or larger. Scales larger than 1:1000 are typical for site plans, whereas small-scale maps are generally used for planning or reconnaissance mapping.

CADD/GIS layers are arranged by theme, which offers convenience in accessing the database or for turning on and off the display of each theme. For this reason, individual layers must be georeferenced

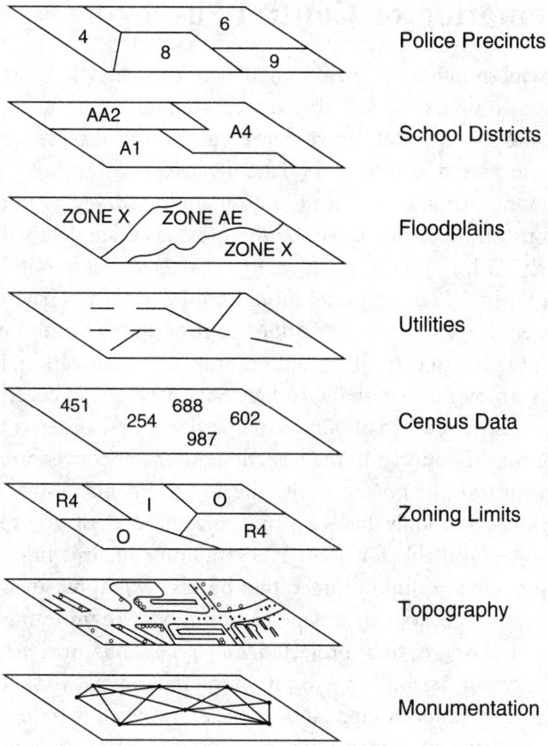

FIGURE 168.1 Simplified GIS layers.

TABLE 168.1 General Types of Imagery and Associated Usage

Imagery Type	Usage
Black and white aerial photography	Topographic and planimetric mapping
Natural color aerial photography	Topographic and planimetric mapping
Infrared aerial photography	Vegetation analysis, land use/land classification
Satellite imagery	Small-scale mapping, vegetation analysis, land use/land classification

to a common ground reference (e.g., state plane, universal transverse mercator, latitude/longitude) so that data from various layers geographically match one another when overlaid. Features are assembled into a database separated into layers, but in a common coordinate system. Not all information is collected to the same degree of horizontal and vertical accuracy and scale. Enlarging a 1:24,000 (1 in = 2000 ft) topographic map to show features at 1 in = 100 ft can misplace features by as much as 0.5 in on the map or 100 ft on the ground. Similarly, for display of topographic relief, contours taken from 1:24,000 quadrangles are considered accurate to within one-half contour interval. These data can be used to create a topographic map showing 2-ft contour intervals far exceeding the accuracy of the source data.

The widespread adoption and availability of digital databases have had a dramatic impact on mapping and GIS/CADD applications. Because data can be manipulated at scales that are inconsistent with the scale at which the data was collected or compiled, care must be taken to fairly represent the data at appropriate scales and contour intervals without falsely conveying more accuracy and precision than is present. Because of the separation of planimetric features and topographic elevations into various layers, and depiction at any scale, problems arise when scales are increased beyond their original values. When merging data collected by traditional survey means, satellite remote sensing, or aerial platforms, maintaining the metadata for each layer in a CADD/GIS system or project becomes as important as the data set itself.

168.2 Monumentation or Control Surveying

This singular subject of establishing a monumentation network is one of the most important issues relating to the surveying applications for a GIS. Selecting the coordinate system and datum is the first step in building a geospatial database. Horizontal and vertical control for field surveys must ultimately be placed within the same coordinate system as other GIS data, or vice versa. Accuracy standards should set the level of accuracy of the monumentation network and remaining surveying information obtained. Most likely, the monumentation control network will be the most accurate data within the GIS. Conversely, the GIS data are most likely the least accurate data set in a CADD drawing containing survey information. In any case, the two data types, CADD and GIS, should not be treated as having equivalent accuracy.

A monumentation system for a GIS is a network of control points or monuments, established in the field, that are referenced not only to each other, but also to a common datum. In the U.S., this common datum is typically established by the National Geodetic Survey (NGS), an agency of the federal government. The NGS has established and maintains a network of first-order monuments with assigned horizontal coordinates across the country. In the past, first order was conventionally thought of as having an accuracy level of 1 part in 100,000; however, with the advancements in surveying equipment technology, first-order monuments commonly have a much higher level of accuracy. In addition, vertical elevation information is often available for many NGS monuments. It is important to understand that these NGS monuments provide a common thread that bonds geographic areas across the country.

Each state has its own coordinate system, as dependent on the map projection used, which is generally a result of the shape of the state or area to be projected. *Map projections* form the basis for planar mapping of geographic areas. A surveyor establishing the monumentation network for a given area must be knowledgeable of the projection and resultant *state plane coordinate system* used in the area.

When we establish a geodetic monumentation network for a GIS, we are typically densifying the existing NGS monumentation system in our area of interest. Often, many NGS monuments are recovered in the field and then used as reference points within the densified monumentation network. The result of the monumentation survey is a network of readily identifiable points throughout an area that can be easily used for land surveying and *aerial photogrammetry.* The fundamental issue at hand is that, once the monumentation network is established, a common datum will exist throughout an area that can be used to reference all subsequent surveying data. Even past surveying information can be moved to this newly established datum via a coordinate transformation.

The actual procedures for establishing a geodetic monumentation network can be divided into several stages that build upon each other:

1. Research existing NGS monuments and field recover the monuments.
2. Design a network layout as referenced to the recovered NGS monuments.
3. Plan the field survey activities to establish the monuments.
4. Execute the field survey activities.
5. Mathematically adjust the network and calculate the coordinates of the monuments.

Recovering the existing NGS monuments in the field can be very time consuming and, at times, frustrating. Many monuments have not been recovered or used for several years; thus, they are often difficult to find. Once recovered, these NGS monuments and their previously established coordinates will be used to reference and constrain the new monuments to be established.

Once the positions of the recovered NGS monuments are known in relation to the area of interest to serve as the geographic region supported by a GIS, a monumentation network is designed that provides the layout of the new monuments. This layout is composed of the existing NGS monuments, the new monuments to be established, and the baselines connecting all of the monuments. The strength and accuracy of a monumentation network is dependent on the design of the layout and, specifically, the baselines within the layout. Today, the most common means used to perform the survey for the establishment of a geodetic monumentation network is the *global positioning system* (GPS). GPS surveying is a method of using satellites and portable field receivers to accurately determine the coordinates of a point.

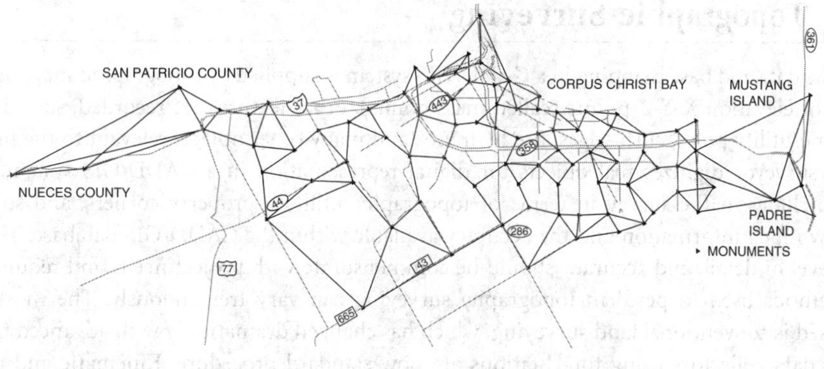

FIGURE 168.2 Baseline and monument layout for a municipal GIS application.

(GPS surveying is discussed in detail in Chapter 167) A baseline between two monuments is created when separate GPS receivers are set on both points and they receive credible satellite data simultaneously. Figure 168.2 illustrates a designed and successfully implemented baseline and monument layout for a municipal GIS application. In the design of the network layout, consideration must be given to the field procedures required to create the needed baselines throughout the monumentation network.

Once the desired monumentation layout is designed, the field activities themselves must be carefully planned. A poorly planned GPS field survey can result in no useful data collected. Assuming that all goes well during the GPS field surveys and that good data were received from the satellites, the resultant coordinates of the new monuments can be calculated. This entire process is often called *balancing*, *adjusting*, or *constraining* the network. During this process, the entire monumentation network is referenced to the NGS datum, and state plane coordinates are assigned to each monument. State plane coordinates differ from surface coordinates as would be needed for conventional plane table land surveying. Note that state plane coordinates are used in map projections to account for the curvature of the earth. Every point has a scale factor that must be applied to the state plane coordinates (northing and easting) of that point to convert the point's state plane coordinates to surface coordinates at sea level. To account for the difference of the terrain's elevation that normally differs from sea level, a sea level correction factor must be applied as well. The combination of a scale factor and a sea level correction factor is commonly termed a *combined scale factor.* The following equations are used to convert state plane coordinates to surface coordinates:

$$\text{Combined scale factor} = \text{Scale factor} \times \text{sea level correction factor}$$

$$\text{Surface coordinate} = \text{State plane coordinate/combined scale factor}$$

It is extremely important that the relationship between surface and state plane coordinates is understood. A conventional land surveyor does not measure distances, for example, along the map projection utilized, but, rather, along the true terrain surface being surveyed. Because of this, features must be scaled accordingly.

After the horizontal coordinates of the monuments are established, vertical elevations can be determined for the points, such as to support surveys that are in need of relational elevations. Once again, a similar method of balancing or adjusting the network is utilized as based upon the known elevations of points within the monumentation network. Much advancement has been made in accurately calculating vertical elevations using GPS surveying. This is mostly a result of the ability to more accurately account for the true shape of and anomalies in the earth's surface. The accuracy achieved by GPS surveying has resulted in reduced errors in location of features, both horizontally and vertically. Integration of these data into a CADD/GIS system requires proper coordinate transformations and adjustment with the combined scale factor.

168.3 Topographic Surveying

Much of the data and base mapping in a CADD/GIS system is supplied by topographic surveying. Besides collection of elevation X-Y-Z points (called mass points), other features are recorded, such as locations of telephone/utility poles, curb edges, fire hydrants, and many other objects relevant to the project. In a sense, the surveyor digitizes real objects for digital representation in a CADD/GIS system. What the surveyor includes in the survey in terms of topography, utilities, property corners, and so forth will dictate how much information and the accuracy available within the CADD/GIS database. The types of features, level of detail, and accuracy should be commensurate with project needs and requirements.

The methods used to perform topography surveying can vary tremendously. The most common method used is conventional land surveying, which has changed dramatically with advanced technology. Electronic data collectors using total stations are now standard procedure. Kinematic and differential GPS receivers are used routinely to gather elevation and other features for topographic surveys. In conjunction with conventional land surveys, aerial photogrammetry is a common means of obtaining contours for planning purposes. Although conventional land surveying and aerial photogrammetry are some of the most popular means of collecting topographic data for GIS applications. Other methods, such as *satellite imagery*, may be applied as well. Care must be taken to ensure that the vertical and horizontal resolution and accuracy are adequate for the intended purpose. While satellite-derived topographic information is useful for planning purposes over large spatial extents, the vertical accuracy may limit its use for site development, construction, or other purposes that require more accurate elevations. Similarly, the horizontal location of planimetric features from satellite imagery is less accurately known than data collected by photogrammetry, *Light detection and ranging* (LIDAR), electronic total station, GPS, or other field survey techniques. Evidence of misalignment becomes apparent when contours or other features such as stream channels derived from other sources of differing accuracy are overlaid on the imagery.

In performing a conventional topographic survey, a surveyor will traverse between the geodetic network monuments, obtaining as much topographic information as desired. Again, the information sought is dependent on the demands of the GIS that dictate the detail needed in terms of the survey data obtained. The traversed line of the survey between the monuments is balanced in terms of angles and distances as based upon the known coordinates of the monuments; this will ensure that coordinates of the obtained topographic features are properly determined in relation to the datum used for the GIS. Figure 168.3 represents a series of points obtained for a GIS using conventional topographic surveying techniques. Figure 168.4 illustrates the same data in their final form, as would be input graphically into the GIS. Every point has its own unique address and attribute. A fire hydrant, for example, will have its own specific coordinates. A relational database can be linked to the fire hydrant graphic element within the GIS, which would provide such information as when the hydrant was installed and when last checked to ensure its proper operating condition. Assigning codes in the field to represent features such as fire hydrants using an electronic data collector and total station simplifies the plotting and database development in the office.

When acquiring aerial photogrammetry for topographic mapping and other purposes, control points must be established within a coordinate system consistent with other mapped features. Linking the aerial photography to the monumentation network is achieved by referencing the aerial photographs to known points on the ground that can be clearly identified in the photographs. This is typically done by *painting*, that is, laying down of panels (or in many cases, taping) white or contrasting colored marks on the ground in the shape of an "X" or a "V" that can be clearly identified in the photograph. These marks are then tied to the monument system conventionally with ground surveys. The end result is an aerial photograph, or series of photographs, rectified to accurately illustrate the existing topography on the proper coordinate system. Once the aerial photographs are produced, they are digitized into graphic elements that are directly incorporated into a GIS. With the horizontal position of the topographic features, the elevation contours and many specific point elevations can be determined as well.

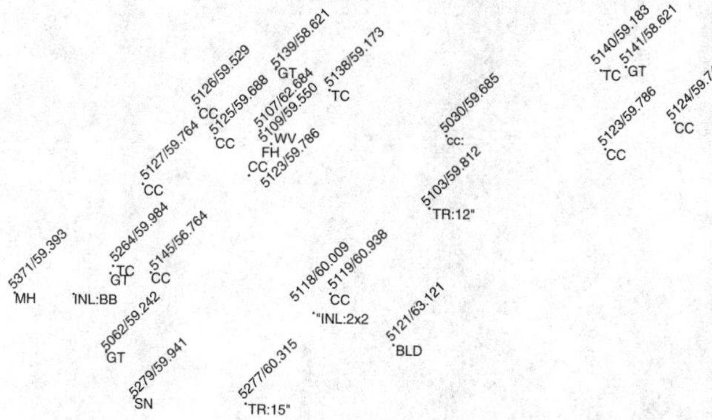

FIGURE 168.3 Series of points obtained using conventional topographic surveying techniques.

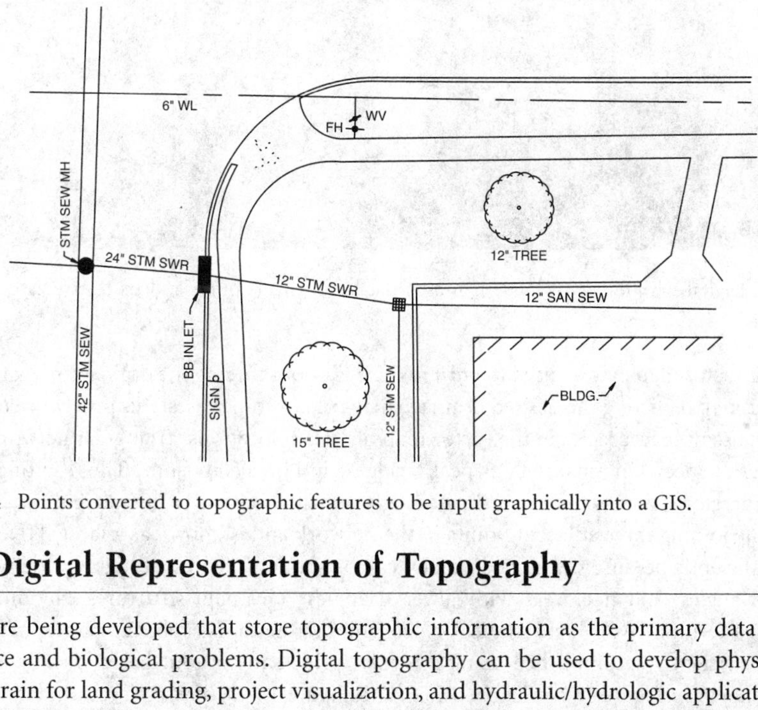

FIGURE 168.4 Points converted to topographic features to be input graphically into a GIS.

168.4 Digital Representation of Topography

Many GISs are being developed that store topographic information as the primary data for analyzing water resource and biological problems. Digital topography can be used to develop physically realistic models of terrain for land grading, project visualization, and hydraulic/hydrologic applications. A *digital elevation model* (DEM) is an ordered array of numbers that represents the spatial distribution of elevations above some arbitrary datum in a landscape. It usually consists of elevations sampled at discrete points on an irregular spacing over the landscape or on regular intervals. Field surveys may be used to collect elevations and then transform them into a DEM, also called a *digital terrain model* (DTM). When discussing the use of DEMs, it is important to consider the way in which the surface representation is to be used and how it was generated. The ideal data structure for a DEM depends on the project purpose. There are three principal ways of structuring digital elevation data: (1) grid based or raster; (2) triangular irregular network (TIN); and (3) contours.

Grid-based methods use regularly spaced elements or cells to represent the terrain. Grid DEMs are useful for automated watershed delineation. Preprocessing and quality control of the DEM are necessary to insure accurate results for applications such as hydrological modeling. No automatically produced drainage network is likely to be very accurate in flat areas because drainage directions assigned across flat areas are not assigned using information directly held in the DEM. This is particularly true where elevation differences are less than the contour interval of the original map used to create the DEM. A

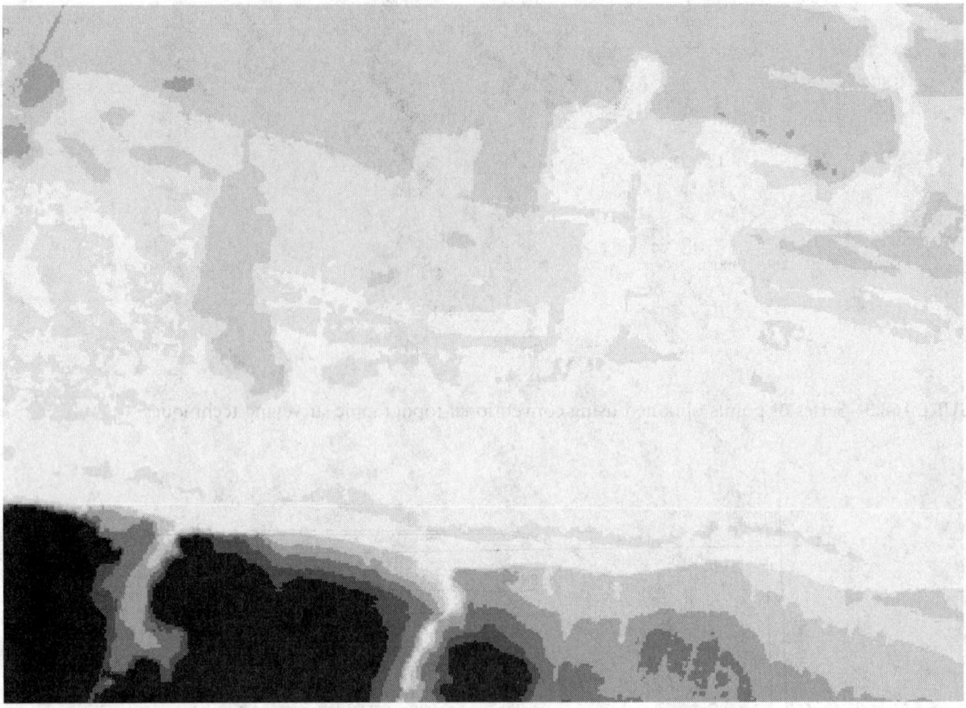

FIGURE 168.5 Digital elevation map based on a raster data structure. (Courtesy Vieux & Associates, Inc., Norman, OK.)

river or stream vector map may be used to *burn in* the elevations forcing the drainage network to coincide with the vector map depicting the desired drainage network. In some cases, this procedure may introduce anomalies because of inaccuracies in the vector map of stream locations. Triangular networks are usually sample surfaces at specific points, such as peaks, ridges, and breaks in slope. The resulting *TIN* formed consists of points stored as x, y, and z coordinates together with pointers to their neighbors. The elemental area is the plane joining three adjacent points in the network and is known as a facet. TIN elevations are stored more efficiently because the facets in a TIN can be varied to represent highly variable terrain with many small triangles, and flat areas with fewer triangles. TIN data structures can provide striking visualizations of terrain. Hydrologic studies or determining the topographic attributes of the landscape may be more efficiently accomplished using a regular grid than contour-based or TIN DEMs. In the three images shown in Figure 168.5, Figure 168.6, and Figure 168.7, digital elevation data are portrayed as raster, TIN, and contours. The data contained in this 20-ft resolution DEM were developed from LIDAR data acquired in January though March 2001. Cell values in the DEM were derived from a TIN produced from the bare earth mass points measured by LIDAR. Individual cell values are based on an average of 25 points inside each 20-ft DEM cell. These data were collected and derived by the North Carolina Floodplain Mapping Program as part of its effort to modernize flood insurance rate maps statewide. Although created for specific use in the engineering aspects of floodplain delineation, the data were also developed to address the elevation data requirements of the broader geospatial data user community. The North Carolina Floodplain Mapping Program (NCFMP, 2003) was established in response to the extensive damage caused by Hurricane Floyd in 1999. The vertical accuracy of the elevation data is less than or equal to 25 cm root mean square error (RMSE). LIDAR DEMs are becoming standard means for topographic mapping at very high spatial resolutions. Because of the high resolution of measurement, automated and manual procedures are required to remove points that do not belong to the terrain surface, such as water, vegetation, buildings, and system noise.

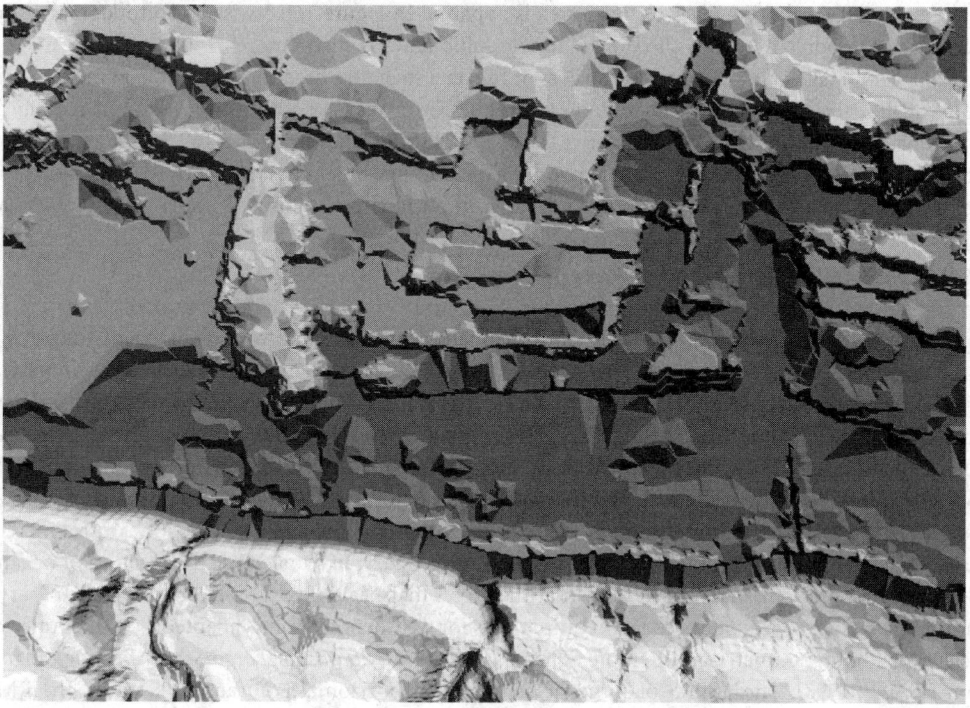

FIGURE 168.6 Digital elevation model using the TIN data structure.

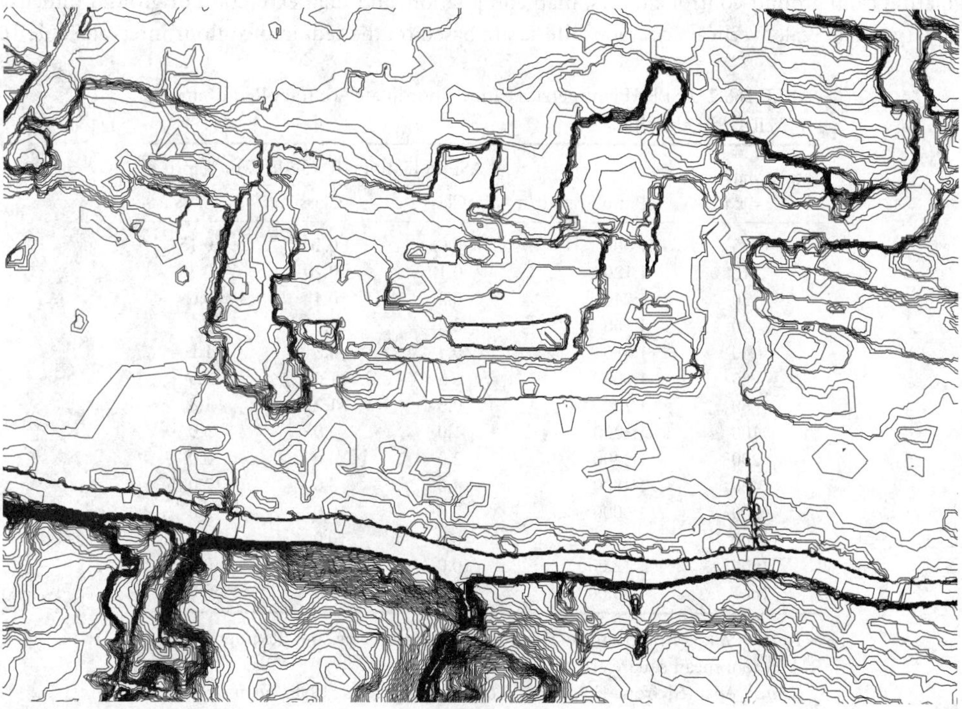

FIGURE 168.7 Digital elevation model using contours.

From the DEMs, the Tar River channel area is apparent in Figure 168.5 with steep topography south and a flatter floodplain area to the north. Topographic relief is made more visible in Figure 168.6 by using the TIN image data structure. Regardless of the data structure used to represent terrain, the analyst should realize that the DEM likely contains anomalies or artifacts resulting from data processing that can distort accurate representation of the surface. The contour lines in Figure 168.7 were derived from the TIN at 2-ft contour intervals. The field checking of contours, breaklines, and derived drainage features such as stream locations and watershed boundaries is important just as it is with paper topographic maps.

168.5 GIS and Survey Specifications

Development of engineering plans often relies on spatial data collected from a variety of sources with mixed accuracies. Defining accuracy standards for mixed data makes metadata all the more important for a project. The purpose of the collected data defines the horizontal accuracy, target map scale, and vertical relief. Horizontal and vertical control survey accuracy should be selected that are consistent with the intended purpose of the map or analysis. Usually 1:5000 third-order control procedures (horizontal and vertical) will provide sufficient accuracy for most engineering work. In most small-scale or GIS mapping, third-order, class II methods and fourth-order topographic and construction control methods are adequate. Higher order control surveys may not be necessary. In some cases, the location of a feature is known more precisely than can be shown at the target map scale. Table 168.2 provides a summary of planimetric feature coordinate accuracy requirements for well-defined points adapted from the American Society for Photogrammetry and Remote Sensing (ASPRS) *Standards for Large-Scale Mapping* (adapted from USACE, 1998). The ASPRS planimetric standard for horizontal accuracy is based on the RMSE between map coordinates at the specified map scale and field locations of well-defined features determined by an independent check survey of higher accuracy. The term "limiting RMSE" is the minimum error allowed for a percentage of points or confidence interval. The RMSE is the cumulative result of all errors originating from ground-control surveys, map compilation, and final extraction of ground dimensions at the target map scale. Vertical accuracy criteria are based on the required contour interval. Table 168.3

TABLE 168.2 ASPRS Planimetric Feature Coordinate Accuracy Requirements for Well-Defined Points

Target Map Scale		ASPRS Limiting RMSE in X or Y in Feet		
1" = x ft	Ratio Scale ft/ft	Class 1	Class 2	Class 3
5	1:60	0.05	0.10	0.15
10	1:120	0.10	0.20	0.30
20	1:240	0.2	0.4	0.6
30	1:360	0.3	0.6	0.9
40	1:480	0.4	0.8	1.2
50	1:600	0.5	1.0	1.5
60	1:720	0.6	1.2	1.8
100	1:1200	1.0	2.0	3.0
200	1:2400	2.0	4.0	6.0
400	1:4800	4.0	8.0	12.0
500	1:6000	5.0	10.0	15.0
800	1:9600	8.0	16.0	24.0
1000	1:12,000	10.0	20.0	30.0
1667	1:20,000	16.7	33.3	50.0

Note: ASPRS, American Society for Photogrammetry and Remote Sensing; RMSE, root mean square error.

Source: Adapted from U.S. Army Corps of Engineers. 1998. *Engineering and Design: Geospatial Data and Systems.* EM 1110-1-2909, CEIM-IM-PD. USACE Publication Depot, Hyattsville, MD.

TABLE 168.3 ASPRS Topographic Elevation Accuracy Requirement for Well-Defined Points

	ASPRS Limiting RMSE in Feet					
	Topographic Feature Points			Spot or Digital Terrain Model Elevation Points		
Target Contour Interval (ft)	Class 1	Class 2	Class 3	Class 1	Class 2	Class 3
0.5	0.17	0.33	0.50	0.08	0.16	0.2
1	0.33	0.66	1.0	0.17	0.33	0.5
2	0.67	1.33	2.0	0.33	0.67	1.0
4	1.33	2.67	4.0	0.67	1.33	2.0
5	1.67	3.33	5.0	0.83	1.67	2.5

Note: ASPRS, American Society for Photogrammetry and Remote Sensing; RMSE, root mean square error.

Source: Adapted from U.S. Army Corps of Engineers. 1998. *Engineering and Design: Geospatial Data and Systems.* EM 1110-1-2909, CEIM-IM-PD. USACE Publication Depot, Hyattsville, MD.

shows specifications for topographic elevation accuracy requirements (RMSE) for well-defined points. When a DTM is being generated, then an equivalent contour interval may be specified. The RMSE of vertical elevation of well-defined points are determined by differential leveling or other methods of higher accuracy. The equivalent contour interval of the DTM used in defining the accuracy standard is an important component of the metadata because contours at any interval may subsequently be derived in a GIS/CADD mapping system. Both tables are guides for selecting target map scale, contour interval, and vertical/horizontal accuracy for planimetric and topographic data acquisition.

The scale and accuracy of all data collected, whether by field survey instruments, photogrammetry, LIDAR, satellite remote sensing, or GPS should be consistent with project goals. *Over*-specifying the precision of data collection can lead to costly data acquisition that wastes project resources. *Under*-specifying precision can lead to erroneous planning and engineering design that results in costly construction modifications. Care should be exercised to meet project goals with *adequate* scale and accuracy, and to use metadata to document the source and processing of each data layer in a CADD/GIS system.

Defining Terms

Aerial photogrammetry — The surveying of surface features through the use of photographs, as taken from an aerial perspective (i.e., an airplane).

Combined scale factor — A factor that can be used to convert state plane coordinates to surface terrain coordinates, considering the map projection used and the relative elevation of the point of interest.

Digital elevation model (DEM) — An ordered array of numbers that represent the spatial distribution of elevations above some arbitrary datum in a landscape.

Digital terrain model (DTM) — An ordered array of numbers that represent attributes related to terrain that include, but not limited to, the spatial distribution of elevations above some arbitrary datum. Often used synonymously with a DEM.

Geographic information system (GIS) — A computer-based system that stores and manages information and data as related to a geographic area.

Global positioning system (GPS) — A method of surveying that uses transmitting satellites and portable field receivers to accurately determine the coordinates of a point.

Light detection and ranging (LIDAR) — An active sensory system that uses laser light to measure distances between an airborne platform and points on the ground, including trees, buildings, and other features, to collect and generate densely spaced and highly accurate elevation data.

Map projection — The projection of the Earth's surface to a two-dimensional plane (or map), considering the curvature of the Earth.

Metadata — Data *about* the data including origin, lineage, and other aspects of the data including documentation of its resolution, scale, accuracy, projection, and datum.

Monumentation surveying — The establishment or recovery of a horizontal and/or vertical coordinate control network as based upon a layout of monuments or control points.

Photogrammetric mapping — A method for acquiring topographic features, planimetric features, orthophotography, and land use by aerial photography using overlapping stereo imagery.

Raster data — A data set composed of regularly spaced grid cells with a single attribute assigned to each cell.

Satellite imagery — Images of the Earth's surface as obtained from orbiting satellites.

State plane coordinate system — A coordinate system used throughout the United States that is based on the map projection of one or more zones within each state.

Topographic surveying — The surveying of topographic features through conventional land surveying, aerial photogrammetry, or other means.

Vector data — A vector data set is composed of points, lines, and polylines/polygons that may be assigned multiple attributes within a database.

References

NCFMP, State of North Carolina. 2003. North Carolina Floodplain Mapping Program. Available at: *www.ncfloodmaps.com/pubdocs/NCFPMPHndOut.pdf.* Accessed on 2 May 2003.

U.S. Army Corps of Engineers. 1998. *Engineering and Design: Geospatial Data and Systems.* EM 1110-1-2909, CEIM-IM-PD. USACE Publication Depot, Hyattsville, MD.

ASPRS. 1990. ASPRS Interim Accuracy Standards for Large-Scale Maps, ASPRS, March 1990. Available on the web at: *http://www.asprs.org/resources.html.*

American Congress on Surveying and Mapping. 1993. *Surveying and Land Information Systems.* 53(4). Bethesda, MD.

Davis, R. E., Foote, F. S., Anderson, J. M., and Mikhail, E. M. 1981. *Surveying Theory and Practice.* McGraw-Hill, New York.

FGDC. 1998. Content Standards for Digital Geospatial Metadata, Federal Geographic Data Committee, version 2.0, FGDC-STD-001-1998.

NCFMP. 2003. North Carolina Floodplain Mapping Program. Available on the Internet at http://www.ncfloodmaps.com/pubdocs/NCFPMPHndOut.pdf Last accessed 2 May 2003.

USACE. 1998. *Engineering and Design - Geospatial Data and Systems,* EM 1110-1-2909, USACE Publication Depot, CEIM-IM-PD, Hyattsville, MD.

Further Information

One of the largest sources of information pertaining to surveying applications for geographic information systems is the American Congress of Surveying and Mapping (ACSM), 5410 Grosvenor Lane, Bethesda, MD 20814-2122. The ACSM publishes journals relating to surveying, mapping, and land information.

The NGS Information Center maintains a tremendous amount of valuable information on geodetic surveying, horizontal and vertical coordinate listings, and other applicable publications. This information may be obtained from NOAA, National Geodetic Survey, N/CG17, 1315 East-West Highway, Room 9202, Silver Spring, MD 20910-3282.

The "American Society for Photogrammetry and Remote Sensing (ASPRS) Standards for Large-Scale Mapping" (ASPRS, 1990) and the Federal Geographic Data Committee (FGDC), "Geospatial Positioning Accuracy Standards, Part 4: Standards for Architecture, Engineering, Construction (A/E/C) and Facility Management" (FGDC, 1998), provide accuracy standards for geospatial (CADD/GIS) data.

169

Remote Sensing

Jonathan W. Chipman
University of Wisconsin, Madison

Ralph W. Kiefer
University of Wisconsin, Madison

Thomas M. Lillesand
University of Wisconsin, Madison

Remote sensing involves the use of airborne and space-imaging systems to inventory and monitor Earth resources. Broadly defined, remote sensing is any methodology employed to study the characteristics of objects from a distance. Using various remote sensing devices, we remotely collect *data* that can be analyzed to obtain *information* about the objects, areas, or phenomena of interest. This chapter discusses sensor systems that record energy over a broad range of the *electromagnetic spectrum*, from ultraviolet to microwave wavelengths.

Remote sensing affords a practical means for frequent and accurate monitoring of the Earth's resources from a site-specific to global basis. This technology is aiding in assessing the impact of a range of human activities on our planet's air, water, and land. Data obtained from remote sensors have provided information necessary for making sound decisions and formulating policy in a host of resource development and land use applications. Remote sensing techniques have also been used in numerous special applications. Expediting petroleum and mineral exploration, locating forest fires, providing information for hydrologic modeling, aiding in global crop production estimates, monitoring population growth and distribution, and determining the location and extent of oil spills and other water pollutants are but a few of the many and varied applications of remote sensing that benefit humankind on a daily basis. It should be pointed out that these applications almost always involve some use of *ground truth* or on-the-ground observation. That is, remote sensing is typically a means of extrapolating from, not replacing, conventional field observation.

169.1 Electromagnetic Energy

The sun and various other sources radiate electromagnetic energy over a range of wavelengths. Light is a particular type of *electromagnetic radiation* that can be seen or sensed by the human eye. All electromagnetic energy, whether visible or invisible, travels in the form of sinusoidal waves. Wavelength ranges of special interest in remote sensing are shown in Table 169.1.

When electromagnetic energy is incident upon an object on the Earth's surface, it can interact with the object in any or all of three distinct ways. The incident energy can be reflected, transmitted, or

0-8493-1586-7/05/$0.00+$1.50

TABLE 169.1 Components Electromagnetic Spectrum

Wavelength	Spectral Region
0.3 to 0.4 μm	Ultraviolet
0.4 to 0.7 μm	Visible: 0.4 to 0.5 μm = blue
	0.5 to 0.6 μm = green
	0.6 to 0.7 μm = red
0.7 to 1.3 μm	Near infrared
1.3 to 3.0 μm	Mid-infrared
3 to 14 μm	Thermal infrared
1 mm to 1 m	Microwave

absorbed. The absorbed component goes into heating the body and is subsequently re-emitted from the object. The particular mix of these three possible interactions is dependent on the physical nature of objects. For example, healthy vegetation normally appears green because the blue and red components of the incident light are absorbed by chlorophyll present in plant leaves. In contrast, concrete surfaces strongly reflect blue, green, and red wavelengths nearly equally and appear light gray. Remote sensors record such variations in energy interaction (both in visible and invisible wavelengths) in order to discriminate among Earth surface features and to assist in quantifying their condition.

All objects at a temperature greater than absolute zero radiate energy according to the formula

$$M = \sigma \varepsilon T^4 \tag{169.1}$$

where M is the total radiant exitance (radiated energy) from the surface of a material (W m^{-2}), σ is the *Stefan-Boltzmann constant* (5.6697 · 10^{-8} W m^{-2}K^{-4}), ε is the emissivity (efficiency of radiation) of the material, and T is the absolute temperature (K) of the emitting material. The general shape of the resulting curves of *emitted energy* (radiant exitance) versus wavelength is shown in Figure 169.1(a). The wavelength at which the amount of emitted energy is a maximum is related to its temperature by *Wien's displacement law*, which states that

$$\lambda_m = \frac{A}{T} \tag{169.2}$$

where λ_m is the wavelength of maximum spectral radiant exitance (μm), A = 2898 μm K, and T = temperature (K). Note that the sun's energy has a peak wavelength of 0.48 μm (in the visible part of the spectrum), whereas the Earth's energy has a peak wavelength of 9.67 μm (invisible to the human eye, but able to be sensed by a thermal scanner).

Reflected energy is recorded when objects are sensed in sunlight in the ultraviolet, visible, nearinfrared or midinfrared portions of the spectrum. Radiated energy is recorded in the thermal infrared portion of the electromagnetic spectrum using radiometers and thermal scanners. This allows, for example, the detection and recording of the heated effluent from a power plant as it flows into a lake at a cooler temperature.

Remote sensing systems can be active or passive systems. The examples cited earlier are passive systems in that they record reflected sunlight or emitted radiation. Radar systems are called *active systems* because they supply their own energy. Pulses of microwave energy are transmitted from radar systems toward objects on the ground, and the backscattered energy is then received by radar antennas and used to form images of the strength of the radar return from various objects. Lidar, like radar, is an active remote sensing technique. Such systems use pulses of laser energy directed toward the ground and measure the time of pulse return. The return time for each pulse back to the sensor is processed to calculate the variable distances between the sensor and the various surfaces present on (or above) the ground.

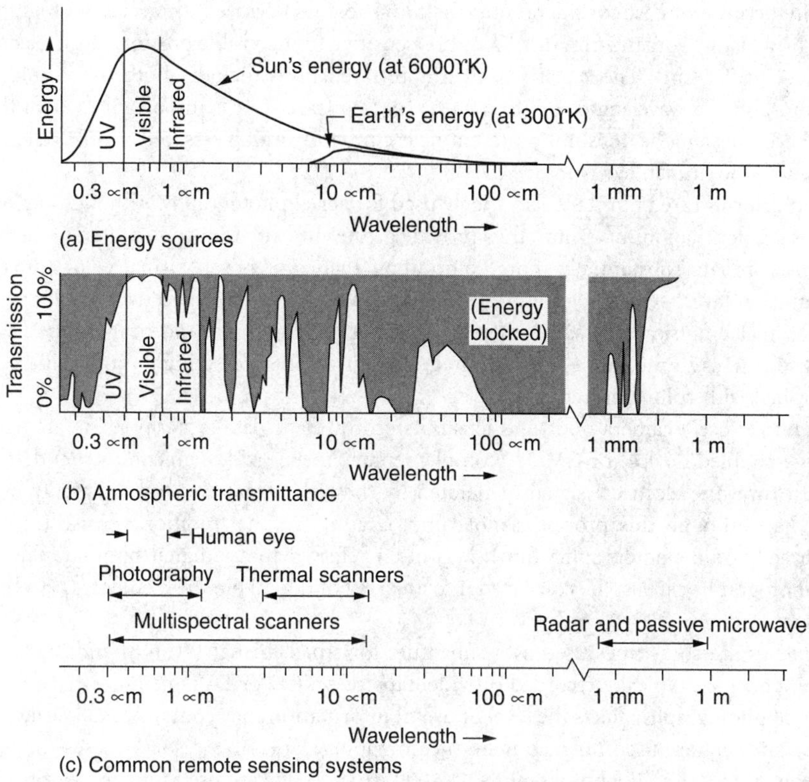

FIGURE 169.1 Spectral characteristics of (a) energy sources, (b) atmospheric effects, and (c) sensing systems. Note that the wavelength scale is logarithmic. (*Source:* Lillesand, T. M., Kiefer, R. W., and Chipman, J. W., *Remote Sensing and Image Interpretation,* 5th ed., John Wiley & Sons, New York, 2004. With permission.)

169.2 Atmospheric Effects

Because the atmosphere contains a wide variety of suspended particles, it offers energy interaction capabilities just as "ground" objects do. The extent to which the atmosphere transmits electromagnetic energy is dependent upon wavelength, as shown in Figure 169.1(b). The sensing systems typically used in various wavelength ranges are shown in Figure 169.1(c). Energy in the ultraviolet wavelengths is scattered greatly, which limits the use of ultraviolet wavelengths from aerial or space platforms. The atmosphere is transparent enough in the visible, nearinfrared, and midinfrared wavelengths to permit aerial photography and multispectral sensing in these wavelengths. In this region the blue wavelengths are scattered the most and the midinfrared wavelengths are scattered the least. In the thermal infrared region, there are two "windows" where the atmosphere is relatively transparent: 3–5 μm and 8–14 μm wavelength (most aerial thermal scanning is done in the 8–14 μm band). At microwave wavelengths, the atmosphere is extremely transparent, and many radar systems can be operated in virtually all weather conditions.

169.3 Remote Sensing Systems

Aerial Photography

Aerial photographs can be taken on any of several film types, from a variety of altitudes. Mapping cameras typically use an image size of 230 by 230 mm. Smaller-format cameras (70 mm and 35 mm) can be used where large-area coverage with great geometric fidelity is not required. The interpretability of aerial photographs is highly dependent on the selection of film type and image scale.

Principally because of cost considerations, the film most widely used for aerial photography is black and white (b/w). Panchromatic films are b/w films sensitive to the visible portion of the electromagnetic spectrum (0.4 to 0.7 μm). The sensitivity of b/w infrared films includes both the visible part of the spectrum and also the wavelengths 0.7 to 0.9 μm (near infrared). It is important to note that infrared energy of these wavelengths does not represent heat emitted from objects, but simply reflected infrared energy to which the human eye is insensitive.

Color and color infrared films are also widely used for aerial photography. Although the cost of using these films is greater than for b/w films, they provide greater information content due to the human eye's ability to discriminate substantially more colors than shades of gray. Normal color films have three separate emulsion layers sensitive to blue, green, and red wavelengths, respectively. Color infrared films have three similar emulsion layers, but they are sensitive to green, red, and near-infrared wavelengths, respectively. Again, as in the case of b/w infrared films, it is reflected sunlight, not emitted energy, that is photographed with color infrared film.

Digital cameras use a camera body and lens but record image data typically using an array of either CCD (charge-coupled device) or CMOS (complementary metal oxide semiconductor) detectors rather than film. In turn, the electrical signals generated by these detectors are stored digitally using various storage media. Although this process is not "photography" in the traditional sense (images are not recorded directly onto photographic film), it is often referred to as "digital photography." Of course, hardcopy photographs can also be converted into an array of digital picture elements (*pixels*) using some form of image scanner.

Video cameras are sometimes used as a substitute for small-format (70 mm and 35 mm) cameras. Video camera data are typically recorded on videotape or on CD or DVD media.

The scale of photographs affects the level of useful information they contain. Small-scale photographs (1:50,000 or smaller) are used for reconnaissance mapping, large-area resource assessment, and large-area planning. Medium-scale photographs (1:12,000 to 1:50,000) are used for the identification, classification, and mapping of such items as tree species, agricultural crop types, vegetation communities, and soil types. Large-scale photographs (larger than 1:12,000) are used for the intensive monitoring of specific items such as surveys of the damage caused by plant diseases, insects, or tree blow-downs. Applications such as hazardous waste site assessment often require very large-scale photographs.

The principles of *photogrammetry* can be used to obtain approximate distances and ground elevations from aerial photographs using relatively unsophisticated equipment and simple geometric concepts, as well as to obtain extremely precise maps and measurements using "softcopy" photogrammetric procedures. The latter approach involves the use of digital, rather than hardcopy, images. The digital images are either acquired by a digital camera designed for photogrammetric applications or by scanning airphotos taken with a precision mapping camera.

Softcopy photogrammetric procedures are implemented in an integrated hardware and software environment providing for end-to-end spatial data capture, manipulation, analysis, storage, display, and output of softcopy images. Such systems enable automated generation of digital elevation models (DEMs), and the capture of two-dimensional and three-dimensional data for direct entry into a geographic information system (GIS). They also enable the production of perspective views (singly or in a fly-through series) and *digital orthophotos*. Digital orthophotos are images that result from pixel-by-pixel mathematical correction for the various geometric distortions present in conventional aerial photographs.

Multispectral, Thermal, and Hyperspectral Scanners

Multispectral scanners are electro-optical devices that sense selectively in multiple spectral bands using electronic detectors rather than film. They sense one small area on the ground at a time and, through scanning, build up two-dimensional images of the terrain for a swath beneath an aircraft or spacecraft. Through this process, they collect rows and columns of image data that can be computer processed. As shown in Figure 169.1, multispectral scanners can sense in a much broader spectral range than film.

Multispectral scanners are the sensing devices used in most Earth resource satellite systems operating in the optical portion of the electromagnetic spectrum (discussed later).

Thermal scanners are electro-optical devices that sense in the thermal infrared portion of the electromagnetic spectrum. They do not record the true internal temperature of objects (kinetic temperature), but rather their apparent temperature based on the radiation from their top surfaces (radiant temperature). Because they sense energy emitted (rather than reflected) from objects, thermal scanning systems can operate day or night. Multiple thermal bands can be sensed simultaneously, as in the case of the National Aeronautic and Space Administration's (NASA) *thermal infrared multispectral scanner.* Successful interpretations of thermal imagery have been made in such diverse tasks as determining rock type and structure, locating geological faults, mapping soil type and moisture conditions, determining the thermal characteristics of volcanoes, studying evapotranspiration from vegetation, locating cold water springs, determining the extent and characteristics of thermal plumes in lakes and rivers, delineating the extent of active forest fires, and locating underground coal mine fires.

Often, thermal scanners are designed and operated to acquire oblique views of the terrain ahead of an aircraft. These *forward-looking infrared* (*FLIR*) systems can be operated on a wide variety of fixed-wing aircraft and helicopters as well as from ground-based mobile platforms. FLIR systems have been used extensively in military applications. Civilian use includes such applications as fire fighting, electrical transmission line maintenance, law enforcement activities, and nighttime vision systems for automobiles.

Hyperspectral scanners acquire multispectral images in many very narrow, contiguous spectral bands throughout the visible, nearinfrared, and midinfrared portions of the electromagnetic spectrum. These systems typically collect 200 or more channels of data, which enables the construction of an effectively continuous reflectance spectrum for every pixel in the scene (as opposed to the four to six broad spectral bands commonly used by satellite systems). Research continues in the development of *ultraspectral scanners,* which employ channels numbering in the thousands.

Imaging Radar

An increasing amount of valuable environmental and resource information is being acquired by active radar systems that operate in the microwave portion of the spectrum. Microwaves are capable of penetrating the atmosphere under virtually all conditions. Depending on the specific wavelengths involved, microwave energy can penetrate clouds, fog, light rain, and smoke. Imaging radar uses an antenna pointed to the side of the aircraft or spacecraft. Because the sensor is mounted on a moving platform, it is able to produce continuous strips of imagery depicting very large ground areas located adjacent to the flight line.

Microwave reflections from Earth materials bear no direct relationship to their counterparts in the visible portion of the spectrum. For example, many surfaces that appear rough in the visible portion of the spectrum may appear smooth as seen by microwaves (e.g., a white sand beach). The appearance of various objects on radar images depends principally on the orientation of the terrain relative to the aircraft or spacecraft (important because this is a side-looking sensor), the object's surface roughness, its moisture content, and its metallic content.

Radar image interpretation has been successful in applications as varied as mapping major rock units and surficial materials, mapping geologic structure (folds, faults, and joints), discriminating vegetation types (natural vegetation and crops), determining sea ice type and condition, and mapping surface drainage patterns (streams and lakes). Imaging radar has also been used extensively as a means of topographic mapping. *Imaging radar interferometry* is the basis for this application. It is based on analysis of the phase difference of the radar signals as received by two antennas located at different positions during imaging. If the angle and distance between the two antennas is known with a high degree of accuracy, the phase difference associated with any point on the terrain can be used to determine the elevation of the point. The *Shuttle Radar Topography Mission,* conducted in 2000, used this procedure to map over 119 million km^2 of the Earth's surface, including over 99.9% of the land area between 60° N and 56° S latitude.

Radar images can also be used to study Earth surface changes that occur between two passes of a radar system. This procedure is known as *differential interferometry* and usually involves the acquisition of one interferometric image pair from the period before the surface change occurs and a third image after the change. The phase difference between the "before" and "after" images can be corrected to account for topography, with the residual phase differences then representing changes in the position of features on the surface (often measured to within 1 cm or better). This procedure has been used in such applications as monitoring ground uplift near volcanoes, measuring the displacements across faults after earthquakes, studying the movement of glaciers and ice sheets, and detecting land subsidence due to the pumping of groundwater, oil extraction, mining, and other activities.

Lidar

Lidar (which stands for *light detection and ranging*) is yet another very effective means for determining the elevation of objects on or above the Earth's surface. Early lidar systems were profiling devices that obtained elevation data only directly under the path of an aircraft. The application of these systems was limited to such activities as bathymetric mapping.

Modern lidar systems are scanning devices that are typically operated in combination with an airborne global positioning system (GPS) for x, y, z sensor location and an inertial measurement unit (IMU) for measuring the angular orientation of the sensor with respect to the ground. This enables the determination of the three-dimensional position of all points "hit" by the lidar pulses.

Most lidar systems are capable of recording multiple returns associated with each outgoing pulse. For example, in a forested area lidar can be used to discriminate not only such features as the top of the forest canopy and bare ground but also surfaces in between (such as the intermediate forest structure and the vegetation growing under the tree canopy). In urban areas, the first return of lidar data typically measures the elevations of tree canopies, building roofs, and other unobstructed surfaces. These data sets are easily acquired, processed, and made quickly available to support high-resolution contour production, and bare-earth surface elevation determination for DEM construction. A distinct advantage to lidar is that all the data are georeferenced from inception, making them inherently compatible with GIS applications.

169.4 Remote Sensing from Earth Orbit

The use of satellites as sensor platforms has made possible the routine acquisition of remotely sensed data of the Earth's surface on a global basis. The field of space remote sensing is a rapidly changing one, with numerous countries and commercial firms developing and launching new systems on a regular basis. The application of satellite image interpretation has already been demonstrated in many fields, such as agriculture, botany, cartography, civil engineering, environmental modeling and monitoring, forestry, geography, geology, geophysics, habitat assessment, land resource analysis, land use planning, oceanography, range management, and water resources.

Land remote sensing from space began on an operational basis in the civilian sector with the launch of Landsat-1 in 1972. At the time of this writing (2003), the most recently launched satellite in the Landsat series is Landsat-7. This satellite images each spot on the Earth's surface once each 16 days, providing for frequent, synoptic, repetitive, global coverage. The multispectral scanner onboard is the *Enhanced Thematic Mapper Plus* (*ETM+*), which senses in six bands of the spectrum, from the blue through the midinfrared, with a ground resolution cell size of 30 m. A seventh band senses in the thermal infrared with a ground resolution cell size of 60 m. The system also incorporates an eighth b/w or panchromatic band having a resolution of 15 m.

Launched in 1986, SPOT-1 was the first in a series of Earth resource satellites conceived and designed by the French Centre National d'Etudes Spatiales (CNES). The most recently launched SPOT satellite, SPOT-5, carries multiple sensing systems. Together, these systems provide data ranging from panchromatic images with 2.5-m resolution and a 60-km swath, to multispectral data with a resolution of 1 km and a 2250-km swath. The SPOT-5 system also features fore-and-aft stereo image acquisition and across-track pointable imaging at angles up to 31° either side of vertical. This allows both across-track stereoscopic imaging from side-by-side orbits, and more frequent opportunities to record images of any ground location.

While Landsat and SPOT data have the longest history of availability and use, many other Earth observation satellites have been operated or are planned for launch in the not-too-distant future. Those systems operating in the optical portion of the electromagnetic spectrum can be generally classified into the categories of moderate-resolution systems, high-resolution systems, and hyperspectral systems. These are complemented by existing or planned radar and lidar systems.

Moderate resolution systems include the Indian Remote Sensing (IRS) series of satellites, the RESURS series initiated by the former Soviet Union and continued by Russia, several systems launched by Japan, and the CBERS satellite systems jointly developed by China and Brazil. A partial list of countries (or agencies) operating, planning, or developing such systems includes Argentina, Australia, Brazil, Canada, China, the European Space Agency, France, Germany, India, Israel, Italy, Japan, Korea, Malaysia, Russia, South Africa, Spain, Taiwan, Thailand, United Kingdom, Ukraine, and the United States.

High-resolution systems have principally been developed and launched commercially. To date, those systems planned or launched by U.S. firms include Space Imaging's IKONOS system, the QuickBird satellite operated by EarthWatch, Inc., and ORBIMAGE's OrbView-3 system. ImageSat International has also launched the first of a planned series of EROS high-resolution satellites. Table 169.2 summarizes these systems. Except for EROS-A, all feature a panchromatic band with a spatial resolution of 1 m or better and four multispectral bands (blue, green, red, near infrared) with spatial resolution of 4 m or better. It should also be noted that it is possible to merge the finer-resolution panchromatic data with the coarser multispectral data digitally to synthesize a multispectral composite having the effective spatial resolution of the panchromatic data. This process is often called "pan-sharpening" and can be used to generate color composites very similar in appearance and utility to color and color infrared photographic images. In fact, such products are often used as a complement to, or replacement for, aerial photography in many applications.

Hyperspectral systems that have been operated from space include the Hyperion instrument carried on the EO-1 spacecraft launched by NASA. The Hyperion provides 242 spectral bands of data over the 0.36- to 2.6-μm range, each of which has a width of 0.010 to 0.011 μm. The spatial resolution of this experimental sensor is 30 m, and the swath width is 7.5 km.

TABLE 169.2 High-Resolution Earth Observation Satellite Systems

Satellite	Launch Date	Spatial Resolution (m)	Spectral Bands (μm)	Swath Width (km)	Altitude (km)
IKONOS	Sept. 24, 1999	1.0	Pan: 0.45–0.90	11.0	681
		4.0	1: 0.45–0.52		
			2: 0.52–0.60		
			3: 0.63–0.69		
			4: 0.76–0.90		
EROS-A	Dec. 5, 2000	1.8	Pan: 0.50–0.90	13.5	480
QuickBird	Oct. 18, 2001	0.6	Pan: 0.45–0.90	16.5	450
		2.4	1: 0.45–0.52		
			2: 0.52–0.60		
			3: 0.63–0.69		
			4: 0.76–0.90		
OrbView-3	2003 (planned)	1.0	Pan: 0.45–0.90	8.0	470
		4.0	1: 0.45–0.52		
			2: 0.52–0.60		
			3: 0.63–0.69		
			4: 0.76–0.90		
EROS-B1	2004 (planned)	0.82	Pan: 0.50–0.90	13.0	600
		3.48	Four multispectral bands planned		

Source: Lillesand, T. M., Kiefer, R. W., and Chipman, J. W., *Remote Sensing and Image Interpretation*, 5th ed., Wiley & Sons, New York, 2004. With permission.

Several other hyperspectral systems have been proposed or are in development. These include, among others, the U.S. Navy's Naval EarthMap Observer (NEMO) satellite and the Australian Resource Information and Environmental Satellite (ARIES). The latter system is designed to include a hyperspectral imager with up to 105 bands located between 0.4 and 2.5 μm, 30-m resolution, a 15-km swath, up to 30° cross-track pointing, and a 7-day repetitive coverage period. The future for these and similar hyperspectral satellite systems is an extremely bright one.

Radar systems operated from space have included the Almaz-1 system of the former Soviet Union, the ERS-1, ERS-2, and Envisat systems of the European Space Agency, Japan's JERS-1 system, and Canada's Radarsat system. Japan also plans to launch the PALSAR system in the not-too-distant future.

Among the precursors to these systems were the experimental spaceborne systems Seasat-1 and three Shuttle Imaging Radar systems (SIR-A, SIR-B, and SIR-C). As mentioned earlier, the Shuttle Radar Topography Mission (SRTM) was a brief but highly productive operational program to map global topography using radar interferometry.

Present and future radar systems afford a wide range of data acquisition options. Among these are various wavelength bands, the plane of polarization (vertical or horizontal) of signal transmission and reception, swath width, viewing angle, and spatial resolution. For example, the planned PALSAR system will provide multiple polarization modes at look angles (relative to vertical) of 10 to 51°, at swath widths as large as 350 km, and resolutions of 10 to 100 m. Such systems will continue the application of radar data to such tasks as sea ice reconnaissance, coastal surveillance, land cover mapping, agriculture and forest monitoring, oil slick detection, and geologic mapping.

Lidar systems are still in their infancy in terms of their application from space. The first such system, the Ice, Cloud, and Land Elevation Satellite (ICESat) includes lidars operating both in the visible and near infrared, at 0.532 μm and 1.064 μm, respectively. These systems transmit pulses about 40 times per second and measure the timing of the received returns to within 1 nanosecond. The laser footprints at ground level are approximately 70 m in diameter, and are spaced 170 m apart along-track.

ICESat is a component of NASA's Earth Observing System Program (discussed in the next section). The primary purpose of ICESat is to collect precise measurements of the mass balance of polar ice sheets, and to study how the Earth's climate affects ice sheets and sea level. The ICESat lidar can measure changes of as small as 1.5 cm in the elevation of ice sheets, over areas of 100 km by 100 km. This should permit scientists to determine the contribution of the ice sheets of Greenland and Antarctica to global sea level change to within 0.1 cm per decade. In addition to measurements of surface elevation ICESat is designed to collect vertical profiles of clouds and aerosols within the atmosphere, using sensitive detectors to record backscatter from the 0.532 μm wavelength lasers.

169.5 Earth Observing System

No discussion of space remote sensing is complete without mention of the Earth Observing System (EOS), which is part of the NASA-initiated Earth Science Enterprise (ESE). The ESE is an international earth science program aimed at providing the observations, understanding, and modeling capabilities needed to assess the impacts of natural events and human-induced activities on the Earth's environment. The program incorporates both space- and ground-based measurement systems to provide the basis for documenting and understanding global change with an initial emphasis on climate change. The program also focuses on the necessary data and information systems to acquire, archive, and distribute the data and information collected about the Earth. The intent is to further international understanding of the Earth as a system.

The EOS component of the ESE includes currently operational observing systems (beginning with Landsat-7), new programs under development, and planned programs for the future. Clearly, programs of this magnitude, cost, and complexity are subject to change. Also, the EOS program includes numerous platforms and sensors that are outside the realm of land-oriented remote sensing. We make no attempt to describe the overall program, but only wish to highlight the Moderate Resolution Imaging Spectro-Radiometer (MODIS) included on the first two EOS-dedicated platforms, the Terra and Aqua spacecraft.

Terra and Aqua were launched in late 1999 and mid-2002, respectively. MODIS is a sensor that is intended to provide comprehensive data about land, ocean, and atmospheric processes simultaneously. MODIS provides 2-day repeat global coverage with moderate spatial resolution (250, 500, or 1000 m, depending on wavelength) in 36 carefully chosen spectral bands. The total field of view of MODIS is ±55°, providing a swath width of 2330 km.

A large variety of data products can be derived from MODIS data. Among the principal data products available are the following:

- *Cloud mask* at 250 and 1000 m resolution during the day and 1000-m resolution at night.
- *Aerosol concentration and optical properties* at 5-km resolution over oceans and 10-km resolution over land during the day.
- *Cloud properties* (optical thickness, effective particle radius, thermodynamic phase, cloud top altitude, cloud top temperature) at 1- to 5-km resolution during the day and 5-km resolution at night.
- *Vegetation and land surface cover*, conditions, and productivity, defined as vegetation indices corrected for atmospheric effects; soil, polarization, and directional effects; surface reflectance; land cover type; and net primary productivity, leaf area index, and intercepted photosynthetically active radiation.
- *Snow and sea-ice cover and reflectance.*
- *Surface temperature* with 1-km resolution, day and night, with absolute accuracy goals of 0.3 to 0.5° over oceans and 1° over land.
- *Ocean color* (ocean-leaving spectral radiance measured to 5%), based on data acquired from the MODIS visible and nearinfrared channels.
- *Concentration of chlorophyll a* (within 35%) from 0.05 to 50 mg/m^3 for case 1 (deep ocean) waters.
- *Chlorophyll fluorescence* (within 50%) at surface water concentration of 0.5 mg/m^3 of chlorophyll *a*.

169.6 Digital Image Processing

The digital data acquired by the systems described in this chapter are typically computer processed to produce images through *digital image processing.* Through various image-processing techniques, digital images can be enhanced for viewing and human image interpretation. Digital data can also be processed using computer-based image classification techniques to prepare various thematic maps, such as land cover maps. Digital image processing procedures are normally integrated with the functions of a GIS.

Defining Terms

Electromagnetic radiation — The transmission of energy in the form of waves having both an electric and a magnetic component.

Electromagnetic spectrum — Electromagnetic radiation is most simply characterized by its frequency or wavelength. The resulting array is called the *electromagnetic spectrum.* The spectrum is normally considered to be bounded by cosmic rays at short wavelengths and by radio waves at long wavelength.

Emitted energy — The energy radiated by an object resulting from its internal molecular motion (heat). All objects above "absolute zero" in temperature radiate energy.

Ground truth (or reference data) — Field observations or other information used to aid or verify the interpretation of remotely sensed data.

Photogrammetry — The science, art, and technology of obtaining reliable measurements, maps, digital elevation models, thematic geographic-information-system data, and other derived products from photographs.

Pixel — The cell representing each combination of row and column (picture element) in a digital image data set.

Reflected energy — That component of incident energy that is reflected from an object.

Remote sensing — Studying the characteristics of objects from a distance by recording and analyzing electromagnetic energy, typically from ultraviolet to microwave wavelengths.

References

American Society of Photogrammetry (ASP), *Manual of Photogrammetry,* 4th ed., ASP, Falls Church, VA, 1980.

American Society of Photogrammetry (ASP), *Manual of Remote Sensing,* 2nd ed., ASP, Falls Church, VA, 1983.

American Society for Photogrammetry and Remote Sensing, *Glossary of the Mapping Sciences,* American Congress on Surveying and Mapping and American Society of Civil Engineers, Bethesda, MD, 1994.

American Society for Photogrammetry and Remote Sensing (ASPRS), *Manual of Photographic Interpretation*, 2nd ed., ASPRS, Bethesda, MD, 1995.

American Society for Photogrammetry and Remote Sensing (ASPRS), *Principles and Applications of Imaging Radar, Manual of Remote Sensing,* 3rd ed., Vol. 2, ASPRS, Bethesda, MD, 1998.

Avery, T. E. and Berlin, G. L. 1992. *Fundamentals of Remote Sensing and Airphoto Interpretation,* 5th ed., Macmillan, New York.

Barrett, E. C. and Curtis, L. F., *Introduction to Environmental Remote Sensing,* Stanley Thornes, Cheltenham, England, 1999.

Campbell, J. B., *Introduction to Remote Sensing,* 3rd ed., Guilford Press, New York, 2002.

Elachi, C., *Introduction to the Physics and Techniques of Remote Sensing.* Wiley & Sons, New York, 1987.

Jensen, J. R., *Introductory Digital Image Processing: A Remote Sensing Perspective,* 2nd ed., Prentice Hall, New York, 1996.

Jensen, J. R., *Remote Sensing of the Environment: An Earth Resource Perspective,* Prentice Hall, Upper Saddle River, NJ, 2000.

Lillesand, T. M., Kiefer, R. W., and Chipman, J. W., *Remote Sensing and Image Interpretation,* 5th ed., Wiley & Sons, New York, 2004.

Sabins, F. F., Jr., *Remote Sensing: Principles and Interpretation,* 3rd ed., Freeman, New York, 1997.

Wolf, P. R., and Dewitt, B. A., *Elements of Photogrammetry,* 3rd ed., McGraw-Hill, New York, 2000.

Further Information

Two of the leading professional societies dealing with remote sensing and photogrammetry are the American Society for Photogrammetry and Remote Sensing (ASPRS, *www.asprs.org*) and the Canadian Remote Sensing Society (CRSS, *www.casi.ca*). The ASPRS publishes the monthly journal *Photogrammetric Engineering & Remote Sensing,* as well as many books and special publications. The CRSS publishes the *Canadian Journal of Remote Sensing.*

In contrast with the ASPRS and CRRS, the International Society for Photogrammetry and Remote Sensing (ISPRS) has no individual members. It has a membership of 50 sustaining members, such as ASPRS and CRRS. The ISPRS has seven technical commissions, each of which has many working groups. The ISPRS holds numerous meetings throughout the world, and publishes many journals, proceedings, and technical reports, such as the *International Archives of Photogrammetry and Remote Sensing* and the *ISPRS Journal of Photogrammetry and Remote Sensing.* For further information, see *www.isprs.org.*

Information on the availability of cartographic and image data throughout the United States, including aerial photographs and satellite images, can be obtained from the U.S. Geological Survey, Earth Science Information Center (*http://ask.usgs.gov*).

For information on the availability of Landsat data on a worldwide basis, see *http://edc.usgs.gov* or *www.landsat.org.*

For information on the availability of SPOT data on a worldwide basis, see *www.spotimage.fr/home* or *www.spot.com.*

Information on the availability of data from various other sensing systems can be found by using an Internet search engine.

XXIV

Control Systems

170

Principles of Feedback Control

Hitay Özbay
Bilkent University, Ankara Turkey on leave from The Ohio State University

In this chapter, fundamental properties of feedback control are discussed with engineering applications in mind. However, feedback appears in many other disciplines, such as agriculture, biology, medicine, pharmaceutics, economics, business management, political science, etc. A typical example of feedback control is steering of an automobile: we would like to keep the vehicle on the road (in the cruising lane). The road curvature determines the desired path to be followed. On the other hand, unpredictable gusting winds, bumps, or potholes on the road may move the vehicle off the cruising lane, unless corrective steering action is taken. As the driver (or autopilot) detects, using vision or other sensors, deviation from the desired path he/she can take the corrective steering action. Clearly, without the feedback from vision (or other sensory mechanism), the automobile cannot be kept on the road for a long time, even if we know the desired path *a priori*. In this example, the position of the vehicle relative to the center of the lane is the system *output* to be controlled, vision or another sensor provide a measured value of the output to the controller (human driver or automatic steering mechanism). This information is processed together with the desired path to be followed (i.e., *reference input*), and then steering angle (i.e., the *control input*) is determined by the controller. The immeasurable variables such as wind, bumps, potholes, as well as measurement errors due to imprecise sensing, represent the *disturbances* in this feedback system. The main goal of feedback control is to reduce the effect of uncertainty (i.e., disturbances in the above example) on the output of the system. In fact, it is widely accepted by control engineers that the only reason to use feedback is to cope with uncertainty.

170.1 Feedback Control in Engineering

Generalizing the steering example given above, a feedback control system contains a *process* (a cause–effect relation) whose operation depends on one or more variables (inputs) that cause changes in some other variables of interest (outputs). If an input can be manipulated, then it is said to be a control input; otherwise it is considered as a disturbance, or noise. The *controller* compares the measured output (which is obtained by a *sensor*) with the desired output (reference) and produces the control input for the process; hence the feedback loop is closed (Figure 170.1). In typical engineering applications, the controller is a computer, or a human interfacing with a computer. In this case, the control signal generated by a computer

0-8493-1586-7/05/$0.00+$1.50

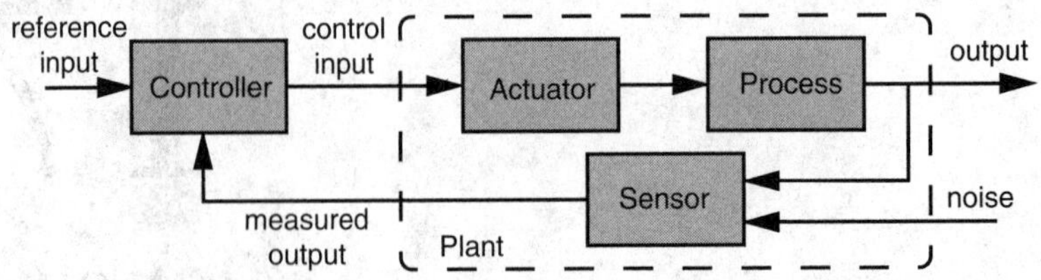

FIGURE 170.1 Feedback control system.

needs to be converted to an input that the process takes. For example, in automatic steering applications discussed above, the computer generates a small electrical signal (in the form of a voltage or current) that needs to be converted to an angle of the steering column; typically, a DC motor does the job. Such a device is called an *actuator.* The sensors and the actuators are interface devices between the physical process and the controller. They are selected by the process engineer, depending on the physical application in mind, design specifications, and economic considerations. The physical process together with the actuators and sensors form the *plant* to be controlled.

In order to design a controller, a mathematical model of the plant (describing the dynamical behavior of the system) is derived. In the next section, I discuss systems represented by ordinary differential equations with constant coefficients, that is, finite-dimensional (lumped parameter) linear, time-invariant systems. Other types of mathematical models include infinite dimensional linear systems (e.g., systems with time delays, and spatially distributed systems described by partial differential equations), nonlinear dynamical systems, and fuzzy systems described by a set of logical rules. Control design techniques used depend on the type of dynamical equations representing the plant.

Typical goals in feedback controller design include:

- Stability of the closed loop system
- Reduction of sensitivity to modeling uncertainty
- Disturbance attenuation
- Tracking of a reference input

Clearly, in controller design the biggest concern is the first item in the above list. If the controller is not designed properly, a stable open-loop system (i.e., stable controller and stable plant connected in cascade with no feedback) may become unstable when the loop is closed with feedback. This is the main reason why feedback control is an important research area in engineering and applied mathematics.

In practice, after a controller is designed for a physical system, it is tested under different operating conditions by performing numerical simulations and laboratory experiments. If the results are satisfactory, then the controller is deployed and the feedback loop is formed around the actual physical system. Otherwise, the mathematical model is refined, and/or some of the design specifications are relaxed in order to satisfy all the design requirements. A summary of this process is shown in Figure 170.2.

170.2 Fundamentals of Feedback for Linear-Time-Invariant Systems

In this section, to illustrate the benefits of feedback, we consider a simple control system, where the plant and the controller are single input–single output (SISO), linear, time-invariant systems whose transfer functions are rational functions. To further simplify the notation and discussion, we assume that the sensor is perfect (measured output is the same as the output of the process). Then, the feedback system shown in Figure 170.1 reduces to the system shown in Figure 170.3, where $C(s)$ and $P(s)$ are the transfer functions of the controller and the plant, respectively. (Plant transfer function is defined as $P(s) = Y(s)/U(s)$, where $Y(s)$ and $U(s)$ represent the Laplace transforms of the output $y(t)$, and the input $u(t)$, respectively.) The tracking error, e, is defined as the difference between the reference input r and the

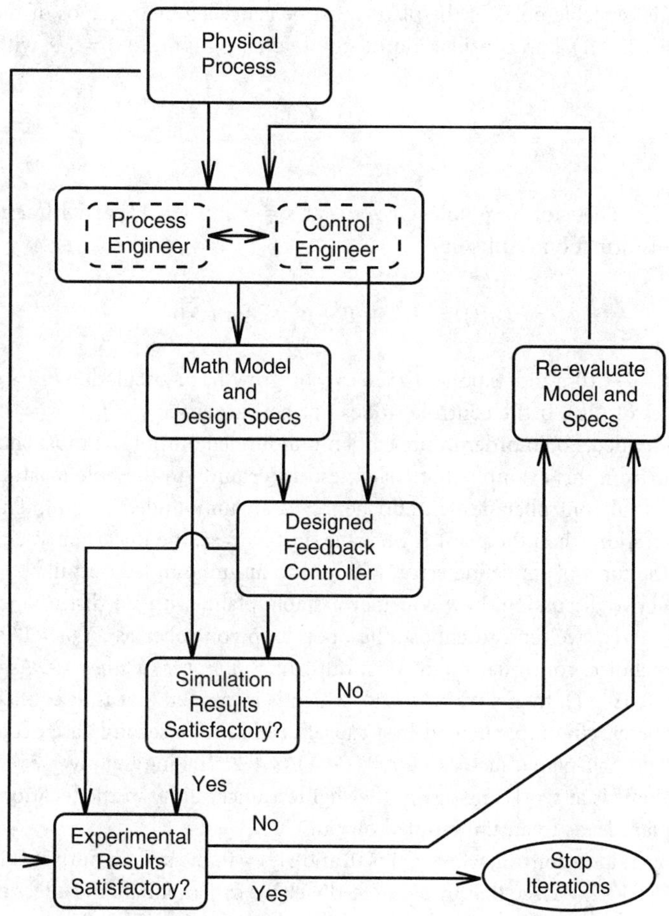

FIGURE 170.2 Controller design iterations.

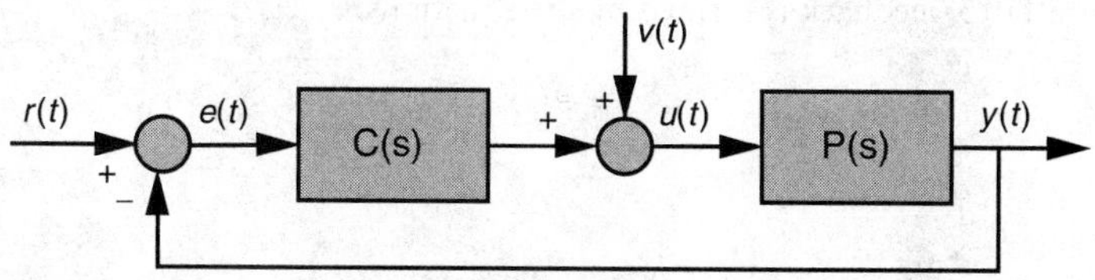

FIGURE 170.3 Feedback system with controller *C* and plant *P*.

output y, that is, $e(t) = r(t) - y(t)$. In Figure 170.3, the control input is u, and the disturbance is v. We want to make the tracking error "small" in the presence of disturbance and modeling errors in the plant.

Open Loop Control

First, consider the open loop control, where the Laplace transform of the output is given by

$$Y(s) = P(s)C(s)R(s) + P(s)V(s)$$

Clearly, when the plant is unstable, a small disturbance may result in an unbounded output. (A system is stable if its transfer function has no poles in $\{s : \text{Re}(s) \geq 0\}$.) Even if we assume that $v(t) = 0$, to get a bounded output, the unstable poles of the plant must be cancelled by the zeros of the controller. As an example, let $P(s) = 1/(s - 1)$, and consider a unit step reference input, $R(s) = 1/s$, with $V(s) = 0$. Then,

$$Y(s) = \frac{C(c)}{s(s-1)}$$

Assuming that $C(s)$ does not have poles or zeros at $s = 0$ and $s = 1$, partial fraction expansion and inverse Laplace transformation results in

$$y(t) = Ae^t + B + y_c(t), \text{ for } t > 0$$

where $A = C(1)$, $B = -C(0)$; and exponential decay, or growth, or oscillations in $y_c(t)$ depend on the location of the poles of $C(s)$. If the controller does not have a zero at $s = 1$, that is, if $C(1)$ is not equal to zero, $y(t)$ is unbounded. So, in order to prevent an unbounded output, we must choose $C(s)$ such that $C(1) = 0$. It is clear from this example that a slightest uncertainty in the pole location of the plant will lead to an unsuccessful controller design, and hence to an unbounded output. Thus, the open loop control is out of question when the plant is unstable; in this case, the uncertainty, either in the form of disturbance or in the form of modeling error, will lead to an unbounded output.

In light of the above discussion, now consider a stable plant, with no disturbance, $v = 0$. Then, in order to make $y(t) \approx r(t)$, we tend to choose the open loop controller as $C(s) = 1/P(s)$. There are two problems with this choice: controller might turn out to be *improper* or *unstable.* As an example of the first case, let $P(s) = 1/(s + 1)$, then $C(s) = (s + 1)$, which is improper, that is, it contains a differentiator, and hence cannot be exactly implemented in a causal fashion. The second case occurs when the plant has a zero in the right half-plane, such as $P(s) = (s - 1)/(s + 2)$. Taking the inverse of this plant leads to a controller whose pole is at $s = 1$, and again, the slightest uncertainty in the location of this right half-plane zero of the plant leads to an unbounded output.

Finally, for the open loop control, when $r(t) = 0$, and $v(t) \neq 0$, the system output is independent of the controller: $Y(s) = P(s)V(s)$. We see that by using feedback we can attenuate the effect of this disturbance.

Feedback Control

As shown in Figure 170.3, feedback is formed by the equations

$$E(s) = R(s) - Y(s)$$

and

$$Y(s) = P(s)C(s)E(s) + P(s)V(s)$$

Therefore, the output is determined from the identity

$$(1 + P(s)C(s))Y(s) = P(s)C(s)R(s) + P(s)V(s)$$

Note that the right side of the above equality corresponds to the output of the open-loop control system discussed before. In order to determine $Y(s)$ uniquely, the operator $(1 + P(s)C(s))$ must be causally invertible; otherwise the feedback system is said to be ill-posed. For example, when $P(s) = K_p$ and $C(s) = -1/K_p$, the feedback system is ill-posed. Another example of an ill-posed systems is $P(s) = 2s/(s + 1)$, $C(s) = -1/2$, in which case the inverse of $(1 + P(s)C(s))$ is $(s + 1)$, an improper transfer function. In general, the *feedback system becomes ill-posed when*

$$1 + P(\infty)C(\infty) = 0$$

From now on we assume that the feedback system is well-posed; and we define the *sensitivity function*, $S(s)$, as the inverse of $(1 + P(s)C(s))$. The word sensitivity will be defined shortly, and hence this notation will be made clear.

In a well-posed feedback system the internal signals of interests are

$$E(s) = S(s)R(s) - S(s)P(s)V(s)$$

$$U(s) = C(s)S(s)R(s) + S(s)V(s)$$

Stabilization

We say that a feedback system is bounded input/bounded output (BIBO) stable if every bounded input pair $(r(t), v(t))$ leads to bounded internal signals $(e(t), u(t))$. (A signal $f(t)$ is said to be bounded if there exists a finite number M_f satisfying $|f(t)| < M_f$.)

It can be shown that the feedback system is BIBO stable if and only if transfer functions $S(s)$, $P(s)S(s)$, and $C(s)S(s)$, are stable, that is, they have no poles in the closed right half-plane, $\{s : \text{Re}(s) \geq 0\}$. It is important to note that stability of the plant and/or controller does not imply stability of the feedback system; conversely, stability of the feedback system does not imply stability of the plant nor the controller. In fact, one of the reasons to use feedback is to stabilize an unstable plant. For example, let $P(s) = 1/(s - p)$, with $p > 0$, and $C(s) = K_c$, it is a simple exercise to show that closed loop system poles, that is, the poles of S, PS, and CS, are the roots of

$$s - p + K_c = 0$$

Thus, with a simple gain $K_c > p$, we can stabilize the system using feedback. In order to have stability robustness against a possible uncertainty in p, we might want to choose K_c much greater than the maximum known value of p. Recall that with an open loop control scheme it is impossible to stabilize this system in the presence of uncertainty in p.

In general, feedback system stability can be determined by finding the closed loop system poles. Let $P = N_p/D_p$, and $C = N_c/D_c$, where N_p, D_p, N_c, D_c are polynomials, and the pairs (N_p, D_p) and (N_c, D_c) are coprime. (A pair of polynomials (N, D) is said to be coprime if $N(s)$ and $D(s)$ do not have common roots.) Then, it is easy to see that all the poles of S, PS, and CS are precisely the roots of the characteristic equation

$$D_p(s)D_c(s) + N_p(s)N_c(s) = 0$$

Once this equation is constructed, numerical root-finding tools can be used to calculate the closed-loop system poles. More discussion on stability analysis and controller design can be found in subsequent chapters of this handbook, where different design/analysis techniques are studied for more general classes of systems. In the remaining sections of this chapter on the fundamental properties of feedback, it is assumed that the feedback system is stable.

Tracking

One of the goals of feedback control is to make the output, $y(t)$, track the reference input, $r(t)$. Therefore, we would like to design the controller in such a way that the tracking error, $e(t) = r(t) - y(t)$, is "small." When $v(t) \equiv 0$, the Fourier transform of the error is $E(j\omega) = S(j\omega)R(j\omega)$. Unless $S \equiv 0$, the error cannot be zero. So, the best we can do is make $|S(j\omega)|$ "small" whenever $|R(j\omega)|$ is not. For example, consider a unit step reference, $R(s) = 1/s$, and let $P(s) = 1/(s - 1)$. Since $R(0) = \infty$, we need to have $S(0) = 0$, so that $E(0) \neq \infty$. Since the plant does not have a pole at $s = 0$, in order to make $S(0) = 0$ the controller needs to have a pole at $s = 0$. Accordingly, we choose a PI (proportional plus integral) controller whose transfer function is in the form $C(s) = K(s + z)/s$. The characteristic equation is

$$s^2 + (K - 1)s + Kz = 0$$

For stability we need $K > 1$, and $z > 0$. Moreover, we can place the closed loop poles anywhere in the complex plane by properly selecting K and z. The tracking error is given by

$$E(s) = \frac{s-1}{s^2 + (K-1)s + Kz}$$

If we place one of the closed loop poles at $s = -1$, and the other one at $s = -r$, then we have

$$|E(j\omega)| = \left|\frac{1}{j\omega + r}\right| \leq 1/r$$

This is achieved when $K = r + 2$, and $z = r/(r + 2)$. Note that the right side of the above inequality can be made arbitrarily small by selecting large values of r; hence, the tracking error can be made small. On the other hand, it is a simple exercise to check that for the above design

$$u(t) = -1 + A \exp(-rt) + B \exp(-t)$$

where $A = K(z - r)(r + 1)/(r(r - 1))$, and $B = 2K(1 - z)/(r - 1)$, and when r is large we have $A \approx K = r + 2$. That leads to large values of $|u(t)|$ in the transient response, possibly resulting in severe actuator damage or saturations. To summarize the discussion, in order to obtain good tracking performance we need use high gain controllers. However, the gain of the controller should not exceed a certain limit, determined by the actuator's capacity to handle large inputs.

Disturbance Attenuation

In this section we assume that $r(t) = 0$, and $v(t) \neq 0$. Recall that in open loop control the output is $Y(s) = P(s)V(s)$, and hence the controller does not play a role in reducing the effect of disturbance on the output. However, in feedback control, the output is $Y(s) = S(s)P(s)V(s)$. Compared to the open loop response the output is scaled by a factor of $|S(j\omega)|$ in the frequency domain. Therefore, as in the tracking discussion, whenever $|V(j\omega)|$ is "large" we need to make $|S(j\omega)|$ "small" by a proper choice of the controller $C(s)$. This way, the effect of the disturbance on the output can be reduced significantly. As an example, consider a unit step disturbance $V(s) = 1/s$, for the unstable plant $P(s) = 1/(s - 1)$, and let the feedback controller be the same PI controller defined above $C(s) = K(s + z)/s$. Then, the output due to disturbance v is obtained as $Y(s) = 1/(s^2 + (K - 1)s + Kz)$. Again, choosing $K = r + 2$ and $z = r/(r + 2)$, we have

$$|Y(j\omega)| = \left|\frac{1}{(j\omega + r)(j\omega + 1)}\right| \leq \frac{1}{r}$$

Hence the output can be made as small as desired, provided that the actuator is not pushed to its limits.

Sensitivity

Consider a function F, which depends on a parameter α. Sensitivity of F with respect to variations of α is defined as

$$S_F^\alpha = \frac{\Delta F / F}{\Delta\alpha / \alpha} = \frac{\alpha}{F}\frac{\partial F}{\partial \alpha}$$

where ΔF represents variations in F due to variations in α by an amount $\Delta\alpha$, at the nominal value of α. Therefore, the functions defined above need to be evaluated at the nominal value of α. Now let F to be the transfer function from r to y, denoted by T, and the parameter α be the plant transfer function, P. In the open loop case we have $T = PC$, and in the feedback control case $T = 1 - S = PC/(1 + PC)$. Since P is simply a "model" of the plant, it does not capture the "true" behavior of the actual plant. Accordingly, assume that the "true" plant has a transfer function in the form $P + \Delta P$, where ΔP represents the modeling uncertainty. By applying the above definition of the sensitivity, we find that sensitivity of T with respect to variations in P is 1 for the open loop control, and it is equal to $1/(1 + PC)$ for feedback control. Recall that $1/(1 + PC) = S$, which we called the "sensitivity" function for the now obvious reason. Having unity sensitivity (open loop case) means that, any percentage variation of P will result in the same percentage variation in T. On the other hand, a properly designed controller can reduce the sensitivity (i.e., smaller than unity) in the frequency region of interest. Note that, by definition of the sensitivity, we have

$$\frac{\Delta T}{T} = S\frac{\Delta P}{P}$$

and hence

$$\Delta T = S\ \Delta P\ C\ S$$

Thus, we need to make the sensitivity $|S(j\omega)|$ "small" whenever $|\Delta P(j\omega)C(j\omega)|$ is "large." This property is also needed for stability robustness in the presence of plant uncertainty. For details, see subsequent chapters in this volume, and sources listed below.

References and Further Reading

Principles of feedback control and basic feedback system analysis and design techniques can be found in all undergraduate level textbooks on feedback control systems. Some of the newer texts include detailed discussions on the links between sensitivity and robustness, a topic that has been traditionally left for a second course on feedback control. A partial list of introductory textbooks on feedback control systems is given below.

P. R. Bélanger, *Control Engineering: A Modern Approach*, Saunders College Publishing, Fort Worth, TX, 1995.

R. C. Dorf and R. H. Bishop, *Modern Control Systems*, 9th ed., Prentice Hall, New York, 2001.

J. C. Doyle, B. A. Francis, and A. R. Tannenbaum, *Feedback Control Theory*, Macmillan, New York, 1992.

G. F. Franklin, J. D. Powell, and A. Emami-Naeini, *Feedback Control of Dynamic Systems*, 4th ed., Prentice Hall, New York, 2002.

G. C. Goodwin, S. F. Graebe, and M. E. Salgado, *Control System Design*, Prentice Hall, New York, 2001.

B. C. Kuo, *Automatic Control Systems*, 7th ed., Prentice Hall, New York, 1995.

K. Ogata, *Modern Control Engineering*, 4th ed., Prentice Hall, New York, 2002.

H. Özbay, *Introduction to Feedback Control Theory*, CRC Press, Boca Raton, FL, 2000.

K. Morris, *Introduction to Feedback Control*, Harcourt/Academic Press, New York, 2001.

C. E. Rohrs, J. L. Melsa, and D. G. Schultz, *Linear Control Systems*, McGraw-Hill, New York, 1993.

New developments in feedback control theory and their applications in many engineering disciplines are reported regularly in academic journals. The most prominent journals in this area, most of which are published monthly, are listed below.

IEEE Transactions on Automatic Control
IEEE Transactions on Control Systems Technology
Automatica, journal of IFAC

Control Engineering Practice, journal of IFAC
SIAM Journal on Control and Optimization
Systems & Control Letters
Mathematics of Control Signals and Systems
International Journal of Control
International Journal of Robust and Nonlinear Control
International Journal of Adaptive Control and Signal Processing
European Journal of Control
AIAA Journal of Guidance Control and Dynamics
ASME Journal of Dynamic Systems Measurement and Control

The two largest professional societies active in the control systems area are:

Control Systems Society of IEEE (Institute of Electrical and Electronics Engineers): *www.ieeecss.org*
IFAC, International Federation of Automatic Control: www.ifac-control.org

Besides publishing some of the academic journals listed above, these societies regularly hold meetings devoted to the latest developments in the control systems area. Some of the major control conferences are:

Conference on Decision and Control, held once a year in December
American Control Conference, held once a year in early summer
Conference on Control Applications, held once a year, typically in the summer
European Control Conference, held once a year late summer
IFAC World Congress, held once every 3 years, typically in the summer.

171

Root Locus

Desineni Subbaram Naidu
Idaho State University

171.1 Introduction

In general, the root locus is the locus of the roots of an algebraic equation with constant coefficients when the value of a parameter of the equation is changed between two limits. With particular reference to the field of control system, the root locus was invented by Walter Evans during the 1950s [1–3] for which he was honored with the prestigious Bellman Control Heritage Award in 1988 by American Automatic Control Council (AACC) and the paper [2] was reprinted in the recent volume *Control Theory: Twenty-Five Seminal Papers (1932–1981)* [4]. Basically, the root locus method gives the information regarding closed-loop stability by showing the location of the closed-loop poles or the roots of characteristic equation of a closed-loop control system for variation of the values of a parameter of the system. The root locus technique was developed for both continuous-time systems described by ordinary differential equations and Laplace transforms with s as the complex variable and discrete-time systems described by ordinary difference equations and Z- transforms with z as the complex variable. This article is based on the author's teaching experience in the area of control systems and the well-known texts such as those found in References 5 through 7.

171.2 Concept and Definition of Root Locus

We first address root locus for continuous-time systems in detail and then briefly deal with the root locus for discrete-time system. Consider a negative feedback control system shown in Figure 171.1. Here, a standard notation is used with $s = \sigma + j\omega$ as Laplace complex variable, $R(s)$ and $Y(s)$ are reference input and the actual output, respectively. Further, K is any parameter associated with the forward transfer function $G(s)$ and $H(s)$ is the feedback transfer function. It is well-known that

$$\frac{Y(s)}{R(s)} = M(s) = \frac{KG(s)}{1+KG(s)H(s)} \tag{171.1}$$

0-8493-1586-7/05/$0.00+$1.50

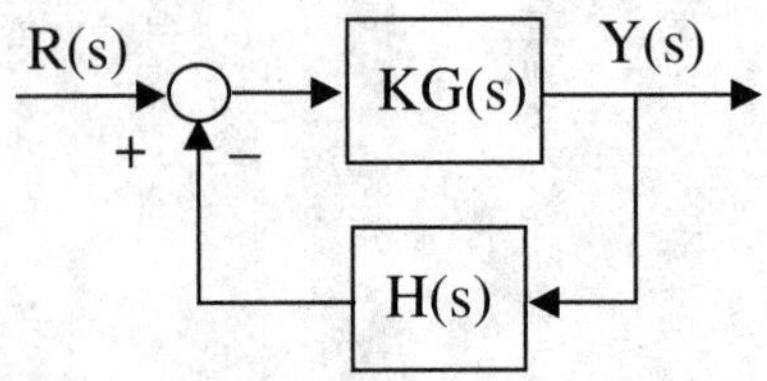

FIGURE 171.1 Negative feedback control system.

The stability of this closed-loop system requires that the location of the poles of the transfer function $M(s)$ or the roots of the characteristic equation (C.E.)

$$F(s) = 1 + KG(s)H(s) = 0 \tag{171.2}$$

lie in the left half s plane. Let us note that if the original open-loop transfer function, say $G_1(s)H_1(s)$, is embedded with a parameter K, it can always be rewritten as $G_1(s)H_1(s) = KG(s)H(s)$. Also, if $KG(s)H(s)$ can be written as a rational function as

$$KG(s)H(s) = \frac{KP(s)}{Q(s)} \tag{171.3}$$

then, the C.E. (171.2) can be written as

$$Q(s) + KP(s) = 0. \tag{171.4}$$

Concept of Root Locus

In the context of systems and control, the root locus is *defined* as the locus (or trajectory) of the roots of the characteristic equation of a negative feedback control system for all positive (and/or negative) values of a parameter associated with the system transfer function.

Thus, the root locus for the negative-feedback control system (171.1) is the locus of the roots of the characteristic equation (171.2) for all positive (and/or negative) values of the parameter K. To illustrate the concept of the root locus, let us consider a simple second order system $G(s) = 1/s(s + 2)$ with open-loop poles at $s = 0, -2$ and $H(s) = 1$. Then the C.E. (171.2) becomes

$$F(s) = s^2 + 2s + K = 0. \tag{171.5}$$

The roots s_1 and s_2 of (171.5) can be easily found as

$$s_{1,2} = -1 \pm \sqrt{1 - K} \tag{171.6}$$

for various positive values of K tabulated as shown. The locus of the roots shown in Table 171.1 is sketched in Figure 171.2. It is clear from Figure 171.2 that the locus (trajectory) of the roots s_1 and s_2 (171.6) of the C.E. (171.5) for values of $K = 0$ to ∞ starts (corresponding to $K = 0$) at the open-loop poles located at $s = 0$ and $s = -2$ and travels through the values corresponding to $K = 0.5, 1, 2, 3, \infty$. Since

TABLE 171.1 Roots of the Characteristic Equation (171.5)

K	0	0.5	1	2	3	∞
s_1	0	−0.293 + j0	−1 + j0	−1 + j1	−1 + j1.414	−1 + j∞
s_2	−2	−1.707 − j0	−1 − j0	−1 − j1	−1 − j1.414	−1 − j∞

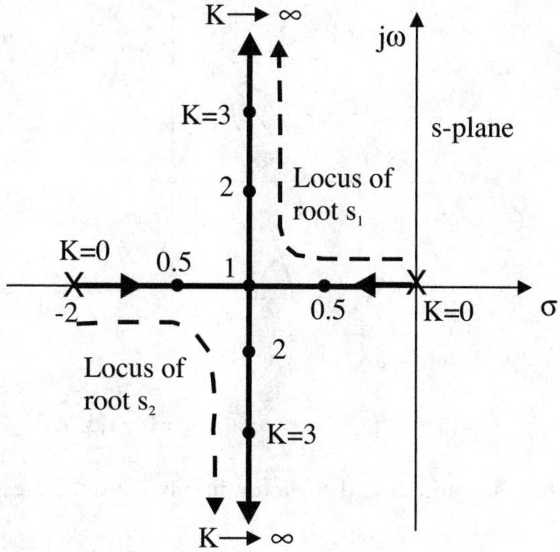

FIGURE 171.2 Root locus concept: Second order system.

the roots of the C.E. for all values of K lie in the left-half of the s-plane, the second-order, negative-feedback control system (171.1) is stable for all positive values of K. Thus, one of the most important useful characteristics of the root locus is to find the stability of the closed-loop system *once for all* for all values of $K = 0$ to ∞.

Definition of Root Locus

The root locus is *defined* as the locus (trajectory) of the roots of the characteristic equation of a feedback control system. Note that the above procedure is given only to get the concept of the root locus and the root locus for any general feedback control system is not obtained *directly* by finding the roots of the C.E. and then plotting or sketching the same in the s-plane as shown above, but is done *indirectly* by developing some rules/steps for constructing or sketching the root locus as given next.

Rules/Steps for Constructing Root Locus

Let us reconsider the C.E. (171.2) written as

$$G(s)H(s) = -\frac{1}{K}. \tag{171.7}$$

and focus only on positive values of $0 \le K \le \infty$ noting that a similar treatment (called *complementary root locus*) can be developed for negative values $-\infty \le K < 0$. Note that $G(s)H(s)$ is a complex number and as such has a *magnitude* and *angle* or *phase*. This leads us to write the *one* relation (171.7) in terms of *two* relations as [8] (where, $|\,|$ indicates *magnitude* and $\angle$ indicates the *angle*)

$$|G(s)H(s)| = \left|-\frac{1}{K}\right| = \frac{1}{K}, \qquad 0 \le K < \infty, \qquad \text{Magnitude Condition}$$

$$\angle G(s)H(s) = \angle\left\{-\frac{1}{K}\right\} = \angle\{-1 + j0\}, \quad 0 \le K < \infty, \qquad \text{Angle Condition} \tag{171.8}$$

$$= \text{odd multiple of } \pi,$$

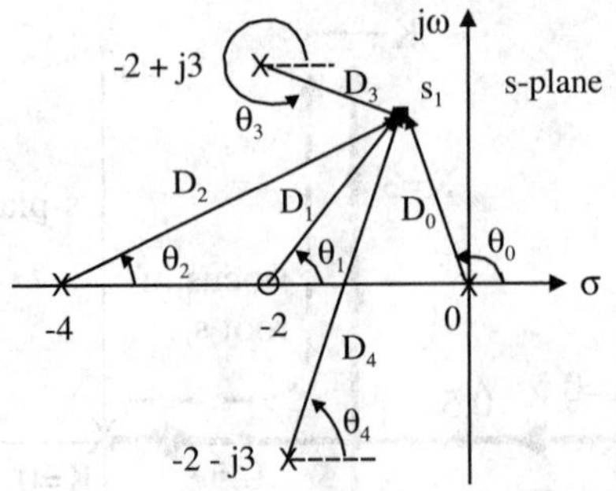

FIGURE 171.3 Magnitude and angle conditions of root locus.

$$= (2r+1)\pi, \qquad r = 0, \pm 1, \pm 2, \ldots \tag{171.9}$$

To further illustrate the above magnitude and angle conditions, consider a feedback control system

$$KG(s)H(s) = \frac{K(s+2)}{s(s+4)(s^2+4s+13)} = \frac{K(s+2)}{s(s+4)(s+2-j3)(s+2+j3)} \tag{171.10}$$

and any point s_1 on the root locus in the s-plane as shown in Figure 171.3. Note that the distances from s_1 to various open-loop poles and zero are given as $D_0 = s_1$, $D_1 = s_1 + 2$, $D_2 = s_1 + 4$, $D_3 = s_1 + 2 - j3$, $D_4 = s_1 + 2 + j3$ and the corresponding angles are shown as $\theta_0, \theta_1, \ldots, \theta_4$. Then the magnitude condition (171.8) implies that

$$\left|\frac{1}{K}\right| = \left|\frac{(s_1+2)}{s_1(s_1+4)(s_1+2-j3)(s_1+2+j3)}\right| \rightarrow$$

$$K = \left|\frac{s_1(s_1+4)(s_1+2-j3)(s_1+2+j3)}{(s_1+2)}\right| = \frac{D_0 D_2 D_3 D_4}{D_1} \tag{171.11}$$

$$= \frac{\text{Product of the distances from all the poles of } G(s)H(s) \text{ to the point } s_1}{\text{Product of the distances from all the zeros of } G(s)H(s) \text{ to the point } s_1}$$

Similarly, with respect to the Figure 171.3 the angle condition (171.9) implies that

$$\angle G(s_1)H(s_1) = \angle\left\{\frac{(s_1+2)}{s_1(s_1+4)(s_1+2-j3)(s_1+2+j3)}\right\}$$

$$= \theta_1 - [\theta_0 + \theta_2 + \theta_3 + \theta_4] \tag{171.12}$$

$$= \text{odd multiple of } \pi$$

Next, we develop various rules/steps required for constructing the root locus. Also, we first consider an illustrative (or running) example described by

$$KG(s)H(s) = \frac{K(s+2)}{s(s+4)(s+6)(s^2+6s+34)} = \frac{K(s+2)}{s(s+4)(s+6)(s+3-j5)(s+3-j5)} \tag{171.13}$$

which will be used to demonstrate the various rules/steps involved in the construction of the root locus. The reason for selecting a high-order system (171.13) is that this example illustrates most of the rules, steps, or properties of root locus. Also, note that the root locus and root loci will be used interchangeably.

Rule/Step 1: Starting Points of Root Locus

From the magnitude condition (171.8), the starting points of the root locus which correspond to the value of $K = 0$ are given by

$$|G(s)H(s)| = \frac{1}{K=0} = \infty \qquad (171.14)$$

meaning that the starting points of the root loci are those values of s for which $G(s)H(s)$ becomes infinity or the *poles* of $G(s)H(s)$.

In the illustrative example (171.13), the starting points of the root locus are at the poles of $G(s)H(s)$ or at $s = 0, -4, -6, -3 + j5, -3 - j5$, which also means that there are 5 (as many as there are poles) branches of root loci. A branch of root locus is the locus of one root of the C.E. as K changes from zero to infinity.

Statement of Step 1: The root locus starts at the poles of the open-loop transfer function KG(s)H(s).

Rule/Step 2: Ending Points of Root Locus

From the magnitude condition (171.8), the ending points which correspond to the value of $K = \infty$ are given by

$$|G(s)H(s)| = \frac{1}{K=\infty} = 0 \qquad (171.15)$$

meaning that the ending points of the root loci are those values of s for which $G(s)H(s)$ becomes zero or the *zeros* of $G(s)H(s)$. In the illustrative example (171.13), the ending points of the root locus are at the zeros of $G(s)H(s)$ or at $s = 2$. Note that if there are 5 branches of root loci as per (171.14), then all these 5 branches need to end at 5 zeros. This means that if there is only 1 finite zero (at $s = 2$) for the system (171.13), then the remaining 4 branches of root loci starting at the poles of $G(s)H(s)$ end at 4 zeros located at infinity.

Statement of Step 2: The root locus ends at the zeros (both finite and infinite) of the open-loop transfer function KG(s)H(s).

Rule/Step 3: Number of Branches of Root Locus

Since the number of branches must equal the number of roots or the order of the C.E., the number of branches of root locus much equal the *order* of the C.E. or the number of open-loop poles or zeros, whichever is higher.

In this particular example (171.13), the total number of branches is 5 the number of open-loop poles.

Statement of Step 3: The total number of branches of the root locus is equal to the number of open-loop poles or zeros, whichever is higher.

Step/Rule 4: Root Locus on Real Axis

Considering any point s_1 on the real axis of the s-plane, it can be easily seen that the angle condition (171.9) or (171.12) is satisfied only for that section on the real axis for which the *total* number of poles and zeros of $G(s)H(s)$ to the *right* of that section is *odd*.

In the present example (171.13), the root locus exists on the real axis on the sections between $s = 0$ and $s = -2$ and $s = -4$ and $s = -6$ as shown in Figure 171.4.

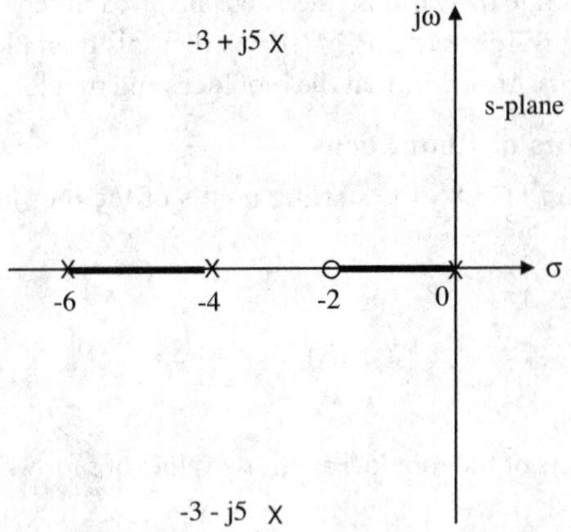

FIGURE 171.4 Root locus on real axis.

Statement of Step 4: The root locus exists on a section of the real axis if the total number of poles and zeros of KG(s)H(s) to the right of the section is odd.

Rule/Step 5: Symmetry of Root Locus

Since the root locus is drawn in complex (*s*-) plane, and the roots not on the real axis are always complex-conjugates, the root locus is symmetrical w.r.t. the real axis of the *s*-plane.

Statement of Step 5: The root locus is symmetrical w.r.t. the real axis in s plane.

Rule/Step 6: Asymptotes of Root Locus

The nature of the root locus near infinity in the *s*-plane is determined by the asymptotes. There are two properties of the asymptotes: (i) angles of asymptotes and (ii) intersection of asymptotes with real axis.

(i) Angles of Asymptotes

These angles for $s \to \infty$ are given by

$$\theta_r = \frac{(2r+1)\pi}{|n-m|}, \qquad r = 0,1,2,\ldots,|n-m|-1, \tag{171.16}$$

where, *n* and *m* refer to the number of finite poles and zeros respectively of *KG*(*s*)*H*(*s*).

In the present example (171.13), we have $n = 5$ and $m = 1$ and $n - m - 1 = 3$, so that the angles for asymptotes (171.16) are given by (stopping at $\theta_r = \theta_3$)

$$\theta_0 = \frac{\pi}{4} = 45^0, \theta_1 = \frac{3\pi}{4} = 135^0, \theta_2 = \frac{5\pi}{4} = 225^0 = -135^0, \theta_3 = \frac{7\pi}{4} = 315^0 = -45^0 \tag{171.17}$$

(ii) Intersection of Asymptotes with Real Axis

The intersection σ_a of the asymptotes on the real axis of the *s*-plane is given by

$$\sigma_a = \frac{\text{Sum of finite poles of } G(s)H(s) - \text{Sum of finite zeros of } G(s)H(s)}{n-m} \tag{171.18}$$

For the running example (171.13), we have

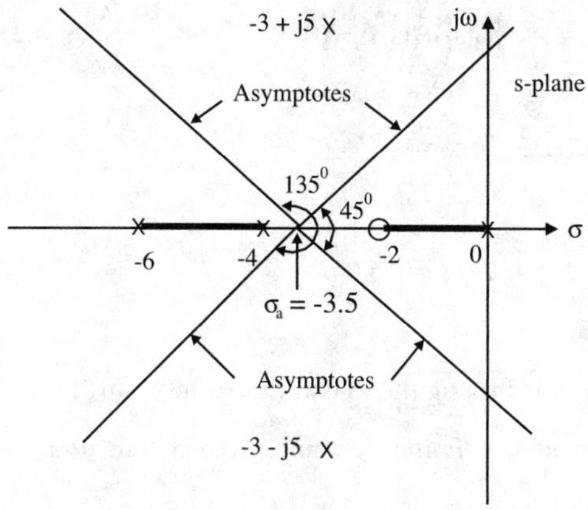

FIGURE 171.5 Asymptotes of root locus.

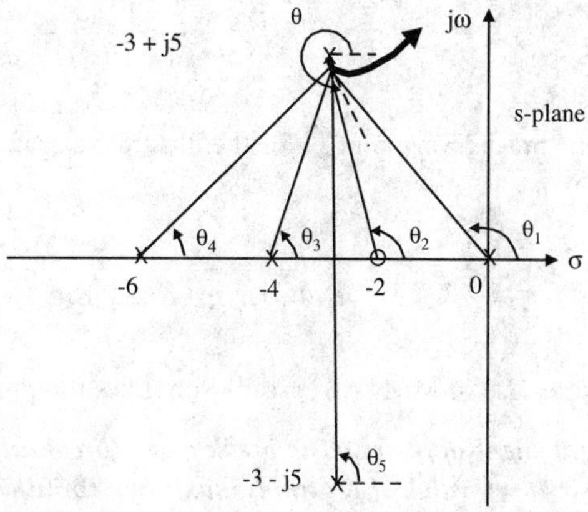

FIGURE 171.6 Angles of departure for root locus.

$$\sigma_a = \frac{[0-4-6-3+j5-3-j5]-[-2]}{5-1} = -3.5 \tag{171.19}$$

The situation at this stage is depicted in Figure 171.5.

Statement of Step 6: The angles and the intersection of asymptotes are given by (171.16) and (171.18).

Rule/Step 7: Angles of Departure and Arrival

The method of finding the starting and ending points of the root locus on the real axis or from the real poles and zeros is straight forward and is clearly found from the earlier Steps 1–2, whereas the method of finding starting and ending points for the root locus from the complex-conjugate poles and zeros is not that straightforward. The angles of departure (from the complex poles) and the angles of arrival (at complex zeros) are found by considering a point s_1 very close to these complex-conjugate poles/zeros and applying the angle condition (171.9). Thus, consider a point s_1 very close to the complex pole at $-3 + j5$ as shown in Figure 171.6. Applying the angle condition (171.9) to the point s_1, we have

$$\begin{aligned}
&\angle(s_1+2)-\left[\angle s_1+\angle(s_1+4)+\angle(s_1+6)+\theta+\angle(s_1+3+j5)\right]=\text{odd multiple of }\pi,\\
&\theta_2-\left[\theta_1+\theta_3+\theta_4+\theta+\theta_5\right]=\text{odd multiple of }\pi,\\
&180-tan^{-1}\left(\frac{5}{3-2}\right)-\left[180-tan^{-1}\left(\frac{5}{3}\right)+tan^{-1}\left(\frac{5}{4-3}\right)+tan^{-1}\left(\frac{5}{6-3}\right)+\theta+90^0\right]=\pi,\\
&\theta=-67.4^0
\end{aligned} \tag{171.20}$$

Statement of Step 7: The angles of departure from the complex poles and the angles of arrival at complex zeros are found using the angle condition (171.9).

Rule/Step 8: Break-Away or Break-In Points on Real Axis

The root locus, starting from a pair of poles (ending at a pair of zeros) on the real axis, for some values of K, will break away (break in) from (to) a point s_1. This is found using the fact that the value of K is maximum (or minimum) at the break away (or break in) point. This is easily found first by rewriting the C.E. (171.2) in terms of K and then using simple calculus of finding the maximum (or minimum) of K at the point s_1 as

$$K = -\frac{1}{G(s_1)H(s_1)} \rightarrow \frac{dK}{ds_1} = 0 \rightarrow \frac{d}{ds_1}\{G(s_1)H(s_1)\} = 0 \tag{171.21}$$

The break away point s_1, for the illustrative example (171.13) becomes

$$\frac{d}{ds_1}\left\{\frac{(s_1+2)}{s_1(s_1+4)(s_1+6)(s_1^2+6s_1+34)}\right\} = 0 \rightarrow s_1 = -5.0024. \tag{171.22}$$

Note: Use of MATLAB© greatly simplifies the procedure for finding the maximum (minimum) at s_1.

Statement of Step 8: The break away (break in) point at s_1 is found by using the fact that at s_1, the value of K attains maximum (minimum).

Rule/Step 9: Intersection of Root Locus with $j\omega$ Axis

The value of $K = Kc$ corresponding to the point of intersection of the root locus with the $j\omega$ axis in the s-plane indicates the critical point for stability of the closed-loop system. Beyond this value of Kc (or for some range of values of Kc), the roots of the C.E. lie in the right-half of the s plane leading to instability. This critical point is easily found using the well-known Routh-Hurwitz test for the C.E. (171.2). For this example (171.13), the C.E. (171.2) is

$$s^5 + 15s^4 + 118s^3 + 484s^2 + (816 + K)s + 2K = 0 \tag{171.23}$$

Using the Routh-Hurwitz [6], we get the following three conditions

$$K > 0;\ K < 2101.07;\ K < 1356.16 \text{ or } K < -1444.83 \tag{171.24}$$

yielding the result as

$$0 < K_c < 1356.16. \tag{171.25}$$

Also, using this value of $Kc = 1356.16$ in the row above the row determining the critical value of Kc, the points of intersection with the $j\omega$ axis are found to be $\pm j4.7773$.

Statement of Step 9: The intersection of root locus with the $j\omega$ axis in the s plane is found using the Routh-Hurwitz test.

Rule/Step 10: Calculation of the value of K on Root Locus

The value of K on the root locus is found using the basic magnitude condition (171.8). Let us calculate the value of K at the crossing of the root locus with the $j\omega$ axis so that the result can be easily verified with that obtained by using Routh-Hurwitz test used in Step 9. Thus, using (171.8) or (171.11) at the point $s_1 = j4.7773$, we have

$$K = \frac{D_0D_2D_3D_4D_5}{D_1} = \frac{4.7773x6.231x7.67x3.0083x10.2272}{5.179} = 1356.34 \tag{171.26}$$

The small difference between this value of K and the value of Kc calculated under Step 9 may be attributed to numerical approximation.

Statement of Step 10: The value of K at any point on the root locus is found using the magnitude condition (171.8).

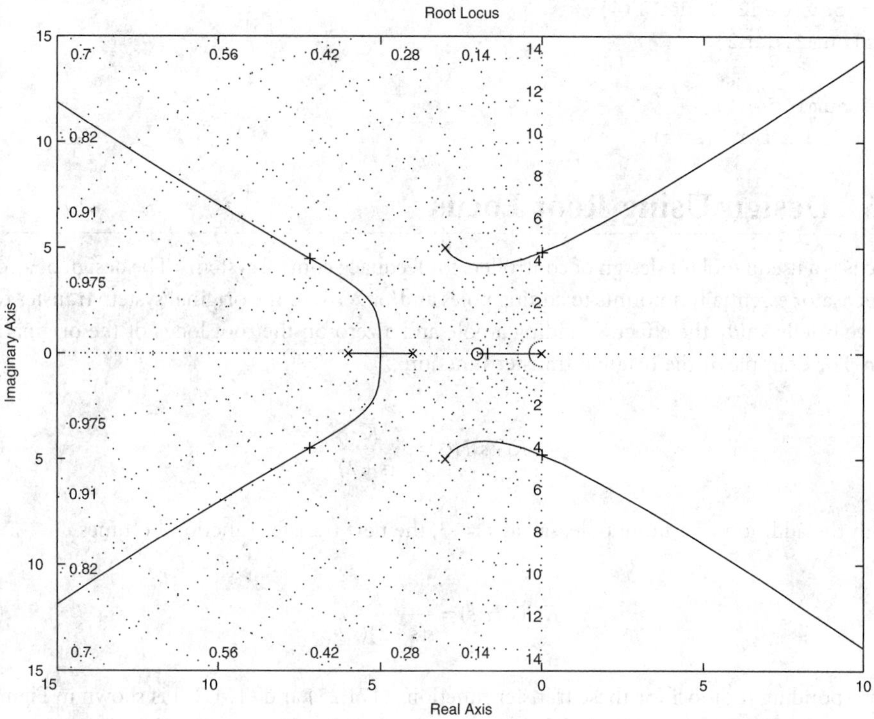

FIGURE 171.7 Root locus diagram using MATLAB©.

171.5 Software for Root Locus

The first semi-automatic plotting device for root locus was called "Spirule", invented by The Spirule Company in 1950s [3] for measuring and adding/subtracting angles and for measuring and multiplying/dividing magnitudes. However, with the introduction of computers, the root locus can now be easily obtained using various software packages available [9, 10], notable among them being MATLAB© and SIMULINK©. For this example, the following MATLAB (*m*) file is used to generate the Figure 171.7.

```
%% TO DRAW THE ROOT LOCUS FOR CONTINUOUS-TIME SYSTEMS
%% OPEN LOOP TRANSFER FUNCTION IS
%% G(s)H(s) = K(s+2)/s(s+4)(s+6)(s^2+6s+34)
%%THE ZERO IS AT -2 AND THE POLES ARE AT 0; -4; -6; -3+5*j; -3-5*j
%% FEEDBACK TRANSFER FUNCTION IS H(s)=1
%%
num1=poly([-2]);
den1=poly([0;-4;-6;-3+5*j;-3-5*j]);
%%rlocus(num1, den1);
%% ALTERNATIVELY WE CAN FIND THE NUM POLYNOMIAL AND
%% DEN POLYNOMIAL AS SHOWN BELOW
 num2=[1 2];
 den21=[1 0];
 den22=[1 4];
 den212 = conv(den21,den22);
 den23=[1 6];
 den24=[1 6 34];
 den234 = conv(den23,den24);
```

```
 den2=conv(den212,den234);
rlocus(num2,den2);
grid
sys=tf(num1,den1);
findK = rlocfind(sys)
```

171.6 Design Using Root Locus

Root locus is a useful tool for design of controllers for feedback control systems. The design of a controller or compensator essentially amounts to adding poles and/or zeros to the original system transfer function. Hence, we briefly study the effect of adding a pole and a zero on the root locus of the original transfer function. For example, if the original transfer function

$$KG(s)H(s) = \frac{K}{s(s+2)} \tag{171.27}$$

then with the addition of a finite pole, say at s = -3, the new transfer function becomes

$$KG(s)H(s) = \frac{K}{s(s+2)(s+3)} \tag{171.28}$$

The corresponding root loci for these transfer functions (171.27) and (171.28) is shown in Figure 171.8. On the other hand, with the addition of a zero, say at s = -3, the new transfer function becomes

$$KG(s)H(s) = \frac{K(s+3)}{s(s+2)} \tag{171.29}$$

and the corresponding root loci for (171.27) and (171.29) is shown in Figure 171.9. By proper choice of the pole and/or zero at s = –3, we can have a desired shape for the root locus to satisfy the design specifications.

171.7 Root Locus for Discrete-Time Systems

The root locus procedure for discrete-time systems described by Z transforms is similar to the above procedure for continuous-time systems described by Laplace transforms. The one main difference between the two procedures is the stability boundary, that is, the stable region for a discrete-time system in z domain is the unit circle. For example, for a discrete-time system described by the open-loop transfer function

$$KG(z)H(z) = \frac{K(z+0.5)}{(z^2-1.4z+0.48)} \tag{171.20}$$

the root locus plot is shown in Figure 171.10.

171.8 Conclusions

The root locus technique is a very powerful tool for analysis and design of automatic control systems and other systems where one is interested in finding the effect of variation of a particular parameter. Even after more than half a century since its invention, the root locus is still an interesting area of research [11–17].

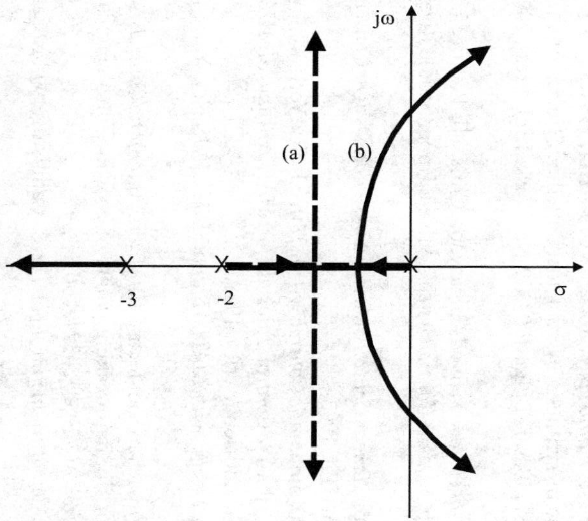

FIGURE 171.8 Root locus diagram for adding poles: (a) original system with poles at $s = 0$ and $s = -2$, and (b) original system with added pole at $s = -3$.

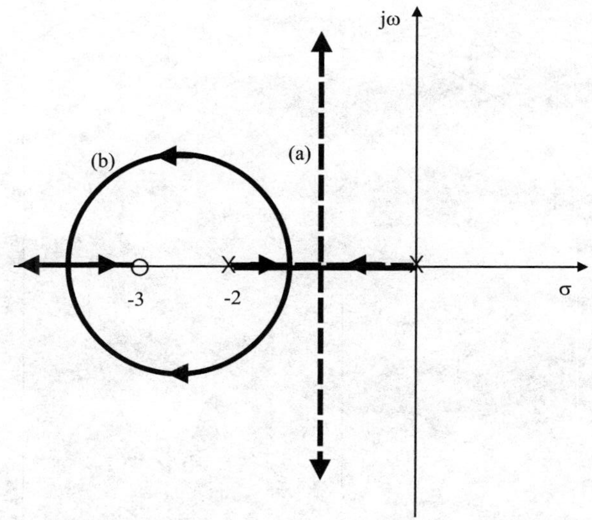

FIGURE 171.9 Root locus diagram for adding zeros: (a) original system with poles at $s = 0$ and $s = -2$ and (b) original system with added zero at $s = -3$.

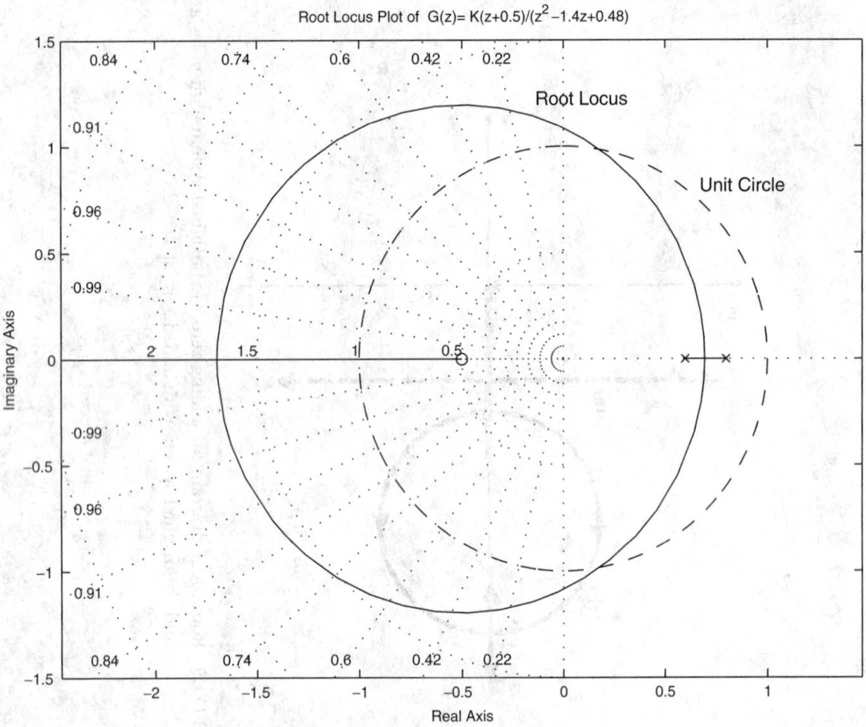

FIGURE 171.10 Root locus diagram for discrete system using MATLAB.

References

[1] W.R. Evans, "Graphical analysis of control systems," *AIEE Trans.*, 67, 547–551, 1948.

[2] W.R. Evans, "Control system synthesis by root locus method," *AIEE Trans.*, 69, 66–69, 1950.

[3] W.R. Evans, *Control System Dynamics*. New York: McGraw-Hill, 1954.

[4] W.R. Evans, "Control system synthesis by root locus method," in *Control Theory: Twenty-Five Seminar Papers (1932–1981)*, Basar, T., Ed. 109–112 New York: IEEE Press, 2001. (Reprinted from *AIEE Trans.*, 1950.)

[5] B.C. Kuo, *Automatic Control Systems*, 7th ed. Englewood Cliffs, NJ: Prentice Hall, 1995.

[6] R.C. Dorf and R.H. Bishop, *Modern Control Systems*, 9th ed. Upper Saddle River, NJ: Prentice Hall, 2001.

[7] J. J. D'Azzo and C.H. Houpis, *Linear Control System Analysis and Design: Conventional and Modern*, 4th ed. New York: McGraw-Hill, 1995.

[8] B.C. Kuo and F. Golnaraghi, *Automatic Control Systems*, 8th ed. New York: John Wiley & Sons, 2003.

[9] W.C. Messner and D.M. Tilbury, *Control Tutorials for MATLAB and SIMULINK: A Web-Based Approach*. Menlo Park, CA: Addison-Wesley, 1998.

[10] A.Tewari, *Modern Control Design with MATLAB and SIMULINK*. New York: John Wiley & Sons, 2002.

[11] K. Steiglitz, "Analytical approach to root locus," *IRE Trans. Automatic Control*, AC-6, 326–332, 1961.

[12] C.F Chen, "A new rule for finding breaking points of root loci involving complex roots," *IEEE Trans. On Automatic Control*, AC-10, 373–374, 1965.

[13] P. Dransfield and D.F. Haber, *Introducting Root Locus*. Cambridge, U.K.: Cambridge University Press, 1973.

[14] A. de Paor, "The root locus method: famous curves, control designs and non-control applications," *Intl. J. Eletric. Eng. Edu.*, 37, 344–356, October 2000.

[15] A.M. Eydgahi and M. Ghavamzadeh, "Complementary root locus revisited," *IEEE Trans. Edu.*, 44, 137–143, 2001.

[16] A. Balestrino, A. Landi, and L. Sani, "Complete root contours for ciricle criteria and relay autotune implementation," *IEEE Control Systems Mag.*, 22, 82–91, 2002.

[17] T.J. Cavecchi, "Phase margin revisited: phase root locus, Bode plots, and phase shifters," *IEEE Trans. Edu.*, 46, 168–174, 2003.

172

Nyquist Criterion and Stability[1]

Norman S. Nise

California State Polytechnic University, Pomona

Frequency response methods for the analysis and design of control systems were developed by H. Nyquist and H. W. Bode in the 1930s. These methods are older than, but not as intuitive as, the root locus, which was discovered by W. R. Evans in 1948. Frequency response yields a new vantage point from which to view feedback control systems. This technique possesses distinct advantages (1) when modeling transfer functions from physical data, (2) when designing lead compensators to meet a steady-state error requirement and a transient response requirement, (3) when determining the stability of nonlinear systems, and (4) in setting ambiguities when sketching a root locus. This chapter introduces frequency response concepts and the determination of stability using the Nyquist criterion.

172.1 Concept and Definition of Frequency Response

In the steady state, sinusoidal inputs to a linear system generate sinusoidal responses of the same frequency. Even though these responses are of the same frequency as the input, they differ in amplitude and phase angle from the input. These differences are a function of frequency.

Sinusoids can be represented as complex numbers, or vectors, called *phasors*. The magnitude of the complex number is the amplitude of the sinusoid and the angle of the complex number is the phase angle of the sinusoid. Thus, $M_1\cos(\omega t + \phi_1)$ can be represented as $M_1 \angle \phi_1$, where the frequency, ω, is implicit.

Since a system causes both the amplitude and phase angle of the input to be changed, we can therefore think of the system itself as represented by a complex number defined so that the product of the input phasor and the system function yields the phasor representation of the output.

Consider the system in Figure 172.1. Assume that the system is represented by the complex number, $M(\omega) \angle \phi(\omega)$. The output steady-state sinusoid is found by multiplying the complex number representation of the input by the complex number representation of the system. Thus, the steady-state output sinusoid is

[1]Adapted from Nise, N. S. 2000. *Control Systems Engineering*, 3rd ed. Wiley, Hoboken, NJ. Copyright © 2000 by John Wiley & Sons, Inc. this material is used by permission of John Wiley & Sons, Inc.

0-8493-1586-7/05/$0.00+$1.50

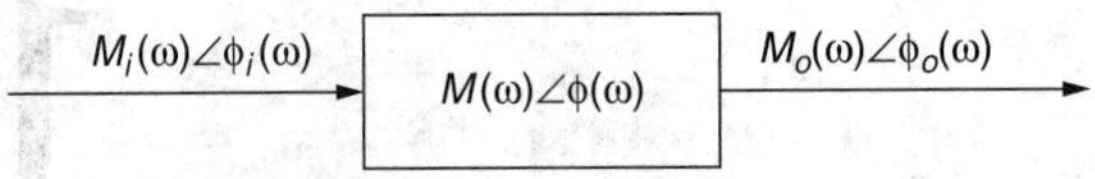

FIGURE 172.1 Steady-state sinusoidal frequency response function.

$$M_o(\omega)\angle\phi_o(\omega) = M_i(\omega)M(\omega)\angle(\phi_i(\omega)+\phi(\omega)) \tag{172.1}$$

From Equation(172.1), we see that the system function is given by

$$M(\omega) = \frac{M_o(\omega)}{M_i(\omega)} \tag{172.2}$$

and

$$\phi(\omega) = \phi_o(\omega) - \phi_i(\omega) \tag{172.3}$$

Equations (172.2) and (172.3) form the definition of frequency response. We call $M(\omega)$ the *magnitude frequency response*, and $\phi(\omega)$ the *phase frequency response*. The combination of the magnitude and phase frequency responses is called the *frequency response* and is $M(\omega) \angle \phi(\omega) = G(j\omega)$.

If we know the transfer function, $G(s)$, of a system, we can find $G(j\omega)$ by using the relationship [Nilsson, 1990].

$$G(j\omega) = G(s)|_{s\to+j\omega} = M(\omega)\angle\phi(\omega) \tag{172.4}$$

172.2 Plotting Frequency Response

$G(j\omega) = M(\omega) \angle \phi(\omega)$ can be plotted in several ways. Two of these ways are (1) as a function of frequency with separate magnitude and phase plots, or (2) as a polar plot, where the phasor length is the magnitude and the phasor angle is the phase. When plotting separate magnitude and phase plots, the magnitude curve can be plotted in decibels (dB) vs. log ω, where dB = 20 log M. The phase curve is plotted as phase angle vs. log ω. Plots that use dB and log frequency are called *Bode plots*. Bode plots can be easily drawn using asymptotic approximations [Nise, 2000].

As an example, find the analytical expression for the frequency response of the system, $G(s) = 1/(s + 2)$. Then, plot the separate magnitude and phase diagrams, as well as the polar plot.

First, substitute $s = j\omega$ in the system function and obtain $G(j\omega) = 1/(j\omega + 2) = (2 - j\omega)/(\omega^2 + 4)$. The magnitude of this complex function, $|G(j\omega)| = M(\omega) = 1/\sqrt{\omega^2+4}$, is the magnitude frequency response. The phase angle of $G(j\omega)$, $\phi(\omega) = -\tan^{-1}(\omega/2)$, is the phase frequency response.

$G(j\omega)$ can be plotted in two ways — Bode plots and a polar plot. The Bode plots are shown in Figure 172.2(a), where the magnitude diagram is $20 \log M(\omega) = 20 \log(1/\sqrt{\omega^2+4})$ vs. log ω, and the phase diagram is $\phi(\omega) = -\tan^{-1}(\omega/2)$ vs. log ω. The polar plot, shown in Figure 172.2(b), is a plot of $M(\omega) \angle \phi(\omega) = 1/\sqrt{\omega^2+4} \angle -\tan^{-1}(\omega/2)$ for different ω.

172.3 Stability

A linear, time-invariant system is *stable* if the natural response approaches zero as time approaches infinity. A linear, time-invariant system is unstable if the natural response grows without bound as time approaches infinity. Finally, a linear, time-invariant system is marginally stable if the natural response neither decays nor grows, but remains constant or oscillates as time approaches infinity.

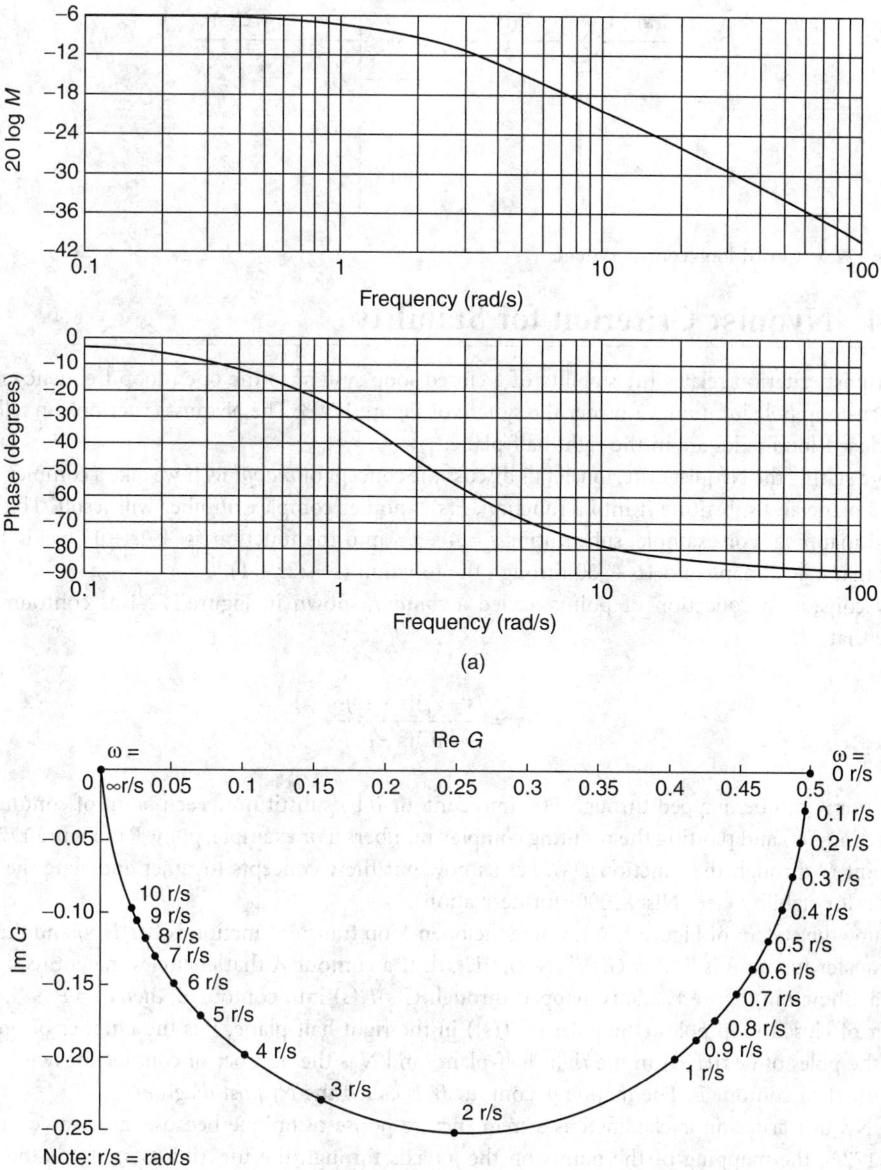

FIGURE 172.2 Frequency response plots for $G(s) = 1/(s + 2)$: (a) Bode plots, and (b) polar plot.

From the point of view of the transfer function, stable systems have closed-loop transfer functions with only left half-plane *poles*, where a pole is defined as a value of s that causes $F(s)$ to be infinite, such as a root of the denominator of a transfer function. Unstable systems have closed-loop transfer functions with at least one right half-plane pole and/or poles of multiplicity greater than one on the imaginary axis. Marginally stable systems have closed-loop transfer functions with only imaginary axis poles of multiplicity one and left half-plane poles. Stability is the most important system specification. An unstable system cannot be designed for a specific transient response or steady-state error requirement. Physically, instability can cause damage to a system, adjacent property, and human life. Many times, systems are designed with limit stops to prevent total runaway.

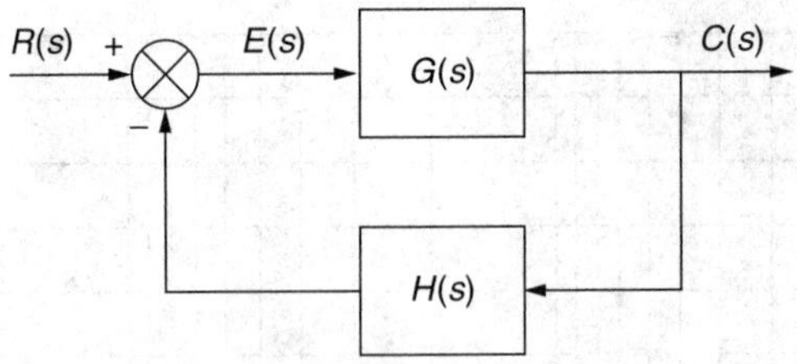

FIGURE 172.3 Closed-loop control system.

172.4 Nyquist Criterion for Stability

The Nyquist criterion relates the stability of a closed-loop system to the open-loop frequency response and open-loop pole location. Consider the system of Figure 172.3. The Nyquist criterion can tell us how many closed-loop poles are in the right half-plane.

Before stating the Nyquist criterion, let us discuss the concept of *mapping*. If we take a complex number on the s plane and substitute it into a function, $F(s)$, another complex number will result. This process is called mapping. For example, substituting $s = 4 + j\,3$ into the function $(s^2 + 2s + 1)$ yields $16 + j30$. We say that $4 + j3$ maps into $16 + j30$ through the function $(s^2 + 2s + 1)$.

Now consider a collection of points, called a *contour*, shown in Figure 172.4 as contour A. Also, assume that

$$F(s) = \frac{(s - z_1)(s - z_2)\cdots}{(s - p_1)(s - p_2)\cdots} \tag{172.5}$$

Contour A can be mapped through $F(s)$ into contour B by substituting each point of contour A into the function $F(s)$ and plotting the resulting complex numbers. For example, point P in Figure 172.4 maps into point Q through the function $F(s)$. Let us now put these concepts together and state the Nyquist criterion for stability (see [Nise, 2000] for derivation).

Assume the system of Figure 172.3, where the open-loop transfer function is $G(s)H(s)$ and the closed-loop transfer function is $T(s) = G(s)/[1 + G(s)H(s)]$. If a contour A that encircles the entire right half-plane, as shown in Figure 172.5, is mapped through $G(s)H(s)$ into contour B, then $Z = P - N$. Z is the number of closed-loop poles (the poles of $T(s)$) in the right half-plane, P is the number of open-loop poles (the poles of $G(s)H(s)$) in the right half-plane, and N is the number of counterclockwise encirclements of -1 of contour B. The mapping, contour B, is called the *Nyquist diagram*.

The Nyquist criterion is classified as a frequency response technique because, around contour A in Figure 172.5, the mapping of the points on the $j\omega$ axis through the function $G(s)H(s)$ is the same as substituting $s = j\omega$ into $G(s)H(s)$ forming the frequency response function, $G(j\omega)H(j\omega)$. Thus, part of the Nyquist diagram is the polar plot of the frequency response of $G(s)H(s)$. Let us look at two examples that illustrate the application of the Nyquist criterion.

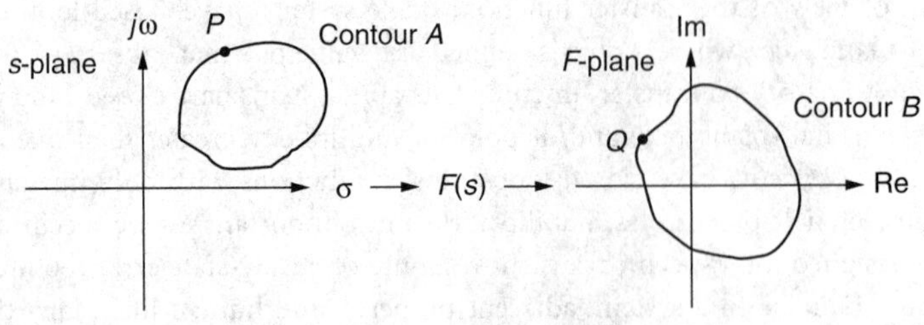

FIGURE 172.4 Mapping contour A through function $F(s)$ to contour B.

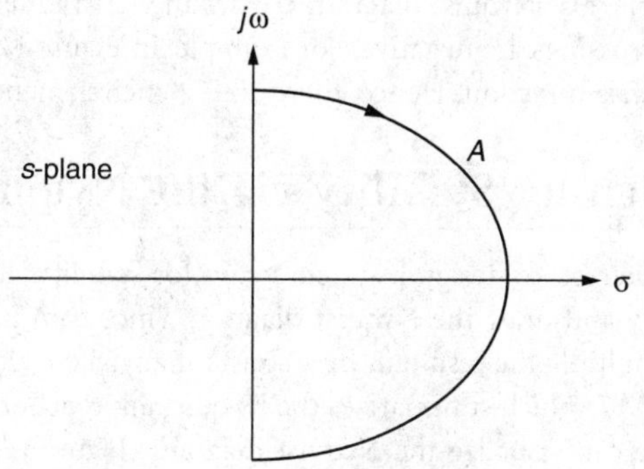

FIGURE 172.5 Contour enclosing right half-plane to determine stability.

In Figure 172.6(a), contour A maps through $G(s)H(s)$ into a Nyquist diagram that does not encircle -1. Hence, $P = 0$, $N = 0$, and $Z = P - N = 0$. Because $Z = 0$ is the number of closed-loop poles inside contour A, which encircles the right half-plane, this system does not have any right half-plane poles and is stable.

On the other hand, Figure 172.6(b) shows a contour A that generates two clockwise encirclements of -1 when mapped through the function $G(s)H(s)$. Thus, $P = 0$, $N = -2$, and $Z = P - N = 2$. The system is unstable because it has two closed-loop poles in the right half-plane ($Z = 2$). The two closed-loop poles are shown inside contour A in Figure 172.6(b) as zeros of $1 + G(s)H(s)$. The reader should keep in mind that the existence of these poles is not known *a priori*.

In this example, note that clockwise encirclements imply a negative value for N. The number of encirclements can be determined by drawing a test radius from -1 in any convenient direction and

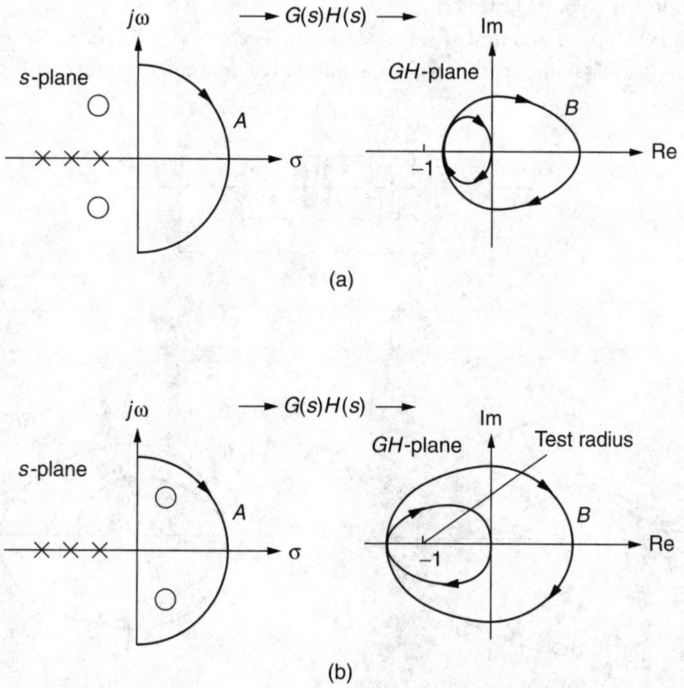

FIGURE 172.6 Examples of mapping for the Nyquist criterion: (a) contour A does not enclose closed-loop poles; and (b) contour A encloses closed-loop poles.

counting the number of times the Nyquist diagram crosses the test radius. Counterclockwise crossings are positive and clockwise crossings are negative. For example, in Figure 172.6(b), contour B crosses the test radius twice in a clockwise direction. Hence, there are −2 encirclements of the point −1.

172.5 Gain Design for Stability via the Nyquist Criterion

We now use the Nyquist criterion to design a system's gain for stability. The general approach is to set the loop gain equal to unity and draw the Nyquist diagram. Since gain is simply a multiplying factor, the effect of the gain is to multiply the resultant by a constant anywhere along the Nyquist diagram. For example, consider Figure 172.7, which summarizes the Nyquist approach for a system with variable gain K. As the gain is varied, we can visualize the Nyquist diagram (Figure 172.7(c)) expanding (increased gain) or shrinking (decreased gain) like a balloon. This motion could move the Nyquist diagram past the −1 point and change the stability picture. For this system, since $P = 2$, the critical point must be encircled two times in the counterclockwise direction by the Nyquist diagram to yield $N = 2$ and a stable system. A reduction in gain would place the critical point outside the Nyquist diagram where $N = 0$ yielding $Z = 2$, an unstable system.

From another perspective, we can think of the Nyquist diagram as remaining stationary and the −1 point moving along the real axis. In order to do this, we set the gain to unity and position the critical point at $-1/K$ rather than −1. Thus, the critical point appears to move closer to the origin as K increases.

Finally, if the Nyquist diagram intersects the real axis at −1, then $G(j\omega)H(j\omega) = -1$. From the root locus, when $G(s)H(s) = -1$, the variables s is a closed-loop pole of the system. Thus, the frequency at which the Nyquist diagram intersects −1 is the same frequency at which the root locus crosses the $j\omega$ axis. Hence, the system is marginally stable if the Nyquist diagram intersects the real axis at −1.

In summary, if the open-loop system contains a variable gain K, set $K = 1$ and sketch the Nyquist diagram. Consider the critical point to be at $-1/K$ rather than at −1. Adjust the value of gain K to yield stability based upon the Nyquist criterion.

Let us look at an example. For a unity feedback system, where $G(s) = K/[s(s + 3)(s + 5)]$, find the range of gain K for stability, instability, and marginal stability. For marginal stability, also find the frequency of oscillation.

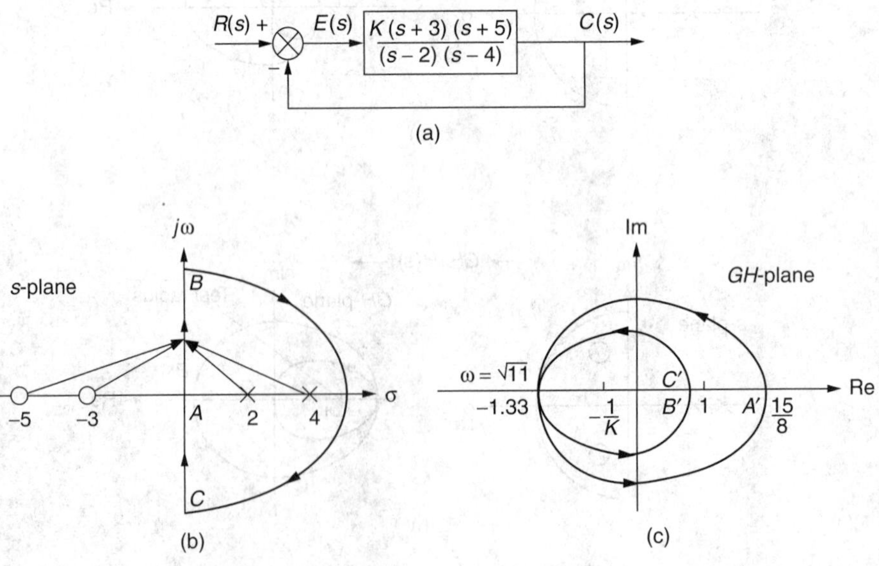

FIGURE 172.7 Feedback control system to demonstrate Nyquist stability: (a) system, (b) contour, and (c) Nyquist diagram.

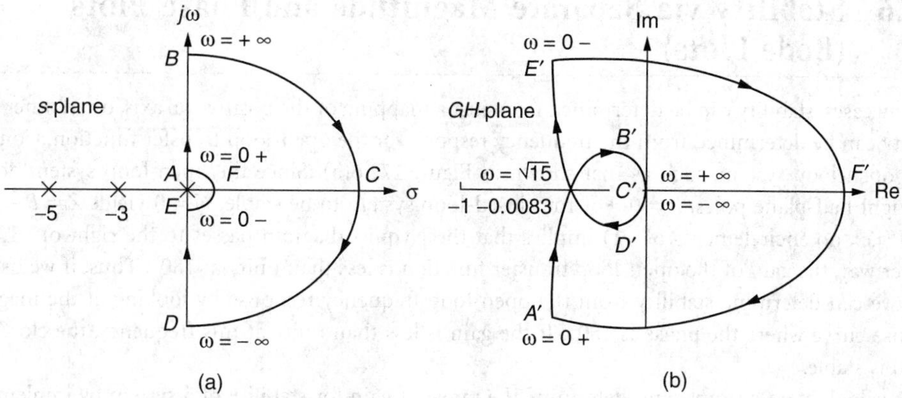

FIGURE 172.8 (a) Contour for example; and (b) Nyquist diagram.

First, set $K = 1$ and sketch the Nyquist diagram for the system using the contour shown in Figure 172.8(a). Conceptually, the Nyquist diagram is plotted by substituting the points of the contour shown in Figure 172.8(a) into $G(s) = 1/[s(s + 3)(s + 5)]$. The contour, as shown, must detour around open-loop imaginary axis poles in order to plot a continuous Nyquist diagram. The detour, however, is epsilon close to the open-loop poles to ensure that any closed-loop right half-plane poles close to the imaginary open-loop poles are still inside the contour and are counted.

From A to B, we use

$$G(j\omega) = \left.\frac{1}{s(s+3)(s+5)}\right|_{s\to j\omega} = \frac{-8\omega^2 - j\omega(15-\omega^2)}{64\omega^4 + \omega^2(15-\omega^2)^2} \tag{172.6}$$

and let ω vary from $0+$ to ∞. The mapping in Figure 172.8(b) goes from A' at ∞ to B' at the origin.

Around the infinite circle from B through C to D, the mapping is found by replacing each complex factor in $G(s)$ by its polar form. Thus,

$$G(s) = \frac{1}{(R_0\angle\theta_0)(R_3\angle\theta_3)(R_5\angle\theta_5)} \tag{172.7}$$

The angles are drawn from the respective poles to a point on the infinite circle. The R_i's are the vector lengths (in this case, infinite). Hence, all points on the infinite circle map into the origin.

The negative imaginary axis from D to E maps into a mirror image of he mapping from A to B, because $G(j\omega)$ has an even real part and an odd imaginary part. From E through F to A, we can again use Equation(172.7). R_0 is zero. Thus, the resultant magnitude is infinite. At E, the angles add to –90°. Hence, the resultant is +90°. Similar reasoning yields the mapping of F and A to F' and A', respectively.

Finally, let us find the point where the Nyquist diagram intersects the negative real axis. We set the imaginary part of Equation(172.6) equal to zero using $\omega = \sqrt{15}$. Then we substitute this value of ω back into Equation(172.6) and find that the real part equals –0.0083.

From the contour of Figure 172.8(a), $P = 0$ and, for stability, N then must be equal to zero. From Figure 172.8(b), the system is stable if the critical point lies outside the contour ($N = 0$) so that $Z = P - N = 0$. Thus, K can be increased by 1/0.0083 = 120.48 before the Nyquist diagram encircles –1. Hence, for stability, $K < 120.48$. For marginal stability, $K = 120.48$. At this gain, the Nyquist diagram intersects –1, and the frequency of oscillation is $\sqrt{15}$ rad/s.

Now that we have used the Nyquist diagram to determine stability, we can develop a simplified approach that uses only the mapping of the positive $j\omega$ axis.

172.6 Stability via Separate Magnitude and Phase Plots (Bode Plots)

In many cases, stability can be determined from just a mapping of the positive $j\omega$ axis, or, in other words, stability can be determined from the frequency response of the open-loop transfer function. Consider a stable open-loop system, such as that shown in Figure 172.6(a). Since the open-loop system does not have right half-plane poles, $P = 0$. For the closed-loop system to be stable, $N = 0$ yields $Z = P - N = 0$. $N = 0$ (i.e., no encirclements of -1) implies that the Nyquist diagram passes to the right of -1. Stated another way, the gain of the open-loop transfer function is less than unity at 180°. Thus, if we use Bode plots, we can determine stability from the open-loop frequency response by looking at the magnitude response curve where the phase is 180°. If the gain is less than unity at this frequency, the closed-loop system is stable.

Let us look at an example and determine the range of gain for stability of a system by implementing the Nyquist stability criterion using asymptotic log-magnitude and phase plots. From the log-magnitude plot, we will determine the value of gain that ensures that the magnitude is less than 0 dB (unity gain) at the frequency where the phase is ±180°.

Shown in Figure 172.9 are the asymptotic approximations to the frequency response of $G(s) = K/[(s + 2)(s + 4)(s + 5)]$, scaled to $K = 40$. Since this system has all of its open-loop poles in the left half-plane, the open-loop system is stable. Hence, the closed-loop system will be stable if the open-loop frequency response has a gain less than unity when the phase is 180°. Accordingly, we see that at a frequency of approximately 7 rad/s, when the phase plot is 180°, the magnitude plot is −20 dB. Therefore, an increase in gain of +20 dB is possible before the system becomes unstable. Since the gain plot was scaled for a gain of 40, +20 dB (a gain of 10) represents the required increase in gain above 40. Hence, the gain for instability is $40 \times 10 = 400$. The final result is $0 < K < 400$ for stability. This result, obtained by

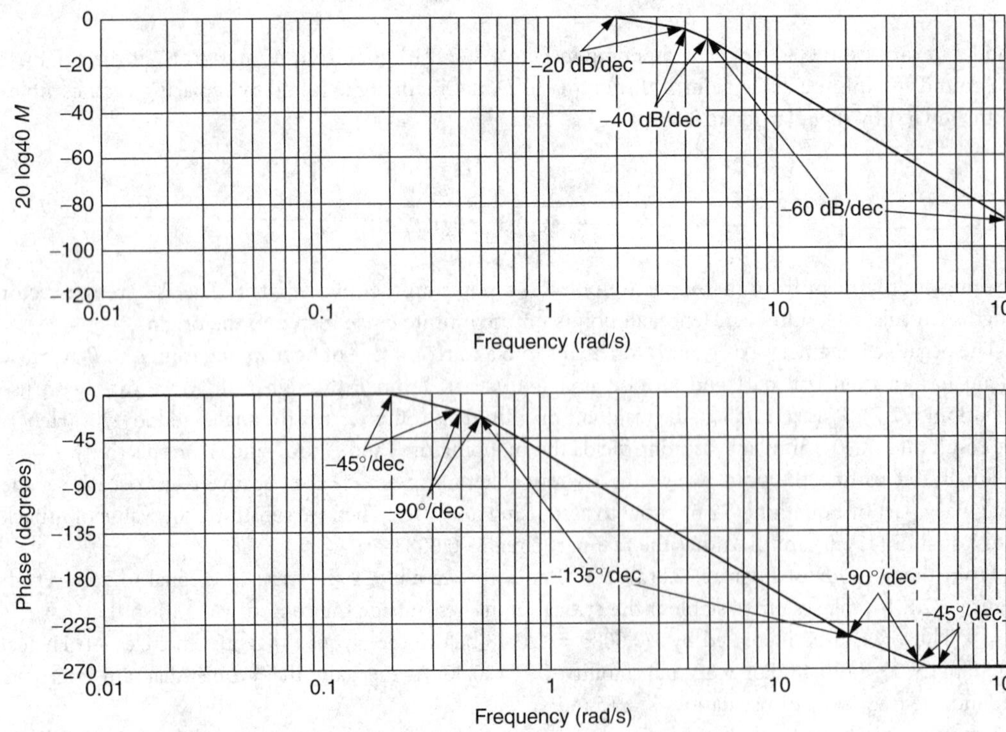

FIGURE 172.9 Log-magnitude and phase plots.

approximating the frequency response by Bode asymptotes, can be compared to the result obtained from the actual frequency response, which yields a gain of 378 at a frequency of 6.16 rad/s.

In this chapter, we discussed the concept of frequency response, including the Nyquist criterion and its application to the stability of linear feedback control systems. The concept of frequency response can be further applied to transient response analysis and design. Frequency response techniques can also be applied to digital and nonlinear control systems.

Defining Terms

Bode plot — A sinusoidal frequency response plot where the magnitude response is plotted separately from the phase response. The magnitude plot is decibels vs. log frequency, and the phase plot is phase vs. log frequency. In control systems, the Bode plot is usually made for the open-loop transfer function.

Frequency response — The combination of magnitude and phase frequency responses expressed as separate magnitude and phase responses, or as a complex function of frequency.

Magnitude frequency response — The ratio of the magnitude of the steady-state sinusoidal response to the magnitude of the sinusoidal input as a function of frequency. The ratio can be expressed in decibels.

Nyquist criterion for stability — Assume a feedback system where the open-loop transfer function is $G(s)H(s)$ and the closed-loop transfer function is $T(s) = G(s)/[1 + G(s)H(s)]$. If a contour A that encircles the entire right half-plane is mapped through $G(s)H(s)$ into contour B, then $Z = P - N$. Z is the number of closed-loop poles (the poles of $T(s)$) in the right half-plane, P is the number of open-loop poles (the poles of $G(s)H(s)$) in the right half-plane, and N is the number of counterclockwise encirclements of -1 of contour B. The mapping, contour B, is called the Nyquist diagram.

Nyquist diagram — A mapping of a closed contour on the s plane that encloses the right half-plane.

Phase frequency response — The difference between the phase angle of the steady-state sinusoidal response and the phase angle of the input sinusoid as a function of frequency.

Phasor — A complex number or vector representing a sinusoid.

Pole — A value of s that causes $F(s)$ to be infinite, such as a root of the denominator of a transfer function.

Stability of a linear, time-invariant system — That characteristic of a linear, time-invariant system defined by a natural response that decays to zero as time approaches infinity.

References

Nilsson, J. W. 1990. *Electric Circuits,* 3rd ed. Addison-Wesley, Reading, MA.

Nise, N. S. 2000. *Control Systems Engineering*, 3rd ed. John Wiley & Sons, New York.

Further Information

Bode, H. W. 1945. *Network Analysis and Feedback Amplifier Design.* Van Nostrand Reinhold, New York.

Dorf, R. C. and Bishop, R. H. 1995. *Modern Control Systems*, 7th ed. Addison-Wesley, Reading, MA.

Kuo, B. C. 1995. *Automatic Control Systems,* 7th ed. Prentice Hall, Englewood Cliffs, NJ.

Kuo, F. F. 1966. *Network Analysis and Synthesis.* John Wiley & Sons, New York.

Nyquist, H. 1932. Regeneration theory. *Bell Syst. Tech. J.*, January, 126–147.

173

System Compensation

Francis H. Raven
University of Notre Dame

The frequency response of a control system may be determined experimentally. Since this is the frequency response for the actual system, correlation criteria that relate transient response to frequency response may then be used to ascertain the transient behavior of the system. Various system **compensation** techniques are available for changing the frequency response so as to improve the transient behavior of the system.

173.1 Correlation between Transient and Frequency Response

Figure 173.1 shows a second-order, type 1 system. For this system, the **natural frequency** is $\omega_n = \sqrt{K_1/\tau}$ and the **damping ratio** is $\zeta = 1/(2\sqrt{K_1\tau})$. The transient behavior is completely described by ζ and ω_n [Raven, 1995]. For a sinusoidal input $r(t) = r_o \sin \omega t$, after the initial transients have died out, the response will be $c(t) = c_o \sin(\omega t + \phi)$. The ratio of the amplitude of the output sinusoidal c_o to the input sinusoidal r_o is $M = c_o/r_o$. The value of ω at which M is a maximum is

$$\omega_p = \omega_n\sqrt{1-\zeta^2} \qquad 0 \le \zeta \le 0.707 \tag{173.1}$$

where ω_p is the value of ω at which M attains its peak or maximum value. The corresponding peak or maximum value of M which is designated M_p is

$$M_p = \frac{1}{2\zeta\sqrt{1-\varsigma^2}} \qquad 0 \le \zeta \le 0.707 \tag{173.2}$$

The preceding result has significance only for $0 \le \zeta \le 0.707$, in which case, $M_p \ge 1$. Solving Equation (173.2) for the damping ratio ζ gives

0-8493-1586-7/05/$0.00+$1.50

$$\zeta=[(1-\sqrt{1-1/M_p^2})/2]^{1/2} \quad M_p \geq 1 \tag{173.3}$$

The transient behavior of higher-order, type 1 systems is closely approximated by the preceding correlation criteria. For the case in which $M_p < 1$, the transient response is described by a first-order, type 1 system whose time constant τ_c is

$$\tau_c = 1/\omega_c \quad M_p < 1 \tag{173.4}$$

where ω_c is the value of ω at which the $G(j\omega)\,H(j\omega)$ plot crosses the unit circle. That is, $|G(j\omega)H(j\omega)| = 1$.

Figure 173.2 shows a second-order, type 0 system. For this system, the natural frequency is $\omega_n = \sqrt{(1+K_o)/b}$, and the damping ratio is $\zeta = a/(2\sqrt{(1+K_o)b}$. The value of ω at which M is a maximum is the same as that given by Equation (173.1). The corresponding peak or maximum value of M is

$$\frac{M_p}{K_o/(1+K_o)} = \frac{1}{2\zeta\sqrt{1-\zeta^2}} \quad 0 \leq \zeta \leq 0.707 \tag{173.5}$$

For large values of K_o, then $K_o/(1+K_o) \approx 1$, and this criterion becomes the same as that for a type 1 system. Solving Equation (173.5) for the damping ratio ζ gives

$$\zeta = \left\{\left[1-\sqrt{1-(K/M_p)^2}\right]/2\right\}^{1/2} \quad M_p/K \geq 1 \tag{173.6}$$

where

$$K = K_o/(1+K_o)$$

For the case in which $M_p/K = M_p/[K_o/(1+K_o)] < 1$, the transient response is described by a first-order, type 0 system whose time constant τ_c is

$$\tau_c = 1/\omega_c \tag{173.7}$$

where ω_c is the value of ω at which the $G(j\omega)H(j\omega)$ plot crosses the circle whose radius is

$$r = \frac{K_o}{\sqrt{1+(1+K_o)^2}} \tag{173.8}$$

For large values of K_o, the radius r approaches the unit circle and the preceding criterion becomes the same as that for a type 1 system.

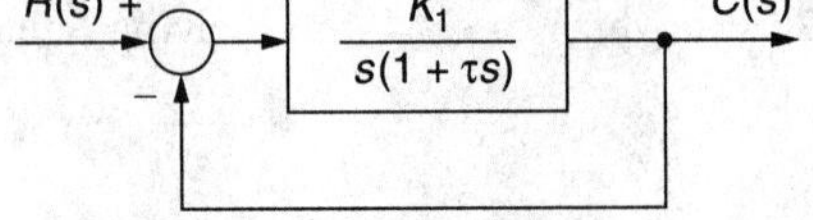

FIGURE 173.1 Second order, type 1 system.

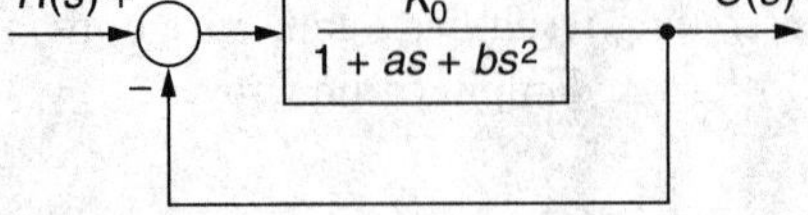

FIGURE 173.2 Second order, type 0 system.

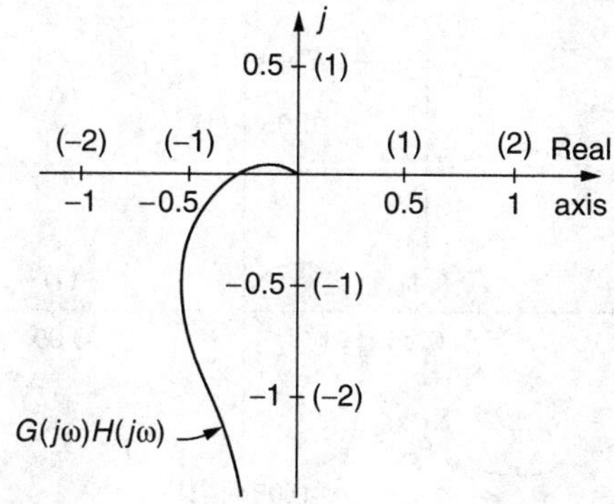

FIGURE 173.3 Typical polar plot.

173.2 Determining K to Yield a Desired M_p

Figure 173.3 shows a typical polar plot of $G(j\omega)H(j\omega)$. If the gain K of the system is doubled, the value $G(j\omega)H(j\omega)$ is doubled at every point. Multiplying the old scale by a factor of 2 yields the polar plot for the system in which the gain K has been doubled. Values of this new scale are shown in parentheses. Changing the gain K does not affect the shape of the polar plot.

An M circle is shown in Figure 173.4. The center of the circle is located on the real axis at $-M^2/(M^2 - 1)$, and the radius is $M/(M^2 - 1)$. The line drawn from the origin, tangent to the M circle at the point P, has an included angle of ψ. The value of $\sin\psi$ is

$$\sin\psi = 1/M \tag{173.9}$$

A characteristic feature of the point of tangency P is that a line drawn from the point P perpendicular to the negative axis intersects this axis at the −1 point.

The procedure for determining the gain K so that the $G(j\omega)H(j\omega)$ plot will have a desired value of M_p is as follows.

1. Draw the polar plot for $G(j\omega)H(j\omega)/K$.
2. Draw the tangent line to the desired M_p circle [$\psi = \sin^{-1}(1/M_p)$].
3. Draw the circle with center on the negative real axis that is tangent to both the $G(j\omega)H(j\omega)/K$ plot and the tangent line, as is shown in Figure 173.5.
4. Erect the perpendicular to the negative real axis from point P. This perpendicular intersects the negative real axis at the point $-A = -0.05$.

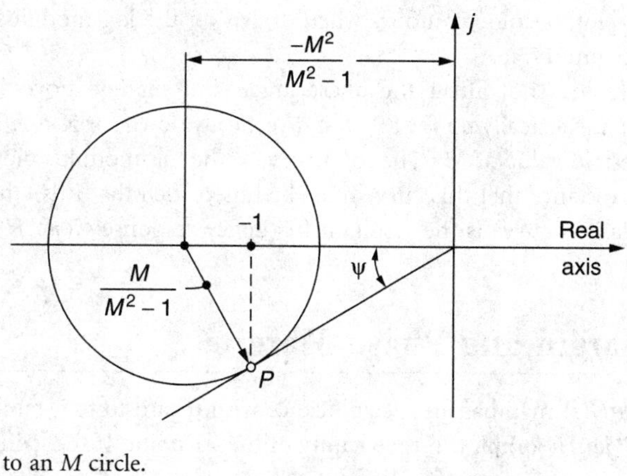

FIGURE 173.4 Tangent to an M circle.

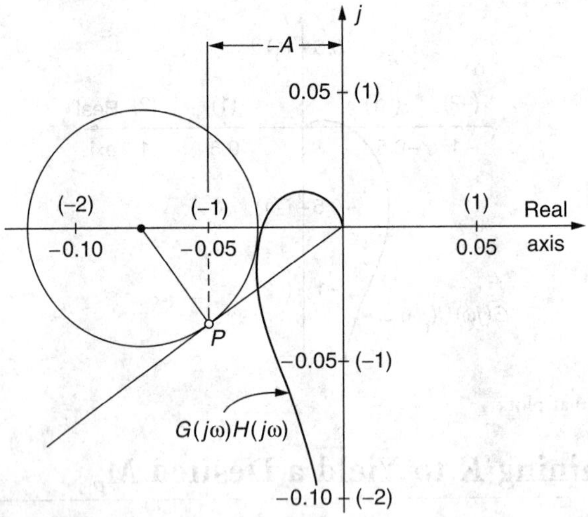

FIGURE 173.5 Determination of K to yield a desired M_p.

5. In order that the circle drawn in Step 3 is the desired M_p circle, this point should be –1 rather than $-A$. The required gain is that value of K which changes the scale so that this becomes the –1 point. Thus $K(-A) = -1$ or $K = 1/A$.

As illustrated in Figure 173.5, the perpendicular drawn from point P to the negative real axis intersects the axis at the value $-A = -0.05$. Multiplication of the scale by a factor of 20, as is shown in Figure 173.5 by the numbers in parentheses, converts this point to the –1 point. Thus, the required value of the gain K is 20, such that the polar plot $G(j\omega)H(j\omega)$ is tangent to the desired M_p circle.

The plot $G(j\omega)H(j\omega)/K$ is the plot for the case in which K is 1. By constructing the plot for $G(j\omega)H(j\omega)$ rather than $G(j\omega)H(j\omega)/K$, the resulting value of $1/A$ represents the required factor K_c by which the gain should be changed such that the resulting polar plot will be tangent to the desired M_p circle. That is,

$$\text{New gain} = K_c \text{ (original gain)} \tag{173.10}$$

Another method for representing frequency response information is the log-modulus or Nichols plot [Nichols, 1947]. A log-modulus plot is a plot of log $|G(j\omega)H(j\omega)|$ versus $\measuredangle G(j\omega)H(j\omega)$. Lines of constant M and constant α, where

$$M = |C(j\omega)/R(j\omega)| \text{ and } \alpha = \measuredangle C(j\omega)/R(\omega) \tag{173.11}$$

are circles on the polar plot, become contours when drawn on the log-modulus plot. These M and α contours are shown in Figure 173.6.

Changing the gain K does not affect the phase angle, but merely moves the log-modulus plot $G(j\omega)H(j\omega)/K$ for a system vertically up for $K > 1$ and vertically down for $K < 1$. The M_p contour shown in Figure 173.7 is the desired value of M_p. The solid curve is the log-modulus plot of $G(j\omega)H(j\omega)/K$ for the system. The vertical distance that this curve must be raised such that it is tangent to the desired M_p contour is log K. The dashed curve is the resultant frequency response $G(j\omega)H(j\omega)$. The frequency at the point of tangency is ω_p.

173.3 Gain Margin and Phase Margin

The –1 point of the $G(s)H(s)$ map has great significance with regard to the stability of a system. Figure 173.8 shows a typical $G(j\omega)H(j\omega)$ plot in the vicinity of the –1 point. If the gain were multiplied by an

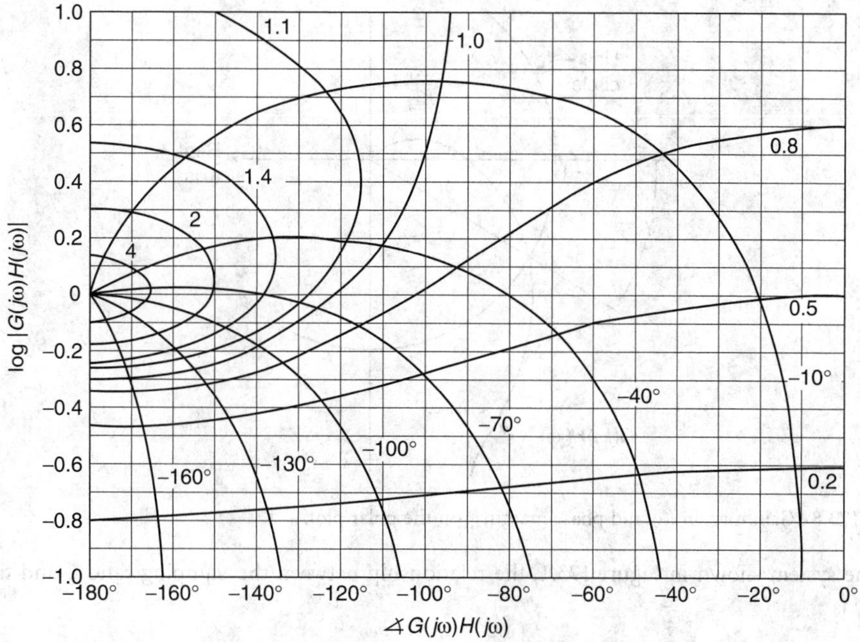

FIGURE 173.6 Log-modulus representation for lines of constant M and lines of constant α.

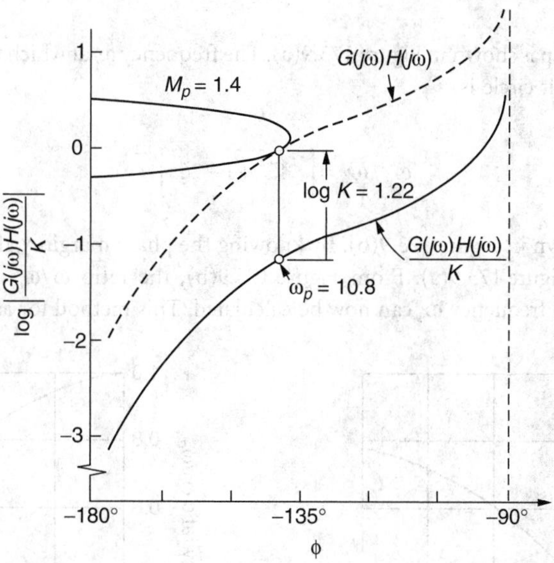

FIGURE 173.7 Determination of K for a desired M_p.

amount K_M, called the **gain margin**, the $G(j\omega)H(j\omega)$ plot would go through the −1 point. Thus, the gain margin is an indication of how much the gain can be increased before the $G(j\omega)H(j\omega)$ plot goes through the critical point.

The angle γ in Figure 173.8 is the angle measured from the negative real axis to the radial line through the point where the polar plot crosses the unit circle. If the angle γ were zero, the polar plot would go through the −1 point. The angle γ, called the **phase margin**, is a measure of the closeness of the polar plot to the critical point. A positive phase margin indicates a stable system, as does a gain margin greater than one.

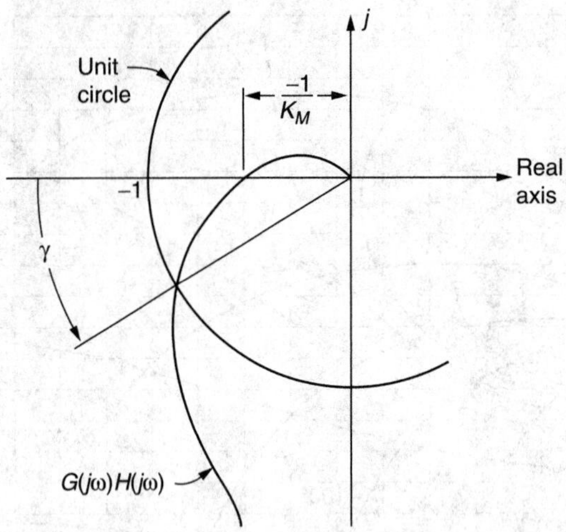

FIGURE 173.8 Gain margin K_M and phase margin γ on the polar plot.

For the system shown in Figure 173.1, the relationship between the damping ratio ζ and the phase margin γ is

$$\zeta = \frac{\tan\gamma\sqrt{\cos\gamma}}{2} \tag{173.12}$$

A plot of this relationship is shown in Figure 173.9(a). The frequency ω_c at which the open-loop frequency response crosses the unit circle is

$$\omega_c / \omega_n = \left[\sqrt{4\zeta^4 + 1} - 2\zeta^2\right]^{1/2} \tag{173.13}$$

The relationship is shown in Figure 173.9(b). By knowing the phase margin γ, the damping ratio ζ may be determined from Figure 173.9(a). From Figure 173.9(b), the ratio ω_c/ω_n can be found. Thus, by knowing ω_c, the natural frequency ω_n can now be calculated. This method for ascertaining the damping

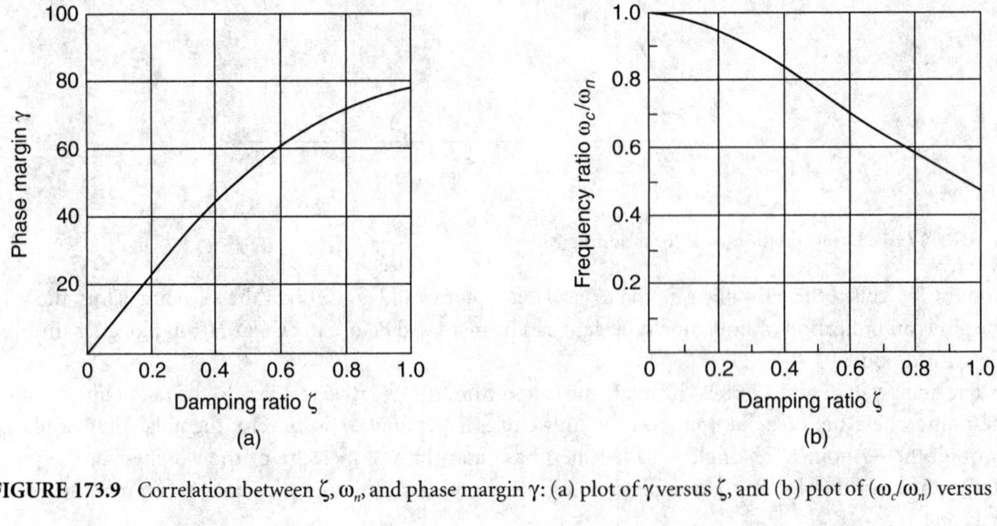

FIGURE 173.9 Correlation between ζ, ω_n, and phase margin γ: (a) plot of γ versus ζ, and (b) plot of (ω_c/ω_n) versus ζ.

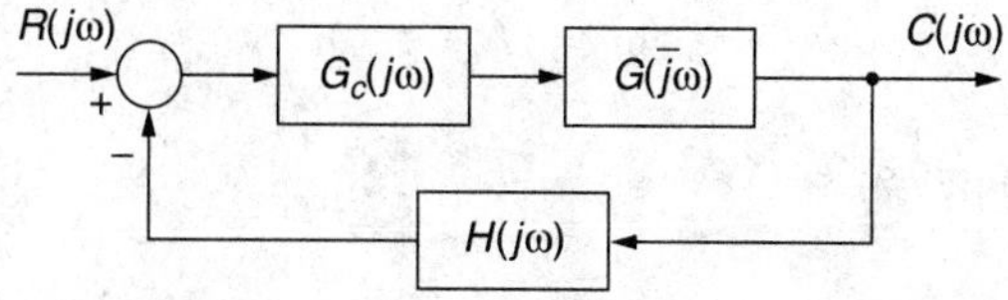

FIGURE 173.10 Series compensator $G_c(j\omega)$.

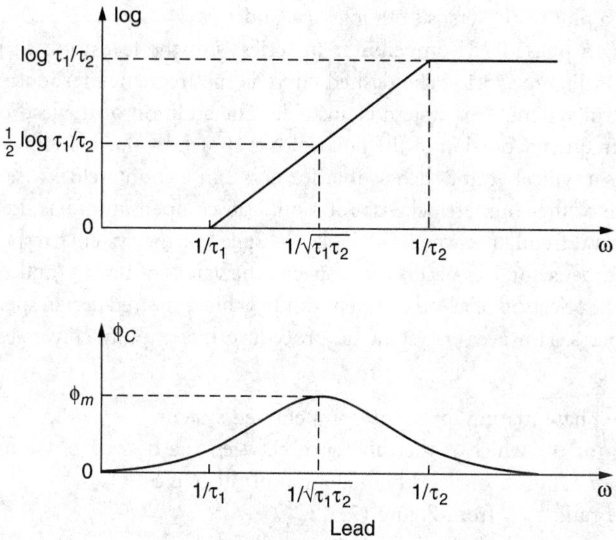

FIGURE 173.11 Bode diagram for a phase lead compensator.

ratio ζ and natural frequency from the phase margin γ and the frequency ω_c at which the open-loop frequency response crosses the unit circle yields good approximations for systems other than that shown in Figure 173.1.

173.4 Series Compensation

A change in the gain K changes the scale factor of the polar plot, but does not affect the basic shape of the plot. In the design of control systems, it is often necessary to change the shape of the polar plot in order to achieve the desired dynamic performance. A common way of doing this is to insert elements in series with the feed-forward portion of the control, as is illustrated by the elements $G_c(j\omega)$ shown in Figure 173.10. This method of compensating the performance of a control system is called **cascade** or **series compensation.**

Lead Compensation

The frequency response of a phase lead compensator is

$$\frac{1+j\tau_1\omega}{1+j\tau_2\omega} \quad \tau_1 > \tau_2 \tag{173.14}$$

The asymptotic approximation to the Bode diagram [Bode, 1945] for $(1 + j\tau_1\omega)/(1 + j\tau_2\omega)$ is shown in Figure 173.11. The frequency at which the maximum phase lead occurs is at

$$\omega_m = 1/\sqrt{\tau_1\tau_2} \tag{173.15}$$

The value of the maximum phase lead is

$$\phi_m = \tan^{-1}\frac{(\tau_1/\tau_2)-1}{2\sqrt{\tau_1/\tau_2}} \tag{173.16}$$

Figure 173.12 shows a plot of ϕ_m versus both log τ_1/τ_2 and τ_1/τ_2.

The effect of using a phase lead compensator in series with the feed-forward portion of a control system is illustrated in Figure 173.13. The dashed curve is the frequency response $G(j\omega)H(j\omega)$ for the uncompensated control system. This system is unstable. The addition of the lead compensation $G_c(j\omega)$ to reshape the high-frequency portion of the polar plot is shown by the solid line curve. Note that lead compensation rotates a typical vector such as that for $\omega = 2$ in a counterclockwise direction away from the −1 point. Because of the counterclockwise rotation, lead compensation has the very desirable effect of increasing the natural frequency, which increases the speed of the system's response.

To select a lead compensator, it is necessary to specify the values of both τ_1 and τ_2. Because of the two unknowns τ_1 and τ_2, the selection of a lead compensator to achieve desired design specifications is basically a trial-and-error process. However, a systematic procedure that rapidly converges is described in the following steps:

1. Determine the phase margin for the uncompensated system.
2. Select a value for ϕ_m, which is the difference between the desired phase margin and the value obtained in step 1, plus a small additional amount, such as 5°.
3. Determine the ratio τ_1/τ_2 from Figure 173.12.
4. Determine the frequency where the log magnitude for the uncompensated system is −0.5 log(τ_1/τ_2). Use this frequency for ω_m.
5. Because the phase lead compensator provides a gain of 0.5 log(τ_1/τ_2) at $\omega_m = 1/\sqrt{\tau_1\tau_2}$, this will be the frequency where the compensated system crosses the unit circle. Determine the resulting phase margin for the compensated system.

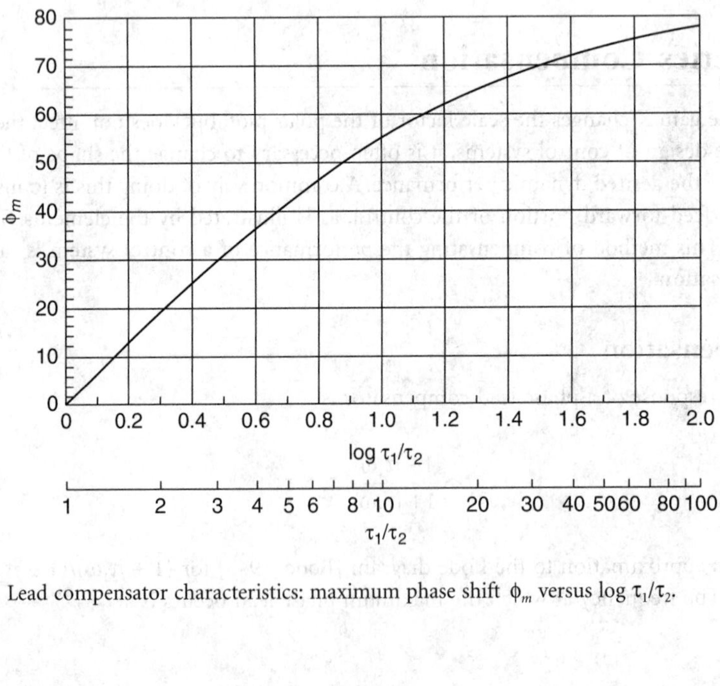

FIGURE 173.12 Lead compensator characteristics: maximum phase shift ϕ_m versus log τ_1/τ_2.

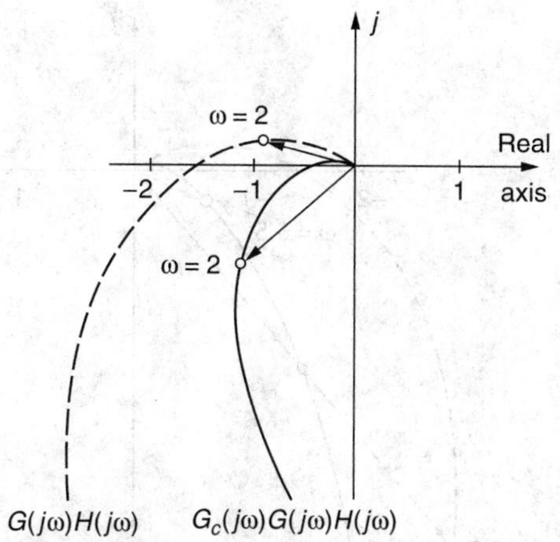

FIGURE 173.13 Use of phase lead to reshape a polar plot.

If the phase margin is too small, increase the value τ_1/τ_2 and, if it is too large, decrease the ratio τ_1/τ_2 and then repeat the steps.

Lag Compensation

The frequency response for a phase lag compensator is

$$\frac{1 + j\tau_2\omega}{1 + j\tau_1\omega} \qquad \tau_1 > \tau_2 \tag{173.17}$$

The Bode diagram for $(1 + j\tau_2\omega)/(1 + j\tau_1\omega)$ is shown in Figure 173.14. Note that when $\omega = 10/\tau_2$, the negative phase shift ϕ_c is very small. The negative phase shift associated with lag compensation is usually undesirable. The effectiveness of lag compensation is attributed to the attenuation ($-\log \tau_1/\tau_2$) that occurs at high frequencies.

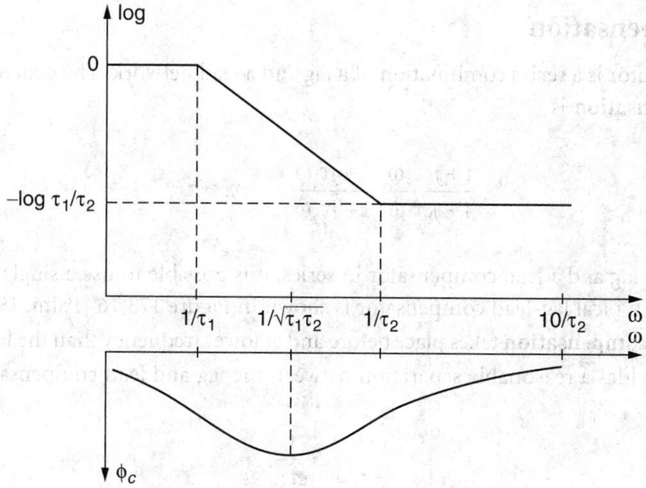

FIGURE 173.14 Bode diagram for a lag compensator.

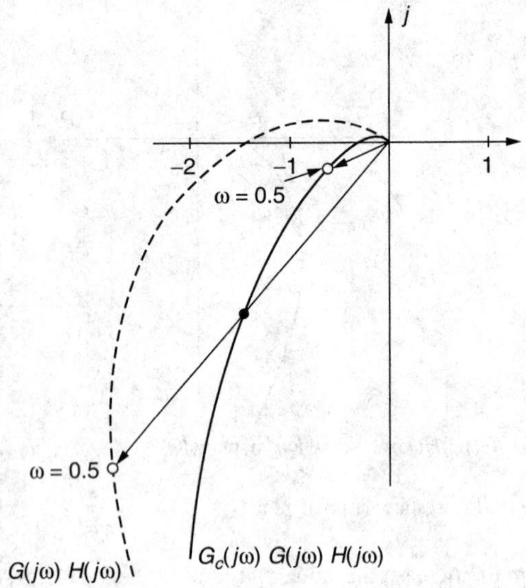

FIGURE 173.15 Use of phase lag to reshape a polar plot.

The effect of using a phase lag compensator in series with the feed-forward portion of a control system is illustrated in Figure 173.15. The dashed curve is the frequency response of the uncompensated system. Note that the effect of lag compensation is to shorten a typical vector, such as that for $\omega = 0.5$, and to rotate it slightly in a counterclockwise direction. A procedure for determining the lag compensator for obtaining a desired dynamic performance is described in the following steps.

1. Add 5° to the desired phase margin and then subtract 180° from this result.
2. For the uncompensated system, determine the value of the $\log|G(j\omega)H(j\omega)|$ at the angle determined in step 1.
3. Set this value of $\log|G(j\omega)H(j\omega)|$ equal to $\log \tau_1/\tau_2$. When the lag compensator is added to the uncompensated system, this will be the point where the resultant system crosses the unit circle (i.e., $\log|G(j\omega)H(j\omega)| - \log \tau_1/\tau_2 = 0$). This will be the $10/\tau_2$ frequency.

Lag-lead Compensation

A lag-lead compensator is a series combination of a lag and a lead network. The general transfer function for **lag-lead compensation** is

$$\frac{1+jc\tau_2\omega}{1+jc\tau_1\omega}\,\frac{1+j\tau_1\omega}{1+j\tau_2\omega} \qquad \tau_1 > \tau_2 \tag{173.18}$$

Rather than using a lag and a lead compensator in series, it is possible to use a single compensator. The Bode diagram for a typical lag-lead compensator is shown in Figure 173.16 [Palm, 1986]. Because $c\tau_1 > c\tau_2 > \tau_1 > \tau_2$, the **lag compensation** takes place before and at lower frequency than the **lead compensation.** A factor of five provides a reasonable separation between the lag and lead compensations:

$$\frac{1}{\tau_1} = 5\frac{1}{c\tau_2}$$

or

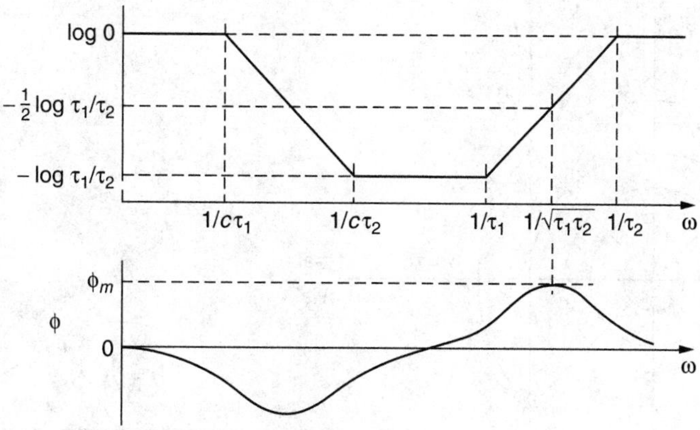

FIGURE 173.16 Bode diagram for a lag-lead compensator.

$$c = 5\frac{\tau_1}{\tau_2} \tag{173.19}$$

The maximum phase shift ϕ_m occurs at $\omega = 1/\sqrt{\tau_1\tau_2}$, and the corresponding gain is $-0.5\log\tau_1/\tau_2$. This is the same as for a lead compensator, except that the sign of the gain $0.5\log\tau_1/\tau_2$ is negative. This feature makes the lag-lead compensator considerably more effective than the lead compensator only.

173.5 Internal Feedback

Another method commonly used to alter frequency response characteristics is that of providing a separate **internal feedback** path about certain components [D'Souza, 1988]. A plot of the function $G^{-1}(j\omega) = 1/G(j\omega)$ is called an inverse polar plot. Figure 173.17 shows a typical inverse polar plot for the function $G^{-1}(j\omega)$. At any frequency ω, the vector from the origin to a point on the plot defines the vector $G^{-1}(j\omega)$ for that frequency. The length of the vector is $|G^{-1}(j\omega)| = 1/|G^{-1}(j\omega)|$, and the angle is

$$\measuredangle G^{-1}(j\omega) = \measuredangle \frac{1}{G(j\omega)} = -\measuredangle G(j\omega) \tag{173.20}$$

On the inverse plane, lines of constant M are circles of radius $1/M$. The center of these concentric M circles is at the point $x = -1$ and $y = 0$ (e.g., the -1 point). A plot of the M circles on the inverse plane is shown in Figure 173.18. Because the reciprocal of -1 is still -1, this point has the same significance for an inverse polar plot as for a direct polar plot. The lines of constant $\alpha = \measuredangle[C(j\omega)/R(j\omega)] = -\measuredangle[R(j\omega)/C(j\omega)]$ are radial straight lines (rays) which pass through the -1 point.

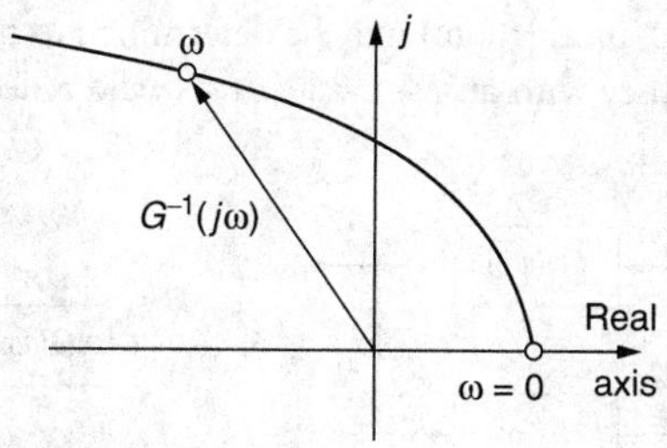

FIGURE 173.17 Typical inverse polar plot $G^{-1}(j\omega)$.

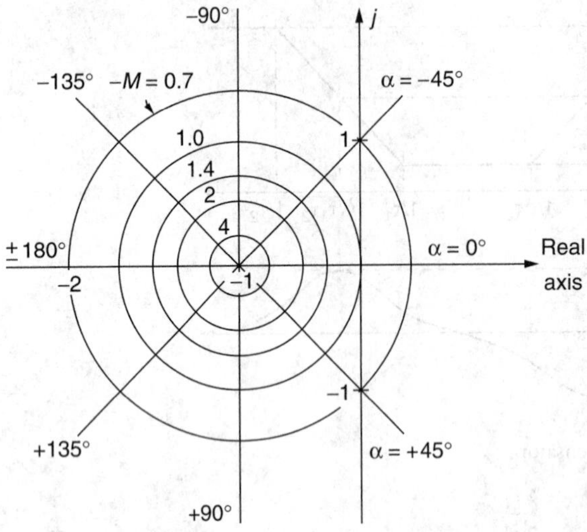

FIGURE 173.18 M circles and α rays on the inverse plane.

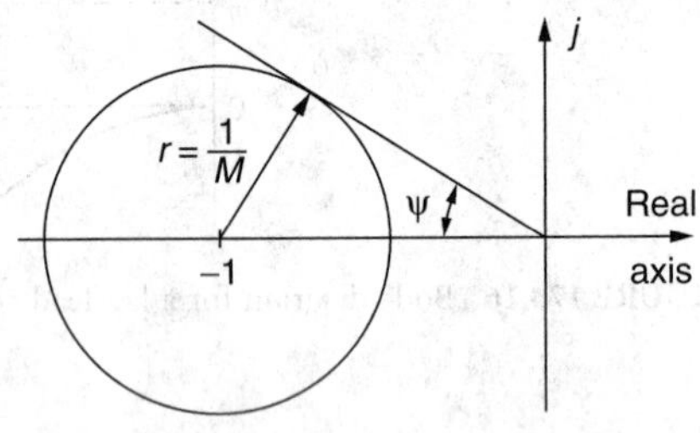

FIGURE 173.19 Tangent line to an M circle.

As illustrated in Figure 173.19, the angle ψ of a radial line drawn from the origin and tangent to any M circle is

$$\sin\psi = 1/M \tag{173.21}$$

The general procedure for determining the required gain K to yield a desired M_p is as follows:

1. Plot the inverse function $KG^{-1}(j\omega)H^{-1}(j\omega)$.
2. Construct the tangent line at the angle $\psi = \sin^{-1}(1/M_p)$.
3. Construct the circle which is tangent to both the $KG^{-1}(j\omega)H^{-1}(j\omega)$ plot and the tangent line.
4. The center of the circle is at the point $-A$. The desired gain is $K = A$.

When the function $G^{-1}(j\omega)H^{-1}(j\omega)$ is plotted rather than $KG^{-1}(j\omega)H^{-1}(j\omega)$, then A is equal to the factor K_c by which the gain should be changed to yield the desired M_p.

The major advantage of using the inverse plane is realized for systems with internal feedback. For the system of Figure 173.20,

$$G^{-1}(j\omega) = \frac{1+G_1(j\omega)H_1(j\omega)}{G_1(j\omega)} = G_1^{-1}(j\omega) + H_1(j\omega) \tag{173.22}$$

As illustrated in Figure 173.21, the vectors $G_1^{-1}(j\omega)$ and $H_1(j\omega)$ may be added as vector quantities to yield $G^{-1}(j\omega)$. For a given $G_1^{-1}(j\omega)$ plot, $H_1(j\omega)$ may be determined to move any given point to a desired location, such as a point of tangency with an M_p circle, or to yield a desired phase margin.

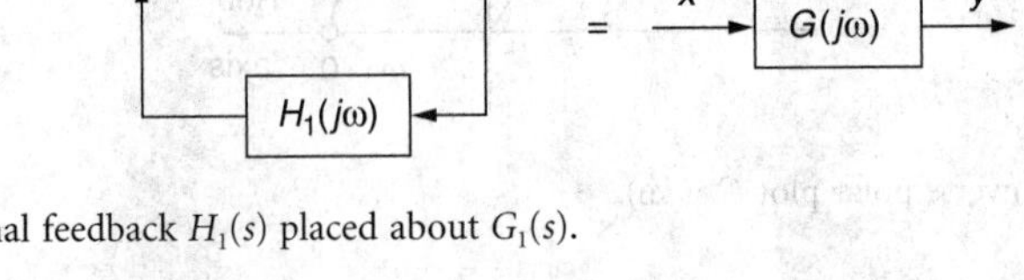

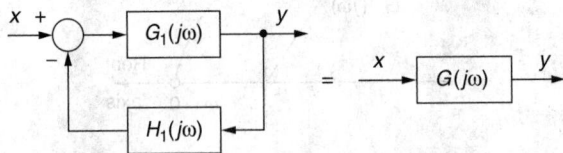

FIGURE 173.20 Internal feedback $H_1(s)$ placed about $G_1(s)$.

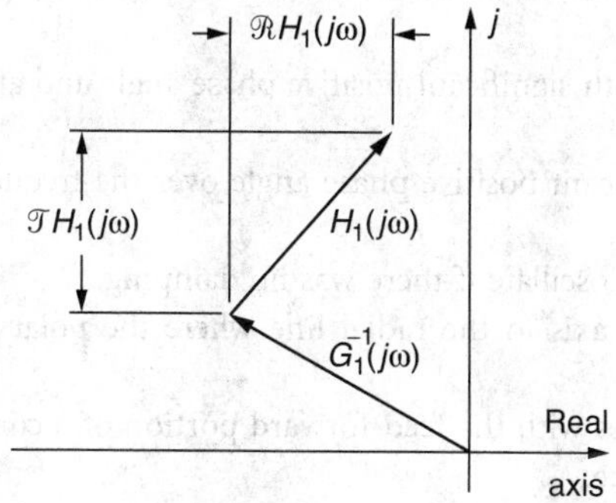

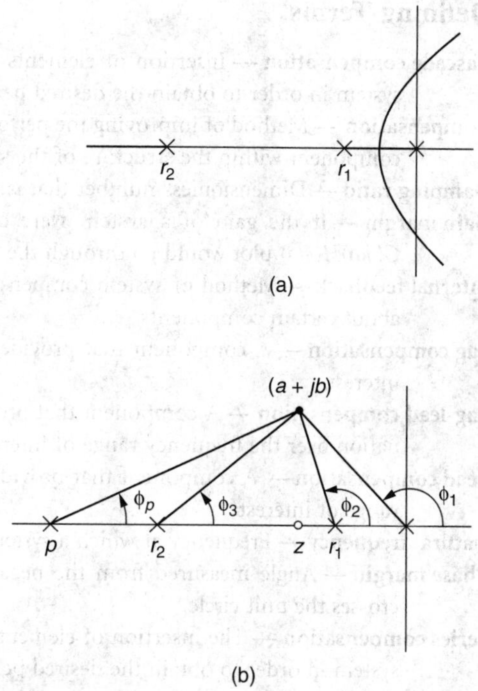

FIGURE 173.21 Vector addition of G_1^{-1} $(j\omega)$ and $H_1(j\omega)$ to yield $G^{-1}(j\omega)$.

FIGURE 173.22 (a) Root locus plot for $s(s - r_1)$ $(s - r_2)$ $+ K = 0$, and (b) addition of phase lead compensation.

If $H_1(s) = \alpha$, a constant, then the $G^{-1}(j\omega)$ plot is shifted to the right by the distance α. If $H_1(s) = \beta s$, then $H_1(j\omega) = j\beta\omega$, and the $G^{-1}(j\omega)$ plot is shifted upwards vertically in proportion to the frequency ω. If $H(s) = \alpha + \beta s$, then the $G^{-1}(j\omega)$ plot is shifted both to the right and upwards vertically.

173.6 Compensation on the *S* Plane

The transfer function for a phase lead compensator may be written in the form

$$\frac{1+\tau_1 s}{1+\tau_2 s} = \frac{s+1/\tau_1}{s+1/\tau_2} = \frac{s-z}{s-p} \quad \tau_1 > \tau_2 \tag{173.23}$$

where $z = -1/\tau_1$ is a zero, and $p = -1/\tau_2$ is a pole. The root locus plot for the system whose characteristic equation is $s(s - r_1)$ $(s - r_2) + K = 0$ is shown in Figure 173.22(a). Assume that the root locus plot goes through the root location for the dominant roots (point $a \pm jb$) shown in Figure 173.22(b). The zero is drawn directly under the point $a + jb$. Application of the angle condition gives

$$\phi_1 + \phi_2 + \phi_3 + \phi_p - 90° = 180°$$

Hence,

$$\phi_p = 270° - (\phi_1 + \phi_2 + \phi_3)$$

The angle ϕ_p determines the location of the pole. Similarly, phase lag and lag-lead compensators [Dorf, 1989] may be designed on the *s* plane.

Defining Terms

Cascade compensation — Insertion of elements in series with the feed-forward portion of a control system in order to obtain the desired performance.

Compensation — Method of improving the performance of a control system by inserting an additional component within the structure of the system.

Damping ratio — Dimensionless number that is a measure of the amount of damping in a system.

Gain margin — If the gain of a system were changed by this factor, called the gain margin, the $G(j\omega)H(j\omega)$ plot would go through the −1 point.

Internal feedback — Method of system compensation in which an internal feedback path is provided about certain components.

Lag compensation — A component that provides significant attenuation over the frequency range of interest.

Lag-lead compensation — A component that provides both significant positive phase angle and attenuation over the frequency range of interest.

Lead compensation — A component that provides significant positive phase angle over the frequency range of interest.

Natural frequency — Frequency at which a system would oscillate if there was no damping.

Phase margin — Angle measured from the negative real axis to the radial line where the polar plot crosses the unit circle.

Series compensation — The insertion of elements in series with the feed-forward portion of a control system in order to obtain the desired performance.

References

Bode, H. W. 1945. *Network Analysis and Feedback Amplifier Design.* Van Nostrand Reinhold, New York.

Dorf, R. C. 1995. *Modern Control Systems,* 7th ed. Addison-Wesley, Reading, MA.

D'Souza, A. F. 1988. *Design of Control Systems.* Prentice Hall, Englewood Cliffs, NJ.

Nichols, J. H. and Phillips, R. S. 1947. *Theory of Servomechanisms.* McGraw-Hill, New York.

Palm III, W. J. 1986. *Control Systems Engineering.* John Wiley & Sons, New York.

Raven, F. H. 1995. *Automatic Control Engineering,* 5th ed. McGraw-Hill, New York.

174

Process Control

Thomas E. Marlin
McMaster University

Process control addresses the application of automatic control theory to the process industries. These industries typically involve the continuous processing of fluids and slurries in chemical reactors, physical separation units, combustion processes, and heat exchangers in industries such as chemicals, petroleum, pulp and paper, food, steel, and electrical power generation. Generally, the goal of process control is to reduce the variability of key process variables that influence safety, equipment protection, product quality, and production rate. To achieve reduced variability in these key variables, we must adjust selected manipulated variables. As a result, the "*total variability*" is not reduced, but it is transferred from important to less important areas of the process, such as cooling water, steam systems, and fuel distribution systems.

174.1 Defining the Process Control Design Problem

Process control design involves the following decisions: (1) selecting sensors that reflect process performance, (2) selecting final elements with good dynamic responses and sufficient range, (3) ensuring that the feedback process dynamics are favorable for good control performance, (4) selecting a structure for linking controlled and manipulated variables, and (5) selecting algorithms and tuning parameters to give robust performance. Note that key decisions such as process flexibility and locations of sensors and final elements involve the structure of the process, and the linking of measured and manipulated variables involves the structure of the control system. If these structural decisions are made poorly, even the best control algorithms will not provide acceptable performance.

Before beginning the design process, the engineer must thoroughly define the control objectives using the following categories:

- Safety
- Environmental protection
- Equipment protection
- Smooth, easy operation
- Product quality

0-8493-1586-7/05/$0.00+$1.50

- Efficiency and profit
- Monitoring and diagnosis

Each of these objectives should be defined as quantitatively as possible; for example, the temperature of the reactor will not exceed 400 K for a feed rate change of ±20% or a feed temperature step change of 20 K. Clearly, the thorough understanding of the process, including disturbed situations, is required to define the design problem. A simplified statement of objectives for distillation control is given in Table 174.1 with sensors and control designs. Several more complete examples are available in Marlin (2000), and methods for estimating economic benefits from control are presented by Marlin et al. (1991).

An important step is the definition of the desired dynamic performance of the key variables. This definition will be used in the initial design and can also be used to monitor and evaluate the achieved performance of the operating plant. In evaluating control performance, many factors are relevant, as shown in Figure 174.2. First, the controlled variable should have a small deviation from its set point (or

TABLE 174.1 Simplified Control Design Form for Distillation Process in Figure 174.1

Control Objectives	Control Designs
1. Safety	**1. Safety**
a. Prevent high pressure in closed vessel	a. Pressure controller PC-1 Pressure safety relief valve PV-3
b. Draw operator's attention to unusual conditions	b. Provide alarms for variables that strongly affect operation.
2. Environmental Protection	**2. Environmental Protection**
a. Prevent hydrocarbon release to environment	a. Relief valve piped to flare to combust hydrocarbons
3. Equipment Protection	**3. Equipment Protection**
a. Ensure that liquid flows through pumps continuously	a. Liquid level control for feeds to pumps
b. Prevent corrosion due to water condensation in distillation tower	b. Monitor the top tray temperature and alarm on low value
4. Smooth Operation	**4. Smooth Operation**
a. Attenuate upstream disturbances before affecting the distillation tower	a. Include feed drum with averaging level control (LC-2) to reduce amplitude of variations in flow rate and composition
b. Moderate flow variations to downstream units	b. Averaging level control for liquid inventories (LC-1 and LC-3). Pressure control must be tightly tuned for safety
5. Product Quality	**5. Product Quality**
a. Control the distillate (top) quality to 4% ± 0.5% heavy key	a. Measure the composition onstream and control it using feedback (AC-1) and control by adjusting the reflux
6. Efficiency and Profit	**6. Efficiency and Profit**
a. Control the bottoms product near its desired value of 3% ± 1.5% to reduce energy consumption	a. Measure a tray temperature (T7) that gives a reasonable inference of the bottom's composition and control by adjusting the reboiler duty. This could be tuned slowly if unfavorable interactions occur with AC-1
b. Control the pressure of the tower to its lower bound to reduce energy consumption	b. Adjust the pressure (PC-1) set point to reduce the liquid level in the flooded condenser (L4) near 0%
7. Monitoring and Diagnosis	**7. Monitoring and Diagnosis**
a. Monitor the tray operation and effectiveness	a. Measure the pressure difference across sections of the tower to indicate hydraulic behavior and corrosion. Also, measure selected tray temperatures to indicate the composition profile
b. Monitor the material balance on the unit	b. Measure flows into and out of the unit to provide redundancy
c. Monitor the energy consumption	c. Measure the reboiler flow rate (F9)
d. Monitor the product quality	d. Calculate the standard deviation and amount of violations of product quality specifications
e. Monitor the performance of key instrumentation	e. Provide duplicate sensors for key measurements, the pressure (PC-1 & P3) and the control tray temperature (TC-7 & T10)

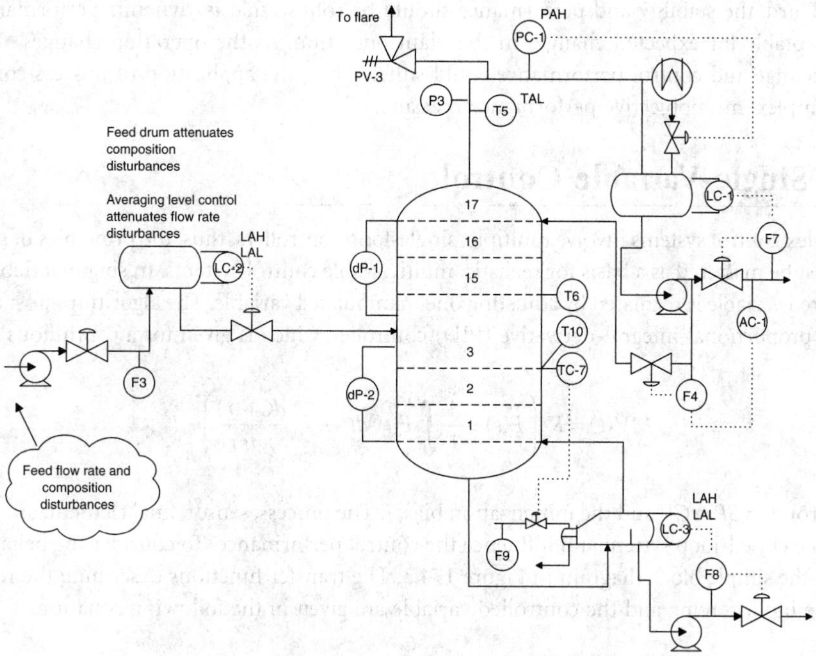

FIGURE 174.1 Two-product distillation process with instrumentation and multiloop control.

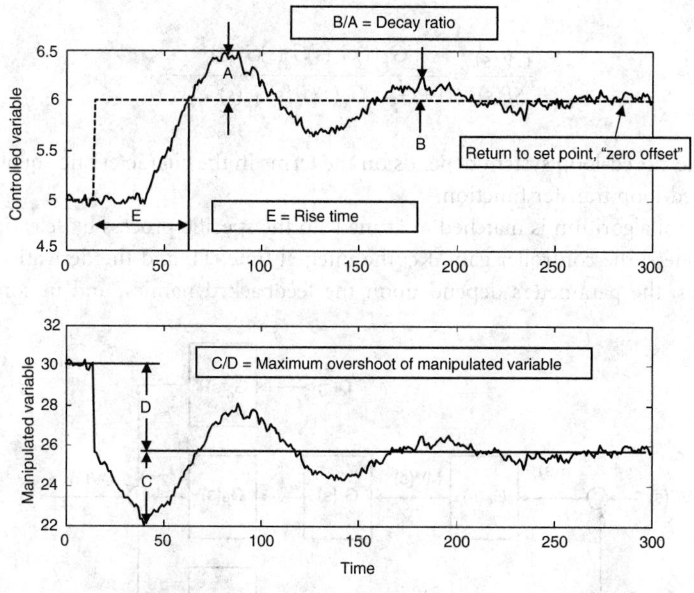

FIGURE 174.2 Example of control-loop dynamic response.

reference value); this can be measured by the integral of the error squared for short data sets and by the standard deviation of the error for long data sets. Note that the contribution of this deviation due to sensor noise cannot be eliminated by control. Second, the behavior of the manipulated variable is crucial in many processes. Large rapid changes in the manipulated variable could lead to damage of process equipment; an example is adjustments to the fuel to a fired heater, where large, long-term fuel variations could lead to excessive thermal stresses. In other cases, disturbances could be propagated to other important units; an example is tight level control that could lead to excessive flow fluctuations thorough the plant. Third, the stability and performance should be robust, that is, dynamic performance should remain acceptable for expected changes in the plant operation. As the operation changes, the process dynamics change and control performance could suffer. Thus, the application of process control must satisfy a complex, multiobjective performance measure.

174.2 Single-Variable Control

Most complex control systems involve multiple, single-loop controllers; thus, the principles of single-loop control must be mastered as a basis for realistic, multivariable control systems. In single-variable systems, one measured variable is regulated by adjusting one manipulated variable. The algorithm most commonly used is the proportional-integral-derivative (PID) controller, which is given for a continuous system:

$$MV(t) = K_c\left(E(t) + \frac{1}{T_i}\int_0^t E(t')dt' - T_D\frac{dCV(t)}{dt}\right) + I \tag{174.1}$$

with the error, $E = SP - CV$ and the initialization bias, I. The process, sensor, final element, and controller appear in the closed-loop system and influence the control performance. To consider the behavior of the system, see the simple block diagram in Figure 174.3. The transfer functions describing the relationship between the input forcing and the controlled variable are given in the following equations.

Disturbance response:
$$\frac{CV(s)}{D(s)} = \frac{G_D(s)}{1+G_p(s)G_v(s)G_c(s)G_f(s)G_s(s)} \tag{174.2}$$

Set point response:
$$\frac{CV(s)}{SP(s)} = \frac{G_p(s)G_v(s)G_c(s)G_f(s)}{1+G_p(s)G_v(s)G_c(s)G_f(s)G_s(s)} \tag{174.3}$$

The stability of the closed-loop system depends on the terms in the characteristic equation, the denominator of the closed-loop transfer function.

The single control algorithm is matched or "tuned" to the specific process by selecting values for the adjustable parameters: the controller gain, Kc; the integral time, TI; and the derivative time, Td. In all tuning approaches, the parameters depend upon the feedback dynamics, and in some methods, the

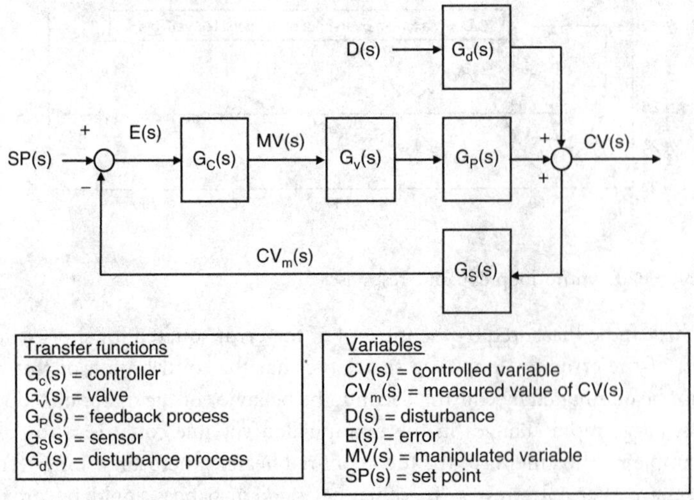

FIGURE 174.3 Single-loop block diagram.

disturbance dynamics are also considered. The method based on a stability margin developed by Ziegler-Nichols (1942) was used for many years; this method established a stability margin for proportional-only feedback and estimated tuning constant values based on heuristics. When digital computing was available to optimize tuning, methods based on closed-loop dynamic performance were developed. The success and generality of these methods depended on the definition of control performance measures and scenarios considered. Initial methods considered only controlled variable performance for perfect models (Lopez et al., 1969), while later methods included sensor noise (Fertik, 1975).

Newer tuning methods are available, such as Morari and Zafiriou (1989) and Marlin (2000), which provide a better balance of performance and robustness. The tuning correlations presented here are restricted to processes that have stable, monotonic step responses, so that they can be adequately modeled using first-order with dead-time models (Brosilow and Joseph, 2002).

Feedback process:
$$\frac{CV(s)}{MV(s)} = \frac{K_p e^{-\theta s}}{\tau s + 1} \tag{174.4}$$

Disturbance process:
$$\frac{CV(s)}{D(s)} = \frac{K_D e^{-\theta_D s}}{\tau_D s + 1} \tag{174.5}$$

The following IMC tuning correlations provide tuning values to achieve a user-specified closed-loop response time, τ_{CL}.

$$K_p K_c = \frac{2\tau + \theta}{2\tau_{CL}} \qquad T_I = \tau + \frac{\theta}{2} \tag{174.6}$$

The smaller the value of the closed-loop time constant (τ_{CL}), the more aggressive the controller. As guidelines, the closed-loop time constant should be greater than 1.7θ and always greater than 0.2τ. Details on the development of IMC tuning rules and more general correlations are provided in Rivera et al. (1986) and Morari and Zafiriou (1989).

The tuning charts in Figure 174.4 give PI tuning that provides robust control with moderate adjustments to the manipulated variable. Details on their development and more general correlations are provided in the tuning charts in Marlin (2000). Regardless of the tuning correlation, the resulting values should be considered as *initial estimates* to be modified to achieve the desired performance for a specific process.

The same single-loop PI controller can be successfully applied to level control, or more generally, inventory control. However, because of the unique process dynamics and control objectives, special tuning approaches are required. The dynamic response for the level in Figure 174.5 is an integrator, which is open-loop unstable. In addition, we usually select from two different objectives for control. Tight level control seeks to maintain the level close to its set point; it is appropriate for controlling small levels and for processes in which the inventory affects process performance, such as chemical reactors. Loose or averaging level control seeks to achieve moderate changes to the manipulated flow rate, while maintaining the level within acceptable limits; it is appropriate for most tank and drum levels, when preventing downstream disturbances is important. Tuning for linear PI level control is given in Figure 174.5. Derivations for these expressions and an approach for nonlinear PI control are available in Marlin (2000), and more background and correlations on level control are given by Cheung and Luyben (1978).

Single-Loop Control Performance

Using a well-tuned feedback algorithm does not ensure acceptable control performance because control performance depends on all elements in the closed-loop system, as seen in Equation (174.2). Important insights can be obtained by considering the frequency response of Equation (174.2), $|CV(j\omega)|/|D(j\omega)|$, shown in Figure 174.6. For disturbances with very low frequencies, feedback control is much faster than the

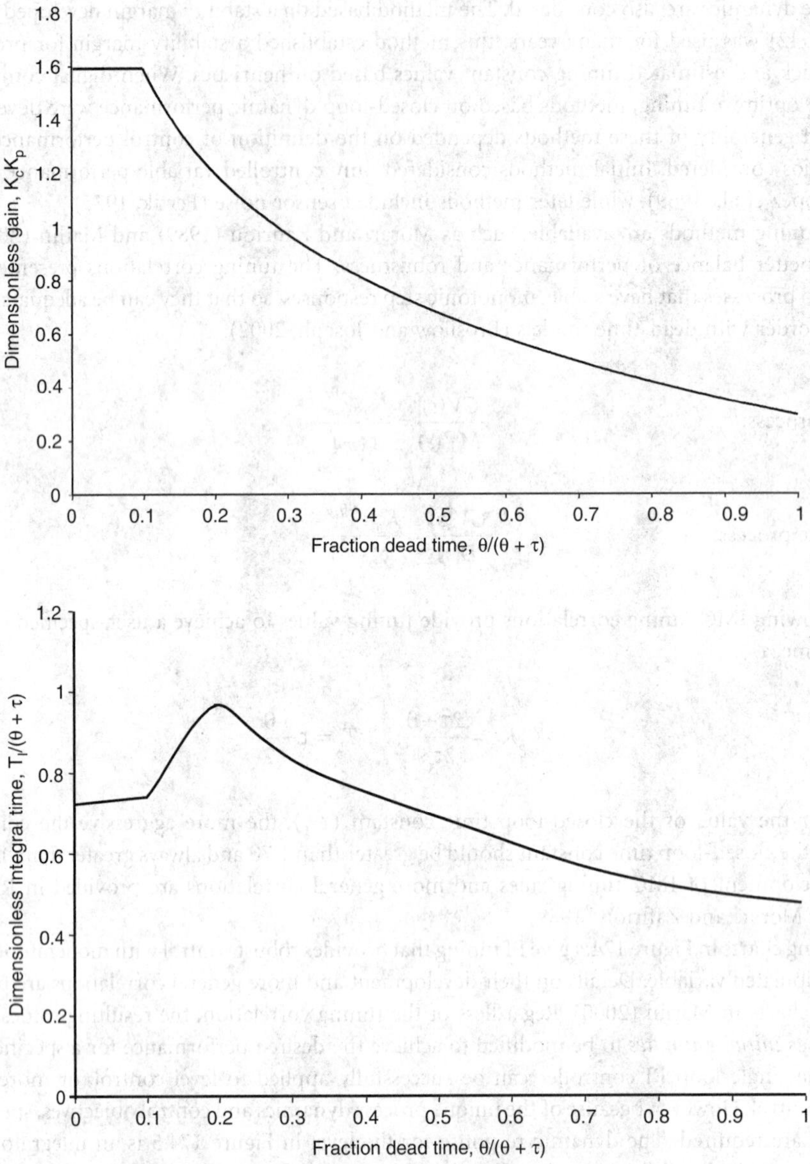

FIGURE 174.4 Proportional integral tuning correlations. (*Source:* Marlin, T. 2000. *Process control: designing processes and control systems for dynamic performance.* McGraw-Hill, New York. With permission.)

disturbance, and is thus quite effective. For very high frequencies, the disturbance time constant attenuates the effect of the disturbance on the controlled variable, and the performance is quite good, although not due to feedback compensation. In some intermediate range of frequencies, resonance occurs and feedback control performance is poor. The location of the resonance peak depends primarily on the feedback dynamics for a (stable, well-tuned) feedback control system. Therefore, the dynamics of the feedback system (process, sensor, and final element) impose a fundamental limitation to feedback performance.

The effects of various elements in the feedback loop on the performance of single-loop control are summarized in Table 174.2. This table provides the basis for design decisions to improve control performance. It is important to note that large dead times and time constants in the feedback loop are always detrimental, whereas large time constants in the disturbance path, which is not in the feedback loop, can improve control performance.

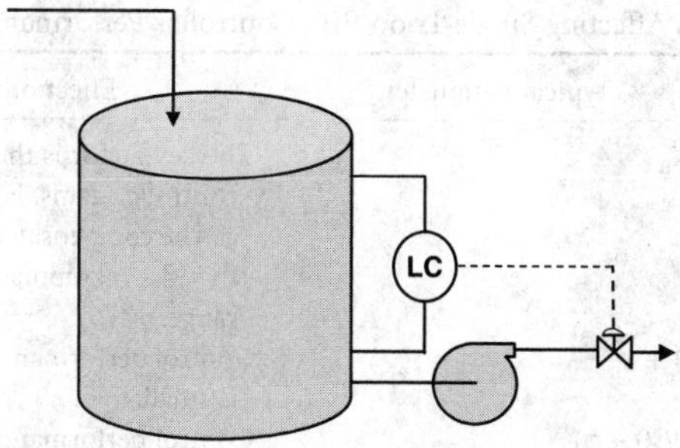

$$K_c = \frac{-0.737\Delta F_{max}}{\Delta L_{max}}$$

$$T_I = \frac{4A}{-K_c}$$

ΔL = maximum allowable change in level (m)

ΔF = maximum flow disturbance magnitude (m^3/min)

A = cross sectional area (m^2)

For both controls, set point = 50% of range

For averaging level control, ΔL ≈ 40% of range

For tight level control, ΔL ≈ 5% of range

FIGURE 174.5 Level control design with tuning for averaging and tight level control.

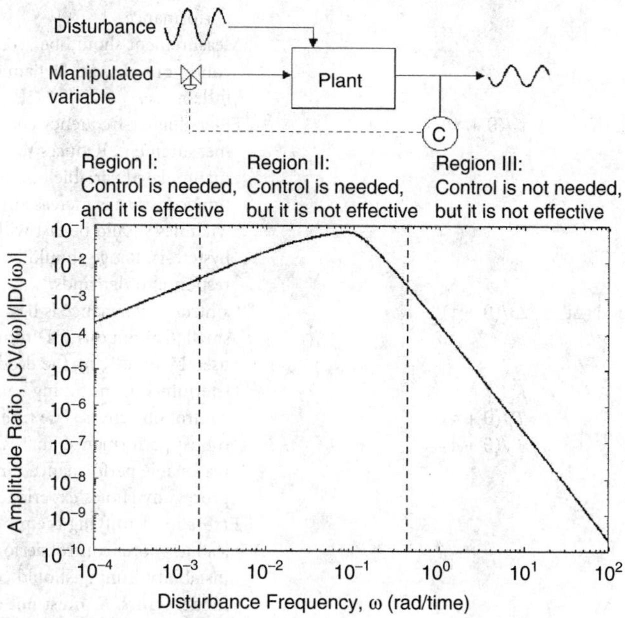

FIGURE 174.6 Typical closed-loop frequency response.

Instrumentation provides the essential measurements and final elements for process control. Important features of the sensor include accuracy, reproducibility, reliability, and dynamics; these issues are discussed in Hughes (2002). Important features of the final element, which is usually a valve, include stiction, hysteresis, characteristic, dynamics, tight shutoff, and failure position; these issues are discussed in Driskell (1983).

Single-Loop Control Improvement Through Process Modifications

The principles in Table 174.2 can be used in designing high-performance single-variable control loops by modifying the process dynamics, either feedback or disturbance as appropriate. In the heat exchanger example in Figure 174.7a, the initial design has slow feedback dynamics because of the heat exchanger

TABLE 174.2 Summary of Factors Affecting Single-Loop PID Controller Performance

Key Factor	Typical Parameter	Effect on Control Performance
Feedback process gain	K_p	The key factor is the product of the process and controller gains. For example, a small process gain can be compensated by a larger controller gain. Note that the manipulated variable must have sufficient range.
Feedback process "speed"	$\theta + \tau$	Control performance is always better when this term is small.
Feedback fraction process dead time	$\theta/(\theta + \tau)$	Control performance is always better when this term is small.
Inverse response	Numerator term in transfer function, $(\tau s + 1)$ with $\tau < 0$	Control performance degrades for large inverse response.
Magnitude of disturbance effect	$K_d\|(\Delta D\|$	Control performance is always better when this term is small.
Disturbance dynamics	τ_D	Control performance is best when the disturbance is slow, that is, the time constant is large.
	ω_D	Feedback control is effective for low-frequency disturbances and is least effective at the resonant frequency.
	θ_D	Disturbance dead time does not influence performance.
Sensor		Measurement should be accurate as operating conditions change. Dynamics should be fast with little noise.
Filter, $1/(1 + \tau_f s)$	$\tau_f/(\theta + \tau)$	Filters higher-frequency components of measurement. Reduces the variability of the manipulated variable but degrades control as filter time constant is increased.
Final element		Dynamics should be fast without sticking or hysteresis. Range should be large enough for response to demands.
Controller execution period (Δt)	$\Delta t/(\theta + \tau)$	Control performance is best when this parameter is small. Continuous PID tuning correlations can be used by modifying the dead time, $\theta' = \theta + \Delta t/2$.
Controller tuning	$K_c K_d$ $T_I/(\theta + \tau)$ $T_d/(\theta + \tau)$	Determined from tuning correlations based on control objectives. The tuning should provide robust performance, that is, should provide reasonable performance for the expected range of process dynamics experienced in the plant.
Modeling errors		Errors in identifying the process model parameters lead to poorer control performance and, potentially, instability. Tuning should consider the estimate of model errors. K_p must not change sign.
Limitations on manipulated variable	min < mv(t) < max	Limitations on manipulated variables reduce the operating window, that is, the range of achievable conditions. An active limit would cause steady-state offset from the set point.

Source: Marlin, T. 2000. *Process control: designing processes and control systems for dynamic performance.* McGraw-Hill, New York. With permission.

being in the feedback loop. The improved design in Figure 174.7b involves rapid mixing of a bypass stream to shorten the feedback dynamics, thus providing better performance. In addition, the cooling stream flow rate in Figure 174.7b can be manipulated independently, which is essential when it is a process stream from another section of the plant.

In the distillation example in Figure 174.1, the feed flow rate and composition experience variation at frequencies near the resonant frequency of the composition control loop. The design includes features

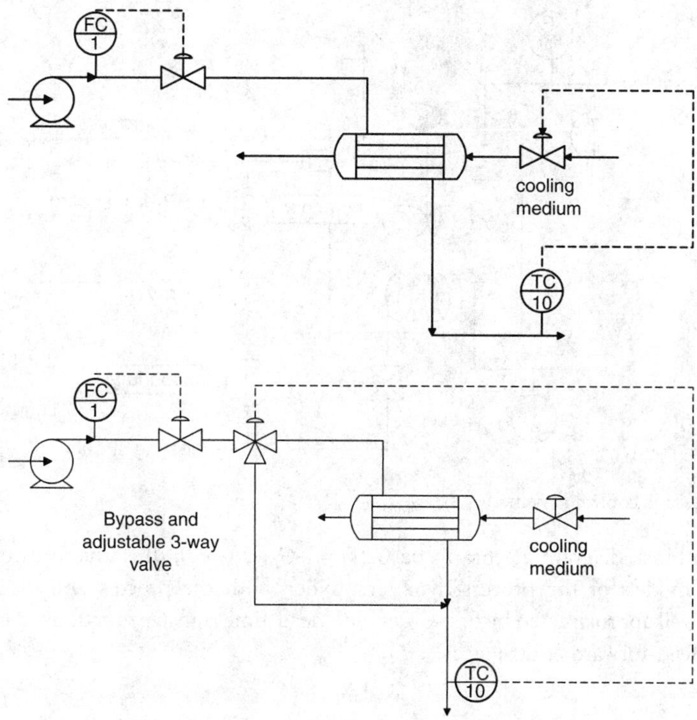

FIGURE 174.7 Heat exchanger control with (a) slow and (b) fast feedback dynamics.

to moderate the effects of disturbances. The feed flow rate disturbance has been attenuated by the addition of a feed storage drum with averaging level control, and the feed composition disturbance is attenuated by mixing in the feed storage drum. As a result, the disturbances experienced by the distillation tower are smaller in magnitude, and the control performance is substantially improved.

Single-Loop Control Improvements Through Control Structure Modifications

The control performance of a single variable can be improved, without changing the process, by modifying the control structure. Four common control structure modifications are presented here. The first new structure is cascade control, which uses a secondary measured variable to provide an early indication of selected disturbances. As shown in Figure 174.8, the secondary variable, the measured heating medium flow rate is controlled using feedback principles, and the set point of the flow controller is adjusted via feedback to achieve the desired outlet temperature. Each of the controllers can use the standard PID controller algorithm, and the inner or secondary controller must be tuned first. The cascade design provides good performance for only some disturbances; for example, the design in Figure 174.8 gives good performance for disturbances in the pressure of the heating medium source, but it does not improve performance for feed temperature disturbances.

Cascade control must conform to specific design criteria. In particular, the secondary variable must (1) be measured, (2) indicate the occurrence of an important disturbance, (3) be causally affected by the final element, and (4) have feedback dynamics much faster than the primary variable (Marlin, 2000). A multiple-layer cascade control design is common to compensate for several disturbances; however, each layer must conform to the four design criteria above.

The second control structure provides an alternative to feedback by measuring the disturbance before it influences the process. This *feed-forward* modification, shown in Figure 174.9, eliminates the feedback dynamics from the control compensation for the measured disturbance and can theoretically provide excellent control. The feed-forward control algorithm to give perfect feed-forward compensation can be

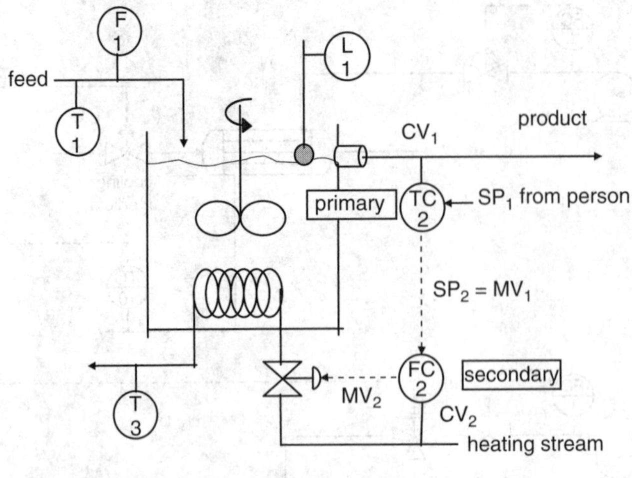

FIGURE 174.8 Cascade control system.

determined from block diagram algebra to be $G_{ff}(s) = -G_D(s)/Gp(s)$; thus, the feed-forward algorithm depends on the models of the process dynamics. When both the process and disturbance transfer functions can be well approximated by first-order with dead-time transfer functions in Equations (174.4) and (174.5), the feed-forward controller is

$$\text{Feedforward controller:} \qquad G_{ff}(s) = \frac{MV_{ff}(s)}{D(s)} = -\frac{K_d}{K_p}\left(\frac{\tau s+1}{\tau_D s+1}\right)e^{-(\theta_D-\theta)s} \qquad (174.7)$$

Equation (174.7) is often used for feed-forward control and can be realized using standard gain, dead-time, and lead/lag algorithms.

Feed-forward control must conform to specific design criteria. In particular, the measured disturbance variable must (1) be measured, (2) indicate the occurrence of an important disturbance, (3) not be causally affected by the final element, and (4) have disturbance dynamics not faster than the compensation dynamics (Marlin, 2000). Usually, feedback control is retained because feed-forward compensates for only the measured disturbance(s) and because the feed-forward control is not perfect due to model mismatch. The feed-forward and feedback adjustments can be added, as in Figure 174.9, using the property of linearity.

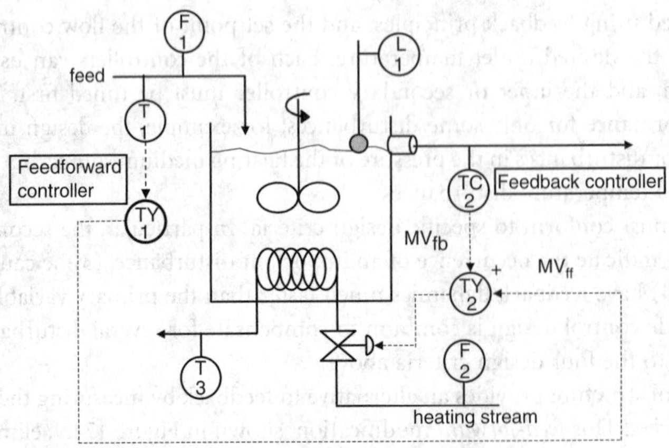

FIGURE 174.9 Feed-forward feedback control system.

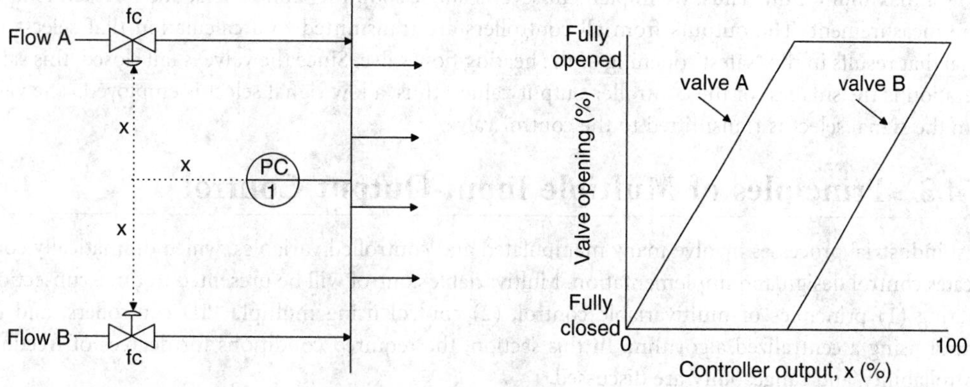

FIGURE 174.10 Pressure control using a split range strategy.

The third control structure is split range control, which enables one controller to adjust more than one final element. Split range can be used to (1) improve the speed of response, (2) expand the range of the manipulation, or (3) utilize the most economically attractive manipulated variable. An application involving an economic objective is shown in Figure 174.10, in which a pressure controller adjusts two sources of material to satisfy demands in the plant. The controller calculates only one output (x) that is transmitted to two final elements. The final elements are calibrated to respond to the controller output as shown in the figure, so that only one valve is moving for any value of x. Under normal operation, v_A is adjusted and v_B is fully closed; this represents the case in which flow A is less expensive. When the controller requests a large flow of material and the controller output (x) is over 50%, v_A is fully opened and v_B is adjusted. In general, the feedback dynamics, at least the process gain, can change when the valves are switched, and the engineer should investigate the need for adjusting the controller tuning based on the controller output value.

The fourth control structure is *signal select*, which is explained here for an application in which three controllers adjust only one final element, as shown for a chemical reactor in Figure 174.11. We want to

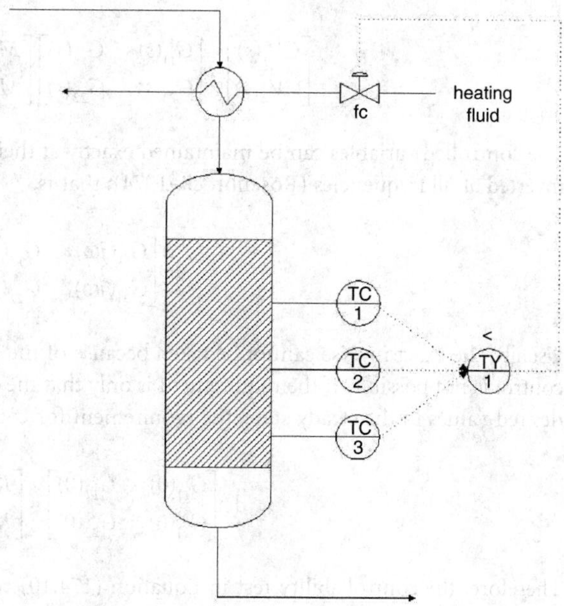

FIGURE 174.11 Reactor temperature control using a signal select strategy.

control the temperature but also ensure that the hottest temperature in the packed bed reactor remains below a maximum limit. Thus, we implement several single-loop PID controllers, one for each temperature measurement. The outputs from all controllers are transmitted to a calculation that selects the signal that results in the "safest" opening of the heating flow valve. Since the valve is fail closed, this safest operation is the smallest of the controller output values; thus, a low signal select is employed. The value from the signal select is transmitted to the control valve.

174.3 Principles of Multiple Input-Output Control

Most industrial processes involve many manipulated and controlled variables, which dramatically complicates control design and implementation. Multivariable control will be presented in three subsections covering (1) principles of multivariable control, (2) control using multiple PID controllers, and (3) control using a centralized algorithm. In this section, the required conditions for degrees of freedom, controllability, and rangeability are discussed.

The selection of variables to be controlled is based on the control objectives; that is, we control what is required for safety, product quality, production rate, and so forth; refer again to Table 174.1 for an example of the relationships between control objectives and control variable selection. After the controlled and manipulated variables have been defined, we must ensure that control is possible, that is, that we have sufficient *degrees of freedom.* A simple test is to compare the number of manipulated and controlled variables. Clearly, to achieve desired (independent) values of the controlled variables, the plant must have at least as many manipulated (final elements) as controlled variables. If this condition is not satisfied, the process design must be modified to include additional flexibility and final elements.

The final elements must also be able to be controlled, which is referred to as *controllability.* Because of the complex nature of chemical processes, final elements do not influence only one controlled variable. Thus, *interaction* exists, with interaction resulting from the effect of a manipulated variable on many controlled variables. Because of interaction, the engineer cannot determine whether control is possible by observation. Several methods are available to determine whether a selection of controlled and manipulated variables is controllable; the method should be matched to the specific application and dynamic behavior required, continuous tracking, continuous regulation, batch control, and so on. Rosenbrock (1974) provides an excellent discussion with calculation methods. For continuous processes, a relevant controllability test is given in the following for the two input/two output system shown in Figure 174.12a.

$$\begin{bmatrix} CV_1(s) \\ CV_2(s) \end{bmatrix} = \begin{bmatrix} G_{11}(s) & G_{12}(s) \\ G_{21}(s) & G_{22}(s) \end{bmatrix} \begin{bmatrix} MV_1(s) \\ MV_2(s) \end{bmatrix} + \begin{bmatrix} G_{d1}(s) \\ G_{d2}(s) \end{bmatrix} D(s) \tag{174.8}$$

The controlled variables can be maintained exactly at their desired values if the feedback process can be inverted at all frequencies (Rosenbrock, 1974); that is,

$$\det \begin{bmatrix} G_{11}(j\omega) & G_{12}(j\omega) \\ G_{21}(j\omega) & G_{22}(j\omega) \end{bmatrix} \neq 0 \tag{174.9}$$

Usually, the exact inverse cannot be taken because of the process dynamics (e.g., dead times), so perfect control is not possible. If the requirement is only that the controlled variables can be maintained at their desired values in the steady state, the requirement for controllability is

$$\det \begin{bmatrix} G_{11}(0) & G_{12}(0) \\ G_{21}(0) & G_{22}(0) \end{bmatrix} = \begin{bmatrix} K_{11} & K_{12} \\ K_{21} & K_{22} \end{bmatrix} \neq 0 \tag{174.10}$$

Therefore, the controllability test in Equation (174.10) requires that the steady-state gain matrix must be invertable, that is, its determinant must not be zero. This serves as a useful controllability test for

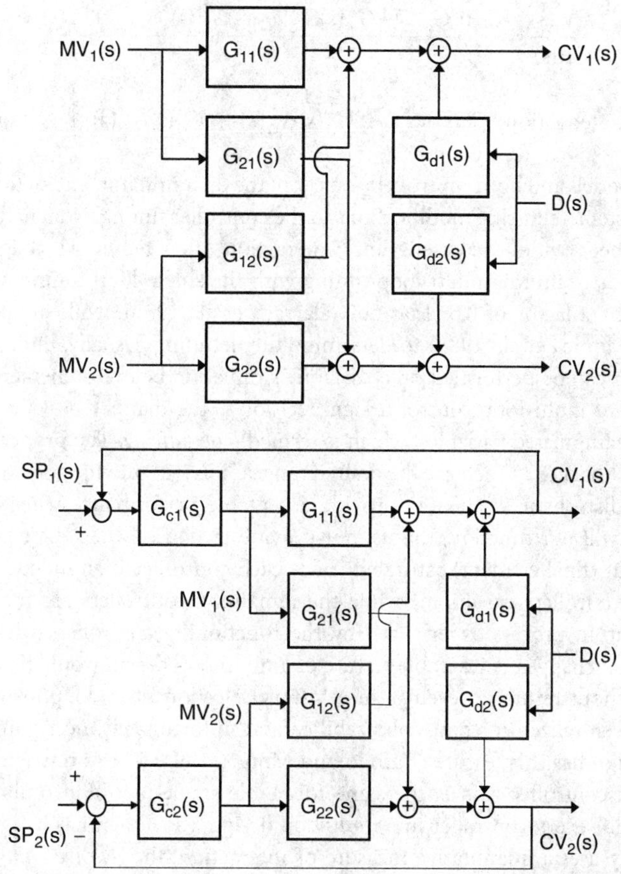

FIGURE 174.12 Multiloop (2 × 2) block diagram for (a) process only and (b) multiloop control.

continuous processes that must be satisfied before the design of the control system begins. Note that this test is only valid at the linearization point and does not ensure that the dynamic performance is satisfactory at all operating conditions.

A final process requirement involves the ability of the system to maintain controlled variables at their set points as the values of the set points and process disturbances vary within expected limits. The acceptable range of plant operation is often referred to as the *operating window* or the *feasible region*. Qualitatively, the process must be able to adjust the manipulated variables over a wide enough range of values to reach the set points and reject the expected range of disturbances.

When a process design satisfies the three essential requirements (degrees of freedom, controllability, and rangeability) the design procedure can continue to address the feedback strategy to achieve good dynamic performance. If the process is deficient in any of these areas, the process must be modified by, for example, increasing the final element capacity or adding an additional manipulated variable. In the following subsections, the two major multivariable control approaches are introduced: multiloop control and centralized multivariable control.

174.4 Multiloop Multivariable Control

Often, multiple single-loop controllers are used to automate a multivariable process, but the existence of *interactions* between loops dramatically increases the challenges in analysis and design. For the simple two input/two output system in Figure 174.12b, each input can influence both outputs (when $G_{ij}(s) \neq 0$ for $i \neq j$). As a result the closed-loop transfer function is

$$\frac{CV_1(s)}{SP_1(s)} = \frac{G_{c1}(s)G_{11}(s) + G_{c1}(s)G_{c2}[G_{11}(s)G_{22}(s) - G_{12}(s)G_{21}(s)]}{CE(s)} \tag{174.11}$$

with the characteristic equation, $CE(s) = 1 + G_{c1}(s)G_{11}(s) + G_{c2}(s)G_{22}(s) + G_{c1}(s)G_{c2}(s)[G_{11}(s)G_{22}(s) - G_{12}(s)G_{21}(s)]$.

Since all process models and both controllers appear in the denominator, interaction affects the stability of the closed-loop system, requiring modifications to the controller tuning. No reliable, general multiloop tuning method has been widely accepted. Since interaction often requires less aggressive tuning, one approach recommends beginning multi-loop tuning with the single-loop tuning modified by reducing the controller gains by a factor of 0.5 (Marino-Galarraga et al., 1987), while an alternative uses a log-modulus criterion (Monica et al., 1988) to determine the detuning. Usually, multiloop tuning requires trial and error, which can be performed on a dynamic simulation before implementation in a plant.

The most important multi-loop control design decision is the manner in which the controllers link the controlled and manipulated variables, which is termed *loop pairing*. The proper loop pairing should yield a control system that can (1) achieve the desired rangeability, (2) provide good dynamic performance for the most likely disturbances, and (3) provide acceptable performance when some loops are not functioning. A loop will not function, that is, not provide feedback, when the controller is in manual (off) or has its output (final element) saturated. Since each controller is an independent algorithm, any one or a group of controllers can be off while the remaining controllers are functioning. When this occurs, the input-output process experienced by the functioning controllers will change, that is, the characteristic equation changes. A desirable feature of a multiloop design would be that a control system would be stable (without retuning) even if an arbitrary selection of its loop(s) were turned off. This feature is termed decentralized integral stabilizability, and unfortunately, no rigorous method exists to ensure a control design has this feature (Campo and Morari, 1994). A less restrictive feature is that the sign of the feedback controller gain is the same for stable single-loop and multiloop control. If this criterion is satisfied for every controller in a multiloop design, the design has integrity.

A test for integrity uses a quantitative measure of interaction, the *relative gain array* (Bristol, 1966; McAvoy, 1983). The relative gain is defined as

$$\lambda_{ij} = \frac{\left\{\dfrac{\partial CV_i}{\partial MV_j}\right\}_{MV_k=\text{const}k\neq j}}{\left\{\dfrac{\partial CV_i}{\partial MV_j}\right\}_{CV_k,\text{const}k\neq i}} = \frac{\left\{\dfrac{\partial CV_i}{\partial MV_j}\right\}_{\text{other loops open}}}{\left\{\dfrac{\partial CV_i}{\partial MV_j}\right\}_{\text{other loops closed}}} \tag{174.12}$$

The relative gain array is formed as a two-dimensional matrix of elements, λ_{ij}. Each relative gain element indicates how interaction influences the steady-state gain of a specific input-output pairing, with a value of 1.0 indicating no influence and large deviations from 1.0 indicating strong influence. The relative gain can be evaluated using only the open-loop steady-state gains (McAvoy, 1983).

Clearly, control is simpler when the process gain "seen" by every controller never changes sign; this good property is characterized by positive relative gains. A multiloop control design with an input-output ($M_j - CV_i$) pairings having a negative relative gain (λ_{ij}) will have one of the following properties (McAvoy, 1983; Morari and Zafiriou, 1989):

- Unstable with all controllers in automatic
- Unstable when only one controller ($MV_j - CV_i$) is in automatic
- Unstable when one controller ($MV_j - CV_i$) is in manual and all others in automatic

Each of these situations is undesirable. Thus, if a design has a negative relative gain pairing, its is said to have *poor integrity*. Note that the results presented here are for stable processes; modified results are available for open-loop unstable processes (Hovd and Skogestad, 1994).

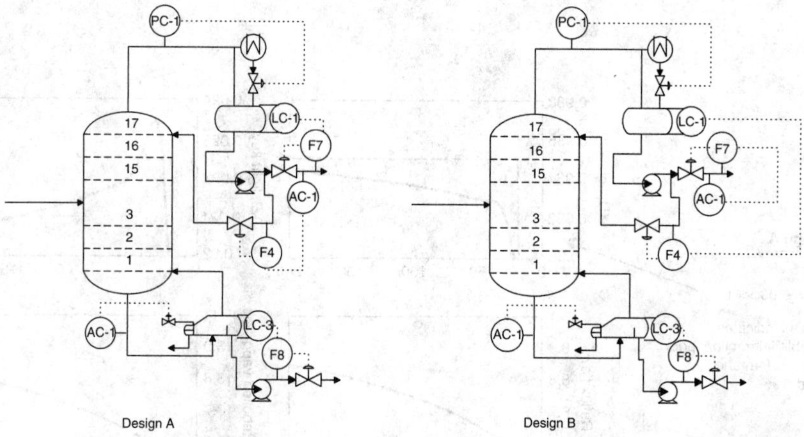

Relative Gain Array (Design A):

	Reflux	Reboil
XD	6.09	−5.09
XB	−5.09	6.09

Relative Gain Array (Design B):

	Distillate	Reboil
XD	0.39	0.61
XB	0.61	0.39

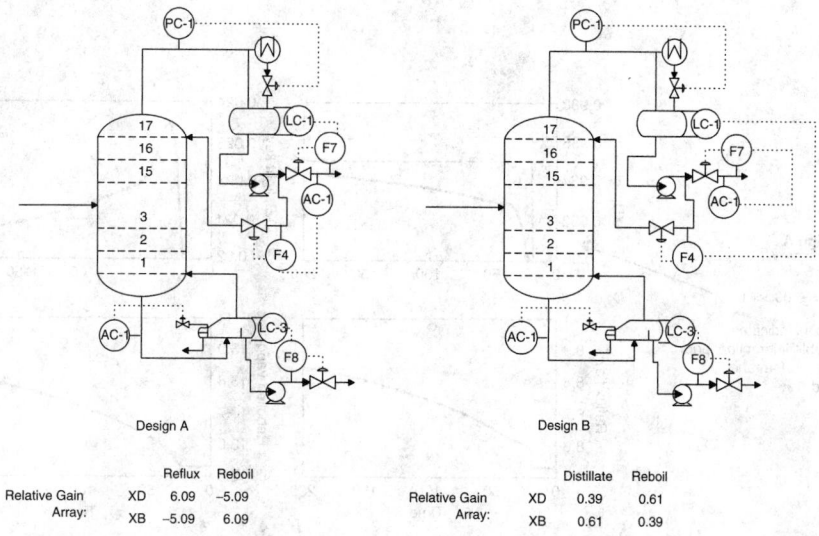

FIGURE 174.13 Alternative multiloop distillation control designs with relative gain arrays.

Naturally, good dynamic performance is the overall goal; unfortunately, no rigorous shortcut test is available to ensure good dynamic control performance. However, several useful guidelines exist. First, the most important variables should be paired with manipulated variables that have sufficient range, that is, variables that can be adjusted to compensate for all expected disturbances. Second, the most important variables should be paired with manipulated variables that provide fast feedback compensation. Third, the process equipment and control structures should be selected to attenuate disturbances; this could include tankage before sensitive units and enhancements such as cascade and feed-forward control. Also, inventories (liquid level and solid mass) should be controlled in a manner that attenuates flow variation to integrated sections of the plant.

Fourth, the loop pairing should not have strongly unfavorable interaction. The effects of favorable and unfavorable interaction will be demonstrated through a two-product distillation tower example. Distillation affects a separation of feed components based on differences in volatility, and the reboiler consumes a large amount of energy at high temperature. The tower in Figure 174.13 is considered; this tower has a conventional pressure control design (Sloley, 2001). The two designs considered differ in the variables adjusted to control the product qualities. Data regarding this case study are available (Marlin, 2000:161). Both designs have loop pairings on positive values of relative gain. In spite of the seemingly small changes in loop pairings, the relative gains are very different, with Design B having the relative gain closer to 1.0.

The control performances of the two designs are shown in Figures 174.14 for changes in one set point and for a feed composition disturbance. Clearly, Design B is better for the set point change, while Design A is better for the disturbance response. An important observation is that control design rules based solely on relative gains being near 1.0, as has been suggested, are not reliable. Since the feedback dynamics are similar for the two cases, the very different performances are not due to dynamics.

The key principle demonstrated by this example is that the directionality of interactions and disturbances strongly affects multiloop control performance. Favorable interaction can occur where one loop, while controlling a variable, introduces adjustments that also tend to reduce the variability of other interacting controlled variables. Both favorable and unfavorable interactions are possible; thus, the proper pairing selection is crucial. Short-cut methods are available for analyzing the likely effects of interaction (Stanley et al., 1985; Skogestad and Morari, 1987). The methods calculate an estimate of the effect of a specific disturbance on closed-loop performance using some simplifying assumption about the dynamics. The following equations summarize the key results for the relative disturbance gain method, which calculates an estimate of the multiloop performance, $IE_i\ dt$, for the system in Figure 174.12b.

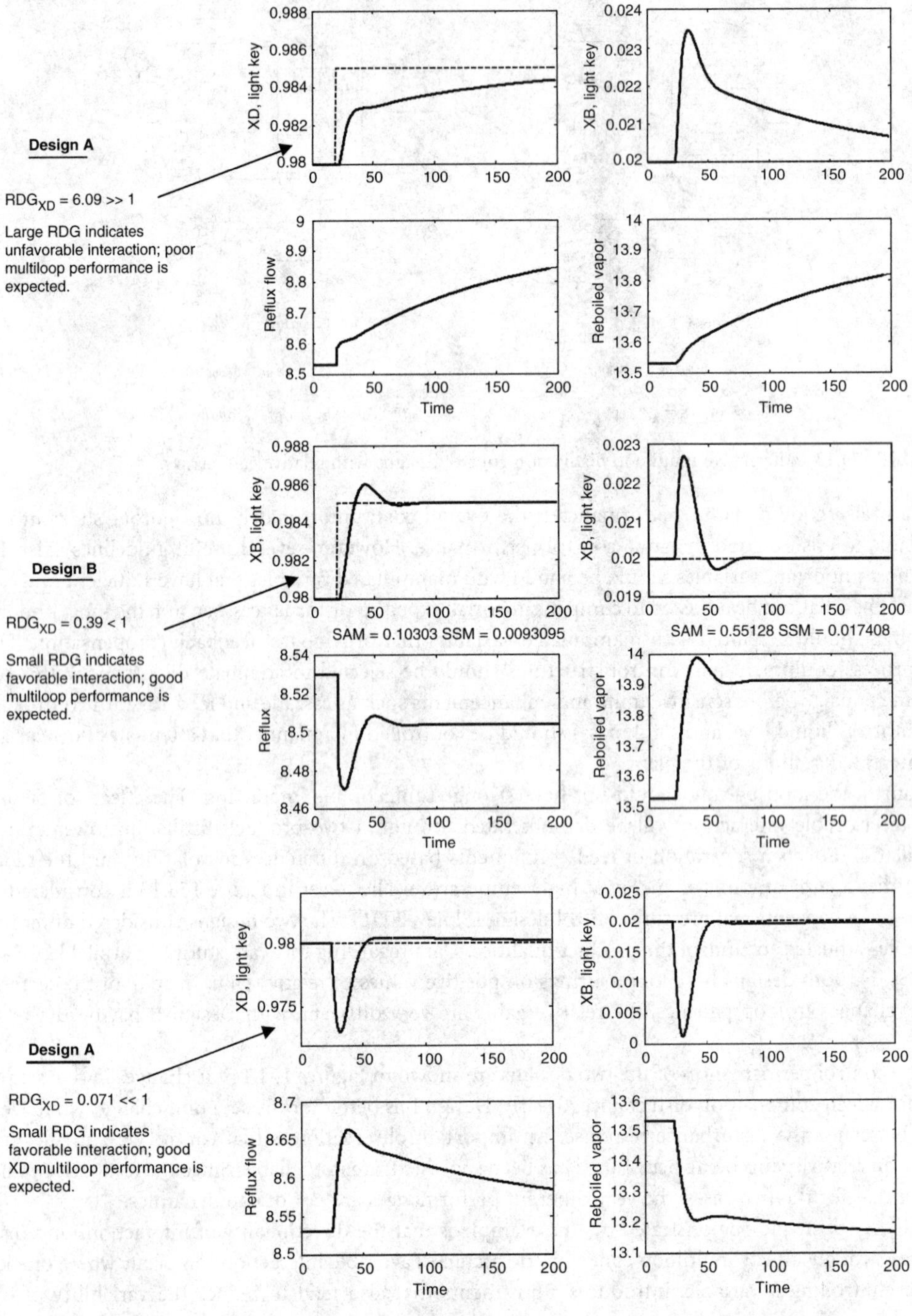

FIGURE 174.14 Dynamic performance of multiloop distillation control for (a) set point change and (b) feed composition disturbance.

Continued on facing page.

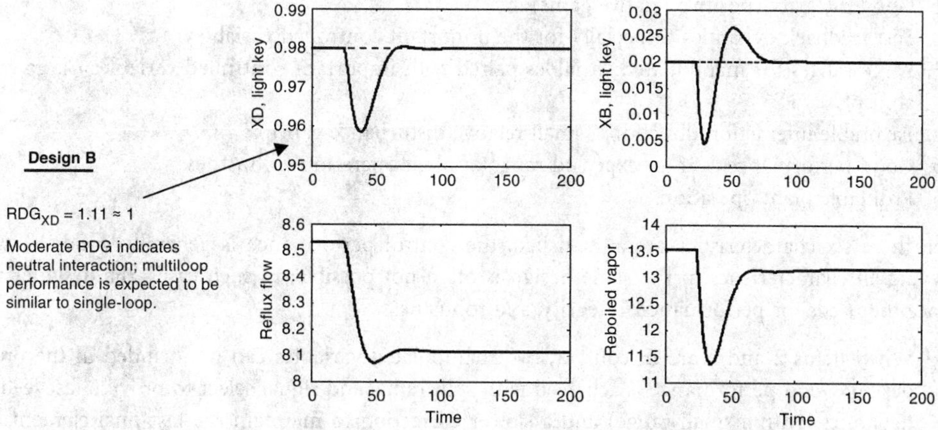

FIGURE 174.14 *Continued.*

Multiloop control performance: $$\int_0^{\infty}(E_1(t))_{ML}\,dt = RDG\, f_{tune}\int_0^{\infty}(E_1(t))_{SL}\,dt \tag{174.13}$$

Relative disturbance gain: $$RDG = \left(1 - \frac{K_{d2}K_{12}}{K_{d1}K_{22}}\right)\lambda_{11} \tag{174.14}$$

Detuning factor: $$f_1\left(\frac{K_{c1}}{T_{I1}}\right)_{SL}\left(\frac{K_{c1}}{T_{I1}}\right)_{ML}^{-1} \tag{174.15}$$

Single-loop performance: $$(E_1)_{SL} = \frac{K_D T_{I1}}{K_{11}K_{c1}} \tag{174.16}$$

Some important results demonstrated by this method (Marino et al., 1985) follow:

- The performance is estimated using only steady-state gains.
- The ratio of the disturbance gains (direction) has a strong effect on performance.
- The effect of the process gains is captured by the relative disturbance gain, which includes the relative gain and the disturbance direction.
- Controller tuning has a relatively small effect on the controller performance, because the detuning factor (f_i) often has a relatively small range (Marino et al., 1987).
- Multiloop performance is proportional to single-loop performance. If single-loop is poor, it is unlikely that multiloop will be acceptable.

Much can be deduced from the relative disturbance gain. If it is much greater than 1.0, interaction degrades control performance; if it is much smaller than 1.0, interaction likely improves performance (with the caveat that plus/minus error cancellation can also lead to a small integral error).

Note that the values calculated for the distillation performance in Figure 174.14 predict the relative control performances between Designs A and B. These shortcut methods are generally reliable for indicating the relative performance of multiloop systems, which is very useful when selecting a good design from many candidates.

In summary, multiloop designs must satisfy key criteria of degrees of freedom, controllability, and rangeability. From candidates that satisfy these requirements, a good design is selected to achieve the following characteristics:

1. Good integrity (positive relative gains)
2. Fast feedback dynamics (especially for the important controlled variables)
3. Large ranges for manipulated variables paired with important controlled variables (large rangeability)
4. Favorable interaction directions (small relative disturbance gain)
5. Good performance over an expected range of plant operation (robustness)
6. Profitable plant operation.

When these six characteristics are all satisfied, the control performance is generally good. However, achieving all characteristics in the same design is often not possible; in such cases, the designer must balance the needs for performance, integrity, and robustness.

- When items 2 and 3 are in conflict, two manipulated variables can be included in the process design. Good performance is achieved with split range and signal select to provide fast feedback dynamics (with a small range) and a slower correction to maintain the fast final element away from saturation.
- When items 1 and 2 are in conflict, some industrial systems achieve good performance by designing control systems paired on zero or negative relative gains (McAvoy, 1983b; Arbel et al., 1996). When a pairing on a zero or negative relative gain is selected, the possibility exists for unstable control to occur when a loop is off. Therefore, the plant personnel should be thoroughly trained and the automation system should have interlock features to disable controllers when interacting loops are not functioning.
- When item 6 is not consistent with the other characteristics, the control system is usually designed to achieve good dynamic operation and provide slow compensation to approach high profitability. The concept of constraint control is often used to improve profitability.

Unfortunately, no systematic procedure is available for designing the structure for multiloop control that includes all options; the skill and experience of the engineer strongly influences the design. The final design should be checked by dynamic simulation.

174.5 Centralized Multivariable Control

Multiloop control involves independent, decentralized controllers applied to a process with interactions. It seems reasonable to provide more information to the controller by allowing it to calculate all manipulated variables using a model of the entire process, with all input-output dynamics. Such a controller cannot be achieved by extending the PID algorithm, but several algorithms are available for centralized, multivariable feedback control. Here, model predictive control (MPC) is presented. This method addresses many practical issues in industrial applications, can be extended to a high-order system (many manipulated and controlled variables), and involves reasonable real-time computing. As a result, it is the dominant centralized control method used in the process industries.

As with most centralized controllers, the algorithm optimizes the dynamic response. The optimization is based on a model of the open-loop process, so that the (estimated) process interactions are considered explicitly in the controller algorithm. Thus, at each controller execution, the adjustments in the manipulated variables are determined that optimize an appropriate objective function measuring closed-loop control performance. Some of the important objectives in centralized process control include the following:

- The controlled variables should quickly return to their set points.
- The manipulated variable adjustments should not be too rapid. This objective protects process equipment; for example, large, rapid adjustments to the fuel to a fired heater could damage the metal tubes and brick refractory.
- The manipulated variables must observe hard constraints on their allowed values. For example, the adjustment of valve openings must remain between 0% and 100%, even if values outside of this range would result in a (predicted) performance improvement.

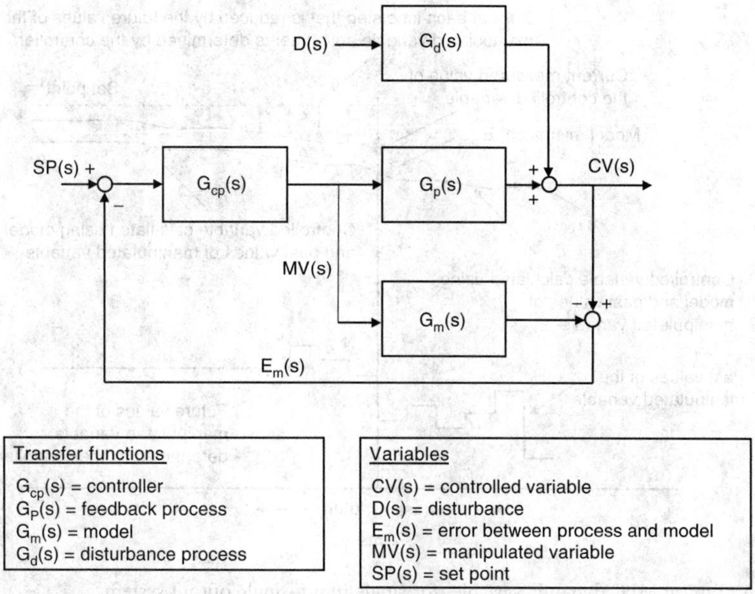

FIGURE 174.15 Block diagram of internal model control (IMC).

- The controlled variables should observe soft constraints on their allowed values. Note that feedback control cannot eliminate constraint violations in the dependent variables in all circumstances, because of slow feedback dynamics, potentially fast, large disturbances, and limitations to the adjustments in manipulated variables.
- The feedback system should perform well, that is, be stable and provide reasonable dynamic performance, over a range of process dynamics. This robustness requirement is needed because of plant-model mismatch, even at a nominal operation, and these errors increase due to changes in operation of non-linear plants.
- The preceding objectives should be achieved as the degrees of freedom for control change. For example, one or more of the manipulated variables can be fixed at their current values because an operator places the variables on "manual," that is, removes the variable from the control problem.

These objectives, especially the need to observe limitations in a problem with changing variable limitations, require solving an optimization every time the controller is executed. The controller to achieve this performance will be explained using the IMC (internal model control) structure shown in Figure 174.15.

At each execution, the controller performs an open-loop optimization to calculate future controller outputs that provide good performance for the controlled and manipulated variables, as shown in Figure 174.16. Only the first controller output is transmitted to the plant. This adjustment is also used as the input to a linear model of the process. Since the controller bases its results on a dynamic plant model, the difference between the measured plant outputs and the model predictions is the feedback signal, which is used to correct the model prediction. All controller and model calculations are performed at each execution, and the dynamic transient is optimized using a rolling horizon approach.

The calculations of the controller are based on the following problem, which is solved at every controller execution:

$$\min_{\Delta MV_j^c} \sum_{i=1}^{I} w_i \sum_{n=1}^{N} (SP_{i,n} - CV_{i,n} - (E_{mm})_i)^2 + \sum_{i=1}^{I} \rho_i \sum_{n=1}^{N} (V_{i,n}^{+} + V_{i,n}^{-})^2 + \sum_{j=1}^{J} q_j \sum_{m=1}^{M} (\Delta MV_{j,m}^{c})^2 \quad (174.17)$$

The problem is subject to the dynamic plant model (given here for CV_i and MV_j)

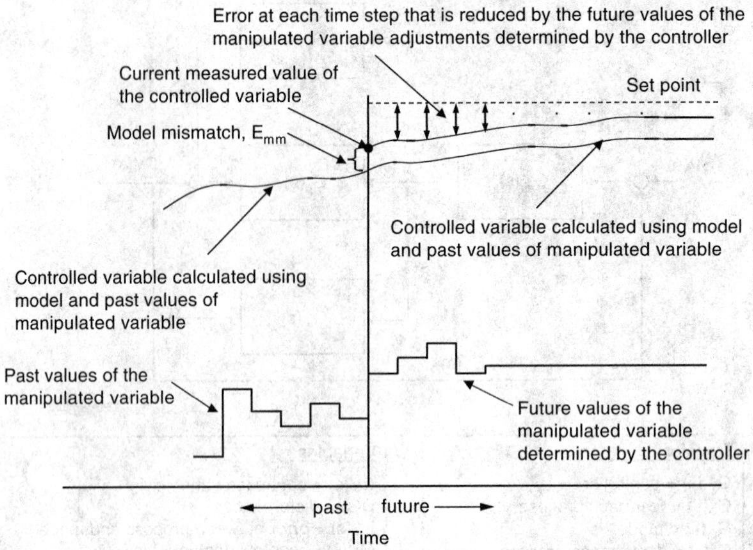

FIGURE 174.16 Typical MPC dynamic variables for single input–single output system.

$$\begin{bmatrix} CV_{i,1} \\ CV_{i,2} \\ \cdots \\ \cdots \\ CV_{i,N} \end{bmatrix} = \begin{bmatrix} a_{11} & a_{12} & \cdots & \cdots & a_{1M} \\ \cdots & \cdots & \cdots & \cdots & \cdots \\ \cdots & \cdots & \cdots & \cdots & \cdots \\ \cdots & \cdots & \cdots & \cdots & \cdots \\ a_{N1} & \cdots & \cdots & \cdots & a_{NM} \end{bmatrix} \begin{bmatrix} \Delta MV^{c}_{j,1} \\ \Delta MV^{c}_{j,2} \\ \cdots \\ \cdots \\ \Delta MV^{c}_{j,M} \end{bmatrix} + \begin{bmatrix} a_{1,M+1} & a_{1,M+2} & \cdots & \cdots & \cdots \\ \cdots & \cdots & \cdots & \cdots & \cdots \\ \cdots & \cdots & \cdots & \cdots & \cdots \\ \cdots & \cdots & \cdots & \cdots & \cdots \\ a_{M,M+1} & \cdots & \cdots & \cdots & \cdots \end{bmatrix} \begin{bmatrix} \Delta MV^{past}_{j,M+1} \\ \Delta MV^{past}_{j,M+2} \\ \cdots \\ \cdots \\ \cdots \end{bmatrix} \tag{174.18}$$

and feedback error:

$$\begin{bmatrix} (E_{mm})_1 \\ (E_{mm})_2 \\ \cdots \\ (E_{mm})_I \end{bmatrix} = \begin{bmatrix} (CV_{measured} - CV_{model})_1 \\ (CV_{measured} - CV_{model})_2 \\ \cdots \\ (CV_{measured} - CV_{model})_I \end{bmatrix} \tag{174.19}$$

The bounds on manipulated variables (change per execution and value at each execution) are

$$(\Delta MV_j)_{\min} \le \Delta MV^{c}_{j,m} \le (\Delta MV_j)_{\max} \tag{174.20}$$

$$(MV_j)_{\min} \le MV_{current} + \sum_{m=1}^{K} \Delta MV^{c}_{j,m} \le (MV_j)_{\max} \qquad \text{for } K = 1, M \tag{174.21}$$

Violations of controlled variable bounds follow:

$$(CV_i)_{\max} - CV_{i,n} + V^+_{i,n} \geq 0 \qquad \text{for } V^+_{i,n} \geq 0 \tag{174.22}$$

$$(CV_i)_{\min} - CV_{i,n} - V^-_{i,n} \leq 0 \qquad \text{for } V^-_{i,n} \geq 0 \tag{174.23}$$

The objective function in Equation (174.17) has three terms that are minimized for good performance. The first term sums the deviations between the future controlled variables and their set points. The second term represents the violations of controlled variable limits that are penalized; the values for the violations are determined in Equations (174.22) and (174.23). A penalty is used in place of strict bounds because imposing strict bounds could lead to infeasibilities. The third term involves the changes in the manipulated variables that are penalized for two reasons. First, large changes might be undesirable in the plant; second, more moderate adjustments lead to a robust controller that provides acceptable performance over a range of plant-model mismatch. This third term in the objective is often referred to as "move suppression."

The dynamic model in Equation (174.18) represents the effects of all past and future controller output adjustments calculated by the controller over the input horizon. Naturally, the past adjustments (ΔMV^{past}) are not affected by the current controller calculation, which determines only the future adjustments (ΔMV^C). A step-weight model represents the effects of the manipulated variable adjustments on the controlled variables. An example of the model for one $MV_j - CV_i$ pair is given in the following for changes in manipulated variables from an initial steady state:

$$CV_{i,n} = [a_{n,1} \quad a_{n,2} \quad \cdots \quad a_{n,M}] \begin{bmatrix} \Delta MV^c_{j,1} \\ \cdots \\ \Delta MV^c_{j,M} \end{bmatrix} \tag{174.24}$$

The step weights (a_{nm}) are determined from empirical modeling (e.g., Cryor, 1986). Note that the separate values for the model step weights (a_{nm}) exist for every input-output ($MV_j - CV_i$) model.

Feedback is included by the $(E_{mm})_i$ corrections in Equation (174.19). Since the controller adjustments are based on a dynamic model, the feedback is the mismatch between the model and the plant. A common assumption is that the feedback (E_{mm}) for all samples in the future output horizon has the same value as the value at the time of the controller execution.

Bounds on the future manipulated variable values can be included; recall that the past manipulated variable values are known but not adjustable. Bounds can be enforced on the change per execution period (ΔMV^c_{jm}) and on the total value of the manipulated variable.

The adjustments for the manipulated variables at the current time are implemented; the remainder of the calculation is discarded, and the dynamic model is integrated one time step. The controller then waits until its next execution.

This formulation has several important features. First, it involves an open-loop control calculation that is updated through feedback; thus, the control involves a rolling horizon that looks ahead to the steady state achieved by the closed-loop system. Second, it is a convex quadratic program that can be solved efficiently to a global optimum (Reklaitis et al., 1983). Third, the controller ensures that bounds on the manipulated variables are strictly observed. Fourth, although bounds on the controlled variables cannot be strictly observed with assurance, the bounds on the controlled variables are expressed using goal programming (Williams, 1999) to penalize violations. Fifth, the controller is easily extended to high-dimensional, multivariable systems, including systems with unequal number of manipulated and controlled variables. Sixth and finally, feed-forward control is easily implemented for multivariable systems.

Further details on the widely applied QDMC algorithm are provided by Garcia and Morshedi (1986). A thorough presentation of model-predictive control is given in Maciejowski (2003). The historical development of model-predictive control along with industrial experiences is reported in Qin and Badgwell (1997), and the classic introduction of model predictive control was presented by Cutler and Raymaker (1979).

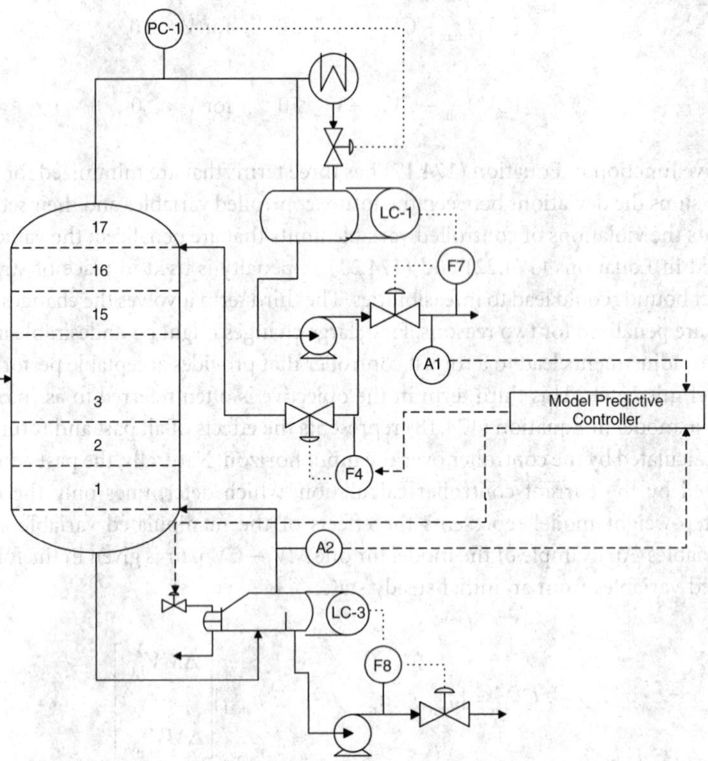

FIGURE 174.17 Distillation tower composition control with model predictive control.

The dynamic performance possible with model-predictive control is demonstrated with an example of distillation control shown in Figure 174.17. Model-predictive control is applied to the distillation tower for situations with constraints.

- *Manipulated variable constraint:* In this situation, the reboiler duty is limited because of a maximum possible heating medium flow rate. The maximum amount of reboiled vapor is 14.1 kmol/min. The results are given in Figure 174.18a for a set point change in the X_D controlled variable. Because one of the manipulated variables encounters a constraint, both controlled variables cannot be maintained at their set points. Since the controller objective (Equation (174.17)) considers both controlled variables, the controller adjusts the one remaining, unconstrained manipulated variable to minimize the sum of the (squared) errors for the distillate and bottoms compositions. Neither controlled variable achieves its set point, but each is maintained "close" to its set point. Modifications can be made to implement a priority ranking for controlled variables so that the more important can be returned to their set points when all controller variables cannot be returned to their set points (Swartz, 1995).
- *Controlled variable constraint:* Again, the set point response for a change in the X_D set is considered. In this situation, the light key in the bottoms should be maintained below a specified limit or costly economic penalties would occur. The maximum value for X_B is 0.0205, and this limitation is included in the controller through a very severe penalty on any X_B values that exceed the limit. The results are given in Figure 174.18b. To reduce the disturbance to X_B due to interactions, the controller has slowed the adjustment to the manipulated variables slightly. Therefore, slightly more time is required to change the distillate composition, X_D. However, the controller achieves the dual goals of reasonably fast X_D response while X_B is maintained within its specified upper limit. This excellent performance is due to the capability of the controller and the perfect model used in this simulation example. Such excellent performance would not be expected for a realistic nonlinear

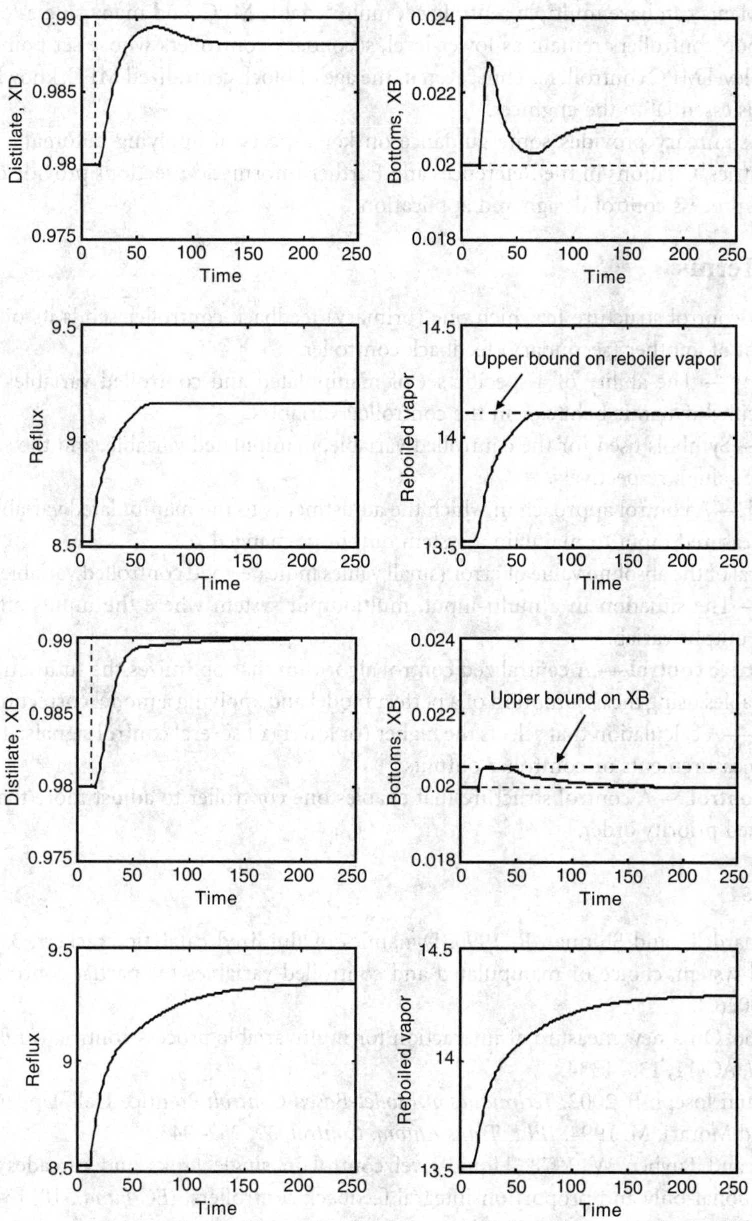

FIGURE 174.18 Dynamic performance of MPC distillation control designs for constraints on (a) reboiler and (b) bottoms product composition. (*Source:* Marlin, T. 2000. *Process control: designing processes and control systems for dynamic performance.* McGraw-Hill, New York. With permission.)

process with dynamics changing due to alterations in operating conditions, but quite good performance can be achieved using centralized model predictive control.

Given the success of centralized control, the reader may wonder about using this technology for centralized control of large plants having hundreds to thousands of variables. Although theoretically possible, such large MPC controllers are not now used because of (1) the difficulty in building the models, (2) the computation time for solving the optimization problem, and (3) the challenge to the plant personnel in understanding the controller results. Typically, centralized MPC is applied to blocks of variables that have substantial interaction among themselves and weak effects on the remainder of the plant. Thus, plants can have multiple centralized, multivariable MPC, and many single-loop controllers. Also single-loop controllers remain as lower-level, secondary controllers whose set points are adjusted by the higher level MPC controllers. Thus, even in the age of block centralized MPC, knowledge of single-loop control is essential to the engineer.

This brief summary provides some guidance on key aspects of applying automatic control in the process industries. Citations in the References and Further Information sections provide details essential for successful process control design and application.

Defining Terms

Cascade — A control structure in which one (primary) feedback controller sends its output to the set point of another (secondary) feedback controller.

Controllability — The ability of a specific set of manipulated and controlled variables to achieve the specified dynamic behavior in the controlled variables.

CV, MV, SP — Symbols used for the controlled variable, manipulated variable, and the set point (reference value), respectively.

Feed forward — A control approach in which the adjustments to the manipulated variable are based on a measured input to maintain a system output unchanged.

IAE — Integral of the absolute value of error (small values indicate good controlled variable performance).

Interaction — The situation in a multi-input, multioutput system where the inputs affect more than one output variable.

Model-predictive control — A centralized control algorithm that optimizes the future trajectory of the variables using the assumption of a perfect model and applying a model correction via feedback.

Signal select — A calculation that selects the higher (or lower) of several control signals; it can be applied to measurements or controller outputs.

Split-range control — A control structure that enables one controller to adjust more than one valve in a fixed priority order.

References

Arbel, A., Rinard, I., and Shinnar, R. 1996. Dynamics of fluidized catalytic crackers. 3. Designing the control system, choice of manipulated and controlled variables for partial control, *IEC Res.*, 34, 3014–3026.

Bristol, E. 1966. On a new measure of interaction for multivariable process control, *IEEE Trans. Autom. Control*, AC-11, 133–1334.

Brosilow, C. and Joseph B. 2002. *Techniques of Model-Based Control*, Prentice Hall, Upper Saddle River.

Campo, P. and Morari, M. 1994. *IEEE Trans Autom. Control*, 39, 952–943.

Cheung, T.F. and Luyben W. 1978. Liquid-level control in single tanks and cascades of tanks with proportional-only and proportion-integral feedback controllers, *IEC Fund.*, 18, 15–21.

Cryor, J. 1986, *Times Series Analysis*, Duxbury Press, Boston.

Cutler, C. and Raymaker B., Dynamic matrix control — a computer control algorithm, paper presented at AIChE national meetings, April 1979.

Driskell, L. 1983. *Control Valve Selection and Sizing*, Instrument Society of America, Research Triangle Park, NC.

Fertik, H. 1975. Tuning controllers for noisy processes. *ISA Trans.*, 14:292–304.

Galarraga, M., McAvoy, T., and Marlin, T. 1987. Short-cut operability analysis 2. estimation of detuning parameter for classical control systems, *IEC Res.*, 26, 511–521.

Garcia, C. and Morshedi, M. 1986, Quadratic programming solution of dynamic matrix control (QDMC), *Chem. Eng. Commun.*, 46:73–87.

Hovd, M. and Skogestad, S. 1994. Pairing criteria for decentralized control of unstable plants, *IEC Res.*, 33, 2134–2139.

Hughes, T. 2002. *Measurement and Control Basics*, 3rd ed., Instrument Society of America, Research Triangle Park, NC.

Lopez, A., Murrill, P., and Smith, C. 1969. Tuning PI and PID digital controllers, *Instrum. Control Sys.*, 42, 89–95.

Maciejowski, J. 2002. *Predictive Control with Constraints*, Pearson Education, Harlow, England.

Marlin, T., Perkins, J., Barton, G., and Brisk, M. 1991. Process control benefits, a report on a joint industry-university study, *J. Process Control*, 1, 68–83.

Marlin, T. 2000. *Process Control: Designing Processes and Control Systems for Dynamic Performance*, 2nd ed., McGraw-Hill, New York.

McAvoy, T. 1983a. *Interaction Analysis*. Instrument Society of America, Research Triangle Park, NC.

McAvoy, T. 1983b. Some results on dynamic interaction analysis of complex control systems, *IEC PDD*, 22, 42–49.

Monica, T., Yu, C., and Luyben, W. 1988. Improved multiloop controllers for multivariable processes, *IEC Res.*, 27, 969–973.

Morari, M. and Zafiriou, E. 1989. *Robust Process Control*. Prentice Hall, Englewood Cliffs, NJ.

Qin, S. J. and Badgwell, T. 1997. An overview of MPC technology, in Kantor, J., Garcia, C., and Carnahan, B. Eds., *Chemical Process Control-V: Assessment And New Directions for Research*, Proceedings of the Fifth International Conference on Chemical Process Control, Tahoe City, California, January 7–12, 1996, American Institute of Chemical Engineers, New York, pp. 232–256.

Reklaitis, G., Ravindran, A., and Rgasdell, K. 1983. *Engineering Optimization, Methods and Applications*, Wiley, New York.

Rivera, D., Skogestad S., and Morari, M. 1986. Internal model control, 4. PID controller design, *IEC PDD*, 25, 252–265.

Rosenbrock, H. 1974. *Computer-Aided Control System Design*. Academic Press, New York.

Skogestad, S. and Morari, M. 1987. Effect of disturbance directions on closed-loop performance. *IEC Res.*, 26, 2323–2330.

Sloley, A. 2001. Effectively control distillation pressure, *Chem. Eng. Prog.*, 38–48.

Stanley, G., Marino-Galarraga, M., and McAvoy, T. 1985. Short-cut operability analysis 1. Relative disturbance gain, *IEC PDD*, 24, 1181–1189.

Swartz. C. 1995. An algorithm for hierarchical supervisory control, *Comp. Chem. Eng.*, 19, 1173–1180.

Williams, H. 1999. *Model Building In Mathematical Programming*, 4th ed., Wiley, New York.

Ziegler, J. and Nichols, N. 1942. Optimum settings for automatic controllers. *Trans. ASME*, 64:759–768.

Further Information

Many useful standards for industrial practice in process control are documented in *Standards and Practices for Instrumentation and Control*, 11th edition, 1992, published by the Instrument Society of America.

A crucial contribution of process control is improved plant safety. An introduction to approaches for designing safe process control is presented in *Guidelines for the Safe Automation of Chemical Processes*, 1993, published by the American Institute of Chemical Engineers.

A topic of growing interest in process automation that complements automatic control is statistical process control; a good introduction to issues and methods is given in Box and Kramer, 1992, Statistical process monitoring and feedback adjustment: a discussion, *Technometrics*, 34:251–267, with further discussions on pages 268–285.

The following references provide drawings and explanations for many industrial control designs:

Duckelow, S. 1991. *The Control of Boilers*, 2nd ed., Instrument Society of America Press, Research Triangle Park, NC.

Liptak, B. 1995. *Instrument Engineers Handbook: Process Control*, 3rd ed., CRC Press, Boca Raton, FL.

Luyben, W., Tyreus, B., and Luyben, M. 1998. *Plantwide Process Control*, McGraw-Hill, New York.

Shinskey, F.G., 1978. *Energy Conservation Through Control*, Academic Press, New York.

175

Digital Control

Michael J. Piovoso
Penn State University

Control is defined as the art and science of transferring sources of variations that would otherwise result in deviation from a desired dynamic response to the input of the system. As an example, consider the speed control of an automobile set to keep the car moving at 60 miles per hour. If the automobile encounters a hill, the automobile would slow down if the controller did not make an appropriate adjustment. If the device providing the manipulating inputs to the dynamic system is digital in nature, we say that the system is being controlled digitally.

Figure 175.1 illustrates a digital control system. The objective of the control system is to choose inputs to the system to be controlled so that controlled output, y(t) matches some reference input, r(t). The analog-to-digital converter, A/D, takes analog inputs and converts them to a sequence of sampled values every clock tick or T_s seconds. The signal marked r(t) is the desired response of the dynamic system output, y(t). The digital processor computes the input to the dynamic system based on the sampled values of r(t) and y(t) so as to correct for any deviations between the two. The processor output is then converted back to an analog signal through the digital-to-analog converter, D/A, that is then used to drive the dynamic system to be controlled, often referred to as the plant. A clock drives the A/D, the processor and the D/A. At each clock tick, the sampled output is compared to the desired output, and a corrective input is applied to the system. Nothing happens again until the next clock tick, leaving the system blind between sampling times.

Digital control systems are both nonlinear and time varying even if the plant to be controlled is linear and time invariant. The nonlinearity results from the fact that A/D converter must quantize the input signal into a finite number of bits, and hence multiple analog signal values convert into the same digital signal. The time-varying behavior is due to the clock. Imagine that the reference input changes instantaneously at some time t. Depending on when this change occurs, different behaviors result. If the change occurs at the beginning of the sampling interval, it is caught immediately; whereas, if it occurs in the middle of the interval, a different time response of the output would eventually results because the input would not be seen until the next clock tick.

Despite these difficulties, a digital controller is almost always used in practice. There are many reasons for this:

0-8493-1586-7/05/$0.00+$1.50

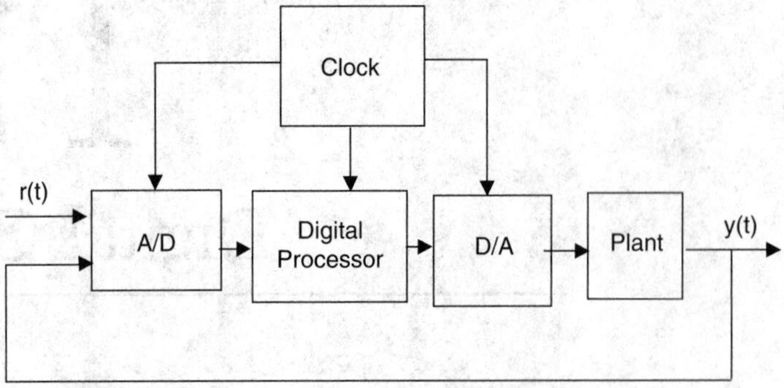

FIGURE 175.1 Digital control system.

1. Since we are dealing with digital hardware, there are none of the drifting effects in either age or temperature that occurs with analog hardware. The same input sequence operated on by an algorithm produces the same output sequence independent of time and temperature.
2. Exact reproducibility is possible for multiple applications since the same inputs always produce the same outputs.
3. There is a high degree of accuracy. More accuracy is always possible by using more bits in the processor or if necessary in the A/D and/or the D/A.
4. There is flexibility in that one processor can control multiple loops or perform additional tasks beside control.
5. Significant performance enhancement is possible by using digital control. Model-based control approaches such as optimal control, adaptive control and model predictive control can only be effectively implemented digitally.

Digital control systems can be developed two different ways. In the first, we recognize that the dynamic system to be controlled is analog in nature, design an analog controller, and then mimic its performance digitally. The second way is to view the world from the perspective of the digital processor. The digital processor observes a sequence of numbers that corresponds to the sampled values of the output of the plant and produces a sequence of numbers that corresponds to the eventual inputs to the plants. Suppose that we were to find a discrete-time system whose output is the same sequence of numbers as that generated from sampled values of y(t) when the analog system input is the output of the D/A due to the sequence of numbers computed by the digital processor. Such a discrete-time system would provide the sampled values for the output, and the algorithm that controls how that sequence is generated, would be the digital controller shown in Figure 175.1 used to control the plant.

This chapter is organized as follows. Section 175.1 is a review of the representation and analysis of discrete time system. Section 175.2 discusses the use of digital simulation as a tool for implementing an analog controller digitally. Section 175.3 focuses on the design of digital controller directly using root locus methodology for which a short review is provided. The pulse transfer function is developed as a means of generating the digital system representation of the analog plant that is the model for designing the controller.

175.1 Discrete Time Systems

Digital controller design is based on the behavior of linear, shift-variant, discrete-time systems. An understanding as to how these systems behave is crucial to the design of digital controller. In many ways, there are a number of similarities between the behavior of linear time-invariant analog systems and linear shift-invariant, discrete-time systems. The behavior of such systems depends on the solution of the describing difference equation or equivalently, on the location of the poles and zeros of its transfer function.

Difference Equations

Linear, time or shift-invariant, discrete-time systems are characterized by linear difference equations with constant coefficients. The general equation that describes any linear, shift-invariant system is

$$y_k = a_{n-1}y_{k-1} + \cdots + a_0 y_{k-n} + b_m x_k + b_{m-1}x_{k-1} + \cdots + b_0 x_{k-m} \tag{175.1}$$

The solution of this equation defines the behavior of the system for a given input. There are two components to the total solution: the homogeneous and particular parts. The homogeneous solution is the solution to the difference equation with zero input and defines the underlying characteristic behavior of the system. If the input is zero, Equation (175.1) reduces to:

$$y_{h,k} = a_{n-1}y_{h,k-1} + \cdots + a_0 y_{h,k-n} \tag{175.2}$$

where the subscript h refers to the homogeneous part. Assume that solution is of the form of $y_{h,k} = Kp^k$. Substituting into Equation (175.2) results in the characteristic equation that defines the necessary condition that p must satisfy.

$$p^n - a_{n-1}p^{n-1} - \cdots - a_0 = 0 \tag{175.3}$$

If the n roots of Equation (175.3) are $p_1, p_2, \cdots p_n$, then the homogenous solution is

$$y_{h,k} = K_1 p_1^k + K_2 p_2^k + \cdots + K_n p_n^k \tag{175.4}$$

Note that $y_{h,n}$ will tend to zero if magnitude of each root of the characteristic equation is less than 1. If any p_i has magnitude greater than 1, the homogeneous solution will grow without bound, resulting in an *unstable* system.

The particular solution, $y_{p,k}$, is the component of y_k that is due solely to the forcing input x_k. For a linear shift-invariant system, the particular solution will take the same form as the input provided that the input is not of the same form as one of the components of the homogeneous solution. Thus, if the input is a sinusoidal, the particular solution is a sinusoidal of the same frequency with only the amplitude and phase changing. If the input is a polynomial in time, the particular solution is a polynomial of the same order as the input.

From a control system perspective, we desire the particular solution to match the reference input because the objective of the system is to force the plant output to match the reference. What keeps that from happening is the homogeneous solution. The faster the homogeneous solution tends to zero, the faster the controlled output approaches the reference input. This implies that the smaller p_i is in magnitude, the faster that contribution to the homogeneous solution tends to zero.

Transfer Functions

An alternative way of representing discrete-time systems represented by equation is developed through Z-transforms. The Z-transform is a one-to-one mapping of a sequence of numbers into a function of the complex variable z. The one-sided Z-transform is defined as

$$F(z) = \sum_{n=0}^{\infty} f_n z^{-n} \tag{175.5}$$

Knowledge of the Z-transform is equivalent to knowledge of the sequence of numbers and vice versa.

Note that the Z-transform of a sequence delayed by m samples is z^{-m} times the Z-transform of undelayed sequence provided that the sampled values of the sequence are zero for negative arguments. Applying the Z-transform to Equation (175.1) results in

$$\frac{Y(z)}{X(z)} = \frac{b_m + b_{m-1}z^{-1} + \cdots + b_0 z^{-m}}{1 - a_{n-1}z^{-1} - \cdots - a_0 z^{-n}} \tag{175.6}$$

Note that this representation of a system is completely equivalent to the difference equation representation of Equation (175.1). This is clear from the fact if input sequence and the system are known, then the Z-transform of the output sequence can be computed from Equation (175.6). A knowledge of $Y(z)$ is equivalent to a knowledge of the output sequence y(k).

The system represented by Equation (175.6) is causal. That is, all outputs are computable from knowledge of present and past inputs and past outputs only. If we multiply the right side of equation (175.6) by z^n, the numerator and denominator become polynomials in z instead of z^{-1} provided that $n \geq m$:

$$H(z) = \frac{b_m z^n + b_{m-1}z^{n-1} + \cdots + b_0 z^{n-m}}{z^n - a_{n-1}z^{n-1} - \cdots - a_0} = \frac{B(z)}{A(z)} \tag{175.7}$$

Equation (175.6) and Equation (175.7) are equivalent transfer functions of the system defined by the difference equation (175.1). Note that by inspection, the transfer function can be written from knowledge of the difference equation and vice versa.

Also, Equation (175.3), the characteristic equation of the system defined by Equation (175.1), is the same equation that define the roots of the denominator polynomial, $A(z)$. The roots of the denominator polynomial are the poles of the system. The roots of the numerator polynomial are the zeros. Knowledge of the poles provides the same information as to the homogeneous solution of the difference equation as does the roots of the characteristic equation. For the system to be stable, the magnitude of the poles must be less than one. This implies that the poles must be located inside a circle of radius 1 in the z-plane. The z-plane is the representation of the complex values of z that are candidates for substitution into Equation (175.7).

Using transfer functions to represent discrete-time systems has many advantages. Difference equations need not be solved, and the transfer function of the interconnections between dynamic systems can be easily computed. As an example, consider the closed-loop system shown in Figure 175.2 with $G_c(z)$ and $G_p(z)$ being the controller and plant transfer function, respectively. The transfer function between the reference input and controlled output, called the closed-loop transfer function, is

$$\frac{Y(z)}{R(z)} = G_{CL}(z) = \frac{G_c(z)G_p(z)}{1 + G_c(z)G_p(z)} \tag{175.8}$$

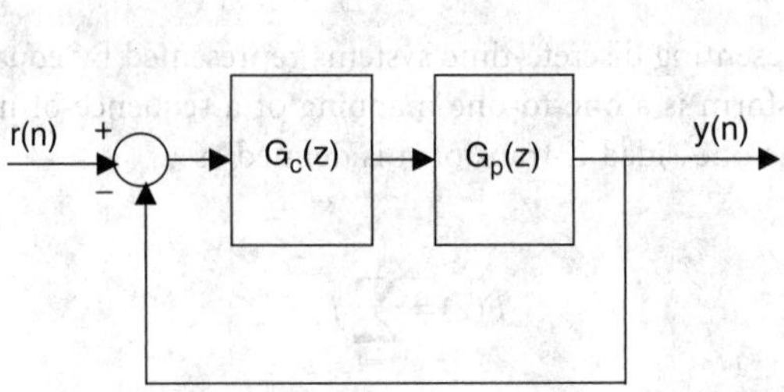

FIGURE 175.2 Closed-loop system.

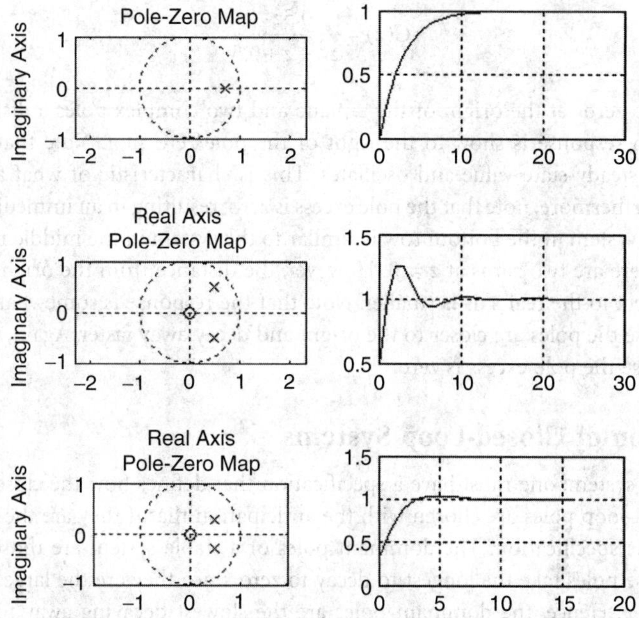

FIGURE 175.3 Relationship between pole and zeros and the corresponding step responses.

System Response

To design a controller, one must understand the dynamic response of both the plant and the closed-loop system — that is, the system response from reference input to output. The response to any input is the sum of the particular and homogeneous responses. The form of the homogeneous response is known from the pole location, and that provides an understanding as to how the homogeneous solution for a stable system decays to zero.

For example, consider a system with a pole at $z = 0.5$. The homogenous solution would be proportional to 0.5^k. If the pole were located at $z = 0.2$ instead of $z = 0.5$, the system would settle faster to the particular solution because the homogeneous part of the response decays away faster. For $z = 0.8$, the system would be slower. In general, the response of a system with poles closer to origin of the Z-plane is faster than ones with poles farther away.

Figure 175.3 illustrates several examples of the effect of poles and zeros on system response. The left column gives the pole and zero location in the z-plane of three systems, while the right column shows the step response of each system. The system in the top row is a first order with transfer function

$$G(z) = \frac{0.3}{z - 0.7} \tag{175.9}$$

Shown for reference in the figures on the left is a circle of radius 1, called the unit circle. Since the poles are inside this circle, the system is stable. For this example, there is no zero and a pole at $z = 0.7$. This will produce a homogeneous solution of 0.7^k, which, as it dies away, the output reaches the particular solution or the steady-state output value of 1. Note that the first output of the system response is 0. Mathematically, this results from having more poles than zeros. The difference between the number of poles and the number of zeros is the pole excess. In this case, there is one pole and no zeros. If there had been three poles and one zero, the first two values, or the pole excess, of the step response would have been zero.

The middle and bottom rows show two related second-order systems. The system in the middle row has the transfer function:

$$G(z) = \frac{0.5z^2}{z^2 - z + 0.5} \tag{175.10}$$

This system has two zeros at the origin of the z-plane and two complex poles a distance of 0.707 from the origin. The step response is show to the right of the pole/zero plot. Note that the output of this system exceeds the steady-state value and oscillates. This is characteristic of what are called an *under-damped* systems. Furthermore, note that the pole excess is zero, resulting in an immediate output response to a step input. The system in the bottom row is similar to the system in the middle row in that the poles are complex and there are two zeros at $z = 0$. However, the distance from the origin is now 0.5774 and the angle with respect to the real axis is smaller. Note that the response is somewhat faster in settling to the 1. This is because the poles are closer to the origin and decay away faster. Again, the system response is immediate because the pole excess is zero.

Characterization of Closed-Loop Systems

To design a control system, one must have a specification that defines how the closed-loop system is to respond. The closed-loop poles are chosen with the anticipation that if they are the dominant ones, the system will meet the specification. The dominant poles of a stable system are those that are closest to the unit circle. These poles take the longest to decay to zero since they are the largest distance from the origin of the z-plane. Hence, the dominant poles are the slowest decaying away to zero and have the biggest influence on determining the time for settling to the steady-state solution.

Specifications for the closed-loop system typically center on the speed and amount of overshoot of the system response. A system may be fast enough in terms of reaching the target value or in settling to a desired result, but if the overshoot or the amount by which the output exceeds the target value is too large, then the overall response may be unacceptable. If the overshoot is acceptable but the system takes too long to reach the desired level or to settle, then the overall response will be unacceptable. The typical performance specifications usually center on the closed-loop response to a step input. The specifications that relate to speed of response are:

1. *Rise time, T_R:* Typically defined as the time required going from 10 to 90% of the final value. The 10% lower limit is used so that any time delay in the system is not directly counted as part of the system response. Time delays cannot be controlled away, and hence they do not reflect the speed of response of the system
2. *Time to peak, T_P:* The time to peak is the time at which the output peaks. This is the time to reach the first overshoot peak.
3. *Settling time, T_S:* The time required for the output to be within a guaranteed level of the final value. Typically, this is measured as being within 5% or 2% of the final value.

The overshoot, M_p, is the difference between the peak and the steady-state values divided by the steady-state value multiplied by 100 to make it a percent. Typical values are 3 to 20%. Most often, some overshoot is desired. The reason is that the speed in getting to the target value can be greatly improved if the system output is allowed to overshoot the steady-state value by a small amount.

Given a set of specifications, the question becomes where the dominant poles of the closed-loop should be located so that the system is likely to meet that specification. The location is determined by approximating the closed loop as a second-order system without any finite zeros and defining the poles of that system so that the specifications are meet. If the assumption that such a second-order approximation is a good model for the closed loop is valid, then the results would be nearly as predicted. If it is not valid, then the approximation may or may not completely satisfy the system requirement. If the specifications are not met, the closed-loop poles need to be altered, a new controller designed and the system retested.

To understand how a second-order system behaves, consider the step response of the analog system:

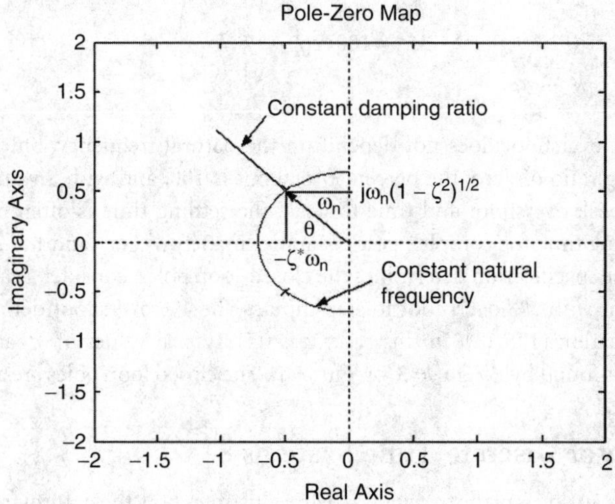

FIGURE 175.4 Complex poles in the s-plane.

$$G(s) = \frac{\omega_n^2}{s^2 + 2\zeta\omega_n s + \omega_n^2} \tag{175.11}$$

where ζ is the damping ratio and ω_n is the natural frequency. If the damping ratio is less than 1, the poles of this system are located at $s_{1,2} = -\zeta\omega_n \pm j\omega_n\sqrt{1-\zeta^2}$. Figure 175.4 illustrates the pole location in the s-plane. Note that the complex pole defines a right triangle with the two arms of length $\zeta\omega_n$ and $\omega_n\sqrt{1-\zeta^2}$. The hypotenuse is of length ω_n, and the cosine of the angle θ is just the damping ratio, ς. Note that the locus of points in the s-plane corresponding to constant damping ratio between 0 and 1 (ω_n varies) is a ray emanating from the origin for which the cosine of the angle of the ray with respect to the negative real axis is the damping ratio. Likewise, the locus of points for fixed natural frequency, ω_n, (damping ratio varying from 0 to 1) is a semicircle in the left s-plane or radius, ω_n.

The step response of the second-order system of Equation (175.11) is given by

$$y(t) = 1 - \frac{e^{-\varsigma\omega_n t}}{\sqrt{1-\varsigma^2}}\sin(\omega_n\sqrt{1-\varsigma^2}t + \theta) \tag{175.12}$$

$$\text{with } \theta = \tan^{-1}\left(\frac{\sqrt{1-\varsigma^2}}{\varsigma}\right)$$

Given this response, the time to peak and percent overshoot are calculated by taking the derivative of Equation (175.12), setting it equal to zero and solving for the values of time for which $y(t)$ has local maxima and minima. Putting the time for the peak into Equation (175.12), the percent overshoot can be found. The values of the time to peak and the percent overshoot are, respectively,

$$T_p = \frac{\pi}{\omega_n\sqrt{1-\varsigma^2}} \tag{175.13}$$

and

$$M_p = 100\exp\left(-\frac{\pi\varsigma}{\sqrt{1-\varsigma^2}}\right) \tag{175.14}$$

Note that the peak overshoot does not depend on the natural frequency, but only on the damping ratio. With a damping ratio of zero, the percent overshoot is 100, and with a value of 1 or greater, it is zero. In addition to peak overshoot and time to peak, the settling time is often of interest. For analog systems, the 5% settling time for complex poles is approximately $3/\zeta\omega_n$ and the 2% is $4/\zeta\omega_n$.

To illustrate how the specification determines the closed-loop poles, consider a case where a 5% settling time of 3 sec is sought with 5% overshoot to step inputs. The 4% overshoot defines the damping ratio found by solving Equation (175.13). In this case, ς is 0.7. Typical values of ς are between 0.5 and 1. The 5% settling time is found by $3/\zeta\omega_n = 3$ or $\zeta\omega_n = 1$. The closed loop poles are approximately $-1 \pm j$.

Characterization of Discrete Time Systems

The mapping from the analog s-plane to digital z-plane illustrates how the natural frequency and damping ratio in the analog world manifest itself in the digital world. A pole in the s-plane at $s = s_0$ produces a transient of the form $e^{s_0 t}$. Sampling every T seconds generates a sequence whose Z-transform has to a pole at $z = z_0 = e^{s_0 T}$ where T is the sampling time. Figure 175.5 illustrate the z-plane that is used to locate the pole and zeros. The circle of radius one is known as the unit circle. For stability, all the poles of the discrete time system must be within this circle. The heart-shaped curves are the ones that correspond to constant damping ratio. For a damping ratio close to 1, the curves are close to the real axis. If the points corresponding to a constant damping ratio are mapped into the digital world as shown above, they produce this set of heart-shaped curves. As the damping ratio decreases towards 0, the constant damping ratio curves move toward the unit circle. Alternatively, the points on the locus of constant natural frequency are the curves that begin and end on the unit circle. The values of the natural frequency are given in steps of the 0.1 times pi divided by the sampling time.

Figure 175.5 can be used to define the desired closed-loop pole locations for a given specification. A specification, such as the peak overshoot, specifies only one parameter, namely the damping ratio. Another constraint is needed to specify the other parameter ω_n. The specification that is most often given for a digital control system is the settling time. For discrete time systems, the distance from the origin to the pole location determines this as was noted earlier. For a 5% settling time, the homogeneous solution must

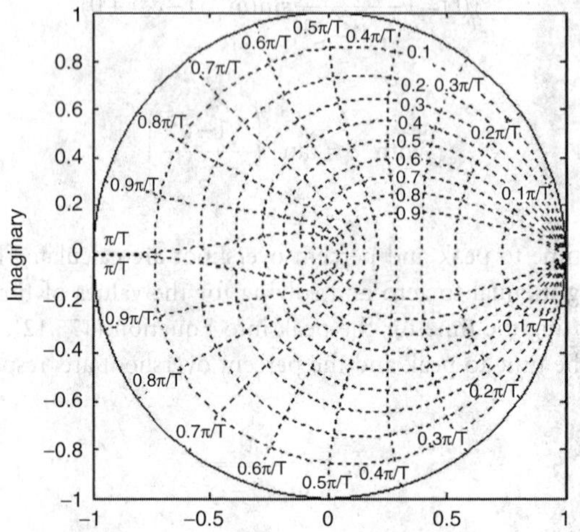

FIGURE 175.5 Grid with curves of constant damping ratio and constant natural frequency.

decay away 95% of its maximum effect. To determine an approximate settling time, take the dominant pole (the one closest to the unit circle) and solve for the smallest sample number, N, that satisfies $R^N \le 0.05$ where R is the distance from the origin to the pole location. The actual settling time is closer to the value N plus the pole excess because the number of samples of response delay is the pole excess.

In the design problem, the inequality above must be solved for R knowing the desired settling time. For example, suppose that we want 2-sec settling and have a sampling time 0.25 sec. This corresponds to eight samples, and if there is a pole excess of two, we need to have the dominant poles of the closed-loop system be located a distance R from the center of the z-plane so that the transient due to that pole settles in six samples after the pole-excess delay. Solving $R^6 = 0.05$, yields desired closed-loop poles at a distance of approximately 0.6 from the center of the z-plane. The actual location for the dominant poles should be at those points corresponding to a distance of 0.6 from the origin and intersecting the desired damping ratio. If the desired damping ratio was 0.7, the location of the dominant poles should be at $z = 0.52 \pm j0.3$.

175.2 Digital Simulation of Analog Controllers

Analog systems can be implemented using only integrators together with appropriate feedback. To approximate such systems digitally means finding an appropriate digital simulation of the analog integrator which will, in turn, define a specific transformation from the s-domain to the z-domain. There are a number of different ways of approximating an integrator or measuring the area under a function. Consider the one shown in Figure 175.6. An approximation to the area under a curve can be estimated by taking the sampled value at time kT and holding it constant until a new sample is available at time $(k + 1) * T$. This produces a set of rectangles and forms what is called forward Euler integration.

The difference equation describing this approximation to an analog integrator is found by noting that the area under the curve at time nT is the area under the curve at time $(n - 1)T$ plus the area under the rectangle formed at time $(n - 1)T$.

$$S_n = S_{n-1} + T^* x((n-1)T) \tag{175.15}$$

where S_n is the approximate area under the curve at time nT. The transfer function of the system corresponding to Equation (175.15) is $T/(z - 1)$. Since the transfer of an integrator is $1/s$, a comparison of the two suggest that a substitution of s by $(z - 1)/T$ would convert an analog system transfer function into a digital system that corresponds to the simulation with a forward Euler integration scheme.

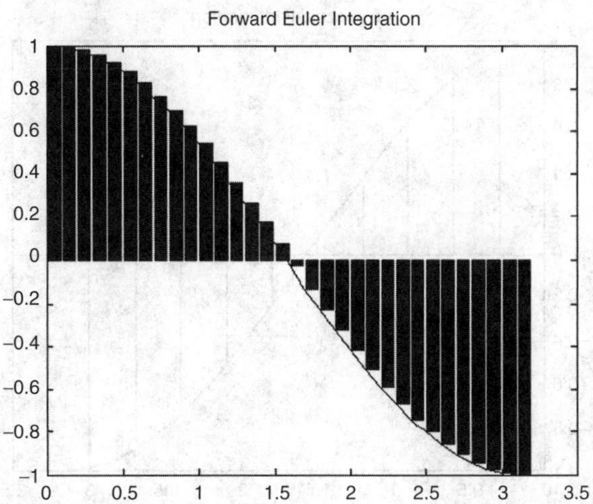

FIGURE 175.6 Forward Euler integration.

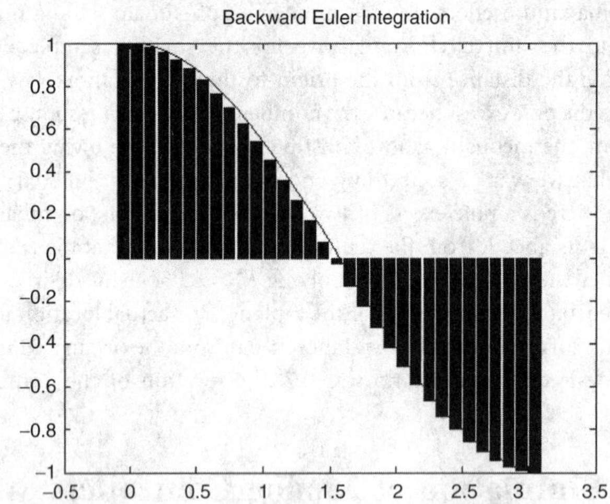

FIGURE 175.7 Backward Euler integration.

Alternatively, the rectangles could be generated by taking samples at time nT and projecting them backwards one sample interval forming a backward Euler integration approximation. Figure 175.7 illustrates this, and in this case, the approximate area under the curve is given by

$$S_n = S_{n-1} + T * x(nT) \tag{175.16}$$

The transfer function of this integrator is $Tz/(z-1)$. The substitution for s that would correspond to backward Euler integration is $(z-1)/Tz$.

A better approximation would be to join samples, $x(nT)$ and $x((n-1)T)$ by a straight line and measure the areas under the resulting trapezoids, leading to trapezoidal integration. This is also known as Tustin approximation and is illustrated in Figure 175.8. The equation describing this approximation is

$$S_n = S_{n-1} + \frac{T}{2} * (x(nT) + x((n-1)T)) \tag{175.17}$$

This corresponds to a substitution for s by $2/T[(z-1)/(z+1)]$.

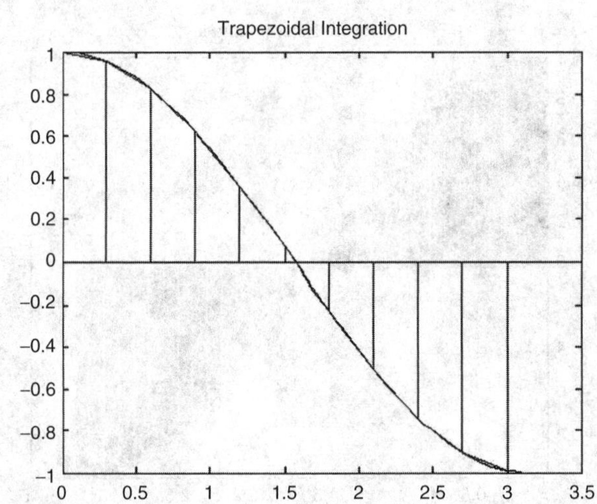

FIGURE 175.8 Trapezoidal integration or Tustin approximation.

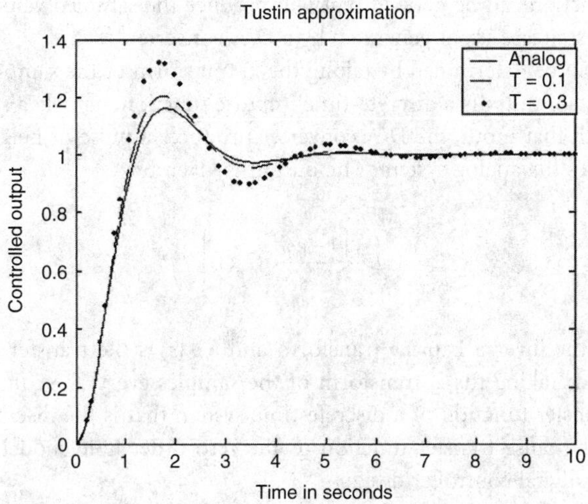

FIGURE 175.9 Tustin approximation with two different sampling times.

Each of these methods provides an adequate approximation provided that the sampling time is chosen sufficiently small. The sampling time should be chosen as large as possible without sacrificing performance too much. For very small sampling time, the changes in the controller output becomes small, creating computational accuracy problems. Choosing a good sampling time is important. A rule of thumb suggested by Astrom and Wittenmark (1997) is to choose the sampling time so that the product of the sampling time and the gain crossover frequency (in radians per second), $T\omega_c$, of the loop gain (i.e., the product of the controller and plant transfer function), is between 0.15 and 0.5. As the sampling time increases, the trapezoidal method tends to offer the most robustness because of its improved accuracy. In fact, the Tustin approximation has the property of mapping the left or the stable region of the s-plane onto the area inside the unit circle, the stable region in the digital world. The frequency axis in the analog domain, the $j\omega$ axis, is mapped onto the frequency axis in the digital world, the unit circle.

The procedure is very simple. Given an analog controller and its transfer function, convert it to a discrete time system by choosing a sampling time T and substituting for s the relationship corresponding to which approximation is chosen: forward, backward, or Tustin integrator approximations. This then gives a discrete transfer function from which a difference equation can be extracted and implemented.

As an example, consider a lead compensator with transfer function $4[(s + 1)/(s + 2)]$ for a system with a transfer function $1/s(s + 1)$. The gain crossover frequency is approximately 1.6 radians per second. Thus an appropriate sampling time is between 0.1 to 0.3 sec. Figure 175.9 illustrates the results of using a Tustin approximation with the two extremes in the choice of the sampling time. The solid curve is the result with an analog controller, the dashed line corresponds to a sampling time of 0.1 sec and the dotted line corresponds to 0.3 sec.

175.3 Design of Digital Controllers

Digital controllers can be generated either by designing an analog controller and simulating it or by designing a digital controller directly. Designing a digital controller directly requires that the plant be discrete time, not analog. Fortunately, there is a way to convert an analog system directly into a meaningful digital one. Figure 175.1 illustrates that the output of the processor goes directly to a D/A converter. The output of the D/A is held to its input value until a new input is applied. This produces a staircase-like function to the input of the plant. The output of the plant is sampled by the A/D and inputted to the processor. From the perspective of the processor, a digital value is being applied to some system and another digital value is being returned as an output of that system. This looks like a digital system, and in fact, the transfer function can be defined that will produce the sampled values of the analog output when it is excited by a staircase input generated by a D/A converter.

The digital system can be determined by taking the Z-transform of the sampled values of the analog output when the processor outputs a discrete-time impulse function, namely a sequence that is 1 at $n =$ 0 and 0 elsewhere. With that input, the D/A converter produces a pulse of height 1 and duration of T seconds that then excites the analog system. The output is given by

$$L^{-1}\left\{\frac{1-e^{-sT}}{s}G_p(s)\right\}$$

where L^{-1} represents the inverse Laplace transform and $G_p(s)$ is the transfer function of the analog system being controlled. Taking the Z-transform of the samples every T seconds of the resulting time signal produces the transfer function of a discrete-time system that is the discrete model for the plant. This system is called the pulse transfer function or the zero-order hold model of the analog plant. It forms the basis for the digital controller design.

Note that the transfer function depends on the sample time T. The poles of the discrete transfer function are located at $e^{s_0 T}$ where s_0 is the location of the analog plant pole. If T is chosen very small, all the poles of the analog system map into values very close to 1. This creates problems from a design perspective. The desired closed-loop poles are also going to be near 1. The reason can be seen by the discussion above concerning settling time. If we desire a system with a certain settling, the number of samples that this corresponds to is that time divided by the sampling time. If the sampling time is small, the number of samples is large. A large number of samples imply that the distance from the origin to the pole location, R, must be near 1 so that R^N is 0.05, where N is the number of samples desired for the settling time.

Clearly, the choice of the sampling time, T, is an important consideration for the design of a digital controller. If T is very large, then the controller would not be able to respond in a timely fashion to unmeasured disturbances. Astrom and Wittenmark (1997) point out several rules of thumb that one can employ in choosing T. If T_r is the 10 to 90% rise time, the sampling time can be chosen so that there are four to ten samples per rise time. Alternatively, the choice can be related to the closed-loop bandwidth. Choose the sampling rate so that the sampling frequency is 10 to 30 times the closed-loop bandwidth in hertz.

Relatively slow sampling rates are tolerable in control systems because the closed loop is a low-pass filter. Thus, the higher frequencies are attenuated, eliminating the need for a higher sampling frequency. Furthermore, the nature of the game in control is the performance of the closed-loop system, not information. Thus, we can tolerate slower sampling rates than might be suggested from a consideration of the issues centered on digital signal processing.

Root Locus Design

The pulse transfer function provides us with a model for designing a controller. There are a number of different techniques that can be employed. Here, we will discuss the use of root locus. Root locus is an algebraic methodology for plotting the location of the roots of the denominator polynomial, or the poles, of the closed-loop transfer function as a function of a multiplicative parameter K. These roots are given by the characteristic equation,

$$1+KG(z)=0 \tag{175.18}$$

where $G(z) = N(z)/D(z)$ and $N(z)$ and $D(z)$ are the numerator and denominator polynomials, respectively. We seek the solution for Equation (175.18) as a function of positive K. We will not examine negative K because in the vast majority of control problems, K will be positive. In order to find a solution, $G(z)$ must be negative real. If it is, there is always a positive K to satisfy Equation (175.18). Thus, the necessary and sufficient condition for a point to be on the root locus and a solution to Equation (175.18) is that

$G(z)$ be negative real or that the angle of $G(z)$ be an odd integer of 180°. Every point on the root locus satisfies this condition.

The art of designing a controller is to find dynamics in the controller such that the desired closed-loop poles are dominant (closest to the unit circle) and on a root locus. This means that the angle of $G(z)$ is an odd integer multiple of 180° at the desired location of the z-plane. This angle can be computed from the difference of the angle associated with $N(z)$ and the angle associated with $D(z)$. The angle of a polynomial is the sum of angles from each root to the desired location. This angle is measured with respect to the positive real axis. The gain that K needed, so that $z = z_0$ is a closed-loop pole, is

$$K = \frac{1}{|G(z_0)|} = \frac{|D(z_0)|}{|N(z_0)|} \tag{175.19}$$

where z_0 is a point on the root locus. The magnitude of a polynomial evaluated at a complex value z_0 is the product of the distances from each root to the point in question, z_0.

Controller Design Using Root Locus

Controller design using root locus is an art and must be often done iteratively. First, one needs a model of the system to be the controller. The specifications imply the locations for the dominant closed-loop poles. A design is then developed and simulated to determine if specifications are met. If specifications are not met, then a modification of the design is required to achieve the desired requirements. When the specifications are met in the simulation, a controller can be implemented and tested. If this fails to meet specifications, then a more accurate model is required for design. Figure 175.10 presents a flow diagram that illustrates this concept.

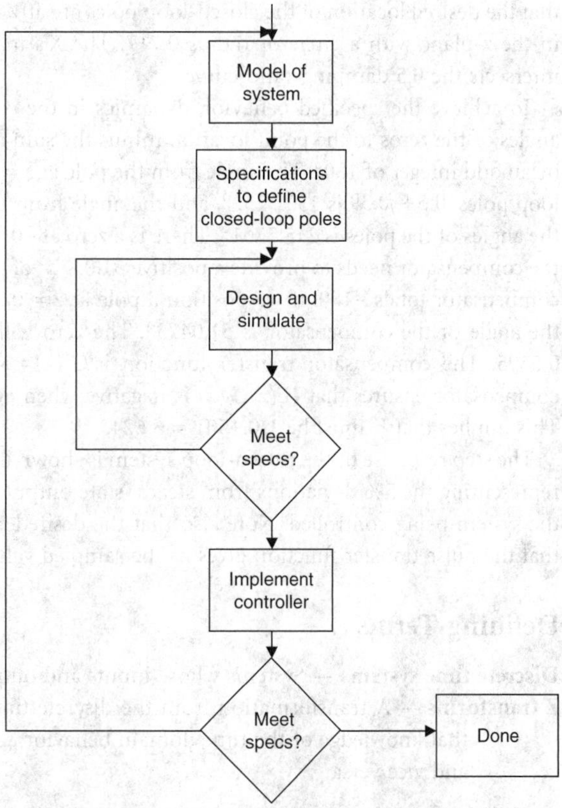

FIGURE 175.10 Flow diagram for controller design.

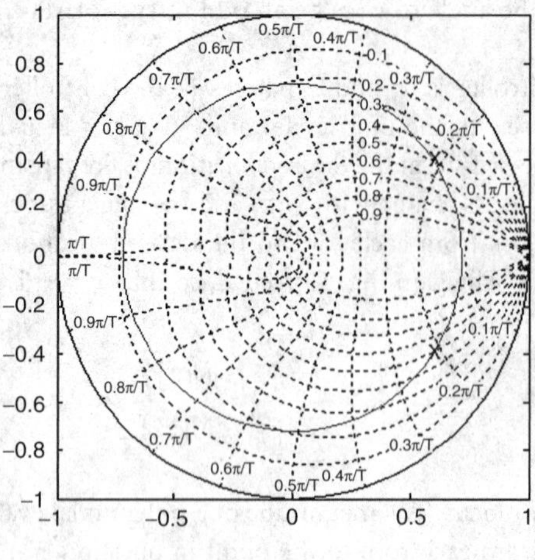

FIGURE 175.11 Closed-loop pole location for design example.

Assume that we have a DC motor to control with transfer function of $G_p(s) = 1/s(s+1)$. The specifications are that the 5% settling time is no more than 3 sec, with a damping ratio of 0.5. The 5% settling is given by $3/\zeta\omega_n$, yielding $\zeta\omega_n = 1$, from Equation (175.14). A sampling rate of 0.3 would meet the criteria set by the rules of thumb defined earlier. The pulse transfer function model for the analog system with sampling time of 0.3 is $G_z(z) = 0.0408(z+0.9049)/z^2 - 1.7408z + 0.7408$. Note that the pole excess is 1. The number of samples required to have a 3-sec settling time is 10. However, because of the pole excess, we need to design for nine samples to achieve 3 sec of settling time. Solving $R^9 = 0.05$ for R gives R = 0.717. If we draw a circle on the grid with the unit circle of radius 0.717 on the grid of Figure 175.5 and seek the intersection of the damping ratio of 0.5 with a circle of radius 0.717, we find that the desired location of the closed-loop poles are at $z = 0.6 \pm j0.39$. Figure 175.11 shows the unit circle in the z-plane with a circle of radius 0.717. The X's mark the spot on the circle of radius 0.717 that intersects the 0.5 damping-ratio curve.

To achieve the specified behavior, dynamics in the compensator must be such that the sum of the angles of the zeros to the point location minus the sum of the angles of the poles to pole location must be an odd integer of 180°. The angle from the pole at z = 1 to the location of one of the complex closed-loop poles, $0.6 + j0.39$ is 135.7252°, and the angle from the pole at z = 0.7408 is 109.8509. The sum of the angles of the poles is 245.5761°. There is a zero at –0.9049 that contributes a phase of 14.5288°. Thus the compensator needs to provide a positive 51.0473° of phase so that the sum of angles of the plant and compensator totals –180. If we position a pole at z = 0, we can compute the position of a zero so that the angle of the compensator is 51.0473°. The zero will need to provide 84.0712°, and its location is 0.5595. The compensator transfer function is $G_c(z) = K(z - 0.5595/z)$. Recall that our choice of the compensator ensures that $G_z(z)G_c(z)$ is negative when evaluated at 0.6 + j0.39, and is in fact, –0.1501. This implies that K must be 1/0.1501 = 6.624.

The step response of the closed-loop system is shown in Figure 175.12. Superimposed are the two lines representing the 5% deviations from steady-state output. The dashed line is the actual analog output of the system being controlled. Note also that the desired settling time is achieved. Furthermore, observe that the pulse transfer function gives us the sampled values of the analog output.

Defining Terms

Discrete time systems — Systems whose inputs and outputs change only at discrete points in time.

Z transforms — A transformation from the discrete time domain to a complex variable domain such that knowledge of the time domain behavior completely defines the complex variable domain and vice versa.

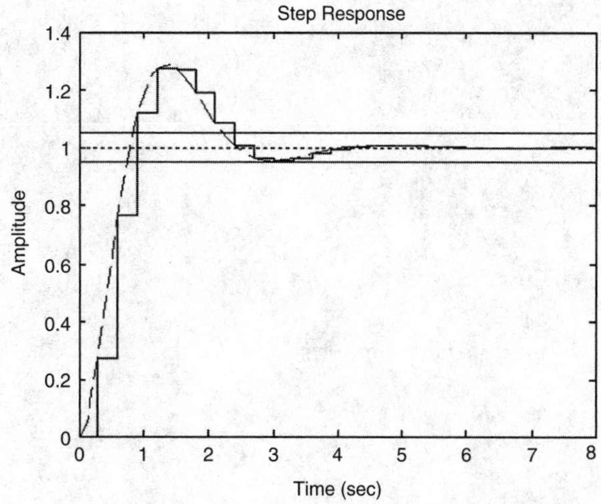

FIGURE 175.12 Step response of the digitally controlled system.

Simulation of analog controllers — A digital controller design technique in which an analog controller is designed and then converted to a discrete time system whose behavior approximates the analog controller.

Pulse transfer function — The discrete time system whose time outputs are the samples at a prespecified rate of the output of a corresponding analog system that is being driven by a zero-order hold analog-to-digital converter.

Root locus design — A technique for design controllers based on positioning the dominant poles of the closed-loop system at a prescribed location.

References

Astrom, K. J. and Whittenmark, B. 1997. *Computer Controlled Systems: Theory and Design*, 3rd ed. Prentice Hall, Englewood Cliffs, NJ.

Dorsey, J. 2001. *Continuous and Discrete Control Systems,* McGraw-Hill, New York.

Franklin, G. F., Powell, J. D., and Workman, M. L. 1998. *Digital Control of Dynamic Systems*, 3rd ed. Addison-Wesley, Reading, MA.

Houpis, C. H. and Lamont, G. B. 1992. *Digital Control Systems: Theory, Hardware, Software*, 2nd ed. McGraw-Hill, New York.

Kuo, B. C. 1995. *Digital Control Systems*, 2nd ed. Oxford University Press, London.

Phillips, C. L. and Nagle, H. T. 1995. *Digital Control System Analysis and Design*, 3rd ed. Prentice Hall, Englewood Cliffs, NJ.

176

Robots and Controls

Thomas R. Kurfess
Georgia Institute of Technology

Mark L. Nagurka
Marquette University

Credit for bringing robots to life is given to two individuals, Karel Capek and Isaac Asimov. In 1922, Capek wrote a play called *Rossum's Universal Robots* (RUR) describing how robots would turn on mankind and eventually take over the world. In the 1940s, Asimov, who is credited with coining the term "robotics," freed us from a view of robots as malevolent beings, painting a view of robots as our helpmates, improving our lives and making us more productive.

176.1 Robot Definition

The term "robot" is defined by The Robotics Industry Association as a "reprogrammable multifunctional manipulator designed to move material, parts, tools, or specialized devices, through variable programmed motions for the performance of a variety of tasks." It consists of mechanical links, often in a serial chain with one link grounded or attached to a frame, that are connected via revolute (i.e., hinge) or prismatic (i.e., sliding) joints and actuated via gear trains or directly by electric motors or hydraulic or pneumatic drives. A robot will generally include position sensors (such as potentiometers or optical encoders) and may include contact, tactile, force/torque, proximity, or vision sensors. At its distal end, a robotic manipulator is typically fitted with an end-effector, such as a gripper, enabling it to accomplish desired tasks. An example of an industrial robot is the Case Packer from CAMotion, Inc. shown in Figure 176.1.

176.2 Robot Control Problem

The fundamental control problem in robotics is to determine the actuator signals required to achieve the desired motion and specified performance criteria. If the robot is to perform a task while in contact with a surface, it is also necessary to control the contact force applied by the manipulator. Although the control problem can be stated simply, its solution may be quite complicated due to robot nonlinearities. For example, because of coupling among links, the robot dynamics in a serial link robot design are described mathematically by a set of coupled nonlinear differential equations, making the controls problem challenging. The controls problem becomes even more difficult if the links exhibit flexibility, and hence cannot be modeled as rigid.

0-8493-1586-7/05/$0.00+$1.50

FIGURE 176.1 Case Packer Cartesian Robot. (Courtesy of CAMotion, Inc.)

To achieve the desired motion and possibly contact force characteristics, the planning of the manipulator trajectory is integrally linked to the control problem. The position of the robot can be described by a set of joint coordinates in *joint space* or by the position and orientation of the end-effector using coordinates along orthogonal axes in *Cartesian or task space*. The two representations are related, that is, the Cartesian position and orientation can be computed from the joint positions via a mapping (or function) known as *forward kinematics*. The motion required to realize the desired task is generally specified in Cartesian space. The joint positions required to achieve the desired end-effector position and orientation can be found by a mapping known as the *inverse kinematics*. This inverse kinematics problem may have more than one solution, and a closed-form solution may not be possible, depending on the geometric configuration of the robot. The desired motion may be specified as point to point, in which the end-effector moves from one point to another without regard to the path, or it may be specified as a continuous path, in which the end-effector follows a desired path between the points. A trajectory planner generally interpolates the desired path and generates a sequence of set points for the controller. The interpolation may be done in joint or Cartesian space.

Some of the robot control schemes in use today include independent joint control [Luh, 1983]; Cartesian-space control [Luh et al., 1980]; and force control strategies such as hybrid position/force control [Raibert and Craig, 1981] and impedance control [Hogan, 1985]. In independent joint control, each joint is considered as a separate system and the coupling effects between the links are treated as disturbances to be rejected by the controller. Performance can be enhanced by compensating for robot nonlinearities and interlink coupling using the method of computed torque or inverse dynamics. In Cartesian-space control, the error signals are computed in Cartesian space and the inverse kinematics problem need not be solved. In this chapter, independent joint control is analyzed in depth to provide insight into the use of various controllers for joint position control. Information on force control schemes can be found in the references cited above as well in Bonitz [1995].

176.3 Basic Joint Position Dynamic Model

Many industrial robots operate at slow speeds, employing large gear reductions that significantly reduce the coupling effects between the links. For slowly varying command inputs, the drive system dominates

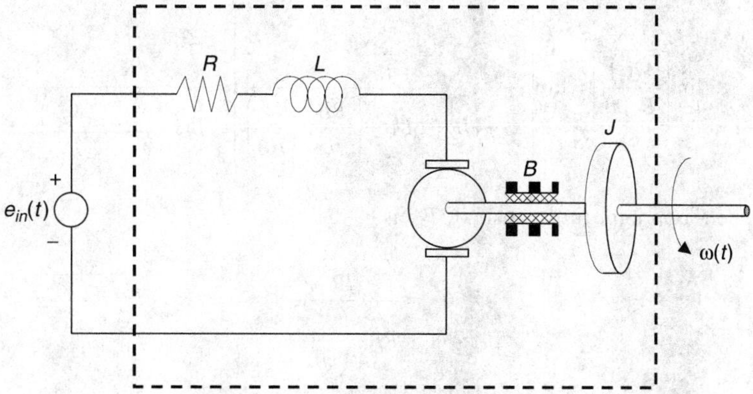

FIGURE 176.2 Simple actuator model for a single-link robot.

the dynamics of each joint and each link can be considered rigid. Under these conditions, each joint can be controlled as an independent system using linear system control techniques. Figure 176.2 is a model of a single actuator-link model, representative of a single link robot. The model includes an effective inertia, J,

$$J = J_m + r^2 J_l \tag{176.1}$$

where r is the gear ratio of the joint, J_m is the motor inertia, and J_l is the link inertia. The model also includes an effective damping coefficient, B, and a motor armature inductance and resistance, L and R, respectively. The relationship between the motor input voltage and the rotational speed is given by the second-order differential equation,

$$\ddot{\omega}(t) + (JR + BL)\dot{\omega}(t) + \left(\frac{RB}{JL} + \frac{K_t^2}{JL}\right)\omega(t) = \frac{K_t}{JL} e_{in}(t) \tag{176.2}$$

where $e_{in}(t)$ is the motor input voltage, $\omega(t)$ is the link angular velocity or rotational speed, and K_t, is a motor torque constant (torque per current ratio). Assuming zero initial conditions and defining the Laplace transformations,

$$\Im\{\omega(t)\} \equiv \Omega(s) \qquad \Im\{e_{in}(t)\} \equiv E_{in}(s)$$

the transfer function between the input voltage, $E_{in}(s)$ and the link rotational velocity, $\Omega(s)$, can be written as

$$\frac{\Omega(s)}{E_{in}(s)} = \frac{\dfrac{K_t}{JL}}{s^2 + (JR + BL)s + \left(\dfrac{RB}{JL} + \dfrac{K_t^2}{JL}\right)} \tag{176.3}$$

For most motor systems, the overall system dynamics are dominated by the mechanical dynamics rather than the electrical "dynamics," that is, the effect of L is small in comparison to the other system parameters. Assuming negligible L in equation (176.3), the following first-order transfer function can be developed:

$$\lim_{L\to 0}\left[\frac{\Omega(s)}{E_{in}(s)}\right]=\lim_{L\to 0}\left[\frac{\frac{K_t}{JL}}{s^2+(JR+BL)s+\left(\frac{RB}{JL}+\frac{K_t^2}{JL}\right)}\right]=\frac{K_t}{JRs+\left(RB+\frac{K_t^2}{JL}\right)} \tag{176.4}$$

Defining the open-loop joint time constant, T_m, as

$$T_m=\frac{JR}{RB+\left(\frac{K_t^2}{JL}\right)} \tag{176.5}$$

and the open-loop link gain as

$$K_m\equiv\frac{K_t}{RB+\left(\frac{K_t^2}{JL}\right)} \tag{176.6}$$

Equation (176.4) can be rewritten as

$$\frac{\Omega(s)}{E_{in}(s)}=\frac{K_m}{T_m s+1} \tag{176.7}$$

The assumption of the motor inductance being small, resulting in negligible electrical dynamics is considered reasonable if the negative real part of the single pole from the electrical dynamics is approximately three times larger than the negative real part of the mechanical dynamics. This is known as the dominant pole theory, and is valid for most motors. However, for small motors, such as those used in micro-electromechanical systems (MEMS) devices, this assumption may not be valid.

Equation (176.7) provides a relationship between joint angular velocity and input motor voltage, and is a useful first-order model for velocity control. However, for most robot applications, position control rather than velocity control is desired. For the typical case of lower velocities, the dynamics of the individual joints can be considered to be decoupled. Recognizing that the position of the joint is the integral of the joint velocity, a transfer function between position, $\Theta(s)$, which is the Laplace transform of $\theta(t)$, and the input voltage can be developed:

$$G(s)=\frac{\Theta(s)}{E_{in}(s)}=\frac{\frac{1}{s}\Omega(s)}{E_{in}(s)}=\left(\frac{1}{s}\right)\frac{K_m}{T_m s+1}=\frac{K_m}{s(T_m s+1)} \tag{178.8}$$

where $G(s)$ is the plant transfer function for the single link. Equation (176.8) is a model for position control, and the same simple dynamics can be used to model a variety of systems including high precision machine tools, which are essentially multiaxis Cartesian robots [Kurfess and Jenkins, 2000].

176.4 Independent Joint Position Control

For many applications, the assumption of negligible link coupling is reasonable and independent joint position control can be a successful strategy. Two classical control methods — proportional derivative (PD) control and proportional integral derivative (PID) control — are typically implemented. This

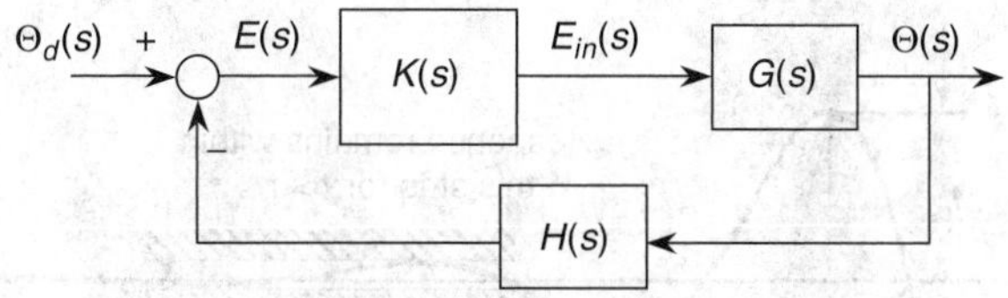

FIGURE 176.3 Generalized closed-loop system configuration.

section develops design strategies for both types, and, for completeness, considers the even simpler proportional (P) controller. Figure 176.3 represents the closed-loop feedback configuration for the single link model, where $G(s)$ represents the link or plant dynamics given by Equation (176.8), $K(s)$ represents the controller that is to be designed, and $H(s)$ represents the feedback sensor dynamics. In general, the relationship between the desired angular position, $\Theta_d(s)$, and the actual angular position of the link, $\Theta(s)$, is given by Black's Law (1934a, 1934b),

$$G_{CL}(s) = \frac{\Theta(s)}{\Theta_d(s)} = \frac{K(s)G(s)}{1+K(s)G(s)H(s)} \tag{176.9}$$

The angular error, $E(s)$, is given by the difference between the desired angle and the actual angle. The transfer function between $\Theta_d(s)$ and $E(s)$ is given by

$$\frac{E(s)}{\Theta_d(s)} = \frac{\Theta_d(s)-\Theta(s)}{\Theta_d(s)} = \frac{1}{1+K(s)G(s)H(s)} \tag{176.10}$$

The controller typically includes a power amplifier that provides the power to drive the system. In this example, the output of the controller is a voltage, that is, the power amplifier's function is to provide sufficient current at the desired voltage to drive the system. Saturation occurs when the desired input voltage exceeds the maximum amplifier voltage output capacity, or when the product of the voltage and current exceeds the amplifier's rated power capacity. The effect of saturation is ignored in the following examples. Finally, the controller output voltage can be related to the input command signal via the following transfer function,

$$\frac{E_{in}(s)}{\Theta_d(s)} = \frac{E(s)}{\Theta_d(s)}K(s) = \frac{\Theta_d(s)-\Theta(s)}{\Theta_d(s)}K(s) = \frac{K(s)}{1+K(s)G(s)H(s)} \tag{176.11}$$

Equation (176.11) can be used to determine the motor command voltage as a function of the desired angular trajectory.

Definition of Specifications

Specifications are generally provided to ensure that the desired system behavior is achieved. The specifications define the closed-loop system characteristics, and the usual practice is to tune the controller gains to achieve the desired performance. For this exercise, a 2% settling time, t_s, and a maximum percent overshoot, M_P, are specified. These two quantities (and others) are visualized in the typical second-order response shown in Figure 176.4. They define a desired dynamic response of the closed-loop system described by a second-order model of the form,

$$G_{CL}(s) = \frac{\omega_{nd}^2}{s^2 + 2\zeta_d \omega_{nd} s + \omega_{nd}^2} \tag{176.12}$$

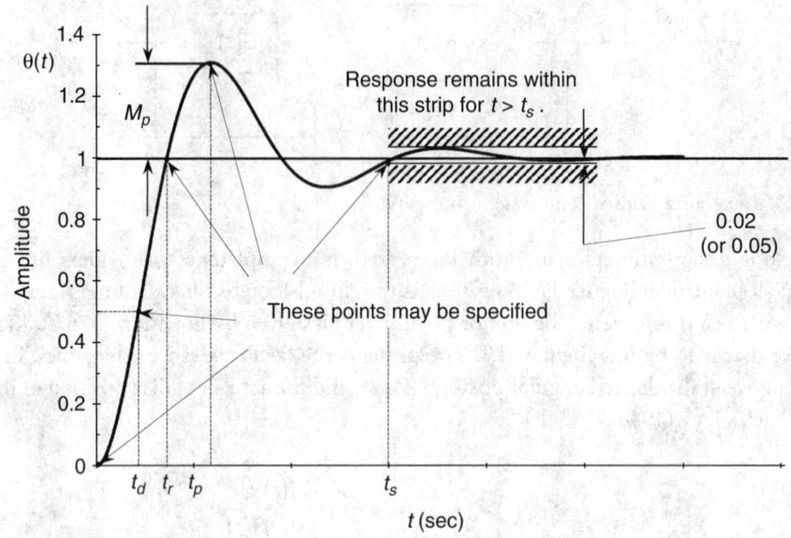

FIGURE 176.4 A typical second-order response.

where ζ_d and ω_{nd} are the desired damping ratio and (undamped) natural frequency, respectively, and are related to the desired overshoot M_{Pd} and desired settling time t_{sd}, of the target closed-loop system by the expressions,

$$M_{Pd} = e^{-\left(\zeta_d / \sqrt{1-\zeta_d^2}\right)\pi} \tag{176.13}$$

$$t_{sd} \approx \frac{4}{\zeta_d \omega_{nd}} \tag{176.14}$$

where the latter is a conservative approximation of the settling time. Solving for ζ_d in Equation (176.13) yields

$$\zeta_d = \sqrt{\frac{\ln(2M_{Pd})}{\pi^2 + \ln(2M_{Pd})}} = \frac{\ln(M_{Pd})}{\sqrt{\pi^2 + \ln(2M_{Pd})}} \tag{176.15}$$

From Equations (176.14) and (176.15), ω_{nd} can be expressed as

$$\omega_{nd} \approx \frac{4\sqrt{\pi^2 + \ln(2M_P)}}{t_{sd} \ln(M_P)} \tag{176.16}$$

From Equation (176.12), the characteristic equation that defines the system response is given by

$$s^2 + 2\zeta_d \omega_{nd} s + \omega_{nd}^2 = 0 \tag{176.17}$$

This equation can be used for the design of various controllers. The poles of the closed-loop system are shown in Figure 176.5 for a variety of desired damping ratios, ζ_d. Nominally, for joint position control a damping ratio of 1, corresponding to a critically damped system, is desired. This places both poles in

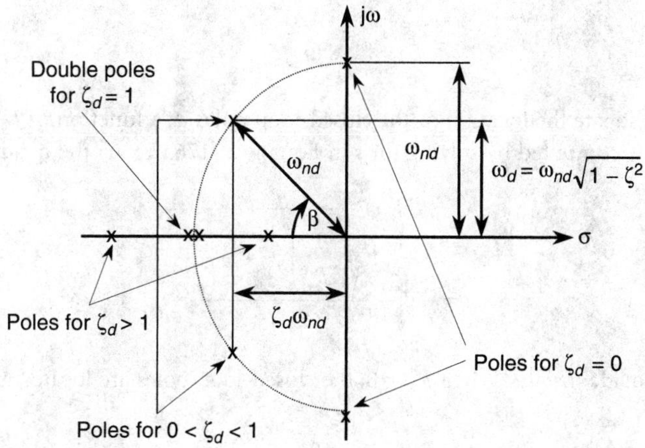

FIGURE 176.5 Pole locations for desired closed-loop system dynamics.

the same location on the real axis, and yields the fastest system response without overshoot. For generality in the control design examples, the variables ζ_d and ω_{nd} are used rather than specific values for these variables [Nagurka and Kurfess, 1992].

Proportional (P) Control

In P control, the simplest analog control algorithm, the control signal is proportionally related to the error signal by a constant gain, K_P. Thus, the transfer function for a proportional controller is a constant,

$$K(s) = K_P \tag{176.18}$$

The closed-loop transfer function of the single actuator-link system with a proportional controller is

$$G_{CL}(s) = \frac{K_P K_m / T_m}{s^2 + (1/T_m)s + K_P K_m / T_m} \tag{176.19}$$

from Equations (176.8), (176.9), and (176.18). The characteristic equation for the transfer function given in Equation (176.19) is

$$s^2 + (1/T_m)s + K_P K_m / T_m = 0 \tag{176.20}$$

With one free design parameter, the proportional gain, K_P, the target closed-loop system dynamics will be virtually impossible to achieve. By equating the actual closed-loop characteristic Equation (176.20) to the desired closed-loop characteristic Equation (176.17),

$$s^2 + (1/T_m)s + K_P K_m / T_m = 0 = s^2 + 2\zeta_d \omega_{nd} s + \omega_{nd}^2 \tag{176.21}$$

only ω_{nd} can be specified directly, giving the value of K_P by

$$K_P = \omega_{nd}^2 \frac{T_m}{K_m} \tag{176.22}$$

(found by equating the coefficients of s^0). Once ω_{nd} has been chosen, ζ_d is fixed to be

$$\zeta_d = \frac{1}{2\omega_{nd}T_m} \tag{176.23}$$

It is useful to investigate the location of the closed-loop poles as a function of K_P. The pole locations on the s-plane can be computed by solving for s in Equation (176.21) via the quadratic formula,

$$s = \frac{-\frac{1}{T_m} \pm \sqrt{\left(\frac{1}{T_m}\right)^2 - 4\frac{K_P K_m}{T_m}}}{2} \tag{176.24}$$

Two values of K_P provide insight. When $K_P = 0$, the closed-loop poles are located at

$$s = \begin{cases} 0 \\ -\frac{1}{T_m} \end{cases} \tag{176.25}$$

corresponding to the open-loop pole locations. When K_P is selected such that the radical in Equation 176.24 vanishes,

$$K_P = \frac{1}{4K_m T_m} \tag{176.26}$$

the two roots are located at

$$s = \begin{cases} -\frac{1}{2T_m} \\ -\frac{1}{2T_m} \end{cases} \tag{176.27}$$

When $0 \le K_P \le 1/(4K_m T_m)$, the closed-loop poles of the system are purely real. When $K_P > 1/(4K_m T_m)$, the closed-loop poles of the system are complex conjugates with a constant real part of $-1/(2T_m)$. Increasing the proportional gain beyond $1/(4K_m T_m)$ only increases the imaginary component of the poles; the real part does not change. A graphical portrayal of the pole locations of the system under proportional control is shown in the root locus plot of Figure 176.6. The poles start at their open-loop positions when $K_P = 0$, and transition to a break point at $-1/(2T_m)$ when $K_P = 1/(4K_m T_m)$. After the break point, they travel parallel to the imaginary axis with only their imaginary parts increasing while their real parts remain constant at $-1/(2T_m)$. It is clear from Figure 176.6 that only specific combinations of ω_{nd} and ζ_d can be achieved with the proportional control configuration. Such relationships can be observed directly via a variety of controls design tools for both single-variable and multivariable systems [Kurfess and Nagurka, 1993, 1994].

Proportional Derivative (PD) Control

To improve performance beyond that available by proportional control, PD control can be employed. The dynamics for the PD controller are given by

$$K(s) = K_D s + K_P \tag{176.28}$$

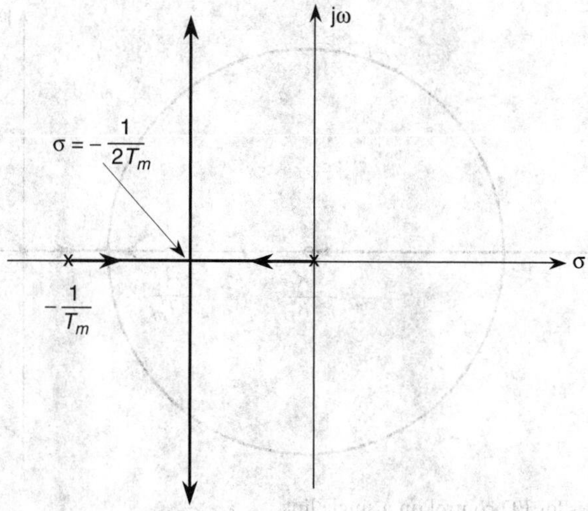

FIGURE 176.6 Root locus for proportional control on a single link.

where the constants K_P and K_D are the proportional and derivative gains, respectively. Employing Black's Law with the PD controller, the closed-loop transfer function for the single actuator-link model is

$$G_{CL}(s) = \frac{\frac{K_m}{T_m}(K_D s + K_P)}{s^2 + \left(\frac{K_m K_D + 1}{T_m}\right)s + K_m K_P / T_m} \tag{176.29}$$

There are now additional dynamics associated with the zero (root of the transfer function numerator) located at

$$s = -\frac{K_P}{K_D} \tag{176.30}$$

These dynamics can affect the final closed-loop response. Ideally, the zero given in Equation (176.30) will be far enough to the left of the system poles on the s-plane that the dominant pole theory applies. (This would mean the dynamics corresponding to the zero are sufficiently fast in comparison to the closed-loop pole dynamics that their effect is negligible.)

Expressions for K_P and K_D can be found by equating the closed-loop characteristic Equation (176.29),

$$s^2 + \left(\frac{K_m K_D + 1}{T_m}\right)s + K_m K_P / T_m = 0 \tag{176.31}$$

to the target closed-loop characteristic equation that possesses the desired natural frequency and damping ratio, Equation (176.17). Equating the coefficients of s^0 and solving for K_P yields

$$K_P = \omega_{nd}^2 \frac{T_m}{K_m} \tag{176.32}$$

Similarly, equating the coefficients of s^1 and solving for K_D yields

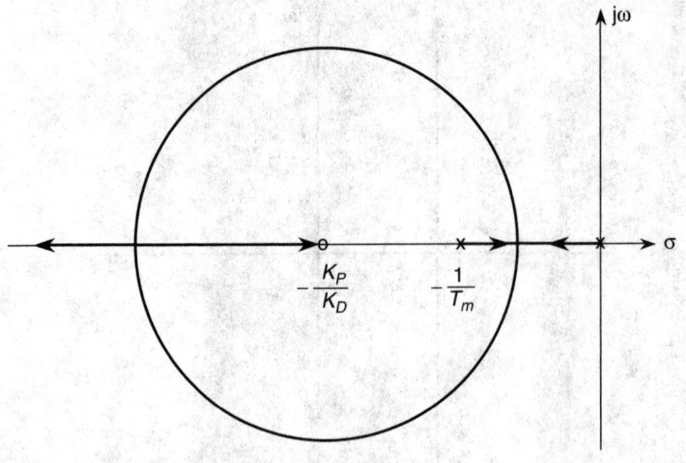

FIGURE 176.7 Root locus for PD control on a single link.

$$K_D = \frac{2\zeta_d \omega_{nd} T_m - 1}{K_m} \tag{176.33}$$

This approach of solving for control gains is known as pole placement. As with all control design methods, the gains derived using Equations (176.32) and (176.33) should be checked to ensure that they are realistic.

Figure 176.7 presents a typical root locus plot for a single actuator-link model using a PD controller where the forward loop gain is varied. The location of the single zero ($s = -K_P/K_D$) can be seen on the real axis in the left of the s-plane. For this design, the zero dynamics will be insignificant in comparison to the pole dynamics. It is noted that the break point for the root locus presented in Figure 176.7 is slightly to the left of the break point in Figure 176.6 ($s = -1/2T_m$). Furthermore, the locus of the complex conjugates poles is circular and centered at the zero.

Classical control tools such as the root locus are critical in control design. Without such tools the designer could use the relationships given by Equations (176.32) and (176.33) to place the poles at any location. However, the resulting location of the zero may have been too close to the closed-loop poles resulting in system dynamics that do not behave as desired. This is described in detail in the following section.

Effect of Zero Dynamics

To demonstrate the effect of a zero, consider two systems that have the same damping ratios (ζ) and natural frequencies (ω_n). The first system is given by

$$G_1(s) = \frac{R_1(s)}{C(s)} = \frac{\omega_n^2}{s^2 + 2\zeta\omega_n s + \omega_n^2} \tag{176.34}$$

and the second system is

$$G_2(s) = \frac{R_2(s)}{C(s)} = \frac{s + \omega_n^2}{s^2 + 2\zeta\omega_n s + \omega_n^2} \tag{176.35}$$

where the input is $c(t)$ and the output is $r_i(t)$ (i = 1,2). The difference between the two systems is the zero located at $s = -\omega_n^2$ for the system described by Equation (176.35). For a unit step input,

$$C(s) = \frac{1}{s} \tag{176.36}$$

the unit step responses of the two systems are

$$R_1(s) = \frac{\omega_n^2}{s^2 + 2\zeta\omega_n s + \omega_n^2}\left(\frac{1}{s}\right) \tag{176.37}$$

and

$$R_2(s) = \frac{s + \omega_n^2}{s^2 + 2\zeta\omega_n s + \omega_n^2}\left(\frac{1}{s}\right) = \frac{s}{s^2 + 2\zeta\omega_n s + \omega_n^2}\left(\frac{1}{s}\right) + \frac{\omega_n^2}{s^2 + 2\zeta\omega_n s + \omega_n^2}\left(\frac{1}{s}\right) \tag{176.38}$$

The difference between the two responses is the derivative term, corresponding to the first term on the right side of Equation (176.38):

$$\frac{s}{s^2 + 2\zeta\omega_n s + \omega_n^2}\left(\frac{1}{s}\right)$$

The time domain responses for Equations (176.37) and (176.38), respectively, are

$$r_1(t) = 1 - e^{-\zeta\omega_n t}\left[\left(\frac{\zeta}{\sqrt{1-\zeta^2}}\right)\sin(\omega_d t) + \cos(\omega_d t)\right] \tag{176.39}$$

and

$$r_2(t) = \left\{1 - e^{-\zeta\omega_n t}\left[\left(\frac{\zeta}{\sqrt{1-\zeta^2}}\right)\sin(\omega_d t) + \cos(\omega_d t)\right]\right\} + \left\{e^{-\zeta\omega_n t}\left(\frac{1}{\omega_n\sqrt{1-\zeta^2}}\right)\sin(\omega_d t)\right\} \tag{176.40}$$

where ω_d, the damped natural frequency, is given by

$$\omega_d = \omega_n\sqrt{1-\zeta^2} \tag{176.41}$$

Several items are worth noting. First, Equations (176.39) and (176.40) are valid for $0 \le \zeta < 1$. Second, the difference between the two responses is

$$\left\{e^{-\zeta\omega_n t}\left(\frac{1}{\omega_n\sqrt{1-\zeta^2}}\right)\sin(\omega_d t)\right\} \tag{176.42}$$

which is the impulse response of the system given by Equation (176.34). For the initial transient of the step response, the impulse response is positive and, therefore, adds to the overall system response. This indicates that the overshoot may be increased, which is the case for the example shown in Figure 176.8 for $\omega_n = 1$ rad/s and $\zeta = 0.707$. In Figure 176.8, the proportional term is the step response given by Equation (176.39), the derivative term is given by expression (176.42), and the total response is given by the sum of the two expressions in Equation (176.40).

From this example, the addition of the zero results in increased overshoot. The real part of the poles is

$$s = \zeta\omega_n = 0.707 \tag{176.43}$$

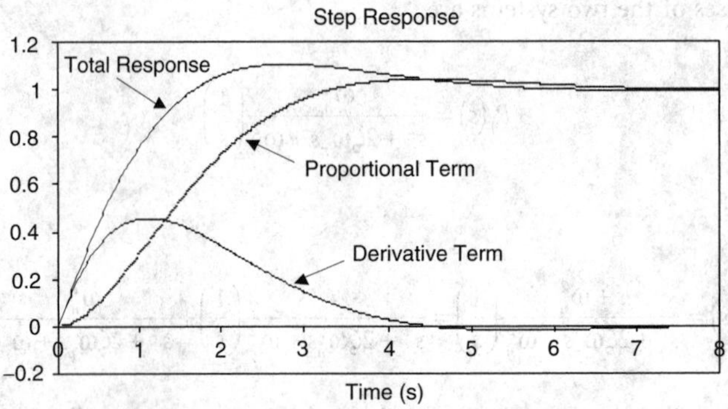

FIGURE 176.8 Step response of two systems.

and the location of the zero is

$$s = 1 \tag{176.44}$$

Thus, the dominant pole theorem does not hold and the dynamics of the zero cannot be ignored. Again, this is demonstrated by the increased overshoot illustrated in Figure 176.8.

Proportional Integral Derivative (PID) Control

A PID controller can be designed to provide better steady-state disturbance rejection capabilities compared to PD or P controllers. The form of the PID controller is given by

$$K(s) = K_D s + K_P + \frac{K_I}{s} = \frac{K_D s^2 + K_P s + K_I}{s} \tag{176.45}$$

For the single actuator-link model, the closed-loop transfer function using a PID controller is given by

$$G_{CL}(s) = \frac{\frac{K_m}{T_m}\left(K_D s^2 + K_P s + K_I\right)}{s^3 + \left(\frac{K_m K_D + 1}{T_m}\right)s^2 + \frac{K_m K_P}{T_m}s + \frac{K_m K_I}{T_m}} \tag{176.46}$$

The closed-loop dynamics are third order and, in comparison to the second-order PD case, the control design has an extra gain that must be determined and an additional pole, located at $s = -d$, added to the target dynamics. The closed-loop characteristic equation,

$$s^3 + \left(\frac{K_m K_D + 1}{T_m}\right)s^2 + \frac{K_m K_P}{T_m}s + \frac{K_m K_I}{T_m} = 0, \tag{176.47}$$

is equated to the new target closed-loop characteristic equation,

$$\left(s^2 + 2\zeta_d \omega_{nd} s + \omega_{nd}^2\right)(s + d) = 0 \tag{176.48}$$

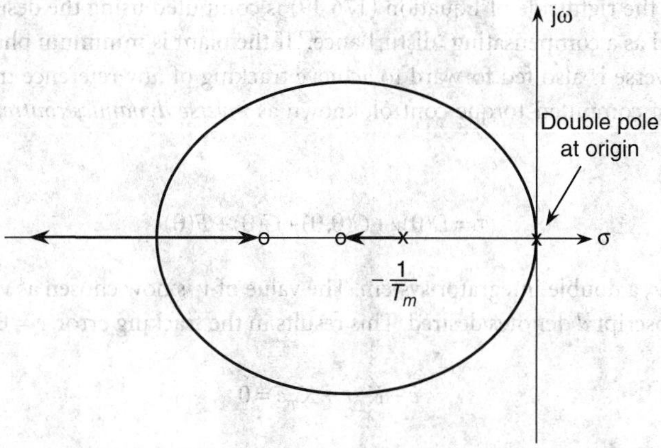

FIGURE 176.9 Root locus for PID control on a single link.

As with the PD controller, the location of the zeros in Equation (176.46) as well as the extra pole at $s = -d$, should have a negative real parts that are at least three times as negative as the target closed-loop pole locations.

Figure 176.9 presents a possible root locus plot for the single actuator-link model with a PID controller where the forward loop gain is varied. The two zeros and extra pole can be seen on the real axis to the left in the s-plane. As with the PD control design, the gains used for achieving the target dynamics must be well understood. It is easy to pick gains that move both zeros and the extra pole far to the left in the s-plane. However, it should be noted that the further left the dynamics are shifted in the s-plane, the faster they become, influencing less the overall response of the system. However, to increase the response speed of these dynamics requires higher gains that ultimately result in higher power requirements (and the possibility of instability). As such, while designing for higher gains is rather straightforward mathematically, the implementation may be limited by the physical constraints of the system.

Unlike the locus of the complex conjugate poles for the model employing PD control (Figure 176.6), the locus here need not be circular. Also, given the increased number of closed-loop system parameters (including the control gains and design parameters such as d, the sensitivity of the closed-loop dynamics to changes of these parameters may be important and should be investigated. For example, if the proportional gain changes a small amount (which is bound to happen in a real system), the impact in the overall system performance must be understood. Mathematically, sensitivity can formally be shown to be a complex quantity (having both magnitude and angle), and a variety of techniques can be used to explore sensitivity to parameter changes [Kurfess and Nagurka, 1994].

176.5 Method of Computed Torque

If the robot is a direct-drive type without gear reduction or if the command inputs are not slowly varying, the control scheme of the previous section may exhibit poor performance characteristics and instability may even result. One method of compensating for the effects of link coupling is to use feed-forward disturbance cancellation. The disturbance torque is computed from the robot dynamic equation, summarized in vector-matrix form as

$$\tau = D(\theta)\ddot{\theta} + C(\theta,\dot{\theta}) + G(\theta) + F(\dot{\theta}) \quad (176.49)$$

where τ is the $n \times 1$ vector of joint torques (forces), $D(\theta)$ is the $n \times n$ inertia matrix, $C(\theta,\dot{\theta})$ is the $n \times 1$ vector of Coriolis and centrifugal torques (forces), $G(\theta)$ is the $n \times 1$ vector of torques (forces) due to gravity, and $F(\theta)$ is the $n \times 1$ vector of torques (forces) due to friction. In the feed-forward disturbance cancellation scheme, the right side of Equation (176.49) is computed using the desired value of the joint variables and injected as a compensating "disturbance." If the plant is minimum phase (has no right-half s-plane zeros), its inverse is also fed forward to achieve tracking of any reference trajectory.

Another version of computed-torque control, known as *inverse dynamics control,* involves setting the control torque to

$$\tau = D(\theta)v + C(\theta,\dot{\theta}) + G(\theta) + F(\dot{\theta}) \tag{176.50}$$

which results in $\ddot{\theta} = v$, a double integrator system. The value of v is now chosen as $v = \ddot{\theta}_d + K_D(\dot{\theta}_d - \dot{\theta}) + K_P(\theta_d - \theta)$ where subscript d denotes desired. This results in the tracking error, $e = \theta_d - \theta$, which satisfies

$$\ddot{e} + K_D\dot{e} + K_P e = 0 \tag{176.51}$$

The gains K_P and K_D can be chosen for the desired error dynamics (damping and natural frequency). In general, computed-torque control schemes are computationally intensive due to the complicated nature of Equation (176.49), and require accurate knowledge of the robot model [Bonitz, 1995].

176.6 Cartesian Space Control

The basic concept of Cartesian space control is that the error signals used in the control algorithm are computed in Cartesian space, obviating the solution of the inverse kinematics. The position and orientation of the robot end-effector can be described by a 3 × 1 position vector, p, and the three orthogonal axes of an imaginary frame attached to the end effector. (The axes are known as the normal (n), sliding (s), and approach (a) vectors.) The control torque is computed from

$$\tau = D(\theta)\ddot{\theta} + C(\theta,\dot{\theta}) + G(\theta) + F(\dot{\theta}) \tag{176.52}$$

$$\ddot{\theta} = J(\theta)^{-1}[\ddot{x}_d + K_D\dot{e} + K_P e - \dot{J}(\theta,\dot{\theta})\dot{\theta}] \tag{176.53}$$

where $J(\theta)$ is the manipulator *Jacobian* that maps the joint velocity vector to the Cartesian velocity vector, $\ddot{x}_d$ is the 6 × 1 desired acceleration vector, $e = [e_p^T e_o^T]^T$, e_p is the 3 × 1 position error vector, e_o is the 3 × 1 orientation error vector, K_D is the 6 × 6 positive-definite matrix of velocity gains, and K_P is the 6 × 6 positive-definite matrix of position gains. The actual position and orientation of the end-effector is computed from the joint positions via the forward kinematics. The position error is computed from $e_p = \theta_d - \theta$ and, for small error, the orientation error is computed from $e_o = \frac{1}{2}[n \times n_d + s \times s_d + a \times a_d]$. The control law of Equations (176.52) and (176.53) results in the Cartesian error equation,

$$\ddot{e} + K_D\dot{e} + K_P e = 0 \tag{176.54}$$

The gain matrices K_D and K_P can be chosen to be diagonal to achieve the desired error dynamics along each Cartesian direction [Spong and Vidyasagar, 1989].

The Cartesian space controller has the disadvantage that the inverse of the Jacobian is required, which does not exist at *singular configurations.* The planned trajectory must avoid singularities, or alternative methods such as the SR pseudoinverse [Nakamura, 1991] must be used to compute the Jacobian inverse [Bonitz, 1995].

Defining Terms

Cartesian or task space — The set of vectors describing the position and orientation of the end-effector using coordinates along orthogonal axes. The position is specified by a 3 × 1 vector of the coordinates of the end-effector frame origin. The orientation is specified by the 3×1 normal, sliding, and approach vectors describing the directions of the orthogonal axes of the frame. Alternately, the orientation may be described by Euler angles, roll/pitch/yaw angles, or an axis/angle representation.

Forward kinematics — The function that maps the position of the joints to the Cartesian position and orientation of the end-effector. It maps the joint space of the manipulator to Cartesian space.

Inverse kinematics — The function that maps the Cartesian position and orientation of the end-effector to the joint positions. It is generally a one-to-many mapping, and a closed-form solution may not always be possible.

Jacobian — The function that maps the joint velocity vector to the Cartesian translational and angular velocity vector of the end-effector $\dot{x} = J(\theta)\dot{\theta}$.

Singular configuration — A configuration of the manipulator in which the manipulator Jacobian loses full rank. It represents configurations from which certain directions of motion are not possible or when two or more joint axes line up and there is an infinity of solutions to the inverse kinematics problem.

References

Black, H. S., Stabilized feedback amplifiers, *Bell Syst. Tech. J.*, January 1934a.

Black, H. S., Stabilized Feedback Amplifiers, U.S. Patent No. 2,102,671, 1934b.

Bonitz, R. G., Robots and controls. In *The Engineering Handbook*, Dorf, R. D., Ed., CRC Press, Boca Raton, FL, 1995.

Hogan, N., Impedance control: an approach to manipulation, parts I, II, and III. *ASME J. Dynamic Syst. Meas. Control*, 107, 124, 1985.

Kurfess, T. R. and Nagurka, M. L., A geometric representation of root sensitivity, *ASME J. Dynamic Syst. Meas. Control*, 116, 305–309, 1994.

Kurfess, T. R. and Nagurka, M. L., Foundations of classical control theory with reference to eigenvalue geometry, *J. Franklin Inst.*, 330, 213–227, 1993.

Kurfess, T. R. and Jenkins, H. E., Ultra-high precision control, in *Control Systems Applications*, Levine, W., Ed., , CRC Press, Boca Raton, FL, 2000, p. 212.

Kurfess, T. R., Precision manufacturing, in *Mechanical System Design Handbook*, Nwokah, O. and Hurmuzlu, Y., Eds., CRC Press, Boca Raton, FL, 2002, p. 151.

Luh, J. Y. S., Conventional controller design for industrial robots — a tutorial, *IEEE Trans. Syst. Man. Cybern.*, SMC-13, 298–316, 1983.

Luh, J. Y. S., Walker, M. W., and Paul, R. P., Resolved-acceleration control of mechanical manipulators, *IEEE Trans. Automat. Control*, AC-25, 464, 1980.

Nagurka, M. L. and Kurfess, T. R., An alternate geometric perspective on MIMO systems, *ASME J. Dynamic Syst. Meas. Control*, 115, 538, 1993.

Nagurka, M. L. and Kurfess, T. R., A Unified Classical/Modern Approach for Undergraduate Control Education, lecture notes from National Science Foundation Workshop, 1992.

Nakamura, Y., *Advanced Robotics, Redundancy, and Optimization*, Addison-Wesley, Reading, MA, 1991.

Raibert, M. H. and Craig, J. J., Hybrid position/force control of manipulators, *ASME J. Dynamic Syst. Meas. Control*, 103, 126, 1981.

Spong, M. W. and Vidyasagar, M. *Robot Dynamics and Control*, John Wiley & Sons, New York, 1989.

177

State Variable Feedback

Thomas L. Vincent
University of Arizona

The fundamental concept of control system design deals with creating a feedback loop between the output and the input of a dynamical system. This is done to improve the stability characteristics of the system. State variable feedback is associated with the idea of using every state variable in the feedback loop, and is used in both nonlinear and linear systems. However, it is only with linear systems that a relatively complete theory is available, with some of the results given in this chapter. The advantage of using state variable feedback for linear control system design is its simplicity. However, in order to use state variable feedback, every state variable must be either measured directly or estimated. Since measuring every state variable is impractical for many control applications, a state estimator must usually be included as a part of the state variable feedback control system design.

177.1 Linear State Space Control Systems

A linear state space model of a control system is given by

$$\dot{\mathbf{x}} = \mathbf{Ax} + \mathbf{Bu} \tag{177.1}$$

$$\mathbf{y} = \mathbf{Cx} + \mathbf{Du} \tag{177.2}$$

where the dot denotes differentiation with respect to time, x is an $N_x \times 1$ dimensional state vector, u is an $N_u \times 1$ dimensional control vector, y is an $N_y \times 1$ dimensional output vector, A is a square $N_x \times N_x$ matrix, B is $N_x \times N_u$, C is $N_y \times Nx$, and D is $N_y \times N_u$. The matrices A, B, C, and D are composed of system constants. Nonlinear systems are often approximated by this same linear system in the neighborhood of some steady-state equilibrium point.

The state vector

$$\mathbf{x} = [x_1 \cdots x_{N_x}]^{\mathrm{T}} \tag{177.3}$$

where $()^{\mathrm{T}}$ denotes transpose, is defined by the solution to the differential Equation (177.1). All control inputs to the system are given by the vector

$$\mathbf{u} = [u_1 \cdots u_{N_u}]^{\mathrm{T}} . \tag{177.4}$$

0-8493-1586-7/05/$0.00+$1.50

Control inputs include both open-loop command inputs external to the system and closed-loop control inputs internal to the system. Control is ultimately transmitted to the system through the use of actuators that may or may not be modeled in Equation (177.1). It follows from Equation (177.1) that the state of the system cannot be determined until all inputs, along with initial conditions for the state, have been specified. All of the system quantities measured by means of sensors are given by the output vector

$$\mathbf{y} = [y_1 \cdots y_{N_y}]^{\mathrm{T}} \tag{177.5}$$

The sensor output may be of the general form of Equation (177.2), where it is a function of both the state and control inputs (e.g., accelerometer) or it may be simply one or more of the state variables (e.g., position sensor, tachometer). The control system design problem is to determine an automatic control algorithm such that the relationship between a command input r(t) and the output y(t) yields acceptable performance in terms of tracking, stability, uncertain inputs, and so forth. These qualitative performance criteria are determined by the controlled system eigenvalues.

177.2 Controllability and Observability

The concepts of controllability and observability are fundamental to state space design. If a system is controllable, then it is always possible, using state variable feedback, to design a stable controlled system. In fact, the controlled systems eigenvalues may be arbitrarily placed. If a system is observable, then it will always be possible to design an estimator for the state.

The system in Equation (177.1) is controllable if for every initial state x(0) there exists a control u(t), $t \in [0,T]$, where T is some finite time interval, that will drive the system to any other point in state space. The system in Equations (177.1) and (177.2) is observable if, given a control u(t), $t \in [0,T]$, the initial state x(0) can be determined from the observation history y(t), $t \in [0, T]$.

Controllability and observability for linear systems in the form of Equations (177.1) and (177.2) may be checked using the Kalman controllability and observability criteria. The Kalman controllability criterion is that the system in Equation (177.1) is controllable if and only if

$$\mathrm{rank}[\mathbf{P}] = N_x \tag{177.6}$$

where P is the controllability matrix

$$\mathbf{P} = [\mathbf{B}, \mathbf{AB}, \cdots, \mathbf{A}^{N_x - 1}\mathbf{B}]. \tag{177.7}$$

According to the Kalman observability criterion, the system in Equations (177.1) and (177.2) is observable if and only if

$$\mathrm{rank}[\mathbf{Q}] = N_x \tag{177.8}$$

where Q is the observability matrix,

$$\mathbf{Q} = \begin{bmatrix} \mathbf{C} \\ \mathbf{CA} \\ \mathbf{CA}^2 \\ \vdots \\ \mathbf{CA}^{N_x - 1} \end{bmatrix} \tag{177.9}$$

State variable feedback design requires that both the controllability and observability criteria be satisfied, which will be assumed in the following discussion. State space design may be broken into two parts: eigenvalue placement (assuming full state information is available) and observer design to supply any missing state information.

177.3 Eigenvalue Placement

If every component of the state x is measured then C = I and D = 0 so that y = x. Given that the desired controlled system eigenvalues are known, the design problem is reduced to finding an $N_u \times N_x$ feedback gain matrix K such that Equation (177.1) under state variable feedback of the form

$$\mathbf{u} = \mathbf{Fr}(t) - \mathbf{Kx}(t) \tag{177.10}$$

will have the prescribed eigenvalues. Here, r(t) is an $N_x \times 1$ vector of command inputs and F is an $N_x \times N_r$ input matrix. The F matrix will in general be required, since the dimension of the command inputs may not equal the dimension of the control vector. It also provides for scaling of the command input. Substituting Equation (177.10) into Equation (177.1) yields

$$\dot{x} = \overline{\mathbf{A}}\mathbf{x} + \overline{\mathbf{B}}\mathbf{r} \tag{177.11}$$

where

$$\overline{\mathbf{A}} = \mathbf{A} - \mathbf{BK} \tag{177.12}$$

$$\overline{\mathbf{B}} = \mathbf{BF} \tag{177.13}$$

The controlled system of Equation (177.11) is of the same form as the original system in Equation (177.1), except the matrix $\overline{\mathbf{A}}$ now depends on the matrix of feedback gains K, and the control input, u, is now the command input r. For a constant command input, the equilibrium solutions to Equation (177.11) depend on both F and K. However, K alone affects the eigenvalues of the controlled system. If Equation (177.1) is completely controllable, then any desired set of eigenvalues for $\overline{\mathbf{A}}$ can be obtained through an appropriate choice for K [Davison, 1968; Wonham, 1967].

For single-input systems (u is a scalar) of low dimensions, determining K for direct eigenvalue placement is easy to do. For example, consider a second-order system of the form of Equation (177.1) with

$$A = \begin{bmatrix} 0 & 1 \\ a_1 & a_2 \end{bmatrix} \quad B = \begin{bmatrix} 0 \\ b \end{bmatrix} \tag{177.14}$$

This system satisfies Equation (177.6) so that state space design is possible. Under state variable feedback,

$$\overline{A} = \begin{bmatrix} 0 & 1 \\ a_1 - bk_1 & a_2 - bk_2 \end{bmatrix} \tag{177.15}$$

that has the characteristic equation

$$\lambda^2 + (bk_2 - a_2)\lambda + (bk_1 - a_1) = 0 \tag{177.16}$$

This system can now be made to behave like a second-order system with eigenvalues $\overline{\lambda}_1$ and $\overline{\lambda}_2$ satisfying the characteristic equation

$$(\lambda-\bar{\lambda}_1)(\lambda-\bar{\lambda}_2)=\lambda^2+2\zeta\omega_n\lambda+\omega_n^2=0 \qquad (177.17)$$

by matching coefficients to yield

$$k_1=(a_1+\omega_n^2)/b; \quad k_2=(a_2+2\zeta\omega_n)/b \qquad (177.18)$$

This procedure is also applicable when u is a vector, but it is complicated by the fact that the gain matrix K is not unique. However, a general procedure to handle this situation is available [Wonham, 1985]. The procedure can also be readily adapted for numerical solution so that higher dimensional problems can be handled as well.

Unless one is certain about what eigenvalues to use, the question of where to place the controlled system eigenvalues remains a fundamental design problem. There are methods available for handling this aspect of state variable feedback design as well. Some of the more commonly used methods are the linear-quadratic-regulator design [Bryson and Ho, 1975] and robust control design [Green and Limebeer, 1994]. These methods focus on other system performance criteria for determining the feedback gain matrix K, which, in turn, indirectly determine the controlled system eigenvalues. Software is available, based on these methods, for doing control design on a PC [Shahian and Hassul, 1993].

177.4 Observer Design

Even if every component of the state vector x is not directly measured, it is possible to build a device called an *observer* [Luenberger, 1971] that will approximate x. An observer that reconstructs the entire state vector is called an *identity observer.* One that only reconstructs states not directly measured is called a *reduced-order observer.* Identity observers are of a simpler form.

An observer uses the system model with an additional output feedback term to generate an estimate for the state. If we let $\hat{\mathbf{x}}$ denote the estimate of x, then an identity observer is obtained from

$$\dot{\hat{\mathbf{x}}}=\mathbf{A}\hat{\mathbf{x}}+\mathbf{B}\mathbf{u}-\mathbf{G}[\hat{\mathbf{y}}-\mathbf{y}] \qquad (177.19)$$

where $\hat{\mathbf{x}}(0)=0$, G is an $N_x \times N_y$ matrix to be determined, and

$$\hat{\mathbf{y}}=\mathbf{C}\hat{\mathbf{x}}+\mathbf{D}\mathbf{u} \qquad (177.20)$$

is the predicted measurement in accordance with Equation (177.2). The structure of the state estimator from Equations (177.19) and (177.20) is the same as a Kalman filter [Kalman, 1960], which is used for state estimation in the presence of Gaussian random noise inputs.

Substituting (177.20) into (177.19) and rearranging terms yields the identity observer,

$$\dot{\hat{\mathbf{x}}}=[\mathbf{A}-\mathbf{G}\mathbf{C}]\hat{\mathbf{x}}+[\mathbf{B}-\mathbf{G}\mathbf{D}]\mathbf{u}+\mathbf{G}\mathbf{y} \qquad (177.21)$$

The components of the G matrix are chosen so that the error equation

$$\dot{\mathbf{e}}=\dot{\hat{\mathbf{x}}}-\dot{\mathbf{x}}=[\mathbf{A}-\mathbf{G}\mathbf{C}]\mathbf{e} \qquad (177.22)$$

is stable. The *return time* of the error equation should be chosen so that it is faster than the return time of the controlled system. The stability properties of the error equation are completely under the designer's control, provided that the system in Equations (177.1) and (177.2) is observable. For controllable and observable linear systems, observers do not change the closed-loop eigenvalues of the controlled system.

Rather, they simply adjoin their own eigenvalues. That is, the eigenvalues for the controlled system using an identity observer are those associated with [A – BK] and [A – GC], respectively.

As an example, consider the second-order system given by Equation (177.14), in which the output is the first state variable x_1, that is,

$$\mathbf{C} = [1 \quad 0] \quad D = 0 \tag{177.23}$$

The observability criterion of Equation (177.8) is satisfied by this system so that an observer can be built. In this case,

$$A - GC = \begin{bmatrix} -g_1 & 1 \\ a_1 - g_2 & a_2 \end{bmatrix} \tag{177.24}$$

which has the characteristic equation,

$$\lambda^2 + (g_1 - a_2)\lambda + (g_2 - a_2 g_1 - a_1) = 0 \tag{177.25}$$

By choosing the constants g_1 and g_2, the observer can now be made to behave like a second-order system with eigenvalues $\hat{\lambda}_1$ and $\hat{\lambda}_2$ satisfying the characteristic equation,

$$(\lambda - \hat{\lambda}_1)(\lambda - \hat{\lambda}_2) = \lambda^2 + 2\hat{\zeta}\hat{\omega}_n\lambda + \hat{\omega}_n^2 = 0 \tag{177.26}$$

Equating coefficients between Equations (177.25) and (177.26) yields

$$g_1 = a_2 + 2\hat{\zeta}\hat{\omega}_n; \qquad g_2 = a_1 + a_2 g_1 + \hat{\omega}_n^2 \tag{177.27}$$

The observer in this case is of the form

$$\dot{\hat{x}}_1 = \hat{x}_2 + g_1(y - \hat{x}_1) \tag{177.28}$$

$$\dot{\hat{x}}_2 = a_1\hat{x}_1 + a_2\hat{x}_2 + g_2(y - \hat{x}_1) + bu \tag{177.29}$$

These equations must be solved on-line to yield $\hat{x}_1$ and $\hat{x}_2$. The resulting controller is

$$u = r - k_1\hat{x}_1 - k_2\hat{x}_2 \tag{177.30}$$

In this case the actual measurement $y = x_1$ could be used in Equation (177.30) instead of $\hat{x}_1$. However, if the measurement of y contains noise, this noise can be filtered through the observer by using the full state estimate $\hat{x}$ as used in Equation (177.30).

The observer differential equations may be solved on-line by either digital or analog methods. The initial conditions for solving these equations will generally not be known. However, in most situations, this does not present difficulties with a properly designed observer. Any error resulting from setting all the initial conditions equal to zero will rapidly become small according to the error equation.

Defining Terms

Closed-loop control inputs — Inputs to a control system, given as a function of the output, that are determined by an automatic control algorithm within the system (usually computed by an analog or digital device).

Feedback loop — Any connection between the input and the output of a dynamical system.

Input — The inputs to a dynamical system are quantities that can affect the evolution of the state of the system.

Open-loop command inputs — Inputs to a control system, given as a function of time, that are initiated by a human operator or some other external device used to specify a desired output.

Output — Those functions of the state and control, possibly the states themselves, that can be measured.

Return time — For asymptotically stable systems, with eigenvalues λ_i, the return time is defined by $T = 1/\min|\mathrm{Re}(\lambda_i)|$, $i = 1, \ldots, N_x$ and the eigenvalue(s) corresponding to T is called the *dominant eigenvalue(s)*. The dominate eigenvalue(s) corresponds to the slowest time constant in the system.

State variables — Those variables that identify the state of a system. The states of a dynamical system are those dynamical components of a system that completely identify it at any moment in time.

References

Bryson, A. E. and Ho, Y. C. 1975. *Applied Optimal Control.* Halsted, New York.

Davison, E. J. 1968. On pole assignment on multivariable linear systems. *IEEE Trans. Autom. Control,* AC-13:747–748.

Green, M. and Limebeer, D. 1994. *Linear Robust Control.* Prentice Hall, Englewood Cliffs, NJ.

Kalman, R. E. 1960. A new approach to linear filtering and prediction problems. *Trans. ASME J. Basic Eng.*, 82:34–45.

Luenberger, D. G. 1971. An introduction to observers. *IEEE Trans. Autom. Control,* AC-16:596–602.

Shahian, B. and Hassul, M. 1993. *Control System Design Using MATLAB.* Prentice-Hall, Englewood Cliffs, NJ.

Wonham, W. M. 1967. On pole assignment in multi-input controllable linear systems. *IEEE Trans. Autom. Control,* AC-12:660–665.

Wonham, W. M. 1985. *Linear Multivariable Control,* 3rd ed. Springer-Verlag, New York.

Further Information

Brogan, W. L. 1985. *Modern Control Theory.* Prentice-Hall, Englewood Cliffs, NJ.

Dorf, R. C. 1995. *Modern Control Systems.* Addison-Wesley, Reading, MA.

Grantham, W. J. and Vincent, T. L. 1993. *Modern Control Systems Analysis and Design.* John Wiley & Sons, New York.

Kalman, R. E., Ho, Y. C., and Narendra, K. S. 1963. Controllability of linear dynamical systems. *Contrib. Differential Equations,* 1:189–213.

Kuo, B. C. 1991. *Automatic Control Systems.* Prentice Hall, Englewood Cliffs, NJ.

178

Nonlinear Control Systems

Andrea Serrani
The Ohio State University

In the last 20 years nonlinear control has reached the level of a mature discipline, both in its theoretical developments and in engineering applications. Examples of successful application of modern nonlinear control theory are widespread, and range from aerospace to robotics, and from electrical and mechanical to biomedical engineering. The advantage of nonlinear control systems design versus more conventional (i.e., linear) design methodologies lies in the fact that the fundamentally nonlinear nature of the plant to be controlled is taken directly into account (and, sometimes, exploited) rather than neglected or ignored. The price to pay, however, is the lack of a general design methodology that is applicable to *all* nonlinear systems, and a substantial complexity of the mathematical tools required for the analysis and the synthesis of nonlinear control systems. The first drawback is intrinsic to the nature of the problem, and it is alleviated by the fact that most nonlinear design techniques indeed apply to entire *classes* of nonlinear systems sharing a common structure and properties. The second requires the modern control engineer to acquire the mathematical tool necessary to master the discipline, and meet the challenges for more demanding applications.

178.1 Differential Geometric Methods

Feedback Linearization

The application of concepts and methodologies from differential geometry lies at the foundation of modern nonlinear control theory. The widely popular nonlinear design technique known as *linearization by feedback* is one of the most important results due to the differential geometric approach. Given a nonlinear system in the form

$$\dot{x} = f(x) + \sum_{i=1}^{m} g_i(x)u_i = f(x) + g(x)u \tag{178.1}$$

with state $x \in \Re^n$ and control input $u \in \Re^m$, the problem of *feedback linearization* amounts to finding a state feedback control

0-8493-1586-7/05/$0.00+$1.50

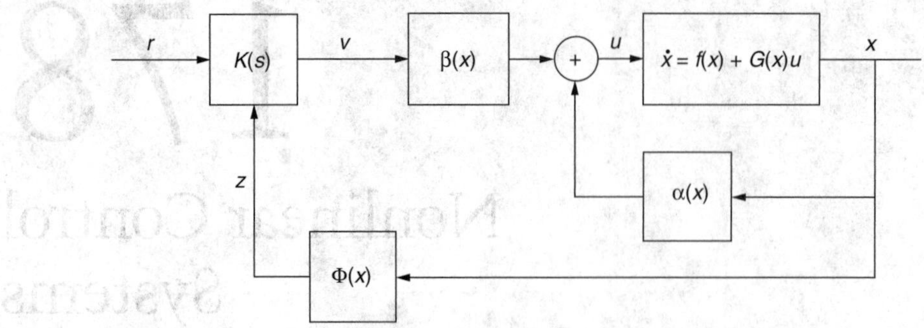

FIGURE 178.1 Inner-loop and outer-loop control based on feedback linearization.

$$u = \alpha(x) + \beta(x)v \tag{178.2}$$

and a change of coordinates in the state space $z = \Phi(x)$ such that the resulting closed-loop system $\dot{x} = f(x) + g(x)\alpha(x) + g(x)\beta(x)v$, when expressed in the new set of coordinates $z(t)$, coincides with the minimal state-space realization of m decoupled linear systems with transfer function

$$G_i(s) = \frac{1}{s^{r_i}}, \qquad i = 1,\ldots,m \tag{178.3}$$

and $r_1 + r_2 + \ldots + r_m = n$. If this is the case, a linear controller $K(s)$ can then be design for Equation (178.3) to obtain the desired response, with the feedback linearizing control Equation (178.2) playing the role of an *inner loop controller* in a multiloop controller design (see Figure 178.1).

In general, the functions $\alpha(x)$, $\beta(x)$, and the transformation $\Phi(x)$ may be defined or remain valid only in a neighborhood $W \subset \Re^n$ of a given point $x_0 \in \Re^n$. If this is the case, the control Equation (178.2) and the change of coordinates $x \mapsto z$ are said to solve the *local* feedback linearization problem; otherwise, if $W \equiv \Re^n$, the Equation (178.1) system is said to be *globally linearizable by state feedback*. Given any nonlinear systems of the form in Equation (178.1), the solvability of the feedback linearization problem (both local and global) can be given a complete characterization in terms of properties of certain *distributions* generated by the vector fields $f(x)$, $g_1(x),\ldots,g_m(x)$, that is, properties of nonlinear equivalents of the invariant and controlled-invariant subspaces found in geometric linear systems theory. It is worth noting that the condition of local (global) existence of a linearizing state-feedback transformation does not depend on the choice of the coordinates , that is, on the choice of the state-space representation of the system model. Crucial, in this regard, is the concept of *nonlinear relative degree*. Given m output functions $y_i = h_i(x)$, the Equation (178.1) system is said to have (vector) relative degrees at a point x_0 with respect to the output $y = (y_1 \ldots y_m)'$ if there exists an m-tuple of integers $\{r_1, r_2, \ldots r_m\}$ such that

1. $L_{g_j} L_f^k h_i(x) = 0$ for all $j = 1,\ldots,m$, for all $k = 1,\ldots,r_i - 2$, for all $i = 1,\ldots,m$, and for all x in a neighborhood of x_0.
2. The matrix $A(x) = \begin{pmatrix} L_{g_1} L_f^{r_1-1} h_1(x) & \cdots & L_{g_m} L_f^{r_1-1} h_1(x) \\ \vdots & \ddots & \vdots \\ L_{g_1} L_f^{r_m-1} h_m(x) & \cdots & L_{g_m} L_f^{r_m-1} h_m(x) \end{pmatrix}$ is nonsingular at x_0

Then, the system in Equation (178.1) is feedback linearizable in a neighborhood of x_0 if it is possible to find m functions such that, when equipped with the "dummy output" $\phi_1(x), \phi_2(x), \ldots, \phi_m(x)$, the system in Equation (178.1) has vector relative degree at $y = (\phi_1 \cdots \phi_m)'$ and, in addition,

3. $r_1 + r_2 + r_m = n$.

Note: The notation $L_f \lambda(x) = \frac{\partial \lambda}{\partial x} f(x)$ stands for the Lie derivative of the smooth function $\lambda(x)$ along the vector field $f(x)$, with $L_f^k \lambda(x) = \frac{\partial (L_f^{k-1}\lambda)}{\partial x} f(x)$.

Although their existence can be completely determined a priori, finding such functions is in general a very difficult task, since it amounts in solving a system of nonlinear partial differential equations. However, important classes of nonlinear systems can be easily shown to be feedback linearizable, and the linearizing feedback computed quite effortlessly. Among these latter, the most notable example is the class of fully actuated *Euler-Lagrange systems*, that is, mechanical systems described by equations of the form

$$\frac{d}{dt}\frac{\partial L(q,\dot{q})}{\partial \dot{q}_i} - \frac{\partial L(q,\dot{q})}{\partial \dot{q}_i} = u_i, \qquad i = 1,\dots,n$$

in which each degree of freedom q_i is controlled independently. In the specific case of a fully actuated rigid robotic manipulator, the Euler-Lagrange equations read as

$$M(q)\ddot{q} + C(q,\dot{q})\dot{q} + g(q) = u \tag{178.4}$$

where $q \in \Re^n$ is the vector of joint positions, $M(q)$ is the $n \times n$ inertia matrix, $C(q,\dot{q})$ accounts for the centripetal and Coriolis forces, $g(q)$ is the gravitational force, and $u \in \Re^n$ is the control torque at each joint. It is easy to check this since the inertia matrix is nonsingular for any $q \in \Re^n$, Equation (178.4) has a (global) vector relative degree $\{2,2,\dots 2\}$ with respect to the output $y = q$, and the linearizing feedback $u = M(q)[v - C(q,\dot{q})\dot{q} - g(q)]$ yields the closed-loop system $\ddot{q} = v$, that is, a system of n decoupled double integrators. The application of a feedback-linearizing loop in the control of rigid manipulators is commonly referred to as a *computed-torque control.*

Input-Output Linearization

In many cases, it may be impossible to find a set of functions $\phi_1(x), \phi_2(x), \dots, \phi_m(x)$ satisfying the above conditions 1 and 2 for which the additional condition 3 holds, or the system may already be endowed with an output function $y = h(x) \in \Re^m$ for which the system has vector relative degree $\{r_1, r_2, \dots r_m\}$, but $r_1 + r_2 + r_m < n$. In this case, it is still possible to simplify the nonlinear system to its maximum extent, applying a state feedback of the kind in Equation (178.2) and a coordinate transformation $z = \Phi(x)$ such that, in a closed loop, the resulting input/output map $v \mapsto y$ is that of a linear system, but the closed-loop system itself is not a minimal realization of this map. Specifically, consider for simplicity the case of a SISO nonlinear system in the form

$$\begin{aligned} \dot{x} &= f(x) + g(x)u \\ y &= h(x) \qquad x \in \Re^n \end{aligned} \tag{178.5}$$

and assume that the system has relative degree $r < n$ at a point x_0, that is, $L_g L_f^k h(x) = 0$ for all x in a neighborhood of x_0 and all $k < r - 1$, and $L_g L_f^{r-1} h(x_0) \neq 0$. Then, it is possible to find a change of coordinates $\Phi(x): x \mapsto z = (\xi, \eta)$, with $\xi \in \Re^r, \eta \in \Re^{n-r}$, such that in the new coordinates the systems has the form

$$\begin{aligned}
\dot{\xi}_1 &= \xi_2 \\
\dot{\xi}_2 &= \xi_3 \\
&\cdots \\
\dot{\xi}_r &= a(z) + b(z)u \\
\dot{\eta}_1 &= q_1(z) + p_1(z)u \\
&\cdots \\
\dot{\eta}_{n-r} &= q_{n-r}(z) + p_{n-r}(z)u \\
y &= \xi_1
\end{aligned} \tag{178.6}$$

with $b(z) \neq 0$ in a neighborhood of $z_0 = \Phi(x_0)$. Systems of the kind in Equation (178.6) are said to be in *normal* form.

The feedback law

$$u = \frac{v - a(z)}{b(z)} = \frac{v - L_f^r h(x)}{L_g L_f^{r-1} h(x)} \tag{178.7}$$

yields a closed-loop system in which the input-output map is that of a chain of r integrators, $y(s) = \frac{1}{s^r} v(s)$, which is precisely a system of the form

$$\begin{aligned}
\dot{\xi}_1 &= \xi_2 \\
\dot{\xi}_2 &= \xi_3 \\
&\cdots \\
\dot{\xi}_r &= \nu \\
\dot{\eta}_1 &= \bar{q}_1(\eta, \xi) + \bar{p}_1(\eta, \xi)\nu \\
&\cdots \\
\dot{\eta}_{n-r} &= \bar{q}_{n-r}(\eta, \xi) + \bar{p}_{n-r}(\eta, \xi)\nu \\
y &= \xi_1
\end{aligned} \tag{178.8}$$

The application of the linearizing control renders the n-r states $\eta_1, \ldots, \eta_{n-r}$ *unobservable* from the output y. If the system in Equation (178.8) is to be controlled closing an outer-loop controller between y and v, particular care must be taken in ensuring that the internal dynamics $\dot{\eta}(t) = \bar{q}(\eta(t), \xi(t)) + \bar{p}(\eta(t), \xi(t))\nu(t)$ remains well behaved, that is, the unobservable trajectory $\eta(t)$ remains bounded for all time t.

Example

The kinematic model of a cart-like mobile robot is given by

$$\begin{aligned}
\dot{x} &= \nu \cos(\theta) \\
\dot{y} &= \nu \sin(\theta) \\
\dot{\theta} &= \omega
\end{aligned}$$

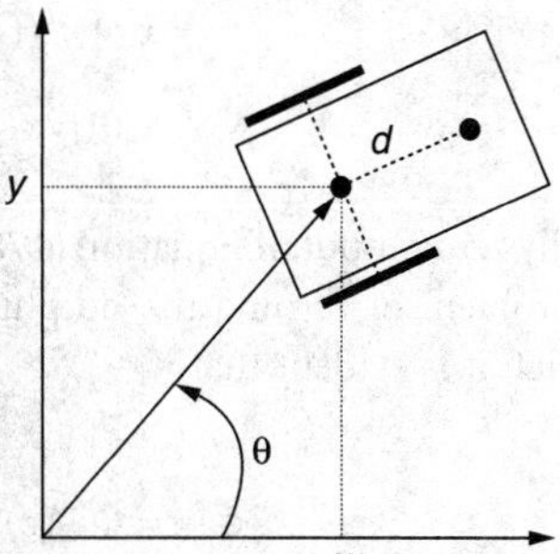

FIGURE 178.2 Cart-like mobile robot.

where (x,y) are the Cartesian coordinates of the center of the vehicle's center of rotation, θ is the vehicle attitude with respect to a reference frame, and (ν,ω) are the linear and angular velocities respectively, playing the role of control inputs (Figure 178.2). It can be easily verified that the system does not have relative degree with respect to the output function $h = (x,y)'$; hence the input-output map between the control input and the Cartesian coordinates cannot be rendered linear. However, the system has relative degree with respect to the output

$$\bar{h} = \begin{pmatrix} \xi_1 \\ \xi_2 \end{pmatrix} = \begin{pmatrix} x + d\cos(\theta) \\ y + d\sin(\theta) \end{pmatrix}$$

which corresponds to the position of some point on the cart away from the center of rotation. Application of the globally defined feedback law

$$\begin{pmatrix} \nu \\ \omega \end{pmatrix} = \begin{pmatrix} \cos(\theta) & \sin(\theta) \\ -\dfrac{\sin(\theta)}{d} & \dfrac{\cos(\theta)}{d} \end{pmatrix} \begin{pmatrix} u_1 \\ u_2 \end{pmatrix}$$

yields the system

$$\dot{\xi}_1 = u_1$$

$$\dot{\xi}_2 = u_2$$

$$\phi = -\frac{\sin(\theta)}{d} u_1 + \frac{\cos(\theta)}{d} u_2$$

in which the input-output relationship, in the Laplace domain, is precisely $\bar{h}(s) = \begin{pmatrix} \frac{1}{s} & 0 \\ 0 & \frac{1}{s} \end{pmatrix} \begin{pmatrix} u_1(s) \\ u_2(s) \end{pmatrix}$.

Zero Dynamics

The notion of *zero dynamics* plays a fundamental role in the design of feedback control laws aiming at rendering asymptotically stable an equilibrium point for a nonlinear system. Its importance stems from the fact that it constitutes a nonlinear equivalent of the familiar notion of a transmission zero of a linear system, and, to a certain extent, accounts for the definition of a nonlinear version of the celebrated Evans root-locus design technique. Consider again the SISO nonlinear system in Equation (178.5) and assume that the origin $x = 0$ is an equilibrium of the unforced system, that is, $f(0) = 0$. Without loss of generality, assume also that $h(0) = 0$. The problem of *zeroing the output* is that of finding all possible pairs $(x_*, u_*(\cdot))$ consisting of initial conditions and input functions other than the trivial pair $(x_*, u_*(\cdot)) = (0,0)$, such that the trajectory of the system

$$\dot{x}(t) = f(x(t)) + g(x(t))u_*(t)$$
$$x(0) = x_*$$

yields an identically zero output. If Equation (178.5) has relative degree r at the origin, the characterization of the solution problem of zeroing the output is derived directly from the normal form of Equation (178.6), from which it is evident that

$$x_* \in \{x = \Phi^{-1}(z): \xi = 0\}, \qquad u_*(t) = -\frac{a(z)}{b(z)}$$

and the resulting nontrivial system dynamics are given by

$$\begin{aligned}\dot{\eta}_1 &= \overline{q}_1(\eta, 0)\\ &\cdots\\ \dot{\eta}_{n-r} &= \overline{q}_{n-r}(\eta, 0)\end{aligned} \tag{178.9}$$

Equation (178.9), which characterizes the dynamics of systems when the output is forced to be identically zero, is referred to as *the zero dynamics* of Equation (178.5). Note that the zero dynamics is not intrinsic to the unforced system, rather it is imposed by a particular choice of an input function $u(t)$ and an initial condition $x(0)$ forcing the trajectories of the systems to evolve on an invariant hypersurface of the state space.

The role played by the zero dynamics of Equation (178.9) in the local stabilization of the equilibrium at the origin can be explained as follows. Since, without loss of generality, the change of coordinate $(\xi, \eta) = \Phi(x)$ can be always chosen to preserve the origin, Equation (178.9) has an equilibrium at $\eta = 0$. If the equilibrium in question is known to be asymptotically stable, a simple strategy to stabilize the origin of Equation (178.5) is that of designing a feedback control law that drives asymptotically the trajectories of the system to the zero dynamics of Equation (178.9). In this regard, application of the linearizing control of Equation (178.7) yields Equation (178.8), which reads as

$$\dot{\xi} = A_c\xi + B_c\nu$$
$$\dot{\eta} = \overline{q}(\eta, \xi) + \overline{p}(\eta, \xi)\nu$$

where (A_c, B_c) are in controllable companion form. Then, choosing the additional control $n = F\xi$, where $F = (-c_0\ -c_1\ \ldots\ -c_{r-1})$ is a feedback matrix such that $A_c + B_cF$ has all eigenvalues with a negative real part, yields the desired asymptotic convergence of $\xi(t)$ to the origin. Trajectories of the system will then be attracted to the zero dynamics of the systems, and convergence to the origin is implied by asymptotic stability of $\eta = 0$. It is worth noting that the method guarantees only *local* stabilization, valid in general only for trajectory originating from initial conditions in a neighborhood of the origin. On the other hand, asymptotic stability of $\eta = 0$ for Equation (178.9) is not required to be exponential, that is, stabilizability of the Jacobian linearization of the Equation (178.5) is not assumed. The method resembles the classic root-locus design technique for minimum-phase linear systems, where stabilization of the closed-loop system is achieved forcing the closed-loop poles to approach the stable open loop transmission zeros of the plant by means of dynamic output feedback.

Local Output Tracking

The problem of letting the output of a system to track a specified reference trajectory occupies a central role in nonlinear control theory. One of the most common scenarios occurs when the reference trajectory

$y_R(t)$ to be tracked by the output $y(t)$ of Equation (178.5) is generated by a *reference model*, which is usually given as a stable linear system of the form

$$\dot{x}_R = Ax_R + Bu_R$$
$$y_R = Cx_R$$

where $u_R(t)$ is an exogenous input. The role of the reference model is that of shaping the commanded variables so that a desired response in terms of settling time and overshoot during transient is achieved. Again, the role of relative degree and zero dynamics of the plant are crucial in finding a constructive solution. Assume that the plant Equation (178.5) has relative degree r, and that the reference model is chosen in such a way that the relative degree between u_R and y_R is greater or equal to r, that is,

$$CB = CAB = \cdots = CA^{r-2}B = 0$$

It is now easy to see that a control which forces the tracking error $e(t) = y(t) - y_R(t)$ to converge asymptotically to zero is given by the multiloop dynamic feedback law

$$\begin{aligned}\dot{\zeta} &= A\zeta + Bu_R \\ u &= \frac{\nu - L_f^r h(x)}{L_g L_f^{r-1} h(x)} \\ \nu &= CA^r\zeta - \sum_{i=1}^{r} c_{i-1}(\xi_i - CA^{i-1}\zeta) + CA^{r-1}Bu_R\end{aligned} \tag{178.10}$$

where c_i, $i = 0,\ldots,r-1$ are the coefficients of the gain matrix F defined in the previous section. To see why the control in Equation (178.10) achieves local tracking of the reference trajectory, note that the inner-loop feedbackof Equation (178.7), together with the change of coordinates $z = (\xi,\eta)$, puts the system into the form of Equation (178.8). Then, the additional change of coordinates $\tilde{\xi}_i = \xi_i - CA^{i-1}\zeta$, $i = 1,\ldots,r$, and the application of the outer-loop control ν defined in Equation (178.10) yields a closed-loop system of the form

$$\begin{aligned}\dot{\tilde{\xi}} &= (A_c + FB_c)\tilde{\xi} \\ \dot{\eta} &= \varphi(\eta, \tilde{\xi}, \zeta, u_R) \\ \dot{\zeta} &= A\zeta + Bu_R \\ e &= \tilde{\xi}_1\end{aligned}$$

in which the trajectory $\tilde{\xi}(t)$ converges asymptotically to the origin, as the matrix $A_c + B_cF$ has, by design, all eigenvalues with a negative real part. Again, particular care must be taken in ensuring that the trajectories of the "unobservable subsystem"

$$\dot{\eta}(t) = \varphi(\eta(t), \tilde{\xi}(t), \zeta(t), u_R(t))$$

remain bounded in the operating range. The closed-loop system is shown in Figure 178.3.

A somewhat different situation occurs when the trajectory to be tracked by the plant of Equation (178.5) is generated by an *autonomous* system of the form

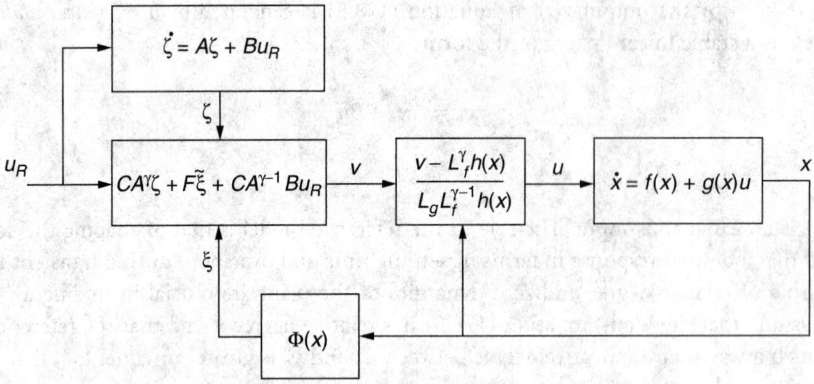

FIGURE 178.3 Model-reference tracking controller.

$$\dot{w} = s(w)$$
$$y_R = r(w) \tag{178.11}$$

with state $w \in \Re^q$, and such that the equilibrium at $w = 0$ is stable, and the Jacobian linearization at the origin $S = \left.\frac{\partial s}{\partial w}\right|_{w=0}$ has all eigenvalues on the imaginary axis. This ensures that the reference Equation (178.11), for initial conditions in a neighborhood of the origin, generates trajectories which are bounded, but do not vanish asymptotically. (If, for some initial conditions, the reference trajectory was such that $y_R(t) = r(w(t)) \to 0$ as $t \to \infty$, the problem of asymptotic output tracking would be reduced to the problem of zeroing the output of the plant.) This setup of the problem includes common situations in which the reference signals to be tracked are constant or sinusoidal signals. If the equilibrium at the origin $x = 0$ for the plant in Equation (178.5) is stabilizable in the first approximation, that is, the pair (F,G) defined as

$$F = \left.\frac{\partial f}{\partial x}\right|_{x=0}, \qquad G = \left.\frac{\partial g}{\partial x}\right|_{x=0}$$

is stabilizable, then asymptotic stability of the origin of the unforced closed-loop system (i.e., when $w = 0$) is usually imposed as an additional control objective. If this is the case, the solution of the problem of local output tracking with local exponential stability is obtained without requiring the existence of a relative degree for the plant, or even asymptotic stability of its zero dynamics. As a matter of fact, it can be shown that the solvability of the problem is equivalent to the existence of a controlled-invariant submanifold of the extended state space $X = \Re^q \times \Re^n$, which can be locally described by the graph of a mapping, on which the tracking error vanishes identically. In other words, the problem is solvable if and only if there exists mappings $x = \pi(w)$ and $u = c(w)$, defined in a neighborhood of $(w,x) = (0,0)$, which solve the *nonlinear regulator equation*

$$\frac{\partial x}{\partial w} s(w) = f(\pi(w)) + g(\pi(w))c(w)$$

$$h(\pi(w)) = r(w)$$

The set $\{(w,x): x = \pi(w)\}$ is called *an error-zeroing manifold*, since at each point in this set the error $e = h(x) - r(w)$ is identically zero. The pair $x = \pi(w)$, $u = c(w)$ can therefore be viewed as a solution of the problem of zeroing the output of the augmented system

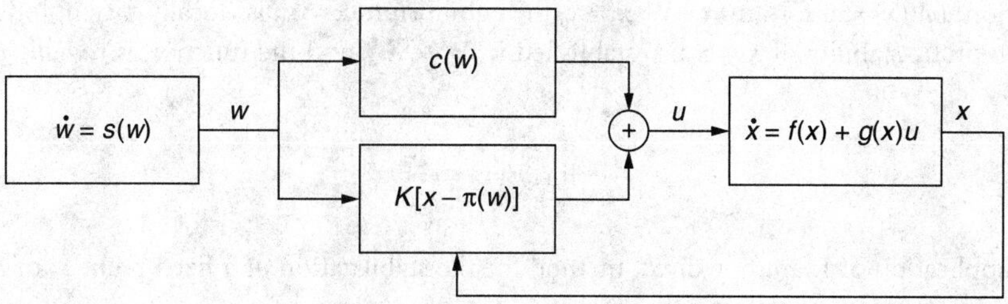

FIGURE 178.4 Tracking controller with autonomous reference model.

$$\dot{w} = s(w)$$
$$\dot{x} = f(x) + g(x)u$$
$$e = h(x) - r(w)$$

whose zero dynamics are precisely given by the trajectories of Equation (178.11). If the solution of the regulator equation is available, the control law which solves the problem of local output tracking is obtained as a combination of the "feed-forward" control $u_{ff} = c(w)$, required to render the error-zeroing manifold invariant under the trajectories of the closed-loop system, and of a "feedback" term of the form $u_{fb} = K[x - \pi(w)]$, which has the role of rendering the error-zeroing manifold locally exponentially attractive. The feedback gain K must be selected in such a way that $F + GK$ has all eigenvalues with a negative real part. The resulting control-loop system is shown in Figure 178.4.

178.2 Lyapunov Techniques

Control Lyapunov Functions

A fundamental tool for the analysis of the stability of a fixed point of a nonlinear system is the celebrated *direct method of Lyapunov.* A fixed point x_0 of a nonlinear system of the form $\dot{x} = f(x)$, $x \in \Re^n$, is stable if there exists a continuously differentiable function $V\colon \Re^n \to \Re_{\geq 0}$, called a Lyapunov function, that is positive definite in a neighborhood W of x_0, that is,

$$V(x_0) = 0$$

and

$$V(x) > 0$$

for each

$$x \in W, x \neq x_0$$

and such that the Lie derivative of V along the vector field $f(x)$ is negative semi-definite in W, that is,

$$L_f V(x) \leq 0$$

for any

$$x \in W$$

If, in addition, $L_f V(x) < 0$ for any $x \in W$, $x \neq x_0$ the equilibrium $x = x_0$ is (locally) asymptotically stable. Global asymptotic stability of $x = x_0$ is established if $W \equiv \Re^n$, and the function is *radially unbounded*, that is,

$$\lim_{\|x\| \to \infty} V(x) = \infty$$

A natural application of Lyapunov direct method for the stabilization of a fixed point x_0 of a nonlinear control system by means of state-feedback

$$\dot{x} = f(x) + g(x)u$$

consists of choosing a *candidate Lyapunov function* $V(x)$ and finding a feedback control $u = \alpha(x)$, with $\alpha(x_0) = 0$, such that

$$\dot{V}(x) := L_f V(x) + L_g V(x)\alpha(x) < 0$$

for all

$$x \in W, x \neq x_0$$

Again, if global asymptotic stability is the property of interest, $V(x)$ must be chosen to be radially unbounded and $u = \alpha(x)$ must guarantee that $\dot{V}(x) < 0$ on $\Re^n - \{x_0\}$. Clearly, the existence of such a feedback law depends on the choice of the candidate Lyapunov function $V(x)$. If the fixed point x_0 is stabilizable by means of state feedback, that is, if there exists $u = \alpha(x)$ such that $x = x_0$ is an asymptotically stable equilibrium of

$$\dot{x} = f(x) + g(x)\alpha(x)$$

then, existence of a Lyapunov function $V(x)$ is guaranteed by the *converse Lyapunov theorem*. On the other hand, in solving the feedback stabilization problem, it is of primary importance to recognize *a priori* whether the choice of a candidate Lyapunov function is appropriate or not, in the sense that a corresponding stabilizing law $u = \alpha(x)$ exists for a given $V(x)$. A candidate Lyapunov function is a *control Lyapunov function* (*CLF*) if

$$L_f V(x) < 0$$

for each

$$x \neq x_0$$

such that

$$L_g V(x) = 0$$

It is quite obvious that functions $V(x)$, which are such that

$$L_f V(x) > 0$$

for some

$$x \neq x_0$$

such that

$$L_g V(x) = 0$$

cannot be used at all for feedback stabilization by Lyapunov direct method. Note that the CLF property can be checked *a priori*. Once a CLF is found, there are several methods to design the stabilizing control $u = \alpha(x)$, the most famous being the *Artstein–Sontag universal formula*,

$$\alpha(x) = \begin{cases} -\dfrac{L_f V(x) + \sqrt{[L_f V(x)]^2 + [L_g V(x)]^4}}{L_g V(x)} & \text{if } L_g V(x) \neq 0 \\ 0 & \text{if } L_g V(x) = 0 \end{cases}$$

In some cases, when a Lyapunov function $V(x)$ is known such that

$$L_f V(x) \leq 0$$

for any

$$x \in W$$

which means that $x = x_0$ is a stable equilibrium for the uncontrolled system, a control law yielding asymptotic stability of x_0 can still be found, even if $V(x)$ fails to be a CLF. Assume, in addition, that the set

$$Q = \{x \in \Re^n : L_f V(x) = 0 \text{ and } L_f^k L_g V(x) = 0 \text{ for all } k = 0, 1, 2, \ldots\}$$

is such that

$$Q = \{x_0\}$$

Then, it can be shown by means of a simple application of *La Salle's invariance principle* (see Khalil [2002], section 4.2), that the so-called *Jurdjevic–Quinn* control

$$\alpha(x) = -L_g V(x)$$

succeeds in stabilizing asymptotically the equilibrium $x = x_0$. A typical example of the application of the Jurdjevic–Quinn control is given by damping control of mechanical systems, in which $V(x)$ is an energy function, and $\alpha(x) = -L_g V(x)$ plays the role of adding a dissipation term. Consider, for example, the model of the rotational dynamics of a fully actuated rigid body

$$I\dot{\omega} = -\omega \times I\omega + u$$

where $I \in \Re^{3\times 3}$ is the positive definite inertia matrix, $\omega \in \Re^3$ is the angular velocity about the body-fixed axes, and $u \in \Re^3$ is the control torque. Choosing the Lyapunov function candidate $V(\omega) = \frac{1}{2}\omega' I \omega$, one obtains

$$L_f V(\omega) = 0 \text{ for any } \omega \in \Re^3$$

Hence, $V(\omega)$ is not a CLF. Application of the Jurdjevic–Quinn control $u = -\omega$ yields

$$\dot{V}(\omega) = -\|\omega^2\|$$

which implies global asymptotic stability of the origin.

Backstepping

Lyapunov-based techniques are instrumental in the design of stabilizing control law for interconnected systems. Typically, the design is accomplished in a recursive fashion, building a new CLF at each step of the recursion, and finding the corresponding control. One such method is commonly known as *backstepping*, and it is applicable to systems exhibiting a triangular structure of the form

$$\begin{aligned}
\dot{z} &= f_0(z) + g_0(z)x_1 \\
\dot{x}_1 &= x_2 + f_1(z, x_1) \\
\dot{x}_2 &= x_3 + f_2(z, x_1, x_2) \\
&\cdots \\
\dot{x}_r &= u + f_r(z, x_1, \ldots, x_r)
\end{aligned}$$

which are said to be in *strict-feedback form*. Let $z \in \mathfrak{R}^{n-r}$ and $x = (x_1, x_2, \ldots, x_r)'$, and assume that the equilibrium $z = 0$ of the system

$$\dot{z} = f_0(z) + g_0(z)\nu$$

is stabilizable by means of a smooth state feedback law $\nu = \alpha_1(z)$; that is, assume that a radially unbounded CLF $V_0(z)$ is known such that

$$L_{f_0}V(z) + L_{g_0}V(z)\alpha(z) < 0$$

for all

$$z \neq 0$$

Then, it can be shown that the function

$$V_1(z, x_1) = V_0(z) + \frac{1}{2}(x_1 - \alpha_1(x))^2$$

is a CLF for the system

$$\begin{aligned}
\dot{z} &= f_0(z) + g_0(z)x_1 \\
\dot{x}_1 &= \nu + f_1(z, x_1)
\end{aligned} \tag{178.12}$$

with corresponding globally stabilizing control

$$\alpha_2(z, x_1) = -(x_1 - \alpha_0(z)) - f_1(z, x_1) - L_{g_0}V_0(z)(x_1 - \alpha_0(z)) + L_{f_0}\alpha_1(z) + L_{g_0}\alpha_1(z)x_1 \tag{178.13}$$

The procedure can obviously be iterated, until a CLF and a corresponding stabilizing feedback for the original system is found. It is worth noting that the method has an interesting interpretation related to the concept of zero dynamics discussed previously. As a matter of fact, the zero dynamics of Equation (178.12) with respect to the "dummy output" $y_1 = x_1 - \alpha_1(z)$ that was obtained choosing the initial conditions $x_1(0) = \alpha_1(z(0))$ and the control input $v_* = -f_1(z,x_1) + L_{f_0}\alpha_1(z) + L_{g_0}\alpha_1(z)x_1$, reads as

$$\dot{z} = f_0(z) + g_0(z)\alpha(z)$$

and, by assumption, possesses a globally asymptotically stable equilibrium at the origin. The control law of Equation (178.13) is completed by the term $-L_{g_0}V_0(z)(x_1 - \alpha_0(z))$, ensuring that the trajectories of the closed-loop systems of Equations (178.12) and (178.13) remain bounded, while the term $-(x_1 - \alpha_0(z))$ drives asymptotically the trajectories toward the zero dynamics.

Adaptive Control

Concepts and techniques from Lyapunov theory are also central in adaptive control of uncertain nonlinear systems. Given a parameter-dependent single-input nonlinear systems in the form

$$\dot{x} = f(x,\theta) + g(x,\theta)u$$

where $f(0,\theta) = 0$ regardless of the value of the unknown parameter vector $\theta \in \Re^p$, we consider the problem of designing a parameter-independent dynamic feedback law of the form

$$\dot{\xi} = \varphi(\xi, x)$$
$$u = \beta(\xi, x)$$

with state $\xi \in \Re^v$, such that the trajectories of the closed-loop system

$$\dot{x} = f(x,\theta) + g(x,\theta)\alpha(\xi, x)$$
$$\dot{\xi} = \varphi(\xi, x)$$

are globally bounded, and $x(t) \to 0$, as $t \to \infty$. Assume that a parameter-dependent CLF $V(x,\theta)$ is given, and that the following hold:

$L_gV(x,\theta)$ is independent of θ.
The stabilizing control $u = \alpha(x,\theta)$ associated with the CLF $V(x,\theta)$ is *affine* in θ, that is, $\alpha(x,\theta) = \alpha_0(x) + \alpha_1(x)\theta$ for some *known* functions $\alpha_1(x)$, $\alpha_1(x)$.

Then, the stabilizing dynamic feedback can be found by appealing to the principle of *certainty-equivalence*, replacing the unknown parameter vector in the control $u = \alpha(x,\theta)$ with an estimate $\hat{\theta}(t)$ generated by the update law

$$\dot{\hat{\theta}} = -(L_gV(x)\alpha_1(x))'$$

In this case, evaluation of the Lyapunov function candidate

$$W(x,\hat{\theta},\theta) = V(x,\theta) + \frac{1}{2}\|\hat{\theta} - \theta\|^2$$

along with solutions of the closed-loop system

$$\dot{x} = f(x,\theta) + g(x,\theta)\alpha_0(x) + g(x,\theta)\alpha_1(x)\hat{\theta}$$

$$\dot{\hat{\theta}} = -(L_g V(x)\alpha_1(x))'$$

yields

$$\dot{W}(x,\hat{\theta},\theta) = \frac{\partial V}{\partial x}[f(x,\theta) + g(x,\theta)\alpha(x,\theta)] \leq 0$$

Thus, application of La Salle's invariance theorem yields global boundedness of all trajectories and regulation of $x(t)$ to the origin. If the first property above does not hold, the principle of certainty equivalence cannot be applied, as the update law cannot be computed. If this is the case, a Lyapunov function that depends on the estimate $\hat{\theta}$ rather than on θ must necessarily be employed in analyzing the asymptotic properties of the closed-loop system. In the simpler case in which the plant model depends linearly on the unknown parameter, that is, for systems in the form

$$\dot{x} = f(x) + F(x)\theta + g(x)u \tag{178.13}$$

a solution to the adaptive stabilization problem follows from the existence of an *adaptive control Lyapunov function* (*ACLF*). A Lyapunov function candidate $V(x,\theta)$ is an ACLF for Equation (178.13) if there exists a parameter-dependent control $u = \alpha(x,\theta)$ such that

$$\frac{\partial V}{\partial x}\left[f(x) + F(x)\left(\theta + \frac{\partial V'}{\partial \theta}\right) + g(x)\alpha(x,\theta)\right] < 0$$

for all $x \neq 0$ and for all θ. Then, an adaptive controller is obtained as

$$\dot{\hat{\theta}} = F'(x)\frac{\partial V'(x,\hat{\theta})}{\partial x}$$

$$u = \alpha(x,\hat{\theta})$$

The asymptotic properties of the closed-loop systems are evaluated with the aid of the Lyapunov function

$$W(x,\hat{\theta}) = V(x,\hat{\theta}) + \frac{1}{2}\|\hat{\theta} - \theta\|^2$$

which is independent of the actual unknown parameter vector θ. A remarkable feature of the ACLF-based method is that it is amenable to backstepping design. As a matter of fact, for the augmented system

$$\begin{aligned} \dot{x} &= f(x) + F(x)\theta + g(x)\xi \\ \dot{\xi} &= u \end{aligned} \tag{178.14}$$

an ACLF is readily available as

$$V_2(x,\xi,\theta) = V(x,\theta) + \frac{1}{2}(\xi - \alpha(x,\theta))^2$$

for which a control $u = \beta(x, \xi, \theta)$, yielding

$$\frac{\partial V_2}{\partial(x,\xi)} \left[f(x) + F(x) \left(\theta + \frac{\partial V_2'}{\partial \theta} \right) + g(x)\xi \atop \beta(x,\xi,\theta) \right] < 0$$

for all

$$x \neq 0, \qquad \xi \neq 0 \text{ and for all } \theta$$

can be easily computed. Then, an adaptive controller for Equation (178.14) is given by

$$\dot{\hat{\theta}} = F'(x)\frac{\partial V'(x,\hat{\theta})}{\partial x} - \frac{\partial \alpha'}{\partial x}(\xi - \alpha(x,\hat{\theta}))$$

$$u = \beta(x, \xi, \hat{\theta})$$

178.3 Input-Output and Dissipative Systems Techniques

Dissipative Systems

A useful tool for the analysis and design of nonlinear control systems is offered by the theory of *dissipative systems.* A nonlinear system of the form

$$\begin{aligned} \dot{x} &= f(x) + g(x)u \\ y &= h(x) \end{aligned} \tag{178.15}$$

with $u \in \Re^m$, $y \in \Re^m$ is said to be *dissipative* with respect to a *supply rate* if there exists a continuous positive definite function $S(x)$, called a *storage function,* such that along solutions of Equation (178.15) the following *dissipation inequality*

$$S(x(t;x_0)) - S(x_0) \leq \int_0^t q(u(s), h(x(s)))ds, \qquad t \geq 0$$

holds for any initial condition x_0 and all input functions $u(\cdot)$. (Although in principle this is not required for the definition of dissipativity, assumption of positive definiteness of the storage function $S(x)$ helps to simplify the statement of the stability results described in this section.) Of particular interest is the case in which $S(x)$ is continuously differentiable, so that the dissipation inequality can be given the infinitesimal form

$$\dot{S}(x) \leq q(u, h(x))$$

for all

$$x \in \Re^n, u \in \Re$$

In addition, a system is said to be *strictly dissipative* if there exists a positive definite function $a(x)$ such that

$$\dot{S}(x) \leq -\alpha(x) + q(u, h(x))$$

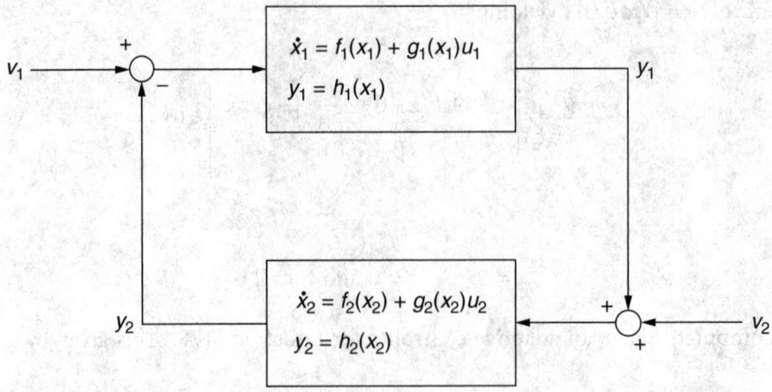

FIGURE 178.5 Feedback loop interconnection.

for all

$$x \in \Re^n, u \in \Re$$

Loosely speaking, if we interpret the storage function as some sort of "energy" of the system, and the supply rate as the power supplied to the system by an external source, the dissipation inequality establishes the property that the rate of change of the energy stored by the system does not exceed the supplied power at any time.

Passive Systems

Among dissipative systems, an important role is played by *passive* systems, for which $q(u,y) = u'y$, and by *strictly output passive* systems, for which $q(u,y) = u'y - \varepsilon y'y$, for some $\varepsilon > 0$. Clearly, a passive system can be rendered strictly output passive through application of the output feedback $u = v - \varepsilon y$, where $v \in \Re^m$ is a new input. Passive systems possess a stable equilibrium at the origin when the input is identically zero, while strictly output passive systems satisfy, in addition,

$$u(t) \equiv 0 \Rightarrow \lim_{t \to 0} y(t) = 0$$

as a result of La Salle's invariance theorem. If, in addition to being strictly output passive, a system is *zero-state detectable*, that is, if

$$u(t) \equiv 0 \text{ and } y(t) \equiv 0 \Rightarrow \lim_{t \to 0} x(t) = 0$$

then the origin is an asymptotically stable equilibrium of the systems. Strict output passivity, in combination with zero-state detectability, is a powerful tool for the analysis of the stability of complex interconnected systems. As a matter of fact, the properties of passivity and strict output passivity are preserved under feedback interconnection: given the system interconnection depicted in Figure 178.5, with

$$\dot{x}_1 = f_1(x_1) + g_1(x_1)u_1$$
$$y_1 = h_1(x_1)$$
$$\dot{x}_2 = f_2(x_2) + g_2(x_2)u_2$$
$$y_2 = h_2(x_2)$$

and $u_1 = v_1 - y_2$, $u_2 = v_2 + y_1$, if the two systems are passive or strictly output passive with storage functions $S_1(x_1)$ and $S_2(x_2)$ respectively, the overall system with input $v = (v_1,v_2)'$ and output $y = (y_1,y_2)'$ is passive with storage function $S_1(x_1) + S_2(x_2)$, and strictly output passive if both systems are strictly output passive. Strictly output passive systems enjoy the property of having a *finite* L_2*-gain*, which means that, if the input is a square-integrable signal, then the output of the system is defined for all time, is square-integrable, and satisfies $\|y\|_{L_2} \le \gamma \|u\|_{L_2}$, for some $\gamma > 0$. It is easy to see that finite L_2*-gain* systems are dissipative with respect to the supply rate $q(u,y) = \gamma^2\|u\|^2 - \|y\|^2$. Stability of feedback interconnections of finite L_2*-gain* systems (not necessarily strictly output passive) can be determined by means of the celebrated *small gain theorem.* Suppose that the systems in Figure 178.4 have finite L_2*-gain* γ_1 and γ_2, respectively, and that $\gamma_1\gamma_2 < 1$. Then, the interconnection is stable, and all trajectories converge to the origin if the overall system is zero-state detectable.

Passivity-based techniques are also instrumental in solving the problem of stabilizing cascade systems of the form

$$\begin{aligned} \dot{z} &= f(z,\xi) \\ \dot{\xi} &= a(\xi) + b(\xi)u \end{aligned} \tag{178.16}$$

where, for the sake of simplicity, we take $u \in \Re$. Let $V(z)$ be a Lyapunov function such that

$$\frac{\partial V}{\partial z} f(z,0) \le 0, \qquad \text{for all } z \in \Re^n$$

which implies that the subsystem

$$\dot{z} = f(z,0)$$

has a stable, but not necessarily asymptotically stable, equilibrium at the origin. Assume that there exists a "dummy output" $y = h(\xi)$ such that the driven subsystem can be written as

$$\dot{z} = f(z,0) + p(z,\xi)y$$

and such that the driving subsystem with output $y = h(\xi)$ is passive with storage function $S(\xi)$, and zero-state detectable. Then, the feedback

$$u = -\frac{\partial V}{\partial z} p(z,\xi) + v$$

transforms Equation (178.16) into a passive system with input v, output y, and storage function $V(z) + S(\xi)$. Application of the control $v = -h(\xi)$ yields convergence of all trajectories $\xi(t)$ to zero, while trajectories $z(t)$ are guaranteed to remain bounded. Convergence of $z(t)$ to the origin is, in turn, established if the driven subsystem is zero-state detectable with respect to y, or if

$$\frac{\partial V}{\partial z} f(z,0) < 0, \qquad \text{for all } z \in \Re^n,\ z \ne 0$$

The one described above is an example of a general methodology called *feedback passivation design*, which aims at rendering a given systems passive with respect to a certain output by means of feedback. The concepts of zero dynamics and relative degree play again a fundamental role in feedback passivation design. As a matter of fact, a system of the form of Equation (178.15) can by rendered passive with respect to a given output $y = h(x)$ only if the system has a vector relative degree $\{1,1,\ldots,1\}$ with respect to y, and the corresponding zero dynamics have a stable equilibrium at the origin. The next example (taken from Van der Schaft [2000]) helps clarify the matter.

Example

The attitude dynamics of a rigid body can described by the equations

$$\dot{\eta} = -\frac{1}{2}\varepsilon'\omega$$
$$\dot{\varepsilon} = \frac{1}{2}\eta\omega + \varepsilon \times \omega \tag{178.17}$$
$$I\dot{\omega} = -\omega \times I\omega + u$$

where $\eta \in \Re$, $\varepsilon \in \Re^3$ are the *Euler parameters*

$$\eta = \cos\left(\frac{\varphi}{2}\right), \qquad \varepsilon = \sin\left(\frac{\varphi}{2}\right)k$$

corresponding to a rotation of an angle φ about the axis $k \in \Re^3$, $\|k\| = 1$. The (η,ε) dynamics is passive with respect to the input/output pair (ω,ε), with storage function $S(\eta) = 2(1 - \eta)$, while the angular velocity dynamics is passive with respect to the input-output pair (u,ω), with the storage function given by the kinetic energy $T(\omega) = \frac{1}{2}\omega' I\omega$. The attitude dynamics is therefore the cascade of two passive systems. Choosing the control $u = -\varepsilon + \nu$, the cascade system is rendered passive with respect to the new input ν and the output ω, with storage function $V(\eta,\omega) = S(\eta) + T(\omega)$. Note that the system has relative degree $\{1,1,\ldots,1\}$ with respect to the output $y = \omega$, and that the corresponding zero-dynamics is

$$\dot{\eta} = 0$$
$$\dot{\varepsilon} = 0$$

which has a stable equilibrium at the origin. Application of the additional output-feedback control $\nu = -\omega$ renders $\omega = 0$ attractive, yielding $(\varepsilon(t),\omega(t)) \to (0,0)$ by virtue of zero-state detectability of ε with respect to the output ω.

Input-to-State Stability

A nonlinear system of the form

$$\dot{x} = f(x,u) \tag{178.18}$$

is said to be input-to-state stable (ISS), if there exists a continuously differentiable function $V(x)\colon \Re^n \to \Re_{\geq 0}$, class-$K_\infty$ functions $\alpha_1(\cdot)$, $\alpha_2(\cdot)$, $\sigma(\cdot)$, and a class-K function $\theta(\cdot)$ such that

$$\alpha_1(\|x\|) \leq V(x) \leq \alpha_2(\|x\|)$$
$$\frac{\partial V}{\partial x} f(x,u) \leq -\sigma(\|x\|) + \theta(\|u\|)$$

for all $x \in \Re^n$ and all $u \in \Re^m$. (A class-K function $\lambda\colon \Re_{\geq 0} \to \Re_{\geq 0}$ is a continuous and strictly increasing function satisfying $\lambda(0) = 0$. A class-K_∞ function is a class-K function such that $\lim_{s\to\infty} \lambda(s) = +\infty$.)

Therefore, an ISS system is dissipative with respect to a supply rate of the form $q(u,y) = \theta(\|u\|) - \sigma(\|x\|)$, with a continuously differentiable, positive, definite, and radially unbounded storage function. An ISS system clearly possesses a globally asymptotically stable equilibrium at the origin, when the input is identically zero. Conversely, global asymptotic stability of the origin for the unforced system (in short, 0-GAS) does not, in general, imply ISS. However, 0-GAS systems are feedback equivalent to ISS systems, in the sense that if Equation (178.18) is 0-GAS, then there exists a feedback law of the form

$$u = \alpha(x) + \beta(x)v$$

with $\beta(x)$ invertible everywhere, such that the closed-loop system with input v

$$\dot{x} = f(x, \alpha(x) + \beta(x)v)$$

is ISS. As a result, it is in principle possible, by means of state feedback, to enforce the desirable property of ISS on a system that is globally asymptotically *stabilizable.* Since the ISS property is preserved under series interconnection, enforcing ISS by feedback proves to be a powerful tool for the *global* stabilization of cascade systems. Consider again the cascade system of Equation (178.16), and assume that there exists a function $\xi = \kappa(z)$ such that the system

$$\dot{z} = f(z, \kappa(z))$$

has a globally asymptotically stable equilibrium at the origin. Then, a "feedback law" $\xi = \alpha(z) + \beta(z)v$ exists such that the system with input v

$$\dot{z} = f(z, \alpha(z) + \beta(z)v)$$

is ISS. The change of coordinates $\beta^{-1}(z)[\xi - \alpha(z)]$ transforms Equation (178.16) into a system of the form

$$\dot{z} = f(z, \alpha(z) + \beta(z)v)$$

$$\dot{v} = \bar{a}(z, v) + \bar{b}(z, v)u$$

Since the cascade of two ISS systems is ISS, necessarily the cascade of an ISS system with a 0-GAS system is 0-GAS. Therefore, to globally stabilize Equation (178.16), it suffices to globally stabilize the origin of the driving subsystem $\dot{v} = \bar{a}(z, v) + \bar{b}(z, v)u.$

The ISS property is also instrumental in analyzing stability of loop interconnections of systems. As a matter of fact, ISS systems possess an *asymptotic gain* with respect to the input, in the sense that there exists a class-*K* function $\gamma(\cdot)$ such that

$$\limsup_{t\to\infty} \|x(t)\| \le \gamma(\|u\|_\infty)$$

for any

$$u(\cdot) \in L_\infty$$

A version of the *small-gain* theorem for ISS systems involving asymptotic gains states that the loop interconnection of two ISS systems

$$\dot{x}_1 = f_1(x_1, u_1)$$
$$\dot{x}_2 = f_2(x_2, u_2)$$
$$u_1 = x_2$$
$$u_2 = x_1 + \nu$$

each having asymptotic gain $\gamma_1(\cdot)$ and $\gamma_2(\cdot)$, respectively, is ISS with respect to the overall input ν if the composition $\gamma_1 \times \gamma_2(\cdot)$ is a simple contraction, that is, if $\gamma_1 \times \gamma_2(s) < s$, for any $s \geq 0$. Several versions, characterizations, extensions, and applications of the ISS have flourished in the last decade, and have contributed to the success of the ISS framework as a powerful analysis and design tool for nonlinear systems. For a survey, see Sontag [1995].

References

H. H. Khalil. *Nonlinear Systems*, 3rd ed. Prentice Hall, New York, 2002.

E. D. Sontag, On the input-to-state stability property, *Eur. J. Control,* 1, 24, 1995.

A. van der Schaft. *L2-Gain and Passivity Techniques in Nonlinear Control*, 2nd ed. Springer-Verlag, Heidelberg, 2000.

Further Information

General Theory

1. M. Vidyasagar. *Nonlinear Systems Analysis*, 2nd ed. Prentice Hall, Englewood Cliffs, NJ, 1993.
2. S. S. Sastry. *Nonlinear Systems: Analysis, Stability, and Control.* Springer-Verlag, New York, 1999.
3. H. H. Khalil. *Nonlinear Systems*, 3rd ed. Prentice Hall, Upper Saddle River, NJ, 2002.

Differential-Geometric and Algebraic Methods

4. A. Isidori. *Nonlinear Control Systems*, 3rd ed. Springer-Verlag, London, 1989.
5. H. Nijmeijer, A. van der Schaft. *Nonlinear Dynamical Control Systems.* Springer-Verlag, New York, 1990.
6. V. Jurdjevi. *Geometric Control Theory.* Cambridge University Press, Cambridge, U.K., 1997.
7. G. Conte, C. H. Moog, and A. M. Perdon. *Nonlinear Control Systems: an Algebraic Setting.* Springer-Verlag, London, 1999.

Lyapunov-Based Techniques

8. M. Krstic, I. Kanellakopoulos, and P. Kokotovi. *Nonlinear Adaptive Control Design.* John Wiley & Sons, New York, 1995.
9. R. Marino and P. Tomei. *Nonlinear Control Design: Geometric, Adaptive, and Robust.* Prentice Hall, Englewood Cliffs, NJ, 1995.
10. A. Isidori. *Nonlinear Control Systems II.* Springer-Verlag, London, 1999.

Dissipative Systems

11. R. Sepulchre, M. Jankovi, and P. Kokotovi. *Constructive Nonlinear Control.* Springer-Verlag, London, 1997.
12. A. van der Schaft. *L2-Gain and Passivity Techniques in Nonlinear Control, 2nd ed.* Springer-Verlag, London, 2000.
13. E. D. Sontag, On the input-to-state stability property, *Eur. J. Control,* 1, 24, 1995.

179

Introduction to Mechatronics

Robert H. Bishop
University of Texas at Austin

Mechatronics is a natural stage in the evolutionary process of modern engineering design. The development of the computer, and then the microcomputer, embedded computers, and associated information technologies and software advances, made mechatronics an imperative in the latter part of the 20th century. Standing on the threshold of the 21st century, with expected advances in integrated bio-electro-mechanical systems, quantum computers, nano- and pico-systems, and other unforeseen developments, the future of mechatronics is full of potential and bright possibilities. The *Handbook of Mechatronics* [1] contains a complete description of the field of mechatronics and is an excellent resource for further reading.

179.1 Basic Definitions

The definition of mechatronics has evolved since the original definition by the Yasakawa Electric Company. In trademark application documents, Yasakawa defined mechatronics in this way [2,3]: "The word, Mechatronics, is composed of 'mecha' from mechanism and the 'tronics' from electronics. In other words, technologies and developed products will be incorporating electronics more and more into mechanisms, intimately and organically, and making it impossible to tell where one ends and the other begins."

The definition of mechatronics continued to evolve after Yasakawa suggested the original definition. One oft-quoted definition of mechatronics was presented by Harashima, Tomizuka, and Fukada in 1996 [4]. In their words, mechatronics is defined as "the synergistic integration of mechanical engineering, with electronics and intelligent computer control in the design and manufacturing of industrial products and processes."

That same year, another definition was suggested by Auslander and Kempf [5]: "Mechatronics is the application of complex decision making to the operation of physical systems." Yet another definition due to Shetty and Kolk appeared in 1997 [6]: Mechatronics is a methodology used for the optimal design of electromechanical products. More recently, we find the suggestion by Bolton [7] that "[a] mechatronic system is not just a marriage of electrical and mechanical systems and is more than just a control system; it is a complete integration of all of them."

All of these definitions and statements about mechatronics are accurate and informative, yet each one in and of itself fails to capture the totality of mechatronics. Despite continuing efforts to define mechatronics, to classify mechatronic products, and to develop a standard mechatronics curriculum, a consensus

0-8493-1586-7/05/$0.00+$1.50

opinion on an all-encompassing description of "what is mechatronics" eludes us. This lack of consensus is a healthy sign. It says that the field is alive, that it is a youthful subject. Even without an unarguably definitive description of mechatronics, engineers understand from the definitions given above and from their own personal experiences the essence of the *philosophy* of mechatronics.

For many practicing engineers on the front line of engineering design, mechatronics is nothing new. Many engineering products of the last 25 years integrated mechanical, electrical, and computer systems, yet were designed by engineers that were never formally trained in mechatronics *per se*. It appears that modern concurrent engineering design practices, now formally viewed as part of the mechatronics specialty, are natural design processes. What is evident is that the study of mechatronics provides a mechanism for scholars interested in understanding and explaining the engineering design process to define, classify, organize, and integrate the many aspects of product design into a coherent package. As the historical divisions between mechanical, electrical, aerospace, chemical, civil, and computer engineering become less clearly defined, we should take comfort in the existence of mechatronics as a field of study in academia. The mechatronics specialty provides an educational path, that is, a road map, for engineering students studying within the traditional structure of most engineering colleges. Mechatronics is generally recognized worldwide as a vibrant area of study. Undergraduate and graduate programs in mechatronic engineering are now offered in many universities. Refereed journals are being published and dedicated conferences are being organized and are generally highly attended.

It should be understood that mechatronics is not just a convenient structure for investigative studies by academicians; it is a way of life in modern engineering practice. The introduction of the microprocessor in the early 1980s and the ever increasing desired performance to cost ratio revolutionized the paradigm of engineering design. The number of new products being developed at the intersection of traditional disciplines of engineering, computer science, and the natural sciences is ever increasing. New developments in these traditional disciplines are being absorbed into mechatronics design at an ever-increasing pace. The ongoing information technology revolution, advances in wireless communication, smart sensors design (enabled by MEMS technology), and embedded systems engineering insures that the engineering design paradigm will continue to evolve in the early 21st century.

179.2 Key Elements of Mechatronics

The study of mechatronic systems can be divided into the following areas of specialty:

- Physical systems modeling
- Sensors and actuators
- Signals and systems
- Computers and logic systems
- Software and data acquisition

The key elements of mechatronics are illustrated in Figure 179.1. As the field of mechatronics continues to mature, the list of relevant topics associated with the area will most certainly expand and evolve.

Central to mechatronics is the integration of physical systems (e.g., mechanical and electrical systems) utilizing various sensors and actuators connected to computers and software. The connections are illustrated in Figure 179.2. In the design process, it is necessary to represent the physical world utilizing mathematical models; hence physical system modeling is essential to the mechatronic design process. Fundamental principles of science and engineering (such as the dynamical principles of Newton, Maxwell, and Kirchoff) are employed in the development of physical system models of mechanical and dynamical systems, electrical systems, electromechanical systems, fluid systems, thermodynamical systems, and microelectromechanical (MEMS) systems. Mechatronic systems utilize a variety of sensors to measure the external environment and actuators to manipulate the physical systems to achieve the desired goals. The sensors measure linear and rotational motion, acceleration, force, torque and power, flow, temperature, distance and proximity, and light and images. Sensors are becoming increasingly smaller, lighter, and in many instances, less expensive. The trend to micro- and nano-scales supports the continued

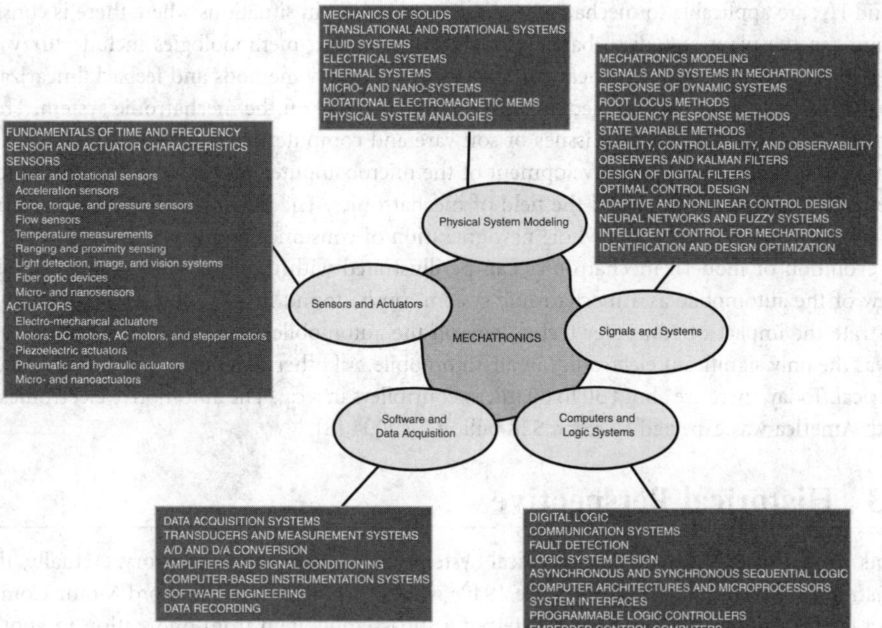

FIGURE 179.1 The key elements of mechatronics.

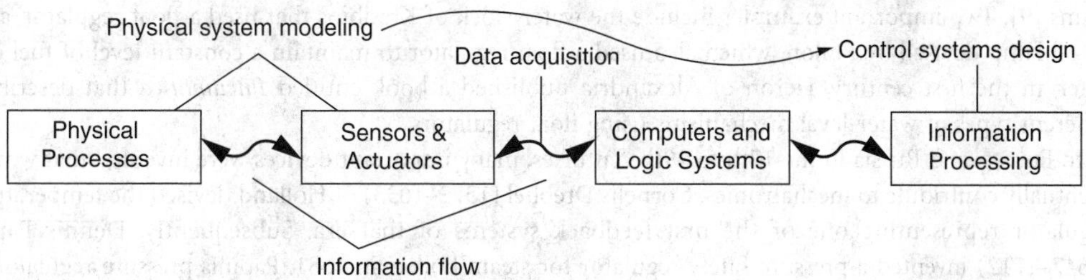

FIGURE 179.2 Mechatronic system connections between the physical processes and the information processing through the sensors and actuators.

evolution of mechatronic system design. Actuators are following the same trends as sensors in reduced size and cost, and can be grouped into electromechanical actuators, electrical actuators, piezoelectric actuators, hydraulic and pneumatic actuators, and microelectromechanical MEMS (microtransducers).

As noted, an important step in the design process is to accurately represent the system with mathematical models. What are the mathematical models used for? They are employed in the design and analysis of appropriate control systems. The historic subject of systems and controls is central to mechatronic system design. The application of control system design to mechatronic systems runs the gamut from classical frequency domain design (using linear, time-invariant system models) to modern multi-input, multioutput (MIMO) state space methods (again assuming linear, time-invariant models) to nonlinear, time-varying methods. The classical design methods use transfer function models in conjunction with root locus methods or frequency-response methods, such as Bode, Nyquist, and Nichols. Although the transfer function approach generally uses single-input, single-output (SISO) models, it is possible to perform MIMO analysis in the frequency domain. Modern state space analysis and design techniques can be readily applied to SISO and MIMO systems. Pole placement techniques are used to design the controllers. Optimal control methods, such as linear quadratic regulators (LQR), have been discussed in the literature since the 1960s and are in common use today. Robust optimal control design strategies, such as H_2 and H_∞, are applicable to mechatronic systems, especially in situations where there is considerable uncertainty in the plant and disturbance models. Other design methodologies include fuzzy control, control, adaptive control, and nonlinear control (using Lyapunov methods and feeback linearization).

Once the control system is designed, it must be implemented on the mechatronic system. This is the stage in the design process wherein issues of software and computer hardware, logic systems, and data acquisition take center stage. The development of the microcomputer, and associated information technologies and software, has impacted the field of mechatronics. The integration of computers and electromechanical systems has led to a whole new gneration of consumer products.

The evolution of modern mechatronics can be illustrated with the example of the automobile. An overview of the automobile as a mechatronic system can be found in *The Mechatronics Handbook* [1]. To illustrate the impact of computer technology on the automobile, consider that until the 1960s, the radio was the only significant electronics in an automobile. All other functions were entirely mechanical or electrical. Today, there are about 30 to 60 micro controllers in a car. The automotive electronics market in North America was expected to reach $28 billion by 2004 [8].

179.3 Historical Perspective

Attempts to construct automated mechanical systems have an interesting history. Actually, the term "automation" was not popularized until the 1940s when it was coined by the Ford Motor Company to denote a process in which a machine transferred a subassembly item from one station to another and then positioned the item precisely for additional assembly operations. But successful development of automated mechanical systems occurred long before then. For example, early applications of automatic control systems appeared in Greece from 300 to 1 B.C. with the development of float regulator mechanisms [9]. Two important examples include the water clock of Ktesibios that used a float regulator, and an oil lamp devised by Philon, which also used a float regulator to maintain a constant level of fuel oil. Later, in the first century, Heron of Alexandria published a book entitled *Pneumatica* that described different types of water-level mechanisms using float regulators.

In Europe and Russia in the 17th to 19th centuries, many important devices were invented that would eventually contribute to mechatronics. Cornelis Drebbel (1572–1633) of Holland devised the temperature regulator representing one of the first feedback systems of that era. Subsequently, Dennis Papin (1647–1712) invented a pressure safety regulator for steam boilers in 1681. Papinís pressure regulator is similar to a modern-day pressure-cooker valve. The first mechanical calculating machine was invented by Pascal in 1642 [10].The first historical feedback system claimed by Russia was developed by Polzunov in 1765 [11]. Polzunov's water-level float regulator, illustrated in Figure 179.3, employs a float that rises and lowers in relation to the water level, thereby controlling the valve that covers the water inlet in the boiler.

Further evolution in automation was enabled by advancements in control theory traced back to the Watt flyball governor of 1769. The flyball governor, illustrated in Figure 179.4, was used to control the speed of a steam engine [12]. Employing a measurement of the speed of the output shaft and utilizing the motion of the flyball to control the valve, the amount of steam entering the engine is controlled. As

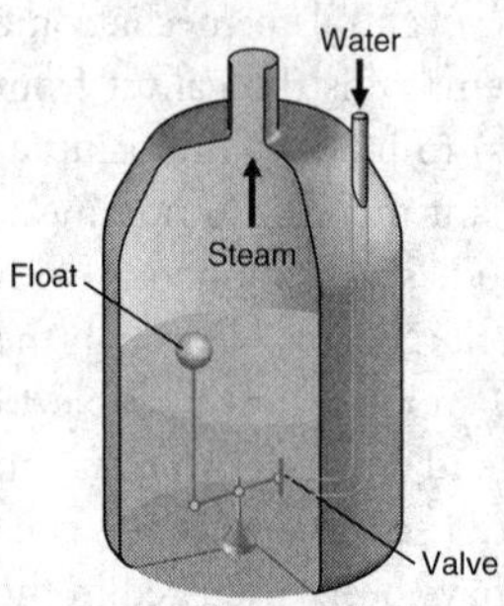

FIGURE 179.3 Water-level float regulator. (From R. C. Dorf and R. H. Bishop, Eds., *Modern Control Systems*, 9th ed., Prentice Hall, New York, 2000. With permission.)

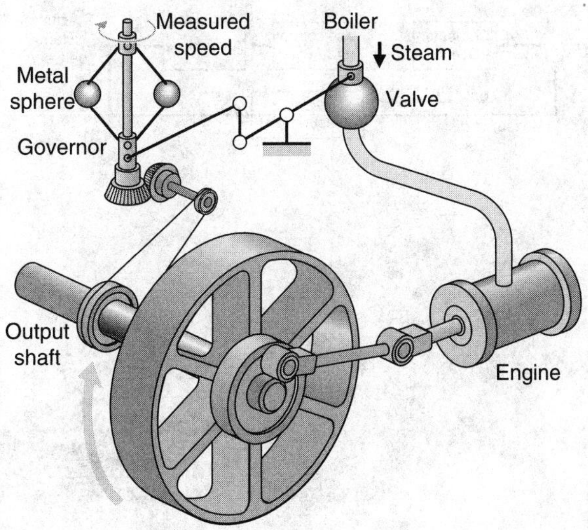

FIGURE 179.4 Watt's flyball governor. (From R. C. Dorf and R. H. Bishop, Eds., *Modern Control Systems*, 9th ed., Prentice Hall, New York, 2000. With permission.)

the speed of the engine increases, the metal spheres on the governor apparatus rise and extend away from the shaft axis, thereby closing the valve. This is an example of a feedback control system where the feedback signal and the control actuation are completely coupled in the mechanical hardware.

These early successful automation developments were achieved through intuition, application of practical skills, and persistence. The next step in the evolution of automation required a *theory* of automatic control. The precursor to the numerically controlled (NC) machines for automated manufacturing (to be developed in the 1950s and 1960s at the Massachusetts Institute of Technology) appeared in the early 1800s with the invention of feed-forward control of weaving looms by Joseph Jacquard of France. In the late 1800s, the subject now known as control theory was initiated by J. C. Maxwell through analysis of the set of differential equations describing the flyball governor [13]. Maxwell investigated the effect that various system parameters had on the system performance. At about the same time, Vyshnegradskii formulated a mathematical theory of regulators [14]. In the 1830s, Michael Faraday described the law of induction that would form the basis of the electric motor and the electric dynamo. Subsequently, in the late 1880s, Nikola Tesla invented the alternating-current induction motor. The basic idea of controlling a mechanical system automatically was firmly established by the end of the 1800s. The evolution of automation would accelerate significantly in the 20th century.

The development of pneumatic control elements in the 1930s matured to a point of finding applications in the process industries. However, before 1940, the design of control systems remained an art generally characterized by trial-and-error methods. During the 1940s, continued advances in mathematical and analytical methods solidified the notion of control engineering as an independent engineering discipline. In the United States, the development of the telephone system and electronic feedback amplifiers spurred the use of feedback by Bode, Nyquist, and Black at Bell Telephone Laboratories [15–19]. The operation of the feedback amplifiers was described in the frequency domain and the ensuing design and analysis practices are now generally classified as "classical control." During the same time period, control theory was also developing in Russia and Eastern Europe. Mathematicians and applied mechanicians in the former Soviet Union dominated the field of controls and concentrated on time domain formulations and differential equation models of systems. Further developments of time domain formulations using state variable system representations occurred in the 1960s and led to design and analysis practices now generally classified as "modern control."

The World War II mobilization led to further advances in the theory and practice of automatic control in an effort to design and construct automatic airplane pilots, gun-positioning systems, radar antenna

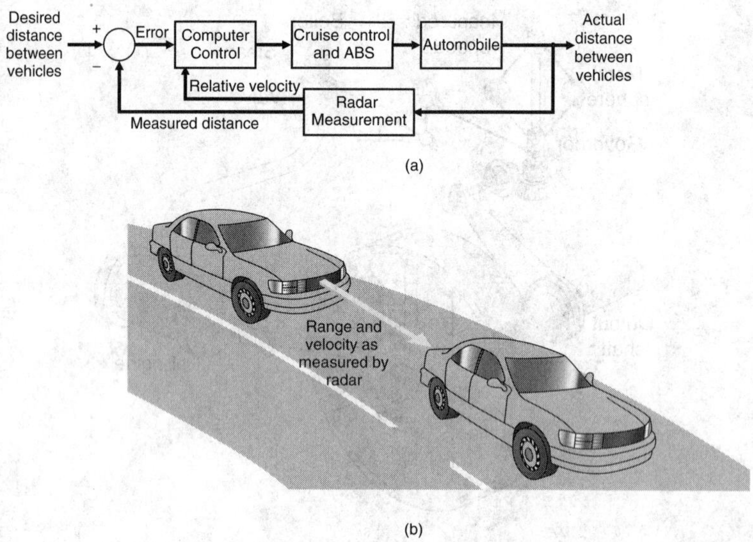

FIGURE 179.5 Using a radar to measure distance and velocity to autonomously maintain desired distance between vehicles. Adapted from R. C. Dorf and R. H. Bishop, Eds., *Modern Control Systems*, 9th ed., Prentice Hall, New York, 2000. With permission.)

control systems, and other military systems. The complexity and expected performance of these military systems necessitated an extension of the available control techniques and fostered interest in control systems and the development of new insights and methods. Frequency domain techniques continued to dominate the field of controls following World War II with the increased use of the Laplace transform, and the use of the so-called *s*-plane methods, such as designing control systems using root locus.

On the commercial side, driven by cost savings achieved through mass production, automation of the production process was a high priority beginning in the 1940s. During the 1950s, the invention of the cam, linkages, and chain drives became the major enabling technologies for the invention of new products and high-speed precision manufacturing and assembly. Examples include textile and printing machines, paper converting machinery, and sewing machines. High-volume precision manufacturing became a reality during this period. An example of high-volume automated manufacturing of paperboard containers for packaging [20]. The automated paperboard container-manufacturing machine employs a sheet-fed process wherein the paperboard is cut into a fan shape to form the tapered sidewall, and wrapped around a mandrel. The seam is heat sealed and cured while another sheet-fed source of paperboard is used to cut out the plate to form the bottom of the paperboard container, formed into a shallow dish and assembled to the cup shell. The lower edge of the cup shell is bent inwards over the edge of the bottom plate sidewall. A heat-sealing process is used to prevent leaks and provide a level edge for standup while the brim is formed on the top to provide the required stiffness. All of these operations are carried out while the work piece undergoes a precision transfer from one turret to another and is then ejected. The production rate of a typical machine averages over 200 cups per minute. Automated paperboard-container manufacturing did not involve any nonmechanical system except an electric motor for driving the line shaft. These machines are typical of paper-converting and textile machinery and represent automated systems significantly more complex than their predecessors.

The development of the microprocessor in the late 1960s led to early forms of computer control in process and product design. Examples include NC machines and aircraft control systems. Yet the manufacturing processes were still entirely mechanical in nature and the automation and control systems were implemented only as an afterthought. The launch of Sputnik and the advent of the space age provided yet another impetus to the continued development of controlled mechanical systems. Missiles and space probes necessitated the development of complex, highly accurate control systems. Furthermore, the need to minimize satellite mass (i.e., to minimize the amount of fuel required for the mission) while providing accurate control encouraged advancements in the important field of optimal control. Time domain methods developed by Liapunov, Minorsky, and others, as well as the theories of optimal control developed by Pontryagin in the former Soviet Union and Bellman in the United States, were well matched with the increasing availability of high-speed computers and new programming languages for scientific use.

Advancements in semiconductor and integrated circuits manufacturing led to the development of a new class of products that incorporated mechanical and electronics in the system and required the two together for their functionality. The term *mechatronics* was introduced by Yasakawa Electric in 1969 to represent such systems. Yasakawa was granted a trademark in 1972, but after widespread usage of the term, released its trademark rights in 1982 [2–4]. Initially, mechatronics referred to systems with only mechanical systems and electrical components — no computation was involved. Examples of such systems include the automatic sliding door, vending machines, and garage door openers.

In the late 1970s, the Japan Society for the Promotion of Machine Industry (JSPMI) classified mechatronics products into four categories [2]:

Class I: Primarily mechanical products with electronics incorporated to enhance functionality. Examples include numerically controlled machine tools and variable-speed drives in manufacturing machines.

Class II: Traditional mechanical systems with significantly updated internal devices incorporating electronics. The external user interfaces are unaltered. Examples include the modern sewing machine and automated manufacturing systems.

Class III: Systems that retain the functionality of the traditional mechanical system, but the internal mechanisms are replaced by electronics. An example is the digital watch.

Class IV: Products designed with mechanical and electronic technologies through synergistic integration. Examples include photocopiers, intelligent washers and dryers, rice cookers, and automatic ovens.

The enabling technologies for each mechatronic product class illustrate the progression of electromechanical products in stride with developments in control theory, computation technologies, and microprocessors. Class I products were enabled by servo technology, power electronics, and control theory. Class II products were enabled by the availability of early computational and memory devices and custom circuit design capabilities. Class III products relied heavily on the microprocessor and integrated circuits to replace mechanical systems. Finally, Class IV products marked the beginning of true mechatronic systems, through integration of mechanical systems and electronics. It was not until the 1970s with the development of the microprocessor by the Intel Corporation that integration of computational systems with mechanical systems became practical.

The divide between classical control and modern control was significantly reduced in the 1980s with the advent of "robust control" theory. It is now generally accepted that control engineering must consider both the time domain and the frequency domain approaches simultaneously in the analysis and design of control systems. In addition, during the 1980s the utilization of digital computers as integral components of control systems became routine. There are literally hundreds of thousands of digital process-control computers installed worldwide [21–22]. Whatever definition of mechatronics one chooses to adopt, it is evident that odern mechatronics involves computation as the central element. In fact, the incorporation of the microprocessor to precisely modulate mechanical power and to adapt to changes in environment are the essence of modern mechatronics and smart products.

179.4 Future of Mechatronics

Mechatronics, the term coined in Japan in the 1970s, has evolved over the past 25 years and has led to a special breed of intelligent products. What is mechatronics? It is a natural stage in the evolutionary process of modern engineering design. For some engineers, mechatronics is nothing new, and for others it is a philosophical approach to design that serves as a guide for their activities. Certainly, mechatronics is an evolutionary process, not a revolutionary one. It is clear that an all-encompassing definition of mechatronics does not exist, but in reality, one is not needed. It is understood that mechatronics is about the synergistic integration of mechanical, electrical, and computer systems. One can understand the extent that mechatronics reaches into various disciplines by characterizing the constituent components comprising mechatronics, which include (1) physical systems modeling, (2) sensors and actuators, (3) signals and systems, (4) computers and logic systems, and (5) software and data acquisition. Engineers and scientists from all walks of life and fields of study can contribute to mechatronics. As engineering and science boundaries become less well defined, more students will seek a multidisciplinary education with a strong design component. Academia should be moving toward a curriculum that includes coverage of mechatronic systems.

In the future, growth in mechatronic systems will be fueled by growth in constituent areas. Advancements in traditional disciplines fuel the growth of mechatronics systems by providing "enabling technologies." For example, the invention of the microprocessor had a profound effect on the redesign of mechanical systems and design of new mechatronics systems. We should expect continued advancements in cost-effective microprocessors and micro controllers, sensor and actuator development enabled by advancements in applications of MEMS, adaptive control methodologies and real-time programming methods, networking and wireless technologies, mature CAE technologies for advanced system modeling, virtual prototyping, and testing. The continued rapid development in these areas will only accelerate the pace of smart product development. The Internet is a technology that, when utilized in combination with wireless technology, may also lead to new mechatronic products. While developments in automotives provide vivid examples of mechatronics development, there are numerous examples of intelligent systems in all walks of life, including smart home appliances such as dishwashers, vacuum cleaners, microwaves, and wireless network enabled devices. In the area of "human-friendly machines" (a term used by H. Kobayashi [23]), we can expect advances in robot-assisted surgery, and implantable sensors and actuators. Other areas that will benefit from mechatronic advances may include robotics, manufacturing, space technology, and transportation. The future of mechatronics is wide open.

References

1. Bishop, R. H., Ed., *The Mechatronics Handbook*, CRC Press, Boca Raton, FL, 2002.
2. N. Kyura and H. Oho, "Mechatronics—an industrial perspective," *IEEE/ASME Trans. Mechatronics*, 1, 10–15, 1996.
3. T. Mori, "Mecha-tronics," Yasakawa Internal Trademark Application Memo 21.131.01, July 12, 1969.
4. F. Harshama, M. Tomizuka, and T. Fukuda, "Mechatronics: what is it, why, and how? An editorial," *IEEE/ASME Trans. Mechatronics,* 1, 1–4, 1996.
5. D. M. Auslander and C. J. Kempf, *Mechatronics: Mechanical System Interfacing*, Prentice Hall, Upper Saddle River, NJ, 1996.
6. D. Shetty and R. A. Kolk, *Mechatronic System Design*, PWS Publishing, Boston, 1997.
7. W. Bolton, *Mechatronics: Electrical Control Systems in Mechanical and Electrical Engineering*, 2nd ed., Addison-Wesley Longman, Harlow, England, 1999.
8. B. Jorgensen, "Shifting gears," *Auto Electron. Electron. Bus.*, February. 2001.
9. I. O. Mayr, *The Origins of Feedback Control*, MIT Press, Cambridge, MA, 1970.
10. D. Tomkinson and J. Horne, *Mechatronics Engineering*, McGraw-Hill, New York, 1996.
11. E. P. Popov, *The Dynamics of Automatic Control Systems*, Gostekhizdat, Moscow, 1956; Addison-Wesley, Reading, MA, 1962.
12. R. C. Dorf and R. H. Bishop, *Modern Control Systems*, 9th ed., Prentice Hall, Upper Saddle River, NJ, 2000.
13. J. C. Maxwell, "On governors," Proceedings of the Royal Society of London, 16, 1868, in *Selected Papers on Mathematical Trends in Control Theory*, R. Bellman and R. Kalaba, Eds., Dover, New York, 1964, pp. 270–283.
14. I. A. Vyshnegradskii, "On controllers of direct action,"*Izv. SPB Tekhnotog. Inst.*, 1877.

15. H. W. Bode, "Feedback: the history of an idea," in *Selected Papers on Mathematical Trends in Control Theory*, R. Bellman and R. Kalaba, Eds., Dover, New York, 1964, pp. 106–123.
16. H. S. Black, "Inventing the negative feedback amplifier," *IEEE Spectrum*, December 1977, pp. 55–60.
17. J. E. Brittain, *Turning Points in American Electrical History*, IEEE Press, New York, 1977.
18. M. D. Fagen, *A History of Engineering and Science on the Bell Systems*, Bell Telephone Laboratories, 1978.
19. G. Newton, L. Gould, and J. Kaiser, *Analytical Design of Linear Feedback Control*, John Wiley & Sons, New York, 1957.
20. M. K. Ramasubramanian, *Mechatronics — the Future of Mechanical Engineering: Past, Present, and a Vision for the Future,* SPIE Press, 4334, 2001.
21. R. C. Dorf and A. Kusiak, *Handbook of Automation and Manufacturing*, John Wiley & Sons, New York, 1994.
22. R. C. Dorf, *The Encyclopedia of Robotics*, John Wiley & Sons, New York, 1988.
23. H. Kobayashi, Guest editorial, *IEEE/ASME Trans. Mechatronics,* 2, 217, 1997.

XXV

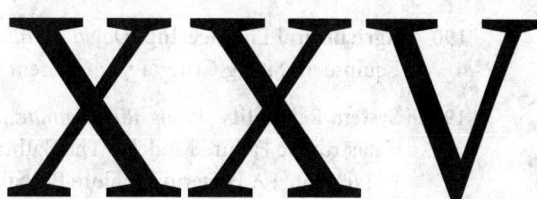

Manufacturing

180

Types of Manufacturing

Richard J. Schonberger
Schonberger & Associates, Inc.

Although there are many ways to categorize manufacturing, three general categories stand out (which probably have emerged from production planning and control lines of thought):

1. *Job-shop production.* A job shop produces in small lots or batches.
2. *Mass production.* Mass production involves machines or assembly lines that manufacture discrete units repetitively.
3. *Continuous production.* The process industries produce in a continuous flow.

Primary differences among the three types center on output volume and variety and process flexibility. Table 180.1 matches these characteristics with the types of manufacturing and gives examples of each type. The following discussion begins by elaborating on Table 180.1. Next are comments on hybrid and uncertain types of manufacturing. Finally, five secondary characteristics of the three manufacturing types are presented.

180.1 Job-Shop and Batch Production

As Table 180.1 shows, job-shop manufacturing is very low in volume but is highest in output variety and process flexibility. In this mode, the processes — a set of resources including labor and equipment — are reset intermittently to make a variety of products. (Product variety requires flexibility to frequently reset the process.)

In tool and die making, the first example, the volume is generally one unit — for example, a single die set or mold. Since every job is different, output variety is at a maximum, and operators continually reset the equipment for the next job.

Casting in a foundry has the same characteristics, except that the volume is sometimes more than one. That is, a given job order may be to cast one, five, ten, or more pieces. The multipiece jobs are sometimes called lots or batches.

A bakery makes a variety of products, each requiring a new series of steps to set up the process — for example, mixing and baking a batch of sourdough bread, followed by a batch of cinnamon rolls.

0-8493-1586-7/05/$0.00+$1.50

TABLE 180.1 Types of Manufacturing — Characteristics and Examples

	Job-Shop Production	Mass Production	Continuous Production
Volume	Very low	High	Highest
Variety	Highest	Low	Lowest
Flexibility	Highest	Low	Lowest
Examples	Tool and die making, casting (foundry), baking (bakery)	Auto assembly, bottling, apparel manufacturing	Paper milling, refining, extrusion

180.2 Mass Production

Second in Table 180.1 is mass production. Output volume, in discrete units, a high. Product variety is low, entailing low flexibility to reset the process.

Mass production of automobiles is an example. A typical automobile plant will assemble two or three hundred thousand cars a year. In some plants just one model is made per assembly line; variety is low (except for option packages). In other plants, assembly lines produce mixed models. Still, this is considered mass production since assembly continues without interruption for model changes.

In bottling, volumes are much higher, sometimes in the millions per year. Changing from one bottled product to another requires a line stoppage, but between changeovers production volumes are high (e.g., thousands). Flexibility, such as changing from small to large bottles, is low; more commonly, large and small bottles are filled on different lines.

Similarly, mass production of apparel can employ production lines, with stoppages for pattern changes. More conventionally, the industry has used a very different version of mass production: Cutters, sewers, and others in separate departments each work independently, and material handlers move components from department to department to completion. Thus, existence of an assembly line or production line is not a necessary characteristic of mass production.

180.3 Continuous Production

Products that flow — liquids, gases, powders, grains, slurries — are continuously produced, the third type in Table 180.1. In continuous process plants, product volumes are very high (relative to, for example, a job-shop method of making the same product). Because of designed-in process limitations (pumps, pipes, valves, etc.) product variety and process flexibility are very low.

In a paper mill, a meshed belt begins pulp on its journey though a high-speed multistage paper-making machine. The last stage puts the paper on reels holding thousands of linear meters. Since a major product changeover can take hours, plants often limit themselves to incremental product changes. Special-purpose equipment design also poses limitations. For example, a tissue machine cannot produce newsprint, and a newsprint machine cannot produce stationery. Thus, in paper making, flexibility and product variety for a given machine are very low.

Whereas a paper mill produces a solid product, a refinery keeps the substance in a liquid (or sometimes gaseous) state. Continuous refining of fats, for example, involves centrifuging to remove undesirable properties to yield industrial or food oils. As in paper making, specialized equipment design and lengthy product changeovers (including cleaning of pipes, tanks, and vessels) limit process flexibility; product volumes between changeovers are very high, sometimes filling multiple massive tanks in a tank farm.

Extrusion, the third example of continuous processing in Table 180.1, yields such products as polyvinyl chloride (PVC) pipe, polyethylene film, and reels of wire. High process speeds produce high product volumes, such as multiple racks of pipe, rolls of film, or reels of wire per day. Stoppages for changing extrusion heads and many other adjustments limit process flexibility and lead to long production runs between changeovers. Equipment limitations (e.g., physical dimensions of equipment components) keep product variety low.

180.4 Mixtures and Gray Areas

Many plants contain a mixture of manufacturing types. A prominent example can be found in the process industries, where production usually is only partially continuous. Batch mixing of pulp, fats, or plastic granules precedes continuous paper making, refining of oils, and extrusion of pipe. Further processing may be in the job-shop mode: slitting and length-cutting paper to customer order, secondary mixing and drumming of basic oils to order, and length cutting and packing of pipe to order.

Mixed production also often occurs in mass production factories. An assembly line (e.g., assembling cars or trucks) may be fed by parts, such as axles, machined in the job-shop mode from castings that are also job-shop produced. Uniform mass-made products (e.g., molded plastic hard hats) may go to storage where they await a customer order for final finishing (e.g., decals) in the job-shop mode. An apparel plant may mass produce sportswear on the one hand, and produce custom uniforms for professional sports figures in the job-shop mode on the other.

More than one type of manufacturing in the same plant requires more than one type of production planning, scheduling, and production. The added complexity in management may be offset, however, by demand-side advantages of offering a fuller range of products.

Sometimes a manufacturing process does not fit neatly into one of the three basic categories. One gray area occurs between mass production and continuous production. Some very small products — screws, nuts, paper clips, toothpicks — are made in discrete units. But because of small size, high volumes, and uniformity of output, production may be scheduled and controlled not in discrete units but by volume, thus approximating continuous manufacturing. Production of cookies, crackers, potato chips, and candy resembles continuous forming or extrusion of sheet stock on wide belts, except that the process includes die cutting or other separation into discrete units — like mass production. Link sausages are physically continuous, but links are countable in whole units.

Another common gray area is between mass and job-shop production. A notable example is high-volume production of highly configured products made to order. Products made for industrial uses — such as specialty motors, pumps, hydraulics, controllers, test equipment, and work tables — are usually made in volumes that would qualify as mass production, except that end-product variety is high, not low.

These types of manufacturing with unclear categories do not necessarily create extra complexity in production planning and control. The difficulty and ambiguity are mainly terminological.

180.5 Capital Investment, Automation, Advanced Technology, Skills, and Layout

The three characteristics used to categorize manufacturing — volume, variety, and flexibility — are dominant but not exhaustive. To some extent, the manufacturing categories also differ with respect to capital investment, automation, technology, skills, and layout.

Typically, continuous production is highly capital intensive, whereas mass production is often labor intensive. The trend toward automated, robotic assembly, however, is more capital intensive and less labor intensive, which erodes the distinction. Job-shop production on conventional machines is intermediate as to capital investment and labor intensiveness. However, computer numerically controlled (CNC) machines and related advanced technology in the job shop erode this distinction as well.

As technology distinctions blur, so do skill levels of factory operatives. In conventional high-volume assembly, skill levels are relatively low, whereas those of machine operators in job shops — such as machinists and welders — tend to be high. But in automated assembly, skill levels of employees tending the production lines elevate toward technician levels — more like machinists and welders. In continuous production, skill levels range widely — from low-skilled carton handlers and magazine fillers to highly skilled process technicians and troubleshooters.

Layout of equipment and related resources is also becoming less of a distinction than it once was. The classical job shop is laid out by type of equipment: all milling machines in one area, all grinding machines in another. Mass and continuous production have been laid out by the way the product flows: serially and linearly. Many job shops, however, have been converted to cellular layouts — groupings of diverse machines that produce a family of similar products. In most work cells, the flow pattern is serial from machine to machine, but the shape of the cell is not linear; it is U-shaped or, for some larger cells, serpentine. Compact U and serpentine shapes are thought to provide advantages in teamwork, material handling, and labor flexibility.

To some degree, such thinking has carried over to mass production. That is, the trend is to lay out assembly and production lines in U and serpentine shapes instead of straight lines, which was the nearly universal practice in the past. In continuous production of fluids, the tendency has always been toward compact facilities interconnected by serpentine networks of pipes. Continuous production of solid and semisolid products (wide sheets, extrusions, etc.), on the other hand, generally must move in straight lines, in view of the technical difficulties in making direction changes.

Defining Terms

Batch — A quantity (lot) of a single item.

Changeover (setup) — Setting up or resetting a process (equipment) for a new product or batch.

Continuous production — Perpetual production of goods that flow and are measured by area or volume; usually very high in product volume, very low in product variety, and very low in process flexibility.

Job-shop production — Intermittent production with frequent resetting of the process for a different product or batch; usually low in product volume, high in product variety, and high in process flexibility.

Mass production — Repetitive production of discrete units on an assembly line or production line; usually high in product volume, low in product variety, and low in process flexibility.

Process — A set of resources and procedures that produces a definable product (or service).

Process industry — Manufacturing sector involved in continuous production.

Further Information

Industrial Engineering. Published monthly by the Institute of Industrial Engineers.

Manufacturing Engineering. Published monthly by the Society of Manufacturing Engineers.

Knod, E. M. and Schonberger, R. J. 2001. *Operations Management: Meeting Customers' Demands*, 7th ed. McGraw-Hill Irwin, Boston. See, especially, chapters 14, 15, and 16.

181
Quality

Matthew P. Stephens
Purdue University

Joseph F. Kmec
Purdue University

Although no universally accepted definition of *quality* exists, in its broadest sense quality has been described as "conformance to requirements," "freedom from deficiencies," or "the degree of excellence which a thing possesses." Taken within the context of the manufacturing enterprise, quality — or, more specifically, manufacturing quality — shall be defined as "conformance to requirements." This chapter focuses on the evaluation of product quality, with particular emphasis directed at statistical methods used in the measurement, control, and tolerances needed to achieve the desired quality. Factors that define product quality are ultimately determined by the customer and include such traits as reliability, affordability or cost, availability, user friendliness, and ease of repair and disposal. To ensure that quality goals are met, manufacturing firms have initiated a variety of measures that go beyond traditional product inspection and record keeping, which, by and large, were the mainstays of quality control departments for decades. One such initiative is total quality management (TQM) [Saylor, 1992], which focuses on the customer, both inside and outside the firm. It consists of a disciplined approach using quantitative methods to continuously improve all functions within an organization. Another initiative is registration under the ISO 9000 series [Lamprecht, 1993], which provides a basis for U.S. manufacturing firms to qualify their finished products and processes to specified requirements. More recently, the U.S. government has formally recognized outstanding firms through the coveted Malcolm Baldrige Award [ASQC, 1994] for top quality among U.S. manufacturing companies. One of the stipulations of the award is that recipient companies share information on successful quality strategies with their manufacturing counterparts.

181.1 Measurement

The inherent nature of the manufacturing process is variation. Variation is present due to any one or a combination of factors including materials, equipment, operators, or the environment. Controlling variation is an essential step in realizing product quality. To successfully control variation, manufacturing firms rely on the measurement of carefully chosen parameters. Because measurement of the entire population of products or components is seldom possible or desirable, *samples* from the *population* are chosen. The extent to which sample data represent the population depends largely on such items as sample size, method of sampling, and time-dependent variations.

Measured data from samples taken during a manufacturing process can be plotted in order to determine the shape of the *frequency distribution*. The frequency distribution can give a visual clue to the process average and dispersion. The latter is referred to as *standard deviation*. Figure 181.1 shows a

0-8493-1586-7/05/$0.00+$1.50

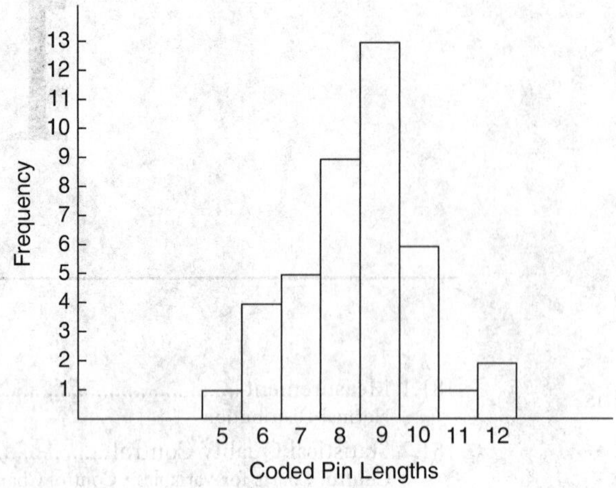

FIGURE 181.1 Coded pin lengths.

frequency distribution plot of 40 coded pin lengths expressed in thousands of an inch greater than 1 in. Thus, the coded length 6 represents an actual length of 1.006 in. For the data shown, average coded pin length is 8.475 and standard deviation is 1.585.

Normal Distribution

Although there is an infinite variety of frequency distributions, the variation of measured parameters typically found in the manufacturing industry follows that of the normal curve. The normal distribution is a continuous bell-shaped plot of frequency versus some parameter of interest, and is an extension of a histogram whose basis is a large population of data points. Figure 181.2 shows a normal distribution plot superimposed on a histogram. Some important properties of the normal distribution curve are:

- The distribution is symmetrical about the population mean μ.
- The curve can be described by a specific mathematical function of population mean μ and population standard deviation σ.

An important relationship exists between standard deviation and area under the normal distribution curve. Such a relationship is shown in Figure 181.3 and may be interpreted as follows: 68.26% of the

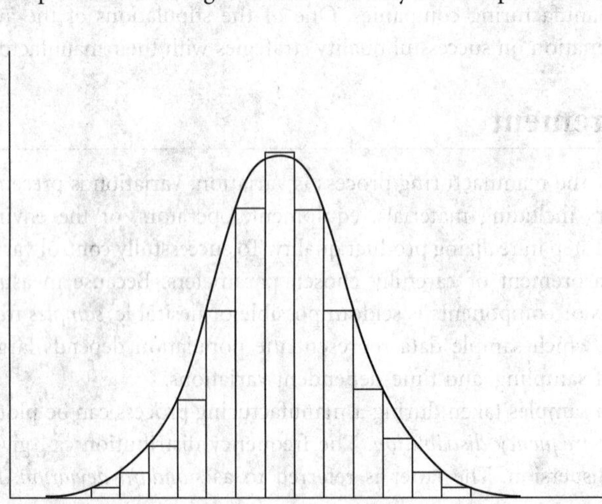

FIGURE 181.2 Normal distribution curve.

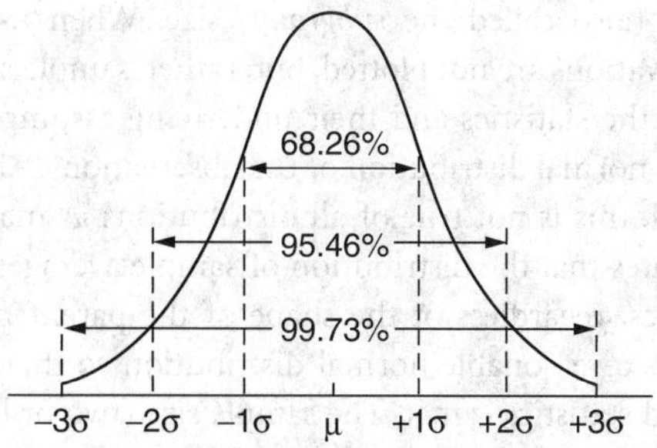

FIGURE 181.3 Percentages under the normal curve.

readings (or area under the curve) will be between ±1σ limits, 95.46% of the readings will be between ±2σ limits, and 99.73% of the readings will be between ±3σ limits. The significance of this relationship is that the standard deviation can be used to calculate the percentage of the population that falls between any two given values in the distribution.

181.2 Statistical Quality Control

Statistical quality control (SQC) deals with collection, analysis, and interpretation of data to monitor a particular manufacturing or service process and ensure that the process remains within its capacity. In order to understand process capability, it is necessary to realize that variation is a natural phenomenon that will occur in any process. Parts will appear identical only due to the limitation of the inspection or measurement instrument. The sources of these variations may be the material, process, operator, time of the operation, or any other significant variables. When these factors are kept constant, the minor variations inherent in the process are called *natural* (or *chance*) *variations*, as opposed to variations due to *assignable causes*.

Control charts are utilized to determine when a given process variation is within the expected or natural limits. When the magnitude of variation exceeds these predetermined limits, the process is said to be *out of control*. The causes for out-of-control conditions are investigated and the process is brought back in control. Control charts or the control limits for the natural or chance-cause variations are constructed based on the relationship between the normal distribution and the standard deviation of the distribution. The property of normal distribution and its relationship to the standard deviation of the distribution was discussed in Chapter 180. As stated earlier, since approximately 99.73% of a normal distribution is expected to fall between ±3σ of the distribution, control limits are established at $\bar{X} \pm 3\sigma$ for the process. Therefore, any sample taken from the process is expected to fall between the control limits or the $\bar{X} \pm 3\sigma$ of the process 99.73% of the time. Any sample not within these limits is assumed to indicate an out-of-control condition for which an assignable cause is suspected.

Control charts can be divided into two major categories: control charts for variables (measurable quality characteristics, i.e., dimension, weight, hardness, etc.), and control charts for attributes (quality characteristics not easily measurable and therefore classified as conforming or not conforming, good or bad, etc.).

Control Charts for Variables

The most common charts used for variables are the $\bar{X}$ and R charts. The charts are used as a pair for a given quality characteristic. In order to construct control charts for variables, the following steps may be followed:

1. Define the quality characteristic that is of interest. Control charts for variables deal with only one quality characteristic; therefore, if multiple properties of the product of the process are to be monitored, multiple charts should be constructed.

2. Determine the sample (also called the *subgroup*) size. When using control charts, individual measurements or observations are not plotted, but, rather, sample averages are utilized. One major reason is the nature of the statistics and their underlying assumptions. Normal statistics, as the term implies, assumes a normal distribution of the observations. Although many phenomena may be normally distributed, this is not true of all distributions. A major statistical theory called the *central limit theorem* states that the distribution of sample averages will tend toward normality as the sample size increases, regardless of the shape of the parent population. Therefore, plotting sample averages ensures a reasonable normal distribution so that the underlying assumption of normality of the applied statistics is met. The sample size (two or larger) is a function of cost and other considerations, such as ease of measurement, whether the test is destructive, and the required sensitivity of the control charts. As the sample size increases, the standard deviation decreases; therefore, the control limits will become tighter and more sensitive to process variation.
3. For each sample calculate the sample average, $\bar{X}$, and the sample **range**. For each sample, record any unusual settings (e.g., new operator, problem with raw material) that may cause an out-of-control condition.
4. After about 20 to 30 subgroups have been collected, calculate

$$\bar{\bar{X}} = \frac{\sum \bar{X}}{g}; \quad \bar{R} = \frac{\sum R}{g}$$

where $\bar{\bar{X}}$ is the average of averages, $\bar{R}$ is the average of range, and g is the number of samples or subgroups.
5. Trial upper and lower control limits for the $\bar{X}$ and R chart are calculated as follows:

$$\mathrm{UCL}_{\bar{X}} = \bar{\bar{X}} + A_2\bar{R}; \quad \mathrm{UCL}_R = D_4\bar{R}$$
$$\mathrm{LCL}_{\bar{X}} = \bar{\bar{X}} + A_2\bar{R}; \quad \mathrm{LCL}_R = D_3\bar{R}$$

A_2, D_3, and D_4 are constants and are functions of sample sizes used. These constants are used to approximate process standard deviation from the range. Tables of these constants are provided in Banks [1989], De Vor et al. [1992], Grant and Leavenworth [1988], and Montgomery [1991].
6. Plot the sample averages and ranges on the $\bar{X}$ and the R chart, respectively. Any out-of-control point that has an assignable cause (new operator, etc.) is discarded.
7. Calculate the revised control limits as follows:

$$\bar{X}_o = \frac{\Sigma\bar{X} - \Sigma\bar{X}_d}{g - g_d}; \quad R_o = \frac{\Sigma R - \Sigma R_d}{g - g_d}; \quad \sigma_o = \frac{R_o}{D_2}$$

$$\mathrm{UCL}_{\bar{X}_o} = \bar{X}_o + A\sigma_o \quad \mathrm{UCL}_R = D_2\sigma_o$$

$$\mathrm{LCL}_{\bar{X}_o} = \bar{X}_o - A\sigma_o \quad \mathrm{LCL}_R = D_1\sigma_o$$

The subscripts o and d stand for revised and discarded terms, respectively. The revised control charts will be used for the next production period by taking samples of the same size and plotting the sample average and sample range on the appropriate chart. The control limits will remain in effect until one or more factors in the process change. Figure 181.4 shows control charts of $\bar{X}$ and R values for ten subgroups. Each subgroup contained five observations because none of the ten data points lie outside of either upper and lower control limits; the process is designated "in control."

The control charts can be used to monitor the out-of-control conditions of the process. It is imperative to realize that patterns of variation as plotted on the charts should give clear indications to a process that

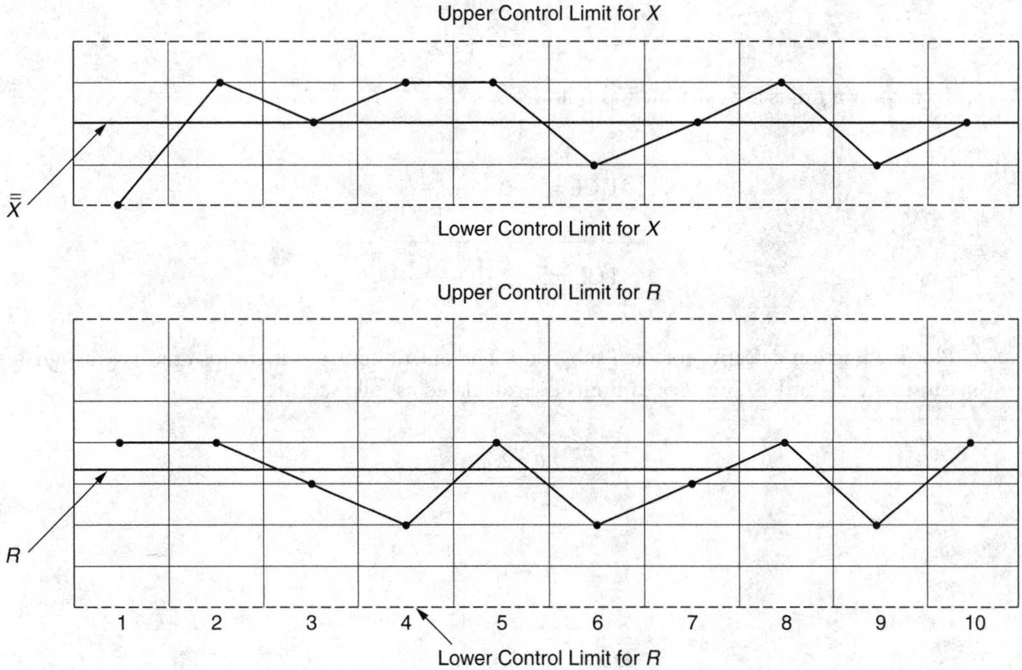

FIGURE 181.4 Control charts for $\bar{X}$ and R.

is headed for an out-of-control condition or one that displays an abnormal pattern of variations. Whereas no point may actually fall out of the limits, variation patterns can often point to some unusual process behavior that requires careful study of the process.

Control Charts for Attributes

For those quality characteristics that are not easily measured — or in such cases where count of defects of defective items are involved or go-no-go gages are used — control charts for attributes are used. These charts can be grouped into two major categories:

- Charts for defectives or nonconforming items
- Charts for defects or nonconformities

Charts for Nonconforming Items

The basic charts in this group are the fraction **nonconforming** chart (*p* chart), percent nonconforming chart (100*p* chart), and count of nonconforming chart (*np* chart). The procedure for the construction, revision, and the interpretation of control charts for attributes is similar to that for $\bar{X}$ and *R* charts. The following steps may be used to construct a *p* chart:

1. Once sample size has been established, fraction nonconforming, *p*, is determined for each sample by

$$p = \frac{np}{n}$$

   where *n* is the sample size and *np* is the count of defectives or nonconforming items in the sample.
2. After roughly 20 to 30 subgroups have been collected, calculate $\bar{p}$, the value of the central line, or the average fraction defective.

$$\bar{p} = \frac{\Sigma np}{\Sigma n}$$

3. Trial control limits are calculated as follows:

$$\text{UCL} = \bar{p} + 3\sqrt{\frac{\bar{p}(1-\bar{p})}{n}}$$

$$\text{LCL} = \bar{p} - 3\sqrt{\frac{\bar{p}(1-\bar{p})}{n}}$$

4. Plot the fraction defective for each subgroup. The out-of-control subgroups that have assignable causes are discarded, and revised limits are calculated as follows:

$$p_o = \frac{\Sigma np - \Sigma np_d}{\Sigma n - n_d}$$

$$\text{UCL} = p_o + 3\sqrt{\frac{p_o(1-p_o)}{n}}$$

$$\text{LCL} = p_o - 3\sqrt{\frac{p_o(1-p_o)}{n}}$$

5. If the lower control limit is a negative number, it is set to zero. Sample points that fall above the upper limit indicate a process that is out of control. However, samples that fall below the lower limit, when the lower control limit is greater than zero, indicate a product that is better than expected. In other words, if a sample contains fewer nonconforming items than the process is capable of producing, the sample fraction defective will fall below the lower control limit. For this reason, some practitioners may choose to set the lower limit of the attribute charts to zero. This practice, however, may mask other problems or potentials for process improvements.

Other charts for nonconforming items are simple variations of the p chart. In the case of the 100p chart, all values of the p chart are expressed as percentages. In the case of the np chart, instead of plotting fraction or percent defectives, actual counts of nonconforming or defective items are plotted. See Banks [1989], De Vor et al. [1992], Grant and Leavenworth [1988], and Montgomery [1991] for greater detail. The formulas for the central line and the control limits for an np chart are given below. It is assumed that the revised value for universe fraction defective, p, is known. If p is not known, then the procedure for the p chart must be carried out to determine the revised value for the universe fraction defective:

$$\text{Central line} = np_o$$

$$\text{Control limits} = np_o \pm 3\sqrt{np_o(1-p_o)}$$

where n is the sample size and p_o is the universe fraction defective.

Charts for Defects or Nonconformities

Whereas the charts for defective or nonconforming items are concerned with the overall quality of an item or sample, charts for defects look at each defect (i.e., blemish, scratch, etc.) in each item or sample. One may consider an item a nonconforming item based on its overall condition. A defect or **nonconformity** is that condition that makes an item a nonconforming or defective item.

In this category are *c* charts and *u* charts. The basic difference between the two is the sample size. The sample size, *n*, for a *c* chart is equal to one. In this case the number of nonconformities or defects are counted per a single item. For a *u* chart, however, $n > 1$. See Banks [1989], De Vor et al. [1992], Grant and Leavenworth [1988], and Montgomery [1991] for the formulas and construction procedures.

181.3 Tolerances and Capability

As stated earlier, *process capability* refers to the range of process variation that is due to chance or natural process deviations. This was defined as $\bar{X} \pm 3\sigma$ (also referred to as 6σ), which is the expected or natural process variation. *Specifications* or *tolerances* are dictated by design engineering and are the maximum amount of acceptable variation. These specifications are often stated without regard to process spread. The relationships between the process spread or the natural process variation and the engineering specifications or requirements are the subject of process capability studies. Process capability can be expressed as:

$$C_p = \frac{\text{US} - \text{LS}}{6\sigma}$$

where

C_p = process capability index
US = upper engineering specification value
LS = lower engineering specification value

A companion index, C_{pk}, is also used to describe process capability, where

$$C_{pk} = \frac{\text{US} - \bar{X}}{3\sigma}$$

or

$$C_{pk} = \frac{\bar{X} - \text{LS}}{3\sigma}$$

The lesser of the two values indicates the process capability. The C_{pk} ratio is used to indicate whether a process is capable of meeting engineering tolerances and whether the process is centered around the target value $\bar{X}$. If the process is centered between the upper and the lower specifications, C_p and C_{pk} are equal. However, if the process is not centered, C_{pk} will be lower than C_p and is the true process capability index. See De Vor et al. [1992], Grant and Leavenworth [1988], and Montgomery [1991] for additional information.

A capability index less than one indicates that the specification limits are much tighter than the process spread. Hence, although the process may be in control, the parts may well be out of specification. Thus, the process does not meet engineering requirements. A capability index of one means that as long as the process is in control, parts are also within specifications. The most desirable situation is to have a process capability index greater than one. In such cases, not only are approximately 99.73% of the parts within specifications when the process is in control, but even if the process should go out of control, the product may still be within the engineering specifications. Process improvement efforts are often concerned with reducing the process spread and, therefore, increasing the process capability indices.

An extremely powerful tool for isolating and determining those factors that significantly contribute to process variation is statistical design and analysis of experiments. Referred to as "design of experiments," the methodology enables the researcher to examine the factors and determine how to control these factors in order to reduce process variation and therefore increase process capability index. For greater detail, see Box et al. [1978].

Defining Terms

Assignable causes — Any element that can cause a significant variation in a process.

Frequency distribution — Usually a graphical or tabular representation of data. When scores or measurements are arranged, usually in an ascending order, and the occurrence (frequency) of each score or measurement is also indicated, a frequency distribution results. The frequency distribution in the following table indicates that ten samples were found containing zero defectives.

No. of Defectives	Frequency
0	10
1	8
2	7
3	8
4	6
5	4
6	2
7	1
8	1
9	0

Nonconforming — A condition in which a part does not meet all specifications or customer requirements. This term can be used interchangeably with *defective.*

Nonconformity — Any deviation from standards, specifications, or expectation; also called a *defect.* Defects or nonconformities are classified into three major categories critical, major, and minor. A critical nonconformity renders a product inoperable or dangerous to operate. A major nonconformity may affect the operation of the unit, whereas a minor defect does not affect the operation of the product.

Population — An entire group of people, objects, or phenomena having at least one common characteristic. For example, all registered voters constitute a population.

Quality — Quality within the framework of manufacturing is defined as conformance to requirements.

Range — A measure of variability or spread in a data set. The range of a data set, R, is the difference between the highest and the lowest values in the set.

Sample — A small segment or subgroup taken from a complete population. Because of the large size of most populations, it is impossible or impractical to measure, examine, or test every member of a given population.

Specifications — Expected part dimensions as stated on engineering drawings.

Standard deviation — A measure of dispersion or variation in the data. Given a set of numbers, all of equal value, the standard deviation of the data set would be equal to zero.

Tolerances — Allowable variations in part dimension as stated on engineering drawings.

References

American Society for Quality Control (ASQC). 1994. *Malcolm Baldrige National Quality Award — 1994 Award Criteria*. ASQC, Milwaukee, WI.

Banks, J. 1989. *Principles of Quality Control.* John Wiley & Sons, New York.

Box, G. E. P., Hunter, W. G., and Hunter, J. S. 1978. *Statistics for Experimenters.* John Wiley & Sons, New York.

De Vor, R. E., Chang, T. H., and Sutherland, J. W. 1992. *Statistical Quality Design and Control.* Macmillan, New York.

Grant, E. L. and Leavenworth, R. S. 1988. *Statistical Quality Control*, 6th ed. McGraw-Hill, New York.
Lamprecht, J. L. 1993. *Implementing the ISO 9000 Series.* Marcel Dekker, New York.
Montgomery, D. C. 1991. *Statistical Quality Control*, 2nd ed. John Wiley & Sons, New York.
Saylor, J. H. 1992. *TQM Field Manual.* McGraw-Hill, New York.

Further Information

Statistical Quality Control, by Eugene Grant and Richard Leavenworth, offers an in-depth discussion of various control charts and sampling plans.

Statistics for Experimenters, by George Box, William Hunter, and Stewart Hunter, offers an excellent and in-depth treatment of design and analysis of design of experiments for quality improvements.

Most textbooks on statistics offer detailed discussions of the central limit theorem. *Introduction to Probability and Statistics for Engineers and Scientists*, written by Sheldon Ross (John Wiley & Sons, New York, 1987) is recommended.

The American Society for Quality Control (P.O. Box 3005, Milwaukee, WI 53201-3005, phone: (800)248-1946) is an excellent source for reference material, including books and journals, on various aspects of quality.

182

Flexible Manufacturing

Andrew Kusiak
University of Iowa
Chang-Xue (Jack) Feng
Bradley University

Flexible manufacturing systems (FMSs) have emerged to meet frequently changing market demands for products. The distinguishing feature of an FMS is in its software (e.g., numerical control programs) rather than the hardware (using, for example, relays and position switches). Fixed automation, programmable automation, and flexible automation are the three different forms of automation [Groover, 2001]. Numerical control (NC) and computer numerical control (CNC) technologies are essential in FMSs. Developments in robotics, automated guided vehicles (AGVs), programmable controllers (PCs), computer vision, group technology (GT), and statistical quality control (SQC) have accelerated the applications of FMSs. The computer technology (hardware and software) is a critical factor in determining the performance of flexible manufacturing systems. An FMS is frequently defined as a set of CNC machine tools and other equipment that are connected by an automated material-handling system, all controlled by a computer system [Askin and Standridge, 1993].

The FMS concept has been applied to the following manufacturing areas [Kusiak, 1986]: (1) fabrication, (2) machining, and (3) assembly. An FMS can be divided into three subsystems [Kusiak, 1985]: (1) management system, (2) production system, and (3) material-handling system. The management system incorporates computer control at various levels; the production system includes CNC machine tools and other equipment (e.g., inspection stations, washing stations); the material handling system includes AGVs, robots, and automated storage/retrieval systems (AS/RSs).

The planning, design, modeling, and control of an FMS differs from the classical manufacturing system. The structure and taxonomy of an FMS has been studied in Kusiak [1985]. Some planning tools — for example, IDEF (the U.S. Air Force ICAM definition language) [U.S. Air Force, 1981] — have been proposed for modeling of material flow, information flow, and dynamic (simulation) modeling of an FMS. Numerous mathematical models of FMSs are discussed in Askin and Standridge [1993]. An FMS can be structured as a five-level hierarchical control model [Jones and McLean, 1986] (see Table 182.1).

Manufacturing concepts such as just-in-time (JIT) manufacturing, lean manufacturing, time-based manufacturing, quick response manufacturing, synchronous manufacturing, and agile manufacturing relate to the flexible manufacturing approach. Here the term *flexible* describes the system's ability to adjust to customers' preferences; JIT reduces the throughput time and the level of inventory; "lean" stresses efficiency and cost reduction; and "agile," "time-based," "quick response," and "synchronous" all address the speed of a manufacturing system in responding to frequently changing market demands. Other terms such as flexible manufacturing cells (FMCs), computer-aided manufacturing (CAM), and computer-integrated manufacturing (CIM) are frequently used to describe flexible manufacturing.

0-8493-1586-7/05/$0.00+$1.50

TABLE 182.1 FMS Control Architecture (National Institute of Standards and Technology)

Level	Planning Horizon	Functions
1. Facility	Months to years	Information management Manufacturing engineering Production management
2. Shop	Weeks to months	Task management Resource allocation
3. Cell	Hours to weeks	Task analysis Batch management Scheduling Dispatching Monitoring
4. Workstation	Minutes to hours	Setup Equipment sequencing In-process inspection
5. Equipment	Milliseconds to minutes	Machining Measurement Handling Transport Storage

Source: Jones, A. and McLean, C. 1986. A proposed hierarchical control model for automated manufacturing systems. *J. Manufacturing Syst.*, 5:15–25.

182.1 Flexible Machining

The basic features of a flexible machining system (FMS), as well as the differences between a flexible machining system and the equivalent classical machining system, are presented in Kusiak [1990] in the form of eight observations. These observations are based on the analysis of over 50 flexible machining systems.

An Example of FMS

Figure 182.1 illustrates one type of flexible machining system (FMS). The FMS produces parts for a family of DC motors used in the FANUC CNC system. The parts are of two types: rotational (e.g., a motor shaft) and prismatic (e.g., a supporting house). The production system includes a CNC turning machine tool, a CNC grinding machine tool, a vertical machining center, and a horizontal machining center, as well as two coordinate-measuring machines (CMMs). Programmable controllers (PCs) are also used in the CNC control system to perform some auxiliary functions. Quite frequently, a PC can be used as an independent CNC control device in a position control mode (as opposed to a path control mode), for example, in a CNC drilling machine to directly drive the servo motor. A CMM measures automatically the precision of the machined parts, including their nominal values and tolerances. It includes a moving head, a probe, a microcomputer, and input/output (I/O) devices. It is controlled by an NC program, and it feeds the measured data to the corresponding CNC system and records them for the purpose of statistical quality control (SQC). The essential function of the SQC system in an FMS is to analyze the data from the on-line measuring of the CNC machine or CMM and to construct a set of X-bar and R charts, which shows the measured values in their means and ranges based on a predetermined sample size and grouping. These mean and range values are then compared to the designed nominal value and control limits, respectively, to determine whether the process is under statistical control.

The automated material handling system in Figure 182.1 includes an AGV, three robots, two part buffers, and the automated storage and retrieval system (AS/RS). The function of the AGV in Figure

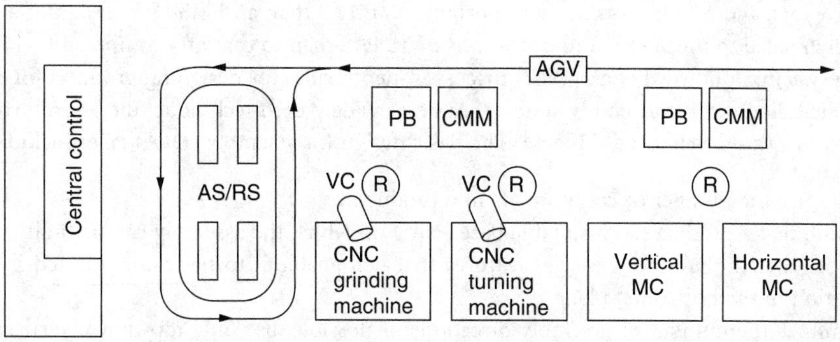

FIGURE 182.1 Physical layout of a flexible machining system. AS/RS: automated storage and retrieval system; PB: part buffer; CMM: coordinate-measuring machine; R: robot; VC: video camera; MC: machining center.

182.1 is to transfer loads to remote locations following a guided path. For the long-distance handling in this FMS, a robot cannot provide the mobility of an AGV, and a conveyor does not offer the flexibility desired. The AGV sends the raw material from the AS/RS to the part buffer of each workstation and brings back the finished products from the part buffers to the AS/RS. An industrial robot is a reprogrammable, multifunctional manipulator designed to move materials, parts, tools, or special devices through variable programmed motions for the performance of a variety of tasks (Robotics Industries Association). In the system in Figure 182.1, the three robots load parts from part buffers to each CNC machine tool, transfer parts from the CNC machine tools to the CMMs, and unload parts from machines to the part buffers. The AS/RS is used to temporarily store the raw material and finished product. Although the AGV and the robots in the system in Figure 182.1 function separately, it is possible for them to work jointly to provide both mobility and flexibility. For example, an AGV can be interfaced with a robot to provide more flexible loading and unloading capability, and a robot can be mounted on an AGV to achieve mobility.

The control and management system includes the central control minicomputer, the microcomputer resided in each CNC machine and robot, and two industrial video cameras. The computer control system performs two fundamental functions: monitoring and control. For example, it stores process plans and NC programs; dispatches programs to the individual pieces of equipment; controls the production, traffic, tools, and quality; monitors the material handling system; and monitors and reports the system performance. The video camera (VC) monitors the critical machining processes. A VC may or may not have the capability to analyze and process the data collected. In the former case, it can process the image based on computer vision technology and inform the CNC system; in the latter case, a human expert monitors the image taken by the VC and then manually informs the corresponding CNC system. Other types of monitoring might be used in an FMS, for example, measuring the cutting force, torque, or power. In some types of monitoring, a measuring head may be placed in the tool magazine and used to measure the machining quality of certain operations as required. The use of the manufacturing automation protocol (MAP) in manufacturing is reaching a momentum. The equipment purchased from different vendors is linked through a shop local-area network (LAN), as the vendors tend to adopt the MAP standard. Tool management is one of the functions of the computer control system [Kusiak, 1990].

182.2 Flexible Assembly

Most of the observations regarding the flexible machining systems discussed in Kusiak [1990] apply to the flexible assembly systems (FASs) as well. Any differences that may arise are of terminology rather than concept. The earliest FASs were designed for assembling printed circuit boards. More recent FASs have been developed for mechanical assembly. The automated material-handling system — including

AGVs, conveyors, and robots — is more important in an FAS than an FMS. The design of a product is more closely related to the design and operations of an FAS than to that of a component with a flexible machining system. Boothroyd et al. [1982] discuss in-depth rules for designing products for automated assembly, including flexible assembly systems. Another widely used technique, the assembly evaluation method (AEM), was developed at Hitachi. The basic design for assembly (DFA) rules include:

1. Minimize the number of components in a product.
2. Ensure that a product has a suitable base (part) on which the assembly can be built.
3. Ensure that the product base has features that will enable it to be readily located in a suitable position in the horizontal plane.
4. Minimize the number of assembly directions; if possible, use only top-down, vertical assembly motions.
5. If possible, design the product so that it can be assembled in layers.
6. Provide chamfers or tapers to help the component be easily positioned.
7. Avoid expensive and time-consuming fastening operations and avoid using assembly tools (e.g., screwdrivers).
8. Design the product so that the assembly does not have to be lifted or rotated, since the lifting and rotating results in complicated fixtures and grippers, more degrees of freedom of the robot, and increased cycle time.
9. Design the product to simplify packaging. The more uniform and simple the product packaging is, the easier it will be to apply a robotic packaging station that uses standard carts and pallets.
10. Use symmetric features whenever possible; otherwise, exaggerate asymmetry to facilitate identification and orientation of components.
11. Design a component to avoid tangling when feeding.
12. Design product variations to allow common handling and assembly methods.

An example of mechanical product design for flexible assembly is presented in Elmaraghy and Knoll [1989]. The authors discuss design for flexible assembly of a family of DC motors and the design of the flexible assembly system.

The IBM Proprinter is a frequently cited example of an electronic product that was designed for flexible automated assembly. The goals for the printer design were relatively simple:

- Develop a modular design.
- Minimize the total number of parts.
- All parts are to be self-aligning for ease of assembly.

Other principles of design for assembly that were used in this example are as follows:

- No fasteners
- No springs
- No pulleys
- No excess modules
- No adjustment
- No labels
- No paint
- No extra parts
- No multidirectional assembly
- No alignment/location tools or fixtures
- No external tests
- No engineering change notices (ECNs)
- No custom builds or options
- No manual assembly

182.3 The Economic Justification of Flexibility

Flexibility is a critical objective in design, planning, and operating an FMS. An FMS is intended to accommodate a large number of types of products and allow for more changes of products and higher equipment utilization, whereas in lean and agile manufacturing, flexibility means not only more changes and more product types, but also quicker changes, plus a reasonable rather than ultimately high utilization. Although none of the definitions of manufacturing flexibility have been generally accepted, the following types of flexibility are frequently considered: machine flexibility, process flexibility, product flexibility, routing flexibility, volume flexibility, expansion flexibility, process sequence flexibility, operation flexibility, and production flexibility. How to price the product produced in an FMS causes difficulties and often creates conflicts between accounting and engineering departments. There is some evidence that the U.S. and Japan operate FMSs in radically different ways. Jaikumar [1986] shows that Japanese users had achieved at the time greater flexibility and yet shorter development time than users in the U.S.

References

Askin, R. and Standridge, C. 1993. *Modeling and Analysis of Manufacturing Systems.* John Wiley & Sons, New York.

Boothroyd, G., Poli, C., and Murch, L. 1982. *Automatic Assembly.* Marcel Dekker, New York.

Elmaraghy, H. and Knoll, L. 1989. Flexible assembly of a family of DC motors. *Manuf. Rev.*, 2:250–256.

Groover, M., 2001. *Automation, Production Systems, and Computer-Integrated Manufacturing.* Prentice Hall, Upper Saddle River, New Jersey.

Jaikumar, R. 1986. Post-industrial manufacturing. *Harv. Bus. Rev.*, 86:69–76.

Jones, A. and McLean, C. 1986. A proposed hierarchical control model for automated manufacturing systems. *J. Manuf. Syst.*, 5:15–25.

Kusiak, A. 1985. Flexible manufacturing systems: a structural approach. *Int. J. Prod. Res.*, 231057–1073.

Kusiak, A. 1986. Parts and tools handling systems. In *Modelling and Design of Flexible Manufacturing Systems*, A. Kusiak, Ed. Elsevier, New York, pp. 99–110.

Kusiak, A. 1990. *Intelligent Manufacturing Systems.* Prentice Hall, Englewood Cliffs, NJ.

U.S. Air Force. 1981. *U.S. Air Force Integrated Computer Aided Manufacturing (ICAM) Architecture Part II, Volume IV — Functional Modeling Manual (IDEF0).* IFWAL-tr-81-4023. Air Force Materials Laboratory, Wright-Patterson AFB, OH.

Further Information

The *Handbook of Flexible Manufacturing Systems* published by Academic Press in 1991 provides a good overview of various aspects of FMSs from theoretical and application point of view.

The *International Journal of Flexible Manufacturing Systems* (Kluwer) publishes papers in design, analysis, and operations of flexible fabrication and assembly systems.

The *Journal of Intelligent Manufacturing* (Chapman and Hall) emphasizes the theory and applications of artificial intelligence in design and manufacturing.

A comprehensive review of existing FMS models and future directions is provided in A. Gunasekran, T. Martikainen, and P. Yli-Olli. 1993. Flexible manufacturing systems: an investigation for research and applications. *Eur. J. Operational Res.*, 661–26.

For up-to-date information on flexible manufacturing, see R. Dorf and A. Kusiak, Eds., *Handbook of Design, Manufacturing and Automation* (John Wiley & Sons, New York, 1994).

Classic readings in just-in-time manufacturing include: (1) *Toyota Production Systems, Beyond Large-Scale Production*, by Taiichi Ohno, former vice president of Toyota Motors Company (Productivity Press, Portland, OR, 1988); (2) *Toyota Production Systems: An Integrated Approach to Just-in-Time*, 3rd ed., by Yasuhiro Monden (IIE Engineering & Management Press, Norcross, GA, 1996); (3) *Non-Stock Production: The Shingo System for Continuous Improvement*, by Shigeo Shingo (Productivity Press, Portland, OR,

1988); and (4) *Japanese Manufacturing Techniques: Nine Hidden Lessons in Simplicity*, by Richard J. Schonberger (Free Press, New York, 1982).

The *Machine that Changed the World — The Story of Lean Production* (HarperCollins, New York, 1991), by J. Womack, D. Jones and D. Roos of the then MIT International Automobile Program, coined the term *lean production*, and discusses the differences between the mass production and lean production. James Womack and Daniel Jones then followed up with *Lean Thinking: Banish Waste and Create Wealth in Your Corporation* (Simon & Schuster, New York, 1996), to address how to implement lean manufacturing and services. *Agile Competitors and Virtual Organization* (Van Nostrand Reinhold, New York, 1995) by S. Goldman, R. Nagel, and K. Preiss of the Lehigh University Lee Iacocca Institute, coined the term *agile manufacturing* and *mass customization*.

Competing against Time: How Time-Based Competition Is Reshaping Global Markets (Free Press, New York, 1990), by George Stalk, Jr. and Thomas M. Hout of the Boston Consulting Group, coins the term *time-based manufacturing*. *Quick Response Manufacturing: A Company Wide Approach to Reducing Lead Times*, by Rajan Suri (Productivity Press, Portland, OR, 1998), coined the term *quick response manufacturing*. *The Goal* by Eliyahu M. Goldratt and Jeff Cox (North River Press, Great Borrington, MA, 1984), has set the stage for synchronous manufacturing, and *The Theory of Constraints* by Eliyahu M. Goldratt (North River Press, Great Borrington, MA, 1990), coined the term *theory of constraints*. The Big Three automobile companies in Detroit later termed Goldratt's theory of constraints *synchronous manufacturing*.

In 1987, the Society of Manufacturing Engineers (SME) published *Automated Guided Vehicles and Automated Manufacturing* by R. Miller. It discusses AGV principles and their applications in FMSs.

Flexible Assembly Systems by A. Owen was published in 1984 (Plenum Press, New York). In 1987, G. Boothroyd and P. Dewhurst published *Product Design for Assembly Handbook* (Boothroyd Dewhurst, Inc.), which presents comprehensive design for assembly rules and examples. A commercial software package is also available. For design of electronic products assembly, see S. Shina, *Concurrent Engineering and Design for Manufacture of Electronic Products* (Van Nostrand Reinhold, New York, 1991).

183

Managing for Value

Edward M. Knod, Jr.
Western Illinois University

As careers progress, engineers assume an increasingly diverse set of managerial responsibilities. Success in meeting those responsibilities demands — at the personal level — two things: First, the engineer-manager must possess an appropriate set of managerial skills and requisite implementation abilities. Such talents are broadly applicable. With periodic refresher training and situation-specific fine-tuning, they are useful in nearly any setting. Second, the individual must understand the road to achieving organizational success. Defining characteristics mark well-managed companies. These are not so much *outcome* characteristics as they are *process* characteristics; they describe the ways that managers and employees view things and get things done — the manner in which successful outcomes are achieved. In a nutshell, the focus is on *value*; specifically, the creation of value-adding operations in value-adding companies that create and deliver goods and services valued by society.

This chapter begins with an overview of contemporary management thought and practice, continues with a discussion of value as a core concept that underlies management actions in commercial ventures, presents and discusses a set of proven principles to guide managing-for-value activities, and concludes with suggested sources of additional information.

183.1 Management: Fundamental Concepts

Engineers often experience a sense of frustration with the imprecise or inexact nature of the "science" that they encounter during an initial foray into management studies. Indeed, *management* — in theory as well as in practice — has two "halves." One is more objective, measurable, or tangible; often, its prescriptions have met the rigors of repeated field testing. The other is "softer," more subjective, and less established, perhaps; it is less tangible and less measur*able*. Engineers must not (*can*not) avoid the latter, but they might need to adopt a different perspective. On the one hand, they gain a deeper appreciation for the challenges faced by social scientists who must venture into the murky waters of research on human behavior; perhaps new cross-professional alliances emerge. On the other hand, the engineer must cling to a healthy skepticism until knowledge can be separated from conjecture.

0-8493-1586-7/05/$0.00+$1.50

An overview of management in contemporary competitive manufacturing and service organizations might first consider managers themselves. Managerial duties and activities, types and effects of management roles, requisite managerial skills and attributes, and leadership are key dimensions.

Duties and Activities

Briefly, the primary duty of management is to ensure organizational success in the creation and delivery of goods and services. Provide outputs that customers value and provide them in ways that keep customers coming back for more. Popular definitions of management often employ lists of activities that describe what managers do. Some activities — decision making, motivating, and communicating, for example — are quite generic; their importance is recognized by the widespread availability of training programs on those topics. One's ability to perform these activities is highly transferable from one job or company to another.

Other management activities tend to be more setting specific; analysis and description within the job or organizational context can be more revealing. These activities fall into one or more of three general, and overlapping, duties: creating, implementing, and improving.

- *Creating*. Activities such as planning, designing, staffing, budgeting, and organizing accomplish the creativity required to build and maintain customer-serving capacity. Product and process design, facility planning and layout, workforce acquisition and training, and materials and component sourcing are among the tasks that have a substantial creative component.
- *Implementing*. When managers authorize, allocate, assign, schedule, or direct, emphasis shifts from creating to implementing — putting a plan into action. A frequent observation is that the biggest obstacle to successful implementation is lack of commitment to the plan. Therefore, teaching, motivating, and mentoring are common activities exhibited by managers as plans are put into action. During implementation, managers also perform controlling activities — that is, they monitor performance and make necessary adjustments.
- *Improving*. Environmental changes (e.g., new or revised customer demands, challenges from competitors, and social and regulatory pressures) necessitate improvements in output goods and services. In response to — or better yet, in anticipation of — those changes, managers re-create; they start the cycle again with revised plans, better designs, new budgets, and so forth.

Outcomes, desirable or otherwise, stem from these activities. A goal-oriented manager might describe a broad aim as "faster product development," or perhaps a more immediate and focused objective as "reduce routing mis-reads by 20%," but he or she will try to attain that goal by employing various activities aimed at creating, implementing, and improving.

Management Roles

During the 20th century, theorists developed various models of management; their efforts usually driven by some perceived deficiencies with previous models. Quinn et al. [2003] summarize four of the more popular of those models (rational goal model, internal process model, human relations model, and open systems model) and argue that all have a place in contemporary management practice. They suggest movement away from an "either-or" mentality toward a more accommodating recognition that managers do face competing values and must employ the best of various approaches as they strive to attain what might appear to be conflicting objectives. In meeting those aims, managers may expect to assume the roles of mentor, facilitator, monitor, coordinator, director, producer, broker, and innovator; each role comes with its own set of required competencies.

A great many scholars cast managerial role expectations in a less generic fashion; they prefer to address roles within the context of the task at hand. Often they note potential negative as well as positive role expectations. For example, Nicholas [1998] gives specific attention to managerial roles in such activities as implementing cellular manufacturing, just-in-time (JIT), total preventive maintenance (TPM), and total quality management (TQM). Other writers point out unhealthy effects — stress and burnout, for example — that might be tied to role-related aspects of managers' jobs. Role overload, role underload, role ambiguity, and role conflict have drawn attention [Aldag and Kuzuhara, 2002: chapter 9].

Requisite Skills and Attributes

Exact requirements are hard to pin down, but any skill or attribute that helps a manager make better decisions is desirable. When various lists of generic skills are compared, considerable (and not unexpected) similarities emerge. Some skills and traits (e.g., good time-management habits, effective communications skills, and pleasant personality) are obviously useful and serve managers in any job. Categorization of skills is helpful, but often reflects the background or discipline of the list maker. Aldag and Kuzuhara [2002], for example, categorize needed management skills as personal, interpersonal, or managerial. Bateman and Zeithaml [1993] propose a slightly more revealing classification; they suggest that managers need technical skills, interpersonal and communications skills, and conceptual and decision skills.

Extension of these broad categories into job-specific lists is perhaps unwarranted given the current emphasis on cross-functional career migration and assignments to interdisciplinary project teams or product groups. On the other hand, recent innovations, business trends, and newsworthy events have served to bring newer types of skills into play and to add renewed emphasis to a few "older" ones. Advancements in electronic business (e-business) or commerce (e-commerce) and the impact on integrated supply chain management (SCM) efforts, for instance, bump up the need for managers to possess greater computer and information system literacy [Chopra and Meindl, 2001: chapter 14]. In a similar fashion, increases in global business ventures place emphasis on foreign language skills and knowledge of foreign cultures. And, recent turmoil in stock markets and revelations of large-scale improprieties serve to focus attention on specific attributes that relate to personal and business ethics and on skills that suggest possession of sound business acumen.

The lasting power of the importance of these newly emphasized skills is unknown. The mandate for managers to possess and exhibit leadership skills, however, emerged early in the 20th century and has proven resilient.

Leadership

Leadership is one of the most heavily studied topics in management and organizational research. But noted leadership scholar Bernard M. Bass [1990] admits that, "There are almost as many different definitions of leadership as there are persons who have attempted to define the concept." Such a statement cannot be expected to sit well with engineers who are accustomed to much closer agreement on definitions of key concepts. Bass goes on to argue, however, that sufficient similarity (among definitions) exists to permit a tentative classification scheme, and he proceeds to enumerate 11 separate conceptual foundation definitions of leadership along with numerous hybrids [Bass, 1990]. Clearly, much work remains before leadership moves very far from the realm of conjecture. (Leadership is one of those "soft" topics mentioned at the beginning of this section.)

We *do* know that effective leaders are both task oriented and people oriented; they take the mission seriously and they take care of their people. We also know that managers are expected to be leaders, and that successful managers are often described — by peers as well as by subordinates — as being good leaders. But, leadership researchers maintain that leadership and management are not the same things. Stephen R. Covey summarizes the distinction made by this camp by noting, "Leadership focuses on doing the right things; management focuses on doing things right" [Covey, 1996]. A fair critique might suggest that beyond the two traits noted at the outset of this paragraph, effective leadership skills are largely situation specific.

Engineers who wish to delve into the subject of leadership will find a rich assortment of materials. Wren [1995] provides insight into the history of leadership study; Kouzes and Posner [2003] offer a practical guide on leadership for everyday application; and Hesselbein, Goldsmith, and Beckhard [1996] offer a collection of essays on leadership composed by leading authorities.

183.2 Value: Evolution of a Concept

Value is not a new concept, nor is the recognition of value enhancement as a worthy goal. Value analysis and value engineering programs keep value as a primary focus; efforts to build world-class excellence also place a high premium on value; and value-chain analysis and the study of value migration also, as their names imply, revolve around the concept of value. These and related activities have resulted in a set of characteristics that serve to describe value-oriented organizations.

Value Analysis and Value Engineering

Value analysis (VA) programs originated in the late 1940s. Originally limited to purchased items, value analysis proved its merit and was subsequently applied to in-house designs under the title value engineering (VE). The buyer or customer assigns value to a good, a service, an activity or a process; value is the worth the customer imputes. Value, therefore, is a good indicator of what the customer is willing to pay, and as such, serves as a practical (market) limit on the provider's costs. VA/VE programs seek to increase value by increasing the performance-to-cost ratio.

Function statements lie at the heart of VA/VE. In standard verb–noun format, a vessel may be said to "contain fluid." If pressure requirements exist, the provider may add the function "sustain pressure." A customer might also want the vessel to "insulate fluid" and could be expected to pay more for the addition of that third performance function — recognition of value added. (Each performance function statement, of course, is clarified with operational performance specifics.) Formal VA/VE strives to ensure that the proposed vessel provides requested functions and can be delivered at a low price. As Nicholas [1998: chapter 2] notes, VA/VE often accompanies concurrent engineering and design for manufacture/assembly (DFMA) efforts. Two caveats close the VA/VE discussion:

- Customers hesitate to pay for unwanted functions; even marketing will not open some pocketbooks.
- Customer desire for performance functions can be unstable; a current requirement may disappear tomorrow. In that case, the first caveat applies.

World-Class Management

Notions about value and about how companies ought to go about providing value changed substantially during the 1980s and 1990s. By the end of the 1970s, an eclectic set of factors had combined to force managers to critically assess old assumptions about the overall capacity required to provide continuous improvement in meeting evolving customer needs. Conference room briefings dealt with emerging markets, shifting patterns of global competitiveness and regional dominance in key industries, the spread of what might be called Japanese management and manufacturing technologies, and the philosophy and tools of the total quality movement. Leading industrial engineering and operations management experts released novel (at that time) prescriptions for how managers ought to establish and maintain world-class excellence in their organizations [Schonberger, 1982; Hall, 1983; Shingo, 1985; Suzaki, 1987]. The principles and implementation techniques first appeared in manufacturing and continue to have profound influence in that sector. At this writing, however, considerable inroads into the service sector exist as well.

The Malcolm Baldrige National Quality Award, first awarded in 1988, helped give widespread verification to some of the beneficial aspects of total quality management. It also served as an impetus for North American managers to examine other quality-oriented programs and awards — Japan's Deming Prize, The European Quality Model, The Shingo Prize, and ISO 9000/QS 9000 registration efforts, for example. Customer-centered, multidimensional definitions of quality became the rule; training programs addressed simple but effective quality-improvement tools; and vigorous attacks on variation proliferated as meaningful performance indicators (e.g., capability indexes) grew in popularity. Perhaps most noteworthy, improvement programs expanded to incorporate quality of *process* in addition to quality of *product.*

Improved quality was the first but not the sole objective of the world-class management movement. Blackburn, for instance, provided a reminder that *time*, in addition to quality, could be a formidable competitive weapon [Blackburn, 1991]. Just-in-time (JIT) performance has received considerable attention [Schonberger, 1982; Schneiderjans, 1993], but the pervasiveness of time-related factors in the development of value also extends into the reduction of such factors as throughput times (cycle or lead time in some settings), queue times, setup or changeover times, and transport times as well [Knod and Schonberger, 2001: chapter 11]. Time–quality interdependencies emerged; valued performance traits such as responsiveness and flexibility — frequently cited as quality dimensions — hinge on effective time management. Better *and* faster are not mutually exclusive.

Better quality and timing outcomes address the numerator of the performance-to-cost ratio. But world-class management has been equally focused on the denominator — costs. Simply stated, any cost incurred in performing an activity that was not valued by customers (immediate or subsequent) could be recovered by elimination of the activity. As world-class ideas began to spread, respected voices from the accounting profession called for revisions in managerial accounting so that more accurate statements of *true* costs for both value-adding and non-value adding activities might be obtained [Johnson and Kaplan, 1987].

Unquestionably, the world-class excellence movement has resulted in much improved identification and delivery of outcomes that customers value. Moreover, it has prompted scrutiny of the value-adding chain itself.

Value Chain Analysis

Competition drives value chain analysis. The technique is applicable at the firm and industry level, at the product or product-family level, and at the activity (or operation) and process level. Porter [1985] suggests that a company may add value through both primary and support activities. Primary activities include logistics (inbound and outbound), operations, marketing, sales, and service. Procurement, technology, human resources, and administration are notable supporting activities in Porter's model. Value is reflected by total revenue, or what customers are willing to pay for the company's goods and services. So long as the company's overall business model remains viable, value exceeds costs, and profit results.

At the product or product group level, supply chain analysis is the popular term [Simchi-Levi et al., 2000; Chopra and Meindl, 2001]. In manufacturing, the role of suppliers as contributors to (or detractors from) value was quickly acknowledged as world-class ideas spread. Comparison of competing suppliers' functional performance levels, of course, dictates the degree to which potential customers acknowledge product or service differentiation and are willing to pay for it. And, of course, the pressure to contain costs lurks at each stage of the supply chain.

Value-chain mapping as a technique for removing waste from activities and processes receives strong endorsement from the Lean Enterprises Institute (*www.lean.org*). The technique draws heavily from process flow-charting, but also mandates a critical assessment of the value-added — from the perspectives of the next-process customer as well as each downstream customer — at each step (activity, operation, workstation, etc.) of the value stream. The formula is simple: steps that add no value are waste; eliminate them. Streamline remaining steps by cutting as much waste as possible. The result is a leaner process.

To sum up, analyses of value chains at all levels have revealed common features

1. Value chain elements (operations, processes, products, suppliers, and so forth) that add no value are targets for elimination. Customers are unwilling to cover the costs required to maintain them.
2. Those elements that do add value are opportunities for competitors. The possibility for those competitors to encroach with superior performance-to-cost ratios demands a vigilant program of continuous improvement.
3. Elements that are unable to maintain some degree of distinction in the eyes of customers will not survive. A distinctive or core competency — superior value delivered on one or more dimensions — is a must.
4. Regardless of initial business plan or value-chain design, value will migrate.

Value Migration

At the broadest level, the concept of value migration applies to business design competitiveness [Slywotzky, 1996]. What a company is worth, its *market value,* is dependent on how well the company's business design is perceived to be capable of supplying what its customers want. Value flows into companies with superior designs, remains stable when the design roughly matches customers' expectations, and flows from the company if competitors (old or new) present customers with better options.

Value *can* migrate out of a company when customers change what they want (value), or when they change where they expect to find that value along the value chain. A company may lose (or voluntarily surrender) a product or component that costs too much to make, a process that becomes obsolete, or a function (e.g., design, manufacturing, or distribution) that can be performed better, faster, and cheaper elsewhere, for instance. If the change, the loss of a component in this case, does not result in a lower perception of the company's worth, outflow value migration has not occurred. In fact, deliberate handing off (outsourcing) can be a strategic plus when accomplished, say, under a broader program of honing distinctive competencies. Schonberger [2001: chapter 10] provides a thorough discussion and examples. However, when a company is stripped of value, perhaps due to its own complacency, someone needs to answer.

Prevention of *all* unwanted value migration *may* not be possible. However, a company (or plant, or office, etc.) must maintain a critical competency level. One solution is to attain success at value inflow to offset value outflow. New processing competencies, first-to-market products, engineering innovations leading to significant product and service improvements, and new strategic alliances are possible avenues for increased value inflow. To increase likelihood of successful inflow of value, value-focused management is a plus.

Value-Focused Management

In the 1970s North American managers allowed production scheduling and control systems to become too complicated, cumbersome, and costly. Costs skyrocketed as back-office and staff personnel served increasingly cumbersome systems, often paying allegiance to mandated reports and controls to the detriment of keeping an eye on what customers valued. "Lean" was clearly missing.

A more competitive approach lies in the simplification of the production environments themselves [Knod and Schonberger, 2001: chapter 16; Steudel and Desruelle, 1992: chapter 8; Schneiderjans, 1993: chapter 6]. Although simplicity has proven an elusive goal, there is merit in its continued pursuit. An array of publications, seminars, and other vehicles for disseminating "how and why" advice has bolstered the spread of contemporary world-class management theory and research. The information bounty can be reduced, tentatively, to a set of core items that might be said to define worthy aims for value-focused managers.

Table 183.1 presents those core items as a short list of six broad operating guidelines. They reflect dominant themes in world-class management, require leadership to implement, and help define what today's successful value-focused managers are all about.

183.3 Managing-for-Value Principles

Table 183.2 presents the final and most detailed component of this general-to-specific look at contemporary management. It is a set of action-oriented, prescriptive principles for managing for value. Originally crafted by Schonberger [1986] in a previous version aimed more specifically at manufacturing, the principles have evolved, stood the test of time, and apply to service settings as well. The principles apply to managers at any level and define ways for increasing competitiveness — by taking direct action to enhance value — in ongoing operations. Brief supporting rationale and techniques or procedures that exemplify each principle appear in the right column; Knod and Schonberger [2001: chapter 3] provide

TABLE 183.1 Characteristics of Value-Focused Management

Customers at center stage. The customer is the next person or process — the destination of one's work. The provider–customer chain extends, process to process, on to final consumers. Whatever a firm produces, customers would like it to be better, faster, and cheaper; prudent managers therefore embrace procedures that provide *total quality, quick responses*, and *waste-free* (economical) *operations*. These three aims are mutually supportive and form the core rationale for many of the new principles that guide managers.
Focus on improvement. Managers have a duty to embrace improvement. A central theme of the total quality (TQ) movement is constant improvement in output goods and services and in the processes that provide them. Sweeping change over the short run, exemplified by business process reengineering [Hammer and Champy, 1993], anchors one end of the improvement continuum; the rationale is to discard unsalvageable processes and start over so as not to waste resources in fruitless repair efforts. The continuum's other end is described as incremental continuous improvement and is employed to fine-tune already sound processes for even better results.
Revised "laws" of economics. Examples of contemporary logic include the following: Quality costs less, not more. Costs should be allocated to the activities that cause their occurrence. Prevention (of errors) is more cost-effective than discovery and rework. Training is an investment rather than an expense. Automation can be beneficial *after* process waste is removed and further reduction in variation is needed. Value counts more than price (e.g., in purchasing). Desired market price should define (target) manufacturing cost, not the reverse.
Elimination of waste. Waste is anything that doesn't add value; it adds cost, however, and should be eliminated. Waste detection begins with two questions: "Are we doing the right things?" and "Are we doing those things in the right way?" Toyota identifies seven general categories of waste [Suzaki, 1987: chapter 1], each with several subcategories. Schonberger [1990: chapter 7] adds opportunities for further waste reduction by broadening the targets to include nonobvious waste. Simplification or elimination of indirect and support activities (e.g., production planning, scheduling, and control activities; inventory control; costing and reporting, etc.) is a prime arena for contemporary waste-reduction programs [Steudel and Desruelle, 1992: chapter 8].
Quick-response techniques. Just-in-time (JIT) management, queue limiters, reduced setups, better maintenance, operator-led problem solving, and other procedures increase the velocity of material flows, reduce throughput times, and eliminate the need for many interdepartmental control transactions. Less tracking and reporting (which add no value) reduces overhead. Collectively, quick-response programs directly support faster customer service [Blackburn, 1991: chapter 11].
Human resources management. Increased reliance on self-directed teams (e.g., in cells or product groups and/or on-line operators for assumption of a larger share of traditional management responsibilities is a product of the management revolution of the 1980s that had noticeable impact in the 1990s. Generally, line or shop employees have favored those changes; they get more control over their workplaces. There is a flipside: As employee empowerment shifts decision-making authority, as JIT reduces the need for many reporting and control activities, and as certain supervisory and support-staff jobs are judged to be non-value-adding, many organizations downsize. Lower- and mid-level managers and support staff often bear the job-loss burden.

a more detailed discussion. Schonberger [2001] has compiled an extensive report of widespread successful applications of these principles in his international benchmarking study, World Class by Principles.

Whether an engineer is thrust into a new position with management responsibilities or simply desires to gain a greater understanding of what successful management is all about in today's competitive arena, he or she will encounter the concept of value quite early in the process. Customers all along the value chain have new sources of information — data about performance levels, prices, options, and availability are much more available today. Value-based buying is, increasingly, not an option; it is a mandate.

Defining Terms

Management — Activities that have the goal of ensuring an organization's competitiveness by creating, implementing, and improving capacity required to provide goods and services that customers want.

Process — A particular combination of resource elements and conditions that collectively cause a given outcome or set of results. Process performance depends on how the process components were designed, built or installed, operated, and maintained.

Value — Perceived worth; the amount that a buyer is willing to pay for a good or service.

Value analysis/value engineering (VA/VE) — Formal examination of existing design or method with the aim of increasing the performance-to-cost ratio.

TABLE 183.2 Managing-for-Value Principles

Principle	Rationale and Examples
Get to know customers; team up to form partnerships and share process knowledge.	Providers are responsible for getting to know their customers' processes and operations. By so doing, they offer better and faster service, perhaps as a member of customers' teams.
Become dedicated to rapid and continual increases in quality, flexibility, and service; and decreases in costs, response or lead time, and variation from target.	The logic of continuous improvement, or *kaizen* [Imai, 1986], rejects the "if it ain't broke. . ." philosophy; seeks discovery and then prevention of current and potential problems; and anticipates new or next-level standards of excellence.
Achieve unified purpose through shared information and cross-functional teams for planning/design, implementation, and improvement efforts.	Information sharing keeps all parties informed. Early manufacturing/supplier involvement (EMI/ESI), and concurrent or simultaneous product and process design are components of the general cross-functional team design concept.
Get to know the competition and world-class leaders.	Benchmarking [Camp, 1989] elevates the older notion of "reverse engineering" to a more formal yet efficient means of keeping up with technology and anticipating what competitors might do. Search for best practices.
Cut the number of products (e.g., types or models), components, or operations; reduce supplier base to a few good ones and form strong relationships with them.	Product line trimming removes nonperformers; component reduction cuts lead times by promoting simplification and streamlining. Supplier certifications and registrations (e.g., ISO 9000) lend confidence, allow closer partnering with few suppliers (e.g., via EDI), and reduce overall buying costs.
Organize resources into multiple chains of customers, each focused on a family of products or services; create cells, flow lines, plants-in-a-plant.	Traditional functional organization by departments increases throughput times, inhibits information flow, and can lead to "turf battles." Flow lines and cells promote focus, aid scheduling, and employ cross-functional expertise.
Continually invest in human resources through cross-training for mastery of multiple skills, education, job and career path rotation, health and safety, and security.	Employee involvement programs, team-based activities, and decentralized decision responsibility depend on top-quality human resources. Cross-training and education are keys to competitiveness. Scheduling — indeed, all capacity management — is easier when the work force is flexible.
Maintain and improve present equipment and human work before acquiring new equipment; then automate incrementally when process variability cannot otherwise be reduced.	*TPM*, total productive maintenance [Nakajima, 1988], helps keep resources in a ready state and facilitates scheduling by decreasing unplanned downtime, thus increasing capacity. Also, process improvements must precede automation; get rid of wasteful steps or dubious processes first.
Look for simple, flexible, movable, and low-cost equipment that can be acquired in multiple copies — each assignable to a focused cell, flow line, or plant-in-a-plant.	Larger, faster, general-purpose equipment can detract from responsive customer service, especially over the longer run. A single fast process is not necessarily divisible across multiple customer needs. Simple, dedicated equipment is an economical solution; setup elimination is an added benefit.
Make it easier to make/provide goods and services without error or process variation.	The aim is to prevent problems or defects from occurring — the fail-safing (*pokayoke*) idea — rather than rely on elaborate control systems for error detection and the ensuing rework. Strive to do it right the first time, every time.
Cut cycle times, flow time, distance, and inventory all along the chain of customers.	Time compression provides competitive advantage [Blackburn, 1991]. Removal of excess distance and inventory aids quick response to customers. Less inventory also permits quicker detection and correction of process problems.
Cut setup, changeover, get-ready, and startup times.	Setup (or changeover) time had been the standard excuse for large-lot operations prior to directed attention at reduction of these time-consuming activities [Shingo, 1985]. Mixed-model processing demands quick changeovers.
Operate at the customer's rate of use (or a smoothed representation of it); decrease cycle interval and lot size.	Pull-mode operations put the customer in charge and help identify bottlenecks. Aim to synchronize production to meet period-by-period demand rather than rely on large lots and long cycle intervals.

TABLE 183.2 Managing-for-Value Principles (*Continued*)

Principle	Rationale and Examples
Record and *own* quality, process, and problem data at the workplace. Ensure that front-line associates get first chance at problem solving — before staff experts.	When employees are empowered to make decisions and solve problems, they need appropriate tools and process data. Transfer of point-of-problem data away from operators and to back-office staff inhibits responsive, operator-centered cures.
Cut transactions and reporting; control *causes*, not symptoms.	Transactions and reports often address problem symptoms (e.g., time or cost variances) and delay action. Quick-response teams, using data-driven logic, directly attack problem causes and eliminate the need for expensive reporting.
Market each improvement; share results with employees, suppliers, and customers.	Aside from the obvious benefits to company marketing, this principle fosters supply-chain development and nourishing. Also, company bulletins and displays allow employees — the ones responsible for the improvements — to keep abreast of progress and share in the credit.

Source: Knod, E.M. and Schonberger, R.J. 2001. *Operations Management: Meeting Customers' Demands*, 7th ed. Irwin/McGraw-Hill, Burr Ridge, IL. Adapted with permission.

Value chain — Series of transformation processes that move a good or service from inception to final consumer. Ideally, each step adds worth (value) in the eyes of both next-stage and ensuing customers.

Value migration — Changes in either (1) priorities for what a customer requires, or deems worthy, from time to time, or (2) points along the supply chain at which the customer expects to find what is sought.

References

Aldag, R. J. and Kuzuhara, L. W. 2002. *Organizational Behavior and Management*. South-Western, Cincinnati, OH.

Bass, B. M. 1990. *Bass & Stogdill's Handbook of Leadership*, 3rd ed. Free Press, New York.

Bateman, T. S. and Zeithaml, C. P. 1993. *Management: Function and Strategy*, 2nd ed. Richard D. Irwin, Burr Ridge, IL.

Blackburn, J. D. 1991. *Time-Based Competition*. Business One-Irwin, Homewood, IL.

Camp, R. C. 1989. *Benchmarking*. ASQC Quality Press, Milwaukee, WI.

Chopra, S. and Meindl, P. 2001. *Supply Chain Management*. Prentice Hall, Inc., Upper Saddle River, NJ.

Covey, S. R., Three roles of the leader in the new paradigm, in *The Leader of the Future*, Hesselbein, F., Goldsmith, M., and Beckhard, R., Eds., Jossey-Bass, San Francisco, 1996, chapter 16.

Hall, R. 1983. *Zero Inventories*. Dow Jones-Irwin, Homewood, IL.

Hammer, M. and Champy, J. 1993. *Reengineering the Corporation*. HarperCollins, New York.

Hesselbein, F., Goldsmith, M., and Beckhard, R., Eds. 1996. *The Leader of the Future*. Jossey-Bass, San Francisco.

Imai, M. 1986. *Kaizen: The Key to Japan's Competitive Success*. Random House, New York.

Johnson, H. T. and Kaplan, R. S. 1987. *Relevance Lost: The Rise and Fall of Management Accounting*. Harvard Business School Press, Boston.

Knod, E. M. and Schonberger, R. J. 2001. *Operations Management: Meeting Customers' Demands*, 7th ed. Irwin/McGraw-Hill, Burr Ridge, IL.

Kouzes, J. M. and Posner, B. Z. 2003. *The Leadership Challenge*. John Wiley & Sons, New York.

Nakajima, S. 1988. *Introduction to TPM: Total Productive Maintenance*. Productivity Press, Cambridge, MA.

Nicholas, J. M., 1998. *Competitive Manufacturing Management*. Irwin/McGraw-Hill, Burr Ridge, IL.

Quinn, R. E., Faerman, S. R., Thompson, M. P., and McGrath, M. R. 2003. *Becoming a Master Manager*, 3rd ed. John Wiley & Sons, New York.

Schneiderjans, M. J. 1993. *Topics in Just-in-Time Management.* Allyn and Bacon, Needham Heights, MA.

Schonberger, R. J. 1982. *Japanese Manufacturing Techniques: Nine Hidden Lessons in Simplicity.* Free Press, New York.

Schonberger, R. J. 1986. *World Class Manufacturing: The Lessons of Simplicity Applied.* Free Press, New York.

Schonberger, R. J. 1990. *Building a Chain of Customers.* Free Press, New York.

Schonberger, R. J. 2001. *Let's Fix It! Overcoming the Crisis in Manufacturing.* Free Press. New York.

Simchi-Levi, D., Kaminsky, P., and Simchi-Levi, E. 2000. *Designing and Managing the Supply Chain.* Irwin/McGraw-Hill, Burr Ridge, IL.

Shingo, S. 1985. *A Revolution in Manufacturing: The SMED [Single-Minute Exchange of Die] System.* Productivity Press, Cambridge, MA.

Slywotzky, A. J. 1996. *Value Migration.* Harvard Business School Press, Boston, MA.

Steudel, H. J. and Desruelle, P. 1992. *Manufacturing in the Nineties.* Van Nostrand Reinhold, New York.

Suzaki, K. 1987. The *New Manufacturing Challenge: Techniques for Continuous Improvement.* Free Press, New York.

Wren, J.T., Ed. 1995. *The Leader's Companion.* Free Press, New York.

Further Information

Periodicals

Industrial Engineering, Institute of Industrial Engineers.

Industrial Management, Society for Engineering and Management Systems, a society of the Institute of Industrial Engineers.

Journal of Operations Management, American Production and Inventory Control Society.

Production and Inventory Management Journal, American Production and Inventory Control Society.

Production and Operations Management, an international journal of the Production and Operations Management Society.

Quality Management Journal, American Society for Quality.

The Journal of Supply Chain Management, Institute for Supply Management, Inc.

Websites

American Society for Quality (*www.asq.org*)

Educational Society for Resource Management (APICS) (*www.apics.org*)

Institute for Industrial Engineers (IIE) (*www.iienet.org*)

Lean Enterprises Institute (*www.lean.org*)

184

Design, Modeling, and Prototyping

William L. Chapman
Hughes Aircraft Company

A. Terry Bahill
University of Arizona

To create a product and the processes that will be used to manufacture it, an engineer must follow a defined system design process. This process is an iterative one that requires refining the requirements, products, and processes of each successive design generation. These intermediate designs, before the final product is delivered, are called **models** or **prototypes**.

A model is an abstract representation of what the final system will be. As such, it can take on the form of a mathematical equation, such as $f = m \times a$. This is a deterministic model used to predict the expected force for a given mass and acceleration. This model only works for some systems and fails both at the atomic level, where quantum mechanics is used, and at the speed of light, where the theory of relativity is used. Models are developed and used within fixed boundaries.

Prototypes are physical implementations of the system design. They are not the final design, but are portions of the system built to validate a subset of the requirements. For example, the first version of a new car is created in a shop by technicians. This prototype can then be used to test for aerodynamic performance, fit, drivetrain performance, and so forth. Another example is airborne radar design. The prototype of the antenna, platform, and waveguide conforms closely to the final system; however, the prototype of the electronics needed to process the signal often comprises huge computers carried in the back of the test aircraft. Their packaging in no way reflects the final fit or form of the unit.

184.1 The System Design Process

The system design process consists of the following steps:

1. Specify the requirements provided by the customer and the producer.
2. Create alternative system design concepts that might satisfy these requirements.
3. Build, validate, and simulate a model of each system design concept.
4. Select the best concept by doing a trade-off analysis.
5. Update the customer requirements based on experience with the models.
6. Build and test a prototype of the system.
7. Update the customer requirements based on experience with the prototype.
8. Build and test a preproduction version of the system and validate the manufacturing processes.
9. Update the customer requirements based on experience with the preproduction analysis.

0-8493-1586-7/05/$0.00+$1.50

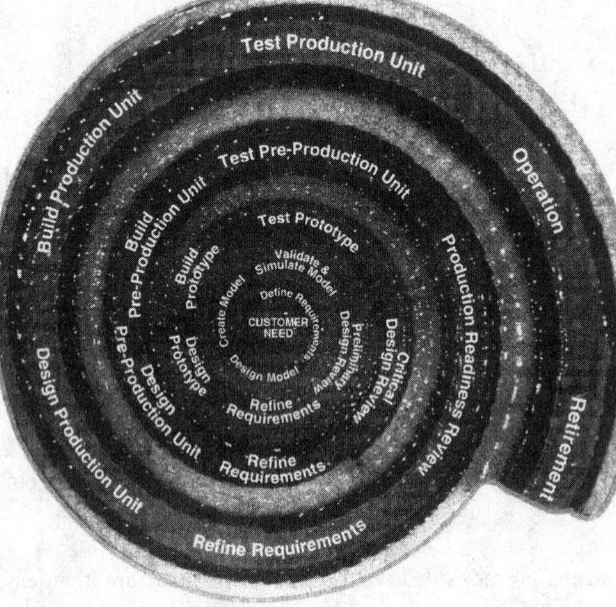

FIGURE 184.1 The system design process.

10. Build and test a production version of the system.
11. Deliver and support the product.

This can be depicted graphically on a spiral diagram as shown in Figure 184.1.

The process always begins with defining and documenting the customer's needs. A useful tool for doing this is quality function deployment (QFD). QFD has been used by many Japanese and American corporations to document the voice of the customer. It consists of a chart called the "house of quality." On the left is listed what the customer wants. Across the top is how the product will be developed. These are often referred to as *quality characteristics*. The "whats" on the left are then related to the "hows" across the top, providing a means of determining which quality characteristics are the most important to the customer [Akao, 1990; Bahill and Chapman, 1993].

After the customer's needs are determined the design goes through successive generations as the design cycle is repeated. The requirements are set and a model or prototype is created. Each validation of a model or test of a prototype provides key information for refining the requirements.

For example, when designing and producing a new airborne missile, the initial task is to develop a model of the expected performance. Using this model, the systems engineers make initial estimates for the partition and allocation of system requirements. The next step is to build a demonstration unit of the most critical functions. This unit does not conform to the form and fit requirements but is used to show that the system requirements are valid and that an actual missile can be produced. Requirements are again updated and modified, and the final partitioning is completed. The next version is called the *proof-of-design* missile. This is a fully functioning prototype. Its purpose is to demonstrate that the design is within specifications and meets all form, fit, and function requirements of the final product. This prototype is custom-made and costs much more than the final production unit will cost. This unit is often built partly in the production factory and partly in the laboratory. Manufacturing capability is an issue and needs to be addressed before the design is complete. More changes are made and the manufacturing processes for full production readied. The next version is the proof of manufacturing or the preproduction unit. This device will be a fully functioning missile. The goal is to prove the capability of the factory for full-rate production and to ensure that the manufacturing processes are optimum. If the factory cannot meet the quality or rate production requirements, more design changes are made before the drawings are released for full-rate production. Not only the designers but the entire design and production team must take responsibility for the design of the product and processes so that the customer's requirements are optimized [Chapman et al., 1992]. Also see Chapters 80, 180, and 181 of this book for additional information on design and production.

Most designs require a model upon which analysis can be done. The analysis should include a measure of all the characteristics the customer wants in the finished product. The concept selection will be based on the measurements done on the models. See Chapter 209 for more information on selection of alternatives. The models are created by first partitioning each conceptual design into functions. This decomposition often occurs at the same time that major physical components are selected. For example, when designing a new car, we could select a mechanical or electronic ignition system. These are two separate concepts. The top-level function — firing the spark plugs — is the same, but when the physical components are considered the functions break down differently. The firing of the spark plugs is directed by a microprocessor in one design and a camshaft in the other. Both perform the same function, but with different devices. Determining which is superior will be based on the requirements given by the customer, such as cost, performance, and reliability. These characteristics are measured based on the test criteria.

When the model is of exceptional quality, a prototype can be skipped. Advances in computer-aided design (CAD) systems have made wire-wrapped prototype circuit boards obsolete. CAD models are so good at predicting the performance of the final device that no prototype is built. Simulation is repeated use of a model to predict the performance of a design. Any design can be modeled as a complex finite-state machine. This is exactly what the CAD model of the circuit does. To truly validate the model, each state must be exercised. Selecting a minimum number of test scenarios that will maximize the number of states entered is the key to successful simulation. If the simulation is inexpensive, then multiple runs of this model should be done. See Chapters 13, 29, 95, 96, 97, and 159 for more information on simulations. The more iterations of the design process there are, the closer the final product will be to the customer's optimum requirements.

For other systems, modeling works poorly and prototypes are better. Three-dimensional solids modeling CAD systems are a new development. Their ability to display the model is good, but their ability to manipulate and predict the results of fit, force, thermal stresses, and so forth is still weak. The CAD system has difficulty simulating the fit of multiple parts (such as a fender and car frame) because the complex surfaces are almost impossible to model mathematically. Therefore, fit is still a question that prototypes, rather than models, are best able to answer. A casting is usually used to verify that mechanical system requirements are met.

Computer-aided manufacturing (CAM) systems use the CAD database to create the tools needed for the manufacture of the product. Numerical control (NC) machine instructions can be simulated using these systems. Before a prototype is built, the system can be used to simulate the layout of the parts, the movement of the cutting tool, and the cut of the bar stock on a milling machine. This saves costly material and machine expenses.

Virtual reality models are the ultimate in modeling. Here the human is put into the loop to guide the model's progress. Aircraft simulators are the most common type of this product. Another example was demonstrated when the astronauts had to use the space shuttle to fix the mirrors on the Hubble telescope. The designers created a model to ensure that the new parts would properly fit with the existing design. They then manipulated the model interactively to try various repair techniques. The designers were able to verify fit with this model and catch several design errors early in the process. After this, the entire system was built into a prototype and the repair rehearsed in a water tank [Hancock, 1993].

Computer systems have also proved to be poor simulators of chemical processes. Most factories rely on design-of-experiments (DOE) techniques, rather than a mathematical model, to optimize chemical processes. DOE provides a means of selecting the best combination of possible parameters to alter when building the prototypes. Various chemical processes are used to create the prototypes that are then tested. The mathematical techniques of DOE are used to select the best parameters based on measurements of the prototypes. Models are used to hypothesize possible parameter settings, but the prototypes are necessary to optimize the process [Taguchi, 1976].

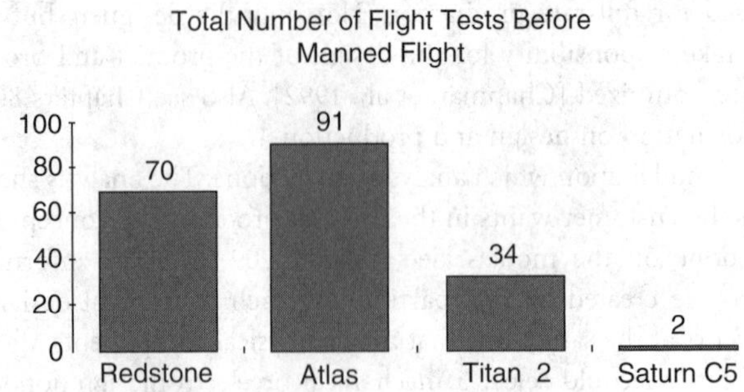

FIGURE 184.2 Flight tests for the manned space program rockets.

The progressive push is to replace prototypes with models, because an accurate fully developed model is inexpensively simulated on a computer compared to the cost and time of developing a prototype. A classic example is the development of manned rockets. Figure 184.2 shows the number of test flights before manned use of the rockets.

The necessity for prototypes diminished rapidly as confidence in computer models developed. Initially, many of the rockets exploded in their first attempts at launch. As more was learned about rocketry, sophisticated models were developed that predicted performance of the rocket based on real-time measurements of valves, temperatures, fuel levels, and so forth. Using modern computers, the entire model could be evaluated in seconds and a launch decision made. This eliminated the need for many flight tests and reduced the cost of the entire Apollo moon-landing program.

184.2 Rapid Prototyping

Rapid prototyping is the key to reducing design time for parts and processes. Design is an iterative process. By creating prototypes quickly, the design can be completed faster. Japanese automobile manufacturers develop prototypes of their products in 6 months, whereas U.S. companies take 13 months [Womack et al., 1990]. This advantage allows the Japanese companies to get to the market faster, or if they choose they can iterate their design one more time to improve their design's conformance to the customer's requirements. Japanese automakers' ability to create the prototype quickly is due in part to better coordination with their suppliers, but also to the exacting use of design models to ensure that the design is producible.

As seen in Figure 184.3, the lead in prototype development accounts for 44% of the advantage that Japanese producers have in product development time. The rest of the advantage is from rapid creation of the huge dies needed to stamp out the metal forms of the automobiles. Design of these important manufacturing tools is given as much attention as the final product. By creating flexible designs and ensuring that the teams who will produce the dies are involved in the design process, the die development time is cut from 25 months in the U.S to 13.8 months in Japan. The rapid creation of the tooling is a key to fast market response.

Stereolithography is a new method of creating simple prototype castings. The stereolithography system creates a prototype by extracting the geometric coordinates from a CAD system and creating a plastic prototype. The solids model in the CAD system is extracted by "slicing" each layer in the *z* axis into a plane. Each layer is imaged by a laser into a bath of liquid photopolymer resin that polymerizes with the energy from the laser. Each plane is added, one on top of the other, to build up the prototype. The final part is cured and polished to give a plastic representation of the solids model. It illustrates the exact shape of the part in the CAD system. This technique will not, however, verify the function of the final product because the plastic material used will not meet strength or thermal requirements [Jacobs, 1992].

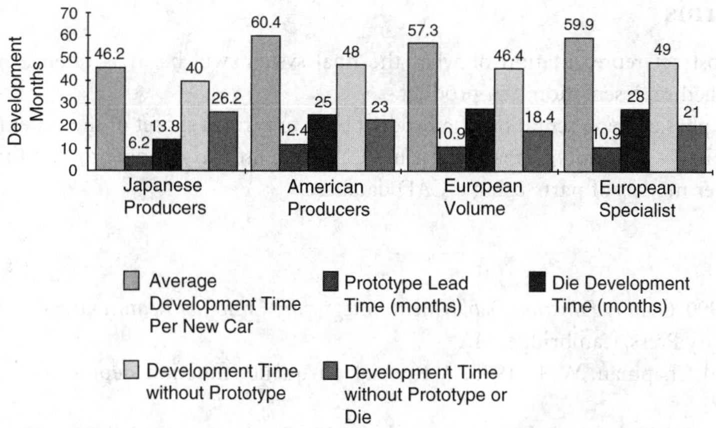

FIGURE 184.3 Comparison of Japanese, American, and European car producers. (Based on Womack, J. P., Jones, D. T., and Roos, D. 1990. *The Machine that Changed the World.* Rawson Associates, New York.

Software developers also use rapid prototyping. This technique is used to get a fast, although barely functional, version of the product into the customer's hands early in the design cycle. The prototype is created using the easiest method to simulate functionality to the viewer. The customer comments on what is seen and the developers modify their design requirements. For example, when developing expert systems, models are almost never used. One of the driving rules is to show a prototype to the customer as soon as possible and afterward throw it away! The purpose of building the prototype was to find out what the knowledge that needs to be represented is like, so that the appropriate tool can be selected to build the product. If, as is usually the case, a nonoptimal tool was used for the prototype, then the prototype is thrown away and a new one is developed using better tools and based on better understanding of the customer's requirements. Beware, though — often a key function displayed in the prototype is forgotten when the prototype is abandoned. Be certain to get all the information from the prototype [Maude and Willis, 1991]. The fault with this technique is that the requirements are not written down in detail. They are incorporated into the code as the code is written, and they can be overlooked or omitted when transferred to a new system.

184.3 When to Use Modeling and Prototyping

When should modeling versus prototyping be used? The key difference is the value of the information obtained. Ultimately, the final product must be created. The prototype or model is used strictly to improve the final product. Costs associated with the prototype or model will be amortized over the number of units built. The major problem with models is lack of confidence in the results. Sophisticated models are too complex for any single engineer to analyze. In fact, most models are now sold as proprietary software packages. The actual algorithms, precision, and number of iterations are rarely provided. The only way to validate the algorithm (not the model) is by repeated use and comparison to actual prototypes. It is easier to have confidence in prototypes. Actual parts can be measured and tested repeatedly, and the components and processes can be examined. Prototypes are used more often than models once the complexity of the device exceeds the ability of the computer to accurately reflect the part or process. During the initial design phases, models must be used because a prototype is meaningless until a concept has been more firmly defined. At the other extreme, modeling is of limited benefit to the factory until the configuration of the part is well known. A general rule is to build a model, then a prototype, and then a production unit. Create even a simple mathematical model if possible so that the physics can be better understood. If the prototype is to be skipped, confidence in the model must be extremely high. If there is little confidence in the model, then a minimum of two or three prototype iterations will have to be done.

Defining Terms

Model — An abstract representation of what the final system will be. It is often a mathematical or simplified representation of a product.

Prototype — A physical representation of a product built to verify a subset of the system's requirements.

Stereolithography — A prototype-manufacturing technique used to rapidly produce three-dimensional polymer models of parts using a CAD database.

References

Akao, Y. (Ed.) 1990. *Quality Function Deployment: Integrating Customer Requirements into Product Design.* Productivity Press, Cambridge, MA.

Bahill, A. T. and Chapman, W. L. 1993. A tutorial on quality function deployment. *Eng. Manage. J.* 5(3):24–35.

Chapman, W. L., Bahill, A. T., and Wymore, A. W. 1992. *Engineering Modeling and Design.* CRC Press, Boca Raton, FL.

Hancock, D. 1993. Prototyping the Hubble fix. *IEEE Spectrum.* 30(10):34–39.

Jacobs, P.F. 1992. *Rapid Prototyping & Manufacturing: Fundamentals of StereoLithography.* McGraw-Hill, New York.

Maude, T. and Willis, G. 1991. *Rapid Prototyping: The Management of Software Risk.* Pitman, London.

Taguchi, G. 1976. *Experimental Designs,* 3rd ed., vols. 1 and 2. Maruzen, Tokyo.

Womack, J. P., Jones, D. T., and Roos, D. 1990. *The Machine that Changed the World.* Rawson Associates, New York.

Further Information

Pugh, S. 1990. *Total Design: Integrated Methods for Successful Product Engineering.* Addison-Wesley, London.

Suh, N.P. 1990. *The Principles of Design.* Oxford University Press, New York.

Wymore, A. W. 1993. *Model-Based Systems Engineering.* CRC Press, Boca Raton, FL.

185

Materials Processing and Manufacturing Methods

Chang-Xue Jack Feng
Bradley University

One of the most critical aspects of engineering applications involves the selection, shaping, and processing of raw materials obtained either directly from the Earth's crust or by special synthesis and purification methods. The history of successful cost competitive technologies and the development of modern societies is closely linked to one of two factors: (1) a new processing method for manufacturing a material with broad engineering utility, or (2) an altogether new discovery of a material with a composition that makes possible dramatically improved engineering properties. A good example of a processing method that can revolutionize engineering technologies is the now well-known discovery of zone refining and crystal growth of semiconductors, such as silicon and germanium, which are at the heart of the present-day microelectronic industries. Similarly, the discovery of new materials is best exemplified by the finding in the mid-1980s that certain ceramics (mixtures of yttrium-barium-copper oxides) — long known to be ferroelectric — can actually lose all resistance to electric current and become superconductors at a relatively warm temperature of about 90 K. This discovery of a new material is likely to have a major impact in many fields of technology.

Engineers and scientists continue the rich historical trend of seeking new materials or an unusual method for shaping/processing a material. Particularly striking evidence of this is found in many of the current worldwide activities in the fast-growing field of nanotechnology. Thus, spectacular engineering properties are attainable by making extremely small structures that are many times smaller than the

0-8493-1586-7/05/$0.00+$1.50

thickness of human hair. The principal goal is to synthesize nanometer-grained bulk materials or composite films in which at least one functional material has dimensions somewhere between 1 and 50 nm. Alternately, methods are being considered to carve out of bulk materials nanosize structures that promise to have a revolutionary impact on engineering and the physical and biological sciences. Examples of such materials and methods include the nanotechnology of making quantum wells and dots for laser optics, micromachined gears, ductile ceramics, and molecular motors. This chapter intends to provide a "road-runner's" view of methods for shaping and processing materials both by the well-established techniques (e.g., casting, welding, powder sintering) and the more recent evolving methods. For a more detailed discussion of the topics in this chapter, refer to Groover (2002) and Kalpakjian and Schmid (2003).

185.1 Processing Metals and Alloys

The major methods for making engineering parts from metallic alloys consist of casting from the molten (liquid) state, followed by cold or hot working of the casting by processes such as rolling, forging, drawing, extrusion, and spinning. Powders of metals and alloys are also made into shapes by pressing and sintering or hot pressing (HP) and hot isostatic pressing ("hipping"). Grinding, polishing, and welding are some of the final steps in the production of an engineering component. There is a growing interest in encouraging near-net-shape manufacturing methods for engineering components to minimize the wastage in machining and grinding operations. Powder metallurgy methods provide an important advantage in this regard.

Casting Methods

Casting of liquid metals and alloys is one of the oldest and best-known processes for making ingots, or shapes in final form determined by the mold or die cavity in which the liquid is poured. The solidification of the liquid metal takes place after the melt is allowed to flow to all the areas of the mold. Shrinkage during solidification makes for a smaller casting size than the mold size, and this is accounted for in the design of the mold. Molds can vary from simple cavities to multipart molds clamped together and containing ceramic cores to create hollow spaces in some parts of the casting.

Sand casting is a metal casting process where sand is mixed with clay, the best-known traditional mold material, and is packed around a wooden pattern that is subsequently removed when the sand casting mold is ready for pouring of liquid metal. A small channel called a *gate* provides the liquid metal access to the mold cavity, while a riser column becomes the reservoir of liquid metal that feeds the shrinkage of the liquid metal during solidification. The sand-casting process is an inexpensive method for making a small number of parts; a range of ferrous and nonferrous alloys can be made by this method. The rough finish of sand-cast parts usually requires further machining and grinding.

Die casting is performed by pouring liquid alloys into preshaped metal molds machined to good accuracy, tolerances, and surface finish. Thus, the finished product is of good surface quality in die casting, and complex shapes can be made. Gravity die casting involves the flowing of the liquid metal under natural gravity conditions, whereas pressure die casting involves forcing the melt into the die under pressure. Centrifugal casting relies on a rotating mold to distribute the melt in the mold cavity by the action of centrifugal forces, a method especially suitable for making large-diameter metal pipes. Aluminum, zinc, copper, and magnesium alloys are often made by die-casting methods. Because machined molds are required, die casting is more expensive than sand castings, although molds are reused, and thus the process is good for large-scale production.

Investment casting or lost-wax casting is used particularly for alloys with high-melting temperatures and when high-dimensional accuracy and precision are needed. The high-melting metals attack the dies used in normal die casting, and thus patterns are produced from wax using a metal mold. The wax patterns are coated with a ceramic slurry and heated to melt the wax, whereupon a dense ceramic mold is created. Molten alloys are then forced into the ceramic mold cavity by use of pressure or by a centrifugal

TABLE 185.1 Summary of Metalworking Processes

Process	Economic Quantity	Typical Materials	Optimum Size
Sand casting	Small–large	No limit	1–100 kg
Die casting — gravity	Large	Al, Cu, Mg, Zn alloys	1–50 kg
Die casting — pressure	Large	Al, Cu, Mg, Zn alloys	50 g–5 kg
Centrifugal casting	Large	No limit	30 min–1 m diam.
Investment casting	Small–large	No limit	50 g–50 kg
Closed-die forging	Large	No limit	3000 cm^3
Hot extrusion	Large	No limit	500 mm diam.
Hot rolling	Large	No limit	—
Cold rolling	Large	No limit	—
Drawing	Small–large	Al, Cu, Zn, mild steel	3 mm–6 mm diam.
Spinning	One-off–large	Al, Cu, Zn, mild steel	6 mm–4.5 m diam.
Impact extrusion	Large	Al, Pb, Zn, Mg, Sn	6 mm–100 mm diam.
Sintering	Large	Fe, W, bronze	80 g–4 kg
Machining	One-off–large	No limit	—

Source: Bolton, W. 1981. *Materials Technology.* Butterworth, London. With permission.

process. Complicated shapes (e.g., aerospace turbine blades) or shapes in which fine detail is sought (e.g., figurines) are made by investment or lost-wax casting methods.

Metalworking Methods

Hot working of metal alloys is normally the next step in the shaping of the cast alloy into a component, and is also used to induce some refinement of the grain structure of the material to obtain the necessary ductility or strength. In hot working, the temperature is high enough that recrystallization occurs at a rate faster than work hardening due to deformation. Forging, rolling, and extrusion are some examples of hot-working processes.

Hot forging is a process of simply pressing the hot metal between two surfaces in open, closed, or flat-faced dies. This process is closest to the ancient art of the blacksmith, but modern manufacturing plants employ high-power hydraulic presses to deliver impact blows to achieve forging of large steel crankshafts, aerospace propellers, and so forth. Hot extrusion consists of taking a cylindrical billet from the cast ingot and placing the hot billet in a die of a slightly larger size. A piston or ram extrudes the hot metal through appropriately cut orifices or openings in the die. The process is akin to squeezing toothpaste out of a tube under pressure. Direct and inverted extrusion are the two most common variations of the extrusion process. Hot rolling is a popular method of reducing the cast ingots into billets, slabs, plates, or sheets by pressing the hot metal between rotating rollers. Four high-rolling mills consist of smaller-diameter work rolls backed by larger diameter backup rolls. Many other types of cluster mills also exist, including planetary mills and pendulum mills.

Cold working of metals and alloys has the advantage of producing a clean, smooth surface finish and a harder material. Cold rolling of aluminum foil or steel strips are some examples of products in daily use that have surfaces that can vary from a standard bright finish to matte and plated finishes, depending on the application. Drawing is another cold-working process in which cylindrical rods can be drawn through a die to make wires or sheets of metal can be shaped into cup-shaped objects by deep drawing the sheet into a die by pressure from the punch. By clamping the sheet around the edges, the pressing operation can produce domestic cookware, automobile bodies, and a variety of cup-like shapes. Circular cross sections can be made by spinning a metal blank in a lathe while applying pressure to the blank with a tool to shape it. Explosive forming is a special process used for shaping large-area sheets of metal into contoured panels such as those needed in communication reflectors. The wave generated by detonation of an explosive charge is used to convert a sheet blank into the shape of the die. Table 185.1 is a summary of the processes used with a listing of the advantages and limitations of each process.

TABLE 185.2 Typical Tolerance Limits, Based on Process Capability, for Various Processes

Process	Typical Tolerance Limits mm	(inches)	Process	Typical Tolerance Limits mm	(inches)
Sand casting			**Abrasive**		
Cast iron	±1.3	(±0.050)	Grinding	±0.008	(±0.0003)
Steel	±1.5	(±0.060)	Lapping	±0.005	(±0.0002)
Aluminum	±0.5	(±0.020)	Honing	±0.005	(±0.0002)
Die casting	±0.12	(±0.005)	**Nontraditional processes**		
Plastic molding			Chemical machining	±0.008	(±0.003)
Polyethylene	±0.3	(±0.005)	Electrical discharge	±0.0025	(±0.001)
Polystyrene	±0.15	(±0.005)	Electrochem. grind	±0.0025	(±0.001)
Machining			Electrochem. machine	±0.005	(±0.002)
Drilling, diameter			Electron beam cutting	±0.008	(±0.003)
6 mm (0.250 in.)	+0.08, −0.03	(±0.003, −0.001)	Laser beam cutting	±0.008	(±0.003)
25 mm (1.000 in.)	+0.13, −0.05	(±0.006, −0.002)	Plasma arc cutting	±1.3	(±0.050)
Milling	±0.08	(±0.003)			
Turning	±0.05	(±0.002)			

Source: Groover, Mikell P. 2002. *Fundamentals of Modern Manufacturing*, 2nd ed. John Wiley & Sons, New York.

Machining and Finishing

The cast, hot- or cold-worked metallic alloys are often subjected to some type of machining operation, which encompasses cutting, grinding, polishing, shearing, drilling, or other metal-removal methods. Chemical etching or polishing methods can also be used as finishing steps in the production of components from metals. Planing, turning, milling, and drilling are the main methods of removing chips from a metal part during machining.

Surface finish and dimensional accuracy of the component are the main goals of machining. Some recent studies in tolerance design were reviewed and reported in Feng and Kusiak (2000) and Feng et al. (2001). Table 185.2 illustrates the typical tolerances that can be achieved by common manufacturing processes. Figure 185.1 illustrates the typical process chains that a designer and a manufacturing engineer or technician should be aware of. Some recent turning surface roughness studies are reviewed in Feng and

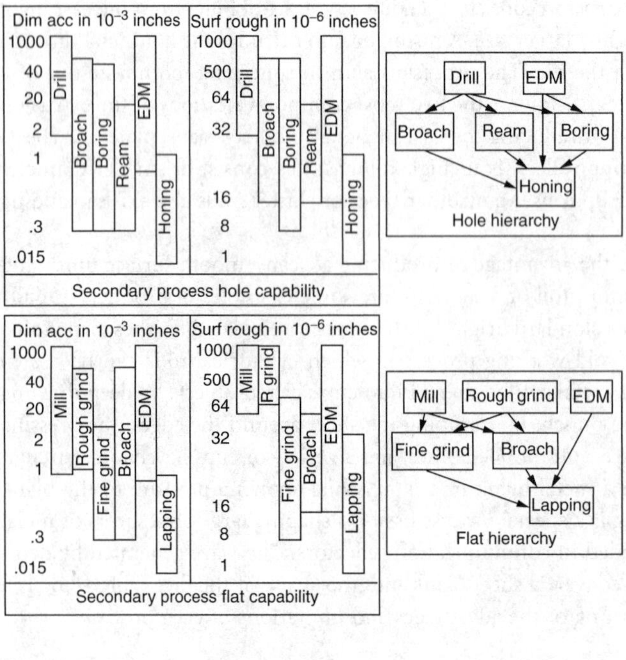

FIGURE 185.1 Process chains with tolerance levels. (*Source:* Wright, P. K. 2001. *21st Century Manufacturing*. Prentice Hall, Upper Saddle River, NJ. With permission.)

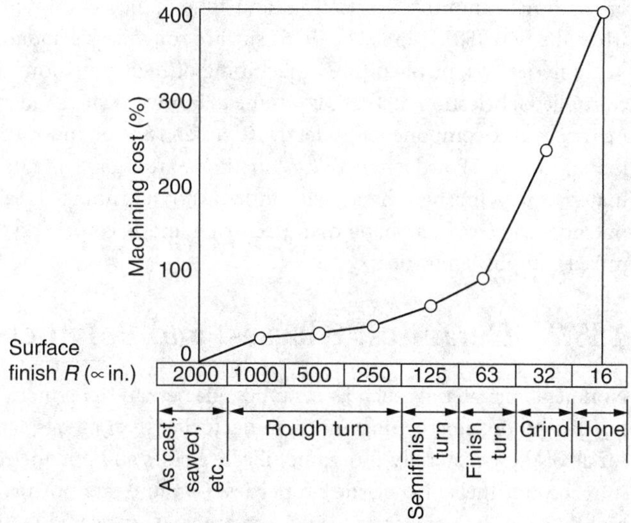

FIGURE 185.2 Relative machining cost with respect to surface roughness specifications. (*Source:* Kalpakjian, S. and Schmid, S. R. 2003. *Manufacturing Processes for Engineering Materials*, 4th ed. Prentice Hall, Upper Saddle River, NJ. With permission.)

Wang (2003) and honing surface roughness studies in Feng et al. (2002). Table185.3 shows typical surface roughness values produced by the various manufacturing processes, and Figure 185.2 shows the relative costs caused by additional manufacturing operations with respect to surface roughness specifications.

Cutting and grinding tools required in machining are usually made from hard metal (carbide) or ceramic materials; a *machinability index* measures the ease with which a particular material can be machined with conventional methods. The higher the machinability index, the easier it is to machine the metal or alloy. In addition to these methods, metal removal can also be achieved by ultrasonic abrasion, laser or electron beam cutting and drilling, electrical discharge or spark machining, arc milling, and chemical milling. Honing has been and will remain to be the only process that will provide both the surface roughness and crosshatched lay directions for the internal surface of engine cylinder liners in

TABLE 185.3 Surface Roughness Values Produced by Various Manufacturing Processes

Process	Typical Surface Finish	Range of Roughness	Process	Typical Surface Finish	Range of Roughness
Casting			**Abrasive**		
Die casting	Good	1.2 (30–65)	Grinding	Very good	0.1–2 (5–75)
Investment	Good	1.5–3 (50–100)	Honing	Very good	0.1–1 (4–30)
Sand casting	Poor	12–25 (500–1000)	Lapping	Excellent	0.05–0.5 (2–15)
Metal forming			Polishing	Excellent	0.1–0.5 (5–15)
Cold rolling	Good	1–3 (25–125)	Superfinishing	Excellent	0.02–0.3 (1–10)
Sheet metal drawing	Good	1–3 (25–125)	**Nontraditional**		
Cold extrusion	Good	1–4 (30–150)	Chemical milling	Medium	1.5–5 (50–200)
Hot rolling	Poor	12–25 (500–1000)	Electrochemical	Good	0.2–2 (10–100)
Machining			Electric discharge	Medium	1.5–1.5 (50–500)
Boring	Good	0.5–6 (15–250)	Electric beam	Medium	1.5–1.5 (50–500)
Drilling	Medium	1.5–6 (20–250)	Laser beam	Medium	1.5–1.5 (50–500)
Milling	Good	1–6 (30–250)	**Thermal**		
Planing	Medium	1.5–12 (60–500)	Arc welding	Poor	5–25 (250–1000)
Reaming	Good	1–3 (30–125)	Flame cutting	Poor	12–25 (500–1000)
Shaping	Medium	1.5–12 (60–500)	Plasma arc cutting	Poor	12–25 (500–1000)
Sawing	Poor	3–25 (100–1000)			
Turning	Good	0.5–6 (15–250)			

Source: Groover, Mikell P. 2002. *Fundamentals of Modern Manufacturing*, 2nd ed. Wiley & Sons, New York.

order to help improve lubrication, and thus reduce wear, reduce emission, and increase oil efficiency when the new IS013565 (ISO, 1996) surface roughness standard is implemented (Feng et al., 2002).

Numerically controlled (NC) machining of high-precision components and sophisticated robotically controlled fabrication methods continue to be investigated and applied in the manufacture of engineering materials and components. Refer to Chapter 182 for more discussions on NC, robotics, and flexible manufacturing. Powder metallurgy methods are most suitable for directly converting powders of raw materials into finished shapes with little or no machining. This method, discussed later, has the greatest potential for near-net-shape manufacturing and is commonly used to consolidate metal, ceramic, and mixed composite powders.

185.2 Ceramics, Glasses, and Polymers

Nonmetallic materials such as ceramics, glasses, and polymeric organic materials require generally different methods for shaping and manufacturing than metals, although some common methods do exist for all of these materials. Inorganic silicate glasses and polymer melts are processed by methods that take into account their viscoelastic properties. Ceramics are normally refractory and high-melting materials and thus require powder sintering, hot pressing, or related methods for production.

Powder Processing and Sintering

Powder sintering of metal alloy powders and ceramics is a versatile method of producing near-final shapes of a component. Ceramic powders are typically blended with solid or liquid additives, binders, and so forth, and the batched system is treated by a series of mixing, chemical dissolution, washing, de-airing, or filtration to produce a slurry, paste, or suspension of solid particles in a liquid, often referred to as the *slip*. Depending on the consistency (viscosity of the syrup) of the suspension, forming a shape is achieved by tape casting onto plastic sheets or by pouring the slip into a plaster mold, followed by drying to remove the residual liquids in the batch. The shaped porous ceramic at this stage is called a *green ceramic body*. Green machining, surface grinding, or application of coatings or glazes is done at this stage. The final step in the process is consolidation to a nearly nonporous solid by firing or sintering at a high temperature. The sintered microstructure is usually a multiphase combination of grains, secondary phases, and some degree of remaining porosity.

Another common way of sintering ceramics, and particularly metal powders, is by direct pressing of the powders with some solid additives (called *sintering aids*). The component shape is determined by a die cavity and the powders are cold pressed at relatively high pressures, followed by heat treatment to create the sintered microstructure. In a combination of these steps, pressure and heat can be applied simultaneously in hot pressing or hot isostatic pressing ("hipping"). In these methods the powder fills in the die cavity, and, as the pressure is applied by hydraulic rams, the die is heated simultaneously to consolidate the powder to a solid. The evolution of a microstructure from powders to dense ceramics with grains and grain boundaries is now a well-developed science-based technology. Thus, deliberate process control is needed, including awareness of atmospheres used in firing, particle size distribution, chemistry of additives, control over grain boundary mobility, and elimination of exaggerated grain growth. The recent development of fast sintering methods includes the use of sparks, microwaves, and plasmas to activate the surfaces of very fine powders (micrometers to as small as a few nanometer particles) and cause consolidation to occur in just a few minutes. Microwave sintering and plasma-activated sintering are examples of promising new methods on the horizon for very fast sintering to achieve small grain sizes and high-purity materials in both metallic and ceramic systems. For an in-depth discussion of the above topics, refer to German (1994).

Glasses and Glass-Ceramics

The manufacture of glasses from mixtures of silicon-dioxide (silica) and a number of other oxides (e.g., sodium, calcium, aluminum oxides) is based on melting the mixtures in a tank-like furnace to create a homogeneous bubble-free viscous melt. The composition of the glass is chosen such that, on cooling the melt during shaping, formation of the thermodynamically favored crystalline material does not occur. To make glass plates or sheets (e.g., car windshields), the molten liquid is allowed to float onto the surface of a liquid bath of low-melting metal that is immiscible with the glass (the lighter glass floats on the higher-density metal melt). This float glass process is but one of a variety of ways of converting molten silicate liquids to glassy shapes. Viscous glass streaming out of a furnace can be trimmed into "gobs" that drop into a preshaped die while blowing air under pressure to force the mushy viscous gob into the final shape (e.g., a beer bottle). Molten glass compositions can also flow through bushings containing small orifices to create fiberglass materials or optical fibers. Optical fibers of high purity are also processed in drawing towers using preforms of carefully engineered clad-and-core compositions. Thousands of glass compositions with specific chemical ingredients are known for generating color, inducing a refractive index gradient, permitting selective transparency, or resisting weathering. The common window glass composition is based on a mixture of the oxides of sodium, calcium, and silicon, and is popularly called soda-lime-silica glass.

Glass-ceramics are materials made by the heat treatment of a shaped piece of glass so as to allow the formation of copious amounts of small crystallites without changing the macroscopic dimensions of the object. The optical transparency of the glass changes to a translucent appearance typical of most glass-ceramics. It is best known in the form of domestic cookware sold under the trade name Corelle or Corningware, by Corning Glass Company. The glass-ceramic process is a useful method of obtaining fully dense ceramics from the glassy state and requires good control of the nucleating agents that are used to catalyze the crystallite formation in the glass-ceramic microstructure. The uniformity of the crystal grains in the structure and the small size are factors that give glass-ceramics higher strength and better mechanical properties compared to glasses. The number of glass-ceramics available for use in engineering practice has grown enormously in the last 25 years; applications of glass-ceramics in electrical, magnetic, structural, and optical fields are becoming more common. Machinable glass-ceramics based on a mica-like crystalline phase have made it easier to shape glass-ceramics by conventional machine shop tools.

Sol-Gel Methods

Sol-gel processes make it possible to use organic-polymeric synthesis methods to produce ceramics and glasses at relatively low temperatures. The major advantage of sol-gel techniques is that all the chemicals needed in a glass or ceramic can be blended into a solution to create a "sol" at close to room temperature. This avoids the high-temperature processing normally used in making a ceramic or a glass. The sol is then made to gel by appropriate hydrolysis reactions, and the gel is then fired into a monolithic body or made into coatings, powders, films, and so forth. The calcination and sintering steps remove water and vestiges of organic bonds to form an inorganic ceramic product. Not all ceramics can be made by sol-gel techniques — not because of any fundamental limitations, but due mainly to the lack of well-developed connections between a suitable starting precursor and the synthesis protocol to make the final ceramic or glass. For making bulk monoliths or coatings and films, the use of drying control agents in sol-gel processing is crucial to attain crack-free objects.

Polymer Processing

The forming of organic polymeric materials essentially consists of making powders or granules of the requisite blend of polymers with additives such as carbon, chalk, and paper pulp. In the next step, the granules are heated to soften the polymer mixture, and the soft material is viscoelastically shaped into the final component. The processing methods mainly involve extruding, injection molding, casting, and calendering. The particular process used depends on the quantity, size, and form in which the polymer object is desired (e.g., sheets of plastic or bulk rods). Thermosetting polymers go through an irreversible chemical change upon heating and they are mostly made by molding or casting. Examples include thermosetting resins such as urea, melamine or phenol formaldehydes, and polyesters. Thermoplastic polymers are processible by repeated softening as long as the temperature is not too high to cause

TABLE 185.4 Comparison of Various Polymer-Processing Methods

Process	Production Rate	Material Type	Optimum Size
Extrusion	Fast	Th, plastic Th, set	Few mm–1.8 m
Blow molding	Fast	Th, plastic	10^{-6}–$2m^3$
Injection molding	Fast	Th, set	15 g–6 kg
Compression molding	Fast	Th, set	Few mm–0.4 m
Transfer molding	Fast	Th, set	
Layup techniques	Slow	Th, set and filler	Few mm–0.4 m, and 0.01–400 m^2
Relational molding	Medium	Th, plastic	10^{-3}–30 m^3
Thermoforming	Fast	Th, plastic	10^{-3}–20 m^2

Source: Bolton, W. 1981. *Materials Technology.* Butterworth, London. With permission.

TABLE 185.5 Comparison of Costs for Polymer-Processing Methods

	Capital Cost	Mold Cost	Cycle Time	Wall Thickness	Detrimental Factors
Injection molding	High	High	Very fast	Very accurate	Flash lines
Blow molding	High/medium	Medium	Fast	Uneven thinness	Local damage
Rotational casting	Low	Low	Slow	Good (if thick)	Inhomogeneity
Vacuum forming	Low/medium	Low	Slow	Poor	Local thinning

Source: Alexander, J. M., Brewer, R. C., and Rowe, G. W. *Engineering Processes*, vol. 2 of *Manufacturing Technology.* Halsted Press, New York. With permission.

decomposition and charring. Ease of flow makes these attractive for processing by injection molding and extrusion methods. Examples include polycarbonate, polyethylene, polystyrene, polyamide (nylon), and polyvinyl chloride (PVC). Inclusion of gas gives rise to foamed plastics such as polystyrene foam (for thermal insulation), whereas polymers are reinforced by fibers (typically glass fibers) to form composites with better stiffness and strength.

The glass fiber polymer composites are shaped by injection molding and are useful in environments where moisture-resistant, high-fracture tough materials are needed. Continuous sheets of plastic are made from PVC or polyethylene polymers by calendering. This process consists of feeding the heated polymer granules or powder through a set of heated rollers. The first roller converts the granules to a sheet that is subsequently reduced to the appropriate thickness and surface finish. Laminates and coatings of polymers are formed by pressing, casting, and roll-coating technologies. Large polymer pieces such as boat hulls are put together manually by pasting several layers of gum-fiber-reinforced polyester. For a large number of parts (automobile bodies, etc.), vacuum forming of PVC or polypropylene-type polymers is the desired method. A comparison of some polymer-processing methods is shown in Tables 185.4 and 185.5.

185.3 Joining and Fastening Processes

Temporary joints between the shaped finished components made from several materials are easily achieved by mechanical means such as nuts, bolts, pins, and so forth. Some permanent joining of two materials requires physical or chemical changes at the joining surfaces.

Mechanical Fastening

Mechanical fastening may be preferred over other joining methods for its ease of manufacturing; ease of assembly and transportation; ease of parts replacement, maintenance, and repair; ease in creating designs that require movable points; and lower overall cost of manufacturing the product. Mechanical fastening methods can be classified into two major groups: (1) those that allow for disassembly, and (2) those that create a permanent joint. Threaded fasteners including screws, bolts, and nuts are examples of the first group, and rivets and eyelets illustrate the second. Other methods include (1) interference fit, including press fitting, shrink and expansion fits, snap fits, and retaining rings; (2) stitching, stapling, sewing, and cotter pins; and (3) inserts in moldings and castings, integral fasteners (e.g., lanced tabs, embossed protrusions, beading, dimpling), and crimping. Refer to Groover (2002) and Kalpakjian and Schmid (2003) for more discussion.

Welding

For joining metals and alloys, welding, brazing, and soldering are the most common methods. Fusion welding consists of melting a part of the metal at the joint interface and allowing the joint to form by flow of the liquid metal, usually with another filler metal. When the same process is assisted by compressive forces applied by clamping the two pieces, the process is called *pressure welding*. Brazing and soldering use a lower-melting filler metal alloy to join the components. The parent pieces do not undergo melting and the liquid filler acts as the glue that makes the joint. The heat required to cause melting in the welding process classifies the process. Thus, in arc welding, heat is provided by creating an electric arc between electrodes often in the presence of a shielding atmosphere. Gas welding is performed using a gas mixture fuel to create a high-temperature flame that is targeted at the joining interface. The most popular gas-welding device is the oxyacetylene welding torch commonly seen in metalworking shops. Thermite welding is a well-known process for joining large sections in remote locations where gas or electric welding are difficult to perform. The thermite process relies on the exothermic heat generated by the reaction of iron oxide with aluminum. The molten product of the exothermic reaction is brought in contact with the joining interface and causes local melting of the parent pieces, followed by cooling, to form the weld. Electron and laser beam welding are some of the modern methods for joining metals and ceramics. Other methods include electroslag welding, friction welding, and explosive welding, which are made possible by a transient wave generated by detonating an explosive charge.

Brazing and Soldering

Brazing and soldering both use filler metals to join and bond two (or more) metal parts to provide a permanent joint. They are attractive compared to welding in the following conditions: (1) the metals have poor weldability, (2) dissimilar metals are to be joined, (3) the intense heat of welding may damage the components being joined, (4) the geometry of the joints does not lend itself to any of the welding methods, and/or (5) high strength is not a requirement (Groover, 2002). Brazing is a joining process in which a filler metal is melted and distributed by capillary action between the faying surfaces of the metal parts being joined. No melting of the base metals occurs in brazing; only the filler melts, as opposed to fusion welding. Brazing can be classified into torch brazing, furnace brazing, induction brazing, resistance brazing, dip brazing, infrared brazing, and braze welding. Soldering is similar to brazing and can be defined as a joining process in which a filler metal with a melting point not exceeding 4500°C (840°F) is melted and distributed by capillary action between the faying surfaces of the metals parts being joined. As in brazing, no melting of the base metals occurs, but the filler metal wets and combines with the base metal to form metallurgical bond. Common soldering methods include hand soldering, wave soldering, and reflow soldering.

Joining of Plastics, Ceramics, and Glasses

The joining of ceramics and glasses involves some methods common to metals and some that are specific to preparing glass-metal seals. Ceramic joining and bulk ceramic processing are closely related; the main joining methods consist of adhesive bonding, brazing with metals or silicates, diffusion bonding, fusion welding, and cementitous bonding. Brazing of metals to ceramics has attracted much recent attention because of its applications in heat engines, where high-temperature ceramic components (e.g., silicon nitride based) are in use. Thus, metallic alloys based on Ti, Zr, and Fe-Ni-Co are used to braze silicon nitride parts designed for high-temperature service. Thermal expansion mismatch between joining materials is an important criterion in the choice of those materials. Studies of glass-metal, ceramic-metal, ceramic-ceramic, and ceramic-polymer interfaces have led to a sophisticated understanding of joints, coatings, and enamels used in ceramic, electronic, and thermal engineering applications.

185.4 Fabrication of Microelectronic and Micromechanical Devices

Electronics Assembly and Packaging

Integrated circuits (ICs) constitute the heart of an electronic system, but the complete system consists of much more than packaged ICs. These ICs and other components are mounted and interconnected on printed circuit boards (PCBs), which in turn are interconnected and contained in a chassis or cabinet. Chip packaging is only part of the total electronic package. The rest of electronics packaging is briefly introduced here. A well-designed electronic package serves the following functions: power distribution and signal interconnections, structure support, circuit protection from physical and chemical hazards in the environment, dissipation of treatment generated by the circuits, and minimum delays will signal transmission within the system (Groover, 2002). A complex electromechanical system is organized into levels, which are often called the packaging hierarchy: (1) Level 0, intraconnections on the chip; (2) Level I, chip to package interconnections to form IC package; (3) Level 2, IC package to circuit board interconnections; (4) Level 3, circuit board to rack, or card-to-board packaging; and (5) Level 4, wiring and cabling, connections to the cabinet.

Microfabrication and Nanofabrication

Miniaturization is an important trend in materials processing. Microoelectromechanical systems (MEMS) illustrate this trend. These MEMS can potentially power microrobots to repair human cells, and produce microknives for surgery and camera shutters for precise photography. Some are now widely used in sensors, ink-jet printing mechanisms, and magnetic storage devices. The more recent trend is the development of nanoelectromechanical systems (NEMS), which operate on the same scale as biological molecules. Figure 185.3 indicates the relative sizes and other factors usually associated with these and traditional manufacturing terms. MEMS products can be classified by type of device or by application. The first classification includes microsensors, microactuators, microstructures and microcomponents, and microsystems and microinstruments. The second classification includes ink-jet heads, thin-film magnetic heads, compact disks, automotive, medical, chemical and environmental, and other applications in scanning probe microscopes, biotechnology, and electronics.

Typical microfabrication technologies include silicon layer processes, LIGA processes, photochemical machining, electrodischarge machining, electron-beam machining, laser beam machining, ultrasonic machining, wire electrodischarge machining, ultra-high precision machining using single-crystal diamond cutting tools and control systems with resolutions as fine as 0.01 μm, electrochemical fabrication, microstereolithograpliy, and photofabrication (Madou, 1997). Two good references on MEMS are Madou (1997) and Kovacs (1998).

Nanostructures consist of physical features whose dimensions are in the range of 1 to 100 nanometers (nm). Structures of this scale can almost be thought of as purposely arranged collections of individual atoms and molecules. Two alternative processing technologies may be used to fabricate these NEMS items: (1) additive molecular processes that build the nanostructure from individual atoms, and (2) nanofabrication technologies similar to microfabrication processes, only performed on a small scale. A good reference in NEMS is Goddard et al. (2003).

Two kinds of new nanomaterials have been widely covered in the media in terms of their science, discovery, and potential uses: fullerenes and carbon nanotubes. Fullerenes can be used for applications

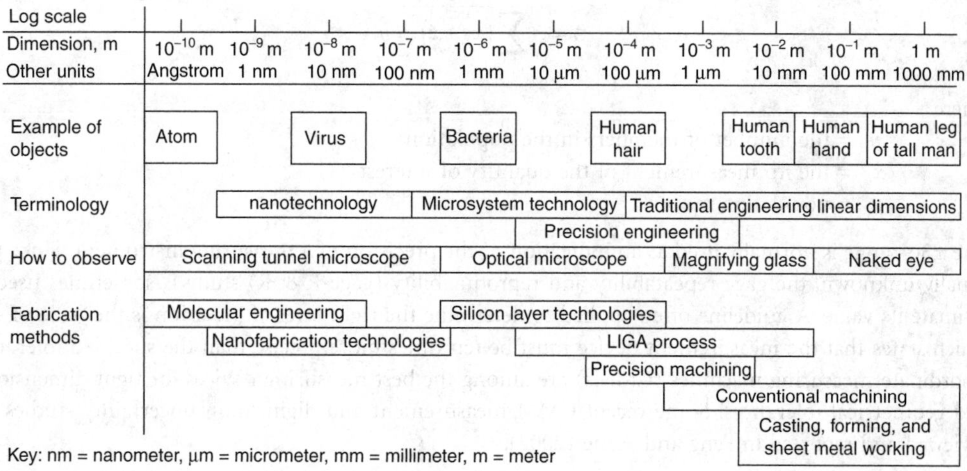

FIGURE 185.3 Terminology and relative sizes for microsystems and related technologies. (*Source:* Groover, M.P., *Fundamentals of Modern Manufacturing*, 2nd ed., p. 846, Wiley, New York. With permission.)

such as medical imaging and drug delivery. Carbon nanotubes look like cylinders of rolled-up chicken wire, but they are ten times stronger than steels, one-sixth the weight of steels, and more conductive than copper. Labs have been able to produce them for more than 10 years, but only with recent advances in equipment and instruments have the prices begun to drop to the point where commercial uses could be feasible. In 1999, the cost of buckyballs was around $600 per gram, compared to only $10 at the end of 2003.

Near-term commercial applications of nanomaterials will be largely found in nanoparticle-reinforced materials and nanoparticle-filled coatings. Nanoparticle-reinforced materials provide additional strength or some other desired property to polymers. Toyota, for example, uses layered silicates, in its cars on fan belt covers. It has also combined materials at the nanoscale to make bumpers as strong as traditional ones, but 60 times thinner. Nanoparticle-filled coatings can be used to create scratch resistant surfaces, antireflective coatings, and corrosion-resistant coatings for various metals.

185.5 Measurement and Inspection

A basic requirement in materials processing is to meet the design specifications in dimensions, tolerances, hardness, and surface finishes. Measurement is a procedure in which an unknown quantity is compared with a known standard. Inspection is a procedure in which a part or product characteristic, such as a dimension or hardness, is examined to determine whether it conforms to design specifications. Many inspection procedures rely on measurement techniques, while others using gaging methods. The inspection by measurement produces variables or quantitative data while inspection by gaging results in attributes or qualitative data. Gaging (also spelled as *gauging*) determines simply whether the part characteristic meets or does not meet the design specification. It is usually faster, but scant information is provided about the actual value of the characteristic of interest.

Some concepts and principles apply virtually to all measurements. The most important are accuracy and precision. Accuracy is the degree to which the measured value agrees with the true value of the quantity of interest. A measurement system is accurate when it is absent of systematic errors. Systematic errors are deviations from the true value that are consistent from one measurement to the next. Precision is the degree of repeatability in the measurement process. A measurement procedure is precise when the random errors in this procedure are minimized. Random errors are usually associated with human participation in the measurement process. They are assumed to obey a normal distribution whose mean is zero and estimated standard deviation is σ given by

$$\sigma = \sqrt{\sum (x_i - \bar{x})^2 / n}$$

where

n = the number of members in the population
x_i = the ith measurement of the quantity of interest
μ = population mean

The $\pm 3\sigma$ range is normally used as an indication of the precision of a measuring instrument. Since μ is usually unknown, the gage repeatability and reproducibility (gage R & R) study is sometimes used to estimate its value. A guideline often applied to determine the right level of precision is the "rule of 10," which states that the measurement device must be ten times more precise than the specified tolerance. Coordinate measuring machines (CMMs) are among the best measuring devices for tight dimensional and geometrical tolerances. Some recent CMM measurement and digitization uncertainty studies are reviewed and reported in Feng and Wang (2002).

Acknowledgments

This revision is based on the same chapter contributed originally by Dr. Subhash H. Risbus of the University of California at Davis to the first edition of this book.

Defining Terms

Casting — The shaping of a metal alloy into a solid piece by pouring liquid metal into the cavity inside a sand mold.

Machining — Final steps in the manufacture of a component; may include cutting, polishing, and grinding operations.

Metalworking — Process steps to shape the cast material into components by cold or hot work. Examples of processes include rolling, forging, extrusion, and drawing.

Powder processing — The conversion of powdered material into a solid piece by cold pressing and heating (sintering) or simultaneous heating and pressure application (hot pressing).

Sol-gel process — The use of organic or organometallic solutions to make the component into a bulk piece or coating at lower temperatures than in conventional melting and casting. Used for oxide ceramics and glasses at present.

References

Benedict, G. F. 1987. *Nontraditional Manufacturing Processes*. Marcel Dekker, New York.

Feng, C-X. and Kusiak, A. 2000. Robust tolerance design with the design of experiments approach, *Trans. ASME, J. Manuf. Sci. Eng.*, 122:520–528.

Feng, C-X. and Wang, X-F. 2003. Surface roughness predictive modeling: neural networks vs. regression, *IIE Trans.*, 35:11–27.

Feng, C-X. and Wang, X-F. 2002. Subset selection in predictive modeling of the CMM digitization uncertainty, *SME J. Manuf. Syst.*, 21:419–439.

Feng, C-X., Wang, J., and Wang, J-S. 2001. An optimization model for concurrent selection of tolerances and suppliers, *Comput. Ind. Eng.*, 40:15–33.

Feng, C-X., Wang, X-F. and Yu, Z. 2002. Neural networks modeling of engine cylinder liner honing surface roughness parameters defined by ISO13565, *SME J. Manuf. Syst.*, 21:395–408.

German, R. M. 1994. *Powder Metallurgy Science*, 2nd ed. Metal Powder Industries Federation, Princeton, NJ.

Goddard, W. A. III, Brener, D. W., Lyshevski, S. E., and Iafrate, G. J. , Eds., 2003. *Handbook of Nanoscience, Engineering and Technology.* CRC Press, Boca Raton, FL.

Groover, M. P. 2002. *Fundamentals of Modern Manufacturing,* 2nd ed. John Wiley & Sons, New York.

International Standards Organization. 1996. Geometrical Product Specifications (GPS). Surface Texture: Profile Method; Surface Having Stratified Functional Properties. Part 2: Height Characterization Using Linear Material Ratio, ISO 13565-2. Geneva, Switzerland.

Kalpakjian, S. and Schmid, S. R. 2003. *Manufacturing Processes for Engineering Materials,* 4th ed. Prentice Hall, Upper Saddle River, NJ.

Kovacs, C. T. A. 1998. *Micromachined Transducers Handbook.* McGraw-Hill, New York.

Madou, M. 1997. *Fundamentals of Microfabrication.* CRC Press, Boca Raton, FL.

Wright, P. K. 2001. *21st Century Manufacturing.* Prentice Hall, Upper Saddle River, NJ.

Further Information

Sources of current information on materials and manufacturing methods include the following professional journals:

ASME Journal of Manufacturing Science and Engineering
IEEE Transactions on Components, Packaging, and Manufacturing Technology
IIE Transactions on Design and Manufacturing
SME Journal of Manufacturing Systems
SME Journal of Manufacturing Processes
Journal of Materials Engineering and Performance
American Ceramic Society Bulletin
Powder Metallurgy
Materials Science and Engineering A and *B*
Transactions of the NAMRI/SME

186

Machine Tools and Processes

Yung C. Shin
Purdue University

Machine tools are the machinery used to process various materials in order to get desired shapes and properties. Machine tools typically consist of a base structure that supports various components and provides overall rigidity; kinematic mechanisms and actuators, which generate necessary motions of the tools and tables; and various feedback sensors. Machine tools are often used in conjunction with jigs and fixtures, which would hold the part in position, and processing tools. An example of a machine tool used in material removal processing is shown in Figure 186.1.

The first generation of modern machine tools was introduced during the industrial revolution in the 18th century with the invention of steam engines. Machine tools opened an era of automation by providing the means to replace human work with mechanical work. In the early days, automation with machine tools was performed using various kinematic mechanisms, but it evolved into programmable automation with the use of computer numerical control. Today, many machine tools are operated with electrical or hydraulic power and controlled by a computer.

The major application of automation using machine tools in the early years was in transfer lines. Many special-purpose machines were grouped together with a part-moving system, such as a conveyor, providing the transfer of parts from one machine to another. A specific operation was performed on each station, and the part was moved to the next station until the finished product was obtained. Recently, however, **flexible automation** has been actively pursued to cope with continuous and rapid change of product designs and cycles. In such systems, several numerically controlled machines are clustered together to perform a variety of jobs. Programs are used, instead of inflexible mechanisms, to control the machines, so that the system can easily adapt to different job requirements.

186.1 Economic Impact

In most industrialized countries, manufacturing contributes about 20 to 30% of the nation's GNP. Since most manufacturing processes are performed with machine tools, the production and consumption of machine tools play a significant role in the nation's economy. Machine tool companies in the U.S. directly employ about 350,000 people. If the surrounding ancillary industries are included — such as the tooling industry, material-handling industry, and sensors and controller manufacturers — the total

0-8493-1586-7/05/$0.00+$1.50

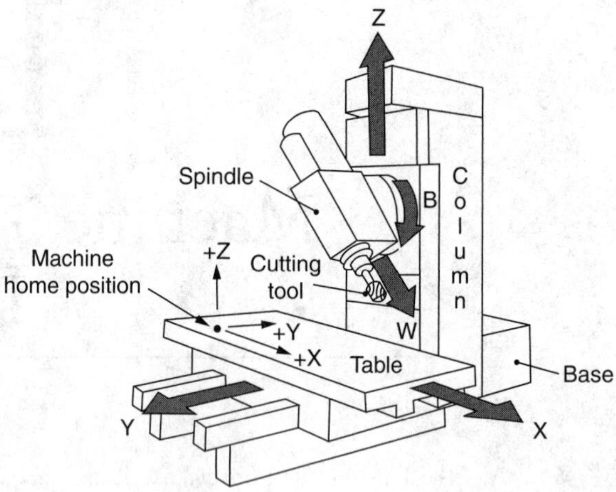

FIGURE 186.1 An example of machine tools.

employment would double that figure. More importantly, the machine tool industry affects a much larger user community.

186.2 Types of Machine Tools

Various manufacturing processes led to development of a variety of machine tools. Therefore, machine tools can be categorized by manufacturing processes. Manufacturing processes can be grouped into four main categories: casting, forming, material removing, and joining.

Casting machines are used to generate various unfinished or finished parts of defined shapes from shapeless materials. Casting is typically performed under high temperatures with molten material. The material in liquid state is poured into a mold or a die that has the shape of the final part, and is subsequently solidified. Examples of casting processes entail sand casting, centrifugal casting, die casting, and continuous casting. No processing tools are used on this type of machine, and high accuracy is not usually required. Casting is usually followed by forming or machining processes in order to attain a final shape or to improve the dimensional accuracy.

Forming machines are used to change the shape of materials in the solid state. During the process, no significant change of volume or weight in the material occurs, but the mechanical properties are altered due to the large amount of bulk deformation and high stresses. Forming processes are performed at either high or low temperature. The required force and power are usually large. Examples of forming machine tools include presses; forging machines; rolling mills; and stamping, drawing, extrusion, and bending machines.

Material removal processes are used to obtain an accurate shape of a part that has been obtained by casting or forming processes. During the material removal process, the volume as well as the shape is changed. Machine tools used for material removal processes often require the highest accuracy. Typical processes under this category include turning, drilling, milling, boring, grinding, and so forth. Material removal processes have been used mostly for metals but have recently found new applications in non-metallic materials such as ceramics and composites.

Machine tools for material removal processes are designed to provide multiaxis motions. They usually consist of a table used for mounting a workpiece, a spindle to hold tools, and a motor and gears to provide various speeds and feeds. In order to obtain a precision motion, axes are often controlled by microprocessor-based computer numerical controllers.

In addition to the processes already mentioned, many nontraditional machining processes are commonly used these days. Nontraditional machining processes include any that use another energy source with or without mechanical power. Typical nontraditional machining processes are electrodischarge machining (EDM), electrochemical grinding, laser machining, and electron-beam machining. These processes utilize either electric or heat energy to remove the material.

Joining is an operation to assemble components. Examples are welding, soldering, adhesion, and riveting. These processes are usually performed at the final stage to fabricate a product from various finished components. The types of machine tools in this category are friction-welding, arc-welding, electron-beam–welding, and laser-welding machines. In addition, robots are commonly used for repetitive assembly operations.

186.3 Control of Machine Tools

In the early days, machine tools were controlled automatically by means of mechanical systems, such as cams, gears, and levers, which provided repetitive or sequential motions of machine tools. These mechanical devices shifted the operator's role from turning wheels or moving levers to supervising overall operation.

Later, hydraulic control replaced the mechanical control systems. Hydraulic systems provided more precise actions with higher power. The hydraulic control later combined with electronic switches were able to generate much more complex operations than mechanical control systems, but were eventually replaced by electronic controls.

Numerically controlled (NC) machines were developed to achieve automatic operation of machine tools with a program. The NC machine was first conceived by John Parson in 1948, and the first prototype was introduced by Massachusetts Institute of Technology (MIT) in 1952. In early NC machines, electric relays were grouped and hard-wired so that users could generate a complex sequence of operations. In NC machines, programs to generate commands for the sequences of operations were stored into punched tapes, which were subsequently read into a computer which interpreted the program and generated pulses to drive mechanisms.

In the beginning mainframes were used as computers to process the data. However, microprocessors and magnetic media replaced numerical control in the 1970s and opened the era of computer numerical controllers (CNCs). Instead of being connected to the mainframe computer and sequentially controlled, CNC machines were designed with individual microprocessor-based controllers. CNC machines are usually equipped with a tool magazine, which holds multiple tools. These machines can perform many different operations on a single machine with automatic tool change and hence are called *machining centers.*

Computer programs used to operate the CNC machines are called *part programs*, which contain a sequence of commands. The commands used in a part program include preparatory G codes and M codes defining miscellaneous functions, as well as specific codes such as S, F, T, or X codes. The G codes are universally used regardless of the type of CNC and are defined in both ISO (International Standards Organization) and EIA (Electronic Industry Association) standards (RS-273-A). Some examples of these codes are shown in Table 186.1. With NC or CNC machine tools, part programs can be generated at a remote site. The English-like languages, such as automatic programming tool (APT), are commonly used for the ease of generating part programs. Nowadays, various computer-aided manufacturing (CAM) packages provide means to graphically generate part programs from design drawings created by computer-aided design (CAD) software. The part programs generated with CAM software must be postpro-

TABLE 186.1 Examples of G and M Codes for Part Programs

G00	Point-to-point positioning	M03	Spindle rotation, CW
G01	Linear interpolation	M04	Spindle rotation, CCW
G02	Circular interpolation, CW	M05	Spindle stop
G03	Circular interpolation, CCW	M06	Tool change
G06	Spline interpolation	M08	Coolant No. 1 on
G17	*xy*-plane selection	M09	Coolant stop
G41	Tool radius compensation, left		

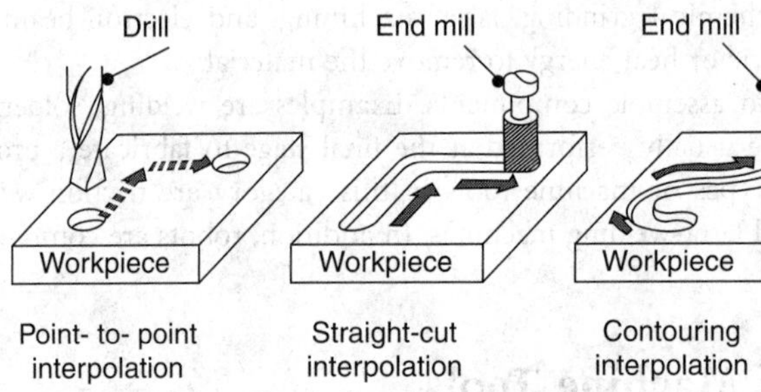

FIGURE 186.2 Illustration of interpolations.

cessed into the format of the specific CNC to be used and downloaded to the machine, where it can be edited if necessary.

Interpolation

Most CNC controllers have the capability of performing various types of interpolation. The simplest type is point-to-point interpolation. This scheme, in which beginning and final positions are of greatest importance, is usually used for rapid movement of tools or workpieces. The second type is the straight-line interpolation, which is used to generate a motion following a straight line with a specified speed. Accurate positioning and maintenance of a constant speed are required. The most complicated type of interpolation is contouring. Trajectory commands for contouring are typically generated using circular or cubic spline functions. Figure 186.2 shows the graphical illustration of these interpolations.

Feedback Control

The generated commands are sent to the actuators of each axis to provide the necessary motion of the pertinent component in either open-loop or closed-loop configuration. In early CNC machines, stepper motors were often used as actuators, but their usage is at present limited to inexpensive, small machine tools. Either DC (direct current) or AC (alternating current) servomotors are popularly used for modern machine tools with feedback control. Figure 186.3 shows the open-loop control system with a stepper motor, whereas Figure 186.4 depicts the closed-loop configuration.

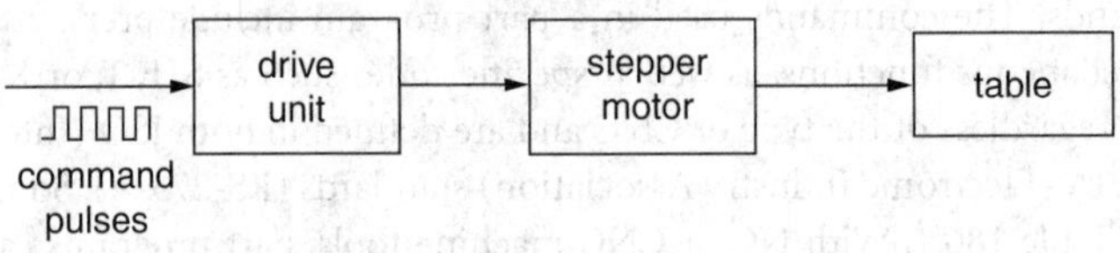

FIGURE 186.3 Open-loop control system with stepper motors.

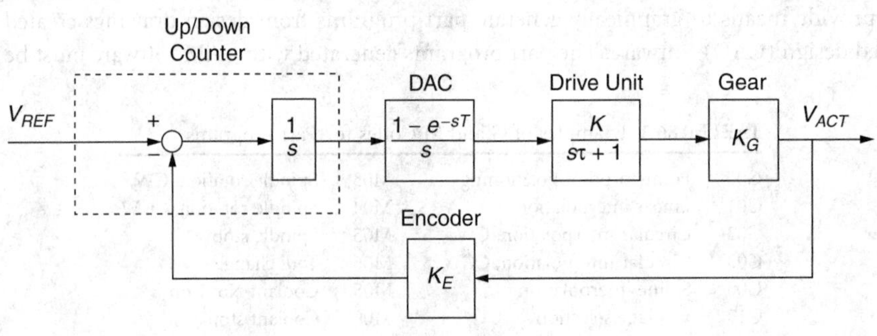

FIGURE 186.4 Closed-loop control system with stepper motors.

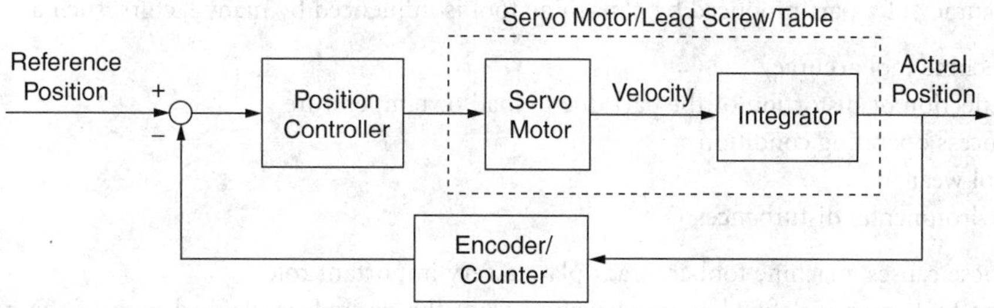

FIGURE 186.5 Digital position control system.

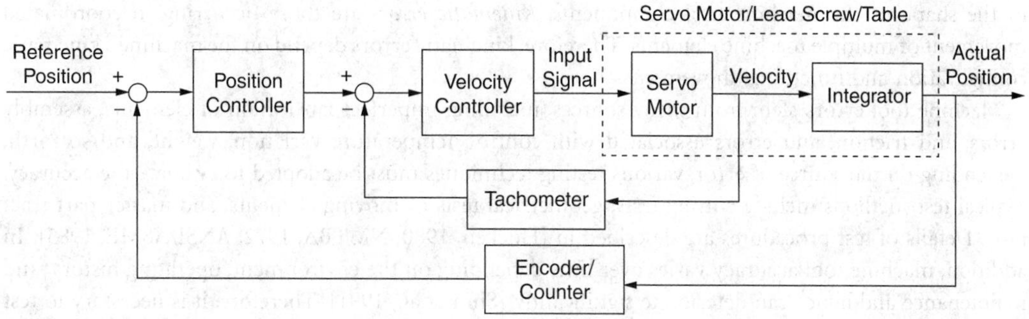

FIGURE 186.6 Digital control system with position and velocity feedback.

Early NC systems used a control scheme based on pulse reference and counting. In this scheme, positional reference commands are input to the system as a series of pulses, and the position feedback is obtained through an encoder. The command pulses are generated by the NC controller, with each pulse corresponding to a fixed **basic length unit** (BLU), which determines the system's accuracy. The frequency of the pulses determines the velocity of the axis, and positional accuracy is achieved by comparing the feedback pulses with the input pulse through the up-down counter.

In more recent CNC systems, command generation and control are performed digitally using a digital controller with servomotors. Proportional-integral-derivative (PID) controllers are most popularly used as feedback controllers with either single- or double-feedback loops. An example of a position control system with a proportional controller is shown in Figure 186.5. The more advanced system shown in Figure 186.6 has two feedback loops, which perform velocity control in the inner loop and position tracking in the outer loop. The most common method uses a proportional type for the position control and a PI controller for the velocity loop. These controllers are designed for the nominal operating condition of the machine.

Recently, adaptive control (AC) has opened a new era for machine tool control. Adaptive control provides the machine tool with the ability to adapt itself to the dynamic changes of the process condition to maintain optimum performance. Without adaptive control, operating conditions might have to be chosen too conservatively to avoid failure, thereby resulting in the loss of productivity. In addition, adaptive control can improve the accuracy of an existing machine, particularly in contouring. Adaptive control is particularly important for the realization of untended operation in the factory of the future.

186.4 Machine Tool Accuracy

The standard of accuracy has been increasing steadily over the last few decades. With the advent of high-precision machine tool technology and measurement techniques, manufacturers can presently attain accuracy that was almost impossible to achieve a few decades ago. For example, machining accuracy of 5 to 10 μm from regular machine tools and 0.1 μm from precision machine tools is routinely achieved. It is projected that the accuracy will improve about tenfold every 2 decades.

The accuracy of a part produced by a machine tool is influenced by many factors, such as

- Machine tool accuracy
- Deflection or distortion of the part due to load or temperature
- Process operating condition
- Tool wear
- Environmental disturbances

Among these causes, machine tool accuracy plays a very important role.

Machine tool error is defined as the deviation from the desired or planned path of the movement between the total and the workpiece. Machine tool errors are typically grouped into geometric and kinematic errors. *Geometric error* includes positional inaccuracies of the individual link and the errors in the shape of the machine tool components. *Kinematic errors* are those occurring in coordinated movement of multiple machine elements. Therefore, kinematic errors depend on the machine's kinematic configuration and functional movements.

Machine tool errors stem from many sources, including imperfect fabrication of elements, assembly errors, and friction, and errors associated with control, temperature variation, weight, and so forth. Depending on the source of error, various testing techniques must be adopted to evaluate the accuracy. Typical test methods include cutting tests, geometrical tests of moving elements, and master-part tract tests. Details of test procedures are described in [Hocken, 1980; NMTBA, 1972; ANSI/ASME, 1985]. In addition, machine tool accuracy varies over time depending on the environment, operating history, and maintenance and hence can deteriorate significantly [Shin et al., 1991]. Therefore, it is necessary to test the accuracy of the machine tools periodically and recalibrate the axes.

Machine tool error is divided into two categories: repeatable (deterministic) and nonrepeatable (random) errors. Typical measurements of positional errors are shown in Figure 186.7. Due to backlash and friction, errors often exhibit hysteresis if the measurements are performed in both directions. Generally, accuracy is defined as the worst possible error at a particular position, and is represented in terms of a deterministic value defined as the mean deviation from the target position and the probabilistic distribution of random variation. Mathematically, positional accuracy is defined as

$$\varepsilon(x, y, z) = \max(|\mu + 3\sigma|, |\mu - 3\sigma|)$$

where μ is the mean positional accuracy and σ is the variance of the random part. The nonrepeatable portion is used to specify the repeatability. Typically, three sigma (variation) values are used to determine the repeatability.

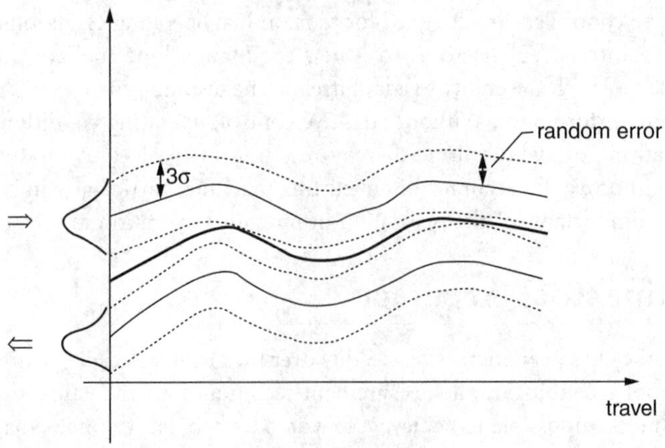

FIGURE 186.7 Presentation of errors of a machine tool axis moving in two directions.

Defining Terms

Basic length unit — Smallest resolution attainable in positioning.

Flexible automation — Automation used to produce a variety of parts without converting the hardware setup.

References

American National Standards Institute/American Society of Mechanical Engineers (ANSI/ASME). 1985. *Axes of Rotation: Methods for Specifying and Testing*. ANSI/ASME B89.3.4.M.

Hocken, R. 1980. *Technology of Machine Tools*, vol. 5. Machine Tool Task Force, Lawrence Livermore Laboratory, Livermore, CA.

NMTBA. 1972. *Definition and Evaluation of Accuracy and Repeatability for Numerically Controlled Machine Tools*, 2nd ed. National Machine Tool Builders Association, McLean, VA.

Shin, Y.C., Chin, H., and Brink, M.J. 1991. Characterization of CNC machining centers. *J. Manuf. Syst.*, 10:407–421.

Ferreira, P.M. and Liu, C.R. 1993. A method for estimating and compensating quasistatic errors of machine tools, *ASME J. Eng. Ind.*, February, pp. 149–159.

Franklin, G.F., Powell, J.D. and Emami-Naeini, A. 2002. *Feedback Control of Dynamic Systems*, 4th ed. Prentice Hall, New York.

Jorgensen, B.R. and Shin, Y.C. 1997. Dynamics of machine tool spindle/bearing systems under thermal growth, *Trans. ASME J. Tribology*, 119:875–882.

Shin, Y.C., Chin, H. and Brink, M.J. 1992. Characterization of CNC Machining Centers, *J. Manuf. Syst.*, 10:407–421.

Shin, Y.C. and Wei, Y. 1992. A statistical analysis of positional errors of a multiaxis machine tools, *Precision Eng.*, 14:139–146.

Taniguchi, N. 1982. Current status and future trends of ultraprecision machining processes, *Metalworking*, March, pp. 34–47.

Further Information

A comprehensive treatment of machine tool technologies is presented in the following two books:

Lawrence Livermore National Laboratory. 1980. *Technology of Machine Tools*. LLNL, Livermore, CA.

Weck, M. 1984. *Handbook of Machine Tools*, 4 vols. Wiley & Sons, New York.

187

Ergonomics/Human Factors

Waldemar Karwowski
University of Louisville

According to the International Ergonomics Association (IEA, 2000), ergonomics (or human factors) is the scientific discipline concerned with the understanding of interactions among humans and other elements of a system, and the profession that applies theory, principles, data, and methods to design in order to optimize human well-being and overall system performance. Ergonomists contribute to the design and evaluation of tasks, jobs, products, environments and systems in order to make them compatible with the needs, abilities, and limitations of people.

Ergonomics discipline promotes a holistic approach to work systems design and management that considers the physical, cognitive, social, organizational, environmental, and other relevant factors (Karwowski and Marras, 1999; Karwowski, 2001). *Physical ergonomics* is concerned with human anatomical, anthropometric, physiological, and biomechanical characteristics as they relate to physical activity. *Cognitive ergonomics* is concerned with mental processes, such as perception, memory, reasoning, and motor response, as they affect interactions among humans and other elements of a system. *Organizational ergonomics* (also known as macroergonomics) is concerned with the optimization of sociotechnical systems, including their organizational structures, policies, and processes. Relevant topics include communication, crew resource management, work design, design of working times, teamwork, participatory design, community ergonomics, cooperative work, new work paradigms, virtual organizations, telework, and quality management.

0-8493-1586-7/05/$0.00+$1.50

187.1 Origins of Ergonomics Discipline

The science of ergonomics originated in 1857 when Wojciech Jastrzebowski of Poland defined the term by combining two Greek words: ergon (work) + nomos (laws). This new science signified then the human work, play, thinking, and devotion as reflected in the manner of optimizing the use of four distinct human characteristics: motor (physical), sensory (aesthetic), mental (intellectual), and spiritual or moral (Karwowski, 1991). The term ergonomic was independently reinvented by K. E. H. Murrell in 1949. The contemporary *human factors* discipline the parallel term for this new scientific discipline adopted in the United States, discovers and applies information about human behavior, abilities, limitations, and other characteristics to the design of tools, machines, systems, tasks, jobs, and environments for productive, safe, comfortable, and effective human use (Sanders and Mccormick, 1993). In this context, ergonomics deals with a broad scope of problems relevant to the design and evaluation of work systems, consumer products, and working environments, whereas human–machine interactions affect human performance and product usability. The wide scope of issues addressed by ergonomics is presented in Table 187.1.

Human factors design and engineering aims to optimize the design and functioning of human–machine systems with respect to complex characteristics of people and the relationships between system users, machines, and outside environments. According to the Board of Certification in Professional Ergonomics (BCPE), a practitioner of ergonomics is a person who (1) has a mastery of a body of ergonomics knowledge, (2) has a command of the methodologies used by ergonomists in applying that knowledge to the design of a product, system, job, or environment, and (3) has applied his or her knowledge to the analysis, design testing, and evaluation of products, systems, and environments. The areas of current practice in the field can be best described by examining the focus of technical groups of the Human Factors and Ergonomics Society (2003), as illustrated in Table 187.2.

187.2 Concept of Human–Machine Systems

A human–machine system can be broadly defined as an organization of people and the machines they operate and maintain in order to perform assigned jobs that implement the purpose for which the system was developed (Meister, 1987). The human functioning in such a system is described in terms of perception, information processing, decision making, memory, attention, feedback, and human response processes. Furthermore, the human work taxonomy can be used to describe five distinct levels of human functioning, ranging from primarily physical tasks to cognitive tasks (Karwowski and Rodrick, 2001). These basic, but universal, human activities are (1) tasks that produce force (primarily muscular work); (2) tasks of continuously coordinating sensory-monitor functions (e.g., assembling or tracking tasks); (3) tasks of converting information into motor actions (e.g., inspection tasks); (4) tasks of converting information into output information (e.g., required control tasks); and (5) tasks of producing information (primarily creative work). Any task in a human–machine system requires processing of information that is gathered based on perceived and interpreted relationships between system elements. The processed information may need to be stored by either a human or a machine for later use.

One of the important concepts in the design of human–machine systems is the stimulus–response compatibility paradigm (Wickens and Carswell, 1997). This paradigm relates to the physical relationship (compatibility) between a set of stimuli and a set of responses as this relationship affects the speed of human response. The spatial relations between arrangements of signals and response devices in human–machine systems with respect to direction of movement and adjustments are often ambiguous, with a high degree of uncertainty regarding the effects of intended control actions. It should be noted that the information displayed to the human operator can be arranged along a continuum that defines the degree to which that information is spatial-analog (i.e., information about relative locations, transformations, or continuous motion); or linguistic-symbolic/verbal (i.e., a set of instructions, alphanumeric codes, directions, or logical operations). The scope of ergonomics factors that need to be considered in design, testing, and evaluation of any human–machine system is shown in Table 187.3 in the form of an exemplary ergonomic checklist.

TABLE 187.1 Classification Scheme for Human Factors/Ergonomics

1. General	
	Human Characteristics
2. Psychological aspects	
3. Physiological and anatomical aspects	
4. Group factors	
5. Individual differences	
6. Psychophysiological state variables	
7. Task-related factors	
	Information Presentation and Communication
8. Visual communication	
9. Auditory and other communication modalities	
10. Choice of communication media	
11. Person–machine dialog mode	
12. System feedback	
13. Error prevention and recovery	
14. Design of documents and procedures	
15. User control features	
16. Language design	
17. Database organization and data retrieval	
18. Programming, debugging, editing, and programming aids	
19. Software performance and evaluation	
20. Software design, maintenance, and reliability	
	Display and Control Design
21. Input devices and control	
22. Visual displays	
23. Auditory displays	
24. Other modality displays	
25. Display and control characteristics	
	Workplace and Equipment Design
26. General workplace design and buildings	
27. Workstation design	
28. Equipment design environment	
29. Illumination	
30. Noise	
31. Vibration	
32. Whole body movement	
33. Climate	
34. Atmosphere	
35. Altitude, depth, and space	
36. Other environmental issues	
	System Characteristics
37. General system features	
38. Total system design and evaluation	
39. Hours of work	
40. Job attitudes and job satisfaction	
41. Job design	
42. Payment systems	
43. Selection and screening	
44. Training	
45. Supervision	
46. Use of support	
47. Technological and ergonomic change	
48. General health and safety	

TABLE 187.1 Classification Scheme for Human Factors/Ergonomics (*Continued*)

49. Etiology
50. Injuries and illnesses
51. Prevention
Social and Economic Impact of the System
52. Trade unions
53. Employment, job security, and job sharing
54. Productivity
55. Women and work
56. Organizational design
57. Education
58. Law
59. Privacy
60. Family and home life
61. Quality of working life
62. Political comment and ethical
63. Approaches and methods

Source: Ergonomics Abstracts, Taylor & Francis, London, 2001.

187.3 Human–System Design: Compatibility Requirements

Ergonomics advocates systematic use of the knowledge concerning relevant human characteristics in order to achieve compatibility in the design of interactive systems of people, machines, environments, and devices of all kinds to ensure specific goals (HFES, 2003). Typically, such goals are improved (system) effectiveness, safety, ease of performance, and contribution to overall human well-being. Although a key to the above definition of HFES is "compatibility," this term has been typically used in the past in relation to design of displays and controls, such as the classical spatial (location) compatibility, or the intention-response-stimulus compatibility related to movement of controls (Wickens and Carswell, 1997).

The use of compatibility in a greater context of ergonomics systems has been advocated by my colleagues and I (Karwowski et al., 1988; Karwowski, 1991). Recently, I introduced the term "human-compatible systems" (Karwowski, 1997) in order to focus on the need for comprehensive treatment of compatibility in the ergonomics discipline. The 1978 edition of the *American Heritage Dictionary* defines "compatible" as follows: "1. capable of living or performing in harmonious, agreeable, or congenial combination with another or others; and 2. capable of orderly, efficient integration and operation with other elements in a system." Probing further, the word "congenial" is defined as suited to one's needs, agreeable, while the "harmony" (from Greek *harmos*, meaning joint) is defined as "1. agreement in feeling, approach, action disposition, or the like, sympathy, accord; and 2. the pleasing interaction or appropriate combination of elements in a whole."

Symvatology: The Science of Artifact–Human Compatibility

An *artifact system* is a set of all artifacts (meaning objects made by human work) as well as natural elements of the environment), and their interactions occurring in time and space afforded to us by nature. Due to its interactions, the artifact system is often a dynamic system with a high level of complexity that can exhibit a nonlinear behavior. A human system can be defined as the human (or humans) with all the characteristics (physical, perceptual, cognitive, emotional, etc.) that are relevant to an interaction with the artifact system. The notation of artifact–human system compatibility expresses the premise that it is the artifact system that should be compatible with the human system, and that such compatibility should primarily be ensured by design of the artifact system.

The science of the artifact–human (system) compatibility or "symvatology" (Karwowski, 2000) was proposed as the corroborative discipline to ergonomics in order to build solid foundations for the science

TABLE 187.2 Subject Interests of Technical Groups of Human Factors and Ergonomics Society

Technical Group	Description/Areas of Concern
I. Aerospace systems	Applications of human factors to the development, design, operation, and maintenance of human–machine systems in aviation and space environments (both civilian and military).
II. Aging	Human factors applications appropriate to meeting the emerging needs of older people and special populations in a wide variety of life settings.
III. Communications	All aspects of human-to-human communication, with an emphasis on communication mediated by telecommunications technology, including multimedia and collaborative communications, information services, and interactive broadband applications. Design and evaluation of both enabling technologies and infrastructure technologies in education, medicine, business productivity, and personal quality of life.
IV. Computer systems	Human factors aspects of (1) interactive computer systems, especially user interface design issues; (2) the data-processing environment, including personnel selection, training, and procedures; and (3) software development.
V. Consumer products	Development of consumer products that are useful, usable, safe, and desirable. Application of the principles and methods of human factors, consumer research, and industrial design to ensure market success.
VI. Educators' professional	Education and training of human factors and ergonomics specialists in academia, industry, and government. Focus on both degree-oriented and continuing education needs of those seeking to increase their knowledge and/or skills in this area, accreditation of graduate human factors programs, and professional certification.
VII. Environmental design	Human factors aspects of the constructed physical environment, including architectural and interior design aspects of home, office, and industrial settings. Promotion of the use of human factors principles in environmental design.
VIII. Forensics professional	Application of human factors knowledge and technique to standards of care and accountability established within legislative, regulatory, and judicial systems. The emphasis on providing a scientific basis to issues being interpreted by legal theory.
IX. Industrial ergonomics	Application of ergonomics data and principles for improving safety, productivity, and quality of work in industry. Concentration on service and manufacturing process, operations, and environments.
X. Medical systems and functionally impaired populations	All aspects of the application of human factors principles and techniques toward the improvement of medical systems, medical devices, and the quality of life for functionally impaired user populations.
XI. Organizational design	Improving productivity and the quality of life by an integration of psychosocial, cultural, and technological factors and with user interface factors (performance, acceptance, needs, limitations) in design of jobs, workstations, and related management systems.
XII. Personality and individual differences in human performance	The range of personality and individual difference variables that are believed to mediate performance.
XIII. Safety	Research and applications concerning human factors in safety and injury control in all settings and attendant populations, including transportation, industry, military, office, public building, recreation, and home improvements.
XIV. System development	Concerned with research and exchange of information for integrating human factors into the development of systems. Integration of human factors activities into system development processes in order to provide systems that meet user requirements.
XV. Test and evaluation	A forum for test and evaluation practitioners and developers from all areas of human factors and ergonomics. Concerned with methodologies and techniques that have been developed in their respective areas.
XVI. Training	Fosters information and interchange among people interested in the fields of training and training research.
XVII. Visual performance	The relationship between vision and human performance, including (1) the nature, contents, and quantification of visual information and the context in which it is displayed; (2) the physics and psychophysics of information display; (3) perceptual and cognitive representation and interpretation of displayed information; (4) assessment of workload using visual tasks; and (5) actions and behaviors that are consequences of visually displayed information.

TABLE 187.3 Examples of Factors to be Used in Ergonomics Checklists

I. Anthropometric, biochemical, and physiological factors

1. Are the differences in human body size accounted for by the design?
2. Have the right anthropometric tables been used for specific populations?
3. Are the body joints close to neutral positions?
4. Is the manual work performed close to the body?
5. Are there any forward-bending or twisted trunk postures involved?
6. Are sudden movements and force exertion present?
7. Is there a variation in worker postures and movements?
8. Is the duration of any continuous muscular effort limited?
9. Are the breaks of sufficient length and spread over the duration of the task?
10. Is the energy consumption for each manual task limited?

II. Factors related to posture (sitting and standing)

1. Is sitting/standing alternated with standing/sitting and walking?
2. Is the work height dependent on the task?
3. Is the height of the work table adjustable?
4. Are the height of the seat and backrest of the chair adjustable?
5. Is the number of chair adjustment possibilities limited?
6. Have good seating instructions been provided?
7. Is a footrest used where the work height is fixed?
8. Has the work above shoulder or with hands behind the body been avoided?
9. Are excessive reaches avoided?
10. Is there enough room for the legs and feet?
11. Is there a sloping work surface for reading tasks?
12. Have the combined sit-stand workplaces been introduced?
13. Are handles of tools bent to allow for working with the straight wrists?

III. Factors related to manual materials handling (lifting, carrying, pushing, and pulling loads)

1. Have tasks involving manual displacement of loads been limited?
2. Have optimum lifting conditions been achieved?
3. Is anybody required to lift more than 23 kg?
4. Have lifting tasks been assessed using the National Institute for Occupational Safety and Health (1991) method?
5. Are handgrips fitted to the loads to be lifted?
6. Is more than one person involved in lifting or carrying tasks?
7. Are there mechanical aids for lifting or carrying available and used?
8. Is the weight of the load carried limited according to the recognized guidelines?
9. Is the load held as close to the body as possible?
10. Are pulling and pushing forces limited?
11. Are trolleys fitted with appropriate handles and handgrips?

IV. Factors related to the design of tasks and jobs

1. Does the job consist of more than one task?
2. Has a decision been made about allocating tasks between people and machines?
3. Do workers performing the tasks contribute to problem solving?
4. Are the difficult and easy tasks performed interchangeably?
5. Can workers decide independently on how the tasks are carried out?
6. Are there sufficient possibilities for communication between workers?
7. Is sufficient information provided to control the assigned tasks?
8. Can the group take part in management decisions?
9. Are the shift workers given enough opportunities to recover?

V. Factors related to information and control tasks

Information

1. Has an appropriate method of displaying information been selected?
2. Is the information presentation as simple as possible?
3. Has the potential confusion between characters been avoided?
4. Has the correct character/letter size been chosen?
5. Have tests with capital letters only been avoided?
6. Have familiar typefaces been chosen?
7. Is the text/background contrast good?
8. Are the diagrams easy to understand?
9. Have the pictograms been properly used?
10. Are sound signals reserved for warning purposes?

TABLE 187.3 Examples of Factors to be Used in Ergonomics Checklists (*Continued*)

Controls

1. Is the sense of touch used for feedback from controls?
2. Are differences between controls distinguishable by touch?
3. Is the location of controls consistent and is sufficient spacing provided?
4. Have the requirements for the control–display compatibility been considered?
5. Is the type of cursor control suitable for the intended task?
6. Is the direction of control movements consistent with human expectations?
7. Are the control objectives clear from the position of the controls?
8. Are controls within easy reach of female workers?
9. Are labels or symbols identifying controls properly used?
10. Is the use of color in controls design limited?

Human–computer interaction

1. Is the human–computer dialog suitable for the intended task?
2. Is the dialog self-descriptive and easy to control by the user?
3. Does the dialog conform to the expectations on the part of the user?
4. Is the dialog error-tolerant and suitable for user learning?
5. Has command language been restricted to experienced users?
6. Have detailed menus been used for users with little knowledge and experience?
7. Is the type of help menu fitted to the level of user's ability?
8. Has the QWERTY layout been selected for the keyboard?
9. Has logical layout been chosen for the numerical keypad?
10. Is the number of function keys limited?
11. Have the limitations of speech in human–computer dialogue been considered?
12. Are touch screens used to facilitate operation by inexperienced users?

VI. Environmental factors

Noise and vibration

1. Is the noise level at work below 80 dBA?
2. Is there an adequate separation between workers and source of noise?
3. Is the ceiling used for noise absorption?
4. Are the acoustic screens used?
5. Are hearing conservation measures fitted to the user?
6. Is personal monitoring to noise/vibration used?
7. Are the sources of uncomfortable and damaging body vibration recognized?
8. Is the vibration problem being solved at the source?
9. Are machines regularly maintained?
10. Is the transmission of vibration prevented?

Illumination

1. Is the light intensity for normal activities in the range of 200 to 800 lux?
2. Are large brightness differences in the visual field avoided?
3. Are the brightness differences between task area, close surroundings, and wider surroundings limited?
4. Is the information easily legible?
5. Is ambient lighting combined with localized lighting?
6. Are light sources properly screened?
7. Can the light reflections, shadow, or flicker from the fluorescent tubes be prevented?

Climate

1. Are workers able to control the climate themselves?
2. Is the air temperature suited to the physical demands of the task?
3. Is the air prevented from becoming either too dry or too humid?
4. Are draughts prevented?
5. Are the materials/surfaces that have to be touched neither too cold nor too hot?
6. Are the physical demands of the task adjusted to the external climate?
7. Are undesirable hot and cold radiation prevented?
8. Is the time spent in hot or cold environments limited?
9. Is special clothing used when spending long periods in hot or cold environments?

Chemical substances

1. Is the concentration of recognized hazardous chemical substances in the air subject to continuous monitoring and limitation?
2. Is the exposure to carcinogenic substances avoided or limited?
3. Does the labeling on packages of chemicals provide information on the nature of any hazard due to their contents?

TABLE 187.3 Examples of Factors to be Used in Ergonomics Checklists (*Continued*)

4. Can the source of chemical hazards be removed, isolated, or their releases from the source reduced?
5. Are there adequate exhaust and ventilation systems in use?
6. Are protective equipment and clothing (including gas and dust masks for emergencies and gloves) available at any time if necessary?

Based on Dul, J. and Weerdmeester, B., 1993, *Ergonomics for Beginners: A Quick Reference Guide*, Taylor & Francis, London.

of ergonomics. To optimize both human and system well-being and their performance, the system–human compatibility should be considered at all levels, including the physical, perceptual, cognitive, emotional, social, organizational, environmental, and other relevant system characteristics. This requires a way to measure the inputs and outputs that characterize the set of system–human interactions (Karwowski, 1991), and the ability to identify, evaluate, and measure the artifact–human compatibility.

The Complexity-Incompatibility Principle

As discussed by Karwowski et al. (1988), the lack of compatibility (ergonomics incompatibility), defined as degradation (disintegration) of the artifact–human system, is reflected in the system's measurable inefficiency and/or losses in human and system performance. In order to express the innate relationships between the system's complexity and compatibility, Karwowski, Marek, and Noworol (1988, 1994) proposed the complexity-incompatibility principle, which can be stated as follows: "As the (artifact–human) system complexity increases, the incompatibility between the system elements, as expressed through their interactions at all system levels, also increases, leading to greater ergonomic (non-reducible) entropy of the system, and decreasing the potential for effective ergonomic intervention."

The above principle was illustrated by Karwowski (1995), using as an example of design of the chair and the design of a computer display, two common problems in the area of human–computer interaction. In addition, Karwowski (1992) also discussed the complexity-compatibility paradigm in the context of organizational design. It should be noted here that the above principle reflects the natural phenomena that others in the field have described in terms of difficulties encountered in humans interacting with consumer products and technology in general. For example, according to Norman (1993), the paradox of technology is that functionality added to an artifact typically comes with the trade-off of increased complexity. These added complexities often lead to increased human difficulty and frustration when interacting with these artifacts. One of the possible reasons for the above is that a technology that has more features also has less feedback. Moreover, Norman argued that the added complexity cannot be avoided when functions are added, and can only be minimized with good design that follows natural mapping between the system elements (i.e., the control–display compatibility).

187.4 Human–Computer Interaction

HCI definition

The subfield of human–computer interaction (HCI) focuses on creating usable computer systems (Karwowski, Rizzo, and Rodrick, 2002). The Curriculum Development Group of the Association for Computer Machinery (ACM) Special Interest Group on Computer–Human Interaction (SIGCHI) defines HCI as a discipline concerned with the design, evaluation, and implementation of interactive computing systems for human use, and with the study of major phenomena surrounding them (Hewett et al., 1992). According to Dix et al. (1993), the "human–computer interaction (or HCI) is the study of the people, computer technology, and the way these influence each other." We study HCI to determine how we can make this computer technology usable by people. Both definitions stress two main dimensions of HCI. The first one points out that in order to build usable and human-centered technologies, it is necessary to study the users, roles, needs, previous experiences, and the real context of use. The second one considers

HCI to be the discipline that focuses on human cognition and behavior, and applies relevant knowledge to the development of diverse cognitive artifacts.

User-Centered Design Paradigm

The main principles for the user-centered design approach in HCI formulated by Gould and Lewis (1983) follow:

- Understanding users. It is important to have an explicit representation about users' cognitive competences and attitudes and an explicit representation of the cognitive nature of the work that people have to perform.
- Interaction design. A consistent sample of the end users must be engaged in working with the design team.
- Evaluation. Since the beginning of the design process, the system has to be evaluated. It is important to evaluate together the human and machine components of the human-machine system (user evaluation).
- Iterative design. Design-evaluation-redesign until system performance attains the usability goal established for the system.

User involvement in system development can lead to better-designed products from the perspective of customers (Nielsen, 2000). Designers should work in conjunction with users, enrolling the users early in the developing process, when their contributions to the system design are thought to be fundamental. User involvement is often critical to ensure sufficient information about initial system requirements; assess whether the product meets end user requirements and needs, and to gather data for the next version of the design. The evaluation phase is fundamental in order to measure whether user requirements are being satisfied. The main evaluation methods suggested by the HCI field are cognitive walkthrough, usability tasting, and heuristic evaluation (Helander et al., 1997).

User interface adaptation has gained considerable recognition in recent years to ensure quality of use of interactive computer systems with respect to end-user abilities, requirements, preferences, and interests. Interactive computer systems can be classified into adaptable and adaptive systems (Akoumianakis and Stephanidis, 1997). A system is called adaptable if it provides tools that make it possible for the end user to change the system's characteristics (Opperman, 1994).

Usability Engineering

Usability engineering (Nielsen, 2000), a set of activities that take place throughout the life cycle of the product, focuses on the multidimensional properties of a user interface. The following five attributes constitute the concept of usability (Nielsen, 1997):

- Learnability
- Efficiency
- Memorability
- Errors]
- Satisfaction (Nielsen, 1997)

The usability engineering process (Good et al., 1986) includes the following basic steps:

1. Definition of measurable usability attributes.
2. Setting of the quantitative levels of desired usability for each attribute. Together, an attribute and a desired level constitute a usability goal.
3. Testing the product against the usability goals. If the goals are met, no further design is needed.
4. If further design work is needed, analysis of the problems that emerge.
5. Analysis of the impact of possible design solutions.

6. Incorporation of user-derived feedback in product design.
7. Return to Step 3 to repeat the test, analysis, and design cycle.

187.5 Ergonomics in Industry

The knowledge and expertise offered by ergonomics as applied to industrial environments can be used to provide engineering guidelines regarding redesign of tools, machines, and work layouts; evaluate the demands placed on the workers by current jobs; simulate alternative work methods and determine potential for reducing physical job demands if new methods are implemented; and provide a basis for employee selection and placement procedures. The basic foundations for ergonomics design are based on two components of industrial ergonomics: engineering anthropometry and biomechanics. Occupational biomechanics can be defined as the application of mechanics to the study of the human body in motion or at rest (Chaffin and Anderson, 1993). Occupational biomechanics provides the criteria for application of anthropometric data to the problems of workplace design.

Engineering anthropometry is an empirical science branching from physical anthropology that (l) deals with physical measurements for the human body (such as body size, form or shape, and body composition), including, for example, the location and distribution of center of mass, weights, body links, or range of joint motions; and (2) applies these measures to develop specific engineering design requirements. The recommendations for workplace design with respect to anthropometric criteria can be established by the principle of design for the extreme, also known as the method of limits (Pheasant, 1986). The basic idea behind this concept is to establish specific boundary conditions (percentile value of the relevant human characteristic), which, if satisfied, will also accommodate the rest of the expected user population. The main anthropometric criteria for workplace design are clearance, reach, and posture. Typically, clearance problems involve the design of space needed for the legs or safe passageways around and between equipment. If the clearance problems are disregarded, they may lead to poor working postures and hazardous work layouts. Consideration of clearance requires designing for the largest user, typically by adapting the 95th percentile values of the relevant characteristics for male workers. Typical reach problems in industry include consideration of the location of controls and accessibility of control panels in the workplace. The procedure for solving the reach problems is usually based upon the fifth percentile value of the relevant characteristic for female workers (smaller members of the population). When anthropometric requirements of the workplace are not met, biomechanical stresses that manifest themselves in postural discomfort, lower-back pain, and overexertion injury are likely to occur.

187.6 The Role of Ergonomics in Prevention of Occupational Musculoskeletal Injury

Lack of attention to ergonomic design principles at work and its consequences have been linked to occupational musculoskeletal injuries and disorders. Work-related musculoskeletal disorders (WRMDs), such as upper-extremity cumulative trauma and lower-back disorders (LBDs), affect several million workers each year, with total costs exceeding $100 billion annually. For example, the upper-extremity cumulative trauma disorders account today for about 11% of all occupational injuries reported in the United States and has resulted in a prevalence of work-related disability in a wide range of occupations.

Work-related musculoskeletal disorders can be linked to several factors, including increased production rates leading to thousands of repetitive movements every day, widespread use of computer keyboards, higher percentage of women and older workers in the workforce, and better record keeping of employers due to better reporting requirements by the Occupational Safety and Health Administration (OSHA). Other factors include greater employee awareness of these disorders and their relation to working conditions, as well as a marked shift in the social policy regarding recognition of compensation for work-related disorders. Given the significance of the reported WRMDs, the federal government increases its efforts to reduce the frequency and severity of these disorders. For more information about these efforts, see Waters et al. (1993); for NIOSH guidelines for manual lifting, see http://www.cdc.gov/niosh/94-110.html.

The current state of knowledge indicates that chronic muscle, tendon, and nerve disorders may have multiple work-related and non-work-related causes. Therefore, WRMDs are not classified as occupational diseases but rather as work-related disorders, where a number of factors may contribute significantly to the disorder, including work environment and human performance at work. Frequently cited risk factors of WRMDs are listed below (Karwowski and Rodrick, 2001):

- Repetitive exertions
- Posture
  - Shoulder (elbow above mid-torso reaching down and behind)
  - Forearm (inward or outward rotation with a bent wrist)
  - Wrist (palmar flexion or full extensions)
  - Hand (pinching)
- Mechanical stress concentrations over the base of palm, on the palmar surface of the fingers, and on the sides of the fingers
- Vibration
- Cold
- Use of gloves

A risk factor is defined as an attribute or exposure that increases the probability of disease or disorder. Work-related musculoskeletal disorders at work are typically associated with repetitive manual tasks that impose repeated stresses to the upper body, that is, the muscles, tendons, ligaments, nerves, tissues, and neurovascular structures. For example, the three main types of disorders to the arm are tendon disorders (e.g., tendonitis); nerve disorders (e.g., carpal tunnel syndrome); and neurovascular disorders (e.g., thoracic outlet syndrome or vibration, Raynaud's syndrome).

From the occupational safety and health perspective, the current state of ergonomics knowledge allows for management of WRMDs in order to minimize human suffering, potential for disability, and related worker's compensation costs. Ergonomics can help to identify working conditions under which WRMDs might occur, develop engineering design measures aimed at elimination or reduction of the known job risk factors, and identify the affected worker population and target it for early medical and work intervention efforts. The ergonomic intervention should allow management to perform a thorough job analysis to determine the nature of specific problems; evaluate and select the most appropriate intervention(s); develop and apply conservative treatment (implement the intervention); on a limited scale if possible; monitor progress; and adjust or refine the intervention as needed.

Most of the current guidelines for control of the WRMDs at work aim to (1) reduce the extent of movements at the joints, (2) reduce excessive force levels, and (3) reduce exposure to highly repetitive and stereotyped movements (Putz-Anderson, 1988). Workplace design to prevent the onset of WRMDs should be directed toward fulfilling the following recommendations:

- Permit several different working postures.
- Place controls, tools, and materials between waist and shoulder heights for ease of reach and operation.
- Use jigs and fixtures for holding purposes.
- Re-sequence jobs to reduce the repetition.
- Automate highly repetitive operations.
- Allow self-pacing of work whenever feasible.A
- Allow frequent (voluntary and mandatory) rest breaks.

For example, some of the common methods to control the wrist posture are (l) altering the geometry of tool or controls (e.g., bending the tool or handle), (2) changing the location or positioning of the part, or (3) changing the position of the worker in relation to the work object. In order to control the extent of force required to perform a task, one can (1) reduce the force required through tool and fixture redesign, (2) distribute the application of the force, or (3) increase the mechanical advantage of the (muscle) lever system.

187.7 Fitting the Work Environment to Workers

Ergonomics job redesign focuses on fitting jobs and tasks to workers' capabilities of workers — for example, designing out unnatural postures at work, reducing excessive strength requirements, improving work layout, introducing appropriate designs of hand tools, and addressing the problem of work rest requirements. As widely recognized, ergonomics must be seen as a vital component of the value-adding activities of the company. Even in strictly financial terms, the costs of an ergonomics management program will be far outweighed by the costs of not having one. A company must be prepared to accept a participative culture and to utilize participative techniques. Ergonomics-related problems and consequent intervention should go beyond engineering solutions and must include design for manufacturability, total quality management, and work organization alongside workplace redesign or worker training.

An important component of the management efforts to control musculoskeletal disorders in industry is the development of a well-structured and comprehensive ergonomics program. The basic components of such a program should include the following: (1) health and risk factor surveillance, (2) job analysis and improvement, (3) medical management, (4) training, and (5) program evaluation. Such a program must include participation of all levels of management; medical, safety, and health personnel; labor unions; engineering; facility planners; and workers.

The expected benefits of ergonomically designed jobs, equipment, products, and workplaces include the improved quality and productivity with error reduction, enhanced safety and health performance, heightened employee morale, and accommodation of people with disabilities to meet recommendations of the Americans with Disabilities Act and Affirmative Action programs. However, the recommendations offered by ergonomics can be successfully implemented in practice only with full understanding of the production processes, plant layouts, quality requirements, and total commitment from all management levels and workers in the company. Furthermore, these efforts can only be effective through participatory cooperation between management and labor through development of in-plant ergonomics committees and programs. Ergonomics must be treated at the same level of attention and significance as other business functions of the plant — for example, the quality management — and should be accepted as a cost of doing business, rather than an add-on activity calling for action only when the problems arise.

187.8 Human and Safety Aspects of Advanced Technologies

Advances in the global economy induce a need to implement adaptive organizational structures and utilize people-oriented, rather than technology-centered, approaches to the design and operation of advanced, information-based manufacturing systems. Information-based companies often apply organizational structures and procedures that enhance communication and cooperation between departments in order to increase market competitiveness, and ensure safe and comfortable working environments. Such structures should enable people to do what they can do best through an improvement of their technical qualifications, decision-making abilities, and utilization of advanced tools for information and knowledge processing, and feedback systems (Karwowski and Marras, 1999). Application of the science of ergonomics facilitates reaching these objectives.

System Integration Paradigm

The contemporary manufacturing paradigm stipulates the systems approach to integration of technology, people, and organization (Karwowski, Salvendy, Badham et al., 1994). Appropriate technology needs to be implemented with due consideration of requirements for selection and training of human resources, allocation of decision making, and control over technology by the people charged with its use. Organizational structures allowing persons to manage available resources and adapt quickly to a changing marketplace also are needed. While many different concepts were developed in the recent past to realize this new manufacturing paradigm, they all attempt to capture critical requirements of the global economy and changing market conditions (i.e., cost, quality, flexibility, and time). Among others, these concepts include lean manufacturing, concurrent engineering, and agile manufacturing, as the more recent manufacturing philosophy and business approach.

The term *agile manufacturing* was first introduced with the publication of the *21st Century Manufacturing Enterprise Strategy* by the Iacocca Institute (1991), which argued that a new competitive environment is emerging, which is acting as a driving force for change in manufacturing, and that competitive advantage will accrue to those enterprises that develop the capability to rapidly respond to the demand for high-quality, highly customized products. In order to achieve the agility required to respond to these driving forces and to develop the required capability, it is necessary to integrate flexible technologies with a highly skilled, knowledgeable, motivated, and empowered workforce. This must be done within organization and management structures that stimulate cooperation both within and between firms.

According to the Iacocca Institute (1991), the agile manufacturing enterprise exhibits the following characteristics:

- Concurrency in all activities
- Continuing education for all employees
- Customer responsiveness
- Dynamic multiventuring capabilities
- Employees valued as intellectual assets
- Empowered individuals working in teams
- Environmental concern and proactive approach
- Accessible and usable knowledge
- Skilled and knowledgeable employees
- Open system architectures
- Right-first-time designs
- Total quality philosophy
- Short cycle times
- Technology awareness and leadership
- Enterprise integration
- Vision-based management.

Agile manufacturing can be viewed as the business concept that brings together many ideas in order to develop an appropriate response to global market opportunities. The agility comes from integration of organization, people and information-based technology into a coordinated and interdependent (often virtual) system.

Management Integration Framework for Quality, Ergonomics, and Occupational Safety Issues

Advanced technologies constitute complex systems that require a high level of integration from both the *design and management* perspectives. *Design integration* focuses on the interactions between hardware (computer-based technology), organization (organizational structure), information system, and people (human skills, training, and expertise). *Management integration* refers to the interactions between various system elements across the process and product quality, workplace and work system design, occupational safety and health programs, and corporate environmental protection polices. All design (integration) factors describing these subsystems must be considered at all levels of management system integration, that is, quality, occupational safety and health, and ergonomics. As illustrated in Table 187.4, system integration is the complex function of design subsystems (design factors) and management subsystems (management factors). The above framework can be expressed as follows:

TABLE 187.4 Framework for Integration of Design and Management Subsystems

System Integration Design Subsystems	Management (Integration) Subsystems			
	Safety and Health	Ergonomics	Environmental Protection	Quality
Organization				
Technology				
Information Systems				
People				

SYSTEM INTEGRATION = f (DESIGN INTEGRATION, MANAGEMENT INTEGRATION)

Many elements of quality, ergonomics, and safety and health issues directly or indirectly influence the integration efforts and the related corporate strategy with respect to planning for integration (Dzissah et al., 2001). In general, system performance can be improved by minimizing process deficiencies, accidents, environmental pollution, and increasing general well-being of all employees. For example, working conditions influence product quality, while productivity improvement is dependent on quality, macroergonomic and safety (occupational and health, natural environmental) management activities.

As discussed by Karwowski et al. (2002), due to inherent interrelationships between product quality defects, workplace design defects, and operational defects, both conceptual and operational models for the management system that integrates quality, ergonomics, and occupational safety and health are needed. The concepts of quality, ergonomics, and a broadly defined safety area reveal several interacting dimensions that need to be considered by the integration methods. The multidimensionality of factors involved in the integration calls for a methodology that is capable of establishing relationships between different dimensions to identify various directions for simultaneous system improvement.

Defining Terms

Ergonomics (or human factors) — The scientific discipline concerned with the understanding of interactions among humans and other elements of a system, and the profession that applies theory, principles, data, and methods to design in order to optimize human well-being and overall system performance. Ergonomists contribute to the design and evaluation of tasks, jobs, products, environments, and systems in order to make them compatible with the needs, abilities and limitations of people (International Ergonomics Association 2000; *www.iea.cc*).

Work-related musculoskeletal disorders (WRMDs) — These disorders are typically associated with repetitive manual tasks with forceful exertions, performed with fixed body postures that deviate from neutral, such as those at assembly lines and those using hand tools, computer keyboards, mice, and other devices. These tasks impose repeated stresses to the soft tissues of the arm, shoulder, and back, including the muscles, tendons, ligament, nerve tissues, and neurovascular structures, which may lead to tendon and/or joint inflammation, discomfort, pain, and potential work disability.

Occupational biomechanics — The application of mechanics to the study of the human body in motion or at rest, focusing on the physical interaction of workers with their tools, machines, and materials in order to optimize human performance and minimize the risk of musculoskeletal disorders.

Engineering anthropometry — An empirical science branching from physical anthropology that deals with physical measurements for the human body, such as body size, form (shape), and body composition, and application of such measures to the design problems.

References

Akoumianakis, D. and Stephanidis, C., 1997, Supporting user-adapted interface design: USE-IT system, *Interacting with Computers*, 9, 73–104.

Chaffin, D. B. and Anderson, G. B. J., 1993, *Occupational Biomechanics*, 2nd ed., John Wiley & Sons, New York.

Dix, A., Finlay, J., Abowd, G., and Beale, R., 1993. *Human Computer Interaction*, Prentice Hall, New York.

Dul, J. and Weerdmeester, B., 1993, *Ergonomics for Beginners: A Quick Reference Guide*, Taylor & Francis, London.

Dzissah, J., Karwowski, W., and Yang, Y. N., 2001, "Integration of quality, ergonomics, and safety management systems," in *International Encyclopedia of Ergonomics and Human Factors*, Karwowski, W., Ed., Taylor & Francis, London, pp. 1129–1135.

Goldman, S. L., Nagel, R. N., and Preiss, K., 1995, *Agile Competitors and Virtual Organizations*, Van Nostrand Reinhold, New York, 1995.

Good, M., Spine, T. M., Whiteside, J., and George, P., 1986. User-derived impact analysis as a tool for usability engineering, *Proceedings of Human Factors in Computing Systems. CHI'86*, Association of Computing Machinery, New York.

Gould, J. D. and Lewis, C., 1983, Designing for usability: key principles and what designers think, in *Proceedings of the CHI'83 Conference on Human Factors in Computing Systems*, Association for Computing Machinery, New York, pp. 50–53.

Helander, M. G., Landaur, T. K., and Prabhu, P. V., Eds., 1997, *Handbook of Human–Computer Interaction*, Elsevier: Amsterdam.

Hewett, T., Baecker, R., Card, S., Cary, T., Gasen, J., Mantiel, M., Perlman, G., Strong, G., and Verplank, W. 1992, ACM SIGCHI Curricula for Human–Computer Interaction, Report of the ACM SIGCHI Curriculum Development Group, Association for Computing Machinery, New York.

Human Factors and Ergonomics Society (HFES), 2003, *Directory and Yearbook*, HFES, Santa Monica, CA.

Iacocca Institute, 1991, *21st Century Manufacturing Enterprise Strategy. An Industry-Led View*, Vols. 1 and 2, Iacocca Institute, Bethlehem, PA.

International Ergonomics Association, 2000, The Discipline of Ergonomics, available at: *www.iea.cc/ergonomics/*.

Karwowski, W., 1991, Complexity, fuzziness and ergonomic incompatibility issues in the control of dynamic work environments, *Ergonomics*, 34: 671–686.

Karwowski, W., 1992, The complexity-compatibility paradigm in the context of organizational design of human–machine systems, in *Human Factors in Organizational Design and Management*, Brown, V. O. and Hendrick, H., Eds., Elsevier, Amsterdam, pp. 469–474.

Karwowski, W., 1995, A general modeling framework for the human–computer interaction based on the principles of ergonomic compatibility requirements and human entropy, in *Work with Display Units 94*, Grieco, A., Molteni, G., Occhipinti, E. et al., Eds., North-Holland, Amsterdam, pp. 473–478.

Karwowski, W., 1997, Ancient wisdom and future technology: the old tradition and the new science of human factors/ergonomics, in Proceedings of the Human Factors and Ergonomics Society 4th Annual Meeting, Albuquerque, New Mexico, Human Factors and Ergonomics Society, Santa Monica, CA, pp. 875–877.

Karwowski, W., 2000, Symvatology: the science of an artifact–human compatibility, 2000, *Theor. Issues Ergonomics Sci.*, 1:76–91.

Karwowski, W., Ed., 2001, *International Encyclopedia of Ergonomics and Human Factors*, Taylor & Francis, London.

Karwowski, W., Kantola, J., Rodrick, D., et al., 2002, Macroergonomics aspects of manufacturing, in *Macroegonomcis: An Introduction to Work System Design*, Hendrick, H. W. and Kleiner, B. M., Eds., Lawrence Erlbaum Associates, Mahwah, NJ, pp. 223–248.

Karwowski, W., Marek, T. and Noworol, C., 1988, Theoretical Basis of the Science of Ergonomics, in *Proceedings of the 10th Congress of the International Ergonomics Association*, Sydney, Australia, Taylor & Francis, Ltd., August, pp. 756–758.

Karwowski, W., Marek, T. and Noworol, C., 1994, The complexity-incompatibility principle in the science of ergonomics, in *Advances in Industrial Ergonomics & Safety*, vol. 6, Aghazadeh, F., Ed., Taylor & Francis, London, pp. 37–40.

Karwowski, W. and Marras, W. S., Eds., 1999, *The Occupational Ergonomics Handbook*, CRC Press, Boca Raton, FL.

Karwowski, W., Rizzo, F., and Rodrick, D., 2002, Ergonomics in information systems, in *Encyclopedia of Information Systems,* H. Bidgoli, Ed., Academic Press, New York, pp. 185–201.

Karwowski, W. and Rodrick, D., 2001, Physical tasks: analysis, design and operation, in *Handbook of Industrial Engineering,* 3rd ed., Salvendy, G., Ed., John Wiley & Sons, New York, pp. 1041–1110.

Karwowski, W. and Salvendy, G., Eds., 1994, *Organization and Management of Advanced Manufacturing*, John Wiley & Sons, New York.

Karwowski, W. G. Salvendy, R. Badham, P. Brodner, C. Clegg,, L. Hwang, J. Iwasawa, P. T. Kidd, N. Kobayashi, R. Koubek, J. Lamarsh, M. Nagamachi, M. Naniwada, H. Salzman, P. Seppälä, B. Schallock, T. Sheridan, and J. Warschat, 1994, Integrating people, organization and technology in advanced manufacturing, *Hum. Ergonomics Manuf.*, 4:1–19.

Kidd, P. T., 1994, *Agile Manufacturing: Forging New Frontiers.* Addison-Wesley, Reading, MA.

Meister, D., 1987, Systems design, development and testing, in *Handbook of Human Factors*, Salvendy, G., Ed., John Wiley & Sons, New York, pp. 17–42.

Nielsen, J., 1997, Usability engineering, in *The Computer Science and Engineering Handbook*, Tucker, A. B. Jr., Ed., CRC Press, Boca Raton, FL, pp. 1440–1460.

Nielsen, J., 2000, *Designing Web Usability: The Practice of Simplicity*, New Riders: Indianapolis, IN.

Norman, D. A., 1993, *Things that Make Us Smart*, Addison-Wesley, Reading, MA.

Opperman, R., 1994, Adaptively supported adaptability, *Int. J. Hum. Comput. Stud.*, 40:455–472.

Pheasant, S., 1986, *Bodyspace: Anthropometry, Ergonomics and Design*, Taylor & Francis, London.

Putz-Anderson, V., Ed., 1988, *Cumulative Trauma Disorders: A Manual for Musculoskeletal Diseases of the Upper I Limbs*, Taylor & Francis, London.

Sanders, M. S. and Mccormick, E. J., 1993, *Human Factors in Engineering and Design*, 6th ed., McGraw-Hill, New York.

Waters, T. R., Putz-Anderson, V., Garg, A., et al., 1993, Revised NIOSH equation for the design and evaluation of manual-lifting tasks, *Ergonomics*, 36:749–776.

Wickens, C. D. and Carswell, C. M., 1997, Information processing, in *Handbook of Human Factors*, Salvendy, G., Ed., John Wiley & Sons, New York, pp. 89–129.

Womack, J., Jones D., and Roos, D., 1990, *The Machine that Changed the World*, Rawson Associates, New York.

Further information

Grandjean, E., 1988, *Fitting the Task to the Man*, 4th ed., Taylor & Francis, London.

Helander, M., Ed., 1988, *Handbook of Human–Computer Interaction*, North-Holland, Amsterdam.

Kroemer, K. H. E., Kroemer, H. B., and Kroemer-Elbert, K. E., 1994, *Ergonomics: How to Design for Ease and Efficiency*, Prentice Hall, Englewood Cliffs, NJ, 1994.

Salvendy, G., Ed., 1987, *Handbook of Human Factors*, John Wiley & Sons, New York.

Salvendy, G. and Karwowski, W., Eds., 1994, *Design of Work and Development of Personnel in Advanced Manufacturing*, John Wiley & Sons, New York.

Wilson, J. R. and Corlett, E. N., Eds., 1990, *Evaluation of Human Work: A Practical Methodology*, Taylor & Francis, London.

Woodson, W. E., 1981, *Human Factors Design Handbook*, McGraw-Hill, New York.

Ergonomics Information Sources and Professional Societies

International Ergonomics Association (IEA), *www.iea.cc*

Crew System Ergonomics Information Analysis Center (CSERIAC), Wright Patterson AFB, Dayton, OH 45433-6573, phone (513) 255-4842, fax (513) 255-4823

Ergonomics Information Analysis Centre (EIAC), School of Manufacturing and Mechanical Engineering, The University of Birmingham, Birmingham B15 2TT England, phone +44-21-414-4239, fax +44-21-414-3476

Journals

Ergonomics, published by Taylor & Francis, London, United Kingdom

Ergonomics Abstracts, published by Taylor & Francis, London, United Kingdom

Human Factors, published by The Human Factors and Ergonomics Society, Santa Monica, CA

International Journal of Human Factors and Ergonomics in Manufacturing, published by John Wiley & Sons, New York

Applied Ergonomics, published by Elsevier, Amsterdam

International Journal of Human–Computer Interaction, published by Lawrence Erlbaum Associates Inc., Mahwah, NJ

International Journal of Industrial Ergonomics, published by Elsevier, Amsterdam

International Journal of Occupational Safety and Ergonomics, published by the Central Institute for Labour Protection, Warsaw

Theoretical Issues in Ergonomics Science, published by Taylor & Francis, London

188

Pressure and Vacuum

Peter Biltoft
University of California, Lawrence Livermore National Laboratory

Charles Borziler
University of California, Lawrence Livermore National Laboratory

Dave Holten
University of California, Lawrence Livermore National Laboratory

Matt Traini
University of California, Lawrence Livermore National Laboratory

188.1 Pressure

Pressure (P) can be defined as a force (F) acting on a given area (A); $P = F/A$. It is measured using metric (SI) units, pascals (Pa), or English units, pounds per square inch (psi). Pressure can also be indicated in absolute (abs) or gage (g) (as measured with a gage) units. Relationships between metric and English units of measure are as follows:

- 100 kPa = 1 bar = 14.7 psia ≈ 1 atmosphere (atm).
- 100 kPa gage (or 100 kPa) =14.7 psig (or 14.7 psi).
- 1 MPa abs = 147 psia.
- 1 MPa gage (or 1 MPa) = 147 psig (or 147 psi).

These units can be used when measuring seven kinds of pressure:

- *Absolute pressure:* sum of atmospheric and gage pressures.
- *Atmospheric pressure:* ambient pressure at local elevations. The atmospheric pressure at sea level is approximately 100 kPa or 14.7 psia.
- *Burst pressure:* theoretical or actual pressure at which a vessel/component will fail.
- *Gage pressure:* pressure above atmospheric pressure as measured by a gage.
- *Maximum allowable working pressure (MAWP):* maximum pressure at which component is safe to operate. MAWP is the maximum permissible setting for relief devices. Service, rated, design, and working pressure (WP) are the same as MAWP.
- *Maximum operating pressure (MOP):* maximum pressure for continuous operation. MOP should be 10 to 20% below MAWP.
- *Pressure test:* a test to ensure that a vessel or system will not fail or permanently deform and will operate reliably at a specified pressure.

0-8493-1586-7/05/$0.00+$1.50

TABLE 188.1 Comparison of Temperature Scales[a,b]

Criterion	Fahrenheit (°F)	Rankine abs (R)	Celsius or Centigrade (°C)	Kelvin abs (K)
Water boils	212	672	100.0	373.0
Water freezes	32	492	0.0	273.0
Zero °F	0	460	−17.7	255.3
Absolute zero	−460	0	−273.0	0.0

[a] Conversion factors: $°F = \frac{9}{5}°C + 32$; $°C = \frac{5}{9}(°F - 32)$; $R = °F + 460$; $K = °C + 273$.

[b] Use absolute temperatures, either Rankine or Kelvin, in all calculations involving the ideal gas law, except when temperature is constant.

Source: This work was performed under the auspices of the U.S. Department of Energy by Lawrence Livermore National Laboratory under contract N. W-7405-Eng-48.

Temperature

Temperature (T) is the intensity of heat in a substance and may be measured on the graduated scale of a thermometer. It is usually measured in degrees Fahrenheit (°F) or degrees Celsius (°C). The two absolute scales are Rankine (R) and Kelvin (K) (see Table 188.1).

Volume

Volume (V) refers to a space occupied in three dimensions. It is generally measured in cubic meters (m^3), cubic centimeters (cm^3), cubic inches ($in.^3$), or cubic feet (ft^3). Volume may also be measured in units of liters (L) or gallons (gal).

Basic Pressure Equations

General Gas Law

The general (or ideal) gas law is the basic equation that relates pressure, volume, and temperature of a gas. It is expressed mathematically as

$$\frac{PV}{T} = \text{constant} \tag{188.1}$$

where

P = pressure in kilopascals absolute (kPa abs) or pounds per square inch absolute (psia). (Add atmospheric pressure of 100 kPa or 15 psia to gage pressure.)
V = volume in any convenient cubic unit such as liters, cubic feet, etc.
T = temperature in absolute units of Rankine or Kelvin.

When the conditions of a gas system change, they do so according to this general gas law, at least where moderate pressures are concerned. At higher pressures ($P > 200$ atm) gases do not obey the general gas law exactly; this problem is discussed later.

The following checklist will assist in solving general gas law problems:

1. Sketch the problem and mark all conditions.
2. List initial and final values of P, V, and T. Use absolute units for P and T.
3. Write the general gas law for initial and final conditions. Cross out conditions that remain constant.
4. Substitute known values and solve for unknowns.
5. Convert your answer if it is in units different from those given in the problem.

The general gas law expressed for initial (subscript 1) and final (subscript 2) conditions is

$$\frac{P_1V_1}{T_1} = \frac{P_2V_2}{T_2} \tag{188.2}$$

Any one of these quantities can be solved for by simply cross-multiplying. Following is an example to show how easily the general gas law can be applied and how much can be determined about the behavior of gases.

Example

A cylinder of dry nitrogen is received from a vendor. The temperature of the nitrogen is 70°F, and the pressure within the cylinder is 2250 psig. An employee inadvertently stores the cylinder too close to a radiator so that within 8 hours the nitrogen is heated to 180°F. What is the new pressure within the cylinder?

Initial	Final
P_1 = 2250 psig = 2265 psia	P_2 = (find)
$V_1 = V_2$ = constant	$V_2 = V_1$ = constant
T_1 = 70°F + 460° = 530 R	T_2 = 180°F + 460° = 640 R

$$\frac{P_1\cancel{V_1}}{T_1} = \frac{P_2\cancel{V_2}}{T_2} \quad \text{cross-multiplying,} \quad P_2 = \frac{P_1T_2}{T_1}$$

$$P_2 = \frac{2265 \times 640}{530} = 2735 \text{ psia} = 2720 \text{ psig} \tag{188.3}$$

Pascal's Law

According to Pascal's law, pressure exerted on a confined fluid will act in all directions with equal force.

$$\text{Pressure} = \frac{\text{Force}}{\text{Area}} \quad \text{or} \quad \text{Force} = \text{Pressure} \times \text{Area} \tag{188.4}$$

Example

Find the force on the end of a cylinder (see Figure 188.1).

$$F = AP \quad \text{or} \quad F = 450 \text{ in.}^2 \times 90 \text{ lb/in.}^2 = 40\ 500 \text{ lb of force}$$

Safety Factor

The ratio of the calculated failure pressure (or actual failure pressure if known) to the MAWP is the safety factor.

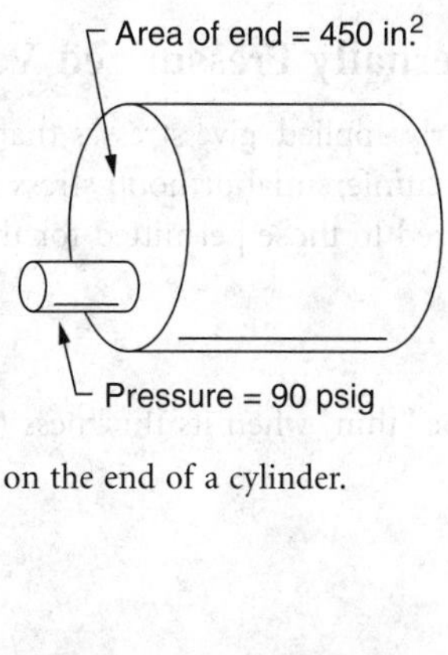

FIGURE 188.1 Example problem: force on the end of a cylinder.

$$\mathrm{SF} = \frac{\text{Burst pressure}}{\text{MAWP}} \tag{188.5}$$

A safety factor related to a value other than the failure pressure should be so identified with an appropriate subscript, such as SF_y for a safety factor based on the yield strength, or SF_u for a safety factor based on the ultimate strength of the material.

Stored Energy in a Pressurized Systems

Gas confined under high pressure is like a hungry tiger in a cage. If it should suddenly escape in a populated area, someone would probably be injured. When a gas-pressure vessel fails, it propels jagged vessel fragments in all directions. The energy of the gas, assuming isentropic expansion, is expressed as

$$U_{\text{gas}} = \frac{P_1 V}{k-1}\left[1-\left(\frac{P_2}{P_1}\right)^{(k-1)/k}\right] \tag{188.6}$$

where

- U_{gas} = stored energy
- P_1 = container pressure
- P_2 = atmospheric pressure
- V = volume of the pressure vessel
- k = c_p/c_v = 1.41 for N_2, H_2, O_2, and air, and 1.66 for He (see Baumeister, 1958, chapter 4)
- c_p = specific heat at constant pressure
- c_v = specific heat at constant volume

This expression as written will yield an energy value in joules if pressure and volume units are in megapascals (MPa) and cubic centimeters (cc), respectively.

A liquid confined under high pressure is also a potential safety hazard. However, for systems of comparable volume and pressure, the amount of stored energy in the liquid case will be considerably less than that contained in the gas because liquids are essentially noncompressible. The energy involved in the sudden failure of a liquid-filled vessel may be conservatively determined from

$$U_{\text{liq}} = \frac{1}{2}\left(\frac{P_1^2 V}{B}\right) \tag{188.7}$$

where B is the liquid bulk modulus. Some typical bulk moduli (B) are 300,000 psi for water, 225,000 psi for oil, and 630,000 psi for glycerin.

An additional and significant potential source of released energy for confined gases is chemical reactions, such as hydrogen and oxygen. For these special situations, a separate analysis addressing any potential chemical energy release should also be made (see Baumeister, 1958, chapter 4).

Basic Design Equations (Internally Pressurized Vessels)

The following equations, when properly applied, give stresses that can be expected to exist in a design — longitudinal or axial stress (s_1), circumferential or hoop stress (s_2), and radial stress (s_3) (see Figure 188.2). These stresses must be compared to those permitted for the vessel material after application of the appropriate safety factor.

Thin Cylinders

A cylinder (or any vessel) is said to be "thin" when its thickness (t) and internal radius (r_i) satisfy the following relationship:

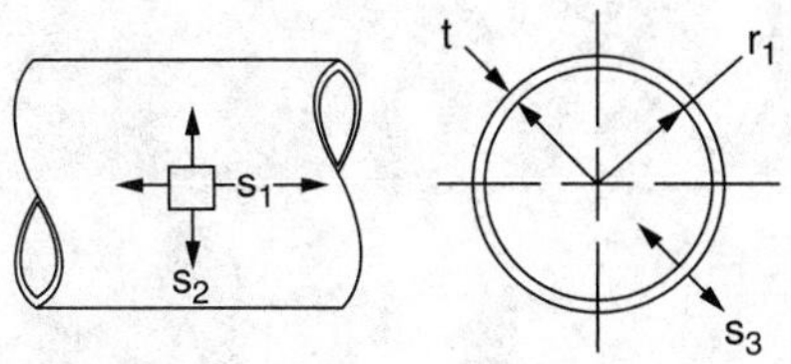

FIGURE 188.2 Stress in thin cylinders.

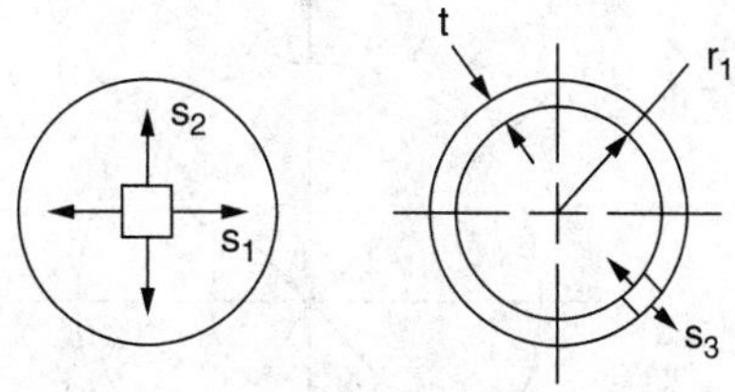

FIGURE 188.3 Stress in thin spheres.

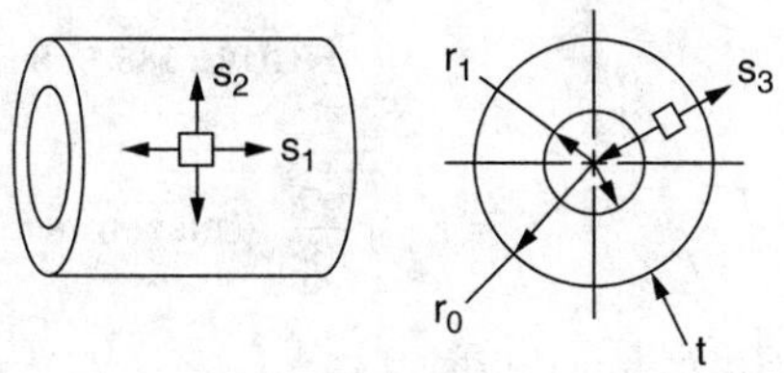

FIGURE 188.4 Stress in thick cylinders.

$$\frac{t}{r_i} \le 0.1 \qquad (188.8)$$

The following stresses exist in a thin cylinder:

$$s_1 = \frac{P}{2} \times \frac{r_i}{t}, \quad s_2 = P \times \frac{r_i}{t}, \quad s_3 = -P \qquad (188.9)$$

Given $P = 800$ psig, $r_i = 5$ in., and $t = 0.25$, then $s_1 = 8000$ psi, $s_2 = 16{,}000$ psi, and $s_3 = 800$ psi.

Thin Spheres

The following stresses exist in a thin sphere (see Figure 188.3):

$$s_1 = s_2, \quad s_2 = \frac{P}{2} \times \frac{r_i}{t}, \quad s_3 = -P \qquad (188.10)$$

Given

$$\frac{t}{r_i} = 0.05 \ (\le 0.1), \quad \frac{r_i}{t} = 20, \quad s_1 = s_2 \qquad (188.11)$$

then $s_1 = 800/2 \times 30 = 8000$ psi and $s_3 = 800$ psi. Note that the hoop stress of a thin sphere is half the hoop stress of a corresponding thin cylinder.

Thick Cylinders (Roark)

A cylinder or sphere is "thick" when its dimensions satisfy the following relationship [Roark, 1989]:

$$\frac{t}{r_i} > 0.1 \qquad (188.12)$$

The following stresses exist in a thick cylinder (see Figure 188.4):

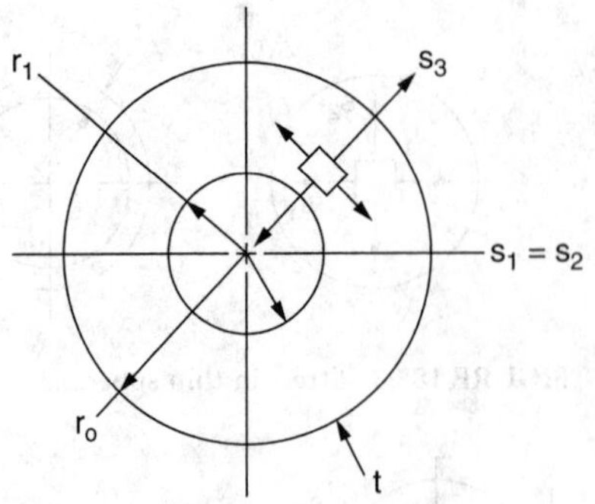

FIGURE 188.5 Stresses in thick spheres.

$$S_1 = P\frac{r_i^2}{(r_o^2 - r_i^2)}, \qquad S_2 = P\frac{r_o^2 + r_i^2}{(r_o^2 - r_i^2)} \text{ (max. inner surface)}, \tag{188.13}$$

$$S_3 = -P \text{ (max. inner surface)}$$

Thick Cylinders (Lamé)

Compare Roark's equation with that of Lamé [Baumeister, 1958, chap. 5]. Use the same vessel and conditions as above, with hoop stress maximum at the inner surface.

$$S_1 = P\frac{R^2 + 1}{R^2 - 1}, \qquad R = \frac{r_o}{r_i} \tag{188.14}$$

Note that the Lamé equation agrees with Roark's. Actually, it is the same equation.

Thick Spheres

Figure 188.5 shows the stress relationship in thick spheres, where $t/r_i > 0.1$.

$$s_1 \text{ (max. inner surface)} = s_2,$$

$$s_2 = \frac{P(r_o^3 + 2r_i^3)}{2E(r_o^3 - r_i^3)} \quad \text{where } E \text{ is the joint efficiency} \tag{188.15}$$

$$s_3 \text{ (max. inner surface)} = -P$$

188.2 The Vacuum Environment

As a practical definition, any environment in which the pressure is lower than atmospheric may be referred to as a vacuum. It is useful to express this subatmospheric pressure condition in units which reference zero pressure. Atmospheric pressure is approximately equal to 760 torr, 101 300 Pa, or 14.7 psi. Table 188.2 provides conversion factors for some basic units.

TABLE 188.2 Conversion Factors for Selected Pressure Units Used in Vacuum Technology

	Pa	torr	atm	mbar	psi
Pa	1	0.0075	$9.78 \cdot 10^{-6}$	0.01	$1.45 \cdot 10^{-4}$
torr	133	1	$1.32 \cdot 10^{-3}$	1.333	0.0193
atm	$1.01 \cdot 10^{5}$	760	1	1013	14.7
mbar	100	0.75	$9.87 \cdot 10^{-4}$	1	0.0145
psi	6890	51.71	$6.80 \cdot 10^{-2}$	68.9	1

The *ideal gas law* is generally valid when pressure is below atmospheric

$$PV = nRT \tag{188.16}$$

where

P = pressure [atm]
V = volume [liters]
n = amount of material [moles]
R = gas law constant [0.082 liter-atm/K-mole]
T = temperature [K]

The rate at which molecules strike a surface of unit area per unit time is referred to as the *impingement rate* and is given by

$$I = 3.5 \times 10^{22} \left(\frac{P}{MW \times T} \right) \tag{188.17}$$

where

P = pressure [torr]
MW = molecular weight [g/mole]
T = temperature [K]

The distance, on average, that a molecule can travel without colliding with another molecule is the *mean free path* and is given by

$$\lambda = \frac{1}{\sqrt{2}\pi d_o^2 n} \tag{188.18}$$

where

λ = mean free path [m]
d_o = molecular diameter [m]
n = gas density (molecules per cubic meter) [m^{--}]

The *average velocity* of molecules in the gas phase is a function of the molecular weight of the gaseous species and the average temperature:

$$\bar{V} = 145.5 \sqrt{\frac{T}{MW}} \tag{188.19}$$

where V = average velocity [m/s].

Methods for Measuring Subatmospheric Pressures

A wide variety of gages are used in vacuum applications. Gage selection is based upon the vacuum system operating pressure range, the presence of corrosive gases, the requirement for computer interface, and unit cost.

Subatmospheric pressure gages (Table 188.3), which infer pressure (gas density) from measurement of some physical property of the rarefied gas, have the unfortunate characteristic of being gas species–sensitive. Most commercial gages are calibrated for nitrogen; calibration curves for other gas species are available. Other factors which influence the readings of pressure gages used in vacuum technology include: location of the gage on the system, cleanliness of the gage elements, orientation of the gage, and environment of the gage (temperature, stray electric, and magnetic fields). Gage readings are typically accurate to ±10% of the reported value.

TABLE 188.3 Subatmospheric Pressure Gages

Type	Principle of Operation	Range [torr]
U tube manometer	Liquid surface displacement	1–760
Bourdon tube	Solid surface displacement	1–760
McLeod gage	Liquid surface displacement	10^{-4}–10^{-1}
Diaphragm gage	Solid surface displacement	10^{-1}–760
Capacitance manometer	Solid surface displacement	10^{-3}–760
Thermocouple gage	Thermal conductivity of gas	10^{-3}–1
Piraini gage	Thermal conductivity of gas	10^{-3}–760
Spinning rotor gage	Viscosity of gas	10^{-7}–10^{-2}
Hot cathode ion gage	Gas ionization cross section	10^{-2}–10^{-4}

TABLE 188.4 Vacuum Pumps

Type	Principle of Operation	Range [torr]
Rotary vane	Positive gas displacement	10^{-2}–760
Rotary piston	Positive gas displacement	10^{-2}–760
Cryo-sorption	Gas capture	10^{-3}–760
Oil vapor diffusion	Momentum transfer	10^{-9}–10^{-3}
Turbomolecular	Momentum transfer	10^{-9}–10^{-3}
Sputter-ion	Gas capture	10^{-0}–10^{-3}
He cryogenic	Gas capture	10^{-0}–10^{-3}

Methods for Reducing Pressure in a Vacuum Vessel

Vacuum pumps (Table 188.4) generally fall into one of two categories: primary (roughing) pumps and secondary (high-vacuum) pumps. Primary pumps are used to evacuate a vacuum vessel from atmospheric pressure to a pressure of about 10 mtorr. Secondary pumps typically operate in the pressure range of from 10^{-10} to 10^{-3} torr.

Pump speed curves such as those presented in Figures 188.6 and 188.7 are generally supplied by the pump manufacturer and reflect optimal performance for pumping air.

Vacuum System Design

The following relationships provide a first-order method for calculating conductances, delivered pump speed, and time to achieve a specified pressure. These relationships assume that conductance elements are of circular cross-section and have smooth internal surfaces, and that the gas being pumped is air.

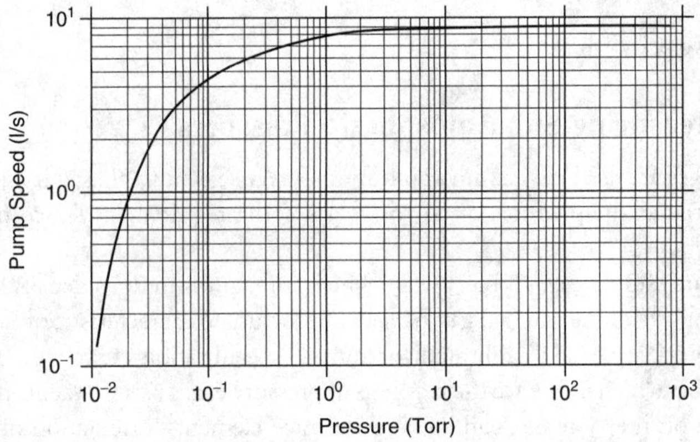

FIGURE 188.6 Pump speed curve typical of a two-stage oil-sealed rotary vane pump.

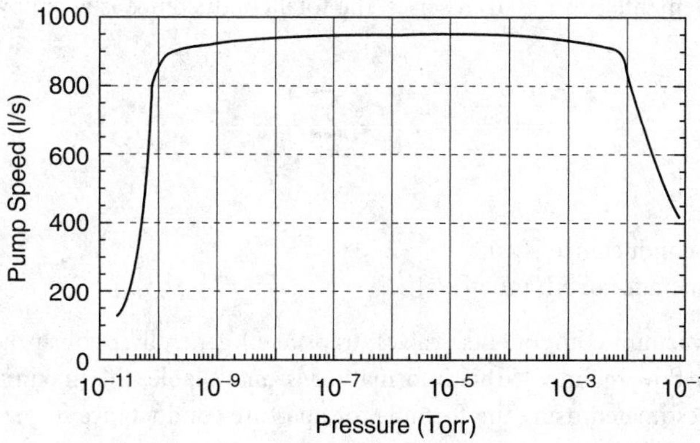

FIGURE 188.7 Pump speed curve typical of a fractionating oil-vapor diffusion pump equipped with a liquid nitrogen cold trap.

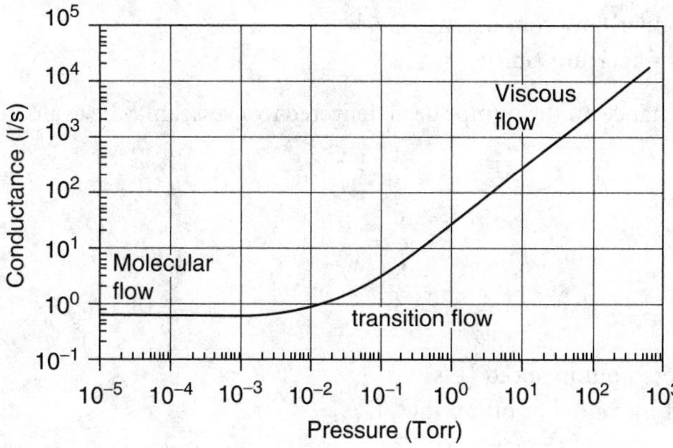

FIGURE 188.8 Conductance as a function of pressure for a 1.0" I.D. tube 10 ft long.

The conductance of tubes and orifices is a function of the flow mode (see Figure 188.8), which in turn is determined by the average pressure and the conductance element inside diameter. Flow modes may be determined by the following method:

$$D\bar{P} > 0.18 \Rightarrow \text{ turbulent flow}$$
$$D\bar{P} < 0.004 \Rightarrow \text{ molecular flow} \tag{188.20}$$

where

D = tube inside diameter [in.]
P = average pressure [torr]

$$C = \frac{3000\bar{P}D^4 + 80D^3}{L} \tag{188.21}$$

where

C = conductance [L/s]
D = tube inside diameter [in.]
$\bar{P}$ = average pressure [torr]
L = tube length [in.]

For conductance elements connected in series, the total conductance is given by

$$\frac{1}{C_{tot}} = \sum_{i=1}^{n} \frac{1}{C_i} \tag{188.22}$$

where

C_{tot} = total conductance [L/s]
C_i = conductance of ith tube [L/s]

Manufacturers of vacuum components (valves, traps, etc.) generally publish the conductance of the unit in the molecular flow regime. If this information is unavailable, the maximum conductance of a valve or trap may be estimated using the formula for aperture conductance in molecular flow:

$$C_m^{ap} = 75A \tag{188.23}$$

where

C_m^{ap} = molecular flow conductance [L/s]
A = area of aperture [in.^{2}]

The effect of conductance on the pump speed delivered to a vessel may be evaluated using the following relationship:

$$\frac{1}{S_t} = \frac{1}{C_{tot}} + \frac{1}{S_p} \tag{188.24}$$

where

S_t = delivered pump speed [L/s]
S_p = speed measured at pump inlet [L/s]

The amount of time required to evacuate a vessel using a primary pump is given by:

$$t = \frac{V}{S_t} \ln\left(\frac{P_1}{P_2}\right) \tag{188.25}$$

where

t = time to pump from P_1 to P_2 [s]
V = volume of vessel [L]
S_t = delivered pump speed [L/s]
P_1 = initial pressure [torr]
P_2 = final pressure [torr]

The lowest pressure (ultimate pressure) that a vacuum system can attain is a function of the total gas load (which is the algebraic sum of gas loads due to leaks, outgassing [see Table 188.5], permeation, and process gas) and the pumping speed of the secondary pump.

$$Q_{tot} = S_t P_u \tag{188.26}$$

where

Q_{tot} = total gas load [torr-L/s]
P_u = ultimate pressure [torr]

TABLE 188.5 Outgassing Data for Materials Used in Vacuum Vessels

	Treatment	q [W/m²]	q [T · l/s · cm²)
		Metals	
Aluminum	15 h at 250°C	$5.3 \cdot 10^{-10}$	$3.98 \cdot 10^{-13}$
Aluminum	100 h at 100°C	$5.3 \cdot 10^{-11}$	$3.98 \cdot 10^{-14}$
6061 aluminum	Glow discharge and 200°C bake	$1.3 \cdot 10^{-11}$	$9.75 \cdot 10^{-15}$
Copper	20 h at 100°C	$1.46 \cdot 10^{-9}$	$1.10 \cdot 10^{-12}$
304 stainless steel	30 h at 250°C	$4.0 \cdot 10^{-9}$	$3.00 \cdot 10^{-12}$
316L stainless steel	2 h at 800°C	$4.6 \cdot 10^{-10}$	$3.45 \cdot 10^{-13}$
U15C stainless steel	3 h at 1000°C and 25 h at 360°C	$2.1 \cdot 10^{-11}$	$1.58 \cdot 10^{-14}$
		Glasses	
Pyrex glass	Fresh	$9.8 \cdot 10^{-7}$	$7.35 \cdot 10^{-9}$
Pyrex glass	1 month in air	$1.55 \cdot 10^{-6}$	$1.16 \cdot 10^{-9}$
		Elastomers	
Viton-A	Fresh	$1.52 \cdot 10^{-3}$	$1.14 \cdot 10^{-6}$
Neoprene	—	$4.0 \cdot 10^{-2}$	$3.00 \cdot 10^{-5}$

Source: O'Hanlon, J. F. 1989. *A User's Guide to Vacuum Technology,* 2nd ed. John Wiley & Sons, New York.

Defining Terms

Atmosphere — The gaseous environment in a defined space.
Conductance — Mass throughput divided by the pressure drop across a tube.
Gas — The state of matter distinguished from the solid and liquid states by relatively low density and viscosity, relatively great expansion and contraction with changes in pressure and temperature, the ability to diffuse readily, and the spontaneous tendency to become distributed uniformly throughout any container.
Mole — Amount of material, 6.02E + 23 particles; 1 mole of carbon weighs 12 grams.
Pressure — Force per unit area (expressed in units such as psi, Torr, and mBar).
Pressure gauge — An instrument for measuring pressure.
Outgassing — Evolution of gas from a solid or liquid in a vacuum environment.
Vacuum — An environment with pressure below atmospheric pressure (760 Torr).

References

Baumeister, T. 1958. *Marks' Mechanical Engineers' Handbook,* 9th ed. McGraw-Hill, New York.
Borzileri, C. V. 1993. DOE pressure safety: pressure calculations. Rev. 10, Appendix C. Lawrence Livermore National Laboratory, Livermore, CA.
Hoffman, D. M., Singh, B., and Thomas, J. H. 1998. *Handbook of Vacuum Science and Technology.* Academic Press, San Diego, CA.
Lafferty, J. M. 1998. *Foundations of Vacuum Science and Technology.* John Wiley & Sons, New York.
O'Hanlon, J. F. 1989. *A User's Guide to Vacuum Technology,* 2nd ed. John Wiley & Sons, New York.
Roark, R. J. 1989. Pressure vessels: pipes. In *Roark's Formulas for Stress and Strain,* 6th ed. McGraw-Hill, New York, p. 363.

Further Information

Beavis, L. C. and Harwood, V. J. 1979. *Vacuum Hazards Manual.* American Vacuum Society, New York.
Drinkwine, M. J. and Lichtman, D. 1979. *Partial Pressure Analyzers and Analysis.* American Vacuum Society, New York.

Dushman, S. and Lafferty, J. M. 1976. *The Scientific Foundations of Vacuum Technique*. John Wiley & Sons, New York.

Faupel, J. H. 1956. *Trans. Am. Soc. Mech. Eng.*, 78:1031–1064.

Hablanian, M. H. 1990. *High Vacuum Technology: A Practical Guide*. Marcel Dekker, New York.

Harris, N. S. 1989. *Modern Vacuum Practice*. McGraw-Hill, New York.

Madey, T. E. and Brown, W. C. 1984. *History of Vacuum Science and Technology*. American Vacuum Society, New York.

Roth, A. 1966. *Vacuum Sealing Techniques*. Pergamon Press, New York.

Roth, A. 1982. *Vacuum Technology*. Elsevier, New York.

American Society of Mechanical Engineers (ASME). Sec. VIII: pressure vessels, division-1. In: *ASME Boiler and Pressure Vessel Code*. ASME, New York.

Vossen J. L. and Kern, W. 1978. *Thin Film Processes*. Academic Press, New York.

Weissler, G. L. and Carlson, R. W. 1979. *Vacuum Physics and Technology*. Academic Press, New York.

Wilson, N. G. and Beavis, L. C. 1979. *Handbook of Leak Detection*. American Vacuum Society, New York.

Websites

Lawrence Livermore Laboratory, safety information relevant to vacuum and pressure systems: *www.llnl.gov/es_and_h/esh-manual.html.*

American Vacuum Society home page: *http://avs.org*

American Vacuum Equipment Manufacturers International home page: *www.avem.org*

American Society of Mechanical Engineers home page: *www.asme.org*

189

Food Engineering

R. Paul Singh
University of California, Davis

The food industry relies on a wide range of unit operations to manufacture a myriad of processed foods. In order to design food processes, a practitioner in the food industry uses fundamental principles of chemistry, microbiology, and engineering. During the last 30 years, the food engineering discipline has evolved to encompass several aspects of food processing. The diversity of processes typically employed in a food processing plant is illustrated in Figure 189.1. Typical food processes may include sorting and size reduction, transport of liquid foods in pipes, heat transfer processes carried out using heat exchangers, separation processes using membranes, simultaneous heat and mass transfer processes important in drying, and processes that may involve a phase change such as freezing. A food engineer uses the concepts common to the fields of chemical, mechanical, civil, and electrical engineering in addition to food sciences to design engineering systems that interact with foods. When foods are used as raw materials they offer unique challenges. Perhaps the most important concern in food processing is the variability in the raw material. To achieve consistency in the final quality of a processed food, the processes must be carefully designed to minimize variations caused by processing. In this chapter, some of the unique engineering issues encountered in food processing will be presented. For more details on the engineering design of food processing systems, see Singh and Heldman [2001], Toledo [1991], Brennan et al. [1990], Heldman and Singh [1981], Loncin and Merson [1979], and Charm [1978].

189.1 Liquid Transport Systems

In a food processing plant, one of the common operations is the transport of liquid foods from one piece of processing equipment to the next. Characteristics of the liquid food must be known before a liquid transport system can be designed. As seen in Figure 189.2, a linear relationship exists between the shear stress and shear rate for a Newtonian liquid, such as water, orange juice, milk, and honey. The viscosity of a Newtonian liquid is determined from the slope of the straight line. Viscosity is an important property needed in many flow-related calculations. For example, viscosity of a liquid must be known before the volumetric flow rate in a pipe, under laminar conditions, can be calculated from the following equation:

$$\dot{V} = \frac{\pi \Delta P R^4}{8 \mu L} \tag{189.1}$$

For non-Newtonian liquids, the relationship between shear stress and shear rate is nonlinear. Non-Newtonian liquids are classified as shear thinning, shear thickening, Bingham, and plastic fluids. Both

0-8493-1586-7/05/$0.00+$1.50

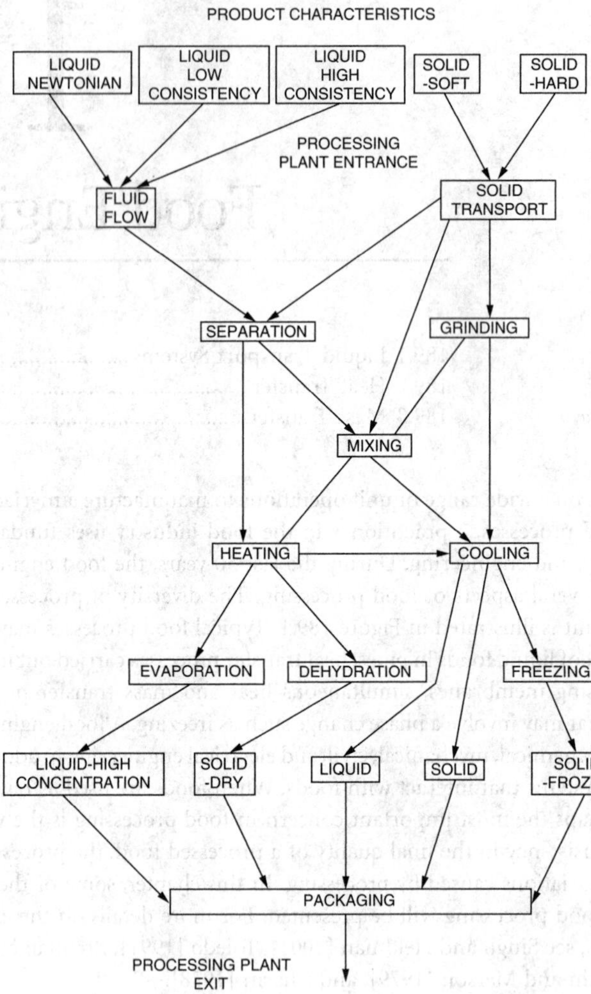

FIGURE 189.1 Generalized flow in a typical food manufacturing plant. (*Source:* Heldman, D. R. and Singh, R. P. 1981. *Food Process Engineering*, 2nd ed. AVI, Westport, CT.)

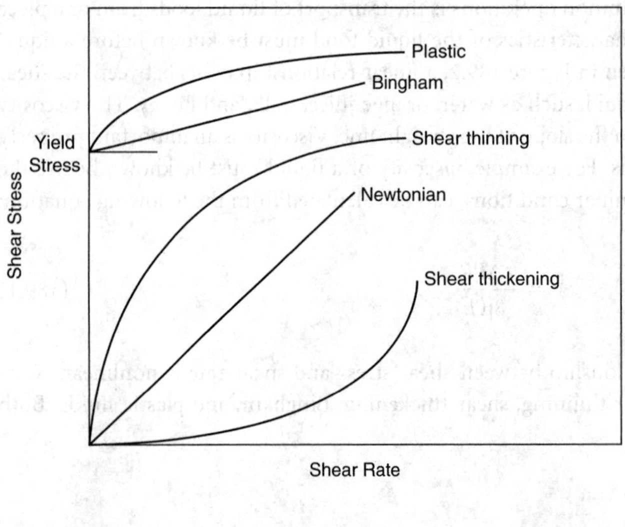

FIGURE 189.2 Shear stress–shear rate relationships for Newtonian and non-Newtonian liquids. (*Source:* Singh, R. P. and Heldman, D. R. 2001. *Introduction to Food Engineering*, 3rd ed. Academic Press, San Diego, CA.)

TABLE 189.1 Values of Coefficients in the Herschel-Bulkley Fluid Model

Fluid	K	n	σ_0	Typical Examples
Herschel–Bulkley	>0	$0 < n < 1$	>0	Minced fish paste, raisin paste
Newtonian	>0	1	0	Water, fruit juice, honey, milk, vegetable oil
Shear-thinning (pseudoplastic)	>0	$0 < n < 1$	0	Applesauce, banana puree, orange juice concentrate
Shear-thickening	>0	$0 < n < \infty$	0	Some types of honey, 40% raw corn starch solution
Bingham plastic	>0	1	>0	Toothpaste, tomato paste

Source: Steffe, J. F. 1992. *Rheological Methods in Food Process Engineering*. Freeman Press, East Lansing, MI.

Newtonian and non-Newtonian characteristics can be described by the Herschel–Bulkley model [Herschel and Bulkley, 1926].

$$\sigma = K\left[\frac{du}{dy}\right]^n + \sigma_0 \tag{189.2}$$

The values of the consistency coefficient, K, and the flow behavior index, n, in Equation (189.2) are used to differentiate between different types of non-Newtonian liquids (Table 189.1). A few measured value of K and n using capillary and rotational viscometers are shown in Table 189.2.

TABLE 189.2 Rheological Properties of Selected Liquid Foods

Product	Temperature (°C)	Composition	Consistency Coefficient (m) (Pa·s^n)	Flow Behavior Index (n)	Measurement Method
Apple juice	27	20° Brix	0.0021	1.0	Capillary tube
Apple juice	27	60° Brix	0.03	1.0	Capillary tube
Applesauce	25	31.7% T.S.	22.0	0.4	Coaxial cylinder
Applesauce	27	11.6% T.S.	12.7	0.28	Capillary tube
Apricot puree	21	17.7% T.S.	5.4	0.29	Coaxial cylinder
Apricot puree	25	19% T.S.	20.0	0.3	Coaxial cylinder
Banana puree	24	Unknown	6.5	0.458	Coaxial cylinder
Banana puree	24	Unknown	10.7	0.333	Capillary tube
Corn syrup	27	48.4% T.S.	0.053	1.0	Coaxial cylinder
Cream	3	20% fat	0.0062	1.0	Unknown
Cream	3	30% fat	0.0138	1.0	Unknown
Grape juice	27	20° Brix	0.0025	1.0	Capillary tube
Grape juice	27	60° Brix	0.11	1.0	Capillary tube
Honey	24	Normal	5.6	1.0	Capillary tube
Honey	24	Normal	6.18	1.0	Single cylinder
Olive oil	20	Normal	0.084	1.0	Unknown
Peach puree	27	10.0% T.S.	4.5	0.34	Capillary tube
Pear puree	27	14.6% T.S.	5.3	0.38	Capillary tube
Pear puree	27	15.2% T.S.	4.25	0.35	Coaxial cylinder
Pear puree	32	18.31% T.S.	2.25	0.486	Coaxial cylinder
Pear puree	32	45.75% T.S.	35.5	0.479	Coaxial cylinder
Skim milk	25	Normal	0.0014	1.0	Unknown
Soy beam oil	30	Normal	0.04	1.0	Unknown
Tomato concentrate	32	5.8% T.S.	0.223	0.59	Coaxial cylinder
Tomato concentrate	32	30% T.S.	18.7	0.4	Coaxial cylinder
Whole milk	20	Normal	0.0212	1.0	Unknown

Adapted from Heldman, D. R. and Singh R. P. 1981. *Food Process Engineering*, 2nd ed. AVI, Westport, CT.

When designing transport systems for water and other Newtonian liquids, the flow regime is determined by calculating the Reynolds number. For example, a flow with Reynolds number smaller than 2,100 is considered to be laminar, whereas a Reynolds number greater than 10,000 indicates turbulent flow. In the case of non-Newtonian liquid foods, a generalized Reynolds number is calculated using both the consistency coefficient K and the flow behavior index n. The generalized Reynolds number (N_{GRe}) is defined as

$$N_{GRe} = \frac{\rho \bar{u}^{2-n} D^n}{2^{n-3} K \left[\frac{3n+1}{n}\right]^n} \tag{189.3}$$

It is well known that for Newtonian liquids flowing in a pipe, the ratio between the mean velocity, $\bar{u}$, and the maximum velocity, u_{max}, under laminar conditions is given by

$$\frac{\bar{u}}{u_{max}} = 0.5 \tag{189.4}$$

Similarly, in the case of turbulent flow, for a Newtonian liquid the velocity ratio is

$$\frac{\bar{u}}{u_{max}} = 0.82 \tag{189.5}$$

In the case of non-Newtonian liquids, the velocity ratio is influenced by the flow behavior index, n, and is determined as

$$\frac{\bar{u}}{u_{max}} = \frac{n+1}{3n+1} \tag{189.6}$$

The velocity ratio for non-Newtonian liquids plotted as a function of generalized Reynolds number is shown in Figure 189.3 [Palmer and Jones, 1976]. Knowledge of such velocity profiles is important in the design of aseptic processing systems for liquid foods.

189.2 Heat Transfer

Heating and cooling processes are widely used in the food industry. The three common modes of heat transfer — conduction, convection, and radiation — play an important role in the processing of foods.

The governing equation for heat transfer in the conductive mode is

$$\frac{\partial T}{\partial t} = \frac{\partial}{\partial x}\left[\alpha \frac{\partial T}{\partial x}\right] \tag{189.7}$$

Under steady state conditions, Equation (189.7) reduces to the well-known Fourier law:

$$q_x = -kA\frac{dT}{dx} \tag{189.8}$$

The solution of both transient and steady-state equations (Equations (189.7) and (189.8)) requires a knowledge of the physical and thermal properties of foods. Considerable research has been done to measure food properties such as thermal conductivity, density, specific heat, and thermal diffusivity [Rao

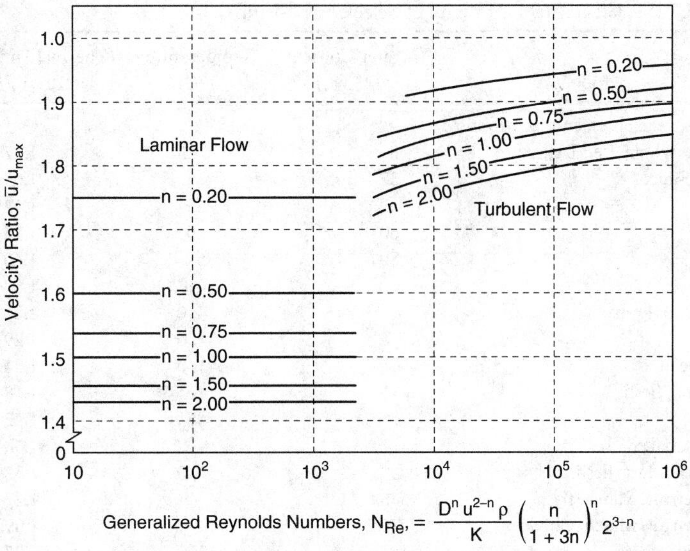

FIGURE 189.3 Plot of velocity ratio versus generalized Reynolds numbers. (*Source:* Palmer, J. and Jones, V. 1976. Prediction of holding times for continuous thermal processing of power law fluids. *J. Food Sci.*, 41:1233.)

and Rizvi, 1995]. A computerized database developed by Singh [1993] contains over 2,500 food-property combinations, along with literature citations. Some selected food property values are shown in Tables 189.3, 189.4, and 189.5. It is evident that for most high-moisture foods, the property values are greatly influenced by the presence of water. In fact, many of the empirical models used for predicting thermal properties are based on the amount of water present in a food material.

Thermal conductivity of foods can be estimated from an empirical equation developed by Sweat [1995] using 430 data points:

$$k = 0.25m_c + 0.155m_p + 0.16m_f + 0.135m_a + 0.58m_m \tag{189.9}$$

A more rigorous model of thermal conductivity has been suggested by Kopelman [1966] to account for the nonisotropic nature of foods. For example, for a food material containing fibers, when the fibers in a food are parallel to the applied heat transfer, the thermal conductivity may be calculated as

$$k_{\parallel} = k_L\left[1 - N^2(1 - k_S/k_L)\right] \tag{189.10}$$

If the food fibers are perpendicular to the direction of heat transfer,

$$k_{\perp} = k_L \frac{1 - Q''}{1 - Q''(1 - N)} \tag{189.11}$$

$$Q'' = \frac{N}{(1 - k_S/k_L)} \tag{189.12}$$

The specific heat of a food material is mainly influenced by the components of the food. Thus, specific heat may be estimated from the knowledge of food composition using the following equation:

TABLE 189.3 Thermal Diffusivity of Selected Foodstuffs

Product	Water Content (% wt.)	Temperature[a] (°C)	Thermal Diffusivity ($\times 10^{-7}$ m²/sec)
Fruits, vegetables, and by-products			
Apple, whole, Red Delicious	85	0–30	1.37
Applesauce	37	5	1.05
	37	65	1.12
	80	5	1.22
	80	65	1.40
	—	26–129	1.67
Avocado, flesh	—	24, 0	1.24
Seed	—	24, 0	1.29
Whole	—	41, 0	1.54
Banana, flesh	76	5	1.18
	76	65	1.42
Beans, baked	—	4–122	1.68
Cherries, tart, flesh	—	30, 0	1.32
Grapefruit, Marsh, flesh	88.8	—	1.27
Grapefruit, Marsh, albedo	72.2	—	1.09
Lemon, whole	—	40, 0	1.07
Lima beans, pureed	—	26–122	1.80
Pea, pureed	—	26–128	1.82
Peach, whole	—	27, 4	1.39
Potato, flesh	—	25	1.70
Potato, mashed, cooked	78	5	1.23
	78	65	1.45
Rutabaga	—	48, 0	1.34
Squash, whole	—	47, 0	1.71
Strawberry, flesh	92	5	1.27
Sugarbeet	—	14, 60	1.26
Sweet potato, whole	—	35	1.06
	—	55	1.39
	—	70	1.91
Tomato, pulp	—	4, 26	1.48
Fish and meat products			
Codfish	81	5	1.22
	81	65	1.42
Corned beef	65	5	1.32
	65	65	1.18
Beef, chuck[b]	66	40–65	1.23
Beef, round[b]	71	40–65	1.33
Beef, tongue[b]	68	40–65	1.32
Halibut	76	40–65	1.47
Ham, smoked	64	5	1.18
Ham, smoked	64	40–65	1.38
Water	—	30	1.48
	—	65	1.60
Ice	—	0	11.82

[a] Where two temperatures separated by a comma are given, the first is the initial temperature of the sample and the second is that of the surroundings.
[b] Data are applicable only where juices that exuded during heating remain in the food samples.
Source: Singh, R. P. 1982. Thermal diffusivity in food processing. *Food Technol.*, 36:87–91.

TABLE 189.4 Thermal Conductivity of Selected Food Products

Product	Moisture Content (%)	Temperature (°C)	Thermal Conductivity (W/m·K)
Apple	85.6	2–36	0.393
Applesauce	78.8	2–36	0.516
Beef, freeze dried			
1,000 mm Hg pressure	—	0	0.065
0.001 mm Hg pressure	—	0	0.037
Beef, lean			
Perpendicular to fibers	78.9	7	0.476
Perpendicular to fibers	78.9	62	0.485
Parallel to fibers	78.7	8	0.431
Parallel to fibers	78.7	61	0.447
Butter	15	46	0.197
Cod	83	2.8	0.544
Egg, white	—	36	0.577
Egg, yolk	—	33	0.338
Fish muscle	—	0–10	0.557
Grapefruit, whole	—	30	0.450
Honey	12.6	2	0.502
	80	2	0.344
Juice, apple	87.4	20	0.559
	36.0	20	0.389
Milk	—	37	0.530
Milk, condensed	90	24	0.571
Milk, skimmed	—	1.5	0.538
Milk, nonfat dry	4.2	39	0.419
Olive oil	—	15	0.189
	—	100	0.163
Pork			
Perpendicular to fibers	75.1	6	0.488
		60	0.540
Parallel to fibers	75.9	4	0.443
		61	0.489
Pork fat	—	25	0.152
Potato, raw flesh	81.5	1–32	0.554
Potato, starch gel	—	1–67	0.040
Poultry, broiler muscle	69.1–74.9	4–27	0.412
Salmon			
Perpendicular	73	4	0.502
Salt	—	87	0.247
Strawberries	—	14 to 25	0.675
Sugars	—	29–62	0.087–0.22
Turkey, breast			
Perpendicular to fibers	74	3	0.502
Parallel to fibers	74	3	0.523
Vegetable and animal oils	—	4–187	0.169
Wheat flour	8.8	43	0.45
		65.5	0.689
		1.7	0.542
Whey		80	0.641

Source: Reidy, G. A. 1968. Thermal Properties of Foods and Methods of Their Determination. M.S. thesis, Michigan State University, East Lansing, MI.

TABLE 189.5 Specific Heats of Foods

Product	Water	Experimental
Beef	68.3	3.52
Butter	15.5	2.051–2.135
Milk, whole	87.0	3.852
Skim milk	90.5	3.977–4.019
Beef, lean	71.7	3.433
Potato	79.8	3.517
Apple, raw	84.4	3.726–4.019
Bacon	49.9	2.01
Cucumber	96.1	4.103
Blackberry, syrup pack	76.0	
Potato	75.0	3.517
Veal	68.0	3.223
Fish	80.0	3.60
Cheese, cottage	65.0	3.265
Shrimp	66.2	3.014
Sardines	57.4	3.014
Beef, roast	60.0	3.056
Carrot, fresh	88.2	3.81–3.935

Source: Adapted from Reidy, G. A. 1968. Thermal Properties of Foods and Methods of Their Determination. M.S. thesis, Michigan State University, East Lansing, MI.

$$c_p = 1.424m_c + 1.549m_p + 1.675m_f + 0.837m_a + 4.187m_m \tag{189.13}$$

Food composition values are available in a computerized database [Singh, 1993] and *Agricultural Handbook Number 8* [Watt and Merrill, 1975]. Composition of selected foods is given in Table 189.6.

Newton's law of cooling is used when the mode of heat transfer is by convection. Thus,

$$q = h_c A(T_p - T_\infty) \tag{189.14}$$

The convective heat transfer coefficient is determined using the lumped heat capacity method, as illustrated by Singh and Heldman [2001]. One of the key pieces of processing equipment used in the heating and cooling of foods is a heat exchanger. In the case of Newtonian liquids, both plate- and tubular-type heat exchangers are commonly used. For example, milk pasteurization is done using plate-type heat exchangers. When working with non-Newtonian liquids such as fruit purees and pastes, or liquid foods that contain food particulates such as beef chunks in a gravy, the scraped-surface heat exchangers are used. To calculate the rate of heat transfer in Equation (189.14), a value for the convective heat transfer coefficient must be known. For a scraped-surface heat exchanger, the following relationship is recommended:

$$h = 1.2\left(\frac{k}{D}\right) N_{\mathrm{Re}}^{0.5} N_{\mathrm{Pr}}^{0.33} N_m^{0.26} \tag{189.15}$$

A more complete and rigorous analysis of non-Newtonian heat transfer in scraped-surface heat exchangers is given by Harrod [1987].

Unsteady-state heat transfer is important in many food processing applications. Reactions occurring in the initial stages of a heating or cooling process can significantly affect the quality attributes of foods. For example, in the food canning process, knowledge of temperature and time inside the can during the heating and cooling processes is important when designing thermal processes that ensure required

TABLE 189.6 Composition of Selected Foods

Food	Water (%)	Protein (%)	Fat (%)	Carbohydrate (%)	Ash (%)
Apples, fresh	84.4	0.2	0.6	14.5	0.3
Applesauce	88.5	0.2	0.2	10.8	0.6
Asparagus	91.7	2.5	0.2	5.0	0.6
Beans, lima	67.5	8.4	0.5	22.1	1.5
Beef, hamburger, raw	68.3	20.7	10.0	0.0	1.0
Bread, white	35.8	8.7	3.2	50.4	1.9
Butter	15.5	0.6	81.0	0.4	2.5
Cod	81.2	17.6	0.3	0.0	1.2
Corn, sweet, raw	72.7	3.5	1.0	22.1	0.7
Cream, half-and-half	79.7	3.2	11.7	4.6	0.6
Eggs	73.7	12.9	11.5	0.9	1.0
Garlic	61.3	6.2	0.2	30.8	1.5
Lettuce, iceburg	95.5	0.9	0.1	2.9	0.6
Milk, whole	87.4	3.5	3.5	4.9	0.7
Orange juice	88.3	0.7	0.2	10.4	0.4
Peaches	89.1	0.6	0.1	9.7	0.5
Peanuts, raw	5.6	26.0	47.5	18.6	2.3
Peas, raw	78.0	6.3	0.4	14.4	0.9
Pineapple, raw	85.3	0.4	0.2	13.7	0.4
Potatoes, raw	79.8	2.1	0.1	17.1	0.9
Rice, white	12.0	6.7	0.4	80.4	0.5
Spinach	90.7	3.2	0.3	4.3	1.5
Tomatoes	93.5	1.1	0.2	4.7	0.5
Turkey	64.2	20.1	14.7	0.0	1.0
Turnips	91.5	1.0	0.2	6.6	0.7
Yogurt, whole milk	88.0	3.0	3.4	4.9	0.7

microbial sterilization. Standard methods involving Heisler charts may be used to determine the temperature history of conduction heating of foods [Holman, 1990]. Due to the low thermal diffusivities of foods, obtaining values of Fourier number and temperature ratios from the Heisler charts is cumbersome. Numerical methods are often employed to solve the governing heat-transfer equation [Heldman and Singh, 1981].

For heating or cooling processes without phase change, the thermal diffusivity is usually a weak function of temperature. However, when a phase change occurs during a process, such as in freezing or boiling, the properties of the food material change, often dramatically.

$$\alpha(T) = \frac{k(T)}{\rho(T) c_{pa}(T)} \tag{189.16}$$

Due to the nonlinearities introduced by the variable properties, more rigorous approaches are used to calculate heat transfer. In the case of freezing, food properties are influenced by the state of water. For example, a dramatic change in the specific heat of sweet cherries as a function of temperature in the freezing zone is shown in Figure 189.4. The freezing time, t_F, may be estimated using the following equation:

$$t_F = \frac{\rho H_L}{T_F - T_\infty}\left(\frac{P'a}{h_c} + \frac{R'a^2}{k}\right) \tag{189.17}$$

In the freezing of foods, there is a phase change of water into ice. Other processes where such complexities are encountered include thawing, dehydration, and evaporation. Design considerations of such processes are elaborated in Heldman and Lund [1992] and Singh and Mannapperuma [1990].

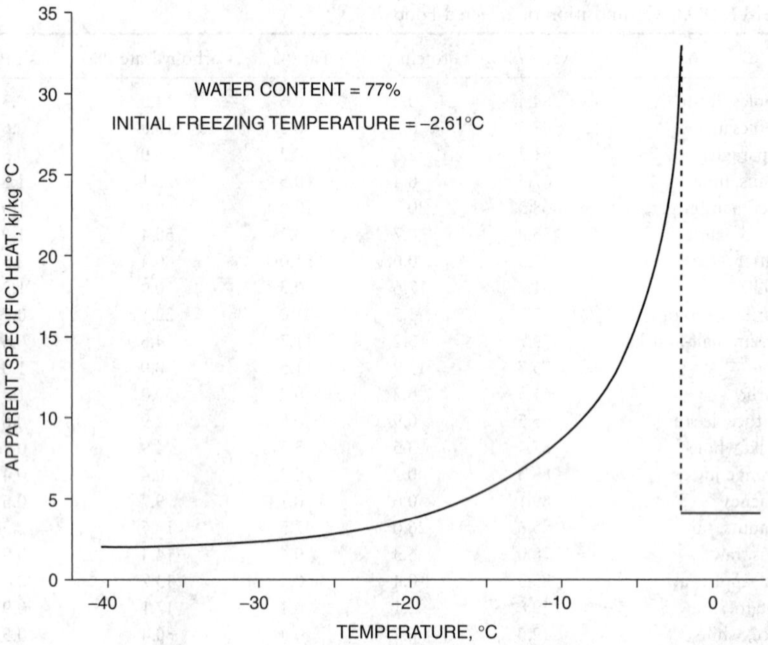

FIGURE 189.4 Predicted apparent specific heat of frozen sweet cherries as a function of temperature. (*Source:* Heldman, D. R. and Lund, D. B. 1992. *Handbook of Food Engineering*. Marcel Dekker, New York.)

TABLE 189.7 Permittivities of Fresh Fruits and Vegetables at 2.45 GHz

Product	Moisture Content (% wet basis)	Tissue Density (g/cm^3)	Relative Dielectric Constant ε′	Relative Dielectric Loss Constant ε″
Apple	88	0.76	54	10
Banana	78	0.94	60	18
Carrot	87	0.99	56	15
Grapefruit	91	0.83	73	15
Lemon	91	0.88	71	14
Mango	86	0.96	61	14
Papaya	88	0.96	67	14
Peach	90	0.92	67	14
Potato	79	1.03	57	17
Strawberry	92	0.76	71	14
Sweet potato	80	0.95	52	14

Note: All values measured at 23°C.
Source: Datta, A. K. and Anantheswaran, R. C. 2001. *Handbook of Microwave Technology for Food Applications*. Marcel Dekker, New York.

Microwave heating of foods has become popular in domestic cooking. However, industrial applications have been limited largely due to nonuniformity of raw materials that causes difficulties in obtaining consistent heating rates. Nevertheless, manufacturing of microwaveable foods for cooking in domestic ovens has increased dramatically during the last few decades. In preparing foods for microwave applications, the dielectric properties of foods are required. Table 189.7 gives the relative dielectric constant, ε′, and relative dielectric loss constant, ε″, for selected foods at 2.45 GHz, the frequency employed in domestic ovens. The conversion of the microwave energy to heat can be approximated by the following equation (given by Copson [1975]):

$$P_D = 55.61\times10^{-14}\, E^2 f \varepsilon' \tan\delta \tag{189.18}$$

and the penetration of microwave energy into a food may be obtained by using an equation developed by von Hippel [1954]. Accordingly, the depth Z, below the surface of the food material at which the electrical strength is 1/e of the electrical field in the free space, is given by the following expression:

$$Z = \frac{\lambda}{2\pi}\left[\frac{2}{\varepsilon'(\sqrt{1+\tan^2\delta}-1)}\right]^{1/2} \tag{189.19}$$

189.3 Mass Transfer

Analogous to heat transfer, mass transfer in a food is described by equations similar to Equation 189.7, as follows:

$$\frac{\partial c}{\partial t} = D_e\left(\frac{\partial^2 c}{\partial x^2}\right) \tag{189.20}$$

Solution of the preceding equation provides information on the rate of change in concentration of a substance in a food material, such as how moisture may be distributed in a food, or how during salting operation, the salt concentration may appear in a food. The unsteady-state mass transfer equation is widely used in a variety of food processing applications such as for determining drying rates, diffusion of solutes, and design of packaging materials for foods. A steady-state solution of Equation (189.20) is used to determine transport of a gas through a polymeric material used in food packaging:

$$\dot{m}_B = \frac{P_B A(p_{B1} - p_{B2})}{(x_2 - x_1)} \tag{189.21}$$

The preceding discussion provides ample evidence that food processing offers complex situations that require the use of advanced mathematical skills to develop appropriate solutions. The data on food properties presented in this chapter show the diversity of input conditions prevalent in the design and analysis of food processing operations.

Defining Terms

Newtonian liquids — Newtonian liquids exhibit a linear relationship between shear stress and rate of shear. The slope of the straight line passing through the origin of the axis is used to calculate viscosity of the liquid. Examples of Newtonian liquids include water, milk, and orange juice.

Non-Newtonian liquids — Non-Newtonian liquids exhibit a nonlinear relationship between shear stress and rate of shear. Furthermore, these liquids may be classified as time-independent or time-dependent non-Newtonian liquids. The rheological properties that describe non-Newtonian liquids are consistency coefficient, flow behavior index, and yield stress. Examples of non-Newtonian liquids are minced fish paste, apple sauce, banana puree, and tomato paste.

Physical and thermal properties — Physical and thermal properties of foods include thermal diffusivity, thermal conductivity, specific heat, density, and viscosity. A knowledge of these properties is essential in designing food processes and equipment.

Nomenclature

Symbol	Quantity	Unit
A	Area (normal to x direction) through which heat flows	(m²)
a	Product thickness for an infinite slab, diameter for an infinite cylinder, and diameter for a sphere	

Symbol	Quantity	Unit
α	Thermal diffusivity	(m^2/s)
c	Concentration	(kg/m^3)
c_p	Specific heat capacity	(kJ/kg°C)
c_{pa}	Apparent specific heat capacity	(kJ/kg°C)
D	Pipe diameter	(m)
D_e	Diffusivity	(m^2/s)
E	Electrical field strength	(V/m)
$\frac{du}{dy}$	Strain	s^{-1}
ε'	Relative dielectric constant	
ε''	Relative dielectric loss constant	
f'	Frequency	(Hz)
h_c	Convective heat-transfer coefficient	($W/m^{2}°C$)
H_L	Latent heat of fusion	(kJ/kg)
K	Consistency coefficient	($Pa{\cdot}s^n$)
k	Thermal conductivity	(W/m·°C)
k_L	Thermal conductivity of liquid component	(W/m°C)
k_S	Thermal conductivity of solid component	(W/m°C)
L	Length of pipe	(m)
μ	Viscosity of liquid	(Pa·s)
$\dot{m}_B$	Mass flow rate	(kg/s)
m_a	Mass fraction of ash	
m_c	Mass fraction of carbohydrate	
m_f	Mass fraction of fat	
m_m	Mass fraction of water	
m_p	Mass fraction of protein	
N^2	Volume fraction of solids or discontinuous phase in the fibrous product	
N_m	Number of blades on the mutator	
N_{Nu}	Nusselt number	
N_{Pr}	Prandtl number	
N_{GRe}	Generalized Reynolds number	
n	Flow behavior index (dimensionless)	
ΔP	Pressure drop	(Pa)
P	Power at the penetration depth	(W)
P_B	Permeability coefficient of component B	
P_D	Power dissipation	(W/cm^3)
P'	Coefficient in Equation (189.17)	
p_B	Partial pressure of gas	Pa
Q''	Coefficient defined in Equation (189.12)	
R	Radius of pipe	(m)
R'	Coefficient in Equation (189.17)	
q	Rate of heat transfer	(W)
q_x	Rate of heat flow in x direction by conduction	(W)
R	Radius of pipe	(m)
r	Density	(kg/m^3)
s	Shear stress	(Pa)
σ_0	Yield stress	(Pa)
$tan\delta$	Loss tangent, $\varepsilon''/\varepsilon'$	
T	Temperature	(°C)
T	Surrounding temperature	(°C)
T_F	Freezing point	(°C)
T_p	Surface temperature	(°C)
t_F	Freezing time	(s)
t	Time	(s)
$\bar{u}$	Mean velocity	(m/s)
$\dot{V}$	Volumetric flow rate	(m^3/s)
x	Length	(m)
y	Length	(m)
Z	Penetration depth	(m)

References

Brennan, J. G., Butters, J. R., Cowell, N. D., and Lilly, A. E. V. 1990. *Food Engineering Operations*, 3rd ed. Elsevier, New York.

Chandra, P. K. and Singh R. P., 1994. *Applied Numerical Methods in Food and Agricultural Engineering*. CRC Press, Boca Raton, FL.

Charm, S. E. 1978. *The Fundamentals of Food Engineering*, 3rd ed. AVI, Westport, CT.

Copson, D. A. 1975. *Microwave Heating*. AVI Publishing Co., Westport, CT.

Datta, A. K. and Anantheswaran, R. C. 2001. *Handbook of Microwave Technology for Food Applications*. Marcel Dekker, New York.

Harrod, M. 1987. Scraped surface heat exchanger: a literature survey of flow patterns, mixing effects, residence time distribution, heat transfer and power requirements. *J. Food Proc. Eng.* 9(1):1–62.

Heldman, D. R. and Lund, D. B. 1992. *Handbook of Food Engineering*. Marcel Dekker, New York.

Heldman, D. R. and Singh, R. P. 1981. *Food Process Engineering*, 2nd ed. AVI, Westport, CT.

Herschel, W. H. and Bulkley, R. 1926. Konsistenzmessungen von gummi-benzollusungen. *Kolloid-Zeitschr.*, 39:291.

Holman, J. P. 1990. *Heat Transfer*, 7th ed. McGraw-Hill, New York.

Kopelman, I. J. 1966. Transient Heat Transfer and Thermal Properties in Food Systems. Ph.D. thesis, Michigan State University, East Lansing, MI.

Loncin, M. and Merson, R. L. 1979. *Food Engineering: Principles and Selected Applications*. Academic Press, New York.

Palmer, J. and Jones, V. 1976. Prediction of holding times for continuous thermal processing of power law fluids. *J. Food Sci.*, 41(5):1233.

Rao, M. A. and Rizvi, S. S. H. 1995. *Engineering Properties of Foods*, 2nd ed. Marcel Dekker, New York.

Reidy, G. A. 1968. Thermal Properties of Foods and Methods of Their Determination. M.S. thesis, Michigan State University, East Lansing, MI.

Singh, R. P. 1993. *A Computerized Database of Food Properties*. CRC Press, Boca Raton, FL.

Singh, R. P. and Heldman, D. R. 2001. *Introduction to Food Engineering*, 3rd ed. Academic Press, San Diego, CA.

Singh, R. P. and Mannapperuma, J. D. 1990. Developments in food freezing. In: *Biotechnology and Food Process Engineering*, H. G. Schwartzberg and M. A. Rao, Eds. Marcel Dekker, New York, pp. 309–358.

Steffe, J. F. 1992. *Rheological Methods in Food Process Engineering*. Freeman Press, East Lansing, MI.

Sweat, V. E. 1995. Thermal properties of foods. In: *Engineering Properties of Foods*, 2nd ed., M. A. Rao and S. S. H. Rizvi, Eds. Marcel Dekker, New York, pp. 99–138.

Toledo, R. T. 1991. *Fundamentals of Food Process Engineering*, 2nd ed. Van Nostrand Reinhold, New York.

von Hippel, A. R. 1954. *Dielectrics and Waves*. MIT Press, Cambridge, MA.

Watt, B. K. and Merrill, A. L. 1975. *Composition of Foods. Agriculture Handbook No. 8*. U.S. Department of Agriculture, Washington, DC.

Further Information

The following references contain a wealth of information from proceedings of recent conferences and collaborative projects on engineering and food:

Fito, P., Ortega-Rodriquez, E., and Barbosa-Canovas, G. V. 1996. *Food Engineering 2000*. International Thompson Publishing, New York.

Jowitt, R., Escher, F., Hallstrom, B., Meffert, H. F. T., Spiess, W. E. L., and Gilbert, V. Eds. 1983. *Physical Properties of Foods*. Applied Science, London.

Jowitt, R., Escher, F., Kent, M., McKenna, B., and Roques, M. Eds. 1987. *Physical Properties of Foods — 2*. Elsevier Applied Science, London.

Le Maguer, M. and Jelen, P., Eds. 1986. *Transport Phenomena*, vol. 1 of *Food Engineering and Process Applications*. Elsevier Applied Science, London.

Le Maguer, M. and Jelen, P., Eds. 1986. *Unit Operations*, vol. 2 of *Food Engineering and Process Applications.* Elsevier Applied Science, London.

Lozano, J. E., Anon, C., Parada-Arias, E., Barbosa-Canovas, G. V. 2000. *Trends in Food Engineering*, Technomic Publishing, Lancaster, PA.

Singh, R. P. and Medina, A. G., Eds. 1989. *Food Properties and Computer-Aided Engineering of Food Processing Systems.* Kluwer, Dordrecht, The Netherlands.

Singh, R. P. and Oliveira, F., Eds. 1994. *Minimal Processing of Foods and Process Optimization.* CRC Press, Boca Raton, FL.

Singh R. P. and Wirakartakusumah, M. A., Eds. 1992. *Advances in Food Engineering.* CRC Press, Boca Raton, FL.

Singh, R. P., Ed. 1986. *Energy in Food Processing.* Elsevier, Amsterdam, The Netherlands.

Spiess, W. E. L. and Schubert, H., Eds. 1990. *Physical Properties and Process Control*, vol. 1 of *Engineering and Food.* Elsevier Applied Science, London.

Spiess, W. E. L. and Schubert, H., Eds. 1990. *Preservation Processes and Related Techniques*, vol. 2 of *Engineering and Food.* Elsevier Applied Science, London.

Spiess, W. E. L. and Schubert, H., Eds. 1990. *Advanced Techniques*, vol. 3 of *Engineering and Food.* Elsevier Applied Science, London.

Yano, T., Matsuno, R., and Nakamura, K., Eds. 1994. *Developments in Food Engineering. Part 1.* Blackie Academic and Professional, London.

Yano, T., Matsuno, R., and Nakamura, K., Eds. 1994. *Developments in Food Engineering. Part 2.* Blackie Academic and Professional, London.

Welti-Chanes, J., Barbosa-Canovas, G. V., and Aguilera, J. M. 2002. *Engineering and Food for the 21st Century.* CRC Press, Boca Raton, FL.

In addition to the above references, the following journals contain research papers on food engineering topics: *Journal of Food Engineering, Journal of Food Process Engineering, Journal of Food Processing and Preservation, International Journal of Food Properties, Journal of Texture Studies,* and *Journal of Food Science.*

190

Agricultural Engineering

David J. Hills
University of California, Davis

Agricultural engineering is the application of engineering principles and animal/plant biology to the production, handling, processing, packaging, and use of agricultural materials. Although agricultural engineering encompasses a broad area, this chapter focuses only on crop production and specifically on equipment associated with cultural practices. The three paramount operations in crop production are soil tillage, planting, and harvesting. Additionally, judicious attention paid to water, nutriment, and plant protection can maximize crop yields.

Selection of agricultural equipment requires the consideration of three parameters that can lead to high performance: machine, power, and operation. Measures of machine performance are the rate and quality at which the operations are accomplished. Rate is an important measure because agriculture is sensitive to changeable weather and crop ripeness. Most biological materials are fragile and many are perishable, so the amount of product damage or reduction in product quality caused by a machine's operation is important. The second performance parameter that must be considered is the effectiveness with which power is applied to accomplish an agricultural operation. Selecting the proper implement for a certain task and matching the implement to the engine power are critical for obtaining high power efficiency. The final performance parameter refers to the machine operator. Knowledge of the machine's mechanisms and of the overall agricultural enterprise is needed by the operator. Sensors and automatic controls are incorporated in modern equipment for additionally improving operator efficiency.

190.1 Equipment Sizing Criteria

Field Capacity

Field capacity refers to the amount of processing that a machine can accomplish per hour on either an area or a material basis. These two capacities are expressed as follows:

$$C_f = \frac{SWE}{10} \tag{190.1}$$

0-8493-1586-7/05/$0.00+$1.50

$$C_m = \frac{SWYE}{10} \tag{190.2}$$

where

C_f = field capacity on an area basis (ha/h)
C_m = field capacity on a material basis (t/h)
S = travel speed (km/h)
W = machine working width (m)
Y = crop yield (t/ha)
E = field efficiency (decimal)

Theoretical field capacity is used to describe a machine's capacity when the field efficiency term is equal to 1.0. This capacity implies that the machine is utilizing its full width and assumes no interruption for turns or other idle time. For cultivators and many harvesters, which work in rows, the machine working width is equal to the row spacing times the number of rows processed in each pass. Operator performance is not perfect, however, so less than the full width of such machines is used in order to ensure coverage of the entire land area; that is, there is some overlapping on each pass. A range of efficiency values for various field operations is provided in Table 190.1. The actual values depend on operator skill, equipment condition, and field, crop, and environmental conditions.

Typical operating speeds for various machines are listed in Table 190.1. Travel speeds for harvesters and other machines that process a product are limited by their materials-handling capacity. For machines that do not process a product, such as tillage machines, the speed is limited by other factors, such as available power, quality of the work, and safety.

Power Requirements

The power required for an agricultural machine is determined by its intended use. Tractors, for example, typically provide power to implements in three forms: drawbar, rotary, and hydraulic. Pulled or towed implements are powered through the traction of drive wheels and the pull, or **draft**, from the drawbar. Rotary power is obtained from the power takeoff (PTO) shaft. Either linear or rotary power can be produced by a tractor's hydraulic system. These three power terms are defined in Equations (190.3), (190.4), (190.5), respectively:

$$P_{db} = \frac{D_u WS}{3.6} \tag{190.3}$$

where

P_{db} = **drawbar power** (kW)
D_u = unit draft of the implement (kN)
W = machine working width (m)
S = travel speed (km/h)

$$P_{pto} = \frac{TN}{9.5} \tag{190.4}$$

where

P_{pto} = **PTO power** (kW)
T = torque (N-m)
N = revolutions per minute (rpm)

TABLE 190.1 Typical Field Equipment Efficiencies and Operating Speeds

Machine	Field Efficiency (%) Range	Field Efficiency (%) Typical	Speed (km/h) Range	Speed (km/h) Typical
		Tillage		
Cultivator (field)	70–90	85	5.0–13.0	9.0
Cultivator (row crop)	70–90	80	4.0–8.0	5.5
Harrow (disk)	70–90	80	5.0–9.9	6.5
Harrow (spiketooth)	70–90	85	5.0–9.5	8.0
Harrow (springtooth)	70–90	85	5.0–9.5	8.0
Landplane	70–90	85	3.0–8.0	6.5
Plow (chisel)	70–90	85	6.5–10.5	10.5
Plow (disk)	70–90	85	3.5–9.5	7.0
Plow (moldboard)	70–90	80	5.0–9.5	7.0
Rotary hoe	70–85	80	8.0–15.5	11.0
Rotary tiller	70–90	85	1.5–7.0	5.0
Subsoiler/ripper	70–85	80	3.0–5.0	4.0
		Planters		
Grain drill	65–80	70	4.9–9.0	6.5
Row crop planter	50–75	60	3.0–6.5	4.0
		Harvesters		
Combine	65–80	70	3.0–11.0	5.0
Corn picker	60–75	65	3.0–6.5	4.0
Harvester — cotton	60–75	70	3.0–6.5	5.0
Harvester — potato	55–70	60	2.5–6.5	3.0
Harvester — tomato	55–70	60	2.5–6.5	4.0
Hay baler	60–85	75	4.0–8.0	5.5
Mower	75–85	80	6.5–11.0	8.0
		Miscellaneous		
Mower — flail	75–90	85	5.0–7.0	5.5
Sprayer — boom type	50–80	65	5.0–11.0	10.5
Spreader — fertilizer	70–70	70	5.0–8.0	7.0

Sources: John Deere. 1993. *Fundamentals of Machine Operation — Tillage*, John Deere Service Publications, Moline, IL; Kepner, R. A., et al. 1978. *Principles of Farm Machinery*, 3rd ed. AVI, Westport, CT; Srivastava, A. K., et al. 1993. *Engineering Principles of Agricultural Machines*, American Society of Agricultural Engineers, St. Joseph, MI; and American Society of Agricultural Engineers (ASAE). 2003. *Standards 2003 — Standards, Engineering Practices and Data*, 50th ed. ASAE, St. Joseph, MI.

$$P_{\text{hyd}} = \frac{pQ}{1000} \tag{190.5}$$

where

P_{hyd} = hydraulic power (kW)
P = pressure of pumped oil (kP)
Q = oil flow rate (L/s)

The total power requirement for operating implements is the sum of implement power components converted to equivalent PTO power.

$$P_T = 1.15\left[\frac{P_{\text{db}}}{K_t} + P_{\text{pto}} + P_{\text{hyd}}\right] \quad (190.6)$$

where

P_T = total implement power requirement (kW)
1.15 = factor that adds 15% to total power for acceleration, slope, and so on
K_t = tractive and transmission efficiency (decimal)

Values for the equation variables can be obtained directly from the implement manufacturer or can be estimated from typical values as listed in Table 190.2. The major factors influencing draft on tillage tools are soil characteristics, forward speed, and crop resistance. The draft for most pull-type, nontillage implements is in the form of rolling resistance.

Approximate values for tractive and transmission efficiency for two- and four-wheel-drive tractors and crawler tractors are provided in Table 190.3. As shown, tractive efficiencies for four-wheel-drive tractors are somewhat higher than those for two-wheel-drive tractors, especially for soft soils. Crawler-type tractors seldom have more than 5% slip, even on soft soils.

Fuel requirements for tractors can be estimated from Table 190.4, provided that the maximum PTO power rating and the actual PTO power requirement (as calculated by Equation (190.6)) are known. The drawbar power is always less than PTO power because of drive-wheel slippage, tractor rolling resistance, and friction losses in the drive between the engine and the wheels. The sum of these losses forms the basis of the tractive and transmission coefficients listed in Table 190.3. These coefficients are essentially the ratios of drawbar power to PTO power.

190.2 Equipment Selection

Soil Tillage

Tillage is used for seedbed preparation, weed control, incorporation of crop residues and fertilizer materials, breaking soil crusts and hardpans to improve water penetration and aeration, and shaping the soil for irrigation and erosion control. The tillage requirement is determined by the type of crop, soil type, and field conditions. A tillage implement consists of a single tool or a group of tools, together with the associated frame, wheels, hitch, control and protection devices, and power transmission components.

Tillage operations for seedbed preparation are often classified as primary or secondary. A primary tillage operation constitutes the initial, major soil-working operation after harvest of the previous crop. It is normally designed to reduce soil strength, cover plant materials, and rearrange soil aggregates. The main objective of secondary tillage is to break down large clods and to prepare a seedbed ideal for planting. An ideal seedbed provides for good seed-to-soil contact, conserves moisture needed for germination, and allows for vigorous and uninhibited root and shoot growth.

Implements used for primary tillage are moldboard plows, disk plows and tillers, heavy disk harrows, chisel plows, subsoilers, rotary plows, listers, and bedders. Moldboard plows and heavy disk harrows are the most commonly used primary tillage tools. Implements used for secondary tillage are disk harrows, cultivators, spike and spring tooth harrows, and rotary hoes and cultivators. The most common implement used for secondary cultivation is the disk harrow. Generally, several tillage operations are performed before the field is ready for planting. In dry climates, culti-packers are also used as the final tillage operation before planting. Increasing the soil density in the top few centimeters helps retain soil moisture.

Example 190.1

What is the engine power requirement for a four-wheel drive tractor pulling a three-bottom, 400-mm moldboard plow to a depth of 18 cm and at a speed of 6.0 km/h on medium textured soil?

TABLE 190.2 Draft and Energy Requirements for Selected Field Equipment Operated at 5 km/h

Machine	Unit Draft (kN/m)	Energy or Work (kW-h/ha)
	Tillage	
Cultivator (field)	0.9–4.4	2.4–12.0
Cultivator (row crop)	0.6–1.2	1.6–3.3
Harrow (disk)	0.7–1.5	2.0–4.0
Harrow (spiketooth)	0.3–0.9	0.7–2.4
Harrow (springtooth)	1.0–4.4	2.1–12.2
Landplane	4.4–11.7	12.2–31.3
Plow (chisel — 18 to 23 cm)	2.9–13.1	8.1–36.9
Plow (moldboard or disk)		
Light soils — 18-cm depth	3.2–6.3	8.7–17.5
Medium soils — 18-cm depth	5.3–9.5	14.6–25.8
Heavy soils — 18-cm depth	8.5–16.6	22.1–46.1
Rotary hoe	0.4–0.9	1.3–2.4
Rotary tiller — 10-cm forward slice	12.2–24.5	25.8–51.6
Subsoiler/ripper 2-m spacing		
Light soils — 40-cm depth	16.0–26.3/unit	7.2–12.0
Medium soils — 40-cm depth	23.3–36.5/unit	10.1–15.7
	Planters	
Grain drill	0.4–1.5	1.1–3.9
Row crop planter — 1-m spacing	0.5–0.8/row	1.1–2.4
	PTO Power (kW/m or kW/row)	
	Harvesters	
Combine — small grain	3.6–11.0	7.2–22
Corn picker	1.5–3.7/row	4.4–8.8
Harvester — cotton (spindle)	7.5–11.2/row	12.9–18.4
Harvester — potato	0.7–1.5/row	
Hay baler		8.3–13.8
Rotary mower (grass, legumes)	7.3–19.6	1.2–2.0
		9.2–24
	Miscellaneous	
Mower — flail		0.9–2.0
Sprayer — boom type	0.2 kW	0.02–0.4
Spreader — fertilizer	0.3–1.2	0.9–3.1

Sources: John Deere. 1993. *Fundamentals of Machine Operation — Tillage*, John Deere Service Publications, Moline, IL; Kepner, R. A., et al. 1978. *Principles of Farm Machinery*, 3rd ed. AVI, Westport, CT; Srivastava, A. K., et al. 1993. *Engineering Principles of Agricultural Machines*, American Society of Agricultural Engineers, St. Joseph, MI; and American Society of Agricultural Engineers (ASAE). 2003. *Standards 2003 — Standards, Engineering Practices and Data*, 50th ed. ASAE, St. Joseph, MI.

TABLE 190.3 Tractor Tractive and Transmission Coefficients, K_t

Soil Condition/Tractive Condition	Two-Wheel Drive	Four-Wheel Drive	Crawler
Concrete	0.87	0.88	—
Firm, untilled	0.72	0.78	0.82
Tilled, reasonably firm	0.67	0.75	0.80
Freshly plowed, soft	0.55	0.70	0.78

Source: Zoz, F. M., Turner, R. J. and Shell, L. R. 2002. Power delivery efficiency: a valid measure of belt and tire tractor performance. *Trans. ASAE*, 45:509–518.

TABLE 190.4 Tractor Fuel Conversion, PTO kW-h/L

Loading, % of Maximum PTO Power	Gasoline	Diesel	LP Gas
100	1.90	2.57	1.57
80	1.74	2.50	1.49
60	1.50	2.26	1.34
40	1.16	1.87	1.09
20	0.76	1.27	0.74

Sources: Kepner, R. A. et al. 1978. *Principles of Farm Machinery*, 3rd ed. AVI, Westport, CT; and Hunt, D. R. 2001. *Farm Power and Machinery Management*, 10th ed. Iowa State University Press, Ames.

Solution. Data from Tables 190.2 and 190.3 are incorporated into Equations (190.3) and (190.6), respectively.

$$P_{\mathrm{db}} = \frac{(7.4)(3 \cdot 400/1000)(6)}{3.6} = 14.8\ \mathrm{kW}$$

$$P_T = 1.15\frac{14.8}{0.75} = 22.7\ \mathrm{kW}$$

Crop Planting

Agricultural plants usually begin from either seeds or seedling transplants. Important factors affecting seed germination and emergence include seed viability, soil temperature, availability of moisture and air to the seeds, and soil strength and resistance to seedling emergence. The planter can exert a strong influence on the rate of germination and emergence of seeds through control of planting depth and firming of soil around the seeds or roots of seedlings. In addition, the planter must meter seeds at the proper rate and, in some cases, must control the horizontal down-the-row placement of seeds in a desired pattern.

Equipment is available for three seeding practices: broadcasting, drilling, and precision planting. Broadcasting refers to random placement of seeds on the soil surface. Seed is metered from a hopper through a variable orifice onto a spinning disk, which accelerates the seed and distributes it. Drilling is the random down-the-row or horizontal placement of seeds in furrows that are then covered. In a seed drill, seeds are metered from a series of hoppers, typically by fluted wheels, into small furrows that are subsequently covered and pressed. High-density plantings, high costs of hand thinning, and erratic performance of mechanical thinners have resulted in the development of precision seeding techniques. In precision planting, the seeds are planted in rows at uniform spacing. Precision planters are similar in operation to press drills; however, the metering hardware is more complex, allowing for seed placement at precise depths and locations.

Table 190.5 provides data on typical seeding rates and practices for selected crops. The wide ranges of seed spacing reflect the dependence on climate, season, and type of market. For agronomic crops, the lower seeding rates and wider spacings are more typical for nonirrigated conditions. For vegetable crops, spacing is determined by type of market (fresh versus processor), desired fruit size, harvesting equipment dimensions, and time of year. Depth of seeding is determined by antecedent soil moisture, soil type, and soil temperature. These data can be used to establish general criteria for planting and cultural equipment.

Example 190.2

A 3.4-m wide grain drill is used to plant a 50-ha field in wheat. Approximately how many hours are required for this planting operation?

TABLE 190.5 Traditional Seed Depth and Rate for Selected Crops

Crop	Depth to Plant Seed (cm)	Spacing between Plants in Row (cm)	Spacing between Rows (cm)	No. Seeds per Gram	Seeding Rate (kg/ha)
Alfalfa	0.6–1.3	Drilled	Drilled	480	9–22
Barley	1.3–2.5	Drilled	Drilled	28	81–108
Bean, snap	2.5–3.8	10–30	46–107	4–5	75–100
Broccoli	0.6	36–91	61–91	320	0.5–1.5
Cabbage	0.6	36–91	61–91	320	0.5–1.5
Carrot	0.6	3–8	38–61	820	2–4
Cauliflower	0.6	36–61	61–122	320	0.5–1.5
Celery	0.6	15–30	46–91	2500	1–2
Corn, field	2.5–5.0	13–30	61–91	4–6	12–18
Corn, sweet	2.5–5.0	15–25	61–91	4–6	10–17
Cotton	2.5–5.0	3–8	61–97	2–4	18–25
Cucumber	2.5	30–91	91–122	40	3–6
Lettuce, head	0.6–1.3	15–25	46–91	900	1–3
Oat	1.3–2.5	Drilled	Drilled	28	54–143
Onion	1.3	8–10	38–91	305	3–5
Pea, English	5.0	3–5	46–91	3–6	100–250
Potato (tubers)	10.0	23–38	76–107	—	1000–2000
Rice	0.6–1.9	Drilled	Drilled	25	75–179
Sorghum	1.3–5.0	Drilled	Drilled	62	17–50
Soybean	2.5–3.8	3–15	30–91	6–12	20–50
Squash, bush	2.5	46–122	91–152	4–14	2–7
Sugar beet	0.7–1.5	10–15	60–91	120	0.5–1.5
Tomato	1.3	30–91	61–152	250–430	0.5–1.5
Watermelon	2.5	61–244	183–244	1–3	1–3
Wheat	1.5–2.5	Drilled	Drilled	26	67–101

Sources: Doane-Western. 1981. *Facts and Figures for Farmers*, 4th ed. Doane Agricultural Services, St. Louis, MO; Maynard, D. N. and Hochmuth, G. J. 1996. *Knott's Handbook for Vegetable Growers*, 4th ed. John Wiley & Sons, New York; and Treadgill, E. D. 1988. Plant growth data. In: *Handbook of Engineering in Agriculture*, R. H. Brown, Ed. CRC Press, Boca Raton, FL, pp. 129–131.

Solution. Data for Table 190.1 are used in Equation (190.1) for determining the field capacity, and then the operation time is calculated.

$$C_f = \frac{(6.5)(3.4)(0.7)}{10} = 1.55 \text{ ha/h}$$

$$\text{Time} = \frac{50 \text{ ha}}{1.55 \text{ ha/h}} = 32 \text{ h or four 8 h days}$$

Crop Harvest

The final crop production operation is the harvesting of the plant parts that have economic value to the grower. In some cases, more than one plant part may have economic value. Many crops are highly perishable products that must be harvested within a very narrow time range, handled carefully, and either processed, properly stored, or consumed fresh soon after harvesting. Mechanical harvesters are available for a number of crops. The general groups are hay and forage harvesters, grain harvesters, and fruit, nut, and vegetable harvesters.

Hay and forage harvesters are used in producing animal feed as ensilage or hay. Ensilage involves cutting the forage at 70 to 80% moisture, allowing it to field dry to 50 to 60% moisture, chopping it into short lengths to obtain adequate packing, and preserving it by fermentation in an airtight chamber. Equipment for the following steps are required: cut/condition, windrow, wilt, chop, transport, and store. For hay production, the forage is cut and allowed to dry to a moisture content of 15 to 23% before storage. Hay production requires equipment for the following steps: cut/condition/swath, rake into windrows, dry, bale or chop, transport, and store.

The harvesting of cereal grains, grasses, and legumes is accomplished almost entirely with the combine. A combine has five general mechanical functions: cutting, feeding, threshing, separating, and cleaning. The cutting operation is accomplished with a cutter bar and reel. The feeding mechanism distributes and delivers the crop material to the threshing cylinder in a steady uniform flow. The threshing operation is accomplished with a cylinder working against iron bar concaves. The separating mechanism extracts the straw. Cleaning of the grain, the final step, is accomplished with screening devices and blowers.

Harvesters for fruits, nuts, and vegetables are usually crop specific and may be equipped with special sensors for product selection according to maturity and size. Although these harvesters are designed for a broad range of specialty crops, they may be broadly classified according to the physical location in which the harvestable portion of the crop is located. These four crop zones are root (e.g., for sugar beets, potatoes); surface (e.g., for beans, tomatoes); bush and trellis (e.g., for boysenberries, grapes); and tree (e.g., for olives, almonds). Successful harvest mechanization requires a total systems approach, which includes varietal breeding, cultural practices, materials handling, grading and sorting, and ultimate processing. Harvester selection is therefore based on these factors, as well as such factors as economics and available labor.

Example 190.3

A tomato harvester is observed to travel at 4.0 km/h with a design width of 1.5 m. The average yield for the field is 80 t/ha. What is the material capacity of the machine?

Solution. Data from Table 190.1 are used in Equation (190.2) for determining the material handling capacity.

$$C_m = \frac{(4.0)(1.5)(80)(0.60)}{10} = 28.8 \text{ t/h}$$

Defining Terms

Draft — The horizontal force required to propel an implement in the direction of travel.

Drawbar power — The power to pull or move an implement at a uniform speed. It is chiefly a function of implement draft and forward speed.

Field capacity — The amount of processing that a machine can accomplish per hour, expressed on either an area or a material basis.

PTO power — The power to operate an implement from the power-takeoff shaft. It is chiefly a function of torque and rotational speed.

References

American Society of Agricultural Engineers (ASAE). 2003. *Standards 2003 — Standards, Engineering Practices and Data*, 50th ed. ASAE, St. Joseph, MI.

Doane-Western. 1990. *Facts and Figures for Farmers*, 4th ed. Doane Agricultural Services, St. Louis, MO.

Hunt, D. R. 2001. *Farm Power and Machinery Management*, 10th ed. Iowa State University Press, Ames.

John Deere. 1993. *Fundamentals of Machine Operation — Tillage*, John Deere Service Publications, Moline, IL.

Kepner, R. A., Bainer, R., and Barger, E. L. 1978. *Principles of Farm Machinery*, 3rd ed. AVI. West-port, CT.

Maynard, D. N. and Hochmuth, G. J. 1996. *Knott's Handbook for Vegetable Growers*, 4th ed. John Wiley & Sons, New York.

Srivastava, A. K., Goering, C. E., and Rohrbach, R. P. 1993. *Engineering Principles of Agricultural Machines.* American Society of Agricultural Engineers, St. Joseph, MI.

Treadgill, E. D. 1988. Plant growth data. In: *Handbook of Engineering in Agriculture,* R. H. Brown, Ed. CRC Press, Boca Raton, FL, pp. 129–131.

Zoz, F. M., Turner, R. J. and Shell, L. R. 2002. power delivery efficiency: a valid measure of belt and tire tractor performance. *Trans. ASAE,* 45:509–518.

Further Information

Applied Engineering in Agriculture. Published bimonthly by the American Society for Engineering in Agricultural, Food, and Biological Systems.

Brown, R. H., Ed. 1988. *Handbook of Engineering in Agriculture.* CRC Press, Boca Raton, FL.

Culpin, C. 1992. *Farm Machinery,* 12th ed. Blackwell Scientific, London.

Goering, C.E., Stone, M.L., Smith, D.W., and Turnquest, P.K. 2003. *Off-Road Vehicle Engineering Principles,* American Society of Agricultural Engineers, St. Joseph, MI.

Transactions of the ASAE. Published bimonthly by ASAE, the American Society for Engineering in Agricultural, Food, and Biological Systems.

191

System Reliability

Rama Ramakumar
Oklahoma State University

Application of system reliability evaluation techniques is gaining importance because of its effectiveness in the detection, prevention, and correction of failures in the design, manufacturing, and operational phases of a product. Increasing emphasis on the reliability and quality of products and systems, coupled with pressures to minimize cost, further emphasize the need to study and quantify reliability and arrive at innovative designs.

Reliability engineering has grown significantly during the past 5 decades (since World War II) to encompass many subareas, such as reliability analysis, failure theory and modeling, reliability allocation and optimization, reliability growth and modeling, reliability testing (including accelerated testing), data analysis and plotting, quality control and acceptance sampling, maintenance engineering, software reliability, system safety analysis, Bayesian analysis, reliability management, simulation, Monte Carlo techniques, and economic aspects of reliability.

The objectives of this chapter are to introduce the reader to the fundamentals and applications of classical reliability concepts and bring out the importance and benefits of reliability considerations.

0-8493-1586-7/05/$0.00+$1.50

191.1 Catastrophic Failure Models

Catastrophic failure refers to the case in which repair of the component is not possible, not available, or of no value to the successful completion of the mission originally planned. Modeling such failures is typically based on life test results. We can consider the "lifetime" or "time to failure" T as a continuous random variable. Then,

$$P\,(\text{survival up to time } t) = P\,(T > t) \equiv R\,(t) \tag{191.1}$$

where $R(t)$ is the *reliability* function. Obviously, as $t \to \infty$, $R(t) \to 0$ because the probability of failure increases with time of operation. Moreover,

$$P\,(\text{failure at } t) = P\,(T \le t) \equiv Q\,(t) \tag{191.2}$$

where $Q(t)$ is the unreliability function. From the definition of the distribution function of a continuous random variable, it is clear that $Q(t)$ is indeed the distribution function for T. Therefore, the failure density function $f(t)$ can be obtained as

$$f(t) = \frac{d}{dt}Q(t) \tag{191.3}$$

The *hazard rate function* $\lambda(t)$ is defined as

$$\lambda(t) \equiv \lim_{\Delta t \to 0} \frac{1}{\Delta t}\begin{bmatrix}\text{probability of failure in } (t, t+\Delta t),\\ \text{given survival up to } t\end{bmatrix} \tag{191.4}$$

It can be shown that

$$\lambda(t) = \frac{f(t)}{R(t)} \tag{191.5}$$

The four functions $f(t)$, $Q(t)$, $R(t)$, and $\lambda(t)$ constitute the set of functions used in basic reliability analysis. The relationships between these functions are given in Table 191.1.

TABLE 191.1 Relationships between Different Reliability Functions

$f(t)$	$\lambda(t)$	$Q(t)$	$R(t)$
$f(t) = f(t)$	$\lambda(t)\exp\left[-\int_0^t \lambda(\xi)d\xi\right]$	$\frac{d}{dt}Q(t)$	$-\frac{d}{dt}R(t)$
$\lambda(t) = \frac{f(t)}{1-\int_0^t f(\xi)d\xi}$	$\lambda(t)$	$\frac{1}{1-Q(t)}\frac{d}{dt}(Q(t))$	$-\frac{d}{dt}[\ln R(t)]$
$Q(t) = \int_0^t f(\xi)d\xi$	$1-\exp\left[-\int_0^t \lambda(\xi)d\xi\right]$	$Q(t)$	$1 - R(t)$
$R(t) = 1-\int_0^t f(\xi)d\xi$	$\exp\left[-\int_0^t \lambda(\xi)d\xi\right]$	$1 - Q(t)$	$R(t)$

Source: Ramakumar, R. 1993. *Engineering Reliability: Fundamentals and Applications.* Prentice Hall, Englewood Cliffs, NJ. With permission.

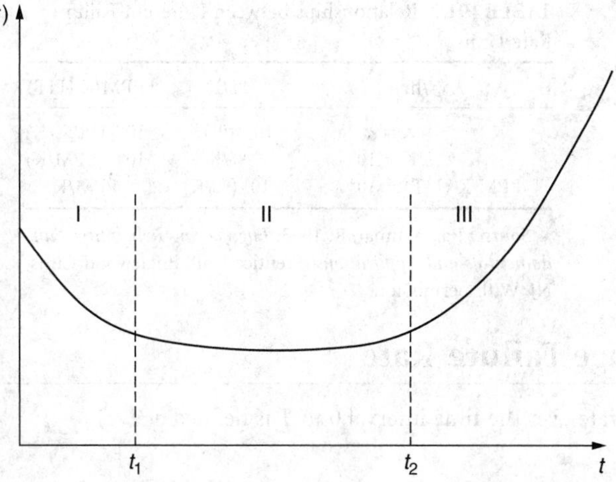

FIGURE 191.1 Bathtub-shaped hazard function. (*Source:* Ramakumar, R. 1993. *Engineering Reliability: Fundamentals and Applications*. Prentice Hall, Englewood Cliffs, NJ. With permission.)

191.2 The Bathtub Curve

Of the four functions discussed, the hazard rate function $\lambda(t)$ displays the different stages during the lifetime of a component most clearly. In fact, typical $\lambda(t)$ plots have the general shape of a bathtub curve as shown in Figure 191.1. The first region corresponds to *wear-in* (infant mortality) or early failures during debugging. The hazard rate goes down as debugging continues. The second region corresponds to an essentially constant and low failure rate — failures can be considered to be nearly random. This is the useful lifetime of the component. The third region corresponds to the *wear-out* or fatigue phase with a sharply increased hazard rate.

Burn-in refers to the practice of subjecting components to an initial operating period of t_1 (see Figure 191.1) before delivering them to the customer. This eliminates all the initial failures from occurring after delivery to customers requiring high-reliability components. Moreover, it is prudent to replace a component as it approaches the wear-out region (i.e., after an operating period of $[t_2 - t_1]$). Electronic components tend to have a long useful life (constant hazard) period. The wear-out region tends to dominate in the case of mechanical components.

191.3 Mean Time to Failure

The mean or expected value of the continuous random variable *time to failure* is the *mean time to failure* (MTTF). This is a very useful parameter that is often used to assess the suitability of components. It can be obtained using either the failure density function $f(t)$ or the reliability function $R(t)$ as follows:

$$\text{MTFF} = \int_0^\infty t f(t)\,dt \quad \text{or} \quad \int_0^\infty R(t)\,dt \tag{191.6}$$

In the case of repairable components, the repair time can also be considered as a continuous random variable with an expected value of MTTR. The mean time between failures, MTBF, is the sum of MTTF and MTTR. For well-designed components, MTTR << MTTF. Thus, MTBF and MTTF are often used interchangeably.

TABLE 191.2 Relationships between Different Failure Rate Units

λ (#/hr)	%/K	PPM/K (FIT)
$\lambda = \lambda$	10^{-5} (%/K)	10^{-9} (PPM/K)
%/K = $10^5\ \lambda$	%/K	10^{-4} (PPM/K)
PPM/K (FIT) = $10^9\ \lambda$	10^4 (%/K)	PPM/K

Source: Ramakumar, R. 1993. *Engineering Reliability: Fundamentals and Applications.* Prentice Hall, Englewood Cliffs, NJ. With permission.

191.4 Average Failure Rate

The average failure rate over the time interval 0 to T is defined as

$$\mathrm{AFR}(0,T) \equiv \mathrm{AFR}(T) = -\frac{\ln R(T)}{T} \tag{191.7}$$

191.5 A Posteriori Failure Probability

When components are subjected to a burn-in (or wear-in) period of duration T, and if the component survives during $(0, T)$, the probability of failure during $(T, T + t)$ is called the *a posteriori failure probability* $Q_c(t)$. It can be found using

$$Q_c(t) = \frac{\int_T^{T+t} f(\xi)d\xi}{\int_T^{\infty} f(\xi)d\xi} \tag{191.8}$$

The probability of survival during $(T, T + t)$ is

$$R(t|T) = 1 - Q_c(t) = \frac{\int_{T+t}^{\infty} f(\xi)d\xi}{\int_T^{\infty} f(\xi)d\xi} = \frac{R(T+t)}{R(t)} = \exp\left[-\int_T^{T+t} \lambda(\xi)d\xi\right] \tag{191.9}$$

191.6 Units for Failure Rates

Several units are used to express failure rates. In addition to $\lambda(t)$, which is usually in number per hour, *%/K* is used to denote failure rates in percent per thousand hours, and *PPM/K* is used to express failure rate in parts per million per thousand hours. The last unit is also known as FIT for "fails in time." The relationships between these units are given in Table 191.2.

191.7 Application of the Binomial Distribution

In an experiment consisting of n identical independent trials, with each trial resulting in success or failure with probabilities of p and q, the probability P_r or r successes and $(n - r)$ failures is

$$P_r = {}_nC_r p^r (1-p)^{n-r} \tag{191.10}$$

If X denotes the number of successes in n trials, then it is a discrete random variable with a mean value of (np) and variance of (npq).

In a system consisting of a collection of n identical components with a probability p that a component is defective, the probability of finding r defects out of n is given by the P_r in Equation (191.10). If p is the probability of success of one component and if at least r of them must be good for system success, then the system reliability (probability of system success) is given by

$$R = \sum_{k=r}^{n} {}_nC_k p^k (1-p)^{n-k} \tag{191.11}$$

For a system with redundancy, $r < n$.

191.8 Application of Poisson Distribution

For events that occur *in time* at an average rate of λ occurrences per unit of time, the probability $P_x(t)$ of exactly x occurrences during the time interval (0, t) is given by

$$P_x(t) = \frac{(\lambda t)^x e^{-\lambda t}}{x!} \tag{191.12}$$

The number of occurrences X in (0, t) is a discrete random variable with a mean value of μ of (λt), and a standard deviation σ of $\sqrt{\lambda t}$. By setting $X = 0$ in Equation (191.12), we obtain the probability of no occurrence in (0, t) as $e^{-\lambda t}$. If the event is failure, then no occurrence means success and $e^{-\lambda t}$ is the probability of success or system reliability. This is the well-known and often used exponential distribution, also known as the constant hazard model.

191.9 Exponential Distribution

A constant hazard rate (constant λ) corresponding to the useful lifetime of components leads to the single parameter exponential distribution. The functions of interest associated with a constant λ are

$$f(t) = \lambda e^{-\lambda t}, \qquad t > 0 \tag{191.13}$$

$$R(t) = e^{-\lambda t} \tag{191.14}$$

$$Q(t) = Q_c(t) = 1 - e^{-\lambda t} \tag{191.15}$$

The a posteriori failure probability $Q_c(t)$ is independent of the prior operating time T, indicating that the component does not degrade no matter how long it operates. Obviously, such a scenario is valid only during the useful lifetime (horizontal portion of the bathtub curve) of the component.

The mean and standard deviation of the random variable *lifetime* are

$$\mu \equiv \text{MTFF} = \frac{1}{\lambda} \quad \text{and} \quad \sigma = \frac{1}{\lambda} \tag{191.16}$$

191.10 Weibull Distribution

The Weibull distribution has two parameters — a scale parameter α and a shape parameter β. By adjusting these two parameters, a wide range of experimental data can be modeled in system reliability studies. The associated functions are

$$\lambda(t) = \frac{\beta t^{\beta-1}}{\alpha^{\beta}}; \quad \alpha > 0, \beta > 0, t \ge 0 \tag{191.17}$$

$$f(t) = \frac{\beta t^{\beta-1}}{\alpha^{\beta}} \exp\left[-\left(\frac{t}{\alpha}\right)^{\beta}\right] \tag{191.18}$$

$$R(t) = \exp\left[-\left(\frac{t}{\alpha}\right)^{\beta}\right] \tag{191.19}$$

With $\beta = 1$, the Weibull distribution reduces to the constant hazard model with $\lambda = (1/\alpha)$. With $\beta = 2$, the Weibull distribution reduces to the Rayleigh distribution.

The associated MTTF is

$$\text{MTFF} = \mu = \alpha\Gamma\left(1 + \frac{1}{\beta}\right) \tag{191.20}$$

where Γ denotes the gamma function.

191.11 Combinatorial Aspects

Analysis of complex systems is facilitated by decomposition into functional entities consisting of subsystems or units and by the application of combinatorial considerations and network modeling techniques.

A *series structure* (or chain structure) consisting of n units is shown in Figure 191.2. From the reliability point of view, the system will succeed only if all the units succeed. The units may or may not be physically in series. If R_i is the probability of success of the ith unit, then the series system reliability R_s is given as

$$R_s = \prod_{i=1}^{n} R_i \tag{191.21}$$

if the units do not interact with each other. If they do, then the conditional probabilities must be carefully evaluated.

If each of the units has a constant hazard, then

$$R_s(t) = \prod_{i=1}^{n} \exp(-\lambda_i t) \tag{191.22}$$

where λ_i is the constant failure rate for the ith unit or component. This enables us to replace the n components in series by an equivalent component with a constant hazard λ_s where

$$\lambda_s = \sum_{i=1}^{n} \lambda_i \tag{191.23}$$

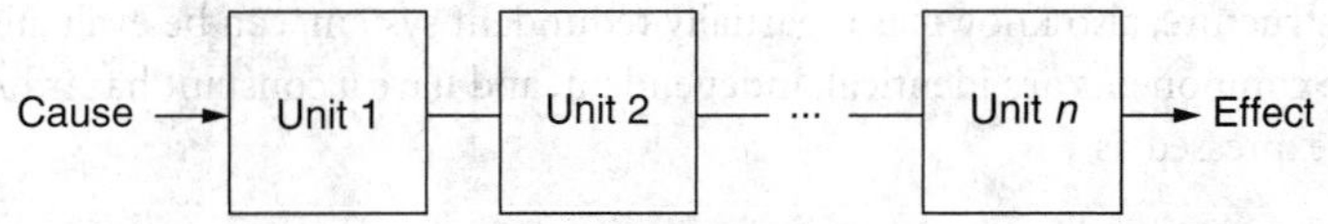

FIGURE 191.2 Series or chain structure. (*Source:* Ramakumar, R. 1993. *Engineering Reliability: Fundamentals and Applications*. Prentice Hall, Englewood Cliffs, NJ. With permission.)

If the components are identical, the $\lambda_s = n\lambda$ and the MTTF for the equivalent component is $(1/n)$ of the MTTF of one component.

A *parallel structure* consisting of n units is shown in Figure 191.3. From the reliability point of view, the system will succeed if any one of the n units succeeds. Once again, the units may or may not be physically or topologically in parallel. If Q_i is the probability of failure of the ith unit, then the parallel system reliability R_p is given as

$$R_p = 1 - \prod_{i=1}^{n} Q_i \tag{191.24}$$

if the units do not interact with each other (i.e., are independent).

If each of the units has a constant hazard, then

$$R_p(t) = 1 - \prod_{i=1}^{n} [1 - \exp(-\lambda_i t)] \tag{191.25}$$

and we do not have the luxury of being able to replace the parallel system by an equivalent component with a constant hazard. The parallel system does not exhibit constant hazard even though each of the units has constant hazard.

The MTTF of the parallel system can be obtained by using Equation (191.25) in Equation (191.6). The results for the case of components with identical hazards λ are: $(1.5/\lambda)$, $(1.833/\lambda)$, and $(2.083/\lambda)$ for $n = 2$, 3, and 4, respectively. The largest gain in MTTF is obtained by going from one component to two components is parallel. It is uncommon to have more than two or three components in a truly parallel configuration because of the cost involved. For two nonidentical components in parallel with hazard rates λ_1 and λ_2, the MTTF is given as

$$\text{MTTF} = \frac{1}{\lambda_1} + \frac{1}{\lambda_2} - \frac{1}{\lambda_1 + \lambda_2} \tag{191.26}$$

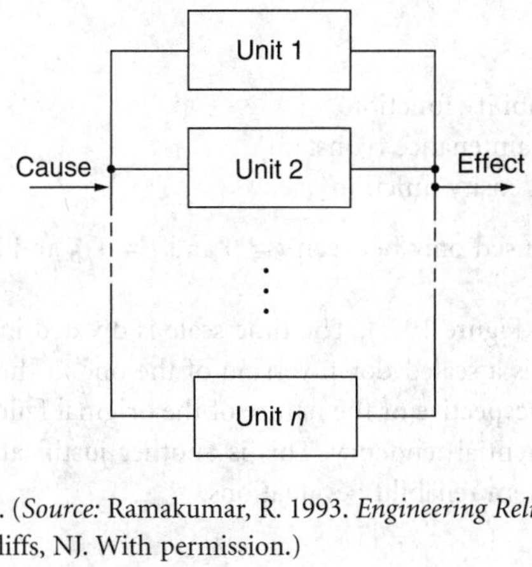

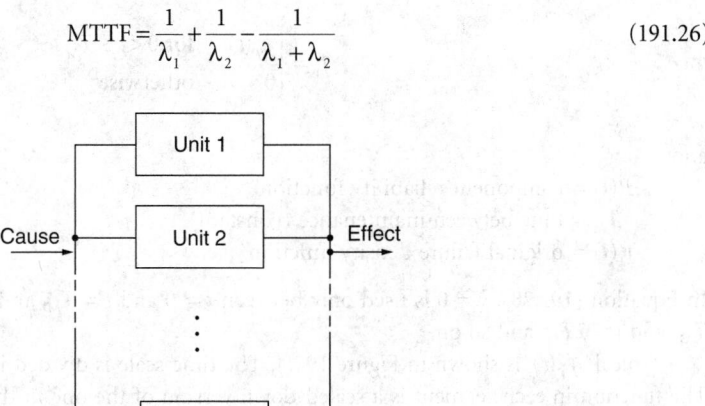

FIGURE 191.3 Parallel structure. (*Source:* Ramakumar, R. 1993. *Engineering Reliability: Fundamentals and Applications.* Prentice Hall, Englewood Cliffs, NJ. With permission.)

An r-out-of-n structure, also known as a partially redundant system, can be evaluated using Equation (191.11). If all the components are identical, independent, and have a constant hazard λ, then the system reliability can be expressed as

$$R(t) = \sum_{k=r}^{n} {}_nC_k e^{-k\lambda t}(1-e^{-\lambda t})^{n-k} \tag{191.27}$$

For $r = 1$, the structure becomes a parallel system. For $r = n$, it becomes a series system.

Series-parallel systems are evaluated by repeated application of the expressions derived for series and parallel configurations by employing well-known network reduction techniques.

Several general techniques are available for evaluating the reliability of complex structures that do not come under purely series or parallel or series-parallel. They range from inspection to cut-set and tie-set methods and connection matrix techniques that are amenable to computer programming.

191.12 Modeling Maintenance

Maintenance of a component could be a scheduled (preventive) one or a forced (corrective) one. The latter follows in-service failure and can be handled using Markov models discussed later. Scheduled maintenance is conducted at fixed intervals of time, irrespective of the system continuing to operate satisfactorily.

Scheduled maintenance, under ideal conditions, takes very little time (compared to the time between scheduled maintenance events) and the component is restored to an "as new" condition. Even if the component is irreparable, scheduled maintenance postpones failure and prolongs the life of the component. Scheduled maintenance makes sense only for those components with increasing hazard rates. Most mechanical systems come under this category. It can be shown that the density function $f_T^*(t)$, with scheduled maintenance included, can be expressed as

$$f_T^*(t) = \sum_{k=0}^{\infty} f_1(t - KT_M)R^k(T_M) \tag{191.28}$$

where

$$f_1 = \begin{cases} f_T(t) & \text{for } 0 < t \le T_M \\ 0 & \text{otherwise} \end{cases} \tag{191.29}$$

and

$R(t)$ = component reliability function
T_M = time between maintenance (constant)
$f_T(t)$ = original failure density function

In Equation (191.28), $k = 0$ is used only between $t = 0$ and $t = T_M$, and $k = 1$ is used only between $t = T_M$ and $t = 2T_M$ and so on.

A typical $f_T^*(t)$ is shown in Figure 191.4. The time scale is divided into equal intervals of T_M each. The function in each segment is a scaled-down version of the one in the previous segment, the scaling factor being equal to $R(T_M)$. Irrespective of the nature of the original failure density function, scheduled maintenance gives it an exponential tendency. This is another justification for the widespread use of exponential distribution in system reliability evaluations.

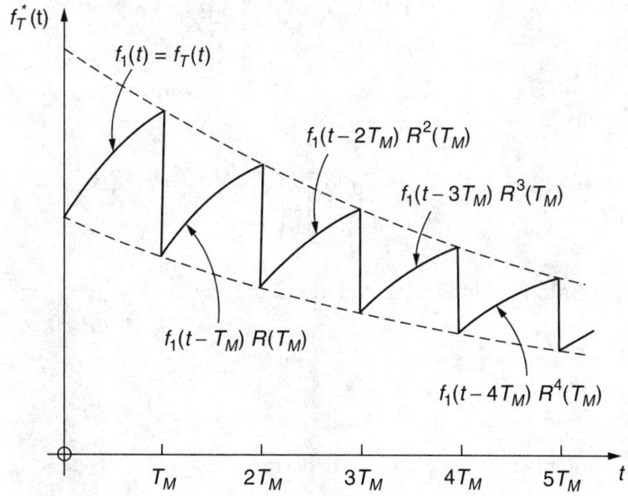

FIGURE 191.4 Density function with ideal scheduled maintenance incorporated. (*Source:* Ramakumar, R. 1993. *Engineering Reliability: Fundamentals and Applications.* Prentice Hall, Englewood Cliffs, NJ. With permission.)

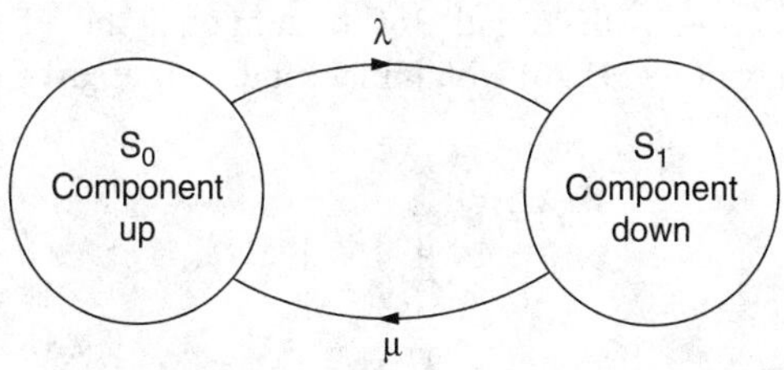

FIGURE 191.5 State-space diagram for a single reparable component. (*Source:* Ramakumar, R. 1993. *Engineering Reliability: Fundamentals and Applications.* Prentice Hall, Englewood Cliffs, NJ. With permission.)

191.13 Markov Models

Of the various Markov models available, the discrete-state, continuous-time Markov process has found many applications in system reliability evaluation, including the modeling of reparable systems. The model consists of a set of discrete states (called the state space) in which the system can reside and a set of transition rates between appropriate states. Using these, a set of first-order differential equations is derived in the standard vector-matrix form for the time-dependent probabilities of the various states. Solution of these equations incorporating proper initial conditions gives the probabilities of the system residing in different states as functions of time. Several useful results can be gleaned from these functions.

191.14 Binary Model for Reparable Component

The binary model for a reparable component assumes that the component can exist in one of two states — the *up* state or the *down* state. The transition rates between these two states, S_0 and S_1, are assumed to be constant and equal to λ and μ. These transition rates are the constant failure and repair rates implied in the modeling process and their reciprocals are the MTTF and MTTR, respectively. Figure 191.5 illustrates the binary model.

The associated Markov differential equations are

$$\begin{bmatrix} P_0'(t) \\ P_1'(t) \end{bmatrix} = \begin{bmatrix} -\lambda & \mu \\ \lambda & -\mu \end{bmatrix} \begin{bmatrix} P_0(t) \\ P_1(t) \end{bmatrix} \tag{191.30}$$

with the initial conditions

$$\begin{bmatrix} P_0(0) \\ P_1(0) \end{bmatrix} = \begin{bmatrix} 1 \\ 0 \end{bmatrix} \tag{191.31}$$

The coefficient matrix of Markov differential equations, namely

$$\begin{bmatrix} -\lambda & \mu \\ \lambda & -\mu \end{bmatrix}$$

is obtained by transposing the matrix of rates of departures

$$\begin{bmatrix} 0 & \lambda \\ \mu & 0 \end{bmatrix}$$

and replacing the diagonal entries by the negative of the sum of all the other entries in their respective columns. The solution of Equation (191.30) with initial conditions as given by Equation (191.31) yields

$$P_0(t) = \frac{\mu}{\lambda+\mu} + \frac{\lambda}{\lambda+\mu} e^{-(\lambda+\mu)t} \tag{191.32}$$

$$P_1(t) = \frac{\lambda}{\lambda+\mu}[1 - e^{-(\lambda+\mu)t}] \tag{191.33}$$

The limiting, or steady-state, probabilities are found by letting $t \to \infty$. They are also known as limiting availability A and limiting unavailability U and they are

$$P_0 = \frac{\mu}{\lambda+\mu} \equiv A \quad \text{and} \quad P_1 = \frac{\lambda}{\lambda+\mu} \equiv U \tag{191.34}$$

The time-dependent $A(t)$ and $U(t)$ are simply $P_0(t)$ and $P_1(t)$, respectively.

Referring back to Equation (191.14) for a constant hazard component and comparing it with Equation (191.32) which incorporates repair, the difference between $R(t)$ and $A(t)$ becomes obvious. Availability $A(t)$ is the probability that the component is up at time t, and reliability $R(t)$ is the probability that the system has continuously operated from 0 to t. Thus, $R(t)$ is much more stringent than $A(t)$. While both $R(0)$ and $A(0)$ are unity, $R(t)$ drops off rapidly as compared to $A(t)$ as time progresses. With a small value of MTTR (or large value of μ), it is possible to realize a very high availability for a reparable component.

191.15 Two Dissimilar Reparable Components

Irrespective of whether the two components are in series or in parallel, the state space consists of four possible states: S_1 (1 up, 2 up), S_2 (1 down, 2 up), S_3 (1 up, 2 down), and S_4 (1 down, 2 down). The actual system configuration will determine which of these four states correspond to system success and failure. The associated state-space diagram is shown in Figure 191.6. Analysis of this system results in the following steady-state probabilities:

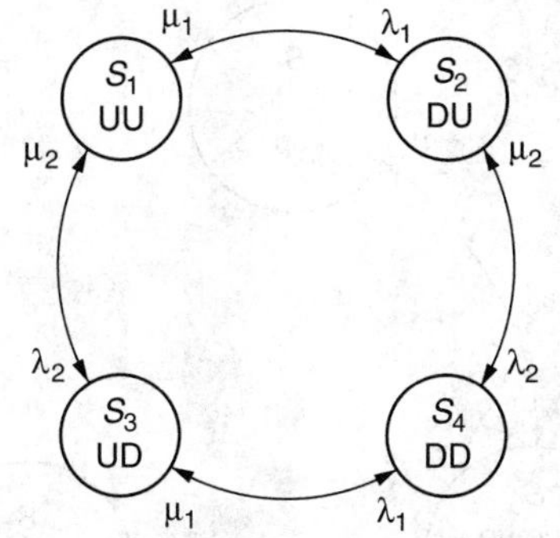

FIGURE 191.6 State-space diagram for two dissimilar reparable components. (*Source:* Ramakumar, R. 1993. *Engineering Reliability: Fundamentals and Applications.* Prentice Hall, Englewood Cliffs, NJ. With permission.)

$$P_1 = \frac{\mu_1\mu_2}{\text{Denom}};\quad P_2 = \frac{\lambda_1\mu_2}{\text{Denom}};\quad P_3 = \frac{\lambda_2\mu_1}{\text{Denom}};\quad P_4 = \frac{\lambda_1\lambda_2}{\text{Denom}} \tag{191.35}$$

where

$$\text{Denom} \equiv (\lambda_1 + \mu_1)(\lambda_2 + \mu_2) \tag{191.36}$$

For components in series, $A = P_1$, $U = (P_2 ++ P_3 ++ P_4)$, and the two components can be replaced by an equivalent component with a failure rate of $\lambda_s = (\lambda_1 ++ \lambda_2)$ and a mean repair duration of r_s, where

$$r_s \cong \frac{\lambda_1 r_1 + \lambda_2 r_2}{\lambda_s} \tag{191.37}$$

Extending this to n components in series, the equivalent system will have

$$\lambda_s = \sum_{i=1}^{n} \lambda_i \quad \text{and} \quad r_s \cong \frac{1}{\lambda_s}\sum_{i=1}^{n} \lambda_i r_i \tag{191.38}$$

$$\text{and system unavailability} = U_s \cong \lambda_s r_s = \sum_{i=1}^{n} \lambda_i r_i \tag{191.39}$$

For components in parallel, $A = (P_1 + P_2 + P_3)$, $U = P_4$, and the two components can be replaced by an equivalent component with

$$\lambda_p \cong \lambda_1(\lambda_2 r_1) + \lambda_2(\lambda_1 r_2) \quad \text{and} \quad \mu_p = \mu_1 + \mu_2 \tag{191.40}$$

$$\text{and system unavailability} = U_p = \lambda_p(1/\mu_p) \tag{191.41}$$

Extension to more than two components in parallel follows similar lines. For three components in parallel,

$$\mu_p = (\mu_1 + \mu_2 + \mu_3) \quad \text{and} \quad U_p = \lambda_1\lambda_2\lambda_3 r_1 r_2 r_3 \tag{191.42}$$

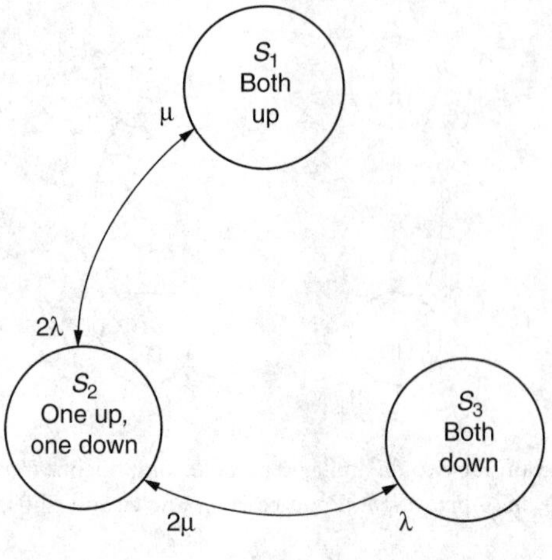

FIGURE 191.7 State-space diagram for two identical reparable components. (*Source:* Ramakumar, R. 1993. *Engineering Reliability: Fundamentals and Applications*. Prentice Hall, Englewood Cliffs, NJ. With permission.)

191.16 Two Identical Reparable Components

In this case, only three states are needed to complete the state space: S_1 (both up), S_2 (one up and one down), and S_3 (both down). The corresponding state-space diagram is shown in Figure 191.7. Analysis of this system results in the following steady-state probabilities:

$$P_1 = \left(\frac{\mu}{\lambda+\mu}\right)^2; \quad P_2 = \frac{2\lambda}{\mu}\left(\frac{\mu}{\lambda+\mu}\right)^2; \quad P_3 = \left(\frac{\lambda}{\lambda+\mu}\right)^2 \tag{191.43}$$

191.17 Frequency and Duration Techniques

The expected residence time in a state is the mean value of the passage time from the state in question to any other state. Cycle time is the time required to complete an *in* and *not in* cycle for that state. Frequency of occurrence (or encounter) for a state is the reciprocal of its cycle time. It can be shown that the frequency of occurrence of a state is equal to the steady-state probability of being in that state multiplied by the total rate of departure from it. Also, the expected value of the residence time is equal to the reciprocal of the total rate of departure from that state.

Under steady-state conditions, the expected frequency of entering a state must be equal to the expected frequency of leaving that state (this assumes that the system is *ergodic*, which will not be elaborated for lack of space). Using this principle, frequency balance equations can be easily written (one for each state) and solved in conjunction with the fact that the sum of the steady-state probabilities of all the states must be equal to unity to obtain the steady-state probabilities. This procedure is much simpler than solving the Markov differential equations and letting $t \to \infty$.

191.18 Applications of Markov Process

Once the different states are identified and a state-space diagram is developed, Markov analysis can proceed systematically (probably with the help of a computer in the case of large systems) to yield a wealth of results used in system reliability evaluation. Inclusion of installation time after repair, maintenance, spare, and stand-by systems, and limitations imposed by restricted repair facilities are some of the many problems that can be studied.

191.19 Some Useful Approximations

For an r-out-of-n structure with failure and repair rates of λ and μ for each, the equivalent MTTR and MTTF can be approximated as

$$\text{MTTR}_{\text{eq}} = \frac{\text{MTTR of one component}}{n-r+1} \quad (191.44)$$

$$\text{MTTR}_{\text{eq}} = \left(\begin{matrix}\text{MTTF}\\ \text{of one component}\end{matrix}\right)\left(\frac{\text{MTTF}}{\text{MTTR}}\right)^{n-r}\left[\frac{(n-r)!(r-1)!}{n!}\right] \quad (191.45)$$

The influence of weather must be considered for components operating in an outdoor environment. If λ and λ' are the normal weather and stormy weather failures rates, λ' will be much greater than λ, and the average failure rate λ_f can be approximated as

$$\lambda_f \cong \left(\frac{N}{N+S}\right)\lambda + \left(\frac{S}{N+S}\right)\lambda' \quad (191.46)$$

where N and S are the expected durations of normal and stormy weather. For well-designed, high-reliability components, the failure rate λ will be very small and $\lambda t << 1$. Then, for a single component,

$$R(t) \cong 1 - \lambda t \quad \text{and} \quad Q(t) \cong \lambda t \quad (191.47)$$

and for n dissimilar components in series,

$$R(t) \cong 1 - \sum_{i=1}^{n} \lambda_i t \quad \text{and} \quad Q(t) \cong \sum_{i=1}^{n} \lambda_i t \quad (191.48)$$

For the case of n identical components in parallel,

$$R(t) \cong 1 - (\lambda t)^n \quad \text{and} \quad Q(t) \cong (\lambda t)^n \quad (191.49)$$

For the case of an r-out-of-n configuration,

$$Q(t) \cong {}_nC_{(n-r+1)}(\lambda t)^{n-r+1} \quad (191.50)$$

Equations (191.47) through (191.50) are called rare-event approximations.

191.20 Reliability and Economics

There are two fundamental facts involved in the interaction of reliability and economics. First of all, it is impossible to realize 100% reliability, and second, the closer we want to approach this ideal value, the more expensive the system will become. Then, how much should be spent on an item (or system) and what should be the design reliability? These are very difficult questions since the answers are highly case specific, and often depend on how nonmonetary items are quantified in monetary terms. No matter how this is done, it will evoke controversy. To add to the complexity, resources spent on improving reliability are certain and near term, while the cost of failure is uncertain and far removed in the time frame.

There are three basic categories of options for improving reliability:

- Provide additional redundancy
- Design with environmental capability
- Conduct environmental testing and redesign based on results

Ultimately, an integrated cost-benefit tradeoff will have to be used, keeping in mind the less precise and more subjective nature of the world we live in.

Recognizing the vast scope of this topic, only one aspect, namely the economics of redundancy, will be considered here. There are two basic redundancy configurations:

- System redundancy (parallel-series configuration)
- Unit redundancy (series-parallel configuration)

System Redundancy

Figure 191.8 illustrates these two redundancy configurations. In system redundancy, the basic or original system consisting of a number of units in series is completely replicated one or more times and are operated in parallel. In unit redundancy, each of the units in the basic system is replicated and connected in parallel. These parallel units are operated in series to form the redundant system. All the units in series

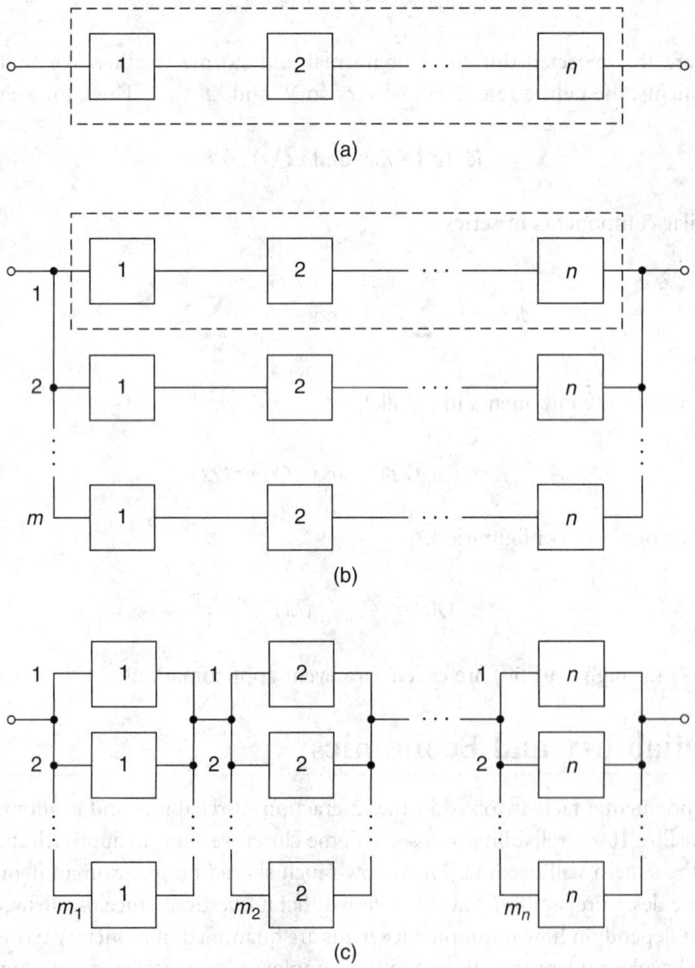

FIGURE 191.8 Redundancy configurations. (a) Basic (original) system. (b) System redundancy; parallel-series configuration. (c) Unit redundancy; series-parallel configuration. (*Source:* Ramakumar, R. 1993. *Engineering Reliability: Fundamentals and Applications.* Prentice Hall, Englewood Cliffs, NJ. With permission.)

in the original system may or may not be identical. However, replication of any one unit always involves identical units. The number of units in parallel with each basic unit in unit redundancy may or may not be equal. A simple economic analysis is considered next.

For the basic system,

$$\text{Total Cost} = C_0 = \sum_{i=1}^{n} c_i \tag{191.51}$$

$$\text{Reliability} = R_0 = \prod_{i=1}^{n} p_i \tag{191.52}$$

where c_i and p_i are the unit cost and reliability of the *i*th unit and *n* is the number of units in series in the basic system.

For the system redundancy configuration,

$$\text{Total Cost} = C_s = mC_0 \tag{191.53}$$

$$\text{Reliability} = R_s = 1-(1-R_0)^m \tag{191.54}$$

where m is the number of replicated systems operating in parallel.

For the unit redundancy configuration,

$$\text{Total Cost} = C_u = \sum_{i=1}^{m} c_i m_i \tag{191.55}$$

$$\text{Reliability} = R_u = \prod_{i=1}^{n} [1-(1-p_i)^{m_i}] \tag{191.56}$$

where m_i is the number of replicated units for the *i*th basic unit.

In general, system redundancy configuration is costly and wasteful as compared to the unit redundancy configuration.

Unit Redundancy

Starting with a basic system as shown in Figure 191.8(a) with a total cost C_0 and reliability R_0, we seek to improve the reliability to *R* by employing unit redundancy at minimum cost. By applying simple optimization techniques, we can derive the values of m_i, $i = 1,2,\ldots,n$, that will minimize the total cost and achieve the desired system reliability value of *R*. The results are summarized in the following equations.

$$m_i = \frac{\ln(1-R^{\alpha i})}{\ln(1-p_i)} \tag{191.57}$$

in which

$$\alpha_i = \frac{\left[\dfrac{c_i}{\ln(1-p_i)}\right]}{\left[\displaystyle\sum_{j=1}^{n} \frac{c_j}{\ln(1-p_i)}\right]} \qquad \text{for } i = 1,2,\ldots,n \tag{191.58}$$

The m_i values calculated using Equation (191.58) should be rounded up to the nearest integer for obvious reasons. Once the m_i values are known, the minimum cost figure can be easily calculated using Equation (191.55).

Defining Terms

Availability — Availability $A(t)$ is the probability that a system is performing its required function successfully at time t. The steady-state availability A is the fraction of time that an item, system, or component is able to perform its specified or required function.

Bathtub curve — For most physical components and living entities, the plot of failure (or hazard) rate versus time has the shape of the longitudinal cross section of a bathtub.

Hazard rate function — The plot of instantaneous failure rate versus time is called the hazard function. It clearly and distinctly exhibits the various life cycles of components.

Mean time to failure (MTTF) — Mean time to failure is the mean or expected value of time to failure.

Parallel structure — Also known as a completely redundant system, it describes a system that can succeed when at least one of two or more components succeeds.

Redundancy — Refers to the existence of more than one means, identical or otherwise, for accomplishing a task or mission.

Reliability — Reliability $R(t)$ of an item or system is the probability that it has performed successfully over the time interval from 0 to t. In the case of irreparable systems, $R(t) = A(t)$. With repair, $R(t) \leq A(t)$.

Series structure — Also known as a chain structure or nonredundant system, it describes a system whose success depends on the success of all its components.

References

Billinton, R. and Allan, R. N. 1992. *Reliability Evaluation of Engineering Systems: Concepts and Techniques*, 2nd ed. Plenum, New York.

Lewis, E. E. 1987. *Introduction to Reliability Engineering*. John Wiley & Sons, New York.

Ramakumar, R. 1993. *Engineering Reliability: Fundamentals and Applications*. Prentice Hall, Englewood Cliffs, NJ.

Shooman, M. L. 1990. *Probabilistic Reliability: An Engineering Approach*, 2nd ed. R.E. Krieger, Malabar, FL.

Further Information

Green, A. E. and Bourne, A. J. 1972. *Reliability Technology*. Wiley-Interscience, New York.

Henley, E. J. and Kumamoto, H. 1991. *Probabilistic Risk Assessment — Reliability Engineering, Design, and Analysis*. IEEE Press, New York.

IEEE Transactions on Reliability. Institute of Electrical and Electronics Engineers, New York.

O'Connor, P. D. T. 1985. *Practical Reliability Engineering*, 3rd ed. John Wiley & Sons, New York.

Proceedings: Annual Reliability and Maintainability Symposium. Institute of Electrical and Electronics Engineers, New York.

Siewiorek, D. P. and Swarz, R. S. 1982. *The Theory and Practice of Reliable System Design*. Digital Press, Bedford, MA.

Trivedi, K. S. 1982. *Probability and Statistics with Reliability, Queuing, and Computer Science Applications*. Prentice Hall, Englewood Cliffs, NJ.

Villemeur, A. 1992. *Reliability, Availability, Maintainability and Safety Assessment*, vols. 1 and 2. John Wiley & Sons, New York.

192

Computer Integrated Manufacturing: A Data Mining Approach

Bruno Agard
École Polytechnique de Montréal

Andrew Kusiak
The University of Iowa

To remain competitive a company has to respond to market changes. For a company to quickly react to these changes, integration of manufacturing activities becomes a necessity. Integration implies that necessary information can be accessed at any time and location, which in turn means that a robust and up-to-date information system containing all information necessary for any stage of the manufacturing process is necessary. Having the information stored in a database is not sufficient; rather the information should be transformed into knowledge. This chapter identifies areas of potential applications of data mining in computer-integrated manufacturing. An attempt will be made to answer the following questions:

1. How can data mining technology be employed beyond prediction and modeling?
2. How can several quality, process-planning, and maintenance problems in manufacturing environment be overcome?
3. How can data mining results be used among different applications?

192.1 Computer Integrated Manufacturing

Definition and Rationale

Computer integrated manufacturing (CIM) involves integration of manufacturing activities by a computer system. This integration makes it possible for various functional areas (e.g., planning, manufacturing, design, and control) to exchange information.

According to Kalpakjian and Schmid [7], advances in automating manufacturing processes have been driven by several competitive forces, such as the need to continuously improve productivity and product

0-8493-1586-7/05/$0.00+$1.50

quality and to reduce manufacturing costs. In this way, automation has become a key factor. Automating manufacturing facilities, agile manufacturing, and integration into a single system has led to CIM.

The goals for CIM are to improve productivity, increase product quality and uniformity, minimize cycle time, and reduce labor costs [7]. CIM leads to:

- Timely response to rapid changes in market demands and product changes
- Better use of materials, machinery, personnel, and reduction of inventory
- Better production control and management of manufacturing operations
- Manufacture of high-quality products at low cost

Computer simulation of manufacturing processes and systems considers process variability and process optimization.

CIM Requirements

To fully realize the benefits of integration, CIM has to be a global system that allows all users to share the same up-to-date information. CIM calls a large database that should be shared by the entire organization. The database must contain up-to-date, detailed, and accurate data. The data relate to products, designs, machines, processes, materials, production, marketing, inventory, and so on. Ideally the data should be collected in real time using automated data acquisition systems.

The data acquisition system should collect and transfer information to the database as needed by different users. The database information should be easily accessible by users, such as designers, manufacturing engineers, process planners, financial officers, and managers. Moreover, different users may require different access modes and tools.

Due to a variety of access modes, the database should be protected against failure or unauthorized use. Management of such a database is critical.

For real-time management and online data analysis, many sensors that provide information about the system are necessary. The data acquisition system populates the database with the following types of data:

- Products and processes
  - Product data: part shapes, dimensions, specifications
  - Data management attributes: owner, revision level, part number
  - Production data: manufacturing process involved
  - Operational data: scheduling, lot size, assembly requirements
  - Resources data: capital, machines, equipment tooling, personnel and their capabilities
- Current manufacturing
  - Number of parts being produced per unit of time
  - Part dimensional accuracy
  - Part surface finish
  - Part weight

The database should make it possible to observe the state of the production system, including the number of parts that have been produced, stock levels, requirements, production rates, and so on. This will help in the decision making and control of the production system. Communication and exchange of data, information, and knowledge among different systems calls for use of standards, such as STEP (Standard for the Exchange of Product Model Data) [6]. STEP facilitates the transfer of information from the design stage to planning, and manufacture without reentering product-related data.

192.2 Data Mining

The volume and scope of manufacturing data are growing at an unprecedented rate, making data analysis and decision making based on the data increasingly difficult for engineers and managers. Numerous tools, including statistics, are used for data analysis. Statistics makes it possible for engineers and managers to plan products and evaluate processes. They provide population-based answers using various models (e.g., regression) and parameters (e.g., mean, confidence interval).

Data mining offers algorithms and tools for discovery of nontrivial, implicit, previously unknown, and potentially useful and understandable patterns from large data sets [1]. It concentrates on discovery of models applicable to an individual or group of individuals (subpopulation) rather than entire populations of objects (e.g., parts).

Data Mining Tasks

Data mining realizes two kinds of tasks: *prediction* produces a model that can be used for classification or estimation; and *description* produces understandable and useful patterns and relationships that permit users to understand a complex database. Westphal and Blaxton [14] described data mining tasks as follows:

- Classification involves assigning labels to data records. The labels come from a small predefined set (good/bad or red/white/yellow). The job of the data miner is to build a model that will successfully classify the records.
- Estimation is the task of filling in a missing value in a particular field of an incoming record as a function of other fields in the record. The usual statistical regression techniques are most often employed for estimation. Estimation is also a popular application of neural networks.
- Segmentation or clustering breaks the population into smaller subpopulations having similar behavior. Clustering methods achieve maximum homogeneity with groups and maximum heterogeneity between groups. Clustering is an unsupervised learning tool aiming at finding natural groupings (clusters). Cluster detection, affinity grouping, and link analysis are mostly used for clustering.
- The description task focuses on explaining the relationships among the data. Among others, link analysis and visualization techniques can be used. Link analysis determines associations between the variables, and visualization techniques are used to simplify data comprehension with adapted representations.

Data Mining Algorithms

An algorithm determines a function (model) linking an output with inputs. Various classes of algorithms are used to mine data [2]. Two general model types are available:

- If the output variable is discrete valued, classification modeling is employed.
- If the output variable is continuous valued, prediction modeling is employed.

Common *predictive modeling* (also called classification or supervised learning) techniques include neural networks, decision trees, and decision rules algorithms.

In *clustering modeling* (called segmentation or unsupervised learning), an algorithm identifies sets of similar examples in some optimal fashion. To obtain these sets, various methods are used, such as k-means algorithm, hierarchical algorithms, pattern recognition, Bayesian statistics, and neural networks.

In *frequent pattern extraction models* (called association rule mining), algorithms extract combinations of variables that exist in the data with some predefined level of regularity. For instance, an association A $\Rightarrow$ B can be provided with two statistical measures: support and confidence. *Support* measures the number of times that A exists as a fraction of the total data. *Confidence* measures the fraction of times that B exists in the data when A is present. An association with high confidence and support is provided to the user.

General Methodology

Figure 192.1 illustrates the data mining process. According to Fayyad et al. [5], this process includes the following steps:

1. Develop an understanding of the application domain, relevant prior knowledge, and goals of the end user.

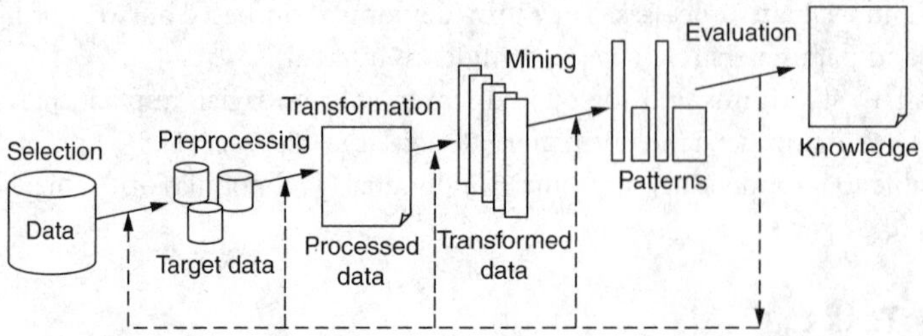

FIGURE 192.1 The data mining process. (*Source:* Fayyad U. M., Piatetsky-Shapiro G., Smyth P. et al., *Advances in Knowledge Discovery and Data Mining*, MIT Press, Cambridge, MA, 1996.)

2. Create a target data set.
3. Clean and preprocess data.
4. Reduce and project data, that is, reduce the effective number of variables under consideration.
5. Select data mining task (classification, regression, clustering).
6. Select data mining algorithm(s), that is, determine which models and parameters are appropriate.
7. Mine data — search for patterns of interest in a particular representational form or a set of such representations, including classification rules, decision trees, regression, clustering, and so on.
8. Interpret mined patterns. Return to any of Steps 1 through 7 for further iteration as necessary.
9. Consolidate discovered knowledge.

192.3 CIM and Data Mining

The information system of an integrated enterprise disposes the information in the areas depicted in Figure 192.2. With the growing number of applications of enterprise requirement planning (ERP) and material and manufacturing requirement planning (MRP and MRP II) in companies of various sizes, more information about products, production capacity, and means of production is becoming available. Along the product life cycle, plenty of information is generated by product data management (PDM) systems that define products, data from marketing studies to define market shares, design form information, and so on.

The data reflect the experience of the designer and manufacturing process planner. All this information has not been sufficiently used. The latter leads to high expectations for applications of data mining in a CIM environment. These data can be mined to meet numerous goals, including understanding the manufacturing system — specific studies of a machine, a set of machines, a product, or a set of products; focus on a time interval, process, or any other subset of the database. Data mining in a CIM environment can help managers to understand complex systems by:

- Presenting high-probability rules for a better understanding of the system
- Presenting complete information to enhance decision making
- Providing information to predict performance or response to disturbances

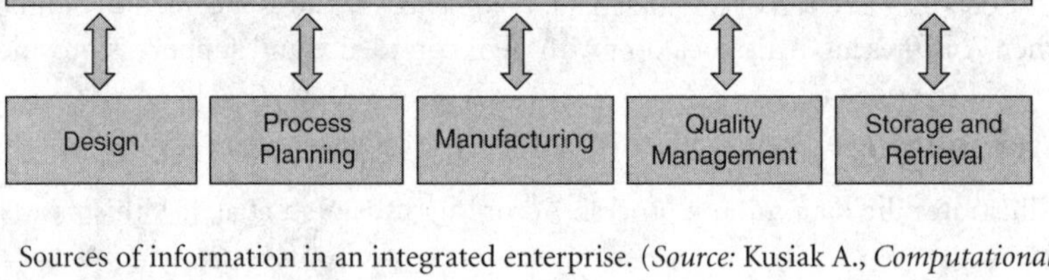

FIGURE 192.2 Sources of information in an integrated enterprise. (*Source:* Kusiak A., *Computational Intelligence in Design and Manufacturing*, Wiley-Interscience, New York, 2000.)

- Estimation of production costs
- Providing information for fast detection of defects in a production or in supply chain

Benefits of data mining follow:

- Accelerate development of a new product/process by analogy to existing designs.
- Recommending manufacturing processes to accelerate process definition or to assist inexperienced engineers.
- Identification of the best set of process parameters from the experience of historical projects.
- Data mining can be used for group technology applications. Classification can set a group to a new part and extract all relevant processes, times, and costs from the database.
- Group technology with data mining makes possible standardization of part design and minimization of design duplication, taking advantage of the design and processing similarities (clustering).
- Process plans can be standardized and developed more efficiently.
- Parts can be managed as families. Data mining can suggest the use of a similar part, product, process, and design stored in the database.
- The best sets of manufacturing parameters can be extracted.

In the following, we discuss how to extract useful information for product manufacturing from databases created independently at different points in the product life cycle.

Data Mining Process in Engineering

Büchner et al. [4] proposed a data mining process in a manufacturing environment (Figure 192.3). The process begins with the identification of a problem by management. The application of data mining techniques focuses on resolution of a specified problem. Steps discussed by Büchner et al. [4] are summarized below.

1. *Human resource identification.* To resolve an identified manufacturing problem, the following three experts are recommended:
   A. Domain expert. This expert could belong to an engineering unit. S/he will join the study to provide information on the manufacturing process, and is able to analyze the results.
   B. Data expert. This individual comes from the company's informatics technologies department. The data expert is necessary to locate in the database the data of interest to the problem at hand.
   C. Data mining expert. This person, typically an outsider, is in charge of the processing of the data. S/he will select the best algorithm and parameters to perform the analysis.

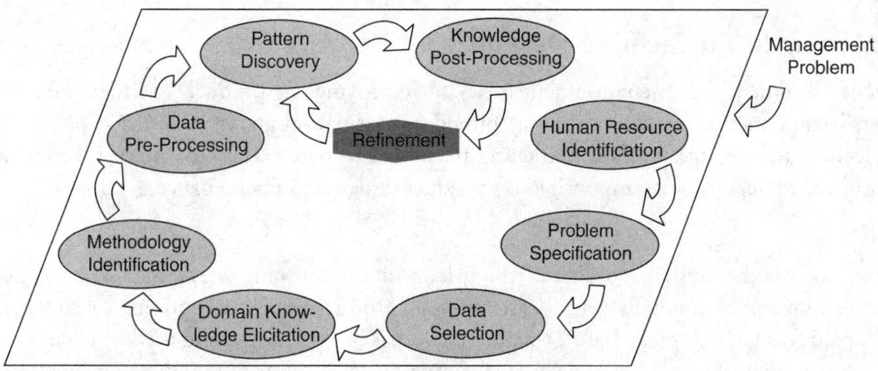

FIGURE 192.3 Data mining process in manufacturing environment. (*Source:* Büchner A. G., Anand S. S., and Hugues J. G., *Stud. Informatics Control,* 6, 319–328, 1997.)

2. *Problem specification.* After initial analysis and discussion, the experts develop a better comprehension of the problem. The problem is decomposed into subproblems that are easier to understand and manage. The subproblems that can be resolved with data mining techniques are identified, and the most suitable data mining approach is selected (association rules, classification, estimation, cluster analysis, frequent pattern extraction, regression, deviation detection, or visualization techniques; Fayyad et al. [5]).
3. *Data selection.* This step involves analyzing the state of the data required to solve the problem at hand. Four main characteristics are considered:
   A. Identification of relevant attributes
   B. Accessibility of data
   C. Population of required data attributes
   D. Distribution and heterogeneity of data
4. *Domain knowledge elicitation.* During this stage, the data mining expert attempts to elicit any domain knowledge of interest that could be incorporated in the discovery process:
   A. Specific constraints on the search space
   B. Hierarchical generalization
5. *Methodology identification.* The best data mining methodology for solving the specified problem is selected.
6. *Data preprocessing.* This step is comprised of the following components:
   A. Adaptation of the database to a format that the data mining algorithm can understand.
   B. Predict/fill in the missing data in the database as needed by the data mining algorithm.
   C. Reduction in problem size.
7. *Pattern discovery.* This step involves using algorithms that automatically discover patterns in the preprocessed data. Selection of the algorithm depends on the data mining goal. Pattern discovery is an iterative process that calls for several refinements. When the volume of data is too large for computation in a reasonable time, decomposition of the data set could be considered. Kusiak [9] proposed two forms:
   A. Feature set decomposition: partitioning a data set on features (e.g., columns of a spreadsheet)
   B. Object set decomposition: partitioning a data set on objects (e.g., rows of the spreadsheet)
8. *Knowledge post-processing.* Trivial and obsolete information must be filtered out and discovered knowledge must be presented in a user-readable way. Visualization techniques may help to facilitate the comprehension of extracted knowledge. The new knowledge must be validated; the domain and data mining experts should work together at this stage. Often the knowledge post-processing provides supplemental knowledge to refine pattern discovery, and may necessitate a return to data preprocessing as well.

Data Mining in Engineering Design

Analysis of a product and its environment is key to improving the product. In order to better match customers' requirements, marketing analysis provides information about customer expectations. Data mining techniques are able to assist marketing personnel to better learn the market; results are then communicated to departments responsible for product design and maintenance.

Marketing

Marketing can use data mining to discover knowledge about customers. The data to be mined could include the number of products sold; description of product contents (options, variants); how the customer paid (cash, check, credit card); whether credit has been arranged and with which bank; services obtained (warranties, delivery dates, delivery at home, installation at home); whether instructions for using the product were requested; customer's age, employment, hobbies, marital status; whether s/he has children; and whether product was purchased for self or as gift.

Informal data could be provided by salespeople, such as the time needed by a customer to make a decision, and whether s/he considered competing products.

From these types of data, the following knowledge can be extracted:

1. Customer requirements can be separated into product clusters. This will allow producing customized products for each cluster instead of a generic product for all consumers. The type and number of products needed can be determined.
2. Predict customers who are likely to buy a particular product/service. With this information we are able to size the production ranges that will help in design (select materials and processes).
3. Anticipate future requirements and wishes for new functionality, so as to work on new products or adapt existing products.
4. Discovery of links between options, as in product/customer, customer/service, and product/service that could be useful to create new products/services or to improve existing designs.
5. Improve the services offered around a product. Propose different ways of buying products — for example, with a customer-specific credit. An agreement with a bank or a credit organization could be necessary. Other services could be organized around delivery (with a transportation company), installation (with hotlines or specialists), teaching (courses or web pages), and so on.

These results, combined with national or local statistics, could be helpful in determining whether a new product/service has to be created, when, for how many customers, at what price, with what parameters and options, and so on. With this knowledge, the design team will be able to develop the new product/service. Berry and Linoff [3] present many examples and applications of data mining in marketing, sales, and customer support.

Yeo et al. [15] used data mining for the prediction of claim costs for automobile insurance industry. Insurance companies are confronted with a dilemma. When claim costs are high, there is a high profit from each consumer but a small number of consumers. On the other hand, if claim costs are low, the profit per consumer is small but there is a large number of consumers. In this situation, policyholders have been classified according to their perceived risk, and claim costs within each risk group have been modeled.

Engineering Design

The product development process generates a large volume of information. The information comes from PDM systems, computer-aided design (CAD) systems, simulations, computation, tests, and so on. In this section we discuss how data mining techniques can extract information from product and design process databases.

The goal of data mining in product development is to learn from past designs in order to assist in decision making of new designs. For instance, Leu et al. [11] applied data mining for prediction of tunnel support stability. In this case study, a company had to design a tunnel through rock for a railroad. Designing such a project requires a large amount of surveying and experience. A crucial task is to estimate the causes of instability of tunnel support that depend on many parameters such as rock characteristics and construction characteristics. Tunnel project engineers had incomplete understanding of rock mechanisms and the impact of construction methods and procedures.

To solve this complex problem, traditional tools, discriminant analysis, and multiple nonlinear regression methods did not prove to be successful. The company had a large database of previous tunnel descriptions and decided to apply data mining techniques. After classification of the historical projects according to rock mass, neural networks were trained and tested on filtered data. The study demonstrated the ability of data mining to deliver new project insights in areas where traditional techniques had not been successful.

To conduct a design project, several meetings are necessary. Reports from these meetings contain all ideas that emerged, which ideas were developed, which decisions have been taken, and which solutions were discontinued and why, as well as the set of tasks that must be done before the next meeting. At the same time, e-mails are exchanged between the members of the team, some tasks are accomplished, others are cancelled, decisions are made, and so on.

In an integrated company, all this information is stored in a database. PDM systems are designed to collect such information. In an ISO-9001 certified company, the PDM system is a central information system for quality data. When a project is finished, members of the team can find in the database how and why a decision was made. However, PDM systems are often used for storage when they could be used for knowledge extraction. Mining such data could provide knowledge about product design development, how decisions are made, and how ideas and concepts are used and developed. In a repetitive design environment, data mining can be used for learning from previous designs. The data mining results can be used to provide the right information to the right person at the right time.

For example, in a product design team, some tasks are always done in the same order. For each task, the person who has to perform it needs a certain amount of data. All data are shared in the database. Data mining would save time by providing the necessary knowledge. Moreover, data mining will lead to understanding repetitive design activities and design reuse. The discovered knowledge will accomplish the following:

- Identify valid, novel, useful, and understandable patterns in design (about methods and processes). This knowledge is useful for repetitive designs or repetitive elements of designs.
- Provide adequate information when needed.
- Learning relations between customer needs and design specifications.
- Enabling extraction of relations between design requirements and manufacturing specifications.
- Understanding how requirements are translated into product specifications.
- Project evaluation is facilitated by comparisons with previous projects.
- Understanding of team's decisions and better team coordination.

Simoff and Maher [13] used text analysis to analyze different aspects of participation in a collaborative design project (in an educational design context). Their goal was to understand the style of collaboration and the potential of different environments in a computer-supported collaborative design. The authors learned that different environments facilitate different types of collaboration. After extraction of related and unrelated activities, they reconstructed and ordered task-related activities to understand how the tasks evolved and evaluated individual contributions.

After Sales

After-sales service follows the relationship that was initiated in sales services. It is an important source of information for developing a market overview, and for improving the product or service provided to the consumer. For instance, at the time of product/service delivery and/or after delivery, the consumer may be asked questions about use and preferences, among many other possibilities.

Sforma [12] analyzed the relations between a power company and its consumers. Each consumer was considered as a unique individual rather than as one of many similar consumers. The goal was to discover changes in their habits and then quickly respond with suggestions and appropriate services. The customer database was analyzed according to the following process:

1. Extraction of only the data that was considered useful for the analysis.
2. Application of simple statistics to aggregate values and calculation of new meaningful variables.
3. Organization of the data in a format suitable for the application of a classification technique.

Data mining allowed the company to contact each consumer individually when a change was observed in the energy consumption pattern. A better knowledge of the consumer by the company implies a stronger relationship and loyalty.

Berry and Linnof [3] studied a company manufacturing diesel engines. During the warranty period, the manufacturer received complaints from independent distributors that maintained engines covered by the manufacturer's warranty. Each complaint was studied by an expert to determine if the manpower and the parts used looked acceptable. A set of existing rules was used to analyze the problems without great success. Data mining was applied to a large set of complaints and new rules were discovered. The new rules were linked to an automatic routing of claims implied that faster service would save the company millions of dollars per year.

The knowledge extracted from after-sales services must be transmitted to different departments and be integrated with the product and service improvement process. To accomplish this integration, the following are needed:

- Time to fix a maintenance problem should be provided to the ERP system.
- Delivery delays should be automatically transmitted to sales services.
- Recurrent breakdowns should be identified and a corrective task should be automatically suggested to the design department and be integrated as a future design task.
- Recurrent breakdowns should be signaled to the maintenance department in order to develop a specific answer suitable to each of these specific breakdowns.
- Recurrent requirements (about products and services) should be transmitted to the marketing and project department to evaluate market evolution and possibilities for improvement.

Data Mining for Quality Management

Data mining can increase product quality by optimizing product quality and by isolating and deleting quality detractors. Two options can be considered in optimizing product quality: increasing average product quality or reduction of the product variability (variance reduction).

In the following example, a particular product characteristic is measured and represented as shown in Figure 192.4. Three levels of quality have been defined: low, medium, and high. The objective is to identify and understand the conditions that contribute to varying quality.

1. Discrimination techniques are applied to discriminate between different classes of objects. These discrimination techniques can be levels, limits, a set of rules, or an application of clustering methods to find models that are able to separate different quality levels.
2. Once the groups have been identified, the products are grouped into low-, medium-, and high-quality clusters.
3. For each quality class, select the objects.
4. For each group of objects selected, extract the conditions that have led to these different states.

After these steps have been completed, we know which conditions led to each level of quality, and then we will be able to select the best set of parameters that lead to a particular quality.

The second way to improve the quality of the manufacturing system is to isolate quality detractors and/or underline reasons why a product may not pass a quality test. The data mining process is based on discrimination of good and bad results and provides the conditions that lead to each quality level. Kusiak [10] performed a fault analysis using the data mining approach to identify the causes of solder defects in a circuit board.

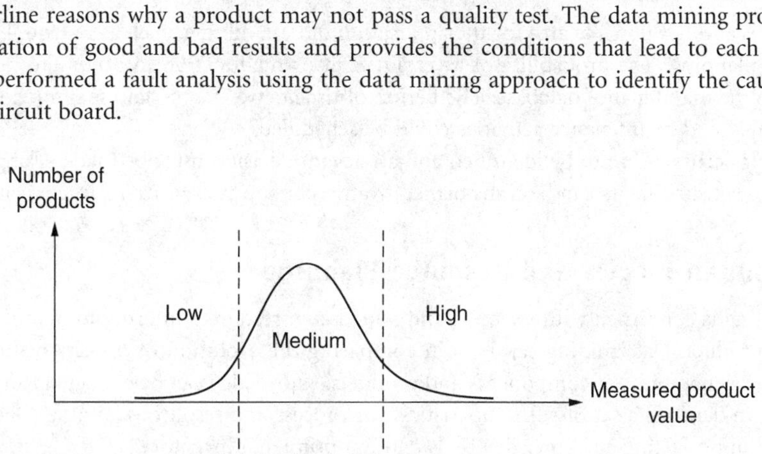

FIGURE 192.4 Product quality.

Data Mining in Manufacturing

Data mining can be used to obtain various types of information aimed at the following objectives:

- Capturing the best manufacturing practices
- Reducing operations expenses
- Predict and prevent failure of components
- Identifying conditions leading to critical situations in order to correct the problem and prevent future failures

Sensitivity Analysis

Sensitivity analysis can be applied to determine which parameters (of the manufacturing system or the process) impact a particular characteristic (productivity, quality, efficiency). Once this knowledge is obtained, it will be possible to optimize the parameters.

A guided data mining analysis can be performed. The parameters that lead to different values of output can be extracted from the database. For instance, if there is an acceptable value of the output, we will have the parameters that lead to it from the model. Next, if the output has to be in a particular range, we can select a range for the inputs or optimize the value of each input in order to reduce manufacturing cost. For example, a parameter that has a low influence on output and a high cost could be modified.

What-If Analysis

A what-if analysis determines what the output (e.g., productivity, quality, efficiency) will be according to new entries or parameters (e.g., time, hardness, temperature). This type of predictive analysis is easy to realize and can be beneficial. The hypothesis is that the new parameters will have the same impact on the output as the former parameters. The steps of a what-if analysis follow:

1. Generate a predictive model from the database (regression analysis or neural networks).
2. Provide a new parameters to the model.
3. Evaluate the new output.
4. Accept the output, and go to Step 2.

This type of analysis determines parameter sensitivity to the output. One can predict the output for a certain combination of input parameters, and then select the set of parameters that optimizes a goal, such as minimizing costs and risks.

Preventive Maintenance

Maintenance is another possible area for the application of data mining, such as in troubleshooting and maintenance planning. The probability of a certain fault or malfunction to occur can be predicted in advance from the maintenance database. The period of time between a certain malfunction or problem can be evaluated and maintenance activities could be scheduled.

Once a fault occurs, it has to be identified and a relevant decision must be made. Using data mining techniques, unexpected faults can be transformed from repairs to preventive maintenance.

Data Mining in Process and Resource Planning

Process planning is concerned with selecting and sequencing resources in order to manufacture a component or a product. Data mining can help in comparing the evolution of process planning activities across different generations of components and products. Knowledge can be extracted from the production database in defining a set of rules (heuristics) for process and resource planning. Data mining can also be useful in predicting customer demands and in optimizing inventory.

192.4 Conclusion

An integrated company generates a large volume of data stored in transactional databases and data warehouses. This chapter discussed possible applications of data mining techniques in an integrated company. Data mining allows to profit from the data collected across many departments. Moreover, data mining provides different forms of knowledge that can be used to meet different goals. Marketing, engineering design, after-market sales service, quality management, manufacturing, and planning activities can all benefit from data mining tools.

References

[1] Anand S. S. and Büchner A. G., *Decision Support Using Data Mining*, Financial Times Pitman Publishers, London, 1998.

[2] Apté C., Data mining: an industrial research perspective, *IEEE Comput. Sci. Eng.*, April–June, 6–9, 1997.

[3] Berry A. J. A. and Linoff G., *Data Mining Techniques: For Marketing, Sales, and Customer Support*, Wiley, New York, 1997.

[4] Büchner A. G., Anand S. S., and Hugues J. G., Data mining in manufacturing environments: goals, techniques and applications, *Stud. Informatics Control,* 6, 319–328, 1997.

[5] Fayyad U. M., Piatetsky-Shapiro G., Smyth P. et al., *Advances in Knowledge Discovery and Data Mining*, MIT Press, Cambridge, MA, 1996.

[6] International Standards Organization (ISO), ISO 10303-1. Industrial Automation Systems and Integration. Product Data Representation and Exchange. Part 1. Overview and Fundamental Principles, ISO, Geneva, 1994.

[7] Kalpakjian S. and Schmid R.-R., *Manufacturing Engineering and Technology,* 4th ed., Prentice Hall, Upper Saddle River, NJ, 2000.

[8] Kusiak A., *Computational Intelligence in Design and Manufacturing*, Wiley-Interscience, New York, 2000.

[9] Kusiak A., Decomposition in data mining: an industrial case study, *IEEE Trans. Electron. Packaging Manuf.*, 23, 345–353, 2000.

[10] Kusiak A., Rough set theory: a data mining tool for semiconductor manufacturing, *IEEE Trans.Electron. Packaging Manuf.*, 24, 44–50, 2001.

[11] Leu S.-S., Chen C.-N., and Chang S.-L., Data mining for tunnel support stability: neural network approach, *Autom. Constr.*, 10, 429–441, 2001.

[12] Sforma M., Data mining in a power company customer database, *Electric Power Sys. Res.*, 55, 201–209, 2000.

[13] Simoff S. J. and Maher M. L., Analysing participation in collaborative design environments, *Design Stud.*, 21, 119–144, 2000.

[14] Westphal C. and Blaxton T., *Data Mining Solutions*, John Wiley & Sons, New York, 1998.

[15] Yeo A.-C., Smith K. A., Willis, R. J., and Brooks M., Clustering technique for risk classification of claim costs in the automotive insurance industry, *Int. J. Intelligent Sys. Accounting Finance Manage.*, 10, 39–50, 2001.

XXVI

Aeronautical and Aerospace

193

Aerodynamics

John F. Donovan
McDonnell Douglas Corporation

Aerodynamics is a subset of fluid dynamics that deals with the flow of air about objects, typically aircraft, missiles, or their components. Much of the work in aerodynamics focuses on the generation of forces and moments on a body due to the air flowing over and through the body. Aerodynamics deals with theoretical and numerical predictions of performance characteristics, experimental determination of performance characteristics, and the design of new and improved geometries using this information. This chapter provides the basic understanding and equations to enable the engineer to calculate the performance of many aerodynamic configurations and interpret aerodynamic data.

193.1 Background

Understanding aerodynamics requires a knowledge of the basic principles of fluid mechanics. These principles are covered in Section 6. However, several key points required for the understanding of aerodynamics are provided in this section.

To understand how lift is generated, it is important to know the relationship between velocity and pressure in a fluid flow. For flows where the effects of viscosity are negligible, the density is constant, and gravity does not play a role, *Bernoulli's equation* describes this relationship:

$$p+\rho\frac{V^2}{2}=p_0=\text{constant along a streamline} \tag{193.1}$$

where p is the static pressure, ρ is the density of the fluid medium, and V is the fluid velocity. The constant in Equation (193.1) is the total pressure, denoted p_0, and is the sum of the static pressure, p, and the dynamic pressure, $\rho V^2/2$. A *streamline* is the path that a fluid particle makes in the flow, assuming the flow is steady, and Equation (193.1) indicates that the total pressure is constant along this line. If the velocity and pressure are known at one point on the streamline, the flow quantities can be computed at another point on the streamline by rewriting Equation (193.1) as

$$p_2+\rho\frac{V_2^2}{2}=p_1+\rho\frac{V_1^2}{2} \tag{193.2}$$

0-8493-1586-7/05/$0.00+$1.50

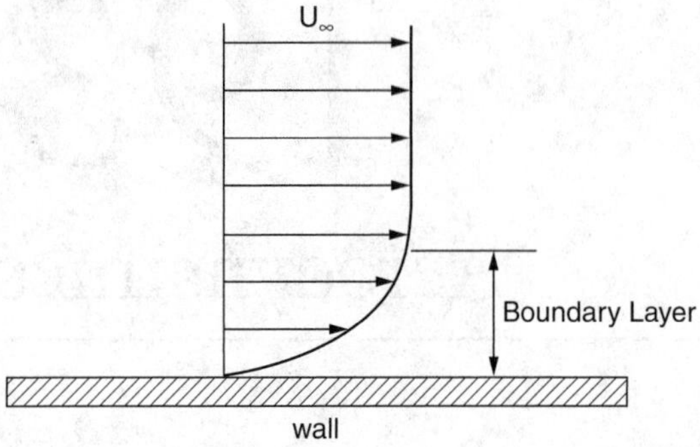

FIGURE 193.1 Velocity profile in a boundary layer.

where the subscripts refer to two points along the streamline. The total pressure is a measure of the energy in the flow and Equation (193.1) indicates that this energy is shared between the static pressure and the dynamic pressure (kinetic energy). If the velocity goes to zero at a point in the flowfield, this point is referred to as a *stagnation point* and the pressure there is the stagnation pressure, which is equal to the total pressure.

The effects of *viscosity* and the *no-slip* condition are also important concepts for understanding aerodynamics. When a fluid moves over the surface of a body, it actually "sticks" to the surface of the body so that there is no relative velocity between the fluid and the surface. This is called the no-slip condition and is caused by intermolecular forces and molecular–scale surface roughness. Due to the effects of viscosity, a *boundary layer* is formed near the surface where the velocity increases from zero at the surface to the freestream value, U_∞, far away (see Figure 193.1). Viscous effects are important only in the boundary layer for most streamlined bodies. Since viscous effects are important in the boundary layer, Bernoulli's equation cannot be applied there. In fact, in the absence of strong curvature, the pressure at the wall where the velocity is zero is equal to the static pressure in the outer flow and not the total pressure as predicted by Bernoulli's equation.

Reynolds number is a measure of the importance of viscous effects and is defined as the ratio of momentum forces to viscous forces:

$$\mathrm{Re} = \frac{U_\infty \rho l}{\mu} \tag{193.3}$$

where l is a length scale of the flowfield and μ is the viscosity of the fluid. As the Reynolds number is increased, the effects of viscosity become less important.

193.2 Flow about a Body

This section describes the forces developed in a constant–density flow about general bodies. Figure 193.2 illustrates the flow about an arbitrary two-dimensional body. The body is traveling through still air at a

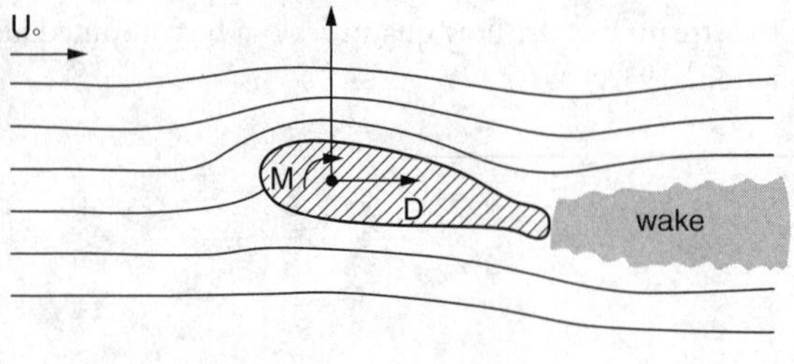

FIGURE 193.2 Flow about an arbitrary two-dimensional body illustrating the aerodynamic forces and moment exerted by the flowfield. The reference frame is fixed to the body.

velocity of U_∞ and the figure is drawn in the reference frame of the body so the fluid appears to move past the body. Two forces are generated — *lift, L,* and *drag, D.* Lift is the force perpendicular to the incoming velocity and drag is the force parallel to the incoming velocity. A *moment, M,* is also exerted on the body. As the freestream velocity changes, the forces on the body change even if the flowfield remains approximately similar. It is therefore useful to define nondimensional force coefficients, which vary much less with flow speed than the forces themselves. The forces are nondimensionalized by the dynamic pressure times an area associated with the body, S (in the case of a wing, it is usually the wing area). The lift and drag coefficients are defined as

$$C_L \equiv \frac{L}{\frac{1}{2}\rho U_\infty^2 S} \tag{193.4}$$

$$C_D \equiv \frac{D}{\frac{1}{2}\rho U_\infty^2 S} \tag{193.5}$$

The moment coefficient is defined using an additional parameter, l, a length scale of the body (in the case of a wing, it is usually the streamwise length of the wing), as

$$C_M \equiv \frac{M}{\frac{1}{2}\rho U_\infty^2 Sl} \tag{193.6}$$

The particular choice of the reference area and length is not critical, but when using published data it is important to know which reference quantities the coefficients are based upon.

Two types of forces act on the surface of a body due to the motion of the fluid — *pressure forces* and *viscous forces.* Pressure forces are simply due to the pressure, p, and act perpendicular to the surface. Viscous forces, or friction forces, are caused by the flow "rubbing" against the surface as it passes over the body. This force acting on a unit area of the surface is termed the *wall shear stress,* τ_w, and acts tangentially to the surface. It is proportional to the velocity gradient normal to the wall at the wall, and is defined as

$$\tau_w = \mu \left.\frac{\partial U}{\partial y}\right|_w \tag{193.7}$$

where μ is the viscosity and the subscript w denotes that the derivative is evaluated at the wall. No matter how complex the surface geometric or the flowfield, the only way that aerodynamic forces are transmitted to a body is through these two mechanisms.

Lift

A lift force is generated on a body by pressure differences between the top and bottom of the body. If the average pressure over the bottom of the body is higher than the average pressure over the top of the body, the lift force points upward. If we consider the flow over a two-dimensional cylinder without viscosity, the streamline pattern is symmetric, as shown in Figure 193.3(a). The presence of the cylinder in the flow requires the same amount of fluid to pass in a smaller area. Thus, the fluid must speed up in some regions of the flowfield. A fluid element moving along the centerline slows down to zero velocity at the stagnation point A, where the pressure is equal to the total pressure. As the fluid moves away from the center of the cylinder, it speeds up, and a minimum pressure occurs in the two regions marked B. This reduction in pressure can be seen using Bernoulli's equation (Equation (193.1)) because, as the flow speeds up, the static pressure drops in order for the total pressure to remain constant. The fluid slows

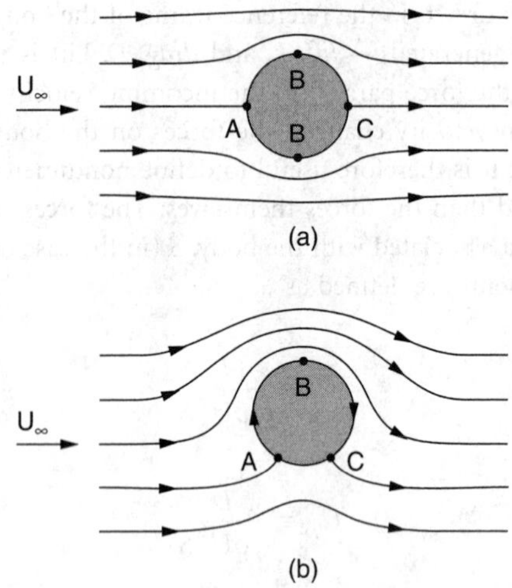

FIGURE 193.3 Flow about a cylinder without spin (a) and with spin (b).

down again toward the rear of the cylinder as the streamlines come together and another stagnation point is formed at *C*. The total lift force is the integral of the pressure over the entire surface in a direction perpendicular to the freestream flow:

$$L = \int p dA_y \tag{193.8}$$

where A_y is the component of the area perpendicular to the freestream flow. Thus, on the upper surface of the cylinder, there is an upward force, but there is an equal and opposite downward force on the lower surface because the flow is symmetric, and the net lift is zero.

If the cylinder is spun about its axis, as shown in Figure 193.3(b), the flow pattern becomes asymmetric. The stagnation points at *A* and *C* have moved below the centerline and the minimum pressure occurs only on the upper surface. In this case, it can be seen that the average pressure over the lower half of the cylinder will be higher than that over the top because the stagnation points are both on the bottom. In this case, there is a net lift force pointing up. The same effect is used in the aerodynamics of sport balls. For example, in tennis, a top spin on the ball causes it to drop much more quickly than a ball without top spin because of the downward-pointing lift vector.

By spinning the cylinder, a *circulation* has been introduced about the body. Circulation, Γ, is related to the average velocity tending to move fluid elements around the body. This average is made up of velocities on a path encircling the body. However, a nonzero value of circulation does not indicate that the fluid elements are actually moving around the cylinder. As the cylinder is spun faster, and Γ is thus increased, the lift force increases.

The relationship between lift and circulation is termed the *Kutta-Joukowski theorem* after the two people who independently discovered it, and is expressed as

$$L' = \rho U_\infty \Gamma \tag{193.9}$$

where L' is the lift per unit span of the cylinder. This theorem applies to any two-dimensional body about which a circulation exists, not just a cylinder. Circulation about a body can be represented as a **vortex**. A vortex is a circular flowfield in which the fluid rotates about a common axis, and the farther the fluid element is from the axis, the slower it moves around the axis. In the case of lift generation, the vortex is not attached to the same fluid elements, but is "bound" to the body and is often called a *bound vortex*.

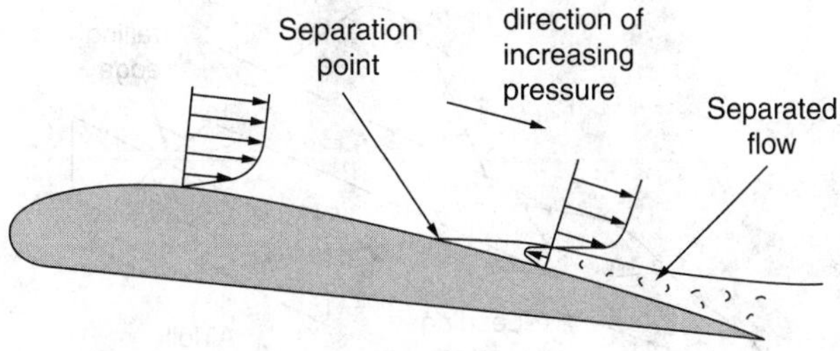

FIGURE 193.4 Diagram illustrating boundary layer separation.

Drag

For a general body, the total drag force results from two sources — *pressure drag* and *viscous drag*. Pressure drag is simply the component of the total pressure force on the body that is in the streamwise direction (parallel to the direction of the freestream). In many flows, *boundary layer separation* causes significant pressure drag. Boundary layer separation can occur when a boundary layer flows into a region of increasing pressure. For fluid elements outside the boundary layer, this increase in pressure is traded off as a decrease in velocity, as can be seen by Bernoulli's equation. Because the pressure is approximately constant across the boundary layer, the fluid near the wall must also undergo an increase in pressure. The fluid near the wall has less energy than the flow at the edge of the boundary layer because it is moving slower. If the pressure increases too rapidly in the downstream direction, the velocity near the wall can reverse. If this happens, the boundary layer is said to have separated, as shown in Figure 193.4. The boundary layer leaves the surface and a recirculating region is developed between the airfoil surface and the boundary layer fluid that left the surface. As the flow is not attached to the surface, it does not continue on a path which leads to a higher pressure in that region as it would if the flow had not separated. The result is a lower pressure acting over the surface. This surface usually has a component of area facing downstream and the low pressure acting on this area is seen as drag (pressure drag). Pressure drag, D_p, is calculated as

$$D_p = \int p dA_x \tag{193.10}$$

where A_x is the streamwise component of the area. For this reason, boundary layer separation is usually considered to have a negative effect on a flow.

Viscous drag, D_v, is present in all vehicles that move through a fluid. It is caused by the wall shear stress, which was discussed earlier in this section. Wall shear stress acts tangentially to the surface, so the total viscous drag is determined by integrating the streamwise component of τ_w over the surface:

$$D_v = \int \tau_w dA_x \tag{193.11}$$

193.3 Two-Dimensional Airfoils

Typically, wings have been analyzed by splitting the problem into two parts: (1) analysis of a wing section (*airfoil*), and (2) modification of the airfoil properties to account for the effects of the complete, finite wing. An airfoil is the cross section of a wing in a plane parallel to the freestream velocity and perpendicular to the wing, as shown in Figure 193.5.

Airfoils are usually described using the nomenclature indicated in Figure 193.6. The longest dimension of the airfoil section is termed the *chord*, and the line connecting the *leading edge* to the *trailing edge* is the *chord line*. Airfoil *thickness* is the maximum distance between the upper and lower surfaces along a

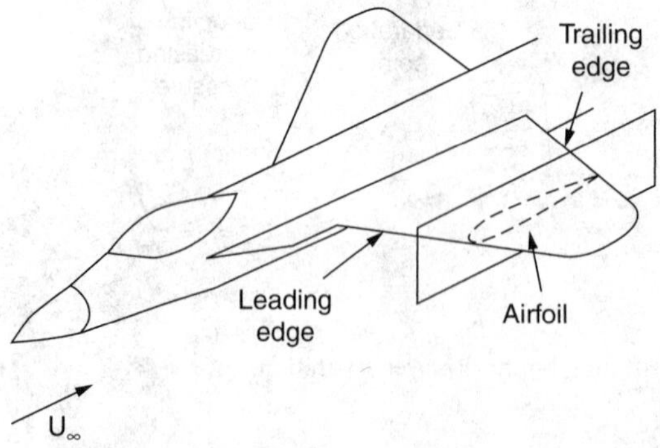

FIGURE 193.5 Diagram illustrating the relationship between an aircraft wing and an airfoil.

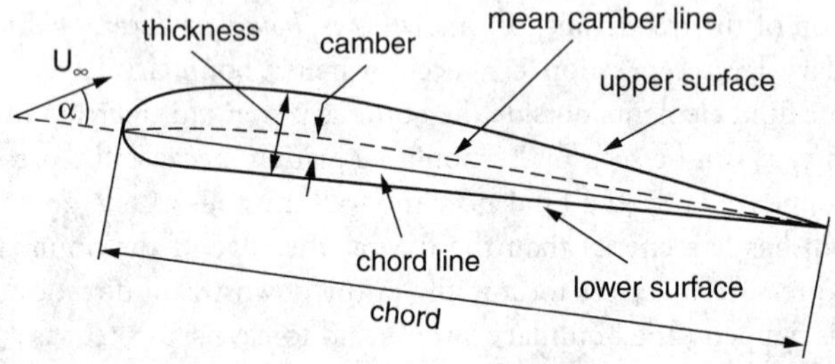

FIGURE 193.6 Airfoil terminology.

line perpendicular to the chord line. Thickness is usually expressed as a percentage of the chord (e.g., a 12% thick airfoil has a maximum thickness of 12% of the chord). The *mean camber line* is the locus of points halfway between the upper and lower surfaces. The variation of the thickness along the mean camber line is termed the *thickness distribution*. *Camber* of the airfoil is defined as the maximum distance between the mean camber line and the chord line in a direction perpendicular to the chord line. When a fluid is moving past the airfoil, the angle of attack is defined as the angle between the freestream velocity and the chord line, denoted as α in Figure 193.6.

Airfoil designs that cover a wide range of thickness, thickness distributions, and camber have been designed and tested by the National Advisory Committee for Aeronautics (NACA) (the predecessor of the current NASA). These include the *four-digit* series (e.g., the NACA 2412); the *five-digit* series (e.g., the NACA 23012); and the *six-digit* series (e.g., the NACA 65-218). The numbers indicate various characteristics of the airfoils. A complete description of the numbering system can be found in Abbott and von Doenhoff [1959]. These and other airfoil designs and their performance characteristics can be found in the books in the Further Information section.

In selecting an airfoil for a particular application, the variation of lift, drag, and moment with angle of attack is necessary. A typical variation of lift coefficient with angle of attack is shown in Figure 193.7. Recall from the earlier discussion that this data represents the performance of a two-dimensional airfoil and is sometimes referred to as *infinite wing data*. The value of α when the lift is zero is termed $\alpha_{L=0}$, the zero-lift angle of attack. As the angle of attack is increased from $\alpha_{L=0}$, the lift increases linearly over a wide range of α. The slope of this line is the *lift slope*. As α becomes large, the pressure gradients on the upper surface of the airfoil also become large, and separation of the boundary layer occurs. Separation causes a large reduction in lift and the airfoil is said to be stalled. The maximum lift generated by the airfoil is indicated by $C_{L,max}$. Before the decrease in lift occurs, the curve becomes nonlinear because of viscous effects.

In the linear region of the curve where viscous effects have a small impact on the lift, *thin airfoil theory* can be used to predict the airfoil behavior. If the airfoil thickness is small in comparison to its chord,

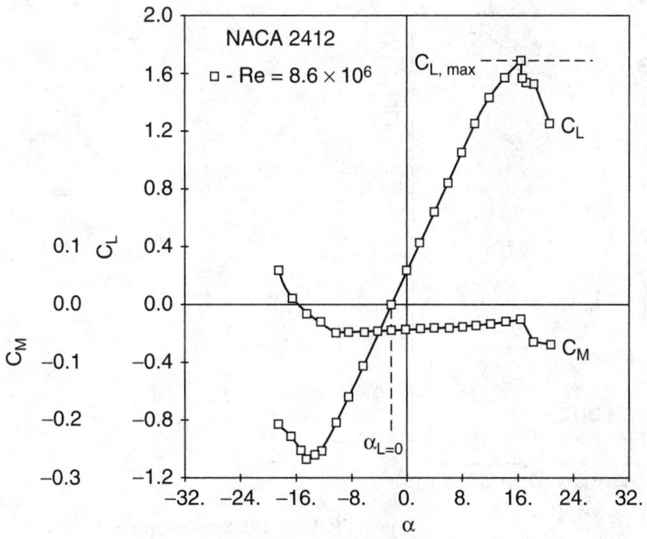

FIGURE 193.7 Lift and moment coefficients as a function of angle of attack for the NACA 2412 airfoil.

the flowfield about the airfoil can be represented by distributing vortices along the mean camber line. An infinite number of solutions are possible for the distribution of the strengths of the vortices as a function of chord. All but one of these solutions is eliminated by applying the *Kutta condition*. The Kutta condition essentially requires that the flow be physically realistic in that the fluid cannot flow around a sharp trailing edge. More specifically, it states that the following be true: (1) the value of the circulation, Γ, is such that the flow leaves the trailing edge smoothly; (2) if the included angle at the trailing edge is finite, then the trailing edge is a stagnation point; and (3) if the trailing edge is cusped (zero included angle), then the velocities from the upper and lower surfaces are equal in magnitude and direction at the trailing edge. Thin airfoil theory then yields a lift slope of 2π, so in the case of symmetric airfoil $C_L = 2\pi\alpha$. For relatively thin airfoils, less than about 12%, this prediction is accurate to within several percent of experimentally determined values.

Camber controls the amount of lift generated at zero angle of attack. A symmetric airfoil (zero camber) generates no lift at $\alpha = 0$. An airfoil with positive camber generates lift at $\alpha = 0$, and zero lift occurs at a negative angle of attack, as shown in Figure 193.7.

Figure 193.7 also shows the variation of moment coefficient with angle of attack for a typical cambered airfoil. The moment is usually taken about a point one quarter of the way along the chord line from the leading edge. This point is known as the *quarter-chord point*. Note that the value of the moment is negative, indicating that the moment acting on the airfoil tends to rotate it in the counterclockwise direction for the airfoil depicted in Figure 193.6. Camber also affects the moment; a symmetric airfoil produces no moment and an increase in the camber causes an increase in the magnitude of moment.

In Figure 193.7, the moment coefficient is approximately constant over a wide range of angle of attack. The quarter-chord point was chosen as the point about which to determine moments for specifically this reason — it is the *center of pressure* for an airfoil. The center of pressure is defined as the point about which the moment is constant, independent of angle of attack. Another useful point in aerodynamics is the *aerodynamic center*, which is the point about which the moment is zero. For a symmetric airfoil, this point happens to be also at the quarter-chord point. In general, the aerodynamic center may or may not lie on the body itself. As viscous effects become important, the moment varies from its ideal behavior, as did the lift behavior. Near stall, the magnitude of the moment increases because of boundary layer separation altering the static pressure distribution.

Drag data are usually presented as a function of lift coefficient, instead of angle of attack, in what is termed a *drag polar*. An example of a drag polar is shown in Figure 193.8 for a cambered airfoil. This

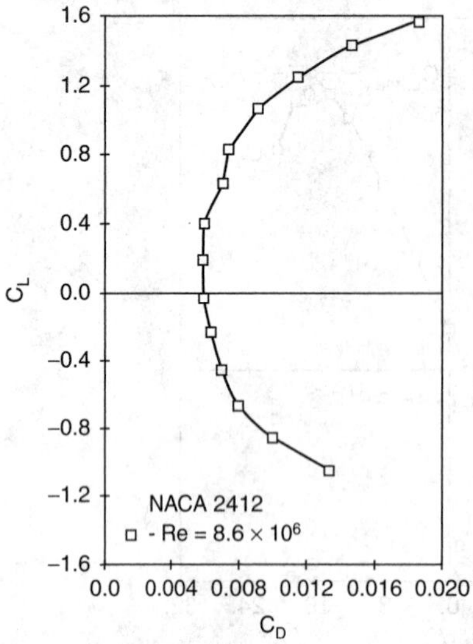

FIGURE 193.8 Drag polar for the NACA 2412 airfoil.

drag coefficient includes both the viscous and pressure drags. The penalty in drag for generating high lift coefficients is clear from Figure 193.8. In addition, when the airfoil approaches stall, and the upper surface boundary layer separates, the drag increases dramatically. This increase is caused by the large pressure drag associated with separation, as indicated in Section 193.2. In many applications, the ratio of lift to drag, *L*/*D*, is an important performance parameter. The *L*/*D*s of different airfoils can easily be compared using a drag polar simply by looking at the slope of a line drawn between a point on the polar and the origin of the plot.

193.4 Finite Wing Effects

Two-dimensional, or infinite wing, performance was discussed in the previous section. Of course, all real aircraft have finite wings. In this section, the modification of infinite wing performance due to the finite nature of real wings is presented. The primary difference between the two cases is that, in the finite wing, the flow can be three-dimensional (i.e., flow can occur into and out of the paper in Figure 193.4 as well as in the plane of the paper). If a wing is generating lift, then the average pressure on the lower surface must be higher than that on the upper surface. Near the wing tips, this pressure difference causes flow from the lower surface to the upper surface, as shown in Figure 193.9. This leakage around the wing tips causes the flow to rotate about a streamwise axis as it leaves the wing tip, forming *wing tip vortices*, as shown in Figure 193.10. Inboard of the wing tips, the flow has a spanwise component due to the leakage around the wing tips. As shown in Figure 193.9, the flow over the lower surface tends to move out toward the wing tip, and the flow over the upper surface tends to move inboard. Clearly, the flow is not two-dimensional, as was assumed in the previous section.

Wing tip vortices induce flow about their axes. In the region of the wing, it can be seen from Figure 193.9 that the resulting induced flow will be downward. This induced downward flow is termed *downwash* and locally rotates the incoming velocity vector, as shown in Figure 193.11. Locally, the angle of attack is reduced by α_i. Thus, the effective angle of attack is $\alpha_{eff} = \alpha - \alpha_i$. Lift is generated perpendicular to the local freestream velocity, and as it has been rotated by α_i, then the lift vector is also rotated by α_i. The lift vector now has a component parallel to the freestream that is far from the wing, D_i— the *induced drag*. In summary, the downwash induced in the region of the wing by the wing tip vortices reduces the effective angle of attack of the airfoil section. This tilts the lift vector downstream, resulting in a component of lift acting in the drag direction. Typical airfoil sections might have a *L*/*D* ratio of about 100. Thus, only a small rotation of the large lift vector can cause a significant increase in drag, which appears as pressure drag.

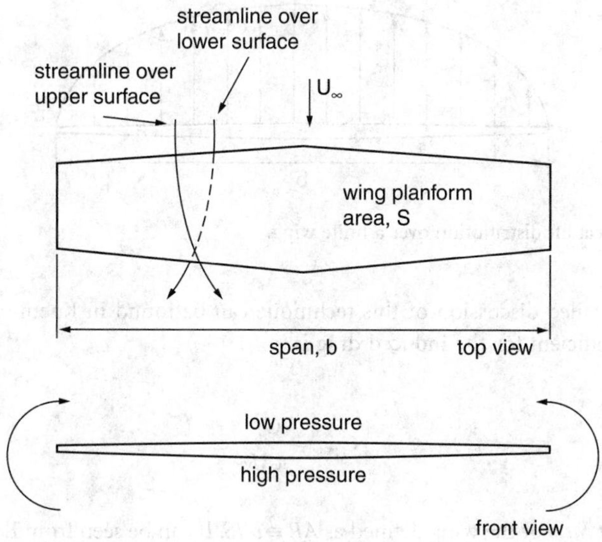

FIGURE 193.9 Finite wing effects.

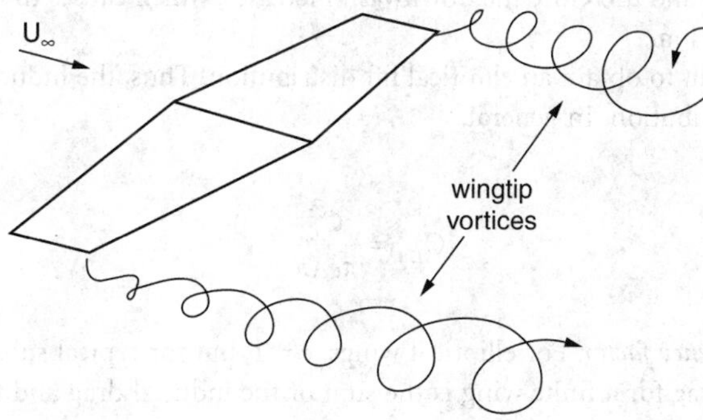

FIGURE 193.10 Wing tip vortices formed by a finite wing.

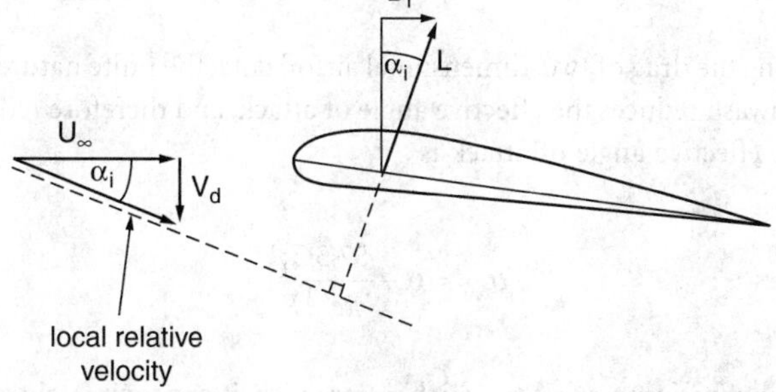

FIGURE 193.11 Diagram illustrating the tilting of the lift vector due to the downwash.

The induced drag of a wing depends on the downwash distribution along its span. It can be shown that the downwash distribution depends on the lift distribution across the wing span. Lift per unit span can vary for several reasons: (1) changes in chord, (2) changes in angle of attack, and (3) variations in airfoil shape. An elliptical distribution of lift across the span, as shown in Figure 193.12, provides a minimum induced drag. Lifting-line theory can be used to predict the downwash for an elliptical lift

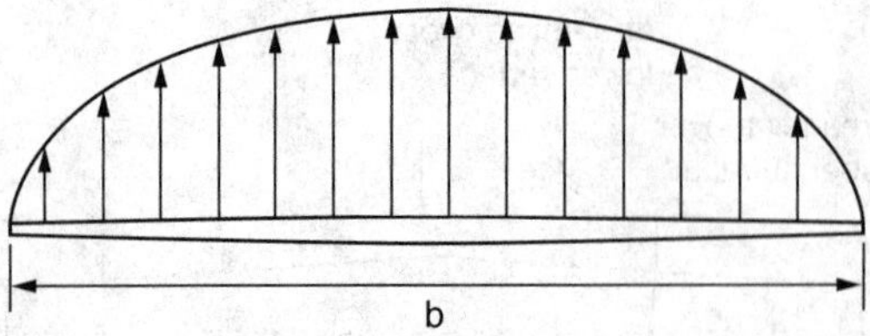

FIGURE 193.12 Elliptical lift distribution over a finite wing.

distribution and a detailed discussion of this technique can be found in Kuethe and Chow [1986]. In this case, the drag coefficient for the induced drag is

$$C_{D,i} = \frac{C_L^2}{\pi AR} \tag{193.12}$$

where AR is the *aspect ratio* of the wing defined as $AR \equiv b^2/S$. It can be seen from Equation (193.12) that, as the aspect ratio increases (i.e., the wing becomes longer and more slender), the induced drag is reduced. This equation also indicates the dependence of induced drag on lift. As the lift increases, the strength of the tip vortices increases, and therefore the downwash increases, which causes the lift vector to tilt more in the downstream direction.

In practice, it is difficult to obtain an elliptical lift distribution. Thus, the induced drag is higher than that for an elliptical distribution. In general,

$$C_{D,i} = \frac{C_L^2}{\pi eAR} \tag{193.13}$$

where e is the *span efficiency factor.* For elliptical wings, $e = 1$, but for typical subsonic aircraft, $0.85 < e < 0.95$. Thus, the total drag for a finite wing is the sum of the induced drag and the infinite wing drag:

$$C_{D,\text{tot}} = C_{D,2\text{–}D} + \frac{C_L^2}{\pi eAR} \tag{193.14}$$

In addition to affecting the drag of two-dimensional airfoil data, the finite nature of a wing also reduces the lift slope. The downwash reduces the effective angle of attack, and therefore reduces the lift generated. For a general wing, the effective angle of attack is

$$\alpha_{\text{eff}} = \alpha - \frac{57.3C_L}{\pi e' AR} \tag{193.15}$$

where e' is another span effectiveness factor, which, in practice, is approximately equal to e. The reduced lift for a given angle of attack indicates a reduction in the lift slope, a, which becomes

$$a = \frac{a_0}{1 + 57.3a_0/(\pi e' AR)} \tag{193.16}$$

where a_0 is the two-dimensional lift slope and a is the lift slope including finite wing effects.

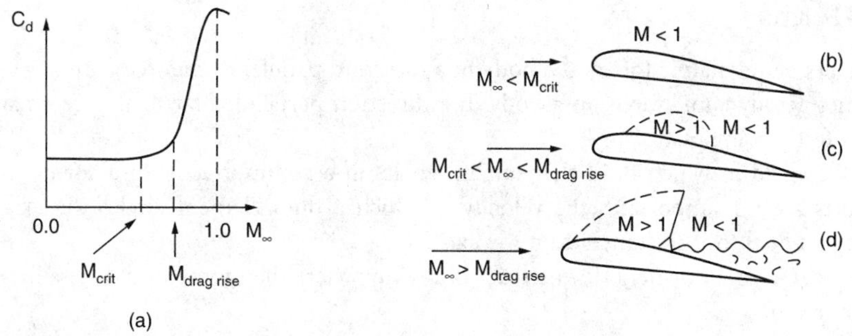

FIGURE 193.13 Drag as a function of Mach number (a), and airfoil flowfields associated with the three Mach number ranges (b–d).

193.5 Effects of Compressibility

It has been assumed to this point that the flow speed has been low enough that the effects of compressibility of the fluid have been negligible. (Section 6 describes compressibility in more detail and should be reviewed at this time if the reader is unfamiliar with this topic). However, as the freestream speed and thus the freestream Mach number (M_∞) are increased, compressibility effects are seen. Only a brief discussion of some of these effects is presented here.

If the local Mach number increases above 1.0, the possibility exists for the formation of shock waves. This effect is one of the major differences between compressible and incompressible flows and is depicted in Figure 193.13. Over the upper surface of an airfoil, the flow moves faster than the freestream. Thus, the Mach number in a region above the upper surface can be supersonic even if the freestream Mach number is less than one (Figure 193.13(c)). The freestream Mach number at which the flow over the upper surface reaches Mach 1.0 is termed the *critical Mach number*, M_{crit}. As the freestream Mach number increases, a shock wave forms on the upper surface, causing boundary layer separation because of the strong adverse pressure gradient (Figure 193.13(d)). In this case, the drag of the airfoil increases dramatically, as shown in Figure 193.13(a). The Mach number where the drag begins to rise is the *drag divergence Mach number*, $M_{\text{drag rise}}$.

High-speed subsonic aircraft often have swept wings. The purpose of wing sweep is to increase $M_{\text{drag rise}}$ so the aircraft can fly faster. By sweeping the wing back, the airfoil "sees" only the component of the freestream Mach number normal to the leading edge, which is lower than the total freestream Mach number, as shown in Figure 193.14. Thus, the critical Mach number with sweep should be $M_{\text{cr,sweep}} = M_{\text{cr}}/\cos\theta$. In practice, because of the complex flowfield induced by wing sweep, the benefit of sweep is less than this. A more complete discussion of compressibility effects in aerodynamics can be found in Anderson [1984].

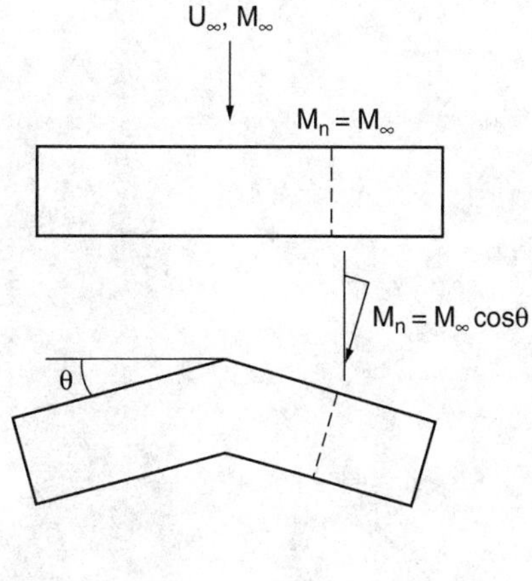

FIGURE 193.14 The effect of wing sweep on the normal Mach number.

Defining Terms

Drag — The net aerodynamic force on a body in a direction parallel to the freestream.

Lift — The net aerodynamic force on a body in a direction perpendicular to the freestream direction and to the wing platform.

Vortex — A flowfield in which fluid elements rotate about a common axis. In a vortex where viscous effects are not important, the velocity at which a fluid element circles the axis is inversely proportional to its distance from the axis.

Wake — The region of a flowfield downstream of a body where the momentum is less than that of the freestream.

References

Abbott, I. H. and von Doenhoff, A. E. 1959. *Theory of Wing Sections*. McGraw-Hill, New York.

Anderson, J. D. 1984. *Fundamentals of Aerodynamics*. McGraw-Hill, New York.

Kuethe, A. M. and Chow, C. Y. 1986. *Foundations of Aerodynamics*, 4th ed. John Wiley & Sons, New York.

Further Information

A much more complete discussion of aerodynamics can be found in *Fundamentals of Aerodynamics* by Anderson, and in *Foundations of Aerodynamics* by Kuethe and Chow. A good introductory text is *Introduction to Flight* (McGraw-Hill, New York, 1978) by J. D. Anderson.

Detailed lift, drag, and moment data for a wide variety of aerodynamic shapes over a wide speed range can be found in *Fluid Dynamic Drag* by Hoerner. A good source of low-Reynolds number airfoil data is *Airfoils at Low Speeds* (Stokely, Virginia Beach, VA, 1989) by M. S. Selig, J. F. Donovan, and D. B. Fraser.

Boundary layers were only mentioned briefly, but for further information, *Boundary Layer Theory* (McGraw-Hill, New York, 1968) by H. Schlichting is a good source.

194

Response to Atmospheric Disturbances

Ronald A. Hess
University of California

The aerodynamic forces that allow the flight of heavier-than-air vehicles are created by the relative motion between the aircraft and the air mass through which it moves. In the case of a nonquiescent atmosphere, that is, one that is moving relative to the Earth, the motion of the air mass itself contributes to this relative motion. Since the atmosphere is rarely quiescent, the subject of aircraft stability and control in a moving atmosphere is of considerable importance to aerospace engineers from the standpoint of performance, passenger comfort, and safety.

194.1 Descriptions of Atmospheric Motion

The fundamental approach that physicists and engineers adopt in describing the motion of a continuum like the atmosphere is a *field description.* In a field description, the velocity of the atmosphere (relative to the Earth) is considered to be a continuous function of location and time. Consider Figure 194.1 showing an orthogonal axis system fixed relative to the Earth. At some point, x_0, y_0, and z_0, and at some time instant t_0, the three components of the velocity of the atmosphere are $u_a(x_0,y_0,z_0,t_0)$, $v_a(x_0,y_0,z_0,t_0)$, and $w_a(x_0,y_0,z_0,t_0)$. The field description allows a convenient means of categorizing atmospheric motion.

Mean Wind

If, for example, $u_a = U$, $v_a = V$, and $w_a = W$, where U, V, and W are constants, one has defined a mean wind. Obviously, this is an idealized description, but nonetheless, a useful one. For example, airliner flight in the jet stream can often be adequately described by such a mean wind condition.

Wind Shear

If, for example, $u_a = u_a(x)$, $v_a = V$, and $w_a = W$, one has defined a simplified wind shear, in which the velocity field varies with position in the x direction, but not with time. In applications to aircraft stability

0-8493-1586-7/05/$0.00+$1.50

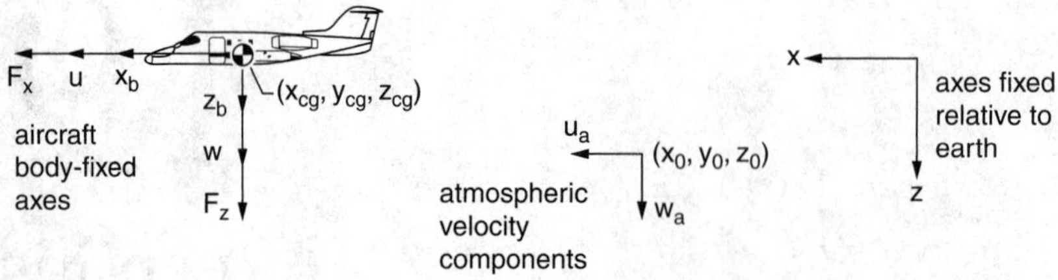

FIGURE 194.1 Axis systems for description of atmospheric turbulence.

and control, changes in $x(\Delta x)$ over which significant changes in u_a occur are typically assumed to be such that $\Delta x >> l_f$ where l_f represents the fuselage length.

More complex wind shears are possible. If, for example, $u_a = u_a(x)$, $v_a = 0$, and $w_a = w_a(x)$, and

$$u_a(x) = -U\left[\sin\left(\frac{2\pi(x-x_0)}{X_h}\right)\right] \quad \text{for } x_0 \le x \le X_h + x_0 \tag{194.1}$$

$$u_a(x) = 0 \text{ for } x < x_0 \text{ and } x > X_h + x_0 \tag{194.2}$$

$$w_a(x) = \frac{W}{2}\left[1 - \cos\left(\frac{2\pi\left[x - x_0 - \frac{(X_h - X_v)}{2}\right]}{X_v}\right)\right] \tag{194.3}$$

$$\text{for } x_0 + \frac{(X_h - X_v)}{2} \le x \le x_0 + \frac{(X_h + X_v)}{2}$$

$$w_a(x) = 0 \text{ for } x < x_0 + \frac{(X_h - X_v)}{2} \text{ and } x > x_0 + \frac{(X_h + X_v)}{2} \tag{194.4}$$

one has defined a simplified version of a meteorological phenomenon called a *microburst*. A microburst is a downburst that is confined to a relatively small area, and is characterized by a massive downward air mass movement that disperses radially near the ground. An intense microburst can be extremely hazardous to aircraft when landing or taking off [Wingrove and Bach, 1989]. Figure 194.2 represents a vertical

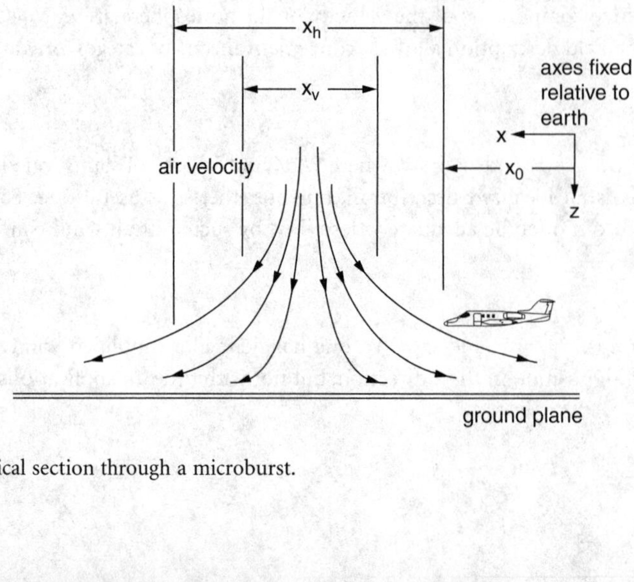

FIGURE 194.2 Vertical section through a microburst.

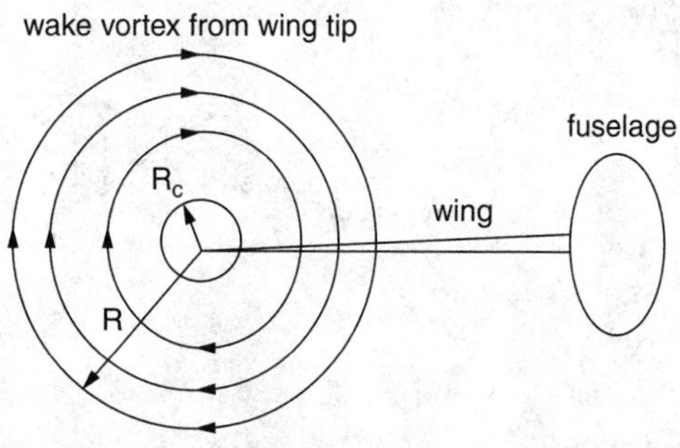

FIGURE 194.3 A wake vortex from aircraft wing tip; aircraft viewed from behind.

cross-section through the center of a microburst velocity field with the arrows representing wind direction. Equations 194.1 and 194.2 represent the "headwind and tailwind" air velocities as the microburst is traversed. Equations 194.3 and 194.4 represent the "down-draft" air velocities during the same encounter. The parameters X_h and X_v represent the distances over which the particular microburst velocity fields extend. The x_0 is the Earth-fixed coordinate x at which the head and tailwind components of the downburst become important. These are obviously a function of altitude. The following parameter values model a particular microburst that occurred at Philadelphia International Airport in 1976 and resulted in the loss of an aircraft [Psiaki and Stengel, 1985]:

$$X_h = 10{,}000 \text{ ft } U = 35 \text{ ft/sec } X_v = 3{,}000 \text{ ft } W = 20 \text{ ft/sec} \tag{194.5}$$

Wake Vortices

If, for example, $u_a = 0$, and we consider a tangential velocity,

$$V_t = \sqrt{V^2(y,z)+W^2(y,z)} = \Gamma_\infty/2\pi\left[\frac{R}{R_c^2+R^2}\right] \tag{194.6}$$

one has a simple model for a wake vortex. In Equation 194.6, Γ_∞ is referred to as the circulation (ft²/sec), R_c is the radius of the core of the vortex (ft), and R is the distance from the axis of the vortex (ft). Wake vortices are created when a wing generates lift. As the name implies, they are characterized by a swirling or vortical flow, emanating from each wingtip of an aircraft as shown in Figure 194.3. For large commercial aircraft, the wake vortices can be quite severe and can persist for some distance behind the aircraft generating the vortices. They can also pose serious dangers for other aircraft encountering them. The following parameter values describe the wake vortex created by a Boeing 747. This vortex was implicated in an aviation accident involving another aircraft that encountered it after takeoff [Fiorino, 2002]:

$$\Gamma_\infty = 5{,}328 \text{ ft}^2\text{/sec } R_c = 14.8 \text{ ft} \tag{194.7}$$

Thus, if an aircraft were following the Boeing aircraft with $R = 20$ ft, tangential air velocities on the order of 30 ft/sec would be experienced at that radial location.

Turbulence

If, for example, $u_a = U$, $V_a = V$ and $w_a = w_a(x,t)$, and if the changes Δx over which significant changes in w_a occur are such that Δx is the same order of magnitude as l_f, one has defined a simplified atmospheric *turbulence field* or *gust field*. If, as in Figure 194.1, the direction of the w_a axis is vertical, the field is referred to as a vertical turbulence field or a one-dimensional "upwash" field [Etkin, 1972].

As opposed to the velocity fields that have been introduced thus far, turbulence cannot be adequately described in deterministic fashion. That is, the air velocity at any point (x,y,z) and time (t) is a random variable. Given this fact, the intensity of turbulence is typically quantified by the root mean square (RMS)

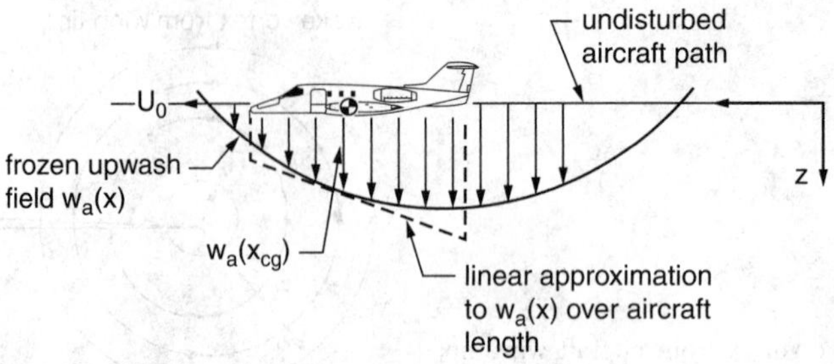

FIGURE 194.4 Aircraft in a one-dimensional, frozen upwash field.

value of each of the velocity components. Thus, for example, at some point x_0 in the one-dimensional upwash field just described,

$$\sigma_{w_a} = RMS\ value\ of\ w_a(x_0,t) = \sqrt{\lim_{T\to\infty} \frac{1}{2T} \int_{-T}^{T} w_a^2(x_0,t)dt} \tag{194.8}$$

Simplified Description

In aircraft stability and control analyses, it is often assumed that the statistical description of the turbulence velocity field, such as the RMS velocity at a point, is not a function of time or the origin or orientation of the axis system used to describe the field. These assumptions are referred to as stationarity, homogeneity, and isotropy, respectively. A final simplifying assumption is employed if, as is often the case, the aircraft is moving through the turbulence field with a velocity that is significantly larger in magnitude that the RMS value of the turbulence. In this case, the spatial and temporal variation of the turbulence field may be replaced by the spatial variation alone. Thus, $w_a(x,t) = w_a(x)$. This simplification is referred to as the frozen turbulence field hypothesis or Taylor's Hypothesis, after the British researcher G. I. Taylor [Houbolt, 1973].

194.2 Turbulence and Aircraft Dynamics

Figure 194.4 is a representation of an aircraft flying through the one-dimensional, frozen upwash field described in the preceding section. The velocity profile of the field is represented in the figure, with the vertical arrows indicating the magnitude and direction of the air velocity at each point on the idealized linear flight path. The problem now becomes one of determining the aerodynamic forces and moments that are created on the aircraft by the indicated turbulence velocities. To accomplish this, a mathematical description of the frozen turbulence field, $w_a(x)$, is needed.

Fourier Integral Representation

The spatially dependent $w_a(x)$ can be represented as a sample function from an ergodic random process [Bendat and Piersol, 1966]. One measure associated with such sample functions is the autocorrelation function defined as

$$\phi_{w_a w_a}(\xi) = \lim_{X\to\infty} \frac{1}{2X} \int_{-X}^{X} w_a(x) w_a(x-\xi)dx \tag{194.9}$$

The autocorrelation function has a Fourier integral or Fourier transform representation

$$\phi_{w_aw_a}(\xi) = \frac{1}{2\pi}\int_{-\infty}^{\infty} \varphi_{w_aw_a}(\varpi)e^{j\varpi\xi}d\varpi \tag{194.10}$$

and

$$\varphi_{w_aw_a}(\varpi) = \int_{-\infty}^{\infty} \phi_{w_aw_a}(\xi)e^{-j\varpi\xi}d\xi$$

Here ϖ represents spatial frequency with units of radians per foot. $\varphi_{w_aw_a}(\varpi)$ is referred to as the *power spectral density* of $w_a(x)$. Power spectral density representation constitutes one of the primary ways in which the characteristics of atmospheric turbulence have been measured and reported [Houbolt, 1973].

Two power spectral forms widely used in turbulence analyses are the von Karman and Dryden spectra [Etkin, 1972]. For example, the von Karman spectrum can be given by

$$\varphi_{w_aw_a}(\varpi) = \sigma^2_{w_a} L \frac{1+\frac{8}{3}(1.339L\varpi)^2}{[1+(1.339L\varpi)^2]^{\frac{11}{6}}} \tag{194.11}$$

with the Dryden form given by

$$\varphi_{w_aw_a}(\varpi) = \sigma^2_{w_a} L \frac{1+3(L\varpi)^2}{[1+(L\varpi)^2]^2} \tag{194.12}$$

where L is referred to as the *turbulence scale length*. Depending on altitude and turbulence intensity, scale lengths from 200 to 5000 ft have been measured [Houbolt, 1973]. Often in flight control system analysis, a simpler spectrum than that given by Equations 194.11 and 194.12 is used. For example,

$$\varphi_{w_aw_a}(\varpi) = 2\sigma^2_{w_a} L \frac{1}{1+(L\varpi)^2} \tag{194.13}$$

The simplicity of the spectrum of Equation 194.13, its ability to closely approximate the spectra of Equations 194.11 and 194.12 and the preservation of the parameters σ_w and L, suggest its use.

If one limits the spatial "duration" of $w_a(x)$ to a large but finite value, $X = X_M$, such that for $X > X_M$, $w_a(x) = 0$, then $w_a(x)$, itself, is amenable to representation by a Fourier integral or Fourier transform as

$$w_a(x) = \frac{1}{2\pi}\int_{-\infty}^{\infty} W_a(j\varpi)e^{j\varpi x}dx \tag{194.14}$$

and

$$W_a(j\varpi) = \int_{-\infty}^{\infty} w_a(x)e^{-j\varpi x}dx$$

The first of Equations 194.14 can be used to demonstrate that $w_a(x)$ may be envisioned as the sum of an infinite number of sinusoids, each of infinitesimal amplitude, with the frequency of each constituent sinusoid differing only infinitesimally from that of its neighbor [McRuer et al., 1973].

Aerodynamic Force/Moment Prediction

The problem of determining the aerodynamic forces and moments exerted on an aircraft as it moves through an upwash field can be further simplified as follows: Consider the case where $w_a(x)$ (envisioned as the infinite sum of infinitesimal sinusoids just described) has the majority of its power below a frequency ϖ_0. If the fuselage length l_f is such that

$$l_f \leq \frac{1}{8}\left(\frac{2\pi}{\varpi_0}\right) \tag{194.15}$$

where $(2\pi/\varpi_0)$ is the spatial wavelength (ft) of the constituent sinusoid with frequency ϖ_0, then the variation in $w_a(x)$ over the length of the aircraft can be considered to be approximately linear. The inequality in Equation (194.15) is derived from the fact that a straight line (of appropriate slope) can be considered an adequate approximation to a sinusoid if the length of the line is less than one-eighth of the period (wavelength) of the sinusoid.

Figure 194.4 suggests that, at the instant shown, the relative motion of the aircraft and upwash field is aerodynamically equivalent to that which would be in evidence if the aircraft were moving through quiescent air but with (1) an instantaneous vertical velocity component given by

$$w(t) = -w_a(x_{cg}) \tag{194.16}$$

and (2) an instantaneous angular pitch rate of

$$q(t) = \frac{dw_a(x_{cg})}{dx} = \frac{1}{U_0}\frac{dw_a(x_{cg})}{dt}\bigg|_{x_{cg}=U_0 t} \tag{194.17}$$

As Figure 194.1 indicates, $w(t)$ is the z_b body-axis component of the velocity of the aircraft center of gravity (cg) in a quiescent atmosphere, and x_{cg} is the x-coordinate of the aircraft center of gravity in the Earth-fixed axis system used to describe the upwash field. U_0 is the x_b body-axis component of the aircraft velocity in steady flight in a quiescent atmosphere.

The manner in which turbulence effects are included in the aircraft equations of motion can now be described. The force equations can be written in the familiar form:

$$\bar{F} = m\frac{d\bar{v}}{dt} \tag{194.18}$$

where $\bar{F}$ represents the vector sum of the aerodynamic and propulsive forces acting on the aircraft with mass m, and $\bar{v}$ represents the velocity of the aircraft center of gravity. For aircraft applications, Equation 194.18 is typically linearized about a condition of steady, wings-level flight [Etkin, 1972; McRuer et al., 1973]. The components of $\bar{F}$ in this equation are expressed as linear functions of the aircraft's linear and angular velocity perturbations and their derivatives. Thus, for example, referring to Figure 194.1, the x_b and z_b body-axis components of $\bar{F}$ can often be simplified to

$$\begin{aligned} F_x &= m[-g\theta + X_u u + X_w w + X_\delta \delta] \\ F_z &= m[Z_u u + Z_{\dot{w}}\dot{w} + Z_w w + Z_q q + Z_\delta \delta] \end{aligned} \tag{194.19}$$

where θ, u, w, and δ represent, respectively, perturbations in the aircraft pitch attitude, perturbations in the x_b and z_b body-axis components of the velocity of the aircraft center of gravity, and perturbations in control/propulsive inputs. X and Z represent the components of the aerodynamic and propulsive forces in the x_b and z_b body-axis directions. The quantities X_u, Z_u, and so on are referred to as mass-normalized stability derivatives [McRuer et al., 1973]. Equation 194.19 can now be modified to include the effects of the one-dimensional upwash field by simply replacing w by $(w - w_a)$, and q by $\left(q+\frac{\dot{w}_a}{U_0}\right)$ on the right sides of these equations. Note that since the left sides of the equations are describing inertial as opposed to aerodynamic terms in the equations, they are not included in these substitutions. The result of these substitutions will be the addition of "disturbance" terms to each of Equations 194.19.

Extension to More Complex Fields

The preceding treatment describing modification to the X and Z-force equations to account for a one-dimensional upwash field can be extended to more complex turbulence fields, such as those involving consideration of u_a and v_a terms in the gust field. In such treatments, it is typical to consider the entire aircraft submersed in the u_a component existing at the aircraft center of gravity (equivalent to considering the aircraft as a point). The v_a terms in the lateral-directional equations are handled in a manner similar to that for the w_a terms in the longitudinal equations, that is, treating the variation in $v_a(x)$ over the length of the aircraft as approximately linear. The equations of motion so derived can also be used to study aircraft response to the deterministic velocity fields discussed earlier, such as the microburst and wake vortex descriptions.

Computer Simulation

Computer simulation of aircraft encountering moving air masses such as turbulent upwash fields requires that one create the desired time history of the disturbance inputs as a function of time rather than position. This can be accomplished through spectrum factorization of the power spectral density of the upwash field. For example, converting the spectrum of Equation 194.13 from spatial frequency ϖ (rad/ft) to temporal frequency ω (rad/sec) can be accomplished as follows:

$$\varphi_{w_a w_a}(\omega) = \frac{1}{U_0}\varphi_{w_a w_a}(\varpi)\Big|_{\varpi=\frac{\omega}{U_0}} \tag{194.20}$$

Assume that the spatial spectrum had parameters $\sigma = 5$ ft/sec and $L = 1{,}000$ ft, and that the aircraft was flying at a speed of 500 ft/sec. The temporal spectrum corresponding to Equation 194.13 would take the form:

$$\varphi_{w_a w_a}(\omega) = \frac{100}{[1+(2\omega)^2]} \quad (\text{ft}^2/\text{sec}^2)/\text{rad} \tag{194.21}$$

The spectrum of Equation 194.21 can be spectrum factorized as

$$\varphi_{w_a w_a}(\omega) = \frac{10}{[1+(2j\omega)]}\cdot\frac{10}{[1+(-2j\omega)]} \tag{194.22}$$

Finally, a random function of time (sample function from an ergodic random process) with the power spectral density of Equation 194.21 can be obtained by passing white noise with unity covariance through a linear transfer function given by

$$G(s) = \frac{10}{1+2(j\omega)}\bigg|_{j\omega=s} \tag{194.23}$$

In this manner, $w_a(t)$ can be created for use in computer simulation.

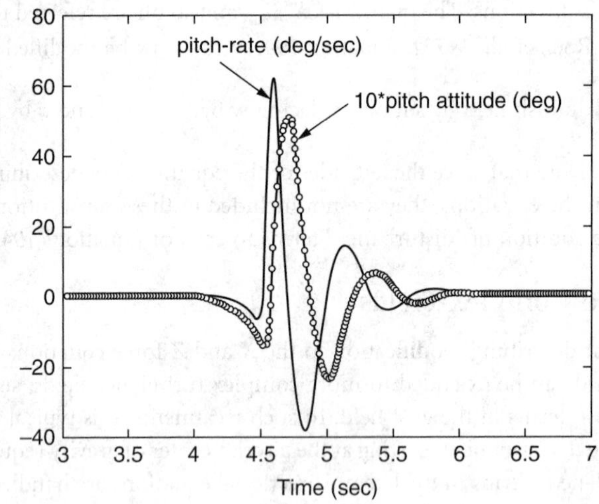

FIGURE 194.5 Pitch-rate and pitch-attitude time histories for light aircraft traversing wake vortex at 90°.

Examples

Two computer simulation examples of aircraft encountering moving air masses are discussed. The first involves a light aircraft encountering the tip vortex described by Equations 194.6 and 194.7. The flight condition is an altitude of 5,000 ft with an aircraft velocity (U_0) of 219 ft/sec. The aircraft crosses the tip vortex emanating from the left wing of the generating aircraft at an angle of 90°. A simulated pilot controlling pitch attitude is included. Only the effect of the vertical velocity component of the vortex is considered. It should be noted that the length of the aircraft (approximately 24 ft) will not invalidate the linear approximation to the upwash field used in creating the vehicle equations of motion.

The second example involves a small executive jet aircraft encountering an upwash field with a spatial power spectral density given by Equation 194.13 with $\sigma = 10$ ft/sec and $L = 1{,}000$ ft. The flight condition is an altitude of 40,000 ft and an aircraft velocity (U_0) of 677 ft/sec. Again a simulated pilot controlling pitch attitude is included. The approximate length of the aircraft in question is 46 ft.

The spatial "bandwidth" ϖ_0 of the turbulence spectrum, defined here as the value of ϖ_0 where the spectrum $\varphi_{w_a w_a}(\varpi_0)$ has dropped to 5% of its zero frequency value, is 0.00437 rad/ft. These values mean that the constraint of the inequality in Equation 194.15 is clearly satisfied.

Figure 194.5 shows the pitch rate and scaled pitch attitude for the light aircraft. Note the very large pitch rates that occur as the aircraft crosses the tip vortex center. Figure 194.6 shows the pitch rate and scaled pitch attitude for the executive jet for a 10-sec period.

Defining Terms

Field description — A representation of a scalar or vector quantity as a continuous function of position and time.

Frozen turbulence field (Taylor's hypothesis) — Considering the time and spatial variation of turbulent velocity field as a spatial variation only.

Mean wind — A constant air velocity, independent of location and time.

Microburst — An atmospheric phenomenon that is confined to a relatively small area, characterized by a massive downward air mass movement that typically disperses radially near the ground.

Power spectral density — A means of describing the distribution of the power in a signal with frequency.

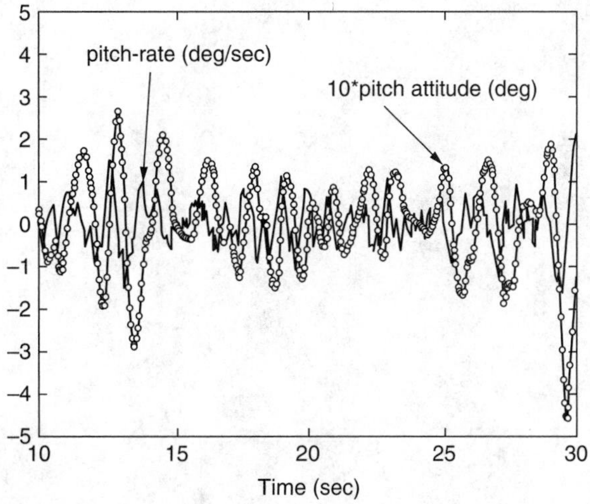

FIGURE 194.6 Pitch-rate and pitch-attitude time histories for executive jet traversing one-dimensional upwash field.

Root mean square (RMS) value — The square root of the integral of a squared function, divided by the integration interval.

Turbulence (gust) field — A field representation of the rapid, random variations of air velocity components at a point in space.

Turbulence scale length — A parameter appearing in the power spectral density representation of a frozen turbulence field. It is related inversely to the break frequency of the power spectral density, that is, the larger the characteristic length, the smaller the spatial or temporal break frequency.

Wake vortex — The swirling air flow emanating from the tip of any wing that generates lift.

Wind shear — The variation in air velocity as a function of position, often associated with microbursts.

References

Bendat, J. S. and Piersol, A. G. 1966. *Measurement and Analysis of Random Data,* Wiley, New York.

Etkin, B., 1972, *Dynamics of Atmospheric Flight,* Wiley, New York.

Fiorino, F., 2002, Upset events, wake vortices at center of Flight 587 probe, *Aviation Week Space Technol.,* 156:39.

Houbolt, J. C., 1973, Atmospheric turbulence, *AIAA J,* 11:421–437.

McRuer, D. T., Ashkenas, I., and Graham, D., 1973, *Aircraft Dynamics and Automatic Control,* Princeton University Press, Princeton, NJ.

Psiaki, M. L. and Stengel, R. G., 1985, Analysis of aircraft control strategies for microburst encounter, *J. Guidance, Control Dyn.,* 8:553–559.

Wingrove, R. C. and Bach, R. E., 1989, Severe winds in the Dallas/Ft. Worth micorburst measured from two aircraft, *J. Aircraft,* 26:221–224.

Further Information

A thorough overview of the topic of turbulence and flight is presented in the 1981 Wright Brothers Lectureship in Aeronautics: Etkin, B., 1981, Turbulent wind and its effect on flight, *J. Aircraft,* 18:327–345.

195

Computational Fluid Dynamics

Ramesh K. Agarwal
Wichita State University

In the past 2 decades, enormous progress has been made in computational methods for the study of fluid flow phenomena in a broad variety of scientific and engineering disciplines. Examples include aircraft, ship, and automobile design; oceanography; meteorology; and astrophysics. The focus of this chapter is, however, on computational aerodynamics, which is rapidly emerging as a crucial enabling technology for the analysis and design of flight vehicles. Computational simulations, coupled with targeted experimental testing, have proven to be a cost-effective way of developing flight vehicles. Such combined evaluations are performed to ensure that the final product will efficiently meet target performance characteristics, using few design cycles. On the other hand, for some flight regimes, such as hypersonic flows at high altitude, experimental testing may not be feasible, and in such cases even greater reliance is placed on computational simulations.

In aeronautical applications, the computational analysis of the aerodynamic performance of a flight vehicle — be it an aircraft, a helicopter, or a launch vehicle — is a multistep process. First, a geometric description of the configuration is obtained in discretized form. Second, a grid is generated around the object to provide a finite set of points over which the flow field solution is calculated. The flow is then solved, and enormous quantities of data for pressure, temperature, and velocity variables are processed to obtain the aerodynamic quantities of interest — namely, the lift, drag, moment coefficients, and other parameters required to assess the flight vehicle performance.

In this respect, recent developments in computational fluid dynamics primarily fall into two broad categories: (1) computational prediction of the flow field about increasingly complex configurations, and (2) the simulation of flow physics leading to a better understanding of such flow phenomena as transition and turbulence. This chapter covers only the first category: the computational tools employed in aerodynamic analysis and their application in analyzing the aerodynamic performance of complex configurations.

195.1 Geometry Modeling and Grid Generation

In the aerospace industry, the geometric description of the configuration in discretized form is routinely created using computer-aided design (CAD) systems. However, the generation of a suitable three-dimen-

0-8493-1586-7/05/$0.00+$1.50

sional grid about complete aircraft configurations remains a formidable task. In this respect, there have been significant developments in the past decade, with both the algebraic and partial differential equations–based techniques fairly well developed for the generation of structured global grids about complex three-dimensional shapes. The recent thrust has been toward zonal grids and unstructured grid technologies since they are thought to be more effective for treating flows over complex three-dimensional objects. In addition, there have been major advances in adaptive grid and grid embedding techniques, for optimal distribution of grid points around an object in order to resolve sharp gradients and discontinuities, at the least cost possible. A review article by Steinbrenner and Anderson [1990] appropriately describes these developments.

195.2 Flow Simulation Algorithms

For flow field solutions, mathematical models vary in complexity from the simple Laplace equation to the unsteady compressible Navier–Stokes equations. Figure 195.1 describes the equations of fluid dynamics for mathematical models of varying complexity.

During the late 1960s, panel methods were developed for calculating inviscid subsonic flow about complex aerodynamic shapes by solving the Laplace equation. Also, boundary-layer solution techniques were developed to allow for viscous effects, primarily for attached flows. During the early 1970s, a major breakthrough was achieved by Murman and Cole [1971], who solved the mixed elliptic-hyperbolic transonic, small-disturbance equation. Later, Jameson [1974] extended the Murman–Cole technique to the solution of the transonic full-potential equation. Such codes are routinely used in transport aircraft design and have proven to be an effective tool, when combined with boundary-layer corrections, for predicting the aerodynamic performance of a variety of flight vehicles in the transonic range.

Most of the recent progress in the development of numerical algorithms has been for the solution of the Euler and Navier–Stokes equations, especially the development of efficient and accurate shock-capturing algorithms. Central-difference algorithms with artificial viscosity (dissipation), characteristics-based upwind schemes, and total-variation-diminishing schemes have been developed for hyperbolic conservation laws. Strong discontinuities (shocks) are captured by these schemes with little oscillation and minimal dispersive and dissipative errors. It is beyond the scope of this chapter to provide details of these techniques, and the reader is referred, for example, to a survey paper by Jameson [1990].

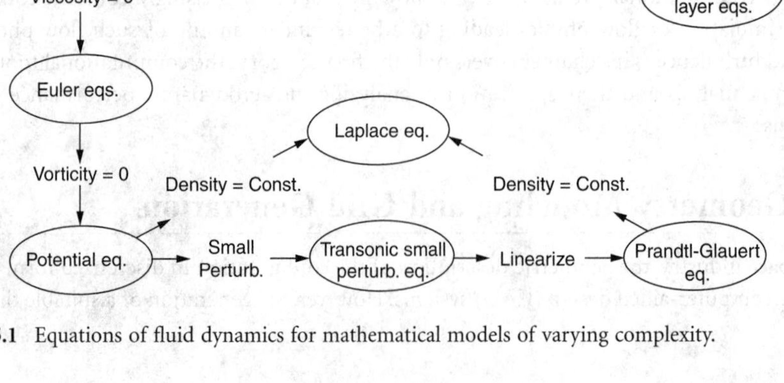

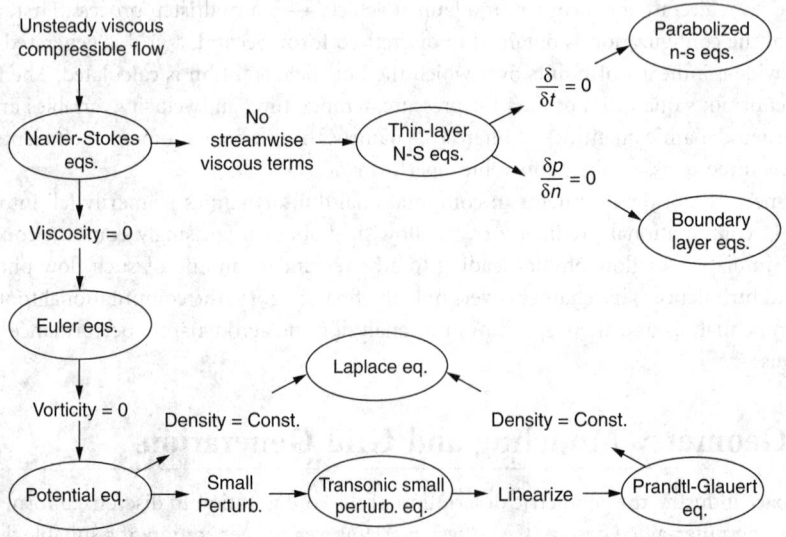

FIGURE 195.1 Equations of fluid dynamics for mathematical models of varying complexity.

195.3 Turbulence Modeling

The currently intractable problem of accurately and practically modeling turbulence effects must be addressed before computational aerodynamics becomes a reliable routine prediction tool for flow situations where viscous effects are dominant. Examples of such situations include separated flows over control surfaces at high angles of attack and aerodynamic heating on high-speed vehicles. The computation of the effect of turbulence from first principles remains an elusive goal at the present time, even for simple two-dimensional shapes such as airfoils, and using maximum available computing power.

In current aerodynamic analyses, turbulence effects are accounted for by phenomenological models. The time-averaged form of the Navier–Stokes equations (known as the Reynolds-averaged Navier–Stokes equations) are solved. Time-averaging introduces the turbulent Reynolds stresses in the Navier–Stokes equations, and these are calculated by multiplying an eddy-viscosity coefficient with the strain-rate tensor. The phenomenological description of eddy viscosity is known as "turbulence modeling." At present, there is no universal turbulence model that works for all flow situations. Generally, turbulence models are developed by validating them against experimental data for simple flow situations, and are then used for calculation of complex flow fields. This approach introduces an element of uncertainty into the prediction of complex flows.

A variety of turbulence models of varying complexity have been developed over the last half-century. The underlying theoretical framework behind these models and their range of applicability are not given here, but an interested reader can find a comprehensive discussion in a review paper by Rubesin [1987].

195.4 Flow Simulation Examples

The selected flow simulation examples given here basically draw upon the author's own work and that of co-workers at the McDonnell Douglas Research Laboratories. The purpose of these examples is to illustrate the current state in computational aerodynamics for the prediction of aircraft, launch vehicle, and helicopter flow fields, from a manufacturer's point of view.

Figure 195.2 shows the surface grid on a complete twin-jet transport aircraft. Figure 195.3 shows the pressure distribution on the aircraft at a cruise Mach number of 0.76 and angle of attack of 2° obtained with the Euler code described by Deese and Agarwal [1988]. Figure 195.4 shows the comparison of the computed pressure distribution, at various spanwise locations of the wing, with experimental data. The inviscid Euler solution tends to predict shocks that are too strong and too far downstream on the wing upper surface. Addition of the viscous terms moves the shock location and strength toward

FIGURE 195.2 Surface grid over a complete twin-jet aircraft.

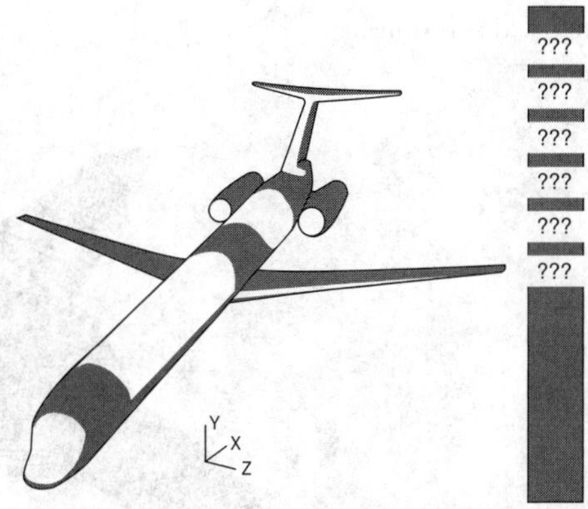

FIGURE 195.3 Euler solution predictions of surface pressure on a generic twin-jet transport aircraft; $M_\infty = 0.76$, $\alpha = 2.0°$.

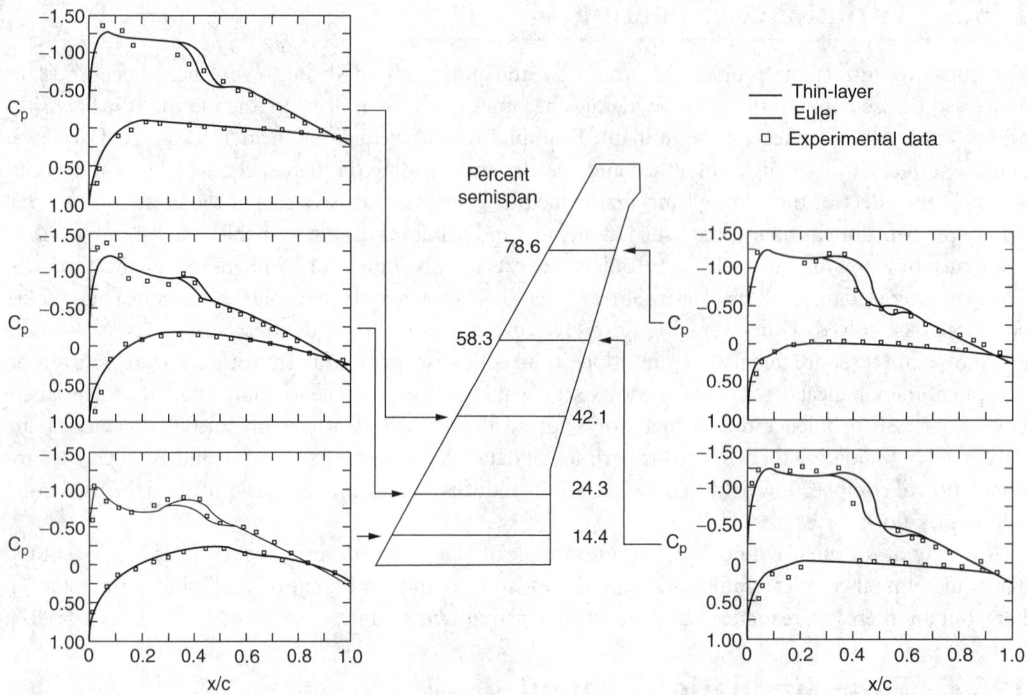

FIGURE 195.4 Comparison of Euler and thin-layer Navier–Stokes pressure distribution with experimental data for typical transport wing body at $M_\infty = 0.76$, Re = 6.39×10^6, $\alpha = 2.0°$, $160 \times 34 \times 42$ mesh.

better agreement with the experimental data, except in the region near the suction peak, where better grid resolution is needed to capture the high-flow field gradients. Figure 195.5 shows the surface grid on a complete fighter aircraft with faired inlets. Figure 195.6 shows the pressure distribution on the aircraft at a Mach number of 0.90 and an angle of attack of 4.84°, obtained with an Euler code described in Deese and Agarwal [1988]. Figure 195.7 shows the comparison of the computed and experimental pressure distributions at various spanwise locations of the fighter wing. The Euler solution tends to predict a suction peak higher than that observed experimentally. Better resolution of the relatively small-radius fighter wing should improve the agreement with data. The shocks on the wing upper surface are predicted to be stronger and slightly downstream of the measured shocks, as is typical of inviscid solutions.

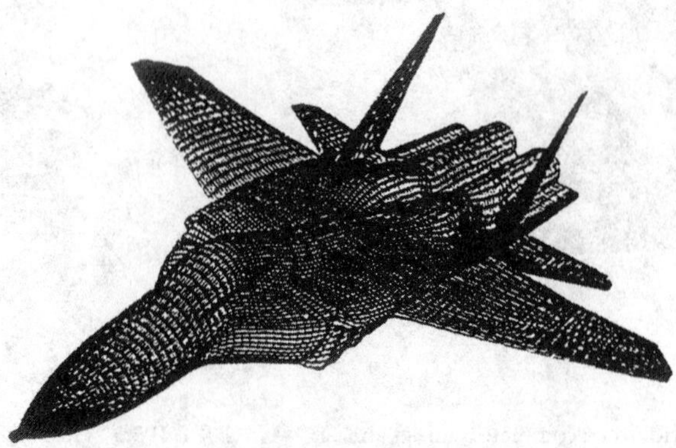

FIGURE 195.5 Surface grid on a generic fighter with faired inlets.

FIGURE 195.6 Surface pressure distribution on the generic fighter, Mach 0.9, α = 4.84°.

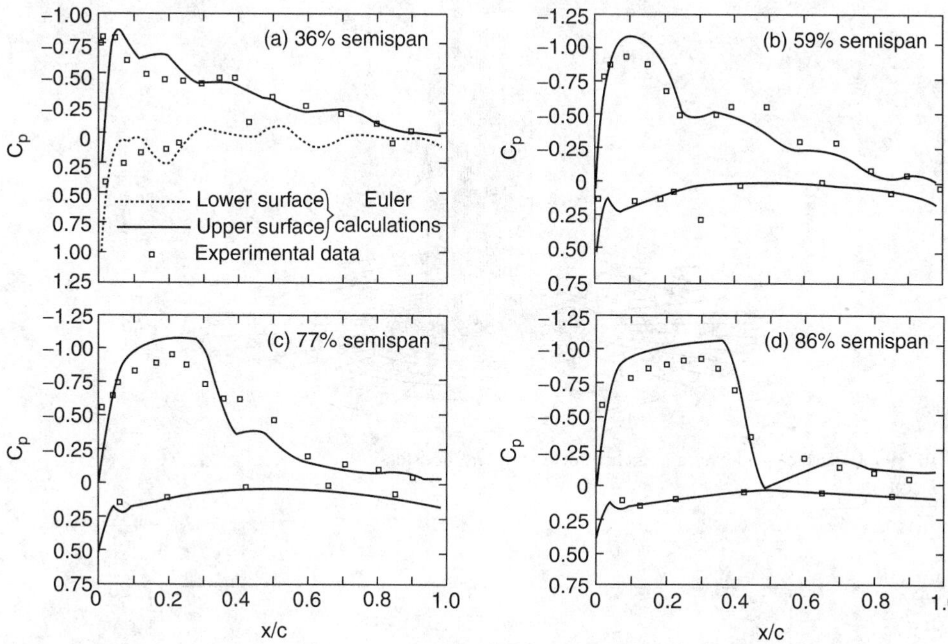

FIGURE 195.7 Surface pressure distribution on the wing of the generic fighter; $M = 0.9$, α = 4.84°.

Figure 195.8 shows the surface grid on a launch vehicle with two boosters. Figure 195.9 shows the comparison of the computed pressure distribution on the main vehicle of a two-booster configuration with experimental data. The predictions are in good agreement with the data, except for the region just upstream of the booster nose. Inclusion of viscous effects should improve the agreement between the calculations and the experimental data.

Figure 195.10 shows the grid distribution in the symmetry plane for a generic helicopter fuselage. Figure 195.11 shows the pressure distribution on an isolated helicopter fuselage obtained with the Navier–Stokes code described by Deese and Agarwal [1988]. Figure 195.12 shows the pressure distribution on an ONERA rotor blade in forward flight, obtained with the rotary-wing Euler code described by Agarwal and Deese [1987]. Figure 195.13 compares computed and experimental pressure distributions on the ONERA rotor in forward flight.

As a final example, Figure 195.14 shows the unstructured grid about an ONERA M6 wing. Figure 195.15 shows the computed pressure distribution on the wing at various spanwise locations and its

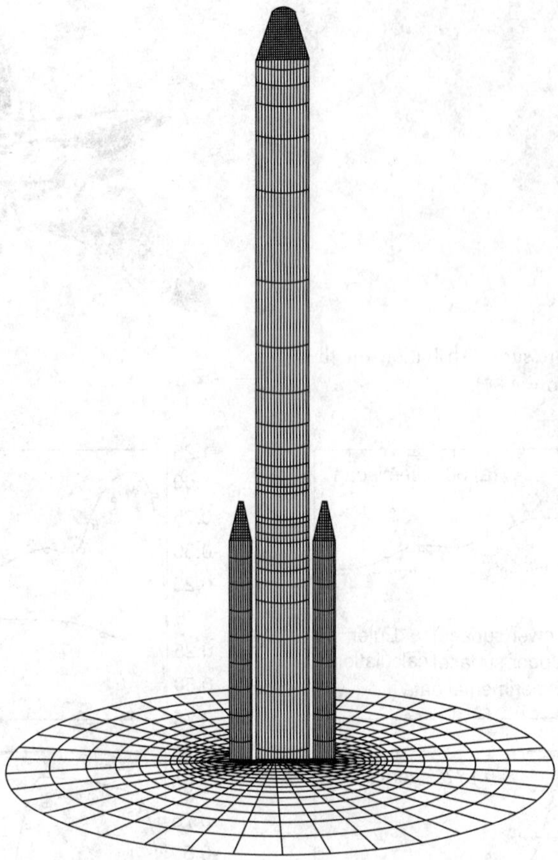

FIGURE 195.8 Surface grid over a launch vehicle with two boosters.

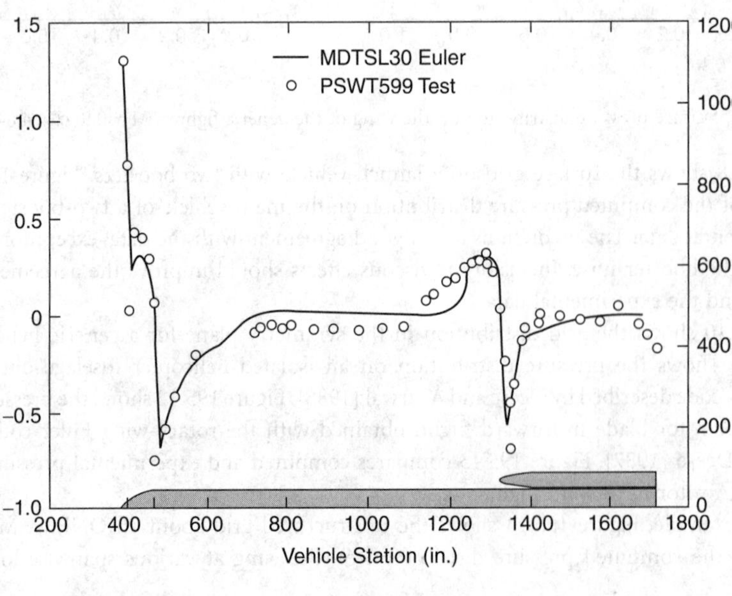

FIGURE 195.9 Surface pressure distribution on the core of a launch vehicle with two boosters; $M_\infty = 1.05$, $\alpha = 0.0$.

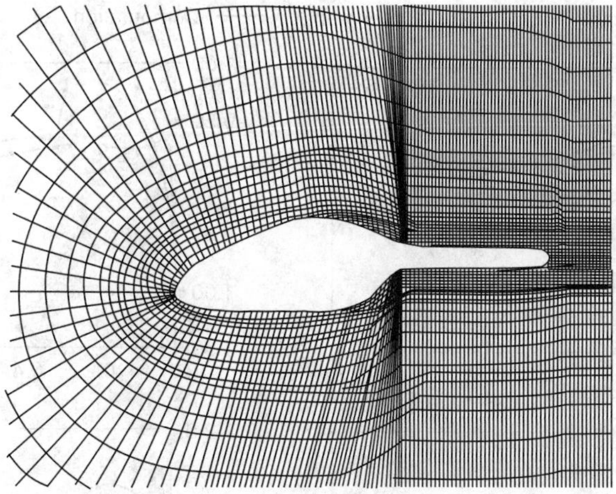

FIGURE 195.10 Grid distribution in the symmetry plane for a generic helicopter fuselage.

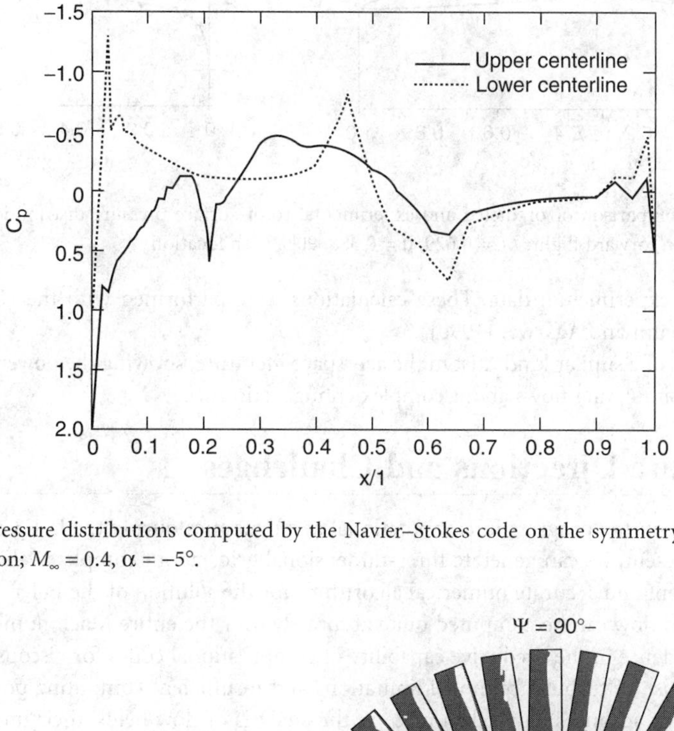

FIGURE 195.11 Pressure distributions computed by the Navier–Stokes code on the symmetry plane of a generic fuselage configuration; $M_\infty = 0.4$, $\alpha = -5°$.

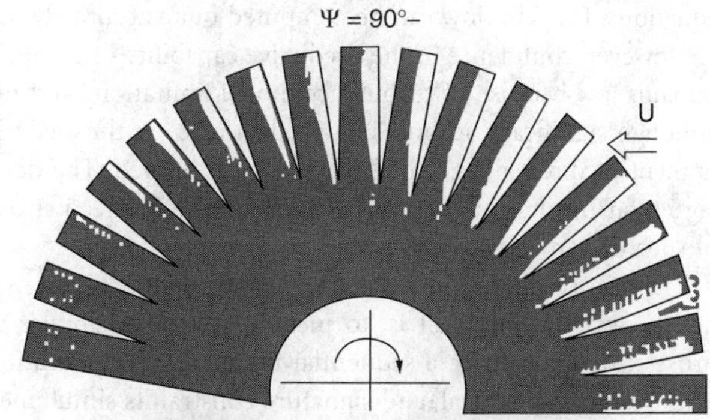

FIGURE 195.12 Euler solutions for flow about an ONERA three-bladed rotor in forward flight; $M_t = 0.629$, $\mu = 0.388$.

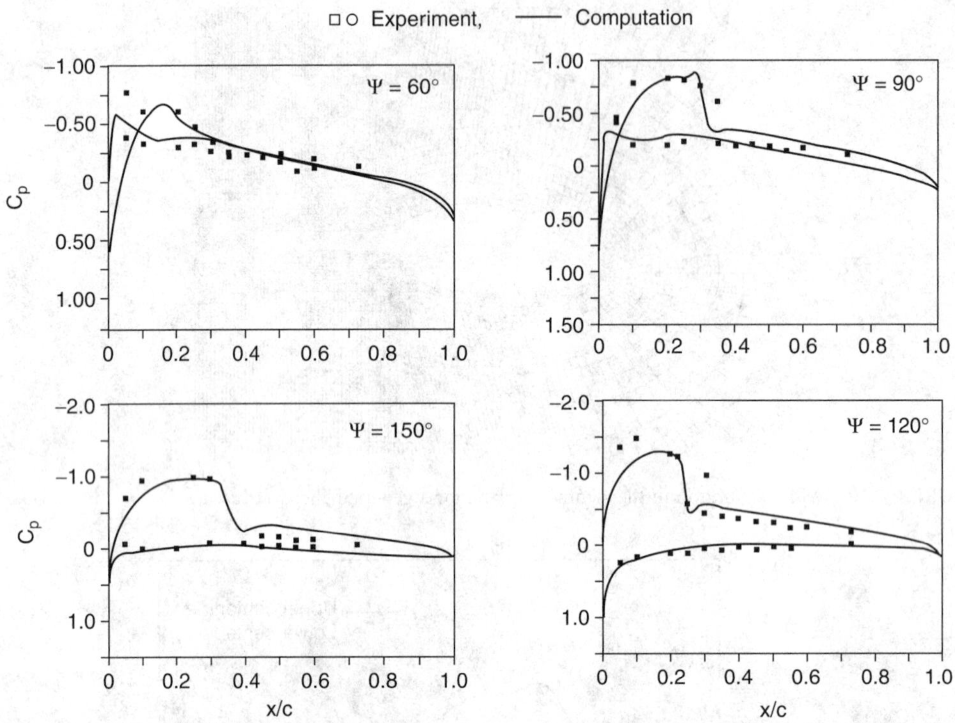

FIGURE 195.13 Comparison of predicted and experimental rotor surface pressure distributions for the ONERA three-bladed rotor in forward flight; $M_t = 0.629$, $\mu = 0.388$, 90% span location.

comparison with experimental data. These calculations were performed with the Navier–Stokes code described by Marcum and Agarwal [1990].

Many examples of a similar kind exist in the aerospace literature, showing the power of computational aerodynamics in simulating flows about complex configurations.

195.5 Future Directions and Challenges

We can look back on the progress in computational aerodynamics during the last 3 decades with great satisfaction. At present, we can generate three-dimensional grids about complete vehicle configurations and employ efficient and accurate numerical algorithms for the solution of the Euler and Navier–Stokes equations. Inviscid flow can be computed quite accurately over the entire Mach number range.

However, confidence in the predictive capability of computational codes for viscous-dominated flows remains low because of turbulence model limitations and insufficient computing power. Furthermore, although significant advances have taken place in the analysis of flow fields, their impact on the development of direct design methods has been limited. The development of inverse design techniques has lagged far behind that of analysis tools. In the future, a strong emphasis is needed on the development of such design methods.

The next computational challenge will be in developing methods and techniques for multidisciplinary design optimization, that is, to include several disciplines simultaneously in the design optimization process, rather than in a sequential manner. A configuration could be optimized taking into account aerodynamic, structural, and signature constraints simultaneously.

Finally, two of the emerging computer-related technologies, artificial intelligence and massively parallel technology, are likely to have major impacts on flight vehicle design. Expert system technology will have a substantial impact on automating the design process, thus reducing the cost and time of a design cycle.

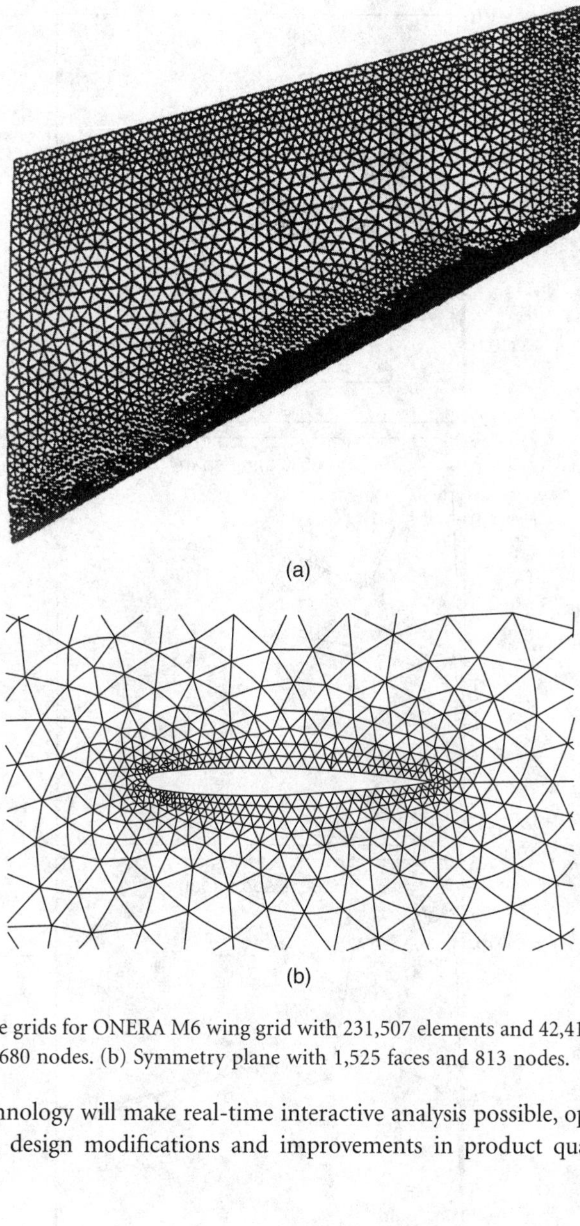

FIGURE 195.14 Surface grids for ONERA M6 wing grid with 231,507 elements and 42,410 nodes. (a) Wing surface with 15,279 faces and 7,680 nodes. (b) Symmetry plane with 1,525 faces and 813 nodes.

Massively parallel technology will make real-time interactive analysis possible, opening up great opportunities for real-time design modifications and improvements in product quality, at a substantially reduced cost.

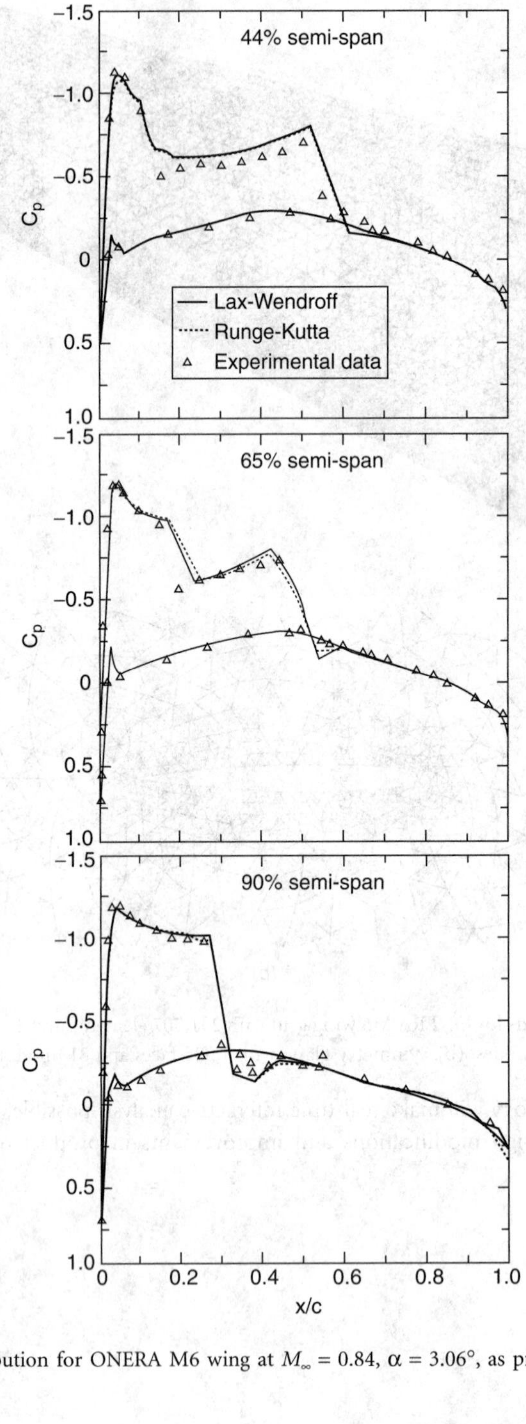

FIGURE 195.15 C_p distribution for ONERA M6 wing at $M_\infty = 0.84$, $\alpha = 3.06°$, as predicted by the unstructured grid code.

References

Agarwal, R. K. and Deese, J. E. 1987. Euler calculations for the flow field of a helicopter rotor in hover. *J. Aircraft*, 24:231–238.

Deese, J. E. and Agarwal, R. K. 1988. Navier–Stokes calculations of transonic flow about wing/body configurations. *J. Aircraft*, 25:1106–1112.

Jameson, A. 1974. Interactive solution of transonic flows over airfoils and wings, including flows at Mach 1. *Commun. Pure Appl. Math.*, 27:283–309.

Jameson, A. 1990. Full potential, Euler and Navier–Stokes schemes. In: *Applied Computational Aerodynamics: Progress in Astronautics and Aeronautics*, A. R. Seebass, Ed. American Institute of Aeronautics and Astronautics, Washington, DC, pp. 39–88.

Marcum, D. L. and Agarwal, R. K. 1990. Finite element Navier–Stokes solver for unstructured grids. *AIAA J.*, 30:648–654.

Murman, E. M. and Cole, J. D. 1971. Calculation of plane steady transonic flows. *AIAA J.*, 9:114–121.

Rubesin, M. W. 1987. *Turbulence Modeling: Supercomputing in Aerospace*. CP-254. National Aeronautics and Space Administration, Washington, DC.

Steinbrenner, J. P. and Anderson, D. A. 1990. Grid generation methodology. In: *Applied Computational Aerodynamics: Progress in Astronautics and Aeronautics*, A. R. Seebass, Ed. American Institute of Aeronautics and Astronautics, Washington, DC, pp. 91–130.

196

Aeronautical and Space Engineering Materials

Nesrin Sarigul-Klijn
University of California at Davis

196.1 System and Material Requirements

Aeronautical and space engineering materials requirements are much different than other engineering disciplines due to the exposure of flight vehicles to complex dynamic loading conditions during missions. This environment, which influences material selection, will be somewhat different depending on the operating altitude of the vehicle, as depicted in Figure 196.1. Flight systems are classified in terms of whether the system will operate only within the atmosphere ("aeronautical"), in space ("space"), or within the atmosphere as well as in space ("aerospace"). Structural failures in-flight can often result in loss of life and vehicles; hence, the engineer faces a major design requirement of a high degree of structural integrity against failure and a low weight requirement to achieve stringent performance goals. A flight vehicle structure must be designed to meet a number of conflicting requirements that include adequate strength to meet maximum expected loads with a suitable factor of safety, adequate stiffness so that deformations are kept within acceptable levels, good in-service properties such as fatigue and corrosion resistance, and good tolerance to atmospheric or space environment conditions.

Flight vehicle engineers have a wide range of materials to choose from for their systems including the commonly used ones described below. In aeronautical and space applications, materials must have both high strength-to-weight and high stiffness-to-weight ratios. Considering the degradation of mechanical properties of materials at elevated temperatures, care must be used in selecting materials for thermal environment applications. Additional properties and parameters to be considered are the strength of the material throughout the entire temperature range, fatigue, stiffness, thermal stress, toughness, stability, cost, availability, producibility, formability, and corrosion. This chapter is intended to direct readers to proper materials groups for their application and to provide typical approximate values of properties for representative materials; readers must refer to material data from the provided sources list at the end of this chapter or from material data sheets to obtain exact numerical properties to be used for their specific application.

In general, materials can be grouped as nonferrous metals, ferrous metals, plastics, ceramics, and composite materials. Although there have been new advances in materials, aluminum is still the most commonly used material for flight systems owing to its ease in forming, ready availability, and low

0-8493-1586-7/05/$0.00+$1.50

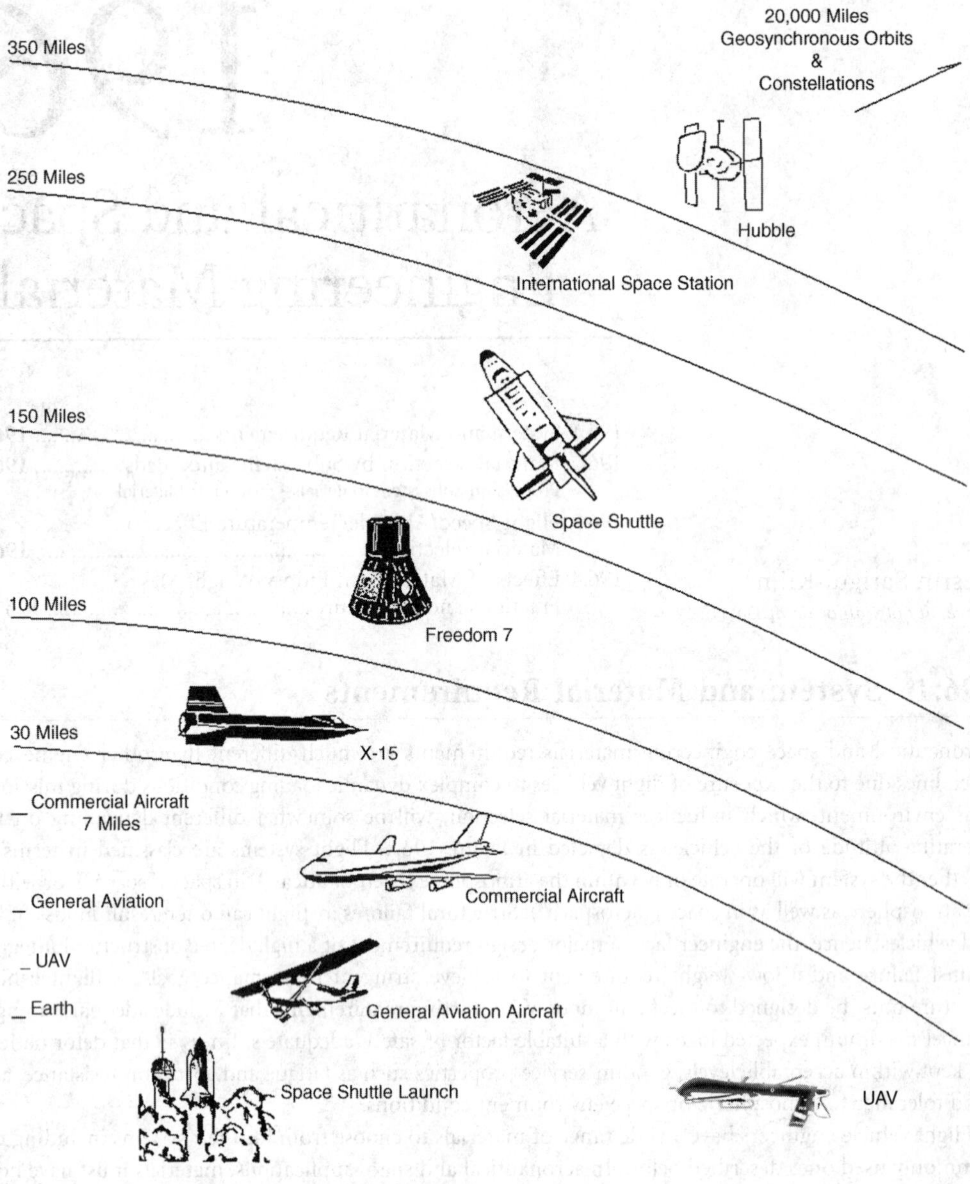

FIGURE 196.1 Aeronautical and space vehicles and their design environment within the atmosphere and beyond.

recycling cost. On the other hand, composites are preferred for their easy tailorability. Metal matrix composites and ceramic matrix composites are expected to be the next-generation advanced composite materials if their low toughness can be improved. Steel, which is a ferrous metal, is used for strength-critical applications. Titanium is used for higher-temperature applications and has good corrosion resistance. As temperature requirements increase beyond 1,500°F (815°C), nickel is used, and beyond 2,000°F (1,093°C), carbon-carbon composites and refractory metals such as beryllium and zirconium in the form of thermal coating are used. Three of the most commonly used flight system material groups are described below: light alloys, used in low-temperature applications; titanium alloys, used in high-temperature applications; and carbon fiber composites, used in higher-temperature applications.

Light alloys are usually based on aluminum and are used in the majority of aeronautical and space applications. Their primary advantages are high strength- and stiffness-to-weight ratios, relatively low cost, and a range of available fabrication processes. They depend on relatively low-temperature heat-treatment processes in obtaining desired material properties, which implies poor strength and stiffness at elevated temperatures and relatively poor fatigue properties. Heat-treatable aluminum alloys are designated as 2xxx, 6xxx, and 7xxx series based on the primary alloying elements. Heat treatment in general can greatly increase the strength of these alloys. For stiffness-critical applications, 6061-T6 is used owing to its weldability and low cost. The most common 2xxx-series alloy used in aerospace is the 2219. Retrogression, a relatively new type of heat treatment is applied to the 7xxx series to improve the combination of strength and stress-cracking resistance. Identified as the T77xx, these alloys are used in primary structures, such as the Boeing 777 upper-wing skin plate, as spar and stringers materials of 7055-T77 and 7150-T77. The higher-strength alloys in general are poorer in fatigue, and their use is restricted to compression-sensitive areas. Tensile loads tend to produce more fatigue problems. The maximum temperature at which light alloys can be used for continuous operations is about 266°F (130°C). This temperature value is what occurs when a flight vehicle is traveling at twice the speed of sound due to kinetic heating when allowance is made for heat dissipation by the structure. Aluminum alloys are often thought of as corrosion resistant, but aluminum is a reactive metal and is susceptible to various forms of corrosion. In order to improve resistance to corrosion, aluminum alloys need to be coated or anodized. Despite some of the abovementioned limitations, aeronautical and space systems that do not use light alloys are rare.

A more recent light alloy that offers a lower-density, higher modulus of elasticity and higher strength than the 2xxx, 6xxx, and 7xxx aluminum alloy series is aluminum-lithium. Alcoa develops the most mature form, and properties are available in sources given at the end of this chapter. The disadvantages of these newer alloys are welding difficulty and directional dependence in modulus. Known applications of aluminum-lithium alloys include the Centour upper stage and payload adaptors for the Titan-IV launch vehicle.

Magnesium alloys, on the other hand, are about two-thirds as dense as aluminum alloys and have similar strength-to-weight ratios. They have poor resistance to corrosion and low wear resistance. The biggest disadvantage is that magnesium is not formable at room temperature.

Beryllium is highly attractive for stiffness-critical structures owing to its very high specific stiffness, which is the ratio of elastic modulus to density, with a value on the order of six times greater than most metals. In addition, it has very low coefficient of thermal expansion and high thermal conductivity. Beryllium in powder form becomes toxic, and it has low fracture toughness. It is commonly used in space structures.

Titanium alloys have greater density than light alloys, but they have higher strength-to-weight and stiffness-to-weight ratios and they maintain these values even at elevated temperatures. Despite their inherently high cost and difficulty to manufacture, the use of titanium alloys has become common especially when weight reduction is required, or strength is required at temperatures higher than that can be sustained by aluminum alloys or polymer composites. The wide-chord fan blade, full moving stabilizer surfaces, and internal structure of combat aircraft provide good examples of titanium alloy applications. Their biggest use is in engine structures for discs, blades, shafts, high-pressure compressors, and nozzle assemblies. Titanium alloy comprises almost 10% of the Boeing 777 aircraft's weight. Ti-6Al-4V is the most general-purpose, high-strength alloy used in aerospace applications.

Carbon fiber composites have high specific stiffness, specific strength, and excellent resistance to fatigue when compared with metallic alloys. Due to their high cost, composites are not replacing metals on commercial aircraft, and the focus on technical developments is changing to find cheaper methods of manufacture with less emphasis on improving properties. Carbon fiber composites have excellent fatigue properties, although these vary depending on the fiber lay-up, matrix type, and loading pattern. Composite matrices include polymers (epoxies), metals, carbon, and ceramics. In advanced composites, the fibers provides the strength and stiffness, and the matrix (resin) holds the fibers together as well as transferring loads to allow the composite to act as a whole. Advanced composites can be categorized with respect to their matrix types as polymer matrix composites (PMCs), metal matrix composites (MMCs), intermetallic matrix composites (IMCs), and ceramic matrix composites (CMCs). The PMCs are further grouped according to their polymer matrices as thermosets (epoxy) or thermoplastics (*polyethylene*).

TABLE 196.1 Specific Modulus and Strength of Typical Fiber Composites and Bulk Metals

Material	Specific Modulus Msi-in³/lb (GPa-m³/kg)	Specific Strength Ksi-in³/lb (MPa-m³/kg)
Graphite	513 (0.13)	4610 (1.15)
Glass	137 (0.034)	2489 (0.62)
Unidirectional graphite/epoxy	454 (0.113)	3764 (0.94)
Unidirectional glass/epoxy	86 (0.021)	2368 (0.59)
Cross-ply graphite/epoxy	240 (0.06)	936 (0.233)
Cross-ply glass/epoxy	53 (0.013)	197 (0.04)
Steel	107 (0.026)	334 (0.083)
Aluminum	107 (0.026)	426 (0.11)

TABLE 196.2 Density, Elastic Modulus, and Poisson's Ratio Rough Values of Common Aerospace Materials (Room Temperature)

Material	Density lb/ft³ (kg/m³)	Elastic Modulus lb/in² (MPa)	Poisson's Ratio
Aluminum alloy	173 (2771)	10×10^6 (69×10^3)	0.33
Titanium alloy	276 (4422)	16×10^6 (110×10^3)	0.34
Copper	556 (8907)	15×10^6 (103×10^3)	0.34
Nickel	556 (8907)	29×10^6 (204×10^3)	0.31
Steel	480 (7689)	30×10^6 (207×10^3)	0.30
Pyrex	144 (2307)	6.2 lb/ft³	0.24
Quartz	160 (2563)	7.9 lb/ft³	0.33
Rubber	1100–950 lb/ft³	0.23–0.0005 lb/ft³	0.4–0.5
Cork	240 (3845)	—	—
Wood	38 (609)	Grain direction: 1.5×10^6 (11×10^3) Orthogonal to grain direction: 0.1×10^6 (0.6×10^3)	—
Carbon fiber — epoxy matrix (60% volume fraction carbon fiber)	105 (1682)	Longitudinal: 32×10^6 (220×10^3) Transverse: 1×10^6 (6.9×10^3)	0.25

The strength- and stiffness-to-weight ratios of composite materials can best be illustrated by "specific" properties. These are simply the result of dividing the mechanical properties of a material by its density. Generally, there is a spread of specific properties and it must be taken into account in analysis. The specific tensile strength and specific tensile modulus values allow the engineer to compare materials on the basis of mechanical properties per unit density. The higher the number, the better the material is for the structural efficiency and the lower the weight. Specific modulus and strength values of the most common aerospace composite and metal materials are presented in Table 196.1. For ready comparison, Table 196.2 lists a wider range of materials with their density, elastic modulus, and Poisson's ratio at room temperature.

196.2 Material Selection by Subsystems and Loads

The driving force behind the advancement of aviation has always been the desire to fly higher and faster. Within the atmosphere, the motion of vehicles takes place under the influence of gravity, aerodynamic forces, and possibly some type of air-breathing propulsion force. In space, however, the mode of propulsion must be entirely independent of air for its thrust. Therefore, system loading conditions must be predicted correctly. In general, loads acting on the flight systems are complex and include static, dynamic,

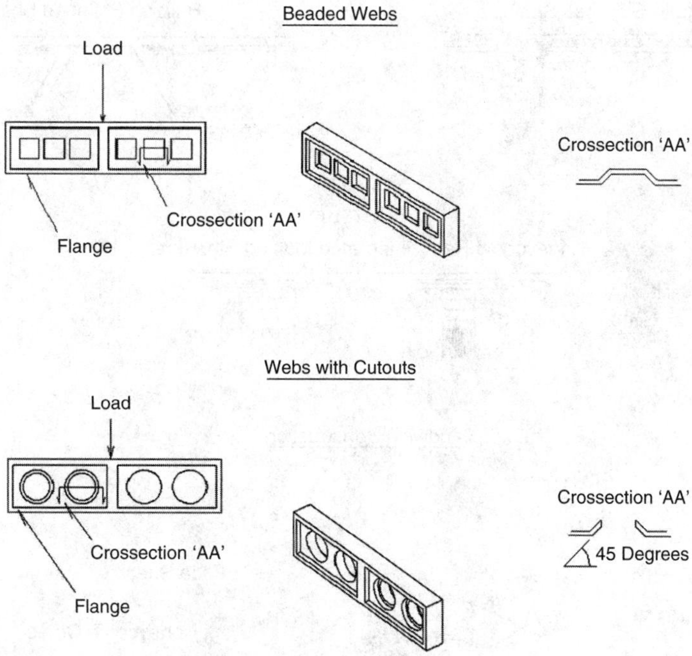

FIGURE 196.2 Lightweight aerospace primary structural members. (a) Web forms with beads and cutouts. (b) Skin shells integrated, built-up and honeycomb forms. *Continued on facing page.*

and repeated (cyclic) types, and may act under a wide range of temperatures. Therefore, the engineer needs to understand the behavior of materials under complex loading. For space transportation vehicles, the loads could even get more complicated, which introduces new constraints to the design. For example, for space transportation vehicles, additional loading requirements are imposed to avoid coupling of characteristic launch vehicle frequencies to critical spacecraft control system frequencies.

The load-carrying structural elements of flight vehicles are the "primary structure." The other structures such as control surfaces are called the "secondary structure." Flight vehicle structural forms are evolved from the need for minimum-weight design to achieve stringent performance goals. As new materials are introduced, structural forms change to take full advantage of the materials' load-carrying characteristics, thereby reducing weight. Lightweight structural forms are depicted in Figure 196.2 for both metal and composite materials. Figure 196.2(a) shows two lightweight web forms, and Figure 196.2(b) details various lightweight shells, including an internally stiffened shell to reduce part count, a built-up, a honeycomb-resin-integrated molded structure, and a sandwich construction for high dynamic-pressure applications. An all-metal, lifting-surface structural form partially skinned to show cutouts on the webs and a carbon fiber/epoxy matrix composite, skinned foam-core wing, lightweight lifting-surface structural form are depicted in Figure 196.3. In a weight-reduction effort at the Scaled Model Aerospace Research and Testing Laboratory (SMARTLAB) at the University of California-Davis, kevlar, carbon, and kevlar/carbon fibers and epoxy matrix materials are used to manufacture various fuselages for an uninhabited aerospace vehicle (UAV) (Figure 196.4). This UAV was flight tested and successfully carried 5.5 times its empty weight as payload. The lightweight structure was achieved using aeroelastic tailoring and manufacturing techniques for advanced composites in design.

The flight vehicle design process starts with an overall configuration that meets the defined vehicle-fairing geometry constraints. Using the acceleration load factors and safety factors applied to the estimated limit loads, the structure is sized using the structural properties for a given material. It is then necessary to estimate the stiffness characteristics of the structure and compare these with minimum requirements. If the stiffness is inadequate, the structure is resized to increase cross-section thickness or inertia, or stiffer materials are applied, and the analysis is repeated. Advanced methods such as aeroelastic tailoring

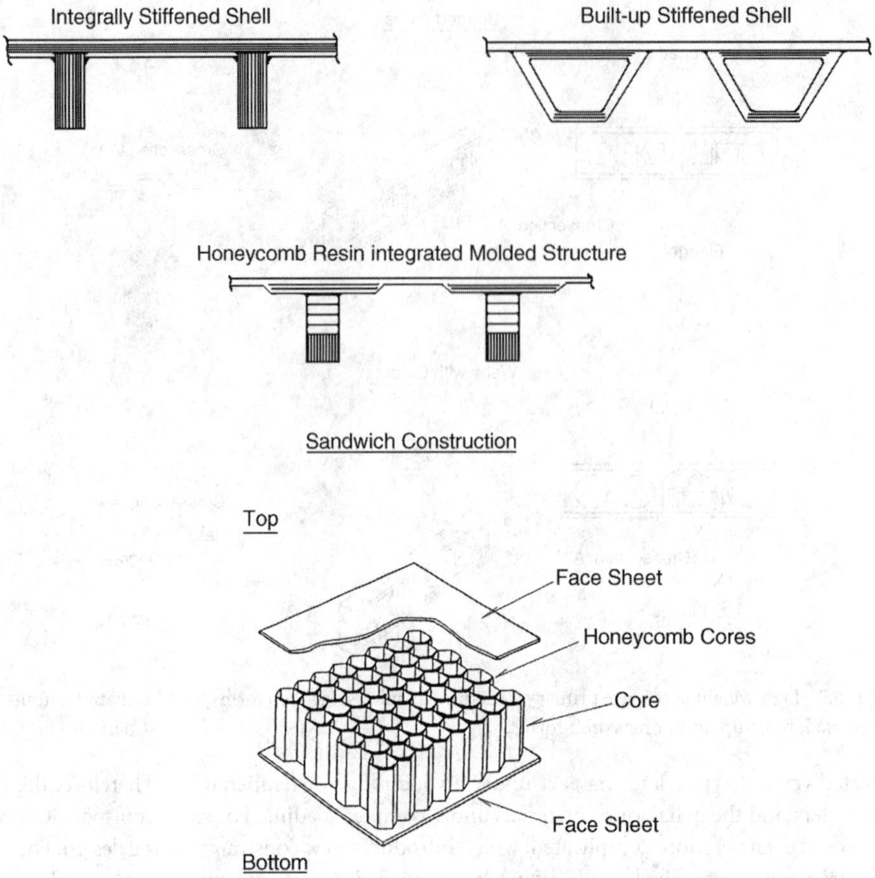

FIGURE 196.2 *Continued.*

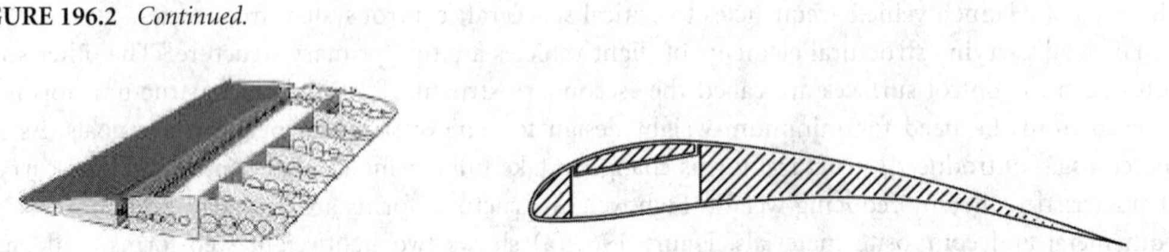

FIGURE 196.3 Representative lifting surfaces. (a) Metal partially skinned showing lightening cutouts on the rib webs. (b) Composite wing cross-section with foam core and 80/20 carbon fiber/epoxy skin.

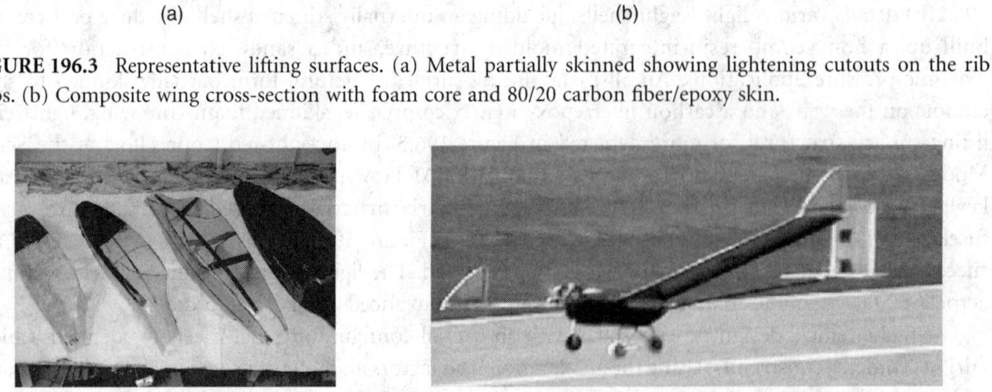

FIGURE 196.4 Fuselages using kevlar/carbon/glass fibers, epoxy matrix material for an uninhabited aerospace vehicle. (Courtesy of Sarigul-Klijn, N., SMARTLAB, University of California-Davis, Davis, CA.)

TABLE 196.3 Spacecraft Subsystem Temperature Requirements

Subsystem	Temperatures (°F)	Temperatures (°C)
Solar panels	–200 to +170	–129 to 76
Batteries	30 to 50	–1 to 10
Electronics	Zero to 140	–17 to 60
Propulsion tanks	40 to 120	4 to 48
Optical systems	70	21

to take into account the effects of stiffness together with strength in the preliminary design phase are new methods used for lightweight designs. Limit loads to be expected during each critical period are based on rational predictions with allowance for statistical variation when possible. The factor of safety is an arbitrary number on type and use of structure to account for variations in materials properties from batch to batch, manufacturing quality, load redistribution, and possible strength degradation due to service treatment. Factor of safety values vary between 1.1 to 1.5, depending on whether the vehicle system is going to carry humans or not and whether it is for commercial or military usage.

In addition to mechanical loads, the structure should resist vibration and thermal effects without excessive deformations. As an example, the typical temperature environment for spacecraft by subcomponent is listed in Table 196.3. Stiffness requirements for launch vehicles include fundamental frequencies required being above minimum values to avoid excessive acceleration during launch and ascent. In addition, the deployed frequencies in space must be acceptable for spacecraft control system requirements.

In the process of the optimum material selection, parameters that must be considered are payload type, atmospheric or space environment, speed, and life cycle, measured in terms of flight hours or years. Once the constraints are determined, then the designers must consider the tradeoff between strength and toughness when selecting alloys and their heat treatments. For minimum weight, high specific strength is desired. For damage tolerance, high toughness is needed; however, as strength increases, usually toughness decreases. The best engineering material is strong enough and tough enough, and provides the minimum weight design at lowest cost. Depending on the operating environment and the system and design requirements, one property or another may be critical, and the material choice is made accordingly. Tables 196.3 and 196.4 provide typical properties of metals and composites commonly used in aeronautical and space applications.

Propulsion Subsystem Materials

All flight vehicle systems except gliders and balloons require a propulsion system for sustaining flight. Propulsion systems have different material requirements depending on whether air breathing, combined cycle, or rocket propulsion types exist. Among propulsion subsystem elements, material selection for blades will be focused for propeller blade and compressor blade cases.

A propeller blade of a piston-engine aircraft experiences vibratory excitations. As a result of vibratory excitation, the blade has to be designed in such a way as to ensure that no natural frequencies are excited in the propeller operating range. The first generation of propeller-driven aircraft used blades made by the natural composites (wood). However, the propeller blade is in many ways a very suitable application for carbon fiber composites. Composites are a good choice as blade materials owing to their tailorability in lay-up and fiber orientations to avoid harmful resonance conditions without significantly impacting the strength or weight of the blades. In addition to the structural requirements of the blades, there are environmental requirements to resist erosion from foreign objects such as stones during ground operations.

Aluminum alloy blades have to be repaired frequently to remove notches, which can significantly reduce fatigue life. Composite blades, however, when suitably treated, require little maintenance, are easy to repair, and last significantly longer than metal blades. For example, aluminum blades are worn after only 10,000 flight hours, while composite blades can last more than three times longer. The cost of ownership of composite blades is therefore lower and the life-cycle cost of running turbo pumps can be significantly reduced.

Advanced high-temperature propulsion materials applications include PMCs for fans; MMCs, IMCs, and superalloys for compressor and turbine materials; and CMCs for turbine applications.

Spacecraft Materials

Prior to selection of materials for spacecraft applications, their strength, corrosion resistance, electrochemical interaction, toxicity, flammability and outgassing must be considered. The latter four properties of toxicity, odor flammability, and outgassing properties are specific to human flight missions and relate to safety aspects. Outgassing is one of the major problems associated with using materials in a vacuum. One example criteria is that its outgassing shall not exceed 1% of its total mass.

All materials, whether structural or not, are flight qualified through extreme tests before commitment to production. For example, interior materials for commercial aircraft must satisfy federal requirements for low smoke and toxic emissions in a fire. In addition, these materials must be durable and cleanable. Cockpit canopies or windshields must meet standards for bird strikes, and compliance is demonstrated by analysis and testing. Wiring and electronic components must meet standards for reliability in the service environment. Paints must be environmentally friendly and strippable to enable vehicle refurbishment without damage to the basic structure. Sealants must have good in-service properties for application, and must perform in the service environment for the design life of the system.

196.3 Flight Speed/Altitude/Temperature Effects on Material Selection

The speed of aerospace vehicles has increased over time. The speed of sound in gases depends on the nature of the particular gas and its temperature. For air at standard sea level under standard temperatures, the speed of sound is 340 m/sec or 1,115 ft/sec or 1,224 km/h or 761 mph. In the stratosphere, where the airliners cruise, the speed of sound is 1,060 km/h (659 mph). Current transport aircraft, with the exception of the Concord, cruise at a speed below the speed of sound in stratosphere. The flight vehicle speed divided by the speed of sound is defined as the Mach number, M. Most airliners then cruise at speeds of $M < 1$. A typical flight vehicle traveling twice the speed of sound would see an equilibrium upper-surface skin temperature of 130°C and a nose temperature of 160°C, as sketched in Figure 196.5. The material choice is greatly affected by dynamic pressure, which is one-half the air density, ρ times the flight speed, U, squared:

$$q = \frac{1}{2}(\rho U^2)$$

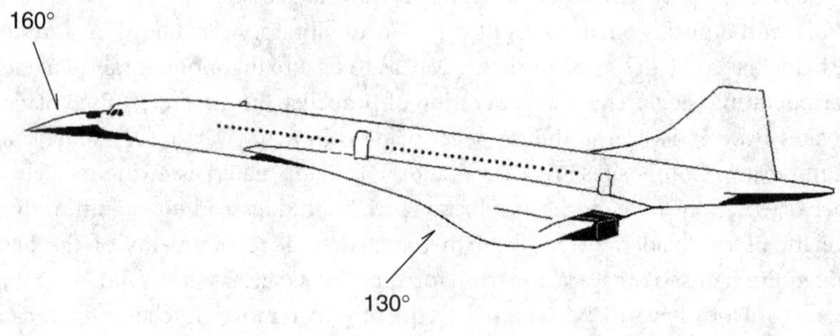

FIGURE 196.5 Representative high-speed aerospace structure's equilibrium skin temperatures at supersonic flight.

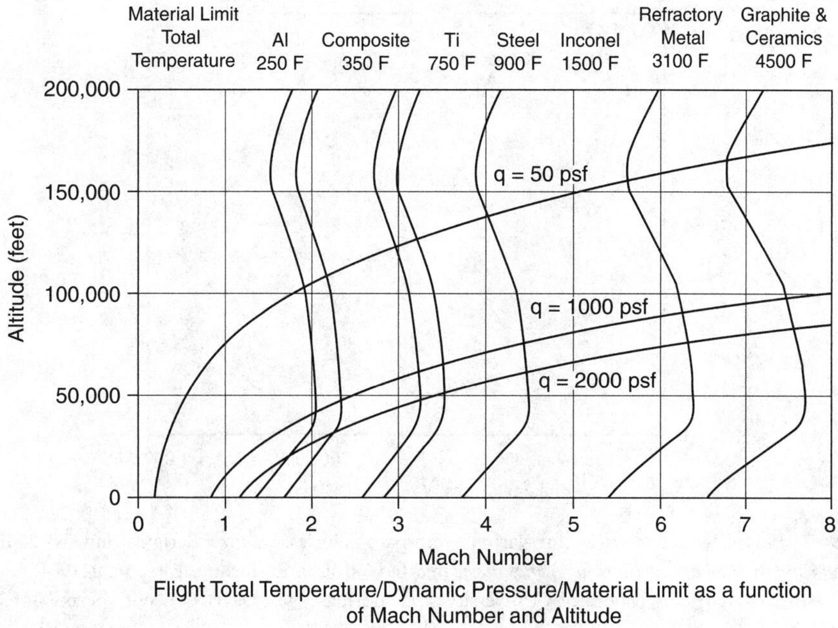

FIGURE 196.6 Flight total temperature and dynamic pressure material limit as a function of mach number and altitude. (*Source:* Sarigul-Klijn, N. 2002. *Material Corridors and Empty Weight Fractions for Atmospheric and Orbital Vehicles*, Technical Report 021001. Scaled Model Aerospace Research and Testing Laboratory, University of California-Davis, Davis, CA. With permission.)

As the flight Mach number and altitude change, the values of total temperature and dynamic pressure change and will limit the types of materials available for that application. Figure 196.6 shows material limits for aerospace systems. In this figure, the total temperature, and dynamic pressure limits are plotted as a function of Mach number and altitude. As seen in Figure 196.6, heating at high Mach numbers and altitudes can generate extreme skin temperatures. Unfortunately, materials able to withstand high temperatures and flight loads for extended periods of time are not presently available, especially at low densities. Therefore, a commercial hypersonic vehicle to carry passengers has not yet been developed.

Aerodynamically heated structures are one of the very recent challenges to the field of engineering design and materials research and development. There are many materials that have application to the design of aerodynamically heated structures. Titanium and high-strength steel are the first two candidate materials. The third high-strength material for consideration in primary structures subject to aerodynamic heating is aluminum. The fourth is beryllium and it has poor ductility and high cost, followed by molybdenum, and columbium materials combined with proper coating. They must be coated to be satisfactory in service. Heat-sink materials should have higher thermal conductivity, high specific heat and melting, and for aerospace applications they must have low specific gravity. Metals can resist temperatures higher than their melting points by convection cooling, which is heat exchange with a coolant. Metal alloys suitable for elevated temperatures are usually nickel and cobalt based, as they exhibit high strength and modulus and maintain their properties at high temperatures. Densities are similar to steel values. The most commonly used elevated-temperature alloys are A-286, Inconel 718, and Invar.

In extreme heat conditions such as during the reentry phase, ablative materials are used. The heat is carried off with material removal. The less material is lost, the more efficient the ablative material. Ablative materials must have a low thermal conductivity so that the heat may remain concentrated in the thin surface layer.

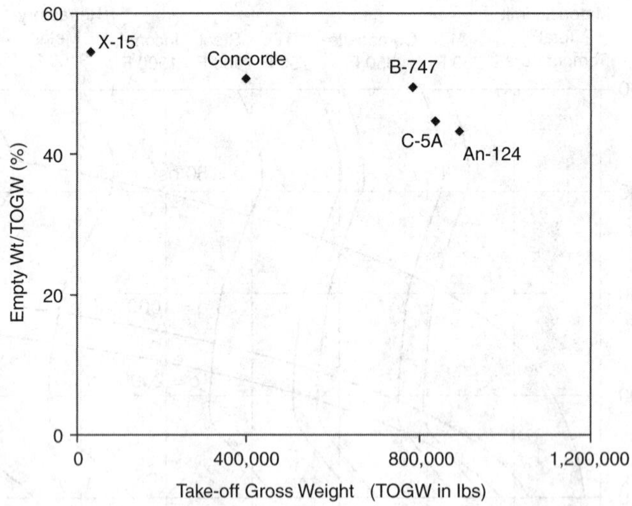

FIGURE 196.7 Empty weight fractions for various aerospace vehicles. (*Source:* Sarigul-Klijn, N. 2002. *Material Corridors and Empty Weight Fractions for Atmospheric and Orbital Vehicles*, Technical Report 021001. Scaled Model Aerospace Research and Testing Laboratory, University of California-Davis, Davis, CA. With permission.)

196.4 Effects of Materials on Empty-Weight Mass Fractions and the Future

Tremendous progress in aeronatics, space, and aerospace has been made in the 100 years since the first flight of the Wright Flyer in 1903. The aerospace systems of the future must be affordable and even lighter in weight in order to go beyond Planet Earth.

Potential materials and process contributions to affordability include lower costs for raw materials, fabrication, assembly, and maintenance. Innovative design concepts and materials and process improvements will help achieve the goal of more affordable systems.

The empty-weight to maximum gross-weight ratios of aeronautical and space vehicles are critical parameters in their performance and material selection plays a big role in this value. The empty-weight fraction for a flight system is defined as the empty weight, We, divided by the take-off gross weight, Wo. It differs from the dry mass fraction in that it does not include payload since payloads are on the order of 5% to 35% of take-off gross weight. The take-off gross weight (or the maximum gross weight) includes the aircraft's empty weight, aircraft fuel, and payload.

$$\text{Empty weight fraction} = \text{Empty weight/Maximum gross weight}$$

The empty-weight fractions versus maximum take-off gross weight for various current aeronautical and space vehicles are plotted in Figure 196.7. Compared to the wide variation in the maximum gross-weight values of the current fleet of aeronautical and space vehicles, their empty-weight fractions are much less variable. The maximum gross weight of a Boeing 747 is about 850,000 lbs; the same value for the space shuttle system is 4.5 million lbs, of which 80% is fuel and oxidizer. The X-15, on the other hand, carried propellant equal to 55% of its launch mass. This value is similar to other supersonic aircraft, such as the Concorde.

Increased concern for the environment has had a significant influence on how materials are produced, processed, and disposed of. More environment-friendly materials that meet all performance requirements are replacing some materials. The downward trend of toxic chemical release into the environment is expected to continue. The American Society for Testing and Materials (ASTM) has developed standard test-specimen geometries, and the test techniques for materials and material data provided in this chapter

TABLE 196.4 Room Temperature and Elevated Temperature Approximate Properties for Metals Commonly Used in Aeronautical and Space Applications

Material	Density lb/in^3 (g/cm^3)	Modulus of Elasticity 10^3 [ksi (Mpa)]	Ultimate Tensile Stress ksi (Mpa)	Compressive Yield Stress ksi (Mpa)	Fracture Toughness ksi$\sqrt{in}$ (mPa $\sqrt{cm}$)	Coefficient of Thermal Expansion $F^{-1} \times 10^{-6}$ ($C^{-1} \times 10^{-6}$)	Poisson's Ratio	Elongation (%)
Aluminum								
2219-T8511 Sheet	0.103 (2.85)	10.6 (73.1)	58 (400)	43 (300)	33 (360)	12.3 (22.1)	0.33	6
6061-T62 Plate	0.098 (2.71)	10 (69)	42 (290)	35 (340)	26 (290)	12.7 (22.9)	0.33	6
7075-T73 Sheet	0.101 (2.80)	10.4 (71.4)	67 (460)	55 (380)	36 (400)	12.3 (22.1)	0.33	10
Steel								
17-4PH H1150 Bar	0.284 (7.86)	29.2 (202)	125 (862)	90 (620)	90 (990)	6.2 (1.11)	0.27	16
A286 Bar	0.287 (7.95)	29.1 (201)	130 (896)	85 (590)	100 (1100)	9.1 (16.4)	0.31	15
Titanium								
Ti-6Al-4V Bar	0.16 (4.43)	16 (110)	134 (923)	131 (903)	85 (930)	4.7 (8.5)	0.31	10
Ti-5Al-2.5Sn Sheet	0.162 (4.49)	15.3 (107)	123 (848)	118 (814)	65 (710)	5.2 (9.4)	—	10
Magnesium and Beryllium								
Mg-AZ31B-H24 Sheet	0.064 (1.77)	6.5 (44.8)	39 (270)	24 (170)	20 (220)	14.1 (25.4)	0.35	8
Beryllium Plate	0.067 (1.86)	42.5 (293)	65 (450)	—	9.5 (104)	6.4 (11.5)	—	—
Inconel and Invar								
Inconel 718 Sheet	0.297 (8.22)	29.4 (203)	180 (1240)	155 (1070)	85 (930)	12.8 (23.0)	0.29	12
Invar	0.291 (8.06)	21 (145)	90 (620)	—	—	1.1 (2.0)	0.29	—

TABLE 196.5 Properties for Representative Composite Materials Commonly Used in Aeronautical and Space Applications

	Range of Mechanical Properties						
Material Fiber/Matrix	Specific Gravity	Modulus of Elasticity Msi (GPa)	Ultimate Tensile Strength Ksi (Mpa)	Coefficient of Thermal Expansion μ in/in/°F (μm/m/°C)	Service Temperature Limits °F (°C)	Composite Type
Graphite/Epoxy	1.6	26 (181)	181 (1500)	0.0111 (0.02)	<750 (400)	PMC
Glass/Epoxy	1.8	6 (39)	154 (1062)	5 (9)	<750 (400)	PMC
Silicon-Carbide/Aluminum (SiC/Al)	2.6	17 (117)	175 (1200)	7 (12)	<1800 (1000)	MMC
Graphite/Aluminum	2.2	18 (124)	65 (448)	10 (18)	<1800 (1000)	MMC
Silicon-Carbide/Calcium-Alumino-Silicate (SiC/CAS)	2.5	17 (121)	58 (400)	2.5 (4.5)	<2700 (1500)	CMC
Carbon/Carbon	1.68	2 (13)	5 (36)	1 (2)	<6000 (3315)	CCC

Note: Density depends on volume fraction and density of constituents. Design strengths and modulus depend on constituent property values, fiber volume fraction, and fiber orientation.

largely used their published values. Underlying principles of materials testing include assurances that test techniques do not influence test results, that test data are reproducible, and that the tests are in fact providing reliable information that can be used in making sound engineering design decisions.

In addition, technological availability of a material for a specific mission is important. A good example is the materials factors that have been influencing the pace of developing human space vehicles. Space launch vehicles have empty weight fractions of 5 to 20% of their maximum gross weight. Such low fractions are acheived by eliminating wings, landing gear, and aerodynamic control surfaces. Designers should be cautioned on assuming that future materials will show significant improvements in empty weight fractions.

Historical data on structural material weight percentages for aeronautical vehicles show that the empty-weight fraction values achieved varied from 60 to 40%, covering the last 100 years of flight vehicle design improvements and material-related developments. There has been significant progress in the understanding of fundamental materials behavior, enabling improved prediction of materials performance in service. Research must continue to ensure higher standards of excellence in aerospace materials together with improved specific properties.

Defining Terms

Ablative material — Materials used for dissipation of heat by mass removal.

Advanced composite material — Structural material made by combining high-performance reinforcements in a matrix material such as epoxy or metal.

Aeroelastic tailoring — Use of stiffness constraints together with strength constraints in early stages of a design to achieve minimum-weight components.

Brittleness — Property of breaking without noticeable warning or deformation.

Creep — Slow deformation of a material under stress that results in permanent change in geometry.

Damping — Dissipation of energy during oscillations.

Elastic limit — Maximum stress level that a material can withstand without deforming permanently.

Factor of safety — Ratio of ultimate strength to the maximum design stress.

Fatigue strength — Load-carrying ability without failure of a material subjected to a loading that is repeated a definite number of times.

Material allowable — Statistically derived material-mechanical property from coupon tests that is used in developing design allowable. An A-basis material allowable is a value that is equaled or exceeded by 99% of the population of values with a confidence of 95%. A B-basis material allowable is a value that is equaled or exceeded by 90% of the population of values with a confidence of 95%.

Natural frequency — Frequency in cycles per second (Hz) at which a system will oscillate when excited by a time-varying load and then the load is removed.

Stiffness — Measure of load that is required to cause a unit deformation.

Strength — Amount of load that a structure can carry without collapse or extreme deformation.

Toughness — Relative degree of resistance to impact loads without fracture.

Ultimate strength — Stress level that causes fracture of material that is calculated based on maximum value of the force and the undeformed cross-section area.

Reference

Sarigul-Klijn, N. 2002. *Material Corridors and Empty Weight Fractions for Atmospheric and Orbital Vehicles*, Technical Report 021001. Scaled Model Aerospace Research and Testing Laboratory, University of California-Davis, Davis, CA.

Further Information

American Society for Metals International, *Engineered Materials Handbook* (ASM-HDBK), Vols. 1 and 2, ASM International, Materials Park, OH.

American Society for Metals International, ASM Specialty-HDBK *Heat-Resistant Materials,* Davis, J. R., Ed., 1997.

American Society for Testing and Materials (ASTM) is one of the largest nonprofit voluntary standards development organizations in the world.

Federal Aviation Regulations (FAR) provides standards that commercial airplanes must satisfy. Parts 23/25/27/29/31/33/35/39.

MIL-STD-1530. Defines the Air Force Structural Integrity Program (ASIP) to ensure structural integrity of USAF airplanes.

Handbook of Metallic Materials (MIL-HDBK 5). Provides standardized design values and related design information for metallic materials and structural elements used in aerospace structures.

The Composite Materials Handbook (MIL-HDBK 17). 2002. Provides property data on current and emerging polymeric matrix composite materials.

197

Propulsion Systems

Jan C. Monk
National Aeronautics and Space Administration

Rocket propulsion is an application of Newton's first, second, and third laws of motion. Newton's first law of motion states that a particle not subjected to external forces remains at rest or moves with constant velocity in a straight line. A rocket lifting off the launch pad goes from a state of rest to a state of motion. Newton's second law of motion states that force equals mass times acceleration. Force in the equation is the rocket thrust, where mass is the amount of rocket fuel being burned and converted into gas, which expands and then escapes from the rocket. As the gas exits the combustion chamber through a nozzle, it picks up speed. Newton's third law of motion states that for every action, there is an equal and opposite reaction. With rockets, the action is the expelling of gas out of the engine; the reaction is the force or thrust of the rocket in the opposite direction.

197.1 Performance Characteristics

In the process of producing thrust, rocket engines generate more power per unit weight than any other engine. To enable a rocket to climb into low-Earth orbit, it is necessary to achieve velocities in excess of 28,000 kph. Escape velocity is a speed of about 40,250 kph. To achieve these velocities, the rocket engine must burn a large amount of fuel and push the resulting gas out of the engine as rapidly as possible. Containing and controlling this power is the basic challenge in the development of these devices. For example, the power density produced by liquid hydrogen (LH_2) turbomachinery utilized by the space shuttle main engine (SSME) is approximately 83 horsepower per pound of turbopump weight.

Rocket propulsion system design solutions are quite varied: thrust levels from ounces to millions of pounds force; liquid and solid *propellants*; and liquid systems with pressures that are maintained by turbopumps or pressurized tanks. Liquid system applications vary from small pressure-fed, storable monopropellant thrusters for keeping satellites stationary, to large turbopump-fed, cryogenic bipropellant engines for boost propulsion. *Combustion chamber* pressures vary from a few pounds per square inch (psi) to several thousand psi. Generally, liquid propulsion systems consist of a propellant feed system, and *injector*, a combustion chamber, and a nozzle. The propellant feed system includes ducting and valves for controlling flows and, in the case of pump-fed systems, turbomachinery that draws propellants from lightweight propellant tanks and increases the pressure to the level necessary to support the desired combustion chamber pressure.

0-8493-1586-7/05/$0.00+$1.50

The ideal rocket propulsion equation is

$$\Delta V_{\text{ideal}} = g_0 I_{\text{sp}} \ln \frac{M_0}{M_1} \tag{197.1}$$

where ΔV_{ideal} is the ideal delta velocity imparted on a vehicle, g_0 is the gravitational constant, I_{sp} is the propulsion system's specific impulse, M_0 is the initial mass of the vehicle, and M_1 is the final or burnout mass of the vehicle. This equation provides two important performance parameters: specific impulse, which is a measure of propulsion system efficiency expressed in seconds, and vehicle burnout mass, which includes all structures (tankage, thrust structure, etc.), residual propellants, engine systems, feed systems, pressurization systems, auxiliary systems, electronic systems, upper stages, payload supporting structures, and the payload itself.

One of the more important internal rocket engine parameters is characteristic exhaust velocity, commonly referred to as C-star (C^*), which relates combustion chamber pressure, chamber throat area, and propellant flow rate. Theoretical characteristic exhaust velocity C^* is computed as follows:

$$C^* = \frac{P_{ns} A_t g_0}{\dot{w}_{tc}} \tag{197.2}$$

where P_{ns} is *nozzle* stagnation pressure in psi, A_t is throat area in square inches, and $\dot{w}_{tc}$ is chamber propellant mass flow rate in pounds-mass per second. A number of losses will reduce the actual C^* realized. These losses are generally a function of injector design and are related to mixture ratio maldistribution, mixing, and so on. The actual C^* realized by a given design is

$$C^*_{\text{act}} = \eta_{c^*} C^* \tag{197.3}$$

where η_{c^*} is C^* efficiency, typically between 0.80 and 0.99.

Another useful parameter is thrust coefficient, which relates thrust F, chamber pressure, and throat area as follows:

$$F = C_F P_{ns} A_t \tag{197.4}$$

where CF is the thrust coefficient, P_{ns} is nozzle stagnation pressure, and A_t is throat area. Once again, additional parameters must be added to reflect actual values. This yields the following:

$$F = \eta_{CF} C_F P_{ns} A_t - P_a A_e \tag{197.5}$$

where η_{CF} is thrust coefficient efficiency, typically between 0.90 and 0.97, P_a is local atmospheric pressure in psi, and A_e is exit area in square inches. This equation yields thrust at any point between *sea level* and *vacuum* conditions.

Specific impulse I_{sp} is an overall efficiency term and is defined as

$$I_{\text{sp}} = \frac{F}{\dot{w}_t} \tag{197.6}$$

where F is thrust level in pounds-force and $\dot{w}_t$ is the total mass flow rate in pounds-mass per second. Specific impulse can be computed for the engine or thrust chamber by utilizing either engine thrust and flow rate or thrust chamber thrust and flow rate, as appropriate. Specific impulse can also be computed if C^* and the thrust coefficient are known. This relationship is expressed as

$$I_{sp} = \frac{C^* C_F}{g_0} \tag{197.7}$$

Again, one must maintain consistency between theoretical values and actual values.

Thrust and specific impulse are commonly calculated at either sea level or vacuum conditions for reference or comparative purposes. Later discussions will refer to sea level thrust (F_{sl}), vacuum thrust (F_{vac}), sea-level-specific impulse (Isp_{sl}), and vacuum-specific impulse (Isp_{vac}).

Mixture ratio is the ratio between the oxidizer and fuel flow rates, and is expressed in equation form as

$$\text{MR} = \frac{\dot{w}_o}{\dot{w}_F} \tag{197.8}$$

where $\dot{w}_o$ is oxidizer flow rate in pounds per second and $\dot{w}_F$ is fuel flow rate in pounds per second. Mixture ratio can be computed for the engine or thrust chamber by utilizing either engine flow rates or thrust chamber flow rates, as appropriate.

Expansion ratio ε is a ratio of the thrust chamber nozzle exit area, A_e, and the thrust chamber throat area, A_t:

$$\varepsilon = \frac{A_e}{A_t} \tag{197.9}$$

A more complete definition of these and other rocket engine equations, including solid propellant systems, can be found in *Rocket Propulsion Elements* [Sutton, 1992].

197.2 Liquid Rocket Engine Cycles

A number of power cycles are available for liquid propellant systems. These include pressure-fed, expander, gas generator, and staged combustion cycles. Each cycle has advantages and disadvantages; the one selected for a given application is determined after a series of system trade studies. A description of a number of engine systems is given in Table 197.1. A brief description of each of these power cycles follows.

Pressure Fed

This system consists of a *thrust chamber assembly,* associated ducting and valves necessary for control, pressurized tankage, and the pressurization system for the tankage. This system is widely utilized for satellite attitude control, orbital transfer, and as auxiliary propulsion for most major launch vehicles. Pressure-fed systems are perhaps the simplest of all propulsion systems, but are performance limited because of the weight penalty associated with increasing chamber pressures. As pressures increase, tank wall thickness and the mass of the gases needed to maintain tank pressures increase. Tank pressures are set by chamber pressure plus pressure losses in the cooling circuit (if any), injector, valves, and ducting. In most pressure-fed applications, combustion chambers are passively cooled (i.e., film cooled or radiative/*ablative*). The space shuttle utilizes pressure-fed systems for the orbital maneuvering system and the reaction control system. A schematic of a simple pressure-fed system is given in Figure 197.1.

Expander

This is the simplest of the turbopump-fed systems primarily because the power source for the turbines is the thrust chamber cooling circuit. Only the thrust chamber requires an ignition system. Pump discharge pressures are set by chamber pressure plus pressure losses in the cooling circuit, turbine, injector, valves, and ducting. The combustion chamber is *regeneratively cooled.* In some applications, extensible

TABLE 197.1 Liquid Rocket Engine Characteristics

Engine	SSME	RL-10-A4	LR87	LR91	RS-27A	F-1	H-1	J-2
Producer	Rocketdyne	P&W	Aerojet	Aerojet	Rocketdyne	Rocketdyne	Rocketdyne	Rocketdyne
Launch vehicle	Space shuttle	Centaur	Titan II/IV first stage	Titan II/IV second stage	Delta	Saturn V first stage	Saturn IB first stage	Saturn IB second stage/ Saturn V second stage/ Saturn V third stage
Status	Active	Active	Active	Active	Active	Inactive	Inactive	Inactive
Propellants	LO_2/LH_2	LO_2/LH_2	N204/Aerozine 50	N204/Aerozine 50	LO_2/RP-1	LO_2/RP-1	LO_2/RP-1	LO_2/LH_2
Thrust, sea level, lbf	394,000	—	446,000 (dual-thrust chamber system)	—	200,000	1,522,000	206,145	—
Thrust, vacuum, lbf	488,800	20,800	529,000	100,000	237,000	1,748,000	230,170	230,000
Specific, impulse, sea level, s	365.1	—	254.0	263.3	255.0	265.4	264.9	
Specific impulse, vacuum, s	452.9	449.0	302.0	318.0	302.0	304.1	295.8	427.0
Mixture ratio	6.00	5.50	1.90	1.86	2.25	2.27	2.23	5.50
Total flow rate, lb/sec	1,079.3	46.3	1,751.7	314.5	784.8	5,734.7	778.1	538.6
Fuel flow rate, lb/sec	154.2	7.1	604.0	110.0	241.8	1,753.7	240.9	82.9
Oxidizer flow rate, lb/sec	925.1	39.2	1,147.6	204.5	542.9	3,981.0	537.2	455.8
Nozzle stagnation, psia	3,100	575	827	827	700	982	652	763
C-star (engine), ft/sec	7,507	7,696	5,611 (each)	5,597	10,707	5,297	5,509	
Area ratio	77.5	84	15	49.2	12	16	8	27.5
Throat area, in.^{2}	81.2	19.3	182.0 (each)	65	373.08	961.4	204.35	
Length, in.	168	70/90	150	110.62	149	220.4	101.61	133
Exit diameter, in.	96.00	46.00	86.25 (each)	64.00	76.00	143.50	45.62	80.50
Exit area, in.^{2}		1,618.7	5,842.6	5,842.6	5,842.6	15,400.0	1,634.8	
Powerhead diameter, in.	99.8		85	64		104.75		
Dry weight, lb	7,004	370	4,583	1,284	2,444	18,616	2,003	3,480
Turbine drive	Fuel-rich preburner	Heated hydrogen (expander)	Gas generator	Gas generator	Gas generator	Gas generator	Gas generator	Gas generator
Start method	Tank head	Tank head	Solid propellant cartridge	Solid propellant cartridge	Solid propellant cartridge	Tank head	Solid propellant cartridge	Gas bottle

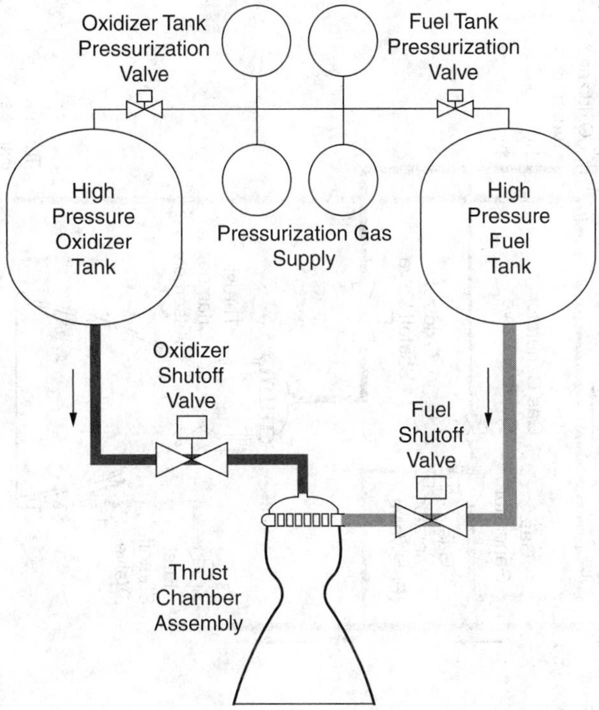

FIGURE 197.1 Pressure-fed propulsion system schematic.

radiation-cooled nozzle extensions are used to increase area ratio while maintaining a short stowed length. Expander cycles are limited in the combustion chamber pressure that can be attained because the energy available to drive the turbine(s) is obtained from the combustion chamber cooling circuit. For applications that require operation at sea level, this reduces the area ratio that can be achieved without *side loads*. Nozzle flow separation is discussed later. The RL10 engine is utilized by the Centaur upper stage for the Atlas-Centaur and Titan-Centaur launch vehicles. A schematic of a simple expander system is given in Figure 197.2.

Gas Generator

This is the most common engine cycle in use today. Turbine power is derived from a separate combustor or gas generator (GG) that utilizes the same propellants as the main system. This hot gas is routed through the turbopump turbines and is dumped overboard. Pump discharge pressures are set by chamber pressure plus pressure losses in the cooling circuit, injector, valves, and ducting. Because the gas generator is parallel to the main chamber, turbine pressure losses do not impact pump discharge pressure in most designs. This highlights one of the disadvantages of this cycle. The gas generator propellants are not used in the main chamber to produce thrust. Some concepts use GG gases for cooling nozzle extensions, but the thrust added is minimal. Gas generators are operated at relatively low mixture ratios because turbine temperatures must be maintained in the 1,000 to 2,000° Rankine range. The main combustor mixture ratio is biased higher to offset this parasitic flow. The combination of poor thrust efficiency of the GG gases and main chamber mixture ratio bias results in a specific impulse penalty. The gas generator cycle was utilized on the F-1 and J-2 engines of the Saturn V launch vehicle and is currently in use on the Delta, Atlas, and Titan launch vehicles. A schematic of a simple gas generator system is given in Figure 197.3.

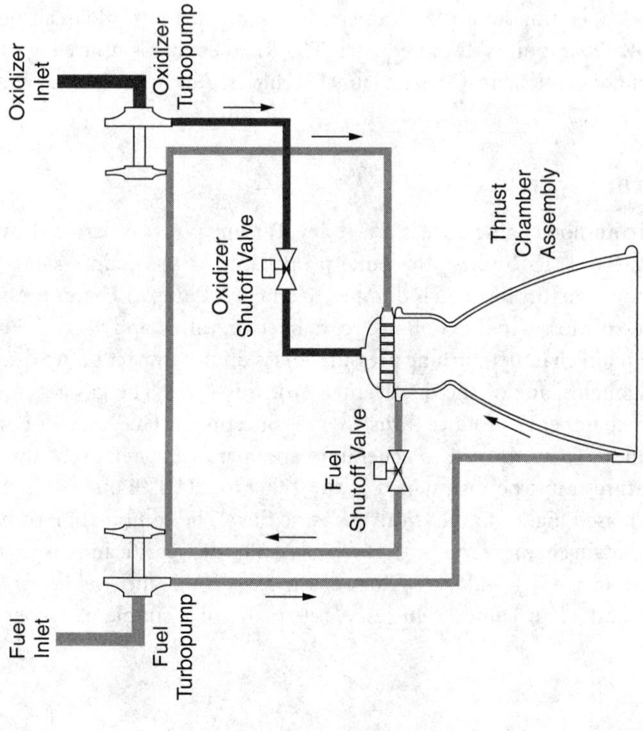

FIGURE 197.2 Expander engine system schematic.

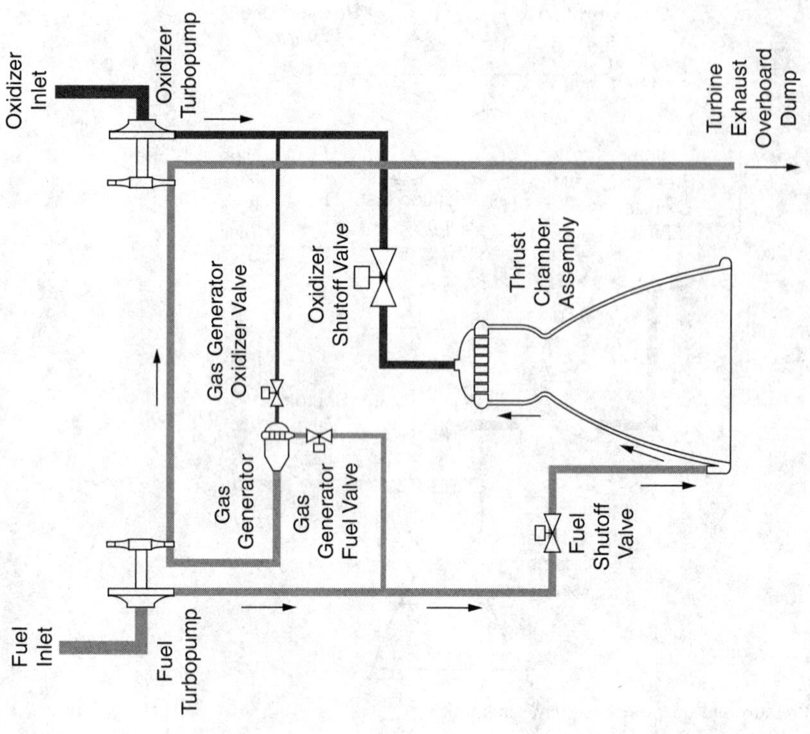

FIGURE 197.3 Gas generator engine system schematic.

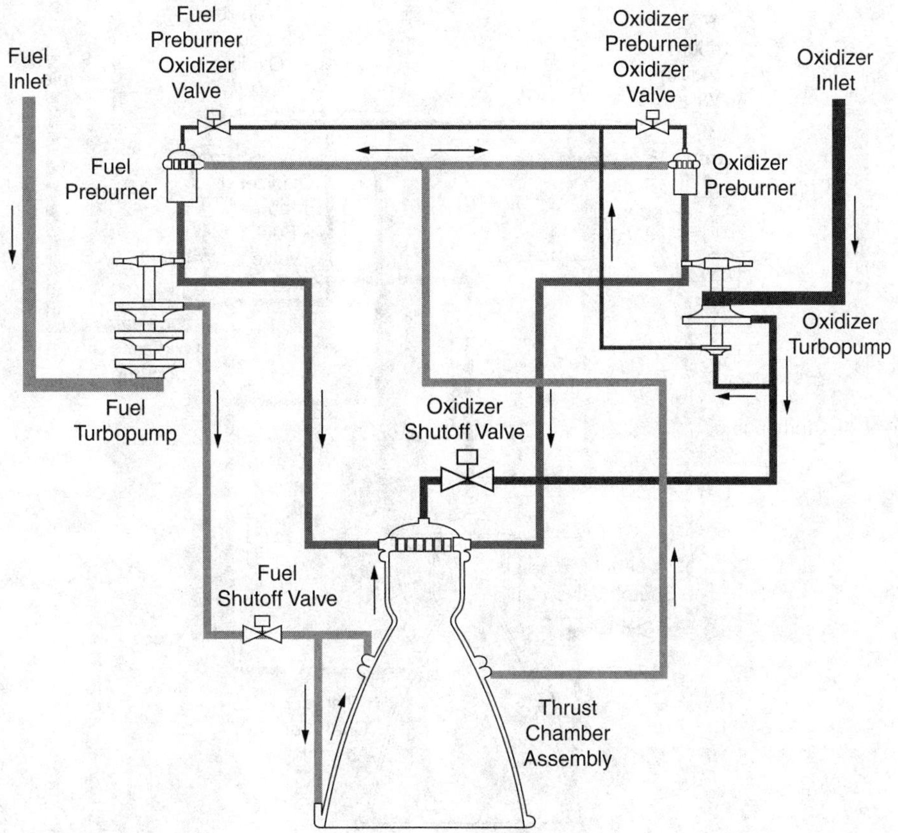

FIGURE 197.4 Staged combustion cycle engine system schematic.

Staged Combustion

The staged combustion cycle provides the highest performance of conventional chemical rocket engines. Turbine power is derived from a separate combustor or preburner which also utilizes the same propellants as the main system. In bipropellant systems, the hot gas is routed through the turbopump turbines to the main injector where it is mixed with the other propellant and is combusted in the main chamber. Pump discharge pressures are set by chamber pressure plus pressure losses in the cooling circuit, turbine, injector, valves, and ducting. Thrust chambers are regeneratively cooled. Staged combustion cycle engines developed in the U.S. have utilized a fuel-rich preburner. Several rocket engine systems developed in Russia have utilized an oxidizer-rich preburner. In the former case, the fuel-rich hot gases are mixed with oxidizer in the main chamber. In the latter, oxidizer-rich hot gases are mixed with fuel in the main chamber. The staged combustion cycle utilizes all propellants in the main combustion chamber, which provides maximum performance. A schematic of a simple staged combustion system is given in Figure 197.4.

A variant of the staged combustion cycle is the full-flow cycle, in which the oxidizer pump turbine is driven with an oxidizer-rich preburner and the fuel pump is driven with a fuel-rich preburner. This cycle offers some simplification in turbomachinery design because of the simplified seal design between the pump end and the turbine end, and a significant reduction in turbine temperatures because all the propellants can be utilized in the turbine drive circuits. This concept is currently under study as a candidate engine system for the next generation launch vehicle. A schematic of a simple full flow staged combustion system is given in Figure 197.5.

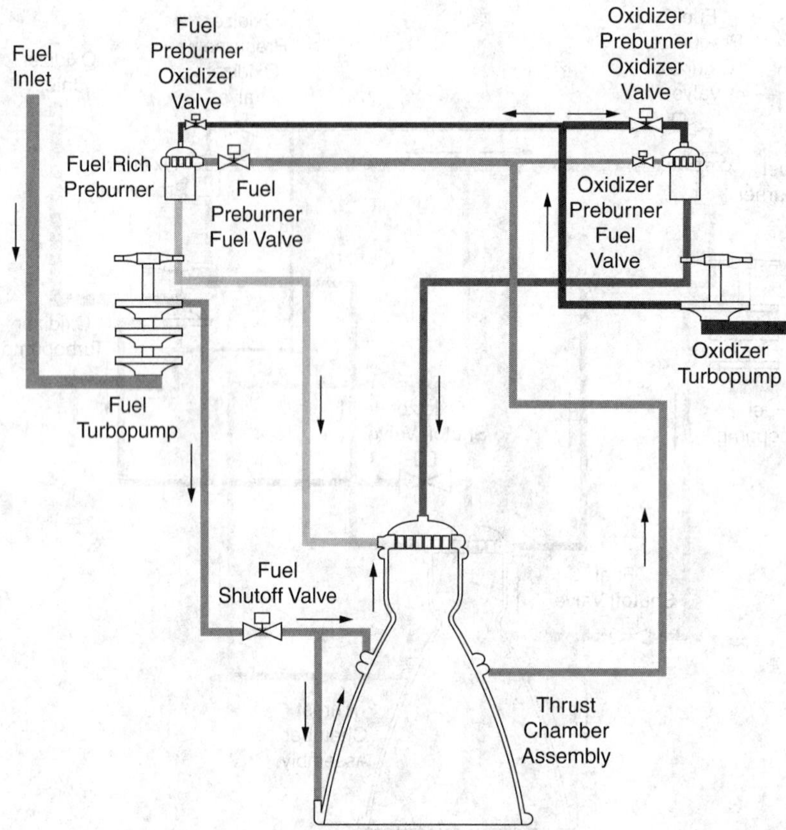

FIGURE 197.5 Schematic of full-flow, staged combustion-cycle engine system.

197.3 Major Components

Main Injector

The purpose of the injector is to introduce propellants into the combustion chamber in a controlled manner, to atomize the propellants, and to mix the propellants at the proper mixture ratio in a homogenous manner. Mixture ratio variations across the injector face are one of the most common problems that the designer will encounter, and these maldistributions lead to combustion efficiency losses. In some cases, maldistributions are deliberately introduced. To enhance the durability of combustion chambers, a film of fuel is injected at the outer circumference of the injector. In order to produce a stable combustion process, baffle elements which are cooled with fuel are commonly used. The most common injector concepts are coaxial, showerhead, and impinging. These are illustrated in Figure 197.6.

The coaxial injector consists of a series of concentric tubes into which the oxidizer is introduced through a center tube and the fuel introduced through the annular area formed by a second tube. This type of injector is commonly used in oxygen-hydrogen engines. Impinging and showerhead injectors, which are commonly used in oxygen-kerosene and storable propellant engines, consist of a series of two sets of orifices. One injects the oxidizer, and the other the fuel. The number of orifices, injection velocities, and injection angles are selected to provide consistent atomization and mixing of propellants. Impinging injectors slant the orifices to impinge the two propellant streams against each, enhancing mixing. The design utilized by the conventional impinging injector consists of a series of concentric copper rings containing the injection orifices. The rings alternate between oxidizer and fuel. The outer ring is generally

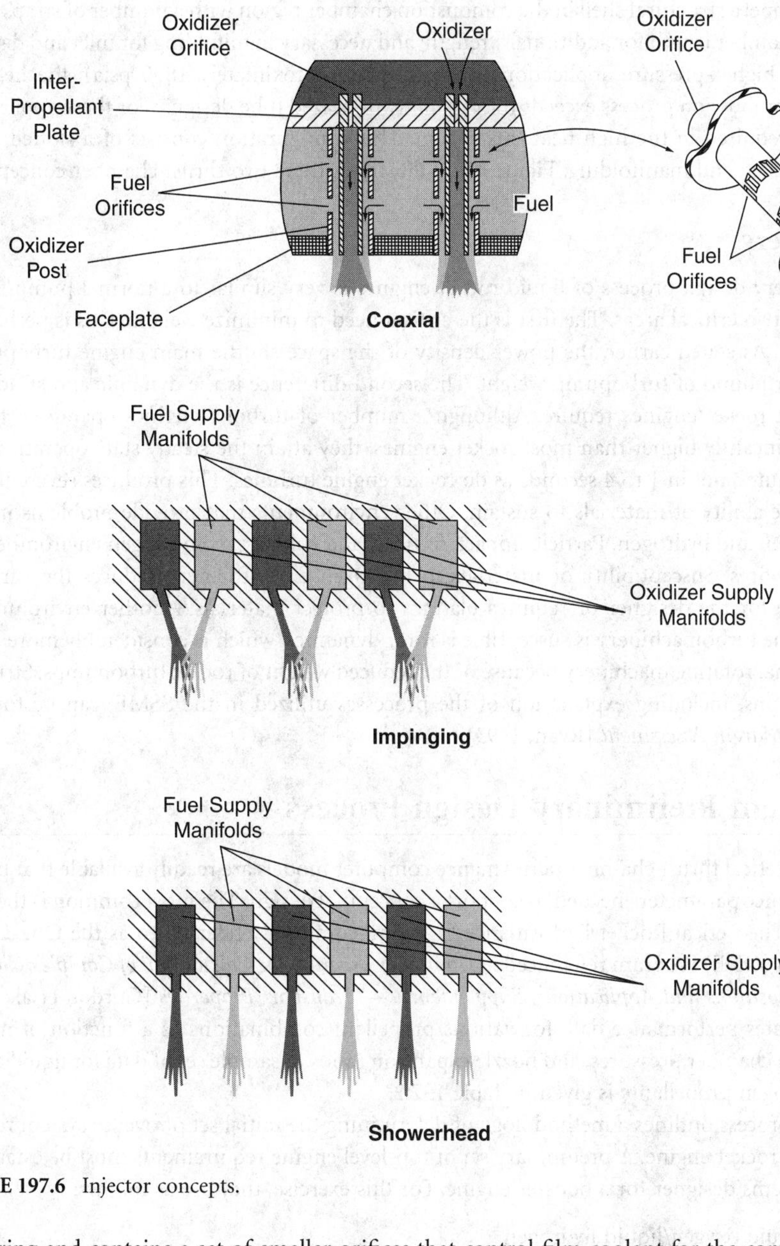

FIGURE 197.6 Injector concepts.

a fuel ring and contains a set of smaller orifices that control film coolant for the combustion chamber. Separate manifolding routes propellants to each set of orifices.

Thrust Chamber

A number of design solutions have been utilized in thrust chambers, varying from passively cooled ablatives to a number of regeneratively cooled concepts. In some applications, the thrust chamber is composed of two separate components. The upper portion — including the throat region and a portion of the expansion region — is commonly called a combustion chamber. The lower portion — consisting of the remainder of the expansion region — is called a nozzle. Regeneratively cooled thrust chamber designs include brazed tube bundles, copper with milled channels, and steel with milled channels. The bundled tube concept utilizes: steel tubes, pressed to vary the shape necessary for formation of the overall thrust chamber shape; a structural shell in the combustion chamber region with a number of straps spaced along the thrust chamber length for additional strength; and necessary manifolding for inlet and discharge coolant flow. For higher-pressure applications (greater than approximately 1800 psia), the heat load produced by the combustion process exceeds the capability of brazed tube designs. For these applications, a copper liner is required in the high heat flux region. This configuration consists of a slotted copper liner, structural jacket, and manifolding. Figure 197.7 illustrates these two thrust chamber concepts.

Turbomachinery

The turbomachinery design process of liquid rocket engines is very similar to a normal pump/turbine design, except for two critical areas. The first is the critical need to minimize weight. This is perhaps the greatest difference. As stated earlier, the power density of the space shuttle main engine turbopump is 83 horsepower per pound of turbopump weight. The second difference is the dynamic and steady state environments that rocket engines require. Although a number of turbojet engines operate at turbine temperatures significantly higher than most rocket engines, they attain the steady state operating point in a matter of minutes, not in 1 to 4 seconds as do rocket engine turbines. This produces severe thermal strains that tax the ability of materials to sustain. Other environments that provide problems in some materials are oxygen and hydrogen. Particle impact, fretting, and rubbing in an oxygen environment can lead to disastrous fires. Susceptibility of materials to hydrogen embrittlement reduces the variety of materials available for the designer or requires platings to protect materials. Another environment to which rocket engine turbomachinery is susceptible is rotor dynamics, which is considerably more critical than in conventional rotating machinery because of the reduced weight of rocket turbopumps. Structural design considerations, including explanation of the processes utilized in the SSME, can be found in *Structural Design/Margin Assessment* [Ryan, 1993].

197.4 System Preliminary Design Process

A number of theoretical thrust chamber performance computer models are readily available that provide the basic performance parameters needed to support a conceptual design. The most common is the Finite Area Combustor Theoretical Rocket Performance Program, commonly referred to as the One-Dimensional Equilibrium (ODE) Program referenced in *Computer Program for Calculation of Complex Chemical Equilibrium Compositions and Applications, Supplement I — Transport Properties* [Gordon et al., 1984]. This model generates performance data for various propellant combinations as a function of mixture ratios, combustion chamber pressures, and nozzle expansion ratios. A sample set of data for liquid oxygen (LO_2)/liquid hydrogen propellants is given in Table 197.2.

The following process outlines a methodology of determining the initial set of overall system requirements for a liquid rocket engine. A preliminary set of top-level engine requirements must be established by the vehicle systems designer for a booster engine. For this exercise, they are as follows:

Propellants: Liquid oxygen/liquid hydrogen
Sea-level thrust: 400,000 pounds force
Mixture ratio: 6.0:1.0

In some instances, a minimum specific impulse value is specified. However, in most cases, the design value should be selected as the result of vehicle-engine trade studies, along with engine weight, recurring costs, and nonrecurring costs.

One of the first choices to be made by the engine designer is the value of combustion chamber pressure. This choice is also the result of a series of trades. For this exercise, 2,000 psia has been chosen and is a good first approximation for the optimum value when recurring costs are one of the more important parameters. This value will provide good performance, while simplifying turbomachinery to two pump stages with moderate turbine temperatures. An initial assumption needs to be made as to engine cycle. For a booster application, either a gas generator cycle or a staged combustion cycle usually provides

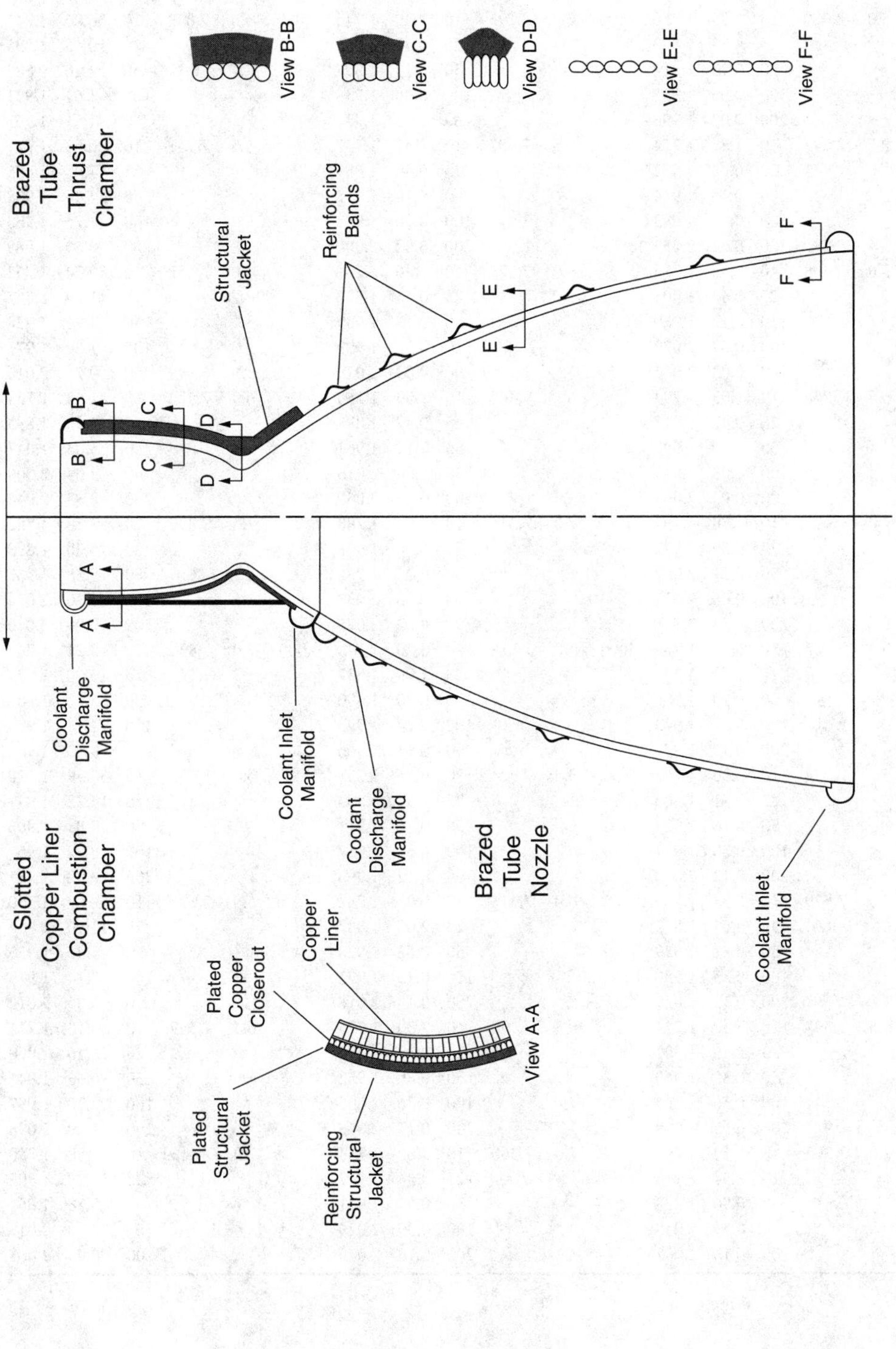

FIGURE 197.7 Two thrust chamber concepts.

TABLE 197.2 Theoretical Performance of Oxygen-Hydrogen Combustor

P_c	MR	C-star	ε	P_e	C_f	P_c	MR	C-star	ε	P_e	C_f	P_c	MR	C-star	ε	P_e	C_f
100.0	5.0	7,680	10	1.29	1.760	1000	5.0	7,795	10	1.29	1.745	3,000	5.0	7,829	10	36.26	1.741
			25	0.37	1.860				25	0.37	1.841				25	10.45	1.835
			50	0.15	1.917				50	0.15	1.895				50	4.10	1.889
			100	0.06	1.962				100	0.06	1.938				100	1.61	1.931
			200	0.02	1.996				200	0.02	1.971				200	1.61	1.963
	5.5	7,547	10	1.37	1.774		5.5	7,692	10	1.37	1.757		5.5	7,741	10	0.63	1.751
			25	0.40	1.882				25	0.40	1.859				25	11.08	1.851
			50	0.16	1.944				50	0.16	1.917				50	4.40	1.908
			100	0.06	1.993				100	0.06	1.963				100	1.75	1.953
			200	0.02	2.032				200	0.02	2.000				200	0.69	1.989
	6.0	7,408	10	1.46	1.786		6.0	7,575	10	1.46	1.769		6.0	7,637	10	39.74	1.762
			25	0.43	1.901				25	0.43	1.877				25	11.74	1.867
			50	0.17	1.969				50	0.17	1.939				50	4.71	1.928
			100	0.07	2.022				100	0.07	1.989				100	1.89	1.977
			200	0.03	2.066				200	0.03	2.029				200	0.76	2.016
	6.5	7,266	10	1.53	1.794		6.5	7,448	10	1.53	1.780		6.5	7,522	10	41.63	1.733
			25	0.47	1.917				25	0.47	1.894				25	12.43	1.883
			50	0.19	1.990				50	0.19	1.961				50	5.03	1.947
			100	0.08	2.049				100	0.08	2.015				100	2.04	2.000
			200	0.03	2.097				200	0.03	2.059				200	0.83	2.042
	7.0	7,127	10	1.58	1.799		7.0	7,316	10	1.58	1.788		7.0	7,396	10	43.58	1.782
			25	0.50	1.928				25	0.50	1.908				25	13.20	1.898
			50	0.21	2.007				50	0.21	1.980				50	5.38	1.967
			100	0.08	2.071				100	0.08	2.038				100	2.20	2.024
			200	0.03	2.124				200	0.03	2.086				200	0.90	2.070
500.0	5.0	7,767	10	6.20	1.749	2000	5.0	7,818	10	6.20	1.742	4,000	5.0	7,836	10	48.21	1.740
			25	1.79	1.845				25	1.79	1.837				25	13.90	1.834
			50	0.70	1.900				50	0.70	1.890				50	5.45	1.887
			100	0.27	1.943				100	0.27	1.933				100	2.14	1.929
			200	0.11	1.977				200	0.11	1.966				200	0.83	1.962
	5.5	7,654	10	6.56	1.762		5.5	7,724	10	6.56	1.753		5.5	7,751	10	50.39	1.750
			25	1.91	1.865				25	1.91	1.853				25	14.72	1.849
			50	0.76	1.924				50	0.76	1.911				50	5.84	1.906
			100	0.30	1.971				100	0.30	1.957				100	2.32	1.951
			200	0.12	2.008				200	0.12	1.993				200	0.92	1.987
	6.0	7,529	10	6.93	1.769		6.0	7,616	10	6.93	1.762		6.0	7,652	10	52.67	1.760
			25	2.05	1.877				25	2.05	1.867				25	13.90	1.864
			50	0.82	1.939				50	0.82	1.928				50	5.45	1.925
			100	0.33	1.989				100	0.33	1.977				100	2.14	1.973
			200	0.13	2.029				200	0.13	2.016				200	0.83	2.012
	6.5	7,397	10	7.30	1.784		6.5	7,496	10	7.30	1.775		6.5	7,539	10	55.10	1.771
			25	2.19	1.901				25	2.19	1.887				25	16.46	1.880
			50	0.89	1.969				50	0.89	1.952				50	6.66	1.945
			100	0.36	2.025				100	0.36	2.005				100	2.70	1.997
			200	0.15	2.069				200	0.15	2.048				200	1.09	2.038
	7.0	7,262	10	7.62	1.791		7.0	7,368	10	7.62	1.784		7.0	7,416	10	43.58	1.780
			25	2.34	1.915				25	2.34	1.902				25	13.20	1.895
			50	0.96	1.988				50	0.96	1.971				50	5.38	1.964
			100	0.39	2.048				100	0.39	2.029				100	2.20	2.019
			200	0.16	2.097				200	0.16	2.075				200	0.90	2.065

optimum performance. For this exercise, the staged combustion cycle is selected. This simplifies the initial set of calculations in that the engine flow rate and thrust chamber flow rate are approximately identical. Table 197.3 provides a typical ODE output for a thrust chamber operating at a chamber pressure of 2,000 psia and a mixture ratio of 6.0. The first parameter to select is area ratio. From previous vehicle trade studies, the optimum nozzle exit pressure (P_e) for a first stage or booster vehicle is approximately 6.5 psia, the optimum for a single-stage-to-orbit vehicle is approximately 4.0 psia, and the optimum for a parallel burn core stage (booster and core stages ignite at sea level) is 2.5 psia. Standard atmospheric conditions for sea level and various altitudes can be found in *Terrestrial Environment (Climatic) Criteria Guidelines for Use in Aerospace Vehicle Development, 1993 Revision* [Johnson, 1993].

Another consideration is side loads on the nozzle. For an overexpanded nozzle ($P_e < P_a$), unsymmetrical flow separation results if significant dynamic loads are applied to the nozzle (pressure times surface area forces). These side loads can ultimately destroy a nozzle and, at a minimum, result in significant weight increases. Side loads are computed using both empirical techniques and detailed nozzle fluid flow with computational fluid dynamics techniques. Calculating the locations within the nozzle where flow separation will occur is very difficult and side loads can usually be quantified accurately with test data.

As the application for this exercise is a booster, an exit pressure of approximately 6.5 psia is desired. Using Table 197.3, an expansion ratio of 30 yields an exit pressure of 6.2 psia, which is close enough for a first approximation of the engine characteristics. Using the initial vehicle thrust and mixture ratio requirements, theoretical values of C^* and C_F obtained from Table 197.3, and assumed values of C^* efficiency of 0.99 (readily obtainable with oxygen/hydrogen coaxial tube injectors) and C_F efficiency of 0.97, various engine parameters can be computed using Equations (197.3), (197.5), (197.6), (197.8), and (197.9) and are shown below in order of computation:

F_{vac}	400,,000 lb	A_t	109.22 in.^{2}
Propellants	Oxygen/hydrogen	A_e	3276.7 in.^{2}
MR	6.0	$\dot{w}_t$	932.27 lbm/sec
Cycle	Staged combustion	$\dot{w}_f$	133.18 lbm/sec
P_{ns}	2000 psia	$\dot{w}_o$	799.09 lbm/sec
e	30:1	Isp_{vac}	429.1 s
C_F	1.8311	F_{sl}	351 846 lbf
C^*	7539.8 ft/sec	Isp_{Sl}	377.41 s

197.5 Conclusion

The science of rocketry has enabled some of humankind's greatest achievements, ranging from instantaneous global communications and accurate weather forecasting via geostationary satellites to trips to the moon. The National Aeronautics and Space Administration's (NASA) space shuttle is one of the most complex flying machines ever built and is the only partially reusable launch vehicle. While several countries have rockets capable of carrying a variety of payloads, the U.S. and Russia are the only countries with spacecraft that can transport a crew to and from orbit. France and China have expendable launch vehicles now in use, and Japan is developing another.

The U.S. is on the threshold of a next-generation launch system. NASA aerospace engineers and industry experts are exploring new concepts, including the first fully reusable launch vehicle. Developing rockets for 21st-century missions promotes enhanced technologies to meet new challenges, including balancing design requirements between operability, performance, weight, and cost. The next generation of spaceships will open new doors to the space frontier.

TABLE 197.3 Theoretical Performance of Oxygen-Hydrogen Combustor at a Chamber Pressure of 2,000 psia and a Mixture Ratio of 6.0

	Chamber	Throat	Exit	Exit	Exit	Exit	Exit	Exit	Exit	Exit	Exit	Exit
PINF/P	1.00	1.74	74.83	128.69	188.56	253.32	322.26	394.88	470.81	631.58	715.98	802.86
P, atm	136.09	78.262	1.8186	1.0575	0.72172	0.53723	0.42231	0.34464	0.28906	0.21548	0.19008	0.16951
T, K	3571.28	3362.47	1992.35	1810.36	1688.69	1598.55	1527.58	1469.4	1420.31	1340.93	1308	1278.43
ρ, g/cc	6.2908-3	3.8815-3	1.5692-4	1.0044-4	7.3493-5	5.7792-5	4.7541-5	4.0334-5	3.4998-5	2.7634-5	2.4990-5	2.2801-5
H, cal/g	−235.74	−515.64	−1948.34	−2093.47	−2187.58	−2255.9	−2308.88					
U, cal/g	−759.64	−1003.92	−2229.01	−2348.45	−2425.4	−2481.02	−2524	−2558.69	−2587.56	−2633.49	−2652.24	−2668.94
G, cal/g	−15092.1	−14503.3	−10236.4	−9624.47	−9212.44	−8905.8	−8663.53	−8464.38	−8295.98	−8022.83	−7909.25	−7807.12
S, cal/(g)(K)	4.16	4.16	4.16	4.16	4.16	4.16	4.16	4.16	4.16	4.16	4.16	4.16
M, mol wt	13.546	13.685	14.106	14.11	14.111	14.111	14.111	14.111	14.111	14.111	14.111	14.111
(DLV/DLP)T	−1.02183	−1.01646	−1.00018	−1.00006	−1.00002	−1.00001	−1.00001	−1.00000	−1.00000	−1.00000	−1.00000	−1.00000
(DLV/DLP)P	1.3823	1.3057	1.0054	1.0018	1.0007	1.0004	1.0002	1.0001	1.0001	1.0000	1.0000	1.0000
CP, cal/(g)(K)	1.8834	1.7168	0.8179	0.784	0.7651	0.7518	0.7414	0.7328	0.7254	0.7131	0.7079	0.7031
GAMMA ($)	1.1455	1.1465	1.2106	1.2199	1.226	1.2307	1.2346	1.2379	1.2409	1.2461	1.2484	1.2505
Son vel, m/sec	1584.6	1530.4	1192.3	1140.8	1104.5	1076.7	1054.1	1035.3	1019.1	992.2	980.9	970.5
Mach number	0.000	1.000	3.175	3.456	3.659	3.819	3.951	4.065	4.164	4.333	4.406	4.474
						Performance Parameters						
AE/AT		1	10	15	20	25	30	35	40	50	55	60
CSTAR, ft/sec		7616	7616	7616	7616	7616	7616	7616	7616	7616	7616	7616
CF		0.659	1.631	1.699	1.741	1.771	1.794	1.813	1.828	1.852	1.862	1.871
CF–VAC		1.235	1.765	1.815	1.847	1.870	1.888	1.902	1.913	1.931	1.939	1.946
IVAC, lb–s/lb		292.2	417.7	429.6	437.2	442.6	446.8	450.1	452.8	457.1	458.9	460.5
ISP, lb–s/lb		156.1	386	402.1	412.1	419.3	424.7	429.1	432.7	438.4	440.7	442.8
P_e, psia		1150.1	26.7	15.5	10.6	7.9	6.2	5.1	4.2	3.2	2.8	2.5

Note: PINF = 2000.0 psia; O/F = 6.0000.

Defining Terms

Ablation — A passive cooling technique in which heat is carried away from a vital part by absorption into a nonvital part, which may melt or vaporize and then fall away, taking the heat with it.

Combustion chamber — A device — which includes a throat region — to mix, burn, and control propellants.

Injector — A device to distribute and inject propellants into the combustion chamber.

Nozzle — A device used to accelerate the combusted gases.

Propellant — Fuel [the chemical(s) the rocket burns] and an oxidizer (oxygen compounds) to ignite the fuel.

Sea level — Standard atmospheric conditions at an altitude of zero feet.

Side loads — Unsymmetrical loads put on a nozzle because of internal flow separation of overexpanded gases.

Regeneratively cooled — A cooling technique in which propellants, usually fuel, are utilized to remove heat from the inner wall of a combustor in a heat exchange process.

Thrust chamber assembly — An assembly consisting of the main injector, combustion chamber, and nozzle. Depending upon fabrication techniques, the combustion chamber and nozzle can be separate components or combined into a single component.

Vacuum — Conditions where atmospheric pressure can be considered to be 0.0 psia.

References

Gordon, S., McBride, B., and Zeleznik, F. 1984. Computer Program for Calculation of Complex Chemical Equilibrium Compositions and Application, Supplement I — Transport Properties. NASA Technical Memorandum 86885, National Aeronautics and Space Administration, Office of Management, Scientific and Technical Information Program.

Johnson, D. 1993. Terrestrial Environment (Climatic) Criteria Guidelines for Use in Aerospace Vehicle Development, 1993 Revision. NASA Technical Memorandum 4511, National Aeronautics and Space Administration, Office of Management, Scientific and Technical Information Program.

Ryan, R. 1993. Structural Design/Margin Assessment. NASA Technical Paper 3410, National Aeronautics and Space Administration, Office of Management, Scientific and Technical Information Program.

Sutton, G. P. 1992. *Rocket Propulsion Elements*, 6th ed. John Wiley & Sons, New York.

Further Information

- American Institute of Aeronautics and Astronautics (AIAA)
- American Society of Mechanical Engineers (ASME)
- National Space Society
- Marshall Space Flight Center, Central Technical Library (phone: 205-544-4524)

198

Aircraft Performance and Design

Francis Joseph Hale
North Carolina State University

Two major considerations in the performance and design of an aircraft are its range and maneuverability. *Range* deals with how far the aircraft can fly with a given fuel load and given payload under specified flight conditions. *Maneuverability* treats turning rates, rates of climb, and acceleration. A performance analysis starts with knowledge of the physical characteristics of an aircraft, either actual or hypothetical, and determines how the aircraft will fly under various flight conditions. The converse of the performance analysis is the design of an aircraft to meet a set of operational requirements and constraints. Design, however, is an iterative process, and the first few tries, which yield major aircraft characteristics, may be referred to as *conceptual* or *feasibility* designs. The performance analysis and preliminary design processes are often intertwined in a series of iterations and it may be difficult to distinguish between the two.

The flight path and behavior of an aircraft are determined by the interaction between its characteristics and those of the environment in which it is operating — namely, the atmosphere. The aircraft characteristics can be categorized as its physical characteristics, such as shape, mass, volume, and surface area; the characteristics of the propulsion, guidance, and control subsystems; and the structural characteristics, such as loading and temperature limitations and the stiffness (rigidity) of the structure. The environment affects the flight of an aircraft through the field forces, surface forces, and inertia forces. The only field force that needs to be considered is gravity, which appears as the *weight* and is a function of the mass of the aircraft. The surface forces are the aerodynamic forces (lift, drag, and side forces), which are strongly dependent on the shape and surface area of the aircraft, especially of the wings, and the properties of the atmosphere. The inertia forces result from non-equilibrium processes and are very important in dynamic and stress analyses of aircraft.

Aircraft operational units are still principally English and will be given preference over SI units in this article with distance in nautical miles (nmi), airspeed in knots (kt) and feet per second (ft/sec), force and weight in pounds (lb), and pressure in lb/ft^2, where 1 nmi = 1.15 mi = 1.84 km; 1 kt = 1 nmi/h = 1.15 mi/h = 1.84 km/h = 0.51 m/sec; 1 ft/sec = 0.59 kt; 1 lb = 4.448 N = 0.4535 kg; and 1 lb/ft^2 = 47.87 Pa.

0-8493-1586-7/05/$0.00+$1.50

TABLE 198.1 Standard Atmosphere Ratios

Altitude (1000 ft)	0	10	20	30	36	40	50	60
Density ratio σ	1.000	0.738	0.533	0.374	0.297	0.246	0.152	0.094
Sonic ratio a*	1.000	0.965	0.929	0.821	0.867	0.867	0.867	0.867

Note: $\rho_o = 23.769 \cdot 10^{-4}$ lb-s/ft²; $a_o = 1116$ ft/sec.

198.1 Aircraft Forces and Subsystems

Atmosphere

Although the standard atmosphere comprises a number of concentric layers surrounding the Earth, only the first two layers of the lower atmosphere are of interest. The *troposphere* starts at sea level and is characterized by a decreasing ambient temperature. At 36,089 ft (11,000 m) above sea level (the *tropopause*), the temperature becomes and remains essentially constant in the *stratosphere* until reaching an altitude of 82,021 ft (25,000 m). In the troposphere, as the atmospheric temperature decreases so does the local speed of sound a, thus affecting the Mach number M; in the isothermal stratosphere the sonic velocity remains constant. Table 198.1 is an abbreviated listing, as a function of the altitude h, of the two atmospheric ratios of most interest — namely, the *density ratio* σ (ρ/ρ_o), where ρ is the atmospheric density at altitude h and ρ_p is the sea level density ($\rho = \rho_o\sigma$), and the *sonic ratio a** (a/a_o). The standard-day sea level values of ρ_o and a_o are given below Table 198.1.

A commonly used exponential approximation for the density ratio σ is

$$\sigma = \rho/\rho_o \cong \exp(-h/\beta) \tag{198.1}$$

where β is an empirical factor with a generally accepted value of 23,800 ft (7254 m). Because the atmosphere is assumed to be an ideal gas, variations in the actual temperature and pressure will produce appropriate changes in the actual density. Consequently, on a hot day or with a below-standard barometric pressure, the density will be lower than the standard value for that altitude and the aircraft and propulsion system will perform as though at a higher altitude.

Aerodynamic Forces

In coordinated flight, the two principal aerodynamic forces are the *lift L* and the *drag D.* The lift is perpendicular to the velocity (*airspeed*) of the aircraft and has the primary function of compensating for the weight of the aircraft. The drag is parallel to the airspeed and resists the motion of the aircraft. The wing is the major source of lift and drag, and is designed to maximize the lift while minimizing the drag.

Wing parameters of importance with respect to performance are (1) the wing span b, the distance from wing tip to wing tip; (2) the average chord c_w, the distance from the front (leading edge) of the wing to the back (trailing edge); and (3) the wing area S, the area of one side of the wing to include the area included by the fuselage. The ratio of the wing span to the average wind chord is the aspect ratio AR of the wing, a measure of the narrowness of the wing. The following relationships are useful:

$$S = bc_w;\quad AR = b/c_w = b^2/S \tag{198.2}$$

The lift and drag forces can be obtained from the following expressions:

$$L = \tfrac{1}{2}\rho_o\sigma V^2 SC_L = qSC_L;\quad D = \tfrac{1}{2}\rho_o\sigma V^2 SC_D = qSC_D \tag{198.3}$$

where q is the dynamic pressure and C_L and C_D are the dimensionless lift and drag coefficients.

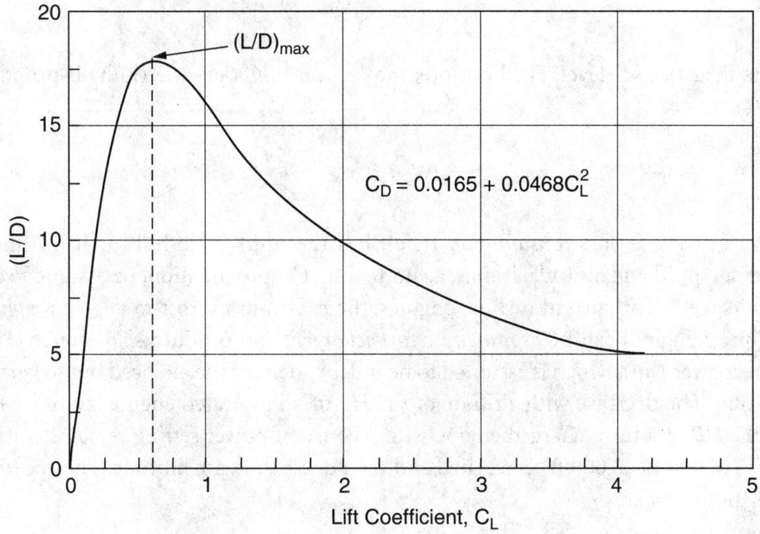

FIGURE 198.1 Typical variation of L/D with C_L for a specified drag polar.

The drag polar is an expression for C_D as a function of C_L and is an important performance and design parameter. The *parabolic drag polar* is an approximation that can be used with many subsonic (and thin-winged supersonic) aircraft configurations and can be written as

$$C_D = C_{DO} + KC_L^2 \quad \text{where } K = 1/(\pi eAR) \tag{198.4}$$

C_{DO} is the *zero-lift drag coefficient* of the aircraft, KC_L^2 represents the *drag-due-to-lift coefficient*, and e is the *Oswald span efficiency* (on the order of unity and less). The lift-to-drag ratio (L/D) or (C_L/C_D) represents the *aerodynamic efficiency* of an aircraft. It has a maximum value that is a design characteristic of the aircraft. This maximum value $(L/D)_m$ cannot be exceeded, although the aircraft may, and usually does, fly at lower values. The expression for the maximum (L/D) of a parabolic drag polar is

$$(L/D)_m = 1/[2(KC_{DO})^{1/2}] \tag{198.5}$$

A large $(L/D)_m$ calls for low values of K (high AR and large e) and C_{DO}. Figure 198.1, a typical plot of the (L/D) ratio for a specific aircraft and specific parabolic drag polar shows the variation in (L/D) as C_L varies. Typical values of $(L/D)_m$ for several classes of aircraft are: 35 for sailplanes, 18 for Mach 0.8 transports, 7 for supersonic aircraft, and 3 for helicopters.

Propulsion Subsystems

Current aircraft engines are air breathers that produce thrust or power, or a combination thereof. The *turbojet* engine represents a thrust producer, whereas an internal combustion engine in combination with a propeller (a *piston-prop*) represents a power producer. Turbofans, unducted fans, turboprops, and propfans combine the characteristics of both to varying degrees. Turbofans and unducted fans are usually described in turbojet terms and turboprops and propfans in piston-prop terms.

The actual performance and functional relationships of an aircraft engine should be obtained from power plant charts. However, to a first approximation, the *thrust* T (lb) produced by a turbojet can be considered to be independent of the airspeed and, for a given throttle (percent rpm) setting, to be directly proportional to the atmospheric density, so that

$$T(h) \cong T_o \sigma \tag{198.6}$$

where T_o is the thrust at sea level. Furthermore, the *fuel consumption rate* (lb/h) is proportional to the thrust, so that

$$dW_f / dt = cT \tag{198.7}$$

where c, the *thrust specific fuel consumption* (tsfc), has the units of lb/h/lb or h^{-1}. Although c varies somewhat with airspeed and altitude, it may, again to a first approximation, be assumed constant for all altitudes and airspeeds. For current turbine engines, the maximum *thrust-to-engine weight ratio* (T/W_e) $\cong$ 6–8. Piston-prop engines can be *aspirated* or *turbocharged*, and produce *shaft power* (*HP*), measured in units of horsepower (hp), which is assumed to be independent of the airspeed but to vary with altitude and power setting. The decrease with altitude of the *HP* of an aspirated engine is similar to that for the turbojet, but the *HP* of a turbocharged engine with a constant power setting remains constant until the *critical altitude* (15,000 to 20,000 ft) is reached and then decreases with altitude. The approximations for the density relationships are

$$HP \cong (HP)_0 \sigma (\text{aspirated}) \quad \text{and} \quad HP \cong (HP)_0 (\sigma / \sigma_{cr}) \; (\text{turbocharged}) \tag{198.8}$$

where σ_{cr} is the density ratio at the critical altitude. Figure 198.2 shows the qualitative relationship of the *HP* with altitude of the two types of engines. (The variation of turbojet thrust is similar to that shown for the aspirated engine.) The propeller converts the shaft horsepower into the *thrust power P*, which is equal to the product of the thrust and airspeed *TV*. The shaft power, thrust power, and thrust are related by the expressions:

$$P = TV = k\eta_p (HP); \quad T = k\eta_p (HP)/V \tag{198.9}$$

where η_p is the propeller efficiency (on the order of 80 to 85% for a well-designed, constant speed propeller) and k is a conversion factor with a value of 326 when V is in knots. Note that for a given horsepower, the thrust power of a piston-prop is independent of the airspeed. However, the available thrust is inversely proportional to V, decreasing as the airspeed increases, whereas the thrust of a turbojet

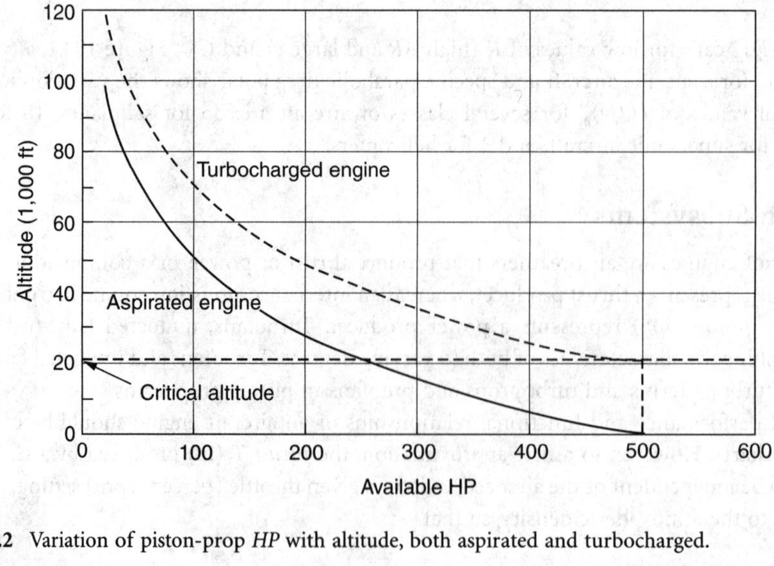

FIGURE 198.2 Variation of piston-prop *HP* with altitude, both aspirated and turbocharged.

is constant and the thrust power (TV) increases with V. Because of these differences, propeller and jet aircraft fly differently for best performance. The fuel consumption rate (lb/h) of a piston-prop is proportional to the HP being delivered, so that

$$dW_f / dt = c^*(HP) = c^* TV / (k\eta_p) \qquad (198.10)$$

where c^*, the *horsepower specific fuel consumption* (hpsfc) with units of lb/h/hp, has the same variations as the tsfc and will also be assumed to be constant. Current piston-prop engines are relatively small (less than 1000 hp) because they are the heaviest of all the engines with $(HP)_m/ W_e \cong 0.5$ hp/lb of engine weight. The specific fuel consumption is an extremely important performance parameter. Some typical values, all expressed as an equivalent tsfc (lb/h/lb), are listed below.

Rocket engines	10
Ramjets	3
Turbojets (with afterburner)	2.5
Turbojets	0.9–1.0
High-bypass turbofans	0.6–0.8
Turboprops	0.5–0.6
Piston-props	0.4–0.5

The order of this list also indicates the relative airspeed regime of the flight vehicles in which these engines are used. Piston-props are used in aircraft with airspeeds on the order of 220 kt or less; turboprops have higher speeds up to Mach 0.7 or so; turbofan engines for airspeeds up to M 0.85; and so on up the speed ladder with turbojets (and very low-bypass turbofans) in supersonic aircraft, ramjets for M 3.0 and higher, and rocket engines in missiles and space boosters. Furthermore, the piston-prop engine is the least expensive and the heaviest of the engines, and the cost increases and weight decreases as the list is ascended.

Weight Fractions

Weight may well be the most important consideration in the design and performance of an aircraft. Every extra pound of weight is accompanied by an increase in the wing area and in the thrust and fuel required, all leading to a further increase in aircraft weight and thus adversely affecting performance and costs (both initial and operating) of the aircraft. Weight fractions are useful in performance and design analyses and are obtained by expressing the gross (total) weight of the aircraft W_o as the sum of the weights of the major components and then dividing each weight by the gross weight. Using a simple weight breakdown made up of the *structural weight* W_s, the *engine weight* W_e, the *fuel weight* W_f, and the *payload weight* W_{PL}, the *gross weight* W_o of the aircraft is the sum of these component weights. The weight fractions are obtained by dividing through by W_o to obtain

$$1 = W_s / W_o + W_e / W_o + W_f / W_o + W_{PL} / W_o \qquad (198.11)$$

With this breakdown, W_s includes not only the weight of the structure, but also the weight of everything not included in the other categories. It includes the weight of all the equipment and landing gear, for example, and even the weight of the flight and cabin crews when appropriate. Aircraft manufacturers often lump W_s and W_e together into the *operational empty weight* and combine W_f and W_{PL} into the *useful load*. In Equation (198.11), the sum of the individual fractions must always be unity. For a turbojet, $W_e / W_o = (T_m / W_o)/(T_m / W_e)$, and, for a piston-prop, $W_e / W_o = (HP_m / W_o)/(HP_m / W_e)$. Values of T_m/W_o are on the order of 0.25 to 0.35 for large high-subsonic transports and on the order of 0.1 hp/lb of aircraft weight for the HP/W_o of the typical piston-prop.

Although the determination of W_s is quite complex, order of magnitude values for W_s/W_o can be used to give a feel for its size and significance. As W_o increases, W_s/W_o decreases because, for example, minimum volume requirements for cabin and cargo space and fixed equipment weights have a higher impact on the structural weight fractions of the smaller aircraft. For large subsonic transports, W_s/W_o values are on the order of 0.45, but smaller aircraft, such as fighters and general aviation aircraft, can have values in excess of 0.55. Note that in this article the structural weight does not include engine weight.

The payload ratio (the payload weight fraction) is of major interest inasmuch as it relates the payload weight to the gross weight of the aircraft. For example, $W_{PL}/W_o = 0.1$ means that 10 pounds of aircraft weight is required for each pound of payload (or excess weight). As W_{PL}/W_o is equal to unity minus the sum of the other weight fractions, any increase in any of these three either decreases the payload weight fraction, thus increasing W_o or decreasing the payload, or decreases W_f/W_o, thus decreasing the range. For example, the increase of T_m/W_o to convert a CTOL aircraft to a V/STOL aircraft can either dramatically reduce the range and payload or increase the gross weight.

198.2 Level Flight

For any aircraft in unaccelerated level flight,

$$L = W; \quad T = D; \quad dX/dt = V \tag{198.12}$$

The weight, however, does not remain constant during flight. As fuel is used, the weight decreases at a rate that is the negative of the fuel consumption rate (Equation (198.7)), so that for a *turbojet* $dW/dt = -cT$. The *instantaneous range* (mileage), the nautical miles per pound of fuel, is equal to dX/dt divided by dW/dt and, with the relationship from Equation (198.13) that the required thrust $T = W/(L/D)$, can be expressed as

$$dX/dW_f = V/cT = V(L/D)/cW \tag{198.13}$$

The last expression of Equation (198.13) shows that the best mileage (maximum range per pound of fuel) of an aircraft is obtained when the product of V and (L/D), which are coupled, is maximized, and c and W are individually minimized. Because W decreases along the flight path, the mileage improves with time and distance.

Although the mileage at a particular point or instant of time is of interest, the range for a given amount of fuel and set of flight conditions and the amount of fuel required to fly a specified range (from point 1 to point 2) are of greater interest. With c assumed constant, integrating the mileage over the cruise flight path while holding V and C_L (and thus L/D) constant, yields the *Breguet range equation* in the form:

$$X = [V(L/D)/c]\ln MR = [V(L/D)/c]\ln[1/(1-\zeta)] \tag{198.14}$$

where the mass ratio $MR = W_1/W_2$ and the *cruise-fuel weight fraction* $\zeta = \Delta W_f/W_1 = (MR-1)/MR$. In order to keep both V and L/D (i.e., C_L) constant along the flight path as fuel is used, the altitude must increase so as to keep W/σ constant (see Equation (198.15)). The required flight path angle, however, is sufficiently small (<1°) so that the level flight approximation is still valid. Such flight is known as *cruise-climb*. It gives the maximum range for a given fuel load and, for long flights in a controlled area, can be approximated by a series of constant altitude legs with appropriate altitude increases (*stepped altitude flight*). Note that the larger the product of V and L/D is, the longer the range.

The airspeed V and the lift coefficient C_L (and thus L/D) are related by expressions from Equation (198.12) and Equation (198.3) — namely,

$$V = \sqrt{\frac{2(W/S)}{\rho_o \sigma C_L}}; \quad C_L = \frac{2(W/S)}{\rho_o \sigma V^2} = \frac{(W/S)}{q} \tag{198.15}$$

Equation (198.15) shows that for a give C_L, a high airspeed V calls for a high *wing loading* (W/S) and a small σ (a high altitude) for good range performance. Consequently, turbojet aircraft *must* fly high and fast to achieve their best range performance; reduced flight times and flight above weather are accompanying bonuses. The *best-range* airspeed (and maximum range) of a particular turbojet is 60% greater at 30,000 ft than at sea level, and 80% greater at 35,000 ft. For long-range subsonic transports, design W/S values are on the order of 120 to 135 lb/ft^2. Shorter-range and fighter aircraft have lower values to meet requirements for shorter runway lengths or for greater maneuverability.

For a *piston-prop*, the relationships of Equation (198.13) still apply, but now the *thrust power TV* is of primary interest and equilibrium requires that the thrust power be equal to the drag power ($TV = DV$). The required thrust power $P \equiv TV = WV/(L/D)$. With $dW/dt = -dW_f/dt = -c^*TV/(k\eta p)$, the mileage (nmi/lb) of a piston-prop is

$$dX/dW_f = k\eta_p V/(cTV) = k\eta_p(L/D)(c^*W) \tag{198.16}$$

Note that the mileage of a piston-prop is explicitly independent of the airspeed, which must, however, be that required to achieve the desired L/D. Good mileage still calls for a high L/D and low values of c^* and W as for the turbojet. Holding C_L (and thus L/D) constant yields the piston-prop version of the *Breguet range equation:*

$$X = [326\eta_p(L/D)/c^*]\ln MR = [326\eta_p(L/D)/c^*][\ln 1/(1-\zeta)] \tag{198.17}$$

with V in kt. Although the range is explicitly independent of V, V must have the value appropriate to the cruise (L/D). Increasing V by increasing the altitude or W/S will not increase the range but will decrease the flight time (and increase the power required). Although the maximum range occurs when (L/D) is at its maximum value, the associated airspeed, especially with an aspirated engine, is usually much lower than the maximum airspeed of the aircraft and the required power for cruise is below the best operating point for the engine. Consequently, the cruise of a piston-prop is customarily at 75% of the maximum power available to reduce flight time and favor the engine. With respect to other power plants, whereas a turbojet is a single-flow engine that produces only jet thrust, a *turbofan* is a multiflow engine that uses a turbine to drive a multibladed ducted fan that produces thrust power in addition to jet thrust. It has characteristics of both turbojet and propeller aircraft, and its performance (and specific fuel consumption) lies in the region between the two but closer to the turbojet. The turbojet range equation is normally used.

A *turboprop*, on the other hand, is primarily a power producer, with little or no jet thrust, that uses the piston-prop equations and is replacing the piston-prop because it is lighter and capable of higher speeds. It has a lower C_{DO}, but a higher sfc, than the piston-prop. (*Derating* a turboprop results in an altitude variation of the power which resembles that of a turbocharged piston-prop.) The maximum airspeed of a turboprop with a conventional propeller is limited by the drop in η_p at the higher airspeeds. The *propfan* uses a double row of multibladed, variable camber propellers to increase the operating airspeed. The *unducted fan* (UDF) with an ultrahigh bypass ratio is still driven directly from the turbine, but resembles the propfan in appearance and approaches its performance. The *absolute ceiling* of an aircraft occurs when the maximum available thrust of a turbojet is equal to the required thrust, $T_m = W/(L/D)_m$, or when the maximum available power of a piston-prop is equal to the required power, $(HP)_m = WV/[k\eta_p(L/D)_m]$. At the ceiling, the rate of climb is zero and the aircraft cannot turn without losing altitude.

198.3 Climbing Flight

The steady-state climbing flight equations are

$$T - D - W\sin\gamma = 0; \quad L - W\cos\gamma = 0; \quad R/C = dh/dt = V\sin\gamma \tag{198.18}$$

where γ, the *climb* (*flight path*) *angle*, is the angle between the horizon and the airspeed vector. The two climb equations of interest are

$$\sin\gamma = (T-D)/W = (T/W) - [\cos\gamma/(L/D)]; \quad R/C = V\sin\gamma = (TV - DV)/W \qquad (198.19)$$

The first expression in Equation (198.19) shows that γ is determined by the *excess thrust* (the thrust not needed to overcome the drag) per unit weight, whereas the *R*/*C* is determined by the amount of *excess power* per unit weight. The two climb programs of special interest are *steepest climb*, γ_m, and *fastest climb*, $(R/C)_m$. Both have their largest values at sea level, where the excess thrust and excess power are at a maximum, and both decrease with altitude until γ and *R*/*C* both go to zero at the absolute ceiling. Using the expressions of Equation (198.19), the climb performance can be obtained either graphically (as sketched in Figure 198.3 for a turbojet) for various altitudes, weights, power settings, and velocities, or by analytic methods.

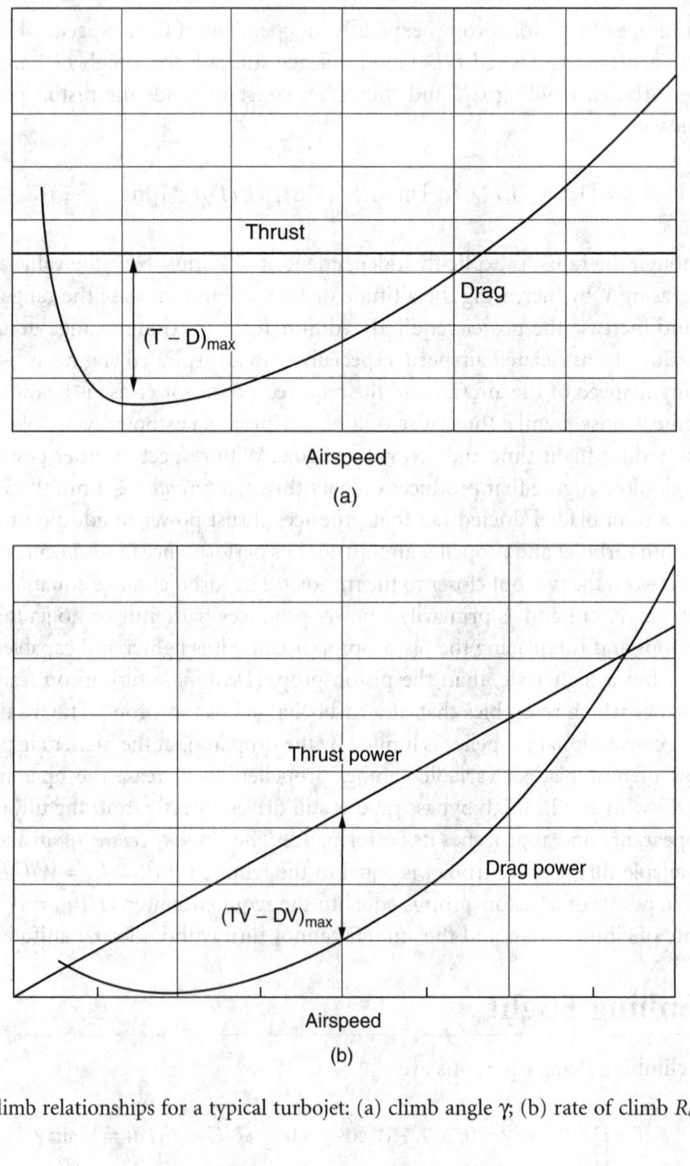

FIGURE 198.3 Climb relationships for a typical turbojet: (a) climb angle γ; (b) rate of climb *R*/*C*.

Climb angles for conventional aircraft are not large, as can be seen by dropping the drag term in Equation (198.20) to obtain the inequality that sin $\gamma_m < (T_m/W)$, which shows that the maximum climb angle for a high-performance turbojet transport, with $T_m/W = 0.33$, will be less than 19°. Obviously, γ for $(R/C)_m$ will be less. As expected, the higher the available thrust for a turbojet and the higher the available power for a piston-prop, the better the climb performance is. In general, the turbojet outclimbs the piston-prop and the climb airspeeds are higher.

198.4 Turning Flight

One aspect of maneuverability is the ability to turn. The equations governing steady-state turns in the horizontal plane are

$$T = D; \quad L = W/\cos\phi; \quad \dot{\chi} = \frac{g\tan\phi}{V} \tag{198.20}$$

where ϕ is the *bank angle* (deg), $\dot{\chi}$ is the *turning rate* (rad/sec), V is the airspeed (ft/sec), and g is the acceleration of gravity (32.2 ft/sec²). The *load factor* is defined as $n = L/W$ (g's). From Equation (198.20) and Equation (198.3),

$$n = \frac{L}{W} = \frac{1}{\cos\phi} = \left(\frac{T}{W}\right)(L/D); \quad V = \sqrt{\frac{2n(W/S)}{\rho_o\sigma C_L}} = \sqrt{\frac{2(W/S)}{\rho_o\sigma C_L\cos\phi}} \tag{198.21}$$

When $n > 1$ g, as in a turn, the airspeed is less than that for wings at level flight with the same power setting, and the stall speed also decreases. It can be shown that the best turning performance (tightest turn and fastest turn) for both types of aircraft occurs at low speeds (on the edge of a stall) and at low altitude while using maximum thrust or power.

Defining Terms

Air breathers — Engines that obtain oxygen from the atmosphere.
Aspect ratio — Wingspan divided by the average chord length.
CTOL — Conventional (horizontal) takeoff and landing aircraft.
Drag polar — Drag coefficient–lift coefficient relationship.
Lift-to-drag ratio (L/D) — Aerodynamic efficiency of an aircraft.
Mach number — Air speed divided by local sonic velocity.
Mass ratio (MR) — Aircraft weight at start of cruise divided by aircraft weight at end of cruise.
Payload ratio — Payload weight divided by gross weight of an aircraft.
V/STOL — Aircraft capable of both vertical and short takeoffs and landings.
Weight fraction — Component weight divided by aircraft weight.
Wing loading (W/S) — Aircraft weight divided by wing area (lb/ft²).

References

Anderson, J. D. 1999. *Aircraft Performance and Design*, McGraw-Hill, New York.
Hale, F. J. 1984. *Introduction to Aircraft Performance, Selection, and Design*. John Wiley & Sons, New York.
Houghton, E. L. and Brock, A. E. 1970. *Aerodynamics for Engineering Students*. Edward Arnold, London.
McCormick, B. W. 1979. *Aerodynamics, Aeronautics, and Flight Mechanics*. John Wiley & Sons, New York.
Raymer, D. P. 1989. *Aircraft Design: A Conceptual Approach*. American Institute of Aeronautics and Astronautics, Washington, DC.

Further Information

Aerospace America is a monthly publication of the American Institute of Aeronautics and Astronautics (AIAA) that presents current technical and political activity as well as a look at the future. For subscription and membership information, call (800) 639-2422.

Aerospace Engineering is a monthly publication of the Society of Automotive Engineers (SAE) that features articles on technical specialties and new technology. For subscription and membership information, call (724) 776-1383; fax (724) 776-1615; or e-mail circulation@sae.org.

Air & Space is a bimonthly publication of the Smithsonian Institution that contains a mix of popular and quasi-technical articles written for the general public, but with items of interest for the more technical reader. For subscription and membership information, contact Air & Space/Smithsonian, P.O. Box 420113, Palm Coast, FL 32142-0013.

Aviation Week & Space Technology is a weekly publication by McGraw-Hill that is considered by many readers to be the most important and authoritative aerospace publication of both broad and detailed interest. For subscription information, call (800) 525-5003.

199

Spacecraft and Mission Design

Wallace T. Fowler
University of Texas, Austin

The design of a spacecraft and the design of its mission are closely related parts of the same iterative process. The mission concept usually comes first, but new missions for existing spacecraft designs have also been developed. The key to any successful design is the development of a good set of requirements. In any case, the mission concept is often changed by design factors, and the spacecraft design is driven by mission requirements. Effort spent in defining and refining requirements is crucial to good design.

The design process starts with either a mission concept or spacecraft and arrives at a complete design. However, every design process is unique, and no design is optimal. Factors will always arise that do not seem to fit. When this happens, the mission requirements or the design process should be modified to accommodate or eliminate the unique factors.

The match between spacecraft and mission is determined by a number of factors. The two most important preliminary design factors are the mass of the mission payload and the amount of velocity change, ΔV, that must be provided to the payload to carry out the mission. These two factors dictate much of the mission and spacecraft design, as they are the primary drivers for the mass of the booster assembly. Other primary design factors are mission duration and destination to which the payload is being sent. A 2-day mission on which the spacecraft stays near the Earth will require a very different hardware/mission plan combination than will a multiyear mission to an outer planet. If a spacecraft is manned, its design and mission planning are greatly complicated. Safety factors that are acceptable for unmanned flight are unacceptable for manned flight. Life support and safety equipment, as well as backups for such systems, dominate the design. One factor that should not be ignored during the design

0-8493-1586-7/05/$0.00+$1.50

of a spacecraft/mission combination is operations. A detailed operations plan is as important to a successful space mission as is good hardware and a good mission plan. Inclusion of operations planning in the initial spacecraft and mission design process, including planning for contingencies, has the potential for lowering mission costs, increasing the chances of mission success, and lowering the potential for problems caused by unforeseen operator/hardware interactions.

199.1 Spacecraft Environments

Every spacecraft must be able to survive in several environments. All those built on earth must be able to survive the earth's atmosphere or be protected from it. The Earth's standard atmosphere is described in Chapter 198, and atmosphere models for other planets are available in NASA literature. The launch environment, characterized by vibration, noise, *g*-loads, aerodynamic loads, transition from air to vacuum, and so on, constitutes a major test for spacecraft. The space environment presents another set of problems for the designer. Hard vacuum, radiation, and temperature extremes are common to all missions. Spacecraft that fly through or beyond the Van Allen radiation belts experience more severe radiation hazards than those spacecraft that stay inside the Van Allen belts. For manned spacecraft, extended flight beyond the protection of the Van Allen belts implies special crew radiation shielding. Orbital debris is also a hazard that must be considered.

Vehicles that reenter the Earth's atmosphere or enter the atmosphere of another planet must contend with entry aerodynamics and accompanying loads and heating. For landers, abrasive dust can be a problem. Finally, if a spacecraft has a crew, its internal environment must support life while its structure protects the habitable volume. Good discussions of the role of environment in spacecraft design can be found in Fortescue and Stark [2003] and Griffin and French [1991].

199.2 Fundamental Principles

The primary factor linking the spacecraft and the mission scenario is linear momentum change, and this is the key driver in overall sizing of the spacecraft and its booster system. Except for missions dominated by attitude-control requirements, the mission scenario that requires the lowest total linear momentum change will be the most efficient in terms of propellant. Thus, in designing mission and spacecraft, it is important to (1) explore many mission scenarios to ensure that the total momentum change required is as low as possible; (2) use staging to discard unneeded mass and reduce the propellant required during each mission phase; (3) employ engine/propellant combinations that produce large momentum changes with relatively small amounts of propellant (i.e., propellants with high exhaust velocities relative to the vehicle); (4) keep all spacecraft component masses as small as possible; and (5) use unmanned spacecraft whenever possible to keep vehicle masses small. It is also important to use space-qualified hardware when available (and appropriate) to reduce cost and risk.

The ability of a propulsion system to impart momentum changes to a spacecraft is the velocity of the exhaust with respect to the spacecraft. The key equation is

$$c = I_{sp}g \tag{199.1}$$

where c is exhaust velocity relative to the vehicle, I_{sp} is specific impulse, and g is the acceleration of gravity at the surface of the Earth. c is in m/sec (ft/sec), g is in m/sec^2 (ft/sec^2), and I_{sp} is in units of seconds.

Note: The value used for g in Equation (199.1) does not change when the spacecraft is in a non-Earth gravity field.

Propellant combinations (fuel and oxidizer) are often listed in propulsion references with a corresponding I_{sp} value. These values imply an appropriate engine in which to combine the propellants efficiently. In spacecraft design studies, trade-offs between propellant storability and I_{sp} are common. I_{sp}

values for most large solid rockets are just under 300 seconds. The I_{sp} value for the space shuttle main engines (liquid oxygen–liquid hydrogen) is more than 400 seconds.

A second basic equation, which relates the mass of the spacecraft assembly at the *i*th stage in the mission (the total mass to be propelled at that point in the mission) to propellant mass that produces the velocity change, is

$$\Delta V_i = c_i \ln(m_{oi} / m_{fi}) \tag{199.2}$$

where ΔV_i is the velocity increment provided to the spacecraft assembly by the propellant in the *i*th stage, c_i is the exhaust velocity of the propulsion system in the *i*th stage relative to the spacecraft assembly, m_{oi} is the mass of the spacecraft assembly prior to burning the propellant in the *i*th stage, and m_{fi} is the mass of the spacecraft assembly after the *i*th stage propellant is expended. The mass of the propellant burned in the *i*th stage is the difference between m_{oi} and m_{fi}.

Sequential staging is a very important concept. Mass that is no longer needed should be discarded as soon as is practical in order to avoid using propellant to change the momentum of useless mass. The following equation shows the total ΔV for a two-stage vehicle.

$$\Delta V = c_1 \ln(m_{o1} / m_{f1}) + c_2 \ln(m_{o2} / m_{f2}) \tag{199.3}$$

When staging occurs, excess hardware (tanks, engines, pumps, structure) is dropped after a stage burns out and before the next stage begins its burn. Thus, m_{f1} will be larger than m_{o2}. This difference is one of the most important factors in space vehicle design and performance. A good discussion of staging can be found in Wertz and Larson [1998].

Equation (199.2) does not apply directly when the two different types of engines with different I_{sp} values are being used simultaneously. The space shuttle first stage (with solid rocket boosters and main engines burning simultaneously) and the Delta booster (with strap-on solid rockets around a liquid rocket core vehicle) are examples of this. A simple modification of Equation (199.2) applies, however. This equation looks like Equation (199.2) but has a significant difference in the definition of the exhaust velocity. For such vehicle mission stages,

$$\Delta V = c^* \ln(m_o / m_f) \tag{199.4}$$

where c^* is a weighted combination of the exhaust velocities of the two types of engines being used. Specifically,

$$c^* = f_1^* c_1 + f_2^* c_2 \tag{199.5}$$

where c_1 and c_2 are the exhaust velocities for the two engine types being used, and f_1 and f_2 are the respective mass flow rate fractions for the two engine types (note that $f_1 + f_2 = 1$). Equations (199.4) and (199.5) allow characteristic ΔV values to be calculated for vehicle stages involving parallel burning of dissimilar engine types.

199.3 Spacecraft/Mission Categories

Launch Vehicles

Launch vehicles and launches are a common element/system for almost all spacecraft. The choice of launch vehicle is determined by the mission, the mass to be boosted to orbit, the required ΔV, the compatibility between launcher and payload, and the overall cost of the launch (launcher plus launch services). The *International Reference Guide to Space Launch Systems* [Isakowitz, 1999] is an excellent source of information on launchers. A good rule of thumb to keep in mind is that for eastward launches into low Earth orbit (LEO) from a latitude of about 30°, between 9300 m/sec (30,500 ft/sec) and 9,450 m/sec (31,000 ft/sec) total ΔV must be supplied by the booster. (This accounts for gravity and drag losses as well as increasing the potential and kinetic energy of the spacecraft.) The time from launch to orbit insertion will depend on the booster acceleration capabilities and constraints, but will generally be less than 8 min.

Sounding Rockets

Sounding rockets are usually small, and their missions are characterized by short mission duration and short times out of the atmosphere of the Earth (or another planet). Their objectives are usually scientific, their equipment is powered by batteries, and they usually possess a recovery system for the instrumentation package (or perhaps the entire vehicle). Sounding rocket payloads usually consist of structure, instrumentation, a control system, a power system, communications, and a recovery system. The mass and ΔV requirement for a sounding rocket depend primarily on the mass to be boosted, the propellant combination to be used, and the altitude to which the payload is to be sent.

Earth-Orbiting Spacecraft

There are many varieties of Earth-orbiting spacecraft. Examples include the original Sputnik and Explorer spacecraft, the space shuttle, and communications satellites such as Intelsat. There are resource mappers, weather satellites, military satellites, and scientific satellites. There are many active satellites in orbit around the Earth today. Useful satellite lifetimes range from a few hours to years. It is likely that the Lageos satellite, a passive laser reflector, will be providing useful data for at least 100 years. Typical Earth-orbiting spacecraft systems are the structure, power system, thermal management system, communications, sensors, computation/data storage, propulsion, and possibly a deorbit/entry system. Launch dates, orbits, and missions of a wide variety of Earth-orbiting spacecraft have been summarized in compact form in the *TRW Space Log* [Thompson, 1987]. Especially useful are the summary documents in this series.

Lunar Spacecraft

A special category of Earth-orbiting spacecraft meriting separate mention is lunar spacecraft. Such spacecraft must climb out of the Earth's gravity well, requiring a ΔV of more than 3100 m/sec (10,200 ft/sec) to reach the moon starting in LEO. Spacecraft that orbit the moon require about 1070 m/sec (3500 ft/sec) to enter an orbit about the moon and 2070 m/sec (6800 ft/sec) to land. Special thermal considerations are necessary for vehicles designed to rest on the lunar surface because of the slow rotation rate of the moon (the lunar day and night are 14 Earth days each). Lunar spacecraft have the systems typical of Earth-orbiting spacecraft, with the possible exception of the deorbit/entry system.

Interplanetary Spacecraft

Interplanetary spacecraft have much in common with lunar vehicles, but they also have many unique characteristics. Spacecraft targeted to the outer planets have longer mission times and different power supplies than those that remain in the inner solar system. The inner planets (Venus and Mercury) present particular problems for the designer of landers. The atmosphere of Venus is hot, dense, and corrosive. The surface of Mercury, except near the poles, sees wide swings in temperature. A point on the equator of Mercury sees 88 Earth days of sunlight followed by 88 Earth days of darkness. Obviously, structure, materials, and thermal systems are important for such spacecraft. Injection onto a minimum-energy transfer to Mars requires 3600 m/sec (11,800 ft/sec). Orbiting Mars will require another 2100 m/sec (7000 ft/sec).

Entry Vehicles

Entry vehicles are spacecraft (or spacecraft segments) designed to use aerodynamic lift and drag to slow or turn the spacecraft, leading to an orbital plane change, lowering of an orbit, or a descent into the planetary atmosphere. Aerothermodynamic considerations (both stagnation point temperature and integrated heat load) and deceleration loading dominate the design of such vehicles. Aerodynamic control surfaces are the primary means of trajectory shaping, which in turn controls the heating and deceleration. Survivability is a primary design consideration. An entry vehicle is a special type of hypervelocity vehicle, usually a glider (see Chapter 198 for additional information).

Landers

Landers are a special class of spacecraft. Some landers are also entry vehicles and some are not, depending on whether the body on which the landing is to take place has an atmosphere. Landings on targets with atmospheres can involve parachutes, cushions, shock absorbers, and so on. Landings on airless bodies usually involve either rocket braking or braking via penetration of the surface of the target body.

Returning Spacecraft

Spacecraft that return to the planet of origin are a very special case. Vehicles that fly to Earth orbit and then deorbit and land on Earth, although complex, are much simpler than vehicles that fly from the Earth to a target, land operate, launch, return to Earth, enter the atmosphere, and are recovered. For such vehicles, mass is by far the most important driver for all design decisions, and repeated staging is the primary feature of the overall design. Often the redesign of the staging process can greatly simplify an otherwise complex vehicle and mission (the lunar orbit rendezvous used in the Apollo program is an excellent example).

199.4 Spacecraft Subsystems

Once the overall mission and spacecraft requirements have been defined, the process of defining requirements for various spacecraft subsystems can begin. The integration of the various subsystems into a harmoniously functioning spacecraft can be very difficult. One way to reduce the difficulties in spacecraft integration is to spend sufficient time and effort *early* in carefully and completely defining the interfaces between the subsystems of the spacecraft. Time spent in defining interfaces will repay itself many times as the design progresses. The following is a brief discussion of the major design considerations of the principal spacecraft subsystems. Chetty [1991], Fortescue and Stark [2003], and Griffin and French [1991] provide excellent discussions of various spacecraft subsystems.

Structure Subsystem

The structure of the spacecraft holds the components together, provides load paths for launch forces, maneuver loads, and so on, and is usually an integral part of the thermal control system and/or the electrical system for the spacecraft. Design constraints that the structural subsystem must usually meet include stiffness requirements, placing principal inertia axes in preferred directions, sustaining mission loads, and serving as a ground for all spacecraft electrical equipment. Spacecraft structures can be metallic, composite, or ceramic. Structural mass is generally 5 to 20% of the overall spacecraft launch mass.

Power Subsystem

The type of power system chosen for a spacecraft is strongly dependent on the mission of the spacecraft and the power demands of the equipment to be carried. Short-duration missions often use batteries. Longer-duration missions that go no farther from the sun that Mars can use photoelectric power (solar cells). Long-duration, low-power missions on which the spacecraft go very far from the sun (e.g., Voyager) use radioisotope thermoelectric generators (RTGs). Other potential power sources are solar thermionic systems, chemical dynamic systems, and nuclear power systems. Limitations on power often place constraints on spacecraft designs and mission operations.

Attitude Control Subsystem

Most spacecraft require attitude stabilization to point antennas, properly orient solar arrays, point sensors, and orient thrusters. Attitude control consists of sensing the current attitude of the spacecraft and applying torques to reorient the spacecraft to a desired attitude.

Three-Axis Stabilization

The most complex and precise spacecraft attitude control is three-axis stabilization accomplished using momentum wheels or reaction wheels. Attitude thrusters are also used for three-axis stabilization, either as the primary attitude control mechanism or to desaturate the momentum wheels. Three-axis stabilization systems can be fast and accurate. For example, the attitude control system on the Hubble Space Telescope, which uses reaction wheels, can point the optical axis of the telescope to within 0.007 arc seconds of the desired direction.

Spin Stabilization

Spin stabilization is much simpler and less precise than three-axis stabilization. Spin-stabilized spacecraft must be rigid and spinning about the principal axis with the largest mass moment of inertia. The first U.S. orbiting satellite, Explorer 1 (launched 31 January 1958), provided an early demonstration of the need for spinning spacecraft to be rigid. Explorer 1 was a long, slender cylinder spinning about its long axis. It possessed flexible whip antennas that dissipated energy and caused the spacecraft to tumble within a day of orbit insertion.

Dual-Spin Spacecraft

Some spacecraft have a spinning section and a despun section. Such spacecraft must have systems to maintain the attitude of the despun section and to control the direction of the spin vector. Reactions between the spinning and despun sections are major design and operational considerations.

Gravity Gradient Stabilization

Spacecraft can use the gradient of the gravitational field of a planet to provide restoring torques and maintain the spacecraft in an orientation with the spacecraft's long axis pointing along the local vertical. Spacecraft designed to employ gravity gradient stabilization usually feature mass concentrations at the ends of a long boom that lies along the local vertical. Gravity gradient stabilization works best with near-circular orbits and would not work on even moderately elliptic orbits. A pointing accuracy of about 1° with some wobble is achievable with gravity gradient stabilization of Earth satellites.

Magnetic Stabilization

Spacecraft attitudes can be stabilized or changed using torques developed by running currents through coils (torque rods) mounted in the spacecraft if the spacecraft is orbiting a body with a significant magnetic field. The primary difficulty with this type of attitude stabilization system is keeping track of the relative orientations of the spacecraft axes and the magnetic field lines as the spacecraft moves along in its orbit. Magnetic stabilization is slow and coarse.

Solar Radiation Stabilization

The pressure of solar radiation falling on a spacecraft can be used to create torques. Spacecraft that employ solar radiation stabilization are large and lightweight.

Telecommunications Subsystem

The primary function of the telecommunications subsystem is to receive, process, and transmit electronic information. Some satellites (communications, intelligence, and weather satellites) are designed to meet one or more telecommunications mission objectives. The telecommunications subsystem often demands a significant amount of electrical power, and this demand often drives the design of the power system. The spacecraft computer/sequencer can be considered part of the telecommunications system or a separate system. Chetty [1991], Fortescue and Stark [2003], and Griffin and French [1991] provide excellent discussions of spacecraft telecommunications.

The telecommunications system often drives the attitude and pointing requirements for the spacecraft. Furthermore, the capabilities of the spacecraft telecommunications antenna(s), coupled with limited power available for signal transmission, often require the use of large and expensive receiving antennas on Earth. This can greatly increase operational costs. Finally, deployable spacecraft antennas carry additional risks. The very successful Galileo mission was severely impacted by the failure of the onboard high-gain antenna to deploy properly.

Propulsion Subsystem

The spacecraft propulsion subsystem will be considered separately from the launch vehicle and upper stages. The propulsion subsystem is responsible for orbit maintenance, small orbit changes, attitude control system desaturation, and so on. Spacecraft onboard propulsion systems use chemically reacting fuels and oxidizers, monopropellants and catalysts, cold gas, and ion thrusters. Chetty [1991], Fortescue and Stark [1992], and Griffin and French [1991] provide excellent discussions of propulsion systems.

Thermal Control Subsystem

Spacecraft thermal control is of utmost importance. Maintaining appropriate thermal environments for the various spacecraft is essential to proper component function and longevity. The primary problem with thermal management of spacecraft is that all excess heat must ultimately be radiated by the spacecraft. For many unmanned spacecraft, passive thermal control involving coatings, insulators, and radiators is ideal. For other spacecraft, active thermal control is necessary. Active systems can employ heaters, radiators, heat pipes, thermal louvers, and flash evaporators. Significant spacecraft mass reductions can sometimes be made if the excess heat from one component is conducted to another component that would otherwise require active heating.

Spacecraft Mechanisms

Some spacecraft are mechanically passive and have no mechanisms in their designs. However, even these spacecraft require mechanisms to deploy them from their boosters. Mechanically active spacecraft can feature a wide variety of mechanisms. There are deployment systems for booms and antennas, docking mechanisms, robot arms, spin/despin systems, sensor pan and tilt mechanisms, gyroscopes, reaction wheels (also considered as part of the attitude control system), landing legs, and airlocks. The types of mechanisms depend on the mission of the spacecraft.

Launch Vehicles

The launch vehicle, or booster, is an integral part of most spacecraft systems. The *International Reference Guide to Space Launch Systems* [Isakowitz, 1999] covers the capabilities of most currently available launch vehicles in detail. The rotation of the Earth has a major effect on launch vehicle performance. The eastward motion of the launch site must be considered when computing the payload that a launch vehicle can place in orbit. The launch azimuth and the latitude-dependent eastward velocity of the launch site can be combined to determine how much velocity assist is provided by Earth's rotation. For westward launches (to retrograde orbits), the rotation of the Earth extracts a penalty, requiring additional velocity to be provided by the booster to make up for the eastward velocity of the launch site.

Upper Stages

Launch vehicles are often combined with upper stages to form a more capable launch vehicle. Commonly used upper stages include the IUS, the Centaur, and several types of PAM (payload assist modules). Upper stages are also carried aloft in the payload bay of the space shuttle to boost spacecraft to higher orbits. Designers should always consider the option of using an available upper stage if the mission requires more ΔV than can be provided by the basic launch vehicle.

Entry/Landing Subsystems

Entry subsystems can be considered either as a separate set of devices or as parts of other subsystems. The heat shield can be reusable or ablative, and could be considered to be part of the thermal control subsystem. The retrorockets, if any, might be considered to be part of the propulsion system. Parachutes, landing legs, cushions, wings, aerosurfaces, a streamlined shape, and landing gear, when needed, belong solely to the category of entry and landing subsystems.

Subsystem Integration and Redundancy

The interplay among spacecraft subsystems is great, and their integration into an efficient, smoothly operating, and reliable spacecraft is an immense challenge. Redundant system elements can provide added reliability if the risk of failure must be lowered. However, redundancy increases initial costs and usually increases overall mass. It is often good to avoid physically redundant elements, choosing physically different but functionally redundant elements instead. System elements that are to be functionally duplicated should be identified as early in the design process as possible.

199.5 Spacecraft/Mission Design Process

The spacecraft/mission design process can be outlined as follows:

1. Beginning with a set of mission goals, develop a set of mission and spacecraft requirements that must be met in order to achieve the goals. An excellent discussion of requirements and how to develop them is found in Wertz and Larson [1998]. Mission goals may change during the design process, and when this happens, the requirements must be reevaluated.
2. Before beginning a methodical approach to the design, it is almost always helpful to bring the design team together and conduct a brainstorming session. For those who are unfamiliar with brainstorming, Adams [1986] is recommended. The brainstorming session should be expected to identify several unorthodox candidate scenarios or components for the mission and spacecraft.
3. Develop a conceptual model for the spacecraft and the mission, identifying major systems and subsystems, and most important, the interfaces between systems, both within the spacecraft itself and between the spacecraft and its support elements on the ground and in space. A preliminary mission scenario and rough timeline should also be developed at this time. These conceptual elements should be developed from the mission and spacecraft requirements. This step will allow the definition of major systems and subsystems and will identify major interactions between mission events and spacecraft hardware elements. Preliminary indications of pointing requirements, required sensor fields of view, ΔV requirements, allowable spacecraft mass, and required and desired component lifetimes should come from these considerations. Brown [1998], Fortescue and Stark [2003], Griffin and French [1991], and Isakowitz [1999] give information appropriate to this topic.

4. The role of heritage in spacecraft design is important. Systems and technologies that have worked well on past spacecraft are often good candidates for new spacecraft with similar missions. Examine the recent past for similar missions, spacecraft with similar requirements, and so on. Although sufficiently detailed systems overviews of recent spacecraft are often difficult to obtain, the effort necessary to obtain them is almost always worth the effort. However, the design team should be careful to avoid early adoption of a candidate system from earlier spacecraft in order to avoid being locked into a system that only marginally meets or does not meet design requirements.
5. Once the first four steps have been taken (some more than once), the design team should develop preliminary spacecraft design candidates (spacecraft/scenario/timeline combinations), concentrating on systems and subsystems and their performance, masses, power requirements, system interactions, thermal input/output, and costs. Several candidate designs should be considered at this stage and their predicted relative characteristics compared. These considerations allow the design team to learn more about the requirements and how best to meet them. It is likely that the design team will recycle through some or all of the design process at this point for one or more of the candidate designs. The possibility of developing a new composite design with the best features of several candidate designs should not be ignored.
6. As candidate designs are refined, the design team should develop criteria for use in choosing among candidate designs. The information on the candidate designs should be organized and presented at a preliminary design review. The feedback from those attending the review is valuable in the identification of strengths and weaknesses of candidate designs.
7. At some point, determined by budget, timeline, technical considerations, and the characteristics of the candidate designs, one design concept must be chosen for development. Once the choice is made, the design team should carry out a complete analysis of the chosen design. This analysis should include a detailed mission scenario and timeline, launch vehicle choice, ΔV requirements for every maneuver, hardware choices for every system and subsystem, system interaction analyses, manufacturability considerations, component and overall cost analyses, and failure mode effects analyses. Any remaining system/subsystem hardware choices should be made, and the design should be presented at a critical design review. The feedback from the critical design review will often result in major improvements in the design. Questions arising at the critical design review will often precipitate another pass through the design process.
8. The effects of real or imagined environmental impacts on the public of mission and/or spacecraft features and resulting political pressures cannot be overemphasized. The Galileo mission to Jupiter provides good examples. First, there were public protests at the Kennedy Space Center in Florida because Galileo carried a "nuclear" power device (an RTG). Most protestors had no idea that RTGs had been launched safely for over 2 decades and that RTGs carried by Apollo were still in operation on the Moon. Also, since Galileo flew by Venus and then used an Earth fly-by to provide a gravity assist to help get it to Jupiter, there were those who predicted that this "nuclear" device might drift off course and hit the Earth. Mission designers must learn to anticipate such protests and provide pertinent information to the public in a straightforward and clear format.

Defining Terms

Mission scenario — A sequence of events and times that, in their entirety, meet the objectives of the mission.

Specific impulse — The amount of thrust force obtained from a fuel/oxidizer/engine combination when a unit weight of fuel is burned in 1 second. The units of specific impulse are seconds because the force and weight units cancel.

Subsystem — A part of an overall system that can be isolated to some degree. Subsystems interact and cannot actually be isolated, but it is convenient to consider subsystems in sequence rather than all at once.

References

Adams, J. L. 1986. *The Care and Feeding of Ideas: A Guide to Encouraging Creativity*, 3rd ed. Addison-Wesley, Reading, MA.

American Institute of Aeronautics and Astronautics. 1998. *AIAA Aerospace Design Engineers Guide*, 4th ed. American Institute of Aeronautics and Astronautics, Washington, DC.

Brown, C. D. 1998. *Spacecraft Mission Design*, 2nd ed. AIAA Education Series, American Institute of Aeronautics and Astronautics, Washington, DC.

Chetty, P. R. K. 1991. *Satellite Technology and Its Applications*. TAB Professional and Reference Books, Blue Ridge Summit, PA.

Fortescue, P. and Stark, J. 2003. *Spacecraft Systems Engineering*, 3rd ed. John Wiley & Sons, New York.

Griffin, M. D. and French, J. R. 1991. *Space Vehicle Design*. AIAA Education Series, American Institute of Aeronautics and Astronautics, Washington, DC.

Isakowitz, S. J., Ed. 1999. *International Reference Guide to Space Launch Systems*, 3rd ed. AIAA Space Transportation Technical Committee, AIAA, Washington, DC.

Sellers, J. J. 1994. *Understanding Space — An Introduction to Astronautics*. McGraw-Hill, New York.

Thompson, T. D., Ed. 1987. *TRW Space Log, 1957–1987*. TRW Space & Technology Group, Redondo Beach, CA.

Wertz, J. R. and Larson, W. J., Eds. 1998. *Space Mission Analysis and Design*, 3rd ed. Space Technology Library, Kluwer Academic, Boston.

Further Information

An excellent and inexpensive source of information on the orbital mechanics necessary for mission planning is Bate, R. R., Mueller, D. D., and White, J. E. 1971. *Fundamentals of Astrodynamics*. Dover, New York.

Details on the locations of the planets, their moons, the orbit of the Earth about the sun, values of astrodynamical constants, and many other items pertinent to mission planning are found in *The Astronomical Almanac*, from the Nautical Almanac Office, Naval Observatory, published yearly by the U.S. Government Printing Office, Washington, DC; and Seidelmann, P. K., Ed. 1992. *Explanatory Supplement to the Astronomical Almanac*. University Science Books, Mill Valley, CA.

An excellent overview of the interplay between spacecraft, launch vehicle, trajectory, and operations is given in Yenne, B., Ed. 1988. *Interplanetary Spacecraft*. Exeter, New York.

XXVII

Safety

200 Hazard Identification and Control

Mansour Rahimi
University of Southern California

Injuries and illnesses costs were estimated at $480 billion to the U.S. economy in 1998 [National Safety Council, 2003]. Yet engineered systems are often designed without an extensive evaluation for their potential hazards. A look at multidisciplinary literature in the field of safety and health shows that a large number of tools and techniques are available for hazard identification, analysis, and control. However, little standardization exists for using these techniques across fields such as engineering, physics, chemistry, sociology, psychology, business, and law. Specifically, design engineers do not make extensive use of safety resources commonly recommended by the safety community, nor are they fully familiar with hazard identification and analysis techniques [Main and Frantz, 1994]. In fact, as the systems under study (e.g., product, machine, production cell, shop floor) become more complex in terms of hardware, software, and human, environmental, and organizational variables, the need to employ a more comprehensive hazard evaluation and control methodology becomes more critical. In addition, the recent developments in environmentally conscious design, industrial ecology, and sustainable development require a deeper understanding of hazard consequences within and across the product life-cycle span (i.e., design, development, production, operation, deployment, use, and disposal). Therefore, successful safety programs require significant effort and resources on the part of a multidisciplinary team of investigators with support from all levels of management within a safety-cultured organizational structure [Hansen, 1993].

200.1 Hazard Identification

It is imperative that engineers become aware of *hazard* prevention principles early in their education. Also, the principles and techniques outlined in this chapter work best when applied early in the design

0-8493-1586-7/05/$0.00+$1.50

of systems. The negative impact of hazards (e.g., accidents, illnesses) will be far greater (sometimes irreversible) in the later stages of the system development.

One way of classifying the sources of hazards is by the type of dominant (hazardous) energy used in the operation of the system. This energy is normally traced back to the root cause of an injury or illness. In this chapter, the following primary hazards are considered: physical, chemical, airborne contaminants, noise, and fire. Other hazards are extensively considered in the reference material, and ergonomic hazards are discussed elsewhere in this handbook. The objective of any hazard control program is to control losses from injuries and damages by identifying and controlling the degree of exposure to these sources of energy.

200.2 Physical Hazards

Physical hazards are usually in the form of kinematic force or lifting or impact by or against an object. More than 50% of compensable work injuries are included in this category.

Human Impact Injuries

The severity of an injury depends on the velocity of impact, magnitude of deceleration, and body size, orientation, and position. The kinetic energy formula used to describe the impact injury is

$$E_{\text{ft-lb}} = (Wv^2)/2g \tag{200.1}$$

where W is the weight of an object or part of the body (lb), v is the velocity (ft/sec), and g is gravity (ft/sec^2). However, if the impacting surface is soft, the kinetic energy for the impact is

$$E_{\text{ft-lb}} = [W(2sA)]/2g \tag{200.2}$$

where s is the stopping distance (ft), and A is the deceleration (ft/sec^2). For example, for both of the above cases, the human skull fracture occurs at 50 ft-lb of kinetic energy. Hard hats are expected to prevent the transfer of this energy to the human skull.

Trip, Slip, and Fall

These injuries comprise 17% of all work-related injuries. Falls are the second largest source of accidental deaths in the U.S. (after motor vehicle accidents). Jobs related to manufacturing, construction, and retail and wholesale activities are the most susceptible to these types of hazards, comprising about 27% of all worker compensation claims in the U.S. These hazards include slipping (on level ground or on a ladder), falling from a higher level to a lower one (or to the ground), falling due to the collapse of a piece of floor or equipment, and failure of a structural support or walkway. Principles of tribology are being used to study the control mechanisms for these accidents. Tribology is the science that deals with the design and analysis of friction, wear, and lubrication of interacting surfaces in relative motion. The basic measure of concern is the coefficient of friction (COF), which in its simplest form is the horizontal force divided by the vertical force at the point of relative motion. A COF greater than 0.5 appears to provide sufficient traction for normal floor surfaces. However, a number of other conditions make this hazard evaluation difficult: unexpectedness of surface friction change versus human gait progression, foreign objects and debris on a path, walkway depression, raised projections (more than 0.25 in.), change in surface slope, wet surfaces, improper carpeting, insufficient lighting, improper stair and ramp design, improper use of ladders, guardrails and handrails, and human visual deficiencies (color weakness, lack of depth perception and field of view, inattention, and distraction).

Mechanical Injuries

The U.S. Occupational Safety and Health Act specifically states that one or more methods of machine guarding shall be provided the operator and other employees to protect them from hazards such as those created by point of operation, ingoing nip points, rotating parts, flying chips, and sparks. Other hazards in this category are cutting by sharp edges, sharp points, poor surface finishes, and splinters from wood and metal parts; shearing by one part of a machine moving across a fixed part (e.g., paper cutters or metal shearers); crushing of skin or tissue caught between two moving parts (e.g., gears, belts, cables on drums); and straining of a muscle (overexertion) in manual lifting, pushing, twisting, or repetitive motions.

Another important category is the hazard caused by pressure vessels and explosions. These hazards are generally divided into two types — boilers that are used to generate heat and steam, and unfired pressure vessels that are used to contain a process fluid, liquid, or gas without direct contact of burning fuel. The American Society of Mechanical Engineers (ASME) covers all facets of the design, manufacture, installation, and testing of most boilers and process pressure vessels in the Boiler and Pressure Vessel Code (total of 11 volumes). The primary safety considerations relate to the presence of emergency relief devices or valves to reduce the possibility of overpressurization or explosion. Explosions can be classified on the basis of the length-to-diameter ratio (L/D) of the container. If the container has an L/D of approximately one, the rise in pressure is relatively slow and the overpressurization will cause the container to rupture. In containers with a large L/D ratio, such as gas transfer pipes and long cylinders, the initial flame propagates, creating turbulence in front of it. This turbulence improves mixing and expansion of the flame area and the speed of travel along the vessel, increasing the pressure by as much as 20 times very rapidly. Other types of explosions are caused by airborne dust particles, boiling liquid and expanding vapor, vessels containing nonreactive materials, deflagration of mists, and runaway chemical reactions.

Explosions may have three types of effects on a human body. First, the blast wave effect that carries kinetic energy in a medium (usually air), which although decaying with distance, can knock a person down (overpressure of 1.0 $lb/in.^2$) or reach a threshold of lung collapse (overpressure of 11 $lb/in.^2$). The second type of effect is thermal, usually resulting from the fire. The amount of heat radiated is related to the size of the fireball and its duration of dispersion. The radiant energy is reduced according to the inverse distance-squared law. Most explosive fireballs reach temperatures about 2400°F at their centers. The third effect is the scattering of the material fragments. All three of these injury effects are multifactorial and they are very difficult to predict.

200.3 Chemical Hazards

A Union Carbide plant in Bhopal, India, accidentally leaked methyl isocyanate gas from its chemical process. It left over 2500 dead and about 20,000 injured. There are over 3 million chemical compounds, and an estimated 1000 new compounds are introduced every year.

Hazardous/Toxic Substances

The health hazards of chemicals can be classified into acute or chronic. The acute ones are corrosives, irritants, sensitizers, and toxic and highly *toxic substances.* The chronic ones are carcinogens, liver, kidney and lung toxins, bloodborne pathogens, nervous system damages, and reproductive hazards. For a reference listing of maximum exposures to these substances, refer to the technical committees of the American Conference of Governmental Industrial Hygienists (ACGIH) and American Industrial Hygiene Association (AIHA). Rules and standards related to manufacture, use, and transportation of these chemicals are promulgated and their *compliance* is enforced by governmental agencies such as the Occupational Safety and Health Administration (OSHA), the Environmental Protection Agency (EPA) and the Department of Transportation (DOT).

Routes of Entry

There are four ways by which toxic substances can enter the human body and cause external or internal injuries or diseases [Shell and Simmons, 1990].

1. Cutaneous (on or through the skin)
   a. Corrosives damage skin by chemical reaction
   b. Dermatitis is caused by irritants such as strong acids; sensitizers, such as gasoline, naphtha, and some polyethylene compounds
   c. Absorbed through skin, but affecting other organs
2. Ocular (into or through the eyes)
   a. Corneal burns due to acids or alkali
   b. Irritation due to abrasion or chemical reaction
3. Respiratory inhalation (explained later in Section 200.4)
4. Ingestion
   a. Toxic substances may be ingested with contaminated food
   b. Fingers contaminated with toxic chemicals may be placed in mouth
   c. Particles in the respiratory system are swallowed with mucus

Mechanisms of Injury

Toxic agents cause injury by one or a combination of the following mechanisms.

1. Asphyxiants
   a. Asphyxia refers to a lack of oxygen in the bloodstream or tissues with a high level of carbon dioxide present in the blood or alveoli
   b. Gas asphyxiants (e.g., carbon dioxide, nitrogen, methane, hydrogen) dilute the air, decreasing oxygen concentration
   c. Chemical asphyxiants make the hemoglobin incapable of carrying oxygen (carbon monoxide) or keep the body's tissues from utilizing oxygen from the bloodstream (hydrogen cyanide)
2. Irritants that can cause inflammation (heat, swelling, and pain)
   a. Mild irritants cause hyperemia (capillaries dilate)
   b. Strong irritants produce blisters
   c. Respiratory irritants can produce pulmonary edema
   d. Secondary irritants can be absorbed and act as systemic poisons
3. Systemic poisons
   a. Poisons may injure the visceral organs, such as the kidney (nephrotoxic agents) or liver (hepatotoxic agents)
   b. Poisons may injure the bone marrow and spleen, interrupting the production of blood (benzene, naphthalene, lead)
   c. Poisons may affect the nervous system, causing inflammation of the nerves, neuritis, pain, paralysis, and blindness (methyl alcohol, mercury)
   d. Poisons may enter the bloodstream and affect organs, bones, and blood throughout the body (usually with prolonged exposure)
4. Anesthetics
   a. May cause loss of sensation
   b. May interfere with involuntary muscle actions causing respiratory failure (halogenated hydrocarbons)
5. Neurotoxins
   a. Neurotics affect the central nervous system
   b. Depressants cause drowsiness and lethargy (alcohol)
   c. Stimulants cause hyperactivity
   d. Hypnotics are sleep-inducing agents (barbiturates, chloral hydrate)
6. Carcinogens
   a. Cancers of the skin at points of contact (tar, bitumen)
   b. Cancers of internal organs and systems have numerous known and suspected causes (labeled as suspected carcinogens)

7. Teratogenic effects. A substance that may cause physical defects in the developing embryo or fetus when a pregnant female is exposed to the substance for a period of time.

200.4 Airborne Contaminants

The greatest hazard exists when contaminants are smaller than 0.5 μm, and thus can be directly introduced into the bloodstream through alveolar sacs (e.g., zinc oxide, silver iodide). Particles larger than 0.5 μm are entrapped by the upper respiratory tract of trachea and bronchial tubes (e.g., insecticide dust, cement and foundry dust, sulfuric acid mist). There are two main forms of airborne contaminants [Brauer, 1990] — particulates (dusts, fumes, smoke, aerosols, and mists), and gases or vapors.

Dusts are airborne solids, typically ranging in size from 0.1 to 25 μm, generated by handling, crushing, grinding, impact, detonation, and so on. Dusts larger than 5 μm settle out in relatively still air due to the force of gravity. Fumes are fine solid particles less than 1 μm generated by the condensation of vapors. For example, heating of lead (in smelters) vaporizes some lead material that quickly condenses to small, solid particles. Smokes are carbon or soot particles less than 0.1 μm resulting from incomplete combustion of carbonaceous material. Mists are fine liquid droplets generated by condensation from the gaseous to liquid state, or by the dispersion of same by splashing, foaming, or atomizing.

Gases are normally formless fluids which occupy space and which can be changed to liquid or solid by a change in pressure and temperature. Vapors are the gaseous form of substances that are normally in a liquid or solid state.

The measures for toxicity of the above substances are given in parts per million (ppm) for gases and vapors, and milligrams per cubic meter (mg/m^3) for other airborne contaminants. The criteria for the degree of toxicity are the threshold limit values (TLVs) based on review of past research and monitory experience. (Early OSHA standards listed permissible exposure limits, or PELs.) TLVs are airborne concentrations of substances which are believed to represent conditions to which nearly all workers may be repeatedly exposed, 8 hours a day, for lifetime employment, without any adverse effects. For acute toxins, short-term exposure levels (STELS) are indicated for a maximum of 15 minutes of exposure, not more than four times a day. Because exposures vary with time, a time-weighted average (TWA) is adopted for calculating TLVs:

$$\text{TLV(TWA)} = \sum (E_i T_i / 8) \tag{200.3}$$

where E_i is the exposure to the substance at concentration level i and T_i is the amount of time for E_i exposure in an 8-hour shift.

In many environments, there are several airborne substances present at the same time. If the effects of these substances are additive and there are no synergistic reactions, the following formula can be used for the combination of TLVs:

$$X = (C_1 / T_1) + (C_2 / T_2) + \cdots + (C_n / T_n) \tag{200.4}$$

where C_i is the atmospheric concentration of a substance and T_i is the TLV for that substance. If $X < 1$, the mixture does not exceed the total TLV; if $X \geq 1$, the mixture exceeds the total TLV.

200.5 Noise

Noise-induced hearing loss has been identified as one of the top ten occupational hazards by the National Institute for Occupational Safety and Health (NIOSH). In addition to hearing loss, exposure to excessive amounts of noise can increase worker stress levels, interfere with communication, disrupt concentration, reduce learning potential, adversely affect job performance, and increase accident potential [Mansdorf, 1993]. Among many types of hearing loss, sensorineural hearing loss is the most common form in occupational environments. Sensorineural hearing loss is usually caused by the loss of ability of the inner ear (cochlea nerve endings) to receive and transmit noise vibrations to the brain. In this case, the middle ear (the bone structures of maleus, incus, and stapes) and the outer ear (ear drum, ear canal, and ear lobe) may be intact.

A comprehensive and effective hearing conservation program can reduce the potential for employee hearing loss, reduce workers compensation costs due to hearing loss claims, and lessen the financial burden of noncompliance with government standards. Current OSHA standards require personal noise dosimetry measurements in areas with high noise levels. Noise dosimeters are instruments which integrate (measure and record) the sound levels over an entire work shift. Noise intensities are measured by the dBA scale, which most closely resembles human hearing sensitivity. For continuous noise levels, OSHA's permissible noise exposure is 90 dBA for an 8-hour shift. If the noise levels are variable, a time-weighted average is computed. For noise levels exceeding the limit values, an employee hearing conservation program must be administered [Mansdorf, 1993:318–320].

200.6 Fire Hazards

Fire is defined as the rapid oxidation of material during which heat and light are emitted. An estimated $10.5 billion in property damage occurred as a result of fires in 2001 [Karter, 2001]. The most frequent causes of industrial fires are electrical (23%), smoking materials (18%), friction surfaces (10%), overheated materials (8%), hot surfaces (7%), and burner flames (7%). The process of combustion is best explained by the existence of four elements — fuel, oxidizer (O_2), heat, and chain reaction. A material with a flash point below 100°F (vapor pressure < 40 lb/in.^{2}) is considered flammable, and higher than 100°F is combustible. In order to extinguish a fire, one or a combination of the following must be performed: the flammable/combustible material is consumed or removed, the oxidant is depleted or below the necessary amount for combustion, heat is removed or prevented from reaching the combustible material not allowing for fuel vaporization, or the flames are chemically inhibited or cooled to stop the oxidation reaction.

Fire extinguishers are classified according to the type of fire present. Class A involves solids that produce glowing embers or char (e.g., wood, paper). Class B involves gases and liquids which must be vaporized for combustion to occur. Class C includes Class A and B fires involving electrical sources of ignition. Finally, Class D involves oxidized metals (e.g., magnesium, aluminum, titanium). In addition to heat, the most dangerous byproducts of fires are hazardous gases and fumes, such as CO, CO_2, acrolein formed by the smoldering of cellulosic materials and pyrolysis of polyethylene, phosgene ($COCl_2$) produced from chlorinated hydrocarbons, sulfur dioxide, oxides of nitrogen (NO_x) resulting from wood products, ammonia (NH_3) when compounds containing nitrogen and hydrogen burn in air, and metal fumes from electronic equipment. For a complete reference to facility design requirements and existing requirements, refer to NFPA Code 101.

200.7 An Engineering Approach to Hazard Control

It is impossible to design an absolutely "safe" engineered system. Nevertheless, the following list is suggested to eliminate the critical hazards and reduce other noncritical hazards to an acceptable level, called *risk control.* It is important to mention that the effectiveness of the control mechanism is relatively reduced by using the lower items on this list.

1. Identify and eliminate the source of hazardous energy. For example, do not use high-voltage electricity. Or consider the use of noncombustible and nontoxic material in environments with fire potentials. This approach is not always practical or cost-effective.
2. Reduce the degree of hazardous energy. For example, use low-voltage solid-state devices to reduce heat buildup in areas with explosion hazards. This approach is practical in some cases, yet costly in other design applications.

3. Isolate the source of hazard. Provide barriers of distance, shields, and personal protective equipment to limit the harmful effects of the hazardous agents. To control the sequence of events in time and space, a lockout/tagout procedure is recommended.
4. Minimize failure. Include constant monitoring of critical safety parameters (e.g., gas concentrations or radiation levels). The monitoring system should detect, measure, understand, and integrate the readings, and respond properly.
5. Install a warning system. Similar to consumer product warnings, all system components should be equipped with warning and proper communication systems. Operators (and the general public) should be warned of the type of hazard present and the means by which information can be obtained in case of an accident. However, too many warning signs and display indicators may confuse the operator.
6. Ensure safe procedures. A common cause of accidents is the inadequacy of procedures and the failure to follow proper procedures.
7. Provide for backout and recovery. In case of an accident, this defensive step is taken to reduce the extent of injury and damage. This step incorporates one or more of the following actions: (a) normal sequence restoring, in which a corrective action must be taken to correct the faulty operation; (b) inactivating only malfunctioning equipment which is applied to redundant components or temporarily substituting a component; (c) stopping the entire operation to prevent further injury and damage; and (d) suppressing the hazard (e.g., spill containment of highly hazardous substances).

When hazard exposure cannot be reduced through engineering controls, an effort should be made to limit the employee's exposure through administrative controls, such as (a) rearranging work schedules and (b) transferring employees who have reached their upper exposure limits to an environment where no additional exposure will be experienced.

200.8 Hazard Analysis and Quantification

A number of *hazard analysis techniques* have been developed to study the hazards associated with a system. For a detailed review, see *Guidelines for Hazard Evaluation Procedures* [American Institute of Chemical Engineers, 1985], Gressel and Gideon [1991], and Vincoli [1993]. These techniques vary in terms of their hazard evaluation approaches and the degree to which hazard exposures can be quantified. A precursor to any of these techniques is the system, which is divided into its small and manageable components, analyzed for causes and consequences of any number of potential hazards, and then synthesized to consider hazard effects on the whole system. Five hazard analysis techniques are briefly presented here.

Preliminary Hazard Analysis

Preliminary hazard analysis (PHA) is the foundation for effective systems hazard analysis. It should begin with an initial collection of raw data dealing with the design, production, and operation of the system. The purpose of this procedure is to identify any possible hazards inherent in the system. One example is energy as the source of hazard to explore the multitude of circumstances by which an accident can occur in a system. Table 200.1 demonstrates an actual use of PHA in the design phase of metal chemical vapor deposition. The four main categories of this table are hazard, cause, main effects, and preventive control. The hazard effects and corrective/preventive measures are only tentative indicators of potential hazards and possible solutions.

Failure Mode, Effects, and Criticality Analysis

While PHA studies hazards in the entire system, failure mode, effects, and criticality analysis (FMECA) analyzes the components of the system and all of the possible failures that can occur. This form of analysis identifies items whose failures have a potential for hazardous consequences. In this analysis, each item's

TABLE 200.1 A Sample for Application of Preliminary Hazard Analysis to Design of Metal Organic Chemical Vapor Deposition (Only Two Hazards Are Listed)

Hazard	Cause	Main Effects	Preventive Control[a]
Toxic gas release	Leak in storage cylinder	Potential for injury and fatality from large release	Develop purge system to remove gas to another tank Minimize on-site storage Provide warning system Develop procedure for tank inspection and maintenance Develop emergency response system
Explosion, fire	Overheat in reactor tube	Potential for fatalities due to toxic release and fire Potential for injuries and fatalities due to flying debris	Design control system to detect overheat and disconnect heater Provide warning system for temperature fluctuation, evacuate reaction tube, shut off input valves, activate cooling system

[a] This column is simplified to show the major categories of hazard control techniques.
Source: Adapted from Kavianian, H. R. and Wentz, C. A. 1990. *Occupational and Environmental Safety Engineering and Management*. Van Norstrand Reinhold, New York.

TABLE 200.2 A Sample for the Application of Failure Mode, Effects, and Criticality Analysis to the Metal Organic Chemical Vapor Deposition Process (Only Three System Components Are Listed)

System Component	Failure Mode	Effects	Criticality Ranking[a]
Reactor tube	Rupture	Release of pyrophoric gas causing fire and release of toxic gases	III
Control on reactor heater	Sensor fails; response control system fails	Reactor overheating beyond design specification	II
Refrigeration equipment	Failure to operate	Increase in vapor pressure; cylinder rupture	IV

[a] Criticality ranks are based on a scale from I to IV.
Source: Adapted from Kavianian, H. R. and Wentz, C. A. 1990. *Occupational and Environmental Safety Engineering and Management*. Van Norstrand Reinhold, New York.

function must be determined. Once this is done, the causes and effects of the failure of the components are indicated. Then, the criticality factor of each failure is determined, and a quantified severity rating is given to the factor. Table 200.2 shows an example for FMECA. Because the frequency of each potential occurrence is also an important factor, a risk assessment matrix (depicted in Table 200.3) can be used to codify the risk assignment. A design team (including the safety engineer) can use this FMECA to redesign components or parts of the system to reduce the criticality rating to predetermined acceptable regions (preferably a hazard risk index between one and five).

Hazard and Operability Study

Hazard and operability study (HAZOP) is one of the most tedious, yet thorough, forms of hazard analysis. It identifies potentially complex hazards in a system. HAZOP examines a combination of every part of the system and analyzes the collected data to locate potentially hazardous areas. The first step is to define the system and all specific areas from which data will be collected. This will help in defining the HAZOP team. Next, a team of experts in these areas is assembled to analyze the collected data. The team usually consists of experts in the fields of engineering, human behavior, hygiene, and organization and management, and other personnel who may have operational expertise related to the specific system being analyzed. Once this is done, an intensive information gathering process begins. All aspects of the system's operation and its human interfaces are documented. The information is then broken down into small information nodes. Each node contains information on the procedure or specific machine being used in

TABLE 200.3 Hazard Risk Assessment Matrix

	Hazard Category			
Frequency of Occurrence	I Catastrophic	II Critical	III Marginal	IV Negligible
Frequent	1	3	7	13
Probable	2	5	9	16
Occasional	4	6	11	18
Remote	8	10	14	19
Improbable	12	15	17	20

Note: The hazard risk index numbers and related suggested criteria are 1–15, unacceptable; 6–9, undesirable; 10–17, acceptable with review; and 18–20, acceptable without review.

TABLE 200.4 HAZOP Data Table for Vacuum Air Vent Node and Reverse Flow Guide Word (Design Intention: To Vent Air into the Sterilizer following a Vacuum Stage)

Guide Word	Cause	Consequence	Recommendation
Reverse flow	Control valve leakage	Ethylene oxide leak into utility room	Air vent should not receive air from utility room, but from exhaust duct to reduce risk from reverse flow leakage of ethylene oxide

Source: Adapted from *Hazard and Operability Study of an Ethylene Oxide Sterilizer.* 1989. National Institute for Occupational Safety and Health, NTIS Publication No. PB-90-168-980. Springfield, VA, p. 27.

the system. Each node is then interconnected logically with other nodes in the system. Each node is also given guide words that help identify its conditions. Table 200.4 gives an example of guide words. Each guide word is a functional representation of subsystem hazard. The team can analyze and determine the criticality or likelihood that this node could produce a hazard. At this point, the HAZOP team will need to determine what course of action to take. This procedure is one of the most comprehensive, yet time consuming, of the analysis tools. It is widely used in large and complex systems with critical safety components, such as petrochemical facilities.

Fault Tree Analysis

Fault tree analysis (FTA) uses deductive reasoning to qualitatively and quantitatively depict possible hazards that occur due to failure of the relationships between the system components (e.g., equipment, plant procedures). FTA uses a pyramid style tree analysis to start from a top undesired event (e.g., accident, injury) down to the initial causes of the hazard (e.g., a joint separation under vibrating forces, erroneous operating procedure). There are four main types of event symbols — a fault event, basic event, undeveloped event, and normal event. A fault event is considered to be an in-between event and never the end event. A basic event is considered to be the final event. An undeveloped event is an event which requires more investigation because of its complexity or lack of analytical data. A normal event is an event which may or may not occur. Each one of these events is joined in the tree by a logic symbol. These gates explain the logic relationship between each of the events. For example, a main event is followed in the tree by two possible basic events. Either event can produce the main event. The logic gate in this case would be an *or* gate, but if both events could have contributed to the main event an *and* gate would be used. FTA's ability to combine causal events together to prevent or investigate accidents makes it a powerful accident investigation and analysis tools.

Event Tree Analysis

Event tree analysis (ETA) is similar to FTA, with the exception that ETA uses inductive reasoning to determine the undesired events which are caused by an earlier event. ETA uses the same pyramid structure as the previous analysis. However, rather than working from top to bottom, ETA works from left to right, dividing the possibility of each event into two outcomes — true (event happening) or false (event not happening). By taking an initial failure of an event, the tree designer tries to incorporate all possible desired and undesired results of the event. The advantage to this system is that it helps predict failures in a step-by-step procedure. This helps the analyst to provide a solution or a countermeasure at each analysis node under consideration. ETA's main weakness lies with its inability to incorporate multiple simultaneous events.

Other techniques such as management oversight, risk tree, "what-if" analysis, software hazard analysis, and sneak circuit analysis are discussed in the sources listed in the Further Information section.

Defining Terms

Compliance — The minimum set of requirements by which an environment conforms to local, state, and federal rules, regulations, and standards. According to the seriousness of the violation, a workplace may be cited by OSHA for imminent danger, serious violation, nonserious hazards, and de minimis violations. In addition to penalties (fines of up to $70,000 per violation), other civil and criminal charges may be brought against responsible supervisors and managers.

Hazard — A set of (or change in) a system's potential and inherent characteristics, conditions, or activities that can produce adverse or harmful consequences, including injury, illness, or property damage (antonym to safety).

Hazard analysis techniques — A number of analytical methods by which the nature and causes of hazards in a product or a system are identified. These methods are generally designed to evaluate the effects of hazards and offer corrective measures or countermeasures.

Mechanical injuries — A type of physical injury caused by excessive forces applied to human body components, such as cutting, crushing, and straining (ergonomic hazards).

Risk control — The process by which the probability, severity, and exposure to hazards (per mission and unit of time) are considered to reduce the potential loss of lives and property.

Toxic substances — Those substances that may, under specific circumstances, cause injury to persons or damage to property because of reactivity, instability, spontaneous decomposition, flammability, or volatility (including those compounds that are explosive, corrosive, or have destructive effects on human body cells and tissues).

References

American Institute of Chemical Engineers. 1985. *Guidelines for Hazard Evaluation Procedures*. American Institute of Chemical Engineers, New York.

Brauer, R. L. 1990. *Safety and Health for Engineers*. Van Nostrand Reinhold, New York.

Gressel, M. G. and Gideon, J. A. 1991. An overview of process hazard evaluation techniques. *Am. Ind. Hyg. Assoc. J.*, 52:158–163.

Hansen, L. 1993. Safety management: a call for (r)evolution. *Prof. Saf.*, 38: 6–21.

Karter, M. J. 2001. *Fire Loss in United States during 2001*. National Fire Protection Association, Quincy, MA.

Main, B. W. and Frantz, J. P. 1994. How design engineers address safety: what the safety community should know. *Prof. Saf.*, 39:33–37.

Mansdorf, S. Z. 1993. *Complete Manual of Industrial Safety*. Prentice Hall, New York.

Occupational Hazards. 1993. Editorial, 1993, November, p. 6.

National Safety Council. 2003. Report on Injuries in America. Available at: www.nsc.org/lrs/statinfo/99report.htm. Accessed on April 23, 2003.

Shell, R. L. and Simmons, R. J. 1990. *An Engineering Approach to Occupational Safety and Health in Business and Industry*. Institute of Industrial Engineers, Norcross, GA.

Vincoli, J. W. 1993. *Basic Guide to System Safety*. Van Nostrand Reinhold, New York.

Further Information

Barbara, A. P., Niland, J., Quinlan P. J., and Plogg, H. 1996. *Fundamentals of Industrial Hygiene.* National Safety Council, Itasca, IL.

Hammer, W. 1993. *Product Safety Management and Engineering,* 2nd ed. American Society of Safety Engineers, Des Plaines, IL.

Hansen, D. J., Ed. 1991. *The Work Environment: Occupational Health Fundamentals.* Lewis Publishers, Chelsea, MI.

Kavianian, H. R. and Wentz, C. A. 1990. *Occupational and Environmental Safety Engineering and Management.* Van Nostrand Reinhold, New York.

Kohn, J. P., Friend, M. A., and Winterberger, C. A. 1996. *Fundamentals of Occupational Safety and Health.* Government Institutes, Inc., Rockville, MD.

National Safety Council. 1997. *Accident Prevention Manual for Business and Industry: Engineering and Technology,* 11th ed. National Safety Council, Itasca, IL.

Timbrell, J. A. 1995. *Introduction to Toxicology.* Taylor and Francis, London.

201

Regulations and Standards

A. Keith Furr

Virginia Polytechnic Institute and Virginia State University

Engineers and corporations in the United States and other industrial nations have long recognized the need for the establishment of basic minimum standards for safe design and construction of structures and equipment. As a result, many technical groups and associations have established guidelines for these minimum standards. However, it generally has been left up to the federal, state, and local governments to decide which of these are mandated. For example, fire codes are usually designed and administered locally. At least one state has virtually no fire codes that apply statewide. In other cases, the application of the codes is not universal. In Virginia, only selected categories of structures are covered by the statewide code, while local building codes vary widely in their scope and interpretation depending on the local officials responsible for their administration. A disaster is often required to reveal the deficiencies in applicable codes and their enforcement. The Hurricane Andrews aftermath is an excellent example of this; as a result, the state has dramatically strengthened its building code standards. Another area needing consideration is the concept of including a design safety factor. The need is well recognized but perhaps the safety factors themselves may need to be reevaluated. Prior to the World Trade Center disaster, no one realistically expected that airliners with virtually full fuel loads would crash into the buildings; the support structure could not withstand the extremely high temperatures created by the burning fuel and the buildings collapsed. Should more consideration of extreme what-ifs be included in building designs? There have been several airline crashes in the past few years that have been blamed on structural failures considered highly unlikely, but which nevertheless occurred.

After World War II, the need was recognized in the United States for "universally" applicable standards, at least in certain fields. One of the earliest defined and adopted national standards was in the field of nuclear energy, a prototype set for many of the agencies that have since been established. The responsible federal agency, the Nuclear Regulatory Commission (NRC) (originally, the Atomic Energy Commission; only current names of regulatory agencies are used in this chapter), set the basic standards by which activities involving radioactive materials created as a result of reactor-related operations were regulated. Note that radiation created by medical devices and natural materials does not fall under the purview of the NRC. States typically have at least partially assumed this responsibility. As more information has been developed, the original standards were modified to accommodate the new data, generally being made more stringent. States were given the option of either allowing the federal agency to supervise the use of these materials or adopting regulations at least as stringent as the federal regulations and administer the same themselves. This model has persisted in many of the other regulatory acts passed since then. In the early 1970s, the Occupational Safety and Health Act (OSHA) was passed, covering a very large number of industries. The standards in the OSHA Act were, for the most part, incorporated directly

0-8493-1586-7/05/$0.00+$1.50

from the standards and guidelines established by industrial associations. Unfortunately, the law does not permit changes to the standards to be made easily, so that the legal OSHA standards often lag behind the industrial standards. Recently, there have been revisions to the regulatory provisions to make them more applicable to current safety requirement levels.

Other major federal safety and health standards are those of the Environmental Protection Agency (EPA), the Department of Transportation, the Food and Drug Administration, the Department of Energy, the Federal Aviation Administration, and a host of other specialized agencies. The scope of regulations is a rapidly evolving area. In recent years, federal operations have been brought under both OSHA and EPA regulations from which they were previously excluded. OSHA has taken a very active role in the operations of medical- and health-related organizations with the Bloodborne Pathogens Act, and the problems of the disabled are now regulated under the Americans with Disabilities Act (ADA). A major area possibly to be changed significantly in the future is noise control. Many engineers agree that the current noise safety program is too lenient. The initial threshold may to be changed from 90 dB to 85 dB and the doubling ratio as a measure of sound level intensity changed from 5 dB to 3 dB as more scientifically sound. These two changes will require major engineering modifications in many cases, and probably additional costs.

Another potential regulatory area that will have a significant impact on engineering design is indoor air quality, an area insufficiently researched until recently. As most buildings are currently being built or retrofitted with heating, ventilation, and air conditioning systems to completely manage the building's internal environment, health problems are developing in building occupants. Typically, only a modest fraction of the personnel are affected, but among these significant health problems may exist, often due to biological contaminants. Building materials are also of concern, which were not considered until fairly recently. Everyone by now should be aware of the problems with asbestos, and the enormous expense of abating associated problems. Similar problems appear to exist with effluents from some interior finish materials, such as those that contain formaldehyde.

201.1 Engineering Practices

Incorporated in the standards for nuclear safety, and now in most other safety and health regulatory standards, was the concept of achieving compliance with safety standards by engineering controls wherever possible. For example, it should, in principle, be impossible to operate a nuclear reactor unsafely. Approaching criticality too rapidly should be impossible by having interlocks that would shut down operations automatically if this were done or if one of a number of other conditions existed, such as loss of coolant, excessively high temperatures, excessively high radiation levels in monitored areas, earthquakes, and loss of information in critical circuits. The control mechanisms are supposed to be designed so that redundant failures have to occur to make the them operate unsafely; no single equipment failure should render operations unsafe (in the sense that an automatic or controlled shutdown should always occur when a failure occurs). This example, taken from the nuclear industry, is reflected in many others. Of course, experience has shown that humans can bypass almost any safety feature if they try. The catwalk in the Kansas City hotel fell due to human error. The Chernobyl incident was due to a combination of poor design a small safety margin, and operations that were lax in maintenance, training, and function. In some cases, failure occurs due to the lack of sufficient knowledge or information. The collapse of the Tacoma Narrow Bridge in 1940, only 4 months after completion, is an excellent example. In any case, safety and health standards today require that, wherever feasible, safety engineering principles as identified by the regulations must be the first option in complying with the standards. Procedural rules as a compliance method may be used on occasion but are definitely a second choice. A fairly recently developing problem that engineers should recognize is the substitution of substandard components that look like identical specified items but do not meet necessary quality standards. This has affected a vast range of manufactured items ranging from cars to medical devices to structural components. The engineer needs to take extra care to prevent these substitutions from occurring.

Many of the earlier industrial standards, including those emphasized in the earliest versions of the OSHA standard, and briefly alluded to in the previous paragraph, are intended to prevent physical injuries. Bending brakes and hydraulic presses are required to have controls making it impossible for the machines, if used properly, to injure any part of the operator's body. Grinding tools and other devices with exposed rotating components are supposed to be designed to prevent hair or clothing from being grabbed by the rotating component or for materials to be thrown by the tool and impact a person, particularly in the face and eye. The design of portable electrical tools is required to make it impossible for them to produce a shock under normal operation. Ladders, scaffolds, and guardrails are defined to prevent employees from slipping or falling. Indeed, one of the major weaknesses in the original standards was that they defined safety in such minute detail that they were seen by many to be concerned with trivia, so that they failed to some extent in their mission to increase safety. Nevertheless, even with the original weaknesses, most of which have now been addressed, the creation of OSHA was a major step forward in occupational safety. The general public benefits, as well as employees, since incidents may have an impact outside the immediate area. If a machine or structure is designed properly, it should be safe for everyone with a legitimate need and training to use the device or facility.

Other areas of the original OSHA Act dealt with fire safety by adopting (in some cases modifying them slightly) standards of the National Fire Prevention Association on the storage of flammable materials, limiting bulk storage, storage within facilities, and sizes of containers for appropriate classes of liquids. Similar restrictions have been established for compressed gas systems, especially gases with inherent safety problems, such as hydrogen, oxygen, and ammonia. With few changes, these regulations have remained in effect and are well understood by engineers charged with designing appropriate facilities. Note that not all of the standards mentioned above were incorporated into OSHA, typically only those applying to employee safety, but other portions of the standards are incorporated into other regulatory standards with which engineers must comply.

Recent changes in the OSHA standard are often more concerned with meeting performance specifications than with specific directives, allowing the engineer, owner, or user to decide how best to meet the compliance goal. Meeting safety and health regulatory standards through engineering practices is still required as the first option, with personal protective devices and procedural controls used as options when engineering controls are not practicable. A major opportunity to implement this concept was created a few years ago when OSHA made a distinction between industrial operations, from the pilot plant scale upward, and those of laboratory-scale operations. It was recognized that the designs appropriate to a large plant where a modest number of chemicals are in use, albeit on a substantial scale, are not appropriate for the laboratory environment, where literally hundreds of different chemicals are used on a rapidly changing basis. Therefore, the OSHA Laboratory Safety Act was added to the basic OSHA Act, preempting the OSHA general industry standards for laboratories. Neither section of the Act is more permissive than the other in terms of meeting the intended goal, a healthy and safe environment. The two different environments do lead to substantially different engineering approaches to many problems, and engineers specializing in the two areas need to understand the differences.

One of the major issues, even in the early versions of the OSHA standards, concerns the levels of specific airborne contaminants in the workplace. For various reasons, such as toxicity, neurological risk, fetal hazards, and corrosiveness, permissible exposure levels (PELs) were set by OSHA. Theoretically, a person in good health could work an 8-hour work day without harm if the exposure levels were less than these, but could not be allowed to work if they exceeded the PELs based on an 8-hour time-weighted average. In some cases, higher short-term exposure limits (STELs) were established, and in others, ceiling (C) limits were established. The initial values were essentially the threshold limit values (TLVs) recommended by the American Conference of Governmental Industrial Hygienists (ACGIH). The latter are reviewed on a yearly basis and have changed significantly for specific substances. OSHA attempted to change their PELs (which are legally enforceable) to be the same in most cases to the more current values of the ACGIH (which are guidelines and are not legally enforceable). However, court actions required OSHA to rescind the changes, so that the OSHA PELs are in many cases out of date the current available knowledge. Some states adopted the updated values on their own, so that now engineers must be aware of the law in their area of expertise. A conservative approach would be to design using the ACGIH guidelines since they normally provide an additional safety level. In neither case should the limits be considered as black and white. Many factors will modify the sensitivity of a specific individual to a given contaminant; for example, natural variations in individual tolerance or tolerance due to current health conditions. A factor receiving increased attention is possible gender and age differences. The PELs and TLVs were based on an 8-hour day, 40-hour work week for a healthy adult male. A good design practice is to maintain an action level no more than 50% of the PEL, and preferably less. Many of the newer OSHA standards now incorporate the 50% action level concept.

Ventilation engineers need to be aware of local factors in designing appropriate systems capable of achieving acceptable levels. The problems of designing ventilation systems to remove air contaminants do not seem to be as highly appreciated by some engineers as might be expected. Routine ventilation designs for offices and homes are geared typically for temperature control. Since hot air rises, inlet and exhaust duct work and openings are typically placed in the ceiling. However most airborne contaminants are heavier than air, so general room exhausts for these contaminants should be located close to the floor, or no lower than waist height. The air flow should not draw or discharge contaminated air indoors through occupants' breathing zone. Various recommendations exist for the amount of fresh air needed depending on the type of occupancy. Typically, in an office environment, a value of 20 cubic feet per minute (cfm) of fresh air is recommended. Relatively few older structures meet this requirement. Even if the original design was satisfactory, if subsequent modifications are not thoroughly reviewed and designed appropriately, they will fall drastically short of meeting employees' needs. In chemically contaminated areas, such as laboratories, the number of air changes per hour is often used as a criterion, with current levels being 10 to 12 per hour. This has serious implications for laboratories since the normal requirement is for this to be 100% fresh air. In office structures, the air is normally partially recycled so that only 15% to 30% may be fresh, but this, while economical, may lead to other problems.

The recycling of air in newer energy-efficient buildings appears to be at least partly responsible for the increasing evidence of health problems in these structures. Recycling air can cause an air quality problem that exists in one part of a building to eventually spread throughout the building. This is especially true when the system is itself the cause of the problem. Most duct systems are not designed for ease of cleaning, and over time become dirty. The dirt and grease that builds up in these systems is an ideal breeding ground for molds and bacteria, to which many individuals are allergic. Not only should high-efficiency filters be used and replaced frequently, but the duct work interior needs to be cleaned periodically. If air quality is brought under regulation, there will be significant engineering requirements on the design of the systems to facilitate cleaning, filter designs, and selection of materials for construction to limit the emission of volatile organic compounds. There is increasing evidence that long-term exposure to some organics, such as formaldehyde, may cause problems to hypersensitive individuals at levels significantly lower than the permissible PELs or TLVs. The National Institutes of Occupational Safety and Health (NIOSH) are already recommending levels substantially below those of both OSHA and the ACGIH for a number of materials due to low, continuous exposures. Unfortunately, our knowledge of health effects in general and especially those related to long-term exposures at low levels leaves a great deal to be desired, although it is improving. Statistically, the problem with low levels of airborne contaminants is very much like the situation with low levels of radioactivity. Health problems may exist, but the experimental evidence is virtually impossible to obtain because of the extremely large number of subjects required for direct measurement. The problems are compounded by the wide variation in individual sensitivities. One individual may be made very ill after exposure to a pollutant while another may be unaffected by it. What if only 1% are affected while the remaining 99% are not? Do you design, at possibly great cost, to prevent problems for the 1%?

The most effective and energy-conserving means of controlling contaminants is to capture them at the source. In a factory, large-scale scavenging systems may be appropriate, while laboratory facilities would be better designed so that all work with the potential to release hazardous vapors, fumes, or gases be done in fume hoods with at least 100 feet per minute (fpm) face velocity. Note that handicapped workers would face difficulties in working in hoods, so specially designed units should be specified where this occurs.

Once contaminated air has left a facility, it is hoped that dilution with the outside air will reduce contaminants to safe levels. Air pollution standards have been established that may or may not be sufficient. Problems such as acid rain and ozone depletion are well documented in the literature and have been extensively publicized in the media. In laboratory buildings, for example, engineers until quite recently typically placed rain caps over fume hood exhausts so that the contaminated air was forced back onto the roof of a building and in many cases was recaptured by the building's air ventilation system. Often, once a laboratory building was completed, subsequent modifications added fume hoods to a duct system that did not have the capacity to handle them. Fire codes typically require laboratories to be at negative pressure with respect to the corridors so that fire within a laboratory cannot spread via the corridors. The negative pressure also prevents contaminated air from spreading via the corridors as well. Exhaust ducts for the newer hoods were often allowed to penetrate floors intended to be fire separations. The fire separation must be maintained. Standards recommended by either the ACGIH, NFPA, or American Society of Heating, Refrigerating and Air Conditioning Engineers (ASHRAE) must still be met.

The Environmental Protection Agency (EPA) has been a major factor in changing engineering practices. For example, one of the most significant standards dealt with underground petroleum storage tanks. A large percentage of organizations with a substantial number of employees, including manufacturing plants, transportation departments, hospitals, and universities, have either buried tanks of heating oil or gasoline. Tanks that have been buried for 20 years or more have an increasing probability of leakage as they age. The EPA regulation required all of these tanks above a certain capacity to be protected from leaking into the environment by 1998. Typically, this involved replacing a tank with a double-shelled unit equipped with appropriate sensors to ensure that leaking does not occur. If environmental contamination has already occurred, the owner was required to clean up the contaminated site. Another example involves the disposal of hazardous medical waste. The EPA set very strict pollution standards for incinerators to burn these wastes that, in effect, caused the practice to be virtually discontinued.

Although addressed in Chapter 93, the related issue of prevention of ground contamination from landfills will be briefly mentioned here. In the last 5 years, a very large number of landfills have been closed due to the enactment of strict standards on landfills. Although no responsible person would deliberately dispose of any kind of hazardous waste routinely at a municipal landfill, it is estimated that 3% to 5% of all municipal solid waste could be characterized as hazardous waste, primarily from households and small business establishments. The engineering problem is to prevent pollutants from seeping from landfills into surrounding groundwater. The landfills must be sealed to prevent leakage and to prevent entry of water. Monitors are required to ensure that the prevention measures are successful. To lengthen the life of existing landfills, and to reduce the cost of new ones, recycling programs have been growing dramatically and have given rise to their own engineering disciplines. Alternative means of hazardous and solid waste disposal have been developed to reduce or eliminate the burial of wastes that could remain hazardous for thousands of years. Various incineration methods have been developed that must comply with EPA's clean air standards (see previous paragraph) as well as being subject to the "not-in-my-back-yard" (NIMBY) syndrome. Dozens of techniques have been developed for disposal of medical wastes, which are regulated by state programs of varying stringency. No matter what engineering approach is involved in any of these environmental issues (and others not mentioned), the net result must be to reduce the level of contaminants in the soil or air below a regulatory-defined level either by destruction or modification. Underground burial has been deemed unacceptable for most hazardous materials, with the notable exception of radioactive isotopes, for which no practical means of destruction exists. The major regulatory act that sets standards for the disposition of materials with hazardous properties is the Resource Conservation and Recovery Act (RCRA). Since most facilities do not process their hazardous waste locally, packaging and transportation of these materials must also meet requirements of the Department of Transportation. These standards have changed significantly within the past several years and now are generally consistent with comparable international requirements. Another major EPA standard is the community-right-to-know standard. Under this act, communities must be made aware of potential industrial chemical hazards in their areas. Several hundred chemicals have been specifically identified for which detailed emergency response plans are required. Localities are required to maintain local emergency planning committees to oversee this regulation.

The original standards for making buildings accessible to the handicapped were less successful than desired, so the ADA was passed in the early 1990s. This standard is much more stringent and carries the potential for much more severe penalties. Not only are the now familiar wheelchair ramps and modified bathroom facilities required, but a number of other engineering changes are specified as well. Safety equipment, such as safety showers and eyewash fountains, must now meet specifications to ensure that a disabled person can use them. Fire alarms must provide a high-intensity visual alarm as well as an audible alarm. Doors must provide means to prevent a disabled person from being trapped or injured by the opening and closing mechanism. Instructions on public signs, including safety signs, must be provided in braille where appropriate. Meetings that could be attended by the disabled must ensure that the information can be received by a disabled person. This could include audio and or visual equipment for the hearing or visually impaired, or possibly a sign-language interpreter. The availability of these resources must be made known at the time the meeting is announced. Places of refuge must be identified to which disabled persons could seek temporary protection while awaiting rescue in the event of a fire or other incident. All new facilities must be designed to comply with the standard, and older ones will have to be brought toward full compliance. There is currently some legal uncertainty as to the interpretation of certain portions of this standard, but it certainly should be considered by engineers and architects in their work.

The Food and Drug Administration (FDA) not only establishes standards for nutritional information and for such things as drugs, pesticides, and residues of these materials on foods, but also regulates or provides guidelines for the design of x-ray machines, lasers, microwave units, and other generators of electromagnetic emissions. Currently cell phone emissions are a matter under consideration. Most FDA safety standards are intended as requirements for the manufacturer and effectively require engineering safeguards to be incorporated into these devices that a user cannot easily bypass. For example, industrial x-ray cameras are required to be interlocked so that the intense radiation areas within the target area cannot be accessed while the beam is on. The radiation levels within this area are such that a serious radiation burn to the hands can occur within 30 seconds. Requirements for the design of x-ray units for diagnostic tests now mandate that the unit define the exposed target area by light so that areas other than the one of interest are not exposed. This has resulted in dramatic reductions for patient exposures. Therapeutic uses of radiation are not regulated and can be quite high but levels are set by prudence and experience. Similar restrictions are imposed on lasers which have sufficient power to cause blindness if a beam were to reach the eye directly or by nondiffuse reflection. The standards also include guidelines for the safe use of the equipment by the operators, although these are recommendations only and are not legally binding. The ACGIH has established TLVs for laser exposures as well as for chemical exposures.

The NRC does not regulate radiation exposures due to x-rays or the use of radioactive materials not related to byproduct materials (materials made radioactive by reactor operations). However, dose rates for x-rays are generally equivalent to those for radiation from radioactive materials. The design of x-ray facilities must take into account radiation levels external to the facility with care since x-rays are normally directed, and hence radiation levels can be high in spots.

Also, states typically regulate the radioactive materials not covered by the NRC in a comparable fashion. In January 1994, the NRC regulations were changed in several significant ways. For example, exposure limits, which had been primarily concerned with external exposures, now must incorporate exposures due to ingestion, inhalation, and possible percutaneous absorption; certain external limits were changed, including those related to the extremities, which now include the area below the knee as well as the areas beyond the elbows, since there are no critical organs within those bodily components; and women who may be pregnant (and so declare) must be limited to 10% of the whole body exposure permissible for other monitored employees, in order to limit exposure to the fetus. The NRC has also stepped up the stringency with which it monitors compliance, and in recent years it has become commonplace for the NRC to impose substantial fines on industrial users, hospitals, and universities.

There are other standards and guidelines that have a significant impact on safety and health. Among these are the OSHA Hazard Communication Act (29 CFR 1910.1200) for industrial users of hazardous materials, and its companion for laboratory employees (OSHA 29 CFR 1450). Both require extensive training of employees and provision for responses to emergencies. The previously mentioned Bloodborne

Pathogens Act (OSHA 29 CFR 1919.1030) is a very strong standard. All of these incorporate significant engineering requirements to protect employees through either provision of emergency measures or uses of protective equipment.

The National Institutes of Health (NIH) Guidelines for Research Involving Recombinant DNA Molecules represent a significant standard which is, in effect, a regulation although it is labeled a guideline. Research programs which do not comply with these guidelines will not receive funding from NIH. Characteristics of the facility in which the research is done, the procedures followed, and the experimental subjects are all carefully defined by the standard. Further, any genetically engineered products of these studies must receive explicit approval prior to being placed on the market since these have implications for genetically modified foods. Even if considered safe, many persons are afraid to eat modified foods. The impact of this standard is still somewhat controversial even if the end product is clearly beneficial, such as making it possible to clone a human fetus for recovery of stem cells that could be used to repair human organs. In a similar vein, the Centers for Disease Control have published standards governing microbiological research practices based on concerns that research workers might contract a contagious disease. This concern is not due to a high probability that contagion may spread beyond the facility, as in fact this has not been known to occur in this country, but is designed to ensure that it does not. A disturbing trend is that modern medicine's disease-fighting capabilities are rapidly decreasing in efficacy as infectious organisms become immune to previously effective treatments.

The preceding material presented a very brief overview of a portion of the regulatory environment. Certain topics that were not mentioned are of comparable importance to those covered. Safety and health programs, once developed only voluntarily by enlightened organizations due to their recognition that they were cost-effective, are now virtually completely defined by regulations and standards applicable to the substantial majority of employees. Engineering subdisciplines have developed to provide the supportive infrastructure for the regulations. This is a rapidly developing field. Many of the regulations cited did not exist a few years ago, and every one of them has undergone significant changes in that time and continue to evolve.

There is a tendency by many to believe that regulations stifle productivity. This may be so if the regulations are applied inflexibly, without common sense. A sensible safety program will without doubt improve productivity. One of the country's most profitable companies is E. I. DuPont, which pioneered many of the safety programs later incorporated into standards. This company is generally regarded as having one of the very best safety and health programs in private industry. Again, DuPont has not undertaken these programs for altruistic reasons, but because they are cost-effective.

References

No date is given in citations of regulatory standards, as these undergo continual revision. The most current version should always be used.

American Conference of Governmental Industrial Hygienists. *Industrial Ventilation: A Manual of Recommended Practice*. American Conference of Governmental Industrial Hygienists, Cincinnati, OH.

American Conference of Governmental Industrial Hygienists. *Threshold Limit Values for Chemical Substances and Physical Agents and Biological Exposure Indices*. American Conference of Governmental Industrial Hygienists, Cincinnati, OH.

Environmental Protection Agency. *Hazardous Waste Regulations*. Title 40, Code of Federal Regulations. U.S. Government Printing Office, Washington, DC.

Nuclear Regulatory Commission. *Radiation Safety Regulations*. Title 10, Code of Federal Regulations. U.S. Government Printing Office, Washington, DC.

Occupational Health and Safety Administration. Title 29, Code of Federal Regulations. U.S. Government Printing Office, Washington, DC.

U.S. Department of Health and Human Services, Centers for Disease Control, and National Institutes of Health. *Biosafety in Microbiological and Biomedical Laboratories*. U.S. Government Printing Office, Washington, DC.

U.S. Department of Transportation. *Hazardous Materials, Substances, and Waste Regulations.* Title 49, Code of Federal Regulations. U.S. Government Printing Office, Washington, DC.

XXVIII

Engineering Economics and Management

202

Present Worth Analysis

Walter D. Short
National Renewable Energy Laboratory

Evaluation of any project or investment is complicated by the fact that there are usually costs and benefits (i.e., cash flows) associated with an investment that occur at different points in time. The typical sequence consists of an initial investment followed by operations and maintenance costs and returns in later years. Present worth analysis is one commonly used method that reduces all cash flows to a single equivalent cash flow or dollar value (Palm and Qayum, 1985). If the investor did not care when the costs and returns occurred, *present worth* (also known as *net present value*) could be easily calculated by simply subtracting all costs from all income or returns. However, to most investors the timing of the *cash flows* is critical due to the time value of money; that is, cash flows in the future are not as valuable as the same cash flow today. Present worth analysis accounts for this difference in values over time by discounting future cash flows to the value in a base year, which is normally the present (hence, the term "present" worth). If the single value that results is positive, the investment is worthwhile from an economic standpoint. If the present worth is negative, the investment will not yield the desired return as represented by the discount rate employed in the present worth calculation.

202.1 Calculation

The present worth (PW), or net present value, of an investment is

$$\mathrm{PW} = \sum_{t=0}^{N} \frac{C_t}{(1+d)^t} \tag{202.1}$$

where N is the length of the analysis period, C_t is the net cash flow in year t, and d is the *discount rate.*

Analysis Period

The analysis period must be defined in terms of its first year and its length, as well as the length of each time increment. The standard approach, as represented in Equation (202.1), is to assume that the present is the beginning of the first year of the analysis period ($t = 0$ in Equation (202.1)), that cash flows will be considered on an annual basis, and that the analysis period will end when there are no more cash flows that result from the investment (e.g., when the project is completed or the investment is retired)

0-8493-1586-7/05/$0.00+$1.50

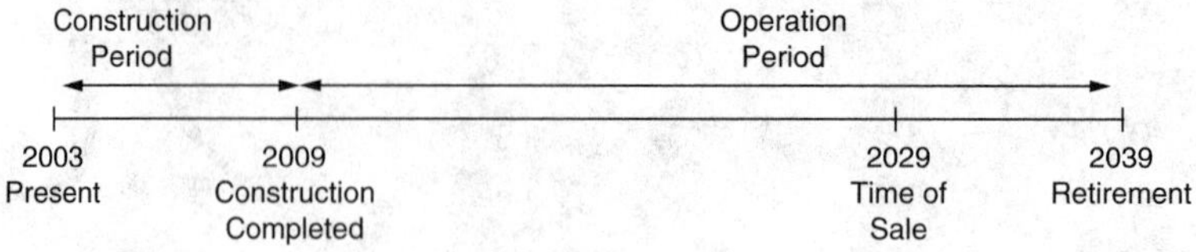

FIGURE 202.1 Comparison of investments.

[Ruegg and Petersen, 1987]. For example, the present worth for a refinery that will require 6 years to construct and then last for 30 years (to 2039) can be represented as

$$\mathrm{PW}(2003) = \sum_{t=0}^{36} \frac{C_t}{(1+d)^t} \tag{202.2}$$

In this case the analysis period covers both the construction period and the full operation period, as shown in Figure 202.1. In Equation (202.2) the parenthetical (2003) explicitly presents the base year for which the present worth is calculated. This designation is generally omitted when the base year is the present.

However, any of these assumptions can be varied. For example, the present can be assumed to be the first year of operation of the investment. In the refinery example, the present worth can be calculated as if the present were 6 years from now (2009) — that is, the future worth in the year 2009 — with all investment costs between now and then accounted for in a single turnkey investment cost (C_0) at that time ($t = 0$). In this case the base year is 2009.

$$\mathrm{PW}(2009) = \sum_{t=0}^{30} \frac{C_t}{(1+d)^t} \tag{202.3}$$

The length of the analysis period is generally established by the point in time at which no further costs or returns can be expected to result from the investment. This is typically the point at which the useful life of the investment has expired. However, a shorter lifetime can be assumed with the cash flow in the last year accounting for all subsequent cash flows. For the refinery example the analysis period could be shortened from the 36-year period to a 26-year period, with the last year capturing the salvage value of the plant after 20 years of operation.

$$\mathrm{PW} = \sum_{t=0}^{26} \frac{C_t}{(1+d)^t} \tag{202.4}$$

where C_{26} is now the sum of the actual cash flows in C_{26} and the salvage value of the refinery after 20 years of operation. One common reason for using a shorter analysis period is to assess the present worth, assuming sale of the investment. In this case, the salvage value would represent the price from the sale minus any taxes paid as a result of the sale.

Finally, the present worth is typically expressed in the dollars of the base year. However, it can be expressed in any year's dollars by adjusting for inflation. For example, in the refinery case, in which the base year is 2009 (Equation (202.3)), the present worth value expressed in the dollars of the year 2009 can be converted to a value expressed in today's (2003) dollars, as follows:

$$\mathrm{PW}_{2003\$}(2009) = \frac{1}{(1+i)^{2009-2003}} \mathrm{PW}_{2009\$}(2009) \tag{202.5}$$

where i is the annual *inflation rate* between 2003 and 2009 and the subscripts indicate the year of the dollars in which the present worth is presented.

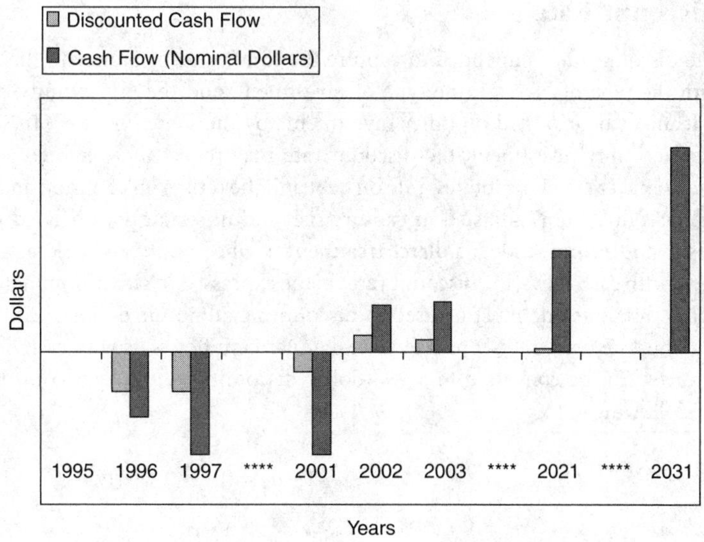

FIGURE 202.2 Analysis period for a refinery.

Cash Flows

Cash flows include all costs and returns where costs are negative and returns or income are positive (see Figure 202.2). The cash flow for each year of the analysis period must include all actual costs and income associated with the investment during that year. Thus, cash flows must include all initial capital costs, all taxes, debt payments, insurance, operations, maintenance, income, and so forth.

Cash flows are typically expressed in terms of the actual dollar bills paid or received (*nominal* or current *dollars*). Alternatively, cash flows can be expressed in terms of the dollars of a base year (*real* or constant *dollars*). Nominal dollar cash flows, *Cn*, can be converted to real dollar cash flows, *Cr* (and vice versa) by accounting for the effects of inflation

$$C_0^r = \frac{C_t^n}{(1+i)^t} \tag{202.6}$$

where i is the annual inflation rate and t is the difference in time between the base year ($t = 0$) and the year of the nominal dollar cash flow.

As shown in Figure 202.2 (which used a nominal discount rate of 0.2 or 20%), discounting can significantly reduce the value of cash flows in later years. A discounted cash flow is the present worth of the individual cash flow. Discounted cash flow in period s is shown in Equation 202.7 where the cash flow is assumed to occur at the end of year s.

$$PW_s = \frac{C_s}{(1+d)^s} \tag{202.7}$$

There are instances where the cash flow occurs continuously throughout the year (e.g., product sales). This can be captured by a continuous discount rate d' (see next section for definition) and an integral discount formula.

$$PW_s = \int_{s-1}^{s} e^{-d't} c(t)dt \tag{202.8}$$

where $c(t)$ is the rate of cash flow throughout the year.

Discount Rate

The discount rate is intended to capture the time value of money to the investor. The value used varies with the type of investor, the type of investment, and the opportunity cost of capital (i.e., the returns that might be expected on other investments by the same investor). In some cases, such as a regulated electric utility investment, the discount rate may reflect the weighted average cost of capital (i.e., the after-tax average of the interest rate on debt and the return on common and preferred stock). Occasionally, the discount rate is adjusted to capture risk and uncertainty associated with the investment. However, this is not recommended; a direct treatment of uncertainty and risk is preferred (discussed later).

As with cash flows, the discount rate can be expressed in either nominal or real dollar terms. A nominal dollar discount rate must be used in discounting all nominal dollar cash flows, whereas a real discount rate must be used in discounting all real dollar cash flows. As with cash flows, a nominal dollar discount rate, *dn*, can be converted to a real dollar discount rate, *dr*, by accounting for inflation with either of the following:

$$1 + d_n = (1 + d_r)(1 + i)$$

$$d_r = \frac{1 + d_n}{1 + i} - 1 \tag{202.9}$$

Since cash flow should include the payment of taxes as a cost, the discount rate used in the present worth calculation should be an after-tax discount rate (e.g., the after-tax opportunity cost of capital).

As noted in the cash flow section above, it is possible to discount continuous cash flows with a continuous discount rate d'. This continuous discount rate is related to the discrete discount rate d as follows:

$$d' = \ln(1 + d)$$

202.2 Application

Present worth analysis can be used to evaluate a single investment or to compare investments. As stated earlier, a single investment is economic if its present worth is equal to or greater than zero but is not economic if the present worth is less than zero.

Present worth > 0. The investment is economic.
Present worth < 0. The investment is not economic.

The use of present worth to compare multiple investments is more complex. There are several possible situations, as shown in Figure 202.3, including investments that impact one another (dependence, point 1 in Figure 202.3), investments that are mutually exclusive (only one investment is possible from a set of investments, point 2 in Figure 202.3), and/or investments for which there is a budget limitation (point 3 in Figure 202.3).

If the investments do have an impact on one another (i.e., they are dependent), then they should be considered together and individually. Investments that impact one another are reevaluated as a package, with new cash flows representing the costs and returns of the combination of investments. The present worth of all individual investments and all combinations of dependent investments should then be compared. This comparison will, by definition, include some mutually exclusive investments (i.e., they cannot all be undertaken because some are subsumed in others). Each set of these mutually exclusive alternatives can be resolved to a single alternative by selecting the combination or individual investment with the highest present worth from each set. For example, if a refinery and a pipeline are being considered as possible investments and the pipeline would serve the refinery as well as other refineries, then the presence of the pipeline might change the value of the refinery products (i.e., the two investments are

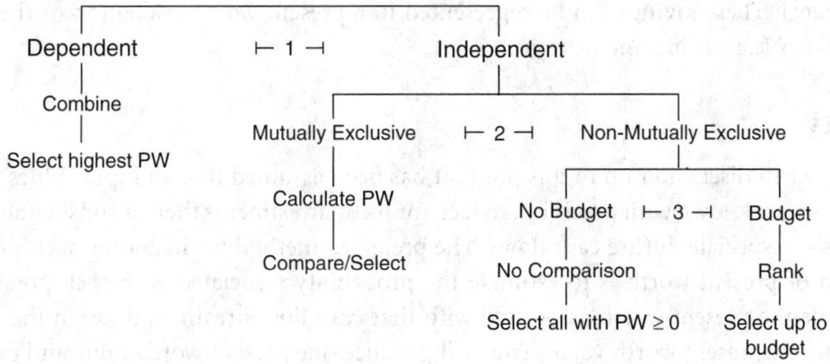

FIGURE 202.3 Cash flows.

not independent). In this case, the present worth of the refinery alone, the present worth of the pipeline alone, and the present worth of the combined pipeline and refinery should be calculated and the one with the highest present worth selected.

Once each set of dependent investments has been resolved to a single independent investment, then the comparison can proceed. If only one investment can be made (i.e., they are mutually exclusive), then the comparison is straightforward; the present worth of each investment is calculated and the investment with the highest present worth is the most economic. This is true even if the investments require significantly different initial investments, have significantly different times at which the returns occur, or have different useful lifetimes. Examples of mutually exclusive investments include different system sizes (e.g., three different refinery sizes are being considered for a single location), different system configurations (e.g., different refinery configurations are being considered for the same site), and so forth.

If the investments are not mutually exclusive, then one must consider whether there is an overall budget limitation that would restrict the number of economic investments that might be undertaken. If there is no budget (i.e., no limitation on the investment funds available) and the investments have no impact on one another, than there is really no comparison to be performed and the investor simply undertakes those investments that have positive present worth and discards those that do not.

If the investments are independent but funds are not available to undertake all of them (i.e., there is a budget), then there are two approaches. The easiest approach is to rank the alternatives, with the best having the highest benefit-to-cost ratio or savings-to-investment ratio. (The investment with the highest present worth will not necessarily be the one with the highest rank, since present worth does not show return per unit investment.) Once ranked, those investments at the top of the priority list are selected until the budget is exhausted. Present worth can be used in the second, less desirable approach by considering the total present worth of each combination of investments whose total initial investment cost is less than the budget. The total present worth of the investment package is simply the sum of the present worth values of all the independent investments in the package. That combination with the greatest present worth is then selected.

202.3 Other Considerations

Three commonly encountered complications to the calculation of present worth are savings versus returns, uncertainty, and externalities.

Savings vs. Income/Returns

In many, if not most, investments, the positive cash flows or benefits are actual cash inflows that result from sales of products produced by the investment. However, there are also a large number of investments in which the positive cash flows or benefits are represented by savings. For example, the benefit of investing in additional insulation for steam pipes within a refinery is savings in fuel costs (less fuel will be required to produce steam). These savings can be represented in a present worth evaluation of the insulation as positive cash flows (i.e., as income or returns).

Uncertainty

In the present worth discussion up to this point, it has been assumed that all input values to the present worth calculation are known with precision. In fact, for most investments there is considerable uncertainty in these values — especially future cash flows. The preferred method for including such uncertainties in the calculation of present worth is to estimate the probability associated with each possible cash flow stream, calculate the present worth associated with that cash flow stream, and assign the probability of the cash flow to the present worth value. This will produce one present worth value and one probability for each possible cash flow stream (i.e., a probability distribution on the present worth of the investment). This distribution can then be used to find statistics such as the expected present worth value, the standard deviation of the present worth value, confidence intervals, and so forth. These statistics, especially the expected present worth value and confidence intervals, can then be used in the decision process.

Externalities

In many cases, not all costs and benefits are included in the cash flows because they are not easily quantified in terms of dollars or because they do not benefit or cost the investor directly. Such costs and benefits are referred to as *externalities* because they are generally considered external to the direct economic evaluation (i.e., they are not included in the present worth calculation). For example, the cost of air emissions from a refinery are often not considered in a present worth calculation, even though they impact the local community as well as the employees of the refinery. Such emissions may result in lost work days and more sick pay, as well as the loss of the local community's goodwill to the refinery, making future refinery expansion more difficult. Likewise, emissions may affect the health and quality of life of local residents.

Externalities can be considered qualitatively, along with measures such as the present worth, when evaluating an investment. Or externalities can be explicitly considered within a present worth calculation by estimating their costs and benefits in terms of dollars and including these dollars in present worth cash flows. In this case, these costs and benefits are said to have been *internalized* and are no longer externalities.

Defining Terms

Analysis period — The period during which the investor will consider the costs and benefits associated with an investment. This period is usually determined by the period during which the investor will develop, operate, and own the investment.

Cash flow — The dollar value of all costs and benefits in a given period. Cash flows are normally expressed in nominal dollars, but can be expressed in real dollars.

Discount rate — The interest rate that represents the time value of money to the investor. For most investors, this is the opportunity cost of capital (i.e., the rate of return that might be expected from other opportunities to which the same capital could be applied). Discount rates can include inflation (nominal discount rate) or exclude inflation (real discount rate).

Inflation rate — The annual rate of increase in a general price level, frequently estimated as the gross domestic product deflator or the gross national product deflator. Estimates of future value are generally provided by macroeconomic forecasting services.

Nominal dollars — Current dollars; dollars in the year that the cost or benefit is incurred.

Present worth — Net present value; the sum of all cash flows during the analysis period discounted to the present.

Real dollars — Constant dollars; value expressed in dollars of the base year. Real dollars are the value excluding inflation after the base year.

References

Palm, T. and Qayum, A. 1985. *Private and Public Investment Analysis.* South-Western, Cincinnati, OH.

Ruegg, R. and Petersen, S. 1987. *Comprehensive Guide for Least-Cost Energy Decisions.* National Bureau of Standards, Gaithersburg, MD.

Further Information

Au, T. and Au, T. P. 1983. *Engineering Economics for Capital Investment Analysis.* Allyn & Bacon, Boston, MA.

Brown, R. J. and Yanuck, R. R. 1980. *Life Cycle Costing: A Practical Guide for Energy Managers.* Fairmont Press, Atlanta, GA.

National Renewable Energy Laboratory. 1993. *A Manual for the Economic Evaluation of Energy Efficiency and Renewable Energy Technologies.* Draft. National Renewable Energy Laboratory, Golden, CO.

Ruegg, R. and Marshall, H. 1990. *Building Economics: Theory and Practice.* Van Nostrand Reinhold, New York.

Samuelson, P. A. and Nordhaus, W. D. 1985. *Economics,* 12th ed. McGraw-Hill, New York.

Stermole, F. J. 1984. *Economic Evaluation and Investment Decision Methods,* 5th ed. Investment Evaluations Corporation, Golden, CO.

Weston, J. F. and Brigham, E. F. 1981. *Managerial Finance,* 7th ed. Dryden Press, Fort Worth, TX.

203

Project Analysis Using Rate-of-Return Criteria

Robert G. Beaves

Robert Morris University

Many decision makers find rate-of-return investment criteria to be more intuitive and therefore easier to interpret than net present value (NPV). As a result the internal rate of return (IRR) continues to be widely used despite its legendary quirks and the well-established superiority of NPV. A second rate-of-return criterion, the modified internal rate of return (MIRR), provides more NPV consistency than does the IRR, but is less widely known and used. This chapter compares NPV, IRR, and MIRR and demonstrates some shortcomings of rate criteria.

203.1 Net Present Value

A project's *net present value* (NPV) represents *the change in the value of the firm* that is expected to result from the implementation of that project. The NPV of a *project* is calculated by summing the present values of all cash flows associated with it while preserving the negative signs of flows into the project and the positive signs of flows from the project. NPV can be defined as follows:

$$\text{NPV} = \sum_{t=0}^{n} a_t (1+k)^{-t} \qquad (203.1)$$

where

a_t = the cash flow at the time t (end of period t)
k = the firm's opportunity cost; the discount or "hurdle" rate
n = the number of periods in the project's life.

Implementation of a project that has a positive NPV increases firm value, whereas implementation of a project having a negative NPV reduces firm value. Because management's goal is to maximize firm value, a project should be rejected if its NPV is negative and accepted otherwise. When forced to choose among competing positive-NPV projects, the project that offers the highest NPV should be preferred because

0-8493-1586-7/05/$0.00+$1.50

it provides the largest increase in firm value. It is well established that the NPV criterion provides theoretically correct accept/reject and project-ranking decisions.

203.2 Internal Rate of Return

A project's *internal rate of return* (IRR) is that discount rate k^* for which that project's NPV is equal to zero. Note that a project's IRR is independent of the firm's *opportunity cost k*. A project's IRR represents *the average periodic rate of return earned on funds while they are invested in that project.* Because IRR is an average rather than a constant rate, the rate of return generated by a project in any one period will generally not equal its IRR. Indeed, the internal return earned by a project in any particular period is usually indeterminable.

Consider project A in Table 203.1, which has an IRR of 12.48%. Project A's rate of return during the first period (time 0 to time 1) is unknown. It is impossible to allocate the $700 payment at time 1 between return of some portion of the initial $1000 investment and return on that investment. Clearly the investment remaining in project A at time 1 need not be $424.80 as would be the case if its 12.48% IRR was a constant rate (i.e., allocating $124.80 of the $700 released at time 1 to return on the initial investment and the remaining $575.20 to return of investment). Although a project's IRR represents the average rate of return earned on funds invested in that project, the amount of funds so invested varies from period to period and is usually indeterminable.

A project should be rejected if its IRR is less than the firm's opportunity cost k and accepted otherwise. NPV and IRR provide the same accept/reject decisions for projects that have unique IRRs. The use of IRR for accept/reject decision is complicated, however, by the fact that some projects have multiple IRRs.

It is well established that ranking competing projects on the basis of their IRRs can provide an incorrect choice from among those projects. Consider projects A and B in Table 203.1. Both projects require the same $1000 initial investment, have been assigned the same 10% opportunity cost and have the same 12.48% IRR. Thus, ranking projects A and B on the basis of their IRRs provides no clear preference. Nonetheless, project A's NPV is only $34.56, whereas project B's is $62.12. Project B has the higher NPV because its entire initial investment and any accumulating returns remain invested in project B and earning an average rate of 12.48% for three periods. On the other hand, much of project A's initial investment and accumulating returns are released from project A prior to time 3. Project A is the economic equivalent of project B only if its intermediate cash flows ($700, $300) can be "reinvested" to earn a 12.48% periodic return from the time they are released by project A to time 3.

Any two projects can be correctly ranked by examining the IRR of an incremental or "difference" project. Project B–A in Table 203.1 is such an incremental project for projects A and B created by subtracting each cash flow of A from the respective cash flow of B. Because the IRR of incremental project

TABLE 203.1 Projects A, B, and C[a]

	Project A	Project B	Project B–A	Project C
Time				
0	−1,000	−1,000	0	−2,000
1	700	0	−700	1,400
2	300	0	−300	600
3	200	1,423	1,223	400
Method				
IRR	12.48%	12.48%	12.48%	12.48%
NPV	$34.56	$69.12		$69.12
IB	$1,000.00	$1,000.00		$2,000.00
MIRR	11.25%	12.48%		11.25%
$MIRR_{2000}$	10.63%	11.25%		11.25%

[a] $k = 10\%$.

B–A exceeds the firm's 10% opportunity cost, project B (the project from which a second project was subtracted) should be preferred over A. If the IRR of the incremental project was less than k, project A would have been preferred. Because incremental IRR analysis can only provide pair-wise rankings, its use becomes tedious when more than two projects are being compared.

203.3 Modified Internal Rate of Return

A project's *modified internal rate of return* (MIRR) represents the *average periodic rate earned on a fixed investment amount* over the expected life of the project. That fixed investment amount is known as the project's *investment base.* A project's MIRR can be defined as

$$\text{MIRR} = \left[\frac{(\text{IB} + \text{NPV}) \times (1+k)^n}{\text{IB}}\right]^{1/n} - 1.0 \tag{203.2}$$

where

NPV = the project's NPV, with k as discount rate
IB = the project's investment base; to be defined in greater detail later.

In calculating a project's MIRR, that project's investment base and any accumulated returns are assumed to earn the project's IRR while invested in the project and to earn the firm's opportunity cost when not retained by the project. Thus a project's MIRR is not really an "internal" rate in the sense that it is based solely on returns earned within the project. MIRR is better described as an "overall" rate that is based on all returns earned over the project's time horizon whether from the project itself or from external "reinvestment" of cash flows. Project B in Table 203.1 has an MIRR of 12.48% because, for its entire 3-period life, that project's $1000 investment base and all accumulated returns remain invested in the project B earning its 12.48% IRR. Project A's MIRR is lower because significant amounts of its investment base and accumulated returns are released at times 1 and 2 and are assumed to earn the firm's 10% opportunity cost for the balance of project A's life.

Projects with MIRRs greater than or equal to the firm's opportunity cost k should be accepted, whereas those with lower MIRRs should be rejected. As Equation (203.2) suggests, the MIRR always provides accept/reject decisions that are consistent with those provided by NPV. In contrast to the IRR, the MIRR is uniquely defined for projects that have at least one negative and one positive cash flow and is generally a function of the firm's opportunity cost k. Further, the MIRR provides an NPV-consistent ranking of competing projects that have the same scale or investment base IB.

When comparing projects having different investment bases, MIRRs adjusted to some common scale will provide an NPV-consistent ranking. Scale adjustments are required because the MIRR, like all rate criteria, does not preserve scale. Consider project C (Table 203.1), which was created by doubling project A's cash flows. Although a project's NPV doubles when its size is doubled, its IRR and MIRR are unchanged. This loss of scale in the calculation of rate of return criteria represents a serious shortcoming for their use in ranking projects.

203.4 Project Investment Base

A project's investment base represents the time 0 value of all external funds (i.e., funds not provided by the project itself) needed to finance the project. All projects considered thus far required a single cash inflow (i.e., negative flow), which occurred at time 0. The investment base of such projects is simply the amount of that initial inflow, $-a_0$. Because some projects have more complex cash flow patterns, it is necessary to establish the following general rule for determining a project's investment base: *A project's investment base is determined by multiplying the minimum cumulative present value associated with that project's stream of cash flows times –1.0.*

TABLE 203.2 Project D[a]

	Cash Flow (*at*)	Present Value	Cumulative PV
Time			
0	−1,000	−1,000.00	−1,000.00
1	300	272.73	−727.27
2	−400	−330.58	−1,057.85
3	700	525.92	−531.93
4	−500	−341.51	−873.44
5	1,500	931.38	57.94
Method			
IRR	11.48%		
NPV	$57.94		
IB	$1,057.85		
MIRR	11.18%		

[a] $k = 10\%$.

Consider project D in Table 203.2, for which the firm's opportunity cost is assumed to be 10%. Using 10% as the discount rate, present values and cumulative present values are calculated for each of project D's cash flows. The minimum cumulative present value associated with project D's cash flows is −$1057.85 Thus, project D's investment base is $1057.85. Assuming a 10% opportunity cost for all funds not currently needed by project D, $1057.85 is the minimum time 0 amount that will fund project D to its termination at time 5.

Note that the final cumulative present value listed in Table 203.2 is necessarily project D's NPV of $57.94. Project D's MIRR is calculated by substituting into Equation (203.2) as follows:

$$\left[\frac{(57.94+1057.85)\times(1.10)^5}{1057.85}\right]^{1/5} -1.0 = .1118$$

In other words, $1057.85 committed to project D at time 0 will earn an average return of 11.18% per period for five periods.

There are other possible definitions of a project's investment base. One common definition is the sum of the present values of the project's negative cash flows times −1.0. That definition, however, overlooks the fact that negative cash flows can be funded from prior-occurring positive cash flows and any external returns earned on those cash flows. As a result, the project's investment base can be overstated and its MIRR understated. For example, the sum of the present values of the negative cash flows of project D times −1.0 is $1672.09. Using that figure as project D's investment base would have produced an MIRR of 10.75%.

203.5 Scale-Adjusted MIRR

MIRR as defined in Equation (203.2) should only be used to compare projects of identical scale (i.e., the same investment base). Assuming projects B and C in Table 203.1 are mutually exclusive, they are not directly comparable on the basis of their MIRRs, because project C's 11.25% MIRR is earned on an investment base of $2000, whereas project B's 12.48% MIRR is earned on an investment base of only $1000. A project's MIRR can be adjusted to any scale by replacing its investment base (IB) in Equation (203.2) with the desired scale as follows:

$$\text{MIRR}_S = \left[\frac{(S+\text{NPV})\times(1+k)^n}{S}\right]^{1/n} -1.0 \qquad (203.3)$$

where MIRRS is the project's MIRR adjusted to scale S and S is the desired scale to which the project's MIRR is being adjusted.

These scale adjustments are based on the assumption that differences in project scales can be invested at the firm's opportunity cost k. Investing at the firm's opportunity cost increases a project's scale but does not affect its NPV. Where two or more competing projects are being compared, the MIRRs of all must be adjusted to a common scale (preferably the largest investment base among the projects). MIRRs adjusted to a common scale of \$2000 were calculated in Table 203.1 for projects A, B, and C. Those scale-adjusted MIRRs reveal that the firm should be indifferent in choosing between projects B and C but should prefer either to project A. These preferences are consistent with the NPV ranking of these three projects.

203.6 Project Life Differences

Comparing projects that have different project lives creates certain problems — no matter what decision criterion is being used. Such comparisons generally require an assumption as to what occurs at the end of each project's life — that is, whether the project is or is not replaced with a similar project. Where all competing projects are one-time expenditures with no replacement, their NPVs can be directly compared even if they have different project lives. Such comparison assumes that funds released by any project earn the firm's opportunity cost k until the end of the longest-lived project. In contrast, the MIRRs of competing one-time projects cannot be directly compared if those projects have different lives. Rate criteria such as MIRR and IRR measure average performance per period over a project's life, whereas NPV measures cumulative performance over a project's life.

Consider projects E and F in Table 203.3. If we assume that both projects are one-time expenditures that will not be replaced at the end of their respective lives, the MIRRs of projects E and F are not directly comparable. Project E earns an average rate of 13.05% per period for three periods, whereas project F earns an average rate of 11.92% per period for five periods. The MIRR of any project can be calculated over a common life z periods longer than its project life as follows, by assuming that all funds can be invested at the firm's opportunity cost k during those z additional periods:

$$\mathrm{MIRR}_{S,n+z} = \left[\frac{(S+\mathrm{NPV})\times(1+k)^{(n+z)}}{S}\right]^{1/(n+z)} - 1.0 \tag{203.4}$$

where $\mathrm{MIRR}_{S,n+z}$ is the project's MIRR adjusted to common scale S and common life $n+z$, and $n+z$ is the common life over which the project's MIRR is being calculated. Adjusted to project F's five-period life, project E's MIRR is 11.82% and is less than project F's MIRR of 11.92%. Thus, MIRRs calculated over a common five-period life provide the same ranking of projects E and F as is provided by their NPVs.

TABLE 203.3 Project E and F[a]

	Project E	Project F
Time		
0	–5,000	–4,000
1	2,500	–1,100
2	2,000	2,000
3	2,000	2,000
4		2,000
5		1,500
Method		
IRR	15.02%	13.15%
NPV	\$428.25	\$452.93
IB	\$5,000.00	\$5,000.00
MIRR	13.05%	11.92%
$\mathrm{MIRR}_{5000,5}$	11.82%	11.92%

[a] $k = 10\%$.

TABLE 203.4 Machines A and B[a]

	Machine A		Machine B	
	Cash Flow	Cumulative PV	Cash Flow	Cumulative PV
Time				
0	−25,000	−2,5000.00	−12,000	−12,000.00
1	7,000	−1,8913.04	−4,000	−15,478.26
2	7,000	−1,3620.04	10,000	−7,916.82
3	9,000	−7,702.39	9,000	−1,999.18
4	10,000	−1,984.86	6,000	1,431.34
5	7,000	1,495.38		
Method				
IRR	17.41%		19.10%	
NPV	$1,495.38		$1,431.34	
IB	$25,000.00		$15,478.26	
MIRR	16.34%		17.57%	
$MIRR_{25000,5}$	16.34%		16.29%	

[a] $k = 15\%$.

Where competing projects are assumed to be replaced at the end of their initial project lives, project rankings for projects having different lives cannot simply be based on the NPVs of the respective projects. Instead, revised projects are generated for each competing project by extending their cash flow streams to some common terminal point. NPVs and MIRRs calculated for such revised projects should provide consistent project rankings.

Example: Project Analyses

A firm has been approached by two different manufacturers who want to sell it machines that offer labor savings. The firm will buy one machine at most because both machines perform virtually the same services and either machine has sufficient capacity to handle the firm's entire needs (i.e., mutually exclusive projects). The firm's cost accounting department has provided management with projected cash flow streams for each of these machines (Table 203.4) and has suggested a 15% opportunity cost for these machines. When one considers of the IRRs and unadjusted MIRRs of machines A and B in Table 203.4, it is clear that both are acceptable, since the IRR and the MIRR of each exceeds the firm's 15% opportunity cost. It may appear that machine B should be favored because both its IRR and its unadjusted MIRR are higher than those of machine A. Note, however, that machines A and B have different investment bases and different lives. Adjusting machine B's MIRR to A's larger scale of $25,000 and to project A's longer five-period life lowers it from 17.57% to 16.29%. When adjusted to a common scale and a common life, the MIRRs of machines A and B provide the same ranking of those two projects as does NPV.

203.7 Conclusion

Rate-of-return criteria can provide correct accept or reject decisions for individual projects and can correctly rank competing projects when used properly. The concept of a rate of return seems easier to understand than NPV. After all, most personal investments (stocks, bonds, CDs, etc.) are ranked according to their rates of return. The NPV criterion, however, is the "gold standard" and its characteristics of preserving project scale and measuring cumulative performance make it more convenient to use when ranking projects.

Rate-of-return criteria should be viewed as complementing the NPV by providing project performance information in a slightly different form. The IRR provides the average rate earned per period on funds invested in the project itself while they remain so invested. In contrast, the MIRR provides the average rate earned per period on the project's investment base over the entire life of the project, assuming that funds earn the firm's opportunity cost when not invested in the project itself.

Defining Terms

Internal rate of return — The average periodic rate earned on funds invested in the project itself.

Investment base — The time 0 value of the funds that must be provided to (invested in) the project during its life.

Modified internal rate of return — The average periodic rate earned on the project's investment base over the life of the project.

Net present value — The expected increase in a firm's value if and when it implements the project.

Opportunity cost — The rate of return the firm believes it can earn on projects of similar risk. Also referred to as the *minimum required return,* the *discount rate,* or the project's *hurdle rate.*

Project — A potential investment represented by a stream of cash flows of which at least one cash flow is positive and at least one is negative.

References

Bailey, M. J. 1959. Formal criteria for investment decisions. *J. Polit. Econ.,* 67:476–488.

Beaves, R. G. 1993. The case for a generalized net present value formula. *Eng. Econ.,* 38:119–133.

Bernhard, R. H. 1989. Base selection for modified rates of return and its irrelevance for optimal project choice. *Eng. Econ.,* 35:55–65.

Hirshleifer, J. 1958. On the theory of optimal investment decision. *J. Polit. Econ.,* 66:329–352.

Lin, S. A. Y. 1976. The modified rate of return and investment criterion. *Eng. Econ.,* 21:237–247.

Mao, J. T. 1966. The internal rate of return as a ranking criterion. *Eng. Econ.,* 11:1–13.

Shull, D. M. 1992. Efficient capital project selection through a yield-based capital budgeting technique, *Eng. Econ.,* 38:1–18.

Shull, D. M. 1993. Interpreting rates of return: a modified rate-of-return approach. *Financial Pract. Educ.,* 3:67–71.

Solomon, E. 1956. The arithmetic of capital budgeting decisions. *J. Bus.,* 29:124–129.

Further Information

Au, T. and Au, T. P. 1992. *Engineering Economics for Capital Investment Analysis,* 2nd ed. Prentice Hall, Englewood Cliffs, NJ.

204

Project Selection from Alternatives

Chris Hendrickson
Carnegie Mellon University

Sue McNeil
University of Illinois, Chicago

Practical engineering and management requires choices among competing alternatives. Which boiler should be used in a plant? What kind of instrumentation should be planned for a new multistory building? Which financing scheme would be most desirable for a new facility? These are practical questions that arise in the ordinary course of engineering design, organizational management, and even personal finances. This chapter is intended to present methods for choosing the best among distinct alternatives.

204.1 Problem Statement for Project Selection

The economic project selection problem is to identify the best from a set of possible alternatives. Selection is made on the basis of a systematic analysis of expected revenues and costs over time for each project alternative. Usually, the objective is to maximize the overall net benefits of the selected project or projects, although other objectives such as minimizing environmental impact might be pursued. Here, we focus on the overall net benefits.

Project selection falls into three general classes of problems. Accept-reject problems (also known as a *determination of feasibility*) require an assessment of whether an investment is worthwhile. For example, the hiring of an additional engineer in a design office is an accept-reject decision. Selection of the best project from a set of mutually exclusive projects is required when there are several competing projects or options and only one project can be built or purchased. For example, three different configurations of a new sewage treatment plant may be considered, but only one configuration will be built. Finally, capital budgeting problems are concerned with the selection of a set of projects when there is a budget constraint and many, not necessarily competing, options. For example, a state highway agency will consider many different highway rehabilitation projects for a particular year, but generally the budget is insufficient to allow all to be undertaken, although they may all be feasible.

204.2 Steps in Carrying Out Project Selection

A systematic approach for economic evaluation of projects includes the following major steps [Hendrickson and Au, 1989]:

0-8493-1586-7/05/$0.00+$1.50

1. Generate a set of project or purchase *alternatives* for consideration. Each alternative represents a distinct component or combination of components constituting a purchase or project decision. We shall denote project alternatives by the subscript x, where $x = 1, 2, \ldots$ refers to projects 1, 2, and so on.
2. Establish a *planning horizon* for economic analysis. The planning horizon is the set of future periods used in the economic analysis. It could be very short or long. The planning horizon may be set by organizational policy (e.g., 5 years for new computers or 50 years for new buildings), by the expected economic life of the alternatives, or by the period over which reasonable forecasts of operating conditions may be made. The planning horizon is divided into discrete periods — usually years, but sometimes shorter units such as months. We shall denote the planning horizon as a set of $t = 0, 1, 2, 3, \ldots, n$, where t indicates different periods, with $t = 0$ being the present, $t = 1$, the first period, and $t = n$ representing the end of the planning horizon.
3. Estimate the *cash flow profile* for each alternative. The cash flow profile should include the revenues and costs for the alternative being considered during each period in the planning horizon. For public projects, revenues may be replaced by estimates of benefits for the public as a whole. In some cases, revenues may be assumed to be constant for all alternatives, so only costs in each period are estimated and the problem simplifies to one of cost minimization. Cash flow profiles should be specific to each alternative, so the costs avoided by not selecting one alternative (say, $x = 5$) are not included in the cash flow profile of the alternatives ($x = 1, 2$, and so on). Revenues for an alternative x in period t are denoted $B(t, x)$, and costs are denoted $C(t, x)$. Revenues and costs should initially be in *base-year* or constant dollars. Base-year dollars do not change with inflation or deflation. For tax-exempt organizations and government agencies, there is no need to speculate on inflation if the cash flows are expressed in terms of base-year dollars and a MARR without an inflation component is used in computing the net present value (NPV). For private corporations that pay taxes on the basis of then-current dollars, some modification should be made to reflect the projected inflation rates when considering depreciation and corporate taxes.
4. Specify the *minimum attractive rate of return (MARR)* for discounting. Revenues and costs incurred at various times in the future are generally not valued equally to revenues and costs occurring in the present. After all, money received in the present can be invested to obtain interest income over time. The MARR represents the tradeoff between monetary amounts in different periods and does not include inflation. The MARR is usually expressed as a percentage change per year. For many public projects, the MARR is based on the guidelines provided by the Office of Management and Budget (OMB 1992, OMB 2003). In 2003, the recommended MARR on 30-year projects is 3.2%. The value of MARR is usually set for an entire organization based upon the opportunity cost of investing funds internally rather than externally in the financial markets. For public projects the value of MARR is a political decision, so MARR is often called the *social rate of discount* in such cases. The equivalent value of a dollar in a following period is calculated as (1 + MARR), and the equivalent value two periods in the future is $(1 + \text{MARR})(1 + \text{MARR}) = (1 + \text{MARR})^2$. In general, if you have Y dollars in the present — denoted $Y(0)$, then the future value in time t — denoted $Y(t)$ is

$$Y(t) = Y(0)(1 + \text{MARR})^t \tag{204.1}$$

or the present value, $Y(0)$, of a future dollar amount $Y(t)$ is

$$Y(0) = Y(t)/(1 + \text{MARR})^t \tag{204.2}$$

5. Establish the criterion for accepting or rejecting an alternative and for selecting the best among a group of mutually exclusive alternatives. The most widely used and simplest criterion is the *net present value* criterion. Projects with a positive NPV are acceptable. From a set of mutually exclusive alternatives (from which only one alternative may be selected), the alternative with the highest

NPV is best. The next section details the calculation steps for the NPV and also some other criterion for selection.

6. Perform sensitivity and uncertainty analysis. Calculation of NPVs assumes that cash flow profiles and the value of MARR are reasonably accurate. In many cases, assumptions are made in developing cash flow profile forecasts. Sensitivity analysis can be performed by testing a variety of such assumptions, such as different values of MARR, to see how alternative selection might change. Formally treating cash flow profiles and MARR values as stochastic variables can be done with probabilistic and statistical methods.

204.3 Selection Criteria

Net Present Value

Calculation of NPVs to select projects is commonly performed on electronic calculators, on commercial spreadsheet software, or by hand. The easiest calculation approach is to compute the net revenue in each period for each alternative, denoted $A(t, x)$:

$$A(t,x) = B(t,x) - C(t,x) \qquad (204.3)$$

where $A(t, x)$ may be positive or negative in any period. Then, the NPV of the alternative, NPV(x), is calculated as the sum over the entire planning horizon of the discounted values of $A(t, x)$:

$$\text{NPV}(x) = \sum_{t=0}^{n} A(t,x)/(1+\text{MARR})^t \qquad (204.4)$$

Other Methods

Several other criteria may be used to select projects. Other discounted flow methods include *net future value*, denoted NFV(x), and *equivalent uniform annual value*, denoted EUAV(x). It can be shown [Au and Au, 1992] that these criteria are equivalent where

$$\text{NFV}(x) = \text{NPV}(x)(1+\text{MARR})^n \qquad (204.5)$$

$$\text{EUAV}(x) = \frac{\text{NPV}(x)\cdot \mathit{MARR}\cdot(1+\text{MARR})^n}{[(1+\text{MARR})^n - 1]} \qquad (204.6)$$

The net future value is the equivalent value of the project at the end of the planning horizon. The equivalent uniform annual value is the equivalent series in each year of the planning horizon.

Alternatively, benefit-to-cost ratio (the ratio of the discounted benefits to discounted costs) and the internal rate of return (the equivalent MARR at which NPV[x] = 0) are merit measures, each of which may be used to formulate a decision. For accept-reject decisions, the benefit-to-cost ratio must be greater than one and the internal rate of return greater than the MARR. However, these measures must be used in connection with incremental analyses of alternatives to provide consistent results for selecting among mutually exclusive alternatives (see, for instance, Au and Au [1992]).

Similarly, the payback period provides an indication of the time it takes to recoup an investment but does not indicate the best project in terms of expected net revenues over the entire planning horizon. Algebraically, the payback period is calculated by summing up cash flows until the sum becomes positive:

$$PBP(x) = \textit{lowest } k \textit{ for which} \sum_{t=0}^{k} B(t,x) \geq \sum_{t=0}^{k} C(t,x)$$

204.4 Applications

To illustrate the application of these techniques and the calculations involved, two examples are presented.

Example 204.1: Alternative Bridge Designs

A state highway agency is planning to build a new bridge and is considering two distinct configurations. The initial costs and annual costs and benefits for each bridge are shown in the following table. The bridges are each expected to last 30 years.

	Alternative 1	Alternative 2
Initial cost	\$15,000,000	\$25,000,000
Annual maintenance and operating costs	\$15,000	\$10,000
Annual benefits	\$1,200,000	\$1,900,000
Annual benefits less costs	\$1,185,000	\$1,890,000

Solution. The NPVs for a MARR of 0.05 (5%) are given as follows:

$$\mathrm{NPV}(1) = (-15\,000\,000) + (1\,185\,000)/(1+0.05) + (1\,185\,000)/(1+0.05)^2$$

$$+(1\,185\,000)/(1+0.05)^3 + \cdots + (1\,185\,000)/(1+0.05)^{30}$$

$$= \$3\,216\,354$$

$$\mathrm{NPV}(2) = (-15\,000\,000) + (1\,890\,000)/(1+0.05) + (1\,890\,000)/(1+0.05)^2$$

$$+(1\,890\,000)/(1+0.05)^3 + \cdots + (1\,890\,000)/(1+0.05)^{30}$$

$$= \$4\,053\,932$$

Therefore, the department of transportation should select the second alternative, which has the largest NPV. Both alternatives are acceptable since their NPVs are positive, but the second alternative has a higher net benefit.

Example 204.2: Equipment Purchase

Consider two alternative systems for providing side collision warnings for transit buses [McNeil et al., 2002]. The first method is a manual one based on the alertness of the driver; and the second is a driver aid using specialized sensors. Which method should be used? We shall solve this problem by analyzing whether the new driver aid has benefits in excess of the existing manual method minus the additional costs.

Solution. Following the steps outlined earlier, the problem is solved as follows:

1. The alternatives for consideration are the existing driver-based method, and the sensor-based warning system. The alternatives are mutually exclusive because the adoption of one precludes the other.
2. The planning horizon is assumed to be 3 years to coincide with the expected life of the sensors.
3. A transit agency in a medium-sized city has approximately 900 buses.
4. The side collision warning systems is expected to either eliminate or reduce the severity of crashes, saving \$4 million per year
5. The cash flow profile for alternative 2 is given in the following table.

System acquisition costs	\$5,000 per bus
Annual maintenance and operating costs	\$1,000 per bus
Annual crash savings	\$4,000,000
Annual savings over costs	\$3,100,000

TABLE 204.1 Relative Net Present Value of New Equipment with Different Assumptions

Acquisition Cost (per bus) ($)	Annual Crash Savings ($)	Annual Maintenance and Operation (per bus) ($)	MARR		
			0.02	0.05	0.08
3,000	2,500,000	1,000	$1,914,213	$1,657,197	$1,423,355
5,000	2,500,000	1,000	$114,213	($142,803)	($376,645)
7,000	2,500,000	1,000	($1,685,787)	($1,942,803)	($2,176,645)
3,000	4,000,000	1,000	$6,240,038	$5,742,069	$5,289,001
5,000	4,000,000	1,000	$4,440,038	$3,942,069	$3,489,001
7,000	4,000,000	1,000	$2,640,038	$2,142,069	$1,689,001

The values are estimated using engineering judgment, an analysis of historical crash records, and widely accepted estimates of the costs of crashes.. The MARR is assumed to be 0.05 (5%). The NPV is computed as follows.

$$\text{NPV}(2) = -4{,}500{,}000 + 3{,}100{,}000/(1 + 0.05) + 3{,}100{,}000/(1 + 0.05)^2 + 3{,}100{,}000/(1 + 0.05)^3 = \$3{,}942{,}069 \qquad (204.7)$$

6. Using the criterion NPV(2) > 0, so alternative 2 is selected.
7. To determine the sensitivity of the result to some of the assumptions, consider Table 204.1. The table indicates that if the annual saving in crash costs is reduced to $2,500,000 per year, the additional investment in the side collision warning system is justifiable at the MARR of 0.02 (2%) but not at a MARR of 0.05 (5%.)

This example illustrates the use of the NPV criteria for an incremental analysis, which assumes that the benefits are constant for both alternatives and examines incremental costs for one project over another.

204.5 Conclusion

This chapter has presented the basic steps for assessing economic feasibility and selecting the best project from a set of mutually exclusive projects, with NPV as a criterion for making the selection.

Defining Terms

Alternatives — A distinct option for a purchase or project decision.
Base year — The year used as the baseline of price measurement of an investment project.
Cash flow profile — Revenues and costs for each period in the planning horizon.
Equivalent uniform annual value — Series of cash flows with a discounted value equivalent to the net present value.
Minimum attractive rate of return (MARR) — Percentage change representing the time value of money.
Net future value — Algebraic sum of the computed cash flows at the end of the planning horizon.
Net present value — Algebraic sum of the discounted cash flows over the life of an investment project to the present.
Payback period — The first period for which the sum of revenues exceeds the sum of costs.
Planning horizon — Set of time periods from the beginning to the end of the project; used for economic analysis.

References

Au, T. and Au, T. P. 1992. *Engineering Economics for Capital Investment Analysis,* 2nd ed. Prentice Hall, Englewood Cliffs, NJ.

Hendrickson, C. and Au, T. 1989. *Project Management for Construction.* Prentice Hall, Englewood Cliffs, NJ. Available at: *www.ce.cmu.edu/pmbook.*

McNeil, S., Duggins, D., Mertz, C., Suppé, A., and Thorpe, C. 2002. A Performance Specification for Transit Bus Side Collision Warning System. Proceedings of 9th World Congress on Intelligent Transport Systems, Chicago, IL, October 14–18.

Office of Management and Budget (OMB). 1992. Circular No. A-94, Revised (Transmittal Memo No. 64), October 29, Guidelines and Discount Rates for Benefit-Cost Analysis of Federal Programs. OMB, Washington, DC.

Office of Management and Budget (OMB). 2003. 2003 Discount Rates for OMB Circular No. A-94. Available at: www.whitehouse.gov/omb/memoranda/m03-08.html. Accessed on March 15, 2003.

Park, C. S. 2002. *Contemporary Engineering Economics, 3rd ed.* Addison-Wesley, Reading, MA.

Further Information

A thorough treatment of project selection is found in Au and Au, *Engineering Economics for Capital Investment Analysis.* Many examples are presented in Park, *Contemporary Engineering Economics.*

205

Depreciation and Corporate Taxes

Chris Hendrickson
Carnegie Mellon University

Tung Au
Carnegie Mellon University

The government levies taxes on corporations and individuals to meet its cost of operations. Regulations on depreciation allowances are part of taxation policy. The tax laws promulgated by the federal government profoundly influence capital investments undertaken by private corporations. Economic valuation of the after-tax cash flows of an investment project based on projected tax rates and inflation effects provides a rational basis for accepting or rejecting investment projects.

205.1 Depreciation as Tax Deduction

Depreciation refers to the decline in value of physical assets over their estimated useful lives. In the context of tax liability, *depreciation allowance* refers to the amount allowed as a deduction in computing taxable income, and *depreciable life* refers to the estimated useful life over which depreciation allowances are computed. The depreciation allowance is a systematic allocation of the cost of a physical asset over time. A *fully depreciated* asset has received depreciation allowances equal to its original purchase price.

The methods of computing depreciation and the estimated useful lives for various classes of physical assets are specified by government regulations as a part of the tax code, which is subject to periodic revisions. Different methods of computing depreciation lead to different annual depreciation allowances and hence have different effects on taxable income and the taxes paid.

Let P be the purchase cost of an asset, S its estimated salvage value, and N the depreciable life in years. Let D_t denote the depreciable allowance in year t, and T_t denote the accumulated depreciation up to and including year t. Then for $t = 1, 2, \ldots, N$,

$$T_t = D_1 + D_2 + \cdots + D_t \tag{205.1}$$

An asset's book value B_t is simply its historical cost less any accumulated depreciation. Then

$$B_t = P - T_t \tag{205.2}$$

0-8493-1586-7/05/$0.00+$1.50

TABLE 205.1 Modified Accelerated Cost Reduction System (MACRS) Depreciation for Personal Property

Recovery Year	3-Year Period	5-Year Period	7-Year Period	10-Year Period
1	33.33	20.00	14.29	10.00
2	44.45	32.00	24.49	18.00
3	14.81	19.20	17.49	14.40
4	7.41	11.52	12.49	11.52
5		11.52	8.93	9.22
6		5.76	8.92	7.37
7			8.93	6.55
8			4.46	6.55
9				6.56
10				6.55
11				3.28

Source: Internal Revenue Service. 2002. *Instructions for Form 4562 Depreciation and Amortization.* U.S. Department of the Treasury, Washington, DC.

or

$$B_t = B_{t-1} - D_t \tag{205.3}$$

Among the depreciation methods acceptable under the tax regulations, the straight-line method is the simplest. Using this method, the uniform annual allowance in each year is

$$D_t = (P - S)/N \tag{205.4}$$

Other acceptable methods, known as *accelerated depreciation methods,* yield higher depreciation allowances in the earlier years of an asset and less in the later years than those obtained by the straight-line method. Examples of such methods are sum-of-the-years'-digits depreciation and double-declining-balance depreciation [Au and Au, 1992]. For example, the double-declining-balance depreciation allowance in any year t is:

$$D_t = \frac{2P}{N}(1 - 2/N)^{t-1} \tag{205.5}$$

Under the current IRS regulations on depreciation, known as the Modified Accelerated Cost Reduction System (MACRS), the estimated useful life of an asset is determined by its characteristics that fit one of the eight specified categories. Furthermore, the salvage value S for all categories is assumed to be zero and all assets with a life of 10 years or less are assumed to be purchased and sold at mid-year. The MACRS schedules are generally double-declining balance switching to straight-line depreciation or straight-line depreciation alone for the longer property life periods. The MACRS depreciation schedules for 3-, 5-, 7- and 10- year recovery periods are shown in Table 205.1, where each entry in the table is the allowable percentage of the purchase cost depreciation allowance in that year. Table 205.2 shows some example assets for the 3- to 10-year recovery schedules. Note that the MACRS schedule can be changed at any time, so accessing up-to-date information for computing allowable depreciation amounts is advisable.

205.2 Tax Laws and Tax Planning

Capital projects are long-lived physical assets for which the promulgation and revisions of tax laws may affect tax liability. For the purpose of planning and evaluating capital projects, it is important to under-

TABLE 205.2 Recovery Period and Examples of Property Class for MACRS Depreciation

Recovery Period	Examples of Depreciable Assets
3 years	Qualified rent-to-own property, race horse
5 years	Automobiles, office equipment
7 years	Office furniture, railroad track
10 years	Vessels, barges, and tugs, fruit- or nut-bearing trees or vines

Source: Internal Revenue Service. 2002. *Instructions for Form 4562 Depreciation and Amortization.* U.S. Department of the Treasury, Washington, DC.

stand the underlying principles, including adjustments for the transition period after each revision and for multiyear "carry-back" or "carry-forward" of profits and losses.

The federal income tax is important to business operations because profits are taxed annually at substantial rates on a graduated basis. Except for small businesses, the corporate taxes on ordinary income may be estimated with sufficient accuracy by using the marginal tax rate. *Capital gain,* which represents the difference between the sale price and the book value of an asset, is taxed at a rate lower than on ordinary income for private individuals if it is held longer than a period specified by tax laws. For corporations in 2003, capital gains are taxed at the same rate as other taxable income.

Some state and/or local governments also levy income taxes on corporations. Generally, such taxes are deductible for federal income tax to avoid double taxation. The computation of income taxes can be simplified by using a combined marginal tax rate to cover the federal, state, and local income taxes. If F is the federal marginal rate on taxable income, R is the state marginal tax rate and X is the combined rate, then:

$$X = F + R - F \cdot R \tag{205.6}$$

Tax planning is an important element of private capital investment analysis because the economic feasibility of a project is affected by the taxation of corporate profits. In making estimates of tax liability, several factors deserve attention: (1) number of years for retaining the asset, (2) depreciation method used, (3) method of financing, including purchase versus lease, (4) capital gain upon the sale of the asset, and (5) effects of inflation. Appropriate assumptions should be made to reflect these factors realistically.

205.3 Decision Criteria for Project Selection

The economic evaluation of an investment project is based on the merit of the *net present value (NPV),* which is the algebraic sum of the discounted net cash flows over the life of the project to the present. The discount rate is the minimum attractive rate of return specified by the corporation.

The evaluation of proposed investment projects is based on NPV criteria, which specify the following: (1) an independent project should be accepted if the NPV is positive and rejected otherwise; and (2) among all acceptable projects that are mutually exclusive, the one with the highest positive NPV should be selected.

205.4 Inflation Consideration

Consideration of the effects of inflation on economic evaluation of a capital project is necessary because taxes are based on then-current dollars in future years. The year in which the useful life of a project begins is usually used as the baseline of price measurement and is referred to as the *base year.* A *price index* is the ratio of the price of a predefined package of goods and service at a given year to the price of the same package in the base year. The common price indices used to measure inflation include the consumer price index, published by the Department of Labor, and the gross domestic product price deflator, compiled by the Department of Commerce.

For the purpose of economic evaluation, it is generally sufficient to project the future inflation trend by using an average annual inflation rate j. Let A_t be the cash flow in year t, expressed in terms of base-year (year 0) dollars, and A'_t be the cash flow in year t, expressed in terms of then-current dollars. Then

$$A'_t = A_t(1+j)^t \tag{205.7}$$

$$A_t = A'_t(1+j)^{-t} \tag{205.8}$$

In the economic evaluation of investment proposals in an inflationary environment, two approaches may be used to offset the effects of inflation. Each approach leads to the same result if the discount rate i, excluding inflation, and the rate i', including inflation, are related as follows:

$$i' = (1+i)(1+j) - 1 = i + j + ij \tag{205.9}$$

$$i = (i' - j)/(1+j) \tag{205.10}$$

The NPV of an investment project over a planning horizon of n years can be obtained by using the constant price approach as follows:

$$\text{NPV} = \sum_{t=0}^{n} A_t(1+i)^{-t} \tag{205.11}$$

Similarly, the NPV obtained by using the then-current price approach is

$$\text{NPV} = \sum_{t=0}^{n} A'_t(1+i')^{-t} \tag{205.12}$$

In some situations the prices of certain key items affecting the estimates of future incomes and/or costs are expected to escalate faster than the general inflation. For such cases the differential inflation for those items can be included in the estimation of the cash flows for the project.

205.5 After-Tax Cash Flows

The economic performance of a corporation over time is measured by the net cash flows after tax. Consequently, after-tax cash flows are needed for economic evaluation of an investment project. Since interest on debts is tax deductible according to the federal tax laws, the method of financing an investment project could affect net profits. Although the projected net cash flows over the years must be based on then-current dollars for computing taxes, the depreciation allowances over those years are not indexed for inflation under the current tax laws.

It is possible to separate the cash flows of a project into an operating component and a financing component for the purpose of evaluation. Such separation will provide better insight to the tax advantage of borrowing to finance a project, and the combined effect of the two is consistent with the computation based on a single combined net cash flow. The following notations are introduced to denote various items in year t over a planning horizon of n years:

A_t = net cash flow of operation (excluding financing cost) before tax
$\mathcal{A}_t$ = net cash flow of financing before tax

$\mathbf{A}_t = \mathrm{A}_t + A_t$ = combined net cash flow before tax
Y_t = net cash flow of operation (excluding financing cost) after tax
Y_t = net cash flow of financing after tax
$\mathbf{Y}_t = \mathrm{Y}_t + Y_t$ = combined net cash flow after tax
D_t = annual depreciation allowance
I_t = annual interest on the unpaid balance of a loan
Q_t = annual payment to reduce the unpaid balance of a loan
W_t = annual taxable income
X_t = annual marginal income tax rate
K_t = annual income tax

Thus, for operation in year $t = 0, 1, 2, \ldots, n$,

$$W_t = \mathrm{A}_t - D_t \tag{205.13}$$

$$K_t = X_t W_t \tag{205.14}$$

$$\mathrm{Y}_t = \mathrm{A}_t - X_t(\mathrm{A}_t - D_t) \tag{205.15}$$

For financing in year $t = 0, 1, 2, \ldots, n$,

$$I_t = Q_t - A_t \tag{205.16}$$

$$Y_t = A_t + X_t I_t \tag{205.17}$$

where the term $X_t I_t$ is referred to as the *tax shield* because it represents a gain from debt financing due to the deductibility of interest in computing the income tax.

Alternately, the combined net cash flows after tax may be obtained directly by noting that both depreciation allowance and interest are tax deductible. Then,

$$W_t = \mathrm{A}_t - D_t - I_t \tag{205.18}$$

$$\mathbf{Y}_t = \mathbf{A}_t - X_t(\mathrm{A}_t - D_t - I_t) \tag{205.19}$$

It can be verified that Equation (205.19) can also be obtained by adding Equations (205.15) and (205.17), while noting $A_t = \mathrm{A}_t + At$ and $Y_t = \mathrm{Y}_t + Yt$.

When an asset is sold, capital gains taxes must be paid on the difference between the sale price and the book value (as calculated by Equations (205.1) and (205.2)). If the sale price is less than the book value, a capital loss is incurred. Capital losses can be used to offset capital gains.

205.6 Evaluation of After-Tax Cash Flows

For private corporations, the decision to invest in a capital project may have side effects on the financial decisions of the firm, such as taking out loans or issuing new stock. These financial decisions will influence the overall equity-debt mix of the entire corporation, depending on the size of the project and the risk involved.

Traditionally, many firms have used an adjusted cost of capital, which reflects the opportunity cost of capital and the financing side effects, including tax shields. Thus, only the net cash from operation Y_t obtained by Equation (205.15) is used when the NPV is computed. The after-tax net cash flows of a proposed project are discounted by substituting Y_t for A_t in Equation (205.11), using after-tax adjusted cost of capital of the corporation as the discount rate. If inflation is anticipated, Y_t' can first be obtained in then-current dollars and then substituted into Equation (205.12). The selection of the project will be based on the NPV thus obtained without further consideration of tax shields, even if debt financing is involved. This approach, which is based on the adjusted cost of capital for discounting, is adequate for small projects such as equipment purchase.

In recent years another approach, which separates the investment and financial decisions of a firm, is sometimes used for evaluation of large capital projects. In this approach, the net cash flows of operation are discounted at a risk-adjusted rate reflecting the risk for the class of assets representing the proposed project, whereas tax shields and other financial side effects are discounted at a risk-free rate corresponding to the yield of government bonds. An adjusted NPV reflecting the combined effects of both decisions is then used as the basis for project selection. Detailed discussion of this approach may be found elsewhere [Brealey and Myers, 2000].

205.7 Effects of Various Factors

Various depreciation methods will produce different effects on the after-tax cash flows of an investment. Since the accelerated depreciation methods generate larger depreciation allowances during the early years, the NPV of the after-tax cash flows using one of the accelerated depreciation methods is expected to be more favorable than that obtained by using the straight-line method.

If a firm lacks the necessary funds to acquire a physical asset that is deemed desirable for operation, it can lease the asset by entering into a contract with another party, which will legally obligate the firm to make payments for a well-defined period of time. The payments for leasing are expenses that can be deducted in full from the gross revenue in computing taxable income. The purchase-or-lease options can be compared after their respective NPVs are computed.

When an asset is held for more than a required holding period under tax laws, the capital gain is regarded as long-term capital gain. In a period of inflation the sale price of an asset in then-current dollars increases, but the book value is not allowed to be indexed to reflect the inflation. Consequently, capital gain tax increases with the surge in sale price resulting from inflation.

Example 205.1

Suppose a light, general-purpose truck is purchased for $25,000 in February. This truck is expected to generate a before-tax uniform annual revenue of $7,000 over the next 6 years, with no salvage value at the end of 6 years. According to the current IRS regulations, this truck is assigned to a 5-year property class with no salvage value after the depreciation period. The MACRS schedule in Table 205.1 is used to compute the annual depreciation allowance. The combined federal and state income tax rate is 38%. Assuming no inflation, the after-tax discount rate of 8%, based on the adjusted cost of capital of the corporation, is used. Determine whether this investment proposal should be accepted.

Solution. Using the 5-year recovery period schedule in Table (205.1), the annual depreciation allowance D_t is found to be $25,000 multiplied by the allowable percentage depreciation allowance in each year. We assume that the tax loss in year 2 may be offset by income elsewhere, so the loss results in a tax shield of $380. Tax amounts are rounded to the nearest dollar.

Period T	Capital Investment	Before-Tax Cash Flow, A_t	Depreciation Allowance Dt	Taxable Cash Flow $A_t - D_t$	Taxes Kt	After-Tax Cash Flow Y_t
0	−25,000		—	—	—	−25,000
1		7,000	5,000	2,000	760	6,240
2		7,000	8,000	−1,000	−380	7,380
3		7,000	4,800	2,200	836	6,164
4		7,000	2,880	4,120	1,566	5,434

Period T	Capital Investment	Before-Tax Cash Flow, A_t	Depreciation Allowance Dt	Taxable Cash Flow $A_t - D_t$	Taxes Kt	After-Tax Cash Flow Y_t
5		7,000	2,880	4,120	1,566	5,434
6		7,000	1,440	5,560	2,113	4,887

Using the adjusted cost of capital approach, the NPV of the after-tax net cash flows discounted at 8% is obtained by substituting Y_t for A_t in Equation (205.11),

$$NPV = -25\,000 + (6240)(P|F, 8\%, 1) + (7380)(P|F, 8\%, 2) + (6194)(P|F, 8\%, 3)$$

$$+ (5434)(P|F, 8\%, 4) + (5434)(P|F, 8\%, 5) + (4887)(P|F, 8\%, 6) = 2794$$

in which $(P|F, 8\%, t) = 1/(1.08)^t$ is the present worth factor of a future amount discounted at an 8% rate for t years. Since NPV = $2,794 is positive, the proposed investment should be accepted.

Example 205.2

Consider a proposal for the purchase of a computer workstation that costs $20,000 and has no salvage value at disposal after 4 years. This investment is expected to generate a before-tax uniform annual revenue of $7,000 in base-year dollars over the next 4 years. An average annual inflation rate of 5% is assumed. The MACRS 5-year property depreciation schedule is used to compute the annual depreciation allowance. The combined federal and state income tax rate is 38%. Based on the adjusted cost of capital of the corporation, the after-tax discount rate, including inflation, is 10%. Determine whether this investment proposal should be accepted.

Solution. Depreciation in each year is calculated by multiplying the 5-year depreciation percentage in each year by $20,000. This annual depreciation allowance will not be indexed for inflation, according to the IRS regulations. At the end of the 4 years, the computer has a book value of $3,456 (as calculated from Equation 205.1 and 205.2). When the computer is disposed of with no salvage value, this amount would represent a capital loss that could be used to offset capital gains accumulated elsewhere in the business. Taxes in year 4 would then be $3,509 multiplied by the combined marginal tax rate (0.38) less the capital-gain tax shield of $3,456 multiplied by the combined marginal tax rate (0.38).

The annual before-tax revenue of $,7000 in base-year dollars must be expressed in then-current dollars before computing the income taxes. From Equation (205.7),

$$A'_t = (7000)(1 + 0.05)^t$$

where $t = 1$ to 4 refers to each of the next 4 years. The after-tax cash flow Y'_t for each year can be computed by Equation (205.15). The step-by-step tabulation of the computation for each year is shown in the following table.

Period T	Capital Investment	Before Tax Cash Flow (Base Year \$) A_t	Before Tax Cash Flow (Current Year \$) A'_t	Depreciation D_t	Taxable Income $A'_t - Dt$	Capital Loss	Taxes Kt	After Tax Cash Flow Y'_t
0	−20,000							−20,000
1		7,000	7,350	4,000	3,350		1,273	6,077
2		7,000	7,718	6,400	1,318		501	7,217
3		7,000	8,013	3,840	4,123		1,586	6,427
4		7,000	8,509	2,304	6,205	3,456	1,045	7,464

Using the adjusted cost of capital approach, the NPV of the after-tax cash flows discounted at $i' =$ 10%, including inflation, can be obtained by substituting the value of Y_t' for A_t' in Equation (205.10) as follows:

$$\text{NPV} = -20\,000 + (6077)(1.1)^{-1} + (7217)(1.1)^{-2} + (6427)(1.1)^{-3} + (7464)(1.1)^{-4}$$

$$= 1416$$

Since NPV = \$1416 is positive, the investment proposal should be accepted.

Example 205.3

A developer bought a plot of land for \$100,000 and spent \$1.6 million to construct an apartment building on the site for a total price of \$1.7 million. The before-tax annual rental income after the deduction of maintenance expenses is expected to be \$300,000 in the next 6 years, assuming no inflation. The developer plans to sell this building at the end of 6 years when the property is expected to appreciate to \$2.1 million, including land. Suppose the entire cost of construction can be depreciated over 32 years based on the straight-line depreciation method, whereas the original cost of land may be treated as the salvage value at the end. The tax rates are 34% for ordinary income and 28% for capital gain, respectively. Based on the adjusted cost of capital, the developer specifies an after-tax discount rate of 10%.

Solution. Using Equation (205.4), the annual depreciation allowance D_t over 32 years is found to be \$50,000. Noting that $P = 1\ 700{,}000$ and $T_t = (6)\ (50\ 000) = 300{,}000$, the book value of the property after 6 years is found from Equation (205.2) to be \$1.4 million.

Ignoring the assumption of mid-year purchase to simplify the calculation and assuming no inflation, the after-tax annual net income in the next 6 years is given by Equation (205.13):

$$Y_t = 300\,000 - (34\%)(300\,000 - 50\,000) = 215\,000$$

The capital gain tax for the property at the end of 6 years is

$$(28\%)(2\,100\,000 - 1\,400\,000) = 196\,000$$

Using the adjusted cost of capital approach, the NPV of after-tax net cash flows in the next 6 years, including the capital gain tax paid at the end of 6 years discounted at 10%, is

$$\text{NPV} = -1\,700\,000 + (215\,000)(P \mid U, 10\%, 6)$$

$$+(2\,100\,000 - 196\,000)(P \mid F, 10\%, 6)$$

$$= 311\,198$$

in which $(P \mid U, 10\%, 6)$ is the discount factor to present at 10% for a uniform series over 6 years, and $(P \mid F, 10\%, 6)$ is the discount factor to present at 10% for a future sum at the end of 6 years. Since NPV = \$311,1 98 is positive, the proposed investment should be accepted.

Defining Terms

Base year — The year used as the baseline of price measurement of an investment project.
Capital gain — Difference between the sale price and the book value of an asset.
Depreciable life — Estimated useful life over which depreciation allowances are computed.
Depreciation — Decline in value of physical assets over their estimated useful lives.

Depreciation allowance — Amount of depreciation allowed in a systematic allocation of the cost of a physical asset between the time it is acquired and the time it is disposed of.

Net present value — Algebraic sum of the discounted cash flows over the life of an investment project to the present.

Price index — Ratio of the price of a predefined package of goods and service at a given year to the price of the same package in the base year.

Tax shield — Gain from debt financing due to deductibility of interest in computing the income tax.

References

Au, T. and Au, T. P. 1992. *Engineering Economics for Capital Investment Analysis,* 2nd ed. Prentice Hall, Englewood Cliffs, NJ.

Brealey, R. and Myers, S. 2000. *Principles of Corporate Finance,* 6th ed. McGraw-Hill, New York.

Internal Revenue Service. 2002. *Instructions for Form 4562 Depreciation and Amortization.* U.S. Department of the Treasury, Washington, DC.

Park, Chan S. 2002. *Contemporary Engineering Economics,* 3rd ed. Prentice Hall, New York.

Further Information

Up-to-date information on the tax code may be found at websites for individual state taxation departments and the following websites:

U.S. Internal Revenue Service, *www.irs.gov*
Federation of Tax Administrators, *www.taxaadmin.org*

206

Financing and Leasing

Charles Fazzi
Saint Vincent College

Financing activities, broadly speaking, relate to a firm's decision as to where to obtain the funds necessary to operate the business. The objective of this article is to describe the different approaches that a firm considers when making this decision as it relates to investment alternatives. The market for funds is divided into the money market and the capital market. The money market includes short-term debt securities, that is, securities that will mature in 1 year or less. The capital market is the market for longer-term borrowings, that is, sources of cash with a time horizon of more than 1 year. The capital market is further subdivided into an intermediate capital market that includes debt securities with a maturity of more than 1 but less than 10 years in duration; and a long-term capital market which includes debt securities which generally have a maturity of more than 10 years. Financing activities for capital projects rarely utilizes money market funds.

The capital market also encompasses the market for equity securities. These sources of funds include the issuance of both preferred stock and common stock. They have the longest time horizon since these securities are normally issued for the life of the corporation. Thus they may be thought of as sources of financing with an infinite time horizon.

This chapter deals with the major sources and forms of capital market funds. In addition, leasing will be discussed. Leasing is a specific type of economic transaction that blends both the acquisition of an asset (an investment decision) with a long-term loan commitment (a financing decision). Under the right conditions, leasing can represent a most attractive approach to the financing process for a firm.

206.1 Debt Financing

There are three primary sources of intermediate and long-term debt financing: term loans, mortgage loans, and bonds.

Term Loans

A **term loan** is simply a loan that is paid off over some number of years called the term of the loan. Term loans may or may not be secured by the assets of the firm. These loans are usually negotiated with commercial banks, insurance companies, or some other financial institution. They can generally be

0-8493-1586-7/05/$0.00+$1.50

negotiated fairly quickly and at a low administrative cost. The amount borrowed (principal) and the interest is paid off in installments over the life of the loan. The rate of interest on the term loan can be fixed over the life of the loan but usually it is a variable interest rate that is linked to the prime rate. Thus the rate of interest for a term loan can be set at "2% over the prime rate," the borrower may pay 8% when the prime rate is 6% in the first year, and 9% in the second year when the prime rate is 7%. Variable rate loans of this type can include a "collar," which sets upper and lower limits on the interest rate that can be charged, or a "cap," which sets an upper limit only. The prime rate of interest is the rate charged on short-term business loans to financially sound companies. Term loans can often carry restrictive provisions or constraints on the financial activities of the borrower such as a dividend restriction.

Mortgage Loans

Mortgage loans are essentially term loans that are secured by the real estate that was purchased with the cash proceeds from the loan. They infrequently extend past a time period of 20 years and may be repaid only over 5 years.

Bonds

Bonds are intermediate to long-term debt agreements issued by corporations, governments, or other organizations, generally in units of $1,000 principal value per bond. A bond represents two commitments on the part of the issuing organization, the promise to pay the stated interest rate periodically and the commitment to repay the $1,000 principal when the bond matures. Most bonds pay interest semiannually at one-half the annual stated or coupon rate of interest. The term *coupon rate* refers to the fact that bonds often have coupons attached that may be detached and redeemed for each interest payment.

A company that floats a bond issue may have the bonds sold directly to the public by placing the bond issue with an investment banker. A bond issue can also be privately placed with a financial institution such as a commercial bank, insurance company, corporate pension fund, or a university endowment fund. The contract between the issuer and the lender is called the bond indenture. If the bond is publicly marketed, a trustee is named to ensure compliance with the bond indenture. In most cases the trustee is a commercial bank or an investment banker. In a private placement, the purchasing institution normally acts as it own trustee.

Bonds are issued in a wide variety of circumstances. A debenture is an unsecured bond that is backed by the full faith and credit of the issuer. No specific assets are pledged as collateral. In the event of default, debenture holders become general creditors of the issuer.

A mortgage bond, like a mortgage loan, is collateralized by a mortgage on some type of asset such as land, building, or equipment.

Convertible bonds are bonds that may be converted into shares of common stock, at a predetermined conversion rate. The decision to convert the bond to common stock is at the option of the bondholder. The conversion feature is attractive to investors and therefore lowers the cost of borrowing.

Most bond issues include a call feature among the provisions. This feature gives the issuer a chance to buy back the bonds at a stated price prior to maturity. Callable bonds give the issuer flexibility in financing their activities. The callable feature favors the borrower and therefore increases the effective cost of borrowing.

A sinking fund requirement establishes a procedure for the orderly retirement of a bond issue over its life. This provision requires the periodic repurchase of a stated percentage of the outstanding bonds. The issuing company can accomplish this through purchases of their bonds in the open market or through use of the call provision previously described.

206.2 Equity Financing

Equity financing utilizes a corporation's ability to sell ownership interest in the organization through the sale of capital stock in the corporation. This stock can be issued as either preferred stock or common stock. While a corporation can choose to issue preferred stock, it must issue common stock as a representation of the individual ownership interest of the investors.

Preferred Stock

Preferred stock is legally an equity security that represents an ownership interest in a corporation. The term preferred arises from the fact that preferred stock has two important preferences over common stock: preference as to payment of dividends and preference as to stockholders' claims on the assets of the business in the event of bankruptcy.

Preferred stock is often characterized as a hybrid security that possesses some of the features of bonds and some of the features of common stock. Preferred dividends are fixed in amount similar to bond interest, and must be paid before common dividends are paid. Like bondholders, preferred stockholders do not participate in the growth of corporate earnings and collect only the dividends that are stated on the stock certificate. Like common stockholders, each dividend payment must be voted on and approved by the board of directors of the corporation. Most preferred stock issuances are cumulative, meaning that missed dividends accumulate as "dividends in arrears" which must be paid before any common dividends can be paid. Timely payment of preferred dividends is very important to the corporation. Corporations with preferred dividends in arrears find it very difficult to raise other forms of capital. Therefore, payment of preferred dividends is almost as important as timely payment of bond interest to the corporation. Preferred stock can also be both convertible and callable which was discussed earlier in conjunction with bonds.

Common Stock

The common stockholders are the owners of the corporation, as evidenced by the voting rights that are normally associated with **common stock**. Each share of common stock normally has one vote in electing the corporation's board of directors. The board is the ultimate corporate authority and is the group to whom the president of the corporation reports. The board selects the president and approves the appointment of all corporate officers. The board members are responsible to the stockholders, and the stockholders have the authority to elect a new board if their performance is deemed unsatisfactory.

Common stock is often referred to as the residual equity of the corporation. Money paid to the corporation for common stock does not have to be repaid, and dividends on common stock are only paid when approved by the board of directors. Common dividends are never in arrears, and common stockholders never have a claim to any specified dividend level. If a corporation prospers and grows the board will normally vote to increase the dividend level to reflect this growth. As dividends grow, the value of the stock increases and the stockholders also benefit from capital gains that accrue from the increased value. Of course, all of this can also work in reverse if the corporation does not prosper. The shareholders could ultimately lose their investment if the corporation goes bankrupt.

Corporations sometimes issue two classes of common stock, one of which does not include any voting rights. Thus, voting rights are retained by one class of common stock, and the other class will not have any voting rights but will have the ability to participate in earnings growth. This will allow one group of owners to retain voting control, but allow another group to participate in earnings and dividends without giving up control. This is a way to raise additional equity capital without losing control of the corporation. Large publicly held corporations rarely have multiple classes of common stock.

206.3 Leasing

A **lease** can be defined as a contractual relationship in which the owner of the property or asset (lessor) conveys to a business or person (lessee) the right to use the property or asset for a specified period of time in exchange for a series of payments. Thus, in a lease contract the lessee is able to use the leased assets without assuming ownership. Leasing has become a very popular way for many businesses to acquire the necessary resources to run their operations. The leasing event is a hybrid transaction which combines the acquisition of necessary resources with a commitment to finance the acquisition. In general two types of leases are offered in the market today: operating leases and financial or capital leases.

Operating Lease

An **operating lease** is a short-term lease written usually for a period of time that is substantially shorter than the asset's useful life. The lessor assumes most of the risks of ownership including maintenance, service, insurance, liability, and property taxes. The lessee can cancel an operating lease on short notice. Thus, the operating lease does not involve the long-term fixed future commitment of financial resources and is similar to renting. There are no balance sheet effects recorded with an operating lease and the rental payments are treated as period expenses.

Financial Lease

A **financial lease** or capital lease is a contract by which the lessee agrees to pay the lessor a series of payments whose sum equals or exceeds the purchase price of the asset. Typically, the total cash flows from the lease payments, the tax savings, and the residual value of the asset at the end of the lease will be sufficient to pay back the lessor's investment and provide a profit. Most financial leases are "net" leases, in that the fundamental ownership responsibilities such as maintenance, insurance, and property and sales taxes are placed upon the lessee. The lease agreement is a long-term agreement between both parties and is noncancellable. In the case of an unforeseen event, the contract may be cancellable, but the lessor will typically impose a substantial prepayment penalty. The decision to use financial leasing as a means of acquiring an asset is often viewed as an alternative to a purchase transaction using long-term debt financing to generate the necessary cash.

Sale and Leaseback

A sale and leaseback is a fairly common arrangement in which a firm sells an asset to a lender/lessor and then immediately leases back the property. The advantage to the seller/lessee is that the selling firm receives a large infusion of cash that may be used to finance other business activities. In return for this cash, the selling firm issues a long-term lease obligation to the lessor in order to make economic use of the asset during the lease period. Title to the asset is transferred to the lessor and the lessor will realize any residual value the asset may have at the end of the lease. The lessee may realize a tax advantage in this situation if the lease involves a building on owned land. Land is not depreciable if owned outright, but the full amount of the lease payment may be tax deductible. This allows the lessee to indirectly depreciate the cost of the land through the deductibility of the lease payment.

Accounting Treatment of Leases

Accounting for leases has changed dramatically over time. Prior to 1977, lease financing was attractive to some firms because the lease obligation did not appear on the company's balance sheet. As a result leasing was regarded as a form of "off-balance-sheet financing." In 1977, the Financial Accounting Standards Board issued an accounting standard that clearly set out criteria that distinguish a capital lease from an operating lease. Any lease that does not meet the criteria for a capital lease must be classified as an operating lease. A lease is considered to be a capital lease if it meets any one of the following conditions:

1. The title of the asset being leased is transferred to the lessee at the end of the lease term.
2. The lease contains an option for the lessee to purchase the asset at a very low price.
3. The term of the lease is greater than or equal to 75% of the economic life of the asset.
4. The present value of the lease payments is greater than or equal to 90% of the fair value of the leased property.

A capital lease is recorded on the lessee's balance sheet as a lease asset with an associated lease liability. The amount of the asset and liability is equal to the present value of the minimum future lease payments. Thus, for capital leases, leasing no longer provides a source of off-balance sheet financing. Operating leases must be fully disclosed in the footnotes to the financial statements.

Lessees like the operating lease method for lease accounting. One obvious reason is that the operating method never reflects the cumulative liability for all future lease payments. The term off-balance sheet financing means the lessee has financed the acquisition of asset services without recognizing an asset or liability on the financial statements. This leads to more favorable financial performance measures for the company.

Tax Treatment of Leases

For tax purposes, the lessee can deduct the full amount of the lease payment in a properly structured lease. The Internal Revenue Service (IRS) wants to be sure that the lease contract truly represents a lease and not an installment purchase of the asset. To ensure itself that a true lease is present, the IRS will look for a meaningful residual value at the end of the lease term. This is usually construed to mean that the term of the lease cannot exceed 90% of the economic life of the asset. In addition, the lessee should not be given the option to purchase the asset for any thing less than the fair market value of the leased asset at the end of the lease term. The lease payments should be reasonable in that they provide the lessor not only a return of principal, but also a reasonable interest return as well. In addition, the lease term must be for less than 30 years; otherwise it will be construed as an installment purchase of the asset. The IRS wants to assure itself that the transaction is not a disguised installment purchase that has more rapid payments that will lead to deductions than would be allowed from the depreciation deduction under an outright purchase. With leasing, the cost of any land is amortized in the lease payments. In an outright purchase of land, depreciation is not allowable. When the value of a lease includes land, lease financing can offer a tax advantage to the firm. Offsetting this tax advantage is the likely residual value of the land at the end of the lease period.

Evaluating Lease Financing

To evaluate whether or not a lease financing proposal makes economic sense, it should be compared with financing the asset with debt. Whether leasing or borrowing is best will depend on the patterns of cash flow for each financing method and on the opportunity cost of funds. The method of analysis will require the calculation of the present values for the lease alternative and the present value of the borrowing alternative. The alternative with the lower present value of the cash outflows will be the better choice. The calculations can be rather extensive and complicating. Moreover, the calculations deal with uncertainty that must be considered in making the final decision.

Defining Terms

Bond — A long-term debt instrument issued by a corporation or government.

Common stock — The stock representing the most basic rights to ownership in the corporation.

Financial lease — A long-term lease that is not cancellable.

Lease — A contract under which one party, the owner of an asset (lessor), agrees to grant the use of the asset to another party (lessee), in exchange for periodic payments.

Operating lease — A short-term lease that is often cancellable.

Preferred stock — A type of stock that usually promises a fixed dividend, with the approval of the board of directors. It has preference over common stock in the payment of dividends and claims on assets.

Term loan — Debt originally scheduled for repayment in more than 1 year, but generally less than 10 years.

References

Clay, R. H. and Holder, W. W. A practitioner's guide to accounting for leases, *J. Accountancy,* August, 61–68, 1977.

Financial Accounting Standards Board. Statement of Financial Accounting Standards No. 13, Accounting for Leases, Financial Accounting Standards Board, Stamford, CT, November 1976.

Revsine, L., Collins, D. W., and Johnson, W. B., *Financial Reporting and Analysis,* 2nd ed., Prentice Hall, Upper Saddle River, NJ.

Ross, S. A., Westerfield R. W., and Jordan, B. D., *Essentials of Corporate Finance,* 3rd ed., McGraw-Hill, New York.

Van Horne, J. C. and Wachowicz, J. M., 2001 *Fundamentals of Financial Management,* 11th ed., Prentice Hall, Upper Saddle River, NJ.

Further Information

The Institute of Management Accountants offers a variety of publications and educational programs on finance and accounting topics. They can be reached at 10 Paragon Drive, Montvale, NJ, 07645, telephone 1-800-638-4427.

207

Risk Analysis and Management

Bilal M. Ayyub
University of Maryland

Risk is associated with all projects and business ventures undertaken by individuals and organizations regardless of their size, nature, and time and place of execution and use. Risk is present in various forms and levels even in small domestic projects such as adding a deck to a residence, and in large multibillion-dollar projects such as developing and producing a space shuttle. These risks could result in significant budget overruns, delivery delays, failures, financial losses, environmental damage, and even injuries and loss of life. Risks are taken even though they could lead to devastating consequences because of potential benefits, rewards, survival, and future return on investment. Risk taking is viewed as an expression of higher levels of intelligence. This chapter defines and discusses risk and its dimensions, risk assessment processes and their fundamental analytical tools, risk management and control, and risk communication.

207.1 Risk Terminology

Definitions necessary for presenting risk-based technology methods and analytical tools are presented in this section.

Hazards

A hazard is an act or phenomenon posing potential harm to some person(s) or thing(s), that is, a source of harm and its potential consequences. For example, uncontrolled fire is a hazard, water can be a hazard, and strong wind is a hazard. In order for the hazard to cause harm, it needs to interact with person(s) or thing(s) in a harmful manner. The magnitude of the hazard is the amount of harm that might result,

0-8493-1586-7/05/$0.00+$1.50

including the seriousness and exposure levels of people and the environment. Hazards need to be identified and considered in project lifecycle analyses since they could pose threats and lead to project failures.

The interaction between a person (or a system) and a hazard can be voluntary or involuntary. For example, exposing a marine vessel to a sea environment might lead to its interaction with extreme waves in an uncontrollable manner, that is, an involuntary manner. Although the decision of a navigator of the vessel to go through a storm system that is developing can be viewed as a voluntary act in nature, and might be needed to meet schedule or other constraints, the potential rewards of delivery of shipment or avoidance of delay charges offer an incentive that warrants such action. Other examples can be constructed where individuals interact with hazards for potential financial rewards, fame, and self-fulfillment and satisfaction, ranging from undertaking investments to climbing cliffs.

Reliability

Reliability can be defined for a system or a component as its ability to fulfill its design functions under designated operating or environmental conditions for a specified time period. This ability is commonly measured using probabilities. Reliability is, therefore, the occurrence probability of the complementary event to failure as provided in the following expression:

$$\text{Reliability} = 1 - \text{Failure Probability} \tag{207.1}$$

Event Consequences

For an event of failure, consequences can be defined as the degree of damage or loss from some failure. Each failure of a system has some consequence(s). A failure could cause economic damage, environmental damage, injury or loss of human life, or other possible events. Consequences need to be quantified in terms of failure-consequence severities using relative or absolute measures for various consequence types to facilitate risk analysis.

For a successful event, consequences can be defined as the degree of reward or return or benefits from success. Such an event could cause economic outcomes, environmental effects, or other possible events. Consequences need to be quantified using relative or absolute measures for various consequence types to facilitate risk analysis.

Risk

The concept of risk can be linked to uncertainties associated with events. Within the context of projects, risk is commonly associated with an uncertain event or condition that, if it occurs, has a positive or a negative effect on a project 's objectives.

Risk originates from the Latin term *risicum* meaning the challenge presented by a barrier reef to a sailor. The Oxford dictionary defines risk as the chance of hazard, bad consequence, loss, and so on. Also, risk is the chance of a negative outcome. To measure risk we must accordingly assess both of its defining components, including the chance for negativity and potential rewards or benefits. Estimation of risk is usually based on the expected result of the conditional probability of the event occurring times the consequence of the event given that it has occurred.

A risk results from an event or sequence of events called a scenario. The event or scenario can be viewed as a cause and, if it occurs, results in consequences with severities. For example, an event or cause may be shortage of personnel needed to perform a task needed to produce a project. The event in this case of personnel shortage for the task will lead to a consequence on project cost, schedule, and/or quality. The events can reside in the project environment that may contribute to project success or failure, such as project management practices, or external partners or subcontractors.

Risk has certain characteristics that should be used in the risk assessment process. Risk is a characteristic of an uncertain future, and is neither a characteristic of the present nor the past. Once uncertainties are resolved and/or the future is attained, the risk becomes nonexistent. Therefore, we cannot describe risks for historical events or risks for events that are currently being realized. Similarly, risks cannot be directly associated with a success. Although risk management through risk mitigation of selected events could result in project success leading to rewards and benefits, these rewards and benefits cannot be considered as outcomes only of the nonoccurrence of these events associated with the risks. The occurrence of risk events leads to adverse consequences that are clearly associated with their occurrence; however, their nonoccurrences are partial contributors to the project success that lead to rewards and benefits. The credit in the form of rewards and benefits cannot be given solely to the nonoccurrence of these risk events. Some risk assessment literature defines risk to include both potential losses and rewards. They need to be treated separately as (1) risks leading to adverse consequences, and (2) risks that contribute to benefits or rewards in tradeoff analyses. An appropriate risk definition is this context is a threat (or opportunity) that could affect adversely (or favorably) achievement of the objectives of a project and its outcomes.

Developing an economic, analytical framework for a decision situation involving risks requires examining the economic and finance environments of a project. This environment could have significant impacts on the occurrence probabilities of events associated with risks. This complexity might be needed for certain projects in order to obtain justifiable and realistic results. The role of such an environment in risk analysis is discussed in subsequent sections.

Formally, risk can be defined as the potential of losses and rewards resulting from an exposure to a hazard or as a result of a risk event. Risk should be based on identified risk events or event scenarios. Risk can be viewed to be a multidimensional quantity that includes event-occurrence probability, event-occurrence consequences, consequence significance, and the population at risk; however, it is commonly measured as a pair of the probability of occurrence of an event, and the outcomes or consequences associated with the event's occurrence. This pairing can be represented by the following equation:

$$Risk \equiv \left[(p_1,c_1),(p_2,c_2),\ldots,(p_i,c_i),\ldots,(p_n,c_n)\right] \tag{207.2}$$

where *pi* is the occurrence probability of an outcome or event *i* out of *n* possible events, and *ci* is the occurrence consequences or outcomes of the event. A generalized definition of risk can be expressed as

$$Risk \equiv \left[(l_1,o_1,u_1,cs_1,po_1),(l_2,o_2,u_2,cs_2,po_2),\ldots,(l_n,o_n,u_n,cs_n,po_n)\right] \tag{207.3}$$

where l is likelihood, o is outcome, u is utility (or significance), cs is causal scenario, po is population affected by the outcome, and n is the number of outcomes. The definition according to Equation 207.3 covers all attributes measured in risk assessment that are described in this chapter, and offers a complete description of risk, from the causing event to the affected population and consequences. The population-size effect should be considered in risk studies since society responds differently for risks associated with a large population in comparison to a small population. For example, a fatality rate of 1 in 100,000 per event for an affected population of 10 results in an expected fatality of 10^{-4} per event whereas the same fatality rate per event for an affected population of 10 million results in an expected fatality of 100 per event. Although, the impact of the two scenarios might be the same on the society (same risk value), the total number of fatalities per event/accident is a factor in risk acceptance. Plane travel may be "safer" than for example recreational boating, but 200 to 300 injuries per accident are less acceptable to society. Therefore, the size of the population at risk and the number of fatalities per event should be considered as factors in setting acceptable risk.

Risk is commonly evaluated as the product of likelihood of occurrence and the impact severity of occurrence of the event:

$$\text{RISK}\left(\frac{\text{Consequence}}{\text{Time}}\right) = \text{LIKELIHOOD}\left(\frac{\text{Event}}{\text{Time}}\right) \times \text{IMPACT}\left(\frac{\text{Consequence}}{\text{Event}}\right) \tag{207.4}$$

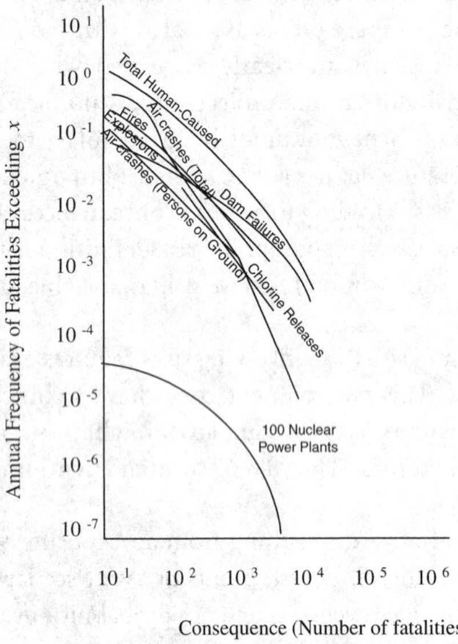

FIGURE 207.1 Example risk profile.

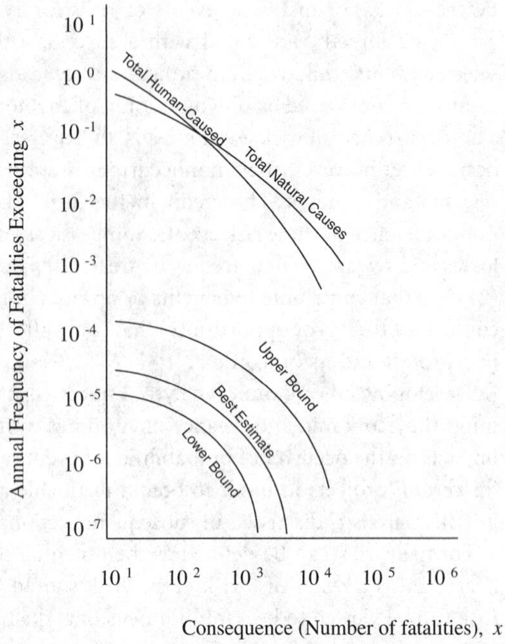

FIGURE 207.2 Uncertain risk profile.

In Equation 207.4, the likelihood can also be expressed as a probability. Equation 207.4 presents risk as an expected value of loss or an average loss. A plot of occurrence probabilities and consequences is called a risk profile or a Farmer curve. An example Farmer curve is given in Figure 207.1 based on a nuclear case study, provided herein for illustration purposes. It should be noted that the abscissa provides the number of fatalities, and the ordinate provides the annual frequency of exceedence for the corresponding number of fatalities. These curves are sometimes constructed using probabilities instead of frequencies. The curves represent median average values. Sometimes, bands or ranges are provided to represent uncertainty in these curves. They represent confidence intervals for the average curve or for the risk curve. Figure 207.2 shows example curves with uncertainty bands. This uncertainty is sometimes called meta-uncertainty.

The occurrence probability (p) of an outcome (o) can be decomposed into an occurrence probability of an event or threat (t), and the outcome-occurrence probability given the occurrence of the event ($o \mid t$). The occurrence probability of an outcome can be expressed as follows using conditional probability concepts:

$$p(o) = p(t)p(o \mid t) \tag{207.5}$$

In this context, threat is defined as a hazard or the capability and intention of an adversary to undertake actions that are detrimental to a system or an organization's interest. In this case, threat is a function of only the adversary or competitor, and usually cannot be controlled by the owner or user of the system. However, the adversary's intention to exploit his capability may be encouraged by vulnerability of the system or discouraged by an owner's countermeasures. The probability ($p(o \mid t)$) can be interpreted as the vulnerability of the system in case of this threat occurrence. Vulnerability is a result of any weakness in the system or countermeasure that can be exploited by an adversary or competitor to cause damage to the system.

Performance

The performance of a system or component can be defined as its ability to meet functional requirements. The performance of an item can be described by various elements including such items as speed, power,

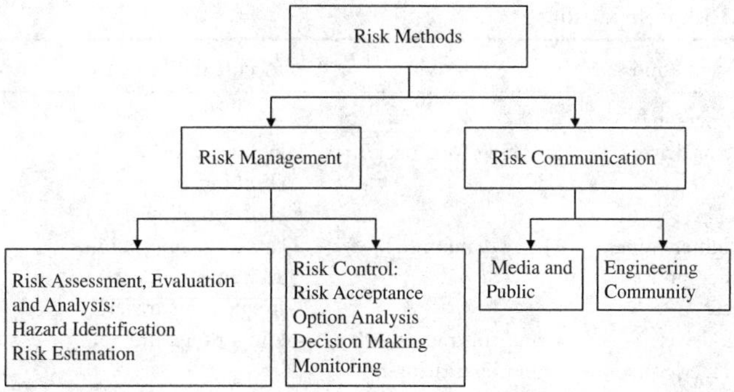

FIGURE 207.3 Risk-based technology methods.

reliability, capability, efficiency, and maintainability. The design and operation of the product or system influence performance.

Risk-Based Technology

Risk-based technologies (RBTs) are methods or tools and processes used to assess and manage the risks of a component or system. RBT methods can be classified into risk management, which includes risk assessment/risk analysis and risk control using failure prevention and consequence mitigation, and risk communication as shown in Figure 207.3.

Risk assessment consists of hazard identification, event-probability assessment, and consequence assessment. Risk control requires the definition of acceptable risk and comparative evaluation of options and/or alternatives through monitoring and decision analysis. Risk control also includes failure prevention and consequence mitigation. Risk communication involves perceptions of risk, which depends on the audience targeted, and thus is classified into risk communication to the media, the public, and to the engineering community.

Safety

Safety can be defined as the judgment of risk acceptability for the system. Safety is a relative term since the decision of risk acceptance may vary depending on the individual making the judgment. Different people are willing to accept different risks as demonstrated by different factors such as location, method or system type, occupation, and lifestyle. The selection of these different activities demonstrates an individual's safety preference despite a wide range of risk values. Table 207.1 identifies varying annual risks for different activities based on typical exposure times for these activities. Also Figure 207.4, from Imperial Chemical Industries, Ltd. shows the variation of risk exposure during a typical day that starts by waking up in the morning from sleep and getting ready to go to work, then commuting and working during morning hours, followed by a lunch break, then additional work hours followed by commuting back to having dinner, and a round trip on a motorcycle to a local pub. The ordinate in this figure is the fatal accident frequency rate (FAFR) with a FAFR of 1.0 corresponding to one fatality in 11,415 years, or 87.6 fatalities per 1 million years. The figure is based on an average number of deaths in 10^8 hours of exposure to a particular activity.

Risk perceptions of safety may not reflect the actual level of risk in some activity. Table 207.2 shows the differences in risk perception by the League of Women Voters, college students, and experts regarding 29 risk items. Only the top items are listed in the table. Risk associated with nuclear power was ranked the highest by League members and college students, whereas experts ranked it as the 20th. Experts placed motor vehicles as the highest risk. Public perception of risk and safety varies by age, gender, education, attitudes, and culture, among other factors. Individuals sometimes do not recognize uncertainties associated with a risk event or activity, which causes an unwarranted confidence in an individual's

TABLE 207.1 Relative Risk Comparisons

Risk of Death	Occupation	Lifestyle	Accidents/Recreation	Environmental Risk
1 in 100	Stunt person			
1 in 1,000	Race car driver	Smoking (one pack/ day)	Skydiving, rock climbing, snowmobiling	
1 in 10,000	Fire fighter, miner, farmer, police officer	Heavy drinking	Canoeing, automobile, all home accidents, frequent air travel	
1 in 100,000	Truck driver, engineer, banker, insurance agent	Using contraceptive pills, light drinking	Skiing, home fire	Substance in drinking water, living downstream of a dam
1 in 1,000,000		Diagnostic X-rays, smallpox vaccination (per occasion)	Fishing, poisoning, occasional air travel (one flight per year)	Natural background radiation, living at boundary of nuclear power
1 in 10,000,000		Eating charcoal-broiled steak (once a week)		Hurricane, tornado, lightning, animal bite, or insect sting

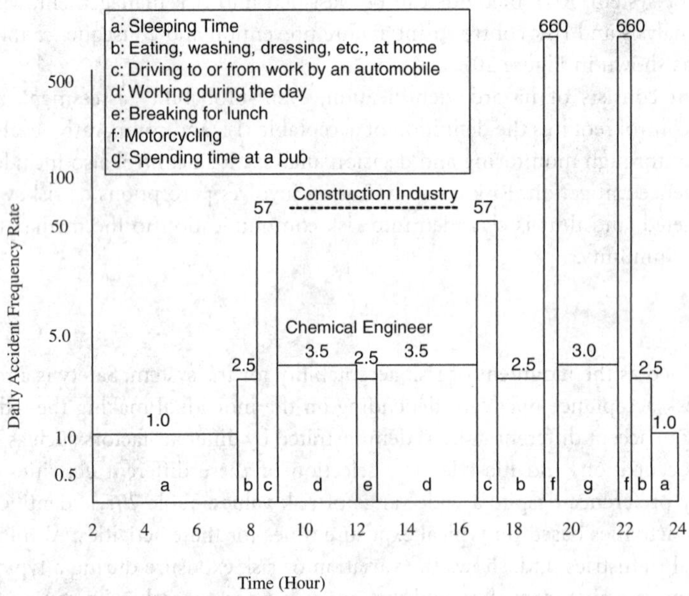

FIGURE 207.4 Daily death-risk exposure for a working healthy adult.

perception of risk or safety. Rare causes of death are often overestimated and common causes of death are often underestimated. Perceived risk is often biased by the familiarity of the hazard. The significance or the impact of safety perceptions stems from that decisions are often made on subjective judgments. If the judgments hold misconceptions about reality, this bias affects the decision. For example, the choice of a transportation mode — train, automobile, motorcycle, bus, bicycle, and so on — results in a decision based on many criteria including such items as cost, speed, convenience, and safety. The weight and evaluation of the decision criteria in selecting a mode of transportation rely on the individual's perception of safety that may deviate sometimes significantly from the actual values of risks. Understanding these differences in risk and safety perceptions is vital to performing risk management decisions and risk communications as provided in subsequent sections on risk management and control.

TABLE 207.2 Risk Perception

Activity or Technology	League of Women Voters	College Students	Experts
Nuclear power	1	1	20
Motor vehicles	2	5	1
Handguns	3	2	4
Smoking	4	3	2
Motorcycles	5	6	6
Alcoholic beverages	6	7	3
General aviation	7	15	12
Police work	8	8	17
Pesticides	9	4	8
Surgery	10	11	5
Fire fighting	11	10	18
Large construction	12	14	13
Hunting	13	18	23
Spray cans	14	13	25
Mountain climbing	15	22	28
Bicycles	16	24	15
Commercial aviation	17	16	16
Electric (non-nuclear) power	18	19	9
Swimming	19	29	10
Contraceptives	20	9	11
Skiing	21	25	29
X-rays	22	17	7
High school or college sports	23	26	26
Railroads	24	23	19
Food preservatives	25	12	14
Food coloring	26	20	21
Power mowers	27	28	27
Prescription antibiotics	28	21	24
Home applications	29	27	22

207.2 Risk Assessment

Risk Assessment Methodologies

Risk studies require the use of analytical methods at the system level that considers subsystems and components in assessing their failure probabilities and consequences. Systematic, quantitative, qualitative, or semiquantitative approaches for assessing the failure probabilities and consequences of engineering systems are used for this purpose. A systematic approach allows an analyst to evaluate expediently and easily complex systems for safety and risk under different operational and extreme conditions. The ability to quantitatively evaluate these systems helps cut the cost of unnecessary and often expensive redesign, repair, and strengthening or replacement of components, subsystems, and systems. The results of risk analysis can also be utilized in decision analysis methods that are based on cost-benefit tradeoffs.

Risk assessment is a technical and scientific process by which the risks of a given situation for a system are modeled and quantified. Risk assessment can require and/or provide both qualitative and quantitative data to decision makers for use in risk management.

Risk assessment or risk analysis provides the process for identifying hazards, event probability assessment, and consequence assessment. The risk assessment process answers three basic questions: (1) What can go wrong? (2) What is the likelihood that it will go wrong? (3) What are the consequences if it does go wrong? Answering these questions requires the utilization of various risk methods as discussed in this chapter.

A risk assessment process should utilize experiences gathered from project personnel including managers, other similar projects and data sources, previous risk assessment models, experiences from other

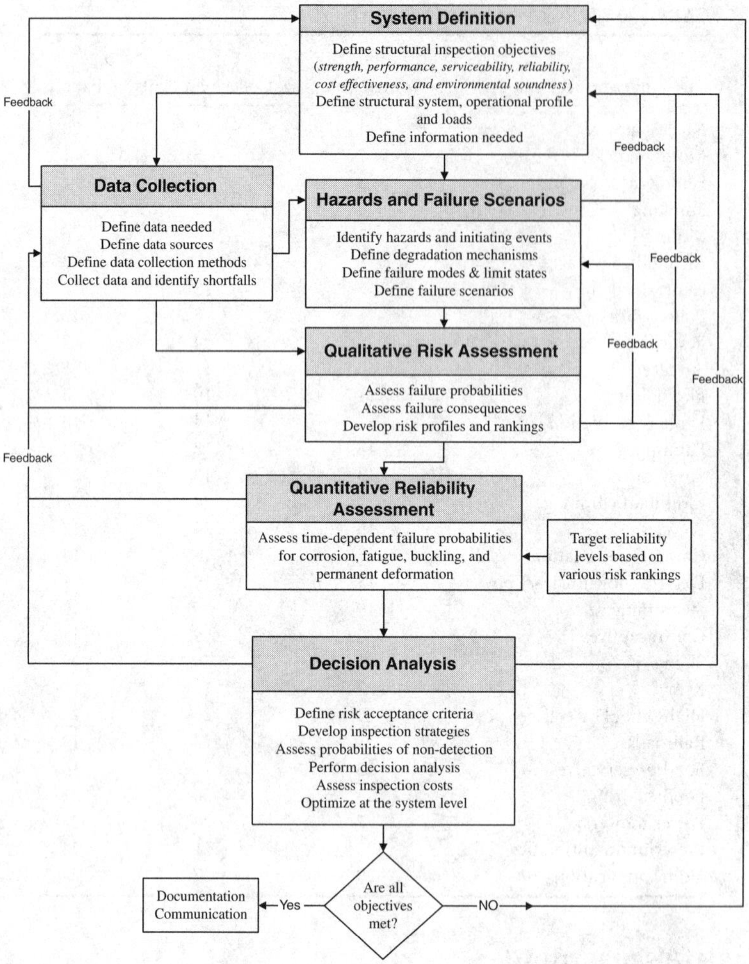

FIGURE 207.5 Methodology for risk-based, life-cycle management of structural systems.

industries and experts, in conjunction with analysis and damage evaluation/prediction tools. A risk assessment process is commonly part of a risk-based or risk-informed methodology that should be constructed as a synergistic combination of decision models, advanced probabilistic reliability analysis algorithms, failure-consequence assessment methods, and conventional performance assessment methodologies that have been employed in related industry for performance evaluation and management. The methodology should realistically account for the various sources and types of uncertainty involved in the decision-making process.

In this section, a typical overall methodology is provided in the form of a workflow or block diagram. The various components of the methodology are described in subsequent sections. Figure 207.5 provides an overall description of a methodology for risk-based management of structural systems for the purpose of demonstration. The methodology consists of the following primary steps:

1. Definition of analysis objectives and systems
2. Hazard analysis, definition of failure scenarios, and hazardous sources and their terms
3. Collection of data in a life-cycle framework
4. Qualitative risk assessment
5. Quantitative risk assessment
6. Management of system integrity through failure prevention and consequence mitigation using risk-based decision making

These steps are briefly described below with additional background materials provided in subsequent sections.

The first step of the methodology is to define the system. This definition should be based on a goal that is broken down into a set of analysis objectives. A system can be defined as an assemblage or combination of elements of various levels and/or details that act together for a specific purpose. Defining the system provides the risk-based methodology with the information that it needs to achieve the analysis objectives. The system definition phase of the proposed methodology has four main activities. The activities are to

1. Define the goal and objectives of the analysis
2. Define the system boundaries
3. Define the success criteria in terms of measurable performances
4. Collect information for assessing failure likelihood
5. Collect information for assessing failure consequences

For example, structural systems require a structural integrity goal that can include objectives stated in terms of strength, performance, serviceability, reliability, cost-effectiveness, and environmental soundness. The objectives can be broken down further to include other structural integrity attributes, such as alignment and watertightness in case of marine vessels. A system can be defined based on a stated set of objectives. The same system can be defined differently depending on these stated objectives. A marine vessel structural system can be considered to contain individual structural elements such as plates, stiffened panels, stiffeners, longitudinals, and so on. These elements could be further separated into individual components and/or details. Identifying all of the elements, components and details allows an analysis team to collect the necessary operational, maintenance, and repair information throughout the life cycle on each item so that failure rates, repair frequencies, and failure consequences can be estimated. The system definition might need to include nonstructural subsystems and components that would be affected in case of failure. The subsystems and components are needed to assess the consequences.

In order to understand failure and the consequences of failure, the states of success need to be defined. For the system to be successful, it must be able to perform its designed functions by meeting measurable performance requirements. But the system may be capable of various levels of performance, all of which might not be considered a successful performance. While a marine vessel may be able to get from point A to point B only at a reduced speed due to a fatigue failure that results in excessive vibration at the engine room, its performance would probably not be considered successful. The same concept can be applied to individual elements, components, and details. It is clear from this example that the vessel's success and failure impacts should be based on the overall vessel performance that can easily extend beyond the structural systems.

With development of the definition of success, one can begin to assess the likelihood of occurrence and causes of failures. Most of the information required to develop an estimate of the likelihood of failure might exist in maintenance and operating histories available on the systems and equipment, and based on judgment and expert opinion. This information might not be readily accessible, and its extraction from its current source might be difficult. Also, assembling it in a manner that is suitable for the risk-based methodology might be a challenge.

Operation, maintenance, engineering, and corporate information on failure history needs to be collected and analyzed for the purpose of assessing the consequences of failures. The consequence information might not be available from the same sources as the information on the failure itself. Typically there are documentations of repair costs, reinspection or recertification costs, lost person-hours of labor, and possibly even lost opportunity costs due to system failure. Much more difficult to find and assess are costs associated with the effects on other systems, the cost of shifting resources to cover lost production, and things like environmental, safety loss, or public relations costs. These may be attained through carefully organized discussions and interviews with cognizant personnel including the use of expert opinion elicitation.

Risk Events and Scenarios

In order to adequately assess all risks associated with a project, the process of identifying risk events and scenarios is an important stage in risk assessment. Risk events and scenarios can be categorized as follows:

- Technical, technological, quality, or performance risks, such as unproven or complex technology, unrealistic performance goals, and changes to the technology used or to the industry standards during the project
- Project-management risks, such as poor allocation of time and resources, inadequate quality of the project plan, and poor use of project management disciplines
- Organizational risks, such as cost, time, and scope objectives that are internally inconsistent, lack of prioritization of projects, inadequacy or interruption of funding, resource conflicts with other projects in the organization, errors by individuals or an organization, and inadequate expertise and experience by project personnel
- External risks, such as shifting legal or regulatory environment, labor issues, changing owner priorities, country risk, and weather
- Natural hazards, such as earthquakes, floods, strong wind, and waves generally require disaster recovery actions in addition to risk management.

Within these categories, several risk types can be identified.

Identification of Risk Events and Scenarios

The risk assessment process starts with the question, "What can go wrong?" The identification of what wrong can go entails defining hazards, risk events, and risk scenarios. The previous section provided categories of risk events and scenarios. Risk identification involves determining which risks might affect the project and documenting their characteristics. Risk identification generally requires participation from a project team, risk management team, subject matter experts from other parts of the company, customers, end users, other project managers, stakeholders, and outside experts on an as needed basis. Risk identification can be an iterative process. The first iteration may be performed by selected members of the project team or by the risk management team. The entire project team and primary stakeholders may take a second iteration. To achieve an unbiased analysis, persons who are not involved in the project may perform the final iteration. Risk identification can be a difficult task because it is often highly subjective, and there are no unerring procedures that may be used to identify risk events and scenarios other than relying heavily on the experience and insight of key project personnel.

The development of the scenarios for risk evaluation can be created deductively (e.g., fault tree) or inductively (e.g., failure mode and effect analysis (FMEA)) as provided in Table 207.3. The table shows methods of multiple uses including likelihood or frequency estimation expressed either deterministically or probabilistically. In addition, they can be used to assess varying consequence categories including items such as economic loss, loss of life, or injuries.

The risk identification process and risk assessment requires the utilization of these formal methods as shown in Table 207.3. These different methods contain similar approaches to answer the basic risk assessment questions; however, some techniques may be more appropriate than others for risk analysis depending on the situation.

Risk Breakdown Structure

Risk sources for a project can be organized and structured to provide a standard presentation that would facilitate understanding, communication, and management. The previously presented methods can be viewed as simple linear lists of potential sources of risk, providing a set of headings under which risks can be arranged. These lists are sometimes called risk taxonomy. A simple list of risk sources might not provide the richness needed for some decision situations since it only presents a single level of organization. Some applications might require a full hierarchical approach to define the risk sources, with as

TABLE 207.3 Risk Assessment Methods

Method	Scope
Safety/review Audit	Identifies equipment conditions or operating procedures that could lead to a casualty, or result in property damage or environmental impacts.
Checklist	Ensures that organizations are complying with standard practices.
What if	Identifies hazards, hazardous situations, or specific accident events that could result in undesirable consequences.
Hazard and operability study (HAZOP)	Identifies system deviations and their causes that can lead to undesirable consequences and determine recommended actions to reduce the frequency and/or consequences of the deviations.
Preliminary hazard analysis (PrHA)	Identifies and prioritizes hazards leading to undesirable consequences early in the life of a system. It determines recommended actions to reduce the frequency and/or consequences of the prioritized hazards. This is an inductive modeling approach.
Probabilistic risk analysis (PRA)	Methodology for quantitative risk assessment developed by the nuclear engineering community for risk assessment. This comprehensive process may use a combination of risk assessment methods.
Failure modes and effects analysis (FMEA)	Identifies the components (equipment) failure modes and the impacts on the surrounding components and the system. This is an inductive modeling approach.
Fault tree analysis (FTA)	Identifies combinations of equipment failures and human errors that can result in an accident. This is a deductive modeling approach.
Event tree analysis (ETA)	Identifies various sequences of events, including both failures and successes that can lead to an accident. This is an inductive modeling approach.
Delphi technique	Assists in reaching consensus of experts on a subject such as project risk while maintaining anonymity by soliciting ideas about the important project risks that are collected and circulated to the experts for further comment. Consensus on the main project risks may be reached in a few rounds of this process.
Interviewing	Identifies risk events by interviews of experienced project managers or subject matter experts. The interviewees identify risk events based on experience and project information.
Experience-based identification	Identifies risk events based on experience including implicit assumptions.
Brainstorming	Identifies risk events using facilitated sessions with stakeholders, project team members, and infrastructure support staff.

many levels as are required to provide the necessary understanding of risk exposure. Defining risk sources in such a hierarchical structure is called a risk breakdown structure (RBS). The RBS is defined as a source-oriented grouping of project risks organized to define the total risk exposure of a project of interest. Each descending level represents an increasingly detailed definition of risk sources for the project. The value of the RBS can be in aiding an analyst to understand the risks faced by the project.

An example RBS is provided in Table 207.4. In this example, four risk levels are defined as shown in the table. The project's risks are viewed as Level 0. Three types of Level-1 risks are provided in the table for the purpose of demonstration. The number of risk sources in each level varies and depends on the application at hand. The subsequent Level 2 risks are provided in groups that are detailed further in Level 3. The RBS provides a means to systematically and completely identify all relevant risk sources for a project.

The risk breakdown structure should not be treated as a list of independent risk sources since commonly they have interrelations and common risk drivers. Identifying causes behind the risk sources is a key step toward an effective risk management plan including mitigation actions. A process of risk interrelation assessment and root cause identification can be utilized to potentially lead to identifying credible scenarios that could lead to snowball effects for risk management purposes.

System Definition for Risk Assessment

Defining the system is an important first step in performing a risk assessment. A system can be defined as a deterministic entity comprising an interacting collection of discrete elements that is commonly defined using deterministic models. The word *deterministic* implies that the system is identifiable and

TABLE 207.4 Risk Breakdown Structure for a Project

Level 0	Level 1	Level 2	Level 3
			History, experiences, culture, personnel
		Corporate	Organization structure, stability, communication
			Finances conditions
			Other projects
			...
	Management		History, experiences, culture, personnel
			Contracts and agreements
		Customers and stakeholders	Requirement definition
			Finances and credit
			...
			Physical environment
		Natural environment	Facilities, site, equipment, materials
			Local services
			...
			Political
			Legal, regulatory
Project risks	External	Cultural	Interest groups
			Society and communities
			...
			Labor market, conditions, competition
		Economic	Financial markets
			...
			Scope and objectives
		Requirements	Conditions of use, users
			Complexity
			...
			Technology maturity
	Technology		Technology limitations
		Performance	New technologies
			New hazards or threats
			...
			Organizational experience
		Application	Personnel skill sets and experience
			Physical resources
			...

not uncertain in its architecture. The definition of the system is based on analyzing its functional and/or performance requirements. A description of a system may be a combination of functional and physical elements. Usually functional descriptions are used to identify high information levels on a system. A system may be divided into subsystems that interact. Additional detail leads to a description of the physical elements, components, and various aspects of the system.

The examination of a system needs to be made in a well-organized and repeatable fashion so that risk analysis can be consistently performed, which ensures that important elements of a system are defined and extraneous information is omitted. The formation of system boundaries is based on the objectives of the risk analysis.

The establishment of system boundaries can assist in developing the system definition. The system boundary decision is partially based on what aspects of the system's performance are of concern. The selection of items to include within the external boundary region also relies on the goal of the analysis. Beyond the established system boundary is the external environment of the system.

Boundaries beyond the physical/functional system can also be established. For example, time may also be a boundary since an overall system model may change as a product is further along in its life cycle. The life cycle of a system is important because some potential hazards can change throughout the life cycle. For example, material failure due to corrosion or fatigue may not be a problem early in the life of a system; however, this may be an important concern later in the life cycle of the system.

Along with identifying the boundaries, it is also important to establish a resolution limit for the system. Resolution is important since it limits analysis detail. Providing too little detail might not provide enough information for the problem. Too much information may make the analysis more difficult and costly due to the added complexity. The depth of the system model needs to be sufficient for the specific problem. Resolution is also limited by the feasibility of determining the required information for the specific problem. For failure analysis, the resolution should be to the components level where failure data are available. Further resolution is not necessary and would only complicate the analysis.

The system breakdown structure is the top-down division of a system into subsystems and components. This architecture provides internal boundaries for the system. Often the systems/subsystems are identified as functional requirements that eventually lead to the component level of detail. The functional level of a system identifies the function(s) that must be performed for the operation of the system. Further system decomposition into "discrete elements" leads to the physical level of a system definition identifying the hardware within the system. By organizing a system hierarchy using a top-down approach rather than fragmentation of specific systems, a rational, repeatable, and systematic approach to risk analysis can be achieved.

Further system analysis detail is addressed from modeling the system using some of the risk assessment methods described in Table 207.3. These techniques develop processes that can assist in decision making about the system. The logic of modeling based on the interaction of a system's components can be divided into induction and deduction. This difference in the technique of modeling and decision making is significant. Induction logic provides the reasoning of a general conclusion from individual cases. This logic is used when analyzing the effect of a fault or condition on a systems operation. Inductive analysis answers the question, "What are the system states due to some event?" In reliability and risk studies, this "event" is some fault in the system. Several approaches using the inductive approach include: PrHA, FMEA, and ETA. Deductive approaches provide reasoning for a specific conclusion from general conditions. For system analysis, this technique attempts to identify what modes of a system/subsystem/component failure can be used to contribute to the failure of the system. This technique answers the question, "How a system state can occur?" Inductive reasoning provides the techniques for FTA or its complement success tree analysis (STA).

Selected Risk Assessment Methods

Qualitative versus Quantitative Risk Assessment

Risk assessment methods can be categorized according to how the risk is determined by quantitative or qualitative analysis. Qualitative risk analysis uses judgment and sometimes "expert" opinion to evaluate the probability and consequence values. This subjective approach may be sufficient to assess the risk of a system, depending on available resources.

Quantitative analysis relies on probabilistic and statistical methods, and databases that identify numerical probability values and consequence values for risk assessment. This objective approach examines the system in greater detail to assess risks.

The selection of a quantitative or qualitative method depends on the availability of data for evaluating the hazard and the level of analysis needed to make a confident decision. Qualitative methods offer analyses without detailed information, but the intuitive and subjective processes may result in differences in outcomes by those who use them. Quantitative analysis generally provides a more uniform understanding among different individuals, but requires quality data for accurate results. A combination of both qualitative and quantitative analyses can be used depending on the situation.

Risk assessment requires estimates of the failure likelihood at some identified levels of decision-making. The failure likelihood can be estimated in the form of lifetime failure likelihood, annual failure likelihood, mean time between failures, or failure rate. The estimates can be in numeric or non-numeric form. An example numeric form for an annual failure probability is 0.00015, and for a mean time between failures

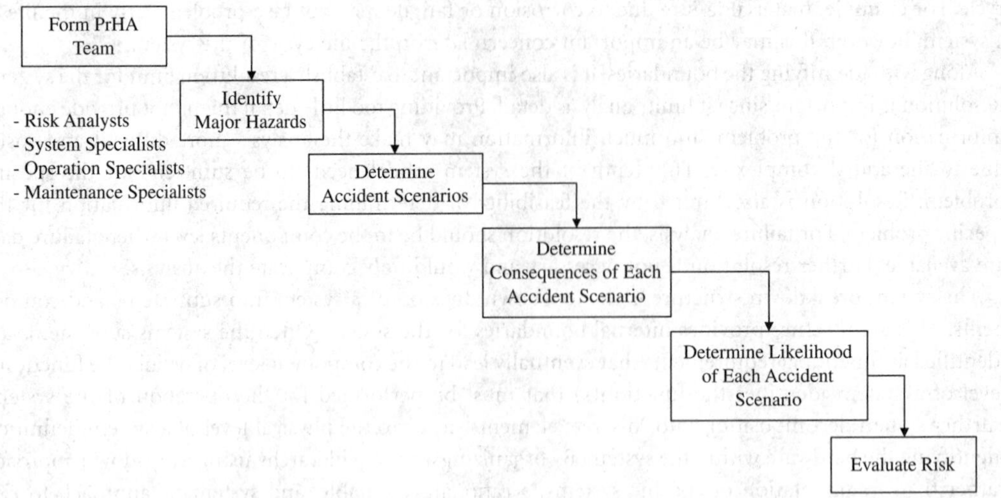

FIGURE 207.6 Preliminary hazard analysis process.

is 10 years. An example non-numeric form for "an annual failure likelihood" is large, and for a "mean time between failures" is medium. In the latter non-numeric form, guidance needs to be provided regarding the meaning of terms such as large, medium, small, very large, very small, and so on. The selection of the form should be based on the availability of information, the ability of the personnel providing the needed information to express it in one form or another, and the importance of having numeric versus non-numeric information in formulating the final decisions.

The types of failure consequences that should be considered in a study need to be selected. They can include production loss, property damage, environmental damage, and safety loss in the form of human injury and death. Approximate estimates of failure consequences at the identified levels of decision making need to be determined. The estimates can be in numeric or non-numeric form. An example numeric form for production loss is 1,000 units. An example non-numeric form for production loss is large. In the latter non-numeric form, guidance needs to be provided regarding the meaning of terms such as large, medium, small, very large, very small, and so on. The selection of the form should be based on the availability of information, the ability of the personnel providing the needed information to express it in one form or another, and the importance of having numeric versus non-numeric information in formulating the final decisions.

Risk estimates can be determined as a pair of the likelihood and consequences, and computed as the arithmetic multiplication of the respective failure likelihood and consequences for the equipment, components and details. Alternatively, for all cases, plots of failure likelihood versus consequences can be developed. Then, approximate ranking of them as groups according to risk estimates, failure likelihood, and/or failure consequences can be developed.

Preliminary Hazard Analysis

Preliminary hazard analysis (PrHA) is a common risk-based technology tool with many applications. The general process is shown in Figure 207.6. This technique requires experts to identify and rank the possible accident scenarios that may occur. It is frequently used as a preliminary method to identify and reduce the risks associated with major hazards of a system.

Failure Mode and Effects Analysis

Failure mode and effects analysis (FMEA) is another popular risk-based technology tool as shown in Figure 207.7. This technique has been introduced both in the national and international regulations for the aerospace (US MIL-STD-1629A), processing plant, and marine industries. The Society of Automotive Engineers in its recommended practice introduces two types of FMEA: design and process FMEA. This

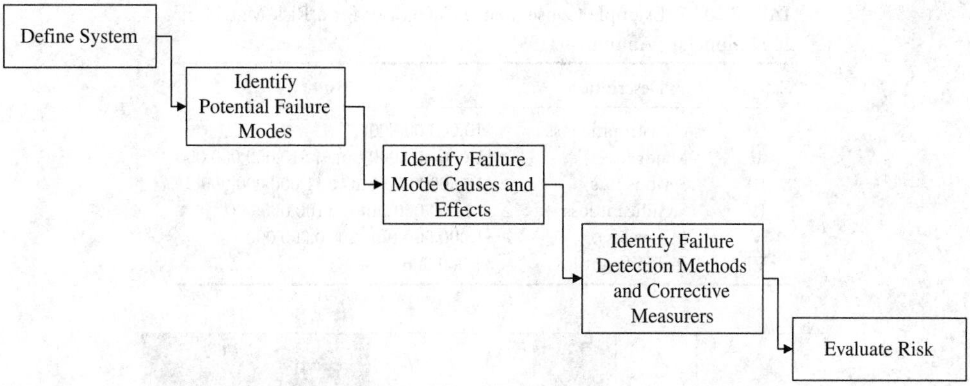

FIGURE 207.7 Failure mode and effects analysis process.

analysis tool assumes that a failure mode occurs in a system/component through some failure mechanism; the effect of this failure on other systems is then evaluated. A risk ranking can be developed for each failure mode for the effect on the overall performance of the system.

The various terms used in FMEA with examples based on the manufacturing of personal flotation devices (PFDs) are provided under subsequent headings to include failure mode, failure effect, severity rating, causes, occurrence rating, controls, detection rating, and risk priority number.

Risk Matrices

Risk can be assessed and presented using matrices for preliminary screening by subjectively estimating probabilities and consequences in a qualitative manner. A risk matrix is a two-dimensional presentation of likelihood and consequences using qualitative metrics for both dimensions. According to this method, risk is characterized by categorizing probability and consequence on the two axes of a matrix. Risk matrices have been used extensively for screening various risks. They may be used alone or as a first step in a quantitative analysis. Regardless of the approach used, risk analysis should be a dynamic process, that is, a living process where risk assessments are reexamined and adjusted. Actions or inactions in one area can affect risk in another; therefore continuous updating is necessary.

The likelihood metric can be constructed using the categories shown in Table 207.5, whereas the consequences metric can be constructed using the categories shown in Table 207.6 with an example

TABLE 207.5 Likelihood Categories for a Risk Matrix

Category	Description	Annual Probability Range
A	Likely	≥ 0.1 (1 in 10)
B	Unlikely	≥ 0.01 (1 in 100) but < 0.1
C	Very unlikely	≥ 0.001 (1 in 1,000) but < 0.01
D	Doubtful	≥ 0.0001 (1 in 10,000) but < 0.001
E	Highly unlikely	≥ 0.00001 (1 in 100,000) but < 0.0001
F	Extremely unlikely	< 0.00001 (1 in 100,000)

TABLE 207.6 Consequence Categories for a Risk Matrix

Category	Description	Examples
I	Catastrophic	Large number of fatalities, and/or major long-term environmental impact.
II	Major	Fatalities, and/or major short-term environmental impact.
III	Serious	Serious injuries, and/or significant environmental impact.
IV	Significant	Minor injuries, and/or short-term environmental impact.
V	Minor	First aid injuries only, and/or minimal environmental impact.
VI	None	No significant consequence.

TABLE 207.7 Example Consequence Categories for a Risk Matrix in 2003 Monetary Amounts (US$)

Category	Description	Cost
I	Catastrophic loss	≥ $10,000,000,000
II	Major loss	≥ $1,000,000,000 but < $10,000,000,000
III	Serious loss	≥ $100,000,000 but < $1,000,000,000
IV	Significant loss	≥ $10,000,000 but < $100,000,000
V	Minor loss	≥ $1,000,000 but < $10,000,000
VI	Insignificant loss	< $1,000,000

Probability Category							
	A	L	M	M	H	H	H
	B	L	L	M	M	H	H
	C	L	L	L	M	M	H
	D	L	L	L	L	M	M
	E	L	L	L	L	L	M
	F	L	L	L	L	L	L
		VI	V	IV	III	II	I
	Consequence Category						

FIGURE 207.8 Example risk matrix.

provided in Table 207.7. The consequence categories of Table 207.6 focus on the health and environmental aspects of consequences. The consequence categories of Table 207.7 focus on the economic impact, and should be adjusted to meet specific needs of industry and/or applications. An example risk matrix is shown in Figure 207.8. In the figure, each boxed area is shaded depending on a subjectively assessed risk level. Three risk levels are used herein for illustration purposes of low (L), medium (M), and high (H). Other risk levels may be added using a scale of five levels instead of three levels if needed. These risk levels are also called *severity factors*. The high (H) level can be considered as unacceptable risk level, the medium (M) level can be treated as either undesirable or as acceptable with review, and the low (L) level can be treated as acceptable without review.

Event Modeling: Event, Success Trees, and Fault Trees

Event modeling is a systematic, and often the most complete, way to identify accident scenarios and quantify risk for risk assessment. This risk-based technology tool provides a framework for identifying scenarios to evaluate the performance of a system or component through system modeling. The combination of event tree analysis (ETA), success tree analysis (STA), and fault tree analysis (FTA) can provide a structured analysis of system safety.

Event tree analysis is often used if the successful operation of a component/system depends on a discrete (chronological) set of events. The initiating event is first followed by other events leading to an overall result (consequence). The ability to address a complete set of scenarios is developed since all combinations of both the success and failure of the main events are included in the analysis. The probability of occurrence of the main events in the event tree can be determined using a fault tree or its complement, the success tree. The scope of analysis for event trees and fault trees depends on the objective of the analysis.

Event tree analysis is appropriate if the operation of some system/component depends on a successive group of events. Event trees identify the various combinations of event successes and failures as a result of an initiating event to determine all possible scenarios. The event tree starts with an initiating event followed by some reactionary event. This reaction can be a success or failure. If the event succeeds, the most commonly used indication is the upward movement of the path branch. A downward branch of the event tree marks the failure of an event. The remaining events are evaluated to determine the different possible scenarios. The scope of the events can be functions/systems that can provide some reduction to

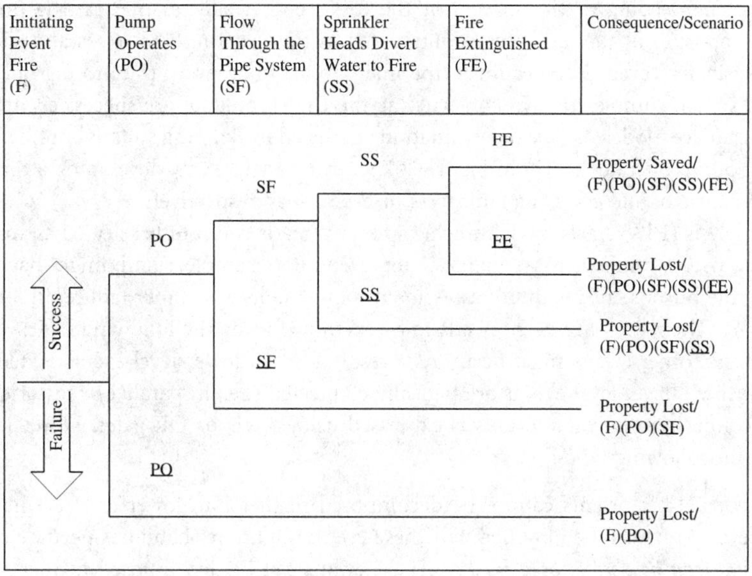

FIGURE 207.9 Event tree example for sprinkler system.

the possible hazards from the initiating event. The final outcome of a sequence of events identifies the overall state resulting from the scenario of events. Each path represents a failure scenario with varying levels of probability and risk. Different event trees can be created for different event initiators. Figure 207.9 shows an example event tree for the basic elements of a sprinkler system that might be critical for maintaining the integrity of a marine vessel.

Based on the occurrence of an initiating event, event tree analysis examines possible system outcomes or consequences. This analysis tool is particularly effective in showing the interdependence of system components that is important in identifying events and that at first might appear insignificant, but due to interdependency result in devastating results. Event tree analysis is similar to fault tree analysis because both methods use probabilistic reliability data of the individual components and events along each path to compute the likelihood of each outcome.

A quantitative evaluation of event tree probability values can be used for each event to evaluate the probability of the overall system state. Probability values for the success or failure of the events can be used to identify the probability for a specific event tree sequence. The probabilities of the events in a sequence can be provided as an input to the model or evaluated using fault trees. These probabilities for various events in a sequence can be viewed as conditional probabilities and therefore can be multiplied to obtain the occurrence probability of the sequence. The probabilities of various sequences can be summed up to determine the overall probability of a certain outcome. The addition of consequence evaluation for a scenario allows generation of a risk value. For example, the occurrence probability of the top branch scenario in Figure 207.9 is computed as the product of the probabilities of the events that comprise this scenario, that is,

$$F \cap PO \cap SF \cap SS \cap FE$$

or (F)(PO)(SF)(SS)(FE) for short.

Complex systems are often difficult to visualize and the effect of individual components on the system as a whole is difficult to evaluate without an analytical tool. Two modeling methods that have greatly improved the ease of assessing system reliability/risk are fault trees (FT) and success trees (ST). A fault tree is a graphical model created by deductive reasoning that leads to various combinations of events, which in turn leads to the occurrence of some top event failure. A success tree shows the combinations of successful events leading to the success of the top event. A success tree can be produced as the complement (opposite) of the fault tree as illustrated in this section. Fault trees and success trees are used to further analyze event tree headings (the main events in an event tree) to provide further detail to understand system complexities. In constructing the FT/ST, only failure/success events that are considered significant are modeled. This determination is assisted by defining system boundaries. For example, the event "pump operates" (PO) in Figure 207.9 can be analyzed by developing a top-down logical breakdown of failure or success using fault tress or event trees, respectively.

Fault tree analysis (FTA) starts by defining a top event that is commonly selected as an adverse event. An engineering system can have more than one top event. For example, a ship might have the following top events for the purpose of reliability assessment: power failure, stability failure, mobility failure, or structural failure. Then, each top event needs to be examined using the following logic: in order for the top event to occur, other events must occur. As a result, a set of lower-level events is defined. Also, the form in which these lower level events are logically connected (i.e., in parallel or in series) needs to be defined. The connectivity of these events is expressed using AND or OR gates. Lower-level events are classified into the following types:

- *Basic events.* These events cannot be decomposed further into lower level events. They are the lowest events that can be obtained. For these events, failure probabilities need be obtained.
- *Events that can be decomposed further.* These events can be decomposed further to lower levels. Therefore, they should be decomposed until basic events are obtained.
- *Undeveloped events.* These events are not basic and can be decomposed further. However, because they are not important, they are not developed further. Usually, the probabilities of these events are very small or the effect of their occurrence on the system is negligible, or can be controlled or mediated.
- *Switch (or house) events.* These events are not random, and can be turned on or off with full control.

The symbols shown in Figure 207.10 are used for these events. Also, a continuation symbol is shown, which is used to break up a fault tree into several parts for the purpose of fitting it in several pages.

FTA requires the development of a tree-looking diagram for the system that shows failure paths and scenarios that can result in the occurrence of a top event. The construction of the tree should be based on the building blocks and the Boolean logic gates.

The outcome of interest from the fault tree analysis is the occurrence probability of the top event. Since the top event was decomposed into basic events, its occurrence can be stated in the form of AND and OR of the basic events. The resulting statement can be restated by replacing the AND with the intersection of the corresponding basic events, and the OR with the union of the corresponding basic events. Then, the occurrence probability of the top event can be computed by evaluating the probabilities of the unions and intersections of the basic events. The dependence between these events also affects the resulting probability of the system.

For large fault trees, the computation of the occurrence probability of the top event can be difficult because of their size. In this case, a more efficient approach is needed for assessing the reliability of a system, such as the minimal cut set approach. According to this approach, each cut set is defined as a set of basic events where the joint occurrence of these basic events results in the occurrence of the top event. A minimal cut set is a cut set with the condition that the non-occurrence of any one basic event from this set results in the non-occurrence of the top event. Therefore, a minimal cut set can be viewed as a subsystem in parallel. In general, systems have more than one minimal cut sets. The occurrence of the top event of the system can, therefore, be due to any one of these minimal cut sets. As a result, the system can be viewed as the union of all the minimal cut sets for the system. If probability values are assigned to the cut sets, a probability for the top event can be determined.

A simple example of this type of modeling is shown in Figure 207.11 for a pipe system using a reliability block diagram. If the goal of the system is to maintain water flow from one end of the system to the other, then the individual pipes can be related with a Boolean *logic*. Both pipe (a) and pipe (d) and pipe (b) or pipe (c) must function for the system to meet its goal as shown in the success tree Figure 207.12(a).

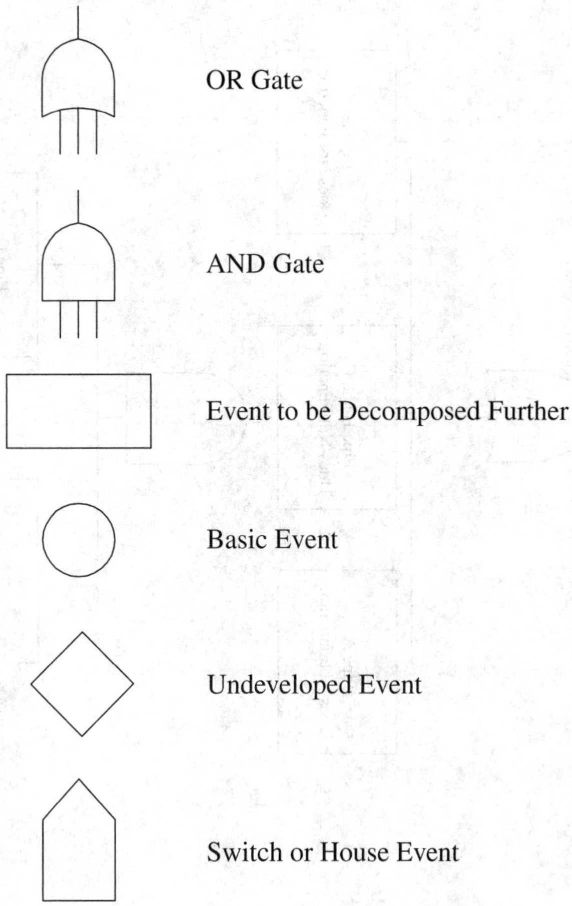

FIGURE 207.10 Symbols used in fault tree analysis.

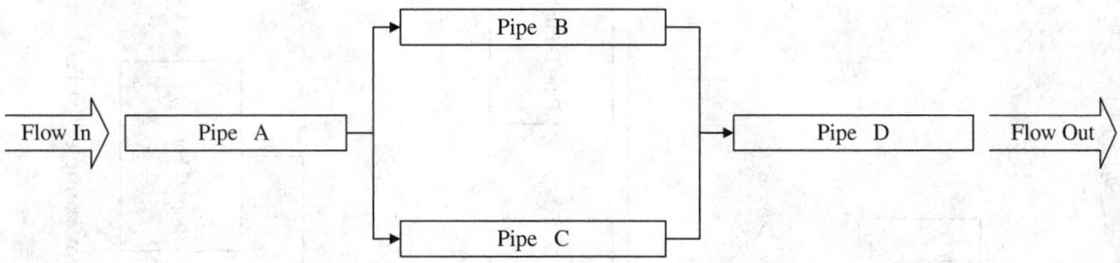

FIGURE 207.11 A reliability block diagram for a piping system.

The complement of the success tree is the fault tree. The goal of the fault tree model is to construct the logic for system failure as shown in Figure 207.12(b). Once these tree elements have been defined, possible failure scenarios of a system can be defined.

As was previously described, a failure path is often referred to as a cut set. One objective of the analysis is to determine the entire minimal cut sets, where a minimal cut set is defined as a failure combination of all essential events that can result in the failure top event. A minimal cut set includes in its combination all essential events, that is, the nonoccurrence of any of these essential events in the combination of a minimal cut set results in the nonoccurrence of the minimal cut set. These failure combinations are used to compute the failure probability of the top event. The concept of the minimal cut sets applies only to the fault trees. A similar concept can be developed in the complementary space of the success trees, and is called the minimal pass set. In this case, a minimal pass set is defined as a survival (or success)

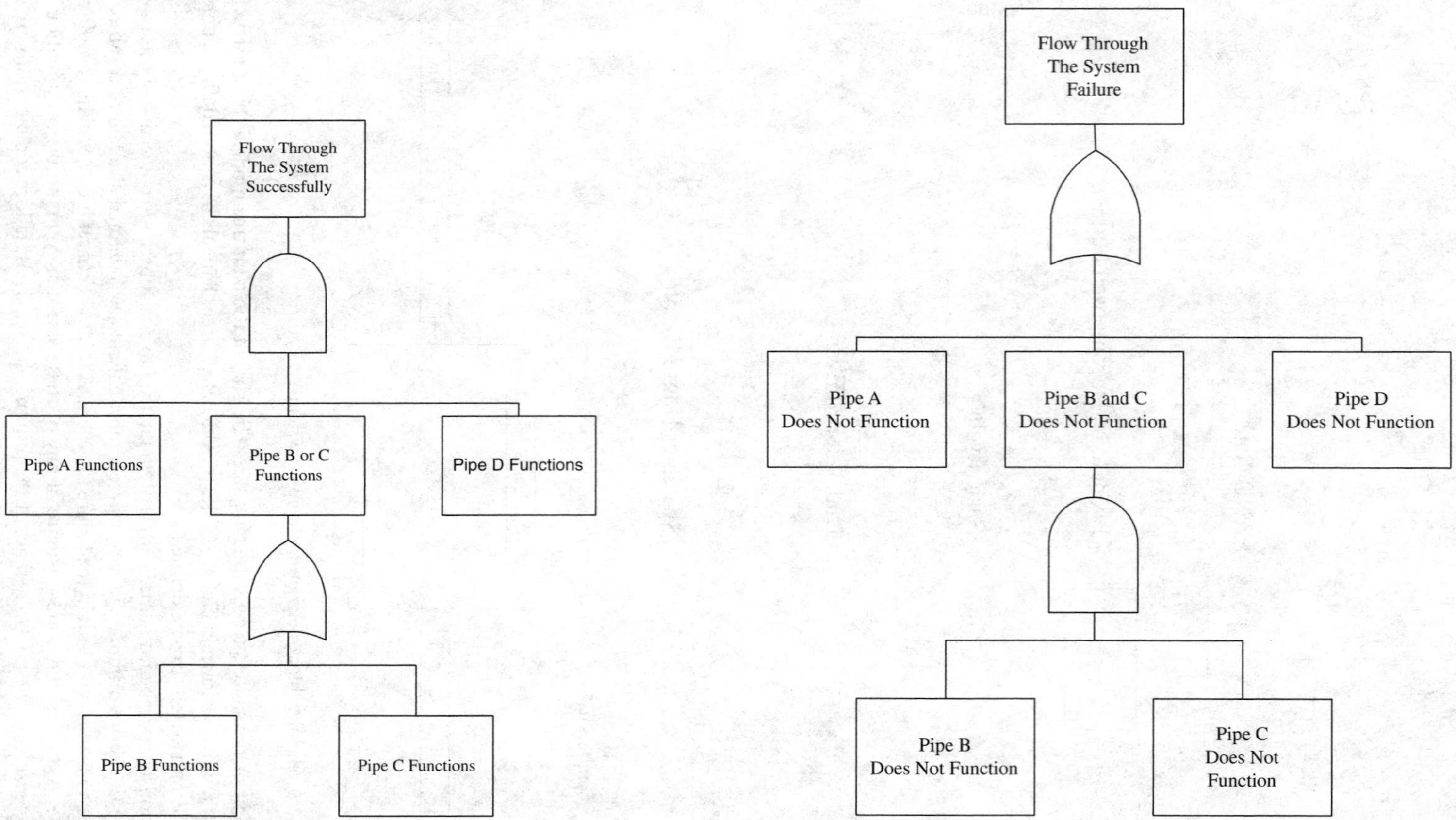

FIGURE 207.12 (a) Success tree for pipe system example. (b) Fault tree for pipe system example.

combination of all essential success events that can result in success as defined by the top event of the success tree. For the piping example, the minimal cut sets are

$$A \tag{207.6a}$$

$$D \tag{207.6b}$$

$$B \text{ and } C \tag{207.6c}$$

A minimal cut set includes events that are all necessary for the occurrence of the top event. For example, the following cut set is not a minimal cut set:

$$A \text{ and } B \tag{207.7}$$

Example 207.1: Trends in Fault Tree Models and Cut Sets

This example demonstrates how the cut sets can be identified and constructed for different arrangements of OR and AND gates logically defining a top-event occurrence. Generally, the number of cut sets increases by increasing the number of OR gates in the tree. For example, Figure 207.13 shows this trend by comparing cases a, b, and d. On the other hand, increasing the number of AND gates results in increasing the number of events included in the cut sets as shown in case c in Figure 207.13.

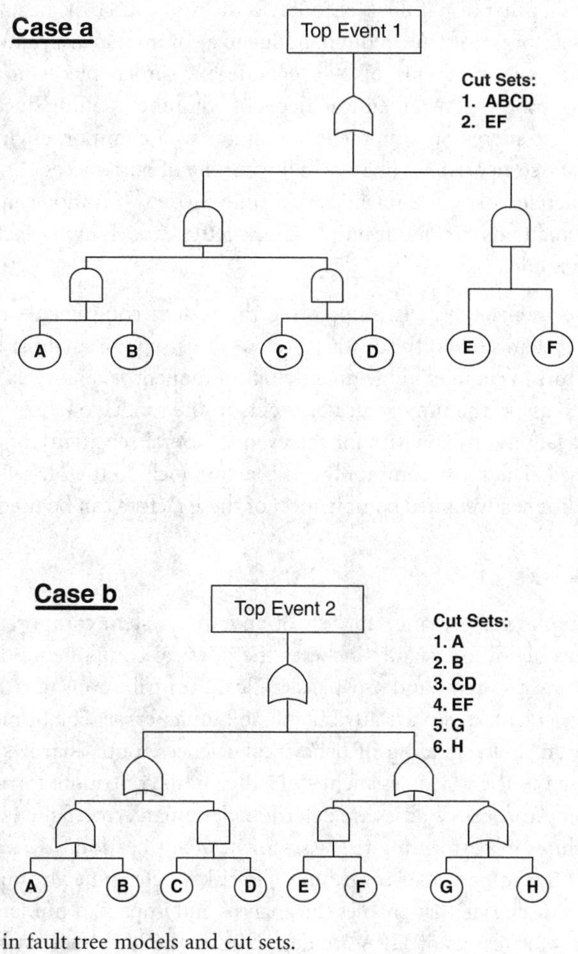

FIGURE 207.13 Trends in fault tree models and cut sets. *Continued on next page.*

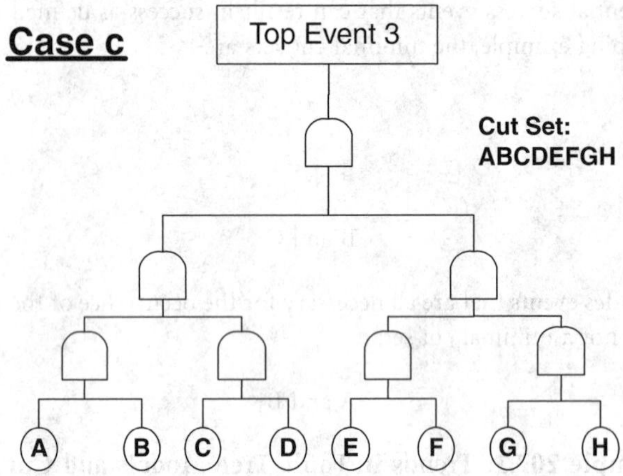

FIGURE 207.13 *Continued.*

Common-cause scenarios are events or conditions that result in the failure of seemingly separate systems or components. Common cause failures complicate the process of conducting risk analysis because a seemingly redundant system can be rendered ineffective by a common cause failure. For example, an emergency diesel generator fed by the same fuel supply as the main diesel engine will fail with the main diesel generator, if the fuel supply is the root source of the failure. The redundant emergency diesel generator is not truly redundant due to a common cause failure. Another example of common cause events is the failure of two separate but similar pieces of machinery due to a common maintenance problem, two identical pieces of equipment failing due to a common manufacturing defect, or two pieces of equipment failing due to a common environmental condition such as the flooding of a compartment or a fire in the vicinity of both pieces of machinery. A method for calculating the reliability of a system while taking into account common cause effects is the beta-factor model. Other methods include multiple Greek-letter model, alpha factor model, and beta binomial failure-rate model.

Part of risk-based decision analysis is pinpointing the system components that result in high-risk scenarios. Commercial system reliability software provides this type of analysis in the form of system-reliability sensitivity factors to changes in the underlying component reliability values. In performing risk analysis, it is desirable to assess the importance of events in the model, or the sensitivity of final results to changes in the input failure probabilities for the events. Several sensitivity or importance factors are available and can be used. The most commonly used factors include the Fussell–Vesely factor, and the Birnbaum factor. In addition, a weighted combination of these factors can be used as an overall measure.

Human-Related Risks

Risk assessment requires the performance analysis of an entire system composed of a diverse group of components. The system definition readily includes the physical components of the system; however, humans are also part of most systems and provide significant contributions to risk. It has been estimated that nearly 90% of the accidents at sea are attributable to human error. The human contribution to risk can be estimated from an understanding of behavioral sciences. Both "hardware failure" and human error should be addressed in the risk assessment since they both contribute to risks associated with the system. Once the human error probabilities are determined, human error/failures are treated in the same fashion as hardware failures in performing risk assessment quantification.

The determination of the human error contribution to risk is determined by human reliability analysis (HRA) tools. HRA is the discipline that enables the analysis and impact of humans on the reliability and safety of systems. Important results of HRA are determining the likelihood of human error as well as ways in which human errors can be reduced. When combined with system risk analysis, HRA methods provide an assessment of the detrimental effects of humans on system performance. HRA is generally considered to be composed of three basic steps: error identification, modeling, and quantification.

Other sources of human-related risks are in the form of deliberate sabotage of a system from within the system or threat from outside the system, such as a computer hacker or a terrorist. The hazard in this case is not simply random but intelligent. The methods introduced in earlier sections might not be fully applicable for this risk type. The threat scenarios to the system in this case have a dynamic nature that are affected by the defense or risk mitigation and management scenarios that would be implemented by an analyst. The use of game theory methods might be needed in this case in combination with other risk analysis and management methods. Game theory is introduced later in this section.

Human Error Identification

Human errors are unwanted circumstances caused by humans that result in deviations from expected norms that place systems at risk. It is important to identify the relevant errors that are necessary to render a complete and accurate risk assessment. Human error identification techniques should provide a comprehensive structure for determining significant human errors within a system. Quality HRA allows for accuracy in both the HRA assessment and overall system risk assessment.

Identification of human errors requires knowledge about the interactions of humans with other humans or machines (physical world). It is the study of these interfaces that allows for the understanding of human errors. Potential sources of information for identifying human error may be determined from task analysis, expert judgment, laboratory studies, simulation, and reports. Human errors may be considered active or latent depending on the time delay between when the error occurs and when the system fails.

It is important to note the distinction between human errors and human factors. Human errors are generally considered separately from human factors in the application of information about human behavior, abilities, limitations, and other characteristics in the design of tools, machines, system tasks, jobs, and environments for productive, safe, comfortable, and effective human use. Human factors are determined from performing descriptive studies to characterize populations and experimental research. However, human factors analysis may contribute to human reliability analysis.

Human Error Modeling

Once human errors have been identified, they must be represented in a logical and quantifiable framework along with other components that contribute to system risk. This framework can be determined from development of a risk model. Currently, there is no consensus on how to model human reliably. Many of these models use human event trees and fault trees to predict human reliability values. The identifications of human failure events can also be identified using FMEA. The human-error rate estimates are often based on simulation tests, models, and expert estimation.

Human Error Quantification

Quantification of human error reliability promotes the inclusion of the human element in risk analysis. This is still a developing science requiring understanding of human performance, cognitive processing, and human perceptions. Since an exact model for human cognition has not been developed, much of the current human reliability data are based on accident databases and simulation and other empirical approaches. Many of the existing data sources were developed from specific industry data, such as the nuclear and aviation industries. The application of these data sources for a specific problem should be thoroughly examined prior to application for a specific model. The result of the quantification of human reliability in terms of probability of occurrence is typically called a human error probability (HEP). There are many techniques that have been developed to help predict the HEP values. The technique for human error rate prediction (THERP) is one of the most widely used methods for HEP. This technique is based on data gathered from the nuclear and chemical processing industries. THERP relies on HRA event tree modeling to identify the events of concern. Quantification is performed from data tables of basic HEP for specific tasks that may be modified based on the circumstances affecting performance.

The degree of human reliability is influenced by many factors often called performance-shaping factors (PSF). PSFs are those factors that affect the ability of people to carry out required tasks. For example, the knowledge that people have on how to don/activate a personal flotation device (PFD) will affect the performance of this task. Training (another PSF) in donning PFDs can also assist in the ability to perform this task. Another example is the training that is given to passengers on airplanes before takeoff on using seatbelts, emergency breathing devices, and flotation devices. Often the quantitative estimates of reliability are generated from a base error rate that is then altered based on the PSFs of the particular circumstances. Internal PSFs are an individual's own attributes (experience, training, skills, abilities, attitudes) that affect the ability of the person to perform certain tasks. External PSFs are the dynamic aspects of situation, tasks, and system that affect the ability to perform certain tasks. Typical external factors include environmental stress factors (such as heat, cold, noise, situational stress, time of day), management, procedures, time limitations, and quality of the person–machine interface. With these PSFs, it is easy to see the dynamic nature of HEP evaluation based on the circumstances of the analysis.

Reducing Human Errors

Error reduction is concerned with lowering the likelihood for error in an attempt to reduce risk. The reduction of human errors may be achieved by human factors interventions or by engineering means. Human factors interventions include improving training or improving the human–machine interface (such as alarms, codes, and so on) based on an understanding of the causes of error. Engineering means of error reduction may include automated safety systems or interlocks. Selection of corrective actions to take can be done through decision analysis considering cost-benefit criteria.

Game Theory for Intelligent Threats

Game theory can be used to model human behavior, herein considered as a threat to a system. Generally, game theory utilizes mathematics, economics, and the other social and behavioral sciences to model human behavior.

An example of intelligent threats is terrorism and sabotage as an ongoing battle between coordinated opponents representing a two-party game, where each opponent seeks to achieve their own objectives within a system. In the case of terrorism, the game involves a well-established political system versus an emerging organization that uses terrorism to achieve partial or complete dominance. Each player in this game seeks a utility, or benefit, that is a function of the desired state of the system. In this case, maintaining system survival is the desired state for the government; whereas the opponent seeks a utility based on the failure state of the system. The government, as an opponent, is engaged in risk mitigation, whose actions seek to reduce the threat, reduce system vulnerability, and/or mitigate the consequences of successful attacks. The terrorists, as an opponent, can be viewed as the aggressor who strives to alter or damage their opponent's desired system state. This game involves an intelligent threat, and is dynamic. The game is ongoing until the probability of a successful disruptive attempt of the aggressor reaches an acceptable level of risk, a stage where risk is considered under control, and the game is brought to an end. Classical game theory can be used in conjunction with probabilistic risk analysis to determine optimal mitigation actions that maximize benefits.

A classical example used to introduce game theory is called the prisoner's dilemma and is based on two suspects captured near the scene of a crime and are questioned separately by authorities such as the police. Each has to choose whether or not to confess and implicate the other. If neither person confesses, then both will serve, say, 1 year on a charge of carrying a concealed weapon. If each confesses and implicates the other, both will go to prison for, say, 10 years. However, if one person confesses and implicates the other, and the other person does not confess, the one who has collaborated with the police will go free, while the other person will go to prison for, say, 20 years under the maximum penalty. The strategies in this case are confess or don't confess. The payoffs (penalties here) are the sentences served. The problem can be expressed compactly in a payoff table of a kind that has become fairly standard in game theory, illustrated in Table 207.8. The entries in this table mean that each prisoner chooses one of the two strategies, that is, the first suspect chooses a row and the second suspect chooses a column. The

TABLE 207.8 Payoff Table in Years for Prisoner's Dilemma Game

		Second Suspect	
		Confess	Don't Confess
First	Confess	(10, 10)	(0, 20)
Suspect	Don't confess	(20, 0)	(1, 1)

two numbers in each cell of the table provide the outcomes for the two suspects for the corresponding pair of strategies chosen by the suspects as an ordered pair. The number to the left of the comma is the payoff to the person who chooses the rows, that is, the first suspect. The number to the right of the comma is the payoff to the person who chooses the columns, that is, the second suspect. Thus, reading down the first column, if they both confess, each gets 10 years, but if the second suspect confesses and first suspect does not, the first suspect gets 20 years and second suspect goes free. This example is not a zero-sum game since the payoffs are all losses. However, many problems can be cast with losses (negative numbers) and gains (positive numbers) with a total for each cell in the payoff table. A zero-sum game occurs when the payoffs in each cell add up to zero.

The solution to this problem needs to be based on identifying rational strategies that can be based on both persons wanting to minimize the time they spend in jail. One suspect might reason that "two things can happen: either the other suspect confesses or keeps quiet. Suppose the second suspect confesses, I will get 20 years if I don't confess, 10 years if I do; therefore in this case it's best to confess. On the other hand, if the other suspect doesn't confess, and I don't either, I get a year; but in that case, if I confess I can go free. Either way, it's best if I confess. Therefore, I'll confess." But the other suspect can and presumably will reason in the same way. In this case, they both confess and go to prison for 10 years each, although if they had acted irrationally, and kept quiet, they each could have gotten off with 1 year each. The rational strategies of the two suspects have fallen into something called dominant-strategy equilibrium. The meaning of the term dominant-strategy equilibrium requires defining the term *dominant strategy*, which results from an individual player (suspect in this case) in a game evaluating separately each of the strategy combinations being faced, and, for each combination, choosing from these strategies the one that gives the greatest payoff. If the same strategy is chosen for each of the different combinations of strategies that the player might face, that strategy is called a dominant strategy for that player in that game. The dominant strategy equilibrium occurs if, in a game, each player has a dominant strategy, and each player plays the dominant strategy; then that combination of (dominant) strategies and the corresponding payoffs are said to constitute the dominant strategy equilibrium for that game. In the prisoner's dilemma game, to confess is a dominant strategy, and when both suspects confess, a dominant-strategy equilibrium is achieved. The dominant-strategy equilibrium is also called Nash equilibrium. When no player can benefit by changing his/her strategy while the other players keep their strategies unchanged, that set of strategies and corresponding payoffs constitute the Nash equilibrium.

The prisoner's dilemma game is based on two strategies per suspect that can be viewed as deterministic in nature, that are, nonrandom. In general, many games, especially those that permit repeatability in choosing strategies by players, can be constructed with strategies that have associated probabilities. For example, strategies can be constructed based on probabilities of 0.4 and 0.6 that add up to 1. Such strategies with probabilities are called mixed strategies as opposed to pure strategies that do not involve probabilities of the prisoner's dilemma game. A mixed strategy occurs in a game if a player chooses among two or more strategies at random according to specific probabilities.

In general, gaming could involve more than two players. In the prisoner's dilemma game, a third player that could be identified is the authority and its strategies. The solution might change as a result of adding the strategies of this third player. The use of these concepts in risk analysis and mitigation needs further development and exploration.

Economic and Financial Risks

Economic and financial risks can be grouped into categories that include market risks, credit risks, operation risks, and reputation risks. These four categories are described in subsequent sections.

Market Risks

Governments and corporations operate in economic and financial environments with some levels of uncertainty and instability. A primary contributor to defining this environment is interest rates. Interest rates can have a significant impact on the costs of financing a project, corporate cash flows, and asset values. For example, interest rates in the U.S. shot up in 1979 and peaked in 1981, followed by gradual decline with some fluctuations until 2002.

For projects that target global markets, exchange rate instability can be a major risk source. Exchange rates have been volatile ever since the breakdown of the Bretton Woods system to fixed exchange rates in the early 1970s. An example of a bust-up in exchange rates is plummeting value of the British pound and Italian lira due to failure of the exchange-rate mechanism in September 1992.

Many projects are dependent on availability of venture capital and the stock performance of a corporation, thereby introducing another risk source related to stock market volatility. Stock prices rose significantly in the inflationary booms of the early 1970s, and then fell considerably a little later. They recovered afterward, and fell again in the early 1981. The market rose to a peak until it crashed in 1987, followed by an increase with some swings until reaching a new peak fueled by Internet technologies until its collapse in 2001.

Other contributing factors to economic and financial instability is commodity prices in general and energy prices in particular, primarily crude oil. The hikes in oil prices in 1973 to 1974 affected commodity prices greatly and posed series challenges to countries and corporations.

Another contributing source to volatility is derivatives for commodities, foreign currency exchange rates, and stock prices and indices, among others. Derivatives are defined as contracts whose values or payoffs depend on those of other assets, such as options to buy commodities in the future or options to sell commodities in the future. They offer not only opportunities for hedging positions and managing risks that can be stabilizing, but also speculative opportunities to others that can be destabilizing and a contributor to volatility.

Credit Risks

Credit risks are associated with potential defaults on notes or bonds by corporations including subcontractors. Credit risks can be associated with market sentiments that determine a company's likelihood of default that could affect its bond rating, ability to borrow money, and maintain projects and operations.

Operational Risks

Operational risks are associated with several sources that include out-of-control operations risk that could occur when a corporate branch undertake significant risk exposure that is not accounted for by corporate headquarters, leading potentially to its collapse. An example is the British Barings Bank that collapsed primarily as a result of its failure to control market exposure created within a small overseas branch of the bank.

Another category here is liquidity risk in which a corporation needs more funding than it can arrange. It could also include money transfer risks and breaches of agreement.

Operational risks include model risks. Model risks are associated with the models and underlying assumptions used to incorrectly value financial instruments and cash flows.

Reputation Risks

The loss of business attributable to loss in a corporation's reputation can pose another risk source. This risk source can affect its credit rating, and ability to maintain clients, workforce, and so on. This risk source usually occurs at a slow attrition rate, and may be an outcome of poor management decisions and business practices.

Data Needs for Risk Assessment

In risk assessment, the methods of probability theory are used to represent engineering uncertainties. In this context, it refers to event occurrence likelihoods that occur with periodic frequency, such as weather, yet also to conditions that exist but are unknown, such as the probability of an extreme wave. It applies to the magnitude of an engineering parameter, yet also to the structure of a model. By contrast, probability is a precise mathematical concept with an explicit definition. We use the mathematics of probability theory to represent uncertainties, despite the fact that those uncertainties take many forms.

Although the term *probability* has a precise mathematical definition, its meaning when applied to the representation of uncertainties is subject to differing interpretations. The *frequentist* view holds that probability is the propensity of a physical system in a theoretically infinite number of repetitions; that is, the frequency of occurrence of an outcome in a long series of similar trials (e.g., the frequency of a coin landing heads-up in an infinite number of flips is the probability of that event). In contrast, the *Bayesian* view holds that probability is the rational degree of belief that one holds in the occurrence of an event or the truth of a proposition; probability is manifest in the willingness of an observer to take action upon this belief. This latter view of probability, which has gained wide acceptance in many engineering applications, permits the use of quantified professional judgment in the form of subjective probabilities. Mathematically, such subjective probabilities can be combined or operated on as any other probability.

Data are needed to perform quantitative risk assessment or provide information to support qualitative risk assessment. Information may be available if data have been maintained on a system and components of interest. The relevant information for risk assessment includes possible failures, failure probabilities, failure rates, failure modes, possible causes, and failure consequences. In the case of a new system, data may be used from similar systems if this information is available. Surveys are a common tool used to provide data. Statistical analysis can be used to assess confidence intervals and uncertainties in estimated parameters of interest. Expert judgment may also be used as another source of data. Uncertainties regarding data quality should be identified to assist in the decision-making process.

Data can be classified to include generic and project- or plant-specific types. Generic data are information from similar systems and components. This information may be the only information available in the initial stages of system design. Therefore, potential differences due to design or uncertainty may result from using generic data on a specific system. Plant-specific data are specific to the system being analyzed. This information is often developed after the operation of a system. Relevant data need to be identified and collected, as data collection can be costly. The data collected can then be used to update the risk assessment. Bayesian techniques can be used to combine objective and subjective data.

Data can be classified as failure probability data and failure consequence data. Failure probability data can include failure rates, hazard functions, times between failures, results from reliability studies, and any influencing factors and their effects. Failure-consequence data include loss reports, damages, litigation outcomes, repair costs, injuries, and human losses, as well as influencing factors and effects of failure-prevention and consequence-mitigation plans. Areas of deficiency in terms of data availability should be identified, and sometimes failure databases need to be constructed. Data deficiency can be used as a basis for data collection and expert-opinion elicitation.

207.3 Risk Management and Control

Adding risk control to risk assessment produces risk management. Risk management is the process by which system operators, managers, and owners make safety decisions, regulatory changes, and choose different system configurations based on the data generated in the risk assessment. Risk management involves using information from the previously described risk assessment stage to make educated decisions about system safety. Risk control includes failure prevention and consequence mitigation.

Risk management requires the optimal allocation of available resources in support of group goals. Therefore, it requires the definition of acceptable risk, and comparative evaluation of options and/or alternatives for decision making. The goals of risk management are to reduce risk to an acceptable level and/or prioritize resources based on comparative analysis. Risk reduction is accomplished by preventing an unfavorable scenario, and reducing the frequency and/or reducing the consequence. A graph showing the risk relationship is shown in Figure 207.14 as linear contours of constant risk, although due to risk aversion these lines are commonly estimated as nonlinear curves and should be treated as nonlinear

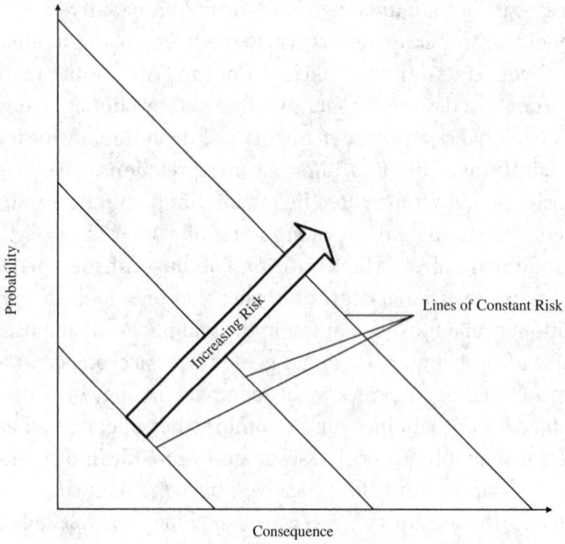

FIGURE 207.14 Risk graph.

curves. Moreover, the vertical axis is termed as probability whereas it is commonly expressed as an annual exceedence probability or frequency as shown in Figure 207.1. In cases involving qualitative assessment, a matrix presentation can be used as shown in Figure 207.8. The figure shows probability categories, severity categories, and risk ratings. A project's base value is commonly assumed as zero. Each risk rating value requires a different mitigation plan.

Risk Acceptance

Risk acceptance constitutes a definition of safety as discussed in previous sections. Therefore, risk acceptance is considered a complex subject that is often subject to controversial debate. The determination of acceptable levels of risk is important to determine the risk performance that a system needs to achieve to be considered safe. If a system has a risk value above the risk acceptance level, actions should be taken to address safety concerns and improve the system through risk reduction measures. One difficulty with this process is defining acceptable safety levels for activities, industries, structures, and so on. Since the acceptance of risk depends on society perceptions, acceptance criteria do not depend on the risk value alone. This section describes several methods that have been developed to assist in determining acceptable risk values as summarized in Table 207.9.

Risk managers make decisions based on risk assessment and other considerations including economic, political, environmental, legal, reliability, producibility, safety, and other factors. The answer to the question "How safe is safe enough?" is difficult and constantly changes due to varying perceptions and understandings of risk. To determine "acceptable risk," managers need to analyze alternatives for the best choice. In some industries, an acceptable risk has been defined by consensus. For example, the U.S. Nuclear Regulatory Commission requires that reactors be designed such that the probability of a large radioactive release to the environment from a reactor incident shall be less than 1×10^{-6} per year. Risk levels for certain carcinogens and pollutants have also been given acceptable concentration levels based on some assessment of acceptable risk. However, risk acceptance for many other activities are not stated.

For example, qualitative implications for risk acceptance are identified in several existing maritime regulations. The International Maritime Organization High Speed Craft Code [2002] and the U.S. Coast Guard Navigation and Vessel Inspection Circular [1993] for passenger submersible guidance both state that if the end effect is hazardous or catastrophic, a backup system and a corrective operating procedure are required. These references also state that a single failure must not result in a catastrophic event, unless the likelihood is extremely remote.

TABLE 207.9 Methods for Determining Risk Acceptance

Risk Acceptance Method	Summary
Risk conversion factors	This method addresses the attitudes of the public about risk through comparisons of risk categories. It also provides an estimate for converting risk acceptance values between different risk categories.
Farmer's curve	It provides an estimated curve for a cumulative probability risk profile for certain consequences (e.g., deaths). It demonstrates graphical regions of risk acceptance/ nonacceptance.
Revealed preferences	Through comparisons of risk and benefit for different activities, this method categorizes societal preferences for voluntary and involuntary exposure to risk.
Evaluation of magnitude of consequences	This technique compares the probability of risks to the consequence magnitude for different industries to determine acceptable risk levels based on consequences.
Risk effectiveness	This method provides a ratio for the comparison of cost to the magnitude of risk reduction. Using cost-benefit decision criteria, a risk reduction effort should not be pursued if the costs outweigh the benefits. This may not coincide with societal values about safety.
Risk comparison	The risk acceptance method provides a comparison between various activities, industries, and so on, and is best suited to comparing risks of the same type.

Often the level of risk acceptance with various activities is implied. Society has reacted to risks through the developed level of balance between risk and potential benefits. Measuring this balance of accepted safety levels for various risks provides a means for assessing society values. These threshold values of acceptable risk depend on a variety of issues including the activity type, industry, and users, and society as a whole.

Target risk or reliability levels are required for developing procedures and rules for ship structures. For example, the selected reliability levels determine the probability of failure of structural components. The following three methods were used to select target reliability values:

- Agreeing on a reasonable value in cases of novel structures without prior history
- Calibrating reliability levels implied in current successfully used design codes
- Choosing a target reliability level that minimizes total expected costs over the service life of the structure in dealing with a design for which failure results in only economic losses and consequences

The first approach can be based on expert opinion elicitation. The second approach called *code calibration* is the most commonly used approach, as it provides the means to build on previous experience. For example, rules provided by classification and industry societies can be used to determine the implied reliability and risk levels in respective rules and codes, and then target risk levels can be set in a consistent manner, and new rules and codes can be developed to produce future designs and vessels that are of similar levels that offer reliability and/or risk consistency. The third approach can be based on economic and tradeoff analysis. In subsequent sections, the methods of Table 207.9 for determining risk acceptance are discussed.

Risk Conversion Factors

Risk analysis shows that there are various taxonomies that demonstrate risk categories often called "risk factors." These categories can be used to analyze risks on a dichotomous scale that compares risks that invoke the same perceptions in society. For example, the severity category may be used to describe both ordinary and catastrophic events. Grouping events that could be classified as ordinary and comparing the distribution of risk to a similar grouping of catastrophic categories yields a ratio describing the degree of risk acceptance of ordinary events as compared to catastrophic events. Comparison of various categories determined the risk conversion values as provided in Table 207.10. These factors are useful in comparing the risk acceptance for various activities, industries, and so on. By computing the acceptable risk in one activity, an estimate of acceptable risk in other activities can be calculated based on risk conversion factors. A comparison of several common risks based on origin and volition is shown in Table 207.11.

TABLE 207.10 Risk Conversion Values for Various Risk Factors

Risk Factors	Risk Conversion Factor	Computed RF Value
Origin	Natural/human-made	20
Severity	Ordinary/catastrophic	30
Volition	Voluntary/involuntary	100
Effect	Delayed/immediate	30
Controllability	Controlled/uncontrolled	5 to 10
Familiarity	Old/new	10
Necessity	Necessary/luxury	1
Costs	Monetary/nonmonetary	NA
Origin	Industrial/regulatory	NA
Media	Low profile/high profile	NA

NA = not available.

TABLE 207.11 Classification of Common Risks

		Voluntary		Involuntary	
Source	Size	Immediate	Delayed	Immediate	Delayed
Human made	Catastrophic	Aviation		Dam failure, building fire, nuclear accident	Pollution, building fire
	Ordinary	Sports, boating, automobiles	Smoking, occupation, carcinogens	Homicide	
Natural	Catastrophic			Earthquakes, hurricanes, tornadoes, epidemics	
	Ordinary			Lightning, animal bites	Disease

Farmer's Curve

The farmer's curve is a graph of the cumulative probability versus consequence for some activity, industry, or design as shown in Figures 207.1 and 207.2. This curve introduces a probabilistic approach in determining acceptable safety limits. The probability (or frequency) and consequence values are calculated for each level of risk generating a curve that is unique to the hazard of concern. The area to the right (outside) of the curve is generally considered unacceptable since the probability and consequence values are higher than the average value delineated by the curve. The area to the left (inside) of the curve is considered acceptable since probability and consequence values are less than the estimated valve of the curve.

Method of Revealed Preferences

The method of revealed preferences provides a comparison of risk versus benefit and categorization for different risk types. The basis for this relationship is that risks are not taken unless there is some form of benefit. Benefit may be monetary or some other item of worth such as pleasure. The different risk types are for the risk category of voluntary versus involuntary actions as shown in Figure 207.15.

Magnitudes of Risk Consequence

Another factor affecting the acceptance of risk is the magnitude of consequence of the event that can result from some failure. In general, the larger the consequence, the less the likelihood that this event may occur. This technique has been used in several industries to demonstrate the location of the industry within society's risk acceptance levels based on consequence magnitude as shown in Figure 207.16. Further evaluation has resulted in several estimates for the relationship between the accepted probability of failure and the magnitude of consequence for failure as provided by Allen in 1981 and called herein the CIRIA (Construction Industry Research and Information Association) equation:

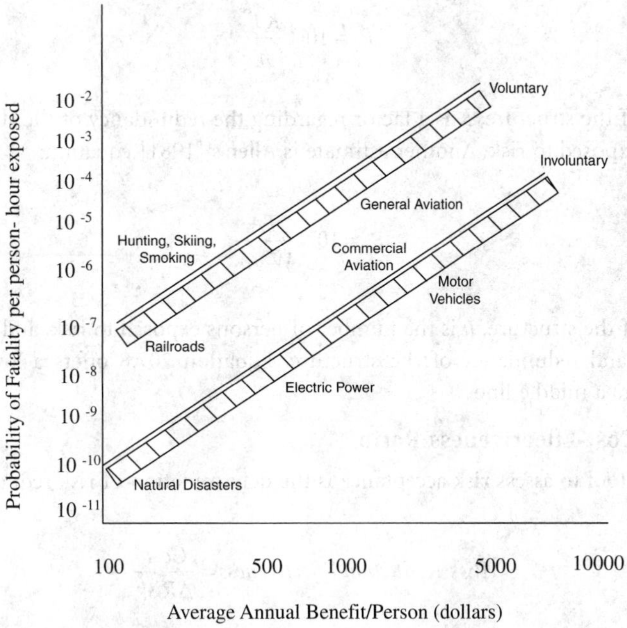

FIGURE 207.15 Accepted risk of voluntary and involuntary activities.

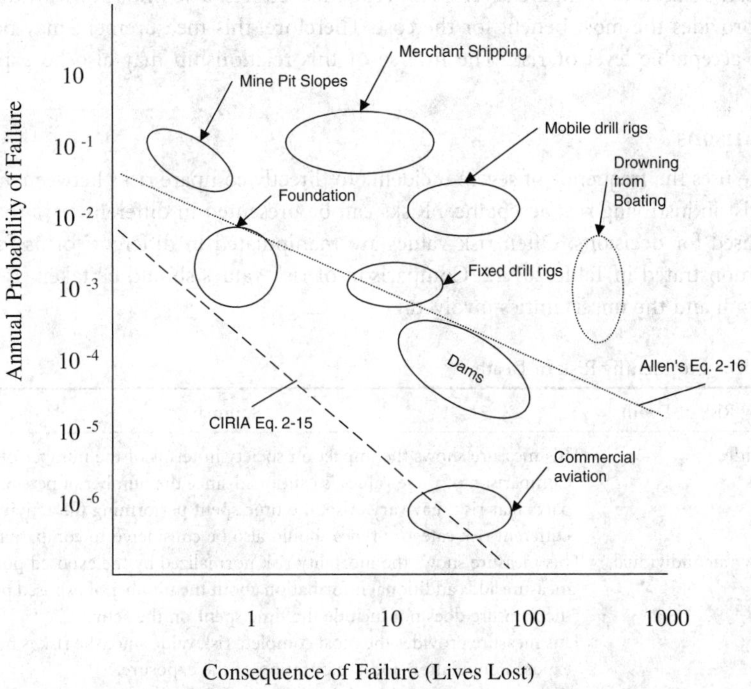

FIGURE 207.16 Target risk based on consequence of failure for industries.

$$P_f = 10^{-4} \frac{KT}{n} \tag{207.8}$$

where T is the life of the structure, K is a factor regarding the redundancy of the structure, and n is the number of people exposed to risk. Another estimate is Allen's [1981] equation:

$$P_f = 10^{-5} \frac{TA}{W\sqrt{n}} \tag{207.9}$$

where T is the life of the structure, n is the number of persons exposed to risk, and A and W are factors regarding the type and redundancy of the structure. Equation 207.8 offers a lower bound, whereas Equation 207.9 offers a middle line.

Risk Reduction Cost-Effectiveness Ratio

Another measuring tool to assess risk acceptance is the determination of risk reduction effectiveness:

$$\textit{Risk}\ \text{Reduction}\ \textit{Effectiveness} = \frac{\textit{Cost}}{\Delta \textit{Risk}} \tag{207.10}$$

where the cost should be attributed to risk reduction, and $\Delta Risk$ is the level of risk reduction as follows:

$$\Delta \textit{Risk} = (\text{Risk before mitigation action}) - (\text{Risk after mitigation action}) \tag{207.11}$$

The difference in Equation 207.11 is also called the benefit attributed to a risk reduction action. Risk effectiveness can be used to compare several risk reduction efforts. The initiative with the smallest risk effectiveness provides the most benefit for the cost. Therefore, this measurement may be used to help determine an acceptable level of risk. The inverse of this relationship may also be expressed as cost effectiveness.

Risk Comparisons

This technique uses the frequency of severe incidents to directly compare risks between various areas of interest to assist in justifying risk acceptance. Risks can be presented in different ways that impact how the data are used for decisions. Often risk values are manipulated in different forms for comparison reasons as demonstrated in Table 207.12. Comparison of risk values should be taken in the context of the values' origin and the uncertainties involved.

TABLE 207.12 Ways to Identify Risk of Death

Ways to Identify Risk of Death	Summary
Number of fatalities	This measure shows the impact on society in terms of the number of fatalities. Comparison of these values is cautioned since the number of persons exposed to the particular risk may vary. Also, the time spent performing the activity may vary. Different risk category types should also be considered to compare fatality rates.
Annual mortality rate/individual	This measure shows the mortality risk normalized by the exposed population. This measure adds additional information about the number of exposed persons; however, the measure does not include the time spent on the activity.
Annual mortality	This measure provides the most complete risk value since the risk is normalized by the exposed population and the duration of the exposure.
Loss-of-life exposure (LLE)	This measure converts a risk into a reduction in the expected life of an individual. It provides a good means of communicating risks beyond probability values.
Odds	This measure is a layperson format for communicating probability, for example, 1 in 4.

This technique is most effective for comparing risks that invoke the same human perceptions and consequence categories. Comparing risks of different categories is cautioned since the differences between risk and perceived safety may not provide an objective analysis of risk acceptance. The use of risk conversion factors may assist in transforming different risk categories. Conservative guidelines for determining risk acceptance criteria can be established for voluntary risks to the public from the involuntary risk of natural causes.

Rankings Based on Risk Results

Another tool for risk management is the development of risk ranking. The elements of a system within the objective of analysis can be analyzed for risk and consequently ranked. This relative ranking may be based on failure probabilities, failure consequences, risks, or other alternatives with concern toward risk. Generally risk items ranked highly should be given high levels of priority; however, risk management decisions may consider other factors such as costs, benefits, and effectiveness of risk reduction measures. The risk ranking results may be presented graphically as needed.

Decision Analysis

Decision analysis provides a means for systematically dealing with complex problems to arrive at a decision. Information is gathered in a structured manner to provide the best answer to the problem. A decision generally deals with three elements: alternatives, consequences, and preferences. The alternatives are the possible choices for consideration. The consequences are the potential outcomes of a decision. Decision analysis provides methods for quantifying preference tradeoffs for performance along multiple decision attributes while taking into account risk objectives. Decision attributes are the performance scales that measure the degree to which objectives are satisfied. For example, one possible attribute is reducing lives lost for the objective of increasing safety. Additional examples of objectives may include minimize cost, maximize utility, maximize reliability, and maximize profit. The decision outcomes may be affected by uncertainty; however, the goal is to choose the best alternative with the proper consideration of uncertainty. The analytical depth and rigor for decision analysis depend on the desired detail in making the decision. Cost-benefit analysis, decision trees, influence diagrams, and the analytic hierarchy process are some of the tools to assist in decision analysis. Also, decision analysis should consider constraints, such as availability of the system for inspection, availability of inspectors, preferences of certain inspectors, and availability of inspection equipment.

Benefit-Cost Analysis

Risk managers commonly weigh various factors including cost and risk. The analysis of three different alternatives is shown graphically in Figure 207.17 as an example. The graph shows that alternative (C) is the best choice since the levels of risk and cost are less than for alternatives (A) and (B). However, if the only alternatives were A and B, the decision would be more difficult. Alternative (A) has higher cost and lower risk than alternative (B); alternative (B) has higher risk but lower cost than alternative (A). A risk manager needs to weigh the importance of risk and cost in making decisions, as well as resource availability, and to make use of risk-based decision analysis.

Risk-benefit analysis can also be used for risk management. Economic efficiency is important to determine the most effective means of expending resources. At some point the costs for risk reduction do not provide adequate benefits. This process compares the costs and risk to determine where the optimal risk value is on a cost basis. This optimal value occurs, as shown in Figure 207.18, when costs to control risk are equal to the risk cost due to the consequence (loss). Investing resources to reduce risks below this equilibrium point does not provide a financial benefit. This technique may be used when cost values can be attributed to risks. This analysis might be difficult to perform for certain risk such as risk to human health and environmental risks since the monetary values are difficult to estimate for human life and the environment.

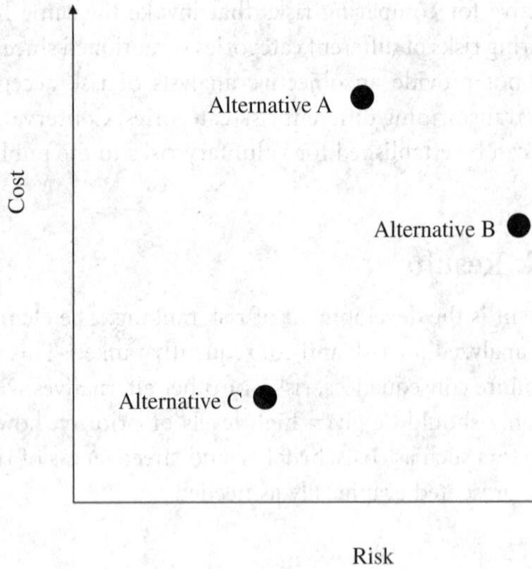

FIGURE 207.17 Risk benefit for three alternatives.

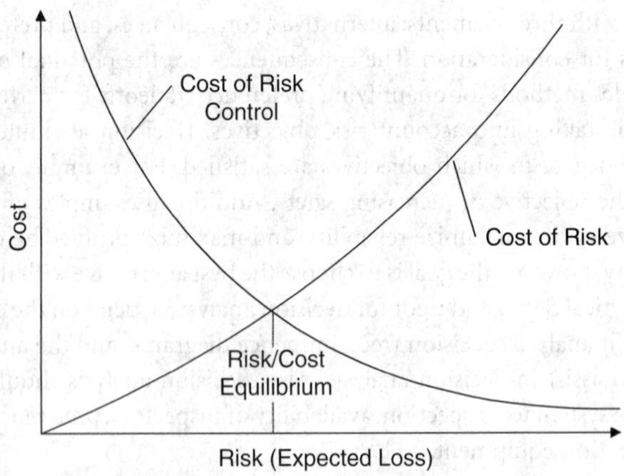

FIGURE 207.18 Comparison of risk and control costs.

The present value of incremental costs and benefits can be assessed and compared among alternatives that are available for risk mitigation or system design. Several methods are available to determine which, if any, option is most worth pursuing. In some cases, no alternative will generate a net benefit relative to the base case. Such a finding would be used to argue for pursuit of the base case scenario. The following are the most widely used present-value comparison methods: (1) net present value (NPV), (2) benefit-cost ratio, (3) internal rate of return, and (4) payback period. The NPV method requires that each alternative need to meet the following criteria to warrant investment of funds: positive NPV and the highest NPV of all alternatives considered. The first condition ensures that the alternative is worth undertaking relative to the base case; for example, it contributes more in incremental benefits than it absorbs in incremental costs. The second condition ensures that maximum benefits are obtained in a situation of unrestricted access to capital funds. The NPV can be calculated as follows:

$$NPV = \sum_{t=0}^{k} \frac{(B-C)_t}{(1+r)^t} = \sum_{t=0}^{k} \frac{B_t}{(1+r)^t} - \sum_{t=0}^{k} \frac{C_t}{(1+r)^t} \tag{207.12}$$

where B is future annual benefits in constant dollars, C is future annual costs in constant dollars, r is annual real discount rate, k is number of years from the base year over which the project will be evaluated, and t is an index running from 0 to k representing the year under consideration.

The benefit of a risk mitigation action can be assessed as follows:

$$\text{Benefit} = \text{unmitigated risk} - \text{mitigated risk} \tag{207.13}$$

The cost in Equation 207.13 is the cost of the mitigation action. The benefit minus the cost of mitigation can be used to justify the allocation of resources. The benefit-to-cost ratio can be computed, and may also be helpful in decision making. The benefit-to-cost ratio can be computed as

$$\text{Benefit-to-Cost Ratio } (B/C) = \frac{\text{Benefit}}{\text{Cost}} = \frac{\text{Unmitigated Risk - Mitigated Risk}}{\text{Cost of Mitigation Action}} \tag{207.14}$$

Ratios greater than 1 are desirable. In general, the larger the ratio, the better the mitigation action.

Accounting for the time value of money would require defining the benefit-cost ratio as the present value of benefits divided by the present value of costs. The benefit-cost ratio can be calculated as follows:

$$B/C = \frac{\displaystyle\sum_{t=0}^{k} \frac{B_t}{(1+r)^t}}{\displaystyle\sum_{t=0}^{k} \frac{C_t}{(1+r)^t}} \tag{207.15}$$

where B_t is future annual benefits in constant dollars, C_t is future annual costs inconstant dollars, r is annual real discount rate, and t is an index running from 0 to k representing the year under consideration. A proposed activity with a B/C ratio of discounted benefits to costs of 1 or more is expected to return at least as much in benefits as it costs to undertake, indicating that the activity is worth undertaking.

The internal rate of return (*IRR*) is defined as the discount rate that makes the present value of the stream of expected benefits in excess of expected costs zero. In other words, it is the highest discount rate at which the project will not have a negative *NPV*. To apply the *IRR* criterion, it is necessary to compute the IRR and then compare it with a base rate of, say, a 7% discount rate. If the real *IRR* is less than 7%, the project would be worth undertaking relative to the base case. The *IRR* method is effective in deciding whether or not a project is superior to the base case; however it is difficult to utilize it for ranking projects and deciding among mutually exclusive alternatives. Project rankings established by the *IRR* method might be inconsistent with those of the *NPV* criterion. Moreover, a project might have more than one *IRR* value, particularly when it entails major final costs, such as cleanup costs. Solutions to these limitations exist in capital budgeting procedures and practices that are often complicated or difficult to employ in practice and present opportunities for error.

The payback period measures the number of years required for net undiscounted benefits to recover the initial investment in a project. This evaluation method favors projects with near-term and more certain benefits, and fails to consider benefits beyond the payback period. The method does not provide information on whether an investment is worth undertaking in the first place.

The previous models for benefit-cost analysis presented in this section do not account for the full probabilistic characteristics of B and C in their treatment. Concepts from reliability assessment 4 can be used for this purpose. Assuming B and C to normally distributed, a benefit-cost index ($\beta B/C$) can be defined as follows:

$$\beta_{B/C} = \frac{\mu_B - \mu_C}{\sqrt{\sigma_B^2 + \sigma_C^2}} \tag{207.16}$$

where μ and σ are the mean and standard deviation. The failure probability can be computed as

$$P_{f,B/C} = P(C > B) = 1 - \Phi(\beta) \tag{207.17}$$

In the case of log-normally distributed B and C, the benefit-cost index ($\beta B/C$) can be computed as

$$\beta_{B/C} = \frac{\ln\left(\dfrac{\mu_B}{\mu_C}\sqrt{\dfrac{\delta_C^2+1}{\delta_B^2+1}}\right)}{\sqrt{\ln[(\delta_B^2+1)(\delta_C^2+1)]}} \tag{207.18}$$

where δ is the coefficient of variation. Equation 207.18 also holds for the case of log-normally distributed B and C. In the case of mixed distributions or cases involving basic random variables of B and C, the advanced second moment method or simulation method can be used. In cases where benefit is computed as revenue minus cost, benefit might be correlated with cost, requiring the use of other methods.

Example 207.2: Protection of Critical Infrastructure

This example is used to illustrate the cost of cost-benefit analysis using a simplified decision situation. As an illustration, assume that there is a 0.01 probability of an attack on a facility containing hazardous material during the next year. If the attack occurs, the probability of a serious release to the public is 0.01 with a total consequence of \$100 billion. The total consequence of an unsuccessful attack is negligible. The unmitigated risk can therefore be computed as

$$\text{Unmitigated risk} = 0.01(0.01)(\$100\text{ billion}) = \$10\text{ million}$$

If armed guards are deployed at each facility, the probability of attack can be reduced to 0.001 and the probability of serious release if an attack occurs can be reduced to 0.001. The cost of the guards for all plants is assumed to be \$100 million per year. The mitigated risk can therefore be computed as

$$\text{Mitigated risk} = 0.001(0.001)(\$100\text{ billion}) = \$0.10\text{ million}$$

The benefit in this case is

$$\text{Benefit} = \$10\text{ million} - \$0.1\text{ million or} \sim\$10\text{ million}$$

The benefit-to-cost ratio is about 0.1. Therefore, the \$100 million cost might be difficult to justify.

Risk Mitigation

A risk mitigation strategy can be presented from a financial point of view. Risk mitigation in this context can be defined as an action to either reduce the probability of an adverse event occurring or to reduce the adverse consequences if it does occur. This definition captures the essence of an effective management process of risk. If implemented correctly, a successful risk mitigation strategy should reduce any adverse (or downside) variations in the financial returns from a project, which are usually measured by either (1) the NPV, defined as the difference between the present value of the cash flows generated by a project and its capital cost and calculated as part of the process of assessing and appraising investments; or (2) the IRR, defined as the return that can be earned on the capital invested in the project, that is, the discount rate that gives an NPV of zero, in the form of the rate that is equivalent to the yield on the investment.

Risk mitigation involves direct costs like increased capital expenditure or the payment of insurance premiums; hence, risk mitigation might reduce the average overall financial returns from a project. This reduction is often a perfectly acceptable outcome, given the risk aversion of many investors and lenders.

A risk mitigation strategy is the replacement of an uncertain and volatile future with one where there is less exposure to adverse risks and so less variability in the return, although the expected NPV or IRR may be reduced. These two aspects are not necessarily mutually exclusive. Increasing risk efficiency by simultaneously improving the expected NPV or IRR and simultaneously reducing the adverse volatility is sometimes possible and should be sought. Risk mitigation should cover all phases of a project from inception to closedown or disposal.

Four primary ways are available to deal with risk within the context of a risk management strategy as follows:

- Risk reduction or elimination
- Risk transfer, such as to a contractor or an insurance company
- Risk avoidance
- Risk absorbance or pooling

These four methods are described in subsequent sections.

Risk Reduction or Elimination

Risk reduction or elimination is often the most fruitful form for exploration. For example, could a system design be amended so as to reduce or eliminate either the probability of occurrence of a particular risk event or adverse consequences if it occurs? Alternatively, could the risks be reduced or eliminated by retaining the same design but using different materials or a different method of assembly? Other possible risk mitigation options in this category include a better labor relations policy to minimize the risk of stoppages, training of staff to avoid hazards, better site security to avoid theft and vandalism, preliminary investigation of possible site pollution, advance ordering of key components, noise abatement measures, good sign posting, and liaisons with the local community.

Risk Transfer

A general principle of an effective risk management strategy is that commercial risks in projects and other business ventures should be borne wherever possible by the party that is best able to manage them, and thus mitigate the risks. Contracts and financial agreements are the principal forms to transfer risks. Companies specializing in risk transfer can be consulted that could appropriately meet the needs of a project. Risks can be transferred alternately to an insurance company which, in return for a payment (i.e., premium) linked to the probability of occurrence and severity associated with the risk, is obliged by the contract to offer compensation to the party affected by the risk. Insurance coverage can range from straight insurance for expensive risks with a low probability, such as fire, through performance bonds, which ensure that the project will be completed if the contractor defaults; to sophisticated financial derivatives such as hedge contracts to avoid risks such as unanticipated losses in foreign exchange markets.

Risk Avoidance

A most intuitive way of avoiding a risk is to avoid undertaking the project in a way that involves that risk. For example, if the objective is to generate electricity but a nuclear power source, although cost-efficient, is considered to have a high risk due to potentially catastrophic consequences, even after taking all reasonable precautions, the practical solution is to turn to other forms of fuel to avoid that risk. Another example would be the risk that a particularly small contractor would go bankrupt. In this case, the risk could be avoided by using a well-established contractor instead for that particular job.

Risk Absorbance and Pooling

In cases where risks cannot economically (or at all) be eliminated, transferred, or avoided, they must be absorbed if the project is to proceed. Normally, a sufficient margin in the project's finances needs to be created to cover the risk event should it occur. However, it is not always essential for one party alone to bear all these absorbed risks. Risks can be reduced through pooling, possibly through participation in a consortium of contractors, when two or more parties are able to exercise partial control over the incidence and impact of risk. Joint ventures and partnerships are other examples of organizational forms for pooling risks.

Uncertainty Characterization

Risk can be mitigated through proper uncertainty characterization. The presence of improperly characterized uncertainty could lead to the higher likelihood of adverse event occurrence and consequences. Also, it could result in increasing estimated cost margins as a means of compensation. Therefore, risk can be reduced by a proper characterization of uncertainty. The uncertainty characterization can be achieved through data collection and knowledge construction.

Example 207.3: Cost-Benefit Analysis for Selecting a Transport Method

Table 207.13 shows four transportation methods considered by the warehouse owner discussed in previous examples in this chapter to supply components from the warehouse to one of its major customers in a foreign country. The available alternatives for the modes of transport are (1) air, (2) sea, (3) road and ferry, and (4) rail and ferry. The company management team has identified four relevant attributes for this decision situation: (1) punctuality, (2) safety of cargo, (3) convenience, and (4) costs. The first three attributes are considered to be a benefits parameter against the fourth one that is the cost of transportation. The weights of importance scores allocated to the three benefits attributes are 30 for punctuality, 60 for safety of cargo, and 10 for convenience. After a brainstorming session by the management team, the performance of each transportation mode was assessed according to these attributes. The assessment results are shown in Table 207.13. The optimal alternative can be selected by applying the concept of cost-benefit analysis. The results are shown in Table 207.14. Inspection of the table reveals that the fourth alternative gives the highest ratio of 1.16. Therefore, the rail and ferry transportation mode can be selected as the best alternative. The plot in Figure 207.19 of the value of benefit against the value of cost for each alternative can be constructed to show that alternative 4 is the best option with the highest weighted benefit of 81 against a cost of $70,000, confirming previous weighted-benefit-to-cost-ratio computations. Alternative 3 comes as second option using the weighted benefit-to-cost ratio but at a low-cost value of only $40,000. A cost-benefit tradeoff analysis can be made between alternatives 4 and 3. A cost-

TABLE 207.13 Assessments of Modes of Transportation for Delivery to Foreign Clients

Alternatives	Cost (in thousands of dollars)	Attribute (0100)		
		Punctuality	Safety	Convenience
A1: Air	150	100	70	60
A2: Sea	90	0	60	80
A3: Road and ferry	40	60	0	100
A4: Rail and ferry	70	70	100	0
	Weight of importance	30	60	10

TABLE 207.14 Benefit-to-Cost Ratio Computations for Modes of Transportation

Alternatives	Cost (in thousands of dollars)	Benefit cores 0–100			Weighted Benefit	(Weighted Benefit)/Cost	Rank
		Punctuality	Safety	Convenience			
A1: Air	150	100	70	60	78	0.52	3
A2: Sea	90	0	60	80	44	0.49	4
A3: Road and ferry	40	60	0	100	28	0.70	2
A4: Rail and ferry	70	70	100	0	81	1.16	1
	Normalized weight of importance	0.30	0.60	0.10			

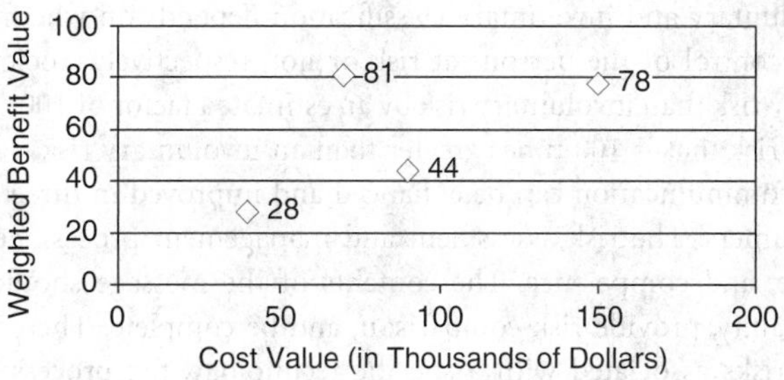

FIGURE 207.19 Cost-benefit analysis of transportation modes.

conscious decision maker might choose alternative 3, whereas a benefit-driven decision maker might select alternative 4. If a decision maker is concerned with both, the weighted benefit-to-cost ratio of 1.16 of alternative 4 makes it the optimal choice.

207.4 Risk Communication

Risk communication can be defined as an interactive process of exchange of information and opinion among stakeholders such as individuals, groups, and institutions. It often involves multiple messages about the nature of risk, or expresses concerns, opinions, or reactions to risk managers or to legal and institutional arrangements for risk management. Risk communication greatly affects risk acceptance and defines the acceptance criteria for safety.

Risk communication provides the vital link between risk assessors, risk managers, and the public to understand risk. However, this does not necessarily mean that risk communication will always lead to agreement among different parties. An accurate perception of risk provides for rational decision making. The Titanic was deemed the unsinkable ship, yet was lost on its maiden voyage. Space shuttle flights were perceived to be safe enough for civilian travel until the Space Shuttle Challenger disaster. These modes of transport obviously had risks that were not perceived as significant until after the disaster. Risk communication is a dynamic process that must be considered prior to management decisions.

The communication process deals with technical information about controversial issues. Therefore, it needs to be skillfully performed by risk managers and communicators who might be viewed as adversaries by the public. Risk communication between risk assessors and risk managers is necessary to effectively apply risk assessments in decision making. Risk managers must participate in determining the criteria for determining what type and level of risk are acceptable and unacceptable. This communication between the risk managers and risk assessors is necessary for a better understanding of risk analysis in making decisions.

Risk communication also provides the means for risk managers to gain acceptance and understanding by the public. Risk managers need to go beyond the risk assessment results and consider other factors in making decisions. One of these concerns is politics, which is, of course, largely influenced by the public. Risk managers often fail to convince the public that risks can be kept to acceptable levels. Problems here are shown by the public's perception of toxic waste disposal and nuclear power plant operation safety. As a result of public perceptions (fears), risk managers may make decisions that are conservative to appease the public.

The value of risk calculated from risk assessment is not the only consideration for risk managers. All risks are not created equal, and society has established risk preferences based on public preferences. Decision makers should take these preferences into consideration when making decisions concerning risk.

To establish a means of comparing risks based on the societal preferences, risk conversion factors (RCFs) may be used. An RCF expresses the relative importance of different attributes concerning risk. An example of possible RCFs is shown in Table 207.10. These values were determined by inferences of public preferences from statistical data with the consequence of death considered.

For example, the voluntary and involuntary classification depends on whether the events leading to the risk are under the control of the persons at risk or not, respectively. Society, in general, accepts a higher level of voluntary risk than involuntary risk by an estimated factor of 100. Therefore, an individual will accept a voluntary risk that is 100 times greater than an involuntary risk.

The process of risk communication can be enhanced and improved in three aspects: (1) process, (2) message, and (3) consumers. The risk assessment and management process needs to have clear goals with openness, balance, and competence. The contents of the message should account for audience orientation and uncertainty, provide risk comparison, and be complete. There is a need for consumer guides that introduce risks associated with a specific technology, the process of risk assessment and management, acceptable risk, decision making, uncertainty, costs and benefits, and feedback mechanisms. Improving risk literacy of consumers is an essential component of the risk communication process.

Listed below are considerations for communicating risk recommended by the U.S. Army Corps of Engineers [1992].

- Risk communication must be free of jargon.
- Consensus of expert needs to be established.
- Materials cited and their sources must be credible.
- Materials must be tailored to audience.
- The information must be personalized to the extent possible.
- Motivation discussion should stress a positive approach and the likelihood of success.
- Risk data must be presented in a meaningful manner.

Defining Terms

Risk — The assessment of potential consequences.
Safety — The judgment of risk acceptability.
Reliability — The ability of a system to meet performance requirements for a specified time period and operation/environmental conditions.

References

Allen, D. E. 1981. Criteria for design safety factors and quality assurance expenditure, in *Structural Safety and Reliability*, Elsevier, 667–678.

International Maritime Organization (IMO). 2002. High Speed Craft Code, London, U.K.

U.S. Army Corps of Engineers (USACE). 1993. Guidebook for Risk Perception and Communication in Water Resources, IWR Report, USACE, Report 93-R-13, Alexandria, VA.

U.S. Coast Guard. 1993. Navigation and Vessel Inspection Circular, 5-93, Washington, D.C.

Further Information

Ang, A. H-S. and Tang, W. H. 1990. *Probability Concepts in Engineering Planning and Design*. Vol. 2 of *Decision, Risk, and Reliability*. John Wiley & Sons. New York.

Ayyub, B. M. 2003. *Risk Analysis in Engineering and Economics*. Chapman & Hall/CRC Press, Boca Raton, FL.

Ayyub, B. M. 2001. *Elicitation of Expert Opinions for Uncertainty and Risks*. CRC Press, Boca Raton, FL.

Ayyub, B. M. and McCuen, R. 2003. *Probability, Statistics and Reliability for Engineers and Scientists*, 2nd ed. Chapman & Hall/CRC Press, Boca Raton, FL.

Kumamoto, H., and Henley, E. J. 1996. *Probabilistic Risk Assessment and Management for Engineers and Scientists*, 2nd ed. IEEE Press, New York.

Modarres, M. 1993. *What Every Engineer Should Know About Reliability and Analysis*. Marcel Dekker, New York.

Modarres, M., Kaminskiy, M., and Krivstov, V. 1999. *Reliability Engineering and Risk Analysis: A Practical Guide*. Marcel Decker, New York.

208

Sensitivity Analysis

Harold E. Marshall
National Institute of Standards and Technology

Sensitivity analysis measures the impact on project outcomes of changing one or more key input values about which there is uncertainty. For example, a pessimistic, expected, and optimistic value might be chosen for an uncertain variable. Then an analysis could be performed to see how the outcome changes as each of the three chosen values is considered in turn, with other things held the same.

In engineering economics, sensitivity analysis measures the economic impact resulting from alternative values of uncertain variables that affect the economics of the project. When computing **measures of project worth**, for example, sensitivity analysis shows just how sensitive the economic payoff is to uncertain values of a critical input, such as the **discount rate** or project maintenance costs expected to be incurred over the project's **study period**. Sensitivity analysis reveals how profitable or unprofitable the project might be if input values to the analysis turn out to be different from what is assumed in a single-answer approach to measuring project worth.

Sensitivity analysis can also be performed on different combinations of input values. That is, several variables are altered at once and then a measure of worth is computed. For example, one scenario might include a combination of all pessimistic values, another all expected values, and a third all optimistic values. Note, however, that sensitivity analysis can in fact be misleading [Hillier, 1969] if all pessimistic assumptions or all optimistic assumptions are combined in calculating economic measures. Such combinations of inputs would be unlikely in the real world.

Sensitivity analysis can be performed for any measure of worth. And since it is easy to use and understand, it is widely used in the economic evaluation of government and private-sector projects. Office of Management and Budget [1992] Circular A-94 recommends sensitivity analysis to federal agencies as one technique for treating uncertainty in input variables. The American Society for Testing and Materials [2002], in its *Standards on Buildings Economics*, describes sensitivity analysis for use in government and private-sector applications.

208.1 Sensitivity Analysis Applications

How to use sensitivity analysis in engineering economics is best illustrated with examples of applications. Three applications are discussed. The first two focus on changes in project worth as a function of the change in one variable only. The third allows for changes in more than one uncertain variable.

The results of sensitivity analysis can be presented in text, tables, or graphs. The following discussion of sensitivity analysis applied to a programmable control system uses text and a simple table. Subsequent

0-8493-1586-7/05/$0.00+$1.50

TABLE 208.1 Energy Price Escalation Rates

Energy Price Escalation Rate	Net Savings
Low	$15,000
Moderate	20,000
High	50,000

analyses use graphs. The advantage of using a graph comes from being able to show in one picture the outcome possibilities over a range of input variations for one or several input factors.

Sensitivity Table for Programmable Control System

Consider a decision on whether to install a programmable time clock to control heating, ventilating, and air conditioning (HVAC) equipment in a commercial building. The time clock would reduce electricity consumption by turning off that part of the HVAC equipment that is not needed during hours when the building is unoccupied.

Using **net savings** (NS) as the measure of project worth, the time clock is acceptable on economic grounds if its NS is positive — that is, if its **present value** savings exceed present value costs. The control system purchase and maintenance costs are felt to be relatively certain. The savings from energy reductions resulting from the time clock, however, are not certain. They are a function of three factors: the initial price of energy, the rate of change in energy prices over the life cycle of the time clock, and the number of kilowatt hours (kWh) saved. Two of these, the initial price of energy and the number of kWh saved, are relatively certain. But future energy prices are not.

To test the sensitivity of NS to possible energy price changes, three values of energy price change are considered: a low rate of energy price escalation (slowly increasing benefits from energy savings), a moderate rate of escalation (moderately increasing benefits), and a high rate of escalation (rapidly increasing benefits). Table 208.1 shows three NS estimates that result from repeating the NS computation for each of the three energy price escalation rates.

To appreciate the significance of these findings, it is helpful to consider what extra information is gained over the conventional single-answer approach, where, say, a single NS estimate of $20,000 was computed. Table 208.1 shows that the project could return up to $50,000 in NS if future energy prices escalated at a high rate. On the other hand, it is evident that the project could lose as much as $15,000. This is considerably less than **breakeven**, where the project would at least pay for itself. It is also $35,000 less than what was calculated with the single-answer approach. Thus, sensitivity analysis reveals that accepting the time clock could lead to an uneconomic outcome.

There is no explicit measure of the likelihood that any one of the NS outcomes will happen. The analysis simply shows what the outcomes will be under alternative conditions. However, if there is reason to expect energy prices to rise, at least at a moderate rate, then the project very likely will make money, other factors remaining the same. This adds helpful information over the traditional single-answer approach to measures of project worth.

Sensitivity Graph for Gas Heating Systems

Figure 208.1 shows how sensitive NS is to the time over which two competing gas heating systems might be used in a building. The sensitivity graph helps you decide which system to choose on economic grounds.

Assume that you have an old electric heating system that you are considering replacing with a gas furnace. You have a choice between a high-efficiency or low-efficiency gas furnace. You expect both to last at least 15 to 20 years, and you do not expect any significant difference in building resale value or salvage value from selecting one system over the other. So you compute the NS for each gas furnace as compared to the old electric system. You will not be able to say which system is more economical until you decide how long you will hold the building before selling it. This is where the sensitivity graph is particularly helpful.

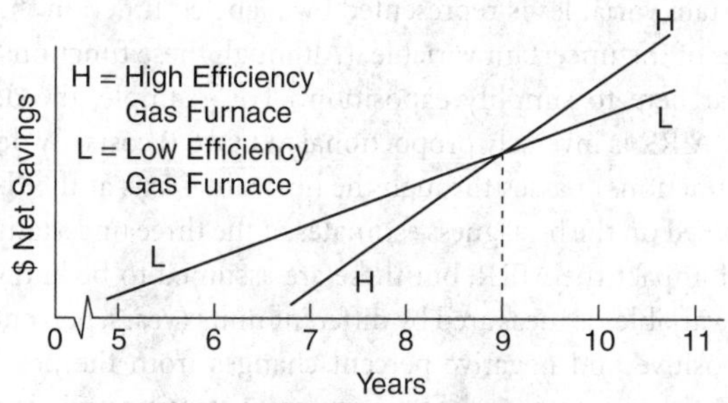

FIGURE 208.1 Sensitivity of net savings to holding period.

Net savings are measured on the vertical axis, and time on the horizontal axis. The longer you hold the building, the greater will be the present value of NS from installing either of the new systems, up to the estimated life of the systems. But note what happens in the 9th year. One line crosses over another. This means that the low-efficiency system is more **cost-effective** than the high-efficiency system for any holding period up to 9 years. To the left of the crossover point, NS values are higher for the low-efficiency system than for the high-efficiency system. But for longer holding periods, the high-efficiency system is more cost-effective than the low-efficiency system. This is shown to the right of the crossover point.

How does the sensitivity graph help you decide which system to install? First, it shows that neither system is more cost-effective than the other for all holding periods. Second, it shows that the economic choice between systems is sensitive to the uncertainty of how long you hold the building. You would be economically indifferent between the two systems only if you plan to hold the building 9 years. If you plan to hold the building longer than 9 years, for example, then install the high-efficiency unit. But if you plan to hold it less than 9 years, then the low-efficiency unit is the better economic choice.

Spider Diagram for a Commercial Building Investment

Another useful graph for sensitivity analysis is the **spider diagram.** It presents a snapshot of the potential impact of uncertain input variables on project outcomes. Figure 208.2 shows — for a prospective commercial building investment — the sensitivity of the **adjusted internal rate of return (AIRR)** to three uncertain variables: project life (PL), reinvestment rate (RR), and operation, maintenance, and replacement costs (OM&R). The spider diagram helps the investor decide if the building is likely to be a profitable investment.

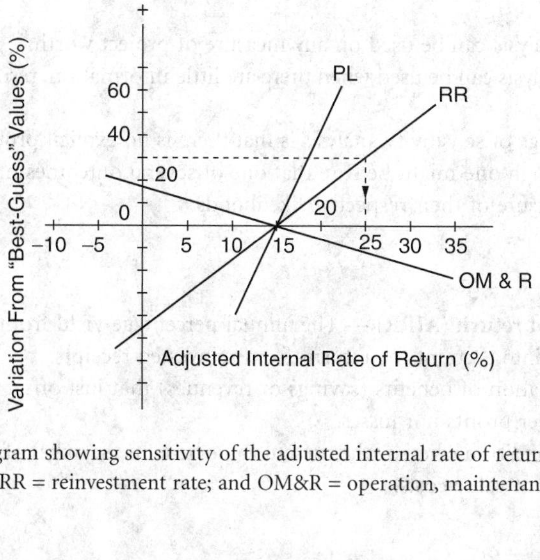

FIGURE 208.2 Spider diagram showing sensitivity of the adjusted internal rate of return to variations in uncertain variables. PL = project life; RR = reinvestment rate; and OM&R = operation, maintenance, and replacement costs.

Each of the three uncertain variables is represented by a labeled function that shows what AIRR value results from various values of the uncertain variable. (Although these functions are not necessarily linear, they are depicted as linear here to simplify exposition.) For example, the downward-sloping OM&R function indicates that the AIRR is inversely proportional to OM&R costs. By design, the OM&R function (as well as the other two functions) passes through the horizontal axis at the "best-guess" estimate of the AIRR (15% in this case), based on the best-guess estimates of the three uncertain variables. Other variables (e.g., occupancy rate) will impact the AIRR, but these are assumed to be known for the purpose of this analysis. Since each of the variables is measured by different units (years, percent, and dollars), the vertical axis is denominated in positive and negative percent changes from the best-guess values fixed at the horizontal axis. The AIRR value corresponding to any given percent variation indicated by a point on the function is found by extending a line perpendicular to the horizontal axis and directly reading the AIRR value. Thus a 30% increase in the best-guess reinvestment rate would yield a 25% AIRR, assuming that other values remain unchanged. Note that if the measure of AIRR were also given in percent differences, then the best-guess AIRR would be at the origin.

The spider diagram's contribution to decision making is its instant picture of the relative importance of several uncertain variables. In this case, the lesser the slope of a function is, the more sensitive is the AIRR to that variable. For example, any given percent change in OM&R will have a greater impact on the AIRR than will an equal percent change in RR or PL, and a percentage change in RR will have a greater impact than an equal percentage change in PL. Thus an investor will want to know as much as possible about likely OM&R costs for this project, because a relatively small variation in estimated costs could make the project a loser.

208.2 Advantages and Disadvantages

There are several advantages of using sensitivity analysis in engineering economics. First, it shows how significant any given input variable is in determining a project's economic worth. It does this by displaying the range of possible project outcomes for a range of input values, which shows decision makers the input values that would make the project a loser or a winner. Sensitivity analysis also helps identify critical inputs in order to facilitate choosing where to spend extra resources in data collection and in improving data estimates.

Second, sensitivity analysis is an excellent technique to help in anticipating and preparing for the "what-if" questions that are asked in presenting and defending a project. For instance, if asked what the outcome will be if operating costs are 50% more expensive than expected, an answer is forthcoming. Generating answers to "what-if" questions will help assess how well a proposal will stand up to scrutiny.

Third, sensitivity analysis does not require the use of probabilities, as do many techniques for treating uncertainty.

Fourth, sensitivity analysis can be used on any measure of project worth.

Finally, sensitivity analysis can be used when there are little information, resources, and time for more sophisticated techniques.

The major disadvantage of sensitivity analysis is that there is no explicit probabilistic measure of **risk exposure**. That is, although one might be sure that one of several outcomes might happen, the analysis contains no explicit measure of their respective likelihoods.

Defining Terms

Adjusted internal rate of return (AIRR) — The annual percentage yield from a project over the study period, taking into account the returns from reinvested receipts.

Breakeven — A combination of benefits (savings or revenues) that just offset costs, such that a project generates neither profits nor losses.

Cost-effective — The condition whereby the present value benefits (savings) of an investment alternative exceed its present value costs.

Discount rate — The minimum acceptable rate of return used in converting benefits and costs occurring at different times to their equivalent values at a common time. Discount rates reflect the investor's time value of money (or opportunity cost). "Real" discount rates reflect time value apart from changes in the purchasing power of the dollar (i.e., exclude inflation or deflation) and are used to discount constant dollar cash flows. "Nominal" or "market" discount rates include changes in the purchasing power of the dollar (i.e., include inflation or deflation) and are used to discount current dollar cash flows.

Measures of project worth — Economic methods that combine project benefits (savings) and costs in various ways to evaluate the economic value of a project. Examples are life-cycle costs, net benefits or net savings, benefit-to-cost ratio or savings-to-investment ratio, and adjusted internal rate of return.

Net savings — The difference between savings and costs, where both are discounted to present or annual values. The net savings method is used to measure project worth.

Present value — The time-equivalent value at a specified base time (the present) of past, present, and future cash flows.

Risk exposure — The probability that a project's economic outcome is different from what is desired (the target) or what is acceptable.

Sensitivity analysis — A technique for measuring the impact on project outcomes of changing one or more key input values about which there is uncertainty.

Spider diagram — A graph that compares the potential impact, taking one input at a time, of several uncertain input variables on project outcomes.

Study period — The length of time over which an investment is evaluated.

References

American Society for Testing and Materials (ASTM). 2002. Standard guide for selecting techniques for treating uncertainty and risk in the economic evaluation of buildings and building systems, E1369-02, in *ASTM Standards on Buildings Economics,* 5th ed. ASTM, West Conshohocken, PA.

Hillier, F. 1963. The derivation of probabilistic information for the evaluation of risky investments. *Manage. Sci.,* April, p. 444.

Office of Management and Budget (OMB). 1992. *Guidelines and Discount Rates for Benefit-Cost Analysis of Federal Programs.* Circular A-94, 29 October. OMB, Washington, DC.

Further Information

National Institute of Standards and Technology. 1992. *Uncertainty and Risk,* part II in a series on least-cost energy decisions for buildings. National Institute of Standards and Technology, Gaithersburg, MD. VHS tape and companion workbook are available from Video Transfer, Inc., 5709-B Arundel Avenue, Rockville, MD 20852. Phone: (301)881-0270.

Marshall, H. E. 1988. *Techniques for Treating Uncertainty and Risk in the Economic Evaluation of Building Investments.* Special Publication 757. National Institute of Standards and Technology, Gaithersburg, MD.

Ruegg, R. T. and Marshall, H. E. 1990. *Building Economics: Theory and Practice.* Chapman & Hall, New York.

209

Life-Cycle Costing*

Wolter J. Fabrycky
Virginia Polytechnic Institute and State University

Benjamin S. Blanchard
Virginia Polytechnic Institute and State University

A major portion of the projected **life-cycle cost** (**LCC**) for a specific product, system, or structure is traceable to decisions made during conceptual and preliminary design. These decisions pertain to operational requirements, performance and effectiveness factors, the design configuration, the maintenance concept, production methods and quantity, utilization factors, logistic support, phaseout planning, and disposal. Such decisions guide subsequent design and production activities, product distribution functions, and aspects of sustaining system support. Accordingly, if the final life-cycle cost is to be minimized, it is essential that considerable cost emphasis be applied during the early stages of system design and development.

209.1 Life-Cycle Costing Situation

The combination of inflation, cost growth, reduction in purchasing power, budget limitations, increased competition, and related factors has created an awareness and interest in the total cost of products, systems, and structures. Not only are the acquisition costs associated with new systems rising, but the costs of operating and maintaining systems already in use are also increasing. This is due primarily to a combination of inflation and cost growth factors traceable to the following:

1. Unsatisfactory quality of products, systems, and structures in use
2. Engineering changes mandated during design and development
3. Changing suppliers in the procurement of system components
4. System production and/or construction changes
5. Changes in logistic support capability
6. Estimating and forecasting errors
7. Unforeseen events and problems

*Material presented in this chapter is adapted from Chapter 6 in W. J. Fabrycky and B. S. Blanchard, *Life-Cycle Cost and Economic Analysis,* Prentice Hall, Englewood Cliffs, NJ, 1991.

0-8493-1586-7/05/$0.00+$1.50

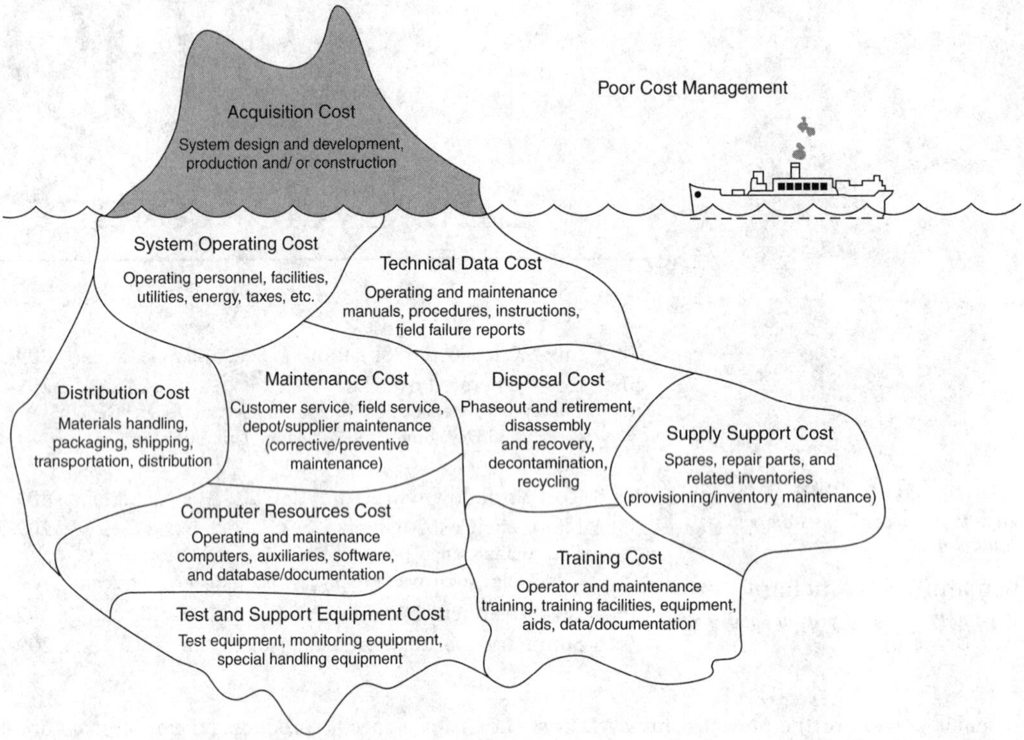

FIGURE 209.1 The problem of total cost visibility.

Experience over the past several decades indicates that cost growth due to various causes has ranged from five to ten times the rate of inflation. At the same time, budget allocations for many programs are decreasing from year to year. The result is that fewer resources are available for acquiring and operating new systems or products and for maintaining and supporting existing systems. Available funds for new projects, when inflation and cost growth are considered, are decreasing rapidly.

The current economic situation is further complicated by some additional problems related to the actual determination of system and/or product cost. Some of these are listed below.

1. Total system cost is not fully visible, particularly those costs associated with operation and support. The cost visibility problem is due to an "iceberg" effect, as is illustrated in Figure 209.1.
2. Individual cost factors are often improperly applied. Costs are identified and frequently included in the wrong category: variable costs are treated as fixed (and vice versa), indirect costs are treated as direct costs, and so on.
3. Existing accounting procedures do not always permit a realistic and timely assessment of total cost. In addition, it is often difficult (if not impossible) to determine costs on a functional basis.
4. Budgeting practices are often inflexible regarding the shift in funds from one category to another, or from year to year, to facilitate cost improvements in system acquisition and utilization.

Current trends of inflation and cost growth, combined with the problems listed above, have led to inefficiencies in the utilization of valuable resources. Systems and products have been developed that are not cost-effective. It is anticipated that these conditions will worsen unless an increased degree of cost consciousness is assumed by engineers and project managers.

LCC is determined by identifying the applicable functions in each phase of the life cycle, costing these functions, applying the appropriate costs by function on a year-to-year basis, and then accumulating the costs over the entire span of the life cycle. To be complete, LCC must include all producer and consumer costs from origin of concept to phase-out and disposal.

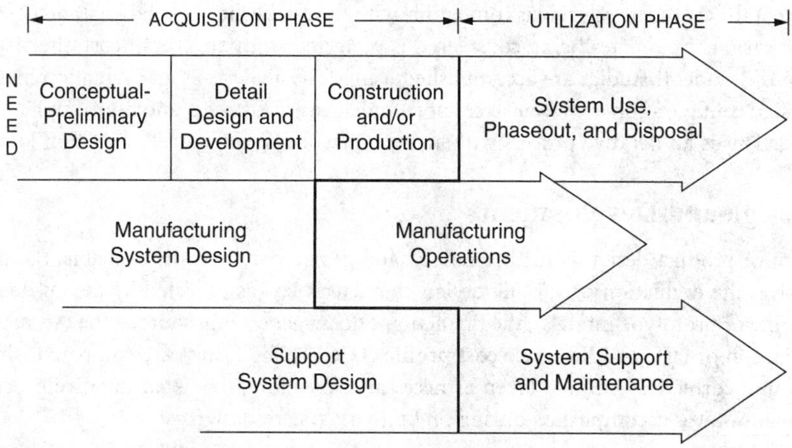

FIGURE 209.2 Product, process, and support life cycles.

209.2 Cost Generated over Life Cycle

LCC includes all costs associated with the product, system, or structure that are anticipated over the defined life cycle. The life cycle and major functions associated with each phase are illustrated in Figure 209.2. Life-cycle costing is employed in the evaluation of alternative system design configurations, alternative production schemes, alternative logistic support policies, alternative disposal concepts, and so on. The life-cycle concept, tailored to the specific system being addressed, forms the basis for life-cycle costing.

There are many technical and nontechnical decisions and actions required throughout the product or system life cycle. Most actions, particularly those in the earlier phases, have life-cycle implications and directly affect LCC. The analysis constitutes a step-by-step approach employing LCC figures of merit as a criterion to arrive at a cost-effective solution. This analysis process is iterative in nature and can be applied to any phase of the life cycle of the product, system, or structure. Cost considerations over the system/product life cycle are summarized in the following sections.

Conceptual System Design

In the early stages of planning and conceptual design (when requirements are being defined), quantitative cost figures of merit should be established to which the system or product is to be designed, tested, produced (or constructed), and supported. A **design-to-cost (DTC)** goal may be adopted to establish cost as a system or product design constraint, along with performance, effectiveness, capacity, accuracy, weight, reliability, maintainability, supportability, and so on. Cost must be an active rather than a resultant factor throughout the system design process.

Preliminary System Design

With quantitative cost requirements established, the next step includes an iterative process of synthesis, tradeoff and optimization, and system/product definition. The criteria defined in the conceptual system design are initially allocated (or apportioned) to various segments of the system to establish guidelines for the design and/or the procurement of needed element(s). Allocation is accomplished from the system level down to the level necessary to provide an input to design and to ensure adequate control. The factors projected reflect the target cost per individual unit (i.e., a single equipment unit or product in a deployed population) and are based on system operational requirements, the system maintenance concept, and the plan for disposal.

As system development evolves, various approaches are considered that may lead to a preferred configuration. LCC analyses are accomplished in (1) evaluating each possible candidate, with the objective of ensuring that the candidate selected is compatible with the established cost targets, and (2) determining which of the various candidates being considered is preferred from an overall cost-effectiveness standpoint. Numerous tradeoff studies are accomplished using LCC analysis as an evaluation method, until a preferred design configuration is chosen. Areas of compliance are justified, and noncompliant approaches are discarded. This is an iterative process with an active-feedback and corrective-action loop.

Detail Design and Development

As the system or product design is further refined and design data become available, the LCC analysis process involves the evaluation of specific design characteristics (as reflected by design documentation and engineering or prototype models), the prediction of cost-generating sources, the estimation of costs, and the projection of LCC as a **life-cycle cost profile (LCCP)**. The results are compared with the initial requirement and corrective action is taken as necessary. As before, this is an iterative process, but at a lower level than what is accomplished during preliminary system design.

Production, Utilization, and Support

Cost concerns in the production, utilization, support, and disposal stages of the system or product life cycle are addressed through data collection, analysis, and an assessment function. High-cost contributors are identified, cause-and-effect relationships are defined, and valuable information is gained and utilized for the purposes of product improvement through redesign or reengineering.

209.3 Cost Breakdown Structure

In general, costs over the life cycle fall into categories based on organizational activity needed to bring a system into being. These categories and their constituent elements constitute a **cost breakdown structure (CBS)**, as illustrated in Figure 209.3. The main CBS categories are as follows:

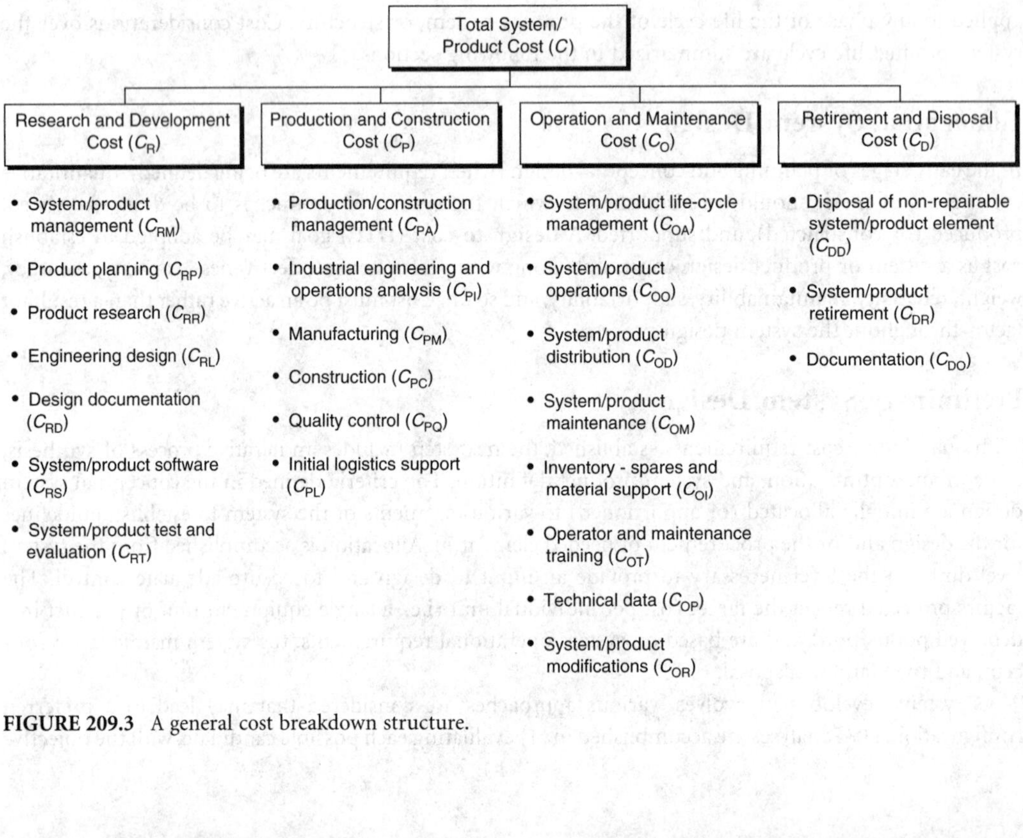

FIGURE 209.3 A general cost breakdown structure.

1. *Research and development cost.* Initial planning, market analysis, feasibility studies, product research, requirements analysis, engineering design, design data and documentation, software, testing and evaluation of engineering models, and associated management functions.
2. *Production and construction cost.* Industrial engineering and operations analysis, manufacturing (fabrication, assembly, and test), facility construction, process development, production operations, quality control, and initial logistic support requirements (e.g., initial customer support, the manufacture of spare parts, the production of test and support equipment, etc.).
3. *Operation and support cost.* Customer or user operation of the system or product in the field, product distribution (marketing and sales, transportation, and traffic management), and sustaining maintenance and logistic support throughout the system or product life cycle (e.g., customer service, maintenance activities, supply support, test and support equipment, transportation and handling, technical data, facilities, system modifications, etc.).
4. *Retirement and disposal cost.* Disposal of nonrepairable items throughout the life cycle, system/product retirement, material recycling, and applicable logistic support requirements.

The cost breakdown structure links objectives and activities with organizational resource requirements. It constitutes a logical subdivision of cost by functional activity area, major system elements, and/or one or more discrete classes of common or like items. The CBS provides a means for initial resource allocation, cost monitoring, and cost control.

209.4 Life-Cycle Cost Analysis

The application of life-cycle costing methods during product and system design and development is realized through the accomplishment of **life-cycle cost analyses (LCCAs)**. An LCCA may be defined as a systematic analytical process of evaluating various designs or alternative courses of action with the objective of choosing the best way to employ scarce resources.

Where feasible alternative solutions exist for a specific problem and a decision is required for the selection of a preferred approach, there is a formal analysis process that should be followed. Specifically, the analyst should define the need for analysis, establish the analysis approach, select a model to facilitate the evaluation process, generate the appropriate information for each alternative being considered, evaluate each alternative, and recommend a proposed solution that is responsive to the opportunity.

Cost Analysis Goals

There are many questions that the decision maker might wish to address. There may be a single overall analysis goal (e.g., design to minimum LCC) and any number of subgoals. The primary question should be as follows: What is the purpose of the analysis and what is to be learned through the analysis effort?

In many cases the nature of the problem appears to be obvious, but its precise definition may be the most difficult part of the entire process. The design problem must be defined clearly and precisely and presented in such a manner as to be easily understood by all concerned. Otherwise, it is doubtful whether an analysis of any type will be meaningful. The analyst must be careful to ensure that realistic goals are established at the start of the analysis process and that these goals remain in sight as the process unfolds.

Analysis Guidelines and Constraints

Subsequent to definition of the problem and the goals, the cost analyst must define the guidelines and constraints (or bounds) within which the analysis is to be accomplished. Guidelines are composed of information concerning such factors as the resources available for conducting the analysis (e.g., necessary technical skills, availability of appropriate software, etc.), the time schedule allowed for completion of the analysis, and/or related management policy or direction that may affect the analysis.

In some instances, a decision maker or manager may not completely understand the problem or the analysis process and may direct that certain tasks be accomplished in a prescribed manner or time frame that may not be compatible with the analysis objectives. On other occasions, a manager may have a preconceived idea as to a given decision outcome and direct that the analysis support the decision. Also, there could be external inhibiting factors that may affect the validity of the analysis effort. In such cases the cost analyst should make every effort to alleviate the problem by educating the manager and documenting unresolved issues.

Relative to the technical characteristics of a system or product, the analysis output may be constrained by bounds (or limits) that are established through the definition of system performance factors, operational requirements, the maintenance concept, and/or through advanced program planning. For example, there may be a maximum weight requirement for a given product, a minimum reliability requirement, a maximum allowable first cost per unit, a minimum rated capacity, and so on. These various bounds, or constraints, should provide the bases for tradeoffs in the evaluation of alternatives. Candidates that fall outside these bounds are not allowable.

Identifying Alternatives

Within established bounds and constraints, there may be any number of approaches leading to a possible solution. All possible alternatives should be considered, with the most likely candidates selected for further evaluation. Alternatives are frequently proposed for analysis even though there seems to be little likelihood that they will prove feasible. This is done with the thought that it is better to consider many alternatives than to overlook one that may be very good. Alternatives not considered cannot be adopted, no matter how desirable they may actually prove to be.

Applying Cost Breakdown Structure

Applying the cost breakdown structure is one of the most significant steps in life-cycle costing. The CBS constitutes the framework for defining LCC categories and provides the communications link for cost reporting, analysis, and ultimate cost control.

In developing the CBS one needs to proceed to the depth required to provide the necessary information for a true and valid assessment of the system or product LCC, identify high-cost contributors and enable determination of the cause-and-effect relationships, and illustrate the various cost parameters and their application in the analysis. Traceability is required from the system-level LCC figure of merit to each specific input factor.

209.5 Cost Treatment over Life Cycle

With the system/product cost breakdown structure defined and cost-estimating approaches established, it is appropriate to apply the resultant data to the system life cycle. To accomplish this, the cost analyst needs to understand the steps required in developing cost profiles that include aspects of inflation, the effects of learning curves, the time value of money, and so on.

In developing a cost profile, there are different procedures that may be used. The following steps are suggested:

1. Identify all activities throughout the life cycle that will generate costs of one type or another. This includes functions associated with planning, research and development, testing and evaluation, production/construction, product distribution, system/product operational use, maintenance and logistic support, and so on.
2. Relate each activity identified in step 1 to a specific cost category in the cost breakdown structure. All program activities should fall into one or more of the CBS categories.
3. Establish the appropriate cost factors in constant dollars for each activity in the CBS, where constant dollars reflect the general purchasing power of the dollar at the time of decision (i.e., today). Relating costs in terms of constant dollars will allow for a direct comparison of activity levels from year to year prior to the introduction of inflationary cost factors, changes in price

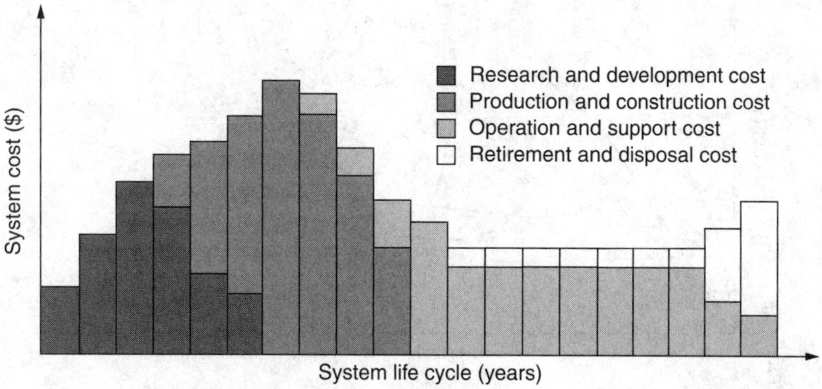

FIGURE 209.4 Development of life-cycle cost profiles.

levels, economic effects of contractual agreements with suppliers, and so on, which can often cause some confusion in the evaluation of alternatives.

4. Within each cost category in the CBS, the individual cost elements are projected into the future on a year-to-year basis over the life cycle as applicable. The result should be a cost stream in constant dollars for the activities that are included.
5. For each cost category in the CBS and for each applicable year in the life cycle, introduce the appropriate inflationary factors, economic effects of learning curves, changes in price levels, and so on. The modified values constitute a new cost stream and reflect realistic costs as they are anticipated for each year of the life cycle (i.e., expected 2004 costs in 2004, 2005 costs in 2005, etc.). These costs may be used directly in the preparation of future budget requests, since they reflect the actual dollar needs anticipated for each year over the life cycle.
6. Summarize the individual cost streams by major categories in the CBS and develop a top-level cost profile.

Results from the sequence of steps above are presented in Figure 209.4. First, it is possible and often beneficial to evaluate the cost stream for individual activities of the life cycle such as research and development, production, operation and support, and so on. Second, these individual cost streams may be shown in the context of the total cost spectrum. Finally, the total cost profile may be viewed from the standpoint of the logical flow of activities and the proper level and timely expenditure of dollars. The profile in Figure 209.4 represents a budgetary estimate of future resource needs in monetary terms.

When dealing with two or more alternative system configurations, each will include different levels of activity, different design approaches, different logistic support requirements, and so on. No two systems alternatives will be identical. Thus, individual profiles will be developed for each alternative and ultimately compared on an equivalent basis utilizing the economic analysis techniques found in textbooks on engineering economics. Figure 209.5 illustrates LCC profiles for several alternatives.

209.6 Summary

Life-cycle costing is applicable in all phases of system design, development, production, construction, operational use, and logistic support. Cost emphasis is created early in the life cycle by establishing quantitative cost factors as "design to" requirements. As the life cycle progresses, cost is employed as a major parameter in the evaluation of alternative design configurations and in the selection of a preferred approach. Subsequently, cost data are generated based on established design and production characteristics and used in the development of LCC projections. These projections, in turn, are compared with the initial requirements to determine the degree of compliance and the necessity for corrective action. In essence, LCC evolves from a series of rough estimates to a relatively refined methodology and is employed as a management tool for decision-making purposes.

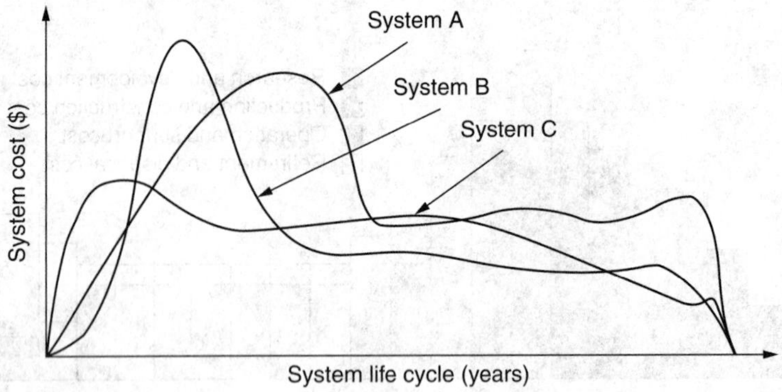

FIGURE 209.5 Life-cycle cost profiles of alternatives.

Defining Terms

Cost breakdown structure (CBS) — A framework for defining life-cycle costs, it provides the communications link for cost reporting, analysis, and ultimate cost control.

Design-to-cost (DTC) — A concept that may be adopted to establish cost as a system or product design constraint, along with performance, effectiveness, capacity, accuracy, size, weight, reliability, maintainability, supportability, and others.

Life-cycle cost (LCC) — All costs associated with the product or system as anticipated over the defined life cycle.

Life-cycle cost analysis (LCCA) — A systematic analytical process for evaluating various alternative courses of action with the objective of choosing the best way to employ scarce resources.

Life-cycle cost profile (LCCP) — A budgetary estimate of future resource needs over the life cycle.

Further Information

For a complete view of life-cycle cost and economic analysis, see Fabrycky, W. J. and Blanchard, B. S. 1991. *Life-Cycle Cost and Economic Analysis*. Prentice Hall, New York. Further information may be obtained from the following:

Blanchard, B. S. and Fabrycky, W. J. 1998. *Systems Engineering and Analysis,* 3rd ed., Prentice Hall, Upper Saddle River, NJ.

Canada, J. R. and Sullivan, W. G. 1989. *Economic and Multiattribute Evaluation of Advanced Manufacturing Systems,* Prentice Hall, Upper Saddle River, NJ.

Dhillon, B. S. 1989. *Life Cycle Costing: Techniques, Models and Applications,* Gordon and Breach Science Publishers, New York.

Fabrycky, W. J., Thuesen, G. J., and Verma, D. 1998. *Economic Decision Analysis,* 3rd ed., Prentice Hall, Upper Saddle River, NJ.

Ostwald, P. F. 1992. *Engineering Cost Estimating,* 3rd ed., Prentice Hall, Upper Saddle River, NJ.

Thuesen, G. J. and Fabrycky, W. J. 2001. *Engineering Economy,* 9th ed., Prentice Hall, Upper Saddle River, NJ.

210

Project Evaluation and Selection

Hans J. Thamhain
Bentley College

Few decisions are more fundamental to business viability than resource allocations for new projects. Virtually every organization selects and implements projects, ranging from product developments to organizational improvements, and from customer contracts to R&D activities and bid proposals. Pursuing the "wrong" project not only (1) drains a company's resources, but also causes the enterprise to (2) miss alternative opportunities, (3) operate less flexibly and responsively in the marketplace, and (4) miss opportunities for leveraging core competencies. Project opportunities must be analyzed relative to their potential value, viability, and importance to the enterprise. Four major dimensions should be considered: (1) added value of the new project, (2) cost of the project, (3) readiness of the enterprise to execute the project, and (4) managerial desire. A well-organized *project evaluation and selection process* provides the framework for systematic data gathering and informed decision making toward resource allocation. Typically, these decisions can be broken into four principal categories:

1. *Deciding initial feasibility.* Screening and filtering, quick decision on the viability of an emerging project for further evaluation
2. *Deciding strategic value to enterprise.* Identifying alternatives and options to proposed project
3. *Deciding detailed feasibility.* Determining the chances of success for a proposed project.
4. *Deciding project go/no-go.* Committing resources for a project implementation.

While these decisions look logical and straightforward, developing meaningful support data is a complex process. It is also expensive, time consuming, and often highly eclectic. Typically, decision making requires the following inputs:

- Specific resource requirements
- Specific implementation risks

0-8493-1586-7/05/$0.00+$1.50

TABLE 210.1 Typical Criteria for Project Evaluation and Selection

The criteria relevant to the evaluation and selection of a particular project depend on the specific project type and business situation, such as project development, custom project, process development, industry, and market. Typically, evaluation procedures include the following criteria:

- Development cost
- Development time
- Technical complexity
- Risk
- Return on investment
- Cost-benefit
- Product life cycle
- Sales volume
- Market share
- Project business follow-on
- Organizational readiness and strength
- Consistency with business plan
- Resource availability
- Cash flow, revenue, and profit
- Impact on other business activities

Note: Each criterion is based on a complex set of parameters and variables.

- Specific benefits (economics, technology, markets, etc.)
- Benchmarking and comparative analysis
- Strategic perspective

While there are plenty of challenges in evaluating project opportunities in terms of cost, time, risks, and benefits, such as those shown in Table 210.1, it is relatively straightforward in comparison to *predicting project success.* The difficulty here is in defining a meaningful *aggregate measure for rating project value and success.* Methods range from purely intuitive to highly analytical. No method is seen as truly reliable in predicting success, especially for more complex types of projects. Yet, some companies have a better track record in selecting "winning" projects than others, which seems to be related to the ability to create an integrated picture of the potential benefits, costs, and risks for the proposed project relative to the company's strengths and strategic objectives. Producing such a composite is both a science and an art. Traditionally, managers have used predominately *rational selection processes* to support project selections. However, purely rational/analytical processes apply to only a limited number of business situations. Many of today's complex project scenarios require the integration of both analytical and judgmental techniques to evaluate projects in a meaningful way toward conclusions on *right, successful, or best choice.*

Yet, in spite of the intricacies involved in project selection, systematic information gathering and standardized methods are at the heart of any project evaluation process, and provide the best assurance for reliably predicting project outcome and repeatability of the decision process. Approaches to project evaluation and selection fall into one of three principal classes:

- Primarily *quantitative* and *rational* approaches
- Primarily *qualitative* and *intuitive* approaches
- *Mixed approaches,* combining both quantitative and qualitative methods

Because of the interdisciplinary complexities involved, analyzing a new project opportunity is a highly interactive effort among various resource groups of the enterprise and its partners. Often, many meetings are needed before (1) a clear picture emerges of potential benefits, costs, and risks involved, and (2) data emerge that are useful for the project evaluation and selection process, regardless of their quantitative, qualitative, or combined nature.

TABLE 210.2 Description of Four Project Proposals

Project option P1	Do not accept any new project proposal; hence, no investment capital is required, nor is revenue generated.
Project option P2	This opportunity requires a $1,000 investment at the beginning of the first year, and generates a $200 revenue at the end of *each* of the following 5 years.
Project option P3	This opportunity requires a $2,000 investment at the beginning of the first year, and generates a variable stream of net revenues at the end of *each* of the next 5 years as follows: $1,500, $1,000, $800, $900, $1,200.
Project option P4	This opportunity requires a $5,000 investment at the beginning of the first year, and generates a variable stream of net revenues at the end of *each* of the next 5 years as follows: $1,000, $1,500, $2,000, $3,000, $4,000.

TABLE 210.3 Cash Flow of Four Project Options or Proposals, Assuming MARR of i = 10%

End of Year	Do-Nothing Option P1	Project Option P2	Project Option P3	Project Option P4
0	0	−1,000	−2,000	−5,000
1	0	200	1,500	1,000
2	0	200	1,000	1,500
3	0	200	800	2,000
4	0	200	900	3,000
5	0	200	1,200	4,000
Net cash flow	0	0	+3,400	+7,500
NPV | N = 5	0	−242	+2,153	+3,192
NPV | N =	0	+1,000	+9,904	+28,030
ROI | N = 5	0	20%	54%	46%
$CB{=}ROI_{NPV\|N=5}$	0	76%	108%	164%
$N_{PBP}\,\|\,i{=}0$	0	5	1.5	3.3
$N_{NPV}\,\|\,i$	0	7.3	5	3.8

Note: CB, cost-benefit; NPV, net present value; PBP, payback period; ROI, return on investment.

210.1 Quantitative Approaches to Project Evaluation and Selection

Quantitative approaches are often favored to support project evaluation and selection if the decision requires economic justification. They are also commonly used to support judgment-based project selection. One of the features of quantitative approaches is the generation of numeric measures for simple and effective comparison, ranking, and selection. These approaches also help to establish quantifiable norms and standards, and lead to repeatable processes. Yet, the ultimate usefulness of these methods depends on the assumption that the decision parameters, such as cash flow, risks, and the underlying economic, social, political, and market factors can actually be quantified and reliably estimated over the project life cycle. Therefore, quantitative techniques are effective and powerful decision support tools, if meaningful estimates of cost-benefits, such as capital expenditures and future revenues, can be obtained and converted into net present values for comparison. As an example, Table 210.2 describes four project options that will be evaluated in this chapter using various quantitative methods, with results summarized in Table 210.3.

Net Present Value Comparison

The net present value (NPV) method uses discounted cash flow as the basis for comparing the relative merit of alternative project opportunities. It assumes that all investment costs and revenues are known, and that economic analysis is a valid basis for project selection.

We can determine the NPV of a single revenue, stream of future revenues, or costs expected in the future. Two types of presentations are common: present worth (PW) and net present value.

PW is the single revenue or cost (also called annuity A) that occurs at the end of period *n*, subject to the prevailing interest rate *i*. Depending on management philosophy and enterprise policies, this interest rate can be *internal rate of return (IRR)* realized by the company on similar investments, **minimum attractive rate of return (MARR)** acceptable to company management, or the prevailing discount rate. PW is calculated as follows:

$$PW(A|\ i,\ n) = PW_n = A\,\frac{1}{(1+i)^n}$$

For the examples used in this chapter, we consider the IRR as the prevailing interest rate.

The NPV is defined a series of revenues or costs, A_n, over *N* periods of time, at a prevailing interest rate *i*:

$$NPV(A_n|i,N) = \sum_{n=1}^{N} A_n\frac{1}{(1+i)^n} = \sum_{n=1}^{N} PW_n$$

Three special cases exist for NPV calculation: (1) *for a uniform series of revenues or costs* over N periods, $NPV(A_n\,|\,i, N) = A[(1+i)^{N-1}]/i(1+i)^N$; (2) *for an annuity or interest rate* i *approaching zero,* $NPV = A^*N$; and (3) *for the revenue or cost series to continue forever,* $NPV = A/i$.

Table 210.3 applies these formulas to four project alternatives described in Table 210.2, showing the most favorable 5-year *NPV* of $3,192 for project option P3.

Return on Investment Comparison

Perhaps one of the most popular measures for project evaluation is the *return on investment (ROI)*:

$$ROI = \frac{Revenue(R) - Cost(C)}{Investment(I)}$$

ROI calculates the ratio of net revenue over investment. In its simplest form the stream of cash flow is *not* discounted. One can look at the revenue on a year-by-year basis, relative to the initial investment. For example, project option 1 in Table 210.3 would produce a 20% ROI each year, while project option 2 would produce a 75% ROI during the first year, 50% during the second year, and so on. Although this is a popular measure, it does not permit a meaningful comparative analysis of alternative projects with fluctuating costs and revenues. Furthermore, it does not consider the time value of money. In a more sophisticated way, we can calculate the *average ROI per year*:

$$\overline{ROI}(A_n, I_n|N) = \left[\sum_{n=1}^{N}\frac{A_n}{I_n}\right]\Big/N$$

We can then *compare the average ROI to the minimum attractive rate of return, MARR*. Given a MARR of 10% for our project environment, all three project options P1, P2, and P3 compare favorably, with project P2 yielding the highest average return on investment of 54%.

Cost-Benefit

Alternatively, we can calculate the NPV of the total ROI over the project lifecycle. This measure, known as *cost-benefit (CB)*, is calculated as the present-value stream of net revenues divided by the present-value stream of investments. It is an effective measure for comparing project alternatives with fluctuating cash flows:

$$CB = ROI_{NPV}(A_n, I_n|i,N) = \left[\sum_{n=1}^{N} NPV(A_n|i,N)\right] \bigg/ \left[\sum_{n=1}^{N} NPV(I_n|i,N)\right]$$

In our example of four project options (Table 210.3), project proposal P2 produces the highest cost-benefit of 206% under the given assumption of i = MARR = 10%.

Payback Period Comparison

Another popular figure of merit for comparing project alternatives is the *payback period (PBP)*. It indicates the time period of net revenues required to return the capital investment made on the project. For simplicity, *undiscounted* cash flows are often used to calculate a quick figure for comparison, which is quite meaningful if we deal with an initial investment and a steady stream of net revenue. However, for fluctuating revenue and/or cost steams, the *NPV* must be *calculated for each period individually* and cumulatively added up to the "break-even point" in time, N_{PBP}, when the *NPV* of revenue equals the investment. Mathematically,

$$N_{PBP} \ldots \triangleright\triangleright\triangleright \text{ when } \sum_{n=1}^{N} NPV(A_n|i) \geq \sum_{n=1}^{N} NPV(I_n|i)$$

In our example of four project options (Table 210.3), project proposal P2 produces the shortest, most favorable payback period of 1.9 years under the given assumption of i = MARR = 10%.

Pacifico and Sobelman Project Ratings

The previously discussed methods of evaluating projects rely heavily on the assumption that technical and commercial success is ensured, and all costs and revenues are predictable. Because these assumptions do not always hold, many companies have developed their own special procedures and formulas for comparing project alternatives. Two examples illustrate this special category of project evaluation metrics. The *project rating factor (PR)* was originally developed by Carl Pacifico for assessing chemical products and predicting commercial success:

$$PR = \frac{pT * pC * R}{TC}$$

Pacifico's formula is in essence an ROI calculation adjusted for risk. It includes the probability of technical success ($0.1 < pT < 1.0$], probability of commercial success ($0.1 < pC < 1.0$), total net revenue over project lifecycle (R), and total capital investment for product development, manufacturing setup, marketing, and related overheads (TC).

The second example shows a formula developed by Sobelman:

$$z = (P * T_{LC}) - (C * T_D)$$

It represents a modified cost-benefit measure that takes into account both the development time and the commercial life cycle of the product. It also includes average profit per year (P), estimated product lifecycle (T_{LC}), average development cost per year (C), and years of development (T_D).

TABLE 210.4 Comparison of Quantitative and Qualitative Approaches to Project Evaluation

Quantitative Methods	Qualitative Methods
Benefits	Benefits
Simple comparison, ranking, selection	Search for meaningful evaluation metrics
Repeatable process	Broad-based organizational involvement
Encourages data gathering and measurability	Understanding of problems, benefits, opportunities
Benchmarking opportunities	Problem solving as part of selection process
Programmable	Broad knowledge base
Useful input to sensitivity analysis and simulation	Multiple solutions and alternatives
	Multifunctional involvement leads to buy-in
	Risk sharing
Limitations	Limitations
Many success factors are nonquantifiable	Complex, time-consuming process
Probabilities and weights may change	Biases via power and politics
True measures do not exist	Difficult to proceduralize or repeat
Analyses and conclusions are often misleading	Conflict and energy intensive
Methods mask unique problems and opportunities	Does not fit conventional decision processes
Stifle innovative decision making	Intuition and emotion dominates over facts
Lack people involvement, buy-in, commitment	Justify wants over needs
Do not deal well with multifunctional issues and dynamic situations	Lead to more fact finding than decision making
May mask hidden costs and benefits	
Pressure to act quickly and possibly prematurely	

Going Beyond Simple Formulas

While quantitative methods of project evaluation have the benefit of producing relatively quickly a measure of merit for simple comparison and ranking, they also have many limitations, as summarized in Table 210.4. Because of these limitations, alternatives to these strictly quantitative methods have been developed that use a broader set of measures in determining the long-range cost and benefits of a project proposal to the enterprise. These methods rely to a large degree on *qualitative, judgmental decision making*. They cast a broad data-gathering net and consider a wide spectrum of factors that are often difficult to quantify or even to describe. Yet, in spite of the limitations of quantitative evaluation and the increased use of qualitative approaches, virtually every organization supports its project selections with some form of quantitative measures, the most popular being cost-benefit and payback period.

210.2 Qualitative Approaches to Project Evaluation and Selection

Especially for project evaluations that involve complex sets of business criteria, the narrowly focused quantitative methods must often be supplemented by broad-scanning, intuitive processes and collective, multifunctional decision making such as *Delphi, nominal group technology, brainstorming, focus groups, sensitivity analysis*, and *benchmarking.* Each of these techniques can either be used by itself to determine the *best, most successful, or most valuable* option. Or, these techniques can be integrated into an analytical framework for *collective multifunctional decision making.*

Collective Multifunctional Evaluations

This process relies on subject experts from various functional areas for collectively defining and evaluating broad **project success** criteria, while employing both quantitative and qualitative methods. The first step is to define the specific organizational areas critical to project success and to assign expert evaluators. For a typical product development project, these organizations may include R&D, engineering, testing, manufacturing, marketing, product assurance, and customer services. These function experts should be given the time necessary for the evaluation. They also should have a commitment from senior management for full organizational support. Ideally, these evaluators should have the responsibility of implementing the project, should it be selected.

The next step is for the evaluation team to define the factors that appear critical to the ultimate success of the projects under evaluation and arrange them into a concise list that includes both quantitative and qualitative factors. A mutually acceptable scale must be worked out for scoring the evaluation criteria. Studies of collective multifunctional assessment practices show that simplicity of scales is crucial to a workable team solution. Three types of scale have produced the most favorable results in field studies: (1) 10-point scale, ranging from +5 = most favorable to −5 = most unfavorable; (2) 3-point scale, +1 = favorable, 0 = neutral or cannot judge, and −1 = unfavorable; and (3) 5-point scale, A = highly favorable, B = favorable, C = marginally favorable, D = most likely unfavorable, and F = definitely unfavorable. **Weighing of criteria** is not recommended for most applications as it complicates and often distorts the collective evaluation.

Evaluators first score individually all factors that they feel qualified to make an expert judgement on. Collective discussions follow. Initial discussions of project alternatives, their markets, business opportunities, and technologies involved, are usually beneficial, but not necessary for the first round of the evaluation process. The objective of this first round of expert judgments is to get calibrated on the opportunities and challenges presented. Further, each evaluator has the opportunity to recommend (1) actions needed for better project assessment; (2) whether additional data are needed; and (3) suggestions that would enhance project success and the evaluation score. Before meeting at the next group session, agreed-on action items and activities for improving the decision process should be completed. With each iteration, the function expert meetings are enhanced with more refined project data. Typically, between three and five iterations are required before a project selection can be finalized.

210.3 Recommendations for Effective Project Evaluation and Selection

Effective evaluation and selection of project opportunities involves many variables of the organizational and technological environment, often reaching far beyond cost and revenue measures. While economic models provide an important dimension of the project selection process, most situations are too complex to use simple quantitative methods as the sole basis for decision making. Many of today's project evaluation procedures include a broad spectrum of variables and rely on a combination of rational and intuitive processes for defining the value of a new project venture to the enterprise. The better a firm understands its business processes, markets, customers, and technologies, the better it will be able to evaluate the value of a new project venture. Further, manageability of the evaluation process is critical to its results, especially in complex situations. The process must have a certain degree of structure, discipline, and measurability to be conducive to the intricate multivariable analysis. One method of achieving structure and manageability of the process calls for grouping the evaluation variables into four categories: (1) consistency and strength of the project with the business mission, strategy, and plan; (2) multifunctional ability to produce the project results, including technical, cost, and time factors; (3) success in the customer environment; and (4) economics, including profitability. Modern **phase management** and **stage-gate processes** provide managers with the tools for organizing and conducting project evaluations in a systematic way. The following section summarizes suggestions that can help managers in effectively evaluating projects for successful implementation.

Seek out relevant information. Meaningful project evaluations require relevant quality information. The four categories of variables identified above can provide a framework for establishing the proper metrics and detailed data gathering.

Ensure competence and relevancy. Ensure that the right people become involved in the data collection and evaluation processes.

Take top-down look first; detail comes later. Detail is less important than information relevancy and evaluator expertise. Do not get hung up on lack of data during the early phases of project evaluation.

Evaluation processes should iterate. It does not make sense to spend a lot of time and resources in gathering perfect data to justify a no-go decision.

Select and match the right people. Whether the project evaluation consists of a simple economic analysis or a complex multifunctional assessment, competent people from functions critical to the overall success of the project should be involved.

Define success criteria. Deciding on a single project or choosing among alternatives, evaluation criteria must be defined. They can be quantitative, such as ROI, or qualitative, such as the chances of winning a contract. In either case, these evaluation criteria should cover the true spectrum of factors affecting success and failure of the project(s). Only functional experts, discussed previously, are qualified to identify these success criteria. Often, people from outside the company, such as vendors, subcontractors, or customers, must be included in this expert group.

Strictly quantitative criteria can be misleading. Be aware of evaluation procedures based on quantitative criteria only (ROI, cost, market share, MARR, etc.). The input data used to calculate these criteria are likely to be based on rough estimates and are often unreliable. Evaluations based on predominately quantitative criteria should at least be augmented with some expert judgment as a "sanity check."

Condense criteria list. Combine evaluation criteria, especially among the judgment categories, to keep the list manageable. As a goal, try to stay within 12 criteria for each category.

Gain broad perspective. The inputs to the project selection process should include the broadest possible spectrum of data from the business environment that affect success, failure, and limitations of the new project opportunity. Assumptions should be carefully examined.

Communicate. Facilitate communications among evaluators and functional support groups. Define the process for organizing the team and conducting the evaluation and selection process.

Ensure cross-functional representation and cooperation. People on the evaluation team must share a strategic vision across organizational lines. They also must sense the desire of their host organizations to support the project if selected for implantation. The purpose, goals, objectives, and relationships of the project to the business mission should be clear to all parties involved in the evaluation/selection process.

Do not lose the big picture. As discussions go into detail during the evaluation, the team should maintain a broad perspective. Two global judgment factors can help to focus on the big picture of project success: (1) overall benefit-to-cost perception, and (2) overall risk of failure perception. These factors can be recorded on a 10-point scale, −5 to +5. This also leads to an effective two-dimensional graphic display of competing project proposals.

Do your homework between iterations. As project evaluations are most likely conducted progressively, action items for more information, clarification, and further analysis surface. These action items should be properly assigned and followed up, thereby enhancing the quality of the evaluation with each consecutive iteration.

Take a project-oriented approach. Plan, organize, and manage your project evaluation/selection process as a *project.*

Resource availability and timing. Do not forget to include in your selection criteria the availability and timing of resources. Many otherwise successful projects fail because they cannot be completed within a required time period.

Use red team reviews. Set up a special review team of senior personnel. This is especially useful for large and complex projects with major impact on overall business performance. This review team examines the decision parameters, qualitative measures, and assumption used in the evaluation process. Limitations, biases, and misinterpretations that may otherwise remain hidden can often be identified and dealt with.

Stimulate creativity and candor. Senior management should foster an innovative ambience for the evaluation team. Evaluating complex project situations for potential success or failure involves intricate sets of variables that are linked among the organization, technology, and business environment. It also involves dealing with risks and uncertainty. Innovative approaches are required to evaluate the true potential of success for these projects. Risk sharing by senior management, recognition, visibility, and a favorable image in terms of high priority, interesting work, and importance of the project to the organization, have been found to be strong drivers toward attracting and holding quality people on the evaluation team, and to gain their active and innovative participation in the process.

Manage and lead. The evaluation team should be chaired by someone who has trust, respect, and leadership credibility with team members. Senior management can positively influence the work environment and the process by providing guidelines, charters, visibility, resources, and active support to the project evaluation team.

In summary, effective project evaluation and selection require a broad-scanning process across all segments of the enterprise and its environment to deal with the risks, uncertainties, ambiguities, and imperfections of data available for assessing the value of a new project venture relative to other opportunities. No single set of broad guidelines exist that guarantees the selection of successful projects. However, the process is not random! A better understanding of the organizational dynamics that affects project performance and the factors that drive cost, revenue, and other benefits, can help in gaining a better, more meaningful insight into the future value of a prospective new project. Seeking out both quantitative and qualitative measures incorporated into a combined rational/judgmental evaluation process often yields the most reliable predictor of future project value and desirability. Equally important, the process requires managerial leadership and skills in planning, organizing, and communicating. Above all, leaders of the project evaluation team must be social architects who can unify the multifunctional process and its people. They must share risks and foster an environment that is professionally stimulating and strongly linked with the support organizations eventually needed for project implementation. This is an environment that is conducive to cross-functional communication, cooperation, and integration of the intricate variables needed for effective project evaluation and selection.

Defining Terms

Annuity (A) — Present worth of a revenue or cost at the end of period *n*.

Cost-benefit (CB) — Net present value of all returns on investment in dollars.

Cross-functional — Actions that span organizational boundaries.

Internal rate of return (IRR) — Average return on investment realized by a firm on its investment capital.

Minimum attractive rate of return (MARR) — The minimum rate of return on new investments acceptable to an organization.

Net present value (NPV) — Net present value of a stream of future revenues or costs.

Payback period (PBP) — Time period needed to recover original investment.

Phase management — Projects are broken into natural implementation phases, such as development, production, and marketing, as a basis for project planning, integration, and control. Phase management also provides the framework for *concurrent engineering* and *stage-gate processes*.

Project rating (PR) factor — Measure developed by Carlo Pacifico for predicting project success.

Project success — A comprehensive measure, defined in both quantitative and qualitative terms that include economic, market, and strategic objectives.

Present worth (PW) — Present value of a revenue or cost at the end of period *n* (also called annuity).

Stage-gate process — Framework for executing projects within predefined stages (see also *phase management*) with measurable deliverables (*gate*) at the end of each stage. The gates provide the review metrics for ensuring successful transition and integration of the project into the next stage.

Weighing of criteria — A multiplier associated with specific evaluation criteria.

z — Project rating factor, a measure developed by Sobelman for predicting project success.

References

Blackburn, S. 2002. The ABCs of evaluation, *Int. J. Project Manage.*, 20 (6), 488.

Brenner, M. 1994. Practical R&D project prioritization, *Res. Technol. Manage.*, 37(5), 38–42.

Cooper, R. 2002. Optimizing the stage-gate process: What best-practice companies do, *Res. Technol. Manage.*, 45 (6), 43–50.

Daniel, H. Z. 2003. Project selection: a process analysis, *Ind. Marketing Manage.*, 32(1), 39.

Dickinson, M. W. 2001. Technology portfolio management: optimizing interdependent projects over multiple time periods, *IEEE Trans. Eng. Manage.*, 48(4), 518–530.

Graves S., Rinquest J., and Medaglia, A. 2002. *Models and Methods for Project Selection.* Amsterdam: Kluwer.

Jacob, W. F. 2003. In search of innovative techniques to evaluate pharmaceutical R&D projects, *Technovation*, 23 (4), 291.

Martino, J. 1995. *R&D Project Selection*, John Wiley & Sons, New York.

Meade, L. M. 2002. R&D project selection using the analytic network process, *IEEE Trans. Eng. Manage.*, 49 (1), 59–68.

Menke, M. M. 1994. Improving R&D decisions and execution, *Res. Technol. Manage.*, 37(5), 25–32.

Miller, J. B. 2002. Engineering project appraisal: the evaluation of alternative development schemes, *Construction Manage. Econ.*, 20 (4), 380.

Obradovitch, M. M. and Stephanou, S. E. 1990. *Project Management: Risk and Productivity.* Daniel Spencer Publishers, Bend, OR.

Ojanen, V. 2002. Applying quality award criteria in R&D project assessment, *Int. J. Prod. Econ.*, 80(1), 119.

Phillips, J., Bothell, T., and Snead, L. 2002. *The Project Management Scorecard: Measuring Success of Project management Solutions,* Butterworth-Heinemann, New York.

Rad, P. F. 2002. A model to quantify the success of projects, *AACE Int. Trans.*, 68(1), 51–55.

Remer, D. S., Stokdyk, S. B., and Van Driel, M. 1993. Survey of project evaluation techniques currently used in industry. *Int. J. Prod. Econ.*, 32(1), 103–115.

Sage, A. 1983. *Economic System Analysis: Microeconomics for Systems Engineering, Engineering Management and Project Selection,* Elsevier, Amsterdam.

Say, T. 2003. Is your firm's tech portfolio aligned with its business strategy? *Res. Technol. Manage.*, 46(1), 32–39.

Seymour S. H. Jr. 1992. *The Capital Budgeting Decision: Economic Analysis of Investment Projects,* Pearson, New York.

Shtub, A., Bard, J. F., and Globerson, S. 1994. *Project Management: Engineering, Technology, and Implementation*, Prentice Hall, New York.

Skelton, M. T. and Thamhain, H. J. 1993. Concurrent project management: a tool for technology transfer, *Project Manage. J.*, 26(4), 41–48.

Sounder, W. E. 1984. *Project Selection and Economic Appraisal,* Van Nostrand Reinhold, New York.

Worstell, J. 2002. Identifying, justifying and prioritizing technical projects, *Chem. Eng. Progress,* 98(3), 72–80.

Further Information

The following journals are good sources of further information: *Engineering Management Journal* (ASEM), *Engineering Management Review* (IEEE), *Industrial Management* (IIE), *Journal of Engineering and Technology Management* (JETM), *Journal of Product Innovation Management* (PDMA), *Project Management Journal* (PMI), and *Transactions on Engineering Management* (IEEE).

Further, the following professional societies present annual conferences and specialty publications that include discussions on project evaluation and selection: American Society for Engineering Management (ASEM), Rolla, MO 65401, (314) 341-2101; Institute of Electrical and Electronic Engineers (IEEE); Engineering Management Society (EMS), East 47 St., New York, NY 10017-2394; Product Development and Management Association (PDMA), 17000 Commerce Pkwy, Mount Laurel, NJ 08054, (856) 439-9052; Project Management Institute (PMI), Upper Darby, PA 19082, (610) 734-3330; Project World, 600 Worcester Road, Natick, MA 01760, (508) 628-9652.

211

Critical Path Method

John L. Richards
University of Pittsburgh

The purpose of this chapter is to describe the three-step, iterative decision-making process of planning, scheduling, and controlling with the critical path method (CPM). CPM is a network-based analytical tool that models a project's activities and their predecessor/successor interrelationships. **Planning** is the development of a **work breakdown structure (WBS)** of the project's activities. **Scheduling** is the calculation of **activity parameters** by doing a forward and a backward pass through the network. **Controlling** is the monitoring of the schedule during project execution by **updating** and **upgrading**, as well as the modifying of the schedule to achieve feasibility and optimality using cost duration analysis and critical resource analysis.

211.1 Planning the Project

Project planning requires the development of a work breakdown structure, which then becomes the basis for a network model of the project. This model can then be used to evaluate the project by comparing regular measures of performance.

Project Performance Measures

The three common performance measures in project management are *time*, *cost*, and *quality*. The overall objective is to accomplish the project in the least time, at the least cost, with the highest quality. Individually, these objectives conflict with each other. Thus, the manager must seek an overall solution by trading off among them. Further, since the overall project is defined by the activities that must be done, the overall project duration, cost, and quality will be determined by individual activity times, cost, and quality levels.

0-8493-1586-7/05/$0.00+$1.50

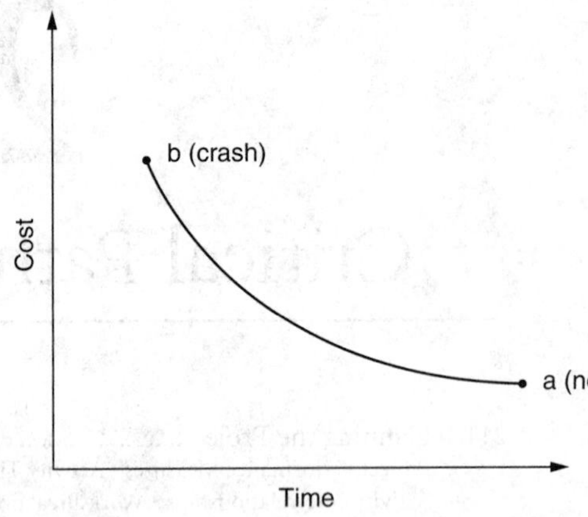

FIGURE 211.1 Activity time-cost tradeoff curve. The normal time (point a) is the least cost/longest activity duration. The crash time (point b) is the least activity duration/highest cost.

Activity Time-Cost Tradeoff

For a specified quality level for a given activity, the manager initially selects that combination of resources (labor, equipment, and material) that accomplishes that particular activity at the least cost. This is the *normal* duration on an *activity time-cost tradeoff curve* (Figure 211.1). Thus, since each activity is to be done at its least cost, the overall project will be done at the least total cost. However, in order to reduce an activity's duration, the activity cost must increase. For example, one can work overtime at premium rates or use more expensive equipment, which increases cost, in order to reduce an activity's duration. *Crash* is the shortest possible activity duration, no matter how high the cost. The inverse relationship between time and cost yields curves with negative slopes.

Activity Interrelationships

There are two possible relationships between a pair of activities in a project network: (1) one must immediately precede the other (*predecessor*), or (2) one must immediately follow the other (*successor*). If there is no predecessor/successor relationship, the activities may be done simultaneously. These predecessor/successor relationships are derived from absolute constraints such as physical/technological, safety, and legal factors; or imposed constraints such as the selection of resources, methods, and financing. The manager should initially incorporate relationships derived only from absolute constraints. Relationships derived from imposed constraints should be added only as necessary to achieve feasibility. This approach to predecessor/successor relationships yields the least constrained project network initially.

The basic predecessor/successor relationship is finish to start with no lead or lag. However, more sophisticated models allow three other types: start to finish, finish to finish, and start to start. In addition, each of the four could have a lead time or a lag time. Thus there are twelve possible ways to describe a particular predecessor/successor relationship.

Work Breakdown Structure

The WBS of a project is the listing of all the individual activities that make up the project, their durations, and their predecessor/successor relationships. It should be the least costly and least constrained method of executing the project, that is, normal durations and absolute constraints only. Therefore, if the schedule resulting from this initial WBS is feasible, then it is also optimal. If, however, the schedule is infeasible because of time and/or resource considerations, then the manager would want to achieve feasibility with the least additional cost. (Scheduling and feasibility/optimality are discussed in later sections.)

TABLE 211.1 WBS of Swimming Pool Construction

Activity ID	Duration	Description	Immediate Predecessors
A101	10	Order and deliver filtration equipment	—
A202	5	Order and deliver liner/piping	—
B301	4	Excavate for pool	—
B202	3	Install liner/piping	A202, B301
B102	2	Install filtration equipment	A101
C301	2	Fill pool	B202
B401	5	Construct deck	B202
C302	2	Connect and test system	C301, B102
B501	3	Landscape area	B401

There are three approaches to developing a WBS: (1) by physical components, (2) by process components, and (3) by spatial components. *Physical components* model a constructed product (e.g., build wall). *Process components* model a construction process (e.g., mix concrete). *Spatial components* model a use of time or space (e.g., order steel or cure concrete). No matter which of the three approaches is used to define a particular activity in a WBS, each should be described by an action verb to distinguish an activity from an event. (Note: There can be special activities in a WBS that involve time only and no cost, such as curing concrete. There can also be dummy activities for logic only that have no time or cost.)

A project's WBS must be developed to an appropriate level of detail. This means that activities must be broken down sufficiently to model interrelationships among them. Also, a standard time period (hour, shift, day, week, etc.) must be chosen for all activities. An appropriate WBS will have a reasonable number of activities and reasonable activity durations.

Example 211.1: Work Breakdown Structure

Table 211.1 shows a WBS for constructing a backyard in-ground swimming pool with a vinyl liner. Note that a time period of days was selected for all activities.

211.2 Scheduling the Project

The critical path method of project scheduling is based upon a network model that can be analyzed by performing a forward and a backward pass to calculate activity parameters.

CPM Network Models

There are two types of network models: activity oriented and event oriented. Both types have nodes connected by arrows that model events (points in time) and activities (processes over time).

Activity-Oriented Diagram

The activity-oriented diagram is also called an *arrow diagram* (ADM) or *activity on arrow* (AOA). The activities are the arrows, and the events are the nodes. Dummy activities (depicted as dashed arrows) may be required to correctly model the project for logic only. Activity identification is by node pairs (node *i* to node *j*). This was the original diagramming method, which is easy to visualize but difficult to draw.

Event-Oriented Diagram

The event-oriented diagram is also called a *precedence diagram* (PDM), *activity on node* (AON), or *circle network*. The activities are the nodes, and the events are the ends of arrows. There are no dummies. All arrows are for logic only, which also allows for modeling the twelve types of activity interrelationships

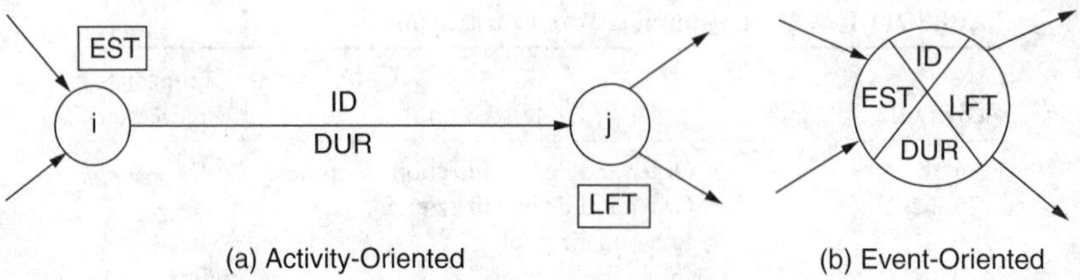

FIGURE 211.2 Activity graphical models. The activity identification (ID) and duration (DUR) are from the WBS. The EST and LFT are calculated from the forward and backward passes, respectively.

discussed earlier. This diagramming method is easier to draw and well suited to computer use. Although developed after AOA, the AON has become the preferred method.

CPM Network Calculations

There are three steps in the manual analysis of a CPM network that determine the activity parameters. The forward pass determines the *earliest start time* (EST) of each activity. The backward pass determines the *latest finish time* (LFT) of each activity. The other times, the *earliest finish time* (EFT) and the *latest start time* (LST), and the floats, *total float* (TF), and *free float* (FF), are then determined from a table. The calculation process is the same for either type of network (ADM or PDM). Before beginning the process, one needs to establish a time convention — beginning of time period or end of time period. The example in this chapter uses beginning of time period, thus the first day is day one. (The end-of-time-period convention would begin with day zero.)

Forward Pass

To determine the EST of an activity, one compares all the incoming arrows — that is, the *heads* of arrows — choosing the *largest* earliest event time. The comparison is made among all the immediately preceding activities by adding their ESTs and respective durations. The process begins at the start node (the EST being zero or one) and proceeds to the end node, taking care that all incoming arrows get evaluated. At the completion of the forward pass, one has determined the overall project duration. The ESTs can be placed on diagrams as shown in Figure 211.2.

Backward Pass

The backward pass begins at the end node with the overall project duration (which is the LFT) and proceeds to the start node. This process determines the LFT for each activity by comparing the outgoing arrows — that is, the *tails* of arrows — choosing the *smallest* latest event time. The comparison is made among all the immediately succeeding activities by subtracting their durations from their respective LFTs. At the completion of the backward pass one should calculate the original project start time. The LFTs can be placed on diagrams as shown in Figure 211.2.

Floats

The other two times (EFT and LST) and floats (TF and FF) are determined in a tabular format using the following relationships: (1) EFT = EST + duration, (2) LST = LFT – duration, (3) TF = LFT – EFT or TF = LST – EST, and (4) FF = EST (of following activities) – EFT. Activities with a TF = 0 are on a **critical path**. There may be more than one critical path in a network. If the duration of any critical path activity is increased, the overall project duration will increase. Activities with TF > 0 may be increased without affecting the overall project duration. On the other hand, free float is that amount of total float that can be used by the activity without affecting any other activities. If TF equals 0, then FF equals 0, necessarily. Free float may comprise some, all, or no portion of total float.

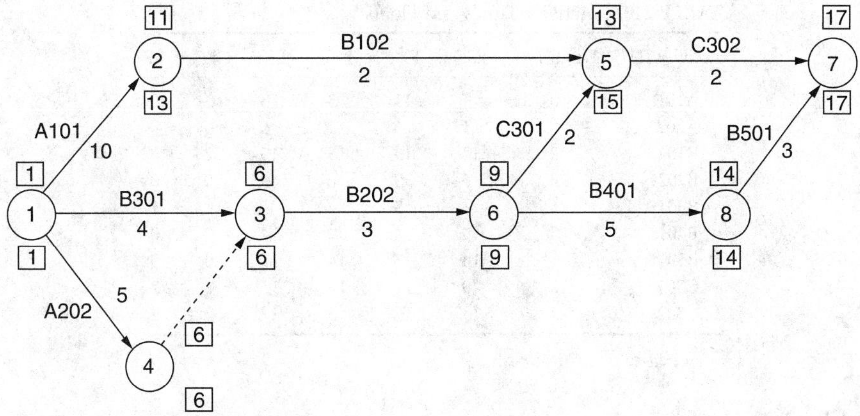

(a) Activity-Oriented Network (ADM or AOA)
Note that activity 4-3 is a dummy and is also on the critical path.

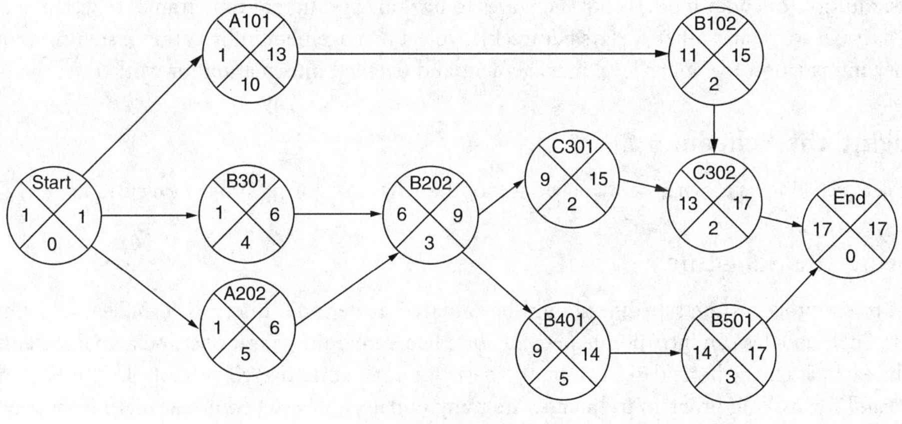

(b) Event-Oriented Network (PDM or AON)
Note that implied start and end nodes are shown with zero durations.

FIGURE 211.3 CPM network calculations (forward and backward passes). The overall project duration was determined to be 16 days (17 – 1 = 16).

Example 211.2: CPM Scheduling

This example continues with the project introduced in Table 211.1. The CPM calculations for the forward and backward passes are shown in Figure 211.3 for both network types. Table 211.2 lists all times and floats.

211.3 Controlling the Project

CPM-based project management provides the tools to control time and money in a dynamic and hierarchical project environment during project execution.

Managing Time and Money

Computerized project management systems can provide a variety of informational outputs for use in managing a project. These include network diagrams showing activity interrelationships, bar charts showing activity durations, tabular listings showing activity parameters, and profiles showing cash flows or resource utilization during the project.

TABLE 211.2 Activity Times and Floats

Activity ID	Duration	EST	EFT	LST	LFT	TF	FF
A101	10	1	11	3	13	2	0
A202[a]	5	1	6	1	6	0	0
B102	2	11	13	13	15	2	0
B202[a]	3	6	9	6	9	0	0
B301	4	1	5	2	6	1	1
B401[a]	5	9	14	9	14	0	0
B501[a]	3	14	17	14	17	0	0
C301	2	9	11	13	15	4	2
C302	2	13	15	15	17	2	2

[a] Activities on critical path.

Hierarchical Management

Project management generally occurs in a multiproject environment with multiple time parameters and multiple managerial levels. Multiple project models integrate cash flow and resource profiles of several projects. Multiple calendar models allow activities to be done at different time frames (e.g., some may be done 5 days per week, and others 7 days per week). Project management information systems can provide summary information for upper-level management and detailed information for workers in the field.

Managing the Schedule

The project schedule is a dynamic managerial tool that changes during project execution.

Updating the Schedule

As the project proceeds the schedule should be **updated** at periodic intervals to reflect actual activity progress. Such updates can incorporate percent complete, remaining durations, and actual start and end dates for each activity. The updates can be the basis for evaluating overall project objectives (time and money) and for making progress payments. After any update, a new schedule calculation must be done to determine new times and floats.

Upgrading the Schedule

Any change to an existing schedule either by changing a planned duration or an activity relationship is a schedule **upgrade.** A schedule upgrade can occur either prior to start of the project, or any time during the project, based on new information. After any upgrade, a new schedule calculation must be done to determine new times and floats.

Managing the Floats

A negative total float indicates that the project will overrun the stated completion date. A positive (>0) total float indicates that the project will be completed earlier than the stated completion date. A particular network will generally have many different total float paths, including negative ones if it is behind schedule, and no zero ones if it is ahead of schedule.

Free float indicates that amount of time an activity can be manipulated without affecting any other activity (and therefore the project as a whole). When managing the floats, one would want to use free float before total float. Once total float is used, the activity becomes part of a new critical path.

The use of the floats can be a difficult contractual issue among the parties. It is a good idea to decide beforehand how the floats can be used. Otherwise the issue may be part of a delay claim at completion of the project.

TABLE 211.3 Activity Times and Floats (Update)

Activity ID	Duration	EST	EFT	LST	LFT	TF	FF
A101[a]	9	6	15	6	15	0	0
A202[a]	2	6	8	6	8	0	0
B102[a]	2	15	17	15	17	0	0
B202[a]	3	8	11	8	11	0	0
B301	0	—	—	—	—	—	—
B401[a]	5	11	16	11	16	0	0
B501[a]	3	16	19	16	19	0	0
C301	2	11	13	15	17	4	4
C302[a]	2	17	19	17	19	0	0

[a] Activities on critical path.

TABLE 211.4 Activity Times and Floats (Upgrade)

Activity ID	Duration	EST	EFT	LST	LFT	TF	FF
A101[a]	9	6	15	6	15	0	0
A202	2	6	8	7	9	1	0
B102[a]	2	15	17	15	17	0	0
B202	3	8	11	9	12	1	0
B301	0	—	—	—	—	—	—
B401	5	11	16	12	17	1	0
B501	3	16	19	17	20	1	1
C301	2	11	13	15	17	4	4
C302[a]	3	17	20	17	20	0	0

[a] Activities on critical path.

Example 211.3: Updating and Upgrading

This example continues with the project introduced in Table 211.1 and originally scheduled in Table 211.2.

1. *Updating*. Assume that it is the beginning of day 6 and that activity B301 is done, and A101 has 9 days left and A202 has 2 days left. These updated durations are used in Figure 211.3 to recalculate activity times and floats as shown in Table 211.3. Note that a second critical path has developed and that the overall project duration has been extended 2 days.
2. *Upgrading*. Now assume that immediately after updating, the duration for activity C302 gets changed to 3 days and B401 must precede it. These upgrades could be incorporated into a revised Figure 211.3 (not shown) to recalculate another set of activity times and floats as shown in Table 211.4. Note that there is now only one critical path and the overall project duration has been extended another day to day 20.

211.4 Modifying the Project Schedule

Project planning, scheduling, and controlling are iterative decision-making processes. It is highly unlikely for an initial schedule to be both feasible and optimal in the first iteration. Likewise, it is highly unlikely that the actual project execution will match the original project plan exactly. Therefore, one must know how to modify the project schedule in order to achieve feasibility and optimality. The modification process involves changing activity duration, changing activity relationships, or both.

Cost Duration Analysis

Cost duration analysis (CDA) utilizes activity time/cost tradeoff curves (discussed earlier) in order to compress the overall project schedule. The objective is to buy back each time unit in the cheapest possible manner until the desired completion date is reached (feasibility). Only activities on a critical path need be reduced. The others with positive float simply have the float reduced. The problem can become very complex in a large network with multiple critical paths where the incremental additional costs for the activities are different.

Critical Resource Analysis

The approach to **critical resource analysis (CRA)** is different from CDA in that it seeks to extend the overall project duration the least amount in order to resolve resource conflicts (i.e., achieve feasibility). CRA can be viewed from one of two perspectives: (1) constrained resources — staying below a specified limit, or (2) resource leveling — selecting a constant limit. The solution approach for either is the same. The problem is one of ordering (predecessor/successor relationships) those activities that have resource conflicts during the same time period. The pairwise comparison of all such activities in a large network with many critical resources presents a huge combinatorial problem. The only viable solution approaches are based upon heuristic decision rules. (A simple rule could be that the predecessor activity should be the one with the smaller LST.)

Combined CDA and CRA

Combining CDA and CRA to achieve a feasible and optimal schedule is virtually impossible for all but the simplest networks. Although the CDA problem does have rigorous mathematical solutions, they are not incorporated in most commercial software. On the other hand, the software generally does incorporate heuristic-based solutions for the CRA problem. Therefore, one should use the software in an interactive decision-making manner.

Example 211.4: CDA and CRA

Assume that after the upgrade as shown in Table 211.4, it is decided that the desired completion date is day 19, and also that activities B102 and B401 cannot be done simultaneously because of insufficient labor. Further, assume that B401 can be reduced from 5 days to 3 days at an additional cost of $200 per day and that C302 can be reduced from 3 days to 2 days at an additional cost of $400.

Solution. The solution approach to this problem is to work two cases: (1) B102 precedes B401, and compress B401 and/or C302 if they lie on a critical path; and (2) B401 precedes B102, and again compress B401 and/or C302. Case 1 would yield an overall project duration of 25 days, and one can readily see that it is impossible to reduce it by 6 days (one can only gain a total of 3 days from activities B401 and C302). Case 2 (B401 precedes B102) yields an overall project duration of 21 days, with both B401 and C302 on the same critical path. One should choose the least expensive method to gain 1 day — that is, change B401 to 4 days for $200. This yields a project duration of 20 days and an additional critical path. Therefore, reducing B401 another day to 3 days does not get the project to 19 days. Instead, the more expensive activity (C302) must be reduced from 3 to 2 days for $400. The answer to the problem, then, is B401 goes from 5 to 4 days for $200, and C302 goes from 3 to 2 days for $400. Thus, the overall project duration is compressed from 21 to 19 days from a total additional cost of $600.

211.5 Project Management Using CPM

CPM was first developed in the late 1950s by the Remington Rand Corporation and the DuPont Chemical Company. Since then, many software manufacturers have developed sophisticated computer-based management information systems using CPM. In addition to performing the CPM calculations discussed in this chapter, such systems can provide data for creating the historical file of an ongoing project, for developing estimating information for a future project, and for performance evaluation of both the project and the participating managers. CPM has even become a well-accepted means for analyzing and resolving construction disputes.

The successful use of CPM as a managerial tool involves not only the analytical aspects discussed in this chapter, but also the attitude that is displayed by those using it in actual practice. If CPM is used improperly as a weapon, rather than as a tool, there will be project management failure. Therefore, successful project management must include positive team building among all project participants, along with the proper application of the critical path method.

Defining Terms

Activity parameters — Activity times (EST, EFT, LFT, and LST) and activity floats (TF and FF) calculated in the scheduling step.

Controlling — Third step in the interactive decision-making process that monitors the accomplishments of the project by updating and upgrading, and seeks feasibility and optimality by cost duration analysis and critical resource analysis.

Cost duration analysis (CDA) — Reducing durations of selected activities in the least costly manner in order to achieve a predetermined project completion date.

Critical path (CP) — String of activities from start to finish that has zero total floats. There may be more than one CP, and the CPs may change after an update or upgrade.

Critical resource analysis (CRA) — Sequencing selected activities in such a manner so as to minimize the increase in project duration in order to resolve resource conflicts among competing activities.

Planning — The first step in the interactive decision-making process that determines the work breakdown structure.

Scheduling — The second step in the interactive decision-making process that determines the activity parameters by a forward and a backward pass through a network.

Update — Changing remaining activity durations due to progress only, and then rescheduling.

Upgrade — Changing activity durations and interrelationships due to new information only, and then rescheduling.

Work breakdown structure (WBS) — Listing of individual activities that make up the project, and their duration and predecessor/successor relationships.

References

Antill, J. M. and Woodhead, R. W. 1990. *Critical Path Methods in Construction Practise,* 4th ed. John Wiley & Sons, New York.

Hendrickson, C. and Au, T. 1989. *Project Management for Construction.* Prentice Hall, Englewood Cliffs, NJ.

Moder, J. J., Philips, C. R., and Davis, E. W. 1983. *Project Management with CPM, PERT and Precedence Diagramming,* 3rd ed. Van Nostrand Reinhold, New York.

Further Information

Journal of Management in Engineering and *Journal of Construction Engineering and Management.* Published by the American Society of Civil Engineers.

Project Management Journal. Published by the Project Management Institute.

The Construction Specifier. Published by the Construction Specifications Institute.

Journal of Industrial Engineering. Published by the American Institute of Industrial Engineers.

212

Intellectual Property: Patents, Trade Secrets, Copyrights, Trademarks, and Licenses

David Rabinowitz
Moses & Singer LLP
Steven M. Hoffberg
Moses & Singer LLP

An engineer's work can be legally protected in a number of different ways. The legal rights to the work (and sometimes the work itself) are referred to as intellectual property.

Because an engineer's work is normally a useful product, patents — which protect useful inventions — are the typical means of obtaining the exclusive legal right to use or distribute the engineer's work. For those inventions or technologies that are unpatentable, difficult to patent, or may be maintained in secrecy despite commercial exploitation, a good practical substitute can be trade secret protection, which requires only that the work product be useful and that it be kept confidential. When the engineer's work product must be shared with another in order to make commercial use of it, license agreements covering such distribution can be used to preserve control over the engineer's work.

Copyrights protect what the Copyright Act calls "works of authorship," which include nonfunctional work products and the nonfunctional aspects of functional works, namely, the parts or aspects that are not necessary to the work's functionality. In addition, copyrights can protect some rights to even such functional things as computer programs, diagrams, textbooks, and architectural drawings, although the courts try not to protect (with copyrights) the underlying ideas and functionality that such things express or contain.

Trademarks protect not work product itself, but the engineer's (or his or her employer's) right to be known as the source of the work product. Trademark law forbids others to falsely claim to be the source, origin or sponsor of the engineer's work, or to falsely claim that the engineer is the source, origin, or sponsor of the work of another.

Thus, a computer program used in manufacturing a new drug can be protected by a patent or trade secret, a copyright, a trademark, and one or more licenses. A patent can be obtained for the logic of the program, the computer readable medium containing the program, or the process of manufacturing and the use of the computer program in that process. A copyright can be obtained in the program's code, which prevents the copying of the program. If the owner of the program does not wish to publicly reveal how it works, the program can be protected by keeping it confidential as a trade secret. If the program

0-8493-1586-7/05/$0.00+$1.50

is to be distributed to others, the code can be protected by a legally enforceable access-control device. A trademark can be used to identify the origin of the computer program. Finally, through license agreements, the program's owner can control further use or distribution of the product.

212.1 Patents

A patent gives the inventor or assignee the right to exclude others from making, using, selling, or offering for sale an invention, for a limited period, typically 20 years from the filing date of an application in a particular country or jurisdiction (e.g., the European Community). In order to be patentable in the U.S., an invention must be filed on behalf of the first and true inventors, be useful, novel, and nonobvious, and the application must include a written description of the invention, an enabling disclosure, a description of the best mode for practicing the invention, and at least one claim that defines the scope of legal protection. Outside the U.S., the owner of the invention is the applicant, and the first to file gains priority. Further, in various countries, the standards for patentability differ, as do the requirements for the application, although a single comprehensive application may be prepared and translated as necessary. A patent is like a contract between the inventor and the government that, in exchange for a disclosure of the invention, the patentee receives a limited period of exclusive use. A patent is not considered an absolute monopoly nor an exception to antitrust law, and therefore the enforcement of patent rights may be limited by other considerations.

The main type of patent is the regular utility patent, which protects a functional invention, which may be a product, composition, or process. Other types of patents include design patents, which protect nonfunctional ornamental features of industrial designs, and plant patents, which protect asexually reproduced plants. Sexually reproduced plants are protected by a parallel system of Plant Variety Protection Act certificates.

In the U.S., a regular utility patent application must be filed within 1 year of the first publication, offer for sale, or use; outside the U.S., an absolute novelty bar is often imposed, without any grace period. Therefore, an applicant seeking international protection should file a priority application, that is, one meeting the requirements of most or all countries in which protection might be sought, before any publication, offer for sale, or use. If protection is desired only within the U.S., the application may be delayed for up to 1 year. This priority application can be a provisional patent application, a regular utility application, or a Patent Cooperation Treaty (International) patent application designating the countries of interest. According to international treaties covering almost all countries, the priority claim from a first-filed regular utility patent application in any one country must be followed up by an International patent application or respective national patent applications in the countries of interest within one year from the priority date. An International application can delay the filing of national patent applications by at least 30 months from the priority date, allowing consolidated initial examination by a single examiner, in a single language prior to prosecution in each country or regional office. Patent applications (including International applications) are typically published 18 months after the priority date, although in the U.S. there are some exceptions. Once a patent application is published, damages based on the published claims may begin to accrue. The granted patent issues after examination, allowance, and payment of an issue or registration fee, and only then can it be enforced. Maintenance fees or annuities may be required.

A provisional patent application in the U.S. is similar to a regular utility application, except that it remains unexamined, and need not have claims. All other requirements are the same, and therefore the description should meet the same standards of completeness. The regular utility patent application or International patent application claiming priority from the provisional application must be filed within 1 year. The utility application may be completely revised and updated without prejudice in the U.S., so long as the provisional application was filed before publication, use or offer for sale. The use of a provisional application potentially extends the life of a patent by 1 year, since the expiration is calculated from the filing of the utility application.

Computer software, either as a general-purpose machine that is programmed to perform specific functions, or as a method performed in accordance with the software, is generally considered patentable in most countries. Algorithms may therefore be protected in context, although laws of science in the abstract are unpatentable. In the U.S., computer-readable media containing code is also patentable, thus allowing computer software *per se* to be protected.

In the U.S., the so-called business method patent has become popular. This type of patent typically has claims that de-emphasize the underlying technology in favor of the application of the technology to the internal business processes. Business method patent applications often encompass an entire business model, including both technologies and their context within the organization, and therefore can be voluminous. Such patents may meet significant resistance outside of the U.S.

The patent examination process reviews both the formalities of the application, as well as the patentability of the claims in view of the prior art. In the U.S., there is a requirement to disclose material prior art known to applicants. During examination, the patent examiner seeks to identify the closest prior art that expressly or inherently meets each and every element of a claim, rendering the claim anticipated, or one or more references that, alone or together, would have rendered the invention obvious to one of ordinary skill in the art. In International applications, such an obviousness rejection is called "lack of inventive step." In response, the applicant can identify errors or inconsistencies in the rejection, or provide additional evidence in support of patentability. Typically, no new matter may be added to the description of the invention, although in some cases it may be possible to overcome a rejection by filing a new application with an amended text. In the U.S., the applicant is typically given two opportunities to overcome the rejection, after which a Request for Continued Examination with fee must be filed, or the rejection appealed. In many cases, the examination in a single application or continuation application is limited to a single invention. A single original patent application may thus form the basis for a number of patents, through filing of division or continuation applications.

A patent does not provide a right to use an invention. Others may have concurrent patent rights that would be infringed by commercializing a patented technology. The publication of the patent or patent application may allow others with earlier priority to draft claims encompassing the disclosed embodiments. When two U.S. applicants claim the same invention, an interference proceeding may be instituted, with the patent awarded to the earlier inventor who has not suppressed or concealed the invention, and has pursued the invention with diligence between conception and reduction to practice or patent application filing. After publication of a claim in a patent or patent application, another seeking to interfere with it has one year to copy the claim in another patent application.

A prior art reference in the U.S., claiming a different invention, may be overcome by documented evidence of conception by the inventor prior to the effective date of the reference. For this reason, and for purposes of interference, inventors are encouraged to document their inventions in notebooks, pages of which are contemporaneously witnessed and dated by two noninventors who understand the contents, to provide evidence of the date of conception and progress toward reduction to practice. Electronic lab notebooks and other evidence might also be acceptable.

U.S. patents are enforced by lawsuits in federal District Courts. Infringement may be found either by literal infringement, that is, an accused method, composition or apparatus directly corresponds to each element of a patent claim, or under the Doctrine of Equivalents, where an equivalent to a claim element will suffice in order to prove infringement. Remedies are also available against those who engage in "contributory infringement" (selling goods that have no substantial noninfringing use in commerce) and those who induce others to infringe.

A prevailing patentee is entitled to proven damages, with an award no less than a reasonable royalty. Patent infringement is not considered criminal, and indeed no intent to infringe need be shown in order for liability to attach. However, in the U.S., intentional infringement is one factor in awarding enhanced damages, which can be up to three times actual damages, and attorneys' fees. A prevailing patentee is entitled to an injunction; that is, an order of court against further infringement of the patent. The injunction is a powerful weapon that may lead to a settlement based on the value of the infringing business, rather than merely the damages to the patentee, which can be less.

An engineer will often be concerned with three types of opinions from patent counsel: patentability or validity, infringement, and clearance. A patentability opinion provides an estimate of the issues that will be encountered in obtaining patent protection, most often focusing on the prior art identified in a search. A patentability opinion is typically requested prior to preparation of a patent application. A validity opinion investigates an issued patent, which has a presumption of validity attached to it. An infringement opinion provides an analysis of patent claims against a product or method, and typically arises in cases where a patent or published application is identified as a concern. Often, this analysis presumes validity, although this may also be investigated. The infringement analysis generally requires review of the proceedings before the Patent Office, called the "file wrapper," since statements made by applicant during prosecution may limit the scope of protection or define claim terms. A clearance opinion seeks to determine the degree of risk that a company will incur by launching a new product or service. Therefore, a search of unexpired patents is made to determine potentially relevant claims, which are then analyzed for scope and validity.

212.2 Trade Secrets

A broad group of things can be legally protected as trade secrets. Anything commercially useful — not just inventions or processes — can be protected as long as it gives its owner a commercial advantage over competitors who do not know the secret.

Trade secret protection can be obtained for things that are not patentable or copyrightable. They need not be novel or original. The sole requirements for trade secret protection are that the information be useful and secret.

Trade secret protection is most useful when, whether or not a patent or copyright can be obtained, the commercial value of the work or information can best be preserved by keeping it confidential. For example, the structure or design of a computer program may be valuable but insufficiently novel to be patentable and too functional to be copyrightable. By distributing the program solely in object code and maintaining the secrecy of the source code and documentation, however, the program can be protected from the use by another who learns of the program structure or design in violation of an individual's agreement or duty of confidentiality.

The legal essence of a trade secret is the obligations of people privy to the secret to keep it secret. Those who learn a trade secret without violating an obligation to keep it secret and without someone else violating such an obligation are free to use it and publicly disclose it. Consequently, it is advisable for a company intending to maintain its trade secrets to promulgate secrecy policies and to obtain written agreements from its personnel and from third parties who become privy to the secret, such as distributors or licensees, to adhere to those policies. Even inside the company, it is useful, in later persuading a court that the trade secret is indeed secret, to distribute information related to trade secrets on a restricted basis and to institute security procedures to keep trade secrets restricted to company premises and personnel. While employees who leave for another job cannot be prevented from remembering what they know of their former employer's trade secrets, they can sometimes be prevented from taking another job where their knowledge will inevitably be used to confer the secret's advantages upon the new employer.

Unlike patents, however, trade secret protection does not protect against reverse engineering. If the secret is contained in a product distributed without an agreement from the recipient not to reverse engineer the product (or to further distribute it to some one not bound by such an agreement), the recipient or its transferee can use the secret if they can figure it out by reverse engineering.

212.3 Licenses

One of the principal ways to exploit intellectual property is to let others use it. An agreement allowing another to use intellectual property, while retaining ownership of it, is referred to as a license.

Licenses are flexible tools. They can, for example, permit (or forbid) commercial use of the intellectual property, permit (or forbid) further disclosure, use, or distribution of the intellectual property, limit or regulate the rights of the licensee in the event of malfunction, and cover any other subject in any other way that the law will permit and enforce.

A license can be granted by the owner of a patent, trade secret, copyright, or trademark. Where the licensee is to make commercial use of the licensed thing, compensation to the licensor may (but need not) take the form of a royalty. The intellectual property rights in a product need not be licensed all together. For, example, a patent owner has the exclusive right to make, use, and sell its invention. The three rights can be licensed separately, in combination with one another or piecemeal. They can be divided in time and territory. The licensee may get the exclusive right to the invention or the nonexclusive right. (An exclusive license prohibits the patent owner from licensing the same right to another, whereas a nonexclusive license does not.)

Licenses binding on consumers have become commonplace in recent years. So-called shrink-wrap licenses and, for products distributed on the Internet, click-wrap licenses, commonly restrict what users can do with the products they buy and what their recourse is if any kind of problem occurs. While the validity of these licenses and some of their terms have been and continue to be challenged, most courts enforce them.

212.4 Copyrights and Anticircumvention Rights

Copyrights protect original works of authorship, a phrase that has come to mean almost any original expression of an idea that is put in a tangible form. Works that can be copyrighted include literary works, musical works, dramatic works, pantomimes and choreographic works, pictorial, graphic, and sculptural works, motion pictures and other audiovisual works, sound recordings, and architectural works. As noted above, however, copyrights only protect nonfunctional features of a work.

Associated with the Copyright Act are recently created protections for certain engineering products. Under the Digital Millennium Copyright Act of 1998, the creator of a software device that prevents unauthorized access to a copyrightable work can legally prevent circumvention of that device. Copyright-like protection has also been extended to mask works for semiconductor chip products and designs of useful articles, such as boat hulls.

The U.S. is a signatory to the Berne Convention, and as a result copyrighted works of U.S. citizens enjoy the same copyright protection as other signatory countries accord their nationals.

A copyright prohibits others from, most importantly, copying and distributing a work. It also prohibits others from adapting the work for other works, from displaying and performing it, where applicable, and from importing copies of it from outside the U.S. Because a copyright also prevents others from authorizing the exercise of rights under copyright, the copyright owner has the exclusive right to license rights to the work.

The author (or authors) of the work owns the copyright when it is created. They are free to transfer all of their rights under copyright from the date of creation, except for some rights-recapture rules that generally become effective many years later. However, if the work is produced during the course of regular employment of the author and within the scope of that employment, the employer is treated as the author and original owner of the copyright. Such works are referred to as works made for hire. Certain categories of works not created by regular employees, including, for example, contributions to collective works, can also be works made for hire, if so agreed in writing.

Unlike patents, no official grant of a copyright is necessary to own a valid copyright. Copyright arises automatically when an original work of authorship is fixed in a tangible medium of expression. It is a procedural prerequisite to suing for copyright infringement that an attempt be made to register the owner's claim to copyright with the United States Copyright Office (if the work was created by U.S. nationals or was first published in the U.S.), but the copyright subsists without registration and can be sold or licensed.

The former rule that a work had to be published with a copyright notice to protect copyright was repealed in 1989, and notice of copyright is no longer required. Nevertheless, the notice has a practical use in warning that copyright is claimed in the work and thereby in deterring infringement, and in preserving the owner's rights to certain categories of damages.

Unless outweighed by other considerations, it is advisable to promptly register any copyrightable work that may be worth protecting against infringement. The principal advantages of prompt registration are (1) an infringer will be liable for a special kind of damages called "statutory damages," which are available whether or not the copyright owner can prove any actual damages; (2) an infringer may, in the court's discretion, be required to pay the copyright owner's attorney fees incurred in pursuing the infringement case; and (3) the copyright owner will be presumed to have a valid copyright, casting the burden on the infringer to disprove it. Considerations that might outweigh registration are the (relatively minimal) cost and, more importantly, the deposit requirement.

To register the copyright in most works, two copies of the best edition of the copyrighted work must be deposited with the Copyright Office. This requirement raises an obvious problem when the work is one that contains trade secrets. A prominent example is a computer program. The Copyright Office has therefore agreed to accept only portions of computer programs as deposit copies, and there are other categories of works where the deposit requirement has been modified for practical reasons.

The term of a copyright depends on its date of publication or creation or the date of the author's death, depending upon the dates of creation and publication and whether the work was made for hire. For new works the most important duration rule is the life of the author plus 70 years, except for works made for hire, in which copyright lasts 95 years from date of publication (or 120 years from creation, if shorter). For works published before 1978, the term of protection is 95 years from first publication, if renewal was made. As a general but not invariable rule, no work predating 1923 is still copyrighted in the U.S.

For an infringement of copyright, the copyright owner is entitled to an injunction, to its damages from the infringement and, to the extent that they do not overlap the owner's damages, the profits made by the infringer from the infringement. If the copyright was registered before the infringement began, the owner has the additional remedies noted above. In extreme cases, all infringing copies and the materials used to make them may be seized and impounded pending final determination of the case. However, the defendant in a frivolous infringement case can be awarded its counsel fees.

212.5 Trademarks

A trademark or service mark (both referred to here as marks) is a word, phrase, picture, or other thing that is used to indicate the source, origin, or sponsor of goods or services. It need not be original or novel, but may not be chosen to suggest a connection with another person or company and may not be the generic name for the goods or services involved. That is, "You've Got Mail," for example, cannot be a trademark for an Internet email delivery service. Ownership of a mark gives the owner the exclusive right to use the mark to indicate the source, origin, or sponsorship of the goods or services with which it is associated.

As with copyrights, marks need not be registered to be valid. Although marks can be registered federally and with state governments, trademark rights are actually obtained by use of the mark in connection with the owner's goods or services. Rights in a mark can be reserved by registering with the USPTO the claimant's intent to use the mark, but the intent to use never ripens into a mark unless the claimant actually uses the mark in interstate commerce.

Unlike copyrights, trademarks do not need be registered or subject to attempted registration as a procedural prerequisite for an infringement lawsuit. Federal registration, however, confers the following important procedural rights and practical benefits:

- Registration prevents someone else who starts using the same mark for the same goods or services after the registration from acquiring any rights in it in the U.S. Otherwise, if the users are geographically separated, it is possible for a so-called junior user to acquire rights in its own territory.
- The registration can be found by another prospective user or registrant, in many cases deterring them without legal action from starting to use or trying to register the same mark. If a prospective

registrant does try to register the same mark, the earlier registration should be found by the trademark examiner and registration of the second mark may then be refused.
- As with a copyright registration, a federal trademark registration is presumed valid, casting on an infringer the burden of disproving registrant's trademark rights.

A trademark registration has a term of 10 years from the date of registration, which can be renewed for 10-year periods indefinitely.

Trademark infringement is established if someone else uses the mark or something similar and that use creates a likelihood of confusion as to the source of the second user's goods or services. That is, there will be infringement if either the second user's goods or services are thought to come from the trademark owner or if the trademark owner's goods or services are thought to come from the second user. The principal factors in determining whether likely confusion, and therefore trademark infringement, has occurred are the similarity of marks and the similarity of goods, although many other factors, such as the sophistication of the purchasers and the good faith of the second user, can be considered. Actual confusion resulting from the second user's use, such as customer complaints to the trademark owner from consumers who bought the second user's goods, is, of course, strong evidence of likely confusion, but is not necessary to a finding of trademark infringement.

Federal remedies for trademark infringement include injunction, the mark owner's damages, and the (nonoverlapping) infringer's profits resulting from the infringement. In exceptional cases, the courts can award the trademark owner its counsel fees and treble damages (i.e., damages three times the trademark owner's actual damages). Seizure of infringing goods can also be ordered. State law may also allow the recovery of punitive damages in extreme cases.

Recently, a federal prohibition against so-called dilution of a trademark was enacted (antidilution laws previously existed in many states). Dilution means the use of a mark by a second user in such a way that the identification of the mark with its owner is blurred or tarnished. Dilution is different from infringement in that it does not require likely confusion — the goods on which the second user puts the mark can be completely different from the owner's goods. Dilution protects only famous trademarks, but courts' determinations of which marks are famous have been inconsistent.

One important note about trademark licenses: because trademark rights are granted to allow the trademark owner to protect its commercial reputation, conversely, for a trademark license to be valid the licensor must control the nature and the quality of the goods or services sold under the mark. If the trademark owner fails to control the quality of licensed goods and services, the mark no longer has any meaning, since it is not telling the public anything about the source of the goods or services. Consequently, a mark owner that grants such so-called "naked" licenses runs the risk of a court declaring that the mark has lost its meaning and therefore has been abandoned.

Defining Terms

Copyright — A set of exclusive rights granted to authors of original works of authorship. Copyright includes the exclusive rights to copy, adapt, distribute, or offer for distribution, import, and, depending on the work, to perform or display the work, including by electronic transmission.

Trademark — One or more words or pictures used by an entity to indicate that it is the source, origin, or sponsor of goods or services. Trademarks may be registered or unregistered and include trade names and service marks.

Patent — A set of exclusive rights to make, use, sell, offer for sale, or import a product made by the patented process. Patents are granted upon examination for useful inventions, defined by a set of claims that are novel and nonobvious. The application for a patent must demonstrate possession of the invention, be enabling to practice the invention, and include a description of the best mode.

Trade secret — A trade secret is any information maintained in confidence that gives its owner a business advantage. Trade secrets can include scientific or engineering information, business information, databases, or any other information meeting the foregoing description.

References

The following are leading legal reference works in their respective fields:

Altman, L. and Callmann, Rudolf, Callmann on Unfair Competition, Trademarks, and Monopolies, 4th ed. Thomson West, St. Paul, MN.

Chisum on Patents, Matthew Bender, New York.

Lipscomb's Walker on Patents, 3rd ed. Lawyers Co-operative, Rochester, NY.

McCarthy on Trademarks and Unfair Competition, 3rd ed. Clark Boardman Callaghan, Deerfield, IL.

Nimmer on Copyright, Matthew Bender, New York.

Further Information

The U.S. Patent and Trademark Office can be accessed at *www.uspto.gov.* The website provides a vast amount of information, including procedures and forms for filing applications, and online searches of existing patents and trademarks. Nevertheless, an application to register either a patent or trademark should not be undertaken without legal counsel, and legal counsel should also be secured if a search is to be conducted in preparation of filing an application.

The U.S. Copyright Office can be accessed at *www.loc.gov/copyright.* Like the USPTO site, it provides a vast amount of information about existing registered copyrights and the procedures and forms used to obtain copyright registrations. As with patents and trademarks, it is highly advisable to consult legal counsel before filing an application to register a copyright.

XXIX

Materials Engineering

213

Properties of Solids

James F. Shackelford
University of California, Davis

The term *materials science and engineering* refers to that branch of engineering dealing with the processing, selection, and evaluation of solid-state materials [Shackelford, 2000]. As such, this is a highly interdisciplinary field. This chapter reflects the fact that engineers outside of the specialized field of materials science and engineering largely need guidance in the selection of materials for their specific applications. A comprehensive source of property data for engineering materials is available in *The CRC Materials Science and Engineering Handbook* [Shackelford and Alexander, 2001]. This brief chapter will be devoted to defining key terms associated with the properties of engineering materials and providing representative tables of such properties. Because the underlying principle of the fundamental understanding of solid-state materials is the fact that atomic- or microscopic-scale structure is responsible for the nature of materials properties, we shall begin with a discussion of structure, which will be followed by a discussion of the importance of specifying the chemical composition of commercial materials. These discussions will be followed by the definition of the main categories of material properties.

213.1 Structure

A central tenet of materials science is that the behavior of materials (represented by their **properties**) is determined by their structure on the atomic and microscopic scales [Shackelford, 2000]. Perhaps the most fundamental aspect of the structure-property relationship is to appreciate the basic skeletal arrangement of atoms in **crystalline** solids. Table 213.1 illustrates the fundamental possibilities, known as the 14 **Bravais lattices.** All crystalline structures of real materials can be produced by "decorating" the unit cell patterns of Table 213.1 with one or more atoms and repetitively stacking the unit cell structure through three-dimensional space.

213.2 Composition

The properties of commercially available materials are determined by chemical composition as well as structure [Shackelford, 2000]. As a result, extensive numbering systems have been developed to label materials, especially metal **alloys**. Table 213.2 gives an example for gray cast irons.

0-8493-1586-7/05/$0.00+$1.50

TABLE 213.1 The 14 Bravais Lattices

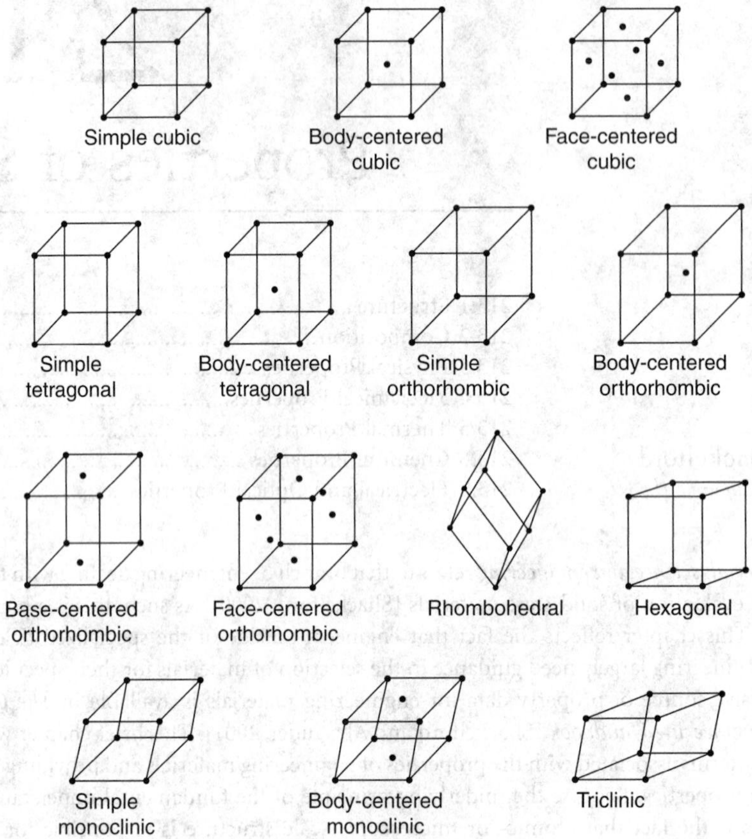

Source: Shackelford, J. F. 2000. *Introduction to Materials Science for Engineers,* 5th ed., Prentice-Hall, Upper Saddle River, NJ. p. 63.

TABLE 213.2 Composition Limits of Selected Gray Cast Irons (%)

UNS	SAE Grade	C	Mn	Si	P	S
F10004	G1800	3.40 to 3.70	0.50 to 0.80	2.80 to 2.30	0.15	0.15
F10005	G2500	3.20 to 3.50	0.60 to 0.90	2.40 to 2.00	0.12	0.15
F10006	G3000	3.10 to 3.40	0.60 to 0.90	2.30 to 1.90	0.10	0.15
F10007	G3500	3.00 to 3.30	0.60 to 0.90	2.20 to 1.80	0.08	0.15
F10008	G4000	3.00 to 3.30	0.70 to 1.00	2.10 to 1.80	0.07	0.15
F10009	G2500	3.40 min	0.60 to 0.90	1.60 to 2.10	0.12	0.12
F10010	G3500	3.40 min	0.60 to 0.90	1.30 to 1.80	0.08	0.12
F10011	G3500	3.50 min	0.60 to 0.90	1.30 to 1.80	0.08	0.12
F10012	G4000	3.10 to 3.60	0.60 to 0.90	1.95 to 2.40	0.07	0.12

Source: Data from ASM International. 1993. *ASM Metals Reference Book,* 3rd ed., p. 263. ASM International, Materials Park, OH, p. 263.

213.3 Physical Properties

Among the most basic and practical characteristics of engineering materials are their physical properties. Table 213.3 gives the **density** of a wide range of materials in units of Mg/m^3 (= g/cm^3), whereas Table 213.4 gives the **melting points** for several common metals and ceramics.

TABLE 213.3 Density of Selected Materials (Mg/m^3)

Metal		Ceramic		Glass		Polymer	
Ag	10.50	Al_2O_3	3.97–3.986	SiO_2	2.20	ABS	1.05–1.07
Al	2.7	BN (cub)	3.49	SiO_2 10 wt% Na_2O	2.291	Acrylic	1.17–1.19
Au	19.28	BeO	3.01–3.03	SiO_2 19.55 wt% Na_2O	2.383	Epoxy	1.80–2.00
Co	8.8	MgO	3.581	SiO_2 29.20 wt% Na_2O	2.459	HDPE	0.96
Cr	7.19	SiC (hex)	3.217	SiO_2 39.66 wt% Na_2O	2.521	Nylon, type 6	1.12–1.14
Cu	8.93	Si_3N_4 (α)	3.184	SiO_2 39.0 wt% CaO	2.746	Nylon 6/6	1.13–1.15
Fe	7.87	Si_3N_4 (β)	3.187			Phenolic	1.32–1.46
Ni	8.91	TiO_2 (rutile)	4.25			Polyacetal	1.425
Pb	11.34	UO_2	10.949–10.97			Polycarbonate	1.2
Pt	21.44	ZrO_2 (CaO)	5.5			Polyester	1.31
Ti	4.51	Al_2O_3 MgO	3.580			Polystyrene	1.04
W	19.25	$3Al_2O_3$ $2SiO_2$	2.6–3.26			PTFE	2.1–2.3

Source: Selected data from Shackelford, J. F., and Alexander, W., Eds. 2001. *CRC Materials Science and Engineering Handbook,* 3rd ed. CRC Press, Boca Raton, FL.

TABLE 213.4 Melting Point of Selected Metals and Ceramics

Metal	M.P.(°C)	Ceramic	M.P. (°C)
Ag	962	Al_2O_3	2049
Al	660	BN	2727
Au	1064	B_2O_3	450
Co	1495	BeO	2452
Cr	1857	NiO	1984
Cu	1083	PbO	886
Fe	1535	SiC	2697
Ni	1453	Si_3N_4	2442
Pb	328	SiO_2	1723
Pt	1772	WC	2627
Ti	1660	ZnO	1975
W	3410	ZrO_2	2850

Source: Selected data from Shackelford, J. F. and Alexander, W., Eds. 2001. *CRC Materials Science and Engineering Handbook,* 3rd ed. CRC Press, Boca Raton, FL.

213.4 Mechanical Properties

Central to the selection of materials for structural applications is their behavior in response to mechanical loads. A wide variety of mechanical properties are available to help guide materials selection [Shackelford and Alexander, 2001]. The most basic of the mechanical properties are defined in terms of the **engineering stress** and the **engineering strain**. The engineering stress, σ, is defined as

$$\sigma = P/A_o \tag{213.1}$$

where P is the load on the sample with an original (zero stress) cross-sectional area A_o. The engineering strain, ε, is defined as

$$\varepsilon = [l - l_o]/l_o = \Delta l/l_o \tag{213.2}$$

where l is the sample length at a given load and l_o is the original (zero stress) length. The maximum engineering stress that can be withstood by the material during its load history is termed the *ultimate*

TABLE 213.5 Tensile Strength of Selected Wrought Aluminum Alloys

Alloy	Temper	TS (MPa)
1050	0	76
1050	H16	130
2024	0	185
2024	T361	495
3003	0	110
3003	H16	180
5050	0	145
5050	H34	195
6061	0	125
6061	T6, T651	310
7075	0	230
7075	T6, T651	570

Source: Selected data from Shackelford, J. F. and Alexander, W., Eds. 2001. *CRC Materials Science and Engineering Handbook*, 3rd ed. CRC Press, Boca Raton, FL.

TABLE 213.6 Young's Modulus of Selected Glasses (GPa)

Type	E
SiO_2	72.76–74.15
SiO_2 20 mol% Na_2O	62.0
SiO_2 30 mol% Na_2O	60.5
SiO_2 35 mol% Na_2O	60.2
SiO_2 24.6 mol% PbO	47.1
SiO_2 50.0 mol% PbO	44.1
SiO_2 65.0 mol% PbO	41.2
SiO_2 60 mol% B_2O_3	23.3
SiO_2 90 mol% B_2O_3	20.9
B_2O_3	17.2–17.7
B_2O_3 10 mol% Na_2O	31.4
B_2O_3 20 mol% Na_2O	43.2

Source: Selected data from Shackelford, J. F. and Alexander, W., Eds. 2001. *CRC Materials Science and Engineering Handbook*, 3rd ed. CRC Press, Boca Raton, FL.

tensile strength, or simple **tensile strength**, TS. An example of the tensile strength for selected wrought (meaning "worked," as opposed to cast) aluminum alloys is given in Table 213.5. The "stiffness" of a material is indicated by the linear relationship between engineering stress and engineering strain for relatively small levels of load application. The **modulus of elasticity**, E, also known as **Young's modulus**, is given by the ratio

$$E = \sigma/\varepsilon \tag{213.3}$$

Table 213.6 gives values of Young's modulus for selected compositions of glass materials. The "ductility" of a material is indicated by the percent elongation at failure ($= 100 \times \varepsilon_{failure}$), representing the general ability of the material to be plastically (i.e., permanently) deformed. The percent elongation at failure for selected polymers is given in Table 213.7.

213.5 Thermal Properties

Many applications of engineering materials depend on their response to a thermal environment. The **thermal conductivity**, k, is defined by **Fourier's law**:

TABLE 213.7 Total Elongation at Failure of Selected Polymers

Polymer	Elongation[a]
ABS	5–20
Acrylic	2–7
Epoxy	4.4
HDPE	700–1000
Nylon, type 6	30–100
Nylon 6/6	15–300
Phenolic	0.4–0.8
Polyacetal	25
Polycarbonate	110
Polyester	300
Polypropylene	100–600
PTFE	250–350

[a] Percentage in 50-mm section.

Source: Selected data from Shackelford, J. F. and Alexander, W., Eds. 2001. *CRC Materials Science and Engineering Handbook*, 3rd ed. CRC Press, Boca Raton, FL.

TABLE 213.8 Thermal Conductivity and Thermal Expansion of Alloy Cast Irons

Alloy	Thermal Conductivity W/(mK)	Thermal Expansion Coefficient 10^{-6} m/(m°C)
Low-C white iron	22[a]	12[b]
Martensitic nickel-chromium iron	30[a]	8–9[b]
High-nickel gray iron	38–40	8.1–19.3
High-nickel ductile iron	13.4	12.6–18.7
Medium-silicon iron	37	10.8
High-chromium iron	20	9.3–9.9
High-nickel iron	37–40	8.1–19.3
Nickel-chromium-silicon iron	30	12.6–16.2
High-nickel (20%) ductile iron	13	18.7

[a] Estimated.

[b] 10 to 260°C.

Source: Data from ASM International. 1993. *ASM Metals Reference Book*, 3rd ed. ASM International, Materials Park, OH, p. 270.

$$k = -[dQ/dt]/[A(dT/dx)] \tag{213.4}$$

where dQ/dt is the rate of heat transfer across an area A due to a temperature gradient dT/dx. It is also important to note that the dimensions of a material will, in general, increase with temperature. Increases in temperature lead to greater thermal vibration of the atoms and an increase in average separation distance of adjacent atoms. The **linear coefficient of thermal expansion**, α, is given by

$$\alpha = dl/l\,dT \tag{213.5}$$

with α having units of mm/(mm°C). Examples of thermal conductivity and thermal expansion coefficient for alloy cast irons are given in Table 213.8.

213.6 Chemical Properties

A wide variety of data are available to characterize the nature of the reaction between engineering materials and their chemical environments [Shackelford and Alexander, 2001]. Perhaps no such data are more fundamental and practical than the **electromotive force series** of metals shown in Table 213.9. The

TABLE 213.9 Electromotive Force Series of Metals

Metal	Potential (V)	Metal	Potential (V)	Metal	Potential (V)
Anodic or Corroded End					
Li	−3.04	Al	−1.70	Pb	−0.13
Rb	−2.93	Mn	−1.04	H	0.00
K	−2.92	Zn	−0.76	Cu	0.52
Ba	−2.90	Cr	−0.60	Ag	0.80
Sr	−2.89	Cd	−0.40	Hg	0.85
Ca	−2.80	Ti	−0.33	Pd	1.0
Na	−2.71	Co	−0.28	Pt	1.2
Mg	−2.37	Ni	−0.23	Au	1.5
Be	−1.70	Sn	−0.14		
				Cathodic or noble metal end	

Source: Selected data from Shackelford, J. F. and Alexander, W., Eds. 2001. *CRC Materials Science and Engineering Handbook*, 3rd ed. CRC Press, Boca Raton, FL.

voltage associated with various half-cell reactions in standard aqueous environments are arranged in order, with the materials associated with more anodic reactions tending to be corroded in the presence of a metal associated with a more cathodic reaction.

213.7 Electrical and Optical Properties

To this point, we have concentrated on various properties dealing largely with the structural applications of engineering materials. In many cases the electromagnetic nature of the materials may determine their engineering applications. Perhaps the most fundamental relationship in this regard is **Ohm's law**, which states that the magnitude of current flow, I, through a circuit with a given resistance R and voltage V is related by:

$$V = IR \tag{213.6}$$

where V is in units of volts, I is in amperes, and R is in ohms. The resistance value depends on the specific sample geometry. In general, R increases with sample length, l, and decreases with sample area, A. As a result, the property more characteristic of a given material and independent of its geometry is **resistivity**, ρ, defined as

$$\rho = [RA]/l \tag{213.7}$$

The units for resistivity are ohm · m (or Ω · m). Table 213.10 gives the values of electrical resistivity for various materials, indicating that metals typically have low resistivities (and correspondingly high electrical conductivities) and ceramics and polymers typically have high resistivities (and correspondingly low conductivities).

An important aspect of the electromagnetic nature of materials is their optical properties. Among the most fundamental optical characteristics of a light-transmitting material is the **index of refraction**, n, defined as

$$n = v_{vac}/v \tag{213.8}$$

where v_{vac} is the speed of light in vacuum (essentially equal to that in air) and v is the speed of light in a transparent material. The index of refraction for a variety of polymers is given in Table 213.11.

TABLE 213.10 Electrical Resistivity of Selected Materials

Metal (Alloy Cast Iron)	ρ (Ω · m)	Ceramic	ρ (Ω · m)	Polymer	ρ (Ω · m)
Low-C white cast iron	$0.53 \cdot 10^{-6}$	Al_2O_3	$>10^{13}$	ABS	$2–4 \cdot 10^{13}$
Martensitic Ni-Cr iron	$0.80 \cdot 10^{-6}$	B_4C	$0.3–0.8 \cdot 10^{-2}$	Acrylic	$>10^{13}$
High-Si iron	$0.50 \cdot 10^{-6}$	BN	$1.7 \cdot 10^{11}$	HDPE	$>10^{13}$
High-Ni iron	$1.4–1.7 \cdot 10^{-6}$	BeO	$>10^{15}$	Nylon 6/6	$10^{12}–10^{13}$
Ni-Cr-Si iron	$1.5–1.7 \cdot 10^{-6}$	MgO	$1.3 \cdot 10^{13}$	Phenolic	$10^{7}–10^{11}$
High-Al iron	$2.4 \cdot 10^{-6}$	SiC	$1–1 \cdot 10^{10}$	Polyacetal	10^{13}
Medium-Si ductile iron	$0.58–0.87 \cdot 10^{-6}$	Si_3N_4	$>10^{11}$	Polypropylene	$>10^{15}$
High-Ni (20%) ductile iron	$1.02 \cdot 10^{-6}$	SiO_2	10^{16}	PTFE	$>10^{16}$

Source: Selected data from Shackelford, J. F. and Alexander, W., Eds. 2001. *CRC Materials Science and Engineering Handbook*, 3rd ed. CRC Press, Boca Raton, FL.

TABLE 213.11 Refractive Index of Selected Polymers

Polymer	*n*
Acrylic	1.485–1.500
Cellulose acetate	1.46–1.50
Epoxy	1.61
HDPE	1.54
Polycarbonate	1.586
PTFE	1.35
Polyester	1.50–1.58
Polystyrene	1.6
SAN	1.565–1.569
Vinylidene chloride	1.60–1.63

Source: Selected data from Shackelford, J. F. and Alexander, W., Eds. 2001. *CRC Materials Science and Engineering Handbook*, 3rd ed. CRC Press, Boca Raton, FL.

Defining Terms

Alloy — Metal composed of more than one element.
Bravais lattice — One of the 14 possible arrangements of points in three-dimensional space.
Crystalline — Having constituent atoms stacked together in a regular, repeating pattern.
Density — Mass per unit volume.
Electromotive force series — Systematic listing of half-cell reaction voltages.
Engineering strain — Increase in sample length at a given load divided by the original (stress-free) length.
Engineering stress — Load on a sample divided by the original (stress-free) area.
Fourier's law — Relationship between rate of heat transfer and temperature gradient.
Index of refraction — Ratio of speed of light in vacuum to that in a transparent material.
Linear coefficient of thermal expansion — Material parameter indicating dimensional change as a function of increasing temperature.
Melting point — Temperature of transformation from solid to liquid upon heating.
Ohm's law — Relationship between voltage, current, and resistance in an electrical circuit.
Property — Observable characteristic of a material.
Resistivity — Electrical resistance normalized for sample geometry.
Tensile strength — Maximum engineering stress during a tensile test.
Thermal conductivity — Proportionality constant in Fourier's law.
Young's modulus (modulus of elasticity) — Ratio of engineering stress to engineering strain for relatively small levels of load application.

References

ASM International. 1993. *ASM Metals Reference Book*, 3rd ed. ASM International, Materials Park, OH.

Shackelford, J. F. 2000. *Introduction to Materials Science for Engineers*, 5th ed. Prentice Hall, Upper Saddle River, NJ.

Shackelford, J. F. and Alexander, W., Eds. 2001. *The CRC Materials Science and Engineering Handbook*, 3rd ed. CRC Press, Boca Raton, FL.

Further Information

A general introduction to the field of materials science and engineering is available from a variety of introductory textbooks. In addition to Shackelford [2000], readily available references include the following:

Ashby, M. F. 1999. *Materials Selection in Mechanical Design*, 2nd ed. Butterworth-Heinemann, Oxford.

Askeland, D. R. and Phule, P. P. 2003. *The Science and Engineering of Materials*, 4th ed. Brooks-Cole, Independence, KY.

Callister, W. D. 2003. *Materials Science and Engineering — An Introduction*, 6th ed. John Wiley & Sons, New York.

Schaffer, J. P., et al. 1999. *The Science and Design of Engineering Materials*, 2nd ed. WCB-McGraw-Hill, Boston.

Smith, W. F. 2004. *Foundations of Materials Science and Engineering*, 3rd ed. McGraw-Hill, New York.

As noted earlier, *The CRC Materials Science and Engineering Handbook* [Shackelford and Alexander, 2001] is available as a comprehensive source of property data for engineering materials. In addition, ASM International published between 1985 and 2002 the *ASM Handbook*, a 21-volume set concentrating on metals and alloys. ASM International has also published a four-volume set entitled the *Engineered Materials Handbook*, covering composites, engineering plastics, adhesives and sealants, and ceramics and glasses.

214

Failure Analysis

James F. Shackelford
University of California, Davis

Failure analysis and prevention are important, practical considerations relative to the applications of materials in engineering design [ASM International, 2002]. This chapter begins by exploring the various types of failures that can occur. There is now a well-established, systematic methodology for the analysis of failures of engineering materials. Beyond **failure analysis** as a "postmortem" examination, the related issue of **failure prevention** is equally important as the basis for avoiding future disasters. Fracture mechanics provides a powerful technique for both analyzing failures and modifying engineering designs for the prevention of future failures. Similarly, nondestructive testing serves as one set of techniques for a comprehensive analysis of failures, as well as a primary strategy for the prevention of future failures, especially in the identification of critical flaws. The two primary techniques on which we will focus are radiographic testing (typically using X rays) and ultrasonic testing.

The issues of failure analysis and failure prevention are taking on increasing levels of importance with the growing awareness of the responsibilities of the professional engineer. Ethical and legal demands for the understanding of engineering failures and the prevention of such future catastrophes are moving the field of materials science and engineering into a central role in the broader topic of engineering design.

214.1 Types of Failures

A wide spectrum of failure modes has been identified. To illustrate the range of these types of failures, we will begin with a brief description of several of the most common failures in structural metals.

Ductile fracture is observed in a large number of the failures occurring in metals due to overload — that is, simply taking a material beyond the elastic limit, beyond its ultimate tensile strength, and, subsequently, to fracture. The microscopic result of ductile fracture is a characteristic "dimpled" fracture surface morphology.

Brittle fracture is characterized by rapid crack propagation without significant plastic deformation on a macroscopic scale. The cleavage texture of a brittle fracture surface comes from both *transgranular* fracture, involving the cleavage of microstructural grains, and *intergranular* fracture, occurring by crack propagation between adjacent grains.

0-8493-1586-7/05/$0.00+$1.50

Fatigue failure is the result of cyclic stresses that would, individually, be well below the yield strength of the material. The fatigue mechanism of slow crack growth gives a distinctive "clamshell" fatigue fracture surface.

Corrosion-fatigue failure is due to the combined actions of a cyclic stress and a corrosive environment. In general, the fatigue strength (or fatigue life at a given stress) of the metal will be decreased in the presence of an aggressive, chemical environment.

Stress-corrosion cracking (SCC) is another combined mechanical and chemical failure mechanism in which a noncyclic tensile stress (below the yield strength of the metal) leads to the initiation and propagation of fracture in a relatively mild chemical environment. Stress-corrosion cracks may be intergranular, transgranular, or a combination thereof.

Wear failure encompasses a broad range of relatively complex surface-related damage phenomena. Both surface damage and wear debris can constitute "failure" of materials intended for sliding contact applications.

Liquid erosion failure is a special form of wear damage in which a liquid, rather than another solid, is responsible for the removal of material. Liquid-erosion damage typically results in a pitted or honeycomb-like surface region.

Liquid-metal embrittlement is another form of material failure caused by a liquid. In this case the solid loses some degree of ductility or fractures below its yield stress in conjunction with its surface being wetted by a lower–melting-point liquid metal. This failure mode occurs in specific solid-liquid metal combinations.

Hydrogen embrittlement is perhaps the most notorious form of catastrophic failure in high-strength steels. A few parts per million of hydrogen dissolved in these materials can lead to fine "hairline" cracks and a loss of ductility. A variety of commercial environments serve as sources of hydrogen gas that, in turn, dissociates into atomic hydrogen, which can readily diffuse into the alloy.

Creep and *stress-rupture failures* can occur above about one-half the absolute melting point of an alloy. *Creep* is defined as plastic deformation over an extended period of time under a fixed load. Failure of this type can occur near room temperature for many polymers and certain low–melting-point metals, such as lead, but may occur above 1,000°C in many ceramics and certain high–melting-point metals, such as the superalloys.

Complex failures are those in which the failure occurs by the sequential operation of two distinct fracture mechanisms. An example would be initial cracking due to stress-corrosion cracking and, then, ultimate failure by fatigue after a cyclic load is introduced simultaneously with the removal of the corrosive environment. The possibility of such sequences should always be considered when conducting a failure analysis.

214.2 Failure Analysis Methodology

A systematic sequence of procedures has been developed for the analysis of the failure of an engineering material. Although the specific methodology will vary with the specific failure, the principal components of the investigation and analysis are given in Table 214.1. This specific set of components is given in the *ASM Handbook* volume on failure analysis and prevention [ASM International, 2002]. Because of the general utility of **nondestructive testing** for failure prevention as well as failure analysis, the subject is covered in detail later in this chapter. **Fracture mechanics** is the general analysis of the failure of structural materials with preexisting flaws. This subject will be discussed separately in the next section. Regarding failure analysis methodology, fracture mechanics has provided a quantitative framework for evaluating structural reliability.

214.3 Fracture Mechanics

The **fracture toughness**, *KIc*, is the most widely used material parameter associated with fracture mechanics [Shackelford, 2000] and is, in general, of the simple form

TABLE 214.1 Principal Components of Failure Analysis Methodology

Collection of background data and samples
Preliminary (visual) examination of the failed part
Preliminary record keeping
Mechanical testing
Chemical analyses (bulk and/or surface)
Macroscopic (1 to 100×) examination of fracture surfaces
Microscopic (>100×) examination of fracture surfaces
Application of stress analysis and/or fracture mechanics
Simulated-service testing
Analyzing the evidence, formulating conclusions, and writing the report

Source: Adapted from ASM International. 2002. *ASM Handbook, Volume 11: Failure Analysis and Prevention*. ASM International, Materials Park, OH.

$$KIc = \sigma f\,(\pi a)^{1/2} \tag{214.1}$$

where σf is the overall applied stress at failure and a is the length of a surface crack (or one-half the length of an internal crack). Fracture toughness has units of $MPa{\cdot}m^{1/2}$. Typical values of fracture toughness for various metals and alloys range between 20 and 200 $MPa{\cdot}m^{1/2}$. Values of fracture toughness for ceramics and glass are typically in the range of 1 to 9 $MPa{\cdot}m^{1/2}$, values for polymers are typically 1 to 4 $MPa{\cdot}m^{1/2}$, and values for composites are typically 10 to 60 $MPa{\cdot}m^{1/2}$. Although Equation (214.1) is widely used for predicting flaw-induced or "fast" fracture, one must keep in mind that the subscript *I* refers to "mode *I*" loading — that is, simple, uniaxial tension. Fortunately, mode *I* conditions predominate in most practical systems, and the wide use of *KIc* is justified. Another aspect of fracture mechanics is the relationship to fatigue and other mechanical phenomena in which flaw size increases incrementally prior to failure.

214.4 Nondestructive Testing

As with fracture mechanics, nondestructive testing can serve to analyze an existing failure or be used to prevent future failures [Bray and Stanley, 1997]. The dominant techniques within this field are radiography and ultrasonics.

Radiographic Testing

X-rays compose a portion of the electromagnetic spectrum. Although diffraction techniques allow dimensions on the order of the x-ray wavelength (typically less than one nanometer) to be determined, x radiography produces a "shadow graph" of the internal structure of a part with a much coarser resolution (typically on the order of 1 mm). The common chest x-ray is a routine example of this technology. In industrial applications, x radiography is widely used for the inspection of castings and weldments. A schematic of the technique is given in Figure 214.1.

A key factor in this test is the thickness of material through which the x-ray beam can penetrate. For a given material being inspected by a given energy x-ray beam, the intensity of the beam, I, transmitted through a thickness of material, x, is given by Beer's law:

$$I = I_0 e^{-\mu x} \tag{214.2}$$

where I_0 is the incident beam intensity and μ is the linear absorption coefficient for the material. The intensity is proportional to the number of photons in the beam and should not be confused with the energy of photons in the beam. The absorption coefficient is a function of the beam energy and of the elemental composition of the material. Experimental values for the linear absorption coefficient of iron as a function of energy are given in Table 214.2. The general trend is a steady drop in the magnitude of

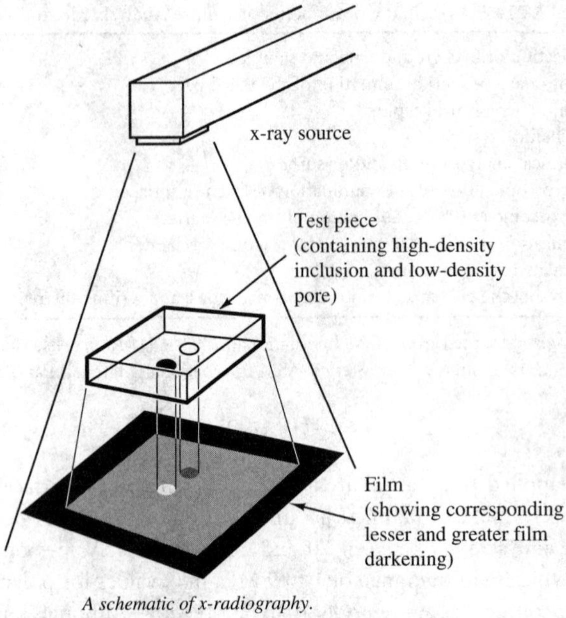

FIGURE 214.1 A schematic of x-radiography. (*Source:* Shackelford, J. F. 2000. *Introduction to Materials Science for Engineers*, 5th ed. Prentice Hall, Upper Saddle River, NJ. With permission.)

TABLE 214.2 Linear Absorption Coefficient of Iron as a Function of x-ray Beam Energy

Energy (MeV)	μ (mm^{-1})
0.05	1.52
0.10	0.293
0.50	0.0662
1.00	0.0417
2.00	0.0334
4.00	0.0260

Source: Selected data from Bray, D. E. and Stanley, R. K. 1997. *Nondestructive Evaluation.* CRC Press, Boca Raton, FL.

μ with increasing beam energy until above 1 MeV, where μ levels off. Below 1 MeV the absorption of the beam is due to mechanisms of photon absorption and photon scattering. Above 1 MeV, however, an additional absorption mechanism comes into play (electron-positron pair production).

The dependence of the linear absorption coefficient on elemental composition is illustrated by the data of Table 214.3. The general trend of data is that μ for a given beam energy increases with atomic number. As a result, low–atomic-number metals such as aluminum are relatively transparent and high–atomic-number metals such as lead are relatively opaque.

Ultrasonic Testing

X radiography was seen to be based on a portion of the electromagnetic spectrum with relatively short wavelengths in comparison to the visible region. Similarly, ultrasonic testing is based on a portion of the acoustic spectrum with frequencies above those of the audible range (20 to 20 000 cycles/second or Hz). Typical ultrasonic inspections are made in the 1- to 25-MHz range. A key distinction between x radiography and ultrasonic testing is that the ultrasonic waves are mechanical in nature and require a transmitting medium, whereas electromagnetic waves can be transmitted in a vacuum.

TABLE 214.3 Linear Absorption Coefficient of Various Elements for an x-ray Beam with Energy = 100 keV (= 0.1 MeV)

Element	Atomic Number	μ (mm^{-1})
Titanium	22	0.124
Iron	26	0.293
Nickel	28	0.396
Copper	29	0.410
Zinc	30	0.356
Tungsten	74	8.15
Lead	82	6.20

Source: Selected data from Bray, D. E. and Stanley, R. K. 1997. *Nondestructive Evaluation.* CRC Press, Boca Raton, FL.

Whereas attenuation of the x-ray beam in the solid was a dominant consideration for x radiography, typical engineering materials are relatively transparent to ultrasonic waves. The key consideration for ultrasonic testing is the reflection of the ultrasonic waves at dissimilar material interfaces. The reflection coefficient, R, defined as the ratio of reflected beam intensity, Ir, to incident beam intensity, Ii, is given by

$$R = Ir/Ii = [(Z_2 - Z_1)/(Z_2 + Z_1)]^2 \qquad (214.3)$$

where Z is the acoustic impedance, defined as the product of the material's density and velocity of sound, with the subscripts 1 and 2 referring to the two dissimilar materials on either side of the interface. The high degree of reflectivity by a typical flaw, such as an internal crack, is the basis for defect inspection. Figure 214.2 illustrates a typical "pulse echo" ultrasonic inspection based on this principle. The oscillations in the figure represent voltage fluctuations in a piezoelectric transducer that is used to both send and detect the ultrasonic signal. The horizontal scale is proportional to time in a time frame of several microseconds. The initial pulse on the left is reflected back from a flaw and seen in the mid-range, with a small pulse on the right representing a reflection from the back side of the sample. The insert is a C-scan in which numerous adjacent A-scan pulses are viewed together to represent the spatial location of the flaw (impact damage in a NASA graphite/epoxy wind tunnel blade). The dashed line corresponds to the displayed A-scan. The limitations of this method include the difficulty of applying the techniques in complex-shaped parts and the loss of information due to microstructural complexities such as porosity and precipitates.

Other Methods of Nondestructive Testing

A wide spectrum of additional methods are available in the field of nondestructive testing. The following four methods are among the most widely used for failure analysis and prevention.

Eddy current testing is a versatile technique for inspecting electrically conductive materials. The impedance of an inspection coil is affected by the presence of an adjacent test piece, in which alternating (eddy) currents have been induced by the coil. The net impedance is a function of the composition or geometry of the test piece. The popularity of eddy current testing is due to the convenient, rapid, and non-contact nature of the method. By varying the test frequency the method can be used for both surface and subsurface flaws. Limitations include the qualitative nature and the need for an electrically conductive test piece.

Magnetic-particle testing is a simple, traditional technique widely used due to its convenience and low cost. Its primary limitation is the restriction to magnetic materials. (This is not as restrictive as one might initially assume, given the enormous volume of structural steels used in engineering.) The basic mechanism of this test involves the attraction of a fine powder of magnetic particles (Fe or Fe_3O_4) to the "leakage flux" around a discontinuity such as a surface or near-surface crack.

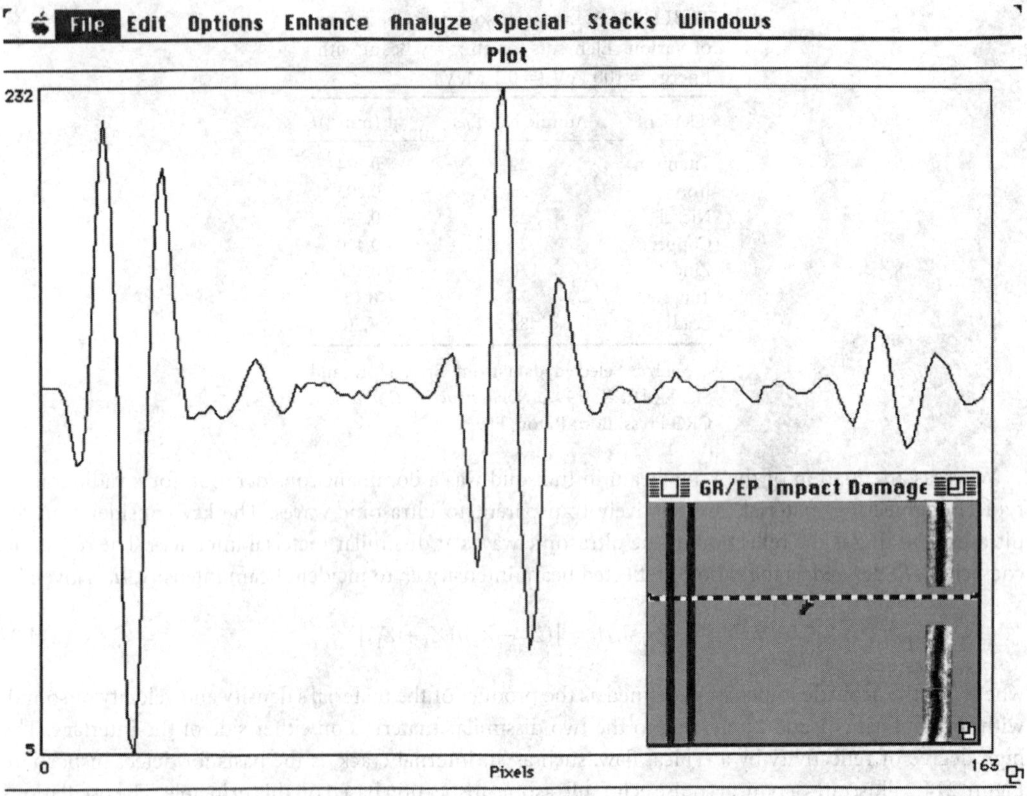

FIGURE 214.2 Typical ultrasonic pulse echo A-scan of a structural defect. (Ultrasonic imaging software provided courtesy of D. Bailey, United States Air Force.)

Liquid-penetrant testing is, like magnetic-particle testing, an inexpensive and convenient technique for surface defect inspection. It is largely used on nonmagnetic materials for which magnetic-particle inspection is not possible. The basic mechanism for this method is the capillary action of a fine powder on the surface of a sample in which a high-visibility liquid has penetrated into surface defects. The limitations of this technique include the inability to inspect subsurface flaws and the loss of resolution on porous materials.

Acoustic-emission testing has assumed a unique role in failure prevention. This nondestructive test, in addition to being able to locate defects, can provide an early warning of impending failure due to those defects. In contrast to conventional ultrasonic testing, in which a transducer provides the source of ultrasound, acoustic emission is the set of ultrasonic waves produced by defects within the microstructure of a material in response to an applied stress. With the material as the source of ultrasound, transducers serve only as receivers. In general, the rate of acoustic-emission events rises sharply just prior to failure. By continuously monitoring these emissions, the structural load can be removed in time to prevent that failure. The primary example of this preventive application is in the continuous surveillance of pressure vessels.

214.5 Engineering Design for Failure Prevention

Finally, it is important to note that engineering designs can be improved as a result of the application of concepts raised in Sections 214.1 and 214.2 — that is, failure analysis can lead to failure prevention. An example of this approach includes the avoidance of structural discontinuities that can serve as stress concentrators.

Defining Terms

Failure analysis — The systematic examination of a failed engineering part to determine the nature of the failure.

Failure prevention — The application of the principles of failure analysis to prevent future catastrophes.

Fracture mechanics — Analysis of failure of structural materials with preexisting flaws.

Fracture toughness — Critical value of the stress-intensity factor at a crack tip necessary to produce catastrophic failure.

Nondestructive testing — A family of techniques for observing material flaws and characteristics without impairing the future usefulness of the part under inspection.

References

ASM International. 2002. *ASM Handbook, Volume 11: Failure Analysis and Prevention.* ASM International, Materials Park, OH.

Bray, D. E. and Stanley, R. K. 1997. *Nondestructive Evaluation.* CRC Press, Boca Raton, FL.

Shackelford, J. F. 2000. *Introduction to Materials Science for Engineers*, 5th ed. Prentice Hall, Upper Saddle River, NJ.

Further Information

In addition to the references, readily available sources of further information include:

ASM International. 1987. *ASM Handbook, Volume 12: Fractography.* ASM International, Materials Park, OH.

ASM International. 1989. *ASM Handbook, Volume 17: Nondestructive Evaluation and Quality Control.* ASM International, Materials Park, OH.

215

Liquids and Gases

Bruce E. Poling
University of Toledo

There are a number of thermodynamic and kinetic properties of gases and liquids that are important in engineering applications. Properties that are discussed in this chapter include viscosity, thermal conductivity, heat capacity, and vapor pressure. In the material that follows, a brief description of each of these four properties is given. However, engineers often do not want a description; rather, they want a number or an equation. The strategy that should be used to accomplish this latter task depends on the particular situation, but many features of this strategy are independent of the desired property. Thus, a general strategy is presented first.

For gases at low pressures and pure liquids, experimental data for many substances are available and have generally been correlated by equations and sometimes by nomographs. Daubert et al. [1997] present references to data, an evaluation of the quality of the data, and correlating equations for all the properties discussed in this section for over 1300 substances. Yaws [1999] presents correlating equations for all the properties discussed in this section for 1700 substances, but unlike Daubert et al. [1997], Yaws' data sources are not so well documented. Daubert et al. [1997] have chosen correlating equations that give the best fit or have some basis in theory; Yaws [1999] uses polynomials in his correlations that have the advantage of mathematical simplicity. Yaws' equation forms especially should not be extrapolated outside the temperature range over which data have been fit. There are many other correlations, summaries of data sources, and tabulations of values. Some of these are summarized in Table 215.1 and others are described in the section that relates to a specific property. If information for a compound is not available, or information at temperatures or pressures outside the range where experimental data have been measured is required, estimation methods such as those described in Poling et al. [2001] may be used. Poling et al. [2001] also present recommended methods for finding properties for gases that cannot be considered ideal (at high pressure) and for mixtures of compounds.

215.1 Viscosity

The *viscosity* of a fluid is a measure of its thickness, or how easily it flows. The viscosity, η, of a substance is defined as the shear stress per unit area divided by the velocity gradient. Fluids for which the viscosity is independent of shear stress are called *Newtonian fluids,* and this class includes gases and most common liquids. Polymer solutions, inks, coatings, and paints are often non-Newtonian — that is, their thickness and flow characteristics change with shear stress. Viscosity has dimensions of mass per length per time. A common viscosity unit is the poise, defined as one gram per centimeter per second. The kinematic

0-8493-1586-7/05/$0.00+$1.50

TABLE 215.1 Information Available in Various References

Reference Number[a]	1	2, 3	4	5, 6	7
Viscosity					
Vapor		V	V, N	E	V
Liquid	V, E	V	V, N	E	V
Thermal Conductivity					
Vapor	V	V	V	E	V
Liquid	V	V	V, N	E	V
Heat Capacity					
Ideal gas	V	V	E	E	V
Liquid		V	E	E	
Vapor pressure	E	V	V	E	V

[a] 1. Dean [1999]; 2. Lide [2002]; 3. Vargaftik et al. [1996]; 4. Perry and Green [1997]; 5. Yaws [1999]; 6. Daubert et al. [1997]; 7. Kaye and Laby [1986].

Note: E = equation, V = values, N = nomograph.

viscosity is defined as the ratio of viscosity to density. A common unit for the kinematic viscosity is the stoke, defined as cm^2/s.

For pure liquids or pure gases at low pressure (near 1 atm) the viscosity is a function of temperature but is insensitive to changes in pressure. Vapor viscosity varies approximately linearly with temperature; the temperature dependence of liquid viscosity is approximately exponential. The viscosity of many common gases at low pressure can be found from the nomograph in Perry and Green [1997]. Viswanath and Natarajan [1989] have compiled liquid-viscosity data for over 900 compounds. Daubert et al. [1997] have fit vapor viscosity, η^V, to

$$\eta^v = \frac{AT^B}{1 + \frac{C}{T} + \frac{D}{T^2}} \tag{215.1}$$

Daubert et al. [1997] correlate liquid viscosity, η^L, with the equation

$$\ln \eta L = A + B/T + C \ln (T) + DTE \tag{215.2}$$

Constants to be used in Equations (215.1) and (215.2) for water are listed in Table 215.2. In both equations, T is in K.

215.2 Thermal Conductivity

The *thermal conductivity* of a fluid, λ, is a measure of the rate at which heat is transferred through the fluid by conduction, that is, in the absence of convection. Units used for thermal conductivity are W/(m K). These may be converted to English or cgs units by

$$\text{W/(m K)} \times 0.5778 = \text{Btu/(h} \cdot \text{ft} \cdot {}^\circ\text{R)}$$

$$\text{W/(m K)} \times 0.8604 = \text{kcal/(cm} \cdot \text{h} \cdot \text{K)}$$

Unlike liquid viscosity which shows a strong temperature dependence, both vapor and liquid thermal conductivity shows a more nearly linear temperature dependence. Vargaftik et al. [1994] have compiled thermal conductivity data for over 700 compounds. Yaws [1999] correlates vapor thermal conductivity, λ^V, with the equation:

TABLE 215.2 Constants to be Used in Chapter Equations for Water

Property	Vapor Viscosity, ηV, Pa · s	Liquid Viscosity, ηL, Pa · s	Vapor Thermal Conductivity, λV, W/m · K	Liquid Thermal Conductivity, λL, W/m · K	Ideal Gas Heat Capacity, C_p^o, J/mol · K	Liquid Heat Capacity, C_p^L, J/mol · K	Vapor Pressure, bar
Equation in text	(215.1)	(215.2)	(215.3)	(215.4)	(215.7)	(215.7)	(215.9)[a]
A	6.1839E-07	−52.843	5.3E-04	−0.432	33.933	276.37	−7.77224
B	0.67779	3703.6	4.7093E-05	5.7255E-03	−8.4186E-03	−2.0901	1.45684
C	847.23	5.866	4.9551E-08	−8.078E-06	2.9906E-05	8.125E-03	−2.71942
D	−73930	−5.879E-29		1.861E-09	−1.7825E-08	−1.4116E-05	−1.41336
E		10			3.6934E-12	9.3701E-09	
T range, K	273–1073	273–646	273–1073	273–633	100–1500	273-533	273–647.3

[a] In Equation (215.9), for water, use Pc = 221.2 bar, and Tc = 647.3 K.

$$\lambda^V = A + BT + CT^2 \tag{215.3}$$

For liquid thermal conductivity, λ^L, Daubert et al. [1997] use the equation:

$$\lambda^L = A + BT + CT^2 + DT^3 + ET^4 \tag{215.4}$$

Constants to be used for water vapor in Equation (215.3) and liquid water in Equation (215.4) appear in Table 215.2. T is in K.

215.3 Heat Capacity

Generally speaking, the *heat capacity* of a fluid is the amount of heat required to increase a unit mass of the substance by one degree. Heat capacities are used to calculate sensible heat effects and are important in the design of heat exchangers. From a historic point of view, the heat capacity of water is of particular significance because, at one time, a calorie was defined as the amount of heat required to heat 1 gram of water 1°C. The term *specific heat* is the ratio of the heat capacity of a substance to that of water. Although the definition of the calorie is now in terms of joules, and although the heat capacity of water varies slightly with temperature, for engineering purposes the heat capacity of liquid water may still be taken as 1 cal/(g · °C) over the range of 0 to 100°C. This is particularly convenient because the four sets of units — cal/(g · °C), cal/(g · K), Btu/(lb · °F), and Btu/(lb · °R) — are all numerically equivalent. Note that the temperature units in these four sets refer to temperature changes, so °C and K, for example, are equivalent. Because the heat capacity of water is 1 cal/(g · °C), the specific heat is numerically equal to the heat capacity when the units are cal/(g · °C). For gases particularly, heat capacities are often given in molar units. Thus, if a heat capacity with units of cal/(g · °C) is multiplied by a substance's molecular weight, the molar heat capacity in cal/(g · mol · °C) is obtained. The units, cal/(g · °C), may be converted to J/(g · °C) by multiplying by 4.186.

From a rigorous thermodynamic point of view, the term *heat capacity* means heat capacity at constant pressure, Cp, and is the change in enthalpy with respect to temperature in a constant pressure experiment. However, other heat capacities can be defined. The heat capacity at constant volume, Cv, is the change in enthalpy with respect to temperature at constant volume, and for liquids and solids these two heat capacities are essentially the same number. For ideal gases

$$Cp = Cv + R \tag{215.5}$$

Two other heat capacities that are sometimes seen are the change in enthalpy of a saturated liquid with respect to temperature (this path is at neither constant P nor V but, rather, along the vapor pressure curve) and the amount of energy required to effect a temperature change while maintaining the liquid in a saturated state. Fortunately, for liquids not close to the critical point and for ideal gases, these latter two heat capacities are essentially the same number as Cp.

For engineering applications, the heat capacities most often needed are *ideal gas heat capacities* and heat capacities of a condensed phase, that is, liquid or solid. The ideal gas heat capacity, C_p^o, for a monatomic gas in the ideal gas state (examples of monatomic compounds include mercury, neon, helium, and argon) is 20.8 J/(g · mol · K). This is 2.5R, where R is the gas constant. For compounds that contain more than one atom, C_p^o is a function of temperature (but not pressure) and its value generally increases with increasing temperature. The heat capacity of a mixture of ideal gases, $C_{p,\text{mix}}^o$ may be calculated with the following equation:

$$C_{p,\text{mix}}^o = \sum_i x_i C_{p,i}^o \tag{215.6}$$

where x_i is the mole fraction of component i and $C_{p,i}^o$ is for pure i.

Liquid heat capacities are weak functions of temperature below the normal boiling point. For some compounds Cp in this temperature range increases slowly with temperature, but for others Cp passes through a shallow minimum. The Cp behavior for water is an example of this latter behavior. Cp for water passes through a minimum at about 40°C; the values at 0 and 100°C are each about 1% higher than this minimum value. Above the normal boiling point, Cp for liquids begins to rise and for all compounds approaches infinity as the critical point is approached.

Both liquid and ideal gas heat capacities are often correlated with temperature by equations such as

$$C_P = A + BT + CT^2 + DT^3 + ET^4 \tag{215.7}$$

Values of the constants in Equation (215.7) for liquid water from Daubert et al. [1997] and for water vapor in the ideal gas state from Yaws [1999] are given in Table 215.2. In Equation (215.7) T is in K. Stull et al. [1969] have compiled ideal gas heat capacity data from approximately 900 compounds and Zábranský et al. [1996] have compiled data for liquid heat capacities for over 1,600 compounds.

215.4 Vapor Pressure

The *vapor pressure*, or saturation pressure, of a pure liquid is the pressure where both liquid and vapor are in equilibrium. Vapor pressure is a function only of temperature, and each substance has a unique vapor pressure curve. Three important points on this curve are the normal boiling point, the triple point, and the critical point. The normal boiling point corresponds to the temperature where the vapor pressure is 1 atm, and the triple point is the one point on the vapor pressure curve where vapor, liquid, and solid can exist in equilibrium. The triple point temperature is essentially the same number as the melting point temperature because pressure has little effect on melting point temperatures. At the critical point, the vapor and liquid phases become identical, and, above the critical temperature, the two phases are no longer distinct. The vapor pressure increases rapidly with temperature, and, in fact, the log of the vapor pressure varies nearly linearly with the reciprocal of the absolute temperature. Thus, if vapor pressures are known at two temperatures, reliable vapor pressures can be determined at other temperatures by interpolating on log P_{vp} versus $1/T$, where P_{vp} is the vapor pressure and T is in kelvins. However, this linear relationship is not exact and extrapolations over large temperature ranges can sometimes lead to unacceptable errors.

A number of equations have been developed for correlating vapor pressures. Perhaps the most common is the Antoine equation:

$$\ln P_{vp} = A + \frac{B}{C+T} \tag{215.8}$$

More recently, the Wagner equation has generally been accepted as one of the best equations for correlating vapor pressures. This equation has the form,

$$\ln(P_{vp}/P_c) = (A\tau + B\tau^{1.5} + C\tau^3 + D\tau^6)/T_r \tag{215.9}$$

In both Equations (215.8) and (215.9), the constants for a particular compound are determined by fitting vapor pressure data. In Equation (215.9), P_c is the critical pressure, P_{vp} is the vapor pressure and will have the same units as P_c, T_r is the reduced temperature, T/T_c, $\tau = 1 - T_r$, and both T and T_c are in kelvins. Some authors have claimed that Equation (215.9) is improved if the exponents 3 and 6 are replaced with 2.5 and 5. Vapor pressure constants are listed for 468 compounds in Poling et al. [2001]; constants to be used in Equation (215.9) for water are listed in Table 215.2. Boublík et al. [1984] and Ohe [1976] list sources for experimental vapor pressure data for approximately 1000 and 2000 substances, respectively.

Defining Terms

Heat capacity — Generally, constant pressure heat capacity, which is defined as the change in enthalpy with respect to temperature at constant pressure.

Ideal gas heat capacity — The heat capacity of a substance at the specified temperature and in the ideal gas state. Unless otherwise stated, this generally means constant pressure heat capacity.

Thermal conductivity — The proportionality constant between the heat transfer rate (by conduction) per unit area and the temperature gradient; a measure of how fast heat is transferred by conduction through a substance.

Vapor pressure — The pressure at which both the liquid and vapor phases of a pure substance are in equilibrium.

Viscosity — The proportionality constant between the shear stress per unit area and the velocity gradient; a measure of the resistance to deformation.

References

Boublík, T. V., Fried, V., and Hála, E. 1984. *The Vapour Pressure of Pure Substances,* 2nd ed. Elsevier, New York.

Daubert, T. E., Danner, R. P., Sibel, H. M., and Stebbins, C. C. 1997. *Physical and Thermodynamic Properties of Pure Chemicals: Data Compilation*, Taylor & Francis, Washington, DC.

Dean, J. A. 1999. *Lange's Handbook of Chemistry,* 15th ed., McGraw-Hill, New York.

Kaye, G. W. C. and Laby, T. H., 1986. *Tables of Physical and Chemical Constants,* 15th ed. Longman, New York.

Lide, D. R. 2002. *CRC Handbook of Chemistry and Physics,* 83rd ed. CRC Press, Boca Raton, FL.

Ohe, S. 1976. *Computer Aided Data Book of Vapor Pressure,* Data Book, Tokyo.

Perry, R. H. and Green, D. 1997. *Perry's Chemical Engineer's Handbook,* 7th ed. McGraw-Hill, New York.

Poling, B. E., Prausnitz, J. M., and O'Connell, J. P. 2001. *The Properties of Gases and Liquids,* 5th ed. McGraw-Hill, New York.

Stull, D. R., Westrum, E. F., Jr., and Sinke, G. C. 1969. *The Chemical Thermodynamics of Organic Compounds,* 2nd ed., John Wiley & Sons, New York.

Vargaftik, N. B., Filippov, L. P., Tarzimanov, A. A., and Totskii, E. E. 1994. *Handbook of Thermal Conductivity of Liquids and Gases*, CRC Press, Boca Raton, FL.

Vargaftik, N. B., Vinogradov, Y. K., and Yargin, V. S. 1996. *Handbook of Physical Properties of Liquids and Gases*, Begell House, New York.

Viswanath, D. S. and Natarajan, G. 1989. *Data Book on the Viscosity of Liquids*, Hemisphere, New York.
Yaws, C. L. 1999. *Chemical Properties Handbook*, McGraw-Hill, New York.
Zábranský, M., Rùzicka, V., Majer, V., and Domalski, E. S. 1996. *Heat Capacity of Liquids: Critical Review and Recommended Values*, American Chemistry Society, and American Institute of Physics, for National Institute of Standards and Technology, Washington, DC.

Further Information

Introductory-level material on heat capacities and vapor pressures can generally be found in thermodynamics textbooks such as *Introduction to Chemical Engineering Thermodynamics* by J. M. Smith, H. C. Van Ness, and M. M. Abbott, 6th ed. (McGraw-Hill, New York, 2001). Introductory-level material on viscosities and thermal conductivities can be found in textbooks dealing with fluid flow and heat transfer, respectively. One such textbook is *Unit Operations of Chemical Engineering*, by W. L. McCabe, 6th ed. (McGraw-Hill, New York, 2001). As mentioned in the text, Table 215.1 lists the type of information available in the books listed in the reference section.

216 Biomaterials

Scott J. Hazelwood
(Second Edition)
University of California, Davis

R. Bruce Martin
(First Edition)
University of California, Davis

The term *biomaterials* usually refers to human-made materials used to construct prosthetic or other medical devices for implantation in human beings. It can also refer to materials that otherwise come in contact with the tissues of internal organs (e.g., tubing to carry blood to and from a heart-lung machine). Despite advances in many areas of materials science, it would be a mistake to assume that modern technology has the ability to replace any part of a living organism with an artificial material (or artificial organ) that will be superior to the original structure. Although it is possible to imagine situations in which this might be true in some limited sense, the synthetic organism as a whole will rarely work better than when the original organ was in place. For example, a segment of bone may be replaced with a similar structure of titanium alloy having greater strength, but the metallic material will lack the ability of bone to adapt to its changing mechanical environment or repair fatigue damage over time.

216.1 History

The history of biomaterials can be divided into three eras. Prior to 1850, nonmetallic materials and common metals were used to fabricate simple prosthetic devices. Wood, ivory, iron, gold, silver, and copper were used to replace items such as teeth and noses, and to hold fractured bones together while they healed. In 1829, Levert experimented with lead, gold, silver, and platinum wire in dogs, but these metals clearly did not have the desired mechanical attributes. Furthermore, without anesthesia, patients could not endure long surgeries in order to implant significant prostheses or fixation devices.

The second era of biomaterials, between 1850 and 1925, was defined by the rapid development of surgery as something other than an emergency procedure. The advent of anesthesia just before the mid-19th century precipitated this development. In addition, x-rays were discovered by Roentgen and found immediate application in orthopedics in the late 1800s, revealing for the first time the true nature of many skeletal problems. Finally, the acceptance of the aseptic surgical procedures propounded by Lister gradually but dramatically reduced the rate of postsurgical infections.

The period from 1925 to the present is the third era, in which the primary advances in the various surgical specialties have resulted from three important developments. The first was the development of cobalt chrome and stainless steel alloys in the 1930s and 1940s, respectively. The second was the

0-8493-1586-7/05/$0.00+$1.50

development of polymer chemistry and plastics in the 1940s and 1950s. The third was the discovery of ways to produce useful quantities of penicillin and other antibiotics. The ability to further reduce surgical infection rates and to fabricate many devices that were compatible with biological tissues significantly advanced the ability of surgeons to treat a great variety of problems. Most of the biomaterials commonly in use today were developed more than 30 years ago; the intervening years have been ones of gradual refinement.

216.2 Problems Associated with Implanted Devices

There are many problems that a biomaterial should ideally overcome. It must be formable into the desired shape using economical methods and without compromising its mechanical properties. It must not corrode in the presence of body fluids. This property entails the avoidance of crevice and fretting corrosion. A biomaterial must not poison the patient. It must either be free of toxic substances or those substances must be adequately locked into the structure of the material. Toxic ions that may be gradually released must not accumulate or lead to a long-term immunological response. A biomaterial must also be easy to sterilize (using steam under pressure, radiation, or ethylene oxide gas) without damaging its properties. The material must not break, either due to an occasional acute overload or due to fatigue from repeated functional loads. The strength and fatigue properties must combine with the shape of the implant to keep stresses within safe limits, particularly where stress concentrations cannot be avoided. Usually, the material must not simply work for a year or two, but for many years. A *connective tissue* implant must not perturb the stresses in adjacent tissues into a state that leads to atrophy of the tissues or prohibits normal repair of fatigue damage. Finally, implants must be made from materials that can be removed and replaced if they fail. Materials that are impervious to attack by the body may be difficult for the surgeon to remove as well.

216.3 Immunology and Biocompatibility

All organisms try to keep out foreign matter. If they fail in this, they work very hard either to destroy the invading object if its molecules look destructible or to isolate it through encapsulation in fibrous tissue if it looks impregnable. The problem of detecting foreign material is very complex, and the problem of overcoming an immune system that successfully copes with this task looms very large. In some cases, *fibrous encapsulation* does not have an adverse effect on an implant, but, as a general rule, biomaterials should be able to avoid the natural inclination of the body to either encapsulate them or break them down. Since doing this by destroying the natural defenses of the body (i.e., the immune system) is a poor tactic, the best way to proceed is to mask the foreign material from the chemical sensors of the host. This is difficult to do, since most materials will be easily recognized as "outsiders" and will not be able to avoid the consequences. One useful principle to remember is that materials with molecules that look like biological molecules will be more readily attacked by destructive cells. For example, nylon and polyethylene both have a core of carbon atoms with hydrogens attached along their lengths. The main difference between the two is that polyethylene has a CH_3 terminal group, whereas nylon has an NH_2 terminal group. Since proteins have the latter kind of terminus, nylon tends to be more susceptible than polyethylene to degradation by cells of the immune system.

On the other hand, it may be very difficult to induce the body to attach tissue to impregnable materials such as polyethylene. In fact, no polymers have been found that are both immune to degradation and amenable to tissue adhesion. Some materials, though, are very attractive to connective tissue cells. For example, hydroxyapatite is nearly identical to bone mineral, and will quickly become integrated with bone when implanted in the skeleton. Unfortunately, hydroxyapatite is not nearly as tough and strong as bone. The commonly used materials that are tough and strong — metals — cannot be destroyed by the immune system, so the body tends to encapsulate them in fibrous tissue. This makes it difficult to rigidly attach them to bone.

TABLE 216.1 Mechanical Properties of Metals and Ceramics Commonly Used in Biomedical Implants[a]

Material	Elastic Modulus (GPa)	Maximum Strain (%)	Tensile Strength (MPa)
Stainless steel	193	10	1000
Cast cobalt-Cr	235	8	670
Wrought cobalt-Cr	235	12	1170
Ti-6Al-4V alloy	117	10	900
Pure titanium	100	15	410
Al_2O_3[b]	380	0	50
ZrO_2[c]	210	0	—
Apatite	62	0	690
C-Si composite	21	0	690

[a] Values are approximations compiled from the references listed at the end of this chapter. Given the variability in manufacturing processes, precise values cannot be given for mechanical properties.
[b] *Compressive* strength = 4 GPa.
[c] *Compressive* strength = 3 GPa.

TABLE 216.2 Mechanical Properties of Polymers Commonly Used in Biomedical Implants[a]

Material	Elastic Modulus (MPa)	Maximum Strain (%)	Tensile Strength (MPa)
Silicone rubber[b]	2.4	700	—
Polyether urethane	—	700	41
PTFE	0.4	300	30
UHMWPE	500	350	35
PMMA	2000	2	30[c]

[a] Values are approximations compiled from the references listed at the end of this chapter. Given the variability in manufacturing processes, precise values cannot be given for mechanical properties.
[b] Modulus at 100% strain.
[c] Compressive strength = 90 MPa.

216.4 Commonly Used Implant Materials

Human-made biomaterials may be broadly categorized into metals, polymers, ceramics, and carbon composites. The materials in each of these categories may contain several forms, such as solids, membranes, fibers, or coatings. They may also serve many purposes, including replacing structural organs or organs that carry out chemical exchanges, housing electronic devices, repairing damaged or congenitally defective cardiovascular or connective tissue, or delivering drugs.

Tables 216.1 and 216.2 show mechanical properties of metals and polymers commonly used as biomaterials. Table 216.3 shows the properties of some connective tissues that these materials are designed to replace.

TABLE 216.3 Mechanical Properties of Major Connective Tissues[a]

Material	Elastic Modulus (MPa)	Extensibility (%)	Tensile Strength (MPa)
Bone	20,000	1.5	150
Cartilage (costal)	14	8.0	1
Collagen (tendon)	1300	9.0	75
Keratin[b]	5000	2.0	50

[a] Values are approximations compiled from the references listed at the end of this chapter. Given the variability of biological tissues, precise values cannot be given for mechanical properties.
[b] For the alpha-form region of tensile tests of wool.

TABLE 216.4 Approximate Compositions of Common Biomaterial Alloys[a]

Alloy	Approximate Composition
Cobalt-chrome	60% Co, 28% Cr, 6% Mo, 3% Ni, 3% other (Fe, C, Mn, Si)
Stainless steel	62% Fe, 18% Cr, 14% Ni, 3% Mo, 0.03% C, 3% other (Mn, P, S, Si)
Titanium	90% Ti, 6% Al, 4% V, 0.4% other (Fe, C, O)

[a] Values are approximations compiled from the references listed at the end of this chapter.

216.5 Metals

Cobalt-chromium alloys were the first corrosion-resistant alloys to be developed and have proven very effective in surgical implants, beginning in 1936 when Venable reported the use of such a material in orthopedics. Most alloys of this material are primarily cobalt. Chromium is added to provide resistance to corrosion. They are usually regarded as the metal of choice in skeletal prostheses. They have historically been available in cast and wrought forms. As with stainless steel, the wrought material is substantially stronger. The mechanical properties and composition of each alloy discussed in this section are shown in Tables 216.1 and 216.4, respectively. Modifications of Co-Cr alloys contain less cobalt and chromium and larger amounts of other alloy materials (35% nickel in one case and 15% tungsten in another). They can be forged and heat treated to obtain tensile strengths significantly above those of stainless steel and the more common Co-Cr alloys — as high as 1800 MPa.

Stainless steel (usually 316L) is a common industrial alloy that has been very successful as a surgical implant material. Both stainless steel and cobalt-chrome alloys owe their corrosion resistance to the formation of a passive ceramic-like CrO_2 coating on the surface, and it is important that this coating not be scratched during implantation. Care must be taken with stainless steel materials in orthopedic applications because of their susceptibility to crevice and stress corrosion. The ductility of these alloys can be increased by heat treatment, and their strength can be increased by cold working of the material. Cast stainless steels are unsuitable for use in weight-bearing situations because of their large grain sizes and low fatigue strengths. Type 316LVM (low carbon [L, typically less than 0.03%], vacuum melt [VM]) material is preferred. The carbon concentration must remain at small amounts in stainless steel materials to maintain resistance to corrosion.

Titanium alloys are primarily Ti-6Al-4V (i.e., 6% aluminum, 4% vanadium, and 90% titanium). Pure titanium is also used, primarily in dental applications. As with cobalt-chrome alloys and stainless steel, titanium and its alloys offer excellent biocompatibility and corrosion resistance. In fact, the titanium oxide (TiO_2) passive layer formed on the surface of these alloys increases the resistance to corrosion compared to that of the other implant metals. Titanium alloys are becoming increasingly popular in orthopedics because their strength is as good as the other metal implant materials, but they are only half as stiff (Table 216.1). This is potentially important because a large elastic modulus mismatch between implant and bone causes stress concentrations in some places and tends to *stress shield* the bone in others. However, the modulus of titanium alloys is still several times greater than that of bone (Tables 216.1 and 216.3), and marked decreases in stress shielding when these alloys are used have not yet been demonstrated. Titanium alloys are also sensitive to notching. A notch or scratch on titanium alloy implants will significantly reduce their fatigue life.

216.6 Polymers

Polymethylmethacrylate (PMMA) is an acrylic plastic from which parts may be machined. It also may be polymerized in the body to fix a hip prosthesis stem into the medullary canal or fill a bony defect. Frequently, this polymer will have an additive: (1) barium to increase its radiographic visualization, or (2) an antibiotic to prevent infection following surgery. The mechanical properties of PMMA are shown in Table 216.2; barium and antibiotic additions do not substantially affect these properties. Typically, PMMA used for orthopedic applications to fix implants is prepared by vacuum or centrifuge mixing systems that reduce the void space in the polymerized form. Treatment of the PMMA in this way increases its ultimate strength compared to that which is mixed by hand. PMMA polymerizes with an exothermic reaction that causes the doughy mass used in hip surgery to reach temperatures in the vicinity of 40°C, although laboratory reports indicate temperatures of 60°C to 90°C may be achieved depending on the amount of cement. In addition, PMMA has the effect of causing blood pressure to drop momentarily when it is implanted during hip surgery. PMMA is also used in the manufacture of hard contact lenses and of intraocular lenses for cataract patients. For this application, polymerization is initiated with heat to produce an extremely clear material with good optical properties. Soft contact lenses are usually hydrogels made of homo- or co-polymers of hydroxyethyl methacrylate; other methacrylates and silicones have also been used, however.

Ultra-high molecular-weight polyethylene (UHMWPE) has a very simple chemical structure, consisting of chains of carbon atoms with hydrogen atoms attached to the sides. In the ultra-high molecular-weight form, these chains achieve a molecular weight of 1 to 4 million. Sterilization of UHMWPE components used as bearing surfaces against one of the metal alloys is mainly accomplished through gamma irradiation, which oxidizes polyethylene and may negatively affect the wear properties of the material. Conducting the sterilization procedure in an environment free of oxygen significantly reduces the wear rate of UHMWPE by inducing cross-linking between polymer chains. Polyethylene has also been used to replace the ossicles of the inner ear.

Polydimethylsiloxane (silicone rubber) is a widely used polymer that was first applied biomedically in a hydrocephalus shunt in 1955. It has a long history of biological compatibility and clinical testing in a great variety of applications, and a "medical grade" (Silastic) has been developed with superior *biocompatibility* and mechanical properties. It can be sterilized by steam, radiation, or ethylene oxide. This polymer can be manufactured with various degrees of cross-linking to adapt its mechanical properties for other purposes as well. Silicone rubber is commonly used for catheters. Rubbery sheets of silicone have been used in hernia repair, and a gel silicone often has been used for breast prostheses or augmentation. (The recent controversy over the safety of the latter application is a good example of the difficulty in assessing the biocompatibility of biomaterials. Different individuals may respond quite differently to the same material.) Harder versions have served as balls in ball-and-cage heart valves. Silicone rubber is widely used in pacemakers and biomedical research to form seals against body fluids for electrical leads. It is used as the functional membrane in both kidney dialysis and extracorporeal blood oxygenator machines. It also has been used in drug delivery implants, where it serves as a membrane through which the drug slowly passes. Silicone also serves as a structural material in prosthetic heart valves and to replace the auricle or the ossicles of the ear. In orthopedics, silicone rubber is formed into soft, flexible "strap hinges" for the replacement of arthritic finger, wrist, or toe joints. However, its fatigue resistance is not as good as it should be for this application.

Dacron has been used for blood vessel prostheses since the pioneering work of DeBakey in 1951. Dacron is thrombogenic, so the pores in the fabric soon become filled with coagulated blood, which is then replaced by a tissue called *neointima* that serves as a biological wall between the Dacron and the blood. However, if the vessel diameter is smaller than 6 mm, the neointima will prevent blood flow in the tube. Dacron has also been used in prosthetic heart valves.

Polytetrafluoroethylene (PTFE, Teflon, or Gore-Tex) has been used to produce blood vessel prostheses smaller than 6 mm in diameter, and the neointima appears to be thinner with this material. However, it too is unsatisfactory for most human blood vessels that are smaller than 3 mm. PTFE is also used to make prosthetic heart valves, ligaments, shunt tips, and artificial ossicles for the ear.

Polyether urethane has been used for years in blood bags and tubing for kidney dialysis machines. It has also been tried as a material for blood vessel replacement. It is frequently found in intra-aortic balloon pumps and in artificial hearts, where it lines the chambers and forms the pumping diaphragm. The primary requirement in the latter application is fatigue resistance; a prosthesis that is to last 10 years must flex about 360 million times. Materials that reportedly have been tried for this purpose and found to have insufficient fatigue properties include several other polymers used in tubing for kidney dialysis machines: polyvinylchloride, silicone rubber, and natural and synthetic rubbers.

Polyalkylsulfones are used in blood oxygenator membranes.

Hexsyn is a brand name of a recently developed elastomer that has good biological compatibility and an extraordinarily high fatigue life (more than 300 million cycles to failure, compared to 600,000 for silicone rubber in an ASTM D430 flexure test). It has been used for heart valves and human finger-joint prostheses.

Epoxy resins have been used to encapsulate electronic implants and with carbon fiber reinforcement for dental implant posts.

Polydepsipeptides, *polylactic acid*, *polyglycolic acid*, *polydioxanone*, and *polycaprolactone* polymers have been used with some success as biodegradable implants. Implant materials that disintegrate gradually in body fluids are useful as sutures, screws, pins, bone plates, to fill defects in bone or other tissues, and for the delivery of embedded drugs.

216.7 Ceramics

Ceramics generally have hydrophilic surfaces amenable to intimate bonding with tissues. They are very biocompatible but brittle relative to biological materials, including bone. One of their primary applications is as a coating (optimally 25 to 30 micrometers thick) for metals to promote attachment to bone. *Hydroxyapatite* [$Ca_{10}(PO_4)_6(OH)_2$] and *tricalcium phosphate* [TCP, $Ca_3(PO_4)_2$] in various forms have proven to be very biocompatible and form intimate bonds with bone. Unfortunately, their tensile strengths are not great. They also may be made in porous or granular forms and used to fill bone defects in lieu of a bone graft. TCP may eventually be resorbed and replaced by bone; hydroxyapatite is less resorbable.

Alumina, a polycrystalline form of Al_2O_3, is a widely used industrial ceramic. It has relatively good strength and excellent wear-resistance characteristics, but is very stiff and brittle compared to bone. It is sintered from alumina powder under pressure at 1600°C to form structures with relatively smooth surfaces. A principal use of this material is to form the heads of total hip prostheses. Alumina has also been used in dental implants, as corneal replacements, and as substitutes for bones in the middle ear.

Zirconia (ZrO_2) is also used in bearing surfaces as the head of femoral joint implants. As with alumina, zirconia has good strength and wear-resistance properties. Because of its low modulus and high strength, zirconia femoral heads can be manufactured into a wide range of diameters and lengths for joint replacement applications.

Bioglass (Na_2O-CaO-P_2O_5-SiO_2) is a glass with soluble additives designed to form a silica gel at its surface and thereby aid chemical bonding to bone. Bioglass implants have been used in many dental applications and may be used to replace several bones in the ear.

216.8 Carbon Materials

Graphite has a weak, anisotropic crystalline structure, but isostatic (or turbostratic) carbon is isotropic and relatively strong. The two most useful forms of turbostratic carbon in biomedical applications are vitreous (or glassy) carbon and pyrolytic low-temperature isotropic (LTI) carbon. These are both isotropic materials, but the latter is more wear resistant and stronger. Turbostratic carbons, carbon-fiber PMMA, and carbon-silicon composites have excellent biocompatibility and can have elastic moduli similar to bone. However, it typically turns out that in order to reduce the stiffness of these materials to match that of bone, the strength must be reduced to less than that available in metals. In the last decade, a number of these materials have been tested for various biomedical applications, but few have been widely used. Carbon-silicon and carbon-coated graphite, however, have been used in heart valves, and pyrolytic carbon coating has been shown to improve bone growth into porous metal surfaces.

Defining Terms

Biocompatibility — The ability to remain in direct contact with the tissues of a living person without significantly affecting those tissues, or being affected by them, other than in a prescribed way. For example, a biocompatible material used in a blood vessel wall would contain the blood under pressure but would not adversely interact with its constituents or other adjacent tissues.

Biomaterial — An engineering material suitable for use in situations where it may come into direct contact with internal tissues of the body. Usually, this means in a surgically implanted device, but it may also include devices that contain or process blood, or other fluids, that will be returned to the body. Sometimes, *biomaterial* is used to refer to a biological structural material.

Connective tissue — Tissue that primarily serves a mechanical rather than a metabolic or chemical function. The volume of the cells in such tissue is small relative to the extracellular components. For example, tendons are composed primarily of extracellular collagen, with a small fraction of their volume occupied by the cells that produce the collagen.

Fibrous encapsulation — The formation of connective tissues around a foreign object in order to isolate it from the rest of the body, both physically and chemically.

Stress shield — To reduce the stress in a region of bone by placing it in parallel with an implant that is stiffer. Bone is able to adjust its structure to the applied loads, and if the imposed load is reduced, the affected bone will atrophy. This may leave the prosthesis inadequately supported and lead to failure.

References

Budinski, K. G. and Budinski, M. K., *Engineering Materials: Properties and Selection*. Prentice Hall, Upper Saddle River, NJ, 1999.

Gebelein, C. G., *Polymeric Materials and Artificial Organs*. American Chemical Society, Washington, DC, 1984.

Hastings, G. W. and Ducheyne, P., *Natural and Living Biomaterials*. CRC Press, Boca Raton, FL, 1984.

Hench, L. L. and Ethridge, E. C., *Biomaterials: An Interfacial Approach*. Academic Press, New York, 1982.

Kossowsky, R. and Kossovsky, N., *Materials Sciences and Implant Orthopaedic Surgery*. Martinus Nijhoff, Boston, MA, 1986.

Lin, O. C. C. and Chao, E. Y. S., Perspectives on biomaterials. *Proceedings of the 1985 International Symposium on Biomaterials*. Taipei, Taiwan, 25–27 February 1985. Elsevier, Amsterdam, 1985.

Nahum, A. M. and Melvin, J., *The Biomechanics of Trauma*. Appleton-Century-Crofts, Norwalk, CT, 1985.

National Research Council. *Internal Structural Prostheses: Report of a Workshop on Fundamental Studies for Internal Structural Prostheses*. National Academy of Sciences, Washington, DC.

Ratner, B. D., et al. *Biomaterials Science: An Introduction to Materials in Medicine*. Academic Press, San Diego, CA, 1996.

Shackelford, J. F., *Bioceramics*. Gordan and Breach Science, Australia, 1999.

Tsuruta, T., et al. *Biomedical Applications of Polymeric Materials*. CRC Press, Boca Raton, FL, 1993.

Wright, T. M. and Li, S., Biomaterials. In: *Orthopaedic Basic Science*, 2nd ed., Buckwalter, J. A., Einhorn, T. A., and Simon, S. R., Eds. American Academy of Orthopaedic Surgeons, Rosemont, IL, 2000.

Further Information

Black, J., *Orthopaedic Biomaterials in Research and Practice*. Churchill Livingstone, London, 1988.

Kambik, H. E. and Toshimitsu, A., *Biomaterials' Mechanical Properties*. ASTM International, Philadelphia, PA, 1994.

Szycher, M., *Biocompatible Polymers, Metals, and Composites*. Technomic, Lancaster, PA, 1983.

Szycher, M., *High Performance Biomaterials: A Comprehensive Guide to Medical and Pharmaceutical Applications*. Technomic, Lancaster, PA, 1991.

Szycher, M., *Szycher's Dictionary of Biomaterials and Medical Devices.* Technomic, Lancaster, PA, 1992.
Von Recum, A. F., *Handbook of Biomaterials Evaluation: Scientific, Technical, and Clinical Testing of Implant Material.* Taylor & Francis, Philadelphia, PA, 1999.

XXX

Mathematics

217

General Mathematics

CONTENTS

217.1 Trigonometry

Triangles

In any triangle (in a plane) with sides *a*, *b*, and *c* and corresponding opposite angles *A*, *B*, and *C*,

$$\frac{a}{\sin A} = \frac{b}{\sin B} = \frac{c}{\sin C} \qquad \text{(Law of sines)}$$

$$a^2 = b^2 + c^2 - 2cb\cos A \qquad \text{(Law of cosines)}$$

$$\frac{a+b}{a-b} = \frac{\tan\frac{1}{2}(A+B)}{\tan\frac{1}{2}(A-B)} \qquad \text{(Law of Tangents)}$$

$$\sin\frac{1}{2}A = \sqrt{\frac{(s-b)(s-c)}{bc}} \quad \text{where } s = \frac{1}{2}(a+b+c)$$

0-8493-1586-7/05/$0.00+$1.50

$$\cos\frac{1}{2}A = \sqrt{\frac{s(s-a)}{bc}}$$

$$\tan\frac{1}{2}A = \sqrt{\frac{(s-b)(s-c)}{s(s-a)}}$$

$$\text{Area} = \frac{1}{2}bc\ \sin A$$

$$= \sqrt{s(s-a)(s-b)(s-c)}$$

If the vertices have coordinates (x_1, y_1), (x_2, y_2), (x_3, y_3), the area is the *absolute value* of the expression

$$\frac{1}{2}\begin{vmatrix} x_1 & y_1 & 1 \\ x_2 & y_2 & 1 \\ x_3 & y_3 & 1 \end{vmatrix}$$

Trigonometric Functions of an Angle

With reference to Figure 217.1, $P(x, y)$ is a point in any one of the four quadrants and A is an angle whose initial side is coincident with the positive x axis and whose terminal side contains the point $P(x, y)$. The distance from the origin $P(x, y)$ is denoted by r and is positive. The trigonometric functions of the angle A are defined as:

$$\sin A = \text{sine } A \quad = y/r$$

$$\cos A = \text{cosine } A \quad = x/r$$

$$\tan A = \text{tangent } A \quad = y/x$$

$$\text{ctn } A = \text{cotangent } A \quad = x/y$$

$$\sec A = \text{secant } A \quad = r/x$$

$$\csc A = \text{cosecant } A \quad = r/y$$

Angles are measured in degrees or radians; 180° = π radians; 1 radian = 180/π degrees.

The trigonometric functions of 0°, 30°, 45°, and integer multiples of these are directly computed.

	0°	30°	45°	60°	90°	120°	135°	150°	180°
sin	0	$\frac{1}{2}$	$\frac{\sqrt{2}}{2}$	$\frac{\sqrt{3}}{2}$	1	$\frac{\sqrt{3}}{2}$	$\frac{\sqrt{2}}{2}$	$\frac{1}{2}$	0
cos	1	$\frac{\sqrt{3}}{2}$	$\frac{\sqrt{2}}{2}$	$\frac{1}{2}$	0	$-\frac{1}{2}$	$-\frac{\sqrt{2}}{2}$	$-\frac{\sqrt{3}}{2}$	−1
tan	0	$\frac{\sqrt{3}}{3}$	1	$\sqrt{3}$	∞	$-\sqrt{3}$	−1	$-\frac{\sqrt{3}}{3}$	0
ctn	∞	$\sqrt{3}$	1	$\frac{\sqrt{3}}{3}$	0	$-\frac{\sqrt{3}}{3}$	−1	$-\sqrt{3}$	∞
sec	1	$\frac{2\sqrt{3}}{3}$	$\sqrt{2}$	2	∞	−2	$-\sqrt{2}$	$-\frac{2\sqrt{3}}{3}$	−1
csc	∞	2	$\sqrt{2}$	$\frac{2\sqrt{3}}{3}$	1	$\frac{2\sqrt{3}}{3}$	$\sqrt{2}$	2	∞

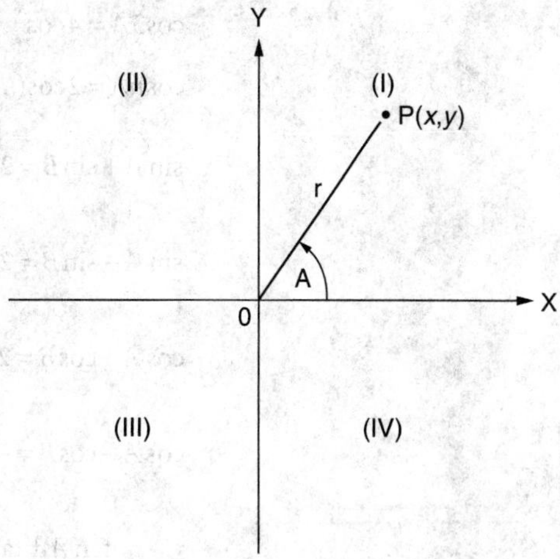

FIGURE 217.1 The trigonometric point. Angle A is taken to be positive when the rotation is counter-clockwise and negative when the rotation is clockwise. The plane is divided into quadrants as shown.

Trigonometric Identities

$$\sin A = \frac{1}{\csc A}$$

$$\cos A = \frac{1}{\sec A}$$

$$\tan A = \frac{1}{\operatorname{ctn} A} = \frac{\sin A}{\cos A}$$

$$\csc A = \frac{1}{\sin A}$$

$$\sec A = \frac{1}{\cos A}$$

$$\operatorname{ctn} A = \frac{1}{\tan A} = \frac{\cos A}{\sin A}$$

$$\sin^2 A + \cos^2 A = 1$$

$$1 + \tan^2 A = \sec^2 A$$

$$1 + \operatorname{ctn}^2 A = \csc^2 A$$

$$\sin(A \pm B) = \sin A \cos B \pm \cos A \sin B$$

$$\cos(A \pm B) = \cos A \cos B \mp \sin A \sin B$$

$$\tan(A \pm B) = \frac{\tan A \pm \tan B}{1 \mp \tan A \tan B}$$

$$\sin 2A = 2 \sin A \cos A$$

$$\sin 3A = 3 \sin A - 4 \sin^3 A$$

$$\sin nA = 2 \sin(n-1)A \cos A - \sin(n-2)A$$

$$\cos 2A = 2\cos^2 A - 1 = 1 - 2\sin^2 A$$

$$\cos 3A = 4\cos^3 A - 3\cos A$$

$$\cos nA = 2\cos(n-1)A\cos A - \cos(n-2)A$$

$$\sin A + \sin B = 2\sin\frac{1}{2}(A+B)\cos\frac{1}{2}(A-B)$$

$$\sin A - \sin B = 2\cos\frac{1}{2}(A+B)\sin\frac{1}{2}(A-B)$$

$$\cos A + \cos B = 2\cos\frac{1}{2}(A+B)\cos\frac{1}{2}(A-B)$$

$$\cos A - \cos B = -2\sin\frac{1}{2}(A+B)\sin\frac{1}{2}(A-B)$$

$$\tan A \pm \tan B = \frac{\sin(A \pm B)}{\cos A \cos B}$$

$$\text{ctn}\, A \pm \text{ctn}\, B = \pm\frac{\sin(A \pm B)}{\sin A \sin B}$$

$$\sin A \sin B = \frac{1}{2}\cos(A-B) - \frac{1}{2}\cos(A+B)$$

$$\cos A \cos B = \frac{1}{2}\cos(A-B) + \frac{1}{2}\cos(A+B)$$

$$\sin A \cos B = \frac{1}{2}\sin(A+B) + \frac{1}{2}\sin(A-B)$$

$$\sin\frac{A}{2} = \pm\sqrt{\frac{1-\cos A}{2}}$$

$$\cos\frac{A}{2} = \pm\sqrt{\frac{1+\cos A}{2}}$$

$$\tan\frac{A}{2} = \frac{1-\cos A}{\sin A} = \frac{\sin A}{1+\cos A} = \pm\sqrt{\frac{1-\cos A}{1+\cos A}}$$

$$\sin^2 A = \frac{1}{2}(1-\cos 2A)$$

$$\cos^2 A = \frac{1}{2}(1+\cos 2A)$$

$$\sin^3 A = \frac{1}{4}(3\sin A - \sin 3A)$$

$$\cos^3 A = \frac{1}{4}(\cos 3A + 3\cos A)$$

$$\sin ix = \frac{1}{2}i(e^x - e^{-x}) = i\sinh x$$

$$\cos ix = \frac{1}{2}(e^x + e^{-x}) = \cosh x$$

$$\tan ix = \frac{i(e^x - e^{-x})}{e^x + e^{-x}} = i \tanh x$$

$$e^{x+iy} = e^x(\cos y + i \sin y)$$

$$(\cos x \pm i \sin x)^n = \cos nx \pm i \sin nx$$

Inverse Trigonometric Functions

The inverse trigonometric functions are multiple valued, and this should be taken into account in the use of the following formulas.

$$\sin^{-1} x = \cos^{-1} \sqrt{1-x^2}$$

$$= \tan^{-1} \frac{x}{\sqrt{1-x^2}} = \text{ctn}^{-1} \frac{\sqrt{1-x^2}}{x}$$

$$= \sec^{-1} \frac{1}{\sqrt{1-x^2}} = \csc^{-1} \frac{1}{x}$$

$$= -\sin^{-1}(-x)$$

$$\cos^{-1} x = \sin^{-1} \sqrt{1-x^2}$$

$$= \tan^{-1} \frac{\sqrt{1-x^2}}{x} = \text{ctn}^{-1} \frac{x}{\sqrt{1-x^2}}$$

$$= \sec^{-1} \frac{1}{x} = \csc^{-1} \frac{1}{\sqrt{1-x^2}}$$

$$= \pi - \cos^{-1}(-x)$$

$$\tan^{-1} x = \text{ctn}^{-1} \frac{1}{x}$$

$$= \sin^{-1} \frac{x}{\sqrt{1+x^2}} = \cos^{-1} \frac{1}{\sqrt{1+x^2}}$$

$$= \sec^{-1} \sqrt{1+x^2} = \csc^{-1} \frac{\sqrt{1+x^2}}{x}$$

$$= -\tan^{-1}(-x)$$

217.2 Series

Bernoulli and Euler Numbers

A set of numbers, $B_1, B_3, \ldots, B_{2n-1}$ (Bernoulli numbers) and $B_2, B_4, \ldots, B_{2n}$ (Euler numbers) appear in the series expansions of many functions. A partial listing follows; these are computed from the following equations:

$$B_{2n} - \frac{2n(2n-1)}{2!} B_{2n-2} + \frac{2n(2n-1)(2n-2)(2n-3)}{4!} B_{2n-4} - \cdots + (-1)^n = 0$$

and

$$\frac{2^{2n}(2^{2n}-1)}{2n}B_{2n-1} = (2n-1)B_{2n-2} - \frac{(2n-1)(2n-2)(2n-3)}{3!}B_{2n-4} + \cdots + (-1)^{n-1}$$

$$\begin{array}{ll} B_1 = 1/6 & B_2 = 1 \\ B_3 = 1/30 & B_4 = 5 \\ B_5 = 1/42 & B_6 = 61 \\ B_7 = 1/30 & B_8 = 1385 \\ B_9 = 5/66 & B_{10} = 50\,521 \\ B_{11} = 691/2730 & B_{12} = 2\,702\,765 \\ B_{13} = 7/6 & B_{14} = 199\,360\,981 \\ \vdots & \vdots \end{array}$$

Series of Functions

In the following, the interval of convergence is indicated; otherwise it is all x. Logarithms are to the base e. Bernoulli and Euler numbers (B_{2n-1} and B_{2n}) appear in certain expressions.

$$(a+x)^n = a^n + na^{n-1}x + \frac{n(n-1)}{2!}a^{n-2}x^2 + \frac{n(n-1)(n-2)}{3!}a^{n-3}x^3 + \cdots$$

$$+ \frac{n!}{(n-j)!\,j!}a^{n-j}x^j + \cdots \qquad [x^2 < a^2]$$

$$(a-bx)^{-1} = \frac{1}{a}\left[1 + \frac{bx}{a} + \frac{b^2x^2}{a^2} + \frac{b^3x^3}{a^3} + \cdots\right] \qquad [b^2x^2 < a^2]$$

$$(1 \pm x)^n = 1 \pm nx + \frac{n(n-1)}{2!}x^2 \pm \frac{n(n-1)(n-2)x^3}{3!} + \cdots \qquad [x^2 < 1]$$

$$(1 \pm x)^{-n} = 1 \mp nx + \frac{n(n+1)}{2!}x^2 \mp \frac{n(n+1)(n+2)}{3!}x^3 + \cdots \qquad [x^2 < 1]$$

$$(1 \pm x)^{1/2} = 1 \pm \frac{1}{2}x - \frac{1}{2\cdot 4}x^2 \pm \frac{1\cdot 3}{2\cdot 4\cdot 6}x^3 - \frac{1\cdot 3\cdot 5}{2\cdot 4\cdot 6\cdot 8}x^4 \pm \cdots \qquad [x^2 < 1]$$

$$(1 \pm x)^{-1/2} = 1 \mp \frac{1}{2}x + \frac{1\cdot 3}{2\cdot 4}x^2 \mp \frac{1\cdot 3\cdot 5}{2\cdot 4\cdot 6}x^3 + \frac{1\cdot 3\cdot 5\cdot 7}{2\cdot 4\cdot 6\cdot 8}x^4 \pm \cdots \qquad [x^2 < 1]$$

$$(1 \pm x^2)^{1/2} = 1 \pm \frac{1}{2}x^2 - \frac{x^4}{2\cdot 4} \pm \frac{1\cdot 3}{2\cdot 4\cdot 6}x^6 - \frac{1\cdot 3\cdot 5}{2\cdot 4\cdot 6\cdot 8}x^8 \pm \cdots \qquad [x^2 < 1]$$

$$(1 \pm x)^{-1} = 1 \mp x + x^2 \mp x^3 + x^4 \mp x^5 + \cdots \qquad [x^2 < 1]$$

$$(1 \pm x)^{-2} = 1 \mp 2x + 3x^2 \mp 4x^3 + 5x^4 \mp \cdots \qquad [x^2 < 1]$$

$$e^x = 1 + x + \frac{x^2}{2!} + \frac{x^3}{3!} + \frac{x^4}{4!} + \cdots$$

$$e^{-x^2} = 1 - x^2 + \frac{x^4}{2!} - \frac{x^6}{3!} + \frac{x^8}{4!} - \cdots$$

$$a^x = 1 + x\log a + \frac{(x\log a)^2}{2!} + \frac{(x\log a)^3}{3!} + \cdots$$

$$\log x = (x-1) - \frac{1}{2}(x-1)^2 + \frac{1}{3}(x-1)^3 - \cdots \qquad [0 < x < 2]$$

$$\log x = \frac{x-1}{x} + \frac{1}{2}\left(\frac{x-1}{x}\right)^2 + \frac{1}{3}\left(\frac{x-1}{x}\right)^3 + \cdots \qquad \left[x > \frac{1}{2}\right]$$

$$\log x = 2\left[\left(\frac{x-1}{x+1}\right) + \frac{1}{3}\left(\frac{x-1}{x+1}\right)^3 + \frac{1}{5}\left(\frac{x-1}{x+1}\right)^5 + \cdots\right] \qquad [x > 0]$$

$$\log(1+x) = x - \frac{1}{2}x^2 + \frac{1}{3}x^3 - \frac{1}{4}x^4 + \cdots \qquad [x^2 < 1]$$

$$\log\left(\frac{1+x}{1-x}\right) = 2\left[x + \frac{1}{3}x^3 + \frac{1}{5}x^5 + \frac{1}{7}x^7 + \cdots\right] \qquad [x^2 < 1]$$

$$\log\left(\frac{x+1}{x-1}\right) = 2\left[\frac{1}{x} + \frac{1}{3}\left(\frac{1}{x}\right)^3 + \frac{1}{5}\left(\frac{1}{x}\right)^5 + \cdots\right] \qquad [x^2 > 1]$$

$$\sin x = x - \frac{x^3}{3!} + \frac{x^5}{5!} - \frac{x^7}{7!} + \cdots$$

$$\cos x = 1 - \frac{x^2}{2!} + \frac{x^4}{4!} - \frac{x^6}{6!} + \cdots$$

$$\tan x = x + \frac{x^3}{3} + \frac{2x^5}{15} + \frac{17x^7}{315}$$

$$+ \cdots + \frac{2^{2n}(2^{2n}-1)B_{2n-1}x^{2n-1}}{(2n)!} \qquad \left[x^2 < \frac{\pi^2}{4}\right]$$

$$\text{ctn}\, x = \frac{1}{x} - \frac{x}{3} - \frac{x^3}{45} - \frac{2x^5}{945} - \cdots - \frac{B_{2n-1}(2x)^{2n}}{(2n)!\,x} - \cdots \qquad [x^2 < \pi^2]$$

$$\sec x = 1 + \frac{x^2}{2!} + \frac{5x^4}{4!} + \frac{61x^6}{6!} + \cdots + \frac{B_{2n}x^{2n}}{(2n)!} + \cdots \qquad \left[x^2 < \frac{\pi^2}{4}\right]$$

$$\csc x = \frac{1}{x} + \frac{x}{3!} + \frac{7x^3}{3\cdot 5!} + \frac{31x^5}{3\cdot 7!} + \cdots$$

$$+ \frac{2(2^{2n+1}-1)}{(2n+2)!}B_{2n+1}x^{2n+1} + \cdots \qquad [x^2 < \pi^2]$$

$$\sin^{-1} x = x + \frac{x^3}{6} + \frac{(1\cdot 3)x^5}{(2\cdot 4)5} + \frac{(1\cdot 3\cdot 5)x^7}{(2\cdot 4\cdot 6)7} + \cdots \qquad [x^2 < 1]$$

$$\tan^{-1}x = x - \frac{1}{3}x^3 + \frac{1}{5}x^5 - \frac{1}{7}x^7 + \cdots \qquad [x^2 < 1]$$

$$\sec^{-1}x = \frac{\pi}{2} - \frac{1}{x} - \frac{1}{6x^3} - \frac{1\cdot 3}{(2\cdot 4)5x^5} - \frac{1\cdot 3\cdot 5}{(2\cdot 4\cdot 6)7x^7} - \cdots \qquad [x^2 > 1]$$

$$\sinh x = x + \frac{x^3}{3!} + \frac{x^5}{5!} + \frac{x^7}{7!} + \cdots$$

$$\cosh x = 1 + \frac{x^2}{2!} + \frac{x^4}{4!} + \frac{x^6}{6!} + \frac{x^8}{8!} + \cdots$$

$$\tanh x = (2^2-1)2^2B_1\frac{x}{2!} - (2^4-1)2^4B_3\frac{x^3}{4!} + (2^6-1)2^6B_5\frac{x^5}{6!} - \cdots \qquad \left[x^2 < \frac{\pi^2}{4}\right]$$

$$\text{ctnh}\, x = \frac{1}{x}\left(1 + \frac{2^2B_1x^2}{2!} - \frac{2^4B_3x^4}{4!} + \frac{2^6B_5x^6}{6!} - \cdots\right) \qquad [x^2 < \pi^2]$$

$$\text{sech}\, x = 1 - \frac{B_2x^2}{2!} + \frac{B_4x^4}{4!} - \frac{B_6x^6}{6!} + \cdots \qquad \left[x^2 < \frac{\pi^2}{4}\right]$$

$$\text{csch}\, x = \frac{1}{x} - (2-1)2B_1\frac{x}{2!} + (2^3-1)2B_3\frac{x^3}{4!} - \cdots \qquad [x^2 < \pi^2]$$

$$\sinh^{-1}x = x - \frac{1}{2}\frac{x^3}{3} + \frac{1\cdot 3}{2\cdot 4}\frac{x^5}{5} - \frac{1\cdot 3\cdot 5}{2\cdot 4\cdot 6}\frac{x^7}{7} + \cdots \qquad [x^2 < 1]$$

$$\tanh^{-1}x = x + \frac{x^3}{3} + \frac{x^5}{5} + \frac{x^7}{7} + \cdots \qquad [x^2 < 1]$$

$$\text{ctnh}^{-1}x = \frac{1}{x} + \frac{1}{3x^3} + \frac{1}{5x^5} + \cdots \qquad [x^2 > 1]$$

$$\text{csch}^{-1}x = \frac{1}{x} - \frac{1}{2\cdot 3x^3} + \frac{1\cdot 3}{2\cdot 4\cdot 5x^5} - \frac{1\cdot 3\cdot 5}{2\cdot 4\cdot 6\cdot 7x^7} + \cdots \qquad [x^2 > 1]$$

$$\int_0^x e^{-t^2}dt = x - \frac{1}{3}x^3 + \frac{x^5}{5\cdot 2!} - \frac{x^7}{7\cdot 3!} + \cdots$$

Error Function

The following function, known as the error function, erf x, arises frequently in applications:

$$\text{erf}\, x = \frac{2}{\sqrt{\pi}}\int_0^x e^{-t^2}dt$$

The integral cannot be represented in terms of a finite number of elementary functions; therefore, values of erf x have been compiled in tables. The following is the series for erf x:

$$\operatorname{erf} x = \frac{2}{\sqrt{\pi}}\left[x - \frac{x^3}{3} + \frac{x^5}{5\cdot 2!} - \frac{x^7}{7\cdot 3!} + \cdots\right]$$

There is a close relation between this function and the area under the standard normal curve. For evaluation it is convenient to use z instead of x; then erf z may be evaluated from the area $F(z)$ by use of the relation

$$\operatorname{erf} z = 2F(\sqrt{2}z)$$

Example

$$\operatorname{erf}(0.5) = 2F[(1.414)(0.5)] = 2F(0.707)$$

By interpolation, $F(0.707) = 0.260$; thus, $\operatorname{erf}(0.5) = 0.520$.

Series Expansion

The expression in parentheses following certain series indicates the region of convergence. If not otherwise indicated, it is understood that the series converges for all finite values of x.

Binomial

$$(x+y)^n = x^n + nx^{n-1}y + \frac{n(n-1)}{2!}x^{n-2}y^2 + \frac{n(n-1)(n-2)}{3!}x^{n-3}y^3 + \cdots \quad [y^2 < x^2]$$

$$(1 \pm x)^n = 1 \pm nx + \frac{n(n-1)x^2}{2!} \pm \frac{n(n-1)(n-2)x^3}{3!} + \cdots \quad [x^2 < 1]$$

$$(1 \pm x)^{-n} = 1 \mp nx + \frac{n(n+1)x^2}{2!} \mp \frac{n(n+1)(n+2)x^3}{3!} + \cdots \quad [x^2 < 1]$$

$$(1 \pm x)^{-1} = 1 \mp x + x^2 \mp x^3 + x^4 \mp x^5 + \cdots \quad [x^2 < 1]$$

$$(1 \pm x)^{-2} = 1 \mp 2x + 3x^2 \mp 4x^3 + 5x^4 \mp 6x^5 + \cdots \quad [x^2 < 1]$$

Reversion of Series

Let a series be represented by

$$y = a_1x + a_2x^2 + a_3x^3 + a_4x^4 + a_5x^5 + a_6x^6 + \cdots \quad (a_1 \neq 0)$$

To find the coefficients of the series

$$x = A_1y + A_2y^2 + A_3y^3 + A_4y^4 + \cdots$$

$$A_1 = \frac{1}{a_1} \quad A_2 = -\frac{a_2}{a_1^3} \quad A_3 = \frac{1}{a_1^5}(2a_2^2 - a_1a_3)$$

$$A_4 = \frac{1}{a_1^7}(5a_1a_2a_3 - a_1^2a_4 - 5a_2^3)$$

$$A_5 = \frac{1}{a_1^9}(6a_1^2 a_2 a_4 + 3a_1^2 a_3^2 + 14a_2^4 - a_1^3 a_5 - 21a_1 a_2^2 a_3)$$

$$A_6 = \frac{1}{a_1^{11}}(7a_1^3 a_2 a_5 + 7a_1^3 a_3 a_4 + 84a_1 a_2^3 a_3 - a_1^4 a_6 - 28a_1^2 a_2^2 a_4 - 28a_1^2 a_2 a_3^2 - 42a_2^5)$$

$$A_7 = \frac{1}{a_1^{13}}(8a_1^4 a_2 a_6 + 8a_1^4 a_3 a_5 + 4a_1^4 a_4^2 + 120a_1^2 a_2^3 a_4 + 180a_1^2 a_2^2 a_3^2 + 132a_2^6 - a_1^5 a_7$$

$$-36a_1^3 a_2^2 a_5 - 72a_1^3 a_2 a_3 a_4 - 12a_1^3 a_3^3 - 330a_1 a_2^4 a_3)$$

Taylor

1. $$f(x) = f(a) + (x-a)f'(a) + \frac{(x-a)^2}{2!}f''(a) + \frac{(x-a)^3}{3!}f'''(a)$$

$$+ \cdots + \frac{(x-a)^n}{n!}f^{(n)}(a) + \cdots \qquad \text{(Taylor's series)}$$

(Increment form)

2. $$f(x+h) = f(x) + hf'(x) + \frac{h^2}{2!}f''(x) + \frac{h^3}{3!}f'''(x) + \cdots$$

$$= f(h) + xf'(h) + \frac{x^2}{2!}f''(h) + \frac{x^3}{3!}f'''(h) + \cdots$$

3. If $f(x)$ is a function possessing derivatives of all orders throughout the interval $a \le x \le b$, then there is a value X, with $a < X < b$, such that

$$f(b) = f(a) + (b-a)f'(a) + \frac{(b-a)^2}{2!}f''(a) + \cdots$$

$$+ \frac{(b-a)^{n-1}}{(n-1)!}f^{(n-1)}(a) + \frac{(b-a)^n}{n!}f^{(n)}(X)$$

$$f(a+h) = f(a) + hf'(a) + \frac{h^2}{2!}f''(a) + \cdots + \frac{h^{n-1}}{(n-1)!}f^{(n-1)}(a)$$

$$+ \frac{h^n}{n!}f^{(n)}(a+\theta h), \quad b = a+h, \quad 0 < \theta < 1$$

or

$$f(x) = f(a) + (x-a)f'(a) + \frac{(x-a)^2}{2!}f''(a) + \cdots + (x-a)^{n-1}\frac{f^{(n-1)}(a)}{(n-1)!} + R_n$$

where

$$R_n = \frac{f^{(n)}[a + \theta \cdot (x-a)]}{n!}(x-a)^n, \quad 0 < \theta < 1.$$

The above forms are known as Taylor's series with the remainder term.

4. Taylor's series for a function of two variables:

$$\text{If}\left(h\frac{\partial}{\partial x}+k\frac{\partial}{\partial x}\right)f(x,y)=h\frac{\partial f(x,y)}{\partial x}+k\frac{\partial f(x,y)}{\partial y};$$

$$\left(h\frac{\partial}{\partial x}+k\frac{\partial}{\partial y}\right)^2 f(x,y)=h^2\frac{\partial^2 f(x,y)}{\partial x^2}+2hk\frac{\partial^2 f(x,y)}{\partial x\,\partial y}+k^2\frac{\partial^2 f(x,y)}{\partial y^2}$$

etc., and if

$$\left(h\frac{\partial}{\partial x}+k\frac{\partial}{\partial y}\right)^n f(x,y)\bigg|_{\substack{x=a\\y=b}}$$

where the bar and subscripts mean that after differentiation we are to replace x by a and y by b,

$$f(a+h,b+k)=f(a,b)+\left(h\frac{\partial}{\partial x}+k\frac{\partial}{\partial y}\right)f(x,y)\bigg|_{\substack{x=a\\y=b}}+\cdots$$

$$+\frac{1}{n!}\left(h\frac{\partial}{\partial x}+k\frac{\partial}{\partial y}\right)^n f(x,y)\bigg|_{\substack{x=a\\y=b}}+\cdots$$

MacLaurin

$$f(x)=f(0)+xf'(0)+\frac{x^2}{2!}f''(0)+\frac{x^3}{3!}f'''(0)+\cdots+x^{n-1}\frac{f^{(n-1)}(0)}{(n-1)!}+R_n$$

where

$$R_n=\frac{x^n f^{(n)}(\theta x)}{n!},\quad 0<\theta<1$$

Exponential

$$e=1+\frac{1}{1!}+\frac{1}{2!}+\frac{1}{3!}+\frac{1}{4!}+\cdots$$

$$e^x=1+x+\frac{x^2}{2!}+\frac{x^3}{3!}+\frac{x^4}{4!}+\cdots \qquad \text{(all real values of } x\text{)}$$

$$a^x=1+x\log_e a+\frac{(x\log_e a)^2}{2!}+\frac{(x\log_e a)^3}{3!}+\cdots$$

$$e^x=e^a\left[1+(x-a)+\frac{(x-a)^2}{2!}+\frac{(x-a)^3}{3!}+\cdots\right]$$

Logarithmic

$$\log_e x = \frac{x-1}{x} + \frac{1}{2}\left(\frac{x-1}{x}\right)^2 + \frac{1}{3}\left(\frac{x-1}{x}\right)^3 + \cdots \qquad (x > \tfrac{1}{2})$$

$$\log_e x = (x-1) - \frac{1}{2}(x-1)^2 + \frac{1}{3}(x-1)^3 - \cdots \qquad (2 \ge x > 0)$$

$$\log_e x = 2\left[\frac{x-1}{x+1} + \frac{1}{3}\left(\frac{x-1}{x+1}\right)^3 + \frac{1}{5}\left(\frac{x-1}{x+1}\right)^5 + \cdots\right] \qquad (x > 0)$$

$$\log_e(1+x) = x - \frac{1}{2}x^2 + \frac{1}{3}x^3 - \frac{1}{4}x^4 + \cdots \qquad (-1 < x \le 1)$$

$$\log_e(n+1) - \log_e(n-1) = 2\left[\frac{1}{n} + \frac{1}{3n^3} + \frac{1}{5n^5} + \cdots\right]$$

$$\log_e(a+x) = \log_e a + 2\left[\frac{x}{2a+x} + \frac{1}{3}\left(\frac{x}{2a+x}\right)^3 + \frac{1}{5}\left(\frac{x}{2a+x}\right)^5 + \cdots\right] \qquad (a > 0, -a < x < +\infty)$$

$$\log_e \frac{1+x}{1-x} = 2\left[x + \frac{x^3}{3} + \frac{x^5}{5} + \cdots + \frac{x^{2n-1}}{2n-1} + \cdots\right] \qquad (-1 < x < 1)$$

$$\log_e x = \log_e a + \frac{(x-a)}{a} + \frac{(x-a)^2}{2a^2} + \frac{(x-a)^3}{3a^3} - \cdots \qquad (0 < x \le 2a)$$

Trigonometric

$$\sin x = x - \frac{x^3}{3!} + \frac{x^5}{5!} - \frac{x^7}{7!} + \cdots \qquad \text{(all real values of } x)$$

$$\cos x = 1 - \frac{x^2}{2!} + \frac{x^4}{4!} - \frac{x^6}{6!} + \cdots \qquad \text{(all real values of } x)$$

$$\tan x = x + \frac{x^3}{3} + \frac{2x^5}{15} + \frac{17x^7}{315} + \frac{62x^9}{2835} + \cdots + \frac{(-1)^{n-1}2^{2n}(2^{2n}-1)B_{2n}}{(2n)!}x^{2n-1} + \cdots$$

($x^2 < \pi^2/4$, and B_n represents the nth Bernoulli number)

$$\cot x = \frac{1}{x} - \frac{x}{3} - \frac{x^2}{45} - \frac{2x^5}{945} - \frac{x^7}{4725} - \cdots - \frac{(-1)^{n+1}2^{2n}}{(2n)!}B_{2n}x^{2n-1} + \cdots$$

($x^2 < \pi^2$, and B_n represents the nth Bernoulli number)

217.3 Differential Calculus

Notation

For the following equations, the symbols $f(x)$, $g(x)$, etc., represent functions of x. The value of a function $f(x)$ at $x = a$ is denoted $f(a)$. For the function $y = f(x)$ the derivative of y with respect to x is denoted by one of the following:

$$\frac{dy}{dx}, \quad f'(x), \quad D_x y, \quad y'$$

Higher derivatives are as follows:

$$\frac{d^2y}{dx^2} = \frac{d}{dx}\left(\frac{dy}{dx}\right) = \frac{d}{dx} f'(x) = f''(x)$$

$$\frac{d^3y}{dx^3} = \frac{d}{dx}\left(\frac{d^2y}{dx^2}\right) = \frac{d}{dx} f''(x) = f'''(x)$$

$$\vdots$$

and values of these at $x = a$ are denoted $f''(a)$, $f'''(a)$, and so on (see Table of Derivatives in Appendix).

Slope of a Curve

The tangent line at point $P(x, y)$ of the curve $y = f(x)$ has a slope $f'(x)$ provided that $f'(x)$ exists at P. The slope at P is defined to be that of the tangent line at P. The tangent line at $P(x_1, y_1)$ is given by

$$y - y_1 = f'(x_1)(x - x_1)$$

The *normal line* to the curve at $P(x_1, y_1)$ has slope $-1/f'(x_1)$ and thus obeys the equation

$$y - y_1 = [-1/f'(x_1)](x - x_1)$$

(The slope of a vertical line is not defined.)

Angle of Intersection of Two Curves

Two curves, $y = f_1(x)$ and $y = f_2(x)$, that intersect at a point $P(X, Y)$ where derivatives $f_1'(X), f_2'(X)$ exist, have an angle (α) of intersection given by

$$\tan\alpha = \frac{f_2'(X) - f_1'(X)}{1 + f_2'(X) \cdot f_1'(X)}$$

If $\tan\alpha > 0$, then α is the acute angle; if $\tan\alpha < 0$, then α is the obtuse angle.

Radius of Curvature

The radius of curvature R of the curve $y = f(x)$ at the point $P(x, y)$ is

$$R = \frac{\left\{1+[f'(x)]^2\right\}^{3/2}}{f''(x)}$$

In polar coordinates (θ, r) the corresponding formula is

$$R = \frac{\left[r^2 + \left(\frac{dr}{d\theta}\right)^2\right]^{3/2}}{r^2 + 2\left(\frac{dr}{d\theta}\right)^2 - r\frac{d^2r}{d\theta^2}}$$

The *curvature* K is $1/R$.

Relative Maxima and Minima

The function f has a relative maximum at $x = a$ if $f(a) \geq f(a+c)$ for all values of c (positive or negative) that are sufficiently near zero. The function f has a relative minimum at $x = b$ if $f(b) \leq f(b + c)$ for all values of c that are sufficiently close to zero. If the function f is defined on the closed interval $x_1 \leq x \leq x_2$ and has a relative maximum or minimum at $x = a$, where $x_1 < a < x_2$, and if the derivative $f'(x)$ exists at $x = a$, then $f'(a) = 0$. It is noteworthy that a relative maximum or minimum may occur at a point where the derivative does not exist. Further, the derivative may vanish at a point that is neither a maximum nor a minimum for the function. Values of x for which $f'(x) = 0$ are called "critical values." To determine whether a critical value of x, say x_c, is a relative maximum or minimum for the function at x_c, one may use the second derivative test:

1. If $f''(x_c)$ is positive, $f(x_c)$ is a minimum.
2. If $f''(x_c)$ is negative, $f(x_c)$ is a maximum.
3. If $f''(x_c)$ is zero, no conclusion may be made.

The sign of the derivative as x advances through x_c may also be used as a test. If $f'(x)$ changes from positive to zero to negative, then a maximum occurs at x_c, whereas a change in $f'(x)$ from negative to zero to positive indicates a minimum. If $f'(x)$ does not change sign as x advances through x_c, then the point is neither a maximum nor a minimum.

Points of Inflection of a Curve

The sign of the second derivative of f indicates whether the graph of $y = f(x)$ is concave upward or concave downward:

$f''(x) > 0$: concave upward
$f''(x) > 0$: concave downward

A point of the curve at which the direction of concavity changes is called a point of inflection (Figure 217.2). Such a point may occur where $f''(x) = 0$ or where $f''(x)$ becomes infinite. More precisely, if

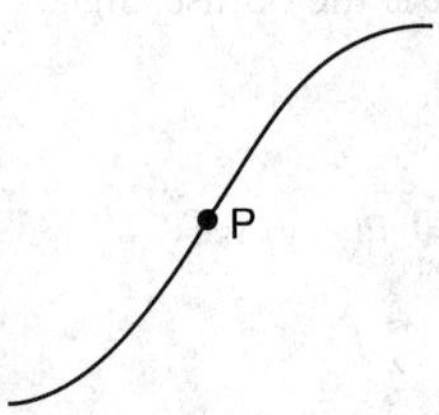

FIGURE 217.2 Point of inflection.

the function $y = f(x)$ and its first derivative $y' = f'(x)$ are continuous in the interval $a \le x \le b$, and if $y'' = f''(x)$ exists in $a < x < b$, then the graph of $y = f(x)$ for $a < x < b$ is concave upward if $f''(x)$ is positive and concave downward if $f''(x)$ is negative.

Taylor's Formula

If f is a function that is continuous on an interval that contains a and x, and if its first $(n + 1)$ derivatives are continuous on this interval, then

$$f(x) = f(a) + f'(a)(x-a) + \frac{f''(a)}{2!}(x-a)^2 + \frac{f'''(a)}{3!}(x-a)^3$$

$$+ \cdots + \frac{f^{(n)}(a)}{n!}(x-a)^n + R$$

where R is called the *remainder*. There are various common forms of the remainder:

Lagrange's Form

$$R = f^{(n+1)}(\beta) \cdot \frac{(x-a)^{n+1}}{(n+1)!}, \quad \beta \text{ between } a \text{ and } x$$

Cauchy's Form

$$R = f^{(n+1)}(\beta) \cdot \frac{(x-B)^n (x-a)}{n!}, \quad \beta \text{ between } a \text{ and } x$$

Integral Form

$$R = \int_a^x \frac{(x-t)^n}{n!} f^{(n+1)}(t)\,dt$$

Indeterminant Forms

If $f(x)$ and $g(x)$ are continuous in an interval that includes $x = a$, and if $f(a) = 0$ and $g(a) = 0$, the limit $\lim_{x \to a} [f(x)/g(x)]$ takes the form "0/0," called an *indeterminant form. L'Hôpital's rule* is

$$\lim_{x \to a} \frac{f(x)}{g(x)} = \lim_{x \to a} \frac{f'(x)}{g'(x)}$$

Similarly, it may be shown that if $f(x) \to \infty$ and $g(x) \to \infty$ as $x \to a$, then

$$\lim_{x \to a} \frac{f(x)}{g(x)} = \lim_{x \to a} \frac{f'(x)}{g'(x)}$$

(The above holds for $x \to \infty$.)

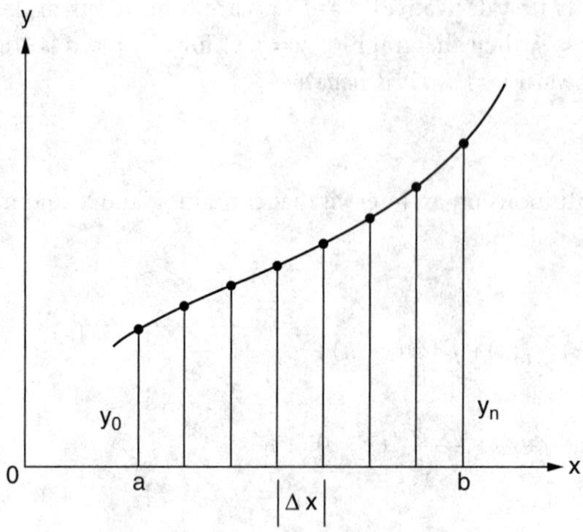

FIGURE 217.3 Trapezoidal rule for area.

Examples

$$\lim_{x\to 0}\frac{\sin x}{x}=\lim_{x\to 0}\frac{\cos x}{1}=1$$

$$\lim_{x\to\infty}\frac{x^2}{e^x}=\lim_{x\to\infty}\frac{2x}{e^x}=\lim_{x\to\infty}\frac{2}{e^x}=0$$

Numerical Methods

1. *Newton's method* for approximating roots of the equation $f(x) = 0$; A first estimate x_1 of the root is made; then, provided that $f'(x_1)\neq 0$, a better approximation is x_2:

$$x_2 = x_1 - \frac{f(x)}{f'(x_1)}$$

The process may be repeated to yield a third approximation, x_3, to the root:

$$x_3 = x_2 - \frac{f(x_2)}{f'(x_2)}$$

provided $f'(x_2)$ exists. The process may be repeated. (In certain rare cases the process will not converge.)

2. *Trapezoidal rule for areas* (Figure 217.3): For the function $y = f(x)$ defined on the interval (a, b) and positive there, take n equal subintervals of width $\Delta x = (b - a)/n$. The area bounded by the curve between $x = a$ and $x = b$ [or definite integral of $f(x)$] is approximately the sum of trapezoidal areas, or

$$A \sim \left(\frac{1}{2}y_0 + y_1 + y_2 + \cdots + y_{n-1} + \frac{1}{2}y_n\right)(\Delta x)$$

Estimation of the error (E) is possible if the second derivative can be obtained:

$$E = \frac{b-a}{12} f''(c)(\Delta x)^2$$

where c is some number between a and b.

Functions of Two Variables

For the function of two variables, denoted $z = f(x, y)$, if y is held constant, say at $y = y_1$, then the resulting function is a function of x only. Similarly, x may be held constant at x_1, to give the resulting function of y.

The Gas Laws

A familiar example is afforded by the ideal gas law relating the pressure p, the volume V, and the absolute temperature T of an ideal gas:

$$pV = nRT$$

where n is the number of moles and R is the gas constant per mole, 8.31 ($J \cdot K^{-1} \cdot mole^{-1}$). By rearrangement, any one of the three variables may be expressed as a function of the other two. Further, either one of these two may be held constant. If T is held constant, then we get the form known as Boyle's law:

$$p = kV^{-1} \qquad \text{(Boyle's law)}$$

where we have denoted nRT by the constant k and, of course, $V > 0$. If the pressure remains constant, we have Charles' law:

$$V = bT \qquad \text{(Charles' law)}$$

where the constant b denotes nR/p. Similarly, volume may be kept constant:

$$p = aT$$

where now the constant, denoted a, is nR/V.

Partial Derivatives

The physical example afforded by the ideal gas law permits clear interpretations of processes in which one of the variables is held constant. More generally, we may consider a function $z = f(x, y)$ defined over some region of the xy plane in which we hold one of the two coordinates, say y, constant. If the resulting function of x is differentiable at a point (x, y), we denote this derivative by one of the notations

$$f_x, \quad \delta f / dx, \quad \delta z / dx$$

called the *partial derivative with respect to x*. Similarly, if x is held constant and the resulting function of y is differentiable, we get the *partial derivative with respect to y*, denoted by one of the following:

$$f_y, \quad \delta f / dy, \quad \delta z / dy$$

Example. Given $z = x^4y^3 - y \sin x + 4y$, then

$$\delta z / dx = 4(xy)^3 - y \cos x$$

$$\delta z / dy = 3x^4 y^2 - \sin x + 4$$

217.4 Integral Calculus

Indefinite Integral

If $F(x)$ is differentiable for all values of x in the interval (a, b) and satisfies the equation $dy/dx = f(x)$, then $F(x)$ is an integral of $f(x)$ with respect to x. The notation is $F(x) = \int f(x)\,dx$ or, in differential form, $dF(x) = f(x)\,dx$.

For any function $F(x)$ that is an integral of $f(x)$, it follows that $F(x) + C$ is also an integral. We thus write

$$\int f(x)\,dx = F(x) + C$$

Definite Integral

Let $f(x)$ be defined on the interval $[a, b]$ which is partitioned by points $x_1, x_2, \ldots, x_j, \ldots, x_{n-1}$ between $a = x_0$ and $b = x_n$. The jth interval has length $\Delta x_j = x_j - x_{j-1}$, which may vary with j. The sum

$$\sum_{j=1}^{n} f(\upsilon_j)\Delta x_j,$$

where υ_j is arbitrarily chosen in the jth subinterval, depends on the numbers $x_0, \ldots, x_n$ and the choice of the v as well as f; but if such sums approach a common value as all Δx approach zero, then this value is the definite integral of f over the interval (a, b) and is denoted $\int_a^b f(x)\,dx$. The *fundamental theorem of integral calculus* states that

$$\int_a^b f(x)dx = F(b) - F(a),$$

where F is any continuous indefinite integral of f in the interval (a, b).

Properties

$$\int_a^b [f_1(x) + f_2(x) + \cdots + f_j(x)]dx = \int_a^b f_1(x)\,dx + \int_a^b f_2(x)dx + \cdots + \int_a^b f_j(x)\,dx$$

$$\int_a^b cf(x)\,dx = c\int_a^b f(x)\,dx, \quad \text{if } c \text{ is a constant}$$

$$\int_a^b f(x)\,dx = -\int_b^a f(x)\,dx$$

$$\int_a^b f(x)\,dx = \int_a^c f(x)\,dx + \int_c^b f(x)\,dx$$

Common Applications of the Definite Integral

Area (Rectangular Coordinates)

Given the function $y = f(x)$ such that $y > 0$ for all x between a and b, the area bounded by the curve $y = f(x)$, the x axis, and the vertical lines $x = a$ and $x = b$ is

$$A = \int_a^b f(x)dx$$

Length of Arc (Rectangular Coordinates)

Given the smooth curve $f(x, y) = 0$ from point (x_1, y_1) to point (x_2, y_2), the length between these points is

$$L = \int_{x_1}^{x_2} \sqrt{1+(dy/dx)^2}\,dx$$

$$L = \int_{y_1}^{y_2} \sqrt{1+(dx/dy)^2}\,dy$$

Mean Value of a Function

The mean value of a function $f(x)$ continuous on $[a, b]$ is

$$\frac{1}{(b-a)}\int_a^b f(x)dx$$

Area (Polar Coordinates)

Given the curve $r = f(\theta)$, continuous and nonnegative for $\theta_1 \le \theta \le \theta_2$, the area enclosed by this curve and the radial lines $\theta = \theta_1$ and $\theta = \theta_2$ is given by

$$A = \int_{\theta_1}^{\theta_2} \frac{1}{2}[f(\theta)]^2 d\theta$$

Length of Arc (Polar Coordinates)

Given the curve $r = f(\theta)$ with continuous derivative $f'(\theta)$ on $\theta_1 \le \theta \le \theta_2$, the length of arc from $\theta = \theta_1$ to $\theta = \theta_2$ is

$$L = \int_{\theta_1}^{\theta_2} \sqrt{[f(\theta)]^2 + [f'(\theta)]^2}\, d\theta$$

Volume of Revolution

Given a function $y = f(x)$ continuous and nonnegative on the interval (a, b), when the region bounded by $f(x)$ between a and b is revolved about the x axis, the volume of revolution is

$$V = \pi\int_a^b [f(x)]^2 dx$$

Surface Area of Revolution (Revolution about the x axis, between *a* and *b*)

If the portion of the curve $y = f(x)$ between $x = a$ and $x = b$ is revolved about the x axis, the area A of the surface generated is given by the following:

$$A = \int_a^b 2\pi f(x)\{1+[f'(x)]^2\}^{1/2} dx$$

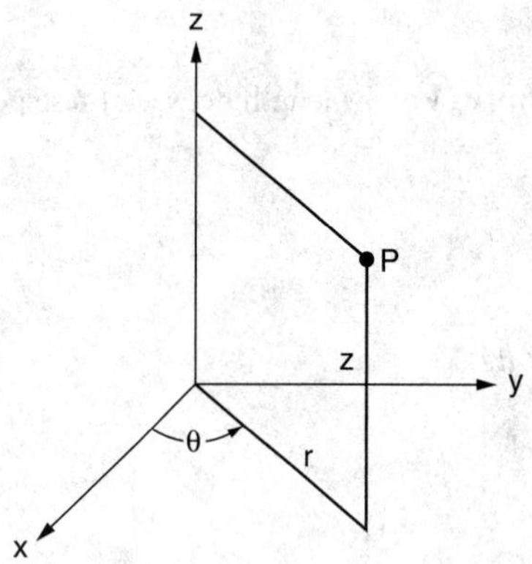

FIGURE 217.4 Cylindrical coordinates.

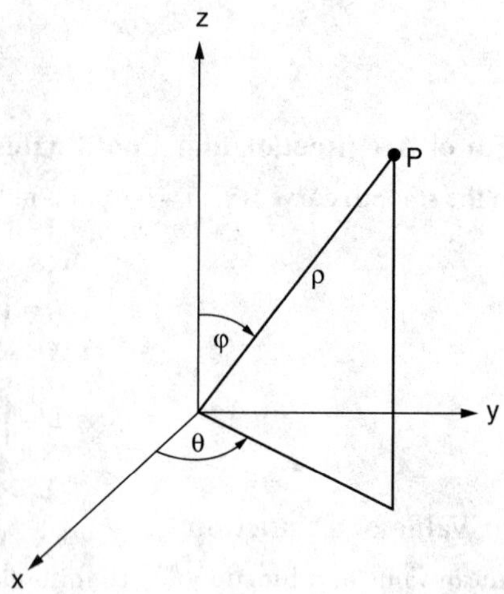

FIGURE 217.5 Spherical coordinates.

Work

If a variable force $f(x)$ is applied to an object in the direction of motion along the x axis between $x = a$ and $x = b$, the work done is

$$W = \int_a^b f(x)\,dx$$

Cylindrical and Spherical Coordinates

1. Cylindrical coordinates (Figure 217.4):

$$x = r\cos\theta$$

$$y = r\sin\theta$$

Element of volume $dV = r\,dr\,d\theta\,dz$.

2. Spherical coordinates (Figure 217.5):

$$x = \rho\sin\phi\cos\theta$$

$$y = \rho\sin\phi\sin\theta$$

$$z = \rho\cos\phi$$

Element of volume $dV = \rho^2 \sin\phi\,d\rho\,d\phi\,d\theta$.

Double Integration

The evaluation of a double integral of $f(x, y)$ over a plane region R,

$$\iint_R f(x, y)\,dA$$

is practically accomplished by iterated (repeated) integration. For example, suppose that a vertical straight line meets the boundary of R in at most two points so that there is an upper boundary, $y = y_2(x)$, and a

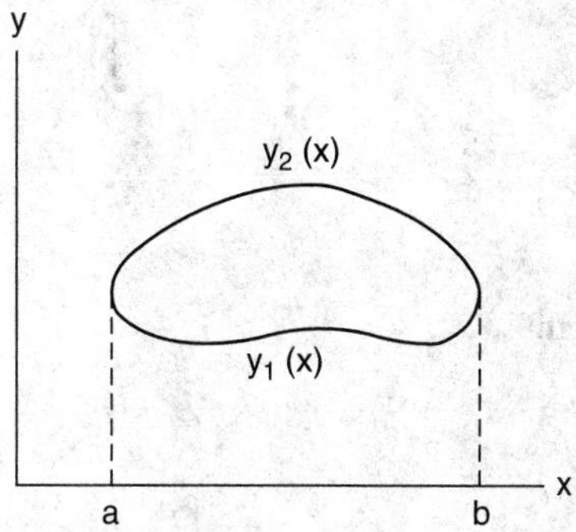

FIGURE 217.6 Region R bounded by $y_2(x)$ and $y_1(x)$.

lower boundary, $y = y_1(x)$. Also, it is assumed that these functions are continuous from a to b (see Figure 217.6). Then

$$\iint_R f(x,y)dA = \int_a^b \left(\int_{y_1(x)}^{y_2(y)} f(x,y)dy \right) dx$$

If R has left-hand boundary, $x = x_1(y)$, and a right-hand boundary, $x = x_2(y)$, which are continuous from c to d (the extreme values of y in R), then

$$\iint_R f(x,y)dA = \int_c^d \left(\int_{x_1(y)}^{x_2(y)} f(x,y)dx \right) dy$$

Such integrations are sometimes more convenient in polar coordinates, $x = r\cos\theta$, $y = r\sin\theta$, $dA = r\,dr\,d\theta$.

Surface Area and Volume by Double Integration

For the surface given by $z = f(x, y)$, which projects onto the closed region R of the xy plane, one may calculate the volume V bounded above by the surface and below by R, and the surface area S by the following:

$$V = \iint_R z\,dA = \iint_R f(x,y)dx\,dy$$

$$S = \iint_R \left[1 + (\delta z/\delta x)^2 + (\delta z/\delta y)^2\right]^{1/2} dx\,dy$$

[In polar coordinates, (r, θ), we replace dA by $r\,dr\,d\theta$.]

Centroid

The centroid of a region R of the xy plane is a point (x', y') where

$$x' = \frac{1}{A}\iint_R x\,dA, \qquad y' = \frac{1}{A}\iint_R y\,dA$$

and A is the area of the region.

Example. For the circular sector of angle 2α and radius R, the area A is αR^2; the integral needed for x', expressed in polar coordinates, is

$$\iint x\,dA = \int_{-\alpha}^{\alpha}\int_{0}^{R}(r\cos\theta)r\,dr\,d\theta$$

$$= \left[\frac{R^3}{3}\sin\theta\right]_{-\alpha}^{+\alpha} = \frac{2}{3}R^3\sin\theta$$

and thus,

$$x' = \frac{\frac{2}{3}R^3\sin\alpha}{\alpha R^2} = \frac{2}{3}R\frac{\sin\alpha}{\alpha}$$

Centroids of some common regions are shown in Table 217.1.

TABLE 217.1 Centroids

	Area	x'	y'
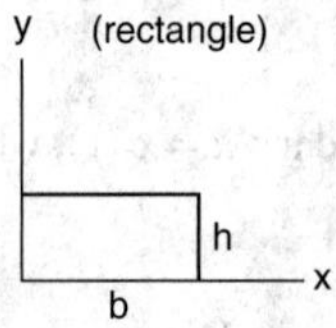	bh	$b/2$	$h/2$
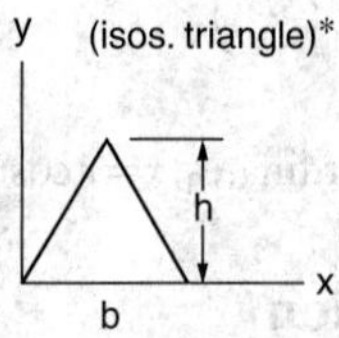	$bh/2$	$b/2$	$h/3$
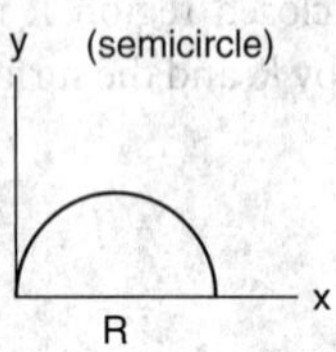	$\pi R^2/2$	R	$4R/3\pi$
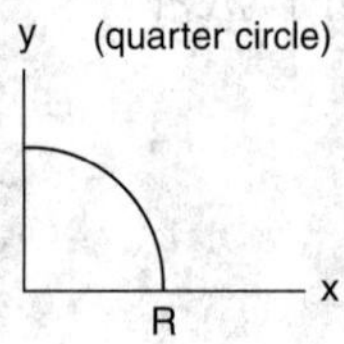	$\pi R^2/4$	$4R/3\pi$	$4R/3\pi$
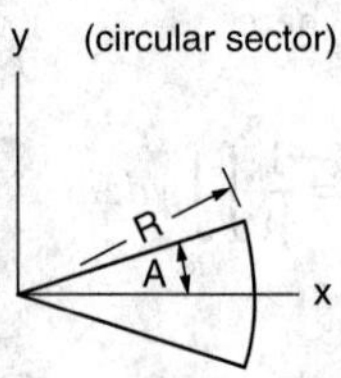	R^2A	$2R\sin A/3A$	0

* $y' = h/3$ for any triangle of altitude h.

217.5 Special Functions

Hyperbolic Functions

$$\sinh x = \frac{e^x - e^{-x}}{2} \qquad \operatorname{csch} x = \frac{1}{\sinh x}$$

$$\cosh x = \frac{e^x + e^{-x}}{2} \qquad \operatorname{sech} x = \frac{1}{\cosh x}$$

$$\tanh x = \frac{e^x - e^{-x}}{e^x + e^{-x}} \qquad \operatorname{ctnh} x = \frac{1}{\tanh x}$$

$$\sinh(-x) = -\sinh x \qquad \operatorname{ctnh}(-x) = -\operatorname{ctnh} x$$

$$\cosh(-x) = \cosh x \qquad \operatorname{sech}(-x) = \operatorname{sech} x$$

$$\tanh(-x) = -\tanh x \qquad \operatorname{csch}(-x) = -\operatorname{csch} x$$

$$\tanh x = \frac{\sinh x}{\cosh x} \qquad \operatorname{ctnh} x = \frac{\cosh x}{\sinh x}$$

$$\cosh^2 x - \sinh^2 x = 1 \qquad \cosh^2 x = \frac{1}{2}(\cosh 2x + 1)$$

$$\sinh^2 x = \frac{1}{2}(\cosh 2x - 1) \qquad \operatorname{ctnh}^2 x - \operatorname{csch}^2 x = 1$$

$$\operatorname{csch}^2 x - \operatorname{sech}^2 x = \operatorname{csch}^2 x \operatorname{sech}^2 x \qquad \tanh^2 x + \operatorname{sech}^2 x = 1$$

$$\sinh(x+y) = \sinh x \cosh y + \cosh x \sinh y$$

$$\cosh(x+y) = \cosh x \cosh y + \sinh x \sinh y$$

$$\sinh(x-y) = \sinh x \cosh y - \cosh x \sinh y$$

$$\cosh(x-y) = \cosh x \cosh y - \sinh x \sinh y$$

$$\tanh(x+y) = \frac{\tanh x + \tanh y}{1 + \tanh x \tanh y}$$

$$\tanh(x-y) = \frac{\tanh x - \tanh y}{1 - \tanh x \tanh y}$$

Bessel Functions

Bessel functions, also called cylindrical functions, arise in many physical problems as solutions of the differential equation

$$x^2 y'' + xy' + (x^2 - n^2)y = 0$$

which is known as Bessel's equation. Certain solutions, known as *Bessel functions of the first kind of order n*, are given by

$$J_n(x)=\sum_{k=0}^{\infty}\frac{(-1)^k}{k!\Gamma(n+k+1)}\left(\frac{x}{2}\right)^{n+2k}$$

$$J_{-n}(x)=\sum_{k=0}^{\infty}\frac{(-1)^k}{k!\Gamma(-n+k+1)}\left(\frac{x}{2}\right)^{-n+2k}$$

In the above it is noteworthy that the gamma function must be defined for the negative argument $q:\Gamma(q) = \Gamma(q+1)/q$, provided that q is not a negative integer. When q is a negative integer, $1/\Gamma(q)$ is defined to be zero. The functions $J_{-n}(x)$ and $J_n(x)$ are solutions of Bessel's equation for all real n. It is seen, for $n =$ 1, 2, 3,…, that

$$J_{-n}(x)=(-1)^n J_n(x)$$

and, therefore, these are not independent; hence, a linear combination of these is not a general solution. When, however, n is not a positive integer, a negative integer, or zero, the linear combination with arbitrary constants c_1 and c_2,

$$y=c_1 J_n(x)+c_2 J_{-n}(x)$$

is the general solution of the Bessel differential equation.

The zero-order function is especially important as it arises in the solution of the heat equation (for a "long" cylinder):

$$J_0(x)=1-\frac{x^2}{2^2}+\frac{x^4}{2^2 4^2}-\frac{x^6}{2^2 4^2 6^2}+\cdots$$

while the following relations show a connection to the trigonometric functions:

$$J_{1/2}(x)=\left[\frac{2}{\pi x}\right]^{1/2}\sin x$$

$$J_{-1/2}(x)=\left[\frac{2}{\pi x}\right]^{1/2}\cos x$$

The following recursion formula gives $J_{n+1}(x)$ for any order in terms of lower-order functions:

$$\frac{2n}{x}J_n(x)=J_{n-1}(x)+J_{n+1}(x)$$

Legendre Polynomials

If Laplace's equation, $\nabla^2 V = 0$, is expressed in spherical coordinates, it is

$$r^2\sin\theta\frac{\delta^2 V}{\delta r^2}+2r\sin\theta\frac{\delta V}{\delta r}+\sin\theta\frac{\delta^2 V}{\delta\theta^2}+\cos\theta\frac{\delta V}{\delta\theta}+\frac{1}{\sin\theta}\frac{\delta^2 V}{\delta\phi^2}=0$$

and any of its solutions, $V(r, \theta, \phi)$, are known as *spherical harmonics*. The solution as a product

$$V(r,\theta,\phi) = R(r)\Theta(\theta)$$

which is independent of ϕ, leads to

$$\sin^2\theta\,\Theta'' + \sin\theta\cos\theta\,\Theta' + [n(n+1)\sin^2\theta]\Theta = 0$$

Rearrangement and substitution of $x = \cos\theta$ leads to

$$(1-x^2)\frac{d^2\Theta}{dx^2} - 2x\frac{d\Theta}{dx} + n(n+1)\Theta = 0$$

known as *Legendre's equation*. Important special cases are those in which n is zero or a positive integer, and, for such cases, Legendre's equation is satisfied by polynomials called Legendre polynomials, $P_n(x)$. A short list of Legendre polynomials, expressed in terms of x and $\cos\theta$, is given below. These are given by the following general formula:

$$P_n(x) = \sum_{j=0}^{L} \frac{(-1)^j(2n-2j)!}{2^n j!(n-j)!(n-2j)!} x^{n-2j}$$

where $L = n/2$ if n is even and $L = (n-1)/2$ if n is odd.

$$P_0(x) = 1$$

$$P_1(x) = x$$

$$P_2(x) = \frac{1}{2}(3x^2 - 1)$$

$$P_3(x) = \frac{1}{2}(5x^3 - 3x)$$

$$P_4(x) = \frac{1}{8}(35x^4 - 30x^2 + 3)$$

$$P_5(x) = \frac{1}{8}(63x^5 - 70x^3 + 15x)$$

$$P_0(\cos\theta) = 1$$

$$P_1(\cos\theta) = \cos\theta$$

$$P_2(\cos\theta) = \frac{1}{4}(3\cos 2\theta + 1)$$

$$P_3(\cos\theta) = \frac{1}{8}(5\cos 3\theta + 3\cos\theta)$$

$$P_4(\cos\theta) = \frac{1}{64}(35\cos 4\theta + 20\cos 2\theta + 9)$$

Additional Legendre polynomials may be determined from the *recursion formula*

$$(n+1)P_{n+1}(x) - (2n+1)xP_n(x) + nP_{n-1}(x) = 0 \quad (n = 1, 2, \ldots)$$

or the *Rodrigues formula*

$$P_n(x) = \frac{1}{2^n n!}\frac{d^n}{dx^n}(x^2-1)^n$$

Laguerre Polynomials

Laguerre polynomials, denoted $L_n(x)$, are solutions of the differential equation

$$xy'' + (1-x)y' + ny = 0$$

and are given by

$$L_n(x) = \sum_{j=0}^{n} \frac{(-1)^j}{j!} C_{(n,j)} x^j (n = 0, 1, 2, \ldots)$$

Thus,

$$L_0(x) = 1$$

$$L_1(x) = 1 - x$$

$$L_2(x) = 1 - 2x + \frac{1}{2}x^2$$

$$L_3(x) = 1 - 3x + \frac{3}{2}x^2 - \frac{1}{6}x^3$$

Additional Laguerre polynomials may be obtained from the recursion formula

$$(n+1)L_{n+1}(x) - (2n+1-x)L_n(x) + nL_{n-1}(x) = 0$$

Hermite Polynomials

The Hermite polynomials, denoted $H_n(x)$, are given by

$$H_0 = 1, \quad H_n(x) = (-1)^n e^{x^2} \frac{d^n e^{-x^2}}{dx^n}, \quad (n = 1, 2, \ldots)$$

and are solutions of the differential equation

$$y'' - 2xy' + 2ny = 0 \quad (n = 0, 1, 2 \ldots)$$

The first few Hermite polynomials are

$$H_0 = 1 \qquad H_1(x) = 2x$$

$$H_2(x) = 4x^2 - 2 \qquad H_3(x) = 8x^3 - 12x$$

$$H_4(x) = 16x^4 - 48x^2 + 12$$

Additional Hermite polynomials may be obtained from the relation

$$H_{n+1}(x) = 2xH_n(x) - H'_n(x)$$

where prime denotes differentiation with respect to x.

Orthogonality

A set of functions $\{f_n(x)\}$ ($n = 1, 2, \ldots$) is orthogonal in an interval (a, b) with respect to a given weight function $w(x)$ if

$$\int_a^b w(x) f_m(x) f_n(x) dx = 0 \quad \text{when } m \neq n$$

The following polynomials are orthogonal on the given interval for the given $w(x)$:

Legendre polynomials:	$P_n(x)$	$w(x) = 1$
		$a = -1, b = 1$
Laguerre polynomials:	$L_n(x)$	$w(x) = \exp(-x)$
		$a = 0, b = \infty$
Hermite polynomials:	$H_n(x)$	$w(x) = \exp(-x^2)$
		$a = -\infty, b = \infty$

The Bessel functions *of order n*, $J_n(\lambda_1 x)$, $J_n(\lambda_2 x), \ldots$, are orthogonal with respect to $w(x) = x$ over the interval $(0, c)$ provided that the λ_i are the positive roots of $J_n(\lambda c) = 0$:

$$\int_0^c xJ_n(\lambda_j x) J_n(\lambda_k x) dx = 0 \quad (j \neq k)$$

where n is fixed and $n \geq 0$.

Functions with $x^2/a^2 \pm y^2/b^2$

Elliptic Paraboloid (Figure 217.7)

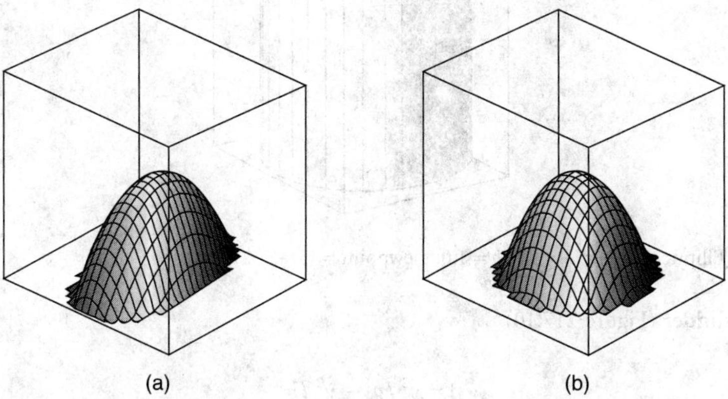

FIGURE 217.7 Elliptic paraboloid. (a) $a = 0.5$, $b = 1.0$, $c = -1.0$; viewpoint = (5, −6, 4). (b) $a = 1.0$, $b = 1.0$, $c = -2.0$; viewpoint = (5, −6, 4).

$$z = c(x^2/a^2 + y^2/b^2)$$

$$x^2/a^2 + y^2/b^2 - z/c = 0$$

Hyperbolic Paraboloid (Commonly Called *Saddle*) (Figure 217.8)

$$z = c(x^2/a^2 - y^2/b^2)$$

$$x^2/a^2 - y^2/b^2 - z/c = 0$$

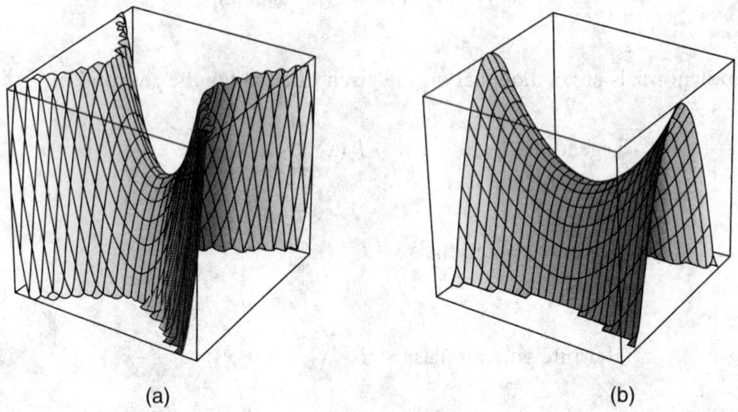

FIGURE 217.8 Hyperbolic paraboloid. (a) $a = 0.50$, $b = 0.5$, $c = 1.0$; viewpoint = (4, –6, 4). (b) $a = 1.00$, $b = 0.5$, $c = 1.0$; viewpoint = (4, –6, 4).

Elliptic Cylinder (Figure 217.9)

$$1 = x^2/a^2 + y^2/b^2$$

$$x^2/a^2 + y^2/b^2 - 1 = 0$$

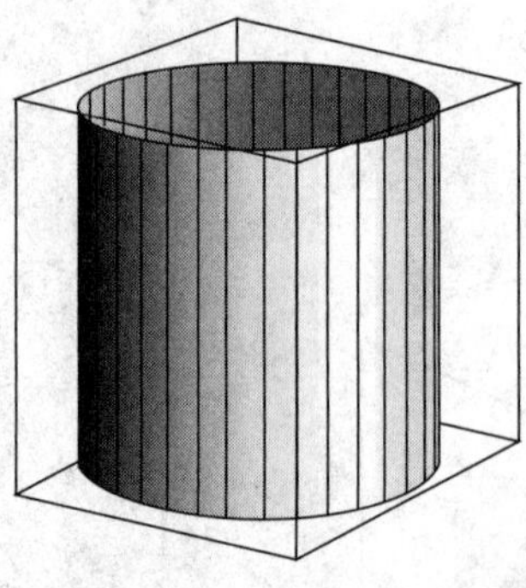

FIGURE 217.9 Elliptic cylinder. $a = 1.0$, $b = 1.0$; viewpoint = (4, –5, 2).

Hyperbolic Cylinder (Figure 217.10)

$$1 = x^2/a^2 - y^2/b^2$$

$$x^2/a^2 - y^2/b^2 - 1 = 0$$

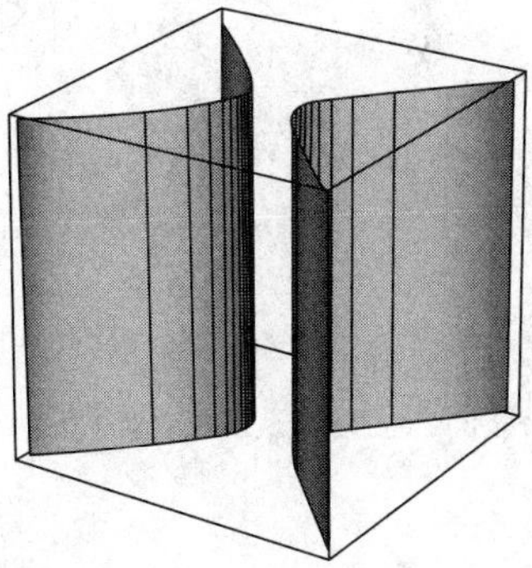

FIGURE 217.10 Hyperbolic cylinder. $a = 1.0$, $b = 1.0$; viewpoint = (4, –6, 3).

Functions with $(x^2/a^2 + y^2/b^2 \pm c^2)^{1/2}$

Sphere (Figure 217.11)

$$z = (1 - x^2 - y^2)^{1/2}$$

$$x^2 + y^2 + z^2 - 1 = 0$$

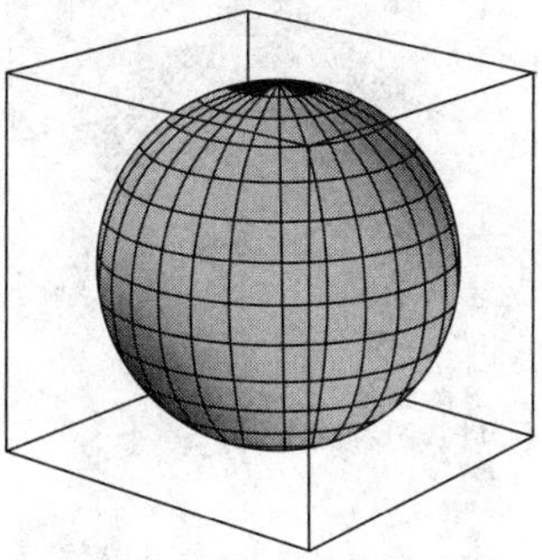

FIGURE 217.11 Sphere. Viewpoint = (4, –5, 2).

Ellipsoid (Figure 217.12)

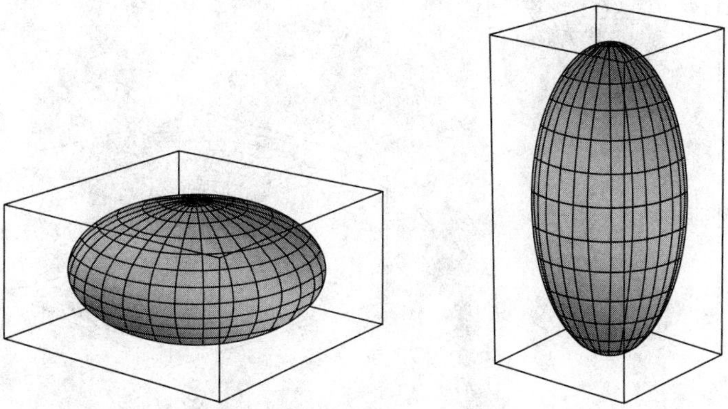

FIGURE 217.12 Ellipsoid. (a) $a = 1.00$, $b = 1.00$, $c = 0.5$; viewpoint = (4, –5, 2). (b) $a = 0.50$, $b = 0.50$, $c = 1.0$; viewpoint = (4, –5, 2).

$$z = c(1 - x^2/a^2 - y^2/b^2)^{1/2}$$

$$x^2/a^2 + y^2/b^2 + z^2/c^2 - 1 = 0$$

Special cases:

$$a = b > c \quad \text{gives oblate spheroid}$$

$$a = b < c \quad \text{gives prolate spheroid}$$

Cone (Figure 217.13)

$$z = (x^2 + y^2)^{1/2}$$

$$x^2 + y^2 - z^2 = 0$$

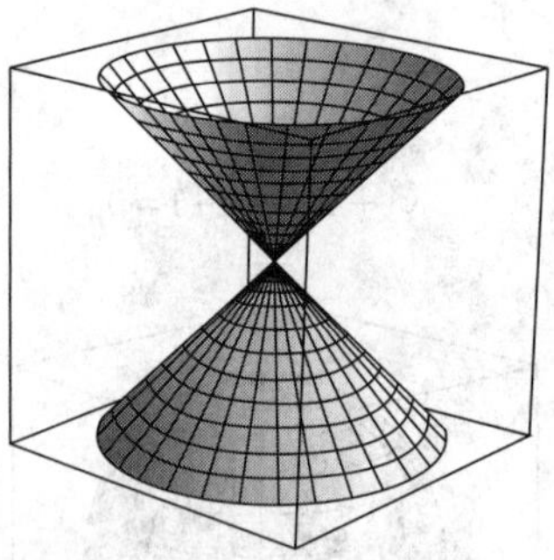

FIGURE 217.13 Cone. Viewpoint = (4, −5, 2).

Elliptic Cone (Circular Cone if $a = b$) (Figure 217.14)

$$z = c(x^2/a^2 + y^2/b^2)^{1/2}$$

$$x^2/a^2 + y^2/b^2 - z^2/c^2 = 0$$

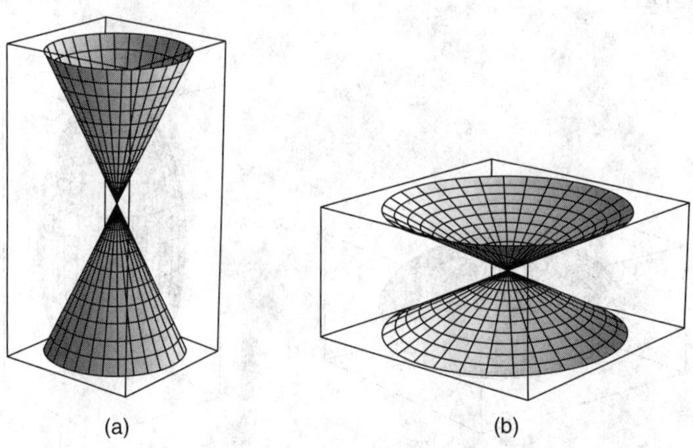

FIGURE 217.14 Elliptic cone. (a) $a = 0.5$, $b = 0.5$, $c = 1.00$; viewpoint = (4, −5, 2). (b) $a = 1.0$, $b = 1.0$, $c = 0.50$; viewpoint = (4, −5, 2).

Hyperboloid of One Sheet (Figure 217.15)

$$z = c(x^2/a^2 + y^2/b^2 - 1)^{1/2}$$

$$x^2/a^2 + y^2/b^2 - z^2/c^2 - 1 = 0$$

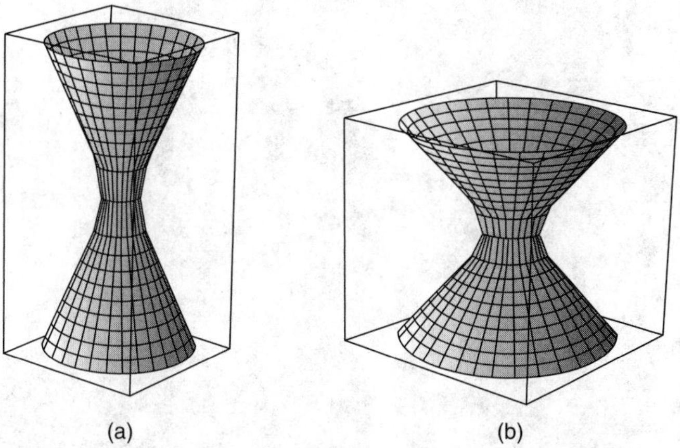

FIGURE 217.15 Hyperboloid of one sheet. (a) $a = 0.1$, $b = 0.1$, $c = 0.2$; $\pm z = c\sqrt{15}$; viewpoint = (4, –5, 2). (b) $a = 0.2$, $b = 0.2$, $c = 0.2$; $\pm z = c\sqrt{15}$; viewpoint = (4, –5, 2).

Hyperboloid of Two Sheets (Figure 217.16)

$$z = c(x^2/a^2 + y^2/b^2 + 1)^{1/2}$$

$$x^2/a^2 + y^2/b^2 - z^2/c^2 + 1 = 0$$

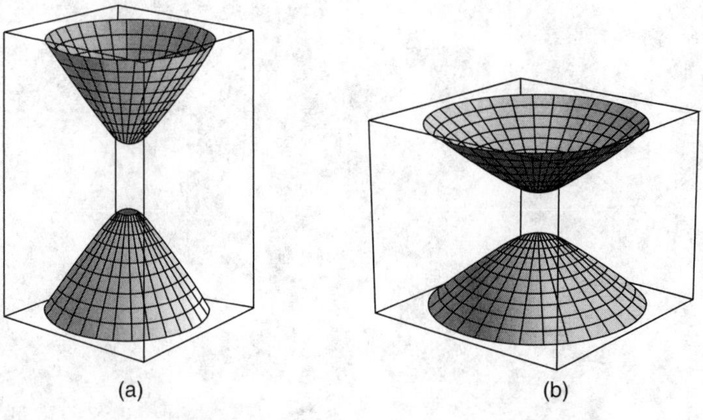

FIGURE 217.16 Hyperboloid of two sheets. (a) $a = 0.125$, $b = 0.125$, $c = 0.2$; $\pm z = c\sqrt{17}$; viewpoint = (4, –5, 2). (b) $a = 0.25$, $b = 0.25$, $c = 0.2$; $\pm z = c\sqrt{17}$; viewpoint = (4, –5, 2).

218

Linear Algebra Matrices

George Cain
Georgia Institute of Technology

218.1 Basic Definitions

A *matrix* **A** is a rectangular array of numbers (real or complex):

$$\mathbf{A} = \begin{bmatrix} a_{11} & a_{12} & \cdots & a_{1m} \\ a_{21} & a_{22} & \cdots & a_{2m} \\ \vdots & & & \\ a_{n1} & a_{n2} & \cdots & a_{nm} \end{bmatrix}$$

The *size* of the matrix is said to be $n \times m$. The $1 \times m$ matrices $[a_{i1}\ a_{i2}\ \ldots\ a_{im}]$ are called *rows* of **A**, and the $n \times 1$ matrices

$$\begin{bmatrix} a_{1j} \\ a_{2j} \\ \vdots \\ a_{nj} \end{bmatrix}$$

are called *columns* of **A**. An $n \times m$ matrix thus consists of n rows and m columns; a_{ij} denotes the *element*, or *entry*, of **A** in the ith row and jth column. A matrix consisting of just one row is called a *row vector*, whereas a matrix of just one column is called a *column vector*. The elements of a vector are frequently called *components* of the vector. When the size of the matrix is clear from the context, we sometimes write $\mathbf{A} = (a_{ij})$.

A matrix with the same number of rows as columns is a *square* matrix, and the number of rows and columns is the *order* of the matrix. The diagonal of an $n \times n$ square matrix **A** from a_{11} to a_{nn} is called the *main*, or *principal*, *diagonal*. The word *diagonal* with no modifier usually means the main diagonal. The *transpose* of a matrix **A** is the matrix that results from interchanging the rows and columns of **A**. It is

0-8493-1586-7/05/$0.00+$1.50

usually denoted by $\mathbf{A}^T$. A matrix $\mathbf{A}$ such that $\mathbf{A} = \mathbf{A}^T$ is said to be *symmetric*. The *conjugate transpose* of $\mathbf{A}$ is the matrix that results from replacing each element of $\mathbf{A}^T$ by its complex conjugate, and is usually denoted by $\mathbf{A}^H$. A matrix such that $\mathbf{A} = \mathbf{A}^H$ is said to be *Hermitian.*

A square matrix $\mathbf{A} = (a_{ij})$ is *lower triangular* if $a_{ij} = 0$ for $j > i$ and is *upper triangular* if $a_{ij} = 0$ for $j < i$. A matrix that is both upper and lower triangular is a *diagonal* matrix. The $n \times n$ *identity matrix* is the $n \times n$ diagonal matrix in which each element of the main diagonal is 1. It is traditionally denoted $\mathbf{I}_n$, or simply $\mathbf{I}$ when the order is clear from the context.

218.2 Algebra of Matrices

The sum and difference of two matrices $\mathbf{A}$ and $\mathbf{B}$ are defined whenever $\mathbf{A}$ and $\mathbf{B}$ have the same size. In that case $\mathbf{C} = \mathbf{A} \pm \mathbf{B}$ is defined by $\mathbf{C} = (c_{ij}) = (a_{ij} \pm b_{ij})$. The product $t\mathbf{A}$ of a scalar t (real or complex number) and a matrix $\mathbf{A}$ is defined by $t\mathbf{A} = (ta_{ij})$. If $\mathbf{A}$ is an $n \times m$ matrix and $\mathbf{B}$ is an $m \times p$ matrix, the product $\mathbf{C} = \mathbf{AB}$ is defined to be the $n \times p$ matrix $\mathbf{C} = (c_{ij})$ given by $c_{ij} = \sum_{k=1}^{m} a_{ik}b_{kj}$. Note that the product of an $n \times m$ matrix and an $m \times p$ matrix is an $n \times p$ matrix, and the product is defined only when the number of columns of the first factor is the same as the number of rows of the second factor. Matrix multiplication is, in general, associative: $\mathbf{A}(\mathbf{BC}) = (\mathbf{AB})\mathbf{C}$. It also distributes over addition (and subtraction):

$$\mathbf{A}(\mathbf{B} + \mathbf{C}) = \mathbf{AB} + \mathbf{AC} \text{ and } (\mathbf{A} + \mathbf{B})\mathbf{C} = \mathbf{AC} + \mathbf{BC}$$

It is, however, not in general true that $\mathbf{AB} = \mathbf{BA}$, even in case both products are defined. It is clear that $(\mathbf{A} + \mathbf{B})^T = \mathbf{A}^T + \mathbf{B}^T$ and $(\mathbf{A} + \mathbf{B})^H = \mathbf{A}^H + \mathbf{B}^H$. It is also true, but not so obvious perhaps, that $(\mathbf{AB})^T = \mathbf{B}^T\mathbf{A}^T$ and $(\mathbf{AB})^H = \mathbf{B}^H\mathbf{A}^H$.

The $n \times n$ identity matrix $\mathbf{I}$ has the property that $\mathbf{IA} = \mathbf{AI} = \mathbf{A}$ for every $n \times n$ matrix $\mathbf{A}$. If $\mathbf{A}$ is square, and if there is a matrix $\mathbf{B}$ such that $\mathbf{AB} = \mathbf{BA} = \mathbf{I}$, then $\mathbf{B}$ is called the *inverse* of A and is denoted $\mathbf{A}^{-1}$. This terminology and notation are justified by the fact that a matrix can have at most one inverse. A matrix having an inverse is said to be *invertible*, or *nonsingular*, while a matrix not having an inverse is said to be *noninvertible*, or *singular*. The product of two invertible matrices is invertible and, in fact, $(\mathbf{AB})^{-1} = \mathbf{B}^{-1}\mathbf{A}^{-1}$. The sum of two invertible matrices is, obviously, not necessarily invertible.

218.3 Systems of Equations

The system of n linear equations in m unknowns

$$\begin{aligned}
a_{11}x_1 + a_{12}x_2 + a_{13}x_3 + \cdots + a_{1m}x_m &= b_1 \\
a_{21}x_1 + a_{22}x_2 + a_{23}x_3 + \cdots + a_{2m}x_m &= b_2 \\
&\vdots \\
a_{n1}x_1 + a_{n2}x_2 + a_{n3}x_3 + \cdots + a_{nm}x_m &= b_n
\end{aligned}$$

may be written $\mathbf{Ax} = \mathbf{b}$, where $\mathbf{A} = (a_{ij})$, $\mathbf{x} = [x_1\, x_2 \ldots x_m]^T$, and $\mathbf{b} = [b_1\, b_2 \ldots b_n]^T$. Thus $\mathbf{A}$ is an $n \times m$ matrix, and $\mathbf{x}$ and $\mathbf{b}$ are column vectors of the appropriate sizes.

The matrix $\mathbf{A}$ is called the *coefficient matrix* of the system. Let us first suppose the coefficient matrix is square; that is, there are an equal number of equations and unknowns. If $\mathbf{A}$ is upper triangular, it is quite easy to find all solutions of the system. The ith equation will contain only the unknowns x_i, x_{i+1}, …, x_n, and one simply solves the equations in reverse order: the last equation is solved for x_n; the result is substituted into the $(n - 1)$st equation, which is then solved for x_{n-1}; these values of x_n and x_{n-1} are substituted in the $(n - 2)$th equation, which is solved for x_{n-1}, and so on. This procedure is known as *back substitution.*

The strategy for solving an arbitrary system is to find an upper-triangular system equivalent with it and solve this upper-triangular system using back substitution. First, suppose the element $a_{11} \neq 0$. We may rearrange the equations to ensure this, unless, of course the first column of **A** is all 0's. In this case, proceed to the next step, to be described later. For each $i \geq 2$ let $m_{i1} = a_{i1}/a_{11}$. Now replace the ith equation by the result of multiplying the first equation by m_{i1} and subtracting the new equation from the ith equation. Thus,

$$a_{i1}x_1 + a_{i2}x_2 + a_{i3}x_3 + \cdots + a_{im}x_m = b_i$$

is replaced by

$$0 \cdot x_1 + (a_{i2} + m_{i1}a_{12})x_2 + (a_{i3} + m_{i1}a_{13})x_3 + \cdots + (a_{im} + m_{i1}a_{1m})x_m = b_i + m_{i1}b_1$$

After this is done for all $i = 2, 3, \ldots, n$, the equivalent system,

$$\begin{aligned} a_{11}x_1 + a_{12}x_2 + a_{13}x_3 + \cdots + a_{1n}x_n &= b_1 \\ 0 \cdot x_1 + a'_{22}x_2 + a'_{23}x_3 + \cdots + a'_{2n}x_n &= b'_2 \\ 0 \cdot x_1 + a'_{32}x_2 + a'_{33}x_3 + \cdots + a'_{3n}x_n &= b'_3 \\ &\vdots \\ 0 \cdot x_1 + a'_{n2}x_2 + a'_{n3}x_3 + \cdots + a'_{nn}x_n &= b'_n \end{aligned}$$

results, in which all entries in the first column below a_{11} are 0. (Note that if all entries in the first column were 0 to begin with, then $a_{11} = 0$ also.) This procedure is now repeated for the $(n-1) \times (n-1)$ system

$$\begin{aligned} a'_{22}x_2 + a'_{23}x_3 + \cdots + a'_{2n}x_n &= b'_2 \\ a'_{32}x_2 + a'_{33}x_3 + \cdots + a'_{3n}x_n &= b'_3 \\ &\vdots \\ a'_{n2}x_2 + a'_{n3}x_3 + \cdots + a'_{nn}x_n &= b'_n \end{aligned}$$

to obtain an equivalent system in which all entries of the coefficient matrix below a'_{22} are 0. Continuing, we obtain an upper-triangular system $\mathbf{Ux} = \mathbf{c}$ equivalent with the original system. This procedure is known as *Gaussian elimination*. The numbers m_{ij} are known as the *multipliers*.

Essentially the same procedure may be used in case the coefficient matrix is not square. If the coefficient matrix is not square, we may make it square by appending either rows or columns of 0's as needed. Appending rows of 0's and appending 0's to make **b** have the appropriate size is equivalent to appending equations 0 = 0 to the system. Clearly the new system has precisely the same solutions as the original system. Appending columns of 0's and adjusting the size of **x** appropriately yields a new system with additional unknowns, each appearing only with coefficient 0, thus not affecting the solutions of the original system. In either case, we may assume the coefficient matrix is square, and apply the Gaussian elimination procedure.

Suppose the matrix **A** is invertible. Then, if there were no row interchanges in carrying out the above Gaussian elimination procedure, we have the *LU factorization* of the matrix **A**:

$$\mathbf{A} = \mathbf{LU}$$

where **U** is the upper-triangular matrix produced by elimination and **L** is the lower-triangular matrix given by

$$\mathbf{L}=\begin{bmatrix} 1 & 0 & \cdots & \cdots 0 \\ m_{21} & 1 & 0 & \cdots 0 \\ \vdots & & \ddots & \\ m_{n1} & m_{n2} & \cdots & 1 \end{bmatrix}$$

A *permutation* $\mathbf{P}_{ij}$ matrix is an $n \times n$ matrix such that $\mathbf{P}_{ij}\mathbf{A}$ is the matrix that results from exchanging row i and j of the matrix **A**. The matrix $\mathbf{P}_{ij}$ is the matrix that results from exchanging rows i and j of the identity matrix. A product **P** of such matrices $\mathbf{P}_{ij}$ is called a *permutation* matrix. If row interchanges are required in the Gaussian elimination procedure, then we have the factorization

$$\mathbf{PA} = \mathbf{LU}$$

where **P** is the permutation matrix giving the required row exchanges.

218.4 Vector Spaces

The collection of all column vectors with n real components is *Euclidean n-space*, and is denoted $\mathbf{R}^n$. The collection of column vectors with n complex components is denoted $\mathbf{C}^n$. We shall use *vector space* to mean either $\mathbf{R}^n$ or $\mathbf{C}^n$. In discussing the space $\mathbf{R}^n$, the word *scalar* will mean a real number, and in discussing the space $\mathbf{C}^n$, it will mean a complex number. A subset **S** of a vector space is a *subspace* such that if **u** and **v** are vectors in **S**, and if c is any scalar, then $\mathbf{u} + \mathbf{v}$ and $c\mathbf{u}$ are in **S**. We shall sometimes use the word *space* to mean a subspace. If $B = [\mathbf{v}_1, \mathbf{v}_2, \ldots, \mathbf{v}_k\}$ is a collection of vectors in a vector space, then the set **S** consisting of all vectors $c_1\mathbf{v}_1 + c_2\mathbf{v}_2 + \ldots + c_m\mathbf{v}_m$ for all scalars $c_1, c_2, \ldots, c_m$ is a subspace, called the *span* of B. A collection $\{\mathbf{v}_1, \mathbf{v}_2, \ldots, \mathbf{v}_m\}$ of vectors $c_1\mathbf{v}_1 + c_2\mathbf{v}_2 + \ldots + c_m\mathbf{v}_m$ is a *linear combination* of B. If **S** is a subspace and $B = \{\mathbf{v}_1, \mathbf{v}_2, \ldots, \mathbf{v}_m\}$ is a subset of **S** such that **S** is the span of B, then B is said to *span* **S**.

A collection $\{\mathbf{v}_1, \mathbf{v}_2, \ldots, \mathbf{v}_m\}$ of n-vectors is *linearly dependent* if there exist scalars $c_1, c_2, \ldots, c_m$, not all zero, such that $c_1\mathbf{v}_1 + c_2\mathbf{v}_2 + \ldots + c_m\mathbf{v}_m = 0$. A collection of vectors that is not linearly dependent is said to be *linearly independent*. The modifier *linearly* is frequently omitted, and we speak simply of dependent and independent collections. A linearly independent collection of vectors in a space **S** that span **S** is a *basis* of **S**. Every basis of a space **S** contains the same number of vectors; this number is the *dimension* of **S**. The dimension of the space consisting of only the zero vector is 0. The collection $B = \{\mathbf{e}_1, \mathbf{e}_2, \ldots, \mathbf{e}_n\}$, where $\mathbf{e}_1 = [1, 0, 0, \ldots, 0]^T$, $\mathbf{e}_2 = [0, 1, 0, \ldots, 0]^T$, and so forth ($\mathbf{e}_i$ has 1 as its ith component and zero for all other components) is a basis for the spaces $\mathbf{R}^n$ and $\mathbf{C}^n$. This is the *standard basis* for these spaces. The dimension of these spaces is thus n. In a space **S** of dimension n, no collection of fewer than n vectors can span **S**, and no collection of more than n vectors in **S** can be independent.

218.5 Rank and Nullity

The *column space* of an $n \times m$ matrix **A** is the subspace of $\mathbf{R}^n$ or $\mathbf{C}^n$ spanned by the columns of **A**. The *row space* is the subspace of $\mathbf{R}^m$ or $\mathbf{C}^m$ spanned by the rows of **A**. Note that for any vector $\mathbf{x} = [x_1\ x_2 \ldots x_m]^T$,

$$\mathbf{Ax} = x_1\begin{bmatrix} a_{11} \\ a_{21} \\ \vdots \\ a_{n1} \end{bmatrix} + x_2\begin{bmatrix} a_{12} \\ a_{22} \\ \vdots \\ a_{n2} \end{bmatrix} + \cdots + x_m\begin{bmatrix} a_{1m} \\ a_{2m} \\ \vdots \\ a_{nm} \end{bmatrix}$$

so that the column space is the collection of all vectors **Ax**, and thus the system $\mathbf{Ax} = \mathbf{b}$ has a solution if and only if **b** is a member of the **A** column space.

The dimension of the column space is the *rank* of **A**. The row space has the same dimension as the column space. The set of all solutions of the system $\mathbf{Ax} = \mathbf{0}$ is a subspace called the *null space* of **A**, and the dimension of this null space is the *nullity* of **A**. A fundamental result in matrix theory is the fact that, for an $n \times m$ matrix **A**,

$$\text{rank } \mathbf{A} + \text{nullity } \mathbf{A} = m$$

The difference of any two solutions of the linear system $\mathbf{Ax} = \mathbf{b}$ is a member of the null space of **A**. Thus this system has at most one solution if and only if the nullity of **A** is zero. If the system is square (i.e., if **A** is $n \times n$), then there will be a solution for every right-hand side **b** if and only if the collection of columns of **A** is linearly independent, which is the same as saying the rank of **A** is n. In this case the nullity must be zero. Thus, for any **b**, the square system $\mathbf{Ax} = \mathbf{b}$ has exactly one solution if and only if rank $\mathbf{A} = n$. In other words, the $n \times n$ matrix **A** is invertible if and only if rank $\mathbf{A} = n$.

218.6 Orthogonality and Length

The *inner product* of two vectors **x** and **y** is the scalar $\mathbf{x}^H\mathbf{y}$. The *length*, or *norm*, $\|\mathbf{x}\|$, of the vector **x** is given by $\|\mathbf{x}\| = \sqrt{\mathbf{x}^H\mathbf{x}}$. A *unit vector* is a vector of norm 1. Two vectors **x** and **y** are *orthogonal* if $\mathbf{x}^H\mathbf{y} = 0$. A collection of vectors $\{\mathbf{v}_1, \mathbf{v}_2, \ldots, \mathbf{v}_m\}$ in a space **S** is said to be an *orthonormal* collection if $\mathbf{v}_i^H\mathbf{v}_j = 0$ for $i \neq j$ and $\mathbf{v}_i^H\mathbf{v}_i = 1$. An orthonormal collection is necessarily linearly independent. If **S** is a subspace (of $\mathbf{R}^n$ or $\mathbf{C}^n$) spanned by the orthonormal collection $\{\mathbf{v}_1, \mathbf{v}_2, \ldots, \mathbf{v}_m\}$, then the *projection* of a vector **x** onto **S** is the vector

$$\text{proj}(\mathbf{x}; \mathbf{S}) = (\mathbf{x}^H\mathbf{v}_1)\mathbf{v}_1 + (\mathbf{x}^H\mathbf{v}_2)\mathbf{v}_2 + \cdots + (\mathbf{x}^H\mathbf{v}_m)\mathbf{v}_m$$

The projection of **x** onto **S** minimizes the function $f(\mathbf{y}) = \|\mathbf{x} - \mathbf{y}\|^2$ for $\mathbf{y} \in \mathbf{S}$. In other words the projection of **x** onto **S** is the vector in **S** that is "closest" to x.

If **b** is a vector and **A** is an $n \times m$ matrix, then a vector **x** minimizes $\|\mathbf{b} - \mathbf{Ax}\|^2$ if and only if it is a solution of $\mathbf{A}^H\mathbf{Ax} = \mathbf{A}^H\mathbf{b}$. This system of equations is called the *system of normal equations* for the least-squares problem of minimizing $\|\mathbf{b} - \mathbf{Ax}\|^2$.

If **A** is an $n \times m$ matrix and rank $\mathbf{A} = k$, then there is a $n \times k$ matrix **Q** whose columns form an orthonormal basis for the column space of **A** and a $k \times m$ upper-triangular matrix **R** of rank k such that

$$\mathbf{A} = \mathbf{QR}$$

This is called the *QR factorization* of **A**. It now follows that **x** minimizes $\|\mathbf{b} - \mathbf{Ax}\|^2$ if and only if it is a solution of the upper-triangular system $\mathbf{Rx} = \mathbf{Q}^H\mathbf{b}$.

If $\{\mathbf{w}_1, \mathbf{w}_2, \ldots, \mathbf{w}_m\}$ is a basis for a space **S**, the following procedure produces an orthonormal basis $\{\mathbf{v}_1, \mathbf{v}_2, \ldots, \mathbf{v}_m\}$ for **S**:

Set $\mathbf{v}_1 = \mathbf{w}_1 / \|\mathbf{w}_1\|$.
Let $\tilde{\mathbf{v}}_2 = \mathbf{w}_2 - \text{proj}(\mathbf{w}_2; \mathbf{S}_1)$, where $\mathbf{S}_1$ is the span of $(\mathbf{v}_1\}$; set $\mathbf{v}_2 = \tilde{\mathbf{v}}_2 / \|\tilde{\mathbf{v}}_2\|$.
Next, let $\tilde{\mathbf{v}}_3 = \mathbf{w}_3 - \text{proj}(\mathbf{w}_3; \mathbf{S}_2)$, where $\mathbf{S}_2$ is the span of $\{\mathbf{v}_1, \mathbf{v}_2\}$; set $\mathbf{v}_3 = \tilde{\mathbf{v}}_3 / \|\tilde{\mathbf{v}}_3\|$.

And so on: $\tilde{\mathbf{v}}_i = \mathbf{w}_i - \text{proj}(\mathbf{w}_i; \mathbf{S}_{i-1})$, where $\mathbf{S}_{i-1}$ is the span of $\{\mathbf{v}_1, \mathbf{v}_2, \ldots, \mathbf{v}_{i-1}\}$; set $\mathbf{v}_i = \tilde{\mathbf{v}}_i / \|\tilde{\mathbf{v}}_i\|$. This is the *Gram–Schmidt procedure.*

If the collection of columns of a square matrix is an orthonormal collection, the matrix is called a *unitary matrix*. In the event that the matrix is a real matrix, it is usually called an *orthogonal matrix*. A unitary matrix **U** is invertible, and $\mathbf{U}^{-1} = \mathbf{U}^H$. (In the real case an orthogonal matrix **Q** is invertible, and $\mathbf{Q}^{-1} = \mathbf{Q}^T$.)

218.7 Determinants

The *determinant* of a square matrix is defined inductively. First, suppose the determinant det **A** has been defined for all square matrices of order < n. Then

$$\det \mathbf{A} = a_{11}\mathbf{C}_{11} + a_{12}\mathbf{C}_{12} + \cdots + a_{1n}\mathbf{C}_{1n}$$

where the numbers $\mathbf{C}_{ij}$ are *cofactors* of the matrix **A**:

$$\mathbf{C}_{ij} = (-1)^{i+j} \det \mathbf{M}_{ij}$$

where $\mathbf{M}_{ij}$ is the $(n-1) \times (n-1)$ matrix obtained by deleting the ith row and jth column of **A**. Now det **A** is defined to be the only entry of a matrix of order 1. Thus, for a matrix of order 2, we have

$$\det \begin{bmatrix} a & b \\ c & d \end{bmatrix} = ad - bc$$

There are many interesting but not obvious properties of determinants. It is true that

$$\det \mathbf{A} = a_{i1}\mathbf{C}_{i1} + a_{i2}\mathbf{C}_{i2} + \cdots + a_{in}\mathbf{C}_{in}$$

for any $1 \le i \le n$. It is also true that det **A** = det $\mathbf{A}^T$, so that we have

$$\det \mathbf{A} = a_{1j}\mathbf{C}_{1j} + a_{2j}\mathbf{C}_{2j} + \cdots + a_{nj}\mathbf{C}_{nj}$$

for any $1 \le j \le n$.

If **A** and **B** are matrices of the same order, then det **AB** = (det **A**)(det **B**), and the determinant of any identity matrix is 1. Perhaps the most important property of the determinant is the fact that a matrix is invertible if and only if its determinant is not zero.

218.8 Eigenvalues and Eigenvectors

If **A** is a square matrix, and $\mathbf{Av} = \lambda\mathbf{v}$ for a scalar λ and a nonzero, **v**, then λ is an *eigenvalue* of **A** and **v** is an *eigenvector* of **A** that *corresponds* to λ. Any nonzero linear combination of eigenvectors corresponding to the same eigenvalue λ is also an eigenvector corresponding to λ. The collection of all eigenvectors corresponding to a given eigenvalue λ is thus a subspace, called an *eigenspace* of **A**. A collection of eigenvectors corresponding to different eigenvalues is necessarily linear-independent. It follows that a matrix of order n can have at most n distinct eigenvectors. In fact, the eigenvalues of **A** are the roots of the nth degree polynomial equation

$$\det(\mathbf{A} - \lambda\mathbf{I}) = 0$$

called the *characteristic equation* of **A**. (Eigenvalues and eigenvectors are frequently called *characteristic values* and *characteristic vectors.*)

If the nth order matrix **A** has an independent collection of n eigenvectors, then **A** is said to have a *full set* of eigenvectors. In this case there is a set of eigenvectors of **A** that is a basis for $\mathbf{R}^n$ or, in the complex case, $\mathbf{C}^n$. In case there are n distinct eigenvalues of **A**, then, of course, **A** has a full set of eigenvectors. If there are fewer than n distinct eigenvalues, then **A** may or may not have a full set of eigenvectors. If there is a full set of eigenvectors, then

$$\mathbf{D} = \mathbf{S}^{-1}\mathbf{A}\mathbf{S} \quad \text{or} \quad \mathbf{A} = \mathbf{S}\mathbf{D}\mathbf{S}^{-1}$$

where **D** is a diagonal matrix with the eigenvalues of **A** on the diagonal, and **S** is a matrix whose columns are the full set of eigenvectors. If **A** is symmetric, there are n real distinct eigenvalues of **A** and the corresponding eigenvectors are orthogonal. There is thus an orthonormal collection of eigenvectors that span $\mathbf{R}^n$, and we have

$$\mathbf{A} = \mathbf{Q}\mathbf{D}\mathbf{Q}^{\mathrm{T}} \quad \text{and} \quad \mathbf{D} = \mathbf{Q}^{\mathrm{T}}\mathbf{A}\mathbf{Q}$$

where **Q** is a real orthogonal matrix and **D** is diagonal. For the complex case, if **A** is Hermitian, we have

$$\mathbf{A} = \mathbf{U}\mathbf{D}\mathbf{U}^{\mathrm{H}} \quad \text{and} \quad \mathbf{D} = \mathbf{U}^{\mathrm{H}}\mathbf{A}\mathbf{U}$$

where **U** is a unitary matrix and **D** is a *real* diagonal matrix. (A Hermitian matrix also has n distinct real eigenvalues.)

References

Danial, J. W. and Noble, B. 1988. *Applied Linear Algebra.* Prentice Hall, Englewood Cliffs, NJ.

Strang, G. 1993. *Introduction to Linear Algebra.* Wellesley-Cambridge Press, Wellesley, MA.

219

Vector Algebra and Calculus

George Cain
Georgia Institute of Technology

219.1 Basic Definitions

A vector is a directed line segment, with two vectors being equal if they have the same length and the same direction. More precisely, a *vector* is an equivalence class of directed line segments, where two directed segments are equivalent if they have the same length and the same direction. The *length* of a vector is the common length of its directed segments, and the *angle between* vectors is the angle between any of their segments. The length of a vector $\mathbf{u}$ is denoted $|\mathbf{u}|$. There is defined a distinguished vector having zero length, which is usually denoted $\mathbf{0}$. It is frequently useful to visualize a directed segment as an arrow; we then speak of the nose and the tail of the segment. The *sum* $\mathbf{u} + \mathbf{v}$ of two vectors $\mathbf{u}$ and $\mathbf{v}$ is defined by taking directed segments from $\mathbf{u}$ and $\mathbf{v}$ and placing the tail of the segment representing $\mathbf{v}$ at the nose of the segment representing $\mathbf{u}$ and defining $\mathbf{u} + \mathbf{v}$ to be the vector determined by the segment from the tail of the $\mathbf{u}$ representative to the nose of the $\mathbf{v}$ representative. It is easy to see that $\mathbf{u} + \mathbf{v}$ is well defined and that $\mathbf{u} + \mathbf{v} = \mathbf{v} + \mathbf{u}$. Subtraction is the inverse operation of addition. Thus, the *difference* $\mathbf{u} - \mathbf{v}$ of two vectors is defined to be the vector that when added to $\mathbf{v}$ gives $\mathbf{u}$. In other words, if we take a segment from $\mathbf{u}$ and a segment from $\mathbf{v}$ and place their tails together, the difference is the segment from the nose of $\mathbf{v}$ to the nose of $\mathbf{u}$. The zero vector behaves as one might expect: $\mathbf{u} + 0 = \mathbf{u}$, and $\mathbf{u} - \mathbf{u} = 0$. Addition is associative: $\mathbf{u} + (\mathbf{v} + \mathbf{w}) = (\mathbf{u} + \mathbf{v}) + \mathbf{w}$.

To distinguish them from vectors, the real numbers are called *scalars*. The product $t\mathbf{u}$ of a scalar t and a vector $\mathbf{u}$ is defined to be the vector having length $|t|\,|\mathbf{u}|$ and direction the same as $\mathbf{u}$ if $t > 0$, and the opposite direction if $t < 0$. If $t = 0$, then $t\mathbf{u}$ is defined to be the zero vector. Note that $t(\mathbf{u} + \mathbf{v}) = t\mathbf{u} + t\mathbf{v}$, and $(t + s)\mathbf{u} = t\mathbf{u} + s\mathbf{u}$. From this it follows that $\mathbf{u} - \mathbf{v} = \mathbf{u} + (-1)\mathbf{v}$.

The *scalar product* $\mathbf{u} \cdot \mathbf{v}$ of two vectors is $|\mathbf{u}||\mathbf{v}| \cos \boldsymbol{\theta}$, where θ is the angle between $\mathbf{u}$ and $\mathbf{v}$. The scalar product is frequently called the *dot product*. The scalar product distributes over addition:

$$\mathbf{u}\cdot(\mathbf{v}+\mathbf{w})=\mathbf{u}\cdot\mathbf{v}+\mathbf{u}\cdot\mathbf{w}$$

0-8493-1586-7/05/$0.00+$1.50

and it is clear that $(t\mathbf{u}) \cdot \mathbf{v} = t(\mathbf{u} \cdot \mathbf{v})$. The *vector product* $\mathbf{u} \times \mathbf{v}$ of two vectors is defined to be the vector perpendicular to both $\mathbf{u}$ and $\mathbf{v}$ and having length $|\mathbf{u}||\mathbf{v}|\sin\theta$, where θ is the angle between $\mathbf{u}$ and $\mathbf{v}$. The direction of $\mathbf{u} \times \mathbf{v}$ is the direction a right-hand threaded bolt advances if the vector $\mathbf{u}$ is rotated to $\mathbf{v}$. The vector product is frequently called the *cross product.* The vector product is both associative and distributive, but not commutative: $\mathbf{u} \times \mathbf{v} = -\mathbf{v} \times \mathbf{u}$.

219.2 Coordinate Systems

Suppose we have a right-handed Cartesian coordinate system in space. For each vector **u**, we associate a point in space by placing the tail of a representative of **u** at the origin and associating with **u** the point at the nose of the segment. Conversely, associated with each point in space is the vector determined by the directed segment from the origin to that point. There is thus a one-to-one correspondence between the points in space and all vectors. The origin corresponds to the zero vector. The coordinates of the point associated with a vector **u** are called *coordinates* of **u**. One frequently refers to the vector **u** and writes $\mathbf{u} = (x, y, z)$, which is, strictly speaking, incorrect, because the left side of this equation is a vector and the right side gives the coordinates of a point in space. What is meant is that (x, y, z) are the coordinates of the point associated with **u** under the correspondence described. In terms of coordinates, for $\mathbf{u} = (u_1, u_2, u_3)$ and $\mathbf{v} = (v_1, v_2, v_3)$, we have

$$\mathbf{u}+\mathbf{v}=(u_1+v_1, u_2+v_2, u_3+v_3)$$

$$t\mathbf{u}=(tu_1, tu_2, tu_3)$$

$$\mathbf{u}\cdot\mathbf{v}=u_1v_1+u_2v_2+u_3v_3$$

$$\mathbf{u}\times\mathbf{v}=(u_2v_3-v_2u_3, u_3v_1-v_3u_1, u_1v_2-v_1u_2)$$

The *coordinate vectors* **i**, **j**, and **k** are the unit vectors $\mathbf{i} = (1, 0, 0)$, $\mathbf{j} = (0, 1, 0)$, and $\mathbf{k} = (0, 0, 1)$. Any vector $\mathbf{u} = (u_1, u_2, u_3)$ is thus a linear combination of these coordinate vectors: $\mathbf{u} = u_1\mathbf{i} + u_2\mathbf{j} + u_3\mathbf{k}$. A convenient form for the vector product is the formal determinant

$$\mathbf{u}\times\mathbf{v}=\det\begin{bmatrix}\mathbf{i} & \mathbf{j} & \mathbf{k}\\ u_1 & u_2 & u_3\\ v_1 & v_2 & v_3\end{bmatrix}$$

219.3 Vector Functions

A *vector function* **F** *of one variable* is a rule that associates a vector $\mathbf{F}(t)$ with each real number t in some set, called the *domain* of **F**. The expression $\lim_{t\to t_0} \mathbf{F}(t) = \mathbf{a}$ means that for any $\varepsilon > 0$, there is a $\delta > 0$ such that $|\mathbf{F}(t) - \mathbf{a}| < \varepsilon$ whenever $0 < |t - t_0| < \delta$. If $\mathbf{F}(t) = [x(t), y(t), z(t)]$ and $\mathbf{a} = (a_1, a_2, a_3)$, then $\lim_{t\to t_0} \mathbf{F}(t) = \mathbf{a}$ if and only if

$$\lim_{t\to t_0} x(t) = a_1$$

$$\lim_{t\to t_0} y(t) = a_2$$

$$\lim_{t\to t_0} z(t) = a_3$$

A vector function **F** is *continuous* at t_0 if $\lim_{t\to t_0} \mathbf{F}(t) = \mathbf{F}(t_0)$. The vector function **F** is continuous at t_0 if and only if each of the coordinates $x(t)$, $y(t)$, and $z(t)$ is continuous at t_0.

The function **F** is *differentiable* at t_0 if the limit

$$\lim_{h\to 0}\frac{1}{h}[\mathbf{F}(t+h)-\mathbf{F}(t)]$$

exists. This limit is called the *derivative* of **F** at t_0 and is usually written $\mathbf{F}'(t_0)$, or $(d\mathbf{F}/dt)(t_0)$. The vector function **F** is differentiable at t_0 if and only if each of its coordinate functions is differentiable at t_0. Moreover, $(d\mathbf{F}/dt)(t_0) = [(dx/dt)(t_0), (dy/dt)(t_0), (dz/dt)(t_0)]$. The usual rules for derivatives of real valued functions all hold for vector functions. Thus, if **F** and **G** are vector functions and s is a scalar function, then

$$\frac{d}{dt}(\mathbf{F}+\mathbf{G})=\frac{d\mathbf{F}}{dt}+\frac{d\mathbf{G}}{dt}$$

$$\frac{d}{dt}(s\mathbf{F})=s\frac{d\mathbf{F}}{dt}+\frac{ds}{dt}\mathbf{F}$$

$$\frac{d}{dt}(\mathbf{F}\cdot\mathbf{G})=\mathbf{F}\cdot\frac{d\mathbf{G}}{dt}+\frac{d\mathbf{F}}{dt}\cdot\mathbf{G}$$

$$\frac{d}{dt}(\mathbf{F}\times\mathbf{G})=\mathbf{F}\times\frac{d\mathbf{G}}{dt}+\frac{d\mathbf{F}}{dt}\times\mathbf{G}$$

If **R** is a vector function defined for t in some interval, then, as t varies, with the tail of **R** at the origin, the nose traces out some object C in space. For nice functions **R**, the object C is a *curve*. If $\mathbf{R}(t) = [x(t), y(t), z(t)]$, then the equations

$$x = x(t)$$

$$y = y(t)$$

$$z = z(t)$$

are called *parametric equations* of C. At points where **R** is differentiable, the derivative $d\mathbf{R}/dt$ is a vector *tangent* to the curve. The unit vector $\mathbf{T} = (d\mathbf{R}/dt)/|d\mathbf{R}/dt|$ is called the *unit tangent vector*. If **R** is differentiable and if the length of the arc of curve described by **R** between $\mathbf{R}(a)$ and $\mathbf{R}(t)$ is given by $s(t)$, then

$$\frac{ds}{dt}=\left|\frac{d\mathbf{R}}{dt}\right|$$

Thus, the length L of the arc from $\mathbf{R}(t_0)$ to $\mathbf{R}(t_1)$ is

$$L=\int_{t_0}^{t_1}\frac{ds}{dt}\,dt=\int_{t_0}^{t_1}\left|\frac{d\mathbf{R}}{dt}\right|dt$$

The vector $d\mathbf{T}/ds = (d\mathbf{T}/dt)/(ds/dt)$ is perpendicular to the unit tangent **T**, and the number $\kappa = |d\mathbf{T}/ds|$ is the *curvature* of C. The unit vector $\mathbf{N} = (1/\kappa)(d\mathbf{T}/ds)$ is the *principal normal*. The vector $\mathbf{B} = \mathbf{T}\times\mathbf{N}$ is the *binormal*, and $d\mathbf{B}/ds = -\tau N$. The number τ is the *torsion*. Note that C is a plane curve if and only if τ is zero for all t.

A *vector function* **F** *of two variables* is a rule that assigns a vector $\mathbf{F}(s, t)$ to each point (s, t) in some subset of the plane, called the *domain* of **F**. If $\mathbf{R}(s, t)$ is defined for all (s, t) in some region D of the plane, then as the point (s, t) varies over D, with its rail at the origin, the nose of $\mathbf{R}(s, t)$ traces out an object in space. For a nice function **R**, this object is a *surface*, S. The partial derivatives $(\partial\mathbf{R}/\partial s)(s, t)$ and $(\partial\mathbf{R}/\partial t)(s, t)$

are tangent to the surface at $\mathbf{R}(s,t)$, and the vector $(\partial\mathbf{R}/\partial s) \times (\partial\mathbf{R}/\partial t)$ is thus *normal* to the surface. Of course, $(\partial\mathbf{R}/\partial t) \times (\partial\mathbf{R}/\partial s) = -(\partial\mathbf{R}/\partial s) \times (\partial\mathbf{R}/\partial t)$ is also normal to the surface and points in the direction opposite that of $(\partial\mathbf{R}/\partial s) \times (\partial\mathbf{R}/\partial t)$. By electing one of these normals, we are choosing an *orientation* of the surface. A surface can be oriented only if it has two sides, and the process of orientation consists of choosing which side is "positive" and which is "negative."

219.4 Gradient, Curl, and Divergence

If $f(x, y, z)$ is a scalar field defined in some region D, the *gradient* of **f** is the vector function

$$\text{grad}\, f = \frac{\partial f}{\partial x}\mathbf{i} + \frac{\partial f}{\partial y}\mathbf{j} + \frac{\partial f}{\partial z}\mathbf{k}$$

If $\mathbf{F}(x, y, z) = F_1(x, y, z)\mathbf{i} + F_2(x, y, z)\mathbf{j} + F_3(x, y, z)\mathbf{k}$ is a vector field defined in some region D, then the *divergence* of **F** is the scalar function

$$\text{div}\, \mathbf{F} = \frac{\partial F_1}{\partial x} + \frac{\partial F_2}{\partial y} + \frac{\partial F_3}{\partial z}$$

The curl is the vector function

$$\text{curl}\, \mathbf{F} = \left(\frac{\partial F_3}{\partial y} - \frac{\partial F_2}{\partial z}\right)\mathbf{i} + \left(\frac{\partial F_1}{\partial z} - \frac{\partial F_3}{\partial x}\right)\mathbf{j} + \left(\frac{\partial F_2}{\partial x} - \frac{\partial F_1}{\partial y}\right)\mathbf{k}$$

In terms of the vector operator *del*, $\nabla = \mathbf{i}(\partial/\partial x) + \mathbf{j}(\partial/\partial y) + \mathbf{k}(\partial/\partial z)$, we can write

$$\text{grad}\, f = \nabla f$$

$$\text{div}\, \mathbf{F} = \nabla \cdot \mathbf{F}$$

$$\text{curl}\, \mathbf{F} = \nabla \times \mathbf{F}$$

The *Laplacian operator* is div (grad) $= \nabla \cdot \nabla = \nabla^2 = (\partial^2/\partial x^2) + (\partial^2/\partial y^2) + (\partial^2/\partial z^2)$.

219.5 Integration

Suppose C is a curve from the point (x_0, y_0, z_0) to the point (x_1, y_1, z_1) and is described by the vector function $\mathbf{R}(t)$ for $t_0 \le t \le t_1$. If f f is a scalar function (sometimes called a *scalar field*) defined on C, then the integral of f over C is

$$\int_C f(x, y, z)\, ds = \int_{t_0}^{t_1} f[\mathbf{R}(t)]\left|\frac{d\mathbf{R}}{dt}\right| dt$$

If **F** is a vector function (sometimes called a *vector field*) defined on C, then the integral of **F** over C is

$$\int_C \mathbf{F}(x, y, z) \cdot d\mathbf{R} = \int_{t_0}^{t_1} \mathbf{F}[\mathbf{R}(t)] \cdot \frac{d\mathbf{R}}{dt} dt$$

These integrals are called *line integrals.*

In case there is a scalar function f such that $\mathbf{F} = \text{grad}\, f$, then the line integral

$$\int_C \mathbf{F}(x,y,z)\cdot d\mathbf{R} = f[\mathbf{R}(t_1)] - f[\mathbf{R}(t_0)]$$

The value of the integral thus depends only on the end points of the curve C and not on the curve C itself. The integral is said to be *path independent*. The function f is called a *potential function* for the vector field **F**, and **F** is said to be a conservative field. A vector field **F** with domain D is conservative if and only if the integral of **F** around every closed curve in D is zero. If the domain D is simply connected (i.e., every closed curve in D can be continuously deformed in D to a point), then **F** is conservative if and only if curl **F** = 0 in D.

Suppose S is a surface described by $\mathbf{R}(s, t)$ for (s, t) in a region D of the plane. If f is a scalar function defined on D, then the integral of f over S is given by

$$\iint_S f(x,y,z)\,dS = \iint_D f[\mathbf{R}(s,t)]\left|\frac{\partial \mathbf{R}}{\partial s}\times\frac{\partial \mathbf{R}}{\partial t}\right| ds\,dt$$

If **F** is a vector function defined on S, and if an orientation for S is chosen, then the integral of **F** over S, sometimes called the **flux** of **F** through S, is

$$\iint_S \mathbf{F}(x,y,z)\cdot d\mathbf{S} = \iint_D \mathbf{F}[\mathbf{R}(s,t)]\cdot\left(\frac{\partial \mathbf{R}}{\partial s}\times\frac{\partial \mathbf{R}}{\partial t}\right) ds\,dt$$

219.6 Integral Theorems

Suppose **F** is a vector field with a closed domain D bounded by the surface S oriented so that the normal points out from D. Then the *divergence theorem* states that

$$\iiint_D \text{div}\,\mathbf{F}\,dV = \iint_S \mathbf{F}\cdot\mathbf{dS}$$

If S is an orientable surface bounded by a closed curve C, the orientation of the closed curve C is chosen to be consistent with the orientation of the surface S. Then we have *Stokes's theorem*:

$$\iint_S (\text{curl }\mathbf{F})\cdot\mathbf{dS} = \oint_C \mathbf{F}\cdot\mathbf{ds}$$

References

Davis, H. F. and Snider, A. D. 1991. *Introduction to Vector Analysis,* 6th ed. Wm. C. Brown, Dubuque, IA.

Wylie, C. R. 1975. *Advanced Engineering Mathematics,* 4th ed. McGraw-Hill, New York.

Further Information

More advanced topics leading into the theory and applications of tensors may be found in J. G. Simmonds, *A Brief on Tensor Analysis* (1982, Springer-Verlag, New York).

220

Complex Variables

George Cain
Georgia Institute of Technology

220.1 Basic Definitions and Arithmetic

A *complex number* is an ordered pair $z = (x, y)$ of real numbers x and y. The sum, difference, product, and quotient of two complex numbers $z = (x, y)$ and $w = (u, v)$ are defined by

$$z \pm w = (x \pm u, y \pm v)$$

$$zw = (xu - yv, xv + yu)$$

$$\frac{w}{z} = \left(\frac{xu + yv}{x^2 + y^2}, \frac{xv - yu}{x^2 + y^2}\right)$$

The *real part* of the complex number $z = (x, y)$ is x, and the *imaginary part* of z is y. If $z_1 = (a, 0)$ and $z_2 = (b, 0)$, then $z_1 \pm z_2 = (a \pm b, 0)$, $z_1 z_2 = (ab, 0)$, and $z_1/z_2 = (a/b, 0)$. Thus, for complex numbers with 0 imaginary part, complex arithmetic coincides with the usual arithmetic for real numbers, and we see that the complex numbers are an extension of the real numbers. In this case, we write simply a for the complex number $(a, 0)$, and so on.

For the complex number $i = (0, 1)$ we have that $i^2 = (-1, 0) \equiv -1$. Now note that for any $z = (x, y)$, it is true that

$$z = (x, y) = (x, 0) + (0, y) = x + iy$$

and the algebra of complex numbers reduces to the usual algebra for real numbers, together with the fact that $i^2 = -1$.

The *conjugate* of a complex number $z = x + iy$ is defined to be $\bar{z} = x - iy$. The *modulus* of z is the real number $|z| = \sqrt{z\bar{z}} = \sqrt{x^2 + y^2}$. We have the following properties of the conjugate and the modulus:

0-8493-1586-7/05/$0.00+$1.50

$$\overline{(z \pm w)} = \overline{z} \pm \overline{w}$$

$$\overline{zw} = \overline{z}\,\overline{w} \quad \text{and} \quad \overline{\left(\frac{w}{z}\right)} = \frac{\overline{w}}{\overline{z}}$$

$$|z+w| \le |z| + |w|$$

$$|zw| = |z||w| \quad \text{and} \quad \left|\frac{w}{z}\right| = \frac{|w|}{|z|}$$

Considering $z = (x, y)$ to be the rectangular coordinates of a point in the plane establishes a one-to-one correspondence between the set of all complex numbers and the points in the plane. We thus speak of points in the plane as being complex numbers. The usual Euclidean distance $d(z, w)$ between the points z and w is then given by $d(z, w) = |z - w|$. Let (r, θ) be polar coordinates of the point $z = (x, y)$. (We always assume $r \ge 0$.) Then r is, of course, simply $|z|$, the modulus of z. The number θ is called an *argument* of z and is denoted arg z. Thus, $z = r(\cos\theta + i\sin\theta)$. If $w = s\,(\cos\vartheta + i\vartheta)$, then

$$zw = rs[\cos(\theta + \vartheta) + i\sin(\theta + \vartheta)]$$

$$\frac{z}{w} = \frac{r}{s}[\cos(\theta - \vartheta) + i\sin(\theta - \vartheta)]$$

$$z^n = r^n[\cos n\theta + \sin n\theta]$$

Note that a complex number has an infinite number of arguments, and arg z denotes any one of them. Any interval $(a, b]$ of reals of length 2π contains exactly one argument of z. The particular value of arg z in the interval $(-\pi, \pi]$ is called the *principal value of the argument* and is denoted Arg z.

220.2 Complex Functions

Let **C** denote the set of all complex numbers (the *complex plane*). A set $\mathbf{S} \subset \mathbf{C}$ is said to be a *neighborhood* of the point z_0 if there is an $r > 0$ such that $z \in \mathbf{S}$ whenever $|z_0 - z| < r$. The point z_0 is an *interior point* of **S** in case **S** is a neighborhood of z_0. A point z_0 is called a *limit point* of a set **A** if every neighborhood of z_0 meets **A**. A set that is a neighborhood of each of its points is called an *open set*, and a set that contains all its limit points is called a *closed set.*

A function f from a subset of the complex numbers into the complex numbers is called a *complex function.* The statement that the *limit* of f at z_0 is w_0 means that for each $\varepsilon > 0$, there is a $\delta > 0$ such that $|f(z) - w_0| < \varepsilon$ whenever z is in the domain of f and $|z - z_0| < \delta$. We write

$$\lim_{z \to z_0} f(z) = w_0$$

Note that it is not required that z_0 be in the domain of f.

The function f is *continuous* at the point z_0 in its domain if $\lim_{z\to z_0} f(z) = f(z_0)$. If z_0 is an interior point of the domain of f, then f is said to be *differentiable* at z_0 if the limit

$$f'(z_0) = \lim_{z \to z_0} \frac{f(z) - f(z_0)}{z - z_0}$$

exists. The number $f'(z_0)$ is the *derivative* of f at z_0.

A complex function of the form $p(z) = a_n z^n + a_{n-1} z^{n-1} + \cdots + a_1 z + a_0$ is called a *polynomial.* A *rational function* r is the quotient of two polynomials: $r(z) = p(z)/q(z)$. A polynomial is continuous and differentiable at every complex number, and a rational function is continuous and differentiable everywhere except at those points z_0 for which $q(z_0) = 0$.

220.3 Analytic Functions

A complex function f is *analytic* on an open set **U** if it has a derivative at each point of **U**. The function f is said to be analytic at the point z_0 if it is analytic on some open set containing z_0. A function analytic on all of the complex plane is an *entire* function. If $f(z) = u(x, y) + iv(x, y)$ is analytic in an open set **U**, then the real and imaginary parts of $f(z)$ satisfy the Cauchy-Riemann equations:

$$\frac{\partial u}{\partial x} = \frac{\partial v}{\partial y} \quad \text{and} \quad \frac{\partial u}{\partial y} = -\frac{\partial v}{\partial x}$$

and both u and v satisfy *Laplace's equation* in **U**:

$$\frac{\partial^2 u}{\partial x^2} + \frac{\partial^2 u}{\partial y^2} = 0$$

A function that satisfies Laplace's equation in a domain **U** is said to be *harmonic* in **U**. Thus, the real and imaginary parts of a function analytic in an open set are harmonic on the same set.

220.4 Integration

An integral of a complex function f is defined on a curve (or *contour*) C in the complex plane and is simply the line integral of f on C. Thus if $z = \gamma(t)$, $a \le t \le b$ defines a contour C from the point $z_0 = \gamma(a)$ to the point $z_1 = \gamma(b)$, the *contour integral* of f along C is given by

$$\int_C f(z)dz = \int_a^b f(\gamma(t))\gamma'(t)dt$$

All of the usual nice linearity properties of integrals hold for complex integrals. Thus,

$$\int_C [c_1 f(z) + c_2 g(z)]dz = c_1 \int_C f(z)dz + c_2 \int_C g(z)dz$$

for any complex numbers c_1 and c_2. If C_1 and C_2 are contours, then $\int_{C_1+C_2} f(z)dz$ is defined by

$$\int_{C_1+C_2} f(z)\,dz = \int_{C_1} f(z)\,dz + \int_{C_2} f(z)\,dz$$

A contour given by $z = \gamma(t)$, $a \le t \le b$ is called a *simple* contour if $\gamma(t) \ne \gamma(s)$ for $a < t, s < b$, and $t \ne s$. A *closed* contour is a contour for which $\gamma(a) = \gamma(b)$. If C is given by $z = \gamma(t)$, $a \le t \le b$, then $-C$ is the contour given by $z = \gamma(a + b - t)$, $a \le t \le b$. It is clear that

$$\int_{-C} f(z)dz = -\int_C f(z)dz$$

Suppose C is a simple closed contour and f is a function analytic at all points on and inside C. Then

$$\int_C f(z)dz = 0$$

This is the celebrated *Cauchy-Goursat theorem.*

An open connected set **U** such that every simple closed contour in **U** encloses only points of **U** is said to be *simply connected.* If z_0 and z_1 are points in a simply connected open set **U**, and if C_1 and C_2 are contours in **U** from z_0 and z_1, then for any f analytic in **U**, it is true that

$$\int_{C_1} f(z)dz = \int_{C_2} f(z)dz$$

A simple closed contour C given by $z = \gamma(t)$, $a \le t \le b$ is *positively oriented* if, for z_0 inside C, the argument $\text{Arg}[\gamma(t) - z_0]$ increases as t goes from a to b. If f is analytic in simply connected region **U**, and if C is a positively oriented simple closed curve in **U**, then for every z_0 in **U** that is not on C, we have

$$f(z_0) = \frac{1}{2\pi i}\int_C \frac{f(z)}{z - z_0}dz$$

This is the celebrated *Cauchy integral formula.* If f is analytic at a point z_0, then it has derivatives of all orders, and

$$f^{(n)}(z_0) = \frac{n!}{2\pi i}\int_C \frac{f(z)}{(z - z_0)^{n+1}}dz$$

If the function f is continuous on an open connected set **U** and if

$$\int_C f(z)dz = 0$$

for every closed contour C in **U**, then f is analytic on **U**. This is *Morera's theorem.*

If f is analytic on an open set **U**, the point z_0 is in **U**, and C_R is a circle in **U** of radius r centered at z_0 on which $|f(z)| \le M$, then

$$\left|f^{(n)}(z_0)\right| \le \frac{n!M}{r^n}$$

If follows from this that a bounded entire function must be constant.

220.5 Series

Let f be a function analytic in an open set **U**, and let z_0 be a point of **U**. Then there is a unique power series $\sum_{k=0}^{\infty} a_k(z - z_0)^k$ such that

$$f(z) = \sum_{k=0}^{\infty} a_k(z - z_0)^k$$

for all z in some neighborhood of z_0. This series is the *Taylor series* of f at z_0 and converges to $f(z)$ for all z inside a circle C, on which is found the singularity of f closest to z_0. The radius of C is called the *radius of convergence* of the series. This series may be differentiated term by term to obtain the Taylor series for the *n*th derivative of f at z_0:

$$f^{(n)}(z)=\sum_{k=n}^{\infty}k(k-1)\cdots(k-n+1)a_k(z-z_0)^{k-n}$$

The radius of convergence of this series is the same as that of the Taylor series for f. It follows easily that, for each $n = 0, 1, 2, \ldots$, we have

$$a_n=\frac{1}{n!}f^{(n)}(z_0)$$

Let C_0 and C_1 be two concentric circles centered at a point z_0, with C_0 having a smaller radius than C_1. Suppose the function f is analytic on an open set containing the two circles and the annular region between them. Then, for each point in the annular region bounded by the circles, we have

$$f(z)=\sum_{k=0}^{\infty}a_k(z-z_0)^k+\sum_{k=1}^{\infty}b_k(z-z_0)^{-k}$$

where

$$a_k=\frac{1}{2\pi i}\int_C\frac{f(z)}{(z-z_0)^{k+1}}dz$$

$$b_k=\frac{1}{2\pi i}\int_C\frac{f(z)}{(z-z_0)^{-k+1}}dz$$

for any positively oriented simply closed contour C around the annular region. This is a *Laurent series.* It is sometimes written

$$f(z)=\sum_{k=-\infty}^{\infty}c_k(z-z_0)^k$$

with

$$c_k=\frac{1}{2\pi i}\int_C\frac{f(z)}{(z-z_0)^{k+1}}dz$$

220.6 Singularities

Suppose that z_0 is a singular point of the function f and f is analytic in a neighborhood of z_0 (except, of course, at z_0). Then z_0 is called an *isolated singular point*, or an *isolated singularity*, of f. In case z_0 is an isolated singular point of f, the Laurent series

$$f(z)=\sum_{k=-\infty}^{\infty}c_k(z-z_0)^k$$

represents f for all z such that $0 < |z - z_0| < r$, for some $r > 0$. If $c_k = 0$ for all $k \leq -1$, then z_0 is said to be a *removable* singularity. (The value of f at z_0 can, in this case, be redefined to be c_0, and the function so defined is analytic at z_0.) If there is a negative integer $N < -1$ such that $c_k = 0$ for all $k \leq N$, then z_0 is a *pole* of order p, where p is the largest positive integer for which $c_{-p} \neq 0$. A pole of order one is called a *simple pole*. An isolated singularity that is neither a removable singularity nor a pole is called an *essential* singularity. Thus, z_0 is an essential singularity if, for every negative N, there is a $k < N$ such that $c_k \neq 0$.

If z_0 is an isolated singularity of the function f, then the coefficient c_{-1} in the Laurent series given is the *residue* of f at z_0. If z_0 is a simple pole, then the residue of f at z_0 is given by

$$\mathrm{Res}(z_0) = \lim_{z \to z_0} (z - z_0) f(z)$$

If z_0 is a pole of order p, then

$$\mathrm{Res}(z_0) = \lim_{z \to z_0} \frac{1}{(p-1)!} \frac{d^{p-1}}{dz^{p-1}} [(z - z_0)^p f(z)]$$

The importance of the idea of the residue comes from *Cauchy's residue theorem*, which says that if C is a positively oriented simple closed contour and the function f is analytic on and inside C except at the points $z_1, z_2, \ldots, z_n$, then

$$\int_C f(z)\,dz = 2\pi i \sum_{k=1}^{n} \mathrm{Res}\,(z_k)$$

220.7 Conformal Mapping

We can regard a one-to-one complex function f on a region **D** as defining a *mapping* from **D** into the plane. In the event that f is analytic, this mapping is *conformal*; that is, angles are preserved. More importantly, such a mapping preserves the harmonic property of a function. Thus, if f is analytic and $w = f(z) = u(x, y) + iv(x, y)$ maps a domain $\mathbf{D}_z$ onto a domain $\mathbf{D}_w$, and if

$$\frac{\partial^2 \Phi(u,v)}{\partial u^2} + \frac{\partial^2 \Phi(u,v)}{\partial v^2} = 0 \text{ in } \mathbf{D}_w$$

then

$$\frac{\partial^2 \Theta(x,y)}{\partial x^2} + \frac{\partial^2 \Theta(x,y)}{\partial y^2} = 0 \text{ in } \mathbf{D}_z$$

where $\Theta(x, y) = \Phi[u(x, y), v(x, y)]$. Conformal mappings are useful in solving boundary value problems involving Laplace's equation in a region of the plane. The given region is mapped conformally to another region in which the solution is either known or easy to find.

References

Ahlfors, L. V. 1979. *Complex Analysis,* 3rd ed. McGraw-Hill, New York.

Churchill, R. V. and Brown J. W. 1990. *Complex Variables and Applications,* 5th ed. McGraw-Hill, New York.

Saff, E. B. and Snider, A. D. 1976. *Fundamentals of Complex Analysis for Mathematics, Science, and Engineering.* Prentice Hall, Englewood Cliffs, NJ.

Further Information

Extensive compilations of conformal mappings are found in *Dictionary of Conformal Representations*, by H. Kober (1957, Dover Publications, New York.), and *Complex Variables and Applications*, by R. V. Churchill and J. W. Brown (1990, McGraw-Hill, New York.)

221

Difference Equations

William F. Ames
Georgia Institute of Technology

Difference equations are equations involving *discrete variables*. They appear as natural descriptions of natural phenomena and in the study of discretization methods for differential equations, which have continuous variables.

Let $y_n = y(nh)$, where n is an integer and h a real number. (One can think of measurements taken at equal intervals, $h, 2h, 3h, \ldots$, and y_n describes these.) A typical equation is that describing the famous Fibonacci sequence — $y_{N+2} - y_{n+1} - y_n = 0$. Another example is the equation $y_{n+2} - 2zy_{n+1} + y_n = 0$, $z \in C$, which describes the Chebyshev polynomials.

221.1 First-Order Equations

The general first-order equation $y_{n+1} = f(y_n)$, $y_0 = y(0)$ is easily solved, for as many terms as are needed, by *iteration*. Then $y_1 = f(y_0)$; $y_2 = f(y_1), \ldots$. An example is the logistic equation $y_{n+1} = ay_n(1 - y_n) = f(y_n)$. The logistic equation has two fixed (critical or equilibrium) points where $y_{n+1} = y_n$. They are 0 and $\bar{y} = (a - 1)/a$. This has physical meaning only for $a > 1$. For $1 < a < 3$ the equilibrium $\bar{y}$ is asymptotically stable, and for $a > 3$ there are two points y_1 and y_2, called a *cycle of period two*, in which $y_2 = f(y_1)$and $y_1 = f(y_2)$. This study leads into chaos, which is outside our interest. By iteration, with $y_0 = 1/2$, we have $y_1 = (a/2)(1/2) = a/2^2$, $y_2 = a(a/2^2)(1 - a/2^2) = (a^2/2^2)(1 - a/2^2), \ldots$.

With a constant, the equation $y_{n+1} = ay_n$ is solved by making the assumption $y_n = A\lambda^n$ and finding λ so that the equation holds. Thus $A\lambda^{n+1} = aA\lambda^n$, and hence $\lambda = 0$ or $\lambda = a$ and A is arbitrary. Discarding the trivial solution 0 we find $y_n = Aa^{n+1}$ is the desired solution. By using a method called the *variation of constants*, the equation $y_{n+1} - ay_n = g_n$ has the solution $y_n = y_0a^n + \sum_{j=0}^{n-1} g_j a^{n-j-1}$, with y_0 arbitrary.

In various applications we find the first-order equation of *Riccati type* $y_n y_{n-1} + ay_n + by_{n-1} + c = 0$ where a, b, and c are real constants. This equation can be transformed to a linear second-order equation by setting $y_n = z_n/z_{n-1} - a$ to obtain $z_{n-1} + (b + a)z_n + (c - ab)z_{n-1} = 0$, which is solvable as described in the next section.

221.2 Second-Order Equations

The second-order linear equation with constant coefficients $y_{n+2} + ay_{n+1} + by_n = f_n$ is solved by first solving the homogeneous equation (with right-hand side zero) and adding to that solution any solution of the

0-8493-1586-7/05/$0.00+$1.50

inhomogeneous equation. The *homogeneous equation* $y_{n+2} + ay_{n+1} + by_n = 0$ is solved by assuming $y_n = \lambda^n$, whereupon $y^{n+2} + a\lambda^{n+1} + b\lambda^n = 0$ or $\lambda = 0$ (rejected) or $\lambda^2 + a\lambda + b = 0$. The roots of this quadratic are $\lambda_1 = \frac{1}{2}(-a+\sqrt{a^2-4b})$, $\lambda_2 = -\ \frac{1}{2}(a+\sqrt{a^2-4b})$ and the solution of the homogeneous equation is $y_n = c_1\lambda_1^n + c_2\lambda_2^n$. As an example consider the Fibonacci equation $y_{n+2} - y_{n+1} - y_n = 0$. The roots of $\lambda^2 - \lambda - 1 = 0$ are $\lambda_1 = \frac{1}{2}(1+\sqrt{5})$, $\lambda_2 = \frac{1}{2}(1-\sqrt{5})$, and the solution $y_n = c_1\,[(1+\sqrt{5})/2]^n + c_2[(1-\sqrt{5})/2]^n$ is known as the *Fibonacci sequence.*

Many of the orthogonal polynomials of differential equations and numerical analysis satisfy a second-order difference equation (recurrence relation) involving a discrete variable, say n, and a continuous variable, say z. One such is the *Chebyshev equation* $y_{n+2} - 2zy_{n+1} + y_n = 0$ with the initial conditions $y_0 = 1$, $y_1 = z$ (*first-kind* Chebyshev polynomials) and $y_{-1} = 0$, $y_0 = 1$ (second-kind Chebyshev polynomials). They are denoted $T_n(z)$ and $V_n(z)$, respectively. By iteration we find

$$T_0(z)=1,\quad T_1(z)=z,\quad T_2(z)=2z^2-1,$$

$$T_3(z)=4z^3-3z,\quad T_4(z)=8z^4-8z^2+1,$$

$$V_0(z)=0,\quad V_1(z)=1,\quad V_2(z)=2z,$$

$$V_3(z)=4z^2-1,\quad V_4(z)=8z^3-4z$$

and the general solution is $y_n(z) = c_1T_n(z) + c_2V_{n-1}(z)$.

221.3 Linear Equations with Constant Coefficients

The general kth-order linear equation with constant coefficients is $\sum_{i=0}^{k} p_i y_{n+k-i} = g_n$, $p_0 = 1$. The solution to the corresponding homogeneous equation (obtained by setting $g_n = 0$) is as follows. (a) $y_n = \sum_{i=1}^{k} c_i\lambda_i^n$ if the λ_i are the distinct roots of the characteristic polynomial $p(\lambda) = \sum_{i=0}^{k} p_i\lambda^{k-i} = 0$. (b) If m_s is the multiplicity of the root λ_s, then the functions $y_{n,s} = u_s(n)\,\lambda_s^n$, where $u_s(n)$ are polynomials in n whose degree does not exceed $m_s - 1$, are solutions of the equation. Then the general solution of the homogeneous equation is $y_n = \sum_{i=1}^{d} a_i u_i(n)\lambda_i^n = \sum_{i=1}^{d} a_i \sum_{j=0}^{m_i-1} c_j n^j \lambda_i^n$. To this solution one adds any particular solution to obtain the general solution of the general equation.

Example 221.1

A model equation for the price p_n of a product, at the nth time, is $p_n + \frac{b}{a}(1+\rho)\,p_{n-1} - \frac{b}{a}\rho p_{n-2} + (s_0 - d_0)/a = 0$. The equilibrium price is obtained by setting $p_n = p_{n-1} = p_{n-2} = p_e$, and one finds $p_e = (d_0 - s_0)/(a+b)$. The homogeneous equation has the characteristic polynomial $\lambda^2 + (b/a)(1+\rho)\lambda - (b/a)\rho = 0$. With λ_1 and λ_2 as the roots the general solution of the full equation is $p_n = c_1\lambda_1^n + c_2\lambda_2^n + p_e$, since p_e is a solution of the full equation. This is one method for finding the solution of the nonhomogeneous equation.

221.4 Generating Function (z Transform)

An elegant way of solving linear difference equations with constant coefficients, among other applications, is by use of *generating functions* or, as an alternative, the z transform. The generating function of a

TABLE 221.1 Important Sequences

y_n	$f(x)$	Convergence Domain
1	$(1-x)^{-1}$	$\lvert x\rvert<1$
n	$x(1-x)^{-2}$	$\lvert x\rvert<1$
n^m	$xp_m(x)(1-x)^{-n-1}$*	$\lvert x\rvert<1$
k^n	$(1-kx)i^{-1}$	$\lvert x\rvert<k^{-1}$
e^{an}	$(1-e^a x)^{-1}$	$\lvert x\rvert<e^{-a}$
$k^n \cos an$	$\dfrac{1-kx\cos a}{1-2kx\cos a+k^2x^2}$	$\lvert x\rvert<k^{-1}$
$k^n \sin an$	$\dfrac{kx\sin a}{1-2kx\cos a+k^2x^2}$	$\lvert x\rvert<k^{-1}$
$\binom{n}{m}$	$x^m(1-x)^{-m-1}$	$\lvert x\rvert<1$
$\binom{k}{n}$	$(1+x)^k$	$\lvert x\rvert<1$

* The term $p_m(z)$ is a polynomial of degree m satisfying $p_{m+1}(z)=(mz+1)p_m(z)+z(1-z)p'_m(z), p_1=1$.

sequence $\{y_n\}$, $n = 0, 1, 2, \ldots$, is the function $f(x)$ given by the formal series $f(x) = \sum_{n=0}^{\infty} y_n x^n$. The z transform of the same sequence is $z(x) = \sum_{n=0}^{\infty} y_n x^{-r}$. Clearly, $z(x) = f(1/x)$. A table of some important sequences is given in Table 221.1.

To solve the linear difference equation $\sum_{i=1}^{k} p_i y_{n+k-i} = 0$, $p_0 = 1$ we associate with it the two formal series $P = p_0 + p_1 x + \cdots + p_k x^k$ and $Y = y_0 + y_1 x + y_2 x^2 + \cdots$. If $p(x)$ is the characteristic polynomial then $P(x) = x^k p(1/x) = \bar{p}(x)$. The *product* of the two series is $Q = YP = q_0 + q_1 x + \cdots + q_{k-1}x^{k-1} + q_k x^k + \cdots$ where $q_n = \sum_{i=0}^{n} p_i y_{n-i}$. Because $p_{k+1} = p_{k+2} = \cdots = 0$, it is obvious that $q_{k+1} = q_{k+2} = \cdots = 0$ — that is, Q is a polynomial (formal series with finite number of terms). Then $Y = p^{-1}Q = q(x)/\bar{p}(x) = q(x)/x^k p(1/x)$, where p is the characteristic polynomial and $q(x) = \sum_{i=0}^{k} q_i x^i$. The roots of $\bar{p}(x)$ are x_i^{-1} where the x_i are the roots of $p(x)$.

Theorem 1. If the roots of $p(x)$ are less than one in absolute value, then $Y(x)$ converges for $|x| < 1$.

Theorem 2. If $p(x)$ has no roots greater than one in absolute value and those on the unit circle are simple roots, then the coefficients y_n of Y are bounded. Now $q_k = g_0$, $q_{n+k} = g_n$, and $Q(x) = Q_1(x) + x^k Q_2(x)$. Hence $\sum_{i=1}^{\infty} y_i x^i = [Q_1(x) + x^k Q_2(x)]/[\bar{p}(x)]$.

Example 221.2

Consider the equation $y_{n+1} + y_n = -(n+1)$, $y_0 = 1$. Here $Q_1 = 1$, $Q_2 = -\sum_{n=0}^{\infty}(n+1)x^n = -1/(1-x)^2$.

$$G(x) = \frac{1 - x/(1-x)^2}{1+x} = \frac{5}{4}\frac{1}{1+x} - \frac{1}{4}\frac{1}{1-x} - \frac{1}{2}\frac{x}{(1-x)^2}$$

Using the table term by term, we find $\sum_{n=0}^{\infty} y_n x^n = \sum_{n=0}^{\infty}[\frac{5}{4}(-1)^n - \frac{1}{4} - \frac{1}{2}n]x^n$, so $y_n = \frac{5}{4}(-1)^n - \frac{1}{4} - \frac{1}{2}n$.

References

Fort, T. 1948. *Finite Differences and Difference Equations in the Real Domain.* Oxford University Press, London.

Jordan, C. 1950. *Calculus of Finite Differences.* Chelsea, New York.

Jury, E. I. 1964. *Theory and Applications of the Z Transform Method.* John Wiley & Sons, New York.

Lakshmikantham, V. and Trigrante, D. 1988. *Theory of Difference Equations.* Academic Press, Boston, MA.

Levy, H. and Lessman, F. 1961. *Finite Difference Equations.* Macmillan, New York.

Miller, K. S. 1968. *Linear Difference Equations.* Benjamin, New York.

Wilf, W. S. 1994. *Generatingfunctionology,* 2nd ed. Academic Press, Boston, MA.

222

Differential Equations

William F. Ames
Georgia Institute of Technology

Any equation involving derivatives is called a *differential equation.* If there is only one independent variable the equation is termed a *total differential equation* or an *ordinary differential equation.* If there is more than one independent variable the equation is called a *partial differential equation.* If the highest-order derivative is the nth, then the equation is said to be nth order. If there is no function of the dependent variable and its derivatives other than the linear one, the equation is said to be *linear.* Otherwise, it is *nonlinear.* Thus $(d^3y/dx^3) + a(dy/dx) + by = 0$ is a *linear* third-order ordinary (total) differential equation. If we replace by with by^3, the equation becomes nonlinear. An example of a second-order linear partial differential equation is the famous wave equation $(\partial^2u/\partial x^2) - a^2(\partial^2u/\partial t^2) = f(x)$. There are two independent variables x and t and $a^2 > 0$ (of course). If we replace $f(x)$ by $f(u)$ (say u^3 or sin u), the equation is nonlinear. Another example of a nonlinear third-order partial differential equation is $u_t + uu_x = au_{xxx}$. This chapter uses the common subscript notation to indicate the partial derivatives.

Now we briefly indicate some methods of solution and the solution of some commonly occurring equations.

222.1 Ordinary Differential Equations

First-Order Equations

The *general* first-order equation is $f(x,y,y') = 0$. Equations capable of being written in either of the forms $y' = f(x)g(y)$ or $f(x)g(y)y' + F(x)G(y) = 0$ are *separable* equations. Their solution is obtained by using $y' = dy/dx$ and writing the equations in differential form as $dy/g(y) = f(x)dx$ or $g(y)[dy/G(y)] = -F(x)[dx/f(x)]$ and integrating. An example is the famous *logistic* equation of inhibited growth $(dy/dt) = ay(1 - y)$. The integral of $dy/y(1 - y) = a\,dt$ is $y = 1/[1 + (y_0^{-1} - 1)e^{-at}]$ for $t \geq 0$ and $y(0) = y_0$ (the initial state called the *initial condition*).

Equations may not have unique solutions. An example is $y' = 2y^{1/2}$ with the initial condition $y(0) = 0$. One solution by separation is $y = x^2$. But there are an *infinity* of others — namely, $y_a(x) = 0$ for $-\infty < x \leq a$, and $(x - a)^2$ for $a \leq x < \infty$.

If the equation $P(x,y)dy + Q(x,y)dx$ is reducible to

0-8493-1586-7/05/$0.00+$1.50

$$\frac{dy}{dx}=f\left(\frac{y}{x}\right) \quad \text{or} \quad \frac{dy}{dx}=f\left(\frac{a_1x+b_1y+c_1}{a_2x+b_2y+c_2}\right)$$

the equation is called *homogeneous* (nearly homogeneous). The first form reduces to the separable equation $u + x(du/dx) = f(u)$ with the substitution $y/x = u$. The nearly homogeneous equation is handled by setting $x = X + \alpha$, $y = Y + \beta$, and choosing α and β so that $a_1\alpha + b_1\beta + c_1 = 0$ and $a_2\alpha + b_2\beta + c_2 = 0$. If $\begin{vmatrix} a_1 & b_1 \\ a_2 & b_2 \end{vmatrix} \neq 0$ are always possible; the equation becomes $dY/dX = [a_1 + b_1(Y/X)]/[a_2 + b_2(Y/X)]$ and the substitution $Y = Xu$ gives a separable equation. If $\begin{vmatrix} a_1 & b_1 \\ a_2 & b_2 \end{vmatrix} = 0$ then $a_2x + b_2y = k(a_1x + b_1y)$ and the equation becomes $du/dx = a_1 + b_1(u + c_1)/(ku + c_2)$, with $u = a_1x + b_1y$. Lastly, any equation of the form $dy/dx = f(ax + by + c)$ transforms into the separable equation $du/dx = a + bf(u)$ using the change of variable $u = ax + by + c$.

The general first-order linear equation is expressible in the form $y' + f(x)y = g(x)$. It has the *general solution* (a solution with an arbitrary constant c)

$$y(x)=\exp[-\int f(x)dx]\{c+\int \exp[f(x)dx]g(x)dx\}$$

Two noteworthy examples of first-order equations are as follows:

1. An often-occurring nonlinear equation is the *Bernoulli equation,* $y' + p(x)y = g(x)y^{\alpha}$, with α real, $\alpha \neq 0$, $\alpha \neq 1$. The transformation $z = y^{1-\alpha}$ converts the equation to the linear first-order equation $z' + (1 - \alpha)p(x)z = (1 - \alpha)q(x)$.
2. The famous *Riccati equation,* $y' = p(x)y^2 + q(x)y + r(x)$, cannot in general be solved by integration. But some transformations are helpful. The substitution $y = y_1 + u$ leads to the equation $u' - (2py_1 + q)u = pu^2$, which is a Bernoulli equation for u. The substitution $y = y_1 + v^{-1}$ leads to the equation $v' + (2py_1 + q)v + p = 0$, which is a linear first-order equation for v. Once either of these equations has been solved, the general solution of the Riccati equation is $y = y_1 + u$ or $y = y_1 + v^{-1}$.

Second-Order Equations

The simplest of the second-order equations is $y'' + ay' + by = 0$ (a, b real), with the initial conditions $y(x_0) = y_0$, $y'(x_0) = y_0'$ or the boundary conditions $y(x_0) = y_0$, $y(x_1) = y_1$. The general solution of the equation is given as follows.

1. $a^2 - 4b > 0,\quad \lambda_1 = \tfrac{1}{2}(-a + \sqrt{a^2 - 4b}),\quad \lambda_2 = \tfrac{1}{2}(-a - \sqrt{a^2 - 4b})$

$$y = c_1\exp(\lambda_1 x) + c_2\exp(\lambda_2 x)$$

2. $a^2 - 4b = 0,\quad \lambda_1 = \lambda_2 = -\tfrac{a}{2},$

$$y = (c_1 + c_2x)\exp(\lambda_1 x)$$

3. $a^2 - 4b < 0,\quad \lambda_1 = \tfrac{1}{2}(-a + i\sqrt{4b - a^2}),\quad \lambda_2 = \tfrac{1}{2}(-a - i\sqrt{4b - a^2}),$

$$i^2 = -1$$

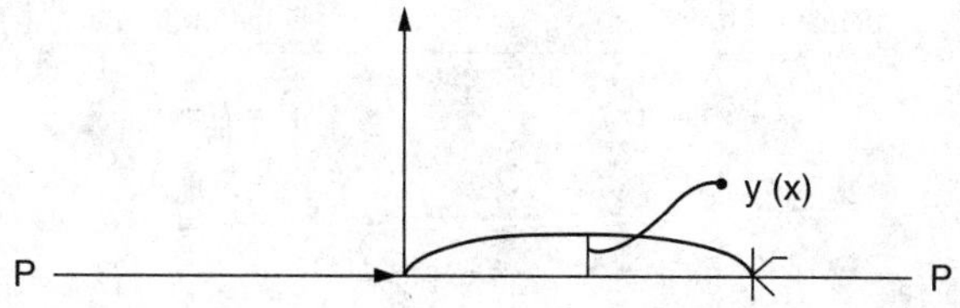

FIGURE 222.1

With $p = -a/2$ and $q = \frac{1}{2}\sqrt{4b - a^2}$,

$$y = c_1 \exp[(p+iq)x] + c_2 \exp[(p-iq)x] = \exp(px)[A \sin qx + B \cos qx]$$

The initial conditions or boundary conditions are used to evaluate the arbitrary constants c_1 and c_2 (or A and B).

Note that a linear problem with specified data may not have a solution. This is especially serious if numerical methods are employed without serious thought.

For example, consider $y'' + y = 0$ with the boundary condition $y(0) = 1$ and $y(\pi) = 1$. The general solution is $y = c_1 \sin x + c_2 \cos x$. The first condition $y(0) = 1$ gives $c_2 = 1$, and the second condition requires $y(\pi) = c_1 \sin \pi + \cos \pi$ or "$1 = -1$," which is a *contradiction.*

Example 222.1: The Euler Strut

When a strut of uniform construction is subjected to a compressive load P it exhibits no transverse displacement until P exceeds some critical value P_1. When this load is exceeded, buckling occurs and large deflections are produced as a result of small load changes. Let the rod of length ℓ be placed as shown in Figure 222.1.

From the linear theory of elasticity (Timoshenko), the transverse displacement $y(x)$ satisfies the linear second-order equation $y'' = Py/EI$, where E is the modulus of elasticity and I is the moment of inertia of the strut. The boundary conditions are $y(0) = 0$ and $y(a) = 0$. With $k^2 = P/EI$ the general solution is $y = c_1 \sin kx + c_2 \cos kx$. The condition $y(0) = 0$ gives $c_2 = 0$. The second condition gives $c_1 \sin ka = 0$. Since $c_1 = 0$ gives only the trivial solution $y = 0$ we must have $\sin ka = 0$. This occurs for $ka = n\pi$, $n = 0,1,2,\ldots$ (these are called *eigenvalues*). The first nontrivial solution occurs for $n = 1$ — that is, $k = \pi/a$ — whereupon $y_1 = c_1 \sin(\pi/a)$, with arbitrary c_1. Since $P = EIk^2$ the critical compressive load is $P_1 = EI\pi^2/a^2$. This is the buckling load. The weakness of the linear theory is its failure to model the situation when buckling occurs.

Example 222.2: — Some Solvable Nonlinear Equations

Many physical phenomena are modeled using nonlinear second-order equations. Some general cases are given here.

1. $y'' = f(y)$, first integral $(y')^2 = 2 \int f(y)dy + c$.
2. $f(x,y',y'')$. Set $p = y'$ and obtain a first-order equation $f(x,p,dp/dx) = 0$. Use first-order methods.
3. $f(y,y',y'') = 0$. Set $p = y'$ and then $y'' = p(dp/dy)$ so that a first-order equation $f[y,p,p(dp/dy)] = 0$ for p as a function of y is obtained.
4. The *Riccati Transformation* $du/dx = yu$ leads to the Riccati chain of equations, which linearize by raising the order. Thus,

	Equation in y	Equation in u
1.	$y' + y^2 = f(x)$	$u'' = f(x)u$
2.	$y'' + 3yy' + y^3 = f(x)$	$u''' = f(x)u$
3.	$y''' + 6y^2y' + 3(y')^2 + 4yy'' = f(x)$	$u^{(iv)} = f(x)u$

This method can be generalized to $u' = a(x)yu$ or $u' = a(x)f(u)y$.

Second-Order Nonhomogeneous Equations

The general solution of $a_0(x)y'' + a_1(x)y' + a_2(x)y = f(x)$ is $y = y_H(x) + y_P(x)$, where $y_H(x)$ is the general solution of the homogeneous equation (with the right side zero) and y_p is the particular integral of the equation. Construction of particular integrals can sometimes be done by the *method of undetermined coefficients* (Table 222.1). This applies only to the linear constant coefficient case in which the function $f(x)$ is a linear combination of a polynomial, exponentials, sines and cosines, and some products of these functions. This method has as its basis the observation that repeated differentiation of such functions gives rise to similar functions.

Example 222.3

Consider the equation $y'' + 3y' + 2y = \sin 2x$. The characteristic equation of the homogeneous equation $\lambda^2 + 3\lambda + 2 = 0$ has the two roots $\lambda_1 = -1$ and $\lambda_2 = -2$. Consequently, $y_H = c_1e^{-x} + c_2e^{-2x}$. Since $\sin 2x$ is not linearly dependent on the exponentials and since $\sin 2x$ repeats after two differentiations, we assume a particular solution with undetermined coefficients of the form $y_P(x) = B\sin 2x + C\cos 2x$. Substituting into the original equation gives $-(2B + 6C)\sin 2x + (6B - 2C)\cos 2x = \sin 2x$. Consequently, $-(2B + 6C) = 1$ and $6B - 2C = 0$ to satisfy the equation. These two equations in two unknowns have the solution $B = -\frac{1}{20}$ and $C = -\frac{3}{20}$. Hence $y_P = -\frac{1}{20}(\sin 2x + 3\cos 2x)$ and $y = c_1e^{-x} + c_2e^{-2X} - \frac{1}{20}(\sin 2x + 3\cos 2x)$.

A general method for finding $y_P(x)$ called *variation of parameters* used as its starting point $y_H(x)$. This method applies to *all* linear differential equations irrespective of whether they have constant coefficients. But it assumes $y_H(x)$ is known. We illustrate the idea for $a(x)y'' + b(x)y' + c(x)y = f(x)$. If the solution of the homogeneous equation is $y_H(x) = c_1\phi_1(x) + c_2\phi_2(x)$, then vary the parameters c_1 and c_2 to seek $y_P(x)$ as $y_P(x) = u_1(x)\phi_1(x) + u_2(x)\phi_2(x)$. Then $y_P' = u_1\phi_1' + u_2\phi_2' + u_1'\phi_1 + u_2'\phi_2$ and choose $u_1'\phi_1 + u_2'\phi_2 = 0$. Calculating y_P'' and setting in the original equation gives $a(x)u_1'\phi_1' + a(x)u_2'\phi_2' = f$. Solving the last two equations for u_1' and u_2' gives $u_1' = -\phi_2 f/wa, u_2' = \phi_1 f/wa$,

TABLE 222.1 Method of Undetermined Coefficients — Equation $L(y) = f(x)$ (Constant Coefficients)

	Terms in $f(x)$		Terms to be Included in $y_P(x)$
1.	Polynomial of degree n	(a)	If $L(y)$ contains y, try $y_P = a_0x^n + a_1x^{n-1} + \cdots + a_n$.
		(b)	If $L(y)$ does not contain y and lowest-order derivative is $y^{(r)}$, try $y_P = a_0x^{n+r} + \cdots + a_nx^r$.
2.	$\sin qx$, $\cos qx$	(a)	$\sin qx$ and/or $\cos qx$ are not in y_H; $y_P = B\sin qx + C\cos qx$.
		(b)	y_H contains terms of form $x^r\sin qx$ and/or $x^r\cos qx$ for $r = 0, 1, \ldots, m$; include in y_P terms of the form $a_0x^{m+1}\sin qx + a_1x^{m+1}\cos qx$.
3.	e^{ax}	(a)	y_H does not contain e^{ax}; include Ae^{ax} in y_P
		(b)	y_H contains $e^{ax}, xe^{ax}, \ldots, x^ne^{ax}$; include in y_p terms of the form $Ax^{n+1}e^{ax}$.
4.	$e^{px}\sin qx$, $e^{px}\cos qx$	(a)	y_H does not contain these terms; in y_p include $Ae^{px}\sin qx + Be^{px}\cos qx$.
		(b)	y_H contains $x^re^{px}\sin qx$ and/or $x^re^{px}\cos qx$; $r = 0, 1, \ldots, m$ include in y_P $Ax^{m+1}e^{px}\sin qx + Bx^{m+1}e^{px}\cos qx$.

where $w = \phi_1\phi_2' - \phi_1'\phi_2 \neq 0$. Integrating the general solution gives $y = c_1\phi_1(x) + c_2\phi_2(x) - \{\int [\phi_2 f(x)]/wa\}\phi_1(x) + [\int (\phi_1 f/wa)dx]\phi_2(x)$.

Example 222.4

Consider the equations $y'' - 4y = \sin x/(1 + x^2)$ and $y_H = c_1e^{2x} + c_2e^{-2x}$. With $\phi_1 = e^{2x}$ and $\phi_2 = e^{-2x}$, $w = 4$, so the general solution is

$$y = c_1e^{2x} + c_2e^{-2x} - \frac{e^{-2x}}{4}\int \frac{e^{2x}\sin x}{1+x^2}dx + \frac{e^{2x}}{4}\int \frac{e^{-2x}\sin x}{1+x^2}dx$$

The method of variation of parameters can be generalized as described in the references.

Higher-order systems of linear equations with constant coefficients are treated in a similar manner. Details can be found in the references.

Series Solution

The solution of differential equations can only be obtained in closed form in special cases. For all others, series or approximate or numerical solutions are necessary. In the simplest case, for an initial value problem, the solution can be developed as a Taylor series expansion about the point where the initial data are specified. The method fails in the *singular case* — that is, a point where the coefficient of the highest-order derivative is zero. The general method of approach is called the *Frobenius method.*

To understand the nonsingular case consider the equation $y'' + xy = x^2$ with $y(2) = 1$ and $y'(2) = 2$ (an initial value problem). We seek a series solution of the form $y(x) = a_0 + a_1(x-2) + a_2(x-2)^2 + \cdots$. To proceed, set $1 = y(2) = a_0$, which evaluates a_0. Next, $y'(x) = a_1 + 2a_2(x-2) + \cdots$, so $2 = y'(2) = a_1$ or $a_1 = 2$. Next $y''(x) = 2a_2 + 6a_3(x-2) + \cdots$ and from the equation, $y'' = x^2 - xy$, so $y''(2) = 4 - 2y(2) = 4 - 2 = 2$. Hence $2 = 2a_2$ or $a_2 = 1$. Thus, to third-order $y(x) = 1 + 2(x-2) + (x-2)^2 + R_2(x)$, where the remainder $R_2(x) = [(x-2)^3/3!]y'''(\xi)$, where $2 < \xi < x$ can be bounded for each x by finding the maximum of $y'''(x) = 2x - y - xy'$. The third term of the series follows by evaluating $y'''(2) = 4 - 1 - 2 \cdot 2 = -1$, so $6a_3 = -1$ or $a_3 = -1/6$.

By now the nonsingular process should be familiar. The algorithm for constructing a series solution about a nonsingular (ordinary) point x_0 of the equation $P(x)y'' + Q(x)y' + R(x)y = f(x)$ (note that $P(x_0) \neq 0$) is as follows:

1. Substitute into the differential equation the following expressions:

$$y(x) = \sum_{n=0}^{\infty} a_n(x-x_0)^n, \quad y'(x) = \sum_{n=1}^{\infty} na_n(x-x_0)^{n-1},$$

$$y''(x) = \sum_{n=2}^{\infty} n(n-1)a_n(x-x_0)^{n-2}$$

2. Expand $P(x)$, $Q(x)$, $R(x)$, and $f(x)$ about the point x_0 in a power series in $(x - x_0)$ and substitute these series into the equation.
3. Gather all terms involving the same power of $(x - x_0)$ to arrive at an identity of the form $\sum_{n=0}^{\infty} A_n(x-x_0)^n \equiv 0$.
4. Equate to zero each coefficient A_n of step 3.
5. Use the expressions of step 4 to determine a_2, a_3, … in terms of a_0, a_1 (we need two arbitrary constants) to arrive at the general solution.
6. With the given initial conditions, determine a_0 and a_1.

If the equation has a regular singular point — that is, a point x_0 at which $P(x)$ vanishes and a series expansion is sought about that point — a solution is sought of the form $y(x) = (x = x_0)^r \sum_{n=0}^{\infty} a_n(x-x_0)^n, a_0 \neq 0$ and the index r and coefficients a_n must be determined from the equation by an algorithm analogous to that already described. The description of this Frobenius method is left for the references.

222.2 Partial Differential Equations

The study of partial differential equations is of continuing interest in applications. It is a vast subject, so the focus in this chapter is on the most commonly occurring equations in the engineering literature: second-order equations in two variables. Most of these are of the three basic elliptic, hyperbolic, and parabolic types.

Elliptic equations are often called *potential equations* since they occur in potential problems where the potential may be temperature, voltage, and so forth. They also give rise to the steady solutions of parabolic equations. They require boundary conditions for the complete determination of their solution.

Hyperbolic equations are often called *wave equations* since they arise in the propagation of waves. For the development of their solutions, initial and boundary conditions are required. In principle, they are solvable by the method of characteristics.

Parabolic equations are usually called *diffusion equations* because they occur in the transfer (diffusion) of heat and chemicals. These equations require initial conditions (e.g., the initial temperature) and boundary conditions for the determination of their solutions.

Partial differential equations (PDEs) of the second order in two independent variables (x,y) are of the form $a(x,y)u_{xx} + b(x,y)u_{xy} + c(x,y)u_{yy} = E(x,y,u,u_x,u_y)$. If $E = E(x,y)$ the equation is linear; if E depends also on u, u_x, and u_y, it is said to be *quasilinear*, and if E depends only on x, y, and u, it is *semilinear*. Such equations are classified as follows: If $b^2 - 4ac$ is less than, equal to, or greater than zero at some point (x, y), then the equation is elliptic, parabolic, or hyperbolic, respectively, at that point. A PDE of this form can be transformed into canonical (standard) forms by use of new variables. These standard forms are most useful in analysis and numerical computations.

For hyperbolic equations the standard form is $u_{\xi\eta} = \phi(u,u_\eta,u_\xi,\eta,\xi)$, where $\xi_x/\xi_y = (-b+\sqrt{b^2-4ac})/2a$, and $\eta_x/\eta_y = (-b-\sqrt{b^2-4ac})/2a$. The right sides of these equations determine the so-called characteristics $(dy/dx)|_+ = (-b+\sqrt{b^2-4ac})/2a, (dy/dx)|_- = (-b-\sqrt{b^2-4ac})/2a.$

Example 222.5

Consider the equation $y^2u_{xx} - x^2u_{yy} = 0$. $\xi_x/\xi_y = -x/y$, $\eta_x/\eta_y = x/y$, so $\xi = y^2 - x^2$ and $\eta = y^2 + x^2$. In these new variables the equation becomes $u_{\xi\eta} = (\xi u_\eta - \eta u_\xi)/2(\xi^2 - \eta^2)$.

For parabolic equations the standard form is $u_{\xi\xi} = \phi(u,u_\eta,u_\xi,\eta,\xi)$ or $u_{\eta\eta} = \phi(u,u_\eta,u_\xi,\xi,\eta)$, depending upon how the variables are defined. In this case $\xi_x/\xi_y = -b/2a$ if $a \neq 0$, and $\xi_x/\xi_y = -b/2c$ if $c \neq 0$. Only ξ must be determined (there is only one characteristic) and η can be chosen as any function that is linearly independent of ξ.

Example 222.6

Consider the equation $y^2u_{xx} - 2xyu_{xy} + x^2u_y + u_y = 0$. Clearly, $b^2 - 4ac = 0$. Neither a nor c is zero so either path can be chosen. With $\xi_x/\xi_y = -b/2a = x/y$, there results $\xi = x^2 + y^2$. With $\eta = x$, the equation becomes $u_{\eta\eta} = [2(\xi + \eta)u_\xi + u_\eta]/(\xi - \eta^2)$.

For *elliptic equations* the standard form is $u_{\alpha\alpha} + u_{\beta\beta} = \phi(u, u_\alpha, u_\beta, \alpha, \beta)$, where ξ and η are determined by solving the ξ and η equations of the hyperbolic system (they are complex) and taking $\alpha = (\eta + \xi)/2$, $\beta = (\eta - \xi)/2i(i^2 = -1)$. Since ξ and η are complex conjugates, both α and β are real.

Example 222.7

Consider the equation $y^2 u_{xx} + x^2 u_{yy} = 0$. Clearly, $b^2 - 4ac < 0$, so the equation is elliptic. Then $\xi_x/\xi_y = -ix/y$, $\eta_x/\eta_y = ix/y$, so $\alpha = (\eta + \xi)/2 = y^2$ and $\beta = (\eta - \xi)/2i = x^2$. The standard form is $u_{\alpha\alpha} + u_{\beta\beta} = -(u_\alpha/2\alpha + u_\beta/2\beta)$.

Methods of Solution

Separation of Variables

Perhaps the most elementary method for solving linear PDEs with homogeneous boundary conditions is the method of *separation of variables*. To illustrate, consider $u_t - u_{xx} = 0$, $u(x,0) = f(x)$ (the initial condition) and $u(0,t) = u(1,t) = 0$ for $t > 0$ (the boundary conditions). A solution is assumed in "separated form" $u(x,t) = X(x)T(t)$. Upon substituting into the equation we find $\dot{T}/T = X''/X$ (where $\dot{T} = dT/dt$ and $X'' = d^2X/dx^2$). Since $T = T(t)$ and $X = X(x)$, the ratio must be constant, and for finiteness in t the constant must be negative, say $-\lambda^2$. The solutions of the separated equations $X'' + \lambda^2 X = 0$ with the boundary conditions $X(0) = 0$, $X(1) = 0$, and $\dot{T} = -\lambda^2 T$ are $X = A \sin \lambda x + B \cos \lambda x$ and $T = Ce^{-\lambda^2 t}$, where A, B, and C are arbitrary constants. To satisfy the boundary condition $X(0) = 0$, $B = 0$. An infinite number of values of λ (eigenvalues), say $\lambda_n = n\pi (n = 1,2,3,\ldots)$, permit all the eigenfunctions $X_n = b_n \sin \lambda_n x$ to satisfy the other boundary condition $X(1) = 0$. The solution of the equation and boundary conditions (not the initial condition) is, by superposition, $u(x,t) = \sum_{n=1}^{\infty} b_n e^{-n^2\pi^2 t} \sin n\pi x$ (a Fourier sine series), where the b_n are arbitrary. These values are obtained from the initial condition using the orthogonality properties of the trigonometric function (e.g., $\int_{-\pi}^{\pi} \sin mx \sin nx \, dx$ is 0 for $m \neq n$ and is π for $m = n \neq 0$) to be $b_n = 2\int_0^1 f(r) \sin n\pi r \, dr$. Then the solution of the problem is $u(x,t) = \sum_{n=1}^{\infty} \left[2\int_0^1 f(r) \sin n\pi r \, dr \right] e^{-n^2\pi^2 t} \sin n\pi x$, which is a Fourier sine series.

If $f(x)$ is a piecewise smooth or a piecewise continuous function defined for $a \le x \le b$, then its Fourier series within $a \le x \le b$ as its fundamental interval (it is extended periodically outside that interval) is

$$f(x) \sim \frac{1}{2} a_0 + \sum_{n=1}^{\infty} a_n \cos[2n\pi x/(b-a)] + b_n \sin[2n\pi x/(b-a)]$$

where

$$a_n = \left[\frac{2}{(b-a)} \right] \int_a^b f(x) \cos[2n\pi x/(b-a)] dx, \quad n = 0, 1, \ldots$$

$$b_n = \left[\frac{2}{(b-a)} \right] \int_a^b f(x) \sin[2n\pi x/(b-a)] dx \quad n = 1, 2, \ldots$$

The Fourier sine series has $a_n \equiv 0$, and the Fourier cosine series has $b_n \equiv 0$. The symbol ~ means that the series converges to $f(x)$ at points of continuity, and at the (allowable) points of finite discontinuity the series converges to the *average value* of the discontinuous values.

Caution: This method *only* applies to linear equations with homogeneous boundary conditions. Linear equations with variable coefficients use other orthogonal functions, such as the Bessel functions, Laguerre functions, Chebyshev functions, and so forth.

Some nonhomogeneous boundary value problems can be transformed into homogeneous ones. Consider the problem $u_t - u_{xx} = 0$, $0 \le x \le 1$, $0 \le t < \infty$ with initial condition $u(x,0) = f(x)$, and boundary conditions $u(0,t) = g(t)$, $u(1,t) = h(t)$. To homogenize the boundary conditions set $u(x,t) = w(x, t) + x[h(t) - g(t)] + g(t)$ and then solve $w_t - w_x = [\dot{g}(t) - \dot{h}(t)] - \dot{g}(t)$ with the initial condition $w(x,0) = f(x) - x[h(0) - g(0)] + g(0)$ and $w(0,t) = w(1,t) = 0$.

Operational Methods

A number of integral transforms are useful for solving a variety of linear problems. To apply the Laplace transform to the problem $u_t - u_{xx} = \delta(x)\delta(t)$, $-\infty < x < \infty$, $0 \le t$ with the initial condition $u(x, 0^-) = 0$, where δ is the Dirac delta function, we multiply by e^{-st} and integrate with respect to t from 0 to ∞. With the Laplace transform of $u(x, t)$ denoted by $U(x, s)$ — that is, $U(x,s) = \int_0^\infty e^{-st}u(x,t)dt$ — we have $sU - U_{xx} = \delta(x)$, which has the solution

$$U(x,s) = A(s)e^{-x\sqrt{s}} + B(s)e^{x\sqrt{s}} \quad \text{for } x > 0$$

$$U(x,s) = C(s)e^{-x\sqrt{s}} + D(s)e^{x\sqrt{s}} \quad \text{for } x < 0$$

Clearly, $B(s) = C(s) = 0$ for bounded solutions as $|x| \to \infty$. Then, from the boundary condition, $U(0^+,s) - U(0^-,s) = 0$ and integration of $sU - U_{xx} = \delta(x)$ from 0^- to 0^+ gives $U_x(0^+,s) - U_x(0^-,s) = -1$, so $A = D = 1/2\sqrt{s}$. Hence, $U(x,s) = (1/2\sqrt{s})e^{-\sqrt{s}|x|}$ and the inverse is $u(x,t) = (1/2\pi i)\int_\Gamma e^{st}U(x,s)ds$, where Γ is a Bromwich path, a vertical line taken to the right of all singularities of U on the sphere.

Similarity (Invariance)

This very useful approach is related to dimensional analysis; both have their foundations in group theory. The three important transformations that play a basic role in Newtonian mechanics are translation, scaling, and rotations. Using two independent variables x and t and one dependent variable $u = u(x,t)$, the *translation group* is $\bar{x} = x + \alpha a$, $\bar{t} = t + \beta a$, $\bar{u} = u + \gamma a$; the *scaling group* is $\bar{x} = a^\alpha x$, $\bar{t} = a^\beta t$, and $\bar{u} = a^\gamma u$; the *rotation group* is $\bar{x} = x \cos a + t \sin a$, $\bar{t} = t \cos a - x \sin a$, $\bar{u} = u$, with a nonnegative real number a. Important in what follows are the *invariants* of these groups. For the translation group there are two $\eta = x - \lambda t$, $\lambda = \alpha/\beta$, $f(\eta) = u - \varepsilon t$, $\varepsilon = \gamma/\beta$ or $f(\eta) = u - \theta x$, $\theta = \gamma/\alpha$; for the scaling group the invariants are $\eta = x/t^{\alpha/\beta}$ (or $t/x^{\beta/\alpha}$) and $f(\eta) = u/t^{\gamma/\beta}$ (or $u/x^{\gamma/\alpha}$); for the rotation group the invariants are $\eta = x^2 + t^2$ and $u = f(\eta) = f(x^2 + t^2)$.

If a PDE and its data (initial and boundary conditions) are left invariant by a transformation group, then similar (invariant) solutions are sought using the invariants. For example, if an equation is left invariant under scaling, then solutions are sought of the form $u(x,t) = t^{\gamma/\beta}f(\eta)$, $\eta = xt^{-\alpha/\beta}$ or $u(x,t) = x^{\gamma/\alpha}f(tx^{-\beta/\alpha})$; invariance under translation gives solutions of the form $u(x,t) = f(x - \lambda t)$; and invariance under rotation gives rise to solutions of the form $u(x,t) = f(x^2 + t^2)$.

Examples of invariance include the following:

1. The equation $u_{xx} + u_{yy} = 0$ is invariant under rotation, so we search for solutions of the form $u = f(x^2 + y^2)$. Substitution gives the ODE $f' + \eta f'' = 0$ or $(\eta f')' = 0$. The solution is $u(x,t) = c \ln \eta = c \ln(x^2 + t^2)$, which is the (so-called) fundamental solution of Laplace's equation.

2. The nonlinear diffusion equation $u_t = (u^n u_x)_x (n > 0)$, $0 \le x$, $0 \le t$, $u(0,t) = ct^n$ is invariant under scaling with the similar form $u(x,t) = t^n f(\eta)$, $\eta = xt^{-(n+1)/2}$. Substituting into the PDE gives the equation $(f''f)' + ((n+1)/2)\eta f' - nf = 0$, with $f(0) = c$ and $f(\infty) = 0$. Note that the equation is an ODE.
3. The wave equation $u_{xx} - u_{tt} = 0$ is invariant under translation. Hence, solutions exist of the form $u = f(x - \lambda t)$. Substitution gives $f''(1 - \lambda^2) = 0$. Hence, $\lambda = \pm 1$ or f is linear. Rejecting the trivial linear solution we see that $u = f(x - t) + g(x + t)$, which is the general (d'Alembert) solution of the wave equation; the quantities $x - t = \alpha$, $x + t = \beta$ are the characteristics of the next section.

The construction of all transformations that leave a PDE invariant is a solved problem left for the references.

The study of "solitons" (solitary traveling waves with special properties) has benefited from symmetry considerations. For example, the nonlinear third-order (Korteweg-de Vries) equation $u_t + uu_x - au_{xxx} = 0$ is invariant under translation. Solutions are sought of the form $u = f(x - \lambda t)$, and f satisfies the ODE, in $\eta = x - \lambda t$, $-\lambda f' + ff' - af''' = 0$.

Characteristics

Using the characteristics of the solution of the hyperbolic problem $u_{tt} - u_{xx} = p(x,t)$, $-\infty < x < \infty$, $0 \le t$, $u(x,0) = f(x)$, $u_t(x,0) = h(x)$ is

$$u(x,t) = \frac{1}{2}\int_0^t d\tau \int_{x-(t-\tau)}^{x+(t-\tau)} p(\xi,\tau)d\xi + \frac{1}{2}\int_{x-t}^{x+t} h(\xi)d\xi + \frac{1}{2}[f(x+t)] + f(x-t)]$$

The solution of $u_{tt} - u_{xx} = 0$, $0 \le x < \infty$, $0 \le t < \infty$, $u(x,0) = 0$, $u_t(x,0) = h(x)$, $u(0,t) = 0$, $t > 0$ is $u(x,t) = \frac{1}{2}\int_{-x+1}^{x+t} h(\xi)d\xi$.

The solution of $u_{tt} - u_{xx} = 0$, $0 \le x < \infty$, $0 \le t < \infty$, $u(x, 0) = 0$, $u_t(x, 0) = 0$, $u(0, t) = g(t)$, $t > 0$ is

$$u(x,t) = \begin{cases} 0 & \text{if } t < x \\ g(t-x) & \text{if } t > x \end{cases}$$

From time to time, lower-order derivatives appear in the PDE in use. To remove these from the equation $u_{tt} - u_{xx} + au_x + bu_t + cu = 0$, where a, b, and c are constants, set $\xi = x + t$, $\mu = t - x$, whereupon $u(x,t) = u[(\xi - \mu)/2, (\xi + \mu)/2] = U(\xi, \mu)$, where $U_{\xi\mu} + [(b + a)/4]U_\xi + [(b - a)/4]U_\mu + (c/4)U = 0$. The transformation $U(\xi, \mu) = W(\xi, \mu)\exp[-(b - a)\xi/4 - (b + a)\mu/4]$ reduces to satisfying $W_{\xi\mu} = \lambda W = 0$, where $\lambda = (a^2 - b^2 + 4c)/16$. If $\lambda \ne 0$, we lose the simple d'Alembert solution. But the equation for W is still easier to handle.

In linear problems, discontinuities propagate along characteristics. In nonlinear problems, the situation is usually different. The characteristics are often used as new coordinates in the numerical method of characteristics.

Green's Function

Consider the diffusion problem $u_t - u_{xx} = \delta(t)\delta(x - \xi)$, $0 \le x < \infty$, $\xi > 0$, $u(0,t) = 0$, $u(x,0) = 0[u(\infty,t) = u(\infty,0) = 0]$, a problem that results from a unit source somewhere in the domain subject to a homogeneous (zero) boundary condition. The solution is called a *Green's function of the first kind.* For this problem there is $G_1(x,\xi,t) = F(x - \xi,t) - F(x + \xi,t)$, where $F(x,t) = e^{-x^2/4t}/\sqrt{4\pi t}$ is the *fundamental* (invariant) *solution.* More generally, the solution of $u_t - u_{xx} = \delta(x - \xi)\delta(t - \tau)$, $\xi > 0$, $\tau > 0$, with the same conditions as before, is the Green's function of the first kind

$$G_1(x,\xi,t-\tau)=\frac{1}{\sqrt{4\pi(t-\tau)}}[e^{-(x+\xi)^2/4(t-\tau)}-e^{-(x+\xi)^2/4(t-\tau)}]$$

for the semifinite interval.

The solution of $u_t - u_{xx} = p(x,t)$, $0 \le x < \infty$, $0 \le t < \infty$, with $u(x,0) = 0$, $u(0,t) = 0$, $t > 0$ is $u(x,t) = \int_0^t d\tau \int_0^\infty p(\xi,\tau)G_1(x,\xi,t-\tau)]d\xi$, which is a superposition. Note that the Green's function and the desired solution must both satisfy a zero boundary condition at the origin for this solution to make sense.

The solution of $u_t - u_{xx} = 0$, $0 \le x < \infty$, $0 \le t < \infty$, $u(x,0) = f(x)$, $u(0,t) = 0$, $t > 0$ is $u(x,t) = \int_0^\infty f(\xi)G_1(x,\xi,t)d\xi$.

The solution of $u_t - u_{xx} = 0$, $0 \le x < \infty$, $0 \le t < \infty$, $u(x,0) = 0$, $u(0,t) = g(t)$, $t > 0$ (nonhomogeneous) is obtained by transforming to a new problem that has a homogeneous boundary condition, Thus, with $w(x,t) = u(x,t) - g(t)$ the equation for w becomes $w_t - w_{xx} = -\dot{g}(t) - g(0)\delta(t)$ and $w(x,0) = 0$, $w(0,t) = 0$. Using G_1, above, we finally obtain $u(x,t)=(x/\sqrt{4\pi})\int_0^t g(t-\tau)e^{-x^2/4t}/\tau^{3/2}d\tau$.

The Green's function approach can also be employed for elliptic and hyperbolic problems.

Equations in Other Spatial Variables

The spherically symmetric wave equation $u_{rr} + 2u_r/r - u_{tt} = 0$ has the general solution $u(r,t) = [f(t-r) + g(t+r)]/r$.

The Poisson-Euler-Darboux equation, arising in gas dynamics,

$$u_{rs} + N(u_r + u_s)/(r+s) = 0$$

where N is a positive integer ≥ 1, has the general solution

$$u(r,s)=k+\frac{\partial^{N-1}}{\partial r^{N-1}}\left[\frac{f(r)}{(r+s)^N}\right]+\frac{\partial^{N-1}}{\partial s^{N-1}}\left[\frac{g(s)}{(r+s)^N}\right]$$

Here, k is an arbitrary constant and f and g are arbitrary functions whose form is determined from the problem initial and boundary conditions.

Conversion to Other Orthogonal Coordinate Systems

Let (x^1, x^2, x^3) be rectangular (Cartesian) coordinates and (u^1, u^2, u^3) be any orthogonal coordinate system related to the rectangular coordinates by $x^i = x^i(u^1, u^2, u^3)$, $i = 1, 2, 3$. With $(ds)^2 = (dx^1)^2 + (dx^2)^2 + (dx^3)^2 = g_{11}(du^1)^2 + g_{22}(du^2)^2 + g_{33}(du^3)^2$, where $g_{ii} = (\partial x^1/\partial u^i)^2 + (\partial x^2/\partial u^i)^2 + (\partial x^3/\partial u^i)^2$. In terms of these "metric" coefficients, the basic operations of applied mathematics are expressible. Thus (with $g = g_{11}g_{22}g_{33}$),

$$dA=(g_{11}g_{22})^{1/2}du^1du^2;\quad dV=(g_{11}g_{22}g_{33})^{1/2}du^1du^2du^3;$$

$$\text{grad }\phi=\frac{\vec{a}_1}{(g_{11})^{1/2}}\frac{\partial\phi}{\partial u^1}+\frac{\vec{a}_2}{(g_{22})^{1/2}}\frac{\partial\phi}{\partial u^2}+\frac{\vec{a}_3}{(g_{33})^{1/2}}\frac{\partial\phi}{\partial u^3}$$

($\vec{a}_i$ are unit vectors in direction i);

TABLE 222.2 Some Coordinate Systems

Coordinate System	Metric Coefficients	
	Circular Cylindrical	
$x = r\cos\theta$	$u^1 = r$	$g_{11} = 1$
$y = r\sin\theta$	$u^2 = \theta$	$g_{22} = r^2$
$z = z$	$u^3 = z$	$g_{33} = 1$
	Spherical	
$x = r\sin\psi\cos\theta$	$u^1 = r$	$g_{11} = 1$
$y = r\sin\psi\sin\theta$	$u^2 = \psi$	$g_{22} = r^2$
$x = r\cos\psi$	$u^3 = \theta$	$g_{33} = r^2\sin^2\psi$
	Parabolic Coordinates	
$x = \mu\nu\cos\theta$	$u^1 = \mu$	$g_{11} = \mu^2 + \nu^2$
$y = \mu\nu\sin\theta$	$u^2 = \nu$	$g_{22} = \mu^2 + \nu^2$
$z = \frac{1}{2}(\mu^2 - \nu^2)$	$u^3 = \theta$	$g_{33} = \mu^2\nu^2$

Note: Other metric coefficients and so forth can be found in Moon, P. and Spencer, D. E. 1961. *Field Theory Handbook*. Springer, Berlin.

$$\text{div}\vec{E} = g^{-1/2}\left\{\frac{\partial}{\partial u^1}[(g_{22}g_{33})^{1/2}E_1] + \frac{\partial}{\partial u^2}[(g_{11}g_{33})^{1/2}E_2] + \frac{\partial}{\partial u^3}[(g_{11}g_{22})^{1/2}E_3]\right\}$$

(here $\vec{E} = (E_1, E_2, E_3)$);

$$\begin{aligned}\text{curl}\vec{E} = g^{-1/2}\Bigg\{&\vec{a}_1(g_{11})^{1/2}\left(\frac{\partial}{\partial u^2}[(g_{33})^{1/2}E_3] - \frac{\partial}{\partial u^3}[(g_{22})^{1/2}E_2]\right)\\ &+\vec{a}_2(g_{22})^{1/2}\left(\frac{\partial}{\partial u^3}[(g_{11})^{1/2}E_1] - \frac{\partial}{\partial u^1}[(g_{33})^{1/2}E_3]\right)\\ &+\vec{a}_3(g_{33})^{1/2}\left(\frac{\partial}{\partial u^1}[(g_{22})^{1/2}E_2] - \frac{\partial}{\partial u^2}[(g_{11})^{1/2}E_1]\right)\Bigg\}\end{aligned}$$

$$\text{div grad}\psi = \nabla^2\psi = \text{Laplacian of }\psi = g^{-1/2}\sum_{i=1}^{3}\frac{\partial}{\partial u^i}\left[\frac{g^{1/2}}{g_{ii}}\frac{\partial\psi}{\partial u^i}\right]$$

References

Ames, W. F. 1965. *Nonlinear Partial Differential Equations in Science and Engineering, Volume 1.* Academic Press, Boston, MA.

Ames, W. F. 1972. *Nonlinear Partial Differential Equations in Science and Engineering, Volume 2.* Academic Press, Boston, MA.

Brauer, F. and Nohel, J. A. 1986. *Introduction to Differential Equations with Applications.* Harper & Row, New York.

Jeffrey, A. 1990. *Linear Algebra and Ordinary Differential Equations.* Blackwell Scientific, Boston, MA.

Kevorkian, J. 1990. *Partial Differential Equations.* Wadsworth and Brooks/Cole, Belmont, CA.

Moon, P. and Spencer, D. E. 1961. *Field Theory Handbook.* Springer, Berlin.

Rogers, C. and Ames, W. F. 1989. *Nonlinear Boundary Value Problems in Science and Engineering.* Academic Press, Boston, MA.

Whitham, G. B. 1974. *Linear and Nonlinear Waves.* John Wiley & Sons, New York.
Zauderer, E. 1983. *Partial Differential Equations of Applied Mathematics.* John Wiley & Sons, New York.
Zwillinger, D. 1992. *Handbook of Differential Equations.* Academic Press, Boston, MA.

Further Information

A collection of solutions for linear and nonlinear problems is found in E. Kamke, *Differential-gleichungen-Lösungsmethoden und Lösungen*, Akad. Verlagsges, Leipzig, 1956. Also see G. M. Murphy, *Ordinary Differential Equations and Their Solutions*, Van Nostrand, Princeton, NJ. 1960 and D. Zwillinger, *Handbook of Differential Equations*, Academic Press, Boston, MA, 1992. Nonlinear problems see

Ames, W. F. 1968. *Nonlinear Ordinary Differential Equations in Transport Phenomena.* Academic Press, Boston, MA.
Cunningham, W. J. 1958. *Introduction to Nonlinear Analysis.* McGraw-Hill, New York.
Jordan, D. N. and Smith, P. 1977. *Nonlinear Ordinary Differential Equations.* Clarendon Press, Oxford, UK.
McLachlan, N. W. 1955. *Ordinary Nonlinear Differential Equations in Engineering and Physical Sciences,* 2nd ed. Oxford University Press, London.
Zwillinger, D. 1992. *Handbook of Differential Equations.* Academic Press, Boston, MA.

223

Integral Equations

William F. Ames
Georgia Institute of Technology

223.1 Classification and Notation

Any equation in which the unknown function $u(x)$ appears under the integral sign is called an *integral equation*. If $f(x)$, $K(x,t)$, a, and b are known then the integral equation for u, $\int_B^A K(x,t)u(t)dt = f(x)$ is called a *linear integral equation of the first kind of Fredholm type*. $K(x,t)$ is called the *kernel function* of the equation. If b is replaced by x (the independent variable) the equation is an equation of *Volterra type of the first kind.*

An equation of the form $u(x) = f(x) + \lambda \int_B^A K(x,t)u(t)dt$ is said to be a linear integral equation of *Fredholm type of the second kind*. If b is replaced by x it is of *Volterra type*. If $f(x)$ is not present, the equation is homogeneous.

The equation $\phi(x)u(x) = f(x) + \lambda \int_a^{b \text{ or } x} K(x,t)u(t)dt$ is the *third kind equation* of Fredholm or Volterra type. If the unknown function u appears in the equation in any way other than to the first power then the integral equation is said to be *nonlinear*. Thus, $u(x) = f(x) + \int_a^b K(x,t) \sin u(t)dt$ is nonlinear. An integral equation is said to be *singular* when either or both of the limits of integration are infinite or if $K(x,t)$ becomes infinite at one or more points of the integration interval.

Example 223.1

Consider the singular equations $u(x) = x + \int_0^\infty \sin (xt)u(t)dt$ and $f(x) = \int_0^x [u(t)/(x-t)^2]dt$.

0-8493-1586-7/05/$0.00+$1.50

223.2 Relation to Differential Equations

The *Leibnitz rule* $(d/d\,x)\int_{a(x)}^{b(x)} F(x,t)dt = \int_{a(x)}^{b(x)} (\partial F/\partial x)\,dt + F[x,b(x)](db/dx) - F[x,a(x)] \times (da/dx)$ is useful for differentiation of an integral involving a parameter (x in this case). With this, one can establish the relation.

$$I_n(x) = \int_a^x (x-t)^{n-1} f(t)dt = (n-1)!\underbrace{\int_a^x \cdots \int_a^x}_{n\text{ times}} f(x)\underbrace{dx\cdots dx}_{n\text{ times}}$$

This result will be used to establish the relation of the second-order initial value problem to a Volterra integral equation.

The second-order differential equation $y''(x) + A(x)y'(x) + B(x)y = f(x)$, $y(a) = y_0$, $y'(a) = y_0'$ is equivalent to the integral equations

$$y(x) = -\int_a^x \{A(t)+(x-t)[B(t)-A'(t)]\}y(t)dt$$
$$+\int_a^x (x-t)f(t)dt + [A(a)y_0 + y_0'](x-a) + y_0$$

which is of the type $y(x) = \int_a^x K(x,t)y(t)dt + F(x)$, where $K(x,t) = (t-x)[B(t) - A'(t)] - A(t)$ and $F(x)$ includes the rest of the terms. Thus, this initial value problem is equivalent to a Volterra integral equation of the second kind.

Example 223.2

Consider the equation $y'' + x^2y' + xy = x$, $y(0) = 1$, $y'(0) = 0$. Here $A(x) = x^2$, $B(x) = x$, $f(x) = x$, $a = 0$, $y_0 = 1$, $y_0' = 0$. The integral equation is $y(x) = \int_0^x t(x-2t)y(t)dt + (x^3/6) + 1$.

The expression for $I_n(x)$ can also be useful in converting boundary value problems to integral equations. For example, the problem $y''(x) + \lambda y = 0$, $y(0) = 0$, $y(a) = 0$ is equivalent to the Fredholm equation $y(x) = \lambda\int_0^a K(x,t)y(t)dt$, where $K(x,t) = (t/a)(a-x)$ when $t < x$ and $K(x,t) = (x/a)(a-t)$ when $t > x$.

In both cases the differential equation can be recovered from the integral equation by using the Leibnitz rule. Nonlinear differential equations can also be transformed into integral equations. In fact, this is one method used to establish properties of the equation and to develop approximate and numerical solutions. For example, the "forced pendulum" equation $y''(x) + a^2 \sin y(x) = f(x)$, $y(0) = y(1) = 0$ transforms into the nonlinear Fredholm equation,

$$y(x) = \int_0^1 K(x,t)[a^2 \sin y(t) - f(t)]dt$$

with $K(x,t) = x(1-y)$ for $0 < x < t$ and $K(x,t) = t(1-x)$ for $t < x < 1$.

223.3 Methods of Solution

Only the simplest integral equations can be solved exactly. Usually approximate or numerical methods are employed. The advantage here is that integration is a "smoothing operation," whereas differentiation is a "roughening operation." A few exact and approximate methods are given in the following sections. The numerical methods are found in Chapter 230.

Convolution Equations

The special convolution equation $y(x) = f(x) + \lambda \int_0^x K(x-t)y(t)dt$ is a special case of the Volterra equation of the second kind. $K(x - t)$ is said to be a *convolution kernel.* The integral part is the convolution integral discussed in Chapter 225. The solution can be accomplished by transforming with the Laplace transform: $L[y(x)] = L[f(x)] + \lambda L[y(x)]L[K(x)]$ or $y(x) = L^{-1}\{L[f(x)]/(1 - \lambda L[K(x)])\}$.

Abel Equation

The Volterra equation $f(x) = \int_0^x (t)/(x - t)^{\alpha}dt$, $0 < \alpha < 1$ is the (singular) Abel equation. Its solution is $y(x) = (\sin \alpha\pi/\pi)(d/dx) \int_0^x F(t)/(x - t)^{1-\alpha}dt.$

Approximate Method (Picard's Method)

This method is one of successive approximations that is described for the equation $y(x) = f(x) + \lambda \int_a^x K(x,t)y(t)dt$. Beginning with an initial guess $y_0(t)$ (often the value at the initial point a) generate the next approximation with $y_1(x) = f(x) + \int_a^x K(x,t)y_0(t)dt$ and continue with the general iteration

$$y_n(x) = f(x) + \lambda \int_a^x K(x,t)y_{n-1}(t)dt$$

Then, by iterating, one studies the convergence of this process, as is described in the literature.

Example 223.3

Let $y(x) = 1 + \int_0^x xt[y(t)]^2dt$, $y(0) = 1$. With $y_0(t) = 1$ we find $y_1(x) = 1 + \int_0^x xtdt = 1 + (x^3/2)$ and $y_2(x) = 1 + \int_0^x xt[1 + (t^3/2)]^2dt$, and so forth.

References

Jerri, A. J. 1985. *Introduction to Integral Equations with Applications.* Marcel Dekker, New York.
Tricomi, F. G. 1958. *Integral Equations.* Wiley-Interscience, New York.
Yosida, K. 1960. *Lectures on Differential and Integral Equations.* Wiley-Interscience, New York.

224

Approximation Methods

William F. Ames
Georgia Institute of Technology

The term *approximation methods* usually refers to an analytical process that generates a symbolic approximation rather than a numerical one. Thus, $1 + x + x^2/2$ is an approximation of e^x for small x. This chapter introduces some techniques for approximating the solution of various operator equations.

224.1 Perturbation

Regular Perturbation

This procedure is applicable to *some* equations in which a small parameter, ε, appears. Use this procedure with care; the procedure involves expansion of the dependent variables and data in a power series in the small parameter. The following example illustrates the procedure.

Example 224.1

Consider the equation $y'' + \varepsilon y' + y = 0$, $y(0) = 1$, $y'(0) = 0$. Write $y(x; \varepsilon) = y_0(x) + \varepsilon y_1(x) + \varepsilon^2 y_2(x) + \dots$, and the initial conditions (data) become

$$y_0(0) + \varepsilon y_1(0) + \varepsilon^2 y_2(0) + \cdots = 1$$

$$y_0'(0) + \varepsilon y_1'(0) + \varepsilon^2 y_2'(0) + \cdots = 0$$

Equating like powers of ε in all three equations yields the sequence of equations

$$O(\varepsilon^0): y_0'' + y_0 = 0, \quad y_0(0) = 1, \quad y_0'(0) = 0$$

$$O(\varepsilon^1): y_1'' + y_1 = -y_0', \quad y_1(0) = 0, \quad y_1'(0) = 0$$

$$\vdots$$

The solution for y_0 is $y_0 = \cos x$ and using this for y_1 we find $y_1(x) = \frac{1}{2}(\sin x - x\cos x)$. So $y(x; \varepsilon) = \cos x + \varepsilon(\sin x - x\cos x)/2 + O(\varepsilon^2)$. Appearance of the term $x \cos x$ indicates a *secular term* that becomes

0-8493-1586-7/05/$0.00+$1.50

arbitrarily large as $x \to \infty$. Hence, this approximation is valid only for $x \ll 1/\varepsilon$ and for small ε. If an approximation is desired over a larger range of x, then the method of multiple scales is required.

Singular Perturbation

The *method of multiple scales* is a singular method that is *sometimes* useful if the regular perturbation method fails. In this case the assumption is made that the solution depends on *two* (or more) different length (or time) scales. By trying various possibilities, one can determine those scales. The scales are treated as dependent variables when transforming the given ordinary differential equation into a partial differential equation, but then the scales are treated as independent variables when solving the equations.

Example 224.2

Consider the equation $\varepsilon y'' + y' = 2$, $y(0) = 0$, $y(1) = 1$. This is singular since (with $\varepsilon = 0$) the resulting first-order equation cannot satisfy both boundary conditions. For the problem the proper length scales are $u = x$ and $v = x/\varepsilon$. The second scale can be ascertained by substituting $\varepsilon^n x$ for x and requiring $\varepsilon y''$ and y' to be of the same order in the transformed equation. Then

$$\frac{d}{dx} = \frac{\partial}{\partial u}\frac{du}{dx} + \frac{\partial}{\partial v}\frac{dv}{dx} = \frac{\partial}{\partial u} + \frac{1}{\varepsilon}\frac{\partial}{\partial v}$$

and the equation becomes

$$\varepsilon\left(\frac{\partial}{\partial u} + \frac{1}{\varepsilon}\frac{\partial}{\partial v}\right)^2 y + \left(\frac{\partial}{\partial u} + \frac{1}{\varepsilon}\frac{\partial}{\partial v}\right) y = 2.$$

With $y(x;\varepsilon) = y_0(u,v) + \varepsilon y_1(u,v) + \varepsilon^2 y_2(u,v) + \ldots$, we have terms

$$O(\varepsilon^{-1}): \frac{\partial^2 y_0}{\partial v^2} + \frac{\partial y_0}{\partial v} = 0 \qquad \text{(actually ODEs with parameter } u\text{)}$$

$$O(\varepsilon^{0}): \frac{\partial^2 y_1}{\partial v^2} + \frac{\partial y_1}{\partial v} = 2 - 2\frac{\partial^2 y_0}{\partial u \partial v} - \frac{\partial y_0}{\partial u}$$

$$O(\varepsilon^{1}): \frac{\partial^2 y_2}{\partial v^2} + \frac{\partial y_2}{\partial v} = -2\frac{\partial^2 y_1}{\partial u \partial v} - \frac{\partial y_1}{\partial u} - \frac{\partial^2 y_0}{\partial u^2}$$

$$\vdots$$

Then $y_0(u,v) = A(u) + B(u)e^{-v}$ and so the second equation becomes $\partial^2 y_1/\partial v^2 + \partial y_1/\partial v = 2 - A'(u) + B'(u)e^{-v}$, with the solution $y_1(u,v) = [2 - A'(u)]v + vB'(u)\,e^{-v} + D(u) + E(u)e^{-v}$. Here A, B, D, and E are still arbitrary. Now the solvability condition — "higher order terms must vanish no slower (as $\varepsilon \to 0$) than the previous term" [Kevorkian and Cole, 1981] — is used. For y_1 to vanish no slower than y_0, we must have $2 - A'(u) = 0$ and $B'(u) = 0$. If this were not true, the terms in y_1 would be larger than those in y_0 ($v \gg 1$). Thus, $y_0(u,v) = (2u + A_0) + B_0 e^{-v}$, or in the original variables $y(x;\varepsilon) \approx (2x + A_0) + B_0 e^{-x/\varepsilon}$ and matching to both boundary conditions gives $y(x;\varepsilon) \approx 2x - (1 - e^{-x/\varepsilon})$.

Boundary Layer Method

The boundary layer method is applicable to regions in which the solution is *rapidly varying*. See the references at the end of the chapter for detailed discussion.

224.2 Iterative Methods

Taylor Series

If it is known that the solution of a differential equation has a power series in the independent variable (t), then we may proceed from the initial data (the easiest problem) to compute the Taylor series by differentiation.

Example 224.3

Consider the equation $(d^2x/dt) = -x - x^2$, $x(0) = 1$, $x'(0) = 1$. From the differential equation, $x''(0) = -2$, and, since $x''' = -x' - 2xx'$, $x'''(0) = -1 - 2 = -3$, so the four term approximation for $x(t) \approx 1 + t - (2t^2/2!) - (3t^3/3!) = 1 + t - t^2 - t^3/2$. An estimate for the error at $t = t_1$ (see a discussion of series methods in any calculus text), is not greater than $|d^4x/dt^4|_{max}[(t_1)^4/4!]$, $0 \le t \le t_1$.

Picard's Method

If the vector differential equation $\mathbf{x}' = f(t,\mathbf{x})$, $\mathbf{x}(0)$ given, is to be approximated by Picard iteration, we begin with an initial guess $\mathbf{x}_0 = \mathbf{x}(0)$ and calculate iteratively $\mathbf{x}_i' = f(t,\mathbf{x}_{i-1})$.

Example 224.4

Consider the equation $x' = x + y^2$, $y' = y - x^3$, $x(0) = 1$, $y(0) = 2$. With $x_0 = 1$, $y_0 = 2$, $x_1' = 5$, $y_1' = 1$, so $x_1 = 5t + 1$, $y_1 = t + 2$, since $x_i(0) = 1$, $y_i(0) = 2$ for $i \ge 0$. To continue, use $x_{i+1}' = x_i + y_i^2$, $y_{i+1}' = y_i - x_i^3$. A modification is the utilization of the first calculated term immediately in the second equation. Thus, the calculated value of $x_1 = 5t + 1$, when used in the second equation, gives $y_1' = y_0 - (5t + 1)^3 = 2 - (125t^3 + 75t^2 + 15t + 1)$, so $y_i = 2t - (125t^4/4) - 25t^3 - (15t^2/2) - t + 2$. Continue with the iteration $x_{i+1}' = x_i + y_i^3$, $y_{i+1}' = y_i - (x_{i+1})^3$.

Another variation would be $x_{i+1}' = x_{i+1} + (y_i)^3$, $y_{i+1}' = y_{i+1} - (x_{i+1})^3$.

References

Ames, W. F. 1965. *Nonlinear Partial Differential Equations in Engineering, Volume I.* Academic Press, Boston, MA.

Ames, W. F. 1968. *Nonlinear Ordinary Differential Equations in Transport Processes.* Academic Press, Boston, MA.

Ames, W. F. 1972. *Nonlinear Partial Differential Equations in Engineering, Volume II.* Academic Press, Boston, MA.

Kevorkian, J. and Cole, J. D. 1981. *Perturbation Methods in Applied Mathematics.* Springer, New York.

Miklin, S. G. and Smolitskiy, K. L. 1967. *Approximate Methods for Solutions of Differential and Integral Equations.* Elsevier, New York.

Nayfeh, A. H. 1973. *Perturbation Methods.* John Wiley & Sons, New York.

Zwillinger, D. 1992. *Handbook of Differential Equations, 2nd ed.* Academic Press, Boston, MA.

225

Integral Transforms

William F. Ames
Georgia Institute of Technology

All of the integral transforms are special cases of the equation $g(s)=\int_a^b K(s,\ t)f(t)dt,$ in which $g(s)$ is said to be the *transform* of $f(t)$, and $K(s,t)$ is called the *kernel* of the transform. Table 225.1 shows the more important kernels and the corresponding intervals $(a,\ b)$.

Details for the first three transforms listed in Table 225.1 are given in this chapter. The details for the others are found in the literature.

225.1 Laplace Transform

The Laplace transform of $f(t)$ is $g(s)=\int_0^\infty e^{-st}f(t)dt$. It may be thought of as transforming one class of functions into another. The advantage in the operation is that under certain circumstances it replaces complicated functions by simpler ones. The notation $L[f(t)] = g(s)$ is called the *direct transform* and $L^{-1}[g(s)]=f(t)$ is called the *inverse transform*. Both the direct and inverse transforms are tabulated for many often-occurring functions. In general $L^{-1}[g(s)]=(1/2\pi i)\int_{\alpha-i\infty}^{\alpha+i\infty} e^{st}g(s)ds,$ and to evaluate this integral requires a knowledge of complex variables, the theory of residues, and contour integration.

Properties of the Laplace Transform

Let $L[f(t)]=g(s), L^{-1}[g(s)]=f(t)$.

1. The Laplace transform may be applied to a function $f(t)$ if $f(t)$ is continuous or piecewise continuous; if $t^n|f(t)|$ is finite for all $t,\ t\to 0, n<1;$ and if $e^{-at}|f(t)|$ is finite as $t\to\infty$ for some value of a, $a>0$.
2. L and L^{-1} are unique.
3. $L[af(t)+bh(t)]=aL[f(t)+bL[h(t)]$ (linearity).

0-8493-1586-7/05/$0.00+$1.50

TABLE 225.1 Kernels and Intervals of Various Integral Transforms

Name of Transform	(a, b)	$K(s, t)$
Laplace	$(0, \infty)$	e^{-st}
Fourier	$(-\infty, \infty)$	$\frac{1}{\sqrt{2\pi}} e^{-ist}$
Fourier cosine	$(0, \infty)$	$\sqrt{\frac{2}{\pi}} \cos st$
Fourier sine	$(0, \infty)$	$\sqrt{\frac{2}{\pi}} \sin st$
Mellin	$(0, \infty)$	t^{s-1}
Hankel	$(0, \infty)$	$tJ_v(st), \quad v \geq -\frac{1}{2}$

4. $L[e^{at} f(t)] = g(s-a)$ (shift theorem).
5. $L[(-t)^k f(t)] = d^k g / ds^k; k$ a positive integer.

Example 225.1

$L[\sin at] = \int_0^\infty e^{-st} \sin a t\, dt = a/(s^2+a^2), s>0.$ By property 5,

$$\int_0^\infty e^{-st} t \sin a t \; dt = L[t \sin a t] = \frac{2as}{s^2+a^2}$$

1. $L[f'(t)] = sL[f(t)] - f(0)$

$$L[f''(t)] = s^2 L[f(t)] - sf(0) - f'(0)$$

$$\vdots$$

$$L[f^{(n)}(t)] = s^n L[f(t)] - s^{n-1} f(0) - \cdots - sf^{(n-2)}(0) - f^{(n-1)}(0)$$

In this property it is apparent that the initial data are automatically brought into the computation.

Example 225.2

Solve $y'' + y = e^t$, $y(0) = 0, y'(0) = 1$. Now $L[y''] = s^2 L[y] - sy(0) - y'(0) = s^2 L[y] - s - 1$. Thus, using the linear property of the transform (property 3), $s^2 L[y] + L[y] - s - 1 = L[e^t] = 1/(s-1)$. Therefore, $L[y] = s^2/[(s-1)(s^2+1)]$.

With the notations $\Gamma(n+1) = \int_0^\infty x^n e^{-x} dx$ (gamma function) and $J_n(t)$ the Bessel function of the first kind of order n, a short table of Laplace transforms is given in Table 204.2.

1. $L[\int_a^t f(t) dt] = \frac{1}{s} L[f(t)] + \frac{1}{s} \int_a^0 f(t) dt].$

Example 225.3

Find $f(t)$ if $L[f(t)] = (1/s^2)[1/(s^2-a^2)]$. $L[1/a \sinh at] = 1/(s^2-a^2)$. Therefore, $f(t) = \int_0^t [\int_0^t \frac{1}{a} \sinh a\, t dt] dt = 1/a^2[(\sinh at)/a - t]$.

TABLE 225.2 Some Laplace Transforms

$f(t)$	$g(s)$	$f(t)$	$g(s)$
1	$\frac{1}{s}$	$e^{-at}(1-at)$	$\frac{s}{(s+a)^2}$
t^n, n is a + integer	$\frac{n!}{s^{n+1}}$	$\frac{t\sin at}{2a}$	$\frac{s}{(s^2+a^2)^2}$
t^n, $n \neq a$ + integer	$\frac{\Gamma(n+1)}{s^{n+1}}$	$\frac{1}{2a^2}\sin at \sinh at$	$\frac{s}{s^4+4a^4}$
$\cos at$	$\frac{s}{s^2+a^2}$	$\cos at \cosh at$	$\frac{s^3}{s^4+4a^4}$
$\sin at$	$\frac{a}{s^2+a^2}$	$\frac{1}{2a}(\sinh at+\sin at)$	$\frac{s^2}{s^4-a^4}$
$\cosh at$	$\frac{s}{s^2-a^2}$	$\frac{1}{2}(\cosh at+\cos at)$	$\frac{s^3}{s^4-a^4}$
$\sinh at$	$\frac{a}{s^2-a^2}$	$\frac{\sin at}{t}$	$\tan^{-1}\frac{a}{s}$
e^{-at}	$\frac{1}{s+a}$	$J_0(at)$	$\frac{1}{\sqrt{s^2+a^2}}$
$e^{-bt}\cos at$	$\frac{s+b}{(s+b)^2+a^2}$	$\frac{n}{a^n}\frac{J_n(at)}{t}$	$\frac{1}{(\sqrt{s^2+a^2}+s)^n}$
$e^{-bt}\sin at$	$\frac{a}{(s+b)^2+a^2}$	$J_0(2\sqrt{at})$	$\frac{1}{s}e^{-a/s}$

1. $L\left[\frac{f(t)}{t}\right]=\int_s^\infty g(s)ds;\quad L\left[\frac{f(t)}{t^k}\right]=\underbrace{\int_s^\infty\cdots\int_s^\infty}_{k\text{ integrals}} g(s)(ds)^k$

Example 225.4

$L[(\sin at)/t]=\int_s^\infty L[\sin at]ds=\int_s^\infty\left[ads/(s^2+a^2)\right]=\cot^{-1}(s/a)$.

1. The *unit step function* $u(t-a)=0$ for $t<a$ and 1 for $t>a. L[u(t-a)]=e^{-as}/s$.
2. The *unit impulse function* is $\delta(a)=u'(t-a)=1$ at $t=a$ and 0 elsewhere. $L[u'(t-a)]=e^{-as}$.
3. $L^{-1}[e^{-as}g(s)]=f(t-a)u(t-a)$ (second shift theorem).
4. If $f(t)$ is *periodic* of period b — that is, $f(t+b)=f(t)$ — then $L[f(t)]=[1/(1-e^{-bs})]\times\int_0^b e^{-st}f(t)dt$.

Example 225.5

The equation $\partial^2 y/(\partial t\partial x)+\partial y/\partial t+\partial y/\partial x=0$ with $(\partial y/x)(0,x)=y(0,x)=0$ and $y(t,0)+(\partial y/\partial t)(t,0)=\delta(0)$ (see property 10) is solved by using the Laplace transform of y with respect to t. With $g(s,x)=\int_0^\infty e^{-st}y(t,x)dt$, the transformed equation becomes

$$s\frac{\partial g}{\partial x}-\frac{\partial y}{\partial x}(0,x)+sg-y(0,x)+\frac{\partial g}{\partial x}=0$$

or

$$(s+1)\frac{\partial g}{\partial x}+sg=\frac{\partial y}{\partial x}(0,x)+y(0,x)=0$$

The second (boundary) condition gives $g(s,\ 0)+sg(s,\ 0)-y(0,\ 0)=1$ or $g(s,\ 0)=1/(1+s)$. A solution of the preceding ordinary differential equation consistent with this condition is $g(s,x)=[1/(s+1)]e^{-sx/(s+1)}$. Inversion of this transform gives $y(t,\ x)=e^{-(t+x)}I_0(2/\sqrt{tx})$, where I_0 is the zero-order Bessel function of an imaginary argument.

225.2 Convolution Integral

The *convolution integral* (*faltung*) of two functions $f(t)$, $r(t)$ is $x(t)=f(t)*r(t)=\int_0^t f(\tau)r(t-\tau)d\tau$.

Example 225.6

$t^* \sin t=\int_0^t \tau\sin(t-\tau)d\tau=t-\sin t$.

1. $L[f(t)]L[h(t)]=L[f(t)*h(t)]$.

225.3 Fourier Transform

The *Fourier transform* is given by $F[f(t)]=(1/\sqrt{2\pi})\int_{-\infty}^{\infty} f(t)e^{-ist}dt=g(s)$ and its *inverse* by $F^{-1}[g(s)]=(1/\sqrt{2\pi})\int_{-\infty}^{\infty} g(s)e^{ist}dt=f(t)$. In brief, the condition for the Fourier transform to exist is that $\int_{-\infty}^{\infty}|f(t)|dt<\infty$, although certain functions may have a Fourier transform even if this is violated.

Example 225.7

The function $f(t)=1$ for $-a\le t\le a$ and $=0$ elsewhere has

$$F[f(t)]=\int_{-a}^{a} e^{-ist}dt=\int_0^a e^{ist}dt+\int_0^a e^{-ist}dt=2\int_0^a \cos st\,dt=\frac{2\sin sa}{s}$$

Properties of the Fourier Transform

Let $F[f(t)]=g(s); F^{-1}[g(s)]=f(t)$.

1. $F[f^{(n)}(t)]=(is)^n F[f(t)]$
2. $F[af(t)+bh(t)]=aF[f(t)]+bF[h(t)]$
3. $F[f(-t)]=g(-s)$
4. $F[f(at)]=1/ag(s/a),\quad a>0$
5. $F[e^{-iwt}f(t)]=g(s+w)$
6. $F[f(t+t_1)]=e^{ist_1}g(s)$
7. $F[f(t)]=G(is)+G(-is)$ if $f(t)=f(-t)(f(t)$ even)
8. $F[f(t)]=G(is)-G(-is)$ if $f(t)=-f(-t)(f$ odd)

where $G(s)=L[f(t)]$. This result allows the use of the Laplace transform tables to obtain the Fourier transforms.

Example 225.8

Find $F[e^{-a|t|}]$ by property 7. The term $e^{-a|t|}$ is even. So $L[e^{-at}]=1/(s+a)$. Therefore, $F[e^{-a|t|}]=1/(is+a)+1/(-is+a)=2a/(s^2+a^2)$.

225.4 Fourier Cosine Transform

The *Fourier cosine transform* is given by $F_c[f(t)]=g(s)=\sqrt{(2/\pi)}\int_{-\infty}^{\infty} f(t)\cos st\,dt$ and its *inverse* by $F_c^{-1}[g(s)]=f(t)=\sqrt{(2/\pi)}\int_{-\infty}^{\infty} g(s)\cos st\,ds$. The *Fourier sine transform* F_s is obtainable by replacing the cosine by the sine in the above integrals.

Example 225.9

$F_c[f(t)], f(t)=1$ for $0<t<a$ and 0 for $a<t<\infty$. $F_c[f(t)]=\sqrt{(2/\pi)}\int_0^a \cos st\,dt=\sqrt{(2/\pi)}(\sin as)/s$.

Properties of the Fourier Cosine Transform

$F_c[f(t)]=g(s)$.

1. $F_c[af(t)+bh(t)]=aF_c[f(t)+bF_c[h(t)]$
2. $F_c[f(at)]=(1/a)g(s/a)$
3. $F_c[f(at)\cos bt]=1/2a[g((s+b)/a)+g((s-b)/a)],a,b>0$
4. $F_c[t^{2n}f(t)]=(-1)^n(d^{2n}g)/(ds^{2n})$
5. $F_c[t^{2n+1}f(t)]=(-1)^n(d^{2n+1})/(ds^{2n+1})F_s[f(t)]$

Table 225.3 presents some Fourier cosine transforms.

Example 225.10

The temperature θ in the semiinfinite rod $0\le x<\infty$ is determined by the differential equation $\partial\theta/\partial t=k(\partial^2\theta/\partial x^2)$ and the condition $\theta=0$ when $t=0, x\ge 0$; $\partial\theta/\partial x=-\mu=$ constant when $x=0$, $t>0$. By using the Fourier cosine transform, a solution may be found as $\theta(x,t)=(2\mu/\pi)\int_0^{\infty}(\cos px/p)(1-e^{-kp2t})dp$.

TABLE 225.3 Fourier Cosine Transforms

$f(t)$	$\frac{g(s)}{\sqrt{2/\pi}}$
$\left.\begin{matrix} t & 0<t<1 \\ 2-t & 1<t<2 \\ 0 & 2<t<\infty \end{matrix}\right\}$	$\frac{1}{s^2}[2\cos s - 1 - \cos 2s]$
$t^{-1/2}$	$\pi^{1/2}(2s)^{-1/2}$
$\left.\begin{matrix} 0 & 0<t<a \\ (t-a)^{-1/2} & a<t<\infty \end{matrix}\right\}$	$\pi^{1/2}(2s)^{-1/2}[\cos as - \sin as]$
$(t^2+a^2)^{-1}$	$\frac{1}{2}\pi a^{-1}e^{-as}$
$e^{-at}, \quad a>0$	$\frac{a}{s^2+a^2}$
$e^{-at^2}, \quad a>0$	$\frac{1}{2}\pi^{1/2}a^{-1/2}e^{-s^2/4a}$
$\frac{\sin at}{t}, \quad a>0$	$\begin{cases} \pi/2 & s<a \\ \pi/4 & s=a \\ 0 & s>a \end{cases}$

References

Churchill, R. V. 1958. *Operational Mathematics*. McGraw-Hill, New York.

Ditkin, B. A. and Proodnikav, A. P. 1965. *Handbook of Operational Mathematics* (in Russian). Nauka, Moscow.

Doetsch, G. 1950–1956. *Handbuch der Laplace Transformation*, vols. I–IV (in German). Birkhauser, Basel.

Nixon, F. E. 1960. *Handbook of Laplace Transforms*. Prentice Hall, Englewood Cliffs, NJ.

Sneddon, I. 1951. *Fourier Transforms*. McGraw-Hill, New York.

Widder, D. 1946. *The Laplace Transform*. Princeton University Press, Princeton, NJ.

Further Information

The references citing G. Doetsch, *Handbuch der Laplace Transformation*, vols. I–IV, Birkhauser, Basel, 1950–1956 (in German) *and* B.A. Ditkin and A. P. Proodnikav, *Handbook of Operational Mathematics*, Moscow, 1965 (in Russian) are the most extensive tables known. The latter reference is 485 pages.

226

Chaos, Fractals, and Julia Sets

Anca Deliu
Georgia Institute of Technology

Chaos refers to the apparently chaotic evolution of certain dynamical systems. *Fractals* are sets with an intriguing structure that reveals itself in shapes with intricate detail recurring on all scales. *Julia sets* are invariant sets for certain family of complex maps. Each Julia set corresponds to one map in the family. Typically the Julia sets are fractals and the dynamics of the corresponding map on the Julia set is chaotic.

Chaotic dynamics (nonlinear science), and fractals (nowhere differentiable curves), and Julia sets (invariant sets) have been known and studied by mathematicians for about one hundred years. Nevertheless it was during the last three decades that the scientific community at large became interested in these concepts and looked at them from the common perspective of their fascinating graphic visualization. In the early eighties many graduate students, and more mature researchers, were spellbound in front of their computers as incredible shapes, unseen before, materialized unexpectedly on the screen out of very simple formulas.

226.1 Chaos

When, in addition to mathematicians, researchers from other fields become aware that simple and deterministic phenomena from physics, engineering, economics, biology, and geology, can lead to unpredictable time evolutions, nonlinear science became very fashionable.

As opposed to classical systems, that eventually settle into periodic motion or steady state, chaotic systems evolve apparently at random in a pattern that can only be described statistically. The complexity of the system is related to the geometry of this pattern. The study of individual trajectories becomes irrelevant and it is replaced by statistical analysis.

The French mathematician Henri Poincaré set the foundations of chaotic dynamics about 100 years ago. He studied the time evolution of three celestial bodies subject to mutual gravitational forces. He showed that irregular orbits (called *chaotic* in today's terminology) could arise due to the nonlinear character of the system. His discovery was surprising since it had been assumed, until then, that all reasonable dynamical systems eventually settle into steady state or periodic motion.

Currently the word *chaos* has a very specific technical meaning that we will describe later in this section. The interesting dynamics of a system occurs on, and around the invariant sets of the system, and thus chaos is understood in terms of the dynamics on and around these sets. A major goal of the study of dynamical systems is to identify these invariant sets and to understand them. An invariant set can be

0-8493-1586-7/05/$0.00+$1.50

attracting or repelling, according to whether it attracts or repels the nearby orbits. The classical invariant sets are the fixed points and the periodic orbits of the system. What the study of chaotic dynamics brought forward during the last three decades was the existence of invariant sets larger than a periodic orbit and (typically) smaller than the whole space. These invariant sets are called attractors or repellers, as the case may be, and the dynamics of the system on these sets is necessarily chaotic. Typically these sets have fractional dimension and assume unusual shapes. Ruelle and Takens referred to these sets as strange attractors, in their work on the onset of turbulence [Ruelle and Takens, 1971].

Let us reemphasize that the technical meaning of the word chaos refers to *deterministic* time evolution that is only perceived as chaotic by a human observer.

Examples of strange attractors discovered in the last decades are the Lorentz attractor, the Henon attractor, and the Smale horseshoe attractor [Falconer, 1990]. In addition to these, the classical, and other, Cantor sets and the Julia sets, are older examples that fit in the same general framework.

The word chaos, with a slightly different meaning than the currently established technical term, was introduced by Li and Yorke in their paper "Period Three Implies Chaos" [Li and Yorke, 1975]. In that paper the authors proved that a map with a periodic point of order three, must necessarily have periodic points of all orders; such a map has orbits with disordered behavior that seems to be chaotic.

But what is, after all a dynamical system? A *dynamical system* is a rule that determines the time evolution of a system, in terms of its initial state. A dynamical system is denoted by a pair (f, X), where $f{:}X \to X$, is a map and X is the phase space of the system. Each point in X represents a specific state of the system and the function f determines the time evolution of the system. In this chapter X is a subset of the Euclidian space R^n or of the complex plane C. In practice a dynamical system describes the time evolution of an observable (a measurement) x, such as temperature, pressure, or population. Depending on whether this time evolution is tracked at all time points, or only at a discrete set of points, a dynamical system is called continuous or discrete.

A *continuous dynamical system* is described by a differential equation of the form

$$\frac{dx}{dt} = f(x) \tag{226.1}$$

where $f{:}X \to X$, is a map with some smoothness. For any initial state $x_0 \in X$ the evolution of the system is described by the solution of Equation (226.1) with initial condition x_0.

A *discrete dynamical system* (f, X) is described by a difference equation of the form

$$x_{n+1} = f(x_n) \tag{226.2}$$

where $f{:}X \to X$, is a mapping with some regularity.

The main question of the theory of dynamical systems is to find out what happens to the system given that it started in some initial state $x_0 \in X$.

For continuous dynamical systems the answer to that question consists in solving Equation (226.2), with initial condition $x_0 \in X$. That solution is a continuous curve called the trajectory of $x_0 \in X$ and it represents the fate of $x_0 \in X$ as time $t \to \pm\infty$.

For discrete dynamical systems the answer consists in solving the difference equation (226.2), with initial condition $x_0 \in X$. That solution is the sequence of iterates of f,

$$x_0, f(x_0), f(f(x_0)), \ldots$$

The sequence of iterates $x_k = f^k(x_0){:}k \geq 0$. is called the orbit of $x_0 \in X$ and x_k represents the state of the system at time $t = k$. Thus a discrete dynamical systems determines the evolution of the system only at a discrete set of points in time t = 0, t = 1, t = 2, …, etc.

Chaos can occur both in the discrete and the continuous dynamical systems. In principle the higher the dimension of the phase space of a dynamical system, the more room and thus opportunity, there is for chaos to occur. In discrete dynamical systems chaos can occur even in dimension one; for continuous dynamical systems chaos can occur only in dimension three or higher. Since discrete dynamical systems are simpler while at the same time they can illustrate all the characteristics of chaotic dynamics we will focus our discussion on them.

Let thus (f, X) be a discrete dynamical system. A point y is called a *fixed point* for f if it is a solution of the equation $f(y) = y$. If the initial state of a dynamical system is a fixed point, then the system will remain in that state at all subsequent time points. Fixed points play an important role in understanding the dynamics of f. A fixed point is called *attractive* if it attracts all the nearby orbits of the system.

Example 226.1

Let $X = \mathbb{R}$ and $f(x) = \sin x$. Then the orbits of all initial states $x_0 \in \mathbb{R}$ converge to the unique fixed point $x = 0$, of $\sin x$, as one can check by pressing repeatedly the sin button on a calculator.

Example 226.2

Let $X = \mathbb{R}$ and $f(x) = e^x$. Then the orbits of all initial states $x_0 \in \mathbb{R}$ diverge to infinity, as one can check by pressing repeatedly the Exp button on a calculator. One will soon get an overflow message.

A dynamical system with a unique fixed point, like the dynamical system of Example 226.1, has a very simple behavior. In the long run all orbits converge asymptotically to the fixed point. Similarly the dynamical system of Example 226.2, has a very simple behavior too since all orbits behave similarly, they diverge to infinity.

Example 226.3

Let $X = \mathbb{R}$ and $f(x) = x\Sigma$. To find the fixed points of f one solves the equation $x\Sigma = x$. Thus the system has two fixed points $x = 0$ and $x = 1$. The fixed point $x = 0$ is attractive, it attracts all initial states $-1 < x_0 < 1$. The fixed point $x = 1$ is repelling.

Notice that if x is a fixed point for a one-dimensional dynamical system, then, if $|f'(x)| < 1$, x is attractive and if $|f'(x)| > 1$, then x is repelling. A point x with $f'(x) = 0$ is called a critical point. Critical points play a special role in the dynamics of a system as we will illustrate later, in the section on Julia sets.

A point y is called a *periodic point* of *period p* for the dynamical system (X,f) if y satisfies the equation $f^p(y) = y$ and $f^q(y) \neq y$ for any $q < p$. If y is periodic, with period p, the orbit of y is a finite set of p points $\{y, f(y), \ldots, f^{p-1}(y)\}$. This orbit is called a *periodic orbit* with period p. Notice that a fixed point is a periodic point of period 0. Periodic points play an important role in understanding the dynamics of f. A periodic orbit is called *attractive* if it attracts the neighboring orbits, in the sense that if the system starts close to that period, then all the subsequent states of the system will converge to that periodic orbit.

Example 226.4

Let $X = \mathbb{R}$ and $f(x) = x\Sigma - 1$. To find the periodic points of f, of order 2, one solves the equation $f(f(x)) = x$, or equivalently $(x\Sigma - 1)\Sigma - 1 = x$. Two solutions of this equation are $x = 0$ and $x = -1$ and one can indeed check that $\{0, -1\}$ is a period for f.

Example 226.5

Let $X = \mathbb{R}$ and $f(x) = -x$. This map has a unique fixed point $x = 0$, and any point $x \neq 0$ is a periodic point of order 2. These periodic orbits $\{x. - x\}$ are neither attractive nor repelling.

The examples above show simple maps with simple dynamics. One can use a calculator to iterate other simple functions $f: X \to X$, similar to the ones above and see what happens. One might assume that all simple functions would behave equally nicely under iteration. That assumption would turn out to be false. Since chaos is understood in terms of dynamics on an invariant set we will give, next, the formal definition of an attractor. Attractors and repellers are similar concepts, in the sense that the attractor of a system is the repeller of the corresponding system, with the reversed arrow of time.

A closed and bounded subset $A \subset X$ is an *attractor* if A is *invariant* — that is, $f(A) = A$—and A is *attracting*—that is, the distance between $f^k(x_0)$ and A converges to zero as $k \to \infty$— for all points $x_0 \in V$ in a neighborhood V of A. The set V is called the *basin of attraction* of A.

We say that f is *chaotic* if the following three conditions hold.

1. There is a point x whose orbit is dense in A — that is, it passes arbitrarily close to any point of A.
2. The periodic point of f are dense in A.
3. The function f has sensitive dependence on initial conditions, i.e., there is a number $\delta > 0$, such that for any $x \in X$ there are points y arbitrarily close to x, such that $|f^k(x) - f^k(y)| \geq \delta$ for some k.

We cannot emphasise enough that chaotic dynamical systems have a *deterministic* evolution, although the three conditions above ensure that the system's evolution seems, and is for practical purposes, unpredictable.

By reversing the arrow of time an attractor becomes a repeller, in other words an attractor for (X, f) is a repeller for (X, f^{-1}) or if f does not have an inverse for $(X, f_1^{-1}, f_2^{-1}, \ldots, f_k^{-1})$ where $f_1^{-1}, f_2^{-1}, \ldots, f_k^{-1}$ are the inverse branches of f. For example 0 is an attracting fixed point for $f(x) = \frac{x}{2}$ and it is a repelling fixed point for $f(x) = 2x$.

In the effort to explain the transition from classical dynamics to chaos three scenarios for the onset of chaos have been proposed [Ruelle, 1989]:

1. The Ruelle — Takens onset of chaos through quasi periodicity.
2. The Feigenbaum onset of chaos through period doubling.
3. The Pomeau — Manneville onset of chaos through intermittency.

The simplest approach is to illustrate the onset of chaos through period doubling. We will introduce the logistic family of maps that provides a simple example of the onset of chaos through periodic doubling. The next example uses the logistic family to illustrate the onset of chaos through the so-called process of periodic doubling.

Example 226.6

Let $X = [0,1]$ and let $f_\lambda: X \to X$ be the logistic family

$$f_\lambda(x) = \lambda x(1-x) \tag{226.3}$$

originally introduced to model the development of certain populations. If the population is x_k at the end of the kth year, it is assumed to be $x_{k+1} = f_\lambda(x_k)$ at the end of the $(k+1)$th year.

As λ increases from 0 to 4, the behavior of f_λ undergoes a sequence of transformations, ranging from the simplest to the most complex [Falconer, 1990].

For $0 < \lambda < 3$, f has a stable fixed point, $x_\lambda = 1 - 1/\lambda$, which attracts all orbits starting at $x_0 \in (0,1)$.

When λ increases through the value $\lambda = \lambda_1 = 3$, the stable fixed point x_λ splits into a stable period 2 orbit, $\{y_\lambda^1, y_\lambda^2\}$. For example, if $\lambda = 3.38$, almost all trajectories settle asymptotically into this periodic cycle.

At $\lambda = \lambda_2 = 1 + \sqrt{6} \approx 3.45$, the period 2 orbit becomes unstable and splits into a stable period 4 orbit. Let λ_k denote the value of lambda at which the kth period doubling occurs.

Then λ_k increases to the critical value $\lambda_\infty \approx 3.57$, where chaos sets in. Feingenbaum discovered that λ_∞ is a universal constant, in the sense that

$$\lambda_\infty - \lambda_k \approx C\delta^k \tag{226.4}$$

for very general families of functions, where λ_k are the points where period doubling occurs.

For $\lambda \in (\lambda_\infty, 4)$ chaos and periodicity alternate, depending on specific window values of λ.

Example 226.7

Let X = (0,1) and let $f(x) = \lambda x(1-x)$ with $\lambda = 3.839$. In other words let $f(x) = 3.839x(1-x)$. Choose any initial state x_0 between 0 and 1 and use a calculator to convince yourself that the orbit of x_0 will settle into repeatedly passing closer and closer to three points $p_1 \approx 0.149$, $p_2 \approx 0.489$ and $p_3 \approx 0.959$.

Although a discussion of continuous dynamical systems is beyond the scope of this article, we will conclude the section on chaos with the example of one of the most famous chaotic systems. Its attractor, called the Lorentz strange attractor is embedded in $\mathbb{R}^3$ and its two-dimensional projections have a characteristic butterfly shape.

Example 226.8

Lorentz [1963], who studied Rayleigh-Benard convection considered the following model. A fluid contained between two rigid plates is subjected to gravity. The top plate is maintained at the temperature T_0 and the bottom plate is maintained at a higher temperature, $T_0 + \Delta T$. Experiments show that, for a certain range of small values of ΔT, the fluid will execute a convective cellular flow. For bigger values of ΔT the flow becomes chaotic. This is similar to what happens in the earth's atmosphere. The specific equations of Lorentz are [Falconer, 1990]:

$$\frac{dx}{dt} = -\sigma x + \sigma y$$

$$\frac{dy}{dt} = -xy + rx - y$$

$$\frac{dz}{dt} = xy - bz$$

where $\sigma = 10$, $r = 28$, and $b = 8/3$.

The system has sensitive dependence on initial conditions. The model above is a very simplified version of the dynamics related to meteorological phenomena. It is thus no surprise that the weather is unpredictable since even this simple model behaves chaotically.

226.2 Fractals

Fractal geometry emerged as a novel area of mathematics in the early eighties, when B. Mandelbrot published his book, *The Fractal Geometry of Nature*. To better highlight some basic features of fractal geometry, we will present them in contrast with their analogues from Euclidian geometry.

Euclidian geometry studies the properties of, and the relationships between, points, lines, angles, curves, surfaces and solids in space. All these shapes are smooth, simple, and regular. Most of them are found in man made objects: highways, parking lots, domes. Some occur in nature: the trajectories and shapes of the planets, or the shapes of crystals. Nevertheless, most objects around us do not have smooth shapes. For example the flames, clouds, turbulent rivers, coastlines and trees all have small intricate detail.

Mathematicians realized only about two decades ago, that these shapes, in their own way, have a mathematical simplicity and beauty that can be described in the language of mathematics. Mandelbrot's remarkable contribution was to realize that the apparent geometrical disorder around us is a universal property of nature, that it is the rule rather than the exception, and to bring that idea before the scientific community. Mandelbrot called these rugged and fuzzy shapes *fractal sets*. The word fractal comes from the Latin fractus, meaning fractured. To be specific, all the sets we talk about in this section are subsets of the Euclidian space R^n.

Fractal geometry studies the properties of, and relationships between fractal sets. Intuitively one can easily recognize a fractal set. A breaking wave is a fractal, a donut is not. The question is how to capture, mathematically, the essence of fractals. Mandelbrot gave a beautiful and simple mathematical definition of a fractal set. His definition involves two different notions of dimension, the *topological dimension* and the *Hausdorff dimension* of a set. We will first state Mandelbrot's definition of a fractal set, and then we will talk about topopological dimension and Hausdorff dimension.

A fractal set is, *by definition,* a set whose topological dimension is *strictly* smaller than its fractal dimension.

The concepts of topological dimension and Hausdorff dimension belong to advanced mathematics and we will discuss them only superficially. We will denote the topological dimension by TDim and the Hausdorff dimension by HDim.

Both the topological and Hausdorff dimensions of a set measure the size of the set, but from different perspectives. The topological dimension of a set can only take integer values, 0, 1, 2, up to n. A subset of R^3 has topological dimension equal to either 0, 1, 2, or 3. The Hausdorff dimension of a set can take any value between 0 and n, including the end points. A subset of R^3 can have any Hausdorff dimension between 0 and 3.

The topological dimension of a set is always less than, or equal to, its Hausdorff dimension. There are sets with TDim = 0 and HDim = n. The consequence that TDim < HDim for a given set is that the set must necessarily explode, in some sense, have infinitely fine detail, be a fractal. Depending on its topological dimension a fractal set can be a fractal dust, a fractal curve, a fractal surface and so on.

The topological dimension of a set measures, one might say, the degree of freedom that an imaginary creature has to navigate within the set. The bigger the dimension, the larger the set. For example a plane a plane is larger than a line since there is more room, freedom of movement, within the plane than along the line.

The topological dimension agrees with our intuitive notion of dimension. A point has topological dimension zero; TDim (point) = 0. A curve has topological dimension one; TDim(curve) = 1. A surface has topological dimension two; TDim(surface) = 2. Moreover, a finite or countable union of sets of a given dimension, has the same dimension. A set consisting of two, three, one hundred points, in fact a dust of points, has topological dimension zero. Similarly, set consisting of two, three or a network of lines has topological dimension one.

One possible way to define the topological dimension of a set $F \subset R^n$ is along the following lines. Cover F by a union of n-dimensional open balls B_i of arbitrarily small radius $F \subseteq \bigcup B_i$. Let n_F be the minimum number of balls that have to overlap, so as to cover the set F, as their radius converges to 0. Then the topological dimension of F is by definition TDim(F) = $n_F - 1$. To see that this definition agrees with our intuition, cover the unit interval F = [0,1] by one dimensional balls of small radius, that is by arbitrarily small open intervals. In order to cover the interval [0,1] in this manner we need that these intervals intersect in at least pairs of $n_F = 2$. Hence according to the definition the topological dimension of the interval is TDim([0,1]) = 2 – 1 =1.

The Hausdorff dimension measures the size of a set, from a different perspective. It measures how well the set fills the ambient space. The bigger the dimension, the larger the set.

A whirl of dust, raised by the wind, may be very thin, or it may so thick as to fill the whole space. The fractal dimension of a dust can assume any value between zero and three. It equals zero, when the dust is extremely sparse, and three when the dust fills the whole three dimensional space. A Cantor set is a dust type of set; it has topological dimension 0, but it has a positive Hausdorff dimension that could be as big as the dimension of the surrounding space.

The graph of a real nowhere differentiable of one real variable may have any Hausdorff dimension between 1 and 2. The trajectory of a particle driven by a Brownian motion has Hausdorff dimension greater than 1. A continuous graph describing the performance of a stock can sometimes fluctuate so much, during periods of market volatility, as to fill whole solid areas on the computer screen. The Hausdorff dimension of this graph can assume any value between one and two.

Next we set some preliminary terminology and notation in order to be able to give the formal definition of the Hausdorff dimension.

The diameter of $U \subset \mathbb{R}^n$ is $|U| = \sup\{|x-y| : x, y \in U\}$. Let $F \subset \mathbb{R}^n$. For $\varepsilon > 0$, an ε-cover of F is any countable collection of sets U_i with $|U_i| < \varepsilon$, and such that $F \subset \cup_i U_i$. Let $s \geq 0$. For any $\varepsilon > 0$, we define

$$H_\varepsilon^S(F) = \inf\left\{ \sum_{i=1}^{\infty} |U_i|^S : \{U_i\} \text{ is an } \varepsilon\text{-cover of } F \right\}$$

As ε decreases to 0, $H_\varepsilon^s(F)$ increases and thus approaches a limit,

$$H^s(F) = \lim_{\varepsilon \to 0} H_\varepsilon^s(F)$$

which we define to be the *s-dimensional Hausdorff measure* of *F*.

For any subset $F \subset R^n$, there is a unique number $D = \text{HDim}(F) \geq 0$ such that for $0 \leq s < D$, $H^s(F) = \infty$ and for $D < s < \infty, H^s(F) = 0$. This number D is called the *Hausdorff dimension* of *F*. If *F* has Hausdorff dimension *D*, the *D*-dimensional Hausdorff measure of *F* may take any value in $[0, \infty]$. The inequality TDim $\leq$ HDim always holds.

The Hausdorff dimension is difficult to estimate, both theoretically and numerically. As a substitute, one often uses a simpler concept, the *fractal dimension*, which conveys similar information about the geometry of a set yet is easier to compute.

Let $F \subset R^n$ be a nonempty bounded subset of some euclidian space, and let $N(F, \varepsilon)$ be the smallest number of cubes of side length equal to ε required to cover *F*. The fractal dimension (also called *box dimension, Kolmogorov entropy*, or *capacity*) is defined as the limit

$$\text{FDim}(F) = \lim_{\varepsilon \to 0} \frac{\log N(F, \varepsilon)}{-\log \varepsilon}$$

If the set F is included in the Euclidian plane, then it can be represented on the computer screen. The number of squares of given side length, necessary to cover such a set can be easily counted, and thus the fractal dimension can be estimated numerically.

For any set, we have the inequalities TDim $\leq$ HDim $\leq$ FDim.

The fractal dimension of the Lorentz attractor was estimated numerically to be around 2.04 and the fractal dimension of the attractor corresponding to λ_∞, from the logistic family, was estimated numerically to be ≈ 0.538 [Falconer, 1990, p. 175].

The key to the inherent simplicity of fractals, despite their complicated shapes, is their *self similar* essence. Think of a fern. It consists of many fronds, each frond a scaled down version of the whole fern. Each frond is itself a set of even smaller frondlets. The original fern is a collection of smaller copies of itself.

The fern looks complicated. The rule according to which it is build is, however, simple. The same observation is true for trees, mountains, clouds.

A map $W : \mathbb{R}^n \to \mathbb{R}^n$ is called a *contractive similitude* if $|W(x) - W(y)| = s|x - y|$, for any $x, y \in \mathbb{R}^n$, for some $0 < s < 1$. The number *s* is the *contraction ratio* of *W*.

A set $A \subset \mathbb{R}^n$ is called *self-similar* if it can be tiled with smaller copies of itself,

$$A = \bigcup_{i=1}^{N} W_i(A) \tag{226.4}$$

where $\{W_i\}$ are contractive similitudes.

The collection of maps $\{W_i\}$ defines a random dynamical system, called an *iterated function system* (Barnsley, 1988). Rather than follow the orbit of a point x_0 under iteration of one map, one defines the nth iterate of this system by $x_n = w_X(x_{n-1})$, where X is a random variable with values 1,2,…n. In other words one starts at some initial condition x_0 and proceeds by applying at random, the maps $\{W_i\}$. This is the, so called *chaos game*, that one can implement, and output on a computer screen. A consequence of the fact that the w_i are contractive maps is that the orbit of any initial condition x_0 is attracted to a compact set A, called, by analogy with the theory of dynamical systems, the attractor of the iterated function system $\{W_i\}$. Typically this attractor is a fractal set, and it can assume almost any shape, depending on the maps $\{W_i\}$. The collection of maps $\{W_i\}$ are called the *fractal code* of the set A.

The self similar sets have many interesting properties. In particular they are the only class of fractal sets whose Hausdorff dimension is (relatively) easy to compute theoretically.

If $\{W_i\}$ is an iterated function system with the non-overlapping condition $W_i(A) \cap W_j(A) = \emptyset$, then

$$\mathrm{HDim}(A) = \mathrm{FDim}(A) = \mathrm{D} \tag{226.5}$$

where D is the unique real solution to the equation $\sum_{i=1}^{N} s_i^D = 1$ and s_i is the contraction ratio of W_i.

The simplest example of a self-similar set is the famous ternary Cantor set C (Figure 226.1). This Cantor set is the attractor of an iterated function system with two maps $W_1 = x/3$ and $W_2 = x/3 + 2/3$, $s_1 = s_2 = 1/3$, hence Equation (226.5) yields HDim(C) = FDim(C) = log 2/log 3. In addition to being the attractor of that iterated function system that Cantor set can be realized by the following construction. From the closed interval [0, 1], remove the middle third interval $(\frac{1}{3}, \frac{2}{3})$. Two closed intervals remain—$[0, \frac{1}{3}]$ and $[\frac{2}{3}, 1]$. Repeat the procedure for all remaining intervals iteratively. The Cantor set C is the fractal "dust" that remains. All points of the form $\{k3^{-n} : n \geq 0, 0 \leq k \leq 3^n - 1\}$ belong to C, but there are many more. Since the ternary Cantor set is included in the unit interval, and hence hard to see on a page, we give an enhanced (two dimensional) version of it in Figure 226.2.

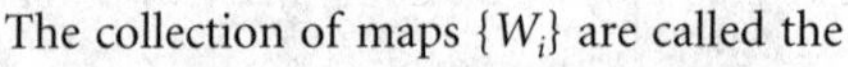

FIGURE 226.1 Ternary Cantor set C.

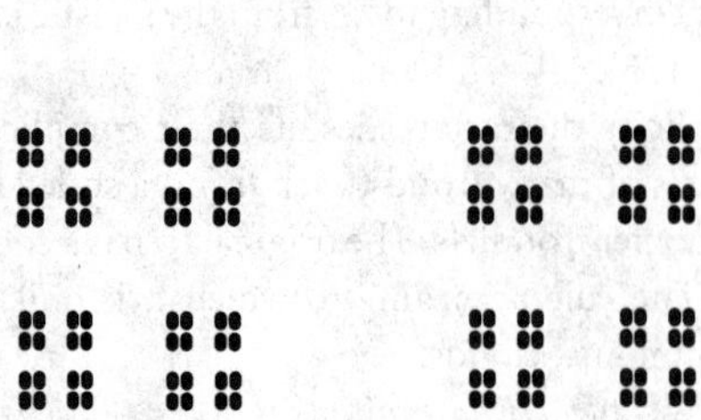

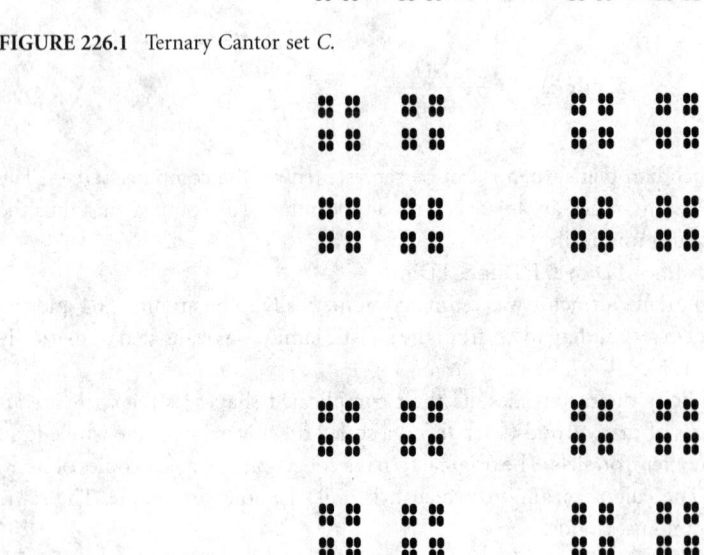

FIGURE 226.2 Two-dimensional construction of Cantor set C.

Another example of a Cantor set is presented in Figure 226.2. For this example $N = 4$, $\{s_i = 1/3 : i = 1,\ldots,4\}$ and the 4 maps $\{W_i\}$ are two dimensional affine maps. Equation (226.5) yields HDim = FDim = log 4/log 3. The topological dimension of both Cantor sets is 0, a feature common to all "dust" (*totally disconnected*) sets.

Michael Barnsley (1988) discovered an ingenious method to realize any shape with fractal character as the attractor of an iterated function system, in other words to determine its fractal code. His method is formalized by the Collage Theorem and it exploits the self similarity property of sets with fractal character. The collage theorem holds in any dimension, but it was originally discovered in connection with two dimensional affine contractions, as a method to encode and generate pictures.

The Collage Theorem states that if a target set T can be covered well enough with smaller copies of itself $w_1(T), \ldots, w_n(T)$, then the set T is close enough to the self similar attractor A of the iterated function system $\{W_i\}$. More formally the Collage Theorem states the following.

Collage Theorem. If $d(T, \bigcup W_i(T)) < s < 1$, then $d(T, A) < \dfrac{d(T, \bigcup W_i(T))}{1-s}$, where A is the attractor of the iterated function system $\{W_i\}$ and d is the Hausdorff distance between sets, that measures, one might say, how well two sets look like one another. Notice that the smaller s is (i.e., the better the collage of the target set T with smaller affine copies of itself), the better the approximation of T by the attractor.

The Collage Theorem gives an easy method to find suitable fractal codes for any set with a self similar shape of nature. A fern or a tree are wonderful candidates for the collage theorem. Once one finds the code of the fern and of the tree, one can represent them on the computer screen by iterating the corresponding iterated function system. The ideas of the Collage Theorem have been exploited by M. Barnsley, and by many other people to develop data compression software. The code of a fern is clearly much smaller, memory wise, than the bit map representation of the same fern.

226.3 Julia Sets

Julia sets are invariant sets for the maps of the quadratic family $f(z) = z^2 + c, z \in \mathbb{C}, c \in \mathbb{C}$ indexed by the complex parameter c.

The Julia sets are repellers for these maps, they are typically fractal curves, or fractal dusts, and they have remarkable shapes. The complement of the Julia set $J(f)$ is called the Fatou set. The dynamics of the map f is very simple on the Fatou set, also called the stable set of f, and the dynamics of f is interesting and chaotic on the Julia set. These quadratic maps, and their invariant sets, were an important subject of study for mathematicians, in particular Julia and Fatou, during the 1920s. The subject became popular again in the early 1980s when it was revisited from the perspective of the emerging science of chaos.

Example 226.9

For $c = 0, f(z) = z^2$ and the corresponding Julia set $J(f)$ is the unit circle $\{|z| = 1\}$.

The dynamics of the complex map above are relatively simple. The nth iterate of the point z is $f^n(z) = z^{2n}$. Hence all the points with $|z|<1$ converge to the fixed point $z = 0$, and similarly all the points with $|z| > 0$ diverge to infinity. The unit circle $|z| = 1$ is invariant under the map $f(z) = z^2$. On the unit circle the map has the three attributes of chaotic behavior, listed in the section on Chaos. Under iteration, a generic point z, with $|z| = 1$, will jump about on the unit circle, and recur infinitely many times close to any given point on the circle. To summarize the unit circle is invariant and all sets that do not intersect the unit circle diverge away from it, either towards 0 or towards infinity, depending on whether the set lies in the bounded, or the unbounded connected component in which the unit circle splits the complex plane.

The unit circle is the Julia set for the map $f(z) = z^2$. Notice that the map $f(z) = z^2$ has periodic points of arbitrarily high order, and all of them lie on the unit circle. These points are repelling since we saw that the whole unit circle is a repelling set. Thus the unit circle is the closure of the repelling periodic points of $f(z)=z^2$. By analogy, this property is used to define all the Julia sets.

The *Julia set* $J_c(f)$ of the map $f(z)=z^2+c, z\in\mathbb{C}, c\in\mathbb{C}$ is, by definition, the closure of the set of repelling periodic points of *f*.

One can show that $J_c(f)$ is nonempty, is closed and bounded, has empty interior, it is either a closed (simple or composite) or a dust, and is generally a fractal. The Julia set $J_c = J_c(f)$ is both forward- and backward-invariant, $J_c = f(J_c) = f^{-1}(J_c)$ and *f* is chaotic on J_c.

Next we will give examples of Julia sets for various values of *c* different from 0. Clearly the value of *c* has a great impact on the dynamics of the map f_c and on the shape of the corresponding Julia set.

The critical points of a map *f* (*z*) are, by definition, the solutions of the equation $f'(z)=0$. All the quadratic complex maps $f(z)=z^2+c$ have the same critical point $z = 0$. It is a well known fact that the orbit of the critical point plays a special role in the dynamics of a system. For all the maps $f(z)=z^2+c$, *c* is the image of 0, $c=f_c(0)$, in other words c is the first iterate of 0. Thus for the quadratic family the study of the orbit of 0 is equivalent to the study of the orbit of *c*.

The dependence of the Julia set of $f(z)=z^2+c$, on the orbit of the critical point 0, or equivalently on the orbit of *c*, can be summarized as follows.

If the orbit of *c* converges to an attracting fixed point, the corresponding Julia set J_c is a simple closed curve. If the orbit of *c* diverges to infinity the corresponding Julia set J_c is a totally disconnected fractal set. If the orbit of *c* is a periodic orbit the corresponding Julia set J_c is a continuous closed curve whose complement has infinitely many connected components. Finally, if the orbit of c is eventually periodic, but not periodic, the Julia set has a dendritic shape (see Falconer [1990, p. 214]).

For complex parameters *c*, with $|c|$ small, the corresponding Julia sets are continuous perturbations of the unit circle. These closed curves, however, do not have any differentiable arc, they are fractal curves.

Example 226.10

For $c\in C$ with $|c|<1/4\,|$ the Julia set of $f(z)=z^2+c$ is a simple closed curve that splits the complex plane in two connected components. The iterates $f^k(z_0)$ of an initial condition z_0 in the bounded component converge to a fixed point of the map $f(z)=z^2+c$, and the iterates $f^k(z_0)$ of an initial condition z_0 in the unbounded component, diverge to ∞. The set of all c with the property that the corresponding Julia set J_c is a simple closed curve is a cardioid that includes strictly the disc $\{c:|c|<\pi\}$.

For complex parameters *c*, with $|c|$ large, the corresponding Julia sets are totally disconnected fractal sets.

Example 226.11

The Julia set of $f(z)=z^2+c$, with $|c|>\frac{1}{4}(5+2\sqrt{6}), J(f)$ is totally disconnected. In other words in this case the Julia set is a fractal dust that repels all nearby orbits. As before the Julia dust J_c is invariant under the forward and backward iterations of f and f is chaotic on J_c.

For values of c so that *c* is a periodic point of f_c, or is eventually a periodic point of f_c, rather than a attracting fixed point, the corresponding Julia set and the dynamics of the system acquire new twists, as seen in the next two examples.

Example 226.12

For c = –1 the map $f_{-1}(z)=z^2-1$ has $f_{-1}(0)=-1$ and $f_{-1}(1)=0$. It turns out that in this case the Julia set J_{-1} is a continuous curve with infinitely many cycles, i.e., the complement of the Julia set of $f_{-1}(z)=z^2-1$ has infinitely many connected components.

Example 226.13

For $c=i$, the map $f_i(z)=z^2+i$, we have $f_i(i)=-1+i$ and the points $-1+i$ and $-i$ form a period of order 2, thus the critical point 0, or equivalently the parameter c = *i*, are eventually periodic points. It turns out that in this case the Julia set J_i is a dendrite.

For proofs of the results stated in the examples see Falconer [1990, p. 214].

There is no better conclusion to a chapter on chaos, fractals and Julia sets than to introduce a concept that blends together features of all these concepts in a beautiful structure known as the Mandelbrot set. The Mandelbrot set is a subset M of the complex plane $M \subset C$ and it consists of all those complex numbers c with the property that the orbit of c (or of the critical point 0) under the quadratic map $f_c(z) = z^2 + c$ is bounded. It is easy to decide, using a computer, whether the iterates of c under f_c stay bounded or diverge. If one checks in this manner a comprehensive array of values of c around the origin, and plots those c values for which the iterates of 0 under f_c are bounded, one discovers an amazing shape. Equivalently the Mandelbrot set is the set of points c, so that the Julia set of f_c is connected. Mandelbrot ran into this set, apparently by accident, and at first he thought that it was the result of some computer mulfunction. The boundary of the set has a fractal look, with intricate detail, and all its geometric features recur on all scales. The main cardioid of the Mandelbrot set corresponds to the set of all c that correspond to a simple closed curve Julia set. The Mandelbrot set is a wonderful discovery that reflects the beauty of fractals and the order of chaos.

References

Barnsley, M. F. 1988. *Fractals Everywhere.* Academic Press, New York.

Devaney, R. L. 1989. *An Introduction to Chaotic Dynamical Systems.* Addison-Wesley, Reading, MA.

Falconer, K. J. 1990. *Fractal Geometry, Mathematical Foundations and Applications.* John Wiley & Sons, New York.

Li, T. Y. and Yorke, J. A. 1975. Period three implies chaos. *Amer. Math. Monthly.* 82:985.

Lorentz, E. N. 1963. Deterministic nonperiodic flow. *J. Atmos. Sci.* 20:130.

Mandelbrot, B. B. 1982. *The Fractal Geometry of Nature.* Freeman, New York.

Ott, E. 1993. *Chaos in Dynamical Systems.* Cambridge University Press, New York.

Penrose, R. 1989. *The Emperor's New Mind.* Oxford University Press, New York.

Ruelle, D. 1989. *Chaotic Evolution and Strange Attractors.* Cambridge University Press,

Ruelle, D. and Takens, F. 1971. On the nature of turbulence. *Comm. Math. Phys.* 20:167.

227

Calculus of Variations

William F. Ames
Georgia Institute of Technology

The basic problem in the *calculus of variations* is to determine a function such that a certain *functional*, often an integral involving that function and certain of its derivatives, takes on *maximum or minimum values*. As an example, find the function $y(x)$ such that $y(x_1) = y_1, y(x_2) = y_2$ and the integral (functional) $I = 2\pi \int_{x_1}^{x_2} y[1+(y')^2]^{1/2}\, dx$ is a minimum. A second example concerns the transverse deformation $u(x,t)$ of a beam. The energy functional $I = \int_{t_1}^{t_2} \int_0^L [\frac{1}{2}\rho(\partial u/\partial t)^2 - \frac{1}{2}EI(\partial^2 u/\partial x^2)^2 + fu] dx\, dt$ is to be minimized.

227.1 The Euler Equation

The elementary part of the theory is concerned with a *necessary* condition (generally in the form of a differential equation with boundary conditions) that the required function must satisfy. To show mathematically that the function obtained actually maximizes (or minimizes) the integral is much more difficult than the corresponding problems of the differential calculus.

The *simplest case* is to determine a function $y(x)$ that makes the integral $I = \int_{x_1}^{x_2} F(x, y, y')\, dx$ stationary and that satisfies the prescribed end conditions $y(x_1) = y_1$ and $y(x_2) = y_2$. Here we suppose F has continuous second partial derivatives with respect to x, y, and $y' = dy/dx$. If $y(x)$ is such a function, then it must satisfy the *Euler equation* $(d/dx)(\partial F/\partial y') - (\partial F/\partial y) = 0$, which is the required necessary condition. The indicated partial derivatives have been formed by treating x, y, and y' as independent variables. Expanding the equation, the equivalent form $F_{y'y'}y'' + F_{y'y}y' + (F_{y'x} - F_y) = 0$ is found. This is second order in y unless $F_{y'y'} = (\partial^2 F)/[(\partial y')^2] = 0$. An alternative form $1/y'[d/dx(F - (\partial F/\partial y')(dy/dx)) - (\partial F/\partial x)] = 0$ is useful. Clearly, if F does not involve x explicitly $[(\partial F/\partial x) = 0]$ a first integral of Euler's equation is $F - y'(\partial F/\partial y') = c$. If F does not involve y explicitly $[(\partial F/\partial y) = 0]$ a first integral is $(\partial F/\partial y') = c$.

The Euler equation for $\sqrt{\frac{2}{\pi}}\sin st$ is $(d/dx)[yy'/[1+(y')^2]^{1/2}] - [1+(y')^2]^{1/2} = 0$ or after reduction $yy'' - (y')^2 - 1 = 0$. The solution is $y = c_1 \cosh(x/c_1 + c_2)$, where c_1 and c_2 are integration constants. Thus the required minimal surface, if it exists, must be obtained by revolving a catenary. Can c_1 and c_2 be chosen so that the solution passes through the assigned points? The answer is found in the solution of a transcendental equation that has two, one, or no solutions, depending on the prescribed values of y_1 and y_2.

0-8493-1586-7/05/\$0.00+\$1.50

227.2 The Variation

If $F = F(x, y, y')$, with x independent and $y = y(x)$, then the *first variation* δF of F is defined to be $\delta F = (\partial F/\partial x)\,\delta y + (\partial F/\partial y)\,\delta y'$ and $\delta y' = \delta(dy/dx) = (d/dx)(\delta y)$ — that is, they commute. Note that the first variation, δF, of a functional is a first-order change from curve to curve, whereas the differential of a function is a first-order approximation to the change in that function along a *particular curve*. The laws of δ are as follows: $\delta(c_1F + c_2G) = c_1\delta F + c_2\delta G; \delta(FG) = F\delta G + G\delta F; \delta(F/G) = (G\delta F - F\delta G)/G^2$; if x is an independent variable, $\delta x = 0$; if $u = u(x, y)$; $(\partial/\partial x)(\delta u) = \delta(\partial u/\partial x), (\partial/\partial y)(\delta u) = \delta(\partial u/\partial y)$.

A necessary condition that the integral $I = \int_{x_1}^{x_2} F(x, y, y')dx$ be stationary is that its (first) variation vanishes — that is, $\delta = \int_{x_1}^{x_2} F(x, y, y')dx = 0$. Carrying out the variation and integrating by parts yields $\delta I = \int_{x_1}^{x_2}[(\partial F/\partial y) - (d/dx(\partial F/\partial y')]\delta y\,dx + [(\partial F/\partial y')\delta y]_{x_1}^{x_2} = 0$. The arbitrary nature of δy means the square bracket must vanish and the last term constitutes the *natural boundary conditions*.

Example

The *Euler equation* of $\int_{x_1}^{x_2} F(x, y, y', y'')dx$ is $(d^2/dx^2)(\partial F/\partial y'') - (d/dx)(\partial F/\partial y') + (\partial F/\partial y) = 0$, with natural boundary conditions $\{[(d/dx)(\partial F/\partial y'') - (\partial F/\partial y')]\delta y\}_{x_1}^{x_2} = 0$ and $(\partial F/\partial y'')\delta y'\big|_{x_1}^{x_2} = 0$. The Euler equation of $\int_{x_1}^{x_2}\int_{y_1}^{y_2} F(x, y, u, u_x, u_y, u_{xx}, u_{xy}, u_{yy})dx\,dy$ is $(\partial^2/\partial x^2)(\partial F/\partial u_{xx}) + (\partial^2/\partial x\partial y)(\partial F/u_{xy}) + (\partial^2/\partial y^2)(\partial F/u_{yy}) - (\partial/\partial x)(\partial F/\partial u_x) - (\partial/\partial y)(\partial F/\partial u_y) + (\partial F/\partial u)$, and the natural boundary conditions are

$$\left[\left(\frac{\partial}{\partial x}\left(\frac{\partial F}{\partial u_{xx}}\right) + \frac{\partial}{\partial y}\left(\frac{\partial F}{\partial u_{xy}}\right) - \frac{\partial F}{\partial u_x}\right)\delta u\right]_{x_1}^{x_2} = 0, \qquad \left[\left(\frac{\partial F}{\partial u_{xx}}\right)\delta u_x\right]_{x_1}^{x_2} = 0$$

$$\left[\left(\frac{\partial}{\partial y}\left(\frac{\partial F}{\partial u_{yy}}\right) + \frac{\partial}{\partial x}\left(\frac{\partial F}{\partial u_{xy}}\right) - \frac{\partial F}{\partial u_y}\right)\delta u\right]_{y_1}^{y_2} = 0, \qquad \left[\left(\frac{\partial F}{\partial u_{yy}}\right)\delta u_y\right]_{y_1}^{y_2} = 0$$

In the more general case of $I = \iint_R F(x, y, u, v, u_x, u_y, v_x, v_y)dx\,dy$, the condition $\delta I = 0$ gives rise to the two Euler equations, $(\partial/\partial x)(\partial F/\partial u_x) + (\partial/\partial y)(\partial F/\partial u_y) - (\partial F/\partial u) = 0$ and $(\partial/\partial x)(\partial F/\partial v_x) + (\partial/\partial y)(\partial F/\partial v_y) - (\partial F/\partial v) = 0$. These are two PDEs in u and v that are linear or quasi-linear in u and v. The Euler equation for $I = \iiint_R (u_x^2 + u_y^2 + u_z^2)\ dx\,dy\,dz$, from $\delta I = 0$, is Laplace's equation $u_{xx} + u_{yy} + u_{zz} = 0$.

Variational problems are easily derived from the differential equation and associated boundary conditions by multiplying by the variation and integrating the appropriate number of times. To illustrate, let $F(x)$, $\rho(x)$, $p(x)$, and w be the tension, the linear mass density, the natural load, and (constant) angular velocity of a rotating string of length L. The equation of motion is $(d/dx)[F(dy/dx)] + \rho w^2 y + p = 0$. To formulate a corresponding variational problem, multiply all terms by a variation δy and integrate over $(0, L)$ to obtain

$$\int_0^L \frac{d}{dx}\left(F\frac{dy}{dx}\right)\delta y dx + \int_0^L \rho w^2 y\delta y dx + \int_0^L p\delta y dx = 0$$

The second and third integrals are the variations of $\frac{1}{2}\rho w^2 y^2$ and py, respectively. To treat the first integral, integrate by parts to obtain

$$\left[F\frac{dy}{dx}\delta y\right]_0^L - \int_0^L F\frac{dy}{dx}\delta\frac{dy}{dx}dx = \left[F\frac{dy}{dx}\delta y\right]_0^L - \int_0^L \frac{1}{2}F\delta\left(\frac{dy}{dx}\right)^2 dx = 0$$

So the variation formulation is

$$\delta\int_0^L\left[\frac{1}{2}\rho w^2y^2+py-\frac{1}{2}F\left(\frac{dy}{dx}\right)^2\right]dx+\left[F\frac{dy}{dx}\delta y\right]_0^L=0$$

The last term represents the *natural boundary conditions.* The term $\frac{1}{2}\rho w^2 y^2$ is the kinetic energy per unit length, the term $-py$ is the potential energy per unit length due to the radial force $p(x)$, and the term $\frac{1}{2}F(dy/dx)^2$ is a first approximation to the potential energy per unit length due to the tension $F(x)$ in the string. Thus, the integral is often called the *energy integral.*

227.3 Constraints

The variations in some cases cannot be arbitrarily assigned because of one or more auxiliary conditions that are usually called *constraints.* A typical case is the functional $\int_{x_1}^{x_2} F(x,u,v,u_x,v_x)dx$ with a constraint $\phi(u,v) = 0$ relating u and v. If the variations of u and v (δu and δv) vanish at the end points, then the variation of the integral becomes

$$\int_{x_1}^{x_2}\left\{\left[\frac{\partial F}{\partial u}-\frac{d}{dx}\left(\frac{\partial F}{\partial u_x}\right)\right]\delta u+\left[\frac{\partial F}{\partial v}-\frac{d}{dx}\left(\frac{\partial F}{\partial v_x}\right)\right]\delta v\right\}dx=0$$

The variation of the constraint $\phi(u,v)=0, \phi_u\delta u+\phi_v\delta v=0$ means that the variations cannot both be assigned arbitrarily inside (x_1,x_2), so their coefficients need not vanish separately. Multiply $\phi_u\delta u+\phi_v\delta v=0$ by a Lagrange multiplier λ (may be a function of x) and integrate to find $\int_{x_1}^{x_2}(\lambda\phi_u\delta u+\lambda\phi_v\delta v)dx=0$. Adding this to the previous result yields

$$\int_{x_1}^{x_2}\left\{\left[\frac{\partial F}{\partial u}-\frac{d}{dx}\left(\frac{\partial F}{\partial u_x}\right)+\lambda\phi_u\right]\delta u+\left[\frac{\partial F}{\partial v}-\frac{d}{dx}\left(\frac{\partial F}{\partial v_x}\right)+\lambda\phi_v\right]\delta v\right\}dx=0$$

which must hold for any λ. Assign λ so the first square bracket vanishes. Then δv can be assigned to vanish inside (x_1,x_2), so the two systems

$$\frac{d}{dx}\left[\frac{\partial F}{\partial u_x}\right]-\frac{\partial F}{\partial u}-\lambda\phi_u=0, \qquad \frac{d}{dx}\left[\frac{\partial F}{\partial v_x}\right]-\frac{\partial F}{\partial v}-\lambda\phi_v=0$$

plus the constraint $\phi(u,v) = 0$ are three equations for u, v and λ.

References

Gelfand, I. M. and Fomin, S. V. 1963. *Calculus of Variations.* Prentice Hall, Englewood Cliffs, NJ.

Lanczos, C. 1949. *The Variational Principles of Mechanics.* University of Toronto Press, Toronto.

Schechter, R. S. 1967. *The Variational Method in Engineering.* McGraw-Hill, New York.

Vujanovic, B. D. and Jones, S. E. 1989. *Variational Methods in Nonconservative Phenomena.* Academic Press, New York.

Weinstock, R. 1952. *Calculus of Variations, with Applications to Physics and Engineering.* McGraw-Hill, New York.

228
Probability and Statistics

Y. L. Tong
Georgia Institute of Technology

In most engineering experiments, the outcomes (and hence the observed data) appear in a random and nondeterministic fashion. For example, the operating time of a system before failure, the tensile strength of a certain type of material, and the number of defective items in a batch of produced items are all subject to random variations from one experiment to another. In engineering statistics, we apply the theory and methods of statistics to develop procedures for summarizing the data and making statistical inference and to obtain useful information with the presence of randomness and uncertainty.

228.1 Elementary Probability

Random Variables and Probability Distributions

Intuitively speaking, a random variable (denoted by X, Y, Z, etc.) takes a numerical value that depends on the outcome of the experiment. Because the outcome of an experiment is subject to random variation, the resulting numerical value is also random. In order to provide a stochastic model for describing the probability distribution of a random variable X, we generally classify random variables into two groups — the discrete type and the continuous type. The discrete random variables are those which, technically speaking, take a finite number or a countably infinite number of possible numerical values (in most engineering applications, they take nonnegative integer values). Continuous random variables involve outcome variables such as time, length, distance, area, and volume. We specify a function $f(x)$, called the

0-8493-1586-7/05/$0.00+$1.50

probability density function (p.d.f.) of a random variable X, such that the probability that the random variable X takes a value in a set A (of real numbers) is given by

$$P[X \in A] = \begin{cases} \sum_{x \in A} f(x) & \text{for all sets } A \text{ if } X \text{ is dicrete} \\ \int_A f(x)dx & \text{for all intervals } A \text{ if } X \text{ is continuous} \end{cases} \tag{228.1}$$

By letting A be the set of all values that are less than or equal to a fixed number t (i.e., $A = [-\infty,t]$), the probability function $P(X \le t)$, denoted by $F(t)$, is called the distribution function of X. We note that, by calculus, if X is a continuous random variable and if $F(x)$ is differentiable, then $f(x) = (d/dx)F(x)$.

Expectations

In many applications, the result of an experiment with a numerical outcome X is a specific function of $X[u(X)$, say]. Because X is a random variable, $u(X)$ itself is also a random variable. We define the expected value of $u(X)$ by

$$E[u(X)] = \begin{cases} \sum_x u(x)f(x) & \text{if } X \text{ is discrete} \\ \int_{-\infty}^{\infty} u(x)f(x)dx & \text{if } X \text{ is continuous} \end{cases} \tag{228.2}$$

provided that, of course, the sum or the integral exists. In particular, if $u(x) = x$, then $E(X) \equiv \mu$ is called the mean of X (of the distribution) and $E(X-\mu)^2 \equiv \sigma^2$ is called the variance of X (the distribution). The mean is a measurement of the central tendency, and the variance is a measurement of dispersion of the distribution.

Some Commonly Used Distributions

There are many well-known distributions that are useful in engineering statistics. Among the discrete distributions, the hypergeometric and binomial distributions have applications in acceptance sampling problems and quality control, and the Poisson distribution is useful for studying queueing theory and other related problems. Among the continuous distributions, the uniform distribution concerns random numbers and can be applied in simulation studies, the exponential and gamma distributions are closely related to the Poisson distribution and, together with the Weibull distribution, have important applications in life testing and reliability studies. All of these distributions involve at least one unknown parameter; hence their means and variances also depend on the parameters. See textbooks in this area for details. For example, Hahn and Shapiro [1967:163–169, 120–134] comprehensively list these and other distributions regarding their p.d.f.'s, graphs, parameters, means, and variances, with discussions and examples of their applications.

The Normal Distribution

Perhaps *the* most important distribution in statistics and probability is the normal distribution (also known as the Gaussian distribution). This distribution involves two parameters, μ and σ^2, and its p.d.f. is given by

$$f(x) = f(x;\ \mu,\ \sigma^2) = \frac{1}{\sqrt{2\pi}\sigma} e^{-\frac{1}{2\sigma^2}(x-\mu)^2} \tag{228.3}$$

for $-\infty < \mu < \infty$, $\sigma^2 > 0$, and $-\infty < x < \infty$. It can be shown that, for a p.d.f. of this form, the values of μ and σ^2 are, respectively, that of the mean and the variance of the distribution. Further, the quantity $\sigma = \sqrt{\sigma^2}$ is called the standard deviation of the distribution. We shall use the symbol $X \sim \mathcal{N}(\mu,\sigma^2)$ to denote that X has a normal distribution with mean μ and variance σ^2.

When plotting the p.d.f. $f(x;\mu,\sigma^2)$ given in Equation (228.3), we see that the resulting graph represents a bell-shaped curve and is symmetric about μ. If a random variable Z has an $\mathcal{N}(0,1)$ distribution, then the p.d.f. of Z is given by (from Equation (228.3)):

$$\phi(z) = \frac{1}{\sqrt{2\pi}} e^{-\frac{1}{2}z^2}, \qquad -\infty < z < \infty \tag{228.4}$$

The distribution function of Z,

$$\Phi(z) = \int_{-\infty}^{z} \phi(u)du, \qquad -\infty < z < \infty \tag{228.5}$$

cannot be given in a closed form; hence it has been tabulated. The table of $\Phi(z)$ can be found in most textbooks in statistics and probability, including those listed in the references at the end of this chapter. (By the symmetry property, $\Phi(z) + \Phi(-z) = 1$ holds for all z.)

228.2 Random Sample and Sampling Distributions

Random Sample and Related Statistics

As noted in Box, Hunter, and Hunter [1978], in the design and analysis of engineering experiments, a study usually involves the following steps:

1. The choice of a suitable stochastic model by assuming that the observations follow a certain distribution. The functional form of the distribution (or the p.d.f.) is assumed to be known except the value(s) of the parameter(s).
2. Design of experiments and collection of data.
3. Summarization of data and computation of certain statistics.
4. Statistical inference (including estimation of the parameters of the underlying distribution and hypothesis-testing problems).

In order to make statistical inferences concerning the parameter(s) of a distribution, it is essential to first study the sampling distributions. We say that $X_1, X_2, \ldots, X_n$ represent a random sample of size n if they are independent random variables and each of them has the same p.d.f. $f(x)$. Due to space limitations, the notion of independence will not be carefully discussed here. But, nevertheless, we say that $X_1, X_2, \ldots, X_n$ are independent if

$$P[X_1 \in A_1, X_2 \in A_2, \ldots, X_n \in A_n] = \prod_{i=1}^{n} P[X_i \in A_i] \tag{228.6}$$

holds for all sets $A_1, A_2, \ldots, A_n$.) Because the parameters of the population are unknown, the population mean μ and the population variance σ^2 are unknown. In most commonly used distributions, μ and σ^2 can be estimated by the sample mean $\overline{X}$ and the sample variance S^2, respectively, which are given by

$$\overline{X} = \frac{1}{n}\sum_{i=1}^{n} X_i, \quad S^2 = \frac{1}{n-1}\sum_{i=1}^{n}(X_i - \overline{X})^2 = \frac{1}{n-1}\left[\sum_{i=1}^{n} X_i^2 - n\overline{X}^2\right] \tag{228.7}$$

(The second equality in the formula for S^2 can be verified algebraically.) Now, because $X_1, X_2, \ldots, X_n$ are random variables, $\bar{X}$ and S^2 are also random variables. Each of them is called a statistic and has a probability distribution that also involves the unknown parameter(s). In probability theory, there are two fundamental results concerning their distributional properties.

Theorem 1 *(Weak Law of Large Numbers).* As the sample size n becomes large, $\bar{X}$ converges to μ in probability and S^2 converges to σ^2 in probability. More precisely, for every fixed positive number $\varepsilon > 0$, we have

$$P[|\bar{X} - \mu| \leq \varepsilon] \to 1, \quad P[|S^2 - \sigma^2| \leq \varepsilon] \to 1 \tag{228.8}$$

as $n \to \infty$.

Theorem 2 *(Central Limit Theorem).* As n becomes large, the distribution of the random variable

$$Z = \frac{\bar{X} - \mu}{\sigma / \sqrt{n}} = \frac{\sqrt{n}(\bar{X} - \mu)}{\sigma} \tag{228.9}$$

has approximately an $\mathcal{N}(0,1)$ distribution. More precisely,

$$P[Z \leq z] \to \Phi(z) \text{ for every fixed } z \text{ as } n \to \infty \tag{228.10}$$

228.3 Normal Distribution–Related Sampling Distributions

One-Sample Case

Additional results exist when the observations come from a normal population. If $X_1, X_2, \ldots, X_n$ represent a random sample of size n from an $\mathcal{N}(\mu,\sigma^2)$ population, then the following sampling distributions are useful.

Fact 1. For every fixed n, the distribution of Z given in Equation (228.9) has exactly an $\mathcal{N}(0,1)$ distribution.

Fact 2. The distribution of the statistic $T = \sqrt{n}(\bar{X} - \mu)/S$, where $S = \sqrt{S^2}$ is the sample standard deviation, is called a Student's t distribution with $\nu = n - 1$ degrees of freedom; in symbols, $t(n - 1)$. This distribution is useful for making inference on μ when σ^2 is unknown. A table of the percentiles can be found in most statistics textbooks.

Fact 3. The distribution of the statistic $W = (n - 1)S^2/\sigma^2$ is called a chi-squared distribution with $\nu = n - 1$ degrees of freedom; in symbols, $\chi^2(\nu)$.

Such a distribution is useful in making inference on σ^2. A table of the percentiles can also be found in most statistics books.

Two-Sample Case

In certain applications, we may be interested in the comparisons of two different treatments. Suppose that independent samples from treatments T_1 and T_2 are to be observed as shown in Table 228.1. The difference of the population means $(\mu_1 - \mu_2)$ and the ratio of the population variances can be estimated, respectively, by $(\bar{X}_1 - \bar{X}_2)$ and S_1^2 / S_2^2. The following facts summarize the distributions of these statistics.

Fact 4. Under the assumption of normality, $(\bar{X}_1 - \bar{X}_2)$ has an $\mathcal{N}(\mu_1 - \mu_2, (\sigma_1^2/n_1) + (\sigma_2^2/n_2))$ distribution; or equivalently, for all n_1 and n_2, the statistic

$$Z = [(\bar{X}_1 - \bar{X}_2) - (\mu_1 - \mu_2)]/(\sigma_1^2/n_1 + \sigma_2^2/n_2)^{1/2} \tag{228.11}$$

TABLE 228.1 Summary of Data for a Two-Sample Problem

Treatment	Observations	Distribution	Sample Size	Sample Mean	Sample Variance
T_1	$X_{11}, X_{12}, \ldots, X_{1n_1}$	$N(\mu_1, \sigma_1^2)$	n_1	$\bar{X}_1$	S_1^2
T_2	$X_{21}, X_{22}, \ldots, X_{2n_2}$	$N(\mu_2, \sigma_2^2)$	n_2	$\bar{X}_2$	S_2^2

has an $\mathcal{N}(0,1)$ distribution.

Fact 5. When $\sigma_1^2 = \sigma_2^2 \equiv \sigma^2$, the common population variance is estimated by

$$S_p^2 = (n_1 + n_2 - 2)^{-1}[(n_1 - 1)S_1^2 + (n_2 - 1)S_2^2] \tag{228.12}$$

and $(n_1 + n_2 - 2)S_p^2 / \sigma^2$ has an $\chi^2(n_1 + n_2 - 2)$ distribution.

Fact 6. When $\sigma_1^2 = \sigma_2^2$, the statistic

$$T = [(\bar{X}_1 - \bar{X}_2) - (\mu_1 - \mu_2)] / S_p(1/n_1 + 1/n_2)^{1/2} \tag{228.13}$$

has a $t(n_1 + n_2 - 2)$ distribution, where $S_p = \sqrt{S_p^2}$.

Fact 7. The distribution of $F = (S_1^2 / \sigma_1^2)/(S_2^2 / \sigma_2^2)$ is called an *F* distribution with degrees of freedom $(n_1 - 1, n_2 - 1)$; in symbols, F $(n_1 - 1, n_2 - 1)$.

The percentiles of this distribution have also been tabulated and can be found in statistics books.

The distributions listed above (normal, Student's *t*, chi-squared, and *F*) form an important part of classical statistical inference theory, and they are developed under the assumption that the observations follow a normal distribution. When the distribution of the population is not normal and inference on the population means is to be made, we conclude that (1) if the sample sizes n_1 and n_2 are large, then the statistic *Z* in Equation (228.11) has an approximate $\mathcal{N}(0,1)$ distribution; and (2) in the small-sample case, the exact distribution of $\bar{X}$ [of $(\bar{X}_1 - \bar{X}_2)$] depends on the population p.d.f. There are several analytical methods for obtaining it, and those methods can be found in statistics textbooks.

228.4 Confidence Intervals

A method for estimating the population parameters based on the sample mean(s) and sample variance(s) involves the confidence intervals for the parameters.

One-Sample Case

Confidence Interval for μ When σ² Is Known

Consider the situation in which a random sample of size *n* is taken from an $\mathcal{N}(\mu,\sigma^2)$ population and σ^2 is known. An interval, I_1, of the form $I_1 = (\bar{X} - d, \bar{X} + d)$ (with width 2*d*) is to be constructed as a *confidence interval* for μ. If we make the assertion that μ is in this interval (i.e., μ is bounded below by $\bar{X} - d$, and bounded above by $\bar{X} + d$), then sometimes this assertion is correct and sometimes it is wrong, depending on the value of $\bar{X}$ in a given experiment. If, for a fixed α value, we would like to have a confidence probability (called confidence coefficient) such that

$$P[\mu \in I_1] = P[\bar{X} - d < \mu < \bar{X} + d] = 1 - \alpha \tag{228.14}$$

then we need to choose the value of *d* to satisfy $d = z_{\alpha/2}\sigma / \sqrt{n}$; that is,

$$I_1 = \left(\overline{X} - z_{\alpha/2}\frac{\sigma}{\sqrt{n}}, \overline{X} + z_{\alpha/2}\frac{\sigma}{\sqrt{n}}\right) \tag{228.15}$$

where $z_{\alpha/2}$ is the $(1-\alpha/2)$th percentile of the $\mathcal{N}(0,1)$ distribution such that $\Phi(z_{\alpha/2}) = 1-\alpha/2$. To see this, we note that from the sampling distribution of $\overline{X}$ (Fact 1), we have

$$\begin{aligned} P\left[\overline{X} - z_{\alpha/2}\frac{\sigma}{\sqrt{n}} < \mu < \overline{X} + z_{\alpha/2}\frac{\sigma}{\sqrt{n}}\right] &= P\left[\frac{|\overline{X}-\mu|}{\sigma/\sqrt{n}} \le z_{\alpha/2}\right] \\ &= \Phi(z_{\alpha/2}) - \Phi(-z_{\alpha/2}) = 1-\alpha. \end{aligned} \tag{228.16}$$

We further note that, even when the original population is not normal, by Theorem 2 the confidence probability is approximately (1 – α) when the sample size is reasonably large.

Confidence Interval for μ When σ² Is Unknown

Assume that the observations are from an $\mathcal{N}(\mu,\sigma^2)$ population. When σ^2 is unknown, by Fact 2 and a similar argument we see that

$$I_2 = \left(\overline{X} - t_{\alpha/2}(n-1)\frac{S}{\sqrt{n}}, \overline{X} + t_{\alpha/2}(n-1)\frac{S}{\sqrt{n}}\right) \tag{228.17}$$

is a confidence interval for μ with confidence probability $1-\alpha$, where $t_{\alpha/2}(n-1)$ is the $(1-\alpha/2)$th percentile of the $t(n-1)$ distribution.

Confidence Interval for σ²

If, under the same assumption of normality, a confidence interval for σ^2 is needed when μ is unknown, then

$$I_3 = ((n-1)S^2/\chi^2_{1-\alpha/2}(n-1), (n-1)S^2/\chi^2_{\alpha/2}(n-1)) \tag{228.18}$$

has a confidence probability $1-\alpha$, when $\chi^2_{1-\alpha/2}(n-1)$ and $\chi^2_{\alpha/2}(n-1)$ are the $(\alpha/2)$th and $(1-\alpha/2)$th percentiles, respectively, of the $\chi^2(n-1)$ distribution.

Two-Sample Case

Confidence Intervals for $\mu_1-\mu_2$ When σ_1^2 and σ_2^2 Are Known

Consider an experiment that involves the comparison of two treatments, T_1 and T_2, as indicated in Table 228.1. If a confidence interval for $\delta = \mu_1 - \mu_2$ is needed when σ_1^2 and σ_2^2 are known, then by Fact 4 and a similar argument, the confidence interval

$$I_4 = \left((\overline{X}_1 - \overline{X}_2) - z_{\alpha/2}\sqrt{\sigma_1^2/n_1 + \sigma_2^2/n_2},\ (\overline{X}_1 - \overline{X}_2) + z_{\alpha/2}\sqrt{\sigma_1^2/n_1 + \sigma_2^2/n_2}\right) \tag{228.19}$$

has a confidence probability 1 – α.

Confidence Interval for $\mu_1-\mu_2$ When σ_1^2 and σ_2^2 Are Unknown but Equal

Under the additional assumption that $\sigma_1^2 = \sigma_2^2$ but the common variance is unknown, then by Fact 6 the confidence interval

$$I_5 = [(\overline{X}_1 - \overline{X}_2) - d, (\overline{X}_1 - \overline{X}_2) + d] \tag{228.20}$$

has a confidence probability 1 – α, where

$$d = t_{\alpha/2}(n_1 + n_2 - 2) S_p (1/n_1 + 1/n_2)^{1/2} \tag{228.21}$$

Confidence Interval for σ_1^2/σ_2^2

A confidence interval for the ratio of the variances σ_2^2/σ_1^2 can be obtained from the *F* distribution (see Fact 7), and the confidence interval

$$I_6 = \left(F_{1-\alpha/2}(n_1 - 1, n_2 - 1)\frac{S_2^2}{S_1^2},\ F_{\alpha/2}(n_1 - 1, n_2 - 1)\frac{S_2^2}{S_1^2} \right) \tag{228.22}$$

has a confidence probability $1-\alpha$, where $F_{1-\alpha/2}(n_1 - 1, n_2 - 1)$ and $F_{\alpha/2}(n_1 - 1, n_2 - 1)$ are, respectively, the $(\alpha/2)$th and $(1-\alpha/2)$th percentiles of the $F(n_1 - 1, n_2 - 1)$ distribution.

228.5 Testing Statistical Hypotheses

A statistical hypothesis concerns a statement or assertion about the true value of the parameter in a given distribution. In the two-hypothesis problems, we deal with a null hypothesis and an alternative hypothesis, denoted by H_0 and H_1, respectively. A decision is to be made, based on the data of the experiment, to either accept H_0 (hence reject H_1) or reject H_0 (hence accept H_1). In such a two-action problem, there are two types of errors that we may commit: the type I error is to reject H_0 when it is true, and the type II error is to accept H_0 when it is false. As a standard practice, we do not reject H_0 unless there is significant evidence indicating that it may be false (in doing so, the burden of proof that H_0 is false is on the experimenter). Thus, we usually choose a small fixed number, α (such as 0.05 or 0.01), such that the probability of committing a type I error is at most α. With such a given α, we can then determine the region in the data space for the rejection of H_0 (called the critical region).

One-Sample Case

Suppose that $X_1, X_2, \ldots, X_n$ represent a random sample of size *n* from an $\mathcal{N}(\mu, \sigma^2)$ population, and $\overline{X}$ and S^2 are, respectively, the sample mean and sample variance.

Test for Mean

In testing

$$H_0 : \mu = \mu_0 \quad \text{vs.} \quad H_1 : \mu = \mu_1 (\mu_1 > \mu_0) \text{ or } H_1 : \mu > \mu_0$$

when σ^2 is known, we reject H_0 when $\overline{X}$ is large. To determine the cut-off point, we note that (by Fact 1) the statistic $Z_0 = (\overline{X} - \mu_0)/(\sigma/\sqrt{n})$ has an $\mathcal{N}(0,1)$ distribution under H_0. Thus if we decide to reject H_0 when $Z_0 > z_\alpha$, then the probability of committing a type I error is α. As a consequence, we apply the following decision rule:

$$d_1 : \text{reject } H_0 \text{ if and only if } \overline{X} > \mu_0 + z_\alpha \frac{\sigma}{\sqrt{n}}$$

Similarly, from the distribution of Z_0 under H_0, we can obtain the critical region for the other types of hypotheses. When σ^2 is unknown, then by Fact 2, $T_0 = \sqrt{n}(\overline{X} - \mu_0)/S$ has a $t(n-1)$ distribution

TABLE 228.2 One-Sample Tests for Mean

Null Hypothesis H_0	Alternative Hypothesis H_1	Critical Region
$\mu=\mu_0$ or $\mu\le\mu_0$	$\mu=\mu_1>\mu_0$ or $\mu>\mu_0$	$\bar{X}>\mu_0+z_\alpha\frac{\sigma}{\sqrt{n}}$ $\bar{X}>\mu_0+t_\alpha\frac{S}{\sqrt{n}}$
$\mu=\mu_0$ or $\mu\ge\mu_0$	$\mu=\mu_1<\mu_0$ or $\mu<\mu_0$	$\bar{X}<\mu_0-z_\alpha\frac{\sigma}{\sqrt{n}}$ $\bar{X}<\mu_0-t_\alpha\frac{S}{\sqrt{n}}$
$\mu=\mu_0$	$\mu\ne\mu_0$	$\lvert\bar{X}-\mu_0\rvert>z_{\alpha/2}\frac{\sigma}{\sqrt{n}}$ $\lvert\bar{X}-\mu_0\rvert>t_{\alpha/2}\frac{S}{\sqrt{n}}$

TABLE 228.3 One-Sample Tests for Variance

Null Hypothesis H_0	Alternative Hypothesis H_1	Critical Region
$\sigma^2=\sigma_0^2$ or $\sigma^2\le\sigma_0^2$	$\sigma^2=\sigma_1^2>\sigma_0^2$ or $\sigma^2>\sigma_0^2$	$(S^2/\sigma_0^2)>\frac{1}{n-1}\chi_\alpha^2$
$\sigma^2=\sigma_0^2$ or $\sigma^2\ge\sigma_0^2$	$\sigma^2=\sigma_1^2<\sigma_0^2$ or $\sigma^2<\sigma_0^2$	$(S^2/\sigma_0^2)<\frac{1}{n-1}\chi_{1-\alpha}^2$
$\sigma^2=\sigma_0^2$	$\sigma^2\ne\sigma_0^2$	$(S^2/\sigma_0^2)>\frac{1}{n-1}\chi_{\alpha/2}^2$ or $(S^2/\sigma_0^2)<\frac{1}{n-1}\chi_{1-\alpha/2}^2$

under H_0. Thus, the corresponding tests can be obtained by substituting $t_\alpha(n-1)$ for z_α and S for σ. The tests for the various one-sided and two-sided hypotheses are summarized in Table 228.2. For each set of hypotheses, the critical region given on the first line is for the case when σ^2 is known, and that given on the second line is for the case when σ^2 is unknown. Furthermore, t_α and $t_{\alpha/2}$ stand for $t_\alpha(n-1)$ and $t_{\alpha/2}(n-1)$, respectively.

Test for Variance

In testing hypotheses concerning the variance σ^2 of a normal distribution, we use Fact 3 to assert that, under $H_0:\sigma^2=\sigma_0^2$, the distribution of $w_0=(n-1)S^2/\sigma_0^2$ is $\chi^2(n-1)$. The corresponding tests and critical regions are summarized in Table 228.3 (χ_α^2 and $\chi_{\alpha/2}^2$ stand for $\chi_\alpha^2(n-1)$ and $\chi_{\alpha/2}^2(n-1)$, respectively).

Two-Sample Case

In comparing the means and variances of two normal populations, we once again refer to Table 228.1 for notation and assumptions.

Test for Difference of Two Means

Let $\delta=\mu_1-\mu_2$ be the difference of the two population means. In testing $H_0:\delta=\delta_0$ versus a one-sided or two-sided alternative hypothesis, we note that, for

TABLE 228.4 Two-Sample Tests for Difference of Two Means

Null Hypothesis H_0	Alternative Hypothesis H_1	Critical Region
$\delta=\delta$ or $\delta\le\delta_0$	$\delta=\delta_1>\delta_0$ or $\delta>\delta_0$	$(\bar{X}_1-\bar{X}_2)>\delta_0+z_\alpha\tau$ $(\bar{X}_1-\bar{X}_2)>\delta_0+t_\alpha\upsilon$
$\delta=\delta_0$ or $\delta\ge\delta_0$	$\delta=\delta_1<\delta_0$ or $\delta<\delta_0$	$(\bar{X}_1-\bar{X}_2)<\delta_0-z_\alpha\tau$ $(\bar{X}_1-\bar{X}_2)<\delta_0-t_\alpha\upsilon$
$\delta=\delta_0$	$\delta\ne\delta_0$	$\lvert(\bar{X}_1-\bar{X}_2)-\delta_0\rvert>z_{\alpha/2}\tau$ $\lvert(\bar{X}_1-\bar{X}_2)-\delta_0\rvert>t_{\alpha/2}\upsilon$

$$\tau=\left(\sigma_1^2/n_1+\sigma_2^2/n_2\right)^{1/2} \tag{228.23}$$

and

$$\nu=S_p(1/n_1+1/n_2)^{1/2} \tag{228.24}$$

$Z_0=[(\bar{X}_1-\bar{X}_2)-\delta_0]/\tau$ has an $\mathcal{N}(0,1)$ distribution under H_0, and $T_0=[(\bar{X}_1-\bar{X}_2)-\delta_0]/\upsilon$ has a $t(n_1+n_2-2)$ distribution under H_0 when $\sigma_1^2=\sigma_2^2$. Using these results, the corresponding critical regions for one-sided and two-sided tests can be obtained, and they are listed in Table 228.4. Note that, as in the one-sample case, the critical region given on the first line for each set of hypotheses is for the case of known variances, and that given on the second line is for the case in which the variances are equal but unknown. Further, t_α and $t_{\alpha/2}$ stand for $t_\alpha(n_1+n_2-2)$ and $t_{\alpha/2}(n_1+n_2-2)$, respectively.

228.6 A Numerical Example

In the following, we provide a numerical example for illustrating the construction of confidence intervals and hypotheses-testing procedures. The example is given along the line of applications in Wadsworth [1990:4.21] with artificial data.

Suppose that two processes, T_1 and T_2, for manufacturing steel pins are in operation, and that a random sample of four pins (of five pins) was taken from the process T_1 (the process T_2) with the following results (in units of inches):

T_1: 0.7608, 0.7596, 0.7622, 0.7638

T_2: 0.7546, 0.7561, 0.7526, 0.7572, 0.7565

Simple calculation shows that the observed values of sample means, sample variances, and sample standard deviations are

$$\bar{X}_1=0.7616,\quad S_1^2=3.178914\cdot 10^{-6},\quad S_1=1.7830\cdot 10^{-3}$$

$$\bar{X}_2=0.7554,\quad S_2^2=3.516674\cdot 10^{-6},\quad S_2=1.8753\cdot 10^{-3}$$

One-Sample Case

Let us first consider confidence intervals for the parameters of the first process, T_1, only.

1. Assume that, based on previous knowledge on processes of this type, the variance is known to be $\sigma_1^2 = 1.80^2 \cdot 10^{-6} (\sigma_1 = 0.0018)$. Then, from the normal table [Ross, 1987:482], we have $z_{0.025} = 1.96$. Thus, a 95% confidence interval for μ_1 is

$$(0.7616 - 1.96 \times 0.0018 / \sqrt{4}, \quad 0.7616 + 1.96 \times 0.0018 / \sqrt{4})$$

or (0.7598, 0.7634) (after rounding off to the fourth decimal place).
2. If σ_1^2 is unknown and a 95% confidence interval for μ_1 is needed, then, for $t_{0.025}(3) = 3.182$ [Ross, 1987:484], the confidence interval is

$$(0.7616 - 3.182 \times 0.001783 / \sqrt{4}, \quad 0.7616 + 3.182 \times 0.001783 / \sqrt{4})$$

or (0.7588, 0.7644).
3. From the chi-squared table with 4 – 1 = 3 degrees of freedom, we have [Ross, 1987:483] $\chi^2_{0.975} = 0.216$, $\chi^2_{0.025} = 9.348$. Thus, a 95% confidence interval for σ_1^2 is $(3 \times 3.178\,914 \cdot 10^{-6}/9.348, 3.178\,914 \cdot 10^{-6}/0.216)$, or $(1.0202 \cdot 10^{-6}, 44.151\,58 \cdot 10^{-6})$.
4. In testing the hypotheses

$$H_0 : \mu_1 = 0.76 \quad \text{vs} \quad H_1 : \mu_1 > 0.76$$

with $\alpha = 0.01$ when σ_1^2 is unknown, the critical region is $\bar{x}_1 > 0.76 + 4.541 \times 0.001783 / \sqrt{4} = 0.7640$. Because the observed value of $\bar{x}_1$ is 0.7616, H_0 is accepted. That is, we assert that there is no significant evidence to call for the rejection of H_0.

Two-Sample Case

If we assume that the two populations have a common unknown variance, we can use the Student's t distribution (with degree of freedom $\nu = 4 + 5 - 2 = 7$) to obtain confidence intervals and to test hypotheses for $\mu_1 - \mu_2$. We first note that the data given above yield

$$S_p^2 = \frac{1}{7}(3 \times 3.178414 + 4 \times 3.516674) \cdot 10^{-6}$$

$$= 3.371920 \cdot 10^{-6},$$

$$S_p = 1.836279 \cdot 10^{-3}, \quad \upsilon = S_p \sqrt{1/4 + 1/5} = 1.231813 \cdot 10^{-3}$$

and

$$\bar{X}_1 - \bar{X}_2 = 0.0062$$

1. A 98% confidence interval for $\mu_1 - \mu_2$ is $(0.0062 - 2.998 \cdot, \upsilon, 0.0062 + 2.998\, \upsilon)$ or (0.0025, 0.0099).
2. In testing the hypotheses $H_0 : \mu_1 = \mu_2$ (i.e., $\mu_1 - \mu_2 = 0$) versus $H_1 : \mu_1 > \mu_2$ with $\alpha = 0.05$, the critical region is $(\bar{X}_1 - \bar{X}_2) > 1.895\upsilon = 2.3344 \cdot 10^{-3}$). Thus, H_0 is rejected (i.e., we concluded that there is significant evidence to indicate that $\mu_1 > \mu_2$ may be true).
3. In testing the hypotheses $H_0 : \mu_1 = \mu_2$ vs $H_1 : \mu_1 \neq \mu_2$ with $\alpha = 0.02$, the critical region is $|X_1 - X_2| > 2.998\upsilon = 3.6930 \cdot 10^{-3}$. Thus, H_0 is rejected. We note this conclusion is consistent with the result that, with confidence probability $1 - \alpha = 0.98$, the confidence interval for $(\mu_1 - \mu_2)$ does not contain the origin.

References

Bowker, A. H. and Lieberman, G. J. 1972. *Engineering Statistics,* 2nd ed. Prentice Hall, Englewood Cliffs, NJ.

Box, G. E. P., Hunter, W. G., and Hunter, J. S. 1978. *Statistics for Experimenters.* John Wiley & Sons, New York.

Hahn, G. J. and Shapiro, S. S. 1967. *Statistical Models in Engineering.* John Wiley & Sons, New York.

Hines, W. W. and Montgomery, D. G. 1980. *Probability and Statistics in Engineering and Management Science.* John Wiley & Sons, New York.

Hogg, R. V. and Ledolter, J. 1992. *Engineering Statistics.* Macmillan, New York.

Ross, S. M. 1987. *Introduction to Probability and Statistics for Engineers and Scientists.* John Wiley & Sons, New York.

Wadsworth, H. M. (Ed.) 1990. *Handbook of Statistical Methods for Engineers and Scientists.* McGraw-Hill, New York.

Further Information

Other important topics in engineering probability and statistics include sampling inspection and quality (process) control, reliability, regression analysis and prediction, design of engineering experiments, and analysis of variance. Due to space limitations, these topics are not treated in this chapter. The reader is referred to textbooks in this area for further information. There are many well-written books that cover most of these topics. The short list of references above consists of a small sample of them.

229
Optimization

George Cain
Georgia Institute of Technology

229.1 Linear Programming

Let **A** be an $m \times n$ matrix, **b** a column vector with m components, and **c** a column vector with n components. Suppose $m < n$, and assume the rank of **A** is m. The standard linear programming problem is to find, among all nonnegative solutions of $\mathbf{Ax} = \mathbf{b}$, one that minimizes

$$\mathbf{c}^{\mathrm{T}}\mathbf{x} = c_1x_1 + c_2x_2 + \cdots + c_nx_n$$

This problem is called a *linear* program. Each solution of the system $\mathbf{Ax} = \mathbf{b}$ is called a *feasible* solution, and the *feasible set* is the collection of all *feasible solutions.* The function $\mathbf{c}^{\mathrm{T}}\mathbf{x} = c_1x_1 + c_2x_2 + \cdots + c_nx_n$ is the *cost function*, or the *objective function.* A solution to the linear program is called an *optimal feasible solution.*

Let **B** be an $m \times m$ submatrix of **A** made up of m linearly independent columns of **A**, and let **C** be the $m \times (n - m)$ matrix made up of the remaining columns of **A**. Let $\mathbf{x}_{\mathrm{B}}$ be the vector consisting of the components of **x** corresponding to the columns of **A** that make up **B**, and let $\mathbf{x}_{\mathrm{C}}$ be the vector of the remaining components of **x**, that is, the components of **x** that correspond to the columns of **C**. Then the equation $\mathbf{Ax} = \mathbf{b}$ may be written $\mathbf{Bx}_{\mathrm{B}} + \mathbf{Cx}_{\mathrm{C}} = \mathbf{b}$. A solution of $\mathbf{Bx}_{\mathrm{B}} = \mathbf{b}$ together with $\mathbf{x}_{\mathrm{C}} = \mathbf{0}$ gives a solution **x** of the system $\mathbf{Ax} = \mathbf{b}$. Such a solution is called a *basic solution*, and if it is, in addition, nonnegative, it is a *basic feasible solution.* If it is also optimal, it is an *optimal basic feasible solution.* The components of a basic solution are called *basic variables.*

The Fundamental Theorem of Linear Programming says that if there is a feasible solution, there is a basic feasible solution, and if there is an optimal feasible solution, there is an optimal basic feasible solution. The linear programming problem is thus reduced to searching among the set of basic solutions for an optimal solution. This set is, of course, finite, containing as many as $n!/[m!(n-m)!]$ points. In practice, this will be a very large number, making it imperative that one use some efficient search procedure in seeking an optimal solution. The most important of such procedures is the *simplex method*, details of which may be found in the references.

The problem of finding a solution of $\mathbf{Ax} \leq \mathbf{b}$ that minimizes $\mathbf{c}^{\mathrm{T}}\mathbf{x}$ can be reduced to the standard problem by appending to the vector **x** an additional m nonnegative components, called *slack variables.* The vector **x** is replaced by **z**, where $\mathbf{z}^{\mathrm{T}} = [x_1x_2 \ldots x_n \quad s_1s_2 \ldots s_m]$, and the matrix **A** is replaced by $\mathbf{B} = [\mathbf{A}\ \mathbf{I}]$, where **I** is the $m \times m$ identity matrix. The equation $\mathbf{Ax} = \mathbf{b}$ is thus replaced by $\mathbf{Bz} = \mathbf{Ax} + \mathbf{s} = \mathbf{b}$, where

0-8493-1586-7/05/$0.00+$1.50

$\mathbf{s}^T = [s_1 s_2 \ldots s_m]$. Similarly, if inequalities are reversed so that we have $\mathbf{Ax} \le \mathbf{b}$, we simply append $-\mathbf{s}$ to the vector $\mathbf{x}$. In this case, the additional variables are called *surplus variables.*

Associated with every linear programming problem is a corresponding dual problem. If the *primal* problem is to minimize $\mathbf{c}^T\mathbf{x}$ subject to $\mathbf{Ax} \ge \mathbf{b}$, and $\mathbf{x} \ge 0$, the corresponding *dual* problem is to maximize $\mathbf{y}^T\mathbf{b}$ subject to $\mathbf{y}^T\mathbf{A} \le \mathbf{c}^T$. If either the primal problem or the dual problem has an optimal solution, so also does the other. Moreover, if $\mathbf{x_p}$ is an optimal solution for the primal problem and $\mathbf{y_d}$ is an optimal solution for the corresponding dual problem, $\mathbf{c}^T\mathbf{x_p} = \mathbf{y}_d^T\mathbf{b}$.

229.2 Unconstrained Nonlinear Programming

The problem of minimizing or maximizing a sufficiently smooth nonlinear function $f(\mathbf{x})$ of n variables, $\mathbf{x}^T = [x_1 x_2 \ldots x_n]$, with no restrictions on $\mathbf{x}$ is essentially an ordinary problem in calculus. At a minimizer or maximizer $\mathbf{x}^*$, it must be true that the gradient of f vanishes:

$$\nabla f(\mathbf{x}^*) = 0$$

Thus, $\mathbf{x}^*$ will be in the set of all solutions of this system of n generally nonlinear equations. The solution of the system can be, of course, a nontrivial undertaking. There are many recipes for solving systems of nonlinear equations. A method specifically designed for minimizing f is the *method of steepest descent.* It is an old and honorable algorithm, and the one on which most other more complicated algorithms for unconstrained optimization are based. The method is based on the fact that at any point $\mathbf{x}$, the direction of maximum decrease of f is in the direction of $-\nabla f(\mathbf{x})$. The algorithm searches in this direction for a minimum, recomputes $-\nabla f(\mathbf{x})$ at this point, and continues iteratively. Explicitly:

1. Choose an initial point $\mathbf{x}_0$.
2. Assume $\mathbf{x}_k$ has been computed; then compute $\mathbf{y}_k = \nabla f(\mathbf{x}_k)$, and let $t_k \ge 0$ be a local minimum of $g(t) = f(\mathbf{x}_k - t\mathbf{y}_k)$. Then $\mathbf{x}_{k+1} = \mathbf{x}_k - t_k\mathbf{y}_k$.
3. Replace k by $k + 1$, and repeat step 2 until t_k is small enough.

Under reasonably general conditions, the sequence $(\mathbf{x}_k)$ converges to a minimum of f.

229.3 Constrained Nonlinear Programming

The problem of finding the maximum or minimum of a function $f(\mathbf{x})$ of n variables subject to the constraints

$$\mathbf{a}(\mathbf{x}) = \begin{bmatrix} a_1(x_1, x_2, \ldots, x_n) \\ a_2(x_1, x_2, \ldots, x_n) \\ \vdots \\ a_m(x_1, x_2, \ldots, x_n) \end{bmatrix} = \begin{bmatrix} b_1 \\ b \\ \vdots \\ b_m \end{bmatrix} = \mathbf{b}$$

is made into an unconstrained problem by introducing the new function $L(\mathbf{x})$:

$$L(\mathbf{x}) = f(\mathbf{x}) + \mathbf{z}^T\mathbf{a}(\mathbf{x})$$

where $\mathbf{z}^T = [\lambda_1 \lambda_2 \ldots \lambda_m]$ is the vector of *Lagrange multipliers.* Now the requirement that $\nabla L(\mathbf{x}) = \mathbf{0}$, together with the constraints $\mathbf{a}(\mathbf{x}) = \mathbf{b}$, give a system of $n + m$ equations

$$\nabla f(\mathbf{x}) + \mathbf{z}^{\mathrm{T}}\nabla\mathbf{a}(\mathbf{x}) = \mathbf{0}$$

$$\mathbf{a}(\mathbf{x}) = \mathbf{b}$$

for the $n + m$ unknowns $x_1, x_2, \ldots, \lambda_1 \lambda_2 \ldots \lambda_m$ that must be satisfied by the minimizer (or maximizer) **x**.

The problem of inequality constraints is significantly more complicated in the nonlinear case than in the linear case. Consider the problem of minimizing $f(x)$ subject to m equality constraints $\mathbf{a}(\mathbf{x}) = \mathbf{b}$, and p inequality constraints $\mathbf{c}(\mathbf{x}) \le \mathbf{d}$ (thus, $\mathbf{a}(\mathbf{x})$ and $\mathbf{b}$ are vectors of m components, and $\mathbf{c}(\mathbf{x})$ and $\mathbf{d}$ are vectors of p components). A point $\mathbf{x}^*$ that satisfies the constraints is a *regular point* if the collection

$$\{\nabla a_1(\mathbf{x}^*),\ \nabla a_2(\mathbf{x}^*), \ldots, \nabla a_m(\mathbf{x}^*)\} \cup \{\nabla c_j(\mathbf{x}^*) : j \in J\}$$

where

$$J = \{j : c_j(\mathbf{x}^*) = d_j\}$$

is linearly independent. If $\mathbf{x}^*$ is a local minimum for the constrained problem and if it is a regular point, there is a vector $\mathbf{z}$ with m components and a vector $\mathbf{w} \ge \mathbf{0}$ with p components such that

$$\nabla f(\mathbf{x}^*) + \mathbf{z}^{\mathrm{T}}\nabla\mathbf{a}(\mathbf{x}^*) + \mathbf{w}^{\mathrm{T}}\mathbf{D}\mathbf{c}(\mathbf{x}^*) = \mathbf{0}$$

$$\mathbf{w}^{\mathrm{T}}(\mathbf{c}(\mathbf{x}^*) - \mathbf{d}) = 0$$

These are the *Kuhn-Tucker conditions.* Note that in order to solve these equations, one needs to know for which j it is true that $c_j(\mathbf{x}^*) = 0$. (Such a constraint is said to be *active.*)

References

Luenberger, D. C. 1984. *Linear and Nonlinear Programming*, 2nd ed. Addison-Wesley, Reading, MA.

Peressini, A. L., Sullivan, F. E., and Uhl, J. J., Jr. 1988. *The Mathematics of Nonlinear Programming.* Springer-Verlag, New York.

230

Numerical Methods

William F. Ames
Georgia Institute of Technology

0-8493-1586-7/05/$0.00+$1.50

Since many mathematical models of physical phenomena are not solvable by available mathematical methods, one must often resort to approximate or numerical methods. These procedures do not yield exact results in the mathematical sense. This inexact nature of numerical results means we must pay attention to the errors. The two errors that concern us here are *round-off errors* and *truncation errors.*

Round-off errors arise as a consequence of using a number specified by m correct digits to approximate a number that requires more than m digits for its exact specification. An example here is using 3.14159 to approximate the irrational number π. Such errors may be especially serious in matrix inversion or in any area where a very large number of numerical operations are required. Certain attempts at handling these errors are called *enclosure methods* [Adams and Kulisch, 1993].

Truncation errors arise from the substitution of a finite number of steps for an infinite sequence of steps (usually an iteration) that would yield the exact result. For example, the iteration $y_n(x) = 1 + \int_0^x xty_{n-1}(t)dt$, $y(0) = 1$ is only carried out for a *few steps*, but it converges in *infinitely* many steps.

The study of some errors in a computation is related to the theory of probability. In what follows, a relation for the error will be given in certain instances.

230.1 Linear Algebra Equations

A problem often encountered is the determination of the solution vector $u = (u_1, u_2, \ldots, u_n)^T$ for the set of linear equations $Au = v$, where A is the $n \times n$ square matrix with coefficients a_{ij} $(i,j = 1,\ldots,n)$ and $v = (v_1,\ldots,v_n)^T$, and i denotes the row index and j the column index. There are many numerical methods for finding the solution, u, of $Au = v$. The direct inversion of A is usually too expensive and is not often carried out unless it is needed elsewhere. Only a few methods will be listed. You can check the literature for the many methods and computer software available. Some of the software is briefly described at the end of this chapter. The methods are usually subdivided into *direct* (once through) or *iterative* (repeated) procedures.

In the following, it will often be convenient to partition the matrix A into the form $A = U + D + L$, where U, D, and L are matrices having the same elements as A, respectively, above the main diagonal, on the main diagonal, and below the main diagonal, and zeros elsewhere. Thus,

$$U = \begin{bmatrix} 0 & a_{12} & \cdots & & a_{1n} \\ 0 & 0 & a_{23} & \cdots & a_{2n} \\ \vdots & \cdots & & \cdots & \\ 0 & 0 & \cdots & \cdots & 0 \end{bmatrix}$$

We also assume the u_j's are not all zero and that $\det A \neq 0$, so the solution is unique.

Direct Methods

Gauss Reduction

This classical method has spawned many variations. It consists of dividing the first equation by a_{11} (if $a_{11} = 0$, reorder the equations to find an $a_{11} \neq 0$) and using the result to eliminate the terms in u_1 from each of the succeeding equations. Next, the modified second equation is divided by a'_{22} (if $a'_{22} = 0$, a reordering of the modified equations may be necessary) and the resulting equation is used to eliminate all terms in u_2 in the succeeding modified equations. This elimination is done n times resulting in a triangular system:

$$\begin{aligned}
u_1 + a'_{12}u_2 \quad &+\cdots \quad +a'_{1n}u_n &= v'_1\\
0 + u_2 \quad &+\cdots \quad +a'_{2n}u_n &= v'_2\\
&\cdots\\
0 + \cdots \quad &+u_{n-1} \quad +a'_{n-1,n}u_n &= v'_{n-1}\\
&u_n &= v'_n
\end{aligned}$$

where a'_{ij} and v'_j represent the specific numerical values obtained by this process. The solution is obtained by working backward from the last equation. Various modifications, such as the Gauss–Jordan reduction, the Gauss–Doolittle reduction, and the Crout reduction, are described in the classical reference authored by Bodewig [1956]. Direct methods prove very useful for sparse matrices and banded matrices that often arise in numerical calculation for differential equations. Many of these are available in computer packages such as IMSL, Maple, Matlab, and Mathematica.

The Tridiagonal Algorithm

When the linear equations are tridiagonal, the system

$$\begin{aligned}
b_1u_1 \quad +c_1u_2 \quad &= d_1\\
a_iu_{i-1} \quad +b_iu_i \quad +c_iu_{i+1} &= d_i\\
a_nu_{n-1} \quad +b_nu_n \quad &= d_n, \quad i = 2,3,\ldots,n-1
\end{aligned}$$

can be solved explicitly for the unknowns, thereby eliminating any matrix operations.

The Gaussian elimination process transforms the system into a simpler one of *upper bidiagonal* form. We designate the coefficients of this new system by a'_i, b'_i, c'_i, and d'_i, and we note that

$$\begin{aligned}
a'_i &= 0, \quad i = 2,3,\ldots,n\\
b'_i &= 1, \quad i = 1,2,\ldots,n
\end{aligned}$$

The coefficients c'_i and d'_i are calculated successively from the relations

$$\begin{aligned}
c'_1 &= \frac{c_1}{b_1} \qquad d'_1 = \frac{d_1}{b_1}\\
c'_{i+1} &= \frac{c_{i+1}}{b_{i+1} - a_{i+1}c'_i}\\
d'_{i+1} &= \frac{d_{i+1} - a_{i+1}d'_i}{b_{i+1} - a_{i+1}c'_i}, \quad i = 1,2,\ldots,n-1
\end{aligned}$$

and, of course, $c_n = 0$.

Having completed the elimination, we examine the new system and see that the nth equation is now

$$u_n = d'_n$$

Substituting this value into the $(n - 1)$st equation,

$$u_{n-1} + c'_{n-1}u_n = d'_{n-1}$$

we have

$$u_{n-1} = d'_{n-1} - c'_{n-1} u_n$$

Thus, starting with u_n, we have successively the solution for u_i as

$$u_i = d'_i - c'_i u_{i+1}, \qquad i = n-1, n-2, \ldots, 1$$

Algorithm for Pentadiagonal Matrix

The equations to be solved are

$$a_i u_{i-2} + b_i u_{i-2} + c_i u_i + d_i u_i + e_i u_{i+2} = f_i$$

for $1 \le i \le R$ with $a_1 = b_1 = a_2 = e_{R-1} = d_R = e_R = 0$.

The algorithm is as follows. First, compute

$$\delta_1 = d_1 / c_1$$
$$\lambda_1 = e_1 / c_1$$
$$\gamma_1 = f_1 / c_1$$

and

$$\mu_2 = c_2 - b_2 \delta_1$$
$$\delta_2 = (d_2 - b_2 \lambda_1) / \mu_2$$
$$\lambda_2 = e_2 / \mu_2$$
$$\gamma_2 = (f - b_2 \gamma_1) / \mu_2$$

Then, for $3 \le i \le (R-2)$, compute

$$\beta_i = b_i - a_i \delta_{i-2}$$
$$\mu_i = c_i - \beta_i \delta_{i-1} - a_i \lambda_{i-2}$$
$$\delta_i = (d_i - \beta_i \lambda_{i-1}) / \mu_i$$
$$\lambda_i = e_i / \mu_i$$
$$\gamma_i = (f_i - \beta_i \gamma_{i-1} - a_i \gamma_{i-2}) / \mu_i$$

Next, compute

$$\beta_{R-1} = b_{R-1} - a_{R-1} \delta_{R-3}$$
$$\mu_{R-1} = c_{R-1} - \beta_{R-1} \delta_{R-2} - a_{R-1} \lambda_{R-3}$$
$$\delta_{R-1} = (d_{R-1} - \beta_{R-1} \lambda_{R-2}) / \mu_{R-1}$$
$$\gamma_{R-1} = (f_{R-1} - \beta_{R-1} \gamma_{R-2} - a_{R-1} \gamma_{R-3}) / \mu_{R-1}$$

and

$$\beta_R = b_R - a_R \delta_{R-2}$$
$$\mu_R = c_R - \beta_R \delta_{R-1} - a_R \lambda_{R-2}$$
$$\gamma_R = (f_R - \beta_R \gamma_{R-1} - a_R \gamma_{R-2}) / \mu_R$$

The β_i and μ_i are used only to compute δ_i, λ_i, and γ_i, and need not be stored after they are computed. The δ_i, λ_i, and γ_i must be stored, as they are used in the back solution. This is

$$u_R = \gamma_R$$

$$u_{R-1} = \gamma_{R-1} - \delta_{R-1} u_R$$

and

$$u_i = \gamma_i - \delta_i u_{i+1} - \lambda_i u_{i+2}$$

for $R - 2 \geq i \geq 1$.

General Band Algorithm

The equations are of the form

$$A_j^{(M)} X_{j-M} + A_j^{(M-1)} X_{j-M+1} + \cdots + A_J^{(2)} X_{J-2} + A_J^{(1)} X_{J-1} + B_j X_j$$
$$+ C_j^{(1)} X_{j+1} + C_j^{(2)} X_{j+2} + \cdots + C_j^{(M-1)} X_{j+M-1} + C_j^{(M)} X_{j+M} = D_j$$

for $1 \leq j \leq N$, $N \geq M$. The algorithm used is as follows:

$$\alpha_j^{(k)} = A_j^{(k)} = 0, \qquad \text{for } k \geq j$$
$$C_j^{(k)} = 0, \qquad \text{for k} \geq N + 1 - j$$

The forward solution ($j = 1,\ldots,N$) is

$$\alpha_j^{(k)} = A_j^{(k)} - \sum_{p=k+1}^{p=M} \alpha_j^{(p)} W_{j-p}^{(p-k)}, \qquad k = M,\ldots,1$$

$$\beta_j = B_j - \sum_{p=1}^{M} \alpha_j^{(p)} W_{j-p}^{(p)}$$

$$W_j^{(k)} = \left(C_j^{(k)} - \sum_{p=k+1}^{p=M} \alpha_j^{(p-k)} W_{j-(p-k)}^{(p)} \right) / \beta_j, \qquad k = 1,\ldots,M$$

$$\gamma_j = \left(D_j - \sum_{p=1}^{M} \alpha_j^{(p)} \gamma_{j-p} \right) / \beta_j$$

The back solution ($j = N,\ldots,1$) is

$$X_j = \gamma_j - \sum_{p=1}^{M} W_j^{(p)} X_{j+p}$$

Cholesky Decomposition

When the matrix A is a symmetric and positive definite, as it is for many discretizations of self-adjoint positive definite boundary value problems, one can improve considerably on the band procedures by using the Cholesky decomposition. For the system $Au = v$, the matrix A can be written in the form

$$A = (I + L)\,D(I + U)$$

where L is lower triangular, U is upper triangular, and D is diagonal. If $A = A'$ (A' represents the transpose of A), then

$$A = A' = (I + U)'D(I + L)'$$

Hence, because of the uniqueness of the decomposition,

$$I + L = (I + U)' = I + U'$$

and therefore,

$$A = (I + U)'D(I + U)$$

that is,

$$A = B'B, \text{ where } B = \sqrt{D}\,(I + U)$$

The system $Au = v$ is then solved by solving the two triangular system

$$B'w = v$$

followed by

$$Bu = w$$

To carry out the decomposition $A = B'B$, all elements of the first row of A, and of the derived system, are divided by the square root of the (positive) leading coefficient. This yields smaller rounding errors than the banded methods because the relative error of $\sqrt{a}$ is only half as large as that of a itself. Also, taking the square root brings numbers nearer to each other (i.e., the new coefficients do not differ as widely as the original ones do). The actual computation of $B = (b_{ij})$, $j > i$, is given in the following:

$$\begin{aligned}
b_{11} &= (a_{11})^{1/2}, & b_{1j} &= a_{ij}/b_{11}, \quad j \geq 2\\
b_{22} &= (a_{22} - b_{12}^2)^{1/2}, & b_{2j} &= (a_{2j} - b_{12}b_{1j})/b_{22}\\
b_{33} &= (a_{33} - b_{13}^2 - b_{23}^2)^{1/2}, & b_{3j} &= (a_{3j} - b_{13}b_{1j} - b_{23}b_{2j})/b_{33}\\
&\vdots\\
b_{ii} &= \left(a_{ii} - \sum_{k=1}^{i-1} b_{ki}^2\right)^{1/2}, & b_{ij} &= \left(a_{ij} - \sum_{k=1}^{i-1} b_{ki}b_{kj}\right)/b_{ii}, \quad i \geq 2, j \geq 2
\end{aligned}$$

Iterative Methods

Iterative methods consist of repeated application of an often simple algorithm. They yield the exact answer only as the limit of a sequence. They can be programmed to take care of zeros in A and are self-correcting. Their structure permits the use of convergence accelerators, such as overrelaxation, Aitkins acceleration, or Chebyshev acceleration.

Let $a_{ii} > 0$ for all i and $\det A \neq 0$. With $A = U + D + L$ as previously described, several iteration methods are described for $(U + D + L)u = v$.

Jacobi Method (Iteration by Total Steps)

Since $u = -D^{-1}[U + L]u + D^{-1}v$, the iteration $u^{(k)}$ is $u^{(k)} = -D^{-1}[U + L]u^{(k-1)} + D^{-1}v$. This procedure has a slow convergent rate designated by R, $0 < R \ll 1$.

Gauss–Seidel Method (Iteration by Single Steps)

$u^{(k)} = -(L + D)^{-1} Uu^{(k-1)} + (L + D)^{-1}v$. The convergence rate is $2R$, twice as fast as that of the Jacobi method.

Gauss–Seidel with Successive Overrelaxation (SOR)

Let $\overline{u}_i^{(k)}$ be the ith components of the Gauss–Seidel iteration. The SOR technique is defined by

$$u_i^{(k)} = (1-\omega)u_i^{(k-1)} + \omega\overline{u}_i^{(k)}$$

where $1 < \omega < 2$ is the overrelaxation parameter. The full iteration is $u^{(k)} = (D + \omega L)^{-1}\{[(1 - \omega)D - \omega U]u^{(k-1)} + \omega v\}$. Optimal values of ω can be computed and depend upon the properties of A [Ames, 1993]. With optimal values of ω, the convergence rate of this method is 2R which is much larger than that for Gauss–Seidel (R is usually much less than one).

For other acceleration techniques, see the literature [Ames, 1993].

230.2 Nonlinear Equations in One Variable

Special Methods for Polynomials

The polynomial $P(x) = a_0x^n + a_1xn^{n-1} + \cdots + a_{n-1} + a_n = 0$, with real coefficients $a_j, j = 0,\ldots,n$, has exactly n roots which may be real or complex.

If all the coefficients of $P(x)$ are integers, then any rational roots, say r/s (r and s are integers with no common factors), of $P(x) = 0$ must be such that r is an integral divisor of a_n and s is an integral division of a_0. Any polynomial with rational coefficients may be converted into one with integral coefficients by multiplying the polynomial by the lowest common multiple of the denominators of the coefficients.

Example

$x^4 - 5x^2/3 + x/5 + 3 = 0$. The lowest common multiple of the denominators is 15. Multiplying by 15, which does not change the roots, gives $15x^4 - 25x^2 + 3x + 45 = 0$. The only possible rational roots r/s are such that r may have the value $\pm 45, \pm 15, \pm 5, \pm 3$, and ± 1, while s may have the values $\pm 15, \pm 5, \pm 3$, and ± 1. All possible rational roots, with no common factors, are formed using all possible quotients.

If $a_0 > 0$, the first negative coefficient is preceded by k coefficients which are positive or zero, and G is the largest of the absolute values of the negative coefficients, then each real root is less than $1 + \sqrt[k]{G/a_0}$ (upper bound on the real roots). For a lower bound to the real roots, apply the criterion to $P(-x) = 0$.

Example

$P(x) = x^5 + 3x^4 - 2x^3 - 12x + 2 = 0$. Here $a_0 = 1$, $G = 12$, and $k = 2$. Thus, the upper bound for the real roots is $1 + \sqrt[2]{12} \approx 4.464$. For the lower bound, $P(-x) = -x^5 + 3x^4 + 2x^3 + 12x + 2 = 0$, which is equivalent to $x^5 - 3x^4 - 2x^3 - 12x - 2 = 0$. Here $k = 1$, $G = 12$, and $a_0 = 1$. A lower bound is $-(1 + 12) = 13$. Hence all real roots lie in $-13 < x < 1 + \sqrt[2]{12}$.

A useful *Descartes rule of signs* for the number of positive or negative real roots is available by observation for polynomials with real coefficients. The number of positive real roots is either equal to the number of sign changes, n, or is less than n by a positive *even* integer. The number of negative real roots is either equal to the number of sign changes, n, of $P(-x)$, or is less than n by a positive even integer.

Example

$P(x) = x^5 - 3x^3 - 2x^2 + x - 1 = 0$. There are three sign changes, so $P(x)$ has either three or one positive roots. Since $P(-x) = -x^5 + 3x^3 - 2x^2 - x - 1 = 0$, there are either two or zero negative roots.

The Graeffe Root-Squaring Technique

This is an iterative method for finding the roots of the algebraic equation

$$f(x) = a_0x^p + a_1x^{p-1} + \cdots + a_{p-1}x + a_p = 0$$

If the roots are $r_1, r_2, r_3, \ldots$, then one can write

$$S_p = r_1^p\left(1 + \frac{r_2^p}{r_1^p} + \frac{r_3^p}{r_1^p} + \cdots\right)$$

and if one root is larger than all the others, say r_1, then for large enough p all terms (other than 1) would become negligible. Thus,

$$S_p \approx r_1^p$$

or

$$\lim_{p\to\infty} S_p^{1/p} = r_1$$

The Graeffe procedure provides an efficient way for computing S_p via a sequence of equations such that the roots of each equation are the squares of the roots of the preceding equations in the sequence. This serves the purpose of ultimately obtaining an equation whose roots are so widely separated in magnitude that they may be read approximately from the equation by inspection. The basic procedure is illustrated for a polynomial of degree 4:

$$f(x) = a_0x^4 + a_1x^3 + a_2x^2 + a_3x + a_4 = 0$$

Rewrite this as

$$a_0x^4 + a_2x^2 + a_4 = -a_1x^3 - a_3x$$

and square both sides so that upon grouping

$$a_0^2x^8 + (2a_0a_2 - a_1^2)x^6 + (2a_0a_4 - 2a_1a_3 + a_2^2)x^4 + (2a_2a_4 - a_3^2)x^2 + a_4^2 = 0$$

Because this involves only even powers of x, we may set $y = x^2$ and rewrite it as

$$a_0^2y^4 + (2a_0a_2 - a_1^2)y^3 + (2a_0a_4 - 2a_1a_3 + a_2^2)y^2 + (2a_2a_4 - a_3^2)y + a_4^2 = 0$$

whose roots are the squares of the original equation. If we repeat this process again, the new equation has roots which are the fourth power, and so on. After p such operations, the roots are 2^p (original roots). If at any stage we write the coefficients of the unknown in sequence

$$a_0^{(p)} \quad a_1^{(p)} \quad a_2^{(p)} \quad a_3^{(p)} \quad a_4^{(p)}$$

then, to get the new sequence $a_i^{(p+1)}$, write $a_i^{(p+1)} = 2a_0^{(p)}$ (times the symmetric coefficient) with respect to $a_i^{(p} - 2a_1^{(p)}$ (times the symmetric coefficient) $- \cdots (-1)^i a_i^{(p)2}$. Now if the roots are r_1, r_2, r_3, and r_4, then $a_1/a_0 = -\sum_{i-1}^{4} r_i, a_i^{(1)} / a_0^{(1)} = -\sum r_i^2, \ldots, a_1^{(p)} a_0^{(p)} = -\sum r_i^{2p}$. If the roots are all distinct and r_1 is the largest in magnitude, then eventually

$$r_1^{2p} \approx -\frac{a_1^{(p)}}{a_0^{(p)}}$$

And, if r_2 is the next largest in magnitude, then

$$r_2^{2p} \approx -\frac{a_2^{(p)}}{a_1^{(p)}}$$

And, in general, $a_n^{(p)} / a_{n-1}^{(p)} \approx -r_n^{2p}$. This procedure is easily generalized to polynomials of arbitrary degree and specialized to the case of multiple and complex roots.

Other methods include Bernoulli iteration, Bairstow iteration, and Lin iteration. These may be found in the cited literature. In addition, the methods given below may be used for the numerical solution of polynomials.

230.3 General Methods for Nonlinear Equations in One Variable

Successive Substitutions

Let $f(x) = 0$ be the nonlinear equation to be solved. If this is rewritten as $x = F(x)$, then an iterative scheme can be set up in the form $x_{k+1} = F(x_k)$. To start the iteration, an initial guess must be obtained graphically or otherwise. The convergence or divergence of the procedure depends upon the method of writing $x = F(x)$, of which there will usually be several forms. A general rule to ensure convergence cannot be given. However, if a is a root of $f(x) = 0$, a necessary condition for convergence is that $|F'(x)| < 1$ in that interval about a in which the iteration proceeds (this means the iteration cannot converge unless $|F'(x)| < 1$, but it does not ensure convergence). This process is called *first order* because the error in x_{k++} is proportional to the first power of the error in x_k.

Example

$f(x) = x^3 - x - 1 = 0$. A rough plot shows a real root of approximately 1.3. The equation can be written in the form $x = F(x)$ in several ways, such as $x = x^3 - 1$, $x = 1/(x^2 - 1)$, and $x = (1 + x)^{1/3}$. In the first case, $F'(x) = 3x^2 = 5.07$ at $x = 1.3$; in the second, $F(1.3) = 5.46$; only in the third case is $F'(1.3) < 1$. Hence, only the third iterative process has a chance to converge. This is illustrated in the iteration table below.

Step k	$x = \frac{1}{x^2 - 1}$	$x = x^3 - 1$	$x = (1 + x)^{1/3}$
0	1.3	1.3	1.3
1	1.4493	1.197	1.32
2	0.9087	0.7150	1.3238
3	−5.737	−0.6345	1.3247
4	…	…	1.3247

230.4 Numerical Solution of Simultaneous Nonlinear Equations

The Eechniques illustrated here will be demonstrated for two simultaneous equations: $f(x,y) = 0$ and $g(x,y) = 0$. They immediately generalize to more than two simultaneous equations.

The Method of Successive Substitutions

The two simultaneous equations can be written in various ways in equivalent form

$$x = F(x, y)$$

$$y = G(x, y)$$

and the method of successive substitutions can be based on

$$x_{k+1} = F(x_k, y_k)$$

$$y_{k+1} = G(x_k, y_k)$$

Again, the procedure is of the first order and a necessary condition for convergence is

$$\left|\frac{\partial F}{\partial x}\right| + \left|\frac{\partial F}{\partial y}\right| < 1 \qquad \left|\frac{\partial G}{\partial x}\right| + \left|\frac{\partial G}{\partial y}\right| < 1$$

in the iteration neighborhood of the true solution.

The Newton–Raphson Procedure

Using the two simultaneous equations, start from an approximation, say (x_0,y_0), obtained graphically or from a two-way table. Then, solve successively the linear equations

$$\Delta x_k \frac{\partial f}{\partial x}(x_k, y_k) + \Delta y_k \frac{\partial f}{\partial y}(x_k, y_k) = -f(x_k, y_k)$$

$$\Delta x_k \frac{\partial g}{\partial x}(x_k, y_k) + \Delta y_k \frac{\partial g}{\partial y}(x_k, y_k) = -g(x_k, y_k)$$

for Δx_k and Δy_k. Then, the $k+1$ approximation is given from $x_{k+1} = x_k + \Delta x_k$, $y_{k+1} = y_k + \Delta y_k$. A modification consists in solving the equations with (x_k,y_k) replaced by (x_0,y_0) (or another suitable pair later on in the iteration) in the derivatives. This means the derivatives (and therefore the coefficients of $\Delta x_k,\Delta y_k$) are independent of k. Hence, the results become

$$\Delta x_k = \frac{-f(x_k, y_k)(\partial g/\partial y)(x_0, y_0) + g(x_k, y_k)(\partial f/\partial y)(x_0, y_0)}{(\partial f/\partial x)(x_0, y_0)(\partial g/\partial y)(x_0, y_0) - (\partial f/\partial y)(x_0, y_0)(\partial g/\partial x)(x_0, y_0)}$$

$$\Delta y_k = \frac{-g(x_k, y_k)(\partial f/\partial x)(x_0, y_0) + f(x_k, y_k)(\partial g/\partial x)(x_0, y_0)}{(\partial f/\partial x)(x_0, y_0)(\partial g/\partial y)(x_0, y_0) - (\partial f/\partial y)(x_0, y_0)(\partial g/\partial x)(x_0, y_0)}$$

and $x_{k+1} = \Delta x_k + x_k$, $y_{k+1} = \Delta y_k + y_k$. Such an alteration of the basic technique reduces the rapidity of convergence.

Example

$$f(x,y) = 4x^2 + 6x - 4xy + 2y^2 - 3$$

$$g(x,y) = 2x^2 - 4xy + y^2$$

By plotting, one of the approximate roots is found to be $x_0 = 0.4$, $y_0 = 0.3$. At this point, there results $\partial f/\partial x = 8$, $\partial f/\partial y = -0.4$, $\partial g/\partial x = 0.4$, and $\partial g/\partial y = -1$. Hence,

$$x_{k+1} = x_k + \Delta x_k = x_k + \frac{-f(x_k, y_k) - 0.4g(x_k, y_k)}{8(-1) - (-0.4)(0.4)}$$

$$= x_k - 0.1275\ 5f(x_k, y_k) - 0.051\ 02g(x_k, y_k)$$

and

$$y_{k+1} = y_k - 0.051\ 02f(x_k, y_k) + 1.020\ 41g(x_k, y_k)$$

The first few iteration steps are shown in the following table.

Step k	x_k	y_k	$f(x_k, y_k)$	$g(x_k, y_k)$
0	0.4	0.3	−0.26	0.07
1	0.43673	0.24184	0.078	0.0175
2	0.42672	0.25573	−0.0170	−0.007
3	0.42925	0.24943	0.0077	0.0010

Methods of Perturbation

Let $f(x) = 0$ be the equation. In general, the iterative relation is

$$x_{k+1} = x_k - \frac{f(x_k)}{\alpha_k}$$

where the iteration begins with x_0 as an initial approximation and α_k is some functional.

The Newton–Raphson Procedure

This variant chooses $\alpha_k = f'(x_k)$ where $f' = df/dx$ and geometrically consists of replacing the graph of $f(x)$ by the tangent line at $x = x_k$ in each successive step. If $f'(x)$ and $f'(x)$ have the same sign throughout an interval $a \le x \le b$ containing the solution, with $f(a)$ and $f(b)$ of opposite signs, then the process converges starting from any x_0 in the interval $a \le x \le b$. The process is second order.

Example

$$f(x) = x - 1 + \frac{(0.5)^x - 0.5}{0.3}$$

$$f'(x) = 1 - 2.3105\,[0.5]^x$$

An approximate root (obtained graphically) is 2.

Step k	x_k	$f(x_k)$	$f'(x_k)$
0	2	0.1667	0.4224
1	1.605	−0.002	0.2655
2	1.6125	−0.0005	...

The Method of False Position

This variant is commenced by finding x_0 and x_1 such that $f(x_0)$ and $f(x_1)$ are of opposite signs. Then, α_1 = slope of secant line joining $[x_0, f(x_0)]$ and $[x_1, f(x_1)]$ so that

$$x_2 = x_1 - \frac{x_1 - x_0}{f(x_1) - f(x_0)} f(x_1)$$

In each following step, α_k is the slope of the line joining $[x_k, f(x_k)]$ to the most recently determined point where $f(x_j)$ has the opposite sign from that of $f(x_k)$. This method is of the first order.

The Method of Wegstein

This is a variant of the method of successive substitutions which forces or accelerates convergence. The iterative procedure $x_{k+1} = F(x_k)$ is revised by setting $\hat{x}_{k+1} = F(x_k)$ and then taking $x_{k+1} = qx_k + (1 - q)\,\hat{x}_{k+1}$. Wegstein found that suitably chosen q's are related to the basic process as follows:

Behavior of Successive Substitution Process	Range of Optimum q
Oscillatory convergence	$0 < q < 1/2$
Oscillatory divergence	$1/2 < q < 1$
Monotonic convergence	$q < 0$
Monotonic divergence	$1 < q$

At each step, q may be calculated to give a locally optimum value by setting

$$q = \frac{x_{k+1} - x_k}{x_{k+1} - 2x_k + x_{k-1}}$$

The Method of Continuity

In the case of n equations in n unknowns, when n is large, determining the approximate solution may involve considerable effort. In such a case, the method of continuity is admirably suited for use on either digital or analog computers. It consists basically of the introduction of an extra variable into the n equations

$$f_i(x_1, x_2, \ldots, x_n) = 0, \quad i = 1, \ldots, n$$

and replacing them by

$$f_i(x_1, x_2, \ldots, x_n, \lambda) = 0, \quad i = 1, \ldots, n$$

where λ is introduced in such a way that the functions depend in a simple way upon λ and reduce to an easily solvable system for $\lambda = 0$ and to the original equations for $\lambda = 1$. A system of ordinary differential equations, with independent variable λ, is then constructed by differentiating with respect to λ. The results are

$$\sum_{j=1}^{n} \frac{\partial f_i}{\partial x_j} \frac{dx_j}{d\lambda} + \frac{\partial f_i}{\partial \lambda} = 0$$

where $x_1, \ldots, x_n$ are considered as functions of λ. The equations are integrated, with initial conditions obtained with $\lambda = 0$, from $\lambda = 0$ to $\lambda = 1$. If the solution can be continued to $\lambda = 1$, the values of $x_1, \ldots, x_n$ for $\lambda = 1$ will be a solution of the original equations. If the integration becomes infinite, the parameter λ must be introduced in a different fashion. Integration of the differential equations (which are usually nonlinear in λ) may be accomplished on an analog computer or by digital means using techniques described in Section 230.8.

Example

$$f(x, y) = 2 + x + y - x^2 + 8xy + y^3 = 0$$

$$g(x, y) = 1 + 2x + 3y + x^2 + xy - ye^x = 0$$

Introduce λ as

$$f(x, y, \lambda) = (2 + x + y) + \lambda(-x^2 + 8xy + y^3) = 0$$

$$g(x, y, \lambda) = (1 + 2x - 3y) + \lambda(x^2 + xy + ye^x) = 0$$

For $\lambda = 1$, these reduce to the original equations, but, for $\lambda = 0$, they are the linear systems

$$x + y = -2$$

$$2x - 3y = -1$$

which has the unique solution $x = -1.4$, $y = -0.6$. The differential equations in this case become

$$\frac{\partial f}{\partial x}\frac{dx}{d\lambda} + \frac{\partial f}{\partial y}\frac{dy}{d\lambda} = -\frac{\partial f}{\partial \lambda}$$

$$\frac{\partial g}{\partial x}\frac{dx}{d\lambda} + \frac{\partial g}{\partial y}\frac{dy}{d\lambda} = -\frac{\partial g}{\partial \lambda}$$

or

$$\frac{dx}{d\lambda} = \frac{\dfrac{\partial f}{\partial y}\dfrac{\partial g}{\partial \lambda} - \dfrac{\partial f}{\partial \lambda}\dfrac{\partial g}{\partial y}}{\dfrac{\partial f}{\partial x}\dfrac{\partial g}{\partial y} - \dfrac{\partial f}{\partial y}\dfrac{\partial g}{\partial x}}$$

$$\frac{dy}{d\lambda} = \frac{\dfrac{\partial f}{\partial \lambda}\dfrac{\partial g}{\partial x} - \dfrac{\partial f}{\partial x}\dfrac{\partial g}{\partial \lambda}}{\dfrac{\partial f}{\partial x}\dfrac{\partial g}{\partial y} - \dfrac{\partial f}{\partial y}\dfrac{\partial g}{\partial x}}$$

Integrating in λ, with initial values $x = -1.4$ and $y = -0.6$ at $\lambda = 0$, from $\lambda = 0$ to $\lambda = 1$ gives the solution.

230.5 Interpolation and Finite Differences

The practicing engineer constantly finds it necessary to refer to tables as sources of information. Consequently, interpolation, or that procedure of "reading between the lines of the table," is a necessary topic in numerical analysis.

Linear Interpolation

If a function $f(x)$ is approximately linear in a certain range, then the ratio $[f(x_1) - f(x_0)]/(x_1 - x_0) = f[x_0,x_1]$ is approximately independent of x_0 and x_1 in the range. The linear approximation to the function $f(x)$, $x_0 < x < x_1$, then leads to the interpolation formula

$$f(x) \approx f(x_0) + (x - x_0) f[x_0, x_1] \approx f(x_0) + \frac{x - x_0}{x_1 - x_0}[f(x_1) - f(x_0)]$$

$$\approx \frac{1}{x_1 - x_0}[(x_1 - x) f(x_0) - (x_0 - x) f(x_1)]$$

Divided Differences of Higher Order and Higher-Order Interpolation

The first-order divided difference $f[x_0,x_1]$ was defined above. Divided differences of second and higher order are defined iteratively by

$$f[x_0, x_1, x_2] = \frac{f[x_1, x_2] - f(x_0, x_1)}{x_2 - x_0}$$

$$\vdots$$

$$f[x_0, x_1 \ldots, x_k] = \frac{f[x_1, \ldots, x_k] - (x_0, x_1, \ldots, x_{k-1}]}{x_k - x_0}$$

and a convenient form for computational purposes is

$$f[x_0, x_1, \ldots, x_k] = \sum_{j=0}^{k}{}' \frac{f(x_j)}{(x_j - x_0)(x_j - x_1)\cdots(x_j - x_k)}$$

for any $k \geq 0$, where the $'$ means the term $(x_j - x_j)$ is omitted in the denominator. For example,

$$f[x_0, x_1, x_2] = \frac{f(x_0)}{(x_0 - x_1)(x_0 - x_2)} + \frac{f(x_1)}{(x_1 - x_0)(x_1 - x_2)} + \frac{f(x_2)}{(x_2 - x_0)(x_2 - x_1)}$$

If the accuracy afforded by a linear approximation is inadequate, a generally more accurate result may be based upon the assumption that $f(x)$ may be approximated by a polynomial of degree 2 or higher over certain ranges. This assumptions leads to *Newton's fundamental interpolation formula* with divided differences:

$$f(x) \approx f(x_0) + (x - x_0) f[x_0, x_1] + (x - x_0)(x - x_1) f[x_0, x_1, x_2]$$
$$+ (x - x_0)(x - x_1)\cdots(x - x_{n-1}) f[x_0, x_1, \ldots, x_n] + E_n(x)$$

where $E_n(x)$ = error = $[1/(n+1)!]f^{(n-1)}(\varepsilon)\pi(x)$ where $\min(x_0,\ldots,x) < \varepsilon < \max(x_0,x_1,\ldots,x_n,x)$ and $\pi(x) = (x - x_0)(x - x_1) \ldots (x - x_n)$. In order to use this most effectively, one may first form a divided-difference table. For example, for third-order interpolation, the difference table is

$$\begin{array}{l|llll} x_0 & f(x_0) & f[x_0,x_1] & & \\ x_1 & f(x_1) & & f[x_0,x_1,x_2] & \\ & & f[x_1,x_2] & & f[x_0,x_1,x_2,x_3] \\ x_2 & f(x_2) & & f[x_1,x_2,x_3] & \\ x_3 & f(x_3) & f[x_2,x_3] & & \end{array}$$

where each entry is given by taking the difference between diagonally adjacent entries to the left, divided by the abscissas corresponding to the ordinates intercepted by the diagonals passing through the calculated entry.

Example

Calculate by third-order interpolation the value of cosh 0.83 given cosh 0.60, cosh 0.80, cosh 0.90, and cosh 1.10.

$$\begin{array}{l|llll} x_0 = 0.60 & 1.185\ 47 & 0.7598 & & \\ x_1 = 0.80 & 1.337\ 43 & & 0.6560 & \\ & & 0.9566 & & 0.1586 \\ x_2 = 0.90 & 1.433\ 09 & & 0.7353 & \\ x_3 = 1.10 & 1.668\ 52 & 1.1772 & & \end{array}$$

With $n = 3$, we have

$$\cosh 0.83 \approx 1.185\ 47 + (0.23)(0.7598) + (0.23)(0.03)(0.6560)$$
$$+ (0.23)(0.03)(-0.07)(0.1586) = 1.364\ 64$$

which varies from the true value by 0.00004.

Lagrange Interpolation Formulas

The Newton formulas are expressed in terms of divided differences. It is often useful to have interpolation formulas expressed explicitly in terms of the ordinates involved. This is accomplished by the Lagrange interpolation polynomial of degree n:

$$y(x) = \sum_{j=0}^{n} \frac{\pi(x)}{(x - x_j)\pi'(x_j)} f(x_j)$$

where

$$\pi(x) = (x - x_0)(x - x_1)\cdots(x - x_n)$$
$$\pi'(x_j) = (x_j - x_0)(x_j - x_1)\cdots(x_j - x_n)$$

where $(x_j - x_j)$ is the omitted factor. Thus,

$$f(x) = y(x) + E_n(x)$$

$$E_n(x) = \frac{1}{(n+1)!}\pi(x) f^{(n+1)}(\varepsilon)$$

Example

The interpolation polynomial of degree 3 is

$$y(x) = \frac{(x-x_1)(x-x_2)(x-x_3)}{(x_0-x_1)(x_0-x_2)(x_0-x_3)} f(x_0) + \frac{(x-x_0)(x-x_2)(x-x_3)}{(x_1-x_0)(x_1-x_2)(x_1-x_3)} f(x_1)$$

$$+ \frac{(x-x_0)(x-x_1)(x-x_3)}{(x_2-x_0)(x_2-x_1)(x_2-x_3)} f(x_2) + \frac{(x-x_0)(x-x_1)(x-x_2)}{(x_3-x_0)(x_3-x_1)(x_3-x_2)} f(x_3)$$

Thus, directly from the data

x	0	1	3	4
$f(x)$	1	1	−1	2

we have as an interpolation polynomial $y(x)$ for $f(x)$:

$$y(x) = 1 \cdot \frac{(x-1)(x-3)(x-4)}{(0-1)(0-3)(0-4)} + 1 \cdot \frac{x(x-3)(x-4)}{(1-0)(1-3)(1-4)}$$

$$-1 \cdot \frac{x(x-1)(x-4)}{(3-0)(3-1)(3-4)} + 2 \cdot \frac{(x-0)(x-1)(x-3)}{(4-0)(4-1)(4-3)}$$

Other Difference Methods (Equally Spaced Ordinates)

Backward Differences

The backward differences denoted by

$$\nabla f(x) = f(x) - f(x-h)$$

$$\nabla^2 f(x) = \nabla f(x) - \nabla f(x-h)$$

$$\ldots$$

$$\nabla f^n(x) = \nabla^{n-1} f(x) - \nabla^{n-1} f(x-h)$$

are useful for calculation near the end of tabulated data.

Central Differences

The central difference denoted by

$$\delta f(x) = f\left(x + \frac{h}{2}\right) - f\left(x - \frac{h}{2}\right)$$

$$\delta^n f(x) = \delta^{n-1} f\left(x + \frac{h}{2}\right) - \delta^{n-1} f\left(x - \frac{h}{2}\right)$$

is useful for calculating at the interior points of tabulated data.

Also to be found in the literature are Gaussian, Stirling, Bessel, Everett, Comrie differences, and so forth.

Inverse Interpolation

This is the process of finding the value of the independent variable or abscissa corresponding to a given value of the function when the latter is between two tabulated values of the abscissa. One method of accomplishing this is to use Lagrange's interpolation formula in the form

$$x = \psi(y) = \sum_{j=0}^{n} \frac{\pi(y)}{(y - y_j)\pi'(y_j)^{x_j}} x_j$$

where x is expressed as a function of y. Other methods resolve about methods of iteration.

230.6 Numerical Differentiation

Numerical differentiation should be avoided wherever possible, particularly when data are empirical and subject to appreciable observation errors. Errors in data can affect numerical derivatives quite strongly (i.e., differentiation is a roughening process). When such a calculation must be made, it is usually desirable first to *smooth* the data to a certain extent.

The Use of Interpolation Formulas

If the data are given over equidistant values of the independent variable x, an interpolation formula, such as the Newton formula, may be used, and the resulting formula differentiated analytically. If the independent variable is not at equidistant values, then Lagrange's formulas must be used. By differentiating three- and five-point Lagrange interpolation formulas, the following differentiation formulas result for equally spaced tabular points.

Three-Point Formulas

Let x_0, x_1, and x_2 be the three points. Thus,

$$f'(x_0) = \frac{1}{2h}[-3f(x_0) + 4f(x_1) - f(x_2)] + \frac{h^2}{3} f'''(\varepsilon)$$

$$f'(x_1) = \frac{1}{2h}[-f(x_0) + f(x_2)] + \frac{h^2}{6} f'''(\varepsilon)$$

$$f'(x_2) = \frac{1}{2h}[f(x_0) - 4f(x_1) + 3f(x_2)] + \frac{h^2}{3} f'''(\varepsilon)$$

where the last term is an error term and $\min_j x_j < \varepsilon < \max_j x_j$.

Five-Point Formulas

Let x_0, x_1, x_2, x_3, and x_4 be the five values of the equally spaced independent variable and $f_j = f(x_j)$.

$$f'(x_0) = \frac{1}{12h}[-25f_0 + 48f_1 - 36f_2 + 16f_3 - 3f_4] + \frac{h^4}{5} f^{(v)}(\varepsilon)$$

$$f'(x_1) = \frac{1}{12h}[-3f_0 - 10f_1 + 18f_2 - 6f_3 + f_4] - \frac{h^4}{20} f^{(v)}(\varepsilon)$$

$$f'(x_2) = \frac{1}{12h}[f_0 - 8f_1 + 8f_3 - f_4] + \frac{h^4}{30} f^{(v)}(\varepsilon)$$

$$f'(x_3) = \frac{1}{12h}[-f_0 + 6f_1 - 18f_2 + 10f_3 + 3f_4] + \frac{h^4}{20} f^{(v)}(\varepsilon)$$

$$f'(x_4) = \frac{1}{12h}[3f_0 - 16f_1 + 36f_2 - 48f_3 + 25f_4] + \frac{h^4}{5} f^{(v)}(\varepsilon)$$

and the last term is again an error term.

Smoothing Techniques

These techniques involve the approximation of the tabular data by a least squares fit of the data using some known functional form, usually a polynomial. In place of approximating $f(x)$ by a single least squares polynomial of degree n over the entire range of the tabulation, it is often desirable to replace each tabulated value by the value taken on by a least squares polynomial of degree n relevant to a subrange of $2M + 1$ points centered, where possible, at the point for which the entry is to be modified. Thus, each smoothed value replaces a tabulated value. Let $f_j = f(x_j)$ to be the tabular points and y_j = smoothed values. A first-degree least squares with three points would be

$$y_0 = \frac{1}{6}[5f_0 + 2f_1 - f_2]$$

$$y_1 = \frac{1}{3}[f_0 + f_1 + f_2]$$

$$y_2 = \frac{1}{6}[-f_0 + 2f_1 + 5f_2]$$

A first-degree least squares with five points would be

$$y_0 = \frac{1}{5}[3f_0 + 2f_1 + f_2 - f_4]$$

$$y_1 = \frac{1}{10}[4f_0 + 3f_1 + 2f_2 + f_3]$$

$$y_2 = \frac{1}{5}[f_0 + f_1 + f_2 + f_3 + f_4]$$

$$y_3 = \frac{1}{10}[f_0 + 2f_1 + 3f_2 + 4f_3]$$

$$y_4 = \frac{1}{5}[-f_0 + f_2 + 2f_3 + 3f_4]$$

Thus, for example, if first-degree, five-point least squares are used, the central formula is used for all values except the first two and the last two, where the off-center formulas are used. A third-degree least squares with seven points would be

$$y_0 = \tfrac{1}{42}[39f_0 + 8f_1 - 4f_2 - 4f_3 + f_4 + 4f_5 - 2f_6]$$

$$y_1 = \tfrac{1}{42}[8f_0 + 19f_1 + 16f_2 + 6f_3 - 4f_4 - 7f_5 + 4f_6]$$

$$y_2 = \tfrac{1}{42}[-4f_0 + 16f_1 + 19f_2 + 12f_3 + 2f_4 - 4f_5 + f_6]$$

$$y_3 = \tfrac{1}{21}[-2f_0 + 3f_1 + 6f_2 + 7f_3 + 6f_4 + 3f_5 - 2f_6]$$

$$y_4 = \tfrac{1}{42}[f_0 - 4f_1 + 2f_2 + 12f_3 + 19f_4 + 16f_5 - 4f_6]$$

$$y_5 = \tfrac{1}{42}[4f_0 - 7f_1 - 4f_2 + 6f_3 + 16f_4 + 19f_5 + 8f_6]$$

$$y_6 = \tfrac{1}{42}[-2f_0 + 4f_1 + f_2 - 4f_3 - 4f_4 + 8f_5 + 39f_6]$$

Additional smoothing formulas may be found in the references. After the data are smoothed, any of the interpolation polynomials, or an appropriate least squares polynomial, may be fitted and the results used to obtain the derivative.

Least Squares Methods

Parabolic

For five evenly spaced neighboring abscissas labeled x_{-2}, x_{-1}, x_0, x_1, and x_2, and their ordinates f_{-2}, f_{-1}, f_0, f_1, and f_2, assume a parabola is fit by least squares. There results for all interior points, except the first and last two points of the data, the formula for the numerical derivative:

$$f_0' = \frac{1}{10h}[-2f_{-2} - f_{-1} + f_1 + 2f_2]$$

For the first two data points designated by 0 and h:

$$f'(0) = \frac{1}{20h}[-21f(0) + 13f(h) + 17f(2h) - 9f(3h)]$$

$$f'(h) = \frac{1}{20h}[-11f(0) + 3f(h) + 7f(2h) + f(3h)]$$

and for the last two given by $\alpha - h$ and α:

$$f'(\alpha - h) = \frac{1}{20h}[-11f(\alpha) + 3f(\alpha - h) + 7f(\alpha - 2h) + f(\alpha - 3h)]$$

$$f'(\alpha) = \frac{1}{20h}[-21f(\alpha) + 13f(\alpha - h) + 17f(\alpha - 2h) - 9f(\alpha - 3h)]$$

Quartic (Douglas–Avakian)

A fourth-degree polynomial $y = a + bx + cx^2 + dx^3 + ex^4$ is fitted to seven adjacent equidistant points (spacing h) after a translation of coordinates has been made so that $x = 0$ corresponds to the central point of the seven. Thus, these may be called $-3h$, $-2h$, $-h$, 0, h, $2h$, and $3h$. Let k = coefficient of h for the seven points. That is, in $-3h$, $k = -3$. Then, the coefficients for the polynomial are

$$a = \frac{524\Sigma f(kh) - 245\Sigma k^2 f(kh) + 21\Sigma k^4 f(kh)}{924}$$

$$b = \frac{397\Sigma kf(kh)}{1512h} - \frac{7\Sigma k^3 f(kh)}{216h}$$

$$c = \frac{-840\Sigma f(kh) + 679\Sigma k^2 f(kh) - 67\Sigma k^4 f(kh)}{3168h^2}$$

$$d = \frac{-7\Sigma kf(kh) + \Sigma k^3 f(kh)}{216h^3}$$

$$e = \frac{72\Sigma f(kh) - 67\Sigma k^2 f(kh) + 7\Sigma k^4 f(kh)}{3168h^4}$$

where all summations run from $k = -3$ to $k = +3$ and $f(kh)$ = tabular value at kh. The slope of the polynomial at $x = 0$ is $dy/dx = b$.

230.7 Numerical Integration

Numerical evaluation of the finite integral $\int_a^b f(x)dx$ is carried out by a variety of methods. A few are given here.

Newton–Cotes Formulas (Equally Spaced Ordinates)

Trapezoidal Rule

This formula consists of subdividing the interval $a \le x \le b$ into n subintervals a to $a + h$, $a + h$ to $a + 2h$, …, and replacing the graph of $f(x)$ by the result of joining the ends of adjacent ordinates by line segments. If $f_j = f(x_j) = f(a + jh)$, $f_0 = f(a)$, and $f_n = f(b)$, the integration formula is

$$\int_a^b f(x)dx = \frac{h}{2}[f_0 + 2f_1 + 2f_2 + \cdots + 2f_{n-1} + f_n] + E_n$$

where

$$|E_n| = (nh^3/12)|f''(\varepsilon)| = [(b-a)^3/12n^2|f''(\varepsilon)|, a < \varepsilon < b$$

This procedure is not of high accuracy. However, if $f''(x)$ is continuous in $a < x < b$, the error goes to zero as $1/n^2$, $n \to \infty$.

Parabolic Rule (Simpson's Rule)

This procedure consists of subdividing the interval $a < x < b$ into $n/2$ subintervals, each of length $2h$, where n is an even integer. Using the notation as above, the integration formula is

$$\int_a^b f(x)dx = \frac{h}{3}[f_0 + 4f_1 + 2f_2 + 4f_3 + \cdots + 4f_{n-3} + 2f_{n-2} + 4f_{n-1} + f_n] + E_n$$

where

$$|E_n| = \frac{nh^5}{180}\left|f^{(iv)}(\varepsilon)\right| = \frac{(b-a)^3}{180n^4}\left|f^{(iv)}(\varepsilon)\right| \quad a<\varepsilon<b$$

This method approximates $f(x)$ by a parabola on each subinterval. This rule is generally more accurate than the trapezoidal rule. It is the most widely used integration formula.

Weddle's Rule

This procedure consists of subdividing the integral $a < x < b$ into $n/6$ subintervals, each of length $6h$, where n is a multiple of 6. Using the notation from the trapezoidal rule, the following results:

$$\int_a^b f(x)dx = \frac{3h}{10}[f_0 + 5f_1 + f_2 + 6f_3 + f_4 + 5f_5 + 2f_6 + 5f_7 + f_8 + \cdots$$

$$+ 6f_{n-3} + f_{n-2} + 5f_{n-1} + f_n] + E_n$$

Note that the coefficients of f_j follow the rule 1, 5, 1, 6, 1, 5, 2, 5, 1, 6, 1, 5, 2, 5, and so on. This procedure consists of approximating $f(x)$ by a polynomial of degree 6 on each subinterval. Here,

$$E_n = \frac{nh^7}{1400}[10f^{(6)}(\varepsilon_1) + 9h^2 f^{(8)}(\varepsilon_2)]$$

Gaussian Integration Formulas (Unequally Spaced Abscissas)

These formulas are capable of yielding comparable accuracy with fewer ordinates than the equally spaced formulas. The ordinates are obtained by optimizing the distribution of the abscissas rather than by arbitrary choice. For the details of these formulas, Hildebrand [1956] is an excellent reference.

Two-Dimensional Formula

Formulas for two-way integration over a rectangle, circle, ellipse, and so forth, may be developed by a double application of one-dimensional integration formulas. The two-dimensional generalization of the parabolic rule is given here. Consider the iterated integral $\int_a^b \int_c^d f(x,y)dxdy$. Subdivide $c < x < d$ into m (even) subintervals of length $h = (d-c)/m$, and $a < y < b$ into n (even) subintervals of length $k = (b-a)/n$. This gives a subdivision of the rectangle $a \le y \le b$ and $c \le x \le d$ into subrectangles. Let $x_j = c + jh$, $y_j = a + jk$, and $f_{i,j} = f(x_i, y_i)$. Then,

$$\int_a^b \int_c^d f(x,\ y)dxdy = \frac{hk}{9}[(f(_{0,0} + 4f_{1,0} + 2f_{2,0} + \cdots + f_{m,0})$$

$$+ 4(f_{0,1} + 4f_{1,1} + 2f_{2,1} + \cdots + f_{m,1}) + \cdots$$

$$+ 2(f_{0,2} + 4f_{1,2} + 2f_{2,2} + \cdots + f_{m,2}) + \cdots$$

$$+ (f_{0,n} + 4f_{1,n} + 2f_{2,n} + \cdots + f_{m,n})] + E_{m,n}$$

where

$$E_{m,n} = -\frac{hk}{90}\left[mh^4 \frac{\partial^4 f(\varepsilon_1,\eta_1)}{\partial x^4} + nk^4 \frac{\partial^4 f(\varepsilon_2,\eta_2)}{\partial x^4}\right]$$

and ε_1 and ε_2 lie in $c < x < d$, and η_1 and η_2 lie in $a < y < b$.

230.8 Numerical Solution of Ordinary Differential Equations

A number of methods have been devised to solve ordinary differential equations numerically. The general references contain some information. A numerical solution of a differential equation means a table of values of the function y and its derivatives over only a limited part of the range of the independent variable. Every differential equation of order n can be rewritten as n first-order differential equations. Therefore, the methods given below will be for first-order equations, and the generalization to simultaneous systems will be developed later.

The Modified Euler Method

This method is simple and yields modest accuracy. If extreme accuracy is desired, a more sophisticated method should be selected. Let the first-order differential equation be $dy/dx = f(x,y)$ with the initial condition (x_0,y_0) (i.e., $y = y_0$ when $x = x_0$). The procedure follows.

Step 1. From the given initial conditions (x_0,y_0), compute $y_0' = f(x_0,y_0)$ and $y_0'' = [\partial f(x_0,y_0)/\partial x] + [\partial f(x_0,y_0)/\partial y]\, y_0'$. Then, determine $y_1 = y_0 + hy_0' + (h^2/2)\, y_0''$, where h = subdivision of the independent variable.

Step 2. Determine $y_1' = f(x_1,y_1)$, where $x_1 = x_0 + h$. These prepare us for the following.

Predictor Steps

Step 3. For $n \geq 1$, calculate $(y_{n+1})_1 = y_{n-1} + 2hy_n'$.

Step 4. Calculate $(y_{n+1}')_1 = f[x_{n+1},(y_{n+1})_1]$.

Corrector Steps

Step 5. Calculate $(y_{n+1})_2 = y_n + (h/2)[(y_{n+1}')_1 + y_n']$, where y_n and y_n' without the subscripts are the previous values obtained by this process (or by steps 1 and 2).

Step 6. $(y_{n+1}')_2 = f[x_{n+1},(y_{n+1})_2]$.

Step 7. Repeat the corrector steps 5 and 6 if necessary until the desired accuracy is produced in y_{n+1}, y_{n+1}'.

Example

Consider the equation $y' = 2y^2 + x$ with the initial conditions $y_0 = 1$ when $x_0 = 0$. Let $h = 0.1$. A few steps of the computation are illustrated.

Step	
1	$y_0' = 2y_0^2 + x_0 = 2$
	$y_0'' = 1 + 4y_0 y_0' = 1 + 8 = 9$
	$y_1 = 1 + (0.1)(2) + [(0.1)^2/2]9 = 1.245$
2	$y_1' = 2y_1^2 + x_1 = 3.100 + 0.1 = 3.200$
3	$(y_2)_1 = y_0 + 2hy_1' = 1 + 2(0.1)3.200 = 1.640$
4	$(y_2')_1 = 2(y_2)_1^2 + x_2 + 5.592$
5	$(y_2)_2 = y_1 + (0.1/2)[(y_2')_1 + y_1'] = 1.685$
6	$(y_2')_2 = 2(y_2)_2^2 + x_2 = 5.878$
5 (repeat)	$(y_2)_3 = y_1 + (0.05)[((y_2')_2 + y_1'] = 1.699$
6 (repeat)	$(y_2')_3 = 2(y_2)_3^2 + x_2 = 5.974$
and so forth...	

This procedure may be programmed for a computer. A discussion of the truncation error of this process may be found in Milne [1953].

Modified Adam's Method

The procedure given here was developed retaining third differences. It can then be considered as a more exact predictor-corrector method than the Euler method. The procedure is as follows for $dy/dx = f(x,y)$ and h = interval size.

Steps 1 and 2 are the same as for the Euler method.

Predictor Steps

Step 3. $(y_{n+1})_1 = y_n + (h/24)[55y'_n - 59y'_{n-1} + 37y'_{n-2} - 9y'_{n-3}$, where y'_n, y'_{n-1}, and so on are calculated in step 1.

Step 4. $(y'_{n-1})_1 = f[x_{n+1},(y_{n+1})_1]$.

Corrector Steps

Step 5. $(y_{n+1})_2 = y_n + (h/24)[9(y'_{n+1})_1 + 19y'_n - 5y'_{n-1} + y'_{n-2}]$.

Step 6. $(y'_{n+1})_2 = f[x_{n+1},(y_{n+1})_2]$.

Step 7. Iterate steps 5 and 6 if necessary.

Runge–Kutta Methods

These methods are self-starting and are inherently stable. Kopal [1955] is a good reference for their derivation and discussion. Third- and fourth-order procedures are given below for $dy/dx = f(x,y)$, where h = interval size.

For third-order procedures (error $\approx h^4$),

$$k_0 = hf(x_n, y_n)$$

$$k_1 = hf(x_n + 1/2\,h, y_n + 1/2\,k_0)$$

$$k_2 = hf(x_n + h, y_n + 2k_1 - k_0)$$

and

$$y_{n+1} = y_n + \tfrac{1}{6}(k_0 + 4k_1 + k_2)$$

for all $n \geq 0$, with initial condition (x_0,y_0).

For fourth-order procedures (error $\approx h^5$),

$$k_0 = hf(x_n, y_n)$$

$$k_1 = hf(x_n + \tfrac{1}{2}h, y_n + \tfrac{1}{2}k_0)$$

$$k_2 = hf(x_n + \tfrac{1}{2}h, y_n + \tfrac{1}{2}k_1)$$

$$k_3 = hf(x_n + h, y_n + k_2)$$

and

$$y_{n+1} = y_n + \tfrac{1}{6}(k_0 + 2k_1 + 2k_2 + k_3)$$

Example

(Third-order) Let $dy/dx = x - 2y$, with initial condition $y_0 = 1$ when $x_0 = 0$, and let $h = 0.1$. Clearly, $x_n = nh$. To calculate y_1, proceed as follows:

$$k_0 = 0.1[x_0 - 2y_0] = -0.2$$

$$k_1 = 0.1[0.05 - 2(1 - 0.1)] = -0.175$$

$$k_2 = 0.1[0.1 - 2(1 - 0.35 + 0.2)] = -0.16$$

$$y_1 = 1 + \tfrac{1}{6}(-0.2 - 0.7 - 0.16) = 0.8234$$

Equations of Higher Order and Simultaneous Differential Equations

Any differential equation of second or higher order can be reduced to a simultaneous system of first-order equations by the introduction of auxiliary variables. Consider the following equations:

$$\frac{d^2x}{dt^2} + xy\frac{dx}{dt} + z = e^x$$

$$\frac{d^2y}{dt^2} + xy\frac{dy}{dt} = 7 + t^2$$

$$\frac{d^2z}{dt^2} + xz\frac{dz}{dt} + x = e^2$$

In the new variables $x_1 = x$, $x_2 = y$, $x_3 = z$, $x_4 = dx_1/dt$, $x_5 = dx_2/dt$, and $x_6 = dx_3/dt$, the equations become

$$\frac{dx_1}{dt} = x_4$$

$$\frac{dx_2}{dt} = x_5$$

$$\frac{dx_3}{dt} = x_6$$

$$\frac{dx_4}{dt} = -x_1x_2x_4 - x_3 + e^{x_3}$$

$$\frac{dx_5}{dt} = -x_3x_2x_5 + 7 + t^2$$

$$\frac{dx_6}{dt} = -x_1x_3x_6 - x_1 + e^{x_1}$$

which is a system of the general form

$$\frac{dx_i}{dt} = f_i(t, x_1, x_2, x_3, \ldots, x_n)$$

where $i = 1,2,\ldots,n$. Such systems may be solved by simultaneous application of any of the above numerical techniques. A Runge–Kutta method for

$$\frac{dx}{dt} = f(t, x, y)$$

$$\frac{dy}{dt} = g(t, x, y)$$

is given below. The fourth-order procedure is shown.

Starting at the initial conditions x_0, y_0, and t_0, the next values x_1 and y_1 are computed via the equations below (where $\Delta t = h$, $t_j = h + t_{j-1}$):

$$k_0 = hf(t_0, x_0, y_0) \qquad l_0 = hg(t_0, x_0, y_0)$$

$$k_1 = hf\left(t_0 + \frac{h}{2}, x_0 + \frac{k_0}{2}, y_0 + \frac{l_0}{2}\right) \qquad l_1 = hg\left(t_0 + \frac{h}{2}, x_0 + \frac{k_0}{2}, y_0 + \frac{l_0}{2}\right)$$

$$k_2 = hf\left(t_0 + \frac{h}{2}, x_0 + \frac{k_1}{2}, y_0 + \frac{l_1}{2}\right) \qquad l_2 = hg\left(t_0 + \frac{h}{2}, x_0 + \frac{k_1}{2}, y_0 + \frac{l_1}{2}\right)$$

$$k_3 = hf(t_0 + h, x_0 + k_2, y_0 + l_2) \qquad l_3 = hg(t_0 + h, x_0 + k_2, y_0 + l_2)$$

and

$$x_1 = x_0 + 1/6(k_0 + 2k_1 + 2k_2 + k_3)$$

$$y_1 = y_0 + 1/6(l_0 + 2l_1 + 2l_2 + l_3)$$

To continue the computation, replace t_0, x_0, and y_0 in the above formulas by $t_1 = t_0 + h$, x_1, and y_1 just calculated. Extension of this method to more than two equations follows precisely this same pattern.

230.9 Numerical Solution of Integral Equations

This section considers a method of numerically solving the Fredholm integral equation of the second kind:

$$u(x) = f(x) + \lambda \int_a^b k(x,t)u(t)dt \quad \text{for } u(x)$$

The method discussed arises because a definite integral can be closely approximated by any of several numerical integration formulas (each of which arises by approximating the function by some polynomial over an interval). Thus, the definite integral can be replaced by an integration formula which becomes

$$u(x) = f(x) + \lambda(b-a)\left[\sum_{i=1}^{n} c_i k(x,t_i)u(t_i)\right]$$

where $t_1,\ldots,t_n$ are points of subdivision of the t axis, $a \le t \le b$, and the cs are coefficients whose values depend upon the type of numerical integration formula used. Now, this must hold for all values of x, where $a \le x \le b$; so it must hold for $x = t_1, x = t_2, \ldots, x = t_n$. Substituting for x successively $t_1, t_2, \ldots, t_n$, and setting $u(t_i) = u_i$ and $f(t_i) = f_i$, we get n linear algebraic equations for the n unknowns $u_1,\ldots,u_n$. That is,

$$u_i = f_i + (b-a)[c_1 k(t_i,t_1)u_1 + c_2 k(t_i,t_2)u_2 + \cdots + c_n k(t_i,t_n)u_n], \qquad i = 1,2,\ldots,n$$

These u_j may be solved for by the methods under the section entitled "Numerical Solution of Linear Equations."

230.10 Numerical Methods for Partial Differential Equations

The ultimate goal of numerical (discrete) methods for partial differential equations (PDEs) is the reduction of continuous systems (projections) to discrete systems that are suitable for high-speed computer solutions. The user must be cautioned regarding the fact that the seeming elementary nature of the techniques holds pitfalls that can be seriously misleading. These approximations often lead to difficult mathematical questions of adequacy, accuracy, convergence, stability, and consistency. Convergence is concerned with the approach of the approximate numerical solution to the exact solution as the number of mesh units increase indefinitely in some sense. Unless the numerical method can be shown to converge to the exact solution, the chosen method is unsatisfactory.

Stability deals in general with error growth in the calculation. As stated before, any numerical method involves truncation and round-off errors. These errors are not serious unless they grow as the computation proceeds (i.e., the method is unstable).

Finite Difference Methods

In these methods, the derivatives are replaced by various finite differences. The methods will be illustrated for problems in two space dimensions (x,y) or (x,t), where t is timelike. Using subdivisions $\Delta x = h$ and $\Delta y = k$ with $u(ih,jk) = u_{i,j}$, approximate $u_x|_{i,j} = [(u_{i+1,j} - u_{i,j})/h] + O(h)$ (forward difference), a first-order $[O(h)]$ method, or $u_x|_{i,j} = [(u_{i+1,j} - u_{i-1,j})/2h] + O(h^2)$ (central difference), a second-order method. The second derivative is usually approximated with the second-order method $[u_{xx}|_{i,j} = [(u_{i+1,j} - 2u_{i,j} + u_{i-1,j})/h] + O(h^2)]$.

Example

Using second-order differences for u_{xx} and u_{yy}, the five-point difference equation (with $h = k$) for Laplace's equation $u_{xx} + u_{yy} = 0$ is $u_{i,j} = \frac{1}{4}[u_{i+1,j} + u_{i-1,j} + u_{i,j+1} + u_{i,j-1}]$. The accuracy is $O(h^2)$. This model is called *implicit* because one must solve for the total number of unknowns at the unknown grid points (i,j) in terms of the given boundary data. In this case, the system of equations is a linear system.

Example

Using a forward-difference approximation for u_t and a second-order approximation for u_{xx}, the diffusion equation $u_t = u_{xx}$ is approximated by the *explicit* formula $u_{i,j+1} = ru_{i-1,j} + (1 - 2r)u_{i,j} + ru_{i+1,j}$. This classic result permits step-by-step advancement in the t direction beginning with the initial data at $t = 0$ $(j = 0)$ and guided by the boundary data. Here, the term $r = \Delta t/(\Delta x)^2 = k/h^2$ is restricted to be less than or equal to 1/2 for stability and the truncation error is $O(k^2 + kh^2)$.

The Crank–Nicolson implicit formula that approximates the diffusion equation $u_t = u_{xx}$ is

$$\begin{aligned}-r\lambda u_{i-1,j+1} + (1+2r\lambda)u_{i,j+1} - r\lambda u_{i+1,j+1} = {} & r(1-\lambda)u_{i-1,j} \\ & + [1-2r(1-\lambda)]u_{i,j} \\ & + r(1-\lambda)u_{i+1,j}\end{aligned}$$

The stability of this numerical method was analyzed by Crandall [Ames, 1993] where the λ, r stability diagram is given.

Approximation of the time derivative in $u_t = u_{xx}$ by a central difference leads to an always unstable approximation, the useless approximation

$$u_{i,j+1} = u_{i,j-1} + 2r(u_{i+1,j} - 2u_{i,j} + u_{i-1,j})$$

which is a warning to be careful.

The foregoing method is *symmetric* with respect to the point (i,j), where the method is centered. Asymmetric methods have some computational advantages, so the Saul'yev method is described [Ames, 1993]. The algorithms ($r = k/h^2$)

$$(1+r)u_{i,j+1} = u_{i,j} + r(u_{i-1,j+1} - u_{i,j} + u_{i+1,j}) \quad \text{(Saul'yev A)}$$

$$(1+r)u_{i,j+1} = u_{i,j} + r(u_{i+1,j+1} - u_{i,j} + u_{i-1,j}) \quad \text{(Saul'yev B)}$$

are used as in any one of the following options:

1. Use Saul'yev A only and proceed line-by-line in the $t(j)$ direction, but *always* from the left boundary on a line.
2. Use Saul'yev B only and proceed line-by-line in the $t(j)$ direction, but *always* from the right boundary to the left on a line.
3. Alternate from line to line by first using Saul'yev A and then B, or the reverse. This is related to *alternating direction methods.*
4. Use Saul'yev A and Saul'yev B on the same line and average the results for the final answer (A first, and then B). This is equivalent to inroducing the dummy variables $P_{i,j}$ and $Q_{i,j}$ such that

$$(1+r)P_{i,j+1} = U_{i,j} + r(P_{i-1,j+1} - U_{i,j} + U_{i+1,j})$$

$$(1+r)Q_{i,j+1} = U_{i,j} + r(Q_{i+1,j+1} - U_{i,j} + U_{i-1,j})$$

and

$$U_{i,j+1} = \frac{1}{2}(P_{i,j+1} + Q_{i,j+1})$$

This averaging method has some computational advantages because of the possibility of truncation error cancellation. As an alternative, one can retrain the $P_{i,j}$ and $Q_{i,j}$ from the previous step and replace $U_{i,j}$ and $U_{i+1,j}$ by $P_{i,j}$ and $P_{i+1,j}$, respectively, and $U_{i,j}$ and $U_{i-1,j}$ by $Q_{i,j}$, and $Q_{i-1,j}$ respectively.

Weighted Residual Methods

To set the stage for the method of finite elements, we briefly describe the weighted residual methods (WRMs), which have several variations — the interior, boundary, and mixed methods. Suppose the equation is $Lu = f$, where L is the partial differential operator and f is known function, of say x and y. The first step in WRM is to select a class of known basis functions b_i (e.g., trigonometric, Bessel, Legendre) to approximate $u(x,y)$ as $\sim\Sigma a_i b_i(x,y) = U(x,y,a)$. Often, the b_i are selected so that $U(x,y,a)$ satisfy the bounary conditions. This is essentially the *interior method.* If the b_i in $U(x,y,a)$ are selected to satisfy the differential equations, but not the boundary conditions, the variant is called the *boundary method.* When neither the equation nor the boundary conditions are satisfied, the method is said to be *mixed.* The least ingenuity is required here. The usual method of choice is the interior method.

The second step is to select an optimal set of constants, a_i, $i = 1,2,\ldots,n$, by using the residual $R_I(U) = LU - f$. This is done here for the interior method. In the boundary method, there are a set of boundary residual R_B, and, in the mixed method, both R_I and R_B. Using the spatial average $(w,v) = \int_V wv dV$, the criterion for selecting the values of a_i is the requirement that the n spatial averages

$$(b_i, R_E(U)) = 0, \quad i = 1,2,\ldots,n$$

These represent n equations (linear if the operator L is linear and nonlinear otherwise) for the a_j.

Particular WRMs differ because of the choice of the b_js. The most common follow.

1. *Subdomain.* The domain V is divided into n smaller, not necessarily disjoint, subdomains V_j with $w_j(x,y) = 1$ if (x,y) is in V_j, and 0 if (x,y) is not in V_j.
2. *Collocation.* Select n points $P_j = (x_j,y_j)$ in V with $w_j(P_j) = \delta(P - P_j)$, where $\int_V \phi(P)\delta(P-P_j)dP = \phi(P_j)$ for all test functions $\phi(P)$ which vanish outside the compact set V. Thus, $(w,R_E) = \int_V \delta(P-P_j)R_E dV = R_E[U(P_j)] \equiv 0$ (i.e., the residual is set equal to zero at the n points P_j).
3. *Least squares.* Here, the functional $I(a) = \int_V R_E^2 dV$, where $a = (a_1,\ldots,a_n)$, is to be made stationary with respect to the a_j. Thus, $0 = \partial I/\partial a_j = 2\int_V R_E(\partial R_E/\partial a_j)dV$, with $j = 1,2,\ldots,n$. The w_j in this case are $\partial R_E/\partial a_j$.
4. *Bubnov–Galerkin.* Choose $w_j(P) = b_j(P)$. This is perhaps the best-known method.
5. *Stationary functional (variational).* With ϕ a variational integral (or other functional), set $\partial\phi[U]/\partial a_j = 0$, where $j = 1,\ldots,n$, to generate the n algebraic equations.

Example

$u_{xx} + u_{yy} = -2$, with $u = 0$ on the boundaries of the square $x = \pm 1, y = \pm 1$. Select an interior method with $U = a_1(1-x^2)(1-y^2) + a_2x^2y^2(1-x^2)(1-y^2)$, whereupon the residual $R_E(U) = -2a_1(2-x^2-y^2) + 2a_2[(1-6x^2)y^2\ (1-y^2) + (1-6y^2)x^2(1-x^2)] + 2$. Collocating at $\left(\frac{1}{3},\frac{1}{3}\right)$ and $\left(\frac{2}{3},\frac{2}{3}\right)$ gives the two linear equations $-32a_1/9 + 32a_2/243 + 2 = 0$ and $-20a_1/9 - 400a_2/243 + 2 = 0$ for a_1 and a_2.

WRM methods can obviously be used as approximate methods. We have now set the stage for *finite elements.*

Finite Elements

The WRM methods are more general than the *finite element* (FE) methods. FE methods require, in addition, that the basis functions be finite elements (i.e., functions that are zero except on a small part of the domain under consideration). A typical example of an often used basis is that of triangular elements. For a triangular element with Cartesian coordinates (x_1,y_1), (x_2,y_2), and (x_3,y_3), define natural coordinates L_1, L_2, and L_3 ($L_i \leftrightarrow (x_i,y_i)$) so that $L_i = A_i/A$ where

$$A = \frac{1}{2}\det\begin{bmatrix} 1 & x_1 & y_1 \\ 1 & x_2 & y_2 \\ 1 & x_3 & y_3 \end{bmatrix}$$

is the area of the triangle and

$$A_1 = \frac{1}{2}\det\begin{bmatrix} 1 & x & y \\ 1 & x_2 & y_2 \\ 1 & x_3 & y_3 \end{bmatrix}$$

$$A_2 = \frac{1}{2}\det\begin{bmatrix} 1 & x_1 & y_1 \\ 1 & x & y \\ 1 & x_3 & y_3 \end{bmatrix}$$

$$A_3 = \frac{1}{2}\det\begin{bmatrix} 1 & x_1 & y_1 \\ 1 & x_2 & y_2 \\ 1 & x & y \end{bmatrix}$$

Clearly $L_1 + L_2 + L_3 = 1$, and the L_i are one at node i and zero at the other nodes. In terms of the Cartesian coordinates,

$$\begin{bmatrix} L_1 \\ L_2 \\ L_3 \end{bmatrix} = \frac{1}{2A}\begin{bmatrix} x_2y_3 - x_3y_2, & y_2 - y_3, & x_3 - x_2 \\ x_3y_1 - x_1y_3, & y_3 - y_1, & x_1 - x_3 \\ x_1y_2 - x_2y_1, & y_1 - y_2, & x_2 - x_1 \end{bmatrix}\begin{bmatrix} 1 \\ x \\ y \end{bmatrix}$$

is the linear triangular element relation.

Tables of linear, quadratic, and cubic basis functions are given in the literature. Note that while the linear basis needs three nodes, the quadratic requires six, and the cubic basis ten. Various modifications, such as the Hermite basis, are described in the literature. Triangular elements are useful in approximating irregular domains.

For rectangular elements, the *chapeau* functions are often used. Let us illustrate with an example. Let $u_{xx} + u_{yy} = Q$, $0 < x < 2$, $0 < y < 2$, $u(x,2) = 1, u(0,y) = 1, u_y(x,0) = 0, u_x(2,y) = 0$, and $Q(x,y) = Q_w\delta(x-1)\delta(y-1)$:

$$\delta(x) = \begin{cases} 0 & x \neq 0 \\ 1 & x = 0 \end{cases}$$

Using four equal rectangular elements, map the element I with vertices at (0,0), (0,1), (1,1), and (1,0) into the local (canonical) coordinates (ξ, η), $-1 \le \xi \le 1, -1 \le \eta \le 1$, by means of $x = \tfrac{1}{2}(\xi + 1)$, $y = \tfrac{1}{2}(\eta + 1)$. This mapping permits one to develop software that standardizes the treatment of all elements. Converting to (ξ, η) coordinates, our problem becomes $u_{\xi\xi} + u_{\eta\eta} = \frac{1}{4}Q$, $-1 \le \xi \le 1, -1 \le \eta \le 1$, $Q = Q_w\delta(\xi - 1)\delta(\eta - 1)$.

First, a trial function $\bar{u}(\xi,\eta)$ is defined as $u(\xi,\eta) \approx \bar{u}(\xi,\eta) = \sum_{j=1}^{4} A_j\phi_j(\xi,\eta)$ (in element I) where the ϕ_j are the two-dimensional chapeau functions

$$\phi_1 = \left[\frac{1}{2}(1-\xi)\frac{1}{2}(1-\eta)\right] \quad \phi_2 = \left[\frac{1}{2}(1+\xi)\frac{1}{2}(1-\eta)\right]$$

$$\phi_3 = \left[\frac{1}{2}(1+\xi)\frac{1}{2}(1+\eta)\right] \quad \phi_4 = \left[\frac{1}{2}(1-\xi)\frac{1}{2}(1+\eta)\right]$$

Clearly ϕ_i take the value one at node *i*, provide a bilinear approximation, and are nonzero only over elements adjacent to node *i*.

Second, the equation residual $R_E = \nabla^2\bar{u} - \frac{1}{4}Q$ is formed and a WRM procedure is selected to formulate the algebraic equations for the A_i. This is indicated using the Galerkin method. Thus, for element *I*, we have

$$\int_{D_I}\int(\bar{u}_{\xi\xi} + \bar{u}_{\eta\eta} - Q)\phi_i(\xi,\eta)d\xi d\eta = 0, \quad i = 1,\ldots,4$$

Applying Green's theorem, this result becomes

$$\int_{D_I}\int\left[\bar{u}_\xi(\phi_i)_\xi + \bar{u}_\eta(\phi_i)_\eta + \frac{1}{4}Q\phi_i\right]d\xi d\eta - \int_{\partial D_I}(\bar{u}_\xi c_\xi + \bar{u}_\eta c_\eta)\phi_i ds = 0, \quad i = 1,2,\ldots,4$$

Using the same procedure in all four elements and recalling the property that the ϕ_i in each element are nonzero only over elements adjacent to node *i* gives the following nine equations:

$$\sum_{e=1}^{4}\left\{\int_{D_e}\int\sum_{j=1}^{9}A_j[(\phi_j)_\xi(\phi_j)_\xi + (\phi_j)_\eta(\phi_i)_\eta] + \frac{1}{4}Q\phi_j\right\}d\xi\, d\eta$$

$$-\sum_{e=1}^{4}\int_{\partial D_e}(\bar{u}_\xi c_\xi + \bar{u}_\eta c_\eta)\phi\, ds = 0, \quad i = 1,2,\ldots,9$$

where the c_ξ and c_η are the direction cosines of the appropriate element (*e*) boundary.

Method of Lines

The *method of lines*, when used on PDEs in two dimensions, reduces the PDE to a system of ordinary differential equations (ODEs), usually by finite difference or finite element techniques. If the original problem is an initial value (boundary value) problem, then the resulting ODEs form an initial value (boundary value) problem. These ODEs are solved by ODE numerical methods.

Example

$u_t = u_{xx} + u^2$, $0 < x < 1$, $0 < t$, with the initial value $u(x,0) = x$, and boundary data $u(0,t) = 0, u(1,t) = \sin t$. A discretization of the space variable (x) is introduced and the time variable is left continuous. The approximation is $\dot{u}_i = (u_{i+1} - 2u_i + u_{i-1})/h^2 + u_i^2$. With $h = 1/5$, the equations become

$$u_0(t) = 0$$

$$\dot{u}_1 = \frac{1}{25}[u_2 - 2u_1] + u_1^2$$

$$\dot{u}_2 = \frac{1}{25}[u_3 - 2u_2 + u_1] + u_2^2$$

$$\dot{u}_3 = \frac{1}{25}[u_4 - 2u_3 + u_2] + u_3^2$$

$$\dot{u}_4 = \frac{1}{25}[\sin t - 2u_4 + u_3] + u_4^2$$

$$u_5 = \sin t$$

and $u_1(0)=0.2$, $u_2(0)=0.4$, $u_3(0)=0.6$, and $u_4(0)=0.8$.

230.11 Discrete and Fast Fourier Transforms

Let $x(n)$ be a sequence that is nonzero only for a finite number of samples in the interval $0 \le n \le (N-1)$. The quantity

$$X(k)=\sum_{n=0}^{N-1} x(n)e^{-i(2\pi/N)nk},\ k=0,1,\ldots,N-1$$

is called the *discrete Fourier transform* (DFT) of the sequence $x(n)$. Its inverse (IDFT) is given by

$$x(n)=\frac{1}{N}\sum_{k=0}^{N-1} X(k)e^{i(2\pi/N)nk},\ n=0,1,\ldots,N-1\ (i^2=-1)$$

Clearly, DFT and IDFT are finite sums and there are N frequency values. Also, $X(k)$ is periodic in k with period N.

Example

$$x(0) = 1, x(1) = 2, x(2) = 3, x(3) = 4$$

$$X(k)=\sum_{n=0}^{3} x(n)e^{-i(2\pi/4)nk},\ k=0,1,2,3,4$$

Thus,

$$X(0)=\sum_{n=0}^{3} x(n)=10$$

and

$$X(1)=x(0)+x(1)e^{-i\pi/2}+x(2)e^{-i\pi}+x(3)e^{-i3\pi/2}=1-2i-3+4i=-2+2i;\ X(2)=-2;\ X(3)=-2-2i.$$

DFT Properties

1. Linearity: If $x_3(n)=ax_1(n)+bx_2(n)$, then $X_3(k)=aX_1(k)+bX_2(k)$.
2. Symmetry: For $x(n)$ real, $\mathrm{Re}[X(k)]=\mathrm{Re}[X(N-k)], \mathrm{Im}[X(k)]=-\mathrm{Im}[X(N-k)]$.
3. Circular shift: By a circular shift of a sequence defined in the interval $0\le n\le N-1$, we mean that, as values *fall off* from one end of the sequence, they are appended to the other end. Denoting this by $x(n\oplus m)$, we see that positive m means shift left and negative m means shift right. Thus, $x_2(n)=x_1(n\oplus m)\Leftrightarrow X_2(k)=X_1(k)e^{i(2\pi/N)km}$
4. Duality: $x(n)\Leftrightarrow X(k)$ implies $(1/N)X(n)\Leftrightarrow x(-k)$.
5. Z-transform relation: $X(k)=X(z)\big|_{z=e^{i(2\pi k/N)}},\ k=0,1,\ldots,N-1$.
6. Circular convolution: $x_3(n)=\sum_{m=0}^{N-1} x_1(m)x_2(n\ominus m)=\sum_{\ell=0}^{N-1} x_1(n\ominus \ell)x_2(\ell)$ where $x_2(n\oplus m)$ corresponds to a circular shift to the right for positive m.

One fast algorithm for calculating DFTs is the radix-2 *fast Fourier transform* developed by J. W. Cooley and J. W. Tucker. Consider the two-point DFT $X(k)=\sum_{n=0}^{1} x(n)e^{-i(2\pi/2)nk}$, $k=0,1$. Clearly, $X(k)=x(0)+ x(1)e^{-i\pi k}$. So $X(0)=x(0)+x(1)$ and $X(1)=x(0)-x(1)$. This process can be extended to DFTs of length $N = 2^r$, where r is a positive integer. For $N = 2^r$, decompose the N-point DFT into *two* $N/2$-point DFTs. Then, decompose each $N/2$-point DFT into *two* $N/4$-point DFTs, and so on until eventually we have $N/2$ *two*-point DFTs. Computing these as indicated above, we recombine them into $N/4$ four-point DFTs and then $N/8$ eight-point DFTs, and so on, until the DFT is computed. The total number of DFT operations (for large N) is $O(N^2)$, and that of the FFT is $O(N \log_2 N)$, quite a saving for large N.

230.12 Software

Some available software packages are listed here.

General Packages

General software packages include Maple, Mathematica, and Matlab. All contain algorithms for handling a large variety of both numerical and symbolic computations.

Special Packages for Linear Systems

In the IMSL Library, there are three complementary linear system packages of note.

LINPACK is a collection of programs concerned with *direct* methods for general (or full) symmetric, symmetric positive definite, triangular, and tridiagonal matrices. There are also programs for least squares problems, along with the QR algorithm for eigensystems and the singular value decompositions of rectangular matrices. The programs are intended to be completely machine independent, fully portable, and run with good efficiency in most computing environments. The LINPACK User's Guide by Dongarra et al. is the basic reference.

ITPACK is a modular set of programs for iterative methods. The package is oriented toward the sparse matrices that arise in the solution of PDEs and other applications. While the programs apply to full matrices, that is rarely profitable. Four basic iteration methods and two convergence acceleration methods are in the package. There is a Jacobi, SOR (with optimum relaxation parameter estimated), symmetric SOR, and reduced system (red-black ordering) iteration, each with semi-iteration and conjugate gradient acceleration. All parameters for these iterations are automatically estimated. The practical and theoretical background for ITPACK is found in Hagemen and Young [1981].

YALEPACK is a substantial collection of programs for sparse matrix computations.

Ordinary Differential Equations Packages

Also in IMSL, one finds such sophisticated software as DVERK, DGEAR, or DREBS for initial value problems. For two-point boundary value problems, one finds DTPTB (use of DVERK and multiple shooting) or DVCPR.

Partial Differential Equations Packages

DISPL was developed and written at Argonne National Laboratory. DISPL is designed for nonlinear second-order PDEs (parabolic, elliptic, hyperbolic [some cases], and parabolic elliptic). Boundary conditions of a general nature and material interfaces are allowed. The spatial dimension can be either one or two and in Cartesian, cylindrical, or spherical (one dimension only) geometry. The PDEs are reduced to ordinary DEs by Galerkin discretization of the spatial variables. The resulting ordinary DEs in the timelike variable are then solved by an ODE software package (such as GEAR). Software features include graphics capabilities, printed output, dump/restart facilities, and free format input. DISPL is intended to be an engineering and scientific tool and is not a finely tuned production code for a small set of problems. DISPL makes no effort to control the spatial discretization errors. It has been used to successfully solve a variety of problems in chemical transport, heat and mass transfer, pipe flow, etc.

PDELIB was developed and written at Los Alamos Scientific Laboratory. PDELIB is a library of subroutines to support the numerical solution of evolution equations with a timelike variable and one or two space variables. The routines are grouped into a dozen independent modules according to their function (i.e., accepting initial data, approximating spatial derivatives, advancing the solution in time). Each task is isolated in a distinct module. Within a module, the basic task is further refined into general-purpose flexible lower-level routines. PDELIB can be understood and used at different levels. Within a small period of time, a large class of problems can be solved by a novice. Moreover, it can provide a wide variety of outputs.

DSS/2 is a differential systems simulator developed at Lehigh University as a transportable numerical method of lines (NMOL) code. See also LEANS.

FORSIM is designed for the automated solution of sets of implicitly coupled PDEs of the form

$$\frac{\partial u_i}{\partial t} = \phi_i(x,t,u_i,u_j,\ldots,(u_i)_x,\ldots,(u_i)_{xx},(u_j)_{xx},\ldots), \quad \text{for } i = 1,\ldots,N$$

The user specifies the ϕ_i in a simple FORTRAN subroutine. Finite difference formulas of any order may be selected for the spatial discretization and the spatial grid need not be equidistant. The resulting system of time-dependent ODEs is solved by the method of lines.

SLDGL is a program package for the self-adaptive solution of nonlinear systems of elliptic and parabolic PDEs in up to three space dimensions. Variable step size and variable order are permitted. The discretization error is estimated and used for the determination of the optimum grid and optimum orders. This is the most general of the codes described here (not for hyperbolic systems, of course). This package has seen extensive use in Europe.

FIDISOL (finite difference solver) is a program package for nonlinear systems of two- or three-dimensional elliptic and parabolic systems in rectangular domains or in domains that can be transformed analytically to rectangular domains. This package is actually a redesign of parts of SLDGL, primarily for the solution of large problems on vector computers. It has been tested on the CYBER 205, CRAY-1M, CRAY X-MP/22, and VP 200. The program vectorizes very well and uses the vector arithmetic efficiently. In addition to the numerical solution, a reliable error estimate is computed.

CAVE is a program package for conduction analysis via eigenvalues for three-dimensional geometries using the method of lines. In many problems, much time is saved because only a few terms suffice.

Many industrial and university computing services subscribe to the IMSL Software Library. Announcements of new software appear in *Directions*, a publication of IMSL. A brief description of some IMSL packages applicable to PDEs and associated problems is now given. In addition to those packages just described, two additional software packages bear mention. The first of these, the ELLPACK system, solves elliptic problems in two dimensions with general domains and in three dimensions with box-shaped domains. The system contains over 30 numerical methods modules, thereby providing a means of evaluating and comparing different methods for solving elliptic problems. ELLPACK has a special high-level language making it easy to use. New algorithms can be added or deleted from the system with ease.

Second, TWODEPEP is IMSL's general finite element system for two-dimensional elliptic, parabolic, and eigenvalue problems. The Galerkin finite elements available are triangles with quadratic, cubic, or quartic basic functions, with one edge curved when adjacent to a curved boundary, according to the isoparametric method. Nonlinear equations are solved by Newton's method, with the resulting linear system solved directly by Gauss elimination. PDE/PROTRAN is also available. It uses triangular elements with piecewise polynomials of degree 2, 3, or 4 to solve quite general steady state, time-dependent, and eigenvalue problems in general two-dimensional regions. There is a simple user input. Additional information may be obtained from IMSL. NASTRAN and STRUDL are two advanced finite element computer systems available from a variety of sources. Another, UNAFEM, has been extensively used.

References

General

Adams, E. and Kulisch, U. (Eds.) 1993. *Scientific Computing with Automatic Result Verification.* Academic Press, Boston, MA.

Gerald, C. F. and Wheatley, P. O. 1984. *Applied Numerical Analysis.* Addison-Wesley, Reading, MA.

Hamming, R. W. 1962. *Numerical Methods for Scientists and Engineers.* McGraw-Hill, New York.

Hildebrand, F. B. 1956. *Introduction to Numerical Analysis.* McGraw-Hill, New York.

Isaacson, E. and Keller, H. B. 1966. *Analysis of Numerical Methods.* John Wiley & Sons, New York.

Kopal, Z. 1955. *Numerical Analysis.* John Wiley & Sons, New York.

Rice, J. R. 1993. *Numerical Methods, Software and Analysis,* 2d ed. Academic Press, Boston, MA.

Stoer, J. and Bulirsch, R. 1976. *Introduction to Numerical Analysis.* Springer, New York.

Linear Equations

Bodewig, E. 1956. *Matrix Calculus.* Wiley (Interscience), New York.

Hageman, L. A. and Young, D. M. 1981. *Applied Iterative Methods.* Academic Press, Boston, MA.

Varga, R. S. 1962. *Matrix Iterative Numerial Analysis.* John Wiley & Sons, New York.

Young, D. M. 1971. *Iterative Solution of Large Linear Systems.* Academic Press, Boston, MA.

Ordinary Differential Equations

Aiken, R. C. 1985. *Stiff Computation.* Oxford University Press, New York.

Gear, C. W. 1971. *Numerical Initial Value Problems in Ordinary Differential Equations.* Prentice Hall, Englewood Cliffs, NJ.

Keller, H. B. 1976. *Numerical Solutions of Two Point Boundary Value Problems.* SIAM, Philadelphia, PA.

Lambert, J. D. 1973. *Computational Methods in Ordinary Differential Equations.* Cambridge University Press, New York.

Milne, W. E. 1953. *Numerical Solution of Differential Equations.* John Wiley & Sons, New York.

Rockey, K. C., Evans, H. R., Griffiths, D. W., and Nethercot, D. A. 1983. *The Finite Element Method — A Basic Introduction for Engineers,* 2d ed. Halstead Press, New York.

Shampine, L. and Gear, C. W. 1979. A User's View of Solving Stiff Ordinary Differential Equations, *SIAM Rev.* 21:1–17.

Partial Differential Equations

Ames, W. F. 1993. *Numerical Methods for Partial Differential Equations,* 3d ed. Academic Press, Boston, MA.

Brebbia, C. A. 1984. *Boundary Element Techniques in Computer Aided Engineering.* Martinus Nijhoff, Boston, MA.

Burnett, D. S. 1987. *Finite Element Analysis.* Addison-Wesley, Reading, MA.

Lapidus, L. and Pinder, G. F. 1982. *Numerical Solution of Partial Differential Equations in Science and Engineering.* John Wiley & Sons, New York.

Roache, P. 1972. *Computational Fluid Dynamics.* Hermosa, Albuquerque, NM.

231
Dimensional Analysis

William F. Ames
Georgia Institute of Technology

231.1 Units and Variables

Dimensional analysis is a mathematical tool whose use will enable the scientist and engineer to save time in planning experiments and correlating results of experiments. The method arose in the science of mechanics. In this science three *units* are regarded as fundamental, namely, *length*, *mass*, and *time*. Other employed units are called *derived units*. For example, velocity and acceleration are derived units defined by reference to the two fundamental units — length and time. The units of force-momentum, mechanical energy, and power are dependent on all three of the fundamental units. The study of electricity, heat transfer, and so forth requires the inclusion of other fundamental units such as charge.

Here the fundamental units adopted are *length* (in centimeters), L; *mass* (in grams), M; and *time* (in seconds), T. This is the CGS system of units. Others are the British, or foot, pound, second (FPS) system and the MKS system, which is closely allied to the CGS system. The dimensional formulas of some basic physical variables in the CGS system are given in Tables 231.1 through 231.3.

Some *variables* are dimensionless. For example, angle is defined as arc/radius; strain as volume/volume; and specific gravity as density/density (dimensions $(L^{-3}\ M)/(L^{-3}\ M) = L^0M^0$). In addition there are dimensional constants: c, the velocity of light in a vacuum, has dimensions LT^{-1}, and g, the gravitational constant, has dimensions $L^3M^{-1}T^{-2}$. Some dimensionless constants also appear. An example is the Reynolds number $N_{Re} = Dv\rho/\mu$ (D is diameter), which has dimensions $(L \cdot LT^{-1} \cdot L^{-3}M)/(L^{-1}MT^{-1}) = L^0M^0T^0$. Some dimensionless groups are listed in Table 231.4.

231.2 Method of Dimensions

The *method of dimensions* is based on the simple observation that the dimensions on both sides of an equation must be the same. To illustrate, suppose we wish to find the distance s traveled in a given time by a particle of mass m falling from rest under the uniform acceleration due to gravity g. Assuming $s = f(g,t,m) = Cg^at^bm^c$, where C is an unknown dimensionless constant and the indices a, b, and c are to be determined. Examining the dimensions in a dimensional formula we have

$$L = (LT^{-2})^aT^bM^c$$

Clearly $a = 1$, $b = 2$, and $c = 0$, so $s = Cgt^2$. It is not within the realm of dimensional analysis to evaluate C.

Another example is the two-body problem of astronomy. Suppose a planet of mass m rotates about the sun (mass $\overline{M}$) in an orbit that is an ellipse of major diameter D. The physical quantities that will affect the sought-for period, t, appear to be the mass of the sun, the mass of the planet, the diameter D,

0-8493-1586-7/05/$0.00+$1.50

TABLE 231.1 Dimensional Physical Quantities

Physical Quantity	Exponents of Dimensions L	M	T	Formula
Volume density	−3	1	0	$L^{-3}M$
Length per unit volume	−2	0	0	L^{-2}
Area density	−2	1	0	$L^{-2}M$
Curvature	−1	0	0	L^{-1}
Linear density	−1	1	0	$L^{-1}M$
Angle	0	0	0	
Mass	0	1	0	M
Length	1	0	0	L
Mass × length	1	1	0	LM
Area	2	0	0	L^2
Moment of inertia	2	1	0	L^2M
Volume	3	0	0	L^3
Speed of density change	−3	1	−1	$L^{-3}M\,T^{-1}$
Velocity per unit volume	−2	0	−1	$L^{-2}\,T^{-1}$
Momentum per unit volume	−2	1	−1	$L^{-2}M\,T^{-1}$
Velocity per unit area	−1	0	−1	$L^{-1}\,T^{-1}$
Viscosity	−1	1	−1	$L^{-1}M\,T^{-1}$
Frequency	0	0	−1	T^{-1}
Mass per second	0	1	−1	$M\,T^{-1}$
Velocity	1	0	−1	$L\,T^{-1}$
Momentum	1	1	−1	$LM\,T^{-1}$
Kinematic viscosity	2	0	−1	$L^2\,T^{-1}$
Action	2	1	−1	$L^2M\,T^{-1}$
Volume per second	3	0	−1	$L^3\,T^{-1}$
Acceleration of density change	−3	1	−2	$L^{-3}M\,T^{-2}$
Acceleration per unit volume	−2	0	−2	$L^{-2}\,T^{-2}$
Force per unit volume	−2	1	−2	$L^{-2}M\,T^{-2}$
Acceleration per unit area	−1	0	−2	$L^{-1}\,T^{-2}$
Pressure	−1	1	−2	$L^{-1}M\,T^{-2}$
Angular acceleration	0	0	−2	T^{-2}
Surface tension	0	1	−2	$M\,T^{-2}$
Acceleration	1	0	−2	$L\,T^{-2}$
Force	1	1	−2	$LM\,T^{-2}$
Temperature	2	0	−2	$L^2\,T^{-2}$
Energy, torque	2	1	−2	$L^2M\,T^{-2}$
Rate of change of volume per second	3	0	−2	$L^3\,T^{-2}$
Power	2	1	−3	$L^2M\,T^{-2}$

TABLE 231.2 Dimensional Thermal Quantities

Physical Quantity	Thermal Units	Dynamic Units
Temperature	θ	θ
Quantity of heat	H	L^2MT^{-2}
Specific heat	Dimensionless	Dimensionless
Heat capacity per unit mass	$L^0\,M^{-1}T^0H\theta^{-1}$	$L^2M^0T^{-2}\theta^{-1}$
Heat capacity per unit volume	$L^{-3}M^0T^0H\theta^{-1}$	$L^{-1}M\,T^{-2}\theta^{-1}$
Temperature gradient	$L^{-1}M^0T^0H^0\theta^{-1}$	$L^{-1}M^0T^0\theta^{-1}$
Conductivity	$L^{-1}M^0T^{-1}H^0\theta^{-1}$	$LMT^{-3}\theta^{1}$
Entropy	$H\theta^{-1}$	$L^2MT^{-2}\theta^{-1}$

TABLE 231.3 Dimensional Magnetic and Electrical Quantities

Physical Quantity	Electromagnetic	Electrostatic
Magnetic pole, P	$L^{3/2}M^{1/2}T^{-1}\mu^{1/2}$	$L^{1/2}M^{1/2}T^{0}\kappa^{-1/2}$
Strength of magnetic field, H	$L^{-1/2}M^{1/2}T^{-1}\mu^{1/2}$	$L^{1/2}M^{1/2}T^{-1}\kappa^{-1/2}$
Magnetic and electric induction, μH, κE	$L^{-1/2}M^{1/2}T^{-1}\mu^{1/2}$	$L^{-1/2}M^{1/2}T^{-1}\kappa^{-1/2}$
Magnetic and electric moments, PL, QL	$L^{5/2}M^{1/2}T^{-1}\mu^{1/2}$	$L^{5/2}M^{1/2}T^{-1}\kappa^{1/2}$
Electric current, I	$L^{1/2}M^{1/2}T^{-1}\mu^{-2}$	$L^{3/2}M^{1/2}T^{-1}\kappa^{1/2}$
Quantity of electricity, Q	$L^{1/2}M^{1/2}T^{0}\mu^{-1/2}$	$L^{3/2}M^{1/2}T^{-1}\kappa^{1/2}$
Potential difference, E	$L^{3/2}M^{1/2}T^{-2}\mu^{1/2}$	$L^{1/2}M^{1/2}T^{-1}/\kappa$
Resistance, R	$LM^{0}T^{-1}\mu$	$L^{-1}M^{0}T\kappa^{-1}$
Capacitance, C	$L^{-1}M^{0}T^{2}\mu^{-1}$	$LM^{0}T^{0}\kappa$
Inductance, ET/I	$LM^{0}T^{0}\mu$	$L^{-1}M^{0}T^{2}\kappa^{-1}$
Permeability, μ	$L^{0}M^{0}T^{0}\mu$	$L^{-2}M^{0}T^{2}\kappa^{-1}$
Permittivity, κ	$L^{-2}M^{0}T^{2}\mu^{-1}$	$L^{0}M^{0}T^{0}\kappa$

TABLE 231.4 Dimensionless Groups in Engineering

Biot number	N_{Bi}	hL/k
Condensation number	N_{Co}	$(h/k)(\mu 2/\rho^2 g)^{1/3}$
Number used in condensation of vapors	N_{Cv}	$L^3\rho^2 g\lambda/k\mu\Delta t$
Euler number	N_{Eu}	$ge(-dp)/\rho V^2$
Fourier number	N_{Fo}	$k\theta/\rho cL^2$
Froude number	N_{Fr}	V^2/Lg
Graetz number	N_{Gz}	wc/kL
Grashof number	N_{Gr}	$\mathrm{L}^3\rho^2\beta g\Delta \mathrm{t}/\mu^2$
Mach number	N_{Ma}	V/V_a
Nusselt number	N_{Nu}	hD/k
Peclet number	N_{Pe}	$DV\rho c/k$
Prandtl number	N_{Pr}	$c\mu/k$
Reynolds number	N_{Re}	$DV\rho/\mu$
Schmidt number	N_{Sc}	$\mu/\rho D_v$
Stanton number	N_{St}	$h/cV\rho$
Weber number	N_{We}	$LV^2\rho/\sigma g_e$

and the gravitational constant G. Thus, we try $t = C\overline{M}^a m^b D^c G^d$ + (similar terms), where the similar terms must have the same dimensions, that is, 0 in length, 0 in mass, and 1 in time. The dimensional equation becomes $T = M^a M^b L^c (L^3 M^{-1}\ T^{-2})^d$. Hence

$$\begin{aligned} &(\text{length}) && 0 = c + 3d \\ &(\text{mass}) && 0 = a + b - d \\ &(\text{time}) && 1 = -2d \end{aligned}$$

The solution of this system is $d = -1/2$, $c = 3/2$, a is arbitrary, and $b = -1/2 - a$. Therefore, $t = C\overline{M}^a m^{(-1/2)-a} D^{3/2} G^{-1/2}$ or $t^2 = C'(\overline{M}/m)^{2a}\,(D^3/Gm)$, which is Kepler's law (the square of the period of a planet is proportional to the cube of the major axis of the orbit). The constants C' and a must be evaluated by experiment.

Many other examples are found in the literature. This section closes with the determination of the height, h, that a liquid rises in a capillary tube. The relevant equation is $h = C \cdot \rho^a r^b s^c g^d \theta^e$, where ρ is the fluid density, r is the radius of the tube, s is the surface tension of the fluid, g is the acceleration of gravity, and θ is the contact angle. The dimensional equation becomes $L = (L^{-3}M)^a L^b (MT^{-2})^c (LT^{-1})^d$, which gives rise to the following three linear equations:

$$\begin{aligned} &\text{(length)} && 1 = -3a + b + d \\ &\text{(mass)} && 0 = a + c \\ &\text{(time)} && 0 = -2c - 2d \end{aligned}$$

in the four unknowns a, b, c, and d. We may express any three of the unknowns in terms of the fourth. Choosing a we find $c = -a$, $d = a$, and $b = 1 + 2a$, so $h = cr(r^2\rho g/s)^a$. Experiments show that h is inversely proportional to r, so that $a = -1$. Thus $h = C(s/r\rho g)$.

References

Bridgeman, P. W. 1937. *Dimensional Analysis*. Yale University Press, New Haven, CT.
Huntley, H. E. 1952. *Dimensional Analysis*. MacDonald, London.

232

Computer Graphics Visualization

Richard S. Gallagher
R. S. Gallagher and Associates

Visualization is the process of seeing something. In the computer graphics field, the term *visualization* has a more specific meaning, and refers to computational techniques to display and understand behavior. The field of computer graphics visualization is still young; much of its fundamental work took place in the late 1980s. Today, it has grown to become one of the key commercial applications of computer graphics. Key technical aspects of computer graphics visualization include the three-dimensional (3-D) display of objects; display of scalar, vector, and tensor quantities; continuum volume display techniques; and animation over time.

232.1 3-D Display of Objects

Most modern computer graphics equipment uses display technology similar to that of television, where a two-dimensional array of dots known as *pixels* have individual dot colors set to produce a composite image. Turning data into such a screen image involves representing the data as geometric components, such as lines or polygons, *transforming* these components to a particular view from the observer, coloring or *rendering* these components, and converting them to dots on the screen through a process known as *scan conversion.*

Transformations are generally performed through matrix multiplication of the coordinates of each component, effecting operations including *rotation, scaling,* and *translation.* The transformation of a point P to its transformed position P' takes the form:

$$[P'] = [T][P] = [r][s][t][P]$$

Scaling and translation matrices can be combined in the form:

0-8493-1586-7/05/$0.00+$1.50

$$\begin{bmatrix} s_x & 0 & 0 & \Delta_x \\ 0 & s_y & 0 & \Delta_y \\ 0 & 0 & s_z & \Delta_z \\ 0 & 0 & 0 & 1 \end{bmatrix}$$

which multiplies the vector $[x, y, z, W]$, with the fourth element being a nonzero *homogenous coordinate* factor of x, y, and z. The rotational transformation is added by multiplying this matrix by matrices corresponding to rotations about the x,y, and z axes:

$$\begin{bmatrix} 1 & 0 & 0 & 0 \\ 0 & \cos\theta & -\sin\theta & 0 \\ 0 & \sin\theta & \cos\theta & 0 \\ 0 & 0 & 0 & 1 \end{bmatrix} \begin{bmatrix} \cos\theta & 0 & \sin\theta & 0 \\ 0 & 1 & 0 & 0 \\ -\sin\theta & 0 & \cos\theta & 0 \\ 0 & 0 & 0 & 1 \end{bmatrix} \begin{bmatrix} \cos\theta & -\sin\theta & 0 & 0 \\ \sin\theta & \cos\theta & 0 & 0 \\ 0 & 0 & 1 & 0 \\ 0 & 0 & 0 & 1 \end{bmatrix}$$

Other transformations include the simulation of *perspective*, which reduces the x and y coordinates in the current view as a function of the z coordinate, and *clipping*, which removes portions of components intersected by either screen boundaries or an arbitrary *clipping plane.*

Rendering techniques generally involve setting a color or color function across each component, often based on the effects of a light source from the observer to the screen. For example, a square polygon facing the observer may be red, while its color may change toward darker shades of red as the polygon is rotated, becoming completely dark when 90° from the observer. Shading techniques may compute color as a continuous function across a component. Gouraud shading, for example, interpolates a color value across a polygon from its corner values, while another technique known as Phong shading computes color from an interpolation of the light source vector itself across the polygon. More advanced rendering techniques include *ray tracing,* which computes the behavior of light rays to simulate effects such as reflectance, shadows, and translucency, and *texture mapping,* which simulates a pattern or image across surfaces of the displayed model.

232.2 Scalar Display Techniques

One of the fundamental operations of visualization is the display of a single variable in space. Some of the visual techniques for displaying scalar field values include:

Color coding. The outside visible surfaces of a model are color-coded with values corresponding to the scalar value (Figure 232.1). For example, a structural model may have elements of high stress colored red, and low stress colored blue.

Isovalue display. An *isovalue* is a region of constant scalar value. In the general case, an isovalue can be idealized as a point in one-dimensional space, a curve in two dimensions, and a surface in three dimensions (Figure 232.2). One particular form of isovalue display, the *contour plot,* displays bands of color on model surfaces corresponding to ranges of isovalues. Isosurfaces are surfaces of constant 3-D scalar value.

Particle displays and implicit isovalues. A color-coded distribution of particles within a 3-D scalar field gives an overview of values throughout the field and can be useful in situations where isosurfaces obscure portions of the interior model. Another application of particle display is to compute and display particles at points corresponding to a particular scalar value. As the density of such particles increases, the resulting image approaches the isosurface of the scalar value in the limit.

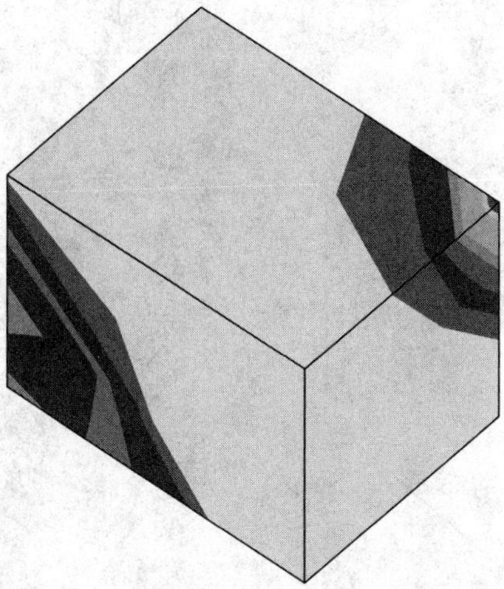

FIGURE 232.1 Contour representation of a scalar value.

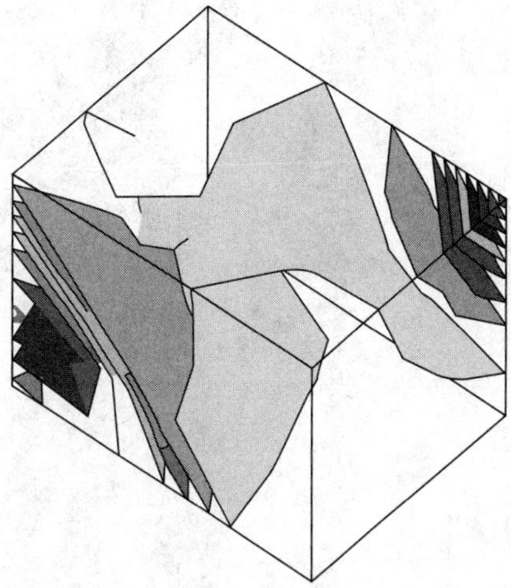

FIGURE 232.2 Isosurfaces of a scalar value.

232.3 Vector and Tensor Field Display

The comprehensible display of multidimensional quantities remains a key area of visualization research. Particular applications include the display of flow fields, multivariate field analysis problems, and examination of derived quantities such as gradients. Some of the current techniques used in vector and tensor field display include the following:

Vector and tensor symbols. Symbols, often known as *glyphs,* use size, shape and color to show multiple values at a point in space. For example, a three-pointed triad may have each point colored and sized according to the value of different quantities at that point. More complex glyphs may vary the shape, direction, and color of multiple components to show complex states of behavior, such as the state of a tensor field.

Vector and tensor field curves. Path curves can be created through a vector or tensor field, generally representing the path of a vector quantity from specified starting locations. Further dimensions of information can be displayed across these path curves by varying quantities such as color, thickness, and cross-sectional geometry.

Particle field display. A distribution of particles within a field can represent a multivariate quantity by varying attributes such as particle density, color, size, and shape. As one example, Figure 232.3 shows a particle distribution whose density varies by the gradient, or rate of change, of a state of stress in the Cartesian *x, y,* and *z* directions.

All of these techniques share a common approach of combining several visual techniques within a single image to display multivariate behavior. In addition, the use of animation (discussed below) adds the dimension of time and motion to evaluating complex behavior visually.

232.4 Continuum Volume Visualization

A 3-D field, particularly a sampled field such as a medical image, can be represented as a field of volume elements or *voxels.* These voxels can be viewed as tiny cubes connecting a regular array of sample points within the volume. Continuum volume visualization techniques generate imagery by performing operations on the voxel field. Such algorithms can allow an entire scalar-valued field to be represented within a single image, using techniques such as translucency and color variation, and can also provide a more accurate means of computing discrete images, such as isosurfaces (Figure 232.4). Above all, a

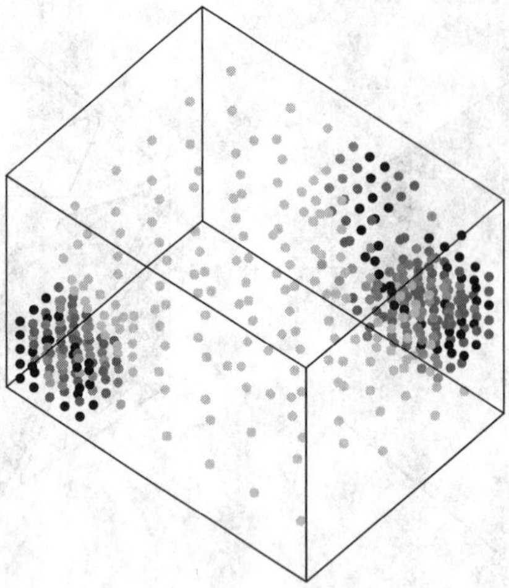

FIGURE 232.3 Particle representation of the gradient of a scalar value. Dot color varies by scalar value, while dot density varies by gradient.

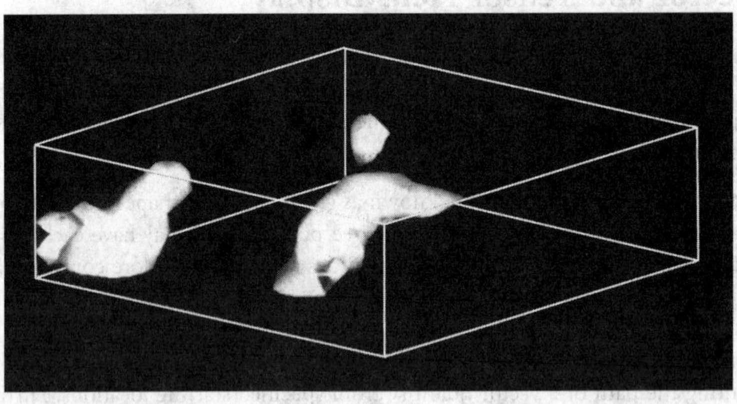

FIGURE 232.4 Isovalues of density within a voxel data set.

voxel representation of a volume keeps an entire volume's data in a form suited to direct visualization and analysis.

Many voxel-based techniques operate in either *image space* or *voxel space*. Image space techniques often involve casting rays through each dot of the screen image into the projected voxel array, computing contributions to the ray's color and opacity from each voxel the ray intersects. Such algorithms may work from front to back, stopping when a ray becomes opaque or exits the volume, or back to front. Preclassification of the voxel space can help optimize these techniques or allow computations to be performed in parallel.

Voxel space techniques evaluate voxels to create polygons which contribute to the image. One such example, the Marching Cubes algorithm patented in 1987 by General Electric, examines which vertices of a voxel are above or below a threshold value, and then orders these binary values into a bit string whose value is used to look up the topology of any isosurface polygons passing through the voxel.

232.5 Animation Over Time

With increasing computing capabilities, a growing area of interest in visualization is the display of behavior which varies with time or motion.

At its most basic level, animation is simply the generation and playback of individual images to produce the sensation of continuous movement, which humans perceive at a rate between 10 and 30 frames per second. The steps involved in producing animated visualization sequences include:

Interpolation of behavior. Within each individual frame, aspects, such as result values or model deformation, can be interpolated between the initial and final positions. While simple linear interpolation is most common, this interpolation function can incorporate issues, such as acceleration and deceleration, as well.

Motion of the object. The position of the model can be interpolated linearly, or along a path, generally using the same function of incrementation as would be used for behavior. Issues in the general case of object motion include continuity across multiple connecting path segments, and the combination of one or more viewing transformations, such as translation and rotation.

Motion of the observer. The observer's position can be likened to the location of a camera within a scene. This camera position can also be interpolated along a path, moving into, out of, or around a scene, while accounting for display characteristics, such as perspective and lighting.

Frame capture and playback. While animated sequences can be directly generated and displayed when adequate computing and display resources exist, animation above a certain level of complexity must be stored frame by frame for later playback. Frame capture can be accomplished in computer memory, as images saved in a given format on computer disk, or by using special-purpose hardware, such as videotape controllers or writable video discs.

232.6 Summary

Techniques such as these share a common purpose of making objects and behavior more comprehensible to the human observer. In particular, they provide engineers with a means to better understand complex 3-D behavior. Trends towards the future in this area include computing performance improvements ranging from better display throughput to massively parallel architectures, a greater degree of interactivity, improved user interfaces, and further development of display algorithms.

Defining Terms

Glyph — A symbol displaying multivariate field values at a point in space.

Isovalue — A region of constant scalar value within a field of geometry. For example, an *isosurface* represents constant-value surfaces within a 3-D volume.

Pixel — A unit two-dimensional screen dot location in a computer graphics image.

Ray tracing — A technique for realistic display of geometry and volumetric data, involving computing the path of a light ray from screen pixel locations to its projected or reflected locations within the 3-D model.

Scan conversion — The process of converting geometric entities, such as lines and polygons, to projected screen dot (pixel) locations on a computer graphics display.

Voxel — A unit volume element within a regular 3-D array, describing a discrete volume space.

References

Foley, J. D., Van Dam, A., Feiner, S. K., and Hughes, J. 1990. *Computer Graphics, Principles and Practice*, 2nd ed. Addison-Wesley, Reading, MA.

Gallagher, R. S., Ed. 1994. *Computer Visualization.* CRC Press, Boca Raton, FL.

Kaufman, A., Ed. 1990. *Volume Visualization.* IEEE Computer Society Press, Los Alamitos, CA.

Lorensen, W. and Cline, H. E. 1987. Marching cubes: a high resolution 3-D surface construction algorithm. *Computer Graphics*, 21(4), 163–169.

Further Information

Association for Computer Machinery (ACM) SIGGRAPH, annual conference proceedings, ACM Press.
IEEE Visualization, annual conference proceedings (1990–present), IEEE Computer Society Press.

Appendix: Mathematical Tables and Formulae

A.1 Greek Alphabet

Greek Letter		Greek Name	English Equivalent	Greek Letter		Greek Name	English Equivalent
Α	α	Alpha	a	Ν	ν	Nu	n
Β	β	Beta	b	Ξ	ξ	Xi	x
Γ	γ	Gamma	g	Ο	ο	Omicron	ŏ
Δ	δ	Delta	d	Π	π	Pi	p
Ε	ε	Epsilon	ĕ	Ρ	ρ	Rho	r
Ζ	ζ	Zeta	z	Σ	σ ς	Sigma	s
Η	η	Eta	ē	Τ	τ	Tau	t
Θ	θ ϑ	Theta	th	Υ	υ	Upsilon	u
Ι	ι	Iota	i	Φ	ϕ φ	Phi	ph
Κ	κ	Kappa	k	Ξ	χ	Chi	ch
Λ	λ	Lambda	l	Ψ	ψ	Psi	ps
Μ	μ	Mu	m	Ω	ω	Omega	ō

A.2 International System of Units (SI)

The International System of units (SI) was adopted by the 11th General Conference on Weights and Measures (CGPM) in 1960. It is a coherent system of units built from seven *SI base units*, one for each of the seven dimensionally independent base quantities: the meter, kilogram, second, ampere, kelvin, mole, and candela, for the dimensions length, mass, time, electric current, thermodynamic temperature, amount of substance, and luminous intensity, respectively. The definitions of the SI base units are given below. The *SI derived units* are expressed as products of powers of the base units, analogous to the corresponding relations between physical quantities but with numerical factors equal to unity.

In the International System there is only one SI unit for each physical quantity. This is either the appropriate SI base unit itself or the appropriate SI derived unit. However, any of the approved decimal prefixes, called *SI prefixes*, may be used to construct decimal multiples or submultiples of SI units.

It is recommended that only SI units be used in science and technology (with SI prefixes where appropriate). Where there are special reasons for making an exception to this rule, it is recommended always to define the units used in terms of SI units. This section is based on information supplied by IUPAC.

Definitions of SI Base Units

Meter: The meter is the length of path traveled by light in vacuum during a time interval of 1/299 792 458 of second (17th CGPM, 1983).

Kilogram: The kilogram is the unit of mass; it is equal to the mass of the international prototype of the kilogram (3rd CGPM, 1901).

Second: The second is the duration of 9 192 631 770 periods of the radiation corresponding to the transition between the two hyperfine levels of the ground state of the cesium-133 atom (13th CGPM, 1967).

Ampere: The ampere is that constant current which, if maintained in two straight parallel conductors of infinite length, of negligible circular cross section, and placed 1 meter apart in vacuum, would produce between these conductors a force equal to 2×10^{-7} newton per meter of length (9th CGPM, 1948).

Kelvin: The kelvin, unit of thermodynamic temperature, is the fraction 1/273.16 of the thermodynamic temperature of the triple point of water (13th CGPM, 1967).

Mole: The mole is the amount of substance of a system which contains as many elementary entities as there are atoms in 0.012 kilogram of carbon-12. When the mole is used, the elementary entities must be specified and may be atoms, molecules, ions, electrons, or other particles, or specified groups of such particles (14th CGPM, 1971). Examples of the use of the mole:

- 1 mol of H_2 contains about 6.022×10^{23} H_2 molecules, or 12.044×10^{23} H atoms.
- 1 mol of HgCl has a mass of 236.04 g.

- 1 mol of Hg_2Cl_2 has a mass of 472.08 g.
- 1 mol of Hg_2^{2+} has a mass of 401.18 g and a charge of 192.97 kC.
- 1 mol of $Fe_{0.91}$ S has a mass of 82.88 g.
- 1 mol of e^- has a mass of 548.60 μg and a charge of −96.49 kC.
- 1 mol of photons whose frequency if 10^{14} Hz has energy of about 39.90 kJ.

Candela: The candela is the luminous intensity, in a given direction, of a source that emits monochromatic radiation of frequency 540 × 10^{12} Hz and that has a radiant intensity in that direction of (1/683) watt per steradian (16th CGPM, 1979).

Names and Symbols for the SI Base Units

Physical Quantity	Name of SI Unit	Symbol for SI Unit
Length	meter	m
Mass	kilogram	kg
Time	second	s
Electric current	ampere	A
Thermodynamic temperature	kelvin	K
Amount of substance	mole	mol
Luminous intensity	candela	cd

SI Derived Units with Special Names and Symbols

Physical Quantity	Name of SI Unit	Symbol for SI Unit	Expression in Terms of SI Base Units	
Frequency[a]	hertz	Hz	s^{-1}	
Force	newton	N	$m{\cdot}kg{\cdot}s^{-2}$	
Pressure, stress	pascal	Pa	$N{\cdot}m^{-2}$	$= m^{-1}{\cdot}kg{\cdot}s^{-2}$
Energy, work, heat	joule	J	$N{\cdot}m$	$= m^2{\cdot}kg{\cdot}s^{-2}$
Power, radiant flux	watt	W	$J{\cdot}s^{-1}$	$= m^2{\cdot}kg{\cdot}s^{-3}$
Electric charge	coulomb	C	$A{\cdot}s$	
Electric potential, electromotive force	volt	V	$J{\cdot}C^{-1}$	$= m^2{\cdot}kg{\cdot}s^{-3}{\cdot}A^{-1}$
Electric resistance	ohm	Ω	$V{\cdot}A^{-1}$	$= m^2{\cdot}kg{\cdot}s^{-3}{\cdot}A^{-2}$
Electric conductance	siemens	S	Ω^{-1}	$= m^{-2}{\cdot}kg^{-1}{\cdot}s^3{\cdot}A^2$
Electric capacitance	farad	F	$C{\cdot}V^{-1}$	$= m^{-2}{\cdot}kg^{-1}{\cdot}s^4{\cdot}A^2$
Magnetic flux density	tesla	T	$V{\cdot}s{\cdot}m^{-2}$	$= kg{\cdot}s^{-2}{\cdot}A^{-1}$
Magnetic flux	weber	Wb	$V{\cdot}s$	$= m^2{\cdot}kg{\cdot}s^{-2}{\cdot}A^{-1}$
Inductance	henry	H	$V{\cdot}A^{-1}{\cdot}s$	$= m^2{\cdot}kg{\cdot}s^{-2}{\cdot}A^{-2}$
Celsius temperature[b]	degree Celsius	°C	K	
Luminous flux	lumen	lm	$cd{\cdot}sr$	
Illuminance	lux	lx	$cd{\cdot}sr{\cdot}m^{-2}$	
Activity (radioactive)	becquerel	Bq	s^{-1}	
Absorbed dose (of radiation)	gray	Gy	$J{\cdot}kg^{-1}$	$= m^2{\cdot}s^{-2}$
Dose equivalent (dose equivalent index)	sievert	Sv	$J{\cdot}kg^{-1}$	$= m^2{\cdot}s^{-2}$
Plane angle	radian	rad	1	$= m{\cdot}m^{-1}$
Solid angle	steradian	sr	1	$= m^2{\cdot}m^{-2}$

[a] For radial (circular) frequency and for angular velocity the unit rad s^{-1}, or simply s^{-1}, should be used, and this may not be simplified to Hz. The unit Hz should be used only for frequency in the sense of cycles per second.

[b] The Celsius temperature θ is defined by the equation

$$\theta / {}^\circ C = T/K - 237.15$$

The SI unit of Celsius temperature interval is the degree Celsius, °C, which is equal to the kelvin, K. °C should be treated as a single symbol, with no space between the ° sign and the letter C. (The symbol °K, and the symbol °, should no longer be used.)

Units in Use Together with the SI

These units are not part of the SI, but it is recognized that they will continue to be used in appropriate contexts. SI prefixes may be attached to some of these units, such as milliliter, ml; millibar, mbar; megaelectronvolt, MeV; and kilotonne, kt.

Physical Quantity	Name of Unit	Symbol for Unit	Value in SI Units
Time	minute	min	60 s
Time	hour	h	3600 s
Time	day	d	86 400 s
Plane angle	degree	°	$(\pi/180)$ rad
Plane angle	minute	′	$(\pi/10\ 800)$ rad
Plane angle	second	″	$(\pi/648\ 000)$ rad
Length	angstrom[a]	Å	10^{-10} m
Area	barn	b	10^{-28} m^2
Volume	liter	l, L	dm^3 = 10^{-3} m^3
Mass	tonne	t	Mg = 10^3 kg
Pressure	bar[a]	bar	10^5 Pa = 10^5 N·m^{-2}
Energy	electronvolt[b]	eV (= $e \times$ V)	$\approx 1.60218 \times 10^{-19}$ J
Mass	unified atomic mass unit[b,c]	u (= $m_a(^{12}C)/12$)	$\approx 1.66054 \times 10^{-27}$ kg

[a] The angstrom and the bar are approved by CIPM for "temporary use with SI units," until CIPM makes a further recommendation. However, they should not be introduced where they are not used at present.

[b] The values of these units in terms of the corresponding SI units are not exact, since they depend on the values of the physical constants e (for the electronvolt) and N_A (for the unified atomic mass unit), which are determined by experiment.

[c] The unified atomic mass unit is also sometimes called the dalton, with symbol Da, although the name and symbol have not been approved by CGPM.

A.3 Conversion Constants and Multipliers

Recommended Decimal Multiples and Submultiples

Multiple for Submultiple	Prefix	Symbol	Multiple for Submultiple	Prefix	Symbol
10^{18}	exa	E	10^{-1}	deci	d
10^{15}	peta	P	10^{-2}	centi	c
10^{12}	tera	T	10^{-3}	milli	m
10^9	giga	G	10^{-6}	micro	μ (Greek mu)
10^6	mega	M	10^{-9}	nano	n
10^3	kilo	k	10^{-12}	pico	p
10^2	hecto	h	10^{-15}	femto	f
10	deca	da	10^{-18}	atto	a

Conversion Factors — Metric to English

To Obtain	Multiply	By
Inches	Centimeters	0.393 700 787 4
Feet	Meters	3.280 839 895
Yards	Meters	1.093 613 298
Miles	Kilometers	0.621 371 192 2
Ounces	Grams	$3.527\ 396\ 195 \times 10^{-2}$
Pounds	Kilograms	2.204 622 622
Gallons (U.S. liquid)	Liters	0.264 172 0524
Fluid ounces	Milliliters (cc)	$3.381\ 402\ 270 \times 10^{-2}$

To Obtain	Multiply	By
Square inches	Square centimeters	0.155 000 310 0
Square feet	Square meters	10.763 910 42
Square yards	Square meters	1.195 990 046
Cubic inches	Milliliters (cc)	$6.102\ 374\ 409 \times 10^{-2}$
Cubic feet	Cubic meters	35.314 666 72
Cubic yards	Cubic meters	1.307 950 619

Conversion Factors — English to Metric

To Obtain	Multiply	By[a]
Microns	Mils	**25.4**
Centimeters	Inches	**2.54**
Meters	Feet	**0.304 8**
Meters	Yards	**0.914 4**
Kilometers	Miles	**1.609 344**
Grams	Ounces	28.349 523 13
Kilograms	Pounds	**0.453 592 37**
Liters	Gallons (U.S. liquid)	**3.785 411 784**
Millimeters (cc)	Fluid ounces	29.573 529 56
Square centimeters	Square inches	**6.451 6**
Square meters	Square feet	**0.092 903 04**
Square meters	Square yards	**0.836 127 36**
Milliliters (cc)	Cubic inches	**16.387 064**
Cubic meters	Cubic feet	$2.831\ 684\ 659 \times 10^{-2}$
Cubic meters	Cubic yards	0.764 554 858

[a] Boldface numbers are exact; others are given to ten significant figures where so indicated by the multiplier factor.

Conversion Factors — General

To Obtain	Multiply	By[a]
Atmospheres	Feet of water @ 4°C	2.950×10^{-2}
Atmospheres	Inches of mercury @ 0°C	3.342×10^{-2}
Atmospheres	Pounds per square inch	6.804×10^{-2}
Btu	Foot-pounds	1.285×10^{-3}
Btu	Joules	9.480×10^{-4}
Cubic feet	Cords	**128**
Degree (angle)	Radians	57.2958
Ergs	Foot-pounds	1.356×10^{7}
Feet	Miles	**5280**
Feet of water @ 4°C	Atmospheres	33.90
Foot-pounds	Horsepower-hours	1.98×10^{6}
Foot-pounds	Kilowatt-hours	2.655×10^{6}
Foot-pounds per minute	Horsepower	3.3×10^{4}
Horsepower	Foot-pounds per second	1.818×10^{-3}
Inches of mercury @ 0°C	Pounds per square inch	2.036
Joules	Btu	1054.8
Joules	Foot-pounds	1.355 82
Kilowatts	Btu per minute	1.758×10^{-2}
Kilowatts	Foot-pounds per minute	2.26×10^{-5}
Kilowatts	Horsepower	0.745 712
Knots	Miles per hour	0.868 976 24
Miles	Feet	1.894×10^{-4}
Nautical miles	Miles	0.868 976 24
Radians	Degrees	1.745×10^{-2}

To Obtain	Multiply	By[a]
Square feet	Acres	**43 560**
Watts	Btu per minute	17.5796

[a] Boldface numbers are exact; others are given to ten significant figures where so indicated by the multiplier factor.

Temperature Factors

$$°F = 9/5(°C) + 32$$

$$\text{Fahrenheit temperature} = 1.8\ (\text{temperature in kelvins}) - 459.67$$

$$°C = 5/9\ [(°F) - 32]$$

$$\text{Celsius temperature} = \text{temperature in kelvins} - 273.15$$

$$\text{Fahrenheit temperature} = 1.8\ (\text{Celsius temperature}) + 32$$

Conversion of Temperatures

From	To		From	To	
Fahrenheit	Celsius	$t_C = \frac{t_F - 32}{1.8}$	Celsius	Fahrenheit	$t_F = (t_C \times 1.8) + 32$
				Kelvin	$T_K = t_C + 273.15$
				Rankine	$T_R = (t_C + 273.15) \times 18$
	Kelvin	$T_k = \frac{t_F - 32}{1.8} + 273.15$	Kelvin	Celsius	$t_C = T_K - 273.15$
				Rankine	$T_R = T_K \times 1.8$
	Rankine	$T_R = t_F + 459.67$	Rankine	Fahrenheit	$t_F = T_R - 459.67$
				Kelvin	$T_K = \frac{T_R}{1.8}$

A.4 Physical Constants

General

Equatorial radius of the earth = 6378.388 km = 3963.34 miles (statute)
Polar radius of the earth = 6356.912 km = 3949.99 miles (statute)
1 degree of latitude at 40° = 69 miles
1 international nautical mile = 1.150 78 miles (statute) = 1852 m = 6076.115 ft
Mean density of the earth = 5.522 g/cm^3 = 344.7 lb/ft^3
Constant of gravitation $(6.673 \pm 0.003) \times 10^{-8} \cdot cm^3 \cdot g^{-1} \cdot s^{-2}$
Acceleration due to gravity at sea level, latitude 45° = 980.6194 cm/s^2 = 32.1726 ft/s^2
Length of seconds pendulum at sea level, latitude 45° = 99.3575 cm = 39.1171 in.
1 knot (international) = 101.269 ft/min = 1.6878 ft/s = 1.1508 miles (statute)/h
1 micron = 10^{-4} cm
1 angstrom = 10^{-8} cm
Mass of hydrogen atom = $(1.673\ 39 \pm 0.0031) \times 10^{-24}$ g
Density of mercury at 0°C = 13.5955 g/mL
Density of water at 3.98°C = 1.000 000 g/mL
Density, maximum, of water, at 3.98°C = 0.999 973 g/cm^3
Density of dry air at 0°C, 760 mm = 1.2929 g/L
Velocity of sound in dry air at 0°C = 331.36 m/s – 1087.1 ft/s
Velocity of light in vacuum = $(2.997\ 925 \pm 0.000\ 002) \times 10^{10}$ cm/s
Heat of fusion of water, 0°C = 79.71 cal/g

Heat of vaporization of water, 100°C = 539.55 cal/g
Electrochemical equivalent of silver 0.001 118 g/s international amp
Absolute wavelength of red cadmium light in air at 15°C, 760 mm pressure = 6438.4696 Å
Wavelength of orange-red line of krypton 86 = 6057.802 Å

π Constants

π = 3.14159 26535 89793 23846 26433 83279 50288 41971 69399 37511
$1/\pi$ = 0.31830 98861 83790 67153 77675 26745 02872 40689 19291 48091
π^2 = 9.8690 44010 89358 61883 44909 99876 15113 53136 99407 24079
$\log_e \pi$ = 1.14472 98858 49400 17414 34273 51353 05871 16472 94812 91531
$\log_{10} \pi$ = 0.49714 98726 94133 85435 12682 88290 89887 36516 78324 38044
$\log_{10} \sqrt{2\pi}$ = 0.39908 99341 79057 52478 25035 91507 69595 02099 34102 92128

Constants Involving *e*

e = 2.71828 18284 59045 23536 02874 71352 66249 77572 47093 69996
$1/e$ = 0.36787 94411 71442 32159 55237 70161 46086 74458 11131 03177
e^2 = 7.38905 60989 30650 22723 04274 60575 00781 31803 15570 55185
$M = \log_{10} e$ = 0.43429 44819 03251 82765 11289 18916 60508 22943 97005 80367
$1/M = \log_e 10$ = 2.30258 50929 94045 68401 79914 54684 36420 76011 01488 62877
$\log_{10} M$ = 9.63778 43113 00536 78912 29674 98645 − 10

Numerical Constants

$\sqrt{2}$ = 1.41421 35623 73095 04880 16887 24209 69807 85696 71875 37695
$\sqrt[3]{2}$ = 1.25992 10498 94873 16476 72106 07278 22835 05702 51464 70151
$\log_e 2$ = 0.69314 71805 59945 30941 72321 21458 17656 80755 00134 36026
$\log_{10} 2$ = 0.30102 99956 63981 19521 37388 94724 49302 67881 89881 46211
$\sqrt{3}$ = 1.73205 08075 68877 29352 74463 41505 87236 69428 05253 81039
$\sqrt[3]{3}$ = 1.44224 95703 07408 38232 16383 10780 10958 83918 69253 49935
$\log_e 3$ = 1.09861 22886 68109 69139 52452 36922 52570 46474 90557 82275
$\log_{10} 3$ = 0.47712 12547 19662 43729 50279 03255 11530 92001 28864 19070

A.5 Symbols and Terminology for Physical and Chemical Quantities

Name	Symbol	Definition	SI Unit
	Classical Mechanics		
Mass	m		kg
Reduced mass	μ	$\mu = m_1 m_2/(m_1 + m_2)$	kg
Density, mass density	ρ	$\rho = m/V$	$kg{\cdot}m^{-3}$
Relative density	d	$d = \rho/\rho^\theta$	1
Surface density	ρ_A, ρ_S	$\rho_a = m/A$	$kg{\cdot}m^{-2}$
Specific volume	v	$v = V/m = 1/\rho$	$m^3{\cdot}kg^{-1}$
Momentum	$\mathbf{p}$	$\mathbf{p} = mv$	$kg{\cdot}m{\cdot}s^{-1}$
Angular momentum, action	$\mathbf{L}$	$\mathbf{L} = \boldsymbol{r} \times \boldsymbol{p}$	J·s
Moment of inertia	I, J	$I = \Sigma m_i r_i^2$	$kg{\cdot}m^2$
Force	$\mathbf{F}$	$\mathbf{F} = d\mathbf{p}/dt = m\mathbf{a}$	N
Torque, moment of a force	$\mathbf{T}$, ($\mathbf{M}$)	$\mathbf{T} = \boldsymbol{r} \times \boldsymbol{F}$	N·m
Energy	E		J
Potential energy	E_p, V, Φ	$E_p = -\int \mathbf{F}{\cdot}ds$	J

Name	Symbol	Definition	SI Unit
Kinetic energy	E_k, T, K	$E_k = (1/2)mv^2$	J
Work	W, w	$W = \int \mathbf{F}\cdot ds$	J
Hamilton function	H	$H(q, p) = T(q, p) + V(q)$	J
Lagrange function	L	$L(q,\dot{q}) = T(q,\dot{q}) - V(q)$	J
Pressure	p, P	$p = F/A$	Pa, $N\cdot m^{-2}$
Surface tension	γ, σ	$\gamma = dW/dA$	$N\cdot m^{-1}$, $J\cdot m^{-2}$
Weight	G (W, P)	$G = mg$	N
Gravitational constant	G	$F = Gm_1m_2/r^2$	$N\cdot m^2\cdot kg^{-2}$
Normal stress	σ	$\sigma = F/A$	Pa
Shear stress	τ	$\tau = F/A$	Pa
Linear strain, relative elongation	ε, e	$\varepsilon = \Delta l/l$	1
Modulus of elasticity, Young's modulus	E	$E = \sigma/\varepsilon$	Pa
Shear strain	γ	$\gamma = \Delta x/d$	1
Shear modulus	G	$G = \tau/\gamma$	Pa
Volume strain, bulk strain	θ	$\theta = \Delta V/V_0$	1
Bulk modulus, compression modulus	K	$K = -V_0(dp/dV)$	Pa
Viscosity, dynamic viscosity	η, μ	$\tau_{x,z} = \eta(dv_x/dz)$	Pa·s
Fluidity	ϕ	$\phi = 1/\eta$	$m\cdot kg^{-1}\cdot s$
Kinematic viscosity	v	$v = \eta/\rho$	$m^2\cdot s^{-1}$
Friction coefficient	μ, (f)	$F_{frict} = \mu F_{norm}$	1
Power	P	$P = dW/dt$	W
Sound energy flux	P, P_a	$P = dE/dt$	W
Acoustic factors			
Reflection factor	ρ	$\rho = P_r/P_0$	1
Acoustic absorption factor	α_a, (α)	$\alpha_a = 1 - \rho$	1
Transmission factor	τ	$\tau = P_{tr}/P_0$	1
Dissipation factor	δ	$\delta = \alpha_a - \tau$	1
		Electricity and Magnetism	
Quantity of electricity, electric charge	Q		C
Charge density	ρ	$\rho = Q/V$	$C\cdot m^{-3}$
Surface charge density	σ	$\sigma = Q/A$	$C\cdot m^{-2}$
Electric potential	V, ϕ	$V = dW/dQ$	V, $J\cdot C^{-1}$
Electric potential difference	U, ΔV, $\Delta\phi$	$U = V_2 - V_1$	V
Electromotive force	E	$E = \int (\mathbf{F}/Q)\cdot ds$	V
Electric field strength	$\mathbf{E}$	$\mathbf{E} = \mathbf{F}/Q = -\text{grad } V$	$V\cdot m^{-1}$
Electric flux	Ψ	$\Psi = \int \mathbf{D}\cdot d\mathbf{A}$	C
Electric displacement	$\mathbf{D}$	$\mathbf{D} = \varepsilon\mathbf{E}$	$C\cdot m^{-2}$
Capacitance	C	$C = Q/U$	F, $C\cdot V^{-1}$
Permittivity	ε	$\mathbf{D} = \varepsilon\mathbf{E}$	$F\cdot m^{-1}$
Permittivity of vacuum	ε_0	$\varepsilon_0 = \mu_0^{-1}c_0^{-2}$	$F\cdot m^{-1}$
Relative permittivity	ε_r	$\varepsilon_r = \varepsilon/\varepsilon_0$	1
Dielectric polarization (dipole moment per volume)	$\mathbf{P}$	$\mathbf{P} = \mathbf{D} - \varepsilon_0\mathbf{E}$	$C\cdot m^{-2}$
Electric susceptibility	χ_e	$\chi_e = \varepsilon_r - 1$	1
Electric dipole moment	p, μ	$\mathbf{p} = Q\mathbf{r}$	C·m
Electric current	I	$I = dQ/dt$	A
Electric current density	$\mathbf{j}$, $\mathbf{J}$	$I = \int \mathbf{j}\cdot d\mathbf{A}$	$A\cdot m^{-2}$
Magnetic flux density, magnetic induction	$\mathbf{B}$	$\mathbf{F} = Qv \times \mathbf{B}$	T
Magnetic flux	Φ	$\Phi = \int \mathbf{B}\cdot d\mathbf{A}$	Wb
Magnetic field strength	$\mathbf{H}$	$\mathbf{B} = \mu\mathbf{H}$	$A\cdot M^{-1}$
Permeability	μ	$\mathbf{B} = \mu\mathbf{H}$	$N\cdot A^{-2}$, $H\cdot m^{-1}$
Permeability of vacuum	μ_0		$H\cdot m^{-1}$
Relative permeability	μ_r	$\mu_r = \mu/\mu_0$	1
Magnetization (magnetic dipole moment per volume)	$\mathbf{M}$	$\mathbf{M} = \mathbf{B}/\mu_0 - \mathbf{H}$	$A\cdot m^{-1}$
Magnetic susceptibility	χ, κ, (χ_m)	$\chi = \mu_r - 1$	1
Molar magnetic susceptibility	χ_m	$\chi_m = V_m\chi$	$m^3\cdot mol^{-1}$
Magnetic dipole moment	$\mathbf{m}$, μ	$E_p = -\mathbf{m}\cdot\mathbf{B}$	$A\cdot m^2$, $J\cdot T^{-1}$
Electrical resistance	R	$R = U/I$	Ω

Name	Symbol	Definition	SI Unit
Conductance	G	$G = 1/R$	S
Loss angle	δ	$\delta = (\pi/2) + \phi_T - \phi_U$	1, rad
Reactance	X	$X = (U/I) \sin \delta$	Ω
Impedance (complex impedance)	Z	$Z = R + iX$	Ω
Admittance (complex admittance)	Y	$Y = 1/Z$	S
Susceptance	B	$Y = G + iB$	S
Resistivity	ρ	$\rho = E/j$	$\Omega \cdot m$
Conductivity	κ, γ, σ	$\kappa = 1/\rho$	$S \cdot m^{-1}$
Self-inductance	L	$E = -L(dI/dt)$	H
Mutual inductance	M, L_{12}	$E_1 = L_{12}(dI_2/dt)$	H
Magnetic vector potential	$\mathbf{A}$	$\mathbf{B} = \nabla \times \mathrm{A}$	$Wb \cdot m^{-1}$
Poynting vector	$\mathbf{S}$	$\mathbf{S} = \mathbf{E} \times \mathbf{H}$	$W \cdot m^{-2}$
	Electromagnetic Radiation		
Wavelength	λ		m
Speed of light			
In vacuum	c_0		$m \cdot s^{-1}$
In a medium	c	$c = c_0/n$	$m \cdot s^{-1}$
Wavenumber in vacuum	$\tilde{v}$	$\tilde{v} = v / c_0 = 1/n\lambda$	m^{-1}
Wavenumber (in a medium)	σ	$\sigma = 1/\lambda$	m^{-1}
Frequency	v	$v = c/\lambda$	Hz
Circular frequency, pulsatance	ω	$\omega = 2\pi v$	s^{-1}, $rad \cdot s^{-1}$
Refractive index	n	$n = c_0/c$	1
Planck constant	h		$J \cdot s$
Planck constant/2π	$\hbar$	$\hbar = h/2\pi$	$J \cdot s$
Radiant energy	Q, W		J
Radiant energy density	ρ, w	$\rho = Q/V$	$J \cdot m^{-3}$
Spectral radiant energy density			
In terms of frequency	ρ_v, w_v	$\rho_v = d\rho/dv$	$J \cdot m^{-3} \cdot Hz^{-1}$
In terms of wavenumber	$\rho_{\tilde{v}}, w_{\tilde{v}}$	$\rho_{\tilde{v}} = d\rho / d\tilde{v}$	$J \cdot m^{-2}$
In terms of wavelength	ρ_λ, w_λ	$\rho_\lambda = d\rho/d\lambda$	$J \cdot m^{-4}$
Einstein transition probabilities			
Spontaneous emission	A_{nm}	$dN_n/dt = -A_{nm}N_n$	s^{-1}
Stimulated emission	B_{nm}	$dN_n / dt = -\rho_{\tilde{v}}(\tilde{v}_{nm}) \times B_{nm} N_n$	$s \cdot kg^{-1}$
Stimulated absorption	B_{mn}	$dN_n / dt = \rho_{\tilde{v}}(\tilde{v}_{nm}) B_{mn} N_m$	$s \cdot kg^{-1}$
Radiant power, radiant energy per time	Φ, P	$\Phi = dQ/dt$	W
Radiant intensity	I	$I = d\Phi/d\Omega$	$W \cdot sr^{-1}$
Radiant excitance (emitted radiant flux)	M	$M = d\Phi/dA_{source}$	$W \cdot m^{-2}$
Irradiance (radiant flux received)	$E, (I)$	$E = d\Phi/dA$	$W \cdot m^{-2}$
Emittance	ε	$\varepsilon = M/M_{bb}$	1
Stefan–Boltzmann constant	σ	$M_{bb} = \sigma T^4$	$W \cdot m^{-2} \cdot K^{-4}$
First radiation constant	c_1	$c_1 = 2\pi hc_0^2$	$W \cdot m^2$
Second radiation constant	c_2	$c_2 = hc_0/k$	$K \cdot m$
Transmittance, transmission factor	τ, T	$\tau = \Phi_{tr}/\Phi_0$	1
Absorptance, absorption factor	α	$\alpha = \Phi_{abs}/\Phi_0$	1
Reflectance, reflection factor	ρ	$\rho = \Phi_{refl}/\Phi_0$	1
(Decadic) absorbance	A	$A = \lg(1 - \alpha_i)$	1
Napierian absorbance	B	$B = \ln(1 - \alpha_i)$	1
Absorption coefficient			
(Linear) decadic	a, K	$a = A/l$	m^{-1}
(Linear) napierian	α	$\alpha = B/l$	m^{-1}
Molar (decadic)	ε	$\varepsilon = a/c = A/cl$	$m^2 \cdot mol^{-1}$
Molar napierian	κ	$\kappa = \alpha/c = B/cl$	$m^2 \cdot mol^{-1}$
Absorption index	k	$k = \alpha / 4\pi\tilde{v}$	1
Complex refractive index	$\hat{n}$	$\hat{n} = n + ik$	1
Molar refraction	R, R_m	$R = \frac{(n^2 - 1)}{(n^2 + 2)} V_m$	$m^3 \cdot mol^{-1}$
Angle of optical rotation	α		1, rad

Name	Symbol	Definition	SI Unit
	Solid State		
Lattice vector	$\mathbf{R}, \mathbf{R}_0$		m
Fundamental translation vectors for the crystal lattice	$\mathbf{a}_1; \mathbf{a}_2; \mathbf{a}_3,$ $\mathbf{a}; \mathbf{b}; \mathbf{c}$	$\mathbf{R} = n_1\mathbf{a}_1 + n_2\mathbf{a}_2 + n_3\mathbf{a}_3$	m
(Circular) reciprocal lattice vector	$\mathbf{G}$	$\mathbf{G}\cdot\mathbf{R} = 2\pi m$	m^{-1}
(Circular) fundamental Translation vectors for the reciprocal lattice	$\mathbf{b}_1; \mathbf{b}_2; \mathbf{b}_3,$ $\mathbf{a}^*; b^*; \mathbf{c}^*$	$\mathbf{a}_1\cdot\mathbf{b}_k = 2\pi\delta_{ik}$	m^{-1}
Lattice plane spacing	d		m
Bragg angle	θ	$n\lambda = 2d \sin\theta$	1,rad
Order of reflection	n		1
Order parameters			
Short range	σ		1
Long range	s		1
Burgers vector	b		m
Particle position vector	$r, \mathbf{R}_j$		m
Equilibrium position vector of an ion	$\mathbf{R}_0$		m
Displacement vector of an ion	$\mathbf{u}$	$\mathbf{u} = \mathbf{R} - \mathbf{R}_0$	m
Debye–Waller factor	$\mathbf{B}, \mathbf{D}$		1
Debye circular wavenumber	q_D		m^{-1}
Debye circular frequency	ω_D		s^{-1}
Grüneisen parameter	γ, Γ	$\gamma = \alpha V/\kappa C_V$	1
Madelung constant	$\alpha, \mathfrak{M}$	$E_{\text{coul}} = \dfrac{\alpha N_A z + z - e^2}{4\pi\varepsilon_0 R_0}$	1
Density of states	N_E	$N_E = dN(E)/dE$	$J^{-1}\cdot m^{-3}$
(Spectral) density of vibrational modes	N_ω, g	$N_\omega = dN(\omega)/d\omega$	$s\cdot m^{-3}$
Resistivity tensor	ρ_{ik}	$E = \rho\cdot\mathbf{j}$	$\Omega\cdot m$
Conductivity tensor	σ_{ik}	$\sigma = \rho^{-1}$	$S\cdot m^{-1}$
Thermal conductivity tensor	λ_{ik}	$J_q = -\lambda\cdot \text{grad}T$	$W\cdot m^{-1}\cdot K^{-1}$
Residual resistivity	ρ_R		$\Omega\cdot m$
Relaxation time	τ	$\tau = l/v_F$	s
Lorenz coefficient	L	$L = \lambda/\sigma T$	$V^2\cdot K^{-2}$
Hall coefficient	A_H, R_H	$\mathbf{E} = \rho\cdot\mathbf{j} + R_H(\mathbf{B}\times\mathbf{j})$	$m^3\cdot C^{-1}$
Thermoelectric force	E		V
Peltier coefficient	Π		V
Thomson coefficient	$\mu, (\tau)$		$V\cdot K^{-1}$
Work function	Φ	$\Phi = E_\infty - E_F$	J
Number density, number concentration	$n, (p)$		m^{-3}
Gap energy	E_g		J
Donor ionization energy	E_d		J
Acceptor ionization energy	E_a		J
Fermi energy	E_F, ε_F		J
Circular wave vector, propagation vector	$\mathbf{k}, \mathbf{q}$	$\mathbf{k} = 2\pi/\lambda$	m^{-1}
Bloch function	$u_k(\mathbf{r})$	$\psi(\mathbf{r}) = u_k(\mathbf{r})\exp(i\mathbf{k}\cdot\mathbf{r})$	$m^{-3/2}$
Charge density of electrons	ρ	$\rho(\mathbf{r}) = -e\psi^*(\mathbf{r})\psi(\mathbf{r})$	$C\cdot m^{-3}$
Effective mass	m^*		kg
Mobility	μ	$\mu = v_{\text{drift}}/E$	$m^2\cdot V^{-1}\cdot s^{-1}$
Mobility ratio	b	$b = \mu_n/\mu_p$	1
Diffusion coefficient	D	$dN/dt = -DA(dn/dx)$	$m^2\cdot s^{-1}$
Diffusion length	L	$L = \sqrt{D\tau}$	m
Characteristic (Weiss) temperature	ϕ, ϕ_w		K
Curie temperature	T_C		K
Néel temperature	T_N		K

A.6 Elementary Algebra and Geometry

Fundamental Properties (Real Numbers)

$a + b = b + a$	Commutative law for addition
$(a + b) + c = a + (b + c)$	Associative law for addition
$a + 0 = 0 + a$	Identity law for addition
$a + (-a) = (-a) + a = 0$	Inverse law for addition
$a(bc) = (ab)c$	Associative law for multiplication
$a\left(\frac{1}{a}\right) = \left(\frac{1}{a}\right)a = 1, a \neq 0$	Inverse law for multiplication
$(a)(1) = (1)(a) = a$	Identity law for multiplication
$ab = ba$	Commutative law for multiplication
$a(b + c) = ab + ac$	Distributive law

Division by zero is not defined.

Exponents

For integers m and n,

$$a^n a^m = a^{n+m}$$

$$a^n / a^m = a^{n-m}$$

$$(a^n)^m = a^{nm}$$

$$(ab)^m = a^m b^m$$

$$(a/b)^m = a^m / b^m$$

Fractional Exponents

$$a^{p/q} = (a^{1/q})^p$$

where $a^{1/q}$ is the positive qth root of a if $a > 0$ and the negative qth root of a if a is negative and q is odd. Accordingly, the five rules of exponents given above (for integers) are also valid if m and n are fractions, provided a and b are positive.

Irrational Exponents

If and exponent is irrational (e.g., $\sqrt{2}$), the quantity, such as $a^{\sqrt{2}}$, is the limit of the sequence $a^{1.4}$, $a^{1.41}$, $a^{1.414}$,….

Operations with Zero

$$0^m = 0 \qquad a^0 = 1$$

Logarithms

If x, y, and b are positive and $b \neq 1$,

$$\log_b(xy) = \log_b x + \log_b y$$

$$\log_b(x/y) = lob_b x - \log_b y$$

$$\log_b x^p = p\log_b x$$

$$\log_b(1/x) = -\log_b x$$

$$\log_b b = 1$$

$$\log_b 1 = 0 \qquad \text{Note: } b^{\log_b x} = x$$

Change of Base ($a \neq 1$)

$$\log_b x = \log_a x \log_b a$$

Factorials

The factorial of a positive integer n is the product of all the positive integers less than or equal to the integer n and is denoted $n!$. Thus,

$$n! = 1 \cdot 2 \cdot 3 \cdot \ldots \cdot n$$

Factorial 0 is defined: $0! = 1$.

Stirling's Approximation

$$\lim_{n\to\infty}(n/e)^n\sqrt{2\pi n} = n!$$

Binomial Theorem

For positive integer n

$$(x+y)^n = x^n + nx^{n-1}y + \frac{n(n-1)}{2!}x^{n-2}y^2 + \frac{n(n-1)(n-2)}{3!}x^{n-3}y^3 + \cdots + nxy^{n-1} + y^n$$

Factors and Expansion

$$(a+b)^2 = a^2 + 2ab + b^2$$

$$(a-b)^2 = a^2 - 2ab + b^2$$

$$(a+b)^3 = a^3 + 3a^2b + 3ab^2 + b^3$$

$$(a-b)^3 = a^3 - 3a^2b + 3ab^2 - b^3$$

$$(a^2 - b^2) = (a-b)(a+b)$$

$$(a^3 - b^3) = (a-b)(a^2 + ab + b^2)$$

$$(a^3 + b^3) = (a+b)(a^2 - ab + b^2)$$

Progression

An *arithmetic progression* is a sequence in which the difference between any term and the preceding term is a constant (d):

$$a, a + d, a + 2d, \ldots, a + (n - 1)d$$

If the last term is denoted l [$= a + (n - 1)d$], then the sum is

$$s = \frac{n}{2}(a + l)$$

A *geometric progression* is a sequence in which the ratio of any term to the preceding term is a constant r. Thus, for n terms,

$$a, ar, ar^2, \ldots, ar^{n-1}$$

The sum is

$$S = \frac{a - ar^n}{1 - r}$$

Complex Numbers

A complex number is an ordered pair of real numbers (a, b).

Equality: $(a, b) = (c, d)$ if an only if $a = c$ and $b = d$
Addition: $(a, b) + (c, d) = (a + c, b + d)$
Multiplication: $(a, b)(c, d) = (ac - bd, ad + bc)$

The first element (a, b) is called the *real* part, the second the *imaginary* part. An alternative notation for (a, b) is $a + bi$, where $i^2 = (-1,0)$, and $i = (0,1)$ or $0 + 1i$ is written for this complex number as a convenience. With this understanding, i behaves as a number, that is, $(2 - 3i)(4 + i) = 8 - 12i + 2i - 3i^2 = 11 - 10i$. The conjugate of $a + bi$ is $a - bi$, and the product of a complex number and its conjugate is $a^2 + b^2$. Thus, *quotients* are computed by multiplying numerator and denominator by the conjugate of the denominator, as illustrated below:

$$\frac{2+3i}{4+2i} = \frac{(4-2i)(2+3i)}{(4-2i)(4+2i)} = \frac{14+8i}{20} = \frac{7+4i}{10}$$

Polar Form

The complex number $x + iy$ may be represented by a plane vector with components x and y:

$$x + iy = r(\cos\theta + i\sin\theta)$$

(See Figure A.1.) Then, given two complex numbers $z_1 = r_1(\cos\theta_1 + i\sin\theta_1)$ and $z_2 = r_2(\cos\theta_2 + i\sin\theta_2)$, the product and quotient are:

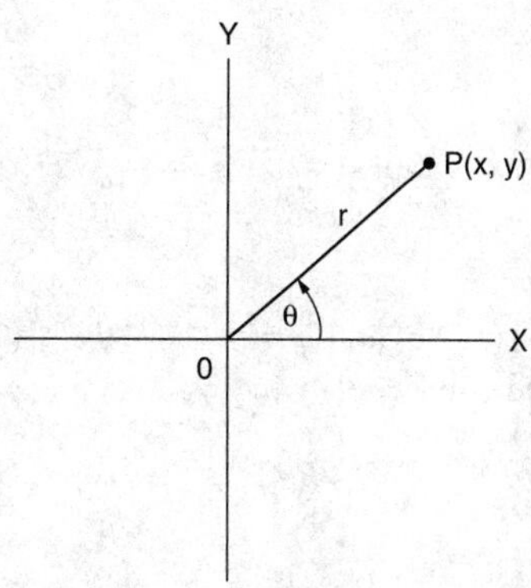

FIGURE A.1 Polar form of complex number.

Product: $z_1 z_2 = r_1 r_2[\cos(\theta_1+\theta_2)+i\sin(\theta_1+\theta_2)]$
Quotient: $z_1/z_2 = (r_1/r_2)[\cos(\theta_1-\theta_2)+i\sin(\theta_1-\theta_2)]$
Powers: $z^n = [r(\cos\theta+i\sin\theta)]^n = r^n[\cos n\theta + i\sin n\theta]$

$$z^{1/n} = [r(\cos\theta+i\sin\theta)]^{1/n}$$

Roots:

$$= r^{1/n}\left[\cos\frac{\theta+k\cdot 360}{n}+i\sin\frac{\theta+k\cdot 360}{n}\right], \qquad k=0,1,2,\ldots,n-1$$

Permutations

A permutation is an ordered arrangement (sequence) of all or part of a set of objects. The number of permutations of n objects taken r at a time is

$$p(n,r) = n(n-1)(n-2)\cdots(n-r+1)$$

$$= \frac{n!}{(n-r)!}$$

A permutation of positive integers is "even" or "odd" if the total number of inversions is an even integer or an odd integer, respectively. Inversions are counted relative to each integer j in the permutation by counting the number of integers that follow j and are less than j. These are summed to give the total number of inversions. For example, the permutation 4132 has four inversions: three relative to 4 and one relative to 3. This permutation is therefore even.

Combinations

A combination is a selection of one or more objects from among a set of objects regardless of order. The number of combinations of n different objects taken r at a time is

$$C(n,r) = \frac{P(n,r)}{r!} = \frac{n!}{r!(n-r)!}$$

Algebraic Equations

Quadratic

If $ax^2 + bx + c = 0$, and $a \neq 0$, then roots are

$$x = \frac{-b \pm \sqrt{b^2-4ac}}{2a}$$

Cubic

To solve $x^3 + bx^2 + cx + d = 0$, let $x = y - b/3$. Then the *reduced cubic* is obtained:

$$y^3 + py + q = 0$$

where $p = c - (1/3)b^2$ and $q = d - (1/3)bc + (2/27)b^3$. Solutions of the original cubic are then in terms of the reduced cubic roots y_1, y_2, y_3:

$$x_1 = y_1 - (1/3)b \quad x_2 = y_2 - (1/3)b \quad x_3 = y_3 - (1/3)b$$

The three roots of the reduced cubic are

$$y_1 = (A)^{1/3} + (B)^{1/3}$$
$$y_2 = W(A)^{1/3} + W^2(B)^{1/3}$$
$$y_3 = W^2(A)^{1/3} + W(B)^{1/3}$$

where

$$A = -\frac{1}{2}q + \sqrt{(1/27)p^3 + \frac{1}{4}q^2}$$
$$B = -\frac{1}{2}q - \sqrt{(1/27)p^3 + \frac{1}{4}q^2}$$
$$W = \frac{-1 + i\sqrt{3}}{2}, \quad W^2 = \frac{-1 - i\sqrt{3}}{2}$$

When $(1/27)p^3 + (1/4)q^2$ is negative, A is complex; in this case A should be expressed in trigonometric form; $A = r(\cos\theta + i\sin\theta)$ where θ is a first or second quadrant angle, as q is negative or positive. The three roots of the reduced cubic are

$$y_1 = 2(r)^{1/3}\cos(\theta/3)$$
$$y_2 = 2(r)^{1/3}\cos\left(\frac{\theta}{3} + 120°\right)$$
$$y_3 = 2(r)^{1/3}\cos\left(\frac{\theta}{3} + 240°\right)$$

Geometry

Figure A.2 to A.12 are a collection of common geometric figures. Area (A), volume (V), and other measurable features are indicated.

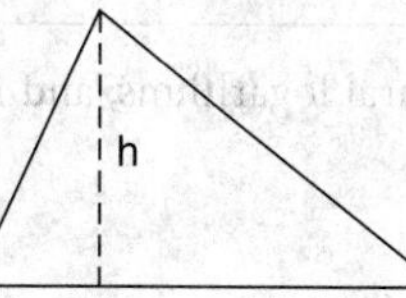

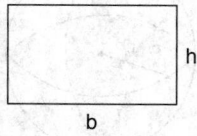

FIGURE A.2 Rectangle. $A = bh$.

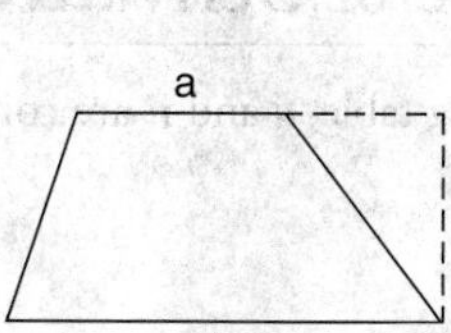

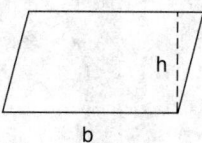

FIGURE A.3 Parallelogram. $A = bh$.

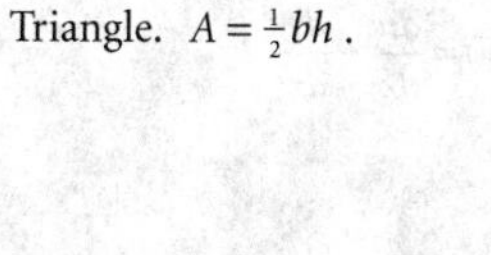

FIGURE A.4 Triangle. $A = \frac{1}{2}bh$.

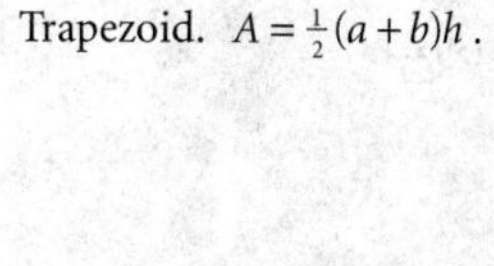

FIGURE A.5 Trapezoid. $A = \frac{1}{2}(a+b)h$.

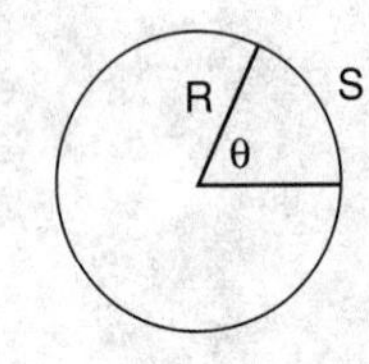

FIGURE A.6 Circle. $A = \pi R^2$; circumference $= 2\pi R$; arc length $S = R\theta$ (θ in radians).

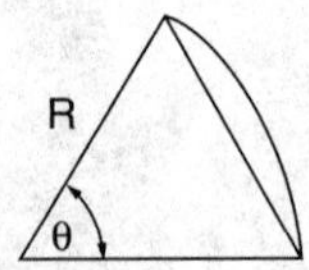

FIGURE A.7 Sector of circle. $A_{\text{sector}} = \frac{1}{2}R^2\theta$; $A_{\text{segment}} = \frac{1}{2}R^2(\theta - \sin\theta)$.

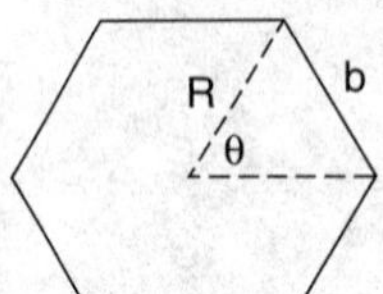

FIGURE A.8 Regular polygon of n sides. $A = (n/4)b^2 \text{ctn}(\pi/n)$; $R = (b/2)\csc(\pi/n)$.

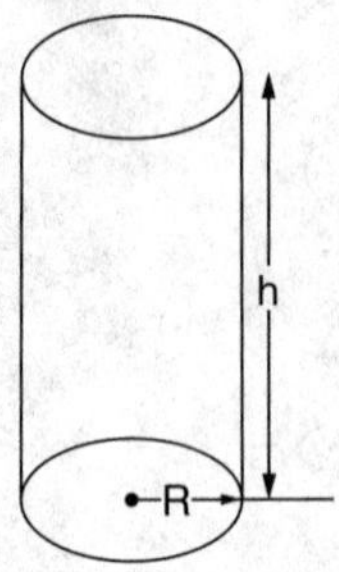

FIGURE A.9 Right circular cylinder. $V = \pi R^2 h$; lateral surface area $= 2\pi Rh$.

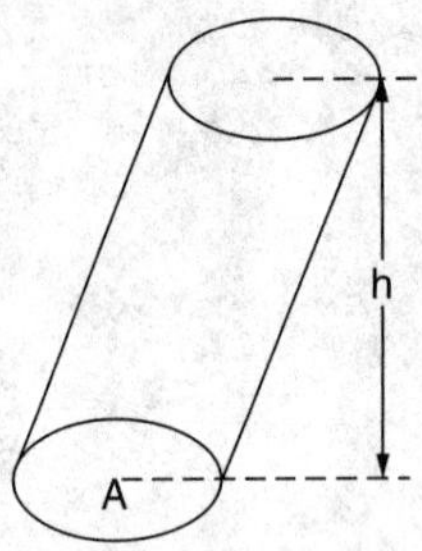

FIGURE A.10 Cylinder (or prism) with parallel bases. $V = Ah$.

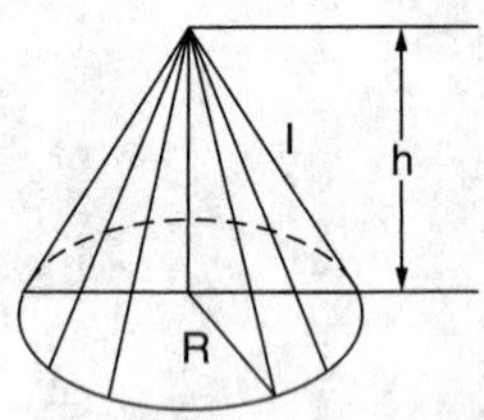

FIGURE A.11 Right circular cone. $V = \frac{1}{3}\pi R^2 h$; lateral surface area $= \pi Rl = \pi R\sqrt{R^2 + h^2}$.

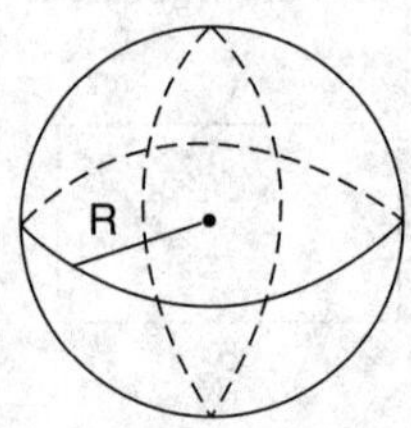

FIGURE A.12 Sphere. $V = \frac{4}{3}\pi R^3$; surface area $= 4\pi R^2$.

A.7 Table of Derivatives

In the following table, a and n are constants, e is the base of the natural logarithms, and u and v denote functions of x.

1. $\frac{d}{dx}(a) = 0$
2. $\frac{d}{dx}(x) = 1$
3. $\frac{d}{dx}(au) = a\frac{du}{dx}$
4. $\frac{d}{dx}(u + v) = \frac{du}{dx} + \frac{dv}{dx}$

5. $\frac{d}{dx}(uv) = u\frac{dv}{dx} + v\frac{du}{dx}$

6. $\frac{d}{dx}(u/v) = \frac{v\frac{du}{dx} - u\frac{dv}{dx}}{v^2}$

7. $\frac{d}{dx}(u^n) = nu^{n-1}\frac{du}{dx}$

8. $\frac{d}{dx}e^u = e^u\frac{du}{dx}$

9. $\frac{d}{dx}a^u = (\log_e a)a^u\frac{du}{dx}$

10. $\frac{d}{dx}\log_e u = (1/u)\frac{du}{dx}$

11. $\frac{d}{dx}\log_a u = (\log_a e)(1/u)\frac{du}{dx}$

12. $\frac{d}{dx}u^v = vu^{v-1}\frac{du}{dx} + u^v(\log_e u)\frac{dv}{dx}$

13. $\frac{d}{dx}\sin u = \cos u\frac{du}{dx}$

14. $\frac{d}{dx}\cos u = -\sin u\frac{du}{dx}$

15. $\frac{d}{dx}\tan u = \sec^2 u\frac{du}{dx}$

16. $\frac{d}{dx}\text{ctn}u = -\csc^2 u\frac{du}{dx}$

17. $\frac{d}{dx}\sec u = \sec u\tan u\frac{du}{dx}$

18. $\frac{d}{dx}\csc u = -\csc u\,\text{ctn}u\frac{du}{dx}$

19. $\frac{d}{dx}\sin^{-1}u = \frac{1}{\sqrt{1-u^2}}\frac{du}{dx}, \quad (-\tfrac{1}{2}\pi \le \sin^{-1}u \le \tfrac{1}{2}\pi)$

20. $\frac{d}{dx}\cos^{-1}u = \frac{-1}{\sqrt{1-u^2}}\frac{du}{dx}, \quad (0 \le \cos^{-1}u \le \pi)$

21. $\frac{d}{dx}\tan^{-1}u = \frac{1}{1+u^2}\frac{du}{dx}$

22. $\frac{d}{dx}\text{ctn}^{-1}u = \frac{-1}{1+u^2}\frac{du}{dx}$

23. $\frac{d}{dx}\sec^{-1}u = \frac{1}{u\sqrt{u^2-1}}\frac{du}{dx},$

$\left(-\pi \le \sec^{-1}u < -\frac{1}{2}\pi;\ 0 \le \sec^{-1}u < \frac{1}{2}\pi\right)$

24. $\frac{d}{dx}\csc^{-1}u = \frac{1}{u\sqrt{u^2-1}}\frac{du}{dx},$

$\left(-\pi < \csc^{-1}u \le -\frac{1}{2}\pi;\ 0 < \csc^{-1}u \le \frac{1}{2}\pi\right)$

25. $\frac{d}{dx}\sinh u = \cosh u\frac{du}{dx}$

26. $\frac{d}{dx}\cosh u = \sinh u\frac{du}{dx}$

27. $\frac{d}{dx}\tanh u = \text{sech}^2 u\frac{du}{dx}$

28. $\frac{d}{dx}\text{ctnh}u = -\text{csch}^2 u\frac{du}{dx}$

29. $\frac{d}{dx}\text{sech}u = -\text{sech}u\tanh u\frac{du}{dx}$

30. $\frac{d}{dx}\text{csch}u = -\text{csch}u\,\text{ctnh}u\frac{du}{dx}$

31. $\frac{d}{dx}\sinh^{-1}u = \frac{1}{\sqrt{u^2+1}}\frac{du}{dx}$

32. $\frac{d}{dx}\cosh^{-1}u = \frac{1}{\sqrt{u^2-1}}\frac{du}{dx}$

33. $\frac{d}{dx}\tanh^{-1}u = \frac{1}{1-u^2}\frac{du}{dx}$

34. $\frac{d}{dx}\text{ctnh}^{-1}u = \frac{-1}{u^2-1}\frac{du}{dx}$

35. $\frac{d}{dx}\text{sech}^{-1}u = \frac{-1}{u\sqrt{1-u^2}}\frac{du}{dx}$

36. $\frac{d}{dx}\text{csch}^{-1}u = \frac{-1}{u\sqrt{u^2+1}}\frac{du}{dx}$

Additional Relations with Derivatives

$$\frac{d}{dx}\int_a^t f(x)dx = f(t) \qquad \frac{d}{dt}\int_t^a f(x)dx = -f(t)$$

If $x = f(y)$, then $\frac{dy}{dx} = \frac{1}{dx/dy}$

If $y = f(u)$ and $u = g(x)$, then $\frac{dy}{dx} = \frac{dy}{du}\cdot\frac{du}{dx}$ (chain rule)

If $x = f(t)$ and $y = g(t)$, then $\frac{dy}{dx} = \frac{g'(t)}{f'(t)}$, and $\frac{d^2y}{dx^2} = \frac{f'(t)g''(t) - g'(t)f''(t)}{[f'(t)]^3}$

(*Note:* Exponent in denominator is 3.)

A.8 Integrals

Elementary Forms

1. $\int a\,dx = ax$

2. $\int a \cdot f(x)dx = a\int f(x)dx$

3. $\int \phi(y)dx = \int \frac{\phi(y)}{y'}dy$, where $y' = \frac{dy}{dx}$

4. $\int (u+v)dx = \int u\,dx + \int v\,dx$, where u and v are any functions of x

5. $\int u\,dv = u\int v\,du = uv - \int v\,du$

6. $\int u\frac{dv}{dx}dx = uv - \int v\frac{du}{dx}dx$

7. $\int x^n dx = \frac{x^{n+1}}{n+1}$, except $n = -1$

8. $\int \frac{f'(x)dx}{f(x)} = \log f(x)$, $[df(x) = f'(x)dx]$

9. $\int \frac{dx}{x} = \log x$

10. $\int \frac{f'(x)dx}{2\sqrt{f(x)}} = \sqrt{f(x)}$, $[df(x) = f'(x)dx]$

11. $\int e^x dx = e^x$

12. $\int e^{ax}dx = e^{ax}/a$

13. $\int b^{ax}dx = \frac{b^{ax}}{a\log b}$, $(b > 0)$

14. $\int \log x\,dx = x\log x - x$

15. $\int a^x \log a\,dx = a^x$, $(a > 0)$

16. $\int \frac{dx}{a^2 + x^2} = \frac{1}{a}\tan^{-1}\frac{x}{a}$

17. $$\int \frac{dx}{a^2 - x^2} = \begin{cases} \frac{1}{a}\tanh^{-1}\frac{x}{a} \\ \text{or} \\ \frac{1}{2a}\log\frac{a+x}{a-x}, \ (a^2 > x^2) \end{cases}$$

18. $$\int \frac{dx}{x^2 - a^2} = \begin{cases} -\frac{1}{a}ctnh^{-1}\frac{x}{a} \\ \text{or} \\ \frac{1}{2a}\log\frac{x-a}{x+a}, \ (x^2 > a^2) \end{cases}$$

19. $$\int \frac{dx}{\sqrt{a^2 - x^2}} = \begin{cases} \sin^{-1}\frac{x}{|a|} \\ \text{or} \\ -\cos^{-1}\frac{x}{|a|}, \ (a^2 > x^2) \end{cases}$$

20. $$\int \frac{dx}{\sqrt{x^2 \pm a^2}} = \log(x + \sqrt{x^2 \pm a^2})$$

21. $$\int \frac{dx}{x\sqrt{x^2 - a^2}} = \frac{1}{|a|}\sec^{-1}\frac{x}{a}$$

22. $$\int \frac{dx}{x\sqrt{a^2 \pm x^2}} = -\frac{1}{a}\log\left(\frac{a + \sqrt{a^2 \pm x^2}}{x}\right)$$

Forms Containing $(a + bx)$

For forms containing $a + bx$, but not listed in the table, the substitution $u = (a + bx)/x$ may prove helpful.

23. $$\int (a+bx)^n dx = \frac{(a+bx)^{n+1}}{(n+1)b}, \ (n \neq -1)$$

24. $$\int x(a+bx)^n dx = \frac{1}{b^2(n+2)}(a+bx)^{n+2} - \frac{a}{b^2(n+1)}(a+bx)^{n+1}, \ (n \neq -1, -2)$$

25. $$\int x^2(a+bx)^n dx = \frac{1}{b^3}\left[\frac{(a+bx)^{n+3}}{n+3} - 2a\frac{(a+bx)^{n+2}}{n+2} + a^2\frac{(a+bx)^{n+1}}{n+1}\right]$$

26. $$\int x^m(a+bx)^n dx = \begin{cases} \frac{x^{m+1}(a+bx)^n}{m+n+1} + \frac{an}{m+n+1}\int x^m(a+bx)^{n-1}dx \\ \text{or} \\ \frac{1}{a(n+1)}\left[-x^{m+1}(a+bx)^{n+1} + (m+n+2)\int x^m(a+bx)^{n+1}dx\right] \\ \text{or} \\ \frac{1}{b(m+n+1)}\left[x^m(a+bx)^{n+1} - ma\int x^{m-1}(a+bx)^n dx\right] \end{cases}$$

27. $$\int \frac{dx}{a+bx} = \frac{1}{b}\log(a+bx)$$

28. $$\int \frac{dx}{(a+bx)^2} = -\frac{1}{b(a+bx)}$$

29. $$\int \frac{dx}{(a+bx)^3} = -\frac{1}{2b(a+bx)^2}$$

30. $$\int \frac{x\,bx}{a+bx} = \begin{cases} \dfrac{1}{b^2}[a+bx-a\log(a+bx)x] \\ \text{or} \\ \dfrac{x}{b}-\dfrac{a}{b^2}\log(a+bx) \end{cases}$$

31. $$\int \frac{xbx}{(a+bx)^2} = \frac{1}{b^2}\left[\log(a+bx)+\frac{a}{a+bx}\right]$$

32. $$\int \frac{xdx}{(a+bx)^n} = \frac{1}{b^2}\left[\frac{-1}{(n-2)(a+bx)^{n-1}}+\frac{a}{(n-1)(a+bx)^{n-1}}\right],\quad n\neq 1,2$$

33. $$\int \frac{x^2dx}{a+bx} = \frac{1}{b^3}\left[\frac{1}{2}(a+bx)^2-2a(a+bx)+a^2\log(a+bx)\right]$$

34. $$\int \frac{x^2dx}{(a+bx)^2} = \frac{1}{b^3}\left[a+bx-2a\log(a+bx)-\frac{a^2}{a+bx}\right]$$

35. $$\int \frac{x^2dx}{(a+bx)^3} = \frac{1}{b^3}\left[\log(a+bx)+\frac{2a}{a+bx}-\frac{a^2}{2(a+bx)^3}\right]$$

36. $$\int \frac{x^2dx}{(a+bx)^n} = \frac{1}{b^3}\left[\frac{-1}{(n-3)(a+bx)^{n-3}}+\frac{2a}{(n-2)(a+bx)^{n-2}}-\frac{a^2}{(n-1)(a+bx)^{n-1}}\right],\quad n\neq 1,2,3$$

37. $$\int \frac{dx}{x(a+bx)} = -\frac{1}{a}\log\frac{a+bx}{x}$$

38. $$\int \frac{dx}{x(a+bx)^2} = \frac{1}{a(a+bx)}-\frac{1}{a^2}\log\frac{a+bx}{x}$$

39. $$\int \frac{dx}{x(a+bx)^3} = \frac{1}{a^3}\left[\frac{1}{2}\left(\frac{2a+bx}{a+bx}\right)^2+\log\frac{x}{a+bx}\right]$$

40. $$\int \frac{dx}{x^2(a+bx)} = -\frac{1}{ax}+\frac{b}{a^2}\log\frac{a+bx}{x}$$

41. $$\int \frac{dx}{x^3(a+bx)} = \frac{2bx-a}{2a^2x^2}+\frac{b^2}{a^3}\log\frac{x}{a+bx}$$

42. $$\int \frac{dx}{x^2(a+bx)^2} = -\frac{a+2bx}{a^2x(a+bx)}+\frac{2b}{a^3}\log\frac{a+bx}{x}$$

A.9 The Fourier Transforms

For a piecewise continuous function $F(x)$ over a finite interval $0 \le x \le \pi$, the *finite Fourier cosine transform* of $F(x)$ is

$$f_c(n) = \int_0^{\pi} f(x)\cos nx\,dx \quad (n = 0,1,2,\ldots) \tag{A.1}$$

If x ranges over the interval $0 \le x \le L$, the substitution $x' = \pi x/L$ allows the use of this definition also. The inverse transform is written

$$\bar{F}(x) = \frac{1}{\pi} f_c(0) + \frac{2}{\pi} \sum_{n=1}^{\infty} f_c(n) \cos nx \quad (0 < x < \pi) \tag{A.2}$$

where $\bar{F}(x) = [F(x+0) + F(x-0)]/2$. We observe that $\bar{F}(x) = F(x)$ at points of continuity. The formula

$$\begin{aligned} f_c^{(2)}(n) &= \int_0^\pi F''(x) \cos nx \, dx \\ &= -n^2 f_c(n) - F'(0) + (-1)^n F'(\pi) \end{aligned} \tag{A.3}$$

makes the finite Fourier cosine transform useful in certain boundary value problems.

Analogously, the *finite Fourier sine transform* of $F(x)$ is

$$f_s(n) = \int_0^\pi F(x) \sin nx \, dx \quad (n = 1, 2, 3, \ldots) \tag{A.4}$$

and

$$\bar{F}(x) = \frac{2}{\pi} \sum_{n=1}^{\infty} f_s(n) \sin nx \quad (0 < x < \pi) \tag{A.5}$$

Corresponding to Eq. (A.6), we have

$$\begin{aligned} f_s^{(2)}(n) &= \int_0^\pi F''(x) \sin nx \, dx \\ &= -n^2 f_s(n) - nF(0) - n(-1)^n F(\pi) \end{aligned} \tag{A.6}$$

Fourier Transforms

If $F(x)$ is defined for $x \ge 0$ and is piecewise continuous over any finite interval, and if

$$\int_0^\infty F(x) dx$$

is absolutely convergent, then

$$f_c(\alpha) = \sqrt{\frac{2}{\pi}} \int_0^\infty f(x) \cos(\alpha x) dx \tag{A.7}$$

is the *Fourier cosine transform* of $F(x)$. Furthermore,

$$\bar{F}(x) = \sqrt{\frac{2}{\pi}} \int_0^\infty f_c(\alpha) \cos(\alpha x) d\alpha \tag{A.8}$$

If $\lim_{x \to \infty} d^n F / dx^n = 0$, an important property of the Fourier cosine transform,

$$f_c^{(2r)}(\alpha)=\sqrt{\frac{2}{\pi}}\int_0^\infty\left(\frac{d^{2r}F}{dx^{2r}}\right)\cos(\alpha x)dx \tag{A.9}$$

where $\lim_{x\to 0} d^rF/dx^r = a_r$, makes it useful in the solution of many problems.

Under the same conditions,

$$f_s(\alpha)=\sqrt{\frac{2}{\pi}}\int_0^\infty F(x)\sin(\alpha x)dx \tag{A.10}$$

defines the *Fourier sine transform* of $F(x)$, and

$$\bar{F}(x)=\sqrt{\frac{2}{\pi}}\int_0^\infty f_s(\alpha)\sin(\alpha x)d\alpha \tag{A.11}$$

Corresponding to Eq. (A.9) we have

$$\begin{aligned} f_s^{(2r)}(\alpha)&=\sqrt{\frac{2}{\pi}}\int_0^\infty \frac{d^{2r}F}{dx^{2r}}\sin(\alpha x)dx \\ &=-\sqrt{\frac{2}{\pi}}\sum_{n=1}^{r}(-1)^n\alpha^{2n-1}a_{2r-2n}+(-1)^{r-1}\alpha^{2r}f_s(\alpha) \end{aligned} \tag{A.12}$$

Similarly, if $F(x)$ is defined for $-\infty < x\ \infty$, and if $\int_{-\infty}^{\infty} F(x)dx$ is absolutely convergent, then

$$f(\alpha)=\frac{1}{\sqrt{2\pi}}\int_{-\infty}^\infty F(x)e^{i\alpha x}dx \tag{A.13}$$

is the *Fourier transform* of $F(x)$, and

$$\bar{F}(x)=\frac{1}{\sqrt{2\pi}}\int_{-\infty}^\infty f(\alpha)e^{-i\alpha x}d\alpha \tag{A.14}$$

Also, if

$$\lim_{|x|\to\infty}\left|\frac{d^nF}{dx^n}\right|=0 \quad (n=1,2,\ldots,r-1)$$

then

$$f^{(r)}(\alpha)=\frac{1}{\sqrt{2\pi}}\int_{-\infty}^\infty F^{(r)}(x)e^{i\alpha x}dx=(-i\alpha)^r f(\alpha) \tag{A.15}$$

Finite Sine Transforms

	$f_s(n)$	$F(x)$
1.	$f_s(n)=\int_0^{\pi} F(x)\sin nx\,dx \quad (n=1,2,\ldots)$	$F(x)$
2.	$(-1)^{n+1} f_s(n)$	$F(\pi-x)$
3.	$\frac{1}{n}$	$\frac{\pi-x}{\pi}$
4.	$\frac{(-1)^{n+1}}{n}$	$\frac{x}{\pi}$
5.	$\frac{1-(-1)^n}{n}$	1
6.	$\frac{2}{n^2}\sin\frac{n\pi}{2}$	$\begin{cases} x & \text{when } 0<x<\pi/2 \\ \pi-x & \text{when } \pi/2<x<\pi \end{cases}$
7.	$\frac{(-1)^{n+1}}{n^3}$	$\frac{x(\pi^2-x^2)}{6\pi}$
8.	$\frac{1-(-1)^n}{n^3}$	$\frac{x(\pi-x)}{2}$
9.	$\frac{\pi^2(-1)^{n-1}}{n}-\frac{2[1-(-1)^n]}{n^3}$	x^2
10.	$\pi(-1)^n\left(\frac{6}{n^3}-\frac{\pi^2}{n}\right)$	x^3
11.	$\frac{n}{n^2+c^2}[1-(-1)^n e^{c\pi}]$	e^{cx}
12.	$\frac{n}{n^2+c^2}$	$\frac{\sinh c(\pi-x)}{\sinh c\pi}$
13.	$\frac{n}{n^2-k^2} \quad (k\neq 0,1,2,\ldots)$	$\frac{\sinh k(\pi-x)}{\sin k\pi}$
14.	$\begin{cases} \frac{\pi}{2} & \text{when } n=m \\ 0 & \text{when } n\neq m \end{cases} \quad (m=1,2,\ldots)$	$\sin mx$
15.	$\frac{n}{n^2-k^2}[1-(-1)^n\cos k\pi] \quad (k\neq 1,2,\ldots)$	$\cos kx$
16.	$\begin{cases} \frac{n}{n^2-m^2}[1-(-1)^{n+m}] & \text{when } n\neq m=1,2,\ldots \\ 0 & \text{when } n=m \end{cases}$	$\cos mx$
17.	$\frac{n}{(n^2-k^2)^2} \quad (k\neq 0,1,2,\ldots)$	$\frac{\pi\sin kx}{2k\sin^2 k\pi}-\frac{x\cos k(\pi-x)}{2k\sin k\pi}$
18.	$\frac{b^n}{n} \quad (\lvert b\rvert\leq 1)$	$\frac{2}{\pi}\arctan\frac{b\sin x}{1-b\cos x}$
19.	$\frac{1-(-1)^n}{n}b^n \quad (\lvert b\rvert\leq 1)$	$\frac{2}{\pi}\arctan\frac{2b\sin x}{1-b^2}$

Finite Cosine Transforms

	$f_c(n)$	$F(x)$
1.	$f_c(n) = \int_0^{\pi} F(x)\cos nx\, dx \quad (n = 0,1,2,\ldots)$	$F(x)$
2.	$(-1)^2 f_c(n)$	$F(\pi - x)$
3.	0 when $n = 1,2,\ldots;\ f_c(0) = \pi$	1
4.	$\frac{2}{n}\sin\frac{n\pi}{2};\ f_c(0) = 0$	$\begin{cases} 1 & \text{when } 0 < x < \pi/2 \\ -1 & \text{when } \pi/2 < x < \pi \end{cases}$
5.	$-\frac{1-(-1)^n}{n^2};\ f_c(0) = \frac{\pi^2}{2}$	x
6.	$\frac{(-1)^n}{n^2};\ f_c(0) = \frac{\pi^2}{6}$	$\frac{x^2}{2\pi}$
7.	$\frac{1}{n^2};\ f_c(0) = 0$	$\frac{(\pi - x)^2}{2\pi} - \frac{\pi}{6}$
8.	$3\pi^2\frac{(-1)^n}{n^2} - 6\frac{1-(-1)^n}{n^4};\ f_c(0) = \frac{\pi^4}{4}$	x^3
9.	$\frac{(-1)^n e^c\pi - 1}{n^2 + c^2}$	$\frac{1}{c}e^{cx}$
10.	$\frac{1}{n^2 + c^2}$	$\frac{\cosh c(\pi - x)}{c\sinh c\pi}$
11.	$\frac{k}{n^2 - k^2}[(-1)^n\cos\pi k - 1] \quad (k \neq 0,1,2,\ldots)$	$\sin kx$
12.	$\frac{(-1)^{n+m} - 1}{n^2 - m^2};\ f_c(m) = 0 \quad (m = 1,2,\ldots)$	$\frac{1}{m}\sin mx$
13.	$\frac{1}{n^2 - k^2} \quad (k \neq 0,1,2,\ldots)$	$-\frac{\cos k(\pi - x)}{k\sin k\pi}$
14.	0 when $n = 1,2,\ldots;\ f_c(m) = \frac{\pi}{2} \quad (m = 1,2,\ldots)$	$\cos mx$

Fourier Sine Transforms

	$F(x)$	$f_s(x)$
1.	$\begin{cases} 1 & (0 < x < a) \\ 0 & (x > a) \end{cases}$	$\sqrt{\frac{2}{\pi}}\left[\frac{1-\cos\alpha}{\alpha}\right]$
2.	$x^{p-1} \quad (0 < p < 1)$	$\sqrt{\frac{2}{\pi}}\frac{\Gamma(p)}{\alpha^p}\sin\frac{p\pi}{2}$
3.	$\begin{cases} \sin x & (0 < x < a) \\ 0 & (x > a) \end{cases}$	$\frac{1}{\sqrt{2\pi}}\left[\frac{\sin[a(1-\alpha)]}{1-\alpha} - \frac{\sin[a(1+\alpha)]}{1+\alpha}\right]$
4.	e^{-x}	$\sqrt{\frac{2}{\pi}}\left[\frac{\alpha}{1+\alpha^2}\right]$
5.	$xe^{-x^2/2}$	$\alpha e^{-\alpha^2/2}$
6.	$\cos\frac{x^2}{2}$	$\sqrt{2}\left[\sin\frac{\alpha^2}{2}C\left(\frac{\alpha^2}{2}\right) - \cos\frac{\alpha^2}{2}S\left(\frac{\alpha^2}{2}\right)\right]^*$

	$F(x)$	$f_s(x)$
7.	$\sin\frac{x^2}{2}$	$\sqrt{2}\left[\cos\frac{\alpha^2}{2}C\left(\frac{\alpha^2}{2}\right)+\sin\frac{\alpha^2}{2}S\left(\frac{\alpha^2}{2}\right)\right]^*$

* $C(y)$ and $S(y)$ are the Fresnel integrals.

$$C(y)=\frac{1}{\sqrt{2\pi}}\int_0^y \frac{1}{\sqrt{t}}\cos t\, dt$$

$$S(y)=\frac{1}{\sqrt{2\pi}}\int_0^y \frac{1}{\sqrt{t}}\sin t\, dt$$

Fourier Cosine Transforms

	$F(x)$	$f_x(\alpha)$
1.	$\begin{cases}1 & (0<x<a)\\ 0 & (x>a)\end{cases}$	$\sqrt{\frac{2}{\pi}}\frac{\sin a\alpha}{\alpha}$
2.	$x^{p-1}\ (0<p<1)$	$\sqrt{\frac{2}{\pi}}\frac{\Gamma(p)}{\alpha^p}\cos\frac{p\pi}{2}$
3.	$\begin{cases}\cos x & (0<x<a)\\ 0 & (x>a)\end{cases}$	$\frac{1}{\sqrt{2\pi}}\left[\frac{\sin[a(1-\alpha)]}{1-\alpha}+\frac{\sin[a(1+\alpha)]}{1+\alpha}\right]$
4.	e^{-x}	$\sqrt{\frac{2}{\pi}}\left(\frac{1}{1+\alpha^2}\right)$
5.	$e^{-x^2/2}$	$e^{-\alpha^2/2}$
6.	$\cos\frac{x^2}{2}$	$\cos\left(\frac{\alpha^2}{2}-\frac{\pi}{4}\right)$
7.	$\sin\frac{x^2}{2}$	$\cos\left(\frac{\alpha^2}{2}-\frac{\pi}{4}\right)$

Fourier Transforms

	$F(x)$	$f(\alpha)$
1.	$\frac{\sin ax}{x}$	$\begin{cases}\sqrt{\frac{\pi}{2}} & \lvert\alpha\rvert<a\\ 0 & \lvert\alpha\rvert>a\end{cases}$
2.	$\begin{cases}e^{iwx} & (p<x<q)\\ 0 & (x<p, x>q)\end{cases}$	$\frac{i}{\sqrt{2\pi}}\frac{e^{ip(w+\alpha)}-e^{iq(w+\alpha)}}{(w+\alpha)}$
3.	$\begin{cases}e^{-cx+iwx} & (x>0)\\ 0 & (x<0)\end{cases}\ (c>0)$	$\frac{i}{\sqrt{2\pi}(w+\alpha+ic)}$
4.	$e^{-px^2}\ R(p)>0$	$\frac{1}{\sqrt{2p}}e^{-\alpha^2/4p}$
5.	$\cos px^2$	$\frac{1}{\sqrt{2p}}\cos\left[\frac{\alpha^2}{4p}-\frac{\pi}{4}\right]$

	$F(x)$	$f(\alpha)$
6.	$\sin px^2$	$\frac{1}{\sqrt{2p}}\cos\left[\frac{a^2}{4p}+\frac{\pi}{4}\right]$
7.	$\lvert x\rvert^{-p} \quad (0<p<1)$	$\sqrt{\frac{2}{\pi}}\frac{\Gamma(1-p)\sin\frac{p\pi}{2}}{\lvert\alpha\rvert^{(1-p)}}$
8.	$\frac{e^{-a\lvert x\rvert}}{\sqrt{\lvert x\rvert}}$	$\frac{\sqrt{\sqrt{(a^2+a^2)}+a}}{\sqrt{a^2+a^2}}$
9.	$\frac{\cosh ax}{\cosh \pi x} \quad (-\pi<a<\pi)$	$\sqrt{\frac{2}{\pi}}\frac{\cos\frac{a}{2}\cosh\frac{\alpha}{2}}{\cosh\alpha+\cos\alpha}$
10.	$\frac{\sinh ax}{\sinh \pi x} \quad (-\pi<a<\pi)$	$\frac{1}{\sqrt{2\pi}}\frac{\sin a}{\cosh\alpha+\cos a}$
11.	$\begin{cases}\frac{1}{\sqrt{a^2-x^2}} & (\lvert x\rvert<a)\\ 0 & (\lvert x\rvert>a)\end{cases}$	$\sqrt{\frac{\pi}{2}}J_0(a\alpha)$
12.	$\frac{\sin[b\sqrt{a^2+x^2}]}{\sqrt{a^2+x^2}}$	$\begin{cases}0 & (\lvert\alpha\rvert>b)\\ \sqrt{\frac{\pi}{2}}J_0(a\sqrt{b^2+\alpha^2}) & (\lvert\alpha\rvert<b)\end{cases}$
13.	$\begin{cases}P_n(x) & (\lvert x\rvert<1)\\ 0 & (\lvert x\rvert>1)\end{cases}$	$\frac{i^n}{\sqrt{\alpha}}J_{n+1/2}(\alpha)$
14.	$\begin{cases}\frac{\cos[b\sqrt{a^2-x^2}]}{\sqrt{a^2-x^2}} & (\lvert x\rvert<a)\\ 0 & (\lvert x\rvert>a)\end{cases}$	$\sqrt{\frac{\pi}{2}}J_0(a\sqrt{a^2+b^2})$
15.	$\begin{cases}\frac{\cosh[b\sqrt{a^2-x^2}]}{\sqrt{a^2-x^2}} & (\lvert x\rvert<a)\\ 0 & (\lvert x\rvert>a)\end{cases}$	$\sqrt{\frac{\pi}{2}}J_0(a\sqrt{a^2-b^2})$

The following functions appear among the entries of the tables on transforms.

Function	Definition	Name
$\mathrm{Ei}(x)$	$\int_{-\infty}^{x}\frac{e^v}{v}dv$; or sometimes defined as $-\mathrm{Ei}(-x)=\int_x^{\infty}\frac{e^{-v}}{v}dv$	Exponential integral function
$\mathrm{Si}(x)$	$\int_0^x\frac{\sin v}{v}dv$	Sine integral function
$\mathrm{Ci}(x)$	$\int_{\infty}^{x}\frac{\cos v}{v}dv$; or sometimes defined as negative of this integral	Cosine integral function
$\mathrm{erf}(x)$	$\frac{2}{\sqrt{\pi}}\int_0^x e^{-v^2}dv$	Error function

Function	Definition	Name
erfc(x)	$1-\text{erf}(x)=\dfrac{2}{\sqrt{\pi}}\int_x^{\infty}e^{-v^2}dv$	Complementary function to error function
$L_n(x)$	$\dfrac{e^x}{n!}\dfrac{d^n}{dx^n}(x^n e^{-x}),\ n=0,1,\ldots$	Laguerre polynomial of degree n

A.10 Bessel Functions

Bessel Functions of the First Kind, $J_n(x)$ (Also Called Simply *Bessel Functions*) (Figure A.13)

Domain: $[x>0]$

Recurrence relation:

$$J_{n+1}(x)=\frac{2n}{x}J_n(x)-J_{n-1}(x),\quad n=0,1,2,\ldots$$

Symmetry: $J_{-n}(x)=(-1)^n J_n(x)$

0. $J_0(20x)$
1. $J_1(20x)$
2. $J_2(20x)$
3. $J_3(20x)$
4. $J_4(20x)$
5. $J_5(20x)$

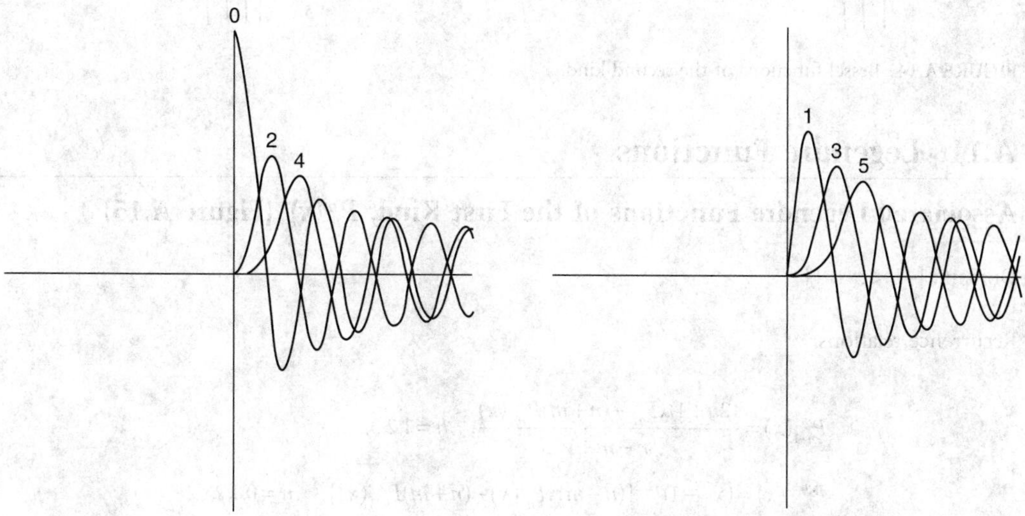

FIGURE A.13 Bessel functions of the first kind.

Bessel Functions of the Second Kind, $Y_n(x)$ (Also Called *Neumann Functions or Weber Functions*) (Figure A.14)

Domain: $[x>0]$

Recurrence relation:

$$Y_{n+1}(x) = \frac{2n}{x} Y_n(x) - Y_{n-1}(x), \quad n = 0, 1, 2, \ldots$$

Symmetry: $Y_{-n}(x) = (-1)^n Y_n(x)$

0. $Y_0(20x)$
1. $Y_1(20x)$
2. $Y_2(20x)$
3. $Y_3(20x)$
4. $Y_4(20x)$
5. $Y_5(20x)$

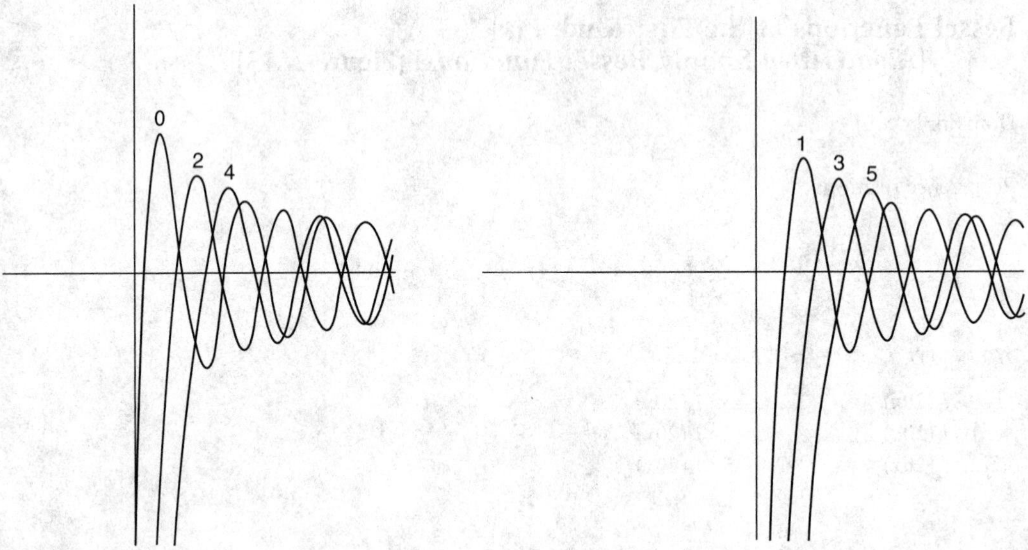

FIGURE A.14 Bessel functions of the second kind.

A.11 Legendre Functions

Associated Legendre Functions of the First Kind, $P_n^m(x)$ (Figure A.15)

Domain: [–1 < x < 1]

Recurrence relations:

$$P_{n+1}^m(x) = \frac{(2n+1)xP_n^m - (n+m)P_{n-1}^m(x)}{n-m+1}, \quad n = 1, 2, 3, \ldots$$

$$P_n^{m+1}(x) = (x^2 - 1)^{-1/2}[(n-m)xP_n^m(x) - (n+m)P_{n-1}^m(x)], \quad m = 0, 1, 2, \ldots$$

with

$$P_0^0 = 1 \qquad P_1^0 = x$$

Special case: P_n^0 Legendre polynomials

1-0. $P_0^0(x)$
1-1. $P_1^0(x)$ 2-1. $0.25P_1^1(x)$

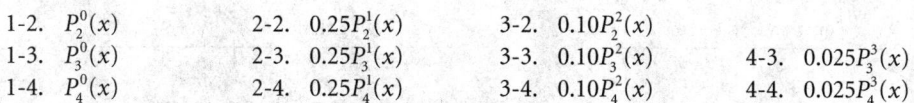

1-2. $P_2^0(x)$	2-2. $0.25P_2^1(x)$	3-2. $0.10P_2^2(x)$	
1-3. $P_3^0(x)$	2-3. $0.25P_3^1(x)$	3-3. $0.10P_3^2(x)$	4-3. $0.025P_3^3(x)$
1-4. $P_4^0(x)$	2-4. $0.25P_4^1(x)$	3-4. $0.10P_4^2(x)$	4-4. $0.025P_4^3(x)$

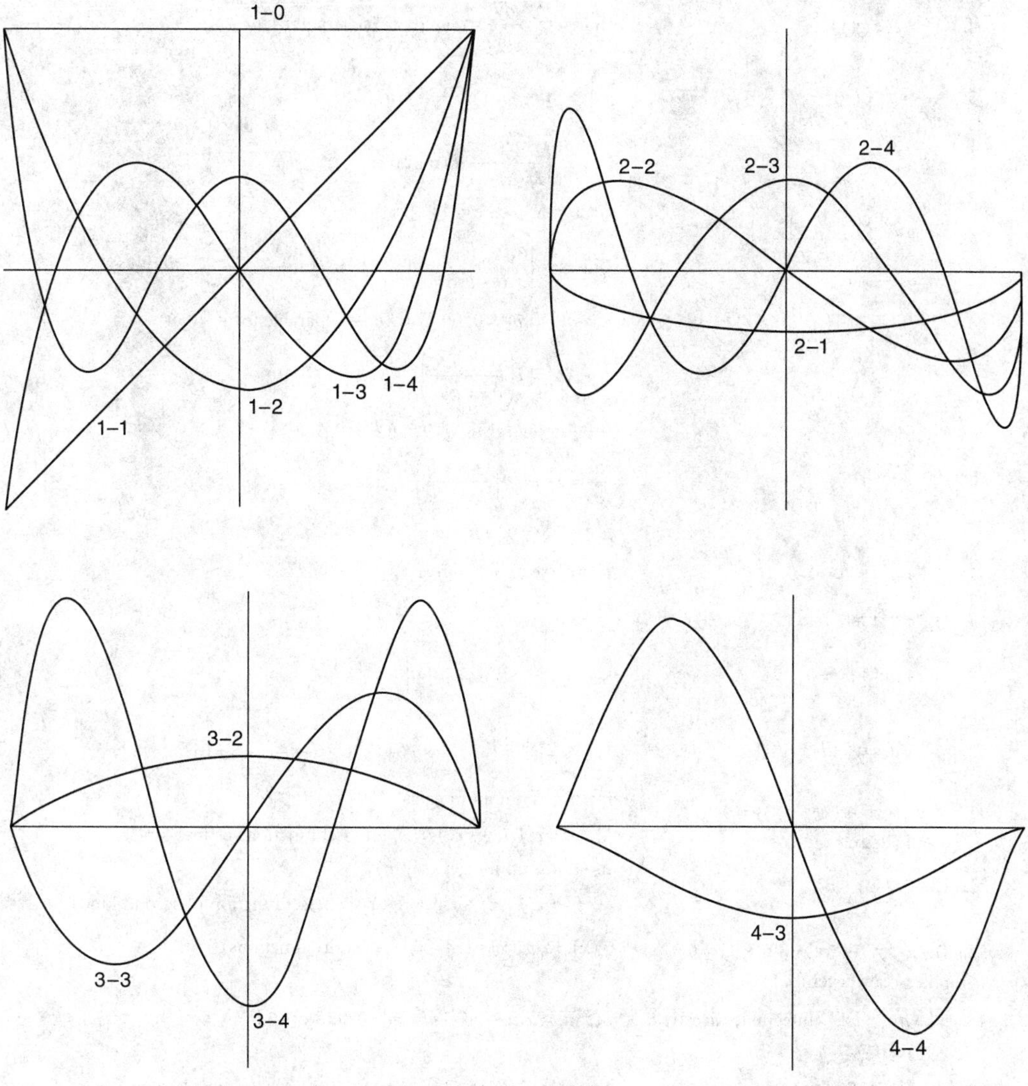

FIGURE A.15 Legendre functions of the first kind.

A.12 Table of Differential Equations

Equation	Solution
1. $y' = \dfrac{dy}{dx} = f(x)$	$y = \int f(x)dx + c$
2. $y' + p(x)y = q(x)$	$y = \exp\left[-\int p(x)dx\right]\left\{c + \int \exp\left[\int p(x)dx\right]q(x)dx\right\}$

Equation	Solution
3. $y' + p(x)y = q(x)y^{\alpha}$ $\alpha \neq 0, \alpha \neq 1$	Set $z = y^{1-\alpha} \to z' + (1-\alpha)p(x)z = (1-\alpha)q(x)$ and use 2
4. $y' = f(x)g(y)$	Integrate $\frac{dy}{g(y)} = f(x)dx$ (separable)
5. $\frac{dy}{dx} = f(x/y)$	Set $= xu \to u + x\frac{du}{dx} = f(u)$ $\int \frac{1}{f(u)-u}du = \ln\lvert x\rvert + c$
6. $y' = f\left(\frac{a_1x+b_1y+c_1}{a_2x+b_2y+c_2}\right)$	Set $x = X + \alpha$, $y = Y + \beta$ Choose $\begin{cases} a_1\alpha + b_1\beta = -c_2 \\ a_2\alpha + b_2\beta = -c_2 \end{cases} \to Y' = f\left(\frac{a_1X+b_1Y}{a_2X+b_2Y}\right)$ If $a_1b_2 - a_2b_1 \neq 0$, set $Y = Xu \to$ separable form $u + Xu' = f\left(\frac{a_1+b_1u}{a_2+b_2u}\right)$ If $a_1b_2 - a_2b_1 = 0$, set $u = a_1x + b_1y \to$ $\frac{du}{dx} = a_1 + b_1f\left(\frac{u+c_1}{ku+c_2}\right)$ $a_2x + b_2y = k(a_1x + a_2y)$
7. $y'' + a^2y = 0$	$y = c_1\cos ax + c_2\sin ax$
8. $y'' - a^2y = 0$	$y = c_1e^{ax} + c_2e^{-ax}$
9. $y'' + ay' + by = 0$	Set $y = e^{-(a/2)x}u \to u'' + \left(b - \frac{a^2}{4}\right)u = 0$
10. $y'' + a(x)y' + b(x)y = 0$	Set $y = e^{-(1/2)\int a(x)dx} \to u'' + \left[b(x) - \frac{a^2}{4} - \frac{a'}{2}\right]u = 0$
11. $x^2y'' + xy' + (x^2 - a^2)y = 0$ $a \geq 0$ (Bessel)	i. If a is not an integer $y = c_1J_a(x) + c_2J_{-a}(x)$ (Bessel functions of first kind) ii. If a is an integer (say, n) $y = c_1J_n(x) + c_2Y_n(x)$ (Y_n is Bessel function of second kind)
12. $(1-x^2)y'' - 2xy' + a(a+1)y = 0$, a is a real (Legendre)	$y(x) = c_1p_a(x) + c_2q_a(x)$ (Legendre functions)
13. $y' + ay^2 = bx^n$ (integrable Riccati) a, b, n real	Set $u' = ayu \to u'' - abx^nu = 0$ and use 14
14. $y'' - ax^{-1}y' + b^2x^{\mu}y = 0$	$y = x^p[c_1J_{\nu}(kx^q) + c_2J_{-\nu}(kx^q)]$ where $p = (a+1)/2$, $\nu = (a+1)/(\mu+2)$, $k = 2b/(\mu+2)$, $q = (\mu+2)/2$

15. Item 13 shows that the Riccati equation is linearized by raising the order of the equation. The *Riccati chain,* which is linearizable by raising the order, is

$$u' = uy,\quad u'' = u[y' + y^2],\quad u'' = u[y'' + 3yy' + y^3],$$

$$u^{(iv)} = u[y''' + 4yy'' + 6y^2y' + 3(y')^2 + y^4],\ldots$$

To use this consider the second-order equation $y'' + 3yy' + y^3 = f(x)$. The Riccati transformation $u' = yu$ transforms this equation to the linear form $u''' = uf(x)!$.

References

Kamke, E. 1956. *Differentialgleichungen* Lösungsmethoden und Lösungen, Vol. I. Akad. Verlagsges, Leipzig.

Murphy, G. M. 1960. *Ordinary Differential Equations and Their Solutions.* Van Nostrand, New York.

Zwillinger, D. 1992. *Handbook of Differential Equations*, 2nd ed. Academic Press, San Diego.

Author Index

G

H

I

J

K

Subject Index

C

D

E

F

G

H

I

O

P

Q

R

S

W

Y

Z